Your Place in the Cosmos
Volume VIII

A layman's book of Astronomy
and
The Mythology
of the eighty-eight celestial
constellations
and
Catalog VIII
by
Dr. James J. Rickard

Mosele & Associates, Inc.
34523 Wilson Rd.
Ingleside, Illinois 60041
U.S.A.

As the International Star Registry approaches it's 30th birthday, we reflect on the origins and development of our service.

The concept's universal appeal is proven by the fact that since our inception we will have named one million stars. Naming a star with ISR remains a creative and positive way to express personal regard and makes a moving gesture of love, affections, adoration or gratitude.

We thank all 20th century star-namers for the enthusiastic way in which they have embraced our service and we look forward with pleasure to continuing the work of the International Star Registry in the Third Millenium.

En momentos en que,ISR ,se aproxima a su 30 aniversario,reflexionamos sobre los orígenes y desarrollo de nuestros Servicios.

El exito de nuestra universal concepción ha quedado probado por el hecho de que hemos dado nombre a mas de un millon de Estrellas.Nombrar una Estrella con ISR continua siendo una creativa y positiva manera de expresar un recuerdo personal,y,es,sin duda,un conmovedor gesto de amor,afecto,adoración o gratitud.

Agradecemos calurosamente a todos aquellos que,en esa parte del Siglo XX, dieron nombre a una Estrella ,por la manera entusiasta en que acogieron nuestro Servicio,y,miramos con una placentera ilusión,el estar continuando nuestro trabajo de ISR en los años que vienen.

Contents

Introduction

Our understanding of the universe today is very different from the universe that astronomers knew just 100 years ago. Astronomers then had no idea how vast and empty space was, nor did they understand the tremendous evolutionary changes that had occurred in the Cosmos.

The Cosmos, the Greek word for universe, is incredibly huge! Because light travels at a finite speed, 186,000 miles per second (300,000 kilometers per second) we are always looking into the past with our telescopes. Some objects are so far away that the light from them has been traveling toward us for 10 billion years. And the sky glows uniformly from photons emitted 15 billion years ago from an early epoch before any stars had formed.

The Cosmos is not eternal; it began about 15 billion years ago in a burst of incredible energy, the Big Bang. This super-hot ball of energy and fundamental particles rapidly expanded and cooled, forming the protons, neutrons, and electrons found in all ordinary matter today. The hot gas eventually cooled even more, allowing the stars and galaxies to form the objects we see today with our telescopes. But these accumulations of matter occupy less than 1/10 of 1 percent of space! The Cosmos is nearly empty due to its expansion.

We are part of this evolutionary history. Our Sun, our planet Earth and all life are made of the atoms composed of the protons and neutrons from the Big Bang. The Earth was cool enough so that the complex molecules of life formed and evolved over billions of years into us, you and me. We are really just another natural stage in the evolution of the Cosmos.

In the description that follows we will examine how ancient astronomers interpreted the sky. We will see how the patterns of bright stars were used for story telling and cultural mythology. And finally, we will close with a look at modern astronomy. The last section of the book contains the personal names of stars assigned by the International Star Registry. This book is part of a series of star names assigned by ISR.

Pre-Historic Star Gazing

Since the beginning of human society, people have looked at the night sky and perceived patterns with the stars. The human brain seems to be genetically wired to look for patterns and regularity—even if none exist. People crane their necks, and use the night sky like a giant Connect-the-Dots game.

> "Mommy, do you see that wolf up there?"

> "Yes, that's to remind us that wolves hunt at night and will carry off small children, if we are not careful!"

In this manner, a mother's concern was made cosmically official and beyond challenge. So it was with stories about bears, serpents, scorpions, and sea monsters.

All societies have connected these stellar dots in various ways, making order out of the chaos above us. Doing so, according to psychologists, makes us feel less afraid because we recognize the repetitive patterns that return year after year. Humans have a sense of time. We have memory that goes beyond the last time we ate. We learn; i.e., develop response patterns, from past experiences and then predict outcomes when faced with similar

circumstances. When there are no surprises, our lives can be tranquil and more productive. Human society likes order. It takes less thought, debate, and decision.

One of the oldest indications that the sky was used to mark the passage of time comes to us from the Caves of Lascaux in southern France. Cro-Magnon artists over 15,000 years ago painted horses, antelope and bulls on the cave walls. And they also drew clusters of dots and squares recently interpreted by researchers as the phases of the Moon—a lunar calendar.

Because the sky patterns are repetitive and predictable, shamans and astrologers developed elaborate fortune-telling beliefs, the precursors to religious faith systems. They kept records of sky events and unusual occurrences such as solar and lunar eclipses, "new" stars, comets, and rare conjunctions of the bright planets. They also named the bright stars. When one knows the name of a person, a relationship can be calm and non-threatening—even friendly. Naming the stars was important because then the stars became familiar friends in the night sky.

We humans have a need to know our place in the Cosmos, our role in its past, present, and future. We see ourselves as a part of an organic universe. We have acted under the belief that our actions are influenced by the Cosmos and our actions influence the Cosmos in return. Hindu and Buddhist religions call this the Law of Karma, the spiritual law of cause and effect. Isaac Newton called this principle the Third Law of Motion—for every action there is an equal and opposite reaction.

Ancient mariners used knowledge of the sky for navigation. They memorized the star pictures and star names. This knowledge was handed down from generation to generation and used to sail the Mediterranean Sea and Indian, and the Pacific oceans. In a very real sense the Diaspora of the human species resulted, in part, from knowledge of the night sky.

What are Constellations?

Constellations are groupings of bright stars in the night sky. The patterns, or "star pictures", as they are called in many languages, do not change from year to year. Those near the celestial poles can be seen sometime during the night all year long; others appear and disappear depending on the seasons.

The bright stars provide a permanent background map against which moving objects and unexpected visitors can be measured. The Sun, Moon, the five bright planets, the unpredictable comets and flashing meteors make up these moving objects. Short-lived bright stars know as novae occasionally appear in the sky as well.

Origin of Names Used Today

Elaborate legends grew around some of the constellations and were passed on from the wise elders to their youngsters. Perhaps the stories were written first and the heavens used as a giant easel to make drawings illustrating the tales. Humans have named these constellations after animals, mythological beasts, boats, rivers, hunters, gods and goddesses, and even craftsmen's tools.

Most of the names western cultures now use for bright stars and the constellations in the Northern Hemisphere came to us from Babylon, Egypt, Greece and Rome. The Bull, the Scorpion and the Lion are known to have been named over 6,000 years ago. The appearance of Orion and the brightest star in the sky, Sirius, rising in the early summer dawn foretold the annual flooding of the Nile.

Oriental cultures have an equally elaborate system of ancient legends and star names. However we use the nomenclature used by professional astronomers, which was derived from the west.

The constellations of the Zodiac are the best known because of their use in astrological horoscopes; but there are even larger, more spectacular star pictures that fill the rest of the sky. About 200 AD, the famous Egyptian astrologer Ptolemy compiled a sky atlas of 48 constellations.

Not all of the stars in the sky were assigned to constellations by ancient astrologers. As modern astronomy began to develop in Europe, more complete sky atlases were prepared. Often new constellations were invented to fill up space between better-known ones. Sometimes they were created for political reasons, to commemorate some obscure event in medieval European history. Most importantly, when European adventurers sailed around the world and viewed the sky from the southern hemisphere for the first time, many constellations were invented out of necessity to complete the jigsaw puzzle of the sky.

In 1603 Johann Bayer, a German astronomer, published a major sky atlas revision named Uranometria. At the same time he introduced the system of bright-star designations still used by professional astronomers today. He

took the letters of the Greek alphabet to name the stars within a constellation in order of their apparent brightness. Thus the star commonly known as Betelgeuse has the Bayer designation Alpha Orionis because it is the brightest star in the constellation Orion. Orionis is the Latin genitive, or possessive case of Orion.

Eventually, astronomers changed some of the traditional boundaries, consolidated some constellations and divided a few very large ones, like Argo Navis, into manageable territories. The astronomers Jakob Bartsch and Johannes Hevelius revised and codified the constellations in 1624 and 1690.

The French astronomers, for example Abbe Nicolas Louis de Lacaille, got into the act in the 18th and 19th centuries. So there was a mixture of names from different times and cultures. Finally in 1930 everyone agreed to the present boundaries of just 88 constellations.

Most professional astronomers know little or nothing about the myth or origin of the constellation names. Researchers are interested in the physical nature of our universe. For them, the constellations are used only to indicate the general direction of objects in the sky. Since the stars in a particular constellation are normally at very different distances along the line of sight, there is usually no physical connection between them. This is similar to the lack of association between individuals who live in Minnesota and Moscow, although both cities start with the letter M.

The Constellations of the Zodiac and the Calendar

The Sun, Moon and the five bright planets are the most obvious of the moving objects in the sky. Like the swallows of Capistrano, the motions of these bodies are repetitive and predictable. And they all move in a narrow band in the sky about 10 degrees wide known as the ecliptic. The ecliptic is really the extension of the Sun's equator. All of the planets revolve around the Sun in the same flat plane and in the same counter-clockwise direction as seen from the Sun's north pole. Of all the constellations, only thirteen provide the background of the ecliptic. Twelve of the 13 are known as the zodiacal constellations.

The ecliptic path is tipped 23.5 degrees with respect to the Earth's equator. It's really the Earth that is tipped, of course. This inclination causes the seasons. As viewed from planet Earth, the Sun stands on the equator at dawn on the first day of spring, known as the vernal equinox. On this day the Sun rises exactly at the East point on the horizon. Then the Sun climbs slowly to 23.5 degrees north of the equator on the first day of summer, the summer solstice. The Sun then reverses its course, slowly returning to the equator on the autumnal equinox when it again appears due east at sunrise. It then continues to its lowest point in the sky, 23.5 degrees south of the equator on the first day of winter or winter solstice.

To pinpoint these seasonal indicators, prehistoric astronomers in many countries constructed calendar circles, such as the world-famous Stonehenge in England.

Although the Sun, Moon and the planets are confined to this ecliptic path, they move at different speeds. The Moon, for example, whizzes around the ecliptic in 29½ days, while Venus, the brightest star-like object in the sky, moves back and forth with respect to the background stars every 224 days. Saturn takes nearly 30 years to complete its circuit.

Ancient Calendars

The Egyptians astrologers noted that the Sun took about 365 days to "circle the sky". They also knew that 365 was not evenly divisible by any small integers except 5. So they chose a more convenient number, 360, and divided the sky into equal parts. This number can be evenly divided by many integers: —2, 3, 4, 5, 6, 8, 9, 10, 12, 15, 18, and 30. In this way they could organize and divide their civil calendar into rational sections. This Egyptian civil calendar, based on this motion of the Sun around the sky, left five extra days to party at the end of the year!

Notice that 7 and 11 are numbers that cannot evenly divide 360. That made them special. The ancients therefore associated greater luck with them, a belief that still influences us at the dice tables of Las Vegas. The number 13, also not on the list, developed the reputation as a bad luck symbol.

The Mayan astrologers of Central America divided the year into four 91-day quarters, with only one extra party day at the end of the year. Obviously the Egyptians had more fun!

Roman astrologers under Julius Caesar realized that there were really 365¼ days in the solar year, and they refined their calendar to more closely represent the solar motions.

However there was still that little problem of the extra six hours per year. So in 46 BC the astronomer Sosigenes devised the plan of adding one extra day each fourth year, the leap year, to keep the Roman social calendar synchronized with the rest of the Universe.

Christianity and the Western world adopted this Julian calendar, with its 12 months of differing numbers of days, until it became apparent that it needed further refining.

The Change From the Julian to the Gregorian Calendar

Over the centuries the spring equinox had been appearing about a day earlier every century, in spite of the leap year correction. By 1582 AD, it had advanced from March 21 to March 11. This worried religious authorities because it affected the dates of important celebrations—Easter, for example. Easter is the Christian feast of the Resurrection that occurred two days after the Jewish Passover. Passover is defined by the Jewish lunar calendar to be the first Sabbath after the first full Moon after the vernal equinox. So Easter is also determined by the spring equinox and the phases of the Moon.

Refining the mathematical calculations used by the Church to determine Easter was one of the goals of Nicolas Copernicus (1474–1547 AD). The Church had used the observations and model of the solar system of Ptolemy from 200 AD. This model placed the Earth in the center of the Universe and uses elaborate geometry to describe and predict the motions of the Sun, Moon and planets, as they appear to move around the Earth.

Debate Over Geocentric and Heliocentric Models

To suggest that the Sun was in the center of the solar system upset Church authorities to such an extent that arrest, torture and death by burning at the stake were common consequences. Copernicus knew this, and delayed the publication of his revolutionary heliocentric theory until after he died.

Galileo Galilei (1564–1642 AD.) got into trouble with Church authorities and the Jesuits during the Inquisition for publicly promoting Copernicus's model of the solar system. Until his time the study of the stars and planets was largely an exercise in philosophy, not observation. Galileo's development of the telescope had allowed him to substantiate the theory with observational evidence. But the religious authorities weren't buying it. He was convicted of heresy and spent the last seven years of his life under house arrest. The Sun-centered model was gradually accepted, however, and following a recent review of his case in Rome, after 350 years his conviction was overturned and Galileo exonerated by the Church.

Regardless of which astronomical model is used, both geocentric and heliocentric models had to be corrected to bring back the spring equinox from March 11 to March 21. So Pope Gregory XIII introduced a new calendar, which he named after himself, as leaders often do, in 1582. By papal decree the day after October 4 would be October 15. And each century year not divisible by 400 would add a "century day" to eliminate equinox creep. In 1900, the century day was added as June 31, as it will be in 2100, 2200 and 2300. But we missed getting the extra holiday in 2000.

It is interesting to note that many countries in Eastern Europe continued to use the older Julian calendar into the 20th century. Even in Great Britain and its colonies the Gregorian calendar was not recognized until 1750! It seems that the rejection of papal authority by Henry VIII in 1530 over his controversial request for a divorce from Catherine of Aragon had calendar consequences for over 200 years.

What happened when British employers were faced with ten fewer days in 1750? Were workers told to skip their vacations that year? And think of the many people who were robbed of their birthday parties.

The Zodiac and Astrology

Today we distinguish between study of astrology and the science of astronomy. Astrology is a belief or faith that somehow events on Earth are influenced by the positions of the planets. Today astronomers look disapprovingly on astrology, calling it a pseudo-science, like fortune telling. But let's remember that the two professions were one and the same until only in the last 200 years. The word astrology is a combination of the Greek words for star and word. In this context astrology means trying to make sense of the motions of celestial bodies, trying to find logical meaning in the sky.

The world model accepted by all ancient cultures placed the Earth in the center of the universe. The heavens were the realm of the gods who exercised power over creatures on Earth. Evil powers inhabited the interior of the Earth. Even today ask a 6-year old child to point at Heaven and at Hell and see how deeply this cosmic error is imbedded in the human psyche.

Faced with no understanding of the causes of earthquakes, volcanoes, hurricanes, floods, pestilence, disease, etc., mankind looked into the night sky for explanations of the catastrophes affecting him. Perhaps he could predict the future by carefully watching the signs.

The Greeks, Romans, and Christian philosophers have argued the question of the existence of free will. Does man really have a choice in moral decisions? Or is he subject to powerful forces beyond his control, maybe beyond his comprehension? Are events predestined? Is fate the overwhelming determinant in the course of our lives? Some people seem to have more 'luck' than others. Astrology is an attempt to correlate events in our individual lives with the relative positions of the Sun, Moon, and planets. It is perhaps the oldest belief system shared by all cultures on Earth.

People of many cultures around the world still believe in astrology, to varying extents. In America, after the comics, the horoscope column is the most read feature in the daily newspaper. The Chinese calendar has its own dos and don'ts depending upon whether its the year of the Snake or the year of the Dog.

Modern Tibetan Buddhist astrologers carefully prepare the yearly calendar looking for days when the planets may have bad positions for certain earthly activities. On some days it is not a good idea to travel, or buy a house. If there is a lot of bad energy, the astrologers just skip that day. If June 7th is a bad day, then the calendar will have a June 6th followed by June 8th!

The New Age spiritual movement of recent years has as one of its central beliefs that nothing happens by chance—it is our job to look for signs around us and adjust our actions to remain in harmony with the Universe. If something bad happens, it may be due to our ignoring the warnings.

The Zodiac

The zodiacal constellations are the only ones of importance to western astrology. Modern western astrology is derived from the Greco-Roman belief system. The sky was divided into 12 constellation zones along the ecliptic, each 30 degrees wide. The Sun spends 30.44 days in each sign. The first sign begins with the vernal equinox. During Roman times, 2,500 years ago, the Sun entered the constellation Aries, the Ram on that date, March 21.

Astrologers divide humans into 12 personality types according to the moment of their birth. Your "Sun sign" is the position of the Sun when you were born. A complete horoscope is a map of the sky showing the relative location of the Sun, Moon and all the planets at a given moment. Conjunctions (apparent closeness) and the angular relationships among the planets add individuality to your persona. According to this system of belief, the compatibility between people is largely based on their horoscopes, which are the major determinant of personalities and characteristics.

Astrologers were highly regarded in ancient civilizations due to the overwhelming influence their advice had over the actions of those in power. Astrologers were often consulted for opinions on important social events. Is next Sunday the best day for crowning the new king? Should we plant the crops this week or next? Should the wedding be on Saturday or Wednesday? In principle an astrologer would be consulted for every important decision.

Precession of the Earth's Axis

Since the Earth wobbles a bit, like a spinning top, the direction of its poles slowly changes. One complete wobble takes 25,800 years. This causes the positions of the equinoxes and solstices to shift slowly against the background of the constellations, and is known as the precession of the equinoxes. The vernal equinox moves from constellation to constellation every 2,150 years. Today the Sun is in the constellation Pisces on March 21, one whole constellation backwards from Aries. Soon the vernal equinox will be in Aquarius. Read more about the Age of Aquarius below in our description of that constellation.

Modern astrologers ignore this shift in the background screen of constellations. They maintain that it is the relative positions of the planets that influence us, not the stars themselves. The permanent star fields behind the planets are just a convenient reference.

Today there are, in fact, 13 constellations that the ecliptic passes through. In 1930 the International Union of Astronomers, an organization of professional astronomers, revised and codified the names and boundaries of the constellations. The boundaries of Sagittarius were officially changed, with part of its former territory being placed in the constellation Ophiuchus. Astrologers ignore this fact, using the original 12 signs of equal length as the map upon which they plot the relationships of influential bodies—Sun, Moon, Mercury, Venus, Mars, Jupiter and Saturn.

The Twelve Constellations of the Astrologers' Zodiac

These 12 constellations are to be found on the Egyptian astrologer Ptolemy's list of 48 constellations. They are organized by the period the Sun was in front of them inPtolemy's time. The traits, aspects and careers listed are those generally accepted by astrologers.

Aries, The Ram. March 21–April 19

Trait: assertive, creative

Positive aspects: creative, brave, direct, pioneering, energetic

Negative aspects: selfish, pugnacious, and impatient

Careers: artist, composer, writer, rock star

Aries was sent by the god Hermes to save the two children of the King of Thessaly from their cruel stepmother. Was killed in the effort. Then the leader of the Argonauts, Jason, using his ship Argo, recovered the Golden Fleece from this mythical ram.

Taurus, The Bull. April 20–May 20

Trait: possessive, permanent

Positive aspects: practical, reliable, ethical in business and moral issues

Negative aspects: self-indulgent, boring, stubborn, opinionated

Careers: bankers, judges

The Greek myth talks about how Zeus, the father of all the gods, fell in love with Europa, the princess of Phoenicia. He disguised himself as a majestic white bull to attract her attention. Once she was on his back he carried her across the sea to the island of Crete. Western civilization developed around cattle husbandry. The cow was a source of milk, meat, clothing, shoes, and field labor, so it was reasonable to see this grouping as a giant bull. Taurus is just east of Orion who some see as a bullfighter or toreador. It is known that the constellation was defined long before the Greeks, but not as a bull. To the Egyptians, this was their god Osiris, while the Chinese saw it as the "White Tiger" or "Great Bridge."

The red giant star Aldebaran marks his eye. Pleiades, the famous star group also called the Seven Sisters, is located within the boundaries of the constellation Taurus.

Gemini, The Twins. May 21–June 21

Trait: communicative, loquacious

Positive aspects: intellectual, talkative, youthful, busy, energetic

Negative aspects: moody, changeable, gossipy, two-faced

Career: TV presenter, communicator

Castor and Pollux were fraternal twin brothers (different fathers, same mother). The brothers were inseparable in life and excelled in the martial arts. Castor was a horseman as well. These Greek heroes were among the men who went with Jason on his voyages on the ship Argo. When Castor died, Pollux went to Zeus to ask that both of them be placed in the sky together.

Cancer, The Crab. June 22–July 22

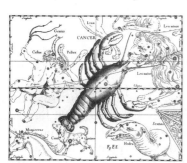

Trait: understanding, obedient

Positive aspects: protective, sensitive, tenacious, cautious, resourceful

Negative aspects: over-protective, self-pitying, unforgiving

Career: doctor, security guard

Crabs are respected inhabitants of the ocean floor. The myth associated with Cancer involves a heroic struggle between Hercules and Hydra, the water snake. Hera sent little Cancer to nip at Hercules's feet to distract him during the fight. Hercules crushed Cancer but at least he was placed in the sky as a reward for his heroic but pitiful effort. This constellation contains no bright stars; but has the beautiful large family of stars M44, known as the Beehive cluster.

Leo, The Lion. July 23–August 22

Trait: impressive

Positive aspects: powerful, generous, creative, showy, dramatic

Negative aspects: bully, pompous, conceited, stubborn

Career: entrepreneur, CEO, politician

The lion is the most feared predator in the savannas of Africa and elsewhere. Lions were prominent in both Egyptian and Hindu astrology. Tribal chiefs often adorn themselves in lion skins as a symbol of their authority. In the Greek myth, Hercules strangled the mighty Leo as one of his twelve labors and then used his skin, which couldn't be penetrated by stone or metal, for his tunic. In the Middle Ages Christians associated Leo with the biblical story of Daniel in the lion's den.

Virgo, The Virgin. August 23–September 22

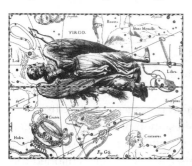

Trait: analytical

Positive aspects: meticulous, organized, discriminating, caring

Negative aspects: workaholic, hypercritical, finicky, a worrier

Careers: scientist, office assistant, customs inspector

Virgo was identified as the goddess of the harvest. In some cultures she is also the goddess of justice, un-swayed by emotional arguments. The Sun is descending to the autumnal equinox at the end of Virgo's month. At the same time the full Moon moves higher in the sky and provides evening light for the harvest. In countries where grapes were grown and harvested, it was traditional to have virgins stomp the grapes for the first crush in wine making.

Libra, The Balance Scales. September 23–October 23

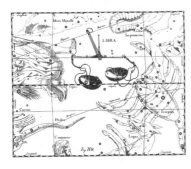

Trait: harmony, balance

Positive aspects: charming, idealist, diplomatic, romantic

Negative aspects: indecisive, oscillating, frivolous

Careers: travel agent, flight attendant, sellers, marketers

The date of equal day-equal night occurs on September 23, the first day of Libra. In some traditions Libra is not a person; but rather, the scales of justice held in the hands of Virgo. Libra also refers to the marketplace, where the fruits of the harvest were sold or traded.

Scorpius, The Scorpion. October 24–November 21

Trait: inventive

Positive aspects: purposeful, subtle, persistent, resourceful

Negative aspects: obstinate, secretive, jealous, crafty

Careers: policeman

The scorpion was sent by Apollo to protect his sister's chastity from the advances of Orion. After many battles, Scorpius finally killed the great hunter. They were then placed in opposite parts of the sky so that they are never up at the same time. The stars in Scorpius actually look like a scorpion with its curved tail and stinger to the south. Astrologers refer to it as Scorpio. Scorpions become more active in the fall, eating insects that are abundant because of the fall harvest.

The new constellation boundaries approved by astronomers in 1930 placed part of Scorpius's previous property in the constellation Ophiuchus.

Sagittarius, The Archer. November 22–December 21

Trait: adventurous

Positive aspects: optimistic, jovial, dependable, freedom-loving

Negative aspects: tactless, capricious, extreme, overly optimistic

Career: lawyer, real estate agent

Sagittarius is the celestial archer and is depicted as a feared centaur, the half-man half-horse creature of Greek mythology. The center of our galaxy, the Milky Way, lies in the direction of Sagittarius. In its boundaries are numerous star cluster and emission nebulae familiar to amateur astronomers worldwide. The arrow in his bow is pointed at Scorpius to keep him at bay.

Capricornus, The Mountain Goat. December 22–January 19

Trait: calculating, cautious

Positive aspects: patient, loyal, disciplined, headstrong

Negative aspects: miserly, rigid, merciless, dominating

Career: engineer, government worker

Capricornus has been recognized since Babylonian times as a goat. Often he is shown with a fish tail, perhaps relating to a manifestation of the Greek god Pan. While fleeing the god Typhon, Pan jumped into the Nile River. The part that was below water took the form of a fish, while his torso, above water, remained that of a goat. Pan is the god of revelry, music, and fine wine.

Aquarius, The Water Carrier. January 20–February 18

Trait: humane, charitable

Positive aspects: friendly, idealistic, intellectual, wise

Negative aspects: tactless, eccentric, iconoclastic

Career: charity worker

The constellation is often drawn as a man pouring a jar of water. The Water Carrier is symbolic of the bearer of life-giving liquids, having altruistic concern for others. Aquarius represents harmony and understanding instead of selfishness and greed. The vernal equinox point will move into Aquarius in the near future. The precession of the Earth's axis takes about 25,800 years to complete its circle. So the vernal equinox remains in a constellation for about 2000 years. Some astrologers ascribe the constellation's characteristics to that "age". Since Roman times we have been living in the Age of Pisces, which has negative characteristics. Soon we will begin the Age of Aquarius with its more positive social aspects. A popular rock musical of the 1970's *Hair*, which embodied the hopes of that generation for a world of harmony, understanding, and peace on Earth.

There is some discrepancy among astrologers as to when the Age of Aquarius actually begins. Some say it began in 1997. Others calculate that the year 2080 will be the beginning. Most agree that we are in the transition period between the two Ages right now. Because of the boundaries adjustment in 1930, the Sun won't enter the modern Aquarius until 2800 AD.

Pisces, The Fishes. February 19–March 20

Trait: impressionable

Positive aspects: compassionate, emotional, receptive, fun

Negative aspects: weak-willed, indecisive, secretive, vague, careless

Career: an actor or actress

Pisces, two fish tied together, has been around since Babylonian times. Ancient Greeks told the story of Venus and her son Cupid being frightened by the monster Typhon while on the banks of the Euphrates River. They escaped with their lives by jumping into the river and taking on the form of fish.

Pisces has been associated with malignant influences. The astrological calendar describes the emblems of this constellation as indicative of violence and death—really bad vibes! The Sun has been in Pisces on the vernal equinox for about 2000 years. This "Age of Pisces" has seen the most destructive wars in human history, as well as the development of weapons of mass destruction.

The Other Constellations Visible In The North

To appreciate and understand the spatial relationships of the constellations, look at the star chart on page…first we list the constellations visible in the northern hemisphere, which carry nearly all of the mythological background. We list them in alphabetical order so, like words in the dictionary, they are easy to find. The origin of the name is listed.

Andromeda, The Chained Princess (In Ptolemy's list of 200 AD)

Daughter of Cepheus and Cassiopeia, she was chained to a rock by the ocean as a sacrifice to the sea monster Cetus. She got into this mess after her mother boasted how beautiful she was. Neptune was angered and decided to torture the girl and her parents. Just as the sea monster was about to grab her, Perseus stepped in and killed him. Cepheus allowed Andromeda and Perseus to be married.

In Andromeda is the nearest large spiral galaxy like our own Milky Way. The Andromeda Galaxy is visible under dark sky conditions as a faint hazy object. With a good pair of binoculars one can make out its elongated spiral shape. The light entering your eyes left that galaxy about two million years ago!

Aquila, The Eagle (In Ptolemy's list of 200 AD)

Aquila is the eagle who carried thunderbolts for Zeus. It lies in a gorgeous section of the Milky Way. The brightest star in Aquila is Altair, which is about 16 light years away, making it a near neighbor. Altair forms one point of the Summer Triangle together with Deneb in Cygnus, and Vega in Lyra.

Auriga, The Charioteer (In Ptolemy's list of 200 AD)

Some stories say that Auriga invented the chariot, or at least the harnesses for attaching two horses to a lightweight war wagon. It was the charioteer's job to manage and maneuver the chariot during battle while the warrior clubbed and sliced his way through the enemy. The chariot was the ancient forerunner of the tank. Remember Ben Hur?

Another myth maintains that Auriga's father, who was crippled, invented this chariot, the original wheel chair, so that his son could move him about more easily.

Bootes, The Bear Driver (In Ptolemy's list of 200 AD)

Bootes is depicted as a hunter carrying a large club in his right hand and the leash to his hunting dogs, Canes Venatici, in the left hand. It is not clear whether Bootes is hunting the bears of the north, Ursa Major and Ursa Minor, just scaring them away, or herding them around the North Star. This large constellation has the bright star Arcturus at its center. To find Bootes in the June sky, extend the arc of the handle of the Big Dipper (Ursa Major) southward. The first bright yellow-white star is Arcturus.

Camelopardalis, The Giraffe (In Hevelius's list of 1690)

Named by Plancius, this constellation has no stars brighter than 4th magnitude. From modern over-lighted cities none of the stars are visible. The giraffe occupies the space between the constellations Auriga and Perseus and the North Pole star Polaris. When you have the opportunity to view the sky under dark conditions, you will find dozens of stars in this constellation.

Canes Venatici, The Hunting Dogs (In Hevelius's list of 1690)

Hunting dogs were highly prized by the ancients. They were not just retrievers, but trackers and sometimes attack dogs. Their skills and loyalty were admired. These two dogs are found close to the handle of the Dipper in Ursa Major. They are the leashed hounds of Bootes, and bear no relation to Orion's dogs, Canis Major and Canis Minor. Hevelius invented this minor constellation in 1687.

Canis Major, The Greater Dog   (In Ptolemy's list of 200 AD)

Canis Major is one of Orion's two hunting dogs. He sits at the feet of his master. The brightest star in this constellation is Sirius, often called the Dog Star. It is also the brightest star in the sky being only about 10 light years away. The Egyptians watched for the early morning rising of Sirius and Orion to forecast the flooding of the Nile River.

Canis Minor, The Lesser Dog (In Ptolemy's list of 200 AD)

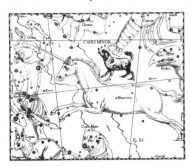

Ancient cultures really loved their dogs! There are no domestic cat constellations. Should we conclude that dogs are better than cats? This small constellation is the second of Orion's dogs. It contains the bright star Procyon. Linking the brightest stars of Orion, Taurus, Auriga, Gemini, Canis Minor and Canis Major makes a large hexagon in the winter sky.

Cassiopeia, The Queen   (In Ptolemy's list of 200 AD)

The familiar W or M, depending of the time of year, this constellation can be seen all year from northern middle latitudes. Her royal family surrounds her. Her husband Cepheus is on her left, her son-in-law Perseus is on the right, and her daughter Andromeda is just beneath her.

Centaurus, The Centaur   (In Ptolemy's list of 200 AD)

The Centaur is the mythical half-horse, half-human monster that embodied the strength and speed of a horse and the intelligence of a human. This centaur is said to be Chiron, who was very wise, and tutored Hercules and Jason. Unfortunately, Hercules wounded him by accident, and he begged to be put out of his misery.

Historically humans developed horse drawn wagons and chariots before attempting to ride on the backs of horses. In battle the hands were used to hold shields and spears. Fighters couldn't hang on or ride bareback and still be effective soldiers. Centaurs didn't have that problem. Eventually we humans invented the technology to stick to the backs of horses—the stirrup (circa 600 AD) changed the nature of cavalry and the history of the western world.

Cepheus, The King (In Ptolemy's list of 200 AD)

He is Cassiopeia's husband and Andromeda's father. He is depicted as being dressed in royal robes with a scepter in his hands. When Cassiopeia boasted that she was more beautiful than the water nymphs, she brought down the wrath of Neptune on her country. An oracle told Cepheus he could save his people if he would sacrifice his daughter, Andromeda, to the monster. Cepheus sadly chained Andromeda to a rock along the shore to await her fate. When Perseus offered to save Andromeda if he could then marry her, Cepheus agreed. In this way, Cepheus saved both his people and his daughter.

Cetus, The Sea Monster, Or Whale (In Ptolemy's list of 200 AD)

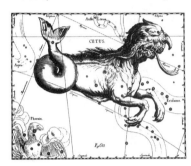

This is an important player in the story of Perseus who rescued the lovely Andromeda, daughter of Cepheus and Cassiopeia. Cetus was the sea monster that just missed getting Andromeda for dinner.

Whales were feared and largely misunderstood by ancient cultures. Because sailors and fishermen ventured out in such small boats, any encounter with a whale was an awesome experience. Cetus is also identified with the whale that swallowed Jonah in the Biblical story.

Coma Berenices, Berenice's Hair (By Tycho Brahe, about 1590)

This cluster of faint stars represents Berenice's hair. She was a beautiful Egyptian princess who offered to Venus to cut off her long flowing hair if her brother, the pharaoh, would return from his battle with the Syrians alive. Well, he did.

Corona Borealis, The Northern Crown (In Ptolemy's list of 200 AD)

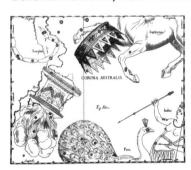

This almost full circle of stars represents a crown of reward given to Ariadne, thedaughter of Minos the second king of Crete. She had saved the life of Theseus, the future king of Athens in his fight against the Minotaurs. Theseus promised to marry Ariadne and take her away from Crete. But things didn't work out, and he never married her. After her suicide, some say that Theseus asked that the crown be placed in the sky to remind him of his shameful behavior. It was the least he could do.

Corvus, The Crow (In Ptolemy's list of 200 AD)

This small constellation looks like a trapezoid. Crows and ravens are clever birds. In some cultures they are seen as signs or omens of bad luck. But they are birds that have adapted to the growth of the human population and are known as resourceful scavengers.

Crater, The Cup (In Ptolemy's list of 200 AD)

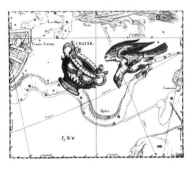

This small constellation represents the drinking goblets used in festivals associated with Bacchus, Apollo, Hercules and others. But sometimes it was associated with the large Hydra grouping. But sometimes it was associated with the large Hydra grouping.

Cygnus, The Swan   (In Ptolemy's list of 200 AD)

Cygnus is depicted as a swan flying, its wings oustreached and is sometimes called the Northern Cross. Its brightest star is Deneb, one of the three stars of the Summer Triangle. Swans are considered especially beautiful, elegant birds. Their dignity and grace are celebrated in many cultures.

Delphinus, The Dolphin   **(In Ptolemy's list of 200 AD)**

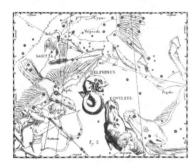

This small compact constellation looks like a little kite. Like whales, dolphins are very intelligent aquatic mammals. There are many stories of shipwrecked mariners being saved by dolphins. Only recently a fisherman in the North Sea was saved from sharks by dolphins that rammed the sharks, driving them away from the fisherman until help arrived.

Draco, The Dragon   (In Ptolemy's list of 200 AD)

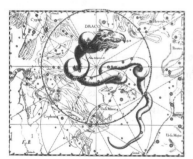

The dragon winds back and forth around 90 degrees of the sky near the North Pole. Some pictures show Draco being restrained by the foot of Hercules immediately to the south. The brightest star in Draco, Thuban, was the North Star four millennia ago. But due to the precession of the Earth's axis, it now points toward Polaris.

Equuleus, The Foal Or Small Horse   (In Ptolemy's list of 200 AD)

This is a small constellation without bright stars just to the west of Pegasus, the winged horse. Some cultures associate the two horses as mother and foal.

Eridanus, The River   (In Ptolemy's list of 200 AD)

Eridanus represents a river in Greek mythology. The river begins near the bright star Rigel, Orion's knee, and continues over 40 degrees to the west under Cetus. The river Eridanus is not only the sixth largest of the constellations, it is also the longest in the sky, continuing into the southern sky and ending with the bright star Achernar above the Small Cloud of Magellan. So only half of this constellation is invisible from Europe and North America.

Hercules, The Champion   (In Ptolemy's list of 200 AD)

Hercules was the greatest hero of mythology. He accomplished so many amazing tasks that some suspect he was given credit for other warriors' feats as well as his own. He killed Leo and clothed himself in the lion's indestructible skin. He destroyed Hydra, the water snake with a hundred hissing heads. He killed the flesh eating birds that had terrorized Arcadia. He conquered and pillaged Troy. He accidentally crushed little Cancer. And he did all that before Tuesday!

Hydra, The Water Serpent   (In Ptolemy's list of 200 AD)

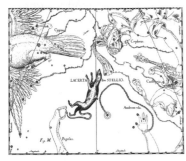

Hydra is a monster with a hundred serpentine heads so frightening it caused most people to die straight away. And if that didn't work, his poisonous breath could do you in. Hercules killed him on Monday.

Hydra, the largest of all the constellations, is another long one that rambles more than 90 degrees through the sky. It used to be even bigger, but was split into four more convenient parts—Sextans, Crater, Corvus and a new, slimmed-down Hydra.

Lacerta, The Lizard   (In Hevelius's list of 1690)

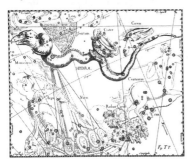

A little constellation, which was created to fill a small vacuum between Cepheus and Pegasus, fairly near the Milky Way. There are no bright stars and there is relatively no myth associated with the common lizard. This is a modern constellation, so sorry no mythology. The history of naming tells us, however, that several astronomers named the area in honor of their favorite king. But the king is dead; long live the lizard!

Leo Minor, The Lion Cub   (In Hevelius's list of 1690)

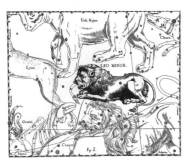

Hevelius named this area north of Leo in 1690. It incorporates a few stars in previously unclaimed territory. He settled on the name of Leo Minor because the stars' proximity to the great grouping known as Leo. This happened long before the birth of Disney's cub Simba, the prince who grew to be the Lion King.

Lepus, The Hare   (In Ptolemy's list of 200 AD)

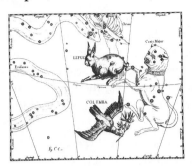

The constellation of the hare, or jack rabbit lies just under Orion and to the west of his hunting dog, Canis Major. Tradition has it that Orion loved to hunt rabbits and he continues hunting the Hare around the celestial circle.

Lupus, The Wolf   (In Ptolemy's list of 200 AD)

It is natural that the wolf would be a feared beast among sheep-herding folk. So the myth of Lupus is not for the weak-of-heart. To make a long, horrid tale short, the cruel king Lycaon sacrificed children on the altar of Zeus. When the god found out, he transformed him into the wolf, Lupus, and placed him in the sky near Centaurus who makes sure the wolf stays in his place.

Lynx, The Cat   (In Hevelius's list of 1690)

Hevelius named this narrow constellation of nondescript stars after the Lynx, a large cat found in the forests of Europe. He said only those with cats eyes could spot it. Perhaps its spots reminded the astronomer of a star cluster. This mysterious and fascinating feline was once widespread, but now has almost entirely disappeared from the European continent. Several countries have adopted measures for its reintroduction, and this extraordinary cat is gradually reappearing.

Lyra, The Harp   (In Ptolemy's list of 200 AD)

The lyre, a type of harp, was a musical instrument highly regarded for its near magical power to soothe the troubled mind and soul. Angels play harps. And so did Orpheus, son of the god Apollo, so sweetly, it is said that trees and flowers followed him, wild animals and birds came to listen, and rivers slowed to hear his music.

This constellation contains Vega, one of the brightest stars in the sky, and a point of the Summer Triangle (along with Deneb and Altair). Because of the precession of the Earth's North Pole, in about 12,000 years Vega will replace Polaris as the North Star.

Monoceros, The Unicorn (In Hevelius's list of 1690)

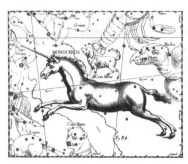

Monoceros is a scattering of faint stars near the celestial equator between Canis Major and Canis Minor. It was named by Plancius, who may have had in mind the Indian Rhinoceros. The unicorn, on the other hand, is a fabulous mythical creature described as a horse with a spiral horn of gold growing from its forehead. Even though it is very shy and rarely seen by mortals, the unicorn can defend itself with that horn. It is considered very lucky to see a Unicorn, which is reportedly visible only to those who search and trust.

Monoceros boasts having two of the most massive stars yet discovered. The two blue-giants are estimated to be some 55 times the size of the Sun, and revolve around each other. Together they are known as Plaskett's Star. The pair is violently unstable and brightens unexpectedly from time to time, giving astronomers a thrill.

Ophiuchus, The Serpent Holder (In Ptolemy's list of 200 AD)

In Greek mythology Ophiuchus is always associated with the first physician, Aesculapius, son of Apollo. He learned the art of healing and restoring life from Chiron, the Centaur. One of his remedies relied upon extracted snake venom, and he holds the nearby constellation Serpens, on his arm.

Ophiuchus has been in fact the "thirteenth constellation" of the zodiac, since 1930, when the constellation boundaries approved by astronomers, sliced off a chunk of Scorpius's boundaries and added it to Ophiuchus. The Sun actually spends the first two weeks of December in front of the serpent holder.

Orion, The Hunter (In Ptolemy's list of 200 AD)

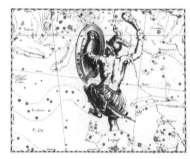

Perhaps the most widely recognized star pattern in the sky is Orion. This giant hunter stands in front of Taurus, like a matador. Four bright stars, among them famous Betelgeuse, enclose the row of three inclined stars known as Orion's Belt. He bears a sword on his belt and carries a lion skin and a huge club. His two faithful dogs are by his side. Another of his prey, Lepus the Hare, is at his feet.

Now if you want to read a gripping story, find some of the ancient stories about Orion, one of the most universal mythological characters. You will find love, thrills, magic healing, bravery, death and pathos. Son of the sea god Neptune, he was handsome and strong and in love first with Merope, one of the Seven Sisters. Later, the beautiful Artemis, goddess of the Moon, also a keen hunter, took his fancy. Finally he was killed by the lowly Scorpion.

With binoculars you can see a hazy patch below the belt stars called the Great Orion Nebula. We know today that it contains hundreds of new stars just formed from the hydrogen gas clouds illuminated by the bright surrounding stars. Some stars are less than 3,000 years old, mere infants! Photos of star birth and protoplanetary systems transmitted from the Hubble Space Telescope are among the most dramatic and revealing of any space photos ever taken.

Pegasus, The Winged Horse (In Ptolemy's list of 200 AD)

Pegasus can be easily identified as a large square next to Andromeda in the autumn sky. Actually one of the stars forming the square is assigned to Andromeda. Let's say the two share the star. But the recurring appearance of this winged horse on the stage of Greek mythology is so much more colorful than the nickname of The Big Square. Pegasus is yet another example of the flights of fancy that Greek mythology could take. Pegasus is said to be the son of Neptune, who sprang to life from the blood of Medusa when Perseus cut off her head. Pegasus carried the warrior Bellerophon to battle against the evil Chimaera. When he really needed an extra boost in speed, he could supercharge using those big wings on his back. Pegasus was enshrined in the night sky as the bearer of Zeus's lightening bolts.

Perseus, The Hero   (In Ptolemy's list of 200 AD)

Perseus set out to bring back the head of Medusa, one of the three snake-haired monsters of old. To equip him for his dangerous journey Pluto loaned him his helmet of invisibility (an ancient cloaking device), Minerva lent him a very shiny shield to blind opponents, and Mercury lent him the famous winged shoes and a diamond sword. With equipment like that he couldn't lose. On his way back with the ugly head, Perseus rescued his future wife Andromeda, who was chained to a rock.

Sagitta, The Arrow   (In Ptolemy's list of 200 AD)

The smallest of all constellations, it could be the emblem of Diane and Apollo, better known as Cupid's Arrow. Arrows were the common missiles of love as well as war in many myths. Sagitta lies in the summer skies near the Milky Way.

Scutum, The Shield   (In Hevelius's list of 1690)

This grouping was called Scutum Sobiescianmum, by Hevelius in 1690, who wanted to honor King Sobieski of Poland. The shield was supposed to represent King Sobieski's Coat-of-Arms. Mercifully, the name was later shortened. The stars making up Scutum are not very bright. The constellation is in the middle of the Milky Way between Sagittarius and Aquila. At a much greater distance than the foreground stars, but in this same direction, is one of the spiral arms of our galaxy, known as the Scutum Arm.

Serpens, The Snake   (In Ptolemy's list of 200 AD)

This constellation is named after the snake Ophiuchus charmed, and from which he extracted medicinal venom. Serpens is another long constellation, extending away from Ophiuchus on both sides. Unique among constellations, Serpens is composed of two separate areas of the sky. The western side is also known as Serpens Caput, the snake's head. The eastern side is sometimes called Serpens Cauda, or snake's tail. To this day, the symbol of the medical profession is the caduceus, which is a winged staff with serpents twined around it. It is from Serpens that the symbol arose.

Sextans, The Sextant (In Hevelius's list of 1690)

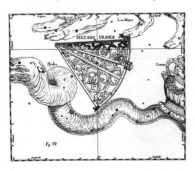

The sextant, or the similar astrolabe, is the star-sighting device used by astrologers to measure the positions of stars. The device was refined in the 15th and 16th centuries for use by navigators at sea. This constellation of faint stars was created by the German astronomer Hevelius in 1687 AD for the device which served him so well in his stellar measurements.

Triangulum, The Triangle   (In Ptolemy's list of 200 AD)

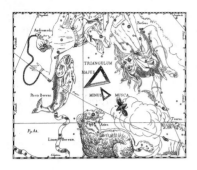

A small constellation of three stars making a triangle just south of Andromeda. In ancient times it was referred to the Delta, the capital letter "D" in the Greek alphabet. It was known to some Greeks as the Nile delta, and to others as the island of Sicily. During Christian times it was sometimes associated with the Trinity, or alternatively with the three-pointed miter hats worn by bishops in the Church.

Ursa Major, The Big Bear Or The Big Dipper   (In Ptolemy's list of 200 AD)

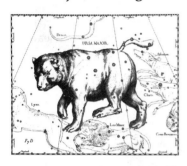

Ursa Major is the best-known constellation in the Northern Hemisphere. Many cultures assigned these stars to the figure of a giant bear. Its seven bright stars are also a physical clustering of stars of common origin. This constellation is circumpolar, meaning it is visible every night of the year slowly circling around the North Star Polaris. Its orientation can be used to tell the time at night.

Many cultures see a figure of a water dipper with a long handle. African-American slaves had folk songs about following the "drinking gourd" at night to travel north to freedom. The two stars on the end of the cup are called the pointer stars, used to locate Polaris. The two stars on the handle side of the cup point south toward the heart of the Lion, the bright star Regulus. The curved handle itself points to the bright star Arcturus in Bootes. If that curve is extended farther it reaches Spica, the bright star in Virgo.

Ursa Minor, The Lesser Bear Or The Little Dipper   (In Ptolemy's list of 200 AD)

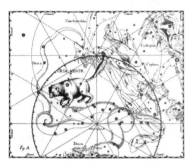

This group is right next to Ursa Major, and is easily remembered by relating the two as either the big and little bears, or the big and little dippers. Also a collection of seven circumpolar stars, the Little Dipper is even closer to the pole. The last star in its handle is in fact the star Polaris that lies less than a degree from the true polar axis of the Earth's rotation.

Astronomers use Ursa Minor a test of sky darkness. There is still too much light pollution if you cannot see all seven stars of the Little Dipper clearly. In most large urban areas with their overuse of street and security lights, the only star visible is often just Polaris itself—and this only with the help of the brighter pointer stars of the Big Dipper.

Vulpecula, The Fox   (In Hevelius's list of 1690)

This is another of the "nouveau" constellations added by the German astronomer Hevelius to his sky atlas. The original name is Vulpecula cum Anser, the Fox and the Goose. It is located next to Cygnus in the summer Milky Way. Perhaps Hevelius meant the Fox with the Swan. Both birds constantly had to be on the lookout for predatory foxes.

Southern Hemisphere Constellations

The constellations listed below are best seen from the Southern Hemisphere since they are below the horizon from latitudes north of 30 degrees. Most of the bright star constellations in the south are located in the southern Milky Way. Argentines and Australians enjoy the full view of the heart of the Milky Way during the longest nights of their winter.

There were several important players in the naming of the southern constellations during the Renaissance.

Petrus Placius was a Dutch astronomer, cartographer and theologian. When the first Dutch expedition set out to explore the East Indies in 1595, he hired the ship's pilot, Pieter Dirkszoorn Keyser, to make observations and fill

in the blank areas around the South Celestial Pole—sort of a moonlighting job. Keyser didn't make it back home, dying in Java, but his catalog of 135 stars was completed and delivered to Placius by his assistant Frederick de Houtman. These early navigators named about ten constellations after exotic birds, reptiles and other animals they had seen on their trip to the Southern seas. Placius drew them on a globe in 1598, and Johann Bayer incorporated the constellation names in his Uranometria in 1603. The rest is history, as they say.

Abbe Nicolas Louis de Lacaille observed and named many of the Southern Hemisphere stars and constellations during an extended stay in South Africa during 1750–1754. He made his observations from the most southerly spot he could get to—the Cape of Good Hope and, amazingly, he did this with a tiny telescope, measuring only one centimeter in diameter. Lacaille was quite a technophile, memorializing about a dozen tools of the day, mostly recent inventions. Many of his names, however, were long and cumbersome. Thankfully, they were shortened to fit better on a sky chart.

Only a few have any attached mythology, since most of this region of the sky is not visible from Greece or Egypt.

Antlia, The Air Pump

(In Lacaille's list of 1752)

This is a small grouping of stars just north of Vela. Originally named Antlia Pneumatica, it honored the latest in mechanical technology, the air pump invented by Robert Boyle. There is no mythology here.

Apus, The Bird Of Paradise

(Named by Dutch navigators Keyser and Houtman on their chart of 1597. In Bayer's list of 1603)

This constellation was named after a tropical bird first encountered by Dutch navigators in Papua New Guinea, who called it Apus Indicus. The birds' skins were first brought to Europe in 1522 by survivors of Magellan's expedition to circumnavigate the world. Apus lies very close to the south celestial pole and is a small circumpolar constellation seen all year from the Southern Hemisphere.

Ara, The Altar

(In Ptolemy's list of 200 AD)

Located south of the stinger of Scorpius, this small group depicts the altar that was erected by the Olympians to give thanks for their victory over the Titans.

Caelum, The Chisel

(In Lacaille's list of 1752)

Named Caela Sculptoris after the tool used by sculptors and stonemasons, this faint star grouping lies between Columba and the southern extent of Eridanus.

Carina, The Keel

(Part of the old Argo Navis, in the IAU list of 1930)

The Greeks named a huge area of the southern sky Argo Navis, tying it to the story of Jason and the ship of his Argonauts. This crown jewel of the spring equatorial skies was unceremoniously dismantled into four parts by practical astronomers—Carina, the keel; Puppis, the ship's stern; Pyxis, the compass and Vela, the sail.

Canopus (Alpha Carinae), the second brightest star in the sky, marks the rudder. Interestingly, Canopus is the only star named in classic times after a historic figure—the pilot of the fleet that sailed the Greek ships to destruction at Troy.

In this direction we are looking along another spiral arm of the Milky Way. Carina contains many interesting objects. The best known is the "star" Eta Carinae. It is an unstable super massive star, about 100 times the mass of the Sun that is shedding it gaseous outer layers. There is a huge cloud of glowing gas that hides our view of the star itself. In the last century Eta Carinae was much brighter than it is today. Because the star is unstable, everyone is waiting for it to go supernova. If it does we will know it. It is only 4,000 light-years away. Nothing this big has exploded so relatively nearby. Some scientists predict a huge flux of lethal X- and Gamma radiation may strike the Earth. Stay tuned!

Chamaeleon, The Chameleon (Named by Dutch navigators Keyser and Houtman on their chart of 1597. In Bayer's list of 1603)

The Dutch navigators Pieter Dirkszoon Keyser and Frederick de Houtman found an unusual family of lizards. These animals had the startling ability to effect a camaflage by changing their colors to match their backgrounds. They were so taken with this strange new reptile that they named a part of the sky for it. The astronomer Johann Bayer approved, and included it in his list, drawing it following the description of the early south sea explorers. The Chameleon is a faint constellation near the South Pole.

Circinus, The Draftman's Compass

(In Lacaille's list of 1752)

The compass is a tool used by surveyors to measure distances, and draftsmen to draw circles. It is next to another tool of the same trades, Norma, the level. This faint constellation lies close to Alpha and Beta Centauri, two of the nearest stars to the Sun and very bright southern Milky Way objects.

Columba, Noah's Dove   (In Royer's list of 1679)

This constellation, first called Columba Noachi, was added by Plancius to honor the homing pigeon or dove that Noah released to find land after the Flood. When the rain stopped, Noah released the pigeon several times until it finally returned with an olive twig from new growth on dry land. The dove is also associated with nearby parts of Argo Navis, which Plancius saw as Noah's Arc, with some seeing it as leading the ship ever westward.

Corona Australis, The Southern Crown   (In Ptolemy's list of 200 AD)

Just as we have a northern crown, so too there is a southern crown. Visible just south of Sagittarius, this crown is named after the olive branch garlands worn by heroes and famous politicians in Greece and Rome.

Crux, The Southern Cross   (In Royer's list of 1679)

This smallest of all constellations makes up in fame what it lacks in size. Crux has bright stars in an easily recognized pattern, making it the best known of the southern groupings. It is incorporated into the flags of several southern countries: Brazil, Australia, and New Zealand to name a few. The entire constellation cannot be seen north of latitude +25 degrees. The Greeks included it as part of Centaurus, but several 16th century explorers viewed it as a distinct group.

The two stars that make the upright post of the cross define a line that points toward the South Pole. Another line from Sirius through Canopus (Alpha Carinae), the second brightest star in the sky, defines a second line that intersects the aforementioned Crux line at about the South Pole. There is no bright star comparable to the northern Polaris at the South Pole. So southern mariners and navigators such as the Polynesians had to develop a more sophisticated knowledge of the sky that northerners.

Crux contains another interesting feature. Australian aborigines and Brazilian Amazon Indians had a distinct name for a round black patch of sky with no stars at all! We call that feature The Coal Sack. It is obvious to the naked eye because it lies right in the middle of the Milky Way. We know today, largely from radio telescopes, that this black patch is really a cold dense cloud of hydrogen gas that blocks our view of the stars behind it. It is located relatively nearby so there are few stars in front of it.

Dorado, The Golden Swordfish

(Named by Dutch navigators Keyser and Houtman on their chart of 1597. In Bayer's list of 1603)

Dorado is supposed to represent a fish called Pompanos, a large, iridescent fish. Hawaiians know it as Mahi-Mahi. It is sometimes called a Dolphin fish.

It also has been depicted as a swordfish. The Spanish name for the large tropical swordfish is in fact Dorado, which is considered a trophy prize by sports fisherman. For astronomers, Dorado is memorable because it contains half of the Large Magellanic Cloud (LMC). The LMC and the Small Magellanic Cloud are distinct star systems, small irregular galaxies that are in orbit around the Milky Way.

In 1987 a giant star in the region of the Tarantula Nebula in the LMC exploded as a supernova. It is the brightest visible supernova in 383 years! It provided unprecedented data to astrophysicists, validating theories of cosmic explosions and nucleosynthesis.

Fornax , The Furnace Or Kiln   (In Lacaille's list of 1752)

A small southern grouping of nondescript stars was named Fornax Chemica, the chemical or laboratory furnace. Tradition has that it honors Antoine Lavoisier inventor of the furnace and one of the pioneers of modern chemistry, who was guillotined in the French Revolution in 1794. Its stars were previously part of the constellation Eridanus.

Grus, The Crane

(Named by Dutch navigators Keyser and Houtman on their chart of 1597. In Bayer's list of 1603)

Grus is a modern grouping, but it is worth mentioning that the crane was a symbol of office of the astronomer in ancient Egypt, perhaps because of its high flight. Also called Phoenicopterus (Flamingo) in the past, its stars were formerly included in Piscis Austrinus.

Horologium, The Pendulum Clock (Named by Dutch navigators Keyser and Houtman on their chart of 1597. In Lacaille's list of 1752)

This is another of the tools of Lacaille. He named it Horologium Oscillitorium, or pendulum clock, in honor of the Dutch physicist Christian Huygens who developed the mechanism for pendulum clocks a century earlier. Huygens had also resolved the rings of Saturn with a telescope.

Hydrus, The Little Water Snake

(Named by Dutch navigators Keyser and Houtman on their chart of 1597. In Bayer's list of 1603)

This triangle of stars lies between the Large and Small Magellanic Cloud near the south celestial pole. It was apparently created as a companion to Hydra, the large female water snake in the northern sky, known since Greek times.

Indus, The American Indian

(Named by Dutch navigators Keyser and Houtman on their chart of 1597. In Bayer's list of 1603)

This is a depiction of an American native holding arrows in both hands. Perhaps the constellation is named after the native peoples of Patagonia and Terra del Fuego, regions which many navigators would have passed on their way to the South Seas.

Upon encountering the natives of the new world, Christopher Columbus mistakenly thought he had landed in the East Indies and called the tribes, Indians. Columbus made several trips to the New World; but apparently never realized that the Indies were still 6,000 miles farther west. Later Italian explorers, like Amerigo Vespucci, eventually proved that the new lands were really distinct continents. That's why they are called the Americas and not the Columbuses.

Mensa, Table Mountain

(In Lacaille's list of 1752)

Named by Lacaille Mons Mensa, or Table Mountain, after the location at which he set up his small 1-cm telescope near Cape Town, South Africa. The stars are not bright; but the constellation includes about half of the Large Magellanic Cloud (LMC), a nearby satellite galaxy of the Milky Way. The other half of the LMC is in Dorado.

Microscopium, The Microscope

(In Lacaille's list of 1752)

This is another of the scientific instruments enshrined in the sky by Lacaille. The microscope, invented in 1590 by a Dutch spectacle maker, allowed scientists to examine the world of the Very Small. More than just a magnifying glass, the microscope reveals the fine structure of nature around us.

Musca, The Fly

(Named by Dutch navigators Keyser and Houtman on their chart of 1597. Renamed by Lacaille on his list of 1752)

These stars were called Apis Australis by Dutch navigators and listed as the Bee (Apus) by the German astronomer Bayer. Lacaille substituted the southern Fly (Musca Australis) for the same stars. This is not an example of miscommunication, rather national rivalries of old.

Norma, The Measuring Square

(In Lacaille's list of 1752)

Just as we have the drawing compass and triangle, so too we have the draughtsman's square right beside it.. Lacaille called it Norma et Regula, or level and square.

Octans, The Octant

(In Lacaille's list of 1752)

The octant was a device for measuring the altitude of a star above the horizon, and thus, one's latitude on the Earth. This tool was invented by John Hadley in 1730 to be used aboard ships at sea as a navigation device. It was superseded by the sextant which is also memorialized in the sky.

Pavo, The Peacock

(Named by Dutch navigators Keyser and Houtman on their chart of 1597. In Bayer's list of 1603)

This is a large constellation near the South Pole. Perhaps the navigators had in mind the myth of the Golden Fleece, since a peacock does appear in it. After the successful voyage of the Argo, its builder, Argus, was transformed into a beautiful peacock, so Pavo has his place in the heavens near to his ship (or at least the pieces of it) in the Southern sky.

Phoenix, The Symbol Of Reincarnation

(Named by Dutch navigators Keyser and Houtman on their chart of 1597. In Bayer's list of 1603)

The Phoenix is part of classical mythology. Many ancient cultures believed that life and history occurred in cycles. The Phoenix was a mythical bird of incredible beauty typifying reincarnation. After living for 500 years the bird built a funeral pyre of twigs from incense bushes and cedar. He then lit the fire, dying in the flames. But from the ashes arose a reborn Phoenix, free from the encumbrances of the past.

Pictor, The Painter

(In Lacaille's list of 1752)

Lacaille's original name was Equuleus Pictoris, the painter's easel (horse, as in sawhorse).

Piscis Austrinus, The Southern Fish

(In Ptolemy's list of 200 AD)

This fish is thought to be the parent of the larger Pisces in the north. Tradition has it that Piscis is drinking from the water pouring out of Aquarius's urn above it. It is distinguished by the bright star, Fomalhaut, Arabic for fish mouth.

Puppis, The Stern

(Part of the old Argo Navis, in the IAU list of 1930)

Formed from the break-up of the old Argo Navis, this is the stern of the ship, the 'poop' deck. The radio source known as Puppis A is the remnant of a supernova from 4,000 years ago.

Pyxis, The Compass

(In Lacaille's list of 1752, part of the old Argo Navis)

This is one of the new constellations fathered by the breakup of Argo Navis. It represents the ship's magnetic compass. Ancient mariners did not have compasses. The navigational uses of the compass were brought back to Europe from China by Marco Polo in the 13th century. Oh well, so much for historical accuracy!

Reticulum, The Crosshairs

(In Lacaille's list of 1752)

Named after the cross hairs, or reticule, in the eyepiece of a telescope.

Sculptor, The Sculptor   (In Lacaille's list of 1752)

Originally named L'Atelier du Sculptor, the sculptor's workshop. The constellation is poorly placed. The sculptor's chisel, Caelum, is not close by.

Telescopium, The Telescope   (In Lacaille's list of 1752)

This unremarkable constellation, first called Tubus Telescopium, is a poor tribute to such an important instrument. The telescope has to be ranked as the most revolutionary instrument of all time. Its use by Galileo caused all sorts of problems for theologians, and Galileo himself. As telescopes became larger, the discoveries multiplied. In the last century the reflecting telescope has grown to 10 meters (400 inches) in diameter. The decade of the 1990's saw the fantastic pictures and data from the Hubble Space Telescope. The impact on astronomy and cosmology of the HST has been equal or greater than Galileo's first instrument.

Triangulum Australe, The Southern Triangle

(Named by Dutch navigators Keyser and Houtman on their chart of 1597. In Bayer's list of 1603)

This triangle is nearly equilateral and has no mythology attached to it.

Tucana, The Toucan

(Named by Dutch navigators Keyser and Houtman on their chart of 1597. In Bayer's list of 1603)

Named after the family of fruit-eating birds with an unbelievably large brightly-colored bill found in Brazil and tropical rainforests in the Americas. This constellation contains within its boundaries the Small Magellanic Cloud and perhaps the most beautiful globular cluster in the sky, 47 Tucanae.

Vela, The Sail

(Part of the old Argo Navis, in the IAU list of 1930)

This ragged circle of stars represents the fully opened sail of the old Argo Navis, billowing as it catches the wind. It has the modern distinction of having the second recognized optical pulsar.

Volans, The Flying Fish

(Named by Dutch navigators Keyser and Houtman on their chart of 1597. In Bayer's list of 1603)

Originally named Piscis Volans after the tropical fish that can jump out of the water and sail in the air for several meters before diving back under. Presumably the Dutch navigators saw schools of these fish in the south seas just as they were racking their brains for a zippy name of a newly-created pattern in the sky.

Like the dove, Columba, it is associated with the ship Argo Navis. Off the coast of western Mexico, sports fishermen try to catch the flying fish as bait for the really big ones, the swordfish, Dorado. Appropriately Volans is right next to Dorado in the southern sky.

The Development of the Science of Astronomy

We can divide the development of the science of astronomy into two historical periods; a Classical and a Modern. The Classical period began with Isaac Newton and the publishing of his book *Philosophiae Naturalis Principia Mathematica* in 1686. The Modern period began with Albert Einstein's seminal paper on Special Relativity in 1905.

The Grandfathers of Astronomy

Marking the beginning of modern astronomy with Isaac Newton is not meant to depreciate the contributions of earlier scientists. Newton himself reflected that his work would not have been possible without the help of his precursors. "I stood upon the shoulders of giants….", he said. Some early astronomers had great insight.

Eratosthenes (276–194 BC) measured correctly the circumference of the Earth. Yes, we have known the Earth was a round globe for some time!

Aristarchos Of Samos (310–230 BC) proposed that the Sun was the center of the universe and that the Earth and planets revolved around it.

Aristotle argued against this model because there was no back-and-forth motion of the stars detected. The argument, a good one, was that if the Earth moved in a circle around the Sun, then the stars would display parallactic motion. This is like looking at a distant object first with one eye and then the other. The object appears to move back and forth with respect to the far horizon. No one guessed how far away the stars are. Stellar parallaxes were not detected until the 19th century.

Aristarchos determined the distance to the Moon using trigonometry. He got the right answer, too! He made a stab at the distance to the Sun by measuring the angle between it and the Moon at First and Third Quarter phases. But without a telescope and an accurate clock, this is a very hard measurement to obtain. His finding was that the Sun was at least 20 times farther away than the Moon and therefore at least 20 times larger as well. The real answer is that the Sun is 400 times farther away. But science is not about being right; it's about performing experiments and honestly reporting the results. If they are wrong, some other scientists will improve upon the measurements and report what they find.

Copernicus (1473–1543) revisited the work of Aristarchos and also proposed the heliocentric model of the solar system. He backed up his theory with hard numbers, derived by applying geometry and mathematics to the best available visual data of the motions of the planets.

Tycho Brahe (1546–1601) lost his nose in a duel over 'honor' or some woman. But no matter. Astronomy takes eyes, not noses. He constructed new sighting instruments and proceeded to record accurate stellar and planetary positions of over 1,000 stars. In 1572 he saw a new bright star appear. That object is known as Tycho's supernova. In 1577 he tracked a comet across the sky and showed that it had to be much farther away than the Moon. Before this most people thought that comets were phenomena within the Earth's atmosphere.

Kepler (1571–1630) was a mathematician and an astrologer. He was a strange mystical character who developed the three laws of planetary motion. Kepler used Brahe's data, particularly the motion of the planet Mars, to derive his theory.

Galileo (1564–1642) did experiments in the physics of motion and built and applied the telescope to celestial observations. He discovered the craters on the Moon, the lunar-like phases of Venus, spots on the Sun, the motions of the four brightest satellites around Jupiter (still called the Galilean moons), and some say the rings of Saturn. He was arrested for heresy because he published his observations in support of the Copernican heliocentric model. He narrowly avoided being burned at the stake!

Classical Astronomy

ISAAC NEWTON (1642–1727) was a brilliant introvert whose insight into both astronomy and mathematics changed the course of human history. He solved long-standing problems in physics and astronomy by inventing calculus. For example:

★ The three laws of motion
★ Law of Universal Gravitation
★ Precession of the Earth
★ The non-spherical shape of the Earth

★ Circular motion
★ Planetary orbits—revised Kepler's laws
★ The tides
★ The orbits of artificial satellites around the Earth

Perhaps the most important was his Law of Gravity. This is the force that holds the Cosmos together. Newton's mathematics showed that if the positions and velocities of every object in the universe were known, then every future state of the universe could be calculated. Now that's scary!

Newton believed that for every Effect there must be a Cause. A consequence of his theory was to impose absolute fatalism upon the Cosmos, and by implication, on the human species as well. Nothing happens by chance.

Newton's *Principia* was actually a compendium of problems he had worked on for more than 20 years. He published his work largely at the urging of the Astronomer Royal, Sir Edmund Halley. Newton was able to show Halley that the comet that now bears Halley's name was a repetitive visitor to the inner solar system with a period of about 76 years.

Because of the mathematical tools developed by Newton, theories and discoveries in physics and astronomy

mushroomed. The planets Uranus, Neptune, and Pluto were discovered by applying gravitational theory.

Newton built the first reflecting telescope, the basic optical design still used in the giant 10-meter mirror telescopes today. The world model inspired by Newton was an infinite universe moving and evolving in absolute time according to the forces of gravity. The universe had no beginning and no end. Time was infinite in extent, as well.

Newton had spent a great part of his life doing alchemy. Alchemists thought that by mixing the correct acids and chemicals that any element could be transmuted into another, such as lead into gold. It turned out to be a little more complicated. Chemical processes only rearrange the outer electrons in atoms. In the 19th century scientists discovered the existence of atoms and organized that knowledge around the Periodic Table of Mendeleev. Not until the 20th century discoveries of the nature of the atomic nucleus did transmutation becomes possible— and then only at the expense of high energy.

The last part of the Classical Period saw the development of electromagnetic theory by scientists like Oersted, Faraday, and James Clark Maxwell. Electrical charge and current flow seems at first sight to be a different phenomenon from magnetism. But the linkage between the two was discovered and finally reduced to a mathematical theory—Maxwell's Four Equations. The ability to generate and transmit electricity was an immediate practical application. Steam engines were replaced by electric motors in the factories. Edison invented the electric light bulb. The theory also predicted the existence of electromagnetic waves of energy that travelled at a rapid but finite speed—186,000 miles (300,000 km) per second. This was the same as the finite speed of light determined by astronomers from observations. Finally we had stumbled onto the nature of light!

The theory predicted the existence of other forms of electromagnetic radiation from the shortest gamma rays to the longest radio waves. The full spectrum is now used to investigate our Cosmos.

The Modern Period

There were some mathematical and philosophical conflicts between Newtonian physics and Maxwellian electromagnetic theory. Scientists don't like discrepancies. If there is even one exception to a theory, then the theory is simply wrong, or at least incomplete. A revised or different theory evolves from new experimental tests. Since the Earth moves in its orbit around the Sun at some 30 km per second, it should be possible to see a change in the speed of light between the direction of motion and the opposite direction. If light travels at the speed X in an unmoving laboratory, then in the direction of orbital motion light should travel at X + 30 km per second. In the direction away from the orbital motion light should travel at X -30 km per second. Like walking in a moving train, one travels faster walking toward the front of the train and slower walking toward the back.

Logical? Yes, but when two American scientists built a device to test this prediction, they found no difference in any direction. Michelson and Morley reported that they had nothing but negative results—nothing, zip, nada. It was the most important null experiment of the 19th century. Why didn't light obey the laws of (Newtonian) physics?

The mathematician and physicist who solved the puzzle was Albert Einstein (1879–1955). While working as an unknown Swiss patent clerk he published a visionary solution known as the Special Theory of Relativity. It was really quite simple. The discrepancies between Maxwell and Newton were due to our definition of Time. Time was not absolute; but relative to, and dependent upon, the relative velocities of two events. Simply put, nothing can travel faster than the speed of light. The speed of light, c, is a fundamental constant in the universe around us. Deal with it. One of the predictions of the theory was that a rapidly moving clock appears to run more slowly as seen by an observer not moving with it. Very non-intuitive! But wait, it gets better.

Suppose you have two cosmonauts on the Earth who are twins. Let one cosmonaut stay on Earth and the other travels off to a star 10 light years away at the speed of light. He explores around that star for a year, and then returns to Earth, again traveling at the speed of light. The two cosmonauts are reunited upon his return to Earth. Amazingly they discover that the twin that went traveling, 21 years previously, is only one year older than when he left. The grounded twin, however, aged 21 years!

Einstein must have had a lot of idle time at the patent office because he also published papers about the photoelectric effect (he got the Nobel prize for this one), how Brownian motion validated the theory of atoms, and the equivalence between matter and energy—the now-famous $E=mc^2$. People noticed. In 1913 Einstein got a real job in Berlin commensurate with his talents. War broke out a year later, but Einstein continued to expand on his theories.

In 1916 Einstein published his most radical theory, the Theory of General Relativity. Because of the difficult math, most scientists didn't understand it. Einstein proposed that gravity, Newtonian gravity, was an illusion. Planets moved in their closed orbits because the Sun's mass curved the space around it. This distortion of Space-Time bent the otherwise straight paths of the planets into closed ellipses. One of the predictions of the General Theory was that light itself would be bent by the same distortion of Space-Time. Many thought he had overdosed on math.

Einstein's theory, in spite of its complexity, made predictions that could be tested observationally. A British team of astronomers set out to see if the bending of light near the Sun could be observed during a total solar eclipse in 1919. When the team analyzed the data, they reported they had detected Einstein's Gravitational Lens effect! The whole story is remarkable given the fact that Einstein was a German scientist and on the loosing side in WW I, just ended in 1918. This one experiment made Einstein an instant world celebrity. He had discovered a fundamental property of the Cosmos that would affect theory in physics and astronomy from then on.

The Size and Contents of the Universe

Einstein's theoretical work attracted attention, but it took decades of detailed work by astronomers to verify it, and fully comprehend the magnitude of its implications. Some built bigger and better telescopes with which to search the skies. Others gathered fundamental knowledge about the birth and death of stars, their lifetimes, and the nuclear processes that provide the energy of stars. Understanding these nuclear processes eventually led to the development of the Atomic and Hydrogen Bombs in 1944 and 1948. Well, no science is totally pure.

In 1914 an American astronomer at the Flagstaff Observatory in Arizona, Vesto Slipher, published a survey of bright galaxies showing that their spectral lines indicated the vast majority were moving away from us. This was mysterious because at that time astronomers believed that galaxies, or nebulae as they were known, were small-ish collections of gas and stars — perhaps new solar systems in formation.

After returning from the army in WW I, Edwin Hubble (1889–1953) began using the largest telescope in the world, the 100-inch atop Mt. Wilson, to study the distances to galaxies. He started with the biggest and brightest, the Andromeda Nebula. By carefully examining photographs taken at different times he saw the brightening and dimming of variable stars known as Cepheid Variables. Bingo! The Andromeda Nebula was really the Andromeda Galaxy located more that a million light years away. It was not inside our Milky Way at all. The Milky Way was only about 100,000 light years across, from edge to edge. Andromeda was a distant and independent galaxy!

Hubble extended his study to as many more galaxies as his 100-inch (2.5 meter) telescope could reach. By 1929 it was obvious that he needed a bigger telescope. George E. Hale, another astronomer at Mt. Wilson and Caltech, obtained a grant from the Rockefeller Foundation to build a 200-inch telescope and place it at Mt. Palomar near San Diego, at that time a much darker site than any in the Los Angeles area.

The Universe—Huge and Still Growing

Hubble had determined the distances to dozens of galaxies. Using those distances and the velocity data from spectral lines, he derived the Hubble Law of Recession — galaxies appear to be racing away from us in proportion to their cosmic distance. The universe was expanding.

The universe was also much bigger than anyone had suspected — hundreds of millions, perhaps billions of light years in extent. Slowly came the realization that we do not see the universe as it exists today! The light that we see from distant galaxies left those sources millions and billions of years ago. The sky is a giant window into the past. Instead of a three-dimensional universe, we must also include the fourth dimension, Time, in the understanding of our place in the Cosmos.

When Einstein met Hubble in 1929, he was very enthusiastic about Hubble's observations. "That's what I've been telling you," Einstein basically said. He tried to explain that General Relativity math predicted that the universe was unstable and therefore must either expand or contract. Hubble really didn't understand what Einstein was talking about. He was a skilled observer, not a theoretician.

Einstein had developed the theory in 1916 when astronomers did not know of the expansion. The universe looked stable and unchanging. To keep it that way Einstein had added a new repulsion force to his equations to counter the pervasive attraction of gravity. When Einstein found out in 1929 that the observations showed an expanding universe, he changed his theory to accommodate Hubble's data.

Radio And Space Astronomy

After WW II many of the scientists involved in the development of Radar and radio communication technology turned their antennas skyward. Radio technology and the development of the digital computer have gone hand in hand. In 1962 the digital radio receiver was developed along with dozens of improvements in low noise amplifiers. Today we have radio telescopes that take observations over the entire spectrum, from submillimeter to meter wavelengths. Those data are used to study objects that are invisible to us, such as large clouds of gas and dust. The astronomical data from these instruments complements optical observations.

In the decades since 1960 we have launched into space scores of astronomical satellites and space probes. Some have taken photographs of distant planets, asteroids and comets. Others have scanned the sky in the high-energy wavelengths that can't be observed from the ground.

The result is that we now have maps of the sky showing X-ray sources, the ultraviolet and the near and far infrared. And this is in addition to the very fine resolution visible light photos taken by the appropriately named Hubble Space Telescope.

With these additional tools astronomers have detected and mapped the Cosmic Microwave Background radiation that comes to us from events nearly 14 billion years ago! We have found stars made of pure neutrons that rotate 1,000 times a second and emit radio wave pulses as they spin. We have discovered Black Holes in the centers of the majority of spiral galaxies. And we have realized that most of the stuff in the universe is dark matter of an as-yet unknown nature.

Black Holes

Using Newton's law of gravity, theoreticians as early as 1783 had constructed a mathematical model of a body so massive that the escape velocity from its surface exceeds the speed of light. Because of this, the object cannot give off anything, even light. Such a body would be extremely compact and the matter inside it impossibly dense. But the physics of super dense bodies attracted no attention until Einstein's General Relativity theory was published in 1916.

General Relativity is a redefinition of gravity, and the math is complex. One of the first solutions to Einstein's equations was made by the German physicist, Swartzschild. His solution described the physical characteristics of a massive body, later to become known as the Black Hole. An American astrophysicist, also became interested in the subject. The father of the atom bomb, J. Robert Oppenheimer, showed in 1939 that matter would fall into a singularity, a dimensionless point in the center of a Black Hole. No known nuclear forces could resist the pressure.

The first indication that Black Holes might be real came from X-ray observations by a satellite telescope in the late 1960's. Here we saw a binary star — one normal star and one invisible companion — in a cosmic dance. The invisible Black Hole is tearing off material from the normal star. As the matter falls into the Black Hole, it emits huge quantities of X-rays. After the Hubble Space Telescope was launched in 1990, its cameras and spectrographs detected the rapid motion of stars near the center of many galaxies. It turns out that Black Holes are common mass concentrations in the centers of most galaxies. We detect them by the tremendous gravitation force they exert on stars and gas in their immediate neighborhoods.

Science fiction writers and movie producers have had a field day with Black Holes. They portray these objects as huge vacuum cleaners sucking in everything around them. Another image is that a Black Hole is a window into another dimension or, perhaps, another universe. In reality the matter collected into the Black Hole stays right there. It doesn't disappear from our universe at all. The gravitational effects of the Black Hole just keep growing and growing and growing.

Astronomy Today and Tomorrow

Our knowledge of the Cosmos is extensive but far from complete. We have a theory of the origin of the universe that is consistent with nearly all observations. The Big Bang theory is detailed and incorporates not only astronomical data but also high-energy particle physics and atomic theories. Simply put, the Cosmos had a beginning; it is not eternal. It came into existence 15 billion years ago and has been expanding and cooling off ever since. Ordinary matter, like the stars and us, is really only a minor fraction of the matter-energy stuff that the universe contains. Some say that ordinary matter is just 5 percent of the universe. Another 25 percent is some sort of Dark Matter; and the biggest component is called Dark Energy and is 70 percent of the Cosmic mix. This is hard to fathom.

The Cosmos is still expanding and seems to be expanding faster today than it did 10 billion years ago! But, the data are still few, so this picture may change. The Hubble Law of Recession is no longer interpreted as distant galaxies actually moving away from us at near the speed of light. Instead, the redshifts we see are really the result of cosmic expansion, the stretching out of the universe in all directions. The universe itself has grown during the billions of years it has taken the light from distant galaxies to reach us. Light waves themselves are stretched out as a consequence.

Eventually, using new and improved telescopes in space and on the ground, we will collect a more complete inventory of the things in the universe. We will build larger optical and radio telescopes. We will send more instruments into space. And we will construct neutrino and gravity wave telescopes. This will certainly bring startling discoveries as well as refinements of established theories. Astronomy is a dynamic and fascinating science.

Possible Life on Other Planets

In our solar system we have found forms of life only on planet Earth. Mars may have rudimentary life forms according to some astrobiologists, since a meteorite from its surface was found to have suspicious signs of fossils. On Europa, one of Jupiter's moons, a giant ocean containing some aquatic life may be present under its ice covering. But nowhere else in our system of planets does there seem to be life today. "Intelligent" life may have only developed on Earth, and it is us.

When one examines the billions of Sun-like stars in our Milky Way we find that many have planetary systems. Over 80 planets have been detected around nearby stars as of the date of this writing. But do any of them harbor intelligent life forms like us, or are we really alone? We have the technology today to send and receive radio messages as far away as 100 light years.

Two important consequences may result if we ever discover intelligent life out there. First of all, these friendly extraterrestrials might be willing to share their knowledge and technology with us. Why is this important? We certainly have problems affecting the well being of the human race on this planet. There may be ways to solve our conflicts without self-destruction. There could also be miracle cures to our diseases.

Secondly, we know that the lifetime of our middle aged Sun is finite. Changes will occur in the future that will make the Earth a hostile environment for our species. There will be long-term global warming as the Sun ages and heats up. But long before that occurs we humans will face self-generated problems of over population, limited food production, air and water pollution, and possibly mass extinction due to a collision with a stray asteroid or comet. Right now all of our eggs are in one basket. We will either become extinct or have to find a new planet or planets elsewhere to live on. Friendly extraterrestrials may have faced the same problem in their past. International Star Registry

International Star Registry®

The following section of this book is a catalog of new star names compiled by the International Star Registry (August 2004 to July 2007). All star names are arranged in alphabetical order and include telescope coordinates. It is much easier to find a star within a constellation than to decipher its Astronomical position in the sky. It was for this reason that during the Astronomical Congress of 1928 eighty-eight constellations were officially recognized. The boundaries are described in Atlas Celeste published at Cambridge, England in 1930. Immediately preceding the catalog is a list of Abbreviations for these various constellations.

In order to locate your personalized star refer to the charts and coordinates which you received in your star package..

In the course of compiling the catalog of names, it is possible that errors or omissions may sometimes occur. If you find any such oversights please notify the International Star Registry and corrections will be included in the next printing.

Andromeda	AND	Cygnus	CYG	Pavo	PAV
Antlia	ANT	Delphinus	DEL	Pegasus	PEG
Apus	APU	Dorado	DOR	Perseus	PER
Aquarius	AQR	Draco	DRA	Phoenix	PHO
Aquila	AQL	Equuleus	EQU	Pictor	PIC
Ara	ARA	Eridanus	ERI	Pisces	PSC
Aries	ARI	Fornax	FOR	Piscis Austrinus	PSA
Auriga	AUR	Gemini	GEM	Puppis	PUP
Bootes	BOO	Grus	GRU	Pyxis	PYX
Caelum	CAE	Hercules	HER	Reticulum	RET
Camelopardalis	CAM	Horologium	HOR	Sagitta	SGE
Cancer	CNC	Hydra	HYA	Sagittarius	SGR
Canes Venatici	CNV	Hydrus	HYS	Scorpius	SCO
Canis Major	CMA	Indus	IND	Sculptor	SCL
Canis Minor	CMI	Lacerta	LAC	Scutum	SCT
Capricornus	CAP	Leo	LEO	Serpens	SER
Carina	CAR	Leo Minor	LMI	Sextans	SEX
Cassiopeia	CAS	Lepus	LEP	Taurus	TAU
Centaurus	CEN	Libra	LIB	Telescopium	TEL
Cepheus	CEP	Lupus	LUP	Triangulum	TRI
Cetus	CET	Lynx	LYN	Triangulum Australe	TRA
Chamaeleon	CHA	Lyra	LYR	Tucana	TUC
Circinus	CIR	Mensa	MEN	Ursa Major	UMA
Columba	COL	Microscopium	MIC	Ursa Minor	UMI
Coma Berenices	COM	Monoceros	MON	Vela	VEL
Corona Australis	CRA	Musca	MUS	Virgo	VIR
Corona Borealis	CRB	Norma	NOR	Volans	VOL
Corvus	CRV	Octans	OCT	Vulpecula	VUL
Crater	CRT	Ophiuchus	OPH		
Crux	CRU	Orion	ORI		

A
Gem 6h52'52" 21d0'40"
A 10
Cnc 8h48'23" 18d42'40"
A & A
Oph 17h0'36" -0d1'55"
A. A. Dooley
Per 4h14'23" 32d6'35"
A. Albert Articus Fantasticus 1922
Lib 15h27'43" -7d8'53"
A and J Always & forever!Happy 19th
Cyg 20h16'30" 50d43'33"
A and R Carvalho
Umi 15h26'37" 72d54'47"
A/B Embree
Ari 2h40'11" 25d43'58"
A Beautiful Blue Jewel
Lyn 6h45'26" 54d44'34"
A Beautiful Mind - A Best Friend
Cyg 19h53'48" 38d57'54"
A Beautiful Star for BBE
Cas 1h29'13" 69d23'52"
A. "Belton"
Ari 2h32'13" 11d15'23"
A Bicycle Built for 2- Tiff and Bev
Cyg 21h23'28" 46d35'52"
A Blessing in A Curse
Dra 19h25'17" 64d22'17"
A Blue Moon in Your Eyes
Her 17h54'48" 26d8'7"
A. Broad Dragonshire
Lib 15h37'15" -13d25'56"
A. Burleigh Oxton
Uma 12h58'49" 56d49'4"
A & C Goodfellow - Forever & Always
Gem 6h20'49" 20d12'7"
A Cat Called Henry
Leo 11h13'32" -0d3'38"
A Charles Miller Star
Ori 6h4'8" 10d23'50"
A Cherished Lady - Doris Watson
Gem 7h4'45" 15d55'47"
A Child's Gift *J.D.R.R.J.M.6*
Leo 11h8'24" -0d27'20"
A Child's Light
Per 3h33'2" 33d34'38"
A Circle Never Ends
Lmi 9h25'55" 33d44'27"
A Clouet & Co Kl Sdn Bhd
Ori 5h36'25" 1d51'56"
A Corini Star
Psc 1h45'24" 8d24'13"
A Cowboy And His Wolf
Per 4h25'46" 40d59'1"
A Cut Above the Rest
Gem 7h42'17" 33d48'27"
A & D Parker
Sge 19h54'40" 16d48'9"
A & D Wolent 05.21
Cyg 20h24'6" 46d55'32"
A Daughters Joy
Cas 1h18'43" 53d23'58"
A Daughter's Love For Her Daddy
Her 17h7'52" 30d39'40"
A. David Amico
Cnc 8h39'30" 30d55'32"
A. David Kornbluh
Her 16h43'55" 6d3'51"
A Day to Remember
Gem 6h31'34" 24d35'56"
A December Dream
Cam 3h52'42" 70d19'41"
A Diamond For Your Ring
Ori 6h13'36" 2d13'40"
A Doug Jones Riff
Uma 9h39'7" 42d15'17"
A Dreamer's One Wish
Cnc 8h24'22" 11d26'0"
A. Druilhet
Psc 1h32'32" 23d42'28"
A E Magee
Umi 14h31'21" 76d56'29"
A F - M B T - OompaBean
Sco 17h38'14" -42d2'8"
A. F. unit
Sgr 18h10'56" -29d35'49"
A F W
Uma 9h19'14" 52d10'2"
A FA LA FA GA PHIGUS
Dra 18h6'12" 70d23'30"
A Families Dream
Sco 16h9'20" -15d43'9"
A Firebird for Liv
Pho 1h43'7" -46d23'13"
A First Wish
Leo 10h8'35" 20d28'16"
a&fo1
Ori 5h43'31" 5d44'57"
A Forever Kind of Love
Cas 1h15'44" 56d31'9"
A. Frances Moore
Ari 3h7'12" 29d39'28"
A Friend Is A Second Self
Peg 22h36'21" 15d35'40"
A. George Ottavino
Aqr 21h47'10" 2d1'34"
a ghrá mo chroí
Tau 5h30'52" 20d33'55"
A ghrá mo chroi
Sge 19h44'40" 17d31'40"
A Ghra' Mo Chroi (Love Of My Heart)
Aqr 21h47'59" 1d25'44"
A Gold'n Diva (Camille Sanon)
Psc 1h51'40" 7d48'43"
A Grigg & Erin O'Reilly 16 Sep 2000
Cru 12h36'13" -58d28'58"
A Guiding Light For Our Marriage
Pyx 8h49'48" -30d2'27"
A. H. "Bunky" Aldred
And 2h32'9" 49d25'10"
A Heart So Big
Uma 11h17'6" 29d23'6"
A Hoffman
Uma 11h57'41" 63d28'37"
A. Hope Stone
Leo 11h21'21" 16d30'4"
A. J.
Cnc 8h20'55" 24d50'7"
A J
Sex 9h45'17" -0d41'26"
A J Augie Duran
Uma 8h25'58" 66d36'13"
A. J. Boyd
Oph 17h33'28" -0d14'24"
A. J. Gould
Aql 19h17'16" 0d22'31"
A. J. Harpring
And 2h24'2" 50d32'11"
A. J. Linville
Uma 8h19'9" 65d45'53"
A. J. Roy
Uma 11h53'2" 59d57'50"
A J 's
And 0h32'19" 39d1'44"
à jamais
Crb 15h46'14" 27d27'56"
A jamais
Ori 6h5'30" 17d11'34"
à jamais aimer
Tau 4h19'29" 3d47'46"
A Jewel in the Sky
Mon 7h21'51" -3d43'4"
A. John Merola M.D.
Boo 15h20'34" 44d55'8"
A. Jordan Doughtie
Leo 11h55'44" 19d2'24"
A. K. Carreiro
Uma 13h53'55" 60d59'52"
A két testvér, Zsozso és Szabolcs
Uma 13h55'41" 50d32'22"
A két testvér, Zsozso és Szabolcs
Uma 13h50'55" 51d21'15"
A Kiss For Brian's Heart
Lib 15h4'6" -23d57'28"
A Kiss for Chrissy
Sgr 18h38'14" -27d10'16"
A Kiss to Build a Dream On
Aqr 23h21'49" -16d43'9"
A Knight & His Damsel
And 0h34'40" 45d43'45"
A L O N I
Eri 3h54'4" -0d43'11"
A L R E N
Ori 5h24'11" -2d45'11"
A&L Rouse
Aql 19h3'16" 8d59'18"
a. La-Verne Mickle
Aqr 22h12'41" 1d9'20"
A leggyönyörubb csillag: Rita
Lib 15h33'43" -10d17'45"
A Light for Ali
Psc 23h58'17" 5d58'34"
A light in our lives JAPollick
Lyn 7h43'28" 46d4'22"
A little bit...
Her 17h19'11" 37d31'20"
A Little Brother's Lasting Love
Lib 15h8'34" -12d24'11"
A Little "Cowboy" Love
Vir 12h23'35" -0d9'49"
A Little Piece Of Heaven
Gem 6h33'40" 21d20'41"
A. Lona & Gertrude Arnold
Cyg 21h10'21" 48d51'46"
A. Louise Wolfe
Cnc 8h52'38" 27d48'17"
A Love That Will Burn Forever L&M
Aql 19h24'29" 2d28'15"
A Love Written in the Stars
Cyg 20h41'47" 46d2'38"
A Lovely Chance
Cam 4h34'51" 67d23'12"
A Lovely Lady
Cas 1h32'25" 60d29'55"
A Lover's Duet
And 0h39'1" 34d8'13"
A Lovers Jupiter
And 0h41'23" 43d32'50"
A Loving Mother:Jacqueline Casselli
Cas 1h15'5" 64d55'15"
A Lucky Star Called Nicole Cichetti
And 0h54'21" 37d56'5"
A lyttle ray of lyght and love
Lyn 9h2'45" 34d59'23"
A&M 090302
Leo 10h21'15" 21d0'36"
A. M. A. T. W
Cru 12h17'51" -57d25'29"
A M - Matt - N S
Uma 10h59'32" 47d42'14"
A&M Maxwell
Cyg 20h37'10" 48d4'5"
a ma vie de coer entier
Uma 8h37'42" 69d25'25"
a ma vie de coer entier N&M
Cyg 19h26'11" 52d33'39"
A Match Made In Heaven
Her 16h49'54" 22d35'29"
A. McCardle Jr.
Sco 17h52'29" -35d47'58"
A mi mami
Lyn 7h33'56" 35d46'5"
A Mi Szüleink: Panni és Pisti
Uma 9h22'8" 72d13'53"
A Milky Way For Mom
Gem 7h19'49" 28d0'30"
A Million I Love You's
Leo 11h45'34" 22d34'55"
A million wishes for Matt
Uma 10h11'41" 56d39'0"
A Monkey On A Rock
Ari 2h43'55" 27d33'43"
A Mothers Endless Love
Psc 1h40'4" 18d33'45"
A Mother's Heart - Ethelyn Baima
Tau 5h16'59" 25d6'48"
A Mothers Infinite Love, Pat Nila
Gem 7h41'4" 25d16'26"
A Mother's Love
Vir 12h8'40" 7d4'34"
A Mothers Love
Lyn 9h21'17" 41d8'40"
A Mother's Love
Uma 9h30'1" 49d15'44"
A Mothers Love
Cas 1h8'21" 63d35'38"
A Mothers Love
Sco 16h45'53" -29d15'28"
A Mother's Love for Stephanie
Sgr 18h41'50" -29d58'58"
A Mother's Love/ Liu-Haller
Cas 0h40'38" 61d46'47"
A Mother's Prayer
And 1h0'36" 38d25'31"
A+MPowell3699
Peg 22h48'50" 33d13'52"
A&N Bear
Uma 10h27'49" 66d48'35"
A N C Y
Sgr 18h18'8" -23d17'56"
A New Beginning
Cas 23h1'13" 57d0'50"
A New Beginning
Uma 8h34'25" 47d1'44"
A New Beginning
Sge 19h42'19" 18d5'12"
A New Beginning (For The Gaia's)
Psc 1h48'25" 9d59'36"
A New Beginning Roberto & Christina
Cep 22h5'43" 71d51'12"
A Newfoundlander's Dream
Aqr 22h41'39" -10d21'22"
A Night with Diane and Brian
Uma 12h49'35" 62d33'25"
A. Pascal 60
Vir 13h48'58" -6d14'56"
A PEACE OF DAD IN THE SKY
Her 16h55'0" 34d14'5"
A Pedra
Uma 9h2'48" 62d50'29"
A Perfect Forever
Ori 6h20'40" 5d52'58"
A Perfect Love
Uma 9h55'12" 66d55'59"
A Perfect World
Dra 14h57'36" 55d36'49"
A. Peter Russillo
Aur 5h46'42" 49d10'19"
A Piece Of Heaven
Ari 2h50'46" 25d25'46"
A Place for Keano Beanos
Uma 11h5'31" 48d7'4"
A place in the sky for Paul &Yvonne
Cyg 19h36'37" 29d32'19"
A Princess and her Soulmate
And 0h23'28" 26d36'36"
. . . a promise made for life
Ori 5h52'41" 21d48'43"
A Promise On Us
Aur 5h36'30" 40d29'25"
A Promise To Love Matthew & Pamela
Cyg 21h26'9" 36d1'32"
A Promise-Maria Joanna Roupen Roland
Uma 9h7'29" 69d7'43"
A. R. C 1935
Cru 12h43'40" -56d54'3"
A. Raiola
Her 17h31'45" 30d53'5"
A. Ralston
Cap 21h42'17" -18d27'18"
A Rietsch for the 21st Century
Ori 5h38'13" 3d9'40"
A & S
Cyg 20h51'39" 43d17'57"
A Scottish Soldier
Cyg 21h40'23" 44d46'57"
A Simple Life
Cap 20h52'58" -25d9'13"
A Simple Wish
Uma 14h26'57" 62d9'22"
A Sirius
Psc 23h53'35" 3d0'44"
A souvenir from above
Cnc 8h17'56" 10d38'26"
A Special Star
Psc 23h45'44" 5d24'2"
A Star As Beautiful As Summer
Aqr 21h47'3" -3d7'26"
A Star for Alex
Cyg 20h39'26" 50d31'12"
A Star for Amy
Psc 0h14'49" 11d13'0"
A Star for an Astronomy Lover Jon
Per 4h49'22" 49d38'5"
A Star for Crystal to gaze upon
Cnc 8h23'31" 8d0'19"
A Star for Elysse - "Elysse Dawn"
Cru 12h8'52" -58d40'14"
A Star for Jill and Annelise
Cyg 21h56'43" 38d33'44"
A Star for Kodi from Mom
Aql 19h36'34" 2d33'51"
A Star for Molly's Dad Mike
And 0h15'23" 29d46'1"
A Star for My Lover Tonya
Lib 15h5'15" -26d55'58"
A star for Peter & Helen: *Eternal*
Lyn 7h16'35" 56d37'51"
A Star for The Prisoners of Hope
Cyg 19h44'41" 38d33'6"
A Star For You, For You Are Mine
Cra 18h22'37" -37d19'49"
A star from Richie for Wendy Rau
Psc 1h11'31" 29d55'18"
A Star in Lara's Heart
Uma 11h26'53" 65d9'21"
A Star In Our Galaxy -Matt and Linz
Cnc 9h9'27" 28d10'7"
"A Star is Baum" for Phyllis & Joe
Cyg 19h45'9" 53d44'30"
A Star is Born - Bob McCoy 1988
And 0h37'13" 42d55'6"
A Star is Born - Jan McCoy 1952
And 0h9'31" 44d49'7"
A Star is Born - Kelly McCoy 1993
And 0h22'26" 44d19'38"
A Star Is Born - Mimi 4/2/67 6:44AM
Ori 5h16'25" 9d10'38"
A Star is Born - Sean McCoy 1984
And 0h45'26" 41d23'54"
A Star IZ Born
Dra 17h25'35" 52d25'11"
A Star Mistaken For A Plane
Uma 11h18'22" 68d22'39"
A Star Named " G " - Gerri Urra
Cyg 21h34'58" 52d44'39"
A Star Named Mavis Boyle
Uma 10h17'22" 67d46'39"
A Star Of Our Own - K P Forever
Cru 12h26'39" -61d13'40"
A Star of Sanford (2005)
Cep 21h23'50" 64d17'24"
A star to dance with
Lyr 18h43'32" 38d40'4"
A star to the left of where we met
Uma 11h36'4" 28d28'5"
A Summers Love
Cnc 8h48'56" 6d45'14"
A Symbol Of Our Immortal Love-D&G05
Ara 17h22'40" -51d56'29"
A Tavernier Star
Uma 12h37'28" 58d13'44"
"A" Team
Uma 9h43'51" 57d38'25"
A Time For Love
Cyg 21h27'18" 51d38'41"
A token of my love for you, Meg
Cap 20h42'8" -14d51'48"
A tout jamais
Cyg 19h46'28" 34d17'57"
"A Treasured Love" Mark & Asha
Pho 1h8'21" -42d20'24"
A True Knight is Trever Peck
Psc 0h53'55" 27d44'37"
A Twinkle in God's Eye for Mama
Cas 0h19'30" 55d21'21"
A V Resa
Lyr 19h13'14" 37d54'40"
A Very Busy Man
Psc 0h21'14" 8d59'47"
A világ legjobb Ancója
Lib 14h54'16" -1d54'30"
A világ Legjobb emberének, Boldinak
Uma 11h57'55" 34d0'7"
A & W
Gru 23h2'14" -40d7'48"
A Ward To The Stars
Per 3h37'32" 45d38'57"
a warlike twin from god
Lmi 10h42'40" 36d0'26"
A Way To Keep Us Close Anywhere.
Sgr 18h48'28" -27d35'7"
A wish come true
Gem 6h50'9" 14d56'53"
A Wish For Margarett
Cnc 7h58'23" 19d8'9"
A Wonderful New Beginning
Ori 5h7'4" 8d58'8"
A Year To Last Forever
Uma 10h26'11" 50d15'59"
A. Yvonne Franklyn
Lib 15h35'48" -4d57'21"
a1m31k04
Aqr 21h27'28" -3d1'4"
A^2
Uma 11h18'36" 51d34'25"
A29 Jenny
And 1h38'42" 49d11'44"
A4 Tech
Lyn 8h30'9" 56d27'30"
a4em~Tyrel and Megan
Sco 16h9'10" -12d45'0"
a5m11k02
Tau 4h7'51" 26d29'10"
AA
Aql 19h11'57" 12d8'47"
AA-7-13-05
Tau 4h25'54" 12d46'26"
AaA UaS
Uma 11h18'3" 68d36'5"
Aada Minea
Tau 3h37'15" 28d45'1"
AADCTS71007
Uma 11h0'3" 66d37'2"
Aadil Patel
Dra 18h49'40" 85d43'31"
Aadyn Patrick
Ori 6h12'17" 19d2'24"
AAGE WOLTER
Umi 14h33'9" 77d40'36"
Aaiden
Sco 16h19'30" -18d48'12"
AAisha BHassan
Crb 16h17'2" 30d5'52"
Aaliah Afia Avettey
Tau 5h2'23" 23d55'10"
Aaliayha Lynn
And 0h9'19" 45d55'17"
Aaliya Yaqub
Gem 7h22'44" 18d55'29"
Aaliyah
Gem 6h40'48" 27d56'36"
aaliyah
Umi 15h30'0" 69d8'30"
Aaliyah
Lyn 7h30'27" 59d43'51"
Aaliyah Elizabeth
Psc 1h13'37" 26d35'26"
Aaliyah Lanee' Lee Love
And 1h24'20" 45d51'12"
Aaliyah Nichols
Lyn 8h46'55" 42d17'37"
Aaliyah Noel Gibbel
Uma 11h7'13" 57d41'25"
Aaliyah's Bright Star
And 0h32'47" 39d0'3"
Aally Boo
Gem 6h44'15" 25d7'38"
Aamanda Sue Brown
Psc 23h50'59" 2d36'17"
Aamayron
Gem 7h7'46" 34d18'49"
Aamira Ardalan
Vul 19h46'37" 25d1'17"
Aanya Parekh
Cnc 9h10'23" 15d49'24"
Aaralyn Rose
Psc 1h17'42" 26d42'48"
Aarhianna Mercierus
Ori 6h0'36" 5d27'19"
Aarick Jeremiah Aguilar
Her 17h36'1" 18d44'23"
Aariene Lockett
Lyr 19h13'13" 26d30'53"
Aarika
Vir 12h31'40" 3d3'37"
aar-jac 86
Ori 6h7'31" 7d14'5"
Aarjon
Uma 9h54'3" 56d36'37"
Aaron
Uma 8h46'14" 56d23'51"
Aaron
Aqr 22h27'36" -12d58'7"
Aaron
Ori 5h32'22" -0d42'57"
Aaron
Ori 5h33'0" -0d24'18"
Aaron
Lib 14h51'29" -3d3'44"
Aaron
Sco 16h34'36" -43d27'43"
Aaron
Cap 21h26'3" -24d16'56"
Aaron
Ori 5h28'22" 1d10'4"
Aaron
Ori 5h4'26" 4d34'14"
Aaron
Psc 23h7'53" 7d38'13"
Aaron
Tau 5h49'6" 17d42'20"
Aaron
Her 17h53'18" 15d28'23"
Aaron
Her 18h40'55" 21d16'17"
Aaron
Boo 14h41'21" 20d17'21"
Aaron
Ori 5h53'50" 21d7'2"
Aaron
Per 3h43'44" 51d44'2"
Aaron A. Paul
Uma 9h53'55" 66d9'28"
Aaron Adamec
Uma 8h36'6" 58d36'52"
Aaron Adams's Everlasting Star
Tau 4h16'45" 26d19'57"
Aaron Agi
Lyn 8h40'24" 34d15'42"
Aaron Alan Malone
Her 17h20'28" 45d58'49"
Aaron Alexander
Lmi 10h3'59" 39d26'8"
Aaron Allan Ward
Leo 9h54'11" 15d4'5"
Aaron & Amanda Faucher
And 0h31'40" 46d14'7"
Aaron and Amy Riede
Uma 11h16'23" 29d56'7"
Aaron and Brooke Wegner
Umi 15h24'58" 74d26'5"
Aaron and Christy Chase Forever
Uma 11h19'5" 47d48'41"
Aaron and Ellen's Star
Lib 15h28'56" -13d3'39"
Aaron and Jennifer Wilker
Cam 4h5'4" 67d25'8"
Aaron and Krystal
Cap 21h12'31" -15d23'55"
Aaron and Nina
Psc 1h21'8" 24d10'44"
Aaron Andrew Wilson
Aqr 22h58'11" -8d12'32"
Aaron Arland
Per 3h2'38" 52d32'4"
Aaron Ash
Umi 13h46'45" 70d52'33"
Aaron Autsen
Lyn 7h3'53" 45d27'35"
Aaron Billy Moon
Gem 6h44'43" 20d5'18"
Aaron Billy Moon
Uma 11h1'25" 64d17'12"
Aaron Black
Boo 14h55'56" 21d28'48"
Aaron Blain
Cep 21h58'28" 62d5'13"
Aaron Blake Sannar
Her 18h9'49" 18d41'33"
Aaron Bradley, My Love, 9 Dec 1980
Uma 11h9'8" 34d55'25"
Aaron & Bridget Fischer
Tau 4h41'46" 17d55'17"
Aaron Bush
Dra 18h59'39" 74d10'28"
Aaron C. Tarantino
Cnc 8h40'52" 32d26'2"
Aaron Carter & Janine Ikue 10/16/04
Ori 6h10'28" 20d15'43"
Aaron Charles Ryan
Ori 5h20'20" 8d22'47"
Aaron Christian Winn
Aqr 21h58'43" 1d39'41"
Aaron Christopher Cosgrove
Cyg 21h44'3" 38d56'42"
Aaron Christopher Jackson
Peg 23h46'59" 27d54'52"
Aaron Conrad Grubbs
Vir 11h53'46" 8d24'35"
Aaron Copeland
Per 3h19'47" 40d58'19"
Aaron Corban Lawson
Per 4h12'32" 34d24'10"
Aaron Cordova
Aql 19h55'40" 11d11'51"
Aaron D'Angelo Garibay
Her 18h35'35" 24d56'55"
Aaron Dannelly Nelson
Aql 19h34'49" -11d12'3"
Aaron David Guiswite
Sco 16h9'1" -19d15'29"
Aaron David Melamed
Uma 11h29'19" 42d21'57"
Aaron David Rapke
Cnc 8h51'14" 26d14'2"
Aaron David Riojas
Ori 6h0'58" 17d17'7"
Aaron David Sudduth
Lyn 7h19'4" 50d47'24"
Aaron Derby
Psc 1h30'19" 24d44'37"
Aaron Donald Hanson
Cyg 20h3'7" 34d3'30"
Aaron Doran Curbo
Cam 4h38'45" 70d40'8"
Aaron Douglas Edmunds
Her 18h19'6" 23d15'56"
Aaron Duffy
Ori 5h5'5" 4d50'35"
Aaron Duran a Chain of Melodies
Cep 22h53'22" 65d58'42"
Aaron E. Campbell
Cyg 21h10'10" 51d10'43"
Aaron E. Fried
Uma 9h33'28" 66d35'16"
Aaron Eduardo
Her 18h8'36" 33d15'12"
Aaron Edward Lopez
Her 18h49'43" 21d24'15"
Aaron Edward Peverley
Lmi 11h4'2" 31d57'28"
Aaron Eighmey
Her 17h44'19" 27d31'16"
Aaron Enoch Teakell
Lib 14h56'57" -24d28'37"
Aaron Evans - Man of Honor
Boo 14h25'28" 45d43'46"
Aaron Ewing
Per 3h24'15" 43d55'29"
Aaron Frank Wallin
Per 4h32'32" 42d39'42"
Aaron & Freya - The Angels' Star
Gem 6h56'37" 29d29'34"
Aaron Fumberger
Sco 16h29'29" -32d6'41"
Aaron G. Dasilva
Cnc 8h38'22" 30d55'37"
Aaron Gabriel
Her 17h19'31" 34d19'29"
Aaron George
Sgr 19h26'15" -13d17'36"
Aaron Gluck-Thaler a.k.a. Ama
Per 3h47'50" 52d2'10"
Aaron H. DuQuesnay
Tau 5h47'37" 17d8'56"
Aaron Haas
Psc 23h5'54" 7d4'9"
Aaron Hammon
Hya 8h49'29" -12d53'17"
Aaron Harpin
Her 16h12'3" 5d20'15"
Aaron Harris
Ari 2h30'1" 27d9'53"
Aaron Henry Meyers
Ori 6h18'20" 14d29'32"
Aaron Hill
Umi 14h24'33" 87d38'48"
Aaron Hockaday's Dream
Ari 2h14'22" 22d54'53"
Aaron Hopkins
Ori 4h59'14" -0d7'6"
Aaron J. Abbarno
Lyn 8h30'3" 39d42'24"
Aaron J. Baird
And 0h44'19" 40d18'47"
Aaron J Castro
Car 10h4'42" -62d24'12"
Aaron J. Lodge
Lib 15h8'7" -1d23'57"
Aaron J. Pond
Cep 22h36'14" 74d16'52"
Aaron Jacob Epel
Aql 18h51'37" -0d5'50"
Aaron James Hatok
And 23h28'49" 38d50'50"
Aaron James Mitchell - 2nd February, 2006
Ori 6h18'46" 13d33'49"
Aaron James Moore
Her 17h25'52" 27d53'52"
Aaron James Zikmund
Ori 6h5'57" 11d13'17"
Aaron & Jess 1 year anniversary
Cru 12h43'5" -56d33'1"
Aaron & Jess 2nd
Her 16h57'49" 33d4'4"
Aaron & Jess's Lucky Star
Uma 8h10'42" 61d55'21"
Aaron & Joelle
Aqr 21h59'38" -14d42'10"

Aaron John Lippy-My Shining Star
Lyn 6h39'3" 59d7'6"
Aaron Jon Bennett
Her 17h47'37" 43d29'37"
Aaron Jon Taylor
Ari 1h54'33" 18d42'33"
Aaron Joseph Maher
Ori 5h53'6" 21d55'8"
Aaron Joseph Morrison
Ori 5h15'54" 12d5'43"
Aaron Kamm Kamin
Aql 19h51'3" -0d18'8"
Aaron & Kathy Partlow
Leo 9h40'43" 27d16'35"
Aaron Keith Young
Uma 8h38'28" 51d52'56"
Aaron Kendall Dillman
Umi 16h49'4" 79d39'11"
Aaron Kovacs Sowd
Sco 17h2'59" -30d9'59"
Aaron & Kristy Burks
Cyg 21h59'26" 50d27'24"
Aaron Kyle Hall
Sco 16h6'44" -18d12'49"
Aaron L. Bell
Cep 22h11'51" 61d8'26"
Aaron Lacy
Per 3h30'27" 39d17'54"
Aaron LaMarr Robinson
Her 17h39'58" 39d34'50"
Aaron Lane
Lib 15h36'38" -25d46'28"
Aaron Lee Preston
Her 17h6'59" 32d3'32"
Aaron Lemuel Nichols
Uma 10h19'31" 42d34'25"
Aaron Lewis Adrian
Cep 21h29'59" 64d6'58"
Aaron Long
Lib 15h20'49" -4d26'1"
Aaron loves Michelle
Umi 13h49'56" 78d40'26"
Aaron Lowell Hamilos
Gem 7h38'22" 24d51'32"
Aaron Mark Otto
Uma 12h6'33" 62d24'13"
Aaron Mathew
Boo 14h48'44" 54d17'53"
Aaron Matthew Becker
Ari 2h41'12" 28d30'46"
Aaron Matus
Her 17h13'54" 26d53'41"
Aaron Meek
Vir 11h57'30" -0d11'26"
Aaron & Melissa Sexstella
Cnc 9h3'28" 28d5'42"
Aaron Michael Bode
Sco 17h1'11" -40d39'10"
Aaron Michael Boyer
Her 17h21'26" 36d35'55"
Aaron Michael Coble
Her 17h57'16" 18d9'58"
Aaron Michael Greenfield
Uma 11h13'1" 72d6'15"
Aaron Michael Jordan Rock
Pho 0h52'24" -56d46'9"
Aaron Michael Kendall
Ori 6h11'50" 0d17'11"
Aaron Michael Krapes
Her 17h33'3" 47d35'6"
Aaron Michael Pieragostini
Aqr 22h44'22" -16d43'49"
Aaron Michael Smith
Lib 15h15'44" -6d26'37"
Aaron Michael Sturm
Cyg 19h46'4" 34d54'44"
Aaron Michael Tapp
Aur 5h45'23" 43d6'55"
Aaron Moeller
Dra 18h11'59" 77d59'35"
Aaron Nathaniel Taylor
And 2h37'27" 40d28'30"
Aaron Nissen
Ari 2h45'52" 26d21'7"
Aaron 'Ogre' Meisner
Uma 13h8'1" 54d30'56"
Aaron Osborne
Dra 20h15'5" 61d58'22"
* Aaron * our love was meant to be
Cnc 8h36'7" 22d56'34"
Aaron P. Torrence
Cep 22h40'53" 68d21'51"
Aaron Patach Jr.
Cep 20h54'31" 63d32'33"
Aaron Paul & Angela Margaret
Cyg 20h11'56" 30d13'30"
Aaron Paul Brewer
Aur 6h54'20" 38d21'42"
Aaron Perl
Gem 6h51'58" 32d5'20"
Aaron Peter Mannion
Per 2h57'3" 56d29'12"
Aaron Philip Doyle
Sgr 18h47'41" -34d10'24"
Aaron Philip Smith
Cep 22h34'26" 64d59'11"
Aaron Phillip Metham
Cru 12h21'10" -59d36'21"
Aaron Phillip Nichols
Per 2h45'10" 53d34'21"
Aaron pookie Tyler
Psc 23h39'57" 3d26'32"
Aaron Principe
Uma 13h23'18" 57d21'42"
Aaron Quidley 1 Creative Son
Ori 6h1'25" -0d50'53"
Aaron Quinn
Sgr 18h20'16" -32d18'25"
Aaron R. Iskowitz
Her 17h27'54" 34d1'27"
Aaron R. Keener
Uma 11h38'22" 32d45'11"
Aaron Rankin
Uma 8h56'19" 57d49'49"
Aaron & Regan Forever
Cyg 20h22'57" 45d43'56"
Aaron Rhys West
Sco 17h12'16" -32d23'14"
Aaron Richard Price
Umi 10h0'19" 86d21'35"
Aaron Robert Bard Nelson
Her 18h2'37" 24d59'35"
Aaron Robert Coates
Ori 6h22'38" 14d58'9"
Aaron Robert Mielke
Ori 5h30'36" 5d9'56"
Aaron Robert Simpson
Lib 14h29'54" -19d1'4"
Aaron Ross Norvell
Uma 12h0'0" 36d6'35"
Aaron Rothstein
Cyg 19h43'13" 33d9'30"
Aaron Samuel Blaes
Aql 19h38'12" -0d1'28"
Aaron & Sarah; Written In The Stars
And 0h47'17" 23d41'14"
Aaron Schneider
Boo 14h53'18" 52d6'52"
Aaron Scott
Dra 18h58'19" 69d53'0"
Aaron Scott & Breanne Marie
Vir 13h24'25" 2d1'8"
Aaron Scott Herman
Ari 2h26'42" 20d43'30"
Aaron shining
Uma 10h50'49" 63d41'39"
Aaron Snider
Leo 11h1'44" 1d9'43"
Aaron & Stephaine Forever
Umi 18h24'32" 87d11'54"
Aaron Stephen Wood
Per 3h32'9" 44d49'26"
Aaron Steven Burns Visit My Grave!
And 1h40'54" 47d5'2"
Aaron Stevenson
Cap 21h2'33" -16d50'1"
Aaron Stuart
Gem 6h9'44" 26d31'13"
Aaron Stuart Hawke istar
Cru 12h17'53" -58d51'59"
Aaron & Suzanne Amstutz
Leo 11h16'15" 6d10'16"
Aaron Taraboletti
Ori 5h26'5" 8d35'13"
Aaron Taylor
Uma 11h59'35" 43d43'46"
Aaron Thomas
Leo 11h5'22" 1d51'40"
Aaron Thomas
Boo 13h47'47" 11d57'58"
Aaron Thomas Berner
Sco 16h47'16" -26d33'50"
Aaron Thomas Ruffner
Dra 10h53'55" 73d41'24"
Aaron Thomas Wickham
Peg 21h28'48" 10d16'34"
Aaron Todd Melchior
Gem 7h42'54" 23d14'0"
Aaron Tolson
Umi 16h14'6" 76d21'34"
Aaron Torres
Vir 13h46'52" 2d1'57"
Aaron Trovato and Lora Lee's Star
Cnc 8h25'37" 26d13'13"
AAron Ushler
Ari 2h20'14" 18d12'34"
Aaron van Boer
Cyg 21h55'11" 36d43'8"
Aaron VanRyne
Ori 5h54'56" 20d50'57"
Aaron Varner
Uma 11h9'52" 62d17'31"
Aaron Victor Rabow
Sco 16h13'25" -23d36'57"
Aaron W Sylvester
Gem 6h46'5" 33d6'17"
Aaron Walton Burgess
Cnc 8h54'30" 10d3'21"
Aaron & Wendy Benes
Gem 7h18'48" 20d38'7"
Aaron William Cowan
Sgr 19h22'28" -23d24'12"
Aaron William Guynn
Uma 12h55'52" 62d52'58"
Aaron William Jones
Sco 16h41'9" -28d3'4"
Aaron Wilson
Leo 10h16'24" 11d41'17"
Aaron Zachery Lyons
Uma 12h50'57" 60d7'50"
Aaron62985
Cnc 8h9'20" 8d11'41"
AaronCherise
Sge 19h30'50" 18d54'48"
AaronEugenePowers03148 2Celestial
Psc 1h36'12" 12d24'13"
Aaron-In My Heart Forever-Love Mel
Lib 16h0'28" -16d51'2"
Aaronius & Adrienneopeia per sempre
Cyg 21h40'13" 52d43'33"
Aaron-Jonius Cuschierius
Vir 12h47'41" 12d53'39"
Aaron-Kasey
Tau 5h9'22" 22d13'20"
Aaron's Aidan
Cap 20h33'34" -14d43'57"
Aarons baby love Bresse&Blythe
Sgr 18h14'48" -16d13'57"
Aaron's Dream
Psc 1h22'10" 27d52'48"
Aaron's Girl
Cyg 20h19'36" 47d34'4"
Aaron's inamorata Kirsten
Cnc 8h47'13" 22d5'42"
Aaron's Light
Lib 14h58'59" -15d48'16"
Aaron's Mom
Cas 1h25'53" 63d46'3"
Aaron's Quest
Umi 14h9'37" 78d4'48"
Aaron's Star
Pho 0h45'12" -48d7'0"
Aarron Rufus
Cep 22h47'25" 75d31'48"
Aarron-Leigh Frank Sawyer - Stargazer
Per 3h38'13" 36d56'22"
Aarta
Sco 17h46'3" -39d45'22"
Aarush's Mommy Hemani
Leo 9h53'43" 22d2'38"
AAS364
Her 16h48'56" 47d48'26"
Aase
Psc 1h23'47" 33d12'41"
AASERA
Vir 13h11'3" -14d23'27"
Aasheeta
Leo 11h35'24" 25d21'12"
Aato
Dra 10h15'56" 79d27'53"
Aatos Koivu
Cep 22h48'55" 68d7'54"
Aavia Lucille Phenoris Williams
Gem 6h44'13" 27d37'0"
Aayushi Dhanak
Cnc 8h10'36" 7d45'42"
AB Bayouth, Jr.
Lyn 7h31'16" 37d56'17"
A.B. Closson
Lib 15h42'37" -29d36'47"
"Ab Imo Pectore"
Uma 11h33'45" 41d24'48"
ab imo pectore
Cyg 21h27'21" 51d21'45"
Ab Romanticus, ab Puella, ab Conivi
Cyg 21h15'53" 47d2'20"
A.B. Smith
Per 3h2'51" 33d14'55"
AB Tower of Strength, Heart of Gold
Cra 18h30'18" -41d8'23"
AB30
Cyg 21h42'11" 46d20'57"
AbaaadY
Psc 1h45'58" 13d23'57"
Abaco's Timberidge Mayaguana
Leo 9h33'38" 27d30'53"
Abagail Lyn Hitson
Uma 10h7'46" 44d0'7"
Abate Assunta "Susy"
Ori 5h25'32" 12d8'26"
Abaton - für Janni
Ori 6h19'12" 9d25'7"
Abayomi Pastor
Her 17h3'24" 43d5'4"
Abba Dabba, Hannia Prado
Mon 7h19'20" -0d55'21"
Abbadabbs
Her 17h11'51" 30d8'11"
Abbana Lil Dot - Love Jane xxoo
Cru 12h42'12" -57d10'7"
Abba's High Horse
Ari 3h1'55" 25d4'14"
Abbas & Karim Hajj Ahmad
Leo 11h3'49" -0d41'58"
Abbas Kashani-Poor
Uma 8h33'23" 69d9'18"
Abbas Meherally
Psc 1h18'54" 16d42'24"
Abbe Rae Zeka
Cap 20h29'57" -18d49'16"
Abbelynn Robertson
Leo 9h54'49" 17d27'1"
Abbe's Adrastea
Her 17h48'59" 45d37'39"
Abbey
Uma 10h19'15" 47d30'51"
Abbey
Lyn 8h7'3" 41d55'58"
Abbey
Peg 22h37'28" 25d30'30"
Abbey
Cmi 7h19'52" 8d55'20"
Abbey Anderson-Rockson
Mon 6h38'37" 10d35'27"
Abbey Brooks Stoker
Cnc 8h50'39" 19d25'16"
Abbey Cameron
Lyn 7h34'0" 56d51'21"
Abbey Katherine Gentry
Crb 16h12'11" 33d14'59"
Abbey Lee
Ari 3h12'31" 27d21'8"
Abbey Lenn
Del 20h33'16" 15d11'36"
Abbey Louise Eastwood - 29.05.1988
Uma 11h54'32" 46d45'50"
Abbey Maree
Cra 19h6'7" -38d59'26"
Abbey O'Brien
And 0h7'20" 44d9'28"
Abbey O'Keefe
Leo 10h25'28" 27d1'1"
Abbey Rae West
And 0h17'2" 38d15'23"
Abbey Rhoades
And 0h18'17" 34d51'38"
Abbey Road
Ori 5h58'39" 18d22'44"
Abbey Road
Sgr 18h59'53" -35d21'48"
Abbey Veronica Boone
And 0h46'20" 25d30'3"
"Abbey's Beautiful Star"
Cam 7h0'57" 71d23'52"
Abbi Grace Barrett
Uma 10h59'40" 33d32'36"
Abbi, Nicola & Gemma
And 0h6'46" 47d45'0"
Abbie
Cas 3h3'2" 64d39'6"
Abbie "Abster" Heisner
Leo 9h39'32" 21d37'47"
Abbie (Bubbles) Edwards
Sco 16h18'41" -14d30'15"
Abbie Catherine Mitchell
And 0h57'24" 44d6'20"
Abbie Charlotte Johnston
Boo 14h13'41" 18d54'51"
Abbie & Charlotte McKean
And 2h21'12" 38d49'50"
Abbie Elizabeth Trone
Cyg 20h28'24" 35d41'58"
Abbie Ford
And 1h50'44" 36d40'40"
Abbie Grace Wheeler
And 1h20'21" 33d56'57"
Abbie Horan
Leo 10h59'6" 10d52'6"
Abbie Jane Currell
And 0h18'32" 35d56'13"
Abbie Jean Freeman
Gem 7h4'30" 33d52'14"
Abbie Lytlle- Artist and Rock Hound
Sco 17h23'39" -42d15'0"
Abbie Margaret
Uma 10h38'39" 47d30'12"
Abbie Marie
Tau 4h31'40" 5d34'17"
Abbie Rae
Sgr 19h39'14" -20d31'26"
Abbie Rose
Sgr 18h33'8" -29d19'33"
Abbie Rose & Randall Scott
Uma 10h52'37" 66d6'21"
Abbie White - Abbie's Star
And 23h59'45" 36d3'54"
Abbie Wright Phillpotts
Umi 13h47'57" 78d17'13"
Abbiepotomus
Vir 13h10'19" -15d25'41"
Abbie's Star
And 23h54'0" 36d58'40"
Abbigail
And 1h11'21" 41d42'44"
Abbigail Ervin
Lmi 10h16'50" 31d22'34"
Abbigail Hulzebos & Stella Hulzebos
Her 18h54'48" 24d6'32"
Abbigail Jay Ludwig
Uma 9h49'48" 49d43'55"
Abbigail Marie Faivre
Lmi 9h58'21" 33d30'43"
Abbigail Marie Sather
Uma 13h52'39" 61d30'22"
Abbigail Nichole
Cnc 8h54'53" 12d54'45"
Abbigailtina
Uma 9h22'13" 52d23'8"
Abbigale Elizibeth Torrey
And 0h35'52" 31d32'2"
Abbigale Feenie Horvath
Leo 9h29'46" 23d19'22"
Abbious Nincompoopious
Cnc 8h21'51" 19d42'5"
Abbott-Elise
And 1h1'42" 41d11'26"
Abboud Family Star
Ori 6h4'38" 20d37'44"
Abboura
Cap 21h44'25" -10d5'34"
Abby
Vir 14h10'54" -18d6'15"
Abby
Lib 14h50'0" -1d57'40"
Abby
Cnc 8h15'18" 9d13'13"
Abby
Ari 3h17'53" 19d36'46"
Abby
And 23h0'50" 37d15'1"
Abby
Cyg 19h37'18" 37d5'24"
Abby
And 23h52'16" 47d13'28"
Abby
And 1h36'18" 47d5'31"
Abby Alane Pelton
Cam 7h26'16" 75d18'47"
Abby Austin
Cas 2h12'35" 63d48'47"
Abby Beckman
Uma 10h42'55" 62d28'29"
Abby Bing
Cmi 7h27'28" 3d34'52"
Abby Biro
Cnc 8h23'27" 14d30'0"
Abby Bunzel
Cas 1h32'20" 65d30'7"
Abby Carpenter
Vir 13h30'54" -20d20'9"
Abby Cat
Lyn 7h13'4" 53d4'3"
Abby Ceppos
Aqr 21h47'16" -6d51'37"
Abby Clay Rutgers
Uma 10h24'43" 47d58'0"
Abby Crittenden Dowdy
And 0h14'1" 27d31'57"
Abby Dougal
Umi 14h29'46" 69d31'18"
Abby Elizabeth Kelley
And 0h45'7" 37d8'49"
Abby Elizabeth Platt
Cyg 21h30'39" 42d7'19"
Abby Fosgate
Ari 2h13'1" 24d2'49"
Abby Freeman-GaMPI Shining Star
And 1h42'30" 41d59'52"
Abby Gaffney
Cas 2h52'44" 58d22'11"
Abby Gailt
Aqr 22h4'40" -16d57'46"
Abby Girly Girl Wehinger
Ari 3h18'55" 19d50'16"
Abby Graham
And 0h11'42" 43d33'1"
Abby Hall
Cas 1h29'52" 67d10'20"
Abby Hannigan
Ori 5h31'47" 10d8'17"
Abby Harmon Johnson
Uma 9h1'30" 70d8'45"
Abby & James - Forever Together
Aqr 20h43'42" -9d30'29"
Abby Jane Tobin
Uma 11h10'37" 40d17'23"
Abby Jo
Dra 16h4'37" 59d3'17"
Abby Jo Rogers Feb. 28th 1933
Cas 0h35'38" 53d52'32"
Abby Johnson
Lmi 10h51'36" 37d59'50"
Abby Joy
Vir 13h56'58" -8d46'28"
Abby Karlsen
Vir 12h24'54" 12d7'7"
Abby Kay
Psc 0h45'14" 16d13'16"
Abby Kynn VonDielingen
Vir 13h49'38" -17d4'39"
Abby Lauren
Leo 11h4'27" -5d6'2"
Abby Lee Holm
And 0h4'46" 39d37'46"
Abby Lee Miller Dance Co 25th year
And 23h11'46" 47d47'23"
Abby Leigh Best
Leo 10h44'50" 22d44'16"
Abby Loudermilk- Love of my life
Gem 7h44'27" 30d24'46"
Abby Louise Caldwell
Lmi 9h26'48" 33d20'49"
Abby Luman
And 1h49'57" 46d43'59"
Abby Marie
Cra 18h12'44" -40d13'39"
Abby Moorehead
Leo 10h44'26" 12d31'39"
Abby Myers
And 23h8'2" 47d21'21"
Abby Nevell
And 23h19'57" 41d42'9"
Abby Nicole Fisher
Ari 1h59'17" 18d23'48"
Abby Norwood
Crb 16h2'33" 32d7'34"
Abby Pierce
Lyn 7h59'19" 49d48'8"
Abby Rae
And 1h42'34" 38d4'26"
Abby Rae Davidson
Umi 14h16'30" 72d35'35"
Abby Rain
And 0h35'18" 43d2'23"
Abby Rebecca Rose
And 2h35'36" 44d0'2"
Abby Rose
Psc 0h40'26" 7d57'7"
Abby Rose
Ari 2h32'57" 25d15'50"
Abby Rose Lee
Peg 22h22'27" 20d53'53"
Abby Rotter
Cnc 9h4'40" 17d2'0"
Abby Schaaf
And 2h32'5" 50d5'50"
Abby Sharon Gaines
And 0h53'6" 36d59'46"
Abby Star
Vir 13h30'41" 11d49'30"
Abby Stember
Cap 21h45'6" -12d38'13"
Abby Szott
Leo 9h39'7" 29d24'30"
Abby Tinidini
Lib 15h5'30" -1d50'31"
Abby Walton
Cas 0h21'26" 54d44'51"
Abby Ytell
Cyg 20h9'0" 58d57'44"
Abby Yunker
Cap 20h32'12" -19d34'25"
Abby Zombro
Leo 10h20'36" 18d59'45"
Abby, 1988-2004
Umi 14h42'7" 79d7'26"
Abby522
Uma 11h12'9" 33d48'35"
Abbyann our God-Daughter
And 2h21'17" 39d12'27"
Abbygail Diana Martinez
And 1h53'59" 42d59'26"
Abbygail Lynne Hetrick
And 1h20'11" 48d33'39"
Abbygail Renee Gwin
Per 2h42'33" 56d35'28"
Abby-Leigh Downing
And 0h58'37" 46d5'57"
Abby's Dad
Tau 3h48'19" 18d32'57"
Abby's Destiny
Uma 11h19'54" 43d48'17"
Abby's Eyes
Mon 8h9'23" -6d0'15"
Abby's Light
Sgr 18h3'50" -26d41'42"
Abby's Star
And 0h10'40" 24d6'49"
Abby's Star
And 0h32'54" 28d16'51"
A.B.C
And 0h19'9" 31d23'27"
ABCD always and forever
Her 18h9'35" 46d38'10"
Abderrazak Cheraït
Ori 4h52'59" -3d15'1"
ABDIINORT, Tibor csillaga
Cas 23h0'49" 59d40'7"
Abdul Al Bayyari
Ori 5h20'51" 1d54'42"
Abdul Aziz A. Karim Al-Khereiji
Gem 6h42'40" 29d0'22"
Abdul Hanan
Cyg 19h45'17" 32d25'52"
Abdul Kareem Ben Abdallah
Gem 6h53'24" 33d3'36"
Abdul Nafie
Uma 9h8'41" 58d37'48"
Abdul Quader Mohammed
Uma 11h29'5" 60d20'53"
Abdul R. Wahid
Umi 0h23'4" 88d52'19"
Abdul Rahim
Aqr 22h13'14" 0d18'38"
Abdulaziz and Shamsa
Cyg 19h55'22" 43d17'17"
Abdul-Aziz Jandali
Ori 6h17'45" 15d2'9"
Abdulkarim & Maryam
And 23h29'14" 36d11'57"
Abdulla and Michelle
Cyg 19h55'28" 35d43'12"
ABDULLAH
Tau 4h2'51" 22d32'18"
Abdullah Dirani
Dra 19h12'18" 65d33'17"
Abdullah Gözener
Psc 0h50'46" 29d51'28"
Abe
Uma 11h58'16" 59d14'30"
Abe Franco
Uma 11h21'11" 58d48'10"
Abe Franco
Boo 14h52'8" 32d22'6"
Abe Little
And 0h26'18" 42d23'26"
Abe M. Shrem
Aur 6h10'35" 46d50'12"
Abe Pederson
Lib 15h6'50" -1d6'45"
Abe Sheldon
Per 3h21'15" 44d52'27"
Abe Wilson Jr (BO)
Umi 17h39'46" 82d29'1"
ABeautifulStar
Lyr 18h49'10" 38d37'39"
Abeccer's Star
Uma 10h27'25" 64d4'36"
Abednego Williams
Umi 16h8'23" 86d15'35"
Abeer
Peg 22h50'5" 4d11'7"
Abeer 31-10-66
Ori 5h53'10" 21d37'52"
Abel A. Cantillo II
Aql 19h6'16" 0d54'41"
Abel Castillon
Leo 10h50'2" 17d31'9"
Abel et Cain
Cap 20h34'37" -11d13'22"
Abel Morales
Uma 9h43'54" 48d15'52"
Abel Star, Katchaz & Aida Dadikozian
Ori 6h12'3" -3d49'8"
Abel, Christian
Uma 11h13'26" 49d6'16"
Abelardo Abby Martinez
Her 16h44'42" 27d37'25"
Abelardo Gama - December 11, 1964
Sgr 18h3'37" -27d52'9"
Abelina Claudette Zelaya Ochoa
Leo 11h50'48" 23d45'57"
Abelito Santiago Guerrero
Lib 15h8'19" -28d4'18"
Abell Courage
Uma 8h19'8" 68d32'11"
Abella
Crb 15h35'17" 32d38'38"
Abel's Breath
Vir 13h22'44" -12d46'47"
Abena Bey
Sgr 18h53'7" -27d26'20"
Abenmarahtau
Cnc 9h4'18" 29d14'47"
Abercrombie
Leo 9h30'9" 26d9'40"
Abercrombie & Pack
Cyg 19h58'51" 32d39'27"
Aberdeen
Cas 23h30'29" 58d39'15"
Aberger, Robert
Uma 11h35'9" 49d44'56"
Aberra Haile
Leo 10h20'28" 10d6'24"
A.BET
Gem 6h27'58" 27d43'48"
ABH04012007
Ari 3h14'24" 22d3'47"
Abha
Umi 17h20'24" 87d14'10"
Abhi Alavilli
Aqr 21h51'14" -1d36'10"
Abhineyt and Svetha
Uma 10h52'51" 64d56'8"
Abhishek Dhayal
Eri 4h29'1" -12d58'41"
abi
Aqr 23h50'33" -10d3'47"
Abi
Cam 3h31'35" 63d49'5"
Abi
Umi 13h32'40" 70d20'19"
Abi
And 23h58'59" 37d1'11"
Abi & Claire's Star
And 0h9'0" 29d20'55"
Abi Lizzy
Cnc 9h15'15" 12d18'59"
Abi Louise Salt
And 23h31'31" 41d35'15"
Abi Megan Fairney
And 0h45'24" 32d39'30"
Abi the Lil' Munchkin
And 2h11'22" 47d27'30"
Abiding Reverence
Sco 17h51'8" -35d45'7"
Abie
Umi 13h18'27" 70d50'33"
Abie
Per 3h20'27" 46d3'8"
Abier Adawi
Ori 5h26'8" 1d52'40"
Abi-G
Lib 15h4'1" -22d23'45"
Abigael Suzanne Rugh
Cnc 8h16'1" 25d57'7"
Abigail
Cnc 8h52'51" 23d1'41"
Abigail
And 0h25'44" 32d4'22"
Abigail
Peg 22h44'11" 23d33'30"
Abigail
Tau 3h38'18" 5d51'22"
Abigail
Ori 5h32'35" 6d34'20"
Abigail
Vir 13h12'17" 4d46'8"
Abigail
And 0h3'12" 45d59'20"

Abigail
And 23h39'51" 43d28'32"
Abigail
Cas 0h20'43" 53d26'0"
Abigail
Cnv 13h35'39" 34d10'21"
Abigail
And 1h26'32" 44d13'8"
Abigail
And 1h36'59" 43d5'40"
Abigail
Uma 11h28'6" 39d40'21"
Abigail
Aqr 21h48'4" -6d5'53"
Abigail
Cas 1h41'17" 63d58'58"
Abigail
Sco 17h33'21" -45d5'57"
Abigail
Cru 12h4'22" -63d8'22"
Abigail *Abby* Sargent Mayo
Ori 5h34'29" -5d23'56"
Abigail (Abby) Verhelst
Uma 10h13'58" 52d53'8"
Abigail 'Abi' Wohlgamuth
Peg 23h12'49" 27d39'6"
Abigail Alice Porter
Sgr 18h8'13" -35d13'39"
Abigail Alova Christine
And 0h15'8" 39d29'23"
Abigail & Andy
Cyg 21h24'47" 34d53'37"
Abigail Angel Novak
Umi 14h34'1" 75d7'54"
Abigail Ann
And 0h51'46" 39d39'23"
Abigail Ann
And 2h10'37" 43d17'30"
Abigail Ann Grant
Aqr 22h29'3" -4d49'1"
Abigail Ann Plessman
Crb 15h44'59" 29d43'14"
Abigail Anne
Ari 2h12'7" 25d0'48"
Abigail Anne
And 2h8'18" 39d3'44"
Abigail Anne Beth
Cyg 21h15'12" 38d45'12"
Abigail Anne Cincilla
Cnc 8h43'35" 6d35'49"
Abigail Anne Rose Ortiz
Gem 7h23'46" 25d5'2"
Abigail Annette Harrahy
Aqr 22h30'29" -19d24'38"
Abigail Anthea Myers
And 0h20'4" 29d9'27"
Abigail Asuncion
Vir 13h55'13" -1d48'15"
Abigail Atkinson
Leo 9h32'40" 28d37'50"
Abigail Aurora
Aqr 22h25'54" -2d1'36"
Abigail Autumn
Her 18h37'5" 19d50'33"
Abigail Autumn Flowers
And 1h17'48" 46d27'45"
Abigail B. Saul - Our Star
And 0h41'36" 39d5'2"
Abigail Bassett & John Kaslewicz
Ori 6h24'53" 17d20'10"
Abigail Bethany & Delaney Cathcart
Umi 11h32'44" 87d54'14"
Abigail Breann Dawson
And 1h13'54" 36d40'15"
Abigail Brianne Tuttle
Psc 23h50'43" 4d37'32"
Abigail Brooke Stauffer
Gem 7h21'51" 24d17'5"
Abigail Caliva Torrente
And 1h42'54" 41d42'11"
Abigail/Caroline
Uma 11h50'22" 31d48'36"
Abigail Catherine
Lib 14h59'4" -6d11'25"
Abigail Catherine Wiser
Sco 16h13'26" -9d42'11"
Abigail Christen
Cnc 8h28'48" 13d7'25"
Abigail Claire Breitzman
Sco 17h18'13" -44d34'48"
Abigail Claire Zschappel
Lyn 6h32'33" 55d48'56"
Abigail Collins
And 1h55'7" 42d26'25"
Abigail D. Wolff
And 0h30'57" 45d58'48"
Abigail Denise Cronin
Dra 20h19'10" 65d7'0"
Abigail Diane Anderson
Cnc 8h41'0" 29d38'27"
Abigail Dickinson Austin
Sgr 18h23'19" -27d14'16"
Abigail Douglas Watkins
Gem 7h42'59" 26d38'13"
Abigail E. McCullough
Cnc 8h35'12" 19d21'52"
Abigail Eden
Cnc 8h56'11" 11d2'12"
Abigail Elaine
Peg 22h27'23" 19d12'17"
Abigail Elizabeth Birkey
And 2h21'14" 39d0'25"
Abigail Elizabeth Greable
And 0h44'31" 46d39'24"
Abigail Elizabeth Puccini
Ari 2h50'5" 25d47'13"
Abigail Elizabeth Ross
Sco 16h8'57" -11d33'24"
Abigail Elizabeth Smith
Gem 7h21'35" 25d23'0"
Abigail Elizabeth Stephens
Lib 14h27'56" -17d26'0"
Abigail Ellenora Thomas
Gem 7h23'36" 25d36'27"
Abigail Esquibel
Tau 3h53'47" 9d15'53"
Abigail Faith Willis M.B.
Lyn 7h42'3" 48d1'35"
Abigail felicitatis
Uma 9h53'1" 48d52'36"
Abigail Florrie Skeet
And 1h29'13" 40d45'11"
Abigail Frances Mohrmann
And 1h21'58" 40d56'34"
Abigail Grace
And 1h26'33" 46d8'31"
Abigail Grace
Gem 6h20'58" 21d9'49"
Abigail Grace
Pyx 8h49'54" -29d58'16"
Abigail Grace Caskey
And 23h48'41" 34d56'46"
Abigail Grace Dobbins
Uma 11h41'2" 46d31'23"
Abigail Grace Holm
Ari 3h14'23" 26d31'16"
Abigail Grace Moores
Umi 15h16'7" 88d56'42"
Abigail Grace Morris
Del 20h33'7" 19d17'20"
Abigail Grace of Tednessa
Ori 5h22'28" 8d1'57"
Abigail Grace Tubbs
Cnc 8h24'0" 28d21'0"
Abigail Isabel Lopez
Ori 5h26'56" 5d11'3"
Abigail Isabela
And 1h29'26" 42d44'36"
Abigail Jane
Vir 13h34'31" 12d50'20"
Abigail Jane
Sco 17h57'9" -39d20'52"
Abigail Jane Hayles
And 23h58'32" 46d46'33"
Abigail Jerrine
And 23h8'23" 48d13'47"
Abigail JoAnn Zamiatowski
Col 6h22'57" -34d43'53"
Abigail Joelle Mallari
Crb 16h13'33" 36d42'23"
Abigail Joy
And 0h27'24" 36d54'44"
Abigail Joy
Leo 11h4'25" 2d21'53"
Abigail Katherine Grzelak-Zehar
Umi 14h28'7" 74d9'23"
Abigail Kathleen
Ari 3h5'13" 17d59'21"
Abigail Kathleen Offenbacher
Vir 13h16'59" -4d10'5"
Abigail Kirby Hartman
Lyn 7h35'41" 45d26'5"
Abigail L Kanellakis
Cnc 9h9'26" 20d30'51"
Abigail Laurel Lambert
Tau 4h23'20" 21d15'32"
Abigail Lily
Peg 22h32'29" 27d55'49"
Abigail Lily Casterton
And 2h37'26" 49d50'57"
Abigail Lorelei May
Gem 6h47'18" 23d46'45"
Abigail Lorn Elane Corea
Cnc 8h10'0" 6d56'56"
Abigail Louise Sloan
Cyg 20h29'28" 34d2'21"
Abigail Lucy O'Brien
Leo 11h10'34" 9d36'19"
Abigail Lyn Sheaffer
Sco 17h53'46" -44d16'34"
Abigail Lynette
And 0h58'16" 39d50'47"
Abigail Lynn
And 2h18'8" 38d19'31"
Abigail Lynn Carswell
And 0h3'3" 45d34'23"
Abigail Lynn Childers
Lmi 10h22'34" 33d55'46"
Abigail Lynn Guicheteau
Cas 1h41'55" 68d23'17"
Abigail Lynn Murphy
Cyg 20h5'3" 32d21'56"
Abigail Lynn Rutherford
Ari 3h14'59" 26d54'49"
Abigail M. De Alba
Psc 1h25'12" 16d15'52"
Abigail M. Hearns (DOB-07/11/2002)
Cnc 8h55'44" 29d20'24"
Abigail Mae
Aqr 22h3'45" 2d50'"
Abigail Maralind Wieland
Uma 9h45'40" 60d3'34"
Abigail Marguerite Muir
Lib 15h4'44" -3d54'25"
Abigail Marie
Aql 19h14'19" 4d27'45"
Abigail Marie
Cyg 21h5'32" 51d53'33"
Abigail Marie Brainerd
And 0h13'47" 32d45'14"
Abigail Marie Brown
Cyg 20h37'33" 43d48'46"
Abigail Marie Carroll
Sgr 19h16'52" -13d39'32"
Abigail Marie Emerson
And 23h43'32" 48d28'2"
Abigail Marie Hornacek
Tau 4h23'58" 16d14'6"
Abigail Marie Kline
Vir 12h22'57" -1d24'17"
Abigail Marie Scheers
And 1h25'40" 45d50'26"
Abigail Mary Martinek
Cnc 8h16'57" 32d56'48"
Abigail Matienzo
Uma 10h15'20" 49d9'45"
Abigail May Ellwood
Aql 19h15'30" 5d47'36"
Abigail McCowan
Cas 1h12'6" 67d44'52"
Abigail Meena
Aql 19h36'32" 8d53'0"
Abigail Morgan
Vul 20h17'9" 24d19'35"
Abigail "Most beautiful in the sky"
And 1h15'2" 36d25'59"
Abigail Nairn
Leo 11h2'26" 4d23'11"
Abigail Naomi
Uma 9h29'22" 46d7'55"
Abigail Negrin
Uma 11h52'38" 35d39'18"
Abigail Niamh Pace
And 2h35'21" 46d15'25"
Abigail Nicole
Lib 15h8'13" -5d14'13"
Abigail Nicole Dennis
Cap 21h39'32" -15d41'58"
Abigail Nicole Wood
Leo 10h51'54" 16d5'9"
Abigail Olajay
And 1h15'22" 45d9'1"
Abigail Ortolano
Sco 17h7'39" -31d55'25"
Abigail Paige Baldwin
And 0h8'39" 28d54'59"
Abigail Pearl
Gem 7h10'30" 34d17'17"
Abigail Peyton Gellman
And 1h19'8" 38d40'58"
Abigail Pritchett
Lyn 7h23'35" 46d51'25"
Abigail Rachel Blanzy
Cas 1h44'44" 65d12'37"
Abigail Rae Schmidt
And 1h50'22" 37d29'53"
Abigail Ray Armijo
Ori 5h38'45" -8d19'58"
Abigail Renee Bentley
Psc 23h23'47" 4d27'28"
Abigail Rhea Carlson
Psc 1h16'13" 17d14'19"
Abigail Rose
Gem 6h43'14" 25d1'36"
Abigail Rose
Tau 3h50'27" 27d32'6"
Abigail Rose Allen
Uma 11h14'32" 59d35'9"
Abigail Rose Barrow
Ari 3h2'34" 14d30'4"
Abigail Rose Dunphy
Lyn 7h49'9" 52d56'11"
Abigail Rose Hapka
Umi 14h27'38" 73d20'33"
Abigail Rose Kern
Ari 2h2'25" 24d15'47"
Abigail Rose Lutz
And 23h43'28" 45d46'7"
Abigail Rose Miles
Gem 6h37'47" 27d42'50"
Abigail Rose "The Little Lion"
Leo 9h29'9" 15d51'51"
Abigail Rose Veres
Uma 10h5'21" 43d36'5"
Abigail Roxanne Davis
Uma 9h6'28" 57d5'42"
Abigail Ruth
Cas 1h1'27" 61d53'3"
Abigail Ruth Catanzaro
Ari 2h7'12" 19d15'1"
Abigail Ruth Fredrickson/Oknick
Vul 19h27'15" 25d21'17"
Abigail Ruth Gay
Gem 7h17'28" 17d37'24"
Abigail Ryder Poole
Uma 10h37'18" 57d57'1"
Abigail Sara Kruger
Crb 15h30'31" 27d21'43"
Abigail Sarah Coyne
And 0h2'24" 33d19'3"
Abigail Sarah Fuller
Leo 10h14'22" 11d16'47"
Abigail Sarah McCauley
And 1h58'57" 37d43'56"
Abigail Soifer
Uma 11h55'12" 38d29'52"
Abigail Sophie
And 1h31'28" 33d47'48"
Abigail Star
And 23h16'29" 48d10'15"
Abigail Susan Carrigan
Vir 14h42'18" -1d21'51"
Abigail Sweet Pea
Uma 9h45'52" 68d20'43"
Abigail Talia Everett
Ari 2h46'32" 25d27'54"
Abigail Taylor Reed
Psc 1h17'47" 28d4'26"
Abigail Theresa Walters
And 2h9'39" 39d32'51"
Abigail Tillina Rabert
Cas 0h16'10" 63d12'11"
Abigail Vail
Umi 16h39'28" 75d29'18"
Abigail Vital - helweh
And 2h19'13" 41d18'3"
Abigail Watson
Gem 7h15'5" 32d40'39"
Abigail Wilkes
And 23h48'26" 33d47'49"
Abigail York Roberts
Psc 0h41'19" 17d36'5"
Abigail, my beloved forever
And 0h32'48" 28d35'20"
Abigail's Angel
Sgr 19h38'9" -17d21'38"
Abigail's Christening Star
And 2h20'51" 42d58'5"
Abigail's Good Morning Star
Ori 6h5'22" 11d40'55"
Abigail's Star
And 1h28'42" 50d5'25"
Abigail's Star
And 2h16'12" 48d47'46"
Abigal Rodrigues Loureiro
Tau 3h38'4" 28d37'52"
Abigale Jean O'Brien
And 0h17'28" 35d25'34"
Abigale Marie Blackley
Cnc 8h9'1" 29d4'51"
Abigale Michaela
Dra 16h0'45" 55d53'32"
Abigale Scarberry Peters
Lyr 18h51'13" 34d41'22"
Abigale Snow Evans
And 2h19'28" 50d5'5"
Abigale Yabut Custodio Toledo
Lib 15h51'42" -10d15'18"
Abigayle Bre-Ann Fullenwider
Vir 13h42'29" -12d31'30"
Abigayle Grace
Lyn 7h7'3" 46d12'43"
Abigayle Rosalind
And 0h1'8" 44d33'38"
Abigeal Ryann McCarthy
Gem 7h28'51" 25d59'14"
Abiola Adeniji-Adele
Aqr 21h9'55" -3d4'12"
Abir El-Achkar
Gem 7h41'56" 19d27'18"
Abird-Ari
Cas 23h9'58" 55d14'54"
Abi's and Noah's Star
And 23h39'43" 41d53'54"
Abi's Star - Ariel
And 0h14'53" 39d14'19"
Abiyoyo
Sgr 19h50'6" -14d32'44"
Abla
Tri 2h25'0" 30d14'40"
A-Block
Uma 13h39'56" 57d13'20"
Abner Ray
Ari 2h56'54" 20d2'55"
Above & Beyond
Cnc 8h48'4" 27d27'38"
Above Earth's Lamentations
Ari 2h40'39" 27d50'44"
ABP-LYP-82107
Ori 6h19'25" 21d15'26"
Abra Elizabeth Karbin
Lib 15h28'39" -6d16'18"
ABRACADABRA
Leo 10h1'58" 7d16'58"
Abracadabra Boomshakadae
Ari 2h47'2" 14d15'8"
Abraham
Uma 13h53'46" 60d23'3"
Abraham Barinholtz & Emily Antila
Lib 15h9'25" -6d59'13"
Abraham Besser
Her 16h40'48" 6d8'40"
Abraham I. Harte
Cep 22h20'24" 64d49'12"
Abraham Jacob Barba
Ori 5h34'22" -1d29'32"
Abraham James Rhodes
Dra 15h39'55" 58d36'40"
Abraham "Kahu" Kawai'i
Aur 5h15'34" 31d33'49"
Abraham Maharba Ruvalcaba
Ari 3h5'30" 28d14'24"
Abraham Michael Sipos
Cap 20h9'12" -21d59'39"
Abraham Nasser
Aur 5h54'51" 36d19'14"
Abraham Neufeld
Ori 4h53'58" 12d14'34"
Abraham Pizarro
Aql 19h8'15" 12d7'15"
Abraham Stephen Middaugh
Per 3h52'58" 49d16'46"
Abraham Trieff
Psc 23h56'58" -0d43'26"
Abraham's Star
Lib 15h24'47" -23d42'2"
Abrahannah
Tau 4h17'52" 17d45'0"
Abrakat, Dietmar Wolfgang
Ori 6h16'44" 16d10'22"
Abram
Her 17h1'2" 24d11'53"
Abram
Lib 15h50'41" -5d4'18"
Abram Ber
Cnc 8h37'40" 18d24'37"
Abram Crawford Harvey
And 1h3'57" 42d16'26"
Abram Samuel
Leo 9h47'55" 8d41'42"
Abram Shain
Gem 7h3'48" 28d52'43"
Abrazos Con Amor
Psc 1h19'46" 31d36'24"
Abrea Sirius One
And 0h8'44" 35d19'45"
Abriana Dorian
Cas 2h46'56" 64d8'29"
Abrianna Delight
Gem 6h46'0" 16d38'32"
Abriel
Dra 17h10'5" 56d30'42"
Abrielle Simone White
Psc 0h28'51" 7d30'27"
Absence makes the heart grow fonder
Sco 16h42'13" -36d51'7"
Absolut Air Traffic Dan 05
Cyg 19h55'32" 58d35'45"
Absolutely Amanda Kassel
And 1h11'34" 42d28'11"
Absolutely Fabio
Her 18h44'13" 20d58'44"
Absolutely Forever! R.S. & J.O.
Uma 13h50'39" 57d2'54"
Absolutio-onis Captiuos-a-um
Aql 19h56'51" -0d30'31"
Absolutus Aeternus Amor
Uma 11h12'58" 30d5'10"
Abu
Cnc 8h51'4" 29d25'7"
Abu
Umi 15h4'48" 72d12'22"
Abu Jesus
Tau 4h23'47" 8d9'31"
Abuela
Tau 4h11'47" 27d52'9"
Abuela Angelita
Del 20h34'55" 14d33'52"
Abueli
Aqr 22h7'31" -19d32'45"
Abuelita Hajek's Special Star!
Cyg 20h14'11" 49d9'17"
Abuelita Leonor J. Jorg de Yampey
Cnc 8h53'41" 15d26'30"
Abuelo Buddy
Aqr 23h2'5" -8d24'34"
Abuelo Elizer
Uma 8h31'35" 65d8'0"
Abumorcos
Sgr 18h8'44" -27d8'44"
Abundantia
Cma 6h59'48" -13d12'55"
Abussimbel
Lyr 19h4'54" 44d19'51"
Abygail Foster - Shining Star
Lyn 7h31'18" 53d26'22"
ABYS (Always By Your Side)
Cyg 19h50'50" 44d29'9"
Abyss
Uma 8h40'14" 67d23'53"
ABZ-25
Uma 11h55'22" 61d27'52"
A.C.
Uma 10h4'29" 57d35'42"
AC & CA forever
Ara 17h30'3" -46d31'10"
AC MUPPIN 143
Cnc 8h11'20" 28d34'40"
ACA
Lib 15h54'28" -17d20'44"
Acabrera
Lyn 6h52'16" 56d42'32"
Acacia
Sgr 17h58'58" -27d29'23"
Academician Veniamin Anilovich
Uma 11h20'20" 35d8'50"
Academy Lake Constance
Ori 6h19'59" 21d26'7"
Academy of Dance 25th Anniversary
Uma 10h49'44" 55d39'35"
Acanday:Brookman. 23.07.60
Aql 19h44'37" 15d58'21"
Acantares
Uma 11h35'7" 45d42'50"
Acarya Lin Baker
Lyn 7h13'43" 56d2'15"
ACB
Uma 13h51'56" 54d42'27"
ACC Nocturne
Gem 7h37'34" 28d36'58"
Acceptance
Ori 6h6'55" 21d13'35"
AccordionStar
Sco 16h13'20" -10d10'26"
Accra Beach Hotel
Crb 15h37'43" 38d45'59"
AccuCrete
Her 16h55'30" 16d59'3"
Ace
Com 13h5'31" 25d14'46"
Ace
Psc 0h44'18" 14d32'16"
Ace
Cnc 8h48'55" 6d59'2"
Ace
Uma 11h5'50" 52d7'56"
Ace
Tri 2h29'0" 37d16'17"
Ace
Cnc 9h10'59" 32d33'9"
ACE
Cyg 20h43'41" 32d25'46"
ACE Distributors Sdn Bhd
Ori 5h30'34" 2d55'18"
Ace Fenyar
Gem 6h59'12" 13d14'32"
Ace Lawson
Leo 10h54'40" 14d12'28"
Ace loves Nicole Goodman
Gem 6h44'23" 17d12'20"
Ace Of Trump
Cep 22h20'46" 57d31'48"
Ace Warren
Uma 10h17'10" 63d50'33"
Acebal
Lib 14h48'49" -3d35'4"
Acena
Cnc 8h22'35" 16d13'40"
Acenath
Peg 23h15'51" 17d4'23"
AceStar
Aur 6h30'47" 36d29'19"
Acey
Del 20h44'57" 15d58'6"
Acey and Pauline Smith
Leo 10h25'11" 6d51'13"
Achaiah
And 2h32'34" 45d0'5"
Achalé une douce graine de vie
Her 18h38'36" 19d46'51"
Achatado Smith
Sgr 18h14'34" -25d8'59"
Achilleus
Cas 0h59'49" 63d23'14"
Achim
Uma 9h23'2" 63d6'1"
Achim Knolle
Uma 8h39'57" 66d43'6"
Achsa
Leo 10h25'29" 20d24'27"
ACI Facility Maintenance
Lyn 8h41'47" 45d0'35"
Acinom
Gem 7h35'29" 26d27'40"
AcireCorley
Lib 14h47'50" -6d57'45"
ACISSEJ RELLIM
Sgr 18h0'20" -26d54'1"
A.C.J. Wheatley - Tony's Star
Uma 12h37'38" 52d30'38"
Ackerman All Stars
Uma 11h10'3" 56d35'0"
ACM 12/02/06
Tau 5h59'44" 28d8'14"
ACNL Presidents' Star
Aqr 23h7'21" -20d3'7"
ACO MAGYARORSZÁG
Lib 15h20'7" -20d7'25"
ACP
Cnc 8h20'1" 32d9'29"
Acqua di Palaemon
Del 20h32'39" 20d3'23"
Acquiesce LCSC220905
Ori 6h3'25" 1d30'10"
Acrelias
Crb 16h7'25" 35d15'25"
Acrelida "Foxy"
Gem 6h21'50" 18d17'50"
Ács Mónika
Lib 15h49'55" -13d54'15"
Act in Spite of Fear
And 0h8'30" 28d56'39"
Actaliz
Uma 11h19'42" 35d28'19"
ACTEMRA
Uma 8h23'57" 62d19'50"
Action Jackson
Peg 22h24'41" 9d33'26"
Actions Speak Louder Than Words
Ari 2h50'46" 29d33'20"
ACTS
Cnc 8h19'19" 13d11'24"
ACV 2601
Ori 5h59'41" -0d10'59"
Aczariah Mischa
Vir 13h3'25" 10d3'0"
AD ASTRA PER ASPERA
Uma 10h27'37" 43d3'7"
Ad astra per aspera
Uma 12h38'6" 59d46'50"
Ad astra, Nemo nisi mors
Cyg 20h14'27" 49d13'39"
Ad Infinitum
Ori 5h11'15" -9d39'28"
Ad infinitum
Sgr 18h4'13" -18d24'25"
Ad Mater Dei Gloriam - JJM
Cam 5h41'18" 63d59'22"
ad nauseam
Vir 14h20'3" 4d16'48"
Ad Vitam
Her 17h23'27" 16d37'3"
Ad Vitam Aeternam
Gem 6h22'0" 18d5'59"
Ad Vitam Aeternam
Mon 6h31'47" 8d25'5"
Ad Vitam Aeternam - For All Time
And 1h34'32" 44d1'2"
Ad Vitam Aeternam of Matthew
Cru 12h27'45" -59d43'3"
Ada
Cas 2h21'15" 66d42'17"
Ada
Cam 4h54'49" 75d57'44"
Ada
Cyg 20h12'40" 48d24'12"
Ada
Cas 1h14'13" 53d47'37"
Ada 1920
Uma 12h37'44" 55d17'35"
Ada Bell
Vir 12h31'13" -0d24'2"
Ada Catherine Dea Leong
Lib 14h53'56" -9d3'33"
Ada Chiara Emilie Mork Finsas
And 23h45'2" 48d27'30"
Ada Crane
Ori 6h7'5" 5d8'4"
Ada D' Orthoux
Cas 23h27'19" 55d19'28"
Ada Eagar & Ella Sue Evertsen
And 0h17'1" 28d0'40"
Ada Elisa
Col 5h48'30" -31d6'28"
Ada Frances Gadoury
Cap 20h24'51" -11d6'57"
Ada Gallager
Lyn 6h22'48" 61d41'40"
Ada Imogene Jones Harless
Cas 1h22'39" 56d12'37"
Ada J. Matti
Cap 20h35'32" -19d18'37"
Ada Julianna
And 2h19'47" 44d51'49"
Ada Light Of My Life
Ari 2h39'3" 17d59'4"
Ada LoveStar
Vir 13h19'14" 4d2'21"
Ada Luz
Crb 15h50'49" 36d19'56"
Ada McGuire
Lib 14h50'10" -5d0'41"
Ada Pompeo
Lyr 19h14'16" 37d20'45"
Ada Shine
Lyr 19h5'42" 42d27'54"
Ada Urquiola "Dream Star"
Lib 15h6'1" -29d22'24"
Ada-Fonsi
Tau 4h50'17" 17d35'20"
adagaue
Pho 0h37'49" -54d12'5"
Adagio
Ori 5h56'1" -0d5'46"
Adah
Ari 2h17'18" 25d6'19"
Adah Godbold Phelps
And 0h58'31" 36d12'47"
Adaira
Ari 3h7'25" 29d1'32"
Adak5051 Carol J. Stankaitis-Young
Cas 1h31'29" 61d11'11"
Adal
Uma 9h36'48" 42d9'5"
Adalaide
Cyg 21h53'7" 38d36'34"
Adaleir
Cnc 8h29'27" 24d24'47"
Adaley Rose
Psc 0h30'6" 8d10'28"
Adalia
Ori 6h5'48" 5d19'19"
Adalia
Cnc 8h12'5" 32d0'23"

Adalia M. Krietemeyer
Cnc 8h44'56" 12d12'11"
Adalida Jean Richardson
Uma 9h3'40" 58d10'49"
Adalie
Uma 9h53'51" 57d53'53"
Adalita Lotus 20.05.2005
Tau 3h24'36" 17d18'22"
Adaliz
Psc 1h42'1" 6d57'45"
Adalyn Grace
Cap 20h40'29" -23d33'43"
Adam
Aqr 22h30'13" -22d59'36"
Adam
Uma 12h9'48" 57d42'33"
Adam
Vir 12h55'32" -0d37'31"
Adam
Sct 18h47'50" -5d39'35"
Adam
Lib 15h28'20" -13d47'43"
Adam
Cap 20h27'31" -13d9'46"
Adam
Ori 6h17'35" 10d47'47"
Adam
Leo 10h24'50" 13d32'49"
Adam
Ori 5h52'56" 2d13'8"
Adam
Aql 19h17'30" 5d38'45"
Adam
Gem 7h32'22" 29d48'31"
adam
Gem 7h32'24" 24d21'39"
Adam
Boo 15h5'16" 21d33'21"
Adam
And 1h50'53" 35d53'33"
Adam
Her 17h14'47" 31d48'15"
Adam
Aur 5h48'58" 46d49'43"
Adam
Per 3h25'21" 46d55'13"
Adam
Per 2h59'58" 46d39'28"
Adam
And 2h25'31" 48d55'6"
Adam 2005
Cap 20h27'35" -27d15'18"
Adam Abdalla
Uma 10h17'50" 69d11'29"
Adam "Ace" Hardin
Her 17h29'42" 22d40'21"
Adam 'Adarni Burnquist' Hill
Her 18h31'35" 21d20'34"
Adam & Adriana Rigano 27-3-2004
Ari 2h6'14" 11d5'49"
Adam Ahmad
Per 2h45'32" 50d12'32"
Adam (AJ) Jacob
Tau 4h47'39" 26d33'52"
Adam Alan Sheldon
Her 17h39'50" 16d43'42"
Adam Alexander Clapp
Lmi 10h47'58" 28d27'57"
Adam Alexander Cole
Lyn 7h17'58" 51d51'26"
Adam Alexander Kraus
Tau 4h32'38" 17d0'15"
Adam & Alisa Keniger
Lib 14h25'31" -19d15'58"
Adam and Chrissy's star
Lib 15h41'35" -4d44'10"
Adam and Danielle
Pho 23h33'27" -40d15'51"
Adam and Haley
Cyg 20h48'20" 35d10'8"
Adam and Jessica
Cra 18h46'28" -39d23'26"
Adam and Kelsey
Cyg 20h35'43" 39d33'12"
Adam and Lindsay Forever
Cyg 21h31'25" 44d38'39"
Adam and Maris
Vir 13h29'16" 8d39'17"
Adam Andrew Long
Lib 14h51'25" -18d1'58"
Adam Angel Craig
Umi 16h10'3" 71d28'16"
Adam & Angela
Pho 0h11'18" -42d29'52"
Adam & Angela Livingston 7/1/2006
Leo 10h11'57" 20d37'1"
Adam Anthony Canturi
Cru 12h2'29" -58d36'59"
Adam Beck and Sonja Brittain
Psc 1h35'42" 21d14'9"
Adam Been
Psc 1h5'44" 9d44'19"
Adam Bell
Leo 10h35'36" 19d47'10"
Adam Benjamin Cohen-Congress
Per 2h38'11" 57d2'56"
Adam Benjamin Rasmussen
Vir 12h17'9" 3d40'58"
Adam Benjamin Spirit of Theseus
Ori 5h36'30" -0d58'29"
Adam Bird
Leo 10h47'22" 22d31'33"
Adam Bonnette
Cnc 8h23'16" 31d27'16"
Adam Boyce Hays
Leo 9h33'9" 11d17'36"
Adam Boyd
And 2h20'41" 48d52'33"
Adam Braginton
Cyg 20h5'21" 48d8'7"
Adam Breest
Lyr 18h55'19" 30d33'0"
Adam Bremen Mondry
Her 17h12'40" 47d49'47"
Adam Bruckner
Tau 4h26'58" 2d17'37"
Adam C. Gillis
Gem 7h25'59" 33d44'3"
Adam C. Raczkowski
Dra 17h10'23" 51d18'31"
Adam C. Weston
Aqr 22h9'31" -1d41'46"
Adam C. Wisor
Her 18h3'10" 17d53'4"
Adam Cardenas Sisk
Leo 11h26'1" 15d9'41"
Adam Carl Witte
Lib 14h53'20" -2d3'19"
Adam Casey Pederson
Ari 2h47'42" 29d2'16"
Adam Charles Zonis Land
Cap 20h41'13" -24d29'48"
Adam Chase Eidson
Per 3h48'34" 40d25'25"
Adam Chauser & Christine Szabo
Lib 14h55'28" -6d53'10"
Adam Cheng
Cru 12h35'19" -64d24'29"
Adam & Cheryl, Together Always
Cep 21h2'19" 57d48'45"
Adam Christian Santucci
Aqr 22h31'17" -0d58'51"
Adam Christopher Frank
Sco 17h25'46" -45d22'22"
Adam Christopher Webster
Ari 3h22'12" 19d49'37"
Adam Churchill Bennett
Gem 6h50'31" 25d8'22"
Adam Cilio The Great
Ari 3h1'14" 22d28'42"
Adam Clark Erb
Sgr 18h54'6" -21d46'52"
Adam Cosmo Daly
Per 2h40'56" 52d30'10"
Adam Craig
Boo 14h59'10" 19d8'52"
Adam Craig Williams
Ori 5h32'46" -1d35'12"
Adam Crain
Ori 5h41'33" -1d25'16"
Adam Crawford
Her 17h0'17" 12d57'32"
Adam Cree
Cep 1h46'52" 81d46'4"
Adam Curtis
Aqr 22h15'38" 1d8'2"
Adam D. Froehlich
Uma 10h31'25" 67d16'43"
Adam D. Wondrely
Lib 15h47'2" -16d56'42"
Adam & Daniel Berkowitz
Ori 5h20'37" 4d50'16"
Adam Daniel Dubowik
Gem 6h46'55" 23d55'31"
Adam David Hunt
Cap 21h35'49" -16d41'22"
Adam David Milsted
Cep 22h18'18" 61d58'41"
Adam David Nixon
Sco 17h15'12" -32d36'25"
Adam David Phillips
Her 17h35'16" 36d48'18"
Adam David's Star
Uma 9h40'57" 59d21'41"
Adam Dealey
Umi 14h41'6" 76d54'28"
Adam Dean Clarke
Vir 12h42'38" 5d0'34"
Adam Demopoulos
Cep 23h35'57" 67d42'22"
Adam Dominic Morris - Adam's Star
Her 17h54'39" 25d33'46"
Adam Dooner
Umi 15h30'14" 72d15'57"
Adam Douglas Boyer Rayzor
Lib 15h20'27" -24d15'47"
Adam Douglas Deppe
Per 3h16'50" 48d21'16"
Adam Douglas Lemonds
Tau 4h34'28" 14d23'40"
Adam Duhon
Vir 13h15'26" -4d58'6"
Adam Dylan Grant
Umi 17h14'13" 83d38'44"
Adam E J Kuhr
Psc 1h59'9" 2d57'20"
Adam E. Kauffman
Her 16h19'24" 19d1'34"
Adam Elliott Rowley
Ari 3h7'17" 25d46'15"
Adam Emilio Bisaccia
Cnc 8h31'13" 28d0'19"
Adam Erceg 24.01.1984 love forever
Cru 12h30'20" -62d24'13"
Adam Evans Vancini Light of My Life
Per 3h53'52" 34d32'29"
Adam & Eve
Ser 15h52'17" 22d18'44"
Adam Evert Newberry
Sgr 18h6'28" -26d57'48"
Adam Felber
Uma 11h14'27" 55d20'12"
Adam Fornataro
Per 3h14'22" 47d20'33"
Adam Frederick Bishop
Lib 15h8'33" -9d50'10"
Adam G Sams
Uma 9h17'36" 58d41'39"
Adam Galexy
Cyg 19h46'33" 37d56'35"
Adam Garff Ethridge
Uma 8h18'59" 69d4'1"
Adam Gaynes: A Sunquam Star
Per 3h30'2" 50d31'55"
Adam Grace
Sco 16h12'52" -14d34'41"
Adam Gregory
Cap 20h31'21" -11d8'9"
Adam Gregory
Ari 2h40'22" 29d31'14"
Adam Hailey
Ori 4h45'53" 4d2'4"
Adam Harpe
Ari 1h53'56" 17d59'47"
Adam Hartley Westbrook
Boo 14h38'19" 26d29'11"
Adam Hawley
Ori 6h5'50" -0d27'52"
Adam Heath
Cnc 8h32'21" 22d49'3"
Adam Hochheiser
Mon 6h53'57" -0d6'2"
Adam Holley
Leo 9h52'23" 30d0'0"
Adam Hommerding
Cnc 8h24'24" 25d51'38"
Adam Ian Gartland
Peg 22h29'13" 25d40'49"
Adam in Orion
Ori 6h5'28" 19d32'8"
Adam J. Carson
Tau 3h47'36" 26d15'19"
Adam J. Hammock
Vir 12h26'0" 3d47'43"
Adam J. Huners
Dra 17h49'52" 64d58'56"
Adam J. Miller
Vir 11h56'31" -0d32'36"
Adam J Pollock
Lyr 19h17'53" 28d31'23"
Adam J Regnier
Leo 10h2'19" 20d18'12"
Adam J Weiner
Cnc 9h19'41" 16d22'30"
Adam Jack King
Her 17h35'3" 23d29'28"
Adam & Jackie
Sge 19h34'33" 18d47'20"
Adam & Jaimie
Uma 10h21'9" 49d42'59"
Adam James Dean
Gem 6h41'30" 31d2'10"
Adam James Deyo
Umi 15h31'6" 72d37'10"
Adam James Dickie
Ori 5h22'9" -5d3'27"
Adam James Facciolla
Lyr 18h15'34" 31d33'52"
Adam & James Gill
Per 4h43'30" 41d58'43"
Adam James Huckins
Umi 14h40'44" 68d8'17"
Adam James Parker
Ori 5h59'36" -2d40'12"
ADAM JANJIS
Cyg 21h14'49" 46d2'25"
Adam Janke's Star
Tau 4h16'31" 28d26'14"
Adam Janusz
Tau 3h37'54" 27d49'13"
Adam Jared Goldberg
Boo 15h19'3" 49d23'39"
Adam & Jean Steinberg
Uma 14h0'50" 57d1'35"
Adam & Jessica
Cyg 19h50'20" 43d2'38"
Adam John Bell
Dor 4h20'4" -51d39'58"
Adam John Boccia
Uma 10h38'3" 69d18'18"
Adam John Charles Bell
Lib 15h0'2" -7d28'11"
Adam John Finkelstein
Aur 5h38'23" 36d56'15"
Adam John Halewood
Per 4h6'31" 34d8'11"
Adam John Oakland
Lyn 7h46'3" 57d7'49"
Adam Jonathan Briggs Wnuk
Ari 2h56'11" 11d27'52"
Adam Joseph
Uma 11h11'11" 50d50'58"
Adam - Joseph - Aziz - Cosman
Cep 22h10'8" 73d40'29"
Adam Joseph Bullock
Leo 10h45'54" 12d57'16"
Adam Joseph Carmody
Leo 9h33'58" 28d56'32"
Adam Joseph Dowell
Aql 19h40'24" -7d18'59"
Adam Joseph Sanders
Per 2h45'41" 44d29'29"
Adam Joseph Southerly
Tau 3h40'3" 8d42'2"
Adam Joseph Waite
Gem 6h40'0" 17d15'3"
Adam Joshua
Ori 6h15'36" 9d4'43"
Adam Joshua Maliborski
Vir 12h36'30" 12d25'22"
Adam Justin Boehling
Her 16h44'32" 33d36'4"
Adam Kaattari
Cep 22h58'46" 70d37'47"
Adam Kalb & Natalia Covarrubias
Cyg 21h15'58" 36d22'55"
Adam Kaos Karlakki
Aqr 21h59'28" -21d6'53"
Adam Keeton
Uma 10h41'1" 64d28'25"
Adam Kenley
Sco 16h13'9" -12d27'43"
Adam Kery
Sgr 17h51'38" -26d4'58"
Adam Kevin Manson
Cru 12h35'56" -59d57'4"
Adam & Kim Greenlee
Psc 1h24'25" 17d19'1"
Adam & Kimberly's Everglow
Cyg 19h50'26" 39d58'14"
Adam Koehler Brown
Psc 23h51'54" 7d50'15"
Adam Koenigsberg
Aur 5h39'0" 35d44'7"
Adam Krzemkowski
Uma 13h13'45" 59d28'31"
Adam L. Tilford
Psc 0h20'52" 21d32'11"
Adam + Lacy Forever
Cap 21h36'56" -14d52'22"
Adam Lane Bartmess
Ori 6h4'8" -0d20'30"
Adam Lardent
Vir 13h20'45" 11d27'34"
Adam Law
Uma 12h24'16" 55d4'30"
Adam Lawrence Booker
Her 17h49'40" 47d0'17"
Adam Lee Bradley
Her 16h57'50" 33d20'37"
Adam Lee Mazza
Uma 10h17'11" 67d32'34"
Adam Lee Morse
Uma 10h59'20" 36d55'7"
Adam Lenard Delk
Boo 14h47'51" 28d21'39"
Adam Lewis Phillpotts
Umi 16h31'16" 83d46'21"
Adam Lorenzo Sunderman
Cep 22h11'22" 60d53'36"
Adam Louis Glazier
Per 4h1'57" 42d6'35"
Adam Louis Morris
Sgr 18h41'15" -21d57'58"
Adam Lucas Farrell
Sco 17h57'42" -32d9'12"
Adam M Inman
Vir 13h25'2" 9d23'16"
Adam & Mandy
Sge 19h37'40" 16d51'24"
Adam Marcus Lensi
Her 16h10'55" 45d47'55"
Adam Marcus Lofft
Gem 7h36'6" 33d32'26"
Adam & Marissa
Cyg 20h2'51" 33d32'10"
Adam Mark Robeck
Cep 21h22'48" 60d7'42"
Adam Marmino, my Hero and my Love
Gem 7h46'11" 26d22'50"
" Adam & Martin - 12.11.2004 "
Cru 12h49'16" -58d6'56"
Adam Martin Weyer
Her 16h53'47" 31d29'25"
Adam Matthew Bird
Aqr 22h10'47" -2d17'51"
Adam Matthew Carbonara
Umi 14h23'57" 77d29'6"
Adam Matthew Kirkland
Ori 5h38'22" -4d49'22"
Adam Mayne Daly
Cnc 8h49'16" 30d31'8"
Adam M.C. Clemons
Vir 13h16'17" -22d15'53"
Adam Meils Foote
Ari 2h18'58" 20d37'41"
Adam & Melissa Dooley's Star
Cyg 21h54'51" 52d51'34"
Adam Micah Starr
Lyn 8h30'57" 51d15'25"
Adam Michael Dupuis
Uma 12h2'18" 41d52'9"
Adam Michael Hale
Her 18h24'22" 16d54'56"
Adam Michael Johansson
Aur 5h18'17" 37d49'40"
Adam Michael Park
Cyg 20h27'4" 30d15'46"
Adam Michael Peralta
Uma 11h42'14" 28d45'41"
Adam Michael Ptak
Uma 11h10'51" 38d4'1"
Adam Michael Sheda
Per 3h30'50" 49d31'25"
Adam Michael Stratton
Psc 23h45'23" 5d50'41"
Adam Michael Tait
Her 18h7'18" 18d23'32"
Adam Michael Vital
Cap 21h25'7" -17d1'9"
Adam Michael Weinstein
Aqr 22h21'15" -0d33'38"
Adam Micheal Messner
Cyg 21h47'38" 38d17'47"
Adam Millward
Cep 22h29'38" 63d37'48"
Adam & Missy
Vir 13h17'51" 11d36'0"
Adam My Loving Son
Uma 11h32'56" 50d5'19"
Adam - My Prince Charming
Leo 11h43'22" 14d11'7"
Adam Navarro
Leo 9h29'51" 14d30'22"
Adam Neil Purvis' Dream
Cyg 19h41'52" 33d6'56"
Adam Nicholas Wilcox
Per 3h32'11" 40d11'52"
Adam Oates
Aqr 21h46'48" -1d32'44"
Adam Oscar
Per 3h20'2" 53d26'13"
Adam - our special shining star
Cru 12h49'55" -64d38'51"
Adam P. Clawson
Ori 5h37'34" -0d53'52"
Adam Pace
Aql 19h43'30" -0d2'54"
Adam Paul Durakovic
Tau 5h45'53" 27d35'41"
Adam Paul Janssen
Gem 7h37'46" 31d20'37"
Adam Peter Johnson
Gem 7h8'8" 26d0'18"
Adam Preston Ryan
Sgr 17h51'55" -18d39'53"
Adam Prime
Aql 20h14'40" -0d51'47"
Adam R Buchnowski
Lib 14h52'9" -1d13'12"
Adam R. Spencer
Uma 14h18'39" 57d33'6"
Adam Raymond
Uma 8h27'41" 67d5'2"
Adam & Rebecca 102281
Cyg 21h21'59" 37d51'29"
Adam {Red Dwarf} Reilly
Dra 10h52'24" 73d24'8"
Adam Richard Laidlow
Leo 10h14'19" 21d44'17"
Adam Richard Moore
Cep 22h38'2" 67d27'59"
Adam Richard Pimenta
Psc 0h52'17" 28d59'32"
Adam Rios
Lyn 6h28'17" 55d0'22"
Adam Robbins
Her 17h22'14" 34d0'4"
Adam Robert Karel
Hya 9h27'9" -0d2'0"
Adam Roberts
Her 17h18'51" 23d24'0"
Adam Roger Wilkins
Boo 14h59'59" 16d10'2"
Adam Rowney, Forever Shining Bright
Cyg 21h49'33" 48d19'5"
Adam Rubin
Umi 16h35'27" 84d21'35"
Adam Ryan
Aur 5h47'42" 46d47'34"
Adam Salim
Tau 4h54'5" 17d30'28"
Adam Samuel Feiring
Psc 0h25'37" 0d5'13"
Adam Sanders
Cap 20h38'49" -13d9'4"
Adam Schneider
Tau 4h54'18" 19d28'3"
Adam Scott Queen
Lib 14h49'9" -0d54'11"
Adam Scott Tinkham
Aql 18h53'1" -0d15'52"
Adam Serge Pennington
Vir 12h45'34" -8d5'30"
Adam Shackleton
Her 18h0'18" 21d28'34"
Adam "Snuggly Bear" Gierczak
Sgr 19h23'4" -14d56'2"
Adam Sobolewski
Uma 9h23'51" 42d1'35"
Adam Squance
Cyg 20h54'40" 41d30'28"
Adam Stary & Ashley Lind
Sge 19h26'16" 18d6'17"
Adam Stein
Her 17h54'44" 32d23'29"
Adam & Steph's Nuzzle Star...iLu
Cyg 19h43'49" 34d50'50"
Adam Sterling
Dra 18h27'13" 53d2'30"
Adam Stevens
Boo 14h50'12" 43d7'7"
Adam Sullivan
Uma 9h37'40" 47d1'17"
Adam Swanepoel
Gem 7h42'33" 33d7'2"
Adam Sydney Joseph
Per 2h24'48" 52d32'17"
Adam T. Sowers
Leo 11h15'23" 24d51'50"
Adam & Taran
Cma 6h47'39" -14d33'30"
Adam Thames
Uma 10h35'15" 58d51'34"
Adam "The Eagle Scout" Jolly
Aql 19h15'35" 4d47'57"
Adam the Rockstar
Cnc 8h0'46" 12d46'16"
Adam Thomas
Her 18h24'40" 22d35'22"
Adam Thomas Artman
Umi 14h38'21" 72d50'21"
Adam Thomas Dodds
Per 4h26'56" 42d28'3"
Adam Thomas Kalwa
Per 4h3'27" 49d27'7"
Adam Thomas Norden
Her 17h16'15" 25d50'14"
Adam Timothy Wright
Uma 10h26'4" 50d27'31"
Adam Tristan Schuetz
Uma 11h21'8" 58d37'30"
Adam Turner
Her 17h35'51" 26d36'53"
Adam Uliassi
Her 17h19'48" 34d27'31"
Adam Vincent Ammirata, Jr.
Vir 12h46'55" -0d5'56"
Adam Vrtkovski's Star 14/05/2004
Cru 12h38'57" -61d22'12"
Adam 'Wade-bong' Wade
Umi 16h14'22" 73d57'2"
Adam Wang Zhe Hofstaedter
Cnc 9h19'27" 13d35'37"
Adam Wayne Deitz
Crb 16h5'28" 37d21'2"
Adam Weeks
Her 16h10'1" 46d46'13"
Adam West
Uma 9h33'18" 49d57'10"
Adam William Chambers
Aql 19h51'9" 16d8'44"
Adam William Purcell
Lib 15h7'13" -10d46'0"
Adam William Zorko
Uma 10h5'53" 65d5'34"
Adam Woda's Star
Her 16h35'4" 48d25'48"
Adam Worth Sherlock
Cep 23h36'45" 72d5'8"
Adam Young Chisholm
And 0h53'51" 39d43'45"
Adam, Candi & Ace Corley's Star
Tri 2h4'17" 31d55'33"
Adam, Jennifer & Bryan Salyer
Uma 10h24'11" 53d47'21"
Adam, Matthias
Uma 8h51'44" 59d15'0"
Adam1991
Her 17h7'25" 33d24'34"
Adamandia Antoniadou
Aqr 20h42'31" 0d24'26"
Adammus Ford
Psc 1h11'57" 27d56'49"
Adamo
Her 17h29'52" 26d9'10"
Adamo
Gem 6h36'46" 24d42'22"
Adamo
Tau 5h26'16" 25d7'46"
Adamo
Cam 3h59'14" 66d7'52"
Adamo Aemilia
Ori 5h55'29" 7d29'38"
Adamo "Girouette" Lafleur
Lac 22h55'57" 50d6'7"
Adamo Immortalis
Cnc 8h25'19" 10d53'19"
Adams
Crb 16h15'31" 30d17'6"
Adams Aster
Leo 10h39'6" 21d39'59"
Adam's Celestial Joie de Vivre
Sco 17h15'15" -44d29'24"
Adam's Diamond in the Sky
Crb 16h23'9" 38d50'45"
Adams Family Star
Dra 16h38'52" 61d12'32"
Adams Family's Shining Star
Tau 4h39'47" 18d29'56"
Adam's Fire
Uma 11h27'24" 51d30'9"
Adam's Hope
Gem 6h49'50" 33d59'13"
Adam's Serenity
Ori 6h11'37" 15d57'47"
Adam's Star
Leo 10h17'45" 27d4'1"
Adam's Star
Cas 0h22'41" 56d36'47"
Adam's Star
Her 17h11'3" 38d28'32"
Adam's Star
Sgr 19h42'15" -15d6'40"
Adams, Manfred Willi
Ori 6h19'4" 13d20'50"
Adamus
Ori 6h17'53" 15d28'38"
Adan
Leo 9h36'3" 28d35'40"
Adan Butkus
Her 16h51'55" 22d34'15"
Adana 746
Cas 0h38'16" 48d52'39"
Adania
Uma 10h36'39" 68d32'46"
Adarsh Dave
Cnc 8h36'29" 23d52'59"
Adasta
Tau 4h19'12" 27d32'43"
ADC 6-10
Aql 19h32'44" 9d5'47"
ADCH HC Jusdandy Duce CDX
Cmi 7h34'15" 5d47'31"
Addam Peter James Warnock
Her 17h36'27" 16d36'43"
Addevalinga
Cyg 20h29'9" 45d6'52"
Addie
Leo 10h50'39" 20d16'22"
Addie
Vir 12h28'43" 2d19'45"
Addie
Peg 0h10'21" 18d6'32"
Addie
Cyg 19h58'33" 57d39'53"
Addie Florence
Uma 11h36'55" 32d59'17"
Addie Peck +
Lmi 9h39'33" 34d53'26"
Addie the Amazon1369 Carebear
Psc 0h40'25" 6d23'39"
Addison
Uma 10h6'17" 49d44'58"
Addison "Addie" Lee Powell
Sco 16h43'33" -42d18'38"
Addison Ann Noone
Uma 10h36'32" 47d40'18"
Addison Brook Perkins
Gem 7h27'8" 30d57'28"
Addison Claire Jones
Aqr 22h7'7" -15d23'58"
Addison Claire Schumacher
Vir 12h0'50" -1d16'0"
Addison Cory 11.27.03
Sgr 20h0'11" -30d14'9"
Addison Diane Miller
And 23h21'45" 43d36'13"
Addison Edmunds Frauenberg
Aqr 21h48'14" 0d44'16"
Addison Elizabeth Rushford
Tau 4h41'22" 5d19'51"
Addison Elizabeth Valentin
And 23h48'45" 46d21'17"
Addison Faith Seminsky
Tau 3h28'53" 13d10'52"
Addison Gayle
Uma 12h5'7" 54d37'57"
Addison Goldberg
And 0h26'53" 37d41'59"
Addison Gorringe
And 2h22'49" 45d56'32"
Addison Grace
And 0h32'27" 39d18'38"
Addison Grace
And 0h9'39" 44d53'36"
Addison Grace
Psc 0h51'59" 31d40'16"
Addison Grace
Apu 14h35'41" -72d36'36"
Addison Grace Masi
Her 17h19'25" 45d28'19"
Addison Grace Oglesby
Sco 16h54'44" -37d3'51"
Addison Jane Warren
Lyr 18h51'23" 43d27'28"
Addison Jean Olsen
Vul 20h29'53" 20d55'6"
Addison Kate Bumbleburg
Cap 21h30'40" -18d41'54"

Addison Kaye McClure
Lyr 19h6'58" 31d29'49"
Addison Knight
Leo 10h47'6" 19d46'22"
Addison LaCole Rush
Peg 0h7'10" 16d32'3"
Addison Lydic
Peg 22h46'33" 6d47'46"
Addison Mae Nolan
Uma 8h59'30" 55d21'51"
Addison Marie Clover
Tau 4h6'28" 11d59'24"
Addison Marie Johnson
Crb 15h47'59" 26d31'56"
Addison Rae Watts
And 23h0'9" 50d36'54"
Addison Raya
Cyg 21h14'37" 47d2'20"
Addison Rose
Cas 23h46'21" 52d56'54"
Addison Rose Mauch
Cyg 21h36'16" 39d0'36"
Addison Victoria Dewar
Vir 12h8'46" 10d33'53"
Addison Wayne Dunn
Uma 12h37'32" 57d59'26"
Addison's Lucky Star
Cam 4h19'7" 69d52'27"
Addison's WISHING Star
Umi 14h43'41" 75d49'59"
Addy Aguilera
Umi 14h24'45" 66d36'1"
AddysCutie
And 0h46'15" 34d36'11"
Addyson L. Bruce
Uma 11h42'34" 44d19'57"
Addyson LaShay Gaillardet
1st B-Day
Mon 7h3'43" -6d13'43"
Ade & Amy's Joy- Adrian
Paul Ledger
Cap 20h43'32" -22d11'25"
Adee Levinstein
Psc 0h41'54" 13d54'0"
Adeel Abbas
Cyg 20h2'14" 35d56'57"
Adel Abdallat
Leo 11h47'9" 21d29'42"
Adel Fakeih - Mayor of
Jeddah 2005
Her 17h34'37" 35d36'43"
Adel Harris
Tau 4h25'58" 23d49'33"
Adel Souto
Peg 21h53'30" 7d14'5"
Adel Zahedi
Gem 6h24'55" 23d42'8"
Adela
And 1h56'33" 39d4'51"
Adela
Uma 10h32'28" 61d35'45"
Adela Wrobel
Cyg 22h1'50" 51d8'10"
Adelaida Vega Melendez
Ari 3h18'13" 29d2'26"
Adelaide
Lyn 7h49'16" 56d50'13"
Adelaide
Ori 6h10'41" -2d6'41"
Adelaide 1951
Per 3h0'10" 51d36'58"
Adelaide A. Arena
And 0h50'32" 35d36'4"
Adelaide Aulich
Lib 15h7'54" -13d59'26"
Adelaide Farah
Lyn 7h54'5" 58d37'14"
Adelaide Janet Coupal
Young
Ori 5h35'30" 6d37'14"
Adelaide Joseph
Sgr 19h38'42" -12d27'33"
Adelaide L. Resendes
Cyg 20h51'40" 43d3'40"
Adelaide T. Bilz
Leo 9h46'56" 26d27'46"
Adelaideblondus minor 20
01 93
Tau 5h51'13" 15d5'16"
Adelais' Hope
And 0h53'52" 42d15'37"
Adele
And 1h48'42" 45d47'42"
Adele
Ori 5h42'56" 2d3'40"
Adele & Armin Elkins
Cyg 20h57'34" 47d6'59"
Adele & David Cox
And 2h23'21" 45d22'6"
Adele Dove
Sco 16h57'42" -39d5'3"
Adele Hymovitz Brown
Uma 8h20'59" 63d1'43"
Adele Jacobs Spitfire
Aur 5h54'38" 54d3'26"
Adèle Martinez
Leo 9h36'21" 13d37'21"
Adele Pate Vanderbilt
Clifford
Ori 6h1'50" 21d29'16"
Adèle Rose Clarisse
Pergay
Her 16h30'40" 22d40'29"
Adele Samantha Larke
And 23h18'19" 50d52'57"
Adele Y. Mussry
Lyn 7h54'35" 39d7'20"
Adele-Cynthia-Catherine
Del 20h36'34" 16d3'41"
Adele's Happiness
Aur 5h12'20" 42d59'49"
Adele's Purple Heaven
And 23h24'59" 38d14'50"
Adelfa "GiGi" Bacallao
Lmi 10h45'5" 30d46'3"
Adelheid Auguste
And 1h21'19" 37d48'38"
Adelheid Fogl
Psc 1h12'32" 28d30'49"
Adelheid Gertrud Kühn
Uma 11h1'15" 47d31'35"
Adelia Gable
Tau 4h25'20" 23d6'32"
Adelia Quillen
Uma 11h50'2" 61d31'9"
Adelia Tomasoni 22
novembre 2003
Boo 14h28'59" 54d20'20"
Adelina
Psc 23h26'23" 7d50'4"
Adelina A. Davila
Ori 5h21'17" 1d44'42"
Adelina Almanza Suira
Ludwig 112770
And 23h49'8" 41d38'35"
Adelina "Della" Marquez
Leo 9h53'44" 22d58'27"
Adelina Jacqueline
Vinciguerra
And 0h18'32" 44d28'49"
Adelina Lora
Lyn 9h2'16" 42d1'6"
Adelina Phoebe Benjamin
Tau 5h46'3" 19d24'26"
Adeline
Uma 13h6'23" 58d50'30"
Adeline
Uma 10h18'35" 55d7'31"
Adeline
Umi 14h46'54" 75d41'19"
Adeline
Umi 14h2'47" 76d49'49"
Adeline Abigail Fish
And 23h51'15" 35d31'52"
Adeline Ambrose Cavicchia
Lib 15h14'55" -5d1'0"
Adeline and Hans' 60th
Anniversary
Cyg 21h5'58" 47d30'48"
Adeline Blaise
Cas 23h59'6" 59d43'8"
Adeline Cora
Leo 10h14'58" 16d11'58"
Adeline Cora
Leo 10h19'25" 16d59'28"
Adeline Elizabeth Moore
And 1h30'17" 37d33'47"
Adeline Fred
Cyg 20h18'53" 55d58'8"
Adeline Grace
Cnc 8h47'32" 15d33'33"
Adeline Juanita Shinnick
Uma 12h15'31" 56d9'2"
Adeline Julianna
Uma 9h5'59" 55d43'5"
Adeline & Julien
Sgr 18h20'42" -23d6'31"
Adeline Laviani
Lyn 7h26'58" 48d38'23"
Adeline LeBlanc
Ari 2h37'19" 29d33'26"
Adeline Marie Harper
Vir 14h42'38" 4d20'54"
Adeline Noland
Aqr 22h17'12" 1d18'43"
Adeline Patrick
And 0h8'18" 39d56'5"
Adeline Rose Kelly
And 1h44'23" 38d30'22"
Adeline Rossiter Haglund
Uma 9h44'12" 45d1'43"
Adeline-Roger 26 Avril
1943
Col 5h23'59" -28d34'2"
Adeline's Christmas Star
Cap 21h6'50" -19d17'26"
Adelio Sarro
Uma 13h50'42" 52d9'59"
Adelita
Leo 10h26'42" 6d30'40"
Adella 04202005
Tau 5h45'24" 19d25'36"
Adellaray
Cam 4h46'22" 62d39'37"
Adelle
Lib 15h43'21" -28d25'35"
Adelle Christian Lilly
Uma 11h39'1" 32d45'32"
Adelle Marie Kasner
Sco 16h5'24" -18d19'31"
adellyn
Sco 16h5'40" -9d41'11"
Adelsia
Uma 8h46'7" 62d10'1"
Adelyn Jean Amiet
Ori 5h17'20" 6d30'5"
Adelyn Rose
Ori 6h13'24" 3d36'50"
Adelyn Ruth Davidson
And 0h12'47" 43d26'33"
Adelynn
And 1h2'23" 42d19'57"
Aden
Cyg 20h22'41" 52d9'24"
Aden Anthony Dorros
Per 2h16'13" 54d35'16"
Aden Cole Hurtado
Aqr 22h16'34" 1d15'11"
ADEN DANIEL
Lyn 8h2'8" 47d12'38"
Aden Isaiah Folkers
Ori 5h55'54" 7d26'19"
Aden John Schmidt
Cep 20h45'2" 65d37'4"
Aden Spirit
Aqr 21h12'16" 0d55'12"
Adena Leigh
Her 17h30'35" 44d28'47"
Adenike Kamilah Ishola
Cnc 9h12'24" 21d15'34"
Adenitran
Uma 12h44'53" 63d8'48"
Adens Victoria
Leo 10h23'19" 26d59'38"
Aden-So You Always
Remember Us-Mia
Psc 1h11'28" 22d49'4"
Adeola
And 1h14'51" 50d22'39"
Aderha Taelynn "Our Little
Angel"
Sgr 19h0'45" -13d52'35"
Adeyemi Adeyeba
Leo 11h25'8" 2d57'50"
ADHOC1227
Ori 4h52'20" 2d12'39"
Adi
Uma 8h39'32" 65d52'51"
Adi & Ruben's Twinkle
Twinkle Star
Boo 14h39'0" 35d58'54"
Adia
Sgr 18h10'7" -19d53'32"
Adian Michael Toomey
Per 4h3'34" 42d35'47"
Adiana
Psc 0h38'48" 7d1'45"
Adianez Enid Cintron
Gem 6h40'55" 20d55'37"
Adiaris
Per 3h16'27" 44d25'16"
Adidas
Psc 23h29'46" 6d52'49"
Adie
Pav 20h59'26" -63d36'3"
Adil
Ori 5h19'2" 1d14'48"
Adila
Vir 14h2'12" -1d16'43"
Adilina
Sgr 18h12'23" -32d0'30"
adilita's light
Uma 10h1'17" 48d51'33"
Adin
Uma 11h7'40" 68d8'36"
Adin & Fiona Jennings
Sgr 20h11'4" -37d50'26"
Adina
Vir 12h20'28" 12d27'5"
Adina Leah
Vul 20h41'6" 26d3'43"
Adina Maya
Psc 0h58'0" 5d4'2"
Adirana
And 1h19'12" 48d45'53"
Adis Kurtovic
Uma 11h41'17" 53d20'30"
Adisa Alic
Uma 10h44'54" 58d26'47"
Adison Grace Bushue
Dra 19h4'11" 56d50'56"
Adison Joan Mullen
Uma 10h45'36" 45d17'3"
Adisynn Jae
Aqr 22h40'29" 1d21'55"
Aditi
Psc 1h26'59" 27d11'10"
Aditi
Gem 7h36'1" 32d58'7"
Aditi Noel
Eri 4h38'58" -0d26'49"
Aditi Patel
Per 4h11'24" 33d54'7"
Adjani Frances DeBellis
And 1h7'54" 46d40'29"
Adjemian Star 5
Dra 17h26'23" 64d9'35"
Ad'ka
Uma 13h49'26" 55d43'27"
Adkazast Astralis Amare
Vel 9h8'10" -43d18'12"
ADL F2282
Psc 1h3'42" 10d22'33"
Adlen Rettinghausen
Lyn 8h20'57" 39d56'39"
Adler
Tau 4h33'0" 29d12'32"
Adlin Barrish
Ori 6h18'51" 3d5'19"
Adlis Brian Wegmann
Psc 22h55'50" 7d38'17"
A.D.M. Brightest Star
Cas 0h56'59" 60d15'30"
Administra
Ori 5h40'48" -2d49'50"
Admiral Optimus Prime
Beta Supreme
Sgr 19h58'58" -33d47'59"
Admiral Steidle
Cnc 8h37'34" 16d31'19"
Admiral Winter's Star
Gem 7h2'44" 24d9'56"
Adnama
Aqr 23h5'8" -8d43'6"
Adnan
Ari 2h46'13" 10d57'18"
Adnan loves Allie
Lyr 19h19'6" 29d50'37"
Adnani, Youcef
Cap 20h38'14" -20d28'33"
Adnashra
Boo 14h14'34" 16d57'48"
Adolf
And 23h3'58" 36d25'59"
Adolf Ehrenhart's STERN
Cep 20h44'4" 58d7'41"
Adolf Eiserlo
Ori 5h13'30" 8d29'9"
Adolf Gampper
Uma 10h12'16" 61d37'26"
Adolf "George" Eunis
Per 3h36'55" 39d32'21"
Adolf und Andrea 28
Uma 13h10'7" 52d38'32"
Adolfo and Kolleen Iglesias
Uma 10h6'33" 45d43'39"
Adolfo Cuadra, MD
Lib 14h36'9" -19d27'39"
Adolfo Flores Vega
Leo 11h31'10" 15d41'54"
Adolfo Garcia
Dra 18h45'53" 50d13'56"
Adolfo Gonzalez
Uma 10h12'1" 44d0'28"
Adolfo Gonzalez
Uma 13h3'0" 54d51'12"
Adolph Calvin & Dorothy
Ann Smith
Cyg 19h39'4" 52d40'27"
Adolphe's Star Eyes
Uma 11h49'29" 53d35'0"
Adonay Segovia
Cep 21h17'57" 60d5'25"
Adonco
Uma 10h53'47" 57d43'10"
Adonia
Cyg 20h59'45" 39d46'49"
Adonijah Masatee
Campbell
Uma 13h35'13" 58d19'31"
Adonis & Aphrodite
And 23h18'55" 52d22'22"
Adonis Jordan
Leo 11h16'21" -1d44'14"
Adonna & Lana Sutton
Cyg 21h14'48" 47d23'10"
Adonness
Cas 23h6'25" 56d8'21"
Adora
And 23h42'0" 42d58'7"
Adora Adora
Cap 21h54'57" -18d22'16"
Adora Kathryn Lee
Ari 3h26'12" 22d48'41"
Adorable Abigayle Taylor
And 0h38'43" 35d49'18"
Adorable Ange Ziad Al
Tanoury
Lyn 7h1'3" 46d13'46"
Adorable Anya
And 23h22'30" 52d8'1"
Adorable Anya
Sgr 18h49'41" -25d56'23"
Adorable Ethan Fine
Vir 11h56'32" -5d34'29"
Adorabuena
Tau 4h44'11" 19d13'17"
Adore
Vir 13h22'2" 11d45'7"
Adore
Psc 1h6'3" 13d20'28"
Adore
Cyg 20h48'16" 42d36'47"
Adorer
Uma 11h38'49" 49d15'7"
Adornetto, Joshua
Uma 8h55'9" 54d24'2"
Adory
Gem 6h55'9" 16d9'32"
Adrain Lee
Uma 11h50'36" 28d31'25"
Adrean
And 23h25'30" 47d53'30"
Adreana Rodriguez
Leo 9h49'58" 16d57'48"
Adreanna's home away
from home
Cnc 9h14'37" 15d31'42"
Adreanna's Star
Sco 16h10'24" -22d20'53"
Adreine, Reneee, and
Gavin
Tri 2h8'45" 33d24'0"
Ad'ren
Crb 16h14'16" 32d33'50"
Adri
Psc 0h40'3" 10d24'25"
ADRI
Lib 15h15'26" -25d30'12"
Adri Solete
Uma 10h0'28" 72d44'5"
Adria
Cas 23h48'8" 58d40'33"
Adria
Sgr 19h15'14" -21d35'43"
Adria
Uma 11h53'35" 38d47'48"
Adria & David Tarleton
Pyx 8h34'35" -31d31'36"
Adria Dawn
Cyg 19h34'55" 44d45'24"
Adria Elise
Cam 4h3'8" 67d37'20"
Adria Kay Bucci
Ori 6h10'17" 18d51'8"
Adria M. Cardaropoli
Uma 9h55'55" 58d21'1"
Adria Rae
Uma 12h3'47" 52d15'58"
Adrian
Aur 6h4'33" 47d50'18"
Adrian
Lyn 8h1'42" 40d4'11"
Adrian
Uma 11h24'36" 30d54'55"
Adrian
Crb 15h36'4" 26d56'19"
Adrian
Leo 10h23'12" 23d34'58"
Adrian
Aql 19h27'49" 7d59'22"
Adrian
Cam 4h11'37" 67d48'51"
Adrian
Uma 9h14'27" 62d8'22"
Adrian
Hya 9h37'48" -21d28'51"
Adrian
Lib 15h1'26" -12d37'17"
Adrian
Dra 9h32'21" 76d9'12"
Adrian
Lib 15h34'51" -25d57'22"
Adrian 23
Cnc 8h43'21" 9d37'24"
Adrian & Alex - Perfectly
Entwined
Cru 12h27'10" -58d14'58"
Adrian & Alexandria
Cnc 8h33'5" 16d29'2"
Adrian Alvarez
Sgr 18h29'49" -31d20'17"
Adrian and Alix
Tau 4h48'7" 22d3'40"
Adrian and Amy forever
Col 6h15'31" -42d49'38"
Adrian and Chanel
Per 4h46'52" 48d54'2"
Adrian Bergin
Cyg 20h35'51" 38d41'42"
Adrian Cox - "A Star for a
Star"
Her 18h17'13" 20d42'18"
Adrian Cronin
Cyg 20h37'28" 59d32'25"
Adrian Dempstar
Cru 12h29'7" -60d21'38"
Adrian Diamond
Gem 7h25'39" 22d6'8"
Adrian & Dini - Happiness
Forever
Cru 12h43'27" -61d37'36"
Adrian Dion - Moesl
Tschach
Aur 5h25'28" 39d26'9"
Adrian Dowlin Bellomo
Tau 3h52'57" 24d9'20"
Adrian Foster
Lyr 18h21'51" 37d47'55"
Adrian G Barr
Dra 10h56'58" 77d41'30"
Adrian Garcia
Leo 10h44'12" 10d12'48"
Adrián González Revilla
Psc 0h59'56" 16d45'50"
Adrian H. Reyes
Uma 11h27'10" 45d57'16"
Adrian & Indy
Her 17h0'43" 13d45'5"
Adrian Izayah
Uma 12h10'50" 60d59'46"
Adrian & Janelle 27
September 2004
Cru 12h33'59" -55d47'20"
Adrian John Wilson's Star
of Love
Cru 12h2'33" -62d46'21"
Adrian Joseph Garcia
Sgr 18h37'0" -29d42'5"
Adrian Kopperud
And 0h22'20" 29d5'19"
Adrian Küffer
Umi 14h56'18" 74d57'6"
Adrian Kyle Torres
Umi 15h20'10" 84d57'28"
Adrian Leighton Hood
Lofaro
Cam 3h50'19" 58d52'50"
Adrian Michael Thulborn
Cyg 20h6'43" 40d34'14"
Adrian Miguel Vega
Sco 16h2'55" -21d53'31"
Adrian Millan
Ari 2h16'8" 14d11'33"
Adrian "My Little Man"
Franco
Lyr 19h18'31" 28d2'24"
Adrian Paenza - Friend and
Humanist
Tau 4h22'26" 23d5'33"
Adrian Percy Dolby
Sgr 18h52'3" -28d42'17"
Adrian Philip Caganap
Lib 15h2'46" -5d32'53"
Adrian Price
Vir 13h18'11" 6d17'1"
Adrian Sam
Sgr 18h57'49" -34d5'41"
Adrian Sanchez
Peg 22h34'31" 11d54'0"
Adrian Scott McCammon
Gem 7h31'19" 24d30'54"
Adrian Sosa
Ari 2h3'12" 23d20'15"
Adrian Susannah
Lyr 19h5'19" 32d2'37"
Adrian & Vanessa
Cyg 21h20'44" 51d0'30"
Adrian Vargas
Lib 15h23'32" -14d23'48"
Adrian Wheeler
Uma 12h5'26" 40d20'39"
Adrian Yurnet
Aqr 21h43'51" 1d9'37"
Adriana
Del 20h45'36" 13d10'3"
Adriana
Peg 23h36'18" 24d57'38"
Adriana
Lyn 9h29'18" 40d22'19"
Adriana
Gem 7h17'16" 32d4'23"
Adriana
Lib 14h42'36" -10d7'44"
Adriana
Sgr 19h20'32" -16d6'3"
Adriana A. Rubalcaba
Cas 3h14'59" 58d14'21"
Adriana Agoglitta
Uma 10h1'15" 50d10'53"
Adriana Ames
Cnc 8h47'8" 21d56'32"
Adriana Anches Yampey
Ori 5h57'53" 6d11'57"
Adriana Boggiani
15/3/1966
Ori 6h7'52" 20d55'23"
Adriana Brigiuti 95
Umi 14h31'0" 71d27'52"
Adriana Brinceanu
Aqr 23h1'26" -10d25'48"
Adriana Calderon
Gem 7h43'11" 23d3'49"
Adriana Chavez-Lopez
Nor 16h6'3" -47d36'18"
Adriana "Chevela" Streit
Uma 8h26'43" 61d27'46"
Adriana Del Vescovo
Sco 16h43'26" -44d8'23"
Adriana Dominique Trotz
Cap 21h57'46" -13d4'38"
Adriana Donnola
Col 5h59'34" -34d56'56"
Adriana D'Urso
Cas 0h31'14" 52d58'20"
Adriana Elizabeth Castro
Hultado
Cnc 8h42'43" 21d27'47"
Adriana F Scala
Vir 12h21'5" 2d19'16"
Adriana Isabel Guadalupe
Mendez A.
Vir 14h28'10" 3d49'22"
Adriana Jose Hipwell
Lyn 6h50'27" 51d34'10"
Adriana Lisette
Tau 5h18'2" 25d40'42"
Adriana Malaver Frederick
Tau 5h50'30" 18d2'10"
Adriana Maria
Lib 14h37'42" -12d24'54"
Adriana Maria Pedroso
Rangel
Ari 2h50'55" 14d20'43"
Adriana Marie Carney
And 23h12'0" 44d49'49"
Adriana Marquez Gonzalez
Sgr 19h18'9" -21d52'28"
Adriana & Matthew Harris
Aql 19h58'17" -0d28'3"
Adriana Mendosa
And 1h44'16" 46d39'15"
Adriana Mercedes
Col 5h42'48" -33d14'29"
Adriana Michele Mirabelli
And 23h9'28" 52d29'26"
Adriana Michelle Vargas
Velazquez
Eri 3h52'46" -0d37'17"
Adriana Milo
Sgr 19h6'34" -31d19'24"
Adriana Monique Ferreira
Vazquez
And 0h16'56" 27d53'22"
Adriana Munera-Reyes
Leo 11h27'33" 5d43'34"
Adriana Negru
Umi 16h20'47" 80d12'33"
Adriana Nicole
Aqr 21h15'56" -8d49'9"
Adriana Oh
Tau 3h37'24" 19d59'32"
Adriana Paredes
Tau 4h55'12" 16d36'43"
Adriana R
Vir 13h49'35" -9d49'55"
Adriana Rodriguez
Crb 16h5'13" 29d31'33"
Adriana Rose Reinersman
Cas 1h39'59" 63d15'41"
Adriana Saraiva
Vir 14h14'6" -18d22'15"
Adriana Sikiric
Cyg 20h56'21" 49d25'15"
Adriana Skyy Wendland
Cnc 8h27'11" 16d42'3"
Adriana Stewart
Lib 15h46'22" -19d27'8"
Adriana Stimola
Lib 15h5'51" -2d50'37"
Adriana Synne Alicea
Sco 16h49'29" -44d43'2"
Adriana Tuma
Aqr 23h27'54" -23d36'26"
Adriana Umana
Sco 17h53'31" -30d38'0"
Adriana Victoria Maldonado
Gem 6h31'39" 25d41'0"
Adriana Victoria RicoToro
Uma 9h9'21" 56d42'51"
Adriana Vikki Dutke
And 1h4'14" 34d5'44"
Adriana Wells
Cas 0h7'56" 53d28'56"
Adriana's
Ari 2h40'54" 13d12'3"
Adriana's Star
Gem 6h28'25" 13d20'54"
AdrianaSulemaRuiz
And 0h46'40" 40d38'21"
Adriane Grace Liedy
Com 12h30'45" 26d46'20"
Adriane Lynn Walker
Gem 7h18'49" 20d6'12"
Adriane S. Giuggio
Lyn 7h47'1" 38d26'50"
Adriane Swalm Durey
Tri 2h9'20" 34d48'6"
Adriane Wilk
Umi 13h54'18" 72d10'56"
Adriane Yvonne Cimino
Cnc 9h13'10" 14d0'14"
Adrianek
Aqr 22h28'15" -1d45'21"
Adriann Alexis Terry
(Duddits)
Cnc 9h4'23" 30d22'8"
Adrianna
And 23h7'41" 35d20'36"
Adrianna
And 23h42'13" 43d48'47"
Adrianna
And 0h42'11" 26d57'50"
Adrianna
Ari 3h24'54" 27d59'28"
Adrianna 7
Tau 4h17'0" 16d41'33"
Adrianna Amelia
Psc 1h12'7" 15d40'40"
Adrianna Christina Healey
Peg 21h35'47" 22d28'56"
Adrianna Christine 2006
Uma 10h42'44" 49d35'9"
Adrianna Ladd LaRose
Lib 14h54'40" -2d2'39"
Adrianna Ladd LaRose
Cas 1h30'30" 62d24'21"
Adrianna Marlene Skelton
Tau 3h45'23" 6d29'10"
Adrianna Mianulli
Crb 15h47'31" 39d31'38"
Adrianna Noelle
Sgr 19h36'36" -27d41'52"
Adrianna Penguin
Sco 16h48'36" -32d17'32"
Adrianna Renee Greco
Lyn 8h16'47" 57d13'29"
Adrianna Renee Ramsey
And 0h41'38" 39d28'26"
Adrianna Revelen
LaMorgese
Aqr 20h40'15" -11d3'40"
Adrianna's Star
Tau 3h41'14" 14d41'34"
Adrianne
Tau 5h28'46" 25d51'31"
Adrianne "AA" Baker
And 23h4'38" 48d27'54"
Adrianne and Tonio
Sge 20h4'32" 16d57'9"
Adrianne Broussard
Vir 12h29'7" 4d25'35"
Adrianne Marie McCoy
Sgr 18h37'26" -23d2'24"
Adrianne Radiance Porter
Aqr 21h14'42" -11d41'8"
Adrianne Rockstar
Gutierrez
And 0h25'51" 38d7'14"
Adrianne Vanderborght
Sgr 18h36'42" -16d59'38"
Adrianne Wade
Aqr 21h49'11" -7d20'1"

Adrianne Wimberley
Leo 11h18'43" 15d22'25"
Adriano
Sex 10h3'4" 2d37'26"
Adriano & Davina
Sge 19h54'0" 17d2'25"
Adriano Geoffrey
Her 18h27'55" 13d6'41"
Adriano Giustiniani
Crb 16h10'23" 25d50'16"
Adriano stella polare
Crb 16h22'41" 31d34'42"
Adrians
Uma 10h49'47" 44d44'13"
Adrian's Ambiguity
Sgr 18h59'6" -32d53'6"
Adrian's Focus
Cru 12h47'59" -59d51'13"
Adrian's Joy
And 2h18'27" 42d41'34"
Adrian's Love
Ari 3h23'32" 30d27'29"
Adrian's Star
Per 3h33'57" 49d50'4"
Adrian's Star
Uma 11h24'24" 57d11'5"
ADRIAN-SHAYEN, 20.01.2003
Uma 12h16'51" 55d57'58"
Adrianus Franke
Cnv 13h30'9" 42d12'16"
Adrideo Beatitudo
Lyr 18h38'34" 30d23'18"
Adriel
Del 20h35'41" 7d9'53"
Adriel Gonzalez
Her 17h5'36" 42d10'13"
Adriel Roni Lubarsky
Sgr 18h4'51" -22d3'36"
Adrielle
And 23h8'57" 41d32'11"
Adrielle
Uma 11h1'47" 50d26'1"
Adrien and Dorothy
Lyr 18h44'7" 39d52'17"
Adrien Oliver Van Doren
Uma 12h32'47" 58d20'28"
Adriene Drucker
Lib 15h52'19" -9d47'23"
Adriene Lynn Celic
And 23h32'10" 48d22'13"
"Adrienn csillaga"
Uma 11h1'14" 64d37'31"
Adrienna Lynn Berg
And 1h41'36" 40d24'34"
Adrienne
And 23h59'20" 34d2'1"
adrienne
Uma 10h35'16" 40d3'59"
Adrienne
Her 16h13'33" 47d44'48"
Adrienne
Ori 4h51'51" 11d27'36"
Adrienne
Com 12h33'52" 18d1'47"
Adrienne
Umi 14h25'28" 65d41'51"
Adrienne
Uma 9h46'44" 57d1'19"
Adrienne
Uma 8h59'50" 54d14'25"
Adrienne
Lyn 6h57'8" 52d31'59"
Adrienne 3952
Cyg 22h0'47" 50d45'25"
Adrienne And Marie
Crb 15h46'54" 27d23'54"
Adrienne and Michael Zarn
Sge 19h38'26" 16d48'57"
Adrienne Anne
Cyg 21h58'37" 48d24'44"
Adrienne Beth
Ari 2h50'36" 16d47'47"
Adrienne Beth Aperghis
Cas 22h57'58" 53d14'36"
Adrienne Brooke
Ori 6h3'45" 6d13'18"
Adrienne Byng
Aql 18h55'6" 9d17'15"
Adrienne Denali
Lib 14h53'51" -3d11'18"
Adrienne Denise Johnson
And 0h48'31" 44d21'47"
Adrienne E. Crippen
Uma 11h35'12" 46d39'44"
Adrienne Elain Walker
Cyg 21h1'57" 46d12'1"
Adrienne Elaine Summers
Com 12h12'27" 28d22'0"
Adrienne Eliazar Gandaoli
Cnc 8h26'29" 27d58'27"
Adrienne Elise
Tau 4h41'27" 17d20'26"
Adrienne Fulton Ryan Johnson
Per 3h45'27" 34d22'48"
Adrienne Grace Borbi
Vir 13h21'47" 13d44'6"
Adrienne Gray McNeill
Leo 11h26'43" 17d14'39"
Adrienne Grierson
Lib 14h23'44" -19d23'54"
Adrienne Gunter
Leo 9h25'25" 10d16'12"

Adrienne Jane
Gem 7h2'59" 19d57'54"
Adrienne Jensen
And 2h34'45" 43d31'34"
Adrienne Jeretta Chittenden Graham
Cas 1h45'48" 65d11'32"
Adrienne L. Cean
Sgr 17h59'3" -17d28'37"
Adrienne L. O'Laughlin
Tau 3h28'8" 11d36'31"
Adrienne Lieberman
Vir 14h33'37" -0d25'33"
Adrienne Marie
And 0h12'23" 43d24'14"
Adrienne Marie
And 1h16'40" 34d32'22"
Adrienne Marie McKay
And 0h14'41" 42d37'30"
Adrienne McKannay Shines Forever
And 1h11'14" 37d39'26"
Adrienne Michelle
Cnc 8h30'40" 26d30'3"
Adrienne Michelle Balut
Com 12h50'21" 26d14'58"
Adrienne & Mike
Cyg 19h46'22" 52d18'23"
Adrienne Mulholland
And 0h10'27" 39d4'24"
Adrienne P. Matteini
Leo 11h18'51" 19d28'18"
Adrienne Pagano
Lyn 6h36'41" 57d43'24"
Adrienne Pasquine
Aqr 22h59'41" -24d27'14"
Adrienne Pearl
Psc 0h51'25" 10d38'13"
Adrienne Poulakakos
Psc 1h0'31" 3d17'44"
Adrienne Randolph
And 2h22'54" 49d28'35"
Adrienne Renee
Cyg 19h53'43" 30d34'2"
Adrienne Riska
Dra 15h36'55" 55d53'38"
Adrienne Rothwell
Tau 4h10'35" 28d57'20"
Adrienne Shelby
Sco 16h56'12" -42d36'5"
Adrienne Ventresca #14
Uma 10h27'7" 66d48'58"
Adrienne Vogt
Cap 20h39'45" -16d12'45"
Adrienne, Scott, Max and Luke
Uma 9h28'14" 64d12'40"
Adrienne, Scott, Max and Luke
Lmi 10h28'38" 36d1'52"
Adrienne, The Light of My Life!
Cnc 8h41'16" 32d25'25"
Adrienne's Hope
Mon 7h15'15" -0d1'26"
Adriennes Pink Star
Sgr 18h6'1" -17d11'52"
Adrienne's Star
Lmi 11h4'4" 28d51'55"
Adriléo
Cnc 8h11'58" 6d54'35"
Adrin Joseph Corelli
Per 2h34'41" 51d33'41"
Adrina Serus
Gem 7h50'17" 27d7'14"
Adrine
Cnc 8h33'32" 7d33'24"
Adrine (Zoween) Salatian
Cas 23h29'29" 58d1'21"
Adriyana Yvonne Lanza
And 23h2'25" 41d9'47"
Adrockaylalapop
Leo 10h5'5" 16d40'2"
Adry
Peg 22h56'12" 25d51'50"
Adry Cepeda-La Estrella en mi cielo
Ari 3h3'50" 13d17'43"
Adryan Elizabeth Shilling
Vir 13h16'44" 7d3'18"
Adryan Mardesich
And 0h34'54" 36d49'20"
Adryann
Sco 16h18'57" -8d33'21"
ADSAJM
Aql 19h29'4" -11d37'20"
ADSL - Delphine et Laurent
Ari 2h48'48" 22d21'48"
Adsum
Uma 8h55'4" 58d22'8"
Adutel
Uma 11h29'1" 41d39'8"
Adventurer Jack
Uma 8h53'49" 51d41'1"
Adviana
Ari 3h14'52" 29d15'36"
adwoa osei
Psc 0h48'24" 16d0'5"
Adymn Calcagni
Uma 10h56'15" 68d11'41"
Adyn Charles Fischer
Ori 5h32'41" 0d17'37"
Ady's Star
Uma 8h35'57" 68d50'4"

Adysha and Nadia
Vul 19h21'51" 21d35'33"
Adyson Mackenzie White
Psc 1h22'48" 7d22'13"
Adz The Maverick 120105
Per 3h20'7" 41d16'2"
Ae Lee
Dor 5h9'24" -67d46'45"
AEB
Dra 16h3'37" 66d20'37"
Aedan Bailey
Her 17h21'35" 39d52'59"
Aedan James Sinhart
Umi 14h37'41" 78d55'21"
Aedan Marshall
Aqr 21h49'55" -3d20'41"
Aedan Micheal Gilman
Her 16h44'59" 44d27'21"
Aegina
Vir 13h2'23" 12d2'50"
AEindigoJL
Cyg 21h46'34" 49d30'1"
Aeiony & Eahnion
Cyg 20h11'4" 41d0'22"
Aelene Hollriegel
Cnc 9h0'40" 14d39'45"
AelizCyna "Cyna" Nickole Gomez
Lib 15h23'33" -6d8'49"
Aelyna Marais
Peg 22h17'45" 12d6'29"
AEMEK
Dra 16h56'48" 52d1'20"
Aemilia Espinoza
Uma 14h14'23" 57d50'44"
AEON
Cyg 20h6'50" 35d26'45"
Aeonea Graystar
Sgr 18h23'16" -24d15'42"
Aeos Tziyon
Tau 4h21'35" 10d23'13"
AERE
And 1h10'15" 37d2'58"
AERI
Sco 17h57'17" -43d12'54"
Aeriana's Guiding Light
Leo 10h26'45" 15d15'58"
Aerielle
Psc 1h23'35" 11d25'52"
Aeris
Uma 10h18'0" 56d29'6"
Aeris Stargirl Jaurique
Peg 0h13'59" 16d58'59"
Aerosmith and Time
Vir 13h10'12" -7d25'2"
Aerro's Heart
Ori 5h44'54" 7d28'12"
Aeryell Dawn Dunnuck
Uma 9h1'35" 72d55'16"
Aescal's Mom
Cru 12h41'9" -56d6'26"
AESMD2K4
Lac 22h25'25" 43d7'33"
Aesmkead
Lyn 8h57'13" 40d36'28"
Aeson Neleus Tripodis
Gem 6h42'48" 25d19'33"
Aesop,Bacchus,Chloe,Daphne & Ella
Cam 4h26'48" 62d58'40"
Aeternitas
Leo 9h30'47" 27d33'38"
Aeternitas
Ari 2h41'34" 15d47'30"
Aeterno Vita
Her 17h0'49" 34d27'10"
Aeternum
Oph 17h17'17" -0d30'45"
Aeternum Amor
Ari 2h44'7" 15d39'13"
Aeternum Amoris
Psc 1h10'43" 17d46'12"
Aeternum Yuki
Lib 15h46'6" -15d2'12"
Aeternus
Oph 17h2'54" -0d41'26"
Aeternus Amo
Ari 2h49'45" 28d25'38"
Aeternus amor
Ori 6h24'19" 17d18'26"
Aeternus Amor
Cyg 21h35'39" 45d23'11"
aeternus amor
Cap 20h20'16" -9d42'17"
Aeternus Amor
Uma 9h20'43" 57d20'55"
Aeternus Eternus
Vir 13h44'37" 2d51'12"
aeternus eternus amor et conventus
Cyg 19h30'17" 29d15'1"
Aeternus Fideliter Vester
Cyg 20h23'23" 34d52'45"
Aeternus Georgiya fabooo
Aqr 23h41'53" -16d1'9"
aeternus letificus
Cyg 19h36'16" 31d43'27"
Aetheria Aqua
Leo 10h22'21" 16d20'37"
Aetr
Cas 0h18'57" 57d42'8"
Aeturnus Furst
Cas 23h10'57" 58d56'12"

A.F. Kosierowski : The Greatest
Cap 20h44'26" -18d29'3"
Afaf Zwein, Foufa
Per 3h31'7" 41d50'56"
Affa Paul
Tau 4h33'17" 28d20'49"
Affe
Per 4h37'17" 48d56'14"
Affinity
Dra 19h43'19" 72d45'2"
Afflatus of Andrea
Cyg 21h46'17" 48d22'0"
Aff-Nim-Way
Uma 8h14'58" 67d41'50"
AFLAC
Uma 12h25'12" 56d42'58"
A.F.-Love-Star
Uma 10h6'14" 46d27'4"
Afonso Alexiades Goncalves Cerejeira
Her 17h10'50" 39d36'21"
Afonso Oliveira Paulino de Noronha
Per 2h58'6" 52d15'13"
Afra & Ted Wade
Cyg 20h38'17" 55d20'0"
Afri Tejero
Lib 15h37'39" -17d44'21"
Africa Zamora Garcia
Vir 12h35'56" -7d44'22"
Afrodinat Heredia
Uma 9h18'2" 60d20'42"
Afsanah Tafreshi
Lyn 7h47'17" 37d20'0"
Afshan Andesha
Tau 4h19'23" 8d35'36"
After Everyone We Have Lost
Her 17h10'9" 30d31'57"
Afterlife
Uma 11h53'37" 55d1'34"
Afton
Aql 19h39'41" 15d33'25"
Afton Allen Eyrich
Cnc 7h56'1" 13d52'57"
Afton Ella Grace
And 1h55'33" 36d29'50"
Afton M. Crewes
Uma 13h34'59" 55d40'22"
Afton Sieting
Gem 7h12'33" 34d39'34"
Afton-05
Leo 9h30'25" 8d27'59"
A.G. Kasselberg
Peg 22h38'56" 27d17'43"
Agadir
Peg 21h56'20" 18d45'31"
Agal
Dra 10h40'11" 77d39'8"
Agapao
Per 3h30'14" 45d11'57"
Agape'
Uma 11h40'17" 52d23'28"
Agape
Cyg 21h32'2" 51d11'23"
Agape
Uma 10h26'38" 44d58'58"
Agape
Gem 7h14'8" 21d44'15"
Agape
Cnc 7h56'8" 16d2'39"
Agape
Gem 7h50'41" 27d21'4"
AGAPE
Cnc 9h17'3" 9d8'49"
AGAPE
Cnc 8h12'52" 11d41'58"
Agape
Gem 6h52'21" 14d41'46"
Agape
Ori 5h13'10" 12d20'51"
AGAPE
Ori 6h3'1" 7d15'10"
Agape
Uma 13h40'30" 58d6'20"
Agape
Cyg 19h25'54" 52d55'16"
Agape
Lyn 6h38'58" 55d24'19"
Agape
Cep 22h58'55" 61d5'27"
Agape/1st Corinthians 13:4-8
Ari 2h51'39" 26d26'48"
Agape Aion - Megan and Doug
Uma 10h29'33" 56d38'47"
Agape Chadash
Tau 3h24'35" 14d52'17"
Agape Oikos
Uma 13h46'57" 48d13'45"
AgapeNova
Uma 10h40'38" 47d24'42"
Agapes Mou
Vir 14h42'30" 6d14'37"
AGAPI
And 23h39'50" 42d12'21"
Agapito
Per 2h30'50" 52d30'30"
Agashkin
Leo 11h4'25" 12d26'31"
Agata
Peg 23h56'25" 27d57'58"

Agata Blaszczynska
Cas 1h55'19" 62d57'3"
Agata Giacobbe
Lyn 8h4'29" 44d9'3"
Agata Moretti
And 2h26'17" 42d18'46"
Agata Turano
Crb 16h5'31" 32d46'6"
Agata's inima
Cep 3h53'11" 84d48'21"
Agatha
Cnv 12h33'43" 48d30'45"
Agatha I. Mattingly
Sgr 17h59'52" -30d49'0"
Agathe
Her 17h37'1" 20d48'16"
Agathe Ancellin
Uma 14h28'4" 57d47'39"
Agathe Delmont
Umi 13h44'42" 72d46'56"
Agathe ENGUEHARD
Cnc 9h19'35" 10d7'24"
Agathe Raciazek
Lib 15h57'6" -17d31'39"
Agatka Opacka
Aql 19h54'1" -0d16'11"
Age
Cap 21h30'5" -20d2'55"
Agelu Cailey
Cnc 8h38'38" 9d27'25"
Agent Frank
Peg 23h19'28" 15d55'23"
ageromnisastrum 20 05 2005
Dra 17h10'43" 55d46'24"
Ageyuhi Unitsi
Sgr 18h12'6" -16d30'36"
Aggie
Cma 6h39'22" -15d47'32"
Aggie Capalino
Sgr 20h24'38" -34d44'19"
Aggie Guy
Cap 20h10'55" -15d58'21"
Aggie Nogas
Gem 6h11'5" 24d37'12"
Aggy
Ari 2h51'3" 27d46'19"
Agica
And 0h8'10" 44d0'47"
Agitarius/ Rocky Jazz
And 0h29'19" 38d2'34"
Aglaia
Cam 4h29'40" 67d13'14"
Aglaothumos Lukeios
Uma 9h31'46" 68d44'16"
AGLEA
Lyn 8h11'4" 33d46'8"
Agnar afi
Cep 22h56'35" 57d48'0"
Agnar afi
Cep 6h48'8" 87d7'41"
Agnelo
Uma 10h47'30" 54d28'52"
Agnes
Cyg 20h31'29" 56d11'10"
Agnes
Leo 10h6'50" 22d26'5"
Agnes A. Brannan Simpson
Col 6h27'27" -36d55'58"
Agnes Adams Kelly Milligan
Uma 12h47'46" 59d0'58"
Agnes (Aggie) L. Plants
And 23h42'50" 47d18'13"
Agnes and Celio da Silveira
Eri 1h37'58" -57d14'8"
Agnes B. Kenmir
Uma 14h10'50" 60d8'35"
Agnes Benedicta
Ori 5h16'13" 7d8'20"
Agnes Boer (Mother,Oma,Friend)
Lib 15h57'58" -14d59'6"
Agnes C. Dattilo
Vir 14h15'45" -14d31'42"
Agnes Carty
Cas 1h10'1" 59d16'13"
Agnes Cecilia Joyce Hunnell
Sgr 19h37'37" -14d48'56"
Agnes Cicha
Lyn 6h18'30" 57d11'41"
Agnès De Fornel
Del 20h56'11" 14d37'51"
Agnes De Gise
Cas 1h24'54" 64d11'33"
Agnes Fontana
Lib 14h48'48" -2d10'30"
Agnes Gannon
Cas 1h44'7" 65d25'35"
Agnes Gerber & Oliver Gurtner
Vul 19h31'42" 24d47'50"
Agnes Grace Whitenite
Cas 3h22'3" 74d25'8"
Agnes Harms
Lib 15h46'39" -24d59'9"
Agnes & Jeremy
Ori 6h14'2" -1d21'1"
Agnes Joy Wright
Cap 20h41'52" -14d48'49"
Agnes K. Shawver
Cap 20h14'53" -23d34'16"
Agnes Koutzun
Lyr 18h45'3" 32d27'47"

Agnes Kreft
Uma 8h42'31" 51d51'2"
Agnes Lawson Clark
Crb 16h12'48" 35d53'31"
Agnes Lorraine Shopshire
Cam 6h2'27" 60d23'3"
Agnes Luella Glass
Crb 16h12'20" 27d37'22"
Agnes Mae Washington
Cas 0h42'20" 53d26'41"
Agnes Makal
Aqr 23h52'6" -17d1'4"
Agnes Nolan Engels
Ari 2h46'13" 29d24'57"
Agnes Norrene Stafford
Cyg 21h38'27" 43d3'17"
Agnes Ormières mon amour
And 23h1'3" 38d59'42"
Agnes Pascual
Sco 16h5'40" -14d38'10"
Agnes Pieprzycka
Umi 14h59'49" 75d52'59"
Agnes Reed ~ October 26, 1998
Cas 1h22'3" 52d15'10"
Agnes Roper
Cyg 19h55'45" 44d22'57"
Agnes Smith - Christmas Star
Cas 1h22'35" 69d42'29"
Agnes Starr DellaSala
Psc 1h25'32" 32d56'28"
Agnes Sze Pui Chan Leung
Gem 6h49'40" 21d57'5"
Agnes Teresa Riggio
Cap 20h18'8" -10d51'38"
Agnès, Ma Comète
Cnc 8h52'21" 26d40'25"
Agnese
Cyg 19h55'29" 35d17'8"
Agnese
Lyr 18h45'47" 41d17'0"
Agnese Atti
Ori 6h9'41" 15d20'46"
Agnese of God
Uma 11h10'26" 44d22'25"
Agnest J. Trott
Umi 15h27'54" 89d41'10"
Agneta Ingrid Tybell
Aql 18h52'25" -1d16'47"
Agniesia
Cap 21h0'27" -14d59'13"
Agnieszka
Sco 17h32'13" -44d23'42"
Agnieszka
Cyg 20h13'28" 51d26'20"
Agnieszka
Ari 2h11'40" 27d35'35"
Agnieszka Bastrzyk
Uma 9h41'9" 54d18'51"
Agnieszka Calka
Psa 22h37'21" -25d40'24"
Agnieszka Ines
Mon 6h31'9" 8d9'19"
Agnieszka Kanik
Cyg 21h43'50" 54d18'15"
Agnieszka Katarzyna
Sco 16h11'52" -15d36'12"
Agnieszka Kolodziejska
Cap 20h38'14" -20d48'4"
Agnieszka Merak
Lib 14h54'0" -21d4'44"
Agnieszka Pilawska
And 2h20'0" 38d43'14"
Agnieszka Rosiak
Sgr 19h22'23" -27d53'59"
Agnieszka, My Beautiful Angel
Cas 1h24'29" 64d35'8"
Agnita Nennig
Crb 15h45'56" 35d6'13"
Agnus and Bertha
Aqr 20h40'49" -10d1'42"
Agnus & Ed Williams
Cyg 20h52'10" 38d48'30"
Agnus Miller *Nancy*
Cyg 20h44'10" 33d2'26"
Agop Tepeli,M.D.
Her 18h4'12" 17d35'10"
AGORASTIA 70
Cnc 8h52'54" 31d26'6"
Agostino Li Calzi
And 23h7'4" 42d40'37"
Agren-Kokorda
Leo 9h32'1" 12d6'53"
Agresta
Cru 12h44'54" -60d12'35"
Agu Axeblood Majajas
Cnc 8h38'2" 12d49'14"
Agu Viivi
Lyn 6h20'43" 53d59'13"
A.G.U.1-Moorley
Uma 8h53'5" 50d30'17"
Aguayo-Gabert
Tri 2h25'31" 35d49'56"
Agueda Antonia Villamán García
Ari 3h27'2" 29d11'5"
AGUFIN
Sco 16h10'32" -16d49'8"
Aguilar
Her 17h31'28" 17d13'34"
AguirrePoole
Cyg 19h56'33" 49d58'10"

Agus & Mike's 1st Anniversary Star
Leo 9h45'14" 28d13'27"
Agusia
Ari 2h15'44" 26d40'34"
Agustin Castro
Umi 14h12'36" 67d20'53"
Agustin Exposito
Per 4h42'12" 39d20'40"
Agustin M. Castellanos, M.D.
Cnc 8h52'46" 14d18'37"
Ah
Uma 11h30'50" 35d1'47"
Ah Busafi
And 0h45'12" 39d50'46"
AH & KG
Tau 3h37'3" 15d53'21"
Aharon
Ori 6h4'33" 11d46'29"
Ahava
And 0h31'23" 40d45'50"
Ahava Be'met
Pho 0h18'3" -43d24'34"
Ahava Le-olam
Aur 5h25'18" 41d13'36"
Ahavah
Ori 5h30'15" 0d20'0"
AHDALAPI
Dra 16h14'21" 54d8'25"
a-heart-s
Cas 0h39'53" 65d23'27"
Ahemia
Lac 22h51'13" 49d0'49"
Aheri
Umi 14h22'13" 75d51'58"
Aheri JARP
Uma 8h44'11" 61d48'48"
Ahh Boo
Sco 17h26'17" -45d0'56"
Ahisha & Jayson
Sco 16h17'43" -16d28'41"
Ahlam Ali Soueid
Mon 7h0'34" 9d13'2"
Ah-La-Voo-22
Vir 12h22'20" -6d38'32"
Ahlea Shining Beautiful
Gem 6h48'15" 26d30'12"
Ahlysha Angela Gopaul
Psc 23h50'0" 1d27'41"
Ahmad Ayed AL Azemi
Uma 12h44'16" 61d24'9"
Ahmad Bitar
Umi 15h51'52" 82d58'15"
Ahmad Lawton
Her 17h59'13" 14d51'36"
Ahmad's Star
Psc 0h47'5" 4d51'54"
Ahmed Amer
Boo 14h43'29" 19d36'46"
Ahmed Anani
Ari 2h54'19" 24d59'14"
Ahmed Elhaddi
Aqr 22h9'17" -15d48'32"
Ahmed Kardous
Boo 14h56'28" 18d14'27"
Ahmed Omran
Uma 11h54'50" 32d55'16"
Ahmed Sjouke Elhusseiny
Leo 10h12'20" 25d4'2"
Ahmet Cemil Baykal
Sco 16h45'1" -31d50'28"
Ahmi's Star
Ori 5h18'51" 15d30'42"
Ahmos Netanel
Eri 4h23'46" -1d12'59"
Ahndrea Moriah McEwan
Lyn 7h49'39" 48d12'38"
Ahnika Littlefield
Cap 21h11'44" -15d28'24"
Ahnlichkeit
Leo 11h10'2" -0d33'38"
ahomea
Uma 10h4'38" 62d21'15"
AHR13004JMC
Lyn 7h21'17" 46d19'36"
AHRE2005
Uma 13h48'0" 56d31'15"
Ahrendt, Michael
Ori 6h9'9" 7d2'1"
Ahrens, Dietrich
Ori 5h0'46" 15d13'19"
Ahrens, Gerhard
Sgr 18h0'25" -28d3'25"
Ahrens's Fly Theory
Uma 11h24'26" 35d28'51"
A.H.S. and A.B.M. 7-08-06
Tau 4h39'38" 28d48'29"
Ahslee E. Justice 1 Cor. 13:4
Her 16h40'38" 33d8'16"
Ah-Son Olivia
Del 20h37'22" 9d22'32"
Ahtramus & Scotia
Uma 10h46'23" 64d39'43"
Ahubek
Uma 8h30'31" 65d45'9"
Ahuva "Pumpkin" Rothschild
Aqr 21h38'1" 1d2'41"
Ahva Afnani
Cap 20h53'19" -25d32'59"
Ahymara
Sco 17h39'20" -35d8'1"

Ai Fei Chong
Crb 15h47'1" 27d37'33"
Ai Minamikubo
Sgr 18h55'52" 20d26'16"
Ai Nhi
Peg 22h37'32" 8d12'51"
Ai Thanh
Tau 5h11'25" 18d12'58"
Aida
Cnc 8h23'39" 14d19'49"
Aida
Lyn 7h49'2" 38d0'24"
Aida
Lyr 18h41'37" 38d31'4"
Aida
Psc 0h26'8" 2d21'56"
Aida
Psc 0h52'56" 5d20'0"
Aida
Agr 22h35'5" -2d5'52"
Aida Ann Vartanian
Per 4h8'46" 35d7'7"
Aida Bresha
Uma 11h4'25" 36d17'5"
Aida Calma
Umi 14h43'12" 79d15'52"
Aida Mautino
Uma 10h54'9" 37d4'20"
Aida Mendiola
Equ 21h17'24" 8d56'11"
Aida Menéndez Oliveros
Uma 11h8'30" 32d0'16"
Aida *Nin*
Psc 0h25'36" 8d2'0"
Aida Stratouri
Uma 10h35'37" 63d17'55"
Aida Tardozzi
Per 3h7'17" 51d42'12"
AidaMartinezFernandez 31.10.80
Ori 6h19'59" 6d8'28"
Aidan
Tau 4h6'54" 21d16'25"
Aidan
And 23h20'27" 44d54'16"
Aidan
Cap 21h32'44" -23d48'11"
Aidan
Uma 13h27'23" 55d41'55"
Aidan
Umi 16h4'14" 85d20'2"
Aidan
Lib 14h52'29" -7d3'19"
Aidan
Sgr 18h47'7" -18d3'44"
Aidan 5
Ori 6h25'11" 10d39'47"
Aidan A. Bernard
Aur 5h41'45" 46d42'7"
Aidan Abraham Nolan
Cas 1h33'48" 64d39'40"
Aidan Anthony
Umi 15h30'22" 69d51'52"
Aidan Asquith
And 0h12'48" 46d26'53"
Aidan Athos Murray
Umi 15h14'55" 68d47'21"
Aidan Barth
Psc 1h9'43" 9d54'26"
Aidan Brianna
Leo 11h45'58" 25d40'29"
Aidan Charles Norvell
Cnc 8h44'34" 18d21'54"
Aidan Christopher Berg
Sgr 19h3'30" -26d48'48"
Aidan Christopher Golden
Lib 15h47'36" -10d3'45"
Aidan Christopher Reidy
Umi 16h18'6" 76d56'13"
Aidan Clare "Angel" Heidinger
And 2h27'50" 49d28'53"
Aidan Clifford Fairchild
Psc 1h58'15" 7d59'1"
Aidan Conner
Leo 11h33'29" 25d19'29"
Aidan Corey Fleischer
Per 3h12'2" 46d8'2"
Aidan Daniel Coates
Umi 14h34'44" 75d46'41"
Aidan Daniel Cooper
Ori 6h4'42" 19d33'24"
Aidan David Ayala
Vir 12h16'56" -11d25'53"
Aidan DeWitt 7.9.99
Uma 11h11'39" 56d2'3"
Aidan E. Juhl
Vir 12h26'55" 5d42'32"
Aidan Edward
Uma 10h1'27" 56d56'22"
Aidan Edward Brady
Per 3h9'8" 56d32'18"
Aidan Edward Mackowiak
Uma 9h43'57" 49d16'44"
Aidan & Ethan Lay
Lib 14h37'37" -25d1'6"
Aidan F. Kane
Cap 20h39'57" -22d4'28"
Aidan Frye
Agl 19h57'21" 13d18'55"
Aidan Geoffery Igler
Umi 16h22'35" 77d0'47"
Aidan Georges Jack Downey
Peg 22h27'0" 25d25'13"
Aidan Gibbons
Cap 21h21'48" -14d58'31"
Aidan Gustavo Leibe
Leo 9h23'41" 11d51'0"
Aidan Harold Ley
Tau 5h36'23" 26d18'44"
Aidan Harry Brand Hosie
Umi 15h4'48" 68d14'7"
Aidan Hogge
Scl 23h37'42" -26d51'15"
Aidan Hunter Burke
Lib 14h52'56" -1d29'28"
Aidan Isaiah Rose
Psc 0h53'20" 32d51'55"
Aidan James
Lyn 8h25'46" 33d42'5"
Aidan James Headley
Vir 12h33'27" 0d52'55"
Aidan James Hollingsworth
Gem 7h31'19" 29d51'30"
Aidan John McGhee
Peg 21h38'52" 23d20'38"
Aidan Joseph "A.J." Wenner
Cep 20h38'44" 65d20'18"
Aidan Kai Cariadus
Gem 6h12'52" 25d3'23"
Aidan Kelley
Ori 5h37'35" 8d20'10"
Aidan Kennedy
Per 3h23'13" 51d51'25"
Aidan L Harp
Her 18h7'31" 48d27'39"
Aidan Lindmark
Aur 5h47'40" 44d23'48"
Aidan Literovich
Uma 11h14'32" 30d9'32"
Aidan M. Baumann
Dra 18h39'6" 65d0'52"
Aidan Major
Ori 6h15'28" 15d21'5"
Aidan Maxwell Lyall
Per 2h48'52" 36d22'50"
Aidan McCann
Lyn 8h23'26" 46d6'30"
Aidan Michael
Her 16h42'49" 29d49'35"
Aidan Michael
Cap 20h35'44" -22d55'13"
Aidan Michael Buckley
Vir 12h49'35" -7d25'44"
Aidan Michael Holland
Umi 16h9'55" 74d12'57"
Aidan Michael Rosa
Leo 11h24'52" 25d47'15"
Aidan Michael Rossi
Psc 23h54'26" 6d58'16"
Aidan Michael Truelove
Cep 22h7'7" 60d24'12"
Aidan Michael-Louis Reyes
Pho 0h29'25" -42d14'52"
Aidan Mitchell Love
Sco 16h33'35" -31d16'53"
Aidan Monaghan Elliot
Umi 17h21'44" 86d7'37"
Aidan Moody Naylor
Cam 3h55'57" 55d29'30"
Aidan Nickolaus Holman Depue
Vir 12h11'31" -10d58'5"
Aidan Ocean
Uma 10h27'11" 65d22'47"
Aidan O'Malley
Ori 5h31'4" 10d53'20"
Aidan Owen Whalen
Cru 12h25'36" -62d58'6"
Aidan Page Hughes
Cap 20h51'20" -20d30'18"
Aidan Patrick
Her 17h39'14" 16d32'42"
Aidan Patrick Healey
Umi 14h20'49" 76d13'15"
Aidan Patrick McGeehan
And 1h38'36" 45d33'16"
Aidan Prescott
Eri 3h45'3" -0d15'13"
Aidan Reilly Garrigan
Per 2h24'14" 51d21'33"
Aidan Reynolds Bonner
Cep 23h41'16" 72d5'37"
Aidan Richard Earlin Whiteaker
Uma 8h41'13" 49d6'17"
Aidan Richard Rioux
Uma 10h12'16" 41d36'11"
Aidan Rosemary Taylor
Tau 5h42'51" 19d15'21"
Aidan Rourk
Dra 18h54'11" 61d29'28"
Aidan Ryan Folmer
Dra 18h0'41" 74d59'18"
Aidan Samuel Korff
Her 17h58'16" 21d14'28"
Aidan Scott Barge
Ori 4h52'32" 1d1'24"
Aidan Sean Burke
Uma 9h59'25" 49d26'27"
Aidan T. Boudreau
Uma 9h8'10" 58d51'28"
Aidan T. Stewart
Uma 12h50'54" 56d55'44"
Aidan Thayer
Aur 5h54'35" 37d20'48"
Aidan Thomas Cole
Leo 11h3'20" 13d2'19"
Aidan Thomas Foley
Cnc 8h23'37" 20d29'57"
Aidan Tooke Pollick
Lyn 9h10'23" 34d43'13"
Aidan Walker Shea
Her 18h37'19" 24d8'43"
Aidan Walter
Per 3h9'47" 53d10'0"
Aidan Wayne Steff
Ori 5h36'57" -1d40'1"
Aidan William Grems
Leo 11h55'42" 23d5'53"
Aidan William Martzloff
Peg 22h29'58" 18d7'55"
Aidan, our Young Man of Courage
Cen 12h38'41" -51d41'15"
Aidan11504
Uma 11h15'35" 31d4'0"
Aidan's Muse
Uma 10h54'57" 68d57'29"
Aidan's Star
Lib 15h24'26" -16d12'40"
Aida's
And 0h20'35" 29d49'9"
Aideen
Peg 22h50'11" 25d1'7"
Aiden
Cam 6h21'43" 69d35'48"
Aiden & Addison
Uma 9h1'40" 69d3'4"
Aiden Allen Beltzer
Sgr 18h22'25" -24d14'15"
Aiden Brent Loots
Cap 21h44'11" -11d50'36"
Aiden Brian Kent
Lib 15h8'29" -13d47'59"
Aiden Christain Haislip
Her 17h34'7" 28d40'57"
Aiden Christopher Gordon Tetterton
Gem 7h10'37" 17d13'24"
Aiden Christopher Hines
Ori 5h45'38" 10d23'14"
Aiden Cole Elliott
Cep 21h11'17" 67d26'10"
Aiden Conner Corvisiero
Leo 10h3'53" 12d50'34"
Aiden Cooper Hall
Lmi 10h43'10" 31d24'12"
Aiden Daniel Black
Lib 15h0'25" -5d9'8"
Aiden Dushan Fongemie
Agr 23h6'51" -10d4'52"
Aiden Frederick Plouffe
Her 16h50'37" 35d42'32"
Aiden Gavin Locht
Leo 11h53'20" 20d45'6"
Aiden Gene Kerr
Leo 9h58'20" 24d2'28"
Aiden George Carroll
Agr 22h7'56" -2d41'49"
Aiden Gilles Lavoie
Lib 15h10'19" -15d48'43"
Aiden Gregory Blum
Cap 20h34'14" -23d29'26"
Aiden Henry Richard Merrett
Umi 15h27'12" 81d51'12"
Aiden James Edward Carella
Ari 2h38'27" 21d56'16"
Aiden Joel Rupp
Umi 13h21'6" 73d48'2"
Aiden John Browne
Boo 14h30'14" 28d33'31"
Aiden John Mike-Lucenti
And 1h30'22" 45d41'46"
Aiden John Schoonmaker
Dra 16h52'47" 63d51'0"
Aiden Jon Abbey
Lib 15h39'54" -5d14'36"
Aiden Joseph Arenstam
Umi 16h35'22" 77d16'10"
Aiden Joseph Donnenberg
Her 16h9'30" 46d14'51"
Aiden Joseph Miller
Her 16h50'29" 24d14'11"
Aiden Joseph Nelligan
Lib 14h50'25" -1d10'28"
Aiden Joshua Newell
Umi 14h32'23" 70d31'6"
Aiden Joshua O'Brian
Cnc 8h44'55" 32d15'12"
Aiden Justin Arizon
Umi 14h35'27" 73d54'33"
Aiden Kyle Baker
Ori 4h56'5" 2d29'53"
Aiden Lee Conger
Vir 13h3'45" 3d8'10"
Aiden Luc
Uma 13h37'2" 55d17'50"
Aiden Luke Scholes
Her 16h12'12" 5d29'40"
Aiden Matthew Short
Umi 14h36'53" 76d55'58"
Aiden May
And 2h17'30" 48d31'32"
Aiden M.D.N. Staples
Ari 2h31'42" 26d23'54"
Aiden Michael Brink
Ori 4h50'43" 11d26'46"
Aiden Michael James
And 1h2'37" 43d35'36"
Aiden Michael Scheller
Agr 22h34'9" 0d42'11"
Aiden Nathaniel Thomas "Aidey-Boo"
Agl 19h27'23" 7d33'2"
Aiden Nicholas Kane
Agr 23h12'34" -24d16'0"
Aiden Noel Lozano
Crb 15h29'12" 29d36'35"
Aiden Parker Bernard
Peg 23h17'36" 17d36'17"
Aiden Parr
Ori 6h9'28" 17d26'48"
Aiden Paul Marshall
Her 17h7'41" 26d17'27"
Aiden Peter Spencer
Cra 18h46'34" -41d3'41"
Aiden Robert Kolsto
Dra 17h15'13" 68d16'14"
Aiden Scott Baker-Stanley
Ori 6h23'4" 15d2'35"
Aiden Shaun Firth
Lac 22h36'27" 41d43'36"
Aiden Thomas
Uma 11h46'29" 58d3'19"
Aiden Thomas Griffiths
Psc 1h11'3" 11d32'9"
Aiden Thomas Pugsley
Her 17h46'33" 47d38'36"
Aiden Tomkins Odell
Gem 7h42'13" 19d29'10"
Aiden Tomlinson
Cnc 9h19'6" 29d3'56"
Aiden Warner Cook's Guardian
Per 3h51'18" 34d52'8"
Aiden Wayne Cukale
Uma 11h18'47" 35d5'49"
Aiden Wesley Barnes
Umi 13h44'34" 77d42'25"
Aiden William Beth
Uma 9h33'20" 56d26'5"
Aiden William Jagels
Col 5h38'1" -28d55'42"
Aiden, Kadence...
Vel 9h15'6" -44d8'47"
Aiden's Nana
Cap 21h33'48" -17d3'42"
Aiden's Star - 12.05.2004
Cru 12h43'51" -63d55'13"
Aidi e Alex
Cas 23h14'27" 54d33'31"
Aidon Travis Seton
Cep 0h7'33" 76d23'11"
Aidrian Jeffrey Salazar
Leo 11h56'2" 23d23'51"
Aidsa
Leo 11h16'46" 15d16'12"
Aidyn Ella Cranmer Plante
Ori 6h6'24" 19d55'31"
Aiemee
Psc 0h42'8" 10d41'42"
Aigan
Vir 14h25'41" 3d13'28"
Aiguille du Midi
Ori 6h0'20" 9d43'21"
Aigul
Uma 9h48'53" 61d59'31"
Aigy, I Love You
Uma 10h17'41" 50d42'22"
Aija
Lyn 9h16'5" 37d15'0"
Aiko
Cam 4h7'41" 63d11'43"
Aiko Shiratori Shikashio
Cas 1h37'46" 64d46'42"
Aikou
Uma 10h23'24" 45d16'22"
Aila Hale
Lmi 9h35'19" 34d45'5"
Ailana Mitchel
Ori 6h7'46" 20d51'4"
Ailani Minda Custodio Toledo
Vir 13h13'8" -4d32'10"
Ailbhe and George
Cyg 19h17'40" 50d47'41"
Ailbhe Hanna Dowling
And 23h31'54" 43d31'50"
Ailee Quay
Umi 14h30'18" 73d47'37"
Aileen
Cas 0h44'26" 65d3'29"
Aileen
Uma 12h21'44" 55d38'6"
Aileen
And 23h36'9" 42d37'29"
Aileen
Cyg 20h44'31" 33d44'58"
Aileen and Mike
Uma 9h2'51" 64d34'5"
Aileen Anderson
Lyn 7h58'19" 35d47'53"
Aileen & Brian
Lyn 8h49'16" 36d7'11"
Aileen Chu
Agr 23h26'20" -18d40'51"
Aileen Collins
Cas 0h58'27" 57d44'31"
Aileen F. Haven
Ori 5h9'4" 7d39'34"
Aileen Faith
Crb 15h42'3" 27d29'47"
Aileen Julia
Agr 22h32'40" 0d3'7"
Aileen M. Agius
Per 3h9'29" 51d48'41"
Aileen Mae Yuzon Achico
Agr 22h46'18" -8d17'33"
Aileen MC18
Cas 1h44'47" 63d27'48"
Aileen McKie Anderson *Ju*
Ori 5h7'21" 12d29'51"
Aileen Santiesteban
And 2h33'42" 37d39'50"
Aileen Schmieder
Leo 9h47'17" 7d21'41"
Aileen's Star
Vir 12h27'35" -6d45'8"
Ailene
Gem 7h16'44" 30d8'59"
Ailene & Destiny Fink
Cyg 19h48'14" 31d12'34"
Ailene Marie Goodwin
Uma 8h48'17" 56d24'43"
Ailesha Ostrich Ringer
And 0h43'57" 25d28'50"
Ailie 21
And 0h48'44" 26d43'42"
Ailie Helen Smith
And 1h46'55" 41d33'15"
Ailiin
Her 16h42'0" 30d22'29"
Ailin Amaba Lopez
Sco 16h18'44" -18d47'58"
Ailin Foshay Kelly
Agr 23h35'49" -7d55'50"
Ailin Hernandez
And 23h0'0" 48d20'2"
Ailisia Alessia
Tau 3h24'29" 18d46'18"
Ailsa Elizabeth
And 0h1'37" 34d43'49"
Ailsa & Ellie Campbell
And 23h40'0" 38d59'37"
Ailsa & Ewan's Star
Gem 6h41'0" 12d2'1"
Ailsa Logan
Tau 4h3'35" 21d29'52"
Ailsa Mary Douglas
And 0h35'34" 30d43'8"
Ailuropoda Melanoleuca
And 23h11'28" 42d37'25"
Aim For the Stars
Cep 22h14'2" 57d39'52"
Aimable
Ari 2h12'41" 24d53'43"
Aime Austin
Sgr 19h43'19" -18d25'51"
aimé Pixie KAP
Tri 2h31'28" 36d17'46"
Aimee
And 0h43'45" 30d56'20"
Aimee
And 0h37'8" 36d45'28"
Aimee
Lyn 8h21'47" 33d45'31"
Aimee
Uma 11h6'47" 38d40'31"
Aimee
And 23h12'39" 49d0'31"
Aimee
Leo 10h10'36" 15d38'1"
Aimee
And 0h20'55" 33d8'6"
Aimee
Leo 10h18'21" 22d49'15"
Aimee
Leo 10h17'33" 26d29'42"
Aimée
Psc 23h21'20" 6d25'51"
Aimee
Agr 22h19'50" -3d52'44"
Aimee
Cas 1h41'3" 64d22'16"
Aimee 35
Agr 21h17'20" -8d11'34"
Aimee Alicia Catherine Marsh
And 1h23'47" 41d3'29"
Aimee and Jeffrey
Cyg 19h28'41" 53d45'33"
Aimee and Scott Forever
Cap 21h35'26" -16d46'54"
Aimee Anderson
Ari 2h12'18" 21d43'57"
Aimee Berwick
Lyn 8h32'53" 51d35'57"
Aimee Charlotte Henderson
And 23h41'14" 38d47'21"
Aimée Dawn
Lib 15h5'25" -1d29'8"
Aimee DeMuro
Lib 15h15'27" -16d12'50"
Aimee Dunsavage
Lyn 7h54'7" 54d28'13"
Aimee Dupree
Ari 3h6'20" 23d4'50"
Aimee Elizabeth
Cnc 8h39'6" 10d13'26"
Aimee Elizabeth Burch
Lyn 8h25'54" 38d4'52"
Aimee Elizabeth Moyer
Uma 12h50'24" 57d45'22"
Aimee Elizabeth Price
Cnc 8h5'45" 14d18'32"
Aimee Gabriel Champagne
Vir 12h31'29" -1d11'7"
Aimee Gibb
And 0h46'31" 27d56'58"
Aimee Gordon
Lib 14h54'59" -3d43'21"
Aimee Grace Blenkinsop
And 23h37'23" 42d50'13"
Aimee Habibi Connolly
Cnc 9h20'42" 6d57'20"
Aimee Jean Gray
Ari 3h25'53" 23d50'11"
Aimee Jill
Cap 21h40'9" -19d38'43"
Aimee Jo Wentworth
Uma 10h36'23" 49d12'21"
Aimee Jourdan
Ari 3h24'55" 29d15'33"
Aimee Joy My Love
And 0h4'34" 33d50'14"
Aimee K. Haller
Cap 20h30'51" -19d10'52"
Aimee Kathleen Marshall
Ari 3h17'36" 27d51'24"
Aimee L. Block
Sgr 17h48'9" -27d59'2"
Aimee L Keenan
Pic 6h31'25" -61d53'26"
Aimee Lea
Sco 16h45'30" -33d36'27"
Aimee Lee
And 1h10'49" 41d57'49"
Aimee Leigh Hackney 28 October 2004
Sco 17h25'12" -45d15'14"
Aimee Louise Liversage
And 23h33'36" 49d38'16"
Aimee Lyn Ray
Cyg 21h31'6" 36d51'35"
Aimee Lynn Fielding
Per 4h34'57" 48d19'30"
Aimee Lynn Schamp
Agr 22h40'28" 0d24'10"
Aimee M. Kalandras
Ari 2h49'54" 20d27'24"
Aimee Marie
And 23h16'57" 40d3'46"
Aimee Marie DiAndrea
Col 5h47'2" -32d25'0"
Aimee Marie Diaz
And 0h31'21" 28d45'15"
Aimee Marie Fournier-Plante
And 23h14'0" 48d25'7"
Aimee Maye Silveira
Tau 5h9'19" 25d24'19"
Aimee My Beloved
Cnc 8h25'0" 25d41'17"
Aimee My Lifelong Love
Cap 20h10'2" -13d12'5"
Aimee Navarrete
Lib 14h43'43" -24d54'12"
Aimee Olivia
And 0h8'50" 29d6'15"
Aimee Patrice Colette Marin
Ari 2h56'57" 22d15'57"
Aimee Renee Harrison
Peg 23h16'5" 18d5'26"
Aimee Rose Maoriello
Mon 7h33'38" -0d45'43"
Aimee S Zelman
Tau 3h30'14" 4d21'5"
Aimee & Sandeep Forever
Cma 6h41'45" -16d16'30"
Aimeé Santos's Star
Mon 6h30'32" 1d37'37"
Aimee (The Light) Morf
Vul 19h56'17" 23d1'2"
Aimee Tucker
Vir 12h0'17" 6d28'52"
Aimee Wyser
Lib 15h40'51" -6d32'37"
Aimee Zagarri
Uma 8h42'58" 56d41'12"
Aimee1015
Lib 14h58'41" -13d23'34"
AimeeLarissaClements
Cas 23h40'32" 52d56'50"
AimeeO loves DanG 2004
Vir 13h58'57" -16d47'55"
Aimee's Christmas Star
Psc 1h6'21" 21d36'49"
Aimee's Star
Lib 14h50'12" -1d18'35"
Aimee's Star!
Dra 19h39'39" 63d34'50"
Aimee's Throne
Vir 12h15'35" 2d15'50"
Aimerick
Lib 15h24'11" -21d50'48"
Aim-G
Sco 17h23'38" -33d54'59"
Aimi Lee Rice
Uma 11h26'47" 54d59'34"
Aimie Costello
Cap 20h34'36" -26d22'48"
Aimie Elizabeth Gonce
Cas 1h34'30" 63d21'38"
Aims
Sco 17h49'43" -40d59'46"
Aimy Caradec
Ari 3h2'3" 18d24'55"
Aina Debatty-Morand
Col 6h20'18" -41d41'57"
Ainas
Umi 14h26'57" 73d36'23"
Aine
Cap 20h28'16" -12d17'9"
Aine (Gaëlle Thomas-Desvaux)
Uma 14h16'56" 56d39'15"
Áine McStravick
Cyg 20h24'35" 53d22'5"
Áine & Steven's Silver Wishing Star
Cru 12h17'27" -61d43'12"
Aine7102004
And 0h27'1" 46d12'11"
Aine's Birthday Star
Cas 23h22'12" 53d14'6"
Aingeal mo chuisle
Col 5h45'21" -32d26'8"
Aingeal Turner
Umi 14h57'22" 74d39'56"
Ainhara
Ori 6h11'51" 7d7'53"
Ainhoa
Uma 11h30'53" 32d25'13"
Ainhoa/Paula
Per 4h15'1" 32d48'46"
Aini e Mimma
Com 12h41'23" 13d46'25"
Aino "The Only One"
Uma 11h27'33" 51d48'0"
AINSLEY
And 0h52'52" 36d10'2"
Ainsley
Agr 22h2'16" -15d10'25"
Ainsley Angelina Heaton
Cnc 8h9'29" 20d39'30"
Ainsley Elizabeth Griggs
And 2h13'11" 45d7'31"
Ainsley Francis Novay
Ari 2h49'59" 21d37'6"
Ainsley G. Blair
Uma 8h36'36" 53d5'28"
Ainsley Kayann Kratochvil 10/15/02
Uma 11h18'57" 53d47'52"
Ainsley Lumsden
Uma 11h48'56" 48d40'28"
Ainsley Marie Siekmann
Leo 9h48'50" 15d44'2"
Ainsley Nichole Stabler
Uma 10h45'38" 59d46'56"
Ainsley Ryan McLain
Agr 21h16'21" -9d54'58"
Ainsley & Samuel's Daddy
Ari 2h18'18" 15d30'48"
Ainsley's First Star
Ori 6h0'31" 21d10'28"
Ainslie T. Fagan
Cam 5h49'0" 60d5'43"
AinuCirdan
Umi 16h15'32" 81d28'58"
Aion
Vir 12h51'43" 6d59'46"
Aionia - Georgia Karnachoriti
Leo 9h46'56" 31d26'33"
Aionios
Lac 22h28'5" 47d10'41"
Aionios Fronima
Agl 19h11'11" 5d48'0"
AIONIOS9897
Peg 22h18'14" 13d59'1"
Air am
Cam 3h59'9" 66d36'49"
Airborne Forever
Ind 20h43'13" -46d19'43"
Airborne Reggie
Vir 12h12'26" 12d11'38"
Airek Dean Borchert
Crb 15h47'19" 30d57'28"
Airen Elizabeth Little
Tau 5h46'12" 17d3'39"
Aires Nichols
Lyr 18h42'59" 38d26'27"
Airman 1st Class Benjamin J. Tamai
Agl 19h46'6" 8d9'50"
Airo
Tau 4h35'0" 7d7'52"
Airy
Ari 3h9'49" 21d47'42"
Airy in the Sky
Ari 2h37'22" 17d39'16"
Airypooh
Uma 8h46'17" 55d28'34"
AIS 07-MWG FRAE
Her 16h35'33" 5d23'48"
Aisha
Sco 17h51'33" -40d46'42"
Aisha Arif
Agr 23h8'48" -6d26'0"
Aisha Brook
Scl 0h2'33" -26d39'37"
Aisha Douaidary
Cas 1h14'8" 61d53'29"
Aisha Hussain
Vir 12h24'38" -8d56'46"
AISHA HUSSAIN
Cam 5h36'15" 59d54'21"
Aisha Lee Castejon
Ori 5h41'49" 2d20'30"
Aisha Savannah - 27 July 2006
Leo 9h56'36" 14d29'48"
Aishiteru Maya
Tra 15h56'51" -61d13'24"

Aisling!
And 1h3'34" 42d45'4"
Aisling Barry Gazzo
Aqr 21h11'54" -7d34'6"
Aisling Behan
Cas 0h52'52" 52d36'24"
Aisling Cooper
Cyg 19h49'19" 39d21'41"
Aisling Daisy Parker
And 2h9'27" 40d55'40"
Aisling Doyle
Cap 20h19'12" -12d47'25"
Aisling Fitzpatrick
Ori 6h19'1" 14d42'11"
Aisling Grace McVeigh
And 0h26'39" 40d32'50"
Aisling KD
Umi 16h10'33" 73d33'3"
Aisling Kelly
And 23h52'3" 47d40'57"
Aislinn
Vir 14h0'34" -14d7'22"
Aislinn Barry
Cap 20h30'58" -12d33'33"
Aislinn bean Columb
Cas 0h29'33" 60d20'52"
Aislinn McCormack
Ori 5h37'10" -2d30'7"
Aislinn Selder Horsfall
Umi 16h14'28" 75d29'22"
Aislyn Abraham
Cam 5h7'46" 69d27'18"
Aissa Iviana
And 0h27'22" 36d39'38"
Aita Ama
Uma 10h14'35" 44d20'17"
Aithel Johnson (Poppy)
Uma 8h26'36" 62d7'44"
Ai-Thuy Thi Do
Tau 5h14'59" 24d59'35"
Aitken College
Cru 12h11'46" -62d2'20"
Aitken's star
Uma 11h31'52" 45d6'58"
Aiyana Lynn
Sco 16h13'26" -32d49'18"
Aiyanna Lynn Miller
Sco 17h37'41" -40d22'38"
Aiyla L Katreec Jenica Shockency
Per 4h47'26" 45d17'27"
AJ
Ari 2h11'43" 25d42'58"
AJ
Ari 2h45'43" 16d18'9"
AJ
Tau 5h29'44" 19d48'6"
AJ
Ori 5h16'17" 0d58'25"
AJ
Her 16h42'47" 6d11'28"
Aj
Uma 10h16'44" 63d52'45"
A.J.
Aqr 22h52'1" -8d53'35"
AJ 3
Psc 0h31'25" 7d58'4"
(AJ) Addison James Manfredi
Tau 4h17'20" 24d18'36"
A.J. & Ali Forever
Ori 4h53'35" -2d59'13"
A.J. and Lyndsay Lane
Vir 13h10'29" -13d10'11"
AJ&BVP
Cyg 20h19'38" 54d35'3"
AJ Cali
Tau 4h6'4" 6d6'21"
A.J. Cech 1
Uma 10h12'55" 43d26'19"
A.J. Chenaille
Uma 12h44'57" 56d16'31"
A.J. Crowley
Uma 11h15'10" 31d2'1"
A.J. & E.J.
Umi 15h22'14" 73d52'43"
AJ HANCOCK
Vir 14h32'11" 4d8'28"
AJ Holman
Ari 1h47'8" 19d17'59"
"A.J." - Infinity & Beyond
Cnv 13h30'14" 32d38'38"
AJ & Jaime
Cet 2h12'11" -23d34'54"
AJ Larson
Tau 4h21'17" 17d24'32"
AJ Politsch, Jr.
Umi 16h16'17" 76d47'29"
AJ Ranalli Always & Forever
Uma 11h17'45" 66d10'21"
AJ Roy
Per 3h3'23" 40d51'27"
A.J. Schulte
Aql 19h18'13" -0d16'58"
A.J. The Magnficent
Leo 11h22'47" 14d37'34"
AJ "The Stud" Amico
Uma 10h25'59" 65d26'6"
AJ Voth
Vir 13h25'56" 12d27'30"
AJ Wenner
Cep 1h20'26" 78d36'42"
AJ, Espresso, and London's Star
Vir 12h40'14" 12d0'1"
AJ21
Ori 5h58'39" 3d18'40"
~~~AJ2AF~~~
Gem 7h44'17" 31d55'7"
Aja - Angel
Lyr 18h41'9" 34d5'56"
Aja Danae Villa Castillo
Sco 17h51'32" -36d29'49"
Aja Leigh
Aqr 23h44'16" -24d21'50"
Aja16
Cap 21h8'56" -15d5'42"
AJAG
Uma 11h22'18" 55d11'34"
Ajai Kamau Gosine
Umi 15h41'52" 85d19'9"
Ajani Marcello Laniyan
Uma 11h58'3" 63d33'41"
Aja's Star*
Leo 9h33'35" 26d37'4"
Ajay & Casey Moran
Uma 10h42'4" 64d50'41"
Ajay Shah
Sgr 18h38'10" -29d6'33"
AJ-Bizzle
Sco 17h29'54" -41d49'13"
AJC & JLP
And 23h19'48" 48d8'22"
Ajeya Sathyagal
Gem 7h9'25" 24d2'7"
Ajit
Umi 15h37'43" 70d5'27"
AJJJ
Lib 15h29'42" -6d4'38"
Ajka
Aql 19h5'51" -10d23'8"
AJL
Umi 1h15'38" 88d52'30"
A.Jo
Cap 20h51'10" -19d4'41"
AJRnWRD
Tri 2h28'20" 31d41'26"
A.J.'s
Aqr 21h10'1" -14d4'40"
AJ's First Love, Lindsay
Cyg 20h48'45" 47d6'37"
AJ's Light
Lib 15h4'6" -24d35'11"
A.J.'s Super Nova
Sco 16h12'50" -16d14'27"
AJ's Wish
Cru 12h44'12" -60d32'21"
ajss-062798
Ori 5h32'19" 5d11'18"
AJUDACHELYN-CONAMORES
Lmi 10h46'51" 33d20'32"
AK
Ori 6h4'3" 19d7'43"
AK Chan
Cyg 20h45'44" 46d5'26"
AK/ZJ 10-08-05...........
Sco 16h17'20" -11d17'22"
aka_olas
Umi 15h10'11" 79d27'48"
Akaleikehe
Sco 17h21'5" -42d23'53"
AKAM
Cnc 8h13'52" 7d31'10"
Akamu Haina
Col 5h50'8" -33d11'28"
Akarin "Crazy Asian" Weatherford
Lib 15h24'13" -7d20'30"
AKatrina14
Cap 21h36'33" -22d30'4"
Akayla Dawn White
And 2h24'26" 50d8'30"
Akbar and Aminas Cute Litte Star
Gem 7h45'7" 31d38'11"
Akbar Sattar
Gem 6h39'21" 22d1'34"
Akdag, Baris
Uma 12h22'49" 61d35'57"
AKE 010686
Uma 11h41'41" 51d14'6"
Ake Aiko
Lyr 18h46'17" 37d31'10"
Akeelah....Shining Brightly Forever
Cnc 8h39'1" 31d1'31"
Akeema
Ari 3h13'38" 29d7'52"
Akemi
Ori 5h40'53" 2d31'38"
Akers' warren
Tau 4h16'51" 2d34'19"
Akestar
Com 12h58'2" 20d8'24"
Akeya
Uma 11h18'42" 45d44'52"
AKH: the Bright and the Beautiful
Cyg 20h11'23" 45d53'55"
Akhil Saklani
Uma 11h39'56" 64d24'20"
Aki Y
Uma 8h37'41" 51d14'19"
Akiko 308
Psc 0h20'26" 2d17'7"
Akiko Humphrey
Uma 8h34'38" 70d42'24"
Akiko Kondo
Lib 14h27'36" -14d58'46"
Akiko Miyagawa
Sgr 19h53'53" -12d44'9"
Akiko Super Star
Uma 13h28'13" 57d52'28"
Akiko's Diamond
Cap 20h55'20" -19d13'6"
Akilah And Travis
Cyg 21h42'11" 30d9'3"
Akilah Howard
Tau 4h36'31" 17d3'24"
A.K.I.L.Y.
Boo 14h42'1" 18d21'24"
Akin
Aqr 21h17'2" -14d16'2"
Akina, The One And Only
Uma 11h47'57" 45d1'32"
Akira
Tau 3h30'43" 29d16'37"
Akira
Tau 4h43'27" 7d20'46"
Akira Luz
And 0h29'56" 27d3'20"
Akira Sinatra
Uma 11h10'6" 43d35'5"
Akira-Yayoi & Tomio-Hideko
Vir 13h16'24" -9d58'24"
Akit Patel
Cnc 8h40'22" 7d1'14"
Akiva Mitzmacher
Ari 2h41'36" 16d59'28"
Akiva Shamar Key
Vir 14h16'3" 0d12'1"
Akly
Sco 17h24'5" -31d30'9"
Ako Celebration "6-15-38"
Cyg 20h35'52" 30d5'43"
Ako - Chan
Eri 3h24'27" -22d31'42"
Akop Gasparyan
Her 18h0'59" 17d30'41"
AKORA
Leo 11h11'41" 16d18'53"
Ákos csillaga, Weigert Zsuzsi
Cas 23h1'8" 58d45'4"
Akosa
Aqr 20h39'49" -12d39'45"
A.K.R.<3J.O. touch God & back.
Umi 14h49'21" 70d0'11"
A.K.R. StarRing Nette
Gem 6h44'18" 27d40'26"
Akram Hanieh
Ari 3h21'9" 29d6'32"
AKS / CFP /Osco 1
Leo 11h24'43" 17d34'15"
AKS012206
Uma 8h28'46" 68d17'18"
AKSEL AND LOLA THOMSEN
Cra 18h36'1" -42d42'44"
Akseli Lintukorpi
Her 17h41'11" 40d27'51"
Aksiniya
Ori 6h5'43" 11d44'52"
akspiritmtn
Sgr 18h36'37" -30d22'25"
Aktra
Cyg 21h42'48" 48d22'6"
Akuna Matata
Aql 19h15'30" 14d46'50"
Akupaxom
Tau 3h31'30" 12d25'19"
Al
Cyg 21h59'51" 47d4'35"
Al
Cyg 19h27'43" 51d24'22"
Al
Per 3h7'9" 47d6'37"
Al
Per 3h14'16" 43d46'56"
Al and Angie Lepse
Per 4h25'32" 34d52'56"
Al and Cheri Lobretto
Uma 10h7'6" 42d11'0"
Al and Christa Unger - October 22
Sco 16h7'46" -11d54'43"
Al and Ginny Malinowski
Lyr 18h51'54" 34d0'29"
Al and Ida Lucas
Cyg 21h44'13" 44d51'16"
Al and Jo forever
Tau 4h2'19" 26d14'1"
Al and Mamie Brink
Uma 8h59'12" 69d0'27"
Al and Melissa "Daddy's Girl"
Peg 23h10'43" 24d33'59"
Al and Nancy Van Why - Love Forever
Aur 5h4'52" 34d30'50"
Al and Ruth Albritton
Ari 2h36'48" 21d10'0"
Al and Syls 50th
Per 2h41'48" 53d53'7"
Al and Zakee Forever
Cyg 21h30'3" 41d11'37"
Al & Angie Fitzgerald
Cyg 20h16'35" 55d26'21"
Al & Anna Bloomquist
Umi 12h42'52" 87d19'9"
Al & Anne (Lucy) Malave
Cyg 21h25'21" 33d16'59"
"Al" Archer
Cmi 7h24'19" 7d39'29"
Al Arzaga
Ari 1h51'53" 19d50'28"
Al B Here
Her 18h56'43" 20d55'54"
Al Baby
Leo 11h38'38" 25d19'3"
Al+Bes
Sgr 18h59'7" -32d11'30"
Al & Betty Colella
Aqr 22h0'28" -2d25'7"
Al Blount
Lyn 6h48'2" 51d50'50"
Al Broxton
Cep 22h13'15" 58d20'38"
Al & Carol Gray Shining Bright
Cyg 21h43'1" 48d38'16"
Al & Cher Plourde Love Burns Bright
Cyg 20h49'50" 38d36'45"
Al Cherry
Her 17h26'2" 47d37'30"
Al & Coreen 40
Cyg 21h55'18" 39d52'45"
Al Czap
Gem 7h50'32" 31d36'41"
Al D'Angelo
Aql 20h1'1" 10d12'53"
Al & Deanne Myers
Aql 19h5'58" 13d50'20"
Al & Diane Richter
Leo 11h22'37" 20d8'48"
Al Emid
Cep 21h45'13" 55d37'46"
Al F Tasch Jr. "1941-2004"
Tau 4h26'44" 15d13'30"
Al Ferrandino
Ori 5h36'4" -0d39'34"
Al Fonseca
Leo 11h21'21" 18d38'15"
Al Franklin
Aql 19h13'2" -2d5'0"
Al Giobbie
Cyg 21h51'16" 50d40'40"
Al & Jeanne Carver
Cyg 21h36'38" 49d19'3"
Al Jordan, Jr.
Ori 6h4'18" -0d19'22"
Al Joseph DeGuzman
Umi 13h20'16" 72d34'32"
Al Kaletta
Crb 15h46'4" 36d23'46"
Al King
Per 2h43'51" 54d56'45"
Al Levy "Superstar!"
Sco 16h41'50" -31d42'19"
Al Livoti
Uma 11h46'2" 53d20'21"
Al Loves Sal
Leo 11h41'34" 25d42'31"
Al Minkoff
Vir 13h18'55" 7d25'3"
al mio papà Albertella Gianni
Cyg 19h54'38" 38d24'14"
Al + MJ = Forever
And 0h24'31" 37d54'26"
Al Morjig
Psc 0h40'15" 12d58'12"
Al Mucho's Wish Star
Ari 2h34'20" 27d35'25"
Al My Forever Love, Teri
Cyg 19h38'30" 27d55'13"
Al n Monica
Sco 16h15'36" -12d40'42"
Al Najmah Jihan Shirbeeni
Aqr 22h32'21" -3d5'4"
Al Navarro aka Baby Star
Lib 14h54'46" -0d59'43"
Al Percy
Del 20h37'3" 15d4'9"
Al Polosky
Her 17h29'7" 26d47'55"
Al Provost
Cep 22h32'17" 57d44'31"
Al Remolde
Psc 1h55'2" 6d20'6"
Al Rose Alligoodius
Ori 6h9'35" 18d15'34"
Al Rosen
Tau 3h24'53" 13d36'49"
Al & Sabine Roma
Uma 8h20'15" 63d15'54"
Al Sammartino
Per 2h51'22" 53d53'24"
Al Sannipoli
Per 3h19'32" 45d41'55"
Al Scalise
Lyn 8h5'29" 40d16'35"
al spaino
Gem 6h27'56" 27d10'29"
Al Valenti
Dra 19h21'8" 65d17'13"
Al & Van 4ever Love
Sco 17h56'39" -31d21'45"
Al Whelan
Cnc 8h22'5" 13d34'6"
AL.12
Uma 8h44'39" 58d7'8"
Alaa Hashem
Cep 2h40'0" 83d12'45"
Alaa Zalzali
Uma 9h40'13" 41d38'3"
"Alack Dreams" never far apart
And 2h29'36" 49d40'58"
Alaedin
Her 17h51'8" 26d36'48"
Alaena Hale
And 2h38'13" 43d57'58"
Alaesa Leigh Hearn
Cnc 8h39'58" 16d44'6"
Alain
Leo 11h0'53" 10d23'36"
Alain
Umi 15h47'14" 81d12'19"
Alain
Cep 22h25'19" 66d42'10"
Alain Amghar
Dra 19h42'5" 67d13'23"
Alain and Diane Lepel
Ori 6h5'25" 13d20'3"
Alain et Janie
Dra 18h34'23" 57d47'49"
Alain Huot
Peg 21h50'41" 21d46'14"
Alain Laforest
Cep 0h42'27" 86d32'6"
Alain Lahana
Uma 12h8'51" 49d2'50"
Alain Landec
Ori 5h53'56" 17d2'8"
Alain Mathieu
Uma 9h5'27" 53d56'23"
Alain Mathieu
Cep 22h32'46" 62d35'24"
Alain Meizoz
Cas 0h32'16" 58d16'47"
Alain "Melt My Heart" Fears
Uma 8h42'10" 63d55'23"
Alain Muller que j'aime
Peg 22h39'53" 24d7'59"
Alain Schirmer
Ori 5h57'5" -0d45'10"
Alain Senges notre Etoile
Her 16h36'21" 28d8'14"
Alain Thierry l'amour éternel
Uma 11h13'47" 64d33'52"
Alain Vincent Ginori
Uma 9h8'52" 47d28'23"
Alaina
Ari 2h38'8" 19d19'31"
Alaina
Uma 9h52'8" 52d40'10"
Alaina
Cas 1h9'7" 64d45'58"
Alaina
Sgr 18h51'42" -18d46'5"
Alaina
Sgr 19h59'27" -30d41'37"
Alaina Ann Anderstrom
And 0h10'17" 37d9'17"
Alaina Ann Mascadri
Cas 1h19'34" 55d2'28"
Alaina Catherine Demalis
And 2h18'23" 45d20'33"
Alaina Christine Ball
Crb 15h25'12" 28d46'25"
Alaina Christine Jevin
Sco 16h11'10" -8d50'5"
Alaina D. Rivellini
And 0h41'41" 40d11'54"
Alaina Heiser
Aqr 21h49'40" -6d28'13"
Alaina Logan
Uma 8h51'20" 48d1'41"
Alaina Maiorano
And 2h19'23" 50d25'49"
Alaina Marie
Peg 21h31'50" 15d50'6"
Alaina Marie Allsworth
And 1h22'56" 41d7'29"
Alaina Marie Brooks
Lib 15h1'20" -10d50'30"
Alaina Marie Rachiele
Aqr 21h42'39" 1d59'15"
Alaina Meseana Fierro
Uma 11h52'31" 36d4'1"
Alaina Rose Montgomery
Sco 17h13'6" -41d58'25"
Alaina Rose Stutzman
Lyn 8h13'28" 45d8'38"
Alaine Garzilli
And 1h17'27" 37d46'27"
Alainn Gra
Sco 17h16'16" -43d9'50"
Alainn Réalta
Gem 6h47'20" 23d47'9"
Alainn Suzanne
Sco 17h38'16" -31d10'48"
Alainna G.
Cas 23h41'14" 53d34'19"
Alainna Lexie
Mon 7h15'9" -0d45'11"
Alainne
Sco 17h57'23" -44d15'56"
Alaire
Cap 20h37'39" -11d6'22"
Alamala
Cru 12h56'45" -64d1'5"
Alamea Pyawasay
Cas 0h53'29" 57d10'26"
Alami
Apu 16h51'4" -79d57'0"
ALAN
Cap 20h29'52" -24d34'22"
Alan
Aqr 23h9'38" -21d4'7"
Alan
Cep 20h47'36" 60d15'25"
Alan
Uma 9h28'53" 72d32'54"
Alan
Uma 10h56'23" 70d4'39"
Alan
Her 16h32'21" 40d54'5"
Alan
Lmi 10h23'16" 31d19'47"
Alan
Crb 15h44'18" 35d28'17"
Alan A Newell
Per 4h34'36" 52d15'38"
Alan A. Rockoff "Rocky's Star"
Sco 17h8'42" -34d4'42"
Alan Adamo
Lib 15h19'46" -9d29'0"
Alan Albert Hough
Cep 22h52'27" 60d44'29"
Alan Alexander Thorne
Per 2h59'3" 57d10'18"
Alan Alfred Manning
Cep 3h21'46" 78d12'17"
Alan and Carolyn Hill
Lyn 9h22'30" 38d6'26"
Alan and Charlene Hutchings
Gem 6h41'29" 29d9'57"
Alan and Cindee's Eternity Star
Cyg 21h12'42" 44d16'14"
Alan and Cindy Burns
Her 18h6'55" 27d26'47"
Alan and Elinor +25
Ori 6h9'56" 8d0'0"
Alan and Sally Cooper
Cyg 19h55'3" 40d3'11"
Alan and Shauna Palmer
Cnc 8h31'16" 24d56'7"
Alan Andrew Dean Miller
Leo 10h17'5" 20d17'54"
Alan & Annette Snyder
Uma 10h44'41" 51d35'36"
Alan Anthony DeVore
Pho 0h45'18" -45d47'41"
Alan Ashenfelter
Lib 15h21'15" -20d20'27"
Alan Axley
Gem 6h53'46" 18d7'14"
Alan B. Steiner
Lyn 7h34'19" 37d13'24"
Alan Blenn Brooks
Cyg 20h56'53" 44d34'23"
Alan & Bobbi Van Reet
Gem 7h4'52" 27d37'21"
Alan Boustead
Leo 11h36'42" 14d44'49"
Alan Brady
Ori 6h16'14" 15d16'29"
Alan Brent Spencer
Her 16h34'15" 14d5'42"
Alan Bridges' RACKSTAR
Dra 18h32'35" 58d40'26"
Alan Bruce Crittenden
Lyn 6h44'45" 56d53'8"
Alan Bruce Morison
Uma 9h27'53" 57d10'51"
Alan (Bud) D. Woods
Cmi 7h30'36" -0d9'33"
Alan C. Crain
Lib 15h30'21" -25d15'17"
Alan C. Fox
Psc 1h16'53" 8d21'32"
Alan C. Hopper
Vir 12h51'8" 1d27'34"
Alan Cameron
Dra 16h1'58" 66d25'3"
Alan Carden
Cep 0h9'2" 80d14'4"
Alan Carrick Douglas
Ori 5h10'44" 8d22'51"
Alan Case
Sco 16h43'30" -25d40'30"
Alan Charles Gross
Ari 3h13'6" 26d44'14"
Alan Charles LeSiege
Cap 21h27'39" -24d5'0"
Alan & Cheryl Pope, "Poopietwo"
Per 3h7'22" 52d36'46"
Alan Clark Batten
Her 17h9'14" 32d2'18"
Alan Cole
Tau 3h55'7" 18d43'40"
Alan Cornfield
Per 2h23'44" 54d36'21"
Alan Craig Mangum
Uma 12h4'11" 54d44'51"
Alan Czak
Per 2h39'28" 53d54'25"
Alan D. Heckel
Lib 14h52'8" -13d57'1"
Alan D. Hutchinson
Sco 17h33'56" -36d49'37"
Alan Dale
Cir 15h15'6" -58d15'51"
Alan David Hodge 3-11-62 - 8-6-04
Psc 0h47'33" 6d21'45"
Alan David Lees
Per 3h0'55" 45d39'19"
Alan "Diamond" Dubs
Ari 3h21'9" 21d52'19"
Alan Donald
Vir 13h14'13" -13d23'52"
Alan E. Butcher
Cen 13h31'1" -36d59'14"
Alan E McPeek
Leo 11h10'48" 11d0'22"
Alan Easley
Aql 19h15'6" -7d33'8"
Alan Edward Colthart
Sgr 17h55'46" -17d34'37"
Alan Edward Colwell
Del 20h36'13" 14d4'44"
Alan Edward Lasday
Aur 5h29'0" 40d39'23"
Alan Edward Perry
Per 4h34'6" 48d11'26"
Alan Edward Seager B.D.,M.B.E.
Lib 15h44'14" -29d32'3"
Alan Franco
Cam 4h4'7" 70d42'33"
Alan Frank Anderson
Leo 9h43'29" 26d21'55"
Alan Fuerstman
Dra 18h26'24" 52d8'16"
Alan Gabstar
Uma 11h43'35" 37d44'6"
Alan Galak
Dra 19h36'32" 69d0'31"
Alan Garett
Uma 11h41'57" 51d24'22"
Alan Geoffrey
Per 4h2'18" 45d15'24"
Alan Geoffrey Furby
Cep 20h43'11" 56d21'34"
Alan Grandpap Stephens
Tau 4h31'4" 28d12'34"
Alan Griffiths
Uma 11h32'58" 55d57'55"
Alan Griffiths
Tri 2h50'32" 31d13'18"
Alan H. Edwards
Cep 22h40'26" 65d45'42"
Alan H. Kenwood
Ari 3h17'10" 29d2'37"
Alan Havern
Uma 11h59'20" 53d7'45"
Alan Holcomb's Shining Star
Uma 11h3'28" 46d21'45"
Alan I. Goldstein, Star Polisher
Her 17h42'29" 22d25'24"
Alan J. Coulter
Cep 4h37'25" 82d44'31"
Alan J. Dressler
Per 3h14'52" 51d58'44"
ALAN J. KOENIG
Cep 22h1'0" 67d31'23"
Alan J. Koenigsberg
Tau 3h25'46" 18d9'27"
Alan J. Robinson
Cnc 9h5'48" 31d44'1"
Alan Jay Fisher
Cap 21h41'0" -9d42'32"
Alan & Joan Theaker
Cyg 21h35'10" 48d58'41"
Alan Johnson
Lib 14h53'56" -1d59'32"
Alan Jon Brusha
Ori 5h56'34" 12d50'12"
Alan Jones
Her 17h12'15" 32d11'46"
Alan & Julie Aeternus
Sco 16h7'11" -12d16'27"
Alan & Julie Haake
Cyg 20h2'21" 37d28'12"
Alan & Juliette Dalhed
Mon 7h21'39" -0d59'12"
Alan K. Schmidt
Lyn 7h22'32" 53d25'43"
Alan Kain
Psc 23h47'18" 5d15'40"
Alan Keith Graber January 30, 1976
Aql 19h25'6" 3d8'50"
Alan Keith Merritt
Per 3h48'8" 43d55'51"
Alan Kiviat
Lib 15h17'24" -10d35'38"
Alan (Kyuss) Simpson
Aqr 23h9'19" -10d10'45"
ALAN L. TINTER
Her 17h53'49" 25d9'0"
Alan L. Trudelle H.I.S. Worldwide
Per 3h34'6" 42d37'28"
Alan Lacey
Cyg 20h42'47" 32d55'52"
Alan Ledbrooke
Per 3h54'20" 33d56'41"
Alan Lee Bolen
Psc 1h21'27" 15d23'52"
Alan Lee Hankin
Vir 13h28'29" 5d14'36"
Alan Lee Hughes
Ori 4h50'28" 12d57'35"
~~~

Alan & Lesley
Cyg 20h34'37" 47d16'35"
Alan Leslie Jones "My True Star"
Sco 17h16'12" -38d57'2"
Alan Levinson
Sgr 18h5'56" -27d48'44"
Alan/Linda's 40th Anniversary Star
Per 3h26'41" 33d58'55"
Alan Marc Keller
Uma 11h19'17" 34d7'42"
Alan Margarito Monroy 2662
Cap 20h15'1" -22d28'59"
Alan Marsh Shoemaker
Her 17h27'29" 35d6'21"
Alan Mattey
Cep 23h10'7" 70d39'58"
Alan Maxwell Jordan
Per 3h19'27" 41d57'17"
Alan McCain My Shining Star
Cyg 20h11'13" 36d9'42"
Alan McMurray
Uma 8h58'21" 56d32'25"
Alan Meindersee
Her 17h48'10" 35d44'50"
Alan & Melissa Binaghi
Cyg 21h40'11" 44d2'36"
Alan Michael Brazzell
Aql 18h57'41" -0d31'7"
Alan Michael Chan Rossi
Aql 19h45'9" -0d29'54"
Alan Michael Simon
Cep 2h14'48" 83d29'27"
Alan Michael Switzer
Umi 15h32'32" 75d29'0"
Alan Mickel Nelson
Leo 11h43'57" 14d25'5"
Alan Neil Breeding
Aql 19h45'34" 7d30'21"
Alan Nelson Osborne
Cma 6h21'27" -30d31'32"
Alan O Connor
Gem 7h30'56" 28d24'9"
Alan Ott
Per 3h18'36" 45d50'43"
Alan P. Gosselin
Lib 15h18'55" -4d48'10"
Alan P. Harris
Her 18h0'11" 21d45'11"
Alan Parker
Uma 9h44'42" 71d4'45"
Alan Pentecost - 30.1.53
Cru 12h17'38" -62d50'50"
Alan Porter
Lyr 18h40'15" 30d51'56"
Alan R. Dickie 4/19/55
Ari 2h26'52" 10d44'12"
Alan R. Dodge
Cra 18h2'56" -37d42'21"
Alan R. Kane (Rappoppop)
Sgr 18h53'39" -16d59'38"
Alan R. Lemery
Uma 11h39'34" 40d52'26"
ALAN R MILLER
Cas 23h59'4" 65d28'39"
Alan Rae
Eri 4h24'11" -35d55'23"
Alan Ramsden
Cep 21h27'7" 68d11'36"
Alan Rawlings
Cep 1h22'23" 80d5'29"
Alan Rendall
Ori 6h6'46" 6d14'59"
Alan Roberts
Gem 6h20'57" 18d57'2"
Alan Roger Bradford
Per 4h25'45" 43d6'33"
Alan Ross Hartley
Lmi 10h37'22" 39d6'56"
Alan Ross Sine
Sco 16h17'26" -15d44'47"
Alan Schorr
Tau 3h45'34" 27d17'52"
Alan Schuyler
Uma 8h45'51" 53d59'33"
Alan Scott Rackliff Barton
Ari 2h27'46" 19d9'36"
Alan Scott Villalobos (Townsend)
Ari 2h15'45" 26d9'34"
Alan Shellard
Cep 22h7'17" 68d25'14"
Alan Skehan
Cap 21h6'19" -16d33'59"
Alan Snagg - Best Dad Ever
Psc 0h7'32" 2d6'44"
Alan Steel
Cep 20h27'47" 61d16'19"
Alan Stuckey
Ari 2h19'47" 25d51'11"
Alan Stumpo
Sgr 19h22'18" -17d57'2"
Alan Sukharev
Cam 4h19'43" 57d26'19"
Alan Swindoll
Cap 23h20'23" 48d22'42"
Alan T. Burleigh Ball of Gas
Leo 9h51'39" 20d32'21"
Alan T L Mak
Cap 21h45'48" -8d49'48"
Alan T. Nicholsen
Ori 5h59'59" 22d31'57"
Alan & Tammy
Sgr 18h0'34" -26d41'54"
Alan Thomas Crook
Cep 22h38'43" 57d34'8"
Alan & Tina Cole
Cyg 21h44'22" 44d52'9"
Alan Todd Stone
Uma 11h28'10" 60d45'1"
Alan Trosino
Cam 4h31'57" 53d14'49"
Alan & Vivian
Cyg 19h42'1" 39d16'2"
Alan & Vivian Turner-Humphrey
Lyr 18h37'38" 39d47'26"
Alan W. Raynes
Ori 6h18'1" 14d54'26"
Alan Wayne Coffey
Ori 5h36'54" -0d59'4"
Alan Webb
Lyr 18h36'36" 39d11'54"
Alan Weeks
Psc 0h32'33" 20d56'40"
Alan Wheeler
Ori 5h35'56" -4d36'31"
Alan William Schaefer
Her 17h13'42" 30d23'15"
Alan William Thielmier
Gem 6h51'35" 12d8'5"
Alan Williams
Uma 10h43'46" 63d37'27"
Alan Wilson
Cep 22h29'46" 73d0'49"
Alan Wittenberg
Per 4h36'21" 48d37'35"
Alan, Jane, Holly, Stevie & Kerry
Cmi 7h45'8" 1d2'11"
Alana
Gem 7h7'10" 28d28'44"
Alana
Gem 6h48'55" 23d18'42"
Alana
Cnc 9h1'38" 28d58'45"
Alana
Lyr 18h53'38" 38d18'30"
Alana
Umi 13h14'58" 72d9'24"
Alana
Uma 13h20'54" 53d28'52"
Alana
Vir 13h50'25" -6d24'12"
Alana
Sgr 18h26'14" -20d54'17"
Alana
Vir 13h52'30" -20d2'1"
Alana Alexandria Anderson
Uma 11h33'56" 29d17'11"
Alana Anderson
Uma 8h37'20" 72d6'31"
Alana Anne Hood
Per 3h8'6" 42d52'15"
"ALANA" Auntie's baby
Gem 7h31'25" 34d35'57"
Alana Bug
Aqr 22h47'35" -6d52'32"
Alana Celeste Cheek
And 0h45'51" 26d18'44"
Alana Cristine Ferreira da Silva
Vir 13h18'4" 12d11'8"
Alana Duguay
Ari 2h20'25" 27d2'43"
Alana Elizabeth
And 0h47'53" 45d45'49"
Alana Geary
Crb 15h47'41" 37d18'1"
Alana Grace Galarza
Crb 15h55'15" 26d47'36"
Alana Ilene
Vir 12h56'45" 3d2'32"
Alana Insley
Sgr 18h13'57" -31d3'49"
Alana Jacy Olsen
Cyg 19h51'0" 31d33'10"
Alana J.J.L. 2-29-1968
Psc 0h8'45" 11d10'13"
Alana Joan Dunn
And 0h24'13" 25d50'27"
Alana Joy Morosky
Uma 9h44'50" 50d23'55"
Alana K. Griffin
Cas 23h33'56" 55d10'42"
Alana Kaye Garza
Aqr 22h50'22" -20d21'5"
Alana Kristina
Lep 5h30'46" -15d52'16"
Alana LaRock
Lyn 7h35'13" 41d16'29"
Alana Louise Broberg
Uma 9h55'18" 54d7'2"
Alana Marie
Ari 3h20'21" 23d18'1"
Alana Marie
Vir 13h2'40" 6d29'34"
Alana Marie Anderson
Per 3h1'53" 48d1'16"
Alana Maureen Sullivan
Lib 15h46'42" -17d27'40"
Alana (Mimi) Faye Kemper
Cnc 8h32'2" 23d48'26"
Alana Montoya
Leo 10h15'31" 17d6'37"
Alana Nicole Maher
Cas 1h12'59" 62d44'35"
Alana Paige Parnell
Lib 14h53'16" -12d24'14"
Alana Patricia Thacker
Leo 10h18'43" 25d31'20"
Alana Ramasar
Lib 14h51'2" -24d28'2"
Alana Rivera
And 23h13'52" 51d50'24"
Alana Rose Wilson
Psc 23h50'8" 6d1'10"
Alana Spencer
Crb 15h35'13" 26d49'3"
Alana Taylor 2187
Aqr 22h44'5" -6d40'39"
Alana Tolentino Honeybunchesofadobo
Gem 7h11'4" 28d0'42"
Alana Vicari
Uma 10h46'46" 56d41'13"
Alana Wong
Leo 11h4'22" 2d20'46"
Alana-Brighton
Sco 17h41'39" -42d28'28"
Alanah Marie Ruffin-Smith
Sgr 19h11'47" -17d26'12"
Alana's Star
Lmi 10h37'34" 27d51'44"
ALANASTAR
Her 18h39'44" 16d26'56"
Alandjen Star
Cyg 19h42'38" 28d15'33"
Alandrea P. Wallace
Cas 1h22'34" 60d21'5"
Alane Gardner
Sco 16h40'23" -37d26'20"
Alane Halverson
Cap 21h54'38" -13d34'16"
Alangail
Tau 3h44'44" 29d21'19"
Alani AION
Psc 1h33'1" 10d28'44"
Alani Indigo Santamarina
Lup 14h47'16" -44d28'58"
Alanis Felina Marsh
Vir 13h9'42" 7d23'48"
Alanis Lynn Larsen
Cas 1h7'15" 57d34'7"
Alankim
Her 17h6'42" 30d57'3"
Alanna Connor
Pho 23h39'38" -39d49'14"
Alanna Cyles
And 0h7'3" 41d32'55"
Alanna Diane Harris
Cnc 9h11'47" 25d57'48"
Alanna G. & Mikhaila N. Reace
Cap 20h35'13" -27d6'54"
Alanna Jane Lazerus
Cyg 20h55'37" 32d37'46"
Alanna Jazelle
Lib 15h8'38" -4d46'14"
Alanna Kayleen Walker
Psc 0h54'23" 6d40'28"
Alanna Kerr
Cap 21h39'25" -19d46'29"
Alanna Kyung Cook
Gem 7h15'29" 20d4'2"
Alanna Lauren Harris
Cam 4h1'0" 53d9'32"
Alanna Lynn Graham
Oph 17h31'57" -0d50'20"
Alanna Marie O'Brien
Cnc 8h48'56" 30d17'9"
Alanna Mathews
And 2h3'6" 46d31'22"
Alanna Meredith Hughes
And 0h1'17" 38d11'30"
Alanna Nichole Penecale
Gem 6h49'39" 31d51'28"
Alanna Noel Lemoi
And 0h1'38" 44d41'37"
Alanna Yorkston
Peg 22h27'19" 4d37'25"
Alannah
Lib 14h24'5" -9d44'3"
Alannah Devinne Daly
And 0h14'37" 28d25'53"
Alannah Grace Cooper
Gem 7h25'28" 15d9'32"
Alannah Roux
Ari 3h11'10" 28d43'32"
Alannah's Star
And 0h24'55" 38d54'13"
Alanna's Allure
Uma 11h31'8" 38d15'21"
Alanni M.
Cyg 21h34'7" 42d6'40"
Alanni Sara Rogers
Lib 14h54'49" -1d51'44"
Alanoofus
Lyn 7h44'12" 41d29'3"
Alan's Heart
Uma 10h59'5" 69d8'58"
Alanzaire Forever & Always
Lib 15h31'4" -19d24'57"
Alapaki Piena
Ari 2h45'14" 13d53'1"
Alar Ruth Brock
Ari 2h37'49" 13d9'15"
Alaric Zane
Sgr 20h0'53" -21d18'1"
A.L.A.R.M.....JM
Lib 15h37'33" -7d0'3"
Alarna Leigh
Ari 2h32'50" 18d22'0"
Alasdair & Eliza - Always & Forever
Cru 12h11'47" -62d13'13"
Alasdair Jack Gibson Owles
Umi 14h40'55" 67d58'17"
Alasdair William Roxburgh
Her 18h8'40" 28d10'55"
alaska sethimus morpheus
Crb 15h24'42" 31d47'3"
Alastair John Stewart
Her 17h41'7" 18d21'21"
Alastair & Lisa Davidson
Peg 22h13'5" 23d50'29"
Alastair Mackay
Aur 6h32'53" 47d34'30"
Alaura Nevaeh Iampieri
Sco 17h30'6" -45d12'47"
Alaxia A.M S. P
Mon 6h51'31" -0d7'52"
Alayna
Uma 11h33'14" 40d31'6"
Alayna Butler
Aqr 21h24'42" 2d1'17"
Alayna Hope Welch
Aql 19h19'54" 6d24'8"
Alayna Kilbane
And 2h16'47" 50d4'31"
Alayna Lynn Thomma
And 2h23'12" 46d43'3"
Alayna Marie
Psc 1h4'42" 21d39'6"
Alayna Marie PCB Lotulelei 21 17 10
Sco 17h53'47" -36d17'23"
Alayna N Christman
Lyn 7h24'39" 53d12'18"
Alayna Rae Barron
And 0h36'7" 29d28'6"
Alayna Reagh Piazza Alton
Cnc 8h44'40" 16d16'23"
Alayna's Star
Cas 0h31'29" 62d31'8"
Alayne
Lyn 8h55'41" 45d49'17"
Alayne & Ray Palacios 03/11/2006
Cyg 19h59'26" 54d45'33"
Alayne, Goddess of Love
Leo 11h19'7" 18d52'38"
Alba
Ari 3h11'30" 23d9'3"
Alba
Cas 0h55'38" 51d39'19"
Alba
Ori 5h49'7" -3d15'7"
Alba
Ori 6h10'39" -2d12'13"
Alba
Oph 17h53'28" -0d33'7"
Alba Aurora
Aqr 22h12'50" 0d54'4"
Alba Di Tommaso
Cap 20h48'27" -25d12'11"
Alba Lopez
And 23h1'59" 46d59'46"
Alba Maria Becker Pombo
Ari 2h11'13" 12d21'50"
Alba Marino
Sgr 18h49'28" -20d52'42"
Alba Roca
Ari 3h9'7" 12d19'59"
Alba t'estimo
Ori 5h14'29" -0d31'7"
Alban
Lmi 9h43'7" 33d19'59"
Alban Trouillard
Peg 22h59'32" 19d35'26"
Albani
Lyn 7h13'6" 49d9'5"
Albania Rosario
Uma 11h44'18" 61d6'24"
AlBara
Cnv 13h22'43" 47d35'36"
Albaybi
Umi 15h26'31" 70d4'3"
Albe 50th- 07/30/55
Leo 11h0'51" 23d12'31"
Albert
Cep 0h4'3" 73d3'18"
Albert
Cep 20h53'0" 59d50'19"
Albert
Uma 12h39'17" 58d8'59"
Albert
Aql 20h11'5" -6d1'55"
Albert
Aql 19h15'24" -9d38'37"
Albert A.and Youvon K. Glover
Cyg 20h21'15" 49d5'49"
Albert Abraham
Ori 6h12'11" 20d59'7"
Albert Alex Varga
Per 2h54'10" 53d49'49"
Albert Amadei
Psc 1h13'13" 27d10'25"
Albert and Christie
Sge 20h2'12" 17d23'2"
Albert Anthony Rophie
Gem 7h16'52" 28d48'34"
Albert Antonacchio
Vir 11h39'44" 2d36'52"
Albert Argudin Jr.
Gem 6h41'18" 18d2'26"
Albert Aston
Per 4h16'30" 50d47'13"
Albert B. Sharpe
Aql 20h0'29" 7d23'41"
Albert Barriga
Boo 15h10'33" 41d10'26"
Albert & Beverly
Cnc 9h6'44" 31d37'8"
Albert C. Johnson
Aqr 22h15'51" 2d14'29"
Albert C. Simidian
Psc 1h30'42" 16d48'9"
Albert C. VandeVelde
Cas 23h18'20" 55d7'45"
Albert Calderon & Gerry Calderon
Ori 6h9'27" 20d45'39"
Albert Campbell Holt
Her 16h28'49" 46d39'44"
Albert Chacon
Cnc 8h32'1" 12d1'30"
Albert Christian WItt
And 23h41'4" 45d27'6"
Albert Clarence Fidalgo Sr.
Ori 6h9'9" 16d16'7"
Albert Clifford Brown
Cnc 8h48'16" 26d34'28"
Albert D. Crisamore, Jr
Ari 3h16'25" 29d34'17"
Albert D. Tront
Boo 15h18'46" 49d22'34"
Albert Daggett Sr.
Uma 11h15'29" 63d49'40"
Albert Damm
Sco 17h0'10" -32d48'32"
Albert & Dawn Engagement Star
Cyg 20h22'26" 43d11'21"
Albert Di Giovanni
Aqr 22h39'58" 0d56'13"
Albert E. Asman
Uma 8h55'42" 49d11'23"
Albert E. Richard III
Vir 13h42'11" -8d49'23"
Albert Earl Figg
Cap 21h28'26" -8d45'47"
Albert Edward Chilcott
Uma 8h23'6" 71d5'10"
Albert Eugene Hiser, Jr.
Sco 16h15'24" -15d14'37"
Albert Feiffer
Boo 14h32'52" 22d13'26"
Albert Frederick Masters
Ori 5h52'36" 7d9'6"
Albert G. Kurtz
Sco 16h17'40" -32d50'36"
Albert Gerard Chiuchiolo "Big AL"
Her 17h52'15" 47d26'52"
Albert Giannetto
Tau 3h47'54" 28d16'32"
Albert Giddo Abraham Jr.
Cnc 8h23'27" 14d29'19"
Albert Gino jammin Tarrillo Jr 1962
Her 18h7'12" 34d19'9"
Albert (Gogo) Castner
Uma 11h35'23" 57d13'0"
Albert Gregory Gonzales Snr.
Lup 15h41'55" -32d21'2"
Albert Guggenberger
Uma 14h4'1" 48d4'59"
Albert Head
Her 17h43'1" 31d18'28"
ALBERT HELDAK
Lac 22h14'49" 49d14'36"
Albert Henri Gutleben
Sco 16h4'9" -22d58'47"
Albert Hernandez
Cep 21h51'19" 55d46'0"
Albert I. Tydings
Leo 10h59'5" 15d56'3"
Albert J. Oleksy Sr.
Ori 5h55'0" 20d56'13"
Albert J. Yonfa
Cyg 19h12'54" 53d29'17"
Albert Jacob Alsdorf
Ori 5h31'45" 0d6'14"
Albert James Patterson
Her 17h35'20" 31d6'12"
Albert James Scullin
Uma 11h33'1" 56d27'54"
Albert & Jennifer Laaksonen
Ari 3h16'18" 27d3'36"
Albert Johnson
And 23h4'40" 51d28'34"
Albert Joseph Frederick Bell
Her 18h21'13" 16d15'29"
Albert Joseph Fris
Psc 1h34'5" 13d53'56"
Albert Joseph Samuel II
Lib 15h24'13" -18d54'32"
Albert Junior Garvin
Her 17h7'23" 17d11'50"
Albert Kenneth "Al" Tolley
Uma 8h50'57" 61d30'23"
Albert Kistinger
Psc 2h3'44" 9d15'53"
Albert L. Auer Jr.
Cep 21h58'45" 58d6'57"
Albert L. Neuhauser
Uma 8h34'9" 46d45'8"
ALBERT LEKO
Her 17h16'40" 35d48'1"
Albert Llambrich Tantan y Laura G.P.
Tau 5h29'30" 22d54'26"
Albert Lopez
Her 17h7'6" 35d50'9"
Albert M. Landry
Ori 5h4'16" 15d23'43"
Albert Mirabal
Ori 5h17'3" 9d1'30"
Albert & Misse Zelouf
Cyg 19h16'16" 50d49'42"
Albert Mistiaen
Cep 3h35'29" 85d34'40"
Albert Monroe Waters
Cas 1h6'0" 58d9'11"
Albert N. Donnantuono
Cep 20h41'41" 61d16'37"
Albert Neill Cameron
Her 17h24'8" 27d23'27"
Albert Newell Pickles IV
Uma 10h6'25" 57d41'51"
Albert Ochroch
Umi 17h40'17" 86d28'47"
Albert Odusanya
Per 3h9'59" 55d42'40"
Albert P. Pellegrino
Per 3h20'55" 51d28'33"
Albert "Pop-Pop" Jesikiewicz
Aqr 22h37'4" -21d7'29"
Albert R. Coates, M.D.
Dra 19h10'53" 69d25'56"
Albert Ray Stoneburner IV
Ori 5h55'33" 18d30'18"
Albert Raymond Trost, Sr.
Her 17h26'24" 37d15'9"
Albert & Rennis Koch's Destiny
Cyg 22h0'20" 50d51'2"
Albert Reutter
Uma 8h21'16" 65d33'24"
Albert "Rick" Larsen
Uma 8h55'25" 71d17'35"
Albert Rocco Girardi Jr.
Lib 15h55'27" -10d30'24"
Albert "Rockstar" Thompson
Ori 4h54'21" 11d55'52"
Albert Roper
Per 3h20'2" 46d59'0"
Albert Rowland, Jr.
Sco 16h55'27" -43d17'27"
Albert Sebastian
Per 4h27'48" 46d4'12"
Albert Shikiar
Aqr 22h14'39" 1d13'35"
Albert Speed
Peg 23h11'23" 27d53'53"
Albert Spiegel
And 2h5'57" 37d31'20"
Albert T. Congemi
Gem 7h0'13" 32d11'17"
Albert T Fiore
Cep 0h35'27" 80d37'35"
Albert T. Willardo MD.
Aur 6h8'33" 50d16'24"
Albert Thomas Ehringer
Vir 14h17'28" -17d38'3"
Albert V Studley
Gem 7h15'22" 32d8'9"
Albert Vincent Manville
And 0h36'5" 42d22'28"
Albert Vito
Ori 5h39'56" 7d5'3"
Albert Wagner
Uma 12h5'21" 32d18'31"
Albert William Goss - 8 August 2004
Vel 8h51'43" -45d53'1"
Albert Wrobel
Cyg 22h1'20" 51d7'45"
Alberta
Vir 12h26'31" 12d19'59"
Alberta
Leo 9h40'3" 28d32'29"
Alberta
Lib 15h0'28" -14d13'5"
Alberta Ann
Cyg 19h26'41" 35d58'59"
Alberta Burke
Ari 3h20'52" 21d31'5"
Alberta Jean
Uma 9h53'36" 72d14'50"
Alberta M. Morett A Star Forever
Cas 23h40'47" 54d20'24"
Alberta Santora
Cas 1h8'27" 63d54'5"
Alberta Smilie Bedell best mom ever
Cas 1h13'26" 55d1'16"
Alberta Spadaccia
Uma 11h42'11" 35d18'59"
Albertje Hendrikje Dekoning Faken
Sco 17h29'28" -40d23'45"
Alberto
Uma 13h40'44" 52d25'30"
Alberto A. Pena
Gem 6h37'0" 12d55'11"
Alberto Basei
Lyn 8h1'7" 52d39'43"
Alberto Cardelle
Uma 10h48'24" 40d55'6"
Alberto Carlos De Jesus
Dra 16h43'46" 52d41'46"
Alberto Dominguez
Aqr 23h50'25" -13d17'37"
Alberto e Cosmina
Ori 5h51'38" 5d10'42"
Alberto Felisari
Her 16h19'0" 5d0'12"
Alberto Geyer Aguinaga
Cap 20h20'3" -23d2'16"
Alberto Hernandez Torres
Uma 9h30'57" 44d24'48"
Alberto Jose Sanchez Martinez
Lib 15h38'8" -29d31'48"
Alberto Medina
Per 3h9'48" 55d9'3"
Alberto My Love
Aur 7h10'37" 41d47'50"
Alberto "Nene" Hernandez
Leo 9h46'30" 32d8'57"
Alberto Pimienta
Aql 20h6'6" 9d58'21"
Alberto Ramos
Cep 20h44'58" 66d48'51"
Alberto Ruiz
Lib 14h49'32" -9d39'30"
Alberto Sena
Her 17h46'52" 16d30'47"
Alberto Uboldi's Star
Uma 10h13'24" 63d48'7"
Alberto y Maribel
Ari 2h50'27" 25d40'45"
Alberto, Claudia y Myla
Sgr 18h38'9" -16d15'18"
Albert's One Wish
Lyn 9h20'11" 35d17'45"
Albertson-Hale Twins
Boo 15h5'29" 40d21'15"
Albi Tolve
Cyg 21h45'49" 45d50'56"
Albialbi
Boo 15h18'3" 48d6'22"
Albie Kaufman
Leo 9h39'22" 26d56'3"
Albie Patten
Dra 17h16'58" 63d6'58"
Albie Stisted
Uma 9h42'12" 50d31'49"
Albin Bilinski
Aqr 22h54'59" -9d28'1"
Albina Maria Picharbo
Cas 1h4'23" 62d1'39"
Albina-Isis
Vir 13h4'42" -13d31'49"
Albo
Uma 9h24'18" 55d47'50"
Albrecht & Elfriede Möller
Uma 10h19'21" 54d11'57"
Albrecht Tuerke
Uma 8h16'6" 71d23'15"
Albrecht, Hans-Hermann
Ori 6h5'3" 6d36'19"
Albrecht, Susanne
Leo 11h11'41" 22d2'14"
Albrecht, Wolfgang
Uma 9h44'34" 57d39'52"
Alby Brown Flaherty
Uma 8h39'37" 49d50'20"
Alby Jake Burge - Alby's Star
Per 3h31'19" 37d24'18"
Alby - My Beautiful Rainbow Maker
Cru 12h1'44" -59d48'19"
Alcaeus-Philotheos
Cam 4h15'4" 63d30'22"
Alchemist
Umi 14h1'46" 72d46'40"
Alchemy
Uma 11h35'43" 47d28'5"
Alchimie
Cas 0h7'6" 57d10'59"
Alchrist
Lyn 7h53'14" 48d53'55"
Alcy
Crb 15h50'17" 28d0'31"
Alcyone
Cas 0h2'38" 56d39'26"
ALD0123PQP
Aqr 21h51'38" -2d8'42"
Alda
Uma 9h41'21" 58d33'4"
Alda Eisele
Ori 5h38'49" -0d40'24"
Alda Moutinho Cyr
Cyg 20h26'28" 43d48'36"
Alda Pilar
Psc 0h54'16" 10d57'18"
Aldag, Klaus Bernhard
Uma 11h54'6" 33d38'24"
Aldana Vanesa Chantres
Ori 5h17'43" -7d1'55"
Aldawnan
Cyg 20h42'27" 45d53'59"
Alde Carlo P. Gavino
Sco 17h15'32" -44d56'8"
Aldea Segond
Ari 3h7'59" 20d27'51"

Aldeen M. Anderson
Ori 5h46'48" 5d56'29"
ALDEFRA
Peg 21h56'27" 21d35'37"
Alden Charles Allen
Her 18h54'43" 23d0'29"
Alden Jace
Aqr 22h5'44" -4d22'51"
Alden Jay Gundy, Jr.
Per 3h27'40" 48d9'53"
Alden John Carlson
Ori 5h40'24" -3d29'40"
Alden Mario
Vir 12h37'43" 5d8'8"
Alden Michael LaBarre
Her 17h31'47" 43d52'8"
Alden Perez Boyajian
Umi 15h27'19" 71d27'30"
Alden Prevost
Per 3h19'19" 47d18'48"
Alden Stewart Grinnell
Vir 12h34'34" -0d4'16"
Alderaan
Cnc 9h20'27" 9d19'18"
Aldie
Uma 14h4'34" 61d22'56"
Aldine & Ardyce Priest
Pho 1h55'2" -45d3'9"
AldJoy
And 1h43'52" 48d48'24"
Aldo
Cas 1h18'2" 69d33'57"
Aldo Alejandro Moreno Hinojosa
Uma 9h35'12" 68d44'43"
Aldo Cavalli-Star And Star Maker
Dra 18h46'25" 71d28'49"
Aldo Coviello Jr.
Umi 16h20'7" 85d15'35"
Aldo de Moor
Her 18h7'38" 34d39'51"
Aldo Kheirallah
Equ 21h6'20" 8d46'57"
Aldo Mario Mignani
Cep 22h7'16" 72d5'25"
Aldo Rinaldo Rosati
Cnc 9h3'27" 17d5'1"
Aldo Troncoso
Cru 11h58'56" -61d17'52"
Aldonka
Lyn 8h18'6" 49d29'46"
Aldoph Butler
Sco 16h14'31" -12d7'38"
AldoRon
Per 3h39'0" 51d14'20"
Aldous
Aur 7h24'42" 40d6'56"
Aldred Abraham Dunbar Jr.
Her 18h35'1" 13d27'47"
Aldy
Gem 6h38'37" 17d59'29"
Ale
Her 16h46'43" 27d46'3"
Ale
Ori 5h9'48" 6d34'42"
Ale...
Per 4h8'39" 44d56'2"
Ale
Umi 17h36'39" 80d14'5"
Ale
Lib 14h55'0" -1d44'45"
Ale Ale
Dra 19h23'9" 72d13'51"
Ale e Miky
Her 17h40'29" 19d8'40"
Ale e Nino
Ori 5h33'4" 10d51'34"
Ale y Juan Fer
Uma 10h52'18" 61d22'46"
Ale, la Dama del Mara
Aql 18h50'20" -1d6'6"
Alea Ann Monaco Kosydar
And 1h19'5" 42d1'36"
Alea Motwane
Ari 2h20'49" 21d27'21"
Aleah Marie
Mon 6h44'27" 8d48'31"
Aleah Marie Begg
Uma 11h17'14" 38d20'10"
Aleah Valentine
Aql 20h16'27" 1d16'16"
Aleane Davis
And 0h38'37" 44d55'10"
Aleasha
And 2h24'27" 49d0'6"
Aleasha
Sgr 19h35'32" -23d42'7"
Alec
Dra 17h47'30" 62d13'42"
Alec
Her 17h25'3" 36d46'23"
Alec ~Biscuit Boy~ Garner
Ari 2h32'21" 22d14'45"
Alec Brian Nygard
Uma 9h22'29" 50d25'49"
Alec Chandlee
Ari 2h19'15" 25d1'50"
Alec Charles Huerta
Vir 13h17'13" -13d15'48"
Alec Eisenbraun
Ori 4h54'19" 2d42'2"
Alec Fredrick Sprouse III
Sgr 19h39'30" -14d4'5"
Alec Gonzales
Aqr 22h32'35" -1d23'35"
Alec Hughes
Cma 6h19'10" -30d37'37"
Alec J. Wickersty
Cap 20h35'29" -14d45'27"
Alec James Harris
Uma 10h42'44" 69d33'51"
Alec John
Tau 3h52'28" 29d14'27"
alec jose subero
Aqr 23h32'18" -17d30'54"
Alec Joseph Riggio
Her 17h52'18" 38d48'56"
Alec Julian Topakas
Uma 8h59'8" 51d54'13"
Alec Lloyd Smith
Ari 2h20'18" 23d55'19"
Alec Martin Sloan
Ori 5h38'57" -0d31'38"
Alec McEwan
Cru 12h38'7" -57d45'2"
Alec Michael Arcara
Sco 17h56'22" -38d51'16"
Alec Michael Mangini
Umi 14h40'27" 74d50'38"
Alec Mothershead
Ori 5h16'15" -5d9'15"
Alec (My Little Man)
Cep 23h16'21" 77d21'5"
Alec R. Plourde
Aql 19h38'2" -7d21'36"
Alec Richard Milkint
Ori 5h27'28" 0d14'10"
Alec Rimer
Aur 5h56'57" 44d55'4"
Alec Rospierski
Uma 11h44'0" 52d10'57"
Alec Shi
Uma 10h42'0" 46d13'21"
Alec Steven Weistreich
Vir 11h39'16" -3d44'2"
Alec: The Great Cuddle Bear
Uma 13h28'51" 58d2'44"
Alec Wade Martin
Ori 5h7'42" 7d37'12"
Alecgator
Gem 6h47'46" 24d57'0"
Alecia 71570
Ori 6h2'18" 21d1'13"
Alecia Alexandra Arnold
And 1h6'0" 45d5'56"
Alecia Alexis
And 0h43'27" 42d1'5"
Alecia B M Ortiz
Aqr 22h57'15" -8d13'17"
Alecia Davis
And 0h13'27" 32d40'18"
Alecia Homer
Cam 4h57'54" 54d23'26"
Alecia Kay
Ari 2h49'27" 24d45'39"
Alecia King
Uma 9h21'48" 46d4'22"
Alecia Lynn Noel Wilson
Del 20h31'21" 12d43'58"
Alecia M. Moore
Cas 2h35'59" 66d48'10"
Alecia Mercier
Tau 5h26'45" 21d8'22"
Alecia Noel
Cap 20h21'50" -16d6'7"
Alecia & Taylor's Star of True Love
Uma 11h25'32" 54d22'1"
Alec's Three Pointer
Cmi 7h14'6" 9d48'6"
Alecsander John Morrison
Vir 13h19'59" -12d45'35"
Alecsandria Cook
Uma 11h23'49" 66d53'22"
Alecsi Russell
And 1h48'5" 45d7'23"
Alecsy
Psc 0h22'58" 9d16'9"
Aled & Sally Daniels
Cas 23h2'24" 54d40'19"
ALEDAN
Boo 14h18'6" 41d13'14"
Aleecelynn (Leece) Thornock
And 2h19'6" 48d36'43"
Aleeda Nilene Creamer
Lyn 7h8'7" 51d37'46"
Aleena
And 1h47'25" 42d40'21"
Aleena 1994
And 22h58'40" 48d22'37"
Aleena McKenna
Lyr 18h30'45" 36d15'16"
Aleena Tompkins
Ori 5h38'12" -1d32'5"
Aleena Voeller
And 0h54'50" 36d11'6"
Aleen's Sweet Sixteen Star
Psc 1h10'28" 27d45'45"
Aleesha
And 0h59'34" 35d42'45"
Aleesha Natalie Presley
Cas 23h46'29" 52d32'9"
Aleesha Srinivasan
Uma 13h27'45" 59d27'28"
Aleeyah, Our Niece - Our Angel
Sco 17h27'14" -40d59'9"
Aleeza
Sgr 18h14'21" -19d8'9"
Alegi
And 1h2'20" 46d21'55"
Alegna
Umi 14h51'6" 74d39'32"
Alegra
Psc 1h35'5" 8d14'22"
ALEGRE
Cas 2h13'50" 62d39'53"
alegria
Cas 2h23'5" 68d49'33"
Alegria Yasha Frank
Gem 7h34'28" 15d0'26"
Aleia Nicole McLeroy
Cam 4h11'7" 64d48'28"
Aleicia Kelley
Cas 2h27'54" 64d17'0"
Aleina Ramirez Gonzalez
Mon 6h48'38" 7d48'23"
Aleisha McCarter
Cam 4h54'8" 59d23'14"
Aleisha Nicole
And 0h47'55" 36d35'44"
Aleister Zin
Psc 0h53'56" 31d17'59"
ALEITHIA ROSE BRUERE
Lyn 8h43'7" 40d24'38"
Alejandra
Lyn 9h23'1" 41d12'46"
Alejandra
Lyn 8h13'55" 42d17'8"
Alejandra
And 23h10'50" 48d15'50"
Alejandra
Gem 7h33'5" 21d52'56"
Alejandra
Ari 3h15'46" 26d53'19"
Alejandra
Sco 16h45'40" -31d41'43"
Alejandra Abarzua
Cet 2h48'1" -0d27'16"
Alejandra Angel
Uma 10h48'58" 58d46'35"
Alejandra Betel Martinez
Cnc 9h5'6" 7d41'3"
Alejandra Bravo Mercado
Cap 20h10'36" -21d24'19"
Alejandra Chavez
Cap 20h50'3" -16d25'10"
Alejandra Gaeta
Lyn 7h44'53" 38d33'29"
Alejandra Hernandez
Cnc 8h23'40" 10d53'26"
Alejandra Kelly Baker-Hughes
Sco 16h50'52" -38d13'16"
Alejandra Restrepo
And 22h59'2" 51d40'14"
Alejandra Rodriguez Lagha
Her 16h34'37" 38d56'41"
Alejandra Santisteban
Lyn 8h43'51" 38d59'22"
Alejandra Tatyana Kloster
Sgr 19h48'28" -13d14'20"
Alejandra Weinberg
Lyn 7h55'54" 35d25'10"
Alejandra's Star
Cnc 8h49'8" 28d10'48"
Alejandra's Star
Sgr 17h57'41" -17d22'9"
Alejandra-Voice Of Enlightenment
Her 16h55'26" 30d26'28"
Alejandra-Voice Of Enlightenment
Her 16h43'29" 30d51'39"
Alejandris
Lmi 10h38'59" 32d23'58"
Alejandro
Sco 16h16'25" -11d57'23"
Alejandro
Lyn 7h35'15" 59d8'54"
Alejandro & Ana-Laura
Sge 19h45'17" 17d8'52"
Alejandro Aragon
Aql 19h53'36" -0d14'27"
Alejandro Barreras Parissakis
Vir 13h18'56" 5d10'48"
Alejandro Christian
Sco 16h48'4" -27d13'11"
Alejandro Craig Collins
Lyn 7h55'8" 38d8'57"
Alejandro "Daddy"
Per 4h42'45" 42d49'31"
Alejandro Daniel Valdez
Per 3h47'23" 41d39'43"
Alejandro David Hernandez
Cas 0h29'1" 50d51'34"
Alejandro Diaz
Uma 12h0'4" 62d30'34"
Alejandro Felix
Ori 5h44'21" -4d7'16"
Alejandro Fernandez
Ori 5h23'19" 0d28'18"
Alejandro Fernandez Abarca
Tau 5h38'15" 18d26'11"
Alejandro Haber
Leo 11h44'2" 13d26'16"
Alejandro Heras
Ori 6h9'29" 5d47'21"
Alejandro Medrano
Her 17h37'5" 26d10'13"
Alejandro Paz
Cyg 20h35'54" 47d15'6"
Alejandro Rodriguez
Cap 21h34'32" -9d29'13"
Alejandro Suarez
Cnc 8h48'50" 14d37'51"
Alejandro y Damien Forever and more
Pho 0h36'14" -48d1'16"
Alejay
Pho 23h59'48" -48d28'48"
Alekos Menelaos
Uma 10h33'51" 45d31'14"
Aleksa, Luka & Nikola Ducic
Vul 19h23'26" 21d42'27"
Aleksandar Ivanovic - 03.05.1998.
Ori 5h58'15" 20d46'28"
Aleksandar Radovanovic
Umi 15h0'51" 86d17'37"
Aleksander and Heather Nicole
Leo 11h30'49" 15d45'45"
Aleksander Daniel
Sco 17h56'25" -38d10'31"
Aleksandr Zhukhovitskiy
Cep 22h0'9" 63d48'40"
Aleksandra
Uma 13h4'3" 56d59'12"
Aleksandra
Tau 5h56'21" 24d18'50"
Aleksandra
And 2h19'5" 41d19'13"
Aleksandra
And 0h51'55" 38d52'43"
Aleksandra Kocic
And 23h44'30" 38d11'58"
Aleksandra Kowalska
Lib 14h53'16" -24d42'35"
Aleksandra Perovic
Per 4h21'13" 44d22'53"
ALEKSANDRA Protector of People
Ori 5h45'13" 0d48'6"
Aleksandra Tesic
Cyg 20h6'11" 37d52'52"
AleksandraBS Pjegica
Peg 22h40'38" 18d30'13"
Aleksei Grigorevich Lonesov
Vir 13h17'35" 12d21'45"
Aleksey Lachev
Uma 10h41'14" 57d55'13"
Aleksi Laiho
Ari 2h50'45" 26d38'37"
Aleksndr A. Luzhnykh
Umi 15h11'19" 82d13'17"
Alekzander Jeremy Mojica
Psc 0h21'56" 6d14'55"
Alekzondra Anastastia
Sgr 18h3'32" -25d3'30"
Aleley
Dra 19h14'13" 67d7'7"
Aleli C. Anacion
And 0h39'58" 27d25'19"
Alella
Ori 6h16'53" 10d5'38"
ALEM
Tau 3h47'48" 6d15'50"
Alemkah Smoke and Laurel (Lilith)
Sco 16h35'25" -37d3'40"
Alen Kontelj
Her 16h18'5" 19d28'51"
Alena
Tau 5h49'49" 26d17'11"
Alëna
Ori 5h39'30" 1d49'0"
Alena
Lyn 6h20'30" 59d58'57"
Alena C. Gasparik
Cap 21h39'16" -8d46'17"
Alena Jae
Lyr 18h33'24" 35d51'34"
Alena Lotte Thomas
Uma 9h15'27" 66d39'47"
Alena Madlmayr
Vir 12h57'46" 7d15'6"
Alena Yurevich
Lib 14h33'56" -12d40'37"
Alena, RN
Uma 13h50'37" 62d4'41"
Alendry Ivan Macalintal
Aur 5h23'54" 43d56'55"
Alene
Psc 1h44'22" 14d36'58"
Alenka
Leo 9h31'19" 6d59'37"
Alenka
Umi 16h45'51" 76d20'27"
Alenka Hianiková
Uma 12h24'20" 54d48'3"
Alenkin
Boo 14h43'49" 17d39'21"
Aleonachka
Ori 5h42'47" -2d18'43"
Alero Samuel-Onyuygo
And 0h46'48" 30d11'56"
Aleron
Her 17h37'59" 37d36'12"
Alesandra Gunn
Ori 5h43'10" -2d53'14"
Alesha
Dra 16h2'50" 61d27'52"
Alesha
Her 17h34'45" 27d8'46"
Alesha R Miller
Sco 16h13'1" -10d44'50"
Alesia
Cas 0h47'38" 58d1'47"
Alesia Candace Perez
Gem 7h40'0" 33d38'39"
Alesia Daughtry
And 0h31'8" 29d14'30"
Aleska
Cam 5h15'33" 66d15'17"
Alessandra
Aqr 23h8'4" -14d22'15"
Alessandra
Umi 13h14'51" 75d40'51"
Alessandra
Per 3h1'10" 51d38'30"
Alessandra
Ari 2h22'7" 23d38'13"
Alessandra
Aql 20h23'34" 6d26'37"
Alessandra
Ori 6h22'8" 11d0'2"
Alessandra
Peg 22h1'37" 14d7'43"
Alessandra
Aur 5h23'49" 48d49'57"
Alessandra Antinori
Ori 6h6'16" -2d12'24"
Alessandra Avitto
Lyr 18h33'49" 27d21'46"
Alessandra Barile
Aur 5h14'10" 45d59'25"
Alessandra Bertazzoni
Aql 20h12'48" -0d22'56"
Alessandra Bonucci
Uma 11h9'40" 61d49'50"
Alessandra Brunetti
Ori 5h10'57" -0d3'43"
Alessandra Bussadori
Uma 14h26'30" 61d0'29"
Alessandra Calabrese
Per 3h24'47" 41d31'40"
Alessandra Carmelita Harpster
Tau 5h10'6" 17d50'13"
Alessandra Ceccarelli
Crb 15h59'12" 25d48'16"
Alessandra D' Andrea
Per 3h31'21" 34d39'59"
Alessandra Désirée Spinelli
Per 3h3'5" 51d39'29"
Alessandra e Maria
Ori 6h12'28" 2d58'10"
Alessandra Foschetti
Cas 0h31'37" 62d28'21"
Alessandra Gamberi
Lyr 18h39'6" 44d49'42"
Alessandra Leggio
And 23h33'36" 38d13'10"
Alessandra Marie
Uma 9h51'43" 49d20'48"
Alessandra Marie Trunzo
Uma 12h4'29" 61d44'9"
Alessandra Militello
Her 16h46'43" 27d46'3"
Alessandra Morabito
Cnc 9h8'6" 31d27'35"
Alessandra Rose
Tri 1h52'49" 31d3'33"
Alessandra Rose Gatta
And 2h14'6" 46d22'15"
Alessandra Savini
And 23h58'5" 39d5'13"
Alessandra Skarshinski-Fred
Cyg 20h40'0" 42d24'32"
Alessandra Solari 9 aprile 1963
Uma 9h4'7" 51d59'47"
Alessandra Steinherr
Cyg 20h14'0" 38d59'40"
Alessandra Tess Lehrer
Tau 3h44'18" 25d36'52"
Alessandra Toreson
Cas 23h45'45" 53d45'7"
Alessandra's Star
Cyg 20h45'38" 33d5'13"
Alessandro
Crb 16h0'21" 26d7'22"
Alessandro
Uma 9h4'7" 51d59'47"
Alessandro
And 23h20'36" 48d46'25"
Alessandro 12.06.1957
Cas 23h4'2" 53d36'31"
Alessandro Adriano
Lyr 18h45'40" 31d42'7"
Alessandro Aebi 10.12.2003
Boo 15h8'28" 20d12'56"
Alessandro Aibel
Vir 13h2'52" -2d40'26"
Alessandro Aleotti
Leo 9h54'21" 7d7'21"
Alessandro Ardrizzi
Ori 6h10'33" 8d45'44"
Alessandro Averone
Her 18h52'38" 22d54'28"
Alessandro Businaro
Her 17h46'55" 16d58'52"
Alessandro Carvani Minetti
Umi 15h36'23" 68d50'38"
Alessandro Charles Scalfi
Psc 0h16'58" 5d13'6"
ALESSANDRO DE PIANTE
Vir 13h15'39" 4d21'36"
Alessandro Fiumidinisi
Cas 1h10'6" 50d46'29"
Alessandro Gigli
Ori 6h19'23" 4d44'41"
Alessandro Il Magnifico
Cep 20h42'48" 66d32'32"
Alessandro La Badessa
Cyg 21h37'21" 53d11'3"
Alessandro Massimo Martini
Lyr 18h36'31" 31d48'1"
Alessandro Miranda
Boo 15h18'36" 43d4'13"
Alessandro Paolo
Boo 14h11'20" 28d6'0"
Alessandro Pesare
Uma 13h41'28" 57d6'59"
Alessandro Pizzi
Ori 6h12'47" 8d29'37"
Alessandro Ricci
Cas 0h31'3" 62d25'24"
Alessandro Rossi
Mon 7h0'43" 7d22'7"
Alessandro Serafini
Lyn 7h38'28" 45d10'8"
Alessandro Sergio Rivero
Leo 9h28'49" 24d47'30"
Alessandro Stelli
Cap 21h26'20" -17d25'18"
Alessandro Traverso
Per 3h3'7" 51d43'48"
Alessandro und Myriam (Muus und Müsli)
Per 2h26'10" 51d44'42"
Alessandro Valentini
Cas 23h44'52" 53d4'55"
Alessandro Volpe
Lib 15h47'48" -18d8'12"
Alessandro William Zipeto
Umi 3h52'25" 89d15'5"
Alessandro, a religião das estrelas
Cyg 21h26'25" 32d36'20"
Alessia
Cyg 21h12'12" 31d10'44"
Alessia
And 1h56'34" 39d12'22"
Alessia
Aur 5h5'55" 40d57'19"
Alessia
Uma 11h29'26" 37d15'17"
ALESSIA
And 23h1'18" 50d52'42"
Alessia
Her 18h12'50" 17d41'51"
Alessia
Mon 7h3'48" 7d26'53"
Alessia
Ori 5h45'26" 1d55'18"
Alessia
Umi 13h33'35" 72d32'43"
Alessia Bertini
Her 17h30'30" 19d13'18"
Alessia Biondi
Cas 0h17'29" 57d5'16"
Alessia Caligari
And 23h2'48" 44d41'39"
Alessia Ciocca
Lyr 18h50'59" 27d55'14"
Alessia Crespi
Her 16h54'12" 16d50'44"
Alessia Donna Novielli
Ari 2h19'3" 23d48'36"
Alessia Feder
Her 18h12'10" 18d59'6"
Alessia Louise Perry-Greene
And 23h51'51" 43d8'43"
Alessia Parro
Cyg 20h0'19" 33d59'12"
Alessia Pellegrini
Cep 22h23'2" 66d9'2"
Alessia per sempre
Per 3h24'56" 50d36'16"
Alessia Rustici Tortorelli
Uma 9h29'8" 44d26'19"
Alessia Sabbadini
Cas 0h5'7" 63d34'37"
Alessia Sarah Jaeger
Tau 3h41'33" 17d17'41"
Alessia Selina Massignani
Psc 1h34'17" 15d0'38"
Alessia & Valerio
Uma 12h19'10" 59d58'33"
Alessia Visintini
Cyg 21h1'29" 39d7'25"
Alessia's
Tau 3h27'44" 10d15'43"
Alessio
Peg 22h41'46" 9d38'18"
alessio cabras (de vico)
Cap 20h17'45" -13d40'14"
Alessio Fino
Cyg 21h7'48" 30d19'35"
Alessio Sassoli
And 23h3'45" 41d44'22"
Alessio "scopeto"
Her 17h39'33" 31d56'43"
ALESSIO STRANIERI
Cam 4h8'15" 58d47'3"
Alessio Vicenzo Tognetti
Her 18h36'10" 13d47'31"
Alessondra Margen "Missy girl"
Tau 5h51'11" 15d36'0"
Aleta
Leo 11h52'26" 21d27'40"
Aletha and Richard McAfee
Aur 5h23'16" 41d15'20"
Aletha Ellen Triola
Her 16h49'24" 32d35'42"
Alethea
Lib 15h14'42" -5d42'31"
Alethea Gayle Stephens
Uma 11h22'45" 71d28'56"
ALETHEO
Sco 16h16'26" -8d39'49"
Alethia Hollis Thompson
Cam 4h11'41" 66d7'30"
Alette
Ori 5h29'49" 7d10'32"
Alevtina V. Lance
Tau 5h55'32" 25d50'12"
Alex
Sge 19h40'12" 18d0'58"
Alex
Sge 19h53'4" 17d31'26"
Alex
Ari 2h38'31" 25d45'42"
Alex
Ori 6h8'37" 7d27'41"
Alex
Tau 4h9'18" 5d11'20"
Alex
Leo 10h58'21" 7d47'14"
Alex
Ori 6h18'55" 9d48'24"
Alex
Crb 15h42'14" 34d14'9"
Alex
Crb 15h32'44" 36d56'54"
Alex
And 0h26'7" 36d56'32"
Alex
Leo 9h43'13" 30d39'33"
Alex
Gem 6h49'49" 32d47'32"
Alex
And 23h23'28" 47d53'40"
ALEX
Cyg 20h41'56" 46d33'50"
Alex
And 23h12'58" 44d25'7"
Alex
And 1h26'57" 49d42'20"
Alex
Uma 11h17'18" 63d9'44"
ALEX
Dra 16h27'36" 62d9'41"
Alex
Dra 17h9'8" 63d32'19"
Alex
Dra 17h25'54" 67d23'49"
ALEX
Uma 13h4'2" 62d47'36"
Alex
Uma 11h34'18" 65d48'11"
Alex
Umi 16h8'59" 72d43'28"
Alex
Cep 22h11'24" 61d44'52"
Alex
Uma 12h26'46" 54d1'30"
Alex
Lyn 8h2'12" 58d34'27"
Alex
Cap 20h21'14" -13d23'28"
Alex
Aqr 23h3'16" -8d9'46"
Alex
Cma 6h46'16" -15d4'47"
Alex
Umi 20h52'14" 89d4'19"
Alex
Ori 5h41'4" -0d22'0"
Alex
Sgr 19h1'14" -35d14'56"
Alex 1989
Uma 9h28'47" 57d35'33"
Alex A
Per 4h26'18" 36d8'29"
Alex Alfaro
Lib 14h54'59" -5d38'24"
Alex and Alyssa Always
Per 3h22'40" 33d16'16"
Alex and Brian
Uma 9h28'56" 66d59'14"
Alex and Danielle, Forever
Leo 10h26'28" 12d20'29"
Alex and Debbye
Lyn 9h13'19" 36d8'23"
Alex and Janelle's Forever Love
Cyg 21h11'7" 44d47'15"
Alex and Jen's Star
Cyg 21h19'19" 38d22'31"
Alex and JoAnne Botko
Uma 12h31'59" 57d11'30"
Alex and Joanne, Forever as one
Cru 12h33'52" -57d19'14"

Alex and Katie
Cyg 20h18'29" 34d37'9"
Alex and Kim Together Forever
Cyg 20h37'27" 41d19'33"
Alex and Kristin
Cyg 20h19'17" 35d20'34"
Alex and Larree
Cru 12h32'57" -59d11'10"
Alex and Lori Caire Forever
And 0h18'57" 28d6'17"
Alex And Priscila Cruz
Cap 20h25'41" -13d19'43"
Alex and Robin's Star
Lib 15h6'49" -5d1'30"
Alex And Sonya Forever
Aur 7h25'29" 40d50'19"
Alex and Sophie
Cyg 20h47'52" 35d19'7"
Alex and Teresa Lopez
Mon 6h48'30" -0d38'28"
Alex and Tiffany Cordaway
Ori 6h15'55" 9d9'51"
Alex Andrea Brown
Uma 9h7'48" 70d35'34"
Alex & Anna Gillham
Sge 19h32'23" 17d20'14"
Alex Anne Budman
Her 18h25'13" 12d27'43"
Alex & Ant's Wishing Star (OPITS)
Cyg 21h9'27" 31d18'57"
Alex Astalos
Uma 11h55'32" 37d23'18"
Alex B.
Lyr 19h10'56" 29d19'36"
Alex Baca
Uma 11h17'13" 47d6'28"
Alex Badger
Cnc 9h16'52" 25d12'46"
Alex Bauer
Gem 6h42'17" 30d30'51"
Alex Beck
Sco 16h37'17" -25d9'18"
Alex Bennet
Cas 1h27'59" 55d47'14"
Alex Bialozor
And 2h4'22" 38d53'11"
Alex Boucher
Cnv 13h30'36" 33d26'15"
Alex Brault - I love you forever
Uma 10h26'19" 58d37'39"
Alex Browning
Gem 7h7'17" 24d10'45"
Alex "Bubba" Hunt-Sampolski
Umi 16h36'47" 77d25'51"
Alex Bun Bun
Cnc 8h42'43" 11d26'46"
Alex Burnette
Cmi 7h25'24" 6d55'9"
Alex Butchie Bojko
Umi 15h35'33" 72d23'46"
Alex & Carly
Uma 9h19'34" 60d40'53"
Alex Carracino
Sgr 18h7'32" -32d40'13"
Alex Carrasquillo
Equ 21h2'6" 8d37'22"
Alex & Catrina
Cyg 20h48'3" 54d21'46"
Alex Chan & Kitty Cheng
Psc 1h51'18" 5d17'46"
Alex Chan & Kitty Cheng
Aql 19h59'41" 15d59'11"
Alex Chase Le'Vasseur
Uma 9h35'36" 53d24'12"
Alex Cherkasky
Per 2h50'26" 48d14'15"
Alex Christopher LaRosa
Ori 5h8'24" 10d54'37"
Alex Christou
Uma 10h59'21" 44d52'48"
"alex & christy"
Ori 5h47'28" 5d51'23"
Alex Cilento
Boo 14h31'34" 42d9'24"
Alex Cody Copeland
Del 20h54'20" 13d14'8"
Alex Costin
Ari 1h51'33" 19d49'37"
Alex Craft - A Star With Us Forever
Uma 8h38'44" 60d56'57"
Alex Cseh
Tau 5h21'2" 18d20'8"
Alex Csillaga
Cap 21h46'48" -18d49'49"
Alex Cudlin
Per 3h37'46" 39d30'26"
Alex D. Petrov
Psc 23h8'9" 8d3'2"
Alex Dalton
Crb 16h12'5" 29d9'52"
Alex David
Leo 11h40'59" 27d4'13"
Alex DellaFranco
Aqr 22h18'59" 0d59'36"
Alex Detore
Uma 11h28'59" 38d51'5"
Alex Dierking
Lmi 10h20'3" 36d1'16"
Alex Do
Leo 11h4'26" 15d9'51"

Alex Dolores Dixon Phelan
Aqr 22h12'36" 1d13'39"
Alex Drake Cigan
Per 3h11'16" 51d32'41"
Alex Durkin
Uma 9h28'56" 45d0'6"
Alex Earl Mcbride
Uma 10h54'28" 66d19'58"
Alex Edgard Allan Edde-Toriel
Cep 22h25'31" 62d45'27"
Alex Ellert
Mon 6h44'21" -1d14'46"
Alex Ellie Louise Scott-Jones
And 23h27'4" 35d51'55"
Alex Falzon & Audree Hashibe
Cas 23h0'55" 57d42'38"
Alex Fernandez
Uma 9h47'9" 64d22'26"
Alex Finney
Psc 1h35'46" 17d47'4"
Alex Francis
Sco 17h37'27" -41d19'23"
Alex Franklin Silvestre
Gem 7h12'24" 34d25'17"
Alex Friend
Uma 9h19'41" 57d53'0"
Alex from Hell
Lyn 6h18'20" 57d28'55"
Alex G. Gradillas
Cnc 8h47'42" 13d18'47"
Alex Gabrielle
And 0h33'25" 42d55'44"
Alex Gale's Star
Cep 22h48'58" 70d49'26"
Alex Garber
Aql 19h54'39" -10d19'8"
Alex Garrett
Lib 15h3'9" -28d37'40"
Alex Gerard Colgan
Her 17h45'4" 28d22'50"
Alex Gorrell
Ori 5h41'38" 5d0'52"
Alex Gould's Star
Uma 9h10'37" 63d18'46"
Alex Grandma Sue Zeke
Uma 11h19'45" 45d44'52"
Alex Grant Wright
Men 4h48'11" -71d10'16"
Alex Gregory Davison
Psc 23h5'9" -0d14'15"
Alex Hanka
Aur 6h7'30" 53d39'31"
Alex Harris Ward
Uma 11h18'18" 35d54'48"
Alex Hatton
Ori 6h12'35" 9d14'6"
Alex Hawksley
Cas 1h54'52" 63d11'13"
Alex Hendriks
Aqr 23h4'37" -10d0'28"
Alex Hunter Lipstein
Cyg 21h17'41" 43d49'23"
Alex & Ian Hardwick
Cyg 21h6'11" 46d17'58"
Alex Jack Kittle
Ori 5h58'14" -0d34'42"
Alex Jacob Wurtz
Uma 10h55'21" 38d18'9"
Alex Jacqueline Robbins
Psc 1h40'57" 23d5'16"
Alex Jade Trolley
And 2h12'9" 50d9'42"
Alex James
Uma 10h33'18" 54d44'27"
Alex James Cringles
Cep 21h44'0" 56d9'40"
Alex & Jennifer '79
Cyg 19h34'37" 53d45'45"
Alex John Ferreira
Vir 13h27'40" 11d47'9"
Alex John Patterson
Umi 16h57'8" 76d56'55"
Alex Johnson
Ori 5h4'4" 15d8'22"
Alex Joseph Masselli
Leo 10h38'29" 22d19'1"
Alex Joseph Rainey
Cyg 19h49'28" 35d56'22"
Alex Joseph's Star
Ori 5h24'38" 5d33'11"
Alex Julian Bissonnette
Ari 3h13'19" 29d1'25"
Alex Kaleokalani Kleissner
Gem 6h14'14" 24d32'54"
Alex Kate - 30th June 1987
Cnc 9h13'41" 26d22'23"
Alex Katz Our Shining Star BMS 2005
Lyn 7h51'47" 56d57'9"
Alex Kaufman
Uma 10h19'48" 42d20'20"
Alex Kronk DuBack
Umi 15h40'31" 77d55'6"
Alex L. Moyers
Cnc 8h20'35" 15d57'37"
Alex Labarces
Sgr 18h47'59" -34d57'39"
Alex & Lara
Uma 10h33'34" 57d36'19"
Alex LaSalle Chase
Vir 12h30'25" 7d57'13"

Alex & Laura's Star of Love
Psc 0h23'22" 15d58'20"
ALEX LEIGH ANN
Sco 16h12'11" -12d14'31"
Alex LeVasseur
Tau 4h55'34" 24d51'46"
Alex Levasseur
Peg 23h16'36" 26d30'27"
*Alex & Liam*s GREAT Birthday Star*
Cyg 19h23'48" 53d54'33"
Alex Lovell
And 1h23'35" 49d21'7"
Alex Loves Caleb Forever
Cyg 19h40'31" 43d45'17"
Alex Luvs His Mom - Valarie Boller
Uma 13h23'7" 57d49'57"
Alex Mackenzie
Psc 0h34'39" 7d37'10"
Alex & Madeline Salustri
Cyg 19h45'11" 34d18'41"
Alex Manno
Leo 10h2'28" 15d49'21"
Alex Manriquez (XMAN)
Sgr 18h21'29" -22d19'56"
Alex Marca
Leo 9h51'9" 27d40'8"
Alex Marco Zeppieri
Cep 23h10'50" 70d52'16"
Alex Marco Zeppieri
Sco 17h56'3" -38d11'51"
Alex Marie
Uma 9h37'36" 53d43'45"
Alex Marshall
Boo 14h36'27" 14d38'24"
Alex Mathew Dennis Pyle
Lyn 8h26'1" 56d39'51"
Alex May Matthews
And 2h30'59" 48d24'20"
Alex McConnell
Cnc 9h13'8" 32d45'53"
Alex McFie Rose
Vir 12h31'8" 12d13'40"
Alex McKay
Uma 9h23'2" 49d15'33"
Alex meine große Liebe
Uma 10h10'23" 49d24'8"
Alex Michael Anderson
Umi 16h12'22" 77d34'19"
Alex Michael Kinnamon
Ori 5h3'22" 9d27'51"
Alex Mireia
Dra 20h25'33" 63d1'24"
Alex Morlan
Uma 11h11'17" 45d18'58"
Alex Moya
Ori 5h35'34" 9d3'43"
Alex Muir
Vir 11h48'11" -0d7'30"
Alex Munteanu
Her 16h50'25" 17d34'34"
Alex My Beloved Son Narcowich
Ori 5h41'2" -0d44'39"
Alex (My Love)
Gem 6h53'13" 27d4'54"
Alex Mykil Flores
Ori 6h19'8" 7d21'48"
Alex Nannoshi
Uma 11h13'18" 49d45'16"
Alex of Ukiah
Aql 19h54'8" 12d19'54"
Alex O'Handley
Uma 11h9'23" 55d15'54"
Alex Olson Fiori
Ari 3h19'40" 24d32'30"
Alex Otto's Light in the sky
Her 17h40'19" 38d22'14"
Alex Papi Oleksiuk
Uma 8h39'14" 60d58'7"
Alex Plaza
Leo 10h17'10" 11d58'21"
Alex Prather
Ori 5h15'11" 5d58'55"
Alex "Pumpkin"
Uma 12h15'25" 60d51'40"
Alex R. Garneau
Cyg 21h28'40" 33d29'15"
Alex Rahimi
Leo 10h35'19" 9d17'42"
Alex Rebecca Bewley
Lmi 10h47'52" 26d17'54"
Alex Reed
Uma 11h46'0" 36d27'10"
Alex Remmel
Uma 9h56'8" 68d25'12"
Alex Richard Doddington
Dra 15h54'52" 62d29'7"
Alex Rivera
Sgr 19h12'2" -18d7'38"
Alex Rodriguez
Sco 16h59'18" -41d55'39"
Alex Ross Olson
Cnc 8h30'34" 31d31'21"
Alex Ryan
Uma 10h41'26" 72d49'2"
Alex Ryan Sims
Her 16h59'45" 18d50'35"
Alex Ryberg
Ori 5h31'37" -3d50'2"
Alex S. Hobbs
Cnc 8h0'26" 17d31'51"
Alex Sandoval
Lyn 6h22'38" 57d14'32"

Alex Scott Hillard, my little nefew
Her 17h31'57" 43d14'39"
Alex Sean Humphrey
Ari 2h19'17" 24d56'36"
Alex Sicignano
Psc 1h20'18" 17d18'39"
Alex Simone McGilvray
Aqr 22h22'0" -14d46'43"
Alex Sky Walker
Cru 12h37'5" -59d34'20"
Alex Sloan Loves Emily White
Dra 17h50'15" 62d7'58"
Alex Smith loves Erica Christensen
Cyg 20h28'47" 48d36'10"
Alex Smith-Wood
Vir 14h12'33" 0d21'52"
Alex Somers
Ori 6h14'39" 15d33'3"
Alex Spick
Per 4h16'19" 46d59'19"
Alex Spleen
Her 17h41'15" 44d45'41"
Alex Suarez
Uma 14h15'43" 56d5'40"
Alex T
Uma 11h29'11" 53d26'24"
Alex Taylor
Leo 9h30'54" 30d58'56"
Alex Thanasiu
Her 17h16'36" 31d50'57"
Alex Thomas
Uma 10h12'19" 47d20'24"
Alex Thomas Carter
Lyr 18h52'25" 35d40'13"
Alex Thomas Gray
Cep 23h0'32" 72d24'51"
Alex Tiernan
Lyn 7h7'20" 57d28'49"
Alex Topping
Sco 17h17'43" -32d26'40"
Alex Trebek
Uma 11h22'23" 59d23'5"
Alex Trombley
Cnc 9h8'22" 10d22'40"
Alex Troy Boyle
Boo 15h12'4" 44d55'25"
Alex "Ty" Tarasovich
Uma 9h36'51" 65d11'49"
Alex & Uncle Gary
Umi 15h20'15" 76d6'12"
Alex und Jörg's Glücksstern
Ori 5h45'12" 7d25'41"
Alex und Patchy
And 2h32'47" 45d59'50"
Alex Valencia
Aqr 20h45'2" -8d55'8"
Alex Vassallo
Sct 18h48'51" -4d27'13"
Alex Viruet
Lyn 7h24'9" 56d36'39"
Alex Walden
Uma 9h43'4" 51d46'57"
Alex Warne Martine MW
Cas 1h26'55" 58d34'18"
Alex Wetherbee
Crb 15h48'34" 36d10'58"
Alex William
Umi 16h29'53" 77d31'25"
Alex Williams
Cru 12h51'0" -64d2'51"
Alex Wozniak
Boo 14h28'52" 53d8'52"
Alex Zumaglino
Boo 14h58'10" 14d33'7"
Alex, Ben & Corrina
Dra 17h3'7" 65d1'10"
Alexa
Uma 12h32'41" 62d21'44"
Alexa
Cap 20h13'58" -15d26'58"
Alexa
Lib 14h44'52" -18d54'14"
Alexa
Cnc 8h25'3" 11d5'44"
Alexa
Tau 4h49'37" 19d5'52"
Alexa
Psc 0h58'18" 19d22'37"
Alexa
Her 16h39'29" 7d2'15"
Alexa
Vir 14h11'49" 5d22'15"
Alexa
Leo 9h23'57" 6d31'23"
Alexa
Cyg 19h51'4" 33d27'29"
Alexa
Uma 10h8'8" 48d17'23"
Alexa
And 23h4'3" 41d51'32"
Alexa Alexander
Vol 8h27'50" -65d32'9"
Alexa Ann
Cas 23h4'22" 57d36'18"
Alexa Ann Kallmeyer
Peg 21h12'36" 19d21'17"
Alexa Aoibh
Sco 17h44'40" -35d42'34"
Alexa Brittany Cherifi
Aql 19h53'37" 10d55'58"

Alexa Catherine Delwiche
And 22h59'48" 48d35'28"
Alexa Cunningham
Gem 6h43'42" 34d33'32"
Alexa D. Singleton
Ari 2h47'37" 15d42'58"
Alexa Desko
Tau 4h12'45" 27d11'17"
Alexa Dorman
And 0h42'22" 30d10'51"
Alexa Dumas
Cnc 9h7'14" 11d36'43"
Alexa Erin Gittleson
Cap 20h14'19" -20d51'1"
Alexa Faith & Alyssa Hope Garbus
Sco 16h7'44" -13d40'56"
Alexa Farver
Dra 20h13'51" 69d37'59"
Alexa Gores
And 23h29'31" 40d12'0"
Alexa Holyn Goldman
Psc 1h11'48" 26d33'22"
Alexa Huff
Lyn 6h27'29" 60d17'8"
Alexa Isabella Marie Dahl
Lyn 9h9'9" 38d34'54"
Alexa Jade
And 23h11'56" 47d27'52"
Alexa Jane Hamelburg
Leo 11h3'49" 20d5'35"
Alexa Jean Bryant
Cyg 19h42'44" 31d48'31"
Alexa Jenna Greenfield
Leo 10h53'13" 8d52'57"
Alexa Joelle
Leo 11h57'9" 24d25'3"
Alexa Joy Didyk
Cap 21h36'15" -10d44'5"
Alexa Katie Wilk
Psc 1h24'4" 25d27'45"
Alexa Krayzelburg
And 1h10'12" 37d14'25"
Alexa Krystal
Sgr 18h44'17" -25d7'27"
Alexa Lambrini Tzaferos
Ori 5h15'11" 0d43'33"
Alexa Leon-Prado
Cnc 8h48'11" 23d49'49"
Alexa Letourneau - superstar
Mon 7h12'58" -0d17'33"
Alexa Lignelli
And 23h34'6" 46d53'29"
ALEXA LYNN
Leo 9h42'55" 27d23'20"
Alexa Lynn Davenport
Cap 21h10'4" -27d11'53"
Alexa Lynn Gandrup
Lyn 6h19'26" 55d25'44"
Alexa Madison
Gem 6h46'29" 20d38'43"
Alexa Maria Thompson
Umi 13h23'49" 72d56'28"
Alexa Marie Marzocca
Gem 7h40'8" 33d3'12"
Alexa Marie Zink
Psc 23h16'32" 1d15'26"
Alexa Mary Jung
Uma 10h12'45" 45d1'53"
Alexa Maureen
And 0h40'34" 39d14'7"
ALEXA MCCOY
Cap 21h6'7" -16d57'27"
Alexa McKenzie Lawrence
And 23h6'45" 48d27'4"
ALEXA MIA MILLS
Sco 17h49'27" -42d40'3"
Alexa Michelle Morrison
Lmi 9h36'33" 36d56'12"
Alexa Miller
Uma 8h34'14" 66d58'19"
Alexa Mohl
Uma 9h57'15" 44d33'11"
Alexa Nicole Bartholomew
Uma 11h51'1" 36d29'25"
Alexa Nicole Marsh
Dra 19h15'22" 73d14'22"
Alexa Olimpia Silva-Papasavas
And 23h14'2" 51d18'34"
Alexa Poulos
Leo 10h22'10" 21d23'10"
Alexa Rae Philapakis
Lyr 18h33'34" 32d34'16"
Alexa Renee
And 0h42'50" 40d39'24"
Alexa Renee Chan
Mon 8h2'11" -0d57'15"
Alexa Renee Moore
Cas 1h29'49" 61d57'48"
Alexa Rose
Cap 20h9'45" -18d5'23"
Alexa Rose
Ari 3h18'34" 28d2'41"
Alexa Rose McDowell Prock
Aqr 22h31'13" -2d7'44"
Alexa Rose Therese
Lib 15h17'31" -14d44'51"
Alexa Rose Zappia
Leo 11h7'20" -2d32'49"
Alexa S. Petruso
Ari 2h51'13" 21d49'56"

Alexa Skyy
Vir 12h9'44" -1d50'36"
Alexa Sophia
Vir 13h42'38" 2d45'42"
Alexa & Taylor Tullmann
Cas 0h45'50" 51d14'49"
Alexa-Allegra
Uma 10h13'34" 42d21'13"
Alexalaurenus
Uma 11h54'18" 51d32'3"
Alexandar Marcellus Jimenez
Per 2h46'49" 53d59'37"
AlexanDawn
Vir 12h13'50" 4d30'36"
Alexander
Ori 5h22'57" 7d45'57"
Alexander
Tau 3h33'30" 8d4'28"
Alexander
Vir 13h22'40" 13d28'52"
Alexander
Lmi 10h45'15" 26d39'16"
Alexander
Her 17h50'30" 20d45'43"
Alexander
Uma 10h11'30" 47d55'30"
Alexander
Her 16h30'5" 48d24'9"
Alexander
Per 3h7'3" 52d5'54"
Alexander
Per 3h38'9" 45d55'47"
Alexander
Umi 4h38'39" 89d17'56"
Alexander
Lib 15h2'32" -0d36'33"
Alexander
Cma 6h57'32" -21d40'46"
Alexander
Cyg 19h48'53" 58d23'57"
Alexander
Pho 1h6'52" -45d3'27"
Alexander
Psc 1h26'7" 3d22'20"
Alexander
Cap 21h24'7" -25d8'47"
Alexander
Cma 6h19'45" -29d0'11"
Alexander A. Amarosa
Uma 12h2'13" 64d26'15"
Alexander A. Rivera
Sco 16h7'13" -11d49'17"
Alexander A. Shenton
Ori 5h32'58" 4d41'48"
Alexander Abramovitch 1
Boo 14h34'18" 52d59'20"
Alexander Adrian
Psc 1h35'3" 15d34'51"
Alexander Agadjanian
Del 20h38'45" 13d32'3"
Alexander Aigen
Boo 14h49'9" 22d41'29"
Alexander and Carlie Spreadbury
Cyg 21h48'10" 52d34'21"
Alexander Anokhin
Aql 19h30'35" 12d5'36"
Alexander Anthony "8-6-99"
Boo 14h18'29" 51d0'26"
Alexander Arthur Agadjanyan
Cnc 9h18'4" 13d29'41"
Alexander Athan Elefteriou
Cru 12h36'12" -56d33'48"
Alexander Athelstan Bushell
Gem 7h28'11" 34d19'55"
Alexander Atwood
Sgr 19h28'37" -41d31'13"
Alexander Azer Joseph Greco
Tau 4h29'2" 21d19'22"
Alexander B. Getka
Uma 11h16'12" 63d50'57"
Alexander B. Mac Laurie III
Her 16h50'4" 34d59'44"
Alexander B. Woolsey
Tau 5h54'19" 25d0'17"
Alexander Bailie - 18.03.1949
And 23h0'11" 52d45'35"
Alexander Berg Piusz
Sco 16h19'14" -12d25'27"
Alexander Bergh
Tau 5h25'21" 21d5'9"
Alexander Blair Wilson
Her 17h40'26" 39d10'57"
Alexander Bonnar
Cep 22h54'55" 72d23'55"
Alexander Booth & Danielle Letarte
Psc 23h2'19" 5d47'44"
Alexander Borland
Her 18h49'13" 22d4'17"
Alexander Brett Smalley
Gem 6h36'35" 12d49'24"
Alexander Brisbois
Umi 14h13'6" 72d38'49"
Alexander Bucevica
Per 3h9'33" 48d45'21"
Alexander C. Scontras
Gem 7h18'36" 31d18'30"
Alexander Canales
Uma 11h30'43" 33d29'39"

Alexander Canty
Cnc 8h18'31" 14d26'49"
Alexander Charles Barger
Per 2h41'59" 53d31'45"
Alexander Charles Bill (Poohbear)
Sco 17h47'4" -44d10'57"
Alexander Charles LaGrotta
Cep 2h19'35" 83d52'44"
Alexander Christian 15092005
Cap 20h13'47" -10d35'50"
Alexander Christian Oxley
Her 18h42'29" 14d1'15"
Alexander Christian Redd
Lib 15h50'20" -11d27'33"
Alexander Christian Renna
Lib 15h30'3" -26d43'27"
Alexander Christopher
Ori 6h13'35" 9d14'10"
Alexander Christopher Buettner
Aur 5h56'19" 52d37'54"
Alexander Christopher Lopez
Psc 0h55'36" 31d16'27"
Alexander Cole
Aur 5h17'25" 45d4'17"
Alexander Cole Wheeler
Uma 11h59'44" 42d39'24"
Alexander Coser
Cyg 19h40'29" 35d30'37"
Alexander Cruz
Her 17h35'25" 33d4'37"
Alexander Curcio
Uma 11h8'42" 39d30'28"
Alexander D. Davis
Her 17h45'45" 43d4'42"
Alexander D. Fairchild
Sco 16h50'25" -26d49'24"
Alexander D. Kaviani
Cnc 8h12'22" 11d40'50"
Alexander David Benedict O Donnell
Ori 6h23'31" 10d54'38"
Alexander David DeLeon
Uma 10h38'19" 41d51'27"
Alexander David Neill
Uma 13h41'25" 57d59'36"
Alexander David Riley
Sgr 19h13'50" -21d39'47"
Alexander Devine
Per 4h47'48" 40d50'40"
Alexander Dieter Bayl
Boo 15h33'10" 46d40'24"
Alexander Doak "Zan" Campbell
Aqr 21h36'58" 1d42'22"
Alexander Dodge
Uma 11h54'57" 37d36'11"
Alexander Dranov
Cnc 8h52'31" 12d39'48"
Alexander Dubitsky
Ari 2h28'27" 22d23'41"
Alexander DuFour
Tau 3h27'57" 15d19'33"
Alexander Dylan
Dra 17h25'33" 64d33'29"
Alexander Earl Merritt
Vir 13h54'33" 0d48'1"
Alexander Edward Altenhofen
Vir 12h10'3" 7d6'49"
Alexander Edward Tessem-Cotton
Per 2h51'40" 46d41'44"
Alexander Elerick
Tau 5h46'13" 25d36'39"
Alexander & Eliot Almeida
Uma 9h31'30" 47d19'41"
Alexander Elizabeth Wahl
Vir 13h23'39" -7d3'44"
Alexander Ells
Dra 16h20'8" 58d27'13"
Alexander Eric Jason Marcy
Umi 17h4'53" 75d25'10"
Alexander Eugene
Leo 11h30'56" 13d9'45"
Alexander Fahlbusch
Ori 5h12'47" 0d44'2"
Alexander Francis Batten
Ori 6h15'36" 8d43'3"
Alexander Francis Ruiz
Tau 4h17'21" 23d4'25"
Alexander Francisco Favata
Uma 9h23'47" 49d25'54"
Alexander Frederick Clavien
Psc 0h42'24" 7d16'42"
Alexander Friedland
Her 17h49'44" 44d48'4"
Alexander G. Granberg
Umi 14h24'12" 77d40'5"
Alexander Gale Mitchell
Her 18h10'28" 28d19'13"
Alexander George Cushing
Umi 16h1'43" 74d4'34"
Alexander George Parker
Uma 11h39'59" 63d39'17"
Alexander George Robertson
Cep 0h16'55" 83d44'40"

Alexander George White
Ori 6h10'12" 2d18'52"
Alexander Gerald Colavecchio
Ari 2h49'56" 15d56'50"
Alexander Gerard Ting Star#1013
Ori 6h6'35" 21d20'51"
Alexander Gerhardt
Ori 6h17'52" -0d13'54"
Alexander Gordon
Lmi 10h7'26" 40d19'52"
Alexander Graham Spence
Aql 19h2'46" 15d30'42"
Alexander Greenlees
Eri 3h48'58" -22d28'20"
Alexander Gregory
Uma 11h38'46" 40d55'30"
Alexander Gregory Magee
Aqr 23h30'47" -8d9'3"
Alexander Griffiphs Scott
Her 17h33'3" 31d38'25"
Alexander Gruber
Per 2h45'49" 55d32'54"
Alexander Gunnard Freeburg
Her 18h55'33" 16d40'27"
Alexander Gutkin
Psc 1h1'38" 13d35'10"
Alexander Hamilton Harley
Per 3h57'31" 33d18'3"
Alexander Hamilton Hess
Umi 15h57'19" 79d34'6"
Alexander Hemphill Denton
Her 17h42'26" 21d54'51"
Alexander Hirjovatij
Lac 22h21'51" 47d29'11"
Alexander Hohnen Tatkovic
Cra 18h0'58" -37d27'25"
Alexander Holt Murray
Psc 1h17'2" 22d31'5"
Alexander Homewood
Her 18h32'17" 12d58'31"
Alexander Ian Johnson
Vir 11h56'52" 2d59'37"
Alexander Irwin
Ori 5h11'46" 12d1'24"
Alexander Ivanov
Per 4h45'48" 46d49'49"
Alexander Ivanov Wilkinson
Cnc 8h54'11" 29d56'56"
Alexander J. Lorenz
Tau 3h37'58" 14d10'19"
Alexander J. Ross
Cma 7h26'14" -23d48'39"
Alexander J. Shane
Cap 21h57'58" -17d30'15"
Alexander J. Thompson
Leo 9h49'17" 21d34'52"
Alexander Jack
Ari 3h20'8" 16d35'53"
Alexander Jack Goulet
Umi 16h16'9" 73d55'14"
Alexander Jackson Merritt
Ori 6h10'43" 15d9'48"
Alexander Jacob (A. J.) Martinez
Gem 7h35'54" 15d37'1"
Alexander Jacob Oberfeld
Lyn 7h47'49" 42d35'27"
Alexander James
Ori 6h7'8" 10d28'46"
Alexander James Ashton
Uma 8h24'40" 70d3'34"
Alexander James Baker
Lib 15h27'58" -25d43'8"
Alexander James Baker
Cnc 8h46'20" 25d2'0"
Alexander James Barrett
Sco 16h50'52" -38d44'20"
Alexander James Bossert
Psc 1h0'47" 4d8'17"
Alexander James DeLong
Sgr 19h2'48" -25d58'54"
Alexander James English
Uma 11h23'48" 60d21'26"
Alexander James Gay
Her 16h50'50" 32d53'16"
Alexander James Hocking AJ
Lib 14h22'59" -10d35'3"
Alexander James Hutchings
Umi 14h47'43" 68d56'32"
Alexander James Kolstad "Alex"
Ori 5h13'18" -1d43'10"
Alexander James Mazak
Cep 22h9'45" 61d6'14"
Alexander James Oeth
Cap 20h30'12" -14d3'58"
Alexander James Pogwist
Dra 18h31'37" 54d11'6"
Alexander James Rees
Her 16h15'32" 24d27'42"
Alexander James Stelly
Her 17h20'45" 16d33'54"
Alexander James Wilson
Ori 5h39'56" 2d19'11"
Alexander Jason Pierce
Uma 9h1'38" 48d54'1"
Alexander Jay Jungen (Zander)
Gem 6h16'52" 21d56'1"
Alexander Jeffrey Zaleski
Umi 16h28'24" 76d50'42"
Alexander Jodush
Psc 1h1'22" 26d27'26"
Alexander Joel Catuogno
Sco 16h7'6" -8d25'36"
Alexander John
Ori 5h14'50" -1d41'27"
Alexander John
Col 5h59'17" -34d42'35"
Alexander John
Vir 14h16'58" 1d38'24"
Alexander John Gribble
Sco 16h43'8" -32d14'22"
Alexander John Oleksinski
Umi 14h37'6" 73d37'34"
Alexander John Porter
Psc 0h26'28" 3d27'59"
Alexander John Tarvardian
Her 17h37'9" 26d48'19"
Alexander Joseph
Tau 3h37'45" 15d49'22"
Alexander Joseph Garner
Ori 5h35'42" -1d45'51"
Alexander Joseph Marek
Her 16h28'34" 44d7'13"
Alexander Joseph Oppedisano
Lib 14h43'6" -9d3'27"
Alexander & Joseph Playfoot
Peg 21h38'32" 24d28'40"
Alexander Joseph Tribe
Uma 10h55'41" 34d16'42"
Alexander Joseph Van Elsis
Her 18h11'0" 48d27'48"
Alexander & Josephine Pope
Cyg 21h58'26" 50d46'6"
Alexander Kamella
And 0h48'5" 41d38'42"
Alexander Kayyal
Cap 21h57'47" -18d21'49"
Alexander Killian Montgomery
Per 3h10'9" 39d6'25"
Alexander Kolin
Boo 14h26'23" 15d43'36"
Alexander Kristian Diaz
Psc 2h1'28" 9d33'52"
Alexander Kritsky
Aur 6h1'13" 36d32'26"
Alexander L Vellozzi
Ori 6h19'54" 15d35'55"
Alexander Lachlan Dickie
Her 17h27'33" 45d29'32"
Alexander Lake Hill
Uma 13h3'18" 57d23'22"
Alexander Landeros Gonzales
Gem 7h21'1" 27d46'44"
Alexander Lawrence Denton
Ori 5h30'5" -0d25'58"
Alexander Lee Faison
Uma 9h21'25" 60d45'17"
Alexander Lemesevski
Ari 2h11'44" 26d4'21"
Alexander Lenihan
Cru 12h3'55" -62d55'7"
Alexander Leto Johnston
Cyg 20h36'26" 31d36'21"
Alexander Leukhin
Cas 2h7'44" 62d27'17"
Alexander Levi
Per 2h59'38" 44d38'3"
Alexander (Lexy) Hennessey
Per 2h18'6" 57d12'18"
Alexander Locke Johnson
Aqr 20h53'29" 2d15'50"
Alexander Logi The Great
Per 2h59'47" 51d49'23"
Alexander Louis Acerenza
Psc 0h20'28" -3d43'57"
Alexander Louis Chaufournier
Her 17h25'8" 44d2'5"
Alexander Louis Vesco
Uma 10h54'36" 38d30'12"
Alexander Lovallo
Ari 2h30'48" 23d32'40"
Alexander Lyons Hill
Cap 20h28'48" -11d39'23"
Alexander M Kendrick
And 1h58'43" 42d31'54"
Alexander Malachi Johnson
Vir 13h24'58" -16d27'2"
Alexander Manuel Lopez
Cyg 21h52'35" 37d16'47"
Alexander Mapley
Aqr 22h28'43" -6d56'55"
Alexander Mark Fischer
Sco 16h19'37" -13d2'53"
Alexander Marshall Gibbons
Aur 6h29'40" 35d21'54"
Alexander Matthew Hawkins
Uma 8h41'19" 57d55'17"
Alexander Matus
Ari 3h2'33" 18d54'57"
Alexander McClure
Her 18h19'9" 22d16'15"
Alexander Meehan
Uma 10h16'40" 49d2'6"
Alexander & Melissa Cross
Eri 3h28'22" -5d12'19"
Alexander Michael
Umi 14h0'9" 77d54'12"
Alexander Michael Garofalo
Lmi 9h23'13" 36d39'47"
Alexander Michael Green
Boo 15h22'10" 44d55'16"
Alexander Michael Iorio
Ori 5h30'22" 5d42'6"
Alexander Michael Lien
Umi 16h21'1" 77d28'0"
Alexander Michael Querin
Pho 2h19'40" -41d18'35"
Alexander Michael Wallace
Her 17h37'7" 30d41'20"
Alexander Michalski
Lib 15h56'0" -17d11'38"
Alexander Miguel Welsh
Uma 9h56'11" 53d43'22"
Alexander & Mira Becker
Per 4h14'46" 36d42'56"
Alexander Mitrofonoff (Alex Mitt)
Ari 2h20'34" 14d36'17"
Alexander Montes
Uma 10h22'25" 45d56'31"
Alexander Morgan, Jr.
Ori 6h19'19" 9d25'31"
Alexander Murko
Cru 12h7'51" -59d17'59"
Alexander Murovanny
Uma 13h54'12" 51d19'52"
Alexander N. Antzoulatos
Uma 9h0'40" 50d50'45"
Alexander N. Handy
Her 16h36'54" 44d17'35"
Alexander Narruhn of San Francisco
Her 17h36'4" 28d50'6"
Alexander Nicholas Carle
Psc 0h27'37" 16d32'31"
Alexander Nicholas Cruz
Crb 15h47'33" 37d3'0"
Alexander Nicholas McLaughlin
Dra 17h6'53" 68d14'50"
Alexander + Nicole
Uma 14h27'32" 62d15'19"
Alexander O. Faigel
Dra 17h43'3" 62d37'3"
Alexander Olivares
Cyg 19h42'19" 44d48'38"
Alexander Ortiz-Knight of the sky
Sco 16h15'8" -18d37'16"
Alexander - Our Beloved Son
Gem 6h53'2" 22d15'11"
Alexander P Riley
Cyg 21h37'27" 45d5'27"
Alexander Pachete
Ari 3h1'47" 21d39'43"
Alexander "Pa-Grandpa" S. Monko
Ori 5h22'32" 2d40'58"
Alexander Parboukov
Cnv 13h44'39" 36d7'59"
Alexander Park
Umi 14h23'9" 74d29'57"
Alexander Paton
Sco 16h9'14" -10d7'2"
Alexander Patrick Fitzgerald
Per 4h36'11" 36d42'3"
Alexander Paul Carson
Psc 1h42'34" 5d31'3"
Alexander Paul Hauth
Pho 0h41'12" -46d50'50"
Alexander Paul Kirchner
Cet 1h26'42" -10d56'27"
Alexander Paul Petrak
Her 16h42'28" 29d31'5"
Alexander Paul Tufto
Aqr 23h12'27" -7d18'40"
Alexander Perry Mincone
Umi 14h41'58" 67d57'29"
Alexander Peskin
Cep 22h18'25" 81d20'42"
Alexander Peskin
Her 17h57'7" 20d9'51"
Alexander Peter Dean
Sgr 18h24'9" -33d39'28"
Alexander Philip Medeiros
Her 16h25'49" 22d44'30"
Alexander Pires
Uma 10h47'31" 61d31'46"
Alexander Puls
Ori 6h21'21" 16d35'44"
Alexander Quinn Hawthorne
Aql 18h53'45" -0d41'44"
Alexander R. Mercer
Psc 1h10'23" 32d25'17"
Alexander Raymond McConeghy
Aur 6h2'8" 47d50'50"
Alexander Re
Aqr 22h30'9" -0d52'55"
Alexander Reese
Ori 5h36'13" -1d39'30"
Alexander Reineck
Psc 1h10'55" 16d32'46"
Alexander Reinhold Herzfeld
Cnc 8h58'16" 11d40'12"
Alexander Restaino
Cma 6h40'3" -13d1'45"
Alexander Richard Taylor
Her 16h37'40" 46d14'4"
Alexander Richard Tuma
Uma 9h38'7" 41d26'33"
Alexander Rimalovski
Gem 7h41'53" 32d25'46"
Alexander Robert
Lib 15h9'0" -7d34'1"
Alexander Robert McCaskie Munn
Umi 17h7'38" 76d2'42"
Alexander Roderick
Her 17h28'13" 25d11'12"
Alexander Rodriguez
Uma 14h27'18" 61d7'15"
Alexander Ruche
Lib 15h29'50" -10d57'43"
Alexander Ruck Keene
Dra 18h33'58" 57d39'48"
Alexander Ryan
Leo 11h48'38" 19d52'56"
Alexander Ryan Astrada
Aur 5h33'32" 41d40'31"
Alexander Ryan Gennaro
Uma 9h22'27" 59d38'17"
Alexander Ryan Suski
Ari 3h8'49" 11d11'13"
Alexander S. Koczak
Cyg 20h57'41" 45d36'38"
Alexander Samuel Melton
Cyg 20h5'19" 33d7'48"
Alexander "Sandy" MacLaren
Cyg 19h38'4" 31d49'20"
Alexander "Sandy" Strachan
Ori 5h39'0" -1d3'4"
Alexander Scott Grube
Her 18h34'29" 14d47'12"
Alexander Scott Michel
Vir 14h38'2" 0d2'57"
Alexander Scott White
Sco 17h25'57" -44d50'31"
Alexander Sebastian
Aur 5h33'31" 46d40'43"
Alexander Shanks
Cep 0h7'59" 77d50'17"
Alexander Sidney Rogers
Uma 12h5'44" 63d50'51"
Alexander Siegfried Emke
Cap 20h26'36" -22d41'29"
Alexander Sklaroff Van Hook
Cep 23h59'58" 69d11'8"
Alexander Star Mendoza
Uma 9h45'35" 44d7'14"
Alexander&Stefanie
Uma 8h22'7" 64d9'35"
Alexander Stephen Lee
Uma 11h48'48" 29d53'18"
Alexander Stephen Tanner
Cnc 8h11'39" 32d27'34"
Alexander Steven Wojtaszek
Sgr 18h13'55" -27d19'55"
Alexander Sutherland
Per 2h52'4" 51d42'28"
Alexander T. Poreda
Aql 18h49'31" -0d33'40"
Alexander & Tania Kalchevy
Cyg 21h14'6" 47d15'36"
Alexander the Craig
Cnc 8h27'22" 21d13'24"
Alexander - The First
Uma 11h0'35" 60d22'53"
Alexander "The Great" Politis
Per 2h19'41" 56d48'55"
Alexander Thomas
Uma 11h27'39" 53d25'1"
Alexander Thomas Brown
Umi 17h6'51" 77d3'1"
Alexander Thomas Casey
Umi 16h24'44" 77d21'12"
Alexander Thomas Curley
Her 16h45'5" 34d11'11"
Alexander Topalov
Uma 8h53'36" 51d34'19"
Alexander Tristan Slone
Gem 6h43'47" 14d32'44"
Alexander Tumolo
Sgr 18h13'23" -19d11'58"
Alexander Tyrone Hazell
Ori 5h26'51" -0d31'42"
Alexander Vincent Merlino
Aql 19h36'0" 13d9'48"
Alexander Von Maltzahn
Her 16h53'34" 17d6'26"
Alexander von Skrobotof Cleves Shaw
Uma 11h41'11" 63d30'39"
Alexander Wade Wallace
Leo 10h31'26" 12d46'11"
Alexander Wainner
Cnc 9h13'37" 10d30'14"
Alexander Walhin
Umi 16h35'17" 81d1'0"
Alexander Warren Kane
Cnc 8h47'12" 28d22'0"
Alexander Willem Prins
Psc 23h29'54" -3d4'27"
Alexander William 11/23/00
Dra 20h24'34" 67d29'7"
Alexander William Bealles
Uma 11h49'10" 55d32'38"
Alexander William Gregor
Tau 5h9'41" 17d55'20"
Alexander William Soteras
Cru 12h46'33" -56d37'57"
Alexander Williams
Cru 11h57'13" -58d41'36"
Alexander Wisniewski
Aql 19h14'45" 5d8'8"
Alexander Wolff Herz
Cnc 8h21'15" 20d1'45"
Alexander Xavier Goss
Ari 2h34'58" 25d4'46"
Alexander Yakovlev
Aur 6h48'45" 44d21'11"
Alexander Yumakaev
Uma 8h50'31" 66d13'42"
Alexander062696
Cnc 8h12'19" 28d47'55"
Alexander-Matei Coroi
Vir 13h13'39" -0d46'49"
Alexander's Dream
Ari 3h22'37" 29d10'47"
Alexander's Glücksstern
Uma 12h16'19" 58d37'3"
Alexander's Star
Dra 17h31'38" 67d27'40"
Alexander's Third Birthday Star
Uma 11h9'31" 68d19'38"
Alexandra
Uma 10h46'24" 67d51'23"
Alexandra
Cas 1h35'32" 69d9'56"
Alexandra
Uma 9h32'13" 64d58'36"
Alexandra
Cam 3h39'2" 63d20'51"
Alexandra
Cas 0h54'1" 60d27'54"
Alexandra
Vir 13h16'34" -4d23'43"
Alexandra
Lib 15h7'15" -5d40'23"
Alexandra
Lib 15h30'29" -6d28'2"
Alexandra
Sgr 18h31'4" -20d2'14"
Alexandra
Sco 17h57'5" -39d10'55"
Alexandra
Sco 17h9'33" -41d3'3"
Alexandra
Lib 14h26'12" -22d55'2"
Alexandra
Sco 16h7'26" -23d45'46"
Alexandra
Gem 7h15'33" 20d30'36"
Alexandra
Vir 12h23'46" 6d33'39"
Alexandra
Ari 3h3'54" 15d25'4"
Alexandra
Leo 11h9'56" 9d51'36"
Alexandra
Vir 13h19'45" 13d48'43"
Alexandra
Per 3h17'4" 41d38'1"
alexandra
Uma 10h13'15" 44d52'10"
Alexandra
Cyg 20h11'43" 37d12'1"
Alexandra
Cyg 21h28'58" 35d15'52"
Alexandra
And 0h44'12" 40d27'52"
Alexandra
Cnv 13h45'20" 30d30'10"
Alexandra
And 1h44'29" 36d7'55"
Alexandra
And 1h48'16" 36d36'54"
Alexandra
Cyg 21h7'50" 47d53'38"
Alexandra
And 22h59'11" 50d56'5"
Alexandra
And 23h49'21" 48d14'17"
Alexandra
And 23h16'1" 49d40'41"
Alexandra
And 23h47'57" 38d38'44"
Alexandra 05
Psc 1h22'48" 29d51'22"
Alexandra 15
Aqr 21h7'33" -10d32'9"
Alexandra Ace
Cru 12h39'4" -58d30'54"
Alexandra Alderman
Cam 7h22'8" 70d5'45"
Alexandra "Alie" Poindexter
And 23h40'50" 34d17'45"
Alexandra "Ally" Elizabeth Zippo
Uma 12h49'31" 58d22'26"
Alexandra Amoret
Uma 10h45'14" 71d15'43"
Alexandra and Brian
Cnv 12h46'43" 38d32'20"
Alexandra and Jacob Terry
Umi 17h3'40" 82d7'42"
Alexandra Angel
Cas 1h29'48" 72d30'49"
Alexandra Angel
Tau 4h21'20" 24d37'42"
Alexandra Ann Pikas
Cap 21h50'8" -14d23'39"
Alexandra Ann Rimoldi
Gem 7h39'20" 21d30'33"
Alexandra Arlene Talbott
Psc 0h57'18" 26d52'51"
Alexandra Aronson
And 23h3'29" 37d15'22"
Alexandra Athena
Psc 0h52'8" 29d14'55"
Alexandra B. Clermont
Uma 9h25'31" 46d35'24"
Alexandra Banes Hopson
Uma 11h57'32" 57d29'56"
Alexandra Barbara McLaren
Aqr 23h50'13" -8d20'13"
Alexandra Beadle-Ryby
And 0h5'5" 35d32'41"
Alexandra Behnke
And 0h31'11" 35d20'27"
Alexandra Bernstein
Leo 10h10'28" 14d54'40"
Alexandra Beyreis-Heim
Aqr 22h30'31" 1d16'51"
Alexandra Brennan Foxhoven
Gem 7h9'6" 33d32'23"
Alexandra Brilliant Beauty
And 23h31'12" 39d9'35"
Alexandra Brittany
And 23h39'47" 47d50'38"
Alexandra Brosch
Uma 10h28'38" 48d16'49"
Alexandra Caitlin Collins
Her 16h8'45" 22d29'51"
Alexandra Cameras
Crb 15h34'16" 38d0'56"
Alexandra Camille
And 1h17'22" 49d53'18"
Alexandra Cason
Sgr 18h5'4" -27d46'14"
Alexandra Catherine Butzirus
Cas 0h54'27" 60d37'58"
Alexandra Cauble
Leo 11h23'7" -5d39'43"
Alexandra Chantie Spencer
Lib 15h5'25" -6d21'0"
Alexandra Chapman
And 1h49'4" 41d29'51"
Alexandra Cheyenne Vopinek
Leo 11h8'7" 15d29'49"
Alexandra Christine Cabrera
And 2h14'53" 48d18'12"
Alexandra Christine Manolis
Lyr 18h46'28" 34d23'41"
Alexandra Clare Burke
And 1h22'12" 46d45'54"
Alexandra Conelli
Lyn 7h31'9" 39d58'29"
Alexandra Dahlkild
And 1h12'4" 41d36'35"
Alexandra - dare to dream
Tau 4h14'10" 21d43'34"
Alexandra Dawn Clark
Tau 4h37'55" 23d49'5"
Alexandra de Vars
Cnc 8h4'19" 25d27'14"
Alexandra & Diana
Uma 9h57'3" 61d17'2"
Alexandra Diane Speaks
Aqr 23h48'7" -15d59'11"
Alexandra Dylana Williams
Lib 15h23'2" -6d1'59"
Alexandra Eileen
Cnc 8h48'33" 13d34'27"
Alexandra Elaine
Mon 6h50'45" 3d24'1"
Alexandra Elaine Morford
And 2h32'48" 44d18'20"
Alexandra Elise Hall
Cyg 20h25'9" 39d37'20"
Alexandra Elise Martyn
Cap 21h47'43" -10d35'21"
Alexandra Elizabeth Hanna
Cam 5h44'39" 59d40'40"
Alexandra Elizabeth Meisinger
Tau 5h59'1" 25d23'3"
Alexandra Ell Adams
Uma 9h41'54" 42d41'47"
Alexandra Emily Biel
Cap 20h35'47" -15d13'12"
Alexandra Erby
Lyn 8h59'37" 45d33'55"
Alexandra Eremia -"Alex"
Lyr 18h56'18" 33d40'26"
Alexandra Evelyn Brown
Uma 8h18'14" 67d34'56"
Alexandra Fernandez
And 1h28'46" 39d46'41"
Alexandra Flora Luongo
Tau 5h48'35" 16d56'35"
Alexandra Flores
Lib 15h26'29" -8d37'43"
Alexandra Flynn McGee
Sco 16h54'26" -32d51'57"
Alexandra Freeman
Cap 21h12'48" -15d28'45"
Alexandra Fuller
Ori 5h20'46" -7d13'59"
Alexandra Fuller
Tau 4h8'56" 13d55'19"
Alexandra Gabrielle Kalfon-Wallace
Cnc 8h41'36" 21d43'54"
Alexandra Gachnang
Aql 19h31'18" 4d34'18"
Alexandra Gale Webster
And 23h8'4" 39d34'59"
Alexandra Gayle Webster
Cyg 20h4'55" 58d9'18"
Alexandra George
Gem 6h27'26" 23d13'4"
Alexandra Gitsi
Uma 11h38'16" 65d33'49"
Alexandra Gorbokon
Cnc 8h16'16" 15d43'10"
Alexandra Grace Madama
And 1h0'42" 46d50'42"
Alexandra Grace Sinclair
Tau 3h59'13" 22d59'40"
Alexandra Grace Weinhardt
And 1h19'39" 49d58'37"
Alexandra Grace Wyman
Psc 1h20'32" 29d17'9"
Alexandra Gretchen Bergman
Crb 16h18'31" 27d29'55"
Alexandra Hängärtner (Lexli)
Umi 16h59'17" 78d26'12"
Alexandra Harbord
And 2h24'37" 48d17'59"
Alexandra Harrison
Tau 5h2'50" 22d6'7"
Alexandra Heydari
Ari 1h47'16" 22d47'21"
Alexandra Hope Cieplinski
And 2h20'31" 47d35'20"
Alexandra Hope Trubee
Lib 15h25'36" -4d52'52"
Alexandra Isabel Edwards
And 0h7'22" 40d3'10"
Alexandra Isoard
Aqr 22h35'13" -23d14'27"
Alexandra J. Cheng
Uma 10h33'52" 54d1'12"
Alexandra Jade
Cas 0h13'34" 51d1'15"
Alexandra Jaine Fielder
And 23h3'6" 52d1'43"
Alexandra Jane West
Cru 12h28'36" -61d57'56"
Alexandra Johanna Nahatis Barrett
Sco 17h7'12" -30d22'41"
Alexandra Jordan
And 1h42'33" 41d9'58"
Alexandra Jordan Thelin
Vir 14h32'58" -2d3'21"
Alexandra K. Pla
Del 20h49'31" 14d3'7"
Alexandra Kabakoff
Lib 15h22'50" -7d44'10"
Alexandra Kate Manalo
And 0h20'54" 33d36'24"
Alexandra Katherine Fiona Farquhar
Cap 20h24'14" -11d10'14"
Alexandra Kress
Ori 6h7'45" 20d8'23"
Alexandra Leann Foulkrod ~ Alli
Cnc 8h32'14" 11d21'41"
Alexandra Lee Kastell
Lyr 18h31'34" 36d1'55"
Alexandra Lee Palmer
Uma 8h55'19" 56d18'47"
Alexandra Leia Nifco
Sgr 18h53'13" -29d13'9"
Alexandra Leigh Everhart
Gem 7h2'53" 33d4'13"
Alexandra Leigh Ferguson
Cas 0h21'54" 50d39'37"
Alexandra "Lexi" Katzman
Tau 4h18'53" 7d31'1"
Alexandra "Lexi" Panzeri
Uma 10h6'9" 57d3'41"
Alexandra Lianna
Cap 21h2'17" -17d31'8"
Alexandra Lily Seaman
And 0h11'18" 33d6'32"
Alexandra Lindsey Parrish
Leo 10h13'38" 23d52'16"
Alexandra Llerena
Crb 15h57'20" 39d25'42"
Alexandra Lody
Ori 5h24'8" 7d59'10"
Alexandra Logan Dorr
Cyg 20h18'35" 58d28'39"
Alexandra Logvy
Cap 21h26'48" -22d30'47"
Alexandra Louise Pickett
Ari 3h11'40" 17d53'55"
Alexandra Love Hulnick
Pho 0h44'14" -42d27'50"
Alexandra Lovell
And 22h58'59" 50d46'34"

Alexandra Lucia Burress
Crb 15h50'14" 31d32'39"
Alexandra Lyn Rallatos
And 1h49'14" 42d39'14"
Alexandra Lyndsay Herzog
Psc 0h40'42" 7d49'3"
Alexandra Lynn
Tau 5h46'40" 21d7'31"
Alexandra Lynn Levick
And 23h32'13" 48d21'58"
Alexandra Lynn Noah
Ori 5h11'53" -1d29'27"
Alexandra Lynn Pellillo
Tau 5h38'15" 21d15'16"
Alexandra Lynn Uridge
Cas 0h37'8" 65d18'4"
Alexandra Lynn Uridge
Cas 1h17'23" 63d6'17"
Alexandra Lynn Uridge
Cas 0h27'7" 60d21'39"
Alexandra Lynn Uridge
Cas 0h31'39" 60d4'34"
Alexandra Lynn Uridge
Cas 0h29'36" 60d18'18"
Alexandra M
Vir 12h50'31" 6d59'3"
Alexandra M Brookes
Sco 16h7'59" -15d30'26"
Alexandra M. Colemere
Mon 6h51'31" -0d11'27"
Alexandra M James
Dra 16h13'43" 55d27'50"
Alexandra Madero
Tau 3h50'36" 20d47'7"
Alexandra Mae Middleton Stauffer
Dra 18h48'10" 52d7'54"
Alexandra Marguerite
Lib 14h55'56" -5d47'3"
Alexandra Maria Jennings
Lib 15h25'19" -20d58'21"
Alexandra Maria Pallisco
Ari 3h21'23" 19d11'49"
Alexandra Marie
Lyn 6h24'5" 55d21'20"
Alexandra Marie Alberta
Dra 18h30'42" 70d3'21"
Alexandra Marie Charlton
And 2h23'44" 49d59'16"
Alexandra Marie Elizabeth Paffie
Dra 19h41'56" 63d11'4"
Alexandra Marie Francis
Tau 4h16'22" 23d41'42"
Alexandra Marie Gavranovic
Sgr 18h24'54" -22d30'20"
Alexandra Marie Herrera
Cnc 8h34'25" 8d51'18"
Alexandra Marie Lauze
Cas 1h49'11" 61d31'14"
Alexandra Marie Parker
Tau 5h59'55" 27d21'42"
Alexandra Marie Prickett
Peg 22h22'57" 15d28'24"
Alexandra Marie Rodriguez
Leo 9h47'8" 13d3'14"
Alexandra Marie Rose Friello
Sgr 19h35'45" -14d14'20"
Alexandra Marie Taylor
Psc 0h0'55" 9d11'11"
Alexandra Marie Valeri
Cap 20h38'35" -19d41'15"
Alexandra Marie Zerega
Lib 15h34'3" -5d44'55"
Alexandra Marina
Lyn 7h34'27" 36d26'58"
Alexandra Marini
Cap 20h26'20" -13d22'30"
Alexandra Marques
Psc 1h19'1" 16d50'23"
Alexandra Martin
Gem 7h21'37" 25d57'28"
Alexandra Mathias
Gem 6h37'53" 20d17'49"
Alexandra McKenzie Beaton
Gem 7h37'23" 19d46'7"
Alexandra McMillan Thorp
Peg 21h59'34" 35d36'11"
Alexandra Merritt
Ori 5h58'5" 6d57'6"
Alexandra Messersmith
Uma 9h35'8" 47d12'21"
Alexandra & Michelle
Cyg 20h12'13" 35d53'59"
Alexandra Miel
And 2h27'5" 47d46'14"
Alexandra Morgan Wucetich
Tau 4h39'10" 6d26'0"
Alexandra Mount 21
Cru 12h49'45" -60d46'25"
Alexandra Muñoz
Mon 7h40'10" -7d19'52"
Alexandra N. Fiust
Lib 15h6'12" -27d19'25"
Alexandra Nicole
Tau 3h46'0" 21d17'54"
Alexandra Nicole
Tau 4h36'16" 27d37'12"
Alexandra Nicole Canfield
Leo 11h17'43" 15d2'4"

Alexandra Nicole D'Onofrio
Tau 3h38'43" 27d40'28"
Alexandra Nicole Eagle
And 1h7'28" 37d6'5"
Alexandra Nicolle Williamson
And 1h22'32" 37d32'15"
Alexandra Noel Neyman
Lyr 18h29'27" 28d58'22"
Alexandra Olivia Billson
Ori 5h1'26" 15d10'29"
Alexandra Paciullo
And 0h53'33" 46d29'8"
Alexandra Pallottie "Shining Star"
Gem 7h9'0" 17d4'45"
Alexandra Parnass
And 0h47'15" 25d23'31"
Alexandra Parness
Equ 21h16'19" 8d58'58"
Alexandra Parsons
Peg 22h18'10" 17d6'33"
Alexandra Penney
Uma 11h29'29" 35d41'29"
Alexandra Piccin
Ori 5h59'29" 21d17'45"
Alexandra Pita
Cam 6h45'11" 68d20'10"
Alexandra Pitta-Chazapi
Uma 9h32'47" 42d34'37"
Alexandra Puchinger
Peg 21h55'34" 13d53'39"
Alexandra Quebec Forbes
Vir 12h8'31" 0d8'10"
Alexandra R. Linville
Tau 3h32'53" 17d16'49"
Alexandra Rae
Tau 5h37'40" 26d17'7"
Alexandra & René
Ori 6h16'37" 10d13'7"
Alexandra Rescher
Cyg 20h51'53" 39d10'24"
Alexandra Richardson
Psc 0h19'11" 18d58'51"
Alexandra RL
Sco 17h27'48" -43d27'58"
Alexandra Rodrigues
Cam 4h2'21" 71d13'0"
Alexandra & Romuald 11 Octobre 2003
Aqr 23h46'5" -4d23'29"
Alexandra Rose Ann Dean
Lyn 9h19'33" 39d10'14"
Alexandra Rose Caleca
Aqr 23h3'3" -8d8'25"
Alexandra Rose Gatelaro
And 1h43'24" 43d16'28"
Alexandra Rose McPhillips
Leo 9h59'56" 13d20'58"
Alexandra Rose Theresa D'Aluisio
Lib 15h25'37" -9d22'32"
Alexandra Roxanna Youmazzo
Cnc 9h5'34" 27d14'24"
Alexandra & Ruedi
Lyr 19h24'33" 38d7'54"
Alexandra Sabine
Uma 8h53'55" 52d52'4"
Alexandra Sacani
Uma 8h38'53" 65d51'42"
Alexandra Sarah Juliette
Cap 21h12'44" -25d32'15"
Alexandra "Sascha" Ilina
Ori 5h59'49" 6d45'21"
Alexandra "Sasha" Averko
Uma 13h49'18" 54d7'32"
Alexandra Scheidell McCulloch
And 1h44'15" 42d35'35"
Alexandra Sensat
Sgr 18h23'34" -28d19'40"
Alexandra Sitlali Patricia
Cas 1h12'46" 50d5'5"
Alexandra Smith
Lib 15h29'4" -4d11'5"
Alexandra Sophia Gheen
And 2h18'30" 45d21'12"
Alexandra Sophia Wellman
And 1h29'41" 42d42'42"
Alexandra Soto
Leo 9h36'44" 27d28'33"
Alexandra Star of Love and Light
Ori 6h6'48" 21d25'3"
Alexandra Stief
Aur 5h18'26" 31d35'31"
Alexandra Stuetze
And 23h16'15" 50d0'19"
Alexandra Taylor Bone
And 23h20'39" 42d2'6"
Alexandra Taylor Buongiovanni
Cnc 8h41'26" 31d2'12"
Alexandra Teresa Suarez
Ori 5h57'4" 18d29'20"
Alexandra Thanh Ha Singer
And 1h2'32" 45d17'43"
Alexandra Theurich
And 0h44'5" 40d28'0"
Alexandra Tiare
Dra 19h33'19" 68d40'49"
Alexandra Tozzi
And 2h36'38" 43d20'11"

Alexandra Tyson
And 1h37'12" 50d28'21"
Alexandra V.
Crb 16h6'47" 37d6'23"
Alexandra Valentina
Sco 16h35'6" -27d32'28"
Alexandra Valta
Cas 0h28'36" 58d47'58"
Alexandra Vecchio
And 1h9'34" 45d45'45"
Alexandra Verstuyft
Hya 9h16'32" -0d12'28"
Alexandra Vickowski
And 2h37'12" 48d7'55"
Alexandra Victoria
And 0h34'24" 44d8'15"
Alexandra Victoria DelCalvo
Leo 10h33'14" 14d18'10"
Alexandra Violet Lahourcade
Uma 12h55'36" 57d8'28"
Alexandra+*Virginie*=199 9
Lib 15h24'43" -23d51'34"
Alexandra Von Eisenhart
Tri 2h20'1" 37d12'18"
Alexandra Walker
Lyr 18h39'52" 37d20'10"
Alexandra Winters
Tau 4h35'55" 8d0'11"
Alexandra Yakovlev
Umi 16h29'38" 77d17'41"
Alexandra Zankl
Sgr 18h25'49" -36d41'38"
Alexandra Zinsmeister
Uma 8h57'35" 52d44'39"
AlexandraJulianaWilliamMiskovich
Tri 2h21'38" 34d47'56"
AlexandraMcAbbott
Lib 15h22'12" -7d21'7"
Alexandra's Always Shining Star
Uma 11h50'40" 28d28'34"
Alexandra's Aura
Gem 6h46'48" 31d43'58"
Alexandra's Aura
Crb 15h54'20" 35d41'40"
Alexandra's Dream
And 1h47'46" 46d18'24"
Alexandra's Dream
Cep 22h32'58" 66d3'35"
Alexandra's Faith
And 23h2'47" 36d53'42"
Alexandra's star
Lyn 6h43'41" 51d41'56"
Alexandra's Wishing Star
Psc 0h22'50" 6d57'52"
Alexandre
Psc 1h55'54" 6d56'13"
Alexandre
Uma 10h3'14" 53d9'17"
Alexandre
Cas 23h36'32" 58d11'50"
Alexandre
Ori 6h20'36" -0d38'37"
Alexandre Bagrintsev
Uma 8h20'15" 65d49'22"
Alexandre Brau-Mouret-ABM
Uma 10h55'10" 38d27'42"
Alexandre et Maeva
Umi 16h18'14" 71d42'52"
Alexandre et Magali
Peg 21h29'34" 15d48'33"
Alexandre Godart
Uma 13h44'35" 52d54'49"
Alexandre Immelé
Leo 11h15'29" 2d14'44"
Alexandre Lachat
Umi 15h58'14" 81d27'12"
Alexandre Le Bouthillier
Ori 6h5'45" 10d19'54"
Alexandre Leroux
Cep 23h56'48" 67d18'7"
Alexandré Marie Easley
Peg 23h42'0" 21d38'15"
Alexandre Mario Ginekis
Ori 4h52'46" 10d52'39"
Alexandre Martin
Cyg 21h5'53" 48d14'38"
Alexandre Patrick Richen
Per 3h33'5" 47d11'3"
Alexandre Pierrot
Peg 21h16'30" 16d19'54"
Alexandre Ramos Puyals
Ori 5h59'39" 20d44'45"
Alexandre Shifman
Psc 1h11'38" 19d3'31"
Alexandre Turcotte
Cap 21h44'38" -10d34'42"
Alexandre Vidal Porto y Matias
Psc 1h17'44" 22d43'49"
Alexandre Witasse
Umi 14h35'10" 88d44'18"
Alexandre, Julie et Frédérick
Cyg 22h2'49" 45d30'17"
Alexandrea Danille Minor
Crb 16h12'0" 36d25'20"
Alexandrea Prevost
And 0h56'40" 40d29'58"

Alexandrew
Uma 9h43'19" 50d12'13"
Alexandria
And 23h40'35" 42d18'19"
Alexandria
Lmi 10h33'34" 37d58'43"
Alexandria
And 0h38'7" 35d47'49"
Alexandria
Psc 0h54'16" 31d17'38"
Alexandria
Gem 6h22'36" 22d17'58"
Alexandria
Uma 11h50'15" 28d23'22"
Alexandria
Gem 6h44'26" 25d30'7"
Alexandria
Tau 4h16'45" 28d34'5"
Alexandria
Tau 5h48'7" 16d48'9"
Alexandria
Ori 5h22'36" 14d46'10"
Alexandria
Vir 14h48'10" 3d15'4"
Alexandria
Sgr 18h3'32" -17d35'27"
Alexandria
Uma 11h2'41" 65d0'13"
Alexandria aka My Toot
And 23h11'9" 35d18'39"
Alexandria Ari
Uma 13h53'57" 47d58'34"
Alexandria Caitlyn Butkovich
Vir 14h40'30" 2d27'18"
Alexandria Carolyn Cameron
And 23h30'53" 48d25'38"
Alexandria Christine
Lib 14h31'46" -24d57'32"
Alexandria Dawn
And 0h20'30" 33d0'47"
Alexandria Dee Neaderhiser
And 0h8'44" 33d58'26"
Alexandria Di Furia Star
Ori 6h21'49" 14d52'3"
Alexandria Elizabeth Hamilton
Cyg 21h46'0" 47d54'9"
Alexandria Elyse McGregor: Soulmate
Ari 2h25'52" 24d40'1"
Alexandria Gabrielle
And 0h57'48" 39d58'59"
Alexandria Gene Lau
Sco 16h13'15" -18d6'54"
Alexandria Grace Cnudde
Sco 17h7'35" -38d15'10"
Alexandria Jenai Rice
And 2h33'30" 47d28'2"
(Alexandria) Jessica Sloate
Uma 9h21'5" 69d20'10"
Alexandria Jo Spaeth
Sco 17h52'5" -36d31'38"
Alexandria Joyce Kohn
And 23h18'7" 44d28'15"
Alexandria Lee
And 2h21'6" 48d57'37"
Alexandria Lee Hall
And 0h44'49" 35d53'12"
Alexandria Lee Nelson
Lib 15h31'22" -12d38'13"
Alexandria "Lexi" Danielle Acquaye
Psc 0h43'25" 14d44'1"
Alexandria Louise Scamehorn
And 0h9'28" 46d38'22"
Alexandria Lynn Caraballo
And 1h37'38" 41d8'33"
Alexandria Lynne Bergmann 3/13/2004
Psc 1h9'29" 17d55'53"
Alexandria Magnus
Vir 13h23'5" -0d40'41"
Alexandria Manitsas Douglas
Lyr 18h25'17" 33d14'26"
Alexandria Marie; Isaiah 26:4
And 0h23'38" 35d39'52"
Alexandria Marie Iseler
And 2h21'53" 46d49'11"
Alexandria Marie McCoy
Ori 6h7'10" 4d57'20"
Alexandria Marie Rizzo
Ari 2h20'57" 14d43'25"
Alexandria Michelle
Sco 16h53'17" -40d51'8"
Alexandria Michelle Stefan
Umi 17h1'30" 79d25'47"
Alexandria Napert
Crb 15h55'38" 37d27'25"
Alexandria Nichole Musick
Tau 4h11'8" 29d10'14"
Alexandria Noel
Cnc 8h41'43" 17d16'12"
Alexandria Noel
And 2h15'46" 44d28'40"
Alexandria Noelle
Tau 4h44'2" 26d53'37"
Alexandria Oliveri Touching Star
And 1h14'35" 46d53'45"

Alexandria Paige VanderVoort
Ori 5h50'49" 20d45'33"
Alexandria Pamela Conner
Cas 0h27'6" 53d53'54"
Alexandria R. Lardent
And 23h40'5" 42d52'45"
Alexandria Rae McKinstry-Lawson
Psc 1h3'12" 10d15'33"
Alexandria Rhea Saba
Ari 2h16'47" 26d53'14"
Alexandria Rose
Psc 0h52'17" 20d57'10"
Alexandria Rose
Uma 12h9'14" 53d2'10"
Alexandria Rose Lukaszek
Tau 4h50'16" 22d28'21"
Alexandria Stanley
Umi 14h27'32" 69d56'11"
Alexandria Taylor Loveless 5-1-94
Gem 7h38'38" 23d57'47"
Alexandria Watson
Peg 0h6'51" 17d44'54"
Alexandria Weston Caterino
Leo 11h54'40" 24d6'31"
Alexandria Zarillo
Lyr 18h41'36" 36d57'43"
Alexandria Zoe
Ori 5h5'27" 4d28'34"
AlexandriaMichelle KristovichWarner
Cnc 8h53'44" 31d9'10"
Alexandrian Joy Casem
Aqr 22h41'23" -24d27'52"
Alexandros Chrysochoidis
Cnc 9h14'43" 18d35'41"
Alexandros Despoinis
Dra 19h39'6" 66d0'20"
Alexane
Leo 9h41'44" 29d51'49"
Alexann Susholtz
And 0h32'31" 31d14'55"
Alexa's Aura
Cap 20h10'8" -26d42'22"
Alexas Faith
Lmi 10h44'35" 37d55'41"
Alexa's Star
Psc 0h17'57" 2d32'36"
Alexa's Wishing Star
And 0h43'31" 37d7'57"
Alexcia Rochelle Gold
Crb 15h38'18" 38d22'49"
Alexe Workman
Peg 22h13'34" 19d56'5"
Alexea Juliano
Crb 16h4'0" 36d7'21"
Alexei 2244
Gem 6h24'7" 19d22'48"
Alexei Rem Kapralov
Vir 13h22'57" 2d3'35"
Alexei Yarosh
Peg 22h20'3" 17d17'23"
Alexej Jalyschko
Uma 11h1'24" 63d31'12"
Alexenburg Family
Uma 8h52'13" 56d29'9"
Alexes Nicol Crespo
Psc 0h16'44" -2d23'10"
Alexess- A Beautiful Shining Star
Peg 23h2'25" 9d55'4"
Alexey Evanovich E
Cnv 13h31'54" 38d24'57"
Alexey Sukharev
Ori 5h24'22" -0d17'4"
Alex-Hubert Gagné
Peg 22h57'28" 18d5'48"
Alexi
Com 13h7'37" 15d24'16"
Alexi Alexis
Peg 21h23'57" 22d33'24"
Alexi Laura Anniuk
And 1h11'58" 36d56'45"
Alexi Soutos
Leo 10h19'46" 24d37'58"
Alexia
Peg 22h36'30" 9d2'35"
Alexia
Uma 11h20'26" 31d57'55"
Alexia
And 0h18'11" 43d38'34"
Alexia
And 2h36'38" 48d54'7"
Alexia
Lib 15h47'17" -26d5'37"
Alexia
Dra 16h34'44" 63d51'43"
Alexia Allyn
And 1h32'57" 50d14'54"
Alexia Audrey Stanley
Cyg 20h25'54" 33d39'36"
Alexia Blicklee Rouquette
Mon 7h15'28" -0d50'20"
Alexia Boynuince
Crb 15h48'56" 27d27'2"
Alexia Da Silva
And 0h25'45" 39d29'9"
Alexia Dee Bates
Lmi 10h37'19" 37d29'26"
Alexia Demi Isaak
Cyg 19h32'51" 29d13'36"

Alexia Denise Grant
Sco 17h19'21" -43d32'21"
Alexia éternité de mon coeur
Uma 9h21'35" 43d20'46"
Alexia Faith
Cap 21h25'0" -18d32'16"
Alexia Flory
And 0h38'21" 37d19'28"
Alexia Hatzimihailidis
Cru 12h27'17" -60d45'42"
Alexia Jasmine Platt
And 23h19'36" 51d48'49"
Alexia & Jeremy
Cyg 19h58'11" 39d20'1"
Alexia Kayla Beck
Vir 13h35'24" -4d37'44"
Alexia Lain Smith
Aqr 22h38'6" 1d17'42"
Alexia Marie
Gem 7h44'45" 23d18'12"
Alexia Morrissey
Crv 11h57'28" -19d37'27"
Alexia Nicole
Ari 3h11'28" 12d0'34"
Alexia Palombo
Aqr 23h36'20" -16d45'48"
Alexia Rémoleux
Cap 20h8'21" -14d30'7"
Alexia Smith
Uma 9h53'4" 46d3'11"
Alexia05
Sco 16h16'19" -16d52'11"
Alexia-Maria
And 0h28'45" 32d40'4"
Alexica
Psc 1h19'3" 17d56'20"
ALEXIO
Cen 13h0'34" -48d14'3"
Alexis
Sco 16h54'27" -35d57'53"
Alexis
Cap 21h10'3" -19d45'15"
Alexis
Lib 15h5'10" -1d14'8"
Alexis
Dra 17h37'34" 63d51'38"
Alexis
Uma 12h0'22" 56d31'31"
Alexis
Uma 13h25'4" 53d11'56"
Alexis
Tau 4h20'27" 18d54'3"
Alexis
Ori 6h5'24" 13d18'41"
ALEXIS
Peg 21h37'2" 26d20'3"
Alexis
Lyr 19h18'29" 29d27'25"
Alexis
And 0h18'22" 27d13'13"
Alexis
Uma 10h4'4" 48d22'34"
Alexis
Cyg 20h18'40" 47d24'12"
Alexis
Cas 0h9'35" 59d41'0"
Alexis
Per 2h46'25" 56d26'47"
Alexis
And 1h30'43" 45d6'44"
Alexis
And 2h31'0" 45d15'10"
Alexis
Leo 9h38'26" 31d11'25"
Alexis
Lyn 9h4'27" 34d36'59"
Alexis
Crb 16h4'1" 32d3'21"
Alexis
Lyr 18h52'53" 36d41'24"
Alexis
And 0h31'50" 38d18'30"
Alexis
And 23h52'24" 35d2'35"
Alexis
And 1h11'37" 41d26'50"
Alexis Adams
Del 20h41'13" 15d23'4"
Alexis Adele Buen - My Shining Star
Ari 3h8'33" 19d11'25"
Alexis Alissa Foster
Cyg 20h22'48" 38d25'1"
Alexis and Grandma's Dragon Star
Cyg 20h0'26" 56d45'17"
Alexis Anderson
Uma 12h17'57" 56d9'59"
Alexis Ann Brisson
Aqr 21h21'52" 2d1'33"
Alexis Ann Olsen
Cap 21h15'30" -17d8'57"
Alexis Ann Warren
And 1h59'44" 42d37'27"
Alexis Anne
Cap 21h23'14" -19d44'1"
Alexis Anne Bailin
Lyr 18h43'52" 40d50'21"
Alexis Anne Jasiecki
And 0h20'23" 44d55'14"
Alexis Anne McCluskey
Umi 12h9'9" 88d12'13"
Alexis Anne Ross
Leo 10h59'57" 15d22'16"

Alexis Anne Weiland
Cap 20h39'26" -17d1'29"
Alexis Ann-Ruiz Alessi
Leo 11h6'28" 9d34'34"
Alexis Ansell
And 23h43'32" 47d34'27"
Alexis Ava Matachek
And 0h32'2" 42d39'56"
Alexis Azaria Kogan
Uma 9h31'27" 60d4'18"
Alexis B. Taylor
Vir 12h28'2" 0d13'14"
Alexis Betancourt
And 2h15'49" 45d54'12"
Alexis Borja
Lyn 6h18'57" 58d42'47"
Alexis Brianna Wiser
Sco 16h4'41" -16d46'19"
Alexis Brianna Wiser
Sco 17h51'18" -30d38'9"
Alexis Brianne
And 1h15'47" 37d56'16"
Alexis Brittany Hayden
Lyr 18h45'11" 32d8'39"
Alexis Brooke Doucette
Lib 15h18'34" -6d3'6"
Alexis Brooklyn
Lyr 19h15'27" 29d43'7"
Alexis Brynn
And 0h23'31" 40d46'51"
Alexis Burns
Uma 8h40'27" 67d52'40"
Alexis Carrasquillo
Uma 13h11'31" 55d14'43"
Alexis Cheyenne
Cap 21h3'23" -18d31'4"
Alexis Chieko Vernon
And 1h19'47" 49d41'23"
Alexis Christine Arnold
And 2h15'13" 40d22'20"
Alexis Christine Barton
Umi 15h51'37" 73d44'7"
Alexis Christine Pickett
Uma 10h28'40" 53d37'59"
Alexis Cincione
Sgr 18h19'12" -32d43'8"
Alexis Consalvo
Sgr 18h54'6" -17d37'5"
Alexis Corbi
Uma 9h43'42" 57d24'41"
Alexis - Cristy
And 0h34'10" 45d6'9"
Alexis D. Robinson
And 0h13'56" 39d50'44"
Alexis Danielle
Sgr 19h5'49" -30d22'28"
Alexis Danielle Thompson
And 1h16'21" 48d14'37"
Alexis Dannielle Durham
And 1h38'52" 39d19'4"
Alexis Delaney Krotz
Sgr 18h36'20" -31d14'52"
Alexis Denise Saunders
And 23h14'20" 43d54'45"
Alexis Diana Carey
And 1h51'53" 36d16'13"
Alexis Domenica Bianchi Monitto
Cra 18h10'37" -39d47'11"
Alexis Elizabeth Forsythe
And 1h34'22" 46d5'21"
Alexis Elizabeth Packer
Aqr 21h5'21" 0d40'11"
Alexis Ellen Kent
Cap 21h18'47" -19d42'37"
Alexis Emily Brown
Aqr 21h11'56" -9d7'36"
Alexis et Myrte
Tau 5h45'48" 18d36'11"
Alexis Faye
Sco 16h10'7" -17d52'22"
Alexis Feldman
And 2h22'13" 46d30'15"
Alexis Ferguson, Will You Marry Me?
Ari 2h25'54" 23d55'6"
Alexis Figacz
Lib 15h40'42" -10d25'50"
Alexis Francisco Penalver
Ori 5h32'9" 2d15'25"
Alexis Ginsberg
Ari 3h6'18" 10d55'20"
Alexis Godfrey
Leo 10h58'4" 8d39'51"
Alexis Goline
And 0h39'30" 36d20'48"
Alexis Gore "Morning Star"
Aqr 20h54'1" -8d27'49"
Alexis Grace
And 0h12'40" 43d15'54"
Alexis Grace Rehorst
Cyg 20h26'46" 35d32'5"
Alexis Greenberg
Sco 16h18'37" -41d18'42"
Alexis Gwen Michel Bougeot
Sco 17h16'51" -41d35'50"
Alexis Hernandez
Gem 7h10'42" 22d4'5"
Alexis Holden
Uma 9h43'45" 50d19'31"
Alexis Hooker
Uma 13h20'38" 56d17'56"

Alexis Hudrisier - 17 Mai 2006 Her 17h46'36" 41d2'9"
Alexis I. Mendelsohn Umi 15h24'32" 81d35'20"
Alexis Isabelle Heuer Vir 14h10'13" -8d12'19"
Alexis J. Colvin Leo 11h25'30" 8d23'24"
Alexis Jacquelyn Gills Cnc 9h9'8" 26d31'14"
Alexis Jade And 23h17'1" 47d26'54"
Alexis Jade Finkel Umi 16h35'27" 77d34'16"
Alexis Jane Lyn 7h39'47" 45d21'5"
Alexis & Jason Aqr 22h25'36" -8d36'39"
Alexis & Jeff Lib 15h40'48" -14d33'1"
Alexis Jones Umi 16h18'25" 71d12'46"
Alexis Jones And 0h14'42" 46d51'20"
Alexis Jordan Lilly Cnc 7h59'29" 16d45'49"
Alexis Kaley Tau 4h6'46" 5d22'5"
Alexis Kate Haney And 1h0'38" 42d34'5"
Alexis Katz Lib 15h11'58" -11d8'41"
Alexis Kodofakas Sgr 19h14'29" -18d22'51"
Alexis Laskaris Uma 9h6'59" 59d7'51"
Alexis Lauren Zeiser And 0h28'15" 44d21'32"
Alexis Lee Finnerty Mon 6h52'28" 8d28'5"
Alexis Lee Jones And 23h47'54" 48d41'27"
Alexis Lee Richards Cmi 7h22'53" -0d4'26"
Alexis Lee Roberts Vir 12h36'12" 3d17'14"
Alexis Leigh Bohls And 2h6'49" 41d15'49"
Alexis Lemon Sgr 19h15'4" -29d48'28"
Alexis Leonie Percy Aqr 21h21'4" -10d23'20"
Alexis "Lexi-Poo" Ontiveros Cas 1h0'11" 61d6'21"
Alexis Louise Merklin And 1h12'59" 37d9'29"
Alexis Luna Ori 6h17'28" 14d4'26"
Alexis Lynea Cerimovic Psc 0h54'49" 31d14'25"
Alexis M. Bisset And 0h48'52" 37d37'23"
Alexis M. Bryant Aqr 23h29'47" -7d59'43"
Alexis M. Wallace Sgr 19h20'26" -20d46'30"
Alexis Mackenzie Ferrell Psc 1h16'40" 32d25'45"
Alexis Madison Post And 1h52'41" 37d31'12"
Alexis Marcael Cep 21h9'57" 71d36'55"
Alexis Maria Uma 9h54'15" 45d18'4"
Alexis Marie Uma 8h40'50" 59d49'14"
Alexis Marie Badolato Lib 14h42'6" -11d17'48"
Alexis Marie Donnelly And 23h0'11" 45d5'45"
Alexis Marie Garcia Lyn 8h59'33" 39d53'10"
Alexis Marie Green And 23h43'50" 45d24'42"
Alexis Marie Leifert Tau 4h28'48" 28d17'41"
Alexis Marie Loretero Lyr 19h10'39" 26d22'52"
Alexis Marie McBride Mon 6h53'9" -0d56'49"
Alexis Marie Peterson Cam 4h5'29" 71d28'30"
Alexis Marie Weaver Sgr 17h58'26" -17d12'15"
Alexis Martins Gem 6h49'43" 32d35'22"
Alexis Mary And 0h20'1" 28d56'9"
Alexis Mason Mon 6h58'2" -1d14'46"
Alexis Mercedes Dasantis Umi 13h45'31" 75d20'14"
Alexis Meredith Jones Cap 20h40'39" -21d18'0"
Alexis Michell Gentry And 0h39'2" 31d30'15"
Alexis Michelle And 2h16'15" 44d35'4"
Alexis Michelle Tau 3h47'29" 29d50'59"
Alexis Michelle Blackburn Cnc 8h54'15" 25d28'21"
Alexis Michelle LoPinto And 2h4'37" 42d29'3"
Alexis Mikayla Sgr 19h39'28" -13d53'10"
Alexis Mishelle Cummings Psc 0h42'39" 2d44'14"
Alexis Moehsmer Lyn 8h52'15" 44d17'47"
Alexis Monquie Clark Vir 14h42'10" 2d36'43"
Alexis Moreno Tau 4h28'32" 19d46'26"
Alexis Morgan Frenda And 1h35'14" 48d27'20"
Alexis Murphy Manning And 1h37'46" 44d7'38"
Alexis Murphy Manning Aqr 22h9'23" 1d24'35"
Alexis - My LiL Doughnut Cap 21h37'41" -14d4'3"
Alexis "My Shining Star" Uma 11h34'18" 53d13'12"
Alexis Myung Paul Cnc 8h33'8" 28d15'25"
Alexis Negron Aur 5h25'43" 48d48'9"
Alexis Niccole Watson Uma 10h40'44" 41d23'53"
Alexis Nichole Aqr 22h19'50" -2d42'47"
Alexis Nicole And 23h43'42" 38d39'36"
Alexis Nicole And 23h17'1" 40d36'14"
Alexis Nicole Cam 5h36'31" 56d33'8"
Alexis Nicole Boudreaux Cap 21h5'16" -18d43'12"
Alexis Nicole Castator Dra 19h7'13" 72d46'50"
Alexis Nicole Gill Ari 2h51'38" 27d48'6"
Alexis Nicole Riley Cyg 20h5'50" 45d54'19"
Alexis Nicole Spungin And 0h45'46" 40d0'29"
Alexis Nicole Titch Cas 1h40'14" 64d54'17"
Alexis Nobilo Campbell Aqr 20h40'52" -7d0'17"
Alexis Olcese Sco 16h15'53" -10d13'21"
Alexis Olivia And 23h18'40" 48d11'21"
Alexis Paige Buzzelli Lyn 7h38'4" 56d23'45"
Alexis Paige Levey Gem 6h31'30" 15d48'0"
Alexis & Philip Cep 4h23'41" 81d53'2"
Alexis Piper Ritchie Tau 5h18'59" 23d30'32"
Alexis Proegler Cas 1h6'18" 48d52'2"
Alexis pudwill Uma 10h14'45" 55d29'18"
Alexis QT Gem 7h5'56" 32d20'43"
Alexis R. Wilson Psc 1h6'28" 33d22'43"
Alexis Rachel Dra 18h48'11" 53d52'50"
Alexis Rachel Markowski Cnc 8h34'38" 24d10'20"
Alexis Rae Amira Vir 11h44'29" 3d57'47"
Alexis Rae Grebe Lmi 10h29'38" 34d50'53"
Alexis Renee George And 23h4'49" 48d27'16"
Alexis Renee McFadyen Uma 12h17'54" 57d53'8"
Alexis Renee Petraro And 1h22'4" 50d2'54"
Alexis Reynolds Cas 0h18'21" 51d0'0"
Alexis Robb Trimas (CHIQ-UITA) Sgr 18h0'41" -27d13'3"
Alexis Rocco and Lily Lyn 9h5'2" 37d37'3"
Alexis Roque Ori 5h37'0" 1d49'6"
Alexis Rose And 2h12'22" 38d42'57"
Alexis Sarshuri Leo 11h6'29" 17d14'56"
Alexis Schuster Ari 3h15'44" 19d53'16"
Alexis Short Leo 9h49'43" 24d28'7"
Alexis Simone DiVasta Gem 6h46'9" 34d34'24"
Alexis Sky Cyg 20h50'40" 32d57'34"
Alexis Skye Knapp Cnc 9h2'21" 11d18'22"
Alexis Skylar Vir 12h32'4" 11d7'50"
Alexis Smith Uma 11h55'50" 31d4'6"
Alexis Smith Aqr 22h2'44" -7d55'58"
Alexis Star Psc 1h11'43" 25d56'46"
Alexis Stoll "Lexi's Light" Cnc 8h58'8" 24d12'27"
Alexis Swearingen Aqr 21h8'15" 1d26'46"
Alexis Thoman Cam 3h48'52" 59d36'27"
Alexis Thompson Cnc 8h43'19" 29d43'55"
Alexis Totsis Leo 10h13'5" 17d53'47"
Alexis Truelson Daughter of Wendy And 0h14'29" 46d5'46"
Alexis Tyler Stott Peg 23h45'21" 22d49'3"
Alexis Victoria Brown/ Cortez Cyg 19h29'57" 31d52'21"
Alexis Victoria French Cnv 12h27'46" 32d29'32"
Alexis Victoria Grant Leo 11h8'55" 22d25'7"
Alexis Zoe Rieger Uma 10h36'57" 56d52'43"
Alexis Anne And 0h12'53" 45d34'8"
Alexisania And 0h48'4" 27d6'57"
Alexis-Diego Lib 15h54'35" -10d36'36"
AlexisLee Aqr 21h51'56" -0d52'27"
Alexis-Marie Cnc 9h4'52" 27d15'38"
Alexisstar Sge 20h13'21" 18d24'30"
Alexius Fortis Optimus Her 17h40'17" 15d12'22"
Alexou Lib 14h48'19" -5d34'18"
Alex's 1st Anniversary Star Cyg 21h22'20" 39d10'18"
Alex's Dream Umi 16h35'59" 76d20'5"
Alex's Dream Catcher Umi 15h21'18" 74d56'0"
Alex's Guiding Star Psc 0h28'32" 3d42'35"
Alex's Star Cru 12h45'27" -60d11'4"
Alex's Star Aqr 21h48'12" -7d40'15"
Alex's Star Her 18h50'39" 22d6'19"
Alex's Toby Uma 12h9'21" 46d12'25"
Alex's Valentine Star And 0h31'14" 27d5'34"
Alexsa Rae Cronise Psc 0h52'19" 10d47'41"
Alexsandra Aquila DiBella Ari 2h18'45" 24d39'51"
Alexsandra Monaspova Vir 13h5'28" -13d57'56"
Alexsandra Owen Del 20h22'25" 9d42'4"
Alexsis the Angel Sco 16h4'20" -16d11'9"
Alexsis Victoria Rubio Psc 1h11'18" 10d54'30"
Alexstar Uma 10h45'27" 49d34'19"
Alexta Ori 5h44'50" 1d52'17"
AlexTuzene And 1h51'0" 41d53'19"
Alexus Aql 19h43'54" -3d55'17"
Alexus "Baby Boo" Serrato Ari 1h55'37" 24d32'6"
Alexus R. Ahlers Vir 12h40'58" 12d30'3"
Alexus Reynee Ochoa And 0h22'53" 32d42'18"
Alexus Rose Malone Tau 3h28'47" 1d46'37"
Alexy Uma 10h14'20" 59d35'36"
Alexy Nicholas Saltekoff Uma 10h32'3" 49d45'6"
Alexyara & Astley Blair Cyg 19h51'45" 32d22'59"
Alexys Mckenzie Biers Cap 20h33'56" -23d28'55"
Alexzander Sgr 18h58'28" -24d9'55"
Alexzander James Per 4h15'38" 50d59'54"
Alexzandria Erica Virinia Crb 15h47'57" 34d0'8"
Alexzandria's Faith Cyg 20h38'52" 49d12'47"
AlexZhannaEthanAdrianMaydanich Umi 13h43'50" 76d57'23"
ALF Lyn 8h15'26" 53d40'18"
Alf Per 1h34'27" 54d28'40"
ALF Crb 16h6'16" 34d59'45"
Alfa Frankie Baby Aql 19h50'3" -0d16'2"
alfalfa Cra 18h14'48" -41d10'27"
Alfalfa Lyn 7h48'33" 50d50'45"
Alfie Cmi 7h23'11" 3d3'23"
Alfie Ori 5h33'59" 14d44'23"
Alfie Aqr 22h18'27" 0d52'49"
Alfie Cma 6h48'50" -14d41'10"
Alfie Lyn 8h13'44" 54d19'2"
Alfie Uma 10h27'30" 65d13'12"
Alfie Benjamin Owen Per 4h2'41" 34d29'18"
Alfie Daniel Green Umi 13h18'31" 76d2'18"
Alfie Fergusson Per 3h1'21" 52d42'26"
Alfie Gee - Alfie's Star Umi 15h8'24" 70d17'51"
Alfie Jack Philip Chisholm Her 16h35'37" 12d34'32"
Alfie James Young Ori 5h25'35" 8d10'1"
Alfie Joseph Per 3h25'50" 40d55'12"
Alfie Kane Mortimer Ori 6h20'16" 15d1'40"
Alfie Liddle Her 18h28'24" 12d17'29"
Alfie Oliver Creese Cep 21h6'14" 83d1'58"
Alfie Raymond Joseph Zaal Ori 5h47'13" 9d0'41"
Alfie Ryan Timothy Laws Dra 18h31'33" 79d41'29"
Alfie Simon Craig Umi 15h43'23" 76d43'11"
Alfie "T" Vir 12h15'38" -10d49'36"
Alfie Tinsley Jarvis Uma 10h24'12" 40d15'47"
Alfie's Guiding Light Umi 14h55'20" 79d5'56"
Alfio-Ti Amo Ori 6h2'34" 13d24'48"
Alfnan Cyg 20h8'27" 53d22'21"
Alfons & Frances Rulis Lyn 7h34'59" 36d17'26"
Alfons Hofer Uma 8h58'9" 64d36'25"
Alfons Hördler Uma 10h12'56" 44d39'3"
Alfons Preisinger Uma 10h57'0" 50d5'59"
Alfons Rose Uma 9h42'2" 47d35'0"
Alfons Wittmann Uma 8h13'36" 62d12'18"
Alfonse Chiulli Per 3h15'39" 51d45'21"
Alfonsino Enrico Pirrone Dra 19h57'42" 79d23'3"
Alfonso Her 16h47'50" 43d55'15"
Alfonso 1974 Lib 15h7'13" -7d13'46"
Alfonso A. Lozada Cnc 8h7'59" 19d36'17"
Alfonso and Antonietta Ciervo Sgr 19h17'41" -35d42'7"
Alfonso Caraos - Rising Star Per 3h12'20" 52d42'36"
Alfonso Forero Lib 14h52'34" -2d2'12"
Alfonso Trotta Aql 19h18'46" -0d56'30"
Alfonso Vasquez Tau 5h16'4" 22d17'53"
Alfonso W. Nardi Aql 19h37'28" -0d32'32"
AlfonsoCarlos Martinez-Conde Ibañez Per 3h15'27" 54d1'24"
Alfred Boo 14h59'44" 17d53'44"
Alfred Uma 13h9'50" 54d31'58"
Alfred Adams Ori 5h13'38" 5d1'19"
Alfred Akbar, Jr. Cnc 9h8'9" 30d55'57"
Alfred "Al" Hulstrunk Uma 10h46'24" 58d40'59"
Alfred Alexander McCarthy Aqr 22h37'23" -1d0'1"
Alfred Anthony Cimino "Gramps" Gem 7h32'58" 26d49'16"
Alfred Anthony Eck Cyg 20h27'21" 48d58'55"
Alfred B. Curtis III Her 16h56'35" 27d37'46"
Alfred Berger Boo 15h4'36" 12d51'19"
Alfred Bertotti Her 18h27'18" 21d36'48"
Alfred "Big Al" Martyn Aur 5h43'2" 49d51'54"
Alfred Bissegger And 2h30'51" 41d38'19"
Alfred C. Bickford Cap 20h23'6" -10d19'54"
Alfred Charles Newey Ori 5h58'54" 11d28'1"
Alfred Christensen Her 16h35'55" 38d17'17"
Alfred Daum Her 17h28'40" 32d4'30"
Alfred & Doris Buss, Christmas 2005 Cyg 21h35'31" 38d32'52"
Alfred E. DeMattia Per 3h18'12" 52d45'49"
Alfred E. McCandless Cnc 8h39'51" 32d3'16"
Alfred Edward Perron Cyg 20h0'8" 40d58'4"
Alfred Erich Degen Ori 4h47'45" 4d30'42"
Alfred F Daugherty Cnc 8h37'32" 13d8'40"
Alfred & Florence Friedrich Umi 15h43'25" 74d41'42"
Alfred Fridlund Uma 10h11'7" 44d54'29"
Alfred Gausling Ori 6h17'58" 15d52'11"
Alfred George Devanny-Rawles Her 18h26'4" 15d39'19"
Alfred Girardot, Dedicated Teacher Cnc 9h10'47" 32d13'34"
Alfred Gomes Santos: Our Nunu Per 3h53'27" 47d24'41"
Alfred Guerin Dra 16h54'13" 57d40'59"
Alfred Hammer Uma 12h1'28" 42d27'43"
Alfred Hawley Smith, Jr. Her 16h43'17" 23d0'6"
Alfred J. Burks And 23h52'39" 42d9'48"
Alfred J. Colacioppo Ori 6h19'29" 16d28'12"
Alfred J. Zahler Uma 8h52'14" 66d47'24"
Alfred Joseph Amidei Aqr 23h9'17" -7d33'58"
Alfred K Pellmann Crb 15h23'21" 27d0'23"
Alfred Knips Ori 4h45'49" 5d10'34"
Alfred L. Hunt Ari 2h33'0" 30d32'43"
Alfred Langguth Tau 4h25'9" 3d57'59"
Alfred Lawecki Ori 5h9'44" -1d28'27"
Alfred M. Miller Jr. Her 17h25'42" 36d28'28"
Alfred Michael Kinnear Her 16h31'51" 34d28'0"
Alfred Michael Windstein Sgr 18h16'28" -19d51'45"
Alfred Nelson And 0h8'51" 29d21'29"
Alfred Patrick Bradley IV Cam 4h2'21" 67d14'38"
Alfred Penn Cnc 8h8'2" 17d26'26"
Alfred Pipitone Leo 10h57'51" 5d20'21"
Alfred (Poppie) Surprenant Aur 5h54'11" 43d2'32"
Alfred Raines Ori 5h16'53" 1d18'30"
Alfred Richard Marochini Ori 6h11'2" -0d14'33"
Alfred S Aqr 20h40'52" -10d29'42"
Alfred S. Rushatz Tau 4h26'37" 27d10'14"
Alfred Stefani Sgr 19h17'24" -21d52'22"
Alfred Stein Uma 9h42'23" 69d25'15"
Alfred W. Blumrosen Uma 13h20'2" 57d9'48"
Alfred W. Parlow Uma 11h2'34" 63d31'49"
Alfred Walter Ori 6h16'20" 15d0'30"
Alfred Wicker Ori 5h14'57" 15d28'12"
Alfreda & Albert's Star Sge 19h50'57" 18d14'44"
Alfreda Elizabeth Emily Stirzaker Vir 13h57'34" 2d42'41"
Alfreda Labadie-Umphress Aqr 22h4'46" -2d18'41"
Alfreda Lavern Henley Chandler Psc 0h57'1" 28d24'14"
Alfreda Roberts Crb 16h17'23" 26d43'50"
Alfredo Alberto Alvarez Leo 11h33'35" 17d39'0"
Alfredo and Jessica Her 17h23'33" 17d29'24"
Alfredo Coleman Her 18h43'35" 21d12'49"
Alfredo "Cubito" Diana Cep 20h28'59" 60d26'7"
Alfredo Fischetti Per 3h0'45" 55d17'20"
Alfredo & Ileana Forever Tau 4h37'48" 18d9'59"
Alfredo Malinis Sgr 19h11'0" -12d32'44"
Alfredo Molino Lmi 10h16'10" 29d53'39"
Alfredo Monzillo Per 3h27'40" 41d34'29"
Alfredo Rafael Ortiz Lyr 18h45'34" 40d58'39"
Alfred's Spirit Ori 5h28'44" 1d5'28"
Alfred's Star of Creativity Per 3h4'57" 41d8'43"
Alfreta Per 3h33'14" 50d20'57"
Alfrieda B Cas 23h4'17" 53d39'28"
Alf's Star Cru 12h35'29" -64d26'55"
ALGEPA Uma 14h18'15" 59d13'21"
Alghero 2001 Cyg 19h54'24" 45d33'41"
algún dia Lyr 19h10'1" 38d35'36"
Al-Hakeem Ori 5h44'57" 12d23'51"
Alhoptex Sco 17h35'0" -45d24'55"
Ali Sgr 18h18'16" -33d41'13"
Ali Lyn 8h6'52" 54d57'28"
Ali Umi 13h47'52" 70d22'3"
Ali Lib 16h1'16" -16d42'43"
Ali Lib 15h3'15" -2d18'18"
Ali And 0h18'33" 32d16'46"
Ali Gem 7h32'27" 20d45'58"
Ali Her 17h38'38" 48d50'6"
Ali Lyr 18h16'6" 32d55'15"
Ali A. Al-Khatib Ori 5h55'43" 17d50'46"
Ali Ahmad Karim Olomi Dra 18h17'58" 58d43'27"
ALI AL Her 17h18'48" 42d58'5"
Ali and Dan And 1h15'53" 36d31'17"
Ali and John Everlong Cyg 20h59'8" 36d19'25"
Ali Atkins Cas 1h31'2" 60d9'33"
Ali Avina Cnc 8h18'2" 8d7'53"
Ali Bin Kassim Al-Angari Uma 8h26'10" 67d11'52"
Ali Bird And 1h44'48" 46d8'33"
Ali Butterfly Umi 14h4'35" 87d10'26"
Ali Caleca And 23h47'46" 47d30'13"
Ali Davis Cnc 9h16'35" 10d52'41"
Ali & Dennis - Eternity +1 Cyg 21h55'28" 37d8'20"
Ali Duff Del 20h50'53" 3d3'5"
Ali Earle Aqr 21h37'22" 2d51'37"
Ali Elizabeth Smith And 22h59'34" 51d34'47"
Ali Eslami Umi 15h50'39" 72d26'29"
Ali Esmaili Cep 20h24'46" 61d17'59"
Ali Finlay Cas 0h55'5" 53d53'50"
Ali - Forever Shine Cru 12h42'45" -64d4'50"
Ali G Cnv 13h39'12" 32d55'17"
Ali G. Nazary Ori 5h56'34" -0d42'25"
Ali Girl Ori 6h25'9" 10d17'59"
Ali Grace Tau 4h33'35" 24d21'6"
Ali Greer Cap 20h53'26" -25d49'41"
Ali Griffin Psc 0h49'37" 13d36'41"
Ali I. Shahzad Ori 5h5'31" -0d54'25"
Ali Jafari Ari 2h29'5" 27d6'7"
Ali Jo Forester Cnc 9h18'27" 8d52'1"
Ali Karahasan Vir 13h17'23" -15d15'42"
Ali & Kate Lyr 19h19'59" 41d54'50"
Ali Keeler Cas 1h18'11" 59d53'39"
Ali Lea Hare Col 5h32'51" -35d5'33"
Ali Lynne Leannah Ori 5h9'49" 1d18'4"
Ali Mahmoud Equ 21h7'50" 11d54'12"
Ali Marie Bramson And 2h23'26" 46d28'29"
Ali Marie Spigiel Lyn 8h59'13" 35d38'23"
Ali Matthews Uma 8h49'52" 46d50'30"
Ali Mayfield Vir 14h15'24" 4d22'4"
Ali McCabe Lyn 8h12'52" 50d43'48"
Ali McDough And 1h6'42" 39d39'6"
Ali Munir Cap 20h32'37" -9d14'12"
Ali my princess Sco 17h56'18" -39d49'47"
Ali Parter Gem 7h13'49" 20d11'52"
Ali Patrick Cap 20h25'48" -9d33'8"
Ali Pessy Lib 15h5'30" -8d9'21"
Ali&Pinar&Yagiz Sozmen Uma 11h13'4" 28d44'57"
Ali Pollack Cas 23h41'54" 56d22'24"
Ali Purcelli "Baby Girl" Cap 21h26'4" -21d37'44"
Ali Raif Dinçkök Umi 16h5'42" 85d9'28"
Ali Thomas Tanveer Leo 10h39'15" 18d4'35"
Ali & Tommy Sge 19h54'32" 18d29'35"
Ali & Tyler Uma 13h30'25" 53d46'35"
Ali Woos Lynn Sgr 19h49'11" -23d51'21"
Alia Ori 6h6'22" 12d14'54"
Alia El-Rifai Gem 6h31'53" 23d12'2"
Alia Pearl And 23h13'41" 40d4'5"
Aliah Lib 15h31'54" -15d4'55"
Aliah LaVanway Mohmand Lyn 7h4'47" 51d33'22"
Alian Cyg 20h43'56" 46d6'33"
Aliandra Tina Pup 7h53'59" -41d11'1"
ALIBABY Psc 1h5'35" 3d30'35"
alibil And 2h34'47" 50d2'5"
Alibruce L'amour Cru 12h32'29" -61d53'17"
Alicanne Cnc 8h27'55" 28d5'30"
Ali-Cat Gem 6h29'5" 23d55'47"
Alice Cnc 8h21'56" 22d45'31"
Alice Ari 3h13'45" 29d14'17"
Alice Leo 9h34'6" 29d38'33"
Alice And 0h19'17" 24d56'54"
Alice Ori 5h56'7" 12d29'3"
Alice Del 20h46'40" 7d4'27"
Alice Ori 6h10'5" 6d31'52"
Alice Aql 19h39'39" 14d17'5"
Alice Cnc 8h48'32" 14d36'0"
Alice Mon 6h28'49" 8d21'11"
Alice And 1h11'3" 45d38'48"
Alice Per 4h23'27" 49d34'23"
Alice Lac 22h54'35" 43d31'56"
Alice And 23h45'42" 48d34'51"
Alice Lyn 8h19'56" 42d19'9"
Alice Cra 18h39'15" -39d26'46"
Alice Lyn 7h49'18" 46d3'24"
Alice Lib 14h32'42" -18d24'40"
Alice Mon 7h36'35" -0d35'12"
Alice A. Beam Ari 2h12'4" 24d59'35"
Alice A. Brent Tau 5h52'20" 17d20'20"
Alice A. Strehan Cnc 9h7'31" 8d48'35"

Alice A. Williams
Cas 0h41'45" 70d23'43"
Alice and Frank Minutello
Uma 11h31'50" 35d24'40"
Alice Ann
Cnc 7h57'31" 11d47'36"
Alice Ann Kisinger
And 1h34'54" 35d39'0"
Alice Ann O'Connell
Uma 12h0'5" 32d45'21"
Alice Ann Teeple
Sco 16h5'16" -13d40'33"
Alice B. Woods (WoodsyLady)
Lib 15h4'0" -26d16'14"
Alice Bagdasarian
Leo 11h16'44" 10d10'11"
Alice Barrett Weidner
Uma 10h47'18" 60d59'37"
Alice Bertha Berky
Crb 15h37'38" 28d10'27"
Alice Bianchi
Peg 22h42'12" 24d3'46"
Alice Billy
Uma 10h24'32" 45d54'32"
Alice Blackwelder
Vir 13h32'19" -8d10'58"
Alice Bradford
Aqr 22h51'10" -6d3'47"
Alice Brennan
Cnc 8h18'42" 28d12'8"
Alice Campbell Boyd
Mon 6h30'56" 9d20'2"
Alice Cary
Lib 14h52'1" -10d34'28"
Alice Catherine Donohoe Ford
Leo 10h9'57" 22d36'40"
Alice Ceri Grey
And 1h33'1" 38d55'54"
Alice Chen - The Light in my Life
Tau 4h25'22" 23d22'34"
Alice & Clemens Urmanowicz
Lyr 18h48'11" 32d42'36"
Alice & Clémentine
Cnc 9h2'37" 25d23'46"
Alice Conolly
Ori 6h15'59" 15d1'3"
Alice Cordero
Aqr 22h41'26" 1d52'50"
Alice Corlito
Per 1h47'57" 50d3'15"
Alice D. Ballard - The Greatest Mom
Lyn 8h0'2" 50d36'12"
Alice Dauterman
And 1h12'43" 35d35'1"
Alice Dawley Toler
Lib 14h23'42" -10d2'24"
Alice Deroo
Uma 10h54'43" 68d15'57"
Alice & Doug
Cas 23h2'33" 54d31'24"
Alice Drayfahl
Aqr 23h43'34" -12d48'55"
Alice Elaine
And 0h9'48" 33d13'16"
Alice Elizabeth
Cas 0h38'28" 52d34'38"
Alice Elizabeth Araiza
Lyr 19h19'4" 35d23'14"
Alice Erlene Manalo
Leo 11h32'26" 27d10'13"
Alice et Najat
Cmi 7h37'28" 3d26'41"
Alice F. Sheftall
Cas 0h25'35" 54d3'3"
Alice Faye
Aqr 23h26'46" -9d52'52"
Alice Fern
Psc 1h24'17" 20d32'59"
Alice Fern Converse
Vir 14h3'10" -16d29'25"
Alice Ferrari
Uma 11h56'50" 33d19'40"
Alice Fong Yu Alternative School
Sgr 19h20'30" -16d26'39"
Alice Foringer
Psc 1h20'48" 31d35'40"
Alice & Frank Nemeth
Lyn 8h6'42" 57d29'16"
Alice Freeman Palmer
Crb 16h7'8" 35d22'47"
Alice Fuentes
Ori 6h20'12" 9d39'1"
Alice Gaser
Lmi 10h35'28" 39d1'5"
Alice Gayle Shackelford
Sgr 19h17'57" -12d19'41"
Alice Gertrude Stevens
Dra 19h15'40" 75d28'56"
Alice Gonzalez
Lyn 7h19'15" 57d38'33"
Alice Grace Bennett
And 0h45'11" 26d3'25"
Alice Grimes
And 0h35'39" 46d5'10"
Alice Harris Meyer
Uma 9h54'43" 54d59'5"
Alice Hart Wertheim
Cam 3h39'44" 64d46'7"
Alice Hawman Sprow
Dra 16h54'41" 54d50'28"
Alice Hill Joining Angels in Heaven
Cas 0h45'7" 63d45'21"
Alice Hollingsworth
Peg 22h26'41" 3d44'7"
Alice Irma Giordano
Cas 1h41'54" 63d33'3"
Alice Isobel McDade
And 23h16'54" 47d51'0"
Alice Jablonski
Uma 12h51'28" 58d18'17"
Alice Jane
Tau 3h38'26" 26d6'33"
Alice Jane Dvorak
Vir 13h52'25" -17d40'11"
Alice Jane Pratt
Cam 4h20'48" 71d49'15"
Alice Jean Leon
And 0h13'24" 36d2'29"
Alice Jeemee Chon
Lyn 7h3'3" 52d2'47"
Alice Jenefer Downs Dart
Cas 0h8'20" 53d50'19"
Alice Jimenez
Crb 16h7'59" 38d41'32"
Alice Johnson
Cnc 8h34'1" 31d59'34"
Alice Katherine Keane
Lmi 10h4'15" 32d50'16"
Alice Kay
Umi 15h48'52" 77d4'18"
Alice Kishori Shah
And 23h3'9" 40d14'43"
Alice Lane Conolly
Ari 2h7'42" 23d34'52"
Alice Laura
And 23h23'33" 44d34'13"
Alice Lavelle
And 23h55'30" 39d27'23"
Alice Lawrence Grinnell
Lyn 7h49'45" 49d14'37"
Alice Lenora
Vir 13h26'4" 5d57'45"
Alice Lily Briffa
And 23h37'12" 40d36'8"
Alice Lindgren
Cas 1h17'19" 59d55'33"
Alice Lost in Wonderland
Sco 16h8'18" -22d14'48"
Alice Louise
Sco 15h53'2" -21d48'50"
Alice Louise~Joseph Thomas Clifford
Psc 23h52'57" 7d32'52"
Alice Lyla Elizabeth Scott
And 0h28'13" 46d31'59"
Alice Lynn Valsecchi
Peg 23h48'31" 22d33'54"
Alice M. Ferreira
Lib 15h4'35" -15d29'25"
Alice M. Hamilton
Lep 5h6'56" -11d15'15"
Alice M. Nolley's Star
Crb 15h52'12" 31d4'20"
Alice Mae Barrow
Peg 22h52'34" 27d0'33"
Alice Mae Halls Roth
Cyg 20h27'39" 39d11'30"
Alice Makin Edgar
Cnc 8h58'1" 23d45'29"
Alice Margaret Running Water Heinz
Cnc 8h33'7" 13d30'43"
Alice Marie
Uma 9h16'52" 49d46'9"
Alice Marie
Mon 7h18'12" -0d13'54"
Alice Marie
Cap 21h29'59" -9d26'46"
Alice Marie Cohen
And 0h32'19" 37d6'36"
Alice Marie Dennison
And 0h59'28" 37d51'52"
Alice Marie Mullins
Crb 15h48'41" 35d5'21"
Alice Marie Patton Dukes
Crb 15h38'6" 30d45'46"
Alice Marie Shelton
Lyr 18h49'44" 35d52'15"
Alice Marina Alyea
Cas 1h9'28" 53d7'7"
Alice Mary Brady
Uma 9h15'54" 62d5'28"
Alice Mary Brew
Cra 19h7'27" -39d20'31"
Alice May Ellard. Forever Fourteen
Cyg 19h42'18" 28d21'27"
Alice May Hubbard-Schaffner
Cyg 20h28'28" 52d57'7"
Alice McCarthy Miller
Uma 11h44'10" 32d10'10"
Alice McCune Dougherty
Gem 7h22'16" 27d37'5"
Alice Meisinger
Cas 1h43'14" 64d2'49"
Alice Millie Southey
And 1h56'38" 40d58'38"
Alice Monbleau
And 2h11'25" 46d49'46"
Alice my love
Her 18h53'16" 20d57'0"
ALICE MY LOVE LINDEMANN
Lyr 19h27'38" 40d51'36"
Alice Niemann
Gem 7h33'23" 26d28'24"
Alice Nuckols Davis
Cas 22h59'23" 53d49'12"
Alice O Lewis
Cnc 8h50'7" 14d4'3"
Alice Olins
And 0h0'21" 35d21'31"
Alice P. Munzo
Lib 14h49'35" -10d4'50"
Alice P Wing
Sgr 19h37'57" -15d59'15"
Alice Page
And 23h45'16" 44d0'21"
Alice Peters' Eastern Star
And 23h28'12" 47d9'35"
Alice Pisani
Cas 0h33'30" 63d44'6"
Alice Poteat "GiGi" God's Gift
Lyn 8h20'36" 47d22'21"
Alice Powell the beautiful
Ari 2h8'6" 26d3'56"
Alice Priscilla
Cas 3h14'51" 58d49'54"
Alice Ravenholt
Tau 5h49'6" 22d18'48"
Alice Rosalie
Psc 0h30'3" 6d16'37"
Alice Rose Vyskocil
Psc 1h39'51" 9d39'51"
Alice Rose Williams
Cnc 8h17'37" 24d36'50"
Alice Rosenbaum
Aqr 23h10'35" -10d25'49"
Alice + Ruedi
Aql 19h31'56" 4d30'16"
Alice Ruth Anne
Sco 16h11'46" -8d43'31"
Alice S. Lowe
Cas 0h13'30" 58d17'1"
Alice Sandberg
Crb 15h52'37" 35d27'18"
Alice Seema Joseph
Leo 9h30'28" 16d18'23"
Alice Selover
Crb 15h36'57" 28d18'12"
Alice Theresa Lessane
Cas 23h7'12" 56d57'33"
Alice und Kasper
Uma 9h5'51" 66d13'15"
Alice V. Pascucci
Cas 23h28'27" 56d1'25"
Alice Van Hoolandt
Cnc 8h21'30" 29d31'25"
Alice Verona Lewis Durnin
Uma 9h28'57" 47d6'5"
Alice W. Anderson
Ari 2h46'0" 26d44'17"
Alice Ward
Cap 21h55'26" -20d37'53"
Alice Waud Thecker
Cyg 20h0'13" 43d32'46"
Alice & Werner
Sco 17h54'28" -35d55'24"
Alice White Berger
Lib 15h6'15" -1d5'52"
Alice Wielkopolska
Uma 11h31'10" 29d55'26"
Alice Yee the love of my life
Lib 15h13'22" -8d30'0"
Alice Zandarski
Gem 7h7'28" 33d51'13"
Alice82DL
Peg 22h54'33" 15d22'29"
Alicea
Sco 16h12'19" -31d49'26"
Alicel
Psc 1h3'49" 32d21'19"
Alicen Johnson
Uma 10h15'14" 45d34'24"
Alice's Star
And 1h53'28" 45d7'6"
Alice's Star
Crb 15h43'29" 36d25'58"
Alice's Star
And 23h28'19" 35d21'32"
Alice's Star
And 0h4'56" 40d34'31"
Alice-unique!
And 0h14'28" 40d18'14"
alichas
Ori 5h40'9" -1d11'14"
Aliche
Cam 4h16'56" 68d19'46"
Alicia
Cam 7h53'25" 70d37'3"
Alicia
Uma 8h19'22" 70d34'7"
Alicia
Uma 8h37'26" 65d21'4"
Alicia
Uma 9h58'53" 61d26'24"
Alicia
Sco 16h9'38" -8d37'2"
Alicia
Cap 21h52'34" -18d19'12"
Alicia
Sco 17h50'21" -38d36'5"
Alicia
Sco 17h22'14" -39d41'58"
Alicia
Sco 16h47'25" -40d55'9"
Alicia
And 0h47'42" 43d21'38"
Alicia
Uma 11h19'40" 31d32'1"
Alicia
Gem 7h37'15" 31d12'20"
Alicia
And 2h29'13" 46d10'10"
Alicia
Cas 0h57'39" 52d43'21"
Alicia
Cas 1h20'58" 53d7'42"
Alicia
And 23h28'26" 48d2'18"
Alicia
And 23h39'0" 48d58'1"
Alicia
Peg 21h45'31" 24d38'1"
Alicia
Vul 19h56'24" 22d42'2"
Alicia
Tau 5h36'56" 22d31'53"
Alicia
Cnc 8h43'46" 15d8'56"
Alicia
Aqr 22h15'5" 0d18'9"
Alicia
Leo 11h27'36" 6d51'31"
Alicia
Tau 4h43'15" 3d9'23"
Alicia
Tau 3h55'19" 5d10'37"
Alicia
Cnc 8h1'44" 13d43'54"
Alicia
Cnc 8h3'53" 7d57'0"
Alicia
Cnc 8h8'41" 8d34'48"
Alicia 10/11/1963
Lib 15h58'41" -20d19'21"
Alicia 18
Sco 17h47'44" -36d5'29"
Alicia and Jamie's
Cyg 19h39'40" 37d39'54"
Alicia and Max's Moon Shine
Lib 15h10'13" -18d56'9"
Alicia Angel with the bent Halo
Leo 11h51'51" 21d25'45"
Alicia Angelica
And 0h19'26" 46d5'36"
Alicia Ann
Gem 6h36'13" 14d38'11"
Alicia Ann Irene Sands
Her 17h40'41" 15d6'16"
Alicia Anna McDowell
And 1h51'24" 45d44'46"
Alicia Anne
Tau 4h30'17" 30d3'42"
Alicia Anne
And 0h49'32" 40d10'38"
Alicia Anne Addison
And 1h26'16" 46d26'8"
Alicia Anne Anderson
Psc 23h30'38" 7d14'9"
Alicia Anne Curley
Psc 1h25'24" 12d32'51"
Alicia & Armando Moa Jr.
Cyg 20h31'5" 48d26'0"
Alicia Avery
Lib 15h24'53" -10d21'24"
Alicia Bacher
Lib 14h58'39" -18d14'1"
Alicia Beall
And 1h41'48" 46d35'49"
Alicia Bear
Cap 20h33'34" -20d11'55"
Alicia Beatriz Pérez Aké
Vir 14h25'47" 3d49'25"
Alicia Borseti
Cap 21h48'39" -13d21'3"
Alicia Brynn Hines
Vir 13h10'3" -5d0'8"
Alicia Bunnell
Sco 16h11'19" -10d55'17"
Alicia Caldwell
Cam 6h43'52" 65d44'20"
Alicia Canipe
Her 17h33'57" 32d7'42"
Alicia Carriero
Tau 5h47'28" 27d50'56"
Alicia Catherine
Ari 2h28'7" 27d4'13"
Alicia Christina Hager
Ari 1h50'18" 18d18'54"
Alicia Christine Butler
And 2h17'8" 49d9'37"
Alicia Clair Burton
Aqr 21h29'29" -0d39'54"
Alicia Claire Scholes
And 0h57'32" 23d30'3"
Alicia Collins
Her 17h18'40" 15d1'16"
Alicia - Come What May
And 1h25'17" 43d43'35"
Alicia D. Stumbo
Her 18h43'55" 20d5'49"
Alicia Dandoy
Leo 11h28'24" 19d5'44"
Alicia Daniele
Cyg 21h39'44" 38d7'2"
Alicia Davis
Tau 3h54'56" 6d30'5"
Alicia Dawn
Sco 16h42'15" -30d15'0"
Alicia Dawn Frazier
Leo 10h9'19" 22d26'27"
Alicia Dawn Parris
Umi 19h32'7" 88d32'11"
Alicia & Dennis
Uma 11h24'20" 62d34'7"
Alicia Diane
Leo 11h20'6" 22d18'20"
Alicia Diane DaSilva
And 2h28'40" 48d5'19"
Alicia Dufifie
Cas 0h39'43" 64d49'50"
Alicia Duschik
Ari 3h5'4" 11d14'1"
Alicia Eaves
Leo 9h38'7" 29d10'49"
Alicia Equestria
And 2h19'35" 46d45'30"
Alicia et Hugo
Lib 15h7'16" -28d46'13"
Alicia Eternal
Ori 5h15'4" -4d8'30"
Alicia Evon
Tau 4h18'19" 26d41'49"
Alicia F. San Diego-Surh
Gem 7h41'45" 16d3'16"
Alicia Faith Elias
Vir 13h17'30" -3d10'51"
Alicia Filotti
Cnc 9h14'10" 30d16'15"
Alicia Fish
Lyn 7h31'43" 37d55'29"
Alicia Foley
Pho 0h17'58" -39d52'37"
Alicia Gabriela
Com 12h46'7" 28d8'35"
Alicia & Gary Pierson
Ari 2h50'3" 26d49'48"
Alicia Gonzalez
Pho 1h1'13" -48d1'29"
Alicia Grace
Uma 9h57'40" 57d29'55"
Alicia Hargroove
And 0h29'11" 32d22'47"
Alicia Hartley
Cas 0h11'0" 56d39'58"
Alicia Isabel Galvan
Leo 9h39'29" 12d11'58"
Alicia J. Bumbarger
Sco 16h10'54" -16d30'40"
Alicia Jane Carter 03.12.1986
Ori 5h12'1" -6d13'34"
Alicia Jane Evans
And 2h30'31" 43d35'34"
Alicia Jane Rose Porile
And 0h47'19" 46d19'11"
Alicia Jane Smith
Gem 6h27'17" 24d27'9"
Alicia & Jeff Schraeder
Sgr 18h11'46" -19d22'7"
Alicia Jennifer Jackson
Cas 1h32'4" 65d32'51"
Alicia Joan Porter
And 23h43'42" 39d41'5"
Alicia Jovita
Leo 9h35'44" 30d28'15"
Alicia Joy
Uma 12h0'29" 48d27'48"
Alicia Joy Brown
Cyg 21h58'15" 45d23'47"
Alicia Joyce McClinton
Ari 3h12'1" 27d8'13"
Alicia June Daly
Aqr 22h59'23" -16d34'34"
Alicia Kae Weitzel
Psc 1h34'28" 17d35'55"
Alicia Kalani
And 0h10'26" 45d8'36"
Alicia Kaplan-Sherman
Cas 2h14'12" 62d29'38"
Alicia Kay
And 23h29'3" 42d2'49"
Alicia Kay
And 1h2'30" 44d29'13"
Alicia Knight
Lyr 18h41'8" 37d31'57"
Alicia Latar
Uma 9h18'33" 50d14'33"
Alicia Leaetta Waligora
Tau 5h19'49" 23d30'19"
Alicia Leah Wilson
Cru 12h33'2" -60d0'57"
Alicia Leanne
Ari 2h36'47" 17d30'14"
Alicia Leigh Barker
Cap 20h51'52" -15d23'24"
Alicia Leigh Jones
Umi 14h35'17" 87d27'57"
Alicia Li Matthews
And 23h55'44" 34d51'16"
Alicia Logan Smith
Cyg 21h50'22" 47d24'39"
Alicia Louise Maberry
Cnc 8h45'22" 15d16'29"
Alicia Loves Kevin Forever
Dra 18h21'20" 52d16'9"
Alicia Lu Chun Mei - Ken's Princess
Aqr 20h41'42" -6d49'35"
Alicia Luther Martin
Cas 1h41'49" 66d0'35"
Alicia Lynn
Leo 11h58'0" 20d5'59"
Alicia Lynn Hooten & Ricky Barnosky
Cyg 20h16'56" 32d26'36"
Alicia Lynn Kobrock
And 2h24'9" 38d46'31"
Alicia Lynn Oakley
And 1h4'4" 43d44'29"
Alicia Lynn Oakley
And 1h0'4" 37d45'34"
Alicia Lynnee Chahine
And 0h29'38" 34d24'9"
Alicia M. Betz
Tri 2h24'43" 30d19'1"
Alicia M. Garcia
Tau 4h21'43" 20d38'55"
Alicia M. Horstmann
Cap 20h44'33" -26d44'51"
Alicia M. Salazar
Cas 0h13'16" 59d22'18"
Alicia M. Vilardo
Ori 5h26'28" -7d35'46"
Alicia M. Williams
Gem 7h53'5" 31d42'24"
Alicia Mae Nikkole
Lyr 18h50'41" 43d39'28"
Alicia Maria Betancourt
Tau 4h0'52" 2d8'53"
Alicia Marie
And 22h58'8" 43d39'20"
Alicia Marie
Cyg 21h46'32" 43d57'36"
Alicia Marie
Lib 15h21'31" -6d0'23"
Alicia Marie Alexander
And 23h17'29" 47d53'11"
Alicia Marie Bergfeld
Cap 20h33'46" -15d10'11"
Alicia Marie Brigham
Tau 5h4'30" 27d4'28"
Alicia Marie Bush
Lib 15h31'6" -7d27'57"
Alicia Marie Diana
Ori 5h53'28" 12d33'43"
Alicia Marie Dixon
Cap 20h36'18" -18d34'23"
Alicia Marie Francine Mauro
And 2h19'42" 46d34'4"
Alicia Marie Garza
Lyr 18h29'40" 36d28'46"
Alicia Marie Hufford
Vir 13h19'34" -13d12'23"
Alicia Marie Jeskey
Cap 21h35'56" -14d39'38"
Alicia Marie June
Cap 20h23'34" -13d43'31"
Alicia Marie Ramirez
Lyn 7h53'37" 49d2'54"
Alicia Marie Smith
Psc 23h56'51" 10d39'55"
Alicia Mary Vultaggio
Vir 13h22'4" -11d23'3"
Alicia Maureen (Punkin) Howe
And 0h46'5" 38d37'3"
Alicia Medrano
And 0h52'53" 41d46'1"
Alicia Mello
Lyn 7h55'18" 56d48'2"
Alicia Michele Greaney
And 23h21'3" 49d25'24"
Alicia Michelle Ramos
Ari 3h7'0" 22d15'43"
Alicia Montalvo Perez
Sco 16h34'22" -37d19'43"
Alicia Morse (My Shining Star)
And 0h45'44" 27d3'53"
Alicia Murphy
Lmi 10h16'39" 38d7'40"
Alicia M.W.
Uma 10h56'40" 52d39'26"
Alicia Nagorski
And 2h29'21" 48d32'17"
Alicia Nez (Shesha)
Vir 11h54'27" 3d54'33"
Alicia Nicole Bihler
Tau 4h43'6" 24d30'6"
Alicia Nicole Cron
Sco 17h52'29" -39d51'22"
Alicia Nicole Hern
Tau 5h57'25" 26d56'57"
Alicia Nicole Kannady
Aqr 23h10'53" -10d26'12"
Alicia Noble Star of the Galaxy
And 1h8'54" 35d0'10"
Alicia Odom
Cap 20h32'17" -8d52'27"
Alicia Orozco
Aqr 21h23'54" -10d13'40"
Alicia Parr
Lyn 6h57'16" 59d50'6"
Alicia Parra
Cap 21h32'58" -12d26'50"
Alicia Pearson Huey
Uma 10h19'11" 52d32'20"
Alicia Perry
Uma 10h21'13" 59d37'49"
Alicia R.
Sgr 18h31'37" -16d4'41"
Alicia R. Escue
Uma 11h31'9" 37d55'28"
Alicia R. Martinez
Psc 0h21'39" -0d10'0"
Alicia Rae's Star
Leo 11h42'58" 24d41'0"
Alicia Rankin
Cas 0h43'59" 64d28'55"
Alicia Rau
And 0h44'16" 38d35'2"
Alicia Regina Assetto
And 0h50'59" 41d1'17"
Alicia Renee
And 1h8'11" 42d7'41"
Alicia Renee Gill
Ori 5h12'35" -8d14'54"
Alicia Rescigno
Cyg 19h54'58" 38d29'34"
Alicia Reyna
Cyg 20h57'41" 36d46'46"
Alicia Riccardi
Cam 7h57'52" 67d24'30"
Alicia Riley Hensel
Sco 16h11'17" -11d12'14"
Alicia Rodrick
Vir 13h3'9" 5d47'19"
Alicia Rose
And 0h37'26" 38d57'22"
Alicia Rosselle
Sgr 20h17'7" -35d14'5"
Alicia Roxanne
Leo 11h18'10" -2d1'59"
Alicia Ruth Segura
And 23h20'14" 49d12'1"
Alicia Sbordone
Crb 15h52'6" 37d37'29"
- Alicia Shum - 21 December 1983 -
Sgr 18h3'26" -27d53'17"
Alicia Terran
Cnc 8h36'26" 18d1'0"
Alicia Tucker's Eye in the Sky
Ari 2h18'47" 19d7'20"
Alicia Underwood Moore
Lib 15h14'45" -15d29'15"
Alicia Userpater VIDA
Cyg 21h52'57" 37d16'3"
Alicia Veonia Dillard
And 1h1'3" 38d17'36"
Alicia Vila
Cap 20h39'2" -22d1'57"
Alicia Villaescusa Karagianes
Cap 21h52'0" -14d50'50"
Alicia Viviana Johnson
Leo 9h34'52" 28d15'5"
Alicia Vurchio's 18th Birthday Star
Psc 0h42'56" 8d29'6"
Alicia Walsh
And 2h23'31" 48d18'27"
Alicia Whitney Knott
Peg 21h38'21" 27d28'2"
Alicia Wilson
Uma 11h29'2" 62d58'31"
Alicia, Danielle, & Kara
Uma 13h55'5" 47d54'3"
Alicia, The Light of my Life
Uma 11h28'52" 38d49'3"
Alicia013
Cas 1h46'22" 64d37'0"
Alicia-Brandon 365
Gem 7h39'57" 22d51'51"
Alicia-Craig
Uma 11h24'16" 38d44'16"
Alicia-Dyan
Psc 23h41'33" 7d5'26"
AliciaKay
Lib 15h37'40" -27d39'31"
Alicia's Eternal Heart
Gem 6h41'13" 26d51'13"
Alicia's Etrinity
Sco 16h12'11" -9d19'41"
Alicia's Heart
Psc 0h19'14" 11d29'14"
Alicia's Jeffrey star
Uma 11h13'10" 46d54'59"
Alicia's Penguin
And 0h12'54" 34d47'37"
Alicia's special star
Sgr 18h10'47" -20d5'56"
Alicia's Star
Ori 5h15'3" 6d18'44"
Alicia's Star
Cnc 9h16'54" 22d17'9"
Alicia's Star
Leo 11h3'14" 20d52'55"
Alicia's Will
Lib 14h43'16" -11d11'0"
Alicja Janosz
Gem 7h9'5" 34d0'2"
Alicja K.
Cap 21h23'29" -18d58'58"
Alicja Stoddard
Crb 15h37'1" 38d53'27"
Alick Halim Campbell
Cru 12h23'47" -61d50'56"
Alida
And 2h22'44" 49d6'17"
Alida
Crb 16h15'4" 26d13'33"
Alida LaVonne Foss
Lib 15h15'46" -12d31'25"

Alidawn Gem 6h32'54" 22d29'24"
Alie Mae Vir 14h40'51" 6d14'43"
AlienQueen Aqr 22h45'4" 0d28'17"
Alijah Jane Peg 22h1'49" 16d46'16"
Alijah Kai Haggins Uma 11h18'35" 37d25'29"
AliJay Forever Ori 5h36'11" 2d16'31"
AliKat Cmi 7h37'27" 4d21'21"
Alikat Cen 13h12'21" -49d16'24"
AliKat13195 Per 3h18'34" 51d21'10"
Alike 49810 Gem 6h49'55" 17d2'50"
Aliki Christodoulidou Sco 16h25'19" -37d36'6"
AliMari Tau 4h39'40" 24d11'27"
Alimata et Sylvain Gravaillac Dra 16h18'22" 65d19'29"
Alina Lyn 8h20'47" 53d50'31"
Alina Vir 11h40'37" -0d54'35"
ALINA Umi 14h55'18" 79d43'28"
Alina Sgr 18h24'18" -33d13'17"
Alina Psc 1h24'32" 11d48'10"
Alina Her 18h46'31" 12d40'5"
Alina Crb 15h47'0" 31d59'49"
Alina Cyg 19h52'27" 33d40'39"
Alina Uma 11h12'21" 41d51'47"
Alina And 0h9'24" 44d58'7"
Alina Bestler Boo 14h14'30" 28d58'44"
Alina Fedorova Leo 9h36'49" 12d38'11"
Alina Gage Cap 20h33'10" -17d3'30"
Alina Gozener Cap 21h27'53" -13d9'3"
Alina Grace Mastrogiovanni Lib 15h31'14" -4d40'29"
Alina Hacker Her 16h11'7" 48d6'6"
Alina I Umi 15h31'11" 69d59'37"
Alina Kheyfets Crb 15h38'1" 31d38'46"
Alina Kim Cnc 9h9'27" 19d8'12"
Alina Kleinschmidt Uma 8h22'38" 60d46'40"
Alina Marie Sisson Gem 7h47'34" 31d10'10"
Alina Marie Torres Gem 6h33'37" 13d12'16"
Alina May Psc 0h29'18" 3d17'24"
Alina Mia Cnc 9h6'37" 20d20'51"
Alina Rebecca Her 18h53'47" 20d40'48"
Alina Rose Stribling Ari 2h12'17" 26d15'53"
Alina Siegenthaler Cam 8h42'15" 77d25'24"
Alina Vanessa Iten And 2h8'18" 46d1'53"
Alina Zadorskaya Cap 21h31'37" -18d0'51"
ALINA, euse Sunneschii Aur 6h34'21" 34d38'34"
ALINACHKA Cnc 8h44'21" 7d59'8"
Alina-Katharina Uma 8h37'18" 63d32'16"
Aline Cas 1h54'59" 61d20'54"
Aline Aqr 21h28'37" 1d17'39"
Aline Leo 9h23'35" 26d41'10"
Aline Cyg 21h38'6" 47d17'25"
Aline and Jane Cap 21h50'14" -14d57'1"
Aline Bellitto Uma 8h46'13" 71d43'58"
Aline "Butch" Louise DeFroy Dowell Cyg 21h27'57" 34d39'11"
Aline et Emmanuel BERNARDIN Umi 14h33'4" 85d44'36"
Aline F Rodriguez Leo 10h43'41" 12d54'8"
Aline Fahed Cyg 19h46'44" 30d3'6"
Aline J. Jones Cyg 20h9'47" 37d51'51"
Aline Layla Toprakjian Cnc 7h58'4" 17d40'22"
Aline Marie (Bella) Tau 5h18'28" 17d10'42"
Aline Mühle Uma 11h38'1" 31d36'41"
Alinka Sco 17h52'50" -36d43'25"
Alinushka Gem 7h24'14" 19d43'48"
Alioscia Bassani Her 16h21'4" 16d59'31"
Aliro Hence Love Cap 21h39'52" -13d27'25"
aliron8 Cas 0h34'15" 47d47'59"
Ali's Bright Eyes Lib 15h18'41" -6d58'14"
Alis Marachelian Tau 4h51'56" 22d20'50"
Ali's Romance Leo 9h47'29" 16d1'45"
Ali's Star Aqr 22h31'38" -5d7'28"
Ali's Star Umi 14h54'40" 72d49'18"
Alis volat propiis Col 5h29'23" -34d50'34"
Alis Volat Propriis Dra 17h19'41" 53d50'42"
Alis Volat Propris Sgr 18h11'5" -19d52'15"
Alisa Lib 15h2'43" -23d12'2"
A-Lisa Lyn 8h6'22" 46d17'16"
Alisa 2000 Uma 12h17'49" 62d20'21"
Alisa Alexandra Valdez Cap 21h32'57" -9d10'30"
Alisa Ann Sullivan Ward Happy 30th And 0h49'33" 37d20'58"
Alisa Bucher Dra 16h28'48" 66d26'0"
Alisa Chantel Dra 10h28'37" 79d35'31"
Alisa Guhl Psc 1h35'53" 20d10'22"
Alisa J. Miller Aqr 22h12'40" -23d30'21"
Alisa Joy McPheron Cnc 8h29'26" 25d51'55"
Alisa Justine Walker Uma 12h27'21" 52d22'15"
Alisa Katherine Buck Dra 18h18'57" 70d26'44"
Alisa Lucile Vance Mon 7h58'25" -0d42'6"
Alisa Marie And 0h48'14" 34d53'53"
Alisa Marie Busch And 2h22'5" 49d19'11"
Alisa Marie Troccia And 2h12'0" 49d9'37"
Alisa Rethemeyer Gem 7h7'1" 32d5'13"
Alisa Salego Psc 1h31'49" 25d42'40"
Alisare G Fitzgerald Uma 9h41'4" 46d26'12"
Alise And 0h52'4" 45d3'31"
Alise Michelle McCall And 0h4'32" 45d14'13"
Alisha Cyg 20h31'14" 42d11'49"
Alisha And 23h11'37" 35d51'34"
ALISHA Her 16h46'18" 12d0'19"
Alisha Cnc 8h1'6" 11d30'46"
Alisha Ori 6h16'1" 13d39'57"
Alisha Psc 1h10'59" 10d1'37"
Alisha Aqr 22h18'18" -14d4'41"
Alisha Cap 20h32'31" -27d33'13"
Alisha Pho 0h38'0" -45d33'28"
Alisha 1983 Cap 20h52'39" -20d20'0"
Alisha Ann Cap 21h37'35" -14d40'32"
Alisha Bader Vir 12h12'6" 2d55'56"
Alisha Bollgen Cmi 8h3'58" -0d18'53"
Alisha Dittbenner And 23h45'57" 45d15'51"
Alisha Everett And 0h20'0" 30d29'55"
Alisha Gail Dettling Aqr 22h37'0" -0d46'7"
Alisha Gaye Hughes Cru 12h34'58" -60d26'12"
Alisha Hat 97 Aur 5h44'32" 52d49'38"
Alisha Hinton Tau 5h43'13" 25d49'41"
Alisha Ibrahim Psc 1h58'13" 7d7'0"
Alisha In The Sky With Diamonds Psc 1h0'39" 32d12'57"
Alisha Izanne Cnc 9h13'2" 32d53'8"
Alisha Jade Jensen - 6.03.1979 Cru 12h19'36" -57d37'52"
Alisha Kay Jones 21 Cru 12h22'40" -57d29'38"
Alisha & Kevin Burge Cyg 21h35'44" 49d24'10"
Alisha Leann Barrett And 2h5'36" 41d41'49"
Alisha Lee Vir 14h56'49" 3d35'11"
Alisha Linn Lyn 7h0'40" 61d49'16"
Alisha Marie Psc 0h15'43" 8d25'37"
Alisha Marie Bohnsack Gem 7h6'20" 24d24'59"
Alisha Marie Dahl Cas 1h36'56" 60d6'29"
Alisha Marie Long And 2h29'34" 37d37'42"
Alisha Mertins Cyg 21h22'32" 50d32'28"
Alisha Ravi And 2h19'2" 48d20'0"
Alisha Renee Zazzi Cap 21h54'19" -24d30'42"
Alisha Repp And 0h9'9" 31d53'40"
Alisha Sandoval Tau 4h30'28" 29d53'4"
Alisha Sharon Psc 0h50'27" 19d5'19"
Alisha Sofia Tau 4h37'8" 22d19'24"
Alisha Starr Leo 11h51'42" 22d19'30"
Alisha Voyanna Reynolds Psc 0h52'20" 18d46'2"
Alisha's Star Del 20h18'45" 15d9'9"
Alison Her 18h34'15" 19d3'36"
Alison Gem 7h50'45" 15d58'56"
Alison Gem 6h49'43" 22d23'37"
Alison Ori 6h12'26" 18d33'49"
Alison Vul 20h28'33" 28d7'40"
Alison Crb 15h44'21" 27d52'0"
Alison Tau 4h16'42" 2d56'8"
Alison Cyg 21h17'58" 42d36'1"
Alison Cyg 21h39'13" 38d35'27"
Alison Uma 10h50'2" 47d48'54"
Alison And 2h19'57" 38d32'59"
Alison And 0h14'0" 40d22'29"
Alison Uma 9h40'46" 43d47'25"
Alison Gem 7h37'19" 34d3'15"
Alison Cap 20h20'37" -23d45'1"
Alison Lyn 6h32'16" 57d33'21"
Alison Lib 14h50'49" -13d32'51"
Alison Vir 14h1'24" -16d25'53"
Alison 50 Psc 0h54'53" 26d31'58"
Alison Ackland Cas 23h50'21" 50d0'47"
Alison (Ali Bear) Miner Vir 12h49'42" 12d26'11"
Alison (Ali) Bodie Aur 5h46'10" 45d55'33"
Alison and Alan 2-14-2004 Cyg 19h39'44" 28d18'21"
Alison and Barry Tau 3h40'3" 6d24'48"
Alison and Jon Angle Del 20h34'0" 4d11'40"
Alison and Mike Gem 7h33'4" 24d2'2"
Alison and Paul's Engagement Cyg 20h20'15" 43d22'52"
Alison Ann Balestra Cyg 20h45'17" 44d49'29"
Alison Ann Holman Vir 11h38'54" 9d34'28"
Alison Bailey Greene And 0h25'33" 33d28'53"
Alison Barrett Cnc 9h3'58" 26d54'7"
Alison Bavin Cas 3h13'59" 59d13'4"
Alison Benninger Uma 11h41'38" 52d17'26"
Alison Bisbee Ori 5h42'0" -9d33'13"
Alison Bresciani Her 17h59'10" 49d41'37"
Alison Brook Crb 15h45'43" 31d13'47"
Alison Bulger Cas 1h59'48" 62d47'57"
Alison C. Goring Gem 6h48'4" 23d15'55"
Alison C Powell Cyg 20h13'35" 42d13'9"
Alison C. Yops And 0h18'36" 46d26'4"
Alison Cali Uma 10h5'14" 70d46'32"
Alison Cardillo Ori 6h9'39" 18d8'15"
Alison Cecilia Clark Cas 1h16'17" 63d45'15"
Alison Christine Leo 11h44'51" 18d59'4"
Alison Christine Leo 10h49'30" 6d21'10"
Alison Christine Pafiolis Knight And 1h10'51" 40d33'39"
Alison & Cliff's Wedding Star. Cyg 21h13'48" 41d21'59"
Alison Cooper Tau 5h35'46" 20d54'4"
Alison Corker And 1h37'34" 38d29'18"
Alison Daniels And 0h46'13" 40d37'39"
Alison & Dave Ari 2h5'4" 24d15'31"
Alison De Los Reyes Mon 6h58'58" -0d56'14"
Alison Denise Reeder Cap 20h19'7" -12d54'33"
Alison Dixon And 23h46'30" 34d24'23"
Alison Dudrey Lyr 18h28'46" 36d16'58"
Alison Dunne And 2h12'39" 49d18'31"
Alison E. Grady Aqr 22h40'25" 1d20'56"
Alison Elizabeth Earnhart Vul 20h26'0" 27d52'4"
Alison Elizabeth Hecht Ori 5h38'57" 4d54'4"
Alison Elizabeth Staples Gem 7h16'49" 32d36'6"
Alison Elizabeth Williams Cyg 19h49'46" 53d10'30"
Alison Estelle Whalen Cas 1h9'34" 62d47'5"
Alison Evans And 2h1'10" 38d15'48"
Alison Eyre Ward And 2h6'8" 46d0'19"
Alison Faith Brown Uma 11h21'37" 57d4'13"
Alison Faye Lyn 8h32'7" 36d25'57"
Alison Gabel, Our Special Star Uma 12h5'35" 62d20'41"
Alison Graeve Crb 16h15'36" 27d54'48"
Alison Greenberg Ari 3h21'45" 28d35'9"
Alison Griffiths Psc 1h11'16" 26d12'56"
Alison Hampton Psc 0h38'21" 6d58'9"
Alison Hannah Barlow Gem 7h15'58" 32d43'41"
Alison Hasha - Cuddlie Bear Ari 2h56'20" 22d16'57"
Alison Hedquist And 1h24'10" 43d40'9"
Alison Helf Gem 6h22'24" 22d5'17"
Alison Herdocia Lyn 8h43'5" 44d28'10"
Alison Herper Lyr 19h8'43" 42d29'6"
Alison Holloway Tau 5h32'11" 26d50'11"
Alison Hope Cap 20h22'29" -11d38'34"
Alison J Sgr 19h20'9" -15d6'57"
Alison Jan Skloot Cap 21h38'39" -9d11'29"
Alison Jane Crippin Lmi 10h14'47" 36d21'24"
Alison Jane Kiefel Cru 12h0'45" -62d54'32"
Alison Jayne Jennett And 1h56'44" 40d42'2"
Alison Jean Murphy Psc 0h15'3" 2d38'13"
Alison Jenkinson And 2h24'26" 44d54'26"
Alison Joan New Cas 1h12'29" 72d36'7"
Alison Jones Cyg 20h0'13" 56d16'52"
Alison June Eber Crb 15h52'49" 36d53'44"
Alison June Grant Leo 11h32'38" 15d2'31"
Alison Kalinowski Giese Leo 10h55'17" -0d10'1"
Alison Karnes Uma 12h13'53" 57d20'24"
Alison Kassel Lyn 8h21'2" 47d21'26"
Alison Kathleen Rush Uma 14h11'40" 59d12'11"
Alison Kathryn Fink Sco 16h5'3" -10d53'37"
Alison Kathryn Pennisi Vir 14h13'50" -17d44'38"
Alison Kruger And 0h35'49" 33d29'56"
Alison Lauren Lyr 18h15'56" 31d30'8"
Alison Lawless Cas 23h46'57" 62d36'15"
Alison Leap Cap 21h9'31" -18d55'43"
Alison Lee Cyg 20h51'57" 46d3'1"
Alison Lee Smythe And 1h32'39" 50d5'33"
Alison Leigh And 2h19'29" 46d23'36"
Alison Levenick Uma 12h44'37" 52d54'28"
Alison Lorenz Cas 2h28'8" 63d32'24"
Alison Lori Cap 21h32'8" -9d50'1"
Alison Lowe Lib 14h51'25" -1d58'2"
Alison Lussier Dyer Uma 10h51'42" 51d17'40"
Alison Lynn Birmingham Uma 10h33'36" 63d53'17"
Alison Lynne Leo 11h31'47" 0d3'58"
Alison M. D'Auria And 0h24'28" 38d11'7"
Alison Marie And 0h8'5" 44d53'36"
Alison Marie Vir 13h54'20" 3d29'32"
Alison Marie Ori 5h8'5" 8d16'24"
Alison Marie Ari 2h13'16" 20d32'13"
Alison Marie Tau 5h55'55" 25d15'16"
Alison Marie King Psc 0h35'17" 8d39'2"
Alison Marie Schoenherr Uma 8h52'39" 66d27'57"
Alison Marjorie Selverstone And 0h54'49" 37d55'19"
Alison Martin Jackson Mon 6h48'20" -0d24'50"
Alison Mary Kreger "Ali Monster" Cas 0h27'50" 61d5'58"
Alison Maureen Tau 4h49'53" 18d44'2"
Alison McLeod Cra 19h1'44" -38d56'14"
Alison McLoone And 23h8'53" 50d34'4"
Alison Michele Rodecker Uma 8h35'4" 46d49'13"
Alison Mode Lyn 8h1'37" 35d16'9"
Alison Moens Cas 1h23'45" 56d33'3"
Alison Montgomery McGovern And 23h44'14" 38d6'20"
Alison Moore And 23h11'51" 52d27'57"
Alison Morgan Clark Uma 11h0'23" 62d7'32"
Alison Nancy Kapner Schwartz Ari 3h21'37" 29d21'12"
Alison Nelson Cas 1h47'51" 60d47'12"
Alison Nicole Norwood Cyg 20h36'33" 41d35'17"
Alison Paige Psc 23h22'55" 7d58'5"
Alison Patterson Gem 6h8'54" 24d17'18"
Alison & Paul Gill Cyg 19h45'26" 40d29'21"
Alison Perkowski Aqr 21h51'3" -6d55'18"
Alison Rae Sansom Crb 15h51'58" 36d47'51"
Alison Renee Egbers Sco 16h43'18" -30d5'27"
Alison Rita Sam Maude Deforest And 0h30'55" 25d13'1"
Alison Robyn Neustein Vir 14h28'44" 3d32'35"
Alison Rossi Umi 15h45'33" 73d21'37"
Alison Ruth Crb 16h17'3" 37d29'30"
Alison Ruth Dunn Crb 16h7'7" 37d20'14"
Alison Ruth Gardner Alexander Cas 22h59'35" 59d15'19"
Alison Sarra Widoff Cyg 19h37'4" 31d59'3"
Alison & Seán Derrig Aur 7h27'46" 39d38'22"
Alison Sipes Sgr 18h42'46" -30d39'22"
Alison Smith Aqr 22h27'30" -19d17'56"
Alison Stamp Cnc 8h1'8" 11d34'3"
Alison Stevens And 1h41'2" 49d44'28"
Alison Sturm Lynch "Ali" Tau 5h48'30" 27d49'39"
Alison Swan Cyg 21h16'48" 54d11'19"
Alison Tara Quine And 1h11'48" 43d0'58"
Alison Vellanoweth And 1h32'27" 40d49'12"
Alison W. Spicer Crb 15h44'21" 37d10'4"
Alison Wadle Corrigan Men 4h46'37" -76d43'32"
Alison York Gem 7h55'13" 19d41'44"
Alison Zeidman Sgr 18h47'56" -34d40'45"
Alison1 Tau 3h24'43" -0d45'59"
AlisonBeckyBrigetteTrevorTaylor Cmi 7h34'44" 6d27'41"
Alison's Heart Leo 10h12'38" 14d37'35"
Alisons Star Tri 2h16'46" 32d20'48"
Alison's Star And 0h38'47" 42d42'1"
Alison's Star Cas 0h56'10" 58d29'23"
Alison's Star Cas 0h31'52" 52d44'36"
Alison's Star Sco 16h14'56" -8d42'59"
Alison's Star Col 6h22'54" -37d36'56"
Alispin Cnc 8h37'44" 30d44'21"
Alissa And 1h7'27" 37d55'22"
Alissa Ari 2h34'44" 22d18'12"
Alissa Cnc 8h32'32" 24d6'47"
Alissa and Nana's Looking Star Ori 5h49'34" -1d54'47"
Alissa B. Haynes And 0h32'35" 36d8'11"
Alissa Bandalene Tau 4h47'56" 26d17'1"
Alissa Bauman Cnc 8h19'17" 13d11'4"
Alissa Beth Powell Ari 3h27'2" 26d34'31"
Alissa Bilger And 23h56'43" 43d22'17"
Alissa Burriss And 2h6'39" 46d29'9"
Alissa Costanza And 1h22'28" 38d19'2"
Alissa Debra Nelson Vir 12h41'4" -7d42'28"
Alissa Dora Stauffer Ser 18h21'7" -14d4'7"
Alissa & Drew Cyg 20h40'43" 51d27'31"
Alissa Dunn Cas 23h42'46" 62d16'13"
Alissa Faye Pierce And 0h56'58" 40d15'27"
Alissa Jennifer Cra 18h0'2" -39d3'20"
Alissa Kinnee Sco 17h19'31" -33d0'9"
Alissa L.1977 Sgr 19h15'28" -16d22'6"
Alissa Marie Costanza Uma 9h34'36" 51d11'50"
Alissa Marie Prather Tau 5h28'52" 24d1'16"
Alissa McClure M.D. Lib 15h14'18" -27d32'1"
Alissa Michelle McElhone Tau 3h51'57" 30d39'39"
Alissa Michelle Moseley Leo 9h32'22" 27d39'48"
Alissa Nichole Gray And 23h48'4" 47d16'39"
Alissa Nicole Cohen And 23h46'34" 46d48'51"
Alissa Scott And 23h3'36" 40d32'35"
Alissa Ursula Sco 17h55'55" -44d1'49"
Alissa V. B. Ori 5h47'25" 11d18'41"
Alissa030381 Psc 1h29'53" 13d37'41"
Alissa's shining star, Love, Daddy Cas 23h53'44" 57d9'49"
Alissa's Starr Lyr 19h22'33" 38d27'19"
Alissia Boo 14h41'15" 10d22'39"
Alistair Cameron Duthie Aqr 22h41'12" 1d46'24"
Alistair Johnston Cyg 21h49'4" 47d13'45"
Alistair Logan McKenzie Her 18h16'59" 15d24'20"
Alistair & Lyn-Ann -Lucidus Levis - Col 5h48'37" -34d18'7"
Alistair Mathers Ori 5h13'6" 7d43'35"
Alistair Thomas Sgr 18h14'6" -29d1'49"
Alistar Psc 1h1'49" 3d50'52"
Alistar J Stannard Vir 12h40'42" -9d16'56"
Alister Asher - Ali's Star Uma 11h20'18" 41d15'52"
AliSue & Buttons Too Pic 5h15'42" -55d40'18"
Aliterikris Cam 4h29'59" 66d56'31"
Alius Veruis Eroticus And 1h20'24" 49d45'4"
a-live! Uma 9h13'44" 56d35'15"
Alivia Tau 5h40'11" 24d11'0"
Alivia Ann Loovis Lyn 8h44'28" 45d22'4"
Alivia Ann Mossler And 1h45'15" 49d4'7"
Alivia Catherine Tagliaferri Ari 3h14'2" 28d48'57"
Alivia Jean Killian Ari 3h16'21" 28d56'42"
Alivia Lea Newsome Uma 10h24'52" 58d48'58"
Alivia Lee White Uma 13h16'0" 52d46'3"
Alivia Lesly Guzman Cas 1h15'52" 58d59'45"
Alivia Marie Peterson Tau 3h46'44" 22d21'38"
Alivia Rae Gem 7h11'44" 33d23'52"
Alivia Rose Uma 10h54'29" 47d11'21"
Alix Aur 5h34'51" 33d20'1"
Alix Del 20h16'17" 9d7'56"
Alix And 0h23'18" 25d40'3"
Alix Andrea Ari 3h20'0" 28d38'27"
Alix Coupé Uma 10h19'37" 56d16'9"
Alix Korapich Cap 20h25'32" -12d43'30"
Alix Light Del 20h25'32" 2d52'3"
Alix Macri Aql 19h38'32" 9d5'38"
Alix Maree Tully Gem 6h58'43" 26d54'58"
Alixandra Cru 12h36'12" -59d53'47"
Alixandra Terese Lyn 6h22'43" 60d12'4"
Aliya Bhabha Lib 15h4'13" -22d15'30"
Aliya Brynne Drake Aqr 21h5'34" -10d43'22"
Aliya Rose Ari 2h6'39" 18d29'59"
Aliyah Brooke Halper Tau 4h39'34" 20d56'54"
Aliyah Griggs Lyn 6h19'52" 56d4'59"
Aliyah Karin Hernandez Leo 10h5'8" 21d30'9"
Aliyah L Lib 15h29'47" -18d48'47"
Aliyah Olivia Sarimazi Uma 9h52'32" 49d8'19"
Aliyah "One In A Million" Ori 5h36'12" -1d54'36"
Aliyah-Rosabelle Ondo-Boxill Oph 17h25'42" 9d45'47"
Aliyah's Joy And 23h46'33" 48d36'27"
Aliyra-Rose Jillian Thurley Cru 12h27'3" -60d47'0"
Aliz csillaga Aqr 22h4'55" -9d34'30"
Alizabeth Marie Stoltenberg Sgr 18h14'25" -22d2'31"
Alizad Forever Aql 19h58'54" 1d49'16"
Alize' Nicolette Harmala And 2h14'7" 41d38'15"
ALJ 1967 Tau 3h29'48" 25d27'52"
Alja Shana Billeter Uma 12h1'20" 64d43'49"

Aljenivaro
Cyg 19h37'53" 39d31'52"

ALJO
Her 16h24'50" 23d49'1"

ALJOSI
Uma 12h4'44" 42d56'6"

Al-Karim Kara
Uma 9h22'47" 72d30'45"

ALKHS
Uma 10h26'57" 63d56'23"

All About Me
Cnc 8h7'19" 12d19'6"

All Emotions 11062005
Uma 8h48'1" 61d21'0"

All Emotions 11062005
Uma 14h17'43" 59d18'17"

All FBL Associates
Uma 12h8'54" 60d22'36"

All I Need
Leo 11h36'39" 18d58'46"

All joking apart!!
Cyg 19h55'36" 37d1'58"

All Lightworkers Worldwide
Cam 4h8'30" 72d0'19"

All my children
Uma 13h15'54" 57d4'10"

All My Love, All My Life
Uma 13h52'22" 57d37'58"

All My Love, All My Life
Cyg 19h55'25" 45d8'35"

All my love, to my angel Jessica
And 2h35'7" 39d20'46"

"All My Sunsets"
Psc 1h25'21" 16d2'12"

All Of My Love
Cas 0h15'56" 58d2'40"

All Of My Love
Sco 16h8'38" -16d30'10"

All Okay? Stuart's Star
Cra 18h48'59" -40d41'17"

All Our Love Always Andi & James
Ori 5h8'54" 8d9'38"

All Saints Eighth Grade-2005
Ori 6h16'24" 10d22'28"

All Star
Her 18h14'59" 23d29'44"

All that I am
Uma 11h39'31" 62d58'58"

All That Is Healing
Lib 15h59'3" -17d50'8"

All the Days of My Life
Her 17h3'8" 47d7'54"

All The Garbage In The World
Per 3h33'14" 47d12'52"

All the Little Things
Ari 2h5'39" 23d3'12"

All the stars
Eri 3h18'52" -41d57'2"

"All The Way"
Gem 7h27'13" 20d28'19"

All Ways and Forever
Leo 9h42'55" 28d58'51"

All4Patti'sStar
And 1h24'45" 44d4'49"

Alla
Uma 11h24'42" 29d24'4"

ALLA
Gem 6h21'53" 21d7'3"

Alla
Lib 15h22'34" -7d48'57"

Alla Aandaxjoon
Cyg 20h0'27" 36d45'17"

Alla Borisovna Pugacheva
Cas 0h55'28" 60d29'48"

Alla J. Chernyavsky
Uma 10h25'9" 40d39'25"

Alla Kogan
Aqr 21h37'24" 1d11'48"

Alla Korot
Cet 1h5'55" 0d46'43"

Alla luce di eterno
Sco 16h45'48" -44d47'49"

Alla nostra Elenina
Aur 6h39'12" 31d34'21"

Alla Reysher
Uma 8h51'20" 52d33'13"

Alla S. Loza
Ser 15h22'6" -0d23'0"

Alla Seifer Shulman
Aqr 22h11'59" -14d51'22"

Allamanda Beach Hotel
Aur 5h45'28" 55d21'23"

Allan
Cyg 19h48'10" 37d45'1"

Allan 300465
Dra 18h58'23" 50d2'39"

Allan and Kelly Kiddy, 12.24.2002
Lib 15h25'19" -16d46'20"

Allan and Moira McPhie
Cyg 20h58'17" 48d32'9"

Allan and Tali's Blue Star
Psc 1h9'16" 32d27'31"

Allan Claros
Dra 15h12'18" 59d9'3"

Allan Cochrane
Her 16h25'22" 45d48'47"

Allan Edward Dean
Uma 9h45'14" 45d19'58"

Allan Eric Rybanic
Ori 5h21'35" -5d11'36"

Allan F. Collier 08/04/1924
Ori 5h57'51" 12d29'39"

Allan F Reddy
Aql 19h32'31" 11d49'50"

Allan Francis Paisley - Happy 50th
Cru 12h38'50" -57d59'45"

Allan Frederic Sauln
Uma 9h7'43" 55d8'58"

Allan G. Raiani
Cyg 20h0'50" 36d39'26"

Allan Giovanni Osuna virgen
Lyn 7h42'59" 57d31'51"

Allan Hayler
Cep 22h55'56" 57d5'2"

Allan & Jackie Forever
Lyn 7h19'56" 56d54'41"

Allan James McNichol, Jr.
Leo 11h6'11" 24d43'1"

Allan & Jean Harvey
Col 6h1'43" -40d40'59"

Allan & Jean Hudson
Cyg 20h27'50" 53d47'29"

Allan John Carter
Cep 22h4'35" 84d13'2"

Allan L. Rowley
Sgr 18h29'46" -16d52'12"

Allan L. Wells
Sco 16h31'25" -37d5'19"

Allan Lehmann
Aqr 23h29'46" -24d29'10"

Allan loves Alison forever
Cyg 21h52'47" 50d2'49"

Allan M. Gerovitz
Cep 21h28'6" 65d7'56"

Allan & Marguerite Raslack
Cyg 21h51'32" 40d30'32"

Allan Michael Ellebracht
Uma 11h47'4" 38d10'41"

Allan - My Eternal Light
Cru 12h30'46" -59d52'40"

Allan Nakata
Equ 21h4'47" 3d53'39"

Allan Prudhomme - New Beginnings
Ari 3h10'44" 23d16'52"

Allan R. Ahrens, Jr.
Aqr 21h19'46" 2d14'16"

Allan Rathbum
Cnc 9h10'18" 32d26'18"

Allan Rosenthal
Ari 2h36'47" 21d46'49"

Allan & Samantha
Ori 6h20'59" 5d49'55"

Allan & Selina Lennon's Eternal Love
Cyg 19h56'24" 35d58'49"

Allan Simpson - Forever Shining
Col 6h18'46" -33d50'44"

Allan Stone
Aqr 21h15'34" -7d46'36"

Allan Torney
Ori 6h5'43" 21d24'51"

Allan Williams
Per 4h47'43" 48d52'43"

Allan - You light up my world - Cinta
Psc 1h50'52" 2d50'48"

Allana and Andrew forever
Cap 21h55'45" -9d18'13"

Allanah Ann Wilmer
And 1h43'59" 48d13'29"

Allanah Wilmer
Umi 13h10'46" 75d14'1"

Allan's Eternal
Gem 7h35'39" 34d43'18"

Allan's Star
Per 4h25'52" 42d4'7"

AllansStar
Aur 6h9'30" 31d27'23"

Allasandra B. Cunningham
Leo 10h15'22" 17d37'38"

Allastar
Dra 18h41'8" 72d54'19"

Allee & Michael
Psc 1h34'12" 20d4'57"

Alleen
Tau 4h42'38" 21d47'32"

Allee's Star
Tri 2h21'31" 32d19'38"

Allegra Glassman
And 23h20'17" 44d46'58"

Allegra Henderson
Psc 0h51'19" 13d41'24"

Allegria
Ari 3h23'7" 28d26'34"

Alleigh Caroline Puckett
Leo 9h23'37" 27d10'23"

Alleigh Rae
Boo 14h57'8" 31d0'55"

Allemanda
Del 20h41'18" 16d19'30"

Alleman-McCoy
Lyr 18h15'11" 34d50'54"

Allen
Vir 13h52'50" -1d32'57"

Allen and Barbara Dreyer "10-19-04"
Umi 13h54'26" 75d42'25"

Allen and Betty Bash
Uma 11h16'52" 30d48'54"

ALLEN AND REA NORGAARD
Lib 15h35'30" -10d7'54"

Allen and Sintija Forever
Vir 13h18'45" 5d52'15"

Allen and Tara Star
Cyg 19h48'4" 30d40'3"

ALLEN ANDREWS GARDNER III
Leo 11h36'48" 1d22'47"

Allen B. Strasburger
Cap 20h28'31" -10d55'52"

Allen B. Theriault
Tau 5h47'7" 22d20'52"

Allen & Barbara Holland
Cyg 21h18'46" 38d17'51"

Allen BCOPTERS Broussard
Gem 7h16'17" 27d25'18"

Allen Bradley
Per 3h3'6" 55d8'0"

Allen Brett's Shining Spirit
Sgr 18h7'28" -27d54'50"

Allen Burke
Crb 16h8'43" 28d0'11"

Allen Carl Wallar
Ori 5h29'13" 2d3'28"

Allen Clark Brock II
Lib 14h43'22" -14d22'43"

Allen Cox and Chris Shoultz Forever
Ari 2h14'36" 23d23'54"

Allen Dale "Flamingo Boy" Landis
Aur 5h50'33" 39d13'1"

Allen Dwane Henry
Ori 5h17'41" -7d40'53"

Allen E. Castillo
Leo 10h10'3" 14d49'20"

Allen Eggar Yeakel
Crb 16h6'17" 36d40'12"

Allen Filion-Rozon
Psc 23h14'57" 7d47'49"

Allen Franklin
Cru 12h18'41" -59d10'36"

Allen Fred Fielding
Ori 6h15'30" 15d34'27"

Allen G. Avery, Jr.
Ori 5h27'20" 0d40'42"

Allen G. "Sparky" Tiefenbach
Aql 19h59'40" -0d18'20"

Allen Henry
Cep 22h3'35" 69d31'31"

Allen Herrington
Ori 6h18'49" 14d45'4"

Allen Hubbert
Nor 16h6'9" -43d38'49"

Allen J. Berliner
Per 4h27'43" 35d2'31"

Allen James Kish
Aqr 20h44'53" 0d27'26"

Allen Jerome Nash
Per 3h29'49" 51d51'45"

Allen & Jodi Stroklund
Lyr 18h50'5" 32d7'18"

Allen Johnson
Aur 6h34'38" 34d37'43"

Allen Joseph Niedzwiecki
Ori 5h43'11" 2d13'37"

Allen K Scribner
Uma 10h59'0" 49d42'42"

Allen Kade Kowalski
Boo 14h23'49" 44d0'54"

Allen Kullervo Luoma
Cru 12h15'48" -57d17'2"

Allen L. McChancey
Cap 21h15'53" -27d13'36"

Allen + Lauren's Lucky Star
Lmi 10h18'32" 36d8'16"

Allen Laverne Lyons
Per 3h16'50" 45d22'45"

Allen Lawrence Pope
Per 2h43'16" 54d9'23"

Allen Lee Cuff
Leo 10h49'19" 17d8'53"

Allen Minster
Sgr 19h53'37" -28d7'32"

Allen Moore
Her 17h12'59" 16d55'53"

Allen Novak
Psc 0h53'9" 10d9'49"

Allen P Slaton
Cnc 8h45'39" 32d9'16"

Allen Prince
Gem 7h6'40" 33d45'38"

Allen R. Burke
Umi 17h30'40" 86d34'47"

Allen R. Cummings
Cnc 8h39'56" 32d57'16"

Allen Robert Paisley
Sgr 19h25'52" -24d15'36"

Allen Roberts Crenshaw
Cru 12h50'38" -58d32'7"

Allen Ron Davis
Ari 2h37'5" 18d0'20"

Allen & Rose Krebs
Psc 22h54'4" 6d37'40"

Allen Rueben Stredwick Jr.
Crb 15h46'31" 29d41'22"

Allen Russell Gallaher
Lyn 7h46'53" 42d40'35"

Allen S. Pack - Husband-Dad-Poppy
Sgr 19h30'45" -15d45'20"

Allen Schelm
Her 17h36'2" 32d29'17"

Allen & Shane Hayes
Umi 14h55'4" 70d8'11"

Allen Stigwandish
Dra 17h41'41" 55d57'35"

Allen Stoddard AAF FAA
Gem 6h46'10" 28d32'51"

Allen Tait Nunnally
Sct 18h48'31" -12d19'14"

Allen Verza
Sco 16h8'41" -13d34'53"

Allen W. Kirby
And 23h9'50" 41d26'30"

Allen W. Parham II
Psc 1h4'35" 7d4'29"

Allen Wayne Strickler
Her 16h41'43" 36d30'58"

Allen William Troutt, Jr.
Vir 11h51'14" 9d52'27"

Allen Woodard
Ori 6h15'35" 10d49'53"

Allena Harnish
Ori 6h19'18" 18d57'12"

Allene
Cap 21h35'25" -15d23'30"

Allene Floyd
Pyx 8h51'8" -25d28'46"

Allen-My One and Only
Umi 15h21'28" 72d8'22"

Allenorama
Uma 10h39'38" 58d23'20"

AllenRoy Paquin
Psc 1h7'13" 19d2'18"

Allen's Star of Glowing Fire
Sco 16h57'13" -32d8'0"

AllenT11Blue
Vir 11h49'59" -3d47'33"

Alles Herz (meaning "All Heart")
Sgr 18h28'2" -16d54'43"

Alle's Star
Crb 15h24'14" 25d35'39"

Allet
Cmi 7h24'13" 8d23'43"

Allevard
Mon 6h46'2" -7d22'28"

Allexa's Wishing Star
Psc 1h23'49" 24d14'48"

Alley
And 1h7'10" 36d49'31"

Alley Allerton
Per 3h13'30" 54d46'36"

Alley Blackford
Uma 8h59'17" 71d18'6"

Alley May Stafinski
Ori 5h27'5" 0d26'19"

Alley97031
And 0h32'35" 41d36'46"

Allgood's Star
Cnc 8h43'19" 28d36'9"

Alli
Leo 11h31'21" 24d50'1"

Alli
Sco 16h13'5" -17d28'37"

Alli - 3232007
Vir 13h32'23" 4d9'8"

Alli and Eric's Little Wishing Star
Psc 0h25'50" 19d19'42"

Alli HaHa
And 23h17'7" 49d11'4"

Alli Nicole
Uma 9h17'50" 53d52'51"

Alli-Adam Forever 082904
Crb 16h17'14" 37d36'37"

Alliah
Tau 5h14'33" 23d0'35"

Alliance
Cyg 20h21'11" 58d58'30"

Allianna Williams, Love of my Life
Cyg 20h43'7" 55d3'49"

Alliarm
Vir 13h7'42" 0d56'6"

Allicyn Amalia Stresen-Reuter
Leo 11h48'8" 16d50'15"

Allie
Cnc 8h41'43" 28d41'47"

Allie
Crb 16h8'3" 37d47'10"

Allie
Her 17h6'0" 39d45'2"

Allie
Per 3h15'38" 57d18'41"

Allie
Uma 11h43'9" 37d48'55"

Allie
Cyg 22h2'24" 52d41'26"

Allie
Cam 4h57'10" 63d36'37"

Allie
Lib 15h37'33" -18d1'51"

Allie
Cma 6h19'41" -16d14'23"

Allie
Vir 12h24'27" -4d40'32"

Allie
Aqr 22h34'51" -2d30'26"

Allie
Sgr 18h31'43" -27d31'9"

Allie
Sco 17h49'11" -41d47'47"

Allie & AJ
Cyg 20h20'18" 53d24'3"

Allie and Matt
Sgr 18h19'17" -22d47'34"

Allie and Richie
Ari 2h54'7" 19d50'49"

Allie B
Lyn 7h54'32" 54d8'54"

Allie Bella (Pork Chop) Bernosky
And 2h13'14" 47d54'2"

Allie Brooke Rosenbloom
Aqr 22h20'55" -1d37'2"

Allie by the Sea
Mon 6h56'43" -8d20'56"

Allie Cat
Vir 13h3'57" 13d36'42"

Allie Christine Jackson
And 23h9'1" 48d10'2"

Allie Darrell Diffin
Uma 11h38'22" 65d18'31"

Allie Ellis
Cas 0h2'34" 59d44'35"

Allie Fahey
Lyn 7h36'46" 53d49'36"

Allie Girl
Ori 6h22'30" 10d22'20"

Allie Grace Pruitt
Lib 15h17'8" -22d7'7"

Allie Hays
Lib 15h20'47" -15d30'5"

Allie Mae
Uma 10h29'58" 52d3'31"

Allie Margaret Ingles
Cas 2h20'44" 64d18'34"

Allie Marie Bloxdorf
And 0h6'2" 33d54'47"

Allie Marie Stockwell
Lyn 8h2'40" 41d32'58"

Allie Mikael Harris
Tau 3h44'36" 21d2'2"

Allie Nicole Puckett
Leo 11h38'19" 19d21'56"

Allie Olivia
Cas 22h58'6" 53d13'1"

Allie Paige
Ori 5h12'4" 1d49'56"

Allie Rae Runge
And 0h31'43" 42d48'43"

Allie Renee' Roberts
Ari 2h8'55" 13d8'33"

Allie Sue
Mon 6h51'22" -0d6'53"

Allie T. Bolch
Cas 0h54'8" 57d27'21"

Allie Yoffee
Uma 10h55'7" 43d35'34"

Allie, homecoming? Love Johnny
And 0h33'32" 43d16'6"

Allie10
Sco 16h13'58" -29d59'53"

Allie-Bubba
And 0h49'14" 42d42'19"

Allie-Kim
Aql 20h0'42" 12d26'28"

Allie's Light
Vir 13h42'16" -11d47'50"

Allie's Star
Lyr 18h21'58" 45d46'8"

Allie's Star
And 23h6'25" 41d56'1"

AllieStar
Ori 5h58'20" 17d41'40"

ALLIEVI
Uma 12h27'9" 61d49'22"

Alligood
Cyg 20h12'46" 36d0'43"

AlliGriff
And 1h56'43" 39d52'15"

Allineare amare un altro
Cyg 20h1'29" 37d44'17"

Alli's shining star
Ari 2h10'19" 25d17'29"

Alli's Way
Cnc 8h14'23" 13d18'30"

Alli's Wish Upon A Star
And 2h15'53" 44d43'29"

Allisa D.C.Sauraan
Vir 13h16'26" -1d40'36"

Allison
Aqr 22h33'39" -2d44'37"

Allison
Aqr 21h45'30" -0d1'34"

Allison
Lib 15h3'54" -17d35'2"

Allison
Cap 21h20'52" -15d12'34"

Allison
Aqr 20h43'55" -10d32'18"

Allison
Uma 9h22'25" 66d32'21"

Allison
Uma 11h29'32" 58d55'44"

Allison
Sco 16h50'51" -26d15'58"

Allison
Sgr 18h41'11" -27d39'20"

Allison
Uma 10h24'34" 41d58'31"

Allison
And 1h43'25" 42d48'19"

Allison
Aur 5h25'59" 33d30'35"

Allison
Crb 16h17'43" 39d14'12"

Allison
And 1h25'52" 47d35'38"

Allison
Lyn 8h2'53" 46d15'37"

Allison
Ari 2h8'24" 21d18'43"

Allison
Ori 5h13'37" 15d1'8"

Allison
Psc 1h8'15" 12d30'9"

Allison
Vir 13h22'16" 1d50'34"

Allison
Cnc 8h54'4" 26d54'46"

Allison
Tau 5h59'20" 28d3'23"

Allison
Cyg 19h31'5" 29d20'50"

Allison
Lyr 19h17'42" 29d4'13"

Allison 1530
And 2h9'10" 41d45'33"

Allison A Brennan Charles J Osborn
Lyn 7h21'46" 50d46'24"

Allison Alane Freeman
Gem 6h28'31" 20d54'22"

Allison "Allie" E. Baker
Aqr 22h17'17" -14d26'8"

Allison and Andy
Tri 1h51'44" 31d16'28"

Allison and Bill's Star
Uma 9h11'8" 56d16'5"

Allison and Henry Foster's Star
Cmi 7h12'33" 2d8'4"

Allison and Jake
Gem 6h35'29" 12d1'26"

Allison and Nate's Star
Cyg 20h49'22" 35d10'25"

Allison Ann Elizabeth Murray Droney
And 23h44'56" 45d59'20"

Allison Anne Armas
Vir 14h25'47" 0d24'14"

Allison Annette Alaimo
And 23h48'7" 37d53'15"

Allison Arbuckle
Cas 1h6'47" 54d11'43"

Allison Artrip
Pup 7h45'53" -14d16'41"

Allison B Affleck
Psc 1h10'5" 18d20'29"

Allison Bates and Michael Stephens
Gem 7h42'35" 21d24'34"

Allison Beauvais
Ori 5h4'53" 4d14'24"

Allison Blair Heaton
Sco 16h10'49" -13d58'47"

Allison Branscombe
Sgr 19h33'37" -37d49'53"

Allison Brown
Mon 7h8'39" -0d43'17"

Allison Bug Reese
Ori 6h3'52" 5d55'47"

Allison Burroughs
And 0h59'59" 37d16'46"

Allison C.
And 2h23'52" 47d7'1"

Allison C. Kirby
Lib 15h27'43" -6d42'41"

Allison C. MautiAmorella
And 23h24'38" 51d45'35"

Allison C. Zack
Psc 1h17'6" 12d27'57"

Allison Cannella
Men 5h21'0" -71d27'18"

Allison Carole Hopkins
Uma 11h20'57" 38d39'52"

Allison Caroline Graiger
Ari 2h43'55" 25d38'17"

Allison Catherine Kalla
Cyg 20h17'26" 39d17'38"

Allison Chansky
And 2h29'44" 48d54'22"

Allison Chappell
Cyg 20h15'19" 34d17'39"

Allison Christie Goldman
Ari 2h17'53" 18d40'15"

Allison Christine Colantuono
Dra 18h43'14" 73d20'42"

Allison Christine James Whitby
Cnc 8h44'2" 31d13'59"

Allison Christine Sperring
Lyn 7h44'13" 47d59'50"

Allison Claire Berends
Uma 11h19'51" 54d18'24"

Allison Clark
Leo 10h37'30" 9d12'43"

Allison Colleen Hartman
Ari 3h9'31" 18d17'41"

Allison Cydney Tysall
Lib 14h39'40" -18d35'51"

Allison Dayle Leong
Ari 2h27'45" 10d31'51"

Allison Dewart
Umi 14h19'26" 69d16'48"

Allison Diaz
Cap 21h29'55" -19d8'37"

Allison Dyche
Lib 15h33'33" -8d13'12"

Allison E. Baker
And 23h45'26" 36d39'44"

Allison Eileen Perry
Gem 6h24'48" 23d6'50"

Allison Elizabeth Adamo
Ori 6h16'5" 15d42'59"

Allison Elizabeth Bartoszek
Leo 9h38'19" 23d17'56"

Allison Elizabeth Kazi
Cam 3h34'36" 53d34'50"

Allison Elizabeth Sheely
Leo 11h2'15" -0d59'13"

Allison Elliot
Lib 15h46'2" -17d21'36"

Allison Emily Franks
Lyn 7h36'33" 47d44'52"

Allison Erica Hopkins
Cas 0h18'36" 51d51'11"

Allison Erin
Cyg 19h37'55" 29d52'3"

Allison Faith Oetinger
And 0h14'44" 44d18'12"

Allison Faith Vance
And 23h48'52" 37d1'13"

Allison Faith, Love, Hope "AM"
And 2h28'45" 44d48'48"

Allison Family
Cru 12h27'27" -60d53'29"

Allison Fannin
And 2h3'3" 41d30'56"

Allison Foelschow
Uma 11h20'4" 29d29'32"

Allison Frangos
And 0h43'11" 40d17'42"

Allison Gaiser
And 1h40'8" 47d1'0"

Allison Gayle Keller
And 0h32'39" 32d17'20"

Allison "G.D." Weinschenk
Psc 0h12'32" -4d40'41"

Allison Godfrey & Mary Corey
Psc 0h56'50" 31d36'30"

Allison Grace
Uma 11h38'27" 30d46'28"

Allison Grace
Cam 3h55'38" 56d29'26"

Allison Grace DePalma
Cas 0h18'6" 62d56'45"

Allison Grace Krilich
Dra 16h6'57" 57d13'29"

Allison & Greg's Love Star
Ori 5h57'2" -0d29'28"

Allison H. Griffith
And 0h49'27" 39d38'17"

Allison Haigney
And 23h11'34" 42d9'39"

Allison Hailey White
Ari 2h42'36" 22d1'9"

Allison Hannahs "Alli's Star"
Lyn 7h34'28" 48d30'28"

Allison Heitzenroeder
Gem 6h59'3" 16d22'56"

Allison Helena Pepper
Leo 10h56'33" 15d34'57"

Allison Helene Hill
Sgr 18h32'26" -24d6'50"

Allison Hendricks
Lep 5h51'13" -11d38'5"

Allison Hillary Gibson
And 23h23'13" 46d59'50"

Allison Hohman
Psc 23h18'6" 7d23'32"

Allison Holt
Gem 7h47'35" 17d19'34"

Allison Hope
Col 5h59'44" -34d48'22"

Allison Huggard
Cas 1h2'32" 49d3'49"

Allison & Hugo
Cyg 21h36'29" 46d43'39"

Allison Jaleen Smith
Tau 5h52'41" 25d20'20"

Allison James Merlino
Tau 4h18'16" 27d1'45"

Allison Janelle Sims
Cmi 7h25'2" -0d2'32"

Allison Jean Lemon
And 0h55'43" 22d4'23"

Allison Jennifer Holz
Psc 0h37'13" 11d6'48"

Allison Jennifer Leonard
Sgr 18h23'50" -29d2'37"

Allison Jill
Lyn 6h23'5" 60d42'13"

Allison Jin Sullivan
Lib 14h53'56" -6d52'10"

Allison Johnson
Leo 10h15'44" 25d56'46"

Allison Joy
Mon 6h48'58" -0d37'31"

Allison Joy Ramos
Sgr 19h43'58" -30d8'54"

Allison Joyce
And 2h36'28" 39d2'33"

Allison K. McDonald
Lyr 19h16'33" 29d36'30"

Allison K. White
And 0h25'55" 30d54'32"

Allison Kate Delmonico
Vir 13h5'4" -3d7'31"

Allison Kate Merkley
Psc 1h46'49" 24d5'2"
Allison Kathryn Leonardo
Leo 11h44'17" 24d39'57"
Allison Kelley
And 23h11'36" 44d5'44"
Allison Keri Lyn Noddin
Lib 15h8'36" -20d38'18"
Allison Kiln
And 2h32'50" 38d4'55"
Allison Kirby
And 0h2'1" 40d59'56"
Allison Kramer
Dra 19h4'59" 69d9'1"
Allison Krispin Waters
Tau 3h42'7" 28d24'29"
Allison L. Fernstrom
And 23h29'16" 46d41'52"
Allison LaDean Nuanes
Cap 21h44'17" -17d12'19"
Allison Lane
Cnc 8h54'0" 10d23'40"
Allison Laurece Babcock
Eri 3h51'51" -0d23'31"
Allison Lauren Byrd
Lac 22h26'21" 48d32'4"
Allison Leah Hail
Ari 3h11'40" 28d34'21"
Allison Lee
Uma 13h36'13" 49d22'43"
Allison Lee Wilson
Leo 9h47'53" 29d23'58"
Allison Leigh
Ori 5h53'54" 18d18'20"
Allison Leigh
Vir 14h7'27" -11d10'56"
Allison Leigh
Lib 15h39'48" -25d5'10"
Allison Leigh Hothersall
Cma 6h51'41" -13d21'50"
Allison Leigh Kuffer
Aqr 22h33'9" 2d30'12"
Allison Lesley Dauer
And 0h48'40" 36d42'51"
Allison Lian
Cnc 8h44'41" 32d28'16"
Allison Lindquist
Leo 9h37'8" 32d39'48"
Allison Livezey
Sco 16h47'50" -29d38'25"
Allison Lizabeth
Tri 2h22'58" 32d2'21"
Allison Lofgren
Cap 21h22'34" -18d9'43"
Allison Lori Hughes
Umi 14h23'47" 75d0'21"
Allison Lorine Melvin
Gem 7h10'49" 33d25'34"
Allison Lussier
Leo 9h27'3" 11d44'48"
Allison Lynn
Lib 14h51'10" -3d47'55"
Allison Lynn
Cap 20h27'41" -18d6'54"
Allison Lynn Austin
Ori 6h8'41" -0d25'38"
Allison Lynn Hawkins
Tau 5h48'17" 25d38'38"
Allison Lynn Tagliere
Mon 6h41'50" -10d7'19"
Allison M. Dwyer
And 1h45'39" 41d53'1"
Allison Mae
Ori 6h16'11" 18d46'37"
Allison Mae Shuler
Cam 4h21'6" 65d54'28"
Allison Mahony
Aql 19h26'50" 4d41'39"
Allison Maria Buttiglieri
And 1h41'41" 44d41'52"
Allison Marie
And 0h44'41" 31d27'18"
Allison Marie
Gem 6h31'22" 18d41'44"
Allison Marie
Cnc 8h34'36" 15d30'54"
Allison Marie
Ari 2h44'4" 26d55'31"
Allison Marie Albrecht
Crb 15h53'6" 36d55'31"
Allison Marie Armour
Aqr 22h57'18" -8d26'20"
Allison Marie Brown
Uma 9h13'54" 58d25'37"
Allison Marie Burianek
Cap 20h25'34" -9d53'0"
Allison Marie Chartowick
And 23h52'26" 47d55'48"
Allison Marie Clay
And 0h36'52" 33d8'13"
Allison Marie Greene
Vir 14h2'16" -16d55'11"
Allison Marie Hirsch
Ari 2h58'18" 20d49'40"
Allison Marie Kroupa
And 0h31'57" 27d13'47"
Allison Marie McCobin
Psc 0h56'38" 28d53'36"
Allison Marie Petrak
Vir 12h16'40" -0d53'18"
Allison Marie Renwick
Tau 5h20'2" 21d20'20"
Allison Marie Scalzi
And 23h30'47" 48d12'22"
Allison Marie Sidlauskas
Cas 0h46'17" 64d29'21"
Allison Marie Williams
Uma 14h7'29" 59d6'20"
Allison Marie Winkler
Vir 12h44'19" 9d9'6"
Allison Marion Faggart
Cap 21h45'18" -9d50'35"
Allison Mary
Com 13h21'58" 25d0'15"
Allison Mary McGovern
Cap 21h50'17" -13d24'51"
Allison May
Lyn 8h28'48" 56d26'2"
Allison McDaniels
Uma 13h25'31" 56d24'5"
Allison McFadden
Lyn 9h10'48" 40d23'27"
Allison Michelle
Christensen
Col 5h31'23" -33d8'57"
Allison Michelle Dunn "Alli"
Tau 5h11'49" 27d30'24"
Allison Michelle Longo
Tau 3h45'39" 18d53'27"
Allison Moeller
And 23h14'42" 35d21'45"
Allison Nicole
Uma 8h29'26" 67d57'14"
Allison Nicole Garieri
Cas 23h40'14" 52d58'46"
Allison Nicole Noel
Cap 20h17'14" -10d16'39"
Allison Nicole Robinson
Ari 2h16'0" 26d9'6"
Allison of Noble Birth
Gem 7h42'33" 34d1'9"
Allison of Venustas
Sco 16h47'13" -42d16'46"
Allison O'Neill Eddy
Leo 9h45'11" 29d2'4"
Allison Oraios
Ari 2h22'56" 19d30'43"
Allison Ostrander
Aqr 22h33'15" -2d49'25"
Allison Paige Beauchamp
Uma 8h42'51" 54d12'36"
Allison Paige Swyt
Gem 6h42'54" 34d43'34"
Allison Perry
Lyr 19h18'22" 34d15'0"
Allison Phillips
Lib 15h6'33" -10d6'38"
Allison Pudwill
Uma 11h58'5" 33d22'27"
Allison R
Cas 0h30'7" 62d59'16"
Allison R. Ladd
Cas 23h20'5" 59d48'29"
Allison Rae Tollestrup
Cyg 19h53'48" 36d22'54"
Allison Rebeccah
Leo 10h41'30" 14d0'53"
Allison Redondo
Sco 17h1'1" -38d13'7"
Allison Renee
Cap 20h53'9" -15d39'29"
Allison Renee Lasher
Lib 15h21'1" -19d14'9"
Allison Rivera
Vir 14h55'55" 1d24'16"
Allison Rose Casale
Aqr 22h8'44" 1d2'21"
Allison Rose Lovata
Per 2h22'47" 54d40'20"
Allison Ruth Burgess
Lyn 8h59'28" 41d7'26"
Allison Ruth Graffius
Sco 17h49'42" -33d23'29"
Allison Sarah Livingstone
Ori 6h23'35" 16d35'13"
Allison Schramm
Leo 11h32'7" 26d44'59"
Allison Schumaker
Gem 7h9'12" 14d26'14"
Allison Sharon Harney AKA
Lava
Lac 22h6'3" 51d30'40"
Allison Shredl
Crb 16h3'53" 35d16'27"
Allison Smith
Cnc 8h15'13" 15d59'34"
Allison Snow
Ari 3h12'32" 11d54'41"
Allison Stull
And 23h43'41" 43d29'15"
Allison Suzanne Thomas
(Alli)
Cas 23h38'2" 56d23'33"
Allison the Lovebird
And 0h41'56" 25d34'50"
Allison Theresa Dina
Leo 10h19'42" 25d44'6"
Allison Tomiko
Lib 15h20'52" -18d59'11"
Allison "Ur beauty shines
bright"
Dra 20h14'23" 63d59'48"
Allison Valerie
And 23h2'6" 36d22'45"
Allison Vanessa
Ori 6h6'51" 11d10'57"
Allison Wanger(Alli Baby)
Tau 3h56'7" 21d11'44"
Allison Waskiewicz
Cap 20h50'17" -21d47'28"
Allison Watson
Cap 21h27'24" -17d37'23"
Allison Whitney
Ori 5h31'16" 10d18'25"
Allison Wolfe & David
Oaster
Ari 2h29'23" 27d43'26"
Allison x
And 2h31'11" 46d55'43"
Allison, my sunshine
And 22h58'13" 48d29'41"
Allison, My Wish!
And 1h16'46" 47d28'57"
Allison212005
Crb 15h42'9" 27d17'44"
AllisonRose
Tau 5h8'41" 18d12'32"
Allison's Aurora
Leo 10h29'22" 14d44'22"
Allison's Dream
Cyg 20h16'21" 55d2'28"
Allison's Light
Lyr 18h42'55" 34d56'43"
Allison's Little Star
And 2h27'59" 46d30'46"
Allison's Love
Vir 13h24'16" 6d27'11"
Allison's Shining Star
Uma 12h6'13" 50d10'16"
Allisons Star
Ari 1h49'2" 20d40'20"
Allison's Star
Tau 5h54'56" 25d2'3"
Allison's Star
Vir 12h46'21" -0d3'1"
Allison's "Stella" Star
Gem 6h55'8" 26d32'1"
Allisonus Werderus
Cma 7h25'13" -16d31'26"
Allistair
Col 6h5'38" -31d48'36"
Allistar
Leo 9h59'54" 27d54'41"
Allister Adel DeNitto
Lyn 8h13'12" 43d16'24"
Allister Raymond Spitznas
Cnc 8h39'20" 30d51'0"
Allister's Star
Cru 12h26'22" -59d57'13"
Allisyn Louise Collins
Vir 15h7'25" 4d5'8"
Allisyn Redman
Sgr 18h29'26" -15d57'52"
Allix Valentine
Lyn 7h31'4" 47d51'33"
Alliza Mari
Lyn 8h38'38" 39d29'38"
Allma-Tadema
Lyr 19h10'45" 27d2'9"
Allochka Sragets
Cap 20h27'15" -13d16'33"
ALLOFYOUWAYS
Dra 15h37'31" 62d57'55"
Alloria
Aqr 22h15'23" -13d30'57"
ALLSTAR
Her 17h21'14" 34d29'44"
Alltaf
Ori 6h5'13" 20d50'45"
allthestarsain'tcloseenough-
foryou
Uma 8h38'44" 47d11'5"
Allumbaugh's Ascendant
Actuality
Ori 5h22'45" 14d4'3"
Alluring Eyes
Sco 16h50'32" -33d1'49"
Allways
Gem 6h20'44" 18d6'26"
Ally
Cnc 8h46'41" 23d40'59"
ALLY
Cnc 9h8'29" 10d44'34"
Ally
And 2h32'9" 49d15'12"
Ally
Cyg 20h3'37" 33d53'45"
Ally and Jason
Ori 6h11'55" 17d49'19"
Ally Brynn Maize
Cam 7h11'43" 78d47'28"
Ally Campo
Sco 17h34'9" -41d59'52"
Ally Cat
Lib 15h56'5" -13d29'56"
Ally Catherine Estrada
And 0h11'59" 36d57'54"
Ally Christine Canepa
Aqr 21h40'59" -0d9'23"
Ally Girl
Uma 9h36'47" 45d7'8"
Ally Josephine Moore
Gem 7h7'28" 20d54'40"
Ally K
And 1h43'30" 44d41'17"
Ally Mae
Mon 6h49'17" 3d23'2"
Ally Maree - My Beautiful
Star
Sco 17h38'49" -32d36'36"
Ally - My Universe
Pyx 8h28'10" -30d5'43"
Ally Purtschert
Uma 10h51'33" 68d41'45"
Ally Roma
Uma 12h49'15" 54d49'21"
Ally Sterman
Ori 5h53'41" 13d6'1"
Ally Wally
Uma 8h33'28" 61d7'13"
Ally Wilson
And 0h40'7" 24d22'50"
Allyce
Cyg 21h42'29" 45d13'2"
Allyce Cobb
Lyn 8h11'13" 50d36'2"
Allyene Brownsburger
Cas 1h26'36" 62d5'15"
Allyl
Ori 4h50'21" 4d15'11"
Allymac
Lyn 8h3'48" 57d15'8"
Allyn & Alfred French
Lyr 18h43'16" 41d14'52"
Allynn
Lmi 10h40'36" 29d29'58"
Ally's Butterfly
And 23h12'31" 48d26'31"
Ally's Light
And 23h8'12" 39d59'20"
Ally's Light
Lmi 10h28'34" 36d27'46"
Ally's Star
Lyn 6h32'17" 55d22'58"
Ally's World
Ari 2h30'51" 20d13'8"
Allyse Virginia
Cnc 8h34'2" 7d47'4"
Allysha
Cyg 20h52'55" 54d52'43"
Allyson
Lib 15h4'58" -2d41'7"
Allyson
Crb 15h46'26" 26d41'38"
Allyson
Gem 7h0'38" 22d14'1"
Allyson
And 1h15'55" 38d48'8"
Allyson
Lyn 7h39'32" 40d45'41"
Allyson Alexander
Ari 1h56'48" 18d43'28"
Allyson and Joshua
Col 6h4'55" -33d14'48"
Allyson Ann Taft
And 0h32'31" 30d27'31"
Allyson B. Hayden
And 23h53'37" 38d17'18"
Allyson Brooke
Lib 15h40'18" -15d11'9"
Allyson Brown~~I love you
baby!
Lyn 7h56'41" 41d53'28"
Allyson C. Borowitz
Leo 10h15'0" 15d10'12"
Allyson Claire
Cap 20h23'48" -11d8'30"
Allyson Gabrielle Exum
Gem 6h22'9" 17d59'46"
Allyson Grace Brown
Cnc 8h43'32" 12d35'9"
Allyson Hannah Keyes
And 2h4'28" 38d46'50"
Allyson Jill Scribner
Uma 11h22'53" 54d54'30"
Allyson Kaylei Altizer
And 0h49'35" 36d18'19"
Allyson Klabough
Lib 15h41'47" -20d30'40"
Allyson Kline- Best Mother
Ever!!!
Leo 9h48'37" 30d45'56"
Allyson L. Schreiber
Cas 1h48'48" 61d11'50"
Allyson Lee Gianlorenzi
Uma 8h37'28" 50d42'4"
Allyson Leigh Gaines
Sco 16h4'21" -26d8'58"
Allyson Mae Danila
Gem 7h0'25" 25d46'53"
Allyson Marie
Tau 4h52'44" 26d48'25"
Allyson Rachel Collins,
"allygator"
Leo 9h31'41" 15d27'21"
Allyson Rachel Tiriolo
Peg 23h13'47" 33d49'8"
Allyson Rose Green
And 23h37'7" 38d34'35"
Allyson Sherwood
Sco 15h52'44" -22d6'3"
Allysonne
Tau 4h47'42" 19d14'28"
Allyson's Gem in the Sky
Psc 23h19'10" 3d58'48"
Allyssa Craig
Peg 22h32'8" 9d16'46"
Allyssa Marie Birth
Gem 6h44'40" 33d21'5"
Allyssa Noel Carless
Ori 5h25'43" -1d13'35"
ALM & BJB
Cas 23h37'12" 51d44'50"
Alma
And 23h29'20" 41d52'19"
Alma
Gem 6h26'49" 25d31'46"
Alma
Cap 20h28'54" -9d59'7"
ALMA
Umi 15h3'31" 86d21'8"
Alma Alpha I
Aur 5h43'58" 42d50'28"
Alma and David
Cap 21h6'36" -24d0'54"
Alma and Manuel
Cyg 20h45'37" 37d13'1"
Alma Azurin DeGuzman
Per 3h42'1" 46d25'32"
Alma Baur-Hänseler
And 0h45'23" 41d34'59"
Alma Blease "English
Rose"
Cap 20h55'13" -24d20'36"
Alma Brown Quien Te
Quierre
Crb 15h39'40" 35d42'12"
Alma Carolina Y Jesus D.
And 2h28'36" 42d35'24"
Alma Cruz
Vir 13h15'0" 0d48'23"
Alma D. Marin
Tau 4h28'45" 19d33'13"
Alma Guerrie
Vir 14h40'49" 3d48'32"
Alma Hammond Carney
Cnc 8h22'30" 28d16'37"
Alma Heindryckx
And 2h17'56" 50d34'56"
Alma Jean Hussey
Per 2h40'26" 42d34'17"
Alma & Jonathan Conway
Wed 27Jun87
Pho 0h47'7" -51d27'48"
Alma June Grimes
And 0h40'43" 40d58'10"
Alma Kabbani, "A.L.F."
Mon 6h51'54" -5d47'55"
Alma Lydia Ugartechea
Psc 0h52'51" 25d56'30"
Alma Marie
Leo 9h34'59" 11d10'39"
Alma Melissa Culverwell
Lib 15h12'13" -6d37'37"
Alma Merry Bradley
Tau 4h19'38" 26d32'38"
Alma Muller
Cas 1h18'5" 72d56'49"
Alma Pauline
Pyx 8h34'25" -18d27'23"
Alma Ramirez
Sgr 19h30'2" -17d34'26"
Alma Reyna
Per 3h27'10" 45d34'49"
Alma Sarro Vairo
Uma 13h41'19" 53d57'58"
Alma Toivonen Hill
Tau 5h26'57" 27d48'51"
Alma Ugartechea
Lyn 8h34'9" 55d11'42"
Alma Umile
Cap 20h52'49" -18d13'47"
Alma Yazbel
Cnc 9h21'23" 19d6'22"
Alma424
Ori 5h32'14" -4d2'39"
Almadal Parsons
Oph 16h39'6" -0d35'44"
Almaguer
Uma 9h43'37" 50d16'35"
Almaleticia Johnson
Per 4h15'17" 45d49'18"
Almaluggia
Lyr 19h12'49" 28d56'48"
AlmanachdeGrillon
Uma 9h58'3" 42d4'2"
Almar
Cam 4h23'28" 62d33'17"
Almaretta
Cam 4h1'6" 56d48'18"
Almar's
Leo 10h58'51" 16d18'11"
Almas Gemelas
Tri 1h59'32" 34d32'49"
Almas Gemelas
Umi 15h27'31" 84d56'15"
ALMAVA
Lac 22h55'18" 46d11'18"
Almend
Cyg 20h5'8" 47d29'19"
Almerindo Sarro
Uma 13h21'8" 54d12'21"
Almina Adele Rodriguez
Col 6h26'45" -35d8'23"
Almir
Cyg 21h30'2" 44d3'38"
Almira Evva La Fond
Umi 16h26'48" 71d40'3"
Almiran Marissa
Psc 1h8'28" 22d49'39"
ALMISOMA
Umi 14h5'9" 70d16'1"
Almita Gemelita
Crb 15h55'17" 33d45'54"
Almon Kissell 06/04/22
Mon 7h11'2" -3d50'18"
Almond Beach Club & Spa
Boo 14h32'41" 52d51'26"
Almond Beach Village
Lyn 6h45'46" 58d3'45"
Almonds
Cam 4h20'7" 68d40'7"
Almost
Ori 5h34'45" 3d2'28"
Almost Home
Psc 1h12'42" 27d27'2"
Almudena
Ori 6h7'57" -3d59'57"
Almudena (Mi Gitana)
And 0h25'43" 39d49'30"
Alnair
Sco 17h53'49" -35d45'51"
Alo Bellus
Aql 19h45'19" -0d1'38"
Alocin & Katie Murphy
Cap 20h48'44" -16d48'51"
Alodé Datodzi
Boo 13h52'9" 11d57'42"
Alofa
Vir 13h25'53" 5d22'34"
Aloha
Uma 10h24'50" 41d43'9"
Aloha au ia oe
Eri 4h26'9" -1d25'26"
Aloha Au Ia Oe T&J
Always
Uma 11h39'8" 48d30'28"
Aloha Jenny & Wolfi
Cas 1h3'17" 49d16'14"
Aloha Wau I'a oe
And 2h16'30" 50d23'13"
Aloha wau 'ia 'oe Adam
Cyg 21h31'18" 38d29'11"
Aloha.Ke.Akua-Satoshi &
Kazuko
Aqr 21h14'15" -6d16'1"
Alois Dischmann
Uma 8h40'53" 54d27'40"
Alois "Louie" Putre
Uma 9h7'59" 63d17'16"
Alois Perzl
Uma 8h15'47" 60d21'15"
alois und katrin
Crb 15h33'49" 37d59'29"
Alojzy Dzladek
Uma 12h42'43" 62d57'11"
Alok
Uma 9h22'10" 57d40'8"
Aloka Bill Godege Gent
Dra 19h7'4" 59d23'4"
Aloma Westby Bruce
Umi 14h6'37" 71d37'24"
Alon Torbiner
Uma 11h47'26" 45d4'20"
"Alona Jai" The Most
Precious Star
Umi 15h21'44" 71d42'52"
Alona Truett
Crb 15h35'32" 26d28'52"
Alondra
Ori 6h0'22" 13d25'22"
Alondra
Umi 14h34'51" 73d20'21"
AlongFiancee
Uma 9h57'15" 49d35'36"
Alonna Nicole Burke-
Monds
Cnc 9h7'45" 14d12'58"
Alonso Papá
Ori 6h19'14" 14d22'56"
Alonso Stornant Collar
Uma 12h35'54" 57d24'7"
Alonushko
Leo 10h58'53" 14d56'27"
Alonzo Berton
Aqr 23h50'51" -14d28'20"
Alonzo Campillo
Cnc 8h9'47" 12d41'52"
Alonzo S. Milligan
Sgr 18h8'0" -22d16'29"
Alora
Cyg 21h49'8" 46d47'48"
Alora
Uma 11h12'35" 35d5'0"
Alora Lynn LaVerne
Sanders
Vir 12h41'21" 6d13'4"
alourskejaime
Per 4h32'22" 39d25'53"
Alowisha
Cyg 20h35'39" 53d3'53"
Alowishus
Tau 5h10'16" 21d37'16"
Aloys A. Stepan
Aql 19h29'26" -0d7'30"
Aloys Verheggen
Tau 3h38'57" 16d24'11"
Aloys Verheggen
Per 2h37'1" 53d55'8"
Aloysius
Ari 3h19'23" 15d14'11"
Alp Uygur
Sco 17h26'23" -41d38'18"
Alpay Ekici
Per 2h59'12" 48d37'39"
Alper Dincer
Vir 13h8'35" 10d55'37"
Alpha
Psa 21h39'10" -32d1'43"
Alpha A.J.E. Omega
Tri 1h49'59" 31d31'27"
Alpha and Omega
Sge 20h2'32" 18d26'12"
Alpha Corvus Gregarious
Gem 7h46'43" 34d20'2"
Alpha Cuniculus
Sco 16h48'9" -26d17'32"
Alpha Delta Pi-Gamma Phi
Leo 11h11'2" 4d28'32"
Alpha Grandpa
Ori 5h8'48" 9d0'1"
Alpha Harrus
Crb 15h45'27" 31d55'55"
Alpha Infinitas
Leo 9h43'36" 13d51'23"
Alpha Jessica Conner
Psc 1h16'34" 10d53'48"
Alpha Kristi
Cnc 8h24'35" 12d42'37"
Alpha Maizie
Dra 17h55'51" 51d40'52"
Alpha Mater Tinka
Cas 23h5'2" 55d2'34"
alpha Maxine
Uma 12h36'31" 61d31'19"
Alpha Megan Orionis
Leo 11h40'10" 15d18'35"
Alpha Namonica
Leo 10h35'43" 8d0'11"
Alpha Nonna 1
Cas 0h50'41" 63d0'8"
Alpha Omega Probst
Ori 5h59'12" 18d26'29"
Alpha Omega Robert
Zaleski Sidius-9
Aql 19h40'55" 16d6'26"
Alpha & Omega - Samuel
O. T. Grey
Cen 13h25'18" -45d19'4"
Alpha Pretty
Vir 13h17'49" -8d17'0"
Alpha Rachaela
Uma 9h37'31" 56d44'3"
Alpha Raphael
Per 3h5'11" 52d19'51"
Alpha S R Y Simon Robert
Young
Ori 5h41'13" -1d39'57"
Alpha Salvatore
Uma 9h26'50" 52d12'29"
Alpha Spottaury
Lyn 8h56'5" 46d10'47"
Alpha Stellius
Lmi 10h8'3" 28d21'30"
Alpha Stelor Cassiopeiae
Cas 0h22'44" 54d54'48"
Alpha Stephanie Anne
Lib 15h21'58" -13d30'32"
Alpha Stride-Noble
Cap 20h52'59" -20d15'42"
Alpha Velezari
Her 17h30'50" 32d1'7"
Alpha WET Majoris
Sgr 19h18'33" -16d8'49"
Alpha Xi Delta Zeta Phi
Chapter
Uma 8h26'53" 62d21'33"
Alpha Xi Delta, Omicron
Chapter
Per 3h18'55" 49d6'49"
Alpha-Fleming
Tau 4h25'39" 9d36'31"
ALPHAGIGI
Psc 1h20'39" 30d40'2"
Alphie
Leo 9h25'8" 24d41'38"
Alphonse Halloway Jr.
Per 2h12'21" 56d45'38"
alphonsepalma
Cep 22h28'47" 73d59'20"
Alpina
Lac 22h10'26" 52d51'7"
Alpna
And 1h33'4" 48d29'30"
Alpo
Aql 19h50'11" -0d31'21"
ALR Eternal Love NYB
Ari 3h12'55" 28d46'18"
AlReta
Cnc 9h5'46" 28d23'1"
Alretta
Cap 20h28'55" -10d31'47"
AL'S HEAVENLY LIGHT
Leo 10h13'24" 21d28'38"
AL's Perpetual Light
Ori 6h3'9" 11d55'3"
Al's Star / Al Everdale
Aql 19h24'42" 12d53'23"
ALS1
Sco 17h57'54" -43d2'19"
Alsamoal E. Abu Ali
Ori 5h17'52" 6d58'1"
alse
Her 17h42'46" 33d57'6"
AlSeb
Uma 11h13'57" 38d27'58"
Alsessio Yasin
Cas 23h6'35" 54d48'51"
Alsia Weir
Per 4h36'35" 43d50'25"
Alsie
Lyn 8h21'26" 40d48'21"
Alsie - You're Still The One
Ara 16h55'39" -49d9'57"
Alsista Dawn Smith
Sgr 18h53'58" -15d54'49"
Alskling Markus
Ori 6h9'34" -1d33'47"
AlSta
Peg 22h18'33" 6d38'48"
ALSTAR
Cep 21h23'51" 60d54'57"

Alta
Tri 2h19'54" 35d1'11"
Alta Gracia
Uma 11h44'25" 55d19'21"
Alta Gram Stanton
Gem 7h27'40" 33d28'48"
Alta Mae Fraley
Ari 2h46'38" 26d58'42"
Alta Mae Newman
Uma 11h11'11" 28d52'24"
Alta Marie Kroplin
Psc 1h52'37" 8d19'59"
Alta Sparks Fangio
Cyg 20h39'45" 31d14'18"
Alta Sultemeier
Peg 22h8'41" 7d35'23"
Altaf Fouad Al Ghanim
Sco 16h9'57" -12d11'48"
Altagracia Itzel Muñoz
Leo 10h58'6" 5d11'26"
Altagracia Lopez
Uma 9h27'13" 49d57'3"
Altair's Magick
Ari 2h3'34" 21d54'29"
ALTAN
Cep 22h48'9" 69d40'7"
Altarion
Cam 7h18'44" 68d21'13"
Al'Tariq Yusef Mosley Jr.
Her 17h28'27" 36d24'18"
Altaris
Aql 18h53'19" 8d41'37"
Altarvia
Ori 5h27'28" 4d10'50"
Alteis Olmesartan
Sco 17h51'19" -37d24'17"
Alter Ego
Fanny&Emmanuel4/9/03
Uma 8h28'42" 65d14'47"
Alter Lee Crittenden, Jr.
Uma 10h34'16" 62d43'44"
Altham 1
Uma 11h52'23" 42d18'29"
Althaus, Tim
Uma 10h29'32" 40d5'42"
Althea
Lmi 10h45'42" 32d40'14"
Althea
And 23h8'20" 43d36'51"
Althea
Cnc 8h13'55" 15d44'55"
Althea
Sgr 17h55'13" -29d40'54"
Althea "Babe" Gillies
Gem 7h31'0" 33d36'55"
Althea D. Ward
Crb 15h38'51" 25d51'25"
Althea (Greek for Healer)
And 0h18'10" 27d26'38"
Althea Holt ~ Brianna ~
Patricia
Leo 11h7'12" 9d50'48"
Althea Jae
Cap 21h47'38" -12d47'13"
Althea Nathan
Lyn 8h0'37" 56d49'9"
Althea Nicole
Leo 9h29'50" 27d59'34"
Alton
Peg 0h10'45" 15d26'15"
Alton Arnall Thomasson
Oph 18h24'5" 10d27'14"
Alton Dean Sargeant
Uma 11h26'12" 40d47'45"
Alton Gillespie - Pinhead
No. 260
Scl 0h59'17" -31d13'14"
Alton Joseph Fountain
Leo 10h5'14" 22d39'6"
Alton Lee Overton, Sr.
Ori 4h56'6" 15d4'53"
Al-Triq Calhoun
Leo 10h29'29" 10d56'0"
Altug
Aqr 21h21'49" -5d8'58"
Alui Altum Spiritus Florence
Psc 1h17'25" 16d54'38"
Aluizio Prata Junior
Ori 5h55'2" -0d46'51"
Alun Bowen
Cep 21h18'41" 65d39'24"
Alusia
Ari 2h5'10" 25d7'34"
Alva
Ori 5h17'58" 0d29'19"
Alva
Cap 20h24'6" -13d40'11"
Alva B. Wilson
Vir 13h23'27" -4d45'14"
Alva & Beverly Wright
Uma 9h18'0" 52d17'30"
Alva Edison Philbrook Sr.
Aur 6h29'53" 32d44'48"
Alva Edison Philbrook, Jr.
Aur 6h57'7" 37d57'9"
Alva Holmberg
Gem 6h42'47" 13d51'16"
Alva O'Neill
Cas 1h20'39" 56d41'14"
Alvard Meloyan
Uma 11h28'25" 59d46'16"
Alvarito Garcia
Aqr 21h40'36" 2d15'33"
Alvaro
Cnc 8h5'41" 14d46'32"
Alvaro
Peg 22h44'24" 26d36'6"
Alvaro
Uma 10h40'31" 71d38'25"
Alvaro Escalera
Per 1h47'51" 48d39'15"
Alvaro Kross
Gem 7h17'54" 15d53'21"
Alvaro Sola
Cra 18h42'46" -40d39'18"
Álvaro y Alex
Cep 21h59'37" 59d34'15"
Alvena-George
Tau 5h39'13" 24d41'17"
Alvera Lorraine Walsh
Aqr 22h39'46" 1d38'28"
Alves
Vir 13h3'16" 12d36'41"
Alvid
Boo 14h48'7" 14d50'12"
Alvin
Cnc 8h40'17" 18d33'14"
Alvin
Per 3h12'11" 56d6'32"
Alvin
Col 6h21'41" -33d54'19"
Alvin and Jermel Always
Aqr 22h42'49" -2d20'46"
Alvin D. Dubin
Per 3h10'7" 54d8'12"
Alvin D. Gilbert
Uma 8h49'9" 46d41'43"
Alvin Francis
Sco 16h44'19" -30d6'30"
Alvin H. Kunze
Cnv 12h59'21" 48d31'15"
Alvin & Jacquelin Kollmann
Cyg 19h33'3" 44d23'20"
Alvin Rivers Sr
Gem 6h49'39" 16d54'38"
Alvin Zonis Land II
Vir 12h0'43" -2d56'26"
Alvina Frances Beinoras
Ori 5h45'54" 8d27'53"
Alvina Yeager
Ori 5h29'2" 3d5'51"
Alvino Grijalva Roybal
Lib 15h25'41" -7d38'12"
Alvin's Pledge
Lib 14h44'40" -16d3'26"
Alvon Blair Sr.
Leo 11h3'27" 1d29'33"
Alvon Clay Keary
Cnc 8h35'36" 22d10'46"
Alvyn Baughn Marable
Sco 17h22'36" -31d58'12"
Alward & Forever
Uma 11h47'37" 42d31'1"
Always
Lmi 9h53'41" 41d16'33"
Always
Cyg 20h51'36" 32d15'10"
Always
Cyg 21h40'57" 41d58'35"
Always!
Cyg 20h15'38" 45d58'13"
Always
Boo 14h44'2" 22d17'33"
"Always"
Sge 19h36'45" 16d43'44"
Always
Sge 19h40'9" 18d8'55"
"Always"
Cnc 8h29'3" 24d32'49"
Always
Leo 9h38'3" 29d0'26"
ALWAYS
Leo 9h37'16" 11d50'31"
Always
Ori 6h25'17" 10d34'34"
Always
Sgr 17h50'43" -25d5'58"
'Always'
Cyg 21h52'40" 53d15'43"
Always
Dra 17h59'6" 53d13'47"
Always
Uma 9h20'58" 57d54'50"
Always
Uma 8h46'10" 53d25'49"
Always
Umi 14h16'10" 70d48'58"
Always Amy
Lib 15h6'41" -7d32'44"
Always and Forever
Crv 11h59'14" -16d49'34"
Always and Forever
Umi 14h39'14" 78d39'23"
Always and Forever
Sco 16h39'12" -32d25'47"
Always and Forever
Cnc 9h19'37" 7d39'15"
Always and Forever
Vir 12h13'59" 12d24'28"
Always and Forever
Psc 1h11'52" 19d11'30"
Always and Forever
Ari 3h3'54" 19d9'6"
Always and Forever
Ari 2h42'28" 16d2'0"
Always and Forever
Cnc 8h15'33" 29d4'20"
always and forever
Aql 19h39'10" 15d56'48"
ALWAYS AND FOREVER
Gem 6h21'8" 18d11'38"
Always and Forever
Cyg 19h43'59" 47d1'2"
Always and Forever
Uma 8h58'3" 50d25'21"
Always and Forever
And 23h4'6" 47d55'56"
always and forever
Lac 22h49'33" 49d5'55"
Always and forever
Cyg 20h8'28" 44d0'36"
Always and Forever
Uma 10h16'47" 42d23'50"
Always and Forever
Her 17h30'34" 33d49'4"
Always And Forever!
AA+MB
Psc 0h14'10" 7d6'56"
Always and Forever C & L
Cyg 21h47'31" 44d23'59"
Always and Forever Durves
Cas 1h52'36" 60d20'28"
Always and Forever Joey
Cep 23h10'24" 72d47'34"
Always and Forever Katlyn
Vir 13h16'15" -2d48'18"
Always and Forever x
Cyg 19h56'0" 40d41'4"
Always and Forever, Angie
Uma 10h57'38" 45d6'12"
Always and Forever, Helen
Leo 10h6'16" 7d29'28"
Always and Never
Aqr 21h15'52" -2d14'18"
Always Anna
Ser 15h59'55" 3d59'49"
Always Becky 143
Ari 2h50'56" 12d43'25"
Always Believe - LDI
Uma 8h52'56" 61d20'42"
Always Bub & Bubba
And 0h26'24" 37d47'46"
Always Burning
Lyn 7h51'54" 59d38'17"
Always Close to your Heart
Cyg 20h12'16" 53d55'56"
Always Diane and Andrew
Cyg 19h39'38" 32d3'24"
Always & Forever
Lmi 10h36'59" 32d34'41"
Always & Forever
Uma 10h29'10" 50d37'23"
Always & Forever
Uma 10h39'6" 48d58'39"
Always & Forever
Mon 6h55'41" 0d13'36"
Always & Forever
Gem 7h26'55" 26d43'7"
Always + Forever
Cas 2h32'41" 61d28'0"
~Always & Forever~
Lib 14h53'17" -5d12'36"
Always & Forever
Sgr 18h5'16" -17d4'7"
Always & Forever
Lib 15h34'0" -14d39'54"
Always & Forever My Luvy
Uma 11h36'50" 41d37'24"
ALWAYS & FOREVER 4-
22-06
Cyg 20h38'25" 44d28'30"
Always & Forever Dana's
Star
And 23h23'55" 42d2'30"
Always & Forever my
Dianne
Lib 14h47'33" -5d40'41"
Always & Forever My
Dreamer Michael
Uma 11h16'37" 31d44'32"
Always & Forever, Andrew
& Malissa
Sco 16h13'28" -16d19'38"
Always Forward 10/15/05
Lyr 18h35'47" 41d26'10"
Always Halfway, Briana
Uma 12h38'50" 60d44'37"
Always Happy Glady's Star
Crb 15h49'30" 37d9'46"
Always Herbie
Umi 16h15'7" 72d29'31"
Always Here
Uma 10h2'25" 55d43'27"
Always In Arduis Fidelis
Bruv
Ori 6h15'43" 15d16'30"
Always In Love
Gem 6h51'20" 33d6'57"
Always Janet 143
Sco 16h6'34" -11d51'3"
Always Jilli
Sgr 18h56'22" -19d8'42"
Always Love
Uma 12h4'25" 32d23'58"
Always Loving Brad
Cas 23h53'6" 53d6'49"
Always Loving You
Uma 9h33'44" 68d26'59"
Always Loving You
Aqr 22h1'25" -1d12'28"
Always Matt and Hanna
Cnv 12h52'49" 43d23'18"
Always N Forever
Aqr 21h6'20" -9d9'2"
Always Our Shining Light -
Our Mum
Vel 9h27'31" -50d42'30"
Always reach for the stars
Matthew
Gem 6h54'11" 22d25'37"
Always Rebecca
Cas 23h29'14" 54d58'23"
Always Renee
Cap 21h50'19" -23d34'22"
Always Rockin Robyn
Cru 12h5'58" -60d34'12"
Always Rose Lee
Ori 4h56'27" 15d4'18"
Always Sammy's Richard
Ari 2h52'44" 10d49'25"
Always shining
Leo 10h40'47" 12d30'28"
Always Shining Roger and
Sandy
Aql 19h46'12" 6d18'45"
Always
Something...Samantha
McAdams
Uma 11h22'41" 36d44'32"
Always - Star of Wayne &
Tremayne
Ori 5h24'8" -9d20'19"
Always T.B.
Cnc 8h29'27" 15d11'16"
Always The Star Ruta
Drobenkaite
Cap 20h27'10" -20d22'30"
Always There
Uma 8h34'47" 71d39'48"
Always Together
Vir 12h41'28" 1d4'34"
always together
Ori 6h10'52" 8d10'29"
Always Together in the Sky
Aql 19h8'11" 1d57'42"
Always Watching
Cap 21h31'41" -13d41'37"
Always with us-Our dia-
mond star-ASH
Aqr 22h24'59" 1d31'36"
Always With You
Uma 11h13'16" 39d48'40"
Always with You
Cyg 20h39'10" 52d56'51"
Always With You Always
With Me
Cyg 20h2'57" 49d55'43"
Always Your Special Lady
Pho 1h24'21" -56d6'44"
Always, I Promise
Ori 4h48'56" 11d12'51"
Always143
Lib 14h49'16" -0d42'41"
Always4EverDK
Uma 11h37'54" 48d52'52"
Always-N-Forever GK
Gem 7h8'7" 34d26'50"
AlwaysOurMama
LoveChristian LoveAna
Cas 0h44'5" 57d53'29"
Alwyn H. Culpepper
Cep 22h48'42" 64d2'13"
Aly
Uma 11h57'54" 47d30'8"
Aly 16
Leo 9h39'4" 24d43'31"
Aly 4 Peter Vassil Jr. 333
Sco 16h15'33" -14d26'28"
Alya G. Reid
Aql 19h12'4" 3d18'23"
Alya Kay Sami
And 2h18'58" 47d57'22"
Alya Sami
Uma 9h15'50" 66d3'42"
Alyâa Kamel
Cep 23h20'40" 79d44'53"
Alyah - you light up our
lives
Tau 4h37'12" 30d24'42"
ALYCAY
Sgr 17h50'16" -28d1'45"
Alyce
Cru 12h49'35" -58d20'44"
Alyce
Per 3h11'25" 52d38'17"
Alyce F. Cabello
Psc 23h31'57" 3d9'26"
Alyce Lynnette Urs
Psc 1h25'15" 21d2'42"
Alycia Aquarius Diamond
Ori 5h29'24" 2d7'15"
Alycia Carney
Cyg 21h0'4" 49d10'5"
Alycia Corrine Chapman
Cam 9h11'18" 73d45'22"
Alycia Elise Fox
Aqr 20h46'8" -2d6'3"
Alycia Faith
Gem 7h1'46" 34d55'6"
AlyDeleoWolfBound
Cam 4h25'16" 67d33'16"
Alyim
Cyg 21h23'1" 39d31'57"
Alyonka
Peg 22h27'33" 4d50'47"
Alys Daly Starlight
Starbright
Sco 17h56'24" -30d12'33"
Alys Marie Loewenstein
Cas 23h0'16" 55d47'5"
Alysa Dawne
And 2h16'0" 46d41'41"
Alysa Gregory
And 2h11'59" 41d36'42"
Alysa Jane Morgan
Tau 4h19'46" 10d27'8"
Alysa K
Sco 16h39'50" -27d27'53"
Alysa O'Neill
Vir 12h56'49" 9d38'48"
Alysa Rose Huffine
Ari 2h4'10" 23d54'4"
Alysa, My guiding Light
Leo 10h38'4" 22d21'58"
Alysandra Marie Darin
Aqr 21h43'56" -7d51'38"
Alyse
And 2h10'15" 42d44'37"
Alyse Ann
And 1h19'14" 47d32'17"
Alyse Anna Elizabeth
Aqr 22h5'17" -1d3'25"
Alyse Cronin
Uma 11h28'36" 59d49'57"
Alyse Grace
And 2h27'16" 48d0'2"
Alyse Jordan Timmons
Uma 11h50'56" 52d37'15"
Alyse Mattea
Aqr 21h2'35" -11d30'6"
Alyse Nicole
Cnc 8h50'27" 32d15'59"
Alyse Susanne Graham-
Martinez
Cap 20h32'53" -12d51'25"
Alysen Hoffman Prettiest
Star Ever
And 0h48'51" 39d50'21"
Alysha
Tau 4h8'22" 18d24'6"
Alysha
Lib 15h30'7" -20d55'22"
Alysha Anderson The Sno
Star
Uma 13h53'9" 61d30'44"
Alysha Castonguay, Teen
America '03
Cas 1h34'30" 68d43'29"
Alysha Eden Abrams
Psc 1h24'41" 27d19'16"
Alysha Henry
Psc 1h5'22" 10d11'39"
Alysha Jessica Seroussi
Cnc 8h1'11" 10d9'58"
Alysha Kay
Tau 5h44'5" 20d9'3"
Alysha Lynn Salt
Sco 16h22'21" -31d50'53"
Alysha Mia Angell
Cyg 20h26'6" 58d17'42"
Alysha (My Wifey)
Leo 11h30'9" 7d26'39"
Alysha You Are My
Starlight
Cru 12h17'10" -56d33'20"
Alyshia Lorraine Vincent
Cyg 19h41'20" 34d21'54"
Alyshia Marie Tam
Ari 2h24'4" 26d34'44"
Alyshia Summer Pate
Vir 13h30'23" -9d7'7"
Alysia
Peg 21h41'3" 25d22'34"
Alysia
And 2h11'26" 49d17'25"
Alysia and Joseph Forever
Ori 6h12'1" 20d45'40"
Alysia Christine
Aql 19h58'44" -0d25'5"
Alysia Marie
Gem 7h1'51" 35d4'30"
Alysia May Brown
Per 3h24'10" 39d57'18"
Alysia Monaghan
Aqr 22h49'59" -16d19'4"
Alysia-Mae Asquith
And 23h28'16" 37d49'47"
Alyson
And 23h3'59" 39d59'53"
Alyson
And 2h21'15" 49d10'18"
Alyson
Cas 1h9'37" 54d42'20"
Alyson
Uma 10h36'21" 51d50'3"
Alyson
Per 3h32'41" 44d46'36"
Alyson
Lib 15h42'10" -17d35'24"
Alyson
Dra 16h29'59" 59d34'0"
Alyson
Eri 4h45'4" -25d47'52"
Alyson Bevan
Gem 6h46'40" 35d2'35"
Alyson *Bickie* Nguyen
Ari 3h8'32" 28d57'1"
Alyson Brooke Bartol
Cnc 8h30'17" 20d56'16"
Alyson Buhonick
Lyn 6h56'26" 50d36'5"
Alyson Christine Harris
Sgr 18h37'21" -29d24'45"
Alyson Claire
Ori 5h9'9" 15d25'46"
Alyson e Camila juntos
para sempre
Crb 16h7'32" 26d31'56"
Alyson Elizabeth
Cyg 19h44'46" 28d52'20"
Alyson Iannitti Annum
Argentum
Aqr 22h33'32" -2d41'5"
Alyson Jean Krauss
Cyg 21h34'44" 38d5'22"
Alyson Kay Dieterle
Cap 21h29'29" -23d21'45"
Alyson M. Heelas
Sco 17h29'52" -45d7'36"
Alyson Marie Simovic
Leo 11h19'18" 21d32'14"
Alyson Mary Teeter
Aql 20h7'44" 15d22'16"
Alyson Morehead
Cnc 9h5'25" 17d50'0"
Alyson My Princess
Cru 12h7'49" -63d44'6"
Alyson Rendall Estes
Tau 4h19'47" 23d38'27"
Alyson Rose Bennett
Sgr 19h20'47" -26d55'5"
Alyson Thea Santoro
Uma 10h2'37" 48d28'40"
Alyson Veronique Holmes
And 0h43'36" 37d55'38"
Alyssa
And 0h52'5" 40d44'58"
Alyssa
And 1h30'27" 42d24'30"
Alyssa
And 1h39'34" 40d52'8"
Alyssa
Gem 7h26'27" 34d26'11"
Alyssa
Leo 9h37'34" 30d13'39"
Alyssa
Uma 10h27'2" 51d48'6"
Alyssa
And 23h13'16" 47d54'23"
Alyssa
And 23h15'58" 46d56'2"
Alyssa
Tau 5h56'11" 24d59'24"
Alyssa
Her 16h41'29" 27d56'50"
Alyssa
Leo 11h58'14" 15d26'53"
Alyssa
Ori 6h6'6" 20d56'28"
Alyssa
Vir 13h26'13" 2d22'3"
Alyssa!
Psa 22h45'44" -27d15'37"
Alyssa
Vir 12h44'19" -7d12'0"
Alyssa
Umi 14h48'38" 82d45'18"
Alyssa
Cap 20h24'21" -20d8'33"
Alyssa
Cap 20h31'47" -19d10'15"
Alyssa
Cap 20h53'26" -16d19'14"
Alyssa
Uma 9h54'12" 57d46'48"
Alyssa
Cas 1h28'12" 66d43'51"
Alyssa Abygail
Psc 23h54'23" 5d52'22"
Alyssa Alina Amorose
Sco 16h59'4" -38d46'7"
Alyssa Alvey
Tau 4h58'3" 25d16'14"
Alyssa Amanda Michelle
Miller
Cnc 9h6'56" 10d9'55"
Alyssa and Ben
Gem 6h42'50" 27d32'45"
Alyssa and Nick
Sco 17h53'27" -36d12'51"
Alyssa and Shawn
Johnston
Cas 0h59'17" 56d51'30"
Alyssa Ann
Uma 9h20'16" 49d40'43"
Alyssa Ann
And 23h31'39" 41d41'56"
Alyssa Ann
Aqr 20h46'13" 1d34'18"
Alyssa Anne
Leo 10h38'39" 19d59'38"
Alyssa Apodaca
And 1h38'50" 41d38'34"
Alyssa B.
And 1h38'25" 42d25'32"
Alyssa Bargerhuff
Mus 12h32'35" -67d16'53"
Alyssa Bella June
Psc 1h16'20" 15d10'18"
Alyssa Beth Ferraioli
And 0h50'12" 38d7'27"
Alyssa Beth Stetkevych
And 23h56'3" 41d54'52"
Alyssa Boettinger
Cnc 8h49'8" 14d44'53"
Alyssa Bonnenfant
Cas 23h59'43" 52d33'33"
Alyssa Briana Zielinski
Tau 5h55'26" 25d22'16"
Alyssa Bruno
Dra 19h15'13" 70d46'5"
Alyssa C
Uma 13h21'32" 60d10'16"
Alyssa Carleigh Mazak
And 0h27'48" 40d29'2"
Alyssa Carmel
Cyg 19h37'18" 32d5'51"
Alyssa Catherine Grace
Hartley
Pav 18h18'12" -59d34'55"
Alyssa Chanel Tomasiello
Psc 1h4'20" 21d27'11"
Alyssa Cherlin
Peg 23h52'21" 23d49'27"
Alyssa Ching
Aqr 23h0'50" -6d15'41"
Alyssa Christine
Ari 3h21'18" 28d20'16"
Alyssa Christine Dobbs
Cap 21h27'16" -8d31'32"
Alyssa Christine Youse
Uma 10h57'34" 45d29'30"
Alyssa Claire Caldini
Leo 10h14'55" 25d5'2"
Alyssa Conrad
Psc 1h33'29" 27d54'44"
Alyssa Corin
Cnc 8h10'47" 11d38'30"
Alyssa Corinne
Vir 14h51'14" 5d56'59"
Alyssa Danielle
Lmi 10h1'15" 32d42'45"
Alyssa Dawn
Leo 9h38'14" 32d0'58"
Alyssa Deanna
Ari 3h4'39" 25d5'12"
Alyssa Decker
And 23h33'17" 36d55'56"
Alyssa Delaney Ward
And 1h25'58" 37d26'10"
Alyssa Dever & Michael
McCarren
Ari 2h55'57" 21d20'47"
Alyssa Dru Harrington
And 23h23'6" 42d44'1"
Alyssa Edwards
Ari 2h23'30" 27d10'55"
Alyssa Einstein
And 0h21'46" 35d26'4"
Alyssa Enticknap
Cru 11h58'4" -62d16'6"
Alyssa Enticknap
Cru 11h57'58" -63d19'16"
Alyssa Erin
And 2h22'21" 45d24'10"
Alyssa Faith
Cas 2h18'32" 65d39'2"
Alyssa Faith (Bear) Reyes
Uma 11h14'3" 41d21'30"
Alyssa G. Miller
Aqr 22h22'4" -24d0'13"
Alyssa Georgiadis
Ari 2h12'22" 23d58'31"
Alyssa Giguere
Ori 5h32'31" 5d52'28"
Alyssa Grace
And 2h21'22" 48d51'18"
Alyssa Grace
And 1h57'10" 45d58'46"
Alyssa Harris
And 0h45'15" 36d52'29"
Alyssa Harvey
Psc 1h13'36" 19d20'12"
Alyssa Hayes
And 1h33'44" 46d0'37"
Alyssa Hertzig
Uma 11h5'48" 42d41'22"
Alyssa Howell
Ari 2h52'33" 20d1'45"
ALYSSA - IL MIO BAMBI-
NO TI AMO
Vir 12h56'2" -4d52'21"
Alyssa J.
Psc 0h5'5" 8d31'39"
Alyssa J. DeGrace
Lib 14h42'35" -10d46'32"
Alyssa Jade Walker
And 1h12'13" 46d10'41"
Alyssa Jai Bulcock
Lyr 19h17'41" 28d15'22"
Alyssa Jane Weishoff
Cas 1h20'46" 65d31'5"
Alyssa Jaylene Raquel
Torres
Dra 17h9'6" 60d38'42"
Alyssa Jean Brown
Cnc 8h50'17" 26d53'10"
Alyssa Jean Gray Mom &
Dad love you
Aqr 23h40'8" -22d46'38"
Alyssa Jean Ward
Tau 4h35'16" 17d20'45"
Alyssa Joi
Cnc 8h34'58" 20d21'47"
Alyssa Jones 1995
Lyr 18h40'2" 38d34'39"
Alyssa Jordan Greenway
Uma 11h1'31" 50d47'50"
Alyssa Jordyn Held
Aql 19h2'29" -0d34'51"
Alyssa & Josh
Cyg 19h36'26" 47d46'29"

Alyssa Joy Marasco
Lib 15h45'17" -18d35'3"
Alyssa Kate Saunders
And 0h50'24" 38d12'31"
Alyssa Kathleen Mason
Psc 0h37'54" 10d47'36"
Alyssa Kay Zepeda
And 1h13'36" 34d6'29"
Alyssa Kaye Hardinger
Psc 0h23'48" 0d25'12"
Alyssa Kee
And 23h48'44" 42d19'36"
Alyssa Korynne Engstrom
Sgr 19h10'51" -22d8'34"
Alyssa Kuwana
Uma 12h50'19" 59d32'23"
Alyssa L. Gagne
Aqr 22h14'4" 2d27'35"
Alyssa Lauren & Nicholas Edward
Psc 1h19'8" 31d40'3"
Alyssa Le'Anne Tuttle
Cyg 20h26'22" 31d52'39"
Alyssa Lee Anne Francis
And 1h30'59" 41d29'14"
Alyssa Lee Fellow
Tau 5h8'33" 25d3'54"
Alyssa Lee Hunter
Ori 6h10'46" 15d9'40"
Alyssa Lee Vana
Leo 11h49'44" 26d39'12"
Alyssa Lin Peterson
Cap 20h7'5" -9d0'4"
Alyssa Lou
And 1h33'37" 47d5'27"
Alyssa (Louie) Strocher
Cyg 20h6'36" 56d17'8"
Alyssa Luchette
Equ 21h15'36" 11d1'3"
Alyssa Lucille
And 1h14'55" 46d32'49"
Alyssa Lynn Alfonso
Peg 22h44'33" 26d0'6"
Alyssa Lynn Johnson
Psc 1h3'31" 31d57'1"
Alyssa Lynn Jursich
And 0h34'30" 45d44'0"
Alyssa Lynn Kracht
Vir 13h8'3" 5d39'44"
Alyssa Lynn Pierson
Lyn 7h35'1" 35d42'40"
Alyssa Lynn Villarde
And 0h32'59" 41d36'15"
Alyssa "Lys" Goldberg
And 1h42'41" 48d32'26"
Alyssa M. Miller
And 1h26'52" 50d14'25"
Alyssa M. Morland
Aqr 22h35'33" 1d17'13"
Alyssa M. Weickert 11-20-1993
Lyn 7h37'51" 40d51'0"
Alyssa M. Zamboldi
Gem 6h43'22" 27d17'50"
Alyssa Mae Wiebe
Uma 8h36'6" 61d14'58"
Alyssa Manske
Sco 17h41'29" -40d14'58"
Alyssa Maria
Dra 19h19'1" 68d22'54"
Alyssa Marie
Sgr 18h3'43" -21d33'13"
Alyssa Marie
Sco 17h43'30" -33d33'30"
Alyssa Marie
And 0h9'54" 31d23'47"
Alyssa Marie
Cnc 8h43'43" 15d19'1"
Alyssa Marie
Vir 13h4'9" 5d30'5"
Alyssa Marie Brand
Vir 13h17'52" 5d59'37"
Alyssa Marie Brunetto
Cnc 8h39'30" 32d45'46"
Alyssa Marie Bucholz
And 1h48'59" 42d14'33"
Alyssa Marie Caruthers
Leo 10h11'10" 14d27'5"
Alyssa Marie Cosentino
Sco 16h16'20" -10d22'35"
Alyssa Marie Gallagher
And 0h52'54" 44d35'51"
Alyssa Marie Hardy
Gem 6h40'56" 13d32'10"
Alyssa Marie Livingston
Uma 9h54'42" 57d52'19"
Alyssa Marie McWilliams
Ori 5h24'26" 2d59'25"
Alyssa Marie Nowlin
And 23h22'9" 51d12'45"
Alyssa Marie Stockwell
Lyn 7h26'19" 45d37'44"
Alyssa Marie Walter
Sco 17h28'4" -32d49'42"
Alyssa Mary Bonness
Uma 11h28'41" 61d22'54"
Alyssa Maureen Pepe
Leo 9h32'59" 26d34'40"
Alyssa May Luhr - Princess
Cru 12h44'11" -60d23'39"
Alyssa May Saavedra
Cap 21h43'28" -18d18'58"
Alyssa Mikhail Manning
Cnc 8h47'15" 15d48'43"

Alyssa Miko
Ori 5h36'41" -1d46'34"
Alyssa Nichole
Mon 6h56'11" -0d59'51"
Alyssa Nichole
And 1h16'23" 49d1'4"
Alyssa Nichole Dillon
Per 1h43'25" 51d55'52"
Alyssa Nicole
And 1h35'54" 48d16'16"
Alyssa Nicole Bois
Lyn 8h15'36" 55d52'11"
Alyssa Nicole Bradley
Cas 0h3'9" 58d18'50"
Alyssa Nicole Casalino
Psc 1h5'44" 9d57'58"
Alyssa Nicole Crummey
Gem 6h42'42" 13d44'15"
Alyssa Nicole Garcia
Lyn 8h49'41" 34d36'1"
Alyssa Nicole Giannasca
And 23h27'42" 43d45'15"
Alyssa Nicole Risickella
Crb 16h13'27" 39d24'31"
Alyssa Nicole's Dreamcatcher
Leo 11h11'38" 11d43'46"
Alyssa Nicoletti
Vir 13h11'44" -18d53'0"
Alyssa Nirel Zimmerman
Umi 16h19'33" 75d25'4"
Alyssa Noel
And 22h58'50" 51d26'57"
Alyssa Noelle
And 0h19'5" 37d11'59"
Alyssa Noelle
And 1h18'31" 38d19'31"
Alyssa Noelle Capuano
Per 3h44'29" 43d46'28"
Alyssa Paige McCue
Sgr 19h12'14" -23d9'31"
Alyssa Paige Paris
Leo 11h8'2" 14d29'52"
Alyssa Prokes
Psc 1h22'46" 31d12'38"
Alyssa R. Moore
And 1h31'2" 47d29'29"
Alyssa R. Patton
And 23h50'6" 42d31'49"
Alyssa Rachelle Bisanz
Lyn 9h10'26" 36d58'8"
Alyssa Rae
Ari 1h58'59" 14d7'41"
Alyssa Rae
Cas 1h44'35" 63d31'46"
Alyssa Rae Dobrzynski
Gem 7h54'36" 15d12'42"
Alyssa Randall my Sugar Bear
Sgr 18h58'2" -25d22'32"
Alyssa Ray
Gem 7h18'56" 19d31'31"
Alyssa Raymond
Cam 4h47'18" 58d59'7"
Alyssa Reid Swan
Cnc 8h52'37" 21d49'6"
Alyssa Renae Extraordinaire
Cas 1h21'17" 61d34'49"
Alyssa Rene Aviles
Sgr 19h4'6" -25d9'45"
Alyssa Renee Rush
And 0h23'13" 31d22'43"
Alyssa Rianne Kimball
Peg 23h50'20" 27d45'51"
Alyssa & Richard Meggison
Cyg 19h49'10" 57d42'43"
Alyssa Rollins
Vir 14h1'16" 4d33'30"
Alyssa Rose
Cap 20h40'50" -22d25'16"
Alyssa Rose
Lib 14h30'13" -24d21'14"
Alyssa Rose Adams
Cep 22h25'19" 65d15'9"
Alyssa Rose Bruno - Peanie's Star
Leo 9h36'0" 28d18'34"
Alyssa Rose Burnett
Tau 5h43'45" 25d56'32"
Alyssa Rose Maioriana
And 0h10'13" 47d6'18"
Alyssa & Santiago
Lyr 18h45'47" 35d50'15"
Alyssa Sapienza Happy Sweet 16
Cnc 9h3'59" 10d35'36"
Alyssa Sharon Angelaccio
Lyn 7h8'21" 53d50'10"
Alyssa Skye Indri
Uma 9h18'44" 53d42'16"
Alyssa Smith
Sco 17h19'49" -33d29'59"
Alyssa St. John Farrell
Gem 7h4'33" 21d33'23"
Alyssa Tao
Leo 10h56'49" 6d22'52"
Alyssa Tavella "Our Shining Star"
And 1h56'50" 47d24'41"
Alyssa Taylor Gore
Leo 10h21'9" 24d26'16"
ALYSSA - The Family Swimming Star
Uma 11h8'39" 53d25'52"

Alyssa Wells Arnold
And 22h58'55" 39d36'6"
Alyssa Wolfensberger (Angel)
And 1h40'15" 44d26'11"
Alyssa Zamora
Gem 6h46'12" 23d38'26"
Alyssa, My Love
Sgr 18h57'32" -29d52'51"
AlyssaJeanM90 a.k.a The BUBBA star!
Ori 5h53'13" 11d43'34"
Alyssa's
Vir 13h14'40" -20d37'54"
Alyssa's Dream
Uma 9h53'25" 59d49'17"
Alyssa's Dream
And 1h3'45" 45d51'9"
Alyssa's Eyes
Cyg 20h32'36" 30d40'7"
Alyssa's Star
Cnc 8h38'50" 31d11'47"
ALYSSA'S STAR
And 23h24'36" 43d10'27"
Alyssa's Star
Ori 6h15'59" 15d8'36"
Alyssa's Star (Luna)
Peg 23h56'38" 22d37'3"
Alyssia
And 0h52'28" 40d26'9"
Alyssia Geraldi
Lyn 8h1'7" 34d17'13"
Alyssia Kaitlin Goodbody Ryan
And 1h39'53" 49d13'11"
Alyssia L
Tau 5h57'31" 25d32'38"
Alyssia Marie
Vir 11h59'17" -9d36'36"
Alyssia Miller
Cap 20h12'3" -18d25'47"
Alysun
And 0h40'20" 31d21'15"
Alyvia Rose Piazza Alton
Sgr 19h43'37" -13d8'1"
Alyxandra Lee Olson
Dra 12h23'16" 72d20'35"
Alyxandra Morgan Soloway
And 1h47'48" 40d46'23"
Alyxandra Stuelcken
Oph 16h52'56" -0d8'36"
Alyxie Marie Angelo
Sco 17h37'29" -44d45'1"
Al-Zaina Shams
Dra 16h36'53" 60d4'45"
Alzina
Lyn 7h15'27" 56d27'2"
AM LL BF4E
Per 2h35'30" 56d3'59"
A.M.A.
Ori 6h11'41" 7d31'19"
Ama
Ori 5h24'44" 13d23'40"
Ama Por Siempre
Cyg 21h12'27" 45d53'38"
"Ama - The Little Angel" - G.A.P. Perera
Ori 5h40'16" -4d39'24"
Amabel Zhang
And 0h25'47" 36d19'46"
Amabilis Angelus
Lib 15h5'54" -3d28'54"
Amachers
Lyr 18h50'12" 36d52'8"
Amada "Maya" Castro
Vul 19h31'3" 23d32'4"
Amada Patricia...siempre contigo..D
Psc 0h56'55" 26d18'48"
AMADACS
Lyn 7h19'20" 44d54'2"
Amadea
Cam 3h47'55" 71d13'11"
Amadeus
Her 16h24'27" 49d44'14"
Amadeus
Peg 23h11'17" 17d57'7"
Amadis Reyes
Peg 21h52'57" 6d59'5"
"Amadora"
Lib 15h12'43" -7d7'33"
Amae Singh
Sgr 18h28'25" -22d39'19"
Amaelia Kate
Cru 12h25'35" -58d3'21"
aMAEzing
Uma 8h50'12" 62d0'2"
Amail
Ori 4h59'53" 13d47'52"
Amaire
Peg 22h29'8" 6d51'10"
Amal
Uma 11h38'50" 61d20'16"
Amal A F Sulaiman
Cnc 8h46'8" 13d54'26"
Amal Da Mama Hassoun
Aqr 22h16'44" -0d7'23"
Amal Hijazi
Cnc 8h43'48" 32d41'47"
Amal Mohamed El Tawansy
Cnc 8h33'59" 8d0'9"
Amal My Hokis
Leo 9h36'29" 21d51'33"

Amal & Nicolas
Her 16h32'9" 29d17'44"
AMALA
Cnc 9h17'31" 10d4'35"
Amali Celeste
Tau 4h3'29" 23d12'48"
Amalia
Del 20h59'39" 14d56'33"
Amália
Cru 12h50'54" -60d52'19"
Amalia
Sco 17h58'41" -42d38'28"
Amalia Aviles
Psc 1h14'45" 23d56'27"
Amalia J. Salais
Cas 23h31'47" 53d31'47"
Amalia Marie
Ari 2h0'13" 12d47'57"
Amalia Nan Goushy
And 23h42'1" 48d26'52"
Amalile Martinez
Psc 1h51'35" 5d20'7"
Aman Khangura
Uma 12h40'17" 57d21'21"
Aman Sala
Sex 10h7'21" 5d1'58"
Amana Nour
Cap 21h10'46" -20d59'28"
Amanada Lynn
Umi 13h44'56" 73d21'43"
amanda
Uma 11h3'8" 69d58'15"
Amanda
Dra 19h5'29" 62d23'1"
"Amanda"
Uma 12h2'28" 65d33'29"
Amanda
Cas 2h32'34" 63d48'11"
Amanda
Cas 2h51'25" 60d15'51"
Amanda
Uma 12h58'40" 57d30'8"
Amanda
Lyn 7h29'19" 53d13'43"
Amanda
Lyn 8h28'41" 56d15'43"
Amanda
Cas 1h34'1" 60d36'54"
Amanda
Cas 1h3'52" 63d34'9"
Amanda
Cap 21h42'22" -18d49'20"
Amanda
Cap 21h56'10" -22d1'57"
Amanda
Vir 13h15'13" -19d34'38"
Amanda
Sco 16h18'6" -12d7'15"
Amanda
Lib 15h32'42" -8d29'46"
Amanda
Cap 21h41'9" -9d42'28"
Amanda
Cap 20h21'35" -12d13'50"
Amanda
Lib 15h53'1" -7d9'16"
Amanda
Vir 14h5'11" -0d7'55"
Amanda
Lib 14h50'20" -0d55'31"
Amanda
Lib 14h51'7" -3d13'22"
Amanda
Mon 6h50'49" -0d40'47"
Amanda
Sco 17h30'57" -42d13'32"
Amanda
Sgr 19h46'58" -44d18'45"
Amanda
Sco 16h44'0" -38d14'55"
Amanda
Sco 16h53'55" -44d17'26"
Amanda
Sgr 18h33'45" -24d3'22"
Amanda
Sco 17h53'5" -35d41'42"
Amanda
Sco 17h54'19" -36d35'48"
Amanda
Ori 5h37'18" 3d35'11"
Amanda
Ari 2h8'0" 11d25'45"
Amanda
Ori 5h8'29" 11d21'10"
Amanda
Ori 6h1'50" 11d4'44"
Amanda
Ori 5h38'34" 12d41'4"
Amanda
Peg 22h29'4" 9d34'16"
Amanda
Psc 23h30'12" 8d3'27"
Amanda
Psc 0h44'23" 16d46'55"
Amanda
Tau 5h47'24" 18d8'24"
Amanda
Tau 5h50'58" 17d23'20"
AMANDA
Tau 4h31'27" 16d58'18"
Amanda
Tau 4h26'58" 20d6'2"
Amanda
Ari 3h23'51" 16d3'21"

Amanda
Ari 3h22'33" 16d22'12"
Amanda
Ari 2h58'16" 21d36'12"
Amanda
Cnc 8h8'17" 7d33'31"
Amanda
Ori 6h21'2" 13d45'4"
Amanda
Leo 10h38'15" 14d4'6"
Amanda
Leo 9h23'44" 10d54'15"
Amanda
Vir 12h42'5" 8d38'28"
Amanda
And 0h34'18" 29d7'7"
Amanda
And 0h33'5" 28d29'52"
Amanda
Ari 2h6'51" 23d13'0"
Amanda
Ari 3h4'59" 25d5'57"
Amanda
Ari 3h6'42" 23d46'30"
Amanda
Del 20h32'43" 19d31'41"
Amanda
Cnc 8h38'43" 15d36'29"
Amanda
Cnc 8h22'14" 19d50'40"
Amanda
Gem 7h42'53" 21d7'57"
Amanda
Gem 7h47'29" 21d18'23"
Amanda
Tau 3h49'15" 28d29'32"
Amanda
Tau 3h39'52" 28d42'1"
Amanda
Tau 3h35'45" 24d41'3"
Amanda
Ari 3h20'40" 29d43'48"
Amanda
Tau 5h53'30" 25d10'24"
Amanda
Tau 5h26'43" 26d22'38"
Amanda
Cnc 8h30'45" 27d57'28"
Amanda
Gem 7h48'14" 24d40'9"
Amanda
And 0h31'57" 33d6'7"
Amanda
Leo 9h41'56" 29d45'42"
Amanda
And 23h20'49" 47d29'46"
Amanda
Cas 23h28'49" 51d42'42"
Amanda
And 23h28'41" 48d48'8"
Amanda
And 23h30'13" 49d3'1"
Amanda
And 23h5'7" 49d32'19"
Amanda
And 23h11'4" 48d11'6"
Amanda
And 23h4'0" 48d26'8"
Amanda
And 23h4'32" 48d1'9"
Amanda
Aur 5h51'21" 53d27'39"
Amanda
And 1h44'30" 45d40'47"
Amanda
Per 2h25'57" 51d37'18"
Amanda
And 23h29'12" 43d38'32"
Amanda
And 0h21'53" 46d19'28"
Amanda
Crb 16h12'3" 39d28'2"
Amanda
Gem 7h17'12" 31d36'59"
Amanda
Gem 7h1'11" 30d9'37"
Amanda
And 0h53'3" 33d47'9"
Amanda
And 0h48'30" 31d37'49"
Amanda
Crb 16h24'25" 34d6'1"
Amanda
Uma 11h31'16" 30d40'35"
Amanda
And 2h37'40" 43d48'39"
Amanda
Lyn 9h13'52" 44d15'39"
Amanda
And 0h51'42" 43d35'21"
Amanda
And 0h44'51" 40d26'42"
Amanda 021689
Aqr 20h41'32" -8d38'42"
Amanda 1
And 0h48'36" 40d9'42"
Amanda 2006
Cnc 8h2'2" 13d20'23"
Amanda 7/26/84
And 1h46'20" 49d19'44"
Amanda 777
Ori 5h36'19" 0d1'7"
Amanda A. Cavanaugh
Cap 20h21'34" -11d54'43"

Amanda Adams
Uma 11h28'50" 58d54'39"
Amanda Aimee Frederick
Pho 0h15'51" -41d52'13"
Amanda aka Beautiful
Cnc 8h53'0" 11d40'41"
Amanda&Alex both angels to someone
Ori 5h19'14" 1d51'2"
Amanda Alexander
Uma 9h26'37" 55d28'47"
Amanda Allen
Cas 23h20'3" 57d22'48"
Amanda Allen
Cnc 8h2'46" 21d52'6"
Amanda Alli
Umi 17h9'47" 75d34'48"
Amanda "Allstar" Converse
Sco 17h35'20" -32d38'0"
Amanda Amabile
Cam 6h8'36" 60d23'22"
Amanda Amanda
And 0h44'37" 43d41'29"
Amanda Anastasia
Cas 1h43'23" 65d20'56"
Amanda and Aimee - BFFL
Lib 15h18'9" -7d26'16"
Amanda and Christina
Cyg 20h43'47" 34d14'8"
Amanda and Chuck
Sco 17h18'49" -40d13'15"
Amanda and Eric's Star
Cnc 8h48'50" 31d49'53"
Amanda and Jessie Willard
Mon 8h1'3" -0d32'49"
Amanda and Joel's Satellite
Cap 20h22'29" -19d35'31"
Amanda and Josh Forever
Gem 7h6'2" 15d35'42"
Amanda and Luke
Cyg 20h23'22" 54d23'40"
Amanda and Ryan
Aqr 23h17'38" -20d52'15"
Amanda and Shaun
And 1h46'55" 38d32'5"
Amanda and Sterling forever!
Lyr 18h52'50" 31d33'50"
Amanda And Zack
Cap 20h23'23" -24d28'22"
Amanda & Andy
Cma 6h33'8" -31d41'19"
Amanda & Angel Michon
Cas 23h44'51" 59d38'22"
Amanda Anjum
Leo 11h43'40" 18d16'25"
Amanda Ann
Gem 6h34'2" 18d22'44"
Amanda Ann
Vir 12h34'12" 11d46'23"
Amanda Ann
Aqr 23h53'15" -10d5'48"
Amanda Anne
And 23h24'21" 49d42'18"
Amanda Anne Gauthier
Gem 6h42'37" 26d28'8"
Amanda Anselmino
And 23h29'57" 49d7'53"
Amanda & Arthur's Untouchable Love
Cyg 19h37'57" 33d48'59"
Amanda Ashley Kelly
Ori 6h9'57" 15d49'59"
Amanda Ashton
Cam 6h52'25" 65d23'7"
Amanda Atkinson
Cas 0h50'40" 65d40'11"
Amanda & Austins Colorado
Ori 5h58'7" 20d42'55"
Amanda Autumn
Aqr 22h22'30" -14d15'28"
Amanda Avene
Leo 11h26'12" -1d29'54"
Amanda Ayn
Crb 16h11'3" 36d48'34"
Amanda B Froneman
Ori 5h52'10" 22d36'27"
Amanda Babson
And 1h5'32" 36d1'36"
Amanda Baker
Lyn 7h44'2" 36d31'27"
Amanda Barbara Pascarella
Gem 7h6'2" 32d33'53"
Amanda Bardak
Cnc 8h12'42" 12d51'36"
Amanda Barfield
Mon 7h18'9" -7d48'30"
Amanda Barkley
Cyg 20h23'58" 34d41'41"
Amanda Barnes
Lyn 7h3'58" 50d19'22"
Amanda Barnes
Her 16h31'36" 25d2'17"
Amanda Bartlebaugh
Cnc 9h2'50" 10d10'13"
Amanda Bastian
Cas 0h2'36" 59d40'7"
Amanda Batson
Cas 2h44'22" 76d1'28"
Amanda Beaudoin
Psc 1h0'20" 15d29'43"

Amanda Beaudry
Cyg 19h25'52" 54d26'31"
Amanda "Beautiful" Culver
And 1h5'31" 35d46'19"
Amanda Beckett
Her 17h46'7" 38d27'58"
Amanda Bedway
Ori 6h5'18" 5d19'3"
Amanda Ben
Cru 12h27'45" -61d8'3"
Amanda Bencivenni
And 23h29'43" 48d36'57"
Amanda Beth
And 1h33'47" 43d28'52"
Amanda Beth
Crb 15h39'32" 27d2'10"
Amanda Beth Burton
Psc 0h34'21" 7d27'29"
Amanda Beth Jordan
Psc 1h23'0" 17d50'6"
Amanda Beth Myers
Cas 0h53'24" 56d56'27"
Amanda Blair Echols
Sco 16h5'43" -21d21'31"
Amanda Blair Waters
Lib 15h41'17" -16d59'4"
Amanda & Blake
Cyg 21h9'44" 31d48'27"
Amanda Bonita
Vul 19h45'55" 25d10'50"
Amanda Bonner
Leo 11h19'36" 16d59'20"
Amanda Botros
Vir 13h11'56" 6d26'31"
Amanda Bouchard
Crb 15h54'46" 28d6'59"
Amanda Bourgeois
Gem 6h32'2" 17d40'46"
Amanda Bourquin
Leo 9h35'56" 31d18'57"
Amanda Boyd
And 1h10'26" 46d16'58"
Amanda Branchaud
Tau 5h57'26" 24d2'3"
Amanda Bre Stevens
Umi 17h4'43" 85d26'9"
Amanda Bree Vuletich
Lyn 7h47'27" 43d56'53"
Amanda Brett
Uma 11h4'53" 71d4'52"
Amanda & Brian Pinto
Crb 15h25'39" 29d36'13"
Amanda Brienzo
Cap 21h45'20" -14d52'54"
Amanda Brooke Adkins
Psc 1h27'1" 27d16'27"
Amanda Brooke Lewis
Aqr 22h41'36" -0d15'44"
Amanda Brooke Louden
Lib 15h14'56" -6d2'47"
Amanda Brown
And 1h50'12" 42d45'25"
Amanda Bryon Blohm
Uma 11h11'23" 33d13'16"
Amanda (bug) Stashenko
Vir 13h26'14" 12d59'12"
Amanda "Bunny" Van Dyke
Lep 5h40'21" -11d23'29"
Amanda "Butterfly Dancer" Parente
And 0h43'21" 44d38'1"
Amanda C Lowe
Lib 15h25'54" -23d52'5"
Amanda C. Walters
Ari 2h23'27" 22d59'57"
Amanda Cain
Crb 16h15'28" 27d26'56"
Amanda Cairo
Cas 23h43'43" 53d55'10"
Amanda Campbell-Forstner
Lyn 6h27'34" 57d32'7"
Amanda Canevari
And 1h0'3" 43d39'23"
Amanda Cardoso
Ari 3h21'9" 28d4'32"
Amanda Carlson
Psc 0h24'51" 9d42'47"
Amanda Carol Jacobson
Leo 9h28'54" 6d57'4"
Amanda Carol Mason
And 1h24'52" 36d6'49"
Amanda Carter
Gem 6h23'16" 18d29'47"
Amanda Casamassa
Cam 3h27'14" 56d21'19"
Amanda Catherine
Crb 16h6'44" 36d36'55"
Amanda Catherine Munn
And 1h44'35" 44d43'11"
Amanda Chapman
Cap 20h35'8" -10d54'57"
Amanda Charity Kisby
Tau 4h0'7" 20d57'21"
Amanda & Charlotte Thomas
Cnc 8h16'6" 9d58'46"
Amanda Charlton's Dancing Star
Pho 23h42'1" -53d32'28"
Amanda Cheri' Dolan-Mandy
Cnc 8h16'5" 15d19'16"
Amanda "Chicken Little"
And 0h3'21" 45d30'58"

Amanda Chircop
Col 5h29'1" -31d15'47"
Amanda & Chris Gallant
Cyg 20h27'17" 57d43'25"
Amanda Christine
And 23h38'44" 46d53'23"
Amanda Christine Bowen
Cyg 21h57'3" 53d42'47"
Amanda Christine Brown
Leo 10h55'7" 19d31'58"
Amanda Christine Marie
Sco 16h31'17" -31d23'28"
Amanda Christine Marshall
And 0h29'19" 32d44'39"
Amanda Christine Meloy
Col 6h6'6" -38d50'43"
Amanda Christine Pinkham
Psc 1h25'1" 20d17'44"
Amanda Claire
And 23h5'12" 52d1'25"
Amanda Claire Weisman
And 1h47'59" 40d27'8"
Amanda Clare
Cnc 8h39'1" 7d36'35"
Amanda Clare Knox
And 0h7'4" 40d43'8"
Amanda Clark
And 1h14'16" 43d18'56"
Amanda Clark
And 0h56'49" 38d32'22"
Amanda Colleen
Lmi 10h27'5" 34d54'26"
Amanda & Conrad
Ari 3h19'59" 12d21'58"
Amanda Constable
Lyn 6h23'37" 56d7'6"
Amanda Cook
Uma 10h13'7" 47d17'59"
Amanda Cool
Ari 2h24'16" 19d43'53"
Amanda Costopoulos
Cas 23h41'42" 57d14'6"
Amanda Cotton
Leo 10h7'46" 22d2'24"
Amanda Counts
And 23h0'10" 52d43'10"
Amanda Crane
And 23h19'21" 43d58'26"
Amanda Cristina Venta
Psc 1h6'8" 7d22'22"
Amanda Crosby
Cas 0h6'16" 57d26'18"
Amanda Crosby's Star
Lyn 8h8'45" 36d5'29"
Amanda Crystal Czarnecki
Lib 14h53'53" -0d46'40"
Amanda Curry
Tau 4h5'43" 5d15'55"
Amanda Cynthia Beisler
Crb 15h46'22" 27d30'33"
Amanda D. Gartside
Tau 4h41'35" 22d43'10"
Amanda Dalene
Uma 10h45'37" 45d5'23"
Amanda Danelle Coyne
Tau 4h22'30" 20d33'25"
Amanda & Daniel Immortally
Ori 6h11'48" 15d31'42"
Amanda Daniele
And 23h13'52" 43d34'47"
Amanda Danielle
Ori 5h57'54" 9d45'27"
Amanda Danielle Anderson
Leo 9h57'52" 16d15'20"
Amanda Danniell Spruce
Cru 11h56'30" -59d3'13"
Amanda Dawn
Ari 3h11'24" 26d17'6"
Amanda Dawn
Crb 15h43'0" 28d2'42"
Amanda Dawn
And 0h45'47" 34d16'30"
Amanda Dawn Henning Harrel
And 23h16'2" 52d45'51"
Amanda Dawn Tagarelli Summa
Cnc 8h7'28" 8d49'57"
Amanda De Guzman
Sgr 19h7'24" -15d29'35"
Amanda DeAne Crist
And 1h7'37" 42d40'31"
Amanda Deanne Smith
Leo 11h13'23" 24d21'59"
Amanda Dearman
And 23h23'44" 42d6'19"
Amanda DeBrot the blonde cheer girl
Lib 15h25'11" -26d39'45"
Amanda Dee Harvey
Lyr 18h45'27" 43d9'10"
Amanda Dee Ramsey
And 23h1'27" 47d28'11"
Amanda Deglman
Lmi 10h35'54" 36d39'56"
Amanda Demarest
Lmi 10h40'18" 31d58'55"
Amanda Denise Butler
Lyn 8h18'14" 33d56'12"
Amanda Diane Thompson
Ari 2h17'31" 23d46'4"
Amanda Dianne Turner
Ari 3h19'54" 27d8'46"

Amanda Dillard Jones
Uma 11h4'12" 64d5'40"
Amanda Dober
Leo 9h27'42" 26d0'38"
Amanda Dukes
And 0h12'27" 28d49'0"
Amanda Durrant
Cap 21h25'50" -26d26'10"
amanda dz
Ari 2h33'29" 25d17'57"
Amanda E Brennan Dublin-Scioto High
Cnc 8h40'39" 17d54'37"
Amanda E Evans/Sesher
Cap 21h46'14" -22d30'49"
Amanda E. Haynes
Vul 20h14'33" 22d51'6"
Amanda E. Taylor
Lmi 9h28'23" 34d21'58"
Amanda Ear
Cnc 8h32'41" 23d32'26"
Amanda— Echantress
And 1h28'58" 43d47'28"
Amanda Edgerton
Leo 11h9'0" 21d40'34"
Amanda Elise Puente
Leo 11h11'49" 16d40'6"
Amanda Elizabeth
Ari 2h7'33" 22d47'36"
Amanda Elizabeth
And 1h27'12" 49d32'42"
Amanda Elizabeth
Cap 20h58'12" -26d14'54"
Amanda Elizabeth Barber
Cap 20h24'19" -26d57'3"
Amanda Elizabeth Barnard
Lmi 9h37'19" 38d55'26"
Amanda Elizabeth Henderson
And 23h42'55" 36d32'43"
Amanda Elizabeth Jones
Lyn 7h10'4" 60d37'1"
Amanda Elizabeth Roscoe
Cnc 9h16'23" 8d35'12"
Amanda Ellen Waters
Sgr 19h2'19" -32d8'59"
Amanda Ellis
Lib 15h39'42" -29d31'58"
Amanda Elyse Bunker
Cas 2h31'35" 66d7'27"
Amanda Emily Bruce
Vir 14h40'38" 3d22'38"
Amanda Engstron
And 23h41'50" 47d45'49"
Amanda&Eric
Vir 14h21'20" 4d54'36"
Amanda & Eric Crompton
Cyg 19h46'40" 33d44'1"
Amanda Euniece
Leo 11h24'41" 6d51'7"
Amanda Evans
And 2h18'12" 41d48'29"
Amanda Eve Brunty
And 0h30'21" 42d48'14"
Amanda Faith Gallant
Gem 7h32'0" 30d21'39"
Amanda Faith Wright
Lib 15h20'37" -19d12'20"
Amanda Faye
Gem 7h48'37" 33d34'41"
Amanda Faye
And 2h27'55" 46d10'1"
Amanda Faye Gilbert
Aqr 20h51'53" -11d22'57"
Amanda Fermina
Gem 6h25'13" 24d18'11"
Amanda Ferrante
Lib 15h43'57" -12d48'46"
Amanda Fildes
Aqr 22h18'52" -19d52'21"
Amanda Finch
And 0h13'41" 28d39'44"
Amanda Follett
And 23h40'47" 33d2'47"
Amanda Frampus
Aqr 23h13'53" -15d14'38"
Amanda Frances
Aqr 22h44'55" -17d4'57"
Amanda Frances Zuccalmaglio
Eri 4h27'15" -0d56'9"
Amanda Francesca Allen "My Angel"
Ori 5h59'38" 17d40'40"
Amanda Francis
Sco 16h17'36" -10d40'5"
Amanda Frayser
And 2h15'23" 46d40'59"
Amanda Fritz
And 1h9'59" 42d13'3"
Amanda Gabriele
Cas 0h47'12" 52d58'10"
Amanda Gail
And 1h20'44" 43d58'12"
Amanda Gail
Ori 5h17'45" 15d15'18"
Amanda Gail
Lib 15h37'14" -17d20'12"
Amanda Gail
Ori 5h24'8" -8d42'27"
Amanda Gainey
And 1h37'47" 42d27'32"
Amanda Garitano
Gem 7h21'34" 13d52'45"

Amanda Gaston
Cnc 8h20'21" 28d50'49"
Amanda Gates Annema
Lyn 7h7'7" 47d29'6"
Amanda Gayle My Beloved
Lyn 8h13'28" 37d13'1"
Amanda & Geoffreys Wishing Star
Eri 3h50'43" -0d45'33"
Amanda Gill
Com 12h20'12" 26d8'27"
Amanda Glodowski
Peg 23h2'47" 16d16'5"
Amanda Glorius
Uma 9h16'22" 63d59'53"
Amanda Gneiting
And 23h32'36" 41d22'23"
Amanda Goodrich
Leo 9h23'40" 12d17'39"
Amanda Grace
Aql 20h16'48" 8d49'49"
Amanda Grace
Sgr 18h35'54" -24d8'35"
Amanda Grace Cordrey
And 0h55'36" 46d23'14"
Amanda Grace Ferraro
Ari 2h56'18" 22d24'51"
Amanda Grace Hunt Neal
And 0h21'17" 34d50'39"
Amanda Grace Rockenbach
Uma 11h48'30" 39d56'47"
Amanda Grace Wallace (Pepsi)
Lyn 7h57'18" 37d35'44"
Amanda Gries
Ari 2h4'20" 18d50'13"
Amanda Hagemeier
Sco 17h9'42" -43d24'14"
Amanda Hairston
Lyn 9h17'53" 34d44'40"
Amanda Harrington
Uma 8h40'47" 57d40'4"
Amanda Hart
And 1h49'21" 46d28'9"
Amanda Hattie Douglas
Cnc 8h13'38" 24d27'56"
Amanda Hazelip
And 0h16'32" 33d54'51"
Amanda Heath
Uma 10h40'14" 44d46'56"
Amanda Herman
Lib 14h52'58" -3d29'18"
Amanda Herndon "Mandy"
Cas 2h57'8" 64d19'18"
Amanda High
Aqr 22h39'22" -0d1'11"
Amanda Hinojosa
Crb 16h8'27" 35d23'49"
Amanda Hinojosa
Com 12h48'48" 27d31'14"
Amanda Hohensee
Vir 13h21'17" 6d50'24"
Amanda Hollan
Lib 15h1'9" -16d28'53"
Amanda Hollis Petruzzi
Gem 7h12'42" 25d49'59"
Amanda Honaker
Lyn 7h35'8" 41d36'0"
Amanda Hope Rueffer
Cap 21h37'51" -11d31'44"
Amanda Hoppensack
Aqr 23h7'52" -9d49'38"
Amanda Huber
Cap 21h55'2" -18d18'50"
Amanda Hutchison
Tau 4h38'48" 23d49'4"
Amanda Irene
Uma 9h57'56" 59d55'6"
Amanda Irene Christensen
Cnc 8h35'40" 15d0'42"
Amanda Irwin Hill
Cnc 8h22'30" 28d47'54"
Amanda J. Dimling
And 0h28'8" 36d45'17"
Amanda Jace Nace
Cyg 20h27'48" 34d30'40"
Amanda Jade Munro
And 0h10'31" 39d41'43"
Amanda & Jaden
Cyg 20h18'38" 42d18'45"
Amanda & James Bynion
Cyg 19h34'22" 28d5'10"
Amanda Jane
Cnc 9h4'8" 12d15'44"
Amanda Jane
Uma 10h58'29" 38d9'6"
Amanda Jane
Sgr 19h19'39" -21d48'32"
Amanda Jane
Sgr 18h11'0" -21d48'32"
Amanda Jane
Eri 4h36'50" -20d29'12"
Amanda Jane
Mon 6h44'50" -0d2'34"
Amanda Jane Fleeman
And 0h10'7" 44d24'14"
Amanda Jane Grek
Aqr 23h33'27" -8d4'32"
Amanda Jane Guenther
Leo 11h25'15" 8d52'31"
Amanda Jane Kooker
And 0h48'52" 39d56'49"
Amanda Jane Newell
Ari 2h59'17" 11d22'46"

Amanda Jane "Savvy"
Her 17h28'26" 15d40'33"
Amanda Jane Smith
Cas 0h12'19" 53d32'19"
Amanda Jane Woollett
Cyg 20h54'30" 30d40'40"
Amanda Jane Wray
And 23h23'16" 41d50'26"
Amanda & Jason Forever
Ara 18h5'12" -53d55'57"
Amanda Jayne McDonald
And 2h27'31" 44d3'32"
Amanda Jayne Tennyson
Cnc 8h14'38" 21d34'30"
Amanda Jean
Lyn 8h48'16" 41d54'0"
Amanda Jean
Psc 1h45'35" 5d52'24"
Amanda Jean
Aqr 22h36'19" -0d16'40"
Amanda Jean Barratt
Leo 11h42'32" 26d13'54"
Amanda Jean Conely
And 0h17'9" 37d33'54"
Amanda Jean Elizabeth Grimmett
Oph 17h27'50" -0d7'42"
Amanda Jean Elsner
Aql 19h23'19" 4d14'53"
Amanda Jean Herman
Sgr 19h43'8" -13d6'44"
Amanda Jean Kielich
Her 17h57'40" 21d22'4"
Amanda Jean Meador
Lyn 7h7'29" 47d36'19"
Amanda Jean Patton
Cap 21h32'5" -13d57'39"
Amanda Jean Taylor
Vir 12h53'44" 10d35'3"
Amanda & Jessica July 17th, 1980
Uma 11h42'55" 58d45'59"
Amanda Jill Hood
Cap 21h1'9" -20d38'45"
Amanda Jill Parker
Sgr 19h9'3" -23d51'1"
Amanda Jillian Cunningham
Ari 3h14'25" 29d29'9"
Amanda Jimenez
And 0h23'4" 40d20'23"
Amanda Jo
Uma 10h5'33" 44d4'19"
Amanda Jo
Cnc 8h19'27" 16d37'43"
Amanda Jo
Sgr 18h21'11" -17d50'44"
Amanda Jo
Vir 11h40'7" -0d36'57"
Amanda Jo Emery
And 0h4'57" 43d59'2"
Amanda Jo Howenstine - "Mandy"
Ori 5h21'31" -0d33'6"
Amanda Jo Lutz
Lib 15h42'29" -17d26'56"
Amanda Jo Newer
Uma 10h1'23" 43d42'54"
Amanda Jo Reeder
Ari 2h20'38" 23d59'52"
Amanda Jo Rice
Cap 20h29'21" -13d3'34"
Amanda Jo Shelow
Sco 17h42'10" -41d18'39"
Amanda Joanne
And 0h32'38" 29d4'12"
Amanda Jones
And 23h33'48" 49d0'40"
Amanda Jonovich
Lyr 18h34'56" 33d40'51"
Amanda Jordan
Crb 16h9'32" 38d17'8"
Amanda Jordan
Gem 6h10'10" 26d6'41"
Amanda Joy
Aqr 21h12'51" -10d7'44"
Amanda Joy
Lyn 8h24'44" 57d7'24"
Amanda Joy Catrillo
Leo 11h51'28" 11d46'1"
Amanda Joy Deibert
Sco 16h8'13" -14d25'56"
Amanda Joy Frankel
Sgr 18h34'13" -22d26'35"
Amanda Joy Haymaker
Lib 16h0'52" -16d20'29"
Amanda Joy Thome
Uma 12h44'0" 56d31'45"
Amanda June
Uma 9h8'47" 68d42'39"
Amanda K. Cohen
Cam 5h29'16" 69d2'52"
Amanda K. Hamill
Cnc 8h2'3" 24d35'50"
Amanda K Houchin
Sco 16h7'29" -9d35'31"
Amanda K. Hutchinson
Del 20h39'27" 19d14'14"
Amanda Kagel
And 23h2'42" 47d34'0"
Amanda Kangas
Gem 7h48'58" 31d33'37"
Amanda Kate
Crb 16h10'33" 36d37'7"

Amanda Kate
Peg 0h11'33" 17d40'35"
Amanda Kate Brown
Lib 15h55'1" -7d3'5"
Amanda Kate Stasonis
And 1h9'27" 34d11'23"
Amanda Kathleen
And 23h46'4" 38d20'13"
Amanda Kathleen Butkovich
Uma 10h51'20" 49d0'31"
Amanda Kathryn
Tau 4h41'43" 7d17'54"
Amanda Kay
Per 4h33'33" 39d9'3"
Amanda Kay
Cyg 19h43'35" 57d2'59"
Amanda Kay Dixon
Psc 0h48'52" 9d52'30"
Amanda Kay Dunbar
Leo 10h32'43" 16d13'44"
Amanda Kay Hedrick
Cnc 9h7'23" 29d33'33"
Amanda Kay Huckleberry
Aql 19h23'35" -0d54'56"
Amanda Kay Reberger
Crb 15h32'34" 29d29'1"
Amanda Kay Smith
Tau 5h30'47" 22d15'24"
Amanda Kelley
Lyn 8h33'55" 34d4'39"
Amanda Kelly Metrick
Cap 21h11'57" -21d2'25"
Amanda Kelly Thompson
Leo 9h54'11" 22d38'12"
Amanda Kelman
Cyg 20h30'40" 44d34'39"
AMANDA KELSEY
Boo 14h44'46" 20d2'25"
Amanda&KendallGarlesky
Lyn 7h12'45" 48d15'11"
Amanda Kianna Litt
Leo 11h12'11" 24d21'0"
Amanda Kountanis
Uma 9h3'24" 69d19'50"
Amanda Kristen Crowner
And 1h4'2" 36d42'31"
Amanda Kristen Kornegay
And 23h14'38" 35d14'9"
Amanda Kristine Alvarez
Uma 10h14'29" 69d10'11"
Amanda Krol
Psc 1h8'22" 7d46'40"
Amanda Kurlan
Lyn 8h32'44" 45d25'29"
Amanda Kuszak
And 2h15'42" 42d18'38"
Amanda Kylie Sayer
Cru 12h40'6" -56d27'21"
Amanda L. Brown
And 0h44'32" 28d5'41"
Amanda L. Garcia
Leo 10h21'29" 26d33'2"
Amanda L. Meli
Tau 5h39'25" 26d45'45"
Amanda L. Price
Lyn 8h2'20" 42d1'23"
Amanda Langerud
Cas 1h17'40" 69d14'13"
Amanda Lanphere
Cyg 20h36'24" 37d29'22"
Amanda Lauren Alves
Crb 15h43'1" 32d51'42"
Amanda Lauren Brzezowski
Vir 12h39'22" -6d36'2"
Amanda Lauren Cowden
Cyg 20h39'59" 55d15'45"
Amanda Lauren Ernst
Lib 15h15'58" -12d32'53"
Amanda Layne Foxwell
And 1h0'45" 40d46'52"
Amanda Le Miller
Uma 9h36'49" 70d38'46"
Amanda Lea Quigley
Com 13h4'42" 18d3'44"
Amanda Leah
And 0h29'49" 33d17'53"
Amanda LeAnn
And 2h21'50" 49d5'0"
Amanda Leavell
Lyn 7h34'31" 38d10'17"
Amanda Lee
Vir 12h34'4" 2d38'38"
Amanda Lee Bardes
Cam 3h58'5" 57d55'50"
Amanda Lee Clark
And 2h32'35" 46d20'46"
Amanda Lee Crowe Vick
Lib 15h47'26" -17d44'52"
Amanda Lee Debons
And 0h16'38" 43d14'10"
Amanda Lee Dougherty
Aqr 22h42'59" 0d36'16"
Amanda Lee Eller
Cnc 8h16'58" 9d59'3"
Amanda Lee Freeman
And 0h49'20" 36d7'11"
Amanda Lee Horne
Lib 15h44'10" -6d14'5"
Amanda Lee Hover
Cnc 8h30'21" 22d46'46"
Amanda Lee Lucchesi
Cam 6h48'3" 62d2'6"

Amanda Lee Luck
Tri 2h11'16" 34d54'3"
Amanda Lee Michell
Leo 10h56'25" 6d43'12"
Amanda Lee Schulenberg
Vir 12h8'52" -1d35'13"
Amanda Lee Stimson
Lmi 10h34'17" 36d58'37"
Amanda Lee Wilkins
Aqr 22h37'9" -15d38'36"
Amanda Lee Willis
Aqr 22h6'0" -3d26'9"
Amanda Leeann Criner
And 1h28'47" 46d16'11"
Amanda Leese
Cyg 19h49'40" 29d47'52"
Amanda Legrand
Cnc 8h34'49" 9d25'53"
Amanda Leigh
Tau 4h37'2" 18d32'39"
Amanda Leigh
And 2h36'33" 40d21'3"
Amanda Leigh
Umi 19h40'45" 86d22'27"
Amanda Leigh Arnett-Romero
Lyn 8h13'8" 35d16'23"
Amanda Leigh Brokate
Per 3h13'36" 40d25'3"
Amanda Leigh Bryan
Umi 14h34'38" 73d19'16"
Amanda Leigh Hunter
Lib 15h43'41" -9d59'3"
Amanda Leigh Johnston
Aqr 20h41'13" -13d19'39"
Amanda Leigh Meiborg
Aqr 22h9'47" 0d18'42"
Amanda Leigh Roberts
Gem 6h59'54" 15d12'53"
Amanda Leigh Roberts
Lib 15h26'29" -23d59'7"
Amanda Leigh Schultz
Gem 6h37'42" 16d5'14"
Amanda Likens
Tau 4h32'58" 19d45'34"
Amanda Liller
Cnc 8h41'21" 24d20'56"
Amanda Lily Snookie Angel Becker
Uma 8h38'25" 50d8'26"
Amanda Linda
Gem 6h48'29" 20d13'14"
Amanda Lipson
Cyg 21h13'21" 31d6'11"
Amanda Long
Uma 13h35'53" 51d50'1"
Amanda Loscinto
Gem 7h30'49" 26d12'57"
Amanda Lostracco
Uma 10h44'37" 47d53'14"
Amanda Lou Montgomery
Ari 2h22'0" 23d57'31"
Amanda Louise
Cas 23h54'58" 50d21'30"
Amanda Louise
And 23h33'14" 42d47'55"
Amanda Louise
Cas 23h11'8" 55d59'49"
Amanda Louise - 21
Aqr 22h55'9" -12d1'53"
Amanda Louise Baldam - Mandy's Star
Cas 0h27'19" 54d30'56"
Amanda Louise Deprez
And 1h13'29" 42d59'27"
Amanda Louise Dupre'
Gem 7h20'21" 16d39'5"
Amanda Louise Galt
Cam 3h48'41" 56d18'37"
Amanda Louise Webb
Ari 2h14'3" 11d36'7"
Amanda Louise-Jean Davis
And 23h16'37" 47d31'38"
Amanda Lovell
Cyg 20h52'11" 44d51'5"
Amanda Loves Barry
Lyr 18h42'26" 30d15'25"
Amanda loves Lyndon forever xoxo
Cru 12h17'3" -59d0'44"
Amanda Luciano
Cap 20h26'36" -12d44'2"
Amanda & Luke's
Ari 3h4'54" 13d55'38"
Amanda Lyn
And 23h44'52" 43d25'28"
Amanda Lyn Evans
Cas 1h39'30" 68d57'35"
Amanda Lynn
Aqr 23h39'47" -13d58'1"
Amanda Lynn
Sgr 18h30'50" -20d10'28"
Amanda Lynn
Lyn 7h49'38" 36d52'1"
Amanda Lynn
Leo 9h55'33" 6d31'9"
Amanda Lynn
Psc 0h57'45" 18d21'34"
Amanda Lynn
And 0h19'0" 26d12'35"
Amanda Lynn
Leo 10h17'29" 24d51'25"
Amanda Lynn Bowman
And 0h34'51" 33d52'2"

Amanda Lynn Brown
Cap 21h31'47" -19d32'53"
Amanda Lynn Burns
Ari 3h15'37" 20d37'6"
Amanda Lynn Dyke
Ori 6h15'42" 10d45'50"
Amanda Lynn Gutierrez
Sco 16h6'11" -13d11'21"
Amanda Lynn Hunt
Cap 20h36'11" -20d43'9"
Amanda Lynn Jarrett
Cas 0h6'44" 57d3'57"
Amanda Lynn Lass
And 23h22'52" 41d46'9"
Amanda Lynn Moore
Per 2h17'21" 51d30'43"
Amanda Lynn Ormsby
Ari 3h17'1" 27d7'23"
Amanda Lynn Perez
And 1h4'40" 38d59'22"
Amanda Lynn Purcell
And 1h8'52" 46d12'2"
Amanda Lynne
And 0h16'34" 38d1'22"
Amanda Lynne
Mon 6h48'52" 6d58'36"
Amanda Lynne David
Cap 21h12'54" -24d38'43"
Amanda Lynne Lyman
And 0h41'0" 43d13'20"
Amanda Lynn's StarDust
And 2h32'18" 47d43'47"
Amanda M.
And 23h4'26" 44d51'7"
Amanda M. Aplet 08/20/85 Lil'Monkey
Leo 9h42'55" 20d34'31"
Amanda M. Ballentine
Dra 17h1'41" 64d31'31"
Amanda M. Frost
And 23h30'43" 46d54'13"
Amanda M. Hill
Lib 15h20'46" -8d5'17"
Amanda M. Magee
Cas 0h40'48" 60d6'5"
Amanda M. (Mandy) Harkins
And 0h51'45" 39d48'9"
Amanda M. Siller
Cas 0h38'57" 58d5'43"
Amanda M Uribe
Tau 5h54'26" 25d23'3"
Amanda MacDowell
Uma 9h33'28" 54d40'44"
Amanda Mae
Sco 17h18'6" -32d42'15"
Amanda Mae Bruno
Sco 16h19'37" -10d41'42"
Amanda Mae Hazel
Lyn 7h1'42" 60d30'35"
Amanda Mae Schweihofer
Cap 21h40'10" -11d48'6"
Amanda MaeJo - Puellas Decoras Meus
And 23h7'47" 43d28'10"
Amanda "Mandapanda" Mesquita
Sco 16h44'7" -44d28'49"
Amanda "Mandy" Lee Zuleta
Uma 11h17'25" 63d51'35"
Amanda Marceaux
Vir 11h44'36" -5d32'11"
Amanda Maria Gibson
Gem 6h52'47" 16d50'2"
Amanda Marie
Ori 6h3'48" 18d4'41"
Amanda Marie
Cnc 8h22'45" 16d31'40"
Amanda Marie
Del 20h33'12" 15d51'28"
Amanda Marie
Cnc 9h11'12" 6d58'16"
Amanda Marie
Vir 12h40'7" 5d15'33"
Amanda Marie
Ori 5h28'17" 13d26'29"
Amanda Marie
Leo 10h35'52" 13d39'45"
Amanda Marie
Vir 13h12'54" 12d56'58"
Amanda Marie
Ori 6h20'30" -1d20'0"
Amanda Marie
Aqr 21h2'48" -6d39'5"
Amanda Marie
Cap 20h41'47" -21d13'52"
Amanda Marie
Lyn 7h5'53" 58d52'24"
Amanda Marie Aguilar
And 0h10'35" 33d0'33"
Amanda Marie Arsenault
Ari 3h4'48" 29d35'19"
Amanda Marie As Sweet As Can Be
Uma 9h41'11" 46d1'23"
Amanda Marie Askwith
Cap 21h49'10" -17d23'46"
Amanda Marie Bentley
And 0h45'15" 41d3'9"
Amanda Marie Boll
Lib 14h24'14" -15d9'33"
Amanda Marie Brown
Vir 14h20'18" 4d11'47"

Amanda Marie Closterman 2006 Vir 14h25'23" -3d20'37"
Amanda Marie Cramer And 0h53'59" 37d35'57"
Amanda Marie DiSilvestro Uma 13h5'35" 53d7'37"
Amanda Marie Dykstra Lib 15h7'55" -4d14'5"
Amanda Marie Garcia Gem 6h54'34" 12d46'58"
Amanda Marie Gileza Leo 9h33'5" 24d0'44"
Amanda Marie Gorham Leo 11h40'42" 21d24'43"
Amanda Marie Gorton Sco 17h27'57" -33d15'39"
Amanda Marie Huff Lyn 7h17'23" 57d26'4"
Amanda Marie Irizarry Gem 6h25'59" 16d34'13"
Amanda Marie Jones And 0h21'47" 28d32'50"
Amanda Marie Julaton Psc 23h53'59" -0d27'22"
Amanda Marie Keefe And 23h11'9" 51d12'15"
Amanda Marie Kidwell Peg 21h40'38" 23d56'36"
Amanda Marie Klein Uma 10h37'47" 71d18'30"
Amanda Marie Kohut Cnc 8h52'6" 28d47'51"
Amanda Marie Liberio And 0h48'45" 45d18'7"
Amanda Marie Logsdon Sgr 20h11'18" -38d49'44"
Amanda Marie Long Uma 11h18'2" 54d16'57"
Amanda Marie Markley Vir 13h19'58" -13d6'33"
Amanda Marie Martin Vir 12h33'7" 3d7'28"
Amanda Marie May Ari 1h59'38" 19d55'51"
Amanda Marie Rodriguez ~ Sweet 16 And 23h16'53" 48d42'46"
Amanda Marie Shiko Cap 21h9'5" -20d52'50"
Amanda Marie Simchick Gem 6h31'19" 25d40'35"
Amanda Marie Smith Her 18h41'46" 20d35'24"
Amanda Marie Sprague Psc 1h40'41" 7d18'20"
Amanda Marie Thiel And 2h5'8" 38d24'33"
Amanda Marie Victor Leo 9h40'20" 29d7'41"
Amanda Marie Zotter And 2h31'23" 44d22'55"
Amanda Mary McNeil Per 4h30'7" 44d16'49"
Amanda Mary Owens Ari 3h9'29" 18d58'12"
Amanda Mary Phillips Cas 23h39'40" 56d12'51"
Amanda Mauree Cas 23h1'14" 59d35'54"
Amanda May Cnc 8h20'43" 26d14'38"
Amanda May and Matthew Philip Cap 21h50'11" -15d49'22"
Amanda May Catherine Powell Uma 11h23'30" 59d58'28"
Amanda May Shipman Vir 13h48'54" -7d54'6"
Amanda Mayer Sgr 19h37'10" -12d15'42"
Amanda Mayne And 1h6'1" 46d43'0"
Amanda Mccord Cnc 8h4'59" 15d45'14"
Amanda McGovern Ari 2h49'25" 29d12'43"
Amanda Medina And 0h11'38" 47d40'5"
Amanda Meholchick Vir 13h54'44" -12d13'15"
Amanda Melancon Lyn 6h29'18" 54d3'55"
Amanda Merrill Aqr 21h51'30" 0d16'13"
Amanda Metell Vir 13h14'55" 12d38'52"
Amanda Meyer And 0h21'11" 29d58'7"
Amanda Meyers Aqr 23h30'52" -13d23'35"
Amanda Mi Amor Para Siempre And 2h18'44" 47d37'19"
amanda: mi puro afphrodita Gem 7h13'23" 17d16'33"
Amanda Micaela Garcia Ari 3h24'7" 7d27'44"
Amanda Michel Parisi Cyg 20h41'41" 30d25'34"
Amanda Michelle Tau 4h37'3" 27d59'40"
Amanda Michelle Gardner Cam 3h36'54" 56d35'19"

Amanda Michelle Hedden Crb 15h50'7" 32d26'59"
Amanda Michelle Hockenbury Lib 14h50'56" -2d11'24"
Amanda Michelle Landis Uma 11h14'17" 29d19'13"
Amanda Michelle McAdams Sco 17h51'16" -35d49'26"
Amanda Michelle Roberts Hor 4h9'33" -41d51'10"
Amanda Michelle Royer Peg 23h26'33" 21d36'46"
Amanda Michelle Shelton Aqr 20h48'39" -10d27'41"
Amanda Michelle Souder Leo 9h22'2" 10d40'29"
Amanda Michelle Underwood Gem 7h32'6" 14d12'44"
Amanda Michelle Wicks Cas 23h1'26" 59d28'15"
Amanda Michelson And 23h37'5" 43d11'47"
Amanda Miller And 1h6'6" 45d57'27"
Amanda Miller And 0h41'9" 25d41'45"
Amanda Miller, mae amorem Sco 17h2'40" -43d57'10"
Amanda Money Leo 10h55'15" 5d32'22"
Amanda Monte And 2h28'15" 46d33'8"
Amanda Moore And 0h13'46" 40d6'32"
Amanda Moore Cas 23h17'15" 59d10'31"
Amanda Morgan Poole And 23h24'56" 41d37'33"
Amanda Morgan Southerly Tau 3h49'33" 7d59'52"
Amanda Morphett And 0h45'41" 45d59'59"
Amanda Morrison And 23h14'26" 35d11'6"
Amanda Morrison Cam 7h23'49" 74d34'30"
Amanda Morse Psc 1h25'14" 32d13'1"
Amanda Mount Uma 10h31'42" 46d19'13"
Amanda Murphy Aqr 22h24'56" -6d53'28"
Amanda - My Babe - Bussey Tau 4h29'39" 20d46'12"
Amanda "My Baby" Lee Uma 8h21'10" 71d17'17"
Amanda My Love Tau 4h18'10" 7d17'4"
Amanda My Love And 1h30'46" 39d55'28"
Amanda My Shining Star Ari 2h19'8" 19d29'52"
Amanda Mychele Rhoades Cas 2h23'23" 62d33'0"
Amanda N Daniels Gem 7h40'45" 32d54'14"
Amanda N Paul Lyr 19h20'8" 28d51'38"
Amanda N. Sadowski Aqr 20h45'53" 0d6'9"
Amanda N Weisner And 1h18'16" 40d24'29"
AMANDA NAIMAN Sco 16h9'56" -11d36'26"
Amanda Naomi Scruggs Uma 8h17'10" 72d56'35"
Amanda Natasia And 1h23'35" 39d5'51"
Amanda & Nathan's Tulip Cru 12h31'0" -60d9'43"
Amanda Newhard Cyg 21h21'52" 51d27'45"
Amanda Nichole Breeden And 1h23'19" 45d3'41"
Amanda Nicole And 2h38'19" 48d56'4"
Amanda Nicole And 23h42'51" 44d12'54"
Amanda Nicole Uma 10h11'6" 52d12'55"
Amanda Nicole Ori 5h40'3" 8d18'38"
Amanda Nicole Cnc 8h47'16" 21d7'26"
Amanda Nicole Cra 19h4'58" -38d10'41"
Amanda Nicole Asher And 0h15'58" 39d9'45"
Amanda Nicole Bogle Psc 0h23'24" 6d31'50"
Amanda Nicole Darvish Lib 15h24'53" -6d24'35"
Amanda Nicole Diggs Lib 15h44'27" -29d44'27"
Amanda Nicole Fritz And 0h46'53" 44d57'5"
Amanda Nicole Jamilla Aqr 22h28'30" -17d59'19"
Amanda Nicole Le Psc 1h0'43" 9d49'46"

Amanda Nicole Maria Leo 10h26'16" 12d22'1"
Amanda Nicole Peters Cas 1h57'9" 63d52'13"
Amanda Nicole Severson Sgr 17h59'16" -28d3'6"
Amanda Nicole Valdez Sgr 18h29'31" -26d50'39"
Amanda Nicole Wyatt Aqr 22h30'56" -0d53'27"
Amanda Nicole Zifchak And 23h7'5" 49d9'59"
Amanda Niu Lib 15h17'38" -26d57'33"
Amanda Noel Blanco, Cara Mia Sgr 19h4'11" -20d0'34"
Amanda Noel Carpenter Leo 10h39'22" 8d3'13"
Amanda Noel Hickson Ori 5h23'58" -4d50'6"
Amanda Noll Lyn 7h42'33" 53d57'6"
Amanda Ogden And 2h14'1" 45d11'1"
Amanda Olfano and son Aql 19h22'7" 14d34'25"
Amanda Olivia Ari 3h25'3" 22d59'21"
Amanda Olivo Uma 11h44'46" 60d54'41"
Amanda Olivo Uma 11h44'33" 62d41'32"
Amanda Orender And 23h23'39" 49d23'40"
Amanda Ortiz Gem 7h48'33" 33d15'1"
Amanda Osterby And 2h17'13" 47d20'2"
AMANDA OSTERBY Apu 14h24'34" -73d58'7"
Amanda Otruba Uma 11h27'9" 38d40'4"
Amanda Paluba Sco 17h50'6" -39d39'23"
Amanda Panda And 1h42'17" 49d8'58"
Amanda *Panda* Parton Umi 13h40'12" 75d0'55"
Amanda Panda Pooh Peg 21h30'48" 16d15'29"
Amanda Panda Shileikis Dancing Star Lyr 18h29'14" 36d12'9"
Amanda Paszk Vir 13h0'26" 6d20'2"
Amanda Patricia Airola And 1h5'12" 44d17'44"
Amanda Pendleton Psc 1h8'44" 26d4'22"
Amanda Perez-Leon Lyn 8h54'6" 46d6'50"
Amanda Perry Uma 11h36'20" 30d28'2"
Amanda Peterson Gem 6h40'36" 20d57'42"
Amanda Pettus And 23h1'7" 39d26'47"
Amanda Pewitt Uma 11h2'2" 62d25'28"
Amanda Pirollo Sex 10h24'49" 1d18'41"
Amanda Pirrone Ori 4h45'59" 12d11'36"
Amanda Platek Uma 9h22'13" 43d28'34"
Amanda Pletcher And 23h18'39" 49d8'58"
Amanda Prisilla Roman Tau 4h29'54" 9d49'36"
Amanda Qualls Cnc 8h29'3" 12d27'7"
Amanda Quinones Cam 4h23'8" 62d45'29"
Amanda Quynn Morrison And 0h38'25" 23d18'1"
Amanda R Cordova Uma 11h11'10" 37d45'59"
Amanda R. Cranmer Vir 13h22'20" 13d28'57"
Amanda R. Heeren Leo 9h55'38" 24d18'43"
Amanda Rae And 0h23'56" 32d45'42"
Amanda Rae Gem 6h57'40" 29d57'27"
Amanda Rae Lyn 8h28'19" 44d8'10"
Amanda Rae Cyg 20h7'24" 35d40'38"
Amanda Rae Gahl Cas 1h21'35" 62d36'17"
Amanda Rae Medlin Cnc 8h31'2" 24d27'47"
Amanda Rae Mudluff Leo 11h37'18" 23d24'22"
Amanda Rae Seltenreich Sco 16h14'48" -18d45'52"
Amanda Rae Sweatman Aqr 22h57'39" -9d31'46"
Amanda Ramo Cap 21h32'38" -13d58'28"
Amanda Rea And 23h24'18" 43d11'41"

Amanda Rebecca Young Oph 16h40'41" -0d47'42"
Amanda "Red" Szuck Cyg 19h39'22" 29d9'3"
Amanda Reilly And 0h16'17" 28d48'4"
Amanda Remsen Lyr 18h48'27" 45d13'18"
Amanda Renae Stine Lyn 7h5'49" 44d22'35"
Amanda Renay Ori 5h54'8" 12d31'50"
Amanda Rene 2007 Leo 10h10'58" 21d18'54"
Amanda Reneé Lib 15h56'9" -18d11'23"
Amanda Renee Ellis Peg 22h31'30" 28d46'48"
Amanda Renee Perna Gem 7h36'7" 21d1'47"
Amanda Renee Pickett And 23h23'5" 38d46'7"
Amanda Renee Walker And 0h46'31" 40d37'15"
Amanda Renee Watson And 1h42'54" 46d14'10"
Amanda Renee Wenzlick And 1h0'8" 35d28'34"
Amanda Rexrode And 2h31'54" 43d59'18"
Amanda Rianne Gellman Com 12h37'15" 18d39'15"
Amanda Richelle Gem 6h49'6" 17d9'16"
Amanda Rieann Bensel And 1h15'38" 36d54'55"
Amanda Rindisbacher Cas 23h37'13" 63d30'49"
Amanda Ritter And 1h49'30" 41d16'28"
Amanda & Robert Filippone Cyg 20h22'5" 54d33'10"
Amanda Robyn Pechousek Aql 19h15'59" 0d31'13"
Amanda Rosa Mazoza Sgr 17h53'11" -29d36'25"
Amanda Rosalie Larner And 23h0'47" 42d25'27"
Amanda Rose Lyn 6h41'20" 51d33'50"
Amanda Rose And 0h42'11" 41d26'28"
Amanda Rose Gem 7h42'51" 33d42'37"
Amanda Rose Ari 3h43'5" 13d51'45"
Amanda Rose Com 13h6'10" 27d58'54"
Amanda Rose Bingham Crb 16h9'57" 38d18'43"
Amanda Rose Carrillo Cas 0h49'2" 58d26'11"
Amanda Rose Connors And 0h34'59" 39d3'30"
Amanda Rose Galanga Umi 15h8'0" 69d29'40"
Amanda Rose Gill Leo 11h37'44" 18d4'18"
Amanda Rose Guile Tau 5h45'0" 23d3'21"
Amanda Rose Guzzo And 1h7'56" 42d19'29"
Amanda Rose Johnson Cas 23h16'36" 59d45'51"
Amanda Rose Kimmel Leo 11h20'24" -1d3'21"
Amanda Rose Lawand Tau 4h47'12" 18d21'21"
Amanda Rose Shores Cap 20h59'55" -26d1'19"
Amanda Rosemary Pultz Crb 15h33'46" 26d21'48"
Amanda Rosenberg Cnc 8h20'17" 9d10'55"
Amanda Russell Uma 9h44'36" 52d22'8"
Amanda Ruyter Psc 1h55'12" 5d51'29"
Amanda Ryan And 1h48'35" 43d46'9"
Amanda Ryan Gem 6h47'34" 16d4'50"
Amanda S. Price And 1h46'18" 41d43'3"
Amanda Sanders Sgr 19h3'49" -13d37'16"
Amanda Scanapico And 1h9'29" 34d10'2"
Amanda Schedler Cnc 9h3'32" 30d13'34"
Amanda Schipper Cyg 19h51'29" 38d23'30"
Amanda Schriock Ari 3h27'13" 26d49'56"
Amanda Scott Ethan Aqr 22h3'3" -22d31'0"
Amanda Seelmann Cas 1h1'56" 54d11'57"
Amanda SF And 23h20'5" 47d52'59"
Amanda Shine Dra 17h14'33" 54d3'12"

Amanda Shirley Elizabeth Chenault Psc 1h25'5" 18d1'34"
Amanda Sis McIntosh Mon 6h52'37" 2d17'20"
Amanda Sketcher Cyg 19h58'32" 47d34'37"
Amanda Smith's Star Cap 21h28'18" -8d32'24"
Amanda Snow Sco 16h16'19" -14d23'7"
Amanda so funnily cute Uma 13h26'8" 55d47'31"
Amanda Sparks Vir 14h24'1" -0d32'26"
Amanda Spoleti Uma 12h26'41" 58d53'21"
Amanda Stapleton God's Angel #4 Gem 6h41'47" 24d34'4"
Amanda Star Leo 10h2'3" 19d24'25"
Amanda Star Evans Leo 10h59'20" 17d44'57"
Amanda Star Smith Tau 5h11'2" 24d11'51"
Amanda Starr Goldasich Battistoni Aqr 21h42'50" -0d23'16"
Amanda Starr Wilkerson And 0h14'6" 43d16'3"
Amanda Stein Ari 3h19'39" 15d45'41"
Amanda Stephenson Tau 5h47'50" 12d36'22"
Amanda Streets Uma 11h30'55" 38d58'7"
Amanda Strmensky Peg 23h10'38" 14d12'54"
Amanda Sue Ferris Cyg 19h57'10" 30d16'55"
Amanda Sue Harvey Aqr 22h7'9" -1d41'52"
Amanda Susan Nolan Ori 5h16'37" 6d15'48"
Amanda Suzanne Richards Cas 1h34'9" 66d38'30"
Amanda Swanson Del 20h33'46" 17d38'4"
Amanda Swindell And 23h46'33" 38d17'52"
Amanda T. Barcelos Uma 8h41'0" 66d33'21"
Amanda Tapiero Uma 12h43'34" 62d35'23"
Amanda Tate And 0h31'18" 45d44'43"
Amanda Teeple Uma 11h15'49" 44d0'41"
Amanda Therese Duggan Gem 6h41'28" 21d0'33"
Amanda Thu Tran Lyn 8h29'10" 44d50'35"
Amanda & Timothy's Star Cyg 19h54'53" 52d2'44"
Amanda "Tinkerbell" Amaral Uma 11h23'18" 57d52'24"
Amanda & Tony's 1 Year Cyg 20h47'18" 43d48'28"
Amanda Torres Aqr 22h17'20" 0d34'0"
Amanda Tugman Cnc 8h20'1" 12d39'31"
Amanda Turley Vir 11h50'52" 3d33'40"
Amanda VanCura Aqr 22h6'30" -0d27'33"
Amanda Vandenburg Vir 13h0'17" -0d8'16"
Amanda & Vasilios Cnc 8h56'50" 15d53'55"
Amanda Victoria Oxford Mon 6h43'45" 7d17'51"
Amanda Waggoner Vir 14h18'5" 4d36'58"
Amanda Walraven And 0h55'47" 41d47'42"
amanda warren Sco 16h55'28" -35d32'37"
Amanda Wendlinger Uma 10h10'35" 45d51'57"
Amanda White Leo 11h23'20" 5d28'12"
Amanda Whitehead And 0h31'57" 27d0'38"
Amanda Wianke Cam 3h44'54" 58d2'54"
Amanda Wiles Lmi 10h17'25" 33d12'21"
Amanda Willis And 1h54'30" 42d25'3"
Amanda Wilson Cas 23h55'37" 53d25'34"
Amanda Woodburn Aqr 22h37'29" -2d6'34"
Amanda Wootton Sco 16h44'37" -25d53'42"
Amanda Zimmerman Psc 1h23'25" 6d31'27"
Amanda Zucker Leo 9h49'9" 13d56'31"
Amanda, 1986-2004 And 0h18'13" 33d0'33"

Amanda, Alex, & Lilian Jarosik Uma 11h47'50" 59d48'42"
Amanda, God will hold us up. Cas 1h15'52" 68d15'35"
Amanda, my angel Gem 6h45'16" 14d55'36"
Amanda619 And 23h17'18" 48d54'33"
Amanda79 Crb 15h45'23" 32d49'32"
Amanda-K Cas 0h17'45" 55d49'6"
AMANDAKERR 1025 Eri 3h47'43" -0d35'14"
Amandalex Sgr 18h12'15" -20d0'58"
Amandalove Dra 19h55'48" 73d6'52"
AmandaLyn941XOXO Sco 17h43'18" -32d38'17"
Amandalynn Leo 11h44'28" 25d4'36"
AmandaNicole Leo 9h29'9" 30d53'40"
Amandarana Sco 16h19'42" -25d34'43"
Amanda-Rose Sco 16h5'25" -29d5'52"
Amanda's Butterfly Mon 7h36'41" -0d34'11"
Amanda's Dream Tau 5h47'50" 12d36'22"
Amanda's eyes Leo 11h58'15" 20d29'20"
Amanda's Faith Uma 10h22'17" 59d22'34"
Amanda's Frog Cap 20h11'24" -9d48'55"
Amanda's Gaze Vir 13h32'35" -18d26'6"
Amanda's Heart Uma 13h33'47" 62d10'52"
Amanda's Immortality And 1h41'29" 49d17'23"
Amanda's Knight Light Cnc 9h13'2" 15d22'59"
Amanda's Light Cap 20h12'43" -17d4'17"
Amandas Love Aqr 21h40'28" -3d11'43"
Amanda's Mongo Psc 0h9'24" 11d53'22"
Amanda's Own Place in the Universe And 1h24'31" 35d23'46"
Amanda's Piece Of Heaven Sgr 18h10'56" -26d36'11"
Amandas' Reflection Lyn 6h45'13" 60d55'26"
Amanda's Shining Star And 0h36'43" 45d7'44"
Amanda's Shining Star And 23h17'48" 47d23'59"
Amanda's Smile Uma 10h45'1" 51d18'42"
Amanda's star And 2h22'9" 46d39'10"
Amanda's Star Psc 1h0'10" 31d50'8"
Amanda's Star Lyn 8h32'34" 43d34'49"
Amanda's Star Psc 0h34'33" 9d11'57"
Amanda's Star Psc 1h7'42" 14d15'35"
Amanda's Star Gem 7h8'5" 26d56'10"
Amanda's Star Ori 5h37'17" -0d12'27"
Amanda's Star Aqr 22h47'48" -21d17'17"
Amanda's Star Vir 14h5'25" -15d9'48"
Amanda's Star Cap 20h17'50" -9d45'19"
Amanda's Star Cap 21h51'51" -10d15'54"
Amanda's Star Sgr 18h36'40" -23d29'27"
Amanda's Star Sgr 18h27'10" -31d59'22"
Amanda's Surprise Vir 11h57'58" 7d28'20"
Amanda's Wishing Star Lyr 19h7'39" 45d9'6"
Amandeep Athwal (Amanie) Vir 12h41'25" 2d53'46"
Amandeep Singh Tau 4h39'25" 23d49'4"
Amandie Gem 7h7'30" 17d1'55"
Amandine Ori 6h18'33" 16d8'7"
Amandine Aqr 22h55'33" -15d22'44"
Amandine Ori 4h51'56" -2d43'24"
Amandine Boussin Cet 2h47'10" 4d15'33"
Amandine Venditti Lib 15h49'29" -12d8'33"

Amandita And 2h27'29" 45d5'12"
Amandos Hodgius Lyn 7h37'48" 56d44'30"
Amandus "Mandy" Fuchs, Jr. Uma 12h35'45" 56d6'25"
Amani Tau 5h19'48" 23d5'41"
Amannufac Leo 11h23'17" 15d39'44"
Amanpreeth Gohal Cnc 8h34'23" 27d56'35"
Amanta Sgr 19h50'39" -15d40'20"
Amantium Irae Amoris Integratio Est Vir 13h39'14" 5d16'47"
Amanya Linda Arnolda Nandiri Eri 4h14'5" -31d55'23"
Amar Ser 15h47'42" -0d3'58"
Amar - prem - Davina Cyg 19h10'10" 50d38'11"
Amara Crb 16h11'8" 33d46'33"
Amara Ari 2h9'43" 16d45'44"
Amara Cnc 9h5'51" 20d8'27"
Amara Deirdre Quinn Col 5h46'32" -35d21'27"
Amara Grace Cnc 8h7'23" 11d54'45"
Amara O'Neill And 2h31'31" 38d51'56"
Amara Sophia Rayno And 1h36'16" 41d32'15"
Amara Willson Cam 4h46'12" 72d8'2"
"Amarante" - Our Happy Place Col 6h3'48" -42d26'11"
Amaranthine Uma 8h34'39" 64d26'29"
AMARANTHINE Leo 11h16'54" -1d47'41"
Amaranthine Gem 6h31'19" 21d25'13"
Amaranthine And 0h21'33" 32d13'36"
Amare per Sempre Uma 10h29'6" 51d42'47"
Amare Pro Infinitio Ori 5h28'11" 14d43'11"
Amare Sempre Megan Peg 22h36'9" 33d22'2"
Amari George Waller-Broom Dra 18h18'21" 72d9'59"
Amari Grace Mcgrath Lib 14h51'24" -2d56'17"
Amari Simone Harris And 1h24'45" 40d38'15"
Amariah Leonor Peterson Cap 21h7'7" -27d14'38"
Amarillo Lmi 10h11'48" 31d43'43"
Amarilys And 2h24'33" 37d26'45"
Amarilys Sgr 18h5'51" -27d11'6"
AmariNati Aqr 23h41'44" -13d28'49"
Amarinder Singh Saini Uma 11h48'16" 60d42'51"
Amaris Ori 5h43'15" -2d35'50"
Amaris & Anthony Cyg 20h55'43" 47d12'57"
Amaris Maria Ramirez Leo 11h18'8" 17d6'14"
Amarla Cir 15h9'42" -57d5'18"
Amarus Uma 9h26'36" 62d24'33"
Amaryah Alyn Seawright Aqr 23h52'15" -11d25'9"
Amaryllis Peg 21h50'17" 8d42'5"
Amaryllis Beach Resort Cam 4h30'30" 74d26'39"
amas star Cnc 9h9'53" 7d16'57"
Ama-Star Ari 2h11'59" 25d32'34"
Amatacotes Ori 5h8'3" 9d18'47"
Amaterasu Dra 14h38'6" 55d49'22"
Amative Per 4h35'22" 39d39'21"
Amato Per 3h8'1" 44d20'46"
Amato P Mayrene Dominic-Romantic Uma 8h13'20" 72d14'26"
Amatorius Aeternus Sgr 17h50'7" -24d39'40"
Amatorius Aeternus Lyn 7h37'37" 49d52'0"
Amatorius Aeternus Ori 6h10'58" 15d54'7"

Amaurilis Vidal
Lyn 9h7'20" 45d55'45"
Amaury - Bouchet
Dra 17h18'39" 57d54'31"
Amaury Lainé
Lib 14h30'53" -22d17'24"
Amaya Isabella
Sco 16h16'20" -29d31'51"
Amaya Jayden
Gem 7h20'15" 28d2'55"
Amaya Lee
And 1h18'59" 49d20'34"
Amaya Lee Donald
Tri 2h37'1" 34d12'47"
Amaya Rae Balentyne
And 1h33'37" 41d22'56"
Amaya Warner
Ari 2h33'10" 11d41'6"
AMayzing Melanie
And 1h5'48" 41d46'18"
Amazed
And 1h12'31" 42d2'46"
Amazed
Uma 11h30'59" 46d24'6"
Amazed
Vir 13h1'6" 9d39'50"
Amazed
Lmi 10h31'47" 29d57'28"
Amazed
Aqr 23h10'14" -9d8'32"
Amazin' Esh
Cyg 20h4'36" 38d50'24"
Amazin' Grace
Vir 12h37'1" -9d22'56"
Amazing
Sco 16h11'56" -14d18'42"
Amazing
Dor 4h56'40" -66d44'37"
Amazing
Lyr 19h9'0" 43d4'25"
Amazing
Cyg 19h46'2" 47d29'37"
Amazing
Tri 2h9'45" 34d27'13"
Amazing
Tau 4h16'52" 26d31'25"
Amazing
Psc 1h27'8" 28d0'58"
Amazing Andreia
Tau 4h2'15" 26d9'21"
Amazing Ashley
And 0h35'18" 34d8'51"
Amazing Aurora
Umi 15h34'0" 78d17'41"
Amazing Dina
Crb 15h49'38" 27d38'14"
Amazing Gammy
Uma 8h56'43" 56d54'33"
Amazing Grace
Sgr 18h7'47" -16d15'30"
Amazing Grace
Sco 16h50'4" -41d58'41"
Amazing Grace
Sco 16h38'10" -31d5'31"
Amazing Grace
Leo 9h54'53" 19d52'47"
Amazing Grace
Tau 5h28'29" 21d59'24"
Amazing Grace and Heart
Lmi 10h24'9" 36d1'26"
Amazing Gracie
And 2h23'38" 46d8'13"
Amazing Hoss
Uma 9h42'31" 71d37'29"
Amazing Husband Andrew
Troeger
Uma 11h38'3" 48d39'17"
Amazing Jake
Umi 16h13'22" 82d49'8"
Amazing K. and M. D.
Ori 5h19'58" 15d8'7"
AMAZING Lauren E.
Spradlin
Cas 0h40'52" 64d5'53"
Amazing Love
Uma 8h45'50" 51d10'7"
Amazing Mae
Cas 23h3'23" 58d53'58"
Amazing Maggie Anderson
Uma 10h58'26" 34d9'6"
Amazing Man
Her 17h16'17" 26d41'58"
Amazing Mom Sue Moore
Peg 21h37'40" 25d10'50"
Amazing P I Love You
XOXOXO
Uma 11h22'46" 35d58'7"
Amazing Polina
Umi 15h9'44" 72d59'52"
Amazing Ray
Uma 9h16'48" 58d47'43"
Amazing Rick
Gem 7h8'2" 33d55'1"
Amazing Sweet Lisa
Vir 12h43'0" 7d21'0"
Amazing Wonderful Reggie
Elliott
Cep 22h0'35" 65d31'53"
AmazingAmy
Mon 6h54'45" -0d13'49"
Amazingly Beautiful Katie
Leo 10h41'29" 17d17'59"
Amazingly Radiant Nicole
(Peapod)
Mon 6h29'40" 11d1'33"
Amazingly Wonderful Titus
Lmi 10h47'4" 27d33'13"
Ambassador Marita
Magpili-Jimenez
Cnc 8h48'52" 28d10'27"
Ambassador Wat T.
Cluverius IV
Per 3h5'29" 51d16'6"
Amber
And 2h25'59" 46d56'52"
Amber
And 1h20'32" 47d38'11"
Amber
And 1h6'20" 45d6'51"
Amber
And 1h39'23" 46d22'46"
Amber
And 23h22'47" 37d30'42"
Amber
And 23h47'13" 47d59'33"
Amber
And 22h59'57" 51d15'45"
Amber
And 23h0'49" 47d29'37"
Amber
Gem 7h27'27" 32d7'38"
Amber
And 2h18'45" 38d42'44"
Amber
And 1h4'5" 42d11'27"
Amber
Cyg 21h34'9" 33d52'34"
Amber
Lyn 7h41'16" 37d43'21"
Amber
Lyn 7h52'14" 42d0'40"
Amber
Lyn 7h33'5" 39d54'6"
Amber
Uma 10h2'44" 41d48'37"
Amber
Ari 3h23'12" 28d56'48"
Amber
Tau 4h6'52" 28d2'17"
Amber
Lyr 19h9'38" 27d11'52"
Amber
Leo 11h18'56" 17d14'29"
Amber
Cnc 8h26'8" 15d20'45"
Amber
Ari 2h49'6" 25d26'14"
Amber
Peg 23h46'8" 18d2'49"
Amber
Cnc 8h55'30" 10d51'1"
Amber
Tau 4h27'33" 16d1'51"
Amber
Vir 14h36'3" 4d9'16"
Amber
Vir 12h48'33" 2d23'11"
Amber
Ori 5h54'43" 12d55'30"
Amber
Aqr 22h36'12" 0d8'43"
Amber
Aqr 22h25'5" -2d0'31"
Amber
Lib 15h23'19" -6d9'37"
Amber
Lib 15h38'6" -19d35'54"
Amber
Lib 14h32'36" -18d21'10"
Amber
Cap 20h9'55" -20d14'30"
Amber
Cma 6h40'48" -20d19'8"
Amber
Cap 21h30'35" -14d2'45"
Amber
Uma 8h45'13" 66d32'35"
Amber
Uma 10h28'35" 66d49'20"
Amber
Uma 12h54'48" 63d1'51"
AMBER
Uma 8h52'45" 57d30'10"
Amber
Lyn 7h29'8" 56d41'42"
Amber
Lyn 7h55'40" 57d39'52"
Amber
Lyn 8h12'57" 54d35'19"
Amber
Uma 12h11'54" 53d34'7"
Amber
Cas 23h43'42" 54d7'29"
Amber
Cas 23h8'51" 59d18'11"
Amber
Cas 0h27'54" 60d2'47"
Amber
Cas 0h32'36" 61d38'58"
Amber
Cap 20h35'17" -23d58'45"
Amber
Sco 16h35'47" -45d31'42"
Amber Acacia
Leo 10h52'32" 17d36'21"
Amber Alexa Gabrielle
Vir 13h3'7" 12d13'36"
Amber Alice
Ori 5h9'55" 15d19'29"
Amber Alice Harvey
And 1h48'41" 46d49'11"
Amber Alicia
Tau 5h33'32" 25d30'10"
Amber "Amburrito" Stewart
Cas 0h12'11" 55d5'54"
Amber and James
Independence Star
Ori 6h19'57" 7d22'3"
Amber and Joseph Kirby
Cyg 21h33'25" 46d57'8"
Amber and Kilei Lacey
Sco 17h22'12" -30d53'0"
Amber and Mike's Magical
Star
Cnc 9h16'50" 10d52'40"
Amber Anderson
Com 13h6'56" 27d58'2"
Amber Ann
And 23h59'34" 40d29'28"
AMBER ANN GROGAN
Aqr 22h0'0" 1d33'12"
Amber Ann Mitchell
Gem 7h21'19" 32d17'31"
Amber Anne D'amore
Cam 6h9'10" 68d48'8"
Amber & Anthony Cavey
Cyg 19h59'39" 47d32'36"
Amber Atkinson
Mon 6h47'24" -3d32'26"
Amber B. Cobb
Sgr 18h49'42" -27d44'45"
Amber Bellamy
Cam 7h20'46" 71d32'36"
Amber Beth Palmer
And 0h42'40" 26d42'10"
Amber Billie
Cas 1h15'52" 54d31'33"
Amber Bray
Psc 1h20'4" 25d36'6"
Amber Brown
Gem 7h39'36" 22d56'50"
Amber Browning
Vir 13h17'10" 5d13'3"
Amber Burris
Cnc 8h29'25" 24d23'54"
Amber Butlers Light
Vir 12h33'42" 6d27'7"
Amber Candice Bass
Sco 16h57'50" -40d19'42"
Amber Caskey
Sco 17h36'59" -41d8'41"
Amber Celeste Cornell
Pho 0h59'54" -56d55'55"
Amber Chiles
Cam 4h14'11" 56d1'19"
Amber Christine
Aqr 21h47'29" -7d6'1"
Amber Clover
Vir 14h17'29" 1d2'51"
Amber Coll
Psc 23h45'51" 6d17'3"
Amber Collette Marum-
Crossan
And 1h55'29" 46d52'26"
Amber D. Johnson
And 1h13'0" 37d18'19"
Amber Danielle Bogart
Psc 0h36'45" 6d26'39"
Amber Darlene Neeves
Sgr 19h49'59" -20d6'13"
Amber & David Reardon
Cyg 20h47'27" 36d46'59"
Amber Dawn
Uma 9h55'8" 50d9'22"
Amber Dawn
Cnc 9h3'12" 14d38'26"
Amber Dawn
Com 12h45'7" 18d24'15"
Amber Dawn
Mon 7h39'29" -0d23'37"
Amber Dawn Bartels
Lyn 7h6'27" 59d16'54"
Amber Dawn Blomme
And 0h29'49" 33d24'28"
Amber Dawn Crabb
Ari 2h11'16" 19d20'48"
Amber Dawn Fox
And 0h42'25" 34d24'26"
Amber Dawn Odom
Uma 10h16'39" 69d3'48"
Amber Dawn Ryzek
Vir 14h11'8" -16d20'58"
Amber Dawn Stark
Lib 15h22'31" -6d40'44"
Amber DeLancey
Vir 12h38'23" -0d25'0"
Amber Dennison-jones
Her 17h0'7" 30d39'43"
Amber DeVondré
Lyn 7h2'2" 51d52'21"
Amber Donoso
And 1h42'42" 39d44'8"
Amber Dubois
Crb 15h20'31" 28d32'26"
Amber E Criss
Gem 6h30'18" 16d35'1"
Amber Eberhardt
Uma 13h38'7" 58d3'12"
Amber Elaine Staley
Psc 0h33'32" 12d10'34"
Amber Elyse Deane
Cap 20h44'25" -18d21'31"
Amber Energy
Lib 15h17'51" -16d24'44"
Amber Erdelyi
Cra 19h2'10" -41d33'40"
Amber Flaim
Tau 5h36'24" 17d7'57"
Amber Forever
Uma 9h26'6" 65d59'8"
Amber & Frank's star forev-
er.
Sco 16h22'38" -26d2'52"
Amber Gail
Mon 6h34'41" 11d32'18"
Amber Garafolo
Uma 9h23'50" 56d18'11"
Amber Gifford
And 23h3'3" 42d30'27"
Amber Golby
And 1h51'16" 37d4'23"
Amber Grace Griffin
And 0h15'24" 35d17'20"
Amber Griffin
And 0h21'59" 27d13'54"
Amber Hale
Lyr 19h12'44" 46d32'0"
Amber Hardy
And 1h5'38" 42d41'13"
Amber Hope
And 0h50'14" 39d47'40"
Amber Ibbetson
And 0h48'4" 45d53'20"
Amber Ivie
Cap 21h29'2" -14d45'30"
Amber Jai Washburn
Sgr 19h50'23" -11d52'16"
Amber Janae Warren
Cam 6h0'5" 61d41'21"
AMBER JAQUIN - BUT-
TERFLY
Tri 2h0'55" 31d32'4"
Amber Jean Kelly
Psc 1h6'2" 25d50'46"
Amber Jean Mielke
Ori 4h55'49" 13d29'29"
Amber Jess
And 0h2'56" 44d0'43"
Amber & Jessie Sutton
Cyg 19h52'44" 52d16'42"
Amber Jo Bulson
And 1h28'59" 49d26'46"
Amber Jones
Cyg 21h15'18" 54d49'14"
Amber Jones
Lib 15h3'0" -7d19'54"
Amber Joy Altman
Vir 12h16'46" 11d14'51"
Amber June 1st mother's
day gift
Cas 1h28'0" 61d20'0"
Amber Jupiter
And 2h32'23" 38d7'28"
Amber Justine Miller
And 22h57'38" 51d34'58"
Amber K. Sulligan
Leo 11h31'12" 25d15'54"
Amber Kate Brown
And 1h14'8" 35d3'16"
Amber Katherine
Uma 11h29'44" 53d35'42"
Amber Kathlene Cox
Cnc 8h16'43" 24d37'56"
Amber Kay Rodzik
Cra 18h46'11" -40d2'9"
Amber Kaye Rickard
And 1h29'11" 44d11'55"
Amber Keeney
Sgr 19h38'52" -16d38'28"
Amber Kessler & Tim
Marcinowski
Uma 10h25'5" 48d54'37"
Amber Ketihi Bowles
Gem 7h6'45" 26d27'1"
Amber Kiesel 1
Cas 1h44'37" 72d21'35"
Amber Kiogima
Lib 15h9'20" -4d20'33"
Amber Kristy Gardner
Psc 1h10'56" 27d24'44"
Amber L. Owens
Sco 17h56'25" -30d54'7"
Amber L. Swartz
Leo 10h31'9" 11d6'21"
Amber Lamberth
And 0h9'37" 45d30'33"
Amber Lanphier
And 23h52'27" 40d18'49"
Amber Lara Eaton
Ari 3h15'11" 27d28'46"
Amber Latham
Tau 3h46'19" 27d43'31"
Amber Lauren
Uma 11h37'59" 37d18'33"
Amber Lauren Harrison
Cnc 8h46'14" 30d28'38"
Amber Layne Nelson
Ari 2h20'0" 22d13'33"
Amber Lea Ruiz (love of
my life)
Uma 13h54'38" 59d20'0"
Amber Leanna Galvan
Cap 21h37'56" -16d10'51"
Amber Lee
Ori 5h36'47" 6d46'39"
Amber Lee
Gem 7h12'3" 26d59'45"
Amber Lee
And 0h21'18" 37d44'5"
Amber Lee Henderson
Gem 7h11'18" 28d25'49"
Amber Leigh
Uma 10h18'59" 61d47'59"
Amber Leigh "0208"
Cap 20h31'56" -26d49'10"
Amber Leigh Jacobs
Cyg 21h54'6" 42d52'45"
Amber Leigh Lockwood
Cas 23h18'40" 55d40'13"
Amber Lenore
And 1h40'4" 48d0'54"
Amber Lenore Fansler
Foster
Sex 10h18'52" -0d25'41"
Amber Leopard
Psc 0h56'58" 32d30'46"
Amber Lescher
Tau 5h1'26" 24d43'45"
Amber Lilly
Tau 5h25'39" 22d30'28"
Amber Lisette Evans
Her 16h52'8" 33d0'4"
Amber Loe
And 1h44'38" 41d35'3"
Amber Louise
Ari 2h19'25" 23d38'45"
Amber Louise Rhoades
And 0h12'6" 46d15'1"
Amber Louise Trepkowski
Uma 9h29'47" 61d50'38"
Amber Luce
Uma 8h44'28" 46d56'52"
Amber Lucille Schlage
Uma 11h18'4" 30d20'49"
Amber Lyn
Cnc 8h36'56" 14d56'4"
Amber Lynn
Tau 5h31'26" 18d32'47"
Amber Lynn
Leo 11h12'21" 0d3'16"
Amber Lynn
Cnc 8h40'23" 7d13'19"
Amber Lynn
Tau 3h59'17" 25d45'59"
Amber Lynn
Lyr 18h36'56" 36d1'53"
Amber Lynn
Psc 1h22'46" 33d30'50"
Amber Lynn
Cap 21h29'59" -23d0'7"
Amber Lynn Abel
Crb 16h2'15" 34d41'53"
Amber Lynn Banks
Cyg 19h39'40" 29d22'7"
Amber Lynn Bradshaw
Cam 4h7'4" 69d27'16"
Amber Lynn DeRosa
And 0h11'12" 25d9'4"
Amber Lynn Foster Is
Loved
Aqr 22h48'55" -8d56'9"
Amber Lynn Gilbert
Cnc 9h4'31" 31d39'45"
Amber Lynn Hall
Leo 11h7'19" 5d25'13"
Amber Lynn Hanes
Sco 16h6'3" -16d0'58"
Amber Lynn Hayes
And 23h42'35" 47d31'39"
Amber Lynn McCormack
Ari 2h50'18" 25d38'35"
Amber Lynn Murphy
Cap 20h18'7" -11d2'21"
Amber Lynn Saltamachia
Lib 14h36'24" -16d45'14"
Amber Lynn Shepherd
Leo 11h51'54" 18d47'8"
Amber Lynn Weaver
Leo 10h44'29" 18d17'33"
Amber Lynne
And 23h12'46" 47d20'16"
Amber Lynnette Wilcox
Uma 10h50'0" 52d30'39"
Amber M Flurry
Cas 23h44'8" 57d45'16"
Amber M Rowell
And 1h41'10" 45d32'8"
Amber Mackay
Cru 11h58'23" -62d16'38"
Amber Marie
Sgr 18h25'21" -32d7'16"
Amber Marie
Aqr 23h28'27" -12d37'41"
Amber Marie
Cap 20h27'16" -10d47'16"
Amber Marie
Lyn 6h51'23" 51d16'36"
Amber Marie
And 0h48'51" 41d23'1"
Amber Marie
Ari 2h3'41" 24d0'28"
Amber Marie
Vul 20h22'52" 28d3'48"
Amber Marie
Tau 4h22'27" 18d3'14"
Amber Marie
Cnc 8h13'34" 8d8'6"
Amber Marie Ballantyne
Uma 10h18'27" 48d38'14"
Amber Marie Beggs
Cap 21h27'26" -19d49'33"
Amber Marie Clark
Sgr 18h10'59" -18d53'33"
Amber Marie Clark
Ari 2h37'18" 25d7'38"
Amber Marie Erikson
Uma 11h22'22" 41d38'6"
Amber Marie Franks
Sco 17h21'34" -41d57'9"
Amber Marie Kuznicki
And 23h37'42" 46d9'39"
Amber Marie Line
Tau 4h30'31" 18d52'56"
Amber Marie Linnins
Mon 7h36'11" -0d54'25"
Amber Marie Rosado
Ari 2h20'34" 18d45'34"
Amber Marie Speciale
Cap 20h24'22" -11d0'36"
Amber Marie Toms
Crb 15h40'10" 38d21'24"
Amber Martinez
And 22h58'9" 50d15'23"
Amber Mary Bratcher
Tau 3h38'5" 29d48'48"
Amber McCoy
Aql 20h23'49" -0d42'19"
Amber McCusker
Uma 11h28'29" 50d24'22"
Amber McDuffie
Sgr 18h55'26" -20d21'49"
Amber Melissa Poe
Tau 3h53'0" 1d0'31"
Amber Michelle Hoefer
Pho 0h4'30" -45d20'26"
Amber Michelle Johnson
Cap 21h14'31" -19d38'17"
Amber Michelle Layne
Crb 15h36'32" 26d35'34"
Amber & Mike's first Star
Umi 16h19'24" 74d34'21"
Amber M.J.M.
Ari 3h20'28" 29d4'7"
Amber Moffatt
Lyn 7h29'52" 53d37'55"
Amber "Mon Petite Solei"
Salcido
Ori 6h19'2" 14d24'27"
Amber Mosley
Lyn 8h13'9" 41d34'59"
Amber My Love
Peg 22h21'41" 7d14'52"
Amber Nadine Van Sciver
And 0h20'42" 29d5'32"
Amber Nichole
And 0h23'47" 41d53'10"
Amber Nichole
Dor 4h19'34" -50d30'41"
Amber Nichole Bloom
Peg 22h25'43" 29d12'12"
Amber Nichole Cooksley
Lib 14h29'28" -13d43'43"
Amber Nichole Tupper
Lib 15h12'28" -11d54'16"
Amber Nichole VanSistine
Lyr 19h1'25" 26d1'47"
Amber Nicole
Tau 5h7'22" 24d24'44"
Amber Nicole
And 1h25'52" 37d45'33"
Amber Nicole
Vir 13h39'15" -2d23'57"
Amber Nicole
Cet 1h8'57" -0d7'39"
Amber Nicole Belmonte
Ari 2h21'54" 25d59'50"
Amber Nicole Canada
Leo 10h4'33" 22d35'23"
Amber Nicole Crawley
Ori 6h7'2" 10d0'44"
Amber Nicole Cronkhite
Uma 9h28'21" 46d59'2"
Amber Nicole Dilbeck
Psc 1h7'45" 21d25'41"
Amber Nicole Doane
Uma 8h32'37" 68d6'34"
Amber Nicole Driggers
Sco 17h9'53" -39d22'28"
Amber Nicole Evans
And 23h52'22" 40d56'22"
Amber Nicole Hannahs
Sgr 18h43'6" -30d20'58"
Amber Nicole Hayden
Tau 5h29'18" 18d49'46"
Amber Nicole Liddell
And 1h28'55" 49d34'39"
Amber Nicole Looney
Cam 4h46'9" 58d19'17"
Amber Nicole Marshall
Aql 19h29'3" 7d45'18"
Amber Nicole Niemiec's
Star
Uma 9h14'19" 56d27'2"
Amber Nicole Peyronel
And 2h9'10" 39d9'51"
Amber Nicole Ross Gowin
And 0h15'46" 28d22'49"
Amber Nicole Smith Sasis
Cas 1h35'46" 60d22'23"
Amber Nicole Staskon
Psc 0h9'49" 12d21'54"
Amber Nicole Stratton
Lyn 7h57'43" 39d48'4"
Amber Nicole Walters
Vir 13h1'6" 2d17'6"
Amber Nicole Ward
Her 17h47'46" 42d20'59"
Amber Niece
And 0h29'33" 34d31'12"
Amber "Nikki" Vigil
Tau 5h37'45" 18d50'10"
Amber Nikkole Pearson
Ori 5h10'22" 7d36'6"
Amber Noel Gargano
Cap 20h8'7" -14d43'25"
Amber Pang Wai Yee
Cnc 8h18'22" 22d48'39"
Amber Pearson
Uma 9h29'4" 62d5'34"
Amber Peck's Shining Star
Vir 12h31'9" 12d51'30"
Amber Phillips
And 0h44'4" 39d12'53"
Amber Pridgen
And 1h19'39" 39d9'12"
Amber Pulley
Cnc 9h2'54" 21d15'14"
Amber R. Piercy
Sgr 19h11'57" -12d51'47"
Amber R. Stuckey
Leo 11h21'31" 17d17'39"
Amber Rackley
And 1h54'55" 41d44'45"
Amber Rae
Tau 4h17'23" 26d46'23"
Amber Rae Bell
Ari 2h8'47" 25d14'12"
Amber Rae Carrell
Psc 1h24'36" 20d55'44"
Amber Rae Silverthorn
Sco 17h20'49" -42d28'12"
Amber Raines
Crb 15h48'53" 34d7'47"
Amber Raven 11/26/73
Lmi 10h10'51" 31d27'57"
Amber Ray
Lyn 7h25'5" 50d16'49"
Amber Reader
And 2h5'47" 42d28'13"
Amber Reann
Aqr 23h40'34" -4d56'42"
Amber Renee Bernd
Cas 1h26'51" 61d5'44"
Amber Renee Ford
And 23h49'43" 35d6'9"
Amber Renee Weaver
Vir 13h51'27" -8d20'35"
Amber Renee Wirth
Lib 14h23'6" -17d41'6"
Amber Richardson
And 23h28'25" 37d11'4"
AMBER & ROBERT-FOR-
EVER LOVE
Ari 2h0'8" 17d32'57"
Amber Romance
Psc 1h6'55" 29d46'44"
Amber Rose
Gem 7h31'3" 24d55'15"
Amber Rose
Lyn 7h54'28" 43d38'28"
Amber Rose
Lyn 8h33'54" 39d3'53"
Amber Rose
Dra 17h13'9" 63d4'42"
Amber Rose Bauta
And 23h32'43" 45d45'0"
Amber Rose Byres
Lib 15h4'51" -18d29'29"
Amber Rose Drews
Leo 9h26'50" 10d1'30"
Amber Rose Hecita
Ari 2h44'8" 13d39'27"
Amber Rose Holly
Sgr 18h31'43" -27d40'49"
Amber Rose Simpson
Cap 20h33'29" -19d40'56"
Amber Ruiz
Lib 14h30'55" -12d50'33"
Amber/Ryan
Cyg 20h15'28" 52d8'40"
Amber & Ryans Love
Tau 5h59'42" 23d10'53"
Amber Sanchez
Lib 15h6'55" -9d33'30"
Amber Sanders
Cas 23h6'59" 56d3'6"
Amber Sandlin
Sco 17h43'2" -42d18'20"
Amber Schiro
Cas 2h18'6" 74d13'50"
Amber & Scott Walker
Uma 11h3'27" 65d26'27"
Amber Shae Roberts
And 0h21'53" 33d38'16"
Amber Shaw
Uma 8h55'59" 68d33'28"
Amber Shelton Davis
Crb 15h30'21" 29d49'51"
Amber Shiree Drake
And 0h15'59" 29d26'6"
Amber Song Buckalew
Aqr 22h9'15" -12d41'24"
Amber Sorsek
Ari 1h47'52" 18d18'47"
Amber Star
Gem 6h41'39" 20d56'49"
Amber Starr Reynolds
Uma 10h34'20" 40d12'5"
Amber Stevenson
Cas 0h34'2" 53d47'4"
Amber Stone
Lyn 7h8'16" 58d28'31"

Amber Sue
Gem 7h23'21" 16d19'5"
Amber Sue Cross
Cas 0h45'42" 61d12'13"
Amber (Sunshine) Urban
And 2h36'40" 46d52'44"
Amber Tai
And 23h14'3" 43d47'19"
Amber Tai
Cap 21h40'24" -10d41'36"
Amber the Angel
And 0h21'31" 32d36'28"
Amber Trackwell "My One Love"
Umi 14h31'7" 85d3'41"
Amber Trautvetter
And 0h13'38" 39d23'13"
Amber_Tseu
Vir 13h0'37" -7d48'41"
Amber Tucker
Psc 2h3'37" 8d24'54"
Amber Valentyn
Ari 2h39'13" 25d52'49"
Amber Vanboxel
Uma 11h21'27" 28d19'55"
Amber Vanlandingham
Gem 6h50'27" 25d25'1"
Amber Walker
And 1h5'52" 43d54'25"
Amber Whatley
And 0h22'9" 32d11'7"
Amber Wiener
And 1h23'14" 48d59'40"
Amber Yarborough
Sgr 18h8'5" -17d26'42"
Amber1128
Ori 6h19'13" 14d36'57"
AMBER2334
And 23h20'34" 38d14'29"
AmberAnne
Aqr 22h28'26" -6d42'9"
AmberChase
Lib 15h13'22" -10d38'40"
AmberDawn
Aqr 22h41'37" 2d4'52"
Amber-Dawn Lamontagne
Umi 13h26'53" 74d27'12"
Amberdoodle
Psc 23h13'6" -0d26'10"
Ambere St. Denis
Lib 15h24'50" -6d45'56"
Amberella
Leo 10h7'40" 26d36'40"
AmberGlow
Uma 11h17'10" 46d5'11"
Amber-Isaiah
Uma 11h36'53" 57d36'28"
AmberK 5000
Leo 11h9'6" 9d6'34"
Amberlea Nicole Westerfield
Lmi 10h37'52" 32d46'52"
Amberlee
Lyn 7h36'33" 35d35'42"
Amberlee C. Toth
Ori 6h2'28" 14d7'22"
Amberley
Leo 10h22'27" 22d22'51"
Amberli Alice Jewell
Gem 6h26'24" 19d22'27"
Amberlicious
Cnc 8h9'12" 13d17'0"
Amberly
Ori 5h13'35" 4d49'5"
amberly
Sco 16h35'59" -26d49'39"
Amberly Ann
Sgr 19h21'29" -13d20'49"
Amberly Desirae Cannon
Cnv 13h39'25" 37d12'44"
Amberly Nichols
Crb 16h13'33" 33d46'50"
Amberlyn Adamo Venustas
Cam 4h22'12" 57d58'40"
Amberlynn Faith Douglas
Ari 2h33'29" 26d25'29"
Amberrose
And 23h26'4" 38d0'3"
Ambers Alaska
Ori 6h18'7" 8d25'22"
Amber's Aura
Sgr 19h0'1" -17d47'2"
Amber's Beauty
Sgr 18h2'48" -26d56'51"
Amber's Dragon
Gem 6h38'56" 15d48'42"
Amber's Everlasting Star
And 0h40'19" 43d40'27"
Amber's Heart and Soul
Sgr 18h3'58" -17d22'23"
Amber's "Hello Kitty"
Lyn 8h1'26" 33d20'47"
Amber's Nightlite
Ari 3h17'3" 16d50'38"
Amber's Smile
Dra 20h12'12" 73d36'36"
Amber's Star
Peg 23h16'45" 28d39'40"
Amber's Star 012406
Psc 0h53'52" 16d34'5"
Ambiguous Butterfly
Umi 14h14'30" 71d26'26"
Ambivalence
Ari 2h15'28" 23d55'8"
Ambix
Ori 5h3'22" 5d37'8"
Ambra
Cam 4h28'44" 53d38'58"
Ambra Casadei
Uma 10h16'15" 72d34'30"
Ambre
Ori 6h16'25" 14d41'37"
Ambre Arnéodau
Umi 10h25'1" 87d22'54"
Ambre Jo
Ori 5h25'5" 14d26'2"
Ambreen
Cas 0h58'4" 58d51'44"
Ambreen Mehkary & Yasir Cheema
Cyg 20h51'37" 48d41'15"
Ambrelee Taylor
Lyn 7h30'12" 58d6'38"
Ambria
Ari 3h1'10" 18d29'31"
Ambria Johns
And 0h53'36" 44d6'30"
Ambrin
Ari 3h5'7" 14d11'55"
Ambrosa
Sco 16h5'55" -17d3'16"
Ambrose LeBlanc
Uma 11h52'45" 39d10'35"
Ambrosia
Ori 6h17'34" 16d22'43"
Ambrosia
Ori 5h59'3" 21d2'35"
Ambrosia
Cam 3h33'55" 61d52'42"
Ambrosina
Cnc 8h53'31" 17d30'6"
Ambrosius, Thorsten
Uma 8h53'55" 61d58'4"
Ambrozia Williams
Uma 13h7'16" 53d15'22"
Ambrus
Uma 11h35'58" 54d21'1"
Ambur & Kyle Love 05
Umi 15h21'20" 73d53'33"
Amby
Sgr 19h27'56" -12d10'58"
AMDL#1
Vir 14h19'5" 5d49'19"
Ame
Crb 15h28'55" 27d43'39"
Ame
Sco 16h41'12" -41d25'14"
Ame Bicho
Uma 8h53'24" 69d49'34"
Ame Marie Kiko
Cap 21h27'2" -10d12'26"
Amedeo Andrea SAVORETTI
Boo 13h57'5" 11d11'11"
Amedeo Ursini
Leo 10h25'19" 9d10'18"
Amee
Sgr 19h15'24" -16d7'8"
Amee DeFriez
Cap 20h33'23" -17d39'41"
Amee the Prettiest of the Pretty
Sco 17h47'37" -41d5'53"
Ameen Arshan
Per 3h41'49" 44d54'40"
Ameen & Nusrat
Cyg 19h41'59" 27d50'37"
Ameena
Uma 12h34'19" 59d5'29"
Ameer
Cap 20h16'22" -19d10'31"
Ameet & Mamta Naik
Cyg 19h36'18" 53d2'3"
Ameika Purcell Minahan
Psc 1h43'8" 15d57'7"
Ameisenköpchen
Uma 9h57'14" 53d7'23"
Amel Chehayeb
Lib 14h50'35" -4d38'45"
Amela Modric
Lib 15h24'9" -27d23'0"
Amelea Raye
Aqr 22h16'6" -1d45'0"
Amelia
Sco 16h12'49" -15d25'13"
Amelia
Sgr 18h8'15" -27d50'59"
Amelia
Leo 10h59'19" 5d6'8"
Amelia
Mon 6h58'3" 3d36'40"
Amelia
Leo 9h45'0" 29d27'8"
Amelia
Ori 6h10'54" 20d56'46"
Amelia
And 0h15'29" 29d30'30"
Amelia
Uma 9h20'11" 44d59'57"
Amelia
And 23h9'4" 39d31'40"
Amelia
And 1h35'35" 50d5'51"
Amelia 12 19 1925
Uma 11h58'58" 36d6'33"
Amelia & Alina Schreck
Ari 2h11'3" 20d43'58"
Amelia (Angel) Cody
Aqr 22h2'39" 0d14'37"
Amelia Anna
Sco 16h6'26" -14d35'9"
Amelia Anne
Ari 2h40'34" 16d11'8"
Amelia Anne Krouse
Aql 18h59'15" 16d47'40"
Amelia Annie Hitchen
And 0h49'19" 22d22'9"
Amelia Bernard
Aqr 22h19'22" -22d20'40"
Amelia Berry
And 2h14'49" 47d41'53"
Amelia Beth
Sco 16h8'15" -16d42'14"
Amelia Betrond Schriver
Ari 3h13'35" 29d30'21"
Amelia Bhani Ramlakhan
Vir 13h5'30" -15d27'37"
Amelia Bibles
Uma 10h43'19" 46d35'55"
Amelia Bo Gebruers
And 23h22'34" 50d42'47"
Amelia Brewer
Peg 22h28'51" 4d13'29"
Amelia C. Whipple
And 0h27'9" 43d54'31"
Amelia Carolann
Vir 12h30'42" 3d18'4"
Amelia Christine Marchese
Aqr 22h31'39" -21d28'4"
Amelia Claire Whitlock
And 0h11'8" 28d13'26"
Amelia Coyle - Nana
Cep 20h46'1" 61d28'24"
Amelia E. Lewis
Leo 10h43'7" 17d1'49"
Amelia Elizabeth
Cru 12h37'5" -59d5'36"
Amelia Elizabeth Austin
And 23h15'0" 48d51'32"
Amelia Elizabeth Lynn Hartmann
And 23h20'10" 48d11'46"
Amelia Elizabeth Spitznas
Tau 5h15'36" 22d2'42"
Amelia Elizabeth Whelan
Cyg 20h14'8" 36d57'1"
Amelia Elizabeth Wiener
Vir 12h26'33" 12d14'20"
Amelia Ella Linnéa Kress
Cyg 19h39'28" 34d40'59"
Amelia Ellen Beale
And 23h40'7" 50d12'0"
Amelia Fotopoulos
Cas 1h17'59" 50d44'7"
Amelia Grace
And 23h3'30" 37d10'25"
Amelia Grace
Cap 21h36'43" -13d36'41"
Amelia Grace Kelly
Cra 18h33'56" -43d48'23"
Amelia Grace Leonard
Umi 13h46'11" 69d35'43"
Amelia Grace Stevens
Cnc 8h6'16" 26d14'19"
Amelia Helena Brown
And 23h24'31" 51d6'30"
Amelia Jacklyn Qirici
And 2h20'23" 50d0'13"
Amelia Jaime Taylor
And 0h10'28" 33d5'4"
Amelia Jane Brubaker
Uma 11h38'29" 35d50'42"
Amelia Jane Robertson
And 23h46'39" 37d30'19"
Amelia Jayne Bartlett
Cyg 21h50'19" 43d52'1"
Amelia Jean
Tau 4h34'24" 14d8'40"
Amelia Jean Drew
And 0h14'4" 39d29'3"
Amelia Jean Fischer
Aqr 21h9'0" 0d53'26"
Amelia Juliette Lesniak
Cap 20h49'47" -20d19'22"
Amelia Kate
And 1h15'12" 42d17'44"
Amelia Kate Richardson
Cas 23h16'50" 56d9'51"
Amelia Kate Talbot
And 2h3'36" 45d30'1"
Amelia Kate Wilson
And 23h40'19" 48d27'49"
Amelia Katie Dunn
And 23h2'21" 49d4'56"
Amelia Kristina Small
And 22h59'18" 52d4'49"
Amelia L. Gouweloos
Vir 14h13'10" -2d49'24"
Amelia L. Ufford
Psc 0h43'30" 6d23'27"
Amelia Lauren Dawn Higgs
And 1h59'21" 42d13'48"
Amelia Lea Brown
Gem 7h24'4" 26d42'0"
Amelia Leigh Davis
Sgr 18h13'34" -20d1'8"
Amelia Louise Long
And 23h15'28" 41d56'28"
Amelia Loves Joe Forever
Aqr 23h42'48" -15d31'2"
Amelia Loves Victor
Aql 19h15'39" 7d12'22"
Amelia Lyn Fager
Tau 4h26'52" 23d23'16"
Amelia Lynn
And 2h12'37" 43d31'51"
Amelia Mae
And 0h41'14" 23d43'35"
Amelia Mae Antonia Mitchell
And 23h15'42" 51d55'18"
Amelia Margaret Pendleton
Uma 12h41'41" 56d58'46"
Amelia Margo Briggs
Leo 10h52'15" -3d20'35"
Amelia Marie
And 0h41'11" 40d17'34"
Amelia Marie Graham
And 2h30'34" 45d42'22"
Amelia Marie Moniz
Cnc 8h54'41" 32d15'52"
Amelia Marie Sheridan
And 2h23'36" 47d23'7"
Amelia Marisa Kendall Klenk
Sco 17h37'34" -42d31'51"
Amelia McKinley
Lib 15h7'51" -10d51'14"
Amelia Nelline Joy Wells
Uma 12h33'38" 56d39'19"
Amelia Olbera
Crb 15h29'9" 31d36'53"
Amelia Ottalie Krausz Ferguson
Vir 13h56'45" -0d18'44"
Amelia Otte Mersiowsky
Cyg 19h47'33" 36d5'14"
Amelia - Our Angel Upon A Star
Col 5h25'49" -31d3'52"
Amelia Paparozzi
Lib 14h49'55" -1d20'8"
Amelia Poppy
And 1h51'56" 40d17'58"
Amelia - Precious Angel
And 1h18'31" 37d54'30"
Amelia Pulcherrima Clarissima
And 23h17'56" 48d33'50"
Amelia Pyros Nevarez
Lyn 6h50'32" 51d34'52"
Amelia Quinn Haley
And 2h30'22" 50d37'21"
Amelia Rae Clark
Uma 11h17'31" 64d15'5"
Amelia Ramsden
Sgr 18h29'49" -36d1'52"
Amelia Randall
Sco 17h22'59" -39d51'7"
Amelia Renee Ramsey
Leo 11h33'10" 20d53'17"
Amelia Renee Read
Ari 2h50'54" 28d5'12"
Amelia Roan
Leo 11h2'33" 6d39'24"
Amelia Rose
Leo 9h27'4" 12d42'19"
Amelia Rose
And 0h56'33" 44d26'51"
Amelia Rose
Dra 9h33'20" 79d41'35"
Amelia Rose Dent
And 2h29'16" 37d32'36"
Amelia Rose Lewis
Uma 9h32'23" 62d41'34"
Amelia Rose Ligot
Cet 1h8'18" -0d53'1"
Amelia Ruth
And 23h55'29" 46d13'49"
Amelia Samantha Grant
Cas 2h45'28" 64d13'13"
Amelia Sargis
Leo 10h43'12" 18d43'38"
Amelia Star Moore
Gem 7h12'56" 26d25'0"
Amelia Victoria McGunigall
And 0h52'57" 22d9'25"
Amelia Westley
And 23h15'16" 35d15'10"
Amelia, Brandy and Tim Beamenderfer
Uma 12h5'5" 51d39'25"
Amelia05111984
And 1h23'4" 34d40'27"
Amelia-Faith Nicole Wood
And 0h18'4" 42d5'15"
Ameliah
Cap 21h41'3" -24d36'3"
Ameliah Marie Nickels
Vir 14h1'50" -17d20'12"
Amelia's Christening Star
Sge 20h20'23" 18d58'15"
Amelia's Guiding Light
Sco 17h57'59" -39d53'0"
Amelia's Star
Ori 6h10'39" 8d50'21"
Amelia's Star
Uma 10h17'50" 42d26'13"
Amelie
And 0h8'52" 28d45'46"
Amélie
Umi 17h41'44" 81d1'52"
Amelie Amber
Dra 16h18'8" 61d8'31"
Amelie and Alexa Copley
Gem 7h46'23" 24d52'52"
Amelie Clare Jeffers
Lyn 8h25'8" 56d29'7"
Amelie Creekmore
Lyr 19h11'8" 26d22'54"
Amelie Daisy Bell
And 2h31'45" 40d0'32"
Amélie Delmi
Uma 10h15'27" 68d57'32"
Amélie Emma Isabel Orgill
And 2h32'45" 49d31'34"
Amélie et Benoit
Cyg 20h0'27" 32d36'25"
Amélie Georgia Moniak
And 0h15'7" 33d13'49"
Amelie Grace Day
And 0h21'11" 45d43'43"
Amelie Ivanka Dobek
Aqr 22h13'14" -1d33'36"
Amelie Joy
Vir 13h33'12" -16d55'15"
Amelie Juliette Lowery
Lmi 10h21'25" 37d13'48"
Amélie Martin St-Pierre
Umi 15h7'59" 78d54'8"
Amelie Michele Hanna
Uma 9h3'12" 64d15'15"
Amélie Nowé
Tau 5h2'46" 19d36'28"
Amelie Rebecca Jane
Cap 20h52'18" -20d27'17"
Amélie Rioux
And 1h47'42" 39d18'17"
Amélie Rosemary Lindt
Vir 13h0'27" 4d45'53"
Amelie Sahara Caines
Uma 11h24'10" 58d44'30"
Amelie Von Koczian
And 23h3'56" 51d23'47"
Amélie, l'étoile de ma vie.
Lib 14h40'19" -9d0'21"
Amelio Richard Tomasiello
Leo 9h51'36" 12d50'7"
Améliyanis
Crv 12h9'42" -13d47'17"
Amely
Uma 11h51'58" 60d6'10"
Amenatave Gutugutuwai
Ari 2h2'49" 19d59'49"
Amerbee
Ori 5h4'0" 15d21'8"
Ameri Lynnette
Ari 2h45'39" 13d14'38"
America
Cas 22h58'14" 55d48'22"
America Arias
Ari 3h13'28" 28d44'31"
America Cano
Lib 15h33'23" -28d43'13"
America G. Guido
Gem 6h48'48" 26d56'14"
American Career Executives
Uma 11h43'18" 34d6'43"
American Hero: Aaron David Bailey
Cap 20h54'26" -19d20'32"
American Heroes Hank & Jennifer
Ori 5h36'48" -0d52'38"
American Pool Construction
Dra 18h20'37" 56d4'47"
America's Heroes
Uma 11h35'32" 57d11'52"
Americo
Leo 11h0'5" 18d9'24"
Americo Anthony Michel, Jr.
Her 17h11'41" 32d49'16"
Americo "Max" Bovio
Uma 11h1'34" 72d44'34"
Americo "Poppie" Venturini
Psc 1h27'18" 13d26'20"
Americool
Uma 9h58'56" 55d44'15"
Americus Lynn Hullet
Dra 17h31'9" 62d9'48"
Amerie Reed Tillett
Cap 21h44'35" -19d43'46"
Amerina Berardinelli
Uma 9h12'55" 50d38'11"
Ameris Elisabetta Molinari
Per 3h12'2" 40d26'21"
AmeriSpa
Uma 11h52'14" 63d8'1"
AMERI-STAR HOMES
Cap 21h25'19" -15d28'8"
Amerith
Cas 23h9'1" 59d36'15"
Amerjit Nicky Mission Sandhu II
Per 3h48'47" 48d46'27"
Amer-Puddin
Vir 11h46'27" 2d30'42"
Ames
Cnc 8h15'42" 10d34'9"
Ames
And 0h2'49" 45d20'1"
Ames&Cate
Cyg 20h20'12" 54d46'13"
Ameshnee Mariemuthoo
Uma 9h24'38" 57d55'43"
Amess
Lyn 7h32'15" 39d12'59"
Amethyst
And 0h55'33" 40d41'30"
Amethyst
Lyr 19h5'10" 31d16'39"
Amethyst Christine Post
Cnc 8h44'20" 19d20'2"
Amethyst Margaret Wilson
Psc 0h40'14" 12d6'45"
Ametropia
Lyn 9h33'43" 40d46'10"
AMEX P2W
Cru 12h48'49" -60d42'20"
Amey Dawson
Cnc 9h17'17" 16d46'10"
AmeyNoelle
Ori 6h19'8" -0d7'24"
Amgad Girgis
Umi 13h41'15" 69d46'19"
AMH
Uma 8h30'47" 66d0'2"
AMI
Lib 15h44'37" -5d26'25"
Ami Adell Glascock
Ari 3h23'7" 19d56'51"
Ami Aimé Pam
Vir 14h58'40" 4d10'32"
Ami and Mikhail
Gem 7h28'21" 31d58'18"
Ami and Tim Thomsen
Boo 15h14'57" 51d35'53"
Ami Jo
Aqr 22h37'22" 0d50'27"
Ami Louise
And 23h21'28" 52d23'5"
Ami Lyhnn Meader
And 1h32'10" 47d27'56"
Ami pour la vie
Lyr 18h38'6" 41d22'42"
Ami Roberts
And 1h24'44" 35d7'42"
Ami Williams
Tau 5h11'33" 17d59'25"
Amia L. Zaide
Umi 13h54'55" 77d10'44"
Amichetti Calabretta
Sco 16h11'56" -11d59'48"
Amici Per Sempre
And 2h25'42" 46d37'37"
Amici speciali
Per 3h44'31" 47d28'16"
Amicitia
Cyg 21h57'2" 39d18'29"
Amicitia
Her 16h41'7" 48d9'38"
Amicitia Catholyn
Cra 17h59'48" -37d20'36"
Amicizia
Uma 12h48'11" 55d38'34"
Amico fedele
Cap 20h7'15" -9d26'20"
AmicosOttimo
Lyr 18h48'46" 38d56'52"
Amiculus
Aqr 21h8'57" -7d18'43"
Amicus
Mon 7h17'58" -0d12'30"
Amicus Charisma
Ori 5h39'27" -3d47'4"
Amidu Froglet Jefferies
Tau 4h1'36" 1d58'59"
Amie
Leo 11h30'1" 5d25'45"
Amie
Gem 7h4'20" 14d26'6"
Amie
And 23h24'9" 41d43'32"
Amie
And 2h32'58" 40d39'45"
Amie
Cra 19h2'43" -41d46'28"
Amie and Adam
Vir 14h2'24" -15d41'14"
Amie Celeste Moore
Cnc 9h6'20" 31d11'17"
Amie Doll
Vir 13h18'41" -8d12'39"
Amie Houser
Umi 13h16'34" 70d39'1"
Amie Jo Jordan
Tau 4h32'36" 17d42'54"
AMIE KATHLEEN RODGERS
And 23h22'7" 42d24'59"
Amie Kolos
Lyn 7h51'48" 58d36'0"
Amie Lance
Lib 15h10'48" -22d29'49"
Amie Lee
Cap 20h14'6" -9d1'18"
Amie Louise McKell B.A. Hons.
Cas 0h23'2" 54d27'28"
Amie Lynn Ewing
Cas 0h36'5" 52d51'55"
Amie Lynn Weis
Ari 2h57'11" 14d16'27"
Amie M Search Hamman
Cnc 8h40'18" 16d36'34"
Amie Marie Brown
Aqr 21h50'42" -2d38'36"
Amie Marrocelli
Sco 16h14'37" -13d13'33"
Amie Michelle Lawrence
Cap 20h40'48" -20d43'15"
Amie Nicole Thurow
Ori 4h48'43" 11d33'4"
Amie Oberndorf Raftus
Ori 5h46'52" 5d27'5"
Amie - Princess Perfection
And 1h56'34" 40d1'35"
Amie Renee
Tau 3h52'37" 25d4'58"
Amie Stacy ~ The Greatest Mom Ever
Cas 1h15'27" 64d20'8"
Amie the Beautiful
Cas 1h16'37" 53d33'0"
Amiee D. K. Hooker
Aqr 21h10'26" -1d58'57"
Amiee Hsueh
Gem 6h18'15" 22d3'38"
Amiee Lynn
Cnc 8h11'18" 28d18'35"
Amiee's Star
And 0h9'18" 44d33'28"
Amie-Joe
Cyg 19h16'34" 52d5'46"
Amier ~ The One I Adore
Ari 2h19'17" 22d1'13"
Amie's Beach
Hya 8h41'3" -11d35'51"
Amie's Star
Del 20h46'33" 19d21'22"
Amie's Star 28
Uma 11h31'48" 46d50'0"
Amie's Wishing Star
Tau 4h39'1" 0d57'19"
Amigo de los pajaros quebrados
Crb 16h16'41" 31d3'13"
Amila Orucevic
Cas 0h21'41" 56d46'43"
Amiliana & Corne
Vir 11h40'31" -6d5'23"
Amilya Charlotte Miller
Peg 22h50'58" 31d36'8"
Amin
Uma 11h40'5" 39d17'47"
Amin Tony Hester
Sgr 19h49'43" -12d40'33"
Amina Hamzoui Khalpey
Sco 17h52'47" -42d1'5"
Amina Néggah
Dra 17h15'4" 52d45'17"
Amino
Ori 6h17'34" 14d5'16"
Aminta
Vir 11h38'30" 4d29'50"
Aminta Flores (Shooting Star)
Peg 23h9'27" 30d26'19"
Amir
Uma 9h46'10" 52d45'17"
Amir Arif Awais
Gem 6h45'22" 34d8'20"
Amir Ashel
Lib 15h4'39" -22d42'40"
Amir Ayoub
Uma 10h2'7" 60d0'17"
Amir Branch
Per 2h46'16" 47d45'6"
Amir Parvin
Uma 12h55'46" 58d3'54"
Amir Rajab
Lyn 9h6'55" 37d57'18"
Amir*** Raza
Gem 7h0'56" 13d32'47"
AMIRA
Vir 12h47'50" 3d46'7"
Amira
Gem 6h1'53" 27d19'23"
Amira
Lyr 18h30'15" 29d19'33"
Amira
And 23h27'38" 35d47'23"
Amira
Cas 1h21'5" 52d55'32"
Amira
Cap 20h38'55" -13d13'25"
Amira
Ori 5h21'41" -8d35'17"
Amira
Aqr 23h21'31" -21d19'51"
Amira Lail
Eri 1h57'29" -53d16'16"
Amira Shagaga
Cap 20h32'25" -17d56'2"
Amira Tova Winograd
And 1h24'30" 44d38'3"
Amis
Uma 10h20'4" 70d12'37"
AMIS-DODASY
Cas 23h33'2" 58d26'51"
Amish Parekh
Sco 16h16'13" -10d59'29"
Amistad Celestial
Dra 16h32'15" 68d43'11"
Amit
Ari 2h48'0" 27d49'32"
Amit
Psc 0h49'31" 7d53'1"
Amitabh Bachchan
Lib 15h15'18" -5d54'37"
Amith & Bridget Forever
Lyr 19h13'44" 27d2'11"
Amitie Lee Gray
Cas 0h11'19" 56d50'35"
Amity Hoenisch
Ori 5h37'16" 8d21'43"
Amity Leora
And 23h8'21" 47d50'20"

Amiya Joy Farris
Crb 15h27'56" 26d49'23"
Amiya Lyn Lacey-Rousou
Cas 23h50'12" 59d51'57"
AMJALLAK
Umi 14h15'9" 76d26'13"
A.M.L. to Patrick
And 1h32'28" 44d5'37"
AML5JMD - YATB
And 23h34'2" 47d43'28"
AMLAMAUCE
Leo 10h58'58" 7d4'0"
Amma
And 0h13'31" 43d18'56"
AmMaLuc3
Ori 5h32'33" 10d19'19"
Ammann, Marie
Uma 9h33'3" 52d31'38"
Ammann, Sophia Mira
Uma 9h37'29" 54d12'23"
Ammari 02.28.05
Uma 12h37'36" 60d20'41"
Ammiel
Ari 2h16'37" 14d27'29"
Ammmore
Uma 13h31'21" 60d9'8"
Ammon, Dr. Karlheinz
Ori 6h21'14" 16d46'15"
Ammorette
Ori 6h18'54" 19d29'4"
Amna
Cnc 8h22'5" 25d39'2"
Amneh and Imran
And 0h18'49" 22d18'48"
Amnon Medan
Cam 4h19'1" 68d4'48"
Amnon's Star
Leo 11h40'15" 25d54'56"
AMO
Lyn 7h8'42" 60d12'35"
Amo enim infinitas infintio
Cyg 19h56'28" 30d42'18"
Amo Eternus
Cyg 20h33'16" 41d12'54"
amo por siempre
Gem 7h4'14" 31d2'5"
Amo Shannon Dirunitas
Pho 1h39'3" -52d3'50"
Amo te Ana
Gem 7h30'3" 27d4'45"
amo una
Umi 15h16'22" 86d54'9"
Amoa Noel Casey, Sr.
Uma 14h19'9" 54d59'40"
Amo-Coniunctio-Astrum
Sco 16h12'42" -14d45'39"
AMODEE8
Ori 5h23'21" -3d56'9"
Amoena Consanguinea
Tau 4h16'36" 10d35'21"
Amoi
Lac 22h44'45" 54d8'41"
Amoi Jesciena Thompson
Lib 15h58'38" -9d46'45"
Amooni
And 0h42'41" 46d8'50"
Amor
Cyg 21h25'36" 47d7'35"
Amor
Cyg 21h32'28" 46d3'22"
Amor
And 1h11'48" 37d2'23"
Amor
Cyg 19h37'32" 30d48'7"
Amor
Vir 12h22'30" 4d29'35"
Amor
Mon 6h57'39" 2d39'39"
Amor
Leo 9h34'42" 28d55'19"
AMOR
Sge 19h26'29" 17d40'17"
amor
Lib 15h10'24" -26d34'37"
Amor a la Distancia
Cyg 21h16'58" 37d8'28"
amor aeternitas
Per 2h26'22" 54d40'41"
Amor Aeternus
And 23h1'42" 50d15'16"
Amor Aeternus
Cap 21h36'59" -21d41'36"
Amor Aeternus "Love
Forever"
Gem 6h47'27" 12d52'25"
Amor Amplus Caelum
Sgr 18h54'21" -29d56'33"
Amor animi arbitrio sumitur,
non ponitur
Uma 9h15'44" 64d6'7"
Amor Asombroso
Aql 20h23'59" 2d47'58"
Amor Atque Studium
Sempiternum
Cyg 20h59'16" 46d41'14"
Amor de Amber
Aqr 21h13'51" -8d58'15"
Amor de Mayo
Gem 7h8'12" 27d9'14"
Amor de me Vida Pattie
Sgr 19h27'30" -29d34'49"
amor de mi alma François
Armand Bernath
Cyg 20h39'57" 42d25'1"
Amor Eri-Gmh
Leo 11h24'11" 21d44'25"
Amor est vitae essentia
And 0h53'15" 44d4'33"
amor eterno
Sge 19h15'3" 18d28'3"
Amor Eterno
Uma 8h20'7" 63d3'55"
Amor Familiae 728
Leo 10h17'3" 13d33'8"
Amor Magnus
Uma 10h54'27" 49d35'37"
Amor (Mario Mader)
Cas 23h49'6" 54d7'41"
AMOR MIO
Leo 11h19'22" 2d37'5"
"Amor Mio" Andres &
Maryela
Sgr 18h14'3" -20d2'12"
Amor Para Siempre
Cha 10h36'47" -76d53'58"
Amor por las Estrellas
Aql 18h51'50" 9d47'46"
Amor quam Amicus
And 1h41'21" 49d23'47"
Amor Suspiro 13
Tau 3h56'10" 23d2'32"
amor verdadero
Ari 2h54'58" 26d59'34"
Amor Verdadero
Per 3h20'21" 47d9'4"
Amor Verdadero
And 0h36'39" 40d26'27"
amor verdadero
Vir 12h37'8" -6d55'31"
Amor Verdadero: Bruce
and Tabitha
Gem 7h53'35" 14d4'31"
Amor Verum
Ari 2h54'23" 26d52'38"
Amor Vincit Omnia
Her 18h11'15" 28d45'15"
Amor Vincit Omnia
Crb 15h51'2" 34d56'16"
Amor Vincit Omnia
Uma 8h55'34" 51d54'20"
Amor Vincit Omnia
Cyg 20h43'48" 45d56'25"
Amor Vincit Omnia
Cep 21h5'5" 55d43'10"
Amor Vincit Omnia
Dra 18h39'26" 69d2'59"
Amor vincit omnia
Pho 1h16'54" -41d9'58"
Amor vincit omnia
Sco 16h41'28" -27d35'33"
Amor vincit omnia.
Promitto.
Leo 11h45'10" 26d52'55"
Amor Vincit Omnia T & S
Cyg 20h31'30" 58d9'52"
Amora
Cyg 21h39'33" 52d21'55"
Amora
Cyg 21h46'22" 40d30'38"
Amora Lia
Sco 16h49'59" -34d8'39"
Amorado
Psc 1h18'45" 31d24'44"
Amorarantxa
Leo 10h50'57" 11d41'26"
amordipaolo
Uma 8h9'4" 59d55'4"
Amore
Uma 9h6'38" 57d43'19"
Amore
Dra 18h21'19" 73d9'32"
Amore
Cas 1h19'8" 66d21'23"
Amore
Lib 15h50'14" -11d2'4"
Amore
Sco 17h53'53" -36d18'35"
Amore
Col 6h15'44" -34d28'31"
amore
Cru 12h47'16" -57d8'42"
Amore
Aql 19h35'25" 12d1'4"
Amore
Aql 20h20'36" 1d49'24"
AMORE
Psc 1h16'6" 14d16'20"
Amore
Ori 5h53'12" 11d55'47"
Amore
Her 18h26'33" 17d6'33"
AMORE
Lmi 10h6'43" 36d56'50"
amore
Cyg 20h59'10" 31d57'15"
Amore
Cyg 20h59'49" 32d51'15"
Amore
And 2h12'11" 38d7'21"
Amore
Lyr 19h1'35" 43d6'51"
Amore 28.11.1969
Lac 22h55'46" 46d34'10"
amore alla luna ed alle
stelle
Cep 23h54'24" 69d42'0"
amore della principessa
Lib 15h46'14" -27d8'21"
amore di aprile
Gem 7h5'53" 34d49'7"
Amore Di Mare
Cyg 21h58'55" 48d0'20"
Amore dolce dolce
Cam 7h32'13" 64d24'4"
Amore Eterna
Crb 15h46'0" 26d48'17"
Amore Eternal
And 1h13'53" 37d21'13"
amore eterno
Vul 21h1'5" 24d3'49"
Amore eterno
Gem 7h34'6" 29d28'52"
Amore Eterno
Sge 20h1'51" 17d46'35"
Amore Eterno
Cap 20h18'30" -11d48'18"
Amore Eterno
Umi 13h46'7" 87d58'14"
Amore Eternus
Sgr 19h41'52" -13d22'39"
Amore gigante
Uma 11h11'45" 44d27'58"
Amore HR
Her 18h27'1" 21d40'5"
amore infinito
Cnc 9h18'43" 9d28'36"
Amore Infinito
Lyr 19h3'31" 31d46'34"
Amore Infinito
Cyg 20h14'10" 48d51'59"
Amore Infinito
Aqr 23h7'42" -8d17'45"
Amore Infinito
Sgr 18h3'28" -27d51'57"
Amore Infrangibile
Lyn 9h12'51" 35d4'50"
Amore Mia Jade
Cra 18h6'41" -37d23'38"
Amore Mio
Eri 3h49'1" -20d53'22"
Amore Mio
Cyg 21h30'47" 51d32'22"
Amore Naturale - D+S
Hor 4h7'41" -41d41'21"
Amore "TTM" Infinito
Umi 15h56'35" 78d37'26"
Amore' Uncontained
Cyg 20h57'32" 44d50'18"
Amoretta
Vir 13h25'41" -4d49'38"
Amorette Anastasia
Muzingo
Lyr 18h36'54" 36d36'1"
Amoris aeterna
Ari 3h4'53" 29d56'11"
AMORNITY 6/11/2004
Ori 5h57'2" 18d5'9"
Amornrat
Uma 9h56'46" 59d52'20"
Amory
Ori 5h50'38" 3d4'4"
Amos Bruce
Aur 5h34'49" 43d30'21"
Amos Eliot
Cas 0h28'20" 61d13'45"
Amos Gordon Cutler
Cep 23h20'43" 76d12'1"
Amos M. Nelson
Her 17h28'38" 37d34'24"
Amos O. Guthrie
Uma 8h38'10" 59d44'54"
Amos Stoll
Ari 2h21'16" 24d13'36"
AmoteJ2
Lep 5h18'13" -11d3'27"
Amour
Aqr 22h36'14" -6d48'24"
Amour
Lib 15h30'13" -12d51'10"
Amour
Dra 19h8'31" 68d6'16"
Amour
Cnc 8h33'7" 22d21'30"
Amour
Vir 12h40'8" 5d20'13"
Amour
Crb 15h38'28" 37d38'26"
Amour
Cyg 20h23'51" 46d57'12"
Amour
Uma 9h18'4" 48d56'27"
Amour
And 1h41'16" 42d32'28"
Amour Asmaa Forever
And 2h30'53" 43d38'56"
Amour d'Angel et Seb
Uma 9h21'17" 63d38'3"
Amour de Florian
Peg 23h14'40" 31d12'23"
Amour de Mélisa
Uma 9h21'24" 43d30'5"
Amour de Stéphanie
Lyn 6h18'33" 56d38'2"
Amour et (Soleil) pour tou-
jours
Cyg 19h47'25" 39d16'29"
Amour Eternel
Lyr 18h36'37" 35d22'35"
amour éternel
Ori 5h30'25" 1d56'54"
amour éternel
Gem 7h21'56" 14d22'45"
Amour Eternel
Ori 5h25'18" -5d13'50"
Amour éternel
Col 5h31'5" -35d6'1"
Amour Nous Avons
Vir 13h20'45" 2d56'55"
Amour Toujours (Love
Always)
Tri 2h16'8" 35d4'55"
Amour Vrai
Psc 1h28'14" 10d43'46"
Amoureux
Gem 7h4'36" 23d17'36"
Amoureux
Lib 15h30'43" -15d26'20"
Amoureux
Lyn 7h53'38" 56d41'38"
amourfameux
Umi 13h21'20" 72d18'10"
Amours
Aur 6h38'59" 48d31'44"
Amours de ma vie
Uma 8h31'10" 61d7'12"
AMP
Uma 11h29'55" 28d48'24"
AMP 41880
Uma 10h21'37" 56d39'18"
Amparo
Gem 7h32'6" 15d50'57"
Amparo Roselló Soria
Ari 2h11'44" 11d40'14"
Amparo Sperry
Lyn 9h6'51" 41d15'57"
Amparo Uribe
Gem 7h52'59" 32d30'18"
Ampy
Psc 1h45'9" 22d17'31"
Amra
Ori 5h25'47" 8d57'27"
Amra B
And 23h28'45" 36d8'24"
Amra Cerenic
Uma 10h0'35" 55d23'22"
Amra Sahbegovic
Cru 12h48'9" -61d0'18"
Amreeta
Cyg 21h36'27" 40d6'43"
Amrit Menda Hathiramani
Ari 2h17'44" 22d21'4"
Amrita
Cnc 8h1'7" 17d47'34"
Amrita
Uma 8h46'36" 59d55'23"
Amrita Petra Rüger
Cas 0h59'38" 58d41'40"
Amrita Ronnachit -one
damn hot star
Pho 2h15'58" -40d21'53"
AMS-50
Gem 7h27'34" 30d54'27"
Amse
Cyg 20h23'56" 52d33'44"
AmStar
Lib 15h37'42" -26d19'33"
Amstel Kersey
Cyg 20h42'39" 30d29'28"
Amtrak
Cmi 7h8'7" 10d17'6"
AMUN-TARI
Uma 11h18'57" 69d40'13"
AMUP
Lyr 18h37'30" 30d21'53"
Amur eterna per Ladina ed
Adrian
Cas 1h28'42" 71d48'10"
Amursmi
Umi 15h32'44" 82d35'49"
Amuwen
Ori 5h56'21" 21d1'16"
AMW aka tafkaa Lil Bear
Umi 14h4'41" 66d13'10"
Amy
Uma 11h51'8" 63d40'14"
Amy
Cas 2h12'6" 64d17'7"
Amy
Cam 3h26'59" 63d6'46"
Amy
Cyg 21h0'35" 54d31'28"
Amy
Cas 23h32'35" 55d52'49"
Amy
Cas 1h22'6" 63d47'56"
Amy
Cas 1h28'43" 60d1'58"
Amy
Aqr 23h32'7" -7d12'32"
Amy
Vir 13h16'39" -4d36'42"
Amy
Aql 18h58'44" -0d59'57"
Amy
Cap 21h30'28" -16d22'7"
Amy
Cap 21h11'4" -19d28'49"
Amy
Lib 15h31'1" -9d33'57"
Amy
Sco 16h3'33" -25d14'31"
Amy
Col 5h35'36" -28d13'20"
Amy
Sgr 18h23'53" -28d22'32"
Amy
Cap 20h8'49" -23d35'4"
Amy
Sco 16h12'46" -30d48'42"
Amy
Sco 17h18'47" -32d39'13"
Amy
Sco 17h4'26" -36d22'2"
Amy
Umi 14h15'7" 66d53'2"
Amy +
Ari 2h48'16" 27d24'37"
Amy
Psc 1h10'45" 24d25'28"
Amy
Del 20h36'23" 15d46'22"
Amy
And 0h21'58" 25d29'19"
Amy
Crb 15h37'51" 26d5'28"
Amy
Tau 5h42'7" 26d37'55"
Amy
Ari 3h12'58" 27d51'47"
Amy
Cnc 8h36'14" 8d14'3"
Amy
Tau 5h26'50" 17d18'47"
Amy
Tau 4h55'42" 17d27'48"
Amy
Aqr 20h46'14" 1d55'9"
Amy
Crb 16h5'49" 33d53'28"
Amy
Gem 7h14'49" 33d1'11"
Amy
And 0h47'18" 37d13'18"
Amy
And 23h11'40" 36d28'4"
Amy
And 0h38'17" 41d45'46"
Amy
And 0h30'39" 42d59'51"
Amy
And 1h36'33" 41d32'48"
Amy
And 1h19'23" 41d50'42"
Amy
And 0h52'41" 42d17'13"
Amy
Per 3h42'26" 42d30'6"
Amy
And 2h36'14" 43d54'54"
Amy
Lyn 6h58'4" 44d47'54"
Amy
Uma 9h28'14" 45d39'27"
Amy
Dra 18h20'43" 48d52'20"
Amy
And 0h0'47" 45d15'41"
Amy
And 23h56'55" 42d59'42"
Amy
And 2h35'21" 45d3'29"
Amy 52
Ori 6h17'38" 10d40'53"
Amy 723
Leo 10h7'45" 27d17'57"
Amy A. Laurent
Her 17h7'9" 32d9'10"
Amy Adams
Tau 5h49'6" 16d58'20"
Amy Aileen Plunkett
And 0h49'13" 22d46'55"
Amy "Aimer" Silar
Leo 11h42'8" 24d18'52"
Amy Akopdjanian
Cnc 9h1'15" 23d49'50"
Amy Alexander
Cam 5h52'8" 61d34'39"
Amy Alexandra Williams
And 0h43'36" 25d52'26"
Amy Alise
Gem 7h36'41" 24d48'40"
Amy "A-mi" Marie Philip
Ori 6h7'55" 11d41'46"
Amy "Amy Beans" Masuga
Uma 10h50'54" 60d21'38"
Amy and Alex Lopez
Uma 11h59'31" 44d52'55"
Amy and Ben's 1st
Anniversary Star
Cyg 21h8'31" 36d55'45"
Amy and Chris Forever
Sgr 18h53'16" -30d39'17"
Amy and Devin's Star
Dra 18h11'8" 72d38'14"
Amy and Dirks Star
Cyg 19h49'29" 39d11'35"
Amy and Eric Palmer
Umi 15h22'26" 71d57'58"
Amy and Kevin Forever
Ori 5h42'17" 11d39'43"
Amy and Liam Brown
Cyg 20h20'38" 58d53'6"
Amy and Matt Forever
Cas 0h14'29" 58d47'7"
Amy and Ryan
Pho 0h5'53" -45d8'5"
Amy and Ryan Forever
Aql 19h45'45" 11d57'41"
amy_and_scott 062493
Cnc 8h45'39" 6d37'13"
Amy and Ted Nation
Her 16h51'19" 30d27'19"
Amy Andersen
Uma 11h22'5" 54d24'15"
Amy Anelle's Way
And 0h38'33" 31d33'34"
Amy "Angel" Christine
Kowieski Star
Sco 17h20'16" -41d17'44"
Amy Angeline Shook
Aqr 23h51'24" -11d47'6"
Amy Anglin
Cap 21h24'44" -19d10'32"
Amy Ann Bauer
Uma 12h42'18" 52d23'40"
Amy Ann Briskie
And 23h19'30" 41d3'13"
Amy Ann Britton
Gem 7h49'59" 29d11'9"
Amy Ann Hall
Cap 21h23'16" -24d20'26"
Amy "Anne Murray"
Gonzales
Ori 6h20'46" 8d38'26"
Amy Anton
Leo 9h29'18" 11d51'32"
Amy Arrison
Sgr 17h47'14" -16d38'57"
Amy B. Maslar
Leo 10h20'52" 8d17'16"
Amy B. Robinson
And 0h49'44" 45d45'50"
Amy B Valbush
Psc 1h5'47" 33d14'42"
Amy Babez
Leo 9h47'39" 12d30'14"
Amy Basso
Tau 5h49'23" 22d5'33"
Amy Bates
Mon 6h55'47" -4d0'49"
Amy Becker
Lib 15h18'4" -21d23'44"
Amy (Bella) Dillmann
Psc 1h7'1" 29d11'21"
Amy Beltran
Mon 7h19'9" -0d13'59"
Amy Bennett
Cas 1h16'40" 67d34'14"
Amy Bergsma Batsakes
Lib 15h5'54" -0d47'37"
Amy Beth
Aqr 22h55'3" -9d9'17"
Amy Beth
Cnc 8h37'11" 7d54'6"
Amy Beth
Aqr 22h19'20" 1d33'27"
Amy Beth
And 0h58'39" 36d26'52"
Amy Beth
And 1h50'25" 46d50'27"
Amy Beth Cutler
Psc 1h32'40" 15d40'23"
Amy Beth Ferrara
Sco 16h11'8" -12d7'2"
Amy Beth Gallagher
Vir 14h13'19" -12d39'26"
Amy Beth Schultz
And 0h7'38" 30d43'45"
Amy Beth Selzer
Psc 0h36'18" 8d57'39"
Amy Blasinski
And 0h27'29" 31d30'49"
Amy Blyda Brown
Sgr 19h50'57" -42d25'4"
Amy Boggioni
And 0h10'33" 40d59'46"
Amy Brake
Ari 2h45'13" 26d4'20"
Amy & Bryan shine togeth-
er forever
Leo 10h16'29" 21d55'8"
amy/bubbles
Lyn 8h32'52" 34d47'2"
Amy C. Carter
Vir 13h53'44" -1d17'7"
Amy C. Hise
And 0h33'8" 28d8'7"
Amy Carney
Tau 4h29'48" 21d5'24"
Amy Carol Easter
Lib 15h9'41" -10d52'10"
Amy Carol Hurst
Tau 4h7'42" 6d8'59"
Amy Cash
Cap 20h54'26" -21d28'1"
Amy Catherine Narquis
Psc 0h25'17" 6d47'24"
Amy Chandler
Lyn 7h36'34" 41d11'12"
Amy Chaplick
Dra 19h16'33" 62d55'40"
Amy Chapman July 6, 1996
Uma 8h15'54" 65d8'26"
Amy Charbonneau
Lyn 7h1'36" 48d25'33"
Amy+Chris LYF
Lyr 18h48'11" 36d45'44"
Amy Christeen
Cap 21h34'41" -17d24'35"
Amy Christina
Psc 0h1'55" 6d26'21"
Amy Christine Carberry
Cap 21h36'50" -13d59'53"
Amy Christine Melendez
Leo 9h57'21" 16d57'35"
Amy Cieciura
Crb 16h2'46" 38d40'49"
Amy Claire
Dra 17h46'24" 58d19'20"
Amy Clark
And 1h43'16" 45d17'15"
Amy Clay - The Eternal
Sunshine
Ari 3h28'54" 27d1'45"
Amy Colleen Haggard
Tau 5h9'16" 24d8'0"
Amy Collins: Bestfriend
Now & Always
Cam 4h45'30" 54d55'56"
Amy Colston
Leo 9h41'5" 29d20'57"
Amy Colucci
Crb 15h34'14" 27d18'5"
Amy Crimmins
Cas 0h14'7" 51d19'49"
Amy Danielle
Tau 4h22'0" 28d25'0"
Amy Danielle Duckworth
Lyr 18h31'0" 37d4'48"
Amy Danielle Tiffany
Uma 11h37'40" 55d56'59"
Amy Danielle Wood
Aqr 21h27'44" -8d1'45"
Amy Darling
Psc 1h2'2" 11d4'23"
Amy Dawn
Mon 6h39'51" -1d10'11"
Amy Dawn Hendricks
And 0h12'29" 43d51'53"
Amy Dee Monson
Leo 11h38'58" 26d18'32"
Amy Denise
Gem 7h16'23" 32d22'41"
Amy Denise Halas
Mon 6h48'5" 8d16'50"
Amy Désirée Goldstein
Tau 5h49'31" 16d21'18"
Amy Diane de los Santos
Aql 19h47'5" -0d48'26"
Amy Diane Rascoe
Gem 6h49'48" 32d47'45"
Amy Dineen
Sex 10h34'59" -0d37'3"
Amy Dollimount
Cas 23h59'24" 59d16'5"
Amy Dosen-Black
Cam 3h38'39" 65d57'28"
Amy E. Anderson
Uma 11h44'36" 49d51'39"
Amy E. Bentz
Del 20h29'32" 5d53'40"
Amy E. Cooper
Cas 1h42'2" 65d52'11"
Amy E Rawls
Aqr 22h33'55" -23d55'35"
Amy E. Whitman
And 23h22'35" 38d2'36"
Amy Ehasz
Sgr 18h36'1" -31d21'17"
Amy Eileen Holston
Lyr 18h47'6" 30d11'32"
Amy Elena Silva
Ari 2h22'53" 13d54'18"
Amy Elizabeth
Ori 5h0'5" 10d19'49"
Amy Elizabeth
Psc 0h25'2" 16d55'34"
Amy Elizabeth
Leo 10h17'12" 16d29'49"
Amy Elizabeth
Cnc 8h58'46" 15d38'13"
Amy Elizabeth
And 0h40'42" 25d13'29"
Amy Elizabeth
Cas 23h31'41" 61d38'34"
Amy Elizabeth
Aqr 22h48'11" -3d28'48"
Amy Elizabeth
Aqr 22h41'52" -8d15'52"
Amy Elizabeth
Vir 13h56'43" -16d15'12"
Amy Elizabeth Aguirre
Lmi 10h37'16" 35d47'46"
Amy Elizabeth Bolen's
Dogstar
Cma 6h50'28" -16d32'32"
AMY elizabeth EVANS
Ori 6h13'49" 15d47'18"
Amy Elizabeth Forsyth
Vir 12h1'11" -0d7'44"
Amy Elizabeth Fulmer
Crb 16h1'56" 29d52'10"
Amy Elizabeth Gaither
Aqr 22h27'31" -1d38'4"
Amy Elizabeth Henderlong
And 23h20'15" 41d44'41"
Amy Elizabeth Jones
And 0h43'7" 36d19'13"
Amy Elizabeth Larmore
Sgr 18h5'16" -27d7'36"
Amy Elizabeth Leanne
Trethewey
And 1h34'32" 49d32'8"
Amy Elizabeth Milner
Cnc 8h55'39" 24d36'3"
Amy Elizabeth Oliver
Cap 20h10'11" -17d48'38"
Amy Elizabeth Roche
Sgr 19h18'47" -21d16'31"
Amy Elizabeth Sinclair
Tau 5h56'29" 23d58'8"

Amy Elizabeth Wolfe Per 4h11'14" 35d27'12"
Amy Ellen Akers (Passolt) Leo 9h48'30" 31d58'42"
Amy Elrod And 0h24'41" 30d56'28"
Amy & Eric June 3, 2006 Cam 7h13'22" 71d43'16"
Amy Esther Chamberlain Crb 15h33'59" 30d43'58"
Amy Faith Ari 3h28'5" 26d1'2"
Amy Faith Robinson Gem 6h33'49" 21d8'27"
Amy Faye Morgan Gem 7h34'56" 19d43'50"
Amy Feingold Bednar Oph 16h56'14" -0d7'47"
Amy Fetrow Cnc 8h10'18" 7d44'52"
Amy Finney And 0h39'53" 27d20'15"
Amy Ford Tau 3h37'43" 9d57'21"
Amy Forever Ari 2h20'18" 24d35'18"
Amy Forever Beautiful Tau 5h49'2" 14d30'25"
Amy Frederick Aqr 21h19'3" -7d46'0"
Amy Freund Lyn 7h7'2" 53d48'32"
Amy Fromsdorf Lmi 10h40'20" 31d45'32"
Amy Fuller Mon 6h44'52" -0d4'49"
Amy Gabriel Peg 21h35'37" 22d25'57"
Amy Gale Gustafson Aql 19h39'18" 8d27'11"
Amy Gedney Vir 13h29'3" -8d2'26"
Amy Gee Sgr 18h48'11" -35d35'33"
Amy Georgia Littler And 0h14'8" 45d26'10"
Amy Gerrans Griess And 1h19'16" 46d49'29"
Amy Gibeley Ogden Sco 16h14'3" -11d19'56"
Amy Gifford And 23h46'33" 34d27'10"
Amy Gillespie Cnc 9h13'44" 13d43'50"
Amy - Goddess of Love Cap 20h9'39" -8d51'58"
Amy Godin Psc 1h28'4" 33d34'37"
Amy God's Star And 1h12'31" 38d40'54"
Amy Goerner And 2h22'44" 45d34'32"
Amy Goline And 0h11'57" 30d37'41"
Amy Greene Aqr 22h55'26" -22d24'51"
Amy Grolnick Lyn 8h6'58" 43d52'34"
Amy Hall Lyn 6h17'18" 57d13'32"
Amy Hapeman And 0h12'34" 25d38'59"
Amy Harris Uma 10h28'3" 45d24'0"
Amy Harrison And 1h18'55" 34d41'47"
Amy Hartman Lib 15h11'54" -20d45'34"
Amy Hass Smith Leo 10h43'26" 15d53'23"
Amy Hatfield Wingate Ori 5h43'39" -5d4'3"
Amy Heath Cen 11h32'29" -47d15'12"
Amy Heinley Cas 23h17'44" 55d44'46"
Amy Higgins Cnc 9h14'29" 17d36'5"
Amy Holter Tau 5h26'14" 26d33'46"
Amy Hook Lyr 18h16'1" 34d54'11"
Amy Hope Ari 2h52'21" 29d28'56"
Amy Huang And 0h10'2" 46d28'50"
Amy Huang - Rising Star Cam 5h40'36" 56d18'8"
Amy Hui Leo 10h17'33" 15d58'57"
Amy Hutto Per 3h30'8" 47d4'50"
Amy Hwai Lin See Tau 3h50'56" 15d22'51"
Amy Ileen Gooch Lyn 7h26'16" 45d23'57"
Amy Isabella Cas 0h8'28" 52d35'58"
Amy & Ivy's little star And 2h36'38" 44d41'16"
Amy Jane Cyg 20h42'18" 49d58'43"
Amy Jane Tau 5h7'57" 18d19'22"
Amy Jane Pennington And 23h40'25" 35d43'12"
Amy Jane Watson - 29/01/2005 Ori 5h36'57" -2d29'25"
Amy Janelle Laughlin Per 3h34'17" 33d14'19"
Amy Janelle Pollard Cyg 21h14'26" 44d35'40"
Amy Jean Crb 16h1'50" 33d53'19"
Amy Jean Cap 21h29'56" -19d5'4"
Amy Jean Ballor Tau 5h55'20" 26d23'3"
Amy Jean Stodola Lyn 8h29'20" 58d17'38"
Amy Jenelle S. Umi 14h13'52" 69d9'17"
Amy Jennifer Wilson And 23h34'36" 38d10'17"
Amy Jo Leo 11h2'59" 4d43'30"
Amy Jo Aqr 22h36'40" -16d5'50"
Amy Jo Angel Heers Gem 7h32'19" 29d44'47"
Amy Jo & Carolyna McConeghy Umi 15h32'55" 72d51'26"
Amy Jo Crabtree Lyr 18h46'4" 37d56'27"
Amy Jo David And 0h28'8" 39d43'56"
Amy Jo Frederica Violet Bushong Aqr 21h6'13" -14d7'50"
Amy Jo Kensinger Cam 7h22'24" 64d31'37"
Amy Jo Lewis Ari 1h50'28" 23d16'50"
Amy Jo McDonald And 0h9'4" 46d15'40"
Amy Jo McNeil Ser 18h19'59" -13d48'33"
Amy Jo Rice Lib 14h30'50" -10d11'41"
Amy & John Forever! Cyg 20h42'3" 36d10'10"
Amy Johnson And 1h18'43" 42d58'47"
Amy Jorgenson Uma 8h48'18" 70d46'52"
Amy Joy McGrail Crb 16h6'20" 36d44'12"
Amy Joy (Trexler) Apgar And 0h54'49" 36d17'59"
Amy Julianna "Little Riding Star" Umi 15h0'52" 72d7'27"
Amy June Lial Ari 3h19'11" 24d5'50"
Amy & Justin Lib 15h26'45" -4d29'32"
Amy K. Miller & Chris A. Miller Cyg 19h45'27" 35d52'5"
Amy K Smithey And 23h47'37" 45d40'8"
Amy Katherine Cas 0h8'17" 53d35'33"
Amy Katherine Radonich Cru 12h1'31" -62d50'20"
Amy Kathleen Morgan Lyn 7h10'7" 52d19'54"
Amy Kathleen Smith And 23h9'26" 45d22'50"
Amy Katrina Winters Uma 11h31'18" 56d8'13"
Amy & Kavicy: Match Made In Heaven Cyg 20h3'46" 44d22'53"
Amy Kay Lyn 7h51'32" 47d2'24"
Amy Kay Baucom Ori 6h20'47" -1d35'8"
Amy Keller And 0h44'13" 31d10'37"
Amy Keller Lyr 19h4'2" 25d58'1"
Amy & Kevin's Wedding Day Star Pup 7h44'21" -23d29'0"
Amy Kifus 9 And 23h7'50" 48d57'55"
Amy Kiley Cap 20h50'2" -25d26'12"
Amy Kimberley Alice Hickey Gem 6h40'30" 30d49'7"
Amy Kizaki Psc 1h52'34" 7d20'10"
Amy Knott McDonald Uma 14h1'2" 54d6'25"
Amy Kuist Cnc 8h31'7" 23d27'47"
Amy L. Borrenpohl Sco 16h41'28" -37d14'0"
Amy L. Cusmano Lib 14h48'10" -15d56'40"
Amy L. Davis - 1436 948 And 0h48'30" 39d33'17"
Amy L. Fisher Aqr 21h41'16" -7d46'8"
Amy L. Scates Leo 10h23'22" 14d12'25"
Amy L. Slusser Cnc 9h12'29" 21d25'5"
Amy L. Stanford Tau 4h12'29" 4d18'30"
Amy L. Webber And 23h31'9" 47d0'52"
Amy La Montagne Leo 10h13'44" 18d29'30"
Amy Ladieu Lyr 18h34'58" 36d11'49"
Amy Lam Gem 7h21'14" 20d36'50"
Amy Lancaster Dra 17h41'28" 59d17'28"
Amy Laura Cubitts Cyg 21h46'21" 37d59'54"
Amy Laura Fontain And 1h12'34" 42d0'2"
Amy Lavada Truelove And 0h36'38" 33d19'36"
Amy Lea Ori 5h54'55" 3d6'7"
Amy Lea Bradford Lyn 8h20'14" 37d56'0"
Amy LeAnn And 2h35'10" 45d6'45"
Amy Leanne Petruskavich Ori 6h16'38" 14d45'50"
Amy Lee Leo 9h29'1" 11d24'34"
Amy Lee Cyg 21h57'25" 47d13'16"
Amy Lee Cam 4h38'7" 54d59'26"
Amy Lee Cyg 20h39'31" 45d45'1"
Amy Lee Gem 7h14'51" 34d30'18"
Amy Lee Lyn 8h6'40" 35d5'34"
Amy Lee Ann Lienhart Psc 0h18'33" 4d15'15"
Amy Lee Boggio And 23h1'28" 48d2'52"
Amy Lee Bond Sco 16h6'26" -8d47'16"
Amy Lee Boots Boo 14h32'33" 16d19'6"
Amy Lee Cronk And 1h40'46" 48d27'31"
Amy Lee Kimbrell And 22h59'25" 47d35'15"
Amy Lee Parker And 2h36'15" 45d12'25"
Amy Lee Pierzchalski And 0h54'7" 37d50'8"
Amy Lee Torla And 1h5'58" 45d21'1"
Amy Leigh Ori 5h17'57" 0d39'52"
Amy Leigh Aqr 22h53'5" -9d26'26"
Amy Leigh Sgr 18h23'3" -22d44'9"
Amy Leigh - Daughter Cnc 8h0'55" 22d29'56"
Amy Leigh Lore Mon 7h21'22" -0d43'22"
Amy Leigh Morriss Uma 9h17'26" 60d20'21"
Amy Leiter Sugerman Uma 9h55'45" 59d43'34"
Amy Linda Robinson And 1h59'58" 42d45'12"
Amy Lineberry Cyg 21h27'13" 33d22'52"
Amy Lisa Gadd And 0h53'38" 43d59'40"
Amy Lisabeth Gloeckner Gem 7h4'55" 17d48'28"
Amy Loomes Pho 0h16'19" -43d4'28"
Amy Lorentz Psc 1h1'8" 9d32'8"
Amy Lorin Lyn 8h21'30" 58d46'58"
Amy Lorraine Faas Ori 5h56'49" 18d28'56"
Amy Lorraine Page Cnc 8h34'35" 17d10'8"
Amy Lou And 1h26'38" 48d2'55"
Amy Lou Pho 0h29'52" -47d59'12"
Amy Lou Friend And 0h41'54" 37d58'42"
Amy Louise Leo 9h23'41" 16d58'11"
Amy Louise Bohlken Aql 19h24'51" 2d24'26"
Amy Louise Bowen Pyx 8h35'55" -25d35'11"
Amy Louise Choquette Sanborn Aqr 22h43'29" -15d27'59"
Amy Louise Good Umi 15h0'20" 70d43'48"
Amy Louise Mainella And 0h17'51" 27d25'16"
Amy Louise Maslin Tau 5h24'41" 17d3'43"
Amy Louise Pepper Tau 5h55'14" 24d46'55"
Amy Louise Schmisseur Crb 15h37'29" 30d25'8"
Amy Louise Stewart And 2h33'26" 49d9'2"
Amy Louise Vanderplank And 23h44'33" 41d52'23"
Amy Loves Ori 6h15'50" 15d20'18"
Amy Loves Chris Forever Tau 5h8'58" 24d9'5"
Amy Loves Jeffrey 4-23-05 TLA Uma 9h42'6" 56d21'46"
Amy Loves Mark Cyg 21h54'42" 47d29'37"
Amy loves Matt Lib 15h26'31" -4d31'22"
Amy Low And 2h31'37" 50d1'4"
Amy Lu And 0h52'15" 40d11'8"
Amy Lucienne Guilmet Gem 7h19'38" 22d23'55"
Amy Lucille Pettit Tau 4h37'3" 6d14'1"
Amy Luna Gem 6h50'8" 15d40'21"
Amy Lynn Ari 2h35'24" 24d43'13"
Amy Lynn Ari 2h10'15" 22d55'39"
Amy Lynn Tau 3h56'31" 21d32'20"
Amy Lynn And 0h20'52" 38d11'25"
Amy Lynn Gem 6h46'58" 32d34'59"
Amy Lynn And 2h22'49" 48d59'58"
Amy Lynn Vir 13h56'24" -18d0'5"
Amy Lynn Cap 20h21'17" -12d28'6"
Amy Lynn Boo 14h49'4" 54d40'55"
Amy Lynn 8/19/81 Leo 10h19'23" 25d55'26"
Amy Lynn Amadio Aqr 22h14'27" -23d24'2"
Amy Lynn Barber Lib 15h3'2" -3d4'32"
Amy Lynn Cerullo And 1h24'42" 49d51'24"
Amy Lynn Curry Del 20h56'41" 13d38'54"
Amy Lynn Davis Aqr 20h41'41" -10d37'44"
Amy Lynn Evans Psc 1h13'1" 32d8'48"
Amy Lynn Feinberg Lyn 8h3'7" 40d42'0"
Amy Lynn Goodpaster And 1h29'50" 49d15'4"
Amy Lynn Hallstrom And 2h2'55" 37d51'37"
Amy Lynn Jursich Cam 3h58'15" 59d33'59"
Amy Lynn McDermott And 0h29'4" 33d22'15"
Amy Lynn Pistininzi Psc 1h54'44" 8d16'31"
Amy Lynn Primm Cen 13h30'27" -37d48'19"
Amy Lynn Ragni Ori 5h52'44" 20d56'12"
Amy Lynn Slone Tau 4h49'30" 19d41'42"
Amy Lynn Stafford Vir 12h48'57" 7d38'4"
Amy Lynn Trudell Psc 0h20'19" 17d15'8"
Amy Lynn Vinson Lib 16h0'15" -16d29'53"
Amy Lynn Vonderheide Sco 16h43'37" -27d13'37"
Amy Lynn Walls Cyg 19h43'8" 36d7'54"
Amy Lynn Zuniga Ori 6h18'49" 15d7'2"
Amy Lynne Ori 5h38'17" -0d20'45"
Amy Lynne Polvani Ari 2h6'57" 20d25'10"
Amy Lynne Stock Cap 20h22'1" -20d0'10"
Amy Lynn's Star - Dolphin's Eye Del 20h51'40" 15d59'0"
Amy M. Lavallee And 0h42'27" 39d47'38"
Amy M. O'Neil Dra 19h48'45" 60d10'19"
Amy M. Ruschak Aqr 21h52'36" 1d42'42"
Amy Mae Richardson Tau 4h41'56" 17d14'32"
Amy Maples Lyn 7h40'17" 45d38'33"
Amy Mareck Lyr 18h46'7" 31d19'29"
Amy Margaret Cyg 21h17'52" 42d52'48"
Amy Margret Mills Aql 18h51'49" -0d49'47"
Amy Marie Mon 7h19'3" -0d46'46"
Amy Marie Lyn 7h29'27" 52d49'5"
Amy Marie And 0h58'24" 45d35'31"
Amy Marie And 0h43'53" 27d45'3"
Amy Marie Cnc 8h37'15" 20d54'55"
Amy Marie Chambers Lib 14h57'12" -6d50'18"
Amy Marie Cornett And 2h33'30" 47d56'50"
Amy Marie Daudelin Crb 16h13'36" 36d58'8"
Amy Marie Ferrell Gem 7h12'58" 32d26'15"
Amy Marie Flack Ari 2h32'15" 25d11'39"
Amy Marie Italiano Cyg 21h37'5" 53d41'41"
Amy Marie Kohle Gem 6h31'48" 26d14'46"
Amy Marie Lamar And 1h6'21" 35d0'9"
Amy Marie Lassila Sco 17h57'8" -43d44'59"
Amy Marie Meacle And 2h12'28" 50d23'29"
Amy Marie Nolte Tau 3h41'25" 19d6'34"
Amy Marie Rizzardi Ari 3h14'44" 29d41'44"
Amy Marie Shriver Psc 0h38'43" 15d44'27"
Amy Marie Smith Peg 21h40'33" 27d31'46"
Amy Marie Van Bibber Aqr 23h4'21" -11d5'7"
Amy Marie Wallace And 1h36'28" 37d18'37"
Amy Marks And 0h22'23" 26d33'13"
Amy Maronet Tau 3h35'6" 12d3'57"
Amy Martin Lib 15h50'43" -20d6'41"
Amy & Matt Cyg 19h41'23" 51d59'42"
Amy May Cnc 8h26'19" 23d11'17"
AMY may our love shine forever Cap 21h53'38" -13d39'9"
Amy Maynard; A Star for a Star And 23h42'6" 36d29'16"
Amy McAtee Cam 4h51'56" 73d7'28"
Amy McCartney Vir 12h17'59" 3d28'45"
Amy McDonald And 2h33'18" 49d35'2"
Amy McErlean Leo 9h53'11" 9d56'11"
Amy & Mel - Baba Nam Kevalam Ori 5h31'34" 2d11'3"
Amy Meldrum's Night Light Psa 21h48'51" -33d19'4"
Amy Melissa Pitts Psc 0h58'9" 26d18'58"
Amy Merckx And 23h17'57" 35d29'34"
Amy Mertz Crb 15h41'24" 35d56'55"
Amy/Michael's Love Star Cam 4h6'26" 69d2'59"
Amy & Michael's Star And 1h10'56" 45d53'16"
Amy Michele Hartley Gem 6h44'58" 28d5'39"
Amy Michele Scarff Umi 13h23'11" 75d57'18"
Amy Michelle Lib 15h26'15" -6d47'3"
Amy Michelle Tau 5h57'43" 25d4'40"
Amy Michelle Davidson And 0h27'20" 37d34'59"
Amy Michelle Dempsay Lib 15h29'29" -15d2'32"
Amy Michelle Gilmore Uma 9h24'56" 59d45'47"
Amy Michelle Miller "My Wiffie" Cas 23h59'43" 60d47'16"
Amy Michelle Spivey Lib 15h44'2" -11d11'20"
Amy Michelle Stannard Tau 3h37'55" 10d46'32"
Amy Michelle Webster Lib 15h29'27" -3d48'37"
Amy Michelle Wei And 2h38'42" 48d59'50"
Amy Migliore-Dest Cyg 20h2'22" 37d27'50"
Amy Miklasiewicz Tau 3h59'47" 9d19'37"
Amy Millicent Brown Psc 1h21'5" 32d56'19"
Amy Milliken Psc 1h14'34" 32d52'56"
Amy Monique And 1h17'12" 34d37'18"
Amy Moore Cam 5h57'45" 74d25'31"
Amy Morin Cyg 20h30'19" 42d22'34"
Amy Mullin Lmi 10h51'42" 28d16'23"
Amy Munger Cas 23h14'20" 55d43'42"
Amy My Eternal Love And Soulmate And 23h9'1" 41d57'30"
Amy my Love Aqr 21h45'35" -0d56'52"
Amy Nannel Linder Leo 11h19'55" 18d14'34"
Amy & Nevil Uma 9h10'34" 72d26'40"
Amy Ngo Sgr 20h26'55" -44d34'32"
Amy Niamh Huckins And 0h2'6" 40d50'36"
Amy Nicholson Tewalt And 23h59'40" 39d26'12"
Amy Nicole Uma 10h22'26" 68d48'23"
Amy Nicole Bozeman Leo 11h11'35" 4d57'45"
Amy Nicole Campbell Cyg 21h59'44" 47d37'0"
Amy Nicole Chapman Leo 9h53'53" 23d30'26"
Amy Nicole Dawes Umi 16h0'18" 80d48'31"
Amy Nicole Evans Vir 13h56'12" 1d49'34"
Amy Nicole Gilbert Aqr 21h36'19" -8d1'50"
Amy Nicole Konieczny Cap 20h19'53" -11d27'54"
Amy Nicole Martonik Aqr 21h49'19" -0d23'34"
Amy Nicole Ramsey Lib 14h49'53" -2d31'53"
Amy Nicole Savell Leo 10h11'46" 16d0'23"
Amy Nina DaSilva Ari 2h25'23" 17d38'6"
Amy Nottridge Cas 1h53'4" 60d12'32"
Amy Noverini And 2h0'15" 45d14'35"
Amy Okonski And 23h11'21" 41d13'40"
Amy Olive Uma 8h12'36" 60d21'59"
Amy Oliver Show Star of KFKA Radio Lyn 6h38'44" 61d16'47"
Amy Olivia Pierce Sco 16h52'39" -26d38'34"
Amy Onozato Sco 16h12'16" -13d48'56"
Amy P Umi 16h12'57" 73d53'43"
Amy Patnode Sco 16h40'33" -37d13'11"
Amy Patricia Cas 1h38'50" 59d9'5"
Amy Pauline Psc 0h46'18" 18d39'58"
Amy Pavia Vir 12h31'9" 2d15'17"
Amy Perkins Lyn 6h58'53" 47d43'25"
Amy Perry Cyg 19h43'9" 37d21'15"
Amy Pickett Cas 1h29'37" 52d15'29"
Amy Poe Ari 2h51'50" 26d37'30"
Amy Powell Aqr 21h48'32" 1d56'55"
Amy Powell Gem 7h17'6" 32d25'24"
Amy Quesnelle Cyg 19h56'30" 46d25'41"
Amy R. Leo 10h47'45" 12d20'2"
Amy R. Durdil Boo 14h31'37" 19d9'20"
Amy R. Outland And 1h44'35" 44d30'36"
Amy R. Polen Cap 21h55'32" -18d31'8"
Amy R. Riddlespurger Sgr 18h18'52" -18d19'21"
Amy Rachelle Turner And 1h30'33" 50d12'39"
Amy Rad Sco 17h6'2" -44d11'29"
Amy Rebecca Ori 5h55'17" 20d40'24"
Amy Rebecca Fleeman Sgr 18h4'34" -24d28'6"
Amy Rebecca Steinmetz Peg 22h22'43" 29d13'19"
Amy Reeder Psc 1h8'3" 19d26'27"
Amy Renae & Gregory Anton Wester Cyg 19h34'26" 28d48'34"
Amy Rene' Cnc 8h10'45" 32d41'46"
Amy Renee Gem 7h16'19" 17d25'5"
Amy Renee Tau 5h19'30" 18d17'53"
Amy Renee Marvin Cap 20h56'45" -21d10'38"
Amy Renee Santimaw Cnc 8h37'17" 7d4'42"
Amy Reza Lyn 7h14'19" 44d52'36"
Amy Rice And 2h31'0" 50d9'45"
Amy Riegelsberger And 0h15'55" 44d44'40"
Amy Rivard Leo 9h23'29" 6d49'9"
Amy Roberts Lyn 8h29'0" 35d11'38"
Amy Robertson Tau 4h21'23" 2d8'14"
Amy Robinson Umi 15h27'31" 72d47'21"
Amy Rosalie Cam 4h46'11" 56d6'35"
Amy Rose Dawkins Lyn 8h1'38" 41d16'4"
Amy Rose Latka Sco 17h28'37" -40d13'3"
Amy Rose Livingstone And 23h23'3" 47d54'56"
Amy Rose McGowan Ari 3h17'26" 26d18'12"
Amy Roue-Robson Mon 6h29'54" 8d32'50"
Amy Rufener Cyg 21h54'11" 41d54'44"
Amy Ruth Gunn Uma 8h18'17" 66d35'20"
Amy Ruth Mitchell Tau 4h33'55" 22d4'48"
Amy Sabrina Gardiner Vir 13h29'31" -19d34'49"
Amy Sarah Vir 13h1'55" -22d16'31"
Amy Scheeringa Cap 21h8'23" -25d23'24"
Amy Schroyer Leo 10h38'25" 16d17'42"
Amy Seaton And 0h18'38" 40d18'38"
Amy Serena Ori 6h18'43" 13d12'31"
Amy Shah Sco 16h12'4" -15d2'57"
Amy Shattuck Mon 6h57'56" 9d52'12"
Amy & Shaun - Eternal Love Psc 1h24'59" 29d13'3"
Amy Shay Tau 5h49'45" 17d41'43"
Amy Sheryl Garber Tau 4h51'7" 26d12'54"
Amy Simpson Cap 20h14'32" -18d1'3"
Amy Smith And 23h48'59" 43d59'2"
Amy 'Snuffles' Kruczek And 23h55'39" 36d11'24"
Amy Staines And 23h10'0" 42d34'0"
Amy Star Cas 0h20'30" 57d6'16"
Amy Stearns Ori 5h35'19" 10d38'19"
Amy Stell And 1h49'30" 43d37'22"
Amy Stenglein And 0h47'21" 36d9'0"
Amy Stephanie And 23h26'56" 37d16'15"
Amy Stratton Vir 12h41'59" -1d24'49"
Amy Stricoff Cas 0h12'3" 62d47'42"
amy sue Lyn 7h57'22" 48d7'18"
Amy Sue Williams Tau 5h11'34" 24d32'2"
Amy Suzanne Sgr 19h6'30" -18d33'36"
Amy Suzanne Mak Aqr 21h2'55" 2d5'11"
Amy Suzanne Osborn Sco 17h28'4" -44d44'0"
Amy Sweetums Miller Ori 6h18'49" 14d42'6"
Amy Synnott Com 12h47'18" 15d11'5"
Amy Synnott Lyn 8h16'18" 37d5'57"
Amy Tamara Byars Cnc 8h58'20" 22d50'51"
Amy Tay Lib 15h22'37" -25d41'57"
Amy Taylor Cas 0h12'17" 63d40'36"
Amy Tegantvoort Uma 10h18'5" 48d38'14"
Amy Teresa Miller Lib 15h28'39" -11d25'10"
Amy the Barefoot Girl And 23h26'39" 48d20'53"
Amy the Beautiful And 1h40'51" 37d56'18"

Amy The Cutest Cat
Lyn 7h10'43" 49d36'17"
Amy the Magnificent
Umi 16h39'47" 88d46'45"
Amy the Muse
Ori 5h57'3" 21d8'47"
Amy Thoman
Cam 7h1'37" 72d24'54"
Amy Thomas
Car 7h44'26" -51d3'22"
Amy Thompson
Leo 11h2'46" 4d57'13"
Amy Ting
Umi 16h52'58" 78d7'1"
Amy T.L. Abbott
Cap 20h37'7" -18d54'59"
Amy Traskowski
Sgr 18h47'40" -16d51'2"
Amy TWinkel
Psc 0h50'19" 9d7'39"
Amy Underwood
Cap 20h36'48" -15d51'58"
Amy V. Heath
Vir 13h16'54" 13d12'46"
Amy Vanderhoop
And 0h34'58" 37d6'4"
Amy Vasquez
Ari 3h15'34" 29d28'36"
Amy Verdi
Psc 1h16'22" 16d46'11"
Amy Verge
And 1h20'22" 47d16'45"
Amy & Victor Lindquist
Love Star
Cyg 21h40'21" 49d40'6"
Amy W. Roberts
Cas 0h23'14" 55d5'14"
Amy Walker
Cas 0h28'18" 63d0'45"
Amy Warinner Mystic Toots
Cnc 9h14'38" 8d7'16"
Amy Waters
And 1h7'20" 35d33'18"
Amy Watson
Cas 0h43'18" 61d59'13"
Amy Whitson
Ari 2h55'46" 19d55'1"
Amy Wiese
Cnc 8h43'0" 9d6'0"
Amy Winters
Leo 11h32'18" 26d46'38"
Amy Woldert
Cyg 20h14'42" 48d46'46"
Amy Woody
Sgr 19h1'44" -35d28'55"
Amy Young, Beauty of the
Heavens
Ori 5h46'51" 6d2'18"
Amy your the Love of my
life
Ori 6h20'54" 2d30'25"
Amy Zolun
Sco 16h41'26" -34d46'9"
Amy, The Most Beautiful
Star Of All
Aqr 22h29'54" -1d50'54"
Amy, to the stars and back!
Jaime
Vir 13h56'48" 6d10'21"
Amy, you will never be in
the dark.
Leo 9h45'23" 25d35'43"
Amy11671
Uma 11h45'7" 29d54'48"
Amya Joy Anderson-
Salaam
Cap 20h39'32" -25d40'23"
Amya Nicole Farquharson
And 0h25'56" 40d15'14"
Amya Patricia
And 1h0'52" 38d8'40"
AmyAndSadieMyers
Ori 6h18'5" 10d33'58"
AmyBeth
Lib 15h25'10" -28d5'57"
Amy......Forever Shining
Brightly
Vir 11h49'8" 8d7'26"
Amy-Jane Cole
And 23h30'28" 47d37'0"
Amy-Kaye Elizabeth
Mitchell
Ori 6h5'59" 18d34'16"
Amykins
Ari 2h12'44" 23d7'45"
Amyliz13
Gem 7h1'39" 29d36'40"
Amy-Louise May
Cru 12h1'18" -60d27'15"
Amylyn Grace Wlazlo
And 0h20'35" 42d48'4"
Amymarie
Aqr 22h7'25" -18d17'42"
Amy-Michelle
Peg 0h12'24" 17d19'30"
Amymichelle amore
Psc 23h35'33" 6d50'19"
Amypie
Uma 12h2'29" 56d30'36"
Amy's
Mon 6h44'26" 10d45'13"
Amy's 18th Birthday Star
And 1h16'27" 49d14'4"
Amy's 21st - 04/01/06
Cap 20h9'42" -14d0'41"
Amy's Angel
Cap 20h41'1" -20d55'52"
Amy's Beloved Star of
Hope
Per 2h56'10" 41d33'7"
Amy's Dragonstar
Sgr 20h19'1" -41d38'12"
Amy's Emma - A Mothers
Day First
Tau 4h28'22" 12d31'44"
Amy's Eye
Ori 5h33'13" 10d13'40"
Amy's Fire AMM 06041974
Ari 2h28'35" 21d36'10"
Amy's Light of Hope
Lyn 7h59'16" 39d11'0"
Amy's Muffin & Supreme
Tobyboy
Umi 14h34'4" 68d28'3"
Amy's Pooh Bear
Psc 1h26'20" 16d43'7"
Amy's Rusty Star
And 0h27'46" 37d31'11"
Amy's Ryan
Uma 11h18'58" 68d45'4"
Amy's Shining Star
Leo 9h25'32" 12d21'19"
Amy's Smile
Aqr 20h47'30" -14d32'9"
Amy's Star
Cap 20h27'36" -11d1'37"
Amy's Star
Vir 13h31'30" -1d8'34"
Amy's Star
Cmi 8h3'28" -0d9'39"
Amy's Star
Cas 23h33'31" 55d16'13"
Amy's Star
Sco 16h35'20" -37d33'39"
Amy's Star
Sgr 20h16'46" -29d36'42"
Amy's Star
Cnc 8h44'28" 7d26'36"
Amy's Star
Leo 9h29'20" 28d8'27"
Amy's star
And 0h35'13" 27d50'18"
Amy's Star
Cnc 8h35'20" 15d13'30"
Amy's Star
And 1h43'26" 37d55'23"
Amy's Star
Gem 7h18'28" 32d46'45"
Amy's Star
And 2h20'26" 47d29'28"
Amy's Star
And 23h31'1" 47d6'14"
Amy's Star
And 23h49'13" 48d22'59"
Amy's Starbright
Gem 6h51'38" 16d7'31"
Amy's Wish
And 0h31'0" 38d2'50"
Amy's Wish Come True
From Her Angel
Ori 5h18'57" -0d38'38"
Amy's Wish Upon A Star!!
And 0h47'58" 27d32'25"
Amy-snuggles
And 2h28'7" 50d26'13"
Amy-Steve Star
Her 17h43'35" 41d53'5"
Amythyst Mae Romero
Hya 9h34'24" -0d39'55"
Amy-ZING in the HEAV-
ENS
Leo 11h0'26" -1d48'33"
Amzo
Cap 20h53'30" -17d2'40"
An
Cas 23h28'38" 51d28'38"
an affirmation of Allison's
beauty
Sco 17h35'58" -39d2'38"
An Amazing Star[t] to Life
Together
Per 3h50'31" 33d35'34"
An Angel called Belle
Ari 2h59'41" 29d8'7"
"An Angel Named Amber"
AmberCollier
And 23h56'58" 33d14'23"
An Angel's Love 3/13/03
Lyn 7h38'55" 52d31'7"
An "Angel's" love is
"Honey" sweet!
Cyg 21h46'38" 49d13'39"
An Angel's Smile
Col 5h50'47" -33d15'13"
an angi von herzen
Cep 21h16'17" 57d40'37"
An Apology for Sharon
Sgr 18h41'4" -23d3'45"
An Asset To My Heart
Uma 8h14'42" 62d57'20"
An dara rós déag
And 1h5'38" 39d14'41"
An Eye in the Sky
Uma 9h9'53" 53d14'40"
An Jing
Uma 11h16'52" 66d18'19"
An Réalta Karen
Vir 13h13'59" 5d59'0"
An Réalta Olive
Mon 6h29'4" -4d36'44"
Ana
Vir 14h17'51" -6d16'53"
Ana
Uma 10h12'1" 70d59'25"
Ana
Uma 9h8'48" 70d56'42"
Ana
And 23h55'38" 35d47'25"
Ana
Cas 0h23'35" 54d17'9"
Ana Alejandra Canseco
Leo 9h38'52" 27d45'34"
Ana Alicia Pena Díaz
Ari 2h35'10" 19d17'8"
Ana Alonso
Peg 22h6'49" 10d59'14"
Ana Alysya Adams
Cas 2h17'24" 73d57'22"
Ana Amalia Rivera de Leal
Cnc 8h40'9" 22d33'24"
Ana Ashmore
And 23h58'31" 36d56'3"
Ana Babes Babic
Cas 23h32'58" 52d56'56"
ANA BAHIBAK HABIBI
BASEL!
Cep 21h51'30" 56d58'5"
Ana Banana
Lyn 7h41'24" 40d9'45"
Ana Barreto
Ant 10h50'40" -36d45'20"
Ana Belen
Vul 19h50'55" 21d55'48"
Ana Bobo Ollie
Sco 16h59'31" -38d37'2"
Ana Bulat
Dra 18h46'30" 59d51'24"
Ana Camelia Ilinca
Aqr 21h59'42" -16d39'41"
Ana Carolina B. Maynart
Cru 12h46'24" -60d29'56"
Ana Carolina Chaves
De'Carli
Crb 15h53'31" 27d35'57"
Ana Carolina Liu
Vir 13h12'39" -13d53'23"
Ana Celsa
Dra 18h59'9" 53d8'1"
Ana Clara Rodrigues
Leo 10h44'7" 15d59'58"
Ana Cristina Forcada
And 0h40'51" 27d13'50"
Ana Cuevas
And 23h48'30" 47d9'30"
Ana del Carmen
Cnc 8h7'57" 27d8'51"
Ana Delgado
Ari 1h58'46" 14d4'43"
Ana Delia Placeres
Pascual
Ori 5h0'31" 5d49'58"
Ana Dorina Pop
Cnc 9h15'59" 31d56'44"
Ana "DUCKIE" Melgar
And 1h20'31" 37d6'7"
Ana Elizabeth
And 2h12'41" 46d8'25"
Ana Elizabeth Moreno
Gem 6h29'37" 25d55'33"
Ana Elizabeth Razo Perez
Cas 23h1'56" 53d44'54"
Ana Elizabeth Warthling
Umi 14h37'5" 70d4'20"
Ana Estrella
Vir 13h3'53" -7d17'53"
Ana Evelyn Juricic
Lyn 7h55'10" 49d37'40"
Ana Felix
Uma 9h35'11" 69d46'23"
Ana G.
Ori 5h1'15" 10d10'48"
Ana Gabriela Cintron
Sgr 19h44'42" -17d48'35"
Ana Garcia
Uma 10h59'58" 54d41'3"
Ana Garcia Clemente
Aql 20h12'48" -0d33'43"
Ana Girithyaveiel
Dra 18h30'3" 54d52'48"
Ana Gladys Zaiduni
Leo 11h47'47" 21d30'15"
Ana Gloria Fernandez de
Condello
Tau 3h54'41" 7d4'3"
Ana Gomez
Com 12h51'6" 26d26'56"
Ana Gonzalez
Com 12h47'21" 26d59'48"
Ana Gracie
Vir 12h10'53" 5d24'41"
Ana Grande
Gem 6h14'23" 26d27'38"
Ana Guerrero
Lyn 9h12'51" 33d14'9"
Ana Huerta
Aqr 21h8'43" -3d8'14"
Ana Iglesias Rodriguez
Her 18h26'21" 23d39'14"
Ana Isabel Coronado
Corchón
And 23h47'19" 43d23'50"
Ana Isabel Locsin
Aur 6h34'20" 35d12'28"
Ana Isabel Mendoza
Lyn 8h4'51" 34d15'13"
Ana Isabel Rocha
Cap 21h27'19" -19d16'7"
Ana Isabel Ruiz Sánchez
Sgr 18h16'47" -28d33'23"
Ana Julia Rivera
Tau 4h33'6" 16d54'5"
Ana Karla Serna
And 1h24'30" 45d35'49"
Ana Kyle, A Bright Star.
Aql 20h6'55" 15d43'57"
Ana L. Bolanos
Ori 6h16'0" 14d9'39"
Ana Latorre Viñes
Aqr 22h14'40" -14d56'46"
Ana Laura Fonseca
Lib 14h29'17" -23d26'47"
Ana Laura Ramos & Yoyo
Hamelius Mtz
Vir 12h6'10" -1d19'14"
Ana Leonor & Maria Teresa
Cas 0h53'4" 62d21'28"
Ana Lilia M
Cas 0h34'42" 69d30'59"
Ana Lilia Tovar
Vir 13h14'4" 1d17'42"
Ana Lise
Psc 1h31'52" 12d45'7"
Ana Lourdes Irwin
Sgr 20h0'50" -42d13'55"
Ana Lucia
Tau 3h53'57" 8d43'57"
Ana Lucia
Com 12h25'2" 24d54'58"
Ana Lucrecia
Eri 4h11'20" -23d11'58"
Ana Luisa
And 0h36'42" 23d57'25"
Ana Luisa
Cnc 9h7'13" 20d48'50"
Ana Luisa Santiago mi vida
mi amor
Lmi 10h46'13" 24d21'4"
Ana M. Frayre
Leo 10h19'47" 14d26'53"
Ana M. Hernandez
Her 17h26'43" 26d46'21"
Ana M. Navas Jara
And 23h53'11" 43d10'16"
Ana M. Prieto
Lib 15h37'56" -7d17'6"
Ana Marcia Amberg de
Castro
Dra 19h39'22" 61d44'50"
Ana Maria
Lib 15h12'15" -6d7'58"
Ana Maria
Cyg 21h20'30" 35d58'45"
Ana Maria
Ari 2h45'49" 16d46'42"
Ana Maria A. M. Soares de
Azevedo
Cas 1h16'24" 55d1'38"
Ana Maria Baldor-Bunn
Mon 7h36'52" -0d50'46"
Ana Maria Hatley
Ori 5h33'13" 0d33'53"
Ana Maria Llambes
And 23h55'40" 39d12'11"
Ana Maria Rivas Aguirre
Tau 4h15'18" 1d29'37"
Ana Maria Rivera
Dra 16h35'47" 68d13'19"
Ana Maria Trujillo Jimenez
Gem 7h49'3" 31d30'52"
Ana Marie
And 1h16'21" 36d34'56"
Ana Marie Antonescu
Leo 10h58'54" 5d18'22"
Ana Marie Natale
Lib 15h29'44" -10d5'25"
ANA MARIJA NANI ANCI
Peg 23h42'1" 14d54'27"
Ana + Mario
Cyg 20h28'55" 58d17'44"
Ana Maurizot (bobita...)
Ori 5h3'13" 5d26'5"
Ana Merino
Leo 10h18'34" 24d18'33"
Ana Moo
Lyn 6h32'34" 57d16'19"
Ana M.Rodriguez
Psc 1h10'40" 27d54'23"
Ana Netanel
Eri 4h29'58" -3d37'22"
Ana Nikolic
Ori 6h16'16" 3d14'10"
Ana & Pau
Ori 5h55'48" 20d48'6"
Ana Paula da Fonseca
Cnc 8h24'47" 30d28'56"
Ana Paula Reis
Cyg 21h32'35" 31d16'48"
Ana Paula Rose
Cas 0h57'48" 50d13'17"
Ana Pena
Cnc 8h40'31" 32d5'10"
Ana Pendic
And 2h14'11" 48d8'28"
Ana & Randy
Lyr 18h50'5" 34d26'16"
Ana Rangel
And 0h41'4" 36d51'37"
Ana Reyes Concepcion
Sgr 20h17'21" -32d53'49"
Ana Rivera
Sco 16h10'16" -10d20'34"
Ana Rosa
Aur 5h9'23" 33d25'6"
Ana Rose
Cnc 8h50'32" 10d54'20"
Ana Ross
Lyn 7h48'31" 52d32'37"
Ana Saa
Cnc 8h50'47" 12d10'12"
Ana Simarro
Aqr 22h39'26" -4d46'56"
Ana Sofia
Lib 15h9'58" -23d45'51"
Ana Starr
Ori 5h27'28" 5d55'3"
Ana The Glow In My Heart
Crb 16h8'40" 36d9'47"
Ana Tosic
Peg 22h21'36" 23d55'7"
Ana Valdivieso
Ari 3h12'5" 11d25'42"
Ana Victoria
And 23h24'43" 51d22'56"
Ana Victoria Murga
Hernandez
Cma 7h26'58" -15d24'54"
Ana Victoria Salazar
Cap 20h34'21" -18d14'7"
Ana Willich
Crb 15h48'48" 27d5'6"
Ana Wright
Cas 0h52'50" 61d11'31"
Ana y Chus 14-02-2004
Cap 21h32'39" -8d31'14"
Ana Yoselin Bugallo
And 0h29'30" 31d20'7"
ana7
Cep 2h42'20" 81d57'28"
Anabel
Aqr 22h57'47" -22d39'42"
Anabel
Cnc 8h43'18" 18d55'14"
Anabel
And 1h24'22" 45d53'39"
Anabel Chartier "Granny
Good Times"
Lyn 7h41'10" 47d52'37"
Anabel Melian
And 2h10'30" 40d58'22"
Anabela
Cam 4h43'21" 61d22'55"
Anabelicious "MH2UH"
Ari 3h8'31" 30d5'10"
Anabell Almazan
Aqr 22h3'33" -3d33'49"
Anabella Garcia
Uma 11h33'59" 63d5'53"
Anabella Haynes Platt
Psc 1h15'27" 16d40'0"
Anabelle
Mon 6h42'58" 4d8'8"
Anabelle
Gem 6h36'37" 24d34'4"
Anabelle
Pav 19h39'54" -58d23'38"
Anabelle J. Miele
Lib 14h59'19" -3d41'57"
Anabelle Mae Blakeley
Peterson
And 23h3'44" 40d26'32"
Anabelle Marie Chemelli
Lib 14h49'14" -3d5'55"
Anabeth Ambriz
Vir 13h42'32" 3d2'17"
Anacelia Martinez
And 1h12'2" 47d2'35"
Anacriz Ramirez
Aqr 21h49'51" -1d52'39"
anaeli
Ant 10h55'50" -35d55'17"
Anaëlle Soracolli
Lib 15h31'54" -5d7'49"
Anaely
And 0h37'57" 25d31'56"
Anahi Stefany Torres
Aql 19h3'8" 8d16'18"
ANAHIT
Aqr 23h13'32" -8d0'34"
Anahita
Per 2h24'39" 55d47'10"
Anahita
Cnc 8h48'44" 32d4'3"
Anahita
Uma 12h6'22" 44d57'44"
Anahita Norouzi
Gem 7h44'11" 33d3'11"
Anahita's Star
Cap 21h4'6" -27d20'44"
Anaili
Leo 9h34'22" 30d51'40"
Anaïs
Ori 5h30'44" 9d30'52"
Anaïs
Cap 20h7'36" -9d58'38"
Anais "Anna" Elizabeth
Gomez
Cas 1h44'0" 66d8'59"
Anais Geraldine Blin
And 2h28'45" 38d9'34"
Anais Grace's Star
Umi 17h13'32" 82d7'25"
Anaïs Leloup
Cap 20h30'6" -10d40'18"
Anais Lucia Christeen
Ledbrook
Pho 1h5'58" -48d28'29"
Anais Rovira
Uma 8h53'19" 65d13'42"
Anais Zaabi
Lyr 18h47'42" 38d20'48"
Anaiya Sage Falk Brown
Tau 5h59'48" 24d2'3"
Anaiyah Elizabeth
Mercurius
Lib 15h0'32" -9d22'54"
Anaja Elaine Parris
Psc 0h21'5" 17d46'0"
Anakila
Sgr 17h54'49" -17d16'38"
Anakristen
Tau 3h31'50" 2d38'26"
Analaiya
Lyn 8h24'26" 34d8'45"
Analays Alvarez Blanco
Leo 11h52'27" 23d0'34"
Analeah Ortiz
Gem 7h39'44" 33d52'45"
Analee
Ori 5h11'36" 15d45'51"
Analee Plew Robinson
Lyr 18h47'55" 39d53'50"
Analia
Gem 7h32'39" 32d36'11"
Analiese O'Leary
Leo 9h43'15" 27d48'33"
Analilia Hernadez
Sco 16h13'31" -16d4'34"
Analisa & Ava
Umi 13h40'11" 75d34'29"
Analisa Rose
And 2h20'18" 43d19'5"
Analise Blanchard
Cas 1h37'27" 68d34'57"
Analise Chatterpaul
Mon 6h53'12" -0d22'3"
Analise Jacqueline Winter
Cas 1h25'26" 61d49'53"
Analise Jacqueline Winter
Psc 1h11'2" 31d7'20"
Analise Winter
Cas 0h59'57" 55d29'59"
ANALU
Lib 15h52'43" -18d12'30"
Analytik J
Sgr 18h36'23" -29d58'46"
Anam Arif Rashid
Umi 14h41'58" 79d3'27"
Anam Cara
Cap 21h20'43" -19d38'56"
Anam Cara
Dra 19h9'53" 64d7'0"
Anam Cara
Cas 0h40'48" 51d18'42"
Anam Cara
Lmi 10h22'43" 30d18'47"
Anam Cara
Lyn 7h49'3" 39d3'35"
Anam Cara
Gem 7h24'50" 21d45'20"
Anam Cara
Ori 5h7'43" 7d6'11"
Anam Cara Irish Shea
Godwin
Cyg 19h57'54" 30d28'23"
Anam Cara Nellie
Mon 7h33'22" -0d17'40"
Anamaria
And 23h28'13" 37d58'31"
Anamaria
Cas 1h28'38" 57d59'32"
AnaMaria Jimenez
And 1h33'57" 44d25'9"
Anamarie Emily Kelly
Lib 15h4'42" -6d57'17"
Anamarie Sangeniti
Psc 0h20'18" 3d16'25"
Anamchara
Sgr 19h1'22" -28d38'57"
Anamchara
Aqr 22h40'10" -2d34'15"
Anamish
Lyn 8h1'52" 39d20'19"
Anand
Ori 4h51'9" 4d3'59"
Anand & Tammy
Gem 7h30'17" 20d27'36"
Ananda Lea Grieser
Cyg 21h38'36" 47d32'36"
Ananda Maeve Wielock
Cas 0h56'0" 53d40'36"
Anando
Men 5h13'22" -70d49'44"
Anannya
Eri 4h18'28" -31d13'19"
Anant Harwalkar
Cas 1h6'31" 64d33'24"
Anaquetinita
And 0h36'53" 27d17'32"
ANARITHA
Uma 13h59'2" 49d36'15"
AnarKahikeE
Lyn 8h29'23" 59d17'56"
Anarkalli J Aakarssha
And 2h28'34" 38d1'2"
Ana-Rosa Bes-Fernandez
Buesso
Cas 1h3'1" 53d23'57"
Anarossa Martinez 57
Umi 15h19'15" 67d52'53"
Anartic Evening Star
Lyn 8h57'20" 34d42'11"
Anas
Cyg 21h20'57" 30d18'30"
Ana's "Arbol" en Las
Estrellas
Sgr 18h24'29" -20d20'54"
Ana's Extra Ordinary
Uma 11h19'27" 32d16'35"
Ana's Star
And 1h25'24" 40d13'42"
Ana's Star ~ 24.1.1986
Cru 12h24'20" -57d13'55"
A-NASA Star
Per 2h51'32" 54d45'1"
Anasazi Ridge
Uma 12h20'3" 55d25'15"
ANASFELIS
Cep 22h1'30" 69d17'46"
Anastacia
Oph 17h34'21" -0d17'30"
Anastacia Alicia-Marie
Price
Sco 16h41'54" -40d7'0"
Anastacia Liana Abell
Gem 7h47'5" 31d24'30"
Anastacia Renee Gonzales
And 2h5'35" 38d48'58"
Anastacija Lee Balic
Lmi 10h53'4" 38d34'13"
Anastashia
And 0h23'7" 42d43'58"
Anastasia
Peg 22h22'52" 33d51'26"
Anastasia
And 2h3'54" 37d58'31"
Anastasia
Aur 6h12'31" 36d7'50"
Anastasia
Gem 7h44'23" 33d14'47"
Anastasia
Cas 0h57'28" 55d54'59"
Anastasia
And 23h19'27" 47d13'46"
Anastasia
Her 17h14'28" 24d23'24"
Anastasia
Tau 5h58'24" 24d59'19"
Anastasia
Cnc 9h0'40" 22d49'1"
Anastasia
Psc 0h25'58" 7d49'57"
Anastasia
Psc 1h17'21" 15d26'19"
Anastasia
Vir 12h7'36" 11d31'21"
Anastasia
Cnc 8h49'40" 8d34'36"
Anastasia
Cra 19h0'30" -38d51'45"
Anastasia
Sco 17h47'13" -41d21'33"
Anastasia
Vir 13h28'39" -18d29'47"
Anastasia
Aqr 23h5'22" -9d13'29"
Anastasia
Cam 5h22'43" 60d19'6"
Anastasia Ashley Monk
Sco 17h26'7" -32d25'21"
Anastasia Blue
Cap 20h32'59" -11d29'59"
Anastasia Chan Choy Yoke
Cas 1h37'12" 64d53'31"
ANASTASIA CHIARA
MILETO
And 0h43'5" 26d44'9"
Anastasia Childs
Cas 23h50'41" 57d16'53"
Anastasia Economides
Lyn 9h21'42" 33d50'18"
Anastasia Eden
Lib 14h53'13" -3d44'33"
Anastasia Hughes
Cas 0h29'29" 52d17'3"
Anastasia Ifandis
Lib 15h43'54" -9d33'53"
Anastasia Ionkina
Cap 21h36'0" -16d3'20"
Anastasia J. Ault
Cas 23h41'49" 53d47'7"
Anastasia J. Checchio
Aqr 22h31'41" 2d29'15"
Anastasia Jude Arnold
Aqr 22h27'20" -1d52'39"
Anastasia Lee Vandiver
Sco 15h51'7" -24d35'58"
Anastasia Marina
And 23h33'7" 48d8'19"
Anastasia McDuffie
Leo 10h19'0" 17d7'45"
Anastasia Papadimitriou
And 1h6'28" 46d51'49"
Anastasia Plonkey
Mon 8h2'50" -0d28'7"
Anastasia Williams
Tau 3h46'34" 21d12'48"
Anastasia-Rosaelia-
Sebastian
Her 16h51'21" 34d50'12"
Anastasios
Uma 9h53'24" 42d45'43"

Anastasios
Lyn 6h34'7" 56d44'23"
Anastasios Michaelides
Uma 10h18'7" 49d55'18"
Anastasios Mousmanis
Cep 21h53'2" 63d24'52"
Anastasios Tsohos
Per 3h10'27" 50d18'47"
Anastasiya
Ari 3h1'54" 29d11'52"
Anastasiya
Tau 3h25'6" -0d26'48"
Anastas'ka
Cap 20h48'15" -16d29'53"
Anastasoff
Lyn 6h31'29" 57d4'6"
Anastassia Andryuschenko
Cnc 8h52'37" 25d53'6"
Anastassia Shapardanova
Aqr 21h37'28" 1d47'49"
Anastassia Syrrakos
Tau 5h36'20" 18d25'9"
Anastastia
Lyr 18h45'35" 44d16'49"
Anat
Uma 10h25'0" 68d43'9"
Anat
Cet 1h9'26" 0d46'34"
Anatoli Tsvetkov Angelov
Cnc 8h40'11" 8d8'1"
Anatolij & Galina
Cas 23h34'50" 56d48'12"
Anatoliy Belitchenko
Aqr 21h14'53" -8d50'25"
Anatoly Garelik
Lyn 8h25'21" 47d11'17"
Anaya Thais
Cyg 20h17'24" 45d48'23"
Anayeli
Leo 10h3'32" 22d12'22"
ANB & NROE 2006
Ari 2h11'44" 22d32'37"
Anbela
Eri 4h34'19" -15d21'19"
ANC -DRE
Sgr 19h2'43" -30d43'10"
Anca
Ari 2h23'30" 18d15'21"
Anca Daniela Dobrea
Cas 23h38'55" 55d13'21"
Anca Emma Iacob
And 23h40'15" 47d11'34"
Anca Lapusan
Psc 1h44'23" 22d32'30"
Anca-Andreea
Cnc 9h7'56" 31d21'41"
Anca-Gabriela
Aqr 22h23'13" -22d3'23"
Ancalimë
Gem 6h40'32" 22d3'53"
Ancestral Fire
Dra 18h20'18" 54d45'1"
An-Ch-El
Uma 10h3'9" 41d33'20"
Ancia & Vale
Cam 3h46'25" 54d36'37"
...and ever
Leo 11h33'7" 17d33'8"
AnDaLaMaDaMi
Cnc 9h3'6" 27d58'44"
Andalerov
Ari 2h3'56" 10d41'52"
Andante Andante
Sgr 18h25'0" -18d48'10"
AndaRik
Cyg 21h26'51" 37d0'16"
Andatura
Sco 16h16'17" -15d41'38"
AndCo21-2-04
Uma 9h28'10" 44d39'54"
Andee Carlson
Aur 6h4'13" 50d0'59"
Andee Lynn Costabile
Aqr 22h43'40" -6d13'43"
Andel
Ori 6h8'25" 18d16'54"
Andel Lucie Vesela
Cas 23h58'33" 59d18'55"
Andeol
Dra 16h12'5" 69d12'0"
Andera e Sandro
Lyr 19h3'38" 26d52'39"
Anders Daddy Cool
Hanneborg
Uma 10h3'21" 58d2'59"
Anders Keady Gustafson
Umi 15h32'32" 71d1'8"
Anders Olaf Darrell
Johnson
Aql 19h25'6" 9d29'20"
Anders Whitman Benson
Cyg 19h59'45" 37d37'17"
Andersen Masimo
Mahoney Breen
Leo 11h1'35" 7d25'24"
Andersenian
Tau 5h49'34" 17d13'6"
Anderson
Lyn 9h19'6" 34d39'30"
Anderson Arnold Sylver
Psc 1h20'50" 29d26'25"
Anderson James
Sgr 18h31'27" -16d42'55"
Anderson McAree
Nightingale
Aur 6h4'24" 50d20'47"
Anderson - Oliver
Umi 15h44'7" 76d9'39"
AndersonGCM
Umi 16h22'44" 71d32'20"
Anderson's Light
Uma 8h26'1" 61d34'25"
Anderson's Love Sweet
Love
Cyg 21h52'34" 49d2'2"
Anderson's Wish
Lyr 18h52'59" 39d36'43"
Andi
Gem 7h8'14" 33d33'39"
Andi
And 0h19'4" 26d9'14"
Andi
Vir 12h46'18" 13d0'43"
Andi
Ori 5h19'44" 0d23'29"
Andi
Sgr 19h20'27" -16d9'23"
Andi Baby
Tau 5h20'7" 28d1'48"
Andi & Bandi forever
Uma 8h24'59" 62d28'40"
ANDI EROS
Uma 10h16'25" 52d57'57"
Andi Ich liebe Dich!
Ori 6h1'2" 7d6'36"
Andi & Jessie
And 0h42'37" 43d41'9"
Andi & Karin
Per 4h16'9" 33d58'51"
Andi Kulkies
Umi 17h6'11" 76d27'14"
Andi Parkhurst
Cnc 8h50'22" 29d55'34"
Andi Roxburgh
And 2h11'29" 44d50'17"
Andi Schatzeli
Aur 5h20'0" 30d0'27"
Andi Warner "Class of
2000"
Ori 4h54'33" -0d55'26"
Andia Abrana Katz
Peg 22h40'51" 24d51'30"
Andida
Lmi 10h7'12" 33d1'39"
AndiDrew
Cyg 19h51'43" 36d28'27"
Andie
And 2h26'14" 44d45'56"
Andie (Hone) Armstrong
Ori 5h31'36" 11d12'53"
Andie James
Cyg 19h48'1" 35d51'17"
Andie & Kevin
Cas 0h54'30" 54d17'18"
Andie Minton
Aqr 22h10'6" 0d43'27"
Andie Szabo
Cap 20h31'5" -11d11'34"
Andigi
Vir 14h15'32" -15d45'16"
Andilea
Cas 1h33'31" 67d11'9"
Andixx
Cap 20h48'23" -20d13'50"
Andjelina
And 23h51'57" 44d36'0"
A.N.D.M Ashton
Gem 7h6'11" 17d26'9"
Andon
Ori 5h37'14" 7d33'28"
Andorah's Mouse
Cyg 21h52'5" 48d15'54"
Andot
Uma 9h11'49" 56d43'52"
Andra
Tau 4h25'27" 24d31'45"
Andra Krick "Love Star"
And 23h10'46" 47d51'8"
Andra Redman
Uma 8h40'59" 59d9'29"
Andra & Stefan
Ori 5h47'4" 6d47'13"
Andralex
Her 17h10'7" 29d47'8"
Andranik & Lusine
Cyg 20h32'7" 39d42'24"
Andranik Mirzoyan
Cap 21h12'54" -17d12'21"
Andraphalees
Vir 12h59'21" -14d32'41"
André
Cap 20h10'41" -21d21'52"
André
Lac 22h47'43" 49d21'18"
André
And 1h17'17" 44d41'53"
Andrè
Uma 10h56'3" 43d43'58"
Andrè
Ori 5h34'47" 9d48'14"
André
Ori 5h7'37" 8d31'11"
Andre Agassi
Her 17h4'55" 24d29'30"
Andre Almeida
Pup 7h45'43" -28d7'56"
André and Vesna - Forever
Uma 9h37'9" 64d22'45"
Andre' Angel White
Tau 4h28'38" 2d45'8"
Andre Antonio Adonnino
Lib 15h46'35" -20d14'0"
André Arcand
Dra 19h18'32" 61d18'20"
Andre Cantiniaux
Peg 21h24'52" 18d30'31"
Andre & Carole Lataille
Lib 15h39'42" -19d50'5"
André Cartier
Apu 16h3'38" -75d9'1"
André Cloutier
Ori 6h9'5" 15d46'58"
André D. Callaway
12/17/72-02/20/04
Sgr 19h37'43" -16d21'27"
Andre David
Uma 12h37'11" 55d5'17"
Andre Dean
Ori 5h18'31" 4d22'45"
André et Rolande Lagassé
23.02.1952
Cyg 20h39'3" 44d59'32"
Andre F. Henry
Dra 17h58'30" 56d20'19"
Andre Frolow
Uma 9h47'41" 49d5'41"
André Hubrich
Ori 6h13'45" 7d32'39"
Andre J. Robin
Her 16h36'34" 5d22'43"
Andre Kian Malkin
Cnc 8h44'42" 16d46'18"
André Körbitz
Uma 9h42'51" 59d49'46"
André Korsnes Brunsæl
Leo 11h6'28" -0d32'15"
André & Kristin Knaebel
Cyg 19h49'20" 53d10'10"
Andre LaFleur
Uma 8h34'31" 67d6'30"
André Lebreton
Peg 21h27'31" 23d30'18"
Andre Lewis Lloyd
Ori 5h24'8" 3d36'4"
Andre Marc Riehle
Uma 9h32'30" 68d59'37"
Andre Marshall
Uma 13h43'7" 58d17'58"
Andre Maurice Moore
Dra 18h42'21" 85d57'30"
Andre Michael Mcgee
Cep 20h40'38" 66d1'54"
Andre My Shining Star
Uma 11h21'15" 64d55'36"
André Olijslager
Uma 11h47'39" 57d38'43"
André Oneil Davis
Vir 13h12'10" -11d41'44"
Andre Parodi 30 ans
Uma 9h16'37" 50d57'13"
Andre S Burnett
Lib 15h58'22" -16d57'1"
Andre Schützner
Uma 10h6'46" 41d36'31"
Andre T
Per 3h7'12" 50d50'49"
Andre Till Schwab
And 23h37'28" 40d1'7"
André Tomas Giguere
Per 3h17'36" 48d47'25"
Andre Young
Uma 9h34'50" 53d14'38"
Andrea
Lyn 8h10'12" 58d23'32"
Andrea
Lyn 7h12'2" 52d35'51"
Andrea
Cyg 21h29'7" 53d27'20"
Andrea
Cas 23h43'48" 53d32'51"
Andrea
Cas 23h2'21" 57d46'45"
andrea
Dra 19h24'26" 61d24'58"
Andrea
Uma 8h48'28" 62d2'46"
Andrea
Uma 9h40'53" 60d12'31"
Andrea
Cas 1h37'45" 66d22'23"
Andrea
Cam 3h21'32" 65d41'35"
Andrea
Cam 5h52'26" 60d16'26"
Andrea
Cam 4h47'43" 66d9'11"
Andrea
Cas 0h42'20" 69d44'44"
Andrea
Cas 1h25'47" 69d8'10"
Andrea
Dra 19h31'47" 69d48'34"
Andrea
Lib 15h55'53" -17d51'52"
Andrea
Sco 16h11'8" -16d19'56"
Andrea
Sgr 19h26'23" -20d18'55"
Andrea
Lep 5h54'37" -16d30'25"
Andrea
Lib 14h39'15" -10d1'38"
Andrea
Vir 12h47'4" -10d20'35"
Andrea
Aqr 23h46'55" -4d2'46"
Andrea
Vir 14h1'6" -2d11'35"
Andrea
Cam 12h28'56" 79d45'6"
Andrea
Ori 5h39'49" -1d38'48"
Andrea
Her 17h51'31" 21d26'3"
ANDREA
Sgr 20h10'58" -38d54'14"
Andrea
Lyn 8h26'47" 48d38'0"
Andrea
And 1h26'8" 49d49'34"
Andrea
And 1h33'43" 47d28'54"
Andrea
And 2h14'8" 50d49'17"
Andrea
Cas 0h33'16" 47d23'52"
Andrea
Uma 9h53'14" 48d41'27"
Andrea
Cas 23h26'52" 52d20'36"
Andrea
Lyn 8h27'36" 38d15'21"
Andrea
Lyn 8h1'36" 38d50'48"
Andrea
Cyg 21h14'55" 30d55'29"
Andrea
And 0h23'33" 42d15'55"
Andrea
And 0h13'45" 42d17'14"
Andrea
And 0h39'47" 41d13'14"
Andrea
Lmi 10h37'43" 34d38'47"
Andrea
Uma 11h21'54" 31d31'46"
Andrea
And 0h32'6" 36d0'3"
Andrea
And 0h16'36" 34d51'33"
Andrea
And 0h2'3" 35d35'8"
Andrea
Cnc 9h5'56" 32d18'22"
Andrea
Ori 5h20'4" 1d15'48"
Andrea
Ori 5h27'32" 5d26'35"
Andrea
Vir 12h32'35" 3d1'22"
Andrea
Mon 6h50'21" 7d15'16"
Andrea
Ori 5h8'26" 13d21'38"
Andrea
Ori 5h56'11" 12d51'2"
Andrea
Psc 23h23'36" 4d15'55"
Andrea
Psc 0h31'47" 14d42'50"
Andrea
Ori 6h3'27" 14d4'35"
Andrea
Leo 10h17'8" 17d42'50"
Andrea
Cnc 9h9'8" 19d53'7"
Andrea
Leo 11h15'36" 20d6'41"
Andrea
Gem 7h6'16" 22d16'22"
Andrea
Peg 21h41'29" 16d55'48"
Andrea
Her 17h51'31" 21d36'3"
Andrea
Her 17h17'49" 15d49'37"
Andrea
Lyr 19h11'2" 27d21'45"
Andrea
Ari 3h14'7" 28d52'44"
Andrea
Ori 5h53'38" 22d38'52"
Andrea
Tau 3h57'43" 27d19'42"
Andrea
Aur 5h59'17" 28d59'2"
Andrea 67-07-01
Lyn 7h19'58" 44d29'3"
Andrea A. Arroyo
Lib 16h0'50" -17d29'13"
Andrea Acheson
Vir 13h0'33" 5d49'17"
Andrea & Adrian
Cyg 21h15'9" 46d6'53"
Andrea Aeternus Devotus
Uma 9h6'0" 62d35'43"
Andrea & Alan Cordray
Lib 15h23'5" -25d51'43"
Andrea Alexis
Uma 11h31'8" 30d39'47"
Andrea Aliberti
And 23h27'48" 48d38'31"
Andrea Almaguer
Psc 1h3'41" 24d40'0"
Andrea and Ally Young
And 23h41'48" 45d11'19"
Andrea and Benji
Sge 19h51'29" 18d38'16"
Andrea and Dan
Vir 13h29'48" 0d1'51"
Andrea and Jason Russell
Pup 7h5'5" -41d18'9"
Andrea and Josh's love
star
Sgr 19h10'18" -31d14'13"
Andrea and Lia...The
brightest light
And 23h20'48" 47d18'57"
Andrea and Mike's Star
Cyg 20h21'21" 52d51'49"
Andrea and Neil Brundage
Uma 13h23'57" 60d27'44"
Andrea and Will's Star
Vir 11h47'16" -4d38'40"
Andrea Anderson
Cas 1h4'22" 48d50'43"
Andrea Anderson
And 0h46'3" 24d39'14"
Andrea Andrews
Lyn 8h31'49" 33d28'19"
Andrea "Andy" Odio
Cam 4h4'1" 53d23'48"
Andrea & Angela
And 0h12'50" 45d59'57"
Andrea Annie Pumpkin E.
Williams
And 1h14'59" 38d52'53"
Andrea Antonio
Cyg 20h4'36" 31d35'59"
Andrea Arredondo
Mon 6h53'6" -0d13'10"
Andrea Avis
And 0h42'21" 33d41'53"
Andrea Badilatti
Sgr 19h32'12" -18d18'57"
Andrea Bakay
Cas 1h38'17" 60d32'50"
Andrea Barbe
And 2h14'42" 45d26'33"
Andrea Bauzulli
Cnc 8h26'19" 17d51'29"
Andrea Begnoni
Psc 1h3'56" 5d58'38"
Andrea Belgau
Cas 1h37'46" 60d4'2"
Andrea Bellavia
Leo 10h14'15" 12d29'20"
Andrea Besancon Kepler
Gem 6h32'39" 26d11'57"
Andrea Biggi
And 1h16'22" 42d43'39"
Andrea Bodel
Leo 11h30'34" -2d27'8"
Andrea Bondurri
Per 3h20'10" 39d57'51"
Andrea Bonk
Sco 17h42'45" -44d21'36"
Andrea Bourne
Sgr 18h56'16" -27d43'37"
Andrea Bowns Fewkes
Dra 18h23'5" 52d23'22"
Andrea Braun Byrne
Peg 22h40'23" 28d37'28"
Andrea Brooke Kenyon
Aqr 23h53'25" -13d14'39"
Andrea Brown
Tau 4h22'19" 14d46'51"
Andrea Brügger
Cas 1h33'23" 72d38'18"
Andrea B's Magic Star!
Peg 22h35'21" 3d8'47"
Andrea Byars
Uma 9h12'11" 56d42'29"
Andrea C Marsh-Acome
Dra 18h52'4" 54d59'38"
Andrea C. Simonelli
Tau 4h19'27" 2d42'35"
Andrea Calanchi
Cyg 21h25'25" 35d0'8"
Andrea Caleiro
Psc 1h38'54" 20d42'39"
Andrea Campbell
Tau 5h39'58" 25d19'51"
Andrea Carlo Chiaselotti
Uma 10h42'0" 40d35'37"
Andrea Carloni
Aqr 20h41'57" -3d5'11"
Andrea Carol
Crb 16h24'7" 29d20'51"
Andrea Carol Eppinette
Mon 6h33'33" -1d23'49"
Andrea Carolina
Tau 4h43'58" 23d51'4"
Andrea Carolina Casillas
Ori 5h19'16" 11d46'35"
Andrea Carolyn
Cyg 20h32'21" 34d11'40"
Andrea Casamassa
Lyn 8h42'16" 36d15'25"
Andrea Casutt
And 23h38'24" 39d12'40"
Andrea Celeste Wright -
BABYDOLL
Ori 4h53'58" 11d14'53"
Andrea Christina
And 0h12'53" 28d44'11"
Andrea Claudia Bader
Vul 19h50'52" 22d31'10"
Andrea Colleen Glaeser
And 1h52'47" 42d12'51"
Andrea Contreras
Gem 7h36'16" 23d58'59"
Andrea Cortes Shines
Forever
And 2h24'0" 49d23'53"
Andrea Crapisi
And 23h54'15" 45d18'37"
Andrea Cristina Fiore
Leo 11h16'49" -6d20'36"
Andrea Cully
Cyg 20h42'24" 46d51'38"
Andrea D. Bucha
And 23h3'19" 48d7'10"
Andrea D. DeMaria
Cas 0h33'49" 57d13'56"
Andrea Dale DeCarlo
Cap 21h20'44" -15d3'13"
Andrea Dämmig
Uma 9h32'56" 51d26'4"
Andrea Danae Moreno
Leo 10h5'13" 19d23'47"
Andrea Daniela Mathis
Cep 21h48'40" 71d26'39"
Andrea Darlene Killion
Aqr 22h44'34" -11d56'43"
Andrea Dawn
And 0h46'58" 45d44'47"
Andrea Dawn
And 2h17'34" 41d19'19"
Andrea Dawn Beaumont
Cas 1h37'48" 72d50'39"
Andrea Dawn Hinkleman
And 0h52'38" 36d55'16"
Andrea De Los Cobos
Crb 16h9'20" 33d32'56"
Andrea "De" Vovchansky
Gem 7h6'26" 15d45'18"
Andrea Decker
Lyn 9h37'15" 40d9'17"
Andrea Demeter
Cap 21h31'3" -19d14'13"
Andrea Demetri Yiallouris
Ori 5h39'42" -1d32'4"
Andrea DiBella
Leo 10h33'51" 17d37'28"
Andrea & Dirk
Ori 6h0'19" 5d39'13"
Andrea Donati
Cas 0h25'15" 62d23'6"
Andrea e Alessandra
Cam 7h22'24" 72d32'25"
Andrea Elizabeth Benson
Gem 6h47'35" 32d19'40"
Andrea Elmore
Crb 15h50'23" 27d45'12"
Andrea & Endre
Szerelemcsillaga
Lib 14h44'32" -18d12'33"
Andrea Ernst
Oph 16h58'12" -0d58'14"
Andrea Estefania
And 0h42'28" 30d41'28"
Andrea Evans
And 2h1'35" 43d47'55"
Andrea Ex Animo
Aqr 22h5'19" -0d48'19"
Andrea Fanella
Ari 2h18'9" 13d10'20"
Andrea Fell's Oasis
And 0h43'16" 42d48'14"
Andrea Ferancová,
03.03.1974
Umi 14h32'24" 84d36'56"
Andrea Ferguson
Lib 14h51'33" -0d43'18"
Andrea Flora Moore
Tau 4h23'31" 24d37'58"
Andrea Fryda Phi Sigma
Chi Alumni
Umi 15h5'30" 70d59'7"
Andrea Gail
Cyg 20h16'32" 38d21'16"
Andrea Gail Gray
Sgr 18h27'28" -16d11'13"
Andrea Gaxiola
Leo 11h4'53" -4d45'54"
Andrea Geissler
Sco 17h41'10" -32d6'8"
Andrea Giannone
24/09/1996
Ori 5h49'1" -8d20'21"
Andrea Giles
Cas 23h28'56" 55d49'54"
Andrea Giordano
Tau 4h22'23" 24d29'28"
Andrea Gomez
Sco 17h23'31" -42d19'47"
Andrea Graydon &
Brennen Haworth
Ari 2h10'2" 22d34'42"
Andrea Gunther
Cas 1h24'57" 52d24'42"
Andrea Hartmann
Tri 2h11'39" 31d55'55"
Andrea Henson
Uma 9h56'56" 49d14'42"
Andrea Hott Stuff Clinton
Lib 14h48'2" -21d57'17"
Andrea Hubert
Gem 7h49'38" 23d20'29"
Andrea I. Cave
Sgr 19h21'26" -20d20'12"
Andrea I Love You Forever
And 23h25'5" 45d22'32"
Andrea Isabela Rodriguez
Peg 21h22'43" 16d29'24"
Andrea Isabelle Bähler
Lmi 10h38'6" 29d9'17"
Andrea J Solberg
Lyn 7h6'41" 54d27'45"
Andrea Jaeger
Lac 22h54'38" 44d37'58"
Andrea Jane
And 23h39'59" 49d22'33"
Andrea Jane Moore
And 0h54'10" 36d46'48"
Andrea Jean Hartman
Uma 11h40'47" 46d32'26"
Andrea Jean Routt
Cyg 21h12'27" 31d41'8"
Andrea Jeanne Scanlon
Leo 9h59'40" 19d22'51"
Andrea & Jeff
Sco 16h47'43" -30d55'31"
Andrea & Jeffrey Bush
Gem 7h28'20" 29d14'8"
Andrea Jo
Lyn 8h27'11" 37d59'55"
Andrea Jo
Uma 10h14'21" 43d19'9"
Andrea Jo Hertel
Aqr 23h36'19" -20d15'17"
Andrea Jo McKee
Psc 23h32'14" 5d46'20"
Andrea & John
Cyg 20h25'51" 39d3'1"
Andrea Johnson
Lib 15h35'1" -24d33'41"
Andrea Jolliffe
Gem 7h55'8" 32d5'29"
Andrea Joy Collins
And 23h28'33" 45d59'35"
Andrea Joy Flannery
Lib 14h53'13" -0d52'6"
Andrea Joy McCarthy
And 0h7'52" 46d14'47"
Andrea + Jürg Schweizer
Lac 22h3'51" 40d32'18"
andrea k.
Psc 0h51'41" 16d26'31"
Andrea K. Sell
Ari 2h15'33" 14d0'33"
Andrea Kadar 11.02.1981
Ori 5h48'51" 9d12'59"
Andrea Kanovsky
And 23h27'5" 39d6'36"
Andrea Katherine
Cap 20h17'7" -20d19'51"
Andrea Kay SAVORETTI
And 1h38'35" 46d27'40"
Andrea Kaye (Knight)
Uma 10h9'0" 65d58'3"
Andrea Keller
Leo 10h25'22" 16d3'20"
Andrea Klein
And 2h30'57" 45d52'47"
Andrea Klemz
And 0h11'35" 43d23'27"
Andrea Klingenberg
Uma 9h59'38" 63d1'13"
Andrea Kowalisyn
And 0h46'4" 37d45'25"
Andrea L. Abbott Webb
Lib 15h29'44" -25d14'20"
Andrea L. Cordts
Leo 11h18'26" 21d58'47"
Andrea L LaPierre
And 0h10'49" 44d23'21"
Andrea L. Parisi, Esq. "You
Did It"
Uma 10h16'27" 41d48'46"
Andrea L. Thom
Cap 20h18'29" -23d55'43"
Andrea Lacy
Cas 1h14'0" 48d56'2"
Andrea Lacy
Ari 3h27'35" 26d2'19"
Andrea Lappas
Her 17h41'21" 16d12'18"
Andrea Lauren Redewill
And 1h10'40" 42d46'48"
Andrea Lea Ebenroth
Aqr 21h38'6" -7d53'26"
Andrea Lee Johnston
Lib 15h43'11" -11d51'8"
Andrea Lee Parker
Per 4h44'39" 46d47'25"
Andrea Lee Tolan
Cas 23h37'38" 54d44'41"
Andrea Leigh
Sgr 18h51'41" -20d23'42"
Andrea Leigh
Aqr 22h53'20" -15d5'24"
Andrea Leigh
Vir 12h37'33" 12d25'52"
Andrea Leigh "bubs" Stein
And 1h44'37" 38d37'14"
Andrea Leigh Colella
Gem 6h46'20" 15d14'2"
Andrea Leigh Ducker
Leo 9h51'15" 22d55'0"
Andrea Len
And 1h56'59" 37d34'29"
Andrea Linden
Ori 6h24'35" 13d37'43"
Andrea Loftus
Vir 13h59'34" -17d17'33"
Andrea Long
Sgr 18h47'5" -18d36'46"

Andrea Long
Ari 2h34'35" 26d12'46"

Andrea Lopaska
Ori 5h12'51" 15d44'8"

Andrea Lopez # 3
Leo 10h10'37" 15d49'55"

Andrea Louise
Ori 4h48'5" 11d58'22"

Andrea Louise
Cap 21h48'30" -18d31'42"

Andrea Louise Guevera
Aqr 21h2'53" 0d36'1"

Andrea Loves Michael
Mon 6h51'0" -0d54'9"

Andrea Loya
Sco 16h8'9" -30d9'8"

Andrea Luca
Boo 14h49'26" 16d12'35"

Andrea Lüthi
Boo 14h59'21" 12d51'10"

Andrea Lyn
Uma 11h12'20" 31d57'29"

Andrea Lynch - Angel - 17.03.1982
And 2h37'6" 48d35'58"

Andrea Lynn
Crb 16h20'1" 38d17'27"

Andrea Lynn
Lib 14h54'42" -1d38'50"

Andrea Lynn
Cas 1h40'2" 60d52'33"

Andrea Lynn Cordaway
Cnc 9h15'25" 18d37'3"

Andrea Lynn Gibson
Uma 13h52'1" 57d7'43"

Andrea Lynn Kahley
Sco 16h2'41" -8d47'18"

Andrea Lynn La Salvia-Hayward
Cnc 8h39'33" 15d9'3"

Andrea Lynn Nathanson Wallick
Vir 12h41'41" 2d19'15"

Andrea Lynn Robinson
Cap 20h29'57" -11d59'13"

Andrea Lynn Spence
And 1h41'11" 43d50'50"

Andrea Lynn Varner
Tau 5h49'47" 16d14'17"

Andrea M Lacy
Vir 13h21'28" 12d41'44"

Andrea M. Page
And 23h22'50" 51d54'12"

Andrea M. Ramirez
Eri 3h53'15" -0d42'20"

Andrea M. Riggs
Cen 11h42'20" -51d33'51"

Andrea M. Yarnall
Sgr 19h54'53" -19d1'0"

Andrea Macauda
Ori 5h8'17" 1d28'7"

Andrea Manasse
Ori 6h0'7" 1d18'34"

Andrea Maria
And 0h11'39" 42d18'10"

Andrea Maria Albin-Platter
Ori 6h11'18" 20d50'29"

Andrea Maria Badiali
And 1h51'5" 46d39'18"

Andrea Maria Otter
Tau 4h12'3" 5d33'22"

Andrea Marie , Robert William
Cyg 20h17'44" 58d38'7"

Andrea Marie Anderson
Lyn 8h45'57" 33d14'37"

Andrea Marie Ciavarro
And 2h20'56" 41d41'19"

Andrea Marie Connolly
Cap 21h27'29" -12d8'32"

Andrea Marie Hall
Cnc 9h5'28" 31d0'54"

Andrea Marie Kulesa
Cas 22h57'18" 54d56'6"

Andrea Marie Sterancsak
Sco 17h53'58" -36d9'28"

Andrea Marie Stinson
Sgr 20h24'7" -32d45'1"

Andrea Marie Valley "The Love Star"
Psc 23h37'27" 2d23'11"

Andrea Mary
And 1h21'18" 48d48'6"

Andrea + Mary
Cas 23h37'39" 54d12'11"

Andrea Mason
Cas 0h10'26" 54d16'16"

Andrea Matia
Cyg 20h19'45" 39d14'53"

Andrea Mauselbärchen Karl
Ori 6h4'12" 10d58'50"

Andrea May Demos
And 23h15'22" 47d6'4"

Andrea McCormick
Cas 0h42'23" 62d38'21"

Andrea McElroy
Mon 6h47'15" -5d37'47"

Andrea Mendez "12-31-2003"
And 1h44'13" 39d7'49"

Andrea Meunier
Crb 15h41'4" 39d11'32"

Andrea Michele Sams
And 0h38'34" 37d4'26"

Andrea Michelle
Cas 0h0'28" 56d16'48"

Andrea Michelle
Lib 15h31'32" -9d11'23"

Andrea Michelle Chandler
Cnc 8h54'0" 27d11'10"

Andrea Michelle Payne
Gem 7h23'54" 19d50'14"

Andrea Michelle Windham
And 0h37'49" 36d59'54"

Andrea Miguel
And 23h18'58" 47d36'21"

Andrea Minoris
Crb 15h27'56" 28d11'20"

Andrea Monique
Tau 5h44'4" 24d13'40"

Andrea Moore
Aqr 22h29'40" -1d20'58"

Andrea Mourin
Vir 13h31'42" -4d34'43"

Andrea Mumford
Leo 11h11'41" -1d36'53"

Andrea Murdaugh
And 1h25'31" 46d31'28"

Andrea Mylove Piazza
Cap 20h28'31" -13d46'27"

Andrea N. Vagnoni
Cnc 9h7'49" 32d21'3"

Andrea Nash
Uma 11h56'27" 52d52'2"

Andrea Neidhart
Peg 21h22'22" 15d50'11"

Andrea Nete Palsgreen Henriksen
Dra 18h30'44" 52d11'11"

Andrea Nichole Tiliakos
Tau 4h6'15" 15d59'12"

Andrea Nicole
Crb 15h41'4" 29d16'58"

Andrea Nicole
Uma 10h16'41" 44d11'25"

Andrea Nicole D'Angelo
And 23h10'39" 50d44'23"

Andrea Nicole Gonzalez "ANG"
Lib 15h52'25" -14d54'19"

Andrea Nicole Turpin
Uma 9h5'22" 62d1'49"

Andrea Noel DiGuardi
Ori 5h7'20" 15d14'32"

Andrea Noncu 69
Peg 23h17'35" 31d5'57"

Andrea Norman "60"
Gem 7h11'11" 33d2'23"

Andrea Nüßemeyer
Ori 6h20'52" 15d10'47"

Andrea Octavia Perez
Gem 7h53'28" 31d27'14"

Andrea Oehm
Sge 19h58'36" 17d0'19"

Andrea Olya
Leo 11h29'30" 10d8'1"

Andrea Ortiz
Cyg 20h5'25" 39d25'19"

Andrea P. Shea
Ori 5h31'50" 1d16'41"

Andrea Paige
And 0h44'44" 35d41'39"

Andrea Pedigo
Cap 21h37'4" -17d5'7"

Andrea Pellegrino
Uma 10h7'27" 56d42'19"

Andrea&Petr
Per 2h24'41" 51d9'38"

Andrea Pierce
Psc 1h3'23" 5d6'59"

Andrea Pimentel Perez Olagaray
And 2h6'41" 41d55'2"

Andrea Piotrowski's Wishing Star
Cas 1h27'57" 55d30'30"

Andrea Preciado
Sex 10h14'6" -0d30'9"

Andrea Price
Sco 16h4'26" -25d43'15"

Andrea Provencio
Psc 1h2'54" 15d4'1"

Andrea Quilli
Per 4h29'56" 31d59'53"

Andrea Quintana
Ori 5h23'0" 0d38'22"

Andrea R. Heintz
Crb 15h37'10" 38d25'0"

Andrea Rachael Hopkins
Uma 13h47'58" 62d2'32"

Andrea Rachiski
Lyn 7h5'36" 48d35'51"

Andrea Rae Abbring
And 1h29'38" 46d21'11"

Andrea Ray Romero
Psc 1h39'14" 14d7'57"

Andrea Reamer
Uma 9h59'25" 62d10'58"

Andrea Renae
And 0h31'38" 45d38'4"

Andrea Renae Rios
Eri 2h50'59" -40d12'54"

Andrea Renato
Per 3h12'44" 42d44'39"

Andrea Rene
Gem 6h7'55" 23d21'12"

Andrea Renee
Ori 6h6'40" 4d21'58"

Andrea Renee Truax
Tau 3h59'21" 17d31'17"

Andrea Renee's Star
Ori 5h35'56" 11d32'40"

Andrea Rezzana
Aql 20h14'49" -0d20'19"

Andrea Rios
Cap 21h29'36" -8d53'43"

Andrea Rizzo
Lyr 18h57'7" 38d14'31"

Andrea Roberts
And 22h58'12" 40d18'47"

Andrea Robyn Herft
Cen 13h27'59" -44d4'38"

Andrea Rodriguez
Lmi 10h34'3" 35d47'59"

Andrea Rosa Rodriguez
Gem 7h16'33" 17d3'28"

Andrea Rose
Cyg 19h47'57" 31d38'52"

Andrea Rose
Lyn 8h24'16" 50d38'2"

Andrea Rose Jacobs
Cas 0h27'44" 67d2'55"

Andrea Rose Rosado
Vir 11h58'57" -6d57'51"

Andrea Rose Stiffelman
And 23h41'7" 45d54'10"

Andrea Rose Weiss
Uma 10h42'59" 50d45'5"

Andrea Rosso
Aql 19h56'4" 15d40'43"

Andrea Rougeron
Vir 13h12'43" -18d9'0"

Andrea Ruth Blaser
Cas 23h49'32" 53d30'13"

Andrea Ryser "Amici per sempre"
Aqr 23h8'36" -19d53'46"

Andrea Sadowy - My Best Friend
Cas 0h46'28" 56d56'55"

Andrea Salena Madison
And 1h17'57" 42d38'27"

Andrea Salleras
Sgr 19h31'38" -13d1'43"

Andrea & Sandro
Uma 11h10'16" 40d15'46"

Andrea Santos
Gem 7h10'12" 32d36'46"

Andrea Santucci
Cas 0h21'50" 50d54'7"

Andrea Sara Hunt
Sco 16h11'19" -28d52'37"

Andrea Schatzstern
Sct 18h29'10" -15d32'11"

Andrea Schmid
Umi 13h47'23" 77d50'40"

Andrea/Sebastian
Ori 4h59'34" 10d56'49"

Andrea Serena
Cas 23h11'8" 55d31'59"

Andrea Silva
Cas 23h19'36" 55d36'30"

Andrea Sirtori
Uma 8h53'38" 46d39'23"

Andrea Sitbon
Cyg 21h36'42" 50d33'20"

Andrea Skeels
And 2h18'2" 39d15'52"

Andrea (spa)
Cam 7h3'14" 69d0'41"

Andrea Spring
Gem 7h40'11" 31d33'38"

Andrea Statman
Cnc 8h44'6" 26d9'57"

Andrea Steimann
And 1h0'41" 44d39'47"

Andrea Steinmann
Cam 5h43'37" 70d41'24"

Andrea Susan Nicholas
Uma 11h56'21" 53d27'57"

Andrea Sylvia
Uma 12h31'6" 54d20'44"

Andrea T.
Leo 11h43'59" 18d11'58"

Andrea T Flores...Por Siempre
And 0h16'37" 29d24'54"

Andrea Telenko
Gem 6h50'7" 21d22'43"

Andrea (The Boss) Antista
Uma 11h25'21" 55d4'33"

Andrea Thiel Kevin Svoboda 1yr Anv
Leo 10h19'12" 22d13'7"

Andrea & Thomas
And 1h51'33" 35d54'13"

Andrea Threrese Trinity Chambers
Psc 1h4'50" 19d45'56"

Andrea Tigris-Minor
Uma 11h56'10" 41d8'37"

Andréa Timpone
Ori 5h28'23" -0d26'36"

Andrea Torvinen's Star
Tau 5h50'50" 27d6'49"

Andrea Triebenbacher
Per 3h39'32" 34d7'20"

Andrea Trinchero Blue 31
And 1h16'4" 42d24'31"

Andrea u. Arnd Pico del Teide 2004
Uma 10h5'51" 45d38'19"

Andrea und Patrick Renggli
Dra 14h42'25" 56d3'19"

Andrea und Yannic
Ori 4h51'45" -0d53'41"

Andrea Upah
Uma 10h15'21" 54d1'13"

Andrea Uribe
Ari 2h3'45" 23d18'11"

Andrea Valenti
Vir 13h18'59" -3d31'18"

Andrea Valle Bayly
Mon 6h34'20" 2d47'57"

Andrea Verónica Aguirre Iglesias
Cap 21h7'9" -19d13'52"

Andrea Vidal
Tau 3h56'53" 23d47'43"

Andrea Vislosky
Cas 0h28'32" 52d46'51"

Andrea Wallace Biggs
Aqr 22h17'25" 2d23'37"

Andrea Walsh Heath
Ari 3h28'22" 26d13'34"

ANDREA WARFIELD
Crb 16h7'26" 38d8'12"

Andrea Weiss
And 0h13'44" 34d12'8"

Andrea Wiedmer
Crb 16h9'12" 26d0'14"

Andrea will you marry me?
Cas 0h32'6" 59d39'23"

Andrea Wirth
And 23h36'8" 37d23'44"

Andrea Yarnall
Sgr 18h54'50" -33d42'27"

Andrea Yelland
Lib 14h48'37" -19d51'12"

Andrea You Had Me From Hello
Aur 7h17'59" 38d15'43"

Andrea Zaccagnino~Salvatore Alfani
Ori 6h4'45" -0d37'16"

Andrea -Zaubermaus-Däster
Dra 20h23'36" 74d51'31"

Andrea Ziemba
And 1h38'44" 42d43'2"

Andréa Zoie
Sco 16h36'55" -26d17'13"

Andrea, 21.08.1985
And 23h11'40" 47d0'52"

Andrea, My First Special Angel
Lib 15h8'21" -8d17'55"

Andrea, My Mother of Strength
Lib 15h34'17" -6d52'19"

Andrea18C
And 2h37'8" 45d18'25"

Andreafrancois
Cam 7h22'54" 61d19'50"

AndreaGaetano
Umi 13h54'20" 69d36'38"

AndreaGeorgePyo- Forever
Per 2h43'34" 54d39'17"

Andreas
Per 1h42'44" 54d35'53"

Andreas
And 2h14'6" 37d46'28"

andreas
Cyg 20h37'44" 31d12'44"

andreas
Uma 9h59'9" 61d27'5"

Andreas
Cap 21h31'7" -19d56'4"

Andreas
Sco 16h40'1" -33d43'44"

Andreas
Sgr 18h45'17" -36d5'25"

Andreas Andy Hanggi
Umi 15h24'4" 84d42'49"

Andreas Aristotle Parkinson
Uma 9h44'19" 42d4'45"

Andreas Arnold
Uma 9h26'59" 49d14'48"

Andrea's Aurora
Ori 5h58'18" -0d4'42"

Andreas Baroso
And 2h34'51" 46d13'38"

Andreas Belmonte
Uma 9h3'47" 52d0'4"

Andreas Bergmann
Uma 11h43'54" 28d51'17"

Andreas Bernardo Dambakakis
Leo 11h11'9" -1d2'34"

Andreas Christopoulos
Umi 16h11'18" 75d57'21"

Andreas De Santa Coloma
Uma 8h35'33" 71d22'1"

Andreas Dembek
Ori 6h10'17" 7d46'45"

Andreas der Grosse
Sct 18h31'59" -15d17'47"

Andreas der Messias
Peg 22h44'57" 3d25'44"

Andreas Dick
Cas 0h9'58" 52d22'38"

Andrea's Dream - Drey's Fantasy
Cnc 8h33'18" 10d49'8"

Andreas Dripke
Ori 5h12'17" 8d36'19"

Andrea's Eternal Light
Tau 5h4'16" 20d56'26"

Andrea's Eye
Uma 14h13'19" 57d43'0"

Andreas *Forever My Star* Rhonda
Leo 9h24'12" 14d3'29"

Andreas Gelhart
Vir 13h59'4" -9d54'43"

Andreas Georg-Schäfer
Uma 14h20'28" 59d1'18"

Andreas Gießmann
Uma 10h57'55" 47d53'8"

Andreas "Hase" Ewigleben
Uma 10h56'1" 33d20'16"

Andreas Haseke
Uma 10h19'32" 61d39'38"

Andrea's Heavenly Light
Gem 7h24'53" 33d18'35"

Andreas Hübler
Uma 13h49'42" 59d35'15"

Andreas Ioannides
Lyr 19h25'25" 37d37'1"

Andreas Kapl
Uma 10h16'19" 60d58'40"

Andreas Keller
Uma 8h37'21" 53d3'28"

Andreas Knapp-Voith
Uma 8h28'19" 64d4'43"

Andreas Kovalski
Uma 8h45'0" 56d45'55"

Andreas Krauß
Uma 11h29'39" 31d50'52"

Andreas Kreidler
Uma 9h21'6" 63d52'10"

Andreas L. Pangerl, a star himself
Uma 10h58'17" 34d6'35"

Andreas Land
Cap 20h59'8" -22d32'21"

Andreas Lenk
Uma 9h4'52" 68d30'19"

Andrea's Light
Gem 6h57'13" 15d6'47"

Andreas Linneous Gubbrealis
Lib 15h9'1" -5d51'22"

Andrea's Love
And 2h30'59" 46d50'36"

Andreas Machner
Ori 6h22'25" 15d7'33"

Andreas Maier
Ori 6h12'26" 21d9'11"

Andreas & Marielle
Uma 10h42'21" 40d31'56"

Andreas Meyer
Her 18h30'8" 21d51'58"

Andreas&Monika
Aql 19h31'47" 4d21'22"

Andreas Mozoras
Del 20h56'34" 13d48'36"

Andreas Mulle
Uma 9h56'57" 55d39'2"

Andreas Nicholas Kounnou
Per 2h13'48" 55d16'28"

Andreas Nicholas Manganas
Ari 2h38'1" 27d41'18"

Andreas Niedrig - "Der Ironman"
Uma 9h51'43" 53d42'42"

Andreas Nikolai Bischoff
Uma 11h24'3" 38d1'58"

Andreas Papandrikos
Ori 5h2'0" 5d28'52"

Andrea's Piece of the Sky
Lyn 8h11'4" 33d52'8"

Andreas Rohner
Per 3h11'23" 42d47'27"

Andreas Rudolf Schweighauser
Vul 20h28'46" 27d37'15"

Andreas Schenk
Her 18h21'24" 12d16'58"

Andreas Schneider
Uma 12h9'43" 58d23'3"

Andrea's Shining Light
Ari 2h18'52" 18d21'26"

Andrea's Shining Star
And 0h25'51" 28d3'5"

Andrea's Shining Star
Vir 12h45'43" -3d22'5"

Andrea's Shinning Star
Leo 11h16'54" 12d34'23"

Andreas Spanner / Tiger
Uma 8h47'57" 72d53'45"

Andreas Stallmann
Uma 9h55'27" 56d43'47"

Andreas Stamatiou
Ori 5h36'17" 14d35'14"

Andrea's Star
Tau 4h36'32" 16d29'6"

Andrea's Star
Ari 3h23'17" 29d22'52"

Andrea's Star
And 1h35'59" 49d17'54"

Andrea's Star
Lib 15h46'7" -5d43'26"

Andrea's Star
Cap 21h40'14" -13d42'31"

Andrea's Star *Lucky 13*
And 23h1'40" 52d18'0"

Andreas Stoiber
Uma 12h18'38" 57d38'48"

ANDREAS - "The bewitched Angel"
Ori 5h27'15" -4d21'50"

Andreas Theodore Warren
Lyn 7h26'47" 59d0'23"

Andreas Theodosis
Sco 16h5'51" -22d21'6"

Andreas Treichl
Uma 8h55'44" 72d29'40"

Andreas und Angela
Uma 9h48'52" 48d37'54"

Andrea's Vision
And 0h8'15" 48d25'16"

Andreas Voigt
Uma 9h32'27" 54d48'3"

Andreas Wortmann
Umi 13h22'56" 72d32'11"

Andreas Yiannakis
Uma 8h22'37" 66d36'52"

andreas-gabriele-kreher
Ori 5h59'12" 12d15'49"

Andreastar
Leo 11h16'15" 1d25'2"

AndreaStarGazer2006
Cyg 19h36'50" 29d42'34"

Andree Ann Lavoie
Pho 1h10'8" -56d32'20"

Andrée Bégin
Vir 13h28'28" 13d39'21"

Andree Fontaine Lauterbach
Uma 9h17'7" 62d4'4"

Andree' Voorhies
Cnc 8h12'12" 17d29'55"

Andreea Elisabeta Merrill
Lib 15h14'34" -10d30'53"

Andreea Ernest
Her 17h55'17" 29d59'15"

Andreea Radu
Sco 16h8'33" -30d35'45"

Andrei
Lib 15h3'16" -18d3'13"

Andrei
Psc 0h26'4" 15d3'13"

Andrei Alexandru Covaciu
Cnc 8h36'52" 14d6'11"

Andrei Antoshin
Dra 18h39'39" 57d10'8"

Andrei Bossov
Dra 15h46'58" 60d25'31"

Andrei V. Shamenko
Her 16h53'53" 36d48'28"

Andreia Brás
And 0h43'30" 22d1'49"

Andreia23
Lib 15h8'43" -6d57'12"

Andreina Diaz
Tau 4h13'7" 16d34'2"

Andrej & Blanka
Uma 8h12'4" 69d31'12"

AndreLana
Lmi 10h23'23" 38d14'2"

André-pépite
Oph 17h21'41" -22d42'21"

Andreriben
Aqr 21h10'36" -13d28'59"

Andrés Alfonzo Olivas
Cep 22h4'23" 64d23'1"

Andres "Aqui Nadie se Rinde"
Psc 0h12'17" -1d31'43"

Andres Chavez
Uma 10h23'8" 69d37'50"

Andres F. Avendanolopez
Lyn 8h1'41" 42d51'34"

Andres Fernandez
Uma 9h50'57" 55d7'40"

Andres Gerardo Guardiola
Ori 5h38'44" -2d32'35"

Andres Glücksstern
Uma 11h29'40" 38d21'33"

André's Islay
Uma 8h35'16" 50d0'40"

Andre's Light
Uma 10h45'41" 56d28'10"

Andres Michael Reyes
Tau 5h44'55" 21d24'6"

Andres Mosquera
Uma 13h47'51" 57d3'11"

Andres Phillip Cabrera
Cap 21h38'43" -20d19'24"

Andres Vargas
Leo 9h30'26" 32d3'33"

Andres Vidal
Crb 16h24'3" 33d54'35"

Andresen, Elke
Uma 8h55'57" 48d31'13"

Andresen, Walter
Uma 11h56'5" 50d16'11"

Andrew
Per 4h27'9" 46d51'12"

Andrew
Per 2h47'44" 50d8'23"

Andrew
Her 16h41'33" 35d35'29"

Andrew
Her 17h22'13" 31d35'35"

Andrew
Uma 11h20'23" 36d42'5"

Andrew
And 0h7'2" 33d59'7"

Andrew
Per 4h38'27" 43d26'10"

Andrew
Per 4h30'49" 44d16'6"

Andrew
Aur 5h35'41" 41d50'51"

Andrew
Cnc 8h23'31" 11d24'47"

Andrew
Ori 6h22'50" 14d52'53"

Andrew
Aql 18h51'46" 9d58'22"

Andrew
Ori 6h6'2" 9d29'19"

Andrew
Ari 3h17'54" 27d28'11"

Andrew
Aur 5h58'36" 29d19'52"

Andrew
Gem 6h32'19" 23d35'31"

ANDREW
Gem 6h52'6" 17d3'59"

Andrew
Cep 2h47'19" 84d43'30"

Andrew
Lyn 7h6'26" 58d13'11"

Andrew
Cep 21h31'15" 64d29'51"

Andrew 22
Leo 10h8'44" 26d3'57"

Andrew 31 Sr
Her 18h12'30" 18d42'48"

Andrew A. R. Gentles
Lib 14h24'30" -11d56'23"

Andrew Albert
Dra 17h10'23" 54d58'37"

Andrew Alex Tolompoiko
Psc 1h20'4" 16d4'47"

Andrew Alexander Daly Jr
Ori 5h29'25" 0d30'43"

Andrew Alexander Grella
Tau 4h11'24" 28d51'22"

Andrew Alexander Tradd 2/11/1991
Uma 9h2'12" 48d15'31"

Andrew Alfred Carlson
Per 3h7'52" 51d13'9"

Andrew & Alix - Together Forever
Col 5h28'26" -35d26'1"

Andrew Alwyn Walls
Cep 21h33'1" 68d19'18"

Andrew and Alannah's Star
Ori 6h6'44" 7d16'17"

Andrew and Avril
Peg 21h35'55" 22d27'27"

Andrew and Catherine Rumble
Uma 8h46'9" 55d8'23"

Andrew and Emma Forever
Umi 17h7'16" 79d46'10"

Andrew and Georgia bound by LOVE
Dra 19h18'52" 60d1'27"

Andrew and Jen
Ori 4h47'23" -0d26'37"

Andrew and Jessica
Psc 1h19'53" 31d46'52"

Andrew and Karen's Star
Pho 0h35'32" -43d8'48"

Andrew and Kate forever
Cru 12h43'27" -57d0'28"

Andrew and Kellie
Gem 7h1'35" 18d46'27"

Andrew and Kelly, Love Forever
Sgr 18h50'21" -29d23'43"

Andrew and Laurie Harrison
Psc 0h7'13" 8d17'26"

Andrew and Margaret Koslosky
Cyg 21h46'30" 46d50'44"

Andrew and Melissa Neal
Crb 16h12'14" 38d4'13"

Andrew and Misty
Crb 15h55'15" 26d3'15"

Andrew and Robin Johnson May 18 02
Uma 11h21'47" 69d46'35"

Andrew and Svea's Hammock Post
Uma 12h39'11" 56d53'14"

Andrew (Andy) Moore
Uma 11h34'18" 41d10'24"

Andrew Anthony Jess Sr.
Uma 9h54'22" 54d1'48"

Andrew Arden Albright
Lib 14h40'18" -17d26'19"

Andrew Armstrong
Cep 1h53'53" 86d25'13"

Andrew Arthur Barclay
Lyn 7h51'6" 54d10'30"

Andrew Arthur Carlson-Doom
Uma 8h46'18" 51d22'56"

Andrew Balettie
Uma 11h55'43" 54d46'45"

Andrew Barbour
Uma 11h26'6" 59d47'7"

Andrew Barry-Pursell
Tau 5h55'33" 25d16'10"

Andrew & Belinda For Eternity
Ara 17h21'5" -50d26'46"

Andrew Bennett
Sco 16h17'12" -16d24'11"

Andrew Benson
Cyg 21h15'48" 42d39'22"

Andrew Biechler King
Ori 6h5'31" 16d17'52"
Andrew & Billy
Ari 2h5'11" 20d35'35"
Andrew Blair Wilson
Per 4h10'30" 32d6'54"
Andrew Blaise Krilich
Leo 11h35'13" 27d30'42"
Andrew Blake O'Steen
Dra 15h59'7" 62d43'58"
Andrew Blake Travis
Per 2h43'33" 55d36'12"
Andrew Blum
Uma 13h50'6" 58d23'51"
Andrew Blunt
Umi 15h11'24" 70d52'10"
Andrew Bombelli
Cnc 8h46'47" 31d10'27"
Andrew Bond
Cep 4h42'1" 80d46'44"
Andrew Bonillo Sr.
Lyn 9h10'18" 34d35'36"
Andrew Bradley Weddle
Her 17h8'41" 31d9'9"
Andrew & Brandie Craig
Cyg 21h25'27" 46d32'39"
Andrew Branham Dyer
Sgr 18h36'7" -27d8'28"
Andrew Bremner & Angie MacPhee
Vir 12h3'7" 1d23'0"
Andrew & Briann Forever & Always
Cyg 19h40'14" 32d22'0"
Andrew Bricker
Boo 14h59'11" 27d28'49"
Andrew Bruno
Sgr 19h20'34" -41d43'27"
Andrew Bryan Kraden
Lyn 7h6'59" 55d7'32"
Andrew Brylee Craighead
Leo 10h38'17" 16d14'52"
Andrew Buchanan Fischl
Her 17h7'15" 30d40'41"
Andrew "Buddy" Larson
Her 17h40'25" 23d37'4"
Andrew Bunker
Uma 9h53'27" 64d47'29"
Andrew Butterworth
Cru 12h37'13" -61d29'34"
Andrew C. Bryniczka
Aqr 23h15'35" -21d41'59"
Andrew C. Dennis
Cnc 9h15'7" 32d0'9"
Andrew C. Gray
Vir 12h58'29" -2d17'31"
Andrew C. Mosteller
Ori 5h51'18" -0d57'13"
Andrew Cameron Shelton
Her 17h43'1" 19d8'31"
Andrew Canales
Uma 11h24'43" 37d24'9"
Andrew Carl Greene
And 1h3'6" 35d31'59"
Andrew Carl Imperiale
Aqr 22h49'24" 2d29'46"
Andrew & Carolyn's Star
Lyr 18h49'23" 43d35'34"
Andrew Carrington Sherwood
Ori 5h30'21" 2d45'34"
Andrew Cattarin
Psc 0h22'53" -1d33'46"
Andrew Chapin
Umi 15h22'51" 72d46'48"
Andrew Charles Jones
Nor 16h18'14" -43d11'36"
Andrew Charles O'Donaghoe 1950-2004
Cru 12h18'41" -56d48'8"
Andrew Charles Pfeiffer Jr
Leo 9h51'23" 28d43'58"
Andrew Charles Pfeiffer Sr
Gem 6h59'21" 14d36'7"
Andrew Charles Samis
And 0h7'39" 45d39'10"
Andrew Charles Wormsbacher
Vir 13h11'4" 6d57'43"
Andrew Christopher
Her 17h20'29" 26d40'56"
Andrew Christopher
Sco 16h15'53" -9d32'29"
Andrew Christopher Blunt
Cyg 21h33'35" 51d5'33"
Andrew Christopher Jones
Gem 7h27'57" 30d22'35"
Andrew Christopher Kleinpeter
Ori 6h6'12" 14d34'7"
Andrew Claridge
Per 3h9'44" 49d36'49"
Andrew Colburn
Cas 1h27'58" 72d35'8"
Andrew Colin Vignaux - Happy 60th
Cru 12h25'11" -61d2'44"
Andrew Collett Forever
Peg 22h21'34" 20d30'7"
Andrew Collins
Cyg 22h0'5" 52d32'53"
Andrew Colverson
Per 4h19'56" 32d24'3"
Andrew Conway
Her 17h20'38" 35d15'37"

Andrew Cordray
Lib 16h0'47" -15d58'15"
Andrew & Cortney
Sge 20h6'50" 18d18'8"
Andrew Craig Diewald
Aur 5h5'30" 35d57'44"
Andrew Craig MacLaughlin
Aur 5h38'35" 43d59'35"
Andrew Cramer
Ari 2h39'51" 30d40'38"
Andrew Crochunis
Cnc 8h24'46" 31d31'0"
Andrew Crooks
Ori 5h13'28" 13d5'46"
Andrew Crump
Dra 18h52'6" 51d9'35"
Andrew Cullen & Emily Page St. John
Cyg 20h32'16" 57d0'31"
Andrew D. Sorensen
Her 16h48'19" 29d57'25"
Andrew D. Stenecker
And 23h20'17" 41d24'17"
Andrew Dale
Dra 19h4'53" 61d12'35"
Andrew Dale George
Uma 11h11'10" 52d37'48"
Andrew Dale Wittenberg
Uma 9h10'8" 63d54'56"
Andrew David
Aur 5h42'6" 44d54'47"
Andrew David Bruscoto
Tau 4h37'33" 28d15'32"
Andrew David Eisner
Per 3h45'29" 51d57'31"
Andrew David Evans
Vul 19h23'47" 26d46'44"
Andrew David & Jennifer Dawn Percy
Cap 21h2'19" -17d40'25"
Andrew David White
Her 17h5'24" 15d8'49"
Andrew David Wilson
Uma 11h11'56" 28d41'41"
Andrew David Woods
Her 16h34'30" 34d27'37"
Andrew Davidson Stargate
Ari 2h45'39" 11d54'29"
Andrew Davis Lindgren
Aql 18h54'9" -0d44'56"
Andrew Dean
Lyn 8h11'53" 41d27'55"
Andrew Dean Dykshorn
Her 17h27'38" 44d42'48"
Andrew Dean Dykshorn
Aql 19h0'38" -0d9'0"
Andrew Dean Moulas
Leo 10h17'29" 15d56'14"
Andrew Decker
Tau 4h32'16" 17d43'23"
Andrew Dell and Holly Anne Joy
Per 3h44'46" 48d48'32"
Andrew DeLuca
And 23h0'7" 51d6'55"
Andrew Di Benedetto
Cep 22h14'36" 82d9'58"
Andrew Diamond
Aqr 20h44'32" -2d12'59"
Andrew Dolinky
Lib 14h53'15" -1d55'5"
Andrew Donald Atchley
Dra 19h49'20" 65d27'4"
Andrew Donis
Ori 5h45'46" 8d2'19"
Andrew Donovan Flores
Her 17h43'28" 48d15'23"
Andrew Douglas Armstrong-Sly
Ori 4h45'12" 0d55'46"
Andrew (Drew) John Cary
Lac 22h48'19" 52d33'1"
Andrew Duncan MacLeod
Cep 22h44'46" 74d34'53"
Andrew Dunn Rhodes
Cnc 8h20'25" 9d30'42"
Andrew E. Anemone
Psc 1h21'12" 13d22'23"
Andrew Edward Hairfield
Dra 18h42'56" 58d55'36"
Andrew Edward John Trolio
Boo 14h52'7" 28d43'54"
Andrew Edward Reed
Del 20h39'37" 17d17'42"
Andrew Edward Ryan
Uma 9h28'5" 63d50'11"
Andrew Ellis
Peg 22h40'50" 24d7'8"
Andrew Ellis
Per 4h24'6" 51d14'32"
Andrew Emmanuel Garcia
Lyn 8h26'43" 41d2'18"
Andrew Ericsson
Ori 5h19'59" -4d58'18"
Andrew & Erin FOREVER!
Aqr 22h32'42" 0d38'48"
Andrew Evan
Cap 21h13'25" -22d2'35"
Andrew F. DiMatteo
Uma 11h57'0" 47d0'28"
Andrew Feher
And 0h26'8" 34d12'55"
Andrew Fertig
Ori 5h58'42" 18d8'32"

Andrew Finch
Uma 9h12'8" 58d43'39"
Andrew Fisher
Uma 14h22'22" 58d22'42"
Andrew Forever Yours Kirstie 121105
Lib 14h32'11" -13d7'49"
Andrew Francis Schneden
Her 17h10'39" 31d31'19"
Andrew Franklin Burroughs
Vir 12h14'7" 1d49'49"
Andrew Franklin Mauk
Her 17h56'34" 23d54'38"
Andrew & Freda Cleary 27th Feb 1965
Cyg 19h35'39" 31d6'31"
Andrew Fujii
Lib 15h51'26" -6d7'15"
Andrew G. Braswell
Per 3h36'48" 44d23'41"
Andrew G. Larkin
Per 2h30'54" 52d27'25"
Andrew G. Makrakis
Psc 1h22'34" 24d48'49"
Andrew & Gabriella
Cas 1h26'7" 60d28'56"
Andrew Gaeddert
Uma 8h41'49" 52d1'41"
Andrew Galea
Ari 1h58'9" 14d47'30"
Andrew Gauthier
Ori 4h51'8" 4d43'13"
Andrew Gentile
Dra 19h4'44" 66d16'35"
Andrew Geoffrey Haldeman
Umi 15h58'54" 76d32'0"
Andrew George Conway - 8 lbs.10 oz.
Tau 5h44'19" 27d2'3"
Andrew George Daly III
Cap 20h24'23" -20d14'19"
Andrew George Donald Burke
Ori 5h49'18" -2d56'12"
Andrew Gerald LaPlante
Aqr 22h56'0" -18d29'47"
Andrew Gonchar
Psc 1h43'49" 17d45'37"
Andrew Gostanian, Jr.
Sgr 18h10'41" -31d2'5"
Andrew Grant
Cep 21h58'29" 69d29'25"
Andrew Gray
Aqr 22h47'43" 0d30'29"
Andrew Grealey
Her 18h27'59" 13d11'15"
Andrew H. "Drew" Starr
Per 2h59'10" 46d22'24"
Andrew Hadley Carter a.k.a. Shaggy
Tau 5h49'22" 17d13'56"
Andrew Hajdukiewicz
Uma 13h40'38" 51d40'4"
Andrew Hale
Cap 21h51'34" -24d30'35"
Andrew Hamilton
Per 3h5'37" 50d44'0"
Andrew Hanna
Uma 9h13'54" 64d26'16"
Andrew Hanson Oeth
Cap 20h20'55" -13d27'52"
Andrew Harvey
Cep 22h10'5" 53d25'23"
Andrew Hastings
Vir 13h29'30" -10d46'12"
Andrew Hatton's 21st Birthday Star
Vir 14h32'12" 3d9'6"
Andrew Haynes
Peg 21h29'58" 14d3'54"
Andrew Heggaton
Leo 10h50'42" 10d29'16"
Andrew Henry Madanci
Psc 1h8'23" 11d3'5"
Andrew Henry Schlick Jr.
Sco 16h10'45" -15d17'1"
Andrew Holtzhausen
Per 2h54'43" 43d31'17"
Andrew Howarth
Lib 15h8'27" -7d57'32"
Andrew Hunt Wetmore
Cas 0h25'29" 61d43'8"
Andrew Ian Martin
Uma 12h20'59" 58d21'30"
Andrew Ian Sherratt
Per 4h31'23" 48d26'42"
Andrew Ian Weiss
Umi 16h28'31" 75d1'43"
Andrew Ioane
Cru 11h58'15" -61d40'47"
Andrew Iskandarian
Ori 5h29'56" 3d27'22"
Andrew J. Balint 20 Years HVB BDB
Uma 10h41'37" 63d53'19"
Andrew J. Camacho
Cas 23h52'6" 53d12'31"
Andrew J. Kivatisky
Cnc 8h54'4" 26d51'36"
Andrew J. Stevens
Psc 1h27'0" 17d43'52"
Andrew J. Trewin
Her 16h32'37" 29d24'30"
Andrew Jack Pryor
Dra 18h32'31" 53d32'54"

Andrew Jacob Allard
Tau 4h27'54" 8d42'21"
Andrew Jacob Deming
Cyg 20h9'22" 41d0'18"
Andrew Jacob Long
Ori 4h53'28" 13d52'37"
Andrew Jacob Redinbaugh
Her 18h45'22" 24d43'16"
Andrew Jacobs
Uma 13h33'48" 55d3'51"
Andrew James
Cap 20h15'28" -11d6'30"
Andrew James
Tau 5h40'22" 23d47'31"
Andrew James Arias
Cep 20h42'34" 62d44'39"
Andrew James Edwards
Cnc 8h48'20" 24d29'17"
Andrew James Flachsmann
Umi 15h33'22" 71d12'58"
Andrew James Hale
Cir 15h18'59" -58d32'8"
Andrew James Harmon
Per 3h25'36" 44d34'25"
Andrew James Jones
Dra 19h22'49" 61d23'7"
Andrew James Oswald
Her 17h38'39" 35d35'2"
Andrew James Renel
Gem 7h30'42" 18d41'38"
Andrew James Rogers
Per 3h29'40" 36d18'51"
Andrew James Swanson
Uma 11h26'57" 35d29'42"
Andrew Jared Scheibner
Per 2h55'31" 32d14'23"
Andrew Jarrett Hill
Per 4h47'19" 44d46'10"
Andrew Jason (A.J.) McGehee
Her 18h44'10" 19d39'44"
Andrew Jason Driggers
Tau 4h23'53" 24d38'24"
Andrew Jay Urion
Cap 21h42'16" -23d58'49"
Andrew Jayden
Umi 16h23'57" 77d6'19"
Andrew Joel Carley
Her 17h36'31" 15d8'25"
Andrew John Arot
Aur 6h24'0" 41d18'30"
Andrew John Burger
Tau 4h35'3" 7d27'27"
Andrew John Coleman
Sgr 18h6'36" -26d55'40"
Andrew John D'Agostino
Umi 14h44'20" 73d40'27"
Andrew John Engel "1977-2005"
Ari 3h9'43" 26d51'58"
Andrew John Hurley
Gem 7h48'58" 33d3'54"
Andrew John Maciog
Per 2h12'19" 53d32'41"
Andrew John Przystawski
Lmi 9h55'7" 36d36'35"
Andrew John Sheeran
Sgr 18h59'55" -34d2'5"
Andrew John Strom
Per 3h8'50" 38d27'45"
Andrew John Utz
Dra 17h23'12" 63d37'9"
Andrew John Walker
Vir 11h44'51" -0d54'17"
Andrew John Wietecha
Tau 4h37'48" 29d21'41"
Andrew Johnston
Cmi 7h39'32" 5d5'47"
Andrew Jonathan Wheeler Long
Lyr 18h28'46" 46d17'6"
Andrew Jordan Crenshaw
Cru 12h52'6" -57d21'46"
Andrew Joseph Akkum
Leo 9h37'58" 25d59'31"
Andrew Joseph Cardone
Per 4h2'8" 45d29'4"
Andrew Joseph Donick
Aur 6h8'28" 52d32'39"
Andrew Joseph Doody
Uma 10h55'44" 58d10'16"
Andrew Joseph Ferrante
Lmi 9h58'34" 36d18'13"
Andrew Joseph Geoghan
Dra 18h49'18" 61d24'27"
Andrew Joseph Gilleran
Ori 6h20'28" 16d14'45"
Andrew Joseph Hess
Lib 15h43'38" -11d48'2"
Andrew Joseph Hill
Ori 5h5'39" 1d45'18"
Andrew Joseph Kitterman
Leo 11h6'40" 5d1'7"
Andrew Joseph Leonhard
Sgr 19h6'11" -21d47'0"
Andrew Joseph Mercado
Uma 8h41'28" 51d33'23"
Andrew Joseph Vasconcelos
Psc 0h33'58" 14d15'36"
Andrew Jung Noh
Uma 9h40'31" 42d20'12"
Andrew Justice Kumar
Lyn 7h15'42" 59d14'12"

Andrew Justin Laine
Leo 11h49'43" 23d36'8"
Andrew Kabelis
Lup 15h42'20" -30d52'29"
Andrew & Kathleen - Dance with you
Nor 16h28'51" -51d20'12"
Andrew & Katie
Uma 9h8'48" 69d7'54"
Andrew Keenan MD, Loving Father
Ori 6h19'51" 7d40'14"
Andrew Keleher
Cru 12h27'58" -60d27'45"
Andrew Kennard
Cep 21h7'56" 55d27'23"
Andrew Kenneth Becerra Jr.
Gem 6h45'0" 26d50'1"
Andrew Kenneth Roy Scott
Per 3h7'29" 55d21'58"
Andrew King-Wah Monkey Wong
Aql 19h47'18" -0d6'47"
Andrew & Kirsty - Love Always & 4eva
Pho 0h36'21" -48d44'7"
Andrew Kronby
Uma 8h37'46" 71d44'38"
Andrew L. Hlebik
Ori 5h34'20" 1d3'58"
Andrew L. Sneen
Umi 17h34'59" 82d49'25"
Andrew L. Young
Crb 15h42'23" 33d9'35"
Andrew L. Young
Crb 15h36'46" 31d24'57"
Andrew Lancaster 27/08/1985
Per 4h27'10" 48d24'3"
Andrew Lane
Ari 2h32'15" 24d56'31"
Andrew Latham
Psc 0h56'3" 24d42'30"
Andrew Lawrence Pearce
Per 2h27'8" 51d26'31"
Andrew Lawrence Walker
Vir 14h14'22" -18d9'56"
Andrew Lawrence Walker Jr.
Cnc 8h29'49" 21d9'44"
Andrew & Leah's First Light Year
Uma 11h50'49" 45d57'48"
Andrew Lederer
Dra 19h43'43" 72d16'6"
Andrew Lee Burgess
Aqr 21h48'59" 0d37'25"
Andrew Lee Coop
Her 16h31'54" 31d35'14"
Andrew Lee King and Kona
Psc 0h44'17" 12d30'33"
Andrew Lee Robson
Cep 21h17'44" 58d26'50"
Andrew Lee Smith
Cnc 8h24'49" 12d11'24"
Andrew Lee Yocom
Ori 5h23'12" -5d12'47"
Andrew Logan Williams
Her 17h28'33" 36d52'54"
Andrew Louis Carroll
Lib 15h6'19" -5d29'49"
Andrew Louis Charles Schneider
Leo 10h14'31" 18d54'55"
Andrew Louis Gostine
Sgr 17h53'26" -16d56'4"
Andrew Louis Martorelli
Ari 2h54'17" 30d1'23"
Andrew loves Johanna
Ori 5h38'7" 7d11'35"
Andrew Lutz
Lyn 7h21'23" 53d10'47"
Andrew Lyon
Per 4h26'48" 39d28'7"
Andrew M Cosello
Her 16h52'25" 28d30'11"
Andrew M. Geiszler
Cep 21h44'27" 61d26'59"
Andrew M. McDonald
And 23h21'44" 42d0'20"
Andrew M Smith
Boo 15h2'58" 40d2'23"
Andrew Macrina
Leo 9h35'34" 29d40'58"
Andrew Maddams
And 23h28'17" 38d2'34"
Andrew Makepeace Ladd III for John
Aqr 20h52'4" 2d25'15"
Andrew Malinchak
Cep 23h14'12" 77d43'52"
Andrew Mallace
Cnc 8h52'19" 11d43'25"
Andrew Manes
Boo 14h58'33" 52d40'0"
Andrew Marshall
Per 3h13'50" 50d37'47"
Andrew Marshall Harhay
Psc 0h23'31" 12d22'41"
Andrew Mason Ward
Ori 4h51'39" 7d27'56"
Andrew Matelwich
Cnc 8h43'41" 32d48'13"

Andrew Matthew Ellis
Aqr 22h39'52" 1d45'39"
Andrew Maxwell Parker
Cnv 13h46'40" 34d0'47"
Andrew Mazur
Cep 21h43'59" 61d57'21"
Andrew Mazza
Vir 13h23'39" 4d1'39"
Andrew McBride - "Forever Young"
Gem 6h50'53" 17d51'21"
Andrew McMahon
Umi 8h43'35" 88d9'15"
Andrew McMahon - "Rock Star"
Uma 9h40'28" 54d35'0"
Andrew McNaughton-Smith
Psc 0h41'37" 8d40'23"
Andrew "Medic 29" Wilson
Her 16h32'32" 33d58'26"
Andrew & Meghan Forever
Lyr 18h25'40" 35d17'0"
Andrew Mellor
Aur 5h54'14" 41d37'31"
Andrew Michael
Per 3h5'45" 54d54'33"
Andrew Michael "Andy" Johnson
Uma 11h36'48" 55d32'39"
Andrew Michael David
Ori 5h45'19" 1d13'13"
Andrew Michael D'Ostilio
Gem 6h42'48" 32d52'49"
Andrew Michael Dowdle
Aqr 22h50'27" -3d36'20"
Andrew Michael Foglesong
Umi 15h3'10" 67d41'28"
Andrew Michael Fritz
Per 4h16'8" 33d38'54"
Andrew Michael Goodman
Her 17h46'30" 44d36'9"
Andrew Michael Karczewski Jr.
Aqr 20h47'10" -13d42'29"
Andrew Michael Klingenberger
Cnc 9h4'11" 23d46'37"
Andrew Michael McCalla
Cap 21h13'12" -15d51'23"
Andrew Michael Onia Montillano
Apu 15h17'19" -71d3'31"
Andrew Michael Woodward
Her 17h44'55" 36d29'28"
Andrew & Michelle - Forever Hotty Love
Cru 12h47'6" -57d34'13"
Andrew Milner
Cep 2h27'23" 80d38'41"
Andrew & Mira Ray - Sept 4, 2004
Cyg 21h19'17" 50d31'4"
Andrew Mittonette
Lib 15h54'53" -5d21'20"
Andrew Montanaro
Her 17h39'2" 31d25'46"
Andrew Moodie & Lisa Larsson
Per 4h46'55" 48d51'48"
Andrew Morenci
Cnc 8h21'22" 23d26'6"
Andrew Motichka
Cnc 8h48'20" 7d22'0"
Andrew Murphy (Murph)
Per 3h10'0" 51d4'26"
Andrew My Love
Uma 12h39'22" 62d27'52"
Andrew Myles Macedo
Lyn 6h36'22" 57d32'14"
Andrew N Cari
Aqr 20h49'35" -12d16'14"
Andrew N Stephanie Martin
Cyg 21h30'21" 51d55'52"
Andrew Nava
Tau 3h44'32" 27d41'5"
Andrew Neal Daufel Sof R N Bardsley
Ari 2h32'56" 24d22'34"
Andrew Netupsky, My Shining Star!
Uma 10h32'9" 50d40'2"
Andrew Nicholas Georgiades
Her 16h18'30" 6d15'27"
Andrew Nicholas Mark Bennett
Ori 5h32'54" -1d39'34"
Andrew Norman Stevenson Dunn
Ori 5h32'4" 0d37'12"
Andrew Novak
Lmi 9h24'44" 37d45'37"
Andrew - Our Stellar Dad 19 May 2007
Tau 4h29'50" 20d59'20"
Andrew P. Alexander
Sgr 17h49'41" -16d29'7"
Andrew P. Bobrowitz
Uma 11h26'11" 69d46'19"
Andrew P Chasar Loves Ashley Nasti
Gem 6h37'0" 21d9'41"
Andrew Patocs 19700728(PattoPatto)
Cru 12h35'9" -58d58'58"

Andrew Patrick Huber
Vir 13h58'41" 3d52'16"
Andrew Patrick McNutt
Aql 19h10'40" 0d16'33"
Andrew Paul
Peg 22h43'58" 32d28'9"
Andrew Paul Cobley - "Hastie"
Uma 13h8'1" 61d31'6"
Andrew Paul Coombs
Cyg 21h54'39" 52d9'11"
Andrew Paul Kopacz
Her 16h28'33" 17d22'33"
Andrew Paul Pettavel
Uma 9h39'45" 41d50'36"
Andrew Paul Wilcock
Leo 9h45'1" 27d34'35"
Andrew Peffall
Uma 13h37'14" 56d4'26"
Andrew Perez - My Best Friend
Ori 5h30'38" 1d49'4"
Andrew Peter Bullock
Psc 0h49'25" 7d41'16"
Andrew Peter James Doyle
Cas 1h13'25" 50d53'35"
Andrew Peter Reimer
Lib 15h14'42" -3d45'52"
Andrew Petros Petralis
Ori 5h40'13" -0d13'12"
Andrew Philipson
Cep 23h7'8" 76d26'39"
Andrew Phythian
Tau 4h26'43" 8d46'22"
Andrew Pirie
Cep 22h57'0" 58d37'33"
Andrew Poling & Denise Carr
Peg 21h36'44" 27d49'18"
Andrew Powell God's Angel #8
Cap 20h24'29" -14d8'35"
Andrew Quinlan Sierra Zero Four
Per 3h1'42" 35d12'20"
Andrew Quinn Martin's Daddy's Star
Psc 0h22'45" 9d58'14"
Andrew R. Barone
Sgr 17h53'58" -19d34'17"
Andrew R Cameron
Cru 12h7'8" -59d16'9"
Andrew R. Lambert
Vir 14h17'52" -20d45'41"
Andrew R. Roe
Per 3h8'24" 54d43'14"
Andrew Ramos
Uma 10h15'45" 47d39'21"
Andrew Reed Roberts
Cep 22h33'16" 65d18'57"
Andrew Renaut
Her 18h12'36" 16d28'9"
Andrew Rhodes and Shigeyo Nagase
Cyg 21h31'17" 44d56'43"
Andrew & Rhonda Webb
Cru 12h19'27" -57d33'23"
Andrew Richard Bealles
Uma 11h38'54" 56d13'41"
Andrew Richard Corona
Cnc 8h50'14" 27d48'49"
Andrew Richard Shiver
Uma 8h53'2" 66d28'50"
Andrew Richard Woods
Uma 9h48'2" 69d22'9"
Andrew Riedel, Lance Corporal
Leo 10h10'12" 12d20'3"
Andrew Robert Bepristis
Per 4h10'44" 50d57'31"
Andrew Robert Bogdan
Pho 0h56'0" -52d47'16"
Andrew Robert Carr
Her 18h23'23" 12d5'0"
Andrew Robert Comi
Uma 9h10'6" 55d31'3"
Andrew Robert Cordaway-Fiore
Lib 15h32'47" -20d17'23"
Andrew Robert Ireland
Uma 11h9'24" 47d49'43"
Andrew Robert Layman
Ari 2h41'28" 30d4'1"
Andrew Robert McClelland
Tau 4h18'12" 27d18'31"
Andrew Robert Miller
Vir 14h34'57" 5d24'14"
Andrew Robert Pierre Marchal
Umi 15h18'54" 80d32'39"
Andrew Robert West
Lmi 9h59'48" 36d45'7"
Andrew Robinson
Aqr 20h42'9" -2d7'30"
Andrew Rock
Her 17h53'41" 28d52'20"
Andrew Ronald 2/15/2006
Uma 10h37'10" 45d24'32"
Andrew Ronald Rennar
And 23h29'1" 48d47'42"
Andrew Rosen
Per 3h18'20" 51d36'48"
Andrew Roy Buckner
Ori 5h27'21" 2d6'53"

Andrew Royal Zimmermann Her 18h47'19" 15d49'50"
Andrew Rudd Tau 3h51'48" 25d18'47"
Andrew Rudick Tau 3h43'28" 27d58'12"
Andrew Russell Weiss Leo 9h48'3" 28d9'55"
Andrew Ryan Canter Ori 5h48'31" 8d45'3"
Andrew Ryan Champagne Ori 5h33'47" 3d42'28"
Andrew Ryan Fidler Gem 7h18'52" 32d53'8"
Andrew Ryan Fidler Gem 7h39'54" 33d33'39"
Andrew S. Combs,Son of Anastasia Boo 14h39'57" 23d36'10"
Andrew S. Gottfried Ori 5h30'51" 1d7'42"
Andrew S. Ogawa, M.D. Uma 9h27'39" 64d59'40"
Andrew S. Sciara Lib 15h34'16" -27d28'16"
Andrew & Sally Turner Now & Forever Vel 10h13'32" -40d23'47"
Andrew & Sara Spells 20/12/2003 Cyg 21h21'27" 50d47'39"
Andrew & Sarah Graves - Love Eternal And 2h18'54" 42d34'40"
Andrew Schaeffer Sgr 18h0'42" -28d17'35"
Andrew Schlessinger Her 16h35'49" 47d50'33"
Andrew Scott Dunham Peg 22h23'47" 26d3'10"
ANDREW SCOTT HAUGHIE Cyg 20h54'49" 31d0'14"
Andrew Scott Karun Uma 14h7'43" 60d17'21"
Andrew Scott Ritchie Her 18h46'39" 13d26'17"
Andrew Scott Robinson Cap 21h14'14" -26d22'4"
Andrew Scott Thompson Sge 20h10'13" 20d54'0"
Andrew Scroggins Tau 3h41'8" 8d24'55"
Andrew Shih Hao Chiu Tau 3h43'26" 28d56'10"
Andrew Sidwell Vir 12h16'38" -0d55'58"
Andrew Simmers Uma 11h50'17" 31d6'34"
Andrew Simon Uma 9h22'33" 42d46'45"
Andrew Sinishtaj Vir 12h45'54" 6d28'8"
Andrew Smalley Lyn 8h17'36" 54d31'11"
Andrew Smith - Love Forever Jen x x Cyg 19h40'52" 30d44'23"
Andrew Smith Ward Lyn 8h27'38" 33d31'49"
Andrew Stephen Napoli Ori 5h28'9" 10d2'37"
Andrew Stephen Werfel Gem 7h18'51" 33d27'17"
Andrew Steven Carty Boo 14h44'22" 34d4'22"
Andrew Steven Hainsworth Vir 13h47'13" -11d53'1"
Andrew Steven Hofinger Ori 5h39'20" -2d27'51"
Andrew Steven Potts Crb 16h13'45" 34d3'25"
Andrew Steven Rollette Her 16h44'6" 36d11'30"
Andrew T. Halbert Dra 17h30'11" 53d26'32"
Andrew T. Hubeli Umi 16h57'50" 82d17'30"
Andrew T. Nill Aql 19h28'5" 3d10'10"
Andrew Tai Her 17h36'47" 26d43'46"
Andrew Takahashi Her 18h55'31" 17d23'43"
Andrew Taylor Engelmann Lac 22h43'34" 50d17'50"
Andrew Taylor Johnson Cap 20h29'23" -15d49'55"
Andrew Tedesco Uma 10h6'40" 50d48'4"
Andrew Teruel Ori 6h25'32" 13d31'40"
Andrew Thad Limer Aql 19h12'30" 12d43'53"
Andrew Thebeault Cnc 8h17'14" 32d41'47"
Andrew Thomas Barnosky Dra 18h37'44" 61d58'13"
Andrew Thomas Cella Aqr 22h45'56" 1d16'11"
Andrew Thomas Constable And 0h6'12" 47d9'24"
Andrew Thomas DeLong Ori 5h30'58" -4d47'26"
Andrew Thomas Duckworth Her 16h29'23" 23d34'24"
Andrew Thomas Duncan Tau 5h46'51" 20d23'47"
Andrew Thomas Kaylor Per 2h40'41" 52d43'2"
Andrew Thomas Maddox Ori 5h16'38" 5d42'37"
Andrew Thomas Meyer Ori 6h12'47" 20d46'46"
andrew thomas oconnor Her 16h53'10" 33d6'35"
Andrew Thomas Pico Verdu Aqr 22h31'23" -9d15'49"
Andrew Thomas Welch Per 3h17'19" 46d21'39"
Andrew Thomas Wilson Lyn 7h8'47" 59d49'22"
Andrew Thomas, Director of Donuts Crb 15h43'30" 26d48'27"
Andrew Truman Case Uma 11h0'21" 58d21'2"
Andrew Tyler Miller to Mars 36 Gem 7h38'56" 17d34'34"
Andrew U. Reyes Sco 17h47'18" -43d5'2"
Andrew Uhrman Vir 13h30'2" -12d4'55"
Andrew Valenzuela Cep 21h7'51" 56d8'29"
Andrew Vernon Peterson Aur 5h51'25" 32d40'19"
Andrew Villa Ori 5h29'8" 6d58'34"
Andrew Vincent James Vir 13h24'33" -9d3'37"
Andrew Vincent Nucolo Tau 3h53'13" 27d15'53"
Andrew Vincent Papucci Gem 7h33'53" 33d57'50"
Andrew W Dmytrasz Lyn 7h7'26" 44d31'47"
Andrew W. Zavetsky Sco 17h53'25" -36d38'28"
Andrew Wade Pate Cnc 8h3'25" 12d11'26"
Andrew Walker Lib 15h42'45" -27d26'25"
Andrew Wallenstein Per 2h15'37" 51d45'11"
Andrew Walsh Her 17h20'51" 28d2'11"
Andrew Warner Russell Psc 1h28'23" 7d2'9"
Andrew Webb Her 17h24'15" 16d7'24"
Andrew Welham Psc 1h52'10" 7d14'1"
Andrew Wilburn And 2h18'18" 41d44'20"
Andrew Wiley Rhodes And 0h43'29" 36d33'46"
Andrew William Burgh Leo 10h26'16" 16d30'49"
Andrew William Fleming Ori 6h13'23" 8d23'32"
Andrew William & Flora Ann Patula Gem 6h3'2" 23d58'5"
Andrew William Harris Cru 12h1'34" -62d52'33"
Andrew William James Walker Dra 16h19'34" 51d33'42"
Andrew William Marshall III Cep 2h27'43" 80d39'46"
Andrew William Niederriter Lyn 7h58'37" 57d26'31"
Andrew William Pierce Uma 10h3'47" 59d9'44"
Andrew Wilson Gem 6h25'9" 19d10'29"
Andrew Wilson Logan Her 16h50'48" 34d22'31"
Andrew Winch Cru 12h36'34" -58d19'46"
Andrew Wingenbach Vir 11h38'22" 0d15'20"
Andrew Wnek Lyn 7h32'53" 50d12'17"
Andrew Woodfin Miller, III Cnc 8h48'50" 12d12'58"
Andrew Woody Gentry-Mace Per 2h14'3" 55d59'25"
Andrew Wooten Leo 10h22'21" 17d24'37"
Andrew- You are my star always- Oma Leo 10h56'41" -1d46'33"
Andrew Youna Cap 21h32'57" -9d25'23"
Andrew, Beloved Super Scottish Hero. Peg 22h23'6" 21d30'50"
Andrew, Kieran, Callum & Megan Bowden Dra 19h8'55" 60d58'5"
Andrew, my darling face starboy! Cas 1h24'32" 67d13'18"
Andrew, my dear sweet Dorky Boy Lac 22h44'53" 52d33'3"
Andrew, Winter Formal '07? Sgr 19h12'48" -16d5'55"
Andrew,Charles,Elijah Sco 16h18'31" -32d18'10"
Andrewboy Uma 11h19'57" 50d2'12"
AndrewJames&DeniseChristineFord Per 3h17'14" 54d7'46"
Andrew-Renee Sgr 18h16'17" -18d49'42"
Andrew's Angel Tonya Leo 11h7'43" 15d9'1"
Andrew's Dream Sgr 19h49'9" -14d33'25"
Andrew's Dream Star Sco 17h9'5" -31d23'39"
Andrew's Eternal Wonder Ori 5h54'9" 9d20'48"
Andrew's FCUoM - June 2005 Pho 1h44'51" -39d59'25"
Andrew's Honey Pet Uma 9h29'52" 46d3'37"
Andrews Memorial UMC Ori 5h49'19" -0d51'55"
andrews place more suited then here Umi 13h59'44" 69d34'6"
Andrew's Star Sgr 19h45'44" -16d19'40"
Andrew's Star Her 17h29'57" 37d53'35"
Andrew's Star Leo 11h27'39" 2d23'11"
Andrew's Star Leo 9h43'42" 14d25'30"
Andrew's Star Ori 5h56'55" 22d27'37"
Andrey Boersma Cep 22h35'43" 76d46'8"
Andrey Pastukh Aqr 22h12'46" -0d26'25"
Andrey Yashin Aur 5h58'8" 29d46'15"
Andria Franqui And 0h14'14" 35d7'50"
Andria Giannis And 0h24'13" 32d36'12"
Andria Humphreys-Ward And 2h19'2" 40d35'45"
Andria Judith Waites And 0h34'27" 28d33'50"
Andria Marie Scianna Uma 11h36'31" 38d12'32"
Andria Michele Leo 11h3'15" -6d31'26"
Andria Michele Johnson And 1h38'3" 42d17'4"
Andria Passaro Uma 9h46'42" 68d17'13"
Andria Rose Spencer Crb 15h30'25" 27d11'53"
Andria Salcedo And 23h5'42" 47d21'12"
Andria58 Lyn 8h31'2" 34d12'57"
Andriana Tau 3h56'33" 21d25'15"
Andriani Uma 13h46'2" 57d0'30"
Andriani Arvanitaki Uma 10h43'53" 39d53'11"
Andria's & Marcus's Place in Heaven Uma 11h33'42" 48d35'17"
Andrijana Trajkovska And 2h21'49" 41d21'56"
Andril Cru 12h14'12" -60d3'43"
Andrin Lmi 10h50'46" 36d16'39"
Andrin Aql 19h36'19" 10d34'34"
Andrin Ori 5h16'37" 0d11'2"
Andrina Goodwin Cyg 20h30'53" 32d21'24"
Andrissas Legacy Ari 2h59'0" 10d38'27"
Andrius Uma 10h59'56" 46d34'31"
Andrius Petras Narkevicius Her 16h11'46" 48d14'58"
Andro H. Kim & Wm. B. Heuring Sgr 17h52'32" -21d15'56"
Androlin Uma 10h33'0" 51d45'4"
Andromeda Cyg 21h20'56" 36d19'26"
Andromeda Cara Hayleigh Vallance And 23h35'48" 36d59'37"
Andromeda Huff And 2h6'28" 42d35'14"
Andromeda Katriona Vail And 0h4'48" 42d47'22"
Andromeda Light Gillespie G Star And 1h23'40" 45d2'55"
Andromeda Pauline And 23h31'2" 38d43'17"
ANDROMEX 1712 Uma 10h19'6" 50d25'41"
Andromonie Leo 10h21'16" 17d32'12"
Androulla Cas 3h13'22" 67d35'33"
Androva, el mundo es tuyo! Ori 5h29'50" 2d32'43"
AnDrumana Sgr 18h39'40" -17d8'24"
Andrusha Cas 1h36'41" 67d40'26"
Andrzej D. T.Q., T.A. M. Sge 19h34'31" 19d0'50"
Andrzej Michalski Cap 21h4'11" -21d20'1"
Andrzej Pisarczyk Uma 11h9'31" 43d7'59"
AnD's Camelot Sgr 19h30'25" -17d13'32"
ANDSAR Del 20h24'58" 9d43'36"
Andshua Cyg 20h17'25" 55d1'51"
Andy Uma 13h4'16" 58d48'6"
Andy Cas 1h19'19" 66d25'39"
Andy Uma 9h46'21" 62d11'31"
Andy Sgr 19h4'57" -33d7'8"
"Andy" Vir 11h38'23" 3d5'39"
Andy Aql 19h47'15" 5d27'37"
Andy Aql 19h34'34" 2d15'6"
Andy Lmi 10h8'0" 39d27'55"
Andy Uma 11h23'12" 49d8'14"
Andy Uma 10h25'37" 46d58'5"
Andy 01-17-62 Aql 19h47'30" -0d54'25"
Andy 3 Per 2h46'56" 40d47'23"
Andy & Abbie Cmi 7h13'4" 10d48'15"
Andy Allen Leo 10h42'19" 8d2'12"
Andy and Alessandra And 0h20'55" 28d38'17"
Andy and Angela Collison's Star Cyg 20h48'22" 37d23'57"
Andy and Carol Sge 19h49'16" 18d52'8"
Andy and Dinah Uma 9h36'34" 48d16'35"
Andy and Emily's Star Umi 15h20'41" 73d53'43"
Andy and Holli Madsen Cas 1h37'56" 63d42'47"
Andy and Janet Cap 20h23'6" -11d5'34"
Andy and Laura McGuire Lyr 19h12'58" 39d32'29"
Andy and Lauren Sgr 18h14'19" -25d10'59"
Andy and Melissa Umi 15h41'17" 72d50'45"
Andy and Melissa Piazza Her 17h16'36" 15d34'23"
Andy and Meredith's Enchantment Dra 17h37'55" 52d34'36"
Andy and Zarah Cap 21h32'57" -15d6'41"
Andy Bate Crb 15h52'35" 29d52'59"
Andy Baxter Sco 17h57'5" -38d35'47"
Andy Beaumont Per 4h15'53" 35d39'21"
Andy & Becky's Love Her 18h3'19" 32d17'29"
Andy Bleifer Cyg 21h53'54" 54d42'11"
Andy Boy Her 16h16'11" 6d13'19"
Andy Brunner Her 17h37'24" 16d50'8"
Andy C. Koepke Sgr 18h18'32" -27d34'0"
Andy & Carolyn Shalosky Cyg 21h17'4" 33d8'17"
Andy & Cassandra 05/20/2004 Aqr 22h23'4" -6d44'52"
Andy & Christina Cyg 20h18'37" 40d9'58"
Andy Cole Aur 5h31'11" 45d58'9"
Andy Cruikshank Lyn 8h6'37" 56d5'53"
Andy & Deirdre Fitzgerald Gilmore Cyg 20h30'13" 53d37'18"
Andy Diete - In Memory Boo 15h5'25" 13d2'45"
Andy & Dom Forever Uma 10h8'38" 42d6'35"
Andy Fankhauser Cep 22h56'58" 71d22'34"
Andy Fedo III Her 16h12'52" 48d9'8"
Andy Foster Per 3h30'19" 43d29'3"
Andy Franzone Aql 19h36'17" 4d23'44"
Andy & Gail Coppock Cyg 19h45'21" 33d48'23"
Andy Gartee Leo 10h25'50" 25d6'44"
Andy Hunter Uma 9h23'58" 60d25'11"
Andy & Jack Sgr 19h2'9" -29d54'16"
Andy Jeffers Vir 12h59'3" -18d23'34"
Andy Jefford Uma 11h55'35" 43d48'38"
Andy Jividen Lac 22h21'20" 45d46'14"
Andy & Jo Harrison Sge 20h20'31" 20d0'24"
Andy John Per 3h7'45" 53d27'27"
Andy Jud Aur 5h57'56" 42d40'7"
Andy & Julie O'Brien Cyg 21h53'55" 48d28'15"
Andy Karoly Crb 16h21'56" 34d38'34"
Andy L. Cnc 8h33'16" 21d12'40"
Andy Lax Uma 8h34'46" 62d29'32"
Andy LeBlanc Ari 3h9'49" 24d7'58"
Andy Lesauvage - Andy's Star Per 2h46'43" 52d33'32"
Andy & Linda Seib Uma 11h0'18" 37d6'21"
Andy Long Van Le Per 4h19'47" 44d34'38"
Andy Loves Emily Forever Star Vir 13h14'14" -17d37'19"
Andy Lynn Uma 13h47'57" 54d43'28"
Andy M. Gleeman Her 18h30'58" 18d23'43"
Andy Main Per 4h24'29" 32d37'1"
Andy Marie Cas 23h41'7" 51d45'25"
Andy Massongill Boo 15h18'27" 32d40'17"
Andy McCabe Uma 8h20'26" 69d22'29"
Andy McEvoy Vir 13h59'18" -21d57'49"
Andy Mein Aql 19h46'13" -0d19'13"
Andy Mitchell Cnc 9h3'38" 15d48'51"
Andy Norbergs Aur 5h38'34" 39d5'4"
Andy Pannebecker Aqr 22h9'23" 0d37'4"
Andy Patric Thomas Boo 14h58'40" 39d2'22"
Andy Peters Her 16h26'10" 6d36'38"
Andy Phillips (One in a Million) Ari 3h21'39" 18d24'14"
Andy Pinson Per 4h25'30" 44d5'55"
Andy Prieto Ori 5h23'4" 2d38'33"
Andy R. Fankell Her 17h24'19" 34d56'19"
Andy Rawling Uma 10h52'2" 67d5'56"
Andy Ristau Uma 11h9'51" 71d21'52"
Andy Schoenig Psc 0h34'27" 17d57'20"
Andy Smith Cap 21h0'12" -18d27'40"
Andy Sniderhan Cyg 20h32'32" 51d32'47"
Andy Squires Ori 5h53'25" -8d35'41"
Andy (Super Nova) Politi Aur 5h43'47" 32d9'40"
Andy The Beloved Umi 9h29'1" 88d45'57"
Andy "The Irish Yankee" Gasser Ori 6h6'56" -3d53'59"
Andy - The Light of My Life Pyx 8h46'32" -27d21'42"
Andy "The Red Giant" Narendra Per 2h48'39" 40d45'41"
Andy Thielking Uma 11h54'26" 36d22'22"
Andy Tremelling Lmi 10h32'54" 35d11'16"
Andy Vu Per 3h52'55" 38d26'52"
Andy White Per 4h10'45" 37d15'3"
Andy&Whitney Keye Anniversary Star Her 16h21'59" 21d7'34"
Andy Workman Lmi 10h9'10" 28d38'37"
Andy Yoder Uma 8h21'46" 67d13'36"
Andy your a Star Cen 13h46'34" -40d22'18"
Andy Z - 60 Tau 5h18'23" 21d39'47"
Andyleron Cma 6h39'36" -14d49'24"
Andyman Umi 14h57'48" 78d54'4"
Andy-n-Lindsay Uma 8h45'28" 61d39'50"
Andyroo Aqr 23h19'5" -13d23'38"
Andy's Flight Tau 4h40'8" 20d46'19"
Andy's My Best Friend Lib 15h46'44" -23d7'31"
Andy's Star Dra 18h21'26" 56d50'27"
Andy's Star Leo 11h50'17" 22d39'29"
Andy's Wave Sgr 19h17'0" -27d59'13"
Andy's Xenion And 0h14'58" 25d14'14"
Andzela Leo 10h0'40" 20d24'17"
Ane Løvstrøm Peg 21h41'43" 22d45'30"
Aneela Nanner Psc 1h21'32" 32d12'19"
Aneeqa (Anki Banki) Col 6h33'48" -39d20'58"
Anees Lyn 6h24'24" 54d3'39"
Anees Per 3h47'49" 46d26'11"
Aneesa Miyoko Chandra Lyn 8h29'33" 42d29'20"
Aneisa's Star Cnc 8h13'32" 21d21'50"
Anel Pasic Umi 14h44'8" 76d33'31"
Anela Ori 5h30'44" -3d21'24"
anela Aql 19h34'38" 10d21'47"
Anela Gem 6h49'6" 13d55'57"
'Anela Jill A Solberg Uma 11h35'28" 44d22'38"
Anela Mendoza Vir 13h23'11" 1d19'20"
Anelalani Castro Peg 23h40'51" 22d52'45"
Aneli Vir 14h21'38" 1d36'35"
Anelia 20 Psc 1h21'56" 33d2'27"
Aneliese Jordan Waldrop And 0h19'58" 31d11'22"
Anelise Adams Cyg 19h44'19" 34d13'52"
Anelivar Leo 10h57'36" 17d55'7"
ANELIZA Gem 7h24'2" 27d9'7"
Anella Ari 3h14'34" 28d2'51"
Anello Di Fidanzamento Uma 12h50'10" 57d54'25"
Anel'yse Renee Anderson Sgr 17h48'1" -29d33'47"
Anemone And 0h39'53" 26d26'36"
anenan Aql 20h8'45" 9d14'41"
Anesa Noelle Acosta And 2h16'51" 47d34'27"
Anesi Cyg 21h39'18" 37d48'25"
Anesty Sco 17h11'48" -33d20'12"
Anet Garcia Lyn 6h36'55" 56d28'31"
Anet N Joshua Always N Forever Cyg 21h25'39" 52d5'6"
Aneta Dukat Lib 15h47'45" -11d13'19"
Aneta Gardias Ori 6h7'50" -0d18'56"
Aneta Joanna Keranen And 2h23'49" 39d18'52"
Anetka Lib 15h4'26" -6d0'20"
Anetka Lib 15h28'15" -18d2'4"
Anett Godo Sgr 18h57'25" -23d33'39"
anetta Aqr 22h34'54" -17d8'27"
Anette Lyn 7h55'39" 33d39'2"
Anette Horlacher And 0h42'14" 37d5'56"
Anette Isaacs Gem 6h10'26" 22d39'44"
Aneurin Cra 18h2'51" -37d14'1"
Aneurysm Cnc 8h48'30" 7d33'46"
Aney Lyn 8h24'29" 35d8'22"
Aney Francis Allen Uma 8h48'42" 65d59'47"
Anezka Nate Alvarez Tau 4h20'4" 25d10'3"
Anfe' Callaway Woodrum Cnc 8h54'33" 13d25'41"
"Ang&El" Tau 5h48'12" 20d10'14"
Ang Kripp Cas 1h33'34" 72d33'16"
Ang & Lyns Lyn 7h31'47" 52d5'14"
Ange Mon 7h3'52" -7d17'32"
Ange Beige And 0h33'26" 32d40'25"
ange bleu Sgr 17h59'32" -24d9'11"
Ange d'avril Uma 11h23'38" 32d1'19"
Ange de Patrick Vir 14h13'57" -8d2'34"
Ange DL Aqr 22h14'0" -8d27'25"
Ange_lilly Ori 5h56'13" 17d54'31"
Ange..... M-M Vir 14h20'37" -19d50'4"
Ange Sans Ailes Ori 5h50'59" 11d47'41"
Angel Ori 5h41'20" 11d24'47"
Angel Ori 4h52'59" 6d50'52"
Angel Ari 3h14'28" 15d40'22"
Angel Tau 4h32'56" 15d32'53"
Angel Tau 4h32'3" 17d2'2"
ANGEL Ari 2h35'58" 20d31'59"
Angel Gem 6h39'19" 20d39'9"
Angel Gem 6h55'0" 22d9'23"
Angel Cnc 8h14'48" 15d49'24"
Angel Cnc 9h11'4" 15d49'10"
Angel Her 18h50'35" 21d22'37"
Angel And 0h34'58" 32d47'35"
Angel Peg 22h27'30" 25d9'58"
Angel Peg 22h10'8" 27d28'57"
Angel Lyr 19h10'22" 26d29'19"
Angel Lyr 18h48'58" 34d49'11"
Angel Lyr 18h39'33" 30d51'41"
Angel Cnc 9h4'9" 31d12'44"
Angel! Gem 6h49'47" 32d29'10"
Angel And 1h57'20" 39d2'48"
Angel Lyn 7h25'27" 49d0'29"
Angel And 23h26'14" 38d18'33"
Angel And 23h53'37" 42d57'34"
"Angel" Lyr 18h31'39" 39d42'36"
Angel Cyg 20h48'32" 38d10'23"
angel Lyr 18h45'11" 44d12'54"
Angel Lyr 18h43'44" 44d9'15"
Angel Lyr 18h50'52" 43d40'12"
Angel Cyg 21h20'11" 51d41'8"
Angel And 23h23'51" 51d40'55"
Angel And 23h59'26" 48d12'45"
Angel Cam 4h44'19" 59d58'53"
Angel Uma 9h37'20" 45d8'17"
Angel Uma 11h23'5" 51d1'34"
Angel Uma 11h51'19" 50d12'59"
Angel Sgr 18h23'30" -20d13'53"
Angel Sco 16h6'26" -15d52'30"
Angel Cap 20h36'33" -20d49'49"

Angel
Cap 21h41'39" -8d55'10"
Angel
Sco 16h12'0" -10d46'18"
Angel
Sgr 19h38'24" -14d58'48"
Angel
Sco 16h12'37" -11d53'34"
Angel
Aql 19h46'21" -0d0'58"
Angel
Aql 19h43'58" -0d7'37"
Angel
Ori 5h36'56" -1d20'41"
Angel
Umi 17h22'32" 78d50'41"
Angel
Umi 15h28'15" 83d34'57"
Angel
Umi 13h8'54" 70d35'33"
Angel
Umi 15h19'7" 68d27'49"
Angel
Umi 15h30'53" 73d54'48"
Angel
Uma 9h35'16" 62d10'54"
Angel
Uma 9h45'14" 61d51'48"
Angel
Uma 11h18'9" 63d51'55"
Angel
Uma 10h42'58" 62d56'4"
Angel
Dra 16h26'0" 63d17'53"
Angel
Uma 13h4'42" 61d36'5"
Angel
Lyn 6h22'32" 55d27'12"
Angel
Uma 8h35'28" 52d37'12"
Angel
Dra 15h40'2" 59d10'18"
Angel
Cyg 20h7'43" 56d31'48"
Angel
Sco 16h43'6" -33d0'42"
Angel
Sco 16h29'25" -34d39'51"
Angel
Sco 17h49'29" -30d19'4"
Angel
Psa 22h33'9" -25d57'58"
ANGEL
Cap 21h47'20" -24d28'12"
Angel
Sco 17h52'31" -42d12'0"
Angel
Sgr 20h13'32" -38d33'6"
Angel
Psc 1h8'45" 3d52'51"
Angel 13
Lyr 18h49'35" 39d18'26"
Angel 14
Umi 16h21'31" 75d5'36"
Angel 1585
Uma 11h7'27" 33d32'11"
Angel Adrian
Col 5h54'39" -37d45'54"
Angel Alberto
Uma 11h26'25" 41d45'28"
Angel Alice
Cap 20h35'13" -11d17'28"
Angel Alicia
Psc 0h24'54" 7d43'29"
Angel Amber
Cas 23h29'56" 51d58'53"
Angel Amber Anson
And 23h48'0" 47d25'34"
Angel Amielia Amanda
Gem 7h52'43" 19d13'56"
Angel Amy
Ari 2h34'32" 21d33'8"
Angel and Lovebug
Peg 22h32'9" 25d4'55"
Angel Andre
Lyr 18h42'44" 38d12'1"
Angel Andrea
And 0h38'45" 41d31'20"
Angel Anja Sweetest Star
And 2h12'44" 48d49'23"
Angel Ann Casey
Peg 22h34'34" 7d0'52"
Angel Avery
Uma 11h29'42" 59d21'51"
Angel Baby
Aql 19h5'7" -0d9'28"
Angel Baby
Cnc 8h47'36" 24d49'58"
Angel Baby
Cyg 20h4'31" 51d1'50"
Angel Baby
And 1h36'20" 41d59'16"
Angel Baby
Lmi 10h6'35" 36d54'57"
Angel Baby Forever in My Heart
Lyr 18h28'28" 32d6'20"
Angel Baby Girl
Vir 13h25'9" 13d27'49"
Angel Baby Girl Dazi
Cap 21h36'58" -16d41'33"
"Angel Baby" Harvey
Ori 6h9'28" 19d50'42"
Angel Bean Dower
Psc 23h25'12" 3d34'17"
Angel Bear
Uma 10h10'11" 43d55'56"
Angel Bear
Cap 20h24'44" -12d28'43"
Angel Beau
Psc 1h5'33" 12d11'48"
Angel Becky Gower Winter Wine
Peg 22h23'5" 16d8'39"
Angel Bella
Lyn 7h59'13" 35d35'39"
Angel Blessings
Psc 1h23'47" 23d30'52"
Angel Boy
Her 18h18'20" 20d52'35"
Angel Brittany
Aqr 22h30'58" -1d2'59"
Angel Buckner
Uma 10h44'39" 59d15'51"
Angel Butterfly Rochon
Psc 1h37'51" 21d45'53"
Angel C. Meade
Ori 5h11'41" 8d52'20"
Angel C. Walker
Cnc 8h8'36" 10d41'39"
Angel Carla Forever
Cep 0h22'40" 72d58'39"
Angel Caroline Whitton
Uma 10h40'22" 71d4'46"
Angel Cartagena
Sco 16h15'54" -10d19'35"
Angel Cathy
And 1h0'52" 41d57'3"
Angel Celeste
Vir 14h16'21" -12d22'39"
Angel Cellilo-Ramirez
Lyr 19h5'51" 36d38'8"
Angel Chang
And 1h4'57" 38d5'2"
Angel Christina Maria
Uma 9h30'46" 44d59'18"
Angel Christina Weir
Cap 21h8'38" -16d46'8"
Angel Clifton
Lib 15h19'20" -5d0'26"
Angel Crouch
Lyr 18h48'51" 33d16'33"
Angel Crystaline Schreib
And 23h45'45" 48d9'36"
Angel Cyssou
Del 20h36'30" 7d18'41"
Angel D. Starcher
Cyg 21h12'47" 44d43'45"
Angel Daddy Koppel (John J. Koppel)
Sgr 18h32'52" -31d16'42"
Angel Dawn
Aqr 22h37'1" -23d39'16"
Angel Dawn's Life Star
And 1h43'12" 50d6'31"
Angel Debbie
Sgr 19h11'28" -13d51'2"
Angel Delgado Batista
Aqr 23h13'28" -9d13'19"
Angel Diana
And 1h39'39" 47d27'1"
Angel Dove
Ori 5h23'14" 7d1'50"
Angel Eddie
Umi 14h21'47" 77d43'48"
Angel Estrella Amanda Lewis
Cyg 19h45'58" 42d39'57"
Angel Eyes
Lyr 18h50'39" 40d40'50"
Angel Eyes
Lyr 18h44'22" 37d55'17"
Angel Eyes
And 1h37'57" 47d24'50"
Angel Eyes
And 23h13'13" 48d23'19"
Angel Eyes
Cyg 20h52'33" 47d36'34"
Angel Eyes
Uma 11h0'6" 50d23'40"
Angel Eyes
Lyn 9h13'50" 34d21'1"
Angel Eyes
Lyn 8h20'27" 44d40'25"
Angel Eyes
Her 18h4'26" 20d39'41"
Angel Eyes
Sge 19h50'48" 17d33'7"
Angel Eyes
Ari 2h14'28" 24d37'44"
Angel Eyes
Lib 15h27'24" -6d55'32"
Angel Eyes
Aqr 22h3'12" -7d50'59"
Angel Eyes
Lib 15h17'53" -20d9'24"
Angel Eyes
Uma 10h12'33" 69d31'16"
Angel Eyes
Dra 17h26'17" 65d31'39"
Angel Eyes
Uma 9h25'4" 60d53'4"
Angel Eyes
Lyn 6h51'22" 57d24'45"
Angel Eyes
Uma 9h28'57" 54d47'38"
Angel Eyes 081906
Aqr 20h52'32" -12d41'18"
Angel Eyes (1)
Psc 1h14'56" 26d23'34"
Angel Eyes Cassidy
Cyg 20h3'35" 39d7'29"
Angel Eyes Jill The Motivator
Cyg 20h43'8" 39d23'42"
Angel Eyes Tasha
Cap 20h12'16" -23d17'4"
Angel Face
Vir 14h6'22" -12d51'49"
Angel & Fancy Face Forever
Cru 12h19'18" -57d53'48"
Angel Fire
Per 4h14'48" 38d57'47"
Angel Flint
Cyg 21h56'39" 37d31'30"
Angel For Debra Lil SS
And 0h21'45" 39d10'41"
Angel - Forever & Always
And 23h9'39" 50d42'57"
Angel Francess
Gem 6h58'34" 23d4'0"
Angel Frank Pomeroy
Cnc 9h1'16" 7d8'54"
Angel from Above: Ashley Fowler
Cnc 8h50'34" 28d29'21"
Angel from Heaven
Uma 9h13'24" 49d32'9"
Angel Galazka
Lyr 18h49'54" 29d47'1"
Angel Girl Herma Bogle
Sco 17h56'8" -39d6'46"
Angel Girl Jeanette
And 0h22'51" 26d42'9"
Angel Gonzalez DeLuz
Sgr 19h41'46" -44d48'35"
Angel Gonzalez III
Her 18h21'14" 17d29'10"
Angel Grace
Lib 14h48'54" -18d47'34"
Angel Granny Fran
Ari 2h57'15" 24d54'49"
Angel Gwen
Uma 8h47'43" 67d44'31"
Angel Hardip Aujla
Cas 1h21'4" 57d7'53"
Angel Harold
Cam 5h52'15" 61d28'0"
Angel Hartley
And 1h15'4" 45d25'28"
Angel Harvey
Lyr 18h50'0" 31d49'42"
Angel Hazel
Dra 15h46'22" 64d7'27"
Angel in Ashes
Lyr 18h30'16" 32d20'21"
Angel In Disguise
Cap 20h10'56" -13d45'23"
Angel in Disguise (The Buddy Star)
Uma 10h26'42" 66d28'50"
Angel In The Sky
Vir 13h22'25" -10d44'41"
Angel is the Next Supergirl
And 0h17'45" 35d58'42"
Angel Ishmael
Sco 17h45'56" -42d24'48"
Angel J
Cap 21h19'26" -15d11'9"
"Angel" J. Araujo
Umi 16h52'47" 81d24'24"
Angel Jacobs
And 0h59'51" 48d13'53"
Angel James
Cep 21h56'38" 59d58'57"
Angel Javier Robles
Cam 4h33'31" 67d46'34"
Angel Jo 08
Cyg 20h8'11" 36d41'8"
Angel JoAnn
Vir 13h15'46" -17d15'27"
Angel Julia
Cnc 9h9'40" 10d30'52"
Angel & Kaban - 24.08.1997
Uma 12h49'23" 61d57'30"
Angel Kae
And 23h15'52" 43d0'3"
Angel Kaszubinski
Cap 21h38'50" -17d56'32"
Angel Katy Anne
Sgr 18h15'17" -20d4'33"
Angel Kay
Leo 11h47'31" 25d38'8"
Angel Kay Hill
Lib 15h10'0" -7d43'3"
Angel Kay Lindsey, My True Love
And 0h48'32" 35d39'29"
Angel Ken has Christies love ALWAYS
Cep 23h13'59" 80d6'41"
Angel Khoury
Cas 22h57'55" 54d37'43"
angel kisses
Cyg 19h53'54" 36d50'41"
Angel L. Torres
Uma 12h36'13" 54d8'28"
Angel L4TM
Psc 1h4'10" 5d30'17"
Angel Lee
Sco 16h14'8" -9d45'34"
Angel Lili
Ori 5h18'20" 12d40'9"
Angel Linda Marie Gaebel
Aqr 23h17'44" -19d30'49"
Angel Lisa
Cap 20h34'49" -19d28'43"
Angel Lisa
Psc 1h3'26" 32d14'15"
Angel Lisa
And 23h25'22" 52d16'24"
Angel "Lite"
Uma 11h47'2" 63d3'29"
Angel Logan Elise
Sco 16h56'33" -45d9'43"
Angel Love Aguilera
Uma 12h5'28" 33d23'2"
Angel Loves Big Poppa
Eri 3h46'29" -0d46'47"
Angel Lucas Alexander Wilson
Gem 6h30'29" 20d1'0"
Angel/Lyn
Leo 11h43'1" 14d48'1"
Angel Maddie
Lyr 18h35'28" 38d26'3"
Angel Mae Nagy
Per 3h17'58" 39d33'55"
Angel Mae Peknik
And 1h47'57" 41d50'29"
Angel - Mallory Root
And 0h46'51" 41d36'42"
Angel Manieri
Uma 12h17'7" 52d54'38"
Angel Manon Martel
Lib 15h55'54" -14d44'48"
Angel Marc Molinsky
Lyr 18h50'14" 42d1'10"
Angel Maria
Crb 16h11'16" 38d25'55"
Angel Maria Byrd
Sco 16h9'53" -16d13'46"
Angel Marie
Cap 20h36'28" -19d9'41"
Angel Marie
Lyn 6h49'10" 56d22'33"
Angel Marie
Cas 23h44'27" 56d18'25"
Angel Melissa Lady
Cnv 13h59'41" 39d48'53"
Angel Meryl
Oph 17h54'42" -0d51'5"
Angel Michael
Lib 15h27'58" -4d33'35"
Angel Michelle
Cnc 9h5'55" 11d36'57"
Angel Michelle
Gem 7h13'32" 20d30'32"
Angel Monique
Tau 4h7'38" 6d33'55"
Angel Morgan
Leo 11h27'15" 16d38'38"
Angel mother
Gem 7h24'33" 26d6'7"
Angel N. Hunt's Star
Lib 15h2'37" -13d52'48"
Angel Nancy
Umi 15h35'30" 74d29'49"
Angel Nanny Sketchley
Lmi 11h4'9" 26d42'6"
Angel Nat
And 0h56'8" 40d29'50"
Angel Nia
Sgr 18h40'29" -28d0'12"
Angel Nicole Gonzalez
Sco 17h17'54" -31d54'55"
angel OF
Sco 16h10'11" -10d46'41"
Angel of Africa, 23.06.2003
Cas 23h6'39" 59d3'49"
Angel Of Beauty and Grace
Cas 1h30'36" 63d49'1"
Angel of Heaven
Psc 23h13'55" 6d19'15"
Angel of Heaven
Psc 0h47'10" 17d9'15"
Angel of Inez
Cnc 8h45'7" 13d17'9"
Angel of Light
Cnc 8h16'52" 16d4'46"
Angel of Love - Kelly Clifton
Sco 16h12'50" -12d8'47"
Angel of Love-Cherished Forever
Mon 6h41'27" -0d20'35"
Angel of Many
Psc 1h18'38" 29d32'47"
Angel of Mary
Cas 1h16'23" 63d18'0"
Angel Of Mine
Aqr 22h23'13" -24d35'0"
Angel Of Mine
Lyr 18h40'39" 39d2'32"
Angel of Mine
Lyr 18h49'0" 44d1'50"
Angel of Mine, Ms. J
Sgr 18h10'19" -31d14'17"
Angel of my heart and now the sky.
Cyg 19h50'0" 33d32'42"
Angel of my heart for ever-more
Cnc 9h12'53" 24d48'4"
Angel of Perfection
Leo 10h26'46" 12d32'48"
Angel of Tabitha
Ari 3h14'12" 15d54'30"
Angel of the Sky Alex Husted
Leo 9h22'20" 13d45'39"
Angel Olmo
Uma 12h4'7" 65d21'40"
Angel on Earth
Cap 21h11'20" -15d59'13"
Angel Payne
Uma 9h52'35" 59d16'6"
Angel Perez
Per 3h6'7" 37d4'1"
Angel Perez
And 23h9'52" 46d29'18"
Angel Pie
Leo 11h8'59" 7d41'25"
Angel Pink
Lyr 18h46'46" 38d24'42"
Angel Pip
Ori 6h9'12" 20d52'57"
Angel Posadas
Psc 1h21'57" 15d36'54"
Angel Princess
And 2h36'19" 44d12'9"
Angel Princess, Judy 11867
Sco 16h58'18" -40d2'34"
Angel Priscella
Cnc 9h21'11" 31d32'6"
Angel Puppy Jaisy
Cma 6h39'45" -12d14'52"
Angel Rachel
Psc 0h24'23" 7d36'7"
Angel Renee
Vir 14h16'15" -12d59'34"
Angel Rivera
Uma 12h12'56" 58d39'31"
Angel Rose
Aqr 21h4'14" -6d44'39"
Angel Rose
Col 5h45'45" -33d47'17"
Angel Rose
And 1h37'58" 42d55'41"
Angel Rose Pellicane
Cap 21h45'43" -12d53'13"
Angel Rutherford
Aqr 21h50'30" 0d56'58"
Angel Ryleigh's Star
Psc 1h19'58" 17d8'59"
Angel Sabie
Cas 0h53'56" 50d19'58"
Angel Samantha
And 0h22'18" 36d28'3"
Angel Sarah
Vir 14h19'11" -20d46'43"
Angel Secundus
Lmi 9h34'15" 38d24'58"
Angel Skye
And 23h17'15" 40d36'43"
Angel Slezak
Lib 15h17'13" -27d18'1"
Angel Star
Lyn 7h9'13" 59d50'28"
Angel Star
Lyr 18h39'59" 39d58'54"
Angel Star
Gem 6h31'4" 14d8'19"
Angel Star "Karissa"
And 0h31'38" 27d44'31"
Angel Suz
Cas 0h29'35" 61d59'10"
Angel (Tanya Leigh Cassandra Bovey)
And 0h38'59" 43d41'42"
Angel Thomas Pomeroy
Cnc 9h2'59" 7d2'51"
Angel Tor
Lyr 18h48'6" 37d27'5"
Angel Tracy
And 0h49'44" 39d10'18"
Angel & Trent
Cyg 20h4'34" 46d31'40"
Angel Valyssa Jaye Hernandez
Cap 21h45'13" -8d51'19"
Angel Victor
Ori 5h32'20" -2d8'14"
Angel Victoria
Peg 21h58'26" 25d21'7"
Angel Vidal Millà
Cnc 8h8'30" 15d8'36"
Angel West
Sgr 18h51'56" -31d49'41"
Angel Z
Uma 12h42'0" 59d33'22"
Angel, Sweet Love of My Life
Vul 20h31'37" 28d13'1"
Angel10
Uma 11h37'28" 62d18'35"
Angel777-1096-7514
Sgr 20h15'48" -33d34'42"
Angela
Sco 17h54'29" -36d21'14"
Angela
Sgr 18h30'45" -28d44'48"
Angela
Sgr 18h1'37" -27d50'57"
Angela
Cyg 21h8'45" 49d6'38"
Angela
Uma 9h46'6" 67d24'48"
Angela
Uma 13h39'19" 62d8'28"
Angela
Cam 5h58'36" 69d50'38"
Angéla
Uma 9h18'51" 71d35'36"
Angela
Lyn 6h39'32" 59d52'42"
Angela
Uma 9h11'9" 53d58'31"
Angela
Sex 9h46'40" -0d38'42"
Angela
Vir 12h35'19" -0d20'29"
Angela
Aqr 20h44'9" -2d45'13"
Angela
Vir 13h45'47" -9d41'18"
Angela
Lib 15h4'40" -2d15'50"
Angela!!
Cap 20h25'29" -13d50'41"
Angela
Lib 14h35'55" -17d56'29"
Angela
Cap 21h44'5" -18d59'17"
Angela
Crb 15h41'37" 28d15'0"
Angela
Boo 14h46'52" 25d11'38"
Angela
Ari 3h13'51" 27d12'16"
Angela
Ari 2h47'44" 26d49'7"
Angela
Tau 5h50'3" 23d55'48"
Angela
Leo 10h12'18" 16d44'20"
Angela
And 0h36'32" 27d52'16"
Angela
And 0h45'14" 27d35'44"
Angela
Ari 2h45'39" 27d13'24"
Angela
Vir 15h0'22" 4d0'37"
Angela
Aqr 20h49'25" 2d16'1"
Angela
Ori 5h39'26" 8d36'21"
Angela
Ori 5h56'34" 11d45'13"
Angela
Gem 6h51'16" 13d38'20"
Angela
Ori 6h15'15" 9d9'41"
Angela
Ori 6h20'14" 9d41'10"
Angela
Cnc 8h24'18" 13d24'58"
Angela
Vir 13h4'38" 13d15'39"
Angela
Tau 4h37'44" 18d41'18"
Angela
Cyg 19h49'7" 50d37'57"
Angela
Cyg 20h26'30" 45d57'15"
Angela
Uma 10h19'19" 46d53'45"
Angela
Uma 11h10'25" 45d13'5"
Angela
Cyg 21h8'45" 49d6'38"
Angela
And 23h44'18" 47d10'41"
Angela
Cas 0h4'39" 55d9'4"
Angela
Cam 4h11'38" 59d56'27"
Angela
Cas 2h8'15" 59d10'12"
angela
Lyr 18h45'57" 39d28'5"
Angela
And 23h19'32" 42d3'12"
Angela
And 0h46'13" 46d34'25"
Angela
And 2h19'49" 49d7'17"
Angela
And 0h49'50" 40d56'57"
Angela
And 2h20'13" 37d58'1"
Angela
And 1h14'4" 42d17'53"
Angela
And 0h43'33" 44d33'58"
Angela
Cyg 19h59'27" 33d19'19"
Angela
Lyn 7h32'26" 37d2'32"
Angela A. Arancio
Lib 15h6'27" -3d32'57"
Angela (Abby) Acosta
Lyn 7h7'17" 61d30'59"
Angela Abeyta
Leo 10h36'54" 23d32'43"
Angela & Alex
And 23h18'51" 48d38'48"
Angela Aliene
Ari 3h10'6" 26d29'24"
Angela Aline Lairmore
Cap 21h4'5" -19d48'47"
Angela Alohalani Midro 7402
Cap 21h24'48" -19d42'0"
Angela Amodeo
And 23h24'35" 48d37'8"
Angela and Laura Coe Infinity Star
Col 5h40'56" -32d30'49"
Angela and Leah
Cyg 20h6'7" 57d0'43"
Angela and Nick
Cyg 21h43'43" 51d46'52"
Angela "Angel" Deles
Ari 3h14'13" 29d30'1"
Angela "Angie" Fowler
Cap 20h26'53" -11d59'58"
Angela Ann DeNeui
Ari 2h23'29" 25d6'50"
Angela Anna Leva
Cyg 21h13'53" 45d50'14"
Angela Anne Johnson
And 0h17'6" 31d31'35"
Angela "Annie" Reid
Cas 23h32'55" 51d57'30"
Angela Antonelli
Vir 12h47'8" 12d44'24"
Angela Aragon
Cnc 8h36'22" 15d37'12"
Angela Armes Special Star
Cap 21h41'4" -17d17'22"
Angela & Austin
Lyr 19h5'22" 42d34'22"
Angela B. Galeana
Cyg 21h53'17" 42d34'59"
Angela Babe
And 1h4'10" 39d45'43"
Angela Barcic
Vir 13h22'13" 12d51'13"
Angela Bauer
And 23h42'46" 38d2'54"
Angela Bibiana Artherton
Tau 4h27'24" 13d42'27"
Angela Bishop
Uma 10h4'32" 52d19'32"
Angela Bordone
Peg 0h2'54" 26d15'37"
Angela Bright Star
Lyn 7h59'27" 38d33'39"
Angela - Brightest Star In My Sky
And 1h35'46" 47d57'43"
Angela Brosche
And 1h44'16" 41d50'37"
Angela Bryce
Gem 6h44'44" 15d38'59"
Angela (BUBBLZ) Kimberly
Cyg 20h9'43" 35d47'24"
Angela Buitendacht
Ari 2h3'12" 22d59'2"
Angela C. Farmer
Lyr 19h3'49" 32d4'1"
Angela C. G. Hoefke
And 23h47'33" 44d14'20"
Angela Cain
Lyn 8h48'23" 44d24'12"
Angela Capadagli
Cap 21h40'20" -14d14'26"
Angela Capstick Carter-Plotkin
Cnc 7h55'50" 18d48'24"
Angela Cardalli Sanford
And 0h35'58" 28d25'41"
Angela Carter Kennedy Johnson
And 1h15'33" 46d46'58"
Angela Celeste Rienstra
And 2h11'8" 46d51'30"
Angela & ChaiLi
Sge 19h51'54" 19d1'17"
Angela & Charles Reed
Cyg 21h27'30" 30d15'17"
Angela Christie - Michael David
Vir 12h53'52" 6d33'58"
Angela Christina Bianco
Cnc 8h59'11" 10d32'57"
Angela Christine
Sco 16h6'48" -8d47'28"
Angela Christine Eberts GWA 04-05
Cas 1h23'3" 52d39'1"
Angela Christine Stafford
Vir 12h41'48" 11d19'31"
Angela Christine Van Duys
And 1h46'16" 41d51'35"
Angela Ciccarelli
Uma 9h23'18" 60d42'11"
Angela "Cinderella" Palazzo
Ari 3h16'52" 16d2'18"
Angela Clare Marshall Baldwin
Cas 0h8'44" 53d38'32"
Angela Colella
Psc 1h20'8" 8d14'6"
Angela Colquitt
Lyr 19h13'28" 45d54'33"
Angela Connelly
Cap 20h53'50" -25d56'43"
Angela Cordner Wilson
Vir 13h12'7" 2d55'21"
Angela Corelli
And 1h10'22" 42d10'22"
Angela Corno
Boo 14h34'18" 24d34'39"
Angela Costilow
Sco 16h56'23" -44d36'8"
Angela Cristina Anaya
And 0h37'11" 32d25'24"

Angela D. Guillen
Lmi 10h47'8" 26d49'6"
Angela D. Harden-GaMPI Shining Star
And 0h52'45" 35d53'29"
Angela D. Owens
Her 18h51'32" 23d37'25"
Angela D. Schroeder
Psc 0h59'44" 7d2'51"
Angela Dahlke
Boo 14h40'30" 30d16'35"
Angela Daile Kellejian
Dra 19h27'21" 66d19'16"
Angela Dallaire
Crb 15h58'2" 34d52'3"
Angela & Dave's New Beginings
Cyg 19h56'14" 37d2'1"
Angela Dawn
And 2h32'3" 43d58'29"
Angela Dawn
Ari 3h17'25" 27d43'2"
Angela Dawn
Ari 2h40'42" 25d17'35"
Angela Dawn
Aqr 22h35'34" 0d30'53"
Angela Dawn
Cas 1h51'39" 60d58'31"
Angela Dawn
Lib 15h33'40" -16d33'31"
Angela Dawn 06
Sgr 20h24'7" -30d33'6"
Angela Dawn Hardesty
Cyg 20h39'22" 42d39'32"
Angela Dawn Lindamood
And 0h43'53" 45d23'14"
Angela Dawn Rikard
Lmi 10h26'58" 33d49'51"
Angela Dawn Schiebout Whitlock
And 1h31'57" 50d27'18"
Angela DeLaBastide
Vir 13h12'55" -15d12'47"
Angela Denise Victoria Maragni
Ari 3h16'21" 28d42'5"
Angela & Derick
Lyn 8h24'59" 51d27'38"
Angela Di Ciaula
Cam 4h33'55" 53d39'57"
Angela Diane Coats
Umi 16h2'39" 86d29'16"
Angela Doughty Apicella
Cas 23h5'29" 57d7'24"
Angela Drake
Ori 5h58'1" 21d24'23"
Angela E. Barreras
Vir 12h54'45" -7d10'2"
Angela E. Ledford
Lyr 18h58'23" 33d33'35"
Angela Elizabeth Guenther
Cyg 21h2'10" 36d53'36"
Angela Elizabeth Roman
And 23h25'44" 38d15'34"
Angela Esson
Psc 0h40'34" 9d9'21"
Angela F. Begazo
Uma 11h8'45" 41d40'0"
Angela Faith
Cam 3h47'23" 55d32'2"
Angela Faith Smith
Dra 20h0'6" 74d23'16"
Angela Favia
Sgr 19h15'15" -15d4'33"
Angela Faye Hambsh
And 23h5'9" 52d16'59"
Angela Feres
Tau 4h42'48" 26d33'32"
Angela Ferguson
Psc 0h16'44" -3d26'22"
Angela Finn
And 23h43'29" 42d20'28"
Angela Fiorillo
Cam 5h34'37" 71d25'37"
Angela Forrest-Burge
Cap 20h39'3" -15d22'56"
Angela Fortunato Smile
Cnc 8h45'52" 31d7'32"
Angela Frances Mackrell
Gem 7h40'37" 34d8'7"
Angela Froemmel
Uma 11h56'3" 30d47'16"
ANGELA GAGNE
Sgr 19h30'54" -16d47'17"
Angela Galante
Lyr 18h58'14" 26d15'26"
Angela Garbett
And 2h21'35" 47d16'41"
Angela Garcia
Tau 4h20'20" 17d30'56"
Angela Geralyn Cincotta
And 2h31'12" 45d25'37"
Angela & Glenn
Gem 7h25'59" 32d3'0"
Angela Grace
And 0h19'38" 29d56'2"
Angela Grace Phillips
And 0h36'7" 42d46'55"
Angela Granillo
Cyg 19h57'26" 44d29'32"
Angela Green
And 23h25'11" 47d44'49"
Angela Groll
Uma 14h19'36" 55d51'59"
Angela H.
Ari 2h10'42" 14d47'42"
Angela Hale
And 1h58'26" 38d5'33"
Angela Hallerud
Uma 9h35'38" 63d12'39"
Angela Harris
Gem 6h54'58" 21d59'46"
Angela Hempling
And 1h37'35" 49d17'14"
Angela Henderson
Gem 6h24'32" 27d30'40"
Angela Hewes
And 2h27'38" 40d53'16"
Angela Holbrook
Ori 5h24'23" 1d0'1"
Angela Holland
Cyg 21h0'6" 54d43'21"
Angela Ieni
Peg 22h2'29" 14d46'1"
Angela Inslee
Lmi 10h5'23" 34d33'13"
Angela Irene Seri
Cyg 20h39'28" 35d9'52"
Angela Isabel Villegas
Tau 5h46'39" 13d33'46"
Angela J. Carland
Lyr 18h49'33" 30d23'28"
Angela J. McCathran
Sco 16h5'41" -16d58'43"
Angela J. Ross
Cyg 19h50'25" 39d15'8"
Angela&Jake
Sco 16h12'54" -14d32'6"
Angela Jane
And 1h26'34" 39d3'21"
Angela Jane Hall
And 23h4'22" 44d34'3"
Angela "Janeth"
Cas 23h47'26" 53d22'57"
Angela Jarod Alpha Prime
Ori 6h15'42" 10d23'39"
Angela Jo
And 23h13'14" 51d50'59"
Angela Joshi
Ori 5h34'48" 12d37'24"
Angela Joy
Cyg 20h14'15" 36d41'13"
Angela Joy Strong
Uma 9h44'34" 51d7'26"
Angela Juliana Wiemeyer
Gem 6h56'20" 27d57'55"
Angela Juliette - beautiful angel
Gem 6h45'6" 18d21'31"
Angela K. 0607
Lyn 8h34'11" 36d39'15"
Angela K. Broadway
And 0h18'1" 39d35'18"
Angela K. Vaden-Williams
Tau 4h55'5" 23d9'22"
Angela Kathleen Hoverter
Cap 21h54'18" -8d53'58"
Angela Kay
Sco 17h51'21" -35d46'44"
Angela Kay
Tau 4h5'34" 6d25'51"
Angela Kay
Cyg 19h38'14" 35d24'34"
Angela Kay Cardozo
Cas 0h27'6" 54d47'18"
Angela Kay Earwood
Psc 0h38'50" 6d57'55"
Angela Kay Palmisano
And 0h57'46" 38d43'25"
Angela Kaye Broadus
And 1h20'38" 37d46'6"
Angela Kaye Hawkins
Gem 6h37'54" 16d55'39"
Angela Kim Renecker
Gem 7h23'52" 25d10'18"
Angela Kindschy
Lib 15h17'32" -20d23'34"
Angela&Klaus
Dra 16h43'35" 59d51'5"
Angela Kortnie Hinton
Lyr 18h30'12" 28d16'50"
Angela Kowalczyk
Lep 5h12'58" -10d52'42"
Angela Kristine
Cap 20h52'18" -18d4'8"
Angela Kristine Burgett
Lib 15h9'21" -7d29'46"
Angela Kristine Field
Lib 15h5'24" -7d28'38"
Angela L. Temple
Leo 11h19'9" 1d57'27"
Angela Lamberton
Umi 15h38'25" 67d42'5"
Angela Laurene Mitts
Lyr 18h31'20" 36d7'51"
Angela Lee Milano
Lyr 18h33'59" 32d51'16"
Angela Lee Verhagen
Ori 6h5'24" 16d55'6"
Angela Leigh
Cas 0h26'28" 65d26'42"
Angela Lenfest
Cnc 8h37'5" 22d42'33"
Angela Lesley Wouters
Cnc 8h7'45" 8d58'3"
Angela & Levon
Lyr 19h8'11" 27d14'37"
Angela Li
And 2h36'43" 37d46'11"
Angela Loraine
Gem 6h1'1" 27d14'0"
Angela "Love of My Life"
Ori 5h47'33" -3d3'1"
Angela Lyn
Sgr 18h14'11" -18d54'19"
Angela Lynn
Gem 7h26'36" 28d7'1"
Angela Lynn Huff
Vir 13h43'16" -6d37'59"
Angela Lynn Martin
Cas 1h23'8" 65d5'58"
Angela Lynn Murphy
Lyn 7h42'28" 38d2'42"
Angela Lynn Temple
Lib 15h53'9" -7d15'57"
Angela Lynn Willebeek-LeMair
Lyn 8h34'22" 34d30'37"
Angela Lynne Lamb
And 23h47'21" 47d49'44"
Angela M. Castelucci
And 1h39'10" 45d52'48"
Angela M Ertzner
And 23h38'3" 42d26'22"
Angela M. Fellows
And 0h43'20" 40d35'17"
Angela M. Ribaudo
Umi 16h8'22" 73d54'40"
Angela M Salerno
Sgr 18h31'9" -16d6'58"
Angela M. Santoro
Leo 11h33'16" 13d44'23"
Angela M. Stancy
And 0h38'7" 33d9'31"
Angela M. Vittiglio
Mon 6h19'58" -9d6'24"
Angela M. Williams
Cap 20h8'0" -9d10'15"
Angela Mae
Cas 1h36'11" 65d34'2"
Angela Mae Mrwik
Del 20h39'50" 15d22'17"
Angela Major
And 23h0'36" 40d29'26"
Angela "Mama" Viglietta
Sgr 19h11'29" -13d13'16"
Angela Manfredi
Sco 16h3'36" -26d3'13"
Angela Mangan Hunt
Uma 10h24'6" 61d50'29"
Angela Manouvelos
Uma 11h20'58" 39d39'21"
Angela Marciantonio
Hya 9h21'14" -2d35'43"
Angela Marey
And 1h5'9" 36d52'16"
Angela Maria
Lib 15h2'22" -1d6'0"
Angela Maria Boyle (10/84-9/05)
And 2h27'7" 42d19'10"
Angela Maria Duda
And 1h42'52" 41d39'34"
Angela Maria Hampton
Cyg 21h20'1" 35d36'14"
Angela Maria Strickland
And 1h39'17" 38d18'20"
Angela Marie
And 0h33'59" 44d8'25"
Angela Marie
Gem 6h45'47" 33d59'14"
Angela Marie
And 1h16'57" 45d54'46"
Angela Marie
And 2h17'3" 49d26'49"
Angela Marie
And 23h11'10" 49d6'30"
Angela Marie
Leo 10h43'22" 15d51'8"
Angela Marie
Ari 2h22'29" 18d18'26"
Angela Marie
Tau 5h26'54" 21d7'4"
Angela Marie
Cas 23h37'43" 56d4'36"
Angela Marie
Cap 21h30'33" -23d55'44"
Angela Marie Anderson - My Angel
Lib 14h52'51" -0d44'25"
Angela Marie Ashfield-Collins
And 0h31'40" 45d40'50"
Angela Marie Boudreaux
Cap 20h36'15" -14d44'4"
Angela Marie Buchanan
Mon 6h54'14" -0d48'33"
Angela Marie Capasso
Uma 10h37'10" 44d47'28"
Angela Marie Cartwright
Aqr 22h13'50" -1d48'25"
Angela Marie Chapin
Ari 2h0'56" 23d36'40"
Angela Marie Cotherman
And 2h27'13" 45d25'18"
Angela Marie Estes
Lib 15h8'9" -5d5'15"
Angela Marie Fimbres
Cnc 8h31'22" 13d0'39"
Angela Marie Foster
Ari 2h36'16" 22d15'26"
Angela Marie Gillingham
Cnc 8h8'20" 8d59'15"
Angela Marie Hilton
Cas 0h37'31" 51d9'44"
Angela Marie Hoagland
Mon 8h1'47" -0d35'35"
Angela Marie Janssen
Cam 5h32'4" 62d57'47"
Angela Marie Kapala
Vir 13h5'4" -19d51'17"
Angela Marie McFadden
Aqr 23h3'34" -9d36'16"
Angela Marie Moore
Lyr 19h8'21" 31d27'44"
Angela Marie Murphy
Psc 23h15'26" 6d25'20"
Angela Marie Murphy
Leo 11h45'6" 20d26'16"
Angela Marie Perreault
Vir 11h40'7" 7d26'0"
Angela Marie Richter
Uma 10h43'26" 61d26'57"
Angela Marie Russo
And 23h20'9" 48d37'32"
Angela Marie Swerlyk
And 0h40'25" 28d42'23"
Angela Marie Towner
Tau 5h44'12" 14d35'23"
Angela Marie Villa
Vir 11h57'52" -9d50'45"
Angela Marie Whitacre
And 23h28'4" 36d37'10"
Angela Marie Zuniga
And 0h19'21" 28d50'25"
Angela Marie LaMas
And 1h48'40" 38d53'58"
Angela Marien
Sgr 19h43'43" -26d45'45"
Angela Marino
Lyr 18h36'43" 36d5'25"
Angela Martha Hurst
Lyn 7h11'6" 47d1'48"
Angela Martinelli
Sgr 19h6'53" -34d42'2"
Angela Mary Coates
Vir 13h25'52" -20d34'8"
Angela Mary Ulrey
Gem 6h30'26" 24d6'20"
Angela Maureen Curry
Uma 8h30'15" 61d24'0"
Angela McDowell
Ari 2h18'44" 19d25'22"
Angela Mech
Ari 3h4'11" 13d6'58"
Angela Meeks
Uma 8h21'24" 59d44'39"
Angela Melkonian - Shining Star
And 2h20'22" 47d48'26"
Angela MG Barr
Cnc 8h15'35" 23d23'18"
Angela Michele Volz
Ari 3h19'18" 18d50'52"
Angela Michelle
And 2h11'25" 49d23'23"
Angela Michelle Hille
Mon 6h32'27" -5d18'33"
Angela Michelle - November 5, 1987
Umi 9h49'43" 88d56'3"
Angela Miller
Vir 12h23'28" -1d0'22"
Angela Miller
Lib 15h5'13" -13d55'44"
Angela Momilani Santiago
Lyn 7h29'53" 55d29'11"
Angela Monique - Mummy's Star
And 23h40'54" 42d36'22"
Angela Moody McKay
Cnc 8h19'35" 11d16'12"
Angela Moore
Sgr 18h29'43" -25d54'18"
Angela Morgan
Cas 0h57'15" 63d36'7"
Angela Mullins
Tau 4h27'57" 17d38'5"
ANGELA (MY ANGEL)
Cas 1h24'24" 64d22'19"
Angela Natale Piatti
Leo 10h23'9" 22d30'56"
Angela Necikowski
Uma 11h0'23" 54d27'23"
Angela Newton
Uma 11h51'4" 55d0'9"
Angela Nichole Silva
Umi 15h19'9" 73d44'30"
Angela&Nick Forever
Sge 19h53'47" 18d48'3"
Angela Nickole Amunategui
Lib 15h6'1" -27d50'8"
Angela Nicole
Psc 1h28'5" 26d55'42"
Angela Nicole Bowersox
Cyg 19h50'46" 31d3'8"
Angela Nicole C.
Lib 14h52'20" -3d46'54"
Angela Nicole Lefler
Cnc 8h37'11" 16d20'8"
Angela Nicole Mendoza
Sgr 19h41'20" -16d2'16"
Angela Nicole Staroba
Uma 12h4'49" 48d54'12"
Angela Niki Christakos
Gem 6h31'10" 14d6'9"
Angela Noelle Strutz
And 23h17'33" 48d36'49"
Angela Nola Palou
Cnc 8h23'7" 9d17'36"
Angela Norris
Psc 22h54'9" -0d4'15"
Angela Offen and Matt Saunders
Crb 16h2'38" 25d43'19"
Angela Ogliari
Lyn 9h9'31" 40d5'30"
Angela P.
Leo 11h29'25" 13d56'50"
Angela Palaggi
Ori 5h53'36" 20d34'54"
Angela Park
Sgr 19h51'1" -12d59'55"
Angela Pineda Hanlon
And 0h16'39" 42d22'51"
Angela Pisanti-Delaney
Peg 21h57'8" 20d13'45"
Angela Porter-Corralez
And 0h34'32" 42d34'17"
Angela Powers
Aqr 22h32'45" 1d40'38"
Angela R. Achille
Cap 21h41'51" -11d11'16"
Angela R Beard
Sco 16h13'53" -13d26'10"
Angela R Billman
Vir 13h20'58" -13d51'34"
Angela R Puente
And 23h37'20" 47d44'0"
Angela R. Smith
Sco 16h19'21" -41d13'18"
Angela & Rachel
Cyg 21h26'41" 30d33'42"
Angela Rae
Cyg 21h21'4" 34d48'49"
Angela Rafaloski
And 23h49'4" 48d22'46"
Angela Reaves
Tau 4h5'49" 8d15'56"
Angela Reeves Southerland
Lyn 7h45'2" 58d31'12"
angela ren'e 12
Uma 12h3'16" 39d59'39"
Angela Rene Roberts
Vir 13h20'51" 3d31'38"
Angela Renea Cowan
Sco 16h52'26" -38d43'41"
Angela Renee
Lyn 7h49'9" 57d49'13"
Angela Renee Coons
And 1h16'10" 36d3'17"
Angela Renee Giddens
Sco 16h44'15" -31d13'31"
Angela Renee Kauffman
Uma 11h42'54" 50d8'6"
Angela Renee Kinner
Vir 13h44'24" 5d11'50"
Angela Riggio
Lyr 19h18'0" 29d52'19"
Angela Riley
Crb 16h12'9" 26d6'36"
Angela Rita Manning
Cas 1h23'52" 68d56'42"
Angela Rivera
Car 10h49'24" -62d10'8"
Angela Rizzo Monte
Cas 1h31'22" 63d19'21"
Angela Roney
And 1h13'12" 37d53'45"
Angela Rosario Salas
Uma 8h32'25" 69d31'59"
Angela Rose
And 1h2'30" 40d27'16"
Angela Rose
Uma 11h13'22" 41d41'26"
Angela Rose Benincase
Lib 14h38'20" -20d24'10"
Angela Rose Charles
Lib 15h48'2" -16d50'33"
Angela Rose Gianfrancesco
Dra 16h13'19" 63d4'4"
Angela Rose Sysol
And 2h23'48" 45d53'45"
Angela Russell
Cyg 19h55'11" 33d22'54"
Angela & Ryan
Mon 6h44'45" -0d3'36"
Angela S. Curran
Cyg 20h5'14" 35d36'10"
Angela Sanders "Jeremy & Tegrett"
Sco 16h9'11" -11d32'19"
Angela Sanders "Jeremy & Tegrett"
Uma 11h18'32" 57d2'20"
Angela Sangu
Cap 21h36'16" -11d0'2"
Angela Sardi
Gem 7h33'50" 22d26'7"
Angela Scheetz
Uma 10h20'55" 49d14'50"
Angela Schmid (Ash)
Crb 15h56'36" 26d26'58"
Angela Schranz
Vir 11h39'23" -0d2'49"
Angela Shaw
Sco 16h54'16" -36d6'28"
Angela Sherie Harmon
Leo 11h20'12" 23d10'0"
Angela Shining Star Christensen
And 23h59'6" 41d45'37"
Angela Solem
Vir 13h19'14" 1d27'47"
Angela Somers
Uma 10h6'3" 45d54'57"
Angèla Soucy
And 1h23'22" 47d19'0"
Angela Sproston my star sister
Cap 20h10'16" -9d27'43"
Angela Steel
Sco 16h52'30" -42d34'33"
Angela Steiner
Cas 0h59'47" 54d22'28"
Angela Sughrue Sweet 16
Tau 4h13'41" 24d4'49"
Angela Susan Thorne
And 1h13'44" 35d11'43"
Angela Swann
Cyg 21h14'39" 47d26'34"
Angela Swearengin
Mon 7h16'38" -0d55'27"
Angela T. Krzyzanowski
Ori 6h4'36" 19d32'38"
Angela Taprogge
Leo 11h45'1" 15d11'38"
Angela Taylor
And 2h21'49" 49d54'7"
Angela Taylor
Uma 8h41'19" 72d35'49"
Angela Templeman
Tau 4h24'22" 12d59'19"
Angela The Beautiful
Sgr 19h8'35" -23d10'34"
Angela - The BEST Mommy
Vir 13h11'42" -6d50'21"
Angela "The Munchkin Star"
Psc 1h11'13" 25d59'14"
Angela Therese LaMonica
Uma 11h33'38" 31d7'24"
Angela Tiby
Gem 7h28'16" 33d7'53"
Angela Trask
Her 17h40'28" 15d56'38"
Angela Turin
Lib 15h8'30" -24d33'31"
Angela+Tyler Forever
Cyg 21h40'0" 35d31'18"
Angela und Manfred
Uma 10h2'29" 46d55'46"
Angela Uzcanga
And 0h11'17" 35d57'33"
Angela Valentino 10-1-1989
Lib 15h9'37" -11d21'19"
Angela Wang
Aqr 23h11'17" -4d58'41"
Angela Weldon
And 23h27'23" 42d50'5"
Angela Wesolowski
Uma 12h7'34" 61d21'52"
Angela Whitmoyer
Vir 15h7'27" 4d2'15"
Angela Wilkerson
Sco 16h14'25" -14d1'37"
Angela Williams
And 0h14'36" 26d27'14"
Angela Williams
Crb 15h36'3" 38d18'19"
Angela Wood
Uma 12h1'7" 38d44'3"
Angela Wright
Uma 9h24'41" 44d19'30"
Angela Xiong
Del 20h50'40" 10d54'18"
Angela Yearby
And 2h28'40" 50d13'44"
Angela Yebra
Cyg 20h9'32" 44d35'31"
Angela & Ysidoro Alvillar
Ari 2h8'6" 15d36'19"
Angela Zulay Goforth
And 2h19'21" 50d29'46"
AnGeLa29
Aqr 22h31'23" 0d22'21"
AngelaAndJustin
Cyg 20h46'49" 42d47'4"
AngelaArtyOAlly
Lyn 6h42'13" 56d54'42"
Angelabelle
Leo 10h30'25" 16d3'4"
Angela-Brynn Burke Lawrence
And 0h30'58" 33d37'54"
Angelaflutist
Gem 6h40'43" 34d30'45"
Angelagary 1970
And 1h1'0" 46d16'3"
Angelalou Spinkii
Dra 17h11'8" 51d45'8"
Angela-Maria Azari
Tri 1h42'30" 31d9'33"
Angela's Alluring Ankh
Aqr 22h44'8" -8d51'46"
Angela's Angel
Lyr 18h35'16" 37d56'18"
Angela's Angels
Aqr 22h31'47" 2d16'13"
Angela's Dream
Leo 10h19'2" 17d57'54"
Angela's Eyes
And 1h8'53" 41d37'55"
Angela's Lucky Sparkle
Cnc 9h15'39" 27d48'57"
Angela's Star
Peg 22h39'5" 7d59'16"
Angela's Star
Tau 4h27'26" 19d37'27"
Angela's Star
Vir 13h22'55" 12d13'5"
Angela's star
And 0h33'29" 38d59'14"
Angelas star
Lyr 18h49'32" 33d50'10"
Angela's World
Uma 9h25'5" 72d10'47"
Angelbaby
And 0h43'49" 43d38'16"
Angelbaby 03.30.86
Ari 2h47'52" 12d43'50"
AngelBear
Lib 15h39'35" -17d59'16"
Angelbear 143
Lib 15h44'50" -13d1'18"
AngelBrat
Lib 15h5'9" -23d30'8"
Angelcake Pollack DL4919
Uma 10h28'56" 65d1'0"
Angeldog777
Cam 4h33'6" 57d44'53"
Angele Baker
Leo 9h41'45" 29d50'17"
Angelé Hartley
Lib 15h20'10" -28d14'31"
Angele Violet Hill Shines For You
Uma 10h5'48" 45d31'36"
Angeleah Grace Rockford
And 23h6'34" 50d32'48"
Angeleena's Star
Uma 11h29'58" 37d43'52"
Angelena Corinne
Leo 11h41'58" 14d58'9"
Angelena McMann Blessing from Above
And 0h46'6" 39d23'46"
Angeles Megias Reinaldo
Ori 5h19'55" 6d32'14"
Angeles Sellers
Psc 0h46'45" 16d54'59"
angeleyes
Cap 20h25'38" -16d0'57"
Angelface
Cnc 9h4'42" 30d41'59"
Angelfish
Lyn 7h17'33" 52d19'17"
AngelGlo
Ari 3h6'39" 29d9'34"
Angelheart
Aur 5h52'10" 53d29'51"
AngelheartLCM
Sco 16h44'43" -30d50'20"
Angelhome
Cnc 8h27'11" 9d53'54"
Angeli
Her 17h16'44" 16d45'17"
Angeli Agrawal
Sgr 17h54'49" -29d21'49"
Angeli P
Cnc 8h6'37" 11d54'13"
Angeli Romana
Lyn 8h28'51" 53d47'39"
Angelia
And 1h4'21" 44d35'49"
Angelia
Lyn 9h3'30" 44d53'18"
Angelia D. Holt
Uma 9h4'32" 55d47'2"
Angelia Estelle Lytle
Psc 0h14'6" 2d26'50"
Angelia Marie Gregory
Cap 21h47'29" -14d37'24"
Angelia Raveneau-Butler
Aqr 22h49'53" -16d48'38"
Angelia Saylor
Sco 16h32'33" -26d17'57"
Angelia's Star
Uma 11h24'35" 38d16'29"
Angelic
Cnc 8h13'39" 16d40'54"
Angelic
Sgr 18h25'39" -32d59'46"
Angelic
Sgr 19h42'25" -12d34'22"
Angelic Diva
Tau 5h52'9" 15d44'52"
Angelic Light
Aqr 21h13'30" -9d14'6"
Angelic Ostrander
Lib 14h54'35" -0d51'9"
angelic spirit
Lyr 19h0'39" 32d23'4"
Angelic Susan
And 2h29'7" 42d0'58"
Angelica
And 0h41'10" 34d36'1"
Angelica
Cas 0h4'56" 54d40'27"
Angelica
And 22h58'28" 47d52'59"
Angelica
Ari 2h47'48" 15d17'9"
Angelica
Psc 1h7'34" 28d56'3"

Angelica
Cnc 9h19'2" 23d35'51"
Angelica
Ori 5h18'33" -6d43'38"
Angelica
Cas 0h14'16" 61d3'22"
Angelica Adamo
Lyn 7h45'28" 38d52'8"
Angelica Ann
Tau 4h19'0" 30d4'5"
Angelica Boccitto
Boo 14h18'29" 40d29'30"
Angelica Carolina Vitola
Tau 4h41'57" 23d50'51"
Angelica Chiari
Ori 5h55'13" -8d20'8"
Angelica Ciervo-Brancato
And 1h2'17" 44d26'44"
Angelica Colmenares
Dra 17h43'18" 54d53'24"
Angelica Constance Cortez
Gem 7h17'23" 25d38'41"
Angelica Dallari
Umi 15h9'52" 76d6'22"
Angelica Dolfi
And 23h42'51" 35d57'33"
Angelica Duval
Tau 5h38'6" 18d44'8"
Angelica Esperanza
Cam 4h14'54" 55d18'49"
Angelica Estrella
And 1h12'30" 45d32'33"
Angelica Hope Eerbeek
And 0h45'7" 39d38'43"
Angelica Juliet Pierce
Aqr 22h22'6" -17d44'1"
Angelica Marie
Lib 15h31'25" -6d41'30"
Angelica Marie Sena
Crb 16h13'16" 34d19'56"
Angelica Marta Virginia Calabrese
Cas 0h49'31" 61d55'36"
Angelica Maude Gethyn Gledhill
And 23h50'40" 40d51'15"
Angelica May Causing
Tau 3h44'6" 4d47'10"
Angelica Mollenhauer
Ari 2h59'0" 30d45'6"
Angelica Moschella
Cas 1h41'36" 64d33'21"
Angelica Nicole
Cnc 8h25'54" 21d40'42"
Angelica Nicole Canales
Gem 6h7'37" 24d6'51"
Angelica Pena
Sgr 18h37'14" -20d31'25"
Angelica Perazzi
Boo 14h58'1" 17d47'7"
Angelica Perez
Sgr 18h36'39" -27d33'30"
Angelica Petty
Cas 0h49'10" 66d40'10"
Angelica Raye Salscheider
Sco 16h13'1" -13d3'40"
Angelica Reyes
And 2h7'23" 38d37'18"
Angelica Rodriguez
And 2h31'39" 44d5'41"
Angelica Rose Lofton
Lyr 18h29'57" 37d9'58"
Angelica Serrano Osorio
Aqr 23h8'56" -6d38'8"
Angelica V. Krader
Vir 13h26'42" -17d15'26"
Angelica Worley
Lyn 7h49'50" 54d36'10"
Angelica Yanneth Ramirez
Lyn 6h17'33" 55d50'51"
Angelica, Puffy & Baby
Aqr 22h5'6" -8d43'42"
AngelicaChristopher
Lyn 8h25'23" 55d41'45"
Angelical Princessita Ekaterina
Sco 16h30'50" -26d38'17"
Angelica's Dawn
Ori 5h53'19" 20d41'30"
Angelicas Eye
Leo 10h7'18" 22d16'29"
Angelica's Star
Mon 6h50'43" -0d9'20"
Angelicus Affectus 02132005
Uma 8h50'47" 65d15'27"
Angelika
Ori 5h54'38" 16d23'52"
Angelika
Leo 10h40'57" 14d11'56"
Angelika
Cyg 20h25'9" 38d29'24"
Angelika Brucksch
Uma 8h51'8" 68d22'3"
Angelika Burghardt
Uma 9h33'57" 43d56'53"
Angelika Eremeeva
Sco 16h38'15" -34d32'19"
Angelika Högele
Uma 9h42'2" 69d14'28"
Angelika Ioannou
Umi 17h24'36" 86d19'38"
Angelika Janowicz
Uma 8h25'9" 67d29'24"

Angelika mein Angel von Rudolph
Uma 11h11'3" 71d14'6"
Angelika Mika
Uma 9h21'27" 63d45'10"
Angelika Nicola Tonoian
Leo 10h28'22" 13d34'14"
Angelika & Patrice
Cas 0h50'19" 47d48'45"
Angelika&Reinhold Stern
Uma 11h48'55" 37d46'34"
Angelika T
Uma 10h28'9" 61d35'14"
ANGELIKI
Cas 0h8'8" 53d30'48"
Angelina
And 23h26'50" 35d40'14"
Angelina
Lyr 18h15'23" 36d50'31"
Angelina
And 0h15'37" 31d39'28"
Angelina
Lib 14h50'29" -1d7'31"
Angelina
Pyx 8h43'34" -31d5'16"
angelina
Sco 17h49'56" -41d48'12"
Angelina Alicia Valenzuela
Uma 10h48'6" 40d21'7"
Angelina Baglini
Sgr 18h16'4" -18d26'52"
Angelina Bregante
Cas 0h23'20" 51d46'20"
Angelina Carmen Lewandowski
Tau 5h56'11" 25d17'49"
Angelina Christina
Vir 12h28'30" 1d49'6"
Angelina Cirignotta
Tri 2h34'34" 34d59'51"
Angelina Elizabeth Solis
And 2h28'39" 50d55'25"
Angelina & Ernest Wiebusch
Cyg 19h58'20" 57d34'13"
Angelina Estrella Vicario
Lmi 10h14'43" 34d41'39"
Angelina G. Cipriano
Cas 0h41'52" 61d13'47"
Angelina Gabriela Navarrete
And 0h36'6" 30d56'43"
Angelina Garner Lee
And 0h50'36" 38d41'0"
Angelina Gromacki
And 0h39'0" 42d44'47"
Angelina Grullon
Cas 2h44'12" 57d33'20"
Angelina H.
Lmi 10h31'43" 30d34'39"
Angelina Isabella Gozzi
Gem 7h19'0" 20d24'10"
Angelina Jean Alberro
Cnc 8h49'56" 27d16'40"
Angelina Jolie
Pho 0h44'0" -41d39'0"
Angelina Joy Pearson
And 0h34'3" 46d30'19"
Angelina Kristina Wellesley Dimmick
Ori 6h8'7" -0d59'48"
Angelina Lilly Bartholomeou
Uma 9h19'7" 54d52'14"
Angelina Lima Coelho
Crb 15h56'34" 28d20'45"
Angelina Lucia Lizardi
And 1h13'12" 38d1'39"
Angelina Luyanda
Lmi 10h31'48" 32d1'28"
Angelina Lynn Fiandaca
Aqr 23h22'41" -11d37'59"
Angelina M. Bruno
Vir 13h18'9" -20d6'3"
Angelina Mae Berkeypile
Crb 16h11'12" 35d56'22"
Angelina Malatesta
Ari 2h47'35" 18d32'11"
Angelina Marianne Buttacavoli
And 0h39'29" 43d0'39"
Angelina Maribito
And 2h34'6" 44d38'17"
Angelina Marie
And 23h31'45" 48d38'34"
Angelina Marie Kamer
Uma 9h30'42" 56d12'35"
Angelina Marie Papariello
And 23h24'14" 43d49'32"
Angelina Maso 29/07/1980
Leo 10h50'52" 14d51'55"
Angelina McDonnell
Leo 9h47'59" 32d53'26"
Angelina Megale Barnecut
And 1h18'52" 44d54'12"
Angelina "mi gordita" Rodriguez
Sgr 19h29'50" -11d59'41"
Angelina Michelle
Leo 10h22'49" 10d55'44"
Angelina Miranda
Leo 10h56'57" 14d3'58"
Angelina Mosca
Cyg 19h55'51" 45d8'18"

Angelina Neis
Ori 6h14'50" 16d12'30"
Angelina Nicole Urana
And 0h1'13" 46d58'7"
Angelina Papingu
Gem 7h20'34" 13d28'36"
Angelina Pilyugina Gisselberg
And 23h16'48" 48d48'10"
Angelina Requal Gutierrez
Lyn 8h17'5" 56d14'9"
Angelina Reyna Ynez Garcia
Gem 7h49'48" 24d58'53"
Angelina Rivera
Cap 21h7'39" -26d37'35"
Angelina Rodriguez
Psc 0h0'34" -3d55'13"
Angelina Ruppert
Lyn 8h8'24" 53d34'11"
Angelina Sasso
Psc 1h6'58" 23d45'26"
Angelina Therese Pietras
Cas 0h54'37" 64d29'33"
Angelina Tlumac
Aqr 22h41'12" -17d47'30"
Angelina Valentine
And 23h31'43" 42d13'26"
Angelina y Cesar
Aqr 21h15'0" -8d26'42"
Angelina Zoe
Leo 11h27'44" 19d28'58"
AngelinaGL3386
And 23h9'12" 40d54'25"
Angeline
Mon 6h57'9" -8d11'30"
Angeline Gainey
Lyn 7h58'9" 43d3'18"
Angeline Gruetzmacher Howard
Cnc 8h52'39" 10d13'57"
Angeline Lau 12-8-1996
And 23h45'35" 39d8'6"
Angeline Mary G. RIvera
Vir 12h1'32" -2d46'57"
Angeline Princess of Bearnagh
And 2h15'20" 50d12'43"
Angeline Provenzano
Sgr 18h5'45" -28d5'4"
Angeline & Shane
Cas 23h3'28" 55d36'47"
Angeline Valianos
Lib 15h24'4" -29d48'17"
Angeline Zacarias Campos
Vir 13h0'2" 0d45'47"
Angelique
Cnc 8h30'2" 6d58'46"
Angelique
Ori 5h19'47" 11d21'41"
Angelique
Psc 23h26'48" 5d18'27"
Angélique
Vir 13h0'34" 11d59'27"
Angelique
Com 12h19'49" 26d49'11"
Angelique
And 22h59'16" 39d59'15"
Angélique
Cyg 20h23'10" 41d39'28"
Angelique
Sco 17h49'39" -33d56'10"
Angelique
Uma 12h24'45" 59d55'33"
Angelique
Uma 11h7'52" 67d26'18"
Angélique Anderlini
Sco 16h9'20" -30d18'35"
Angelique "Angel" Godoy
Aql 19h16'40" -0d43'49"
Angelique Anne Marie Blincow
And 0h55'48" 36d36'47"
Angelique C.
Leo 9h49'11" 29d9'0"
Angelique Clay
Cas 2h34'11" 66d11'29"
Angelique Danielle Murray
Leo 10h10'15" 16d46'12"
Angélique et Mathys Guillemot
Col 6h16'8" -40d17'29"
Angelique & Gavin happily everafter
Her 16h31'17" 34d1'13"
Angelique Gentry
Peg 23h35'40" 19d55'29"
Angelique Gundram
Vir 12h10'36" 7d30'34"
Angelique Krouskos
Cru 12h25'50" -60d48'55"
Angelique Lerdall
Cnc 8h40'0" 16d22'32"
Angelique Louise Brown
Cam 4h13'26" 56d9'48"
Angelique Lyn McCabe
Lyn 9h17'4" 40d30'57"
Angelique Marleen Santana
Lyn 7h44'0" 49d31'24"
Angelique Merold
Psc 0h42'45" 9d45'10"
Angelique Pacem
Vir 13h25'49" 11d14'49"
Angelique Pavan
And 2h18'30" 49d35'14"

Angélique Psila
Cyg 19h47'21" 33d46'31"
Angélique Rameau
Cap 21h36'27" -19d44'34"
Angelique Schmoe
And 23h26'3" 41d21'1"
Angelique Souffle Vital
Vir 12h50'11" 3d50'17"
Angelique, Ballerina of Our Hearts
Cas 0h14'43" 63d38'26"
Angelis
Sco 17h21'1" -42d51'13"
Angelis keh Pavlos
Cyg 21h49'34" 46d19'35"
AngeLisa
Sgr 20h7'21" -36d48'30"
Angelita
Tau 5h24'27" 25d38'47"
Angelita Klostermeier
Uma 11h26'38" 36d36'36"
Angelita Maria Alonso
Uma 9h5'48" 53d46'25"
Angelita Orozco
And 2h4'18" 37d47'47"
Angelita Santos
And 0h31'27" 46d7'51"
Angelita Sierras Duron Grijalva
Ari 2h4'53" 22d59'4"
Angelito D. Powers
And 23h25'21" 47d56'32"
Angelito Riubin Hernandez-Ines 2004
Sco 17h55'26" -37d26'45"
AngelJoy Elena Davis
Peg 21h12'45" 15d1'30"
Angelkitten0204
Aqr 21h43'9" 0d20'19"
Angell 2004
Cnc 8h41'15" 17d34'19"
Angella
Cyg 20h28'6" 57d20'23"
Angella Choi
Psc 1h37'44" 23d31'37"
Angellana
Aql 19h11'42" -0d8'14"
Angellika
Psc 1h18'15" 25d5'41"
AngelMolly
Sgr 19h19'15" -22d3'23"
Angelo
Her 16h43'3" 23d13'25"
Angelo
Peg 23h57'43" 23d18'34"
Angelo
Leo 9h49'23" 7d12'23"
Angelo
Ori 6h20'57" 6d20'59"
Angelo
Uma 8h35'30" 49d56'55"
Angelo
Cnc 9h7'59" 31d0'34"
Angelo Alphonse Celli
Ori 5h27'27" 6d55'11"
Angelo and Arlene Messina
Cyg 20h50'20" 47d31'43"
Angelo and Kaylee's Star
Lib 15h45'37" -5d46'26"
Angelo Bruciato
Umi 13h47'46" 78d30'2"
Angelo Colon- My Seckel Pear
Cyg 20h17'25" 41d15'38"
Angelo & Crystal
Cnc 9h10'0" 24d17'57"
Angelo Cusatis
Per 3h5'22" 56d32'26"
Angelo d' Amore
Cas 1h19'2" 72d42'36"
Angelo DeFendis Jr. "1960-1983"
Vir 13h26'42" 1d5'17"
Angelo DeGregoria, Jr.
Per 3h40'48" 43d11'21"
Angelo DiGiambattista
Cyg 19h29'51" 52d4'19"
Angelo Ernest Lowell
Lib 15h21'11" -5d41'2"
Angelo Ferlini
Cas 0h55'2" 56d41'34"
Angelo Frank Dicharia
Per 4h1'27" 42d37'17"
Angelo Gerald Romano
Cep 22h13'43" 62d7'50"
Angelo Giannone Sr
Aqr 21h14'31" -8d42'19"
Angelo Giorgio Profeta
Per 2h20'50" 56d4'26"
Angelo & Guglielma Carlino
Ori 6h7'21" 21d25'25"
Angelo J. Fazio
Cyg 19h36'2" 34d32'0"
Angelo J Redmond September 22, 2003
Umi 14h50'25" 77d14'5"
Angelo James Magliocco
Sco 16h7'35" -11d34'46"
Angelo John Marino
Psc 0h51'0" 15d35'1"
Angelo John Pino
Umi 14h18'42" 75d7'33"
Angelo Joseph Vaccaro
Uma 11h9'21" 49d25'18"

Angelo Kai Kami-Beveridge
Boo 14h26'21" 24d34'50"
Angelo Krstevski
Cru 12h24'5" -61d15'14"
Angelo Manuzzi
Per 2h48'27" 52d22'20"
Angelo Marino, Sr.
Per 2h37'21" 55d28'49"
Angelo Messina
Aql 19h5'13" -0d43'30"
Angelo Mozilo
Her 16h36'45" 34d21'10"
Angelo Nicholas Gregorio
Ari 3h16'23" 16d42'58"
Angelo Palanti
Sgr 20h11'36" -39d49'52"
Angelo Patire
Cep 20h27'23" 64d14'31"
Angelo Pellicciari
Uma 8h20'57" 65d42'34"
Angelo Perniola
Umi 16h56'37" 83d4'38"
Angelo Petitti Family Legacy
Lyn 7h2'59" 47d21'0"
Angelo Pezzino
Aur 5h47'7" 48d26'10"
Angelo Playa del Carmen 29/8/2001
Boo 14h43'2" 28d47'7"
Angelo Proietti
Per 3h31'41" 36d8'52"
Angelo Ralph Maslotti
Leo 11h8'37" 22d17'8"
Angelo Randazzo
Her 17h16'12" 37d30'0"
Angelo Rodriguez
Uma 11h28'54" 62d1'56"
Angelo Roustas
Aur 5h30'5" 52d47'44"
Angelo Strada
Aur 5h26'5" 40d7'5"
Angelo Valentino
Crb 15h54'22" 29d0'32"
Angelo Vargas
Lib 14h51'26" -0d41'59"
Angelos Kyriacos Tsangarides
Cep 23h26'33" 84d30'12"
Angelo's Star
Per 4h46'30" 46d50'52"
AngelRose
Cap 20h30'36" -22d48'10"
Angels
Ori 5h30'50" -2d44'28"
Angels
And 0h36'36" 42d18'19"
Angels 3, Forever a Triangled Bond.
And 23h16'58" 47d21'49"
Angel's Ash
Per 3h20'13" 41d46'29"
Angels Come From Tennessee
Uma 11h56'57" 64d14'25"
Angel's Eye
Uma 9h42'16" 57d59'0"
Angel's Freedom
Leo 9h22'44" 27d49'28"
Angels' Harps of Gold
Lmi 10h36'0" 32d47'28"
Angel's heaven
Lib 15h12'48" -16d41'58"
Angels in the sky "Arthur & Mary"
Lyn 7h9'44" 58d5'11"
Angel's Jackson
Cmi 7h20'37" 9d22'10"
Angels' King Edward
Lyr 19h3'46" 36d0'28"
Angels Logan and Savion Taylor
Sco 17h58'53" -39d20'10"
Angel's of Flight
Cap 20h52'7" -17d35'19"
Angel's Paradise
Lyr 19h13'11" 27d59'31"
Angel's Rest
And 1h30'38" 40d46'48"
Angel's Rose
Lyn 7h55'2" 38d55'8"
Angel's Shining Star
And 0h36'56" 22d10'26"
Angel's Star
Ari 1h54'56" 18d12'9"
Angel's Star
Leo 11h29'47" 12d33'7"
Angel's Star (Angelica R Wright)
Cyg 20h37'22" 47d27'47"
Angels Star "Jackie"
Cap 21h36'23" -14d7'27"
Angels' Three
Lyn 7h20'32" 58d11'35"
Angelstar
And 0h22'27" 38d44'28"
AngelStarz
Sgr 19h19'7" -20d46'50"
Angeltatt74
Lmi 10h11'56" 37d17'42"
Angelus
Tau 3h50'3" 26d40'1"
Angelus 143
Vir 14h9'39" -14d53'21"

Angelus Adsum
Psc 1h20'15" 17d43'33"
Angelus Seraphim - Angelia Reynolds
Psc 1h16'25" 11d9'9"
Angelyna Deardorff-Hamilton
Leo 10h43'36" 15d51'11"
Angennica
Sgr 18h29'24" -34d25'4"
Ange's Angel
Sco 17h32'57" -35d55'4"
angi
Sgr 18h25'30" -32d41'6"
Angi Hull
And 0h49'17" 38d15'33"
ANGI "ISBAR"
Mon 6h32'7" 5d41'55"
ANGI LOVE GERI
Her 17h41'54" 16d19'44"
Angi, you are forever in my heart
Umi 17h1'45" 75d34'41"
Angie
Cap 21h36'13" -17d5'27"
Angie
Sgr 17h57'10" -17d56'41"
Angie
Dra 19h33'19" 61d2'3"
Angie
Cas 1h57'38" 64d57'59"
Angie
Col 5h51'59" -33d14'28"
Angie
Her 18h5'45" 18d23'46"
Angie
Ari 3h16'33" 22d53'53"
ANGIE
Ari 2h13'43" 23d26'1"
ANGIE
Leo 11h9'6" 16d53'19"
Angie
Leo 10h34'19" 17d55'46"
Angie
Leo 11h9'59" 20d55'40"
Angie
Leo 11h28'28" 19d33'21"
Angie
Aqr 21h59'54" 1d1'0"
aNGie
Ari 2h45'52" 15d34'22"
Angie
Vir 12h40'47" 10d26'12"
~ ANGIE ~
Tau 5h52'42" 14d51'26"
Angie
And 1h3'17" 39d10'57"
Angie
Crb 16h13'27" 33d0'25"
Angie
Crb 16h12'53" 32d38'53"
Angie
Gem 7h10'46" 34d16'49"
Angie
Uma 13h54'9" 48d28'38"
Angie and Brian Sprada
Vir 13h11'38" -11d50'25"
Angie and Dave Forever Love
Cyg 21h11'31" 47d15'34"
Angie and Ray 11/23/03
Cyg 19h30'6" 48d43'12"
Angie and Tim
And 0h17'23" 45d30'3"
Angie "Angus" McWilliams
Cnc 8h25'45" 21d8'51"
Angie - Baby
Sco 16h0'35" -26d5'58"
Angie Baby 4774
Cas 0h20'59" 55d46'14"
Angie Barbara Kirsch
Sco 17h19'50" -44d42'52"
Angie Bear
Lib 15h13'45" -15d22'34"
Angie ~ Bubbles ~ Uhm
Cru 12h42'6" -64d40'32"
Angie Carr
Cap 20h29'5" -26d21'21"
Angie Carter
Peg 23h58'1" 19d11'2"
Angie Castro
And 0h40'35" 33d48'17"
Angie Chambers
And 23h10'40" 51d17'17"
Angie Coleman
And 23h2'39" 51d21'18"
Angie D'Aleo Morelli
Sco 17h6'7" -42d14'55"
Angie Dammen
Uma 10h1'53" 68d3'33"
Angie & Danny Martin Wedding Star
Vir 12h48'51" 3d40'35"
Angie Dowdy (Agnes Helen Griner)
And 23h9'12" 37d14'2"
Angie Ema Deruelle
Ari 2h13'13" 22d48'22"
Angie Evans
And 0h35'53" 31d27'59"
Angie Frantz
And 23h40'35" 41d15'9"
Angie Goss
Uma 11h24'31" 30d21'41"

Angie Guio
Uma 9h2'23" 65d11'15"
Angie Haar
Aqr 23h40'35" -23d13'21"
Angie Hackett's Heavenly Body
Cru 12h26'32" -57d42'39"
Angie HFFH
Crb 15h54'48" 34d26'42"
Angie Hoare - Truly A Star
Umi 16h3'28" 78d21'47"
Angie Hock
Tau 4h43'50" 20d13'11"
Angie Huber Star
Uma 10h17'23" 42d5'30"
Angie & Hugi
Per 4h16'49" 35d1'7"
Angie K
Cas 1h57'55" 61d19'38"
Angie Kate Elvin
Vir 13h21'50" 12d3'30"
Angie Klossner
Uma 9h59'1" 42d5'45"
Angie & Laura Best Friends Forever
Cam 4h9'25" 66d6'25"
Angie M.
Crb 15h49'35" 39d23'3"
Angie M. Andreolo
And 1h28'23" 49d6'19"
Angie M. Clayton
Vul 20h41'27" 24d55'50"
Angie Marie
And 2h18'23" 46d40'30"
Angie Marie Franco
Ori 6h11'26" 15d18'40"
Angie Marie Frizzell
Cam 9h10'13" 79d29'25"
Angie Martin
Cnc 8h39'26" 23d53'22"
Angie Mascitelli
Uma 11h20'15" 52d16'57"
Angie McFadden
Psc 1h22'17" 19d23'43"
Angie Moss
Gem 7h13'5" 19d37'8"
Angie My Shining Star
And 0h22'4" 27d50'1"
Angie & Nelson
Sge 19h26'36" 17d23'44"
Angie Osburn "Sunshine"
Uma 9h50'6" 46d11'17"
Angie Pabon
Leo 11h44'36" 27d39'37"
Angie Padovano
Cam 6h8'30" 60d38'33"
Angie Parris
Del 20h42'9" 15d3'7"
angie pink
And 1h43'44" 49d25'12"
Angie Rachelle
Ari 2h21'53" 10d51'34"
Angie Rae
Mon 6h42'50" -0d17'25"
Angie Ruth White
Tau 4h43'27" 24d35'0"
Angie Salem
Lyn 7h35'36" 49d10'4"
Angie Sardiña
And 1h0'55" 41d25'13"
Angie Sharp
Tau 4h9'35" 26d30'5"
Angie Stakes
Cnv 12h38'33" 39d55'30"
Angie Tarnaski
Ori 6h12'51" 17d30'54"
Angie & Teodoro
Umi 15h19'17" 73d13'22"
Angie Thomas
Cas 2h26'25" 67d50'37"
Angie Turner
Lyn 7h47'2" 38d54'55"
Angie Wason "Angie's Star"
And 0h49'10" 44d24'58"
Angie Will You Marry Me Yes!
And 1h9'7" 36d30'32"
Angie Wiwigacz
Her 17h1'28" 13d41'4"
Angie Worth
Cnc 8h1'13" 15d20'55"
Angie, MSSA
Psc 1h54'41" 8d29'46"
Angie-from-RoBert
Peg 22h27'25" 30d49'5"
Angie.G
Cnc 8h46'31" 32d20'10"
Angielove
Leo 11h41'15" 21d47'4"
Angielyn
Gem 6h3'24" 26d59'56"
AngieM
Cyg 21h50'23" 46d10'37"
Angie's Butterfly
Cas 1h17'45" 54d47'44"
Angies Children
Sgr 18h55'31" -20d6'57"
Angie's Dream
Leo 9h59'44" 18d31'20"
Angie's Little Star
Cru 12h10'54" -62d17'7"
Angie's Love Star
Vir 13h44'0" 2d32'53"
Angie's Piece of Heaven
Cas 1h56'4" 64d0'48"

Angie's Reachable Star
And 1h44'27" 47d51'15"
Angie's Smile
And 1h24'31" 38d39'27"
Angie's Star
Uma 9h35'0" 46d31'0"
Angie's Star
Leo 11h40'39" 23d21'14"
Angie's Star
Cas 23h20'17" 56d6'24"
Angie's Star
Cru 12h47'7" -58d14'12"
Angie's Turtle Star
Gem 7h6'21" 14d35'25"
Angiesarina
Psc 1h21'17" 17d14'40"
Angila Michelle Moore
And 23h28'25" 39d3'5"
AngL 81
Cap 20h54'16" -16d19'54"
Angle Grace
Gem 6h29'43" 25d24'57"
Angler Apts.
And 2h25'17" 47d56'46"
Angolina Triventi-Jovine
Boo 15h23'56" 33d14'37"
Angora
Sco 17h28'27" -39d23'48"
Angry Dragon
Tau 4h27'29" 21d30'30"
Angus
Cma 7h24'3" -24d45'41"
Angus Alan Paul Richardson
Per 4h38'24" 47d26'58"
Angus Alexander Gadd
Cru 12h18'19" -62d44'29"
Angus Arthur Griffith Hayes
Cep 23h10'54" 73d4'7"
Angus David Corvini
Cru 12h26'18" -57d36'11"
Angus - Gus - Smith
Dor 5h38'43" -68d59'16"
Angus Michael MacKay
Ari 2h44'48" 16d36'19"
Angus Paul Gardner
Her 18h56'35" 16d46'3"
Angus 'We luv The Mongrel' Burchall
Cru 12h36'3" -56d12'12"
Angus William Graham - 12 Jan 2000
Cru 12h56'54" -59d29'50"
Angy
Boo 14h28'28" 11d25'53"
Angy
Cyg 20h7'26" 49d5'22"
Angy Christine Shreeve
Lyn 7h55'11" 57d18'2"
Angy Nellen
Cep 2h5'1" 82d48'41"
Anh Thu Do
Uma 10h0'11" 42d26'12"
AnhThi
Psc 1h27'40" 18d13'44"
ANI
Psc 1h11'54" 27d17'24"
Ani
And 1h25'19" 48d24'55"
Ani
Lib 15h5'6" -3d51'10"
Ani
Pho 0h7'7" -46d9'40"
Ani Eblighatian
Com 13h15'5" 26d16'43"
Ani és Nándi szerelemcsillaga
Cas 23h34'29" 55d40'8"
Ani GJVC
Aur 4h50'23" 37d10'5"
Ani L'Dodi v'Dodi Li
Agr 23h11'30" -12d17'30"
Ani ledodi vidodi li Karen
Cap 21h39'3" -9d28'32"
Ani Nalbandian "Our Bright Star"
Leo 11h49'25" 26d22'16"
Ani ohev otach
Uma 11h52'10" 54d2'4"
Ani Skye
Uma 9h41'12" 58d14'43"
Ani Stoyanova Toncheva
Ari 2h38'25" 28d52'45"
Ania
Aqr 20h43'18" -9d5'36"
Ania Camille Ting Star#1016
Ori 5h52'45" 20d48'59"
Ania Frances Manalili
Ari 3h19'17" 21d54'12"
Ania G. Pollack
Leo 11h5'47" 17d25'33"
Ania K
Tau 4h17'3" 2d39'49"
Ania Kubas
Cyg 20h50'19" 43d10'31"
Ania Magdalena Gajlewicz
Ari 2h41'46" 26d25'3"
Ania Zyazeva
Lib 14h59'59" -3d56'52"
Aniah Rose Monae Norfleet's Star
Cas 23h30'26" 53d7'53"
Aniarandall
Lyr 18h52'36" 32d33'33"

Anibal
Cap 21h2'40" -18d13'21"
Anibal Albelo
Uma 8h48'7" 53d41'12"
Anibal Alvarado
Uma 9h9'40" 55d31'16"
Anibal O. Lleras
Lyn 6h55'35" 57d28'25"
Anibal Tavares, Jr.
Leo 9h32'3" 26d41'32"
Anica
And 0h35'7" 27d9'54"
Anica Shaw
And 2h22'34" 38d39'24"
Anicgoa
Cam 4h13'4" 68d52'40"
Anicia Fojas
Cam 5h54'10" 69d31'44"
Aniel Xavion Gonzalez
Peg 21h51'3" 8d55'15"
Aniela Kamienski
And 0h19'52" 45d50'59"
Aniela Koling Sprung
Leo 11h17'32" 15d2'47"
Anielka Maria Rojas Meza
Agr 22h15'21" -3d33'40"
Aniella Katherine Porfilio
Ari 2h50'53" 24d34'49"
Aniello & Clara Garace
Gem 7h1'13" 16d23'22"
Anielski Piekna Zofia
Cas 0h50'29" 63d17'14"
Anie's Star
Ori 5h58'57" 12d12'38"
Aniessa Mae Dull
Sgr 19h33'48" -13d5'34"
Anik
Tri 1h51'42" 27d56'43"
Anika
Tau 4h12'46" 9d16'32"
Anika
Tau 3h28'18" 10d1'16"
Anika
Agr 21h30'39" 2d4'37"
Anika
Vir 13h22'24" 13d33'52"
Anika
Uma 13h12'3" 53d28'23"
Anika
Uma 11h5'42" 55d57'23"
Anika Garza
Uma 8h58'57" 51d45'42"
Anika & Jason
Uma 10h5'22" 42d37'56"
Anika Lynne
Sgr 19h3'27" -30d43'50"
Anika Marie
And 1h9'30" 47d9'28"
Anika Miles
Uma 11h24'31" 28d55'38"
Anika Nagel 7.5.2006
And 23h50'58" 46d8'31"
Anika Novacek
Leo 11h21'11" 16d6'43"
Anika Renee Stone
Sco 16h10'10" -21d11'28"
Anika und Tobias
Uma 10h17'40" 49d45'39"
Anikamartinoff
Uma 11h37'36" 44d42'13"
Aniko
Sco 17h27'16" -31d31'52"
Anikó/Ancsurka
Leo 9h29'48" 12d7'14"
Aniko Elizabeth Roth
Cyg 19h26'4" 46d53'43"
Aniko Suzanne Fekete
Oph 17h43'12" -0d55'29"
Anil Joshua Kainth
Umi 14h52'43" 70d35'8"
Anil Patel
Dra 18h56'57" 59d44'38"
Anima
Uma 9h34'2" 49d13'59"
anima e corpo
Lyn 6h52'19" 56d49'55"
Animal Daddy
Vir 14h21'15" -4d15'2"
Animus
And 1h5'36" 35d59'16"
Anina
And 1h33'56" 41d6'16"
Anina Pasetto
And 1h14'1" 45d53'59"
Anina Reimann
Umi 15h57'24" 83d35'11"
Anina Zvezda
Gem 7h37'17" 14d35'20"
Anioka z Danielek
Leo 10h27'24" 13d6'1"
AniQuay
Vir 12h20'37" 2d32'40"
Anique
Uma 12h0'28" 39d2'24"
ANIQUE
Uma 11h36'17" 63d5'29"
Anis Dupuis
Del 20h46'37" 7d13'55"
Ani's Star
Ari 3h21'47" 28d14'49"
Anisa
Crb 15h42'52" 25d59'56"
anisa &elisa jahaj dy yjetexhaxhit
Psc 1h38'31" 22d0'6"

Anise Colley Johnston
Lyr 19h8'35" 32d31'47"
Anisha Vina Patel
Aql 20h7'51" 15d18'0"
Anishna Zapdes
Leo 11h42'17" 23d5'35"
Anisia
Gem 7h25'59" 32d35'44"
Anissa
And 1h58'50" 41d44'3"
Anissa
Cyg 20h43'47" 38d42'51"
Anissa
Ori 5h40'15" 11d42'13"
Anissa
Lyn 7h20'24" 57d31'57"
Anissa Nicole Magwood
Cas 2h9'6" 64d51'25"
Anissa Robyn Baker
Lib 14h28'39" -18d45'5"
Anissa's Heart
Leo 10h41'52" 17d55'55"
Anita
Del 20h59'31" 15d28'40"
Anita
And 0h32'50" 28d28'29"
ANITA
Vir 12h48'22" 0d19'27"
Anita
Equ 21h5'37" 9d11'40"
Anita
Cnc 8h48'52" 14d7'19"
Anita
And 0h20'28" 46d31'20"
Anita
Cyg 21h8'17" 30d20'45"
Anita
Uma 11h9'12" 42d23'4"
Anita
Psc 1h20'0" 31d53'1"
Anita
Sgr 19h4'47" -17d40'53"
ANITA
Cap 21h37'49" -10d48'8"
Anita
Eri 4h30'19" -0d55'16"
Anita
Lyn 6h58'59" 60d35'22"
Anita
Uma 8h15'59" 65d13'11"
Anita
Umi 14h34'27" 69d51'10"
Anita
Cas 0h37'19" 66d23'48"
Anita *01.02.1944
Uma 8h26'48" 67d39'52"
Anita 10.02
And 0h41'30" 23d27'1"
Anita 11-21
Sco 16h42'57" -32d45'44"
Anita A. Gray
Crb 16h2'42" 36d33'21"
Anita A. Hall
Lib 15h12'16" -10d15'51"
Anita A. Manning
Cru 12h22'31" -61d4'7"
Anita and Alli Friendship Star
Tau 4h31'17" 27d35'5"
Anita and Barry
Cyg 21h22'26" 32d28'10"
Anita and Jack Turtledoves Forever
Cyg 20h57'27" 31d8'45"
Anita and Josh Gurin
Lyr 18h48'19" 39d48'58"
Anita Ann Nunez
Lyn 8h14'19" 41d46'5"
Anita Anne
Tau 4h10'8" 6d5'6"
Anita Aurora Herrera Sneathen
Ari 3h5'32" 22d22'7"
Anita B
Cap 20h25'0" -12d13'58"
Anita Beatriz Hemingway
Cam 5h50'16" 57d55'48"
Anita Berrey
Cnc 8h18'46" 13d59'48"
Anita Bonita
Psc 1h8'53" 25d10'3"
Anita "Bonita" Garcia
And 0h19'17" 32d26'34"
Anita Bosnjak
And 2h27'34" 45d54'39"
Anita Brusa
And 23h43'8" 48d25'40"
Anita Cain
Leo 11h0'33" 2d6'58"
Anita Carnett
Vir 13h55'11" -10d39'47"
Anita Carolla
Mon 7h39'47" -0d43'33"
Anita Chamblee Otero
Ari 1h59'25" 13d42'25"
Anita Crites
Cam 5h24'35" 67d30'5"
Anita D. Fry
Uma 8h51'12" 71d28'52"
ANITA DAGEL
Agr 23h12'9" -6d40'14"
Anita D'Aniello
Cyg 21h7'14" 47d18'0"
Anita D'Aniello
Cyg 21h7'14" 47d18'0"

Anita E. Iezza "Love Star"
Ori 5h17'56" 15d19'21"
Anita Ekenstam Beverly
Lib 15h3'55" -7d0'11"
Anita Ellen Tisdale
Cyg 20h53'1" 33d22'33"
Anita F. Alpern
Her 16h41'31" 6d8'57"
Anita Falcone
Cap 20h11'19" -12d35'36"
Anita Fankhauser
Per 3h19'53" 39d13'51"
Anita Forever
Uma 13h59'37" 60d54'12"
Anita Frentzel
Leo 11h11'5" 25d47'24"
Anita Gonzales
Sco 17h54'5" -35d48'47"
Anita Grace Biggs
Psc 1h29'48" 16d36'44"
Anita H F Wellen
Leo 9h30'22" 22d44'24"
Anita Hafner's Mini Muffin Empire
Dra 15h14'34" 63d11'34"
Anita Heinen
Uma 8h18'54" 59d48'5"
Anita Heinzer
Boo 14h34'19" 19d6'0"
Anita Hillebrandt
Uma 10h27'23" 40d10'38"
Anita Jaimes
Cnc 9h6'53" 14d25'5"
Anita Jean Randolph
Umi 15h43'26" 77d0'43"
Anita Jo Quinn
Lyn 8h27'52" 55d23'31"
Anita & Josi
Lac 22h40'5" 54d11'4"
Anita Joyce Meek
Gem 7h37'51" 14d54'22"
Anita Kanagala
Ari 2h14'16" 25d17'43"
Anita Klein ~ Our Angelic Friend
And 0h16'49" 27d29'40"
Anita Kramer
Uma 10h24'3" 72d38'16"
Anita L.
Mon 6h49'20" -0d41'12"
Anita L. Dunn
Crb 15h43'20" 36d51'13"
Anita L. Falcon (7/26/35-2/26/03)
Uma 10h21'20" 56d15'34"
Anita L. Ricca
Leo 9h38'1" 28d34'11"
Anita L Schnebly
Sgr 20h1'4" -36d29'21"
Anita Lesley Fisk
And 0h6'18" 48d20'30"
Anita Lesley Fisk
And 23h48'11" 39d46'27"
Anita Levine
Tau 5h58'28" 28d19'19"
Anita Lewis-Brown
Cas 23h22'4" 53d13'36"
Anita Lorraine Lasanske
And 0h55'8" 40d6'0"
Anita Louise Lefebvre
Cas 0h31'26" 61d57'34"
Anita Lynn "Gucci" Arndts
Sco 17h36'20" -39d51'40"
Anita Maiorisi
And 1h46'57" 43d4'54"
Anita Maria Johnson Dennis
Uma 10h58'34" 54d56'14"
Anita Marie
Dra 16h18'10" 51d36'5"
Anita Marie Day
Sco 16h7'33" -14d6'44"
Anita Marie Harding
Crb 16h24'41" 33d29'58"
Anita Marie Kurtyak
Cas 23h9'16" 58d33'36"
Anita McNeil
Cam 5h56'46" 66d41'5"
Anita Meier
Cas 0h36'51" 58d51'34"
Anita Mejia Rivera mother of AJR
Tau 4h24'1" 25d22'16"
Anita Menke
And 1h17'38" 37d14'45"
Anita Michelle Means
Sco 17h52'54" -30d24'44"
Anita Mingham
Mon 6h57'42" 0d16'37"
Anita Molly Glass
And 2h6'20" 45d4'51"
Anita & Myron DiVittorio
Ori 5h8'18" -0d45'29"
Anita Natalie Genz Verrilli
Vir 13h23'18" 12d34'56"
Anita "Nini" Lerner
Uma 13h21'51" 55d2'54"
Anita Nocetti
Ori 6h10'41" 7d24'33"
Anita Nyffenegger
Her 17h19'11" 16d33'38"
Anita Oliver
And 2h17'18" 42d1'8"
Anita Palano
Uma 10h15'4" 45d36'24"

Anita Perillo
Ori 5h52'40" 12d29'19"
Anita Piazza
Uma 10h29'26" 45d47'53"
Anita Pigoni
Uma 10h11'32" 64d32'32"
Anita Polizzotto
Agr 21h37'21" 0d39'59"
Anita R. Pugh
Oct 21h28'19" -75d57'15"
Anita Rebecca Shelton
Vir 13h30'25" -8d7'50"
Anita Reichenbach
Peg 21h23'2" 15d32'49"
Anita Romella Rubio
Leo 9h28'38" 27d16'31"
Anita Rose
Crb 16h4'34" 37d14'9"
Anita Ruth Mallory
Cas 23h45'49" 59d53'49"
Anita Salvatore Meeks
Cnc 8h7'45" 16d59'29"
Anita Scelsa
Hya 9h22'26" -0d11'46"
Anita Sienkiewicz
Mon 6h45'12" -0d24'51"
Anita Sigel
Lib 15h59'22" -10d0'20"
Anita "Sparky" Clinton
Lyn 6h52'7" 56d47'57"
Anita Sue Grossbard
Tau 4h3'54" 11d46'45"
Anita - The Light Of Grace
Cru 12h25'55" -61d40'10"
Anita The Magnificent
Cas 23h1'56" 57d31'23"
Anita Toscano
Crb 16h20'49" 30d5'46"
Anita Tracy
Agr 22h10'7" 0d16'14"
Anita Uhrman
Vir 13h18'22" -12d28'56"
Anita und Dani Flury
Uma 11h8'7" 28d33'41"
Anita Varney
Vir 12h53'39" 3d25'43"
Anita W. Guthrie
Cyg 21h28'22" 50d34'16"
Anita Wareham
Leo 11h50'36" 20d42'54"
Anita Weiner
Lib 15h9'25" -11d3'37"
Anita Wright
Cas 2h21'43" 68d47'41"
Anita, 09.01.1985
Lyr 18h30'15" 28d40'1"
Anita-Nanny-Sims
Cap 20h39'36" -17d59'21"
Anita's Dream
Agr 20h42'31" 0d59'50"
Anita's Star
Uma 9h19'30" 66d7'7"
Anita's Star
Cas 23h45'54" 57d55'50"
Anita's Wish Star
Lib 15h6'51" -5d14'54"
Anitastar
Mon 7h15'52" 0d41'26"
Anitha V.P
Agr 22h33'9" 0d56'33"
Aniva
Ori 5h30'53" 8d26'28"
AniWil-Kamaisatononaphani
Lyn 9h10'18" 40d47'54"
Aniya Elaine
And 0h5'42" 47d6'4"
Aniya Renae' Durham
And 1h30'51" 37d38'19"
Aniyah Esperanza Flax
Cap 21h3'50" -21d36'12"
Anja
Umi 15h29'37" 78d10'59"
Anja
Uma 11h15'52" 72d37'43"
ANJA
Cep 1h39'42" 80d48'23"
Anja
Sco 16h38'1" -30d16'45"
Anja
Cap 21h36'51" -24d27'10"
Anja
Sgr 18h7'41" -26d32'5"
Anja
And 22h59'15" 50d17'17"
Anja
Vir 13h37'34" 3d12'42"
Anja
Ori 5h55'0" 15d51'55"
Anja Alaine
And 0h43'35" 38d48'39"
Anja and Leif
Lmi 10h35'58" 38d5'51"
Anja Andersen
Agr 22h9'43" -23d21'7"
Anja e Simone
Her 18h9'31" 27d26'59"
Anja Kaarina Rawlins
And 0h17'19" 38d42'2"
Anja Lea
Vul 21h4'27" 20d57'28"
Anja Ludwig
Ori 6h19'57" -1d27'8"
Anja Lutzka-Prickett
Lib 15h21'25" -29d52'41"

Anja Maria Grabner
Uma 9h5'42" 48d31'13"
Anja Müller
Uma 8h34'25" 50d39'47"
Anja (my lil cup) Rosales
Ori 5h6'7" 13d15'59"
Anja Rickenbach, 24.06.1977
Sge 19h7'27" 16d49'39"
Anja Sarah
And 1h0'10" 41d54'38"
Anja Skeide
Ori 5h17'32" 16d6'5"
Anja van den Hof
Psc 1h27'53" 17d50'1"
Anja W. Lakenberg
Agr 21h45'6" -1d44'1"
Anjali
Agr 22h5'19" -23d14'30"
Anjali
And 23h14'57" 50d30'35"
Anjali Devaraj Tierney
Cam 4h12'11" 69d21'36"
Anjali Kishore Shahani Moreno
And 0h36'53" 28d15'18"
Anjali Mohan
Ari 3h6'11" 15d19'41"
Anjali Ramachandra-Gallacher
Lyn 8h39'37" 44d46'18"
"Anjali" Sara Catelyn Joshi
Ori 5h33'17" 13d41'6"
Anjali Tara Dhillon
Lib 15h31'58" -8d16'48"
Anjana H Patel
Sct 18h46'25" -10d40'12"
Anjana Nirushan
Cru 12h25'44" -58d57'15"
Anjana Prasher-Daniel I Love You
Vir 13h18'49" 12d35'46"
ANJANAN
Uma 9h59'29" 63d58'13"
Anjanett Marie Amore
Com 12h0'4" 21d42'36"
AnJanette and Joe
Agr 22h49'36" -4d23'34"
Anjanette Penley
Agr 23h1'30" -22d38'13"
Anjardi
Lyn 7h45'21" 56d27'33"
Anja's & Steff's Hochzeitsstern
Uma 9h49'27" 65d1'26"
Anjel Iris
Ari 2h7'16" 19d29'24"
Anjelica Gabriella Urciel
Cas 23h43'28" 55d38'26"
Anjelica Marie Hidalgo
Sgr 18h41'10" -35d45'14"
Anjelica Marie Maddox
Ari 2h53'2" 30d53'0"
Anjelika Krychtaleva
Tau 5h56'0" 24d59'59"
Anjelique Camille Augustin
Cas 1h32'15" 62d41'28"
Anjenette
Ori 5h12'54" 5d48'30"
Anjica
And 1h17'1" 48d52'22"
Anjill
Lyn 6h19'22" 59d47'24"
Anji's Star Glow
Ari 2h43'10" 21d16'18"
Anjli Vaswani
Cas 0h47'48" 60d47'10"
Anjna Masani
Cyg 20h36'35" 50d29'46"
Anjolu 5
Uma 8h44'26" 54d22'33"
Anjounette Rose Stelly
Uma 10h53'38" 49d6'20"
Anju Chopra
Cas 1h0'59" 60d15'10"
anjuschka
And 1h58'51" 41d21'17"
Anjuta
Com 12h59'28" 19d41'15"
ANKA
Ari 2h43'59" 22d22'32"
Anka Dawn
And 0h50'9" 35d4'40"
Anke
Uma 9h4'35" 52d4'49"
Anke
And 2h13'39" 48d45'35"
Anke der Stern von Micha
Cnc 8h58'11" 15d32'17"
Anke Gaugele
Uma 8h54'22" 54d31'54"
Anke Krämer - Ergomaus
Uma 9h38'13" 49d3'29"
Anke (M O)
Uma 11h15'24" 43d17'1"
Anke Vinzelberg
And 1h30'19" 48d53'51"
Anker World
Peg 21h50'22" 16d18'48"
Ankerman 230367
Ori 5h55'21" 18d10'17"
Ankhqua
Agr 22h3'57" -14d58'12"
Ankit Sharma
Vir 13h10'13" -22d27'57"

Ankur & Selene's Everlasting Star
Agr 21h9'51" -3d1'50"
Anly Rose 8/14/01
Del 20h43'39" 12d7'49"
AnM05
Lyn 8h0'49" 52d39'27"
Ann
Lyn 7h55'55" 53d22'35"
Ann
Sco 16h11'30" -17d53'18"
Ann
Psc 0h18'36" 6d31'52"
Ann
Psc 0h46'56" 5d22'1"
Ann
Psc 1h13'0" 4d18'29"
Ann
Cyg 19h34'26" 29d40'44"
Ann
Leo 9h34'17" 28d11'5"
Ann
Com 12h31'17" 27d52'25"
Ann
Crb 15h40'19" 38d58'18"
Ann
Uma 9h45'58" 49d32'21"
Ann
And 0h49'50" 40d55'49"
Ann & Alan's Star
Cyg 20h46'45" 33d32'29"
Ann Alford
And 2h21'47" 46d40'47"
Ann and Don Muffly
Tau 5h32'55" 27d27'56"
Ann and John Kozlowski
Cyg 19h36'33" 32d12'41"
Ann and Joseph Cavallaro
Her 16h7'53" 46d19'59"
Ann and J.P. Riturban
Ori 6h11'26" 15d42'26"
Ann and Laurence Mc Guire
Lib 15h7'50" -16d7'54"
Ann and Tom Woodward
Gem 6h20'42" 21d47'17"
Ann Andrews
Uma 8h10'40" 64d55'28"
Ann Areeckal
Lyn 7h40'1" 38d59'7"
Ann & Arpiar Babigian
Cyg 20h1'41" 50d3'42"
Ann Aughinbaugh
Lyn 6h23'55" 54d56'3"
Ann B. Calveric
Cnc 8h11'59" 11d4'12"
Ann B - solismoicroi
Cnc 7h59'29" 12d46'10"
Ann B Trujillo
Psc 1h16'53" 24d2'32"
Ann Baenen
And 1h5'24" 42d25'13"
Ann Barna
Agr 21h56'48" -3d13'47"
Ann & Bart McGlothlin
Cru 12h42'18" -57d6'55"
Ann Berliner Christensen
Cnv 12h34'50" 51d38'25"
Ann Bernadine
And 23h39'21" 35d51'35"
Ann Bessie
Lib 15h34'11" -22d47'18"
Ann + Bill Orloff
Cap 21h34'22" -14d42'43"
"Ann & Bob"
Cyg 20h40'49" 45d30'46"
Ann Bohning Dawe
Cnc 9h11'44" 30d40'10"
Ann "Bunny" Snow
And 23h18'59" 48d8'20"
Ann Burton
And 0h42'7" 25d7'55"
Ann Buyanovsky
Agr 22h4'22" -14d57'11"
Ann C.
Uma 9h27'30" 42d20'43"
Ann C. Stimmel
And 2h38'22" 39d49'29"
Ann Calhoun
Vul 19h45'47" 23d17'28"
Ann Carrick
Ori 5h31'16" -9d9'40"
ANN CARSON
Cnc 8h23'11" 10d21'21"
Ann Catherine Bates née McAnespie
Cas 0h43'3" 69d57'43"
Ann Catherine Foster
Lib 14h51'18" -1d25'27"
Ann Cathrine Campbell
And 1h3'37" 39d56'15"
Ann Chinnis
And 23h58'48" 45d33'57"
Ann Circolone
Cas 0h26'43" 58d2'20"
Ann Clark and Annabelle Terrell
Agr 21h54'27" 0d36'41"
Ann Cody
And 1h41'26" 43d2'51"
Ann Collis of Durham's Star
And 0h0'29" 41d0'33"
Ann Condello
Lyn 7h56'15" 58d5'12"

Ann Conway
Cas 1h33'18" 69d13'37"
Ann Cooper
Uma 11h46'15" 36d20'51"
Ann Cooper Starling
Aql 19h49'12" 6d4'0"
Ann & David DeSimone
Cyg 19h38'6" 35d7'0"
Ann Debo
Aqr 20h42'0" -3d41'47"
Ann Dekay Evans
Tau 4h17'18" 17d59'12"
Ann DiCecco
Uma 8h28'57" 67d44'42"
Ann & Donald Retallick
Cyg 19h26'24" 48d48'9"
Ann Donihe
Psc 1h18'39" 16d35'37"
Ann Dowd Fournier
Tau 4h12'26" 16d5'4"
Ann E. Ferrara
Uma 10h35'31" 62d18'35"
Ann E. Flanagan
Cnc 8h45'0" 7d19'17"
Ann E. Gordon
Psc 0h23'26" 2d27'21"
Ann E. Martin
And 1h57'19" 38d51'12"
Ann E. Rocha
Uma 10h13'14" 67d11'27"
Ann Edwige Jennifer Morrow
Cyg 21h55'10" 50d42'2"
Ann Elaine
Lib 15h23'32" -16d30'0"
Ann Elizabeth
Lib 14h28'59" -16d20'41"
Ann Elizabeth
Leo 11h14'2" 25d30'13"
Ann Elizabeth Carmichael
Vir 12h34'14" 3d7'15"
Ann Elizabeth Clayton
Cyg 21h35'13" 50d13'56"
Ann Elizabeth (Dinks) Frick
Cnc 8h19'1" 19d50'56"
Ann Elizabeth Erdenberger
Vir 14h13'13" -17d41'55"
Ann Elizabeth Fisher
Cas 0h56'0" 54d52'32"
Ann Elizabeth Gordon
Gem 7h36'37" 34d16'58"
Ann Elizabeth Gordon
Gem 6h33'53" 22d37'17"
Ann Elizabeth Hurt
Uma 10h31'1" 66d3'34"
Ann Elizabeth Klasinski
Cas 1h7'23" 58d50'8"
Ann Elizabeth Larroux
And 0h11'36" 34d22'22"
Ann Espinoza
Crb 16h15'53" 36d59'51"
Ann Eterniteum Smith
Cas 0h26'9" 56d56'54"
Ann Evans Star
Cas 0h49'34" 52d8'27"
Ann Fagan Tolbert
Lyn 8h1'52" 53d45'50"
Ann Fala Tedesco
And 22h58'39" 48d5'11"
Ann Floria
Cas 1h39'32" 63d16'18"
Ann Frances Schwartz
Aqr 21h31'18" 0d25'4"
Ann Friederich
Ari 3h13'40" 18d50'19"
Ann Friedman
Aqr 21h55'30" -1d0'16"
Ann Fueshko Nigma
And 1h47'5" 42d7'51"
Ann G.
And 0h13'55" 26d51'47"
Ann Gaasedelen
Ari 2h22'13" 25d0'51"
Ann Ganong Seidler
Lyn 7h26'33" 52d32'9"
Ann Geraghty
Cas 1h54'40" 61d18'2"
Ann Gibson Sluss
Cap 21h9'52" -19d7'4"
Ann Gordon Ferrell
Vir 14h52'48" 3d48'58"
Ann H. Huss
And 1h42'41" 41d50'40"
Ann Haley
Cyg 20h55'17" 40d44'1"
Ann Hall
Cyg 21h39'52" 39d28'50"
Ann Harrison
And 0h20'21" 28d9'30"
Ann Holley
And 23h48'38" 45d9'42"
Ann Hughes
Cas 1h22'20" 69d2'5"
Ann Huitson - Ann's Star
Cas 1h55'48" 62d18'33"
Ann Hyland
Cas 1h17'48" 56d42'13"
Ann Is Goddess
Del 20h26'53" 20d8'9"
Ann Ishenko
Cas 1h24'6" 61d2'57"
Ann & Jackie Mchugh
Uma 10h50'41" 52d13'57"
Ann K Jobson
Cas 0h14'26" 51d27'54"

Ann Kamionka
And 0h53'25" 44d6'2"
Ann Katherine Roberts Morcerf
Sgr 19h17'10" -13d46'44"
Ann Kempster
And 23h40'7" 34d33'59"
Ann & Ken Palmer
Uma 11h43'53" 54d53'14"
Ann Kershner
Crb 16h12'6" 32d57'46"
Ann Knebel My Pumpkin
Crb 15h58'32" 32d13'48"
Ann Kosir Brinkman
Uma 11h31'8" 58d29'19"
Ann Kovace
And 1h42'19" 38d24'46"
Ann Kovach
Gem 7h33'37" 28d40'14"
Ann Kristin Skogen
Gem 8h7'4" 32d42'15"
Ann Kuligowski Hodge
Srp 18h16'26" -0d5'43"
Ann Labbe Merrill
Cas 1h31'16" 62d29'14"
Ann Laurel
Uma 9h45'17" 58d16'21"
Ann & Laurie
Cyg 21h52'21" 40d44'58"
Ann Lawing Miller
Apu 17h12'26" -72d52'42"
Ann Lenner
Cas 1h39'0" 60d44'29"
Ann Leslie Hargrove
Lib 15h7'32" -5d4'59"
ann litwin cashman
Leo 10h11'19" 10d8'2"
Ann Lodewyks
And 2h22'11" 50d33'33"
Ann Loeb
Lyn 7h4'10" 59d52'12"
Ann Loretta Powell
Cam 6h50'58" 66d25'4"
Ann Louise
Cyg 19h42'32" 34d44'0"
Ann Louise A
Lyr 18h54'34" 32d39'59"
Ann Louise Haley
And 0h34'20" 38d13'31"
Ann Louise Morehouse
Vir 15h2'28" 5d15'18"
Ann Louise Williams
Psc 1h13'2" 23d37'40"
Ann Lovelace
Lib 15h30'55" -24d37'55"
Ann Lucy Ragland
Psc 1h25'39" 13d47'46"
Ann M. Crispin
Cas 1h22'36" 54d53'4"
Ann M. Herbst
Cnc 8h50'25" 31d38'56"
Ann M. Peterson
And 0h33'30" 39d50'17"
Ann Margaret Furlong Simmons
Vir 13h3'52" 6d52'56"
Ann & Marianne - The Sparkle Of My Life!
Ara 17h23'45" -52d17'3"
Ann Marie
Uma 8h29'21" 68d31'32"
Ann Marie
Cas 1h31'3" 62d46'17"
Ann Marie
Psc 1h53'6" 9d53'7"
Ann Marie
Psc 1h22'52" 27d25'27"
Ann Marie
Cnc 8h48'6" 28d25'8"
Ann - Marie
And 0h31'50" 31d47'40"
Ann Marie 30
Cas 0h16'37" 56d25'4"
Ann Marie Barnard
Ari 2h36'51" 29d4'25"
Ann Marie Britcher
Tau 3h39'20" 15d44'19"
Ann Marie Britcher
Tau 4h50'40" 22d26'34"
Ann Marie & Carmine
Cyg 21h30'53" 37d8'21"
Ann Marie Contente
Sco 17h28'55" -32d12'56"
Ann Marie Elefante
Cyg 20h56'12" 48d31'19"
Ann Marie Enright
Cas 1h19'59" 63d35'21"
Ann Marie Galente,You are my Star
Cas 0h58'29" 63d37'4"
Ann Marie & Gary Gillum
Psc 1h27'27" 17d49'6"
Ann Marie Hebert
And 23h10'37" 47d40'1"
Ann Marie Hopper
Cas 23h59'43" 56d55'4"
Ann Marie Jolley
Cam 4h28'59" 58d22'16"
Ann Marie Love
Aqr 22h24'10" -19d16'26"
Ann Marie Mangol
Aqr 22h26'15" 2d26'21"
Ann Marie Masson
And 23h56'49" 41d59'9"

Ann Marie Messier
Leo 9h37'27" 27d34'52"
Ann Marie My Love. With Hope.
Dra 18h54'9" 66d34'20"
Ann MArie Onufer Natishan
Uma 11h12'58" 33d8'20"
Ann Marie Pulice
Uma 12h13'45" 55d23'0"
Ann Marie Rossi
Uma 9h32'6" 69d38'16"
Ann Marie Spain
Aqr 21h22'53" 0d9'28"
Ann Marie Travers
Lmi 10h48'1" 37d47'57"
Ann Marie Vehige
Aqr 23h7'18" -19d47'19"
Ann - Marie Wales
Psc 0h5'46" 9d11'6"
Ann Marie Wisniewski
Uma 8h31'54" 73d4'2"
Ann Marie Zeimetz
Uma 10h34'19" 63d51'50"
Ann Marie-Amato
Lyr 18h33'45" 28d49'48"
Ann Marie's Birthday Star
And 0h49'33" 42d29'54"
Ann Marshall
Leo 10h12'14" 14d38'19"
Ann May Ashpole
And 23h44'56" 33d10'36"
Ann Mazzeo
Cnc 8h42'6" 15d56'40"
Ann Mc Laughlin
Cas 23h22'19" 53d30'40"
Ann Meadus
Uma 9h11'9" 63d13'16"
Ann Michaelis
Sco 17h16'41" -32d40'20"
Ann Michele Murphy
Aqr 21h36'2" -0d40'1"
Ann Michele Schytt Christensen
Uma 10h13'42" 66d33'38"
Ann Michelle
Lib 14h59'20" -18d19'11"
Ann Michelle Manner
Cyg 21h43'41" 41d30'44"
Ann Miller
Aql 20h9'50" 4d15'11"
Ann Mimi Kim
Cnc 8h21'58" 14d9'30"
Ann & Monte DeLeo
Lyr 18h42'17" 41d8'57"
Ann Mosher
Boo 13h53'20" 24d45'21"
Ann Moss Jamison
And 1h30'14" 48d49'19"
Ann Mother of Light and Love
Uma 10h11'10" 68d57'24"
Ann Mullen
Cap 20h59'43" -25d34'51"
Ann Murphy Henchey
Cas 0h23'13" 47d0'31"
Ann " My sweet pea" Star
Pup 7h27'56" -34d33'53"
Ann "Nannie Annie" Sheahan
Uma 9h52'48" 69d41'29"
Ann Naomi Mizushima
Gem 6h46'25" 13d29'9"
Ann & Neil
Mon 7h8'27" -0d46'57"
Ann Nelson
Cas 0h41'34" 52d22'57"
Ann Newell - Eye Street Star
Cam 3h28'35" 61d59'55"
Ann Oliva
Lib 15h52'26" -19d43'13"
Ann Ollick
Vir 13h38'46" -13d0'33"
Ann "Onry" Becker
Tau 4h40'14" 19d18'13"
Ann O'Shaughnessy
Ori 6h6'8" 17d43'14"
Ann Patricia Miriello
Aqr 21h30'1" 2d19'52"
Ann Perrone
Uma 11h21'33" 34d48'56"
Ann Phillips
And 1h31'22" 43d19'41"
Ann Pittman
Psc 0h14'57" -4d46'40"
Ann Price Castucci
Sgr 19h13'37" -34d58'19"
Ann Pullaro
Ori 5h43'0" -2d0'1"
Ann Reda
Dra 19h42'29" 60d10'23"
Ann Rhodes 0723
Leo 10h40'52" 8d14'39"
Ann Sakosky
And 1h3'49" 41d53'52"
Ann Salkin
Lyr 19h11'39" 26d30'11"
Ann Santini
And 23h37'19" 40d13'4"
Ann Schilling
Cap 21h48'11" -11d1'15"
Ann Schultz
Crb 15h40'29" 26d16'42"
Ann Schwinghamer
Cas 0h56'32" 57d51'48"

Ann (Sex Kitten)
Lib 15h16'49" -19d59'46"
Ann Sierra
Com 13h6'33" 19d5'26"
Ann S.K. Howlett
Vir 13h44'36" -1d2'44"
Ann Skelly Fulton
Lib 15h8'4" -6d14'49"
Ann Smith Gordon
Tau 5h52'33" 26d56'25"
Ann Spaulding
Cas 0h46'34" 47d51'59"
Ann Speicher
Sgr 19h20'20" -16d7'25"
Ann Stahurski
Cap 20h56'18" -16d2'13"
Ann Stanko
Cas 1h31'7" 63d24'46"
Ann Stedt
Lib 14h30'41" -14d12'6"
Ann Stephenson
Leo 11h32'35" 7d27'33"
Ann Stertzbach
Vir 12h25'18" 6d39'42"
Ann Symington
Cru 12h12'11" -62d2'50"
Ann T. Halvorsen
Gem 6h53'30" 25d17'13"
Ann Thompson
Cap 20h9'12" -9d39'39"
Ann Tinkham
Uma 10h7'15" 53d52'58"
Ann Tobin
Cap 20h23'13" -14d17'50"
Ann V. Doan
Uma 13h48'2" 51d59'19"
Ann Van Ess
Cas 0h31'58" 55d10'43"
Ann Ventrice
Cas 0h35'27" 54d0'24"
Ann Waldersen
Mon 6h44'24" -0d5'44"
Ann Weibel
Cnc 8h51'33" 27d25'58"
Ann Wolverson
Psc 0h18'5" -5d5'4"
Ann Wright
And 0h52'31" 37d58'29"
Ann Zisk
Crb 16h11'35" 35d37'51"
Ann, Loving Mother of Three
And 22h58'38" 50d45'20"
Ann, Mickey, Patricia McCarn
Aql 19h50'12" -0d36'39"
Ann, Mike and Scout
Uma 9h46'49" 54d33'2"
Ann, Mike and Scout
Uma 11h20'0" 37d38'32"
Ann4Dor
Lyr 18h51'4" 31d35'51"
Anna
Cyg 19h54'21" 33d28'21"
Anna
Crb 16h22'12" 31d52'39"
Anna
Lyn 8h48'2" 36d27'31"
Anna
And 0h43'20" 36d36'2"
Anna
And 1h44'11" 41d57'56"
Anna
And 1h1'34" 44d48'45"
Anna
And 0h39'56" 41d39'7"
Anna
Cyg 19h47'8" 46d35'38"
Anna
Uma 9h33'46" 46d18'52"
Anna
And 1h50'18" 45d32'55"
Anna
Per 3h3'7" 51d43'48"
Anna
Boo 15h22'55" 38d12'38"
Anna
Her 17h46'34" 41d53'57"
ANNA
Cnc 9h16'57" 25d32'39"
Anna
Cnc 9h3'0" 20d40'8"
Anna
Ari 3h16'56" 24d36'30"
Anna
Psc 1h1'22" 26d32'54"
Anna
Ori 6h25'4" 10d10'23"
Anna
Psc 0h49'13" 16d19'48"
Anna
Uma 8h57'29" 56d17'31"
Anna
Cas 1h30'24" 60d18'8"
Anna
Boo 14h36'13" 53d51'14"
Anna
Cas 23h2'16" 54d5'20"
Anna
Cas 23h2'59" 53d59'25"
Anna
Lib 14h53'8" -1d37'47"
Anna
Lib 15h39'23" -7d48'27"

Anna
Vir 14h15'0" -17d40'58"
anna
Cru 12h32'53" -61d40'29"
anna
Cru 12h32'40" -61d25'6"
Anna
Sgr 19h24'29" -34d44'52"
Anna
Sco 17h48'25" -39d45'48"
Anna
Sco 16h44'52" -34d37'11"
Anna
Ori 6h5'7" -2d6'54"
Anna 1962
Vir 12h19'46" -9d55'13"
Anna Agnes
Cyg 20h16'14" 35d4'55"
Anna Agnese Staniscia
Cam 7h10'12" 69d21'39"
Anna Agrow - GaMPI Shining Star
Lyn 8h22'31" 55d3'51"
Anna Alexis
Ari 2h30'25" 17d47'10"
Anna and Bella Sennett
Vir 14h16'26" -1d43'41"
Anna and Jacob
Uma 12h23'40" 57d6'18"
Anna and Joe Roth
Cyg 19h47'17" 29d31'46"
Anna And Kyle Everlasting
Cyg 20h28'42" 46d59'55"
Anna and Lisa
Vir 13h45'35" -8d2'4"
Anna Antonella Gloria
And 0h32'11" 41d6'34"
Anna Antonina Bisaga
Umi 8h33'23" 88d8'3"
Anna Aston McArtor
And 23h50'36" 34d24'40"
Anna Azalea Molotsky
And 2h18'19" 50d25'57"
Anna Babe
Lib 15h5'7" -0d45'37"
Anna "Baby Doll" Trucchio
Cyg 20h41'47" 39d36'50"
Anna Banana 25
Gem 7h28'54" 33d49'8"
Anna "Banana" Schenker
Cnc 8h27'2" 21d40'33"
Anna Bannana
Lib 15h7'56" -4d4'40"
Anna Basgier
Lyr 19h14'27" 26d35'51"
Anna Baskeyfield
And 0h2'20" 43d37'33"
Anna Belladonna C. L. D.
Peg 21h37'57" 27d41'35"
Anna Belle Norris
Vir 12h27'0" -2d56'28"
Anna Bergamo Gremillion
Cnc 8h2'4" 21d33'36"
Anna Bernice
And 1h8'16" 36d49'14"
Anna Beth 1
Ari 2h9'11" 27d45'25"
Anna "Blue Eyes"
Gem 7h8'22" 21d3'11"
Anna Bop
Cap 20h26'46" -17d9'32"
Anna Bottros-Yartseva
Lyn 7h48'40" 38d52'33"
Anna Brania & Jason Horne
Cyg 19h46'32" 32d1'26"
Anna Brigid Myers
And 2h11'49" 46d16'8"
Anna Bronwyn Trickey
Cnc 8h51'53" 12d50'39"
Anna Broster
Uma 12h1'27" 35d35'39"
Anna Bruno
Cnc 8h44'34" 14d6'58"
Anna "Bubbles" Montelione
Leo 9h36'29" 26d4'46"
Anna buonomano
Uma 8h31'5" 60d59'3"
Anna Busch
Uma 13h26'25" 53d37'55"
Anna Caroline Menkel
Uma 10h49'52" 44d32'25"
Anna Carolyn Walden, Beloved Mother
Gem 6h56'0" 14d16'5"
Anna Carrillo
Tau 5h57'10" 26d52'14"
Anna Catherine
Umi 4h19'24" 88d44'43"
Anna Catherine McDonough
Com 13h11'54" 17d49'54"
Anna Catherine Riley
And 0h17'52" 29d18'28"
Anna Cathrine Brubaker
Uma 9h41'12" 69d42'52"
Anna Cecelia
Uma 10h56'19" 41d3'25"
Anna Celeste
Sex 10h45'43" -0d1'25"
Anna Celeste
Sgr 18h50'3" -17d39'14"
Anna Cheshire
Ori 6h8'19" -0d13'49"

Anna Christan Davis
Cnc 8h43'0" 14d9'50"
Anna Christian Black
Cas 1h22'3" 68d26'40"
Anna Christina Nabors
Cas 0h37'40" 53d26'49"
Anna Christine
Cas 23h40'20" 52d29'50"
Anna Christine Hernandez
Ari 1h55'57" 17d2'35"
Anna Cinderella Bean
Vir 13h52'32" -6d47'36"
Anna Claire Simendinger
Cas 1h26'51" 65d21'55"
Anna Claire Smith
Sgr 18h12'29" -20d2'1"
Anna Clara
Cnc 8h45'24" 16d13'11"
Anna Cochran "Princess Star"
Gem 6h20'56" 27d52'22"
Anna Collins and Jason Holub
Uma 11h3'48" 36d55'3"
Anna Conforti Rizzo
Lib 15h3'9" -2d12'19"
Anna Costa
Tau 4h31'15" 4d49'35"
Anna Crapanzano
Per 3h43'45" 43d32'26"
Anna Crouse Wrege
Cas 2h49'26" 60d37'52"
Anna Crystal Lindenberg
And 0h39'0" 32d25'58"
Anna D. Greaves
Cnc 8h35'45" 24d51'31"
Anna Daube Freund
Umi 15h14'53" 71d54'46"
Anna Davis George
And 0h33'33" 43d42'56"
Anna Day Funk
Uma 9h43'29" 67d36'20"
Anna Debbie Chris Sydney Thompson
Del 20h55'44" 15d59'53"
Anna Dee
Psc 0h53'11" 28d33'26"
Anna DelMonico
Vir 13h19'52" 12d17'54"
Anna & Didier Bischoff-Jubin
Lyr 19h19'51" 29d45'7"
Anna Dolan
Cyg 19h41'17" 29d0'57"
Anna Douglas Birthday Star
Sco 16h37'54" -40d52'23"
Anna Driganova
Uma 11h15'9" 39d27'43"
Anna D'Simone
Lyr 18h51'42" 30d16'29"
Anna e Giuseppe per sempre
Ori 5h13'33" 2d36'14"
Anna E. M. Bassham
Aql 19h7'55" 13d37'13"
Anna E. McCormick
Cyg 19h54'46" 30d38'26"
Anna Eleytherou
Cas 0h20'46" 61d53'42"
Anna Elise Wright
And 23h4'4" 46d58'39"
Anna Elishava bat Haya TzveFay
Lyn 7h43'6" 41d43'23"
Anna Elizabeth
And 1h31'57" 49d43'56"
Anna Elizabeth
Umi 15h41'17" 74d44'10"
Anna Elizabeth
Lib 15h20'34" -5d19'6"
Anna Elizabeth
Sco 16h48'46" -35d57'26"
Anna Elizabeth Berry
Cas 0h31'18" 61d16'45"
Anna Elizabeth Brezina
Leo 9h54'32" 28d45'25"
Anna Elizabeth Catalanello
And 2h18'55" 46d29'0"
Anna Elizabeth Dunkerly
Uma 11h42'55" 39d17'54"
Anna Elizabeth "Elli" Nunnery
Peg 22h11'28" 7d41'32"
Anna Elizabeth Jarrell Pellegrino
And 23h16'52" 49d18'13"
Anna Elizabeth Ryan
Uma 8h56'24" 47d2'44"
Anna Elizabeth Spence
Psc 2h3'0" 6d41'19"
Anna Elizabeth Vitale
Sco 16h48'21" -39d7'54"
Anna Elizabeth Wojciechowski
Sgr 18h25'48" -16d38'18"
Anna Erin Leet
Uma 11h39'59" 60d52'30"
Anna Ethel Manley "Big Grandma"
Cyg 21h32'6" 47d15'31"
Anna Ettlinger
Dra 19h39'5" 65d1'46"
Anna Fadeley
Leo 11h31'6" -5d30'33"

Anna Fallini Reymore
Cas 1h7'36" 50d38'30"
Anna Flora Spalding
Cam 5h53'47" 58d44'52"
Anna Fox Alosio
Cas 1h3'6" 61d42'0"
Anna Franklin
Lyr 18h31'12" 29d31'11"
Anna Froussakis
And 0h35'9" 32d3'17"
Anna G.
Ori 6h13'17" 8d28'57"
Anna Gabrielle Kraft
And 1h17'42" 38d18'28"
Anna Gaines
Aqr 22h37'47" -7d15'52"
Anna Gainey
Vir 13h15'46" 3d35'47"
Anna Gajb
Sco 17h52'43" -39d47'48"
Anna & Gerardo
Umi 15h38'21" 79d55'21"
Anna Gloria Vega
Tau 5h48'14" 21d34'29"
Anna Glorioso
And 0h35'32" 42d52'5"
Anna Gomah
Aqr 20h44'6" -2d26'8"
Anna Grace
And 0h36'7" 28d23'28"
Anna Grace Bostick
Gem 7h13'38" 28d15'56"
ANNA GRACE MAHORNEY
Lyn 8h27'45" 39d5'13"
Anna Grace Mckelvey
Gem 6h49'18" 18d50'2"
Anna Grace Yukniewicz
Lib 15h5'13" -1d23'3"
Anna Grace Zuckerman
And 0h25'13" 44d42'31"
Anna (Gram) Bischoff
Sgr 19h59'12" -30d27'57"
Anna Greszler Hurd
Sco 16h3'43" -11d0'52"
Anna Grigsby
Cas 23h6'21" 55d23'30"
Anna Gwynnedd Somerville
Cru 12h18'31" -60d38'43"
Anna Habartova
Gem 7h48'5" 32d30'23"
Anna Haugh
Tau 4h15'31" 24d9'8"
Anna Heckman / Granny Reber
Ari 2h9'51" 23d39'43"
Anna Helen
Sco 17h51'56" -39d4'22"
Anna Helleschova
Ori 5h15'53" -2d6'27"
Anna Hofbauer
Uma 8h36'34" 58d48'46"
Anna Hong. My Sister. My Friend.
Cap 20h47'39" -15d42'43"
Anna Irene
And 2h27'27" 43d18'37"
Anna Irons Brennecke
Lyn 8h0'55" 41d55'55"
Anna J. Urbanczyk
Uma 10h8'25" 41d25'40"
Anna ja lisakki
Lyn 7h9'29" 57d23'53"
Anna Jacobs
Ari 2h40'57" 22d1'19"
Anna Jade Lawler Carrington
Uma 11h22'5" 30d22'49"
Anna Jane Lucas
Cas 0h13'11" 61d32'28"
Anna Jane Miller
Cap 20h18'22" -12d10'51"
Anna Jane Miller Woods
Ari 2h7'16" 19d42'49"
Anna Jantos
Cas 2h27'35" 65d37'44"
Anna Jasik
Lib 15h12'38" -5d55'26"
Anna Jayne Capobres
And 0h2'48" 44d9'18"
Anna Jean
Uma 10h25'52" 41d14'8"
Anna Jean
Uma 12h1'16" 49d17'16"
Anna Jean Marsh
Leo 11h21'27" 15d34'12"
Anna Jean Spencer "Star Of Wonder"
Cap 20h40'4" -21d1'11"
Anna Jean Wilson
And 23h7'48" 41d10'58"
Anna Jean's Little Bit of Heaven
Cru 12h34'56" -55d47'3"
Anna Jeremiah
Gem 6h26'24" 20d21'54"
Anna & John Shorts Moon & Pig Star
Tau 5h1'23" 19d34'37"
Anna Johnson
Cas 0h43'2" 66d55'33"
Anna Jordan
Sco 16h9'19" -19d28'11"
Anna Josephine Hill
Cas 1h36'14" 67d42'51"

Anna Josephine Marie Dill Plant
Lyr 18h19'29" 44d59'15"
Anna Julie
Lib 14h53'0" -18d2'8"
Anna K. Cleveland
Lib 15h17'49" -4d18'13"
Anna Kallistei
Cam 5h46'15" 60d54'20"
Anna Kamila Chodon
Cnc 8h30'6" 13d39'41"
Anna Karczewska
Crb 15h53'18" 29d58'6"
Anna Karin Mikulak
Mon 6h50'33" -0d36'6"
Anna Karina
Cap 21h44'30" -18d43'9"
Anna Kate
And 1h32'51" 44d25'42"
Anna Katelyn
Lyn 7h41'53" 53d8'14"
Anna Katharina Grether
Sex 10h44'56" 3d34'6"
Anna Katherine
Aqr 22h40'2" 2d18'18"
Anna Katherine Corr
Lmi 10h50'43" 31d41'13"
Anna Kathleen Plant
Vir 14h33'11" 4d14'50"
Anna Kathleen Ward
Lyn 8h57'46" 43d25'40"
Anna Kathrin Christine Alica
Uma 9h46'34" 55d26'53"
Anna Kathryn Burkett
And 0h11'59" 43d9'34"
Anna Kathryn Hailey
Lyn 6h52'41" 51d11'43"
Anna Kilian Harrison
Ori 5h14'23" 4d23'20"
Anna King
Cnc 8h52'30" 29d24'27"
Anna Kostenko
Cnc 8h8'34" 11d19'34"
Anna Koval Thackray
Cas 0h7'55" 59d44'42"
Anna Koziarski
Lyn 7h53'9" 52d9'20"
Anna Krystyna Mech
Leo 10h45'45" 16d24'44"
Anna L. Hughes
Psc 23h29'51" -3d11'56"
Anna L Kelsey Craddle To Grave
And 0h24'6" 34d22'23"
Anna L. Lyons
Ori 4h50'13" 6d47'17"
ANNA L270480
And 0h11'41" 30d25'32"
Anna La Creativa
And 23h56'16" 48d20'32"
Anna Lake Hill
Uma 13h42'38" 59d25'20"
Anna Lakovetsky
Cap 21h22'34" -19d9'57"
Anna Lannon
Crb 15h41'18" 26d5'53"
Anna Laura
Vir 13h21'19" -3d38'50"
Anna Laura 2k5, Psyche
And 1h18'55" 38d45'42"
Anna Lavina Watkins
Vir 13h4'37" -22d0'45"
Anna Lea
Umi 11h28'0" 88d32'5"
Anna Leary
Lib 14h51'34" -1d0'11"
Anna Lee
And 1h41'46" 49d55'21"
Anna Lee Burris
Crb 15h27'36" 28d9'6"
Anna Lee Fauber
Ari 2h33'33" 24d58'14"
Anna Lee Nelson
Crb 15h52'51" 31d29'42"
Anna Leeze
Lyn 8h5'48" 56d3'43"
Anna Leigh Fisher
Vir 14h39'54" 3d48'5"
Anna Leigh Remsburg
Umi 15h7'46" 76d33'57"
Anna Leigh Rowlands
Lyr 18h46'27" 44d35'29"
Anna Leigh White
Cru 12h49'7" -57d12'24"
Anna Leslie Millstein
And 1h18'12" 35d17'38"
Anna Leticia
Cen 13h34'14" -43d22'29"
Anna Lillian Rechichi
And 1h32'22" 43d45'9"
Anna Liokumovich
Vir 14h1'25" -14d59'2"
Anna Lisa Laucas
Cyg 20h57'17" 31d51'21"
Anna LoBianco
Gem 7h40'50" 33d17'0"
Anna Loechel
Lib 15h54'43" -7d2'7"
Anna Lonas
Sgr 18h59'9" -34d12'47"
Anna Lorraine
Cap 20h20'22" -14d50'15"
Anna Louisa Charnley
Tau 3h41'36" 16d55'37"
Anna Louise
Sgr 19h44'19" -18d21'20"
Anna Louise Newton
Cnc 9h14'57" 11d22'22"
Anna Louise Poorbaugh Stump
Cyg 20h12'33" 41d44'45"
Anna Louise Probelski
Cas 0h26'57" 56d5'21"
Anna Louise Rugen
Psc 0h53'28" 32d28'56"
Anna Louise Schatt Corder
Cas 1h42'2" 60d23'36"
Anna loves John
Uma 13h2'50" 55d16'24"
Anna Loves Mike Forever
Ori 5h34'42" -0d15'47"
Anna Lovett
Cas 0h1'9" 53d43'31"
Anna Lucas
Peg 22h19'41" 11d31'19"
Anna Luisa Dobbs
Tau 3h43'47" 23d40'34"
Anna Luu
Pho 0h33'2" -48d43'4"
Anna Luvisotto
Uma 12h39'27" 61d31'43"
Anna Lynn Gionet
And 0h15'38" 44d32'47"
Anna Lynn Turpin
Cyg 20h1'57" 43d18'32"
Anna Lynne Schneider
Uma 11h44'34" 57d22'1"
Anna M. Balzano
Cas 2h14'39" 66d43'7"
Anna M. Dlugolonski
Aqr 21h32'41" 1d8'59"
Anna M. Klueg
Tau 4h23'35" 19d1'24"
Anna M. Sjoberg
Lyr 19h13'0" 26d32'34"
Anna M. Teres
Uma 12h47'18" 55d32'14"
Anna M Villanueva
Psc 1h3'57" 4d9'51"
Anna Madeline
Ari 3h2'0" 28d37'12"
Anna Mae
Cnc 8h21'27" 24d54'0"
Anna Mae
Com 12h37'26" 27d10'42"
Anna Mae
And 2h30'59" 44d47'11"
Anna Mae
And 0h44'9" 34d32'14"
Anna Mae Bizzozero
And 23h21'29" 48d10'38"
Anna Mae Donlon
Cas 0h58'27" 57d52'50"
Anna Mae Feigh World Best Mother
Cas 0h16'49" 52d32'4"
Anna Mae Goldman -March 12, 1923
Uma 13h41'9" 48d40'0"
Anna Mae Pericak
Ori 5h52'22" -2d50'49"
Anna Magdalena Smith
Cnc 9h17'58" 6d36'3"
Anna Maija
Uma 12h14'26" 61d38'37"
Anna Marangon
Peg 22h12'37" 21d27'54"
Anna Marguerite Parkinson
And 0h21'8" 27d12'47"
Anna Maria
Gem 7h11'51" 28d35'48"
Anna Maria
Uma 10h6'28" 51d24'28"
Anna Maria
Cam 5h52'17" 56d28'51"
Anna Maria
Cas 2h42'6" 57d58'54"
Anna Maria
And 0h28'54" 36d9'52"
Anna Maria
Cyg 19h54'11" 34d1'48"
Anna Maria Borrega
Umi 15h1'28" 69d19'22"
Anna Maria Cass
Lib 15h3'11" -1d19'30"
Anna Maria Coppola
And 23h14'10" 37d13'30"
Anna Maria Di Mascio
Crb 16h24'30" 35d39'55"
Anna Maria Gallo
Psc 1h19'51" 32d12'4"
Anna Maria Howell
Sco 16h47'30" -36d19'5"
Anna Maria Huizar
Cap 21h35'47" -18d10'29"
Anna Maria Milne
Crb 16h24'44" 31d2'57"
Anna Maria Neri
Lyn 8h12'54" 54d14'7"
Anna Maria Nowicka
Cas 23h42'45" 62d16'3"
Anna Maria & Richard DiCasoli
Cyg 21h54'18" 48d11'0"
Anna Maria ROCCO
Psc 0h14'5" 3d16'29"
Anna - Maria Romain
Umi 14h48'12" 79d21'7"
Anna Maria Sollars
Vel 10h20'32" -42d27'7"
Anna Maria, meine große Liebe
Uma 10h9'23" 49d51'9"
Anna Marie
Cam 4h29'54" 58d6'2"
Anna Marie
And 0h32'46" 31d29'19"
Anna Marie
Ori 6h13'11" 19d20'1"
Anna Marie
Vir 13h58'22" 6d20'20"
Anna Marie
Psc 1h37'58" 12d19'23"
Anna Marie
Sgr 18h6'9" -24d53'46"
Anna Marie
Sco 16h8'7" -14d6'46"
Anna Marie
Uma 8h30'45" 61d7'57"
Anna Marie
Lyn 7h20'58" 52d43'42"
Anna Marie
Uma 13h42'23" 55d58'43"
Anna Marie Adams
Aql 19h35'30" 8d7'43"
Anna Marie Affinati Cieslewicz
Cas 0h56'2" 63d40'35"
Anna Marie and James Cullen
Mon 6h49'2" 8d17'55"
Anna Marie Armstrong
Vir 13h25'51" 12d25'53"
Anna Marie Askins
And 2h35'7" 47d30'29"
Anna Marie Ball
And 2h32'9" 39d53'2"
Anna Marie Beach
Ari 3h23'12" 26d21'36"
Anna Marie Beebe
And 23h44'43" 38d11'56"
Anna Marie Bensinger
Cap 20h10'50" -9d8'11"
Anna Marie Bevilacqua
And 23h14'11" 41d36'25"
Anna Marie Curl
Aqr 23h18'14" -17d45'2"
Anna Marie DiCarlo
Cas 1h44'50" 65d22'13"
Anna Marie "Dordy-Nord" C. Brown
Cam 5h42'44" 57d39'34"
Anna Marie Emory
Cas 22h58'54" 54d27'38"
Anna Marie Evers
Leo 11h19'50" 12d36'22"
Anna Marie Genevieve
Sco 17h6'58" -34d52'57"
Anna Marie Guthrie
Psc 0h37'48" 18d9'20"
Anna Marie Halawani
Uma 11h19'49" 39d28'44"
Anna Marie Idleman
Cas 1h36'21" 63d18'56"
Anna Marie Jaeger
Cas 0h48'33" 62d54'26"
Anna Marie Langley
And 1h12'50" 37d47'33"
Anna Marie Lisy
Sgr 18h59'42" -34d59'27"
Anna Marie Overcash
Crb 15h27'0" 26d7'23"
Anna Marie & Rupert
Cyg 21h22'18" 38d2'54"
Anna Marie Sawyer
Sco 16h5'22" -14d27'10"
Anna Marie Schiller
Tau 4h33'15" 4d41'5"
Anna Marie Silveira
Cyg 21h32'14" 37d49'44"
Anna Marie Simiele
Aqr 22h37'39" -0d32'34"
Anna Marie Ward
Gem 6h51'36" 25d33'19"
Anna Marie Washington
Sco 15h55'58" -21d54'54"
Anna Marie Wright
Leo 9h55'17" 18d31'27"
Anna Marie Zingone
Cyg 20h40'10" 39d42'11"
Anna Marie Zouglas
And 0h59'7" 39d59'47"
Anna Marina
Aqr 23h31'57" -19d54'31"
Anna Marina Obremski
Cnc 8h19'29" 16d55'54"
Anna Mary
Cnc 8h16'0" 21d9'12"
Anna Mary Knipp
Uma 11h11'15" 44d11'37"
Anna Mary Sisk Mazany
Cyg 21h59'43" 50d43'12"
Anna Matus
Ari 3h4'40" 19d4'35"
Anna May Hardcastle
Uma 9h8'29" 46d37'37"
anna m.burns
Uma 13h36'32" 48d30'18"
Anna McCabe
And 0h34'28" 27d11'12"
Anna McKinley *Princess Buttercup*
Cnc 8h12'23" 10d51'25"
Anna McLamb
Cam 6h17'13" 68d25'47"
Anna Meeshka Bear
Uma 9h23'9" 64d47'33"
Anna Megan Taylor Ketelsen
And 2h6'32" 46d34'35"
Anna Melchior
Cas 1h14'47" 68d6'23"
Anna Melissa
And 23h29'47" 43d5'24"
Anna Menke
And 1h13'50" 45d51'44"
Anna Michelle Brooks
Cyg 20h57'21" 32d34'59"
Anna Migliaccio
Cap 21h58'8" -22d50'39"
Anna Mina
Ori 6h2'24" 20d42'48"
Anna Mintch
Leo 10h2'53" 13d21'34"
Anna Molino Löf
Cas 0h35'21" 62d14'3"
Anna Mon Amour
Aqr 22h11'41" -22d39'15"
Anna mon amour pour l'éternité
Cnc 8h22'13" 19d35'18"
Anna Montano
Aqr 21h41'8" -6d6'53"
Anna Montgomery Riemer
And 23h23'27" 51d5'43"
Anna Morse
And 0h29'19" 36d53'53"
Anna My Love's Star
Uma 10h49'30" 46d3'58"
Anna My One And Only
Cam 5h47'25" 56d9'41"
Anna Myfanwy Persephone Schütte
And 23h13'19" 47d58'48"
Anna N. Treppiedi
Boo 14h28'6" 53d34'42"
Anna Nalbandyan
Vir 14h21'36" 0d5'14"
Anna Nhi Vo
Gem 6h31'35" 12d4'34"
Anna Nicole Davis
Ori 5h52'14" 21d54'30"
Anna Nicole Denison
Lmi 9h43'18" 38d26'4"
Anna Nicole Weiss
Mon 6h43'1" 11d17'52"
Anna Noelle
Lyr 19h16'24" 28d51'15"
Anna Nunes Flaherty
Aql 19h16'9" 5d42'5"
Anna occhi di cerbiatto
Lib 15h31'58" -7d23'18"
Anna Olinsky
Uma 11h22'48" 58d37'56"
Anna Orlowska
Lib 15h3'14" -3d24'55"
Anna P.
Psc 0h4'7" 1d27'31"
Anna Pakosova
And 0h26'33" 46d25'44"
Anna Pansy Simmons
And 0h20'8" 42d48'3"
Anna Parsons Calloway Charles
Gem 6h51'6" 25d25'38"
Anna "pata" Brichese
Her 17h56'46" 29d37'56"
Anna Patricia
And 23h9'24" 45d27'2"
Anna Pérez Lluis
Ori 5h4'24" -0d47'25"
Anna Perz
Cnc 8h37'6" 15d55'24"
Anna Polatou
And 2h27'7" 42d41'15"
Anna Potiha
Tau 3h52'14" 17d0'40"
Anna Quacky
Cma 7h13'41" -12d47'50"
Anna R. Carbajal
Mon 6h44'45" -0d20'7"
Anna Rachael Dunlap-Hartshorn
Leo 10h17'41" 20d28'13"
Anna Rachel Rodenbaugh
Lib 15h30'15" -26d53'55"
Anna Rebecca Mancuso
Tau 5h26'7" 25d0'42"
Anna Rita Heffel
Leo 9h53'37" 20d18'55"
Anna Rivera (Mame)
Ari 3h4'51" 23d32'5"
Anna & Robert Reissfelder
Cap 20h45'56" -19d36'11"
Anna & Rocky forever, 6-19-2006
Uma 9h32'37" 64d10'39"
ANNA ROG
Gem 7h27'52" 32d18'17"
Anna Romine
And 0h48'19" 42d39'18"
Anna Rosalia Schuler
Uma 11h7'33" 28d32'40"
Anna Rose
Cnc 8h40'54" 32d42'16"
Anna Rose Carda
Vir 13h17'5" 12d56'21"
Anna Rose Carter
And 2h31'59" 48d54'53"
Anna Rose Cornelius
Vir 13h16'43" -3d42'22"
Anna Rose Klein
Aqr 21h46'33" 0d27'3"
Anna Rosella Nason
And 0h43'21" 45d37'14"
Anna Ruth Johanesen
Lib 14h50'34" -18d27'7"
Anna S.
Lyr 19h26'38" 38d37'52"
Anna Sadler's 17th Birthday Star
Cap 20h49'14" -16d12'36"
Anna Scardigli of Italia
And 0h21'14" 41d28'48"
Anna Schneider
And 0h28'30" 42d10'43"
Anna Schuhmacher
Umi 10h43'19" 87d1'13"
Anna+Scottxxxs4eva
Ori 5h52'55" 22d44'35"
Anna Scozzafava Moya Jena
Aqr 22h53'26" -24d34'37"
Anna Sergeeva
Ari 2h51'51" 28d42'59"
Anna Sharon Favata
Uma 9h51'53" 53d11'17"
Anna Sheil
Cas 0h9'31" 53d26'59"
Anna Sherratt Cagle
Lmi 10h43'24" 38d18'13"
Anna Slavinski
Cnc 8h10'58" 30d12'57"
ANNA Smolian Bulgaria
Cap 20h32'11" -13d16'30"
Anna Sophia Bresler
Cyg 20h47'42" 52d23'10"
Anna Sophia Jean
Ari 2h52'12" 17d37'15"
Anna Sophie
Uma 10h25'37" 66d19'38"
Anna Stacher
Cas 2h0'20" 58d46'54"
Anna Stratton
Uma 8h56'16" 66d15'30"
Anna Sue Rowder
Cyg 21h42'25" 42d20'18"
Anna Sullivan
Leo 11h0'46" 9d0'56"
Anna T. Alexander
And 0h13'59" 33d25'28"
Anna T. Cancelliere
And 1h12'37" 45d4'38"
Anna Talis Love Firouz
Tau 4h36'12" 20d28'38"
Anna Teague
Tau 4h38'38" 10d17'4"
Anna Tedeschi
Per 4h18'2" 36d5'37"
Anna Tempel Bell
Psc 0h20'56" 17d23'8"
Anna Tenyotkina
Ori 5h37'50" 8d16'47"
Anna Teresa
And 23h13'24" 43d30'55"
Anna Teresa Venneman
Uma 12h57'33" 58d51'46"
Anna Teresa Viscogliosi
And 23h21'11" 48d17'25"
Anna Theresa Brundige
Lyr 18h33'4" 37d9'51"
Anna Theresa Marsala 10/29/1991
Uma 11h34'41" 58d5'53"
Anna "Tinkerbell" Collins
Leo 9h27'9" 31d43'41"
Anna & Trevor, Forever & For Always
And 0h15'30" 43d21'35"
Anna Truzzolino
Gem 7h16'16" 25d10'35"
Anna Tullio
Cas 23h23'44" 58d2'8"
ANNA und MARC
Uma 8h45'59" 68d7'39"
Anna V
Aqr 21h56'47" 1d9'33"
Anna V. Schiavo
Uma 11h25'26" 53d27'52"
Anna Vassileva
Uma 11h19'18" 44d15'8"
Anna Victoria
Uma 11h10'11" 46d44'25"
Anna Victoria Hilliard
And 1h45'20" 42d51'11"
Anna Villa
Per 4h45'3" 42d25'8"
Anna Vitkovskaia
Her 16h16'44" 44d35'42"
Anna Wagner
Cap 21h52'57" -24d47'52"
Anna Ward
Sco 16h13'27" -9d15'57"
Anna Weiss
Aqr 23h53'32" -12d23'39"
Anna (whale) Maier
Psc 0h36'42" 20d21'52"
Anna Wilson
Cyg 20h50'8" 46d53'47"
Anna Xurong Ji
Umi 15h48'52" 75d54'56"
Anna Yoko Walker
Uma 11h15'45" 36d21'10"
Anna & Zachary Stewart
Psc 1h19'42" 5d53'56"
Anna Zaydenberg
Vir 13h8'28" 13d22'40"
Anna, der hellste Stern am Himmel
Ori 6h12'38" 15d47'33"
Anna, Will You Marry Me Again?
Vir 13h6'59" 10d56'38"
AnnaB
Cam 9h30'22" 82d5'20"
Annababy
Uma 8h13'12" 62d27'22"
AnnaBananaFace
Crb 15h37'47" 25d56'38"
Annabear and Nick
Cyg 20h49'49" 38d16'56"
Annabel Baltazar
And 0h30'17" 42d50'8"
Annabel Claire - 19.09.2005
Cru 12h12'56" -63d49'12"
Annabel Courtney-Henderson
Tau 4h21'56" 17d21'43"
Annabel Jayne
And 2h32'33" 48d59'18"
Annabel K. Webb 14th April 1986
Cas 1h34'50" 68d35'35"
Annabel Lee
Cyg 21h19'15" 42d42'21"
Annabel97
And 1h5'38" 37d36'33"
Annabell Brunschweiler
Uma 10h55'35" 34d1'40"
Annabell Chase Foy
Uma 11h23'3" 41d24'50"
Annabell Edler Becker
And 23h11'18" 48d16'11"
Annabell McDonald
Peg 22h14'59" 12d19'29"
Annabell Paisley Swann
Leo 9h31'20" 12d1'46"
AnnaBella
Vir 12h51'20" 6d20'0"
Annabella
And 0h50'33" 40d19'37"
Annabella
And 1h28'7" 41d3'25"
Annabella
Lib 14h59'23" -19d46'18"
Annabella Grace Ryan
And 2h28'56" 42d10'16"
Annabella Iris Aurora Sipila
Lyn 9h37'49" 40d38'54"
Annabella Jane Granata
Lib 15h5'10" -5d25'24"
Annabella Rose Imbrescia
Ari 2h56'23" 25d30'38"
Annabelle
Ori 6h19'6" 13d36'43"
Annabelle
Crb 15h24'3" 30d27'23"
Annabelle
And 23h21'0" 42d10'28"
Annabelle
And 2h36'34" 50d34'37"
Annabelle
Lib 14h54'5" -1d32'53"
Annabelle
Aqr 21h15'15" -2d24'42"
Annabelle
Sgr 18h16'12" -17d21'48"
Annabelle
Uma 13h26'42" 59d3'22"
Annabelle - 5
Cas 23h38'38" 55d9'15"
Annabelle Cabrigas Hazelton
Gem 6h6'28" 22d35'29"
Annabelle Eberly
Cas 23h15'26" 56d13'41"
Annabelle Grace Glover
And 1h7'56" 46d55'58"
Annabelle Grace Jenkins
Ari 3h14'21" 27d57'28"
Annabelle Grace Sunday
Cyg 19h36'34" 48d5'3"
Annabelle Hope Chaplin
And 23h54'7" 39d31'27"
Annabelle -Hugo- Greg
Psc 0h2'8" 1d42'23"
Annabelle Jane
And 0h47'52" 39d30'20"
Annabelle Jones
And 23h21'32" 35d16'13"
Annabelle Jones
Cam 5h31'5" 66d39'54"
Annabelle Kate Wiedower
Aqr 22h5'59" -12d27'53"
Annabelle Lee Hupp
Ari 2h35'10" 21d44'4"
Annabelle Lisa DelMonte
Vir 13h18'20" 6d6'21"
Annabelle Lusiana Lingga
Ori 5h36'55" -2d3'45"
Annabelle Marie
Cnc 8h47'53" 22d26'28"
Annabelle Marie Sweet Matthews
Her 18h41'18" 20d22'13"
Annabelle - Our Angel
Cru 12h41'43" -59d48'37"
Annabelle Paige Lucas
Leo 10h16'2" 21d34'27"
Annabelle Riley Bryant
Lib 15h12'38" -5d13'16"
Annabelle Rose James
And 0h21'59" 39d36'55"
Annabelle Sedow
Aqr 22h41'11" -3d30'2"
Annabelle Skye
And 0h10'59" 23d18'5"
Annabelleleighton
Aqr 22h12'14" -2d20'33"
Annabelle's Star
Cru 12h2'41" -63d7'25"
Annabel-Merwin
Uma 11h9'59" 60d41'19"
Annabel's Diamond
Aqr 22h46'7" -15d44'26"
Annabel's First Christmas
And 23h38'43" 48d48'45"
Annabel's Star
And 23h40'35" 38d16'40"
Annachiara
Crb 16h21'18" 28d31'19"
Annafaye
Gem 7h22'11" 20d53'50"
AnnaGrabowsky1
And 0h34'34" 28d9'22"
Anna-Grace Glover
And 2h36'30" 45d48'9"
Anna-Greta's 100 År's Stjerna
Uma 11h20'34" 44d24'19"
Anna-Gyöngyike
Cas 0h55'10" 69d26'35"
Annah
Gem 6h50'54" 16d56'20"
Annah
Leo 9h51'40" 20d12'23"
Annah & Carlos Chagas Eternal Star
Cyg 20h12'29" 36d13'57"
Annaha Featherhill
Sgr 18h27'28" -27d14'31"
ANNAIG
Ori 6h20'41" 9d24'39"
Anna-Jason
Cyg 19h58'12" 40d47'32"
Annajayne Leigh Follis
Psc 1h9'23" 25d42'10"
Annakat
Aqr 22h36'44" 0d44'7"
AnnaKramer
Sgr 18h30'30" -16d45'13"
annakrontal
Cyg 21h40'24" 51d52'53"
Annalaura Cadei
Uma 8h23'39" 68d12'6"
AnnaLee
Lyr 18h46'31" 37d53'23"
Annalee
Lyr 19h13'31" 26d22'50"
Annalee Rose
And 0h51'47" 42d47'41"
Annaleigh
Vir 11h59'32" -6d45'24"
Annalena
Cyg 19h41'42" 29d2'5"
Annalena Charlott
Ori 6h14'51" 16d24'24"
Annalene Vorster
Umi 15h9'53" 74d11'47"
Annali
Tri 2h11'33" 32d36'43"
Annalie Rose Escover
Cap 21h54'41" -22d19'15"
Annaliese Amanda Purrington
Uma 14h15'14" 55d34'30"
Annaliese Elizabeth DiMeola
And 23h21'29" 49d14'42"
Annaliese Katerene Weir
Psc 1h10'13" 21d56'38"
Annaliese Sanford
And 0h39'28" 25d41'20"
Annalisa
And 23h20'4" 48d6'30"
Annalisa
Lyr 18h25'52" 30d47'57"
Annalisa
Cyg 19h48'56" 33d12'41"
Anna-Lisa
Cas 0h21'17" 61d15'56"
Annalisa
Uma 10h16'57" 56d25'28"
Annalisa
Umi 16h17'24" 85d7'15"
Annalisa Bianchi Oster
Psc 0h21'6" 15d15'55"
Annalisa Crana
And 1h39'2" 50d8'49"
Annalisa D' Acquisto
Crb 16h20'28" 38d27'9"
Annalisa Lustica 12/28/1980
Cap 20h55'28" -21d17'51"
Annalisa Rose Hertzler
Vir 13h0'36" 3d16'17"
Annalisa Swan Hinckley Jenkins
Ari 3h9'1" 28d49'41"

Annalise Davis Campel
Lyn 7h12'45" 53d30'14"
Annalise Elena Yap
Sco 17h48'32" -38d17'29"
Annalise Grace
Cap 20h19'34" -23d6'59"
Annalise Marie
And 1h27'19" 47d40'41"
Annalise Mikayla D'Auria
Lib 14h39'5" -13d1'35"
annalisebenedict040405
Tau 5h30'1" 22d32'2"
Annaliza Baldemor
Umi 11h56'58" 87d18'38"
Annaly Howland Lakey
Cyg 19h41'29" 35d14'41"
Annalyse
Cru 12h42'30" -56d45'46"
Annamae
And 22h58'43" 47d12'45"
Annamaria
Lac 22h47'30" 50d26'31"
Annamaria
Lac 21h58'21" 37d32'25"
Annamaria
And 1h40'8" 43d12'25"
Anna-Maria
Ori 6h17'57" 13d13'51"
Annamaria
Cam 7h49'13" 77d38'25"
Annamaria*
Cap 21h49'26" -18d8'27"
Anna-Maria
Uma 9h37'12" 54d13'22"
Annamaria Cimorelli
And 2h6'8" 46d13'13"
AnnaMaria Clara Arostegui
Psc 1h13'52" 11d46'21"
Annamaria detta Barbi
Ori 5h58'41" 21d53'39"
Annamaria Di Cicco 28 luglio 1943
Cyg 21h32'2" 34d0'8"
AnnaMaria Julita Sally Barrett Fox
And 23h18'16" 42d24'36"
Annamaria Tracey Reeves
Cas 23h37'52" 55d36'31"
Annamarie
Cyg 22h1'49" 50d1'23"
Annamarie
Cnc 9h21'55" 29d1'23"
Anna-Marie Azaria Nash
Ori 6h7'28" 17d5'31"
Anna-Marie Bardillon
Tau 4h25'25" 22d35'8"
Annamarie Christina Hart
Tau 5h32'19" 22d50'52"
Annamarie Eisen
Gem 7h24'9" 34d6'33"
Annamarie Grace
And 23h27'49" 46d58'53"
Annamarie Musarra
Uma 11h11'50" 49d39'51"
Annamarie Nicole
Cyg 20h36'0" 47d29'31"
Annamarie Nosti
Lib 15h26'18" -21d27'54"
Anna-Marie Tancock 17-5-1976
And 2h30'30" 39d1'57"
Anna-Mimosa
Ori 5h50'45" 5d50'5"
Annangel
Cnc 8h16'42" 20d36'7"
Annanne
Tau 5h5'6" 22d20'31"
Annaratone Paola
Lyn 7h56'51" 42d19'18"
Annare
Lyn 7h34'25" 43d51'56"
ANNARIS
And 2h35'40" 49d15'3"
Anna-Rita Dahl Perkins
Cap 20h53'53" -22d22'17"
Annarita e Virgilio
Her 16h40'24" 6d33'19"
Annarose Lorber Klinger
Lib 15h19'33" -25d7'53"
Anna's Dream Star
Sco 17h10'23" -35d6'20"
Anna's Front Porch Light
Uma 11h7'44" 31d0'4"
Anna's Light
And 0h32'49" 43d5'43"
Anna's Light
Cnc 9h18'13" 12d40'11"
Anna's Light
Lib 15h13'36" -14d47'10"
Anna's Love
Sco 16h13'25" -11d17'28"
Anna's Place In The Heavens
Sco 16h9'17" -18d58'34"
Anna's Rocketship
Tau 5h14'8" 21d59'33"
Anna's Shining Girl Scout Star
And 0h48'38" 44d27'46"
Anna's Star
And 1h20'6" 38d26'19"
Anna's Star
And 0h7'48" 39d54'47"
Anna's Star
Gem 7h7'52" 34d59'38"
Anna's Star
And 0h0'28" 35d30'39"
Anna's Star
Crb 15h54'7" 37d50'40"
Anna's Star
Uma 8h53'44" 64d26'38"
Anastacia Lundine
Ori 5h5'31" 15d59'28"
Annastasia
Sco 16h3'3" -16d20'56"
Annastasia
Aqr 22h33'29" -5d12'40"
Annastasia Tink
Psc 1h40'25" 5d4'16"
Annastazia Sachar
And 1h18'28" 36d46'4"
annatopia
Lyn 7h18'10" 48d7'2"
Annavel Leyva
Psc 1h20'1" 24d18'23"
AnnaZielinska
Uma 10h12'51" 55d32'40"
Ann-Britt Alpha 3
Ari 2h55'22" 22d7'33"
AnnChris
Cyg 19h50'23" 36d27'35"
Anne
And 23h46'35" 34d35'42"
Anne
And 23h42'56" 34d33'35"
Anne
And 2h13'11" 41d56'32"
Anne
And 0h52'49" 37d27'7"
Anne
Uma 11h31'9" 34d47'21"
Anne
Lyn 7h2'14" 47d12'50"
Anne
And 1h33'16" 46d58'49"
Anne
And 0h6'22" 45d47'14"
Anne
Boo 14h27'42" 49d15'9"
Anne
Uma 9h27'4" 47d33'32"
Anne
Vir 14h22'4" 1d13'42"
Anne
Ori 5h54'28" 18d41'27"
ANNE
Gem 7h15'5" 19d50'58"
Anne
Uma 8h22'17" 66d57'43"
Anne
Lib 14h50'46" -1d55'7"
Anne Adler Whitehouse
Eri 4h36'16" -0d43'35"
Anne Alemian
Lib 15h33'17" -16d37'11"
Anne Alinda Kearney
Cas 0h20'59" 62d4'22"
Anne Aliperti
Cas 0h49'21" 56d18'1"
Anne and David Kiddy
Cyg 20h53'0" 36d2'24"
Anne and Harvey Hamff
Cnc 8h31'1" 28d25'37"
Anne and Hilda's Star
Dra 18h52'51" 73d23'34"
Anne and Michael Miller
Aqr 22h23'52" 1d45'13"
Anne and Roy Haugh
Ori 6h4'9" 21d21'26"
Anne Anderson
Psc 0h52'44" 32d19'40"
Anne *Angel* Duggan
Vir 12h52'57" 8d45'26"
Anne Aplington
Cap 21h10'58" -25d6'28"
Anne Arhio
Tau 5h8'52" 19d45'44"
Anne Beegle
Dra 18h34'26" 56d11'34"
Anne Blank
Cas 23h43'19" 54d53'9"
Anne & Bob Peper
Cyg 20h19'5" 54d44'3"
Anne Bowers Smith
Cas 1h10'7" 55d12'21"
Anne Boyer OREOS
Aqr 21h44'3" -0d38'5"
Anne Brophy
Sco 17h18'29" -32d40'46"
Anne C Leonardis
Cas 1h39'23" 62d38'19"
Anne C. Mason
Del 20h45'50" 9d15'4"
Anne Caitlin Gaske
Cap 20h14'46" -9d24'42"
Anne Campanaro
Gem 6h2'1" 24d16'3"
Anne Catherine
Dra 15h23'46" 56d9'7"
Anne Catherine Codd
And 2h23'19" 39d26'13"
Anne Catherine Pennings-Beasley
Ari 1h57'4" 19d22'39"
Anne Cathrine Kanoelehua Rust
Sco 16h28'45" -26d8'1"
Anne Chauvet
Hya 8h29'45" 6d28'23"
Anne Cherry
Cru 12h5'47" -63d22'7"
Anne Christine Jorgensen
And 0h22'56" 32d33'5"
Anne Christine Simmons
Lyn 8h9'29" 45d33'39"
Anne Clarke
Uma 11h25'55" 33d51'17"
Anne Colleen Frei
And 2h29'51" 38d49'26"
Anne Cose
Lib 15h54'5" -10d50'47"
Anne Cottam
Cen 11h41'28" -52d9'30"
Anne D. Pearsall
Ori 6h19'13" 7d26'16"
Anne DeVito Ladd
Ori 6h5'16" 19d48'36"
Anne Didie
And 0h23'18" 36d15'35"
Anne Dimaio
Uma 10h16'14" 62d58'36"
Anne Divers
Equ 21h6'35" 4d11'58"
Anne Dooley
Eri 4h33'20" -20d13'5"
Anne (Dornan) Johnston
Cas 23h37'57" 57d40'43"
Anne Drinnan
Cas 0h51'1" 53d27'47"
Anne Duffy
Ari 2h33'12" 18d48'41"
Anne Duggan
Cas 23h1'47" 59d2'35"
Anne E. Barrett
Uma 10h55'32" 46d6'5"
Anne E. Haislip
Umi 9h46'42" 88d46'33"
Anne E Quilligan
Cap 21h54'30" -22d50'20"
Anne Elizabeth
Sco 17h23'29" -42d43'39"
Anne Elizabeth
Aqr 23h21'42" -12d31'57"
Anne Elizabeth
Leo 10h9'32" 22d28'40"
Anne Elizabeth Costello
Lyn 9h38'59" 41d14'6"
Anne Elizabeth Foster
Mon 6h48'54" -0d52'23"
Anne Elizabeth Fry
Aur 7h18'6" 41d24'17"
Anne Elizabeth Heer-Hale
Lyr 18h45'1" 40d54'59"
Anne Elizabeth Mullen
Uma 13h43'56" 51d39'5"
Anne Elizabeth Pacific
Cas 23h20'50" 56d1'1"
Anne Elizabeth Underwood
Sco 16h11'15" -14d51'57"
Anne Elizabeth Wachenfeld
And 23h50'0" 35d22'23"
Anne et Bertrand
Uma 13h58'49" 56d41'16"
Anne et Olivier
Aqr 23h6'1" -17d23'11"
Anne et Pascal Picard
Uma 12h19'53" 52d19'26"
Anne F. Palumbo
Cas 23h38'52" 51d8'47"
Anne Faherty
Gem 7h26'45" 17d26'13"
Anne Faucher 05-05-1994
Peg 23h26'9" 23d26'39"
Anne Faure - Wartel
Uma 13h37'44" 49d16'36"
Anne Felten
Uma 9h29'24" 47d48'0"
Anne (Firebug) Bomkamp
Cas 23h58'5" 57d54'0"
Anne Fleischman
Cas 0h39'54" 57d26'25"
Anne Florence Westhorpe
Tau 5h42'48" 24d51'36"
Anne Frank
Cnc 8h29'8" 14d57'16"
Anne G. Tanis
Cyg 21h53'48" 49d35'49"
Anne Gainsborough
And 23h19'54" 42d2'34"
Anne Garkani
Lyn 7h42'22" 38d32'0"
Anne Garrison
Crb 15h51'47" 26d1'25"
Anne & George Gennings
Uma 9h6'42" 69d43'57"
Anne Gerulat
Uma 11h30'3" 63d11'33"
Anne H. Holt
Uma 14h5'7" 55d56'7"
Anne Halligan
Lib 15h19'24" -7d25'56"
Anne "Hayes" Stringer
Tau 4h7'17" 15d14'3"
Anne Henderson
And 0h43'25" 45d38'35"
Anne Henning
Peg 23h45'9" 24d10'41"
Anne Herman
And 0h13'37" 27d27'18"
Anne Holland
Cnv 13h47'15" 31d18'43"
Anne Holland
Cas 23h2'1" 54d23'40"
Anne Howard
Ari 3h7'13" 28d13'19"
Anne Hunter
Cap 21h22'18" -24d38'25"
Anne Ian
Uma 12h29'31" 61d27'4"
Anne Irene
Cra 18h21'38" -37d12'25"
Anne Jackson Schultz
Uma 11h38'10" 58d5'32"
Anne Kate Wenzel
Cyg 21h31'23" 33d29'18"
Anne Katherine Borghese
Aqr 22h23'40" -16d40'8"
Anne Katherine Waslin
Sgr 18h5'19" -31d44'14"
Anne Kennell
Sgr 19h23'51" -14d18'50"
Anne Kenyon Cook
And 0h58'10" 34d58'49"
Anne Keyser Peppers
Sco 17h8'10" -42d53'1"
Anne Kieffer
Cas 0h4'52" 54d51'35"
Anne Kosoglow
Ari 2h4'21" 24d26'29"
Anne Kowal
And 1h24'48" 41d38'43"
Anne Kowalesik Dugan 1915-2002
Uma 11h43'16" 33d24'7"
Anne Kvedaras
Per 3h31'40" 47d11'52"
Anne L. Faber
Cnc 8h50'39" 19d40'9"
Anne L Mattie
Uma 11h4'5" 38d56'13"
Anne L. Pomije and Joseph M. Pomije
Mon 6h52'43" -5d38'37"
Anne Labbe
Tau 5h44'57" 21d59'49"
Anne Lamont Heldreich Kukea
Lib 14h49'42" -1d3'29"
Anne - Laure
Cas 23h12'48" 55d10'7"
Anne Laure Mancy
Umi 15h28'46" 82d3'39"
Anne Leffler
Per 4h25'49" 34d47'16"
Anne Leighton "Minxy"
Ari 2h13'14" 22d29'18"
Anne Lerch Rondepierre
Psc 1h13'53" 26d42'4"
Anne Leslie Pallansch
And 23h18'8" 47d30'3"
Anne Lindsey Armbruster
Tau 4h11'56" 10d34'31"
Anne Louise
Aqr 20h40'55" 1d32'34"
Anne Louise Dodt
Lyn 8h1'59" 58d23'1"
Anne Lovejoy
Lyn 7h50'23" 48d9'30"
Anne L.Tate-MacMichael
Lib 15h37'7" -27d13'47"
Anne Lynn DeStasio
Uma 9h29'33" 69d51'50"
Anne Lyon
Cyg 21h14'14" 43d9'42"
Anne M. Christie
Ari 3h9'45" 27d6'2"
Anne Madden Baranski
Vir 13h22'13" 13d22'21"
Anne Madeline Pasquarello Desmond
Ori 6h8'34" 15d19'28"
Anne Magoo Trifiletti
Per 2h50'16" 44d41'15"
Anne Malashevitz
Uma 11h29'21" 46d55'14"
Anne Margaret Bartsch
Tau 4h57'14" 16d10'9"
Anne Margaret LaMont
Lyn 6h28'32" 60d0'50"
Anne Marie
Uma 9h25'58" 68d45'13"
Anne Marie
Cap 21h47'58" -9d14'3"
Anne Marie
Ari 2h48'56" 28d51'52"
Anne Marie
And 0h29'21" 46d21'39"
Anne Marie
And 2h5'9" 39d50'26"
Anne Marie
And 1h14'7" 37d58'32"
Anne Marie and Tim
Cru 12h42'13" -56d43'1"
Anne Marie Balligan
Vir 13h20'22" 11d6'14"
Anne Marie Bathon
Uma 13h13'13" 56d57'14"
Anne Marie Chaille
Mon 6h58'18" 9d38'18"
Anne Marie Dooling
Cas 0h52'49" 64d2'45"
Anne Marie Duffy
Cnc 8h30'29" 16d18'36"
Anne Marie Durfee
Gem 7h22'2" 32d1'21"
Anne Marie Filice
Cap 21h1'34" -22d15'27"
Anne Marie Flynn
And 1h36'13" 43d46'14"
Anne Marie Friedrich
Dra 20h20'5" 70d44'42"
Anne Marie Geswender
Lyr 18h53'0" 32d16'22"
Anne Marie Horn
Ari 2h25'15" 10d49'55"
Anne Marie & James Armstrong
Cyg 19h50'20" 52d47'22"
Anne Marie Johnson
Sgr 17h49'51" -16d39'58"
Anne Marie Ketterle
Umi 14h21'8" 77d24'45"
Anne Marie Kippes
Sgr 19h15'0" -15d51'53"
Anne Marie LaBelle
Cas 23h6'5" 55d54'32"
Anne Marie Loiselle
Ari 2h10'2" 24d11'24"
Anne Marie Otaola
Gem 7h20'22" 33d21'28"
Anne Marie Paul
Sgr 18h58'54" -31d56'47"
Anne Marie Rizzi
Crb 15h58'41" 37d54'7"
Anne Marie Rodgers
Leo 9h28'16" 28d1'48"
Anne Marie Rowse
And 23h14'18" 42d34'3"
Anne Marie Sedney
Psc 0h24'38" -1d31'28"
Anne Marie Shipe
Ari 2h18'18" 26d27'18"
Anne Marie Thibert
Cap 20h32'15" -10d15'23"
Anne Marie Valliere
Uma 10h15'35" 45d59'1"
Anne Marie Vieira-Jagoe
Cyg 20h12'34" 30d11'50"
Anne Marie Weidner
Vir 12h33'16" -0d26'48"
Anne Marie's Palace
Ari 2h20'43" 24d0'50"
Anne Martha Rowe
Aqr 21h18'39" -9d45'56"
Anne Mary Syvulick & Lichacz Girls
Cas 2h22'20" 72d31'7"
Anne Maureen Ashworth
Ori 5h39'59" -0d57'1"
Anne Maureen Baker
Lyn 7h6'55" 60d33'32"
Anne Mayr
Ori 6h19'26" 8d35'13"
Anne McGonagle
Lyn 6h59'54" 48d55'54"
Anne McQueen
Cas 1h13'46" 53d24'42"
Anne Merete og Thomas for evig
Cas 1h36'47" 65d57'18"
Anne Michelle
Tau 5h27'5" 22d13'59"
Anne Michelle Suskind
And 2h37'59" 44d29'41"
Anne & Mike Dosier
Cyg 20h6'20" 55d54'50"
Anne-(mom)-you-(R)-our-Sunshine
Uma 8h25'48" 62d9'39"
Anne Montillo
Her 17h17'10" 46d29'50"
Anne Morgan
And 23h51'50" 46d54'49"
Anne Morton Kimberly
Lyn 8h41'16" 33d38'39"
Anne Murray
And 1h30'6" 50d25'2"
Anne "Nan" Burnham
And 0h48'11" 33d21'41"
Anne Nguyen
Psc 0h46'31" 17d30'11"
Anne Noel
Cir 14h3'13" -66d32'10"
Anne Nunn
Cas 1h0'4" 66d32'16"
Ann-e O'Brien
Gem 7h1'19" 22d15'17"
Anne P. Grant, Light of Ryan's Life
Ori 4h47'44" 11d14'8"
Anne P Sperduto
Uma 9h38'31" 62d13'33"
Anne Pauline
Tau 3h42'56" 10d44'39"
Anne "Precious Angel" Oppermann
Uma 10h1'51" 45d58'16"
Anne Princess
Sgr 19h9'47" -29d43'39"
Anne Priscille Francesca
Cnc 8h6'51" 25d45'25"
Anne Radcliffe
Cyg 20h59'39" 33d4'27"
Anne Regouby
Vir 13h1'51" -4d30'2"
Anne Reith
Lyn 7h34'38" 49d20'2"
Anne & Rex Greenaway
Cyg 21h57'16" 53d32'47"
Anne Rittgers
Ari 2h5'36" 25d20'7"
Anne Roberge de Trettien
Lib 14h25'46" -22d41'31"
Anne & Robin's 50th Anniversary
Gem 7h17'57" 30d54'7"
Anne Rochefort
Cas 23h45'14" 55d52'43"
Anne Ronchetto
Uma 8h57'41" 72d54'52"
Anne Rumrill Culver
Leo 11h5'30" 5d39'15"
Anne Savary
Per 4h27'11" 41d42'57"
Anne Scatto
And 0h29'34" 24d32'41"
Anne Serena Gottschalk
Cap 20h16'56" -10d43'48"
Anne Sheppard
Cas 23h42'54" 51d5'24"
Anne Shorter
Cas 1h39'12" 60d50'37"
Anne Skovgaard
And 0h19'58" 41d2'5"
Anne Smith
Gem 7h6'59" 32d29'31"
Anne Snowdon
Cas 1h37'53" 61d41'41"
Anne Spencer
And 0h52'5" 38d13'39"
Anne Squillante
Uma 10h38'42" 71d6'5"
Anne Stacey Garry Ross Foulkes
Ori 5h11'40" 8d12'9"
Anne Starkey Gordon
And 0h28'28" 38d36'57"
Anne Stucki
Cam 4h3'2" 53d39'14"
Anne Szilva
Del 20h31'31" 20d1'9"
Anne Taylor, Martin Taylor
Sge 19h50'50" 17d42'8"
Anne Theresa
Aqr 20h57'57" -7d39'27"
Anne Thomas
Cas 23h11'37" 54d25'21"
Anne Thomas
Cas 0h15'16" 55d18'35"
Anne Van Tassell
Leo 10h22'43" 26d41'44"
Anne Veronese
And 2h7'56" 41d15'56"
Anne Victoria Lugo-Everlasting Love
And 0h16'19" 28d17'31"
Anne Virginia Lang
Cyg 19h36'1" 31d40'21"
Anne Vrazo Mikel
Srp 18h22'43" -0d8'11"
Anne Waller
Pav 19h16'4" -57d49'54"
Anne Walters
Tau 4h15'45" 26d40'16"
Anne Ward and Darren Doyle
Cyg 19h55'7" 38d14'40"
Anne Watson
Cas 1h12'50" 64d20'20"
Anne Webb
Sco 17h53'43" -35d39'9"
Anne Wert
Ori 6h3'9" 20d37'36"
Anne Williams Chase
Cnc 8h10'17" 16d49'50"
Anne Wurzburger
Lib 15h30'19" -12d30'0"
AnneAlexBoris
Uma 11h14'23" 53d5'38"
Anne-Aurora
Lib 15h1'30" -20d55'47"
Anne-Brigitte & Markus, 06.06.2003
Boo 13h40'16" 21d52'1"
Anne-Caroline Huguenin Walther
Uma 12h9'16" 59d19'8"
Anne-Cathérine
Cnv 14h0'23" 35d31'16"
Anne-Charlotte Drouin
Lib 14h44'21" -17d54'3"
Annechka
Apu 15h30'42" -74d21'47"
Anne-Christine Caputo/Laurent [Fey]
And 23h7'42" 39d57'5"
Annecim
Lyn 7h49'53" 35d32'23"
AnnEdBut
Cyg 21h36'40" 45d40'34"
Annee Stokes
And 1h37'25" 47d35'16"
Anneen Kwe
Uma 11h24'4" 31d44'7"
AnneGelic Grace
Mon 6h45'3" -0d6'23"
Annegret
Crb 16h7'35" 36d2'42"
Annegret
Aur 5h15'29" 42d48'51"
Annegret Bormann
Vir 11h39'42" -4d51'20"
Annegret Wagner
Uma 8h12'52" 67d0'18"
Annehooch
And 1h48'56" 42d31'23"
Anneka Rose
And 23h44'1" 38d3'29"
Anneka's Star
Ori 5h11'54" -6d13'54"
Anneke
And 23h3'8" 40d47'10"
Anneke Blauaugenschatz
Uma 11h4'48" 48d55'21"
Anneke Helen Braam
Uma 9h27'0" 57d1'51"
Anneke, Mother to Madeline 09/18/05
Vir 12h54'33" -9d7'45"
Annekeb
Lyn 8h17'15" 56d1'35"
Anneli
Ori 5h44'22" 3d2'23"
Anne-Lie
Aqr 21h41'19" 2d12'21"
Annelie Damasc
Com 12h23'20" 29d10'24"
Annelie Metrakos
Peg 23h1'52" 28d52'35"
Annelies
And 2h14'43" 45d23'7"
Annelies Malia
Uma 9h50'52" 55d26'8"
Anneliesa
Leo 9h24'7" 11d16'22"
Anneliese
And 0h42'50" 39d10'53"
Anneliese
And 0h18'24" 42d8'21"
Anneliese
Uma 9h37'2" 42d28'32"
Anneliese
Uma 9h24'4" 62d41'46"
Anneliese Bilidas
Lib 15h17'26" -5d52'45"
Anneliese & Helene
Uma 9h55'25" 57d4'31"
Anneliese Hollis Maybach
Ori 5h11'31" 12d40'49"
Anneliese Kusek
Ari 2h43'39" 28d49'12"
Anneliese Laval
Cas 0h18'56" 56d47'48"
Anneliese Regan
And 0h18'46" 30d17'37"
Anneliese Schwarzbach
Uma 11h9'26" 45d29'46"
Annelise Gabrielle Klenk
Lib 14h48'52" -10d41'38"
Annelise Pico
Cmi 8h4'40" -0d2'55"
Annelore Kindler Strauss
Uma 11h18'15" 41d17'14"
Anne-Maree Mitchell
Ari 3h1'47" 26d50'58"
Anne-Marie
Gem 7h45'45" 32d48'16"
Annemarie
Cas 2h41'43" 57d51'52"
Annemarie
Ori 5h28'19" -5d22'47"
Annemarie
Cas 23h18'12" 58d33'56"
Annemarie
Umi 13h33'46" 72d36'21"
Annemarie Caesar
Cas 0h0'7" 53d31'16"
Annemarie Cardamone
Leo 11h16'15" 18d48'7"
Anne-Marie Jahlan
Sgr 18h31'55" -19d10'34"
Anne-Marie Kuttruff
Leo 11h54'45" 25d48'14"
AnneMarie Leigh Phipps
Ori 6h12'2" 5d41'1"
Annemarie "Maus" Baum
Uma 8h14'25" 66d10'54"
Annemarie McRedmond
Lib 14h49'48" -0d58'10"
Anne-Marie/Michele
Ari 3h12'55" 16d20'28"
AnneMarie Morley
Aql 19h52'40" -0d58'46"
Annemarie Nomikos
And 23h22'57" 45d44'30"
Anne-Marie Palynyczak
Uma 10h22'27" 69d36'47"
Anne-Marie Rainbow
And 1h9'44" 42d58'21"
Anne-Marie Reynes
Vir 11h44'49" -1d52'23"
Annemarie - Sparkles Like Your Eyes
And 23h47'8" 35d18'7"
Anne-Marie Stack
Dra 15h16'23" 56d8'52"
Annemarie Stiefvater
Uma 9h58'37" 58d44'7"
Annemarie und Mario Matschl
And 2h37'21" 50d25'56"
Anne-Mette Schou Schadegg
Cas 0h33'38" 61d41'5"
ANNEMEUS-STELLARIS-ANNIMAE III-IV-I
Tau 5h56'3" 26d39'24"
Annemieke Carolien Steen
And 23h34'3" 46d16'18"
ANNEPIE
Vir 13h45'44" -16d28'48"

Anneris0821
Leo 11h56'5" 22d20'13"
Anne's Birthday Star
Sco 16h13'14" -13d25'47"
Annes Do
Lib 15h16'37" -8d46'25"
Anne's Eternal Light
Leo 10h33'3" 8d58'42"
Anne's Gleaming Grace
Lyr 18h33'29" 37d3'56"
Anne's Light
Gem 7h46'3" 31d56'58"
Anne's Magic
Cyg 20h34'17" 35d45'11"
Anne's pEarl
Vir 14h15'9" -3d4'49"
Anne's Shining Light
Cnc 8h48'30" 14d49'9"
Anne's Star
Crv 12h28'14" -14d15'12"
Anne's Star Of Strength
Cas 23h5'47" 59d6'36"
AnneSegStar #1
And 0h16'27" 25d47'29"
Anne-Sophie
Umi 19h32'24" 86d56'32"
Anne-Sophie et Meindert
Col 5h10'55" -28d18'28"
Anne-Sophie Perraud
Ori 5h56'34" 16d53'15"
Annestar
Del 20h34'31" 20d6'40"
Annetate
Psc 0h47'32" 11d13'9"
Anne-Thérèse
Sgr 19h6'39" -22d53'26"
Annett Kleinfeld
And 1h28'17" 45d27'15"
Annett Pohl
Uma 8h39'48" 54d47'15"
Annetta
Ari 2h49'41" 21d19'13"
Annetta Jo (Vaughn) Bannister
Cas 1h47'22" 57d45'9"
Annetta "MiMi" Raye Rinke Baker
Psc 23h25'12" 5d47'16"
Annetta Yeager
Psc 0h17'43" 16d57'17"
Annette
Ori 5h13'8" 11d32'53"
Annette
Ori 5h46'58" 7d2'2"
Annette
Cnc 8h31'41" 24d20'49"
Annette
Uma 11h33'48" 49d31'31"
Annette
And 0h56'19" 37d30'50"
Annette
And 0h58'54" 43d44'45"
Annette
And 0h46'13" 30d48'7"
Annette
Psc 0h55'1" 31d37'9"
Annette
Uma 11h15'13" 35d31'28"
Annette
Lyn 6h42'50" 53d58'7"
Annette
Cas 23h45'39" 53d29'26"
Annette
Cam 4h23'40" 73d18'38"
Annette
Lyn 6h40'15" 60d37'13"
Annette
Aqr 22h36'17" -1d30'35"
Annette
Aqr 22h26'10" -5d7'39"
Annette
Cap 20h30'5" -12d14'10"
Annette Adams-Dzierzba
Tau 5h8'9" 18d10'57"
Annette and Steve
Cyg 20h28'31" 35d39'20"
Annette Bailleux
Ori 6h2'26" 11d24'26"
Annette Barrett
Lmi 9h45'7" 35d33'7"
Annette Beijersbergen
Leo 10h32'28" 27d11'0"
Annette Beker
Uma 9h48'34" 49d14'21"
Annette Brooks
Tau 5h51'52" 17d54'23"
Annette Brown
Cnc 8h43'9" 29d21'49"
Annette "Bunny" Conn
Vir 12h41'58" 11d55'16"
Annette Catelinet
Lmi 10h18'40" 35d44'50"
Annette Cemeona DiToma
Ori 5h37'59" -1d38'22"
Annette Chipman
Ari 3h13'15" 27d17'22"
Annette Clayton Scott
And 0h10'53" 39d34'44"
Annette Crystal Herrera
Ari 3h4'45" 14d46'18"
Annette Denise
Lyn 8h20'27" 37d22'50"
Annette DiBona Hamilton
Cet 0h22'53" -14d32'3"

Annette Dreyfus
Aql 19h17'20" -0d34'36"
Annette E. Waters
Dra 17h23'13" 59d48'36"
Annette - Eine wundervolle Liebe
Uma 11h37'24" 39d40'36"
Annette Elizabeth Cockwell
And 2h37'1" 46d39'15"
Annette Elizabeth Martin Quinn
And 0h6'32" 44d14'47"
Annette Frances Girton Fahnestock
And 23h16'55" 41d42'10"
Annette Giersch
Sco 16h48'50" -43d14'26"
Annette Graf
Uma 10h52'29" 59d40'17"
Annette Guzman
Aqr 22h25'26" -6d57'46"
Annette Hallmark
Cyg 21h37'15" 41d20'42"
Annette Howe
Gem 6h31'53" 24d16'38"
Annette Joy Andersen
Uma 11h52'44" 60d33'49"
Annette Kaminska
Umi 17h37'1" 81d12'8"
Annette & Keith Our Star Forever
Uma 11h26'21" 40d25'56"
Annette L. Luckett
Cnc 8h13'4" 12d49'7"
Annette Leiomi Chalker
Cyg 21h5'46" 36d33'25"
Annette Linkous
Psc 23h19'35" 6d25'35"
Annette Lucence
Cas 23h36'18" 54d30'56"
Annette MacGauley
Vir 11h42'50" -0d19'14"
Annette Malpas
Del 20h38'4" 14d42'34"
Annette Manning Bass
Uma 11h58'23" 37d51'36"
Annette Margaret Nesbitt
Ori 5h42'58" -1d13'16"
Annette Marie Leaf
Del 20h25'39" 5d10'5"
Annette Massman
Vul 19h49'4" 22d26'37"
Annette Nickens
Vul 19h54'35" 22d18'16"
Annette Pape' Walter
And 2h19'27" 49d31'9"
Annette Paun
Uma 10h57'16" 66d0'33"
Annette Quiring
Sco 16h13'39" -38d50'56"
Annette R. Sulzman
Aqr 22h50'18" -0d57'18"
Annette Rene
Cyg 21h44'27" 38d28'48"
Annette Roberts - "Appearing Nightly"
And 23h17'39" 47d25'3"
Annette Saylor
Leo 9h44'9" 23d39'24"
Annette Seagraves
Dra 17h37'52" 53d45'33"
Annette Sidor
And 0h58'53" 43d38'45"
Annette Stucki
Her 17h0'29" 14d31'21"
Annette Sue Elizalde
Leo 10h39'31" 13d5'24"
Annette Sweeney
Vir 11h41'58" -3d18'56"
Annette Synder
Lyr 18h41'11" 34d47'57"
Annette Tauber
Mon 7h18'35" -5d16'1"
Annette Theuerkorn
Leo 11h39'8" 24d14'51"
Annette Trujillo
Uma 11h25'52" 68d27'5"
Annette Valenti Quarrier
And 0h46'38" 42d2'16"
Annette Vielleux
Uma 11h25'52" 41d28'9"
Annette Yvonne
Peg 22h40'30" 3d39'43"
Annette's Amazing Mountian Movers!
Umi 14h1'23" 77d19'59"
Annette's Diamond in the Sky
Leo 9h29'47" 30d3'47"
Annette's Light of Love
Tau 3h39'43" 11d14'7"
Annette's Lite
Vir 15h1'44" 5d30'22"
Annette's Star
Psc 0h59'35" 18d9'10"
Annette's Star
And 2h20'3" 50d15'18"
Annette's Wishing Star
Leo 11h7'41" 11d45'50"
Anney Kincaid Castetter
And 1h41'14" 42d16'34"
Anni
Lib 15h8'42" -4d16'42"
Anni
Lyn 6h47'42" 53d36'6"

Anni&Miki
Cyg 19h42'25" 35d24'7"
Annica Foxcroft
Sgr 18h48'1" -16d49'56"
Annick
Leo 10h24'29" 12d32'0"
Annick
And 23h20'15" 35d28'19"
Annick Laplante
Umi 17h18'42" 81d42'9"
ANNICK MONNERAT
Ori 5h56'4" 3d11'52"
Annie
Ori 6h14'40" 21d9'1"
Annie
Gem 6h22'11" 22d15'32"
Annie
Leo 10h27'19" 16d51'19"
Annie
Cnc 8h19'40" 29d26'15"
Annie
Tau 4h31'55" 21d57'11"
Annie
Vir 13h21'34" 6d26'53"
Annie
Tau 4h19'44" 14d18'47"
Annie
Peg 22h27'42" 4d36'2"
Annie
And 0h32'12" 41d54'14"
Annie
And 1h52'30" 38d56'41"
Annie
Uma 11h52'54" 39d1'1"
Annie
Uma 11h2'12" 38d41'4"
Annie
Uma 10h35'36" 41d40'42"
Annie
Leo 9h30'58" 31d20'49"
Annie
And 1h45'47" 36d13'13"
Annie
And 1h21'25" 49d20'0"
Annie
Cas 0h22'14" 53d1'19"
Annie
Umi 13h25'12" 86d54'51"
Annie
Vir 13h16'50" -22d21'0"
Annie
Uma 13h1'57" 53d35'47"
Annie
Cyg 20h0'43" 59d38'49"
Annie
Cas 23h49'7" 54d35'2"
Annie
Cas 0h13'51" 64d25'16"
Annie
Uma 9h40'31" 69d48'25"
Annie
Sco 17h8'12" -30d4'37"
Annie 25
And 1h5'6" 39d14'53"
Annie 5
Cap 20h17'10" -19d29'4"
Annie Amiant
Sco 17h50'15" -38d31'3"
Annie Amodei
Sco 17h20'19" -42d49'25"
Annie and Matt
Umi 15h18'7" 67d39'11"
Annie and Terry
Sex 10h44'53" -0d2'36"
Annie "Angel Boo"
Psc 1h19'0" 32d28'9"
Annie B
Ari 2h40'6" 27d43'23"
Annie Beaudet
Aqr 21h20'22" -13d28'17"
Annie Bennett
Uma 9h57'8" 50d55'28"
Annie Bernadette Noyes
Cas 23h7'51" 58d32'36"
Annie Blakely
Ori 5h21'54" -7d22'31"
Annie Bodenmiller
Lib 15h33'11" -11d4'42"
Annie Boisclair
Cas 0h7'9" 53d30'19"
Annie Bravard
Cma 6h28'33" -16d13'20"
Annie Buczacki
Cam 6h43'11" 62d30'15"
Annie "Burd" Cramer
Ari 2h32'30" 25d38'25"
Annie Cassidy
Vir 12h28'26" 11d40'50"
Annie Charlotte Gallo Tasserie
Cas 1h2'54" 62d21'21"
Annie Charron-Tenhoff
Cas 1h2'58" 63d23'45"
Annie Cook
And 2h35'55" 50d19'41"
Annie Crowe
Uma 11h23'35" 40d38'35"
Annie Cubeta
Uma 14h9'48" 61d19'0"
Annie Davis
Cru 12h40'44" -56d44'42"
Annie E Campbell
Leo 9h50'26" 29d40'33"
Annie Fox
Col 5h59'42" -34d59'20"
Annie Fox
Sgr 19h51'41" -28d37'3"

Annie Frances Wall Kilian
Cas 23h49'57" 53d21'37"
Annie Garza
Gem 7h40'7" 22d45'30"
Annie Gertrude McClure
Uma 8h51'43" 50d37'29"
Annie Grace Giles Varnell
Lib 15h12'41" -19d36'22"
Annie & Harout-Love, Roupen & Roubina
Ori 5h56'4" 3d11'52"
Annie Ho
Leo 11h33'7" 25d25'14"
Annie Hoke
Gem 6h47'20" 28d26'11"
Annie J
Tau 5h37'13" 26d28'15"
Annie Jane Thompson
And 1h9'47" 46d2'2"
Annie & Jimmy
Ant 9h37'49" -27d17'28"
Annie & Jose
Cyg 20h37'14" 39d58'40"
Annie Joseph
Lyn 9h10'9" 33d15'51"
Annie & Josh
Sco 16h8'58" -9d30'23"
Annie & June
Leo 11h8'20" 21d1'28"
Annie Kearby
Sgr 18h51'22" -32d35'33"
Annie Kelley
Cnc 8h21'39" 15d21'5"
Annie Kellogg
And 2h25'21" 49d3'24"
Annie Klosowicz
Cap 21h37'22" -10d43'15"
Annie Landrum
Ori 5h47'2" 5d53'13"
Annie Lane
And 0h19'0" 38d59'41"
Annie Lara Blehm
Aqr 22h42'18" -2d16'19"
Annie Laughlin
Vir 13h30'24" -2d54'16"
Annie Laurie Bell
And 23h21'7" 41d35'5"
Annie Leake McCabe
Sgr 18h22'49" -23d8'12"
Annie Lee
Peg 22h31'44" 33d28'49"
Annie Lee
Aqr 22h42'48" 0d12'40"
Annie Lee
Cnc 8h44'4" 23d58'1"
Annie Lindskog
Mon 6h26'3" -5d16'58"
Annie Liu
Cyg 19h45'0" 41d16'6"
Annie Locke Scherer
Sco 16h50'29" -27d7'33"
Annie Louise17
And 2h22'8" 48d23'50"
Annie Lundin
Cas 0h50'38" 57d33'46"
Annie Mae Baker
Cas 1h31'12" 63d54'16"
Annie Mae Miller
Cap 20h29'52" -14d51'10"
Annie Mahon
And 23h5'39" 48d16'20"
Annie Malone
Leo 11h54'16" 18d55'48"
Annie Mangan
Vir 13h11'41" 6d31'55"
Annie Mapes
Tau 4h16'34" 8d12'44"
Annie Margaret Kriney
Uma 12h8'12" 47d11'41"
annie marie
Uma 10h10'51" 66d34'37"
Annie Maude Harmon
Cnc 9h12'50" 32d5'36"
Annie Meidl & Adam Radloff
Lyr 18h32'41" 36d58'43"
Annie Michele
And 23h19'48" 42d41'34"
Annie Michele Sims
Her 17h57'13" 24d4'58"
Annie Mini Miney
And 1h39'48" 36d19'34"
Annie Moskofian
Aql 19h50'18" 11d28'4"
Annie Mowlds
And 1h3'17" 45d41'20"
Annie My Red Rose Angel Star
Cap 21h49'12" -14d57'4"
Annie Nguyen, The Princess
And 2h15'43" 49d21'20"
Annie O
And 1h16'42" 38d46'18"
Annie O
Dra 19h20'16" 60d31'40"
Annie O. Wray
Cas 1h18'43" 62d46'54"
Annie Oakley Shooting Star
Crb 15h31'15" 29d58'22"
Annie O'Connell
Lyn 7h33'56" 57d8'42"
Annie Owens
And 1h12'1" 37d38'29"

Annie Pelletier
Cnc 9h13'23" 32d14'20"
Annie Pinto
Sgr 19h31'5" -40d39'13"
Annie Pizzi
Sgr 18h46'30" -27d44'16"
Annie Poo
Aqr 21h15'45" -8d2'59"
Annie Poo
Aqr 21h49'10" -0d15'8"
Annie Prendergast - O' Brien
Cnc 8h20'52" 13d2'26"
Annie Prewitt
Vir 13h25'14" 5d15'19"
Annie Rae Martorana
And 1h14'31" 37d43'56"
Annie Rose Cheatwood Pisch
And 2h10'53" 43d24'55"
Annie Rose - Sweet Sixteen
Cru 12h52'33" -59d50'26"
Annie & Roy Lambert - Shine Forever
Cru 12h22'19" -58d4'4"
Annie Ruth Davidson
Psc 1h20'0" 11d3'7"
Annie Ruth Fritz
Sco 16h55'44" -41d53'54"
Annie Ryan
Aqr 21h37'45" -0d16'8"
Annie Sarah Wills Miller
Peg 21h55'45" 12d26'31"
Annie Soluri
Cma 6h30'49" -24d12'46"
Annie Stoyanov
Ari 2h35'9" 30d49'26"
Annie Syed
Cas 0h5'0" 53d0'3"
Annie & Tadhg
Uma 9h44'10" 55d15'35"
Annie the Bucket
Ori 5h41'55" 0d25'38"
Annie Tsui
Ari 3h16'45" 28d8'59"
Annie Vericker
Uma 8h38'57" 63d56'55"
Annie Vogelpohl "Best Mom Ever"
And 0h32'27" 35d55'21"
Annie Watson
Boo 15h6'10" 47d50'17"
Annie Wylie
Uma 8h14'22" 72d57'36"
Annie Ying Jiang
Cnc 8h44'27" 32d19'39"
Annie Yuran Deng
And 23h26'42" 50d8'36"
Annie, Fred Hicks
Cyg 20h56'10" 41d15'35"
Annie, Warrior Princess
Aqr 23h24'19" -19d22'0"
Annie49
Cap 20h17'44" -13d17'5"
annieandjune
Uma 13h14'43" 56d37'45"
AnnieB
Lib 15h42'48" -26d39'51"
AnnieB
Leo 10h34'5" 8d18'38"
Annie-Claude Brossard
Dra 17h6'4" 64d10'11"
Annie-France Gaudreault
Umi 14h29'17" 82d50'59"
Annie-K
And 0h10'7" 45d52'9"
Annie-Maggs
Lyr 18h38'31" 28d5'16"
Annieree
Cap 20h28'1" -16d44'56"
AnnieRose Di Murro
Sgr 20h18'22" -33d17'47"
Annie's Anikus
Cyg 21h54'51" 38d33'1"
Annie's Baby
Tau 5h38'57" 18d32'31"
Annie's Chelsea
Uma 13h48'25" 50d31'19"
Annie's Clover
And 1h41'52" 43d1'41"
Annie's Eternal Valentine
And 0h20'29" 32d55'38"
Annie's Eyes
Uma 9h54'34" 63d14'13"
Annie's flight
Aqr 21h12'11" -7d41'31"
Annie's Heart
Uma 13h41'40" 55d40'14"
Annie's Knight Star
Ori 4h58'33" -0d33'11"
Annie's Light
Sgr 19h22'7" -23d21'29"
Annie's Normie
Uma 9h34'37" 49d14'51"
Annie's Peace of Heaven
Psc 1h34'4" 21d36'14"
Annie's Star
Cnc 8h34'28" 9d15'15"
Annie's Star
Psc 1h22'6" 29d19'53"
Annie's Star
Cyg 20h48'47" 47d44'36"
Annie's Star
Cam 3h59'13" 57d32'6"

Annie's Star
And 2h15'19" 43d48'49"
Annie's Star
Cap 21h23'22" -24d14'26"
Annie's Star
Cas 1h20'47" 64d7'35"
Annie's Star
Cas 2h29'16" 66d26'29"
Annie's Star - WaVue
Uma 10h9'25" 47d4'48"
Annie's Sunshine
Cas 1h16'31" 62d19'13"
Annie's Very Own Star
Ori 5h56'58" 21d19'27"
AnnieT
Per 3h30'54" 46d11'24"
Annik Viviana
Cyg 20h2'58" 42d30'35"
Annika
Per 1h37'30" 54d25'0"
Annika
Ori 5h8'39" 15d17'29"
Annika
Uma 10h13'19" 58d10'16"
ANNIKA
Sgr 18h16'48" -29d7'2"
Annika Aletta Unke
Vir 14h19'51" 1d12'0"
Annika Beth Heeringa
Cnc 8h45'7" 31d55'39"
Annika & Chloe Moorhead
Sgr 18h17'38" -16d39'34"
Annika Danielsson
Vir 14h49'13" 3d50'30"
Annika Eden Tomlinson Vaus
Lib 15h19'6" -4d23'9"
Annika Grace Wright
And 0h39'18" 31d44'3"
Annika Izobel Beyer
Lib 15h52'58" -12d9'36"
Annika Jaimee
Gem 6h29'57" 18d45'32"
Annika Karin Olsen
And 0h41'6" 45d42'29"
Annika Lam
Tau 3h32'30" 5d51'29"
Annika Marie
Vir 14h8'57" 4d11'0"
Annika Marie Fisler
Lib 14h53'7" -4d24'33"
Annika Mattea's Star
Dra 18h40'21" 50d53'36"
Annika May Little
Leo 9h58'47" 14d11'11"
Annika Michelle Wolfert
Crb 16h8'28" 37d0'23"
Annika Pedd 30.08.2005
Uma 12h27'32" 52d43'24"
Annika Strain
And 0h29'35" 35d24'33"
Annika Tayler Gastelum
And 1h16'17" 38d43'56"
Annika Torelli
Crb 16h4'45" 38d0'5"
Annika Willoughby
Cas 1h44'19" 65d21'52"
Annika16122004Gerteisz
Uma 10h18'41" 39d26'18"
Annika's Lucky Star
Cnc 8h5'36" 16d37'36"
Anniken
Uma 10h42'43" 53d26'34"
Anniken Alsand Skår
Leo 10h17'41" 26d41'48"
Annilove
Sge 19h41'1" 17d44'12"
Annina De Palatis
Per 4h11'8" 35d31'27"
Annina Filomena Pertuso Stewart
Dra 16h13'4" 59d35'19"
Annina Gabrielle Caruso
Vir 13h24'38" 7d17'51"
Annina My Mommy
Uma 13h41'43" 49d3'44"
Annique "Baby Girl" Veenstra
Ori 5h36'50" -2d4'9"
Annique Elizabeth Jordan
Cas 1h42'4" 61d53'9"
Annis Jane Jarrett Lusk
Cyg 21h51'28" 49d44'52"
Annisa Cristel Dueno
Gem 7h11'53" 33d26'12"
Annison Christian
Ari 2h54'17" 29d25'28"
Anniversary of Frank and Chris 9/79
Her 16h35'15" 29d5'18"
Anniversary Star
Lib 14h52'23" -3d52'41"
Anniversary Star #3841
Crb 15h46'3" 33d46'36"
Annkathreen
Sco 17h21'39" -36d45'12"
Annmarie
Cap 21h33'34" -9d51'24"
Annmarie
Lyr 18h49'7" 42d29'7"
AnnMarie Bariexca
Psc 0h43'41" 15d29'28"
Annmarie Carroll
And 22h58'13" 51d56'47"

AnnMarie DelRossi
Leo 10h4'42" 26d34'6"
AnnMarie Duncan
Aql 19h18'33" -3d32'0"
Annmarie Falco
Cas 0h42'8" 62d42'20"
Annmarie Fegeley Shining Angel
Cas 1h26'44" 65d44'8"
Annmarie Forever
Aqr 23h1'41" -6d8'38"
Ann-Marie Gearhart
Lyn 7h33'17" 52d27'14"
Annmarie Katelyn Samuelson
And 2h22'32" 46d46'0"
Annmarie & Mary Catherine O'Donnell
Lyr 19h6'8" 42d40'54"
AnnMarie McGehey
Gem 6h48'7" 33d24'20"
Annmarie Nicole Uribe
And 1h46'2" 50d3'46"
AnnMarie Patton
Lyn 7h40'12" 41d24'2"
AnnMarie Perino
Uma 8h44'23" 56d2'50"
Ann-Marie Porchowsky
Umi 14h22'57" 68d49'28"
AnnMarie Yurick
Sgr 19h8'5" -14d23'17"
Annointed
Uma 8h39'56" 56d56'0"
Annolee Wood Holsti
Ari 2h25'0" 23d37'11"
annoula
Lib 14h53'9" -0d41'26"
Annoying
Uma 9h54'52" 69d53'47"
Ann's Light of Design
And 2h13'16" 46d9'33"
Ann's Star
Cas 0h13'1" 53d30'48"
Ann's Star
Per 4h37'6" 31d29'44"
Ann's Star
And 0h22'38" 36d9'0"
Ann's Star
Ori 5h56'31" 11d58'15"
Annsachd Odo
Per 2h22'10" 57d9'46"
Ann—The Light of My Life
Cap 20h32'3" -16d1'42"
Anntoinette Delores Lindsey
Cnc 9h20'8" 13d52'17"
Annuhs, Karin
Ori 6h14'41" 16d43'54"
Annûniel
Peg 22h17'50" 34d47'35"
annus mirabilis
Ori 5h40'25" 1d0'4"
Annushka
Sgr 20h22'35" -33d48'55"
Anny
Aqr 23h0'32" -7d39'28"
Anny Johanna Ayala
Leo 11h26'26" 11d15'19"
Anny Lau
Vir 13h52'55" -0d10'41"
Anny Q. Lau
Uma 9h4'30" 52d38'45"
Anny-1004
Sgr 18h21'41" -32d15'47"
Annya Walker
Sgr 18h16'27" -27d18'2"
AnointedAngel41479
Lyr 18h27'47" 36d54'6"
Anomaly
Cap 21h31'41" -17d3'56"
Anomie
Leo 10h11'20" 18d10'2"
Anongnat
Ori 5h7'4" 16d7'48"
anoniem
Tri 2h29'55" 31d10'14"
Another Beautiful Thing
Crb 15h45'44" 33d27'28"
Anouc
Her 17h33'14" 26d45'18"
Anouchka
Cas 1h13'56" 49d53'9"
Anouchka Bayard-Blanchard
Leo 11h10'22" 15d53'13"
Anouchka Chenevard Sommaruga
Dra 9h57'54" 77d46'8"
Anouck Maya Benoit
Sco 16h44'19" -31d54'10"
Anouk
Cas 0h25'46" 50d30'35"
Anouk Guilliet
And 1h3'3" 38d17'26"
Anoush Vartanyan
Leo 9h33'8" 22d45'56"
Anouska "Annie" Fitz-Simon
Gem 7h0'32" 19d34'51"
Anouska Fitz-Simon
Uma 11h47'58" 41d12'33"
Anping Hong
Uma 8h52'18" 59d54'40"

A-N-R Straight On Till Morning Cap 20h21'57" -20d3'23"
Anrawel Ari 1h54'57" 13d56'47"
Anruca Lib 15h34'37" -9d33'44"
Ansaka Uma 11h25'35" 62d25'19"
Ansaldi Domenico Cam 6h51'41" 69d33'24"
Ansata Shah Zam Psc 1h22'40" 29d9'45"
Ansha Forever And 1h22'41" 40d42'56"
Anshuman Chandrachud Uma 10h57'41" 39d7'4"
Anshuria Cas 23h48'45" 51d13'29"
Ansias, Nito Ori 6h4'12" 6d18'51"
Ansil Dean Hale Gem 6h50'24" 15d1'39"
Anslee Connor Gothard Cnc 8h46'0" 7d27'30"
Ansleigh Christine Mendoza Lib 14h26'35" -18d28'14"
Ansley Equ 21h6'57" 9d19'59"
Ansley Camille Ori 4h49'28" 12d29'33"
Ansley Christina Boyd Del 20h52'12" 9d26'56"
Ansley Eileen Usher And 0h10'21" 31d8'56"
Ansley Joan Rice Dra 18h39'35" 64d55'55"
Ansley L. Nickell Cnc 8h26'21" 18d20'33"
Ansley Rae Burroughs Lyr 18h51'44" 34d16'22"
Ansley Rooks Tau 3h31'3" 25d6'52"
Anson M. Breen Ari 2h49'34" 29d13'32"
Ansula's Lovestar Sgr 17h53'6" -29d29'49"
Ansumi Uma 12h51'47" 59d30'20"
Ant & Carolyn Stewart - 4 Eternity Cru 12h34'59" -64d36'7"
Ant. Noodle Pho 0h10'6" -44d18'48"
Antal Ibolya Uma 10h6'40" 68d9'11"
Antal István Uma 12h25'30" 55d17'43"
AntandAsh Tau 4h2'54" 28d26'4"
Antara, The Wishing Starr Aqr 21h59'25" 1d45'54"
Antaram Lac 22h27'3" 43d56'34"
Antares II Uma 13h59'28" 49d7'20"
Antaza Leigh Mon 6h51'5" 9d28'11"
Antecco & Trilly Her 16h39'2" 24d17'24"
Anteros Uma 11h30'9" 32d23'32"
Anthea Lyr 19h5'16" 36d28'36"
Anthea Tau 5h21'31" 27d12'21"
Anthea Cap 21h32'32" -9d13'19"
Anthea Cru 12h43'7" -56d54'38"
Anthea Arnol Vir 12h59'18" -21d36'21"
Anthea Maria Cru 12h27'39" -60d44'58"
Anthea the "Light" Sgr 18h22'59" -24d10'46"
Anthelia Desiree' Sgr 19h42'0" -16d38'20"
Anthemos "Mike" Ades Uma 8h47'36" 47d18'41"
Antheny Linzy Lib 15h48'59" -7d28'16"
Anthonette Paboojian Sgr 17h57'1" -21d41'24"
Anthoney Her 17h37'6" 14d47'31"
Anthony Tau 4h47'19" 19d59'58"
Anthony Cnc 8h38'5" 22d31'42"
Anthony Leo 11h51'10" 19d50'32"
Anthony Lyn 7h40'48" 49d2'5"
Anthony Cyg 20h11'54" 40d42'57"
Anthony Per 2h23'49" 56d41'26"
Anthony Lmi 10h19'10" 37d14'5"
Anthony Per 4h9'24" 33d29'47"
Anthony Gem 7h15'59" 33d40'6"
Anthony Sgr 19h50'47" -12d15'9"
Anthony Vir 13h11'52" -1d35'35"
Anthony Sgr 18h26'55" -26d15'3"
Anthony 11 Ori 5h27'52" 3d31'47"
Anthony A. Affa Ori 6h6'43" -0d32'12"
Anthony A. Nasso Lib 14h51'50" -3d39'0"
Anthony A. "Nino" Berni Umi 14h51'59" 73d29'6"
Anthony A. Ruiz Uma 8h26'30" 66d55'37"
Anthony Abril Lerma Vir 14h32'14" 4d7'51"
Anthony Alan Roles Cyg 20h12'41" 32d51'33"
Anthony Alban Gummett Boo 14h52'16" 23d33'17"
Anthony Albert Corelli Her 16h32'10" 20d45'58"
Anthony Alessandro Ciallella Aur 5h22'32" 44d51'38"
Anthony Alexander Gabriel Uma 11h46'1" 40d28'35"
Anthony Alexander Sher Uma 9h29'12" 43d55'53"
Anthony Allen Greer Her 17h57'43" 23d49'40"
Anthony - Always & Forever Cen 13h57'55" -60d18'41"
Anthony & Amanda Cyg 19h35'13" 29d38'51"
Anthony and Antoinette Umi 15h8'39" 70d19'42"
Anthony and Ashleys True Love Ari 2h47'1" 12d34'50"
Anthony and Brandi Lib 15h52'11" -10d24'15"
Anthony and Christina Forever Vir 12h44'9" 8d27'55"
Anthony and Donna Vir 13h4'21" 9d4'17"
Anthony and Ivander Cap 21h5'56" -27d9'13"
Anthony and Jocelyn Crb 15h59'9" 35d47'13"
Anthony and Jodi Potts Tau 4h24'25" 23d23'46"
Anthony and Katie Ari 2h51'7" 18d54'19"
Anthony and Katie Haydock And 2h34'51" 40d37'42"
Anthony and Lauren ~2007~ Uma 9h28'46" 58d8'41"
Anthony and Linda's Wedding Star Dra 19h58'15" 79d3'52"
Anthony and Lynette Hughes Cyg 20h31'20" 30d53'45"
Anthony and Regina Dunleavy Cyg 20h35'12" 38d43'18"
Anthony Anderson Per 3h9'44" 55d51'21"
Anthony Andrew Puglisi Per 4h45'17" 49d15'39"
Anthony Andrew Thompson Lib 15h33'44" -24d43'21"
Anthony Angel Ybarra Tau 4h22'45" 24d8'30"
Anthony Arnold, Jr. Her 17h6'34" 30d0'21"
Anthony Aronson Per 4h16'49" 31d30'59"
Anthony Aya Her 17h10'31" 18d55'17"
Anthony B. Fries's Star Gem 6h45'44" 14d46'52"
Anthony B. Halili Sco 16h12'48" -10d13'58"
Anthony B. Ramdin Ori 5h26'26" 3d14'43"
Anthony "Babe" Tau 4h7'4" 15d59'14"
Anthony & Barbara Jarmusz Cyg 20h12'40" 51d39'57"
Anthony Barone Cas 1h43'52" 63d44'0"
Anthony Barulli Lmi 10h52'19" 25d48'19"
Anthony Beltsky Cap 20h57'34" -16d47'54"
Anthony Benza Ori 5h3'26" 4d24'11"
Anthony Biaggio Joseph Laurino Vir 13h1'38" -0d51'10"
Anthony Blackburn Gem 7h23'8" 14d52'6"
Anthony Blessing & Amanda Mikiten Cyg 19h47'18" 34d42'53"
Anthony Bogard Boo 15h23'42" 40d29'24"
Anthony Bonavita Per 4h17'1" 45d2'47"
Anthony Bongiorno Uma 9h14'11" 62d27'41"
Anthony Booth Dra 18h50'48" 58d37'59"
Anthony Brett Parker Leo 10h24'56" 8d0'21"
Anthony Bryson Whisenant Tau 5h45'11" 15d15'46"
Anthony Brzozowski Aqr 22h17'35" 1d55'58"
Anthony (Bull) Laudicina Cep 3h11'15" 82d48'43"
Anthony Burnett Per 3h31'17" 35d5'8"
Anthony C. Bolds Aqr 22h15'15" 2d9'8"
Anthony C. Chirasello Aur 7h16'12" 42d16'35"
Anthony C. Dijak, Jr. Per 2h11'55" 55d33'9"
Anthony C. Matteo Cep 23h58'1" 86d51'46"
Anthony C. Scaglione Uma 10h9'8" 56d33'11"
Anthony Cabral Porriello Cap 21h34'6" -14d36'41"
Anthony Cajka and Justin Nacarato Ari 2h10'52" 25d56'59"
Anthony Caldera Uma 10h13'14" 50d12'45"
Anthony Calia Uma 10h16'19" 46d53'14"
Anthony Callea Cra 18h0'45" -37d10'32"
Anthony Candell Sr. Lib 15h9'5" -10d7'24"
Anthony Cantalupo Vir 14h2'17" -13d22'26"
Anthony "Cappy" Capozzelli Boo 14h35'56" 27d46'59"
Anthony Carmine Cardinuto Per 3h9'3" 54d5'50"
ANTHONY CARMINE RUZICKA Uma 13h7'41" 59d47'21"
Anthony Carr Her 18h8'1" 32d40'15"
Anthony Cassar Nor 15h42'58" -49d44'46"
Anthony Catapano Cep 23h12'37" 75d13'59"
Anthony Catolina Per 4h9'24" 50d20'2"
Anthony & Chasity Apu 15h25'5" -70d53'51"
Anthony Chicherchia Uma 10h40'32" 40d7'31"
Anthony Christian Robles Sgr 17h54'22" -29d45'15"
Anthony Christopher Monczewski Her 16h38'30" 37d6'43"
Anthony Christopher Rivera Her 17h28'33" 26d43'17"
Anthony Christopher Slayton Her 17h50'52" 41d52'20"
Anthony Colucci - 50 Ori 5h31'15" 5d5'57"
Anthony Connell Memorial Star Nov70 Cru 12h49'56" -64d28'41"
Anthony Connor Sung Cnc 9h8'1" 27d3'43"
Anthony Corbino Psc 23h30'42" 5d21'21"
Anthony D. Harris Per 3h8'15" 45d27'32"
Anthony (Dad) Cota Her 17h31'10" 48d23'34"
Anthony D'Amelio Per 3h18'5" 51d45'0"
Anthony Daniel Massaro Her 18h4'29" 21d16'29"
Anthony Dantuono Dependable Tool Co Her 17h33'28" 31d54'27"
Anthony Darby Dra 19h42'29" 66d11'18"
Anthony David Murtaugh Cnc 8h43'9" 22d9'0"
Anthony David Wall Cep 22h44'24" 72d39'29"
Anthony DeDominicis Cep 22h9'14" 57d1'54"
Anthony Delaney Ser 18h6'4" -12d24'29"
Anthony DeLuca #1 Retirement Plans Her 17h19'18" 29d58'32"
Anthony DeNunzio Ori 5h53'43" 20d53'30"
Anthony DeWitt Cma 7h20'48" -32d47'50"
Anthony DiCola IV Lib 15h4'48" -6d15'44"
Anthony DiFillippo Boo 14h55'46" 25d44'48"
Anthony DiLolle Per 3h20'59" 41d37'57"
Anthony DiPasquale III Leo 9h29'29" 31d16'57"
Anthony DiPietro Per 4h7'54" 32d8'1"
Anthony Dobbs Thompson Ori 5h3'58" 2d29'45"
Anthony Domenico Eduardo Uma 10h23'21" 61d45'56"
Anthony Dominic Colonna Ori 6h20'31" 10d53'0"
Anthony Dominic Corbisiero Del 20h40'29" 15d23'16"
Anthony Doniven Green Gem 6h23'59" 26d42'20"
Anthony Dow Zarrillo Lmi 9h46'6" 35d10'48"
Anthony Downes Per 4h35'6" 31d0'5"
Anthony Dragonetti Dra 20h19'26" 63d32'5"
Anthony Dufour Cyg 21h15'41" 42d15'45"
Anthony E. DePrima "7-30-39" Cep 21h54'33" 63d18'44"
Anthony E Leone Aql 20h23'5" 0d48'36"
Anthony E. Mirti Uma 12h44'38" 56d12'46"
Anthony Earl Barnes Her 17h13'12" 35d27'51"
Anthony Eduardo Villalpando Uma 12h23'54" 57d54'39"
Anthony Edward Califano Tau 4h35'26" 28d45'30"
Anthony Edward Sedlacek Leo 11h20'17" -2d9'39"
Anthony Eliaz Moreno Her 17h43'48" 44d19'56"
Anthony & Emily Her 18h44'33" 25d46'50"
Anthony Eugene Smith Uma 8h38'24" 67d3'14"
Anthony Evan Cucchiaro Her 16h56'25" 32d28'33"
Anthony F. Angello Ari 2h11'12" 24d37'0"
Anthony F Brattoli Uma 10h46'44" 46d3'17"
Anthony F. Bronzo III Pyx 8h41'48" -20d56'26"
Anthony F. Casamassa Uma 11h0'6" 50d47'26"
Anthony F. Fini Cyg 21h29'54" 32d24'44"
Anthony F Stone Boo 14h45'13" 23d44'6"
Anthony Fabrizzo Suttile Rivera Per 2h45'58" 51d21'27"
Anthony Fedorov Tau 4h35'42" 17d52'3"
Anthony Felice Marchione - Mayo Cru 12h38'52" -60d43'13"
Anthony Ferrara Her 16h31'2" 46d55'11"
Anthony Follis Aur 5h39'59" 33d22'25"
Anthony Fragetti Leo 11h43'1" 24d27'3"
Anthony Francis DelVecchio Cyg 19h59'29" 33d31'9"
Anthony Frascone Per 3h34'3" 44d58'12"
Anthony Fugate Her 17h39'15" 26d19'49"
Anthony G. Psc 1h20'44" 3d9'38"
Anthony G. Glenn Cep 3h52'48" 87d44'39"
Anthony G. Mortellito Umi 16h16'54" 73d36'44"
Anthony Galiano Vir 13h19'59" 11d51'34"
Anthony Gasper Anania Aqr 20h47'38" -11d56'45"
Anthony George Bayes Lmi 11h2'26" 28d54'23"
Anthony George Daw - Tony Ori 5h55'45" 16d57'16"
Anthony George DiPietri Psc 1h24'16" 27d15'45"
Anthony George Hudson Ari 3h21'25" 29d25'37"
Anthony Giambruno Tau 4h19'10" 3d15'54"
Anthony Giannuzzi Umi 15h29'43" 76d36'5"
Anthony Giovanni Battista Tau 4h47'9" 28d24'42"
Anthony Giolaj Tri 2h10'12" 33d29'12"
Anthony Gorny - Love Forever Cyg 20h31'35" 42d15'14"
Anthony Graglia Aqr 22h37'40" -16d23'6"
Anthony Guancione III Ori 6h9'56" 15d53'35"
Anthony Gulisano Ori 6h5'55" 20d15'59"
Anthony Hayward Uma 11h27'32" 58d53'4"
Anthony Henry Aur 6h17'42" 37d15'46"
Anthony Hutton Vir 13h7'38" 9d22'36"
Anthony I Love You Natasha Uma 9h50'17" 56d33'24"
Anthony & Irene Patounas Umi 15h56'51" 76d46'13"
Anthony J. Alto Gem 6h36'10" 15d50'37"
Anthony J. Asaro, Sr. Dra 15h7'25" 61d3'52"
Anthony J. Asaro, Sr. Cap 20h53'28" -25d43'46"
Anthony J. Caruso Per 3h13'41" 55d40'34"
Anthony J. Dattilo, Jr. Uma 8h34'39" 53d36'9"
Anthony J. Deford Jr. Vir 14h11'58" -10d45'55"
Anthony J. Delork Cep 21h30'42" 64d25'34"
Anthony J & Dorothy L Kliemann Cyg 20h7'48" 31d48'34"
Anthony J Dukes Aqr 22h33'28" 1d44'36"
Anthony J. Fuentes Gem 7h21'13" 17d35'21"
Anthony J. Grandazza Uma 8h41'16" 70d11'35"
Anthony J. Ian Aql 19h45'46" 13d12'25"
Anthony J. Jannetti Ori 5h48'23" 11d38'58"
Anthony J. Kraus Lib 14h41'25" -10d37'39"
Anthony J. Liberatori Uma 9h21'15" 52d14'4"
Anthony J. Monzo Jr. Uma 9h55'30" 64d56'1"
Anthony J. Nardone Dra 18h30'38" 50d3'1"
Anthony J. Nicoletti Sr Her 18h43'52" 19d48'14"
Anthony J. Nini Cnc 8h43'58" 16d46'36"
Anthony J Perry "My Sweetie" Cep 3h20'23" 84d7'3"
Anthony J. Rusciano Aqr 22h23'10" -16d25'59"
Anthony J. Russo Sgr 18h56'23" -34d29'12"
Anthony J. Scaturro Vir 12h59'9" -16d11'59"
Anthony J. Scesny, Jr. Cep 0h5'15" 76d0'14"
Anthony J. Slavich, Jr. Aur 5h38'24" 48d17'0"
Anthony J. Stokar Ori 6h18'41" 19d34'21"
Anthony J. Surgala Sr. Cas 0h47'2" 65d5'9"
Anthony J. Williams Sco 17h54'44" -36d14'42"
Anthony Jake Miller Boo 15h26'34" 48d39'46"
Anthony Jamal Morgan Sgr 19h17'22" -17d9'23"
Anthony James Dra 19h35'57" 66d56'38"
Anthony James Boo 14h41'31" 43d4'40"
anthony james adams Uma 11h14'37" 47d21'32"
Anthony James Barlow Col 6h1'22" -42d29'48"
Anthony James Brine Cep 20h53'57" 62d19'49"
Anthony James Cook Uma 8h51'4" 62d14'47"
Anthony James Cortese Per 3h9'1" 56d38'49"
Anthony James DiDuro Aqr 23h18'14" -17d53'18"
Anthony James Miller Her 18h35'13" 15d55'19"
Anthony James Posthuma Tau 5h8'36" 25d53'50"
Anthony James Richards Ori 5h35'30" -0d43'6"
Anthony James Villarreal Ori 5h46'20" 11d43'52"
Anthony & Jeannie Forever Sgr 19h33'32" -20d12'59"
Anthony "Jedi" Miguel Hernandez Tau 5h3'12" 27d20'33"
Anthony & Jenai Lyn 7h22'20" 47d12'56"
Anthony & Jennifer Amaral Cyg 19h47'46" 46d19'7"
Anthony & Jenny Lib 15h10'4" -18d28'2"
Anthony Jetland Psc 0h55'49" 28d21'18"
Anthony Jing Rong Zhang Vir 12h36'29" -11d7'9"
Anthony John And 2h24'44" 46d48'24"
Anthony John Boyens Leo 11h33'28" 25d14'5"
Anthony John Carbone Cnc 8h51'51" 14d1'31"
Anthony John Dominello - Junu Cru 12h35'50" -63d56'21"
Anthony John Graham Equ 21h17'20" 7d37'6"
Anthony John Joseph Paholski Uma 11h27'6" 48d52'34"
Anthony John Kunz, Jr. "Lover" Tau 4h19'9" 26d1'24"
Anthony John Lepore Lyn 7h34'55" 47d53'39"
Anthony John Messuri, Sr. Psc 1h22'9" 28d30'41"
Anthony John Morales Leo 10h23'39" 24d28'54"
Anthony John Mosblech Uma 11h21'19" 40d24'5"
Anthony John Neddoff Her 17h31'37" 31d56'7"
Anthony John Paciella "Zipper" Per 4h22'21" 43d21'33"
Anthony John Patrizzo Gem 7h41'7" 22d25'33"
Anthony John Raucstadt Sco 16h14'46" -15d57'54"
Anthony John Vito Taboadela Sgr 18h22'36" -30d31'26"
Anthony John Walker Cep 22h17'47" 66d54'51"
Anthony Joseph Canales Uma 11h8'43" 34d54'53"
Anthony Joseph Casella Cyg 20h24'3" 59d49'10"
Anthony Joseph Fiori III Her 17h55'13" 47d55'38"
Anthony Joseph Gallo, Jr. Sgr 18h21'40" -23d9'24"
Anthony Joseph James Cnc 8h26'1" 18d52'25"
Anthony Joseph LaMacchia IV Vir 13h22'6" 10d51'1"
Anthony Joseph Leone Lyn 6h37'41" 57d44'31"
Anthony Joseph Lombardo Cnc 8h52'7" 32d16'8"
Anthony Joseph Lotito II Umi 16h23'36" 78d46'17"
Anthony Joseph Naccarato Uma 11h30'34" 43d4'52"
Anthony Joseph Noone Boo 14h44'22" 27d27'47"
Anthony Joseph Rex Umi 14h17'20" 76d27'55"
Anthony Joseph Torriero Jr. Umi 16h39'41" 82d28'47"
Anthony Joseph Ulses Aur 5h27'28" 54d10'3"
Anthony Joseph Vega Her 17h37'43" 21d6'42"
Anthony Joseph Vitale Sgr 19h40'32" -15d24'57"
Anthony & Josephine Cyg 19h55'31" 31d48'29"
Anthony & Josephine Forever Cyg 19h56'23" 33d26'29"
Anthony Jude Davis Psc 1h1'48" 27d50'43"
Anthony Justin Burgois Uma 9h16'52" 53d54'38"
Anthony Karlinsky Uma 11h43'1" 60d13'34"
Anthony Keiper's Star Cnc 9h5'28" 26d1'55"
Anthony Kenneth Culver Sgr 18h4'40" -27d31'29"
Anthony Kleiger call sign Tiger Tau 4h24'45" 0d28'47"
Anthony Krokey Umi 15h20'25" 77d28'8"
Anthony Kyle Graviano Leo 9h25'8" 23d59'58"
Anthony L. Laudano, Educator Per 3h13'49" 55d5'32"
Anthony LaConte Her 17h26'4" 30d59'20"
Anthony Lawrence Leo 11h27'45" 18d23'54"
Anthony Lawrence Cap 20h49'37" -15d30'54"
Anthony Leo Lastra Sco 16h48'58" -34d36'41"
Anthony Leonard Scrima Ori 5h39'20" 1d43'56"
Anthony Leonard White Her 16h46'40" 33d40'34"
Anthony Leonardi Her 17h48'49" 42d0'23"
Anthony Leone Lyn 6h24'3" 61d49'23"
Anthony Leone, Our Hero Uma 10h12'45" 66d31'32"
Anthony Liam Frith Dra 19h35'0" 61d28'42"
Anthony Licata Lib 15h25'53" -9d50'30"
Anthony Lil Huero Sco 17h44'42" -42d10'37"
Anthony & Lisa, Happy 25th Anniv. Dor 4h49'32" -53d57'27"
Anthony Liveris Aqr 22h54'36" -7d48'24"
Anthony Lon Mattox, Sr. Uma 10h9'27" 56d37'17"
Anthony Lonnie Dodd Ori 5h48'7" 6d19'15"
Anthony Lopes Lib 15h20'41" -9d56'46"
Anthony Louis Wheeler Gem 7h42'54" 31d34'1"
Anthony Loves Becky Vir 12h0'27" -1d33'9"
Anthony Luaders Cra 18h45'32" -45d12'25"
Anthony Lucero Ori 5h28'24" 0d51'9"
Anthony M. A. Galloway Uma 9h22'12" 62d39'42"
Anthony M. Lovingood Leo 10h35'8" 22d3'58"
Anthony M. Rivellini Per 3h33'6" 44d51'34"
Anthony M. SanGiacomo And 0h14'20" 32d31'53"
Anthony M Stevens Sco 16h13'24" -15d0'28"
Anthony M Valadez Cyg 20h14'17" 41d6'27"
Anthony Macri Cep 21h58'41" 68d59'0"
Anthony Magit Dra 15h55'3" 61d14'7"
Anthony Mancino Jr. Sco 16h11'2" -42d6'21"
Anthony Mangola Her 16h21'43" 20d41'3"
Anthony & Marguerite Raguso Uma 9h30'7" 65d20'42"
Anthony Markoe & Mario Gonzalez Uma 11h5'31" 59d34'30"
Anthony Marrero II Lyn 8h31'20" 50d34'8"
Anthony Marshall Trent Uma 11h40'5" 46d43'56"
Anthony Martin DeJoia Cap 20h32'11" -20d50'45"
Anthony Martinez Per 3h43'5" 45d31'33"
Anthony Masters Aur 5h33'3" 46d23'42"
Anthony Masters Boo 15h37'21" 45d27'46"
Anthony Matthew Leo 9h35'48" 32d21'24"
Anthony Medina Uma 11h46'59" 48d35'17"
Anthony Medinas Jr. 08/30/1946 Vir 13h11'42" -21d58'49"
Anthony Melchiorri Per 3h21'29" 51d36'17"
Anthony & Melissa 24/04/2007 Cru 12h26'17" -59d35'43"
Anthony Mercurio Per 2h38'32" 55d42'55"
Anthony Merschdorf Lyn 9h11'19" 33d47'38"
Anthony Meszaros Uma 10h20'3" 55d46'43"
Anthony Micari Sgr 19h15'27" -13d54'44"
Anthony Michael Cap 21h21'4" -20d3'22"
Anthony Michael Balsamo Psc 1h27'36" 32d36'45"
Anthony Michael Cappasso Per 2h58'30" 55d34'54"
Anthony Michael Conciglio Aqr 21h40'29" 0d10'24"
Anthony Michael DiOrio Cap 21h36'51" -14d35'
Anthony michaEL GlenN Cyg 19h57'36" 34d42'53"
Anthony Michael Grilli Uma 11h6'48" 39d16'5"
Anthony Michael Guldin II Cnc 9h12'44" 31d38'23"
Anthony Michael Leonarduzzi Lyn 8h13'57" 36d52'24"
Anthony Michael Lopez Aqr 22h6'51" -11d9'39"
Anthony Michael Restivo Sgr 19h7'27" -12d23'59"
Anthony Michael Salvatore Uma 11h2'41" 71d27'14"
Anthony Michael Schissler Tau 3h46'9" 19d38'52"

Anthony Michael Tagliaferri Gem 6h41'13" 20d12'24"
Anthony Michael Tringale Leo 11h19'28" 18d23'26"
Anthony & Michaela: A Forever Love Lyr 18h52'17" 31d51'32"
Anthony Micheal Pho 0h56'48" -51d0'27"
Anthony Mitchell Bilan Ori 6h4'48" 5d56'37"
Anthony Moon Cas 23h25'42" 53d32'23"
Anthony "Moon" Delia Cep 21h37'30" 63d4'1"
Anthony Morgan Lyn 6h45'15" 56d18'37"
Anthony Morris Her 18h3'51" 36d48'4"
Anthony Morris Pascal Weinraub Ori 5h38'32" 11d43'35"
Anthony Mosca "The Breadman" Uma 10h13'27" 44d32'21"
Anthony Muti Aqr 22h41'3" 2d26'38"
Anthony "My Dream Boat" Uma 10h15'38" 57d17'9"
Anthony - My Wind-Jannotta Uma 13h49'57" 56d56'56"
Anthony N. Cap 20h47'0" -22d29'17"
Anthony Naples Aql 19h42'14" 3d41'1"
Anthony Neglia Uma 8h39'14" 65d11'47"
Anthony & Netti Distinti Cyg 19h21'37" 53d0'50"
Anthony Neumann Ori 6h9'39" 16d46'52"
Anthony & Nicole Floriano Cyg 21h33'50" 45d5'19"
Anthony Njoroge Ikahu Cen 13h38'38" -42d30'24"
Anthony Nolan Sakhleh Cnc 8h43'43" 32d56'43"
Anthony Occhino And 0h45'8" 38d7'7"
Anthony Orlando And 2h21'49" 50d14'43"
Anthony Orlando Uma 8h38'3" 64d28'47"
Anthony Owen Muscroft Uma 11h34'6" 54d52'39"
Anthony P. Antonelli Per 2h33'45" 53d28'1"
Anthony P. Mulcahy Her 18h32'20" 19d25'58"
Anthony Padovano Lib 14h31'44" -19d9'46"
Anthony Pagan Ra Aur 5h50'47" 51d26'56"
Anthony Pandolfo Umi 16h14'54" 73d18'56"
Anthony Patrick Alston Lyn 7h0'22" 51d13'35"
Anthony Patrick Cormier Psc 23h58'0" 9d4'0"
Anthony Patrick Fortune Cep 22h29'2" 63d44'26"
Anthony Paul Baird Her 17h39'15" 32d11'26"
Anthony Paul Benson Gem 6h10'43" 25d7'3"
Anthony Paul George Anning Uma 9h25'7" 71d16'49"
Anthony Paul Maggiora Tau 4h33'22" 19d43'15"
Anthony Paul O'Brien Her 18h1'35" 17d32'14"
Anthony Paul Pisari Uma 10h42'24" 45d28'20"
Anthony Peter Costen Per 4h41'11" 40d42'46"
Anthony Petrelis Tau 5h37'0" 20d8'2"
Anthony Petroulakis Lyn 9h2'59" 38d29'36"
Anthony Petrovich Her 16h30'26" 34d10'18"
Anthony Philip Estrada Pho 0h35'8" -40d49'54"
Anthony Pinkard Boo 14h52'58" 26d33'8"
Anthony Pronti Sgr 18h21'58" -21d49'11"
Anthony Provenzano Aql 20h11'4" 8d56'6"
Anthony R Auletto Cep 21h32'59" 60d49'53"
Anthony R. David Moreno Her 16h37'3" 27d46'15"
Anthony R. Jones Lyn 8h38'43" 36d48'1"
Anthony R. Povernick Uma 11h48'21" 45d56'28"
Anthony R. San Diego Her 17h21'11" 38d5'39"
Anthony Raber Sgr 18h16'43" -35d50'26"
Anthony Rachel Cap 21h41'43" -13d14'49"
Anthony Rapp Sco 16h43'30" -43d59'31"
Anthony Ray Her 17h19'11" 37d35'52"
Anthony Ray Cortez Sgr 19h53'47" -13d13'29"
Anthony Ray Reyes Uma 11h47'32" 38d13'29"
Anthony Raymond Biondi Lib 15h40'58" -17d41'48"
Anthony Raymond Dacorte Sgr 18h15'18" -19d11'51"
Anthony Raymond Santillan Lyn 8h13'26" 33d46'43"
Anthony Reynolds, Jr. Ari 2h2'13" 23d28'46"
Anthony Riccio Per 4h11'28" 51d35'13"
Anthony Richard Podlaski Cep 21h25'13" 64d42'11"
Anthony Rigby Glass Lyr 18h49'55" 39d47'24"
Anthony Rivelli, My Star! Gem 6h34'24" 25d5'59"
Anthony Robert Masucci "2/15/2006" Umi 14h21'10" 73d39'5"
Anthony Robinson Per 4h6'34" 44d12'51"
Anthony Rodriguez Cnc 8h16'3" 24d51'55"
Anthony Roko Alaga Her 17h15'9" 22d15'24"
Anthony & Rosanna 2006 Cet 2h45'39" 4d50'40"
Anthony Rote Her 17h22'22" 36d45'50"
Anthony Ryan Orr Ori 5h28'51" 1d39'35"
Anthony Ryan Perkins Cru 12h38'8" -56d7'17"
Anthony S. Brayton Her 17h1'41" 46d35'41"
Anthony S. Sciortino Ori 5h28'7" 1d11'50"
Anthony S. Tramonte Vir 13h16'14" 10d50'42"
Anthony Salvatore Gambino Sco 17h40'41" -41d25'11"
Anthony Sands Umi 13h38'10" 89d43'31"
Anthony Savoia Per 2h12'45" 52d42'17"
Anthony Sclafani Tau 5h17'13" 21d50'32"
Anthony Scott Armstrong Dra 18h37'57" 61d32'15"
Anthony Scott Johnson Vir 13h20'48" -3d21'15"
Anthony Scott Shelton Sgr 18h50'59" -27d6'33"
Anthony Sebastian Toro Cap 21h28'59" -19d51'14"
Anthony Sessa Jr. Sco 16h7'23" -15d33'48"
Anthony Sheeran Per 3h5'22" 51d19'4"
Anthony Sherman Nides Cap 20h36'22" -23d19'34"
Anthony Shinn Boo 14h18'52" 40d23'4"
Anthony Signa Uma 10h59'39" 34d31'28"
Anthony Sindoni Uma 10h56'36" 38d27'26"
Anthony Smith Gem 7h47'22" 31d32'7"
Anthony Smith Cyg 21h32'56" 39d10'32"
Anthony Souza Dra 17h18'6" 62d11'52"
Anthony "Spanky" Pensabene Sco 16h5'19" -9d55'1"
Anthony Spatola Cap 21h40'41" -12d48'7"
Anthony Spiegel Boo 15h3'48" 49d4'4"
Anthony Stephen Bossie Sco 16h49'3" -26d56'40"
Anthony Steven Antoniou Ori 6h19'28" 6d39'37"
Anthony & Sylvia Borrelli Lyr 18h34'48" 39d25'31"
Anthony T. Todaro Sco 17h35'46" -39d34'35"
Anthony Tasco Cep 20h47'47" 68d36'56"
Anthony T-Bone McCullough Cap 20h42'24" -20d54'22"
Anthony the Amazing Psc 1h14'31" 14d53'43"
Anthony "The Rock" LaGuerre Per 3h29'10" 52d28'41"
Anthony Thomas Alessi Boo 14h44'32" 25d59'36"
Anthony Thomas Gleason Uma 8h29'22" 68d51'11"
Anthony Thomas Stinson Uma 10h35'4" 41d52'34"
Anthony Thomas Tarolla Sgr 18h6'35" -17d38'1"
Anthony "Tonio" Graham Tau 3h24'25" 11d59'24"
Anthony (Tony) Matulewic Per 3h35'25" 42d5'37"
Anthony "Tony" Rojas Per 2h55'33" 40d39'35"
Anthony (Tony) Sciame Aqr 22h17'51" -14d2'32"
Anthony Tootal Per 3h44'56" 40d55'18"
Anthony Torres Aql 20h12'46" 15d1'24"
Anthony Trabue Hightower Per 4h12'11" 45d22'13"
Anthony Tuccillo Lyn 6h42'13" 50d13'36"
Anthony Tyler Cicerchia Tau 4h1'49" 5d46'51"
Anthony V. Bettencourt "Boob" Her 16h37'47" 21d47'38"
Anthony V. Zehenni : The Skylark Vir 14h15'10" 3d10'44"
Anthony Velez, Jr. Lib 14h51'2" -3d58'20"
Anthony Victor Cincotta Sco 17h37'33" -40d40'58"
Anthony Vincent Gaetano Per 2h24'10" 55d15'36"
Anthony Vincent Silva Uma 11h57'47" 50d39'0"
Anthony Vitaliano Sco 16h18'56" -10d59'20"
Anthony Volpe, Sr. Lib 14h54'1" -12d23'40"
Anthony W Armstrong Per 2h50'35" 53d51'8"
Anthony W. Messina Ori 5h45'5" -0d57'50"
Anthony Wayne Sells For Eternity Cru 12h28'1" -60d48'24"
Anthony & Wendy Moura Eri 4h35'41" -0d53'12"
Anthony Wernimont Leo 10h8'15" 23d14'18"
Anthony White Jr. "I Love You Star" Per 3h12'23" 45d35'0"
Anthony William Lawson Ori 5h10'58" 12d52'6"
Anthony William Sites Aql 19h44'13" -7d19'35"
Anthony William Smallshaw Aqr 22h35'0" -7d53'34"
Anthony & Will's Solar Corner Cyg 20h45'15" 47d38'4"
Anthony Winters Gem 6h39'59" 16d3'3"
Anthony Wolfe Lup 15h27'27" -41d37'45"
Anthony Woosnam Lmi 10h15'24" 39d0'12"
Anthony - Yolanda Cyg 19h28'42" 57d3'54"
Anthony Young Watts Cen 13h2'50" -47d52'37"
Anthony Zaya Sco 17h57'30" -31d3'11"
Anthony, Susan, and Janessa Love Vir 14h10'22" 5d53'48"
Anthony-David Sarah-Kate Sco 17h45'52" -44d38'8"
Anthony's A Train Cnv 13h18'27" 43d14'51"
Anthony's Angel Uma 10h15'34" 45d57'30"
Anthony's Greatest Star Aqr 22h39'54" 2d14'24"
Anthony's Love Umi 17h34'34" 88d11'13"
Anthony's Smile Her 17h47'54" 47d20'36"
Anthonys Star Leo 11h0'12" 8d41'15"
Anthony's Star Gem 7h25'45" 27d19'58"
Anthony's Star Umi 15h50'18" 70d50'18"
Anthony's Star Sgr 19h56'59" -32d12'32"
Anthony's Star Sco 17h10'7" -30d20'27"
Anthony's Way Her 17h35'9" 17d17'2"
Anthonystrong Tau 4h18'23" 1d23'39"
Anthoula And 2h28'22" 40d43'0"
anthoula Sgr 19h14'2" -29d56'41"
Anthy Lakkotripis Chocolate Star Ori 5h28'7" -3d47'49"
Anti Maynard Mon 6h45'43" -0d5'34"
Antia Tau 4h17'20" 9d19'54"
Antigalu Psc 23h22'55" 3d59'14"
Antigone Tau 3h45'32" 28d37'55"
Antiguan Heaven Cyg 20h25'57" 45d0'10"
Antina Peg 21h21'49" 21d29'55"
Antionette 032853 Mon 6h56'44" 3d39'35"
Antionette Sedfawy Oph 17h38'38" -0d1'45"
Antiroutine Sco 17h30'9" -32d4'10"
Antje Lyn 6h44'50" 55d1'27"
Antje Uma 10h19'35" 44d20'30"
Antje Friederike Heeren Uma 8h30'41" 69d28'9"
Antje Hartung Uma 10h11'23" 55d5'33"
Antje Köhler Uma 8h54'59" 60d47'8"
Antje Lewin Psc 23h53'7" 0d20'42"
Antje's Stern Ori 6h12'47" 16d31'38"
Antjuan and Minhy Lib 15h33'56" -12d31'24"
AntoBimba Umi 17h40'4" 81d8'20"
Antohepe I Uma 10h31'10" 49d1'49"
Antoine Uma 10h26'55" 43d44'26"
Antoine Uma 9h25'57" 52d44'58"
Antoine Col 5h43'22" -40d54'5"
Antoine & Aisha Hawkins Uma 8h55'17" 62d19'8"
Antoine Boutros Uma 8h31'38" 65d7'20"
Antoine Cazzone Lib 15h0'50" -3d10'14"
Antoine Courtois Gem 6h59'26" 22d28'52"
Antoine de Boton Per 4h49'21" 41d53'3"
Antoine d'Oiron Mon 6h46'38" -0d46'39"
Antoine Grosjean Cas 1h7'42" 61d3'7"
Antoine Katz Uma 8h35'7" 56d58'1"
Antoine L. Prince Sr. Cap 20h56'52" -16d49'16"
Antoine Legoffic Umi 17h58'51" 89d18'49"
Antoine Merkel Tau 5h19'24" 20d40'4"
Antoine Nessralla Aur 5h53'39" 51d45'49"
Antoine Peleyras Uma 11h26'42" 66d55'51"
Antoine Pierre Baptiste Dupont Cnc 8h39'27" 9d56'38"
Antoine R. Vann Eri 4h30'12" -0d45'18"
Antoine Trayvon Robinson Her 17h13'9" 34d5'20"
Antoinette Lyn 8h23'56" 45d22'15"
Antoinette Crb 15h43'41" 38d47'12"
Antoinette Cyg 21h10'50" 40d54'30"
Antoinette Uma 9h33'41" 63d1'24"
Antoinette Badagliacca And 23h52'37" 48d32'30"
Antoinette Basler Uma 8h24'55" 62d2'4"
Antoinette Camarata Follaco Cyg 21h44'14" 37d59'11"
Antoinette Cangialosi Cas 1h26'53" 63d23'39"
Antoinette Cialdella Vir 12h11'11" 11d29'30"
Antoinette Cincotta Aqr 22h12'57" -23d20'0"
Antoinette Competello Cap 21h23'41" -24d43'51"
Antoinette D'Auria Our Shining Star Cap 20h15'43" -11d50'28"
Antoinette Delano Uma 11h2'32" 52d37'20"
Antoinette DeNise Gem 7h43'17" 31d15'5"
Antoinette DeYonker Gem 7h3'37" 23d34'13"
ANTOINETTE ELIZABETH TOTO And 2h27'20" 43d46'3"
Antoinette Elizabeth Toto Vir 14h40'8" -1d19'22"
Antoinette Gandiello And 23h54'26" 36d47'19"
Antoinette Kindlimann Sgr 18h10'21" -19d41'45"
Antoinette Layla D'Amelio Cyg 20h1'1" 43d24'46"
Antoinette Loftin Psc 1h21'19" 32d27'0"
Antoinette Luise Terrell James Peg 22h41'57" 23d30'37"
Antoinette Marie Harris Sco 16h18'19" -16d30'5"
Antoinette Marie Petrillo 3-11-1924 And 0h19'24" 27d34'34"
Antoinette "Nettie" McPartlan Cyg 19h9'22" 51d19'33"
Antoinette Rebecca DiMashe And 1h43'42" 40d50'37"
Antoinette Roman Pho 1h0'27" -48d6'7"
Antoinette Rose Schwartz Cyg 21h34'46" 50d7'50"
Antoinette Rose Whitworth Tau 3h53'37" 25d43'28"
Antoinette Rowe - Smiles & Sunshine Lib 14h56'15" -4d26'1"
Antoinette Rumolo Cep 22h12'27" 63d21'25"
Antoinette Sblendorio Ori 6h14'8" 16d59'40"
Antoinette & Stacy Cyg 20h14'20" 38d5'5"
Antoinette & Sydney Weitzer Lyr 19h17'55" 42d15'28"
Antoinette "Tinny" D'Amato Uma 10h23'4" 48d30'59"
Antoinette - Where Dreams Come True Lib 15h13'0" -5d33'49"
Antoinette's Light Tau 5h8'14" 26d48'21"
Antoinette's Lucky Star Cas 23h36'20" 57d17'59"
Antoinne D. Fitts Lac 22h47'15" 51d25'25"
Antoinne Dietrich Fitts, Jr. Vir 13h57'11" -17d8'58"
AntolySerendipidy Cnc 9h2'40" 8d41'17"
Anton Ori 5h47'59" -3d46'26"
Anton *17.10.1945 Uma 8h37'20" 49d51'37"
Anton A Brynza Per 3h34'10" 47d47'57"
Anton and Ashley Forever Cyg 20h24'26" 46d30'40"
Anton August Wagenbrenner Peg 22h50'7" 32d5'23"
Anton /Crystal Ritchie Cyg 20h8'29" 44d7'23"
Anton E. Melliger Umi 16h14'41" 74d3'41"
Anton Hartmann And 2h23'53" 42d48'3"
Anton J. Miller Uma 13h39'6" 53d54'9"
Anton Joseph Secero Cyg 21h46'17" 37d24'48"
Anton Konhäuser Uma 13h50'4" 58d38'10"
Anton Scheiber Dra 18h50'20" 59d20'56"
Anton & Schuyler Hornung Lyn 6h54'8" 50d35'37"
Anton Stelios Karashialis Cyg 19h52'58" 56d29'34"
Anton Tran Dang Per 4h24'49" 46d51'47"
Anton William Norman Kiff Umi 14h20'21" 66d12'51"
Antone Carvalho Ori 6h19'2" 16d30'43"
Antonella Gem 6h23'44" 26d16'35"
Antonella Per 3h50'46" 49d4'40"
Antonella Uma 10h7'4" 51d25'55"
Antonella Cas 23h49'22" 50d13'25"
Antonella Cyg 20h12'37" 34d31'8"
Antonella Uma 9h21'56" 43d34'2"
Antonella Boo 15h30'49" 34d43'31"
Antonella Uma 9h24'51" 63d3'51"
Antonella Cam 4h36'11" 58d13'50"
Antonella Cam 7h1'35" 68d59'0"
Antonella Caroli Gem 7h29'32" 19d19'17"
Antonella Di Muro 14 aprile 1959 Peg 22h6'12" 18d13'8"
Antonella Falvo Ori 5h46'44" 7d19'36"
Antonella Gambotto-Burke Vir 13h21'24" -13d49'41"
Antonella LiP - 8 Juin 1971 Gem 7h42'15" 20d7'30"
Antonella Mamo Mon 7h16'59" -8d26'40"
Antonella Mauriello Uma 12h43'53" 57d24'41"
Antonella Montanaro Ari 2h51'14" 29d7'2"
Antonella Petrelli Cam 5h35'10" 58d11'51"
Antonella Petrone Cnc 8h13'6" 28d0'52"
Antonella Sabrina Delorenze Vir 13h19'44" 12d2'15"
Antonella Turrin Uma 13h31'21" 60d7'15"
Antonetta Alvira Przybylowicz Lib 15h48'51" -9d51'36"
Antonetta Presutti Lib 16h1'27" -20d2'47"
Antonette Cas 23h42'32" 56d1'33"
Antonette Raquel Currie Sierra Cyg 20h19'42" 48d12'55"
Antoni Luke Vir 13h9'33" 0d0'12"
Antonia Uma 14h4'9" 50d39'32"
Antonia And 1h36'55" 47d7'25"
Antonia Sco 17h1'1" -34d50'6"
Antonia - Buhitoard Cep 21h24'4" 61d11'33"
Antonia Charlotte Vel 9h27'29" -56d55'15"
Antonia de Belleza e Inteligencia Uma 11h57'30" 35d58'45"
Antonia De Vecchi Umi 15h2'6" 68d50'29"
Antonia Denise Sgr 19h36'58" -14d52'41"
Antonia F M Marriott-Dean Mon 7h4'13" 4d1'43"
Antonia Grace And 23h19'58" 47d8'43"
Antonia M B Waldner Vir 13h21'33" 11d9'59"
Antonia Mannion Cas 1h13'4" 54d16'17"
Antonia Matthaei, 09.05.1988 Lac 22h40'43" 42d35'39"
Antonia Mills Cas 23h29'44" 51d43'32"
Antonia Petrina Tersillo And 23h18'30" 50d16'4"
Antonia R. Lombari Uma 10h35'42" 49d11'46"
Antonia Reichert Uma 10h11'28" 70d1'29"
Antonia Rosalie Psc 0h51'43" 28d35'28"
Antonia S. Harter Cyg 21h36'41" 37d48'39"
Antonia Saladino Uma 13h11'55" 55d47'9"
Antonia Sgier Cam 4h57'22" 79d40'5"
Antonia & Thomas Uma 9h8'34" 63d49'43"
Antonia Whyatt Cyg 20h28'56" 40d41'26"
Antonia Wulff And 23h27'58" 47d30'39"
Antonia Y Antonio Cyg 21h11'29" 34d57'35"
Antonieta Di Longo Uma 11h9'39" 34d47'19"
Antonietta Cappella Lyn 7h22'41" 45d54'30"
Antonietta Smeragliuolo Tau 4h32'45" 19d38'47"
Antoniette And 0h51'52" 46d20'41"
Antonija Stefanovic Gem 7h27'14" 32d22'24"
Antonin Peg 22h19'29" 12d40'51"
Antonin Cas 23h7'59" 59d21'23"
Antonin Simon Per 3h21'27" 42d23'37"
Antonina Cru 12h14'10" -63d37'15"
Antonina A. Dawson And 1h36'39" 38d55'45"
Antonini Béatrice Uma 10h19'8" 58d47'45"
Antonino Her 18h18'33" 27d17'30"
Antonino MANNISI Uma 9h47'49" 52d35'3"
Antonio Ori 5h53'41" 16d50'42"
Antonio Aur 6h30'43" 42d0'21"
Antonio Boo 15h19'35" 42d56'50"
Antonio A. Giannico Her 18h51'33" 24d2'9"
Antonio Abrego Aql 19h40'48" -0d3'0"
Antonio Accardo Aqr 22h53'32" -22d32'40"
Antonio Alessandrini Gem 7h43'24" 32d26'53"
Antonio Alilin Ori 4h49'34" 10d52'13"
Antonio Augello Uma 10h1'6" 59d23'17"
Antonio Bagnato Aur 5h25'5" 39d37'36"
Antonio & Barbara Umi 14h20'9" 72d46'53"
Antonio C Jimenez Her 16h42'36" 46d4'28"
Antonio Calò Peg 22h5'16" 17d8'22"
Antonio Cammarata Lyr 19h13'6" 33d38'34"
Antonio Carmen Zumpano Del 20h30'52" 16d33'37"
Antonio Casssarà Cas 1h18'15" 56d23'46"
Antonio Cernadas Cep 21h47'5" 60d23'0"
Antonio Chmait Leo 11h44'50" 26d20'41"
Antonio Colombo The Best Father Cep 22h30'13" 65d46'31"
Antonio Crupi Uma 8h26'11" 65d40'30"
Antonio D'Alessandro And 0h6'46" 47d23'21"
Antonio David Nickerson Cnc 8h24'19" 10d48'5"
Antonio Davis Vir 11h50'54" 0d23'15"
Antonio Di Cosimo Lyn 6h47'45" 60d49'39"
Antonio Dorien Green Gem 7h20'12" 21d23'52"
Antonio E. Ercolano Cnv 12h44'50" 46d53'50"
Antonio Fiaschetti Crb 15h54'13" 34d27'10"
Antonio Francisco Reyes Lyn 8h16'9" 51d34'9"
Antonio Frank Per 3h5'48" 49d10'32"
Antonio Gatuso Uma 11h41'23" 38d57'59"
Antonio & Gennaro Ruoco Ori 5h21'51" -1d28'5"
Antonio Goncalves Aqr 20h47'6" -8d7'9"
Antonio Guerrera Cap 21h31'51" -11d9'2"
Antonio Hamilton Vir 13h21'57" 6d11'8"
Antonio Hernández Tri 1h46'8" 31d43'40"
Antonio Hernández Uma 11h59'43" 64d3'21"
Antonio Hernandez Cas 1h37'30" 60d47'33"
Antonio Herrera Cep 21h9'44" 66d25'20"
Antonio Iammatteo Umi 17h8'29" 77d36'16"
Antonio J. Santos Lyr 19h19'30" 28d6'54"
António Lares dos Santos Psc 1h12'3" 32d55'19"
Antonio & Letty Sue Albert Uma 9h2'41" 47d46'9"
Antonio Leyva Ori 6h20'8" 13d54'45"
Antonio Lobo Xavier Cra 18h16'46" -40d1'46"
Antonio Luis Austin Sco 16h15'1" -17d6'36"
Antonio M. Caputo Aur 5h49'34" 44d18'4"
Antonio Marruso Her 18h28'9" 20d34'10"
Antonio Mendiola Babauta Sco 16h8'37" -12d38'56"
Antonio Michael Spina Lyn 8h31'51" 47d9'14"
Antonio Michael Zona Lib 14h52'29" -2d27'29"
Antonio Miguel Her 16h31'42" 9d59'48"
Antonio Ojeda Gem 6h52'0" 31d33'19"
Antonio Oppedisano Cep 20h28'54" 62d0'1"
Antonio Pace Lyr 18h59'26" 26d59'45"
Antonio Paul Her 18h0'29" 40d49'4"
Antonio Pavia Per 3h7'24" 40d28'49"
Antonio Peltyn Boo 14h52'1" 26d52'0"
Antonio (Porcellino) Aur 5h56'9" 34d38'32"
Antonio Priori papà unico e meraviglioso Umi 17h29'5" 80d20'10"
Antonio Remington Ori 5h40'46" 3d42'23"
Antonio Rex Barrera Cap 21h22'2" -25d2'42"
Antonio Rocco Masserelli Gem 6h36'34" 26d34'21"

Antonio Tò Aliperti
Umi 17h7'7" 80d15'12"
Antonio Tommaso Caldaroni Jr.
Umi 16h34'45" 79d11'35"
Antonio "Tony" Cristofaro
Cyg 21h49'21" 42d15'13"
Antonio Ugo Afasano
Sgr 20h4'21" -26d4'5"
Antonio V. Atiles
Cep 23h9'2" 79d49'34"
Antonio Venegas
Ori 6h3'54" 15d49'18"
Antonio Vincent DeFrancesco
Umi 15h28'1" 77d50'54"
Antonio William Orta-Taylor
Leo 11h16'46" -1d10'45"
AntonioniAnt
Ori 5h36'45" -3d9'48"
Antonios Charles Kiryakis
Leo 11h7'9" 22d56'34"
Antonios Kyriacos Antoniou
Per 2h16'19" 55d58'13"
Antonios Perdicaris
Sco 16h16'2" -11d48'57"
Antonious "7-17-1935"
Uma 10h23'44" 63d30'48"
Antonioyana
Ari 3h27'32" 27d5'5"
Antonis Chrysochoidis
Gem 6h16'13" 22d53'16"
Antonis Gortzis
Uma 8h54'5" 70d40'16"
Antonius Amicas
Cru 12h25'5" -56d6'49"
Antonius Vogels
Tau 5h51'14" 17d6'12"
Antonius Yeknom Gundisalvus
Lmi 10h28'8" 35d59'32"
Antony
Tau 4h11'45" 14d55'2"
Antony
Cep 0h11'19" 73d20'51"
Antony Aubrey John Tiri
Her 17h38'24" 23d30'43"
Antony & Sheralyn's love is forever
Cyg 20h17'53" 33d40'23"
Antosik
Cas 23h21'3" 54d59'8"
AntRein
Hya 8h48'42" 4d2'38"
Antuan R. Hinson
Per 2h59'27" 49d32'55"
Antwuan Demoshaw Green
Gem 7h46'12" 21d14'34"
Anu
Gem 6h53'12" 16d53'39"
ANU
Lyn 6h30'39" 57d35'41"
Anu?ki!
Cnc 8h48'33" 31d36'35"
Anu Kurichh
Uma 10h32'19" 62d4'33"
Anu-Elina
Leo 10h32'34" 13d51'32"
Anugrah and Eriks eternal love
Ori 6h1'24" 6d22'9"
an-uir
Sco 17h20'40" -32d32'32"
an-uir guren Schnüffel
And 23h10'10" 42d21'48"
Anuj Dua
Cap 20h11'16" -10d30'24"
Anuja Ankola
Cnc 9h8'42" 31d24'50"
Anuluk
Aql 19h57'2" 6d45'6"
Anup Sabharwal
Crb 16h13'10" 26d8'16"
Anuque Perera
Cru 11h56'52" -61d32'25"
Anuradha
Uma 10h20'28" 54d7'22"
Anurag Gupta
Lib 14h57'37" -15d48'47"
Anuschka-Maria John
Boo 14h35'50" 41d14'40"
Anush Platter
Cap 20h16'43" -27d16'27"
Anusha
Lyn 8h32'26" 58d57'44"
Anusia
Ori 5h37'29" 5d1'55"
Anuska & Petr
Aqr 20h55'53" -8d16'58"
Anustida and Glenn's Star
Tau 3h48'58" 3d58'12"
Anutam Rajdeep
Ori 5h56'27" 18d6'34"
Anutik
Per 3h50'0" 32d15'35"
Anvita Chitrapu
Vir 12h24'37" -6d11'6"
Anvue
Peg 22h42'45" 23d48'37"
Anwar & Dalila
Crb 15h51'31" 30d56'2"
Anwar Robinson
Ori 5h26'36" 1d46'55"
Any
Cam 6h39'11" 76d5'56"
Any Monkey Can Do It
Ori 5h34'50" 1d32'41"
Anya
Vir 14h39'34" 3d34'2"
Anya
Ori 5h50'25" 19d26'43"
Anya
Ori 5h55'41" 20d57'35"
Anya
And 23h45'55" 35d6'38"
Anya
And 2h21'22" 47d18'59"
Anya
Uma 11h34'46" 47d56'34"
Anya
Sco 16h8'12" -18d26'46"
Anya
Lyn 7h26'16" 53d46'55"
Anya
Sco 17h16'2" -43d8'30"
[AnYa*Ae]
Cyg 21h38'13" 33d10'7"
Anya Elisabeth Rieder
Cap 21h34'35" -18d9'13"
Anya Elizabeth Butkovich
Vir 14h23'26" 3d22'45"
Anya Emilie Hodgson
And 23h37'29" 42d22'14"
Anya Emily Hotston
And 23h11'30" 44d36'3"
Anya Eve Cooke
And 23h12'37" 42d4'44"
Anya Faith Graham-Turness
And 2h13'0" 41d54'46"
Anya Kamen
Lyn 8h22'47" 53d46'46"
Anya Lucena
Umi 13h18'40" 69d33'23"
Anya Ryan
Lyn 7h7'45" 58d36'25"
Anya Shira Ditkoff
Umi 17h44'3" 80d43'33"
Anya Victoria Hayes 21st Birthday
And 2h29'13" 41d16'18"
Anya Warren
Uma 8h34'51" 46d36'41"
Anyah Brianna
Ori 5h53'32" 18d41'34"
Anya-John
Cyg 21h24'53" 36d24'32"
Anyaliese Tischer Hare
Aqr 23h2'17" -23d22'16"
Anylou
Lmi 10h37'45" 31d41'21"
Anyssa
Lib 15h30'29" -10d20'49"
Anything for you - even the stars
Cru 12h46'51" -64d38'14"
Anything is Possible
Vir 13h19'43" 9d55'32"
Anything is Possible!
Vir 12h53'7" 11d48'7"
Anyuta22
Lyn 7h34'15" 46d31'45"
anywhere
Sco 16h43'32" -31d8'43"
Anzhelika 07-08-1971
Cas 0h52'1" 56d44'19"
Aoi and Shigenori Okumura
Cyg 20h45'19" 38d30'19"
AOI AOYAMA
Vir 13h39'11" 3d14'27"
Aoibh
And 2h3'11" 38d1'56"
Aoibh Katie Tomany
And 0h14'19" 46d37'46"
Aoibheall an Sionnach
Tau 4h43'33" 17d50'57"
Aoibheann
And 1h21'10" 43d9'39"
Aoibheanna Hone
And 2h9'57" 45d34'22"
Aoibhinn Aisling
Dra 17h48'30" 50d49'27"
Aoibhinn Sullivan
Cyg 19h51'2" 31d59'51"
"Aoibhinn" The Beautiful
Uma 12h39'1" 59d0'26"
Aoife
And 0h48'49" 27d54'30"
Aoife
Peg 22h30'4" 29d58'16"
Aoife MacAlasdaire
And 23h7'28" 43d0'55"
Aoife Mae Long
And 2h31'18" 49d56'50"
Aoife Maria Mc Loone
And 23h33'5" 49d24'3"
Aoife McNicholl
Uma 11h31'38" 29d58'1"
Aoife Nora Butler
Tau 4h46'5" 22d41'6"
Aoife Phillips
And 0h15'30" 47d39'22"
Aoife Riley McCarthy
Uma 11h26'1" 41d50'42"
Aoife Trundle
And 1h58'28" 37d56'35"
Aoife's Wishing Star!
Peg 22h42'54" 3d45'35"
Aoileann De Hora
Cnc 9h6'4" 17d25'12"
AOM!!!
Vir 12h19'43" 8d6'46"
Aom
Cap 21h43'2" -17d33'10"
AOP King
Cep 22h23'57" 58d3'49"
Aout
Lyn 7h13'3" 49d42'25"
Aowlee
Cnc 8h0'20" 13d22'42"
AP 420
Cam 4h10'17" 68d42'5"
Apa 50 - Márkus József
Aqr 20h42'36" -2d17'40"
Apa Csillagai, Rebeka és Tamara
Uma 13h33'28" 61d19'51"
APB2002
Vir 13h22'36" -16d54'31"
Ape Face (My Love)
Leo 11h38'41" 23d43'48"
APFANZ05
Aqr 23h26'17" -17d38'39"
Aphrodite
Aqr 21h31'45" 2d22'49"
Aphrodite
Cyg 21h19'27" 45d8'16"
Aphrodite
Uma 11h33'29" 47d29'0"
Aphrodite
And 2h9'30" 43d2'58"
Aphrodite F. Sarelas
Cas 1h43'8" 64d22'38"
"Aphrodite's Heart"
Cyg 19h35'36" 30d41'2"
Api
Uma 14h22'47" 58d2'14"
Api Vargas Rivas
Her 18h47'16" 22d35'22"
Apicius.Com
Uma 10h31'58" 43d6'49"
Apina Solomon's Pathfinder Star
Cyg 20h2'16" 40d34'6"
Apirl Dawn Hall
Ori 5h36'45" 11d45'4"
AplusBequalsLife
Gem 6h46'27" 14d28'2"
Apo
Leo 11h20'49" 1d37'23"
Apolinar "Poli" Rivera
Cap 21h38'56" -13d29'59"
Apollange8
And 1h59'57" 36d52'30"
Apolline
Del 20h38'18" 16d45'40"
Apolline Brichler
Cap 21h32'20" -23d6'10"
Apolline Vergneault
Sgr 18h15'4" -19d18'2"
Apollo
Cma 6h54'16" -13d17'2"
Apollo
Cma 6h22'56" -14d59'12"
Apollo Busleta
Umi 17h22'28" 77d38'24"
Apollo Estrella
Mon 8h3'32" -0d32'17"
Apollo Molinari
Cma 6h53'44" -21d28'43"
Apollo the Apalled
Aur 4h51'50" 36d44'7"
Apollo William Petrarca
Leo 10h48'37" 18d22'18"
Apollos John-Paul Kimbrough
Gem 6h51'24" 27d56'30"
Apollo-Zeppelin
Ori 5h52'41" 21d19'55"
Apostle J. W. Pruitt
Boo 14h11'56" 29d33'24"
Apostolos
Lyn 8h18'11" 38d23'45"
Apostolos Adamopoulos "Paul"
Peg 22h13'58" 33d3'33"
Apostolos Alexakis
Uma 10h9'38" 70d48'11"
Apple of My Eye
Uma 9h38'35" 66d55'27"
apple of my eye
Uma 12h36'12" 55d6'37"
Apple Pie in the Sky
Ari 3h16'43" 28d29'16"
Apples Hyzy
Lyn 7h38'10" 36d7'2"
Appleton - Hough x
Cyg 20h3'0" 58d52'20"
Applicant Services Division
Umi 13h52'0" 76d17'13"
Appling-Smith
Lyn 7h39'1" 35d28'22"
Aprea
Her 17h40'55" 19d9'37"
"Aprian"
Lyn 8h36'22" 38d48'23"
April
And 2h28'45" 41d32'12"
April
And 0h22'22" 37d31'56"
April
And 0h18'52" 44d58'32"
April
Lyn 9h4'38" 36d29'52"
April
Ari 2h47'59" 30d23'22"
April
Lyr 18h51'10" 37d24'27"
April
And 2h22'39" 50d19'0"
April
Cyg 20h58'10" 46d8'3"
April
And 0h21'54" 29d1'59"
April
Cnc 9h17'51" 20d19'4"
April
Com 12h30'1" 28d35'47"
April
Ari 3h6'55" 11d51'1"
April
Psc 0h30'52" 18d53'40"
April
Lib 14h28'27" -17d47'38"
April
Aqr 22h51'12" -16d51'55"
April
Cap 21h23'15" -18d40'8"
April
Lyn 6h27'9" 56d43'9"
April
Umi 13h29'50" 74d27'47"
April 10th
Crb 15h44'53" 32d10'58"
April 4.3.56
Ari 2h45'30" 26d17'54"
April A. Peterson
Vir 15h0'38" 5d4'34"
April Alden
Cam 6h15'9" 60d48'50"
April Amarna
Sgr 18h2'54" -29d49'36"
April and Court's Wedding Star
Vir 13h27'25" -5d30'7"
April and Mikey 2005
Ari 3h13'8" 26d15'30"
April and Ryan Parks
Sge 20h3'34" 16d52'16"
April and Skip
Ari 2h32'39" 30d45'59"
April Andrews
Tau 5h6'42" 27d0'14"
April & Andrew's Wedding Star
Cyg 20h21'17" 47d16'46"
April Andrews-Singh 061474
Gem 7h32'45" 17d55'29"
April Ann
And 0h28'43" 25d21'43"
April Ann
Tau 3h59'11" 2d38'35"
April Ann
And 23h48'40" 36d11'35"
April Ann 84
Sgr 19h12'25" -17d2'27"
April Ann Turner
And 1h37'32" 44d12'39"
April Audine Yoder
Her 18h44'21" 21d8'5"
April B. & Jim T. Together Forever
Cyg 20h22'46" 45d8'47"
April B. Ortega
Ari 2h50'41" 28d8'22"
April Barabash Angel In The Sky
Lyn 7h22'26" 53d14'55"
April Berry's Princess Star
Psc 0h26'3" -0d0'5"
April Blossom Haywood
And 22h58'55" 47d32'36"
April & Bo Forever
Uma 9h7'26" 63d9'37"
April Bradley Clark
Uma 9h31'17" 44d29'31"
April Cassano Rhodes
Gem 6h47'1" 28d32'43"
April Catherine Houten
And 0h15'44" 43d16'30"
April Christine Curtis
Cyg 20h15'26" 51d35'17"
April Christine Jeannie Villagrana
Psc 1h39'57" 23d9'13"
April Christine Willis
Lyr 19h12'38" 27d7'52"
April D. Johnson
And 1h17'58" 45d38'33"
April & Dave's Lucky Star
Psc 0h14'24" 4d10'59"
April Davis
Lib 15h26'21" -12d29'18"
April Dawn
And 0h11'56" 27d54'28"
April Dawn
Cnc 8h3'4" 11d11'22"
April Dawn Dockstader
Uma 11h7'21" 42d9'26"
April Dawn Rush
And 23h26'30" 42d47'30"
April Dawn Zrallack
Uma 11h40'33" 63d40'45"
April Dawn's Star
And 2h11'44" 38d55'43"
April Eck
Uma 13h39'55" 59d28'47"
April Elizabeth
And 0h34'10" 38d30'39"
April Elizabeth Goble
And 1h10'45" 37d40'25"
April Firnhaber
Uma 10h33'51" 68d7'37"
April Fool's Love
Cyg 21h40'10" 42d46'43"
April Golightly
Ari 2h0'15" 19d50'8"
April Gosnell
And 0h29'57" 37d30'58"
April Graber
Oph 16h25'30" -0d11'50"
April Green Eyes
Tau 5h3'43" 16d13'26"
April Gwen Tinsley
Cru 12h12'59" -61d53'53"
April & Hagai
Tau 3h43'35" 23d17'48"
April Hunt
Lyn 7h43'44" 35d16'42"
April I. Cleave
Uma 10h20'39" 43d34'58"
April is Beautiful
Gem 7h4'13" 29d6'10"
April Ivory
Tau 3h42'12" 28d25'36"
April I.W. Horler
Per 3h27'25" 45d25'26"
April J. Hunter
Srp 18h18'25" -0d13'25"
April Jakelsky
Cas 23h45'46" 61d10'15"
April Jaques
Lib 15h4'44" -12d41'12"
April Jean
Lib 15h53'21" -12d55'12"
April Jean Nelson (Earth Angel)
Vir 12h34'11" -6d10'25"
April Joan Matera
Gem 6h31'26" 26d7'5"
April Joy
And 0h37'0" 38d45'2"
April Joy DeWitt
Ari 2h20'42" 10d59'35"
April Kandil
Lyr 18h57'14" 33d19'42"
April L. Gilcreast
Sgr 18h0'45" -18d41'7"
April LaChelle Burke
And 0h41'58" 23d6'36"
April Lauren Wright
Cas 0h44'29" 67d2'10"
April Lea
Tau 4h38'59" 18d40'29"
April Lea Pucel
Tau 4h20'30" 26d30'9"
April Lee
And 1h36'30" 48d4'29"
April Lee Ashby-McGill
Vir 11h38'55" 9d15'45"
April Leigh McRight
Tau 4h31'17" 17d4'16"
April Lena Cortorreal
Ari 2h13'22" 23d46'53"
April Leon's Star
Ari 1h58'14" 22d34'12"
April Liebig
Vir 14h23'38" 4d26'28"
April Louise Altman
Lyn 9h15'1" 41d49'58"
April Louise Conley
Psc 1h19'16" 18d36'47"
April Louise Langlinais
And 0h35'7" 39d8'0"
April Love
Tau 5h3'11" 27d43'43"
April Love
Vir 13h24'19" -17d7'43"
April Lovett
Lib 14h36'25" -12d25'1"
April Lyn Joslyn
Tau 5h29'31" 20d3'21"
April Lynn
Psc 23h16'36" 1d43'47"
April Lynn
Ari 2h7'29" 24d10'6"
April Lynn
Ari 2h18'25" 27d9'27"
April Lynn
And 0h38'15" 41d21'36"
April Lynn
Vir 12h59'56" -15d53'15"
April Lynn
Lyn 6h45'27" 56d22'25"
April Lynn Fillmore
Uma 10h52'15" 44d58'39"
April Lynn Jost
And 0h42'39" 41d5'41"
April Lynn Nielsen
Lyr 19h7'19" 42d25'42"
April Lynne
Uma 11h19'10" 40d58'25"
April Lynne Brabson
Vir 13h14'39" 9d37'10"
April M. Bayinger
Ari 3h17'28" 29d17'56"
April M. Dean
And 2h36'57" 48d23'30"
April M. Mills
Ari 2h17'22" 13d31'14"
April M. Valenti
And 23h25'46" 48d7'14"
April Madison
Crb 15h32'14" 30d43'25"
April Mae Garnish
Ari 2h11'36" 24d54'29"
April Marie
Leo 11h38'50" 17d38'52"
April Marie
Tau 4h31'57" 19d35'59"
April Marie Allen
Umi 15h59'42" 77d38'5"
April Marie Camp
And 23h21'31" 47d19'31"
April Marie Davies
Cnc 9h4'49" 6d59'2"
APRIL MARIE HULTIN
Gem 7h17'36" 19d36'20"
April Marie Meier
Ori 5h40'25" 8d10'40"
April Marie Pedersen
And 0h14'10" 32d51'19"
April Marie Porter
Ari 2h32'14" 25d57'11"
April Marie Scioscio
Psc 1h19'13" 32d42'23"
April Marie Summerlin
Sgr 20h21'52" -31d59'29"
April Martines Sweet Sixteen Star
Sgr 19h8'58" -21d6'33"
April Matthews
Uma 9h36'56" 45d46'28"
April Maxwell
Tau 4h21'47" 17d27'48"
April McBride
Leo 10h39'2" 17d31'11"
April McCants
Cnc 8h29'45" 25d29'22"
April Michelle
Leo 10h43'56" 9d47'28"
April Michelle Bonniville
Ori 5h36'26" 2d41'17"
April Michelle Morton 16 Cents
Gem 6h49'23" 16d32'15"
April Michelle Soudipour
Uma 9h40'31" 45d39'58"
April Monina Zorrilla De Armas
Tau 5h33'41" 19d14'51"
April My Babybear
Ari 2h32'57" 22d6'12"
April N. & Bailey G.
And 1h10'14" 38d8'15"
April Niccole Rider-Clements
Lyn 7h49'37" 57d39'4"
April Nicole Beaty
Ari 2h25'36" 26d37'55"
April Nicole Elkins
Gem 6h54'22" 33d39'38"
April Nicole Hunt
Gem 7h35'5" 27d4'41"
April Niimi-Miller - 2004
Vul 20h17'22" 24d16'4"
April Nikole
Ari 3h10'53" 28d15'28"
April Ovalle
Ari 2h30'8" 24d51'15"
April Panyasithavong
Cap 20h23'45" -10d44'46"
April Perreault
Dra 19h14'53" 79d29'48"
April Perry
Uma 11h33'9" 64d28'13"
April Peters
Ari 3h16'23" 30d11'36"
April Petkosek
And 0h37'10" 37d13'12"
April Pruitt
Cam 4h13'0" 62d10'17"
April Ray, My Heart
Aqr 20h41'4" 0d45'54"
April Rayna Garcia
Gem 6h47'8" 32d14'38"
April Rayne
Dra 18h52'44" 64d10'6"
April René
Aqr 22h26'41" 1d46'29"
April Renea Goins
Per 3h33'0" 44d30'8"
April Renea Hopkins
Cnc 8h11'18" 24d27'24"
April Renee Duarte
Tau 5h8'22" 22d38'58"
April Richardson
Ori 5h24'18" 2d10'41"
April Roberson
Tau 4h31'50" 2d44'20"
April Rodriguez-Rising Star
Cyg 21h14'51" 45d56'8"
April~Rose
Tau 3h56'35" 23d47'54"
April Rose
Tau 5h43'11" 27d11'21"
April Sanchez Woolf
Cas 2h17'15" 65d41'45"
April Scoles
Tau 4h8'12" 24d13'25"
April Showers
Uma 11h19'47" 51d47'7"
April Simone Gordon
And 23h16'57" 41d24'58"
April Spring
Tau 5h48'18" 15d33'26"
April Starr
Cnc 8h31'59" 9d27'2"
April Starr Long
Lib 15h11'42" -4d58'24"
April Sunshine
Cas 0h45'14" 57d43'44"
April Swan of the Eternal Moon
Cyg 22h0'2" 53d23'23"
April Tamiko Sakamoto
And 0h54'54" 36d27'13"
April the love of my life!
Lyn 7h28'9" 59d12'12"
April Trow
And 23h30'47" 38d11'13"
April Wareham
And 23h31'21" 49d29'52"
April Watkins
And 2h7'26" 44d52'52"
April & Will Newman
Cyg 19h58'53" 58d4'21"
Aprilis az én egyetlen lelki társam
Dra 15h41'12" 61d4'13"
April-Kay
And 0h25'1" 38d25'56"
AprilLeeStrait
Psc 1h15'54" 10d41'14"
Aprillio
Umi 15h17'11" 73d24'44"
April's Doorway to Dreams
Cnc 8h36'43" 8d2'53"
April's Heart
Cnc 8h40'32" 17d59'27"
April's Hope Star
Tau 3h47'56" 11d2'19"
Aprils Joy & Light
Lyn 7h4'41" 47d49'46"
April's Peach
Gem 7h19'49" 33d7'25"
April's Star
And 23h41'5" 33d12'56"
AprilShine/LightofmyLife
Aqr 22h30'20" -9d52'33"
aprosdokiti telya agapi
Cas 1h5'12" 49d17'11"
Apryl
And 1h5'40" 46d15'11"
Apryl Dawn Scott
And 0h42'41" 43d0'16"
Apryl Malarcher
Lib 15h28'38" -6d42'31"
Apryl Marie
Vir 13h17'33" 3d49'31"
Apryl Marie Rose
Vir 12h15'55" 11d40'21"
Aptenodytes forsteri, Agape
Sgr 19h35'23" -14d29'59"
Apu, 09.09.1978
Uma 8h40'34" 70d45'54"
Apurva Thakkar
Cyg 20h37'31" 42d34'13"
AQD
Uma 11h26'30" 38d20'28"
Aqiyl
Tau 4h21'2" 2d36'5"
AQUA
Psc 22h55'13" 3d34'16"
Aquarius Emma
Dra 18h16'27" 57d42'57"
Aquarius Ganymede
Aqr 22h5'2" -9d34'16"
Aquarius King
Aqr 21h39'14" 0d20'50"
Aquatica Major
Leo 11h8'56" 2d26'42"
a'–qua-tse'–li sa:-sa
Lyn 7h45'38" 39d11'28"
Aquilla
Lib 15h40'57" -6d38'44"
Aquilla Gail Gorton
Cap 21h31'10" -19d10'1"
Aquinas Bregel Powers
Sgr 19h32'58" -13d36'37"
Ar Amharc
Lmi 10h27'8" 37d34'28"
Ar Brách Grá
Ori 6h12'0" 3d44'12"
ar máthair
Crb 15h41'10" 38d28'21"
a'r ngra'
Lib 14h59'1" -21d3'29"
AR Superstar
Uma 11h36'15" 62d5'52"
Ara
Peg 22h28'10" 5d53'19"
Ara Grace
And 1h11'19" 46d4'27"
Ara Katherine Griffith
Ara 17h30'57" -47d36'50"
Arabella
Sgr 18h59'25" -32d8'6"
Arabella
And 1h9'11" 42d29'47"
Arabella Hodder - Arabella's Star
Cas 0h55'28" 62d59'37"
Arabella Joy
Lyn 6h46'11" 51d52'57"
Arabella Mae
Cyg 20h56'5" 35d54'34"
Arabella Marianna Serafini
Cas 1h5'15" 62d1'37"
Arabella Nelson
Leo 9h35'21" 30d21'11"

Arabella Schraeder
Aqr 22h18'48" -2d58'25"
Arabi
Tau 5h24'4" 17d50'5"
Arabian Horse
Equ 21h9'57" 10d37'13"
Araceli
Psc 23h50'3" 7d59'56"
Araceli
Lyn 7h59'58" 57d38'32"
Araceli Acosta
Gem 6h28'22" 17d12'41"
Araceli Andrade's Shining Star
Sco 17h43'47" -35d53'52"
Araceli Angel Star
Sco 17h0'21" -32d42'6"
Araceli Cheli Villalobos
Tau 4h22'16" 25d19'41"
Araceli Esparza
Cyg 21h35'22" 50d48'4"
Araceli Garcia-Longoria
Sco 16h58'25" -32d7'32"
Araceli Godwin
Ari 2h39'33" 22d12'12"
Aracelis Eneida
Cam 4h35'9" 58d5'10"
Aracely
Tau 3h33'32" 23d52'33"
Aracely Flores Horcasitas
Umi 16h23'43" 73d55'4"
Aradonna
Ori 5h29'54" 3d8'46"
A-Rae Starr
Gem 7h25'47" 33d15'12"
Araf M. Evans-Sykes
Lib 15h47'20" -17d17'12"
Araghie Kanne
Tau 4h33'14" 8d20'51"
ARAGON
Ari 3h15'4" 27d17'26"
Araine
Sgr 18h31'34" -17d46'8"
Arakawa, Takusaburo
Uma 9h22'19" 49d29'19"
Araksik Martirosyan&Arman Mayilyan
Uma 10h3'24" 45d39'37"
Arali
Uma 14h19'49" 59d17'20"
Aralyn Eve Schmidt
Sco 16h17'11" -13d47'53"
Aram Aljammaz
Cnc 8h36'28" 30d48'23"
A-Ram Heart J-Man
Uma 10h14'5" 66d49'24"
Arames Soleil Bowles
Uma 11h27'8" 63d22'4"
Arami
Cap 21h50'55" -8d56'29"
Aramis
Sco 16h11'40" -17d24'1"
Aramis Rodriguez
Sgr 19h18'29" -13d28'1"
Aramis Witcher
Leo 11h12'50" -2d28'40"
Arancha
Leo 9h32'15" 29d5'33"
arancha1319
Cnc 8h47'49" 14d23'10"
Arandini
Cmi 7h8'45" 10d23'12"
Aranea Corryn The Blacque Wydow
Lib 14h52'17" -3d30'46"
Aranel
Ori 5h29'29" 10d13'42"
Aranka (Ari) Bodnar
Leo 11h34'28" 13d31'52"
Aránzazu
Cnc 8h38'36" 19d42'41"
ARAPERI
Leo 9h55'20" 15d9'43"
ARAPERI
Leo 11h58'17" 14d48'48"
Aras Nerual
Ori 4h52'37" 1d24'18"
Aras Rutelionis
Sgr 18h57'58" -34d23'49"
Arasally Rodriguez
Psc 0h43'59" 19d44'27"
Arash 23
Gem 6h49'57" 28d10'37"
Arash-Naseem
Cyg 21h15'48" 42d16'50"
Arastar-Smata
Umi 15h10'37" 78d56'31"
Arathi
Leo 11h44'53" 20d44'4"
Aravaipa
Tau 5h34'51" 27d33'3"
Araven
Cas 0h56'6" 63d41'21"
Aravind & Trinh
Uma 11h10'41" 49d51'17"
Arawyn "hobbit" Francis
Gem 6h25'20" 19d8'10"
Araxie Nadjarian
Per 3h22'22" 42d0'53"
Araya Phuphanich
Leo 11h36'38" 19d57'40"
Arayonelley
Cas 0h41'6" 66d43'57"

ARAZU HOMERA SIGARI
Uma 10h58'44" 59d51'59"
Arbor Lane ~ John & Florence Krebs
Umi 15h29'36" 72d26'35"
Arbular Mathews
Sgr 19h12'41" -16d1'59"
ARC
Cnc 8h47'19" 16d31'49"
ARC-1945 "12-18-45"
Lyn 8h52'20" 39d27'33"
Arcadia
Her 16h50'18" 23d21'56"
Arcadia
Uma 10h45'56" 70d20'56"
Arcadia Color Guard 2006 - 2007
Uma 10h39'46" 55d48'49"
Arcal
And 2h11'17" 48d47'58"
Arcanese
Cam 4h42'34" 70d55'24"
Arcangela
Gem 6h46'14" 32d24'7"
Arcangela Cerrone
Uma 9h25'4" 44d58'35"
Arcangela Lee Gulotta
Leo 9h31'41" 28d35'38"
Arcarius
Ori 5h51'58" 9d10'43"
Arcelia (Arcy) Medrano
Crb 15h32'55" 37d10'58"
Arcelia Garcia
Cam 5h37'4" 68d28'2"
ARCELIA LUNA SESATE
Tau 3h32'1" 17d59'41"
Arcelia Ojeda
Vir 13h13'49" 5d45'46"
Arcenia C. Sana Star
Cap 21h40'2" -18d24'40"
Arcenio Mesa
Lib 14h25'58" -18d36'40"
Arch M. Johnstone
And 23h8'16" 50d45'10"
Archana Raj Gidda
Tau 4h31'55" 21d46'42"
Archana Suresh
Lib 15h10'48" -8d35'50"
Archange Chloée
Del 20h27'28" 15d6'32"
Archangel © 1974
Uma 9h0'59" 53d42'14"
Archangel Raphael
And 23h41'45" 41d12'39"
Archangla
Cap 20h59'4" -16d57'17"
Archbishop Fiorenza
Aql 19h29'40" 5d23'12"
Archer
Per 4h9'59" 41d20'13"
Archibaldus
Her 16h31'44" 44d34'41"
Archie
Ori 6h4'18" 6d35'22"
Archie
Peg 22h26'50" 22d10'13"
Archie and Marion Forever
Cam 4h15'2" 66d41'11"
Archie B
Her 18h15'20" 19d35'46"
Archie Banjo
Per 2h59'1" 42d20'46"
Archie Bickling
Per 4h38'33" 50d56'52"
Archie Biggs' Christening Star
Cep 20h52'40" 59d13'54"
Archie Courtenay Perrott
Uma 12h33'7" 52d21'52"
Archie & Dorothy Muggelberg
Cyg 21h36'42" 38d18'56"
Archie Edwin
Her 16h39'33" 12d44'50"
Archie Finlay Caldwell Barker
Umi 16h4'11" 83d45'40"
Archie Frost
Ari 2h42'9" 29d43'43"
Archie Grahame Edwards
Umi 21h38'11" 89d18'56"
Archie H. Crook
Uma 9h6'30" 56d22'52"
Archie James Birtwhistle
Per 4h14'42" 31d27'22"
Archie John Addis
Dra 17h31'24" 51d34'32"
Archie Kenneth Graham - Archie's Star
Her 18h30'59" 16d31'23"
Archie Lacarte Born: Fri Feb 13/1942
Uma 11h30'47" 56d37'28"
Archie Lowell Jarvis
And 1h56'55" 47d15'8"
Archie Lucas Head
Her 18h22'42" 13d20'14"
Archie Luke Harris - Archie's Star
Her 18h23'18" 20d22'16"
Archie M. Lindsey
Leo 11h14'27" 15d36'30"
Archie Oliver Ames
Her 18h47'55" 15d47'33"

Archie Oscar Morgans
Ori 5h17'20" 7d48'42"
Archie Quinn Smith
Uma 9h19'9" 70d5'1"
Archie Ralfs
Lmi 9h59'38" 37d19'15"
Archie Reginald Lafond - Archie's Star
Her 16h56'20" 15d23'26"
Archie Terence Ward-Hales
Uma 9h21'41" 58d56'38"
Archie Thomas Horner
Her 18h52'34" 21d10'56"
Archie Thomas Webb - Archie's Star
Per 2h40'13" 50d59'44"
Archie William Finlay Poulter
Cep 21h28'52" 67d58'42"
Archienell
Uma 9h49'48" 59d20'46"
Archie's Star
Uma 9h48'5" 62d52'21"
Archilleus
Cnc 8h53'50" 11d43'8"
Archimedes
Gem 6h49'19" 15d41'7"
Archimedes Franklin
Lyr 18h43'23" 31d33'6"
Arciaga Family
Uma 8h53'14" 60d41'3"
Arcie
Her 16h39'44" 11d34'32"
Arcie D. Sapp
Psc 1h39'54" 8d48'17"
Arco-Chrysald
Ori 5h21'46" -4d16'54"
ARCS Foundation Honolulu Chapter
Cet 2h6'35" -23d14'52"
Arctic Fox
Ori 6h19'24" 14d44'49"
Arctic Roz
Gem 6h36'7" 21d38'53"
Ardaz Glendower
Dra 17h26'27" 65d36'6"
Ardell Schneider
And 1h10'1" 41d44'4"
Ardella Oldham
Mon 6h39'20" 6d21'0"
Ardelle
And 0h33'53" 41d48'49"
Arden Joseph Montenaro
Cap 21h41'6" -16d30'56"
Arden Mallory Baner
Umi 15h13'24" 81d10'56"
Ardenia Jewel
Lmi 10h41'59" 34d41'32"
Ardeth
Cas 2h39'34" 67d35'7"
Ardeth Lela
Uma 11h41'42" 34d13'2"
Ardian Kovanxhi
Cam 6h3'36" 62d39'25"
Ardice Russell
Cam 5h33'5" 62d13'11"
Ardie
Aql 20h26'42" -5d48'5"
Ardie Natasha Osias
Aqr 22h56'10" -16d47'43"
Ardie's Perfection From Heaven
Sgr 19h3'19" -17d12'10"
Ardijan "Donny" Gjevukaj
And 23h8'50" 40d30'19"
Ardis Lee
Uma 10h11'38" 64d49'21"
Ardith Sue Carroll
Uma 9h57'58" 53d55'0"
Ardus
Cnc 8h55'41" 24d14'22"
Ardy
Ori 5h29'29" 13d46'15"
Ardyce Rynn
Aur 5h27'16" 32d30'59"
ARDYFAS
Uma 13h49'32" 57d41'27"
Ardys L. Baker
And 0h14'46" 25d50'40"
Ardyth J. Festin
Psc 1h16'33" 21d13'23"
Are we there yet, Daddy Lantz?
And 2h23'38" 45d40'58"
Are You Sleeping
Dra 19h43'19" 68d25'48"
Area 51 Ricardo Hernandez
Lib 15h37'34" -13d31'13"
Areej Dajani's Star
Gem 7h51'7" 27d33'8"
Arek Golas
Per 2h50'2" 42d33'26"
Arek&Julia
Cas 0h1'0" 59d13'47"
Arek's Pearl
Vir 13h3'24" -0d5'42"
Arelia
Uma 12h8'9" 58d57'36"
Arelis
Ori 5h35'2" -3d51'52"
Arelolo
Eri 3h17'48" -42d23'6"
Arely Diaz
Lib 15h2'15" -2d24'49"

Aren James Dickinson
Ori 4h59'25" 13d22'47"
Aren Luo
And 0h55'25" 37d3'32"
ARENA QB #5
Tau 4h10'42" 13d28'5"
Arend Edo Meinders
Dra 20h25'31" 79d36'55"
Arens, Katja
Ori 5h40'8" -0d1'28"
Areola Theodosia
Cas 1h41'44" 67d25'8"
Aretha N. Persaud
Lyn 7h14'5" 44d40'56"
Aretina Perez
Cam 3h51'47" 67d10'41"
Arezoo
Vir 13h44'0" -4d54'15"
Arfie
Uma 9h15'56" 63d49'18"
ARGAM YEGHIAZARYAN
Tau 5h8'38" 25d34'35"
Argean Cook Blackwood
Lib 15h6'23" -4d20'41"
Argelia Nevarez
Lyn 7h53'23" 43d6'44"
Argentina Marie Jordan
And 23h36'49" 48d1'29"
Argentina Shines
Aqr 21h31'24" 2d29'29"
Argentina's Pelota Divina
Gem 7h20'26" 18d6'4"
Argie & Ernest Stanberry, Sr.
Lyn 6h33'13" 60d51'5"
Argo Star
Ari 2h31'53" 27d6'56"
Argun & Lenora's Engagement Star
Cyg 19h34'29" 31d34'41"
Argyle's Angel
Uma 11h18'24" 32d40'30"
Arham Asif
Umi 14h28'14" 75d11'0"
Ari
Sgr 18h57'1" -32d19'32"
Ari
Cap 21h23'8" -25d20'22"
Ari Alexander Flemming
Ori 5h38'18" -2d43'22"
Ari Blue Jones
Ari 2h49'29" 15d52'32"
Ari Bunter
Ori 5h14'5" -6d50'10"
Ari Daniel Wise
Her 17h43'27" 16d13'14"
Ari Furuya
Lyn 7h49'21" 52d39'59"
ARI KAGAN
Aqr 20h55'52" -7d19'7"
Ari Philip Brenton-Melanson
Aqr 22h39'24" -21d18'12"
Ari Rath
Cap 20h57'38" -17d4'31"
Ari Samuel Jaffe
Leo 9h31'1" 25d31'16"
ARI - THE LION OF MY REALM
Leo 11h49'37" 22d45'29"
Aria
Sco 17h13'44" -31d1'24"
Aria Kathleen
Tau 4h21'57" 23d27'45"
Aria Lee DiLoreto
Vir 13h44'59" -4d44'51"
Aria Natasha
Vul 19h49'4" 23d14'24"
Aria Rose Moreaux
Cnc 7h57'45" 18d46'32"
Aria Sovann Maccabee
Lib 15h45'59" -27d9'48"
Aria the Huntress
Cap 20h39'25" -19d47'38"
Ariadna
Crb 16h5'54" 36d52'48"
Ariadna Medina Bota
Ari 3h27'10" 26d30'13"
Ariadna Mora Ferrandis
Uma 10h44'40" 56d37'2"
Ariadne
Uma 12h37'45" 60d18'9"
Ariadne
Sgr 19h39'12" -37d51'29"
Ariadne
Psc 23h29'39" 3d58'50"
Ariadne Dinel Pond
Cnc 8h40'40" 31d5'51"
Ariadne Erika Levada Tannous
Tau 4h26'48" 19d35'58"
Ariah Leanne Sanve
Ari 3h1'22" 31d9'37"
Arial Liana
Aqr 23h16'11" -13d38'0"
Arial Mckenzie Deese
Psc 1h0'31" 9d17'47"
Arián
Per 3h11'37" 45d22'5"
Arian Maree Lopreiato
Tau 4h20'46" 25d44'25"
Ariana
Del 20h26'40" 18d46'21"
Ariana
Ari 2h19'30" 25d22'0"

Ariana
Aqr 21h1'44" 0d7'27"
Ariana
Vir 14h27'47" 3d19'23"
Ariana
Vir 13h2'18" 11d51'46"
Ariana
And 2h21'42" 49d20'58"
Ariana
And 23h39'56" 42d29'24"
Ariana
And 1h21'16" 33d56'40"
Ariana
Cap 20h59'1" -16d6'12"
Ariana
Sgr 18h18'15" -18d18'31"
Ariana Abigail Ramsundar
And 0h15'7" 32d14'13"
Ariana Alexandra
Cas 0h28'31" 60d26'19"
Ariana Alexis Massey
Cam 3h29'36" 61d16'14"
Ariana Aliki
Lib 15h8'16" -4d38'43"
Ariana "Ani" Samaras
Uma 12h9'31" 56d32'4"
Ariana Christina Uriati
Crb 15h36'1" 39d0'13"
Ariana Daniel
And 0h56'28" 38d37'36"
Ariana DeCiuceis
Leo 9h39'37" 27d6'21"
Ariana Eve
Umi 15h2'15" 81d38'38"
Ariana Flores
Aqr 20h39'44" -9d41'23"
Ariana Gabrielle Cecelia Duax
And 0h17'55" 30d32'40"
Ariana Giselle
And 0h41'22" 26d23'8"
Ariana Griffith ~ Forever a Star
Cas 1h15'11" 66d13'36"
Ariana Jean Lopez
Vir 13h14'51" 6d11'15"
Ariana Kamran
And 0h19'38" 42d11'55"
Ariana Lee
And 23h24'58" 41d52'50"
Ariana Lopez
Cap 21h23'46" -26d58'36"
Ariana Luz Martinez
Uma 10h30'54" 57d28'57"
Ariana Lyn
Ari 2h9'55" 24d3'49"
Ariana M. Rivellini
And 0h58'19" 43d37'21"
Ariana Madix
And 23h6'54" 51d28'43"
Ariana Marie
Vir 12h1'19" -0d9'56"
Ariana Marie Huenefeld
Dra 15h19'8" 60d7'34"
Ariana Marie Winter
And 1h13'18" 45d40'19"
Ariana Matias
Cap 20h23'28" -9d24'1"
Ariana Michele Mastrogiannis
And 0h25'34" 25d48'51"
Ariana "Nani" Singleton
Sco 16h10'10" -15d17'17"
Ariana Noel Robbins
Lib 14h57'16" -3d28'35"
Ariana Noelani
Dra 18h43'12" 54d31'31"
Ariana Raynis
Ari 3h6'28" 22d1'40"
Ariana Rodriguez
Cas 0h37'30" 57d13'28"
Ariana Rose
Cam 7h29'21" 61d31'12"
Ariana Rose Miceli
Lib 15h26'56" -20d31'16"
Ariana Ruth Begraft
Lyn 7h36'55" 38d8'24"
Ariana Ruth Sweedler
Per 3h33'17" 34d44'2"
Ariana Skye
And 1h1'17" 41d41'35"
Ariana Skye Blaeuer
Ori 5h50'12" 5d53'4"
Ariana Sofia Rose * Bat Mitzvah
Cyg 20h48'7" 33d7'45"
Ariana Strong
Cyg 20h52'0" 53d40'33"
Ariana Young
Sco 17h20'52" -41d19'35"
Arianah Musique
And 0h34'45" 38d55'44"
ArianaSonsi
Lyr 19h18'20" 29d49'11"
Ariane
Sco 16h57'47" -40d3'10"
Ariane Béatrice Cazaubon
Lyn 7h54'56" 37d47'58"
Ariane Carole Robitaille Guillemette
Vir 11h57'24" 0d6'51"
Ariane Cornelia Daenzer
Ari 3h11'57" 23d18'19"
Ariane Molesso
Uma 11h18'56" 43d10'20"

Ariane Moret 23/01/1936
Aqr 20h40'23" -9d28'41"
Ariane Sauvain
Gem 7h38'55" 14d31'55"
Ariane Schirmer
Peg 22h14'12" 33d19'32"
Ariane Werboff
Uma 10h2'8" 43d35'10"
Ariane's Eyes
Ori 5h51'27" 6d15'37"
Arianna
Vir 13h17'7" 7d16'29"
Arianna
And 0h39'48" 26d56'47"
Arianna
Lyr 18h51'39" 27d18'49"
Arianna
Lyn 7h51'4" 42d34'52"
Arianna
And 2h35'18" 40d39'26"
Arianna
And 2h8'40" 37d41'56"
Arianna
Lyr 18h52'13" 33d16'18"
Arianna
Cyg 21h28'49" 50d3'23"
Arianna
Cap 20h20'29" -10d21'21"
Arianna
Sgr 19h21'29" -13d24'44"
Arianna
Dor 4h19'41" -51d25'32"
Arianna
Ser 18h18'20" -14d28'14"
Arianna ~ 07-05-05-11:04
And 1h41'36" 43d18'40"
Arianna 21-23-5
Lac 22h19'2" 47d28'25"
Arianna Barbero
And 23h13'52" 40d36'38"
Arianna Brielle
Cas 23h58'59" 64d53'18"
Arianna Davis
Uma 9h8'49" 67d40'55"
Arianna Egan-Duwatt
Cam 4h41'35" 68d11'20"
Arianna Elizabeth Cunha
Cas 1h15'48" 62d54'11"
Arianna Giamberini
Her 16h29'45" 30d50'8"
Arianna Gori
Cas 22h59'34" 53d36'57"
Arianna Grace Lewis
Aqr 22h46'50" -11d33'56"
Arianna (Little Bit)
Umi 14h29'18" 83d4'56"
Arianna Lydia Ettel
Vir 12h11'46" -4d57'53"
Arianna M. Cuddy
Lib 15h18'46" -5d5'9"
Arianna M. Garcia
And 0h51'45" 42d31'28"
Arianna Marie Bianculli
Gem 7h39'41" 14d11'32"
Arianna Marie Miller
And 0h19'41" 28d10'48"
Arianna Marie Winston
Lyn 7h30'17" 46d15'52"
Arianna Michelle
And 2h7'15" 45d43'56"
Arianna Morgan Hearing
Ari 2h51'50" 29d7'3"
Arianna Nicole Gonzalez
And 2h37'39" 44d19'56"
Arianna Nikole Foster
Gem 7h45'36" 22d10'3"
Arianna Soraya Lohrer
Vir 14h50'23" -0d10'23"
Arianna Theilen
And 0h32'58" 45d15'37"
Arianna Theresa Evans
Aqr 22h41'26" 2d6'7"
Arianna Travaglini
And 1h2'41" 40d59'39"
Arianna Trevisanut
Umi 15h43'32" 73d42'54"
Arianna Uray Womack
Cas 23h21'41" 56d59'25"
Arianna Zoe Gonzalez-Pacheco
Sco 17h53'56" -30d7'18"
Arianna Zoe Vassallo
Uma 10h55'49" 59d56'10"
Ariannabell Reyes
Lib 14h51'25" -13d14'16"
Arianna's Dad - World's Best Dad
Her 16h31'55" 48d11'1"
Arianna's Grace
Pho 0h35'14" -41d44'27"
Arianne/Andrew
Aql 19h33'24" 7d41'14"
Arianne G.
Uma 13h37'25" 55d24'1"
Arianne Star
Tau 5h9'19" 18d27'1"
Arianrhod
Eri 4h22'52" -24d10'2"
Arias Natalina Angela
Lyr 18h58'7" 33d21'1"
Aria's wolfy
Cma 7h18'7" -12d36'27"
Aric David Sanders
Psc 1h25'24" 28d59'58"

Aric W.
Sco 16h7'16" -9d18'17"
Aric William Pate
Vir 14h42'17" 4d25'26"
Arica
Lyr 18h44'16" 35d19'10"
Arica Elizabeth Williams
Psc 1h43'58" 20d13'34"
Aricca Nicholle
And 23h48'6" 34d49'51"
Aricka Rae
Psc 0h6'32" 1d45'1"
ARIC'S ALASKA
Uma 13h33'9" 57d6'43"
Arie
Cam 3h55'2" 65d26'8"
Arie Nyman
Per 4h38'22" 48d56'11"
Arie Thomas Thuner
Gem 7h49'36" 33d30'9"
Arie van den Berg Star
Peg 22h54'5" 8d22'43"
Arieana
Lmi 10h20'50" 32d29'46"
Arieanna Rombeiro
Umi 5h3'48" 88d24'30"
arief utomo
Crt 11h35'36" -7d52'41"
Ariel
Vir 13h50'59" -11d58'11"
Ariel
Aqr 22h27'4" -0d46'42"
Ariel
Umi 14h37'50" 78d35'19"
Ariel
Sgr 19h9'39" -14d2'6"
Ariel
Uma 12h35'55" 57d46'13"
Ariel
Sgr 18h27'42" -25d54'54"
Ariel
And 1h15'34" 38d49'4"
Ariel
And 2h13'43" 45d38'6"
Ariel
And 2h16'24" 46d3'38"
Ariel
And 23h52'5" 39d51'41"
Ariel
Ari 2h42'49" 21d14'26"
Ariel
Gem 6h56'53" 12d23'15"
Ariel
Tau 4h6'29" 6d57'43"
Ariel
Peg 21h57'33" 26d19'33"
Ariel 3/28/28
Ari 2h15'43" 22d30'28"
Ariel Cooke
Cap 21h35'9" -15d21'22"
Ariel Dubek
Sgr 18h44'57" -36d7'13"
Ariel Hazen Hunley
Uma 11h44'15" 32d2'37"
Ariel Jane Ashwell
Cyg 19h52'55" 31d15'22"
Ariel Jean Townley
And 1h52'0" 38d53'23"
Ariel Joy Eaton-Willson
Ari 2h45'55" 20d55'45"
Ariel Katen
Cnc 9h12'38" 32d27'18"
Ariel & Laura
Uma 9h44'49" 60d48'2"
Ariel Leroux
Sco 17h39'20" -40d26'5"
Ariel Lian
And 23h47'5" 47d31'48"
Ariel Luis
Lyn 7h36'53" 56d28'48"
Ariel Lynn
Aqr 23h4'31" -9d46'2"
Ariel McRae
Lyn 8h14'21" 49d54'51"
Ariel Netanya
And 0h20'39" 33d55'6"
Ariel Niedecken
Vir 14h1'51" -17d20'59"
Ariel Santa Maria
Ori 5h13'54" 1d28'21"
Ariel Sky
Umi 14h23'16" 75d31'29"
Ariel Thao-Oanh Phan
And 23h47'12" 38d0'44"
Ariela
Sco 16h38'41" -25d37'38"
ArielDallas
Lyn 7h50'7" 45d53'56"
ArielHaley
Uma 8h53'17" 56d59'4"
Ariella
Psc 1h14'22" 6d57'39"
Ariella
Cnc 7h58'34" 10d29'5"
Ariella Munsterman
And 23h21'53" 41d17'10"
Ariella Rose Levine
And 0h55'53" 37d22'58"
Arielle
And 1h40'51" 41d47'15"
Arielle
Cnc 7h59'51" 11d58'50"
Arielle
Dra 18h25'46" 54d5'46"

Arielle
Vir 14h14'33" -9d47'30"
Arielle Altimas
Tau 5h47'22" 25d37'34"
Arielle Brittanie Slocum
And 23h36'12" 47d17'17"
Arielle Cara
Vir 15h3'46" 3d2'59"
Arielle Elizabeth Roberts
Ari 3h3'11" 25d34'33"
Arielle Faith Griswold-Wheeler
Cas 23h22'1" 55d43'7"
Arielle Jenae
Cru 12h22'59" -55d58'30"
Arielle Justina Halpern
And 23h31'12" 50d31'56"
Arielle Kebbel
Her 17h55'17" 49d51'53"
Arielle Leila
Vir 12h8'6" 11d8'27"
Arielle Luna
Cyg 21h20'1" 53d49'47"
Arielle Maria Geromin
And 0h52'55" 40d37'58"
Arielle NaRae
Vir 13h8'43" 12d8'35"
Arienne Bertucelli
Psc 1h39'32" 8d49'38"
Aries Angela-AM 2006
Ari 2h19'24" 12d53'49"
Aries Shayna
Ari 2h51'10" 25d25'45"
Arie's Star
Gem 7h28'6" 20d44'32"
Aries2Gems DeFazio
Umi 14h7'6" 75d32'50"
Arietta Christakou
Uma 11h26'3" 29d32'3"
Ariga
Uma 12h6'20" 61d16'30"
Arigatou, John
Sco 17h27'58" -41d57'2"
Arika
Gem 6h31'11" 26d19'50"
Arika Jade Brock
Uma 11h5'28" 59d55'55"
Arika Yamada
Lyn 8h8'3" 52d5'35"
Arikat
Cyg 21h55'35" 53d23'13"
AriKim
Mon 6h33'25" 10d34'21"
Arikius
Aqr 22h22'44" -5d24'51"
Arild Haugland
Per 3h41'51" 45d26'52"
Arimdaun
Tau 3h40'36" 28d6'19"
Arin
Lib 15h10'46" -12d33'4"
Arin & Elijah Senior (UB1 &2)
Ori 6h1'24" 5d35'34"
Arin Elizabeth
Crb 15h52'40" 34d16'46"
Arin Marie Walker
Tau 4h21'53" 16d29'51"
Arin Reeves Pace
Lib 14h29'53" -10d39'44"
Arin Sachiko Murray
Vir 12h20'23" -1d48'7"
Arina
Aql 20h10'35" 13d11'51"
Arina
Vir 14h8'15" 4d1'4"
Arinamos
Cas 3h14'56" 60d35'25"
Arinlee
Pyx 9h10'22" -30d50'48"
Ario ZG1
Cyg 21h7'20" 49d25'55"
Ario ZG1
Cyg 21h7'20" 49d25'55"
Ariola
Lyn 8h45'20" 40d5'13"
Arion Ra
Eri 4h33'15" -21d27'21"
Arioso
Cep 20h52'48" 59d18'18"
Aris
Ori 6h0'32" 4d47'52"
Aris "bisbirikos" Stefanakis
Her 17h36'18" 15d52'9"
Aris Cuadra
And 1h5'52" 46d9'33"
Ari's Silver Star
Cyg 21h50'40" 47d4'38"
Aris Thomas
Uma 8h43'54" 61d14'51"
Arisa
Lib 15h28'36" -23d40'47"
Arisbe - The Love of My Life
Uma 8h41'35" 56d17'53"
Arise 21
Uma 11h28'35" 46d13'6"
Arish'ka
Cyg 20h42'46" 35d34'4"
Ariss - The Star of Happiness
Her 16h25'10" 11d5'44"
Aristaeus
Aqr 23h41'45" -15d20'59"
Aristar
Cmi 7h30'1" 4d14'59"
Aristia
Uma 12h2'58" 61d16'17"
Aristides
Uma 8h25'10" 62d57'31"
Aristophanes: One of Two
Cyg 19h59'47" 35d2'5"
Aristote Niforos
Tau 5h47'23" 19d0'12"
Aristotelis Panteliadis
Uma 11h10'56" 28d32'37"
Aristova Viktoria
Aqr 22h7'25" -16d3'45"
Ariunaa
Cnc 8h17'28" 18d48'27"
Ariza
Leo 11h7'11" -4d5'35"
Arizona Industrial Hardware
Umi 14h57'31" 74d10'2"
Arjan and Christina - Eternity
Her 17h33'19" 49d6'49"
Arjona PR, Si el Norte fuera el Sur
Eri 4h25'43" -31d51'56"
Arjun Madhavan
Aqr 22h34'55" 0d50'17"
Arkady Eydlin
Sco 16h3'32" -12d19'23"
Arkadyr
Sco 16h10'43" -9d46'49"
Arkansas Baptist Children's Homes
Lyn 8h18'54" 34d58'34"
Arkansas Gazer
Cnc 8h40'57" 24d24'15"
Arketa Angel
Vir 14h15'34" -16d27'7"
Arkie Garrett Sheppard
Leo 10h32'48" 15d31'38"
Arlan and Sheila Roeder
Cyg 21h34'7" 53d27'10"
Arle Lommel
Lyr 19h10'51" 39d21'46"
Arleen
And 1h39'55" 39d40'6"
Arleen
Tau 5h55'41" 24d55'37"
Arleen Baker
Lyn 7h41'34" 35d53'13"
Arleen Hersey
Lyr 18h41'0" 43d11'25"
Arleen O'Shea
And 1h24'43" 47d21'1"
Arleena Lovering
Crb 15h46'29" 31d13'37"
Arleicha Patrice
Vir 11h40'12" -3d23'48"
Arleloga
Cam 5h37'40" 73d50'57"
Arlen Keith Bailey
Tau 5h24'13" 27d55'48"
Arlene
Cnc 8h50'24" 28d0'6"
Arlene
Gem 6h36'8" 24d23'22"
Arlene
Leo 11h18'8" 23d28'40"
Arlene
Lyr 19h18'26" 29d23'58"
Arlene
Cyg 19h29'31" 30d49'49"
Arlene
Crb 16h19'45" 33d36'15"
Arlene
And 23h15'57" 48d7'54"
Arlene
Cas 0h23'15" 55d20'11"
Arlene
Cas 2h23'9" 63d4'3"
Arlene
Aqr 22h53'1" -14d41'35"
Arlene A. Davis
Cnc 8h26'58" 15d24'47"
Arlene Andromeda Sparkle
Psc 0h26'6" 8d42'56"
Arlene Anita Beneteau
Dra 20h12'40" 71d27'21"
Arlene Avery
Psc 1h16'34" 24d31'13"
Arlene B. Loyola
Cas 23h5'4" 56d22'34"
Arlene Baumgartner
Lmi 10h7'59" 31d42'59"
Arlene Betty Paulsen
Vir 12h12'41" 12d2'35"
Arlene Boss
Cnc 9h6'53" 30d53'9"
Arlene Bridges Samuels
Lyn 7h37'39" 53d14'21"
Arlene Crespo-Reyes
Cas 1h40'19" 69d4'9"
Arlene Dixon
Leo 11h30'20" 0d23'21"
Arlene Feder
Uma 8h47'37" 72d43'48"
Arlene/Felina
Cap 21h36'15" -8d27'16"
Arlene Fisichelli
Cnc 8h16'58" 15d54'36"
Arlene Floyd
Uma 11h12'6" 48d4'3"
Arlene H. Parsons
Lyr 19h26'59" 41d39'38"
Arlene Harr
Del 20h54'4" 19d0'47"
Arlene Hoffman Hendren
Leo 10h16'25" 17d18'6"
Arlene Hulstrom-Strock
Lib 15h23'9" -10d35'20"
Arlene Katherine Alt
Ori 6h3'18" 1d46'28"
Arlene L Karg
Gem 7h51'38" 33d20'49"
Arlene Lawrence Brumley
Crb 15h42'15" 31d18'34"
Arlene Lee
Cnc 9h18'31" 14d43'22"
Arlene M Iannaccone
Cas 23h50'54" 53d57'22"
Arlene M Ruhnke
Cas 1h24'47" 60d58'2"
Arlene Mae Sand Stone
And 0h50'52" 33d47'56"
Arlene Mae Van Haften
Ari 3h19'39" 26d19'0"
Arlene Marie
Aqr 21h45'18" -1d25'2"
Arlene Medina
Cyg 21h58'0" 51d24'34"
Arlene Meyer Cohen
Crb 15h41'10" 28d27'14"
Arlene Mildred Mary
And 1h25'13" 48d16'47"
Arlene Patricia
Cyg 20h7'25" 38d54'57"
Arlene Patricia Corley
Uma 10h39'49" 48d10'44"
Arlene R. Baluyut
Cyg 19h59'50" 36d10'14"
Arlene R. Bicher
Cap 20h44'47" -21d25'56"
Arlene R. Wolfe
Uma 11h40'23" 47d47'27"
ARLENE RAY
Cap 20h26'19" -26d35'13"
Arlene Richard
Uma 11h42'20" 64d53'25"
Arlene Rome
Gem 6h54'15" 23d26'50"
Arlene Rose Barber
Leo 10h12'42" 17d23'11"
Arlene Ruth Cook
Leo 10h44'43" 7d45'18"
Arlene Ruth Moore Dwyer
Psc 1h9'13" 31d45'10"
Arlene Ruth Shinn
Uma 12h41'20" 60d22'4"
Arlene S. Dailey
Uma 12h27'19" 60d57'29"
Arlene Sabina
Del 20h34'10" 14d21'29"
Arlene Slavelis
And 2h14'40" 38d34'15"
Arlene Stump
Uma 12h27'13" 52d58'41"
Arlene Susan Tribbia 10/24/57
Mon 6h50'2" -1d58'13"
Arlene Tamie Beller Wittels
Cas 1h32'45" 63d39'58"
Arlene Topacio Grinnell
Vir 14h27'3" 0d33'35"
Arlene Torres
Cam 4h48'3" 54d56'53"
Arlene V
Psc 0h57'18" 16d2'22"
Arlene's Joy
Sco 16h10'36" -11d23'36"
Arleny & Juan 4ever
Psc 23h20'40" 2d23'26"
Arleo 23
Cyg 20h59'37" 44d50'4"
Arletis
Gem 6h48'14" 23d43'2"
Arletta Jayne Brown
And 1h10'50" 43d4'57"
Arlette
Uma 9h38'33" 46d11'13"
Arlette du Paty de Clam
Umi 12h31'5" 88d11'45"
Arlette Jacqueline Hunt
Cap 20h9'1" -9d58'59"
Arlette "Sunni" Kerr
Sgr 19h40'54" -14d33'48"
Arlicia Monique Johnson
And 1h19'38" 42d54'54"
Arlie Reece Jones
Lmi 10h29'59" 30d42'6"
Arlin Scott Menke
Vir 13h29'0" 12d22'12"
Arlina Carrillo Aquino
Cap 20h39'39" -20d25'19"
Arlina Marie Carr
Cas 1h27'18" 60d43'0"
Arlinda Joyce
Tau 3h35'20" 26d55'44"
Arlinda Lee Thomas
Tau 4h16'27" 11d37'48"
Arlinda Mae Smith
Aql 19h39'16" 5d51'23"
Arlinda Spriggs
Cyg 20h39'6" 35d59'5"
Arline Baril (Our "nana")
Cas 1h36'14" 62d12'14"
Arline Donn
Cas 0h9'7" 59d20'9"
Arline Pearson
Gem 7h1'1" 14d8'49"
Arlinghaus, Manfred
Uma 13h17'7" 60d49'30"
Arliss Luther Crocker
Aql 19h24'57" 7d14'19"
Arlo Graham Lindahl
Uma 12h0'38" 51d23'18"
Arlo Hettle
Lib 15h59'4" -9d30'34"
Arlo Robert Tilby
Sco 16h25'17" -29d2'18"
Arloa Marcella
Uma 10h25'49" 43d34'57"
Arlos-7
Tau 3h50'57" 24d32'54"
ARlove
Sge 20h5'31" 20d33'25"
ARLYN
Psc 23h54'58" -3d11'7"
Arlyn
Cyg 20h11'15" 56d49'3"
Arlyn Anderson
Peg 22h6'41" 27d23'48"
Arlyne S. Medann
Crb 16h3'15" 36d16'31"
Armaan
Lyr 19h27'15" 42d21'44"
Armaity Fali Singara
Lib 15h32'25" -11d35'34"
Arman Badeyan
Cap 20h22'50" -11d49'2"
Arman Davtyan
Vir 13h43'32" -2d5'51"
Armand
Uma 9h48'58" 64d30'28"
Armand
Uma 8h35'9" 71d16'2"
Armand
Aur 5h19'32" 31d14'28"
Armand & Cybel
Uma 10h33'25" 65d40'1"
Armand Dallaire
Uma 12h41'47" 58d9'7"
Armand "Danny" Ranieri
Per 4h40'53" 44d8'39"
Armand J. Chauvin
Uma 8h44'8" 48d25'43"
Armand John Cruz (8/22/51-7/20/01)
Leo 10h31'32" 18d13'24"
Armand P. Savino
Lib 14h50'19" -15d32'20"
Armand "Pat" Patregnani
Lib 14h57'59" -2d16'15"
Armand Vallès
Leo 10h50'21" 22d46'37"
Armand416229495
Uma 9h39'16" 45d9'56"
Armanda
Cnc 9h6'52" 31d35'38"
Armande
Ori 5h54'10" 17d58'47"
Armandi Muniz
Lib 15h38'42" -20d35'59"
Armando
Ori 5h35'47" -3d8'27"
Armando
Cam 5h22'25" 66d39'0"
Armando
Ori 6h19'39" 14d3'15"
Armando
And 0h22'21" 46d7'56"
Armando Agnitti
Cnv 13h6'31" 48d9'44"
Armando Bianco
Com 13h1'55" 15d0'33"
Armando Borges
Cyg 21h40'46" 46d51'40"
Armando Castillo Samayoa
Eri 4h18'23" -31d23'56"
Armando Corzo 1921-2004
Cam 7h27'0" 69d21'38"
Armando de Jesus Palma Vargas
Ori 4h48'5" -2d32'9"
Armando J. Barone
Aur 5h44'20" 41d2'59"
Armando M. Cledera II
Ari 3h6'6" 12d20'34"
Armando & Nancy
Aql 19h44'26" 12d9'39"
Armando Padilla aka "80 PROOF"
Sgr 20h11'14" -35d45'47"
Armando Rios Jr.
Tau 5h10'44" 23d27'16"
Armando Ruben Vega
Psc 1h7'47" 33d4'4"
Armando Y Angelica
Cyg 20h49'38" 45d15'19"
Armando's Illustrissimus Sideris
Uma 11h17'20" 41d39'21"
Armand's Star
Cnc 9h8'49" 30d16'54"
Armani
Peg 22h34'19" 9d3'19"
Armani M. Eaddy
Uma 9h23'57" 64d58'49"
Armanie Cyr-Miranda
Lyn 7h23'44" 51d21'30"
Armeet
Aur 6h24'48" 40d26'7"
ARMEIER-6200
Lyr 19h3'7" 31d48'40"
Armell Rashaud Macon
Vir 13h14'26" 12d9'59"
Armelle Borredon
Uma 9h18'23" 45d17'16"
Armelle Chassin du Guerny
Vir 11h42'19" -0d4'27"
Armen
Cnc 8h43'43" 23d16'7"
Armen Badeyan
Ori 5h30'39" 12d14'3"
Armen "Grandpa" Vartan Sahagian
Cnc 9h4'4" 32d17'16"
Armen Kalaydjian
Lmi 10h29'47" 37d46'5"
Armen Louis Najarian
Eri 4h58'14" -8d26'37"
Armen Orujyan
Uma 12h21'46" 58d29'20"
Armen's Star
Mon 6h46'57" -0d2'52"
Armetia Joyce Martindale
Lyn 8h37'18" 39d5'47"
Armida
Vir 12h7'15" -7d2'28"
ARMIG
Psc 0h58'55" 29d19'32"
Armik's Fire & Ice
Cmi 7h24'13" 8d35'41"
Armin
Lac 22h55'7" 45d59'36"
Armin
And 23h30'54" 40d49'54"
Armin
Aqr 21h9'50" -14d10'38"
Armin
Lac 22h9'0" 52d50'45"
Armin Bachmann
Cas 1h30'32" 63d4'32"
Armin Finkbeiner
Uma 10h20'44" 71d47'49"
Armin J. Betting
Lib 15h44'1" -12d25'51"
Armin Peter
Uma 12h2'37" 48d9'42"
Armin Roth
Cnc 7h58'35" 16d53'22"
Armina
Aqr 22h10'46" 0d30'54"
Armine Manucharuan
Uma 8h26'58" 61d47'13"
Armineh Hatamian
Per 4h12'50" 39d30'20"
Armisdea
Cas 0h54'24" 51d25'7"
Armita Afsar
Uma 11h31'1" 64d51'14"
Arms of an Angel
Cyg 21h16'39" 31d53'2"
Armstrong/Conn
Cyg 21h55'17" 52d57'15"
Armstrong's Hot Mix
Lib 15h26'48" -9d29'53"
Armstrong's Star
Uma 9h40'42" 45d29'34"
Army Pfc. Harrison J. Meyer
Per 3h21'52" 42d24'15"
Arnaldo Ildebrando SAVORETTI
Lmi 9h23'55" 38d32'3"
Arnaldo Pala - con luminosita stella
Ari 1h53'59" 19d45'9"
Arnaud Boune Premier Fréchette
Cep 0h15'58" 73d20'32"
Arnaud Mabilais
Aqr 20h55'42" 1d33'1"
Arnaud Strulens
Uma 10h34'21" 41d37'22"
Arnd auf'm Kamp
Uma 10h15'23" 55d53'10"
Arne Dietrichs
Ori 6h11'24" 8d20'49"
Arne Oddbjørn Arnesen
Aqr 21h11'41" -3d17'39"
Arnel L. Simmons
Aql 19h49'4" 8d39'42"
Arne's Star
Cnc 8h18'32" 19d52'32"
Arnesia Cooke
Gem 7h36'59" 27d21'31"
Arnett Lieb
Her 18h38'4" 18d23'12"
Arnetta Warren
Uma 10h32'32" 46d8'43"
Arnette
And 0h12'3" 38d27'52"
Arnette~Carter
Cas 1h20'49" 71d47'43"
Arnhold
Lyn 6h57'12" 54d38'14"
Arnie
Uma 12h26'24" 56d38'56"
Arnie and Sara
Sge 19h27'23" 17d10'34"
Arnie Cerio
Leo 10h48'15" 9d52'43"
Arnie & Cheree Boyarsky
Aqr 22h11'53" -1d0'38"
Arnie Hendrickson 5106 - I love you!
Tau 4h21'51" 23d25'4"
Arnie Lipson
Aql 19h5'2" 15d41'18"
Arnie Nishioka
Per 3h19'2" 43d28'41"
Arno Giovagnetti
Uma 11h27'43" 59d59'46"
Arno Kupka
Uma 8h47'26" 56d51'32"
Arno Raehse
Ori 6h1'21" 0d6'3"
Arno Rebernig
And 1h19'43" 42d10'45"
Arnold and Neoma Griffin
Uma 10h24'34" 56d17'32"
Arnold David Kurmin
Tri 2h5'36" 31d50'4"
Arnold Dixon
Per 3h23'43" 52d24'52"
Arnold Feldman
Psc 1h6'31" 3d30'25"
Arnold & Ferol Hackman love always
Lmi 10h41'45" 24d17'38"
Arnold Fuld
Lyn 7h45'7" 47d2'34"
Arnold G.
Psc 0h38'26" 18d2'32"
Arnold Grandis
Her 18h40'2" 21d19'4"
Arnold H. Kassanoff
Cep 21h45'27" 66d32'22"
Arnold Henry Jibben
Cep 21h40'49" 64d15'33"
Arnold Hugh Wilson
Her 18h13'35" 28d33'35"
Arnold/Josephine Bedsworth & family
Aur 5h29'36" 45d49'48"
Arnold Kirby, Jr.
Cap 21h41'6" -12d27'9"
Arnold Lance Emmerling
Uma 11h35'46" 32d2'26"
Arnold Lewis
Uma 8h55'55" 59d10'51"
Arnold Meyer
Ori 6h19'30" 8d16'13"
Arnold & Milagros LaFontaine
Lib 14h34'52" -8d56'9"
Arnold Nauman
Her 17h42'48" 38d36'43"
Arnold P. Eastman
Aur 7h11'49" 42d47'22"
Arnold Paison "5-24-1930"
Per 3h10'40" 50d26'12"
Arnold Piedrahita
Ari 3h4'49" 16d14'48"
Arnold & Sandra Rosen
Cyg 19h59'22" 44d52'31"
Arnold Seefeld
Dra 20h25'42" 63d17'10"
Arnold Soslow
Sco 16h14'13" -10d54'25"
Arnold Tillman
Aqr 23h7'45" -7d55'11"
Arnold, Prince of B & the World
Per 3h14'5" 40d46'1"
Arnold, Wolfgang
Uma 9h50'32" 64d33'16"
ArnoldAnderson
Aqr 22h2'40" -9d38'33"
Arnoldene
Cap 20h27'42" -15d58'34"
Arnoldo Castro Padre De Maria Paola
Tau 4h51'56" 28d13'3"
Arnolds Z K Minders
Aur 6h37'52" 31d22'32"
Arnoud the Rebel Star
Leo 10h12'47" 16d20'24"
Arnstein Lee Stiles
Umi 15h25'4" 69d25'37"
ARON
Vir 12h47'43" 12d18'45"
Aron and Halle Bohlig
Uma 11h55'35" 29d13'15"
Aron J. Wittkamper
Uma 10h21'56" 41d8'37"
Aron Larie
Cap 21h32'44" -13d21'4"
Aron & Margit Kastell
Cyg 21h44'5" 44d48'23"
ARONIMI
Psc 0h8'44" 4d50'51"
Aroone & Chengkou
Pho 0h16'48" -40d41'12"
Aroul Kaliamurthy
Vir 13h24'40" 12d43'25"
aroura lindalice
Sco 16h11'42" -17d22'35"
Arpad Flynn
Uma 8h20'19" 60d34'44"
Arpi
And 2h35'16" 50d6'19"
Arpi
Uma 11h45'40" 44d49'41"
Arpineh Peach
Sco 17h15'4" -33d10'24"
Arran Anthony
Gem 6h51'34" 22d14'50"
Arran Streppa
Sco 16h8'33" -13d8'18"
Arran Thomas James Phipps
Cep 0h29'24" 78d26'2"
Arran's Star
Umi 13h19'51" 71d47'27"
Arran's Star
Per 4h43'24" 45d51'59"
Arren Duggen, I love you
Cru 12h42'54" -57d41'40"
Arrington
Crb 15h34'59" 28d22'51"
Arrive Magnificent
Leo 11h42'7" 24d25'14"
Arrna Elen Dao
Leo 10h38'22" 14d48'35"
Arron Daniel Kinser
Psc 1h20'58" 10d58'39"
Arron Jac Thomas
Dra 17h55'11" 60d11'26"
Arron Jake Scholes
Per 3h38'45" 39d34'6"
Arron Schmitt and Lindsay Shappell
Cnc 8h51'47" 29d29'36"
Arrow Hewitt
Cma 6h44'3" -11d41'24"
Arrow Stannard
Uma 9h2'48" 67d8'15"
Arroyo Carter
Ori 5h12'4" 8d47'33"
Arruda Ohana
Ori 6h4'22" 21d23'59"
Arruffona
Ori 6h5'59" -2d9'30"
AR's Delight
Leo 10h11'53" 16d10'24"
Arsac Edith
Ari 2h23'0" 20d23'43"
Arsam Khoshnevisan
Dra 18h6'26" 75d44'43"
Arsen A. Melkumyan-Marina & Nikolay
Aqr 22h58'41" -6d57'0"
Arseni Sutton
Cep 22h17'16" 64d3'10"
Arsenia Rosa Gibson
And 23h5'19" 48d28'3"
Arsenio Colosinda
Uma 8h47'30" 66d21'8"
Arshi
Lib 15h14'46" -11d5'27"
Arspecio
Ori 6h16'21" 19d20'25"
A.R.T
Her 16h36'37" 45d51'50"
Art and Joan Flynn
Leo 10h10'54" 13d52'24"
Art and Lauren
Cyg 20h4'59" 33d52'26"
Art and Linda Fierro
Cyg 21h53'24" 44d8'3"
Art Aragón
Per 3h50'46" 37d5'40"
Art Bartholomew
Lib 15h48'12" -17d32'58"
Art Berry
Dra 19h48'38" 71d35'43"
Art Boy Zeller
Cyg 19h38'15" 39d25'18"
Art Brisbane, Star Leader
Lib 15h3'53" -12d53'36"
Art Catenacci III
Per 3h8'45" 57d7'0"
Art Cote
Dra 18h20'25" 58d19'43"
Art DeMao
Uma 11h47'26" 50d30'31"
Art & Doris's Star
Cyg 20h46'23" 37d46'8"
Art Doyle
Ori 5h29'49" -2d44'40"
Art & Edith Young
Uma 12h10'11" 49d21'58"
Art & Estel
Cyg 20h22'54" 35d52'0"
Art Fischer
Aur 5h23'22" 37d6'32"
Art Godoy
Cyg 19h58'32" 33d12'53"
Art & Harriet Sturm
Cyg 21h45'36" 53d15'41"
Art Hernandez
Her 17h41'49" 27d58'18"
A.R.T. Hin
And 0h4'35" 38d17'34"
Art Jaehnke
Lyr 18h50'0" 42d53'27"
Art Kaiama
Per 4h2'50" 42d24'52"
Art Krawitz
Aur 5h49'6" 45d53'19"
Art Larson
Uma 12h2'58" 47d1'18"
Art Linko
Leo 11h51'36" 24d13'8"
Art Lopez
Per 3h17'27" 54d38'21"
Art & Louise Feiertag Family
Crb 15h39'58" 38d50'10"
Art & Maxine Jimerson
Lib 15h4'9" -29d0'2"
Art Meisters
Her 16h42'1" 36d23'34"

ART NADEL
Her 17h13'41" 31d52'28"
Art & Nancy
Per 3h41'45" 42d31'59"
ART OSTROM
Sgr 19h56'13" -35d2'44"
Art Paradissis
Boo 14h31'11" 14d53'0"
Art Pisano
Boo 15h38'33" 41d23'6"
Art & Poop
Uma 11h29'59" 34d40'15"
Art Robson, KindlyLoveableFolkHero
Cyg 19h55'26" 59d44'26"
Art Spiro
Cep 21h42'32" 61d34'30"
Art Tompkins
Per 3h33'52" 47d51'37"
Art Wageman
Cep 21h37'14" 61d7'24"
Art Wiedenbein
Per 3h13'45" 41d4'23"
Art Williams
Cep 20h39'30" 55d23'59"
ART1 - Angus' Star
Cen 13h6'49" -47d31'14"
ArtChele
Ari 2h13'1" 18d45'55"
Artemis
Tau 4h25'57" 23d7'55"
Artemis
Ori 5h55'29" 20d52'47"
Artemis
Lib 15h27'40" -17d59'30"
Arthena
Uma 10h45'3" 53d33'9"
Arthol
Leo 11h45'13" 19d53'10"
Arthur
Ori 5h44'22" 0d55'48"
Arthur
Ori 5h21'57" 0d44'38"
Arthur
Per 4h47'29" 42d35'58"
Arthur
Gem 7h14'44" 31d14'54"
Arthur
Uma 11h51'7" 52d31'13"
Arthur
Uma 13h32'43" 56d35'57"
Arthur
Umi 16h49'6" 76d29'54"
Arthur A. Bukowski
Sco 17h51'38" -36d37'59"
Arthur Alexander Moody
And 0h16'10" 43d41'17"
Arthur Alexander Ruble
Aql 19h7'51" 13d43'41"
Arthur and Beryl
Cyg 20h30'25" 59d5'34"
Arthur Andrzejewski
Aql 19h46'17" 7d28'39"
Arthur Anker
Aqr 20h49'33" -11d5'16"
Arthur "Art" John Anderson
Uma 11h27'37" 62d36'58"
Arthur B. Orkwiszewski
Gem 6h49'27" 17d1'31"
Arthur B. Stewart
Uma 8h31'58" 70d58'34"
Arthur Bernard Latin
Uma 10h11'29" 43d38'51"
Arthur Bolter
Ari 2h41'57" 12d43'13"
Arthur Brendon
Her 17h43'42" 15d18'18"
Arthur Brown
Uma 12h2'48" 59d38'24"
Arthur Bruce Norris
Ori 5h55'11" 18d6'30"
Arthur C. Holdsworth
Uma 10h11'59" 42d38'44"
Arthur C. Lemke "King Arthur"
Ari 3h26'1" 23d6'41"
Arthur C. Lyle
Uma 10h16'8" 43d25'7"
Arthur Carl Schafer
Gem 7h21'53" 22d13'28"
Arthur Carven Smith Walker
Per 3h17'58" 44d2'8"
Arthur Charles Gorlick Star
Gem 6h32'45" 24d33'30"
Arthur Chin
Sco 17h47'37" -33d43'41"
Arthur & Claudia Rothstein
Per 3h35'40" 32d8'15"
Arthur Clive Warr
And 0h37'1" 40d59'21"
Arthur Cody Gordon
Uma 10h42'26" 43d42'10"
Arthur Conal O'Hagan
Uma 11h10'41" 45d22'9"
Arthur Dakota Rhea, III "AD"
Ori 5h33'44" -0d58'5"
Arthur DeFusco
Aql 20h10'19" 4d3'15"
Arthur DeVries III
Aql 19h6'20" 15d38'11"
Arthur Donald Anderson
Uma 10h10'25" 50d30'5"
Arthur Douglass
Her 18h34'56" 20d0'22"
Arthur E Goette
Aur 7h5'38" 38d27'5"
Arthur E. Levine, My Daddy
Her 16h36'9" 37d11'12"
Arthur E. Ward, Jr. (Poppy)
Leo 9h36'6" 6d37'43"
Arthur E-Dad Hargett
Uma 10h35'1" 46d57'51"
Arthur Edward George Bumpass
Vir 11h59'27" -9d20'36"
Arthur Edward Stockton
Mon 6h32'27" -5d14'24"
Arthur Edwin Dumont 8/24/1915
Vir 13h50'12" -17d33'45"
Arthur F. McCormick
Sco 16h9'41" -11d3'35"
Arthur Farr Plemons
Aqr 22h36'29" -1d44'44"
Arthur Francis Shea
Uma 11h52'15" 29d30'45"
Arthur G. Cormier
Her 16h43'20" 46d14'58"
Arthur G Davis
Lib 15h47'29" -7d54'7"
Arthur Geaves Rudge
Umi 17h37'10" 84d52'37"
Arthur George Baba Sundar Shay
Cyg 20h39'32" 35d3'17"
Arthur George Schuettinger
Aur 5h50'18" 32d55'7"
Arthur George Stibbs
Per 4h11'34" 43d27'29"
Arthur Hampton Zeitz, Jr.
Uma 10h51'5" 42d56'38"
Arthur Harman
Her 17h7'53" 35d30'3"
Arthur & Helen Pacheco
Lyn 7h59'33" 45d56'51"
Arthur Hillier
Uma 11h23'35" 45d6'36"
Arthur in the sky with Diamonds
Her 17h40'4" 39d41'23"
Arthur Ivan Hall, June 23, 1923
Ori 6h4'42" 21d2'41"
Arthur J. Andrade, Jr.
Cyg 20h41'47" 44d57'35"
Arthur J. Engelbrecht
Ari 3h20'17" 26d27'42"
Arthur J. Lacoste
Aur 5h50'56" 32d17'37"
Arthur J. Long
Aql 19h54'51" -0d29'48"
Arthur J. Rizzi
Aql 19h12'36" 5d48'16"
Arthur J. Stephens 111
Aql 19h15'0" -0d55'36"
Arthur J. Trofe
Cnc 9h0'57" 15d42'18"
Arthur James Ted Chetcuti
Cep 22h5'26" 53d25'52"
Arthur James Wilgus
Lyn 7h57'16" 35d35'57"
Arthur Jefferson Davis Williams
Cep 22h57'18" 66d39'25"
Arthur John Gasson - 03.12.1928
Uma 8h17'55" 69d36'14"
Arthur John Scarpino
Tau 5h39'29" 20d27'1"
Arthur John Wiebe
Cnc 9h4'54" 32d35'8"
Arthur Joseph Benner
Cnc 9h13'27" 9d2'18"
Arthur Joseph Roy 6/21/05-10/02/05
Her 18h19'35" 22d5'25"
Arthur Julius Rice
Ori 5h18'4" 8d8'11"
Arthur K. Testani
Ari 2h50'55" 26d49'21"
Arthur & Karen Bennett
Lyr 18h40'13" 38d35'26"
Arthur Keith Thomassen
Ori 5h54'40" -0d0'54"
Arthur Kramer 80
Cep 0h18'11" 68d37'45"
Arthur L. Curley
Uma 10h50'59" 57d24'27"
Arthur L. Morgan
Vir 11h49'18" 6d7'52"
Arthur LaPaugh
Vir 15h7'2" 6d14'12"
Arthur le Bienvenu
Uma 8h31'55" 69d51'39"
Arthur Lee Conn, Jr.
Psc 1h13'0" 32d4'32"
Arthur Lee Maines
Aur 5h43'5" 43d48'37"
Arthur Lee Nelson
Ori 4h53'13" 13d11'32"
Arthur Leo Bobbitt
Aur 5h45'4" 47d54'26"
Arthur Leon Lauderback
Leo 11h28'33" 22d54'23"
Arthur LePierre
Cyg 19h55'2" 35d51'43"
Arthur Leroux
Tau 4h18'12" 2d53'29"
Arthur Levine M.D.
Uma 11h36'18" 54d7'46"
Arthur M. Brazier
Cep 22h30'12" 65d34'42"
Arthur M. Cardenas
Her 17h33'58" 34d24'43"
Arthur M. Justis, Jr.
Ari 2h17'43" 23d24'55"
Arthur Maggioli
Sgr 18h14'0" -19d6'23"
Arthur Magill Dumican Senior
Cep 20h39'53" 60d27'8"
Arthur Manson Gray
Del 20h47'52" 13d50'29"
Arthur & Marcy Perez
Cyg 20h45'54" 33d28'36"
Arthur Martin Neidhardt
Sco 16h5'9" -13d58'35"
Arthur Mickelson
Del 20h45'50" 14d39'39"
Arthur Mickelson
Del 20h42'24" 15d1'34"
Arthur Miller
Vir 12h10'46" 11d47'36"
Arthur Moies
Leo 9h50'30" 29d38'9"
Arthur "Moose" Fucina
Tau 3h48'58" 21d37'20"
Arthur Mullins
Aql 19h14'30" -0d34'26"
Arthur N. Weatherman Just for Today
Uma 14h5'55" 48d3'27"
Arthur Nadin Rojas
Dra 19h21'27" 65d1'17"
Arthur Norwood
Aur 6h0'57" 47d56'0"
Arthur Ochoa
Tau 4h10'12" 8d12'14"
Arthur Oliver Avis
Vel 10h45'41" -56d22'7"
Arthur Our Brightest Night
Lib 14h46'37" -18d0'47"
Arthur Paul Razo
Ori 6h10'5" 7d4'24"
Arthur & Phyllis Schott
Dra 19h48'21" 60d17'18"
Arthur & Pilar Herrera
Uma 11h50'19" 55d54'22"
Arthur R. Shellhammer
Uma 9h13'13" 61d23'7"
Arthur Richard Husson
Ori 6h5'11" 16d21'38"
Arthur Richard Koogler
Psc 1h1'23" 15d15'1"
Arthur Richard Sutton
Psc 2h3'47" 5d54'33"
Arthur Robert Tatnell
Cap 20h29'21" -10d47'3"
Arthur Robin Ward Christmas
Her 16h59'19" 16d7'26"
Arthur "Rocky" Nester
Dra 17h18'12" 67d53'14"
Arthur Rose
Uma 8h43'50" 72d49'9"
Arthur Roy Dixon III
Tau 4h36'45" 7d39'30"
Arthur Ruvin
Vir 12h3'58" 4d21'23"
Arthur S. De Mario
Vir 13h33'53" -3d52'17"
Arthur Seay Jobe
And 1h19'12" 47d56'1"
Arthur Shorin
Cap 20h38'28" -13d23'29"
Arthur Sicard
Tau 3h58'58" 15d7'21"
Arthur Slimm Jacobs
Ari 2h49'26" 11d32'13"
Arthur Strawbridge
Uma 13h2'30" 62d52'49"
Arthur Thomas Scott (Tom) 7/5/1941
Cnc 8h9'42" 14d16'10"
Arthur Tovmasian
Cnc 8h28'44" 13d57'54"
Arthur Victor
Boo 14h57'42" 17d11'55"
Arthur W. Bisci
Uma 8h42'55" 47d43'47"
Arthur W. Fisher
Uma 8h42'45" 68d6'54"
Arthur W. Gureck Jr.
Ari 1h52'54" 21d43'40"
Arthur Ward Cornwell
Dra 18h35'19" 55d23'19"
Arthur Wiest
Per 3h30'18" 45d32'2"
Arthur William Costa
Umi 14h31'39" 73d53'3"
Arthur Wilson Dennis
Lib 14h32'10" -18d23'9"
Arthur Wilson Dennis II
Vir 12h53'48" 2d15'9"
Arthur1994
Ori 6h14'55" 16d19'19"
ArthurDustinSchocken
Uma 9h53'48" 46d7'20"
Arthurlyn Burleigh
Cas 0h37'20" 53d47'30"
ArthurR-PatriciaS
Sge 20h4'48" 18d28'58"
Arthur's Light
Leo 11h39'30" 26d18'33"
Arthur's Star
Leo 11h43'0" 25d35'13"
Arthur's Star
Umi 17h0'16" 87d37'18"
Arti Priya Prasad
Sco 16h39'1" -41d9'30"
Arti & Redi 5/28/06
Umi 14h58'57" 72d1'1"
Articias Mooshoo
Uma 10h32'19" 46d31'59"
Artie
Per 3h44'23" 48d47'1"
Artie
Her 17h14'29" 17d35'47"
Artie Heilweil
Gem 7h23'31" 15d27'57"
Artie Smartie Star
Cet 0h50'2" -0d57'31"
Artimaña
Vir 12h37'45" -7d43'10"
Artisher Smith
Uma 10h31'40" 68d24'12"
Artistic Angel
Uma 14h1'17" 58d18'51"
ArtmentalisticBrittney
Ari 2h12'36" 24d44'55"
Artricia "Unconventional" Goddess
Cas 0h48'15" 53d31'44"
Art's Astellas Star
Lib 14h53'11" -2d4'46"
Art's Celestial Body #54
Ori 5h36'6" 6d38'51"
Art's Shining Star 06
Uma 9h46'38" 41d53'34"
ARTSandy
Cnv 13h44'21" 31d56'18"
Artur e Valeria
Ari 2h36'54" 27d52'11"
Artur Markov
Sco 17h18'58" -44d52'53"
Artur Wodarczak
Uma 9h40'1" 51d50'51"
Arturis
Sco 17h4'57" -38d29'24"
Arturo
And 0h33'34" 28d2'4"
Arturo
Gem 6h50'27" 22d22'23"
Arturo
Gem 6h47'19" 18d59'22"
Arturo Bustamante
Vir 13h14'20" 5d29'22"
Arturo Chavez
Her 16h43'35" 27d58'55"
Arturo Conrad Gene Tammaro
Cep 22h29'10" 85d11'17"
Arturo & Cordelia Dominguez
Per 4h17'53" 33d25'2"
Arturo Ernesto Hernandez
Vir 13h20'18" 3d35'38"
Arturo Gonzalez
Vir 13h19'24" 1d48'23"
Arturo Joseph Rodriguez
Aql 20h3'35" 11d44'30"
Arturo Paguay
Her 17h2'51" 47d18'14"
Arturo Rosas
Aqr 22h13'1" -2d17'21"
Arturo Souza and Tim Laine
Pho 0h29'37" -57d11'32"
Arty's Love
Cra 18h8'48" -37d29'7"
Arual
Lib 15h9'38" -26d2'58"
Arualkram 980
Cas 0h18'53" 57d7'38"
Aruba "Duba Duba" Sheikh
Psc 22h57'40" 7d54'6"
Arueous Mousimous
Cap 21h47'52" -8d56'26"
arulkens
Pho 0h42'56" -46d12'35"
Arun and Shelly
Uma 9h21'26" 67d28'4"
Arun Kohli
Cru 12h29'33" -61d1'50"
Arun Papa Khanna
Uma 9h27'10" 54d10'9"
Arun Patrick Booth
Cep 21h33'4" 57d14'39"
Arun Prasad
Sco 16h6'5" -10d3'40"
Aruna Angelina
Sgr 18h48'39" -16d16'20"
Arval Melton Holdridge
And 0h52'47" 41d53'47"
Arven
Lac 22h4'34" 51d40'5"
Arvetta M. & Walter W. Wilson
Sge 19h37'25" 17d51'17"
ARVG
Leo 9h32'11" 12d28'58"
Arvidson Family
Mon 7h8'34" -0d6'29"
Arvilla and Jack Williams
Gem 7h30'46" 29d24'34"
Arvin Sharma
Uma 11h47'8" 29d34'38"
Arvinus Prime
Sco 16h9'1" -15d49'7"
Arvon
Her 17h7'59" 45d46'56"
Arwen
And 2h34'28" 50d36'36"
Arwen & Aragon
Leo 10h58'26" 6d54'35"
Arwen Bryana Clafton
And 23h17'6" 41d53'27"
Arwen Elize Tidy
And 23h51'46" 41d5'27"
Arwen Evenstars' Star
Peg 21h40'8" 27d48'0"
Arwen Molly Myne
Ori 5h28'59" 2d4'48"
Arwyn Kathleen Luckett
Lib 15h15'31" -25d35'55"
Arwyn Stiles
Umi 17h36'38" 84d7'20"
Arya Dune
Cnc 8h7'21" 20d59'51"
Arya Grace
And 1h27'33" 47d38'17"
Arya Steele
Tau 4h42'46" 27d18'1"
Aryan
Sco 17h58'55" -38d18'38"
Aryan Yatish Lodhia
Peg 22h39'53" 23d29'50"
Aryana
And 0h50'17" 36d24'43"
Aryana
Lyr 18h25'52" 32d24'38"
Aryana Gandhi
Ari 2h33'53" 25d39'1"
Aryana Huffman
And 2h10'26" 43d9'23"
Aryeh and Sally Edelist
Cyg 21h6'56" 43d23'49"
Aryn Burke Fletcher
Oph 17h19'37" -24d20'32"
Aryn Eliza Bowling
Mon 6h40'26" -9d43'46"
Aryn Nathaniel
Ori 5h30'32" 0d50'1"
Aryn Simmons
Leo 11h37'25" 14d55'37"
Arynjoe
Lmi 10h53'56" 30d9'45"
Aryon Shariati
Cap 21h38'13" -10d27'59"
Arys
Uma 11h33'13" 62d16'12"
Arzalone
Ori 6h10'52" 18d46'50"
Arzberger, Heike
Sgr 18h6'23" -27d51'25"
Arzelle Leighton
Aur 5h41'23" 53d52'33"
Arzenys M. Velez
Lyn 7h31'35" 39d12'22"
Arzu
Sgr 17h57'6" -17d48'44"
Arzzon Hunt
Sge 19h51'46" 17d36'27"
AS COMPANY
Per 2h11'28" 53d8'35"
A.S. Hayles
Cyg 19h49'5" 38d34'29"
As Long As This Star Shines...
Uma 12h3'10" 29d1'9"
AS PM 527
Vir 11h55'34" 2d53'8"
As the Prostate Turns
Cep 22h32'18" 82d9'33"
As You Wish
Hya 9h10'58" -0d15'7"
As You Wish
Lib 14h48'58" -2d39'6"
As you wish
Cyg 21h12'22" 31d56'14"
As you Wish Upon an Alistar
Aqr 22h30'23" 1d52'31"
As You Wish, Barbara
Vir 13h10'28" -1d21'23"
Asa Cuauhtemoc Cruz Duran
Uma 8h55'56" 52d0'18"
Asa Determinare Canis Major
Sgr 18h30'17" -16d9'17"
Asa Rom Pollack
Lyn 8h11'59" 49d38'38"
Asa Sean Villafranco
Dra 18h57'8" 53d7'26"
Asa Star
Sco 16h6'20" -16d6'36"
Asad Ramzanali
Lyn 7h53'17" 34d58'31"
Asad S. Malik
Leo 11h26'41" 26d1'49"
Asako Hidaka Rice
Cas 0h32'56" 53d50'13"
Asato Makiko
Aqr 20h44'54" -1d52'47"
Ascencia
Vir 13h14'24" -5d23'15"
Ascher Yates
Uma 10h32'27" 48d44'52"
A-Scho
Dra 19h1'30" 69d39'23"
A.Scott Anderson
Boo 14h48'30" 43d32'48"
Asean Vue 11/1/96
Umi 13h17'29" 70d33'4"
A.S.G.-"The Mighty"
Uma 13h52'4" 49d39'1"
Ash
Gem 7h25'7" 14d21'24"
Ash
Umi 14h28'54" 73d13'23"
Ash
Uma 13h23'35" 62d55'19"
Ash
Cas 23h43'49" 57d2'20"
ASH
Uma 12h54'21" 59d35'10"
Ash
Lib 15h23'59" -16d8'1"
Ash
Sgr 19h50'56" -28d22'0"
Ash and Andy
Lyn 7h12'11" 56d14'5"
Ash and Ry 42404
Uma 8h13'0" 60d37'39"
Ash Bash
Cnc 8h46'49" 31d22'17"
Ash&Benj
Lyr 19h14'1" 26d30'48"
Ash & Bern Forever
Cyg 21h50'45" 42d0'53"
Ash & Ger
Leo 9h47'2" 17d57'49"
Ash Mac
Cra 18h33'57" -42d24'33"
Ash Mitchell
Vir 14h4'27" 3d43'35"
Ash Rhoades
Uma 10h31'51" 52d9'15"
Ash S
Lyn 9h9'6" 41d52'28"
ASH Senior Musicians 2005
Lyr 18h19'22" 32d8'20"
Ash & Tolb "Forever Promise"
Cyg 19h43'1" 35d56'1"
Ash Tree Meadow
Uma 11h10'20" 33d33'18"
Asha
Uma 11h10'15" 32d24'9"
Asha
Leo 9h58'27" 21d19'57"
Asha
Uma 8h28'40" 67d59'2"
Asha Beena Chadee
Cam 3h28'39" 66d57'0"
Asha Caroline Glenn
Vir 13h41'33" -16d9'42"
Asha Iris
Leo 9h58'23" 16d8'46"
Asha Jaya Hosangadi
Mon 7h19'19" -5d0'48"
Asha Kristina Pappajohn
Ind 20h43'29" -47d0'49"
Asha Laine Pongho
Cra 18h45'45" -39d45'30"
Asha Rene'e Kays
Dra 19h1'38" 63d54'12"
Asha Star
And 0h15'35" 33d16'6"
Asha Wiley
Uma 9h30'9" 45d42'1"
Asha Zoë Ananda Holmes
Ari 2h15'35" 17d11'6"
Ashabee (Elizabeth Ashley Rowe)
Uma 10h47'45" 44d5'51"
Ash-a-lie
Tau 4h33'3" 27d38'18"
Ashamaly
And 23h40'42" 47d14'26"
AshandLee Always
Pho 0h38'19" -41d41'49"
Ashanil
And 1h19'51" 47d43'44"
Ashanti
Crb 15h43'30" 38d48'57"
Ashanti Brianna Hewitt
Umi 16h24'5" 71d30'10"
Asharaf A. Kahn, M.D.
Gem 6h44'36" 23d18'32"
Asha's Star
And 23h10'37" 42d59'8"
ASHBASH2005
Gem 7h3'55" 25d2'9"
Ashbee
Ari 2h42'48" 21d57'24"
ashbugatha
Leo 10h19'38" 16d16'39"
Ashby Johnson
Ori 5h34'56" -0d27'37"
Ashcake
Cnc 8h51'44" 21d5'35"
Asheeka A. Prasad
Sgr 18h12'29" -17d27'42"
Asher
Sgr 19h14'25" -22d10'6"
Asher
Pho 1h6'17" -53d18'53"
Asher Anderson Kelley
Her 17h39'40" 36d40'25"
Asher Claire
Cru 12h15'51" -56d52'31"
Asher Gable
Tau 5h25'47" 21d44'46"
Asher Green
Ori 4h49'11" 10d44'52"
Asher H. J. Laakes
Ori 5h57'14" 20d39'48"
Asher James Walters
Uma 11h37'10" 62d33'31"
Asher John Rataj
Her 17h36'9" 39d2'47"
Asher Keegan Wayne Nicholson
Uma 11h31'24" 46d26'2"
Asher Leaf Kawalek
Leo 11h24'12" 1d45'26"
Asher Max Lipman
Uma 10h55'44" 61d51'44"
Asher Maxwell Wiesen
Cap 20h34'29" -10d21'56"
Asher Roman Johnson
Gem 7h46'38" 21d14'35"
Asher William Kirkwood
Cep 20h40'54" 59d12'14"
Asher William Yearwood
Her 18h43'15" 14d21'6"
Ashera
Gem 6h40'28" 20d37'30"
Ashes (Bobo 42)
Sgr 17h55'2" -29d30'7"
Asheton's True Star
Tau 4h12'48" 16d13'32"
Ashey
Leo 11h3'50" -0d21'2"
AshGeiser
Aql 19h31'7" 5d19'5"
Ashia
Vir 14h42'38" 3d44'49"
Ashis & Neha Roy
Sge 20h4'26" 17d19'21"
Ashish and Ajai
Cyg 21h40'28" 54d23'21"
Ashish Midha
Crb 15h33'39" 37d2'37"
Ashish Sanjay Bambroo
Ari 3h20'53" 24d16'39"
ASHJCB1252000
Lyn 8h0'27" 36d10'25"
Ashkhenochka
Cap 20h12'13" -13d36'7"
Ashlan Stanton
Ari 1h54'42" 24d11'59"
Ashland
Cyg 21h42'8" 38d48'50"
Ashle' Dawson
Cnc 8h21'48" 17d17'17"
Ashle Joann Valasquez
Mon 6h45'16" -4d19'5"
Ashlea
Ari 2h14'38" 23d30'36"
Ashlea Joy Miller
Psc 1h58'29" 3d2'18"
Ashlea Leet
Sco 16h11'26" -13d45'44"
Ashlee
Dra 15h30'40" 56d50'58"
Ashlee
Cnc 8h55'20" 26d0'24"
Ashlee
Leo 10h18'7" 27d21'59"
Ashlee
Psc 1h19'42" 10d21'19"
Ashlee
Uma 9h41'57" 51d35'18"
Ashlee
Lyn 8h35'53" 34d20'16"
Ashlee Anderson
Sco 17h4'52" -38d7'39"
Ashlee Ann
Sgr 18h22'6" -24d53'56"
Ashlee Ann Meyer
Dra 16h35'43" 63d31'38"
Ashlee Boehme
Lyn 7h34'1" 38d4'52"
Ashlee Caitlyn
Cnc 8h32'25" 15d54'29"
Ashlee Crenshaw
Cap 20h21'41" -12d2'37"
Ashlee D. Donovan
Lyn 7h12'50" 48d48'6"
Ashlee Dawn
Lmi 10h39'14" 24d45'20"
Ashlee Diane Lawyer
Cap 20h25'42" -10d9'57"
Ashlee Elizabeth Crackower
And 23h4'21" 51d8'27"
Ashlee Elizabeth Waite
Boo 14h44'40" 31d12'48"
Ashlee Evans
Tau 4h31'4" 29d34'18"
Ashlee Eve Cook
Cas 1h27'26" 52d12'35"
Ashlee Harrison
And 0h43'15" 27d28'13"
Ashlee Hinojosa
Sgr 17h55'36" -17d22'39"
Ashlee Hope Jackson
Uma 10h31'38" 46d21'57"
Ashlee JoRae Shafer
Vir 13h32'29" 3d27'42"
Ashlee & Jordan
Cap 20h55'47" -16d34'24"
Ashlee Kelly
Cnc 8h13'32" 16d25'11"

Ashlee Lauren Rose
Cnc 8h45'11" 12d57'42"
Ashlee Lynn Stigers
Leo 11h5'40" 21d19'25"
Ashlee M Secrest
Leo 11h11'14" 15d1'58"
Ashlee M. Stafford
Cas 0h16'29" 51d15'53"
Ash-Lee Marie Davidson
Cam 3h57'42" 64d45'10"
Ashlee Marie Hester
Uma 8h44'40" 67d0'46"
Ashlee Marie Kimball
Cap 20h33'6" -14d50'0"
Ashlee Melissa Tripp
And 0h18'5" 37d6'42"
Ashlee Mishelle Bazemore
Lyn 9h18'2" 40d41'5"
Ashlee N Starks/Matthew W Rice
Cyg 21h36'1" 43d55'5"
Ashlee Nicol Fiorito
And 0h48'6" 37d42'20"
Ashlee Nicole - My Angel
Ari 2h33'1" 30d29'52"
Ashlee O'Hare
And 23h4'43" 46d21'19"
Ashlee Shaw
And 0h57'10" 37d29'11"
Ashlee Sherman
Lyn 7h47'3" 42d59'5"
Ashlee - Star of Hearts
Cra 18h52'56" -41d34'8"
Ashlee Susanne Hunter
Ori 6h9'46" 15d22'46"
Ashlee the brightest star in my sky
And 23h7'16" 47d36'21"
Ashlee & Travis Tamillo
Sgr 19h44'28" -12d29'34"
Ash-Lee Warren
Cas 0h55'47" 54d16'39"
Ashlee Witt
Gem 6h50'32" 32d34'10"
AshleeAnnSweetSixteen02 2790
Psc 1h9'41" 22d26'37"
Ashlee's star
And 0h47'57" 45d50'15"
Ashlei Anderson
Aql 19h47'20" 13d12'1"
Ashlei Daniell Miller
Lib 15h32'16" -19d45'52"
Ashlei Elizabeth
Ori 5h21'41" -1d7'52"
Ashlei Kate Massad
Per 4h27'31" 49d3'33"
Ashlei L. Philson
Cnc 8h46'42" 31d34'3"
Ashleigh
Gem 7h23'6" 34d26'24"
Ashleigh
And 1h13'42" 42d58'22"
Ashleigh
Cnc 9h12'2" 17d37'36"
Ashleigh
Lyn 7h28'30" 54d12'35"
Ashleigh 11/15/84
Sco 16h6'53" -15d30'27"
Ashleigh 15
Psc 1h25'19" 19d32'50"
Ashleigh B. Rooke
Ari 2h48'57" 15d30'31"
Ashleigh Carter
And 23h23'17" 35d33'44"
Ashleigh Codispot
Psc 0h31'45" 10d27'18"
Ashleigh Crisp
Lyn 7h11'53" 54d22'8"
Ashleigh Davis
Psc 1h33'39" 10d37'28"
Ashleigh Doop
Psc 0h25'44" 9d27'46"
Ashleigh E. Moore
Lyn 8h30'13" 50d20'1"
Ashleigh Elizabeth
Tau 5h59'20" 25d47'23"
Ashleigh Elizabeth Fagone
Lib 15h14'46" -5d13'38"
Ashleigh Elizabeth McKenna
Leo 11h1'41" 16d27'11"
Ashleigh F. Savage
And 0h14'57" 30d31'49"
Ashleigh Georgia Howland
Peg 23h24'47" 18d26'19"
Ashleigh Hatfield
Gem 7h13'7" 32d31'53"
Ashleigh Jean Posey
Gem 6h49'49" 20d45'7"
Ashleigh Joanna Mirabella
Sco 17h20'27" -32d47'2"
Ashleigh Jolly
Cnc 8h9'46" 7d7'20"
Ashleigh Joy Geiger
And 1h37'14" 48d57'37"
Ashleigh Jyl DiGianni
Aqr 22h38'46" 2d5'15"
Ashleigh Kaczmarczyk
Lmi 10h44'20" 35d40'25"
Ashleigh Kressel McClendon
Lib 15h48'49" -14d17'18"
Ashleigh Kristine
And 23h10'29" 42d29'18"
Ashleigh Lauren Bryan
And 1h53'44" 36d8'24"
Ashleigh Lynn Easley
And 1h41'24" 41d56'2"
Ashleigh Marie Catheryn Wilson
Ari 2h43'20" 27d19'30"
Ashleigh Marie Copeland "6-3-93"
And 0h12'27" 25d36'1"
Ashleigh May Swartz
Uma 10h50'44" 65d34'28"
Ashleigh Miranda Farris
And 0h15'30" 41d10'31"
Ashleigh My Lil Lemonade
Umi 15h29'13" 71d12'14"
Ashleigh N. Toth
Vir 12h30'19" 5d50'17"
Ashleigh Nicole
Vir 13h28'18" 13d12'8"
Ashleigh ' Nikki ' Hart
Crb 15h21'34" 28d8'31"
Ashleigh Olivia Cortez
Uma 11h18'16" 42d22'56"
ASHLEIGH ORION
Ori 5h56'17" 8d20'24"
Ashleigh Parker
And 0h11'33" 41d12'51"
Ashleigh Renee Symenski
Lmi 10h20'32" 35d43'47"
Ashleigh Rey Sanders
And 0h37'16" 43d5'37"
Ashleigh Rollins
Sco 16h15'49" -9d13'30"
Ashleigh Stacey Forsyth
And 0h53'38" 46d20'50"
Ashleigh Thompson
And 1h54'42" 37d36'2"
Ashleigh Vallene
Crb 15h35'52" 37d40'40"
Ashleigh Woodfield
Crb 16h13'14" 31d9'42"
Ashleigh-Louise
And 1h50'13" 46d59'58"
AshleighReneéJones
Leo 10h2'47" 24d23'28"
Ashleighs (Ash Buns) Peace
Psc 0h21'25" 10d56'19"
Ashleigh's Smile
Ori 5h51'3" -4d39'24"
Ashleigh's Star
Sgr 18h21'52" -18d22'40"
Ashleigh's Star
Cnc 8h45'3" 27d25'0"
Ashleigh's Sunflower
Gem 7h2'17" 25d59'24"
Ashleigh's Wish
Ori 6h17'29" 10d44'32"
Ashley
Cnc 8h52'56" 11d8'8"
AshLeY
Vir 13h27'2" 11d5'3"
Ashley
Leo 9h59'37" 8d49'40"
Ashley
Tau 5h14'4" 16d37'51"
Ashley
Tau 5h48'0" 16d39'24"
Ashley
Aqr 22h12'1" 2d4'45"
Ashley
Ari 2h21'0" 14d46'6"
Ashley
Vir 13h32'10" 1d12'15"
Ashley
Leo 10h58'48" 7d16'5"
Ashley
Vir 11h39'0" 2d37'27"
Ashley
Gem 7h10'57" 23d49'55"
Ashley
Gem 7h12'21" 27d37'7"
Ashley
Tau 3h58'13" 22d33'42"
Ashley
Crb 15h22'3" 27d46'22"
Ashley
Leo 11h47'32" 25d44'16"
Ashley
And 0h30'31" 30d18'57"
Ashley
Vul 20h40'59" 24d46'58"
Ashley
Ari 2h53'31" 25d14'44"
Ashley
And 0h24'7" 27d32'59"
Ashley
Psc 0h57'52" 25d55'22"
Ashley
Ori 5h55'39" 21d44'22"
Ashley
Gem 6h21'49" 18d44'10"
Ashley
Leo 11h56'42" 22d4'23"
Ashley
Cnc 8h34'44" 15d44'44"
Ashley
And 1h45'9" 47d38'32"
Ashley
And 2h6'53" 45d24'20"
Ashley
And 1h53'19" 46d30'36"
Ashley
And 1h27'32" 49d2'31"
Ashley
And 2h22'58" 47d39'17"
Ashley
Lyn 8h4'11" 47d18'59"
Ashley
Lyr 18h42'49" 39d57'14"
Ashley
And 0h25'51" 45d7'47"
Ashley
And 0h4'5" 47d27'48"
Ashley
And 23h1'52" 44d41'8"
Ashley
And 23h17'0" 43d18'53"
Ashley
And 23h14'33" 46d19'11"
Ashley
And 23h39'41" 47d26'14"
Ashley
And 23h32'5" 48d30'47"
Ashley
Cas 23h34'52" 51d13'7"
Ashley
Uma 9h44'36" 51d8'29"
Ashley
Uma 10h2'18" 45d10'57"
Ashley
Crb 16h10'45" 36d50'16"
Ashley
Crb 15h43'54" 35d33'59"
Ashley
Lmi 10h12'24" 32d6'57"
Ashley
Aur 4h57'2" 32d47'23"
Ashley
And 1h24'5" 35d37'28"
Ashley
Lyn 7h53'53" 35d54'54"
Ashley
Lyn 7h48'25" 37d15'21"
Ashley
Cnc 9h8'13" 30d55'38"
Ashley
And 1h49'47" 39d11'36"
Ashley
And 0h51'3" 43d17'47"
Ashley
And 1h7'47" 44d44'29"
Ashley
And 0h9'36" 39d43'42"
Ashley
Cyg 20h49'55" 31d56'34"
Ashley
Cyg 20h0'22" 33d22'14"
Ashley
Uma 11h23'32" 39d47'52"
Ashley
Aur 6h38'11" 42d35'32"
Ashley
Sco 16h8'42" -18d20'59"
Ashley
Cap 20h18'0" -16d42'42"
Ashley
Cap 21h56'40" -18d36'21"
Ashley
Cap 20h19'11" -11d51'31"
Ashley
Uma 9h47'29" 66d53'25"
Ashley
Lyn 7h6'34" 59d3'10"
Ashley
Uma 10h54'48" 59d22'7"
Ashley
Sgr 17h52'24" -26d57'37"
Ashley
Sgr 18h28'28" -25d19'42"
Ashley
Sgr 18h39'30" -27d56'25"
Ashley
Lib 15h20'9" -23d42'49"
Ashley
Cma 7h10'5" -27d51'2"
Ashley
Sco 17h32'57" -44d3'48"
Ashley
Sgr 18h57'36" -31d51'2"
Ashley
Psc 1h55'55" 6d15'47"
Ashley
Psc 0h52'26" 6d40'51"
Ashley 13
Cyg 19h31'55" 29d22'53"
Ashley 4204
And 1h16'7" 38d51'53"
Ashley 9
Tau 3h29'55" 18d14'26"
Ashley A. Cavanaugh
Cap 20h21'1" -10d46'7"
Ashley A. Hanna
Lib 15h34'2" -25d36'59"
Ashley A. McCormack "Ash"
Sco 17h13'10" -30d58'25"
Ashley A Miller
Leo 10h40'42" 13d37'19"
Ashley A. Ruggiero
Umi 14h57'39" 74d10'7"
Ashley Acres
Uma 10h42'48" 43d20'42"
Ashley Adair
Leo 11h22'37" 14d43'56"
Ashley Adam
Uma 9h9'4" 48d6'11"
Ashley Adams
Tau 5h5'36" 25d37'17"
Ashley Adan
Uma 11h19'10" 59d33'1"
Ashley Addison
And 0h43'33" 25d16'29"
Ashley Alacar
And 23h20'30" 43d10'29"
Ashley Alexander
Cam 6h6'0" 61d44'0"
Ashley Allison Rinaldi
Sgr 19h42'1" -14d54'41"
Ashley Amanda
Lib 14h58'43" -15d56'40"
Ashley Amanda Cline
Sco 16h7'4" -11d36'36"
Ashley Amaral
Gem 6h45'41" 31d24'58"
Ashley Amber Hoff
Per 2h37'29" 52d25'16"
Ashley Ancona
Sco 17h52'10" -35d38'19"
Ashley and Adam Reese
Gem 7h9'25" 28d46'39"
Ashley and Bart
Her 17h10'47" 19d3'22"
Ashley and Blake
Uma 13h15'52" 58d39'41"
Ashley and Brad's Star to Love By
Ori 5h35'53" 4d6'1"
Ashley and Daniel
Cas 1h32'47" 64d40'12"
Ashley and Gregory Humphries
Cnc 9h8'44" 25d25'39"
Ashley and Haley Bowers
Gem 6h55'1" 22d20'14"
Ashley And Jake
Cap 21h39'26" -19d50'29"
Ashley and Jeff's special place
Cyg 19h39'53" 41d27'50"
Ashley and Joey Forever 2005
Her 16h17'15" 5d57'37"
Ashley and Josh
Cyg 20h42'24" 50d34'49"
Ashley and Laura Coe Eternity Star
Cyg 21h21'55" 32d56'53"
Ashley and Matt
Gem 7h41'40" 34d43'24"
Ashley and Matt, together forever.
Uma 11h20'13" 42d36'11"
Ashley and Michael
Cyg 20h40'54" 31d26'41"
Ashley and Miriam 12606
Tau 3h53'5" 23d44'49"
Ashley and Paul
Lmi 10h8'10" 36d39'30"
Ashley and Phil(True Love)
Leo 9h49'35" 25d19'59"
Ashley and Rodie
Cyg 21h25'17" 39d24'13"
Ashley and Rusty
Sco 17h52'36" -34d38'28"
ASHLEY AND STAYTON FOREVER
Uma 9h2'18" 48d27'53"
Ashley and Susan
Lyn 7h11'33" 53d57'59"
Ashley and Thatcher
Cap 21h26'19" -16d32'51"
Ashley and Thereva Forever
Lib 14h34'19" -14d4'22"
Ashley Anderson
Cas 2h35'53" 65d57'51"
Ashley Anderson
Uma 11h27'8" 37d31'16"
Ashley Anderson Owen
Uma 8h39'58" 70d1'54"
Ashley Anfin
Cyg 19h54'40" 31d34'58"
Ashley Angelle 7.7.7
And 0h24'54" 31d40'18"
Ashley & Angus
Lib 15h5'40" -6d51'39"
Ashley Ann
Aqr 22h32'33" -1d11'24"
Ashley Ann
Uma 9h49'46" 71d22'34"
Ashley Ann
And 0h12'4" 31d21'55"
Ashley Ann
Her 16h31'41" 22d26'10"
Ashley Ann
Tau 4h22'6" 9d36'44"
Ashley Ann
And 1h9'9" 38d17'58"
Ashley Ann
And 0h43'3" 39d47'13"
Ashley Ann 16
And 0h50'53" 24d13'10"
Ashley Ann Barker
Uma 8h43'44" 68d41'42"
Ashley Ann Delo
Sgr 18h49'46" -31d58'23"
Ashley Ann Dent
Cnc 8h45'42" 27d9'17"
Ashley Ann Easterday's Star
Aqr 21h11'46" -9d7'0"
Ashley Ann Flynn
And 1h17'37" 41d29'24"
Ashley Ann Gieg
And 0h36'24" 45d36'52"
Ashley Ann Hakala
Lyn 7h40'22" 48d34'17"
Ashley Ann Lane
Leo 10h27'16" 19d23'47"
Ashley Ann Nelson
And 2h24'47" 47d2'7"
Ashley Ann O'Connor
Umi 15h21'26" 68d22'9"
Ashley Ann Reichert
Tau 4h40'54" 23d46'27"
Ashley Ann Riggs
Aqr 22h33'39" -0d53'17"
Ashley Ann Roebken
And 0h41'25" 38d52'18"
Ashley Ann Ryan
Uma 13h16'13" 61d2'52"
Ashley Ann Scott
Sco 16h56'18" -38d21'47"
Ashley Ann Smith
And 1h58'33" 39d1'50"
Ashley Ann Vallis
Per 3h49'48" 43d44'53"
Ashley Ann Voigt
Sco 16h10'42" -11d26'20"
Ashley Anne
And 23h35'10" 37d38'34"
Ashley Anne
Cam 4h0'48" 58d5'26"
Ashley Anne Drewer
Ari 1h48'7" 19d5'35"
Ashley Anne Vannelli
Cnc 8h49'10" 30d53'8"
Ashley Anne Zotta
And 1h43'49" 37d47'34"
Ashley Appell
Sgr 17h55'36" -17d57'42"
Ashley Applegate
Cyg 20h22'37" 45d50'12"
Ashley Applegate 2/1/06
Lyn 9h3'30" 39d20'34"
Ashley Armijo
Lyn 7h0'10" 45d44'30"
Ashley B. Petersen
Cam 4h7'5" 57d49'18"
Ashley (Baby Doll) Gruhn
Cas 1h21'0" 61d32'3"
Ashley Bailey
Psc 1h20'41" 17d39'25"
Ashley Baldauf - Semler
Dra 18h50'20" 60d48'32"
Ashley Bangar
Cam 3h33'13" 58d14'31"
Ashley Banks
Cnc 8h6'33" 13d16'34"
Ashley Barbee
And 0h14'38" 32d7'42"
Ashley Barber
And 23h27'11" 42d41'57"
Ashley Barnett
And 0h16'51" 28d43'15"
Ashley Barrett
Cnc 8h6'38" 25d32'50"
Ashley Bashlor
Cap 20h40'18" -20d28'22"
Ashley Bear
Umi 15h36'37" 71d37'40"
Ashley Beard
Cnc 8h43'46" 31d25'54"
Ashley Beck
Aqr 21h59'25" 0d53'22"
Ashley Bell
Cyg 19h56'21" 51d58'43"
Ashley Beth Carley Ash-ner-nee
Vir 13h25'43" -1d38'37"
Ashley Bialorucki
And 0h35'19" 45d48'55"
Ashley Billings
Gem 7h34'37" 23d54'27"
Ashley "Billy Bad Ass" Michaels
Cam 4h20'47" 68d30'38"
Ashley Bilotta
Cyg 21h38'2" 53d50'53"
Ashley Bishop
Dra 16h32'52" 54d33'55"
Ashley Blount
Ori 5h25'5" 2d42'44"
Ashley Boring
Ari 2h8'45" 25d12'28"
Ashley Bosse
Cam 4h0'52" 56d15'2"
Ashley Bostwick
And 1h13'31" 37d44'17"
Ashley Boyer
Lyr 19h1'11" 42d17'23"
Ashley Braddock Coleman
Tau 3h26'5" 8d46'8"
Ashley Bradley
Umi 16h30'44" 75d32'31"
Ashley Braeuer
And 0h45'29" 36d54'8"
Ashley Brandon
Cas 23h32'1" 53d9'28"
Ashley Brauer
Uma 9h16'23" 49d9'9"
Ashley Breanne Booth
Per 2h31'57" 54d54'4"
Ashley Brianne
And 23h51'40" 32d33'2"
Ashley Brianne Castle
Peg 21h30'32" 15d57'3"
Ashley Bridgefarmer
Tau 4h18'9" 17d30'53"
Ashley Broc Rosser
And 1h39'17" 47d38'57"
Ashley Brooke Dill
Gem 7h44'40" 26d43'25"
Ashley Brooke Glazer
Lyn 7h5'2" 46d45'10"
ashley brooke rogers
Sgr 19h58'31" -20d50'42"
Ashley Brooke Walsky
And 0h54'12" 37d56'16"
Ashley Brooke Wright
And 0h48'15" 40d17'50"
Ashley Brooke's Special Star
Uma 9h8'29" 66d6'9"
Ashley Brooks
Vir 13h49'29" -18d5'37"
Ashley Broski
Uma 9h12'12" 65d26'22"
Ashley & Bryan
And 0h12'53" 44d31'23"
Ashley Brynna Leah Pryce
Crb 15h35'28" 36d24'51"
Ashley Burgin
Vir 13h26'25" -20d44'35"
Ashley Burnham
Uma 10h59'24" 49d58'16"
Ashley Burson
Sgr 18h23'23" -32d1'35"
Ashley Butler
Uma 8h34'52" 60d45'41"
Ashley "Buttercup" Tanaka
Sgr 18h11'21" -32d55'16"
Ashley C. Bolanos
Ari 2h17'1" 13d12'12"
Ashley Camille Cole
Vir 13h23'2" 8d6'3"
Ashley Campbell
Lyr 18h29'25" 36d32'7"
Ashley Campbell
And 0h43'51" 37d8'28"
Ashley Caporusso
Cnc 9h8'54" 30d43'42"
Ashley Carlin
Vir 14h16'2" -12d37'31"
Ashley & Carlos
Crb 15h33'55" 34d42'39"
Ashley Caroline Smith Madden
And 2h8'14" 44d20'33"
Ashley Catherine Alsip
And 23h21'4" 48d38'11"
Ashley Cathleen Hinson
Lib 14h34'57" -8d50'28"
Ashley Cave
Ari 2h35'41" 29d32'3"
Ashley CesarIII CesarJr Maggie Chen
Dra 20h12'36" 65d36'0"
Ashley Chantale Kennewell
Cas 0h58'47" 56d54'34"
Ashley Chapman
Sgr 18h25'17" -36d11'33"
Ashley Charles James Cresswell
Dra 18h43'46" 69d16'39"
Ashley Charney
Vir 13h26'9" 11d49'12"
Ashley Christine
And 1h27'32" 44d29'8"
Ashley Christine Black
And 0h7'58" 45d49'28"
Ashley Christine Brown (Smash)
And 23h21'24" 42d32'44"
Ashley Christine Daniels
Ari 2h40'22" 13d43'23"
Ashley Christine Soper
Leo 10h49'51" 11d5'0"
Ashley Clarke
Lyn 7h20'11" 49d35'12"
Ashley Cleland
Lib 14h29'53" -19d0'56"
Ashley Clementi
Uma 11h15'33" 49d3'57"
Ashley Closser
Lib 15h11'21" -9d26'47"
Ashley Cochran
And 0h20'26" 29d16'27"
Ashley Cogburn
Lyn 8h33'24" 55d34'27"
Ashley Cook
And 2h31'28" 48d59'57"
Ashley & Corey
Cyg 21h26'48" 47d27'11"
Ashley Courtney Atkins
Leo 11h18'46" 14d12'47"
Ashley Curry "My One & Only"
Uma 9h58'9" 46d57'15"
Ashley D. Carlson
Mon 6h55'58" -0d53'28"
Ashley D. Delhomme
Lib 15h24'35" -26d54'2"
Ashley Danae Mazzeo
Lyr 19h19'15" 29d45'45"
Ashley Dang
Cma 6h35'27" -16d51'30"
Ashley Danielle Clark
Lyn 7h9'7" 57d54'30"
Ashley d'Ann Powell
Cnc 8h38'2" 17d41'2"
Ashley Danyeil Damron
Lib 15h25'31" -21d15'33"
Ashley Davis
Lib 14h24'38" -10d57'14"
Ashley Davis
Cru 12h26'24" -59d55'24"
Ashley Dawn
And 1h2'50" 46d25'22"
Ashley Dawn
And 1h55'9" 37d40'21"
Ashley Dawn Ford
Cyg 21h14'6" 40d54'46"
Ashley Dawn Foster
Tau 3h44'46" 16d54'7"
Ashley Dawn Ivy
Uma 9h26'25" 63d29'40"
Ashley Dawn Krolczyk
Cap 21h33'51" -13d56'52"
Ashley Dawn Marie Benson Simons
Gem 6h41'40" 32d31'8"
Ashley Dawn Morland
Tau 4h16'6" 1d6'38"
Ashley Dawn Reitz
Cyg 20h11'38" 46d13'55"
Ashley DeAnn Harvill
Psc 1h26'4" 14d36'37"
Ashley Debrino
Lib 15h2'34" -1d55'55"
Ashley Deluce
Cam 5h32'38" 66d45'49"
Ashley Demopoulos
Lib 14h54'35" -3d40'31"
Ashley Denise Webb
Ari 3h6'13" 15d21'23"
Ashley Diana
And 0h55'50" 37d40'9"
Ashley Diane
Cnc 8h46'31" 18d59'10"
Ashley Dietz
Cas 1h26'38" 68d34'47"
Ashley Dinga
Psc 1h42'3" 25d12'8"
Ashley Donahue Cruz
And 1h27'2" 34d42'15"
Ashley + Donovan Forever Never Ends
Uma 10h56'42" 42d45'8"
Ashley Dorrell
Lyn 6h54'16" 53d17'50"
Ashley Dunham/Closed Heart
And 23h9'29" 43d18'58"
Ashley Dyan Porterfield
Aqr 22h19'34" -3d52'22"
Ashley E.
Cam 5h6'53" 67d45'7"
Ashley E. Giddings
And 1h8'28" 42d15'42"
Ashley E. Tippins
Ori 5h57'25" 22d47'32"
Ashley Eberhard
Her 16h36'42" 12d11'28"
Ashley Eddy
Cas 1h32'13" 64d45'17"
Ashley Eilers
Cyg 19h48'13" 33d44'56"
Ashley Elaine Gramza
Sco 16h5'50" -13d24'36"
Ashley Elisabeth Gambardella
And 0h57'51" 37d0'47"
Ashley Elise Lux
Mon 6h59'28" -0d47'33"
Ashley Elise Steward
And 2h32'23" 48d58'59"
Ashley Elizabeth
And 1h49'10" 46d39'50"
Ashley Elizabeth
Lyr 18h22'46" 38d2'25"
Ashley Elizabeth
Leo 10h18'34" 10d52'40"
Ashley Elizabeth
Aqr 21h17'46" 2d26'50"
Ashley Elizabeth
And 0h28'53" 29d56'30"
Ashley Elizabeth
Ori 6h5'49" 21d21'51"
Ashley Elizabeth
Lib 15h16'1" -19d10'6"
Ashley Elizabeth
Vir 13h35'56" -20d51'50"
Ashley Elizabeth
Sgr 19h10'43" -16d40'46"
Ashley Elizabeth Bell
Ari 2h8'36" 25d4'10"
Ashley Elizabeth Brookover
And 1h49'46" 45d19'58"
Ashley Elizabeth Carter
And 1h6'3" 42d47'41"
Ashley Elizabeth Hampton
And 0h18'48" 27d50'45"
Ashley Elizabeth Howells
Vul 19h50'23" 21d57'54"
Ashley Elizabeth Maloy
Psc 0h0'1" 10d13'37"
Ashley Elizabeth Merrill
Sco 16h57'21" -44d42'25"
Ashley Elizabeth Muscarella
Cas 0h23'52" 52d20'12"
Ashley Elizabeth Nagle
Cnc 9h10'11" 31d35'8"
Ashley Elizabeth Petty
Crb 15h53'37" 34d39'25"

Ashley Elizabeth Pringle
Tau 4h27'35" 3d11'9"
Ashley Elizabeth Reazer
Cas 0h21'48" 52d36'2"
Ashley Elizabeth Scott
Cap 20h14'6" -26d55'54"
Ashley Elizabeth Treco
And 0h7'21" 39d49'4"
Ashley Elizabeth Zierold
And 1h13'21" 46d37'41"
Ashley Ellis
Ari 2h20'10" 17d35'4"
Ashley & Erik Szyluk
Cyg 20h23'38" 55d4'10"
Ashley Erin
Cap 21h39'56" -15d32'20"
Ashley Erin Kelly
Uma 9h45'15" 61d6'15"
Ashley Fago
And 23h39'58" 48d21'30"
Ashley Faith Kronenberg
And 23h21'17" 42d49'18"
Ashley Fallon Duck
Lyn 6h19'58" 56d18'48"
Ashley Farrell Arnold
Cnc 8h13'54" 7d24'5"
Ashley Faye
Dra 19h35'51" 60d24'56"
Ashley Faye King
Cap 21h57'10" -18d27'17"
Ashley Fenwick - Naditch Viola
And 2h23'35" 46d51'47"
Ashley Flaim
Aqr 21h53'21" -4d56'13"
Ashley Foret
Ari 2h40'58" 19d29'2"
Ashley Fraley
And 0h29'44" 32d24'54"
Ashley Franchesca Kay
Lyn 7h44'35" 35d39'32"
Ashley Freed
Boo 14h23'44" 14d0'20"
Ashley Freeland
And 1h36'4" 47d12'26"
Ashley Furgiuele
Ari 2h8'47" 13d15'41"
Ashley Furtado
Crb 16h5'16" 37d52'54"
Ashley Galamba
Cir 14h48'5" -61d40'48"
Ashley Gallagher
Cas 1h20'18" 64d42'22"
Ashley Galloway
Aqr 22h12'9" -13d11'5"
Ashley & Garry Goode Star
Crb 16h13'38" 34d13'59"
Ashley "glitterati" Huffman
Crb 15h53'33" 27d32'12"
Ashley Grace Obel
Lib 15h6'23" -8d52'15"
Ashley Grace Wegner
Vul 20h27'40" 27d44'18"
Ashley Graves
Psc 0h17'19" 16d45'54"
Ashley Grayson 21
Per 2h21'33" 53d51'58"
Ashley Grimm
Gem 7h12'28" 20d16'48"
Ashley Guy
Gem 7h17'8" 23d59'52"
Ashley Gwen Dowdle
Uma 13h34'13" 53d16'21"
Ashley Hagen
And 2h20'35" 50d44'13"
Ashley Hall and Katie Russell
Mon 7h35'32" -0d39'31"
Ashley Halliday's Angelstar
Lib 15h23'32" -20d22'3"
Ashley Halpern
Vir 13h36'11" 6d56'29"
Ashley Hamilton
And 1h15'12" 46d0'49"
Ashley Hanley
And 0h42'25" 28d17'5"
Ashley Hanline Owens
And 0h32'4" 38d40'21"
Ashley Hargrove
Cam 7h38'58" 67d46'18"
Ashley Harris
Gem 7h37'41" 15d43'1"
Ashley Harrison
And 1h24'45" 44d28'31"
Ashley Heart Jeffery
Leo 11h27'37" 15d10'28"
Ashley Heather Wilcox
Ori 5h0'44" -0d49'33"
Ashley Helmer
And 0h47'55" 38d46'19"
Ashley Hickman & Grandmother Lois
And 1h27'0" 45d38'53"
Ashley Hinshaw
Sgr 19h39'35" -36d17'10"
Ashley Hirst
Tau 5h53'33" 24d55'38"
Ashley Hohn
Gem 7h35'59" 22d36'23"
Ashley Holland Carleton
Leo 10h4'19" 11d35'52"
Ashley Hollis
And 0h37'59" 24d17'29"
Ashley Hollobon
Vir 13h8'38" -14d56'1"
Ashley Hood
Cam 6h34'3" 62d38'27"
Ashley House
Aql 20h12'13" 13d46'50"
Ashley I
Her 17h56'52" 21d28'44"
Ashley Ian Smith
Cru 12h20'36" -55d46'51"
Ashley Irene Kronewitter
And 0h35'55" 44d8'29"
Ashley Irene Powers
Com 13h5'11" 26d20'41"
Ashley & Isabella Robinson
Uma 13h51'29" 61d15'53"
Ashley J. Clift
And 0h52'58" 37d16'35"
Ashley & James Forever
Leo 10h37'44" 14d16'30"
Ashley Jane
Cas 1h57'2" 61d25'36"
Ashley Jane My Love 18
Cas 0h16'6" 55d46'30"
Ashley Jane Waters
Uma 14h3'10" 53d15'15"
Ashley / Jared
Ari 3h6'40" 29d58'20"
Ashley & Jason
Sge 19h53'26" 17d39'0"
Ashley & Jason's Love
Vir 11h44'31" 7d34'5"
Ashley Jean
Sgr 19h47'28" -12d23'5"
Ashley Jean Billings
Cyg 19h46'4" 36d1'3"
Ashley Jean Neves
Mon 6h49'23" -0d28'5"
Ashley Jean Purcell
Gem 7h13'42" 28d4'47"
Ashley & Jeff
Lyr 19h2'35" 42d3'56"
Ashley Jeff Sydney Redding
Umi 13h31'15" 69d49'49"
Ashley Jennifer Ahern
And 2h12'10" 46d41'20"
Ashley Jo Nelson
Leo 10h25'28" 12d20'8"
Ashley John Rowe
Cru 12h17'49" -57d14'36"
Ashley Johnson
Sco 16h10'51" -12d34'22"
Ashley Johnson - My Loving Angel
Cru 12h24'14" -58d29'43"
Ashley Johnston
Gem 6h53'46" 20d21'34"
Ashley Joi Millet
Lep 5h42'53" -21d49'5"
Ashley Jordan
Cap 20h30'20" -10d11'3"
Ashley Joy Mindel
Gem 7h20'51" 23d5'15"
Ashley June Fielder
Cas 0h21'49" 51d50'36"
Ashley K Strahan
And 0h9'12" 45d36'0"
Ashley Kackman
Her 17h9'43" 39d42'36"
Ashley Kate
Cas 0h15'57" 57d18'6"
Ashley Kate McGray
Cap 21h19'45" -22d20'55"
Ashley Katelyn Hall
Ari 2h46'43" 19d58'20"
Ashley Kathryn
And 0h52'34" 37d13'0"
Ashley Kay
Crb 16h7'42" 38d44'49"
Ashley Kay Engen
Tau 5h57'59" 28d2'53"
Ashley Kay Merical
Vir 12h10'59" -2d25'9"
Ashley Kay Wood
And 23h47'37" 38d42'27"
Ashley Kayla
Gem 7h43'54" 33d22'57"
Ashley Kelley
Ori 6h11'46" 7d36'7"
Ashley Kennedy
Cas 2h49'54" 61d0'23"
Ashley Kessler
And 2h13'52" 39d3'19"
Ashley Kierzek
Cyg 21h43'24" 45d44'30"
Ashley Kirby - Believe
Gem 6h57'31" 35d19'27"
Ashley Kisslan
And 22h59'33" 49d45'23"
Ashley Klair Lovelace
Ori 5h52'55" 6d55'5"
Ashley Krach
Gem 7h29'48" 28d57'29"
Ashley Krick
Lib 14h51'52" -2d55'5"
Ashley Kristen Vanhorenbeeck
Leo 10h17'24" 18d9'30"
Ashley Kristine
Psc 23h13'14" -2d6'2"
Ashley Kristine Wood
Tau 5h53'58" 23d5'58"
Ashley L. Childers
Lib 15h2'59" -14d58'53"
Ashley L. Martin
Sco 16h58'31" -39d43'21"
Ashley L. Nickerson-Whalen
Leo 11h45'5" 19d36'23"
Ashley L. Van Nortwick
And 1h41'44" 43d12'15"
Ashley La Bar-Slocki
Lyn 7h9'29" 52d35'1"
Ashley La Shelle
Tau 4h37'52" 24d26'34"
Ashley Lane
And 1h10'48" 43d54'32"
Ashley Lang
Cas 0h44'31" 65d18'13"
Ashley Lauren
Aqr 22h8'27" -14d31'28"
Ashley Lauren
Leo 9h22'53" 11d0'5"
Ashley Lauren Alexander
Sgr 18h35'9" -23d31'14"
Ashley Lauren Behar
And 1h36'13" 40d31'48"
Ashley Lauren Clark
Ari 3h27'7" 23d58'6"
Ashley Lauren Cusack
Sco 16h12'59" -10d46'21"
Ashley Lauren Driskell
Sgr 17h52'49" -28d22'40"
Ashley Lauren LaBarre
Uma 11h52'27" 33d26'55"
Ashley Lauren Lynn
Ori 6h15'56" 15d5'20"
Ashley Lauren Perry
Tau 5h16'8" 21d25'25"
Ashley Lauren Ricketts
Uma 11h7'37" 68d32'47"
Ashley Lauren's Wishing Star
Sco 17h54'23" -35d47'34"
Ashley Laveigh
And 1h0'26" 41d25'38"
Ashley Lawson
Lyn 7h24'3" 57d3'55"
Ashley Le Angelle
Cyg 21h40'2" 45d23'51"
Ashley Leann
And 1h33'39" 46d10'36"
Ashley Leann Smith
Tau 4h29'21" 30d34'7"
Ashley Leanne Snedager
Cas 1h20'12" 66d9'0"
Ashley Lee Ulrickson
And 23h13'11" 48d56'40"
Ashley Leigh Bercier
Vir 13h15'16" -22d1'44"
Ashley Leigh Jones
And 0h51'51" 46d40'37"
Ashley Leigh Rayner
Leo 9h39'22" 22d44'10"
Ashley Li
And 0h15'9" 33d42'19"
Ashley Linger
And 0h38'54" 32d45'22"
Ashley "Little Tooter" Inabinet
Psc 0h50'42" 14d17'25"
Ashley Lodema
Cam 3h52'17" 70d10'23"
Ashley Loraine
Sco 16h56'11" -30d25'55"
Ashley Lorraine
Psc 1h23'4" 24d46'57"
Ashley Lorraine Rowe
Cap 20h17'47" -24d45'4"
Ashley Lott's Star
Sgr 18h36'32" -17d39'6"
Ashley Lusby
Uma 10h38'57" 39d42'43"
Ashley Lyn Little
Sgr 18h31'32" -25d23'19"
Ashley Lynn
Sgr 19h2'9" -29d24'3"
Ashley Lynn
Lib 14h53'31" -4d21'51"
Ashley Lynn
Cam 4h27'37" 61d30'54"
Ashley Lynn
Cas 23h39'59" 58d11'45"
Ashley Lynn
Cyg 21h51'4" 45d59'28"
Ashley Lynn
Ari 3h6'8" 25d42'8"
Ashley Lynn
Gem 6h37'39" 24d8'43"
Ashley Lynn 032586
Ari 2h37'47" 30d56'20"
Ashley Lynn Bailey
Vir 12h24'11" 8d47'17"
Ashley Lynn Boswell
Cyg 20h28'49" 31d0'45"
Ashley Lynn Fuoco
And 23h22'52" 42d5'49"
Ashley Lynn Hoover
Cap 20h28'56" -12d35'58"
Ashley Lynn Hughes
Tau 4h37'9" 20d43'6"
Ashley Lynn Jajuga
And 23h11'31" 43d37'46"
Ashley Lynn Kennedy
Cap 20h27'48" -20d23'21"
Ashley Lynn Lemieux Good
Uma 14h23'3" 56d19'39"
Ashley Lynn Leslie
And 23h10'9" 40d34'53"
Ashley Lynn Moffitt
Per 3h25'23" 45d33'21"
Ashley Lynn Morgan
Lib 15h6'34" -16d33'46"
Ashley Lynn Morris
Vir 13h24'44" 11d53'49"
Ashley Lynn Schmidt
Ori 6h20'25" 9d4'45"
Ashley Lynn Wright
Psc 1h40'52" 24d4'22"
Ashley Lynne Kupferman
Aqr 22h30'24" -2d23'51"
Ashley Lynne Witte
Lib 15h5'37" -3d22'18"
Ashley Lynn-Nicole Lewandoski
Sco 17h56'56" -35d15'49"
ASHLEY. M.
Her 16h32'33" 30d24'14"
Ashley M. Johns
Vir 13h49'39" 1d11'58"
Ashley M. Martinez
Tau 4h10'0" 6d38'19"
Ashley M. Morris
Sgr 18h52'48" -21d45'57"
Ashley M. Perfeito
Leo 10h40'6" 16d21'21"
Ashley M. Steensgaard
Per 4h14'23" 52d20'30"
Ashley Mae
Psc 1h16'20" 28d38'17"
Ashley Mae
Leo 11h45'13" 25d0'6"
Ashley Mae Felan
Lyn 8h53'16" 34d55'59"
Ashley Mai
Tau 4h30'18" 14d35'1"
Ashley Major
Umi 16h9'6" 76d58'55"
Ashley Mari Harding
Lib 14h22'27" -11d33'13"
Ashley Maria Perkerson
And 0h13'11" 28d22'56"
Ashley Marie
Ari 2h24'43" 24d1'33"
Ashley Marie
And 0h32'31" 31d46'10"
Ashley Marie
Gem 7h24'25" 26d30'19"
Ashley Marie
Ari 3h19'52" 28d52'43"
Ashley Marie
Tau 4h43'34" 29d1'40"
Ashley Marie
Tau 4h10'51" 27d48'37"
Ashley Marie
Leo 11h13'49" 5d48'45"
Ashley Marie
Cnc 8h28'31" 31d25'3"
Ashley Marie
Gem 6h51'11" 35d5'47"
Ashley Marie
Gem 6h43'48" 33d30'10"
Ashley Marie
Gem 7h58'14" 30d49'16"
Ashley Marie
And 23h46'22" 47d4'21"
Ashley Marie
Lib 15h12'9" -12d37'45"
Ashley Marie
Lib 15h29'4" -8d31'28"
Ashley Marie
Oph 17h38'39" -0d21'51"
Ashley Marie Abreu
Lyn 8h43'38" 40d16'46"
Ashley Marie Adam
And 1h23'29" 36d4'39"
Ashley Marie Aquino
And 23h14'20" 51d47'29"
Ashley Marie Armstrong
Tau 4h22'58" 12d42'27"
Ashley Marie Atkus
Vir 12h52'47" 5d12'54"
Ashley Marie (Baby Girl)
Cap 21h52'37" -9d19'27"
Ashley Marie Baez
And 2h15'38" 49d3'38"
Ashley Marie Baglieri
Ari 2h33'40" 25d6'46"
Ashley Marie Baptista
Ari 3h3'24" 13d42'56"
Ashley Marie Booth
Cap 21h51'28" -19d3'21"
Ashley Marie Boyd
Psc 1h32'16" 15d9'17"
Ashley Marie Burns
And 0h33'54" 45d11'19"
Ashley Marie Bush
Sgr 17h53'3" -28d42'15"
Ashley Marie Cedergren
And 1h22'41" 48d43'1"
Ashley Marie & Christopher Georges
Ori 5h37'11" 11d58'43"
Ashley Marie Coffey
And 1h21'47" 39d12'1"
Ashley Marie Cole
Uma 11h54'5" 37d3'9"
Ashley Marie Collis
Lmi 10h40'29" 31d31'8"
Ashley Marie Cowan
Cnc 8h33'40" 8d4'54"
Ashley Marie Delongs Star
And 0h34'25" 45d43'42"
Ashley Marie Dugan
Cyg 19h50'45" 33d38'38"
Ashley Marie Fillmore
Cap 21h37'20" -13d39'58"
Ashley Marie Finnegan
Cyg 20h49'51" 47d38'29"
Ashley Marie Hampton
Lib 15h2'7" -12d1'51"
Ashley Marie Jett
Uma 10h16'34" 68d38'8"
Ashley Marie Johns
Sgr 18h26'44" -18d25'49"
Ashley Marie Klann
Dra 18h45'20" 56d33'23"
Ashley Marie Lancaster
Lmi 10h29'56" 38d43'17"
Ashley Marie Lopez
Lyn 8h17'20" 57d14'46"
Ashley Marie McKee
Vir 13h26'23" -12d34'24"
Ashley Marie Michalek
Psc 1h15'37" 19d26'44"
Ashley Marie Oates
Eri 4h42'53" -25d4'24"
Ashley Marie Ochoa
Peg 22h55'46" 17d6'33"
Ashley Marie Page
Psc 1h18'26" 24d46'29"
Ashley Marie Parker
Vir 13h9'40" 10d16'59"
Ashley Marie Patterson
And 2h38'13" 46d16'36"
Ashley Marie Petitt
Sgr 18h29'38" -33d7'18"
Ashley Marie Petrone
Lib 14h47'47" -14d38'27"
Ashley Marie R.
Leo 11h38'51" 27d15'23"
Ashley Marie Sodergren
Gem 7h17'12" 17d29'14"
Ashley Marie Temsick "Baby Girl"
Umi 13h26'53" 76d19'18"
Ashley M Yausie
Pho 23h38'22" -51d56'17"
Ashley Marie's Piece of Heaven
Gem 7h12'39" 18d33'17"
Ashley Marilyn "9-26-95"
And 23h18'41" 47d21'21"
Ashley Marilyn "Daddy's Princess"
Cyg 21h15'56" 46d50'42"
Ashley Marisa Watanabe
Ari 2h51'25" 14d36'15"
Ashley Mary Rose Laabs
Cap 21h16'26" -18d39'42"
Ashley Mattson
Tau 3h47'16" 28d58'35"
Ashley Matye
Lib 15h57'53" -12d8'3"
Ashley May
Ari 2h47'59" 25d16'27"
Ashley May Butler
And 1h11'26" 35d39'42"
Ashley McAllister
Tau 5h50'51" 27d23'23"
Ashley McBeth Bloom
Ari 2h59'54" 27d25'30"
Ashley McGrew Cerda Bass
Leo 11h22'57" 6d42'3"
Ashley McLane
Crb 15h28'49" 28d55'35"
Ashley Medders
Cas 23h52'59" 58d24'17"
Ashley Megan Popp the 1st
Sgr 19h45'11" -15d54'55"
Ashley Megan Sonego
And 2h17'32" 50d58'41"
Ashley Mei Chretien
Ori 6h19'17" 8d47'8"
Ashley Melissa Jolly Williams
And 0h51'19" 34d54'12"
Ashley & Michael Forever
Tau 4h12'43" 27d35'58"
Ashley Micheal
Cnc 8h35'5" 22d40'10"
Ashley Michele Ehlert
Aqr 22h31'12" -2d12'41"
Ashley Michele Holbein
Gem 7h3'53" 27d45'19"
Ashley Michelle
Psc 0h7'49" 2d3'37"
Ashley Michelle
Cap 21h57'31" -19d5'2"
Ashley Michelle Johnson
Cnc 8h47'9" 18d29'51"
Ashley Michelle Kalena
And 1h46'1" 45d18'7"
Ashley Michelle Kicklighter
Vir 14h43'41" -1d51'26"
Ashley Michelle Moriarty
Cas 23h0'9" 55d8'8"
Ashley Michelle's Star
Uma 10h10'51" 43d29'8"
Ashley & Mike Trimble
Cnc 8h41'25" 30d15'4"
Ashley Miller
Dra 19h9'11" 59d48'40"
Ashley Miller
Mon 7h17'43" -0d50'42"
Ashley Mitchell
Sgr 18h18'10" -30d54'17"
Ashley Mokres
Oph 17h9'3" -0d49'29"
Ashley Monet Watson
Sco 17h20'36" -42d9'30"
Ashley Monique
Lmi 10h7'59" 35d6'49"
Ashley Montana
Cnc 8h0'18" 11d15'24"
Ashley Moran
Lyn 9h2'5" 36d43'45"
Ashley Morgan
Leo 10h11'57" 11d14'17"
Ashley Moriah Smith
Lib 14h50'38" -5d56'43"
Ashley Morris My Sweetie Budgeetie
And 2h27'25" 49d34'16"
Ashley Mowitz
Sco 17h51'38" -35d46'19"
Ashley Murray
Uma 10h24'28" 45d59'56"
Ashley Mussbacher
Umi 15h33'15" 81d6'20"
Ashley MY ANGEL Scholl
Tau 4h25'53" 17d28'4"
Ashley My Baby
Cap 21h17'56" -21d47'14"
Ashley - My Beloved
Cas 0h10'24" 54d2'24"
Ashley my one and only
Leo 11h54'55" 28d16'36"
Ashley Mykel Borges
Cap 21h34'13" -16d7'0"
Ashley N. Forseth
Sco 17h55'6" -34d39'49"
Ashley N. Goodhue
Uma 11h12'56" 52d2'8"
Ashley ~n~ Karina
Cyg 20h46'6" 41d30'48"
Ashley N McCullough
Tau 5h46'51" 24d31'5"
"Ashley Nacole"
Gem 6h45'12" 31d39'27"
Ashley Neikrie
And 23h11'30" 51d16'25"
Ashley Nelson
And 0h27'20" 39d57'9"
Ashley Nester
Cap 21h44'22" -23d33'37"
Ashley Newbery, Sailor Scout
Mon 6h43'52" -0d21'40"
Ashley Newton
Peg 22h5'46" 6d22'14"
Ashley Nichole
Ori 5h38'58" 3d24'7"
Ashley Nichole
And 0h41'13" 42d35'22"
Ashley Nichole
Aqr 22h39'13" -1d39'28"
Ashley Nichole Adrian
Cap 20h27'37" -10d16'59"
Ashley Nichole Freck
Cnc 8h33'11" 25d57'52"
Ashley Nichole Freck
Cnc 9h20'16" 28d18'3"
Ashley Nichole White
Cnc 8h38'34" 21d25'10"
Ashley Nicole
Gem 6h30'43" 21d59'8"
Ashley Nicole
Ari 2h8'26" 25d11'56"
Ashley Nicole
And 0h49'43" 24d23'38"
Ashley Nicole
Ori 5h29'54" 2d10'33"
Ashley Nicole
Aqr 20h46'20" -12d18'36"
Ashley Nicole
Lib 15h15'52" -8d12'33"
Ashley Nicole
Cap 20h52'53" -22d5'32"
Ashley Nicole
Cap 20h44'33" -20d46'56"
Ashley Nicole
Cas 1h7'34" 64d8'26"
Ashley Nicole
Lib 15h3'51" -25d56'17"
Ashley Nicole
Sco 16h18'30" -28d48'47"
Ashley Nicole
Cap 20h24'14" -23d59'35"
Ashley Nicole
Sco 17h26'32" -41d55'27"
Ashley Nicole Archer
Cam 7h35'24" 63d12'43"
Ashley Nicole Beaver
And 0h38'30" 39d20'59"
Ashley Nicole Bock
Vir 13h4'0" 12d10'17"
Ashley Nicole Cisewski
Gem 7h49'44" 15d57'12"
Ashley Nicole Cope
Lyn 8h6'41" 44d2'45"
Ashley Nicole Cudini
Ari 2h31'52" 25d34'3"
ASHLEY NICOLE CUIFO-LO
Leo 10h30'5" 23d11'55"
Ashley Nicole Daniels
Cyg 19h47'51" 40d16'31"
Ashley Nicole Davis
Vir 12h39'10" -5d2'5"
Ashley Nicole G.
Ari 1h48'15" 18d48'21"
Ashley Nicole Goolsby
Cnc 8h8'10" 24d29'24"
Ashley Nicole Grodecki
Dra 16h16'19" 51d51'39"
Ashley Nicole Grubb
Lyn 7h27'57" 44d59'27"
Ashley Nicole Haggerty
Cnc 8h35'18" 15d15'33"
Ashley Nicole Haug
Aqr 22h10'9" -2d58'51"
Ashley Nicole Houston
Cas 23h27'41" 57d20'51"
Ashley Nicole Howard
Gem 6h34'43" 18d53'17"
Ashley Nicole Johnson
Ari 2h8'57" 22d57'16"
Ashley Nicole Johnson
Vir 12h43'28" 7d51'31"
Ashley Nicole Jones
Psc 1h0'18" 26d42'39"
Ashley Nicole Jones
Ori 5h54'12" 18d30'8"
Ashley Nicole LaDuke
Cap 21h3'32" -25d0'40"
Ashley Nicole Lowery
And 0h24'48" 32d49'13"
Ashley Nicole Mallios
Gem 6h32'52" 15d16'27"
Ashley Nicole Marcum
Tau 4h36'35" 18d53'24"
Ashley Nicole Martin
Tau 5h49'47" 12d42'50"
Ashley Nicole Masterton
Lib 15h8'23" -5d42'37"
Ashley Nicole McBride
Leo 11h37'48" 25d33'18"
Ashley Nicole McVey
Leo 11h4'48" 17d5'3"
Ashley Nicole Meister
And 2h14'0" 41d52'38"
Ashley Nicole Meyer
Lib 15h4'48" -10d56'55"
Ashley Nicole Meza
Cnc 8h11'18" 24d41'3"
Ashley Nicole Moore
Leo 11h45'36" 23d18'56"
Ashley Nicole Moore
Cap 21h14'59" -15d50'46"
Ashley Nicole Olsen
Aqr 23h56'17" -13d57'45"
Ashley Nicole Olson 6.21.2014
Sco 16h14'1" -13d6'11"
Ashley Nicole Pagano
Sgr 18h27'24" -17d52'35"
Ashley Nicole Panelli
Vir 13h15'35" 12d26'11"
Ashley Nicole Penn
Pho 0h35'38" -40d21'31"
Ashley Nicole Podlas
Sco 17h12'48" -39d6'45"
Ashley Nicole Read
Lib 14h30'39" -10d59'13"
Ashley Nicole Rennolds
Sgr 19h27'42" -16d9'14"
Ashley Nicole Reyes
Oph 17h32'38" -0d26'19"
Ashley Nicole Richardson
Psc 1h40'39" 6d40'41"
Ashley Nicole Sellers
Lmi 9h58'19" 40d26'8"
Ashley Nicole Smiley
Umi 17h23'11" 79d45'20"
Ashley Nicole Smith
Cap 20h38'23" -18d55'37"
Ashley Nicole Steadman
Vir 11h54'50" -6d5'36"
Ashley Nicole Stevens
Sco 16h5'58" -17d37'21"
Ashley Nicole Trefero
Ori 5h59'34" 20d56'27"
Ashley Nicole Vargas
Vir 12h39'54" -5d24'7"
Ashley Nicole Waldemar
Uma 10h13'22" 57d6'0"
Ashley Nicole Walton
Tri 2h38'47" 33d48'40"
Ashley Nicole Ward
Ori 5h39'18" 3d37'30"
Ashley Nicole Whisenant
Leo 11h15'58" 6d47'32"
Ashley Nicole Young
Uma 10h5'3" 63d31'35"
Ashley Nikki Dewey
Lib 15h5'4" -1d6'42"
Ashley Nowell Glenz
Leo 10h11'20" 12d2'24"
Ashley Ohlemacher
Uma 9h31'49" 63d26'25"
Ashley One
And 0h21'20" 37d9'3"
Ashley O'reilly
Uma 10h37'45" 55d40'32"
Ashley Owens
Gem 6h20'44" 27d46'25"
Ashley P & Eric G's Star
Cyg 19h36'11" 53d14'28"
Ashley Paige Langmeier
Tau 4h43'16" 27d37'27"
Ashley Parker
Gem 6h50'49" 14d4'47"
Ashley Parton (Ash)
Per 2h54'48" 44d27'8"
Ashley Patrick
Lyn 6h50'50" 57d28'11"
Ashley Penrod
Sco 16h51'38" -34d0'51"

Ashley Penton O'Hagan
And 22h59'8" 51d12'53"
Ashley Pierce
Gem 6h40'20" 26d40'16"
Ashley Pimley
Mon 6h58'35" 0d47'2"
Ashley Poulin
Sco 17h57'25" -42d40'14"
Ashley Powell God's Angel #10
Leo 10h15'38" 20d29'14"
Ashley Prather
Crb 16h10'58" 33d59'16"
Ashley "pré de frène"
Sgr 18h7'7" -26d34'45"
Ashley Prestia
Tau 3h49'4" 29d45'18"
Ashley Prince
Mon 7h5'20" 4d16'51"
Ashley Priscilla Vasquez
Leo 9h59'23" 24d42'18"
Ashley R. Harrison
Cas 0h57'43" 54d31'0"
Ashley R. Mastrangelo
Lib 14h28'37" -10d37'27"
Ashley R. Yerion
Gem 7h18'47" 34d36'48"
Ashley Rachel
Sgr 20h0'2" -25d27'54"
Ashley Rachele Encinas
Vir 13h14'57" 5d24'49"
Ashley Rachelle
And 1h7'7" 45d49'45"
Ashley Rae Byers
Psc 1h45'19" 18d43'53"
Ashley Rae Elliott
And 23h33'41" 47d18'47"
Ashley Rainwater
Sgr 18h10'9" -30d41'22"
Ashley Reavis
Com 12h35'27" 23d22'50"
Ashley Regina Strittmatter
Cas 0h34'37" 52d7'34"
Ashley Rene' Kahn
Crb 16h13'8" 26d46'31"
Ashley Renee
And 1h29'31" 40d5'45"
Ashley Renee
Cap 21h20'2" -15d35'1"
Ashley Renee
Lib 15h27'58" -3d47'39"
Ashley Renee Bailey
Aqr 20h55'48" -11d46'20"
Ashley Renee Biering
And 0h7'19" 32d36'33"
Ashley Reneé Dove
And 23h49'32" 45d11'42"
Ashley Renee Helberg
Sco 16h9'57" -15d30'56"
Ashley Renee Trojanowski
Aqr 21h51'37" -0d41'9"
Ashley Renee-Aunt Niecy's niece
Vir 13h31'57" 3d28'26"
Ashley Rhein
Lmi 9h48'24" 34d23'3"
Ashley Rich
Ari 2h27'3" 25d9'2"
Ashley Robert Millington - Ash's Star
Her 18h40'48" 25d18'9"
Ashley Robin
And 2h25'20" 48d40'39"
Ashley Rosa
Dra 17h29'53" 68d18'3"
Ashley Rose
Uma 8h38'27" 63d13'31"
Ashley Rose
Cas 23h22'48" 58d12'3"
Ashley Rose
Lib 15h14'45" -15d50'22"
Ashley Rose
And 0h24'59" 33d47'55"
Ashley Rose
Tau 4h33'44" 10d45'39"
Ashley Rose Anastasia Bednar
Ari 2h35'16" 27d59'43"
Ashley Rose Bailey
Tau 5h51'2" 17d16'55"
Ashley Rose Czarnecki
Aqr 21h1'32" -11d0'35"
Ashley Rose Denny
Lib 15h10'44" -4d59'14"
Ashley Rose Elizabeth
Gem 6h58'29" 21d17'21"
Ashley Rose Kurcsak
Aqr 20h48'35" 1d37'4"
Ashley Rose Leone
Sgr 18h33'44" -21d35'21"
Ashley Rose Pelner
And 0h43'51" 32d51'12"
Ashley Roya
Vir 12h59'59" -9d17'24"
ASHLEY RUE
Crb 15h36'4" 38d0'43"
Ashley Ruth
Ari 1h59'36" 23d48'53"
Ashley Ryan
And 0h27'27" 31d22'27"
Ashley Ryan Galati-Kangas
Uma 10h39'32" 55d25'27"
Ashley S. Bedward
Aqr 23h55'12" -4d42'1"

Ashley Sabrina Dugle "Sweetheart"
Lyn 7h57'46" 39d36'32"
Ashley Salazar
Leo 9h34'45" 22d31'59"
Ashley Sarah Burnside
Uma 14h1'50" 53d26'29"
Ashley Scenora Simpson
Leo 11h16'11" 1d11'9"
Ashley "Schnookum Wookum" Tubbs
Cyg 20h25'9" 52d3'28"
Ashley & Scott
Aqr 22h34'58" -1d58'58"
Ashley Scott Hamilton
Cas 0h9'13" 53d20'27"
Ashley Seguin
Cyg 21h43'33" 44d17'36"
Ashley Sevilla
Cam 3h20'17" 60d48'59"
Ashley Sheardown
Cas 2h30'39" 72d43'37"
Ashley Shearin
And 0h48'56" 38d25'29"
Ashley Shines On
Leo 10h17'37" 9d19'34"
Ashley Shiu
Cas 1h44'43" 58d2'12"
Ashley Shutske
Cnc 9h7'3" 27d22'3"
Ashley "Silly Goose" Vigil
Psc 1h22'8" 23d13'48"
Ashley Singleton
Gem 6h37'28" 18d27'38"
Ashley Sirabella
Tau 5h33'11" 26d31'49"
Ashley Sitzer
And 0h19'29" 40d37'40"
Ashley Skippy Turbyfill
And 1h51'5" 39d6'34"
Ashley Sorum
Cnc 8h54'17" 28d19'43"
Ashley Soto
And 1h15'40" 49d4'31"
ASHLEY SOTO
Uma 14h15'22" 60d46'22"
Ashley Spearman
Ari 3h9'48" 26d40'31"
Ashley Spoleti
Uma 13h53'51" 57d21'11"
Ashley Stabb
Crb 16h1'54" 29d17'17"
Ashley Steimer
Vir 14h2'13" -16d8'37"
Ashley Stevens
And 0h59'42" 36d45'48"
Ashley Suchoval
Dra 19h13'43" 68d27'26"
Ashley Sue Coblentz
Ari 2h22'20" 26d12'55"
Ashley Talon Spooner
And 0h40'37" 27d43'45"
Ashley Tara Longworth
Cap 21h31'10" -10d8'50"
Ashley Taylor Cleary
Ari 2h54'26" 28d27'28"
Ashley Taylor Murphy
Gem 6h51'35" 32d22'13"
ASHLEY TEMBECK
And 23h7'15" 44d48'22"
Ashley Teresa Hadaway
Vir 14h42'15" 3d18'8"
Ashley the Beautiful Angel
Gem 7h15'2" 23d12'38"
Ashley The Latosa
Ori 5h41'59" 7d45'8"
Ashley - TheSparkleInMyEyes 07.01.05
Lib 15h1'3" -9d23'1"
Ashley Ting Thongsawath
Gem 7h25'53" 26d19'37"
Ashley Tisdale
Cnc 8h37'49" 8d9'45"
Ashley Tomlin
And 0h27'28" 32d51'57"
Ashley Torres
Cha 8h57'52" -78d20'31"
Ashley Totin - Rising Star
Mon 6h37'27" 7d4'52"
Ashley Tozer
Cas 23h16'40" 55d47'11"
Ashley Tranchant
Ari 1h59'24" 23d9'11"
Ashley Treat
Cam 7h8'42" 72d38'41"
Ashley Venters
Tau 4h33'45" 19d39'10"
Ashley Victoria
Lyn 8h14'20" 41d47'33"
ASHLEY VINCI
Leo 11h13'25" 18d9'54"
Ashley- Violeta
Ari 3h23'31" 19d43'46"
Ashley Virginia Brannen
And 23h21'3" 37d56'8"
Ashley Waddell
Uma 10h57'59" 37d9'12"
Ashley Wagner
Leo 11h7'21" 22d12'51"
Ashley Walker
Aqr 23h53'7" -4d7'8"
Ashley Waterman
Tau 4h16'49" 26d11'11"

Ashley Wentworth
Uma 9h29'19" 67d16'5"
Ashley Werner
Aqr 23h11'21" -8d19'50"
Ashley Wildes
Cnc 8h44'46" 31d36'27"
Ashley Willhite
And 0h50'44" 37d13'54"
Ashley Willis
And 1h27'39" 46d33'11"
Ashley Wilson
Lyn 8h27'56" 41d58'10"
Ashley Witt Fertall
Ori 5h12'58" 11d45'57"
Ashley Wood - Miriam's Shining Star
Cru 12h19'38" -56d42'19"
Ashley Yeager
Cam 4h35'54" 79d45'30"
Ashley Yeazel, The Love Of My Life
Psc 1h25'5" 19d38'46"
Ashley Yetzer
Sco 16h38'1" -30d41'57"
Ashley Youngers
Psc 1h17'9" 28d35'18"
Ashley Zalewski
Aqr 23h53'58" -3d56'10"
Ashley Z.J.
Her 18h35'3" 20d17'14"
Ashley, My Princess
Cnc 8h52'19" 8d59'17"
Ashley, the Light in My Heaven
Aqr 21h50'43" -8d7'2"
Ashley, Will You Marry Me?
And 1h8'12" 36d15'3"
Ashley12172001
Ori 6h1'15" 10d16'41"
Ashley229-04
Gem 6h46'55" 25d57'27"
Ashleyalexis Philomena Power
Leo 9h28'10" 10d45'37"
Ashley-Ellen
Tau 4h11'57" 5d13'36"
Ashley-Marie Bodor
And 0h19'16" 45d35'4"
AshleyNBraden1983
And 0h6'56" 43d37'11"
AshleyNicole14
Psc 1h1'3" 10d40'57"
Ashley-n-Rubner ...Love You Always.
Leo 9h57'58" 8d11'4"
Ashleyon
And 1h17'43" 33d58'28"
AshleyPatrice
Cyg 20h27'9" 53d52'3"
Ashley's and John's Star
And 1h26'35" 48d58'45"
Ashley's Angel
Cap 20h55'50" -19d54'23"
Ashley's Angels
Lyn 7h10'47" 57d36'20"
Ashley's Aura
Lyn 7h4'17" 52d20'28"
Ashley's Beauty
Sgr 18h25'46" -30d30'4"
Ashley's Birthday Star From Grayson
Leo 11h3'35" -1d23'31"
Ashley's Brillance
Cyg 21h23'0" 50d4'42"
Ashley's Diamond
Cnc 8h54'1" 26d11'17"
Ashley's Dream
Psc 0h38'39" 19d38'53"
Ashley's Dream
Cas 1h34'1" 65d22'1"
Ashley's Eye
Cap 20h38'38" -15d24'17"
Ashley's Heart
Cas 2h12'52" 71d3'32"
Ashley's Lasting Brilliance
Aqr 22h37'34" 1d59'35"
Ashley's Lola
Tau 4h52'23" 26d55'59"
Ashley's Lucky Star
Psc 0h20'32" 16d44'35"
Ashley's Mom Andrea (Drea)
Gem 7h17'30" 26d23'49"
Ashley's Mulkey Way
Pho 0h21'32" -44d57'17"
Ashley's Paradise
And 23h30'1" 45d35'9"
Ashley's Rising Star
And 0h53'18" 40d10'2"
Ashley's rising star
And 2h35'57" 39d19'44"
Ashley's Rising Star 4.0 Fall 2005
Aqr 21h4'25" -10d50'27"
Ashley's Rosetta Star
And 2h28'26" 42d8'40"
Ashley's Shining Light
Cnc 8h22'6" 28d39'11"
Ashley's Shining Star
Leo 11h19'58" 25d9'32"
Ashley's Shining Star
Lib 15h9'41" -12d5'53"
Ashley's Shooting Star
Vir 12h36'30" 1d11'43"

Ashley's Softball
And 23h10'29" 41d55'18"
Ashley's Song
Cnc 9h4'6" 25d41'46"
Ashley's Star
Cnc 9h12'52" 28d35'21"
Ashley's Star
Cnc 8h23'25" 16d19'20"
Ashley's Star
Mon 7h3'45" 4d33'21"
Ashley's Star
Ori 5h11'3" 11d34'40"
Ashley's Star
Tau 5h46'41" 15d50'41"
Ashley's Star
And 23h22'45" 46d47'43"
Ashley's Star
And 23h29'52" 48d38'34"
Ashley's Star
Uma 9h13'5" 47d47'24"
Ashley's Star
Lyn 7h58'17" 43d57'2"
Ashley's Star
And 1h4'47" 38d25'5"
Ashley's Star
Cyg 19h48'4" 31d48'23"
Ashley's Star
Her 16h31'46" 33d12'50"
Ashley's Star
Lib 14h45'0" -10d49'50"
Ashleys Star
Cap 21h30'50" -16d37'41"
Ashley's Star
Ori 4h56'56" -2d42'18"
Ashley's Star
Ori 5h57'30" -0d36'26"
Ashley's Star
Aqr 22h35'8" -2d4'37"
Ashley's Star
Aqr 22h39'44" -1d59'22"
Ashley's Star
Cam 3h32'21" 67d40'24"
Ashley's Star - "Bean"
Pav 19h52'49" -69d17'31"
Ashley's Sweet Sixteen
Sgr 17h54'12" -26d11'7"
Ashley's Sweet Sixteen Star
Ari 2h9'0" 20d51'0"
Ashley's VW Bus
Cam 5h20'54" 57d36'34"
Ashli
Gem 7h35'47" 24d2'18"
Ashli and Ricky
Lib 15h33'9" -6d14'34"
Ashli Bowman
Cnc 8h40'10" 7d18'33"
Ashli D Rowland
Cnc 8h58'14" 31d50'20"
Ashli Marie Newfield
And 0h14'49" 44d17'57"
Ashli Nicole Dings "10 lil' toes"
And 0h32'59" 45d4'6"
Ashlie
Gem 7h37'12" 32d22'41"
Ashlie
Leo 10h10'52" 23d59'37"
Ashlie and Kellie
Dra 17h29'32" 66d48'56"
Ashlie Chan
Vir 13h21'34" 2d39'7"
Ashlie Chickenhead Lokey
Lib 14h57'6" -18d20'5"
Ashlie D. LeGrand
Aqr 22h32'5" 1d31'47"
Ashlie Lousie
And 0h53'43" 39d51'12"
Ashlie Marie White
Cnc 9h11'11" 20d20'21"
Ashlie McKae
Sco 17h41'40" -40d54'1"
Ashlie Nicole Milne
Ari 2h8'42" 21d52'42"
Ashlie Purvee
And 0h14'3" 29d22'19"
Ashlie S. Russell
Sco 17h23'3" -33d13'6"
Ashlie Taylor Bissell
Lib 15h23'21" -5d23'12"
Ashlie Wyderka
Lyr 19h24'26" 37d48'15"
Ashlies Strength
Lyn 8h1'44" 42d35'34"
AshLilli
Cap 21h42'7" -16d58'48"
Ashling Byrne
Lmi 10h37'32" 38d14'52"
Ashling Fitzgearld
Ari 3h4'14" 18d27'52"
Ashling Lillian O'Brien~Aug.8, 2006
And 1h52'41" 39d9'46"
Ashlinn
Per 3h37'4" 43d32'42"
Ashlinn
Ari 2h22'50" 17d45'41"
Ashlin's Star
Leo 10h18'39" 11d38'42"
Ashly Ann
Gem 7h15'35" 24d0'52"
Ashly Brook Wyllie
Uma 10h33'53" 44d45'41"
Ashly + Jason
Cnc 9h8'21" 31d35'53"

Ashly Lynn Latham
Tau 6h0'0" 24d51'58"
Ashly Marton
Vir 13h46'3" 2d3'31"
Ashly Taylor Mulder
Col 5h45'24" -31d13'6"
Ashlye Christine O'Hara
And 0h48'16" 36d47'35"
Ashlyn
Crb 16h15'1" 39d24'32"
Ashlyn
Sco 16h3'56" -17d45'9"
Ashlyn Alexis
Lib 15h35'57" -16d49'47"
Ashlyn Annis
Cam 3h51'27" 55d55'31"
Ashlyn Bailey Smith
Tau 4h7'51" 28d52'31"
Ashlyn Dawn
Cas 0h29'48" 53d51'23"
Ashlyn Dawn Perrington
And 1h41'59" 49d41'50"
Ashlyn Faith Currie June 29, 2005
And 23h22'20" 45d20'12"
Ashlyn Lark
Cap 20h41'55" -14d40'4"
Ashlyn Mai
Umi 12h48'10" 87d11'35"
Ashlyn Marie
Sgr 18h9'48" -19d6'37"
Ashlyn Marie Curry
And 1h14'53" 49d51'28"
Ashlyn Marie Ferriero
Uma 10h9'31" 51d17'32"
Ashlyn Noelle Baroni
Per 3h27'12" 47d17'55"
Ashlyn Pearl Kratochvil 10/15/02
Uma 10h50'52" 53d41'45"
Ashlyn Rose
Umi 14h27'12" 74d2'44"
Ashlyn Rose Saeger
Umi 14h5'0" 66d49'14"
Ashlyn Satterfield
Mon 6h43'52" 7d7'4"
Ashlyn Taylor Charleston
And 23h42'51" 42d28'49"
Ashlyn Taylor Kreiss
Ari 2h42'58" 28d35'15"
Ashlyne
Ori 6h17'57" 14d41'34"
Ashlyne Meray Manning
Cas 0h43'46" 54d11'2"
Ashlynn Elizabeth Golliher
Lib 15h10'55" -27d12'8"
Ashlynn Hunter
Uma 13h47'50" 51d40'53"
Ashlynn Jordi Buckner
Leo 9h29'37" 24d18'3"
Ashlynn Makayla Delaney Morris
Uma 8h45'27" 55d12'15"
Ashlynn Miriah Parga
Lyn 6h45'14" 56d44'46"
Ashlynn Sklarcik
Lib 15h8'31" -15d38'43"
Ashlynn, Our Smiling Star
Aql 19h6'2" 15d37'35"
Ashlynne Rae
Tau 5h48'37" 22d44'26"
Ashlyn's Star
Peg 23h54'0" 20d56'45"
Ashna
Cap 20h14'48" -17d36'52"
ASHOEL
Cas 23h38'49" 56d10'45"
Ashoford
Vir 12h48'57" 6d48'3"
Ashok Keshavkant Patel
Cam 6h44'19" 75d39'45"
Ashok Sani
Uma 8h11'27" 62d12'56"
Ashot Baghdasaryan
Uma 12h13'15" 55d33'0"
Ashour Youkhanna Chamoun
Cnc 8h32'17" 28d19'50"
Ashra
Her 16h43'38" 41d18'57"
Ash's Anomaly
Cyg 20h47'16" 53d3'0"
Ash's Night Light
Leo 10h12'1" 20d34'5"
~Ash's Star~
Cnc 9h6'18" 26d3'20"
ASH's Syte
Uma 10h25'4" 71d15'52"
Ashtar Commander of Light
Aql 19h42'27" 12d3'21"
Ashten Evo Colucci
Ori 5h32'40" 12d14'0"
Ashti
Leo 11h28'53" -3d11'9"
Ashtin
Uma 11h57'55" 59d27'31"
Ashtin Goldenberg
Ori 6h16'43" 14d53'13"
Ashtin Shayne
And 1h30'31" 46d10'37"
Ashton
Lyn 7h52'30" 49d29'32"
Ashton
Aql 20h13'21" 7d30'55"

Ashton
And 0h35'1" 32d49'35"
Ashton
Uma 11h56'54" 55d24'34"
Ashton
Uma 8h27'34" 72d48'50"
Ashton
Aqr 21h38'11" -1d13'39"
Ashton
Lib 15h27'6" -26d38'16"
Ashton Bryce Sibley
Uma 9h58'42" 60d47'7"
Ashton Conrad Alan McKibbin
Aqr 22h1'21" -9d2'21"
Ashton Curry "Daddy's Little Star"
Uma 9h52'52" 46d3'4"
Ashton Daniel Kelly
Her 16h41'32" 6d30'40"
Ashton Elizabeth Bott
Cnc 8h46'55" 30d40'5"
Ashton Eric Crisp
Ori 5h5'37" 4d32'15"
Ashton Evan
Cnc 8h29'58" 16d0'36"
Ashton Fatula
Aql 19h48'13" -0d30'51"
Ashton Heckermann
Lmi 10h34'51" 37d57'34"
Ashton Hicks
Lyn 7h21'14" 46d7'3"
ASHTON & IAN COWIE
Tau 4h30'43" 17d20'14"
Ashton James Heljenek
Her 16h36'46" 48d19'25"
Ashton Johnathan Newins
Uma 10h59'14" 37d0'39"
Ashton Lee
Leo 11h24'9" 4d7'8"
Ashton Lee Bass
Uma 8h57'27" 64d21'33"
Ashton Lee Gandrup
Her 17h37'38" 40d39'46"
Ashton Lee Mennerich
Cap 20h11'2" -9d54'41"
Ashton Leigh Binford
Del 20h37'56" 3d25'6"
Ashton Mallory McDuff
Lib 14h49'57" -18d41'12"
Ashton Marie's Star * My MoosieBear
Crb 15h25'10" 26d6'2"
Ashton Matthew Delmonico
Umi 14h52'7" 68d25'27"
Ashton Maynard
Uma 10h13'32" 64d47'12"
Ashton Michael Garrett Grimm
Leo 11h7'1" 21d31'50"
Ashton Michael Rose
Leo 9h39'30" 27d59'44"
Ashton Michael, Beloved 05162002
Tau 4h30'7" 28d33'34"
Ashton Milne's Star
Cas 0h14'25" 53d12'52"
Ashton Nicole
Ari 2h7'18" 17d54'0"
Ashton Nicole Stolar
And 0h52'51" 41d58'14"
Ashton Phillips The Amazing One
Sco 16h9'4" -25d13'14"
Ashton Rae
Aql 19h58'36" 10d16'37"
Ashton Reid Dial
Crb 15h32'50" 38d54'59"
Ashton Royea
Cap 20h40'47" -16d58'24"
Ashton Ryan Haupt
Peg 22h44'58" 9d31'17"
Ashton Sloan Woods
Uma 10h56'44" 54d27'46"
Ashton Smith
Gem 7h13'41" 31d27'16"
Ashton Spencer Holdridge
Ori 5h27'43" 2d8'25"
Ashton Taylor Burr
Uma 11h31'42" 42d49'12"
Ashton Thomas Siwald
Lib 14h53'14" -2d15'18"
Ashton Victoria
Gem 7h0'25" 21d28'17"
Ashton's Light
Uma 11h39'20" 42d33'14"
Ashton's Star
Oph 17h1'5" -1d51'29"
Ashtopher
Ori 6h16'39" 10d11'4"
Ashtule
Umi 16h4'24" 76d15'7"
Ash-Ty-Dus Rizzieri Challenger Star
Uma 10h12'4" 54d8'56"
Ashtyn
Leo 9h39'6" 28d10'3"
Ashtyn & Joe
Vir 13h43'47" -3d40'6"
Ashur Kolich Sangari
Lib 15h43'53" -6d18'52"
Ashutosh
Tau 4h2'52" 6d0'1"
Ashwie
Lib 15h8'33" -11d4'2"

Ashworth
Cyg 20h32'44" 37d17'43"
Ashy Boo
Cep 22h29'2" 65d48'55"
Ashy's Fish Slip
Tau 4h11'42" 26d27'18"
Asi
Umi 16h32'28" 82d13'2"
Asia
Cas 23h26'42" 53d12'55"
Asia Elizabeth Nicole Snyder
Peg 22h18'20" 13d51'10"
Asia Gifford
Leo 10h10'42" 17d9'42"
Asia Huk
And 1h49'45" 37d34'16"
Asia Jean Balancier
Uma 9h9'59" 58d6'21"
Asia Ka mierska, Moje Kochanie
Psc 0h1'20" 1d59'9"
Asia Marin
Uma 8h39'32" 59d53'16"
Asia Namiko Thorpe
Agr 23h3'36" -5d59'42"
Asia Neudauer
Cnc 8h16'58" 19d29'2"
Asia Przepalkowska
Cyg 19h44'21" 37d56'54"
Asia Ryan Troche Williams
Uma 11h10'5" 35d34'28"
Asia Saffron McLaughlin
Ari 2h18'22" 11d19'46"
Asia Sitek
Cyg 20h36'56" 35d49'15"
Asia Skora
Cap 20h29'18" -12d9'56"
asiabelcher
Cyg 20h0'24" 34d17'54"
Asia'nae Gonzales
Uma 9h0'35" 51d21'10"
Asicota
Leo 10h27'59" 13d22'1"
Asif and Nidhi
Apu 17h33'37" -70d4'20"
Asim Ansari Alli
Uma 9h9'34" 62d25'45"
Asim Dincer
Leo 11h56'16" 25d9'27"
Asimina L. Demou
Leo 11h29'9" 26d31'23"
Asja Tuco
Sgr 18h15'31" -19d8'19"
ASK Shecky 8
Cap 21h2'14" -22d13'46"
ASK the Laursens
Cas 0h51'54" 61d28'40"
Askia Grant
Sgr 18h17'11" -23d9'12"
Askin & Kaan & Koray Ibrahim
Dra 18h44'52" 55d24'45"
Askold Horajeckyj
Gem 7h44'55" 32d8'53"
Aslan
Cma 6h50'36" -17d13'25"
Aslan
Cma 6h42'51" -14d25'4"
Aslan's Way
Uma 9h14'53" 62d50'44"
Asli Aker & Tugberk Emirzade
Dra 12h12'55" 73d29'57"
Asli Siyahi
Ari 2h47'50" 11d57'41"
Aslihan Piltanoglu
Ari 3h2'47" 14d37'43"
Aslinn Moondragon
Aqr 22h17'27" 1d8'14"
Aslyn Paige Tallulah Ryan
Aqr 21h49'40" -0d57'22"
Asma
Tau 3h53'23" 22d0'19"
Asma Klai
Cnc 9h3'51" 7d41'18"
Asmi
Vir 13h28'13" 7d35'9"
Asmina-Jane
Uma 8h18'58" 65d36'22"
ASN
Uma 11h13'10" 61d9'23"
AsNiDa
Aql 19h20'25" 15d1'24"
Asombroso
Equ 20h57'21" 4d1'46"
ason 22596
Psc 1h28'9" 15d26'1"
Ason and Solomon Hammond
Tri 2h23'27" 33d3'26"
Aspasia
Ari 2h41'57" 21d51'16"
Aspen
Uma 10h59'46" 54d8'30"
Aspen Lights
Sco 17h54'48" -36d35'30"
Aspen Marie Kalkowski
Umi 15h13'26" 68d44'55"
Aspen Mlt.
Uma 11h39'20" 44d23'23"
Aspen Reann Baker
Mon 6h44'14" -0d2'39"
Asphodel
Ori 6h4'59" 5d48'47"

aspiration And 1h21'13" 44d55'9"
Aspiri - The Star "Happy 70th" Gem 7h43'30" 34d33'58"
Aspporo Star Marketing Sdn Bhd Ori 5h30'56" 2d54'19"
Asproulias Periklis Psc 2h4'17" 9d39'18"
Asra The Great Dane Cma 6h45'36" -12d55'15"
Assad Masoud Psc 0h42'54" 12d44'40"
Assaf Eros Aristos Cnc 9h12'18" 10d5'33"
Assembly of Ancestors Vir 13h10'58" 11d50'22"
Assia Aqr 22h33'48" -1d50'19"
Assia Oph 17h13'23" 9d25'48"
Assia Agramelal Gem 7h13'16" 20d7'1"
Assia la dolce Tartarughina Peg 22h39'31" 19d58'37"
Assistant Chief Newell, 2301 Tau 5h7'52" 24d51'25"
Assmahane Baghri And 0h12'36" 39d50'34"
Assumpta Mestres Cyg 20h24'45" 34d26'44"
ASSUNNY Uma 11h42'11" 35d54'8"
Assunta Fusco Campanile Per 2h13'0" 57d15'56"
Assunta Perricone Cyg 19h42'20" 55d16'34"
Assunta Susie Venturino Cas 23h29'46" 53d7'16"
Ast Charlotte Uma 9h26'28" 55d45'7"
ASTA Cma 6h40'9" -15d51'26"
Asta Lise Sequoia Com 12h43'9" 30d0'46"
Astalan Uma 9h6'57" 62d57'22"
ASTAR4GEEKXOXOD-WORKLETTE Lyn 8h22'25" 56d27'12"
astarte's Star Aqr 21h9'16" -1d53'45"
Astellas Uma 11h30'5" 58d2'17"
Astellas Uma 10h57'1" 35d52'0"
Astellas Uma 9h58'48" 43d9'58"
Astellas Ari 2h46'44" 27d51'33"
Aster And 2h35'18" 46d25'34"
Aster de Thelma And 0h56'9" 35d43'13"
Asteri Turbosis Cap 20h58'51" -19d47'21"
Asteria Gratia And 23h1'15" 40d31'56"
Asteria Sophia And 23h13'27" 41d16'14"
Asterias Peg 21h28'22" 15d28'57"
Asterios Geranis Uma 10h16'3" 65d19'25"
Asterisco de Caelan Psc 0h46'12" 16d4'35"
Asterlou Sgr 18h41'14" -32d38'56"
Asteroid B-612 Uma 10h52'8" 42d20'42"
Astéroïde B612 Lyn 6h31'38" 57d18'55"
Asteroide Barbara 612 Uma 11h19'42" 38d10'5"
Asterperious Squadron 219 Uma 9h45'21" 58d18'25"
Astghik Harutyunyan Ari 2h54'7" 21d9'26"
Astghik Melikyan Gem 7h17'50" 20d57'58"
Asti Cma 6h55'40" -14d37'55"
Aston Cooper Burnett Her 18h49'31" 21d59'39"
Astonishing Majestic Megan Smith Psc 23h22'41" 7d16'4"
Astra Bertrada Eternae Cam 6h20'56" 68d5'52"
Astra Shamilae Uma 9h41'54" 50d50'52"
Astra Silveus 253 Eri 3h42'33" -0d49'28"
AstraDawn Umi 14h52'20" 71d14'46"
Astraea Sally Vir 12h14'56" 8d33'55"
astral divinity Cyg 20h23'48" 47d16'25"
Astraleena Del 20h38'32" 15d22'59"
AstralSuzy Uma 11h17'11" 33d40'27"
Astrid Lyr 18h31'46" 36d8'18"
Astrid And 2h7'31" 42d25'4"
Astrid Ari 2h7'5" 16d10'43"
Astrid Cam 3h38'52" 66d54'59"
Astrid Acuna Leo 9h56'2" 15d36'19"
Astrid Bortoluzzi Crb 15h27'49" 25d43'20"
Astrid E. Travasso Sco 17h10'27" -38d56'26"
Astrid Elizabeth Alvaremga Lyn 6h58'46" 51d52'28"
Astrid Elizabeth Liv Ratcliffe Cyg 21h6'7" 37d0'21"
Astrid Huneide Cnc 8h30'34" 14d12'19"
Astrid Isabelle Uma 11h40'22" 38d17'53"
Astrid Kamer & Erich Kreienbühl And 1h12'57" 45d19'43"
Astrid Len Bailey Cap 20h33'43" -23d38'58"
Astrid Marie Tau 5h38'57" 24d34'26"
Astrid & Patrick Umi 17h4'27" 82d35'2"
Astrid Poesch Cru 12h7'3" -61d8'13"
Astrid Reichmuth Umi 16h41'57" 75d41'44"
Astrid + Timo Uma 11h39'24" 54d42'34"
Astrid und Ghiath Uma 9h45'10" 50d12'40"
Astrid Wolf-Hauch, 13.03.1967 Tri 1h32'51" 34d36'59"
Astrid-Aloisius Uma 12h13'49" 59d43'36"
Astrid's Glücksstern Per 4h43'4" 40d25'10"
Astrig Agopian Aqr 23h48'41" -4d43'5"
Astrinakis Car 10h13'32" -59d44'31"
Astriska Cas 1h50'35" 59d37'53"
Astro Amy And 1h12'13" 37d53'18"
Astro Aragao Aur 5h38'10" 48d21'48"
Astro Laporte Cma 7h1'28" -21d44'24"
AstroDan Ari 3h14'22" 15d11'32"
Astrofeggia de Lenny Psc 0h24'44" 8d28'50"
Astroid Leo 9h43'42" 25d31'59"
Astron Ryan Jacob Tau 5h42'58" 21d11'51"
Astronate Her 16h46'16" 8d56'22"
Astrophil and Stella Cyg 20h33'20" 41d13'41"
Astrum ad Heatherthomas Sgr 18h35'3" -23d36'25"
Astrum Aeternus Eternus Diligo And 0h44'32" 44d42'5"
Astrum Amicus Annette Ari 2h45'48" 27d54'13"
Astrum C Esto Perpetua Lyn 8h20'29" 40d43'30"
Astrum Cecelia Crb 15h59'45" 32d19'13"
Astrum de Domina Ali Lyn 6h17'26" 54d17'59"
Astrum decoris: Robert H Lickel Jr. Lyr 19h20'1" 28d37'39"
Astrum Dominic Ori 5h22'4" -1d45'53"
Astrum Eller Her 18h41'44" 19d24'42"
Astrum enim Amor Tau 4h25'54" 3d15'49"
Astrum Fidei Anne et Ben Cyg 21h36'54" 39d26'54"
Astrum Frostic Uma 9h36'57" 59d23'57"
Astrum Lawrentium Cap 20h31'41" -13d30'2"
Astrum Lucidus Copernicus Gem 7h32'48" 32d10'59"
Astrum Maiestas Lyn 8h1'24" 34d31'7"
Astrum Nexae Ori 5h50'26" -0d2'27"
Astrum Patronus Fortis Her 17h37'19" 15d19'44"
astrum peregrinus Leo 9h39'43" 12d30'51"
Astrum Pro Julianus Quod Pium Aur 5h57'46" 29d15'14"
Astrum Taffiensis Gem 7h33'36" 24d35'24"
Astrumi Minutum Deinde Tenus Tu And 2h19'15" 50d7'31"
Astute-Saulute Gem 7h31'31" 32d34'42"
" Astwo Phiwl " Leo 9h55'47" 13d50'4"
Asuman ve Gurhan Tuker And 23h20'48" 48d44'32"
Asusena Vir 12h27'1" 5d15'35"
Asutsascie Cyg 21h12'40" 51d48'17"
Aswin Reimers Uma 12h58'39" 56d4'35"
Asya Uma 11h29'39" 65d12'16"
Asya Ari 2h33'16" 14d12'41"
Asya Aponte Sco 17h56'20" -30d4'7"
Asya Dechen Cam 5h42'12" 65d29'18"
A.T. Das Ori 6h2'23" 19d48'0"
At First Sight Ori 6h1'12" 18d46'36"
At first sight Cyg 20h29'22" 46d16'46"
At Last Umi 15h18'27" 73d58'13"
At Last Uma 13h49'42" 56d47'37"
A.T. Singley Peg 22h13'33" 13d0'24"
Ataahua Pirihoanga Vir 12h28'44" -7d17'34"
Atah va Ani Ori 6h15'37" 8d25'0"
A.T.A.J.A.C.K.S. Uma 9h46'43" 69d11'50"
Atalie Day Justice Tau 4h18'56" 2d49'59"
Atalie McCarthy Ori 6h3'55" 9d51'55"
Atchley Cnc 9h12'32" 14d37'43"
ATChristie Uma 14h16'38" 58d10'38"
Ate & Louise Boyd Leo 10h12'0" 14d36'45"
Atenea Sanchez Cap 20h40'38" -17d48'58"
Atevetu Daniluck Cru 11h58'11" -62d37'31"
Athair Ferrick Sgr 18h56'41" -30d5'34"
Athanas & Carla Eternally Ori 5h36'31" 10d35'30"
Athanasia Francis And 2h19'43" 46d35'40"
Athanasios Gem 6h56'57" 22d14'1"
Athanasopia Cyg 20h41'11" 46d8'47"
Athanassios Vir 15h6'59" 6d11'37"
Athandy Cnc 8h40'54" 6d37'4"
Athea J. Friend Uma 11h38'55" 46d58'7"
Atheena Meling Uma 13h44'20" 56d44'35"
Athelhampton Lmi 9h53'11" 35d24'56"
Athena Vir 12h33'11" 1d41'39"
Athena Vir 13h31'8" 5d53'47"
Athena Tau 4h28'6" 25d9'42"
Athena and Sean Forever I Love You! Cnc 8h7'44" 26d55'10"
Athena Elisha Ori 5h16'49" 12d18'54"
Athena Fessas Her 16h31'40" 24d51'7"
Athena Follower of God Uma 14h4'35" 48d59'34"
Athena Francis Slater And 2h25'41" 49d36'41"
Athena Helga Heath And 1h18'13" 49d55'33"
Athena Janke Mon 6h49'19" -0d3'33"
Athena Jennifer Myers Per 3h44'49" 46d55'50"
Athena Kay Crawford Aqr 22h11'50" -24d5'0"
Athena Lacarte Born: Mon Nov 11/1968 Uma 11h31'57" 57d12'38"
Athena Loukopoulos Sco 17h24'6" -30d21'11"
Athena M Cook Cyg 20h59'52" 47d38'23"
Athena Marso Ori 6h2'44" 11d24'46"
Athena Starwoman Sgr 18h55'0" -16d0'19"
Athena Vokas Lyn 8h12'48" 46d27'27"
Athenapollo 4-27 & 11-8-05 Lyn 7h53'35" 39d16'21"
ATHENE2112 Cyg 19h59'43" 33d12'38"
Atherlicia Uma 10h46'8" 45d59'20"
Athina Cyg 21h6'0" 38d54'42"
Athina Moravec Leo 11h7'8" 22d25'4"
Athina Psylla Uma 9h44'41" 47d29'5"
Athina Seraphine Cyg 21h30'51" 49d8'23"
Athon and Talula Sco 17h8'38" -37d33'30"
Atifa Leo 10h14'29" 26d1'43"
Atilgan, Saner Uma 10h54'15" 55d21'0"
Atinder "Kaka" Singh Boparai Uma 11h23'21" 60d50'9"
Atisha's Star And 2h1'24" 38d7'5"
Atisthan Eri 3h45'29" -0d13'44"
Atiya Bahmanyar And 2h13'2" 42d29'5"
ATIZ Pho 1h23'50" -42d54'55"
ATJ Lib 14h35'32" -12d45'49"
ATK1130040832 Cam 3h20'39" 63d54'23"
Atkham Nguyen Umi 14h41'37" 78d5'8"
Atkins, John Uma 9h47'1" 44d48'10"
Atlanta Marie Feaster And 23h24'36" 42d43'33"
Atlantica Peg 22h57'48" 21d23'2"
Atlantis Ari 2h4'46" 22d13'0"
ATLANTIS And 23h20'49" 50d50'39"
Atlas Gault Tri 2h20'23" 32d19'31"
Atlas-Niehaus Cyg 21h47'33" 46d51'42"
Atom Alexander Brinley Lyn 6h31'5" 57d11'49"
Atom-Adam Uma 9h12'46" 46d44'43"
Atonement Per 3h22'58" 49d29'34"
Ator Uma 11h35'44" 58d39'54"
A-Train Lyn 9h6'13" 34d19'58"
Atrau Cam 4h56'15" 74d42'31"
Atreyu Ser 15h21'36" -0d54'31"
Atreyu Cap 20h36'6" -23d37'1"
Atreyu Xavier Romich Leo 9h44'43" 28d6'1"
ATRIS Her 17h17'54" 16d46'40"
Atropolisa Ori 6h15'6" 15d2'42"
A.T.S.T. Uma 11h38'49" 62d2'44"
"Atsushi" be with you for evermore... Leo 11h9'44" 12d38'27"
ATT Watching Over You Cam 4h12'26" 69d48'45"
Attallah-Shabaz Whatley Cas 1h12'39" 58d17'57"
Attama Family Star Pho 1h34'31" -42d11'17"
Attanasi Cyg 19h40'37" 43d7'9"
Attendente e Pettirosso Cyg 21h58'56" 47d45'34"
Attia Forest Slater Uma 10h33'19" 56d34'55"
Attia, My Love And 0h4'9" 36d46'29"
Atticus Cam 3h57'23" 65d1'9"
Atticus Chance Southwell Cap 21h36'35" -11d49'9"
Attila és Pukri Csillaga Uma 9h13'59" 61d2'3"
Attila Fekete Ori 5h20'11" -0d14'17"
Attila Laszlo Szoboszlai And 23h38'38" 36d5'6"
Attila "Papi" Takacs Cep 2h12'14" 80d6'13"
Attila The Hun Von Rogers Cma 6h42'33" -13d8'50"
Attila Valy Leo 11h47'28" 21d40'46"
Attinger-Morian Uma 11h57'24" 63d1'10"
Atto Cma 6h58'38" -23d42'45"
Attorney Mark B. Peters Uma 10h41'27" 56d20'3"
Atty Per 4h18'18" 33d10'21"
Atul Kumar Singhal Ari 2h47'33" 12d33'56"
Atze Uma 14h18'40" 58d58'22"
Au Renee avec amour Cyg 19h35'39" 32d4'43"
Aubey Cnc 8h7'6" 12d58'51"
Aubin And 0h34'42" 26d56'17"
Aubin's Ladybird Star Umi 15h23'57" 71d25'54"
Aubra Charlene "Shines Forever" Leo 9h58'42" 20d13'2"
Aubre Fouch Sgr 18h30'24" -16d5'19"
Aubree Ari 2h52'34" 13d29'21"
Aubree Ann Johnson And 1h19'33" 42d24'21"
Aubree Greenspun Uma 11h53'29" 63d21'48"
Aubree Jordan Norton Umi 14h33'59" 75d28'37"
Aubree Lynn Gem 7h24'58" 33d15'36"
Aubree Lynn Lyn 7h49'29" 35d30'1"
Aubree Rayne Cap 20h47'51" -15d57'33"
Aubree Rose Cyg 20h19'43" 45d54'26"
Aubrey And 1h17'57" 45d58'2"
Aubrey And 2h26'16" 46d23'50"
Aubrey Leo 10h13'7" 11d54'37"
Aubrey Cap 20h15'10" -17d14'47"
Aubrey Aqr 23h1'3" -10d48'59"
Aubrey Uma 9h55'0" 60d13'36"
Aubrey Amanda Hager Lib 15h58'54" -13d18'14"
Aubrey and Azlynn Abbott Gem 7h35'25" 24d4'15"
Aubrey Anna Carper Psc 1h3'0" 16d51'41"
Aubrey Cobbett Tau 4h35'30" 26d11'27"
Aubrey Davis Cam 5h43'49" 59d54'11"
Aubrey Elise Resech Cas 0h21'16" 55d45'50"
Aubrey Fisher Psc 0h39'38" 14d10'26"
Aubrey Jillian Lib 15h41'10" -7d45'56"
Aubrey Jillian Goldstein Lib 15h44'17" -18d43'53"
Aubrey Joy Uma 8h54'16" 61d24'31"
Aubrey Julianna Brandt And 0h5'33" 39d1'35"
Aubrey June Sco 16h42'24" -30d55'38"
Aubrey LeAnn Johnson Sgr 19h2'33" -15d26'0"
Aubrey Lee Keller Lyn 6h58'49" 47d38'0"
Aubrey & Lenny Tejada Cyg 20h52'37" 31d23'50"
Aubrey Lynn Cap 21h43'13" -20d23'49"
Aubrey Marie Anderson And 0h11'41" 32d46'17"
Aubrey Mikylie Exon Cru 11h58'35" -64d0'26"
Aubrey Nielsen Ori 5h22'39" 14d12'17"
Aubrey Rachel Storbeck Cnc 9h5'19" 6d49'48"
Aubrey Reese Gem 7h18'4" 19d51'48"
Aubrey Rose Conlon Vir 14h53'31" 0d22'49"
Aubrey Thorp Tau 4h8'43" 16d22'57"
AubreyLove And 23h36'57" 48d41'15"
Aubri Brown Uma 13h40'1" 56d0'42"
Aubri Hatzell Aqr 22h29'56" 0d22'31"
Aubri Rhiannon Karp And 1h49'35" 38d52'25"
Aubriana Leo 9h28'24" 25d4'31"
Aubrianna Bright Lep 5h50'35" -26d50'36"
Aubrianne Elise Rote Gem 7h29'14" 33d55'15"
Aubrie Elizabeth Sisk Uma 9h10'12" 49d46'19"
Aubrie Jane Lib 15h4'17" -1d48'43"
Aubrie Lee Sco 17h10'30" -44d37'15"
Aubrie Lee Lyn 7h48'50" 38d0'45"
Aubrie Noel Vul 20h36'51" 27d11'35"
Aubrie Paige Meadows Ori 5h58'28" 20d57'12"
Aubrie Renee Gilman Leo 11h31'34" 23d0'29"
Aubrie Rose Lyn 7h58'36" 41d19'2"
Aubrie Taylor Zalewski Her 16h51'49" 14d46'0"
Aubry Uma 9h24'11" 44d47'41"
Aubry 40 Gem 6h38'53" 27d46'9"
Aubry Lynn Leo 10h26'6" 15d1'38"
Aubry Lynne Naylor And 0h43'26" 41d2'7"
Aubry Star Cep 3h15'5" 82d56'38"
~AUBZ~ Sgr 18h40'36" -36d11'44"
Auchincloss And 1h56'36" 46d0'56"
.:Aud:. Cma 6h50'56" -22d22'53"
Aude Aqr 20h58'46" -10d15'12"
Aude Couturier Ori 5h26'35" -5d35'5"
Aude Lauriol Cma 7h1'9" -20d10'25"
Audentia - Rosalind Davies Cas 1h18'57" 54d52'56"
Audge Ori 5h51'4" 2d39'37"
Audi Cyg 21h40'8" 30d25'24"
Audi Q7 - A Star That Never Follows Lyn 7h57'59" 54d5'2"
Audia Andy Vir 14h17'48" 2d7'32"
Audie Delores Capocelli Gem 6h52'10" 22d31'31"
Audie Eugene Burks Per 3h24'49" 43d31'17"
Audis C. Hill Ori 5h25'17" 2d44'14"
Audi's Dolphin Cnc 8h59'41" 29d45'5"
Audra Vir 14h22'31" 2d22'18"
Audra Uma 9h23'13" 44d11'55"
Audra Cnc 8h52'48" 32d42'30"
Audra Cap 21h35'14" -17d45'32"
Audra Sgr 19h18'2" -23d4'41"
Audra and Chad's Star And 0h33'44" 45d9'38"
Audra B. Rafter Psc 23h17'1" 7d45'39"
Audra Chantel Muntain Cyg 22h2'35" 45d52'0"
Audra Conwell And 0h41'21" 39d47'3"
Audra Dagenais Cyg 20h34'27" 47d54'42"
Audra Florence White And 23h44'59" 44d27'38"
Audra Louise Ware Aqr 22h20'7" -24d17'14"
Audra Lynn Barrier Lib 14h51'19" -6d40'3"
Audra Lynn Otto Cap 21h35'57" -15d7'2"
Audra Marie Dalton Vir 14h27'28" -1d34'26"
Audra Michelle Barker Aql 19h47'57" 16d14'14"
Audra Renee Gapper And 23h1'15" 49d36'10"
Audra & Ryan Forever Cyg 19h37'54" 54d58'23"
Audra the Radiant Sgr 18h4'4" -28d0'14"
Audrabella Cas 0h1'35" 59d4'13"
Audra's Birthday Star And 1h1'12" 37d51'4"
Audrey Gem 7h11'4" 34d41'23"
Audrey Cas 1h0'49" 54d14'19"
Audrey Cas 0h59'33" 56d59'34"
Audrey Cam 3h18'51" 58d12'43"
Audrey And 23h9'42" 48d41'26"
Audrey Uma 14h2'36" 47d57'4"
Audrey And 2h37'41" 49d24'40"
Audrey Gem 6h30'42" 20d53'0"
Audrey Tau 4h41'32" 2d39'53"
Audrey Aqr 22h43'35" 0d45'54"
Audrey Cas 0h57'50" 60d46'56"
Audrey Uma 9h22'23" 63d56'12"
Audrey Cas 1h53'22" 64d53'35"
Audrey Uma 14h5'49" 61d30'5"
Audrey Crv 12h37'10" -13d22'57"
Audrey Cet 1h10'52" -0d21'30"
Audrey Cap 21h49'35" -19d5'51"
AUDREY Aqr 21h14'13" -8d46'6"
Audrey Cap 21h41'50" -10d10'19"
audrey Lib 15h41'5" -11d31'5"
Audrey 112457 Sgr 18h14'55" -19d20'15"
Audrey Agrati Sgr 17h56'11" -18d1'4"
Audrey Aguilan Psc 23h20'25" 1d31'45"
Audrey & Alex Leo 11h8'58" 25d34'36"
Audrey and Mark Garro 10 And 1h39'45" 47d51'0"
Audrey Ann Gem 7h27'7" 27d2'44"
Audrey Ann White Uma 12h50'7" 59d12'57"
Audrey Ann Witko Ori 6h14'21" 8d53'8"
Audrey Anna Boardman Sco 16h53'46" -38d29'9"
Audrey Anne Bronswijk Cyg 19h37'13" 31d37'16"
Audrey Arnaud Peg 21h32'55" 23d30'18"
Audrey Arvin Vir 14h18'53" 3d1'5"
Audrey Barankovich Dzema Lib 15h15'37" -28d54'57"
AUDREY BEHRENS Psc 1h9'34" 28d20'16"
Audrey Bella Leo 11h44'30" 25d4'45"
Audrey Benoît Sco 17h14'37" -41d14'57"
Audrey Bordeau Place Her 16h45'54" 8d51'28"
Audrey Boyza And 2h23'57" 50d21'7"
Audrey Brown Gem 6h40'46" 24d0'40"
Audrey BT Quealey, 1 in 43 Billion Sgr 18h50'2" -27d37'38"
Audrey Campbell Her 18h2'28" 24d49'12"
Audrey Capathia Brown Cas 1h47'33" 64d20'37"
Audrey Cavé Ori 6h20'51" 15d30'17"
Audrey Claire And 0h15'52" 33d27'2"
Audrey Coleman Tau 5h40'34" 26d24'30"
Audrey Copeland Mon 6h13'39" -8d14'47"
Audrey Corinne Huber Uma 8h50'14" 59d42'5"
Audrey Daniel N. Y. Huysmans Vir 14h13'28" -3d44'26"
Audrey Denise Cobb Uma 13h4'50" 52d59'24"
Audrey DeSmith Carey And 0h4'41" 34d58'16"
Audrey Diane Cummings And 1h0'33" 42d30'38"
Audrey Drawbaugh And 2h22'19" 50d24'6"
Audrey Edith Page Crb 16h4'8" 30d47'37"
Audrey Egan Neasham Ari 2h44'58" 25d35'25"
Audrey Elise Spearman Cyg 21h40'53" 41d50'12"
Audrey Elizabeth Crandall Cnc 9h20'10" 17d4'31"
Audrey Elizabeth Grose Cru 12h41'37" -58d50'1"
Audrey Elizabeth Marino Crb 16h5'28" 37d6'47"
Audrey Elizabeth Sytsma Leo 10h18'53" 20d42'24"
Audrey et Xangomossey Cyg 20h26'0" 58d59'34"
Audrey F. and R. Sean Cosper Vir 14h26'11" 4d11'0"
Audrey Faith Alexander Ori 5h25'26" 10d59'18"
Audrey Faye Arnold Sco 16h19'32" -41d39'0"
Audrey Galvin Herr And 2h32'1" 49d6'11"
Audrey & Gene Curry March 15, 1946 Ori 5h11'0" 8d39'50"
Audrey Gilbertson Cap 21h57'46" -10d4'42"
Audrey Gillespie,My Angel Sgr 19h39'21" -30d42'53"

Audrey Gillies - Audbod Star
And 0h17'53" 25d21'19"
Audrey Gonzalez
Cas 0h43'49" 63d31'0"
Audrey Grace
Lyr 18h37'51" 31d23'33"
Audrey Griffin
Gem 6h57'1" 18d11'54"
Audrey Hailey
And 1h11'35" 38d12'55"
Audrey Hamilton Morrissey
Aqr 22h48'48" 0d11'19"
Audrey Helen Brown
Uma 11h54'57" 33d24'36"
Audrey Hillman Fisher
Lib 14h56'30" -10d57'30"
Audrey Hope Brown
Sco 16h57'27" -43d31'27"
Audrey Irwin
Cas 23h13'0" 54d47'20"
Audrey J. Edwards
Cap 21h27'24" -9d20'9"
Audrey JienFen Hanger
Uma 10h30'42" 56d1'4"
Audrey Jo-Bob Louise Renee Ford
And 2h37'41" 40d25'53"
Audrey & John "Jack" Tarbun
Cep 21h47'8" 66d6'9"
Audrey Josephite
Lib 14h54'25" -1d22'20"
Audrey K.
Sco 16h11'26" -17d28'53"
Audrey K Lindvall
Leo 11h34'7" 21d13'16"
Audrey Katherine Magarian
Leo 10h21'9" 17d53'25"
Audrey Kathleen
Peg 22h38'0" 21d52'57"
Audrey Kathleen Thompson
Lyr 19h8'44" 43d17'59"
Audrey & Kelly P.
Uma 9h50'51" 51d38'7"
Audrey Kennedy
Cas 0h56'23" 66d22'45"
Audrey Kindsfather
Crb 16h13'36" 35d52'50"
Audrey Kingsley Fulsi Luminosus
Lyr 18h51'21" 44d52'34"
Audrey Koski
Psc 1h53'54" 5d46'35"
Audrey Kunka
Cas 0h16'10" 54d46'33"
Audrey L. Bogen
And 23h27'56" 41d31'50"
Audrey Leigh Linn
Aqr 22h49'19" -22d17'38"
Audrey Lesage
Cyg 20h18'26" 51d29'52"
Audrey Liebl
Sco 17h24'8" -41d33'26"
Audrey Loder
And 23h10'33" 47d29'31"
Audrey Lorraine
Cas 1h7'52" 62d20'34"
Audrey Louise Long
And 0h37'19" 31d51'51"
Audrey Luckeydoo
Leo 10h8'36" 22d57'2"
Audrey Lyn
Lib 15h6'38" -3d3'26"
Audrey Lynn Arcelli
Lmi 10h33'10" 33d41'1"
Audrey Lynn Lynch
Cap 20h14'25" -15d23'46"
Audrey Lynn Romero
Vir 12h36'37" -8d35'6"
Audrey "Ma" Hanson
Uma 10h49'19" 44d27'0"
Audrey ma Tipounette
Uma 9h31'37" 59d40'34"
Audrey Mae Childers Damron
Lib 14h54'48" -7d13'14"
Audrey Marie
And 1h13'22" 45d25'24"
Audrey Marie Briggs
Lmi 10h50'26" 33d6'14"
Audrey Marie Elisabeth Pepet'lov
Uma 11h26'21" 48d40'33"
Audrey Marie Groarke
Leo 11h44'24" 19d17'51"
Audrey Marroc
Gem 7h19'50" 20d53'10"
Audrey Martin
Crb 15h45'6" 38d45'59"
Audrey Mary Young-Shearing
Cas 0h10'48" 59d30'43"
Audrey Maynard
Cas 0h22'27" 48d38'58"
Audrey Mazzera
Lib 15h50'8" -17d15'18"
Audrey Meadowcroft
Aql 19h40'58" -4d27'47"
Audrey Metral
Cnc 8h4'13" 21d19'13"
Audrey Miguel
Per 4h24'40" 44d38'57"
Audrey Moberly
Dra 18h29'55" 55d57'46"
Audrey Monique Troop
Cap 21h39'41" -20d32'18"
Audrey Mullarky
Lyn 8h26'55" 51d27'3"
Audrey Namour Bergeron
Mon 6h50'13" -7d27'52"
Audrey Nelson
Tau 4h9'3" 16d41'48"
Audrey Nicole Thompson
Cnc 8h17'43" 19d10'6"
Audrey Page
Leo 11h42'59" 21d43'40"
Audrey Perry 0904
Cas 23h51'28" 55d43'34"
Audrey R. Bayless
Cyg 20h5'48" 54d36'1"
Audrey R. Gaskin
Gem 6h23'29" 22d21'20"
Audrey R. Goldstein
Sgr 18h56'43" -28d41'31"
Audrey R. Trubiano
And 23h21'45" 48d1'11"
Audrey Rebecca Boender Borah
Lib 15h56'32" -5d8'30"
Audrey Rebecca Breithaupt
Leo 10h33'32" 14d14'17"
Audrey & Reverend Bill Lawson
Cyg 21h35'39" 38d6'12"
Audrey Rivera
Ari 3h4'57" 19d44'46"
Audrey Roddy
Uma 11h23'37" 29d34'46"
Audrey Roijen
Vir 13h48'25" -9d36'23"
Audrey Rose Foster
Vir 13h26'28" -2d16'55"
Audrey Rose Foster
Cnv 12h41'42" 38d50'40"
Audrey Rose Hewitt
Sgr 18h19'14" -31d17'12"
Audrey Rose Sessums
Leo 9h43'52" 31d47'33"
Audrey Rose Zuckerman
And 23h36'34" 41d50'20"
Audrey S. Garman
Peg 21h40'48" 22d59'32"
Audrey Sheehan
And 1h20'31" 41d0'19"
Audrey Siew Yin Kong
Tau 5h44'51" 22d19'36"
Audrey Star
Cap 21h41'43" -9d14'57"
Audrey Stephanie Aquino Sanchez
Cam 3h37'23" 57d5'37"
Audrey Sue Lanning
Leo 11h4'42" 2d39'53"
Audrey Taylor
Cyg 19h44'20" 32d7'8"
Audrey Tinervia & Mom Ruth Alexis
Umi 14h22'58" 76d17'46"
Audrey Trlida & Sidney Hull
Eri 4h14'39" -0d5'58"
Audrey Vincenza Eisen
And 23h20'55" 38d1'28"
Audrey Yasmine
And 2h16'41" 45d20'41"
Audrey, Grant's Angel
Mon 6h52'16" -0d53'34"
Audrey, 'Princess of Ghana
Cap 20h17'53" -24d22'21"
AudreyDianaSmit
Aqr 23h3'31" -7d45'22"
Audrey"LOGAN"Thomas
Cap 21h13'46" -16d18'49"
Audrey's Aurora
And 1h46'14" 49d0'16"
Audrey's Eye
Cyg 19h41'15" 29d14'1"
Audrey's Guiding Light
Ari 2h14'36" 20d24'57"
Audrey's Heaven
Peg 21h21'43" 17d29'54"
Audrey's Neverland
Uma 8h48'22" 58d12'33"
AUDREY'S STAR
Tau 3h44'51" 22d3'38"
Audrey's Vision
Aqr 21h12'27" -13d15'22"
Audrianna Nicole Fleming
And 23h10'1" 44d1'37"
Audric François Gantaume
Cas 23h0'41" 56d29'47"
Audrie Ann
And 2h36'29" 40d30'22"
Audrie Ann Diamond
Cnc 8h24'5" 9d23'52"
Audrie Jolie Harward
Umi 16h35'13" 77d33'7"
Audrie Lee Fischer
Sgr 19h39'16" -26d6'5"
Audrie Quick
Sco 16h53'59" -34d46'28"
Audry Coddington
Lyr 19h19'23" 29d35'14"
Audry Nichole Neeley
Tau 4h43'3" 23d0'8"
Auduo
Aql 20h21'55" -7d59'35"
AUDY AVIN
Aqr 21h38'11" 1d35'19"
Auey
Cru 12h9'6" -63d6'20"
Augie
Cep 20h43'10" 60d4'33"
Augie
Uma 10h10'5" 62d41'18"
Augie
Tau 4h9'59" 19d58'34"
AUGIE
Uma 11h25'18" 48d10'41"
Augie Bartolone
Cnc 7h56'53" 18d17'27"
Augie Bloomstein: Sei Mio Cuore
Sco 16h17'42" -33d41'2"
Augie Vassiliades
Cnv 12h39'34" 36d9'47"
Augie's Love
Lyn 7h32'50" 40d37'27"
Augie's Star
Cnc 8h19'10" 9d50'55"
August Aloisia Ake
Uma 11h19'43" 49d57'21"
August Andersen Welles
Cap 20h42'41" -20d12'58"
August Anya
Leo 11h42'3" 16d47'5"
August Baker
Gem 7h11'1" 33d6'55"
August Blue Sky
Vir 13h35'38" -12d10'5"
August Charles Krueger
Psc 23h53'38" 2d0'11"
August Eschberger
Uma 12h42'53" 59d44'53"
August Inhofer
Uma 8h39'18" 69d30'0"
August J. Papili
Cnv 13h12'2" 42d27'30"
August J. Wines III
Uma 11h3'29" 49d47'26"
August & Judith Queen
Crb 16h10'42" 30d4'24"
August Liss
Uma 8h58'27" 68d29'31"
August Love
Leo 11h33'52" 22d37'3"
August Oglinsky
Cap 21h53'8" -9d54'57"
August William Panks
Her 17h56'45" 29d7'2"
August Winter
Psa 22h40'56" -27d11'40"
Augusta Brooke Forbes
Cas 0h58'33" 56d57'49"
Augusta Ricciardi
Her 17h55'16" 44d1'18"
AugustFire82086
Leo 9h57'2" 16d23'18"
Augustina Cannella
Leo 10h25'50" 12d1'1"
Augustine
Her 17h44'6" 21d41'40"
Augustine
Uma 10h57'1" 56d7'58"
Augustine Bokano II
Vir 11h53'56" -5d45'42"
Augustine Chiu
And 0h36'5" 35d23'12"
Augustine Franklin Faille Sr.
Her 16h43'37" 46d15'40"
Augustine Garcia
Lyr 18h43'8" 37d59'26"
Augustine & Kimberly
Uma 13h25'42" 62d26'50"
AugustMargaret
Sgr 18h53'36" -16d46'58"
Augusto C. Vargas...Papi Tuto
Cep 21h48'16" 66d19'47"
Augusto J. Hernandez, Jr.
Per 3h3'41" 46d56'15"
Augusto Torres Mor
Ori 4h47'46" 4d29'39"
Augustus Alexander Plemmons
Her 17h28'8" 34d18'48"
Augustus Christmas
Uma 13h30'47" 54d1'22"
Augustus LeeAnn Sirovatka-Thompson
Mon 6h47'43" -0d5'26"
Augustus Lombardi, Jr
Lmi 10h14'15" 36d38'6"
Augustus McGee Konrad
Aqr 21h15'38" -13d41'46"
äUIEA - Dymka
Tau 4h16'20" 19d37'32"
Auika
Ori 5h37'32" 5d43'50"
Auladat
Lib 15h35'57" -26d2'28"
Aulbree Burriss
Lyn 6h43'24" 54d5'41"
Auldon Levern Reeves
Vir 13h17'23" 4d16'31"
Aulich, Sabine
Ori 5h50'17" 6d44'22"
Auman Parrish
Uma 10h47'50" 52d52'49"
Aundraya
And 0h54'8" 42d7'34"
Aundrea
Cyg 21h29'9" 37d11'49"
Aundrea
Ori 6h12'25" 12d43'45"
Aundrea Faye
Cyg 21h46'31" 49d21'27"
Aundrea Mena Rodriguez
Ari 2h5'48" 21d3'9"
Aundrea Shawnee Free 11/11/92
Sco 17h24'45" -33d17'19"
Aundrey Joy Rivera
Lyn 6h21'43" 54d3'14"
Aundria Ann
Uma 10h39'19" 42d5'30"
Aune Elizabeth Wood
Sgr 19h2'46" -19d20'16"
Aunnika
Uma 12h46'18" 59d1'43"
Aunny
Cam 4h36'48" 63d55'27"
Aunt Alice
Lib 14h59'55" -6d42'53"
Aunt Amy, Kate, & Meg's Star "Max"
Cmi 7h46'45" -0d8'11"
Aunt Ann Gates
Cas 0h20'32" 63d6'8"
Aunt Anne
Cas 1h32'22" 67d17'10"
Aunt "B"
Cap 21h44'47" -13d49'52"
Aunt Barbie
Cas 1h1'17" 61d28'52"
Aunt Betsy
Uma 8h27'4" 66d53'20"
Aunt Betty
Cyg 21h31'9" 53d16'35"
Aunt Bootsie
Lyn 8h56'8" 42d27'54"
Aunt Brenda
Cas 0h46'57" 61d11'37"
Aunt Bruce & Aunt Jeanne Star Casey
Del 20h29'39" 16d28'35"
Aunt Carol
Cmi 7h52'13" 1d14'33"
Aunt Cheryl-BDO&LO
Psc 0h11'36" 3d45'59"
Aunt Cindy
Aqr 22h10'22" -23d53'51"
Aunt Colleen
And 2h13'16" 45d0'58"
Aunt Cookie (Carol Cooke)
And 0h58'38" 43d8'40"
Aunt Daisy
Gem 6h41'37" 20d31'17"
Aunt Dee Main 26047
And 0h40'25" 36d39'53"
Aunt Diane
Sco 16h52'5" -26d35'32"
Aunt Dolly Karosick
And 23h7'44" 48d2'11"
Aunt Edith: Star of Love and Grace
Leo 11h4'23" 16d17'2"
Aunt Edna
Cas 22h57'29" 53d59'12"
Aunt Elizabeth's Star
Cyg 20h34'16" 41d39'52"
Aunt Fran
Uma 9h51'37" 68d5'51"
Aunt Gina Smrekar
Cas 0h4'4" 53d56'38"
Aunt Ginny
Cas 2h3'10" 59d13'30"
Aunt Gloria
Uma 10h56'27" 49d12'31"
Aunt Grace
Cas 0h46'19" 60d1'18"
Aunt Helen
Sco 16h15'57" -41d15'3"
Aunt Helen Wunder
Leo 11h1'8" 16d46'10"
Aunt Janet
Cyg 19h40'41" 40d56'36"
Aunt Janet's Star
Lib 15h27'45" -10d28'10"
Aunt Jeanie
And 0h57'37" 40d47'29"
Aunt Jeannie
Vir 12h44'18" 10d55'51"
Aunt Judy
Uma 11h55'29" 63d21'44"
Aunt Judy Zeiser
Vir 13h45'33" -11d10'46"
Aunt Kat
Tau 3h44'11" 15d7'1"
Aunt Kathy and Uncle Bill
Peg 22h18'12" 10d39'30"
Aunt Kay
Lyn 8h21'30" 55d52'9"
Aunt Laura
And 0h41'3" 45d12'19"
Aunt Lillian Reeg
Cas 3h23'32" 74d23'39"
Aunt Lillian's Beauty
Umi 14h43'53" 80d25'23"
Aunt Lillian's Grace
Cas 0h58'48" 56d31'30"
Aunt Linda of Westchester
Sco 16h51'8" -39d9'18"
Aunt Lorraine
Cnc 9h0'49" 8d30'18"
Aunt Mare
Sco 16h14'32" -10d18'34"
Aunt Mare (Marion Melende)
And 0h38'20" 40d6'9"
Aunt Marietta & Uncle Dominic
Uma 11h48'5" 37d26'38"
Aunt Mary
Psc 1h55'47" 7d5'46"
Aunt Mary and Uncle George
Lib 14h51'50" -3d26'37"
Aunt Mary Anne
Cyg 21h23'33" 30d11'45"
Aunt Mary Kenndy
Sco 16h8'15" -23d30'19"
Aunt Mary & Uncle Louis
Uma 10h51'18" 48d30'3"
Aunt Megan and Katy Love Zoe Bane
Ari 2h5'24" 23d49'11"
Aunt Millie
Cyg 22h2'18" 50d47'19"
Aunt Nell
Cyg 21h47'40" 44d33'32"
Aunt Pat
Cas 1h12'12" 58d22'47"
Aunt Peggy
Aqr 22h20'54" -1d13'12"
Aunt Polly
Gem 6h42'49" 17d28'55"
Aunt Rita's star
And 0h26'38" 39d52'40"
Aunt Ruth
Dra 19h12'23" 60d28'30"
Aunt Ruthie's Star
Cas 0h28'49" 51d1'47"
Aunt Sarah
Cas 0h14'20" 52d24'48"
Aunt Sophie & Uncle Chubby~55/75/06
Cyg 19h32'8" 51d51'43"
Aunt Sue (Susan Choolfaian)
And 0h6'56" 44d28'17"
Aunt Sue & Uncle Jack
Uma 9h4'27" 54d12'33"
Aunt Susan
Cas 0h57'40" 54d12'11"
Aunt Sylvia
Cas 1h22'27" 62d2'54"
Aunt Thelma Star
Lib 15h47'10" -17d33'22"
Aunt Tony
And 0h59'28" 34d21'6"
Aunt Tudy
Cnc 8h56'43" 16d3'2"
Aunt VaVa
Cnc 8h44'33" 30d56'40"
Aunte Betty
Psc 23h9'6" 3d57'21"
Aunti Merle Houghton
And 0h44'45" 38d41'13"
Auntie
Lyn 6h59'40" 44d24'13"
Auntie Dianna
Cas 23h13'33" 56d10'34"
Auntie Amy
Del 20h36'53" 7d23'16"
Auntie Babe
Cap 20h24'39" -12d43'8"
Auntie Cheryl
Cnc 8h50'30" 27d45'10"
Auntie Colleen
Lib 14h49'44" -1d25'2"
Auntie Denise
Cas 23h43'41" 56d48'26"
Auntie Denny
Uma 9h35'27" 54d29'21"
Auntie Di
Sgr 19h0'46" -15d45'28"
Auntie Donna
Ari 2h17'39" 24d37'51"
Auntie Doreen
Tau 5h59'48" 24d59'13"
Auntie Hilda
Lyn 9h12'55" 33d55'19"
Auntie Jo Kepler
Cas 1h9'34" 62d29'58"
Auntie Lisa
Sgr 17h53'20" -29d44'43"
Auntie Liz
Per 3h23'47" 35d44'2"
Auntie Lou
Dra 18h11'27" 70d9'11"
Auntie Lynn
And 1h31'17" 46d9'46"
Auntie M - Marlene Amos
Sco 16h12'21" -31d32'0"
Auntie- Mom Jean Roath
Cas 0h57'54" 54d14'16"
Auntie- Mom Shirl Leduc
Ori 6h18'23" 14d18'17"
Auntie Nan
And 0h28'57" 42d31'29"
Auntie Net
Uma 8h59'46" 64d41'15"
Auntie Pam
Psc 0h27'5" 6d14'41"
Auntie Randi Clare Ewing
Cyg 19h42'39" 29d20'26"
Auntie Rita
Cyg 19h50'25" 33d18'43"
Auntie Siobhan
Cas 2h28'45" 61d55'55"
Auntie Teresa
Lib 15h2'41" -3d24'25"
Auntie's Lucky Star Over Vegas
Cas 1h52'48" 60d14'45"
Aunty Faye Katz - Forever with Love
Vir 11h58'15" -11d13'57"
Aunty JoJo
Psc 1h37'6" 20d7'0"
Aunty Mary of Warracknabeal
Cap 20h40'1" -21d48'4"
Aunya Y. Dixon
And 1h30'52" 49d40'18"
Aura
Ori 6h24'37" 10d7'41"
Aura D'Amato
Cas 1h34'4" 62d36'20"
Aura Judith
Lyn 6h57'44" 58d14'33"
Aura of Antonio
Leo 11h33'23" -5d16'44"
Aura of Margaret Kiefer
Cap 21h35'59" -14d12'0"
Aura of romance
Uma 10h34'19" 66d51'46"
Aura Vargas
Uma 9h34'4" 54d2'55"
AuraBella
Vir 13h42'24" -5d57'41"
Auraca Spontane
Ori 6h9'42" 13d13'24"
Aurea
Her 16h39'26" 18d4'36"
Aurea
Vir 12h38'58" -9d47'55"
Aurel
Umi 14h29'39" 69d24'56"
Aurel Joseph
Cet 0h41'56" -1d6'47"
Aurelia
And 1h3'56" 46d51'20"
Aurelia
Lyn 7h19'43" 45d29'51"
Aurelia Alejandra Arias
Del 20h49'14" 5d58'13"
Aurelia Cruz Ureta
Psc 0h14'16" 7d43'6"
Aurelia Drake
Crb 16h7'36" 38d53'35"
Aurelia Grygienc
Cap 21h23'2" -25d29'3"
Aurelia Lynn Roberts
Uma 12h0'58" 45d35'41"
Aurelia Patricia-Marie Rainey
And 0h23'15" 30d7'40"
Aureliano Verità
Ori 6h19'55" 7d25'53"
Aurélie
Tau 3h55'39" 3d58'24"
Aurélie
Mon 6h46'30" -5d44'17"
Aurélie
Umi 15h14'2" 71d5'22"
Aurélie Allain
Uma 10h1'40" 49d56'13"
Aurelie Anne
And 0h35'0" 42d13'34"
Aurélie et Karim
Uma 10h30'9" 56d54'36"
Aurélie Fleur Adam
Cas 0h57'55" 62d30'33"
Aurélie Gauchet
Ari 2h16'2" 13d45'54"
Aurélie Huber
Lib 14h29'26" -19d46'43"
Aurelie McCanles
Cyg 21h48'49" 49d11'12"
Aurélie Mon Amour
Peg 21h41'42" 16d58'41"
Aurélie Pétrequin
Umi 14h5'50" 71d30'30"
Aurélie Waldman
Sco 17h55'9" -37d20'58"
Aurelien and Megan's Star
Cas 3h12'34" 58d56'45"
Aurélien Buisine
Cap 21h49'50" -23d31'3"
Aurelio Campos Jacuinde
Psc 1h2'27" 31d12'20"
Aurelius Cy
Dra 15h6'48" 59d37'6"
Aurelius Lucent
Uma 9h20'27" 57d57'36"
Aureus Venustas
Uma 11h44'19" 47d51'39"
Auri
Crb 16h2'55" 32d58'45"
Auria
Ari 3h2'8" 19d15'35"
Aurialix "Lily" Ureña
Scl 1h6'25" -31d1'29"
Auriel Mary Rankmore
Cas 1h11'51" 51d13'30"
Aurigny
Cru 12h53'33" -60d22'39"
Aurilia Elise
Cyg 21h57'10" 43d38'56"
Aurora
Dra 17h41'47" 51d26'15"
Aurora
Crb 15h44'6" 32d59'33"
Aurora
Uma 10h21'0" 39d34'32"
Aurora
Aql 19h15'49" 14d2'32"
Aurora
Gem 7h8'41" 17d38'52"
Aurora
Cas 1h42'20" 54d41'25"
Aurora
Uma 10h24'34" 54d39'0"
Aurora
Uma 10h15'8" 63d16'54"
Aurora
Lib 15h3'31" -20d44'18"
Aurora
Lib 15h29'51" -5d3'55"
Aurora Abbey and Alex
Uma 10h47'20" 43d26'47"
Aurora Alcaraz Chavez
Lib 15h49'1" -7d23'22"
Aurora Benedetti
Cas 0h26'16" 62d27'5"
Aurora Bonilla
Ori 5h45'7" 6d40'43"
Aurora Borea Lindsay
And 0h4'20" 45d8'9"
Aurora Borealis - Lea Desirée Wulz
Peg 21h34'28" 22d28'45"
Aurora Brightius Kathryn
Crb 16h13'49" 33d17'23"
Aurora chu
Lyr 18h44'34" 35d16'29"
Aurora Danielle Profenno
Gem 7h24'8" 19d15'37"
Aurora Dawn Proper
Crb 15h26'19" 27d23'52"
Aurora G
Cnc 8h15'35" 11d38'31"
AURORA G. VARGAS
Ari 3h4'22" 29d16'6"
Aurora Grace Gray
And 2h21'39" 41d24'33"
Aurora Gray
Crb 15h52'25" 32d41'53"
Aurora Lady Kenzie Marie
Peg 23h25'6" 23d25'16"
Aurora Leigh Ramos
Cas 0h25'5" 51d47'3"
Aurora Louise Scott
Ari 2h4'6" 22d28'47"
Aurora Luna
Sco 17h29'45" -42d21'52"
Aurora Madelyn
Vir 13h5'15" 12d13'47"
Aurora Madrid
Leo 11h54'11" 25d21'44"
Aurora Martinez Duarte "Yoya"
Cam 6h12'11" 60d6'23"
Aurora Mi Amor
Cnc 8h40'57" 14d55'5"
Aurora Moon Girl
Leo 10h26'30" 9d18'22"
Aurora Pena Ramirez
Col 5h36'2" -29d32'53"
Aurora Rain Quick "Rori Rain"
Peg 23h37'25" 12d43'35"
Aurora Rose
Leo 9h40'55" 10d41'39"
Aurora Russi
Vir 12h15'44" 11d44'24"
Aurora Sandro
Her 18h33'27" 18d58'29"
Aurora Sansone
Tri 2h30'16" 31d37'48"
Aurora Shiloh
Aqr 22h33'33" 1d50'57"
Aurora Star Yocum
Umi 14h47'58" 80d23'51"
Aurora Torres
Cas 1h33'17" 65d21'4"
Aurora Yolanda DePalma Mornes
Gem 7h22'28" 15d38'2"
Aurora6
Umi 13h17'47" 74d52'19"
Aurora's
Cyg 20h25'32" 40d48'26"
Aurore
Ori 6h3'46" 0d36'29"
Aurore Fragnière
Cas 23h35'57" 51d33'48"
Aurore Le Blanc
Uma 8h33'12" 68d58'50"
Aus World
Cru 12h57'10" -60d4'0"
Ausha & Kaya Maray Regan
Aqr 22h29'4" -19d9'17"
Auspicious
Dra 17h58'48" 59d40'3"
Ausra
Uma 10h13'15" 44d50'24"
Aussie John
Cen 12h22'54" -44d25'22"
Aussie Peach
Gem 6h7'29" 24d53'14"
Aussie Princess
Leo 9h45'32" 29d37'44"
Austein
Uma 14h6'16" 62d4'30"

Austen Daniel Cnc 8h45'10" 10d53'47"
Austen Ency Kaneshiro Lep 5h31'44" -21d26'42"
Austen Johnson Uma 10h20'17" 46d46'0"
Austen Timothy Wright Her 17h10'28" 30d7'28"
Austen Tosone Vir 13h22'27" 11d56'20"
Austi Gem 6h57'9" 19d51'48"
Austin Her 17h59'51" 16d35'35"
Austin Aqr 20h39'9" 1d5'15"
Austin Lyn 7h48'11" 44d37'26"
Austin Lyr 18h46'49" 43d15'35"
Austin Sco 16h15'28" -12d58'48"
Austin Dra 17h36'3" 67d34'12"
Austin Psc 1h49'48" 5d47'4"
Austin and Abbey Cyg 19h38'31" 31d52'21"
Austin and Alisha Shepperd Aqr 21h49'51" -3d32'45"
Austin and Amanda Uma 12h13'4" 58d9'36"
Austin and Joey Uma 10h8'1" 56d13'35"
Austin and Nicola Anniversary #1 Cyg 20h38'20" 38d28'48"
Austin and Sawyer's Star Tau 5h45'1" 18d7'32"
Austin and Shane Stokely Per 2h22'33" 51d25'50"
Austin Bailey Grantham Tau 3h27'21" -0d16'40"
Austin Bateman Her 16h25'34" 17d51'32"
Austin Bergin Her 16h37'25" 26d13'31"
Austin Bishop Boo 15h29'14" 34d4'2"
Austin Bradley Ari 3h22'36" 25d53'59"
Austin - Brooke Cam 4h15'23" 67d30'51"
Austin & Brooke Wendt Lib 15h9'50" -15d6'4"
Austin Brownlow Dra 16h10'0" 54d12'38"
Austin Burke Uma 11h53'59" 29d17'51"
Austin Cadd Vir 14h48'20" 4d51'45"
Austin Carmody Umi 14h41'26" 75d52'32"
Austin Catanzareti Tau 4h14'24" 11d26'59"
Austin Cern Denin Morioka Leo 10h15'11" 8d15'40"
Austin Chad Hinson Uma 11h27'3" 60d16'43"
Austin Charles Cnv 12h25'38" 41d36'1"
Austin Charles Ward Her 18h50'9" 21d37'23"
Austin Clint Prickett Ari 2h2'13" 24d1'14"
Austin D. Murphy Lyr 19h18'39" 28d7'28"
Austin Daniel Crawford Ori 5h2'25" -0d0'51"
Austin Daniel Hamilton Her 18h42'38" 16d43'35"
Austin David Hawkins Her 17h20'12" 33d43'35"
Austin David Popejoy Gem 7h40'32" 30d3'52"
Austin Dean Kolstad Umi 15h34'51" 78d24'0"
Austin DuBois Sco 16h11'27" -9d31'30"
Austin Emory Macy Walker Cas 1h2'30" 55d38'37"
Austin & Ethan Alexander Sgr 18h50'17" -21d2'38"
Austin Eugene Hurt Aqr 22h38'18" -18d10'49"
Austin Eugene Whitton Uma 8h38'19" 68d49'30"
Austin Evernden Cnv 13h3'36" 46d18'14"
Austin Farrell Quinn Cnc 8h25'37" 11d58'30"
Austin G. LaBarge Umi 14h25'59" 73d59'2"
Austin Garrison Uma 11h44'46" 52d33'23"
Austin Gene Underwood Her 17h18'31" 35d0'45"
Austin Gray Taylor Gem 7h52'11" 15d56'47"
Austin Hamilton Burchett Her 17h50'39" 34d21'32"
Austin Heinke Ori 4h54'41" 3d50'33"
Austin Hill Gem 7h10'32" 17d8'1"
Austin Holliday Cnv 13h19'52" 51d9'29"
Austin Howard Stanchina Ori 5h39'46" 0d47'5"
Austin Hyde Williams I Aur 6h16'29" 30d14'14"
Austin Isaiah Bell Uma 11h32'50" 56d0'47"
Austin J Conlon 052093, My Loving 2 Aql 19h58'29" 6d25'55"
Austin J. Love 071898 Her 18h7'40" 47d19'57"
Austin James Aur 5h53'56" 46d38'11"
Austin James Umi 15h22'13" 71d46'26"
Austin James Cep 23h27'53" 76d59'21"
Austin James Bankuty Cnc 8h18'13" 29d1'29"
Austin James Kruger Cmi 7h34'22" 4d52'14"
Austin James Norris Boo 13h49'32" 17d15'1"
Austin James Pursley And 2h25'33" 47d14'51"
Austin James Quandt Umi 14h0'44" 72d37'56"
Austin Jeffrey Burt Tau 4h40'27" 21d53'47"
Austin Jesus Vazquez Psc 1h16'25" 24d27'59"
Austin John David Triggs Per 4h35'43" 32d18'0"
Austin Joseph Courisky Tau 4h14'9" 15d37'37"
Austin Joseph Guinsler Lib 15h32'0" -26d26'22"
Austin Joseph Perez Per 3h44'22" 48d37'39"
Austin Keith Mocello Tau 4h34'56" 22d6'11"
Austin Kelly Uma 13h26'17" 53d13'52"
Austin L. Fox Ori 6h17'10" 15d14'29"
Austin Lea Crotty Uma 10h26'54" 65d23'31"
Austin Lee Burnworth Her 16h59'25" 19d42'50"
Austin Lee Stephenson Umi 15h4'14" 84d40'32"
Austin "light of my life" Boo 14h7'55" 18d24'4"
Austin Lloyd Smith Uma 13h26'8" 52d46'39"
Austin Longmate Uma 9h46'49" 57d18'4"
Austin M. Schmitt Tau 5h23'16" 26d5'36"
Austin Marie Long Uma 11h52'56" 53d13'36"
Austin Matthew Stevens Her 16h22'29" 19d10'39"
Austin McKay Wright Ninth Birthday Sco 17h6'5" -42d33'24"
Austin Mefford Uma 10h37'5" 61d9'18"
Austin Michael Brooks Umi 16h33'40" 76d51'52"
Austin Michael White Dra 18h50'32" 50d41'40"
Austin Michael Wolfe Leo 10h52'46" 9d47'19"
Austin Mockabee Gem 6h34'54" 22d6'0"
Austin Morgan Aqr 22h39'16" -2d12'19"
Austin- My first and only love Uma 11h43'31" 61d39'49"
Austin Nicholas Dunn Uma 11h38'29" 57d42'57"
Austin Paul Weil Her 18h7'41" 47d58'13"
Austin Place Her 17h33'34" 36d31'13"
Austin Powell Her 16h29'20" 47d21'37"
Austin R. Blanchette Lyn 7h11'53" 59d18'38"
Austin Reid Sumrall Ori 5h38'10" -0d17'20"
Austin Rex Cep 22h26'59" 65d24'41"
Austin Richard Babcock Her 16h34'27" 46d22'14"
Austin & Rusty Healy Sgr 19h15'29" -12d57'25"
Austin & Savannah Gentry Cyg 20h12'22" 40d48'0"
Austin Soul Mate Anspaugh Ori 6h17'49" 10d12'14"
Austin Star 21 Her 17h13'31" 17d33'41"
Austin T Her 17h33'46" 36d28'13"
Austin Tanner Kane Tri 2h32'48" 31d36'23"
Austin Taylor Cortese Boo 14h54'18" 22d17'19"
Austin Taylor White Uma 8h35'20" 58d20'33"
Austin Texas Smith Psc 23h30'9" 8d2'12"
Austin Thomas Tau 5h44'12" 16d49'1"
Austin Thomas Boney Tau 4h42'16" 0d51'4"
Austin Tobias Lauber Psc 1h15'40" 11d15'4"
Austin Tyler Jackson Vir 14h8'1" -16d14'33"
Austin Tyler Jaworski Sgr 19h27'24" -17d30'31"
Austin Tyler Slovin My Shining Star Tau 5h52'5" 16d20'1"
Austin Ulrich Lyn 7h29'23" 54d27'4"
Austin W. Griffin Uma 11h38'30" 39d41'42"
Austin Warden Cap 21h10'53" -17d3'49"
Austin Wayne Crist Del 20h24'42" 6d56'52"
Austin Weeks Uma 10h53'28" 47d17'13"
Austin Weldon Ard Cnc 9h0'2" 27d39'30"
Austin William Griggs Aqr 22h7'31" 0d55'55"
Austin Wilson Barker Equ 21h24'40" 9d28'41"
Austin Younkman Ori 4h51'37" 6d29'26"
Austin Zachary Serrano Leo 10h11'53" 10d1'47"
Austin, Cameron & Haley Bohman Ori 4h55'9" 2d36'20"
Austin, Daltin, and Priscilla Paul Uma 11h43'2" 34d24'33"
austing72195 Cnc 8h26'1" 12d42'2"
Austins Angel Vir 12h10'30" 12d28'42"
Austin's Angel Leo 10h16'9" 26d11'43"
Austin's Awesome SAT-1160 at 12 Aur 5h54'5" 49d28'38"
Austin's North Star Her 17h27'49" 20d4'22"
Austin's reach for your dreams star Pho 1h5'20" -47d12'3"
Austin's Star Cep 0h2'40" 69d59'40"
Auston Boo 14h37'42" 16d26'38"
Australian InternationalHotelSchool Ori 5h31'4" 0d31'59"
Austy Vir 14h21'18" -2d19'31"
Austyn Presley Sirmon Lyr 18h42'42" 31d28'59"
Aut viam inveniam aut faci-am Cnc 9h21'41" 14d7'33"
Authreanna Alondra Schubert Vir 14h17'32" -0d44'7"
Autroy Ewaise Allen Umi 18h19'43" 88d43'38"
Autry L Robertson Umi 16h9'24" 78d30'36"
Autum Uma 9h37'19" 54d47'16"
Autum B. Gilley Uma 9h30'14" 45d11'15"
Autumn Cnc 8h59'40" 11d40'7"
Autumn Mon 6h47'43" 7d45'15"
Autumn Uma 9h13'37" 60d10'22"
Autumn Vir 13h36'15" -1d2'24"
Autumn Sgr 19h30'25" -15d52'35"
Autumn Lib 15h29'30" -18d26'48"
Autumn Car 7h13'31" -51d45'5"
Autumn Aqr 23h8'39" -21d47'15"
Autumn Alexis Mowery Uma 12h52'10" 59d49'30"
Autumn and Adam Raffety Del 20h40'52" 17d8'0"
Autumn Andromeda And 2h32'34" 44d35'2"
Autumn Angel Vir 13h23'30" 13d21'37"
Autumn Angel Cox Lyn 7h25'45" 44d33'2"
Autumn Behnke And 0h27'13" 34d3'3"
Autumn Bird Sgr 19h54'8" -43d5'53"
Autumn Breeze Owen Umi 14h40'18" 73d28'2"
Autumn Bright Mon 8h1'29" -4d21'58"
Autumn Brooke Aqr 22h23'57" -2d31'6"
Autumn Celeste Ari 2h16'39" 25d40'45"
Autumn Claire Aql 19h16'45" 8d6'16"
Autumn Danielle Romines Cas 1h42'51" 60d21'11"
Autumn Dawn Bryson Leo 9h38'20" 25d39'53"
Autumn Dawn Piazza Lib 14h26'40" -12d24'58"
Autumn Elena Uma 9h6'43" 48d28'16"
Autumn Elizabeth Gem 7h45'9" 33d52'2"
Autumn Elizabeth Harvey Sco 17h53'43" -37d42'14"
Autumn Elizabeth Smith And 0h43'25" 42d9'33"
Autumn Grace Wadley Leo 10h11'17" 12d9'28"
Autumn Holt Wilder Cap 21h53'50" -19d32'45"
Autumn Hope Lmi 10h41'54" 26d47'26"
Autumn Jade Cnc 9h9'21" 28d50'45"
Autumn Jade And 0h15'39" 43d55'55"
Autumn Jean Liguori-Bills Ari 2h43'19" 28d5'59"
Autumn Jean Liguori-Bills Ari 2h33'16" 11d10'51"
Autumn Jean Ward And 23h16'22" 48d16'25"
Autumn Jones 1998 Lyn 9h10'47" 42d19'50"
Autumn Lane Cnc 8h44'26" 20d2'44"
Autumn Le Psc 1h2'45" 26d22'54"
Autumn Lea Cseh- Stevens Gem 7h31'47" 33d14'57"
Autumn Leaha Byers And 0h39'28" 26d41'1"
Autumn Leigh Uma 11h38'9" 58d28'15"
Autumn Leigh Bolton Lib 14h26'58" -23d36'4"
Autumn light Leo 10h28'21" 22d39'27"
Autumn Lorae Harper Cas 0h4'29" 50d24'26"
Autumn Louise Wiek Cnc 8h50'55" 14d31'2"
Autumn Luray Burch Aqr 22h34'10" -6d13'10"
Autumn Lynn Her 17h13'2" 20d33'55"
Autumn Lynn And 0h28'11" 42d11'6"
Autumn Lynn Wayson Cas 0h8'24" 57d0'29"
Autumn Marian Longhi Lyr 18h15'17" 35d0'44"
Autumn Marie Bostic And 1h55'30" 43d33'50"
Autumn Marie Perillo And 1h16'50" 37d16'17"
Autumn McBride And 23h42'46" 47d47'5"
Autumn Moon Kitterman Mon 6h34'37" 9d6'37"
Autumn Neichole Cap 21h53'13" -19d11'49"
Autumn Rachelle Swahn Eri 4h29'8" -14d9'47"
Autumn Rain Leo 11h45'44" 27d12'26"
Autumn Raine Winterling Aqr 22h28'16" -23d5'54"
Autumn Rene Angela Cole Leo 11h34'6" 0d7'57"
Autumn Rockett Leo 11h57'16" 14d20'39"
Autumn Rose Jackson Psc 0h55'35" 31d42'5"
Autumn Rose Relyea Sco 16h51'24" -29d32'13"
Autumn Rose Rippl And 1h50'31" 43d6'57"
Autumn Ryan Thomas Cnc 8h51'55" 19d44'56"
Autumn Saleski Lib 15h4'4" -2d59'51"
Autumn Sansom Lib 14h53'40" -9d22'9"
Autumn Shayleigh Jenkins 10-26-2002 Sco 17h41'19" -37d58'58"
Autumn Sherman Uma 11h23'7" 41d15'47"
Autumn Singer and David Califano Gem 6h34'15" 14d22'18"
Autumn Sky McCleary Sco 16h7'8" -10d12'33"
Autumn Skye Candelaria And 2h25'25" 50d24'55"
Autumn Soriano Cnc 8h41'1" 15d21'12"
Autumn Star Vir 13h42'37" 3d38'8"
Autumn Star Sco 17h53'47" -42d3'16"
Autumn "Star Baby" Reed Uma 9h57'16" 47d59'58"
Autumn Swallow Cyg 20h59'53" 49d35'54"
Autumn The Believer Vir 12h41'52" 4d5'33"
Autumn Wood Uma 8h35'45" 47d47'39"
Autumn, the little dancer Tau 4h21'4" 14d25'29"
Autumn's Bliss And 23h17'8" 48d19'59"
Autumn's Grace Sco 17h50'13" -40d5'0"
Autumn's Love Umi 15h7'13" 67d53'22"
Autumns Wonderous Light Crb 16h5'53" 37d54'50"
Autumn-Wright Lib 14h28'4" -17d50'9"
Auveen Victoria Kaveh Ari 3h14'6" 28d46'53"
Aux yeux marrons Déesse de l'amour Cnc 8h52'28" 22d55'58"
Ava Cyg 21h15'24" 41d28'51"
Ava Cas 0h54'48" 54d21'27"
Ava Gem 7h42'54" 30d41'31"
Ava Cas 1h34'18" 61d57'54"
AVA - Ad Vitam Aeternam Leo 10h45'28" 19d46'29"
Ava Adair Kathrine Drafs Cnc 8h27'30" 23d12'38"
Ava Albis Vir 14h12'25" -15d56'28"
Ava Alexis Tzi-wa Trepepi Cyg 20h0'32" 33d27'52"
AVA AMARI Vir 15h4'28" 1d51'25"
Ava Amelia And 2h30'9" 49d5'9"
Ava and Lila Putrino Sco 17h29'12" -37d55'2"
Ava Anley And 1h58'59" 46d22'50"
Ava Anne Rogue And 1h9'6" 37d7'15"
Ava Augusta Uma 10h14'10" 45d20'30"
Ava Beatriz Ceglia And 2h33'18" 42d11'28"
Ava Belle And 23h25'55" 47d24'13"
Ava Blair Lunger Lib 15h43'57" -29d8'39"
Ava Blaire Hounshell Gem 7h14'17" 20d16'23"
Ava Blumenthal Uma 10h28'45" 52d5'59"
Ava Brooke Duckworth Sgr 18h55'29" -34d52'44"
Ava Cameron Ingram And 1h33'42" 48d13'55"
Ava Camille Griego Crb 16h7'53" 36d43'47"
Ava Caroline Lusby Cnc 8h28'44" 26d28'28"
Ava Caroline Yarborough Aqr 22h37'49" -1d55'1"
Ava & Carrie Cyg 19h53'46" 38d43'53"
Ava Catherine Uma 11h57'16" 31d8'57"
Ava Catherine Frechette And 0h22'47" 32d22'12"
Ava Catherine Lenz Aqr 22h13'36" 1d13'20"
Ava Catherine O'Connor And 23h42'28" 47d24'20"
Ava Catherine Whittaker Aql 19h52'36" -10d51'54"
Ava Caylie Sco 16h5'19" -22d43'11"
Ava Childers Leo 10h14'20" 25d0'33"
Ava Christa 23 Août 2004 Vir 12h10'41" -10d5'35"
Ava Christina Tau 3h36'17" 18d2'47"
Ava Christine Fowler Gem 7h16'13" 26d27'39"
Ava Claire Cnc 8h16'11" 10d18'14"
Ava Claire Dempsey Del 20h21'20" 20d36'55"
Ava Constance Burke And 1h57'0" 41d29'47"
Ava Diehl Uma 11h23'12" 56d6'8"
Ava Elisabeth Saper And 23h29'36" 44d26'14"
Ava Elise Cas 1h18'25" 57d41'44"
Ava Elizabeth Vir 13h24'48" -13d12'52"
Ava Elizabeth Bowen And 0h13'36" 45d2'7"
Ava Elizabeth Jones And 0h14'19" 47d59'19"
Ava Elizabeth Phillips Peg 23h47'46" 17d55'6"
Ava Elizabeth Schmiedigen And 1h33'45" 46d48'57"
Ava Elizabeth Schwartz Cyg 20h32'40" 35d13'50"
Ava Elizabeth Zaccour Sco 17h56'48" -38d35'10"
Ava Elle Anderson Cas 1h23'34" 62d49'12"
Ava Gallery Tau 5h42'58" 22d52'48"
Ava Gioconda Barquist Uma 12h2'16" 53d27'47"
Ava Grace Sco 16h6'53" -11d54'22"
Ava Grace Sco 17h41'27" -42d52'38"
Ava Grace Tau 5h12'29" 18d39'21"
Ava Grace Lyn 7h31'30" 40d15'51"
Ava Grace And 2h28'12" 45d3'18"
Ava Grace Ainsworth And 1h53'22" 41d37'36"
Ava Grace Costello Lib 15h33'25" -21d53'14"
Ava Grace Delossantos Psc 1h5'50" 32d15'54"
Ava Grace Jernigan Vir 14h25'1" 5d4'33"
Ava Grace Little Love King Sco 16h42'52" -30d48'18"
Ava Grace Shelbourne Vir 14h17'4" -12d7'18"
Ava Grace Vaughan Gem 7h52'29" 33d56'27"
Ava Grayson Daughters Psc 0h44'39" 7d50'51"
Ava Helena Persephone Killmer Cyg 21h29'43" 38d58'24"
Ava Hubick Uma 13h20'48" 54d50'23"
Ava Irene Rosetti Cnc 8h42'10" 12d38'22"
Ava Isabella Gertrude And 0h29'6" 37d37'39"
Ava Jane Dawson Ori 5h49'40" 2d27'9"
Ava Jane Horvath Leo 9h52'8" 30d26'52"
Ava Jane Jackson Ari 2h16'43" 12d15'19"
Ava Janelle Rush Cas 1h58'3" 62d48'49"
Ava Jean Ari 3h16'23" 29d7'2"
Ava Joell Lytle Psc 1h36'14" 18d46'19"
Ava Jolliemore Lyn 6h47'33" 52d41'40"
Ava Jotautas Uma 10h50'57" 65d13'0"
Ava Juliette Darrin Dra 19h32'16" 79d45'48"
Ava Kaiser Mon 6h47'6" -7d25'49"
Ava Kate Greenlees And 23h41'40" 36d34'40"
Ava Kathryn Gatti And 23h53'29" 45d11'41"
Ava Kaye's Star of Love Eri 4h27'7" -0d7'7"
Ava Kiana Daneshmand Dra 18h6'19" 77d6'26"
Ava Lauren Johnson Psc 0h53'15" 15d6'18"
Ava Layne Copeland And 0h37'8" 35d16'42"
Ava LeeAnn Cook Cap 21h12'35" -17d2'56"
Ava Lewis - Tupperware Tri 1h53'49" 28d21'10"
Ava Liana DiNitto Dra 17h5'39" 61d1'55"
Ava Liberty Cas 0h44'37" 66d23'40"
Ava Lily Lyr 18h48'46" 30d55'59"
Ava Lily Kahn And 2h10'31" 44d45'36"
Ava Lily Mae Aspen Cap 20h58'0" -17d43'10"
Ava Loretta DePrisco Dra 18h28'51" 74d46'3"
Ava Lyn Cnc 8h12'52" 25d21'50"
Ava Mae Kloiber And 0h36'27" 36d54'57"
Ava Margaret Sgr 18h35'21" -22d18'52"
Ava Margaret Cona's Birthday Star Gem 6h53'7" 32d29'39"
Ava Marie Bassak Ari 3h6'9" 12d55'3"
Ava Marie Blanchard Psc 1h6'25" 21d53'52"
Ava Marie Blanchard And 1h40'38" 41d43'48"
Ava Marie Cashman Cas 1h25'43" 56d52'22"
Ava Marie Fama Tau 4h18'38" 29d0'26"
Ava Marie Fedrich Our Shining Star And 1h38'20" 48d14'14"
Ava Marie Hileman Umi 15h22'11" 69d0'20"
Ava Marie Hughes Lib 15h3'43" -0d32'23"
Ava Marie Jaksim Crb 15h40'44" 38d13'31"
Ava Marie Porter And 1h18'49" 37d16'43"
Ava Marie Rossitto Cap 21h13'40" -15d53'57"
Ava Marie Shaw Peg 23h27'1" 16d22'15"
Ava Marie Stureski Umi 14h56'15" 74d0'52"
Ava Marie Victoria Cas 2h7'58" 72d22'54"
Ava Mary Tanceusz Sgr 19h9'21" -14d36'56"
Ava McLaughlin's Christening Star And 2h28'45" 37d42'44"
Ava Meadow Goodman Cap 21h15'55" -27d28'2"
Ava Melanie Beckinsale Cru 12h34'34" -57d32'38"
Ava Michael Gravina Sgr 18h31'13" -27d27'36"
Ava Michele Mohney And 0h9'55" 41d8'55"
Ava Michelle And 23h18'32" 49d38'10"
Ava Montgomery "The Little Tuki" Ari 2h33'4" 26d34'57"
Ava Muffuletto And 1h34'51" 45d26'38"
Ava Nicole DeGeorge Lyn 7h20'59" 46d41'23"
Ava Nicole Hylton And 23h48'57" 34d27'24"
Ava Noel Bruzzio Sco 16h11'28" -9d47'7"
Ava Perrone Crb 15h16'41" 30d26'29"
Ava Petlin And 0h15'33" 44d29'15"
Ava R. Chrostowski "9-28-05" And 1h38'40" 45d10'7"
Ava R. McNamara Cas 1h16'36" 50d49'28"
Ava Raisa And 0h34'1" 25d11'47"
Ava Renee Kora And 0h25'12" 32d20'39"
Ava Robinson Sgr 18h52'17" -15d59'29"
Ava Robson And 2h11'1" 46d18'31"
Ava Rose Uma 9h19'17" 57d47'31"
Ava Rose Burns Gem 6h58'53" 14d17'37"
Ava Rose Czap Sco 16h4'29" -13d26'24"
Ava Rose Palumbo Cnc 8h33'48" 32d23'38"
Ava Rose Preskenis And 0h9'52" 35d40'55"
Ava Rose Rutkowski Lyn 9h17'20" 41d11'40"
Ava Rose Silvius Aqr 23h14'13" -11d58'47"
Ava Rylea And 23h49'1" 45d6'10"
Ava Sade Sample-Moore Lyn 8h28'27" 54d42'40"
Ava Sara LiZi Mayginnes And 1h37'12" 44d46'43"
Ava Shotkoski Uma 10h5'49" 59d3'23"
Ava Sione Hileman Ari 2h18'53" 24d18'30"
Ava Someday Uma 9h43'18" 42d54'5"
Ava Sophia And 0h53'47" 22d50'50"
Ava Sophia Taylor Umi 14h50'35" 72d20'51"
Ava Spina Uma 11h1'42" 45d56'24"
Ava Terese Parrella Lyn 8h47'0" 41d23'56"
Ava Victoria Ari 2h15'56" 23d22'48"
Ava Victoria Kriwinski Lmi 10h8'41" 28d57'2"
Avagnano Andrea Her 16h42'14" 11d19'40"
Avail Spring Downey And 23h19'6" 47d35'35"
Avain Jayde Sco 17h32'59" -34d9'23"
Avais Uma 9h40'43" 45d46'36"

Avalicious
Pyx 8h47'53" -26d22'40"
Ava-Lili Smith - Ava-Lili's Star
And 1h46'58" 39d44'34"
Avalina Anderson
Cyg 20h26'47" 34d10'22"
Avalon
Per 3h18'7" 32d21'0"
Avalon
Psc 0h18'19" 9d43'21"
Avalon
Cep 22h1'28" 57d1'38"
AVALON
Cap 21h18'12" -15d37'19"
Avalon Grace
Lib 14h55'17" -2d3'28"
Avalon Priscilla Kennedy Astles
Psc 0h18'6" 7d48'59"
Avalon Royalty
Peg 21h36'18" 17d41'7"
Avalyne Kalanie Peters
Aqr 22h6'48" -11d3'31"
AVANAG
Lyn 8h56'11" 42d11'9"
Avaneesh
Uma 11h18'28" 65d1'56"
Avani Patel
Tau 5h35'33" 24d0'41"
Avani R. Patel
Tau 3h45'53" 13d45'57"
Avannah Sage
Uma 11h7'35" 39d15'19"
Avarie Grace
Psc 0h3'23" 3d50'41"
Avary Emma Richards
Lib 15h16'1" -18d50'4"
Avary Kathry Taylor
Dra 17h57'21" 56d43'36"
Avary Linden Trump
Lyn 7h23'33" 48d56'3"
Ava's Angel
And 23h5'45" 51d51'51"
Ava's Star
And 1h39'1" 40d7'45"
Ava's Star
And 2h38'16" 37d26'34"
Avatar Meher Baba
Uma 12h16'35" 54d29'5"
AVB Aaronic
Uma 8h48'49" 57d18'16"
Ave Maria
Vir 12h54'9" 1d6'27"
Avea and Milt
Peg 22h26'43" 34d21'8"
avec coeur
Gem 7h28'17" 25d36'51"
Avec Dieux Aime et la Protection
Psc 1h11'34" 27d0'4"
Avec Tout Mon Amour
Col 6h19'37" -34d45'57"
Avec Tout Mon Coeur
Uma 9h26'49" 42d38'23"
Avel Contreras
Cyg 21h31'12" 41d43'22"
Avela McKee
Ari 1h49'12" 24d38'47"
Aven Marie Davis
Cap 21h12'0" -15d41'44"
Aven Zakai 51206
Aql 19h6'0" -8d37'19"
Averett-46
Psc 1h29'45" 28d15'23"
Averi Hatzell
Aqr 22h20'25" 0d14'23"
Averie Eleanor Anfinson
Aqr 22h5'10" -16d50'50"
Averie Megan Rhoades
Cyg 21h58'39" 45d7'31"
Averila '46
Mon 7h0'59" -8d59'27"
Avery
Sco 16h11'22" -11d55'21"
Avery
Cap 21h35'51" -13d34'51"
Avery
Dra 18h26'5" 53d9'34"
Avery
Uma 11h28'18" 63d14'2"
Avery
Lib 15h46'37" -24d26'55"
Avery
Cru 12h30'59" -57d38'6"
Avery
Tau 5h30'26" 27d41'39"
Avery Allsop
Leo 9h31'42" 6d46'9"
Avery Amelia Jane From
Psc 0h26'17" 8d4'6"
Avery Bella Johnson
Sgr 20h3'37" -30d26'33"
Avery Blair Pekar
Tau 5h58'44" 25d23'34"
Avery Blayz Britten
Uma 11h28'20" 60d8'8"
Avery Brian Orzech
Cap 21h43'39" -19d54'36"
Avery Brooke
Uma 9h48'34" 69d8'55"
Avery Byrnes
Mon 6h43'55" 8d54'12"
Avery Cole Morell
Ari 2h5'33" 19d35'16"
Avery D. Warsing
Dra 18h17'42" 59d10'42"
Avery David Adessky
Aqr 20h44'52" -11d42'29"
Avery Diana Hart
And 0h39'50" 45d10'12"
Avery Elisabeth Barlow
Lib 15h55'35" -12d17'16"
Avery Elise
Aql 19h48'13" -0d31'2"
Avery Elizabeth Bender
Lib 14h50'34" -6d38'0"
Avery Elizabeth Estes
Cap 21h26'8" -27d9'57"
Avery Elizabeth Evans
Sgr 19h30'18" -21d47'31"
Avery Elizabeth Keeves
Lyr 19h15'44" 29d29'8"
Avery Elizabeth Rehl
Vir 12h57'16" 14d18'6"
Avery Ellen Owenburg
Sco 16h12'28" -10d15'48"
Avery Grace
Ari 2h38'42" 19d33'23"
Avery Grace Dalesandro
Umi 14h40'35" 77d4'1"
Avery Grace Fierman
And 0h30'7" 31d14'46"
Avery Gunther
Leo 9h40'57" 30d20'6"
Avery Hefner
Ori 6h19'35" 14d21'22"
Avery J Wahl
Uma 9h17'58" 61d30'35"
Avery James Edward Fahrman
Sco 16h14'31" -13d3'28"
Avery Jane
Tau 5h15'28" 16d15'40"
Avery Jean Schapira
Cep 21h51'56" 62d27'14"
Avery Katherine
Gem 6h52'56" 24d59'36"
Avery Katherine Marcomb
Vir 12h26'47" 7d55'8"
Avery Katherine Marcomb
Vir 13h48'16" -5d10'56"
Avery Kathleen Lozano
Crb 15h40'19" 26d16'38"
Avery Laine Woods
Uma 10h46'15" 55d17'35"
Avery Lane Deringer
And 0h41'3" 27d4'27"
Avery Lea Rogers
And 22h59'5" 52d41'44"
Avery Leigh McClinton
Vir 14h6'55" 6d42'45"
Avery Leigh Wehner
Lyn 7h41'34" 39d47'57"
Avery Leigh-Ann
Lib 15h35'46" -26d11'41"
Avery Lynn
Aql 19h42'19" 5d57'57"
Avery Lynn Berry
Uma 9h58'57" 70d24'26"
Avery Marie
And 23h52'34" 42d1'16"
Avery Mcallistaire Dennison
Aql 19h28'35" 5d11'13"
Avery Michael Anne Cunningham
Leo 11h16'9" 19d9'9"
Avery Michael Hornaday
Ori 6h13'4" 13d31'42"
Avery Michelle Malinowski
Uma 9h54'58" 43d8'39"
Avery Nicole
And 1h25'33" 35d58'42"
Avery Nicole Atieno Rabah
Leo 9h59'22" 14d40'24"
Avery Nicole Easton
And 0h19'35" 28d11'11"
Avery Nolan
Lib 15h39'52" -23d38'55"
Avery Olivia Zrno
Aql 20h7'31" 9d55'7"
Avery Parker Spayd
Aqr 23h3'34" -10d34'11"
Avery Paul
Uma 10h57'16" 40d23'47"
Avery Pearl Cabral
Lyn 7h59'25" 55d17'45"
Avery Prem Shankar
Umi 15h26'59" 71d1'28"
Avery Quinn Smith
And 23h55'25" 45d22'43"
Avery Renai Williams
Uma 10h7'58" 46d48'45"
Avery Renea Graham
Ori 6h1'34" 0d31'39"
Avery Rose
Gem 6h54'50" 24d30'44"
Avery Rose Kuhnert
Psc 0h6'26" -1d43'10"
Avery Rose Lewis
And 0h58'2" 43d32'11"
Avery Rose Sagl
Cnc 8h17'44" 30d22'28"
Avery Roxana
Ari 2h5'20" 20d12'57"
Avery Scott Leighton
Aqr 22h44'5" 1d58'0"
Avery Shayne
Aql 18h52'5" -0d26'52"
Avery Simmons
Leo 11h41'59" 15d29'3"
Avery Sophia Miller
Lyn 8h14'16" 50d11'11"
Avery Thomas
Sgr 20h0'59" -30d33'19"
Avery Thomas Joseph Novak
Cnc 8h52'40" 12d30'36"
Avery Toussaint Daly
Cap 21h5'47" -20d27'18"
Avery Weiss
Ari 3h14'9" 27d53'2"
Avery William Barbour
Cep 21h28'53" 64d21'52"
Avery, in his beloved memory
Cap 21h35'47" -19d51'22"
Avi & Chad - This I Swear Is True
Cas 3h14'51" 61d10'34"
Avi Goldyne Star
Lib 15h49'46" -6d45'52"
Avi Ouaknine
Crb 15h56'4" 27d3'56"
Avia Mater Matris Murleen
Psc 1h24'29" 24d48'51"
Avianna Sofia
Cyg 20h12'2" 40d54'33"
Avice Stocken
Ori 5h56'40" 22d8'19"
Avigdor
Sgr 18h8'0" -21d39'51"
Avil Lucille Forbes
Sco 17h15'12" -32d38'9"
Avin Mackenzie Glamp
Cas 1h2'0" 61d15'14"
AVincentVern mon éternel amour
Cas 1h54'40" 67d55'0"
Avis
Cas 1h31'47" 63d56'53"
Avis
And 22h58'58" 47d49'17"
Avis and Al Always
Tri 2h30'21" 35d47'53"
Avis Lillian Watson
Cas 1h37'16" 65d24'40"
Avi's Star
Cma 6h57'0" -19d46'16"
Aviva Barth
Tau 4h12'59" 29d22'20"
Aviva Maytal
Tau 5h33'3" 23d14'51"
Avner Peretz
Lib 16h0'32" -16d26'58"
Avoca Sunrise
Cru 12h56'43" -58d51'37"
Avraham Mordechai Yitzhak
Cam 4h25'20" 72d0'55"
Avram Gelfenzon
Leo 11h56'24" 24d28'11"
Avrie Laura Rose Weighman
Ari 2h54'58" 28d34'38"
Avriel Kimshead
Vir 13h26'11" -8d15'25"
Avril
Vir 13h12'50" -2d9'53"
Avril Angel Eyes
Sgr 18h57'7" -27d18'31"
Avril Lavigne
Lib 15h4'52" -16d11'26"
Avril Lavigne
Lib 15h8'21" -9d18'3"
Avril Lewis
And 23h17'32" 41d27'4"
Avril Louise Johns
Dra 20h23'3" 70d53'27"
Avril P. Beckford, M.D.
Umi 13h47'5" 87d57'13"
AvryAuN
Uma 8h42'13" 69d26'7"
Avtar Singh Bhullar
Her 18h20'27" 24d44'43"
A.W. Faber-Castell (M) Sdn Bhd
Ori 5h33'19" 2d20'12"
AWC 07'
Uma 12h41'49" 54d33'32"
AWENASA HANIA
Uma 12h28'27" 56d55'49"
Awesome Audrey
Cas 23h16'8" 59d33'14"
awesome dude 131
And 0h30'28" 42d23'27"
AwesomeAdam415
Per 3h21'50" 42d23'2"
AwesomeShannon20
And 23h21'36" 46d46'59"
Aweston
Peg 21h31'53" 24d41'16"
AWK-737
Psc 1h8'30" 24d34'48"
AWNE
Gem 6h38'41" 20d40'9"
Awoon
Tau 5h9'19" 23d8'28"
Awuor_Mary-1507
Leo 10h19'33" 10d2'43"
AWYMME
Lyn 7h0'57" 45d26'19"
Axel
Cnc 8h36'14" 19d37'17"
Axel Armand
Gem 6h38'36" 20d54'4"
Axel August Matacia
Lyn 7h24'25" 59d39'10"
Axel Auragnier
Sco 16h40'51" -38d1'15"
Axel Becker
Uma 8h25'1" 65d46'53"
Axel Brenna
Uma 8h53'16" 52d10'12"
Axel Bror Christensen
Ori 5h50'37" -0d4'7"
Axel & Clara Schebesta
Cyg 19h39'21" 52d54'14"
Axel Filliette
Sgr 18h52'47" -18d51'54"
Axel Hillebrand
Ori 5h54'56" 13d0'52"
Axel Potier
Tau 5h48'20" 15d17'21"
Axel Rieger
Uma 10h56'0" 69d36'10"
Axel Säckl
Uma 11h12'57" 63d59'57"
Axel Schönfelder
Uma 9h6'14" 61d26'25"
Axel Stone
Leo 11h20'8" -5d13'9"
Axel + Tom
Ori 5h15'23" -9d12'46"
Axel W. der Frosch + Hunde Schatz
Uma 10h3'0" 69d5'3"
Axel y Tamara
Sge 19h53'10" 16d42'11"
Axelle
Cnc 8h17'54" 25d9'12"
Axelle
Psc 0h44'51" 18d39'53"
Axelle
And 1h29'59" 35d31'51"
Axelle Guegan
Cap 20h33'12" -10d0'4"
Axelle Guerra
Vir 14h19'28" 5d21'11"
Axelle Rollet
Lib 14h47'39" -4d55'32"
Axellou
Ari 3h10'5" 15d50'6"
Axi Nue
Pho 0h32'15" -41d22'44"
Axilleonde
Cas 23h34'0" 54d37'50"
Axl
Uma 11h39'11" 61d51'6"
Axl
Uma 11h42'37" 39d31'57"
Axle Braveheart
Leo 9h58'25" 14d10'2"
AXPERUTYUN
Uma 11h34'14" 47d55'35"
Axsis
Lyr 18h43'32" 34d52'56"
Axylom
Uma 11h38'24" 59d56'12"
Aya and Kenji
Sco 17h29'49" -31d55'29"
Ayah Nasif
And 1h14'18" 45d34'11"
Ayakabodi
Lup 15h41'57" -35d35'21"
Ayako and Takuo Akiyama
Cyg 20h10'41" 34d56'14"
Ayako Hosomizu
Ara 17h6'16" -58d14'19"
Ayako Kari
Tau 4h47'39" 26d56'3"
Ayala Yasmin Erez
Tau 4h28'19" 12d16'20"
Ayana Heaven Nixon
Psc 0h18'59" 15d35'28"
Ayanna
Aql 19h57'26" -1d56'22"
Ayanna Renee
Sgr 18h36'2" -23d2'53"
Ayato & Nagisa
Ari 2h30'36" 13d21'8"
Ayaz, Ayten
Uma 10h11'6" 69d10'19"
Ayça
Umi 14h23'38" 78d22'49"
Aycayia Virgo
And 23h44'6" 46d59'32"
Ayche
Sco 17h44'4" -39d56'20"
Ayçiçek
Lib 14h35'33" -10d59'59"
Ayda
Cap 20h32'34" -21d41'25"
Ayda Shakeri
Uma 11h33'46" 52d43'44"
Aydan Gerald Quinn
Gem 7h25'49" 14d59'22"
Aydan James Crowe
Cru 12h36'47" -59d29'19"
Aydan Margaret Rudolph
Leo 9h47'41" 28d53'39"
Aydan Uppal
Vir 14h3'21" -6d37'16"
Ayde the Care Bear
Ori 5h45'44" 6d23'32"
Ayden
Lyn 7h36'16" 43d20'28"
Ayden Christopher Frey
Vir 14h22'16" -4d2'7"
Ayden Mathew Cook
Aur 6h37'52" 34d57'12"
Ayden Paul Ressa
Cyg 20h10'20" 44d40'48"
Ayden Phillips
Aqr 22h39'46" -4d16'46"
Ayden Riley Cheek
Tau 4h29'15" 3d10'11"
Ayden Seipel
Dra 20h12'58" 70d28'54"
Ayden Teabo
Tau 4h16'24" 27d8'46"
Ayden Wells Peterson
Uma 10h37'48" 44d50'2"
Ayden05
Psc 1h30'17" 15d5'21"
Ayden's
And 0h36'37" 32d16'38"
Ayden's 3rd Birthday Star
Lac 22h25'10" 50d23'29"
Ayden's Aura
Lib 15h6'59" -8d13'34"
Ayden's Star
Tau 4h54'39" 17d45'14"
Aydia Lee Karas
Psc 1h6'54" 30d0'55"
Aydin Shamsuddin Khwaja
Her 17h39'35" 34d57'58"
Ayelet
Lyn 6h50'4" 57d31'7"
Ayeletush & Basush
Uma 10h58'38" 39d37'3"
Ayerim's Sweetdream
Ari 3h5'1" 16d18'16"
Ayers
Dra 19h30'53" 64d54'52"
Ayesha Bhabha
Sgr 18h20'21" -22d41'37"
Ayesha Taliah
Cnc 9h7'19" 27d42'0"
A.Y.F.
Aqr 21h11'19" -8d42'44"
Ayham Abu Ammar
Lmi 9h49'15" 35d27'10"
Ayhan Erdem
Uma 11h8'40" 41d19'57"
Ayin Fuentes
Cnc 8h37'37" 13d34'42"
Ayisha
Cas 0h46'30" 52d44'16"
Ayk Saakyan
Uma 11h26'57" 58d32'14"
AYLA
Cas 23h33'49" 55d29'12"
Ayla
Uma 10h43'57" 45d5'13"
Ayla
And 2h35'24" 42d18'0"
Ayla Allyssa Bartels
Gem 7h10'43" 32d17'43"
Ayla Ann McCarron
Dra 17h31'23" 55d37'45"
Ayla Beth
Uma 11h47'28" 36d34'4"
Ayla Hope
Aqr 22h43'36" 0d55'51"
Ayla Jayne
Uma 10h29'49" 43d48'0"
Ayla Levin
Tau 4h45'32" 19d40'26"
Ayla Levine
Peg 21h45'54" 11d7'32"
Ayleen Shaker
Psc 0h50'5" 11d51'5"
Aylin
Ori 5h55'19" 17d9'38"
Aylin Hardwiger
And 22h59'34" 50d45'41"
Aylin Kinaci
Lib 15h15'19" -8d50'14"
Aylin Thibault
Sgr 18h17'32" -29d0'39"
Aylin Türksever
Crb 15h57'5" 27d12'12"
Aylissa
Lib 15h40'55" -8d1'46"
ayman
Ari 3h11'22" 23d13'35"
- Ayman Chahine -
Pho 2h4'2" -40d43'21"
Ayme Plasencia
Vir 13h51'17" -5d34'23"
Aymee E Thurston
And 23h26'0" 45d18'59"
Aymee Melinda Sells
Cyg 20h30'9" 33d57'57"
AYN'S
Cap 21h55'11" -24d38'28"
Ayn's Manifesto
Aqr 22h15'16" -2d37'28"
Aynslie Claire
Sgr 18h36'24" -20d52'29"
Ayodhya Prasad Gupta
Cnc 8h45'9" 24d33'32"
Ayren Jackson-Cannady
Com 13h13'56" 24d37'58"
Ayren's Star of Sam
Sgr 18h37'20" -31d22'45"
Ayrista
Cnc 8h24'44" 14d43'53"
ayse*
Vir 13h11'20" 12d42'0"
AYSE&EMRECAN ATAK : BIZ
Cru 12h45'41" -57d6'57"
AYSEGUL
Sgr 19h40'46" -14d9'55"
Aysem
And 1h41'48" 42d23'46"
Aysenur
Uma 9h7'8" 51d42'33"
Aysha Khatun
Cas 2h1'55" 62d43'5"
Aysin Dasdemir
Del 20h46'22" 11d2'8"
Ayu
Cas 1h39'33" 64d53'30"
Ayub Shamsia-Zulqernain
Uma 9h38'31" 43d12'0"
Ayumi Taguchi
Uma 12h46'21" 58d23'58"
Ayven Skye Drake
Umi 16h58'37" 79d30'15"
AZ
Uma 11h0'36" 57d1'50"
Aza
Cas 1h6'23" 75d43'27"
Aza Salih
Uma 13h42'43" 57d13'5"
Azadeh
Psc 1h8'52" 12d18'14"
Azadeh Ommanian
Tau 5h1'28" 18d50'12"
Azalea Lucille Holt Wick
And 1h35'17" 46d32'36"
Azan
Uma 9h37'1" 48d19'5"
Azar
And 1h58'26" 42d13'21"
Azar
Crb 16h18'38" 27d26'16"
AZAR Doctor aus Liebe
Her 17h28'29" 18d9'55"
A.Zebadiah Toujours
Lyn 7h6'25" 49d34'13"
Azhane C. Salmon "August 19, 2004"
And 23h9'16" 46d1'54"
AziaMarie Theresa Beckhorn
Ari 2h18'41" 23d36'36"
Azie L. Schone
Tau 3h37'13" 25d3'22"
Azim & Liza
Uma 9h43'22" 45d16'4"
Azima Pendeka
Aqr 21h26'26" -0d56'35"
Azin (luces hermosas)
Lyn 7h4'16" 45d14'25"
Aziz
Psc 0h59'42" 31d37'54"
Aziz Bucater: Star of Bonnie's Life
Psc 1h6'53" 13d21'3"
Aziza Salameh
Ari 3h4'27" 22d59'40"
Aziza Tiffani
Cnc 9h1'57" 28d16'1"
Aziza Zainool
Psc 1h10'37" 32d37'51"
Azlan Khalsom
Ari 2h8'42" 26d21'15"
Azpeitia
And 1h1'2" 45d40'46"
Azreil
Cru 12h28'29" -58d46'41"
Azri Shenkar
Leo 10h14'0" 23d25'46"
Azriel
Cnc 8h41'24" 15d34'41"
Azriel
Cra 18h49'28" -39d20'3"
Azron
Uma 11h27'53" 34d51'20"
Azucena
Cnc 9h7'50" 15d18'57"
Azucena Magnolia
Psc 1h3'25" 3d59'35"
Azucena Sanchez
Cnc 8h11'35" 15d8'7"
Azur Modric
Ari 2h33'2" 27d1'25"
azure
Ori 4h52'35" 10d47'22"
Azuriah (Helped By God)
Aur 5h57'32" 53d58'44"
Azusa
Psc 23h36'59" 4d18'18"
Azzim Williams
Lac 22h26'5" 43d27'5"
Azzoni Adriana
Per 3h25'16" 41d34'26"
Azzue
Leo 9h38'30" 26d41'42"
Azzurra
Aur 5h33'52" 32d33'39"
Azzurra
Umi 17h33'52" 80d0'53"
Azzurra Ammirati
Umi 17h27'14" 80d42'23"
Azzurra Maria Mazzante
Lyr 19h6'40" 26d51'54"
B
Psc 0h51'44" 24d45'52"
B
Umi 16h24'16" 78d17'1"
B...
Lib 15h26'7" -15d2'34"
B 612
Gem 6h52'43" 19d27'9"
B A B 3 (Star of Barbie & Daniel)
Cru 12h34'58" -61d56'38"
B and E 61
Ari 1h59'4" 18d43'37"
B. and E. Binder
Cyg 20h43'54" 51d39'28"
B and J
Lib 15h7'19" -6d56'5"
B and L
Cnc 8h51'7" 30d32'4"
B. Aric Griffin
Aql 19h36'37" 5d33'6"
B - B
Cyg 20h6'31" 31d0'19"
B&B
Cyg 19h39'53" 55d17'25"
"B. B." Punster
Com 12h50'41" 22d15'53"
B & C Forever
Crb 15h25'37" 25d48'5"
B. Conan Gaius
Her 18h17'18" 15d7'59"
B & D 429
Sge 19h36'0" 18d51'4"
B & D Allen 50
Cyg 21h19'52" 47d18'18"
B D Blitzer
Psc 0h59'3" 16d50'18"
B. Dale Reding
Cyg 21h58'20" 49d10'58"
B Denise Harrison
Tau 3h59'56" 29d32'51"
B. Douglas Smith
Gem 7h13'3" 20d11'13"
B & E
Uma 10h37'26" 57d50'1"
B. Earle Edgerton
Uma 12h3'6" 41d5'17"
B&G Forever
Psc 1h12'32" 28d11'43"
B/GEN SARAH P. WELLS
Vir 12h37'3" 7d2'32"
B. Golden
Cap 21h22'20" -23d41'18"
"B. J."
Tau 3h47'49" 27d38'18"
B & J Seitel
And 1h38'9" 45d38'48"
B. J. Teddy
Vir 14h16'42" -7d1'47"
B. J. Traymar
Cap 21h19'54" -25d51'31"
B L B
Cap 20h30'3" -14d14'59"
B&L Laddin's Lamp
Ori 5h27'19" 6d53'11"
B L Roper
Ori 6h4'11" 13d18'55"
B. L. & The Beans
Cap 20h59'5" -21d3'31"
B. Mark Seabrooks
Her 16h49'38" 8d39'47"
B - My Beautiful Shining Star
Cru 12h41'8" -61d29'45"
B. Neil Carr
Ori 5h27'28" 5d45'25"
B. Reszketo Márta "L"
Cnc 9h17'21" 10d46'47"
B. Robert and E. Imogene Tolin
Cyg 20h28'32" 40d44'58"
B R's Bodacious Bounty
Cas 1h21'50" 56d33'7"
B. Samantha Alterman
Cnc 8h19'47" 17d20'46"
B Squared
Umi 16h14'9" 78d37'3"
B. Tóth László
Uma 9h20'40" 70d13'28"
B W + S H Forever
Tau 3h52'41" 8d9'8"
B Wolf
Lib 15h7'53" -27d16'50"
B. Yvonne Richardson
Uma 9h9'43" 49d51'15"
B,B
Cap 20h22'28" -12d11'47"
B-2
Ori 5h4'31" 4d10'14"
B4L
Sgr 18h40'57" -24d30'24"
B612
Tau 4h45'19" 23d59'11"
B612 - Simona Zanetti
And 2h36'41" 45d45'44"
Ba and Nidit
Cru 12h43'37" -57d48'23"
BA & DA
Cyg 20h49'42" 35d48'43"
BA Sixty Year Star
Aqr 23h5'31" -21d33'36"
Ba Xa Ong Xa Star
Cyg 19h35'50" 53d42'1"
Baabee
Aqr 22h15'44" -3d28'18"
Baack, Michael
Gem 7h37'41" 16d34'17"
Baba
Per 4h40'16" 38d35'18"
Baba
Umi 15h24'55" 71d5'54"
baba
Col 5h51'0" -28d51'53"

"Baba"- Dafina Jelesijevic-Dacovic
Ori 6h11'0" 13d26'25"
Baba femi Ogunlade
Uma 10h1'25" 72d4'47"
Baba & Gedo McGee
Uma 13h46'9" 50d9'18"
Baba & JG
Uma 13h36'44" 50d16'36"
Baba Mändli
Sge 20h16'10" 19d12'55"
BABA + Pat Pat
Per 3h39'34" 32d39'56"
"Babae Filia" - (Mother & Daughter)
Ori 6h13'6" 15d7'13"
Babak
Aql 19h17'15" 15d52'12"
Babalou 30
Ori 6h0'9" 22d24'8"
Babatunde Balogun
Gem 7h8'15" 31d17'33"
Babayan
Her 18h39'59" 19d29'59"
babbalove
Cru 12h42'33" -55d58'50"
Babba's Wish
Gem 6h28'51" 18d20'1"
babci ordyna
Dra 16h21'2" 64d0'1"
Babcia
Uma 10h30'16" 66d56'34"
Babcia Helen
Cyg 21h41'56" 43d0'9"
Babcia Stasia
Uma 8h40'32" 49d36'28"
Babcie
Cap 21h56'17" -8d52'46"
Babe
Aql 18h51'57" -1d17'56"
Babe
Pho 1h34'49" -43d40'47"
Babe
Cyg 21h29'27" 51d38'21"
Babe
Lyn 7h34'52" 35d34'24"
"Babe"
Lyr 18h35'4" 32d21'58"
Babe
Crb 15h32'20" 29d17'20"
Babe
Peg 23h16'37" 27d36'23"
Babe
Tau 5h46'30" 17d13'0"
Babe
Cnc 8h44'52" 14d35'41"
Babe
Ori 6h8'38" 14d48'27"
babe
Vir 12h2'13" 1d47'51"
Babe Doll
Gem 7h40'46" 26d47'13"
Babe & Duke
Cep 3h37'15" 81d4'10"
Babe In Total Control Here <\@> 50
Cas 23h0'10" 56d21'40"
Babe (Mélanie Baudoin)
Aqr 23h28'31" -10d43'43"
Babe One More Chance
Psc 0h39'49" 10d24'59"
Babe Star
Cyg 19h37'54" 34d28'31"
BaBe the star
Aqr 22h29'8" -23d57'29"
Babe & Toots Til The End Of Time
Leo 11h24'39" 17d2'44"
Babe Voorheis
Dra 18h18'41" 73d8'2"
Babe7
Per 3h52'44" 36d30'30"
Babeau
Cnc 8h49'5" 31d45'30"
Babelulu
Uma 11h23'4" 58d49'29"
BaBe.MuFFiN.DeViN.POLAR BEAR!!
Leo 10h17'2" 25d6'6"
Babeola
Her 17h48'37" 41d33'10"
Babes
Crb 16h12'36" 38d30'12"
Babe's Beauty
Gem 7h35'43" 14d50'50"
Babe's Eternal Light
And 0h38'31" 33d30'32"
Babe's Tea Time
Ari 3h17'33" 29d3'50"
Babess Forever
And 0h27'55" 32d26'24"
Babeth
And 2h21'56" 38d46'30"
Babeth
Lyn 6h43'34" 61d9'42"
Babette "My Love"
Ori 6h11'12" 15d10'15"
Babette's Paradise
Umi 16h47'45" 80d26'35"
Babez
Uma 10h37'57" 46d20'11"
Babez Star
Umi 19h59'16" 87d10'46"
Babi
Umi 17h29'5" 80d14'31"
Babi
Aqr 22h17'52" -10d51'47"
Babi
Cyg 19h38'3" 31d42'35"
Babi
Ori 5h16'0" 0d50'0"
Babi
Peg 23h21'1" 22d57'55"
Babi Joo
Cap 20h15'37" -25d12'3"
Babi Lidka - Our Star, Our Light
Pho 1h23'10" -50d32'10"
Babi Linh Sawa
Cru 12h41'3" -61d6'14"
Babi Mica
Cas 23h36'17" 51d35'29"
Babiczki
Cap 21h56'52" -10d2'59"
Babie
Ori 5h26'7" 10d15'35"
Babie Jenni
Leo 10h15'47" 25d6'49"
Babies
Lib 15h2'23" -23d23'46"
Babies, Helmut
Ori 6h3'19" 9d36'24"
Babis
And 23h18'15" 41d47'32"
Babki Stella
Tau 5h44'54" 20d45'13"
Babo
Vir 13h22'46" 5d18'11"
Babou + Babe
Psa 22h56'19" -27d37'9"
Babou - Mo's definition of Love
Leo 10h29'57" 19d21'41"
Babr aka "Abigail"
Dra 17h29'43" 58d31'8"
Bab's
Cas 1h32'15" 60d14'17"
Babs
Ori 5h57'53" 17d17'29"
babs
Ori 5h24'16" 14d23'30"
Babs
And 1h56'31" 46d0'0"
Babs Adams
Crb 16h13'28" 26d13'34"
Babs Deeney Labranche
Sgr 18h27'57" -16d26'44"
Babs King
Cnc 8h10'46" 11d25'27"
Babs Lord Powell
Cas 23h25'17" 56d21'3"
Babs Warren
Uma 11h39'39" 31d31'0"
BABSN2
Uma 11h8'21" 40d22'28"
babspieknadziewczyna
Cet 3h9'49" 7d19'3"
Babu
Leo 10h26'23" 22d43'34"
Babu
Ari 2h18'48" 23d46'36"
Babu
Her 17h57'11" 32d52'21"
Babuka Katya
Aqr 20h54'6" -11d34'33"
Babushka Liliya
Psc 0h19'39" -2d41'43"
Baby
Umi 14h36'1" 75d48'22"
baby
Cap 21h52'7" -14d9'20"
"Baby"
Sgr 19h6'47" -14d3'49"
B-A-B-Y
Uma 12h36'22" 57d11'59"
Baby
Per 3h13'25" 37d11'40"
Baby
Gem 7h18'37" 31d12'20"
Baby
And 1h54'50" 37d56'13"
Baby
Peg 23h18'54" 17d48'28"
Baby
Gem 6h53'14" 22d23'26"
Baby
Cnc 9h5'36" 27d39'27"
Baby
Leo 11h41'28" 25d15'34"
Baby
Ori 6h5'5" 14d25'28"
BABY 1
Leo 11h53'14" 24d43'15"
Baby 1Z
Uma 9h56'47" 42d32'46"
Baby # 2
Ari 2h45'6" 28d44'7"
Baby Abbie
And 23h12'54" 45d27'6"
Baby Abrigo
Cyg 20h18'8" 34d20'43"
Baby Aidan Sugino
And 0h55'27" 36d15'42"
Baby Ale
Tau 4h8'44" 4d44'19"
Baby Allison
Psc 23h31'54" 2d7'42"
Baby Always Star
Lyr 18h41'54" 32d17'31"
Baby Amanda Rene
And 23h49'5" 37d37'44"
Baby Angel
Vir 12h40'59" 7d14'13"
Baby Angel
Cnc 8h44'34" 16d14'46"
Baby Angel
Lib 14h43'25" -11d38'38"
Baby Angel Eith
Dra 19h2'17" 64d33'53"
Baby Angel Laura
Cyg 21h46'55" 39d6'57"
"Baby Angel" Mitchell
Umi 14h31'43" 75d0'18"
Baby Angel Morales
Umi 16h52'14" 80d3'4"
Baby Angel Rolle
Ari 2h11'47" 18d59'15"
Baby Angel-Aidan*Sweet Daddy-Juston
Cyg 20h5'55" 31d2'44"
Baby Angus
Gem 7h13'47" 23d5'55"
Baby Anna Renee Joseph
And 0h58'5" 37d30'52"
Baby Anthony
Dra 18h42'32" 63d4'37"
Baby April
Ari 2h44'23" 28d8'19"
Baby Arielluv
Tau 4h10'41" 2d43'22"
Baby "Aroha" Star 22903
Her 16h41'27" 16d23'48"
Baby Ashley!
Ori 5h49'40" 3d44'29"
Baby Austin
Tau 4h36'14" 14d18'14"
Baby "B"
Gem 6h20'9" 22d6'4"
Baby B
Cyg 21h37'50" 46d0'46"
Baby B
Cas 23h4'53" 58d11'50"
Baby B Doll
Umi 15h37'38" 71d40'31"
Baby B. W. Self
Uma 11h43'49" 55d53'37"
Baby Baby Baby
Uma 8h43'20" 60d48'20"
Baby Bailey
Umi 14h37'22" 74d55'2"
Baby Bako
Umi 16h7'53" 77d13'24"
Baby Balooga
Vir 13h38'48" 2d26'53"
Baby Bathmaker
Umi 15h39'57" 68d3'39"
Baby Bauer
Ari 2h39'18" 28d1'30"
Baby Beagan
Peg 23h12'46" 18d25'46"
Baby Bean Stanley
Ser 15h29'0" 11d35'5"
baby bear
Ari 3h0'1" 18d26'7"
Baby Bear
Her 18h47'10" 20d12'48"
Baby Bear
Cnc 8h16'58" 17d39'14"
Baby Bear
Lib 15h17'26" -18d39'5"
Baby Bear
Cap 21h11'14" -22d58'35"
Baby Bear Russell
Umi 15h21'12" 73d19'53"
Baby Bear's Star
Psc 1h59'5" 7d20'15"
Baby Bella Alicia
Vir 13h12'40" -22d38'31"
Baby Beresford
Lib 15h47'35" -10d42'14"
Baby Bernstone's Star
Lmi 10h9'29" 40d47'11"
Baby Billy Bostock
Umi 8h16'15" 88d12'21"
Baby Binks
Lyn 7h39'7" 38d12'0"
Baby Bird
And 2h28'23" 44d42'29"
Baby Bird
Cap 20h48'29" -25d45'30"
Baby Biscardi
Umi 14h42'54" 75d2'29"
Baby Blue Eyes
Lib 15h33'51" -10d1'23"
Baby Boo
Umi 16h23'27" 86d19'6"
Baby Boo
Uma 11h26'36" 46d3'26"
Baby Boo
Psc 0h50'59" 28d36'43"
Baby Boo
Tau 5h44'34" 16d31'4"
Baby Boo and Ninny
Tau 4h2'0" 27d12'33"
Baby Boo-Boo Bear
Aqr 22h12'9" -2d43'7"
Baby Bosher
Umi 14h36'20" 75d12'46"
Baby Boy
Umi 14h25'2" 75d34'25"
Baby Boy
Peg 23h24'5" 21d8'53"
Baby Boy and Baby Girl
Aqr 23h46'10" -18d33'27"
Baby Boy Baptista
Cap 21h15'56" -15d40'59"
Baby Boy Biscardi
Umi 14h22'47" 76d17'52"
Baby Boy Delao
Uma 8h21'32" 61d59'25"
Baby Boy Duffy
Cyg 19h43'47" 39d1'48"
Baby Boy Hechtman
Umi 15h30'27" 72d25'41"
Baby Boy Hippo & Baby Girl Hippo
Uma 12h46'45" 52d59'37"
Baby Boy Jacob
Vir 12h53'55" -5d57'30"
Baby Boy Lipe
Lib 15h17'28" -19d44'57"
(Baby Boy) Michael J Burgess, Sr
Cnc 8h33'11" 24d9'47"
Baby Boy Padilla-MacGillivray
Umi 14h42'58" 75d23'28"
Baby Boy Provo
Umi 13h15'16" 72d31'47"
Baby Boyle
Ori 5h55'48" 19d49'37"
Baby Brack
Umi 15h27'12" 72d30'48"
" Baby Brayanna "
Psc 1h9'59" 27d0'1"
Baby Brown Angels
Cam 4h10'13" 54d10'40"
Baby Bryan, Our Angel Among Stars
Umi 13h22'20" 70d10'24"
Baby Bubba
Umi 14h38'27" 70d35'40"
Baby Burnett
Umi 15h31'57" 75d34'50"
Baby Cade
Tau 4h18'55" 14d32'35"
Baby Caitlin
Tau 5h23'53" 26d47'43"
Baby Cakes
Crb 15h44'39" 27d10'8"
Baby Cakes
Lib 15h7'30" -21d16'25"
Baby Cakes
Aqr 23h17'49" -7d31'11"
Baby Cakes Bob
Per 3h29'58" 49d50'48"
Baby Cameron
Col 5h49'19" -32d8'33"
Baby Caro
Leo 11h32'4" -1d33'19"
Baby Casey Drew McAuliffe
Umi 15h33'24" 72d9'46"
Baby Chad Donald Naphegyi
Ori 6h2'49" 18d1'13"
Baby Champs
Aqr 22h23'42" -22d6'43"
Baby Chavez
Uma 10h52'29" 55d24'7"
Baby Chipmunk
Uma 12h2'7" 60d31'35"
Baby Christian
Umi 13h48'54" 73d14'37"
Baby Christopher
Umi 14h51'20" 67d34'20"
Baby Cian
Umi 13h24'1" 71d55'47"
Baby Coobs
Sgr 18h4'10" -27d28'5"
Baby Corrigan
Leo 9h42'4" 25d5'58"
Baby Courtney
Peg 23h12'19" 17d29'33"
Baby Cousin Dawn Gail
And 0h27'47" 34d20'39"
Baby Crazy
Cnc 7h59'52" 12d46'53"
Baby Curry
Umi 14h34'45" 74d22'55"
Baby Curry
Umi 14h30'16" 75d8'2"
Baby Cusati
Lyn 6h22'57" 59d55'30"
Baby Danielczyk
Uma 10h14'19" 44d16'46"
Baby & Davide
Lyn 8h14'48" 40d17'36"
Baby Dawkins
Uma 11h46'22" 52d1'47"
baby Desire Amosa
Ari 2h37'32" 22d8'27"
Baby Deutsch
Psc 1h11'10" 27d28'15"
Baby Disney
Aur 5h25'59" 46d15'7"
Baby Doll
Lyr 18h48'43" 38d0'47"
Baby Doll
And 23h11'50" 47d8'46"
Baby Doll
And 2h2'10" 37d34'55"
Baby Doll
Leo 10h20'55" 24d22'45"
Baby Doll
Uma 13h40'2" 56d38'38"
Baby Doll
Cas 1h57'33" 65d47'7"
Baby Doll
Sgr 18h7'44" -20d11'9"
Baby Doll
Cap 21h40'47" -13d26'39"
Baby Doll Deb
Ari 2h11'7" 23d11'0"
Baby Doll the Angel
Aqr 22h22'19" -23d5'12"
Baby Doodles
Uma 10h38'22" 56d31'17"
Baby Duck
Tau 3h38'41" 20d54'15"
"Baby Dukes"
Leo 11h16'39" 22d30'4"
Baby Earl
Her 17h34'29" 43d30'50"
Baby Eden Lucas Tyler
Cen 12h24'26" -49d22'54"
Baby Edward
Cyg 19h46'2" 29d19'9"
baby elefant
Lmi 10h0'10" 40d14'55"
Baby Ellie Beth McKee
And 1h12'44" 45d33'15"
Baby Emma
And 1h11'6" 42d32'55"
Baby "Ernie"
Psc 1h7'41" 25d51'25"
Baby Face
Psa 21h39'25" -30d10'54"
Baby Felix Dobson
And 23h3'55" 41d9'4"
Baby Flynn E. Zakarewicz
Uma 10h12'57" 69d53'58"
Baby Frenth
Uma 11h38'10" 56d50'30"
Baby Fresca
Lmi 10h31'9" 36d38'19"
Baby Frudd
Umi 15h37'37" 82d41'36"
Baby Gabriel
Leo 11h2'20" 3d36'15"
Baby Gabriella Traverse
Aql 19h42'29" 6d36'14"
Baby Gabrielle
And 0h2'10" 46d25'5"
Baby Gabrielle
Cas 0h2'2" 56d22'35"
Baby Gadwapus
Tau 4h0'50" 18d27'17"
Baby Giampaolo
Sco 16h36'6" -29d19'48"
Baby Girl
Sco 16h42'58" -30d46'14"
Baby Girl
Sgr 18h17'50" -33d49'38"
Baby Girl
Sco 17h3'45" -37d44'39"
Baby Girl
Car 6h23'41" -50d52'39"
Baby Girl
Umi 14h50'31" 80d3'32"
Baby Girl
Ori 5h36'35" -1d9'36"
Baby Girl
Lib 14h50'0" -2d15'15"
Baby Girl
Vir 13h43'33" -11d20'21"
Baby Girl
Sco 16h5'52" -11d19'11"
Baby Girl
And 23h16'26" 52d59'9"
Baby Girl
Uma 8h34'15" 62d36'31"
Baby Girl
Dra 14h36'54" 63d20'36"
Baby Girl
Tau 3h37'23" 16d47'35"
Baby girl
Ari 2h16'34" 17d49'17"
Baby Girl
Mon 6h46'46" 8d2'12"
Baby Girl
Vir 13h10'59" 13d31'55"
Baby Girl
Leo 9h41'50" 13d57'37"
Baby Girl
Leo 11h30'0" 9d40'37"
Baby Girl
Aqr 21h42'32" 0d41'39"
Baby Girl
Psc 0h44'13" 14d30'28"
Baby Girl
Vir 12h41'31" 0d58'18"
Baby girl
And 0h42'12" 28d49'2"
Baby Girl
And 0h56'5" 23d13'27"
Baby Girl
Gem 6h42'39" 22d3'46"
Baby Girl
Cnc 8h14'13" 18d14'21"
Baby Girl
Gem 7h1'7" 23d12'27"
Baby Girl
And 23h19'37" 49d0'45"
Baby Girl
And 23h27'17" 47d28'39"
Baby Girl
Cyg 20h9'19" 51d17'19"
Baby Girl
And 0h45'59" 45d39'28"
Baby Girl
And 1h6'11" 45d56'52"
Baby Girl
And 1h16'24" 46d56'1"
Baby Girl
Crb 15h47'16" 36d4'13"
Baby Girl
Lyr 18h38'34" 30d45'46"
Baby Girl
Lyn 8h24'31" 34d22'40"
Baby Girl
Cnc 8h31'23" 31d28'53"
Baby Girl
Cnc 8h40'40" 31d30'50"
Baby Girl
And 1h39'8" 44d54'20"
Baby Girl
And 0h32'28" 38d14'16"
Baby Girl
Uma 9h18'36" 43d51'42"
Baby Girl and 808
Tau 4h25'15" 10d32'34"
Baby Girl Ashley
Leo 10h22'48" 21d49'13"
Baby Girl Barsosky
Cmi 7h26'27" 8d40'37"
Baby Girl Buck
Cyg 19h58'53" 41d34'39"
Baby Girl Colarusso
Lyn 7h59'1" 48d14'40"
Baby Girl Duffy
Cyg 19h44'52" 39d1'34"
Baby Girl Furman
Cap 21h37'2" -16d51'1"
Baby Girl Gutierrez
Vir 13h14'51" 7d20'13"
Baby girl Hayes
And 0h42'33" 43d55'40"
Baby Girl Jennifer Marie Torres
Lib 15h15'21" -20d21'24"
Baby Girl Kimi
Cas 1h18'1" 63d43'27"
Baby Girl Liu
Psc 23h51'24" 1d48'6"
Baby Girl (Marisa) your my light !
Her 16h55'41" 37d20'30"
Baby Girl Michelle
Her 16h55'41" 35d35'20"
Baby Girl Nicole Mussara
Lib 15h40'37" -25d45'50"
Baby Girl Owens
And 23h49'56" 37d8'11"
Baby Girl~Samantha Lane Mundy
And 1h23'2" 39d58'21"
Baby Girl - Star of Eternal Love
And 0h52'7" 42d35'11"
Baby Girl Yakel
Psc 1h30'50" 13d9'2"
Baby Girl's Shining Light
Sgr 18h55'2" -16d3'28"
Baby Glenna
Ori 5h37'49" -3d7'48"
Baby Grace Ann Tolbert
Tau 4h20'55" 23d22'11"
Baby Grayden
Her 17h2'20" 35d46'10"
Baby Grayson Thomas Stone
Uma 10h26'27" 52d9'33"
Baby Grill
And 1h49'59" 45d57'39"
Baby Gurl & Jellybean
Psc 22h57'45" 4d12'24"
Baby Guy
Uma 11h44'26" 39d13'31"
Baby Hamels
Uma 12h54'19" 60d18'10"
Baby Hannah
Leo 10h11'34" 20d57'27"
Baby Harper
Umi 14h36'2" 73d43'5"
Baby Hayden
Ori 5h31'44" -3d8'47"
Baby Hayes
Sco 16h9'19" -12d52'59"
Baby Hebert
Sco 16h17'52" -8d32'58"
Baby Hen
Aql 19h48'46" -0d31'15"
Baby Howell
Umi 15h17'33" 68d11'27"
Baby Hunter Lovell
Umi 15h34'25" 78d0'31"
Baby Hussey
Umi 15h36'6" 72d53'7"
Baby Israel
Umi 16h21'39" 76d38'56"
Baby J
Aql 20h1'39" 15d28'2"
Baby Jacaruso
Mon 6h56'58" 10d40'26"
Baby Jack
Cnc 8h11'56" 10d23'40"
Baby Jack
Lib 14h48'57" -5d14'46"
Baby 'Jack Jack' Pelcic
Uma 8h19'27" 68d59'12"
Baby Jack Joseph Nelson's Star
Uma 11h17'55" 72d24'51"
Baby Jack Michael Wood
Ori 5h32'49" -1d28'59"
Baby Jacksyn Haley
Uma 9h23'36" 53d36'7"
Baby Jade
Cas 0h31'17" 61d30'45"
Baby Jake
Sgr 19h51'39" -28d27'14"
Baby Jake
Lmi 10h19'45" 36d52'35"
Baby James Joseph Gardner
Uma 11h45'41" 60d8'50"
Baby jan & Mane jan
Her 16h45'0" 42d32'46"
Baby Jane
Dra 19h10'45" 64d52'46"
Baby Jenn
Psc 1h11'39" 27d19'9"
Baby Jo
Vir 12h48'41" 4d39'8"
Baby Jo
Ori 5h30'41" 4d7'28"
Baby Joachim Jazmines
Psc 0h47'40" 12d24'40"
Baby Joel David Andrews
Umi 13h48'22" 78d7'21"
Baby Joel Perez
Cap 20h47'56" -20d11'0"
Baby JoJo
Cap 21h46'57" -14d23'34"
Baby JoJo Martinez
Peg 21h42'46" 27d35'17"
Baby Jorden
Cyg 20h51'30" 44d18'58"
Baby Joseph Alfred Kennedy 6-29-03
Umi 15h31'31" 74d46'20"
Baby Joseph Angelo Pecora
Gem 6h51'1" 33d8'28"
Baby Julia Mary P
Tau 4h12'53" 25d8'11"
Baby Julian
Her 17h2'55" 29d3'59"
Baby Kaitaia
Cyg 20h8'32" 58d50'33"
Baby Kate, aka Kate Sullivan
Mon 7h19'48" -0d35'35"
Baby Katia
And 1h34'2" 48d21'46"
Baby Kayden
Umi 16h14'53" 70d52'0"
Baby Kelly
Vul 19h45'44" 25d57'54"
Baby Kendele
Cap 21h43'48" -16d54'33"
Baby Kiki
Lib 15h6'18" -6d31'20"
Baby Kim
And 1h4'34" 43d53'50"
Baby Kleckner
Umi 16h42'38" 82d3'13"
Baby Kostka
Umi 14h27'26" 75d13'24"
Baby Kruse
Psc 23h8'17" 6d59'14"
Baby Kuppins
Umi 14h36'43" 79d53'8"
Baby Lally
Umi 17h7'44" 76d43'44"
Baby Lara Horan Falsey
And 23h34'37" 48d20'33"
Baby Laura
Leo 11h34'25" 14d19'25"
Baby Leah
And 1h27'41" 45d57'31"
Baby Lewis Beasley
Uma 11h56'48" 32d55'17"
Baby Lo
Uma 10h0'56" 49d37'33"
Baby Love
Leo 10h37'3" 17d54'27"
Baby Love
Aqr 22h47'57" -2d47'9"
Baby Love
Umi 15h38'4" 72d28'26"
Baby Love
Cyg 20h26'41" 55d48'45"
Baby Magic
And 0h48'7" 40d28'12"
Baby Makenna Rae Jackson
Sco 16h5'14" -17d11'27"
Baby Man and Squirrely Bird Forever
Sge 19h51'31" 18d7'24"
Baby Manzi Roldan
Sco 16h45'36" -30d28'35"
Baby Marina
And 0h13'28" 45d9'19"
Baby Matthew O'Keeffe
Umi 14h11'52" 79d15'45"
Baby Max
Her 17h42'33" 21d7'31"
Baby Max Andrew Miller
Umi 14h42'35" 75d21'19"
Baby Max Lewis Baldwin
Umi 16h7'37" 74d35'25"
Baby Max Santos
Uma 10h21'2" 68d6'6"
Baby McConnell
Uma 11h41'13" 58d0'3"
Baby McCoy
Gem 7h13'59" 19d11'11"

Baby Minsterman
Cnc 8h23'8" 21d41'54"
Baby Morgan
Cap 21h0'45" -23d47'54"
Baby Muccigrosso
Tau 4h16'12" 27d23'42"
Baby Muffin
Aqr 23h26'41" -18d37'2"
Baby Mynx
Leo 9h59'20" 23d11'49"
Baby Nephew Watson
Sco 16h16'12" -11d22'6"
Baby Niamh Critchley
Cyg 21h45'20" 32d20'30"
Baby Nini
Cnc 9h11'40" 26d1'49"
Baby Noah
Lyr 18h33'28" 36d34'47"
Baby Noll
Sco 17h34'7" -34d10'58"
Baby O'Connor
Umi 15h30'46" 76d46'24"
Baby & Ocsi
Cas 0h45'17" 60d37'8"
Baby of Harjit and Maninder Pabla
Umi 16h1'44" 76d58'2"
Baby Page Dalby
Per 2h55'48" 53d42'43"
"Baby Pam"
Tau 5h32'59" 19d58'53"
Baby Patrick Jay
Umi 16h37'44" 81d34'30"
Baby Pelagatti Angel
Umi 15h50'48" 72d44'59"
Baby Pelan
Aqr 22h59'0" -6d58'25"
Baby Penguino
Gem 7h47'1" 25d3'47"
Baby Phat
Cas 1h37'13" 67d18'27"
Baby Pie
Uma 9h29'50" 68d37'21"
Baby Pie - I will always love you!
Cnc 8h52'26" 26d52'27"
Baby Princess Bear
Uma 10h43'23" 47d51'18"
Baby Puffin
Uma 8h42'14" 64d43'19"
Baby Raine
Umi 14h33'52" 74d26'14"
Baby Rajala
Uma 9h59'31" 52d44'55"
Baby Rappazzo
Uma 12h47'31" 62d52'58"
Baby Raub & Max
Umi 14h0'55" 71d45'21"
Baby Raykip
Uma 9h56'40" 58d2'49"
Baby Reggie
Aqr 20h47'32" 1d4'33"
Baby Richard (Richard Stamp)
Cep 2h25'40" 81d12'19"
Baby Robert Booth - Our Angel
Umi 15h31'56" 75d44'41"
Baby Robertson
Aqr 22h39'45" 0d53'26"
Baby Roland
Umi 17h23'28" 75d34'57"
Baby Ron
And 23h9'19" 42d16'13"
Baby Rose
Vir 11h53'1" 8d0'27"
Baby Ross Hoyman
Lmi 10h54'26" 32d55'55"
Baby Roth
Sco 16h42'54" -29d5'35"
Baby R's Star
Her 17h33'53" 19d49'5"
Baby Rucker Yung
Vul 19h36'50" 25d55'49"
Baby Rudy
Umi 15h24'51" 72d11'32"
Baby Rue
Ori 6h21'50" 14d21'35"
Baby Ryan's Star
Uma 12h7'48" 61d28'50"
Baby Sal Rios
And 2h10'57" 38d9'57"
Baby Sam Boland
Cru 12h37'52" -58d30'35"
Baby Samantha
Cnc 8h38'16" 18d59'20"
Baby Sarah
Dra 17h8'1" 68d26'54"
Baby Saville
Ori 5h45'6" 6d49'32"
Baby Sister
And 1h21'44" 47d45'37"
Baby Sister Anna
Dra 19h19'58" 83d35'23"
Baby Squash
Her 16h22'4" 46d7'41"
Baby Squeesh
Cnc 9h20'13" 9d11'32"
Baby Star
Peg 22h20'35" 6d16'37"
Baby Star
Cma 6h14'59" -18d55'59"
Baby Star- Gina Marie Comminello
And 0h33'51" 27d18'56"
Baby Star Reyes Baert
And 1h14'8" 38d33'14"
Baby Ste
Uma 13h12'1" 58d51'47"
Baby Steinbeck
Sgr 18h24'22" -33d1'10"
Baby Steph
And 0h12'37" 26d15'8"
Baby Steven Ciccarelli
Sgr 17h52'40" -28d36'58"
Baby Stigliano
Cam 4h2'30" 67d19'6"
Baby Sue
Umi 13h34'48" 74d38'24"
Baby Tang
Umi 15h24'24" 74d39'35"
Baby Tedesco
Umi 13h20'46" 72d50'3"
Baby Timothy
Uma 12h6'34" 58d51'41"
Baby T-K "The Angel Star"
Umi 16h55'23" 76d39'29"
Baby Tolbert's Shining Star
Ori 6h14'26" 15d11'14"
Baby Tommy Keane
Umi 4h24'37" 88d36'33"
Baby Traylor In My Heart
Cas 1h27'14" 52d15'10"
Baby Trenton Douglas
Umi 15h19'54" 71d18'32"
Baby Twinkle Wink
Gem 7h20'14" 32d8'4"
Baby Vito Carriel
Umi 15h18'40" 78d9'22"
Baby W.
Vir 13h22'46" -3d54'6"
Baby Walker
Uma 10h9'28" 54d45'36"
Baby Watson 2006
Umi 13h59'22" 75d45'4"
BABY WETH
Ori 5h33'38" 0d33'46"
Baby Wheatman
Aqr 22h18'57" -0d1'52"
Baby Whiles
Umi 14h7'30" 68d11'20"
Baby White Tiger
Leo 9h44'30" 24d58'57"
Baby Will
Lyn 8h47'12" 34d31'25"
Baby Williams
Peg 22h19'4" 19d9'32"
Baby Williams II
Umi 14h32'23" 73d41'32"
Baby Windebank
Umi 15h38'11" 76d34'9"
Baby Z
Uma 13h47'41" 56d37'18"
Baby Zachary
Tau 4h9'42" 29d10'54"
Baby, may we always shine together
Aqr 23h0'2" -7d7'30"
BabyBinx
Aqr 22h59'39" -9d43'6"
"Babybird"
Vir 13h41'38" 2d26'48"
Babyblue-Cliff Henry Mason II
Vir 13h50'56" -21d12'31"
Babyboi
Per 3h48'57" 49d0'51"
BabyBoo: Robert "Dumbo" Bashaw
Uma 10h42'9" 58d15'10"
babyboy
Tau 4h25'23" 17d51'38"
Babycakes
Cnc 8h50'3" 9d50'19"
Babycakes
Ori 5h4'53" 13d10'33"
Baby-Cakes
Cap 21h54'3" -17d55'56"
babycakes
Lib 15h15'0" -9d51'47"
Babycakes Alma
And 2h38'33" 51d0'23"
BabyCakes O'Brien
Lib 15h5'11" -2d8'32"
Babycakes's Star
Uma 9h27'22" 72d1'51"
BabyD
Del 20h24'18" 9d39'21"
Babydog
Cmi 7h56'53" 10d50'6"
Babydoll
Sgr 18h1'49" -30d32'13"
Babydoll
Uma 10h25'18" 60d7'24"
babydoll
Aqr 22h24'8" -0d38'25"
baby-doll
Sco 16h17'44" -14d6'37"
Babydoll
Sgr 19h35'24" -15d54'43"
Babydoll
Lib 15h7'25" -16d51'30"
Babydoll
Leo 11h12'33" 14d38'27"
BABYDOLL
Cnc 8h44'33" 7d27'22"
Babydoll
Ori 5h35'50" 3d4'1"
Babydoll
Tau 3h34'29" 2d36'0"
Babydoll
Gem 6h16'39" 24d0'36"
BabyDoll
Ari 2h37'19" 25d56'58"
Babydoll
And 1h45'15" 50d27'33"
BabyDoll
Uma 11h38'59" 47d0'57"
Babydoll
And 23h13'0" 48d35'31"
Babydoll
Lyn 8h3'43" 38d39'0"
Babydoll Brandi
Cas 22h57'41" 55d16'19"
Babydoll Christi
Sco 17h57'13" -31d22'40"
Babydoll Princess
Leo 10h7'21" 25d21'51"
BabyDoll1
Ori 6h10'12" 5d38'52"
babydoll921
Vir 12h28'52" -6d32'5"
Babydoll's Twinkle
Uma 8h59'26" 72d5'5"
Babyface
Cnc 7h58'10" 12d6'50"
Babyface Tweety
And 22h58'23" 47d31'36"
Babygirl
And 23h4'56" 48d58'39"
Babygirl
And 1h15'53" 47d36'54"
Babygirl
And 23h46'47" 43d58'9"
Babygirl
Lmi 10h21'5" 36d23'38"
Babygirl
Ari 2h24'45" 22d26'10"
Babygirl
Tau 4h24'39" 17d43'3"
Babygirl
Vir 14h42'33" 2d33'10"
babygirl
Ari 2h11'12" 25d54'5"
BabyGirl
Leo 10h11'12" 16d48'30"
BabyGirl
Mon 7h31'13" -0d24'0"
babygirl
Ori 5h54'10" -2d44'38"
Babygirl
Sgr 19h33'32" -21d20'45"
Babygirl
Aqr 23h20'39" -22d23'18"
Babygirl
Lib 14h49'48" -24d39'41"
babygirl
Psc 1h39'7" 4d44'37"
BaByGiRL**AJA**
Sgr 17h56'38" -20d26'48"
"BabyGirl" Brittney I Love You
And 2h12'38" 43d53'1"
Babygirl Holly
Cyg 19h13'14" 52d23'45"
Babygirl Jessica
And 0h53'57" 36d19'43"
BabyGirl Jessica
Cnc 8h40'44" 18d39'23"
BabyGirl(Melissa)
Peg 22h45'45" 17d39'53"
Babygirl Michelle
Cnc 9h6'39" 18d3'43"
Babygirl Sarah
Cap 21h58'39" -11d12'14"
Baby-girl to the moon the..and back
Aqr 22h38'59" 1d1'30"
Babygirl, Chassidy's Star
And 2h36'11" 44d20'29"
Babygirl's Star
Tau 4h23'52" 14d34'23"
Babyhead
Cam 7h34'41" 74d17'29"
Baby-Lillie
Umi 13h35'39" 75d2'22"
BabyLove
Cnc 8h18'58" 19d50'36"
Babyman
Vir 13h29'6" 12d50'40"
BabyNuGz
Lib 15h48'19" -17d44'43"
BabyPowder
Sco 17h27'17" -41d30'44"
Baby's shining star
Aqr 22h18'15" -0d38'36"
Baby's Star (Barbara Hendricks)
Cap 21h52'40" -18d54'36"
babysandy
Gem 6h36'17" 13d11'4"
Babyshoes
Ori 5h33'16" -9d1'25"
BabySnakes
Cap 20h26'23" -11d9'46"
Bacardi
Sgr 18h53'41" -20d23'26"
Baccaba
Tau 5h37'8" 24d53'36"
Bacchi
Cyg 21h20'22" 39d20'34"
BACH
Uma 10h42'51" 46d4'41"
Bacha
Cru 12h29'17" -60d35'2"
Bachirou Brame
Tau 5h48'43" 17d11'35"
Bachmann, Elke
Ori 5h26'40" 10d58'26"
Bacio
Uma 9h55'0" 53d34'1"
Backasch, Kay
Uma 13h25'48" 58d55'7"
Background Matrix 3
Cep 0h5'49" 82d47'20"
Bad Ass Michelman
Her 18h5'51" 28d27'3"
Bad Bear
Uma 11h35'2" 54d51'3"
"Bad Kitty" Krystal Lauren
Leo 11h5'43" 10d14'20"
BADADI2UDE
Uma 10h33'31" 48d55'18"
BadBowie
Cma 7h24'5" -24d44'6"
Badger
Sgr 19h11'54" -34d23'21"
Badger
Leo 11h29'32" 16d0'15"
Badger
Gem 6h39'38" 19d57'30"
Badgley Love
Cnc 8h12'41" 32d36'26"
Badi, 09.02.1961
Peg 22h38'53" 23d54'7"
Badiou Nicole
Cnc 8h26'41" 16d27'7"
Badman
Aqr 22h47'16" -23d39'7"
Badmylin45H43H12A4
Uma 14h23'45" 55d10'38"
Bado
Her 17h36'2" 27d52'51"
Bados
Psc 1h30'46" 2d53'39"
Badouna Christina - Principessa
Lyr 18h39'11" 37d7'11"
Badri Zahra Adnani
Lib 14h52'22" -7d38'47"
Badria Moustapha
Gem 7h13'10" 25d0'21"
Badril Bedour
Cas 23h26'2" 56d0'48"
Badriya
Cas 0h49'37" 53d46'34"
Badriya Motalaq Al Shamas
Aqr 23h25'53" -8d16'29"
Badu
Del 20h51'19" 14d4'50"
BAEDMNDWSHGSTR
Lib 14h50'1" -1d13'8"
Baerbel
Gem 6h9'25" 24d27'56"
Baerbel e Lucio
Aur 4h55'38" 32d34'14"
BaF Lorenz
Cyg 21h38'5" 45d44'45"
Bafaver
Lib 15h54'31" -13d59'3"
Baffy Viero Traversa
Aur 6h37'59" 30d54'55"
Baftir
Cyg 21h20'4" 46d45'52"
Bagel
Ori 6h5'29" 10d55'44"
Bagel and Briz
Umi 14h47'35" 76d15'15"
Bagels
Lyn 8h50'53" 40d32'43"
Baggage Bob
Per 3h16'52" 53d34'44"
Baggeler, Thomas Michael
Uma 11h49'51" 60d32'27"
Bagheera - Baby Baggins - Mottice
And 23h19'28" 52d8'42"
Bah
Sgr 19h46'30" -12d17'49"
Bahama Burns
Crb 15h36'52" 31d42'37"
Bahama Star
Cyg 21h38'42" 32d46'53"
Bahar
Cas 2h24'51" 71d59'24"
bahinca
Ori 5h35'50" 1d49'45"
Bahr, Nadine
Ori 6h15'46" 10d52'46"
Bahr, Petra
Ori 5h11'30" 15d49'28"
Bahrawi Firas
Tau 5h51'43" 13d7'17"
Bahrsch, Roland
Gem 7h17'37" 20d24'23"
Bai Vincent
Dra 14h34'19" 56d55'36"
Baierl, Günther
Uma 9h41'6" 67d21'29"
Baiileia Katarraktes
Ari 2h33'9" 26d40'48"
Baila Ruchel
Ori 5h52'1" 11d30'7"
Bailee Lane Huckleberry
Psc 1h10'27" 27d15'0"
Bailee Nicole Bowman
And 2h15'28" 45d48'4"
Bailee Riann Jamison
Vir 14h14'28" -14d29'23"
Bailee Rose
Gem 6h53'55" 16d54'4"
Bailee Sklavos
Uma 9h32'59" 42d22'24"
Bailee-Jade Angel
Leo 9h24'23" 32d16'38"
Bailei Alexandra Saari
Uma 10h35'49" 50d31'57"
Bailey
Dra 18h28'11" 50d59'46"
Bailey
Per 2h13'18" 56d20'24"
Bailey
Crb 16h13'47" 33d48'18"
Bailey
Gem 6h53'33" 16d36'4"
Bailey
And 0h33'4" 28d15'9"
Bailey
Cnc 8h46'25" 23d27'14"
Bailey
Tau 4h23'13" 23d2'31"
Bailey
Aql 18h51'30" 8d41'37"
Bailey
Lep 5h11'4" -11d42'48"
Bailey
Umi 16h29'49" 77d40'41"
Bailey
Umi 17h39'38" 83d39'40"
Bailey
Lib 15h5'14" -7d47'29"
Bailey
Uma 11h52'17" 63d11'0"
Bailey
Pho 0h46'49" -49d23'52"
bailey 21898
Aqr 22h33'38" -2d15'14"
Bailey Ace
Uma 9h38'54" 46d40'52"
Bailey Aileen Stephens
Umi 15h5'59" 71d0'54"
Bailey and Sydney Miller
Ori 5h38'31" -0d56'36"
Bailey Ann Hughes
And 23h25'53" 46d29'55"
Bailey Anne Golub
Ari 3h6'3" 29d22'4"
Bailey Aspen Thompson
Psc 0h24'34" 15d58'51"
Bailey Blackwell
Uma 9h4'13" 71d20'25"
Bailey Bob
Sco 16h11'57" -15d47'36"
Bailey Bylciw
Lyr 18h41'49" 38d14'10"
Bailey Caroline
And 0h37'47" 31d27'21"
Bailey Charles Reynolds
Uma 13h47'4" 53d30'45"
Bailey Christine Bennett
Sgr 18h13'2" -28d9'43"
Bailey Danielle Brame
And 1h37'17" 42d50'46"
Bailey David
And 1h16'30" 43d9'51"
Bailey Day
Lyn 8h11'3" 35d6'54"
Bailey Dexter
Uma 13h41'37" 55d41'55"
Bailey Diffley Lansing
Uma 10h28'28" 68d53'38"
Bailey Dimond Whistler
Psc 1h30'8" 13d0'45"
Bailey Elizabeth
Gem 7h6'8" 29d15'31"
Bailey Elizabeth Ann
Vir 12h57'53" 11d39'52"
Bailey Elizabeth Blackmon
And 0h36'27" 45d5'11"
Bailey Elizabeth Jayne
Vir 12h52'45" -7d25'46"
Bailey Erin Schmied
Leo 11h28'46" 2d9'18"
Bailey Evan Daniels
Cnc 8h20'16" 10d49'50"
Bailey Evan Gordon
Uma 9h31'4" 45d38'57"
Bailey Fink
Per 3h19'22" 47d41'0"
Bailey Genevieve
Vir 12h49'55" 10d52'29"
Bailey Gibart
Lyn 6h55'18" 57d42'35"
Bailey Grace Poort
Crb 15h47'22" 36d17'8"
Bailey Holloway
Uma 11h39'43" 44d57'58"
BAILEY HUNT
Cnc 8h32'53" 17d41'49"
Bailey Irene Drager Santana
Uma 10h32'57" 49d38'14"
Bailey Jane's Star
And 23h5'19" 36d1'14"
Bailey Jay
Cap 21h13'35" -15d26'49"
Bailey Jay Smith
Cyg 19h57'31" 38d48'16"
Bailey Jean
Lib 14h51'0" -5d1'46"
Bailey Joan Nitz
Leo 10h14'36" 8d15'42"
Bailey John Edward Alton Wright
Aqr 22h33'26" -22d36'52"
Bailey Karl J.C. Jordan
Cep 21h12'14" 65d8'6"
Bailey Katherine Fitzgerald
And 2h14'36" 47d2'58"
Bailey Kay
Aqr 22h56'27" -6d31'53"
Bailey Laura
Cas 23h37'42" 55d23'45"
Bailey Leona Radivan
Ori 5h55'18" 18d21'5"
Bailey Lucas John
Her 17h16'14" 14d37'5"
Bailey Lynn Knehr
Lyn 7h10'29" 53d3'23"
Bailey Lynn Roberts
And 1h15'18" 49d10'12"
Bailey M. Glidden
Aur 7h26'50" 40d22'57"
Bailey Mackenzie Natalis
Pho 1h4'19" -46d16'43"
Bailey Makenna White
Cas 3h16'0" 70d17'21"
Bailey Makenzie Maguire
Psc 0h13'41" -3d33'7"
Bailey Mariah
And 0h26'53" 32d33'59"
Bailey Marie Barbounis
Leo 9h29'21" 7d11'16"
Bailey Master of the Games
Leo 11h21'3" 20d57'34"
Bailey McDill
Uma 10h26'15" 46d7'19"
Bailey Michelle Whitworth
Lyn 8h30'24" 55d39'18"
Bailey Morgan
Lyn 8h20'23" 52d38'36"
Bailey Nicholas Mills - 18.06.05
Gem 7h3'45" 14d48'46"
Bailey Nicole Fowler
And 0h14'50" 45d24'42"
Bailey Nicole Gastelum
Lyr 18h42'17" 37d30'42"
Bailey Nicole Haro
Peg 22h36'19" 26d50'30"
Bailey Olivia Soileau
Cas 0h22'49" 52d11'30"
Bailey Paige Curtis
Gem 6h58'52" 29d16'22"
Bailey Patricia Gillette
Lib 15h20'40" -10d58'11"
Bailey Rae
Umi 16h19'49" 76d1'28"
Bailey Raeann Gray
Mon 6h48'24" 3d29'2"
Bailey Ray Hofmeester
Gem 8h5'14" 31d26'51"
Bailey Reale
Aqr 20h44'3" 1d13'38"
Bailey Robertson
Lyn 7h20'47" 44d47'49"
Bailey Rose
Gem 6h29'53" 20d11'22"
Bailey Rose Antal
Cma 6h43'55" -13d33'29"
Bailey Roth
Crb 15h44'2" 27d41'57"
Bailey Sage Hosterman
Lyn 8h40'22" 33d27'32"
Bailey Savannah Dozier
Vir 12h41'21" -8d43'1"
Bailey & Shannon's Eternal Light
Cyg 21h16'15" 46d30'36"
Bailey Terkel
Uma 13h12'38" 58d10'48"
Bailey Thomas Flaskett
Ori 6h18'10" 15d13'51"
Bailey Thomas Graham
Cnc 7h56'6" 18d19'47"
Bailey Tony Pearson
Per 3h36'25" 42d9'38"
Bailey Willow Cummings
Lyn 8h41'1" 45d42'28"
Bailey Woodward
Aql 19h28'57" 3d23'28"
Bailey Yvonne Larson
And 2h26'19" 37d34'50"
Bailey-Elliott Promise
And 23h29'8" 48d36'58"
Bailey-Rae Emily Cogan
And 23h1'11" 50d16'41"
Baileys Comet:Pa's Punkin Jaime Sue
And 2h19'41" 42d14'9"
Bailey's kiss to marriage bliss
Cyg 21h15'43" 46d48'55"
Bailey's PaPa
Tau 3h35'34" 25d47'32"
Bailey's Sadalbari
Uma 11h45'51" 40d27'29"
Bailey's Shining Star 22-7-04
Cra 18h4'9" -37d27'28"
Bailey's Star
Dra 19h31'2" 67d10'43"
Bailey's Star 12/9/2005
Vir 14h15'7" -5d11'52"
bailissamo
Lyn 8h2'22" 33d56'54"
Baily Marie Jones
Crb 15h53'59" 34d5'52"
Baird Sister's Crackerjack Star
Ori 4h53'46" 1d53'10"
Baird's K. K. I. S.
Dra 17h10'2" 51d43'13"
BAIR'S DEN
Uma 9h20'28" 47d43'7"
baj
Cas 0h14'29" 57d15'28"
BAJA
Ant 10h51'49" -36d26'2"
Bajáky András Olivér
Sgr 19h25'32" -30d47'2"
Bajigirl
And 0h27'58" 37d23'16"
Bajupa
Lyn 6h24'10" 61d4'7"
Bak Middle School of the Arts
And 0h19'39" 28d51'9"
Baka
Lmi 10h42'58" 31d17'59"
Bakary
Sco 17h1'0" -40d36'57"
Baker 1
Lib 15h7'22" -12d15'25"
Baker Bedgood
Her 17h2'8" 46d5'4"
Baker Jackson Goldsmith
Leo 11h5'10" 12d41'39"
Baker Keevak
Leo 9h40'22" 22d33'51"
Baker Kindergarten 2004-05
Uma 11h29'49" 63d44'44"
Baker-Friederich Shining Light
Cyg 20h26'33" 46d17'2"
Baker's Lovelight
Cap 21h3'49" -26d17'17"
Bakhta
Cnc 8h49'35" 6d51'42"
Baktygul Bulatkali
Psc 1h46'39" 19d36'56"
Balázs 10
Tau 4h38'56" 10d25'21"
Balbeanus The Barbarian
Boo 14h37'35" 51d18'56"
BALCA
Leo 10h27'55" 23d22'15"
bald eagle hem
Uma 9h24'24" 67d33'27"
Baldassini Famiglia
Hya 9h20'44" 3d31'6"
Baldisch
Uma 12h1'21" 32d6'32"
Baldo Contreras
Cnc 8h0'38" 15d29'14"
Baldo Talevi
Uma 10h43'56" 50d12'33"
Baldonion Benjamin 101097
Lib 15h1'49" -1d3'32"
Baldrich, Wolfgang
Uma 10h55'24" 42d10'51"
Baldwin BOE
Her 16h20'30" 18d13'35"
Bali Zsuzsanna
Aqr 21h39'29" -0d24'15"
Balie Dae Villagomez
Cyg 20h30'56" 34d8'19"
Baliga-Ostasiewski
Cma 6h58'10" -16d47'38"
B.A.L.K.S.
Ori 6h4'53" 15d43'21"
Ballalee
Uma 13h22'7" 62d51'17"
Ballard 777
Cnc 8h51'55" 29d29'7"
Ballard & Walker Union
Cyg 22h1'54" 51d17'42"
Ballario Giorgia
And 0h46'12" 45d53'30"
Ballentine Family
Sge 19h43'50" 16d54'48"
Ballentine-McCarthy
Cnc 8h43'36" 16d24'24"
Ballion
Tau 4h35'36" 12d22'57"
Bálló Katalin
Leo 9h23'52" 12d6'39"
Ballsburg
Uma 12h5'31" 39d53'8"
Ballslinger
Her 17h5'25" 33d30'1"
Bally & Shannon
Lyr 19h0'36" 42d14'34"
BaLo Bleck
Uma 10h22'24" 40d27'49"
Balog Zoltán 78
Uma 12h2'45" 55d39'47"
Baloo Bear
Uma 12h0'17" 58d46'20"
Baloo Griffin
Cnc 7h56'9" 18d41'12"
Balszuweit, Bert
Uma 8h48'44" 47d38'41"
Balthasar Weiss
Cyg 19h55'14" 51d44'49"
Baluchon - Luc Boutaud
Ori 6h0'20" 21d12'6"
Baluga
Cet 1h12'18" -0d36'15"

Balyara
Dra 17h38'53" 59d34'20"
Balz, Werner
Uma 8h12'40" 62d33'41"
Balze
Uma 11h20'59" 63d56'40"
Balzer, Jens
Ori 5h33'3" 10d4'1"
BAM*
Lmi 11h6'26" 31d6'47"
BAM
Cap 20h32'5" -21d16'36"
BAM
Cma 6h30'27" -31d7'4"
Bam Bam
Psc 0h0'8" 4d23'8"
Bam Bam
Umi 16h22'18" 70d46'38"
Bam Bam
Lyn 6h51'13" 57d11'52"
Bam Bam
Aqr 21h10'55" 1d15'24"
Bam Bam Sweetpea
Dra 17h6'43" 68d26'31"
BAM Brittany Ann Milroy
And 1h37'5" 41d56'19"
B.A.M. - Brooke And Martin
And 22h58'5" 48d4'8"
BAM (Samantha)
Leo 11h43'8" 25d46'55"
Bama
Cnv 12h36'37" 51d33'4"
Bama Rae
Uma 11h56'15" 60d29'47"
Bama's Star
Tau 4h31'58" 2d4'54"
BamBam and Pookie
Sge 20h2'34" 16d52'59"
Bamba's Star
Cyg 19h36'28" 48d4'40"
Bambi
Cyg 20h50'40" 31d38'32"
Bambi
Uma 11h34'34" 43d42'13"
Bambi
Peg 23h2'45" 29d52'2"
Bambi
Ari 1h53'1" 19d1'31"
Bambi
Umi 13h9'7" 87d34'43"
Bambi Lyn
Umi 17h34'46" 89d10'10"
Bambi My Dear Love
Uma 9h22'25" 65d39'35"
Bambi S. Marzlo
Uma 11h1'2" 46d17'5"
Bambi Westerman
Cnc 8h33'39" 29d23'28"
Bambia & Bouchra pour la vie
Sgr 19h19'21" -20d51'59"
Bambie
Sco 17h20'45" -32d40'0"
Bambina 10
Umi 15h46'38" 77d30'43"
Bambine
Ori 5h23'50" 4d59'43"
Bambino
Del 20h24'33" 9d43'31"
Bambino
Per 3h37'39" 40d35'26"
Bambino 63
Umi 15h29'45" 76d50'42"
Bambolina - (Valentina Bisantino)
Umi 13h5'44" 70d15'3"
Bamboo
Uma 13h55'20" 56d10'44"
Bambrick
Lyn 8h31'40" 33d47'19"
Bamer's Beauty
Uma 9h38'9" 62d29'55"
Bamidele[Dale] Ajiboye
Leo 11h37'54" 24d10'38"
Bamma Phi Bamma
Cap 20h9'43" -25d14'20"
Bammie
Vir 13h26'49" -12d45'8"
Bampi Len
Lmi 10h39'10" 29d43'24"
Bampy
Her 18h17'30" 19d15'2"
BAMrONE
Dra 17h53'6" 57d50'46"
Bana
Cyg 21h40'44" 41d38'3"
Banacek
Cyg 21h37'9" 50d23'52"
Banacki
Ori 4h48'51" -0d23'3"
Banana Bean
Uma 11h59'44" 38d53'7"
Banana Bread Brandon
Tau 4h15'36" 11d10'4"
Bánáti Tibor
Sco 16h11'43" -35d27'1"
Band of Gold
Cap 20h24'10" -27d10'20"
Bandage
Aql 19h17'31" -7d2'51"
Bandalo, Margarete
Uma 10h47'42" 57d8'1"
Bandar April 24-2005
Tau 4h34'23" 6d23'12"
Bandele
Lyn 8h8'53" 48d2'34"
Bandet, Queen of Jackson Avenue
Uma 10h38'16" 40d57'26"
Bandinelli
Aql 19h17'20" 9d40'48"
Bandir, Abdulahat
Ori 6h7'24" 5d57'54"
Bandit
Uma 9h3'36" 67d8'43"
Bandit
Cma 7h26'47" -18d39'45"
Bandit
Lib 15h29'17" -12d32'10"
Banditos 5
And 2h14'6" 49d9'40"
Bandit's Heart
Cnv 12h40'41" 38d4'17"
Bane Valiant Crain
Per 3h6'51" 37d54'3"
Baner's Blazing Ball
Ori 5h24'51" 1d33'16"
Banessa
Gem 6h52'13" 21d50'48"
Banga
Lyr 19h20'16" 29d38'3"
Banger D.
Aur 5h35'41" 43d37'28"
Banger P.
Umi 14h33'25" 82d28'13"
Banjo
Lib 15h49'19" -11d26'10"
Banjo Femi
Aqr 22h5'14" -16d12'3"
Banner Bread Indeed!
Cyg 21h30'38" 33d7'56"
Bannier, Hinrich
Uma 12h15'13" 62d6'44"
Banning Capps
Vir 13h39'54" 3d23'32"
Banno's Shining Star
Tau 4h16'13" 21d4'22"
Banphrionsa
Cap 20h49'33" -15d12'28"
banphriosna lola
Tau 4h10'9" 26d31'5"
BaNsHeE
Lmi 10h42'16" 26d18'32"
Banter in Love
Cyg 21h7'55" 46d43'53"
Banumpkin
Cep 0h8'2" 71d28'29"
Banville
Pho 23h53'41" -45d29'1"
Bao-Chau Nguyen
Lyn 6h27'9" 58d34'12"
Ba-Pa Smurf
Cap 21h48'12" -10d35'38"
Bapi
Cep 21h43'13" 61d49'27"
Bappert, Jürgen
Ori 6h17'32" 16d36'9"
Baptysta
Umi 15h45'58" 85d27'10"
Bar Bra Patocka
Lyn 6h57'54" 60d51'41"
Bar Mitzvah of Jeremy Sichel
Leo 11h34'36" -2d4'59"
Barabara Ann
Cas 2h13'47" 62d23'30"
Barabas, Kurt
Leo 10h24'4" 19d20'55"
Barak
Peg 22h23'22" 7d24'0"
Baralihon
Pho 2h21'14" -40d37'41"
Barancsa
Sgr 18h10'23" -17d10'24"
Baranya Edit
Aqr 22h56'27" -7d48'16"
Barb
Aqr 22h2'57" -19d27'31"
Barb and Ethan
Cyg 19h35'20" 35d32'39"
Barb Aschettino
Cam 7h30'13" 74d29'5"
Barb Breen
Cnc 8h42'46" 27d52'20"
Barb & Brian: Written in the stars
Uma 11h37'6" 47d0'47"
Barb Bubla
Uma 9h39'5" 68d34'26"
Barb Croisant
Ori 4h56'30" -2d43'0"
Barb & Donna's Cowboy
Uma 12h37'46" 54d23'15"
Barb Downey
And 0h27'27" 36d16'11"
Barb Ehr
Lac 22h47'56" 43d35'44"
Barb Enos
Cnc 9h14'24" 16d27'22"
Barb & Ev Cutter's 50th Anniv Star
Cyg 21h40'1" 54d3'12"
Barb - Fred Perkins
Lyr 19h12'37" 26d31'50"
Barb & Glenn Gelman Happy 25th 2004
Cyg 19h30'13" 57d20'48"
Barb "Granny Goodwitch" Ashford
Sco 17h52'38" -35d51'42"
Barb Haverkos
Tri 2h7'56" 35d5'41"
Barb & Joe Forever
Cyg 21h54'46" 49d4'3"
Barb Korak
Leo 9h52'56" 12d55'29"
Barb Lee
Uma 10h26'25" 54d41'16"
Barb Lewandowski
Aqr 22h22'38" -4d24'0"
Barb Mastriana's personal star
Cas 1h23'16" 55d37'0"
Barb Mossop Hamilton
Aqr 23h0'46" -7d59'46"
Barb My Most Shining Star
Per 2h46'22" 43d22'2"
Barb Piek Kaldenberger
Cas 1h39'2" 61d53'27"
Barb Powel
Her 17h15'55" 23d7'21"
Barb Radziewicz
Her 16h39'0" 6d12'16"
Barb Reinhold
Cyg 20h0'28" 33d27'10"
Barb Sciortino
Cas 0h26'12" 63d0'14"
Barb Welshofer
And 0h37'10" 41d3'49"
Barb Wyatt
Vir 12h49'43" 7d11'7"
Barbadan
Cap 21h45'33" -8d34'36"
Barbalicious
Lyr 18h48'17" 34d17'42"
BarbaPapa
Leo 10h55'28" 24d43'54"
Barbar J Roper
Cap 20h33'6" -18d33'42"
Barbara
Lib 14h43'6" -21d2'50"
Barbara
Cap 21h38'57" -14d48'49"
Barbara
Lib 15h0'43" -13d55'21"
Barbara
Aql 20h10'51" -0d20'34"
Barbara
Lib 14h54'55" -5d7'8"
Barbara
Lib 14h50'0" -2d29'42"
Barbara
Lib 14h53'35" -1d17'35"
Barbara
Vir 13h25'45" -3d51'50"
Barbara
Vir 11h45'36" -2d50'48"
Barbara
Cas 0h26'25" 62d22'35"
Barbara
Cas 0h29'2" 62d23'7"
Barbara
Cas 23h32'57" 57d55'26"
Barbara
Dra 17h9'33" 55d40'34"
Barbara
Uma 12h35'51" 57d26'45"
Barbara
Uma 8h19'49" 65d16'53"
Barbara
Umi 14h15'7" 66d53'2"
Barbara
Sgr 18h27'34" -32d23'44"
Barbara
Sgr 18h36'29" -24d4'3"
Barbara
Sco 17h48'14" -39d33'22"
Barbara
Ori 5h55'13" -8d20'8"
Barbara
Her 16h42'2" 28d59'37"
Barbara
Boo 15h7'39" 39d0'7"
Barbara
Crb 16h24'12" 28d28'27"
Barbara
Gem 7h21'32" 18d38'25"
Barbara
Del 20h38'58" 15d13'16"
Barbara
Ori 6h13'15" 5d40'46"
Barbara
Vir 13h26'26" 11d26'40"
Barbara
Lyr 18h43'31" 35d7'58"
Barbara
Aur 6h25'43" 30d21'52"
Barbara
Lyn 8h30'35" 35d48'13"
Barbara
Per 3h5'54" 44d34'3"
Barbara
Per 4h39'24" 41d0'7"
Barbara
And 23h11'57" 48d11'49"
Barbara
And 23h34'12" 38d55'10"
Barbara
And 23h22'51" 41d43'42"
Barbara
Lyr 19h26'54" 37d44'15"
BARBARA 030101
Lac 22h52'47" 53d52'8"
Barbara 1013
Lib 15h22'32" -9d29'21"
Barbara 3-24
Ari 3h25'50" 25d54'42"
Barbara A. Bassler
Uma 10h0'46" 47d52'12"
Barbara *A Beautiful Sister* Handy
Vir 13h8'1" 9d26'26"
Barbara A. Beer
Uma 10h41'40" 63d20'7"
Barbara A. Budd-Klauber
Leo 11h8'3" 15d5'44"
Barbara A. Davis
Vir 13h30'44" -2d4'3"
Barbara A. Del Nero
Peg 21h47'30" 7d54'51"
Barbara A. Gousha
Crb 15h53'18" 38d47'34"
Barbara A. Henninger
Cam 7h9'49" 79d44'34"
Barbara A. Korkan
And 0h12'27" 46d8'47"
Barbara A. Lieto
Tau 3h39'52" 18d57'52"
Barbara A. Lis
Cnc 8h9'36" 18d21'53"
Barbara A Neimeyer
Crb 15h55'45" 28d29'27"
Barbara A Pelchat
Uma 9h11'2" 48d4'49"
Barbara A Robinson
Uma 13h47'46" 56d6'49"
Barbara A. Smith
Aqr 22h27'55" -6d22'9"
Barbara A. Vines
Vir 12h23'26" -5d10'10"
Barbara A Weber
Ari 3h14'51" 30d56'43"
Barbara Ackerly
Aqr 23h54'30" -13d43'29"
Barbara Adamski
Uma 10h29'33" 62d57'5"
Barbara Agricola and Chris Harden
Aqr 22h14'13" -21d49'50"
Barbara and Bruce Reyle
Cyg 22h0'38" 53d38'46"
Barbara and CC White
Leo 11h26'12" 6d14'6"
Barbara and Danny
Her 18h52'10" 24d13'11"
Barbara and Dorsey 50th
Cyg 21h3'29" 50d27'34"
Barbara and George Kanas
Cyg 20h22'27" 55d26'13"
Barbara and Jack McMillan
Lyr 18h48'35" 30d45'37"
Barbara and Lonie Rudd
Sge 19h39'32" 17d6'12"
Barbara and Roberta Friends Forever
Cyg 20h22'25" 59d20'21"
Barbara and Roz
Sgr 18h51'7" -28d42'33"
Barbara and Val Jean Thurmon
Per 4h42'31" 43d47'20"
Barbara and Val Whitnum
And 0h11'55" 29d2'46"
Barbara and Walter Wood
Her 17h51'31" 27d2'11"
Barbara Anderson
Crb 15h43'54" 26d4'31"
Barbara & Andrew Blumberg
Psc 1h13'31" 26d39'21"
Barbara "Angel" Dufresne
And 23h1'39" 50d39'53"
Barbara Angelica
And 0h14'45" 45d44'9"
Barbara Ann
And 2h20'50" 44d22'3"
Barbara Ann
Crb 15h42'57" 33d5'41"
Barbara Ann
Gem 7h1'53" 26d55'19"
Barbara Ann
Peg 22h31'49" 10d15'16"
Barbara Ann
Psc 1h35'56" 20d55'6"
Barbara Ann
Ori 6h3'59" 13d44'38"
Barbara Ann
Cap 20h24'42" -24d10'34"
Barbara Ann
Sgr 18h0'38" -30d8'56"
Barbara Ann
Sco 17h3'43" -38d38'57"
Barbara Ann
Uma 8h56'30" 58d38'28"
Barbara Ann
Cam 5h20'23" 60d44'24"
Barbara Ann
Sgr 18h7'43" -22d1'56"
Barbara Ann 8-21-1946 to 3-23-1993
Dra 12h52'10" 71d9'14"
Barbara Ann Bammer
Gem 6h12'23" 24d35'39"
Barbara Ann Brayman
Cas 1h37'39" 65d22'37"
Barbara Ann Clark
Leo 11h46'6" 21d20'37"
Barbara Ann Crask Godbey
Tau 5h47'19" 21d10'29"
Barbara Ann Cypert
Lyn 8h17'17" 56d51'1"
Barbara Ann Degretta Monborne
Psc 23h6'58" 7d21'17"
Barbara Ann Donetz
Cas 23h26'58" 57d45'20"
Barbara Ann Ducharme
Sgr 18h38'55" -29d16'57"
Barbara Ann Dyas-Albrecht
Sco 16h15'41" -16d58'1"
Barbara Ann Earles
Aqr 22h18'47" -12d57'27"
Barbara Ann Egbert
Crb 15h51'18" 26d29'8"
Barbara Ann Fleming
And 2h33'33" 38d56'14"
Barbara Ann Grausam Lumia
Uma 13h1'0" 53d30'37"
Barbara Ann Haig
Uma 8h17'53" 65d3'21"
Barbara Ann Harris
Cnc 8h40'59" 23d52'34"
Barbara Ann Hurley
Lyr 18h21'8" 45d28'53"
Barbara Ann Kain
Cas 2h42'38" 66d10'9"
Barbara Ann Leonetti
Gem 6h48'17" 18d4'15"
Barbara Ann Lexa
Sco 16h33'9" -37d49'45"
Barbara Ann Lynch O'Connell
Lib 15h2'4" -2d26'31"
Barbara Ann Massey
Ari 2h46'18" 15d42'27"
Barbara Ann Morris Jensen
Gem 6h39'14" 22d11'14"
Barbara Ann Morvel
Lyr 19h9'27" 45d6'20"
Barbara Ann Myers
Gem 6h50'56" 16d0'0"
Barbara Ann Notholt
Mon 6h33'27" 8d13'11"
Barbara Ann Payne
Lmi 9h30'21" 34d52'44"
Barbara Ann Phillips
Psc 1h50'50" 9d48'28"
Barbara Ann (Pooh Bear)
Cnc 8h38'33" 28d59'39"
Barbara Ann Rachar
Psc 1h10'27" 22d28'33"
Barbara Ann Randall
Umi 14h40'19" 88d40'33"
Barbara Ann 'Robin' Hazuka
And 0h53'2" 35d35'44"
Barbara Ann Russell
Uma 10h7'55" 65d1'6"
Barbara Ann Savini
And 2h3'25" 38d46'0"
Barbara Ann Schurmann
Leo 10h52'15" 12d9'27"
Barbara Ann Simon "Ladybug"
Tau 3h49'48" 17d47'24"
Barbara Ann Sposato
Leo 11h5'52" 21d36'38"
Barbara Ann St. Clair
Cas 0h27'26" 53d26'53"
Barbara Ann Sternquist
Sgr 20h22'35" -29d49'25"
Barbara Ann Stevenson
Crb 16h8'44" 39d1'20"
Barbara Ann Wruck Love Petty Family
Leo 9h57'20" 8d1'42"
Barbara Ann Zoglio
Uma 9h13'24" 47d9'10"
Barbara Anne
Cas 0h14'35" 52d57'43"
Barbara Anne
Aqr 22h10'54" 1d19'17"
Barbara Anne
Cas 1h37'50" 64d6'25"
Barbara Anne
Uma 9h8'16" 58d23'13"
Barbara Anne Griffin
Leo 11h25'18" 26d8'6"
Barbara Ann's Eternal Light
Cap 20h36'15" -26d53'18"
Barbara Arciello
Sco 16h5'1" -8d47'16"
Barbara Atkins
Cyg 20h3'57" 37d33'36"
Barbara B. Fawkes
Aqr 22h9'57" -22d44'13"
Barbara B Stafford
Cap 20h29'22" -20d3'5"
Barbara B Turco
Psc 23h21'4" 7d32'6"
Barbara B Williams
Aqr 22h18'41" 2d8'16"
Barbara B.A. Doran
Mon 6h42'12" -0d1'19"
Barbara Baby Bear
Vir 12h50'33" 4d12'18"
Barbara Bailey
Cas 1h15'4" 64d30'19"
Barbara Bailey Johnston
Peg 22h37'52" 27d3'19"
Barbara Bannwart
Cas 1h50'3" 59d10'38"
Barbara "Barb" Driscoll
Gem 6h55'38" 15d54'3"
Barbara & Barry Lewis
Cyg 19h41'49" 39d10'11"
Barbara Bartnicki
Uma 11h4'27" 38d12'32"
Barbara Bates Ross
And 0h43'21" 41d18'35"
Barbara Beck Abell
Ori 5h12'11" 5d55'51"
Barbara Beck,
Dra 19h30'56" 74d9'32"
Barbara Bell Coleman
Lyn 8h7'34" 52d24'27"
Barbara Benjamin-Creel
And 1h51'37" 39d17'1"
Barbara Bennett Augustine
Uma 10h35'43" 54d39'9"
Barbara Benson "Barb's Beauty"
Cap 21h16'34" -16d12'53"
Barbara Bergmann
Umi 15h28'11" 76d37'29"
Barbara Bertolini
Sco 16h44'49" -27d3'8"
Barbara Bessey
Cyg 20h42'5" 47d41'4"
Barbara - Best Mum in the Universe!
Car 10h15'54" -63d17'51"
Barbara Bibbo
Lib 15h0'36" -14d13'30"
Barbara (Bibi) Bonilla
Cas 1h15'6" 51d6'0"
Barbara Binney Dexter Cianelli
Cas 0h27'30" 61d33'43"
Barbara & Bob Renzi
Eri 3h54'26" -11d8'39"
Barbara "Bobbie"
Gem 7h5'17" 28d41'45"
Barbara Bowen
Vir 12h37'46" 10d7'44"
Barbara Bowers
Cas 2h51'17" 60d22'53"
Barbara Boyd Gould
Cap 21h4'49" -20d24'44"
Barbara Brazilian
Uma 9h36'49" 48d26'39"
Barbara Brennan
Cas 0h50'41" 56d25'42"
Barbara Brennan
Lmi 11h6'37" 25d33'29"
Barbara Brockbank
Leo 9h43'43" 27d59'34"
Barbara Bromm Dalton
Lib 15h33'9" -5d42'39"
Barbara Brooks Andersen
Gem 6h42'50" 26d57'14"
Barbara Brown Lemmond
Sco 16h23'51" -31d14'37"
Barbara Brown Skala
Cas 0h0'26" 57d1'20"
Barbara Brzozowski
Cas 0h24'12" 57d10'57"
Barbara Buckle-Saulnier
Sgr 19h39'55" -19d55'30"
Barbara Budd's Star
Cas 1h4'25" 54d4'47"
Barbara Burton Graf
Uma 9h44'46" 54d15'35"
Barbara Bush
Cap 20h57'34" -19d50'16"
Barbara Bush Elem HISD 2005-2006
Peg 21h32'51" 12d18'19"
Barbara Bussani
Aqr 22h5'10" 1d51'13"
Barbara Bustos Ruz
Psc 0h15'15" 16d19'1"
Barbara C.
Lyn 8h36'43" 40d57'18"
Barbara C. Shehorn
And 0h18'59" 43d56'49"
Barbara Cady
Crb 15h47'13" 37d31'33"
Barbara Calhoun Murphey
Cyg 21h25'20" 36d10'14"
Barbara Cameron
Cas 0h22'38" 59d3'18"
Barbara Campbell
And 23h14'47" 50d59'22"
Barbara Carabetta
Lep 5h16'27" -11d3'32"
Barbara Carol
Cas 1h32'9" 59d40'28"
Barbara Chimento
Sco 16h7'23" -12d28'16"
Barbara & Chrigu
Her 18h4'50" 17d41'39"
Barbara Christine Bechler
Ari 2h39'34" 12d31'24"
Barbara Christine Craft
Psc 0h45'0" 20d51'46"
Barbara Clague Best Mother-In-Law
Cyg 19h30'44" 51d34'33"
Barbara Claire
And 0h24'57" 42d10'36"
Barbara Clarke
And 2h11'29" 41d52'43"
Barbara Clayton Gilligan
Umi 14h43'23" 79d13'43"
Barbara Colello
Umi 14h14'46" 76d17'29"
Barbara Colton
Crb 16h14'51" 27d38'11"
Barbara Cray
Uma 8h34'15" 59d41'48"
Barbara Crites
Cas 23h37'28" 54d31'2"
Barbara Crozier Couch
And 1h14'32" 43d21'30"
Barbara & Cyrill forever
Umi 14h54'14" 68d38'8"
Barbara D. Goodman "Herman"
Cas 1h20'9" 54d7'15"
Barbara D. Marino
Umi 15h35'3" 73d44'42"
Barbara D. Rodier
Uma 11h45'1" 63d13'38"
Barbara Daitch
Cam 5h18'10" 71d24'55"
Barbara Darney
Uma 9h51'13" 62d55'14"
Barbara Davidson - 25th September 1941
Lib 14h31'43" -19d32'51"
Barbara Davis
Cas 1h20'49" 61d38'39"
Barbara Day
Vir 12h22'50" 11d13'11"
Barbara Delehunt
Vir 11h53'35" -2d39'37"
Barbara Denise Tolar Gentry
Sgr 19h45'23" -34d37'9"
Barbara DeWaters
Uma 9h42'57" 70d50'35"
Barbara Diane Camune
Cas 0h38'2" 53d51'54"
Barbara Dickens
Gem 6h40'43" 30d34'38"
Barbara Di'Orio
Aqr 21h18'39" 1d12'23"
Barbara Dodd
Cas 1h21'3" 52d10'3"
Barbara Dombrowski Doyle
Aqr 21h30'24" -7d58'55"
Barbara Donker
Her 16h51'7" 17d9'8"
Barbara Dziekan
Her 17h37'20" 41d36'50"
Barbara E. Kerr
Cap 21h36'25" -13d29'37"
Barbara E. Schwinn
Uma 11h53'11" 60d17'10"
Barbara e Stefano
Her 18h37'34" 20d17'47"
Barbara E. Zimmermann
Lac 22h57'29" 44d41'13"
Barbara Eddy Fell
Crb 16h8'10" 26d38'59"
Barbara Eileen Testa
Uma 13h42'8" 61d2'4"
Barbara Elizabeth
Lib 15h21'36" -6d5'28"
Barbara Elizabeth Cahill
Cas 0h11'7" 57d14'57"
Barbara Elizabeth Smith
Cyg 21h41'53" 42d33'41"
Barbara Ellen Hayes, the Motherstar
And 1h15'37" 37d38'32"
Barbara Ellen Jones
Crb 16h5'29" 36d53'15"
Barbara Ellen Voils Scott
Aqr 22h40'20" 0d28'7"
Barbara Ellen Weiner
Leo 10h26'19" 23d30'31"
Barbara Ellender Allen Godwin
Lib 15h24'38" -19d2'8"
Barbara Elway Rottle
Uma 12h21'48" 59d49'25"
Barbara Elzbieta Depta - My Misia
Lyn 8h21'5" 34d41'20"
Barbara & Eric - Aug. 13 to 4ever
And 2h28'7" 43d16'9"
Barbara Estelle Tiska Vagrin
Lib 14h53'9" -2d47'41"
Barbara Eugenia Theresa Joan Biever
Vir 13h44'16" -9d44'33"
Barbara Eunice Allan
Lyn 9h4'12" 37d50'18"
Barbara Evanita Hill "Roni"
Uma 11h35'6" 52d6'49"
Barbara Evelyn Howitt Acey
Cap 20h22'25" -24d23'9"
Barbara Feldberg
Cam 3h33'50" 54d3'55"
Barbara Fern
Ari 2h51'29" 28d5'20"
Barbara Figone Austin
Lyn 8h55'56" 43d1'52"
Barbara Fisher
Cyg 21h34'49" 46d18'44"
Barbara Fletcher
Crb 15h28'46" 28d53'34"

Barbara Follett Schweger
Psc 1h13'16" 22d48'59"
Barbara Fowler 1974
Cru 12h55'1" -63d14'52"
Barbara & Francis Kilroy
Uma 9h12'13" 46d36'36"
Barbara Franco
Cyg 20h15'39" 40d33'51"
Barbara Franco
Lyr 18h53'9" 35d22'48"
Barbara Franklin's
Brilliance
Uma 9h51'25" 56d49'26"
Barbara Friedericke
Masters Schmidt
Crb 15h28'14" 27d39'30"
Barbara Frye Bass
Sgr 19h11'1" -21d8'0"
Barbara Funkhouser
Mon 7h45'38" -8d46'56"
Barbara G. Deatherage
Cas 0h5'16" 56d36'55"
Barbara Gammon
Uma 8h34'10" 47d15'7"
Barbara Gebell Sobel
Cas 1h58'9" 62d7'19"
Barbara Gill 60th
Commemorative Star
Aql 20h13'37" -6d20'11"
Barbara Gillespie
Cyg 19h44'16" 31d53'18"
Barbara Glassett
Cas 0h14'3" 60d59'11"
Barbara Glithero Taylor
Uma 11h31'38" 56d34'50"
Barbara Gnat
Psc 1h23'9" 33d3'39"
Barbara Goodson
Leo 10h6'49" 13d10'40"
Barbara Grace
Lyr 18h45'9" 38d40'12"
Barbara Green
Crb 15h44'49" 31d6'34"
Barbara Grossen
Umi 13h40'59" 70d10'49"
Barbara Guardado_Liliana
Robledo
Leo 11h19'57" 11d23'22"
Barbara Guastini
Tau 4h12'15" 26d56'5"
Barbara & Guilelmus Feci
Miraculum
Ori 6h2'49" 18d40'51"
Barbara H. Zimmerman
Cas 0h0'57" 53d56'58"
Barbara Haberland
Aqr 23h19'53" -13d23'21"
Barbara Hall 0301
Per 2h56'18" 53d39'38"
Barbara Hall Streeter
Cnc 8h9'32" 13d18'40"
Barbara Halprin
Cyg 20h1'41" 48d44'49"
Barbara Hamman
Psc 0h37'33" 17d50'26"
Barbara Hand Roland
Cas 1h2'13" 63d29'14"
Barbara Hansen
Cnc 8h12'54" 24d22'58"
Barbara Hardin
Lib 15h57'29" -6d2'8"
Barbara Häring
Her 16h26'47" 13d39'55"
Barbara Harlow
Lyn 8h2'53" 55d24'22"
Barbara Hasday
Cas 0h8'12" 60d57'8"
Barbara Helen Rothstein
Crb 16h13'17" 38d30'46"
Barbara Helmke A.K.A.
Babs
Psc 0h33'49" 9d53'7"
Barbara Hendrix
Aqr 22h38'41" -10d4'49"
Barbara Hobson Huff
Oph 17h14'4" -1d0'44"
Barbara Holloway Carmack
Aqr 22h18'54" 1d36'4"
Barbara "Honey" Lewis
Cas 1h16'2" 59d29'50"
Barbara Hrbek Zucker
Ari 2h50'37" 29d33'42"
Barbara Hudson "On Your
80th"
Lyn 7h45'7" 56d44'49"
Barbara Hyvert
Ori 5h45'38" 1d53'10"
Barbara Iervolino
Lyn 8h20'43" 35d49'33"
Barbara Irwin
Cyg 19h37'5" 30d21'8"
Barbara J. Baldino
Del 20h41'51" 11d11'42"
Barbara J Coffland
Cyg 21h24'25" 34d15'3"
Barbara J. Conroy
Cyg 20h0'29" 39d14'2"
Barbara J. Epley
Peg 22h37'3" 25d49'41"
Barbara J. Giunta
Sgr 19h19'10" -15d17'8"
Barbara J. Johnson
Sco 16h13'20" -18d58'55"
Barbara J. Johnson-
Brodsho
Leo 11h24'34" 16d44'0"
Barbara J. Jordan
Crb 15h48'2" 35d27'57"
Barbara J. Koirtyohann
Cas 0h17'39" 58d55'27"
Barbara J. Lauer
And 1h51'22" 46d48'21"
Barbara J. Pegues ("Nan")
Uma 13h47'12" 61d53'22"
Barbara J. Randolph
Dra 15h40'24" 58d48'58"
Barbara J. Ryan
Uma 12h1'14" 58d23'5"
Barbara J. Simone
Psc 1h32'9" 3d15'46"
Barbara Jackson Altum
Sgr 19h34'19" -37d39'27"
Barbara James
Vir 12h52'29" -7d27'51"
Barbara & James Grant
Per 4h25'25" 40d6'0"
Barbara JAMG Armenta
Vir 13h40'50" -16d9'51"
Barbara Jane Rush
Lmi 10h42'56" 33d42'6"
Barbara Janedy Foley
Leo 9h43'26" 25d19'5"
Barbara Jaquish
Cas 1h19'4" 63d51'42"
Barbara Jean
Cas 0h37'18" 66d20'11"
Barbara Jean
Cru 12h21'32" -56d53'22"
Barbara Jean
Sgr 19h18'35" -22d35'22"
Barbara Jean
Gem 6h5'22" 24d15'48"
Barbara Jean
Mon 6h49'10" 8d25'59"
Barbara Jean
Aqr 22h14'53" 1d39'20"
Barbara Jean
Aqr 22h10'1" 1d17'21"
Barbara Jean
And 2h14'4" 46d3'34"
Barbara Jean 8
Ari 2h39'28" 27d14'43"
Barbara Jean (AKA
ZooGee)
Cas 0h46'25" 57d32'36"
Barbara Jean Berti
Vir 13h2'10" -3d29'9"
Barbara Jean Bradfield
Lyn 7h53'17" 39d15'36"
Barbara Jean Cannon
Crt 11h17'39" -21d28'10"
Barbara Jean Durso
Lyn 8h16'49" 56d34'58"
Barbara Jean Heath
And 23h10'21" 47d57'45"
Barbara Jean Hendrix
Crb 15h35'51" 38d8'40"
Barbara Jean Hottenstein
Cyg 21h45'29" 53d32'18"
Barbara Jean Howell
Sgr 19h16'47" -22d37'17"
Barbara Jean Knutson
Crb 15h52'12" 26d23'37"
Barbara Jean Kuchen
BarB
And 0h40'33" 44d20'57"
Barbara Jean Lake
03/26/56-03/24/06
Ori 5h25'13" 9d22'37"
Barbara Jean Lanci
Cas 23h37'16" 52d34'58"
Barbara Jean Marie
Psc 0h23'25" 16d6'6"
Barbara Jean Murray
Vir 12h41'33" -0d0'33"
Barbara Jean Price
Psc 22h55'16" 4d36'45"
Barbara Jean Santomieri
Leo 11h49'17" 25d10'41"
Barbara Jean Smith
And 23h12'57" 51d53'37"
Barbara Jean Traver Oliver
L.V.N.
Lyn 6h44'4" 51d18'31"
Barbara Jean Tripodi
Psc 0h28'34" 15d11'23"
Barbara Jean Voedisch
And 1h31'59" 36d45'21"
Barbara Jean Wiley
Leo 10h40'54" 7d10'22"
Barbara Jeanette Johnson
Sgr 18h23'41" -25d30'53"
Barbara Jeanne Humes
Tau 4h23'11" 26d2'0"
Barbara Jeanne Jones
Cyg 20h19'43" 39d7'38"
Barbara Jean's Celebration
Cnc 8h52'47" 14d40'59"
Barbara & Jeffery Ivanhoe
Gem 6h43'25" 30d44'18"
Barbara JM Darling
Vir 13h14'22" 5d48'1"
Barbara Jo Emery
Psc 22h56'58" 0d36'8"
Barbara Jo Tuohy...my
miss barb
Gem 6h52'43" 22d17'55"
Barbara Joan Kalous is
Geno's Star
And 1h15'39" 50d21'5"
Barbara Joan Marder
Sgr 19h19'25" -20d18'25"
Barbara Joan Varley
Cas 23h29'28" 57d47'6"
Barbara Johnson
Aqr 23h20'52" -14d10'35"
Barbara Jones Star
Uma 11h48'26" 33d42'40"
Barbara Jons
Sco 16h10'19" -16d32'19"
Barbara & Joseph Phillips
Cyg 20h16'44" 51d9'24"
Barbara Jovanov
And 1h7'24" 37d29'26"
Barbara Joy Hovermale
Leo 10h44'10" 15d50'19"
Barbara Joy Humphrey
Lmi 10h19'36" 35d55'1"
Barbara Joyce Borland
Uma 8h15'24" 70d40'16"
Barbara Joyce Fernandez
Mon 6h43'55" -0d5'52"
Barbara Joyce Wagnon
Ari 2h36'4" 23d17'6"
Barbara Julhiet
Tau 5h53'51" 25d6'29"
Barbara June
Cnc 8h20'33" 22d36'12"
Barbara June
And 0h49'55" 38d43'30"
Barbara June Bug Herndon
Vir 13h24'9" -11d44'15"
Barbara K, The Brooklyn
Babe
Cyg 19h56'32" 33d5'22"
Barbara Kashner
And 23h35'43" 43d15'42"
Barbara Katz
Vir 14h40'57" -0d44'12"
Barbara Kay Devries
And 0h22'10" 34d29'51"
Barbara Kelly Blosser
Aqr 21h42'8" 2d34'9"
Barbara Kilcullen
Cas 1h34'5" 68d34'23"
Barbara Knopf
Uma 11h46'52" 28d59'22"
Barbara Kodesh
Uma 12h22'59" 53d22'13"
Barbara L. Chrzan
Cas 23h45'4" 58d40'36"
Barbara L Johnson
Leo 10h10'52" 22d28'56"
Barbara L. Mladenoff
Cam 3h35'33" 66d28'59"
Barbara L. Pearsons
And 23h54'15" 47d11'55"
Barbara L. Watkins
Cyg 19h49'51" 36d12'56"
Barbara Lacusch
Dra 18h14'10" 73d44'24"
Barbara Lagana
Ari 2h3'9" 18d10'42"
Barbara Lamoreaux
Lib 14h40'43" -11d22'31"
Barbara & Larry's Star
Uma 11h27'36" 38d18'25"
Barbara Latini
Lyn 8h30'18" 57d7'57"
Barbara Learned McMillan
Ori 5h10'26" 11d52'35"
Barbara LeBlanc
Lib 14h51'14" -1d32'14"
Barbara Lee Czerwinski-
Maertz
Psc 23h15'0" 3d27'18"
Barbara Lee Eckert
Tau 4h16'57" 17d12'4"
Barbara Lee Kramer
Leo 11h29'20" 16d40'26"
Barbara Lee Suarez
Vir 12h41'32" -2d27'24"
Barbara Lee White
Cap 20h19'40" -19d35'4"
Barbara Leone Blair
Tau 3h27'15" 16d56'6"
Barbara Levenson
Com 12h14'51" 32d16'2"
Barbara Lewis
Cas 1h20'7" 52d14'15"
Barbara Lincoln Thompson
Crb 16h10'51" 38d10'36"
Barbara Linda Rowan
Psc 0h7'27" 2d5'4"
Barbara Locascio
Cas 0h34'22" 61d9'55"
Barbara Loise
And 0h50'25" 39d4'59"
Barbara Lopez
Cyg 19h35'46" 44d8'54"
Barbara Louise
Sgr 18h53'18" -30d29'4"
Barbara Louise 062951
Cnc 8h8'21" 18d28'9"
Barbara Louise Brockhoff
Uma 12h12'10" 58d19'21"
Barbara Louise Heywood
Cas 23h10'53" 55d10'18"
Barbara Louise Mickey-
Dickson
Crb 15h29'55" 26d11'25"
Barbara Louise Northrop
Leo 10h58'58" 22d35'59"
Barbara Louise Rodert
And 0h38'45" 43d19'22"
Barbara Louise Stuart
Cas 0h50'15" 70d36'48"
Barbara Louise Wood
Cnc 9h18'47" 16d39'11"
Barbara Lounsbury
Cam 4h58'7" 59d38'59"
Barbara Lowe
Cas 23h59'26" 53d39'45"
Barbara Lucille White
Lib 15h25'33" -19d48'52"
Barbara Lush
Crb 15h35'34" 28d13'24"
Barbara Lynette Dearing
Uma 8h52'54" 54d5'48"
Barbara Lynn
Vul 19h47'32" 27d34'18"
Barbara Lynn
Leo 11h23'37" 15d2'49"
Barbara Lynn Beall
Leo 9h52'0" 28d23'17"
Barbara Lynn Christensen
Uma 10h33'34" 43d34'0"
Barbara Lynn Osgood
Lib 15h3'5" -0d47'26"
Barbara Lynne Coleman
Crb 16h7'28" 33d56'57"
Barbara M. Bloom
And 0h47'40" 42d3'11"
Barbara M. Brodsky
Leo 11h1'43" 14d50'1"
Barbara M. Davis Hendon
Tau 5h39'13" 18d44'19"
Barbara M. Ligatti
Dra 18h42'50" 58d53'56"
Barbara M. Millette
Aqr 21h15'15" 2d28'45"
Barbara M. Totty
Sco 16h9'49" -13d27'19"
Barbara Mabli
And 0h1'4" 43d20'15"
Barbara Mae Bruce
Crb 16h17'58" 35d7'30"
Barbara Mae Strachan
Uma 12h27'38" 62d13'31"
Barbara Malone
Cyg 19h29'43" 55d15'21"
Barbara Malsack
Cap 21h40'21" -9d30'0"
Barbara "Mami" Ernst
Umi 13h59'28" 69d28'59"
Barbara Marano
Cas 0h26'52" 61d21'20"
Barbara Margaret Holmes
Lib 14h44'49" -25d1'10"
Barbara Marie Chackan
Uma 10h47'51" 46d40'24"
Barbara Marie Miller
Uma 10h37'17" 50d17'10"
Barbara Marie
Thierwechter
Aqr 21h41'17" 1d50'4"
Barbara Martin (Mom)
Her 18h36'14" 21d14'38"
Barbara McCaffrey
Cas 0h13'1" 57d32'22"
Barbara McCall
Cap 20h7'36" -18d38'16"
Barbara Melycher -
Ferravante
Gem 6h40'37" 19d54'3"
Barbara Meyers
Cap 21h2'31" -15d43'45"
Barbara & Mike Lepinski, 2-
27-1954
Uma 11h22'14" 32d26'59"
Barbara Millar
Uma 8h58'49" 66d9'20"
Barbara Miller
Cas 2h50'17" 60d32'26"
Barbara Minchin
Gem 7h13'52" 32d19'15"
Barbara Molnar
And 23h41'16" 47d23'39"
Barbara: mom, grandma,
great-gramma
Per 4h44'53" 36d34'47"
Barbara 'Momma' Gerlach
Vir 13h48'10" 6d36'13"
Barbara Moss
Lmi 9h28'11" 33d53'28"
Barbara Murtagh Nash
Lib 15h36'5" -26d11'35"
Barbara Murtaugh
Uma 14h8'13" 60d14'30"
Barbara Musotto
Cam 9h30'25" 81d58'11"
Barbara Musotto
Cyg 19h44'46" 46d35'29"
Barbara N. Kovacs
Leo 10h20'7" 25d19'57"
Barbara "nana" Johnson
Uma 10h25'54" 69d56'56"
Barbara "Nana" Vamos
Psc 1h17'39" 31d8'27"
Barbara Nance
Cas 0h54'43" 60d38'55"
Barbara Neely Gilford
Cnc 8h47'53" 31d5'42"
Barbara Nelson
Com 12h55'35" 27d3'41"
Barbara Neuhaus
Ori 5h52'1" 12d10'37"
Barbara Niccolai
Aqr 23h30'2" -12d44'25"
Barbara Nicole
Lib 15h41'41" -22d54'21"
Barbara Norton Thorn
Crb 15h43'56" 33d21'48"
Barbara Nova
Aqr 23h4'52" -7d42'16"
Barbara Nyberg
Cam 3h32'52" 60d52'12"
Barbara O 1961
Cnc 8h13'28" 16d13'26"
Barbara Olga Gordon
Aqr 22h16'59" 1d38'29"
Barbara Orsini Hemingway
And 1h46'51" 45d28'28"
Barbara Pavelin
And 23h43'52" 43d0'45"
Barbara Peros
Sgr 18h6'28" -27d50'29"
Barbara Pettit Finch Star
Cas 0h26'23" 56d22'18"
Barbara Pezzolla
Per 4h18'43" 35d2'51"
Barbara & Philip Karas
Umi 15h10'29" 68d1'29"
Barbara Pierallini
Aur 6h38'57" 41d45'56"
Barbara Pileggi
Ori 5h51'47" -0d15'29"
Barbara Prendergast
Per 3h12'58" 55d10'47"
Barbara Prestwood
And 23h21'57" 44d46'43"
Barbara Prochazka
Cyg 19h40'50" 33d18'32"
Barbara Pruitt
Gem 6h40'15" 18d18'22"
Barbara Pucci
Lyn 8h52'9" 40d37'45"
Barbara Puscher
Cap 20h23'55" -26d43'46"
Barbara Quantz
And 0h48'40" 44d38'17"
Barbara Quittschalle
Uma 9h41'33" 51d7'59"
Barbara R. Benning
Aqr 22h0'33" -15d34'51"
Barbara R. Bolf
Aqr 23h4'36" -10d57'44"
Barbara R. Maness
And 23h36'51" 47d11'15"
Barbara R. Nichols
Ori 5h21'30" -5d30'52"
Barbara R. Sobel
Tau 4h36'9" 19d30'17"
Barbara Rachelle
And 2h14'29" 50d30'10"
Barbara Rae
And 1h3'49" 38d51'29"
Barbara Rae
Gem 6h48'38" 26d19'21"
Barbara Rae DeLaRue
Sgr 18h24'58" -16d16'52"
Barbara & Raffaele
Cap 21h26'9" -14d41'23"
Barbara Rand
Cap 21h52'56" -17d52'0"
Barbara Rieben
Cas 1h33'11" 67d51'32"
Barbara Ring
Cas 0h20'3" 62d2'30"
Barbara Roe
Crb 15h55'0" 27d17'10"
Barbara Rose
Gem 6h4'14" 27d39'51"
Barbara Ross
Crb 15h36'51" 32d25'0"
Barbara Ruch
Ori 6h17'11" 9d29'45"
Barbara Ruth DeBow
Mon 7h16'57" -0d50'9"
Barbara S B Mumma
Leo 11h27'19" 25d31'43"
Barbara S. Chambers
Tau 4h21'47" 17d48'31"
Barbara S. Dayhuff
And 0h13'47" 37d4'36"
Barbara S. Young
Uma 11h7'27" 29d24'11"
Barbara Salconi (_bibbi)
Boo 15h28'44" 43d46'58"
Barbara Sanchez Gutierrez
Sco 16h11'7" -30d20'11"
Barbara Schachet
Cas 23h21'6" 53d29'38"
Barbara Schramm - Sid die
Fremde
Uma 14h15'53" 60d24'52"
Barbara Scoles
Lib 15h19'45" -16d2'53"
Barbara Scott Holgate
Cas 23h56'28" 50d14'54"
Barbara Seeley
Uma 10h1'40" 51d49'34"
Barbara Senft
Crb 16h15'15" 33d13'46"
Barbara Shanks
And 1h10'59" 36d45'42"
Barbara Singleton
Cap 21h30'2" -9d13'7"
Barbara Skiffington
Leo 9h44'26" 23d30'6"
Barbara Slater Maus
Cas 1h16'12" 62d19'10"
Barbara Smethurst
Cas 23h35'23" 57d23'54"
Barbara Smith
And 2h31'35" 43d13'39"
Barbara Smith of Camelot
Per 3h33'55" 41d29'50"
Barbara Solheim
Lyn 9h6'54" 38d35'38"
Barbara Sophie
Grafczynski
Cas 23h12'24" 59d38'51"
Barbara Sperl Monroe
Uma 10h7'49" 43d17'50"
Barbara Spigner
Cam 7h14'12" 73d33'51"
Barbara Squeri
Cas 1h17'58" 62d12'1"
Barbara Story
Cas 0h33'18" 62d37'42"
Barbara Struck Dodson
Uma 9h22'14" 65d9'39"
Barbara Stuart
Gem 6h37'33" 17d9'49"
Barbara Stull's SAS Star
Uma 9h57'38" 51d37'56"
Barbara Sue Gibson
Leo 11h19'54" 13d19'45"
Barbara Sue Hitchcock
Leo 9h43'8" 13d38'6"
Barbara Sue Macomb
Uma 10h28'3" 44d42'12"
Barbara Super Nova
Uma 9h28'29" 48d39'2"
Barbara Susan Martin
Tau 3h53'20" 20d16'35"
Barbara Swesen
Uma 11h30'22" 33d8'22"
Barbara Swirnow
Lib 15h3'1" -1d32'16"
Barbara T. Curtis
Leo 11h9'46" 6d46'51"
Barbara Taylor Eaton
Cam 6h0'54" 63d55'50"
Barbara Teng
Lib 15h36'12" -7d24'19"
Barbara Tepedino
Mon 8h2'31" -0d35'38"
Barbara Terry Bell
Eri 4h2'11" -15d9'0"
Barbara "The Electrifying
One"
Uma 9h22'49" 41d42'10"
Barbara Thropp
Lyn 7h54'14" 41d21'44"
Barbara Tiffany Ferris
Lyn 8h11'31" 40d56'45"
Barbara Tooker
Dra 20h19'29" 65d24'36"
Barbara Tower Doney
Cnc 8h37'10" 29d2'14"
Barbara Trivigino
Cnc 8h42'40" 12d53'54"
Barbara Truluck
Sco 17h52'56" -30d47'11"
Barbara Tryon
Cas 1h37'19" 61d23'55"
Barbara Tumbridge
Cas 0h9'57" 57d5'16"
Barbara Upmeier
Ori 4h44'59" 11d20'50"
Barbara V.
Cas 2h16'37" 62d50'15"
Barbara V. Turner
Cyg 20h19'29" 33d34'41"
Barbara VanZile Star
And 0h33'36" 45d42'49"
Barbara Villegas
Her 18h0'33" 29d42'42"
Barbara Virginia Dwyer
Sonntag
Lyr 18h26'51" 35d5'59"
Barbara Visconti
Cap 21h7'3" -16d6'41"
Barbara Voorhees
Cas 1h36'57" 58d1'56"
Barbara Wade
Lib 15h7'45" -7d7'20"
Barbara Weiss
Aqr 23h1'27" -7d55'12"
Barbara Weldele
Tau 4h39'22" 21d16'8"
Barbara White
Gem 6h52'22" 33d46'21"
Barbara Williamson Backer
Ori 5h48'48" -0d11'46"
Barbara Winters
Cas 23h25'23" 58d14'21"
Barbara Y Nicolas
Cen 13h9'46" -46d1'53"
Barbara Young
Uma 8h39'29" 49d58'5"
Barbara Young
Psc 1h5'3" 9d36'55"
Barbara Young Dey
Cas 0h8'30" 50d34'6"
Barbara Younger - Gee
Aqr 21h43'24" -3d9'18"
Barbara Zerzour
Del 20h42'32" 9d30'53"
Barbara Zingler
Lib 15h14'29" -24d50'23"
Barbara, Alexis, and Leo
Hayser
Uma 12h18'16" 57d17'24"
Barbara's Augen
Uma 10h18'9" 57d17'12"
Barbara23011965
Per 4h43'54" 40d43'16"
BarbaraAnneSmith
Cas 2h15'57" 67d25'58"
Barbaraillius
Sco 17h18'0" -34d7'19"
Barbara-One in a Million
Mom
Psc 0h1'33" 6d23'34"
BarbaraOrr-StarMom
Gem 6h56'5" 15d10'38"
Barbara's
Accomplishments
Umi 15h29'9" 71d53'41"
Barbara's Aunt Dorothy
Aur 6h27'3" 34d19'57"
Barbara's Ballad
Psc 0h9'24" 8d19'39"
Barbara's Compassion
Uma 11h44'27" 56d37'19"
Barbara's Dream
Cnc 8h12'45" 7d43'32"
Barbara's Eternal Nightlight
Uma 10h7'59" 71d45'22"
Barbara's Faith
And 0h54'32" 40d19'47"
Barbara's Family
Ori 5h38'22" -0d18'11"
Barbara's First Star On The
Right
Cyg 19h55'19" 52d14'34"
Barbara's Grandma Days
Cyg 20h17'27" 41d39'16"
Barbara's Guiding Light
Powell
Gem 7h38'8" 26d30'21"
Barbara's Hope
Per 3h16'28" 50d43'25"
Barbara's Humor
Crb 15h30'28" 27d27'31"
Barbara's Love
Lyr 18h46'8" 36d36'56"
Barbara's Melodi
Mon 6h51'44" -0d25'35"
Barbara's Prayers
Cas 1h3'47" 59d25'8"
Barbara's Silver Fox
Vul 19h35'57" 22d8'38"
Barbara's Star
Crb 15h29'12" 26d25'13"
Barbara's Star
Cyg 21h39'7" 40d26'25"
Barbara's Star
Lyr 18h16'2" 34d4'20"
Barbara's Star
Psa 22h37'40" -25d25'44"
Barbara-Simos 27
Décembre 2002
Tau 5h46'38" 18d26'7"
Barbarastar
Vul 20h33'18" 26d52'28"
Barbarastar
Lyr 19h26'37" 41d30'4"
Barbarella Di Pietro
And 23h13'0" 52d38'4"
Barbarella mein
Lebenselixier
Cas 23h13'15" 59d13'42"
Barbarina 14-06-1976
Ori 5h9'44" 1d2'1"
Barbaro
Her 18h46'28" 22d20'32"
Barbaro
Peg 22h43'57" 33d4'5"
BarbaStar,Happy Mothers
Day,Luv Jay
Crb 15h54'1" 30d40'0"
Barbchuck
Gem 6h35'7" 15d34'36"
BarbEd
Cnc 8h17'14" 20d2'58"
Bärbel
Uma 8h17'11" 67d10'26"
Bärbel
Uma 12h35'43" 61d39'31"
Bärbel
Uma 12h11'17" 61d30'51"
Bärbel Berta Steiniger
Vir 14h5'22" -10d48'17"
Bärbel Fabian
Uma 11h50'6" 28d34'34"
Bärbel & Heinz
Uma 8h52'8" 49d20'5"
Barbette
Leo 9h25'41" 12d35'23"
Barbi Beegle
Tau 4h35'52" 17d8'51"
Barbi Franklin
And 2h8'25" 42d43'14"
Barbi Vonn
Uma 11h20'41" 66d31'48"
Barbi Wiggins
Vir 15h3'12" 0d57'29"
Barbie
Per 3h22'55" 49d38'38"
Barbie
Cyg 19h38'43" 38d40'52"
Barbie
Umi 16h46'49" 81d38'52"

Barbie and Ken
Cyg 20h24'52" 47d37'26"
Barbie and Ken Culp
Leo 11h30'50" 11d21'50"
Barbie Ann Cromer
Cas 23h12'31" 55d15'24"
Barbie Doll
Uma 11h23'52" 57d47'31"
Barbie Meador
Lib 15h19'34" -7d29'2"
Barbie Mom
Sco 16h50'59" -24d53'44"
Barbie & Mom's Place
Leo 10h7'28" 15d26'52"
Barbina Mello Edwards
Cam 6h32'35" 69d40'24"
BarBo
Ari 2h31'3" 25d43'21"
Barbolute
Ori 5h30'16" 7d58'41"
Barboni Tania
Lyr 18h44'31" 31d4'53"
Barborka
And 0h27'39" 42d50'48"
Barboza Family Star
Uma 11h31'29" 58d32'51"
Barbra Ann Gerhard
Vir 13h16'12" 7d0'22"
Barbra Banavige
Cas 0h52'23" 67d8'43"
Barbra Barker
And 1h13'28" 41d36'14"
Barbra Jean Taber
Lib 15h42'27" -19d1'41"
Barbra-Anne
Lib 15h44'17" -26d57'16"
BarBrooke
And 2h26'4" 41d49'2"
Barb's Darylann
And 23h40'13" 42d18'2"
Barb's Dream "Strike Out A.L.S."
Lyn 7h28'24" 50d4'17"
Barb's Star
Cas 22h57'37" 54d32'10"
Barb's Star
Umi 13h7'57" 74d21'48"
Barby Fredricksen
Gem 7h30'20" 17d56'43"
Barby's Special Star
Cyg 20h2'48" 44d30'12"
Barcelon Family Star
Dra 18h27'47" 73d19'12"
Barclay
And 0h4'17" 41d32'7"
Barda's Cani Maria
Aqr 23h1'34" -7d31'53"
Bardo
Ari 3h2'48" 12d39'14"
Bárdos Ádám
Uma 9h51'6" 45d28'49"
Bare
Uma 9h23'51" 48d0'53"
Bare
Lac 21h58'44" 40d28'2"
Barea Lianka
Cam 4h31'39" 57d28'28"
Bareford's Eternity
Cas 23h38'47" 53d35'16"
Barend Gerhardus Lindeque
Cru 12h8'9" -60d21'26"
Barend Mooibroek
Uma 12h2'34" 62d5'4"
Bäretätzli
Lmi 10h16'3" 33d47'51"
Bargain Hunter John
Uma 9h46'12" 51d19'50"
Barhumbug InMemory Of Peter Godenzi
Cru 12h18'6" -57d4'44"
bari
Sgr 19h21'41" -28d51'56"
Bari
Uma 10h31'32" 46d42'56"
Bari Faith Klein
Cnc 8h57'19" 32d1'36"
Bari & John 7/22/1947
Cyg 20h13'59" 51d11'24"
Bari Leigh
Lmi 9h25'23" 35d3'19"
Bari Mayhew
Psc 1h5'12" 28d32'35"
Barigor
Sge 20h4'19" 17d19'11"
Barkha
And 0h40'31" 25d11'1"
Barkley
Cma 6h31'50" -14d59'51"
Barkley Claire Bennett
Uma 11h23'16" 34d35'7"
Barkley Dulisch-Bilbrey
Dra 16h54'55" 68d2'1"
Barkley's Honor
Uma 11h28'50" 53d22'56"
Barky
Cma 7h8'42" -25d26'44"
Barlby's Brilliant Star
Uma 10h5'9" 51d27'31"
Bärle & Bee
Ori 6h19'30" 15d1'40"
Bärli
Uma 9h42'13" 50d30'6"
Barlow Bob Godfrey
Per 3h15'20" 55d15'39"

Barlux
Uma 10h3'48" 47d13'59"
Barman-Heim
Uma 11h31'48" 61d58'24"
Barmou Karima
Cam 3h44'0" 65d46'59"
Barna
"Husband,Dad,Grandpa" Ari
Per 3h11'24" 54d36'23"
Barnaby Roth
Crb 15h46'32" 27d16'5"
Barnard Hughes
Cep 22h36'19" 65d23'43"
Barner, Berndt
Ori 6h18'26" 15d49'4"
Barnett L. Gershen
Cep 22h32'6" 63d37'41"
Barney
Ori 4h54'17" -0d10'33"
BARNEY
Sgr 19h45'38" -14d6'50"
Barney
Ori 4h54'31" 12d14'16"
Barney and Irene Dial
Uma 10h51'2" 49d28'1"
Barney and Max The Barker Brothers
Cmi 7h25'12" 9d14'58"
Barney Brown
Mon 6h22'23" 7d54'16"
Barney Bruce J.
Uma 9h19'46" 53d8'20"
Barney Buggaboy Jacob
Leo 11h16'51" -2d5'14"
Barney & Dolly Rosenfelder
Umi 14h59'36" 70d7'55"
Barney Durham
Her 16h39'41" 13d31'40"
Barney Lennartson
Ori 6h3'0" 20d53'25"
Barney Reid
Aur 5h52'9" 53d23'53"
Barney Shaw
Lyn 7h53'54" 48d58'42"
Barney's Wishing Star
Per 4h10'26" 47d23'43"
Barnliz
Uma 9h12'47" 61d25'44"
Barnowski, Günter
Uma 10h21'3" 66d9'21"
Barny 13
Psc 1h5'17" 14d12'30"
Baron - April 14, 1981
Peg 0h10'7" 18d2'47"
Baron Blitz Dupont
Sco 16h15'24" -13d42'2"
Baron Boycey of Babeville
Per 4h20'40" 38d51'30"
Baron Brito
Cyg 19h33'35" 52d31'0"
Baron de Kerlébert
Tau 4h28'32" 16d29'52"
Baron Irik Rodbjorn
Uma 14h9'19" 61d0'24"
Baron James Lewis Kirby, Jr.
Uma 11h5'6" 49d14'27"
Baron James Miller - 1st Communion
Umi 15h31'10" 75d20'48"
Baron Loves Amy Forever
Cyg 20h28'17" 53d18'3"
Baron Nicolas
Uma 11h17'39" 48d23'33"
Baronchelli Matteo
Ori 5h55'43" -8d22'22"
Baroness Melba Toast
Uma 9h46'45" 54d25'37"
Baroness Meret Schaukowitsch
Uma 9h15'1" 63d3'17"
Baroni Fabrizio
Aql 20h13'19" -0d55'57"
Baroni Roberta
Cam 6h46'53" 64d12'58"
Baron's World
Cen 13h34'1" -37d40'10"
Baroo
Cam 3h42'54" 63d0'37"
Barr
Uma 11h14'50" 41d32'47"
Barra Brui - Place of many Surtees
Cru 12h26'58" -59d56'4"
Barracuda
Her 17h36'14" 32d25'54"
Barragunda
Crb 15h33'41" -62d26'34"
Barrel Racin Raye
Uma 8h34'1" 47d25'55"
Barret Franklin Snyder
Uma 10h16'20" 44d45'12"
Barrett
Tau 5h30'26" 17d35'55"
Barrett Lea Cummings
Cas 0h41'20" 48d43'47"
Barrett Major
And 23h23'25" 42d29'53"
Barrett Rose Hinojosa
Leo 10h6'20" 12d38'40"
Barrie A. Mack
Lyn 7h12'49" 57d31'42"
Barrie and Doreen Griffiths
Cyg 20h12'58" 33d5'37"

Barrie James Timson
Uma 12h2'30" 65d22'38"
Barrie Mills Eternity
Uma 11h7'56" 66d44'9"
Barrie Ronald Walter Sharpe
Lyn 7h38'16" 36d23'51"
Barrie Thomas
Dra 18h18'2" 78d53'17"
Barrios - 1307
Lyn 8h24'59" 46d27'17"
Barrow -2610
Ori 5h57'17" 17d15'1"
Barrow Cognatio Astrum
Uma 10h8'28" 50d47'39"
Barrow's Bright Eyes
Tau 4h51'47" 23d19'17"
Barrus Patronus
Uma 12h25'0" 54d59'13"
Barry
Ori 5h24'36" 1d10'19"
Barry
And 0h10'50" 45d29'25"
Barry
Her 16h43'23" 43d24'44"
Barry A. Schreier, PH.D.
Lyn 9h12'42" 40d56'26"
Barry and Dawne Waddicor
Per 3h28'26" 53d2'26"
Barry Baker
Psc 1h5'59" 25d18'39"
Barry Ball
Uma 10h18'37" 49d40'49"
Barry Brown
Cyg 20h11'44" 35d35'23"
Barry Coverley
Her 18h2'15" 14d33'31"
Barry Curtis Schellenberg
Umi 13h44'8" 77d14'18"
Barry D Williams
Aqr 22h44'23" -1d4'16"
Barry Darryl
Dra 15h59'53" 56d6'59"
Barry David Hubbard
Ori 5h30'21" 2d10'9"
Barry Denault
Uma 8h48'13" 66d54'51"
Barry & Doreen Rolfe
Cyg 19h41'49" 31d53'53"
Barry E. Page
Lib 14h51'38" -1d4'19"
Barry Emmons
Mon 6h52'35" -0d57'8"
Barry Fred Tolli
Uma 12h58'6" 59d15'27"
Barry Frederick Mitsch
Lib 15h33'11" -6d21'18"
BARRY & GERRY 50TH ANN STAR
Cep 22h27'49" 65d33'30"
Barry Gregory Mattozzi
Uma 8h27'0" 65d19'28"
Barry Grossman
Lib 15h26'47" -26d38'44"
Barry Grylls
Lmi 10h18'13" 36d32'51"
Barry Hamilton
Umi 15h39'6" 72d32'36"
Barry Heather & Chase Barringer
Pyx 8h52'19" -26d58'44"
Barry Hinwood
And 2h6'13" 38d35'17"
Barry Ienni Millennium Star
Aql 19h5'26" -0d17'35"
Barry J Fishman
Tau 4h8'7" 26d39'35"
Barry J. Leonard
Aql 19h23'28" -10d46'49"
Barry J. Weiler, Jr. 1984-2003
Uma 8h54'24" 58d4'14"
Barry & Jenny's Star
Cyg 21h58'54" 50d20'35"
Barry John Richards
Cyg 21h48'15" 50d10'0"
Barry K. George
Her 18h9'17" 18d24'11"
Barry Kenneth Tranckino
Ori 5h33'28" -0d4'0"
Barry Kinya Pollard
Uma 10h38'3" 45d29'32"
Barry Klosak
Her 16h12'13" 14d47'9"
Barry L. Howard 1961
Uma 11h58'18" 53d47'4"
Barry L. Seaman
Gem 7h32'56" 32d36'49"
Barry Lee Smith
Tau 5h21'58" 20d28'35"
Barry loves Tracey
Lib 15h30'12" -7d20'15"
Barry Lynn Clevenger
Uma 11h22'39" 60d21'45"
Barry Maulden
Her 17h7'10" 39d11'55"
Barry & Maura Wallace-Nichols
Cyg 21h36'28" 35d13'56"
Barry McCarthy
Her 18h4'54" 36d54'54"
Barry Medd
Uma 9h6'42" 50d59'2"
Barry Michael Rose
Uma 11h39'1" 63d37'38"

Barry & Natalie Kinder
Uma 10h48'54" 50d22'48"
Barry Patrick Cornish
Lib 14h48'33" -15d26'54"
Barry Peers
Uma 13h5'22" 62d19'5"
Barry Peter Aylett
Cru 12h26'32" -60d24'18"
Barry Reeves
Ori 6h13'14" 8d21'0"
Barry Robert Montgomery
Umi 4h35'21" 88d38'43"
Barry Rodney Goes..."Bazz"
Sgr 17h56'8" -24d2'24"
Barry Salzman's 60th Birthday Star
Vir 13h3'49" -20d13'44"
Barry & Sara
Cyg 20h13'11" 52d5'42"
Barry Schwarzberg
Gem 6h28'54" 25d55'45"
Barry Setser - Sahir
Ari 1h48'8" 23d1'0"
Barry Steven Bachus
Her 16h53'14" 40d53'8"
Barry Steven Draskovich
Boo 14h34'29" 28d35'22"
Barry T. Driscoll
Her 17h38'22" 20d20'1"
Barry the Star of My Heart
Per 3h54'18" 33d16'45"
Barry Todd Bacon
Aqr 22h23'13" 2d35'22"
Barry Underwood Faulkner
Lmi 9h48'50" 37d24'2"
Barry Vincent Smith
Cap 20h26'39" -10d53'53"
Barry William Kiel Jr
Vir 14h5'12" -10d52'12"
Barry Wong
Cru 12h11'31" -60d34'50"
Barry Wood
Her 17h18'48" 27d19'56"
Barry Wyndham Ure
Cap 21h7'14" -15d30'15"
Barry - You're a Star!
Per 4h37'19" 31d28'6"
Barry, the love of my life.
Her 17h47'19" 46d30'21"
Barrymore Keegan Thiergartner
Uma 8h58'21" 66d24'29"
Barry's Bride
Cas 1h21'23" 62d7'52"
Barry's Brilliant Dream
Leo 9h22'26" 25d54'39"
Barry's Shining Star
Tau 4h7'29" 18d29'1"
Barry's Star
Cnc 8h10'55" 15d30'21"
BARSETKA
Uma 8h32'52" 67d27'4"
Bársony Viktória
Uma 12h1'5" 31d42'10"
Bart
Her 17h7'53" 30d46'33"
Bart
Cyg 19h33'32" 28d25'36"
Bart
Tau 4h5'49" 23d35'40"
Bart Allen Caudill (ROCK)
Her 17h36'34" 26d1'1"
Bart and Marilyn's Star
Lib 15h37'44" -9d38'32"
Bart C. Trexler
Leo 9h28'12" 10d55'31"
Bart Hadder
Cep 23h11'22" 70d55'21"
Bart James Hinners
Lyn 9h14'51" 34d30'8"
Bart L. Fischer
Psc 22h59'17" -0d3'54"
Bart Lamens
Her 17h29'6" 41d15'39"
Bart Matthews
Boo 14h24'42" 46d18'1"
BART RITTER
Vir 13h9'36" 13d59'30"
Bart Star I
Cam 4h13'53" 66d36'1"
Bart Stoddard
Uma 11h29'34" 32d57'4"
Bart Thomas Kuehner
Ori 5h41'13" 1d34'37"
Bart Vickers
Cma 6h46'24" -28d10'40"
BartAlliSTAR
Ari 2h52'49" 21d4'49"
Bartelmann, Uwe + Ilona
Uma 11h39'2" 47d36'26"
Bartenjev
Uma 9h9'30" 69d47'32"
Bártfi Adrienn
Psc 22h52'40" 7d45'38"
Barth H. Pemberton
Sco 16h3'35" -14d5'2"
Bartholomew Anthony Colucci
Leo 10h22'12" 26d3'41"
Bartholomew J Murphy
Uma 12h27'21" 57d28'6"
Bartkowiak, Gerd
Ori 5h19'54" 3d22'47"

Bartlett
Uma 9h2'0" 50d13'19"
Bartnik, Bruno
Ori 6h16'26" 10d23'22"
Bartolome Babies
Uma 9h30'21" 44d51'31"
Barton Abel
Ori 5h29'8" 1d56'39"
Barton Clyde Davis
Lyr 18h47'12" 34d29'12"
Barton Cortright
Uma 14h6'31" 49d46'35"
Barton E. Harrison
Uma 10h37'28" 58d45'39"
& Barton Forever
Gem 7h36'23" 22d17'14"
Bartow Wing Riggs
Ari 2h10'48" 25d21'30"
Bart's Star
Dra 16h28'11" 58d59'52"
Baruch Samuel Blumberg
Leo 9h40'11" 30d54'22"
Baryohay&Tamara Davidoff
Berta Star
Umi 14h14'58" 67d0'2"
BAS BANGLE
Tri 1h42'47" 30d41'30"
Basak
Sgr 18h23'13" -32d32'13"
Basan
Cen 13h24'28" -35d12'34"
Basarich
Ori 6h0'41" 10d5'0"
Baschi
Umi 17h31'15" 81d9'9"
baseball bat
Mic 21h14'0" -31d10'12"
Basellandschaftliche Kantonalbank
And 0h43'28" 41d39'31"
BASH
Uma 9h31'19" 68d29'2"
Bashar
Tau 5h31'38" 20d36'8"
Bashar Hasan Al Sayigh
Peg 21h14'5" 19d14'42"
Bashert
Tau 5h54'53" 28d23'40"
Bashkim
Aur 6h32'12" 38d53'40"
Basi Antoun Datak
Gem 6h54'0" 14d21'54"
Basia Bielecka
Cnc 8h14'54" 11d36'49"
Basia i Maciej
Cam 7h50'45" 63d5'53"
Basia Warszawa
Her 17h28'15" 39d44'10"
Basienka
Aur 6h21'22" 31d56'39"
Basil
Cyg 20h10'58" 48d55'34"
Basil
Ari 2h48'15" 27d51'44"
Basil "Bogey" Boeglin
Uma 9h38'42" 42d27'13"
Basil dee Ocho
Cyg 21h30'35" 47d31'25"
Basil Kalkanas love William K.
Cnc 8h47'53" 14d17'51"
BASIL MOHAMMED EL-GHOUL
Ari 2h6'42" 18d51'13"
Basil P. Robertson - Daddy
Lib 15h45'47" -10d58'15"
Basil Sulaka
Gem 7h56'0" 28d27'56"
Basil Wakelin Street
Cru 12h25'46" -56d13'12"
Basile Giulia
Uma 13h31'47" 60d9'56"
BASILIAS
Her 16h46'8" 10d47'44"
Basilio Gonzalez, Sr.
Cap 20h28'52" -20d22'36"
Basium Lee Anne
Uma 12h38'14" 56d12'47"
Basler Wasser
And 0h42'29" 41d29'4"
Basma 7-8-68
Cyg 21h15'57" 54d50'39"
BASOOOO
Uma 11h39'0" 61d53'39"
Bass River's Gus Healy
Sgr 18h13'55" -27d12'24"
Bass Security Services, Inc.
Ori 5h32'45" 0d8'24"
Bassam Hashem, MD
Per 2h55'35" 52d38'2"
Bassam Mohamad Hazime
Ori 4h51'4" 1d10'56"
Basse, Reinhold
Uma 10h42'35" 42d26'5"
Bassel and Malak
Per 2h38'48" 51d14'39"
Bassie
Gem 7h16'29" 25d11'14"
Bassima Mardini
Ori 5h21'42" -8d40'43"
Basti
Tau 5h18'32" 28d37'52"
Bastian Bieler
Cas 1h38'31" 69d9'20"

Bastian Mantey
Uma 9h28'52" 67d14'24"
BASTIEN
Cep 22h58'2" 71d0'41"
Basti's Rio Negro
Uma 12h30'55" 62d23'9"
Basvalakshmi Katepalli
Gem 7h16'23" 25d48'22"
Bat
Uma 8h10'39" 65d36'21"
B.A.T. 08-30-05
Vir 13h26'22" 12d14'12"
Batcheh
Cnc 8h12'29" 25d47'1"
Batel Perez
Aqr 22h23'56" -9d13'27"
Bath Junkie Peeps
Cnc 8h1'28" 10d26'4"
Bathgate-1925
Lyn 7h55'3" 47d38'35"
Bathsheba
Aqr 22h35'29" -17d11'3"
Batis Campillo
Sco 16h52'30" -26d59'3"
Battaini
And 1h7'56" 44d12'47"
Battle
Peg 21h40'29" 20d43'30"
Battman
Uma 8h35'15" 61d36'8"
Batty Bat
Gem 6h32'26" 25d54'48"
Bätzi
Sge 19h55'27" 17d55'37"
Baudin Kathy
Her 16h33'0" 36d38'49"
Bauer Dezso - Apuci
Uma 9h57'3" 50d13'29"
Bauer, Hartmut
Cnc 8h49'49" 26d22'41"
Bauer, Marc
Uma 10h26'49" 52d23'54"
Bauer, Marcus
Ori 6h17'5" 9d39'58"
Bauer, Thomas
Uma 10h59'16" 66d23'37"
Bauhaus NGA
Cra 18h9'57" -39d6'26"
"Bauldie"
Cep 22h31'28" 57d39'35"
Baum
Gem 7h14'0" 20d0'1"
Baumann, Laura
Tau 3h40'4" 28d50'37"
Baumann-Parker Family
Apu 15h48'19" -78d25'48"
Baumbach, Kurt
Uma 8h54'54" 51d10'41"
Baumeister
Uma 11h48'47" 44d23'31"
Baumeister, Christian
Uma 8h16'24" 68d50'10"
Baumgertel, Michael
Ori 6h17'5" 19d10'6"
baumi, Marco Baumgärtner
Cas 0h9'1" 52d25'46"
Baur, Helmut Fritz
Uma 11h39'52" 38d45'57"
Bautz, Paul
Uma 9h0'7" 47d13'55"
Bautz, Ralf
Ori 5h5'15" 6d10'50"
Baxter
Vir 13h27'9" 3d27'48"
Baxter
Cmi 7h40'27" 3d57'42"
Baxter
Psc 1h15'32" 18d31'38"
Baxter An
Lep 5h33'54" -19d43'11"
Baxter Brown
Uma 10h26'22" 72d13'9"
Baxter Jolly & Darren Smith
Gem 6h59'33" 18d38'56"
Baxxxba
Vir 12h41'47" 1d24'19"
Bay
Uma 11h50'8" 36d31'18"
Bay Arthur Cartier
Cam 6h8'29" 69d16'33"
Bay Badillo Hernàndez
Uma 9h13'6" 51d13'34"
Bay Bear
Tau 4h3'43" 5d39'5"
Baya
Dra 18h5'56" 62d13'41"
Baybay Baybay Baybay
Psc 0h50'28" 19d0'43"
Bayden Gilf Parker
Uma 11h37'37" 44d3'35"
Bayer, Ulla
Ori 6h0'25" 14d21'35"
Baygz & Sezz Forever
Ara 17h5'30" -59d44'16"
Bayhead Gables
Uma 11h9'26" 60d37'15"
Baylea Phoenix Miles 130607
Pho 0h38'42" -51d1'39"
Baylee
Dra 15h32'41" 56d34'39"
Baylee Ann Elizabeth Chester
Vir 13h48'43" -5d24'1"

Baylee Curtis
Lmi 10h41'33" 27d20'19"
Baylee Kae Macek
Uma 8h55'1" 55d5'5"
Baylee Magnifique
And 0h39'2" 33d14'33"
Baylee's Star
Ori 5h30'53" -0d9'44"
Bayleigh
Sco 16h9'38" -10d37'38"
Bayleigh Westerlund
Uma 10h31'40" 41d56'33"
Bayley Elizabeth Dawson
And 23h50'11" 45d22'12"
Bayley Forever With Dad
Uma 10h21'23" 69d17'12"
Bayley's Star
Lyn 8h44'34" 42d19'18"
Bayli Raquel Kumbera
Ari 2h20'55" 19d32'14"
Baylie Marie Fyler
And 23h50'8" 46d44'15"
Baylor Elden James
Uma 9h37'17" 60d29'21"
Baylor Mills Wilson
Cap 21h34'25" -17d23'19"
Bayly Ledes
Crb 16h12'6" 33d46'31"
Bayly's Boca Grande
Her 17h6'48" 32d20'0"
Bayram Civilibal 23.04.1958
Uma 8h47'12" 69d15'28"
Bayrische Prinzessin
And 1h57'50" 45d31'48"
Bayyon Williams
Uma 9h52'21" 59d33'6"
Baze Girl
Mon 6h45'0" -0d19'51"
Baze Mpinja
Uma 9h13'6" 67d36'56"
Bazza's Wagon Star
Cru 12h32'13" -62d0'24"
Bazz's Star
Cru 12h47'11" -57d59'55"
BB
Cma 6h54'31" -17d5'30"
B.B.
Lyn 7h4'20" 47d50'9"
BB 03071980
Uma 13h40'25" 59d40'7"
BB Kathy Benson
Cap 21h31'43" -8d52'57"
B-B Music Factory
Tau 4h24'34" 24d1'8"
BB Poppins
Per 4h2'47" 44d2'9"
BB Rubin
Cap 21h14'6" -22d3'22"
BB120
Aqr 22h17'50" -15d32'25"
B.B.C.N
Dra 20h5'10" 74d7'3"
BBD and Sultan forever
Uma 14h17'3" 58d11'37"
BBDK Holeman DLP
And 1h2'57" 39d25'17"
BBG
Ari 3h26'59" 21d6'28"
BBG-2000
Cas 1h23'33" 61d35'27"
BBH64-40
Leo 11h20'43" 17d20'34"
BBHOLT131
Cyg 19h48'2" 33d44'44"
bbHung
Aqr 21h27'22" 1d22'15"
B-Bo
Dra 18h48'27" 52d42'47"
B-Bop
Cnc 9h5'2" 30d55'5"
B-Boy
Psc 0h45'0" 10d8'12"
Bbsox
Lyn 7h33'53" 57d52'46"
BBTT
Cap 21h26'6" -16d29'53"
BBWnetwork
Cnc 8h58'3" 15d59'15"
BC and Tara
And 2h23'57" 46d50'19"
BC Canorum Caelestis
Cyg 20h25'33" 41d13'39"
BC INFINITY +
Vir 13h9'41" -11d44'26"
BcamcolipemliliTSY
Pho 0h31'38" -40d59'45"
B.C.A.V
Tau 5h34'48" 25d48'28"
BCD Star
Lmi 10h43'11" 33d54'1"
BCI30
And 22h59'1" 51d23'51"
BCK 82784
Vir 12h33'30" 1d58'8"
BD Spencer Alfieri
Uma 11h45'0" 59d55'12"
B.D.40
Lib 15h24'27" -7d12'16"
BDRJ Farrell 538
Her 16h49'17" 23d58'53"
bdss1
Ari 2h31'3" 25d57'47"
BE GOOD
Dra 18h47'8" 58d52'31"

Be Hereford Proud Uma 11h22'5" 59d47'29"
Be Mine Uma 11h44'37" 50d4'36"
Be mine Alexandra? And 1h18'13" 37d34'27"
Be My Light Amy Lake Pho 23h52'16" -45d35'34"
Bea Com 12h34'8" 28d10'25"
Bea Peg 0h6'52" 18d55'29"
Bea - Ajan And 1h17'28" 44d16'11"
Bea and Bill's Anniversary Uma 11h25'23" 33d40'24"
Bea Carrick Cas 1h27'27" 66d59'21"
Bea Cook Cas 1h35'50" 62d45'18"
Bea Don loves Weisenhaus McBride Sgr 18h40'33" -28d15'20"
Béa et Rico Cas 0h12'49" 56d34'29"
Bea Provato Uma 11h56'22" 32d32'52"
Bea Thornton Ori 5h49'6" -4d17'21"
Bea Tramuta Lyn 8h12'22" 43d11'40"
Bea Wattenhofer Uma 13h1'31" 58d12'20"
Bea88 Boo 15h21'55" 42d53'28"
Beacon Light RM Harbourd Family 2007 Cru 12h27'27" -55d59'21"
Beagle Ori 5h19'47" 8d13'28"
Beagsters 03 And 1h19'9" 42d59'24"
Bea-John Uma 10h58'58" 72d12'52"
Bealey Cobb Wheeler Cam 5h42'16" 67d19'16"
Beam Me Up Tess Houigan Uma 10h31'11" 41d49'1"
Beam us up Scottie Uma 9h59'28" 67d1'20"
Beamer And 23h3'50" 47d50'15"
Beamer Uma 11h33'13" 48d14'56"
Beamer Uma 9h13'24" 51d30'49"
Beamish Cma 7h27'3" -15d54'36"
Bean Uma 12h38'17" 60d48'15"
BEAN Aqr 23h48'6" -15d21'8"
Bean Peg 22h29'27" 33d0'28"
Bean Aqr 22h47'31" 1d50'7"
Bean Ori 5h23'59" 6d47'20"
Bean And 0h11'37" 25d56'44"
Bean Bean Mathena Tau 5h44'25" 13d12'20"
Bean Land Mon 7h35'21" -0d57'19"
Bean Toolan Sco 17h55'39" -38d21'39"
BeanBink25 Ori 5h4'27" 10d31'39"
Beaner Psc 23h55'53" 0d31'35"
Beaney Ori 5h45'14" 6d4'55"
Beanie's Light Uma 8h40'37" 70d31'42"
Beanie'sBright Vir 14h5'52" 3d37'6"
Beano & Elsie Hamas Together Always Cyg 20h41'58" 37d11'26"
Bean's Lucky Star Pic 5h16'40" -55d38'37"
Bean's Star Cas 0h4'43" 56d4'16"
Beanstar Tri 1h48'55" 29d47'41"
Bear Ari 2h44'41" 12d58'18"
Bear Uma 8h21'0" 70d31'14"
bear Sco 16h14'15" -12d29'41"
Bear and Timber Memorial Star Uma 11h28'20" 63d16'44"
Bear Bear Vir 11h38'58" 2d41'52"
Bear Brown Uma 11h47'5" 64d47'35"
Bear Dempsey White Aqr 22h4'40" -20d58'5"
Bear gets Lucky - 27 January 2007 Ara 17h12'12" -53d49'43"
Bear In The Air Uma 12h44'1" 54d56'14"
Bear Merritt Psc 23h19'48" 6d22'33"
Bear Star Uma 8h58'20" 62d11'38"
Bear Winger Ori 5h4'56" 10d57'1"
Bearah Jean Crb 15h35'41" 37d8'37"
BearBear Her 16h45'32" 29d56'1"
Bear-bear Sco 16h8'47" -11d37'46"
Bearcat Stacey Davis Platypus Ori 6h4'24" 20d31'42"
Bear-cub Jamie (Newton) Umi 16h12'9" 72d28'41"
Beardsley Leo 11h55'55" 21d4'29"
Bearette Uma 8h26'22" 69d12'20"
Bears Baby Girl Umi 15h54'15" 78d16'56"
Bear's Star Vir 12h5'38" 1d49'41"
Bear's Star Cyg 19h55'3" 37d34'16"
Bears Thunder Uma 10h36'57" 53d34'45"
BearsKMM81 Lib 15h49'3" -5d43'14"
Bearystar Uma 9h42'49" 58d45'21"
Bea's Beacon Peg 21h12'5" 15d46'7"
Bea's Special Star And 0h0'16" 41d0'21"
Béa's Sternchen Uma 10h18'29" 63d53'50"
Beasley Uma 8h43'21" 48d38'22"
Beasley Psc 0h52'5" 12d12'3"
BeaStarFruit Aql 19h26'38" 8d17'16"
Beaswy Ari 2h15'51" 25d31'56"
Beat Cyg 21h29'23" 50d34'8"
Beat Boo 14h40'57" 30d0'25"
Beat Umi 15h4'13" 76d10'12"
BEAT Aqr 22h52'0" -16d54'43"
Beat Allemann And 2h29'43" 42d51'51"
Beat Moosmann Cyg 20h13'10" 49d36'29"
Beat Schaffner Cep 22h11'29" 56d49'14"
Beat und Daniela Kieber 31.07.2003 Umi 14h32'22" 68d55'13"
Béata Sco 17h29'1" -41d57'21"
Beata Cyg 20h5'1" 33d10'18"
Beata Malgorzata Kus Sgr 18h29'0" -16d55'40"
Beata My Shining Princess And 0h46'44" 34d24'17"
Beata Paris Vir 13h45'59" 4d14'42"
Beate Uma 11h7'31" 48d44'7"
Beate Böhme Uma 10h1'50" 69d42'33"
Beate & Frank Uma 9h46'50" 66d46'45"
Beate & Ingo Bock -25 Jahre- Uma 10h22'51" 66d35'48"
Beate Matthies-Schulz Uma 9h51'13" 50d17'16"
Beate Querido Te Quiero... Eri 4h29'52" -20d7'0"
Beates Glückssstern Uma 9h27'30" 47d50'13"
Beatka Bala Ari 2h4'9" 20d12'34"
Beatrice And 0h46'52" 25d17'40"
Beatrice Psc 1h3'11" 32d8'46"
Beatrice Lyr 18h42'16" 33d47'24"
Beatrice Cyg 20h7'37" 35d45'47"
Beatrice Sgr 18h11'48" -20d6'21"
Beatrice Uma 8h50'27" 57d29'48"
Beatrice & Angelo Cnc 8h43'29" 32d19'45"
Beatrice Ann Giannotti Lyr 18h49'31" 46d56'17"
Beatrice Babcock Cas 2h16'52" 65d18'28"
Beatrice Barney Uma 11h37'46" 55d53'7"
Beatrice Bastelli Ori 5h17'41" 8d52'44"
Beatrice Bielli Cyg 21h27'48" 51d23'9"
Beatrice C. Owens Gem 7h47'44" 33d51'14"
Béatrice Carli Roger Uma 11h8'47" 41d52'24"
Beatrice Carlotta Meazza Cas 0h26'48" 62d21'12"
Beatrice Caroline Cnc 8h43'37" 28d42'44"
Beatrice Claudine Graf Uma 12h2'16" 55d32'18"
Beatrice Cornett Morgan Peg 21h41'54" 21d44'41"
Beatrice Cowan And 1h57'26" 47d8'24"
Beatrice Cremonese Umi 15h11'4" 78d43'43"
Beatrice D. Young Daughter of IGD Cas 0h46'33" 58d33'48"
Beatrice Daisy Campion Crb 15h54'49" 27d33'48"
Béatrice de Bruyckere Ori 4h56'28" 15d41'52"
Beatrice Dorothy Mixon Psc 1h3'23" 27d9'59"
Beatrice (Druskin) Kukkonen Cas 1h20'49" 54d14'7"
Beatrice Elizabeth Coleman "MaBea" Cas 1h3'7" 49d47'25"
Beatrice Elizabeth Keefe Lib 15h51'2" -18d5'34"
Beatrice Ella Baughman Uma 12h30'51" 53d57'3"
Beatrice Eloisa Vasquez Lyn 8h24'11" 33d51'28"
Béatrice et Marc Cyg 19h28'13" 35d54'17"
Beatrice Francesca Mon 6h33'19" 7d45'5"
Beatrice Friedman And 2h17'22" 43d56'12"
Beatrice Fzalaj Uma 10h13'33" 55d36'43"
Beatrice Gallatin Beuf Lyn 8h31'15" 54d47'35"
Beatrice "Glückstern" Ori 5h52'42" 6d55'37"
Beatrice Grace And 0h16'11" 47d6'27"
Beatrice Gray Uma 11h8'39" 34d32'43"
Beatrice Hileman Mon 6h53'50" -0d29'45"
Beatrice Koenig Cas 1h41'35" 62d34'38"
Beatrice Louisa And 23h10'20" 52d58'55"
Beatrice Mildred Gordon Florimbio Sco 16h17'53" -15d0'19"
Beatrice N. Moran Lyn 7h30'27" 48d16'54"
Beatrice Opera 75 Years ! Uma 11h35'46" 63d2'52"
Beatrice P. Rickson Tau 5h39'49" 21d3'26"
Béatrice Pelus Tau 4h16'35" 9d45'9"
Béatrice Philibert Ori 5h35'35" -1d45'30"
Beatrice Poochie Rodriguez Sco 17h3'59" -36d38'52"
Beatrice Rose Dasch Ari 2h10'3" 15d32'6"
Beatrice Ruby Rankey Cyg 21h18'22" 42d36'32"
Beatrice Ruth Lynch Cam 3h33'29" 62d4'6"
Beatrice Scott-Tuttle Cas 0h34'25" 60d24'6"
Beatrice Silvia Orena Umi 17h28'31" 80d7'23"
Beatrice Steiner Umi 16h1'25" 79d3'47"
Beatrice Zonzini Cam 6h41'31" 62d40'57"
Beatrice Zwirn Rovner Uma 11h16'37" 37d23'3"
Beatrice, 31.05.1972 Com 12h47'30" 13d27'31"
Béatrice,Christian Hélie Uma 10h21'53" 43d6'19"
Beatris Aqr 21h37'23" 1d24'23"
Beatris74 Cnc 8h21'53" 29d5'50"
Beatrix KC Zalavary Cap 20h14'50" -27d31'44"
Beatriz Lyn 8h1'17" 38d53'30"
Beatriz B Garabedian Per 4h5'14" 48d3'16"
Beatriz Blanco Col 5h51'46" -33d18'37"
Beatriz Buenrostro Cap 20h51'9" -14d56'6"
Beatriz Elizabeth Reyes Cas 0h45'43" 65d37'11"
Beatriz Fandos Fontes Aql 19h4'9" -9d17'50"
Beatriz Olvera Stotzer Cnc 9h3'41" 25d7'16"
Beatriz Rodriguez Cas 23h1'30" 53d13'51"
Beatriz Sanchez Perdomo Com 12h49'7" 20d47'20"
Beatriz Susana Dailey Sco 16h8'34" -16d51'48"
Beatriz Vivero And 0h20'48" 45d18'46"
Beatyfull Janicka B. Lyn 8h40'54" 39d58'37"
Beau Per 2h27'49" 56d43'51"
Beau Gem 7h25'46" 21d6'42"
Beau Tau 5h14'58" 26d31'37"
Beau Her 17h1'40" 22d58'3"
Beau Cru 12h46'9" -59d45'51"
Beau Alexander Smith - Love Eternal Cru 12h28'22" -59d39'6"
Beau Allan Gordon Hall Psc 0h7'22" -4d33'22"
Beau Anthony Danos Sco 17h54'28" -41d47'54"
Beau Branch Bragg Tau 4h20'28" 14d31'48"
Beau Bridges Her 17h35'29" 32d27'28"
Beau Cameron Leo 10h57'0" 6d23'44"
Beau Christian Ari 2h19'2" 26d41'32"
Beau Collin Bixler Cap 20h12'7" -21d13'28"
Beau Drenan Dra 18h44'29" 56d46'43"
Beau Henry Brewer Gem 7h56'12" 29d26'52"
Beau James Bixler Cap 21h5'59" -24d51'9"
Beau Joseph Pick Aqr 20h50'14" -11d31'57"
Beau Langille Munn Umi 15h55'0" 73d30'25"
Beau Lilly And 2h30'2" 39d41'11"
Beau Louis Cockayne-Francis Her 17h54'31" 45d8'7"
Beau Oswald Cyg 20h57'48" 47d18'20"
Beau & Renae Davis Forever Psc 1h13'25" 10d40'41"
Beau Russell Boo 14h40'21" 31d58'18"
Beau Ryan Hartwell Ori 6h7'23" 14d22'39"
Beau Serafin Leary Boo 15h6'40" 33d13'14"
Beau Terence Wilder - Beau's Star Ori 6h18'47" 16d14'22"
Beau Tyler Bretz Lyr 19h3'46" 31d8'5"
Beauch Vir 12h26'49" -4d49'29"
Beaudi's Light Cru 12h30'52" -64d18'12"
Beaujean Melbea's Child Cyg 20h5'43" 33d6'17"
Beau's Sparkle Vir 12h0'29" -7d47'0"
Beaut And 2h7'27" 46d23'17"
Beauté de la Brittany Leo 9h50'14" 18d12'57"
beauté outre la société And 1h16'2" 41d42'43"
Beauties and Beasties Uma 9h24'45" 54d28'27"
Beautiful Lyn 7h12'9" 58d14'33"
Beautiful Umi 15h6'45" 71d3'59"
beautiful Uma 11h16'11" 66d30'57"
Beautiful Aqr 21h29'0" -4d16'42"
Beautiful Lib 14h51'33" -3d24'18"
Beautiful Mon 8h3'57" -0d21'34"
Beautiful Aqr 23h35'0" -9d46'49"
beautiful Apu 14h3'52" -80d24'48"
Beautiful Sgr 19h17'55" -32d51'36"
Beautiful Sco 17h55'1" -32d4'27"
Beautiful Sgr 18h35'20" -22d41'40"
Beautiful And 0h50'37" 41d25'39"
Beautiful Lyr 18h53'57" 34d17'52"
Beautiful Gem 7h44'1" 33d56'12"
Beautiful Gem 6h42'17" 32d18'10"
Beautiful Gem 7h15'10" 31d42'43"
Beautiful And 0h58'4" 35d9'49"
Beautiful Tri 2h10'13" 33d10'18"
Beautiful And 2h20'55" 49d30'59"
Beautiful Per 3h15'20" 47d43'15"
Beautiful Cas 0h42'39" 48d38'17"
Beautiful Crb 15h38'21" 38d51'16"
Beautiful Lyr 18h16'32" 39d46'41"
Beautiful Cyg 20h13'59" 49d19'44"
Beautiful Cas 0h23'7" 56d46'46"
Beautiful Cyg 22h0'16" 50d48'48"
Beautiful Leo 10h10'57" 16d24'24"
Beautiful Cnc 8h41'35" 19d43'53"
Beautiful Her 18h36'9" 19d14'38"
Beautiful Gem 7h26'20" 26d15'3"
Beautiful Ari 3h19'37" 19d50'13"
Beautiful Vir 13h18'0" 11d24'47"
Beautiful Leo 10h58'3" 10d49'6"
Beautiful Psc 0h46'15" 11d32'45"
Beautiful Aqr 22h32'52" 1d12'17"
Beautiful Vir 13h46'4" 2d59'8"
Beautiful Ori 5h5'0" 4d38'34"
Beautiful 7Natalie Lyn 7h55'44" 35d23'5"
Beautiful Abigail Lyn 7h39'43" 41d43'49"
" Beautiful Alex " Cnc 8h51'50" 11d44'58"
Beautiful Ali Psc 1h26'50" 15d25'3"
Beautiful Alison And 2h18'57" 48d38'11"
Beautiful Allison Wallis Mon 6h54'45" -0d15'16"
Beautiful Alyssa Marie Sco 16h7'41" -11d41'13"
Beautiful Amanda Cap 21h11'24" -21d55'22"
Beautiful Amber Cyg 20h15'13" 38d48'35"
Beautiful Amber Psc 0h28'10" 11d3'56"
Beautiful Ami Joy Umi 14h6'47" 70d33'25"
Beautiful Amy Aqr 22h51'40" -6d33'39"
Beautiful Amy Ori 6h16'12" 18d45'35"
Beautiful Amy And 1h17'57" 45d8'43"
Beautiful Andrea Sco 16h9'42" -13d21'54"
Beautiful Angel Cap 21h38'20" -12d50'4"
Beautiful Angel Cap 21h57'5" -17d39'17"
Beautiful Angel And 0h33'42" 35d56'57"
Beautiful Angel Lizzie Cas 0h30'29" 61d27'20"
Beautiful Angel of the Sky Gem 6h33'44" 22d18'16"
Beautiful Ann Field And 22h57'48" 51d52'50"
Beautiful Anna Cma 6h39'21" -17d2'36"
Beautiful Anna Rodriquez Lib 15h34'37" -25d3'13"
Beautiful Anne Cas 0h54'21" 73d34'27"
Beautiful Annette And 1h52'15" 38d9'58"
Beautiful Ashlee 21 Cas 23h0'57" 56d57'3"
Beautiful Baby Tau 5h53'22" 26d44'43"
Beautiful Baby Boy Lambright Cap 21h37'3" -13d37'23"
Beautiful Baby Bri Leo 10h53'9" 13d56'49"
Beautiful Baby Gabriel Peg 22h43'13" 14d50'28"
Beautiful Baby Girl And 2h32'1" 46d40'59"
Beautiful Baby Iga Uma 8h39'20" 70d2'9"
Beautiful Barbara Lib 15h13'17" -4d22'26"
Beautiful Barbara Anne Lmi 10h5'44" 35d12'2"
Beautiful Bayley Lib 14h33'36" -19d21'29"
Beautiful Becka Cam 7h50'44" 73d9'18"
Beautiful Becki Cnc 8h48'53" 18d19'35"
Beautiful Becky Ori 5h43'34" 8d7'50"
Beautiful Becky Lyn 8h33'20" 39d14'24"
Beautiful Becky And 2h33'27" 46d44'56"
Beautiful Becky U 22 Tau 5h50'9" 16d20'23"
Beautiful Bernice Wenkle And 22h59'3" 50d40'15"
Beautiful Beth Lib 14h52'49" -16d36'49"
Beautiful Beth Ann And 2h36'40" 44d32'8"
Beautiful Betsy Psc 1h18'16" 16d22'21"
Beautiful Betsy Sgr 18h19'30" -17d53'39"
Beautiful Betty Cnc 8h19'11" 19d46'0"
Beautiful Bianca Walstra Cas 1h20'37" 56d48'22"
Beautiful Billie Mon 7h19'2" -0d46'48"
Beautiful Blair Uma 11h43'37" 60d15'36"
Beautiful Blue Diamond Cyg 21h21'50" 30d9'10"
Beautiful Blue Eyes Aur 6h11'20" 30d2'47"
Beautiful Bonnie Crb 15h57'1" 35d51'58"
Beautiful Boy Cnc 8h51'1" 30d52'59"
Beautiful Brandy Lib 15h5'28" -5d24'56"
Beautiful Brandy Lee Ori 5h11'24" 15d46'21"
Beautiful Brandy Marie And 1h22'19" 49d45'2"
Beautiful Brenda Crb 15h41'54" 38d1'26"
Beautiful Brenda Tau 3h52'42" 0d31'18"
Beautiful Bri Aqr 20h42'51" -10d13'3"
Beautiful Brown Eyed Kathy Vir 13h47'46" 5d13'5"
Beautiful Buckie And 0h53'10" 40d2'43"
Beautiful Butterfly Princess And 0h13'25" 29d57'23"
Beautiful Cailin Cnc 8h22'29" 25d31'20"
Beautiful Cameron Peg 23h23'33" 29d22'19"
Beautiful - Carolyn Cross Dra 17h2'22" 57d23'8"
Beautiful Cassandra Cas 1h14'35" 62d52'13"
Beautiful Charlotte L. Cyg 20h21'29" 38d52'43"
Beautiful Christian's Soulmate Ori 6h1'47" 11d38'39"
Beautiful Christina And 23h8'17" 39d32'28"
Beautiful Claire And 23h3'53" 40d6'6"
Beautiful Claire And 1h5'58" 46d37'6"
Beautiful Crystal And 0h35'21" 37d1'42"
Beautiful Daniela Vir 13h37'45" -15d56'55"
Beautiful Danijela Ari 3h14'23" 29d53'20"
Beautiful Danyelle And 0h50'4" 41d0'58"
Beautiful Dark Goddess Ser 15h17'36" -0d15'5"
Beautiful Debbie O'Toole And 2h20'29" 44d0'44"
Beautiful Deborah Lynn Shine Cas 0h18'44" 54d32'7"
Beautiful Dee Vir 13h21'41" -6d49'27"
Beautiful Dee Cam 7h19'22" 67d55'56"
Beautiful Deloriz Cnc 8h33'52" 17d22'45"
beautiful dice Cnc 7h59'39" 13d11'10"
Beautiful Disaster Lyr 18h47'16" 39d28'11"
Beautiful Disaster Cap 20h27'15" -26d56'10"
Beautiful Donen Ari 3h4'33" 13d59'29"
Beautiful Elizabeth Grace Tipton Peg 21h17'1" 15d50'57"
Beautiful Erin Leo 11h42'42" 26d14'20"
Beautiful Erin Sco 16h47'36" -41d9'19"
Beautiful "ewe" Cap 20h34'15" -11d59'2"
Beautiful Faerie Sco 17h19'42" -43d36'47"
Beautiful For Eternity Per 3h0'17" 44d58'44"
Beautiful Forever Gem 6h43'23" 12d52'56"
Beautiful Freya And 2h30'40" 38d19'13"
Beautiful Genevieve And 0h14'28" 42d41'0"
Beautiful Gina Vir 12h39'18" 6d15'14"
Beautiful Girl Cnc 8h56'30" 9d28'56"
Beautiful Girl And 0h17'30" 26d50'30"
Beautiful Girl!! And 23h51'52" 46d15'44"
Beautiful Glow Gem 6h41'37" 13d14'26"
Beautiful Goddess Cyg 21h54'7" 53d7'35"
Beautiful Gorgeous Vir 14h41'38" 5d8'59"
Beautiful Gosia Blumicz Star. And 1h28'14" 50d7'6"
Beautiful Hazel Karp Cru 12h33'4" -59d55'59"
Beautiful Heather Cyg 21h47'54" 44d5'19"
Beautiful Heidi Ann And 0h24'21" 33d42'3"
Beautiful Helena Theocharous Cyg 21h51'36" 38d53'48"
Beautiful Hiedi Cnc 8h49'29" 18d11'21"
Beautiful Holly Jo Gem 6h27'2" 24d50'2"
Beautiful Hui Liu Lib 14h26'54" -17d4'59"
Beautiful JAM in my Sky Cnc 9h3'52" 18d51'26"
Beautiful Janet Williams Uma 10h6'26" 67d6'39"
Beautiful Janice Ori 5h11'51" 0d53'37"
Beautiful Jaxous Ari 3h20'36" 15d44'34"
Beautiful Jayne And 2h17'19" 37d34'49"
Beautiful Jazmin Gem 6h26'22" 27d4'27"
Beautiful Jeannie Lib 15h4'2" -1d8'39"
Beautiful Jen Lyr 19h10'10" 27d21'2"
Beautiful Jenn Lib 15h20'0" -4d3'52"
Beautiful Jessica Tau 5h7'33" 27d35'24"
Beautiful Jewls Crb 16h3'11" 30d37'3"
Beautiful Jo Hor 3h9'33" -66d21'45"
Beautiful Judy Ori 5h24'19" -4d35'1"
Beautiful Judy Lyn 7h30'54" 45d38'7"
Beautiful Jules Cnc 8h29'23" 18d35'11"
Beautiful Julia And 23h21'14" 43d49'41"
Beautiful Julia Jones Gem 7h48'14" 23d10'32"
Beautiful Julie And 23h22'34" 50d44'32"
Beautiful Karen Ari 3h13'6" 28d51'28"
Beautiful Karina And 0h19'17" 40d44'23"
Beautiful Karyna Martinez And 1h45'32" 38d44'34"
Beautiful/Katie Psc 1h6'2" 32d9'32"
Beautiful Katie And 23h29'8" 42d2'0"
Beautiful Katie Tau 3h44'15" 20d16'47"
Beautiful Katie Cam 12h52'9" 76d38'52"
Beautiful Katie Lynn Psc 1h43'43" 21d11'27"
Beautiful Katie, 21 Today And 23h27'55" 42d32'43"
Beautiful Katy Lmi 10h21'37" 38d12'14"
Beautiful Kelly Leo 9h22'52" 11d39'52"
Beautiful Kelly Barnes Ari 3h18'19" 16d25'45"
Beautiful Kim And 23h13'11" 47d49'30"
Beautiful Kyla And 1h30'17" 49d50'56"
Beautiful Kyle And 0h49'12" 22d15'31"
Beautiful Kyung-Yeon Cap 20h27'29" -13d17'31"
Beautiful Laura And 23h4'5" 51d19'15"
Beautiful Laura Marie And 0h48'22" 36d11'14"

Beautiful Lauren
Cyg 20h35'27" 41d39'46"
Beautiful Leslie
Vir 12h24'35" -5d2'0"
Beautiful Like Nikki
Cnc 9h6'22" 7d57'34"
Beautiful Lily
Aqr 22h32'31" -1d19'18"
Beautiful Linda
Sco 17h54'29" -41d20'19"
Beautiful Lindsey
Cap 20h50'42" -25d31'20"
Beautiful Lisa
Cam 3h40'46" 62d47'34"
Beautiful Lisa Marie Barry
Tau 3h59'45" 22d54'51"
Beautiful Little Girl
Sco 17h5'34" -38d25'22"
Beautiful (Lori) Girl
Tau 5h59'32" 27d41'41"
BEAUTIFUL LOVE
Sgr 18h54'32" -18d22'11"
Beautiful Margarita Karina's Mother
Sgr 18h29'0" -26d37'19"
Beautiful Martha
Cra 18h47'56" -40d19'59"
Beautiful Mary
Lib 15h59'17" -9d31'5"
Beautiful Mary
Vir 12h22'35" 7d34'2"
Beautiful Maryann
Ari 2h14'27" 14d15'42"
Beautiful Meara
Cas 1h38'40" 63d24'15"
Beautiful Mechell
Ori 5h8'9" 7d18'31"
Beautiful Megan Salter
Vir 13h24'55" 2d4'36"
Beautiful Mel
And 2h28'23" 41d18'14"
Beautiful Melanie
Gem 6h40'26" 19d34'27"
Beautiful Melissa La Munyon
Aqr 23h17'45" -14d26'49"
Beautiful Mellie Smid
Cap 21h46'53" -10d6'34"
beautiful memering
Lyr 18h47'12" 37d48'58"
Beautiful Mess
Crb 15h36'24" 27d20'22"
Beautiful Michelle Gordon's Star
Vir 14h0'19" -19d58'26"
Beautiful Michelle Moussa
Leo 9h42'33" 32d31'20"
Beautiful Mirjana Queen of My Heart
Vir 13h19'8" 4d1'33"
beautiful miss bethann telford
Tau 5h50'24" 16d14'59"
beautiful miss michelle melton
Sgr 19h18'10" -14d4'29"
Beautiful Misty
Cas 1h27'20" 52d26'20"
Beautiful Monicka B.
Uma 12h54'12" 59d8'14"
Beautiful Morning
Sgr 18h9'27" -29d42'0"
Beautiful Natalie Brown
Cap 20h54'37" -23d3'15"
Beautiful Nicole
Ari 2h49'1" 15d47'9"
Beautiful Nicole Renee
Sco 16h10'31" -19d7'29"
Beautiful Nicole Renee
Mon 7h17'59" -0d19'24"
Beautiful Oh Seung Youn
And 2h2'35" 43d53'18"
Beautiful One
Cru 12h19'15" -59d1'45"
Beautiful Pat
Cas 1h33'12" 60d37'11"
Beautiful Phoebe Amber
Mon 6h25'18" -10d29'36"
beautiful precious lindsay bender
And 2h16'40" 48d36'24"
Beautiful Princess
And 2h24'35" 49d36'23"
beautiful princess
And 0h45'16" 39d16'41"
Beautiful Princess
Leo 10h37'40" 12d54'57"
Beautiful Princess
Tau 5h34'34" 27d5'24"
Beautiful Princess Jen
And 1h16'54" 43d47'3"
Beautiful Princess Kay
And 23h54'3" 32d22'49"
Beautiful Princess Pip
Car 7h45'22" -53d6'47"
Beautiful PunkRocker Queen Mena
Cap 20h48'57" -19d54'57"
Beautiful Rachel
Psc 1h11'28" 27d23'36"
Beautiful Rebecca
Cyg 19h55'41" 43d2'25"
Beautiful Rebecca
Uma 8h45'31" 57d17'58"
Beautiful Rebecca Nicole Dewater
Cas 23h0'57" 57d59'7"
Beautiful Renee
Aqr 22h10'31" -4d20'16"
Beautiful Renee
And 0h45'8" 40d58'59"
Beautiful Riggy
Tau 4h12'0" 4d51'28"
Beautiful Robin Spread Your Wings
Lyn 8h21'55" 53d4'15"
Beautiful Roxana Queen of Young
Vir 12h29'8" 9d30'59"
beautiful ruttan
Sco 16h14'22" -14d58'29"
Beautiful Sabrina Ann Webster
Cas 2h28'54" 66d4'26"
Beautiful Sara
And 1h31'43" 44d44'46"
Beautiful Sarah
Her 17h17'52" 49d46'13"
Beautiful Sarah
Lyr 19h20'13" 28d22'57"
Beautiful Sarah Zimmerman
Cap 20h58'24" -18d50'17"
Beautiful Sasha
Cnc 9h7'12" 23d30'38"
Beautiful Satinder Kaur
Vir 13h40'54" 5d8'3"
Beautiful Shadow Dog
Cma 6h35'18" -16d56'38"
Beautiful Shanna Q
Cas 23h27'56" 51d33'43"
Beautiful Shannon Renee
Ori 5h56'17" 17d28'29"
Beautiful Sharon Maria
Cas 0h7'31" 56d3'13"
Beautiful Sinead
Cyg 19h44'11" 30d20'22"
Beautiful Slavena
Ari 2h38'52" 24d28'21"
Beautiful Song
Cnc 8h48'59" 16d7'31"
Beautiful Sonya
Cyg 21h52'9" 38d44'11"
Beautiful soul
Her 16h29'36" 49d52'12"
Beautiful Soul
Leo 11h43'18" 24d6'33"
BEAUTIFUL SOUL
Vir 13h22'54" 12d40'23"
Beautiful Stacie
Lib 14h56'21" -0d56'27"
Beautiful Star Cheryl
Lib 15h34'58" -27d30'35"
Beautiful Star, Beautiful Girl
Lyr 19h12'17" 37d47'33"
Beautiful Stephanie
Psc 1h31'59" 25d37'17"
Beautiful Stephanie M.
And 23h32'12" 48d11'37"
Beautiful Super Star
Cas 23h34'19" 58d32'34"
Beautiful Tahjma
Ori 4h47'59" 12d7'26"
Beautiful Tammy
Uma 9h3'46" 72d57'46"
Beautiful Tania
Lmi 10h13'24" 28d17'20"
Beautiful Tara
And 2h32'8" 45d26'15"
Beautiful Tara
And 2h29'52" 41d39'26"
Beautiful Taryn Lesley
And 0h21'6" 27d57'12"
Beautiful Terry
Cyg 20h29'15" 58d36'20"
Beautiful Theresa
Ari 3h16'42" 21d46'43"
beautiful things
Vir 14h45'49" 4d52'4"
Beautiful Tiffany, My Love, My Star
Uma 8h20'9" 68d43'56"
Beautiful Tina
Sgr 18h53'6" -26d1'24"
Beautiful Tiph
Psc 1h35'53" 25d39'25"
Beautiful Tori
Vir 12h45'29" -5d28'46"
Beautiful Tracy Anne
Cas 1h45'2" 68d28'26"
Beautiful Tricia
Leo 10h24'43" 24d37'38"
Beautiful Tricia Michelle
Sco 16h12'44" -10d15'28"
Beautiful Turtle
Cnc 9h8'54" 24d41'5"
Beautiful Val
Lyn 9h12'22" 37d34'41"
Beautiful Wanda Fay
Crb 15h58'39" 27d27'10"
Beautiful wife and mother Kate
Cas 23h25'29" 56d35'39"
Beautiful William
Pho 0h35'7" -49d34'34"
Beautiful Yanni
Crb 15h36'24" 28d24'28"
Beautiful You
Cam 5h30'37" 77d42'51"
Beautiful Yvie
Cap 20h29'8" -15d56'55"
Beautiful Yvonne Adarac
Cas 22h57'36" 56d41'9"
Beautiful Zahra
Uma 12h5'38" 63d31'51"
Beautiful Zonajen
Uma 9h16'35" 60d0'45"
Beautiful, Bright Nana Jane Jarrard
Cas 0h48'48" 57d10'29"
Beautifulilie
And 0h46'3" 38d50'48"
Beautifulist
And 0h48'45" 40d59'25"
BeautifulNikki
And 0h47'32" 35d25'1"
BeautifulWhiteMysticalUnicornsLight
Cru 12h35'59" -63d1'14"
Beautimus Princess
And 1h52'40" 45d21'51"
Beauty
Lyn 7h41'15" 36d8'23"
Beauty
Cap 20h39'8" -18d43'56"
Beauty 1
Uma 13h30'50" 55d43'7"
Beauty and the Beast
Cnc 8h34'54" 8d28'7"
Beauty in the sky a.k.a Kristina
Cas 0h26'58" 61d21'29"
Beauty Larissa
Cep 20h44'18" 56d8'35"
Beauty MBK
Lib 15h52'59" -10d22'45"
Beauty n Beast
Ori 5h20'43" -5d31'3"
Beauty of Jennifer
Vir 12h20'20" 12d20'53"
Beauty of VLG
Cyg 21h30'4" 31d5'46"
Beauty Star
Lyn 7h54'40" 50d35'58"
Beauty Unsurpassed,Cristen Juliette
Sgr 19h5'17" -16d7'26"
Beauty.....Kerry K
Mon 7h46'30" -0d50'49"
Beauvais Childress
Ori 5h23'30" 5d50'55"
Beaux Yeux
Umi 16h59'45" 75d58'0"
Beavis and Butthead
Cru 12h25'44" -60d32'44"
Beazer Scott Craft
Uma 11h58'33" 45d28'8"
Beazle "2-14-2003"
Aql 20h7'38" 10d19'12"
Beazy
Psc 1h9'32" 24d6'21"
Bèb
Cnc 7h57'17" 16d49'30"
Beb and Boodha
Tau 3h39'16" 22d9'10"
Beba
Psc 22h54'45" 5d24'13"
Beba
Per 4h38'6" 42d51'11"
Beba
Cyg 19h53'19" 30d58'51"
Beba Sandra
Lib 15h7'56" -6d1'25"
Bebaki & Pouzaki
Cyg 21h49'13" 47d14'18"
BeBe
Uma 9h50'35" 44d26'49"
Bebé
Tau 3h34'41" 12d41'32"
BeBe
Vir 13h25'10" 13d9'11"
Bebe
Gem 6h25'56" 21d36'40"
Bebe
Gem 7h22'37" 15d13'41"
Bebe
Gem 7h2'32" 27d16'24"
BeBe
Sco 16h5'29" -20d3'3"
Bebe
Uma 11h43'54" 58d49'43"
Bebe
Cru 12h14'49" -56d45'38"
bebe
Apu 15h36'47" -72d39'18"
bebe
Sco 16h45'32" -27d44'51"
Bébé d'ange
Lyn 7h45'3" 57d44'4"
Bebe Hennessy
Lib 15h23'14" -6d58'6"
Bébé Lulu
Aqr 22h21'56" -12d31'44"
Bébé Tati
Cyg 20h46'13" 32d21'24"
Bebecakes
Aql 19h31'24" 11d22'4"
Bebek Karagoz
Lmi 10h18'3" 33d48'26"
Bebes
Per 3h16'1" 46d41'52"
Bebe's Love
Mon 6h32'50" 8d45'44"
Bebe's Piece of the Sky
Cas 1h57'22" 65d33'36"
Bebesaurus
Aur 5h33'50" 38d35'57"
BEBESITA
Uma 11h46'15" 59d51'53"
Bebey
Her 17h14'48" 24d53'2"
Bebo
Gem 7h18'56" 31d33'8"
Bebo
And 2h20'39" 45d6'49"
Bebop
Gem 7h23'33" 21d51'19"
Bebop
Tau 3h37'4" 3d23'42"
bebop/shaun
Leo 11h9'24" 27d30'39"
Bebu
Lib 15h3'54" -6d12'58"
Bec & Chad's Lucky Star
Cru 12h22'34" -57d33'12"
Bec n Jack
Cru 12h36'6" -59d53'28"
Beca Baby
Cas 0h1'37" 53d32'26"
Becan Justin Maclean O'Brien
Crb 16h14'52" 38d11'45"
Because I Love You
Cep 20h42'45" 69d56'13"
Because I love you more! xxx
Leo 11h1'39" -4d16'36"
Because of Love
Cap 21h38'50" -17d24'7"
Because You Love Me
Cep 22h14'17" 61d11'34"
Because You Love Me - Henry & Aliya
Tau 4h32'34" 21d45'7"
Because You Return Every Other Gift
Cnc 8h14'18" 11d21'31"
Because You're My It
Aqr 22h21'59" -3d9'39"
Because, Wherever, Whenever...
Her 17h47'36" 41d21'34"
BecauseYour theOne Amanda & Jamie
Ori 5h12'44" 10d38'9"
BecBec
Vir 13h14'9" -3d4'31"
Becca
Cas 1h35'31" 73d32'34"
Becca
Lyn 7h35'27" 57d4'48"
Becca
Ari 2h53'3" 14d4'49"
Becca
Tau 4h8'17" 5d13'9"
Becca
Tau 4h10'42" 5d37'9"
Becca
Peg 22h2'0" 12d29'26"
Becca
And 0h35'6" 27d33'0"
Becca
And 23h44'39" 41d17'4"
Becca
And 1h37'10" 49d44'53"
Becca A. Balash
And 2h5'17" 42d42'16"
Becca and Chris 2-11-04
Lyr 19h11'53" 26d52'47"
Becca Ann
Boo 14h38'59" 50d53'3"
Becca Bell
And 1h20'17" 39d30'53"
Becca Blyth
Lyn 7h45'23" 39d53'7"
Becca - Boo
Ori 6h15'32" 14d44'29"
Becca Bradley
Aqr 22h40'48" -2d37'4"
Becca Centuri
Cnc 8h41'35" 15d22'52"
Becca Chaos
Tau 4h24'56" 4d17'12"
Becca Claire Little
Cap 21h28'33" -16d7'45"
Becca David
Uma 11h6'26" 49d56'25"
Becca Gucciardi
Cyg 19h44'46" 37d33'50"
Becca Hoage
Vir 12h26'52" 1d53'4"
Becca Hughes
Uma 8h43'39" 54d29'58"
Becca King
And 1h44'21" 46d33'54"
Becca Kirchhoff
Cnc 8h48'55" 25d56'50"
'Becca Lee
And 0h17'27" 40d39'41"
Becca Loves Sean
Psc 0h0'26" 7d36'18"
Becca Lyn
Lmi 10h14'14" 35d35'32"
Becca Mae
Tau 5h11'0" 18d38'3"
Becca O'Neill Wilk
Sgr 19h48'25" -13d51'31"
Becca Patita
And 1h28'51" 48d0'58"
Becca Re
Gem 6h22'12" 24d15'55"
Becca Richter 1967
Aqr 21h49'56" -0d35'18"
Becca Rosendahl
Tau 5h46'51" 19d39'57"
Becca Stone
Tau 4h33'1" 15d30'10"
Becca Style
Lyn 7h39'9" 54d49'26"
Becca Turner
Cnc 9h7'31" 21d54'24"
Becca Virella
Uma 10h57'20" 48d34'0"
Becca West
Lib 15h32'1" -15d31'24"
Becca Westmoreland
And 0h37'50" 31d39'37"
BECCA0221
Psc 23h30'36" 7d25'29"
Becca12
Ori 6h3'58" 9d44'4"
Becca46
Lmi 9h32'4" 37d20'47"
beccaboo
And 0h59'17" 40d8'58"
BeccaBoo
Sgr 19h11'19" -22d0'6"
Becca's 18th
Gem 7h3'31" 27d8'0"
Becca's Angel
Ori 5h38'18" 4d22'11"
Becca's Heart
Cam 4h19'33" 68d26'9"
Becca's Star
Umi 17h1'47" 80d3'21"
Becca's Star
And 23h45'45" 44d17'1"
Becci Bookner
Leo 11h23'4" 16d24'3"
Becci Jo McLaughlin
Leo 11h4'14" 22d42'45"
Beccy and Oliver Lytton
Uma 11h31'5" 55d34'4"
BECHEROVKA
Sco 16h8'18" -10d39'3"
BECHN
Cyg 20h42'29" 51d32'12"
Beci Marie Owens
And 2h13'42" 38d51'3"
Beck
Sgr 17h51'42" -18d22'56"
Beck
Aqr 21h8'12" -5d54'59"
Beck
Uma 11h43'0" 59d30'18"
Beck Harris
Vir 14h2'53" -12d52'39"
Beck the Speck! Happy 18th Honey!
Ari 2h23'50" 11d27'7"
Beck, Heidi
Uma 10h4'18" 51d16'12"
Beck, Johann Georg Heinz
Uma 9h0'43" 50d31'16"
Becka
Aql 19h54'3" -0d25'42"
Becka and Noah's
Psc 1h4'24" 30d22'21"
BECKA BAIR
Gem 6h35'4" 18d32'14"
Becka Lynn Reed
Leo 10h14'29" 22d57'21"
Beck-a-Boo
Cas 1h0'11" 63d40'54"
Beckars
Dra 15h42'48" 61d47'40"
Becka's Star
Mon 7h31'32" -0d46'10"
Becker
Del 20h36'48" 14d1'42"
Becker, Frank
Uma 10h28'54" 42d13'51"
Becker, Helmut
Uma 9h4'35" 51d20'13"
Becker, Lisa
Uma 12h13'53" 60d1'44"
Becker, Manfred
Ori 6h8'17" 16d35'6"
Becker, Uwe
Uma 8h41'21" 48d37'16"
Becker, Wolfgang
Uma 8h48'31" 47d53'8"
Beckers, Dieter
Uma 8h33'49" 72d40'17"
Beckers, Wilhelm
Uma 11h11'35" 34d17'37"
Beckerz
Peg 21h54'36" 9d12'0"
Becket
Ori 6h11'2" -3d45'34"
Beckett Johnas Andrew
Sco 16h13'57" -30d36'24"
Beckett Johnson
Uma 11h48'16" 32d27'20"
Beckett Kershaw
Umi 15h16'5" 76d53'53"
Beckett Lee Bennett
Leo 9h33'35" 17d1'48"
Beckett Sullivan Gabriel
Peg 22h34'49" 12d6'47"
Becki
Cyg 20h29'8" 49d58'40"
Becki
Cas 2h9'24" 62d24'50"
Becki Balok
And 2h17'27" 46d6'18"
Becki O'Brien
Cyg 21h43'55" 41d1'5"
Becki Trivison
Aqr 22h53'55" -5d18'16"
Beckia, Bestia Plurima Mirifica
Aur 5h8'29" 36d19'59"
Beckie Harmon
Lib 15h0'6" -6d5'14"
Beckie Keith
Mon 7h17'41" -0d6'25"
Beckie Kettle
Ari 3h7'9" 11d37'22"
Beckie Kilburn
Cyg 19h44'47" 31d3'58"
Beckie Kimbell
And 23h23'53" 38d11'43"
Beckie Regusci
Cas 1h46'58" 64d57'7"
Beckie Wolff
Leo 9h59'9" 17d39'21"
Becki.est.meus.astrum
And 2h9'0" 46d46'9"
Becki's Anniversary Star
Cyg 19h39'7" 32d15'33"
Beckjerm
Cyg 19h30'10" 32d4'11"
Becklet1003
Lib 15h19'48" -6d6'2"
Becko
Uma 13h38'28" 49d27'16"
BecksBeth
Lyn 7h25'56" 55d18'49"
Beckworth
Uma 10h32'37" 49d5'56"
Becky
Per 3h8'59" 56d44'54"
Becky
And 2h36'19" 48d17'39"
Becky
Lyn 7h31'39" 48d57'15"
Becky
And 0h8'54" 34d53'14"
Becky
Gem 7h19'17" 32d33'36"
Becky
And 1h59'43" 40d39'44"
"Becky"
Leo 10h19'19" 16d55'20"
Becky
Ari 2h44'54" 29d40'23"
Becky
Cnc 9h17'49" 25d54'18"
Becky
Tau 3h37'35" 13d10'58"
Becky
Psc 1h6'52" 15d59'8"
Becky
Uma 8h20'11" 71d50'28"
Becky
Cam 7h30'17" 76d23'35"
Becky
Sco 16h7'30" -15d36'24"
Becky
Cap 20h34'29" -17d55'10"
Becky
Sgr 18h44'12" -28d53'56"
Becky Abbott
Vir 13h2'35" 8d14'56"
Becky Ager
Cyg 20h15'50" 41d40'52"
Becky And Eddie
Cyg 21h13'48" 33d15'32"
becky and eric's star
Cnc 8h36'26" 7d30'27"
Becky and Jeff +1's star
Aqr 21h9'15" -12d15'3"
Becky and Pat Wedding Star
Lib 15h1'45" -3d55'18"
Becky and Steve in Love Forever! 24
Umi 17h1'19" 78d31'34"
Becky Anderson
Lib 14h38'39" -9d59'31"
Becky & Andrew 8/20/03
Lmi 10h11'45" 35d18'39"
Becky Ann
Peg 22h38'1" 12d0'3"
Becky Ann Bartels
Cyg 21h31'15" 33d3'45"
Becky Appleton Castillo
Cnc 8h37'52" 23d45'35"
Becky Audrey Sander
Cas 3h34'11" 70d21'28"
Becky Augustine
And 0h4'55" 41d50'5"
Becky Bach
Uma 10h51'56" 57d25'9"
Becky Banks Grindle
Vir 12h10'24" 6d46'8"
Becky Becktold's Star
Cyg 21h31'8" 50d37'3"
Becky Boeck
Sgr 18h39'35" -18d25'54"
Becky Boehle Dooley
Cas 23h51'1" 53d57'56"
Becky Boo
And 23h25'1" 41d2'0"
Becky Boo
Lyn 7h37'11" 39d55'56"
Becky (Boo) Weisenberger
Cyg 21h9'3" 46d59'33"
Becky Bordner's Star
Cyg 21h17'8" 45d51'59"
Becky & Bud Mazzant
Ori 5h57'56" 11d31'25"
Becky Bullard
Tau 4h39'46" 22d13'45"
Becky Choi
Cnc 8h21'8" 7d29'0"
Becky Drake
Ari 2h42'47" 27d35'42"
Becky Eads
Aqr 22h37'25" -1d17'44"
Becky Fallis
Uma 10h54'18" 58d47'38"
Becky Fogle
Umi 15h39'54" 76d5'49"
Becky Fritsch
Lyn 8h0'26" 37d17'35"
Becky George
Uma 11h43'22" 50d48'49"
Becky Grove Phi Sigma Chi Alumni
Umi 14h19'33" 73d55'11"
Becky Harris
Sgr 17h55'21" -29d20'4"
Becky Hensley
Cas 0h5'39" 54d12'14"
Becky House
Leo 10h21'5" 16d49'59"
Becky in the Sky
Uma 9h59'2" 70d56'24"
Becky Irene Meyer
Sgr 18h40'59" -27d3'6"
Becky & Jack Cashion: Two as One
And 0h44'5" 42d20'3"
Becky Jane
Ari 3h16'15" 28d33'51"
Becky Jayne Hurford
And 23h13'58" 42d5'56"
Becky Jo Mitchell
Uma 12h24'33" 60d51'25"
Becky Johnson
Crb 15h31'38" 29d4'8"
Becky Joy
Crb 15h41'33" 36d49'39"
Becky Kappel
Sgr 18h34'22" -17d3'6"
Becky Keaton
Tau 5h26'46" 21d30'49"
Becky Keenan
Peg 23h31'49" 17d47'11"
Becky & Kevin Lacey's 25th
Her 17h28'13" 20d41'42"
Becky King
Uma 11h4'14" 65d45'38"
Becky Klein
Uma 9h53'53" 49d48'26"
Becky Kolakowski
Vir 12h19'48" -0d58'50"
Becky L. Black
Cyg 21h45'41" 50d33'40"
Becky Lane Larson
Cnc 8h14'19" 15d16'47"
Becky Lee Stewart
Psc 1h25'17" 7d7'19"
Becky Lynn
And 1h16'48" 36d55'13"
Becky Lynn
And 0h36'36" 43d11'32"
Becky Lynn Hollingsworth
And 1h23'18" 43d38'59"
Becky Mae Holt
Uma 9h53'44" 54d30'3"
Becky McClain
Cnc 9h18'1" 10d56'12"
Becky Mesch
Vir 12h43'26" 3d52'45"
Becky Michelle Kalish
And 0h38'31" 31d26'15"
Becky**Mike**Debbie
Pyx 9h1'30" -26d23'2"
Becky "My Beloved Hummingbird"
And 0h1'55" 45d7'43"
Becky Olsen
Sgr 18h6'1" -28d1'56"
Becky * Our Shining Star - 25/04/83
Col 6h33'14" -33d36'6"
Becky Ozkul
Ori 5h56'18" 18d19'56"
Becky P
And 23h22'0" 51d37'48"
Becky P Wong
Ori 5h30'37" 9d24'23"
Becky Pitt - The Cat With Character
Vir 13h10'1" 7d25'14"
Becky Rainer
And 2h38'22" 45d25'28"
Becky Raines Hall
Sco 17h27'52" -33d23'23"
BECKY - Rebecca Margareta
Ori 6h4'9" 13d34'29"
Becky Star of my Heart
And 0h38'15" 41d34'1"
Becky Sue
Psc 1h9'20" 28d59'15"
Becky Sue Sunshine
Cnc 8h39'12" 14d45'50"

Becky Sue, Will You Marry Me?
Crb 15h36'34" 29d1'52"
Becky (Sugarbritches)
Lmi 10h27'28" 35d28'50"
Becky "The Star You Wished Upon"
Uma 9h27'45" 59d43'54"
Becky Thompson
Uma 8h49'50" 72d49'57"
Becky Vickers
And 1h51'28" 45d6'7"
Becky Wade
Aqr 21h45'55" -1d48'29"
Becky Wait
Psc 1h13'5" 10d13'18"
Becky Watts
Cam 4h13'48" 55d56'7"
Becky Woldert
And 0h16'12" 29d34'22"
Becky, Roy Jr.(Buck), Larry, & Gary
Leo 10h22'39" 20d51'59"
Becky-Jane
Ori 5h52'7" 18d20'37"
BeckyJoeBrave
Lyn 6h59'9" 47d18'55"
Beckylou
Cam 4h28'50" 66d55'30"
Beckylu
Aqr 23h19'56" -12d44'10"
Becky's Beacon
Uma 11h4'59" 35d6'55"
Becky's Butterfly
Vir 14h37'44" -0d7'7"
Becky's Lucky 7
Gem 7h46'24" 26d4'38"
Becky's one in a billion star
Cyg 21h19'24" 43d50'3"
Becky's Star
Leo 9h29'45" 31d13'16"
Beckys Star
Com 12h37'15" 22d32'57"
Becky's Star
Cnc 8h38'32" 8d3'45"
Beckys star of stars.
Crb 16h19'14" 39d30'9"
Becky's Sweet Gregory
Boo 14h23'57" 19d33'29"
Bec's Light of Earendil
Pho 0h57'58" -53d54'33"
BECTON MORGAN
Pho 0h47'13" -56d54'44"
Bectra Hughes
Del 20h43'17" 15d7'6"
Bed of Roses
Sco 16h3'24" -27d12'53"
BEDA ZIEGLER
Cnv 13h32'15" 31d32'4"
BEDAR
Psc 0h55'1" 27d46'16"
Bedorf, Hedi Marie
Ori 6h12'43" 8d52'26"
Bedrana Duka
Lyn 7h54'29" 57d46'5"
Bedros Michael Hagop Kevorkian
Cap 21h20'48" -25d16'29"
BEDROS UND MAJA
Lac 22h23'36" 48d19'30"
bedwyr42
Sgr 17h58'40" -28d8'31"
Bee Goby
Per 3h27'9" 44d32'18"
Bee Haug
Cyg 19h36'34" 38d17'32"
Beeb
Aql 19h52'31" 12d31'16"
Beebee
Umi 16h24'5" 75d24'37"
Beebo
Sco 16h10'46" -19d18'2"
Bee-Bop
Cnc 8h8'27" 11d14'28"
Beebs
Uma 11h26'11" 52d7'30"
Beedgen, Heinrich
Uma 10h42'0" 44d11'54"
Beedlebee
Lac 22h52'27" 52d41'1"
BeeF
Umi 15h45'58" 81d40'5"
Beefcake Stu 30
Per 4h33'22" 48d49'19"
Beefkins
Dra 17h36'37" 61d11'22"
Beefy
Uma 11h24'35" 48d37'5"
Beej
Lyr 19h13'48" 27d30'18"
beejoy4ever
Cyg 19h42'9" 43d39'50"
BEELAD
Aqr 22h46'22" -16d27'19"
Beemo
Per 4h48'21" 39d17'52"
Been Bag Partridge
Cyg 21h34'50" 49d46'53"
"Been There, Done That"
Sge 19h49'40" 18d19'51"
Beena
Leo 11h33'33" 26d3'57"
Beena Patel
Ari 2h33'34" 19d7'29"
Beenken-Stern
Uma 11h8'55" 63d51'5"
Beep Beep Beep
Psc 0h31'55" 8d57'19"
Beeper Buzzell
Sco 16h20'52" -25d26'54"
Beer, Christine
Uma 12h47'23" 61d53'18"
Beerschot
Ori 5h12'17" 5d52'44"
Beeschknobb
Uma 9h2'39" 48d55'19"
Beeshow Michelle
Uma 8h44'8" 50d23'12"
Beesinger23
Vir 14h6'15" -16d29'53"
Beetle's Sparkle
Gem 7h40'0" 17d13'17"
Beez311
Cep 21h52'54" 82d13'2"
Beezenpup Sweetness
Lib 15h24'23" -5d8'36"
Beezer
Sco 16h51'19" -41d58'14"
Beezer
Gem 6h3'36" 27d14'12"
Beezl
Per 3h27'54" 48d54'44"
Beezy
Uma 8h50'47" 48d19'22"
Beg 114 GBC 21
Cap 21h47'30" -11d22'38"
Bega
Vir 14h14'42" -20d30'28"
Begali Claudia
Umi 15h11'46" 73d27'27"
Begerow, Melanie
Lib 15h17'38" -10d46'16"
Beggars Canyon Band
Leo 10h16'35" 9d20'18"
beghemot
Lmi 10h24'26" 38d17'13"
BEHDAR
Lyn 7h6'43" 51d52'55"
Behn, Edelgard
Ori 5h21'10" -8d3'26"
Behnaz Almasi Koupaei
Aqr 21h43'24" 2d13'11"
Behold A PaleHorse
Leo 10h29'3" 13d25'3"
Behounek
Her 16h34'41" 36d58'46"
Behrend, Christina
Uma 9h14'13" 59d4'0"
Behrens X2
Eri 4h16'47" -1d12'19"
Behzod Rostam
Cep 21h41'33" 61d32'54"
Beidanet
Sgr 19h20'16" -13d44'10"
Beijos
Uma 10h27'22" 47d5'16"
Beilah's dad
Uma 13h50'23" 55d24'24"
Beisel, Tatjana
Uma 9h38'53" 48d37'32"
Beita
Dra 15h53'26" 55d32'34"
BeJaBa
Psc 1h25'10" 18d43'12"
Bek
Cru 12h25'12" -56d28'31"
Bek Loves Andrew Forever
Cru 12h27'59" -57d37'3"
Beka
Cru 12h32'39" -59d22'37"
Beka
Cnc 8h53'17" 6d41'31"
Beka Bean Soup
Tau 5h30'9" 26d52'50"
Beka Bedoshvili
Cam 5h27'9" 68d41'5"
Beka Boo
And 2h16'47" 46d23'8"
Bekah Baby
Cap 20h11'22" -13d32'7"
Bekah Boo Sondreggers Heavenly Star
Peg 23h12'50" 17d7'5"
Bekah Jensen
Ori 5h45'28" 7d21'4"
Bekah Lee
Tau 5h44'31" 15d22'51"
Bekah Lynne Gronberg A Rising Star
Leo 11h47'9" 22d9'40"
Bekah Marie Bevins
Sco 16h42'17" -26d5'48"
Beke Ildikó
Tau 5h33'9" 24d3'29"
Békési Mihály Tanár Úr
Lib 15h41'17" -19d5'23"
Békésy-Varga Barnabás
Uma 8h53'50" 47d56'33"
Beki Clark
Mon 7h56'36" -0d46'22"
Bekie
Aqr 22h28'8" -12d38'51"
Bekieey
And 1h42'11" 44d3'27"
Beki's Smile
Gem 6h55'1" 21d12'30"
BEKKI BRIDGE 04
Uma 10h39'28" 54d27'16"
Bekki & Shawn - 1st Anniversary
Vir 12h28'47" -5d26'4"
Bel
And 1h40'28" 45d51'53"
Bel Ami
Psc 1h14'57" 31d29'25"
Bel Ange
Cam 4h25'50" 57d29'51"
bel ange
Ori 4h46'58" -0d12'25"
Bela
Cap 20h52'28" -17d44'24"
Bela
Lyr 18h29'55" 36d33'49"
Bela Ashley
Cnc 8h45'51" 18d46'47"
Bela Ashley
Tau 4h13'59" 16d48'41"
Bela & Beth Bobis
Aur 5h57'52" 42d19'6"
Béla Janos Suhajda
Uma 9h28'43" 66d33'48"
Bela & Sonia's Eternal Love
Cru 12h31'18" -62d57'34"
Beladedic, Milos
Ori 6h13'50" 6d59'39"
BelBerth3.24.01
Lyn 8h42'39" 40d53'33"
Beleevid
Vir 13h0'21" -20d21'42"
Belen " la Princesa "
And 1h46'25" 36d17'5"
Belen Nanette
Ari 1h51'31" 12d28'42"
Belen Navarro
Sgr 17h44'7" -21d38'24"
Beleza do Amore
Lyn 7h38'9" 47d46'39"
Belgica del Carmen
Uma 8h59'8" 49d52'8"
BelGray
Vir 14h6'1" -0d18'39"
Beli
Vir 13h24'16" 12d57'57"
Belice
Leo 9h22'31" 11d53'41"
Belides20050409
Per 2h28'16" 53d51'56"
Belief
Cyg 21h52'47" 49d24'32"
Belief of Pi
Cyg 20h26'44" 36d38'3"
believe
Cyg 21h20'42" 32d47'24"
Believe
Peg 21h16'3" 13d45'18"
Believe
Aqr 23h17'33" -13d11'8"
Believe
Cam 3h56'3" 64d34'28"
Believe
Uma 13h25'34" 56d32'4"
Believe in Magic
Pho 0h34'14" -42d14'29"
Believe My Love
Sgr 18h7'5" -31d40'24"
Believe- William G. Schacht
Per 2h52'34" 32d35'33"
Believing Makes Miracles Happen
Uma 9h6'18" 47d12'41"
Belinda
And 0h1'23" 45d34'13"
Belinda
Cap 20h8'55" -25d15'42"
Belinda
Sgr 18h29'35" -27d23'16"
Belinda
Cas 1h7'30" 67d9'14"
Belinda
Cas 3h3'5" 61d46'36"
Belinda
Dra 15h42'15" 62d38'31"
Belinda
Vir 13h49'14" -3d11'0"
Belinda
Psc 0h11'16" -2d11'5"
Belinda Allgood
Sgr 17h55'51" -27d39'31"
Belinda and Justin Forever
Cru 12h22'32" -62d43'34"
Belinda Ann
Crb 15h32'40" 27d54'58"
Belinda Beth Campbell
Lib 15h48'1" -4d23'29"
Belinda boo boo
And 1h40'40" 49d55'33"
Belinda Cassandra Pitt
Cap 21h22'25" -16d36'55"
Belinda Cristina Valdes
Cap 21h40'32" -14d30'23"
BeLinda EsQuer
And 0h38'16" 25d45'3"
Belinda & Gerhard forever
Uma 12h31'48" 54d12'55"
Belinda Guthrie
Aqr 22h28'53" -6d52'54"
Belinda Harrison
Cyg 19h50'7" 36d35'22"
Belinda Jean Merkelis
Cap 21h21'56" -16d47'55"
Belinda Kate Moore
Cen 11h59'59" -52d11'55"
Belinda Kay
And 2h6'33" 37d35'59"
Belinda Kay 11-11-1982
Sco 16h14'12" -29d48'16"
Belinda Lee Green - Princess Bella
Cru 12h42'44" -64d8'25"
Belinda May Purcill (Billy)
Vir 13h20'33" -0d17'2"
BELINDA my W.W. Dragon Lady.
And 2h23'54" 38d41'2"
Belinda Plutz
Cas 0h40'59" 60d45'1"
Belinda Reser
Cyg 20h43'47" 47d27'19"
Belinda Richard
Aqr 21h26'47" -0d26'10"
Belinda S. Stefl
Cyg 20h14'19" 39d1'35"
Belinda Sachvie
Vir 14h5'28" -16d11'59"
Belinda & Scott's Love Star 2006
Cru 12h38'42" -59d55'22"
Belinda Storla
Leo 9h33'41" 27d12'54"
Belinda Strother-Alston
Nor 16h18'47" -43d24'28"
Belinda Teresa Dominguez-Gross
Aqr 21h47'56" -1d7'53"
Belinda Welsh - My Beautiful Wife
Cra 18h57'17" -40d40'11"
Belinda, Tricia + 5
Uma 9h8'11" 59d47'54"
* Belinda's Shining Light *
Cru 12h33'33" -62d31'54"
belita sole
Cnc 8h12'11" 20d47'10"
Belitz, Lutz
Lib 15h17'15" -8d14'22"
Belky B
Ari 3h21'29" 27d11'10"
Bell
Cnv 13h47'41" 36d48'49"
Bell Boy's Guiding Star
Cru 12h34'21" -64d2'44"
Bell G 1.
Cap 20h55'34" -20d17'16"
Bella
Cap 20h58'16" -19d52'58"
Bella
Cap 20h58'8" -22d25'28"
Bella
Lyn 6h40'25" 56d34'1"
Bella
Uma 14h4'20" 57d26'5"
Bella
Uma 8h19'20" 69d8'36"
Bella
Sco 17h27'5" -43d24'40"
Bella
Aqr 22h46'0" -24d45'15"
Bella
And 23h45'5" 33d2'41"
Bella
Lyr 18h57'26" 39d25'25"
Bella
And 0h10'22" 45d26'15"
Bella
And 2h2'24" 47d29'33"
Bella
And 1h41'32" 45d15'43"
Bella
And 1h44'14" 45d56'49"
Bella
And 2h34'46" 48d39'10"
Bella
And 2h33'10" 49d0'58"
Bella
Lyn 7h41'9" 47d58'4"
Bella
Aur 5h42'29" 50d27'30"
Bella
Cas 0h47'15" 57d40'31"
Bella
Tau 3h41'24" 27d3'8"
Bella
Ari 3h20'21" 29d23'36"
Bella
Ari 3h16'28" 28d50'29"
Bella
Cnc 8h50'19" 26d49'7"
Bella
Leo 10h28'34" 15d21'9"
Bella
Leo 9h49'6" 19d39'45"
Bella
Psc 1h15'19" 23d51'36"
Bella
Ari 2h5'45" 21d55'48"
Bella
Ori 6h4'10" 16d34'8"
Bella
Leo 11h36'30" 7d46'16"
Bella
Tau 3h27'11" 8d24'56"
Bella
Ori 4h51'14" 11d11'41"
Bella
Psc 23h23'35" 3d14'18"
Bella Adriana
Cas 1h1'58" 61d46'34"
Bella & Ally Precious Carol
Cyg 19h55'45" 38d57'2"
Bella_Amor
Cyg 21h32'52" 54d15'50"
Bella Anima
Uma 11h25'25" 64d59'30"
Bella Babygirl
Mon 8h0'9" -0d52'50"
Bella Barbra
Uma 10h53'2" 71d27'40"
Bella Blue Skies
Aqr 20h46'8" -9d18'48"
Bella Brenda
Sco 15h57'18" -22d13'59"
Bella Brittany
Sgr 18h13'43" -27d5'17"
Bella Butterfly
And 0h59'41" 46d47'17"
Bella Capella
Vir 14h23'59" -2d27'33"
Bella Christina
Peg 23h49'52" 10d53'54"
Bella Christine
And 1h36'41" 45d57'16"
Bella Cody
Her 17h2'51" 29d46'20"
Bella Corser
Peg 0h6'56" 23d28'20"
Bella Cristiana
Cnc 8h36'52" 18d36'40"
Bella Daniella
Lac 22h28'43" 50d59'36"
Bella Danielle Nicole
And 1h35'23" 47d8'30"
Bella Dawna
Pho 23h51'3" -45d21'59"
Bella Del Amor
Ori 5h44'44" 11d43'54"
Bella Donna
Psc 1h12'49" 15d44'15"
Bella Donna
Cyg 19h40'22" 29d16'52"
Bella Donna
And 0h53'37" 41d4'44"
Bella Donna Lee
Cap 21h37'14" -20d7'37"
Bella Donna Tracy
Gem 7h19'44" 30d54'18"
Bella donne
Uma 8h33'55" 65d16'4"
Bella Dora Hsu
And 23h33'58" 37d50'0"
Bella Duarte
Peg 21h22'50" 18d40'52"
bella e splendida
Sco 16h9'57" -9d19'41"
Bella Elena
Lib 14h54'39" -14d31'52"
Bella Elisabetta
Ori 6h20'28" -0d29'12"
"Bella" Emily Valentino
And 1h0'31" 34d58'19"
Bella Erika 27
Sgr 18h48'38" -29d32'50"
BELLA ERIN
Cap 20h25'55" -26d35'39"
Bella Erin
Cas 22h59'14" 57d26'0"
Bella Farfalla
Cas 0h58'38" 60d18'12"
Bella For Michael
Ari 3h28'53" 23d51'45"
Bella Gabrielle Winn
Ori 6h10'6" 6d52'50"
Bella Galli
And 23h44'3" 41d33'12"
Bella Gina
Uma 11h49'16" 43d38'37"
Bella Grace
And 1h14'37" 38d7'43"
Bella Gramble
Lyn 8h40'32" 33d41'9"
Bella Gretta
Crb 15h38'56" 32d7'48"
Bella Heather
Ori 5h24'40" -0d54'26"
Bella Hula
Lib 15h41'25" -18d46'26"
Bella Isabelle
Lib 14h46'24" -8d52'44"
Bella Jeannettia
Leo 11h26'35" 27d22'31"
Bella Jennifer
Cyg 20h21'51" 50d59'56"
Bella Jess Whitelock
Ori 5h40'31" -2d6'21"
Bella Josie
Gem 7h9'8" 26d34'9"
Bella Juanita
Uma 10h5'30" 48d47'12"
Bella Kathryn Ellena
Vir 13h3'8" -15d27'2"
Bella Kimberly
Cyg 20h37'37" 41d52'36"
(Bella) Lindsay Susanne
Sgr 19h43'25" -14d58'35"
Bella Lisa
And 23h33'1" 48d14'59"
Bella Luce
Vir 12h29'55" 1d36'43"
Bella Luce
Uma 10h23'38" 58d30'10"
Bella Lucia
Crb 15h20'43" 31d28'3"
Bella MacKenzie
Ori 5h34'56" 9d31'23"
Bella Mackenzie Pitman-Page
Col 5h59'54" -30d0'35"
Bella Maggy
Lib 15h31'47" -5d9'1"
Bella Makenzie - Gram's Star
Cru 12h39'50" -59d18'52"
Bella Maria
Cas 23h26'49" 53d1'8"
Bella Maria
Ari 3h22'27" 28d51'13"
Bella Maria
Leo 11h41'9" 27d10'40"
Bella Maria
Lmi 9h46'24" 39d55'15"
Bella Maria Girod
Sgr 19h10'14" -33d5'27"
Bella Maria Sawdy
Cnv 13h29'11" 31d24'7"
Bella Mente
Vel 9h27'38" -40d37'52"
Bella Mia
Dra 17h52'35" 53d33'23"
Bella Mia
Lib 15h13'17" -14d45'56"
Bella Mia
Vir 12h19'48" 5d46'52"
"Bella" - Mia Bella Rosa Bahamonde
Ari 3h16'48" 28d16'13"
Bella Michelle
And 23h10'20" 35d44'12"
Bella Milyavskaya
And 1h52'41" 38d22'40"
bella monkey's star
Lib 15h14'21" -4d55'49"
Bella Mylissa Luce Di Mio Vita
Gem 6h47'52" 32d47'0"
Bella Nicole
Tau 4h6'17" 6d43'0"
Bella Note
Tau 4h26'44" 15d58'44"
Bella Notte
Crb 16h7'4" 35d18'43"
Bella Pescatello
Cnv 12h27'38" 43d17'7"
bella principessa
Psc 22h54'16" 7d59'46"
"Bella" Priscilla
And 0h47'29" 35d34'59"
Bella Ragazza Addie
Gem 7h27'43" 30d13'16"
Bella Rasdal
And 0h43'19" 42d4'0"
Bella Robyn Romaine
Umi 16h22'10" 76d51'18"
Bella Rosa
Vir 13h34'34" 8d39'25"
Bella Rosalia
Ori 4h52'24" 11d13'46"
Bella Rose
Ori 6h18'3" 9d10'53"
Bella Rose
Mon 7h32'42" -4d3'10"
Bella Rose Baker
And 23h38'20" 48d48'5"
Bella Rose Lemieux
Crb 15h33'16" 34d35'59"
Bella Rosemary
Ari 2h9'28" 22d39'44"
Bella Soleil Nairn
Leo 10h42'21" 7d14'55"
Bella Stallion
Equ 21h13'35" 11d41'7"
Bella star of Linda
Uma 8h54'15" 70d2'37"
Bella Tina
Lyn 7h43'0" 51d34'25"
Bella "Tinkles" Carriere
Vir 13h49'0" 2d59'5"
Bella Vella, Louise
Crb 15h45'38" 29d8'10"
Bella Wexler
Uma 12h25'45" 60d59'10"
Bella, segue sempre il suo cuore
Lyr 18h40'25" 36d26'31"
Bella-Boo
Mon 6h26'3" -0d45'39"
bellacamellita
Ori 5h35'1" 0d51'24"
BellaCip
Aqr 23h8'53" -8d35'29"
Bellakathy Centauri
And 1h47'44" 49d27'58"
bellamia
Lyr 19h14'17" 34d28'3"
Bellamore
Ori 6h7'32" 17d19'7"
BellaNatalia
Aqr 22h16'35" 2d15'57"
Bellanca Family Star
Sge 19h54'39" 18d52'53"
Bellanna
Uma 8h19'15" 70d2'46"
Bellarachie's Heliosmicros
Gem 7h6'28" 28d58'24"
Bellarose Violet Jardin
Tau 5h26'45" 22d15'50"
Bella's Daddy
Ori 5h41'43" -2d33'14"
Bella's Hope
And 23h13'20" 45d17'29"
Bella's Star
Gem 6h51'23" 28d10'7"
Bella's Stella
Umi 15h14'11" 69d15'7"
Bella's tiny little thing
Dra 16h55'36" 57d8'9"
Bella-XNMSSO-Un Amor para Siempre
Uma 12h38'45" 62d10'3"
Bella-Zemer
Cyg 19h38'38" 51d45'49"
Belle
Cyg 19h50'58" 51d37'59"
Belle
Cam 4h56'10" 59d6'53"
"Belle"
Uma 8h44'16" 50d7'53"
Belle
And 1h29'32" 34d54'6"
Belle
And 1h17'8" 37d50'52"
Belle
And 0h14'45" 33d24'53"
Belle
Leo 11h12'29" 24d11'14"
Belle
Cyg 20h21'7" 55d49'59"
Belle
Lib 14h59'49" -24d5'6"
Belle Ajolaise
Peg 21h50'40" 17d27'55"
Belle Amore
Leo 10h42'53" 14d5'49"
Belle Baker
Cyg 21h13'3" 46d22'32"
Belle Céline
Lib 15h50'19" -4d11'30"
Belle D&G 9-3-2005
Cyg 20h3'0" 43d41'50"
Belle d'Aiguebelle
Cyg 20h31'36" 41d48'40"
Belle Donne
Uma 13h50'7" 61d48'12"
Belle Emeline
Cam 5h26'7" 69d21'18"
Belle Fille
Sco 17h43'53" -41d47'53"
Belle Heather Wilson
Her 18h4'12" 18d34'24"
Belle Isabelle, mon Isamour
Gem 7h52'16" 15d32'34"
Belle Kendall
Uma 8h20'23" 69d38'14"
Belle L. Bulis
Crb 15h37'47" 38d1'12"
Bell'e Luminoso come una stella
Vir 13h25'41" 5d48'21"
Belle (Maire Bugler Nunan)
Lib 14h51'6" -2d37'17"
Belle McCormick
Tau 4h35'48" 22d29'32"
Belle Olivia
And 1h59'59" 40d1'4"
Belle Renteria
Uma 11h27'34" 33d20'23"
Belle Sarah
And 0h34'27" 43d21'40"
Belle Scott
And 0h30'45" 32d20'12"
Belle & Seth Osborne
Cyg 22h2'5" 45d14'17"
Belle Star
Gem 7h28'44" 32d50'5"
Belle Star
Ari 2h43'29" 27d52'27"
Belle Vita
Sco 17h16'7" -38d51'42"
Bellea Brianna Driver
Uma 9h28'45" 56d53'35"
Belle's In The West Wing
Umi 13h40'35" 71d51'57"
Belle's Light
Sgr 19h44'58" -12d44'10"
Belle's Star
Cyg 20h38'55" 38d20'14"
Belleza
Uma 11h34'56" 38d23'48"
Bellezza
And 0h18'9" 45d57'3"
Bellezza
Uma 9h32'37" 66d24'54"
Bellisimo
Cas 1h22'23" 60d46'18"
Bellissimo Daidra
And 0h35'48" 36d21'2"
Bello
Lyn 7h50'1" 49d15'14"
Bello Accendere di Andrew Michael
Per 2h49'55" 53d34'34"
Bello Amor
Cnc 8h25'56" 23d39'29"
Bello Principessa Michelle
Per 2h17'5" 54d52'4"
Bello Rachel
Cap 20h11'10" -11d2'56"
Bells
Vir 13h23'12" 13d12'34"

Bell's Celestial Eden Aql 19h2'17" -0d56'33"
Bell-Style Tau 3h41'47" 24d52'27"
Bellus Ame Welsh Cnc 9h2'55" 26d5'52"
Bellus Astrum de Linda Cnc 8h31'31" 11d16'48"
Bellus Brooke Ramsey Lib 15h26'5" -12d16'44"
Bellus Bullae Lyn 6h58'16" 60d0'58"
Bellus Femina Cas 22h57'19" 57d9'47"
Bellus Martine Leo 11h13'56" 15d16'13"
Bellus Rebecca Carroll Sco 17h32'45" -37d6'4"
Bellus Sophia And 23h50'57" 39d27'47"
bellus Vicki Cam 3h28'7" 60d33'23"
Bellus Viscus Boo 14h46'10" 26d7'17"
BellusAmor Lyn 8h28'37" 34d17'59"
Belly Button Star Tau 5h26'35" 18d55'58"
Bellyboo Cam 4h14'16" 72d10'19"
Bellybutton Her 16h53'34" 33d18'14"
Belma moje srce And 2h29'13" 37d40'48"
Belovarac / Thorn Sge 19h26'20" 17d28'46"
Beloved Sge 20h6'0" 16d37'27"
Beloved Cyg 20h6'6" 35d34'9"
Beloved Cyg 21h20'1" 31d20'29"
Beloved Crb 15h19'37" 31d33'22"
Beloved Cyg 20h28'33" 40d3'14"
Beloved Cyg 19h37'2" 39d52'10"
Beloved Her 16h37'43" 46d53'18"
Beloved Umi 15h15'40" 72d43'17"
Beloved Sgr 19h46'34" -24d32'21"
Beloved Aleksandr Miheikin Vir 12h14'56" -0d51'14"
Beloved Andy Schipul Her 17h5'28" 32d15'1"
Beloved Baby Bahr Umi 15h32'32" 73d55'47"
Beloved Baby Olivia Grace Queenan Lib 15h12'11" -10d9'32"
Beloved Becky And 0h35'33" 34d8'27"
Beloved Blanche Cmi 7h29'1" 6d55'11"
Beloved Brad Aragon Aqr 23h52'40" -3d43'50"
Beloved Bubs Cyg 20h19'37" 51d22'14"
Beloved Büsi Ori 6h5'18" 18d59'19"
Beloved Carolie Lyn 8h41'10" 41d41'27"
Beloved Cecilia Mon 7h30'49" -0d52'40"
Beloved Chase Uma 8h43'26" 58d36'39"
Beloved Christina Cas 2h3'38" 63d53'44"
Beloved Commander Colonel Dohnke And 23h9'36" 51d27'44"
Beloved Constance Cas 1h33'35" 63d44'42"
"Beloved" - David and Amy Forever Her 17h22'25" 37d9'56"
Beloved Faye Vita Cas 1h17'59" 63d29'8"
Beloved Geogog Del 20h27'1" 20d9'39"
Beloved Husband Smitty,dad ,pop-pop Cep 22h29'20" 62d25'45"
Beloved James Harold Wayne Summers Ori 5h30'3" -0d2'33"
Beloved Jun Sco 16h4'34" -15d39'0"
Beloved Karla Pilavdzic Cnc 8h40'27" 18d44'38"
Beloved Lady Selket Sco 17h51'23" -38d53'30"
Beloved LaMoyne Ari 3h6'58" 14d42'14"
Beloved Lisa Sgr 19h20'38" -23d20'34"
Beloved Mana Aqr 23h26'54" -15d49'27"
Beloved Maureen Aqr 22h47'9" -5d43'47"
Beloved Meghan Ori 5h23'51" -0d59'3"
Beloved Mijita - SS 4:7 Cas 0h34'9" 51d45'26"
Beloved Mommy"Edie"Edith Chiarenza Leo 9h45'15" 27d33'32"
Beloved Mother & Best Friend, Leisa Crb 15h40'8" 26d3'17"
Beloved Mother Colleen Kaye Sco 16h56'43" -32d46'29"
Beloved Mother Dr. Tere Cap 21h47'25" -13d21'31"
Beloved Mother Efstathia Skoulikas Lyn 6h57'8" 50d43'22"
Beloved Mother Norma Trovitch Vir 13h20'24" 12d14'14"
Beloved Noel Cap 21h35'42" -11d13'4"
Beloved Patricia Crb 15h34'30" 26d15'26"
Beloved Pepper Roo Gem 6h36'13" 12d51'42"
Beloved Poppy Pho 23h43'37" -41d17'54"
Beloved Raine Tau 4h25'4" 23d50'3"
Beloved Rita Sgr 19h5'24" -17d42'13"
Beloved s Vir 12h25'53" -6d11'33"
Beloved Sanctuary Choir Lyr 18h25'15" 34d30'35"
Beloved Sandra Aql 18h53'21" -0d55'33"
Beloved Sheri, Light in My Darkness Psc 0h41'49" 3d51'56"
Beloved Steffany Ori 6h15'3" 10d38'19"
Beloved Suzette Uma 13h36'27" 54d38'28"
Beloved Vicky And 0h15'58" 45d53'51"
Belshandor Oph 17h34'4" 7d28'9"
Beltran Family Uma 9h19'4" 50d23'38"
Beluah Cnc 8h39'16" 22d3'22"
Bélus age 30 Uma 13h57'24" 52d17'40"
Belva J George Ari 2h45'37" 28d38'10"
Belvedere Bubba Jenkins Lyn 7h49'33" 47d49'57"
Bempsey Co Uma 9h2'54" 49d55'34"
Ben Uma 10h28'44" 47d59'34"
Ben Her 16h4'51" 48d52'18"
Ben Lyn 7h47'52" 51d11'16"
Ben Per 3h23'6" 35d59'17"
Ben Cnc 8h48'31" 28d22'53"
Ben Gem 6h35'40" 26d15'5"
Ben Ori 5h8'41" 9d18'31"
Ben Psc 1h11'47" 17d57'26"
Ben Ori 6h15'30" -0d47'16"
Ben Umi 16h56'46" 78d11'26"
Ben Aqr 23h30'7" -22d35'43"
Ben Adairs way Gem 6h32'46" 20d52'54"
Ben Adam's Star Uma 8h44'7" 56d14'50"
Ben Aguirre Per 4h23'26" 48d48'40"
Ben Ali Boudoukara Cap 21h42'43" -11d1'36"
Ben & Amanda Goodness Lib 15h23'23" -6d4'1"
Ben and Cam Cyg 19h42'11" 55d48'38"
Ben and Carolyn Cyg 20h46'34" 39d0'34"
Ben and Daina Ari 2h53'54" 24d46'4"
Ben and Georgianna Forever Love Lyr 19h1'47" 43d1'0"
Ben and Jason Garcia And 0h41'55" 42d59'1"
Ben and Kaley's Star Lib 14h52'33" -2d12'52"
Ben and Katie Cyg 20h43'35" 49d11'48"
Ben and Kelly Cra 18h18'29" -39d48'18"
Ben and Lisa Uma 9h29'3" 47d1'3"
Ben and Lisa - Forever Us Lmi 10h6'2" 37d33'47"
Ben and Lorraine Marszalak Sge 19h44'23" 18d28'13"
Ben and Martha Munoz Psc 1h35'57" 20d34'34"
Ben and Melissa Cyg 19h39'23" 56d19'24"
Ben and Monica And 1h15'45" 45d8'5"
Ben and Rin Cru 12h34'36" -56d12'19"
Ben and Trina Ara 16h50'23" -62d9'20"
Ben & Andrea Ozretich Cyg 19h33'15" 31d53'4"
Ben Andrew Jackson Peg 21h26'24" 15d41'29"
Ben Armour Uma 10h36'10" 39d39'49"
Ben & Aubrie24 Pho 0h10'16" -44d14'13"
Ben Bach Dra 18h21'52" 76d30'28"
Ben Bacon Her 18h19'28" 15d9'38"
Ben Banks Sco 16h26'24" -26d33'36"
Ben Becker Uma 11h0'38" 45d19'52"
Ben & Becky Pernol Cyg 20h44'43" 36d29'48"
Ben Benjumangi Eastment Cru 12h15'58" -62d18'2"
Ben Blevins Cep 22h14'52" 76d56'16"
Ben Bomlitz Cnc 7h57'30" 12d42'1"
Ben Boo Per 4h10'57" 46d18'9"
Ben Brosca "My Shining Star" Gem 7h17'57" 24d2'7"
Ben Browning Uma 13h33'10" 55d12'15"
Ben Castle Uma 9h19'33" 49d10'31"
Ben Cook Cnc 9h6'37" 28d34'57"
Ben Crabtree Per 3h27'31" 44d57'49"
Ben Crosby Cru 12h37'13" -61d13'36"
Ben Darin Lukosky Her 18h48'26" 13d59'55"
Ben David Bryant Cnc 7h56'9" 17d40'56"
Ben Doughty Per 2h52'19" 33d25'43"
Ben E. Workman Aql 19h18'19" 15d36'28"
Ben Edward Thompson Per 2h56'26" 45d55'56"
Ben & Emma - 4.03.2005 Cru 12h32'4" -63d29'20"
Ben & Emma's Star Cru 12h37'37" -62d33'31"
Ben & Fer Uma 10h45'45" 44d49'31"
Ben Finley and Kayla Masters Eri 3h48'2" -0d59'28"
Ben Flynt Ori 5h35'31" -0d20'29"
Ben G. Grant Gem 7h34'3" 16d3'38"
Ben Gardner Hamby Sco 16h16'39" -41d27'56"
Ben George Tattersall - Ben's Star Peg 21h30'9" 7d58'53"
Ben & Gosia , A Light born from Love Lib 14h59'15" -24d35'21"
Ben Gray Her 16h29'59" 38d8'16"
Ben & Hampton's Big Bang Black Hole Cyg 20h20'50" 45d15'45"
Ben Hillan Aql 19h41'3" 11d55'25"
Ben Honey Tau 3h47'9" 24d52'17"
Ben Hood Tau 4h2'13" 15d43'0"
Ben Howard Root Sco 16h6'35" -12d51'44"
Ben J. Honaker Her 18h39'21" 15d21'25"
Ben James O'Neill Peg 21h30'49" 3d1'16"
Ben Jenkins Ori 5h49'53" 2d31'38"
Ben & John - warrior brothers Uma 9h14'0" 53d54'25"
Ben Joseph Dolan Per 4h48'57" 49d34'47"
Ben Kannenberg - Ben's Star Cen 11h36'45" -64d18'51"
Ben Kenney Lmi 10h11'49" 31d20'3"
Ben Kersey - Ramius The Dark Ori 6h11'3" 15d14'26"
Ben Khay Sco 17h44'17" -35d45'26"
Ben Kimstra - Light of my life Cru 12h43'44" -56d59'15"
Ben King as The Cheshire Cat Gem 6h53'16" 25d9'22"
Ben Koretz Her 17h5'49" 35d17'4"
Ben & Kourtney's Star Cyg 20h16'55" 48d7'22"
Ben L. Zucker Dra 17h12'26" 64d18'8"
Ben Liam Scheerder Umi 14h42'39" 83d29'10"
Ben Lorusso Lac 22h14'36" 37d44'56"
Ben loves Serah forever Cnc 8h38'43" 15d43'56"
Ben Maher Superstar Cnc 8h46'35" 13d56'56"
Ben & Marika Cra 18h3'2" -37d28'56"
Ben Martin Dra 19h32'24" 65d38'45"
Ben & Mary Cyg 20h46'28" 36d37'32"
Ben Mayer Cnc 8h44'53" 26d42'9"
Ben McAllister Uma 11h9'29" 28d44'24"
Ben McCarron Per 3h14'3" 51d45'6"
Ben McGinn Per 3h38'1" 39d3'43"
Ben & Mel - Eternal Love Cru 12h18'34" -62d40'18"
Ben Merced And 23h23'22" 51d32'7"
Ben Michael Osborne Lmi 9h44'0" 37d27'35"
Ben Michael Wallace Ori 6h18'59" 10d12'23"
Ben Millesen Her 17h30'27" 44d10'51"
Ben Mills Per 2h52'7" 42d2'2"
Ben Miro Bühler Cyg 21h15'20" 30d6'27"
Ben My Knight in Shining Armor Lac 22h16'12" 37d42'11"
Ben Newsome - SUPER-STAR Her 16h40'45" 25d7'21"
Ben no Hoshi Psc 1h24'16" 16d37'35"
Ben Nowak True Blue Hero Per 4h28'8" 43d59'59"
Ben Oliver Berberich Uma 12h5'12" 35d45'8"
Ben Orchard Psc 0h8'48" -2d57'7"
Ben - Our Star - Nat Cru 12h36'9" -64d13'47"
Ben Panasevich Cma 6h46'54" -12d23'28"
Ben Parke Cru 12h4'38" -63d36'35"
Ben & Paula Forever Cra 19h7'2" -37d45'6"
Ben per aspera ad astra Per 3h5'51" 32d17'0"
Ben Perez Uma 10h51'30" 59d42'43"
Ben Ross Cra 18h47'28" -39d56'34"
Ben & Shannon For Eternity Psc 1h53'30" 2d57'15"
Ben Singletary Cep 23h20'8" 77d18'23"
Ben Spies Ori 5h41'59" 1d39'23"
Ben Stewart Uma 9h28'15" 55d0'27"
Ben & Susan Benavidez Lyn 7h1'46" 60d27'31"
Ben T. Wood Per 2h11'52" 54d52'55"
Ben & Tanya's Star Cyg 19h55'8" 39d26'1"
Ben the Boy Uma 12h1'5" 32d34'11"
Ben Thompson's Star Cra 18h55'52" -40d56'28"
Ben & Tiffany Ori 5h34'55" 10d47'42"
Ben Trefilek Her 17h52'52" 23d15'57"
Ben Tyler Hilgendorff Her 17h59'18" 29d11'28"
Ben Vickers Umi 14h41'24" 69d46'29"
Ben Vogelsang Uma 8h46'52" 65d23'21"
Ben Watson And 23h3'26" 41d29'13"
Ben Weatherman Lib 15h1'33" -5d31'28"
Ben Westpoint Ori 5h8'15" 3d31'20"
Ben & Whee Ling Leo 10h22'33" 20d29'16"
Ben William Mangarelli Lib 15h3'58" -2d14'9"
Ben Williams Aqr 22h57'52" -6d49'22"
Ben Yarwood Uma 11h22'29" 55d40'49"
Benali Mon 8h1'2" -0d52'0"
BenAmy Uma 12h20'22" 62d55'21"
BenAnitaWesleyAustynMaeganCourtney Cet 0h56'34" -0d34'26"
BenAnna Gem 6h58'44" 16d12'4"
Benard & Luca Bocconcelli Sgr 19h35'29" -16d59'13"
Benard Peter White Jr. Gem 7h41'13" 32d32'35"
Benaud Boo 14h44'2" 23d57'44"
Benay is our Star Gem 6h11'53" 24d48'55"
Bencks Cnc 9h8'59" 31d22'26"
Bend Family Star Ori 6h4'18" 17d10'34"
Bendaverobin Ori 5h29'36" 9d39'0"
Bender Uma 10h33'46" 46d0'21"
Bender Uma 11h5'6" 53d16'11"
Bender 47 Cam 3h41'27" 71d22'58"
Bender, Eva & Steffen Uma 13h0'12" 53d42'30"
Bender's Bright Sgr 18h28'8" -18d52'57"
Bendixen Cam 3h56'57" 61d40'12"
Bendixen, Nico Uma 11h23'59" 34d17'24"
Bendora Tau 4h21'3" 13d43'47"
Bendrahma Ori 5h31'48" 10d22'7"
Bendschick, Andreas Ori 5h31'5" 10d49'55"
Bene Psc 2h1'8" 6d13'29"
Bene and Sarah forever Lyr 18h49'41" 31d33'27"
BeNe BEST ar Sgr 19h0'17" -15d8'7"
BENEATH Vir 13h30'50" -13d28'9"
Benecio Michael Critelli Sgr 18h8'5" -29d7'59"
Benedetta Tau 4h12'20" 5d57'37"
Benedetta Botta Dra 20h39'18" 70d7'50"
Benedetta Contino Her 17h51'22" 21d2'46"
Benedetta Porcasi Uma 11h23'5" 58d5'37"
Benedetta Regina Cas 1h19'15" 63d48'41"
Benedetto "Ben" DiNardo Her 16h39'3" 20d52'42"
Benedetto C. Esposito Per 3h30'0" 51d28'53"
Benedetto Giovanni Cas 0h14'40" 52d45'50"
Benedetto Simonelli Sco 16h28'37" -26d32'31"
Benedict Affiyie Owusu Her 17h37'8" 37d21'57"
Benedict Brady Cep 22h55'46" 65d10'3"
Benedict J. Figler Ari 3h25'6" 28d4'32"
Benedict & Julia...For Eternity Psc 0h47'48" 13d7'28"
Bénédicte Brassart Uma 9h19'34" 56d20'59"
Bénédicte et Bertrand Psc 1h29'42" 3d14'16"
Bénédicte Marie Umi 15h9'8" 70d49'47"
Benedikt Uma 10h17'50" 68d10'14"
Benedikt Alexander Ori 5h8'51" 15d58'51"
Benedikt Moritz Uma 12h33'13" 56d7'7"
benedikt & sabrina schnuppe Lmi 10h32'39" 37d36'9"
Benedix, Claus Cnc 8h23'40" 18d59'53"
BeNeely Star Uma 10h54'23" 66d50'31"
Benet Hardin Leo 11h18'3" 18d4'25"
Benevolence Psc 23h28'32" 1d33'49"
Benevolent One Sgr 18h52'54" -19d47'59"
Benezet-Toulze Per 3h7'12" 49d49'43"
Bengela Lyn 7h7'41" 46d40'45"
Bengt Åke Grunditz Lyn 8h0'44" 33d46'31"
Bengt Fischer Apu 16h17'19" -76d42'41"
Bengt Robert Asplund III Ari 2h36'9" 14d8'59"
Benicia Plumbing Inc. Aur 6h7'26" 30d0'47"
Benicio Jauregui Cnc 8h55'28" 7d6'13"
BeNicole Abramovitz Uma 10h40'24" 42d43'32"
Benihime Dra 16h7'20" 57d46'12"
Benilde Arreola Lib 15h27'47" -6d3'18"
Benini Stella Cam 5h0'22" 64d21'51"
Benita Peg 22h13'40" 12d2'12"
Benita Knight Dinzeo Petruck Pho 23h48'38" -40d7'33"
Benita Lynn Cucurella And 2h30'41" 45d32'40"
Benita McFadzen Cra 19h1'3" -39d29'54"
Benita Rivera Eri 4h21'35" -32d28'59"
Benita Southworth Gem 6h47'11" 33d28'44"
Benita "Together Forever" Cas 0h4'24" 53d49'16"
Benita's Angel Cap 20h25'35" -14d41'31"
Benito B. Rish Cyg 21h38'34" 36d2'12"
Benito Mesisca Leo 11h14'50" -2d50'39"
Benj Per 3h20'47" 51d14'2"
Benja Tau 5h45'54" 14d19'30"
Benjaman Roy Her 16h22'4" 8d14'12"
Benjami Rodrigo i Aleixandre Aqr 22h14'19" -7d44'49"
Benjamin Lib 15h49'40" -9d24'14"
Benjamin Umi 14h0'54" 88d18'19"
BENJAMIN Sgr 17h53'20" -28d46'30"
Benjamin Ari 2h6'5" 19d37'30"
Benjamin Cnc 8h13'1" 25d48'33"
BENJAMIN Aur 5h41'42" 49d22'58"
Benjamin Crb 16h13'38" 30d57'59"
Benjamin A. Chamberlain Leo 11h31'50" 11d16'26"
Benjamin A. Clark Cnc 8h37'45" 32d10'8"
Benjamin Aaron Davies Peg 22h33'37" 14d11'15"
Benjamin Adam "Bubba" Uma 9h2'8" 51d28'0"
Benjamin Adam Dorman, our buddy Her 16h38'29" 25d17'30"
Benjamin Alan James Catling Uma 11h25'23" 39d32'9"
Benjamin Alexander Boes Uma 10h6'26" 48d28'47"
Benjamin Alexander Dyckerhoff Aqr 21h39'0" -7d40'6"
Benjamin Alexander Mode Ori 5h31'58" 0d43'3"
Benjamin Alexander Trimble Per 4h44'24" 41d47'51"
Benjamin Alfred Rich Sco 16h0'20" -24d8'19"
Benjamin Allan Murray Cap 20h19'20" -23d14'55"
Benjamin Almeida Psc 1h56'13" 3d3'35"
Benjamin and Annette Potter Ori 5h41'2" 3d29'55"
Benjamin and Bridget Downs Umi 14h1'38" 71d39'38"
Benjamin and Christa Ori 5h19'2" -9d4'27"
Benjamin and Elizabeth Zinckgraf Cyg 19h25'4" 54d14'37"
Benjamin and Francis Baker Leo 11h32'7" 16d29'49"
Benjamin Andrew Her 18h26'1" 18d25'48"
Benjamin Andrew Crespo Lyn 8h55'4" 43d26'14"
Benjamin Andrew Kratz Sgr 18h19'28" -32d18'50"
Benjamin Anthony Parr Ori 6h3'20" 7d15'46"
Benjamin Anthony Waer Leo 11h25'37" 9d59'45"
Benjamin Anton Hilmer Vir 12h54'20" 12d39'56"
Benjamin Archie Akers Cyg 19h55'49" 51d21'34"
Benjamin Arthur Blatchford Aql 19h41'19" -0d46'32"
Benjamin Asher MacDonald Tau 4h8'49" 29d35'30"
Benjamin Babcock And Pamela Karst Cyg 20h3'47" 38d6'35"
Benjamin "Benji" Rivers Her 17h36'16" 16d55'32"
Benjamin Bentley Ori 6h24'53" 14d57'16"
Benjamin Brett McLain Aql 18h58'38" -0d2'14"
Benjamin "Bucket" Rodriguez Her 17h49'20" 46d14'25"
Benjamin Bullard Sgr 18h35'0" -22d35'44"
Benjamin Byrick Umi 4h57'36" 89d30'59"
Benjamin C. Meredith Her 18h42'3" 18d18'39"
Benjamin Calvert Gem 6h48'15" 26d44'48"
Benjamin Camisa Davis Uma 12h15'1" 57d21'47"
Benjamin Carl Palmer Tau 5h54'36" 24d19'30"
Benjamin Cayton Adovasio Psc 0h37'14" 11d40'27"
Benjamin Chaniago Cru 11h59'17" -59d52'12"
Benjamin Chappel Per 3h24'17" 51d40'22"
Benjamin Charles Fisk Ori 5h32'15" 7d12'2"
Benjamin Charles Otto Leo 11h6'3" 8d50'58"
Benjamin Charles Shepherd Ari 3h19'3" 24d5'15"
Benjamin Charles Small Uma 10h30'38" 52d50'16"
Benjamin Charles Ward Her 18h40'20" 20d57'56"
Benjamin Chase Culver Vir 13h16'12" -17d30'19"
Benjamin Chase Plyler Uma 13h56'24" 61d39'15"
Benjamin Colak Sgr 19h1'13" -20d8'2"
Benjamin Cole Lyman Her 18h2'2" 25d7'9"
Benjamin Cord Melton Her 18h32'26" 14d21'16"
Benjamin Corkhill Cru 12h1'58" -62d52'48"
Benjamin Corley Rausch Sgr 19h53'29" -12d39'47"
Benjamin Craig Andreas Andersen Cnc 8h28'33" 21d40'55"
Benjamin Crocker Ori 6h2'0" 9d52'3"
Benjamin D. Thomas Her 17h20'13" 36d29'11"
Benjamin Daniel Slaughter Lib 15h8'21" -6d10'26"
Benjamin Daniel Snyder Cep 22h0'52" 62d2'21"
Benjamin David Briggs Cru 12h18'40" -57d15'29"
Benjamin David Forster Leo 11h5'12" -0d51'59"
Benjamin David Glick Aqr 22h38'28" 1d20'40"
Benjamin David Kaufman Lac 22h22'32" 51d12'33"
Benjamin David Tellman Boo 14h30'20" 22d41'41"
Benjamin Diffenderfer Her 17h15'32" 32d25'56"
Benjamin Dolberg "Ben's Star" Vir 13h24'44" 4d15'56"
Benjamin Donald MacKenzie Ari 1h53'46" 23d44'10"
Benjamin Donohoe Cru 12h2'1" -63d12'24"
Benjamin "Dover" Worchel Ori 6h5'51" 10d0'53"
Benjamin Driscoll Per 2h55'11" 48d56'23"
Benjamin E. Fireman Aqr 22h13'1" -1d4'33"
Benjamin E. King's Star Cep 22h6'21" 83d4'59"
Benjamin Eddings Aur 5h37'57" 50d33'42"
Benjamin Edward DeHaven Sco 16h14'30" -14d3'12"
Benjamin Edward Pestovic Her 18h12'32" 22d48'5"
Benjamin Edward Sebastian Hornby Per 2h53'35" 51d2'15"

Benjamin Edward Thomas (Ben)
Her 17h4'13" 16d2'12"
Benjamin Edward Wyatt
Per 2h44'59" 38d30'3"
Benjamin Emerson Malmsten
Uma 10h20'4" 51d53'45"
Benjamin Emmanuel Kerkow
Dra 20h40'38" 71d31'12"
Benjamin Eric Mason
Sco 16h7'39" -14d30'36"
Benjamin Espinosa
Aql 19h45'42" -0d5'56"
Benjamin Ethan Peligri
Lib 15h33'19" -26d15'28"
Benjamin Ethan Peligri
Per 3h16'47" 53d33'1"
Benjamin F. Harrison
Cnc 8h27'35" 23d25'7"
Benjamin Faust
Umi 14h31'46" 69d0'34"
Benjamin Ferguson
Cma 7h11'29" -31d19'51"
Benjamin Flanders
Aqr 21h51'54" -0d31'57"
Benjamin Forrer
Cep 1h40'34" 83d26'7"
Benjamin Fox Willem
Vul 19h54'53" 27d57'40"
Benjamin Francis Becker
And 1h25'33" 47d24'2"
Benjamin Franklin DeBolt
Vir 12h39'27" -8d7'35"
Benjamin Franklin Reinauer II
Cyg 19h33'55" 51d28'20"
Benjamin Freeman Johnson
Cru 12h28'22" -55d48'15"
Benjamin French
Boo 15h45'28" 39d59'12"
Benjamin Gadeau
Cnc 8h24'59" 13d52'32"
Benjamin Galileo Hron
Ari 2h49'11" 14d9'7"
Benjamin Garol
Psc 22h57'15" 7d0'11"
Benjamin George
Per 3h11'36" 50d52'39"
Benjamin (gifta)
Uma 14h20'14" 59d24'9"
Benjamin H. Ramsey
Sgr 18h18'38" -32d39'27"
Benjamin Haight
Per 3h29'37" 51d17'24"
Benjamin Harry Berman
Aqr 23h44'46" -13d25'18"
Benjamin Hartley
Tel 18h28'29" -46d6'46"
Benjamin Haskin (UB 3)
Ori 5h59'14" 11d51'44"
Benjamin Hilmer Kolseph
Uma 13h50'42" 60d45'29"
Benjamin Hogan
Umi 16h16'55" 71d10'56"
Benjamin Howell Palmer
Her 18h25'57" 27d40'17"
Benjamin Hunter
Ori 6h16'42" 10d36'57"
Benjamin Hutchison
Psc 23h28'36" -3d3'34"
Benjamin Imbriano
Sco 16h3'13" -29d44'44"
Benjamin Inoshita Goldstein
Cnc 8h52'58" 10d15'53"
Benjamin Isaac Longden
Aql 19h28'27" 4d25'26"
Benjamin J. and Erica D. Walsh
Cyg 19h40'28" 54d15'20"
Benjamin J. Bond
Lib 15h45'56" -26d57'50"
Benjamin Jack Green
Umi 16h44'58" 75d21'46"
Benjamin Jack Read - Ben's Star
Umi 17h7'57" 76d23'10"
Benjamin Jacob
Her 17h15'15" 46d23'29"
Benjamin Jacob Ellis
Lac 22h21'3" 48d10'19"
Benjamin Jacques Gautreaux
Leo 9h23'22" 23d29'38"
Benjamin James Blackburn
Cep 21h57'20" 65d11'56"
Benjamin James Comer
Aql 19h52'27" -0d29'36"
Benjamin James Iredale
Per 2h56'6" 47d17'22"
Benjamin James Johovich
Her 16h56'0" 34d1'36"
Benjamin James Loydall
Cnc 9h19'37" 18d21'25"
Benjamin James Misterka
Lyr 18h55'33" 34d21'32"
Benjamin James Rubin
Uma 13h6'48" 53d2'49"
Benjamin James Umlor
Cep 22h31'19" 72d6'14"
Benjamin Joel Kleinman
Aur 5h57'25" 42d4'22"
Benjamin John Clemenzi
Her 18h2'47" 21d34'31"
Benjamin John Comley
Aql 19h31'45" 7d54'19"
Benjamin John Murray
Cru 12h39'48" -59d15'54"
Benjamin John Sillitoe
Psc 0h46'31" 16d5'44"
Benjamin Jon Carlson
Sco 16h11'19" -18d28'29"
Benjamin Jordan Enggas
Cep 22h2'23" 72d52'15"
Benjamin Joseph Deary
Her 17h35'38" 32d19'32"
Benjamin Joseph Dinardo
Her 16h5'49" 46d58'54"
Benjamin Joseph Foley
Pho 0h37'13" -54d22'8"
Benjamin Joseph Gibson
Cep 21h20'15" 66d45'11"
Benjamin Joseph Moreland
Vir 12h41'11" 11d21'0"
Benjamin Joseph Strand
Dra 9h31'48" 73d15'0"
Benjamin Jude Steinberg
Her 17h36'24" 35d55'23"
Benjamin K. Davis
Lyn 8h20'11" 47d33'9"
Benjamin & Katie
Psc 23h36'49" 4d28'3"
Benjamin Katz
Lyr 18h53'26" 44d4'38"
Benjamin Kenneth Hamilton
Cam 4h9'16" 53d36'23"
Benjamin Kestenbaum Klafter
Ari 2h18'5" 26d52'44"
Benjamin Klein
Lyn 6h53'33" 52d24'3"
Benjamin Kohlberg
Boo 14h2'25" 9d29'42"
Benjamin Kojo Akyereko
Aqr 22h27'35" -21d14'54"
Benjamin Kolin
Boo 14h31'51" 14d51'12"
Benjamin Kowalis
Her 17h36'48" 38d2'4"
Benjamin & Kristy
Ori 5h41'27" -3d5'10"
Benjamin L. Sojka
Cap 21h23'20" -23d43'33"
Benjamin L Stern
Per 2h46'40" 52d57'9"
Benjamin LaCroix
Ari 3h22'45" 27d22'8"
Benjamin Langer
Per 3h3'34" 56d6'3"
Benjamin Layton Harvey
Her 17h25'26" 32d22'44"
Benjamin Lee Carico
Ari 2h22'48" 14d40'28"
Benjamin Lee Prevedelli
Psc 1h29'26" 13d8'37"
Benjamin Lee Worsham
Uma 11h37'36" 53d16'42"
Benjamin Leo Gutierrez
Aql 19h56'54" 12d46'10"
Benjamin Leonard Flores
Lib 15h4'48" -1d27'17"
Benjamin Lewis Adkins
Her 17h39'52" 42d3'57"
Benjamin Litchfield Ross
Aur 6h0'34" 47d53'25"
Benjamin Lizotte
Cmi 7h30'48" 5d31'38"
Benjamin M. Kochendorfer
Ori 5h45'55" 9d16'48"
Benjamin M. Studer
And 23h57'9" 48d15'10"
Benjamin M. Thompson
Cep 22h30'35" 58d5'41"
Benjamin Mahon
Cru 12h2'17" -63d5'48"
Benjamin Malouf - 26 November 2003
Sgr 18h7'7" -27d25'55"
Benjamin Marc Day
Cma 6h52'4" -16d17'58"
Benjamin Matthew Cook
Uma 13h25'29" 62d38'33"
Benjamin Max
Sco 17h31'49" -36d37'24"
Benjamin Max Polonsky
Her 17h6'42" 30d51'12"
Benjamin Maxim Huntley
Per 4h13'55" 42d55'55"
Benjamin Maxwell Gordon-Pound
Per 2h14'2" 53d21'10"
Benjamin Mayginnes
Her 17h21'19" 36d58'4"
Benjamin McCoy Deines
Ari 3h19'33" 28d45'28"
Benjamin Mezi
Uma 9h3'20" 62d58'50"
Benjamin Michael Bellomo
Ari 2h29'27" 19d7'24"
Benjamin Michael Garrahy
Cru 12h40'57" -62d41'53"
Benjamin Michael Lofft
Gem 7h47'56" 28d48'51"
Benjamin Michael Teasdale
Ari 3h17'47" 29d20'53"
Benjamin Michael Vogel
Aur 5h39'34" 32d43'47"
Benjamin & Michaela
Sgr 18h26'32" -16d32'10"
Benjamin Micheal Madison Denesha
Tau 3h37'42" 28d0'49"
Benjamin & Mommy 7:15 p.m.
Vir 13h26'25" -0d5'48"
Benjamin Morris Jones
Per 3h14'38" 47d24'51"
Benjamin Muller Lyon
Ari 3h5'11" 26d52'15"
Benjamin My Love
Vir 12h59'23" -18d15'38"
Benjamin Myers
Gem 7h6'31" 33d36'9"
Benjamin N. Chappel
Her 16h28'1" 44d53'6"
Benjamin Nelson Brunner
Vir 12h29'40" -0d9'44"
Benjamin Oliver Collison
Per 4h27'11" 42d23'39"
Benjamin Oliver Edward Lancashire
Uma 10h38'11" 41d53'46"
Benjamin Oliver Sankey
Per 3h15'44" 44d55'8"
Benjamin Oliver Skjodstrup Hansen
Uma 10h44'21" 47d47'17"
Benjamin Orellano
Her 18h2'49" 22d15'17"
Benjamin Patrick O'Neill
Oph 17h14'5" -0d20'38"
Benjamin Patrick Sullivan
Cnc 8h9'50" 17d10'43"
Benjamin Paul Konger March 28, 1984
Uma 10h28'52" 50d54'8"
Benjamin Paul Lawrence
Ori 4h53'3" 3d33'31"
Benjamin Paul Mellett - Ben's Star
Her 17h37'51" 20d26'5"
Benjamin Paul Trivelli
Boo 15h9'41" 42d22'25"
Benjamin Perlson
Her 17h19'56" 16d59'8"
Benjamin Phillip Kotta
Ori 6h14'54" 7d50'1"
Benjamin Pollak
Umi 14h21'48" 78d23'42"
Benjamin Ponce Mendez
Cap 21h38'36" -14d37'1"
Benjamin Pospisil Eubanks - 9:01 am
Cam 3h53'46" 64d40'3"
Benjamin R. Roth
Ori 5h1'33" 9d41'40"
Benjamin Ralph Rophie
Gem 7h29'55" 25d50'57"
Benjamin Randolph Brekke
Uma 10h30'31" 67d49'6"
Benjamin Rausch
Uma 11h58'3" 54d50'2"
Benjamin Reily Smith
Sco 17h19'2" -44d10'33"
Benjamin Rhys Kirkham
Per 3h36'52" 39d23'54"
Benjamin Richard Bamford
Psc 1h38'14" 18d27'30"
Benjamin Richard Ewers
Vir 12h13'3" -0d55'14"
Benjamin Riley Singer
Uma 12h7'35" 58d26'6"
Benjamin Roan Demorest
Psc 1h5'50" 28d0'57"
Benjamin Robert Chester
Psc 0h34'48" 21d0'39"
Benjamin Robert Peate
Umi 14h54'3" 81d19'8"
Benjamin Robert Peters
Boo 15h14'41" 41d20'41"
Benjamin Robert Terrence Cridland
Ori 5h19'55" -0d24'31"
Benjamin Rohinton Ludlow Bathurst
Lyn 8h38'33" 37d49'24"
Benjamin Rush Webster
Ori 5h13'1" 7d26'38"
Benjamin Ryan Molnar
Dra 18h45'37" 71d46'28"
Benjamin S Babcock
Gem 7h25'7" 33d20'4"
Benjamin S. Ginsparg
Aqr 22h0'41" 2d11'57"
Benjamin Salsburg
Ari 3h15'42" 24d50'23"
Benjamin Samuel Easton
Her 18h49'55" 21d31'50"
Benjamin & Sandy
Ori 5h55'24" 7d5'17"
Benjamin Santiago Meza
Sgr 19h6'10" -31d9'9"
Benjamin & Sarah's One Year Star
Cyg 21h52'56" 47d33'26"
Benjamin Scott Gregorio
Lib 15h48'52" -10d57'5"
Benjamin Scott Harrison
Aqr 20h50'53" 2d13'36"
Benjamin Scroggins
Lib 15h36'1" -7d46'6"
Benjamin Shack Sackler
Lac 22h41'28" 50d10'31"
Benjamin & Shanice Hall -I Love You
Pho 2h14'27" -39d48'45"
Benjamin Shannon Sample
Leo 9h40'23" 19d36'56"
Benjamin Sidney Powell
Psc 1h17'4" 11d56'12"
Benjamin Skurbe
Sco 16h11'51" -8d41'44"
Benjamin Skyler Smith
Her 17h29'41" 37d49'29"
Benjamin Steck Whetstone
Uma 10h21'15" 60d56'23"
Benjamin Stephen Bassford
Lmi 10h10'41" 30d31'21"
Benjamin Stevens Coccaro
Aqr 22h36'54" -2d3'6"
Benjamin Stewart Barfoot
Her 18h0'43" 19d31'24"
Benjamin Stiles
Umi 15h37'43" 67d54'42"
Benjamin Stimson
Per 3h9'14" 54d18'30"
Benjamin Stricoff
Uma 9h10'39" 72d2'3"
Benjamin Stuart Mullikin
Leo 9h27'33" 28d12'9"
Benjamin Stuart Sutton
Aqr 23h46'45" -18d1'33"
Benjamin T. & Michelle J. Weaver
Cyg 19h35'21" 32d16'8"
Benjamin Theodore Selle
Lib 14h30'8" -11d35'32"
Benjamin Thomas Blue
Uma 8h35'54" 47d5'11"
Benjamin Thomas Butzke
Tau 5h13'45" 17d34'52"
Benjamin Thomas Huyett
Gem 7h0'45" 19d20'21"
Benjamin Thomas Kaufman
Her 17h11'4" 42d1'9"
Benjamin Thomas Meyers
Tau 5h9'34" 27d57'4"
Benjamin Thomas Ramsey
Per 3h12'34" 47d6'41"
Benjamin Thomas Taylor
Lmi 10h3'56" 38d0'42"
Benjamin Thorndike
Cyg 20h38'22" 52d47'5"
Benjamin Todd Cook
Aqr 21h52'2" -2d12'15"
Benjamin Tonson Berkley
Umi 14h18'59" 77d23'25"
Benjamin Trost
Umi 14h26'19" 76d19'53"
Benjamin Troy Staton
Her 17h33'27" 35d35'23"
Benjamin Turoczi
Per 4h22'3" 51d47'20"
Benjamin Vallejos
Cam 4h0'15" 70d37'19"
Benjamin Vance Beaubien
Uma 11h19'30" 45d33'49"
Benjamin Vazquez
Sco 17h43'26" -33d58'6"
Benjamin Vega - 1992
Aqr 22h37'13" -24d19'3"
Benjamin Victor Hieke - Ben's Star
Per 2h53'12" 47d3'35"
Benjamin Viegas
Eri 4h34'28" -0d31'31"
Benjamin W Hamilton
Aur 5h53'19" 54d1'25"
Benjamin Wade West
Tau 3h41'15" 18d16'19"
Benjamin Wagstaff
Dra 16h22'37" 52d46'33"
Benjamin Walid Yono
Aqr 22h9'44" -4d41'27"
Benjamin Walter Hicks
Ori 5h30'28" 14d30'16"
Benjamin Willen 14/11/1985
Sco 16h34'20" -42d57'0"
Benjamin William Buchanan Axon
Cep 23h22'31" 70d37'54"
Benjamin William Jocelyn
Vul 21h8'51" 27d31'17"
Benjamin Wilson
Uma 11h34'48" 46d59'24"
Benjamin Young of Shetland
Aql 19h28'41" 3d59'4"
Benjamin, Our Heavenly Angel
Aqr 22h46'54" 0d16'43"
Benjamin, Sr. & Janet Rideout
Cyg 21h39'12" 36d53'12"
Benjamin052601
Gem 7h30'45" 34d15'27"
BenjaminCourtney
Crb 15h17'36" 29d7'36"
Benjamine Jerome Firtag
Per 2h39'59" 53d28'4"
Benjamin's Rainbow Bright Star
Ara 17h11'53" -55d12'42"
Benjamin's Star
Lmi 10h11'41" 29d0'11"
Benjamin's Valor
Cnc 8h43'14" 22d38'50"
BenjaminSioui/BenoîtBouthillette
Cep 21h43'51" 64d33'2"
Benjemen3-3-99
Umi 14h24'48" 75d35'59"
Benji
Cap 20h50'46" -25d14'44"
Benji Bear
Aql 19h23'53" 8d4'56"
Benji Cantwell
Lib 14h57'20" -24d20'29"
Benji Rose
Umi 14h24'10" 82d2'6"
BenjiAnne
Sge 20h6'0" 17d46'41"
Benjiman T. Blue
Aqr 22h8'6" -0d29'41"
Benjoran
Lyn 8h59'56" 42d15'31"
BenLovesCrystal
Uma 11h2'15" 48d18'22"
Benly M Duran
Leo 11h17'10" 16d58'23"
Ben'n'Al
Lib 14h52'49" -1d56'42"
Benneditha
Ori 5h56'38" 21d41'58"
Bennet Anthony Warren
Aur 5h46'16" 55d1'58"
Bennett
Tri 2h22'50" 32d19'0"
Bennett
Uma 11h37'44" 61d58'12"
Bennett Andrews & Megan O'Brien
And 23h25'33" 36d58'48"
Bennett Brown
Umi 14h39'17" 77d58'52"
Bennett Christopher Wagner
Aqr 22h29'9" -0d14'43"
Bennett George Collins
Psc 0h36'16" 7d15'30"
Bennett James Biscotti
Ori 6h16'51" 9d34'33"
Bennett McDonald Pearson
Sgr 18h31'29" -18d58'50"
Bennett R.
Uma 9h50'46" 51d24'40"
Bennett Reeves Douglas
Uma 11h22'7" 46d11'36"
Bennett Richard Meoli
Sgr 18h14'12" -19d24'30"
Bennett Spencer Weickert
Cap 20h52'38" -23d26'0"
Bennett Taylor Munday
Lib 15h2'55" -2d11'12"
Bennett Wachman Farina
Psc 1h25'48" 26d34'11"
Bennett William Philip Bidwell
Uma 8h13'53" 65d55'25"
Bennett50
Cyg 20h22'46" 48d23'37"
Bennette Andrew Maruca
Umi 16h23'15" 79d5'27"
Bennett's Place
Lib 15h3'54" -9d13'5"
Bennicium Ursus
Uma 10h34'33" 44d5'34"
Bennie & Cora Lee Jew
Cyg 20h30'19" 31d26'58"
Bennie Dunn Thompson
Aqr 20h47'38" -12d25'24"
Bennie Gene Robinson, Jr.
Hya 9h34'6" -0d55'40"
Bennie L. Perkins
Cnc 9h17'15" 20d10'26"
Bennie Lawyers Go Home
Ser 18h18'30" -13d36'25"
Bennie Muller
Cyg 21h43'9" 31d30'45"
Benno
Dra 18h28'4" 50d28'0"
Benny
Her 17h4'39" 31d29'53"
Benny Angel Mendez
Her 17h10'28" 45d48'15"
Benny Boy
Per 2h45'43" 55d58'59"
Benny Boy
Per 4h6'30" 35d16'34"
Benny Boy
Cep 21h3'6" 58d22'11"
Benny Brewster
Lyr 18h49'28" 37d53'49"
Benny Chavez (El Toston)
Sgr 18h4'29" -24d14'3"
Benny Dale Branton
Aql 19h46'58" -0d12'8"
Benny & Debbie
And 0h38'54" 42d47'46"
Benny Fok
Sco 16h30'50" -37d20'9"
Benny Jonah Kraus
Sgr 19h26'31" -15d13'45"
Benny Montalvo
Cam 5h37'57" 58d55'27"
Benny Palmentere
Lib 14h55'55" -0d51'37"
Benny "Papaw" Snyder
Uma 10h11'59" 43d53'24"
Benny Starling
Ori 5h38'53" -1d58'52"
Benny und Julia
Uma 9h23'5" 41d54'38"
Benny & Verena
Uma 11h57'58" 38d10'45"
Benny, my saving grace.
Aql 18h46'54" 7d45'32"
Benny89
And 2h36'28" 50d21'37"
Benny-Brenda Vonfeldt
Cyg 20h29'43" 46d52'39"
BennyD
Uma 11h22'56" 44d14'59"
Benny's Legacy
Lyn 7h22'41" 53d31'52"
Benoit Billet
Peg 21h55'4" 7d39'5"
Benoît Lebihan 09-04-1981
Umi 14h54'39" 79d5'51"
Ben's Bar Mitzvah Star
Cep 22h37'24" 74d14'45"
Ben's Legacy (18.09.06)
Per 2h50'36" 53d9'37"
Ben's Scarlett
Cru 12h43'20" -64d7'11"
Ben's Sheba
Cnc 8h38'7" 7d9'28"
Ben's Special Star!
Dra 17h48'22" 62d11'57"
Bensen Michael Peden
Vir 12h58'27" -19d32'35"
Benson
Cma 7h9'27" -23d22'4"
Benson Gary Bass III
Cap 21h37'41" -19d37'46"
BensonHolmes
Ori 5h45'58" 2d6'29"
Bensye
Lyn 7h28'6" 48d57'52"
Bensye
Lyn 9h12'17" 38d32'30"
Bentarrow
Ari 3h8'49" 24d9'36"
Bente
Lib 15h21'20" -7d59'38"
Bente Fjermestad
And 1h12'1" 37d47'33"
Bente van der Goot
Tau 4h18'40" 12d56'9"
BenTee's Love
Uma 11h27'37" 30d18'23"
Benthin-Preihs
Per 4h6'24" 34d8'34"
BENTINE
Crb 15h40'46" 27d3'4"
Bentley
Vir 11h42'20" 7d11'18"
Bentley
Per 3h50'56" 35d36'24"
Bentley
Psc 1h8'22" 31d38'1"
Bentley J. Blum
Lyn 6h51'17" 56d41'44"
Benton - the 4th FBB
Aql 19h47'29" -0d27'35"
Benton Turner
Aqr 23h44'54" -24d37'17"
Benton's "Piece of Heaven"
Umi 13h44'53" 78d35'51"
Benu
Pho 0h16'37" -56d25'47"
BENU
Pho 1h2'30" -49d47'21"
Benvegnù Barbara
Ori 5h48'10" -4d14'17"
Benvie Robert Fleming
Umi 15h18'6" 80d38'40"
Benymon's Blessing
Dra 17h50'44" 60d49'49"
Benz
Lac 22h54'11" 50d40'41"
Benz, Hartmut
Uma 11h39'30" 33d23'17"
BENZAK STAR
Uma 13h45'21" 52d47'8"
Benzan
Cru 12h26'42" -59d19'51"
BeogrAD festival
Ori 6h1'56" 6d37'19"
BEP Star
And 1h38'37" 40d41'38"
~Ber~
Umi 15h16'18" 67d36'37"
BERA 007
Cap 21h55'24" -14d51'25"
Beracuda
Uma 10h28'2" 53d5'52"
Berangere
Lib 15h46'5" -4d7'13"
Bérangère
Gem 6h25'28" 27d21'36"
Bérangère-Celle et Soleil
Uma 9h25'48" 49d0'36"
BERBSBrightstar
Psc 0h39'39" 13d9'37"
Berdino's Baby
Umi 17h22'48" 78d42'38"
Beren Argetsinger
Umi 17h44'59" 81d5'34"
Bere-Nice Star
Com 12h40'31" 15d16'16"
Berenize Torres
Lep 5h7'3" -12d7'36"
Berens, Christian
Ari 2h22'25" 19d9'54"
Beretta, the Loving Labrador
Uma 10h39'33" 44d10'56"
Berfemir
Lib 15h25'24" -11d19'28"
Berg, Ilse
Uma 13h44'2" 56d25'25"
Berg, Peter
Ori 6h15'37" 10d50'41"
Berger
Per 2h36'20" 55d27'37"
Berger, Manfred
Uma 11h30'35" 38d16'50"
Berger, Wolfgang
Uma 10h37'53" 47d27'46"
Bergeron Rutledge
Leo 10h17'43" 11d6'34"
Berghaus, Gerhard
Uma 10h30'51" 40d6'41"
Bergur og Maria
Uma 11h46'3" 42d44'24"
Berihu
And 23h10'0" 47d2'51"
berilbebe
Tau 4h38'14" 6d5'23"
Beringer Child "The Devoted One"
Uma 11h5'33" 44d13'27"
Berington Brooks
Lyr 18h48'33" 31d57'33"
Berit ahkku
Uma 11h8'57" 34d56'57"
Berkeley Elizabeth Sumner
And 0h5'47" 43d59'28"
Berkem
Sco 16h9'31" -13d41'43"
Berkley Lisa Silverthorne
Sgr 19h0'50" -22d40'50"
Berko
Pho 0h32'36" -42d20'52"
Berlene Martin
And 2h28'54" 45d32'19"
Berlin
Uma 10h16'15" 46d36'45"
Berlin
Lyn 8h18'43" 43d3'59"
Berlin
Lyn 9h35'50" 39d35'35"
Berlin, Manfred
Ori 5h27'31" 13d31'21"
BERLO
Tau 4h6'48" 21d54'52"
Berly
Dra 11h7'25" 78d23'34"
Berly's Jewel
Cap 21h25'45" -18d11'13"
BERMUDA GIRL
Lyn 8h16'4" 33d26'36"
Bermuda-Rude
Aql 19h51'36" -0d39'33"
Berna Akyoney Gurlen
Tau 5h58'24" 24d59'39"
Berna and Miguel
Uma 11h35'2" 33d16'49"
Berna & Evren Kumcu
And 0h34'9" 32d20'16"
Berna User
Uma 11h7'53" 32d50'0"
Bernadette
Leo 10h8'8" 27d46'26"
Bernadette
Ori 5h36'49" 10d15'13"
Bernadette
Tau 4h31'28" 4d16'53"
Bernadette
Eri 4h17'53" -1d11'20"
Bernadette
Sgr 19h41'19" -13d50'44"
Bernadette
Cam 5h0'41" 73d18'56"
Bernadette
Uma 8h58'53" 61d31'55"
Bernadette Alfonso
Uma 11h26'55" 61d30'51"
Bernadette Ann Cosmano
Com 12h25'50" 26d57'17"
Bernadette... Bern Baby Bern
Aqr 22h11'1" -2d39'39"
Bernadette Burns
Crb 16h21'47" 34d5'53"
Bernadette Corsey Graham
Cas 23h35'44" 59d32'41"
Bernadette D. Wallace
Dra 19h3'12" 64d23'1"
Bernadette et Jean-Louis
Uma 9h16'59" 60d43'44"
Bernadette et Louis Stevens
Cas 23h19'24" 55d54'17"
Bernadette - Eternal Sunshine
Cap 21h13'23" -15d45'37"
Bernadette Fredericka Trenholm
And 23h47'28" 47d36'21"
Bernadette Fritz
Psc 1h8'2" 27d1'41"

Bernadette Grace
And 23h57'7" 47d54'53"
Bernadette Holland
Her 17h23'21" 35d5'33"
Bernadette J. Powers
Cam 4h23'16" 56d24'48"
Bernadette Joyce Opperman
Dra 18h3'43" 71d52'16"
Bernadette Keahi Kea Nu'uanu
Lib 15h1'46" -1d5'55"
Bernadette Kelly Wells
Mon 8h5'17" -0d34'56"
Bernadette Lio
Tau 5h51'31" 26d53'44"
Bernadette Maitre
Sco 16h20'44" -42d12'6"
Bernadette Marie Becker
Sco 16h49'8" -26d58'13"
Bernadette Marie Schweiger
Lib 15h47'30" -17d8'10"
Bernadette Marino Dalton
Sgr 18h23'39" -32d37'59"
Bernadette Mary Finn
Psc 1h6'44" 31d56'49"
Bernadette Mary Rice
Cas 23h13'59" 54d49'37"
Bernadette McKinney
Cas 0h13'6" 56d26'15"
Bernadette N. Reyes
Psc 0h37'30" 7d25'10"
Bernadette Petti
Cas 2h55'0" 66d46'21"
Bernadette Poulos
Vir 12h58'22" 10d50'20"
Bernadette Sy
Ori 5h10'41" 12d41'30"
Bernadette Van Wyk
Uma 8h53'36" 48d31'3"
Bernadette Vivian
Uma 12h41'33" 62d25'43"
Bernadette Viviano
Uma 9h29'31" 63d34'50"
Bernadette Warren Collins
And 23h22'34" 43d39'3"
Bernadette,ich liebe Dich für immer
Uma 11h33'45" 30d9'22"
Bernadette-Jeanne-Carla-Maria
Uma 12h52'52" 62d53'47"
BernadettePeacetoallwhoseethislight
Cas 1h27'46" 61d20'58"
Bernadette's Bunny
Lep 5h32'19" -12d41'5"
Bernadette's Rest
Cyg 19h34'59" 29d8'39"
Bernadina
Sco 16h9'56" -40d27'17"
Bernadine
Leo 11h32'29" 26d23'14"
Bernadine
Ari 3h12'12" 28d6'6"
Bernadine M. Gaines
Lib 15h3'5" -1d15'37"
Bernadine Nightwind
Cas 0h26'50" 62d1'46"
Bernadine Rita White
Cap 21h8'5" -16d35'0"
Bernadine Zarcone
Peg 21h24'39" 20d0'58"
Bernard
Cen 13h23'36" -55d18'21"
Bernard and Dorothea Thompson
Lyr 19h24'6" 37d45'19"
Bernard and Kristy Bircher
Uma 8h51'17" 58d5'59"
Bernard Anthone
Aur 5h25'10" 40d42'1"
Bernard Aubard
Vir 13h42'2" -6d19'52"
Bernard B. Brown
Uma 9h18'50" 53d14'33"
BERNARD BAPST
Uma 11h20'54" 29d12'21"
Bernard Barale
Uma 9h19'6" 42d40'9"
Bernard (Bud) E. Willett
Cap 20h47'58" -27d22'28"
Bernard "Buddy" Fox
Leo 11h10'33" -5d6'29"
Bernard C. Stevens, Jr.
Uma 11h36'21" 38d35'11"
Bernard Colin
Peg 22h52'40" 24d48'45"
Bernard Debisschop
Aqr 21h46'46" -6d12'40"
Bernard Dudemaine
Sgr 19h3'7" -23d43'48"
Bernard Dupuis
Cep 21h15'41" 65d45'17"
Bernard E Foley Sr
Vir 12h37'18" -8d5'57"
Bernard et Christine
Cyg 21h31'14" 42d6'22"
Bernard Eugene Smith
Aql 19h10'15" -7d58'14"
Bernard Flobert mon amour
Ori 5h42'26" -2d6'7"
Bernard Foster
Peg 22h19'35" 17d1'42"
Bernard Francis
Lib 15h39'7" -28d37'50"
Bernard Francis, the Keenable Star
Cyg 20h29'1" 52d57'8"
Bernard Gadeau
Leo 9h33'58" 15d32'14"
Bernard Grenier
Her 17h6'1" 17d33'58"
Bernard James Wilford
Her 18h28'15" 25d33'13"
Bernard & Jean Barnes' Star
Uma 11h50'15" 28d31'16"
Bernard Kenny's Star
Ori 5h9'3" 15d26'43"
Bernard Klein
Her 17h21'20" 46d18'23"
Bernard Kyle
Cap 20h32'44" -25d20'33"
Bernard L. Sager
Cap 21h27'12" -16d28'10"
Bernard Lanne
Uma 9h8'55" 56d8'41"
Bernard Lavelle 50
Cas 23h34'5" 51d30'17"
Bernard Luna
Uma 9h53'2" 51d1'19"
Bernard Luxon
Cep 2h15'19" 82d30'57"
Bernard M. Dubb
Tau 5h32'53" 27d44'9"
Bernard Michael Croghan
Cyg 20h8'34" 33d55'30"
Bernard Morin
Cnc 8h38'2" 11d12'0"
Bernard Morris Cooper
Crb 15h46'29" 27d19'58"
Bernard Mrozielski
Uma 9h11'18" 52d54'8"
Bernard Naumoff
Cep 22h23'35" 62d19'44"
Bernard R. Veader
Uma 9h31'37" 64d31'49"
Bernard Robinson
Per 3h13'33" 54d38'28"
Bernard S. Blake Jr.
Her 15h59'35" 44d51'18"
Bernard S. Feldman
Tau 3h39'55" 7d28'2"
Bernard Sellars
Per 4h30'23" 31d7'57"
Bernard Shusman
Per 3h11'57" 53d18'13"
Bernard Tourny
Ari 2h55'37" 21d32'48"
Bernard William Boudrow
Gem 7h42'13" 32d14'5"
Bernard Wrixon At Southern Cross
Cru 12h36'59" -58d42'14"
Bernard X. Gassaway
Per 2h39'46" 57d1'28"
Bernard, Coryphaeus Boonie, Venny
Gem 7h42'31" 34d38'18"
Bernarda-Strong as a Bear
Uma 11h31'41" 33d15'7"
Bernardino Ciuffetelli
Aqr 21h5'28" -5d56'39"
Bernardino R. & Gilda J. Mr. Magoo
Cap 20h36'18" -13d15'28"
Bernáth Tamás
Cap 21h42'24" -18d53'17"
Bernd Böhme
Uma 10h1'59" 42d7'7"
Bernd Burkhard Freytag
Her 18h39'24" 21d24'24"
Bernd Cerny
Ori 5h56'31" 17d14'52"
Bernd Drebing
Uma 8h46'29" 63d59'51"
Bernd Heider
Uma 11h7'28" 58d48'4"
Bernd Hörmann
Ori 5h54'29" 13d5'11"
Bernd Laskowski
Uma 8h39'30" 65d54'54"
Bernd Lenzen
Ori 5h8'20" 0d53'38"
Bernd Mix
Uma 10h56'38" 58d12'46"
Bernd Morhoff
Uma 11h24'29" 30d47'57"
Bernd Müller
Ori 5h42'52" -0d19'24"
Bernd Ohmstede
Uma 11h45'0" 38d24'44"
Bernd Rabus
Uma 10h59'22" 33d37'41"
Bernd Rethemeier
Uma 14h2'31" 62d14'6"
Bernd Schaminski
Ori 5h45'15" -3d37'58"
Bernd Schlubeck
Boo 13h56'24" 22d57'58"
Bernd Schütte
Uma 10h9'44" 60d30'5"
Bernd Schweitzer
Ori 6h20'14" -0d17'19"
Bernd Sonnenberg
Uma 12h1'8" 48d36'10"
Bernd Wenzel 09. April 1960
Uma 13h32'51" 56d1'37"
Berndt Hülsmann
Ori 5h17'14" 8d36'22"
Berndt, Mandy
Psc 1h11'20" 2d51'22"
Bernélis
Umi 13h35'50" 75d4'30"
Bernell
Lyn 8h56'29" 38d2'5"
Bernell Samuel
Leo 11h39'15" 21d4'3"
Bernetta
Crb 15h42'52" 35d54'38"
Bernhard Arnold Merz * 24.08.1955 D
Ori 5h11'2" 15d38'50"
Bernhard Drechsler
Aur 5h27'48" 40d28'48"
Bernhard Klemm
Uma 9h55'19" 42d11'34"
Bernhard Koch
Uma 11h45'55" 30d49'55"
Bernhard Lenherr
Uma 9h44'5" 67d41'6"
Bernhard Manfred Zinke
Uma 11h20'32" 34d23'0"
Bernhard Miksch
Umi 16h22'35" 85d21'56"
Bernhard Moosbrugger
Uma 9h56'13" 62d55'37"
Bernhard Oestereich
Uma 10h34'32" 47d0'6"
Bernhard Schneemann
Uma 9h16'38" 55d31'52"
Bernhard Ypey
Ori 5h9'44" 0d20'0"
Bernhard-Paul Knopp
Uma 10h1'52" 55d42'46"
Bernhardt Schmidt
Cap 20h10'18" -17d52'52"
Bernice
Uma 11h38'50" 62d35'33"
Bernice
Cas 1h35'10" 57d59'38"
Bernice
And 0h17'3" 44d44'28"
Bernice and Brian Wedding Day
Cyg 19h55'30" 41d37'17"
Bernice Ann Geiger
Cnc 9h12'59" 28d5'53"
Bernice Ashburn Young
Com 13h5'4" 27d35'17"
Bernice Baldwin
Uma 11h38'27" 51d32'19"
Bernice Baron
Sgr 18h0'51" -27d49'0"
Bernice "Bee" Nagel
Tau 3h30'59" 13d8'9"
Bernice Brady
Crb 16h13'34" 37d24'18"
Bernice C. Kiefer
Gem 7h9'23" 27d51'41"
Bernice C. Minesci
Cas 2h17'38" 66d46'7"
Bernice C. Rieger
Cas 1h51'35" 61d49'58"
Bernice Christner
Cas 0h0'10" 54d10'37"
Bernice Clark
And 0h20'31" 31d3'37"
Bernice Cox Flowers
Per 4h18'45" 46d3'26"
Bernice Eileen Lucas
Sct 18h46'41" -5d26'7"
Bernice Emily Henkel Lasky
Cap 21h40'46" -11d50'46"
Bernice Etha Sharp
Sgr 20h12'58" -35d48'7"
Bernice Evelyn Saltzman
Uma 10h52'14" 64d54'31"
Bernice Forester
Com 13h5'25" 26d0'1"
Bernice Gay Hamilton
Ori 6h14'50" 15d14'2"
Bernice Hoeper
Cas 1h24'56" 52d13'47"
Bernice J. Ballard
Com 12h0'45" 21d49'35"
Bernice Kimberly
Cas 23h29'17" 53d17'57"
Bernice Koerper
Cas 2h10'33" 64d10'15"
Bernice M. Crowder
Ari 3h7'58" 28d36'31"
Bernice M. Zimmerman
Lmi 10h44'15" 28d6'39"
Bernice McGrath
Cas 0h17'33" 56d11'5"
Bernice "Niecie" Quillen
Tau 4h31'19" 14d45'33"
Bernice O'Donnell
Cam 7h58'37" 61d43'24"
Bernice Pocino
Uma 9h16'31" 71d39'13"
Bernice Righthand
Cyg 20h59'11" 32d39'52"
Bernice Seawood
Cas 0h56'36" 69d45'39"
Bernice V. E. B. Cannon
Col 5h43'11" -33d33'18"
Bernice Weisberg
Ari 3h7'56" 28d53'20"
Bernie
Ari 3h15'8" 27d58'8"
Bernie
Leo 9h49'32" 21d55'17"
Bernie
Ari 3h17'17" 12d6'45"
Bernie
Gem 6h54'15" 31d5'4"
Bernie
Dra 15h55'17" 52d26'30"
Bernie
Cas 0h57'23" 66d33'43"
Bernie
Ori 5h34'11" -0d45'8"
Bernie and Mary Fagersten
Cyg 20h14'6" 51d50'27"
Bernie & Ashley Rowe
Sge 19h53'4" 17d40'5"
Bernie Barc
Cep 0h14'45" 75d3'14"
Bernie Barrie
Lyn 8h16'37" 42d9'25"
Bernie Beard
Aql 19h50'49" 15d20'32"
Bernie & Bill our Eternal Light
Uma 12h12'1" 61d22'44"
Bernie Bird
Ori 5h22'39" -3d52'7"
Bernie & Blair
Ari 2h50'10" 28d35'22"
Bernie Bright
Uma 9h53'45" 52d12'42"
Bernie Burghardt
Ori 6h14'26" -1d22'37"
Bernie Crawford
Cas 0h8'45" 53d44'8"
Bernie & Delma Gonzalez
Cyg 20h36'15" 40d48'12"
Bernie & Fuzzy-Married 60 Years!!
And 0h14'11" 45d21'51"
Bernie Gorman, Jr.
Uma 9h45'43" 59d24'33"
Bernie & Jenn Stabinski
Cyg 19h14'45" 53d59'27"
Bernie Jett Starr Hubner
Gem 7h22'34" 15d19'15"
Bernie & Joan Eternal Love
Uma 8h57'30" 63d18'39"
Bernie Margolis - 75th Birthday
Cep 22h12'21" 70d10'48"
Bernie Millage
Gem 7h0'38" 20d34'49"
Bernie "My Angel on Earth"
Her 17h58'28" 22d4'55"
Bernie Progin
Ari 1h57'49" 13d52'32"
Bernie Redlawsk
Aql 20h5'7" 13d39'22"
Bernie Shaeffer
Her 16h9'50" 49d16'3"
Bernie Sherry
Her 17h15'5" 29d39'3"
Bernie Snelnick
Cnc 8h20'29" 23d52'25"
Bernie Sue Addleman Sones
Cam 5h9'14" 71d28'32"
Bernie The Blue
Dra 18h58'12" 61d36'23"
Bernie The Butch
Cnc 9h10'3" 16d53'14"
Berniece
Crb 16h16'30" 33d22'25"
Berniece Anderson
Ari 3h22'58" 29d21'38"
Berniece Minert
Cas 0h10'20" 56d44'55"
Bernie's Brilliance
Cma 7h13'26" -25d26'54"
Bernie's Light
Ori 5h35'49" 2d53'14"
Bernie's Shining Star
Ari 2h12'10" 24d1'57"
Bernie-Steve
Cyg 21h59'7" 45d7'21"
Bernis & Pezis Kuschelbärenstern
Uma 9h30'38" 47d32'27"
berniste
And 1h43'45" 40d12'59"
Bernita Callahan
Umi 13h24'13" 70d56'51"
Bernita Dodge
Cas 1h12'31" 49d2'38"
BerNor
Com 12h27'57" 26d24'22"
Bernritter
Uma 11h52'38" 35d19'59"
Berrer, Oliver
Uma 10h37'38" 46d44'16"
Berrie Lee
Uma 9h27'50" 56d22'52"
Berry's Life by Tsunekazu & Yuki
Sgr 18h41'54" -20d22'34"
Bersagliere Modesto Verderio
Per 4h41'53" 46d55'3"
Bersaida
And 0h37'17" 22d23'14"
Bert
Vir 13h1'56" -0d39'31"
Bert A. Spencer Jr.
Dra 18h51'1" 60d2'28"
Bert and Jane Salvati
Lyr 19h20'20" 29d59'28"
Bert Brooks, Jr. aka Punkin Wunkin
Per 3h5'51" 55d34'48"
Bert & Diane Kurland
Eri 3h48'13" -0d42'38"
Bert Frick
Ori 6h12'2" 7d50'10"
Bert Gieseke
Uma 8h17'36" 70d30'36"
Bert Goldberg
Cep 22h37'22" 67d6'43"
Bert & Helen Whitten
Uma 11h17'24" 30d30'21"
Bert Jason Coday Feb. 5, 2001
Uma 10h53'35" 60d25'33"
Bert John Lockyer
Per 4h10'0" 43d27'33"
Bert Lee Rabenn
Cep 22h18'1" 61d59'29"
Bert M Brooks Jr.
Aqr 22h43'23" -15d49'39"
Bert "Oompa" Bruce
Aql 19h0'54" -0d48'12"
Bert "Puka" Faiola
Uma 9h59'58" 45d49'53"
Bert Sharp
Sgr 18h15'32" -18d44'42"
Bert Sutherland
Sgr 18h4'20" -16d57'16"
Bert Topping
Uma 10h33'20" 64d35'52"
Berta
Ori 5h40'47" 6d21'34"
Berta Gegner
Uma 11h55'1" 57d7'7"
Berta Herrera Amador
Uma 9h44'3" 62d46'28"
Berta Lidia Menendez Guzman
Cap 21h40'50" -9d8'34"
Berta Magriña
Sgr 18h14'42" -29d25'20"
Bertemma
Cmi 8h1'52" 1d26'3"
Bertenburg, Maria
Vir 13h25'23" 5d41'27"
Bertha
Uma 11h0'17" 45d57'38"
BERTHA
And 0h41'11" 36d40'5"
Bertha and Albert Wishner
Cyg 21h56'44" 38d13'34"
Bertha Duryea Kahler
Leo 10h22'37" 25d19'50"
Bertha E. Kettelle
Cas 0h8'41" 56d48'31"
Bertha & Earl Haines
Cyg 20h25'16" 46d53'8"
Bertha Elizabeth Estep
Leo 9h24'22" 25d17'55"
Bertha Goldberg
Cas 1h24'55" 71d55'45"
Bertha Guevara
Lyr 18h42'47" 37d39'38"
Bertha Hollie
Cas 1h20'13" 64d32'33"
Bertha Horn
Ari 3h6'46" 17d50'54"
Bertha Irene Woody
Tau 4h9'46" 7d10'45"
Bertha Kuma Dartey
Cas 22h59'23" 53d49'12"
Bertha Leticia
Lib 15h8'47" -25d19'46"
Bertha M. Koerper
And 0h43'6" 44d27'3"
Bertha Mae Pehrson
Cas 1h29'53" 61d13'16"
Bertha Mason
Cap 20h20'56" -10d57'42"
Bertha May Wolfe (Bertie)
Gem 6h58'21" 25d47'32"
Bertha "MOM" Miller
Uma 11h8'54" 38d16'7"
Bertha Ramona Rosario Rodriguez
Crb 15h30'30" 29d20'24"
Bertha Resendez
Sco 17h37'53" -30d46'22"
Bertha Vedder
Cas 0h43'30" 52d40'14"
Bertha Williams Immel
Cnc 8h22'17" 22d40'54"
Bertha y Chuy
Sge 20h4'51" 18d7'43"
Bertie
Psc 1h21'6" 25d55'54"
Bertie
Ari 2h30'8" 20d11'18"
Bertie
Cas 0h51'26" 56d32'54"
Bertie
Umi 16h17'34" 79d28'44"
BERTIE
Uma 10h10'27" 65d50'36"
Bertie Bungler
Cep 23h35'18" 80d13'52"
Bertie Crossin
Uma 10h51'41" 41d56'34"
Bertie Hudson
Uma 10h11'31" 59d49'54"
Bertie Lubner
Cru 12h32'47" -60d22'57"
Bertil Sara
And 1h26'35" 47d33'47"
Bertny "Puddin" Gilmore
Peg 22h12'45" 8d9'42"
Berton G. Kanne
Leo 11h7'10" 22d53'43"
Berton Karl Lax
Cnc 8h49'29" 31d24'1"
Bertram
Umi 14h46'56" 75d55'22"
Bertram August Dahl
Del 20h48'25" 17d29'18"
Bertram, Ulfert
Ori 6h14'37" 9d13'5"
Bertrand & Alexandra
Uma 9h39'0" 46d17'18"
Bertrand de Dietrich
Uma 13h31'19" 55d12'35"
Bertrand Kosturi III
Ori 5h49'35" 6d14'55"
Bertrand Ricou
Aqr 22h7'45" -24d10'51"
Bertrand Saugnac
Ori 5h54'23" 1d57'57"
Bert's Star
Sco 16h5'23" -12d17'12"
Bertsch, Manuel
Uma 9h39'28" 64d6'4"
Bertukan
Leo 10h15'34" 16d57'15"
BERUSKA
Crb 15h22'51" 29d2'55"
Beruta - My Little Girl
Cru 12h18'6" -57d7'47"
Berutiful
Tau 4h41'55" 5d45'55"
Beryl
Lyn 7h44'20" 40d6'53"
Beryl and Bill Allen
Cyg 20h28'20" 36d28'35"
Beryl and Ron
Cyg 21h54'41" 52d32'8"
Beryl Edwards Trawick
And 2h28'23" 45d21'8"
Beryl Jean Coles
Sgr 19h57'26" -33d1'31"
Beryl Mary
Uma 9h42'25" 41d47'40"
Beryl Moss
Per 3h26'52" 47d45'38"
Beryl - The Guardian Angel
Uma 12h1'20" 41d23'43"
Beryl Turner Choi
Tau 4h27'33" 24d8'54"
berylalex diamond
Cru 12h44'20" -58d1'45"
BES Stars - Class of 2006
Per 3h3'39" 47d21'26"
Besa
Cyg 20h50'29" 42d59'47"
Besa82
Crb 15h44'6" 29d49'4"
Besame
Uma 10h9'0" 49d42'31"
Besando a Matilde
Sgr 18h53'52" -24d30'30"
Beschen Cove
Cru 12h36'21" -57d59'34"
Bescher
Uma 11h32'2" 61d29'10"
Beschermengel
Cyg 20h21'8" 55d14'41"
Beshala
Aqr 21h35'3" -0d37'11"
Besherta
Leo 10h16'19" 12d23'27"
Besiana
Gem 6h32'43" 21d50'4"
Besiana
Her 18h21'8" 16d10'58"
Beso Multiplique Tres.
Sgr 19h13'11" -27d21'26"
BeSo, 16.11.2003
Cep 22h45'5" 66d21'20"
Besonder Ein
Her 16h15'14" 48d6'42"
BESOS
Ari 2h46'58" 27d46'59"
Bess
Aur 6h1'17" 48d10'12"
Bess Ann
Cap 21h50'10" -10d2'54"
Bess Ann Fagan
Sgr 18h29'50" -32d23'36"
Bess Chakravarty
Lyn 7h34'59" 49d21'22"
Bess England
Cap 21h32'54" -17d5'38"
Bess Marie Hopmans
Mon 6h52'4" -0d19'39"
Bess the Beautiful
Cyg 19h57'45" 49d40'39"
Bessie
Gem 6h48'37" 31d25'28"
Bessie
Crb 15h33'52" 32d3'17"
Bessie
Ori 6h4'8" 15d3'3"
Bessie
Aql 19h49'32" -0d13'6"
Bessie Ann
Cap 20h10'15" -16d14'32"
Bessie Boutsioukos
Cas 2h29'19" 65d43'41"
Bessie Isabella Griffin Blankenship
Sco 16h39'20" -28d40'39"
Bessie Londrata Martines
Cyg 21h57'1" 48d56'12"
Bessie Marr
Crb 15h47'13" 32d56'5"
Bessie Ramsey
Crb 15h25'51" 26d28'42"
Bessie's Glowing Heart
Sco 17h22'19" -30d48'36"
Bessie's Secret
Cyg 21h12'37" 46d56'16"
Bessy Fisseha
Gem 7h22'37" 25d1'23"
Bessy Karina Quintanilla
Lyn 8h22'23" 35d52'33"
Bessy Mate Kerry
Cas 2h2'8" 74d34'16"
Bessy's Wishing Star
Cnc 8h57'56" 23d43'29"
Best Big Sis Star! Gail x 21 x
Crb 15h52'12" 30d25'50"
Best Coast Kim
Cap 20h44'25" -17d9'56"
Best Dad & Doc: David O. Elliott
Pho 0h4'24" -44d42'17"
Best Dad In The Universe
Uma 10h16'57" 57d35'35"
Best Dad KCG
Leo 10h46'36" 9d16'18"
Best Daddy Ever! J.M.L.
Cap 21h39'24" -24d11'1"
Best Doctor in the Cosmos
Per 3h4'4" 47d5'16"
Best Double "D" Frens, Dar & Dezi
Tau 4h2'11" 26d4'8"
BEST FOOT FORWARD
Crb 16h14'46" 39d30'40"
Best Friend
Tau 5h10'58" 20d36'55"
Best Friend Stacey
Cas 1h57'29" 61d20'39"
Best Friends
Her 16h31'8" 46d24'14"
Best Friends: Alyssa and Stacie
Lib 15h34'58" -15d29'36"
Best Friends Forever!
Cma 6h14'22" -20d8'7"
Best Friends Forever
Ari 2h7'49" 25d58'54"
Best Friends Forever- Jess & Ang
Sgr 19h44'59" -12d23'26"
Best Friends Forever Kerri & Jenn
Mon 7h20'58" -0d58'58"
"Best friends" Mary & Marco, 05.10.1998
Uma 8h49'48" 69d7'48"
Best Friends Means Friends FOREVER*
Cnc 9h5'7" 7d52'3"
Best Friends, Soulmates
Lib 14h41'21" -11d58'14"
best grandmother
Per 3h29'13" 46d47'23"
Best Honey Ever
Lib 15h39'5" -6d32'28"
Best Mami & Papi Star
Cas 1h0'24" 63d48'9"
Best Man ~ MWS
Tau 5h27'55" 21d19'9"
Best Mom Ever
Leo 9h48'30" 26d0'18"
Best Mom ever Karen Mandel
Leo 10h21'38" 25d19'45"
Best Mom in the Universe! Neecie!
Uma 8h51'40" 72d43'7"
Best Mommy in the Universe Love LCM
Uma 11h59'31" 53d34'34"
Best Mommy in the World
Sco 16h21'11" -12d16'22"
Best Nana in the World
Psc 1h26'0" 23d8'48"
BeST Star Denise
And 23h39'5" 42d43'53"
Best Teacher's Assistant-March 2005
Umi 14h41'8" 81d41'5"
Best Wife and Mother Dara Howell
Cas 1h56'17" 61d32'46"
Besta
Uma 10h47'52" 59d55'12"
BestDaddyEver
Her 17h38'36" 15d56'15"
Bestest Best Friends F&A I&B
Cyg 20h33'1" 53d34'2"
Bestest Buddy
Cap 20h19'18" -23d11'52"

Bestest Diane Gonzales
Per 2h13'21" 55d13'21"
Bestest Raymond Tam
Per 2h14'30" 55d24'49"
bestfriends 4 ever brother & sister
Sco 16h9'36" -29d48'28"
bestloans4u.com
Del 20h23'34" 15d3'44"
Bet
Boo 15h18'3" 40d30'41"
BETABUNNY
Vir 13h40'13" -7d24'57"
Betandian Artus
Cyg 20h19'6" 49d44'51"
Betbeze
Lyr 19h3'4" 32d32'5"
Bet-c Perry Marjorie Elizabeth Head
Boo 14h42'11" 15d26'44"
Bételgeuse de Constantin d'Octodure
Ori 6h9'3" 21d6'31"
Beth
Gem 6h16'16" 22d13'9"
Beth
Ari 2h38'0" 25d8'22"
Beth
Com 13h5'56" 26d25'37"
Beth
Leo 11h45'27" 26d13'48"
Beth
Tau 5h52'55" 23d2'14"
Beth
Psc 0h39'54" 16d58'31"
Beth
Psc 0h44'5" 18d5'4"
Beth
Lyr 18h35'55" 31d18'8"
Beth
And 1h27'54" 37d0'35"
Beth
And 0h19'23" 42d54'13"
Beth
Uma 11h57'28" 47d10'22"
Beth
And 23h22'3" 47d44'5"
Beth
Sgr 18h17'13" -16d20'44"
Beth
Dra 15h4'17" 62d17'50"
Beth A. Martone
And 1h23'19" 37d7'32"
Beth A Michielsen
Crb 16h1'43" 34d4'30"
Beth Ackerson
Ori 6h22'21" 10d3'12"
Beth AKA Cat
Lyn 7h31'6" 57d10'18"
Beth aka Irish Angel
Gem 7h1'56" 22d11'41"
Beth & Alan Haspel
Cap 20h19'40" -11d42'47"
Beth Alice
Cyg 21h26'27" 31d1'51"
Beth and Alan Nichols
Sge 19h44'21" 17d0'12"
Beth and Alek
Cyg 21h31'55" 47d44'58"
Beth and Duane (Dog) Chapman
Cyg 20h46'23" 35d14'50"
Beth and Jack
And 0h42'30" 36d6'30"
Beth and Joe Fahrner
And 23h35'21" 41d19'40"
Beth and Morgan's Star
Peg 21h22'36" 19d51'25"
Beth Angel
Ret 3h42'36" -53d5'33"
Beth Ann
And 0h40'44" 25d20'12"
Beth Ann
Gem 7h4'18" 23d54'6"
Beth Ann
Leo 9h48'18" 23d26'33"
Beth Ann
Vul 19h47'25" 27d6'30"
Beth Ann
Cas 0h19'15" 51d21'55"
Beth Ann Buck
Tau 4h28'21" 15d18'30"
Beth Ann Bullard
Dra 15h50'14" 54d35'57"
Beth Ann Crotzer James
Cas 1h18'43" 69d2'36"
Beth Ann De Zinna
Sgr 19h53'53" -39d21'19"
Beth Ann "Face's Starlight"
Ori 5h48'19" 11d23'13"
Beth Ann & Garrett Lim
Lep 5h29'32" -14d51'11"
Beth Ann Gee The Little Biffer
Uma 13h33'21" 62d16'58"
Beth Ann Holton
Ori 5h4'23" 6d39'53"
Beth Ann Jones
Crb 15h53'21" 27d9'45"
Beth Ann Oertel
Psc 1h43'29" 5d45'34"
Beth Ann Raudenbush
Cas 23h2'19" 55d17'15"
Beth Ann Sharp
Aqr 23h51'31" -11d15'2"
Beth Ann Weachter
Psc 0h18'6" 4d29'48"
Beth Ann Weldon
Gem 7h10'57" 23d39'11"
Beth Ann Woodward / Johnson
Leo 10h26'58" 22d10'13"
Beth Anne
Psc 0h37'5" 21d26'46"
Beth Anne Keates
Cas 0h12'27" 59d26'18"
Beth Anne Oldershaw
Cyg 19h52'39" 33d53'26"
Beth Anne Petcash
Cyg 19h35'2" 32d0'30"
Beth Anne Purdy
Lmi 10h31'4" 35d14'59"
Beth Barker
Lyn 7h3'41" 44d53'32"
Beth Beasley
Cam 3h41'8" 57d42'24"
Beth "Bethy Pop" Renick
Uma 13h54'2" 51d30'1"
Beth Bouvier and Ed DiGeronimo
Tau 5h56'52" 24d1'6"
Beth & Brett Cavalieri 1999
Cyg 21h20'48" 38d20'9"
Beth Campbell
Mon 6h29'34" 0d10'5"
Beth Carbonaro
Sco 16h19'51" -9d55'19"
Beth Choate Wortmann
Aqr 21h8'10" -5d56'24"
Beth Christine Ramirez
Leo 11h39'23" 18d52'59"
Beth Corning
Vir 12h57'41" -15d19'9"
Beth & Danny Joyce
Cyg 21h14'56" 31d56'31"
Beth E Stanton
Lyr 19h23'56" 38d24'12"
Beth Ellen
Cnc 8h44'17" 15d21'27"
Beth Ellen Home 01/10/1934
Lib 15h49'22" -14d14'43"
Beth Enid Allbaugh
Psc 0h46'43" 17d9'25"
Beth Faulhaber
Lyn 8h7'3" 49d13'30"
Beth Felton-Marks
Boo 14h15'34" 48d34'9"
Beth Firewood
Sgr 18h46'45" -30d51'15"
Beth Gerstenberger Zucker
And 1h53'35" 38d57'27"
Beth Gross
Lib 15h15'27" -12d46'27"
Beth Hallman
Cap 21h13'55" -16d6'33"
Beth Harbor
Cas 0h53'39" 71d1'57"
Beth Hardison* flygirl
And 1h42'38" 43d47'42"
Beth Heger
Boo 15h3'17" 49d46'47"
Beth Helme
Ari 2h32'51" 30d35'0"
Beth "Hunky"
Cas 0h4'53" 54d17'19"
Beth Hunter
Leo 11h10'52" 22d51'2"
Beth Janes
Peg 23h46'1" 13d15'35"
Beth & Jason
Cnv 12h34'55" 43d30'6"
Beth & John's Star
Lyn 9h9'45" 39d6'35"
Beth Jones
Lib 15h49'33" -12d14'20"
Beth Kelly
And 22h59'37" 40d18'27"
Beth Kolman Mueller
Leo 9h48'23" 17d39'49"
Beth L DuVall
Uma 8h34'5" 68d16'41"
Beth Lauren
Her 17h26'8" 34d14'11"
Beth Layne McIntosh Shreve
Vir 12h44'15" 3d16'0"
Beth Lindsay
Sgr 18h51'32" -27d9'58"
Beth Lively
Uma 9h41'29" 46d7'22"
Beth Loves Chris
Cyg 20h3'38" 57d52'21"
Beth Lyons
Tau 3h42'36" 23d44'16"
Beth Malloy
Lyn 7h6'18" 51d45'35"
Beth Marie Lee
And 23h44'2" 39d50'46"
Beth Marie Rivas
Lib 14h32'40" -9d24'28"
Beth Marie Waligorski
And 0h35'39" 29d36'24"
Beth McLean mamma oooooh
Gem 6h47'15" 13d47'12"
Beth "Meisty"
Psc 23h43'56" 1d6'4"
Beth Milne - "Tinkerbell"
Cas 23h39'0" 53d51'57"
Beth Moore's Can-Do Star
Psc 0h48'5" 8d59'21"
Beth Mulligan
Leo 11h41'32" 26d20'18"
Beth my beautiful loveliness
Cas 23h33'4" 58d33'19"
Beth My Love
Cyg 21h38'23" 38d54'43"
Beth Neil
Leo 11h13'17" 9d43'47"
Beth Oxley
Aqr 21h12'32" -9d25'5"
Beth Paula Silverman
Lyn 7h52'12" 35d19'26"
Beth Pedevill
And 1h54'36" 41d21'4"
Beth Pond
Cas 23h21'22" 59d23'35"
Beth S. Rutenber
Per 3h32'38" 43d1'48"
Beth Scanlan 7403 TML
Uma 10h23'6" 57d9'51"
Beth Sepko
Vir 13h38'52" -1d28'20"
Beth & Sid Paly Married 9-3-1951
Tau 5h56'32" 25d0'21"
Beth Simone Eckhaus
Sco 16h10'35" -10d25'26"
Beth Sisco
Ari 2h45'28" 16d28'10"
Beth Starr Thomson
Uma 11h23'3" 29d27'40"
Beth Susan Goldstein - A Great Mom
Her 16h44'43" 33d3'40"
Beth Suzanne Simionescu
Per 4h4'44" 45d17'39"
Beth Tedesco
Sgr 18h19'27" -21d0'49"
Beth - The Dancer
Lib 15h35'25" -15d21'21"
Beth Tilley McClees Kantner
Aur 5h19'43" 44d59'23"
Beth & Toby - Mummy's Star
And 2h13'46" 49d7'6"
Beth Tucker Showell
Cas 0h23'6" 65d5'44"
Beth Uehlein Bradley
Her 18h26'26" 22d56'56"
Beth Vande Hey
Cyg 19h53'43" 36d46'52"
Beth Vanderstar
And 23h26'33" 42d12'7"
Beth Ware
Leo 10h10'52" 27d17'29"
Beth Yabsley
And 23h47'23" 46d54'55"
Beth Young
Cap 20h8'49" -10d8'23"
Beth Ziegler
Cas 1h45'19" 61d7'50"
Beth365
Tau 3h37'26" 16d34'29"
BethAdam
Leo 11h27'58" 21d57'42"
Bethalle Russell
Uma 8h50'35" 52d58'29"
Bethan and Gaffa Morris
Uma 9h28'39" 66d31'37"
Bethan Cole
Umi 13h41'4" 75d19'40"
Bethan Jones
Cyg 19h35'18" 44d7'12"
Bethan Louise Hanson - Star of Hope
And 2h16'19" 37d28'57"
Bethan Roberts
And 23h38'38" 49d35'36"
Bethanie
And 1h3'23" 38d4'25"
Bethanie Lynn Norwood
Mon 7h17'3" -9d29'46"
Bethanie Rosella
Aur 5h22'30" 32d10'10"
Bethann
Tau 3h47'41" 29d21'9"
Bethann C. Mitchell
Lyn 9h25'4" 39d56'45"
BethAnn Corbett
And 1h25'35" 34d48'29"
Bethann M Lawall
Lyn 8h24'7" 34d25'44"
BethAnn Sulborsky
And 0h38'21" 38d57'49"
BethAnne Hoover
Uma 13h54'52" 60d16'15"
Bethannshan Amour
Ori 6h24'40" 10d31'35"
Bethany
Aqr 21h45'20" 0d46'45"
Bethany
Crb 15h25'27" 25d36'48"
Bethany
And 0h26'3" 28d17'27"
Bethany
And 0h21'17" 29d5'44"
Bethany
And 0h13'54" 43d37'7"
Bethany
Lmi 10h6'56" 35d14'59"
Bethany
And 23h17'13" 47d35'17"
Bethany
Cas 1h25'31" 59d12'31"
Bethany
Uma 9h58'47" 48d51'14"
Bethany
Uma 9h56'21" 51d53'46"
Bethany
And 2h15'34" 50d26'13"
Bethany
Cam 4h35'56" 67d6'15"
Bethany
Cas 1h26'6" 61d3'18"
Bethany
Aqr 21h51'49" -1d39'20"
Bethany
Vir 12h44'26" -6d27'40"
Bethany
Leo 11h9'6" -0d12'42"
Bethany
Ori 5h35'19" -2d18'50"
Bethany
Sco 16h13'50" -10d45'12"
Bethany
Sgr 19h11'10" -25d21'44"
Bethany
Sco 16h8'15" -35d48'22"
Bethany A Corbett
Uma 11h0'8" 64d40'29"
Bethany Alexandrea
And 23h5'53" 40d20'17"
Bethany Alice Stopher
Pav 21h5'3" -62d10'55"
Bethany Amanda Monk
Gem 6h52'29" 21d52'17"
Bethany and Derek Forever
Sco 16h39'33" -39d54'59"
Bethany and Hope
Uma 13h58'32" 61d13'6"
Bethany and Martin Gillott
Cyg 21h39'31" 50d14'26"
Bethany and Matt's Star
Lyr 19h7'16" 35d13'27"
Bethany and Vic's Star
Lib 15h19'31" -6d24'11"
Bethany Ann Atwell
Uma 9h47'49" 43d12'14"
Bethany Ann Hanson
And 0h45'29" 42d41'42"
Bethany Ann Johnson
Leo 11h19'13" 24d1'12"
Bethany Ann McKernan
Ori 6h15'26" 8d47'37"
Bethany Ann Robinson
And 2h29'41" 42d2'55"
Bethany Anne Clark
Cas 3h1'19" 61d26'21"
Bethany Anne Peterson
Uma 12h2'13" 51d24'42"
Bethany April Keylock
And 2h13'34" 41d45'7"
Bethany Avery
Mon 7h49'16" -0d32'35"
Bethany Bailey
Lmi 10h5'51" 38d8'5"
Bethany Blakely
And 23h5'17" 40d58'6"
Bethany Brooke
Lyn 8h14'30" 34d43'45"
Bethany Buzynski
And 0h44'16" 25d35'31"
Bethany & Carli Margolis
Leo 10h12'52" 17d16'58"
Bethany Chandler and Chase Bradley
Sco 16h12'49" -9d38'5"
Bethany Christine Hammond
Ari 3h17'6" 21d21'47"
Bethany Christine Norlander
Lyn 7h35'15" 48d15'57"
Bethany Clair Lovell 18
And 23h47'39" 47d40'47"
Bethany Dawn Blaine
Cam 3h30'54" 67d55'53"
Bethany Dawn Miller
Oph 17h50'55" -0d46'22"
Bethany Elaine Slaven 12-30-2000
And 23h5'40" 47d5'10"
Bethany Ellen Hayward
Tau 4h10'3" 16d24'35"
Bethany Emma Woods
Cas 23h35'15" 55d47'34"
Bethany Estee
Uma 9h35'36" 69d20'41"
Bethany Esther Kopsa
Her 17h53'59" 22d16'24"
Bethany Force
Cap 21h40'19" -17d3'14"
Bethany Goodman Stroyne
Lyn 7h22'3" 45d22'9"
Bethany Grace
And 2h27'17" 38d30'18"
Bethany Grace Howarth
Cyg 20h17'23" 40d55'15"
Bethany Hope
Lyn 7h43'53" 42d36'37"
Bethany Jackson
Cyg 21h40'49" 38d16'49"
Bethany & Joshua
Uma 9h38'58" 47d4'9"
Bethany Joy Wolfswinkel
Lyr 19h18'53" 29d27'44"
Bethany K Adamczyk
And 1h18'53" 35d35'33"
Bethany Kay Boardman
Aqr 20h44'21" -9d21'40"
Bethany Kristin Stiefel
And 0h59'30" 43d11'41"
Bethany L. Buck
And 0h13'49" 45d17'47"
Bethany L. Cone
Lib 14h24'22" -18d45'13"
Bethany L. DiPadua
Vir 12h48'21" 6d12'53"
Bethany Lewis
Psc 0h14'42" 11d1'26"
Bethany Lynn
Tau 4h39'10" 20d4'2"
Bethany Lynn
Uma 10h9'6" 42d35'35"
Bethany Lynn Babbitt
Vir 13h38'9" -19d44'49"
Bethany Lynn Strothman
Gem 7h7'44" 21d53'6"
Bethany M Alabek - Higher and First
Gem 6h46'42" 32d17'24"
Bethany Mae Anderson
Uma 9h4'8" 62d24'49"
Bethany Mae Schmidt
Cnc 8h24'56" 26d10'14"
Bethany Marae Leiker
Cnc 8h40'34" 18d40'28"
Bethany Marie
And 2h35'57" 43d46'4"
Bethany Marie
Lib 15h50'17" -11d44'45"
Bethany Marie Durst
Gem 7h3'2" 20d47'57"
Bethany Marie Ferguson "6/13/1972"
Uma 13h28'55" 61d18'29"
Bethany Megan Hauser
And 2h8'35" 39d43'51"
Bethany Michele
Ori 5h23'58" 13d27'3"
Bethany Mulloy Shock
Leo 9h46'28" 13d35'30"
Bethany Powell
Lib 15h0'33" -19d35'53"
Bethany R. Diaz
And 0h41'23" 34d3'54"
Bethany Rachel Kroeger
Lyn 7h49'47" 36d49'30"
Bethany Redfurn of Shetland
And 1h20'36" 49d30'52"
Bethany Reeves
Uma 12h0'5" 36d1'37"
Bethany Renee Barcroft
And 1h26'9" 44d56'25"
Bethany Rimpila's Star
And 2h33'4" 45d31'58"
Bethany Rose
And 23h0'23" 39d23'18"
Bethany Rose
And 2h22'4" 41d52'5"
Bethany Rose Sands Wender
And 1h27'32" 48d15'29"
Bethany Rose's Name Day Star
Cyg 20h29'33" 45d33'17"
Bethany Ruggio
Cas 1h19'10" 53d39'12"
Bethany Shay Green
And 2h21'3" 48d21'22"
Bethany Tucker
And 0h32'57" 43d29'59"
Bethany Tucker
Sco 17h48'44" -45d6'28"
Bethany Walters
Lib 14h47'42" -3d32'8"
Bethany Wehler
Cas 1h36'20" 67d7'22"
Bethany Wireman
Gem 7h27'18" 14d10'21"
Bethany Z
Uma 13h14'28" 58d14'52"
Bethany-Laura
Tau 5h5'48" 18d14'12"
Bethany's Blue Rose
Lyr 18h51'14" 33d3'39"
Bethany's Eyes
Uma 10h59'23" 44d33'45"
Bethany's Wish
Cnc 8h35'19" 21d56'7"
Bethel McKenna Gomes
Cas 0h32'9" 53d37'34"
Bethi
And 1h59'58" 47d27'53"
Bethi <><
Umi 15h36'41" 79d37'43"
Bethie Boo
Uma 9h39'15" 43d19'49"
Bethiel
Dra 18h34'10" 60d5'2"
Bethikiss
And 0h21'38" 26d54'48"
Bethina Guadamuz
Dra 16h47'39" 53d21'43"
Beth-Kahlil
Cyg 19h36'0" 30d14'13"
Bethke's Brightest
Lyn 9h2'23" 45d31'30"
Bethlehem A. Wray
Sgr 19h42'32" -11d54'40"
Bethni
And 1h51'20" 39d4'36"
BethNita
Gem 7h11'7" 25d54'55"
Beth-Oryan
Pyx 8h43'52" -27d15'30"
Beth's Alicorn
Leo 10h17'5" 24d23'26"
Beth's Christening Star
Umi 15h43'0" 79d59'36"
Beth's Getaway
Cyg 19h37'6" 52d0'17"
Beth's Little Star
Aql 19h23'18" -0d3'4"
Beth's Shining Light
Cas 23h1'50" 54d26'57"
Beth's Star
Sgr 18h56'46" -29d22'41"
Beth's Star
Col 6h25'39" -35d7'44"
Beth's Star
And 2h14'0" 49d48'6"
Beth's Star
Cnc 8h20'45" 13d26'36"
BethsyMorganeCohen
Sgr 17h49'6" -23d38'37"
Bethy
Cam 5h9'56" 68d53'44"
Bethy van de Velde
Uma 9h55'25" 41d28'21"
Bethy's Bob-oh
Aur 5h53'40" 52d31'3"
Bethy-You Are My Heart-Gallagher
Lyn 6h31'22" 57d58'41"
Beti,la reina de mi corazon
Cep 23h6'27" 59d59'55"
Betina
Umi 15h47'54" 86d40'41"
Betker, Helmut
Uma 9h21'14" 64d38'41"
Beto
Cap 20h9'46" -17d1'40"
Beto
Gem 7h1'58" 28d10'49"
Beto & Rocio Forever
Oph 17h33'10" 5d3'4"
Beto Villa Leon
Lmi 10h19'36" 28d6'41"
Betony & Bryan
Cyg 21h59'23" 50d59'21"
Betrayed Guitarist
Vir 14h10'12" -18d28'11"
Bet's Star
Cas 0h22'28" 58d13'36"
Betse
Uma 10h56'17" 50d32'42"
Betselann Jurick
Uma 8h25'49" 63d16'21"
Betsey and Dennis
Peg 22h28'19" 29d35'27"
Betsi Brooks Krumm
Vir 15h0'1" 1d45'16"
Betssy
Lib 14h47'7" -9d12'44"
Betsy
Aqr 22h29'40" -0d3'58"
Betsy
Lyn 6h28'6" 60d31'42"
Betsy
Cyg 21h7'59" 54d3'17"
Betsy
Sgr 18h55'24" -25d56'55"
Betsy
Ari 2h36'1" 12d55'4"
Betsy
Tau 5h33'7" 22d32'42"
Betsy
Per 3h20'51" 43d31'25"
Betsy & Alec Forever - Happy 25th!
Cyg 21h33'14" 32d38'23"
Betsy and Darian
Vir 12h17'39" -0d59'42"
Betsy Ann Martinez
Leo 11h2'42" 9d17'33"
Betsy Anne Rouviere
Cas 0h40'51" 60d51'50"
Betsy Berriman
Psc 1h20'15" 15d41'44"
Betsy Boruchoff
Cnc 9h7'21" 32d33'48"
Betsy Charlotte Leighton
And 2h11'2" 49d8'42"
Betsy Coniglio
Psc 23h45'33" 0d32'45"
Betsy Cooper
Aqr 22h12'1" -22d19'59"
Betsy D. Armstrong
Lmi 10h53'26" 28d27'5"
Betsy et Yoni
Uma 9h21'5" 60d20'4"
Betsy Foote
Psc 23h7'49" 2d18'45"
Betsy Hamilton (S.G.)
Lyr 18h37'19" 26d27'23"
Betsy Jane Jr. III
Cap 20h23'12" -12d13'10"
Betsy Jean Colman
Lib 14h35'19" -10d49'50"
Betsy Jo Osborne 143
Crb 15h41'54" 36d49'52"
Betsy Kay Bly
Aqr 22h38'30" 0d38'3"
Betsy Lee
Leo 11h18'9" 14d14'21"
Betsy Lopez
Cet 2h23'56" -20d7'22"
Betsy & Luke Wunderlin
Aqr 22h1'50" -24d15'49"
Betsy Lyman
Lib 15h0'16" -0d41'55"
Betsy M. Leonard
Cas 0h32'25" 53d28'14"
Betsy Mann
And 1h5'25" 39d33'50"
Betsy Margo Weill
And 1h37'43" 42d54'57"
Betsy Mizelle
Cas 23h32'50" 58d56'15"
Betsy Peters
Crb 16h21'10" 34d45'32"
Betsy Rouviere
Boo 14h31'53" 53d45'29"
Betsy Stephens
Psc 22h59'42" 7d49'30"
Betsy Tipton Grise
Cyg 19h43'2" 34d20'20"
Betsy Tuttle
Vir 14h53'31" 1d20'13"
Betsy Tyler Carolyn Voter, I
Umi 14h29'41" 73d6'51"
Betsy Welsh
Lib 15h7'2" -1d14'3"
Betsy-Liz Thomas-Parker
Ori 5h35'16" 11d7'22"
Betsy's Great Ball of Fire
Vir 12h8'54" -8d38'36"
Betsy's Messy House
Cyg 20h15'58" 35d29'36"
Betsy's Neverland
And 23h5'46" 40d57'37"
Betsy's Star
Psc 1h24'59" 11d38'6"
Betsy's Star
Leo 10h22'52" 10d34'7"
Betsy's Star
Vir 13h32'37" -1d36'13"
Betsy's Star
Uma 12h19'6" 55d5'39"
Betsy's Star
Uma 10h44'47" 59d42'12"
Bett Kniseley Marriott
Sco 16h9'27" -16d34'53"
Bettal 45
Cyg 21h14'1" 45d39'19"
Bettcher
Cmi 7h21'43" -0d5'27"
Bette
Cas 2h36'7" 65d1'44"
Bette
Crb 15h48'49" 34d27'59"
Bette
Aqr 21h31'7" 1d31'36"
BETTE
Ori 6h7'55" 17d25'26"
Bette and Terry Lomme
Cyg 20h48'15" 47d45'38"
Bette Brizendine
Crb 15h41'51" 32d42'23"
Bette "Bubbles" Schmid
Cam 5h41'41" 61d12'11"
Bette Gould Cook
Crb 15h28'6" 28d6'22"
Bette J. Dodd
Aqr 20h40'14" -8d35'31"
Bette Jane Daul
And 1h46'54" 42d53'33"
Bette Jane Gerula
Uma 8h41'37" 67d3'53"
Bette Joan Sanson
Uma 9h59'20" 52d4'52"
Bette Louise Bachand Foster
Cnc 8h49'53" 16d30'8"
Bette Rapa
Psc 1h38'43" 25d16'18"
Bette Simon
And 2h6'12" 42d42'13"
Bettejane60
Aqr 20h39'22" -11d18'23"
Better Business Bureau
Cas 0h41'11" 61d27'24"
Better Together
Dra 10h58'51" 75d23'37"
Better Together
Cyg 19h44'21" 42d32'43"
Better Together: Rich and Lexi
Gem 6h52'55" 23d34'39"
Betteridge
Cap 20h27'24" -16d43'9"
Betti
Uma 10h48'11" 59d25'4"
Betti Coffey
Per 3h17'47" 50d37'45"
Betti & Ray Avitia
Cyg 21h21'50" 35d48'10"
Bettie
Aqr 22h3'47" 0d10'45"
Bettie Ahlheit Simmons
And 0h19'37" 43d30'54"
Bettie Henry
Cma 6h22'33" -32d13'49"
Bettie Hopkins
Tau 4h11'49" 10d41'31"

Bettie Jane Trimm
Ori 6h8'18" 7d30'59"
Bettie L Schutter
Gem 7h22'2" 32d7'37"
Bettie L. Troutman
Uma 11h38'36" 59d41'29"
Bettie Rage
Tau 3h39'14" 29d4'45"
Bettie-Ann Hauth
Cnc 8h53'32" 30d54'53"
Bettina
Lyn 7h11'54" 49d39'51"
Bettina
Uma 9h8'55" 55d27'41"
Bettina Ann Schlenker
Psc 0h45'51" 16d57'10"
Bettina Bauser
Cam 6h54'5" 65d11'36"
Bettina Biermann
Uma 10h1'26" 42d19'34"
Bettina Calderazzo
Lib 15h36'17" -18d46'2"
Bettina Cinderella Nolden
And 0h23'23" 31d43'47"
Bettina (Dicke)
Uma 11h17'0" 70d1'24"
Bettina Hauser
Cep 20h45'19" 68d49'46"
Bettina Pfäffli
Cas 23h31'28" 59d55'50"
Bettina Saalfrank
Uma 11h35'19" 29d4'47"
Bettina & Thomas
Mon 7h31'39" -1d14'13"
Bettina & Uli
Uma 8h41'54" 53d4'0"
Bettina Westhelle
Uma 8h35'52" 47d44'36"
Bettinas Welt
Uma 13h36'9" 49d14'23"
Bettinelli & Chapman - Sempre Amore
Cru 12h47'8" -60d54'5"
Betts
Lib 15h47'47" -11d13'4"
Betty
Lib 14h37'46" -18d4'40"
Betty
Psc 0h20'16" -3d29'27"
Betty
Vir 12h25'55" -9d45'14"
Betty
Col 5h36'28" -37d25'20"
Betty
Cyg 21h3'53" 45d11'27"
Betty
Cyg 21h3'53" 45d11'27"
Betty
Uma 9h55'40" 49d46'48"
Betty
And 1h43'4" 45d47'16"
Betty
Cas 0h20'48" 50d40'3"
Betty
And 0h46'50" 37d16'40"
Betty
Crb 16h2'46" 36d4'0"
Betty 02/22/1941
Cyg 19h44'0" 35d56'47"
Betty 1
Cyg 19h49'43" 32d19'34"
Betty A. Hart
And 0h26'22" 40d3'44"
Betty Ackerman
And 1h26'10" 49d19'45"
Betty Ahlstrom Simmons
Cnc 9h5'14" 27d32'49"
Betty Alice+ Barry Frederick Codell
Sgr 18h11'21" -25d52'5"
Betty and Al
Crb 16h2'40" 32d56'3"
Betty and Barney
Psc 1h23'21" 5d34'51"
Betty and Bob Kampfman
Leo 9h31'41" 26d5'9"
Betty and Carmen Yarber
And 0h18'13" 29d16'58"
Betty and Frank Nicholas
Lyn 7h22'49" 53d3'9"
Betty and Lloyd "Luke" McGinnis
Aqr 22h1'34" -23d54'9"
Betty and Mickey Rodin
Cyg 20h18'0" 43d33'10"
Betty and Roy Bulkley
Lyn 9h38'44" 40d9'19"
Betty Ann 111743
Sco 16h23'45" -38d33'30"
Betty Ann Ashley
Crb 16h3'26" 34d20'6"
Betty Ann Bateman
Uma 11h27'54" 60d13'46"
Betty Ann Billings Strodel
Lib 15h17'53" -16d48'33"
Betty Ann Desmond
Crb 16h5'25" 38d40'35"
Betty Ann Fitzgerald
Gem 6h46'48" 26d58'6"
Betty Ann Flickinger
Sco 17h53'51" -36d6'27"
Betty Ann Gibson
Leo 9h56'26" 14d21'54"
Betty Ann Robey
Ori 6h13'30" 2d29'0"
Betty Ann Vinci
Sgr 19h7'12" -23d37'19"
Betty Ann Young
Aqr 22h9'1" -0d25'2"
Betty Anne Miller
And 2h22'40" 49d10'6"
Betty * Aunt Betty
Umi 14h35'38" 73d31'53"
Betty "Aunt Mattie" Martin
Uma 12h19'17" 59d56'12"
Betty "Babe" Bailey
Uma 9h17'22" 67d18'8"
Betty Barnett Counter
Ari 2h57'50" 21d45'41"
Betty & Barney
Uma 11h35'35" 57d12'39"
Betty Bates Bakewell
Cas 23h38'49" 52d45'19"
Betty Beatrice
Cyg 21h54'46" 47d29'51"
Betty Billings
Uma 10h53'15" 45d18'28"
Betty Birle Shining Star
Crb 15h35'19" 28d2'21"
Betty BJ Humphries
Peg 21h54'11" 35d10'41"
Betty Blackler
Per 2h55'52" 51d13'25"
Betty Bodmer Anastasio
Ori 6h11'18" 17d13'9"
Betty Boo
Cyg 19h50'52" 35d40'32"
Betty Boop
Ori 5h33'41" -0d16'13"
Betty "Boopster"
And 2h30'2" 43d22'30"
Betty Botz
Lyr 18h41'19" 34d44'54"
Betty Brauchle's Star
Gem 7h9'5" 29d57'18"
Betty Breiner
Sco 16h4'0" -11d36'46"
Betty Brown
Uma 8h50'5" 51d45'21"
Betty Burt
Lyn 6h54'58" 51d33'35"
Betty Buzek
Lyn 6h23'23" 57d10'35"
Betty C. Wiley
Cap 20h16'56" -10d11'42"
Betty Cadwell
Leo 10h16'50" 13d43'30"
Betty Campos
Umi 15h59'20" 72d54'38"
Betty Collins
Cnc 9h18'39" 16d42'22"
Betty Collins
Cas 1h11'34" 58d58'29"
Betty Connolly
Gem 6h50'30" 31d41'23"
Betty Consiglio
Uma 10h51'1" 63d36'29"
Betty Coy
Leo 10h1'56" 12d30'31"
Betty Davi
Lmi 10h31'30" 37d49'31"
Betty & David Tilton
Cyg 19h8'14" 54d7'18"
Betty Demos
Cas 23h26'42" 57d40'25"
Betty Douglas Offermann
Lib 15h45'22" -8d40'16"
Betty Duffy
Uma 10h16'45" 63d30'49"
Betty Dugal
Lib 15h10'53" -12d29'16"
Betty Duong
Sgr 18h26'16" -20d37'36"
Betty - Eternal Light 12/12/1924
Sgr 18h3'18" -27d57'55"
Betty Evans
Lyr 19h5'4" 45d38'34"
Betty F.
Cas 0h36'28" 54d38'22"
Betty Feri Chang
And 23h42'29" 45d39'43"
Betty Fields
Cyg 19h53'17" 39d34'25"
Betty Forgie
Cas 1h3'27" 49d43'11"
Betty Fox Middlecoff
Leo 11h16'17" 15d1'48"
Betty Gable
Cas 1h7'35" 61d36'7"
Betty Gaschen
Ari 3h20'11" 27d53'30"
Betty George Ward Gulick
Leo 11h3'59" -0d11'47"
Betty Gilliam
Cnc 8h10'57" 25d54'21"
Betty & Gordon's Star of Love
Ori 6h8'54" 16d7'40"
Betty Grace Dunn
And 0h38'8" 38d18'32"
BETTY GRAFFEO
And 0h55'58" 36d35'10"
Betty "Gran" Stewart
Cas 0h37'47" 50d16'31"
Betty Hawblitzel Ikuma
Gem 6h54'3" 29d3'55"
Betty Heinze
Cas 0h14'45" 58d19'33"
Betty Helen Tate
Cas 1h34'42" 66d55'45"
Betty Henning
Pup 8h0'12" -34d58'14"
Betty Hicks
Cas 0h0'46" 56d12'50"
Betty Hill
Gem 6h29'22" 22d40'19"
Betty Hintze
Tau 4h32'33" 27d27'8"
Betty Howell
Uma 12h18'56" 59d4'31"
Betty Hundley
Lyn 8h10'2" 43d51'25"
Betty Hunsley
Lyn 7h8'42" 55d48'15"
Betty & Irving
Aur 5h48'10" 38d17'45"
Betty J
Cnc 8h13'9" 15d42'29"
Betty J. Anderson
Tau 5h52'55" 25d6'56"
Betty J. Bosdell
Sgr 19h36'35" -14d52'49"
Betty J. Burts
Tau 4h33'4" 6d16'38"
Betty J Cottrell
Cap 20h38'20" -10d2'57"
Betty J. Davis
Lib 14h58'34" -9d25'27"
Betty J Kelley
Srp 18h20'2" -0d15'54"
Betty J. Kelly
Cas 1h14'45" 58d51'47"
Betty J. Lee
Ari 2h46'54" 24d25'26"
Betty J. Melchi Terveer
Gem 7h15'3" 25d58'50"
Betty J. Naugles
Sgr 19h15'57" -20d46'50"
Betty J. Teghtmeyer
Aql 19h5'11" -0d51'40"
Betty Jane
Cnc 9h8'15" 15d9'26"
Betty Jane
And 23h42'4" 39d0'43"
Betty Jane and Donald Roseberry, Jr
Uma 10h46'28" 45d50'13"
Betty Jane Cameron
Gem 6h39'13" 13d37'53"
Betty Jane Markland
Cas 1h41'22" 62d58'14"
Betty Jane Sheats
Vir 12h36'20" 2d51'47"
Betty Jane Tschudy
Sgr 20h14'22" -35d57'57"
Betty Jane Wereski
And 2h14'30" 49d3'57"
Betty Jane White
Psc 0h4'39" 3d5'56"
Betty Jean
Cas 23h27'29" 53d3'37"
Betty Jean
Ari 3h15'55" 25d6'33"
Betty Jean Arnold
And 2h23'4" 47d49'35"
Betty Jean Becker
Lyn 7h21'36" 56d35'53"
Betty Jean (Elizabeth) Kay
Pho 1h18'54" -42d36'59"
Betty Jean Fairbrother
Aqr 21h24'20" 2d25'19"
Betty Jean Finklea
Vir 12h37'18" 4d44'52"
Betty Jean Harris
Crb 16h16'36" 36d17'14"
Betty Jean Hess
Uma 11h20'31" 63d45'21"
Betty Jean Howes
Sgr 19h40'28" -13d31'47"
Betty Jean Melton
And 1h50'13" 45d55'9"
Betty Jean Mosher
Lib 14h55'21" -23d24'28"
Betty Jean Odgen Johnson
Uma 10h23'42" 61d0'3"
Betty Jean Runion
Uma 14h14'57" 57d3'22"
Betty Jean Stafford
Leo 10h16'8" 25d38'16"
Betty Jean Varone
And 0h49'54" 38d18'47"
Betty Jepson
Gem 7h26'25" 16d58'37"
Betty Jewel
Sgr 18h28'54" -26d40'22"
Betty Jo & Bobby Dee Jones
Cyg 20h3'30" 40d14'31"
Betty Jo Bradley
Psc 0h29'26" 9d41'50"
Betty Jo Conner
And 23h37'23" 41d54'39"
Betty Jo Heaton Leitz
Crb 15h31'57" 27d49'10"
Betty Jo Hixson
Lyr 18h42'20" 38d30'6"
Betty Jo Hogland
Cas 0h39'19" 51d28'45"
Betty Jo Maxwell
Ori 5h40'14" 4d39'24"
Betty Jo Stubblebine
And 1h42'4" 37d36'25"
Betty Joan Dodson
Lyn 7h2'4" 47d45'33"
Betty "Jody Byers" Yarmchuk
Cas 0h54'32" 59d40'46"
Betty Joe
Psc 0h19'6" 21d18'28"
Betty Joe Allen
Umi 16h12'11" 72d8'12"
Betty & Joe Marshall
Uma 11h6'19" 51d40'57"
Betty Jo's Star
Vir 11h59'25" -6d44'13"
Betty Jo's Yellow Rose
Ori 5h28'40" -4d22'9"
Betty June
Gem 7h18'12" 19d58'42"
Betty Kathleen Lawson
Cas 1h7'36" 51d26'14"
Betty Kelly
Crb 16h7'36" 36d0'5"
Betty Kennard
Cas 1h38'54" 63d23'58"
Betty Kiser
And 23h58'32" 39d39'17"
Betty Kleinman
Uma 13h33'58" 59d41'58"
Betty Kolar
Cas 0h10'12" 57d4'17"
Betty Kramp
Her 17h40'27" 20d0'14"
Betty L. Kenney-Lewis
Leo 11h33'39" 27d8'27"
Betty L. Smith
Cnc 8h35'2" 17d27'55"
Betty L. Torner
Uma 13h43'15" 53d42'46"
Betty L. Woelfel
And 1h1'15" 38d35'33"
Betty Laird
And 2h19'1" 50d29'23"
Betty (Lion Heart) Nubgaard
Uma 12h30'28" 62d5'20"
Betty Lou
Cnc 8h33'56" 31d57'9"
Betty Lou Adams
Cas 0h20'5" 61d38'52"
Betty Lou Brooks
And 2h20'20" 42d50'43"
Betty Lou Dorn
Mon 6h50'38" 7d1'59"
Betty Lou Fortin
And 1h47'43" 42d25'19"
Betty Lou Gerrard Schwan
Tau 4h24'43" 8d39'35"
Betty Lou Shultz
Cyg 20h23'0" 45d55'37"
Betty Lund
Cas 0h9'12" 56d43'46"
Betty Luongo
Leo 9h46'45" 31d39'6"
Betty M. Geroge
Lib 15h34'54" -25d11'54"
Betty Madeline
Cru 12h28'37" -56d8'43"
Betty Mae Duncan
Cnc 8h18'6" 15d59'31"
Betty Majka-Siensa
Uma 10h20'54" 55d7'34"
Betty Manz
Lyn 8h27'11" 33d13'32"
Betty Marie Morrison
Crb 15h55'54" 36d6'2"
Betty Marksheffel
Ari 3h12'59" 30d45'23"
Betty Maurice
Tau 3h46'17" 29d15'16"
Betty Mauzy
Uma 12h53'31" 53d48'27"
Betty Maxwell Cummings
Leo 11h50'49" 26d24'32"
Betty McCrumb
Cas 0h30'12" 54d9'25"
Betty & Mike Gordon
Cyg 19h40'44" 37d21'28"
Betty Miller
Uma 8h43'11" 53d34'2"
Betty Miller
Uma 10h8'31" 53d36'27"
Betty Myrlene
Uma 11h38'16" 35d59'39"
Betty Nell Cornish Hamilton
And 0h51'19" 37d45'15"
Betty Nohe Colson - 80
And 23h46'38" 42d38'20"
Betty & Norman James
Cru 12h13'42" -60d47'53"
Betty Nuzzo-11/02/04
Cyg 21h42'40" 44d31'45"
Betty Oliver
Cyg 20h54'42" 47d9'54"
Betty Organt
Cam 5h3'12" 69d23'23"
Betty Pearson
Uma 8h32'10" 60d19'8"
Betty Peeler-Cross
Cas 23h13'42" 55d28'49"
Betty Piontek
Aqr 22h43'35" 1d14'36"
Betty Principe
Her 18h24'55" 29d47'30"
Betty Quan - Young
Lib 15h41'38" -4d3'0"
Betty Quigley
Cyg 19h35'58" 30d49'47"
Betty R Karr
Umi 14h26'52" 65d50'32"
Betty Rae Puma
Gem 7h49'12" 22d2'52"
Betty Reilly
Lib 14h56'57" -15d25'7"
Betty Ruth Baird Angel In The Sky
Lyr 18h36'53" 31d47'21"
Betty Ruth Coffman
Uma 10h12'5" 54d42'29"
Betty Ruth Gray
Crb 16h5'12" 35d52'57"
Betty Ryan
And 1h15'55" 45d20'33"
Betty S. Fulcher
Uma 11h43'32" 36d15'3"
Betty S. Reynolds
Cas 23h19'40" 56d4'1"
Betty & Sammy Got Engaged!
Gem 6h33'19" 24d33'27"
Betty Schoefernacker
Lyr 18h44'34" 31d52'22"
Betty Scruggs
Uma 10h31'39" 51d34'50"
Betty Shanley- Chestnut Street Elem
Uma 13h16'57" 56d22'10"
Betty Sims
Vir 12h41'7" 8d10'25"
Betty Skelton
Cnc 8h50'12" 12d9'39"
Betty Smith
Gem 7h24'43" 32d5'36"
Betty Spangler
Psc 22h57'4" -0d3'1"
Betty Speechley - Forever With Mark
Pho 23h28'20" -44d23'6"
Betty Stamm
Uma 10h34'7" 72d0'4"
Betty Sue McQueen
Del 21h6'12" 19d39'11"
Betty Sue's Star
Uma 13h58'59" 53d30'2"
Betty Taru
Cyg 19h40'29" 31d4'25"
Betty Taylor Memorial
Crb 15h49'52" 34d44'53"
Betty " The Mama" Smith
Cap 21h17'11" -17d0'30"
Betty Tullipano
Dra 20h21'58" 64d15'12"
Betty Venus McGlothlin
Leo 11h20'14" 2d41'11"
Betty White
And 1h6'55" 35d50'39"
Betty Wiediger
Lib 15h15'54" -21d41'28"
Betty Willams
Crb 15h54'48" 37d36'48"
Betty Willock
Cas 23h54'28" 50d17'9"
Betty Willy Graham
Uma 10h10'51" 44d0'27"
Betty Wilson
Cas 1h19'3" 61d5'37"
Betty Wood
And 1h53'16" 44d2'41"
Betty Wooten Smith
Gem 7h19'54" 24d52'27"
Betty Young
Cas 23h52'48" 54d2'2"
Betty, Kayla, Shea, Nathan, David
Cma 6h16'24" -12d34'48"
Betty,Susan,Laur,M.Ann,Bev,Deb,Reb
Uma 10h9'7" 43d7'22"
BettyAnn Lee
Cas 0h15'18" 57d31'25"
Betty-Boo
Cyg 20h22'53" 55d51'43"
Betty-Dawn
Cas 1h17'29" 69d5'44"
Bettye
Lyr 19h12'44" 26d30'21"
Bettye & Gary Howard
Cyg 20h7'33" 38d3'56"
Bettye J. Harmel
Uma 13h44'28" 57d51'18"
Bettye Jean Retzlaff
Cyg 21h44'44" 32d22'41"
Bettye Scrutchin
Crb 15h47'48" 33d42'29"
Betty-Jack
Cyg 21h34'10" 39d37'39"
Bettyjane Mantor
Aqr 22h41'35" 0d28'29"
BettyLou
Her 18h29'55" 12d8'26"
Bettylou and Palmer Rockswold
Ori 5h15'43" -9d14'18"
BettyLouTaradash
Sgr 19h3'8" -17d31'49"
Betty-May Smith
Cam 3h46'56" 58d55'35"
bettymelin
Uma 11h46'50" 36d55'22"
Bettyplum
And 0h15'23" 47d33'9"
Betty's Best Joel & Kyla
Cap 21h51'38" -14d24'6"
Betty's Dream
Tau 4h32'7" 6d32'4"
Betty's Light
Lyr 18h52'16" 32d4'38"
Betty's Love
Cyg 20h7'24" 40d1'40"
Betty's Shining Star
Cam 5h17'46" 68d24'19"
Betty's Special Star
Ari 3h25'27" 22d1'47"
Betty's Star
Uma 11h13'37" 29d11'9"
Betty's Star
Sco 17h0'51" -30d24'4"
Betty's Sunshine
Vir 13h21'34" 12d48'29"
Bettyus
Uma 11h17'25" 29d20'20"
Between Floors 2004
And 0h43'31" 39d37'38"
Between the Moon and Where You Are
Uma 10h6'42" 53d37'42"
Beulah
Cas 23h31'49" 52d56'7"
Beulah Feldman
Vir 12h29'53" -0d52'32"
Beulah Ford McEntire
Srp 18h21'13" -0d1'42"
Beulah Mathews
Cyg 20h59'12" 35d16'27"
Beulah Wilson
Lyr 19h11'32" 38d8'27"
Beullah Turner
Psc 1h12'12" 25d24'18"
Beuna Mae Henry
Vir 13h0'10" -7d43'18"
Beutel, Waltraud
Uma 12h3'0" 61d6'10"
Bev
And 0h18'56" 26d29'15"
Bev and Paul
Cap 20h21'35" -25d23'0"
Bev and Shawna
Cyg 19h55'53" 31d45'44"
Bev&BillHutchinson#25LoveStar
Cyg 20h21'33" 42d57'6"
Bev Gregory
Aql 19h55'14" 12d32'52"
Bev Hail
Cru 12h43'13" -56d57'21"
Bev & Ron Freestone, Happy 35th!
Gem 6h59'23" 22d16'52"
Bev Walsh
Cyg 20h21'19" 58d16'39"
Bev Wells
Per 2h51'27" 40d8'50"
Bev Wing
Aqr 21h36'39" 1d13'6"
Beverelie June
And 23h25'30" 39d11'8"
Beverely Howard Rosen
Ari 2h46'12" 20d49'45"
Beverley
Cap 20h32'34" -12d32'6"
Beverley
Sgr 19h2'18" -14d51'32"
Beverley Anne Wilen
Crb 15h50'12" 27d11'0"
Beverley B
Peg 21h32'26" 25d46'24"
Beverley Billingsley
Vir 13h41'50" -18d0'9"
Beverley Blewitt
Ori 6h0'26" 21d42'43"
Beverley Carol Jamieson Ralph
Sgr 18h57'14" -34d23'35"
Beverley Colón - 2004
Crb 16h22'6" 32d3'27"
Beverley Dawn - 13.12.1945
Cru 12h14'34" -59d40'25"
Beverley & Graham Jamieson-Ralph
Cru 11h59'58" -59d54'58"
Beverley Jane
Cas 23h34'36" 56d58'33"
Beverley Jayne Hill
Uma 8h35'23" 56d16'27"
Beverley & Jon Loyd
Leo 11h32'30" 21d39'18"
Beverley Lowden
Cas 1h15'29" 53d59'42"
Beverley - My 30 Year Star - Chris
Gem 7h17'19" 19d36'55"
Beverley V. Semm
Vir 15h8'22" 5d3'39"
Beverley Viljoen
Oph 16h22'37" -0d7'50"
Beverley's Watcher
Cnc 8h3'7" 6d42'21"
Beverli J. Hougardy
Cas 1h35'51" 71d1'52"
Beverly
Vir 13h27'58" -1d3'38"
Beverly
Vir 13h58'43" -3d24'45"
Beverly
Ind 21h24'36" -55d17'43"
Beverly
Sco 16h55'56" -42d16'50"
Beverly
Ori 5h54'48" 11d53'17"
Beverly
Peg 21h53'2" 11d1'58"
Beverly
Leo 11h36'36" 10d41'45"
Beverly
Ari 2h49'16" 25d38'16"
Beverly
Tau 5h5'2" 26d51'49"
Beverly
And 23h33'46" 37d33'41"
Beverly
Lyn 7h9'24" 47d21'43"
Beverly
Cnc 9h11'46" 31d15'30"
Beverly 12842
Sgr 18h13'34" -21d2'0"
Beverly A. Bolten
Tri 1h40'53" 28d54'40"
Beverly A. Coker 10-15-1937
Cas 0h34'1" 47d42'47"
Beverly A. Gurkin
Leo 11h17'57" 18d16'1"
Beverly A Hayes
Aqr 22h2'6" -2d38'12"
Beverly A Morrissey
Cap 21h56'0" -8d41'12"
Beverly A. Spicher
Ori 5h25'33" 7d24'30"
Beverly Abdalla Johnson
Crb 15h30'51" 28d45'51"
Beverly Afshari
Cap 21h5'48" -16d39'52"
Beverly Aitken
And 0h55'34" 36d55'4"
Beverly Allen
Gem 7h30'44" 31d3'39"
Beverly and James Evans
Cyg 20h24'31" 47d1'23"
Beverly and Joseph Dance
Aqr 21h49'32" -6d7'43"
Beverly and Lawrence J. Lesko
Per 2h48'6" 54d17'1"
Beverly And Ross Privitera
Cyg 20h6'50" 40d37'15"
Beverly Aney Postema
Uma 11h14'9" 52d35'22"
Beverly Ann
Ori 5h1'59" 5d31'4"
Beverly Ann
Ari 3h11'27" 11d50'26"
Beverly Ann "BAnn" Luffel
Aqr 23h21'16" -7d32'38"
Beverly Ann Buxton
Ori 6h4'42" -0d55'33"
Beverly Ann Jackson
Eri 3h52'44" -0d7'1"
Beverly Ann Jameison
Umi 12h4'49" 89d55'42"
Beverly Ann McNew Thompson
Lib 14h44'55" -17d36'33"
Beverly Ann Mingilino
Leo 11h4'42" 15d3'33"
Beverly "Ann" Person
And 0h18'48" 40d2'33"
Beverly Ann Pettit Twilight
Vir 14h25'51" -4d37'20"
Beverly Ann Pickett
Sco 16h12'56" -42d16'55"
Beverly Ann Shockley
Uma 9h37'36" 50d33'44"
Beverly Ann Sweat
Ari 3h1'42" 24d56'35"
Beverly Ann Thompson Brown
Uma 12h45'40" 58d59'7"
Beverly Ann Ward
Sco 16h8'57" -31d15'59"
Beverly Anne Boden Rogers
Mon 6h45'46" -0d47'25"
Beverly B. North *60th*
Cyg 19h59'29" 35d51'30"
Beverly B. Simpson
Uma 8h49'29" 56d45'14"
Beverly Baumann
Ori 5h50'21" 9d9'43"
Beverly Bernstrom
Tau 4h42'1" 7d32'50"
Beverly Bertagnolli
Leo 11h31'2" 26d54'56"
Beverly Blaha Hand
Gem 7h14'26" 22d55'12"
Beverly Bryant
Lyn 7h40'35" 42d1'15"
Beverly Chamberlain
Leo 11h16'25" 4d44'41"
Beverly Cornelius
Leo 9h30'14" 17d23'1"
Beverly D. Wilson 11-2-1971
And 2h15'53" 49d31'41"
Beverly Dawn Ruocco
Tau 5h10'5" 17d47'14"
Beverly Day 17-09-1955
Vir 12h0'37" -0d4'40"
Beverly Defontes
Sgr 18h13'49" -20d1'47"

Beverly Diane
And 23h28'58" 48d52'52"
Beverly Dodge
Cyg 21h19'41" 46d37'51"
Beverly Dove Slagle
Umi 17h23'10" 83d27'25"
Beverly Elaine McLean
And 0h12'15" 26d35'57"
Beverly F. Dohrmann
Tau 3h32'10" 27d5'11"
Beverly F. Goodlin
Lyr 19h18'4" 28d23'16"
Beverly Figart
Vir 13h27'3" 12d4'49"
Beverly Glass Klinestiver
Cas 0h27'53" 50d24'55"
Beverly Hedrick
Cas 0h42'59" 53d29'7"
Beverly Helen Compton
Ari 2h8'1" 24d26'25"
Beverly Hollandsworth
Sco 16h44'16" -44d7'57"
Beverly Hoover
Sex 10h40'32" -0d56'42"
Beverly J. Cox
Uma 13h43'20" 51d48'19"
Beverly J. Downey
And 23h22'17" 42d18'54"
Beverly J. Sowers
And 0h42'47" 43d33'12"
Beverly Jane
Gem 6h46'1" 21d12'10"
Beverly Jane Ball
Psc 0h16'14" 19d51'50"
Beverly Jane Raisor
Cas 0h15'5" 51d56'8"
Beverly Jean
Ari 2h58'40" 26d59'9"
Beverly Jean
Cam 6h13'15" 75d47'55"
Beverly Jean Duroy
Sgr 18h14'58" -19d44'18"
Beverly Jean Guerrant Stahl
Leo 9h44'31" 14d28'9"
Beverly Jean Morris (Juice)
Mon 7h1'10" 9d50'32"
Beverly Jeanne
And 23h20'57" 35d19'44"
Beverly & Jerry Fisher
Uma 9h41'49" 53d53'51"
Beverly Jo
Tau 4h44'16" 27d7'37"
Beverly Jo Baudino
Cyg 21h26'0" 31d49'47"
Beverly Jo Turkowski
Uma 11h51'19" 57d57'59"
Beverly Jones
Lyn 7h15'0" 52d13'6"
Beverly Joyce McBride Fowler
Cyg 19h55'25" 30d48'46"
Beverly June Atkinson
Uma 8h35'28" 64d14'27"
Beverly Kay Blalock
Dra 18h36'26" 52d6'15"
Beverly Kay Stanford
Cra 18h5'3" -37d10'16"
Beverly Kim Lewis
Leo 10h22'3" 21d26'28"
Beverly Lois (viestenz) Potter
Cap 21h43'2" -10d42'2"
Beverly Lynn
Vul 20h15'37" 23d33'23"
Beverly Lynn Shown
Sco 16h9'51" -13d28'10"
Beverly Lynne Fisher Marshall
Lyn 7h40'13" 56d50'46"
Beverly Mae Anunson
Cas 0h24'16" 61d34'52"
Beverly Mae Umbrianna Sunshine
And 0h47'22" 40d32'54"
Beverly Margarita Hicks
Aqr 22h52'50" -9d11'27"
Beverly Marie Morris
Cnc 8h20'8" 8d22'26"
Beverly Mauro
Cas 2h25'48" 67d7'0"
Beverly Morrison-Moyer
Uma 11h17'0" 32d35'51"
Beverly "My Shining Star"
Cas 0h19'50" 48d10'46"
Beverly Newton's Eastern Star
Mon 7h39'37" -4d3'36"
Beverly Nguyen
Ari 3h5'29" 28d2'54"
Beverly Noreen Tilton
Tau 4h59'42" 24d8'38"
Beverly Oliver
And 1h38'25" 39d26'3"
Beverly Page
Crb 16h10'10" 35d45'20"
Beverly Radley
And 0h31'21" 29d42'15"
Beverly Roberts Cronkite
Aqr 20h55'2" -10d4'30"
Beverly Rose Crane
Lib 15h24'43" -19d53'44"
Beverly Ruth Broome
Mon 6h51'15" 7d47'38"
Beverly S.
Uma 9h23'44" 47d6'53"
Beverly Sanders
Sco 16h55'56" -38d55'48"
Beverly Sanford
Cas 23h21'0" 54d55'4"
Beverly Schindler
Crb 15h48'59" 29d17'29"
Beverly Schmidt
Gem 7h36'41" 33d2'23"
Beverly Schmidt Our Guiding Light
Crb 16h6'0" 35d58'40"
Beverly Schumann
Cas 1h31'12" 64d56'56"
Beverly Smith
Uma 14h22'38" 60d31'19"
Beverly & Steve
Cyg 21h31'31" 34d8'40"
Beverly Sue Barker
Aqr 23h47'9" -19d23'20"
Beverly Sue Clay
Lyn 8h0'17" 43d49'29"
Beverly Sue Joy
And 0h32'6" 41d58'32"
Beverly Taylor
Umi 15h11'24" 81d16'8"
Beverly Taylor Garrett
Mon 6h47'41" -0d8'8"
Beverly the Runner 7942
And 1h1'40" 39d44'58"
Beverly Valladares
Cam 4h5'44" 58d6'41"
Beverly Voron
Uma 12h11'32" 54d36'19"
Beverly Walters
Cas 1h13'50" 69d12'3"
Beverly Wehrmann
Lyn 8h37'3" 40d11'38"
Beverly Wesner-Hoehn
Lyr 18h40'36" 36d28'25"
Beverly White
Lmi 10h13'43" 33d41'8"
Beverly Wierbinski
Uma 10h44'58" 39d32'16"
Beverly Wilson Smith
And 0h20'0" 46d11'40"
Beverly Yaeger
Psc 1h34'30" 18d21'16"
Beverly Zychowka
Ori 5h41'45" 5d55'48"
Beverly1001
Leo 9h51'31" 9d56'56"
Beverly-Jane Yuenger
Cyg 20h38'54" 54d10'48"
BeverlyPugh
And 23h2'6" 50d12'59"
Beverly's Light
Psc 1h18'56" 25d45'30"
Beverly's Love Always
And 0h35'55" 39d10'48"
Beverly's Star
Uma 8h33'40" 63d55'41"
Bevie
Vir 12h17'16" 0d10'39"
Bevilishous
Psc 1h13'55" 21d12'25"
Bevnan1
Lyn 7h47'57" 52d1'32"
Bex 11-70
Cam 3h33'25" 64d8'1"
Bex Soanes
And 23h11'49" 37d48'50"
Bexant Thomas
Ori 6h0'43" 5d40'23"
Bexley Madine Avinash Dallape
Ari 3h26'5" 19d52'10"
BEYBIRISSA
Vir 13h31'20" -1d36'55"
Beyeler Optik
And 0h7'44" 41d17'31"
beyhan
Aqr 22h42'44" -5d48'41"
Beyond Expectation's
Uma 10h14'11" 43d11'37"
Beyond Imagination
Vir 13h17'44" 5d41'10"
Beyreuther, Rainer
Uma 9h49'22" 42d38'26"
Bez
Aql 19h7'50" 12d43'25"
Bezalel
Cnc 8h52'25" 10d47'16"
bezaubernder Prinz (Stephan Diethelm)
Cep 1h49'28" 81d22'47"
Bezh and Kate
Cyg 20h12'2" 32d22'34"
"BF" Eric Franklin Wilmoth
Uma 10h21'34" 53d8'16"
BF n°1 Barbara
Uma 10h22'55" 41d56'50"
BFH Turtle Girl
Psc 1h34'28" 10d6'15"
BFUG
Lyn 8h38'12" 39d53'27"
Bfunk
Hya 9h22'6" 5d51'19"
BFW Commando Star
Gem 6h2'13" 22d52'4"
BG John E. Cornelius
Uma 11h40'52" 63d2'9"
BG Lohnes
Per 2h58'27" 49d29'1"
BG MARSHALL H SCANTLIN
Ari 1h54'6" 22d5'9"
B.G.Dinah and Titch.
Dra 15h7'6" 59d32'5"
BGen Jerry C. McAbee USMC
Per 3h25'0" 43d33'39"
BGreenACarrillo
Uma 11h10'4" 52d20'45"
BGS61JEP92MAP94AMP97CRP99
Ori 5h20'28" 9d16'5"
Bhaggo
Ori 5h20'57" -8d26'27"
Bhaile
Uma 9h18'3" 57d21'32"
Bhakti
Ari 3h19'27" 28d23'33"
Bhanu - "Mumblemouse"
Cep 21h22'22" 66d33'56"
Bharti Lakhaney
Aqr 21h49'5" -7d40'27"
Bhatta-Vespa
Sco 16h42'46" -29d45'59"
Bhavin Prajapati
Sgr 19h47'43" -14d25'56"
Bhavna
And 22h58'40" 51d35'53"
BHnCS
Ori 6h13'3" 2d43'33"
Bhrett Robert Long
Uma 9h2'57" 56d39'6"
BHUMI JOSHI
Uma 11h39'30" 60d39'45"
Bhumika Vyas
Tau 5h26'23" 26d42'22"
Bhupinder Kataria
Psc 1h10'38" 29d7'11"
Bia Bop Bop
And 23h0'29" 51d26'11"
Biagio Butta
And 23h25'8" 41d43'51"
Biagio Savarino
Tau 3h29'29" 24d27'2"
Bianca
Peg 21h39'14" 5d58'42"
Bianca
Ari 2h54'59" 21d57'45"
Bianca
And 23h29'42" 42d11'39"
Bianca
And 1h16'11" 49d28'39"
BIANCA
Uma 11h40'55" 46d10'26"
Bianca
Uma 9h48'23" 46d8'3"
Bianca
Cyg 20h33'49" 48d1'12"
Bianca
Lyn 8h38'35" 36d7'48"
Bianca
And 0h53'33" 36d17'5"
Bianca
Uma 9h12'36" 59d5'41"
Bianca
Uma 9h12'6" 59d55'7"
BiAnCA
Lyn 7h59'55" 55d6'42"
Bianca
Aqr 21h26'47" -0d44'59"
Bianca
Aqr 22h28'54" -3d33'16"
Bianca
Umi 15h55'13" 85d31'46"
Bianca 1608
Leo 11h28'0" -3d50'0"
Bianca Alarcon
Sco 17h54'55" -45d5'0"
Bianca (Bankoo) Cagnana
Eri 2h39'52" -43d50'0"
Bianca Berardi
Uma 8h10'57" 60d47'38"
Bianca Bronwen (Angel Girl)
Cyg 21h43'29" 50d53'41"
Bianca Chan
Psc 1h12'51" 12d4'34"
Bianca Chouls
And 2h1'4" 37d39'51"
Bianca Cooley
Sgr 18h3'3" -27d37'51"
Bianca Corti
Umi 15h16'55" 68d16'25"
Bianca deSouza
Lib 15h17'19" -10d14'11"
Bianca Dilorio
And 23h6'30" 51d20'48"
Bianca D-more
Peg 22h35'40" 29d18'38"
Bianca Enriquez
Vir 12h16'32" 3d52'38"
Bianca Esposito
Cnc 8h14'26" 15d4'7"
Bianca Fawn Starnes Whitmire
Lyr 19h9'32" 32d54'46"
Bianca Ferrara
Lyn 6h49'57" 51d2'26"
Bianca Forte
Cas 0h34'56" 63d18'5"
Bianca Isabella Alexander Thomas
Lyn 6h32'47" 56d21'20"
Bianca Jade Rutsch Cigan
Gem 6h43'39" 23d29'15"
Bianca Jenasace Antolini
Cap 20h35'24" -14d32'45"
Bianca Kassab
Psc 1h41'14" 15d33'4"
Bianca Lamas
And 23h25'17" 38d52'34"
Bianca Lilian Bergh
Leo 11h30'45" 11d35'20"
Bianca M. Dancy
Aqr 22h19'32" -1d44'4"
Bianca Maria
Uma 10h25'26" 58d50'3"
Bianca Marie
Vir 12h18'45" 2d42'22"
Bianca Marie Onorato
Psc 0h55'16" 32d39'15"
Bianca & Martin
Uma 10h49'46" 43d7'14"
Bianca Metoyer
Leo 9h58'23" 31d11'11"
Bianca Michelle Bethany 10/13/1993
Lib 15h17'54" -6d40'23"
Bianca & Miles in the stars forever
Aqr 22h22'41" -7d14'16"
Bianca Milov
Psc 23h22'12" 6d21'46"
Bianca Nicole Perales
Uma 11h34'2" 44d59'9"
Bianca Nobilo
And 1h43'29" 36d18'43"
Bianca Parr
Uma 10h18'21" 57d13'10"
Bianca Patricia I. Zabala
Sgr 18h10'39" -19d29'23"
Bianca Renee
Lmi 10h36'59" 31d38'15"
Bianca Rosabella Smith
Ori 5h33'49" 6d42'47"
Bianca Rose Cannetti
Psc 0h25'56" 8d46'0"
Bianca Sabatini
Umi 14h33'54" 73d43'14"
Bianca Star Wilson
Cyg 19h47'0" 37d17'15"
Bianca Star's star
Cru 12h25'16" -62d0'54"
Bianca Sykes
And 23h37'6" 46d57'4"
Bianca Toni Jamil
Gem 7h24'32" 26d5'43"
Bianca Varano
And 0h28'47" 42d16'54"
Bianca Witaszek
Gem 6h34'40" 26d7'21"
Bianca-Alexandre
Uma 10h14'13" 59d44'33"
Biancarelli
Uma 10h18'51" 53d20'13"
Bianca's Bright Light
Pyx 9h23'18" -29d50'19"
Bianca's Guiding Star
Psc 1h6'20" 10d3'13"
Bianca's Light
Peg 23h38'20" 15d34'14"
Bianka
Vir 14h42'57" -2d3'41"
Bianka Bommarito
Umi 14h36'5" 84d10'46"
Biasio Vittorio
Ori 6h4'12" -2d11'32"
Biatriu
Lyn 9h16'26" 37d57'25"
Biba
Cas 1h25'40" 59d25'37"
Biba Belle Blyth
And 2h5'0" 37d3'14"
Bibby
Uma 13h23'32" 55d56'6"
Bibeau
Cyg 19h37'58" 37d31'53"
BIBELI
Ori 6h4'54" 19d25'57"
Bibi
Her 17h49'34" 38d48'40"
Bibi Fernandez
Aqr 22h26'57" -23d29'52"
Bibi Mayes and Jack Zonghetti
Sge 20h5'49" 18d15'13"
Bibi Z. Rahaman
Gem 7h35'42" 26d37'34"
Bibiana & Marlon
Lyr 18h51'53" 33d5'16"
Bibiana Moreno
Uma 8h36'8" 50d39'7"
Bibiane, guide de lumière
Cas 22h58'20" 59d5'8"
Bibiliolà
Cyg 20h14'10" 36d3'35"
Bibou
Vir 12h44'37" -8d44'4"
Bibou Guy Lavoie
Cep 22h45'35" 66d47'0"
BibouStar
Lac 22h47'43" 42d15'32"
Bice La Pegna
Cas 0h27'39" 60d13'40"
Bichi Christiane
Cyg 19h31'50" 32d13'41"
Bichlmaier, Evelyn Michaela
Uma 10h30'59" 47d29'31"
Bick
Per 1h42'52" 50d52'30"
Bick's Beautiful Life Celebration
Gem 7h12'13" 16d24'21"
Bicky Gal's Star
Ori 5h20'8" -4d13'0"
Biddi
Cap 20h50'25" -25d22'59"
Biebl, Stefan
Uma 10h11'33" 46d17'14"
Bien Nice
Uma 11h45'38" 33d32'54"
Biendli
Umi 17h10'24" 75d57'9"
Biene
Cam 6h27'49" 73d22'31"
Bieneli
Mon 7h1'31" -7d24'40"
Bieneman Steve Viv Stephen Brittany
Aql 19h32'1" 11d9'54"
Bierstedt, Brigitte
Uma 8h56'41" 62d4'42"
Biesold, Heinz
Uma 8h53'28" 49d21'26"
Biff
Cnc 8h4'21" 10d51'3"
Biff
Aql 19h53'57" -0d29'21"
Biff "Buddha" Holland
Gem 6h32'13" 18d46'53"
BIG
Boo 14h43'56" 34d52'17"
big 3 0 rach - United In Sisterhood!
Cap 20h58'45" -20d10'25"
Big A Lil a
Per 4h15'41" 50d50'11"
BIG AIR AL
Tau 4h13'31" 25d46'32"
Big Al
Her 18h33'45" 23d7'33"
Big Al
Psc 0h40'59" 7d45'37"
Big Al
Aur 5h49'51" 47d18'18"
Big Al
And 0h50'11" 40d32'15"
Big Al
Uma 11h52'35" 37d40'20"
Big Al
Vir 13h2'3" -8d36'10"
Big Al
Psc 23h56'52" -0d45'11"
Big Al
Cra 18h3'47" -37d11'38"
"Big Al" Alex Koehne
Uma 10h27'57" 48d46'48"
Big Al Fliegel
Ori 6h12'57" 15d6'30"
"Big" Al Flowers
Her 18h12'35" 15d29'35"
Big Al Kadden
Her 18h23'41" 21d32'0"
Big Al & Little Cait
Gem 7h0'44" 14d46'40"
"Big Al" - Rosenstrauss
Cru 12h26'19" -62d13'48"
Big Al's Comet
Gem 7h18'26" 13d43'54"
Big Ani / Albert V. Gonong
Leo 11h36'6" 18d34'49"
Big Bear
Psc 1h7'42" 28d44'24"
Big Bear - Mark Brian Hagan
Uma 8h37'47" 53d30'12"
Big Betty
Sco 16h12'56" -9d50'28"
Big Bill O'Shea
Cep 20h16'28" 61d43'2"
Big Bird a.k.a Mr Wonderful
Uma 11h40'20" 48d29'48"
Big Blaze
Aqr 23h1'20" -9d26'57"
Big Blue Diamond
And 2h37'53" 39d7'
Big Bofie
Gem 6h44'12" 24d29'30"
Big Boss Kyle
Leo 11h5'18" 24d49'7"
Big Boy, Francis J. Palmer
Dra 17h8'56" 51d6'41"
Big Bro
Per 3h35'12" 32d23'57"
Big Brother
Uma 11h45'7" 51d29'45"
Big Brother
Her 17h11'40" 18d29'10"
Big Brother Doug
Her 17h15'6" 37d50'48"
Big Brother Ross
Cnc 8h11'25" 12d33'44"
Big Brother S. Star
Crb 16h11'31" 36d33'34"
Big Bruce
Aqr 22h20'50" 1d15'45"
Big Buford
Leo 11h24'10" 17d42'13"
Big Buford
Sco 16h6'50" -16d36'32"
Big Bugs - Daniel's Star - 14.10.84
Lib 15h56'22" -10d51'29"
Big Bull
Uma 10h22'4" 58d55'57"
Big Bunz
Sge 19h4'5" 20d28'10"
Big Bushie
Uma 11h37'40" 63d18'47"
Big C
Sco 17h19'11" -44d19'20"
Big C
Her 18h1'15" 19d55'39"
Big D
Uma 9h32'31" 41d57'54"
Big D
Uma 11h1'44" 47d29'5"
Big D
Sco 17h11'0" -44d21'18"
Big D
Sgr 19h57'1" -28d20'48"
Big D
Cam 3h50'40" 71d42'19"
Big D
Uma 14h14'57" 55d3'23"
Big D 5
Leo 10h10'31" 25d56'19"
Big Dad
Uma 11h27'53" 51d48'40"
Big Daddy
Uma 13h35'14" 48d46'44"
Big Daddy
Her 17h34'40" 38d0'5"
Big Daddy
Aur 5h23'41" 39d31'14"
Big Daddy
Her 17h39'57" 22d27'27"
Big Daddy
Vir 12h17'11" 11d23'29"
Big Daddy
Uma 13h36'20" 54d7'39"
Big Daddy
Cep 20h57'25" 69d5'0"
Big Daddy
Aqr 21h46'15" -7d4'21"
Big Daddy 47
Psc 23h47'0" 4d54'52"
Big Daddy and Patti
Cyg 21h35'16" 45d52'28"
Big Daddy Barron
And 22h59'30" 50d52'2"
Big Daddy Donovan Stringfellow
Sco 16h12'58" -19d43'21"
Big Daddy Flex 2/21/1998-4/12/2005
Cep 22h30'20" 73d33'47"
Big Daddy Jay Krue
Cap 21h16'6" -15d3'59"
Big Daddy John deMolitor
Uma 11h22'57" 44d3'12"
Big Daddy Johnson
Leo 11h17'4" 19d3'57"
Big Daddy Julio
Gem 6h45'0" 15d22'30"
Big Daddy Matt
Uma 11h6'54" 64d17'28"
Big Daddy Patrick
Cnc 8h0'8" 11d59'6"
Big Daddy Phil
Her 16h44'37" 25d38'40"
Big Daddy Roy
Cmi 7h26'18" 4d56'18"
Big Daddy Trav
Her 16h48'20" 47d1'55"
Big Daddy, Dandy Don
Psc 1h17'13" 8d14'59"
Big Dave
Cnc 8h17'11" 8d48'41"
Big Dawg Gates
Dra 18h44'20" 54d32'32"
Big Dee
Leo 9h51'14" 12d45'23"
big dee
Gem 7h29'53" 20d35'26"
Big Dog Borealis
Gem 7h29'17" 31d0'24"
Big Dog Jason B Burnett
Cma 7h7'58" -24d27'24"
Big Dog Weishaus
Cnc 9h4'25" 31d26'41"
Big E
Uma 11h1'22" 50d49'12"
Big E
Cam 4h36'13" 66d7'25"
"Big Ed"
Cap 21h47'25" -24d9'20"
Big Ed
Aql 19h43'10" 13d8'5"
Big Ed
Aqr 20h56'11" 1d24'27"
Big Ed & Josie Jenny
Cyg 20h35'26" 56d36'27"
Big Fisher
Del 20h35'46" 15d29'55"
Big Frank
Cep 21h29'0" 61d38'49"
Big Game Dad
Sgr 18h14'39" -27d46'46"
big girl
Uma 10h29'49" 45d41'4"
Big Girl and Mayhore
Lib 15h15'51" -8d43'21"
Big Gorilla and Polar Bear
Her 16h47'37" 37d23'41"
Big Grandma
Cas 1h44'31" 65d4'36"
Big Grandma
Uma 11h3'23" 52d38'55"
"Big Guy"
Uma 10h31'12" 43d35'19"
Big Guy
Uma 8h47'59" 46d58'59"
Big Guy Miller
Cep 22h31'2" 57d52'7"
Big Hairy Roman
Aqr 22h50'11" -7d42'57"
Big Henry
Uma 10h16'41" 52d7'21"
Big Hitter
Sco 17h42'23" -42d7'29"
Big House
Cma 7h10'58" -28d2'27"
Big Hunk
Ori 5h47'38" -1d57'23"
Big J
Umi 14h50'10" 70d14'6"
Big Jack
Cep 2h15'50" 80d36'27"
Big Jack
Gem 6h45'52" 16d9'22"
Big Jen Duncan
Uma 13h36'50" 51d59'21"
Big Jer
Cnc 8h45'42" 15d41'12"
Big Jim
Her 17h28'37" 48d27'35"
Big Jim Bear
Uma 8h52'36" 55d56'30"
Big John
Dra 19h6'1" 73d17'21"
Big John
Uma 11h36'40" 62d18'7"
Big John
Cap 20h26'18" -9d51'58"
Big John
Her 17h49'54" 39d5'18"
Big John
Aur 6h57'26" 41d46'24"
Big John
Aur 6h14'3" 31d33'8"
Big John
Leo 11h42'45" 19d0'41"
Big John 1936
Cas 0h37'21" 47d45'16"
Big John and Metka
Leo 11h21'24" 14d21'30"
Big John Earl & Edith Rogers
Uma 11h24'20" 45d34'39"
Big John, Little John
Uma 11h34'18" 53d10'17"
Big Juicy
Ori 5h58'1" 12d26'56"
Big Kahona
Cyg 21h10'46" 44d45'9"
Big Ken
Her 16h50'13" 17d55'48"
Big Lad Ruggy
Uma 10h54'10" 47d52'42"
Big Little Sister
Sco 16h8'53" -19d30'57"
Big Llama
Uma 10h37'36" 71d48'23"
"Big Lou" in honor of Coach Farrar
Per 4h5'17" 36d36'20"
"Big Lou" Stachiotti
Vir 14h4'27" -22d10'44"
BIG MACK DADDY LOVE
Tau 4h26'47" 15d37'11"
Big Mamoo
Lib 15h32'38" -26d16'49"
Big Mary
Mon 5h59'42" -6d38'41"
Big Massimo
Uma 9h38'23" 60d26'35"
Big Meng
Psc 0h3'31" 4d49'48"
BIG MIKE
Sco 16h51'59" -34d25'56"
Big Mike
Cep 3h42'37" 80d33'35"
Big Mike
And 2h24'24" 50d10'38"
Big Mike's Italian DeStallion
Apu 15h41'48" -71d22'9"
Big Mom
Uma 11h13'2" 52d26'5"
Big Momoo
Cnc 8h16'13" 11d44'8"
Big Nick
Cyg 21h38'49" 49d40'4"
Big Nico
Uma 11h32'32" 58d33'37"
Big Nutbrown Hare
Ori 5h56'30" 17d32'35"
Big O
Dra 15h53'11" 57d21'12"
Big Ol
Cap 21h29'56" -20d23'26"
Big Ozzie
Ori 6h19'1" 20d17'59"
Big P
Cyg 21h22'30" 35d9'31"
Big Papa
Sco 16h2'9" -13d55'56"
Big Papa Wilson
Cep 0h59'57" 83d28'40"

Big Pappa
Uma 11h24'28" 60d11'41"
Big Pappa Steve's Star The Freebird
Cep 22h17'48" 61d50'25"
Big Penny & Little Penny
Vir 14h37'34" 4d21'39"
Big Pig
Aql 19h15'57" 10d48'50"
Big Pop
Cep 21h12'40" 60d9'28"
Big Poppa
Aqr 23h52'28" -10d7'41"
Big Poppa.....Steven Lee Hernandez
Tau 4h21'37" 26d32'47"
Big Red
Tau 4h33'10" 20d51'16"
Big Red
Sgr 20h25'36" -38d5'36"
Big Rich's Shining Star
Aqr 22h28'43" 1d30'41"
Big Sav
Sge 19h52'24" 17d0'58"
Big Sexy
Sco 16h15'33" -17d12'46"
Big Sis
Per 4h2'39" 45d14'53"
Big Sister
And 1h28'22" 47d9'12"
Big Sister
Cas 1h52'7" 69d0'53"
big sister star
Tau 5h35'7" 25d8'58"
Big Spender "46"
Uma 11h25'15" 63d26'16"
Big Starfire
Uma 10h44'22" 69d14'30"
Big Steve
Aur 5h45'42" 29d50'12"
Big Swede
Lib 14h56'48" -4d54'36"
Big Syd
Vir 12h10'33" 0d40'38"
Big T
Cap 20h24'38" -13d47'33"
Big " T " (Taylor Whitehead)
Psc 0h8'2" 6d47'22"
Big Ted
Uma 9h6'3" 64d47'27"
BIG TED-Theodore S. Ardelean Jr.
Per 2h19'54" 57d22'11"
Big Tim
Dra 14h34'23" 58d44'53"
Big Tone
Ari 2h38'24" 26d40'53"
Big Vin
Psc 0h33'16" 10d42'8"
Big Willie
Cep 22h48'5" 64d4'29"
Big Willy
Ari 3h16'12" 20d25'24"
Big Willy
Cyg 20h37'7" 36d43'58"
Big YiaYiá
Cas 23h41'38" 57d25'2"
BigDeals/Spndalot Lampert Bros Star
Uma 11h21'51" 35d26'55"
BigDude
Cma 6h44'23" -16d28'5"
BigE22Always
Cyg 19h58'36" 47d37'36"
Biggie B
Aqr 21h13'17" -9d49'47"
Biggie Star Regalia
Sco 16h49'54" -27d27'36"
Biggs's Twinkler
Uma 10h31'57" 40d55'1"
BigGuy
Her 17h43'28" 46d42'34"
BigHoney6402
Gem 7h27'15" 23d53'41"
Bigjerm Locoarium
Ori 6h6'25" 16d33'46"
BigOlWhompinB612 Etoile de Densmore
Uma 10h23'48" 65d21'19"
Bigs
Ari 2h48'4" 12d48'37"
Bigs n Smalls
Her 16h54'35" 32d44'22"
BigWil Archer2007
Uma 10h28'30" 65d36'5"
BIJIN
Cnc 7h57'16" 14d27'4"
Bijou
Cap 21h50'19" -13d27'26"
Bijou (My Boy) Chelune
Uma 10h18'58" 62d36'26"
Bijou Sea Star
Uma 11h16'51" 49d22'17"
Bijoux
Uma 13h53'22" 51d23'21"
Biker David Bark
Uma 9h33'15" 60d14'6"
Bikini Boesch-Krygier
Peg 23h13'31" 17d52'33"
BIL
Leo 10h10'33" 26d48'18"
Bilal Chami
Gem 7h1'37" 21d55'5"

Bilan Simone Barbadaes
Ori 6h12'59" 6d10'42"
Bilands Dee
Leo 11h43'15" 23d55'8"
Bilanie
Cyg 21h24'38" 46d25'15"
Bilanka Pia Kinaszczuk
Per 3h50'45" 32d13'37"
Bilbo
Uma 11h43'9" 33d33'47"
Biliczki Sandor
Leo 10h42'23" 14d2'20"
Bilinski
Aur 5h52'7" 37d29'22"
Bilinz Star
Cmi 7h28'55" 7d28'12"
Biljana
Tau 4h33'42" 13d18'4"
Bilko
Cyg 20h10'45" 38d2'20"
Bill
Her 17h9'0" 31d42'4"
Bill
Uma 9h23'56" 44d16'37"
Bill
Per 4h14'38" 41d21'0"
Bill
Lib 15h27'28" -9d42'20"
Bill
Lib 15h49'29" -6d15'58"
Bill
Vir 12h13'1" -10d5'27"
Bill
Sco 16h54'9" -38d39'38"
Bill
Sco 16h36'17" -38d14'22"
Bill 12-22-05 Marisol
Cyg 20h33'38" 40d53'41"
Bill Ackerman
Per 2h54'0" 48d10'58"
Bill & Addie Tschinkel
Cyg 21h27'41" 36d55'40"
Bill (amazing leo) Frenkel
Lmi 10h51'27" 24d33'55"
Bill and Amy together for eternity.
Leo 10h15'15" 14d54'1"
Bill and Barb's 50th. Anniversary
Umi 14h37'4" 74d7'13"
Bill and Elaine HAWKINS
Cyg 20h32'35" 41d56'53"
Bill and Eli Hanks-2006 C.E.
And 23h1'13" 47d57'4"
Bill and Flo - 8th August 1942
Cyg 21h32'57" 48d37'29"
Bill and Helen Andrews
Peg 22h55'52" 27d38'31"
Bill and Jackie 10 years
Cyg 19h51'13" 33d40'25"
bill and jean kays
Psc 1h22'28" 17d9'3"
Bill and Judy Kampa
Cyg 20h46'9" 47d35'56"
Bill and Karen
Cyg 20h18'52" 52d56'8"
Bill and Lisa Neal
Leo 9h44'20" 27d18'32"
Bill and Lou Rasher
Vir 12h26'51" -2d37'36"
Bill and Marian Kilgore
Tau 4h30'41" 17d8'23"
Bill and Mary Ann (McGuire) Philmon
Cyg 20h5'50" 51d48'26"
Bill And Mary Carroll
Ori 5h41'55" 3d46'55"
Bill and Meredith Keeler
Leo 11h3'12" 20d8'12"
Bill and Michelle
Crb 15h45'4" 35d12'47"
Bill and Pam Comerford - 11/11/2006
Cyg 19h42'37" 42d36'6"
Bill and Rita Asher
Cyg 20h15'18" 37d31'43"
Bill and Teri Reed
Crb 15h49'59" 39d6'23"
Bill Anderson
Aql 19h8'7" 11d31'49"
Bill & April's True Love Star
Cyg 20h38'11" 36d50'29"
Bill Arno
Tau 3h36'48" 21d47'30"
Bill Bagwell
Uma 11h38'37" 48d8'32"
Bill Baldwin
Uma 11h40'55" 38d0'12"
Bill Banks - Star Hunter
Lyn 7h10'29" 50d39'27"
Bill Barger
Per 4h27'19" 45d26'9"
Bill Bates
And 2h26'4" 41d25'33"
Bill Beaumont
And 23h17'4" 35d20'51"
Bill Bellody
Dra 14h42'2" 59d36'33"
Bill Bernardez
Uma 11h23'20" 31d17'45"
Bill & Betty DeWitt
Cyg 20h32'54" 32d21'17"

Bill & Betty Harding 9/12/1955
Cyg 20h9'46" 41d9'39"
Bill & Betty Schwartz's Lighthouse
Aql 19h48'4" -0d30'12"
Bill & Beverly McKown's 50th
Uma 8h35'51" 50d24'50"
Bill Biddlecombe
Per 4h5'45" 33d3'0"
Bill "Big Kahuna" Joslin
Tau 5h45'32" 20d57'5"
Bill Bishop
Umi 12h33'7" 88d5'45"
Bill Blailock
Uma 9h27'33" 72d55'8"
Bill Blount
Uma 9h40'41" 71d6'52"
Bill Brunz
Ori 5h24'43" -4d55'28"
Bill Burke - Star of my universe
Psc 0h21'38" 0d39'59"
Bill Cantwell
Cep 20h49'57" 60d26'23"
Bill & Carole Fink, 2/4/61
Equ 21h16'14" 9d2'51"
Bill & Carole Holsinger
Lib 15h24'33" -8d30'41"
Bill & Carolyn's Star
Uma 10h10'17" 41d49'38"
Bill Carriger
Aql 19h46'48" -0d0'15"
Bill Carson
Umi 13h59'18" 74d46'34"
Bill Casto's Lucky Star aka Linda
Del 20h39'38" 7d36'4"
Bill Chillianis
Aqr 22h28'52" 1d53'54"
Bill Clinton Lyles
Uma 9h59'3" 58d15'50"
Bill Conboy
Umi 14h43'9" 83d2'54"
Bill Cooney
Cru 12h8'3" -62d13'48"
Bill Cozzie
Lac 22h32'22" 48d24'10"
Bill Davidson
Her 17h10'48" 27d39'14"
Bill Davis
Boo 14h28'40" 48d17'32"
Bill Davis
Lyr 19h13'0" 37d55'16"
Bill Day
Ori 5h47'46" 2d4'11"
Bill Degen
Uma 11h46'24" 36d22'27"
Bill DeRicco
Gem 7h25'5" 32d47'13"
Bill DeSalvo
Uma 10h30'17" 68d0'45"
Bill Dickerson's Hope Crutch Star
Cep 21h30'20" 63d51'8"
Bill Dodge
Her 18h51'5" 23d30'48"
Bill & Dorothy Fowler
Lyn 9h7'58" 39d10'15"
Bill Dulson
Cep 22h14'31" 60d4'34"
Bill Dunne
Ori 4h48'49" -3d35'41"
Bill & Eileen
Uma 11h22'12" 62d28'2"
Bill Engle
Sgr 20h26'46" -42d3'51"
Bill Enslin
Psc 1h25'12" 28d2'44"
Bill Epifanio
Lmi 10h12'56" 40d37'42"
Bill Estabrook
Aqr 22h28'55" -6d38'3"
Bill Ewing Sparks Seventy
Uma 11h26'38" 30d16'42"
Bill Fickel
Cyg 19h50'18" 37d38'52"
Bill Finnegan
Gem 7h38'21" 21d31'42"
Bill Fischahs: Dad, Grandpa, Friend
Ori 5h31'2" 2d7'3"
Bill Fisher 2006 Pro Bono All Star
Dra 15h52'57" 61d18'53"
Bill Flaherty
Lyr 18h53'21" 45d7'13"
Bill Flanigan
Aur 5h51'6" 51d45'43"
Bill Foeller
Crb 15h21'47" 30d38'54"
Bill Fornek
Aur 5h35'18" 33d37'30"
Bill France, Jr.
Uma 11h49'18" 36d21'47"
Bill Frantz Nr1
Cnc 8h40'52" 16d46'12"
Bill Fred Kump
Leo 11h48'32" 11d30'34"
Bill Fry
Vir 12h30'11" -0d7'28"
Bill Gaddie
Cep 22h34'59" 68d35'40"

Bill Gallauer
Lib 15h42'21" -17d12'0"
Bill Garvey
Aqr 22h29'9" -1d40'9"
Bill Garwood
Tau 5h32'2" 27d2'7"
Bill Gavin
Ori 6h5'32" 19d46'55"
Bill Gebhart
Uma 11h20'12" 35d49'56"
Bill Gene Bauer
Uma 11h44'32" 55d45'14"
Bill Gerken
Uma 9h26'5" 42d31'11"
Bill Gibson
Ori 4h50'57" 2d16'47"
Bill Gimbel, IV
Cap 21h2'27" -26d18'15"
Bill Green
Sco 17h48'3" -33d33'29"
Bill Gregg
Sco 17h36'54" -32d26'5"
Bill Griffen 7-6-1939
Cnc 8h8'29" 17d13'17"
Bill H. & Veldena L. Breedlove
Umi 16h29'32" 82d34'35"
Bill "Haerry" Haerr
Leo 11h48'19" 10d25'29"
Bill Hancock - My Eternal Soulmate
Ori 5h40'45" -2d11'51"
Bill Hardin
Boo 14h56'39" 23d31'37"
Bill & Heidi Eden
Cyg 21h22'18" 34d48'50"
Bill & Helen Holt
Cyg 19h48'56" 38d31'55"
Bill Henkelman Lake Waynewood
Uma 8h24'22" 66d15'6"
Bill & Hilarie
Cnc 8h52'58" 19d6'51"
Bill Hill
Gem 7h25'17" 21d23'45"
Bill Humphreys 71
Leo 10h21'27" 26d49'15"
Bill Hutchins Of 42nd Street
Cap 20h19'49" -16d51'3"
Bill & Irene Humphreys
Cyg 20h6'24" 39d51'10"
Bill & Jacquie Edwards
Cyg 19h21'51" 28d9'23"
Bill Janesh
Sco 16h41'55" -25d39'2"
Bill Jarosz
Cnc 8h55'4" 7d58'19"
Bill Jay Isley
Aqr 22h48'3" -11d17'24"
Bill & Jeannine McKee
Uma 8h47'15" 51d22'1"
Bill & Joan Becker
Cyg 20h51'11" 32d23'50"
Bill & Joan Wilson
Cyg 19h34'2" 44d43'16"
Bill & Joan's Golden Star
Cyg 19h39'23" 35d14'48"
Bill Jolles
Per 4h0'55" 35d15'47"
Bill Julian
Uma 9h1'12" 57d58'51"
Bill & Karen Thompson
Cyg 21h57'48" 47d35'55"
Bill Keener
Cep 21h20'52" 56d16'6"
Bill Keeton III
Umi 14h34'44" 73d50'49"
Bill Kenny
Ori 5h6'45" 9d13'47"
Bill Kirkland
Vir 13h40'5" 4d58'38"
Bill Kist Ph.D., hunter, & fishman
Gem 6h3'37" 27d22'27"
Bill Kraus
Per 3h13'59" 31d11'34"
Bill Lancaster
Her 18h50'6" 22d33'34"
Bill Lawrence
Cep 23h27'11" 71d26'5"
Bill Lehr Star
Sgr 17h52'36" -28d29'44"
Bill & Leslie Riviere
Cyg 19h38'42" 35d39'2"
Bill Lifland
Leo 11h7'18" 11d41'0"
Bill & Linda Brackeen
Aql 19h20'52" -8d27'54"
Bill Ludwig
Uma 8h25'24" 64d21'35"
Bill M. Sinclair
Aqr 22h17'41" -1d19'12"
Bill Manning
Uma 12h4'47" 54d2'28"
Bill Marchetti
Uma 9h33'32" 60d54'53"
Bill Maresh
Lac 22h49'28" 53d57'49"
Bill & Margo Bowers
Cyg 21h17'8" 36d22'40"
Bill & Marlene's Star
Cam 4h2'26" 70d3'21"
Bill McKee
Ari 3h5'56" 26d18'38"

Bill MD Cameron
Sco 16h17'31" -9d6'25"
Bill Merola III - Billy's Light
Uma 12h37'33" 55d55'15"
Bill Michell
Psc 1h38'35" 20d21'21"
Bill Mike
Aur 5h25'40" 40d55'16"
Bill & Mill Garner
Gem 7h15'50" 28d18'8"
Bill Muirhead
Cnc 8h51'34" 7d56'13"
Bill & Nada Arnold
Psc 1h28'40" 17d4'13"
Bill Nichols
Ori 6h1'18" 5d41'15"
Bill Nuckols
Lmi 11h3'18" 25d35'24"
Bill O. Pennell
Ori 5h22'51" 2d16'6"
Bill O'Brien
Ori 6h6'54" 12d15'39"
Bill Orfeo
Vir 13h58'8" -1d28'2"
Bill Orwin
Uma 9h35'37" 45d43'4"
Bill Papageorgiou
Her 17h40'53" 42d35'24"
Bill & Pat Robert's Now & Forever
Pho 0h56'3" -49d23'29"
Bill & Pat Warren
Vir 14h9'19" 3d1'12"
Bill & Peg Lawson
Ori 5h19'57" 0d53'30"
Bill & Penny McLendon
Uma 10h26'27" 49d25'22"
Bill Perfitt (My Perfect Star)
Per 2h57'49" 47d13'53"
Bill Petersen
Her 17h45'58" 32d20'30"
Bill Petter BP 21
Tau 4h21'50" 21d19'46"
Bill Phillips
Lib 15h18'11" -23d22'49"
Bill Plumbe
Tau 4h36'48" 10d43'42"
Bill Powers
Ori 5h33'3" -2d49'14"
Bill Price
Per 2h46'5" 51d29'8"
Bill Provence
Boo 14h32'12" 41d33'51"
Bill R Williams
Sco 16h8'27" -11d6'15"
Bill & Rachel
Lyr 19h7'54" 27d18'29"
Bill Rauch
Cep 22h12'33" 64d1'22"
Bill Ray
Lib 15h9'26" -7d40'48"
Bill 'Redcloud' Manning
Aql 19h50'47" -0d18'23"
Bill Rider
Lyr 18h51'18" 37d27'45"
Bill & Rina Swinburn Golden Wedding
Cas 1h26'27" 65d50'59"
Bill & Robin Shaw
Cyg 19h49'27" 32d58'26"
Bill Robinson
Cyg 19h47'54" 57d51'19"
Bill Rondeau
Lib 14h51'54" -0d42'50"
Bill & Sandee
Cyg 20h20'24" 51d29'50"
Bill & Sarah
Per 4h31'53" 39d33'14"
Bill Scott
Lac 22h39'50" 37d41'50"
Bill & Shae's Star
Cyg 20h36'11" 38d10'32"
Bill & Shirley McLendon
Uma 10h30'7" 49d48'57"
Bill Simpson
Cyg 19h44'42" 42d27'26"
Bill Smith's Real Estate Pelican
Ari 3h21'45" 18d48'57"
Bill Stewart - Stellar Dad
Leo 10h21'47" 16d57'3"
Bill Stinner-Trimble
Lmi 10h19'26" 29d21'9"
Bill Stout
Vir 13h25'0" 1d49'24"
Bill Strickley
Uma 12h0'21" 40d31'36"
Bill & Sue Hull 50th Anniversary
Psc 1h26'44" 18d36'15"
Bill & Susan McKissick 0202
Sge 19h41'3" 17d48'54"
Bill & Susan Nash's Wedding Star
Cyg 20h10'0" 39d53'1"
Bill T. Miller (altar boy's gang)
Psc 1h30'59" 11d58'16"
Bill Tamayo
Cap 20h43'45" -21d59'8"
Bill & Tammy Lapp
Gem 7h8'14" 14d53'33"

Bill & Teresa - Friends Forever
Sco 16h8'31" -17d44'1"
Bill & Tess Capuano
Eri 3h52'25" -39d23'53"
Bill Tewell (Toolie)
Cep 22h16'24" 65d54'29"
Bill — the cowboy
Aur 5h37'16" 42d5'36"
Bill the Monster Destroyer Crossey
Her 16h14'45" 48d57'59"
Bill the Thrill
Her 17h58'57" 23d55'38"
Bill Thomas Greatest Dad Ever
Tau 4h5'12" 4d39'14"
Bill Tidwell, Jr.
Ori 5h23'46" 4d6'22"
Bill & Tiny
Lib 14h50'19" -1d27'43"
Bill Tomkiewicz
Cap 20h40'6" -16d52'56"
Bill Trimper
Uma 9h41'10" 63d36'3"
Bill & Trish Miller
Psc 23h38'5" 4d59'11"
Bill Twaddle
Cep 21h12'1" 65d36'30"
Bill Utterback
Boo 14h47'46" 19d34'7"
Bill Vandivier
Cap 21h21'26" -18d10'52"
Bill Vasas
Uma 11h24'12" 63d4'14"
Bill Vedder
Cep 21h3'47" 57d32'29"
Bill Vickery "Vickery's Lucky Star"
Cep 22h42'4" 77d46'16"
Bill Walker
Sgr 18h46'22" -28d50'45"
Bill Walker
Her 18h50'49" 22d13'55"
Bill Walter Memorial
Per 3h8'57" 50d27'1"
Bill Webb
Ori 6h20'18" 10d24'50"
Bill Weitschat
Aur 6h24'32" 41d41'31"
Bill Wilson, Sr.
Sgr 19h15'40" -34d46'6"
Bill Zanker
Her 17h9'46" 32d38'47"
Bill Ziemann "My Blessing Star"
Dra 20h10'40" 62d6'58"
Bill Zwecker
Cep 21h18'39" 61d12'21"
Bill, Jennifer Mourafetis
Cyg 19h55'36" 37d23'34"
Bill, Maxie, & Patty Cowan
Eri 3h55'14" -0d16'53"
BillandChristinesLuckyStar 29Jan2004
Umi 16h14'51" 75d18'57"
BillandNuzzyPacker
Lib 15h43'23" -12d32'38"
billchrissy
Her 17h5'8" 43d29'31"
BillDiverDownGustafson
Vir 11h50'15" 8d11'56"
Billdoe
Cap 21h58'10" -22d25'33"
Bill-E
Leo 11h25'16" 14d57'36"
Billee Alexis
And 23h2'57" 48d0'15"
Billee Mara Krupnick
Oph 17h7'18" -0d44'29"
BillGlor
Her 18h32'2" 18d57'57"
Billi Elizabeth Lovell
Uma 11h44'22" 28d25'50"
Billiam
Cap 21h26'42" -25d33'20"
Billiblue
Cnc 9h18'57" 10d10'15"
Billie
Aqr 22h42'45" 2d0'34"
Billie A. Becker
And 0h11'23" 46d44'31"
Billie A. Jones
Lib 15h28'15" -11d13'17"
Billie Ann Branum
And 0h54'8" 41d20'23"
Billie Buster
Sgr 18h30'30" -21d30'0"
Billie Carstens
Lyn 8h9'13" 34d28'25"
Billie Curtis
Uma 10h40'17" 44d29'35"
Billie D. Patton
Cnc 8h21'15" 12d8'49"
Billie Frances Sharrer
Per 2h54'23" 48d4'33"
Billie Jean
Cas 0h12'54" 54d14'36"
Billie Jean , Our Star
Ari 3h7'6" 13d51'59"
Billie Jean Brown "Billie Bob"
And 0h16'46" 28d14'58"
Billie Jean Davis
Uma 9h32'45" 58d5'54"

Billie Jean Hilsher
Sco 17h33'34" -39d59'48"
Billie Jean Sheehan
Ori 6h20'14" 14d22'14"
Billie Jean Wagoner
Ari 3h21'13" 28d58'56"
Billie Jeane Jones
Sgr 17h53'51" -29d50'33"
Billie Jo
Lib 15h36'39" -7d40'33"
Billie Jo (B.J.) Haynes Lankford
Cyg 21h24'35" 39d12'2"
Billie Jo Poynter
Sgr 18h39'32" -18d6'10"
Billie Jo & Wilbur Morgan
Cyg 19h46'32" 55d46'8"
Billie Joe
Aqr 21h5'21" -13d36'53"
Billie Jorja Whelan
Peg 22h24'29" 22d34'39"
Billie Josephine Infield
Aql 19h45'8" -0d4'0"
Billie Kay Coles
Sco 17h39'10" -39d34'43"
Billie Kay Welch
Cnc 8h46'46" 23d39'12"
Billie Kaye
Aqr 23h45'18" -13d43'54"
Billie Laverne Wagner
Lyn 6h25'36" 59d16'33"
Billie Linda Sigler Taylor
Uma 11h42'6" 29d22'18"
Billie Marilyn McPherson
And 0h42'46" 45d5'16"
Billie McNiven
Cep 23h7'37" 71d12'3"
Billie Miller
Uma 11h1'21" 47d53'23"
Billie Miller Pittman
Uma 11h16'37" 54d23'22"
Billie Nicole Mussman
Cap 21h55'55" -13d39'50"
Billie Rayanna Scott
Cap 21h36'56" -12d8'30"
Billie Rose
Ari 2h35'13" 25d35'44"
Billie Rowe Scruggs
Uma 11h0'10" 68d45'3"
Billie York
Sco 17h3'46" -32d43'43"
Billie-Jade's star forever
Peg 22h31'24" 9d21'6"
Billie-Jean Frage
Uma 9h27'42" 65d1'50"
Billiejean Locklear
And 0h23'14" 41d49'40"
Billie's Rose
Leo 11h40'18" 26d10'9"
Billie's Star
Umi 15h0'29" 70d32'15"
billjeaninetcws
Lib 15h22'44" -21d56'54"
BillKari
Tau 5h39'52" 25d6'47"
BillMagsDean
Cnc 9h3'14" 7d43'27"
BillnSylviaHawhee50
Tau 5h59'0" 23d15'26"
Billori
Lib 14h46'56" -4d22'42"
Bill's 60th Birthday Star
Vel 9h30'6" -40d31'5"
Bill's Ballybunion
Ori 5h25'9" 0d44'50"
Bill's Beloved
Sgr 18h54'40" -23d50'45"
Bill's Big Surf
Cnc 8h20'40" 8d15'46"
Bill's HighLight
Per 2h47'26" 54d14'31"
Bill's Light
Uma 10h30'24" 68d22'59"
Bill's Place
Cep 23h2'54" 70d53'46"
Bill's Sharr
Psc 1h7'12" 33d16'8"
Bill's Sparkling Shepherd ~ Jean
Uma 9h18'12" 54d18'23"
Bill's Star
Sco 16h4'32" -18d59'34"
Bill's Star
Tau 4h24'27" 26d51'46"
Bill's Sumpin From Sumter
Uma 11h6'4" 37d31'55"
Billy
Aur 5h25'44" 40d33'35"
Billy
Cnc 8h43'4" 31d46'40"
Billy
Uma 10h40'50" 52d1'6"
Billy
Per 3h20'37" 50d7'7"
Billy
Lmi 10h54'24" 25d55'13"
Billy
Sco 16h11'42" -17d25'3"
Billy
Aqr 21h19'5" -7d24'34"
Billy
Uma 12h18'6" 56d28'12"
Billy
Col 5h27'7" -28d45'58"

Billy "50"
Cep 20h37'43" 64d6'50"
Billy Aitken
Per 4h3'21" 44d59'18"
Billy Alexander
Lmi 9h42'33" 34d10'26"
Billy and Ally Birks
Gem 6h58'45" 27d59'54"
Billy and Alyse
Cyg 20h19'7" 45d15'36"
Billy and Dandrea
Cnv 12h42'1" 41d0'1"
Billy and Gia
Uma 10h40'23" 52d7'40"
Billy and Jackie Forever
Cyg 20h20'43" 56d1'7"
Billy And Jen For Ten
Cyg 21h9'58" 30d10'24"
Billy And Jenn
Lib 15h55'0" -17d19'30"
Billy and Jerry Jo Forever
Tau 4h27'17" 17d14'31"
Billy and Kerrin
Sco 17h10'0" -32d59'35"
Billy and Lisa McLain
And 0h34'22" 37d59'10"
Billy and Lolly's Star
Per 4h49'28" 47d41'30"
Billy and Nancy Pritchett
Aql 19h6'36" 7d59'18"
Billy Anne
Cyg 21h57'20" 47d35'11"
Billy (big bruv) Skinner
Ori 6h14'59" 2d0'16"
Billy Birken
Her 17h18'56" 35d47'15"
Billy Bleifer
Cyg 21h32'26" 54d13'53"
Billy Blizzard
Ori 5h25'32" 2d24'14"
Billy Bob Butler
Sgr 19h7'58" -18d41'32"
Billy Bobbert
Lyn 8h1'3" 40d6'11"
Billy & Bobby Schiffman
Uma 11h1'44" 51d1'57"
Billy Boughey
Her 17h17'53" 45d56'7"
Billy Brookshire Taer 2007 Shine On
Sgr 18h18'21" -31d33'30"
Billy Bryan
Umi 13h12'11" 69d36'25"
Billy Burnett
Her 16h37'47" 7d39'11"
Billy Buster - Class of 2007
Vir 14h34'14" 4d28'12"
Billy Carrigan
Uma 9h39'31" 51d33'54"
Billy Corathers
Sco 17h53'31" -38d6'15"
Billy Coulson
Uma 10h53'58" 51d42'45"
Billy Crawford
Uma 10h32'14" 48d25'57"
Billy D
Tau 3h26'34" 19d7'54"
Billy "Daddy" Pirozzi
Cep 20h41'10" 60d54'14"
Billy David Greenacre
Cep 0h23'19" 74d4'57"
Billy Dean Garoutte
Cap 20h12'30" -9d17'3"
Billy Drew
Her 16h22'13" 19d34'51"
Billy E. Hanes
Aur 5h15'34" 42d41'5"
Billy & Edda
Eri 4h32'25" -14d37'13"
Billy Eugene Hopgood
And 0h20'9" 42d31'35"
Billy Eugeney
Uma 10h24'20" 44d52'30"
Billy Farmer
Uma 9h54'40" 42d42'53"
Billy Fotis - Daddy to Little Goose
Her 17h20'18" 35d46'38"
Billy Fulco
Ori 6h15'46" 9d26'37"
Billy G "My Inspiration"
Her 17h43'46" 16d0'19"
Billy Gainey
Uma 12h54'40" 58d20'46"
Billy Gareth Standing
Umi 14h59'51" 78d30'53"
Billy Gene Bridges
Her 17h0'19" 13d25'34"
Billy Gene Henson
Cnc 8h27'25" 22d27'54"
Billy & Gina Rominek
Cyg 20h38'56" 42d34'18"
Billy Hegg-James
Aur 5h8'22" 40d7'41"
Billy Heussner & Beth Schirott
Cyg 20h13'49" 32d41'3"
Billy Howard Limmer
Ori 5h47'5" 3d1'30"
Billy Hughes
Per 2h13'19" 55d58'36"
Billy J. Rea
Ori 5h26'17" 0d54'22"
Billy Jack- Angel Forever
Vir 12h48'6" 6d49'27"
Billy Jack Frizzell
Aql 20h17'55" 6d55'14"
Billy Jack Green
Ori 6h3'0" 6d28'37"
Billy Jack Sullivan
Cep 20h34'42" 60d33'50"
Billy Jack Williams
Uma 12h2'9" 42d27'41"
Billy James Schaffer, Jr.
Uma 8h34'11" 72d56'23"
Billy Jo
Lib 15h11'5" -15d36'33"
Billy Joe and Barbra Dean Goodwin
And 0h43'55" 37d10'6"
Billy Joe Screamer
Vir 12h32'19" 12d10'24"
Billy Jones
Gem 6h47'47" 28d12'50"
Billy Joseph Hayes
Her 16h30'49" 34d39'3"
Billy K.
Aql 19h29'32" 0d23'40"
Billy Kallas
Leo 11h10'3" 26d35'38"
Billy Kilkenny
Lyr 18h34'30" 27d34'21"
Billy&Kim Richardson*Endless Love*
Crb 15h29'25" 25d56'27"
Billy & Kitty's Star
Ori 5h42'38" -2d3'4"
Billy Lamb
Dra 17h35'5" 57d7'56"
Billy Lane
Cep 20h39'26" 61d47'1"
Billy Lasater's Star
Per 2h47'40" 54d17'1"
Billy LeRoy Marshall
Leo 10h27'56" 20d33'25"
Billy Lewis
Lyn 7h46'58" 36d59'17"
Billy Luyendyk
Ori 5h24'1" 13d48'15"
Billy "Maestro" Cleary
Cep 21h46'43" 60d9'11"
Billy & Michele
Cyg 20h44'55" 32d9'7"
Billy Monahan
Uma 10h46'55" 51d4'15"
Billy Moore
Uma 8h58'54" 58d41'28"
Billy Moore
Cru 12h38'31" -64d3'19"
Billy Murdy
Cep 20h45'58" 57d5'18"
Billy Murray
Gem 6h46'39" 24d46'55"
Billy Nash
Lyn 7h5'31" 50d43'19"
Billy Navarra
Dra 20h23'28" 71d58'41"
Billy Nickerson III
Cep 2h33'58" 82d48'17"
Billy Orzame
Ori 5h43'47" -3d37'48"
Billy P. and Marti Robinson
Cnv 12h30'23" 32d16'37"
Billy Pat
Per 3h34'34" 35d30'32"
Billy Pettit
Umi 14h53'19" 70d50'32"
Billy Plunkert
Leo 11h18'29" 17d39'30"
Billy Ponce
Her 16h37'53" 46d32'57"
Billy R. Bailey
Sco 17h46'56" -45d31'3"
Billy Ray Bragg
Psc 1h6'33" 27d44'51"
Billy Ray Sacket, Jr.
Per 4h35'52" 32d18'27"
Billy Ray Sellers
Uma 10h58'24" 46d57'42"
Billy Ray Walls
Gem 6h41'47" 25d25'42"
Billy Reese
Sgr 19h34'35" -38d6'15"
Billy Ron
Ori 5h54'31" -0d3'46"
Billy Rosario
Cnc 8h2'22" 25d47'59"
Billy & RoseMarie Haglund
Lyn 6h24'49" 56d4'56"
Billy Rutzen
Ori 5h27'35" 4d59'5"
Billy S. Flowers
Per 3h49'10" 46d34'53"
Billy Samuel
Her 17h8'2" 39d58'57"
Billy Sands "Infinite b"
Sgr 19h35'5" -12d25'21"
Billy Shields
Uma 10h22'59" 62d37'12"
Billy Shine Down
Uma 8h22'6" 60d33'41"
Billy Shkrabek
Her 17h45'5" 43d18'24"
Billy Steele
Cma 6h18'44" -30d55'54"
Billy Talbert
Uma 9h41'9" 58d45'9"
Billy the Cave Bear
Uma 8h15'1" 68d57'56"
Billy the kid
Ari 2h16'26" 23d36'16"
Billy Thomas Dalton 7/6/41 - 9/3/06
Gem 6h56'46" 18d22'25"
Billy Thompson
Ori 4h56'19" -0d21'28"
Billy Thorpe's Rock Star
Ari 3h19'0" 15d14'4"
Billy Thrush
Cnc 8h47'13" 32d3'11"
Billy Toomey's Star
Umi 16h25'18" 80d41'18"
Billy Trimble
Vir 14h9'16" -17d33'50"
Billy Twomey
Uma 11h9'58" 71d41'23"
Billy und Argus
Apu 14h32'39" -73d46'33"
Billy Urquiza
Aql 20h10'22" 15d58'30"
Billy Weimer
Sgr 17h59'41" -17d1'52"
Billy Whizz
Uma 8h21'57" 61d49'36"
Billy Youmans
Ori 5h43'17" -0d38'43"
Billy1
Cyg 20h17'7" 47d52'29"
Billy21, My Eternal Love
Lyn 8h0'24" 39d17'21"
Billy-Bob
Lyn 7h9'24" 49d32'43"
BillyDaymondMullins
Tau 5h40'39" 25d41'9"
Billye Thompson
Umi 15h16'30" 88d4'2"
Billy's Blazing Sun
Vir 12h34'44" -2d14'16"
Billy's Boomer Sooner
Boo 14h37'3" 31d28'8"
Billys Place
Cyg 20h51'44" 33d28'51"
Billy's ROCKSTAR
Lyn 7h24'57" 46d24'53"
Billy's Star
Cnc 8h7'12" 11d35'14"
Biloxi ToJo
Aql 19h46'29" 7d51'7"
Bilwati Salles Eric 16/06/1961
Cas 0h20'17" 53d32'9"
Bim
Tau 5h12'42" 21d28'57"
Bima Padilla - Love now & forever, Daniel.
Cyg 20h38'47" 51d56'40"
Bimba e Bobolo
Lyr 18h36'24" 31d22'32"
Bimbaria
Sco 17h13'0" -34d9'49"
Bimini Bill's SnyderMania
Tau 4h5'4" 6d51'32"
bimper bear
Lib 15h47'28" -8d52'29"
Bim-Tooty
Ori 5h22'5" 4d52'4"
Bina
Cyg 21h46'46" 43d51'19"
bina
Lib 15h42'23" -3d58'37"
Bina Patel
Leo 11h23'9" -4d41'59"
Bina Shahid
Ari 2h53'17" 28d30'3"
BinBin
Gem 7h22'41" 14d3'22"
Binder, Michael
Uma 13h36'29" 56d16'19"
Bindi
Cru 12h54'7" -55d51'51"
Bindu - Mariam
And 2h24'40" 45d44'18"
Bine
Ori 5h49'39" 12d10'38"
BINE
Uma 10h52'2" 57d29'22"
Bine & Stefan-Stern
Uma 11h37'42" 37d24'6"
Bines & Peters Glücksstern
Ori 6h14'46" 6d31'32"
Bing
Uma 10h0'43" 42d6'51"
Bing J. Carbone
Ari 3h5'57" 23d49'53"
Bingger, Dorothea
Uma 10h42'8" 62d39'13"
Bingham Brightness
Aqr 22h49'18" -19d43'30"
Bingo
Umi 13h22'43" 74d13'8"
Binh C Tran
Vir 12h39'58" -9d22'29"
bink bink
Ara 17h37'2" -52d38'30"
BINKI
Lyn 8h28'56" 49d18'55"
Binki 1976
Lib 15h7'10" -0d38'18"
Binkstar
Uma 11h48'39" 33d50'27"
Binky
Gem 7h38'46" 32d5'13"
BINKY
Her 16h59'17" 13d9'57"
Binky Sue Hudec
Psc 0h31'20" 11d31'35"
BINKZ
Tau 3h44'37" 29d16'32"
Binnie and Elaine May 18, 1974
Lyr 19h12'39" 42d27'27"
Binnur
Uma 12h37'32" 62d56'43"
Binny
Uma 11h41'20" 51d11'14"
Bino
Uma 9h14'44" 64d7'25"
bio
Uma 11h39'53" 51d44'19"
BiochemBob
Del 20h48'31" 15d43'58"
Biolato Marco
Her 18h56'17" 23d32'27"
Bionda Lita
Ari 2h36'47" 21d28'30"
Bip 528
Uma 9h46'56" 61d45'29"
Bippy's Star
Lyn 7h58'54" 35d28'49"
Bipsy
Vir 12h49'8" 11d57'56"
Bird Love
And 23h21'41" 47d34'26"
Bird Star
Cnc 8h45'28" 14d53'25"
bird72
Per 3h36'6" 45d34'2"
Birdi
Aql 19h59'16" 4d59'21"
Birdie
Psc 1h33'0" 7d58'59"
birdie
Ori 6h7'58" 15d2'44"
"Birdie" (Birk)
Cyg 20h15'50" 52d22'36"
Birdie Louise Welborn Larner
Lyr 19h19'15" 28d14'40"
Birdie Love
Cnc 8h32'35" 17d36'48"
Birds Star
Ari 3h11'28" 10d48'30"
Birdsong
Uma 10h7'41" 47d18'8"
Birdy
Ari 2h16'24" 11d39'50"
Birgit
Uma 9h29'42" 48d1'3"
Birgit
Uma 11h28'20" 31d4'48"
Birgit 16.4
Uma 12h39'17" 53d49'0"
Birgit Aschenbach
Uma 10h18'19" 55d5'16"
Birgit Fery
Uma 11h40'18" 65d25'9"
Birgit Gebauer
Uma 11h13'1" 43d50'57"
Birgit Henkis
Uma 10h9'12" 70d54'38"
Birgit Küster-Culic
Ori 5h48'19" 11d59'34"
Birgit Pokladeck
Ori 6h14'38" 8d13'4"
Birgit Stadge
Ori 6h17'51" 7d19'15"
Birgit Tillingen
Uma 11h39'4" 41d14'27"
Birgitt
Uma 10h0'7" 42d15'43"
Birgitta - an Amazing Friend
Sco 17h54'0" -40d11'27"
Birgitta Hallgren MNP
Cap 21h9'13" -18d49'29"
Birgitta Olsson-Spellerberg
Boo 14h35'49" 36d53'29"
Birgitte Apel
Gem 7h30'17" 20d13'8"
Birin and Michelle Yucesan
Cnc 8h37'11" 16d56'14"
Birklie Goedde
Gem 6h36'21" 12d27'53"
Birl Orion Shultz
Ori 5h46'9" -2d28'0"
Birmingham
Crb 15h18'10" 32d23'22"
Birmoser, Robert
Ori 6h8'2" 7d0'59"
Biró Bálint a mi csillagunk 50 éves
Cap 20h38'30" -12d10'46"
Birringer, Markus
Cnc 8h53'23" 25d17'9"
Birth of Jinta Yamaki
Lib 15h34'0" -18d58'38"
Birthas Big Bang
Ori 6h6'18" 5d56'59"
Birthday Star of Sharon Whittington
Cyg 21h37'9" 39d15'10"
Biscottino ama nocciolina
Cnv 12h26'20" 33d27'40"
Biscotto d'amore
Cyg 21h40'4" 43d51'1"
Bishop Alfred A. Owens Jr.
Uma 10h23'43" 64d22'8"
Bishop Angela Denise Davis
And 1h16'32" 37d26'41"
Bishop Cameron Audino
Uma 10h17'52" 64d55'24"
Bishop Carrington Edward Morgan, Jr
Cep 23h11'47" 70d0'17"
Bishop Donald J. McCoid
Gem 6h32'41" 20d47'10"
Bishop F. Joseph Gossman
Aql 19h3'7" -0d27'52"
Bishop Mike
Aur 5h46'8" 37d17'1"
Bishop Wayne T. Jackson
Vir 12h6'21" -0d41'0"
Bishwanath Nevatia
Col 5h41'28" -31d33'25"
Bisket Singer
Cyg 19h55'25" 51d52'28"
Bisquettes
Sco 17h41'12" -36d17'32"
Biss Glory
Peg 22h37'33" 11d26'3"
Bissi
Aqr 22h26'37" -8d31'33"
Bistro Al Vino : Al,Brenda,Alana
Umi 14h32'47" 72d53'4"
B.I.S.Wilson3rdGradeClass 2004-2005
Uma 11h14'10" 57d27'23"
BIT
Lib 15h29'21" -26d5'38"
Bit A' Tl E Gonaai Na'n Bas
Cma 6h48'27" -13d58'11"
Bitanem
Uma 12h7'52" 47d59'50"
Bitanem
Leo 9h23'8" 32d28'45"
Bite Falls
Lyn 8h13'0" 45d31'4"
Bite Size
Sgr 18h14'41" -19d13'34"
Bitsy
Cap 21h16'52" -15d44'50"
Bitsy Betsy
Cas 23h23'23" 58d52'50"
Bitsy Johnson
Lyn 7h1'6" 54d2'5"
Bittie
Gem 6h30'45" 21d59'4"
Bitty
And 22h59'47" 48d16'54"
Bitty
Umi 15h18'40" 73d8'10"
Bitty
Cap 21h6'50" -26d34'44"
Bitzer
Hya 9h33'37" -0d6'42"
Bivins
Lib 15h26'9" -6d20'15"
Bixby's Brillance
Sgr 19h44'19" -25d23'19"
Bixente Barandiarán
Cnc 9h7'54" 30d58'15"
Bixiño
Tau 5h55'48" 28d16'43"
Bizzle
And 0h24'32" 25d45'20"
bizzo
Lyn 7h55'15" 56d51'56"
bj
Vul 21h6'54" 27d43'19"
B.J. Baines, Jr.
Per 2h8'20" 54d51'27"
BJ 'Bunny' Wakefield
Uma 13h42'56" 56d2'0"
BJ - If ya knew him, ya loved him 9584
Cru 12h25'16" -61d58'50"
BJ Jenkins
Her 18h23'33" 18d11'55"
BJ Nelson
Uma 9h22'14" 51d54'54"
BJ Norris
Leo 11h26'26" 11d36'44"
BJ "Sunshine" Sayre
Crb 15h31'0" 35d18'14"
BJAllen25LDForde
Cas 1h43'37" 62d19'55"
Bjaloncik
Per 4h6'14" 33d59'28"
Bjanikka Mareine Ben
Sgr 19h8'29" -30d36'16"
B.Jay & Kelly Weiss
Lyr 19h0'8" 27d22'50"
BJBD + JBB Forever!
Aql 19h37'17" 16d20'2"
BjKarePrittThankfulForFamilyFriends
Uma 8h56'2" 46d55'36"
BJOELLE
Lyn 8h11'46" 35d9'18"
Bjorn Andersson
Sgr 18h5'58" -24d25'42"
Bjorn Andrew Mathay
Lib 15h16'20" -7d7'56"
Bjorn Heen Loufman
Uma 12h38'10" 61d45'57"
Bjorn Martin Rettberg
Sgr 18h30'55" -19d8'55"
Bjorn & Misha "Bear"
Uma 9h28'54" 71d54'2"
Björn Nickel
Ori 5h12'46" 8d33'41"
Björn + Sabine
Uma 10h22'31" 55d51'33"
Björn + Saskia
Boo 14h57'17" 18d54'35"
Björn Schmidt/In unändlichär Liäbi Margrith,
Cas 23h46'23" 57d34'35"
BJ's moms: Verna and Barb
Lib 15h34'21" -19d56'18"
BJ's Shimmy Now and Always
Lyn 7h31'18" 37d0'24"
BJUMSEN
Ari 2h30'53" 17d40'55"
BK Chance
Lyn 6h20'53" 57d17'25"
BK FOREVER
Sco 16h5'25" -14d21'40"
BK-CM-LS-RS-TT-JY-GZ
Uma 11h9'21" 49d30'20"
BKF90159
And 1h42'25" 46d43'17"
BKSP1 Sonya Porter & Brian Karel
Mon 7h16'43" -0d13'36"
BL Concept
Uma 9h52'16" 65d21'15"
Blaalop
Her 16h58'11" 32d10'24"
Black
Cyg 20h35'57" 43d0'12"
Black Bart
Cnc 8h45'6" 19d43'4"
Black Belt Nicholas O'Brien
Her 17h15'54" 36d44'4"
Black Bobby
Cru 12h25'37" -60d29'3"
Black Cat
Pho 0h33'9" -41d38'44"
Black Cody
Tau 5h45'7" 16d54'8"
Black Jesus Star
Uma 9h13'54" 64d36'43"
Black Lady the 3rd
Gem 7h10'50" 34d7'0"
Black Pearl
Cyg 19h51'57" 36d16'57"
Black Rose 1980
Tau 3h38'46" 28d29'41"
Black Sheep and Lily
Pho 1h18'33" -42d17'32"
Black Wednesday
Leo 11h7'26" 20d36'15"
Black Widow Helen
Cas 2h23'37" 73d23'56"
Black Widow Queen
Cas 0h32'49" 48d53'30"
Blackburn
Cyg 19h44'43" 30d19'37"
Blackie
Ori 5h32'2" 2d13'34"
Blackie Drumm
Ori 5h41'35" 7d34'18"
Blackie of the C.C.C.
Lib 15h27'27" -26d9'27"
Blackie Wildfire
Tau 4h21'47" 28d26'34"
Blackpaw
Uma 10h0'20" 46d18'42"
Blackshear Star
Umi 16h20'54" 74d43'42"
blackstentribe
Ori 5h35'45" 2d26'55"
Blackstock
Ori 5h30'13" 6d29'39"
blacksunshine
Dra 19h43'55" 60d23'4"
Blackwater
Dra 16h12'36" 62d44'29"
Blacky
Uma 11h57'8" 50d17'2"
BLADE/SHAI/GIA
Cnv 12h48'4" 43d6'37"
Bladon-Spiers (amor in æternum)
Pho 0h45'22" -42d24'21"
Blagden Wharton
Cep 23h44'38" 72d36'58"
Blaike Lauren Hammel
Uma 11h21'43" 38d7'3"
Blaike Stafferd
Aql 19h22'25" 0d28'8"
Blaikie
Umi 15h57'4" 71d22'19"
Blain Harry
Cyg 21h21'23" 38d32'43"
Blain Miller Donnell
Aqr 21h4'21" -9d4'25"
Blaine
Cap 21h1'45" -18d45'10"
Blaine Alan Runge
Sco 16h43'0" -32d19'29"
Blaine and Cindy Kirkpatrick
Sge 20h4'44" 17d58'35"
Blaine and Roy Voyles
Cam 5h39'47" 69d37'13"
Blaine Andrew
Sco 17h17'35" -33d6'55"
Blaine Holt Finkbiner
Cnc 9h2'46" 31d8'35"
Blaine Lorenz Dustin
Lyn 7h52'8" 38d19'37"
Blaine Master 2000
Aqr 23h25'1" -20d10'38"
Blaine Reeve Grayson
Cam 4h15'20" 68d12'39"
Blaine Thomas Cheng Mayginnes
Her 17h53'6" 38d47'1"
BLAINESTAR
Ari 2h54'32" 18d22'4"
Blair Allen's Sick Star
Ari 2h36'8" 20d1'44"
Blair Arrington Walters
Sco 16h8'31" -38d28'26"
Blair Bear
Uma 9h27'10" 65d54'56"
Blair Boudreau
Lyn 7h29'29" 53d57'42"
Blair Buckler
And 0h39'24" 41d22'48"
Blair Davis
Vir 12h10'58" 3d26'21"
Blair Delaine Holt
Gem 7h32'28" 31d8'13"
Blair Elizabeth Henninger
Psc 1h24'47" 31d55'38"
Blair Elizabeth Self
Dra 17h12'38" 53d17'40"
Blair Elyse
Tau 3h38'31" 29d3'42"
Blair Emerson Shannon, Ambassador
Her 17h47'8" 28d22'58"
Blair Ford
Dra 19h15'41" 60d8'0"
Blair Kimberly Browne
And 0h47'3" 25d30'43"
Blair Lantz Brim 5-31
Gem 7h3'21" 25d0'20"
Blair Madalyn Miller
And 1h36'5" 44d25'19"
Blair Michael Paul
Boo 14h35'14" 50d13'33"
Blair & Miranda Pankratz
Tau 4h38'34" 10d28'13"
Blair Montgomery Brown
Sgr 18h9'57" -20d5'57"
Blair Nicole Carmichael
Del 20h39'6" 13d36'52"
Blair Robert MacKenzie
Sgr 18h9'39" -31d1'34"
Blair Rutledge
Lyn 8h59'59" 34d54'27"
Blair Steven Perry
Ori 6h6'45" 17d54'20"
Blair "Sweet Pee" Taylor Thurman
Pho 0h4'1" -45d15'29"
Blair The Wonderful
Aur 5h27'51" 46d10'34"
Blair Thomas McKendrick
Sco 16h9'24" -10d45'18"
Blair Tracy Shaw
Mon 6h52'2" -0d7'10"
Blair Walker Fuller
Psc 1h20'53" 31d38'34"
Blair Winegarner
Umi 16h10'3" 85d34'14"
BlairBear
Aqr 22h4'9" -0d14'16"
Blaire
And 1h23'41" 46d22'6"
Blaire Bear
Umi 16h20'1" 75d54'31"
Blaire "Bear" Ramsay
Vir 13h19'19" -16d22'55"
Blaire Fritzinger
And 2h32'48" 48d47'37"
Blaire Jackson
Lyr 18h25'38" 34d19'6"
Blaire Nichole Brewster
Ori 6h23'30" 16d31'3"
Blaire Nicole Maltarich
Leo 11h27'29" 13d48'27"
Blaire S.
Uma 9h46'24" 49d9'55"
Blair's Sizzling Sizzle 6/10/01
Umi 14h51'20" 78d53'53"
Blairtny
And 23h31'27" 47d53'35"
Blaise
And 0h6'55" 28d46'32"
Blaise
Cap 21h32'44" -10d40'21"
Blaise Adison Mayer
Cyg 21h25'29" 39d45'37"
Blaise Dewey
Cap 20h11'33" -16d26'45"
Blaise James Fitzpatrick
Cnc 9h19'7" 8d26'8"
Blaise John Giacobbo
Cnc 9h5'36" 30d43'57"
Blaise M. Iacofano
Sgr 19h42'23" -16d42'23"
Blaise Patrick Higgins
Umi 15h17'54" 67d35'53"
Blaise Thomas Sindone
Cap 20h20'36" -18d55'7"
Blaise Virgil Stewart
Sgr 18h53'53" -26d23'57"
BLAISE21
And 23h19'47" 47d58'59"

Blake
Cam 3h46'27" 55d39'31"
Blake
Aur 5h41'30" 46d7'17"
Blake
And 0h8'12" 42d6'6"
Blake
Psc 0h50'12" 13d36'19"
Blake
Cru 12h43'19" -58d6'30"
Blake
Lib 15h57'1" -11d28'20"
Blake Aaron our Angel - 6-10-2005
Cru 12h17'20" -57d20'46"
Blake Alan Lemaster
Aqr 22h39'42" -2d26'56"
Blake Alexander Bain - Blake's Star
Ori 6h11'5" -0d34'14"
Blake and Evelyn forever
Lib 14h25'46" -16d48'43"
Blake and Karisa
Uma 9h21'46" 54d8'10"
Blake and Melissa
Lyn 6h56'1" 53d41'36"
Blake and Melissa Forever
Cyg 20h35'17" 51d48'48"
Blake Andrew Brinkman
Psc 23h14'32" -0d1'56"
Blake Andrew Compton
Peg 23h53'1" 28d18'39"
BLAKE ANDREW MESSICK
Tau 4h8'32" 6d43'25"
Blake Andrews
Cep 20h43'37" 69d59'5"
Blake Angelo Deluca
Vir 12h14'29" 11d25'31"
Blake Anthony Clegg
Tau 4h47'11" 22d16'55"
Blake Ashley
Her 17h34'42" 32d24'15"
Blake Austin Reed
Her 17h39'3" 33d17'37"
Blake Ball
Gem 6h22'3" 24d9'30"
Blake Bartlett
Umi 13h42'32" 69d58'29"
Blake Basgall
Uma 11h41'0" 48d23'6"
Blake Brandenburg
Uma 9h33'43" 47d30'51"
Blake C. Hunt
Her 17h26'19" 14d40'56"
Blake Carter Compton
Cap 21h42'52" -9d0'27"
Blake Cody
Tau 5h34'4" 22d6'20"
Blake Cody Fowler - shine brightly
Pho 0h39'51" -49d30'41"
Blake Colby
Dra 14h44'56" 55d25'3"
Blake Cole
Umi 15h52'17" 71d17'33"
Blake Davis
Lib 15h35'57" -14d17'26"
Blake Davis
Leo 9h33'2" 28d23'52"
Blake Delahoussaye
Lac 22h21'4" 46d29'22"
Blake Dewey Peterson
Per 3h45'24" 51d26'22"
Blake Edward
Gem 7h42'1" 32d41'33"
Blake Edward Carlson
Cep 21h40'5" 66d47'14"
Blake Edward Stephens
Boo 14h40'51" 36d27'38"
Blake Enniss
Vir 13h13'33" 2d3'56"
Blake Eplin Pierce
Cyg 20h9'22" 34d13'24"
Blake Eric Haines
Per 2h27'30" 52d23'13"
Blake Eric Swan
Boo 14h42'3" 26d37'59"
Blake Fowler
Cru 11h58'55" -64d34'35"
Blake Haines
Cyg 20h59'53" 54d40'51"
Blake Harrison Fierman
Uma 11h26'8" 33d8'31"
Blake Hughes
Uma 9h1'54" 48d59'15"
Blake Jeznach, Star of Tomorrow
Cyg 19h52'53" 59d31'42"
Blake Johns Bell
Cyg 19h20'55" 29d6'18"
Blake Joseph Monson
Aqr 22h14'27" -0d30'6"
Blake Joshua Abbott
Aql 19h27'18" -0d20'53"
Blake Keelen Altman
Peg 22h5'46" 33d46'39"
Blake Kenneth Baerwald
Her 17h44'33" 40d1'31"
Blake Kostka
Uma 13h35'39" 56d29'47"
Blake Lawrence Roberts
Her 18h37'4" 21d48'45"
Blake Lee Pearson
Per 4h15'50" 49d56'13"
Blake Lloyd Shepard
Boo 15h38'27" 41d58'25"
Blake Marie Barr
Lib 15h31'43" -18d8'49"
Blake Martin
Dra 16h39'34" 63d41'53"
Blake Mitchell Gandrup
Aql 19h24'32" 5d16'5"
Blake Morgan Jones
Uma 9h40'44" 41d27'49"
Blake Murray
Sgr 18h23'56" -34d20'12"
Blake Nicholas Granata
Lib 14h47'15" -22d4'8"
Blake Nicholas Mancuso
Per 2h36'40" 55d18'8"
Blake of ROG-10
Uma 10h18'46" 62d3'34"
Blake Owen Holukoff
Lyr 18h50'55" 45d39'36"
Blake Patrick Missey
Her 16h37'2" 31d4'53"
Blake Peters
Her 18h19'1" 25d35'8"
Blake Pierce Luehlfing
Vir 12h6'6" 3d10'16"
Blake R. Orndorff
Cep 21h42'32" 63d50'16"
Blake Randall
Her 17h50'3" 48d19'7"
Blake Robert Carpenter
Ari 2h57'47" 24d58'54"
Blake Sheldon Strunk
Ari 2h37'52" 28d7'37"
Blake Shelton
Gem 6h2'32" 25d57'7"
Blake Smola
Uma 11h34'1" 58d2'58"
Blake Stuart Keating
Cyg 20h1'52" 55d23'10"
Blake Tyler
Cas 23h1'16" 55d31'33"
Blake VanDernoot
Gem 7h40'53" 34d42'50"
Blake Vann
Lib 15h20'46" -23d48'57"
Blake Wallace Francis
Gem 7h7'54" 31d54'51"
Blake Wayland
And 23h36'8" 38d59'11"
Blake Whaley
Her 16h58'47" 19d18'9"
Blake William Wickham
Tau 4h40'15" 1d57'17"
Blake Williams
Sgr 19h39'58" -14d7'12"
Blake Wilson Anderson 6775
Boo 14h36'16" 36d7'44"
Blake Zon Guarnieri
Ori 5h22'45" 3d44'31"
Blakelee Paige Speelman
Cas 0h29'27" 59d36'44"
Blakely and Ambrosia Forever
Sge 19h48'42" 18d27'18"
Blakemore E. Johnston Jr.
Sco 16h42'25" -29d24'5"
BLAKERS
Lmi 10h6'50" 37d26'5"
Blake's Light - 14/01/2003
Col 6h9'39" -42d54'39"
Blake's Star
Uma 9h26'23" 67d35'21"
BlakeShelton#1Star
Cnc 8h26'30" 15d3'51"
Blalonna
Cmi 7h34'27" 7d1'53"
Blam, Rüdiger
Ori 5h5'26" -0d58'6"
Blamberg, Oliver
Ori 5h20'45" -7d59'54"
Blami
Tri 2h17'21" 32d57'32"
Blanca
Ari 2h15'7" 23d55'27"
Blanca
Sco 16h45'1" -26d24'4"
Blanca and Reuven Nachlieli
Cam 4h31'40" 74d43'10"
Blanca Diaz
Tau 4h30'29" 22d11'23"
Blanca Dorado
Aqr 22h9'15" -7d51'43"
Blanca Doris Shannon
Ari 3h17'37" 19d38'45"
Blanca E. Ruiz
Uma 10h1'3" 66d4'23"
Blanca Estela
Lib 14h58'28" -4d50'35"
Blanca Estela Romero ~ My Angel
And 1h11'59" 36d1'32"
Blanca & Jose Mesa
Gem 6h28'54" 21d29'15"
Blanca Leyva Ward
Aqr 22h50'57" -3d26'43"
Blanca Neri Arismendi, Mi Bella
Tau 5h20'28" 23d28'4"
Blanca Nieves Hermosa
Cap 20h43'47" -22d28'55"
Blanca Reyes Martinez
Her 18h24'29" 27d41'22"
Blanca Rodriguez
Crb 15h50'26" 27d43'37"
Blanca Rodriguez
Lyn 6h20'32" 60d32'10"
Blanca Rosa Ayala
Cap 20h45'24" -16d39'1"
Blanca Silva-Reyes
Cas 1h15'12" 53d29'39"
Blanca Suarez
And 0h23'58" 46d28'32"
Blanchards Dark Star
Vir 11h40'30" 8d16'25"
Blanche and Bryant Vaughan
Cyg 20h54'28" 39d5'41"
Blanche Arlene Specker
Crb 15h39'41" 26d58'12"
Blanche Billings Smith
Uma 11h19'2" 42d49'29"
Blanche Brown
Uma 10h8'29" 62d34'42"
Blanche Canavan
Cyg 21h12'11" 40d0'30"
Blanche Hedrick
And 1h56'7" 41d48'49"
Blanche K. Cummings
Ari 3h9'33" 28d42'17"
Blanche Kathleen Terry
Aqr 23h8'15" -17d33'49"
Blanche Kesseler-Thilman
Ori 5h19'32" 1d40'31"
Blanche Koff
Uma 11h11'40" 52d57'13"
Blanche M. Harris
Cas 0h33'48" 59d18'20"
Blanche M. Mutters
Tri 2h31'13" 30d51'15"
Blanche Price Doyle
Cas 1h26'17" 55d36'33"
Blanche Van Duesen Baldwin
Crb 16h13'41" 38d19'12"
Blanche, who lived out Romans 8:28
Men 5h56'15" -84d39'4"
BLANDIA
Crb 16h4'26" 33d49'58"
Blandine Häfliger
Uma 12h58'37" 52d57'37"
Bländu
Umi 16h55'12" 86d8'15"
Blane and Ali
And 0h30'42" 40d34'35"
Blane and Annie's Diamond n the Sky
Uma 8h31'49" 67d22'50"
Blane Wallace
Dra 20h9'53" 71d54'39"
Blane's Star
Sco 17h43'44" -35d46'12"
Blank
And 0h24'42" 27d41'2"
Blank, Brigitte
Uma 13h2'11" 53d54'30"
Blankenburg, Wilfried
Uma 11h2'56" 40d43'50"
Blankentiz
Psc 1h17'3" 22d57'29"
Blanquart Virginie Choupette
Sco 17h20'56" -42d55'48"
Blanquart Virginie Choupette
Del 20h44'2" 9d18'27"
Blanquita Marie
Ari 2h24'12" 11d55'9"
Blanthorn
Cyg 21h51'52" 51d57'13"
BlanToku
Ori 4h46'48" 11d12'53"
Blas E. Fiorenzano
Eri 4h32'24" -15d24'8"
Blaschke, Klaus
Ori 5h4'13" 5d57'39"
Blase J. Caranese III
Ari 3h26'45" 28d1'7"
Bläser, Katharina
Uma 14h7'19" 56d17'36"
Blash
Pho 0h37'25" -46d52'27"
Blasius
Tau 4h33'56" 19d45'0"
blasterBaby
Cma 7h2'19" -24d23'32"
BLASTR
Ari 3h12'25" 12d8'22"
Blathnait Crowley
Cyg 21h29'57" 36d0'22"
Blatz
Her 16h40'4" 6d56'2"
Blaudschun, Gerald
Uma 11h13'20" 35d0'33"
Blaylock-Carroll
Lyr 18h38'39" 30d49'14"
Blayne
And 0h16'3" 40d53'13"
Blayne
Sgr 18h43'18" -24d23'17"
Blayne Adeline Perlo I
Leo 11h38'36" 26d51'13"
Blayne and Megan always and forever
Cyg 19h59'27" 35d47'23"
Blaze
Dra 17h55'25" 50d45'31"
Blaze
Ori 6h11'29" 18d11'42"
Blaze Belfiore
Uma 13h54'35" 53d7'45"
Blaze D'Amico
And 23h2'16" 39d23'23"
Blaze M. Hendershot
Psc 1h24'45" 29d16'25"
Blaze Starr
Tau 4h18'30" 23d53'11"
Blazer "BEN" Hadinger
Ori 5h23'31" 4d5'35"
Blazing Hope
Lib 15h23'39" -15d42'38"
Bledi
Per 3h19'38" 52d55'3"
Blendean Crawford RN HSI Nurse
Tau 3h42'28" 16d51'12"
Bless the Broken Road
Aqr 22h34'23" 2d17'58"
Bless the Broken Road
Umi 13h28'32" 74d48'42"
Bless U Rina Benning
Uma 8h52'47" 57d24'37"
Blessed Blake
Sco 17h19'25" -44d22'20"
Blessed By God - Judy & Art
Gem 7h43'28" 34d1'29"
Blessed CRAIG Luv & Miss U
Cyg 21h39'19" 48d45'8"
Blessed Infinity
Uma 8h50'26" 54d23'44"
Blessed John Francis
Aql 19h23'45" 2d43'39"
Blessed Little Heart
Psa 22h36'52" -32d57'17"
Blessed Peacemaker
Col 5h42'22" -31d35'44"
"Blessed" - Taylar & John Patty
Cyg 21h20'47" 35d6'12"
Blessed Union
And 0h30'59" 37d55'1"
Blessing
Cep 21h21'44" 65d24'35"
blessing
Eri 3h50'15" -0d46'58"
Blessings By Robin
Gem 7h14'15" 33d39'43"
Blessy Rose Abraham
Psc 23h46'41" -2d33'56"
Bleu Ashton Carvel
Cep 21h52'48" 59d16'34"
Blewski
Lib 14h23'36" -17d21'55"
Blewski 12
Ori 5h58'22" 21d21'37"
BLF BFF
Lyr 18h38'6" 34d46'15"
Blia Yang
Psc 0h59'27" 9d10'10"
Blickie Kurrus Buzan
Ori 5h33'38" -0d15'57"
Blindin
Cas 1h30'54" 72d11'53"
Blingbling Llama Master
Aur 5h40'34" 43d12'12"
Blingham
Cap 20h59'36" -20d8'52"
Blink
Eri 4h10'38" -0d11'54"
Blink
Lib 15h30'44" -7d16'21"
Blinka
Gem 6h2'59" 25d21'34"
Blischke, Alfred
Ori 5h4'36" 6d46'28"
Blish
Cyg 20h42'53" 35d40'33"
B-Lish
Psc 1h23'41" 7d17'2"
Bliss
Lib 15h12'12" -4d44'15"
Bliss
Cap 20h24'53" -21d41'31"
Bliss
Lyr 18h53'17" 35d45'35"
Bliss
Crb 15h48'38" 34d14'4"
Bliss
Per 2h23'16" 57d8'8"
Bliss
Uma 9h8'7" 48d54'57"
Bliss
Uma 9h26'1" 48d23'0"
Bliss on Park Avenue
Lib 14h52'51" -6d48'6"
Bliss R. Fontenot
Aql 20h13'27" 4d15'24"
Blitz
Ori 5h43'55" 1d47'10"
Blitze
Leo 11h23'58" 16d54'51"
Blitzen Westie
Sgr 18h17'6" -28d45'2"
Blitzkrieg
Uma 11h32'58" 38d49'49"
Bliz
Uma 11h50'27" 62d15'22"
Bloching, Walter
Ori 5h31'18" 9d59'31"
"Blocka" Dean's Dream
Cru 12h20'12" -57d50'18"
Blöink, Sarah
Uma 8h46'48" 57d47'24"
Blomberg, Nicole
Leo 10h54'50" 5d18'44"
Blondesherr
Leo 11h3'29" 14d44'52"
Blondie
Cnc 9h2'38" 17d0'31"
Blondie
Cnc 8h52'14" 28d16'52"
Blondie
Crb 15h43'40" 37d28'18"
Blondie
And 1h28'41" 49d22'14"
Blondie
Umi 16h4'51" 71d2'9"
Blondie
Cap 20h17'57" -24d16'19"
Blondie/Kissy Face
Cyg 20h42'22" 40d26'24"
Blondine Dube
Uma 8h54'58" 54d35'22"
Bloo Dream
Boo 14h27'22" 13d41'37"
Bloodfox V
Vul 19h32'4" 25d35'18"
Bloodhead
Sco 16h7'53" -26d13'50"
Bloomford
Ori 5h36'21" -2d10'58"
Bloopette 77454
Peg 23h16'54" 34d12'59"
Blossom
Aqr 21h7'44" -12d28'58"
Blossom Bluff
Vir 12h38'25" 5d51'23"
Blossom Jubilee
Psc 0h26'29" 15d32'19"
BLS8F
Vir 12h34'46" 6d38'22"
BLT (A.K.A Star Ramrod)
And 0h41'34" 27d35'5"
BLUE
Her 18h57'6" 15d50'12"
Blue
Her 17h8'51" 28d15'4"
Blue
Psc 1h9'14" 12d27'44"
Blue
Leo 11h7'4" 9d11'22"
Blue
Uma 10h33'59" 56d44'52"
Blue
Peg 22h38'38" 20d8'7"
blue angel
Cap 20h32'53" -9d59'34"
Blue Angel
Leo 10h26'40" 13d22'20"
Blue Angel
Peg 23h43'56" 11d28'38"
Blue Ariel
Eri 3h41'43" -0d10'44"
Blue Bayou
Cam 4h35'33" 64d41'50"
Blue Dog
Cma 7h7'48" -14d17'48"
Blue Eyed Guy
Ori 5h28'12" 4d14'43"
Blue Eyed Soul
Lyn 9h12'25" 37d54'14"
Blue Eyed Vickie
Uma 9h8'25" 67d24'42"
Blue Eyes
Lyn 6h44'34" 59d31'41"
Blue Eyes
Lyn 6h40'12" 58d52'18"
Blue Eyes
Cyg 20h20'22" 57d4'30"
Blue eyes
Aqr 20h43'2" -11d48'9"
Blue Eyes
Lyn 8h1'31" 42d10'2"
Blue Eyes
Cnc 8h18'40" 12d13'6"
Blue Eyes(Beautiful Darling Casey)
And 0h16'31" 29d20'12"
Blue Fish, Paper Heart
Peg 21h49'24" 10d52'57"
Blue Gazers
Sco 16h14'29" -10d36'14"
Blue Gusanoso 9KJF
Cnc 8h2'49" 21d43'45"
Blue Horizon Apt. Hotel
Umi 14h17'21" 74d56'42"
blue inn
Psc 0h39'1" 15d20'4"
Blue Knox-Moriarty (Fetch Monster)
Cma 6h24'53" -15d37'21"
Blue Laser
Cap 20h58'58" -15d13'46"
Blue Legend
Ari 3h19'18" 20d59'21"
Blue Mamoune
Her 17h58'7" 23d37'30"
Blue Maus
Uma 13h6'3" 53d48'50"
Blue Orchid Beach Hotel
Uma 14h18'33" 58d27'23"
Blue Rose Princess
Vir 13h42'53" -17d18'6"
Blue September
Aqr 21h7'52" -14d29'14"
Blue Serendipity
Ari 2h24'56" 24d21'14"
Blue Sky
Uma 10h58'49" 40d17'12"
Blue Star Petey
Gem 7h11'20" 28d33'15"
Blueberry
Aqr 22h34'55" -17d45'28"
Blueberry Barron
Uma 11h27'44" 33d12'55"
Blueberry Muffin
Pho 0h12'58" -48d37'4"
Bluebird
Cas 23h28'44" 56d13'10"
Bluebird
Tau 4h35'18" 28d31'17"
Bluebird
Ori 5h31'58" 10d34'21"
Blueblee
Ari 3h6'34" 28d43'30"
BlueeApril
Cyg 20h19'18" 56d3'0"
Blue-Eyed Emma
Uma 10h48'53" 60d44'15"
blue-eyed phoenix
Ori 5h49'17" 8d56'1"
BlueEyes
Vir 13h13'47" -4d45'48"
bluefire242
Ari 2h17'41" 12d28'14"
BlueJ122
Cyg 19h41'6" 28d26'58"
Bluejay Bobby
Ori 5h25'48" 7d21'35"
Bluesea
Cru 12h7'56" -61d45'27"
Bluesman
Lyr 18h48'17" 43d42'19"
Bluestar
Aur 5h14'12" 40d42'30"
Bluestone
Pho 0h51'19" -57d11'18"
BlueSue
And 1h0'41" 45d4'20"
Bluey Troes
And 0h31'59" 43d15'6"
Bluford Putnam
Lyn 8h39'53" 33d12'14"
Blume, Jens Peter
Uma 9h0'58" 50d38'41"
Blummer
Cnc 8h39'53" 14d19'48"
Blust Brilliance
Ori 5h11'29" 11d45'52"
Bluuume
Ori 5h45'36" -3d35'19"
B-luv
Ori 5h55'56" 11d48'40"
Bluved
Cyg 19h24'50" 28d46'24"
Blyss
Leo 10h20'55" 23d16'55"
Blythe
Psc 1h51'2" 7d24'18"
Blythe Allison
Cnc 8h6'42" 17d4'15"
Blythe Desire'e Snyder
Lib 14h30'45" -10d48'35"
Blythe Princess
And 0h49'20" 40d47'6"
B.Martine
Cnc 8h16'22" 19d46'46"
B.M.B & M.R.M 2/10/06
Leo 10h13'13" 25d9'33"
bmc
Her 16h49'25" 17d25'43"
BMC2 Hallet
Lyn 6h27'48" 55d3'1"
BMC3373
Psc 0h35'19" 17d38'16"
BMD311
Uma 9h25'21" 61d33'54"
BMichelle
Cam 3h51'55" 53d47'26"
BMP 1
Cnc 8h52'47" 26d44'25"
B.M.S.
Per 3h37'10" 37d18'43"
BMXR 2 8 68 The Guiding Star
Pyx 8h49'52" -30d2'20"
BNC4EVER
Cyg 19h50'52" 38d44'18"
BnK
Lyn 9h16'3" 44d37'56"
Bo
Cnc 8h45'39" 32d4'43"
BO
Per 3h41'21" 45d24'59"
Bo
Uma 11h46'47" 46d59'43"
Bo
Psc 23h11'7" 1d27'13"
Bo
Ori 6h5'7" -0d19'1"
Bo 33
Sco 16h11'12" -22d57'20"
Bo 831
Ari 1h47'12" 23d36'56"
BO ALETHA
Uma 8h52'7" 64d3'28"
Bo and Ashley's Star
Sge 19h12'53" 19d5'24"
Bo & Billy Joe
Cap 20h49'12" -14d55'54"
Bo Boatman
Vir 12h28'51" -2d3'38"
Bo Drachmann
Her 18h23'49" 12d35'3"
Bo Erhard
Cas 0h7'53" 59d42'59"
Bo Erhard
Per 3h38'30" 47d40'4"
Bo Johnna Zarach
Cyg 21h8'10" 47d27'27"
Bo Miller
Aql 19h8'13" 7d6'40"
Bo My Love
Leo 11h18'9" 16d15'16"
Bo Ortenberg
Aqr 21h37'57" -1d37'52"
Bo Peirce
Cma 6h53'5" -14d17'8"
Bo Riley Hauser
Aqr 22h32'24" -3d25'32"
Bo & Sandra Hamm
Cyg 20h1'57" 31d54'49"
Bo Snitchler
Oph 16h39'18" -0d26'40"
Bo Tkach
Ori 6h24'7" 11d26'8"
Bob
Leo 9h29'50" 11d17'57"
BOB
Ari 2h45'45" 20d43'6"
Bob
Gem 6h50'41" 16d55'22"
BOB
Ari 2h38'7" 25d50'22"
Bob
Vul 19h1'24" 23d50'22"
Bob
Lyr 18h43'6" 31d45'13"
Bob
Uma 13h43'8" 48d19'18"
"Bob"
Aql 19h19'27" -0d20'57"
Bob
Ori 5h29'45" -0d59'5"
Bob
Aqr 23h39'40" -11d21'25"
Bob
Sgr 19h35'44" -14d13'12"
Bob
Cap 20h55'15" -21d12'55"
Bob
Uma 8h37'1" 53d56'32"
Bob
Uma 10h10'34" 52d31'13"
Bob
Uma 13h25'57" 56d40'13"
Bob A. Clarke
Vir 13h22'24" 11d42'5"
Bob Adams
Per 3h6'35" 52d4'20"
Bob Allen
Her 17h6'19" 18d0'37"
Bob Allison
Ser 18h15'35" -14d59'39"
Bob and Amy 10-16-04
Ori 6h4'31" 17d41'44"
Bob and Ash
Cyg 21h32'46" 51d41'48"
Bob and Becca
And 23h16'51" 50d14'27"
Bob and Cathy Latham
Uma 9h16'59" 51d53'21"
Bob and Cheri Grant
Lyr 19h18'58" 28d55'6"
Bob and Diana Gatlin
Lyn 8h31'46" 39d16'15"
Bob and Donna Nay
Cyg 20h0'57" 52d17'10"
Bob and Eloise Hoddy
Leo 10h23'39" 25d37'31"
Bob and Evelyn
Cyg 20h46'18" 33d35'29"
Bob and Jan Caron
Uma 13h37'55" 51d15'51"
Bob and Julie Rackham
Cyg 21h42'33" 35d30'8"
Bob and Kay Feeley
Her 16h53'33" 31d8'51"
Bob and Kimberly's 1st Anniversary
Sge 19h11'37" 18d30'2"
Bob and Laura
Crb 16h11'14" 36d58'25"
Bob and Marie
Cyg 19h34'41" 31d58'0"
Bob and Peg Corniea
Gem 6h11'9" 26d2'4"
Bob and Sara's Star
Uma 11h8'44" 71d4'27"
Bob and Shaunne Harriger
And 0h6'44" 39d33'19"
Bob and Sue Engle Happy Anniversary
Cyg 21h23'30" 32d34'31"
Bob and Sue Sedivec
Tri 2h28'27" 31d40'34"
Bob and Susan Love Eternal
Ori 5h22'9" -1d49'16"
Bob and Thelma Forever
Lyr 18h53'12" 44d5'25"
Bob and Tina
And 1h40'47" 42d4'30"

Bob & Ann Kestranek
Uma 11h11'17" 61d11'29"
BOB APPELGATE
Agr 22h34'43" -0d46'43"
Bob Arensberg
Uma 8h36'10" 64d38'9"
Bob B. Brown "The Banjo Man"
Uma 8h53'13" 50d34'44"
Bob Baklik
Uma 12h24'31" 62d45'55"
Bob & Barta
Uma 8h42'47" 70d37'15"
Bob & Bea Everly
Uma 11h54'2" 43d34'7"
Bob & Betty Broderson
Cas 23h4'50" 58d38'2"
Bob Bob
Ari 1h51'11" 19d41'25"
Bob Bourg
Per 3h29'53" 45d20'21"
Bob Bowers
Aql 19h1'21" -0d55'39"
Bob Brady 8-20-1946
Leo 10h20'35" 24d48'31"
Bob Bredehoft's GLOOM Star of Sage
Ori 6h3'13" 17d6'43"
Bob Breed
Uma 8h52'6" 67d55'31"
Bob Brokaw- East Team Leader
Cyg 19h42'10" 39d58'40"
Bob Brotzman
Lib 14h52'53" -1d36'2"
Bob Brzustoski
Uma 13h46'42" 50d10'36"
Bob Bupp Star
Ori 5h39'4" 6d11'18"
Bob Calandro
Ori 5h17'44" 7d32'55"
Bob Caprio
Aur 5h12'13" 47d50'11"
Bob Carieri
Cyg 21h30'28" 30d13'50"
Bob Carney
Cap 21h51'10" -17d6'7"
Bob Chiaviello, Jr.
Cnc 8h38'56" 14d2'11"
Bob & Chrissy Cain
Uma 9h34'59" 46d27'32"
Bob Clark
Cep 22h8'12" 68d28'17"
Bob Clark's Star
Cap 21h21'40" -18d59'14"
Bob Cox
Per 2h18'50" 52d28'28"
Bob Creamer
Her 17h27'55" 44d31'51"
Bob Daddy
Lup 15h43'41" -35d5'16"
Bob Daubney
Uma 12h16'9" 55d0'24"
Bob Davis
Dra 18h59'21" 57d31'40"
Bob Decker Jr.
Sgr 19h56'22" -28d12'1"
Bob DeMay Loves Jamie Hale Forever
Cap 21h37'43" -8d34'19"
Bob Detrick
Husband,Dad,Grandpa
Her 16h43'5" 34d12'41"
Bob Devine
Boo 15h2'49" 48d20'17"
Bob Devlin 2002
Ser 15h16'3" -0d51'40"
Bob Donahoe
Tau 5h6'42" 24d7'50"
Bob Doucet
Dra 19h34'27" 64d2'38"
Bob Eastman
Uma 10h15'0" 57d57'33"
Bob & Edith Bollacker
Uma 8h56'34" 48d32'57"
Bob & Eleanor Stone
Gem 6h36'37" 19d56'51"
Bob - eternally guiding us
Pyx 8h44'6" -27d56'11"
Bob Fanok~Friend and Neighbor
Per 3h33'25" 35d11'14"
Bob Ferree
Tau 5h53'0" 17d3'22"
Bob Fike's Star
Her 17h12'6" 35d18'23"
Bob Fish
Cyg 22h0'6" 52d28'36"
Bob Forstyth
Aql 19h40'44" 14d24'17"
Bob & Fran - Together Forever
Cru 12h30'37" -60d7'1"
Bob Frantz
Psc 1h4'33" 28d5'33"
Bob Fuerst
Cap 20h27'46" -11d18'34"
Bob Gannon's Star
Aur 5h21'22" 41d50'10"
Bob Gant "1943-2006"
Vir 14h10'14" -3d8'46"
Bob Geist
Cep 22h17'7" 65d58'14"
Bob Giard
Uma 11h57'22" 55d51'11"

Bob & Gloria Cooper
Cyg 20h23'47" 34d26'22"
Bob Goldstein
Psc 23h50'33" 7d21'5"
Bob Gorman
Her 17h13'41" 29d32'35"
Bob "Grandpa Harley" Bradshaw
Sgr 19h18'7" -23d30'6"
Bob & Helen Liguori in Scorpio
Sco 16h9'5" -21d59'48"
Bob Hewartson
Cep 22h50'49" 86d40'43"
Bob Hinzman, Husband and Father
Leo 10h10'9" 20d14'38"
Bob Hite
Leo 10h46'31" 13d42'28"
Bob Hoobler
Lac 22h35'10" 49d33'2"
Bob Hughes
Gem 7h53'35" 17d10'7"
Bob Hummell
Leo 9h37'59" 14d26'30"
Bob & Jackie (Lurch Potentate 2004)
Per 3h21'13" 48d28'37"
Bob & Jackie Rasp
Cyg 21h34'18" 38d35'47"
Bob & Jan Merton's (Jarem's) Star
Sge 19h41'21" 17d59'14"
Bob & Jean Hogarth
Her 18h21'26" 18d20'12"
Bob & Jeanine Johnson
Aql 20h1'0" 9d10'27"
Bob & Jeanne Queen
Cyg 20h23'27" 48d21'7"
Bob Jedi
Her 17h43'42" 34d5'23"
Bob & Jo Ann Taylor
Uma 8h47'37" 61d27'49"
Bob & Judy Armstrong
Cyg 21h58'3" 45d19'16"
Bob K.
Her 16h24'57" 47d46'23"
Bob & Kathy Lees Wedding Star
Cnc 9h4'25" 27d4'32"
Bob & Kay's 50th Anniversary
Cyg 20h49'16" 46d45'12"
Bob Keller's Fighting Illini Star
Lyn 7h47'9" 48d39'44"
Bob Kirkham
Cep 22h46'57" 66d45'13"
Bob & Kiyo Allen (Dad & Mom)
Lyn 8h49'35" 34d54'38"
Bob Klym
Lyr 18h31'57" 36d3'14"
Bob Koch
Cyg 20h36'28" 36d1'21"
Bob Kodzis
Her 16h8'17" 23d15'21"
Bob Krum (Digger)
Per 2h47'46" 53d24'41"
Bob Kyker
Per 3h52'14" 36d57'54"
Bob & Laura Sabol~50th Anniversary
Cyg 21h13'16" 46d51'26"
Bob & Laura Sabol~50th Anniversary
Cyg 21h16'30" 45d32'39"
Bob & Lauri
Cyg 19h49'42" 35d45'1"
Bob Lewis
Ari 2h4'11" 27d37'35"
Bob Lightman
Psc 1h31'8" 26d14'25"
Bob & Lisa's Star of Commitment
Cyg 19h36'22" 34d12'18"
Bob Loves Marti Forever
And 0h18'24" 45d39'30"
BOB&LULU
Dra 16h43'49" 57d56'38"
Bob Mansfield
Psc 1h31'58" 10d50'41"
Bob & Mardell Leeper
Cyg 21h27'22" 36d57'2"
Bob & Marlene White
Uma 11h46'35" 28d44'35"
Bob Marsh "NB50"
Sco 16h9'44" -18d59'48"
Bob & Mary Graves
Cyg 20h19'59" 36d32'5"
Bob & Mary K. Quinn
Uma 12h35'39" 57d42'33"
Bob & Mary Petrucelli
Mon 6h50'30" 3d24'31"
Bob Maurer:The Love of My Life
Tau 3h41'17" 22d14'30"
Bob Melendez
And 1h7'18" 46d50'39"
Bob Menaker
Uma 11h30'32" 29d8'48"
Bob Metler
Hya 10h14'30" -26d10'21"
Bob & Michael
Per 2h52'3" 54d5'35"

Bob & Michele, Forever & Always
Agr 22h35'2" -2d32'27"
Bob Mnichowski
Uma 8h44'9" 67d2'5"
Bob n Robyn Geranis Forever
Psc 23h6'22" 6d59'40"
Bob Nelson
Aql 19h2'47" -0d37'52"
Bob Nesheim
Ori 5h41'37" 2d42'48"
Bob Noble
Boo 14h43'52" 21d42'35"
Bob Noble
Uma 11h42'27" 47d5'12"
Bob Norris
Ari 2h49'40" 27d56'14"
Bob O
Ori 6h6'11" 20d51'58"
Bob Orr
Per 3h4'43" 56d36'47"
Bob Palmer Forever Lights Our Lives
Umi 10h47'2" 88d42'26"
Bob Pascale
Lyn 7h23'52" 47d1'37"
Bob & Patricia Regier
Per 3h37'16" 44d17'50"
Bob Pellicore
Agr 21h35'34" 2d17'0"
Bob Penberthy - Simply the Best
Uma 14h1'30" 57d58'33"
Bob & Petie Eckel
Cnc 8h20'41" 26d53'22"
Bob (Pick-a-Box) Courtney
Sco 17h36'18" -42d20'10"
Bob Quindt
Per 2h19'48" 56d25'8"
Bob Reiner
Lyn 7h51'5" 48d3'57"
Bob Ribaudo
Per 2h51'49" 53d50'3"
Bob Rickabaugh
Uma 10h46'29" 46d36'42"
Bob Ringen
Tau 3h25'35" -0d2'30"
Bob Risch
Vir 13h19'56" -11d27'5"
"Bob" Robert O'Neil Suggs
Aql 19h30'46" -0d50'31"
"Bob" Robert Reginald Field
Ori 6h17'27" 15d1'45"
Bob Rogers, a STAR Leader
Scl 23h37'47" -25d20'47"
Bob Rowson
Cep 21h29'53" 68d8'30"
Bob Russow
Aur 5h37'44" 46d10'14"
Bob Sansone
Her 17h22'21" 17d19'52"
Bob Schwellenbach
Ori 6h19'30" 15d36'35"
Bob Seibel
Cyg 20h58'25" 35d14'37"
Bob & Sheri Forever
Cyg 21h59'16" 50d44'39"
Bob & Sherrie Pfeffer Wedding Star
Lib 15h20'4" -7d58'12"
Bob Sherwood
Lyn 7h50'19" 37d14'10"
"Bob & Shirley" (5-9**ALLOURLOVE)
Eri 4h31'44" -1d19'48"
Bob Shotts
Agr 21h47'24" -0d57'54"
Bob Simoneaux
Her 17h7'48" 31d21'30"
Bob Slabaugh
Umi 13h22'59" 74d0'44"
Bob "Soupy" Wright
And 2h37'42" 44d25'41"
Bob St-Onge *F R E E W A Y*
Cep 20h30'54" 62d35'26"
Bob Strang
Her 15h54'1" 40d0'35"
Bob Stump
Her 17h57'27" 28d49'34"
Bob Suppa
Aql 18h57'54" 17d53'6"
Bob Taft Star Son
Her 17h46'29" 45d40'28"
Bob Tarpley
Her 16h32'0" 28d39'20"
Bob Temple's Shining Star
Tau 4h7'34" 14d45'48"
Bob Tencati's Al di la
Cap 21h11'42" -16d18'47"
Bob & Teresa Bienkowski-Our Heroes
Her 17h55'11" 25d6'52"
Bob & Terri Neudauer
Ari 2h17'26" 24d45'15"
Bob & Tina Berry PBS 19 KCMO
Cyg 20h5'51" 44d41'43"
Bob Todd
Per 3h58'50" 37d46'13"
Bob & Trautchen
Cyg 20h20'6" 40d54'54"

Bob Treese
Ori 6h19'7" 18d59'39"
Bob Van Deven
Lyn 8h34'15" 46d25'8"
Bob Waldock 70
Cep 20h47'59" 60d33'4"
Bob Welch
Lyn 7h13'16" 53d13'33"
Bob & Wendy Welch
Vir 12h16'9" -3d34'13"
Bob Wilkins
Uma 12h30'49" 53d51'10"
Bob Wilson
Cap 21h19'35" -16d4'59"
Bob Woods
Her 17h10'31" 27d0'27"
Bob Woods Rocks!
Uma 12h35'13" 58d50'59"
Bob Young
Cnc 8h18'33" 13d49'41"
Bob Yudin
Per 3h7'50" 53d56'44"
Bob Zimmers
Per 2h14'40" 57d3'39"
Bob, Andrea, Nate and Mia Shaver
Leo 9h51'22" 21d22'27"
Bob, Barb & Paul
Peg 22h42'55" 27d18'11"
BOB0719
Cnc 8h17'21" 11d14'16"
Bobaloo
Cam 3h58'10" 64d2'16"
Bobaloo Starlydoo
Sco 17h52'34" -39d21'25"
Bob-a-Lu
Lyn 6h59'25" 47d11'46"
BobandKarrie
Vir 13h19'1" 11d32'36"
BobAndSuzan35StellarYears
Cyg 19h52'3" 38d48'14"
Bobarius
Sco 16h57'19" -44d19'19"
Bobbert
Psc 23h34'33" 2d3'12"
Bobbet
Lib 14h23'3" -14d22'21"
Bobbette 02/09/1969
Agr 23h43'10" -14d39'57"
Bobbi
Agr 22h2'49" -1d53'52"
Bobbi
Tau 4h39'31" 28d13'10"
Bobbi
Lyr 18h53'17" 31d55'2"
Bobbi and Jason
Ari 2h3'40" 24d7'18"
Bobbi Bacon
Gem 6h47'48" 16d36'2"
Bobbi Carla
Umi 14h59'6" 72d41'33"
Bobbi Cossette
Uma 11h26'39" 53d18'45"
Bobbi Danette Rogers
Peg 22h39'8" 11d51'41"
Bobbi Dodson
Mon 6h52'0" -0d45'46"
Bobbi Jean Mouser-Powell
Gem 7h49'0" 26d6'9"
Bobbi Jo
Lyr 19h7'34" 32d21'41"
Bobbi Jo Balzer
Cyg 21h15'38" 44d36'42"
Bobbi Jo Cox
Tau 5h28'51" 26d28'9"
Bobbi L
Uma 11h31'34" 63d1'5"
Bobbi Lee Gardner
Tau 5h42'54" 25d24'15"
Bobbi Lynn Murphy
Lib 15h2'59" -2d46'43"
Bobbi "My Love"
And 1h7'38" 47d24'44"
Bobbi Reichtell
Cam 5h21'45" 71d31'25"
Bobbi Siccardi
Cap 20h57'44" -20d17'21"
Bobbi Sue
Sgr 17h50'45" -17d49'19"
Bobbi Sue Gillespie
And 1h0'9" 34d22'33"
Bobbie
And 1h59'26" 36d10'3"
BOBBIE
Leo 9h51'22" 18d23'4"
Bobbie
Leo 11h57'59" 20d54'45"
Bobbie
Leo 11h8'16" -0d3'12"
Bobbie
Umi 14h38'14" 78d18'33"
Bobbie
Sgr 18h35'50" -23d32'35"
Bobbie Acklin
Cyg 21h37'55" 43d26'16"
Bobbie & Ahto
Eri 3h55'42" -38d48'2"
Bobbie Alexander
Aql 19h42'36" 13d49'34"
Bobbie Beckert DuBois
Uma 8h46'10" 53d26'4"
Bobbie Benson
Her 18h44'52" 22d2'6"

Bobbie Brummell
Cas 1h6'38" 52d34'35"
Bobbie Etta Mercer
Sco 17h45'1" -40d43'4"
Bobbie & Fritz 50th Anniv. Star
Crb 15h38'56" 28d34'34"
Bobbie Gay-Brooks
Sgr 18h15'57" -23d21'38"
Bobbie Goldenberg
Cnc 8h57'53" 7d1'18"
Bobbie Goldsborough
Uma 11h19'46" 39d53'28"
Bobbie "Gorgeous" Martinez
Uma 11h8'51" 32d15'0"
Bobbie Griffin
Her 16h52'57" 31d14'57"
bobbie guillory
Sgr 18h45'2" -23d38'53"
Bobbie Hedley
Gem 6h47'49" 34d26'31"
Bobbie Henson
Uma 8h57'26" 58d48'22"
Bobbie J. Ohm
Sco 16h26'26" -26d9'24"
Bobbie J. Venter
Aur 6h40'41" 37d25'57"
Bobbie Jean
Vir 14h43'17" 3d46'31"
Bobbie Jean Brown
And 1h2'33" 37d38'1"
Bobbie Jean Estes
And 0h43'18" 32d45'59"
Bobbie Jean Johnson "My Angel"
And 1h18'16" 45d47'28"
Bobbie Jo
And 2h27'46" 48d18'40"
Bobbie Jo
And 23h52'0" 46d4'46"
Bobbie Jo
Cnc 9h4'22" 12d11'42"
Bobbie Jo Rains
Cap 20h19'13" -20d28'33"
Bobbie Jo Welch
Uma 11h18'43" 47d22'44"
Bobbie Joe White
Uma 10h6'58" 48d39'32"
Bobbie L. Owens
Psc 23h40'8" 6d22'6"
Bobbie Leann
Ari 2h45'26" 27d36'26"
Bobbie Leatherwood
Agr 21h10'43" -2d45'26"
Bobbie Lee Fraser
Uma 11h59'33" 39d34'51"
Bobbie & Leonard Humphreys-50/2005
Cyg 21h28'53" 44d43'30"
Bobbie McDonald - Love of my Life
Lyn 8h24'54" 39d2'17"
Bobbie Ramirez
Ori 4h58'4" -0d42'19"
Bobbie White
Uma 9h48'13" 41d31'41"
Bobbie Williams
Tau 3h26'18" -0d48'58"
Bobbiecat
Agr 21h14'36" -3d28'59"
Bobbie's Spirit
Sgr 18h35'59" -22d51'47"
Bobbie's Spirit
Ori 4h45'38" 11d56'7"
Bobbi-Jo Pearl Carroll
Leo 10h26'5" 20d47'33"
Bobbi-Li 'Like No Other'
Gem 7h19'49" 24d5'16"
Bobbi's Star
Psc 0h42'22" 10d8'4"
Bobbit
Ori 6h22'43" 10d30'42"
Bobby
Ari 2h56'30" 11d30'25"
Bobby
Ori 5h27'37" 0d10'29"
Bobby
Gem 7h24'29" 24d44'36"
Bobby
Ari 3h15'41" 29d16'19"
Bobby
Tau 3h52'0" 29d51'56"
BOBBY
Leo 11h12'44" 17d33'40"
Bobby
Ari 3h4'41" 24d47'40"
Bobby
Aur 6h33'46" 37d7'48"
Bobby
Her 17h34'9" 41d9'18"
Bobby
Lyn 8h10'7" 50d22'48"
Bobby
Sgr 18h13'55" -19d30'31"
Bobby
Uma 13h37'44" 56d22'39"
Bobby 9
Sgr 18h42'40" -27d9'10"
Bobby Alvarado
Uma 10h34'33" 51d52'53"
Bobby and Anastasia
And 0h52'31" 39d19'44"
Bobby and Michelle
Cyg 20h9'33" 41d34'42"

Bobby and Nell
Ari 2h6'57" 23d56'20"
Bobby and Nicole 11/02/03
Col 6h20'19" -34d54'52"
Bobby Andrew & Lindsey Marie
Uma 11h10'14" 53d13'35"
Bobby Anthony Smith, M.D.
Oph 16h42'10" -0d51'59"
Bobby & Arty Blitman
Lyr 19h12'35" 26d32'33"
Bobby B.
Her 18h7'35" 27d19'44"
Bobby B. Brown
Agr 22h38'57" 1d36'25"
Bobby Baber
Ori 5h43'41" 0d48'35"
Bobby Boi
Uma 8h59'30" 68d7'38"
Bobby Bowen
Psc 1h31'6" 13d26'19"
Bobby Boy's Shining Star
Leo 9h53'32" 17d5'51"
Bobby Bratter
Umi 0h7'0" 89d12'41"
Bobby Bumps' World
Vir 13h41'27" -13d9'39"
Bobby C.
Peg 22h5'50" 33d17'23"
Bobby Calabrese
Ori 5h43'32" -3d11'41"
Bobby Christensen
Sgr 18h23'56" -32d52'59"
Bobby Clara Smith Mayes
Cas 0h50'2" 64d8'42"
Bobby Clarence Pangle
Agr 22h51'12" -8d26'54"
Bobby Coletti
Per 4h5'4" 39d43'16"
Bobby Coppolla
Uma 13h39'26" 53d23'3"
Bobby Corno
Boo 14h26'42" 27d43'34"
Bobby Cota
Agr 21h40'7" -2d57'21"
Bobby Darin
Tau 5h15'29" 18d43'52"
Bobby & Debby's Nightlight
Cyg 21h18'47" 43d15'24"
Bobby Deemer
Uma 11h30'2" 35d52'58"
Bobby Doane
Her 16h45'32" 33d33'31"
Bobby Duke
Her 18h31'42" 23d37'48"
Bobby E Standard
Tau 5h59'44" 25d47'9"
Bobby Freeman
Cnc 8h58'45" 15d3'9"
Bobby French
Dra 19h16'7" 71d24'50"
Bobby G
Cap 20h22'59" -9d49'47"
Bobby G Davis
Ari 2h19'2" 12d58'16"
Bobby Glade
Cap 20h9'1" -10d6'50"
Bobby Glen
Vir 14h19'21" 3d26'33"
Bobby Hamilton, Sr.
Uma 10h50'30" 70d3'50"
Bobby Hill
Uma 11h17'11" 30d2'5"
Bobby J. Lea
Boo 14h23'56" 13d54'8"
Bobby + Jacky
Sco 16h23'47" -33d22'55"
Bobby James
Umi 17h10'0" 84d34'32"
Bobby & Jeannie
Leo 9h26'57" 15d10'9"
Bobby & Jeannie Waddell
Cep 21h58'35" 62d46'8"
Bobby Jo Burress
Her 17h45'9" 22d40'5"
Bobby Jo Trammell
Lyn 7h42'50" 38d11'38"
Bobby Joe Lamb
Com 13h16'48" 27d20'5"
Bobby Joe Stewart
Per 3h11'55" 56d56'53"
Bobby Joe Willmon
Agr 20h45'58" -13d12'40"
Bobby Johansen
Ori 6h15'9" 7d32'2"
Bobby Jutley
Agr 21h10'13" -12d11'30"
Bobby K. Sutphin
Uma 9h7'45" 50d39'44"
Bobby Kleiman
Lyn 6h38'20" 53d56'48"
Bobby L. Tisdale, Jr.
Aur 7h27'32" 40d28'46"
Bobby Lee
Vir 14h4'6" -8d0'37"
Bobby Lee Derrickson
Lac 22h8'12" 53d3'12"
Bobby Lee Duggins
Lyn 9h21'32" 33d29'56"
Bobby Lee (froboy)
Per 3h8'20" 43d32'58"
Bobby Lee- Jersey Boy
Boo 14h37'22" 18d33'23"
Bobby & Lena
Cyg 19h35'57" 28d1'11"

Bobby Lewis
Ori 6h3'45" 16d27'35"
Bobby Lewis Andrews - Roll Tide
Dra 18h39'30" 52d56'32"
Bobby Liebrandt, Jr.
Uma 11h47'36" 51d59'6"
Bobby Lou Hyer
Srp 18h12'49" -0d42'18"
Bobby (Love You Always) Watson
Ori 6h15'31" 13d36'10"
Bobby Lynn Roberts-My Love&My Life
And 23h48'0" 36d44'21"
Bobby M. Gomez
Vir 14h36'38" 4d57'25"
Bobby Martinez
Ari 2h39'57" 28d32'56"
Bobby - Mi Querido
Per 1h43'40" 54d28'46"
Bobby Norton
Cap 20h34'8" -14d0'28"
Bobby "O"
Uma 11h57'49" 29d21'43"
Bobby Prohn
Uma 12h55'20" 53d44'27"
Bobby Ray
Psc 1h32'54" 4d51'35"
Bobby Riffle
Agr 23h55'56" -23d36'40"
Bobby Roy Booker
Her 18h28'36" 25d8'29"
Bobby Roy Pond
Aql 19h33'14" 6d47'16"
Bobby S. Marshall
Her 17h16'16" 13d8'37"
Bobby & Samantha's Star
Crb 15h47'12" 26d1'20"
Bobby Simpson
Her 16h59'56" 29d48'22"
Bobby Sparks
Vir 12h29'4" 10d17'43"
Bobby Spears
Umi 17h25'7" 79d38'46"
Bobby Stow Jr.
Umi 15h27'28" 71d9'18"
Bobby T
Umi 16h41'50" 75d32'16"
Bobby Terrell Johnson
Uma 11h26'4" 39d4'39"
Bobby Tremendous Terry Star
Uma 9h37'58" 63d13'10"
Bobby Two
Sco 17h11'7" -43d55'31"
Bobby & Vanessa Krieger
Cyg 19h24'3" 28d40'1"
Bobby Vigil
Cnc 9h6'44" 11d54'54"
Bobby Walcott 10/25/1925
Sco 16h12'20" -17d9'46"
Bobby Warns
Ori 6h13'56" 9d14'0"
Bobby Wayne Smith
Mon 6h48'54" -0d1'46"
Bobby Webb
Dra 18h11'37" 76d28'20"
Bobby Willis
Lib 15h50'59" -19d59'16"
Bobby Zoellner
Uma 10h42'36" 68d37'55"
Bobby, you will never be forgotten
Uma 10h52'44" 53d39'47"
Bobbyc
Her 17h59'46" 49d40'49"
Bobbye Clayton
Cyg 21h22'43" 50d35'23"
BobbyeLeighMorgan
Vir 13h45'39" -4d45'31"
Bobbyi
Sgr 18h17'46" -30d27'27"
BobbyKniss
Agr 22h33'4" 2d19'58"
Bobby.....My Hero.....
Vir 11h56'37" 6d23'21"
Bobby's Dream
Uma 11h59'57" 35d32'15"
Bobby's Girl
Ori 5h35'20" -2d22'42"
Bobby's Star
Aql 19h45'16" -0d34'24"
Bobby's Star
Uma 10h18'28" 63d38'48"
Bobby's Star
Lyn 8h35'15" 40d32'25"
Bobby's Star
Per 3h19'51" 51d57'4"
Bobby's Star
Cnc 8h3'57" 15d27'38"
Bobby's Territory
Ari 3h5'55" 13d18'33"
Bob-Dad 1927
Cnc 9h3'22" 29d22'43"
Bobecco
Psc 1h17'59" 10d23'48"
Bobert
Vir 14h27'19" -1d48'18"
bobes
Psc 1h9'16" 5d32'47"
Bo-bever Fo-fever
Agr 22h30'29" -2d6'51"
Bobi Cox
Umi 14h4'15" 65d25'1"

Bobie Chen Bao
Umi 14h21'31" 75d45'56"

Bobillier, Alain
Ori 5h20'51" -4d41'53"

Bobino (Bob Smith)
Her 17h25'35" 47d22'11"

Bobi's Star 22-August-1958
Cas 23h28'37" 56d39'6"

Bobizio
And 1h47'7" 49d3'51"

BobKat
Cas 0h51'59" 52d17'19"

boblouis1956
Lib 15h25'22" -13d6'59"

Bobn'Gail
Tri 1h52'52" 30d9'25"

BobnLewis
Ori 6h17'51" 13d46'4"

Bobo
Lyn 8h0'55" 34d26'37"

Bobo
Ori 5h0'12" -1d25'53"

Bobo
Sct 18h53'22" -5d33'35"

Bobo
Pup 8h10'48" -22d54'39"

Bo-Bo Lane
Cma 7h18'34" -13d22'50"

Bobounette
Vir 11h46'23" 9d11'26"

Bob's Eagle's Nest
Cyg 20h42'13" 31d42'26"

Bob's Eternal Light Of Love
Psc 1h1'55" 23d44'15"

Bob's guiding star
Tau 3h40'39" 16d3'26"

Bob's Lucky Star
Cyg 20h6'18" 36d6'48"

Bob's Star
Aqr 22h32'15" -0d59'45"

Bob's Star - looking down with love
Cen 13h58'46" -61d0'44"

Bobsan60
Cap 20h37'13" -21d9'13"

BobStar
Her 17h5'57" 34d56'49"

bobsyl
Uma 11h27'34" 54d3'6"

Bobunni
Sco 17h51'19" -35d50'31"

Boca
Aql 19h59'17" 6d14'7"

Boca Blaze
Gem 7h2'39" 15d53'26"

Boca Braulio Negra
Her 16h56'6" 28d28'47"

Bocaj Redinsfiar
Lyn 8h20'14" 34d55'0"

Bocciolo
Cyg 20h14'39" 49d20'11"

Bocephus
Cep 23h49'28" 83d14'50"

Bochmann, Carsten
Vir 13h7'13" -3d18'22"

Bock 86
Leo 11h16'47" 19d17'53"

Böckelmann, Wolfgang
Uma 11h1'48" 64d40'39"

Böcker, Ingolf
Ori 5h57'4" 21d41'43"

Bocowood Sdn Bhd
Ori 5h35'13" 2d46'46"

Bodach, Gian Luca
Ori 6h1'33" 10d21'32"

Bodacious
Uma 9h10'8" 62d22'6"

Bode Adam Hirschman
Her 16h21'58" 47d37'36"

Böder, Joachim
Sco 16h50'11" -38d29'21"

BODIE
Umi 10h16'23" 86d46'48"

Bodie and Erika Forever
Cyg 20h42'10" 37d33'17"

Bodie Gray "11:52P.M."
Lib 14h37'7" -11d57'23"

Bodie's Blessing
Lyr 19h1'12" 33d13'22"

Bodo Miklos
Lyn 7h26'41" 49d41'22"

Bodo Prusas
Ori 5h56'0" 11d51'8"

Bodo Puppa
Psc 1h18'18" 18d38'10"

Boebeez
Tau 4h38'11" 18d47'57"

Boegemann Family Dreams
Her 17h1'43" 32d31'32"

Boelling 2-10
Uma 9h24'0" 59d5'33"

Boerner-Gordy
Uma 9h14'54" 72d25'39"

Boex, Cecil
Uma 9h8'51" 54d43'44"

Bogey's Beacon
Cma 6h43'4" -15d29'6"

Bogi és Attila szerelemcsillaga
Uma 11h35'46" 36d26'31"

Bogie-Ann Keays
Gem 7h20'34" 22d15'14"

Bogner
Uma 12h26'11" 53d5'10"

Bogumil Prime
Ari 3h15'36" 29d4'51"

Bogumila Dowbysz
Cap 21h54'58" -23d14'50"

Bogumila Orlowska
Leo 11h30'4" 10d4'18"

Bogunia
Uma 9h20'30" 62d32'8"

Bogush Anniversary Star
Cma 6h47'15" -13d15'9"

Boguth
Cyg 19h49'42" 36d2'26"

Bohannon
Uma 11h43'43" 34d18'31"

Bohdan von der Sanfter Loewe "Odin"
Uma 10h48'46" 42d7'16"

Bohemia
Uma 10h22'49" 51d3'51"

Böhl, Werner
Uma 11h13'49" 64d0'12"

Bohl-Lyons
Lib 14h45'6" -9d33'24"

Bohlsar - M45
Tau 5h13'11" 18d21'3"

Bohm, Wolfgang
Uma 13h6'5" 60d13'17"

Böhme, Rica
Uma 13h18'4" 61d3'10"

Bohn, Hans-Jürgen
Uma 10h29'41" 65d46'27"

Bohnen, Josef
Uma 12h58'56" 53d46'56"

Bohnert, Rudi
Ori 5h55'7" 21d31'31"

Boiso
Per 4h13'26" 48d16'1"

BoJ
Cnc 8h42'33" 10d22'0"

Bojangles
Per 3h30'59" 48d20'34"

Bojangles
Lyn 7h13'36" 59d31'3"

Bojinka Miki - a szerelem csillaga
Leo 9h22'15" 12d21'26"

Boka
Lyr 18h46'27" 35d16'30"

BOKA1018
Apu 17h29'33" -69d30'25"

Boki Dujmovic VALENTI-NOVO 2005
Tau 3h34'7" 25d46'45"

Boland Brooks Enterprise
And 23h21'54" 39d7'31"

Bolar
Uma 11h50'54" 52d12'53"

BOLDT'S STAR
Cnc 8h42'7" 13d4'18"

Böllert, Moritz
Psc 1h55'27" 6d5'8"

Bollmann, Carola
Uma 8h26'35" 64d14'38"

Bollo
Uma 9h27'45" 61d18'18"

Bollocks
And 2h35'42" 49d42'39"

Boltonlan 2004
Umi 14h29'39" 87d13'42"

Boluga-Marie
Umi 16h0'6" 71d42'28"

Boly Taing
Cas 0h51'38" 52d35'13"

Bom17022006
Crb 15h57'40" 26d57'46"

BoMa
Cnc 8h11'40" 12d53'52"

Boman R Irani
Ori 5h19'31" 0d43'25"

Bomba Star
Dra 20h1'41" 74d56'33"

Bombadil
Uma 11h7'49" 38d40'28"

Bomber Bouski Tank-Commander Gibson
Uma 13h35'11" 52d30'55"

Bombi Navarro
Peg 21h50'53" 8d29'44"

Bömbis 1955
Uma 8h29'50" 60d12'24"

BOMELI
Dra 16h12'49" 57d51'18"

BOMO
Ori 6h23'43" 10d2'8"

Bon and Frank Soulmates
Lyr 19h8'48" 42d0'21"

Bon Bon
Lyn 7h46'49" 48d48'24"

bon Dimaree
Vir 13h23'7" 4d14'57"

Bon Jontue
Cyg 21h17'34" 44d1'29"

Bon - Sue Hoback
Cas 23h19'10" 58d32'26"

Bon Temps
Peg 0h4'6" 15d56'14"

Bona Fortuna
Mon 7h5'26" -6d26'42"

Bonae Indolis
Ori 5h36'36" -9d32'59"

Bonamy Margaret Brantley
Psc 1h5'19" 4d11'49"

BoNana
Lyn 8h15'35" 35d22'53"

Bonanza Apts.
Tri 2h36'33" 35d58'11"

Bonato Mara
Umi 16h52'49" 79d6'22"

Bonato Mariella
Umi 16h7'21" 70d30'43"

Bonato Milena
Umi 0h45'51" 89d12'45"

BONC 69
Aqr 21h6'8" 0d42'8"

BONCUK
Ari 2h47'18" 20d56'1"

Bone
Vir 12h41'24" 1d9'54"

Boneca
Uma 10h36'54" 57d39'47"

Bonecutter-Williams, Janet 4
Tau 5h57'11" 26d33'46"

Bonesaw
Aqr 22h36'58" -7d19'46"

Boney Ass
Aqr 22h4'0" 0d13'20"

BONFIRE
Gem 7h23'35" 33d6'11"

Bongo Dumont
Uma 9h40'40" 66d47'28"

BONHEUR
Lac 22h26'38" 41d48'42"

Bonica 12/29/03
Umi 13h54'27" 86d3'9"

Boniqueta Pili
Peg 22h15'18" 23d55'46"

Bonis "Ilwu" Cerealios
Ori 5h55'20" 21d15'21"

Bonita
Peg 23h35'17" 19d25'30"

Bonita
Tau 4h20'17" 22d49'5"

Bonita
Ari 2h39'34" 19d50'16"

Bonita
Cnc 9h5'10" 31d24'43"

Bonita
Cam 6h41'52" 62d52'8"

Bonita
Uma 9h51'17" 58d11'1"

Bonita
Lyn 6h35'40" 57d14'58"

Bonita Anita Kay
Lyn 8h10'24" 50d13'21"

Bonita Bea
And 0h32'13" 29d5'27"

Bonita Blu
Ori 5h41'51" 2d17'29"

Bonita Burroughs
Cas 0h13'26" 52d52'6"

Bonita Erica
Leo 9h27'24" 18d31'31"

Bonita Lynch
Cas 0h3'50" 56d29'27"

Bonita Money
And 0h57'35" 36d20'24"

Bonita Rosalie Emberley 25.04.1941
Cas 23h35'15" 56d20'44"

Bonita Springs Charter School
Lyn 8h7'47" 55d54'59"

Bonita Stafford
Cas 0h18'48" 62d52'16"

Bonita Svardal
Leo 11h37'1" 15d24'11"

BonitaNavidMandyMinleyPninaSandraML
Uma 10h40'6" 72d11'35"

Bonjour Princesse !!
Vir 11h47'25" 4d51'53"

Bonne Brynn Wersel
Crb 15h35'39" 31d54'20"

Bonne Etoile de Julian Carrey
Aqr 21h24'56" 1d32'31"

Bonne Lu Pajac
Sco 17h6'38" -34d31'1"

BonnerStar 5.19.53-em*md*dd*jf*fs
Tau 5h49'21" 21d13'29"

Bonnewitz Valérie
Cyg 19h26'46" 36d23'39"

Bonney Lake Elementary Office Staff
Uma 13h17'9" 56d35'26"

Bonni Erin
Tau 5h47'29" 14d59'2"

Bonnie
Ori 5h32'9" 0d7'37"

Bonnie
Tau 3h55'34" 28d40'28"

Bonnie
Gem 6h26'47" 24d39'50"

Bonnie
Leo 9h52'53" 23d19'31"

Bonnie
Crb 16h16'8" 33d39'34"

Bonnie
Cnv 12h43'42" 39d42'44"

Bonnie
And 1h22'30" 40d35'44"

Bonnie
Cas 0h37'57" 55d54'51"

Bonnie
And 23h8'6" 50d10'50"

Bonnie
Uma 8h37'52" 49d38'29"

Bonnie
And 2h35'55" 49d14'50"

Bonnie
And 1h3'34" 46d20'27"

Bonnie
And 0h11'16" 46d38'52"

Bonnie
Lyr 18h38'52" 38d17'40"

Bonnie
Lyn 8h12'54" 55d40'19"

Bonnie
Lyn 7h15'28" 54d8'19"

Bonnie
Cas 23h46'18" 55d27'52"

Bonnie
Cas 2h22'3" 72d24'47"

Bonnie
Cas 1h43'31" 66d52'4"

Bonnie
Psc 0h0'55" -4d37'7"

Bonnie
Mon 7h31'48" -0d20'47"

Bonnie
Ori 5h32'36" -1d14'27"

BONNIE
Vir 12h56'47" -15d13'59"

Bonnie
Cap 21h47'8" -23d43'5"

Bonnie
Cru 12h38'32" -57d53'16"

Bonnie A. Murling
And 0h17'43" 33d43'57"

Bonnie Alexander Hanson
Cnc 8h37'12" 10d43'27"

Bonnie and Clyde
Cyg 21h35'27" 46d13'35"

Bonnie and Clyde
Lib 14h43'22" -12d30'10"

Bonnie and Kane Dean
Uma 11h35'42" 28d21'4"

Bonnie Ann McAlevy
And 1h26'52" 42d36'8"

Bonnie Arden
Uma 12h33'28" 59d41'47"

Bonnie Armstead Bowman
Cyg 20h4'34" 37d28'42"

Bonnie Arnette Bicklmeier 8.3.1942
Cas 0h9'29" 54d15'5"

Bonnie B.
Cet 1h14'27" -0d13'4"

Bonnie B. Johanson
And 0h31'29" 35d24'17"

Bonnie Baird
Sco 16h37'24" -33d56'42"

Bonnie Bernard
And 0h38'14" 31d49'11"

Bonnie & Bernie Lang
Mon 7h7'56" -0d17'26"

Bonnie Beth Kerley
Uma 11h22'50" 59d42'4"

Bonnie Blangiforti
Lib 15h20'11" -21d49'8"

Bonnie Blue July 2, 2001
Cnc 9h16'46" 13d46'10"

Bonnie & Bob
Lyr 18h46'8" 40d0'12"

Bonnie "Bon Bon" Waggoner
Psc 1h26'0" 7d2'58"

Bonnie Brosius
Psc 0h27'14" 3d30'1"

Bonnie Burrell Setzer
Apu 16h29'28" -74d11'22"

Bonnie Candoso
Cas 1h28'59" 64d5'22"

Bonnie Claire Gannon
Uma 12h26'44" 55d39'0"

Bonnie Claire Shelton
And 1h42'15" 43d52'53"

Bonnie & Clyde
Sge 19h43'3" 17d45'32"

Bonnie Cooker
Sco 16h18'6" -19d3'38"

Bonnie Dale
And 2h8'28" 46d14'59"

Bonnie & Dale Jensen
Lyn 7h43'7" 38d55'3"

Bonnie & Dave
Lmi 10h30'56" 31d54'26"

Bonnie Desautels
Ori 5h37'23" 6d29'32"

Bonnie Diane
Cas 23h13'22" 55d31'43"

Bonnie Dohrn
Uma 10h1'40" 72d23'8"

Bonnie E. Duffy
Uma 11h35'11" 29d52'58"

Bonnie E. Miller
Cyg 19h54'6" 33d36'39"

Bonnie E. Reed
Vir 11h52'52" -0d34'33"

Bonnie E. Saunders
Sgr 18h55'41" -29d8'5"

Bonnie Elaine Miller
Cam 7h57'50" 61d23'34"

Bonnie Elise Thomas
And 0h41'44" 25d43'38"

Bonnie Elizabeth "Beth" King LeRoy
Cas 1h20'53" 57d27'39"

Bonnie Elizabeth Millhous
And 1h18'20" 42d27'56"

Bonnie Fae
Lmi 9h41'17" 35d52'56"

Bonnie/Fay 25th Anniversary Star
Uma 12h5'50" 35d48'40"

Bonnie Faye
Cas 1h31'8" 62d25'0"

Bonnie Filkins
Leo 11h53'4" 25d4'6"

Bonnie Flynn
Cru 12h1'4" -62d0'15"

Bonnie Fonk
Psc 23h55'32" 7d18'47"

Bonnie - Forever Our Star, Love B&K
Cas 1h28'48" 65d9'22"

Bonnie Forst
And 23h9'4" 51d51'31"

Bonnie G. Hoffert
Vir 12h17'49" 2d50'23"

Bonnie G Wilk
Lib 15h20'20" -16d1'28"

Bonnie Gabbert
Cyg 21h46'13" 50d17'37"

Bonnie Gallemore Beissenherz George
Lyr 19h17'40" 32d34'9"

Bonnie Giddings
Cru 12h29'31" -63d5'30"

Bonnie & Gordon Gottlieb
Ari 3h15'52" 28d23'51"

Bonnie Gugleron
Tau 5h48'12" 20d26'53"

Bonnie & Hank
Cyg 20h43'59" 33d23'18"

Bonnie Harrington Farrell
Dra 19h36'23" 61d19'47"

Bonnie & Harry Aiello
Cyg 20h25'19" 43d23'43"

Bonnie Hostetler
Peg 21h45'7" 22d34'38"

Bonnie I Galloway
And 0h37'13" 36d35'2"

Bonnie J. Barnett
Sco 17h51'55" -35d47'14"

Bonnie J Miller
Aqr 23h21'6" -18d50'33"

Bonnie Jane Williams
Leo 11h19'56" 14d0'27"

Bonnie Jayne Kraus
Cam 6h4'32" 61d59'36"

Bonnie Jean
Lyn 6h38'50" 59d43'24"

Bonnie Jean Barington
And 0h22'58" 45d25'39"

Bonnie Jean Brierton
Vir 12h21'27" -10d1'16"

Bonnie Jean Daniels
Cas 0h56'27" 58d18'8"

Bonnie Jean Hoyt
Gem 7h37'46" 16d5'10"

Bonnie Jean Tatge
And 1h30'11" 47d37'51"

Bonnie Jeanne Buckner
Cyg 19h37'6" 30d40'25"

Bonnie Jo
Cas 0h0'31" 57d0'51"

Bonnie Jo Bayle
Cnc 8h21'3" 15d27'53"

Bonnie Joan Bochert
And 0h31'1" 45d25'21"

Bonnie Jobe
Aqr 23h53'52" -3d20'21"

Bonnie & Joe Valenti
Cyg 20h32'59" 41d24'26"

Bonnie Joelle
And 2h26'52" 45d51'58"

Bonnie June Shaw
Ari 2h3'50" 10d38'27"

Bonnie Kay Lester
Vir 13h25'58" -0d36'37"

Bonnie Kay McCormack
Sco 17h48'5" -40d28'6"

Bonnie Kelly
Gem 6h36'20" 26d48'7"

Bonnie Kim
Cnc 8h35'20" 31d6'12"

Bonnie Kristen Cheney
Uma 8h47'1" 56d8'45"

Bonnie Krueger
Cmi 7h23'25" 8d24'17"

Bonnie L. Benbow
And 0h47'49" 41d3'27"

Bonnie L. Toth
Ori 5h59'51" 16d57'35"

Bonnie L Woods
Cas 1h15'48" 59d34'13"

Bonnie Lang Silverman
Tau 3h43'31" 29d32'10"

Bonnie Laurel Jones
Tau 5h42'10" 20d26'47"

Bonnie Lee
Lyn 6h48'19" 51d40'44"

Bonnie Lee
Cap 21h27'9" -15d18'25"

Bonnie Lee
Lib 14h39'59" -9d55'7"

Bonnie Lee Chalfant
Col 5h51'31" -41d13'55"

Bonnie Lee Hummel Mortensen
Per 3h9'37" 31d9'57"

Bonnie Lee James
And 0h5'19" 41d6'4"

Bonnie Lee Messlehner
Ori 5h20'37" 6d30'19"

Bonnie Lee VanHaften
Cyg 21h58'34" 44d39'28"

Bonnie Lewis "Nanny"
Cas 0h37'46" 54d59'31"

Bonnie Linch
Lib 15h13'55" -7d2'25"

Bonnie Lou Williams
Umi 15h25'18" 71d39'56"

Bonnie Louise Davidson's star
Ari 2h35'10" 27d16'45"

Bonnie Louise Wolfe
Leo 11h28'52" 16d26'1"

Bonnie Love
Cnc 9h15'44" 24d14'34"

Bonnie Lynn
Crb 15h53'38" 26d26'48"

Bonnie Lynn
Lib 15h14'14" -3d45'8"

Bonnie Lynn Daniel
Crb 15h27'58" 32d3'23"

Bonnie Lynn Fair
Ari 2h58'28" 17d34'38"

Bonnie Lynn Milner
Umi 11h16'18" 88d45'33"

Bonnie Lynn Petro
Lyn 7h33'34" 36d44'6"

Bonnie Lynn Toups Elliot
Gem 6h41'54" 33d7'45"

Bonnie M. Bohadel
Psc 1h9'39" 26d37'58"

Bonnie Mae
Vir 13h18'1" 2d35'0"

Bonnie Mae
Psc 23h52'3" 0d30'52"

Bonnie Mae Brownell
Ari 3h8'23" 22d13'4"

Bonnie Mae Norman Sims
Lyn 7h55'23" 37d42'11"

Bonnie Mai Pham
And 0h38'47" 36d30'8"

Bonnie Marie Bremers
Vir 11h48'18" 4d20'36"

Bonnie Marie Pittman
Uma 11h8'34" 54d25'29"

Bonnie Martin
Cas 0h35'52" 57d5'12"

Bonnie May
Vir 13h23'11" 5d36'18"

Bonnie Mayne
Gem 6h43'10" 20d41'19"

Bonnie McMahan
Cap 21h17'57" -15d20'3"

Bonnie Megan K
Tau 4h0'32" 27d25'41"

Bonnie Meyer
Lyn 9h10'10" 37d44'40"

Bonnie Michelle
Cam 7h25'32" 63d24'10"

Bonnie/Mom052287&040990
Vir 13h15'45" -21d57'57"

Bonnie My Sister My Friend
And 23h41'32" 47d44'0"

Bonnie Newberg
Umi 14h4'42" 71d11'33"

Bonnie Nielsen
Lyn 7h31'2" 43d0'5"

Bonnie Philpott
Lib 15h7'12" -23d23'7"

Bonnie Poon
Lib 14h32'47" -10d36'53"

Bonnie Prizer A Star Sister-In-Law
Leo 10h56'5" 18d25'15"

Bonnie R. Tucker
And 1h40'7" 43d13'49"

Bonnie R. Williams
Cyg 21h50'58" 43d3'19"

Bonnie Rae "Everything"
Lib 14h26'22" -22d34'36"

Bonnie Rae & Robert Michael
Her 17h19'8" 30d23'33"

Bonnie Reilly
And 0h3'2" 35d0'17"

Bonnie Renae
Uma 10h33'38" 41d7'56"

Bonnie Rhodes
Ori 6h3'37" 20d20'19"

Bonnie Rose Swaim-Tribble
Cmi 7h7'49" 5d26'32"

Bonnie Rust
Sco 17h53'13" -36d1'6"

Bonnie Ryan Berry
Tau 4h47'31" 23d6'10"

Bonnie S. Schuckman
And 1h25'13" 43d50'17"

Bonnie & Sandee Yoder
Uma 10h42'31" 70d33'41"

Bonnie Scheer Kurtz
Cyg 19h51'6" 39d19'37"

Bonnie Sheridan Star
Ori 5h41'2" 3d50'53"

Bonnie Smiling at Ginny
Ori 6h15'50" 9d36'49"

Bonnie Stracener
Lib 14h53'15" -2d26'23"

Bonnie Sue
Leo 10h50'7" 16d46'48"

Bonnie Sue
Lyn 8h52'57" 41d9'32"

Bonnie Sue Potter
Cyg 19h58'16" 32d31'24"

Bonnie Taub
Gem 7h46'3" 21d26'35"

Bonnie the Peacemaker
Cap 21h5'54" -17d9'37"

Bonnie Uda Silverstein
Aqr 22h52'53" -16d51'15"

Bonnie Van Uitert
Cam 3h21'35" 61d22'55"

Bonnie VG
Leo 11h42'49" 13d54'29"

Bonnie Ward Simon
Cas 0h29'56" 57d34'20"

Bonnie Washington
Ori 5h26'24" 7d9'58"

Bonnie Whitehurst
Cas 1h10'57" 66d47'32"

Bonnie Whiting - Superstar Teacher
Lep 5h13'48" -12d19'9"

Bonnie Woyewoda
Cnc 8h28'30" 12d28'23"

Bonnie, 10-23-1979
And 1h47'18" 37d55'46"

Bonnie, Dan, Alysha, Coyle
Del 20h22'7" 20d14'1"

Bonnie051185
Ori 4h53'53" 6d52'54"

Bonnie-Ann Morales McLain
Gem 6h48'56" 31d52'58"

Bonnie-Beth The Beautiful
Aqr 23h44'7" -18d23'12"

Bonniellen-Blinkie-Binkie
Sgr 20h21'16" -42d29'44"

Bonnielu
Cyg 20h48'30" 52d0'24"

Bonnie's Angel
Lyr 18h34'44" 39d59'24"

Bonnie's Beacon
Uma 10h42'53" 55d44'55"

Bonnie's Bright and Precious Star
Vir 13h32'5" -6d58'15"

Bonnie's Rusty Star
Aqr 21h34'6" -7d47'35"

Bonnie's Star
Cas 0h47'28" 57d14'57"

Bonnie's Star
Ori 5h28'52" 14d39'7"

Bonnie's Tamara
And 23h8'42" 51d11'29"

BonnieTed
Lyn 7h41'49" 50d47'48"

Bonny & Bill Barnhill
Crb 15h39'8" 38d3'21"

Bonny Jo's Shining Star
Sgr 19h21'46" -27d55'45"

Bonny Lynn
Uma 9h52'9" 68d28'49"

Bonny Lynn
Cnc 8h25'16" 26d22'22"

Bonny Lynn Christie
Gem 7h26'25" 28d8'21"

Bonobo
Uma 11h16'41" 58d53'37"

Bon's Wishing Star
And 2h36'45" 44d44'6"

Bonus Aliquotiens
Leo 11h10'4" -2d8'52"

Bonvicini
Cyg 19h41'35" 28d59'58"

Bonz Seccombe
Cma 7h8'17" -14d35'31"

Bonzi Carolina
And 1h25'41" 35d3'40"

Bonzo
Cmi 7h39'58" 1d29'24"

Bonzo
Cep 21h29'16" 57d53'27"

Boo
Umi 14h27'39" 68d12'11"

Boo
Dra 19h15'11" 69d19'7"

Boo
Aqr 22h8'56" -6d38'15"

Boo
Ori 6h8'3" -3d49'10"

Boo
Sgr 18h10'50" -18d25'13"

Boo
Vir 14h43'8" 4d8'4"

BOO
Tau 5h51'1" 13d2'34"

Boo
Psc 0h13'44" 10d29'0"

Boo
Aqr 20h41'35" 1d41'25"

boo
Vir 13h12'45" 10d0'56"

Boo
Boo 14h25'46" 19d0'8"

boo
Her 16h41'11" 30d41'18"

Boo
Uma 11h5'21" 44d17'13"

Boo
And 23h43'21" 47d48'22"

Boo and JuJu
Sco 17h4'37" -32d55'0"

Boo and Squishy
Sge 20h4'59" 18d10'58"

Boo Bare
Dra 18h27'46" 51d47'43"

Boo Bear
Lyn 7h36'36" 48d39'21"

Boo Bear
Gem 7h14'27" 22d12'10"
Boo Bear
Ari 3h16'51" 29d39'32"
Boo Bear
Cha 10h33'44" -80d34'19"
Boo Bear and Snuggle Bunny
Umi 14h52'30" 80d5'17"
Boo Bear's Butter Bean
And 1h8'12" 37d10'37"
Boo Bear's Star
Psc 0h56'32" 29d45'43"
Boo Boo
Tau 5h9'26" 22d29'11"
Boo Boo
Psc 0h16'35" 16d11'45"
Boo Boo
Aql 19h33'44" 7d54'23"
Boo Boo
And 0h53'5" 41d0'44"
Boo Boo
Cas 0h2'46" 56d31'51"
Boo Boo
Vir 13h28'43" -0d15'44"
Boo Boo Bear
Uma 13h52'16" 52d10'6"
Boo boo bear
Uma 11h37'34" 43d37'12"
Boo Boo Bear
Ari 2h42'10" 12d45'17"
Boo Boo Bright Star
Cyg 20h35'1" 43d15'54"
Boo Boo Kitten
Lyn 8h33'47" 56d22'44"
Boo Boo Star
And 23h18'13" 47d39'27"
Boo Boo's
And 0h34'16" 27d21'50"
Boo Boo's Star
Uma 9h17'52" 57d47'15"
Boo & Puppy's Magical Moment
Cyg 20h26'21" 58d34'39"
Boo u take me 2 the Clouds love Daz
Sgr 18h22'49" -23d47'9"
Booangel
Boo 13h45'49" 18d24'11"
Boobala
Ari 3h29'21" 25d10'32"
Booballa and CheeseMonkey
Aql 19h52'53" -0d15'25"
Boobalu
Sco 16h38'30" -31d26'25"
Boobchka
Umi 14h30'32" 78d35'55"
Boobear
Sco 16h7'32" -13d56'7"
BOOBEE
Vir 12h35'27" -0d44'13"
Booberry
Cam 4h5'51" 68d31'58"
Boobett
Vir 13h2'34" 11d57'7"
Boobida
Lyn 9h9'56" 45d33'5"
Boobie
Lac 22h28'34" 43d45'38"
Boobie
Uma 9h56'4" 41d31'24"
Boobika
Uma 11h41'14" 62d46'1"
BooBink
Aur 5h57'54" 36d31'23"
Boobis
Psc 1h11'53" 28d22'35"
Boobners
Ari 3h5'59" 24d39'52"
boo-boo
Uma 10h44'41" 69d59'16"
Boo-Boo
Umi 14h4'25" 72d27'31"
BooBoo
Cyg 19h58'14" 54d53'34"
BOO-BOO
Dra 18h28'25" 54d29'29"
Boo-Boo
Aqr 23h33'20" -24d17'51"
BooBoo Aurora
And 0h21'1" 27d45'10"
BooBoo & Jill
Tau 5h11'20" 17d44'1"
BooBoo Otis
Aqr 20h39'41" -10d39'14"
Boo-Boo, Eleanor Ann
Gem 6h49'53" 23d44'59"
BooBoo143
Psc 1h3'23" 4d45'7"
BooBooJax
Aqr 21h52'59" -0d19'24"
Booboos
Uma 9h44'59" 42d2'53"
Boo-Boos
Cyg 21h25'18" 39d5'27"
BooBoo's Christmas Star
Umi 13h37'12" 69d36'36"
BOOBOOSTAR
Uma 11h5'20" 36d11'39"
Booby a.k.a. Carmen
Uma 11h47'26" 62d29'1"
Boo-D Forever
Umi 13h19'3" 71d51'28"
Boodini
Uma 8h54'12" 70d4'53"
Boodle
Leo 11h13'28" 17d31'47"
Boodles
Cma 6h53'29" -20d36'22"
Boodog
Per 2h24'13" 57d17'29"
Boof
Uma 8h54'8" 59d45'29"
Boofa
Crb 15h50'47" 33d53'29"
Boofacina
Lyr 18h53'59" 36d35'57"
Boofala
Tau 4h39'8" 20d58'47"
boog
Uma 11h59'50" 64d59'7"
Booga
Lib 14h52'31" -3d34'7"
Booga
Uma 11h52'55" 47d25'1"
Boogaloo
Peg 22h50'1" 12d16'53"
Booger
Pho 23h36'4" -39d46'41"
Booger
Lup 15h36'6" -35d37'38"
Booger Honey's Star
Ari 2h43'11" 26d50'58"
Booger Ryan's Star
Cma 6h37'57" -16d5'35"
Booger's Mate
Aqr 22h40'4" 1d3'45"
Boogie
Peg 22h25'44" 10d12'12"
Boogie
Ari 3h6'18" 19d33'7"
Boogie
Boo 14h44'53" 52d2'36"
Boogie
And 2h15'7" 39d2'33"
Boogie
And 0h36'21" 39d27'42"
Boogie Bear
Umi 15h32'53" 79d36'55"
Boogie&Bobo
Sco 17h0'53" -33d2'22"
Boogie Boo
Cnc 8h41'47" 17d2'29"
Boogie Boy Bruce Da Star McHenry
Leo 9h27'17" 12d18'2"
Boogie Busa
Lyn 7h37'54" 38d59'7"
Boogie Disco's
Col 6h22'23" -41d22'50"
Boogies
Apu 13h59'33" -71d47'26"
Boogies
Vir 14h17'44" 4d11'3"
boogily woogily bear
Tau 4h4'0" 11d50'25"
Boogles
Lyn 7h42'58" 52d41'36"
Boogs
And 2h26'56" 41d37'25"
Booha's Star
Dra 18h40'35" 54d14'16"
Boohbah
Lac 22h21'12" 51d11'6"
Book Lady, Carolyn Kantor
Cam 4h37'38" 57d9'25"
Booker Boys
Tri 2h20'33" 34d29'36"
Booker Pruitt - Loan Star
Per 3h47'22" 33d30'45"
Bookie
Uma 9h19'0" 50d54'15"
Bookie
And 2h30'27" 50d3'29"
Bookie Chapman
Uma 9h40'38" 63d40'8"
Bookies
Uma 8h53'47" 50d43'42"
'Bookworm'
Dra 17h53'15" 52d2'55"
Booky Ben-Ami (Eiger)
Cam 4h19'33" 68d54'20"
Boolachka
Sgr 18h3'18" -27d48'51"
Boolori
Cap 20h55'23" -19d44'58"
Booma & Boompa
Uma 10h27'29" 45d23'53"
Boomer
Cma 6h41'51" -15d3'9"
Boomer
Uma 10h54'33" 69d34'51"
Boomer
Uma 10h11'34" 64d26'56"
Boomer "11/09/2004"
Sco 16h13'8" -12d13'59"
Boomerang Jansen
Cnc 8h47'28" 26d27'0"
Boomers Love
Sco 17h21'52" -31d56'37"
Boomie
Uma 10h7'16" 49d2'10"
Boompa
Cep 0h25'39" 78d58'23"
Boompa Laxner
Uma 8h50'42" 62d13'42"
Boomshakalaka
Uma 8h42'20" 52d21'15"
Boone Allen Stiltner
Aql 19h5'53" -0d11'13"
Boone McCaulley Hutchison
Aqr 21h13'47" -3d4'15"
BooneOakley
Gem 7h23'11" 15d53'51"
Boop>
Cnc 8h37'42" 21d53'28"
Boop
Peg 21h30'16" 14d20'21"
BOOP
Lyn 7h51'15" 49d58'12"
BOOP! Nunnies
Aur 6h15'8" 30d40'10"
Boopa
Her 16h59'10" 30d59'44"
BOOPALINI
Uma 11h41'41" 41d20'58"
Boopers Rose
Sgr 18h26'38" -17d29'15"
Boopie
Sgr 19h23'9" -15d7'33"
Boopie
Lyr 18h40'26" 34d24'38"
boo's 1st anniversary star
Aur 5h52'44" 53d21'10"
Boo's Dream - Anthony Rye
Aqr 21h43'33" -5d39'35"
Boo's Night Light
Umi 14h30'35" 76d45'41"
Boo's Porch
Ori 5h43'8" 0d54'59"
Boo's Quan
Cas 0h27'17" 53d33'37"
Boo's Star
And 23h33'2" 42d6'45"
Boo's Star
Umi 16h21'1" 78d50'17"
Booshazy
Lmi 10h4'59" 38d40'1"
Booskins
Vir 14h5'47" -15d39'3"
Boot
Cnc 8h27'53" 15d26'53"
Booth Family Star- Eric April Trent
Uma 10h11'10" 47d7'26"
Boots
Boo 14h17'46" 52d50'21"
Boots Bautista
Gem 6h47'6" 14d5'30"
Boots Mimi McKiernan
Gem 6h49'39" 16d53'35"
Bootsie Levinson
Psc 1h2'11" 5d5'10"
Bootsy & Sharky
Vir 12h38'18" 13d6'35"
Boozen
Lib 14h40'31" -24d6'39"
Bop
Leo 9h40'14" 27d42'40"
Bopi
Uma 9h48'21" 72d43'10"
BOPPA
Psc 23h59'20" 6d7'20"
Boppa the Great
Sco 17h24'20" -42d23'26"
Boppa's Star
Cas 0h16'17" 57d29'18"
Boppin' Mel
Cas 0h24'56" 57d5'49"
Boppy
Sgr 18h59'27" -35d38'8"
Bora Chang
Umi 15h3'1" 71d30'45"
Bora Shin
Psc 0h54'33" 29d41'1"
Boran & Sabrina's Star
Tau 4h34'16" 15d52'41"
Borasqo
Cyg 21h58'32" 46d13'26"
Borders 300 Winter Park, Florida
Peg 21h45'4" 13d31'15"
Bordi-OsMami
Uma 10h30'3" 70d27'28"
Borelli Family Shining Star
Leo 10h22'57" 26d5'50"
Borgen Family Star
Cas 0h11'37" 53d56'24"
Borgerson
Cep 21h12'39" 66d4'53"
B.Original
Tau 4h10'21" 27d47'57"
Boris
Uma 13h54'36" 52d4'56"
Boris Aronovich
Psc 1h13'55" 26d15'35"
Boris B Good
Uma 10h57'28" 53d5'53"
Boris Catz
Lyn 7h0'59" 48d21'18"
Boris elye eanya tinwe
Tri 2h18'31" 37d7'44"
Boris Gambarin
Sgr 17h50'28" -25d51'53"
Boris Jan Baberkov
Uma 8h50'31" 56d15'41"
Boris Josch
Uma 10h40'1" 57d53'41"
Boris Kagan
Vir 12h38'51" -6d19'58"
Boris L.P. Sasha
Lyn 7h38'47" 39d14'1"
Boris Mouradov
Cep 20h29'11" 63d44'55"
Boris Schakowski
Uma 10h11'2" 47d51'33"
Boris Shklyarevsky
And 1h34'37" 44d17'12"
Boris Stürmann
Ori 5h58'49" 3d23'31"
Born Royalty
Lyr 19h15'2" 36d15'50"
Borntrager
Lib 15h1'24" -6d32'0"
Boroka
Lyn 8h11'5" 57d52'48"
Borrelli's Star
Sco 16h2'58" -10d8'30"
Borsato's Beacon
Umi 14h8'12" 75d43'38"
Borst, Melanie
Uma 9h52'59" 46d42'18"
Borys Petri
Uma 11h22'45" 59d36'18"
Bo's Star "Gabe's Love"
Tau 4h23'39" 22d11'26"
Bös, Winfried
Uma 11h34'49" 49d4'40"
Bosco
Tau 4h52'22" 29d25'38"
Bosco
Uma 14h23'13" 59d32'39"
Bosco Fais
Cap 20h23'11" -13d42'52"
Bosco - Good Ol' Boy
Cnc 8h39'49" 26d34'40"
Boscolo, Cinzia (Forcola)
Ari 2h31'27" 25d34'57"
Bosco's Cosmos
Cnv 12h41'36" 45d33'21"
Bosco's Mom
Ori 6h3'26" 17d11'12"
Bösener, Rolf
Uma 9h55'57" 65d11'17"
Bos-Gideon
Aur 5h33'46" 41d24'58"
Bosio Giuliano Sindaco di Almese
Ori 5h44'54" 11d15'36"
Bösl
Umi 14h57'57" 73d26'59"
Bosmaclacdady
Per 3h44'49" 41d38'14"
Bosnjakovic, Srecko
Uma 10h5'50" 61d25'45"
BoSoleil
Tau 4h43'44" 17d45'22"
Boss Hall
Boo 14h13'36" 28d34'16"
Boss Walker
Her 17h18'40" 25d30'39"
Bossdyno
Cyg 19h57'53" 33d23'23"
Bossman
Cnc 8h32'50" 32d38'25"
Bossman
Leo 10h28'45" 22d25'25"
Boston Andrew Reilly
Umi 16h20'50" 75d58'26"
Boston Michael Kirby
Cep 22h42'55" 72d29'34"
Boston Sunshine
Uma 8h30'51" 67d18'56"
Boston Whalen-Eynon
Cyg 21h34'0" 38d37'48"
Bostwick Family, Suwanee GA
Uma 11h32'35" 58d42'42"
Boswell
And 0h17'59" 32d37'54"
Boswell
Ori 4h48'22" 5d33'41"
Bota
Psc 23h30'14" 7d1'46"
Both, Andreas
Uma 9h58'33" 53d47'0"
Bott, Andreas Thomas
Uma 11h19'7" 35d6'17"
Boubi
Aqr 20h47'5" -3d51'20"
Boubou
Cnc 8h54'30" 26d49'46"
BOUBOU 7
Gem 7h23'35" 32d24'23"
Boucherie, Frank
Uma 11h36'17" 31d28'24"
Bouffard Major
Uma 11h27'4" 63d0'39"
Bougain Martial
Gem 6h30'26" 27d23'25"
Bougainvillea Beach Resort
Per 3h8'10" 55d23'12"
Bouilh a kared : Cédric
Psc 1h43'50" 13d41'38"
Bouillard Virginie
Cmi 7h27'19" 1d52'51"
Bouke "The Love Of My Life"
Uma 10h20'6" 49d57'11"
Boulder Dash Sweet
Cma 6h19'33" -27d40'21"
Bouleman
Ori 6h4'7" -0d39'18"
Boulos Jabre
Sco 16h48'48" -27d28'34"
Boum & Annie
Cyg 19h43'12" 29d32'25"
Bouncing Warrior from the North
Sgr 19h13'14" -21d19'28"
BOUNDLESS
Sgr 17h59'38" -18d10'46"
Boundless and Unbroken
Uma 9h12'32" 51d30'54"
Bournstein 15
Uma 10h37'51" 53d59'38"
Bouvier Karen
Leo 10h55'19" 8d58'57"
Bouy Bear
Cma 6h57'53" -21d54'58"
Bova
Umi 16h30'25" 83d8'42"
Bovis Puniceus
Ori 4h58'38" 10d25'50"
Bowanna Daddy-Barry Gregory
Aqr 23h3'40" -8d59'55"
Bowen
Uma 11h4'55" 48d34'16"
Bowen Jeffrey Uriel Pickett
Tau 4h30'1" 19d54'27"
Bowen Ohana
Gem 7h7'15" 21d35'22"
Bowers Nest
Aql 19h56'51" -5d46'5"
Bowie
Aql 19h45'55" 8d42'28"
Bowie 69
Ori 5h52'31" 12d24'56"
Bowman
Uma 11h35'3" 57d41'10"
Bowman Clinton Rains, III
Ori 6h2'43" 16d19'25"
Bowser
Uma 10h13'45" 44d33'12"
Bowsie
Uma 11h58'7" 28d19'57"
boxnife
Sex 10h10'1" 0d40'10"
Boy Blu
Pho 1h52'7" -43d5'31"
Boy King DiNapoli
Psc 1h48'22" 6d15'23"
Boy, Dietrich
Cnc 9h9'56" 15d46'18"
Boyce's Bequest
Uma 13h21'14" 52d30'51"
Boyd
Dra 18h24'11" 53d36'37"
Boyd Clark
Ori 6h24'2" 17d5'48"
Boyd Jenkins
Uma 11h28'38" 50d57'17"
Boyd Joseph Pritchard
Lib 14h50'0" -11d37'24"
Boyd Young Whisenant
Aql 18h43'39" -0d36'19"
Boyde Family Star
Cyg 21h58'43" 44d22'25"
Boyenval Stéphanie 28/09/1981
Col 6h26'24" -34d45'10"
Boyle Family Star
Cas 2h19'26" 75d30'47"
Boys Wind
Peg 23h50'59" 27d8'50"
Boz
And 0h24'16" 42d10'43"
"Boz" 1948
Lyn 6h30'4" 56d21'27"
Bozena
Sgr 18h22'32" -32d23'57"
Bozenka
Gem 7h1'58" 32d10'17"
bozica i dejan
Peg 23h7'8" 29d9'33"
Bozidar Antonovic - 22.04.1959.
Per 4h10'17" 41d28'21"
Bozidar, Gordana , Nebojsa , Marina
Cas 23h22'48" 54d27'12"
BOZINA
Tau 3h36'33" 21d52'54"
Bozwix8
Lyn 7h27'5" 57d9'25"
Bozych, Amelia Maria
Uma 10h36'5" 41d6'47"
B.P. Karol-Chik
And 23h20'41" 45d42'21"
bra+ca+
Gem 7h39'8" 22d22'28"
Brace Guy Silvestri III
Boo 14h28'35" 47d19'20"
Brace Harris
Sco 16h4'55" -11d5'20"
Brach
Aqr 22h19'15" -16d2'48"
Brach Anthony Hornbuckle
Uma 13h13'33" 62d50'55"
Bracha Gedalya
Lyn 6h51'42" 50d46'59"
Bracho
Vir 13h2'28" 4d51'17"
Brad
Ori 5h21'59" 3d14'20"
B-RAD
Ori 6h7'3" 5d25'53"
Brad
Tau 4h57'48" 17d34'52"
Brad
Her 16h39'50" 35d22'20"
Brad
Cam 4h7'1" 66d50'5"
Brad & Alaina
Cyg 21h28'20" 53d41'52"
Brad Alexander Still
Her 18h20'0" 21d43'17"
Brad Allen Spearbeck
Lib 15h1'23" -1d40'52"
Brad and Denise Squires
Cyg 20h59'48" 30d35'19"
Brad and Emily Danger
Cyg 20h28'15" 31d26'51"
Brad and Joncilee Davis
Cyg 21h53'17" 46d43'32"
Brad and Julie Squires
Sco 17h24'4" -44d18'57"
Brad and Nicole's Star
Ori 6h14'29" 6d25'22"
Brad and Sherri
Uma 11h32'58" 58d52'54"
Brad and Stacey
Lyn 7h56'58" 34d56'33"
Brad and Sunny Goldberg
Lmi 10h35'6" 37d59'59"
Brad and Tara
Ari 2h42'7" 23d29'7"
Brad and Tori Forever
Lyn 7h7'17" 58d14'26"
Brad Andrew Trenwith
Sgr 18h50'25" -26d57'39"
Brad Bentley
Ori 5h19'14" 6d57'42"
Brad & Bobbie
Cyg 20h16'44" 55d38'19"
Brad Bristowe
Sco 17h25'31" -45d14'18"
Brad C. Dunlap
Her 17h10'5" 16d34'17"
Brad & Carlene
Uma 10h50'18" 71d15'42"
Brad & Charlotte Lamson
Crb 15h37'59" 39d9'55"
Brad Edward Hughes
Vir 12h14'8" -11d31'3"
Brad Gearlds
Sgr 19h1'44" -25d54'44"
Brad Harper
Vir 13h13'21" -5d16'20"
Brad Hill "June 24th. 1956"
Cnc 8h29'33" 17d40'34"
Brad Hynde
Uma 9h45'2" 47d11'42"
Brad J. DePascale
Per 4h4'56" 40d50'25"
Brad & Jackie
Cyg 20h11'30" 45d43'27"
Brad Jacobs aka "The Bunny Maker"
Scl 0h12'10" -26d22'23"
Brad Jay Korngold
Uma 8h58'28" 47d16'4"
Brad & Jenna - Valentines Day '06
And 1h45'29" 42d40'48"
Brad Joseph Austing
Sgr 19h18'25" -22d31'7"
Brad Klatt
Uma 11h29'39" 35d30'53"
Brad Koch
Lyn 7h59'8" 47d29'41"
Brad & Kym Anniversary Star 05/03/94
Cru 12h8'27" -62d20'27"
Brad L Converse
Cnc 8h45'32" 32d18'6"
Brad&Laura~ in the stars forever
Her 16h22'53" 47d43'28"
Brad Lei
Cru 12h27'23" -58d50'30"
Brad Long
Aql 19h25'40" 10d41'8"
Brad Maguth
Her 17h14'47" 33d44'9"
Brad Matthew Ralston
Her 17h36'37" 36d19'43"
Brad McLaughlin
Cas 1h8'6" 61d55'42"
Brad & Meme
Umi 14h53'1" 81d18'24"
Brad Michael Dufrene
Ori 5h27'7" 1d23'8"
Brad Milstead
Per 3h4'41" 52d8'4"
Brad & Nicole *Shine Together*
Leo 10h30'57" 17d19'26"
Brad Paisley
Boo 14h11'15" 27d11'45"
Brad Pidwell-Omniluminous
Aqr 23h33'9" -8d57'26"
Brad Porter
Psc 0h35'48" 20d52'28"
Brad Ray, The Shining Star
Lib 15h12'6" -26d31'10"
Brad & Renée King
Crb 15h59'6" 38d37'52"
Brad Roberts " The Star of My Life"
Vir 11h56'46" -0d23'14"
Brad Sattler
Lib 15h6'27" -2d14'14"
Brad Schofield
Eri 3h59'30" -0d55'20"
Brad & Shana Cohen
Umi 14h45'58" 79d37'45"
Brad & Sharlynn, Always & Forever
Uma 11h26'2" 51d22'30"
Brad Squires Shines Forever
And 1h29'24" 34d22'19"
Brad T. LaVigne
Per 3h18'6" 42d10'2"
Brad: The Keeper of My Heart
Vir 12h19'19" 3d9'33"
Brad the sparkle of my life Louise
Cru 12h24'54" -57d47'12"
Brad Thomas Flewelling
Leo 11h10'45" -4d2'12"
Brad Tooman
Cnc 8h10'9" 7d3'2"
Brad Trevarton Ostash
Cep 22h56'26" 72d23'20"
Brad W Goodwin
Aqr 23h27'26" -8d55'52"
Brad Walker
Her 17h19'57" 32d50'43"
Brad Waxman
Her 16h9'13" 47d36'21"
Brad Weiser
Crb 16h12'9" 28d29'11"
Brad William Prentice
Psc 23h29'36" 7d32'36"
Brad William Wheeler
Leo 10h34'19" 20d54'0"
Brad Wright
Her 17h28'6" 15d51'10"
Bradd Lee Johns
Uma 8h11'44" 69d13'25"
Braden Andrew McMahon
Uma 13h6'58" 62d43'7"
Braden Christie 5
Cap 20h23'3" -11d9'6"
Braden Cooper Bundren
Vul 19h49'15" 27d7'44"
Braden Curtis Bieler
Uma 9h58'25" 46d22'7"
Braden Denzil Miller 1
Ari 2h25'36" 24d30'26"
Braden Hebel - Rising Star
Aur 5h44'34" 33d8'17"
Braden James Long - Mr. Early
Cnc 8h19'15" 7d13'54"
Braden Jay Schenk
Gem 7h59'27" 30d11'30"
Braden Jesser
Tau 5h42'31" 25d45'23"
Braden Joseph Hannon
Ari 2h16'34" 14d43'23"
Braden Keeble
Per 3h59'40" 36d29'3"
Braden Lee May
Sco 17h40'34" -32d10'39"
Braden Louis Blondek
Per 3h55'30" 47d40'36"
Braden Matthew Mason - "Braden's Star"
Psc 0h50'33" 20d2'50"
Braden Michael Ballard
Leo 10h23'38" 10d55'20"
Braden Patrick
Cep 0h13'28" 67d33'4"
Braden Ray Vickery
Umi 18h49'40" 86d47'29"
Braden Ray Vickery
Uma 11h12'11" 29d28'47"
Braden Richard Weiss
Uma 13h49'31" 61d31'35"
Braden S.
Umi 15h46'54" 88d37'15"
Braden Seroski
Cap 21h13'24" -17d12'45"
Braden Shea Baker
Ari 3h13'35" 28d59'43"
Braden Sutton Ginsberg
Umi 14h26'6" 74d56'7"
Braden Wesley Johnson
Ari 2h47'22" 15d24'59"
Braden's Heart
Ori 5h45'44" 6d50'18"
Braden's Star
Per 2h23'43" 55d3'35"
Bradey Hobart
Ori 5h58'28" -0d43'48"
Bradey's Robbie
Sco 17h21'29" -41d41'46"
Bradford
Aur 5h11'19" 42d43'43"
Bradford Adam Coughlan
Tau 3h30'59" 8d10'57"
Bradford B. Lyon
Per 3h20'37" 42d35'39"
Bradford C (Brad) Weschler
Aur 5h52'13" 52d8'28"
Bradford Charles Dunlap
Leo 10h11'35" 12d43'16"
Bradford D. Myers
Gru 22h31'41" -39d41'23"
Bradford Daly
Her 18h34'9" 19d3'22"
Bradford F. Henderlong
Vir 12h29'1" 3d5'54"

Bradford Family
Uma 11h59'54" 47d31'18"
Bradford "J-Dawg" Johnston
Umi 14h29'36" 72d3'12"
Bradford Kelly Neal
Leo 11h30'11" 20d3'44"
Bradford Kent Spaulding * Lovebeads
Gem 6h46'15" 30d52'4"
Bradford Lee Archambeau
Dra 20h19'27" 63d39'51"
Bradford Lee Bulkley
Uma 8h55'46" 46d33'56"
Bradford Neal Louison
Leo 9h44'37" 28d17'17"
Bradford Ryan Boal
Per 3h20'24" 51d50'45"
Bradford Swett
Uma 11h16'23" 49d44'26"
Bradford & Tamara: Soulmates
Psc 23h42'59" 0d35'56"
Bradford & Valerie Nagy
Cyg 21h33'19" 30d14'47"
Bradford W. MacKenzie
Ori 6h13'2" 3d12'54"
Bradi Lang
Tau 5h8'1" 22d8'11"
Bradie13
Gem 7h10'35" 18d11'42"
Bradlee Greco
Aqr 22h55'49" -22d43'29"
BRADLEY
Gem 7h17'56" 33d20'7"
Bradley
Aur 5h42'16" 33d6'2"
Bradley A Krueger
Uma 10h16'48" 60d50'4"
Bradley A McBrayer
Per 4h42'14" 43d41'40"
Bradley A. Scott
Dra 16h14'13" 52d2'40"
Bradley Aaron
Uma 12h3'20" 53d24'31"
Bradley Adney
Ari 3h12'9" 21d51'12"
Bradley & Alivia Schwab
Uma 11h28'57" 34d1'23"
Bradley Allen Krueger
Ari 2h48'39" 26d9'26"
Bradley Allen Morris
Lib 14h49'32" -2d30'17"
Bradley Alton Konopa
Uma 11h37'31" 47d9'46"
Bradley And Jenn Forever
Gem 7h22'23" 28d32'22"
Bradley and Louis Root
Gem 6h36'7" 20d44'37"
Bradley and Stephanie
Sge 20h11'40" 18d27'48"
Bradley Asher Molen
Ori 5h24'44" 3d34'59"
Bradley Ballard
Aqr 22h14'9" -3d4'15"
Bradley Becker
Cep 0h18'21" 73d21'8"
Bradley C. Roach
Ori 5h27'34" 5d58'27"
Bradley Christopher McKendrick
Uma 9h52'41" 45d23'17"
Bradley Cofoid
Ori 5h22'40" 2d10'10"
Bradley Curfman
Uma 8h38'21" 59d56'46"
Bradley D. Keefer
Uma 9h31'31" 43d57'24"
Bradley Dallas Hankins
And 2h30'43" 38d0'34"
Bradley Dario Medrano
Her 17h52'10" 28d27'31"
Bradley Davidson-Kiiveri
Cap 20h29'50" -11d0'28"
Bradley Dean
Tau 5h14'2" 17d20'45"
Bradley E. Morrow
Dra 17h26'4" 58d20'36"
Bradley E Stevens
Aur 5h51'13" 52d1'40"
Bradley G Williams
Uma 13h54'31" 55d33'27"
Bradley George Duke
Her 17h59'47" 42d13'43"
Bradley Gordon Cochrane
Cep 23h33'53" 77d30'35"
Bradley Haggerty
Lac 22h24'4" 53d18'27"
Bradley Henry
Her 17h26'19" 33d57'0"
Bradley Hop & Kelly Hop
Ind 21h42'32" -69d31'13"
Bradley Hunter Bedard
Aql 18h58'19" 9d20'42"
Bradley J. Fowler
Per 2h49'12" 54d2'34"
Bradley Jackson Conrad
Uma 11h38'54" 41d40'22"
Bradley James Lunetta
Ori 5h18'35" 12d2'41"
Bradley James Nyberg
Aqr 20h47'7" 0d17'17"
Bradley James Smith
Aql 19h8'45" 13d21'2"
Bradley James Teeters
Her 17h40'6" 36d46'50"
Bradley Jason Stewart
Aur 5h42'9" 54d49'5"
Bradley Jay Chesla
Crb 15h22'58" 25d48'26"
Bradley Jay Dorland
Boo 14h51'22" 20d37'41"
Bradley & Jennifer
Umi 13h55'59" 75d33'30"
Bradley John Hubbard - 24.10.2005
Pho 1h7'20" -52d52'27"
Bradley John Michael
Psc 1h12'58" 17d38'2"
Bradley Johnston 30-11-1983
Sgr 19h11'1" -23d12'5"
Bradley Jon Hunter
Ori 5h40'15" 1d2'27"
Bradley Joseph Mettler
Uma 13h50'58" 48d19'17"
Bradley Joseph Moore
Cnc 9h10'58" 30d59'3"
Bradley Keith Piper
Cru 12h38'33" -60d48'45"
Bradley L. Anderson
Her 16h31'51" 42d38'7"
Bradley Lee Stacy, Jr.
Uma 11h22'6" 36d1'5"
Bradley Leonie John in Space 4 Ever
Cru 12h13'1" -64d20'4"
Bradley Mario Rigolli
Umi 12h51'44" 87d53'10"
Bradley Marshall
Per 2h47'46" 50d16'8"
Bradley & Mary Miller
Cnc 8h19'54" 9d45'24"
Bradley Michael Strong
Per 3h14'37" 47d16'38"
Bradley Michael The Kid
Vir 14h15'33" -13d13'12"
Bradley & Naomi Williams
Pho 0h32'38" -45d13'56"
Bradley Nelson Wolf
Cyg 21h13'52" 45d10'18"
Bradley Paul Strayer
Vir 13h5'21" 12d56'41"
Bradley Peters
Umi 15h39'55" 69d40'22"
Bradley Ray Smith, Jr.
Cnc 9h7'34" 30d43'36"
Bradley Reid Walker
Vir 13h30'50" 4d20'34"
Bradley Robert Wallin
Per 4h35'26" 31d44'37"
Bradley Ryan
Cnc 8h31'2" 12d25'29"
Bradley Ryan Griffith
Ori 5h47'32" 5d53'46"
Bradley S. Koch
Del 20h53'53" 10d51'38"
Bradley Scott
Cas 0h45'28" 64d30'58"
Bradley Scott Miller
Her 16h31'11" 46d22'54"
Bradley Steven and Katie Ann
Per 4h35'13" 38d48'45"
Bradley Steven Perkins
Ori 5h57'25" 17d15'24"
Bradley T
Cnc 8h51'37" 11d50'5"
Bradley Thomas Gabriel
Ari 2h56'31" 27d57'57"
Bradley Thomas Harper
Lyr 18h41'34" 38d38'46"
Bradley Trowbridge
Sco 17h51'5" -36d3'27"
Bradley W. Baker
Aqr 21h4'47" 0d51'26"
Bradley W. Smith
And 1h43'21" 50d27'38"
BradleyandLaci I Cor. 13:1-13
Sge 20h11'54" 17d58'45"
Bradley's Star
Her 18h39'28" 16d50'32"
Bradley's Star
Aql 18h53'12" -0d38'11"
Bradley's Torch
Cru 12h35'48" -59d24'27"
Bradley's Way
Cma 6h46'40" -29d29'31"
Bradlin Jade
Sgr 17h51'44" -29d3'5"
Bradly Thomas
Lyn 9h13'12" 44d10'13"
Brad's "Peace" of the Heavens
Cas 23h48'3" 58d16'41"
Brad's Place
Ori 5h28'51" 0d19'51"
Brad's Shining Light
Uma 10h10'22" 46d27'56"
Brad's Star
Uma 11h19'33" 31d27'50"
Bradsa Rah Star
Ori 5h38'28" -2d39'35"
Brady
Cma 6h15'59" -27d3'34"
Brady
Aql 20h16'46" 0d33'0"
Brady
Tau 5h22'3" 22d17'26"
Brady Alexander Merritt
Ari 1h49'8" 22d52'53"
Brady Allen Haskell
Pyx 8h45'37" -27d50'31"
Brady and Brenda
Lyn 9h5'31" 37d59'22"
BRADY AUSTIN HACKLEY
Crb 15h29'57" 29d12'0"
Brady Beckham Weaver
Leo 11h43'16" 22d6'18"
Brady Chase Borugian
Aqr 21h0'24" 0d43'29"
Brady & Crystal Forever
Psc 0h59'42" 7d45'2"
Brady Daniel Landuyt
Vir 13h42'3" 3d42'36"
Brady Daniel Vestal
Vir 13h30'53" -17d6'43"
Brady David Garlock
Gem 6h18'43" 27d6'28"
Brady Elizabeth Welch
And 0h15'23" 27d38'52"
Brady Farrel Allen Steele
Gem 7h28'24" 20d26'55"
brady forever toobed
Uma 10h42'49" 47d25'11"
Brady Ingram Wilson
Uma 10h44'26" 68d33'56"
Brady James
Gem 6h53'3" 32d0'21"
Brady James Biron
Her 16h37'1" 46d28'47"
Brady John Hill
Psc 0h59'45" 5d27'35"
Brady John Michael Warford
Per 3h20'34" 42d45'26"
Brady Kemp Miller
Umi 14h27'23" 68d48'34"
Brady Marc Weglowski
Her 17h53'10" 45d49'3"
Brady Marshall Katz
Ori 6h17'12" 13d18'51"
Brady Marshall Newell
Cam 4h24'53" 72d46'32"
Brady Michael Spry
Her 17h7'48" 32d33'44"
Brady Mitchell
Uma 11h14'53" 59d20'52"
Brady Mitchell Remington
Umi 15h51'41" 70d36'52"
Brady N. Wiggins
Aqr 22h8'49" -3d13'55"
Brady Orville Heiner
Uma 11h55'56" 58d13'44"
Brady Patrick Matchett
Lib 15h47'48" -9d58'6"
Brady Patrick Quigley
Gem 6h42'5" 23d54'9"
Brady Peet Miller
Ari 3h9'45" 10d58'44"
Brady Porter Wilson
Ori 5h40'34" 0d23'25"
Brady Rickman
Sgr 19h41'13" -14d45'9"
Brady Ridley Hoffman
Lib 14h50'28" -4d18'1"
Brady Robert Ahern
Leo 11h8'46" 8d51'15"
Brady Robert Brau
Vir 13h10'57" 5d31'52"
Brady Rye
And 0h21'13" 38d39'9"
Brady Scot Sheehan
Her 18h43'10" 19d2'46"
Brady Scott Ackerman
Sco 16h10'4" -12d34'27"
Brady Scott Gustinis
Uma 12h59'21" 56d46'9"
Brady "Tai Zao"
Cru 12h41'15" -59d17'41"
Brady Thomas Kilgore
Aql 20h12'23" 8d40'40"
Brady William Haskell
Aqr 20h47'41" -8d47'1"
Brady William Russell
Per 4h12'42" 50d17'12"
Bradyn Darrick Loughner
Psc 0h52'3" 27d55'39"
Bradyn Nell Labbe
Aqr 21h15'5" -3d6'56"
Brady's Birthday Star
Ari 2h22'32" 14d54'44"
Brady's Bright Boreas
Cap 21h35'29" -15d50'6"
Brady's Bright Light
Gem 7h44'16" 14d26'12"
Brady's Glow
Uma 9h39'38" 55d16'44"
Brady's Star
Umi 13h11'11" 71d58'5"
Brady's Star
Uma 8h26'21" 62d21'34"
Brae Bella Morgan
Gem 7h36'3" 21d57'25"
Braebury Homes
Ori 6h18'37" 14d54'4"
Braeden Aroko Rabah
Cap 21h15'41" -25d37'56"
Braeden Edward Cabral "Brady"
Umi 16h17'57" 76d22'24"
Braeden Gregory Murphy 7lbs. 8oz.
Sgr 17h51'50" -29d19'1"
Braeden Layne Boulden
Psc 1h50'19" 5d31'29"
Braeden Michael Allan Boyd
Her 16h36'55" 34d21'25"
Braeden Nicholas Kleinschmit
Gem 6h51'4" 29d10'34"
Braeden Terence Seculer
Her 16h17'55" 26d2'45"
Braeden Thomas
Ori 6h18'57" 16d47'17"
Braeden Thomas Holland
Gem 7h53'12" 16d35'23"
Braeden William Clay
Uma 9h1'48" 60d34'40"
Braedon Alexander Bartoces
Vir 12h0'57" -5d17'17"
Braedyn Jack Lusch
Cnc 8h53'8" 11d39'56"
Braedyn Patrick Politte
Per 4h46'17" 41d34'12"
Braelyn Duerkop
Cnv 13h54'59" 32d22'18"
Brags
Ori 6h17'32" 14d56'9"
Brahea
Cyg 20h32'29" 48d9'2"
Braiden Dean Swanson
Uma 13h22'6" 54d34'27"
Brain Christopher Jordan
Cap 20h21'26" -13d37'4"
Braken
Uma 9h43'31" 59d49'52"
BraLanyndonaljef - Julickon
Uma 9h58'37" 64d31'35"
Bralex
Dra 19h53'56" 70d29'27"
Bralfucious
Cas 0h47'4" 60d39'59"
Bralie
Per 4h45'14" 44d59'53"
Bram Alexander Broderick
Aqr 21h35'49" 2d5'13"
Bram Bouduin
Lib 14h55'0" -24d35'57"
Bram Bullard
Sgr 19h27'0" -15d15'26"
Bramble Dale
Peg 23h39'44" 18d9'44"
Branar
Peg 23h51'47" 17d31'17"
Branchwater Farm
Aqr 22h37'0" -1d49'23"
Branda and Cameron
Cnc 8h12'35" 28d42'28"
Branda Powell
Lyn 8h17'16" 59d24'30"
Brandan Boyd Magennis
Umi 17h34'26" 86d52'44"
Brandan David Horner
Sgr 18h49'14" -26d11'58"
Brandan Edward Wilkins
Gem 7h4'12" 17d11'6"
Brandan Ryan
Psc 1h16'36" 28d5'51"
Brandan Sullivan
And 1h1'27" 46d38'34"
Brande Erin Akers
Dra 19h39'59" 62d5'35"
Brandee
Uma 11h31'13" 54d26'33"
Brandee
Tau 3h39'35" 8d7'27"
Brandee 2005
Cas 1h34'32" 60d18'50"
Brandee Leger
And 0h53'44" 42d3'22"
Brandee-Lashawn
Cam 4h51'12" 60d13'38"
Brandell Family Empire
Uma 10h33'28" 54d45'46"
Brandelyn Rebecca Tate
Lyn 7h59'44" 57d55'19"
Brandelyn Renne
Aql 19h13'0" 16d1'41"
Branden
Aqr 21h14'58" 2d0'31"
Branden and Amanda Enright Forever
Cyg 20h59'35" 46d11'4"
Branden Been
Cnc 8h47'13" 27d54'56"
Branden Brewer
Sco 16h9'20" -14d47'28"
Branden Clawson
Cnc 8h41'23" 15d7'8"
Branden K. Griffin
Lib 15h52'50" -18d9'33"
Branden M. Jenkins
Her 16h37'23" 36d53'58"
Branden Michael Balentine
Her 17h6'25" 30d53'22"
Branden Petersen
Ori 5h44'43" -0d10'27"
Branden Robert Brayshaw
Per 4h24'0" 47d7'54"
Branden Un'Rai Bivins
Dra 18h3'51" 79d7'45"
Branden Valentino D'Cruz
Uma 11h56'58" 39d54'36"
Brandenburg
Her 17h6'54" 31d58'1"
Brandenstein, Gerhard
Uma 9h56'9" 54d47'57"
Brandi
Cas 0h50'40" 65d53'41"
Brandi
Uma 12h5'57" 60d14'4"
B-randi
Aqr 22h56'0" -12d22'13"
Brandi
Tri 2h37'2" 31d35'52"
Brandi
Lyn 9h42'30" 40d14'13"
Brandi
And 2h22'45" 49d12'36"
Brandi
Gem 6h28'29" 22d13'21"
Brandi
Gem 7h17'37" 23d47'32"
Brandi
Psc 23h17'2" 2d1'3"
Brandi
Ori 5h4'53" 3d55'30"
Brandi
Leo 10h39'28" 14d47'0"
Brandi 21
Lib 14h31'42" -17d47'16"
Brandi // Ad Astra Per Aspera
Sgr 19h8'15" -14d30'37"
Brandi Albert
Dra 17h33'53" 58d31'0"
Brandi and Cory's Star 143 forever
Sco 16h11'10" -10d16'54"
Brandi and Gary
And 23h14'4" 44d36'19"
Brandi and Terry King
Cnc 8h13'46" 18d9'16"
Brandi B.
And 0h50'45" 39d12'31"
Brandi Baker you're my shining star
Lib 15h8'7" -0d44'54"
Brandi Barr "Kupkake"
And 1h5'8" 41d12'32"
Brandi Brown
Cas 0h4'48" 57d7'39"
Brandi Buccieri Shepperack
Peg 22h58'52" 26d46'49"
Brandi Burchett
Sco 16h47'36" -27d21'53"
Brandi Carlile
Gem 7h16'30" 15d58'22"
Brandi Day Norris
And 2h33'33" 45d22'59"
Brandi Eetaow Agape
Pho 23h58'30" -46d7'21"
Brandi & Elias
Cyg 19h49'34" 39d7'47"
Brandi Faith Rodriguez
Uma 12h46'45" 62d23'16"
Brandi Gase
Vir 13h46'49" -5d39'52"
Brandi Gean Eisenmann
Cap 21h14'58" -25d9'21"
Brandi Gollihue
And 23h53'30" 40d38'48"
Brandi Golojuch
Psc 0h37'46" 3d57'10"
Brandi H
Uma 10h37'37" 60d50'27"
Brandi Heller
And 23h40'51" 48d14'19"
Brandi "I Love You"
Uma 11h40'46" 42d24'54"
Brandi I. Sawicki
Tau 4h6'5" 28d15'24"
Brandi & Jeremiah Arnold
Cyg 19h55'34" 38d43'28"
Brandi Kay Edgmon
Gem 7h40'56" 16d36'55"
Brandi Kay Nix
Vir 14h3'31" -15d23'43"
Brandi Kaye Arnold
Mon 6h48'43" -0d18'36"
Brandi L Coll
Cnc 8h58'14" 11d38'46"
Brandi Lealani
Uma 10h22'49" 58d0'20"
Brandi Lee
Lib 15h32'3" -6d7'47"
Brandi Lee
Cap 20h32'16" -9d37'5"
Brandi Lee
Cap 21h45'38" -10d20'4"
Brandi Lee Kunselman
Cnc 8h16'57" 32d7'21"
Brandi Leigh
Sgr 18h17'7" -26d17'20"
Brandi Leigh Jones
And 1h31'3" 35d16'58"
Brandi Lockhart My Penguin
Crb 15h35'58" 28d12'10"
Brandi Lyn
Leo 11h57'51" 22d15'29"
Brandi Lyn
Uma 11h22'26" 44d58'22"
Brandi Lynn
And 1h40'2" 38d26'0"
Brandi Lynn
Ari 2h39'25" 25d5'40"
Brandi Lynn Carman
Dra 15h37'43" 55d56'58"
Brandi Lynn Chambers
Lib 14h50'11" -16d8'31"
Brandi Lynn Eissinger
Sco 17h30'4" -31d6'55"
Brandi Lynn Housh
And 23h10'32" 52d21'19"
Brandi Lynn Rick
Vir 14h27'32" 4d41'24"
Brandi Lynn Thompson
Cnc 9h8'21" 25d15'11"
Brandi Lynn Tucker
Ari 2h56'10" 25d18'57"
Brandi Lynne
And 2h36'55" 43d30'2"
Brandi Lynne Williams
And 23h16'13" 51d48'47"
Brandi M. Anderson Christmas Star
Del 20h51'28" 10d59'10"
Brandi M. Steele
Sco 16h9'58" -8d44'58"
Brandi Maier's Star
Cas 23h57'31" 57d22'29"
Brandi Marie
Lyn 6h51'55" 53d34'16"
Brandi Marie
Tau 5h52'5" 13d49'21"
Brandi Marie Davis
Tau 5h47'42" 17d24'0"
Brandi McCullough
Lac 22h42'39" 49d28'6"
Brandi Michelle
Psc 1h23'28" 16d46'21"
Brandi Michelle Beeson
And 0h25'47" 28d40'11"
Brandi Montgomery
Psc 0h44'38" 16d1'56"
Brandi My Sweet Shining Star Always
Cas 0h33'27" 64d5'25"
Brandi N David 7/24/2005
Leo 11h14'13" 24d16'9"
Brandi Nichole Getz "Ladybug"
Lyn 6h29'54" 56d32'52"
Brandi Nichole Thomas
Aql 19h50'29" 0d22'19"
Brandi Nichole Thomas
Aql 19h46'40" 0d8'10"
Brandi NiCole
Sgr 18h47'51" -16d42'37"
BRANDI NICOLE
Sgr 18h5'21" -29d48'26"
Brandi Nicole Hayes
Cap 20h58'4" -21d2'22"
Brandi Nicole Lugo
Cnc 9h17'55" 10d44'3"
Brandi Nicole Pietras
Ari 2h18'34" 14d1'46"
Brandi R. Gwinn
Tau 3h44'51" 28d49'19"
Brandi Rae
And 0h18'52" 40d45'4"
Brandi Redfern
Aqr 21h41'53" 2d37'28"
Brandi Rene' Little
Aqr 23h20'3" -13d13'51"
Brandi Rene Long
Psc 0h24'25" 18d20'29"
Brandi Renee Reynolds
Sco 16h46'26" -40d34'48"
~ Brandi Ruckel ~
Uma 11h37'2" 42d13'47"
Brandi S
Crb 16h12'31" 36d59'24"
Brandi 'Second Star to the Right'
Cap 20h29'43" -24d20'9"
Brandi Shanee
Aqr 23h34'20" -22d28'41"
Brandi Sky
Sco 17h53'48" -36d46'44"
Brandi "Squaw"
Lyr 18h28'16" 28d17'12"
Brandi Starr
Cnc 8h44'9" 15d35'25"
Brandi Stuart
Del 20h45'52" 13d35'17"
Brandi Sue
Cnc 8h20'43" 9d24'11"
Brandi Sugihara
Ori 5h35'38" -1d55'3"
Brandi Wagener
Leo 10h24'13" 20d1'16"
Brandi Wood
Sgr 19h13'25" -34d7'30"
Brandie
Aql 19h30'31" 14d55'4"
Brandie
Ori 5h20'6" 1d32'15"
Brandie
And 0h56'35" 39d44'47"
Brandie A. Shrader
Cam 5h48'20" 60d3'23"
Brandie D. Viars
Cnc 8h31'42" 16d0'19"
Brandie Gowen
Mon 6h32'58" 1d56'44"
Brandie Lee
Lib 14h52'56" -2d46'57"
Brandie Nichole Davis
Lib 15h45'57" -26d3'31"
Brandie Nicole Tyndall
Tau 4h35'43" 27d12'56"
Brandie Nicole Wood
Tau 5h5'33" 22d20'57"
Brandie Noel Bernadette
Vir 14h4'44" -16d12'56"
Brandiholland0428
And 1h26'5" 36d48'59"
Brandilyn Rachelle Bailey
Ori 5h23'14" 0d39'24"
Brandilynn Keene
Cnc 8h58'43" 11d48'20"
BrandiLynnSweetnessBayHokulani7191
Ari 3h1'38" 18d28'56"
Brandi's Beaglegeuse
Sgr 18h52'51" -23d41'14"
Brandi's Brilliance
Dra 18h49'52" 57d57'9"
Brandi's Star
Leo 11h33'46" 13d51'59"
Brando
Cap 20h20'10" -9d21'32"
Brandoggie
Pic 5h38'25" -50d59'10"
Brandol Major
Ari 2h17'2" 23d58'17"
Brandolyn
And 0h14'41" 45d35'25"
Brandon
Her 17h32'55" 40d43'12"
Brandon
Leo 10h5'59" 13d51'2"
Brandon
Psc 0h42'3" 9d41'10"
Brandon
Lib 14h50'11" -0d31'51"
Brandon
Cet 1h16'35" -7d16'13"
Brandon
Cep 20h42'56" 61d27'39"
Brandon A. Lou Athens Academy 2007
Dra 19h26'33" 64d17'49"
Brandon Adler Yoder
Dra 19h14'39" 70d42'17"
Brandon Alexander Cusick
Ari 2h5'5" 25d16'20"
Brandon Alexander Rodney
Boo 14h47'7" 19d53'1"
Brandon Alexander Schwab
Her 17h38'11" 44d29'43"
Brandon Allen
Uma 12h1'51" 36d52'37"
Brandon Allen Johnson
Peg 21h41'31" 21d4'2"
Brandon Allen Moore
Lib 15h25'41" -16d27'1"
Brandon and Amanda Barr
Crb 16h21'11" 32d53'30"
Brandon And Brit always & forever
Sge 19h51'55" 17d20'7"
Brandon and Chelsea - Forever
Aqr 22h36'36" -3d41'47"
Brandon and Cheyne
Cyg 21h15'9" 37d11'0"
Brandon and Dru's Star
Umi 14h16'48" 67d33'15"
Brandon and Katharine
Leo 11h31'49" 1d15'39"
Brandon and Katrina forever in love
Cyg 19h47'47" 34d37'35"
Brandon and Kimberley Link
Uma 13h14'50" 56d35'14"
Brandon and Leslie
Umi 15h23'22" 75d19'59"
Brandon and Mandy Johnson
Cyg 21h46'38" 42d59'1"
Brandon and Michele Clark
Cyg 19h38'35" 29d30'51"
Brandon and Molly Kujawski
Cyg 19h49'31" 43d21'24"
Brandon and Sarah Norton
Umi 14h16'58" 68d31'50"
Brandon and Tracy
Cyg 20h4'13" 33d51'3"
Brandon and Trisha's Star
Lib 15h35'56" -4d47'9"
Brandon Andrew
Lib 15h4'10" -9d22'44"
Brandon Andrew Welty
Cap 21h51'35" -23d27'24"
Brandon Andrew Whyte
Her 17h32'45" 47d9'38"
Brandon & Ashlee
And 1h5'49" 38d11'9"
Brandon B Loves Elizabeth A B
Sgr 17h54'56" -28d52'11"
Brandon Ball
Aql 19h30'6" 14d8'12"
Brandon Barker
Cnv 13h51'57" 30d18'16"
Brandon Begg
Aqr 22h45'39" -15d10'31"
Brandon Bellamy
Ori 6h14'23" 6d17'37"

Brandon Bernard Krock
Her 16h28'49" 21d25'18"
Brandon Bertges
Cam 3h20'15" 61d4'43"
Brandon Billups
Cyg 19h49'10" 51d47'40"
Brandon Box
And 1h8'41" 45d36'44"
Brandon Boyd's Nirvana
Crb 16h11'28" 37d30'42"
Brandon Branch
Lac 22h17'46" 54d40'50"
Brandon Bushwitz
Boo 14h35'20" 34d37'6"
Brandon Callaway
Aur 5h42'10" 40d28'2"
Brandon Cannon
Boo 15h9'7" 33d44'25"
Brandon Carrell Woods
Aql 19h11'17" 3d16'24"
Brandon Carson Love
Sco 17h57'21" -30d47'16"
Brandon Cervantes
Ori 4h50'8" 2d52'58"
Brandon Charles
Gottesman
Leo 11h32'18" 16d27'19"
Brandon Charles Jones
Ari 2h16'20" 25d9'4"
Brandon Cochran
"Motorcross Star"
Lib 15h3'39" -22d16'33"
Brandon Cole Chaffee
Aql 19h16'12" 5d26'51"
Brandon Colton Merz
Cap 21h13'8" -23d10'28"
Brandon Corey Waits
Uma 10h28'2" 44d50'30"
Brandon Cox Owen
Cyg 19h17'6" 51d21'46"
Brandon D. Knight
Gem 7h2'22" 34d39'7"
Brandon D. Saltiel
Aqr 21h49'34" -6d5'5"
Brandon Damon Davis
Her 16h37'53" 47d18'56"
Brandon Davis
Dra 15h32'48" 61d55'45"
Brandon Dean
Boo 14h34'13" 37d15'22"
Brandon Dorholt
Her 17h39'47" 22d18'29"
Brandon Dougherty
Ori 5h35'55" 9d19'55"
Brandon E Impens
Ori 5h55'57" -0d47'51"
Brandon E. Zobisch
Lib 15h13'43" -6d31'28"
Brandon Edward Brondt
Lyn 8h29'44" 33d12'1"
Brandon Ferguson
Gem 7h40'38" 15d55'29"
Brandon Ferral
Uma 11h6'6" 50d29'51"
Brandon Frank Huang
Her 17h49'1" 47d33'5"
Brandon Gibbons
Uma 9h5'37" 47d8'25"
Brandon + Gillian
Her 18h32'34" 19d5'10"
Brandon Glidewell
Uma 8h12'25" 60d52'10"
Brandon Guy Cheek
Sgr 19h9'15" -18d39'51"
Brandon Halasz
Vir 13h59'22" -16d7'23"
Brandon & Hannah
Her 18h3'25" 36d6'42"
Brandon Harris
Lyn 7h22'21" 48d58'16"
Brandon Heath Lester
Pup 7h52'50" -16d13'38"
Brandon Higgins
Cap 21h37'41" -9d11'37"
Brandon Hodgkins
Ori 6h19'20" 8d14'38"
Brandon Hoff
Her 17h41'24" 45d30'45"
Brandon Hoffmann
Ari 3h22'32" 27d33'13"
Brandon Hotye
Lib 15h23'14" -6d20'43"
Brandon I. Chavez
Umi 11h26'19" 87d53'13"
Brandon J. Schwartz
Umi 15h57'30" 75d24'46"
Brandon Jacob
Dra 20h26'13" 67d10'56"
Brandon Jacob Feldman
And 23h36'49" 47d20'42"
Brandon James Coulter
Ori 5h58'12" 17d39'15"
Brandon James De Cicco
Dra 17h39'20" 65d5'25"
Brandon James Hogan
Vir 14h6'2" 2d56'2"
Brandon James Lindley
09/30/75
Lib 15h9'54" -10d20'46"
Brandon James Parfitt
Her 18h50'57" 25d50'11"
Brandon James Ranalli
Her 17h39'19" 24d37'51"
Brandon James Strout
Aqr 22h36'6" -1d35'48"
Brandon James Van Zoest
Cas 0h36'33" 57d29'28"
Brandon James Whitaker
Aur 5h41'30" 44d16'21"
Brandon Jayce Benton
Cnc 8h19'23" 11d54'21"
Brandon & Jennifer's
Destiny
Aqr 22h7'45" -16d47'12"
Brandon Joel
Ori 5h18'54" 12d36'45"
Brandon & JoHanna 4
Eternity
Cyg 21h21'47" 51d51'26"
Brandon Joseph
Ari 3h5'33" 27d43'14"
Brandon Joseph and
Grandma Pischl
Uma 9h19'28" 48d37'37"
Brandon Joseph Bilbrey
Uma 10h17'25" 42d1'39"
Brandon Joseph Carter
Uma 8h42'30" 60d19'8"
Brandon Joseph Dougherty
Uma 12h5'35" 51d59'31"
Brandon Joseph Nave
Cnc 8h53'0" 30d35'3"
Brandon Joseph Rippl
Umi 15h21'48" 76d10'10"
Brandon Joseph Rudy
Her 17h37'6" 37d4'56"
Brandon Joseph Steele
Her 17h19'49" 36d23'3"
Brandon Joseph Tiller
Psc 0h58'41" 9d53'51"
Brandon Julian
Leo 9h57'28" 13d27'15"
Brandon Junior 1
Ori 6h18'39" 9d53'54"
Brandon K. Nash
Leo 11h9'9" 10d2'28"
Brandon Karolick
Psc 0h25'44" 17d9'40"
Brandon Keith Turner
Ari 2h0'58" 22d32'13"
Brandon & Kelly's
Engagement Star
Psc 1h22'39" 32d56'52"
Brandon & Kristen Hall
Cyg 20h15'56" 56d36'35"
Brandon L. Killingsworth
Cyg 20h55'6" 44d42'53"
Brandon Lamar Titus
Uma 8h17'54" 61d4'0"
Brandon Lane Boudreau
Uma 9h47'21" 64d32'58"
Brandon Langlois
Uma 12h44'51" 59d55'11"
Brandon Lapre My Soul
Mate
Pho 0h38'40" -42d31'52"
Brandon Lavendol
Lyn 8h53'38" 37d41'37"
Brandon Laverne Crusha
Uma 11h51'3" 51d52'7"
Brandon Lee
Sgr 17h53'2" -24d57'20"
Brandon Lee Barfield's Star
Per 3h8'43" 37d43'4"
Brandon Lee Ellis
Cap 20h35'43" -19d50'19"
Brandon Lee Evans
Cam 5h32'39" 66d20'50"
Brandon Lee Garrett
Cas 23h1'19" 54d46'16"
Brandon Lee Kusner
Sgr 18h30'10" -33d15'20"
Brandon Lee Patrick Pflanz
Ori 6h5'3" 13d27'52"
Brandon Lee Thome
Tau 4h23'59" 23d41'9"
Brandon Leigh Hillman
Cyg 21h49'15" 39d13'38"
Brandon Leprail Campbell
Her 17h14'58" 36d8'8"
Brandon Lewis Spears
Ori 5h31'54" 1d48'47"
Brandon & Linda Harris
Lac 22h16'31" 37d29'47"
Brandon Little
Her 17h39'43" 37d43'17"
Brandon Long
Her 18h12'18" 21d25'18"
Brandon Louis
Tri 2h29'3" 36d26'50"
Brandon Loves Hollie
Umi 15h26'46" 73d25'15"
Brandon Lowell Covey
Cnc 8h47'7" 15d16'51"
Brandon Luis Sanchez
Cnc 8h49'12" 30d24'28"
Brandon M. Ruffner
Peg 22h41'8" 32d10'11"
Brandon M. Taber
Dra 16h19'2" 66d57'36"
Brandon M Zadzilka
Ari 2h20'20" 25d46'29"
Brandon Malen Ross
Umi 15h21'6" 74d32'15"
Brandon Mark Whitfield
Aqr 21h28'55" -4d29'47"
Brandon Mason
Cep 20h47'54" 68d3'38"
Brandon Matthew
Gem 6h20'16" 26d38'43"
Brandon Maxwell
McDonough
Cap 20h14'29" -18d8'20"
Brandon McCulloch
Vir 13h29'26" -0d4'44"
Brandon McLaughlin
Cas 1h24'2" 58d40'43"
Brandon Melchior
Her 17h39'35" 32d56'29"
Brandon Michael
Per 4h24'45" 44d9'14"
Brandon Michael
Her 16h45'1" 8d50'7"
Brandon Michael
Her 16h38'0" 9d52'13"
Brandon Michael
Uma 9h54'0" 62d38'9"
Brandon Michael Eberly
Sco 16h13'56" -18d0'55"
Brandon Michael Lewis
Gem 7h48'2" 27d24'25"
Brandon Michael Newton
Her 16h21'26" 43d11'13"
Brandon Michael Piros
Vir 13h50'47" 2d18'24"
Brandon Michael Pleman
Lyn 7h2'11" 45d37'25"
Brandon Michael
Richardson
Lmi 10h55'19" 26d57'26"
Brandon Michael Williams
Dra 16h12'14" 61d28'5"
Brandon Michael Wren
Dra 16h12'49" 67d14'2"
Brandon Michael Wright /
B.M.W.
Psc 1h17'57" 32d35'13"
Brandon Miller Estes
Psc 0h44'58" 5d24'42"
Brandon Mills
Leo 10h57'20" -1d37'28"
Brandon Mungai Phillips
Her 17h57'15" 50d5'3"
Brandon Nash
Vir 13h20'12" 9d11'17"
Brandon Nehorai Elghanian
Uma 13h12'7" 58d32'20"
Brandon Nein ~ My
Valentine 2007
Psc 23h29'43" 6d57'45"
Brandon Nero
Peg 21h45'7" 6d41'56"
Brandon & Nikki Knull "I
Love You"
Cyg 20h26'9" 47d4'49"
Brandon Noble
Cap 21h42'3" -15d0'44"
Brandon Nyte
Cnc 8h28'57" 21d2'17"
Brandon Orsborn
Gem 6h37'55" 24d58'49"
Brandon Ottke
Her 17h11'27" 35d27'38"
Brandon Paci
Cnc 9h9'38" 32d0'52"
Brandon Palazzotto
Umi 15h36'46" 83d58'56"
Brandon Paul Anderson
Dra 20h25'23" 67d12'28"
Brandon Philip Gheen
Lib 14h52'17" -3d11'43"
Brandon Pitchess
Psc 0h59'34" 12d8'39"
Brandon Polley
Per 3h0'53" 47d15'13"
Brandon "Poochie" Burns
Lyn 7h43'36" 54d12'24"
Brandon R. Hendrickson
Ori 6h2'51" 20d9'34"
Brandon R. Kessler
Her 16h22'1" 23d20'6"
Brandon Ramiro
Lmi 10h16'39" 28d32'43"
Brandon Ray
Sgr 18h3'26" -27d57'4"
Brandon Richard Krapes
Her 16h30'23" 45d15'21"
Brandon Robert
Her 17h36'54" 15d56'42"
Brandon Robert Kucera
Ari 2h11'18" 24d26'8"
Brandon Robert Ratay
Her 17h23'53" 31d55'56"
Brandon Robert Roeder
Cnc 8h15'22" 11d14'49"
Brandon Robert Wheeler
Cru 12h42'31" -60d28'8"
Brandon Robinson
Vir 12h59'27" 10d47'39"
Brandon Ross Taylor
Lyn 7h30'42" 56d57'56"
Brandon Salvadore Filardo
Uma 11h57'33" 29d45'59"
Brandon & Samantha Carl
Aqr 22h36'50" -1d44'49"
Brandon & Sarah
5/10/2003
Cyg 20h24'1" 33d37'49"
Brandon Scott
Sco 17h22'6" -33d9'54"
Brandon Scott Bruno
Gem 7h46'48" 34d38'51"
Brandon Scott Campbell
Cap 20h14'1" -10d40'43"
Brandon Scott Crosby
Ari 3h19'37" 23d22'16"
Brandon Scott Crossley
Leo 11h21'13" 6d18'17"
Brandon Scott Morehead
Lib 14h49'44" -1d39'58"
Brandon Scott Robinson
Boo 15h25'30" 40d23'41"
Brandon Scott Sensky
Cyg 19h52'13" 38d33'11"
Brandon Scott Smith
Tau 4h42'23" 23d31'7"
Brandon Sean Patrick
Cook-Jewett
Uma 11h59'56" 59d37'52"
Brandon Shane Johnson
Her 17h45'17" 44d40'28"
Brandon Shane Tennimon
Psc 2h0'13" 5d49'2"
Brandon Skye
Uma 8h20'36" 69d32'23"
Brandon Skyler Hirsch
Vir 14h23'3" 3d56'19"
Brandon Smith
Ser 18h21'46" -13d40'4"
Brandon Statzula
Her 16h42'12" 48d40'56"
Brandon & Stephanie
Baxter
Cyg 19h40'49" 35d52'33"
Brandon Stephen Leon
Psc 1h23'35" 27d32'6"
Brandon Steven Harper
Ori 5h24'51" 3d41'55"
Brandon Stevens
Per 3h18'25" 47d27'9"
Brandon Stuart Holck
Her 16h10'15" 47d26'13"
Brandon T. LeMaire
Psc 1h18'30" 10d31'1"
Brandon T. Opela
Cyg 19h56'22" 56d35'35"
Brandon (Tater) Ward
Tau 3h35'6" 3d51'10"
Brandon The Pumpkinking
Ori 4h58'42" -0d19'33"
Brandon Thomas
Dra 16h38'40" 53d22'10"
Brandon Thomas Jagoe
Cyg 19h44'13" 52d6'58"
Brandon Thomas Patty
Cep 22h55'7" 57d16'29"
Brandon Timothy Schulte
Lib 14h22'29" -19d37'33"
Brandon & Tracey Durham
Umi 15h33'57" 74d18'22"
Brandon (Trish's Freebird)
Ari 2h44'56" 15d28'48"
Brandon Tyler Burchard
Cap 21h38'15" -14d0'45"
Brandon Victor Haist
Umi 16h56'28" 77d36'3"
Brandon W. Schmitt
Sgr 17h50'21" -17d21'30"
Brandon Wade Carruthers
Aur 5h52'39" 46d53'0"
Brandon Wade Hooter
Ari 3h5'3" 16d0'57"
Brandon Walsh
Her 18h4'7" 36d18'29"
Brandon Wayne Keeth
Uma 10h11'36" 64d30'28"
Brandon Wayne Robison
Boo 14h17'56" 19d4'41"
Brandon Zimmerman
Umi 14h23'21" 77d47'44"
Brandonia
Uma 10h12'7" 45d13'6"
Brandon's Destiny
Lib 15h19'31" -5d41'5"
Brandon's Diamond In The
Sky
Cap 21h46'29" -19d5'45"
Brandon's Love
Gem 6h29'13" 21d8'18"
Brandon's Star
Leo 11h6'3" 22d17'0"
Brandon's Star
Her 18h10'39" 23d21'26"
Brandon's Star
Her 17h38'19" 32d31'5"
Brandons very own star.
BVOS1
Cnc 8h2'11" 25d39'5"
Brandt Andrew Dettling
Cap 20h25'3" -13d8'48"
Brandt de Ebert, Barbara
Tau 3h43'23" 29d28'34"
Brandt Michael Koehler
Uma 8h15'2" 59d50'42"
Brandt & Rachel-Love for
Eternity
Leo 9h56'44" 16d31'21"
Brandt Roberts
Lyn 8h8'12" 53d33'38"
Brandt V. Hilsen
Her 18h3'10" 31d53'36"
Brandt Wise
Aql 19h8'22" 7d29'15"
Brandula
Psc 1h24'17" 26d5'59"
Brandy
Gem 6h25'0" 23d48'7"
Brandy
Peg 22h42'32" 23d31'12"
Brandy
Aqr 21h54'5" 1d2'46"
Brandy
Ori 5h21'32" 6d56'12"
Brandy
Gem 6h54'21" 12d59'23"
Brandy
Cyg 19h39'14" 34d18'17"
Brandy
Crb 16h7'45" 35d29'6"
Brandy
Cnv 12h39'48" 44d13'40"
Brandy
Cyg 21h18'4" 43d17'41"
Brandy
Uma 11h25'25" 54d22'20"
Brandy
Cap 21h53'24" -18d34'1"
Brandy
Lib 15h5'40" -19d53'47"
Brandy
Sco 17h54'22" -32d19'14"
Brandy A. Montoya-Gurule
Cyg 20h26'42" 52d48'41"
Brandy Alexandria Corrales
Cmi 7h28'9" 6d55'59"
Brandy and Charles
Umi 15h23'44" 75d43'46"
Brandy Ann
Ori 5h40'54" 7d51'28"
Brandy Ann Brown
Ori 5h13'13" -0d48'58"
Brandy Ann Luoma
Ori 6h5'32" 17d16'50"
Brandy Arnold
Cnc 9h12'57" 31d55'30"
Brandy Baby
Ari 2h20'12" 18d2'59"
Brandy Baye Robidoux
Uma 9h35'47" 43d0'45"
Brandy Boswell
And 23h36'34" 42d33'34"
Brandy "Buddy" Beerman
Lib 14h53'7" -5d51'0"
Brandy C. Norris
Cyg 20h45'30" 38d6'14"
Brandy Cain
Lib 15h28'3" -25d51'30"
Brandy Danielle Stanfill
Vir 13h19'27" 13d6'56"
Brandy Dawn Hendricks
Crb 15h52'50" 26d0'16"
Brandy DeMoss - Rising
Star
Umi 16h24'29" 80d38'23"
Brandy Elizabeth Tucker
And 0h44'33" 36d26'59"
Brandy Facey
Vir 13h59'21" -0d7'8"
Brandy Freeman
Lyn 7h57'54" 38d19'20"
Brandy Gail
Ari 2h17'16" 24d2'32"
Brandy Jill Cossman,my
gift from GOD
Mon 7h36'4" -0d20'27"
Brandy Jo Bateman
Lyn 9h5'58" 35d47'2"
Brandy Joanne Durst
Ari 2h5'33" 26d19'33"
Brandy K. Herbert
Ori 5h16'53" 7d49'51"
Brandy Kay Baker
Mon 6h31'53" 4d57'6"
Brandy Kitten Dyer
Ori 5h21'9" 5d43'44"
Brandy L Campbell
Umi 14h36'5" 75d11'40"
Brandy Lawry
Aql 19h11'0" 5d16'57"
Brandy Lee Bentley
Wheatley
Sgr 19h32'53" -38d56'57"
Brandy Lee Matteson
Cyg 21h42'46" 54d35'13"
Brandy Lovelace
Aur 5h44'2" 44d19'29"
Brandy Loves Ronald:
Happy Birthday
Leo 11h0'59" 1d52'6"
Brandy Lyn Mayton
Sco 15h51'50" -20d56'48"
Brandy Lynn
Gem 7h29'10" 34d26'2"
Brandy Lynn Alexander
Crb 15h39'14" 36d54'46"
Brandy Lynn Anderson
Uma 11h58'47" 57d7'49"
Brandy Lynn Budd
Leo 10h12'32" 16d54'4"
Brandy M. Passmore
Lyn 9h5'34" 42d5'15"
Brandy Mae Halm
Tau 4h41'16" 3d26'41"
Brandy Marie
Lyn 9h11'28" 36d58'9"
Brandy & Marjorie
Uma 10h46'49" 60d6'37"
Brandy Martinez
Vir 14h1'21" -15d3'11"
Brandy McCabe
Cma 6h44'55" -13d47'3"
Brandy Meadows
Leo 10h50'6" 14d28'56"
Brandy & Michael
And 23h23'27" 35d18'47"
Brandy Miles
Sco 17h48'38" -39d5'36"
Brandy Moyer
Lmi 10h58'17" 29d56'13"
Brandy Muffy Andregg
Cma 6h22'50" -17d39'11"
Brandy Necol
Cyg 21h49'29" 45d11'16"
Brandy Nichole
Sco 16h12'44" -20d22'9"
Brandy Nichole Hooks
Connelly
Psc 0h12'11" 5d55'50"
Brandy Nicole
Gem 6h38'6" 18d40'50"
Brandy Parker
Aqr 22h23'51" -6d3'12"
Brandy Parnell
Tau 4h29'50" 18d42'43"
Brandy Peeples
Ori 5h23'37" 3d4'38"
Brandy R. Wolfe
And 0h50'56" 45d48'54"
Brandy Rachelle
And 0h33'18" 38d55'36"
Brandy Rose
Ori 5h58'7" 6d33'42"
Brandy Rose Roberts
Aqr 20h52'44" 0d23'46"
Brandy Shank
Gem 7h30'44" 16d25'59"
Brandy Skye Gregart
Aqr 22h37'7" 1d23'8"
Brandy SuperStar Rice
Lyr 18h21'32" 32d50'43"
Brandy Temple
Tau 5h11'37" 18d1'21"
Brandy The Love of My Life
Umi 15h22'3" 71d13'31"
Brandy "The Love of My
Life"
Psc 1h17'23" 4d41'52"
Brandy Thorne
Aqr 20h55'18" -11d50'5"
Brandy Willard
Peg 22h37'59" 12d43'59"
Brandy, Hendrix and
Grampy Woo Woo
Tau 5h55'36" 23d34'21"
Brandye's
Leo 11h34'35" 20d54'21"
BrandyLyn Ann Asher
Aqr 22h36'8" -2d30'2"
brandymelissafilpi
Sco 17h18'18" -43d12'20"
Brandyn
Cru 12h22'9" -61d48'27"
Brandyn Leigh
Gem 7h48'9" 24d34'33"
Brandyn M. Leeson
Uma 9h55'14" 55d45'37"
Brandyn Miller
Uma 12h38'29" 60d18'24"
Brandy's Hideaway
Mon 6h52'59" -0d4'38"
Brandy's Star
Ori 5h8'36" 15d45'14"
Brandy's Star
Ori 5h54'22" 6d37'13"
Brandy's Star
Ari 2h44'54" 30d47'49"
Branelle
Cnv 12h34'53" 45d42'23"
Branif's
Her 17h31'29" 47d47'55"
Branigan & Zunaira
Cyg 21h32'3" 44d39'30"
Branita
Uma 9h27'26" 57d0'26"
Branka
Uma 12h38'31" 58d57'7"
Branka Nevistic
Cas 23h0'23" 58d29'18"
Branko
Aql 19h48'2" -0d40'33"
Branko
Gem 7h5'17" 22d5'59"
Branko Novakovic
Ori 6h21'21" 10d22'50"
Branlin 924
Uma 11h55'37" 29d32'47"
Branna Leigh Palmer
Vir 13h21'12" 2d46'56"
Brannan Jade
Vir 11h38'2" -4d59'55"
Brannen Jennings
Uma 8h48'54" 69d35'45"
Brannigan's Love
Cyg 21h19'58" 41d27'10"
Brannon
Ori 5h25'9" 0d36'7"
Brannon & Dorthea Forever
Sgr 19h23'23" -27d58'27"
Brannon King
Cep 2h1'27" 83d7'13"
Brannon Knox
Cyg 20h8'0" 48d44'37"
Brannon Reid Tucker
Eri 4h20'10" -0d24'55"
Branson Samuel Wilson
Tau 5h49'55" 26d3'6"
Brant Casford
Uma 11h50'4" 32d14'20"
Brant Marvin Michael
Travis
Dra 18h2'47" 73d57'46"
Brantley Eargle
Lmi 10h19'57" 31d11'5"
Branwen
Uma 12h31'55" 61d59'20"
Branzackdon
Lyr 18h41'57" 39d36'26"
Brardy Renee Sryder
Cam 4h8'14" 69d42'53"
brashton
Lyn 7h54'37" 42d6'59"
Bratetic's Perpetual Love
Sgr 18h58'45" -28d18'22"
Bratton, Edith L.-1
Cas 1h28'28" 61d55'44"
Bratton, Kenneth L.-1
Her 17h21'13" 36d55'25"
Bratz, Inge & Günter
Uma 10h16'42" 70d37'50"
Braumlin
Lyn 9h5'1" 37d31'24"
Braun, Manfred
Ori 5h26'38" 13d55'46"
Braune, Peter
Uma 9h10'58" 50d38'16"
Brave & Honorable Marine
Dale Swink
Uma 9h22'20" 46d16'44"
Brave Li'l Soldier Holly J.
McGeogh
Per 3h24'23" 46d53'1"
Brave Saint Margaret
Peg 22h25'29" 4d0'17"
Braveheart
Uma 13h47'30" 53d9'7"
Braveheart Carraway
Cap 20h51'10" -25d42'51"
braveStar
Uma 8h26'45" 62d46'16"
Bravo, Brian!
Ori 5h37'36" 1d20'4"
Brawn
Cep 23h58'52" 74d54'31"
BRAX
Per 3h10'7" 51d15'44"
Braxton
Her 16h39'51" 30d21'24"
Braxton Lee Russell
Ori 5h37'54" -2d39'30"
Braxton Lee Thompson
Umi 19h22'24" 86d58'17"
BRAXTON PAYNE
Her 17h48'58" 38d50'35"
Braxton's Bright Light
Vir 12h37'54" 0d46'14"
Bray Family Star
Umi 16h26'14" 76d46'44"
Bray & Talene Kelly
Wedding Star
Ori 5h37'4" 3d21'20"
Brayden
Umi 14h26'6" 77d5'1"
Brayden
Sgr 19h14'23" -32d19'53"
Brayden Alec Turnmeyer
Uma 13h10'19" 58d15'51"
Brayden Alexander Mikula
Aql 19h59'37" 8d54'36"
Brayden Alston Miller
Cap 21h33'24" -20d55'28"
Brayden Anderson
Schrewe
Cep 21h25'31" 59d40'52"
Brayden Austin Boortz
Uma 8h43'35" 51d12'21"
Brayden Chay Collins
Tau 3h30'52" 22d16'32"
Brayden Christopher
Valentino
Uma 11h13'53" 53d59'23"
Brayden Edward Yarbrough
Aqr 20h50'11" -12d23'42"
Brayden James Tracy
Uma 8h45'48" 65d54'54"
Brayden Lee Galloway
Her 18h11'16" 23d29'43"
Brayden Lee Galloway
Psc 1h21'9" 22d50'36"
Brayden Lee Harvey
Sco 17h51'36" -35d44'48"
Brayden Michael Ray
Her 17h24'56" 16d59'4"
Brayden Rhys Parker - His
Star
Cru 12h28'28" -63d57'7"
Brayden Samuel Cathey
Ori 6h15'34" 10d50'38"
Brayden Scott
Sco 16h13'24" -15d9'0"
Brayden Stephan Rice
Umi 14h26'57" 78d18'23"
Brayden Taylor Hall
Boo 14h57'42" 25d8'2"
Brayden Timothy Newman
Psc 0h52'12" 9d26'10"
Brayden Tines Leary
Sco 16h45'48" -44d29'19"
Brayden Tony Stavreski
Star 9/03/2006
Cru 12h40'14" -63d14'58"
Braydon Alexander Payeur
Ori 6h3'44" 6d49'4"

Braydon Arthur Alario
Cnc 8h2'58" 13d56'18"
Braydon Arthur Kunz
Ori 6h9'54" 14d31'34"
Braydon Brent Hinds
Aql 19h10'29" 6d56'42"
Braydon Loyal Jenks
Lyn 8h16'52" 59d21'52"
Braydon N. Kelman
Sco 16h6'7" -11d38'20"
Braylen Kody
Uma 11h56'40" 34d4'54"
Braylen Stanley Nystrom
Lib 14h53'50" -3d17'42"
Brayson Robert Wusterbarth
Leo 11h23'57" 23d16'16"
Brayton Joseph's Star
Her 17h25'47" 23d26'7"
Brayvon
Her 16h27'30" 18d2'44"
Braz Garland Arellano
Cnc 8h25'26" 13d40'27"
Brazil Fanchon Reynolds
Lib 14h50'44" -0d56'27"
BRB CJG 9.25.04
Uma 9h20'51" 56d31'3"
BRC711
Lyn 6h30'36" 61d20'59"
BRD Partners for Life
Uma 8h26'24" 71d11'56"
Bre
Ari 3h3'6" 14d23'15"
Bre
Lyn 6h55'11" 51d1'9"
Bre' Anna Mokdad
Gem 7h18'55" 19d4'48"
Bre & Mark
Cyg 21h39'15" 38d32'8"
Brea
Lyn 6h21'31" 61d40'30"
BREA
Cap 21h28'47" -15d53'11"
Brea
Sco 16h55'42" -42d38'13"
Brea Danielle Gates
Aqr 20h43'47" -9d4'42"
Brea Kathleen Walker
Crb 16h17'48" 32d54'14"
Brea Mae
Lib 14h42'55" -18d39'34"
Brea Marie Rinchich
Ori 5h15'5" 0d7'8"
Breaenna Celeste
And 1h40'24" 45d11'6"
Brèagha Cridhe
Uma 11h6'31" 67d33'28"
Brealynn
Vir 14h32'32" 3d34'20"
Brean
And 0h13'27" 45d34'44"
Breana
Her 16h20'13" 20d32'52"
Breana
Psc 1h26'50" 5d29'59"
Breana Ann O'Neill
Gem 7h27'37" 17d47'38"
Breana Ashley Matviak
Ori 6h1'58" 21d1'27"
Breana Barnes
Her 16h34'4" 27d3'20"
Breana Marie
And 2h29'40" 49d15'53"
Breana Willison's Light
Vir 14h22'3" -3d20'7"
Breann Carpenter
Leo 9h24'13" 24d24'12"
Breann Jaworowski
Leo 9h48'15" 28d26'19"
Breann Lynn Wallen
And 2h9'21" 43d53'50"
Breann N. Jones
Lmi 10h8'20" 30d44'21"
Breann, The Goddess of my Heart
Ori 6h6'39" 18d40'37"
Breanna
Leo 10h53'54" 19d59'45"
Breanna
Ori 6h19'4" 10d29'24"
Breanna
Crb 16h9'4" 33d7'15"
Breanna
Uma 11h50'49" 46d54'9"
Breanna
Sco 16h13'20" -11d0'25"
Breanna
Cam 3h26'44" 64d3'43"
BREANNA
Uma 11h44'47" 53d42'6"
Breanna (Buggy-Boo)
And 2h36'23" 38d27'17"
Breanna Dawn
Ori 5h13'30" 7d22'24"
Breanna E. Combs
Aql 19h44'42" -0d8'4"
Breanna Elisabeth Kuczynski
Sgr 18h19'39" -31d43'12"
Breanna Harp
Her 16h39'51" 6d20'52"
Breanna Irene
Psc 1h44'46" 5d30'51"
BreAnna & Jay's Star
Uma 9h29'33" 55d58'0"

Breanna Joy Wallis
Cnc 8h33'12" 24d22'35"
Breanna K Wells
Cyg 21h40'53" 47d11'43"
Breanna Kay Balestra
And 1h18'58" 36d9'29"
Breanna Laureen Leidner
And 0h20'8" 30d3'57"
Breanna Layne Anderson
Boo 14h56'13" 30d58'17"
Breanna Lee
Gem 6h4'49" 25d40'47"
Breanna Leigh Whitney
Uma 11h56'40" 37d24'35"
Breanna Littlepage
Sco 16h57'53" -34d34'36"
Breanna Nichole Breckenridge
And 1h19'11" 38d0'51"
Breanna Nichole Whitacre
Tau 3h48'26" 8d9'38"
Breanna Rozelle
And 1h2'43" 36d58'56"
Breanna SugarBottom
Gem 6h41'18" 20d55'10"
Breanna, My One and Only
Leo 10h20'24" 11d32'13"
Breannah
Cra 18h3'56" -37d28'39"
Breanne
Gem 7h40'21" 33d15'52"
Breanne DeBoor
Her 16h15'55" 6d39'27"
Breanne Elizabeth
Uma 11h19'15" 68d38'41"
Breanne forever
Cyg 21h55'8" 48d13'48"
Breanne Lisa Heitshusen
Lib 14h43'10" -10d22'5"
Breanne Ly Sutherland -
And 2h38'29" 44d23'21"
Breanne Marie Barrios
And 0h42'57" 31d31'44"
Breanne Monica Reeves
And 23h35'33" 42d18'51"
Breanne Noel Daws
And 0h25'42" 28d17'41"
Breanne Page
And 1h3'43" 46d15'52"
Breann's Star
And 0h37'16" 30d32'47"
Bre'Ann's Star
Umi 15h30'17" 76d4'3"
Breath Taking Beauty
And 0h39'48" 22d28'20"
Breathe
Uma 11h22'28" 44d57'42"
Breathe
Lac 22h16'8" 37d11'11"
Breauna Rhyanne Butler
And 1h8'31" 34d58'12"
Breawna Elizabeth Sellers
Vul 19h50'38" 22d5'41"
Breck Anthony Blackwell
Aqr 22h33'30" 0d10'22"
Breck Elizabeth
And 2h16'56" 49d16'57"
Brecke Elizabeth Monaghan
Sgr 18h38'13" -17d55'52"
Brecken John Schmidt
Dra 18h42'57" 56d2'37"
breckum
Dra 18h26'43" 75d26'54"
Brecot
Cru 12h57'14" -56d31'24"
Breda Bellissima
And 1h59'34" 42d10'19"
Bree
Uma 9h24'35" 44d42'8"
Bree
Hya 8h15'15" -3d56'42"
Bree Alice King
Sgr 19h32'2" -34d14'55"
Bree Ana
Tau 4h10'54" 5d46'10"
Bree Anna Laws
Uma 9h21'52" 58d30'19"
Bree Callaghan - 7th September 1984
Vir 13h22'50" -17d7'35"
Bree & Chris
And 0h34'26" 38d7'36"
Bree Eaton
Cap 20h21'31" -12d30'42"
Bree Haxell
Cap 20h56'31" -20d14'4"
Bree Loves Stevie
Crb 16h11'6" 34d49'21"
Bree Nicole
Gem 7h53'21" 19d38'13"
Bree Parker McKinstry-Lawson
Psc 1h2'43" 10d41'32"
Bree Smith
Cas 23h4'56" 55d44'1"
Bree Zehnder
Psc 23h15'59" -1d50'47"
Breeana "Breezie" Freil
Uma 11h31'52" 50d33'6"
Breeann
Cam 4h19'24" 63d29'1"
Breeann Ah Sam
Sco 16h11'49" -14d49'26"

Breeann Lanae Gullo
Vir 13h48'22" 3d9'45"
Breeanna
Mon 7h59'31" -8d52'10"
Breeanna Siana-Rose
Tau 4h21'59" 25d37'54"
Breeds
Leo 9h48'2" 28d7'7"
Bree......Eternal Beauty
Leo 11h17'3" -5d8'48"
Breelle Elizabeth
Col 5h38'1" -40d31'4"
Breelynis Majoris
Cnc 8h52'4" 30d53'52"
Breelyn's Birthday Star
Tau 5h15'15" 21d38'10"
Breen
Aql 19h48'2" 4d19'26"
Breen Lyden
Vir 13h11'47" 6d31'37"
Breenna and Patrick *Always*
Sge 20h3'23" 18d31'5"
Breeon
Cmi 7h8'52" 9d23'13"
Breeon
Vir 12h33'15" -9d55'51"
Bree's Happy Place
Aqr 22h3'37" 0d7'3"
Bree's Star
Sco 16h9'22" -14d22'41"
Breeza
Tau 4h27'25" 21d29'6"
Breeze Anna Kramer
Tau 4h35'5" 23d38'3"
Breeze Suozzo
Cma 7h23'58" -14d11'7"
Breeze Weathers
Sco 17h33'24" -40d32'13"
Breezer
Lyn 8h5'51" 40d51'44"
Breezie Dryden
Tau 4h0'26" 17d26'16"
Breezy
Aqr 21h34'26" 2d23'32"
Breezy
Leo 9h56'13" 16d5'35"
Breezy
Sgr 20h4'15" -38d10'0"
Breezy Furlong 581
Vir 13h34'56" 4d51'51"
Breezy Hamilton
Tau 3h47'38" 6d4'32"
Breezy Marie
Leo 11h14'34" -1d42'56"
Breezy's Star
Cas 0h54'12" 55d59'10"
Breghan Marcella
Tau 3h59'4" 18d20'38"
Brehan Clare Furfey
Leo 11h52'23" 19d35'3"
Breheney
Ari 2h33'40" 18d48'58"
Brehme - Andl Star
Cyg 21h36'1" 52d7'1"
Breianons Light
Mon 6h49'6" 7d6'59"
Breina Mae
And 1h8'26" 35d55'22"
Breister/Naud
Her 16h19'57" 42d28'10"
Breiteneder *Löwe* 05.08.1952, Ilse
Ori 5h58'25" -0d42'47"
Brejas236-43VR
And 0h51'13" 21d54'37"
Brekerbohm1934
Uma 10h6'33" 48d49'25"
BREN
Vir 13h4'22" -2d6'55"
Bren & Jen
Ari 2h51'54" 28d14'48"
Bren Myers-Simmerman
Lyr 18h52'50" 35d30'21"
Bren Sherwood 19th Anniversary Star
Lib 15h45'52" -25d41'10"
Brena Dalmacio
Peg 23h48'4" 22d53'49"
Brenanbri
Crb 16h13'35" 28d6'16"
Brenda
And 0h29'26" 32d43'42"
Brenda
Vul 20h16'59" 24d0'2"
Brenda
Peg 22h34'15" 20d22'24"
Brenda
Ari 3h16'54" 19d15'11"
Brenda
Aqr 22h35'49" 0d58'34"
Brenda
Psc 1h23'1" 13d55'8"
Brenda
And 0h36'18" 36d29'22"
Brenda
Gem 7h8'56" 31d37'55"
Brenda
And 1h43'32" 44d48'28"
Brenda
And 0h46'51" 40d1'14"
Brenda
Cam 3h46'53" 59d4'6"
Brenda
Cyg 20h39'19" 40d48'22"

Brenda
Lib 15h26'15" -23d51'3"
Brenda
Sco 16h27'13" -32d40'29"
Brenda
Sco 16h7'51" -14d22'6"
BRENDA
Lib 15h21'46" -14d15'59"
Brenda
Aqr 22h14'48" -14d12'45"
Brenda
Cyg 21h50'57" 55d14'47"
Brenda
Uma 9h42'37" 54d24'42"
Brenda 3271964
Lyn 9h11'8" 41d39'42"
Brenda A. LaFosse
Leo 9h27'5" 23d15'50"
Brenda A. Milligan
Lib 14h30'43" -10d21'47"
Brenda & Amy's Piece of Sky
Gem 7h14'9" 19d3'26"
Brenda and Fube
Cyg 19h39'35" 44d28'9"
Brenda Ann 6-13
Gem 6h48'32" 27d24'7"
Brenda Ann Campbell
Cyg 19h34'59" 28d29'29"
Brenda Ann Oliva
Leo 11h29'15" 22d20'8"
Brenda Ashley McCarthy
Uma 10h4'30" 41d55'24"
Brenda B. Gates
Tau 4h35'38" 24d46'11"
Brenda Bailey
Tri 2h9'22" 32d31'26"
Brenda Baisley Giannelli
Vir 13h3'1" 11d1'15"
Brenda Bartlet
Cap 20h53'52" -18d40'24"
Brenda Bass Lipscomb
And 1h44'43" 36d8'27"
Brenda (BB)
Vir 14h43'22" 3d2'26"
Brenda Bean
Cyg 22h1'6" 51d4'51"
Brenda Berenice
Gem 6h38'59" 14d49'45"
Brenda Bodner's shining light
Cyg 21h43'49" 52d11'14"
Brenda Bouvier-Pleasant
Vir 14h41'0" 3d49'38"
Brenda Bowers
And 23h6'45" 40d18'46"
Brenda "Bre" Showalter
Leo 11h8'54" -3d57'10"
Brenda "Brendon" DeMichael
Ari 2h39'52" 19d54'12"
Brenda Brittain
And 2h15'21" 50d9'23"
Brenda C. Ranocchia
Gem 6h59'55" 16d41'20"
Brenda C. Upham
Com 12h50'33" 30d40'35"
Brenda C. Zendejas
Per 3h46'48" 32d25'57"
Brenda Candido
Lyn 7h1'35" 52d25'7"
Brenda Couch
Lyn 8h5'47" 35d2'36"
Brenda Coward January 17, 1966
Ori 5h20'21" 6d25'4"
Brenda Cox
Sco 16h44'54" -26d27'56"
Brenda Cunningham
Lib 15h43'25" -10d38'4"
Brenda & Curt Vincent
Lyr 19h1'9" 32d43'47"
Brenda Darlene Musson
Vir 14h7'33" -0d47'13"
Brenda Dawn
Com 12h17'37" 31d26'3"
Brenda De Leon
Lib 15h20'55" -24d16'17"
Brenda DeBoer
Uma 13h36'41" 53d48'30"
Brenda Diane
Cas 1h37'13" 57d53'20"
Brenda Dion
Cyg 21h43'35" 33d11'17"
Brenda Donato
Cnc 8h13'30" 18d26'40"
Brenda Dorkable Sanchez
Lib 14h34'27" -23d6'58"
Brenda Dunn Johnson
Aqr 22h17'56" 0d33'16"
Brenda E. Finch Jackson 8-25-1958
Uma 13h31'25" 54d7'3"
Brenda E. Laust
Vir 13h21'14" 10d58'53"
Brenda Elizabeth Haynes, M.D.
Vir 13h18'57" 12d26'49"
Brenda Elyse Lipton
And 2h18'10" 45d21'21"
Brenda Esqueda "I Love You"
Lyn 8h49'58" 35d49'3"
Brenda Esto
Aqr 22h21'51" -6d17'21"

Brenda F. Wickham
Lmi 10h41'46" 32d37'25"
Brenda Faye Loewendick
Cyg 19h37'40" 41d58'56"
Brenda Fox
Leo 9h24'9" 27d27'57"
Brenda G. Morris
And 23h1'11" 40d43'53"
Brenda Gable
Lib 15h5'56" -0d50'27"
Brenda Gail
Crb 15h27'40" 28d9'28"
Brenda Gay Kelley
Uma 10h6'40" 52d19'45"
Brenda Gill
Ori 5h31'59" -0d6'4"
Brenda Gomez
Sco 16h39'8" -34d12'46"
Brenda Goodenough - Divine Star Lady
Del 20h56'18" 11d52'33"
Brenda Gutierrez
Psc 23h45'58" 7d51'54"
Brenda Heine
Lyr 18h52'51" 36d7'12"
Brenda Hernandez
Lib 15h8'54" -10d52'51"
Brenda Irene 1-4-3
Aqr 22h36'26" 0d31'52"
Brenda Irene Houghtalen
Gem 6h56'3" 22d47'31"
Brenda J Corbett
Uma 12h26'38" 58d18'18"
Brenda J Rondum
Tau 3h32'12" 3d2'52"
Brenda Jackson Polk
Cas 1h32'0" 63d3'21"
Brenda Jane (Clark) Glass
Lib 15h55'3" -20d16'48"
Brenda Jane Ritter Nickson
Aql 18h55'57" -0d20'34"
Brenda Jean Salas
Cas 1h28'16" 59d6'35"
Brenda Jo Hankins
Lyn 7h34'40" 47d50'49"
Brenda Jones - Much Loved Shining Star
Cas 0h20'35" 55d44'24"
Brenda K. Stevens
And 23h1'29" 47d29'41"
Brenda Kallvet
And 0h19'30" 38d19'38"
Brenda Kathleen Wolfskill
Ori 5h35'52" 9d37'51"
Brenda Kay
Aqr 21h36'36" 0d55'22"
Brenda Kay
Tau 4h35'30" 21d6'33"
Brenda Kay
Leo 9h29'46" 10d16'50"
Brenda Kay Hester
Cas 2h10'47" 64d3'47"
Brenda Kay Jarvis
Lmi 10h33'18" 36d34'34"
Brenda Kay Lindsey
Ari 2h46'52" 24d0'55"
BRENDA KAY PRINGLE
Ori 6h17'56" 13d48'18"
Brenda Kay Veazey
Lyn 9h7'27" 39d14'28"
Brenda Kaye Strautman Stacy
Lmi 10h17'0" 28d18'23"
Brenda Kelly
Per 2h16'58" 52d6'39"
Brenda & Kenny - True Love Star
Cyg 20h50'20" 55d0'46"
Brenda Kindregan
Psc 1h46'6" 5d42'37"
Brenda Koehring
Lyn 7h7'58" 50d41'14"
Brenda L. Arnold
Umi 14h51'7" 73d32'33"
Brenda L. Coolbaugh
Sgr 18h8'24" -19d36'17"
Brenda L. Cruz
Psc 0h47'7" 18d5'3"
Brenda L. Farrel
And 0h21'6" 46d16'24"
Brenda L. Lebby
Leo 11h38'54" 24d8'3"
Brenda L. Scholl
Sco 17h42'44" -39d3'14"
Brenda L. Thornhill
Del 20h38'13" 15d8'30"
Brenda Lamb
Psc 1h10'23" 23d17'24"
Brenda Larby West End Actress
Cas 1h28'3" 52d19'4"
Brenda Lawson
Cyg 21h44'8" 44d11'19"
Brenda Lea HAYSTKS ISIS -PPLVTB
Cnc 8h12'43" 28d16'27"
Brenda Leann Verhoeff
Cap 20h53'20" -16d43'1"
Brenda Lee
Uma 9h41'24" 56d51'5"
Brenda Lee
Tau 4h18'19" 6d13'6"
Brenda Lee
Lyr 19h11'25" 46d40'48"

Brenda Lee
Lyn 8h25'48" 36d22'59"
Brenda Lee Baysinger
Tau 3h38'51" 15d19'58"
Brenda Lee Bryant (BABY ANGEL)
Uma 8h59'24" 72d46'17"
Brenda Lee Crimmins
Mon 8h1'38" -0d33'53"
Brenda Lee Dashnaw
Sco 17h52'32" -30d26'38"
Brenda Lee Hutchinson
Uma 11h46'18" 33d12'16"
Brenda Lee Ratkus
Cyg 21h49'39" 46d59'41"
Brenda Lee Stack
Leo 10h13'35" 12d10'57"
Brenda Lee Tetreault
Tau 4h47'15" 19d30'2"
Brenda Lee Walls
Uma 11h46'52" 33d58'9"
Brenda Lewis Trout
Uma 11h11'46" 32d38'52"
Brenda Lillian Cervantes
Ari 1h58'27" 19d11'8"
Brenda Lou Anderson
Leo 10h28'22" 16d32'30"
Brenda Louise
And 23h22'32" 42d28'58"
Brenda Lu
Vir 14h16'7" 1d37'39"
Brenda Lynn Archuleta
Tau 4h29'28" 17d2'16"
Brenda Lynn Hudson
Aqr 22h28'8" 2d26'21"
Brenda MacDonald
Uma 11h16'34" 52d11'25"
Brenda Marcella Grinston
And 2h22'54" 42d13'11"
Brenda Marie
Sco 16h41'54" -31d24'39"
Brenda Marie Glatzmaier
Lyr 18h48'45" 45d4'21"
Brenda Marie Kracht
Gem 6h46'15" 19d30'30"
Brenda Martin
Mon 7h48'39" -10d16'32"
Brenda May Bavington
Vir 14h21'33" 7d11'54"
Brenda McBryde
Aql 19h5'49" 2d5'0"
Brenda McFall
Umi 11h31'14" 88d14'25"
Brenda Mi Amore
Uma 9h30'2" 43d45'56"
Brenda Michelle Hinton
Sco 17h34'4" -45d20'32"
Brenda Miles
Lib 14h52'44" -0d32'25"
Brenda Mills
Cas 23h1'34" 58d32'5"
Brenda (MOM) Sparks
Aqr 22h33'17" -3d11'43"
Brenda Moore
Lyn 7h28'52" 57d14'3"
Brenda Munch
Ari 2h6'28" 11d57'16"
Brenda Murphy
Per 3h40'45" 43d47'56"
Brenda My Guiding Star
Tau 4h8'33" 10d57'7"
Brenda My Love
Cyg 21h45'29" 55d11'39"
Brenda Nicholson
Cas 23h7'12" 58d7'48"
Brenda Nolan
Leo 11h40'4" 17d8'9"
Brenda Norton Hall
Lyr 19h0'13" 42d9'12"
Brenda Odite Mahoney
Ori 5h56'7" 21d9'26"
Brenda Palmer
Cyg 21h44'4" 44d50'23"
Brenda Palmer-Rodriguez
Lib 14h59'10" -5d8'54"
Brenda Patrice Woodland
Cnc 8h36'49" 32d25'58"
Brenda Peacock
Ari 2h10'44" 22d23'40"
Brenda Pequignot
Uma 9h6'19" 69d24'41"
Brenda Perry
Cnc 8h31'0" 22d43'30"
Brenda Pinho
Vir 12h51'56" 11d6'34"
Brenda Porras
Sco 16h54'46" -38d13'56"
Brenda Poulton
Gem 7h10'5" 26d16'54"
Brenda Prime
Tau 4h6'20" 6d21'45"
Brenda Raby
Aqr 23h5'36" -7d2'46"
Brenda Righter
Uma 12h44'8" 62d19'29"
Brenda Robertson's Star
Peg 21h50'46" 26d53'52"
Brenda Rock
Uma 10h11'45" 46d46'29"
Brenda RoJon
Lyr 18h16'5" 32d29'50"
Brenda Rose Galloway
Crb 16h16'56" 30d46'15"
Brenda Ryan
Ari 2h8'14" 23d3'7"

Brenda S.
Tri 2h2'17" 33d22'55"
Brenda Sanchez
Cas 23h3'42" 53d52'54"
Brenda Sayler
Lyn 7h41'34" 42d10'16"
Brenda Schoen - SuperSTAR Mom
Lyn 8h22'41" 39d41'17"
Brenda Scott
Cyg 20h32'56" 30d48'53"
Brenda Sherman
Cnc 8h41'7" 32d23'13"
Brenda Siewicki
Leo 11h14'36" 16d42'20"
Brenda Sileo
Lyn 7h44'37" 55d32'19"
Brenda Simpson
Psc 0h20'45" 4d43'18"
Brenda Smith
Leo 10h9'35" 11d43'26"
Brenda & (Son) Anthony Delbridge
Uma 10h15'19" 72d14'36"
Brenda Sonja
Vir 13h34'19" -13d59'0"
Brenda St. John
Tau 4h21'15" 19d50'43"
Brenda Starbright Hamby
Vir 11h46'17" 5d46'8"
Brenda Stark Baldassano
Cnc 8h52'6" 13d11'45"
Brenda Starlight
Cnc 8h0'53" 13d20'49"
Brenda Sue
Per 4h6'5" 43d33'9"
Brenda Sue Martin
Leo 10h24'37" 14d44'31"
Brenda Sue Stout
Cnc 9h4'44" 23d16'13"
Brenda Thompson Stewart
Cam 4h3'31" 55d37'35"
Brenda Thornbrue
Lmi 10h22'44" 38d19'54"
Brenda Townson's Eastern Star
And 2h28'1" 49d37'29"
Brenda van Rensburg
Cra 19h1'7" -37d48'5"
Brenda Ventura
Vir 13h57'22" -17d18'46"
Brenda Villegas
Sco 17h42'9" -33d33'46"
Brenda W. Self (Mother)
Aqr 21h46'46" -8d17'11"
Brenda Wagner
Sco 16h6'5" -22d44'45"
Brenda Walters Russell
Sco 16h15'33" -28d33'10"
Brenda Williams
Mon 6h50'29" -0d26'19"
Brenda Woodward
Ari 3h11'39" 29d7'50"
Brenda Zuniga
Psc 1h16'22" 14d26'18"
Brenda, Krysta & Nick Ecker
Ori 5h34'39" 1d16'27"
Brenda, you're the light of my life
Ori 5h20'52" -0d53'29"
BrendaAnneBlake
Uma 11h25'0" 34d36'33"
brenda-c
Aqr 22h33'12" 1d16'12"
Brendan
Cyg 21h31'49" 54d52'58"
Brendan and Brigit
Crt 11h36'1" -8d1'58"
Brendan and Colleen
Cyg 21h29'27" 33d38'57"
Brendan and Lelia
Cru 12h15'0" -58d8'27"
Brendan and Maggie, La Vie En Rose.
Cyg 19h30'31" 50d57'38"
Brendan Andrew Chandler
Uma 11h18'57" 67d56'23"
Brendan Barr
Ori 5h27'32" -3d9'39"
Brendan Barrett Lull
Sgr 19h47'8" -12d52'5"
Brendan Chance Hudnall
Tau 4h34'32" 30d44'12"
Brendan Charles Kreiss
Tau 3h39'12" 28d6'53"
Brendan Chase Sard
Uma 12h17'24" 52d50'22"
Brendan & Colleen
Cyg 20h55'58" 41d16'56"
Brendan Corey
Vir 14h22'33" 2d29'22"
Brendan David Bakst
Cap 20h33'37" -16d59'3"
Brendan Donald Schwartz
Ari 3h18'36" 28d49'59"
Brendan Edward Roche
Leo 11h45'44" 26d7'55"
Brendan Emillo Westhoff
Lyn 6h47'17" 61d6'39"
Brendan Garvey
Uma 9h28'34" 67d15'40"
Brendan Hourican
Uma 10h9'32" 42d37'54"

Brendan Indiana
Aqr 22h35'0" 1d42'53"
Brendan J 'Viper' Bell
Psc 1h4'25" 10d41'46"
Brendan Jack Draper
Uma 9h46'46" 57d1'49"
Brendan Jack Shaw
Ari 2h38'2" 27d32'17"
Brendan James
Cra 18h43'23" -39d14'58"
Brendan James Donelan
Ari 1h51'31" 22d41'21"
Brendan James Long
Ori 5h30'38" 2d0'43"
Brendan Jason Nailon
Lib 14h53'28" -6d35'13"
Brendan Joe Crouse
Lib 15h11'22" -27d13'25"
Brendan John
Ari 3h8'54" 29d13'4"
Brendan John Moerman
Ori 5h31'34" -0d25'29"
Brendan Jones Rizzuto
Eri 3h44'33" -0d53'14"
Brendan Jorgensen
Her 18h36'15" 25d59'50"
Brendan Joseph Giblin
Uma 11h27'31" 62d46'49"
Brendan Joseph McCarthy, Jr
Her 17h48'7" 28d14'49"
BRENDAN KUNKEL
Boo 14h24'22" 17d45'37"
Brendan Lee Hellbusch
Boo 14h20'44" 46d1'23"
Brendan Luke Reilly
Her 18h15'16" 14d51'20"
Brendan M Alvord
Ari 3h12'32" 29d30'4"
Brendan Maas
Aql 19h7'48" 4d14'34"
Brendan McConnell Kellogg
Cep 22h36'29" 64d38'31"
Brendan McKee
Per 2h45'25" 53d31'44"
Brendan McNaboe
Aqr 22h36'12" -2d5'38"
Brendan Michael
Aqr 21h7'34" -2d37'22"
Brendan Michael Cazenas
Tau 4h35'29" 28d10'11"
Brendan Michael Smith
Psc 0h6'46" 3d42'22"
Brendan Monigan
Aqr 22h59'8" -10d38'23"
Brendan - My Light, My Star, My Galaxy
Cra 18h14'33" -42d1'18"
Brendan Oakes
Per 4h6'46" 40d43'16"
Brendan Orion Porter
Ori 6h5'25" 17d31'42"
Brendan P. Opdycke
Uma 13h3'48" 52d51'42"
Brendan P. Shalloo
Dra 17h20'54" 67d54'54"
Brendan Patrick McCarthy
Ori 5h48'6" 1d43'46"
Brendan Patrick O'Donnell
Leo 11h18'22" 18d46'19"
Brendan & Ronald Wulf
Gem 7h41'43" 33d58'32"
Brendan T. Bennett 9-26-1995
Cmi 7h32'25" 5d17'39"
Brendan Terence Kenny
Umi 14h45'43" 82d19'43"
Brendan Thomas
Per 3h52'37" 33d55'53"
Brendan Thomas Baker-Stanley
Her 16h28'6" 28d51'44"
Brendan Thomas Brady
Boo 15h0'43" 35d5'17"
Brendan Thompson
Cru 11h56'25" -55d48'37"
Brendan Tyler Gipp
Ari 2h3'59" 23d52'54"
Brendan & Tyler Peterson
Aur 5h32'34" 45d50'32"
Brendan Vose Reagan
Cep 22h21'1" 72d31'10"
Brendan W. Horan
Vir 13h2'4" -11d47'13"
Brendan William
Sgr 18h2'5" -28d6'4"
Brendan X. O'Rourke
Ari 2h26'46" 21d1'51"
Brendan's Workshop
Vir 13h3'8" -8d58'6"
Brendas Briannas
Uma 8h47'40" 71d9'15"
Brenda's Criptinite
Her 16h25'18" 49d36'23"
Brenda's Diamond Star
And 0h10'46" 44d35'2"
Brenda's Gram
Cnv 13h31'20" 36d51'47"
Brenda's Heart
Leo 10h9'25" 15d3'27"
Brendas Little Daisy
And 1h2'41" 36d39'47"
Brenda's Native Star
Ari 2h7'17" 22d58'32"
Brenda's Star
Cnc 8h14'10" 16d9'28"
Brenda's Star
Cnc 8h12'10" 25d13'3"
Brenda's Star
And 0h27'5" 42d59'34"
Brenda's Star
Uma 11h24'4" 40d45'54"
Brenda's Star
Cas 2h42'18" 67d49'41"
Brenda's Star
Uma 11h57'23" 61d19'57"
Brenda's Star
Aqr 22h23'41" -0d56'30"
Brenda's Strawberry
Cnc 8h10'40" 16d3'47"
Brenda's Wish
Cas 0h7'27" 59d51'34"
BrendaVanHoff
Gem 6h9'12" 24d56'11"
Brenden Anthony Schmied
Cnc 8h58'43" 17d31'26"
Brenden Davis
Per 3h11'42" 52d45'44"
Brenden Edgeworth
Ori 5h53'45" 21d5'21"
Brenden Lee Newvine
Leo 10h20'46" 11d10'21"
Brenden Levi Sainz
Cap 20h22'11" -13d32'23"
Brenden Myles Facer
Umi 15h1'21" 69d55'19"
Brenden Taylor Huckelberry
Cap 20h26'50" -14d44'56"
Brenden Thomas Harrower
Her 17h20'26" 45d33'43"
Brenden Thomas Mervin
And 0h32'54" 30d24'30"
Brendon and Carmen Wren
Cra 19h7'20" -37d56'49"
Brendon Avery Hutteball
Cnc 8h23'11" 10d31'18"
Brendon Brent Bubba Cordova
Uma 8h22'48" 70d4'59"
Brendon Breshock
Cep 21h58'48" 60d34'16"
Brendon Eugene Walker
Aqr 23h5'21" -8d34'35"
Brendon Gory "skunkrabbit"
Cyg 19h59'38" 32d26'25"
Brendon John 2974
Per 3h27'40" 51d48'54"
Brendon John Mahoney
Cnc 7h59'29" 16d42'58"
Brendon Joseph McNamara
Dor 4h13'20" -55d57'35"
Brendon Kenneth John Dueck
And 1h39'45" 48d10'38"
Brendon Lee Burlison
Ori 5h54'12" 11d25'19"
Brendon Mahoney
Uma 10h36'0" 57d8'38"
Brendon Thomas Franks
Lyn 6h59'28" 50d59'1"
Brendon Van Allen
Ori 6h11'9" 19d27'54"
Brenen Haynes Pratt
Umi 12h48'46" 87d24'12"
Brenlie Melinda Vargas
Ari 2h15'13" 25d17'2"
Brenna
Ori 5h58'13" 21d57'49"
Brenna
Psc 0h30'53" 13d11'28"
Brenna
Cas 0h21'45" 50d5'44"
Brenna
Cam 6h37'58" 76d49'36"
Brenna
Sco 17h34'3" -41d48'23"
Brenna Ann Benedict
Tau 4h52'43" 16d37'54"
Brenna da Boober
Tau 5h55'30" 25d9'57"
Brenna Engle
And 0h11'50" 31d12'22"
Brenna Faith
Crb 16h13'56" 37d34'14"
Brenna Jenkins
Tau 5h44'1" 25d45'26"
Brenna K. Harris
Leo 11h52'54" 26d12'58"
Brenna L.
Cam 3h33'37" 62d44'13"
Brenna Lapsley
And 2h21'13" 50d26'41"
Brenna Lauren Grant
Uma 11h10'30" 34d29'52"
Brenna Lee Cummings
Umi 14h10'18" 76d47'56"
Brenna LindaFay Pomfrey
Peg 21h28'52" 16d1'11"
Brenna MacMillin
And 2h36'21" 44d5'38"
Brenna Marie
Aqr 20h47'33" -8d46'55"
Brenna Marie
Ara 17h25'38" -48d13'28"
Brenna Marie Gleason
Uma 10h31'29" 62d58'32"
Brenna Marie Ransden
Ari 3h16'10" 28d10'55"
Brenna Marques
Cyg 20h26'40" 45d25'54"
Brenna Williams
Psc 0h26'9" 8d16'15"
Brennan
Uma 11h26'29" 48d4'9"
Brennan
Uma 8h49'15" 47d52'31"
Brennan
Cma 7h9'59" -27d58'8"
Brennan Ammentorp
Her 17h18'20" 35d41'30"
Brennan Coleman
Cnc 8h45'58" 30d33'11"
Brennan Crawford
Uma 8h59'18" 70d9'5"
Brennan Donald Kinnison
Cap 21h58'36" -19d27'2"
Brennan Gregory Coleman
Aur 5h35'4" 45d47'1"
Brennan Harding Erickson
Aur 5h52'50" 42d59'57"
Brennan Hart Lammermann
Leo 9h38'12" 28d28'7"
Brennan Hatch Johnson
Leo 11h24'52" 13d41'44"
Brennan Kindal
Ori 5h35'28" -0d47'43"
Brennan LaFleur
Ari 3h10'37" 18d29'35"
Brennan Linnaeus Bachert
Sco 17h35'46" -36d50'0"
Brennan Macy Walker
Cas 0h5'3" 63d10'41"
Brennan Matthew Speer
Psc 1h41'22" 8d5'10"
Brennan McDonald
Uma 11h23'46" 59d14'56"
Brennan Plummer
Lib 15h23'45" -26d57'47"
Brennan Ray Evans
Aql 19h6'0" 3d4'10"
Brennan S Payne
Cap 20h50'48" -19d17'44"
Brennan Taylor
Ori 5h32'10" 9d54'58"
Brennan Wayne Proffitt
Aqr 21h50'26" -7d7'55"
BrennanRoberts
Lyn 8h33'58" 38d22'0"
Brennan's Star
Uma 9h18'53" 53d35'2"
Brenna's Celestial Dream
And 1h19'21" 49d27'53"
Brenna's Playground
Ari 2h18'18" 20d17'39"
Brennen Burnside
Per 3h11'9" 45d47'2"
Brennen C. Harris
Her 16h48'9" 33d16'42"
Brennen Leah Whitaker
Psc 1h24'10" 16d7'10"
Brenner Jones
Her 16h36'6" 6d47'48"
Brennie Hillan
And 2h31'45" 44d2'35"
BRENNITA
Mon 7h17'36" -0d29'45"
Brenny Wright
Uma 10h52'44" 47d26'23"
Brent
Cmi 7h14'43" 10d11'58"
Brent Allen Wilson
Psc 1h10'20" 25d58'7"
Brent and Emily
Ori 6h9'12" 9d19'11"
Brent and Jessica Forever
And 0h15'27" 42d55'46"
Brent and Katie Halling
Uma 8h23'29" 65d40'12"
Brent and Kristie
And 23h14'41" 42d19'27"
Brent and Mary Forever
Her 17h56'29" 28d58'52"
Brent and Rachel
Uma 11h26'21" 62d49'24"
Brent Blakeley
Uma 10h19'46" 62d45'38"
Brent Buccieri Shepperack
Sco 17h41'4" -40d43'43"
Brent & Charlene
Ori 5h11'28" -1d25'54"
Brent Cooper
Umi 15h11'9" 71d44'17"
Brent Dawson Starcher
Sgr 17h59'57" -17d36'34"
Brent Dawson Starcher
Sgr 18h20'18" -23d55'53"
Brent Donenfeld
And 23h24'9" 38d24'5"
Brent "E" C.
Her 18h3'36" 17d33'14"
Brent Edward Bartlett
Her 17h42'19" 36d58'7"
Brent Faduski
Per 2h39'15" 55d58'50"
Brent "FATE" Holly
Per 4h49'0" 45d27'41"
Brent Graeme Duffin
Cep 22h19'29" 63d41'30"
Brent H. Czajkowski
Umi 14h26'43" 75d52'17"
Brent & Heather Peck
Tau 5h32'59" 18d29'58"
Brent James Thomson
Mon 7h8'55" -8d46'2"
Brent Joseph Darin
Uma 8h47'50" 46d44'10"
Brent Joseph Ulrich
Per 3h48'50" 43d46'5"
Brent King
Cep 22h24'8" 64d18'37"
Brent Lanning
Her 16h35'6" 45d53'25"
Brent Leonard Martin
Aur 5h37'13" 53d42'38"
Brent Leroy Farnsworth
Her 18h0'8" 21d29'12"
Brent Matthew Croix Zajac
Leo 11h54'40" 20d31'0"
Brent McGouirk
Ori 5h29'29" 3d11'3"
Brent McIvor
Aql 19h31'47" 12d36'38"
Brent Michael Edwards
Sgr 17h52'34" -16d39'51"
Brent Michael Green
Aql 19h37'19" 13d36'6"
Brent Mitchell Tyler
Her 18h3'44" 22d39'52"
Brent O'Connor
Uma 9h4'22" 56d32'57"
Brent Rakus
Aur 5h43'40" 49d18'47"
Brent Rich
Ori 5h31'2" -2d47'46"
Brent Rivers
Pic 5h32'54" -45d41'24"
Brent Romain - Our Star Forever
Uma 13h34'53" 56d37'57"
Brent 's Serenity
Vir 14h15'3" 5d42'10"
Brent S. Tompkins
Cep 21h30'11" 64d2'21"
Brent Stonecipher
Leo 11h39'13" 24d27'30"
Brent & Tonya
Cyg 21h18'50" 34d59'7"
Brent Tyler and Sonya Yvonne
Aql 19h48'39" 15d0'35"
Brent Van Rossen
Leo 11h0'38" 2d32'14"
Brent William MacFarland
Uma 9h45'57" 59d19'6"
Brentano Sommer "Happy 50th"
Her 18h37'16" 16d16'32"
BRENTEN
Ori 5h27'21" 0d54'44"
Brentenon
Ara 17h38'26" -47d36'22"
Brentlee Loren Evans
Uma 9h50'20" 50d42'40"
Brenton Acott
Cru 12h28'17" -58d40'50"
Brenton Alan Moland
Cep 23h14'27" 78d49'41"
Brenton James Scott
Cru 12h54'35" -58d1'8"
Brenton Koch
Cru 12h30'17" -60d59'13"
Brenton Robert Paolella
Per 2h53'24" 41d29'49"
Brenton Thomas Slocum
Ari 2h21'48" 11d41'25"
Brenton Towers
Cru 11h56'56" -61d47'19"
Brenton's Star
Gem 6h58'55" 24d26'21"
BrentPatriAlex
Dra 19h20'23" 76d51'41"
Brenty's Dream Star
Cru 12h4'25" -61d18'12"
Breonna Ta'Shan Powell
Uma 13h39'15" 58d15'25"
Breshelle
And 0h37'35" 33d38'55"
Bressay
Lib 14h24'4" -11d20'21"
Bret A. Smith
Pho 0h39'40" -40d13'30"
Bret and Liz
Sgr 17h53'17" -29d57'23"
Bret Coley Harte
Uma 9h36'52" 70d23'47"
Bret Daniel Snouffer
Sco 17h27'31" -41d45'6"
Bret E. Bullock
Aur 6h5'53" 47d22'25"
Bret Edward Fendt
Aqr 22h30'10" -1d50'21"
Bret Gordon Edmonds
Lyn 7h4'21" 53d35'44"
Bret Joseph Kacher
Aur 5h12'57" 41d51'30"
Bret M. Jones
Cam 4h23'54" 68d3'6"
Bret Michael
Lyn 8h1'40" 45d52'17"
Bret Michael Fishkind
Uma 10h58'7" 63d40'56"
Bret Patelski
Cnc 8h34'20" 8d16'5"
Bret Spangler
Aur 6h49'10" 44d22'29"
Bret Stein Lesavoy
Ari 2h31'39" 26d48'32"
Bret Stephen
Per 3h20'42" 48d24'38"
Bret W. Cash
Boo 14h21'50" 47d49'29"
Brett
Her 17h41'37" 47d40'51"
Brett
Aur 6h8'51" 47d54'53"
Brett
And 1h31'12" 44d18'48"
Brett
Sgr 19h56'34" -28d35'8"
Brett A. Hetrick
Sgr 17h59'48" -27d13'3"
Brett Acheson
Ari 2h17'50" 25d8'20"
Brett Alan Swanda
Uma 11h1'37" 46d50'57"
Brett Alan Young
Ori 5h28'24" 5d56'47"
Brett Allan Austin
Vir 13h31'2" -20d31'54"
Brett Allan Penfold
Sco 16h11'59" -8d36'34"
Brett Allen Kleinschmit
Leo 11h17'39" -1d18'13"
Brett and Amber
Sco 16h54'36" -35d9'43"
Brett and Brearnne's Light
Dor 4h55'14" -68d23'34"
Brett and Claudia Forever Shining
Cyg 19h35'30" 44d28'31"
Brett and Courtney's Star
Hya 9h13'5" 5d21'53"
Brett and Gilda Forever
Sgr 17h53'29" -29d27'26"
Brett and Michelle - Eternal Love
Ari 2h49'41" 13d39'38"
Brett and Tammy Herrington
Per 3h35'59" 44d15'45"
Brett Antony Barnard
Per 2h56'32" 51d46'6"
Brett "Baby Love" Camara
Lyn 7h12'52" 44d44'0"
Brett Burning Brightly
Psc 23h24'1" 5d30'5"
Brett+Chelsie
Her 18h7'29" 34d37'9"
Brett Collin Farkas
Vir 11h37'27" -4d41'0"
Brett Connor Wilson
Psc 2h0'59" 3d7'37"
Brett Cosmo Thorngren
Cap 21h45'20" -13d15'21"
Brett Creekmore
Vir 12h49'32" 6d8'5"
Brett Daniel Hodson - Hoddo
Cru 12h47'3" -59d33'17"
Brett Daniel Tice 8/7/2003
Leo 11h48'45" 26d37'32"
Brett Douglas Cervantes
Aql 19h51'51" 16d5'41"
Brett Downey 10-13-92 / 8-19-04
Her 17h9'32" 34d25'38"
Brett Evan Grossman
Ori 6h20'44" 9d45'21"
Brett Everett Rush
Her 17h1'55" 16d11'23"
Brett Fetter
Her 18h2'4" 18d1'30"
Brett Franklin
Tau 3h42'22" 26d0'54"
Brett Graveline
Boo 13h52'15" 10d34'5"
Brett Henkel The Best Teacher Star
Her 17h35'36" 36d1'21"
Brett Jackson
Dra 18h18'56" 48d51'25"
Brett James Francis
Psc 0h40'11" 18d6'49"
Brett James "Tex" Custer
Vir 12h54'0" 6d54'14"
Brett Jared Noel
Aqr 21h0'30" 0d27'24"
Brett & Jason
Uma 10h28'16" 64d2'42"
Brett Jay
Psc 23h3'7" 7d13'52"
Brett Joseph Schnacky
Her 18h14'0" 15d56'31"
Brett Lamar Grant
Sco 16h15'51" -17d16'14"
Brett Loves Jocelyn
Cyg 21h32'52" 49d4'48"
Brett M. Middaugh
Cap 21h27'57" -10d52'31"
Brett M. Tozzo
Uma 12h18'32" 52d24'52"
Brett Marie Hendrick
Leo 11h25'45" -6d0'57"
Brett Matthew
Uma 8h14'54" 71d29'25"
Brett Matthew Downey
Uma 8h33'8" 70d2'59"
Brett Matthew Runge
Tau 4h25'27" 30d8'7"
Brett Michael Bauer
Cap 20h15'25" -9d43'4"
Brett Michael Buzzelli
Lac 22h46'23" 49d55'44"
Brett Michael Gile
Per 2h45'36" 53d40'28"
Brett Michael Peters
Aqr 22h2'7" -14d25'28"
Brett Michael Walden
Ari 2h43'1" 25d53'5"
Brett My Shining Star
Aql 19h50'29" 15d0'53"
Brett N. Smith
Vir 12h27'15" -1d42'18"
Brett O. Fowler
Sgr 19h50'47" -13d15'15"
Brett Parker
Dra 14h29'47" 63d55'16"
Brett Patrick Denninger
Sco 16h59'37" -34d21'38"
Brett Patrick Jones
Uma 13h12'4" 55d20'19"
Brett Patrick Judd
Per 3h37'15" 44d11'47"
Brett Peppers
Pho 1h40'3" -48d4'32"
Brett Petty
Uma 9h11'25" 48d49'27"
Brett Philip Mowrer
Pho 1h3'49" -42d52'40"
Brett Phillip James Logan
Leo 9h26'25" 8d39'24"
Brett Porter
Ori 5h38'12" 8d40'1"
Brett Rale Wittig
Aql 19h31'43" 10d58'13"
Brett & Romina - Per Sempre
Col 5h49'24" -34d49'24"
Brett Ronald Presley
Her 16h45'58" 34d24'48"
Brett Russell
Uma 10h41'30" 47d50'56"
Brett Stafford - You're A Star Always
Ori 5h35'59" -2d28'6"
Brett Star
Her 17h10'25" 34d41'50"
Brett Steele... agent provo-cateur!
Cra 18h14'3" -41d2'4"
Brett Stephen Bowen
Lmi 9h53'13" 36d2'51"
Brett T. Little
Cap 21h7'42" -16d36'0"
BRETT - The Brightest Star
Eri 3h50'22" -0d21'2"
Brett Vern Shelley
Ori 5h56'30" 11d22'16"
Brett Wade Eells
Ori 5h42'8" -0d42'43"
Brett Waibel
Aqr 22h53'42" -8d12'3"
Brett Wales
Cyg 20h39'25" 44d52'46"
Brett Wayne Morris
Gem 6h48'8" 14d54'45"
Brett Weston
Ori 6h11'48" 5d47'4"
Brett William Liversage
Leo 10h28'47" 20d16'23"
Brett Woytowich
Cep 22h44'15" 65d46'17"
Brett Yankauskas
Uma 11h44'12" 46d11'57"
Brett Ziegler- My Angel
Cap 20h30'58" -12d37'35"
Brett Zittel
Cyg 19h51'28" 46d57'58"
Brett, Kristin, Tate and Baby
Cnc 8h38'7" 7d24'14"
BrettandMaggie
Gem 6h41'16" 23d2'45"
Brettonwood Ladies
Cas 22h57'55" 53d12'44"
Brettoria
Peg 21h27'12" 17d53'39"
Brett's 21st Star!
Pho 0h39'38" -50d55'52"
Brett's Destiny
Ori 6h7'20" -0d30'24"
Bretty
Cyg 20h35'44" 54d35'24"
Bretty
Her 17h6'51" 28d50'47"
Breu, Karin Theresia
Ori 6h9'32" -0d39'9"
Brevis Andron Uniter
Uma 13h49'22" 57d56'28"
BREW
Uma 10h1'56" 65d47'1"
Brewstar 1030 "10-30-1957"
Lyn 7h13'50" 48d35'25"
Brewster
Her 17h17'18" 46d50'27"
Breyden Gray Dukette
Her 17h22'19" 17d9'56"
Breyden Shane Jackson
Ari 2h8'24" 24d34'34"
Breykhman, Lev
Ori 6h6'53" 13d42'37"
Bri
Gem 6h59'43" 13d30'32"
Bri
Ari 2h51'45" 12d38'18"
Bri
Psc 0h24'36" 8d29'40"
Bri
Aql 19h29'35" -0d6'38"
Bri Bri
Cnc 8h26'2" 19d14'20"
"Bri Bri"
Crb 15h39'3" 31d20'33"
Bri Geyer
And 1h37'39" 38d51'1"
Bria
Cas 23h51'56" 53d7'20"
Bria Ashlea Gore
Crb 15h51'21" 26d29'14"
Bria Brooks
Lyn 7h37'14" 41d7'26"
Bria Brooks
Lmi 10h35'13" 31d51'29"
Bria Malaejah McCoy
Lyn 8h19'49" 33d50'1"
Bria Neuenschwander
Cyg 21h55'40" 43d20'54"
Briah D. Hodgkin
Cas 23h45'13" 57d49'23"
Briah McKenna Harris
Ari 3h12'42" 19d8'24"
Briahna Plotts
And 23h23'45" 47d42'13"
Briahna Rae Traynor
Crb 16h11'25" 27d37'28"
Brian
Crb 15h46'33" 29d7'37"
Brian
Gem 6h46'4" 23d13'53"
Brian
Leo 9h30'3" 16d59'56"
Brian
Tri 1h53'4" 27d35'17"
Brian
Her 18h43'5" 21d16'35"
Brian
Aqr 22h7'53" 1d21'31"
Brian
Vir 12h33'52" 1d38'4"
Brian
Vir 12h56'28" 6d16'48"
Brian
Vir 13h4'19" 6d16'51"
Brian
Lac 22h55'52" 51d10'21"
Brian
Uma 10h35'32" 48d23'53"
Brian
Per 4h25'35" 50d50'0"
Brian
Per 4h5'23" 31d23'32"
BRIAN
Her 17h20'58" 35d38'30"
Brian
And 2h27'21" 44d10'19"
Brian
Peg 23h27'48" 33d53'6"
Brian
Cep 22h36'11" 56d59'43"
Brian
Uma 13h40'14" 56d24'30"
Brian
Lyn 8h25'58" 54d29'22"
Brian
Uma 8h54'26" 59d41'33"
Brian
Cep 0h16'17" 81d35'7"
Brian
Lib 15h27'58" -5d52'24"
Brian
Aqr 23h0'31" -6d17'49"
Brian
Vir 13h34'14" -22d21'22"
Brian #89
Ori 5h23'44" -0d8'59"
Brian A Barr
Cep 22h24'53" 63d29'56"
Brian A. Cross
Cyg 20h20'56" 47d33'28"
Brian A. Hough
Per 4h0'55" 35d10'9"
Brian A Kolafa
Lyn 8h11'49" 40d43'51"
Brian A. Layne
Aql 19h40'7" 7d28'32"
Brian A. Martin
Leo 9h47'18" 13d54'54"
Brian A. Seamer
Vir 13h10'47" -7d9'47"
Brian & Agnieszka
Cyg 20h24'34" 46d47'32"
Brian Alan Nichols
Uma 8h47'36" 53d20'19"
Brian Alan Sayrs
Aql 19h21'18" 3d56'39"
Brian Alexander
And 1h12'47" 37d37'18"
Brian Alexander O'Connor
Boo 13h50'57" 12d4'57"
Brian Alexander Womack
Aqr 22h40'47" -2d10'22"
Brian Allen
Aqr 22h35'28" -1d13'17"
Brian Allen Hancock
Aur 5h42'55" 37d56'8"
Brian Amos Johnson
Gem 6h30'37" 16d51'33"

Brian & Amy's Family Star
Her 18h27'51" 20d0'35"
Brian and Amber Bordano
Sge 19h42'2" 17d8'25"
Brian and Angie
Uma 9h12'32" 65d21'56"
Brian and Arielle
Psc 23h36'13" -0d41'25"
Brian and Ashley
Cnc 8h23'8" 16d10'8"
Brian and Betsy Ober
Sge 19h42'35" 16d56'12"
Brian and Beverly Fisher
Cas 0h40'4" 53d27'26"
Brian and Casey Erdrich
Cyg 19h49'7" 38d57'22"
Brian and Cheri Harkey
Crb 15h32'46" 32d38'35"
Brian and Cynde Dickey 12/31/04
Eri 4h2'47" -8d27'35"
Brian and Debbye Mac Kenzie
Uma 11h10'30" 30d53'28"
Brian and Destinie
Cyg 19h42'41" 54d14'33"
Brian and Elizabeth
Cyg 21h42'21" 38d0'9"
Brian and Emily
Cyg 19h21'4" 52d17'34"
Brian and Gloria Miller
Gem 7h0'51" 13d44'1"
Brian and Heather Long
Pho 0h42'39" -44d11'44"
Brian and Jenn
Cyg 21h36'49" 52d19'41"
Brian and Jennifer Swan
Her 16h46'53" 48d52'4"
Brian and Jennifer Warren
Uma 8h47'38" 60d48'31"
Brian And Jude's Wandering Star
Gem 6h38'34" 19d12'19"
Brian and Julie
Crb 16h18'1" 34d20'25"
Brian and Kate
Psc 0h18'59" 19d20'36"
Brian and Kerri Case
Sge 20h9'49" 21d16'11"
Brian and Lisa Penzone
And 0h38'27" 43d17'0"
Brian and Lisa's Star
Cyg 19h38'25" 30d12'3"
Brian and Michelle's Star, LOVE
Cyg 20h42'17" 37d27'54"
Brian and Monica
Uma 12h55'47" 62d25'52"
Brian and Monica Weida
Cyg 20h4'0" 34d16'9"
Brian and Nadine's Star
Cyg 20h21'40" 54d59'12"
Brian and Natalie
Lib 15h0'12" -24d27'43"
Brian and Rayells eternal love star
Uma 9h27'13" 47d59'21"
Brian and Rebecca
Ori 6h12'32" 14d50'28"
Brian and Roxenne Smith 06/08/1968
Cyg 20h45'16" 44d33'10"
Brian And Tazia Forever
Aqr 21h26'29" -8d25'27"
Brian & Andrea
Lyr 18h45'16" 41d14'15"
Brian Andrew Hughes
Uma 13h15'46" 55d19'16"
Brian Andrew Montgomery
Ori 5h39'55" 9d9'30"
Brian Anthony Robak
Psc 2h0'11" 3d55'55"
Brian Arendell - Brian's Star
Per 4h17'14" 43d28'38"
Brian Arnold
Cyg 20h19'14" 50d53'55"
Brian Arthur Seastone
Gem 6h41'56" 18d16'42"
Brian Arthurs
Cyg 20h39'55" 44d58'5"
Brian&Ashley Holloway Wedding Star
Pho 0h31'19" -41d29'5"
Brian Avery
Cnc 8h12'45" 12d26'45"
Brian Ayling
Cep 21h21'20" 65d10'51"
Brian Baldwin
Tau 3h30'58" 2d24'4"
Brian Barrette
Cyg 21h35'1" 50d42'5"
Brian Becht
Dra 18h8'4" 72d0'39"
Brian + Beverly - our shining light
Cru 12h28'1" -60d55'30"
Brian Blackwell Bit of Heaven
Cen 13h52'48" -60d27'56"
Brian Bluhm
Lyn 6h45'5" 52d26'11"
Brian Boye
Lac 22h15'3" 49d49'42"
Brian Brendle
Per 3h15'59" 33d35'23"
Brian & Brittany's Star
Cyg 19h47'51" 37d9'35"
Brian Brottish Jr.
Gem 6h45'26" 33d24'49"
Brian "Bubba" Lowry
Per 3h2'58" 49d6'9"
Brian Buehrle
Sco 16h45'19" -29d58'4"
Brian Burnsey Burns
Gem 6h44'0" 32d25'28"
Brian C. Keough
Ari 2h27'19" 19d12'14"
Brian C. Lowrie
Aur 7h27'10" 39d42'7"
Brian C. Lutz
Sgr 18h57'8" -35d10'3"
Brian Cali
Sgr 19h11'8" -14d44'30"
Brian & Carey Allison - 28th May 2005.
Uma 9h8'22" 48d48'30"
Brian Carleton Clifford
Cep 2h20'20" 87d36'49"
Brian Carmain
Lib 14h59'19" -14d39'43"
Brian & Cassie
Cyg 21h16'21" 42d0'44"
Brian & Catherine Forever
Leo 10h18'55" 22d40'23"
Brian Champagne
Cas 1h43'53" 60d12'18"
Brian Charles Dean
Cma 6h57'56" -16d45'26"
Brian Charles Ringshall - Grandad Ringshall
Uma 12h23'10" 54d54'39"
Brian Charles Sanders
Cep 22h49'14" 70d45'23"
Brian Chiew - my Clark Kent
Pho 1h6'42" -39d20'15"
Brian & Christie
Cyg 20h37'7" 49d20'3"
Brian & Christina
Uma 11h20'56" 46d53'16"
Brian Christopher
Uma 13h44'8" 59d51'34"
Brian Christopher
Sco 16h9'13" -14d50'27"
Brian Christopher Anderson
Aur 5h27'24" 41d12'10"
Brian Christopher Fernandez
Ari 3h7'45" 29d16'31"
Brian Christopher Mack
Ari 2h12'29" 10d50'15"
Brian Christopher McCormack
Her 18h35'35" 19d15'16"
Brian Christopher Veitch
Psc 1h1'36" 14d55'54"
Brian Christopher Wooten
Per 3h10'6" 56d7'1"
Brian Clark
Aqr 22h27'3" -2d41'44"
Brian Cobb
Aql 20h8'59" 6d33'21"
Brian Collins
Psc 0h34'49" 14d49'12"
Brian Connors
Uma 11h51'38" 50d7'45"
Brian Cooper Davis
Tau 4h11'2" 27d51'14"
Brian Coughlin
Her 18h18'42" 22d42'48"
Brian Crane
Cep 23h39'59" 77d6'4"
Brian Crowe
Ori 5h38'47" 2d36'15"
Brian Crutchfield
Aql 19h32'23" 13d16'21"
Brian Cullen III
Leo 11h19'1" 24d56'44"
Brian Curcura
Aur 4h54'22" 37d45'35"
Brian D. Ferguson
Aql 19h45'59" -0d1'47"
Brian D. McCarthy
Uma 12h6'3" 53d17'30"
Brian D. Molnar
Ori 6h7'23" 10d53'17"
Brian D. Page
Leo 9h53'37" 8d13'6"
Brian D. Page
Tau 4h13'28" 27d26'42"
Brian D. Percival
Aqr 21h59'47" -8d17'0"
Brian D. Walt - Pierce's Daddy
Sco 16h6'3" -17d54'52"
Brian Dale Callahan
Cyg 20h13'4" 55d43'19"
Brian Daniel Christensen
Cnc 8h54'47" 26d33'6"
Brian Daniel Griffin
Gem 6h1'36" 24d26'59"
Brian Daniel Hiltner
Lib 14h58'4" -24d20'29"
Brian Daniel Patrick Mooney
Sgr 18h28'23" -26d20'53"
Brian & Danielle Truini
Uma 10h50'11" 41d2'45"
Brian David Evans
Uma 10h22'8" 41d45'34"
~*~Brian David Hana~*~
Her 16h52'40" 27d55'1"
Brian David Hanson
Dra 17h44'18" 54d20'58"
Brian David Harp
Uma 9h30'42" 41d46'51"
Brian David Holod
Sco 17h30'37" -40d46'1"
Brian David Leapley
Ori 5h23'16" 7d47'47"
Brian David Lee
Her 17h19'4" 14d44'38"
Brian David Lloyd
And 0h45'16" 37d10'1"
Brian David Tyler
Per 3h19'41" 41d32'56"
Brian Dean George
Her 17h53'11" 23d39'40"
Brian & DeAndra
Cyg 20h31'26" 55d30'35"
Brian DeFranco
Uma 9h37'57" 61d51'10"
Brian Del Rio
Cyg 20h56'10" 54d52'19"
Brian Dempsey
Per 2h7'26" 54d44'20"
Brian & Diana Coleman 5-01-1981
Cyg 20h2'58" 58d0'2"
Brian Dickinson
And 0h54'33" 45d40'48"
Brian Dively
Tau 5h58'41" 27d6'33"
Brian Donadio
Vir 13h16'31" 4d25'4"
Brian Donald Wales
Cnc 8h45'23" 26d1'36"
Brian Douglas Burgos
Vir 11h46'44" 6d16'1"
Brian Drew Smith
Her 17h9'19" 46d32'22"
Brian Duffy
Cnc 8h58'23" 23d20'51"
Brian DuFresne
Per 2h53'55" 51d5'17"
Brian Duquette
Per 2h10'54" 54d59'20"
Brian E. Johnson
Uma 10h28'55" 68d51'53"
Brian E. Pickens, M.D.
Pho 0h16'34" -41d6'41"
Brian Earl Cook
Ori 5h24'53" 2d42'2"
Brian Edward Green
Psc 0h29'25" 16d23'39"
Brian Edward Griffin
Leo 10h0'57" 15d25'36"
Brian Edward Howell
Cap 21h1'5" -26d59'17"
Brian Edwards # 44
Cyg 20h40'45" 53d24'35"
Brian Edwin Bormet
Cyg 20h32'16" 49d51'33"
Brian Elliott - Big '50'
Cep 22h45'57" 62d43'34"
Brian Ellis Churchill
Ori 5h24'17" 7d5'56"
Brian Emanuel Lantigua
Ori 5h2'50" 2d10'42"
Brian & Enid - Eternal Love
Cra 18h48'54" -39d32'58"
Brian Ernest Hurley
Cas 0h1'32" 55d7'9"
Brian Eugene
Gem 7h22'21" 17d10'35"
Brian Eugene Lawson
Tau 4h5'48" 30d4'17"
Brian F. Daley
Leo 9h35'14" 31d33'35"
Brian F. Grant
Aqr 22h37'38" -21d31'5"
Brian Faraj
Vir 14h2'33" -15d49'10"
Brian Faria
Aur 5h11'16" 35d11'30"
Brian Ferguson
Cnc 8h16'36" 19d33'23"
Brian Figgins
Boo 14h32'46" 42d0'44"
Brian Fitzgerald
Ari 2h11'49" 17d2'51"
Brian Ford Briscoe
Lac 22h51'45" 48d54'10"
Brian Forrest
Her 18h18'15" 21d14'30"
Brian Francis Mahoney
Psc 0h20'0" 4d47'35"
Brian Frank
Gem 6h56'5" 22d5'16"
Brian Frank DeMichael
Umi 16h46'36" 77d44'7"
Brian Frank Miller
Uma 9h27'21" 49d42'22"
Brian Franklin Bache
Aql 19h29'30" 14d40'0"
Brian Frederick Nicholas
Uma 9h26'52" 44d29'30"
Brian Fredrick
Cen 13h38'17" -38d10'44"
Brian Fuller
Aqr 22h36'9" 2d6'49"
Brian G. Hulla
Cnc 8h37'41" 32d24'49"
Brian G. Krauth
Sgr 19h17'41" -17d4'39"
Brian Galusha
Uma 10h6'54" 45d52'38"
Brian Galusha
Leo 9h31'7" 12d19'39"
Brian George Martin 70!
Tau 5h11'59" 19d16'55"
Brian George Pond
Per 4h46'39" 39d24'14"
Brian German
Cyg 20h32'12" 58d10'19"
Brian Gilbert Hemple
Cep 21h21'58" 66d10'5"
Brian Gillispie
Aqr 20h46'19" -9d14'6"
Brian & Gina
Lyr 18h57'1" 27d12'29"
Brian Gold
Aqr 21h39'15" 0d44'17"
Brian Golden
Uma 11h35'7" 63d0'59"
Brian Grant
Per 3h50'5" 48d52'32"
Brian Gray
Gem 7h44'40" 31d29'44"
Brian Green
Cnc 8h25'15" 13d7'43"
Brian Gregory Klocke
Her 17h29'2" 36d22'43"
Brian Gregory Rockett
Sgr 19h9'39" -17d30'1"
Brian Groene
Uma 11h0'56" 57d37'41"
Brian Gruender
Her 17h38'41" 23d52'58"
Brian Grunden's 30th Birthday Star
Cnc 8h10'25" 27d4'47"
Brian H Watts
Ori 5h21'28" 8d2'4"
Brian Haag
Cnc 8h51'55" 27d14'54"
Brian Hamilton
Per 2h17'17" 55d49'10"
Brian Hanna
Leo 10h39'39" 14d28'1"
Brian Harold Gill
Vir 12h2'46" 4d21'35"
Brian & Heather Wedding Star
Psc 1h25'16" 25d41'3"
Brian - Hetal 4Ever
Aql 19h49'28" -0d19'8"
Brian Hoefler
Per 4h23'27" 31d36'27"
Brian Hofer
Gem 6h50'26" 33d56'3"
Brian Hudec
Sco 17h32'30" -40d27'8"
Brian Hugh Edward Course
Leo 9h44'55" 13d16'55"
Brian Hughes
Uma 13h51'8" 61d20'28"
Brian Hults
Uma 14h15'27" 62d10'14"
Brian & Ina - Mai Stowell
Cyg 21h34'55" 46d9'9"
Brian Inga
Sco 17h48'25" -42d14'33"
Brian J. Bobillier Jr.
Sgr 18h11'51" -16d10'18"
Brian J. Cudahy
Per 4h49'6" 46d36'42"
Brian J. Doyle
Her 17h53'46" 35d15'55"
Brian J. Garbrecht
Leo 10h54'53" -2d36'26"
Brian J. Germann
Uma 10h37'23" 39d48'24"
Brian J. Hubbell
Cnv 12h36'0" 44d7'26"
Brian J. O'Malley
Her 17h28'26" 37d6'19"
Brian J. Palmowski
Cep 22h20'35" 69d38'15"
Brian J. Wood
Uma 9h41'18" 46d30'56"
Brian James Anderson
Per 3h43'53" 50d41'27"
Brian James Caza
Lib 15h2'21" -1d44'8"
Brian James DiFebo Byrne
Cam 6h38'46" 75d21'54"
Brian James Evan McLaughlin
Cyg 19h44'1" 33d3'31"
Brian James Kelly
Lyr 19h18'48" 29d39'1"
Brian James Lee
Ori 5h37'50" 12d14'48"
Brian James Lynch
Per 3h44'21" 52d24'15"
Brian James Michael
Psc 0h25'8" 8d49'29"
Brian James Richards
Cap 20h57'19" -21d2'3"
Brian James Whiston
Per 4h35'16" 45d31'35"
Brian & Jane's Love Star
Cyg 20h9'3" 30d56'59"
Brian Janke's Star
Tau 4h27'47" 29d9'30"
Brian Jason Kendall
Lib 14h53'53" -13d0'41"
Brian Jay Washington II
Crb 16h11'37" 34d30'47"
Brian Jeffrey Fink
Lib 15h51'50" -18d12'27"
Brian Jeffry Blackford
Sco 16h6'54" -14d36'55"
Brian & Jen
Ari 3h14'59" 27d10'4"
Brian & Jennifer's Everlasting Star
Cyg 21h3'19" 45d21'18"
Brian & Jenny Star
Cyg 21h33'30" 31d21'51"
Brian & Jeri Burris
Sge 19h12'46" 18d16'28"
Brian John Davies
Cap 21h34'45" -12d9'24"
Brian Johnson
Cep 0h10'55" 87d57'50"
Brian Johnson
Cep 20h16'8" 60d58'33"
Brian Jubb
Boo 14h30'2" 40d53'51"
Brian/JuliaDemmert-Green
Aqr 21h7'28" -9d15'24"
Brian K. Henderson
Tau 4h39'54" 24d45'26"
Brian K. Lacy, Sr.
Ori 6h17'30" 10d50'22"
Brian K. McGinnis
Uma 11h6'49" 65d25'9"
Brian K Selfridge
Per 3h12'8" 55d56'12"
Brian K. Thornton
Cap 20h31'14" -10d7'28"
Brian K Willich
Aqr 22h14'35" -1d17'58"
Brian Keith
Ori 5h37'52" 0d53'54"
Brian Keith Edwards
Sgr 18h32'16" -25d58'2"
Brian Keith Fitzhugh
Aur 5h4'25" 34d53'22"
Brian Keith Gimlin
Her 17h21'4" 39d14'0"
Brian Keith Hardwick
Per 3h34'8" 49d54'30"
Brian Keith Martin
Uma 11h38'52" 48d12'25"
Brian Keith Wakefield
Ori 6h7'10" -0d49'53"
Brian Kelsey Smith
Her 17h33'50" 32d8'19"
Brian Kenneth Bull
Psc 1h26'7" 18d53'10"
Brian Kenneth Whatuira
Vir 13h28'46" -10d47'58"
Brian Kerzetski, I love your smile
Aqr 22h21'8" -0d33'23"
Brian Kevin Boothe
Psc 0h50'19" 8d9'13"
Brian Kevin Worker Gibbons **** BKW
And 23h47'7" 42d14'14"
Brian Kieth Fulfrost
Lyn 7h12'37" 44d22'40"
Brian & Kim
Cyg 20h47'15" 37d27'32"
Brian King
Gem 6h59'4" 13d43'7"
Brian Kirkham
Per 4h15'48" 46d42'8"
Brian Kirwin
Psc 1h26'30" 29d42'51"
Brian Klatsky
Tau 5h4'44" 20d13'54"
Brian Knott
Boo 14h51'49" 52d23'26"
Brian Kyle Erickson
Psc 0h0'53" 7d19'2"
Brian L. Comeaux
Lac 22h30'48" 54d34'27"
Brian L. DelFavero
Sco 16h44'5" -31d40'24"
Brian L. Kropp
Aur 5h53'7" 35d31'19"
Brian L. Moore
Ori 4h59'18" -0d23'3"
Brian L. Roether
Aqr 22h33'13" -9d11'3"
Brian L. Sherman
Her 17h27'1" 38d30'39"
Brian Larue
Sge 19h46'41" 18d21'45"
Brian Lasley
Aqr 22h13'41" 1d16'6"
Brian & Laura
Ori 6h20'10" 14d18'29"
Brian & Laurie's New Year's Star
Her 18h45'22" 21d29'42"
Brian Lawrence
Cyg 21h52'52" 47d10'28"
Brian Lawrence
Uma 9h18'37" 57d14'18"
Brian Le
Cep 22h50'54" 72d16'32"
Brian Lee
Psc 1h16'56" 18d11'41"
Brian Lee Grieco
Aqr 22h54'41" -8d4'25"
Brian Lee Looper
Aur 5h22'35" 31d50'43"
Brian Lee McKane
Her 18h56'48" 24d14'41"
Brian Lee Miller
Peg 21h54'25" 12d22'49"
Brian Lee Peterson Shines Forever
Umi 15h29'54" 76d28'59"
Brian Lee Rose
Cap 21h9'24" -20d58'31"
Brian Lee Tabigne Guevarra
Her 17h51'52" 28d2'54"
Brian Lerner
Per 3h1'25" 56d40'35"
Brian Lewis Michaletz
Lib 15h52'3" -15d22'0"
Brian Lint
Cyg 21h14'6" 42d25'2"
Brian Lloyd Dobson
Pup 7h25'27" -35d5'42"
Brian Loftus
Per 2h45'46" 51d50'52"
Brian Loud
Cyg 20h4'42" 34d45'16"
Brian Louis Hunt
Hya 9h11'22" -0d29'36"
Brian Louis Leibowitz 03-11-1947
Psc 23h37'59" 2d15'3"
Brian Loves Alyse... You R My Star
Aqr 21h10'26" -12d3'2"
Brian LUKA Scharp
Vir 13h6'17" 6d26'39"
Brian Luvs Sam 4 Ever
Leo 9h43'28" 28d4'0"
Brian Lynn Hogstrom
Tau 5h30'37" 18d18'31"
Brian M. Hazel
Boo 14h25'1" 19d40'45"
Brian M. Kelly
Her 17h43'7" 38d16'25"
Brian M Kim - Meanings Redefined
Uma 13h45'44" 61d8'33"
Brian M. Maciolek
Psc 1h19'17" 10d14'38"
Brian Machado
Umi 15h50'22" 83d30'9"
Brian Malmeth
Gem 6h5'28" 26d12'48"
Brian Mandy
Aql 19h14'48" 16d10'39"
Brian Marengo
Sgr 18h12'45" -29d43'48"
Brian & Margery Hines
Uma 11h53'40" 28d39'47"
Brian & Marlisa 1 Year then 4ever
Crb 16h1'4" 38d44'33"
Brian Martin Brennan
Aur 5h44'29" 38d30'12"
Brian Mathew Yerkes
Cyg 19h54'33" 39d48'22"
Brian Matthew
Ori 6h18'6" 13d36'4"
Brian Matthew Mollick
Her 16h34'33" 32d58'7"
Brian Matthew Richard
Cnc 8h17'42" 11d10'53"
Brian Matthew Sladewski
Uma 11h43'2" 34d32'2"
Brian Matthew Wemple
Leo 10h21'48" 18d27'59"
Brian McArdle
Lib 14h56'21" -6d51'44"
Brian McClure Mace
Aur 5h12'43" 41d9'11"
Brian McGrory
Cep 22h44'27" 75d53'46"
Brian McGuire and Felicita's Star
Cyg 20h36'20" 41d15'50"
Brian McKnight
Gem 7h28'29" 31d46'9"
Brian McLaughlin
Lac 22h13'39" 44d13'15"
Brian McNicholl
Aqr 22h7'12" -22d36'38"
Brian McSorley
Aur 5h50'30" 53d53'2"
Brian Mead
Cyg 19h47'45" 51d11'2"
Brian Medford
Her 16h50'37" 46d33'12"
Brian Michael Beckley
Aur 5h49'2" 55d43'58"
Brian Michael Burgess
Her 18h25'13" 12d40'51"
Brian Michael Clark
Tau 3h50'51" 15d50'7"
Brian Michael Diehl
Per 4h24'46" 52d45'0"
Brian Michael Jones II
And 0h25'55" 29d49'37"
Brian Michael Manes
Per 3h6'15" 56d40'30"
Brian Michael O'Donnell
Aur 5h50'39" 46d44'53"
Brian Michael Testa
Her 17h16'35" 28d38'27"
Brian & Michelle Bradley
Mon 7h18'46" -0d40'30"
Brian Millett
Cyg 20h48'51" 33d41'21"
Brian Mitchell B.A.M
Her 16h14'56" 43d42'56"
Brian Moore
Tau 5h40'16" 20d5'13"
Brian Morrow
Cma 6h13'35" -18d50'2"
Brian - My Knight in Shining Armor
Her 17h18'1" 17d35'19"
Brian My Love
Boo 14h54'34" 23d1'33"
Brian N Johanson
Ori 5h7'7" 4d4'4"
Brian Napier
Leo 11h18'4" 15d25'40"
Brian/Natalie
Cyg 20h24'44" 45d45'39"
Brian Neiberg
Uma 9h59'6" 43d51'26"
Brian Neil Burg
Lyn 6h52'47" 51d55'47"
Brian Nestor
Mon 7h35'9" -0d33'23"
Brian & Ning
Her 17h18'43" 44d23'19"
Brian O Neill
Ari 1h53'12" 19d19'17"
Brian Og Crowley
Ori 4h46'55" -3d22'22"
Brian Olson
Cap 20h8'48" -9d37'20"
Brian Omar Nguyen
Ari 2h36'53" 12d52'3"
Brian O'Neill
Sco 16h44'6" -40d6'3"
Brian O'Rourke
Uma 13h40'12" 52d19'57"
Brian O'Shea
Eri 4h28'22" -2d0'6"
Brian Our Diamond Star
Per 3h2'31" 50d41'13"
Brian Owen Williams
Ori 5h27'23" 3d4'7"
Brian P Carr
Psc 0h43'19" 17d2'26"
Brian P. Dravis
Uma 13h30'6" 58d46'30"
Brian P. Schwartz
Ari 2h37'0" 21d49'4"
Brian Paddington
Uma 9h40'15" 41d37'12"
Brian Paiva
Gem 7h26'48" 29d8'1"
Brian Panton
Per 3h33'59" 46d7'37"
Brian Papesch
Lib 15h9'12" -5d13'0"
Brian Parks In Memory Of Tom Parks
Aql 19h8'45" 1d24'57"
Brian Parrello
Per 3h34'44" 35d7'39"
Brian Patrick
Per 3h44'36" 37d36'22"
Brian Patrick
Sgr 18h56'34" -34d18'40"
Brian Patrick Alexson
Aur 5h43'54" 35d3'16"
Brian Patrick Bennett
Lac 22h30'25" 48d52'19"
Brian Patrick Gibson
Uma 8h38'33" 48d21'27"
Brian Patrick Kevin Lainoff Always
Uma 10h39'12" 50d2'33"
Brian Patrick Schlappy
Uma 11h44'40" 31d48'3"
Brian Patrick Wells
Per 4h25'54" 43d58'57"
Brian Pauli - Daddy's Star
Aqr 21h12'54" -11d31'10"
Brian Payne
Psc 0h42'7" 6d36'38"
Brian Peaty
Cap 21h44'30" -18d10'45"
Brian Peters
Cas 23h4'46" 54d52'26"
Brian & Phyllis Trimmer
Tau 4h35'50" 21d57'46"
Brian Pierce
Tau 5h43'34" 25d44'1"
Brian Preseault- Daddy
Psc 1h7'24" 32d24'15"
Brian Pugh
Her 16h41'24" 36d25'30"
Brian R. Grant
Cnv 13h21'14" 29d40'37"
Brian R. Hall
Uma 10h5'18" 45d53'55"
Brian R. Hall
Uma 12h10'19" 56d31'34"
Brian R. Landis
Uma 11h56'25" 55d22'51"
Brian R. Lucas
Per 2h24'30" 52d4'25"
Brian R. McClain
Per 3h26'24" 51d54'23"
Brian R. Paul II
Aqr 22h39'21" -0d0'3"
Brian R. Rickman
Her 17h15'48" 16d1'27"
Brian R. Tarasi
Ori 5h46'13" 2d21'51"
Brian R. Walsh
Uma 8h57'20" 51d59'30"

Brian+Rachel=Best Couple Ever
Cyg 19h43'15" 43d17'52"
Brian & Racquel
Aqr 21h25'10" 1d56'17"
Brian & Rae Perih
Cap 20h23'56" -9d44'36"
Brian Randolph Page
Ari 2h17'44" 23d50'52"
Brian Reed
Col 5h29'47" -35d14'34"
Brian Reed Woody
Leo 10h6'19" 13d43'32"
Brian Richard Baker
Ori 6h13'11" 15d10'44"
Brian Richard Corner
Uma 9h24'54" 48d46'17"
Brian Richard Fuller
Gem 7h9'2" 33d28'10"
Brian Robbins
Ari 2h42'53" 27d26'22"
Brian Robert Clark
Tau 4h32'37" 28d35'53"
Brian Robert Litz
Tau 4h21'1" 18d13'33"
Brian Roberts Logan
Sco 16h14'49" -12d8'15"
Brian Ross Baker
Pav 19h46'17" -58d4'46"
Brian Ruperto - Rising Star
Boo 15h20'23" 41d29'26"
Brian Ruperto - Shining Star
Per 3h5'56" 55d27'55"
Brian S. Bicher
Cap 21h0'18" -22d24'57"
Brian S. McLaughlin
Her 18h25'2" 17d1'7"
Brian S. Mercer
Gem 7h33'20" 20d4'47"
Brian Salinas
Aql 19h33'19" 11d50'7"
Brian Salter
Cep 21h55'4" 59d10'13"
Brian Sams
Aur 6h15'19" 31d6'29"
Brian Samson
Crb 15h49'50" 29d2'2"
Brian Santos
And 2h22'10" 47d53'1"
Brian + Sarah
Cyg 20h41'32" 44d27'50"
Brian Sawchuk
Lib 15h28'55" -12d31'58"
Brian Scherr
Sgr 18h15'24" -33d28'24"
Brian Schmedding
Uma 10h47'52" 53d40'44"
Brian Schmidt
Cap 21h6'57" -15d36'20"
Brian Schnicke
Her 18h43'7" 13d58'58"
Brian Scott
And 23h12'44" 52d9'7"
Brian Scott Anderson
Lmi 10h14'51" 31d54'3"
Brian Scott Erickson
And 0h33'3" 38d34'27"
Brian Scott Lessin
Lib 14h26'0" -9d57'57"
Brian Scott Stokes
Sco 17h34'25" -44d30'44"
Brian - Shanie - Aiden
Vir 13h24'21" 6d22'33"
Brian Simpson
Uma 8h44'51" 59d1'37"
Brian Smyly Brown
Dra 16h42'37" 52d26'56"
Brian Steven Clark
Her 16h33'15" 41d33'32"
Brian Stuart Miller
Her 17h38'18" 43d57'23"
Brian Szumowski
Cyg 20h15'23" 49d6'10"
Brian T. Foley, Esquire
Vir 12h4'33" 4d27'56"
Brian T. Maley
Cap 20h12'10" -15d25'0"
Brian Taylor Miller
Lmi 10h37'26" 29d22'40"
Brian Taylor (The Best Dad Ever)
Per 2h30'29" 52d46'40"
Brian TC Waring
Tau 4h19'9" 26d29'28"
Brian Terence McInerney
Uma 10h58'2" 55d44'39"
Brian Thomas Bridges Family Star
Uma 14h1'23" 55d6'28"
Brian Thomas Cavanagh
Sgr 19h46'47" -12d45'8"
Brian Thomas Fenton
Sgr 19h22'38" -19d30'14"
Brian Thomas Sedlacek Rabenold
Sgr 18h1'28" -29d0'47"
Brian Thomas Sheehan
Leo 11h8'32" 0d25'11"
Brian Thorn's Everlasting Star
Psc 0h15'5" 15d7'16"
Brian Tivnan
Cep 21h31'42" 66d30'52"
Brian Todd Ely
Her 17h40'36" 47d36'27"
Brian Todd Oschmann
Sgr 18h19'2" -32d19'58"
Brian Turner's Wishing Star
Uma 11h42'23" 30d45'36"
Brian Twigg
Cyg 21h51'25" 54d49'13"
Brian Twohig
Per 3h10'23" 54d37'51"
Brian V. Cooke's Star
Per 2h22'57" 57d16'6"
Brian Vences
Per 3h8'59" 54d39'3"
Brian Vernau
Cnv 13h44'57" 31d3'22"
Brian Victor Pearce
Dra 10h6'39" 73d40'57"
Brian Vito Spalliero
Psc 1h12'16" 27d14'3"
Brian W. Ahern
Dra 17h58'46" 60d59'9"
Brian W. Coughlin
Uma 8h14'1" 62d7'3"
Brian W. Murphy
Aql 19h52'54" -0d12'52"
Brian Walsh
Ari 3h12'19" 18d53'55"
Brian Walter Ralston
Her 16h42'13" 27d7'42"
Brian Wayde Kester
Sco 16h12'32" -11d0'51"
Brian & Wendy
Umi 14h5'56" 71d22'19"
Brian Wesley Hilliard
Sex 9h46'45" -2d45'40"
Brian Wesley Rujan
Gem 6h48'27" 32d52'45"
Brian Westmore
Per 3h35'5" 50d54'46"
Brian Whitfield
Uma 10h8'14" 68d48'41"
Brian William Doumar
Ari 3h4'30" 28d47'46"
Brian William Foster
Uma 11h27'37" 58d58'36"
Brian William Harriman (Wizzo)
Uma 10h7'48" 49d17'47"
Brian William Knuth
Vir 13h13'44" -13d45'19"
Brian William Manwiller
Per 4h31'36" 32d26'48"
Brian William Metcalfe
Per 3h53'51" 52d42'50"
Brian William Miller
Cru 12h56'27" -59d5'11"
Brian William Moore
Sgr 18h54'23" -20d25'22"
Brian William Theisen
Gem 6h33'42" 23d1'32"
Brian Yummy Mazzini
Her 16h51'26" 27d17'23"
Brian Zahn
Psc 1h21'7" 6d14'2"
Brian, Jaxy, and Daulton Sullivan
Dra 17h37'10" 50d43'1"
Brian, Laura & Jackson
Uma 12h0'30" 44d33'19"
Brian, my angel
Dra 15h31'11" 59d11'27"
Brian, My Best Friend, My Love
Ori 5h23'23" 3d6'34"
Brian, Tammie's love
Uma 13h34'45" 53d17'18"
Briana
Lib 15h5'55" -15d40'56"
Briana
Cam 4h38'4" 58d37'58"
Briana
Cyg 20h22'6" 41d2'48"
Briana and James
Leo 11h10'38" 11d5'34"
Briana Asante
Sco 16h13'35" -18d12'49"
Briana Augusto
And 1h2'5" 37d46'45"
Briana Baugh
Ori 6h22'24" 14d43'46"
Briana Belinda Howard
Cnc 8h51'40" 26d5'33"
Briana (Bri) Cortez
Lyr 18h47'1" 39d12'36"
Briana Brooke
Tau 4h16'36" 21d56'11"
Briana Carney
Lib 15h19'7" -8d19'50"
Briana Conrad
And 23h48'26" 46d18'52"
Briana Dae
Del 20h44'15" 7d16'54"
Briana Dale Tingler
Gem 7h21'40" 33d16'10"
Briana Dawn
Tau 4h15'50" 17d44'43"
Briana DiFrancesco
And 0h1'15" 44d24'9"
Briana Eileen
Ori 4h53'42" 7d28'42"
Briana Faith Wheeler
Cyg 21h22'11" 32d53'33"
Briana Girl Lucky Star 13
And 2h9'15" 43d17'7"
Briana Helene
Ori 5h44'16" 8d14'44"
Briana Joy
And 1h5'40" 33d57'18"
Briana K. Ponce
Lyn 8h21'18" 55d52'25"
Briana Lee Moody
And 0h0'28" 34d52'48"
Briana Lieneck
Uma 11h28'0" 42d46'22"
Briana M. Raiola
Ori 5h13'32" 7d51'32"
Briana Mangiacapra
Aqr 20h41'54" -9d16'43"
Briana Marie
Tau 5h31'49" 23d27'17"
Briana Marie Hopkins
Ori 6h12'58" 17d25'45"
Briana Marie Norton
Lib 15h13'41" -19d24'26"
Briana Marie Panos
Cyg 20h34'29" 34d5'2"
Briana Maye
Psc 1h11'31" 24d21'10"
Briana Michelle
And 23h24'4" 47d27'45"
Briana Michelle Abrams
And 23h35'59" 44d30'33"
Briana & Mike
Cyg 21h44'19" 48d7'38"
Briana Mykalah Hoch
Cap 21h36'22" -14d50'22"
Briana Nash
And 23h1'32" 52d24'3"
Briana Nichell Loo
And 23h38'21" 48d33'10"
Briana Olson
And 0h52'59" 37d40'6"
Briana Patrice Pepe
Cyg 19h47'33" 35d50'44"
Briana Rae Campanelli
Uma 9h8'41" 48d43'19"
Briana Rain Axtell
Uma 11h42'59" 62d54'39"
Briana Renee
Tau 5h36'28" 26d23'30"
Briana Rose
Lmi 10h21'38" 36d56'25"
Briana Rose Mejia
And 23h47'20" 48d32'19"
Briana & Tristan
Cyg 21h35'9" 53d16'51"
Briana Yvonne
Cnc 8h36'52" 26d50'45"
BrianandLindaWiur
Cyg 19h32'40" 28d58'11"
Briana's Angel
Gem 6h48'20" 20d47'32"
Briana's Escape
Uma 11h15'0" 68d58'32"
Briana's Star
And 23h40'18" 48d11'28"
Briana's Star
Crb 15h36'19" 37d32'39"
Brianie <3 XOXO
Crv 12h31'55" -15d33'43"
brian-mac
Tau 5h31'44" 25d32'44"
Briann matris of parvulus
Uma 10h6'11" 51d33'15"
Briann Walters
Cas 2h32'35" 66d7'44"
Brianna
Cas 1h32'31" 65d22'59"
Brianna
Uma 8h55'22" 61d33'14"
Brianna
Umi 15h42'43" 73d45'29"
Brianna
Lyn 7h47'7" 52d31'8"
Brianna
Sgr 18h53'58" -16d30'57"
Brianna
Sgr 18h34'32" -24d9'17"
Brianna
Sgr 18h45'6" -27d11'13"
Brianna
Cam 3h56'46" 56d12'37"
Brianna
Lyn 8h42'15" 44d7'33"
Brianna
Gem 6h30'31" 16d19'9"
Bri'anna
Peg 21h23'30" 19d31'43"
Brianna
Ori 4h51'36" 8d10'17"
Brianna
Ari 2h13'6" 12d28'50"
Brianna
Vir 13h52'35" 5d11'54"
Brianna Adame
Uma 8h16'48" 70d35'33"
Brianna and Daddy
Cyg 21h28'47" 32d45'14"
Brianna and Nels' star
Lyn 7h56'54" 46d38'11"
Brianna And Tani's Princess Star
And 2h33'19" 39d20'8"
Brianna Baby Girl
And 0h38'10" 34d15'48"
Brianna Banks
Tau 6h0'52" 27d45'20"
Brianna Bella Dinardo
Sgr 19h14'29" -21d59'34"
Brianna Bertoli
Lyn 6h56'14" 50d27'49"
BRIANNA BOGART
Crb 15h51'28" 27d47'26"
Brianna Carovillano
Vir 13h9'30" 12d27'16"
Brianna Chavez
Cas 1h38'2" 65d51'58"
Brianna Claire Brula's Golden Star
Ari 2h54'44" 28d35'49"
Brianna Corinne
Her 16h26'46" 46d31'57"
Brianna Cortes
Lyn 8h31'30" 38d58'40"
Brianna Crowley Spierto-Diodato
Uma 8h50'11" 64d13'29"
Brianna Cummings/Feingold
Cyg 20h32'0" 35d49'53"
Brianna Danele Snyder
Cyg 21h37'37" 39d25'30"
Brianna Daniela Graham
Lyr 18h47'10" 35d13'9"
Brianna Danielle
And 0h13'40" 37d43'41"
Brianna Danielle Powell
Peg 21h48'31" 16d24'14"
Brianna Dawn DeVore
Cyg 20h46'17" 52d15'48"
Brianna DeRose
Sgr 18h41'14" -26d53'35"
Brianna Dillavou
And 1h15'32" 45d25'35"
Brianna Elaine Dahl
Cas 1h41'8" 65d16'29"
Brianna Elaine James Herb
Cap 20h23'37" -14d33'47"
Brianna Elise Thompson
And 0h36'25" 46d24'32"
Brianna Elizabeth Laney
Aqr 20h39'12" -11d47'30"
Brianna F. Carey
Crb 16h5'55" 36d50'46"
Brianna Faith Rawlings
Psc 0h41'15" 7d48'32"
Brianna Fellner
Uma 14h23'2" 61d48'33"
Brianna Galla
Peg 21h34'59" 26d0'57"
Brianna Grace
Cas 0h22'9" 61d31'44"
Brianna Grant
And 2h7'41" 41d9'0"
Brianna Guarino
And 2h17'19" 48d35'35"
Brianna Haley
Uma 13h41'23" 54d42'25"
Brianna Hope
Ari 2h36'10" 23d56'46"
Brianna Hope Nusbaum
Peg 22h48'15" 19d54'29"
Brianna Huffman
Cam 6h21'28" 62d51'56"
Brianna Isabella
Psc 1h34'48" 18d44'51"
Brianna Jayne Day - 26.06.2007 - 14:12
Cru 12h27'24" -59d28'46"
Brianna Jeannette
Peg 23h37'0" 14d24'5"
Brianna Jo Roseberry
And 0h25'39" 31d55'39"
Brianna Jones
And 0h45'2" 39d14'27"
Brianna Juliet
Cap 21h24'28" -25d39'47"
Brianna Kate & Allana Lyn Gadrow
Sex 9h48'24" 1d35'5"
Brianna Kay
Uma 8h53'7" 62d1'36"
Brianna Kaye Fielding
Cas 1h48'59" 63d1'38"
Brianna Kelly
And 0h31'7" 42d54'19"
Brianna Kenney
Leo 10h24'36" 18d52'48"
Brianna Kristell Friedman
Psc 1h11'7" 33d26'19"
Brianna Lane Galiher
Lyn 7h0'18" 47d42'19"
Brianna Layne Patrick
And 0h11'36" 40d9'50"
Brianna Lee
Ari 2h59'11" 17d46'5"
Brianna Lee Farris
And 0h32'26" 28d52'7"
Brianna Leger
Umi 13h27'12" 76d16'3"
Brianna Leigh Besette
Vir 13h58'47" -16d16'55"
Brianna Linin
Lyr 18h42'38" 37d31'8"
Brianna Lyn Waits
Ara 17h33'36" -46d48'1"
Brianna Lynn Baker
Cap 20h12'12" -15d32'1"
Brianna Lynn Callahan
Gem 7h14'1" 24d0'24"
Brianna Lynn Lusk
Gem 7h42'22" 17d24'54"
Brianna Lynn Rivers
Lib 14h47'23" -18d3'51"
Brianna Lynn Schepis
And 0h25'32" 30d24'29"
Brianna Lynn Walker
And 1h11'49" 37d6'42"
Brianna Lynne Ney
Cas 0h45'7" 57d38'16"
Brianna Lynne Shaughnessy
Crb 16h15'3" 37d56'7"
Brianna Margaret Knochen
And 23h51'26" 48d35'4"
Brianna Maria Sankar
Lib 15h26'44" -25d29'55"
Brianna Marie
Uma 11h37'8" 54d52'44"
Brianna Marie
Cnc 8h26'52" 12d23'28"
Brianna Marie Andrew
Cnc 9h7'42" 11d19'54"
Brianna Marie Chevez
Uma 10h41'20" 44d7'47"
Brianna Marie Schwab
Cnc 9h2'59" 31d18'5"
Brianna Maylor
Cap 20h15'0" -9d32'36"
Brianna Mollie O'Connor
Cnc 8h32'2" 17d50'9"
Brianna Nicole
Ari 3h16'16" 26d0'41"
Brianna Nicole Episcopo
Sco 16h10'29" -14d4'41"
Brianna Nicole Firminger
Ori 5h35'14" 3d19'53"
Brianna Nicole McKewen
Sco 16h12'22" -21d40'25"
Brianna Nicole Powell
Gem 7h5'1" 10d34'15"
Brianna Noelle
Cap 21h53'18" -13d11'1"
Brianna Paige Sorensen
Leo 11h20'45" 15d9'15"
Brianna Price
Ari 2h49'41" 16d44'3"
Brianna Quarry - Rising Star
Lyr 19h13'41" 27d16'0"
Brianna R Kuehling
Leo 11h45'10" 24d35'40"
Brianna Rae
Gem 7h32'31" 26d31'59"
Brianna Rae Kazee
Leo 10h8'37" 20d13'20"
Brianna Rae Ludeman
Gem 6h5'49" 23d29'5"
Brianna Raub
Aqr 22h18'50" 1d14'25"
Brianna Rene Kubitskey
Ori 5h53'1" 22d14'22"
Brianna Rene Rios
Tau 3h57'54" 24d37'10"
Brianna Rose
And 0h33'27" 29d34'7"
Brianna Rose Fineo
Cas 1h14'13" 69d52'39"
Brianna Rose Thompson
Psc 1h36'27" 19d54'40"
Brianna Rose Vallone
Dra 18h54'46" 77d52'40"
Brianna Sera Benz
And 1h27'52" 46d46'31"
Brianna Sosnowski
Cas 1h24'53" 64d50'53"
Brianna Spagnuolo
Cyg 21h21'18" 46d32'7"
Brianna T. Garner
Cap 20h56'11" -17d2'33"
Brianna T. Olsen
Gem 6h51'47" 27d49'59"
Brianna Taylor Swearingen
Uma 13h55'56" 53d47'22"
Brianna York
Vir 11h52'24" 6d33'11"
Brianna Zdanciewicz
Aqr 21h0'3" 0d53'33"
Briannah Chantae Reed
And 2h12'30" 44d43'6"
Briannah Crummey
Cap 21h55'35" -19d28'40"
Briannah Lois
Vir 12h32'7" 4d4'26"
Briannalee
Cas 0h17'42" 50d31'53"
Brianna's Beauty Star
And 0h41'9" 33d9'53"
Brianna's Peanut
Ori 5h55'26" -0d42'53"
Brianne
Vir 12h22'44" -6d10'18"
Brianne
Aqr 21h39'54" 1d48'25"
Brianne
Ori 5h37'8" 12d7'29"
Brianne
Gem 6h38'43" 20d11'11"
Brianne
Leo 10h24'43" 15d10'9"
Brianne #5
Lyn 7h0'21" 51d42'56"
Brianne Barry Carden
Aqr 20h47'20" -10d8'32"
Brianne Bennett
And 0h48'5" 33d8'6"
Brianne Brookes Kaela Ching
Ori 6h9'47" 6d20'17"
Brianne Carpenter
Lyr 18h48'36" 45d5'16"
Brianne Christine Dubuque
Psc 0h46'7" 10d17'48"
Brianne Deininger
Gem 7h22'54" 26d34'8"
Brianne Elizabeth
And 2h19'53" 47d19'39"
Brianne Elizabeth Reilly
Aqr 23h2'21" -9d20'54"
Brianne Elizabeth Taylor
Peg 22h10'12" 10d26'11"
Brianne Evans
Gem 7h28'14" 25d46'24"
Brianne Isabell
And 23h3'27" 52d23'30"
Brianne Klossner
Cam 3h58'8" 69d0'18"
Brianne Leakey
Men 5h15'20" -71d27'0"
Brianne Linette Hanson
And 23h47'19" 36d36'39"
brianne madsen
Uma 12h2'15" 55d30'30"
Brianne Manton
Peg 21h37'42" 22d42'42"
Brianne Marie
And 2h22'7" 49d27'46"
Brianne Marie Jorgensen
Lmi 10h45'24" 32d5'59"
Brianne Michelle Bartlett
Leo 9h35'56" 28d45'56"
Brianne Nicole Downey
Psc 1h31'20" 19d28'10"
Brianne Richelle Conidi
Umi 15h30'0" 74d8'12"
Brianne Rose
Gem 7h25'15" 26d41'59"
Brianne Valenzuela
Cap 20h21'40" -13d12'1"
Brianne's Star
Lmi 10h47'33" 24d8'53"
Brianne's Twinkler
Cnc 8h15'55" 15d30'19"
Brian's 50th 6607
Gem 6h47'40" 31d30'51"
Brian's daughter Wallis
And 1h3'55" 39d48'32"
Brian's Dream
Cnc 9h8'59" 9d35'41"
Brian's Eros
Cnc 8h37'1" 23d11'6"
Brian's eyes
Ori 5h32'44" -0d27'16"
Brian's Legacy
Umi 16h46'10" 83d4'25"
Brian's Love Star
Per 3h27'17" 53d44'1"
Brian's Lucky
Cnc 8h46'28" 18d3'13"
Brians Shining Nia
Cyg 19h40'57" 28d40'56"
Brian's Song
Cep 3h15'0" 82d37'55"
Brian's Spontanaiety
Vir 13h10'18" -12d5'55"
Brian's Stamina
Boo 14h57'21" 52d3'20"
Brian's Star
Cyg 21h42'35" 39d17'41"
Brian's Star
Psc 1h0'43" 26d39'26"
Brian's Star
Tau 4h25'2" 16d13'53"
Brian's Star
Sgr 19h34'33" -16d28'22"
Brian's Star
Dra 17h5'17" 62d35'4"
Brian's Star
Uma 10h49'10" 53d52'47"
Brian's Star
Sco 17h54'12" -36d11'7"
Brian's Way
Per 2h43'10" 54d0'45"
Brian's Wishes
Gem 6h29'56" 24d41'57"
brianwebbboone-bub
Aqr 22h39'59" 1d26'2"
Briar Molly Parker
Cra 18h2'34" -37d11'28"
Briar Owen Bethel
Boo 14h34'42" 52d31'46"
Briar Rose Gibbons
Crb 15h37'53" 35d57'14"
Bribear Beard
Vir 13h35'0" -9d31'8"
Brice
Uma 8h50'1" 62d56'24"
Brice
And 2h25'41" 40d59'9"
Brice and Amanda Vineyard
Sge 19h49'12" 17d51'56"
Brice and Megan
Gem 6h58'11" 22d0'50"
Brice Andrew Yatsko
Tau 5h52'41" 28d12'33"
Brice Ayden
Sgr 18h36'48" -17d34'56"
Brice Cox Star
Aqr 22h55'55" -6d0'13"
Brice Palmer
Boo 14h42'4" 16d42'26"
Brice Thomas Smith
Gem 6h32'57" 16d23'25"
Brice's Star
Cam 4h14'55" 68d58'7"
Bricia Acosta Izabal Newnam
Cap 20h7'32" -20d1'26"
Briciola
Peg 21h26'43" 21d38'56"
Bricket
Psc 23h15'23" 6d35'32"
Briçou
Cas 0h31'58" 58d29'38"
Brid O'Leary
And 2h12'50" 45d8'48"
Bridge and Cliff Einhorn
Aqr 22h58'48" -9d58'37"
Bridge Betriebsdaten AG
Cep 22h53'22" 74d43'44"
Bridge Star Q3-2006-Tiffany Morris
Uma 9h51'55" 70d41'2"
Bridger
Cam 5h49'27" 65d39'9"
Bridger and Porter Jacobson
Cap 20h24'10" -14d14'38"
Bridger James Nakamura-Koyama
Umi 14h32'0" 85d59'53"
Bridger Stowe Grebe
Lyn 7h37'46" 56d38'55"
Bridger T. Baur
Lyn 7h40'40" 39d49'12"
BridgeStar Q4-2005-Heather Linton
Lyn 8h6'11" 48d0'6"
BridgeStar Q4-2006 -Tobey Brewin
Lyn 6h18'24" 56d38'27"
BridgeStar-2004-Diana Jones
Umi 14h53'35" 74d49'23"
BridgeStar-2004-Hope Crihfield
Cyg 20h4'5" 35d0'59"
BridgeStar-2004-Kristine Kleen
Lyr 18h31'28" 36d46'16"
BridgeStar-2004-Sarah Harris
Crb 16h11'26" 32d35'53"
BridgeStar-Illona Tatriubaviciene
Lyr 18h39'2" 37d30'15"
BridgeStar-Jolanta Kupliauskiene
Crb 15h36'22" 31d22'43"
BridgeStar-Q2-2006-Julia Estrada
And 0h15'24" 43d5'43"
Bridget
And 1h8'29" 40d45'38"
Bridget
And 2h24'25" 44d1'2"
Bridget
And 0h31'42" 36d59'33"
Bridget
Crb 16h3'19" 39d3'49"
Bridget
And 2h29'43" 50d10'25"
Bridget
And 23h2'30" 46d11'34"
Bridget
Ari 2h3'37" 13d17'5"
Bridget
Peg 23h37'50" 13d22'2"
Bridget
Psc 1h22'29" 25d1'58"
Bridget
Ari 2h7'39" 26d14'36"
Bridget
Cnc 9h0'54" 15d49'22"
Bridget
Cnc 8h13'37" 29d4'26"
Bridget
Uma 9h3'2" 57d45'7"
Bridget
Cyg 21h53'13" 53d44'43"
Bridget
Leo 11h27'16" -2d8'11"
Bridget
Aqr 22h25'30" -0d11'19"
Bridget
Lib 14h38'8" -9d51'28"
Bridget
Lib 15h27'36" -9d51'10"
Bridget
Psc 1h26'20" 6d25'19"
Bridget
Cap 20h10'52" -25d33'42"
Bridget Allu Staley
Gem 6h17'49" 22d10'13"
Bridget and Breann's Star
Her 17h56'49" 24d0'50"
Bridget and Hatley BFF
Lyr 19h18'5" 29d2'24"
Bridget and Raven
Uma 10h28'7" 55d58'13"
Bridget " Angel" Cahill
Aqr 20h48'45" 1d55'3"
Bridget - Angel of Our Hearts
Tau 4h38'40" 7d41'26"
Bridget Ann Guadagno
Cas 0h19'11" 55d11'27"

Bridget Ann Nedohon (buttercup)
Cyg 19h43'44" 56d42'40"
Bridget Anne
Lib 15h38'59" -12d57'24"
Bridget Anne Pierce
Ari 2h38'30" 20d26'9"
Bridget Ashley
And 0h6'56" 39d29'28"
Bridget Barr
Cam 3h32'59" 62d59'0"
Bridget Buck
Aqr 21h12'46" -5d54'38"
Bridget C. Whalen
Sco 16h13'20" -28d39'9"
Bridget Cathleen O'Neill
Ori 5h32'9" 1d46'15"
Bridget Celeste Stahl
Cnc 7h59'19" 14d52'45"
Bridget & Chris
Tau 5h26'3" 27d16'26"
Bridget Clare Hansen
Cru 12h12'29" -62d18'12"
Bridget Cook
Cas 0h33'21" 63d31'10"
Bridget Elaine
Gem 6h17'54" 21d48'55"
Bridget Elise
Lib 15h15'51" -15d35'24"
Bridget Elizabeth McGrath
Cam 7h23'40" 67d51'21"
Bridget Elizabeth Reilly
Cas 1h7'1" 49d50'45"
Bridget Elizabeth Shea
Lib 15h53'16" -18d13'9"
Bridget Gille
Ori 6h5'23" 19d32'57"
Bridget Helen Gately
Lyn 7h34'12" 36d49'27"
Bridget Josephine Kelly
Uma 10h0'8" 69d27'17"
Bridget K. Winstanley
Cnc 8h3'41" 6d44'18"
Bridget & Kizito Paganini
Umi 15h33'24" 73d6'26"
Bridget Leah
Uma 8h50'32" 48d16'49"
Bridget Leslie Peeples' Star
Cap 20h26'7" -14d0'12"
Bridget Lijovic
Sgr 19h34'29" -23d19'50"
Bridget Louise Day
Cas 0h11'10" 55d3'2"
Bridget Louise Stasenko
Sgr 19h11'4" -13d47'1"
Bridget Lynn Stine
And 0h49'23" 42d52'34"
Bridget M Brooke
Gem 7h9'28" 33d55'34"
Bridget M. Smith
And 23h5'56" 41d49'47"
Bridget & Marc
Cyg 19h30'26" 28d24'3"
Bridget Marie Neal
Aqr 21h30'57" -8d10'42"
Bridget Mary Miller
Sco 17h57'8" -30d55'9"
Bridget McCabe
Uma 10h26'22" 42d7'31"
Bridget McGrath Moss
Cas 3h15'21" 68d2'0"
Bridget Melanie Anne Thomas
And 0h0'41" 43d49'9"
Bridget mi amor
Psc 23h15'10" 3d28'58"
Bridget & Mike Forever
Cyg 20h5'33" 40d5'29"
Bridget Nichole Prinkey
Sgr 18h15'9" -24d35'39"
Bridget Nicole Cabrera
Vir 12h11'26" 4d45'8"
Bridget Onda
Lib 14h27'12" -14d57'8"
Bridget Page Monson
Cyg 21h36'0" 46d30'16"
Bridget R. Guilford
Gem 6h3'36" 25d53'6"
Bridget Rose Frances Carroll
Sco 16h18'37" -9d37'51"
Bridget Rose Franke 4-11-2003
Cep 22h17'48" 64d27'6"
Bridget Rosemary Hinkley
Cas 0h52'50" 52d38'19"
Bridget Ruth McLellan
Psc 0h41'1" 9d2'47"
Bridget Sullivan
Ari 3h14'16" 27d24'44"
Bridget White
Mon 6h51'21" -0d10'3"
Bridget Zwimpfer
And 1h5'27" 37d15'54"
Bridget, the best MOM - universe
Cnc 8h52'24" 28d17'34"
Bridget1978
Leo 9h41'12" 25d34'23"
Bridget22
Leo 9h42'50" 27d45'16"
Bridget72564
And 1h29'53" 49d2'2"
Bridget's Beautiful Dream
Cas 23h54'19" 61d12'55"
Bridget's Blessing
Tau 4h4'55" 7d53'11"
Bridget's Delight
And 1h21'17" 44d38'5"
Bridget's Everlasting Star
Leo 11h9'37" 20d13'4"
Bridget's Star
Cas 0h51'50" 50d25'44"
Bridget's Star
Mon 6h49'29" -7d1'13"
Bridgets Star
Lib 15h44'13" -28d34'39"
Bridgett Ajike
Cap 21h37'42" -15d28'51"
Bridgett Dee Bowers
Aqr 21h31'31" -7d23'31"
Bridgett Denise Sing
And 0h53'36" 40d28'7"
Bridgett Garza
Gem 7h1'3" 21d47'47"
Bridgett Jackson
And 2h6'7" 42d52'24"
Bridgett McTear
Cap 20h22'22" -13d47'41"
Bridgett V. Rykaczewski
Peg 23h50'8" 19d50'33"
Bridgette
Tau 4h29'33" 21d25'2"
Bridgette Aldrich
Cap 21h46'33" -9d55'36"
Bridgette Bester
Eri 3h33'47" -40d7'0"
Bridgette de Carteret Turner
Cru 11h56'29" -59d45'20"
Bridgette Delay
Crb 15h35'1" 28d5'31"
Bridgette Dickson
Vir 12h14'12" 0d28'17"
Bridgette Ellen Witz
Cas 0h0'34" 57d21'28"
Bridgette Janik
Leo 11h20'50" 16d38'33"
Bridgette Jean McCarthy
And 2h35'28" 44d40'30"
Bridgette leigh wallner
Tau 5h55'35" 28d22'50"
Bridgette M Oeser
Vir 14h15'37" -13d58'29"
Bridgette Marie Killian
Sgr 18h53'56" -36d34'53"
Bridgette Marie Reyes
Uma 10h54'19" 59d10'54"
Bridgette Mundy 2/18/58
Mon 7h33'41" -0d47'31"
Bridgette "My Silly Willy"
Cap 21h34'0" -19d38'38"
Bridgette René
Gem 6h50'59" 17d39'48"
Bridgette's Love
Sco 16h7'47" -12d56'8"
Bridgette's Star
Sgr 18h30'44" -33d34'38"
Bridgett's Love
Leo 9h26'27" 11d30'12"
Bridgid Anne Mullen
Vir 13h54'11" -22d12'26"
Bridgitte & Troy
Cyg 19h37'22" 39d4'25"
Bridie
Lyr 18h49'58" 35d20'39"
Bridie Conroy
Cru 12h1'45" -62d15'44"
Bridie Isabel
Tau 3h48'47" 11d44'22"
Bridie Whyte
Peg 21h38'26" 16d12'23"
Bridie's Star Of Strength - 13.07.85
Cnc 8h4'19" 9d15'39"
Bridjet
Crb 16h6'59" 36d44'26"
Bridjett Ogrizovich
Leo 9h47'24" 16d41'44"
Brido - Brian Langridge 1946-2005
Uma 9h8'21" 51d58'21"
Brie
And 0h47'2" 39d39'50"
Brie Carlson
And 2h16'20" 47d10'28"
Brie J Collier
Cru 12h18'10" -57d25'12"
Brie Lawrence
Her 16h30'12" 47d38'31"
Brie LeClair
Cap 21h39'36" -15d19'47"
Brie Martin
Lep 5h19'59" -11d17'45"
Brie Sausser
Tau 5h47'29" 14d45'38"
Brieana Starr
Lep 5h33'8" -11d56'47"
BrieAnn
Sgr 17h54'34" -16d23'47"
Brieann Annzenette Franco
Umi 13h47'23" 69d59'17"
Briee
Lmi 10h43'3" 28d22'6"
Briege McParland
Uma 13h45'18" 57d6'23"
BriehanAlex
Cap 21h8'31" -15d42'55"
Brieler, Wilhelm
Ori 6h8'36" 8d55'17"
Brielle
Sgr 19h20'35" -22d4'34"
Brielle
Uma 8h9'53" 65d2'24"
Brielle & Adam Waters for eternity
Ara 17h8'14" -56d55'45"
Brielle Ann Neumann
Ari 1h51'44" 18d51'53"
Brielle Janice Digiacco
Leo 9h47'41" 29d32'8"
Brielle Marina
Gem 7h48'48" 30d0'30"
Brielle Renee
Cru 12h39'52" -58d53'14"
Brielyn M.
Vul 20h52'7" 22d18'16"
Brien and Joy
Lyn 8h37'29" 36d7'17"
Brienna Leah Williams
Lyn 7h48'26" 43d14'27"
Brienna Leigh Matson
Lib 14h55'24" -4d18'26"
Brienne
Lib 14h58'14" -2d57'7"
Brienne
Gem 7h46'21" 32d32'41"
Brienne Burosch
Crb 15h58'49" 35d16'36"
Briex's Angel
Leo 11h9'21" 3d20'18"
Brig. Gen. Brian Arnold
Per 3h7'44" 47d52'30"
Brig. General Joseph Reynes, Jr.
Per 4h6'31" 34d11'9"
Brigadier General James W. Swanson
Aql 19h32'41" -0d1'46"
Brigadier General Margrit Farmer
Aql 19h12'24" 3d19'31"
Brigadier General Michael J. Basla
Cnc 8h53'29" 8d8'47"
Brigadier General Wade Farris
Her 18h22'38" 15d30'51"
Brigadier General William Gerety
Crb 15h57'25" 28d22'4"
brigádny generál Ing. Juraj BARANEK
Aql 19h47'8" 11d49'19"
Brigetta
Ari 2h47'52" 29d27'0"
Brigette Michelle Marceau Brie 1984
Ori 6h13'40" 15d30'36"
Brigette T. Murdocco
Ori 5h50'43" 11d30'41"
Brigette's Star
And 0h20'27" 31d50'33"
Briggs Alan Haugh
Lib 14h54'48" -2d56'4"
Briggs-Bresser
Mon 6h38'9" 5d42'25"
Brighid
Pho 0h30'19" -45d46'9"
Bright
Peg 22h43'19" 21d59'6"
Bright Adoring Star, William J Meeker
Aql 19h31'30" -0d48'10"
Bright Afton
Uma 10h52'53" 50d20'55"
Bright as Brooke
Aqr 23h13'2" -14d37'4"
Bright Beautiful Ashley
Ari 3h7'32" 20d29'29"
Bright Beautiful Stephanie
Crb 16h9'4" 36d39'6"
Bright Betty Newman
Cas 2h43'40" 57d51'31"
Bright Billy III Guiding Light
Cma 7h1'7" -22d37'15"
Bright Blue Twinkle
Tau 3h54'41" 22d51'58"
Bright Claudia
Per 4h31'0" 39d23'50"
Bright Eye in the Sky
Leo 9h54'6" 22d23'48"
bright eye tarah
Lib 15h52'1" -10d30'28"
Bright Eyes
Ori 5h33'23" -0d28'19"
Bright Eyes
Sgr 18h21'13" -32d59'47"
Bright Eyes
Cnc 8h39'22" 18d1'50"
Bright Eyes
Ari 2h49'54" 29d32'23"
Bright Eyes
Cnc 8h32'19" 13d29'59"
Bright Eyes One
Cep 23h17'53" 79d39'11"
Bright Julia Elena
Lib 15h37'47" -10d25'43"
bright "lil" shiny star
Uma 10h39'50" 42d9'8"
Bright Rosa
Cas 1h37'48" 72d49'13"
Bright Sparkle
Sco 16h13'33" -12d36'45"
Bright Star for my little Julie!
And 2h0'55" 36d9'24"
Bright Star Irene
Cyg 20h3'50" 35d53'46"
Bright Star Joseph Charles
Per 3h11'10" 41d35'21"
Bright Star of Susan
Ori 5h50'13" 9d6'36"
Bright Starlight Fredrica
Sco 17h35'33" -38d11'55"
Bright Zanic
Eri 3h44'57" -12d25'36"
Bright-Deborah
And 1h38'51" 46d48'20"
Brighter Than Sunshine
Mon 7h6'31" -8d43'26"
Brightest Betty, Phil & Gary's Mom
And 1h23'19" 35d32'34"
Brightest Light In My World
Sco 17h38'34" -39d16'29"
Brighton
Pho 23h41'29" -41d41'29"
Brighton
Peg 23h52'12" 17d30'7"
Brigid
Cyg 21h11'36" 46d50'35"
Brigid Eileen McCleery
Cyg 21h36'28" 50d48'26"
Brigid Elizabeth McKinley
Peg 22h19'38" 15d16'48"
Brigid Fay McNamara
Vir 14h16'27" 1d45'16"
Brigid Glioscarnach Glóire
Dra 10h0'34" 78d36'55"
Brigid & Norm Skulina 35th Anniv.
Uma 11h12'42" 34d57'43"
Brigid Rose Emanuel
Ari 2h41'34" 26d50'33"
Brigida
Her 18h45'52" 13d50'17"
Brigida
Sco 16h10'34" -11d51'31"
Brigida Baur-Casati, 26.10.1956
Uma 9h14'21" 57d41'32"
Brigit Anne
Psc 1h58'50" 7d42'56"
Brigit Irene Christensen
Uma 13h30'44" 52d57'17"
Brigit Robison
Cnc 8h43'23" 28d8'1"
Brigitta Dreni
Per 4h16'43" 34d10'35"
Brigitte
Uma 9h44'43" 51d7'35"
Brigitte
Ori 6h11'52" 10d25'16"
Brigitte
Cas 1h1'5" 61d59'28"
Brigitte
Uma 9h14'41" 58d12'59"
Brigitte
Cas 1h41'48" 67d24'26"
Brigitte and Philip Forever
Sgr 18h11'2" -16d8'48"
Brigitte & André
Umi 13h30'24" 74d9'14"
Brigitte Angelica Howe
Com 12h22'38" 28d30'6"
Brigitte Ashley Schell
Lib 15h38'28" -22d18'28"
Brigitte Diez
Her 17h50'52" 43d1'12"
Brigitte E. Brueneman
Vir 12h50'33" 3d36'25"
Brigitte Elise
Cyg 21h33'12" 38d17'1"
Brigitte Foltin
Uma 9h52'37" 41d38'23"
Brigitte For Love
Uma 9h18'9" 57d13'36"
Brigitte Gliese
Ori 4h52'32" -0d3'44"
Brigitte Gudeste
Sgr 18h15'23" -20d0'9"
Brigitte Hauss
Ori 6h23'47" 15d57'34"
Brigitte Hölzer
Uma 9h26'46" 50d21'48"
Brigitte Hummel
Peg 21h38'57" 15d31'32"
Brigitte Jastant
Umi 13h47'14" 72d55'21"
Brigitte Langheld
Ori 5h12'48" 15d58'0"
Brigitte Maniscalco-Meier
Umi 16h5'18" 71d51'11"
Brigitte Marie Desvaux
Lyn 7h8'46" 48d12'55"
Brigitte Maso Deliège 28/03/1954
Ari 2h51'4" 21d46'2"
Brigitte Natchigall
Leo 9h52'40" 12d50'45"
Brigitte (Poo-Poo Bunny Fart)
Lep 5h53'38" -12d26'51"
Brigitte & René
Umi 13h48'22" 70d16'22"
Brigitte S. Howard
Cap 21h51'10" -12d27'13"
Brigitte Schrock
Gem 6h55'41" 26d14'12"
Brigitte Stenders
Uma 10h40'57" 41d22'53"
Brigitte "the always shining"
Per 3h37'48" 45d14'8"
Brigitte Verschnik
And 0h0'37" 42d59'39"
Brigitte Zimmermann
Uma 8h48'29" 60d6'7"
Brigitte, nuestra estrella
Psc 1h23'28" 10d59'28"
Brigitte's Wish
Cnc 8h29'45" 23d32'17"
BrigitteT27021961
Uma 11h44'58" 41d30'58"
Brihan
Sco 17h20'31" -41d31'39"
Bri-Lee
Cyg 20h36'29" 38d24'39"
Briley Ana
Crb 16h2'15" 33d23'20"
Briley Nicole Clanin
Lib 15h2'21" -7d7'21"
Brilia
Uma 11h36'39" 38d12'19"
Brilie
Uma 12h14'10" 53d45'42"
BriLisa
Dra 18h42'5" 55d35'2"
Brill
Ari 2h5'55" 23d23'8"
Brillant Lee
Aql 19h32'17" 11d47'38"
"Brillante della Dio"
Lyn 8h22'17" 36d52'46"
Brillante fiore
Uma 9h40'23" 57d11'53"
Brilliance of Elliot Tanner5/1/1975
Cap 21h46'49" -20d38'5"
Brilliance of Paul Johnston Mathers
Vel 9h29'52" -42d9'15"
Brilliant
Sgr 19h0'45" -12d22'20"
Brilliant Allison
Sgr 18h41'14" -27d15'45"
Brilliant Barney Hendrix
Lyn 6h25'41" 55d12'53"
Brilliant Baz
Cyg 19h49'12" 38d29'4"
Brilliant Irene Symphony
Cyg 19h45'23" 35d15'1"
Brilliant Jul
Leo 11h21'6" 13d9'10"
Brilliant Star Asa
Lib 15h27'52" -10d46'54"
Brilliantine Lillianus Genicus 1923
Gem 6h43'0" 15d19'51"
Brina
Cam 5h38'21" 69d34'58"
Brina Leigh
Leo 9h35'0" 27d3'48"
Brina Lynn
Ari 2h35'43" 25d14'14"
Briney
Uma 11h8'7" 54d42'50"
Briney Marlin
Tau 5h55'40" 25d24'48"
Bri-n-Jess
Her 16h39'50" 6d14'15"
Brinkle B
Peg 23h18'28" 19d27'35"
Brinley Heckermann
Lyr 18h50'30" 36d29'11"
Briony Grace Francis
Cyg 20h55'21" 49d26'6"
Briony Sixsmith
And 1h46'55" 40d40'39"
Briony's Light
Sco 16h46'1" -37d57'33"
Bri's Heart
Gem 6h40'50" 18d45'21"
Bri's Own Star
Tau 3h52'6" 25d25'52"
Bri's Star
And 1h48'23" 45d16'1"
Brisa de mi Eternidad
Cyg 19h32'20" 28d45'59"
BRISAPP
Dra 16h9'59" 63d55'49"
Brischi
Uma 13h13'12" 52d20'48"
Brisean
Gem 6h25'18" 20d8'12"
Briseis
Uma 8h51'12" 66d27'51"
Briselda
Aqr 20h42'14" -8d57'7"
Brisen Sayer Day
Umi 15h25'15" 72d6'56"
brishas star
Peg 22h5'13" 31d47'26"
Brison Reid George
Dra 19h18'47" 74d27'39"
Bristol-McKinley
Per 4h27'34" 44d46'54"
Briston Rashaun Jackson
Peg 21h43'29" 13d16'20"
'Brit' - Britni Rene Schneider
And 0h50'9" 40d39'20"
Brit n Flip
Gem 6h26'31" 23d25'45"
Brita Martiny Yssel
And 0h45'40" 43d2'9"
Brita Melissa
Leo 11h0'17" 9d31'14"
Brita Reichelt
Crb 15h39'24" 31d45'24"
Brital
Lib 15h27'41" -15d43'17"
Britani & Brandon's Spot in the Sky
Ari 3h12'1" 29d16'10"
Britani Carisa Phommyvong
Psc 0h13'5" 10d38'13"
Britani Jane
And 1h46'30" 39d16'9"
Britanica
Leo 10h27'15" 16d52'59"
Britanica
Ori 5h24'17" -0d13'9"
Britanny B. (Limelight from Above)
Leo 10h39'51" 12d42'24"
Britany Dayan
Cap 20h51'13" -15d41'55"
Britany Elizabeth Warren Coen
Vir 14h39'36" -0d6'56"
Britany Whitaker
Cas 1h13'44" 49d17'15"
Brit-Brit
Lyn 8h6'34" 55d15'45"
Brite Blue Iz
Cyg 20h27'29" 52d37'49"
Brite Eyes
Umi 14h30'17" 72d46'0"
Brite Lite
Cnc 8h52'15" 31d5'12"
Britini Loves Chankakada Ork
Umi 14h56'38" 75d10'29"
Britknee Nguyen
Aur 5h48'42" 46d44'19"
Britnee Ann Bradley
Aqr 21h48'18" -0d12'17"
Britnee LeNae
Uma 9h17'50" 64d46'24"
Britnee Tonille
Cam 7h25'20" 73d39'55"
Britney
Crb 16h6'21" 37d55'35"
Britney
Psc 0h58'4" 15d27'7"
Britney Ann Gardner
Gem 6h49'21" 33d51'41"
Britney Castro December 5th, 2006
Uma 10h20'34" 46d35'23"
Britney Fernandez
And 1h25'11" 40d39'18"
Britney Jan O'Connell
Ori 5h1'49" 9d50'8"
Britney June
Lyn 8h17'37" 33d10'8"
Britney Leigh Sanders
And 2h16'18" 45d49'56"
Britney LeShea Keil
Uma 11h27'53" 53d47'51"
Britney Lucas
Uma 11h39'49" 53d59'3"
Britney Michelle Asher
Ori 5h37'22" 3d33'2"
Britney Mortenson
Lib 15h56'20" -18d9'51"
Britney Nicole Niles
Dra 17h25'57" 50d47'43"
Britney Roberts
Lyr 19h20'6" 28d36'53"
Britney & T.J.
Cyg 19h43'40" 35d14'47"
Britney's Amazing Light
Sco 16h18'34" -19d37'49"
Britni
Cnc 8h29'41" 31d18'40"
Britni Bane
Cas 23h30'49" 55d39'24"
Britni Bardin
Sgr 19h8'57" -21d7'41"
Britny Jenkins
Ari 2h53'19" 18d45'55"
Britoni
Cyg 21h15'30" 40d38'50"
Britshaka
Lib 14h53'36" -3d24'5"
britt
Lib 15h8'59" -7d12'58"
Britt and Jay's Star
Oph 17h32'29" 7d58'58"
Britt and Krista Boughey 11.24.01
Her 16h48'4" 33d13'12"
Britt Beefcake Hess
Ari 3h9'40" 29d25'48"
Britt Coleman
Gem 6h33'0" 21d1'21"
Britt EMBA 2005
Lib 15h22'25" -8d45'39"
Britt Faaren Rosenberg
Psc 1h18'11" 33d4'37"
Britt Floch
Aqr 20h45'17" 1d53'2"
Britt Maxfield
Ori 6h8'43" 19d12'16"
Britt Myers
Uma 12h30'40" 56d37'53"
Britt Sutton
Ori 5h37'11" 3d13'9"
Britt The Beautiful
Cnc 8h43'35" 32d15'40"
Britt Vestby Kristensen
Cam 5h21'36" 77d52'59"
Britt, We Love You. G, S, C B-K
Cra 18h16'14" -41d13'56"
Britta
Cap 20h36'21" -16d52'50"
Britta
Uma 10h22'56" 55d11'41"
Britta
Leo 10h5'0" 15d24'48"
Britta
Leo 10h16'46" 26d11'52"
Britta Hamberg
Cas 0h47'32" 57d52'6"
Britta Katherine Jones
Crb 16h19'54" 33d59'11"
Britta & Martin
Uma 10h28'44" 40d17'44"
Britta Schönen
Uma 10h45'45" 48d28'10"
brittandfos02142007
And 23h16'27" 50d20'30"
Brittanee Michelle Sinner
Sgr 19h23'49" -18d33'3"
Brittaney Erin Essenmacher
Uma 11h9'29" 30d53'54"
Brittaney Johnson
Ari 2h33'52" 17d55'13"
Brittaney Star
Ari 3h11'54" 27d50'54"
BRITTANI
Ari 2h26'48" 26d37'8"
Brittani
And 23h0'52" 47d6'6"
Brittani Adams
Sco 17h10'38" -31d42'16"
Brittani Dania
And 1h30'20" 47d49'24"
Brittani & Giang Forever
Lib 14h26'22" -16d43'48"
Brittani Nichole Osborne
Mon 6h53'12" -0d12'36"
Brittani Royelle Ivery
Gem 7h8'41" 28d17'27"
Brittania Blair
Ori 6h3'33" 10d49'19"
Brittanie Leigh
Lyn 6h23'40" 56d17'21"
Brittanie M. Hart
Vir 13h11'49" -18d38'10"
Brittanie's Hope
Aqr 23h44'49" -14d50'53"
Brittanny & Jason Mertz
Cyg 20h14'54" 56d3'29"
Brittany
Cas 0h24'53" 61d11'1"
Brittany
Cas 0h45'49" 64d12'58"
Brittany
Uma 9h5'0" 55d30'18"
Brittany
Uma 10h26'11" 65d32'14"
Brittany
Cap 21h44'4" -10d34'14"
Brittany
Cap 21h45'33" -10d36'20"
Brittany
Lib 14h53'50" -12d2'50"
Brittany
Sgr 18h12'47" -22d7'3"
Brittany
Lib 15h12'6" -7d22'13"
Brittany
Sco 17h20'19" -32d38'7"
Brittany
Tau 3h36'15" 6d13'45"
Brittany
Vir 12h59'45" 0d51'7"
Brittany
Cmi 7h25'29" 3d56'25"
Brittany
Cnc 8h26'54" 10d46'57"
Brittany
Tau 3h32'29" 22d57'4"
Brittany
Tau 3h48'51" 27d44'58"
Brittany
Cyg 19h49'19" 29d42'59"
Brittany
And 0h25'59" 32d37'2"
Brittany
Ari 2h11'45" 24d55'11"
Brittany
Crb 16h5'28" 39d5'24"
Brittany
And 23h17'5" 47d39'55"
Brittany
Per 4h26'44" 43d34'1"
Brittany
And 1h11'15" 40d24'9"
Brittany
And 0h47'33" 38d42'12"
Brittany A Guidry
And 23h23'52" 52d14'21"
Brittany Adele
Aqr 21h46'11" -1d40'54"

Brittany Allen
Del 20h20'46" 9d42'35"
Brittany Amber Bardin
Ari 2h53'54" 18d24'53"
Brittany Amber Mcmurray
Cam 7h31'9" 70d46'20"
Brittany Amber Ramos
Sgr 17h51'51" -28d41'47"
Brittany and Alex
Vir 12h43'19" 10d33'47"
Brittany and Brad's Star
Cyg 19h45'26" 33d22'12"
Brittany and Erica
Ari 3h22'4" 27d41'18"
Brittany and Jamie
Sco 17h38'27" -39d49'47"
Brittany and Ryan
Uma 11h20'8" 39d34'39"
Brittany Andrews
Gem 6h47'46" 26d24'32"
Brittany Ann
Gem 6h55'53" 19d40'33"
Brittany Ann
Uma 9h3'56" 48d53'13"
Brittany Ann
Cas 2h37'52" 64d33'32"
Brittany Ann Baerwald
And 23h8'6" 52d45'14"
Brittany Ann Bair
Leo 10h13'12" 14d3'35"
Brittany Ann Beattie
Lyn 6h39'50" 61d9'45"
Brittany Ann Binsted
Cas 0h52'29" 57d51'27"
Brittany Ann Bray
Uma 9h35'6" 44d36'12"
Brittany Ann Burch
And 0h38'5" 25d35'41"
Brittany Ann Davidson
Cas 23h24'46" 55d3'16"
Brittany Ann Fabre
Ari 3h10'55" 18d12'53"
Brittany Ann Kaminskis
Dra 17h3'3" 52d31'41"
Brittany Ann McDonald
Cam 3h28'34" 67d50'5"
Brittany Ann Neeb
Tau 4h14'53" 27d45'31"
Brittany Ann Scheffer
And 23h56'15" 33d47'58"
Brittany Ann Sipple
Tau 4h7'51" 19d29'58"
Brittany Anne Cornish
Lyn 9h16'17" 36d15'50"
Brittany Anne Dixon
Ari 3h16'29" 28d27'52"
Brittany Anne Johns
Sgr 17h56'23" -19d56'56"
Brittany Anne Louis
Lyr 18h30'29" 36d40'26"
Brittany Anne Wirtjers
And 0h51'58" 43d23'43"
Brittany Arvilla
And 0h32'25" 43d28'57"
Brittany Atwood
And 1h32'30" 40d33'13"
Brittany Autumn
Uma 9h55'4" 72d19'42"
Brittany Azuelo
Lyn 6h43'25" 61d49'33"
Brittany Bailey
Peg 21h10'52" 15d57'2"
Brittany Baird
Vir 12h42'2" 12d45'5"
Brittany Ballard
Cas 0h1'7" 53d24'1"
Brittany Batson
Ari 3h26'43" 21d36'35"
Brittany (Bee) Peck
Sco 16h45'8" -33d7'14"
Brittany Beyer "Pumpkin"
Vir 12h52'36" 10d48'42"
Brittany Bolton
Her 17h17'46" 32d0'11"
Brittany Boyd
Uma 10h27'22" 62d6'51"
Brittany Brainer
Cnc 8h59'26" 14d10'40"
Brittany & Brandon
Ori 5h47'15" -3d30'16"
Brittany Brody
Lib 14h58'45" -9d59'48"
Brittany Busby
Lmi 10h3'4" 38d15'30"
Brittany "Butt" Dobnikar-Smith
Gem 6h22'2" 24d8'22"
Brittany by Firelight
Cnc 8h22'59" 17d56'3"
Brittany Cara Herth
Agr 22h31'38" -13d33'34"
Brittany Cavalaris
Ari 2h41'38" 13d15'17"
Brittany & Chris' Star Forever
Cyg 19h58'38" 33d17'9"
Brittany Christine Wyatt Hilliker
Tau 4h10'45" 6d34'21"
Brittany Christopher
Lyr 18h50'54" 44d54'25"
Brittany Cianelli
Lep 5h11'51" -12d27'2"
Brittany Cole
Uma 10h40'35" 42d53'39"
Brittany Concilus
Cas 0h27'28" 61d39'29"
Brittany Cook
Sgr 18h28'11" -21d11'20"
Brittany "Cooter"
And 1h10'36" 39d8'39"
Brittany Corrine McDonald
Uma 9h37'46" 61d47'19"
Brittany Cortez
Crb 16h13'45" 26d43'19"
Brittany Cowham
Peg 22h23'7" 34d46'29"
Brittany Crawford
Uma 9h4'53" 71d28'34"
Brittany Dahms
Uma 10h57'13" 57d3'14"
Brittany Daniel Bremer
And 23h30'31" 47d55'47"
Brittany Danielle Seagroves
Sco 17h42'19" -36d42'33"
Brittany Danielle Walker
Vir 13h58'54" -7d49'48"
Brittany Danyel Barrios
Ori 5h46'19" 6d9'11"
Brittany Dawn
Ari 2h12'0" 14d8'35"
Brittany Dawn
Cyg 19h48'39" 32d2'54"
Brittany Dawn Eggins
Cyg 20h25'37" 44d11'56"
Brittany DeNea Stroud
And 1h3'46" 34d2'39"
Brittany Denman
Ari 3h6'22" 20d55'19"
Brittany Diamond
Lyn 8h0'43" 56d32'19"
Brittany Diane Fish
Sco 16h14'59" -11d10'38"
Brittany Dior
Cas 0h24'21" 53d40'58"
Brittany Dixon
Tau 4h36'15" 22d16'56"
Brittany D-Marie Snyder
Agr 23h6'27" -22d26'28"
Brittany Drake Koo
Cas 0h42'16" 61d35'12"
Brittany E. Wood
And 0h12'24" 46d4'10"
Brittany Elise Bonine
Leo 11h42'33" 24d18'17"
Brittany Elizabeth
Ori 5h15'5" 8d47'1"
Brittany Elizabeth
Sco 16h10'44" -14d19'22"
Brittany Elizabeth
Sco 17h55'7" -37d34'53"
Brittany Elizabeth Brooks
Vir 12h53'30" 0d55'44"
Brittany Elizabeth Macleod
And 0h17'0" 25d34'55"
Brittany Elizabeth Prosser
Cnc 8h53'22" 28d21'39"
Brittany Elizabeth Sparks
Lmi 9h36'48" 34d41'46"
Brittany Elyse
And 2h35'59" 43d47'9"
Brittany Elyse Towns
Tau 5h52'16" 27d5'22"
Brittany Erin
Agr 22h32'57" -0d15'30"
Brittany Erneston's Star
And 0h41'17" 39d26'49"
Brittany Faught
Uma 8h35'25" 50d8'36"
Brittany Faulkner-Granite Hills 06
And 1h16'52" 44d42'19"
Brittany Frothingham
Ori 6h14'31" 3d19'50"
Brittany G
Ari 1h52'38" 17d25'46"
Brittany Gayle Barnes
Cam 3h25'30" 67d58'21"
Brittany Genther
And 0h15'40" 30d41'44"
Brittany Giello
Ari 2h7'24" 24d4'4"
Brittany Gilmore
And 0h16'47" 28d59'56"
Brittany Goza
Lyn 7h53'54" 38d3'45"
Brittany Green
Lyn 9h5'5" 34d12'25"
Brittany Greenwood a.k.a. "B"
Vir 13h16'48" -11d46'16"
Brittany Griffin
Her 16h33'21" 13d59'12"
Brittany Gunter
Tau 4h28'18" 29d36'1"
Brittany Hafner
Sco 16h11'6" -9d15'37"
Brittany Hallmark
Vir 13h21'14" -6d49'55"
Brittany Heather
Psc 0h7'19" 6d29'53"
Brittany Henderson 143
Cma 6h16'21" -30d3'25"
Brittany Holden (My Wife and Angel)
Tau 4h0'59" 18d49'31"
Brittany "Hot Stuff" Manfredi
Lyn 8h17'30" 50d59'18"
Brittany Hughes
And 23h26'28" 44d26'19"
Brittany Hughes
Cas 22h59'34" 54d12'36"
Brittany Hulsey
Psc 1h55'13" 7d58'23"
Brittany Irene
Tau 4h36'52" 17d44'32"
Brittany Jade Freeman
Sgr 18h23'24" -19d33'21"
Brittany James
Vir 13h13'35" -22d6'36"
Brittany Jane
Lib 15h2'50" -0d48'0"
Brittany Jane Brock
Leo 11h30'25" 21d4'28"
Brittany Jean DeLatte
Uma 11h24'33" 46d2'13"
Brittany Jeffers
Psc 23h54'19" 2d17'43"
Brittany Jennings
Com 12h37'25" 23d12'39"
Brittany Jo
Uma 9h56'19" 64d28'55"
Brittany Jo Benton
Cap 21h10'5" -26d14'43"
Brittany Johnson
Uma 12h6'35" 59d22'13"
Brittany Johnson
Cyg 21h42'36" 44d32'45"
Brittany Jonn Ledesma
Uma 11h2'43" 55d25'31"
Brittany Jordan Van Buren
Tau 3h48'34" 28d28'11"
Brittany Joyce Accetturo
Uma 9h37'48" 67d13'12"
Brittany Justine Beaudoin
Vir 12h18'45" 12d18'25"
Brittany K. Nabors "My Silver Star"
Cyg 19h47'30" 54d39'54"
Brittany Kappel
Gem 6h52'13" 27d58'55"
Brittany Kate Williamson
And 0h57'47" 46d38'3"
Brittany Kaye Olson
Leo 10h19'25" 16d10'46"
Brittany Klein
Uma 11h20'33" 54d55'25"
Brittany Kleve-Bray
Agr 22h5'31" 0d4'39"
Brittany Klingler
Psc 0h23'35" 9d12'56"
Brittany Knight
Leo 11h38'49" 13d53'32"
Brittany L. Seastrand
Mon 6h49'11" -0d19'2"
Brittany Lane
Cnc 7h59'46" 11d23'47"
Brittany Lascko
Lib 15h27'50" -18d53'23"
Brittany Laura Anne
Ori 6h2'47" 6d47'11"
Brittany LeAnn Blizzard
Uma 10h26'36" 43d15'34"
Brittany Lee
And 23h37'11" 36d21'9"
Brittany Lee
Vir 11h39'49" -1d30'56"
Brittany Lee
Sco 17h29'12" -45d22'56"
Brittany Lee Anne Beeman
Cnc 8h42'11" 14d7'38"
Brittany Lee Bennett
Lib 14h55'33" -11d23'7"
Brittany Lee Castiglione
Lyn 8h10'46" 39d23'12"
Brittany Lee Cornelius
Ari 1h57'50" 14d25'28"
Brittany Lee Garcia
And 0h34'34" 31d33'56"
Brittany Lee Harfst
Gem 7h7'59" 27d23'58"
Brittany Lee Kulwicki
Cyg 20h51'34" 31d32'59"
Brittany Leigh
Cap 20h23'44" -24d37'7"
Brittany Leigh De Stefano
Lib 14h53'36" -2d41'12"
Brittany Leigh De Stefano
And 22h59'5" 44d49'37"
Brittany Leigh DeStefano
Umi 15h5'54" 79d42'54"
Brittany Leigh, My September Star
Vir 14h7'6" -4d45'35"
Brittany Lewis
Vir 14h17'26" -8d33'29"
Brittany "Lil' BG" Guidone
And 23h2'45" 44d7'34"
Brittany Linebaugh
Peg 21h41'50" 21d52'9"
Brittany Liz
Cap 20h26'52" -12d53'58"
Brittany Loves Corey Ratke <3
Gem 7h47'21" 32d18'57"
Brittany Lyanne Shipley
And 0h38'10" 34d33'54"
Brittany Lyn
Leo 10h42'28" 18d11'10"
Brittany Lyn Lake
Agr 23h40'26" -10d22'28"
Brittany Lynell
Lib 15h55'27" -18d23'59"
Brittany Lynn
Agr 20h38'55" -2d19'27"
Brittany Lynn Bonner
Uma 11h14'23" 44d59'8"
Brittany Lynn Forever and Always
Uma 10h11'44" 47d34'8"
Brittany Lynn Hastings
Peg 22h27'57" 28d17'55"
Brittany Lynn Katz
Cas 23h42'17" 54d44'43"
Brittany Lynn Kirk
Psc 0h39'29" 11d31'49"
Brittany Lynn Lorenger
Cas 1h12'23" 67d26'6"
Brittany Lynn McDonald
Her 16h41'20" 22d39'25"
Brittany Lynn Ricketts
Cap 21h31'30" -16d26'21"
Brittany Lynn Ross
Lib 15h6'36" -4d29'5"
Brittany Lynn Santee
Cnc 8h28'22" 25d22'54"
Brittany Lynn Tarantino
And 1h19'0" 47d5'40"
Brittany Lynn Troy
Crb 15h37'31" 28d18'45"
Brittany Lynne
Ori 5h37'31" 8d50'10"
Brittany Lynne Madarang
Hor 4h10'23" -42d36'56"
Brittany Lynn-Iulia Harder
Agr 23h7'14" -8d34'28"
Brittany M. Gibson
Vir 14h40'31" 2d24'38"
Brittany M. Otteson
Cnc 8h30'12" 17d27'4"
Brittany M. Sieglaff
Cnc 8h43'50" 12d23'47"
Brittany Machelle Luck
Gem 6h25'27" 27d36'48"
Brittany MaLinda Wilson
And 0h32'15" 36d33'13"
Brittany Mappus
Aur 5h14'45" 42d32'26"
Brittany Marcott
Tau 4h52'26" 18d59'54"
Brittany Marie
And 0h34'6" 43d36'26"
Brittany Marie
And 23h29'54" 41d43'49"
Brittany Marie
Sgr 18h27'8" -17d20'49"
Brittany Marie Beckhorn
Lyn 9h1'4" 38d35'33"
Brittany Marie Blythe
Lib 15h57'33" -10d39'26"
Brittany Marie Lucio
Ori 6h20'21" 10d26'47"
Brittany Marie Snyder
Sgr 18h8'30" -19d9'37"
Brittany Marie Wajsgras
Peg 22h6'6" 8d27'21"
Brittany Marie Williams
And 23h15'57" 39d34'57"
Brittany Marlowe
Cap 21h31'7" -15d34'18"
Brittany Martin
Crb 16h6'17" 37d59'49"
Brittany & Max
Cyg 21h31'14" 50d32'55"
Brittany McDowell
Leo 10h48'37" 15d42'20"
Brittany McHugh
Lib 15h39'16" -17d2'53"
Brittany Michele Parks
And 0h37'54" 26d51'41"
Brittany Montana Stoulp
And 0h26'23" 44d22'44"
Brittany Musacchio
And 0h22'26" 32d10'27"
Brittany My Only Star Forever
And 1h16'53" 38d42'40"
BRITTANY N
Psc 1h23'41" 15d3'50"
Brittany N. Boone Birthday Star
Cap 21h34'58" -17d40'8"
Brittany N. McCallum
Cas 0h5'13" 57d5'30"
Brittany N. Merriman
Uma 10h14'45" 46d41'24"
Brittany N. Mitchell
Vir 13h28'26" -6d55'46"
Brittany N. Zavala
Uma 9h24'0" 65d34'11"
Brittany Nichole Johnson
Ori 6h7'21" 16d7'4"
Brittany Nichole Posey
Cnc 8h48'38" 21d5'10"
Brittany Nichole Schlau
Ori 5h21'40" 0d24'26"
Brittany Nichole Tindall
And 23h10'32" 35d30'3"
Brittany Nicole
And 0h43'45" 42d35'29"
Brittany Nicole
Agr 22h32'30" 0d4'24"
Brittany Nicole
Cap 21h45'38" -16d53'5"
Brittany Nicole Austin-Goyne
And 0h36'33" 36d0'58"
Brittany Nicole Bodenheimer
Gem 7h0'58" 14d6'7"
Brittany Nicole C.
Vir 12h32'56" 8d45'50"
Brittany Nicole Frye
And 22h59'52" 44d7'33"
Brittany Nicole Halterman
Lyn 8h15'37" 57d13'19"
Brittany Nicole Martin
And 0h51'28" 35d53'22"
Brittany Nicole Martin
And 2h33'58" 44d20'41"
Brittany Nicole Minor
Vul 19h47'56" 28d6'5"
Brittany Nix
Psc 1h2'0" 28d10'40"
Brittany Noel Gonzales
Crb 16h11'40" 35d43'45"
Brittany Noell Mintken
And 23h36'19" 50d40'32"
Brittany Noelle Maybrun
Lib 14h51'16" -7d46'35"
Brittany O'Conner
And 2h14'12" 46d26'11"
Brittany Page
Psc 1h23'32" 7d19'35"
Brittany Paige
And 1h6'36" 38d42'26"
Brittany Paige
And 0h20'22" 44d2'3"
Brittany Paige
Tau 3h47'48" 27d1'35"
Brittany Paige Calhoon
Crb 15h45'21" 28d8'36"
Brittany Paige Colacino
And 23h15'8" 49d14'53"
Brittany Paige (The Princess Star)
Ari 2h7'18" 18d33'34"
Brittany Pedersen 1
And 23h2'55" 49d59'24"
Brittany Prater
Vir 13h19'4" -2d30'54"
Brittany Pratt
Agr 21h36'38" 1d8'27"
Brittany Priscilla
Sgr 18h55'54" -34d19'30"
Brittany R. Robinson
Tau 4h21'9" 22d38'38"
Brittany Rachelle Vanden Berg
Psc 0h2'9" -3d2'9"
Brittany Rae
And 2h21'27" 48d52'6"
Brittany Rae's Star
And 0h38'40" 36d43'51"
Brittany Ramah "Smit Star"
Tau 4h6'21" 11d13'8"
Brittany Raye
Cnc 8h40'21" 18d13'41"
Brittany Renee
And 0h20'50" 28d25'37"
Brittany Resler
And 0h17'1" 26d11'53"
Brittany Rhiannon McMillan
Psc 0h48'49" 8d18'18"
Brittany Rianna
Sco 17h26'12" -38d22'6"
Brittany Robertson
Ari 2h37'27" 26d40'38"
Brittany Rose
Del 20h31'37" 16d48'57"
Brittany Rose Eyler
Tau 4h20'38" 8d46'55"
Brittany Rose Singhas
Crb 15h35'7" 26d25'19"
Brittany Saxon Crane
And 0h31'15" 31d4'13"
Brittany Schexnayder
Sgr 18h43'40" -29d21'38"
Brittany Scott
Agr 22h0'8" -21d34'15"
Brittany Shantell Dolores Davis
Cas 1h12'17" 62d51'19"
Brittany Shiedah Mitchell
And 23h14'56" 50d56'49"
Brittany Shinaberry
And 23h46'34" 48d20'11"
Brittany Sierra TJ Fortuno
Ari 2h49'27" 15d42'59"
Brittany Smith Rankin
Lib 14h51'58" -7d28'16"
Brittany Spada your a STAR Godchild
And 0h51'34" 37d28'42"
Brittany Sparkles
Gem 7h36'48" 24d4'9"
Brittany Springmyer
Cyg 21h0'27" 42d52'20"
Brittany Sydzyik
Vir 11h58'23" -0d43'38"
Brittany Taryn Vandenheuvel
Dra 19h42'46" 60d33'23"
Brittany Taylor Sinnott-Klenotich
And 1h25'41" 44d15'27"
Brittany: The Love of My Life
Gem 7h34'41" 22d14'18"
Brittany - The Shopping Star
Crb 16h17'25" 32d32'10"
Brittany Tokmadjian
Cyg 20h15'33" 47d23'52"
Brittany Virginia Stafford
Cyg 19h52'20" 39d10'23"
Brittany Vollmer
And 0h20'49" 42d12'19"
Brittany W. Parker
And 2h20'48" 46d46'7"
Brittany Ward
And 0h12'22" 30d48'43"
Brittany Washington
Cnc 8h31'11" 16d28'10"
Brittany Weaver
Aql 19h28'25" 16d19'13"
Brittany Wood
Vir 12h45'52" 0d44'24"
Brittany Woods
And 0h46'16" 35d14'19"
Brittany Woods
Lyn 7h22'59" 56d20'35"
Brittany Zambito
Agr 22h44'55" -2d21'9"
Brittany, Grant & Gregor Guempel
Umi 14h24'31" 68d38'18"
Brittany.......Beautifully Blonde
Lyn 9h11'33" 33d33'5"
Brittany-Lauren
Sco 16h13'47" -42d3'58"
Brittany's angel star
Tau 4h28'55" 16d23'2"
Brittany's Brilliance
Leo 11h37'6" 15d41'58"
Brittany's Brilliance
Cyg 21h10'53" 45d30'35"
Brittanys Courage
Crb 16h4'33" 29d26'46"
Brittanys Iris
Vir 13h10'13" -2d50'37"
Brittany's Smile
Vir 14h54'3" 3d59'32"
Brittany's Star
Ori 5h42'45" 11d21'59"
Brittany's star
Cnc 9h12'20" 10d46'38"
Brittany's Star
Cnc 7h56'43" 15d46'14"
Brittany's Star
Del 20h21'32" 19d42'33"
Brittany's Star
Lyr 18h53'51" 35d6'27"
Brittany's Star
And 0h43'38" 43d2'17"
Brittany's Star
Dra 18h31'57" 52d30'47"
Brittanythew
Cyg 20h23'8" 47d47'18"
Britten
Lyn 7h32'50" 47d32'33"
Brittin A. Lowry
Lyn 8h58'3" 40d36'59"
Brittiney Wish Bevin - Oct 19, 1985
Cru 12h48'57" -57d6'56"
Brittingham Borealis
Crb 15h51'17" 27d28'10"
Brittinia the Bright Heavenly Star
Peg 22h37'31" 23d10'50"
Brittiny Marsh
Lib 14h57'48" -6d21'18"
Brittiny4-30-07
Tau 5h52'31" 28d9'20"
Brittle
Umi 15h54'54" 71d1'21"
Britt-Marie Jakobsson
Lyr 19h24'48" 38d0'58"
Brittnee Jean
Gem 6h29'7" 20d8'6"
Brittnee Rose Thayer
Sgr 18h38'34" -30d32'47"
Brittness Mitchell
And 0h42'4" 43d2'27"
Brittney
And 0h46'10" 43d5'35"
Brittney
And 23h59'45" 35d14'53"
Brittney
Ari 1h54'39" 24d27'3"
Brittney
Gem 6h31'32" 27d2'35"
Brittney
Ari 3h6'34" 10d56'25"
Brittney
Lyn 8h27'34" 53d36'46"
Brittney
Agr 22h3'32" -4d11'34"
Brittney
Cma 6h42'32" -16d41'41"
Brittney Alexa
Apu 16h29'28" -70d58'45"
Brittney and Gage
Lib 15h23'35" -20d27'38"
Brittney and Michael
Ori 6h17'33" 14d46'27"
Brittney and Tony's Eternal Love
Sco 16h14'31" -12d52'33"
Brittney Angeline Smith
Leo 11h33'24" 14d26'20"
Brittney Ann
Umi 15h56'54" 71d57'42"
Brittney ann Flores
Leo 11h13'21" 18d17'39"
Brittney Ann Wheeler
And 0h14'1" 38d25'10"
Brittney Annette Ehmer 2004
Sgr 17h54'40" -29d24'32"
Brittney Armstrong
Vir 12h46'54" 3d53'50"
Brittney Ashton Phillips
Uma 10h28'59" 42d54'13"
Brittney Camp
Cyg 21h7'56" 47d13'35"
Brittney Cargill
Mon 6h30'47" 8d57'3"
Brittney Cline
Uma 11h47'14" 49d9'52"
Brittney Elaine Bunch
Tau 4h39'32" 27d35'45"
Brittney Elaine Gump
Leo 10h24'46" 26d26'30"
Brittney Elizabeth Robins
Uma 12h9'54" 47d23'25"
Brittney Ellert
Leo 11h40'31" 16d47'49"
Brittney Haghighat
Gem 6h54'12" 21d35'10"
Brittney Hosey - Star of her Eyes
Tau 4h24'45" 0d58'24"
Brittney Joiner
Gem 7h38'40" 28d30'27"
Brittney&Kasey
Uma 9h43'40" 65d21'6"
Brittney Kay
Agr 21h12'18" 0d39'47"
Brittney Kimmell
Cnc 9h13'46" 16d56'46"
Brittney L
And 0h57'28" 23d18'35"
Brittney L. Tegtmeier
Lyn 7h53'21" 55d59'28"
Brittney Lane
Umi 14h56'14" 80d5'16"
Brittney Lane
Vir 15h10'2" 5d6'34"
Brittney Lee Kusner
Sgr 18h13'1" -31d35'52"
Brittney Lynn
Vir 14h6'19" -1d10'12"
Brittney Lynn
And 23h36'14" 47d16'18"
Brittney Lynn Cottrill
And 1h26'52" 43d58'23"
Brittney Lynn Kane
Lib 14h53'51" -2d29'57"
Brittney Matzkanin
Lmi 10h45'52" 25d57'4"
Brittney Michele Hargrove
Sco 16h54'54" -33d30'30"
Brittney Nicole
Lyn 7h59'23" 42d44'50"
Brittney Nicole Cessna
And 0h7'45" 31d17'26"
Brittney Nicole Moore
Tau 4h2'40" 15d57'47"
Brittney Orlowski
And 0h12'22" 43d26'16"
Brittney Phillips
And 2h25'34" 49d19'46"
Brittney Powell
Uma 11h36'25" 45d43'25"
Brittney Rae Sito
Gem 6h28'8" 12d12'40"
Brittney Rice "Pita"
Lyn 6h56'44" 53d13'40"
Brittney Sheiri Coty
Cma 6h53'11" -18d24'38"
Brittney Sky Stellmaker
Ori 6h10'31" 15d48'31"
Brittney Sue
Lib 15h38'32" -28d34'35"
Brittney Wolfe
Gem 7h9'9" 25d18'55"
Brittney's true love star
Sco 17h45'41" -42d59'6"
Brittni
Uma 9h19'38" 47d0'5"
BRITTNI 129
And 23h12'39" 52d35'49"
Brittni Bear
Cap 20h39'10" -16d22'45"
Brittni E. Reed
Peg 21h45'50" 15d32'38"
Brittni Nikol Wood
Vir 12h46'3" -5d25'35"
Brittnia
Sco 17h52'33" -35d59'48"
Brittnie Clark
And 2h12'12" 45d12'15"
Brittnie Jade Dugas
Uma 9h9'19" 69d34'33"
Brittnie's Star
Ari 2h26'18" 25d25'8"
Brittn'Rick420
Lib 14h51'21" -8d26'23"
Brittny Alyse Zavala
Lyn 6h48'15" 54d0'7"
Brittny LaPrail Breithaupt
And 23h16'19" 49d9'48"
Brittny Rose Palermo
And 0h13'27" 29d53'42"
Brittnye
And 0h47'7" 43d49'21"
Brittny's Twinkle in the Sky
Lib 15h27'37" -21d57'14"

Britton and Tom
Cyg 20h5'25" 34d30'16"
Britton Carter Suber
Dra 16h39'45" 57d16'54"
Britton Elise Tandy
Tri 2h16'25" 32d43'41"
Britton Elizabeth Gomez
Crb 16h11'35" 35d2'1"
Britton Lynn Sloat (Toad)
Cnc 7h59'2" 17d17'12"
Brittopia
Umi 13h7'29" 73d10'26"
Britts Altair
Aql 19h41'18" -0d2'19"
Britt-Star
Cyg 21h28'24" 31d53'24"
BrittVal
And 1h29'31" 47d10'16"
britt-waz
Aqr 23h55'44" -10d32'49"
Britty is Pritty
Gem 7h19'53" 20d15'54"
Britty's Star
Ari 2h52'42" 24d30'27"
BRIXI
Uma 8h20'54" 65d1'6"
Brizelle Desireé Celenia Jacobo
And 0h41'20" 40d51'23"
Bri-Z's Kyrié Zanutto
Leo 11h43'4" 22d52'13"
BRO & ALO - Faith, Hope and Love
Aqr 22h35'58" 0d19'37"
Bro. Bobby Christian
Cyg 19h58'45" 35d57'31"
Broadfoot Sparkle
Col 5h5'2" -36d49'14"
Broadway Babe
Psc 1h19'14" 25d55'34"
Broadway Mike
Per 3h40'39" 49d10'46"
Broc Newman
Oph 17h45'26" -0d32'8"
Broccolee
Tau 5h32'23" 21d9'39"
Broccoli Covered With Cheese
Peg 22h45'23" 32d32'33"
Brochmann, Dieter
Uma 10h31'25" 40d59'43"
Brock
Ari 2h50'24" 25d11'3"
B-ROCK
Her 16h59'10" 28d53'2"
Brock Allen Voyna
Cep 21h56'39" 65d19'45"
Brock Andrew McCullough
Aur 5h35'59" 29d55'15"
Brock Anthony Brison
Cnc 8h45'21" 15d12'33"
Brock Ashton Binford
Sco 16h5'5" -26d45'11"
Brock Bauer
Uma 9h15'34" 59d40'0"
Brock Coleman Salm
Per 3h49'3" 35d58'29"
Brock Holmes
Per 4h14'35" 50d40'37"
Brock Lyndon Baumer
Gem 7h24'58" 26d45'47"
Brock Mason Zellinger
Umi 16h33'41" 82d39'3"
Brock Matthew Lakota Kehler
Cam 4h17'45" 72d23'52"
Brock Michael Guercio
Her 17h19'44" 34d41'4"
Brock Nicholas Andrew Nanton
Her 18h53'26" 26d0'57"
Brock of Ages
Aql 19h20'21" -0d48'18"
Brock Reddish
Her 17h8'43" 35d22'23"
Brock Star
Cap 21h39'32" -9d53'16"
Brock Tittle
Ori 5h31'47" 14d2'22"
Brock Warren Good
Gem 6h50'55" 20d50'44"
BrockChels
Uma 11h50'37" 45d26'6"
Brockington Edward Hundley
Aqr 22h29'29" -0d59'41"
Brockman
Uma 12h25'12" 54d54'30"
Brock's Beacon
Cnc 8h37'10" 7d40'46"
Brockstar75
Lyn 7h50'45" 53d18'33"
Brodee Elizabeth Munchkin of Mayhem
Cnc 8h27'29" 13d39'16"
Brodee Smith
Cru 12h1'59" -63d12'6"
Broder and Virginia Erichsen
Uma 12h55'38" 55d0'47"
Broderick Alpha
Srp 18h36'32" -0d34'9"
Broderick Augustine Moravec
Uma 9h29'31" 53d44'21"

Broderick D. Roth
Ori 5h37'25" -1d7'22"
Broderick Pritchett
Psc 1h16'42" 7d8'33"
Brodericks' Valiant View of Heaven
Psc 22h51'45" 3d43'19"
Brodi Falen Mount
And 2h37'46" 39d0'4"
Brodie Deacon Tolen
Per 2h53'2" 45d56'30"
Brodie Fenix
Ori 5h41'4" -0d38'43"
Brodie Glen Presley
Per 4h10'33" 37d10'41"
Brodie Joe Pucciarelli
Her 16h38'58" 21d14'53"
Brodie & Rachel
Ori 6h0'53" 18d8'42"
Brodie William Dandy - 30.11.2004
Cru 12h16'2" -62d31'5"
Brodin Michael Grady
Vir 13h10'11" 7d11'42"
Brodmann, Werner
Ori 6h18'55" 15d42'0"
Brody Alvin Nelson
Umi 15h20'10" 72d10'4"
Brody and Kayla Williams
Ori 5h58'46" 20d44'23"
Brody Benjamin
Cyg 19h36'39" 29d24'42"
Brody Bret Bourland
Cyg 21h42'55" 47d48'41"
Brody Canciamille
Peg 23h33'17" 19d3'52"
Brody Colin Chin
Ari 3h19'27" 28d21'40"
Brody Eric Lucas
Cnc 8h19'51" 24d7'54"
Brody Ernest Bock
Umi 14h33'32" 79d22'28"
Brody James
Uma 11h24'58" 60d46'18"
Brody Joshua Jacques
Dra 17h35'36" 54d40'52"
Brody Michael Jorgensen
Psc 0h33'51" 14d23'38"
Brody Nathan East
Cru 12h27'48" -58d26'49"
Brody Orion Hoffman
Ori 5h32'55" 6d52'29"
Brody Philip Hunter
Ori 5h12'1" -6d58'53"
Brody Richard Frank
Vir 14h54'50" 3d32'38"
Brody Robot Zabielski
Her 17h15'46" 31d37'28"
Brody Rockwell Raymond
Aqr 22h4'30" -0d33'16"
Brody Shaw
Ari 3h22'44" 20d50'35"
Brody Tushman
Gem 7h23'28" 25d52'52"
Brody W. Cox
Uma 9h47'12" 51d54'9"
Brody William Hill
Uma 10h37'23" 48d17'46"
Brody Winn Koch
Aqr 22h41'53" -1d36'28"
Brody's
Uma 10h35'39" 56d28'1"
Brody's Brilliant-Bright Star
Gem 6h54'16" 32d14'12"
Brogan 13.10.1994
Lib 15h16'19" -5d28'28"
Brogdon's Bunch Class of 2005
Sgr 18h43'39" -23d8'31"
Broghan
And 23h2'35" 48d7'10"
Brognoli Reali Flavio
Uma 13h31'21" 60d7'15"
Bröhan, Christa
Uma 11h40'17" 28d24'48"
Broken Computers & Christmas Lights
Aqr 21h9'37" -10d0'0"
Broken Road
Uma 9h40'4" 54d16'10"
Broken Silence
Umi 13h39'47" 75d44'15"
Broll, Peter
Ari 1h58'32" 15d5'26"
Bromelia Andrews
Cmi 7h38'0" 5d45'46"
Bronagh Davis
And 0h26'32" 37d57'24"
Bronco BruVin
Tau 4h12'52" 21d15'44"
Brone Vainauskiene
Cas 23h16'59" 54d27'58"
Broni Cole & Jake
Cma 7h18'1" -31d8'28"
Bronie
Leo 11h16'14" 15d39'32"
Bronislaw Flis
Uma 11h23'25" 71d26'4"
Bronja
And 0h21'20" 30d38'57"
Bronson Banks Gale
Ari 1h47'5" 20d51'53"
Bronson & Hollyn
Cyg 20h31'31" 30d44'18"

Bronson Jye Hose
Lac 22h17'6" 37d52'59"
Bronson Ray
Her 17h39'8" 33d24'33"
Bronson & Tianna Evertsen
Gem 6h37'9" 13d46'48"
Bronson's Star
Ara 17h21'58" -52d8'9"
Bronté Leigh
And 23h15'28" 48d19'19"
Bronus
Her 17h3'57" 17d2'18"
Bronwem Elizabeth
Gem 7h18'40" 14d34'20"
Bronwen and Alastair Murray
Cru 12h14'46" -61d54'6"
Bronwyn Adelle
Cas 1h38'25" 62d36'51"
Bronwyn Ann Abbott
Vir 13h27'58" -15d30'28"
Bronwyn/Branden
Cyg 19h20'43" 28d15'14"
Bronwyn Droog
Equ 21h6'12" 11d48'9"
Bronwyn Elizabeth
Sgr 18h7'33" -26d41'47"
Broochard
Umi 14h25'11" 74d10'58"
Broocks
Uma 10h59'24" 57d55'54"
Brook and Karen Forever
And 2h18'50" 49d19'42"
Brook and Rachel
Gem 6h43'0" 12d26'26"
Brook Ann
Leo 9h26'41" 11d44'30"
Brook Ariel
And 2h31'55" 45d22'50"
Brook Avery Diaz
Leo 10h22'28" 9d42'57"
Brook Black
Sco 16h39'6" -29d38'30"
Brook Cauvet
Umi 13h29'40" 70d2'54"
Brook Ellen Cannon
Aqr 22h55'59" -16d20'33"
Brook Elynn Gile
And 1h46'48" 45d15'41"
Brook Erin Varner
Gem 6h52'50" 31d46'37"
Brook Hamilton
Cnc 8h16'58" 17d0'59"
BROOK HUSKEY
And 23h46'5" 34d18'13"
Brook J. Risner
Aql 19h24'25" 6d13'16"
Brook Leanne Hardesty-McCormick
Crb 15h35'51" 26d1'53"
Brook Nicole
Aqr 22h10'39" -0d5'42"
Brook Nicole Puleo
Aqr 21h30'20" -7d42'47"
Brook Oneto
Tau 4h35'7" 24d37'45"
Brook Reed Crawford
Umi 15h44'6" 70d51'49"
Brook Solley
Per 4h7'14" 48d35'51"
Brook Taplac
Vir 13h52'57" -7d15'47"
Brooke
Lib 15h5'3" -6d0'40"
Brooke
Lib 15h12'0" -12d56'35"
Brooke
Cas 2h9'20" 66d18'55"
Brooke
Sco 17h53'55" -44d55'47"
Brooke
Uma 10h6'17" 45d47'25"
Brooke
And 2h7'42" 41d48'47"
Brooke
Gem 7h6'36" 27d23'17"
Brooke
Gem 7h13'53" 28d11'35"
Brooke
Lyr 19h13'25" 26d27'51"
Brooke
Crb 15h47'3" 28d50'48"
Brooke
Aqr 21h41'42" 1d20'31"
Brooke
Aqr 22h21'38" 1d50'42"
Brooke
Mon 6h47'50" 5d58'25"
Brooke
Ori 5h37'52" 3d26'28"
Brooke
Tau 4h8'19" 6d58'42"
Brooke
Leo 11h34'46" 10d17'1"
Brooke
Tau 4h46'31" 19d11'37"
Brooke Adair Clinkscales
Oph 17h19'29" -0d7'4"
Brooke Afonso Ferriera, 6 lbs 7 oz
Cap 20h21'50" -9d52'34"
Brooke Alexandria Giles
Sgr 19h14'29" -35d20'30"
Brooke Alexis
Peg 21h43'15" 10d24'6"

Brooke Alisha Mealey
And 2h29'27" 44d29'46"
Brooke Alysia Alfaro
Gem 7h11'13" 16d19'15"
Brooke Alyssa
Gem 6h43'4" 33d12'23"
Brooke Alyssa Roche
And 2h29'17" 38d8'41"
Brooke and Andrew Govenar
Cyg 19h51'18" 31d27'46"
Brooke and Brandon
Lyr 18h47'8" 41d5'54"
Brooke and Joel
Lyn 7h34'15" 40d50'35"
Brooke and Marlin's Star
Sge 19h45'53" 17d10'46"
Brooke Angelina M. Ferreia Robbins
Psc 1h29'17" 10d23'9"
Brooke Ann
And 23h45'31" 45d29'28"
Brooke Ann Baker
Psc 0h59'28" 7d31'12"
Brooke Ann Smith
And 2h19'24" 46d45'58"
Brooke Ann Spankowski
And 1h8'34" 45d4'20"
Brooke Anne
And 1h5'58" 36d37'9"
Brooke Anne Lutwin
And 1h7'6" 35d21'5"
Brooke Ashley Evans
Uma 10h9'17" 43d46'5"
Brooke Ashley Shelby
Dra 15h12'46" 62d35'29"
Brooke Ashlyn Power
Lyn 9h38'11" 40d10'40"
Brooke Atchison
Uma 11h3'23" 70d0'24"
Brooke Belling
Cas 0h36'8" 57d31'25"
Brooke Bodine
And 0h20'59" 33d10'28"
Brooke Briza
Cyg 21h21'40" 36d29'22"
Brooke Brunetta
And 0h16'22" 35d51'1"
Brooke C. Koprivec
Cas 3h21'28" 74d20'14"
Brooke Caitlin Tasi
Lib 14h50'17" -2d7'52"
Brooke Cali
And 0h2'24" 34d40'6"
Brooke Carroll - Brookenshpier
Cyg 21h43'49" 31d21'31"
Brooke Catherine Battle
And 1h13'39" 38d44'38"
Brooke Catherine Granko
Gem 6h48'11" 32d55'6"
Brooke Chambers
Lib 15h9'36" -6d26'39"
Brooke Chappell
Aqr 23h50'35" -10d40'0"
Brooke Charlotte Kennedy
Cas 23h52'7" 54d6'25"
Brooke Cheri Danielsen
And 1h38'54" 48d5'31"
Brooke Colston Maltun
Crb 15h42'43" 30d44'43"
Brooke Danielle Flamion
Tau 5h45'44" 18d45'42"
Brooke DeHaven
Sgr 18h16'21" -20d27'56"
Brooke Delaney Gengler
Umi 14h30'6" 76d22'11"
Brooke Dilly
And 23h53'52" 37d56'35"
Brooke E. Flugel
And 1h43'43" 42d48'11"
Brooke Elaine Pierson
Cam 4h13'10" 60d28'57"
Brooke Elise Algiers
And 2h8'1" 45d28'57"
Brooke Elise Williams - OJLYM
Ari 2h45'45" 24d44'4"
Brooke Eliseious
Cap 20h25'47" -14d31'14"
Brooke Elizabeth
Vir 14h22'32" -2d17'30"
Brooke Elizabeth
Crb 16h8'49" 35d54'32"
Brooke Elizabeth Mann 4/14/05
Uma 11h53'32" 50d35'10"
Brooke Elizabeth Michel
Dra 16h42'40" 61d16'56"
Brooke Elizabeth Morrison
Ari 2h20'51" 11d5'12"
Brooke Elizabeth Tracey
Cas 23h48'58" 61d51'26"
Brooke Ellen
Gem 6h49'51" 23d5'41"
Brooke Ellison
Ari 2h21'37" 25d29'5"
Brooke Flutie
Ari 2h3'2" 19d26'18"
Brooke Frances Grace Snider-Brookie
Uma 11h10'28" 57d30'42"
Brooke Frances Williams
Sgr 19h51'17" -35d20'3"

Brooke Gaines
And 1h10'24" 36d45'4"
Brooke Gelbart
Lyn 6h36'29" 59d57'16"
Brooke Hardin
Leo 11h37'47" 15d54'47"
Brooke Hartley
Vir 13h48'29" 3d23'50"
Brooke Haynes
Gem 6h20'41" 21d21'35"
Brooke Hope
Ari 2h7'51" 18d12'58"
Brooke Hotaling
Psc 1h0'38" 14d25'30"
Brooke I Love You
Cyg 21h37'39" 39d10'45"
Brooke Ilene Hansen
Psc 23h24'15" 5d5'34"
Brooke Isabel Hartzell
Aqr 22h20'51" 1d19'5"
Brooke & John
Sge 20h5'0" 17d17'27"
Brooke Joy Rozeboom
And 0h5'15" 44d16'42"
Brooke Julie Walker
Psc 2h0'28" 5d46'25"
Brooke & Justin
Uma 9h56'25" 52d55'35"
Brooke Katherine Dyer's 21st Star
Cra 18h41'8" -38d51'56"
Brooke Kayline Buen - Light of Life
Leo 10h59'6" -2d48'24"
Brooke Krug
Lyr 18h44'41" 35d21'11"
Brooke Lafferty
And 0h44'4" 39d5'58"
Brooke Lashea Atchley
Psc 1h7'52" 11d6'57"
Brooke Lauryn Bourgeois
Cas 23h27'51" 57d38'28"
Brooke Leah Beaudet
And 0h5'25" 46d19'45"
Brooke Leane Pillars
Cnc 8h50'55" 15d46'53"
Brooke Lee
Tau 3h36'15" 15d44'20"
Brooke Lee Hargrove
Tau 3h49'47" 11d53'56"
Brooke Leslie's Light
Cap 21h44'1" -9d56'38"
Brooke Logan Brigdon
And 0h44'55" 41d37'31"
Brooke Lombardo
Gem 7h30'14" 15d7'31"
Brooke Lovell Arnold
Cap 20h32'18" -22d4'58"
Brooke Lowery
Cnc 8h37'33" 7d47'24"
Brooke Lynn Scarchilli
Uma 10h45'2" 48d30'17"
Brooke Lynne Turner
Sgr 18h16'50" -18d22'13"
Brooke M Weed
Cap 20h43'56" -17d20'19"
Brooke Madison App
Ori 6h2'22" 11d24'2"
Brooke Marie Betting
Aqr 22h4'5" -15d20'13"
Brooke Marie Crase
Vir 14h34'31" -0d17'46"
Brooke Marie Riglin
Uma 11h58'42" 61d0'47"
Brooke Marie Stodghill Love Star
And 23h4'6" 36d20'14"
Brooke Marie Yoder
Ori 5h29'8" -0d42'8"
Brooke Marion Catherine Nachtway
Ori 6h2'44" 10d51'51"
Brooke Mary O'Hanlon
Ori 5h13'21" 15d27'9"
Brooke Maxine Barak
Tau 5h13'49" 16d53'4"
Brooke McKenna Radel
Cap 20h34'14" -10d57'52"
Brooke McKenzie Farnum
And 0h13'33" 44d54'55"
Brooke Mehus
And 1h11'42" 41d29'17"
Brooke Michelle Macke's Star
Lmi 10h14'10" 36d59'17"
Brooke Miller Zook
Uma 8h53'33" 68d6'19"
Brooke Moon Kay
Cap 21h1'20" -16d37'38"
Brooke Morgan Edick 1/19/81-8/19/04
Umi 13h50'49" 76d3'36"
Brooke Morgan Heaney
And 0h38'58" 39d26'55"
Brooke: My Angel
Aqr 22h36'19" -1d53'4"
Brooke Nesbitt
Lyn 8h10'42" 39d22'59"
Brooke Nichole Rogers
And 2h34'34" 44d40'1"
Brooke Nicole Cosgrove
Ari 2h4'27" 24d30'44"
Brooke Nicole Hunter
And 2h35'14" 38d42'41"

Brooke Nicole Ogan
And 0h9'34" 35d11'16"
Brooke Nicole Stewart Johnson
And 2h22'51" 48d53'16"
Brooke Noel Barker
Uma 9h42'11" 54d10'7"
Brooke Noelle Bosley
Vul 19h56'34" 29d13'27"
Brooke Noelle-Luv-u-3
Sgr 19h45'41" -31d14'47"
Brooke Olivia Kubiak
Aqr 23h5'3" -13d30'12"
Brooke Olivia Pothoff
And 0h47'35" 39d44'23"
Brooke Pescud-Fletcher
Gem 7h20'9" 20d10'11"
Brooke Powers
And 2h24'12" 42d8'0"
Brooke R Bell
Vir 14h26'46" -4d49'33"
Brooke Rae DeSnyder
Lmi 10h29'54" 35d53'31"
Brooke Ramirez
Psc 1h21'56" 18d54'14"
Brooke Reagan Young
Uma 11h21'47" 53d44'56"
Brooke Reid
Lyn 8h26'36" 42d17'11"
Brooke Renee Brewer
Gem 6h23'26" 27d34'3"
Brooke Rinik
Cyg 19h40'43" 32d7'35"
Brooke Ritchie & Lisa Probst Star
Cas 0h37'45" 61d2'33"
Brooke Robeca
Ori 5h31'46" 4d19'28"
Brooke Roberts
Leo 10h6'50" 23d57'42"
Brooke Roberts
Uma 9h31'9" 55d16'59"
Brooke Robin Wilensky
And 0h18'5" 29d43'17"
Brooke Rose Hansen
And 0h42'32" 37d53'58"
Brooke Rullo
Tau 4h41'13" 5d6'28"
Brooke Ryan Peterson
Lyr 18h43'18" 38d8'47"
Brooke Sawyer
And 23h10'36" 42d7'17"
Brooke Schroeder
Cap 21h28'44" -13d18'28"
Brooke Scott Mullins
Del 20h40'48" 16d40'44"
Brooke Shands
Cas 1h3'28" 50d32'10"
Brooke Sonia Evans
Sgr 18h1'26" -26d17'23"
Brooke Stainton
And 0h18'25" 39d19'25"
Brooke Stalnecker ~ "My Love"
Ori 5h56'40" 12d7'39"
Brooke Strachan
Eri 3h37'31" -41d50'4"
Brooke Sylviane Jennine Capelouto
And 0h46'14" 37d14'55"
Brooke Taylor Gray
Vir 12h47'9" 4d35'55"
Brooke Thomas
Cen 13h10'59" -49d13'28"
Brooke Toni Ostwald
Cru 12h38'51" -58d50'54"
Brooke Tripp
Sgr 18h27'22" -21d14'46"
Brooke Victoria Cabezas
Leo 10h7'29" 22d21'40"
Brooke Wagner
Gem 6h40'30" 20d32'55"
Brooke Weed
Cap 21h56'36" -22d27'23"
Brooke Wright
And 23h57'26" 40d43'51"
Brooke, My Only Love
Aqr 23h0'54" -8d43'14"
Brookelyn Leann Walke
And 2h19'43" 41d48'30"
Brooke-lynn
Sco 17h57'7" -37d59'33"
Brooke-McCarragher
Uma 13h32'12" 56d11'1"
Brooke's
And 23h25'23" 49d42'30"
Brooke's Brightness
Psc 0h41'25" 13d39'10"
Brooke's Goldie
Uma 11h53'32" 47d0'23"
Brooke's Light
Lyn 8h57'40" 36d19'44"
Brooke's Light
Aqr 20h39'39" -9d1'11"
Brooke's Star
Lyn 7h34'3" 56d44'7"
Brooke's Star
Cru 12h31'15" -59d35'16"
Brooke's Star
Cyg 21h30'59" 36d49'42"
Brooke's Star 12.1.83-22.10.04
Col 5h13'59" -39d9'35"
Brooke's Star 15
Aqr 22h44'3" -19d19'34"

Brookie
Gem 7h16'46" 31d57'4"
Brookie
Lyr 18h36'57" 31d44'23"
Brookie and Jay Jay
Gem 6h57'6" 18d19'35"
Brookie Bear
Gem 6h41'52" 21d52'13"
Brookie Cookie
Cap 21h33'27" -14d30'50"
Brookius Prime
Uma 10h12'43" 65d16'24"
Brooklen Nicole Stuart
And 1h4'53" 45d6'50"
Brookley, Zoe & Maxamillian
Dra 9h49'4" 74d8'29"
Brooklyn
Tau 4h22'8" 23d31'17"
Brooklyn Ann Zomer
Uma 10h41'10" 60d6'21"
Brooklyn Ava Young
Umi 16h37'49" 75d14'54"
Brooklyn Denise Brown
Uma 11h36'10" 36d41'10"
Brooklyn Grace Wellborn
Peg 23h48'58" 26d56'12"
Brooklyn Marie
Leo 11h39'47" 13d36'56"
Brooklyn Marie
And 1h38'49" 45d1'13"
Brooklyn Michelle Buttacavoli
Sco 17h48'22" -37d48'10"
Brooklyn Morgan Chong
Vir 11h47'36" -5d46'51"
Brooklyn Olivia Ensley
And 1h17'23" 38d55'22"
Brooklyn Rae
And 0h34'34" 26d13'36"
Brooklyn Rae Mancuso
Crb 16h6'59" 38d33'41"
Brooklyn & Raven MacKay
Gem 7h12'3" 15d29'32"
Brooklyn Rose
Sco 17h3'31" -41d8'49"
Brooklyn & Roslyn
Cyg 21h27'1" 46d35'6"
Brooklyne Alisz
Cnc 9h7'14" 12d55'50"
Brooklynn Ann Bourland
Cnv 12h16'59" 33d18'33"
Brooklynn Bright
Lyn 7h46'59" 41d39'53"
Brooklynn Parker
Uma 9h53'10" 69d19'40"
Brooklynn/Twinkles
Psc 1h34'16" 27d35'40"
Brooklynn's Dreams
Ari 3h15'24" 16d18'37"
Brooklyns Estrella
Vir 12h36'21" 11d7'3"
Brooklyn's star
Vir 13h6'33" 6d55'15"
Brooklyn's Steve Slavin
Leo 11h30'47" 27d3'52"
Brookney Paige Morrell
Per 2h57'33" 51d36'29"
Brooks
Cnc 8h16'13" 7d33'48"
Brooks
Psc 1h29'4" 16d34'35"
Brooks A. Parks, Jr.
Cap 21h5'22" -17d21'44"
Brooks Beacon
Umi 20h0'1" 86d53'1"
Brooks Biagino Winer
Uma 8h11'13" 59d57'59"
Brooks Bryant Stavitzski
Umi 15h29'11" 70d52'26"
Brooks Carney
Ari 2h7'25" 21d40'57"
Brooks' Family Star
Ori 6h19'26" 15d28'13"
Brooks Harmony McKay
And 1h7'28" 46d42'53"
Brooks Martinez
Vir 14h18'5" 1d50'24"
Brooks Towne
Cnc 8h34'34" 7d41'16"
Brooks & Whitney Altizer
Cyg 20h7'35" 52d49'58"
Brooks Wisdom
Her 16h17'40" 24d31'34"
Brookse's Star
Aqr 22h37'3" 0d6'12"
Brooksies Babies
Aql 19h45'48" -0d4'7"
Brooksym
Psc 1h2'55" 33d1'43"
Brother
Peg 22h19'15" 33d32'23"
Brother Branny
Her 18h0'21" 21d29'4"
Brother Charles
Uma 11h13'36" 49d28'38"
Brother Chris
Gem 7h7'51" 26d57'21"
Brother du Pape
Vir 13h4'26" -3d45'34"
Brother Jayson
Hya 9h52'59" -16d9'22"
"Brother John"
Vir 12h11'28" 10d20'33"

Brother Of Light
Uma 12h41'56" 60d22'46"
Brother Tom
Per 3h11'17" 43d48'40"
Brother-Jer510
Per 4h19'8" 47d45'21"
Brothers of Earth and Sky - James
Ori 5h40'25" -0d39'3"
Brothers of Earth and Sky - John
Ori 5h40'24" -0d32'17"
Brower 52
Cas 0h38'24" 47d48'33"
Brown Cougar Eyes
Cyg 19h55'21" 36d43'1"
Brown Deer Owl
Gem 6h42'51" 14d28'24"
Brown Eyed Girl
And 1h51'25" 37d35'22"
Brown Eyed Girl
Lib 15h15'55" -15d56'29"
Brown Eyed Girl
Sco 16h5'32" -24d53'4"
Brown Eyes
Uma 9h3'36" 68d18'33"
Brown Eyes
Ori 5h5'16" 9d42'46"
BrownBear
Crb 15h52'22" 37d54'20"
Browne Ping
Aqr 21h56'44" 0d9'26"
Brownell Legacy
Crb 16h2'25" 32d13'52"
Brownell-Buttich Legacy
Crb 16h8'13" 32d9'46"
Browneyes
Lyn 9h13'1" 38d32'12"
Browneyes
Sgr 18h20'54" -17d35'46"
Brownie
Cap 21h15'44" -16d38'1"
Brownie
Ori 6h1'5" 13d51'4"
Brownie Bailey
Uma 11h40'46" 44d47'27"
Brownie's Joyous Guiding Light
And 1h13'48" 36d36'33"
Brownyn and Slater
Cep 21h34'35" 61d45'4"
Broxton David Gantt
Umi 15h25'53" 71d49'31"
Brr-Double O-Kay-Eee
And 0h55'22" 35d45'45"
BRU
Leo 10h59'33" 15d5'39"
Brub
Psc 22h57'37" -0d3'29"
Brubie
Uma 9h30'9" 47d55'10"
BRUCE
Gem 6h21'59" 18d2'36"
Bruce
Uma 8h29'8" 65d27'19"
Bruce A. Carbo
Cap 21h41'2" -15d38'4"
Bruce A. Quay
Cam 3h56'4" 67d16'56"
Bruce A. Richardson
Oph 17h27'51" 1d47'46"
Bruce A Shining Star
Leo 11h6'29" 16d27'24"
Bruce & Aaron
Lib 15h13'17" -6d12'34"
Bruce Ackermann
Aur 5h58'38" 41d47'19"
Bruce Alan Hamblin
Ori 5h56'37" 18d13'49"
Bruce Alan Ridgway
Per 3h25'34" 46d53'11"
Bruce Alex Royce
Uma 11h32'18" 53d28'18"
Bruce Alfred Ritchie
Ori 5h41'58" -1d54'18"
Bruce Allan Bauer
Her 16h9'32" 45d28'26"
Bruce Allan Goldmacher
Psc 0h43'5" 5d47'39"
Bruce Allen
Uma 10h24'15" 69d33'8"
Bruce Allen Wilson
Uma 9h48'35" 65d47'18"
Bruce and Cara
Sco 16h36'8" -27d24'34"
Bruce and Heidi Whear
Cyg 19h41'5" 48d40'44"
Bruce and Jo
Cyg 19h47'21" 38d35'29"
Bruce and Lynn's Stella Luna
Lyn 6h50'4" 52d19'5"
Bruce and Susan Cohoon
Cyg 21h46'22" 54d10'59"
Bruce Andrew Jones
Uma 10h15'12" 68d55'46"
Bruce Bartlett
Her 17h31'24" 31d23'26"
Bruce Blazejewski Born Aug 29/1970
Uma 11h31'13" 56d35'37"
Bruce Bochy's Gold Star #600
Aql 19h7'55" -0d1'20"
Bruce Borkosky
Sco 17h22'50" -42d32'56"
Bruce Boyd McGlothlin
Cnc 8h48'26" 26d40'44"
Bruce Buddy
Boo 14h43'20" 26d47'12"
Bruce Buehler
Lib 14h49'53" -9d50'57"
Bruce Bunch
Umi 16h22'50" 70d59'0"
Bruce C Woodward
Uma 11h52'7" 50d25'14"
Bruce Carley
Ari 3h11'8" 28d48'18"
Bruce & Carolyn Nason
Cyg 21h53'29" 45d11'9"
Bruce Charles Lowe
Uma 11h53'10" 32d13'36"
Bruce Charles MAC Donald
Leo 11h31'8" 16d36'22"
Bruce Charles Welch
Ari 1h48'24" 20d13'19"
Bruce Colby
Her 18h12'5" 26d15'13"
Bruce Cotton
Uma 9h9'15" 49d30'1"
Bruce David Anderson
Uma 10h31'41" 65d24'56"
Bruce (DNA) Roe
Her 17h18'51" 46d45'40"
Bruce Douglas O'Connor
Vir 12h23'58" 12d26'29"
Bruce E. Kenney III
Aqr 22h35'2" -0d10'52"
Bruce E. Sitler
Aqr 22h3'30" -9d19'9"
Bruce Edward
Aur 5h16'18" 37d1'10"
Bruce Edward Wentworth
Cap 20h51'28" -18d49'51"
Bruce Elroy Enman "4-25-53"
Her 17h25'32" 44d35'46"
Bruce Engstrom
Boo 15h26'57" 46d48'42"
Bruce Ericksonstar 1946
Sgr 19h55'38" -29d10'33"
Bruce Eugene Barton
Ori 6h16'20" 14d50'23"
Bruce F. Boring
Aqr 22h22'58" -22d35'47"
* Bruce F. Kay *
Tau 3h41'49" 14d55'57"
Bruce F Majeski
Ori 5h13'28" 0d14'28"
Bruce G. Bronson
Tau 3h41'53" 26d11'2"
Bruce Gilley
Cap 20h23'32" -9d49'7"
Bruce H. Fornell
Aql 19h56'38" 14d42'26"
Bruce H. Rones
Cyg 19h37'10" 48d42'3"
Bruce H. Valero
Lmi 10h42'45" 29d51'44"
Bruce Hamblet
Boo 15h33'0" 45d11'20"
Bruce Harold Duncan
Crb 15h35'3" 30d59'33"
Bruce Harold Williams
Lib 15h32'46" -11d37'4"
Bruce Holladay - Rising Star
Boo 15h35'59" 49d39'2"
Bruce Hugh Wilkins
Uma 9h44'38" 54d13'10"
Bruce Hunter
Sco 17h53'29" -35d52'12"
Bruce I. Bickle - Light of my life
Leo 11h54'24" 21d9'37"
Bruce J. Bassett Sr.
Cyg 21h46'11" 39d34'26"
Bruce J Bennett
Cnc 8h36'21" 30d37'13"
Bruce J. Olson
Tau 4h33'42" 29d55'3"
Bruce J. Wilson
Cnc 8h31'18" 20d57'40"
Bruce Jacob France
Leo 11h16'31" -0d44'39"
Bruce Jenkins M.D.
Uma 11h51'49" 35d39'23"
Bruce + Jo Allison
Ori 6h21'34" 14d24'29"
Bruce Johns
Per 2h42'56" 53d47'7"
Bruce & Kathryn Pittsley
Per 3h19'24" 48d31'22"
Bruce Kevin Sterns I
Aql 19h55'11" 12d5'10"
Bruce Kiyoshi Natsuhara
Her 18h45'17" 18d0'54"
Bruce Kovner
Cru 12h27'16" -60d47'37"
Bruce L. and Josephine A. Archer
And 2h27'5" 42d17'34"
Bruce Labedis
Dra 17h18'52" 53d9'1"
Bruce Laravuso
Per 3h27'4" 35d14'25"
Bruce Larison
Cnc 8h33'27" 32d10'23"
Bruce Larsen
Her 17h31'49" 44d24'57"
Bruce Lauerman
Psc 23h49'12" 2d2'9"
Bruce Lawrence Moore, Jr.
Ori 4h45'30" 11d41'23"
Bruce Lee Ernsthausen
Tau 5h50'24" 17d8'34"
Bruce Lipman
Ori 6h3'27" 6d39'18"
Bruce & Liz - Soul Mates
Cru 12h37'59" -59d16'39"
Bruce McKinnon & Jean Kim
Cyg 19h55'36" 32d9'40"
Bruce McLachlan LLAP
Cen 13h46'47" -41d14'31"
Bruce Menzies
Psc 0h57'27" 9d31'26"
Bruce Michael Ortman "Baby Michael"
Per 2h31'14" 51d8'28"
Bruce & Michele Pankauski
Sgr 19h20'0" -28d48'22"
Bruce Mook
Uma 9h53'27" 68d50'29"
Bruce Moreno
Uma 9h38'59" 44d47'4"
Bruce Murray Celebrates At 70
Ori 5h23'5" -0d21'58"
Bruce Nankivell
Ori 6h18'47" 8d52'33"
Bruce Pendleton
Lib 15h33'24" -28d17'8"
Bruce "Pop" Sands
Tau 3h43'48" 8d12'42"
Bruce Profsky
Gem 7h33'11" 19d0'26"
Bruce Raymond Wood
Her 16h40'29" 22d53'53"
Bruce Reisinger
Her 17h28'52" 46d7'41"
Bruce Relander
Cyg 20h28'45" 47d24'23"
Bruce Robin Whincup
Tau 4h2'14" 21d23'40"
Bruce Robinson
Uma 10h16'17" 59d20'17"
Bruce Rogers
Her 17h15'22" 25d1'14"
Bruce Rogers
Crb 16h7'0" 37d59'6"
Bruce S. Whitmarsh
Cep 0h41'45" 80d27'22"
Bruce & Sabrina
Tau 5h18'33" 24d40'32"
Bruce Schmidt
Her 17h40'5" 32d16'16"
Bruce Schneider
Vir 13h22'39" -10d20'29"
Bruce Schweiger
Her 18h55'14" 23d51'31"
Bruce & Shirley Schollmeyer
Cyg 21h9'34" 31d32'34"
Bruce Short
Aur 6h22'54" 39d35'37"
Bruce Swede Nelson Sr.
Psc 0h39'43" 6d36'37"
Bruce The Balbarrick
Cep 22h16'49" 58d2'54"
Bruce Timothy Norris
Cnc 8h47'40" 14d51'58"
Bruce Townsend
Lyn 8h2'30" 42d2'37"
Bruce Tyler
Aqr 21h36'8" 1d49'50"
Bruce W. DeKalb 50th Birthday Star
Tau 4h20'25" 27d22'51"
Bruce Walker Swann
Cyg 21h18'53" 53d1'57"
Bruce Wayne Heydorn
Sge 20h17'21" 17d32'28"
Bruce Wayne Holland
Cru 12h7'29" -56d5'18"
Bruce Wayne Weeden
Gem 6h50'35" 30d12'28"
Bruce William Johnson
Tau 5h57'56" 23d57'43"
Bruce William Meachen
Gem 6h59'50" 22d10'13"
"Bruce" William Turner
Cap 21h47'24" -15d25'31"
Bruce Wm. Blount M-12-16-83
Vir 14h3'47" -0d54'56"
Bruce, Bruce
Cap 20h25'7" -22d54'51"
Brucedouglas Hartwell
Umi 15h32'33" 71d6'25"
Brucee
Per 2h56'26" 54d33'52"
"Bruce"-ious Maxiumus "Hueners"-ium
Her 16h28'45" 26d49'36"
Bruce's 50th shining star
Aql 19h44'5" -0d4'57"
Bruce's Big Five O
Her 17h25'22" 44d50'21"
Bruce's Birthday Star
Uma 10h18'26" 43d20'12"
Bruce's Bubby
Ori 5h40'40" -2d17'8"
Bruce's Disco Ball
Lib 15h59'14" -12d20'25"
Bruce's Star - Danny's Dad
Leo 10h15'45" 21d46'42"
Brucetta
Lyn 7h59'18" 35d50'12"
Brucie
Ori 6h12'59" 17d57'21"
Brucie
Ori 6h14'0" 15d6'39"
Bruciegeuse
Aqr 22h48'44" -6d48'48"
Brucifer
Cap 21h4'13" -21d12'0"
Brucifer
Ari 1h54'44" 18d2'49"
Brück, Eckhard
Uma 13h51'1" 55d41'47"
Brückner, Christel
Ori 5h28'19" 14d49'44"
Brud
Vir 13h33'49" 6d54'20"
Brud
Psc 0h33'25" 20d37'40"
Bru-Hay
Per 3h22'11" 47d59'47"
Brujis Superstar
Her 16h57'8" 32d7'14"
Bruley Mathieu
Cap 20h14'22" -9d49'37"
Brumers Roost-John,Cecelia & Family
Uma 9h22'4" 64d33'29"
Brummel
Uma 14h19'17" 55d47'33"
Brümmer, Uwe
Uma 11h32'23" 38d34'36"
Bruna
And 0h7'12" 37d35'4"
Bruna
Mon 7h1'44" 4d47'34"
Bruna Martinengo
Lyr 18h49'16" 41d22'51"
Brunhilde Mathilde Witte *Hilla*
Ori 6h20'6" 19d7'0"
Bruni
Uma 10h47'18" 57d6'51"
Brunia Axmini
Umi 5h19'28" 88d21'4"
Brüning, Uwe
Uma 14h4'31" 55d18'49"
Brunivette Ramirez
Leo 11h52'12" 20d32'19"
Brunner, Max
Uma 12h7'57" 50d16'26"
Brunner, Urs Remo
Uma 8h43'7" 59d54'21"
Bruno
Cep 23h9'40" 79d36'17"
BRUNO
Uma 12h2'56" 50d32'26"
Bruno
Boo 15h34'32" 45d44'56"
Bruno
Leo 10h15'44" 18d3'33"
Bruno
Ori 5h50'20" 3d7'41"
Bruno and Sue Mikos
Umi 13h28'18" 70d26'55"
Bruno Boggs
Uma 14h28'21" 58d42'5"
Bruno Brabant
Cas 0h15'13" 55d29'6"
Bruno Caliciuri dit "cali"
Uma 12h2'47" 59d59'34"
Bruno & Catherine since 2002
Ori 5h42'38" 15d0'22"
Bruno Doublier
Psc 22h57'54" 2d53'24"
Bruno Dutly
Uma 9h26'19" 62d23'5"
Bruno Eddie
Her 17h22'58" 23d31'12"
Bruno et Jérémy
Uma 9h44'30" 66d15'35"
Bruno Ferranti
Lyr 18h32'8" 36d8'16"
Bruno Giacomuzzi
Cep 21h31'14" 64d4'37"
Bruno Gri Di Maiaroff-Valvasone
Aql 19h11'10" -7d11'43"
Bruno Hanlon Bakel
Tau 5h9'9" 24d50'29"
Bruno Hoster-Sacre
Uma 8h44'7" 54d41'38"
Bruno Kattan Rosset
Vir 14h1'3" -18d12'13"
Bruno*Love*
Ari 3h24'24" 20d25'47"
Bruno Naumann
Uma 10h11'15" 49d11'29"
Bruno Pelletier
Leo 11h8'6" 10d32'48"
Bruno Rua
Uma 10h10'28" 60d22'33"
Bruno Ryback
Uma 9h22'9" 48d34'20"
Bruno Schumacher
Uma 9h36'39" 49d1'28"
Bruno Tani
Uma 8h25'47" 61d52'30"
Bruno "Todd" Longhi
Her 17h16'7" 31d24'55"
Bruno Vipulis
Crb 15h35'36" 29d48'25"
Bruno Wagner
Umi 16h14'55" 77d5'42"
Bruno Walter Hauschild
Uma 11h40'13" 33d51'3"
Bruno-George Curtis
Her 16h28'59" 5d25'59"
Bruns, Eduard
Ori 6h15'53" 9d58'22"
BRURIC
Uma 11h0'2" 46d22'21"
Bruser
Mon 8h10'28" -0d43'7"
Brust-Schuster, Mia Zoe
Tau 4h21'29" 25d29'52"
Bruth Rock-12
Ori 5h9'14" 5d39'16"
Brutus
Psc 0h25'35" -3d39'54"
Brutus Bear
Uma 11h44'30" 53d53'23"
Brutus Shira-Ciliberti
Uma 11h19'4" 37d14'46"
BRV
Ori 5h32'23" 3d20'40"
Bry
Cep 21h31'46" 66d20'56"
Bry Bry
Sgr 19h15'14" -12d37'35"
Bry Meloni
Aqr 21h46'21" 0d18'57"
Bryan
Aql 19h30'46" 5d5'3"
Bryan
Leo 10h9'16" 26d1'36"
Bryan
Her 17h58'39" 22d50'3"
Bryan
Her 17h58'0" 25d33'22"
Bryan
Cnc 8h35'25" 30d41'50"
Bryan
Lib 15h30'2" -19d8'43"
Bryan
Vir 13h6'9" -5d12'50"
Bryan
Sco 17h52'6" -31d7'43"
Bryan A. Fitch (Parkville, MD, USA)
Uma 10h3'40" 54d43'17"
Bryan Alexander Maximilian Karsky
Umi 14h24'45" 75d20'22"
Bryan Allen
Ari 3h4'25" 17d55'13"
Bryan Allen Hughes
Ori 6h1'6" -0d48'25"
Bryan Alton Parker
Aqr 21h19'11" 0d37'28"
Bryan and Aaron's Mom
Sgr 18h46'47" -24d42'43"
Bryan and Amanda
Dra 18h55'4" 50d11'44"
Bryan and Angelica's star
Peg 22h49'18" 16d14'47"
Bryan and Darla
Cyg 21h39'7" 48d15'15"
Bryan and Janine
Cyg 19h25'52" 52d23'7"
Bryan and Jill's Star
And 1h1'28" 46d19'34"
Bryan and Lauren's Star
Tau 4h6'24" 4d59'25"
Bryan and Ranelle Harhai
Pyx 8h35'17" -20d57'10"
Bryan and Stacie Gumm
Lyr 18h55'47" 34d59'17"
Bryan Andrew Azer
Cas 0h16'54" 52d14'34"
Bryan Angelo Bacci
Sco 16h3'5" -12d17'22"
Bryan Anthony King
Ori 5h36'44" 8d4'36"
Bryan Bell
Lib 15h13'48" -29d40'29"
Bryan Benson
Eri 4h31'10" -0d6'47"
Bryan & Betty Treharne
Cyg 21h43'7" 51d50'5"
Bryan Bryce Douglas Graves
Aur 5h13'4" 41d42'16"
Bryan C. "EZB" Baker
Uma 10h29'12" 43d14'16"
Bryan Christian Luckey
Vir 13h41'41" 2d35'1"
Bryan & Christina Stephens
Ori 5h35'15" -0d31'42"
Bryan Cochran "Sports Star"
Psc 23h38'42" 0d42'28"
Bryan Crowley
Aql 19h33'18" -9d9'15"
Bryan D. Shiloh
Tau 5h38'36" 21d42'41"
Bryan D. Wright
Gem 7h17'38" 34d6'28"
Bryan David Dworkin
Lyr 18h56'1" 26d30'38"
Bryan David Scheidt
Psc 0h8'27" 6d42'10"
Bryan Donald Abston
Vir 13h14'9" 5d8'29"
Bryan & Emily
Cyg 21h43'34" 43d14'4"
Bryan F. Morales
Boo 15h27'24" 42d28'28"
Bryan Fair
Cep 23h22'33" 73d5'12"
Bryan Foster Myers
And 2h34'37" 40d22'38"
Bryan & Francesca
Cas 0h49'30" 60d10'53"
Bryan Frink
Aql 19h11'5" -0d56'49"
Bryan & Ghislaine Wharton - 25 Years
Cru 12h43'14" -57d2'31"
Bryan Gilmore Hollingsworth
Boo 14h32'55" 23d26'24"
Bryan Gregory Ennis
Ori 5h9'39" 1d15'2"
Bryan Harvey (Dad)
Her 18h51'23" 22d47'59"
Bryan hearts Tobi forever
Ori 6h9'8" 3d39'30"
Bryan Heffalump Sherman
Uma 11h8'16" 62d5'36"
Bryan Highland
Cyg 19h49'57" 36d46'8"
Bryan Holloway
Uma 12h15'10" 56d55'31"
Bryan Ingram Wilson
Uma 9h12'8" 65d11'41"
Bryan Jacob Alvarez
Boo 15h32'59" 45d24'6"
Bryan Jacob Lewis
Ori 5h26'10" 1d22'4"
Bryan James Tapella
Cnc 8h27'53" 25d59'13"
Bryan James Whitcomb
Leo 10h38'28" 18d20'24"
Bryan Joseph Botma
Her 16h18'31" 17d45'46"
Bryan Joseph Bower
Sgr 19h15'54" -34d24'7"
Bryan Jr.
Aur 5h16'47" 42d43'38"
Bryan K. Castleberry
Aur 6h31'33" 35d1'42"
Bryan K. Swartwood III
Her 18h32'48" 18d47'15"
Bryan Keith Kimes
Aqr 21h50'28" -7d56'4"
Bryan Keith Wilson
Sgr 18h53'6" -35d40'54"
Bryan Keller Lane
Psc 0h3'4" 7d21'30"
Bryan L Regina A Isabella D Wagner
Crb 15h37'13" 25d53'22"
Bryan L Smith, Sr & Nadirah A Smith
Cyg 20h54'19" 35d0'35"
Bryan Lamb
Uma 10h12'58" 60d37'57"
Bryan Lawson
Uma 10h49'9" 47d7'23"
Bryan Lee Smith
Sco 17h28'25" -38d12'11"
Bryan Levi Denham
Uma 10h36'29" 47d39'12"
Bryan Mack Nylander
Psc 0h47'25" 5d11'40"
Bryan * Mahasani * Angie
Cyg 20h53'8" 30d44'20"
Bryan Marcus
Per 4h25'58" 51d54'41"
bryan Marr
Ori 5h32'40" 2d8'25"
Bryan Martin
Lyn 7h34'4" 38d6'25"
Bryan Matthew Tutor
Leo 9h22'31" 11d49'44"
Bryan Matthew White
Ori 5h48'17" 3d37'56"
Bryan Michael Hagemeyer
Ori 5h54'18" 1d30'44"
Bryan Miles
Ari 3h17'29" 15d51'18"
Bryan - My Evening Star
Per 4h35'12" 33d3'29"
Bryan overlooking
Her 17h56'18" 29d48'46"
Bryan Parks
Ori 5h56'29" 11d36'5"
Bryan Patrick Dougherty
Per 3h9'7" 51d31'44"
Bryan Patrick Kain
Uma 11h15'56" 60d39'26"
Bryan Patrick Nickerson
Uma 11h26'44" 63d56'31"
Bryan Patrick Thompson
Per 2h44'47" 51d53'20"
Bryan Pfeifer
Her 17h23'17" 27d53'1"
Bryan Poling
Uma 10h10'55" 63d48'21"
Bryan Powell God's Angel #7
Leo 10h5'12" 16d27'11"
Bryan Reed
Cap 20h38'39" -18d39'23"
Bryan Reed Tourner
Cap 20h10'59" -15d58'24"
Bryan Reichenbach
Aqr 21h32'59" 0d47'8"
Bryan Richard Shaw
Lyn 7h46'19" 37d38'54"
Bryan Rickard
Ori 5h55'26" 18d27'28"
Bryan Robert Moser
Uma 9h39'24" 59d47'24"
Bryan Robert Shewmake
Sco 16h30'59" -26d19'32"
Bryan S. Jackson, Super Star
Uma 13h19'55" 53d15'24"
Bryan Salo
Psc 1h34'59" 15d35'47"
Bryan Schenkman
Uma 10h2'12" 71d23'14"
Bryan Scott James
Ori 5h30'52" 0d43'22"
Bryan Simms
Psc 1h55'17" 6d16'12"
Bryan Soule
Cnc 9h12'51" 8d3'18"
Bryan Spencer Hall
Her 17h18'17" 27d5'3"
Bryan T. Wood
Her 17h18'38" 14d17'42"
Bryan "Teach" Lewis
Ori 4h58'7" 15d17'47"
Bryan ~THE ACE~ Morris
Cap 21h36'46" -8d58'0"
Bryan Thomas Moore
Per 4h20'17" 34d22'26"
Bryan Thomas Palmer
Aqr 23h4'3" -8d46'27"
Bryan Thomas Palmer
Uma 11h35'6" 60d28'26"
Bryan Tyler Carmody
Ori 5h39'38" 3d22'4"
Bryan Tyler Rogala
Tau 4h25'8" 21d50'25"
Bryan Wayne Burgess
Cap 21h39'0" -14d46'2"
Bryan Wilcoxen, My Shining Star
Dra 18h38'8" 54d12'36"
Bryan William Snowden
Ori 6h19'13" 14d19'47"
Bryan Wong
Cam 5h27'15" 65d40'14"
Bryan Woodruff
Ari 1h57'30" 14d59'29"
Bryan y Amor por vida
Pho 0h10'40" -42d52'37"
Bryan Yan Ping Lee
Gem 7h44'42" 25d39'15"
Bryana
Crb 15h46'10" 25d51'42"
Bryana 3/2/99
Peg 23h42'52" 15d43'26"
Bryana Miyuki Wong
Leo 11h51'27" 22d1'41"
Bryana Shea 16
Tau 5h27'49" 27d21'45"
BryAnn
And 0h39'49" 39d58'58"
Bryanna
And 1h35'38" 46d55'21"
Bryanna
Leo 10h11'30" 22d7'13"
Bryanna Elizabeth Geisbush
And 2h19'3" 48d45'39"
Bryanna Gonzales
Tau 3h43'17" 17d34'22"
Bryanna Lynn Leake-Rowder
And 0h43'45" 29d19'27"
Bryanna Rita Teneyuca
Ari 3h24'25" 19d38'32"
bryannabenedict040405
Tau 4h24'43" 9d33'35"
Bryanne
Ori 4h59'6" 15d11'40"
Bryanne
Uma 8h58'55" 47d14'37"
Bryanne
And 0h29'16" 41d54'34"
Bryan's Birthday Star
Cep 21h7'31" 59d43'2"
bryan's 'imo 'imo
Tau 4h33'31" 21d7'10"
Bryan's Shining Star
Aqr 21h46'46" -7d28'46"
Bryan's Star
Dra 18h20'30" 52d39'16"
Bryan's Supergiant
Uma 9h24'21" 61d22'5"
Bryanstar
Leo 10h30'10" 18d11'13"
Bryant Davis Shelton
Leo 10h44'20" 22d32'7"
Bryant & Desire
Cyg 19h44'5" 55d28'32"
Bryant Groom
Ori 5h35'32" -0d35'13"
Bryant James Danny Poland Jr.
Gem 6h46'51" 31d58'26"
Bryant John Maybe
Dra 15h7'15" 60d26'23"
Bryant Joseph Miguel
Boo 14h38'18" 29d21'23"
Bryant L. Sampson
Umi 14h37'28" 69d18'52"

Bryant Mabee
Dra 11h40'23" 72d31'1"
Bryant Mayo
Lib 15h16'57" -4d32'18"
Bryant & Megan
Tau 4h8'36" 6d36'23"
Bryant Reid Gilchrist
Uma 12h45'46" 58d40'2"
Bryant Schick
Per 2h12'29" 55d31'17"
Bryant's Brilliance
Her 17h59'37" 23d42'56"
Bryant's Wit
Aqr 22h41'43" -7d32'6"
Bryany
Gem 7h28'14" 25d8'13"
BRYAR ONE
Ari 2h20'33" 26d49'17"
Bryaura
Dra 19h21'29" 65d46'23"
BryBaby<3
Sco 17h17'11" -34d4'40"
Bryce
Sco 17h15'0" -32d37'52"
Bryce
Gem 6h49'43" 17d22'18"
Bryce
Uma 11h36'34" 29d23'30"
Bryce
Psc 1h17'0" 15d41'16"
Bryce
Gem 6h59'53" 14d8'35"
Bryce A. Buckwalter
Tau 3h38'48" 18d28'41"
Bryce Alen Staggs
Her 18h25'0" 12d33'20"
Bryce Alexander Earhart
Gem 7h11'11" 34d46'16"
Bryce Alexander Meder
Aqr 21h3'18" 1d13'13"
Bryce Allen Merritt
Aqr 21h53'24" 1d1'0"
Bryce & Angie
Uma 10h30'40" 67d8'34"
Bryce Ashton
Her 17h15'21" 16d16'12"
Bryce Beau Donnell
Cnc 8h44'28" 6d57'4"
Bryce Cameron Botelho
Peg 22h3'47" 30d56'49"
Bryce Christopher Boutte
Her 17h15'58" 16d54'14"
Bryce Christopher Casimino
Cap 20h48'0" -25d17'21"
Bryce Cole Fagg
Ori 5h44'1" 1d9'17"
Bryce Davis
Cap 20h25'47" -10d29'44"
Bryce Dixon
And 0h31'29" 31d52'34"
Bryce E. Maly
Cam 3h50'14" 68d40'50"
Bryce Edward Hermann
Her 17h0'11" 33d26'21"
Bryce FitzPatrick Maloy
Sgr 18h38'56" -36d16'20"
Bryce Gabriel Mickel
Uma 12h9'56" 45d51'17"
Bryce J. Bernardine
Psc 0h33'35" 4d40'12"
Bryce James Lesinski
Cnc 8h33'39" 32d13'59"
Bryce Jared Scharnhorst
Uma 12h53'4" 54d43'46"
Bryce Jeffrey Russell
Dra 16h29'10" 53d21'0"
Bryce Johnson
Uma 11h41'0" 39d35'0"
Bryce Keaton Manning
Umi 15h38'10" 85d43'47"
Bryce Keeler
Per 3h23'50" 48d39'5"
Bryce Looney
Cnc 8h43'46" 6d40'18"
Bryce MacMillan
Uma 12h2'43" 65d30'57"
Bryce Mahon Dawkins
Ari 2h40'36" 30d1'39"
Bryce McCulley
Crb 16h12'50" 32d55'37"
Bryce Michael Agaman
Per 3h28'19" 33d51'18"
Bryce Michael Godfrey
Lyn 7h58'16" 47d48'24"
Bryce Mitchell
Cyg 19h46'52" 52d24'5"
Bryce Montplaisir Ainslie
Cyg 20h9'30" 39d6'58"
Bryce Moon
And 23h20'45" 47d59'21"
Bryce Munger
Vir 13h52'57" 2d48'51"
Bryce Olivia Armstrong
And 23h26'14" 43d40'45"
Bryce Preston Connally
Gem 6h34'33" 17d55'43"
Bryce Register
Cyg 20h19'3" 37d58'44"
Bryce Victor Robbins
Per 3h11'30" 45d52'50"
Bryce Williams
Ori 5h32'37" -3d47'17"
Brycee Webster
Aqr 23h13'10" -23d37'21"
Brycen Carter
Uma 10h43'11" 40d39'31"
Bryda's Star
Cap 20h14'13" -23d39'52"
Bryen Willis' Lil Fish
Psc 1h30'35" 17d15'18"
Bryleigh Madison Graham
Tau 4h6'5" 25d1'18"
Brylie Greter
Peg 22h42'20" 21d38'4"
Brymi
Ori 5h26'55" 13d18'28"
Bryn Chase
Crb 15h45'13" 31d14'53"
Bryn Kenny
Vul 21h16'39" 24d54'56"
Bryn Magnificus
Sco 17h48'52" -43d0'58"
Bryn Mauldwyn Norman Thomas
Leo 9h40'48" 32d7'38"
Bryn Patrick Mosing
Hya 9h32'33" -0d38'57"
Bryn Quincy Auville-Parks
Uma 10h38'16" 53d19'12"
Bryn Taylor Alsworth
Tau 5h38'59" 27d16'36"
Bryna
Lyn 6h42'55" 53d43'33"
Brynden Curtis Beck
Her 18h56'1" 14d47'9"
Bryn-Drew
Cyg 20h15'13" 54d25'4"
Brynley LaRhea
And 1h22'52" 45d7'24"
Brynn
Vir 12h57'29" 6d15'36"
Brynn
Ori 5h22'50" 3d42'18"
Brynn
Lib 15h41'40" -20d51'3"
Brynn Alexis Dangl
And 0h43'45" 40d24'25"
Brynn Burton's Star
Cnc 8h20'45" 11d31'55"
Brynn Carhart
Cyg 21h39'25" 50d27'27"
Brynn Cheri Edmiston
Lib 15h45'29" -12d56'37"
Brynn E.
And 1h46'31" 36d2'11"
Brynn Elisha Copman
Lyn 7h29'53" 52d52'25"
Brynn Elizabeth Baskin
Uma 9h3'54" 66d16'40"
Brynn Jordan Edwards
Eri 4h18'12" -4d42'51"
Brynn Lynda Skinner
Aql 19h2'7" 15d25'36"
Brynn Nicole
Peg 22h30'40" 11d33'59"
Brynn Patricia May Turner
Cra 18h36'41" -40d45'3"
Brynn Raquel
Ari 2h47'9" 22d1'25"
Brynn Tara Champagne
Cas 0h34'1" 65d31'1"
Brynn Woolf
Gem 7h30'39" 33d46'13"
Brynn555
Uma 10h41'27" 65d23'34"
Brynna
And 2h17'50" 42d17'40"
Brynne and Dave G.
Psc 1h15'21" 24d20'58"
Brynne K. Adams
Ori 5h58'40" -0d56'25"
Brynne Schwartz
Ori 6h7'58" -1d43'59"
Brynne Yvonne Holman 8/17/03
Leo 10h33'0" 27d3'11"
Bryon and Joeleen
Cyg 20h12'19" 46d59'57"
Bryon Curtis Rankin
Aql 20h3'57" 2d34'54"
Bryony Lennon LLB Hons
And 23h37'55" 47d11'43"
Bryony P. Simms
Sgr 18h26'57" -17d38'34"
Bryony's Jewel
Cru 12h37'18" -64d35'35"
Brys Keith Stanley McCaw
Sgr 19h29'31" -18d32'24"
Bryson
Umi 13h10'49" 70d16'59"
Bryson
Cyg 19h37'22" 45d12'1"
Bryson
Vir 12h40'39" 0d12'32"
Bryson Allen McFALL
Vir 13h8'57" -0d53'19"
Bryson Connor Tate
Uma 9h27'37" 57d45'30"
Bryson James Ellis
Aql 19h50'6" -0d57'41"
Bryson James Sullivan
Sgr 18h21'28" -24d0'2"
Bryson Keene Gath
Vir 11h55'26" -4d3'1"
Bryson Matthew Reibel
Gem 7h45'28" 32d47'55"
Bryson Ray Elliott
Vir 13h42'35" 7d12'52"
Bryson Sterling Gordon
Dra 18h35'37" 57d13'32"
Bryson18
Psc 1h30'30" 10d31'51"
BRYSTE
Uma 10h34'39" 40d28'48"
Bryte Star
Lyn 8h15'19" 40d40'15"
Brytnee LeRae Christensen
Lib 15h17'27" -4d50'28"
Brytney Allen
And 23h6'43" 48d47'25"
Bryttany Hull
And 1h1'39" 42d42'4"
Bryttany Leigh
Cnc 9h16'8" 28d48'17"
Bryun Kelley
Ori 6h16'7" 13d15'21"
Brzycki-Sokol 1981
Ari 2h49'15" 28d9'18"
B's Sparkling Ray of Light
Vir 13h39'25" -9d5'36"
Bsa (Bill, Staci, Alyssa)
Leo 10h23'34" 22d56'6"
BSE the Lawyer
Cep 21h42'12" 65d13'45"
BSG Faithfully
Uma 11h44'13" 52d25'7"
Bsie & Carisa
Ori 5h27'58" 5d57'7"
Bsomdotorg
Dra 17h8'18" 54d30'39"
BSQ3-2005-Kathy Frame
Lyn 9h1'5" 33d6'34"
BSVH-CC4
Sge 19h45'22" 16d55'1"
BT Munchkin Baby
Umi 16h18'10" 77d2'8"
BT4EVR
Uma 11h22'59" 46d11'9"
btfl_Jen_07302007
Leo 9h42'58" 24d51'56"
B...the love of my life
Cru 12h2'47" -59d17'12"
BTMott
Aqr 20h51'47" -8d50'27"
BtOrBuBtYh
Uma 11h52'50" 41d5'2"
BTWhite101505
Aqr 21h49'11" 1d55'35"
Bu
And 2h38'2" 50d34'43"
Buang Robert
Lib 15h2'47" -15d14'49"
Bub
Sgr 18h23'42" -23d26'29"
Bub
Cas 0h12'41" 54d10'48"
Bub
Aqr 22h0'23" 0d19'19"
Bub
Ori 5h31'45" 10d32'34"
Bub+WombleForever LubYouInfinity
Gem 6h55'1" 25d6'11"
B-U-B-A 575
Sco 16h14'20" -16d58'51"
buba and booboo
Umi 15h41'17" 79d5'5"
Buba & Papa's Love
Pup 7h33'10" -27d38'6"
Bubachu
Uma 9h47'13" 62d58'2"
Bubala
Vir 13h30'45" 12d59'4"
Bubalah
Uma 13h47'18" 56d48'32"
Bubay
And 1h34'48" 49d9'57"
Bubba
Per 3h24'25" 45d17'28"
Bubba
Lyr 19h9'7" 38d59'35"
Bubba
Cyg 20h25'29" 30d45'7"
Bubba
Gem 7h35'51" 32d57'8"
Bubba
Tri 2h13'6" 32d56'33"
Bubba
Cnv 12h46'56" 37d11'26"
Bubba
Uma 11h35'49" 33d29'22"
Bubba
Tau 3h42'22" 15d49'8"
Bubba
Ori 6h17'31" 16d44'11"
Bubba
Cmi 7h33'40" 6d2'32"
"Bubba"
Leo 11h7'19" 17d15'16"
Bubba
Com 12h41'1" 18d46'27"
"Bubba"
Cep 20h44'36" 55d49'49"
Bubba
Uma 9h27'4" 65d41'55"
Bubba
Mon 6h54'3" -0d59'33"
Bubba and Monkey
And 0h24'7" 42d49'3"
Bubba Brunelle
Peg 22h40'32" 8d2'29"
Bubba Bunny
Umi 16h16'9" 76d45'17"
Bubba Guerrera
Uma 8h36'14" 56d55'38"
Bubba Hammarstrom - P.S.
Lyn 9h4'36" 38d14'31"
Bubba Hudnall
Lib 15h40'22" -22d6'31"
Bubba Jade Ryder
And 0h8'41" 36d12'7"
Bubba Lee
Ori 6h7'10" 6d32'0"
"bubba - munch"
Per 2h21'22" 56d57'54"
Bubba Patenaude
Gem 6h48'27" 33d55'2"
Bubba & To-To
Leo 10h34'34" 14d23'23"
Bubba Weller
Aqr 21h8'3" -14d0'45"
bubbaausten
Leo 9h24'47" 31d55'11"
Bubbabuck
Lyn 8h42'9" 45d4'57"
Bubba's House
Leo 11h29'30" 9d21'14"
Bubba's Legacy
Ori 4h48'47" 12d30'20"
Bubba's Light
Tau 5h44'34" 22d42'35"
Bubba's Lil' One
Ori 6h5'2" 6d22'24"
Bubbas Princess
Her 17h15'26" 15d41'10"
Bubba's Star
Vir 12h54'21" 11d33'50"
Bubba's Star
Cnc 8h21'18" 14d16'3"
BubberAngel&AngelButter& ButterAngel
Umi 15h35'10" 83d18'42"
Bubbie
Lib 15h24'41" -5d16'27"
Bubbie
Tau 4h12'37" 15d59'3"
BUBBIEDOLL
Sco 16h55'28" -41d54'55"
Bubbies
Umi 15h38'21" 71d17'53"
Bubble 06/21/02
Aqr 23h2'17" -10d39'4"
bubbleplugs
Cet 2h26'12" 8d22'31"
Bubbles
Psc 1h26'58" 19d18'3"
Bubbles
Leo 11h41'39" 26d2'52"
Bubbles
Cyg 19h47'57" 29d29'44"
Bubbles
And 23h20'11" 51d31'33"
Bubbles
Uma 9h30'39" 49d40'17"
Bubbles
Uma 11h17'29" 36d57'53"
Bubbles
Eri 4h22'18" -25d15'5"
Bubbles
Lib 15h39'52" -24d3'44"
bubbles
Lib 15h5'32" -26d55'24"
Bubbles Brackett
Ori 5h0'35" 9d37'50"
Bubbles - My Guiding Light
Aqr 22h54'15" -10d57'36"
Bubbly Toes
Ari 2h40'45" 16d59'21"
Bubblypbear
Vir 12h37'50" 2d41'7"
Bubbs
And 22h59'10" 49d0'3"
Bubbs
Uma 11h37'47" 60d49'41"
Bubbs & Babydoll
Umi 15h36'11" 73d42'0"
Bubbu Pashi Manjunath Ganesh Kamath
Aqr 22h39'29" -7d56'47"
Bubby
Cep 21h21'37" 67d52'34"
Bubby
Pyx 8h53'48" -32d9'25"
Bubby
Lmi 11h5'17" 30d57'40"
Bubby & Poppy's Very Special Star
Ori 6h24'3" 14d22'50"
Bubby's Star
Del 20h37'38" 15d38'21"
Bubie, Jimmy Smith
Per 3h33'20" 49d45'58"
Bubina Zvezda
And 0h31'30" 36d42'23"
bubots
Sgr 19h20'36" -34d49'54"
Bub's Star
Cas 0h10'34" 58d42'58"
BUBSTAR 1
Ori 4h56'25" -3d0'14"
Bubsy
Leo 9h33'56" 16d20'12"
BUBU
Ari 2h55'54" 20d38'46"
Bubu
Aqr 21h41'44" 2d32'49"
Bubu
Psc 0h17'7" 11d38'4"
bubu
Umi 16h1'35" 88d0'29"
BuBu
Lib 15h10'49" -11d23'12"
Bubu
Apu 15h4'28" -73d35'36"
Bubu Strasnovova
Tau 4h48'36" 18d48'13"
Bubulina
Aqr 22h35'39" -0d44'52"
Buc Vickers
Uma 9h50'37" 43d45'33"
Buch, Mon coeur
Her 18h41'24" 20d49'12"
Buchanan's Bear
Uma 10h31'1" 61d44'32"
Buchholz, Joachim Julius
Aqr 21h1'16" 2d20'43"
Buchholz, Ludwig
Uma 9h8'32" 52d16'20"
Buchichi
Ori 5h37'28" -1d10'6"
Büchner, Werner
Uma 10h11'31" 50d7'9"
Buck
Leo 11h34'52" 24d6'18"
Buck
Uma 9h30'0" 72d22'47"
Buck (1991 - 2006)
Cma 6h25'33" -17d18'6"
Buck and Trickett 31/05/2002
Cyg 20h11'18" 35d57'20"
Buck Family Star
Uma 10h43'48" 44d57'36"
Buck Futter
Uma 12h26'22" 60d36'15"
Buck Hunziker
Per 3h47'11" 43d46'57"
Buck Johnson
Uma 11h41'3" 32d19'38"
Buck Lovett
Lib 15h5'38" -6d29'37"
Buck Snyder of the 21st Century
And 2h26'53" 48d40'57"
Buck Stokes
Ori 5h36'43" 2d20'3"
Buck, Peter
Uma 13h58'49" 62d14'6"
Bucket Crusher
Cyg 20h59'57" 35d50'34"
Buckethed
Tau 5h50'2" 23d46'29"
Buckeye
Lib 15h51'34" -17d44'8"
Buckeye Beauty (Sara Hudson)
Vir 11h42'49" -2d45'8"
Buckeye Jennabug
Leo 10h41'45" 6d35'35"
Buckk88
Tau 4h27'34" 19d16'4"
Bucko
Boo 14h37'48" 37d33'43"
Bucko and Toots
Peg 22h45'49" 25d37'19"
Bucko Cognat
Her 18h54'57" 24d0'47"
Bucks
Uma 12h39'35" 58d37'44"
Bucks Bad Boys
Peg 21h48'51" 24d57'6"
Bucksteeg, Helga
Uma 10h41'15" 64d18'56"
Bucky
Psc 1h13'14" 24d43'19"
Bucky and Rachie
Ori 4h47'46" 11d13'31"
Bucky Kimmel
Per 3h53'43" 39d20'51"
Bucky-Peanut
Aur 5h46'54" 49d23'22"
Bucky's Star
Uma 9h59'38" 46d29'0"
Bucolic Buffalo
Lib 16h0'22" -17d35'55"
Bud
Sco 16h11'14" -17d13'27"
Bud
Uma 8h56'32" 47d49'5"
Bud
Her 17h42'19" 38d27'23"
Bud
Her 16h57'52" 32d7'21"
BUD
Ori 6h9'56" 14d50'42"
Bud and Atsuko Harlow
And 0h36'17" 28d27'7"
Bud and Barbara50
Cyg 21h19'20" 43d43'4"
Bud and Dorothy Minton
Cas 0h9'55" 54d12'30"
Bud and June Dillashaw
Cyg 19h53'19" 31d6'34"
Bud Anderson
Umi 14h15'30" 69d38'50"
Bud Augustine
Cep 22h11'41" 71d38'44"
Bud Baxley
Ori 5h32'53" 3d41'23"
Bud De Matteo
Per 3h45'13" 38d26'22"
Bud & Helen Hauser
Crb 15h23'5" 27d26'10"
"BUD" Hugh Allison Radford, Jr.
Uma 8h34'31" 69d55'48"
Bud & Jamie Lynn Forever
Cyg 21h51'51" 46d43'7"
Bud Larkin
Lyn 8h8'50" 44d21'1"
Bud & Martha Robertson
Uma 8h12'29" 68d16'43"
Bud N Sis
Tau 5h46'47" 22d22'6"
Bud "Pap Pap" Furrer
Cap 21h49'16" -14d36'10"
Bud Shaffer
Cep 20h48'57" 63d50'10"
Bud Stone
Aur 5h24'47" 32d41'18"
Bud the Musician
Her 16h34'8" 25d46'9"
Bud & Trish Miller
Lyr 19h17'19" 35d34'32"
Bud & Wilma Smith's Piece of Heaven
Vir 13h6'12" -2d9'39"
Bud, Donna, & Colleen
Uma 12h33'29" 55d5'22"
Buda
Ori 5h54'3" -0d20'10"
Buda
Ori 6h0'22" 21d31'57"
~~ Budcastle ~~
Uma 10h26'26" 46d31'17"
Budda Wiley
Cru 12h51'54" -57d57'23"
Budde, Wolfgang
Lib 15h17'54" -8d17'10"
BUDDEEE
Vir 13h26'9" 9d2'25"
Buddha
Uma 11h27'44" 59d38'53"
Buddha <3 4 E
Umi 15h22'16" 68d45'44"
Buddha Bunny
Vir 12h14'40" -10d3'58"
Buddha's Kingdom
Dra 9h56'16" 73d0'56"
Buddi
Aqr 22h24'54" 2d30'32"
Buddie
Ori 6h21'1" 10d14'6"
Buddimathi Kulatunga
Cyg 20h28'59" 56d51'6"
Buddy
Dra 18h24'26" 53d58'16"
Buddy
Uma 9h20'9" 56d27'14"
Buddy
Lib 15h4'36" -7d50'1"
Buddy
Sco 16h5'17" -10d56'47"
Buddy
Aqr 22h6'12" -24d26'4"
Buddy
Cma 6h22'53" -27d21'53"
Buddy
Del 20h35'12" 13d9'59"
Buddy
Vir 12h18'34" 4d14'25"
Buddy
Lmi 10h22'48" 30d57'34"
Buddy
Per 3h44'20" 37d27'27"
Buddy
Psc 1h21'52" 30d47'43"
Buddy
Lyn 9h13'57" 34d39'55"
Buddy
Leo 9h42'14" 31d40'12"
Buddy
Per 3h32'34" 44d10'43"
Buddy
Per 3h9'18" 42d58'20"
Buddy " A Constant Light In Life "
Cap 21h10'39" -22d16'10"
Buddy and Lina Forever
Cen 13h29'52" -37d2'46"
Buddy Baker's Ball 'O Fire
Her 16h44'41" 45d21'14"
Buddy Bone
Cep 1h50'10" 78d10'46"
Buddy Boy
Cap 20h17'5" -8d52'55"
Buddy E.
Aur 5h47'28" 52d59'31"
Buddy Forester
Ori 5h7'24" 3d59'25"
Buddy Graf
Vir 13h25'20" -5d55'41"
Buddy Guy
Vir 12h18'43" 11d23'13"
Buddy & Hallie's Wishing Star
Cap 21h48'35" -8d35'22"
Buddy Langton
Per 2h23'5" 57d6'59"
Buddy Love
Eri 3h57'34" -11d56'33"
buddy melton dog star
Sgr 19h33'38" -14d49'54"
Buddy Number 1
And 1h19'41" 33d56'5"
Buddy & Sarah Johnson Forever
Lyn 6h56'38" 51d13'24"
Buddy Stingo
Uma 13h23'19" 56d6'43"
Buddy Wiegelman
Tau 3h45'59" 22d22'10"
Buddy Young
Lib 15h58'20" -3d53'14"
Buddy12632
Cep 20h31'28" 60d47'3"
BuddyLove's Bonnie
Sco 17h27'1" -45d19'29"
BuddyMelissa
Vir 12h18'34" 9d8'15"
Buddy's Bright Star
Ori 5h43'41" 8d13'35"
Buddy's Light
Cep 20h31'0" 60d48'52"
Buddy's Star
Lib 15h9'2" -14d4'42"
Budge
And 23h52'6" 40d31'56"
Budley
Gem 7h12'35" 19d9'47"
Budnix
Her 16h33'28" 47d9'30"
BUDS
Uma 12h47'6" 55d15'48"
Bud's Dream
Per 3h43'47" 36d38'48"
Bud's Halo
Cyg 20h8'36" 51d9'23"
Budstipe
Cap 21h37'44" -12d12'10"
Bue
Cyg 21h20'24" 33d5'9"
Buela
Psc 1h4'20" 32d15'53"
Buela Jenkins
Uma 10h11'47" 41d37'21"
Buenafe Minon Holbrook
Crb 15h46'4" 29d48'27"
Buffalo
Uma 12h23'35" 61d15'57"
Buffalo Man Adams
Cnc 9h19'27" 30d53'34"
Buffel + Buffalina, 20.12.2000
Uma 9h45'56" 72d34'23"
Büffelchen - ILD - H
Uma 10h11'5" 48d27'44"
Buffi Marie Novotney
Cas 1h4'26" 64d19'41"
Buffi's Hopes and Dreams
Mon 6h50'50" -0d23'34"
Buffy
Cnc 8h13'40" 25d25'37"
Buffy Lea
Cap 20h34'26" -14d25'54"
Buffy Tout
Cas 23h11'26" 59d8'8"
Buford's Way
Uma 9h26'24" 49d6'8"
Bug
Her 17h52'48" 22d14'48"
Bug
Tau 4h36'48" 17d4'7"
Bug
Peg 21h49'19" 14d39'46"
Bug Boat
Uma 9h9'25" 69d44'38"
Bug Eyes
Uma 10h5'45" 57d23'36"
"Bug" Jacob Donald Soley, Jr
Uma 11h25'46" 48d43'33"
Bugatti & Cookie's Star-Our Angels
Sge 19h51'52" 18d35'6"
bugga and ol' jim
Aqr 23h20'9" -14d51'53"
Buggs
Vir 13h13'44" -21d13'57"
Bughi
Cyg 21h37'47" 49d49'25"
Bugie
Uma 12h36'27" 57d25'53"
Bugie's Eyes
Umi 15h31'0" 81d43'10"
Bugs, Loves, and Wonderful Memories
Tau 3h58'31" 22d52'12"
Bugsy
Aur 5h47'11" 49d55'58"
Bugsy
Uma 12h31'28" 54d42'0"
BUI LE NA
Cnc 8h31'2" 21d16'49"
Buick
Leo 10h7'45" 17d16'48"
Buket&Deniz
Lib 15h37'20" -12d58'19"
Buket Torres
And 23h5'37" 51d37'28"
Bukky
Leo 10h54'25" 5d58'19"
Bukwat
And 1h6'20" 35d40'52"
Bula - Treasure Island, Fiji
Uma 10h23'29" 53d17'12"
Bull
Uma 9h27'29" 43d26'10"
Bull - Bisham - Fabre
Peg 23h51'34" 28d18'44"
Bull Frog
Uma 10h21'21" 48d6'15"

BULLDOG
Sco 16h9'15" -13d15'58"
Bulldog Mama
Cma 6h39'44" -15d12'37"
Bulldozer of the IV
Her 17h35'6" 27d17'15"
Bulldozer Sargeant 007
Aql 19h10'18" 9d43'18"
Bulle
Dra 19h51'53" 62d9'29"
Bullet, Poncho, Elmo, Daisy
And 0h26'6" 25d5'58"
Bullfrog
Uma 8h9'21" 61d58'40"
BullMaster Robert S. Marshall
And 22h59'55" 37d10'57"
Bulloch 01/05
Ori 6h1'47" 11d7'13"
Bullock - Headlight In The Sky
Gem 6h21'34" 20d37'53"
Bullvii
Ori 4h53'10" 12d52'0"
Bully
Psc 1h27'52" 28d30'54"
Bully Brophy
Cma 6h53'7" -13d8'52"
BUM
Uma 10h15'50" 53d20'15"
Bumbee
And 2h28'15" 49d46'2"
Bumpa Tim Fugman
Lyn 7h21'34" 57d20'34"
Bumper
Uma 11h34'53" 62d55'24"
bumpkin
Umi 13h53'20" 74d58'48"
Bumpkin
Cyg 21h44'34" 45d6'5"
Bumpy and Phoffey together forever
Cas 23h47'32" 52d58'5"
Bun
Cap 20h8'49" -21d18'26"
Buna's Wishing Star
Leo 10h12'28" 15d27'22"
Bun-Bun Star
Ari 2h3'48" 23d5'10"
Bunca Leonard
Uma 11h33'54" 44d23'42"
Bunchy
Cas 0h35'0" 47d28'53"
Bunjamin
Sgr 18h29'30" -17d8'57"
Bunk and Pat Forever
Mon 7h35'58" -0d32'58"
Bunk Ridenour
Uma 10h45'8" 47d18'52"
Bunker
And 0h0'33" 45d34'15"
Bunker Snyder
Ari 2h16'29" 25d3'34"
Bunkums
Aqr 22h22'16" 1d53'14"
Bunky and Lizzard
Per 4h6'10" 44d9'54"
Bunky and Terrianne
Tau 4h1'19" 19d11'59"
Bunky Boo
Cnc 8h42'38" 12d7'52"
Bunn
Leo 11h31'30" 2d16'10"
Bunni
Peg 23h25'35" 29d25'18"
Bunni
Sco 17h4'41" -33d10'53"
Bunnie
Cyg 19h36'39" 35d34'34"
Bunnie & Kowala
Lyn 6h49'18" 51d6'4"
Bunnie & Stoney Pony - "ALWAYS"
Cnc 8h54'25" 14d5'35"
Bunnies in LoVe
Lep 5h38'2" -11d42'50"
bunns
Cnc 9h21'28" 15d18'27"
Bunny
And 2h30'59" 46d51'33"
Bunny
Umi 13h47'27" 77d3'19"
Bunny
Uma 11h7'6" 54d33'48"
Bunny
Uma 9h23'36" 64d15'20"
Bunny 11-9
Lep 5h17'3" -11d30'55"
Bunny 223
Sco 16h43'34" -25d11'17"
Bunny and Bill
Gem 7h45'7" 21d11'37"
Bunny and Joe Yonkers
Uma 10h57'10" 44d29'5"
Bunny B
Lmi 10h5'24" 40d22'6"
Bunny Bunny
Cnc 8h15'57" 15d17'12"
bunny - bunny
Lep 5h36'58" -16d41'16"
Bunny Crazy
Ori 6h18'14" 15d34'2"
"Bunny" - Dana Marie Ita Bahamonde
Cap 20h27'18" -10d54'21"
Bunny Miranda
Cas 0h7'44" 59d58'39"
Bunny of the Sir
Psc 1h2'15" 27d21'29"
Bunny & The Beast
Psc 23h22'25" 3d5'9"
Bunny Wiener
Lep 6h1'22" -26d23'26"
Bunny & Wifey's Star
Cyg 20h51'57" 45d17'12"
BunnyFish
Gem 6h24'11" 20d46'57"
Bunnykins P.J.K.
Leo 9h23'4" 11d19'57"
Bunny-Kraken
Cas 3h13'7" 63d24'45"
Bunnypoo
Cnc 8h38'32" 7d11'12"
Bunny's Everlasting Light
Cas 0h21'27" 52d58'58"
Buns
Aql 20h37'16" -2d7'27"
Bunter (Quebex Forever Young)
Cma 6h44'1" -16d2'40"
Buntie
Ori 5h45'48" 2d44'57"
Bunty Kidwai
Her 16h50'18" 12d31'15"
Bunz
Aqr 23h55'27" -15d3'25"
Buon' aniversario amore Ti amo, sempre
Cru 12h26'3" -58d47'34"
Buon Anniversario -Love CKK
Cyg 19h40'47" 53d12'13"
Buona Notte
Lib 14h50'26" -2d15'20"
BUONA STELLA TREDICI
Uma 8h38'32" 69d14'16"
Buonafede Alice Luisa Maria
Her 17h29'10" 30d23'11"
Buonavita Carmine
Boo 14h43'2" 25d7'13"
Bupa
Per 3h44'19" 44d3'32"
Buppy
Uma 11h22'35" 46d8'22"
Buraian '05
Uma 9h0'7" 47d25'17"
Burak Tirtir
Crb 16h13'5" 37d1'26"
Burak Yildiz
Uma 11h2'53" 57d11'30"
Burbank
Vir 13h20'44" 13d25'7"
Burbuja
Col 6h5'24" -28d30'47"
Burd Star
Her 18h13'31" 26d13'4"
Burdman
Ori 6h17'13" 14d22'35"
Burga G.
Uma 8h30'21" 69d30'4"
Burgandy
Aqr 20h39'15" -12d39'7"
Burger Bell
Lib 15h5'3" -2d48'39"
Burgern
Lyr 19h19'58" 29d22'3"
Burgess Beginning
Crb 16h2'44" 37d53'32"
Burgos Family
Uma 11h16'34" 56d22'0"
Burgundy
And 2h35'55" 46d46'42"
Burhan
Ori 6h0'1" 13d57'39"
Burk & Jennifer
Lyr 18h48'50" 31d59'2"
Burke William Reynolds
Tau 5h32'4" 26d29'32"
Burkhard Kölbis
Uma 8h42'40" 54d54'7"
Burkhard Schmitz
Ori 5h29'36" 9d35'13"
Burkhard Weppner
Uma 10h29'14" 56d8'40"
Burkhart & Dagmar
Uma 9h42'47" 50d39'2"
Burkley Elizabeth Ayres
Cas 23h51'18" 50d6'51"
Burlesque Barbie Doll
Cru 12h44'50" -60d47'56"
Burley Day
Uma 9h31'54" 57d9'50"
Burmeister, Manfred
Sco 16h55'19" -38d55'45"
Burn
Vir 13h35'47" -7d23'12"
Burnell & Marilyn Cater
Cyg 21h15'38" 40d25'39"
Burnett
Ori 6h8'7" 15d16'10"
Burnie Burnett Shines Forever
Peg 21h54'14" 33d58'36"
Burning Love
Ari 2h46'40" 27d19'58"
Burns Gibson Clark
Ori 5h34'22" 5d4'53"
Burns Newfoundland
Dra 18h7'25" 72d10'4"
Burnt Pumpkin Belle
And 0h46'31" 27d23'7"
Burr Dexter Holstrom
Leo 11h24'26" 11d34'5"
Burrell 50th Anniversary
Cyg 21h14'14" 44d30'11"
Burrell-Carson
Sge 20h9'44" 20d31'10"
Burt "Big Dog" Burchette
Leo 11h14'42" 6d58'39"
Burt Brent 3/15/38
Psc 1h13'22" 13d2'56"
Burt Jaquith's Family
Psc 23h13'58" 6d20'19"
Burt Rubens
Aqr 21h48'30" -2d12'53"
Burt Shepard
Dra 17h7'10" 66d57'54"
Burt Stebbins
Umi 18h9'29" 87d22'34"
Burton Bliss
Cyg 21h22'20" 39d54'51"
Burton David Brent 3/15/38
Psc 1h19'24" 11d49'6"
Burton J
Tau 5h17'39" 17d15'6"
Burton J. Johnson
Tau 3h34'38" 13d6'46"
Burton Scott Price Jr.
Gem 6h28'20" 14d25'21"
Burton Zemansky
Uma 12h6'18" 53d38'2"
Burt's
Uma 12h0'21" 44d54'19"
Burt's Inspiration
Uma 11h39'37" 56d16'41"
Buruda
Peg 22h7'35" 8d36'48"
Busby's Star
Uma 12h19'40" 54d51'14"
Buschmann, Hans
Ori 6h17'42" 10d25'25"
Bushbyville
Cas 1h7'42" 61d23'8"
Bushiwhawha
Uma 10h49'21" 42d32'55"
bushkas star
Vir 12h29'11" 6d40'25"
Bushra
Leo 11h2'24" 0d56'5"
Bushra Chaudhry
Uma 13h49'18" 54d39'47"
Büsi's Zauberstern
Cep 21h47'26" 69d11'42"
Buskirk's Class of 07'
Uma 11h16'53" 58d5'56"
Buss
Cnv 13h59'59" 39d20'40"
Busse, Stefan
Ori 5h1'6" 6d55'48"
Bussie
Peg 22h36'25" 29d18'52"
Bust a ...
Uma 9h45'41" 58d30'50"
Busta Drule
Uma 12h37'23" 55d45'12"
Bustar
Uma 11h35'53" 62d40'52"
Buster
Uma 11h3'2" 61d6'9"
Buster
Uma 8h37'45" 73d3'12"
Buster
Uma 12h17'10" 58d48'24"
Buster
Cma 6h25'2" -29d51'14"
Buster
Cmi 7h42'55" 5d19'22"
Buster
Cmi 7h50'37" 5d25'56"
Buster
Per 3h7'5" 52d15'59"
Buster - A Courageous Superstar
Aql 20h14'15" 9d52'27"
Buster Brownie
Psc 1h40'58" 7d20'3"
Buster D'Andrea
Vir 12h20'33" -3d18'58"
Buster Roper -50- F E & F A
Her 17h28'6" 27d38'40"
Buster Strong, Our Special Boy
Leo 9h37'22" 7d11'22"
Buster VanOver
Ori 5h3'21" 3d51'5"
Busyizzy Jung Yeon Yoo
Leo 10h13'21" 25d38'32"
Butch
Tau 5h44'56" 24d43'54"
Butch
Uma 13h41'25" 51d18'50"
"Butch"
Cas 2h22'32" 68d51'21"
Butch & Amy's Destiny
And 2h3'30" 36d3'48"
Butch and Barbara Queer
Cyg 20h9'6" 32d55'25"
Butch And The Soaring Eagle
Cap 21h19'21" -23d2'13"
Butch & Betty White
Umi 16h45'2" 82d48'56"
Butch Bowen
Ari 1h50'7" 21d23'39"
Butch Dunyon
Her 16h42'29" 47d39'13"
Butch Hickman
Uma 10h52'3" 44d34'14"
Butch Hinchey
Lib 15h2'33" -2d41'27"
Butch Money
Cnc 8h14'32" 25d24'35"
Butch Piechowski
Her 16h45'25" 28d3'13"
Butchie
Per 3h20'25" 53d46'47"
Butchie Leach
Uma 8h43'52" 71d44'21"
Butch's Terrestrial Sphere
Ari 1h58'25" 19d14'52"
Butchy
Cnc 9h10'46" 11d44'44"
Butchy and Whit
Uma 10h42'46" 70d19'9"
Butchy Crispens
Sgr 18h55'27" -16d8'8"
Butchy George
Aur 5h46'43" 54d2'54"
Buthaina N.S. Qawar
Sgr 18h30'4" -18d13'41"
Butichoruyaco Venus Love
Psc 23h56'58" 1d34'16"
Butler Daddy Gene
Cyg 20h58'45" 42d48'22"
Butler van Sanden
And 2h28'38" 43d12'32"
Butter and Waffles
Dra 12h24'27" 71d40'44"
Butter Anderson
Psc 1h28'7" 25d40'49"
Butter Bean
Vir 13h18'51" 13d18'26"
Butter Cup
Mon 7h17'14" -0d34'43"
Butter Dog
Cam 9h13'12" 74d29'57"
Butter Fish
Uma 11h39'18" 44d43'34"
butterballomeda
Lyn 7h7'32" 47d38'40"
Buttercake06
And 0h47'26" 44d54'24"
Buttercup
And 0h56'52" 38d32'40"
"Buttercup"
And 0h36'57" 37d31'32"
Buttercup
Cnc 9h8'3" 30d52'57"
buttercup
Lmi 10h28'46" 34d48'22"
Buttercup
Cas 1h12'7" 54d25'52"
Buttercup
Ori 5h59'52" 6d41'21"
Buttercup
Ori 6h0'5" 18d29'5"
Buttercup
Uma 9h10'32" 55d18'41"
Buttercup
Cmi 8h0'17" -0d5'24"
Buttercup (Jared)
Cap 20h13'42" -9d9'17"
Butterfly
Lib 14h52'28" -16d1'1"
Butterfly
Aqr 22h24'58" -22d38'42"
Butterfly
Tau 4h32'16" 29d36'59"
Butterfly
Uma 11h18'55" 45d53'6"
Butterfly
And 23h42'8" 36d38'38"
Butterfly Amy
Psc 23h49'26" 5d48'35"
Butterfly Beach Hotel
Peg 23h19'34" 34d20'1"
Butterfly Butterfly
Gem 6h31'59" 24d31'57"
Butterfly Girl
And 2h28'24" 44d6'18"
Butterfly Jay
Cyg 20h37'31" 49d14'39"
Butterfly Jenn
Cyg 19h36'37" 34d32'11"
Butterfly Kisses
Lyr 18h43'3" 39d43'7"
Butterfly Kisses
Mon 6h54'55" -0d53'25"
Butterfly Natasha Sozoeva
Ari 2h34'12" 27d44'35"
Butterfly Princess
And 2h10'46" 47d43'17"
Butterfly Rock Cross
Leo 11h53'2" 21d2'29"
Butterfly Wings
Pho 0h41'25" -44d37'13"
Butterfly's Love
Cap 21h2'57" -22d54'48"
Butters
Uma 8h31'18" 62d52'24"
Butters Cotch
Dra 19h53'47" 74d37'11"
Buttface Kevin Ly
Hya 9h24'51" 5d44'19"
Butthead
Leo 10h37'41" 14d36'1"
Butthead Cardenas
Ori 4h44'10" 4d16'37"
Bütti der Glückliche
Uma 8h55'19" 48d52'5"
ButtMunchkin
Ori 5h28'50" -1d37'41"
Büttner, Gabriela
Uma 9h15'5" 64d26'50"
Button
Ser 15h17'34" -0d41'35"
Button
Leo 9h45'47" 26d11'49"
Button Nose
Uma 12h26'39" 55d54'23"
Buttons
Cam 5h21'24" 79d5'51"
Buttons
Sco 16h54'58" -34d6'5"
Buttons
Uma 8h38'30" 51d55'23"
Bütün, Tanja
Uma 8h45'15" 59d46'48"
Buu, 14.11.1986
Aur 6h34'3" 39d4'29"
Buud
Dra 15h40'9" 60d17'12"
buvette
Vir 13h38'2" -0d29'15"
Buxxi
Ori 5h53'17" 18d12'44"
Buyi
Dra 16h55'36" 68d3'34"
Buz 61
Lyn 8h28'44" 45d27'10"
Buz Boisvert, Jr.
Per 3h34'27" 31d52'11"
Buz Simons
Uma 13h3'3" 60d20'21"
Buzga/Hollenbeck/Jones/Warren
Ori 5h2'10" -0d2'21"
Büzu
Crb 16h11'53" 35d17'10"
Buzz
Umi 14h30'17" 74d45'49"
Buzz
Dra 14h52'2" 59d7'40"
Buzz Dicken
Per 2h17'27" 55d20'36"
Buzz York
Aql 18h50'24" 9d47'12"
BuzzBob RoundPants
Boo 14h59'27" 22d59'27"
Buzzie Siegfriedt
Psc 0h19'48" -2d1'48"
Buzzy
Umi 18h51'51" 86d32'10"
BVS - Greatest Dad In The Universe
Gem 7h26'0" 28d18'9"
BWagstaffMackie
Tau 5h49'0" 19d41'50"
Bwana Simba
Eri 4h11'59" -22d57'18"
B.Wayland Wooton
Lyn 7h45'28" 56d42'54"
BWD Group
Sge 20h6'3" 20d44'26"
B-Wear
Crb 16h0'54" 33d50'21"
By George
Cnv 12h41'29" 42d27'1"
ByEaByEa
And 0h43'59" 41d38'53"
Byrdie & Joshie 06
Cyg 20h45'8" 36d42'35"
Byrnes' Family
Uma 10h42'8" 56d42'3"
Byron
Uma 11h28'41" 60d25'56"
Byron
Aqr 22h3'43" -9d31'33"
Byron
Ari 2h23'40" 24d34'16"
Byron
Her 18h27'45" 12d12'3"
Byron Ashley Johnson
Tau 4h28'28" 20d49'21"
Byron "Barney" Beard
Per 3h7'5" 45d40'15"
Byron Beets *Grandpa*
Uma 11h7'51" 35d5'12"
Byron Clithero
Her 16h57'34" 16d28'9"
Byron Dittrich
Per 3h7'4" 54d30'43"
Byron Drew Hanks, Senior
Ori 5h28'51" 3d53'50"
Byron Hassen
Tau 3h29'19" 22d17'31"
Byron J. Leisek
Uma 10h43'18" 44d53'53"
Byron L. Richardson
Vir 13h7'7" -6d58'4"
Byron Lane
Aur 5h50'36" 44d16'39"
Byron & Laura's Soul
Cyg 20h56'22" 36d3'51"
Byron Luke Cheviot
Psc 1h46'34" 18d51'6"
Byron Novem Uncia
Pup 8h1'43" -42d6'32"
Byron P. York, Jr.
Aql 19h25'15" -10d41'51"
Byron Parker Leonard Whitt
Umi 14h40'46" 75d57'9"
Byron Preston Manfull
Psc 1h45'29" 8d48'11"
Byron R. Martinez
Sge 19h55'15" 16d44'53"
Byron Robert Renner
Gem 7h30'2" 25d22'50"
Byron Scott Anderson
Lib 14h50'27" -3d36'17"
Byron (Sonny) Warner
Her 17h14'38" 34d13'23"
Byron Trevor Lewis Didcock
Cru 12h0'26" -62d1'56"
Byron Vladimir
Sco 16h8'14" -21d30'4"
Byron Watkins
Lyr 18h36'49" 39d51'34"
Byron-Wyvonne Grissett
And 23h14'50" 43d36'38"
Bzafata
Gem 7h15'57" 32d44'12"
BZR20
Sgr 18h34'31" -23d36'34"
C
Cas 1h31'24" 65d50'22"
C A T
Gem 7h30'12" 22d11'4"
C. A. TRUDEAU VI-II
Cap 20h17'11" -11d26'26"
C. Adja Diop (Sunflower)
Cnc 8h32'46" 24d55'15"
C and K Cozort Jr.
Ori 5h45'46" 7d19'35"
C&B Ambrose
Apu 14h38'21" -73d48'6"
C B C
Uma 9h26'59" 71d55'22"
C B J Shawanda
Crb 15h50'8" 36d30'41"
C Branson Jr Beloved Dad & Husband
Cyg 20h19'8" 46d4'40"
C + C always and forever
Ari 2h22'59" 24d52'52"
C & C Beasey
Uma 9h52'19" 49d39'48"
C. C (christophe Cordier)
Cnc 8h16'16" 31d1'55"
C&C Colin and Catherine Forever
Lyr 18h51'1" 31d51'48"
C&C Magical Summer '06
Leo 10h24'14" 10d35'6"
C & C's Light of Eternal Love
Uma 9h26'16" 44d18'28"
C. D. Charles Drue Scull
Lep 5h7'24" -15d52'23"
C & D Tilford
Sgr 20h4'2" -25d30'10"
C. Dain Land
Lyr 18h53'27" 31d44'39"
C. Duane Dauner
Lyn 7h24'3" 45d44'15"
C. E. Merchant
Lib 14h23'5" -16d1'56"
C. Edward Starkey
Lmi 10h28'24" 36d30'34"
C. G. G. D.
Tau 5h9'6" 24d30'13"
C H Neocles Radiar Estrela UDX WWD
Lyn 7h3'25" 49d42'57"
C H S Golden Spurs 2007
Boo 15h7'44" 34d40'43"
C. Harvey
Cnc 8h40'26" 14d26'24"
C hasity's Light
And 1h14'53" 45d24'5"
C. I. DeShong
Lyr 18h47'44" 39d32'59"
C. I. Hanson
Sco 17h19'39" -32d12'35"
C. I. T. 03753
Sgr 18h22'38" -35d11'47"
C J Blevins
Uma 10h24'48" 62d1'51"
C&J Borskey
Ori 5h11'30" -0d23'3"
C J Callina
Lep 6h7'22" -11d11'39"
C + J Forever & Always
Umi 15h14'2" 76d8'30"
C. J. Mills
Ori 5h42'2" 7d58'55"
C+J+P+L+D Cox
Lyn 6h36'27" 54d6'19"
C. J. Robey
Aqr 21h39'1" -0d50'59"
C. James McCallar
Sgr 17h48'0" -21d26'10"
C John Singletary
Lyn 8h40'37" 41d1'10"
C. J.'s Star
Umi 14h20'38" 76d5'29"
C. Justin Steckbeck
Uma 10h34'35" 71d32'32"
C K One
Uma 12h9'58" 52d28'26"
C. L. Brown
Lyr 18h28'12" 28d19'4"
C. Lauren Schoen Ed.D
Gem 6h37'47" 14d39'26"
C. Lee a.k.a. Gregarious Recluse
Uma 9h59'14" 47d41'48"
C. Luke Hartmann
Per 2h15'46" 56d6'10"
C. M. Strickland
Aqr 22h10'57" 1d35'21"
C. M. v/o Murmeli
Mon 6h23'16" 8d31'0"
C & M Wright
Cyg 20h49'38" 37d12'48"
C. Marcus Thornburgh
Per 4h48'25" 44d56'38"
C. Marotta "Poppy"
Aqr 22h17'50" -13d53'21"
C. Nicholas Watson
Gem 7h47'18" 14d54'19"
C o l e e n
Cnc 8h2'54" 25d39'36"
C. P. D. Beckwith
Aur 7h18'9" 41d48'49"
C. Palmer
Uma 11h21'12" 34d10'26"
C. Paul Robinson
Uma 9h50'26" 70d45'54"
C. R. Oister
Cap 21h33'58" -17d5'3"
C. R. Torrento
Psc 1h15'11" 27d1'13"
C. & R. Valdez aka VAMGH
Umi 17h43'59" 81d6'13"
C. Raymond & Marilyn LaMarche
Sco 16h49'55" -39d7'49"
C. Richard Morris
Boo 13h49'39" 23d28'15"
C Robert Young
Ori 6h4'22" 21d15'33"
C. Royal
Ori 6h7'58" 16d0'32"
C & S Always
Uma 11h24'36" 58d11'10"
C & S Jackson
Dra 19h40'34" 61d14'12"
C S Keeler Boys
Boo 14h32'48" 17d48'19"
C. Shawn Stahlman
Uma 9h42'53" 56d16'38"
C. Simmers
Psc 1h3'30" 32d19'16"
C. Stuart Hunt
Uma 11h19'16" 28d55'39"
C & T Dancing Pearl
Oph 17h15'53" -0d28'37"
C. Tom Hamilton
Psc 0h36'49" 16d56'17"
C & V Collins
Cyg 21h45'26" 32d52'53"
C Wilson
Uma 9h2'21" 55d31'16"
C. W.'s Soul Tree
Lyn 8h58'20" 41d57'27"
C,K,B,C McGirr
Uma 11h42'39" 62d9'56"
C1 Sam Tx
Crb 15h29'31" 28d0'2"
C202 Ms Pallatroni Class 2004-2005
Umi 14h57'2" 72d54'6"
C202 Ms Pallatroni Class 2005-2006
Umi 15h46'13" 87d17'51"
C2A0R1O7L
Uma 10h9'30" 47d45'56"
C4S4EVA
Uma 9h17'44" 46d46'9"
C92-C94-J98
Psc 1h28'23" 25d46'30"
CA
Leo 10h9'43" 20d58'29"
C.A. Lopez
Del 20h33'57" 14d0'19"
CA. Shaw-Dal.Tex
Leo 11h20'46" 15d15'11"
CAAFE
Cyg 21h16'6" 46d18'4"
Caal 143
Cap 21h42'16" -16d56'17"
CAB
Uma 8h12'27" 63d31'51"
CAB13
Uma 12h17'35" 59d21'29"
CABA
Psc 0h15'34" 5d38'58"
Caballero/Gitana's Magic Aurora
Uma 13h22'24" 54d23'23"
Cabbagehead Joanne
Ori 5h21'43" -7d25'0"
Cabezon
Peg 22h42'3" 24d8'37"
Cabinet Guy
Her 17h17'30" 36d30'4"
Cabo Gonzalez Caterina
Uma 13h57'1" 57d41'14"
Cabo Gonzalez Matias
Uma 13h56'4" 57d33'57"

Cabral
And 23h53'16" 35d19'49"
Cabrie Teresa
Lib 14h45'18" -18d18'39"
Cacaro's Love
Uma 11h45'44" 45d32'50"
Cacau
Cnc 8h37'31" 8d32'3"
Cachetico
Cnc 8h51'44" 11d46'25"
Cachetona
Lyn 7h54'27" 59d13'11"
Cackle Annie
Cam 4h23'17" 67d34'54"
Caco
Crb 15h48'4" 33d10'41"
CAC's Scary
Psc 1h32'33" 13d49'59"
CAC's Star
Psc 1h25'42" 18d33'13"
CAD4466
Del 20h45'19" 13d11'34"
Cadallac Dog
Cma 6h55'11" -14d37'48"
Cadan Bolich
Lib 14h52'56" -3d21'28"
Cadance
Lyr 19h17'57" 29d18'22"
Cadance Jay McDowell
And 0h38'31" 31d17'44"
Cadbury Confectionery
Sales (M) Sdn Bhd
Ori 5h32'8" 2d46'54"
Caddy
Col 5h51'24" -38d44'28"
Cade
Cam 3h58'29" 63d12'0"
Cade Albert Poma
Uma 8h38'54" 68d34'51"
Cade Andrew Pack Moore
Cap 21h9'6" -15d4'37"
Cade Austin Mann
And 23h24'50" 41d33'17"
Cade Ellis
And 1h58'11" 40d10'17"
Cade Jeffrey Sawyer
Her 16h44'12" 19d9'56"
Cade Julian Gray Whitlock
Cnc 8h52'35" 11d14'13"
Cade Kerr
Umi 15h20'37" 75d32'4"
Cade Nicholas Flynn
Uma 11h18'50" 44d51'50"
Cade Patrick Montalbano
Sgr 18h56'4" -33d56'2"
Cade Randall
Leo 11h22'33" 24d38'3"
Cade Thomas George
Guthzeit
Her 16h20'54" 4d17'15"
Cadell Ryan Walton
Vir 13h7'57" 7d49'49"
Caden
Tau 4h44'7" 24d4'5"
Caden
Lyn 8h23'32" 54d14'47"
Caden Alexander Davis
Her 18h25'13" 27d57'47"
Caden Allen Schroeder
Aqr 21h38'30" 1d59'33"
Caden Chase
Pho 0h8'5" -44d26'15"
Caden Christan Meegan
Nicole Vickie
Del 20h41'11" 13d17'46"
Caden Christopher Galenti
Uma 12h16'52" 54d10'59"
Caden Duncan Johns
Umi 13h55'59" 69d34'14"
Caden Frederick McHenry
Cnc 8h36'30" 32d9'43"
Caden Grant Patrick
Birkley
Psc 23h8'26" 7d11'4"
Caden Jacob Custer
Eri 3h44'36" -0d45'36"
Caden Jameson Lanzillo
Sgr 19h39'54" -23d3'39"
Caden Kuchera
Psc 1h59'34" 3d4'34"
Caden M. Spader
Uma 11h16'31" 62d4'2"
Caden Michael Davidson
Her 17h12'58" 31d12'0"
Caden Michael Ryskoski
Aqr 22h13'17" -13d36'6"
Caden Michael Tacoronte
Ari 2h43'17" 24d57'15"
Caden Mitchell Brooks
Uma 10h32'16" 71d27'21"
Caden Nickolas
And 0h40'7" 25d59'17"
Caden Riley McGuire
Leo 11h44'20" 21d25'53"
Caden Ross
Psc 0h11'28" 9d18'10"
Caden Ryan Aiello
Lib 15h37'8" -19d37'42"
Caden Scott
Psc 22h53'11" 5d33'34"
Caden Vaughan Weiland
Cnc 8h11'39" 9d41'25"
Caden William Jungwirth
Ori 6h18'13" 14d16'23"
Cadence
Lyn 9h14'22" 37d39'17"
Cadence
Lyn 6h26'18" 60d0'37"
Cadence Elizabeth Gordon
Vir 12h8'38" 8d7'20"
Cadence Faith McDonald
And 23h47'16" 42d41'37"
Cadence Faye Crull
Uma 8h34'40" 46d48'0"
Cadence Nichole Strickland
Ori 5h43'19" 0d4'39"
Cadence Pearl Link
Uma 9h39'52" 49d42'18"
Cadence Skye
Leo 10h27'27" 13d23'29"
Cadence's Celestial
Sweetness
And 0h41'29" 41d33'31"
Caden's 1st
Cnc 8h53'56" 32d26'31"
Caden's Light - Caden's 1st
Birthday
Cru 12h21'22" -57d55'32"
Cade's Star
Tau 4h38'39" 1d22'7"
Cadi Yee Gum Fung
Vir 11h54'8" 6d10'36"
Cadie Siona Gillespie
Sgr 19h44'5" -23d44'15"
Cadience Abagail
Aqr 20h43'23" -13d5'15"
CADIPE
Ori 6h5'57" 14d16'13"
Cadman Frederick
Her 18h16'41" 14d49'14"
Caducus Angela
Psc 23h12'28" 3d6'27"
Cady
Cas 23h44'19" 53d51'8"
Cady Brock
Cnc 8h40'43" 15d29'46"
Cady Warenda
Cnc 9h6'6" 31d0'58"
Caecus Amor
Uma 10h49'56" 69d37'44"
Caeden
Cru 12h43'26" -60d9'25"
Caeden Paster Fox
Dra 17h26'36" 59d58'44"
CAEL SAOR
Tau 5h28'11" 18d14'14"
Caelan Dean Wagner
Peg 23h50'33" 29d23'43"
Caelan Lewis John
Her 17h20'58" 22d15'48"
Caelestis Jana Cum Amor
Circumvenio
Lac 22h41'42" 49d23'32"
Caeli Enarrant Gloriam Dei
And 1h36'47" 42d45'20"
Caeli Shea
Tau 3h54'45" 9d20'26"
Caelum Aidan Murray
Gem 6h47'26" 17d37'49"
Caely Louise
Ari 2h56'51" 30d30'43"
Caely Louise Welsch
Cas 0h22'41" 63d20'40"
Caelyn Navaroli Hoffman
Cap 20h9'53" -22d27'32"
Caeruleus machinator
Lyr 18h35'8" 27d15'26"
Caesar
Cap 21h42'55" -9d31'10"
Caesar
Cma 7h18'7" -30d9'37"
Caesar and Jackie Parente
Uma 8h21'49" 66d30'20"
Caesar Blue Knight
Cma 7h1'1" -24d43'49"
Caesar Martell
Uma 9h16'48" 60d46'19"
Caetlynn Raye Gardiner
Peg 22h21'31" 25d11'43"
Caety & Dan - Love That
Sparkles!
Psc 1h5'30" 9d36'53"
Cafe
Ori 4h51'10" -0d51'49"
cafe a gogo
Lyn 7h28'8" 47d33'22"
Caffeineboi
Ari 2h17'43" 23d28'37"
CAG-55
Aql 19h2'49" -10d30'31"
Cagan Anthony Sullivan
Per 3h3'23" 43d9'10"
Cage Camron Setzer
Aql 19h32'30" 0d54'33"
cagin
Psc 0h25'33" 4d53'47"
CAHRAH
Cnv 13h14'6" 42d21'33"
Cai Leif
Sgr 18h42'31" -22d8'48"
Cai Xin Zhang
Lib 15h26'9" -11d12'36"
Caidan Sean Murray
And 23h26'21" 48d23'27"
Caiden
Ori 5h18'59" 6d47'45"
Caiden
Ori 5h18'50" 6d21'46"
Caiden
Psc 0h35'49" 6d41'39"
Caiden Alise Donabedian
Gem 7h43'28" 33d1'31"
Caiden Britt-Hart
Lyn 9h0'35" 38d8'45"
Caiden Reid
Uma 10h32'47" 53d7'40"
Caiden Scott Hunt
Peg 22h21'40" 16d47'19"
Caidence Ann
Cap 20h54'4" -16d35'36"
Caidyn Makena Poulter
Uma 8h28'2" 70d58'39"
Caila 9-29-01
Lyr 19h16'39" 41d43'28"
Caila and Travis
Uma 10h6'18" 57d41'3"
Caila Coughlin
Ari 2h21'45" 14d47'26"
Cailean Walker - Our
Buddy
Her 17h28'5" 27d2'16"
Cailee Rae
Cnc 8h41'23" 18d41'45"
Caileigh and Barry
Cyg 20h3'2" 42d47'18"
Caileigh Morgan Young
Leo 10h16'46" 6d57'13"
Caileigh Rebecca
And 0h11'41" 37d28'54"
Cailen Messersmith
Crb 15h47'45" 28d49'42"
Cailey
Crb 15h48'27" 29d45'33"
Cailey
Cam 3h33'37" 61d37'13"
Cailey
Uma 11h34'22" 64d30'48"
Cailey Layne
Lyn 8h40'17" 40d53'21"
Cailin
Uma 10h23'45" 43d14'22"
Cailin
Cap 21h43'56" -12d29'43"
Cailin Bailey Regan
Lyn 7h13'41" 54d14'39"
Cailin Denton's Star
Cnc 8h51'16" 27d50'31"
Cailin Elizabeth Marie
Bentley
Umi 6h31'19" 88d6'20"
Cailin Lea Davis
Ori 5h25'36" 2d13'56"
Cailin Marie White
And 0h56'0" 40d24'18"
Cailin Quinn Egan
Ori 6h6'8" 18d34'34"
Cailleach & the Greenman
Lyr 18h51'19" 35d10'47"
Cailyn Dorothy Goodfriend
Gem 6h41'37" 20d4'50"
Cailyn Joan
Lib 14h28'18" -21d29'5"
CAILYN KK BLAIR
Lyr 19h16'30" 28d39'55"
Cain Foate
Ori 5h6'7" 14d13'0"
Cain Guilfoyle
Per 4h22'49" 46d30'50"
Cain Matthew Smith
Per 3h33'36" 47d5'36"
Cain Salt
Uma 11h57'46" 36d51'1"
Cainan Michael Tucker
Psc 1h10'50" 21d59'35"
Caine Bruce Allen
Col 6h6'4" -28d29'39"
Caine Nakata
Peg 21h31'50" 5d9'24"
C'Aira Brooke Dye
Leo 11h5'6" 8d59'3"
Cairo Antonio Salvatierra
Del 20h18'52" 9d39'54"
Cai's Star
Dra 18h54'4" 60d13'39"
Caison Christopher Holland
Tau 5h36'25" 25d36'29"
Cait Anabel Maguire
Cnc 8h19'48" 9d5'1"
Cait Elizabeth
Dra 18h41'51" 51d49'1"
Caiti
Sgr 19h13'15" -22d15'38"
Caitie Ulle
Ari 2h22'42" 24d27'59"
Caitie's moonstone in the
sky
Dra 17h32'26" 54d22'19"
Caitlan Johnson
Vir 13h6'55" -12d41'44"
Caitlin
Vir 14h2'46" -8d37'38"
Caitlin
Aql 19h50'7" -0d34'43"
Caitlin
Aqr 23h15'35" -8d44'30"
Caitlin
Lyn 8h31'25" 52d36'55"
Caitlin
Cnc 8h19'53" 15d10'33"
Caitlin
Tau 5h45'42" 18d14'33"
Caitlin
Ori 5h23'28" 3d5'47"
Caitlin
Vir 13h59'33" 0d41'33"
Caitlin
And 2h30'44" 50d12'29"
Caitlin
And 2h30'18" 48d13'9"
Caitlin
Crb 15h35'29" 37d25'22"
Caitlin
Cyg 20h35'2" 35d14'53"
Caitlin Alex Duffy
And 23h50'9" 35d18'3"
Caitlin Ali Slater
Psc 1h32'3" 14d49'39"
Caitlin Alice
And 23h14'21" 40d38'11"
Caitlin Alice Foley
Ari 2h34'55" 21d21'49"
Caitlin Andrews
Ori 6h2'55" 15d48'21"
Caitlin Ann Boyle
Sco 16h14'34" -11d56'8"
Caitlin Ann Shannahan
Ori 6h9'58" 20d5'12"
Caitlin Anne Francis
Cyg 19h38'45" 30d55'15"
Caitlin Anne Runnman
LaMar
And 1h38'46" 39d36'19"
Caitlin April Gregg
Lyn 6h49'51" 53d22'28"
Caitlin "Baby" O'Shea
Gem 6h43'36" 19d44'0"
Caitlin Bailey Elizabeth
McCann
And 23h10'21" 43d48'36"
Caitlin Baro
Gem 7h22'5" 19d48'3"
Caitlin Bernadette
Cru 12h19'24" -57d11'30"
Caitlin Blanchard
Cap 21h39'33" -11d50'21"
Caitlin Bonick
Cnc 8h47'56" 17d37'3"
Caitlin Brianne Ford
Ori 6h18'42" 16d35'31"
Caitlin Campbell Gager
Sco 16h29'46" -26d33'1"
Caitlin Chapman
Cyg 19h37'46" 31d3'2"
Caitlin Conroy Hope Cass
Gem 6h44'1" 33d24'21"
Caitlin Cooper
Peg 21h35'31" 10d59'18"
Caitlin Cupp
Uma 11h30'26" 47d43'12"
Caitlin Davis
Ori 5h21'14" 6d37'43"
Caitlin de las Estrellas
Per 2h40'10" 39d56'39"
Caitlin Dee Ann
McCampbell
Mon 6h32'10" 8d29'44"
Caitlin (Dimples)
Sco 16h17'33" -10d6'41"
Caitlin E. Hamill
And 2h18'56" 49d36'19"
Caitlin Eileen
Crb 15h35'57" 38d32'13"
Caitlin Elaine Small
Ari 3h14'2" 26d11'33"
Caitlin Elise Curran
Cnc 8h28'39" 31d9'17"
~Caitlin Elizabeth~
Vir 14h20'34" 3d30'1"
Caitlin Elizabeth
Dra 16h39'7" 63d22'28"
Caitlin Elizabeth Durkin
Peg 22h45'50" 4d41'17"
Caitlin Elizabeth Fitzpatrick
And 23h9'10" 51d42'5"
Caitlin Elizabeth Macy
And 1h15'16" 45d44'45"
Caitlin Elizabeth McGowan
Cap 21h35'23" -13d17'45"
Caitlin Elizabeth Schaub
Tau 4h45'40" 19d16'18"
Caitlin Elizabeth Vickers
And 0h22'9" 31d30'7"
Caitlin Elizabeth Weeks
Lyr 19h19'16" 29d46'47"
Caitlin Elyse Deneen
Cap 20h27'27" -14d27'14"
Caitlin Erin Shoemaker
Sco 16h51'41" -41d56'12"
Caitlin Eva
And 23h3'0" 36d5'36"
Caitlin Faye
Cru 12h29'39" -62d34'39"
Caitlin Fields
Umi 14h57'23" 73d56'39"
Caitlin Foley
Uma 10h14'33" 44d51'21"
Caitlin Frances Lee
Psc 0h0'1" -0d7'0"
Caitlin G. Hoffman
Cep 22h57'18" 68d36'26"
Caitlin Gaffey
Peg 21h35'12" 24d46'55"
Caitlin Garner Dykstra
And 23h4'29" 51d25'53"
Caitlin Grainne Iona
Griffiths
And 1h20'21" 34d57'20"
Caitlin Green
Cru 11h58'1" -62d45'23"
Caitlin Hickman
Cap 21h5'53" -16d47'45"
Caitlin Howden
Cru 11h58'41" -60d42'10"
Caitlin Ilene Redman
Aqr 23h52'3" -16d29'54"
Caitlin Iona Grainne
Griffiths
Gem 6h21'45" 25d9'18"
Caitlin Jade Nitschke
Ari 2h23'20" 12d16'15"
Caitlin Johanna Wolf
Ari 3h13'3" 25d57'29"
Caitlin Joy Guyette
Tau 5h47'52" 22d15'34"
Caitlin Keating Bruce
Umi 16h53'57" 80d20'50"
Caitlin Kelsey Murray
Ari 2h34'50" 24d54'31"
Caitlin Kennelly
Cnc 8h14'18" 15d26'47"
Caitlin Kinsman Simmons
Dra 17h28'32" 54d37'38"
Caitlin Klane
And 1h2'31" 42d10'24"
Caitlin Kroll
Cnc 8h15'44" 9d12'16"
Caitlin Kujawski
Vir 13h21'0" 11d57'15"
Caitlin Leigh Burton
And 23h25'15" 52d24'13"
Caitlin Leigh O'Neill
Psc 0h28'49" 4d23'20"
Caitlin Lock
Sco 17h47'7" -41d42'11"
Caitlin Loesel
Uma 12h23'17" 58d39'35"
Caitlin Lorraine
Cnc 9h6'37" 32d33'43"
Caitlin Louise Wigg
Cnc 8h22'24" 29d21'4"
Caitlin M. Cook
Cnc 8h34'2" 22d57'2"
Caitlin M. Whittle
Cnc 8h5'20" 10d25'5"
Caitlin Mackenzie Query
Ari 2h37'49" 20d5'7"
Caitlin Mae Chakaus
Welcome
Vir 12h9'26" -10d44'34"
Caitlin Mae Llewellyn
Sgr 19h22'11" -27d7'22"
Caitlin Maiette
Sgr 19h5'42" -11d52'8"
Caitlin Marcinkowski
Uma 11h31'15" 59d29'23"
Caitlin Maree D'Souza
Lib 15h17'49" -8d58'15"
Caitlin Margaret Kolb
Umi 14h47'35" 79d21'33"
Caitlin Maria Kainth
And 23h15'56" 35d15'16"
Caitlin Marie
Psc 0h43'13" 15d23'12"
Caitlin Marie
Ori 5h56'26" 17d51'17"
Caitlin Marie
Cap 20h28'39" -20d1'32"
Caitlin Marie Breitenbach
Aqr 22h34'10" -20d42'50"
Caitlin Marie Cowling
Lmi 10h43'28" 28d7'28"
Caitlin Marie Elizabeth
Bogucki
Ari 2h1'24" 20d40'52"
Caitlin Marie Herway
Aqr 22h15'53" -14d17'58"
Caitlin Marie Sky Mallory
Aqr 23h37'1" -16d43'2"
Caitlin Marion Yi Zhong
Larrigan
Umi 14h14'39" 65d46'34"
Caitlin Marissa Murray
Ari 2h10'3" 27d18'0"
Caitlin Marshall Grey
Gem 7h36'3" 23d55'34"
Caitlin May
Her 16h18'4" 42d23'0"
Caitlin May - Our Star In
Heaven
Cru 12h18'47" -57d21'37"
Caitlin Mayes
And 0h12'38" 28d39'6"
Caitlin McCann Davison
Cnc 8h22'38" 27d51'23"
Caitlin McGraw
And 1h42'36" 49d4'47"
Caitlin McKenna
Uma 11h27'57" 60d24'15"
Caitlin McKenna Juenger
Lib 15h8'12" -22d46'40"
Caitlin Melissa Romano
Leo 11h13'17" 8d57'31"
Caitlin Monro
Cru 11h57'7" -60d34'19"
Caitlin Moser
Cap 20h13'33" -13d49'17"
Caitlin Murphy
Mon 6h32'55" -5d17'1"
Caitlin Nancy Lowe
Psa 22h30'40" -31d3'57"
Caitlin Nicholas
Sco 16h6'2" -27d51'15"
Caitlin Nicole
Gem 7h49'8" 23d46'11"
Caitlin Nicole
And 0h12'56" 28d18'42"
Caitlin Nicole Lamar
Lyn 8h24'2" 45d42'8"
Caitlin Nicole Monk
Gem 6h25'17" 25d7'2"
Caitlin O'Connor
Lyn 9h1'27" 36d44'18"
Caitlin Olivia Doherty
Ari 2h32'2" 20d46'18"
Caitlin Paige Schreider
Cas 0h26'19" 56d47'28"
Caitlin Patricia Allen
Cyg 20h14'1" 37d25'23"
Caitlin Phillips
And 2h27'30" 46d9'53"
Caitlin Poppy Shanaghey
And 0h11'32" 45d0'51"
Caitlin Quinlan
And 23h36'2" 49d42'15"
Caitlin Rae Miller
Aqr 22h38'47" -2d50'38"
Caitlin Rebecca Patton
Peg 23h46'51" 26d18'41"
Caitlin Rebecca Williams
And 2h13'24" 50d14'9"
Caitlin Rhian Struk
And 0h50'31" 44d55'19"
Caitlin Riley
Cnc 9h4'55" 30d37'53"
Caitlin Robinson
Gem 6h9'48" 21d48'15"
Caitlin Rosa
Sgr 17h57'15" -23d46'0"
Caitlin Rose
And 0h28'15" 32d29'7"
Caitlin Rose Benfield
And 1h5'53" 45d33'1"
Caitlin Rose Brideau
And 0h30'36" 44d53'30"
Caitlin Rose Burlett
Cnc 9h10'48" 30d35'37"
Caitlin Rose Kraus
Cnc 8h24'21" 14d30'7"
Caitlin Rose Ramirez
Aqr 22h9'13" -14d39'31"
Caitlin Rose Rogers
Vir 12h51'33" -1d26'55"
Caitlin S. Armstrong
Ari 3h29'6" 26d37'18"
Caitlin Shae Henderson
Ari 3h22'50" 28d55'54"
Caitlin Shea
Leo 11h21'30" 12d1'0"
Caitlin Sophie
And 1h14'48" 35d32'28"
Caitlin Starr
Lyr 18h42'43" 38d38'5"
Caitlin Tanner
Aqr 20h46'50" -8d30'30"
Caitlin Taylor Lambon
Uma 11h6'4" 35d5'33"
Caitlin Teresa Buonomo
Gem 7h18'55" 33d47'22"
Caitlin the hottie
Psc 0h58'15" 4d24'8"
Caitlin Tiedeman
Peg 22h7'6" 19d43'6"
Caitlin Trucksess
Cnc 8h29'19" 31d35'17"
Caitlin Veronica 'Muffin'
Krause
Ari 2h32'15" 27d45'29"
Caitlin Veverka
Psc 0h51'29" 25d30'9"
Caitlin Victoria Hathaway
Com 12h18'19" 14d48'31"
Caitlin Victoria McCowan
Gem 6h52'46" 31d15'58"
Caitlin Wai Mallin
Cnc 8h57'3" 13d8'49"
Caitlin Walsh & her Spirit
Guides
Cnc 8h40'55" 10d0'1"
Caitlin's Eye
And 0h26'34" 26d45'40"
Caitlin's Light
And 0h57'43" 45d7'49"
Caitlin's Light
Lib 14h26'21" -18d3'31"
Caitlin's love
Tau 4h22'25" 29d33'35"
Caitlin's Star
Tau 4h22'2" 26d9'1"
Caitlin's Star
Leo 9h38'20" 25d4'51"
Caitlin's Teiko
Cru 12h45'33" -61d20'15"
Caitlyn Alison Gilliland
And 2h21'38" 43d2'30"
Caitlyn and Cory's Star
Dra 15h17'3" 61d0'28"
Caitlyn Anne Cusic
Cyg 21h56'57" 50d16'48"
Caitlyn Augusta Weller
Uma 10h21'53" 49d30'32"
Caitlyn Bandy
Lyr 18h49'1" 34d8'2"
Caitlyn (Caitie) Wesley
Uma 10h0'26" 42d37'3"
Caitlyn "Caity Girl" Breann
Perrin
Aqr 21h3'55" -7d9'3"
Caitlyn Crotty
Lyn 8h20'16" 57d26'37"
Caitlyn Elizabeth James
Vir 14h0'37" -14d57'27"
Caitlyn Elizabeth Kozlowski
Ari 3h21'42" 27d17'53"
Caitlyn Elizabeth Ostfeld
Psc 1h19'22" 27d28'14"
Caitlyn Eve Taylor
And 23h6'56" 51d20'37"
Caitlyn Grace Dresselhaus
Cam 4h50'22" 54d40'20"
Caitlyn Grace Ford
And 2h7'3" 44d27'23"
Caitlyn Hannah Geller
And 2h7'7" 38d51'38"
Caitlyn Hansen
Uma 13h36'47" 54d7'40"
Caitlyn Joy Clarke
Lyr 18h26'59" 32d44'34"
Caitlyn June
And 23h38'22" 34d40'6"
Caitlyn Lee Carter
Leo 10h31'32" 18d1'26"
Caitlyn Louise Holland
Ori 6h24'21" 14d38'23"
Caitlyn Marie Shenkosky
Boo 14h42'20" 52d8'47"
Caitlyn Marie Silevicz
And 23h50'4" 47d49'27"
Caitlyn May Whetstone
Tau 4h14'54" 17d15'34"
Caitlyn Michelle Driggers
Cap 20h22'48" -11d38'48"
Caitlyn Olivia Kline
Sco 17h4'1" -38d46'33"
Caitlyn P. Taylor
And 23h27'52" 42d6'55"
Caitlyn Rae
Vul 19h39'1" 27d22'20"
Caitlyn Renee Haynes
Lyn 7h53'11" 33d59'0"
Caitlyn Rose Schultz
Lyn 7h3'55" 51d1'53"
Caitlyn Shelby
Tau 5h34'10" 27d31'45"
Caitlyn Trinity
Sco 17h20'4" -32d56'8"
Caitlyn Victoria
And 23h35'26" 42d30'57"
Caitlyn Williams
And 0h50'1" 36d29'48"
Caitlynn Jewell Stephens
Lep 5h14'15" -12d22'12"
Caitlynn Marcotte
Sgr 18h59'35" -18d22'24"
Caitlyn's Radiance
And 1h17'24" 36d7'34"
Caitlyn's Star!!
And 2h5'51" 45d53'1"
Caitlyn's Star
Psc 22h58'32" 3d55'45"
Caitrín Aideen Hunter
Cap 20h12'30" -10d25'32"
Caitriona Ann Walsh
Eri 3h48'20" -21d40'36"
Caitriona Kenny -
Singer/Songwriter
Lyr 18h49'4" 35d1'58"
Caitriona McGuinness
And 2h22'46" 38d47'13"
Caitriona Sharkey
Cas 23h59'16" 60d9'42"
Caity E. Lennon
Cyg 20h25'7" 36d38'16"
Caity Grace Always My
Little Girl
And 0h17'11" 44d21'49"
Caity - Lynn
Uma 9h54'45" 48d4'47"
Caity May
Gem 6h38'4" 12d3'16"
Caity Michelle.........Earth
Angel
Lyn 8h5'58" 44d30'59"
Caity Thompson
Psc 1h17'46" 31d48'54"
Caitybear
Ori 5h56'23" 17d56'44"
Caius Alexander Acker
Krohnfeldt
Ori 5h35'56" -1d19'24"
Caiya Louise Milson
And 1h7'26" 45d27'35"
C.A.J. 52
Ori 5h0'0" 11d13'14"
Cajamama For Ever
Cru 12h38'35" -59d31'1"
Caju
Umi 15h52'39" 74d12'12"
Cakes and Pie in the Sky
Per 2h47'12" 54d11'31"
Caki Cake
Her 16h21'59" 44d28'4"
Cakie's Love
Leo 11h22'25" -5d5'40"
CAKLMAC
Uma 11h13'48" 30d48'53"
CAL
Leo 10h19'31" 25d34'3"
Cal (Bud) Flores
Uma 12h10'23" 55d34'24"
Cal Herron
Lib 15h11'36" -8d27'15"

Cal James Hodge
Per 3h20'15" 41d34'56"
Cal & Jodie Swan
Cyg 20h55'2" 37d1'52"
Cal Muren
Ori 5h54'40" 14d20'41"
Cal William
Leo 9h29'11" 25d53'4"
Cal William Backowski
Lib 14h40'42" -18d10'55"
Cal Williams
Umi 16h8'0" 75d42'12"
Cala Lyn Dresbach
Cam 7h22'29" 60d8'58"
Calah Jade
Lyn 7h52'5" 38d17'5"
Calais
And 2h20'57" 49d46'32"
Calamita
And 2h17'39" 42d34'9"
Calandra Lawrence
Lyn 7h50'28" 49d17'14"
Caldona Shands
Tau 5h40'21" 19d56'1"
Caldor - Jeremy H. Jackson
Per 3h49'50" 48d48'51"
Cale Kaba
Ori 6h20'49" 7d22'5"
Calea and Julea December 1, 2006
Cyg 20h1'40" 32d28'33"
Calea Cedar
Aqr 22h26'32" -14d25'17"
Caleb
Uma 12h11'1" 52d30'52"
Caleb
Sgr 18h15'58" -32d2'37"
Caleb
Per 3h18'43" 50d46'40"
Caleb
Gem 6h48'50" 22d58'10"
Caleb 2006
Per 4h47'36" 49d43'32"
Caleb Aethyr
Aqr 21h40'36" 1d57'29"
Caleb Albert Brook
Per 4h21'14" 47d15'4"
Caleb Allen
Her 16h53'46" 32d53'29"
Caleb and Ashley
Sge 19h12'22" 18d18'55"
Caleb and Kara - 7/30/06
Uma 9h34'55" 65d44'39"
Caleb and Tori Forever
Cyg 19h57'33" 36d18'43"
Caleb Aric
Vir 13h10'33" 11d24'48"
Caleb Asher O'Daniel
Lib 15h6'28" -3d40'12"
Caleb Aubrey Rader
Ori 5h55'39" 12d23'1"
Caleb Behan
Lib 15h16'47" -4d54'8"
Caleb Christopher Humbert
Leo 11h24'57" 0d24'18"
Caleb Crutchfield Gooch...Together
Cap 20h35'7" -23d3'59"
Caleb Daniel Martinez
Boo 14h55'48" 18d14'6"
Caleb Daniel Olson
Lib 15h24'30" -13d12'55"
Caleb David Thaemlitz
Uma 10h29'16" 54d53'44"
Caleb Dylan Rogart
Sgr 18h46'28" -32d15'44"
Caleb Elias
Her 17h7'1" 31d7'5"
Caleb Elijah Holzinger
Cyg 19h49'12" 37d55'37"
Caleb Eric
Umi 15h21'26" 77d41'32"
Caleb Ethan Lemoine
Vul 19h46'42" 28d51'11"
Caleb Fentress
Uma 12h11'28" 54d5'31"
Caleb Glasgow
Psc 1h29'16" 16d45'16"
Caleb "Heidelrock"
Vir 12h41'24" 6d40'41"
Caleb Isaiah Schuyler
Her 18h11'26" 18d6'51"
Caleb James Lindman
Lmi 10h3'46" 38d6'26"
Caleb James Neuburger
Ori 4h50'29" 13d23'10"
Caleb James Wood
Lmi 10h20'0" 33d19'50"
Caleb Jason Enders
Ori 5h31'24" 7d9'6"
Caleb Jayson Labbe
Dra 15h33'33" 57d35'35"
Caleb John
Cep 22h49'30" 62d48'39"
Caleb Joseph
Vir 12h48'47" 6d43'18"
Caleb Joseph Hancock
Ori 5h8'30" 2d50'25"
Caleb Joseph McKerchie
Lup 15h48'8" -36d1'46"
Caleb Joseph Regenski
Uma 12h8'33" 46d50'9"
Caleb & Josh Norwood/Calandros
Uma 11h26'0" 47d34'24"
Caleb Joshua Ingram
Sgr 18h32'57" -17d26'2"
Caleb Kissel
Vir 13h17'3" -3d44'38"
Caleb L. Smith
Cnc 8h55'45" 25d56'29"
Caleb Lance Rash
Cnc 8h11'14" 9d4'48"
Caleb Larsh
Uma 11h15'44" 52d41'56"
Caleb Lee
Cap 20h59'14" -27d9'28"
Caleb Lucas Brule
Per 3h19'48" 52d42'19"
Caleb Matthew Wood
Ori 5h37'29" 9d13'18"
Caleb Maximus
Pho 1h6'0" -48d35'51"
Caleb Michael Bourgeois
Lib 15h3'46" -1d28'10"
Caleb Michael Brown
Her 17h38'30" 28d33'5"
Caleb Michael Michalec
Ori 5h47'23" 6d2'42"
Caleb Michael Pepper
Her 16h58'35" 23d1'5"
Caleb Michael Westfall
Uma 11h17'31" 53d35'9"
Caleb Patrick D'Alberto
Cru 12h27'26" -56d7'5"
Caleb Paul Stinner Tomai
Ori 5h40'44" -0d55'33"
Caleb Porter Bausman
Ori 5h31'2" -0d8'43"
Caleb Randall Graham
Leo 11h26'35" 28d7'42"
Caleb Randall Munoz
Her 16h43'57" 32d32'38"
Caleb Ray Romo
Ori 5h29'12" 3d0'54"
Caleb Raymond
Dra 16h36'4" 64d6'37"
Caleb Regenski
Uma 9h11'8" 47d42'4"
Caleb Regenski
Uma 8h59'4" 49d0'1"
Caleb Reid Lambert
Vir 14h3'20" -7d40'46"
Caleb Richard Cramer
Leo 10h18'15" 22d12'23"
Caleb Robert Black
Her 18h3'12" 27d38'15"
Caleb Ryan Rossi
Ori 5h3'30" 13d6'16"
Caleb Samuel
Pho 1h7'4" -47d36'56"
Caleb Sheppard Hooper
Gem 7h10'47" 17d31'7"
Caleb W. Stevens
Sgr 18h53'2" -32d6'29"
Caleb West
Her 17h7'33" 33d36'4"
Caleb William Clark
Gem 7h13'44" 26d55'8"
Caleb's Shining Star
Umi 14h34'16" 75d0'23"
Caleb's Star
Her 17h19'59" 26d54'24"
Caleb's Star
Crb 15h37'2" 26d29'23"
Caledonia
Leo 11h27'53" 1d33'0"
Caleigh Anne Balestra
And 1h36'3" 39d46'37"
Caleigh Banks Kreigh
Gem 6h43'26" 23d30'27"
Caleigh Brooks Edgeworth
Ori 5h15'30" 15d48'1"
Caleigh Tully
Cap 21h54'17" -18d38'11"
Caleil
Ori 5h58'13" 21d6'54"
Calene Thurman
Tau 4h15'14" 15d10'54"
Calette Nicole
Uma 9h12'48" 72d16'4"
CALEX
Cyg 19h25'23" 53d46'58"
Calex
Uma 11h53'44" 31d2'36"
Calex And Amanda
Cyg 21h6'53" 53d10'57"
Caley Baby
Cap 21h11'54" -22d11'34"
Caley Grant
Mon 6h49'38" 8d9'40"
Caley Lisbeth Platteter
And 0h13'10" 26d32'52"
Caleyanne Hope Southerland
Lyn 7h41'29" 59d25'3"
Cali
Aqr 22h36'2" -1d26'41"
Cali
Sgr 19h29'41" -43d28'57"
Cali
Vir 13h36'17" 5d47'2"
Cali Ann Leister
Lib 15h6'8" -6d7'33"
Cali Day
And 1h18'57" 36d36'44"
Cali L.
Peg 22h35'17" 6d14'25"
Cali Lehner
Ori 5h37'1" -0d56'46"
Cali Lynn Caffey
Vir 12h47'39" 5d25'4"
Calia
Uma 12h2'49" 46d43'10"
Calicoe Star
And 2h21'40" 47d29'48"
Calicojak7
Lyn 7h57'5" 54d47'40"
CALIFORNIA
Gem 6h47'8" 18d0'7"
California Memories
Cnc 8h47'25" 17d9'6"
California-Armitage
Dra 18h33'16" 70d2'29"
CaLiMax
Ori 6h18'26" 15d12'56"
Calimë Nambatar
Ori 5h56'24" 21d48'25"
Calimero
Cet 2h19'18" 7d35'54"
Calin Gott
Peg 21h46'52" 11d7'27"
Calina Histë
And 1h19'21" 42d43'39"
Calista
Crb 15h36'2" 31d39'32"
Calista Cervantes
Aql 20h15'25" 8d43'20"
Calisto
Tau 4h2'57" 25d25'4"
Cali-Titus
Uma 11h57'26" 33d57'0"
Calixis
Per 3h20'2" 41d15'4"
Call of the Wild
Cma 6h50'54" -16d0'3"
Callaghan-Goodison Wedding 01.10.04
Cru 12h53'5" -57d40'8"
Callahan
Uma 9h47'54" 55d17'55"
Callain Portner MacLeod
Her 16h17'9" 26d3'37"
Callan Jonah Martin
Cnc 8h18'17" 9d54'26"
Callan's Wish
Tau 3h43'1" 23d48'37"
Callaward
Tau 5h53'39" 27d46'27"
Callee Jo Smith
Sco 17h32'11" -41d37'26"
Calleigh Francesca (Princess)
And 0h45'2" 23d44'47"
Calleigh Jane
Uma 9h12'10" 66d53'53"
Callen Lee Howell
Uma 11h37'43" 63d22'19"
Callenmar 2004
Ori 6h10'18" 16d4'54"
Calley's Comet
And 1h18'49" 49d26'48"
Calli
Cam 3h20'30" 61d41'53"
Calli Anne Catenac
Peg 21h49'42" 13d23'45"
Calli G
Lib 15h26'39" -7d8'22"
Calli L Stuppy
Ori 6h19'36" 14d38'29"
Calli Rowley
And 2h37'24" 46d34'43"
Callia
Leo 11h2'21" 22d10'15"
Callidora-"Gift of Beauty"
Lmi 10h40'13" 31d14'46"
Callidorn
Tau 4h38'13" 20d22'47"
Callie
Vir 13h27'16" 2d17'50"
Callie
Tau 4h30'42" 23d11'46"
Callie
Lyr 18h46'18" 36d14'46"
Callie
Aqr 21h55'20" -8d18'27"
Callie Adele Rymoff Rice
Vir 13h18'33" -12d39'39"
Callie Ann Shrope
Apu 16h26'13" -76d20'8"
Callie Anne
Tau 3h48'32" 17d2'40"
Callie (babygirl)
Lyn 8h33'24" 50d59'21"
Callie Brown
Umi 14h43'2" 73d18'33"
Callie Crea
Lep 5h17'39" -12d7'23"
Callie Elizabeth
Uma 8h11'30" 70d18'42"
Callie Elizabeth
Cyg 21h43'38" 46d43'35"
Callie Elizabeth
Cnc 8h48'2" 21d53'24"
Callie Fc
And 1h43'10" 49d13'34"
Callie Frances
Umi 14h26'53" 77d7'23"
Callie Helwig
Leo 11h10'23" 5d20'33"
Callie Kapp
And 0h19'57" 31d13'12"
Callie Korenek
And 2h33'30" 48d2'25"
Callie Laine Fraser
Tri 2h1'27" 31d39'36"
Callie Lynn Brennan
Sco 17h52'30" -39d53'34"
Callie M. Barefield
And 0h12'53" 42d11'14"
Callie Mae
And 0h18'23" 41d25'52"
Callie R. Hilton
Psc 0h25'0" 0d49'38"
Callie Rae
And 23h9'39" 47d56'47"
Callie Renae Monaco "6-8-01"
Crb 16h3'53" 31d5'59"
Callie Riggs
Aqr 22h1'52" -17d34'34"
Callie Rosa Lea
And 0h26'5" 31d15'51"
Callie Ruth Garner
Leo 10h25'37" 8d56'14"
Callie Ruth Stekardis
Cas 1h19'30" 62d18'2"
Callie Ryann Schramm
Lmi 10h43'56" 39d2'11"
Callie Stillie
Uma 11h35'13" 45d55'43"
Callie Vigil
Ser 15h18'36" -0d58'26"
Callie Weaver
And 0h20'29" 37d20'47"
Callie's Star
Cmi 7h30'13" 4d42'17"
Callinfornia
Leo 10h10'53" 9d4'53"
Calliope
Leo 10h9'47" 13d54'14"
Callipygous
Cap 21h41'51" -17d51'10"
Callista Ariel Prosser
Vir 13h17'0" 11d27'33"
Callisto
Uma 10h43'17" 49d1'1"
Callisto
Sgr 19h28'35" -15d19'26"
Callisto
Dra 17h29'59" 60d51'33"
Callisto Priddy
Uma 9h41'57" 41d30'59"
Callisto Ursa Major 40 Jill MacLeod
Uma 10h26'41" 56d6'23"
Callistus
Uma 11h28'1" 32d3'37"
Callon Joel Siegrist
Leo 10h13'40" 6d41'43"
Calloway Yoakum
Sco 16h50'26" -39d8'57"
Callum Christian McKay
Umi 13h48'57" 74d12'7"
Callum Edward Watts
Aql 19h10'55" 15d25'41"
Callum Erik Peter McMillan 21/02/05
Umi 14h28'42" 69d24'5"
Callum Harry
Uma 10h21'54" 60d34'50"
Callum Harry Sonny Payne
Cep 0h0'17" 72d35'12"
Callum Hill - A Special Wee Star
Uma 11h32'0" 57d11'7"
Callum Howe
Per 2h27'55" 51d58'5"
Callum Jack
Uma 9h14'33" 48d59'13"
Callum Jacob Ng
Cen 13h29'45" -38d19'38"
Callum Jake
Her 18h51'44" 21d49'29"
Callum James Bale Croft
Per 3h20'37" 40d11'30"
Callum James Bayliss
Her 18h24'51" 20d35'39"
Callum James Reynolds
Peg 21h58'44" 36d8'10"
Callum James Rutter
Per 4h37'51" 35d56'51"
Callum Jaydn Wright
Cru 12h51'40" -62d59'30"
Callum Joey Faulkner
Dor 4h21'46" -55d57'55"
Callum John Wilson-Irving
Uma 11h10'54" 39d40'26"
Callum Joseph-Elliott McMahon
Umi 14h6'9" 69d47'52"
Callum Lee Davis
Per 2h15'21" 54d53'44"
Callum Logie Hope
Per 3h28'53" 39d50'32"
Callum Louis John
Ori 6h20'42" -1d45'16"
Callum Malcum
Mon 7h33'51" -1d6'16"
Callum Marc Sefton
Dra 18h32'34" 79d26'59"
Callum Miles Hall
Cep 20h38'6" 61d50'54"
Callum Noel Sexton
Leo 9h22'39" 19d48'5"
Callum & Owen
Umi 14h44'59" 69d30'26"
Callum Patrick Goggins
Her 18h45'24" 15d5'35"
Callum Robert Muir
Ori 5h17'3" 0d36'38"
Callum Stuart Aitken
Per 4h46'41" 43d47'18"
Callum Stuart Burgess Hogg
Her 16h37'42" 12d38'42"
Callum Sunny Thorburn
Cru 12h38'36" -59d27'56"
Callum Tay Roberts
Umi 16h36'36" 75d28'36"
Callum the Great Storyteller Bisp.
Ori 5h18'58" -4d35'14"
Callum's Guardian Angel
Uma 10h20'10" 68d47'6"
Callum's Sparkling Gem
Lmi 9h51'22" 35d17'36"
Cally
And 1h1'24" 38d58'13"
Cally and Ian's Star
Ari 2h57'5" 18d34'57"
Cally May Peterson
And 23h3'3" 40d13'4"
Cally Shioban
And 1h57'6" 46d27'5"
Cally Snellgroves
Cam 3h40'43" 56d19'30"
Cally Zachariadis
Vir 13h24'37" 11d28'29"
Callyn Beth Kuhre
Ari 2h3'44" 23d24'10"
Callypso "Cally"
Lyr 18h33'54" 28d6'11"
callywillyoumarryme
Umi 17h23'49" 84d38'30"
Calm Storm
Psc 1h9'29" 27d35'25"
CALMAR
Cyg 20h21'15" 58d59'14"
Calogero Di Maria
Sco 16h5'59" -11d23'38"
Calongne
Peg 22h34'53" 12d27'44"
Calp
Sco 16h46'18" -31d28'8"
calred
Leo 11h48'10" 12d32'47"
caltha luminosis angelus
Cap 21h16'59" -20d5'18"
Calum Alistair Carr
Per 2h29'4" 57d28'7"
Calum Andrew Sweenie - Calum's Star
Umi 15h5'49" 74d38'25"
Calum Astrum
Psc 0h56'33" 4d50'58"
Calv
Cyg 19h26'55" 36d38'10"
Calvin A.Woody Great Dad & Husband
Uma 11h5'47" 45d10'37"
Calvin Baxter Warren
Lib 15h31'2" -15d25'9"
Calvin Cappelle
Crb 15h23'37" 28d14'49"
Calvin Carl Rumpca
Cnc 9h6'40" 32d39'17"
Calvin Christian Watts
Uma 11h54'55" 51d38'59"
Calvin Clarke D.D.S.
Cap 21h44'35" -8d58'47"
Calvin Coolidge Harnden
Ori 5h3'55" 4d43'38"
Calvin Ferns
Cyg 20h3'52" 39d44'54"
Calvin G. Searl
Uma 14h21'57" 55d41'26"
Calvin Galen Johnson - Isaak
Cep 23h25'52" 73d7'19"
Calvin Glass Sr.
Aqr 22h7'47" 0d49'7"
Calvin H. Knowlton
Uma 14h5'58" 60d11'9"
Calvin Hollins
Dra 19h18'49" 62d56'39"
Calvin James
Per 4h27'45" 42d25'48"
Calvin James Cruthirds
Sco 17h49'16" -42d17'24"
Calvin Jeffrey Nary
Psc 1h24'29" 28d5'11"
Calvin Kenlock
Cep 3h35'9" 87d3'55"
Calvin Lee Cole II
Boo 14h17'33" 18d51'0"
Calvin - Lenaïc - Rebaud
Uma 9h32'3" 56d39'27"
Calvin Lukken Sr.
Cep 22h21'9" 69d27'56"
Calvin Michael Lindholm
Umi 16h24'37" 72d29'29"
Calvin Moores
Aql 19h47'15" 3d25'26"
Calvin Oscar Ahre
Uma 11h12'23" 53d43'52"
Calvin Phillip Watts
Cnc 8h34'10" 24d57'38"
Calvin Ray Cox
Uma 9h24'12" 63d16'4"
Calvin Robert Kron
Cnc 8h7'43" 8d27'56"
Calvin Shimko-Lofano
Her 17h21'5" 24d4'16"
Calvin Wade Burton
Tau 3h52'51" 8d55'51"
Calvin Wadsworth VI
Psc 0h11'26" 9d22'50"
Calvin Warren Turner
Per 3h45'37" 41d8'38"
Calvin, Beverly, Gene, Wallace
Dra 17h0'54" 57d14'31"
Calvinatois Splendidus
Aur 5h42'3" 41d3'48"
Calyn
Lyn 7h33'18" 35d59'11"
Calyn E. Jones
Lmi 10h52'55" 36d40'50"
Calyn Elizabeth Kinney
Srp 18h7'54" -0d4'45"
Calypso
Lyr 18h42'7" 38d22'22"
Calypso And Our Love Shining Bright
Ori 5h0'10" 5d30'8"
Calyx
Vir 11h57'23" 4d34'8"
C.A.M.
Uma 11h1'34" 53d29'59"
CAM " 21 "
Crb 15h34'0" 27d14'2"
CAM Bernal
Ari 2h56'28" 20d47'48"
Cam & Mary Davies, Brayden & Tyler
Umi 14h27'7" 76d50'18"
Cam Nguyen
Cap 20h21'39" -12d38'23"
Camala Gayle
Uma 10h16'40" 43d55'1"
Camalea
Cas 1h1'55" 49d7'13"
Camaraderie
Lyn 9h25'51" 40d11'0"
Camarae Nichol Casillas
Uma 11h28'55" 28d20'41"
CaMarie Joy Coopernia Hoffmanem
Cnc 8h22'9" 24d2'30"
Cambra Jame' Koczkur
Uma 13h45'12" 54d59'50"
Cambria
Uma 12h36'2" 61d14'41"
CambrieAndMichaelForever
Cyg 20h29'13" 47d41'7"
Cambron-69
Gem 6h33'58" 14d24'16"
CamCam
Tau 4h24'16" 17d35'22"
Camcy Sweet
Mon 6h21'13" -0d52'56"
Camdara
Aqr 22h26'25" -5d39'3"
Camden
Psc 1h28'7" 27d29'2"
Camden Alexis Tucker
Ori 6h23'57" 14d55'11"
Camden - Baby Cakes - Benadon
Gem 7h30'38" 19d53'32"
Camden Clay Dawson
Vir 13h19'43" 2d14'29"
Camden David Boucher
Uma 14h1'35" 61d44'53"
Camden Joseph Bade
Cyg 19h51'59" 36d25'9"
Camden Norris
Cap 21h40'0" -16d56'25"
Camden Patrick Rentrop
Ori 5h32'21" 2d58'12"
Camden Paul
Cnc 9h4'39" 30d43'18"
Camden Rainier Smith
Psc 0h2'27" 4d40'5"
Camden & Sarah Cook
Cyg 19h41'40" 30d41'15"
Camdyn Glenn Toumi
Psc 0h2'48" 4d53'3"
Camdyn Maren Glover
Cap 20h18'17" -9d35'20"
Camdynne
Psc 1h18'59" 13d7'8"
Camel
Cam 4h25'31" 71d52'24"
Camel
Cas 1h28'35" 64d41'36"
Camella
Lib 14h30'44" -14d41'36"
Camella E. Busnardo
Sco 16h11'43" -29d11'56"
CAMELLE
Cap 20h22'17" -9d35'11"
Camellia
Cnc 8h22'3" 21d14'42"
Camellia Catherine G. Mirabella
Sco 17h51'36" -42d39'18"
Camellia Rose Day Camp 2006
Uma 9h59'52" 58d58'53"
Camelot
Aqr 23h31'57" -14d45'2"
Camelot
Lib 14h52'1" -0d34'32"
Camelot
Leo 11h30'16" 1d22'55"
Camelot's Tiger
Sgr 19h9'14" -23d36'44"
Cameo Martin-Wylie
Dra 15h8'24" 59d20'29"
Cameo Merasia
Uma 11h2'50" 56d1'50"
Cameo of Light
Lyn 7h27'46" 46d4'30"
Cameo Rose Taylor-Dye
Cyg 19h57'36" 33d11'56"
Camera
Ori 5h10'41" 15d7'32"
Cameren Nicole Cook
Mon 7h18'25" -0d15'43"
Camerista Weistler
Tau 4h4'22" 6d16'33"
Cameron
Ori 6h4'17" 13d52'3"
Cameron
Vir 12h51'27" 11d52'54"
Cameron
Leo 9h57'49" 8d18'44"
Cameron
Per 3h26'40" 43d47'24"
Cameron
Uma 11h54'43" 33d27'52"
Cameron
Aql 19h13'3" -0d34'28"
Cameron
Uma 12h7'25" 62d19'1"
Cameron
Sgr 20h0'59" -44d1'58"
Cameron A. Meloun
Vir 12h35'11" -3d9'3"
Cameron Adamiyatt
Umi 16h59'59" 79d14'51"
Cameron Adams
Cma 6h45'16" -15d25'30"
Cameron & Aimee
Cas 0h31'17" 52d20'15"
Cameron Alex Mageski
Dra 18h36'58" 59d12'16"
Cameron Alexa
And 0h19'59" 29d38'49"
Cameron Alexander Lee
Cnc 8h3'36" 12d59'52"
Cameron Allen Kegley
Aur 5h23'11" 45d53'14"
Cameron Amor January 1,2006
Sgr 18h28'51" -23d39'43"
Cameron and Marissa
Peg 23h12'25" 17d35'14"
Cameron and Nicole Kaikis
And 0h30'58" 42d18'21"
Cameron Angela Gezovich
And 0h43'55" 35d5'41"
Cameron Ann Saunders
And 0h25'38" 37d18'12"
Cameron Anthony Dougherty
Vir 12h44'17" 12d9'48"
Cameron Anthony Lynch
Ori 5h35'40" -1d59'1"
Cameron Anthony Taylor
Umi 17h8'23" 76d35'52"
Cameron Anthony Taylor
Uma 13h8'55" 54d16'23"
Cameron Arthur Cook
Uma 8h28'36" 69d36'59"
Cameron Bennett
Uma 11h30'46" 56d10'22"
Cameron Bissett
Her 16h16'35" 43d44'21"
Cameron Blaine Hunt
Eri 3h25'25" -16d14'14"
Cameron Braman
Cnc 8h47'48" 23d18'2"
Cameron Brooke
Ari 2h34'8" 20d42'43"
Cameron Brown
Lyn 8h19'29" 54d23'33"
Cameron Bruce Star
Per 3h27'10" 33d55'15"
Cameron Cailteux Willig
Ori 6h13'8" 20d55'17"
Cameron-(Cam-Bam)-Terrell Johnson
Uma 9h59'41" 65d53'42"
Cameron & Campbell
Peg 23h19'31" 33d49'28"
Cameron Cash
Vir 13h19'52" 5d34'31"
Cameron Chase Lee
Psc 1h11'13" 28d53'19"
"Cameron Chester Hilbert"
Her 18h10'3" 24d9'24"
Cameron Clark Hayes
Uma 10h27'5" 40d54'40"
Cameron Cousins
Per 3h9'49" 46d22'24"
Cameron D. McKinley
Uma 9h56'31" 66d13'17"
Cameron Dale
Cnc 8h16'42" 18d59'33"
Cameron Dantrail
Crb 15h34'7" 37d56'20"
Cameron David
Sco 16h13'43" -16d55'10"
Cameron David Castellarin
Ari 2h7'11" 22d0'19"

Cameron David Kelly
Lib 15h36'8" -10d3'35"
Cameron David Thorn - Cameron's Star
Ori 6h19'29" 10d29'28"
Cameron Dean Schmidt
Uma 11h49'11" 35d12'14"
Cameron Dean Smith
Peg 22h40'46" 11d23'41"
Cameron DeYoung
Umi 14h49'25" 82d39'52"
Cameron Douglas
Sgr 18h33'49" -17d53'57"
Cameron & Dylan Rome
Cyg 19h50'38" 52d49'15"
Cameron E Bolles- My Love To You, D
And 2h14'28" 38d44'34"
Cameron Elise
And 0h29'22" 25d15'2"
Cameron Elizabeth Taylor
Lib 15h5'50" -0d40'57"
Cameron Elon Ford
Crb 15h36'7" 34d41'57"
Cameron Eric & Kyle Arthur Osborne
Sco 16h9'1" -12d30'11"
Cameron Espino
Ari 3h16'20" 26d57'13"
Cameron Eugene Stroeh
Aur 5h24'26" 40d53'57"
Cameron Eva Elizabeth Shults
Oph 17h52'49" -0d3'11"
Cameron George Leach
Umi 14h25'9" 71d41'44"
Cameron Gibbs
Cru 12h56'15" -63d0'2"
Cameron Grace
Aql 19h16'35" -10d29'44"
Cameron Greathouse
Uma 10h35'10" 68d12'53"
Cameron H Custer's Hope of Light
Ori 4h49'25" 5d53'18"
Cameron Hardy
Psc 23h9'48" 5d31'54"
Cameron Hargreaves
Cru 12h1'54" -63d32'55"
Cameron Harris
Aqr 22h42'39" -16d32'41"
Cameron Heleina Castro
Sco 16h8'8" -8d57'46"
Cameron Henderson
Cnc 8h5'30" 20d5'44"
Cameron Henry Keith Marshall
Per 4h37'23" 42d14'2"
Cameron & Ieuan McIlduff
Ori 6h14'36" 8d32'23"
Cameron Imrie
Cyg 19h44'13" 37d55'48"
Cameron Jackson Wells
Her 16h13'33" 24d53'4"
Cameron James Allen
Umi 19h16'11" 88d3'17"
Cameron James Cook
Umi 14h40'52" 75d29'37"
Cameron James Frank Ormond
Umi 15h59'13" 74d40'11"
Cameron James Gregory 9/14/1995
Her 18h17'14" 21d31'54"
Cameron James McDonald
Lac 22h24'53" 41d32'23"
Cameron James Watson
Per 3h13'17" 46d56'25"
Cameron James Yost
Sgr 18h34'58" -35d49'17"
Cameron Jared Olsen
Her 17h18'41" 49d29'36"
Cameron Jason Peters
Boo 14h14'46" 47d51'23"
Cameron Jess
Lib 15h15'20" -19d15'26"
Cameron Jock
Sco 16h57'31" -44d34'50"
Cameron John Ballantyne Hamilton
Per 3h49'43" 40d4'26"
Cameron John Dunn
Crb 15h59'20" 38d8'45"
Cameron Jon Davies
Aqr 23h4'1" -18d50'3"
Cameron Joseph Burek
Cam 4h17'56" 73d12'55"
Cameron Joseph Cravens
Psc 1h45'59" 13d37'30"
Cameron Joseph Overholser
Per 2h45'50" 56d18'55"
Cameron Joshua Goldberg
Cep 22h55'57" 58d23'36"
Cameron Kelly
And 1h21'44" 43d58'3"
Cameron Kenneth Ciani
Psc 1h10'53" 30d48'9"
Cameron Kevin Kopaunik
Cyg 19h26'51" 51d42'12"
Cameron Ladd Dewey
Ari 2h21'41" 23d47'41"
Cameron Lake Strandin
Uma 9h10'14" 69d50'59"
Cameron Layton Davis
Tau 5h17'36" 25d43'49"
Cameron Lee
Mon 6h52'3" -0d27'19"
Cameron Lee Baldwin
Tau 4h29'5" 13d17'43"
Cameron Lee Duryee
Per 3h13'56" 53d27'11"
Cameron Lloyd Powell
Cep 20h46'45" 60d15'56"
Cameron Logan
Lyn 6h54'45" 58d1'23"
Cameron Lorell Mitchell
Her 16h13'46" 18d15'2"
Cameron "Love Star"
Umi 14h5'15" 73d37'34"
Cameron Lucian
Aqr 23h1'19" -6d20'34"
Cameron Luqa
Peg 23h54'55" 25d26'24"
Cameron Mark Crellen
Uma 9h3'37" 57d53'20"
Cameron Mark Durre
Cyg 20h57'44" 55d10'55"
Cameron Mark Temple
Aqr 20h52'51" -8d25'15"
Cameron Mayflower (Cammy)
Peg 22h28'11" 24d44'0"
Cameron Maynard
Cep 20h59'38" 56d1'40"
Cameron McCartney Martin
Cma 6h31'40" -15d4'43"
Cameron Michael Caudill
Her 16h31'3" 33d58'34"
Cameron Michael David
Per 3h41'33" 41d40'56"
Cameron Michael John Lownds
Uma 11h25'30" 49d44'8"
Cameron Michael Kitchen
Ori 5h30'28" 3d17'27"
Cameron Michael Skeens
Aql 19h38'45" 9d49'4"
Cameron Michelle Cruz
Ari 2h24'33" 10d51'38"
Cameron Milani Scherbakov
Per 3h4'22" 44d49'31"
Cameron Miller-Lemus
Tau 4h35'55" 16d27'25"
Cameron Mitchell 10-28-2002
Sco 17h1'48" -39d16'58"
Cameron Mokdad
Ari 2h24'23" 19d8'12"
Cameron Montgomery
Her 18h19'6" 19d4'28"
Cameron Montville
Per 4h42'29" 46d6'3"
Cameron Munchkin O'Reilly
Uma 9h30'39" 44d32'12"
Cameron Nicholas Donnelly
Cnc 7h57'24" 13d18'15"
Cameron & Nicole
Cyg 20h57'48" 47d34'42"
Cameron Oakes' Star
And 23h3'18" 52d14'26"
Cameron OReillys Infinite Light
Her 17h7'47" 31d10'49"
Cameron Patrick Bohan
Lib 14h52'44" -1d39'59"
Cameron Patrick Rice
Lyn 7h59'23" 52d59'55"
Cameron Pedersen
Uma 10h38'2" 66d43'2"
Cameron Philip Zwick
Uma 10h15'30" 46d24'12"
Cameron Ray
Uma 8h44'1" 50d33'56"
Cameron Reimers
Uma 11h1'2" 65d36'29"
Cameron Richard Smith
Cep 22h56'30" 57d22'8"
Cameron Richard Spiers
Cet 2h15'34" 6d26'32"
Cameron Rudolph Bowman
Lyn 6h45'29" 60d20'56"
Cameron Ryan Bruemmer
Lyn 9h21'30" 33d24'16"
Cameron Ryan Courtney & Cherie 4EVA
Cru 11h59'22" -62d53'10"
Cameron Scot Needham 231992 1:18 PM
Ori 6h19'27" 8d57'10"
Cameron Scott Hendryx
Cap 20h22'39" -12d43'8"
Cameron Scott Salt
Per 3h10'17" 53d48'20"
Cameron Sillars
Dra 20h39'19" 68d8'35"
Cameron Spencer
And 0h50'14" 37d6'16"
Cameron Steele
Mon 6h50'55" -0d38'39"
Cameron Steele Kelly
Lib 15h2'26" -20d40'34"
Cameron Stiles Christensen
Ari 2h52'42" 24d37'51"
Cameron Stokes
Ori 6h0'58" 13d38'21"
Cameron Tyler Lange "Star of Life"
Dra 18h21'1" 67d3'16"
Cameron William Merris
Per 4h11'45" 42d19'30"
Cameron Zachary Veith
Cra 18h49'59" -40d35'46"
Cameron, My Love Master
Cap 21h54'57" -11d4'34"
Cameron's First Star
Umi 14h36'22" 76d31'18"
Cameron's Liberty Star
Sco 17h56'50" -30d40'31"
Cameron's Reba
Aql 19h32'39" 8d1'9"
Cameron's Star
Lyr 18h49'5" 35d4'50"
Cameron's Star
Ara 17h25'59" -51d19'39"
Cameron's Star
Cru 12h25'59" -57d52'20"
Cameron's Star
Lib 15h3'8" -3d38'53"
Cameron's Star
Cap 20h28'49" -9d38'35"
Cameron's Star
Umi 13h21'43" 70d0'49"
Cameryn's Star
Uma 14h11'26" 57d37'47"
Cametmat
Aqr 21h27'51" 0d56'53"
Camey "sweetness" Hileman
Gem 7h7'3" 17d50'34"
Cami
Cmi 7h24'5" 9d44'59"
Cami
And 1h26'54" 43d18'26"
Cami Adolf Shines Forever
Ari 1h57'27" 22d54'4"
Cami and Christopher Klohn
Cyg 20h20'24" 58d32'2"
Cami Place
And 2h19'25" 49d26'46"
Cami Waller
Cyg 20h59'27" 47d34'45"
Camie
Cyg 19h49'20" 36d0'22"
Camie & Madeline Soltren
Aqr 23h36'49" -15d37'3"
Camie's Beauty
Ori 5h21'29" -0d40'10"
Camila
Cen 12h16'32" -42d11'12"
Camila
Vul 20h25'36" 27d42'52"
Camila Aramayo
Peg 22h52'33" 26d32'30"
Camila Del Valle Marcellini
Cap 21h23'36" -19d57'7"
Camila Melo Saavedra
Car 9h52'37" -60d10'31"
Camila Sophia Mendoza
Umi 15h16'35" 67d52'7"
Camila SV
Lib 14h26'46" -16d7'44"
Camilia
Cnc 8h50'45" 31d2'38"
Camilinpinpin
Her 16h54'0" 37d31'37"
Camilla
And 1h14'4" 42d12'52"
Camilla
Crb 16h23'11" 29d20'49"
Camilla
Tau 3h39'38" 29d18'46"
Camilla
Cnc 8h54'11" 11d4'1"
Camilla
Ori 5h43'32" 1d54'24"
Camilla
Ori 5h41'48" -2d34'28"
Camilla & André
And 2h28'46" 41d58'49"
Camilla "bimba"
Uma 11h18'31" 55d43'4"
Camilla Caires-Gauthier
And 23h12'27" 52d19'14"
Camilla "Cami" Baunsgard
Cas 23h16'8" 56d6'25"
Camilla Kay
Crb 16h19'26" 29d36'30"
Camilla Lee Purvis
Tau 3h49'58" 11d56'40"
Camilla Leigh Stuart
Cap 21h27'10" -15d40'32"
Camilla Louise Brissenden
Cas 0h37'17" 50d57'43"
Camilla Margaret Mary Canty Shepherd
Cyg 20h58'52" 31d52'1"
Camilla Organ
Cnc 8h32'33" 15d13'31"
Camilla Solstråle
Cap 20h44'44" -21d8'57"
Camilla Stenkvist
Mon 6h51'22" 8d32'8"
Camilla Stull
And 2h9'44" 41d49'24"
Camilla's
Cas 23h43'55" 56d5'28"
Camille
Cas 23h26'59" 59d57'18"
Camille
Lyn 8h13'41" 43d22'57"
Camille
Lyn 8h9'48" 34d29'37"
Camille
Ori 6h6'10" 6d5'1"
Camille
Gem 7h26'3" 15d7'52"
Camille
Ari 2h42'36" 26d44'0"
Camille 14-04-1995
Uma 9h20'20" 43d48'34"
Camille A Gonzalez
Sco 16h23'5" -27d53'42"
Camille A Gonzalez
Sco 16h8'26" -12d40'38"
Camille A. Wilhelm (Cami)
Cas 1h32'25" 56d52'50"
Camille and Dave
Dra 19h10'5" 77d37'30"
Camille and Lawrence Narissi
Vir 13h53'53" -11d48'49"
Camille Ann
Gem 6h56'43" 13d23'0"
Camille B. Lynch
Cas 23h4'36" 54d12'43"
Camille Bondurant
Sge 19h42'17" 17d29'54"
Camille Boyer 31.07.1985
Tau 3h53'45" 26d46'55"
Camille Camilly Camilly Willy
Sgr 17h52'59" -29d43'55"
Camille Christine Pellegrino
And 23h11'40" 50d47'59"
Camille Djakeli
Tau 4h25'5" 22d26'37"
Camille Elizabeth
Per 4h46'9" 41d51'20"
Camille English
Uma 9h20'15" 41d31'41"
Camille Gnolfo
Crb 16h8'58" 26d41'21"
Camille Gnolfo
Leo 11h12'33" -0d49'52"
Camille Gutierrez
Vir 14h32'40" -0d18'19"
Camille Horn Newbern
Ari 2h38'46" 24d47'11"
Camille J. Fiore
Sco 16h15'28" -27d46'39"
CAMILLE JANE SHIELDS STAR MOM
Cas 1h22'5" 57d15'47"
Camille Jordan Williams
Umi 15h22'46" 68d18'20"
Camille Katisha Antar
Cap 21h50'46" -21d34'41"
Camille Labbé Tanguay
Cas 0h8'49" 54d14'44"
Camille Laurent
Crb 15h58'8" 35d3'20"
Camille Licsi
Sgr 18h36'38" -23d53'12"
Camille Louise Paton
Cam 4h30'20" 76d19'32"
Camille Maria DeMattia Barnes
Cnc 8h4'16" 23d8'21"
Camille Marie Lombardo
Psc 1h17'16" 25d47'16"
Camille Morvay Raia
Cas 2h15'0" 64d33'54"
Camille My Love
Uma 11h22'27" 69d58'0"
Camille Necole Vega
Lyr 18h50'0" 36d47'20"
Camille Pâquier
Lib 15h25'52" -21d58'39"
Camille Pasquini
And 2h20'12" 46d19'13"
Camille Renee
Crb 15h40'55" 27d12'1"
CAMILLE ROCCO NOBILE
Uma 10h6'49" 47d39'7"
Camille Rose
And 1h52'37" 42d8'38"
Camille Roxanne Maccarone
And 23h35'12" 44d16'26"
Camille Roy
Cas 0h2'53" 53d58'34"
Camille Selene
Sco 17h39'45" -41d35'6"
Camille Smetana
Sco 17h58'26" -31d27'50"
Camille Sullivan
Cas 1h15'38" 58d55'7"
Camille Thoman
Cam 5h11'33" 69d37'30"
Camille Turtle Berbano
Gem 6h49'29" 29d24'57"
Camille Weston Schaffer
Psc 0h41'2" 14d32'26"
Camille,John Macy & Bristol
Cmi 7h30'59" 5d37'41"
Camille21
Ori 4h52'50" 13d16'48"
Camillegeorges
Gem 7h0'47" 27d59'57"
Camille-la-Bien-Nommée
Dra 14h39'29" 60d23'14"
Camille's Star
Crb 16h19'53" 30d25'32"
Camillo & Gillian
Cyg 21h13'24" 42d0'18"
Camillus A. Powell
Uma 10h1'44" 46d17'22"
Camilly Geigel
Psc 0h44'21" 19d47'33"
Camin Beth Dean
And 23h7'46" 48d49'24"
Camiolo Andrews Family 1978
Ori 5h7'56" 6d6'42"
Cami's Poppy
Ari 2h5'57" 23d56'17"
Camius Juliron
Hya 10h11'22" -26d10'27"
Camlynne Elizabeth
Uma 11h41'14" 37d11'22"
Cam-Mal Star
Lyn 8h51'54" 40d4'5"
Cammi Jolie
Leo 11h13'17" -3d51'46"
Cammie Jo
Sgr 19h25'7" -13d43'55"
Cammie Louise
Cas 23h34'9" 55d22'8"
Cammy And Chris Forever
And 23h21'53" 48d42'43"
Cammy King
Tau 5h47'28" 16d6'7"
Camoule
Ori 5h42'13" 12d11'38"
Camp Beloved
Vir 13h9'33" 12d37'2"
Camp Courageous of Iowa
Uma 13h17'22" 56d47'27"
Camp Morton
Per 3h32'28" 46d57'38"
Camp Tek Love: Alison & Steve
Gem 7h25'25" 14d15'23"
Campagni D'anima
Lyn 7h58'37" 35d34'59"
Campbell
Uma 11h42'37" 44d13'53"
Campbell "Belle" Gillian Brent
Sco 17h25'21" -42d21'9"
CAMPBELL CHRISTOPHER SMITH
And 23h57'52" 46d40'30"
Campbell Club Forever
Cam 5h2'8" 69d3'43"
Campbell Duffy
Vir 11h49'56" 4d31'47"
Campbell Family Star
Ori 6h11'49" 19d22'47"
Campbell Family, Suwanee GA
Aql 19h45'39" 2d22'25"
Campbell George Baker
Cnc 8h25'28" 7d0'55"
Campbell Hadley Deringer
Per 4h20'4" 52d26'18"
Campbell Jonathan Throckmorton
Vul 19h33'33" 25d54'35"
Campbell Kay Davis
Lib 15h4'50" -8d13'22"
Campbell Lee Youens
Aqr 22h13'13" -1d0'52"
Campbell Leiterman
Sgr 18h24'39" -33d7'35"
Campbell Martin William O'Meara
Psc 1h30'21" 11d31'48"
Campbell Newman - 6th August 2003
Cru 12h3'30" -62d24'12"
Campbell NOBC
Ori 5h0'40" 10d37'25"
Campbell Roth
Vir 13h56'51" -15d26'59"
Campbell Rudder
Crb 16h18'58" 29d35'51"
Campbell Sophia Bruening
Vir 13h29'15" -3d59'44"
Campbell Star
Cyg 21h54'53" 48d57'35"
'Campbell's diamond in the sky'
Cru 12h37'38" -60d37'41"
Campbells Rule The Universe
Cru 12h37'17" -60d7'20"
Campbell's Star
Uma 11h24'24" 58d15'2"
Campbell's Star
Sge 20h6'57" 17d19'34"
Campe
Ori 5h48'13" 0d59'53"
Camp-Eros
Sgr 18h23'30" -29d2'46"
Camren Michael George
Uma 10h23'33" 43d29'52"
Camren Skinner
Umi 14h14'3" 75d23'10"
Camron Blaine Howell
Ari 1h58'23" 19d39'19"
Camron McIntyre
Uma 12h8'38" 55d42'59"
Camron McKenzi Ryan
Mon 6h44'43" -0d0'32"
Camron's Star of Heaven
Vir 13h32'14" -4d26'50"
Camryn
Lyn 7h11'13" 52d43'35"
Camryn Alyssa Moschitta
And 0h14'43" 46d49'34"
Camryn Ann Murray
Sgr 18h10'36" -20d1'15"
Camryn Baylee Fryer
Lyn 9h4'3" 42d5'4"
Camryn Jane Golden
Uma 9h7'14" 71d0'29"
Camryn Landree Wells
Lyn 9h10'40" 34d57'9"
Camryn Marie
Vir 13h21'59" 1d7'34"
Camryn Paige Poglajen
Cap 20h37'46" -21d21'5"
Camryn Ronita King
Crb 15h27'45" 28d55'44"
Camurriey
Uma 9h29'56" 47d51'20"
Camy Wilson
Per 3h29'52" 33d1'32"
C.A.N.
Leo 10h28'9" 18d44'51"
Can Basat
Lib 14h46'36" -10d45'45"
Can Do,Do Do
Cam 3h44'9" 68d27'53"
Can I Call You SweetHeart
Sgr 19h21'30" -17d17'10"
Can Kakmaci
Uma 10h18'16" 43d38'29"
Can10
And 23h25'33" 51d37'50"
Cana Falay
Uma 8h17'31" 62d23'42"
Canada
Lyn 7h33'54" 49d9'43"
Canam
Crb 15h42'41" 38d43'45"
Cancellieri
Tau 3h38'0" 27d59'2"
Candace
Gem 6h38'28" 27d33'24"
Candace
Leo 10h52'10" 6d8'6"
Candace
Ori 5h45'11" 3d41'32"
Candace
And 23h52'52" 42d9'20"
Candace
Lyn 8h52'35" 42d10'42"
Candace
And 2h18'49" 39d14'11"
Candace
And 0h46'12" 38d54'38"
Candace
Lyn 9h13'51" 34d12'18"
Candace
Cas 23h41'58" 61d38'25"
Candace
Lyn 7h35'7" 56d38'1"
Candace
Aqr 21h12'50" -14d5'16"
Candace
Cra 18h37'52" -40d35'47"
Candace Ann Burke
And 2h27'26" 37d57'36"
Candace Barr Breen-Warren
Cap 21h39'58" -14d18'30"
Candace... Be Strong
Her 16h49'26" 29d44'49"
Candace Bernice Scott
Peg 22h51'46" 11d33'47"
Candace Berry
Aqr 22h19'8" -23d33'47"
Candace Beth Patrick
Uma 10h51'17" 43d8'45"
Candace Blake Kirkpatrick
Lyn 8h43'35" 45d29'22"
Candace "Candy" Eck
Psc 0h26'37" 18d45'21"
Candace Claburn
Ari 2h18'40" 18d2'30"
Candace & Cody
Leo 9h58'4" 23d29'40"
Candace Cole
And 1h17'12" 45d58'36"
Candace Darlene Durham
Leo 10h16'15" 15d50'10"
Candace Dawn Steele
Lmi 10h14'14" 34d0'28"
Candace Delicia Parham
Lib 15h19'39" -20d40'31"
Candace Grosman
Dra 18h25'39" 56d40'31"
Candace... Have Faith
Her 16h41'35" 30d23'11"
Candace Holly Nix
Gem 7h3'30" 14d27'10"
Candace Hope Lindsey
Lyn 7h55'23" 38d39'52"
Candace Ilona Katherine Bakay
Lib 15h4'35" -11d17'36"
Candace Ima Duncan
And 0h4'38" 43d58'57"
Candace Jackson
Sco 17h57'54" -35d22'4"
Candace Jade Meyers
Leo 10h14'21" 25d7'49"
Candace K. Gray
And 0h19'45" 43d44'24"
Candace L. Beseau
And 0h56'31" 39d47'20"
Candace L. Beseau
Cas 0h34'54" 55d17'16"
Candace L. Bohlman
Cas 23h31'11" 52d7'18"
Candace L. Fox
Cap 20h15'17" -8d43'4"
Candace Lee Ann Hayes
And 0h43'57" 37d40'59"
Candace Lee Carnell
Tau 4h41'27" 8d28'6"
Candace Lorraine Smith
Gem 6h25'42" 26d33'38"
Candace M. Davis
Crb 16h22'32" 38d19'3"
Candace MacNevin
Her 17h56'39" 27d48'31"
Candace Marie
And 0h21'54" 44d51'11"
Candace Marie Bartsch
Gem 6h47'20" 26d46'31"
Candace Marie Smith
Ori 6h12'7" 15d47'39"
Candace Marie Turner
Crb 15h48'9" 35d11'57"
Candace Marie-Baxley Stephenson
Uma 11h48'20" 44d15'32"
Candace Meehan
Uma 10h8'55" 57d14'22"
Candace Morgan
Lyn 9h19'37" 40d10'11"
Candace & Nathalie
Del 20h47'23" 5d18'49"
Candace Nicole PPool
Lyr 18h46'53" 35d11'50"
Candace Nicole Tuggle
Eri 5h9'54" -8d46'37"
Candace Pirritano
Per 3h23'8" 53d28'3"
Candace Prevette
Gem 6h6'12" 27d39'26"
Candace Quirarte
Gem 6h27'12" 27d13'16"
Candace Raye
And 1h0'29" 38d21'56"
Candace... Remember... I Love You
Her 16h54'44" 28d48'31"
Candace Renee Shepley
And 1h33'24" 40d24'5"
Candace Rosato - Shining Star
Lyn 6h58'51" 48d33'54"
Candace Sharp
Cas 1h1'32" 49d21'54"
Candace Shoemaker
Crb 16h4'29" 36d33'53"
Candace The Beautiful
Uma 10h40'34" 47d49'6"
Candace & Trever "Wedding Star"
Cyg 20h58'48" 49d17'6"
Candace Vizachero Leon
Uma 11h37'46" 49d20'29"
Candace Watkins
Cam 4h6'36" 53d51'17"
Candace Welch
Ori 6h12'30" 9d16'37"
CandaceandBrian's
Gem 7h6'27" 15d10'49"
CandAmaya
Uma 9h17'13" 49d38'4"
Candee A. Swenson, Simply the Best
Gem 6h46'17" 15d17'14"
Candelaria
Aqr 21h58'45" -3d38'33"
Candelaria
Vir 12h25'20" -5d15'21"
Candelaria Villarreal
Sco 17h54'52" -41d58'41"
Candence Lynn
Lib 14h34'34" -10d48'0"
Candeo Ferocia
Cap 20h50'5" -16d28'35"
Candes Zona
Lib 15h6'31" -26d52'24"
Candesco
Col 5h51'44" -33d51'59"
Candi and Dallas Forever One Heart
Cyg 19h35'33" 36d55'17"
Candi & Andrea Eternal Friends
Gem 7h40'28" 30d43'54"
Candi Freil
Uma 13h8'2" 61d50'25"
Candi Huff
Aqr 22h38'21" -0d43'47"
Candi & Keith Olson
Gem 7h46'52" 31d25'19"
Candi Lynn Urich
Aqr 21h20'42" 1d1'43"
Candi Michelle
And 1h16'3" 45d5'55"
Candi Moraga
Cyg 19h38'45" 28d6'25"
Candi Uselman
Aqr 21h48'16" -8d21'18"

Candi Virginia Wagmen
Uma 11h50'7" 40d53'57"
Candi Wahl
Psc 0h38'7" 17d3'40"
Candi Woodford
Lyr 18h50'37" 36d35'28"
Candice
Crb 15h38'6" 31d1'14"
Candice
Cyg 20h12'18" 40d34'53"
Candice
Cnc 8h33'18" 14d30'28"
Candice
Lyr 19h7'5" 26d24'0"
Candice
Leo 10h36'33" 22d34'11"
Candice
Sgr 19h12'41" -17d1'36"
Candice
Lib 14h54'16" -0d41'0"
Candice
Vir 11h53'3" -0d45'46"
Candice
Mon 7h11'23" -0d6'43"
Candice
Sco 16h51'7" -38d56'13"
Candice 1107
And 0h59'23" 41d7'53"
Candice Bailey
And 22h58'49" 39d27'33"
Candice Bergey
And 23h34'30" 45d46'55"
Candice Brooke Randall
Crb 16h19'32" 29d57'53"
Candice Brown
Lmi 10h30'46" 32d19'18"
Candice Brown Johnson
Cet 0h55'39" -0d7'44"
Candice 'Candie' Cesario
Cap 21h38'25" -18d8'8"
Candice Carlo - Eternal
Love
Cnc 8h42'18" 28d18'6"
Candice Clarke the Blue
Rose
Ori 5h35'44" -2d28'45"
Candice Crossfield-GaMPI
Star
Lyr 18h54'14" 31d52'27"
Candice Dawn
And 1h46'54" 38d3'10"
Candice Debraekeleer
Cas 1h18'7" 69d20'51"
Candice Denise
Aqr 22h15'36" -14d41'53"
Candice Diane Caudill
And 0h43'11" 42d46'17"
Candice Dwan Ross
And 0h28'45" 32d53'14"
Candice Ellen Clark
Cas 1h28'12" 62d2'37"
Candice Finlayson
Sco 17h55'22" -37d23'36"
Candice Gabriel Phillips
Vir 13h12'43" -17d54'35"
Candice Grace Spotti
Dra 16h8'31" 56d46'36"
Candice Grey eternity
sweetheart
Pho 1h6'44" -42d36'5"
Candice Hancock
Cnc 8h47'9" 16d27'50"
Candice Heller
Sco 16h55'14" -38d57'28"
Candice Hope
And 0h42'9" 33d16'58"
Candice Jean Golds
Psc 0h18'36" 6d40'22"
Candice Jeanit Davis
Sco 17h5'2" -36d38'35"
Candice Johnson
Sco 17h56'38" -42d26'15"
Candice & Josh
Gem 6h59'57" 27d26'10"
Candice Kathleen Shelley
Leiter
And 1h4'9" 39d51'29"
Candice L. Rogers
Gem 7h37'40" 15d40'53"
Candice L. Weispfenning
Cnc 8h33'45" 30d39'17"
Candice Laurn Elliott
And 0h36'32" 37d47'12"
Candice Leann Hickle
Aqr 22h33'28" -3d34'3"
Candice Lee Fredrics
Tau 4h10'23" 3d49'42"
Candice Lee Johnson
And 23h38'1" 42d18'32"
Candice Lee Sykes
Ari 3h15'5" 21d23'39"
Candice Ling
Tau 5h57'24" 23d17'37"
Candice Lynn Metz
Lib 15h47'5" -3d53'0"
Candice Lynn Mobley
Cyg 19h51'4" 32d30'37"
Candice Lynn Seidel
Vir 14h41'57" 3d53'18"
Candice Lytle
And 0h36'10" 42d13'35"
Candice Maria
Uma 11h18'36" 58d55'41"
Candice Marie
Sgr 19h31'17" -18d7'26"
Candice Marie
Cnc 8h25'29" 13d44'12"
Candice Marie Angel
Carling
Sco 17h53'5" -41d34'51"
Candice Marie Freeman -
14 Oct 1983
Lib 15h55'31" -14d57'7"
Candice Marie Goodman
Sco 16h44'28" -29d54'54"
Candice Marie Sparks
Leo 11h19'48" 15d34'40"
Candice Mary
Lyn 9h29'12" 39d55'24"
Candice Mc.
And 0h27'42" 42d25'31"
Candice Michelle wood
Tau 5h51'0" 16d57'14"
Candice Mock
Cnc 9h14'47" 22d33'41"
Candice Moench
Mon 6h44'55" -0d17'23"
Candice Nicole Thompson
Vir 13h53'59" -18d26'48"
Candice - Our Shining Star
Cru 12h51'29" -57d24'3"
Candice R. Walker
And 1h49'58" 41d0'16"
Candice Rainey
Lyn 8h23'15" 33d10'4"
Candice Renee
And 1h2'26" 38d3'51"
Candice Richards
Sco 17h10'25" -44d57'37"
Candice Rose
And 2h5'55" 42d50'45"
Candice Russell
Aqr 22h26'22" -6d23'12"
Candice Steck
Gem 7h31'2" 34d53'15"
Candice "Sunshine" Beam
Leo 11h26'54" 9d40'47"
Candice Sutnick
And 23h37'37" 42d30'20"
Candice T.
Psc 0h42'54" 17d13'43"
Candice Taylor
Lyn 6h46'18" 54d8'4"
Candice Tipton
And 0h52'25" 33d41'33"
Candice True
Cyg 21h12'50" 32d6'38"
Candice Typhaine Cornu
Her 17h27'3" 20d13'38"
Candice Wanca
Lib 15h38'46" -29d29'4"
Candice Young
Tau 4h18'1" 9d30'42"
Candice Zipler
Cap 21h11'55" -27d27'37"
Candice-Nicholas
Cas 23h39'29" 53d37'53"
Candida
Sgr 17h52'33" -17d48'13"
Candida
And 1h50'18" 45d32'55"
Candida and Russell
McBride
Cas 0h57'36" 50d11'13"
Candida Camile Laine
Uma 9h40'45" 51d45'30"
Candida Elena
And 0h21'15" 25d59'51"
Candida Paulina Choles
Curiel
Peg 22h11'44" 10d14'0"
Candie Carmel
Uma 8h29'32" 63d31'58"
Candie Renee Westbrook
And 1h21'22" 45d41'51"
Candis
Umi 13h33'13" 72d56'53"
Candis Falcon - Mom
Sco 16h28'10" -30d29'45"
Candi's K-9 Angels
Cma 6h46'32" -13d44'28"
Candis Lee Amdahl
Psc 1h29'58" 27d30'28"
Candis Marie Falisi
Cnc 9h7'39" 30d40'11"
Candoche
Lib 15h59'44" -18d22'41"
Candolim
And 1h27'21" 40d6'43"
Candour
Peg 21h50'15" 24d14'31"
Candra Beth
And 23h36'40" 42d28'18"
Candy
Cnc 7h58'36" 11d34'6"
Candy
Vir 13h41'7" 1d38'49"
Candy
Lib 14h26'30" -17d56'4"
Candy
Uma 9h10'2" 59d48'49"
Candy
Sco 16h42'50" -42d9'18"
Candy and Mikey
Cap 20h25'42" -24d26'4"
Candy Beam 2005
Cru 12h0'15" -62d26'51"
Candy Bell
Uma 8h40'14" 65d18'12"
Candy Cane
Ori 6h10'29" 18d8'11"
Candy Clark
Cnc 8h43'27" 15d56'15"
Candy Crack
And 2h8'27" 42d20'33"
Candy Criger Jennings
Sco 17h21'9" -38d20'33"
Candy David
Tau 4h11'38" 6d43'59"
Candy Eaugenia Teso
Ariaga
Tau 4h8'25" 16d21'23"
Candy Haynes
Lib 15h17'21" -20d12'26"
Candy Hoekstra 01171969
Cap 21h11'13" -18d36'23"
Candy Jack McGregor
Memorial Star
Cep 21h13'18" 66d11'7"
Candy Jo Bizelli
Lib 14h24'50" -12d41'41"
Candy Jo Gambrill
Ari 3h7'21" 23d24'48"
Candy Kamens
Cyg 20h6'10" 37d4'15"
Candy L Sego
Ari 2h18'51" 19d49'12"
Candy Lau and Ryan Chiu
Lib 15h30'38" -12d16'38"
Candy Liew
Vir 13h58'29" -16d30'10"
Candy loves Nelson
Sco 16h18'0" -10d1'41"
Candy Mena
Cyg 19h41'40" 36d22'9"
Candy Nerge
Leo 11h11'56" 17d43'14"
Candy Nicole Dykes
And 0h16'22" 46d7'32"
Candy Player's Star
Ori 4h55'14" -0d16'25"
Candy Queen
Cas 0h32'35" 54d33'56"
Candy Ramdass
Cnc 8h19'52" 10d25'6"
Candy Schmal
Peg 22h59'42" 9d48'38"
Candy SWEETS Diaz
Cnc 8h20'16" 14d13'57"
Candy Tar
Lib 15h48'45" -20d4'39"
candyland
Vir 14h16'36" 0d36'45"
Candy's Kisses in Heaven
Ari 3h16'1" 29d35'8"
Candy's Radiance
And 1h22'4" 48d14'26"
Candy's Star
Uma 12h39'4" 56d1'1"
Canel Perret Dme
Ori 5h47'2" 5d48'34"
Canelle & Benjamin Abouaf
Com 13h0'53" 29d2'9"
CANES
Sge 19h51'45" 18d37'11"
Canet
Cru 12h18'27" -59d21'10"
Caney Jeremiah
Uma 10h16'30" 67d4'14"
Cange
Aur 5h32'3" 55d43'18"
Canito's Star
Sge 19h50'32" 18d6'13"
Cannassa
Uma 12h51'30" 59d24'34"
Cannon
Uma 9h33'55" 54d15'40"
Cannon
Sgr 19h32'24" -14d29'7"
Canon Basil O'Sullivan
Uma 11h53'19" 32d2'1"
Canopy
Lyn 7h36'55" 48d11'24"
Canrob
Gem 7h5'5" 32d44'29"
Cant & Cridland
Cyg 20h3'49" 33d46'0"
Can't Take My Eyes Off Of
You
Tau 4h48'47" 16d2'48"
Cantarito
Uma 11h16'33" 45d59'26"
Canterbury Children's
Center
Uma 11h39'55" 58d25'46"
Canterbury Tales
Crb 15h32'49" 32d2'44"
Canti, Loving Memory of
Nick Grieco
Vir 13h26'4" -6d13'17"
Canto Hondo for my Love
Colleen
Vir 12h11'8" 11d13'36"
Cantwell Splendor
Uma 10h28'35" 57d21'32"
Canus Lupus
Cma 7h15'39" -31d17'28"
Canus Sensim
Gem 7h34'5" 33d51'12"
Canyon Alan
Sco 16h11'47" -15d1'6"
Canyon & Danielle
Uma 11h25'19" 52d50'1"
Canyon Hughes
Cas 2h25'22" 67d15'10"
Caoimhe
And 0h54'23" 46d31'28"
Caoimhe Clare Haughey
Uma 12h16'30" 55d11'26"
Caoimhe McDonagh
Uma 9h49'0" 62d26'2"
Caoimhin
Leo 9h46'48" 29d28'22"
Caolan Thomas Farry
Cap 21h19'29" -21d16'49"
"Cap."
Umi 15h59'35" 72d47'17"
Cap Kaylor
Cas 1h9'19" 69d46'9"
C.A.P. Love Always Mom
Ari 3h1'10" 18d54'50"
Cap pour toujours
Gem 6h3'43" 26d16'42"
Cape Gray: Peter and Beth
Gray
Aqr 21h18'50" -14d4'12"
Capella Castlemartyr
Aur 5h12'31" 46d24'17"
Capen
Lyn 8h31'53" 45d8'56"
Capie
And 2h23'48" 46d40'6"
Capi's Star
Uma 13h52'23" 60d23'8"
Capitan Estrella
Psc 1h50'18" 9d12'23"
Capitola Ritchie
Ori 5h26'25" 0d15'4"
Capizzi 70
Uma 10h48'55" 47d45'42"
CapLeoSavGre 1984
Aql 19h51'37" -0d36'4"
Capn Jack Searles
Cep 22h20'25" 62d37'58"
Cap'n Tilly Reynolds of the
Pacific
Umi 15h15'55" 73d30'39"
Capo Billy
Uma 13h8'28" 56d12'56"
CapoKate
Lyn 7h30'19" 36d10'9"
Capoo's Star
Psc 1h28'51" 13d4'32"
Capperon Muriel
Ori 6h21'20" 15d55'43"
Cappotto Giorgia
Her 18h27'40" 22d10'35"
cappuccino
Ari 2h52'10" 28d34'17"
Cappuccino
Ori 6h8'44" 14d1'35"
Cappucine Eva Ferguson
Cas 0h29'9" 61d24'51"
Cappy's Light 40
Vir 14h2'50" -0d53'41"
Capri
Tau 4h43'32" 18d35'40"
Capri Serrie
Vir 11h49'24" -3d52'40"
Caprice Jeaneen Sweet
Cook
Cam 5h15'36" 61d31'18"
Capricorn of Domenica
Cap 20h33'2" -13d16'47"
Capriotti Valentina
And 0h36'19" 46d19'32"
Capritarius
And 0h19'7" 28d28'6"
Capriuolo Roberto
Peg 22h9'7" 6d44'39"
Capt. Burgandy
Cep 22h11'3" 69d42'20"
Capt. Eddie
Lib 14h51'12" -5d9'27"
Capt. Leslie Highley
Cas 23h43'42" 52d46'27"
Capt. Michael Timothy
Thompson
Aql 19h42'45" 9d3'20"
Capt. Victor Borst, CAP
Forever
Uma 11h44'22" 58d39'45"
Captain
Gem 6h41'42" 17d25'18"
Captain B.Feen
Leo 9h49'37" 27d39'28"
Captain Biggers
Sgr 18h17'48" -23d52'50"
Captain Bill
Tau 4h28'11" 18d9'18"
"Captain Bill" Cowan
Gem 7h10'55" 26d36'23"
Captain Billy
Cnc 8h29'31" 12d53'23"
Captain Blue Eyes
Aur 5h48'50" 29d42'23"
Captain Bob Gordon
Uma 8h49'12" 66d4'20"
Captain Caballero
Cnc 9h19'19" 12d55'46"
Captain Carl Mitchell
Leo 9h37'32" 28d38'12"
Captain Chemo
Aql 20h4'45" 14d58'52"
Captain Costar
Sgr 19h18'36" -18d52'36"
Captain Craig Lee Miller
Uma 10h14'25" 51d15'35"
Captain Critter Rick
Lib 15h18'1" -9d31'21"
Captain David Moore
Her 18h19'45" 22d34'50"
"Captain Do Right"
C.T.Kilgore"
Uma 11h22'44" 46d5'40"
Captain Eric Thomas
Paliwoda
Psc 0h43'27" 7d35'22"
Captain Eric Wilder
Pyx 8h56'38" -34d37'42"
Captain Fantastic
Cru 12h39'20" -59d54'52"
Captain Fantastic
Per 3h1'25" 51d0'51"
Captain Floyd T. Holt
Gem 7h15'36" 17d50'34"
Captain Fuzzy Star
Psc 0h14'59" 18d49'17"
Captain Garnet John
Godfrey
Umi 4h53'42" 89d7'31"
Captain Gary Robert
Marville
Aql 19h3'41" 14d53'12"
Captain George L. Jackson
Cep 20h6'4" 60d48'35"
Captain Gerry Nichols
Per 2h21'43" 55d18'12"
Captain Gilbert H. Barker
Aql 19h42'42" 6d42'10"
Captain Griffin Moline
Aql 19h10'49" 15d10'40"
Captain "H"
Hya 8h32'16" -15d44'10"
Captain Hook
Aur 6h25'57" 31d38'11"
Captain Howard Riggs
Per 3h36'58" 45d58'4"
Captain Insano
Boo 14h33'48" 19d51'49"
Captain J. Michael Leeds
Sco 16h35'13" -43d13'37"
Captain Jack
Dra 17h43'35" 54d55'27"
Captain Jane Jasper and
Blackbead
Sgr 18h33'41" -19d54'31"
Captain Jimmy Hayden
Lyr 18h47'20" 29d58'2"
Captain Joe's Polish Star
Aql 19h22'46" 15d14'52"
Captain John P. Fay
Tri 2h22'39" 34d49'8"
Captain John Yu, RS-175
Psc 0h49'27" 16d24'56"
Captain Jonathan Archer
Aql 19h40'3" 14d29'50"
Captain Joshua and his
mate Claire
Uma 10h34'2" 65d35'33"
Captain Kathy Mazza
Tau 4h21'14" 19d50'25"
"Captain" Keith Carpenter
Per 3h18'45" 51d5'21"
Captain Kevin Toto Corrao
Ori 6h7'27" 18d37'4"
Captain Kevy
Cap 21h41'34" -11d9'58"
Captain Kirk Langness
Del 20h42'7" 15d8'32"
Captain Margot
Leo 11h5'44" -1d13'45"
Captain Mark Frank Vedere
Her 17h31'22" 19d18'55"
Captain Mark G. O'Conner
Aql 19h43'7" 11d54'9"
Captain McCoy
Vir 12h53'15" 4d27'48"
Captain Midnight
Her 17h24'38" 39d6'32"
Captain Morgan
Aur 5h23'14" 30d55'15"
Captain Pappy
Vir 13h50'2" -2d30'57"
Captain Phil Bear
Uma 11h20'28" 67d25'45"
Captain Pinky
Tau 5h2'9" 25d29'42"
Captain R. Bruce Allison
Leo 11h26'27" 8d30'15"
Captain Richard Taylor
Boo 14h45'27" 37d20'26"
Captain Rob
Uma 11h12'41" 40d7'58"
Captain Robbo
Vel 10h21'43" -41d32'48"
Captain Rock Sparrow
Leo 10h56'41" 6d23'50"
Captain Ron's Jaeger Star
Psc 1h9'0" 32d21'3"
Captain Sameer Fanous
Leo 11h39'20" 21d28'55"
Captain Squeeky
Cnc 9h15'38" 7d37'12"
Captain Thomas Leigh
McGrath
Gem 7h19'35" 22d15'29"
Captain Trimmer Newell
Aql 19h43'21" 9d29'6"
Captain Walter W. Olsen
Tau 4h32'56" 12d21'30"
Captain Yarr McCune
Vir 14h43'26" 4d52'32"
CapTar
Lyn 9h0'31" 37d37'3"
CAPTJOE
Gem 7h15'34" 28d9'8"
Capt'n Harry
Aql 19h48'26" 6d25'52"
Capture the Dragon
Ori 5h51'1" 9d7'58"
Capuchi
Lyn 7h38'18" 53d31'30"
Capuchica's Delight
And 0h49'26" 43d39'8"
Capucine
Leo 10h8'52" 25d54'58"
Capucine Dauphin
Del 20h37'24" 4d49'8"
Car Car
Cep 22h24'6" 72d46'18"
Cara
Sgr 18h50'8" -16d28'52"
Cara
Psc 23h5'36" 0d11'17"
Cara
Ari 2h16'48" 12d18'53"
~Cara~
Leo 9h33'36" 6d56'39"
Cara
Gem 6h27'26" 16d32'23"
Cara
And 0h42'20" 41d34'1"
cara
And 0h3'53" 41d3'50"
Cara
And 2h19'16" 49d1'53"
Cara Alexandria
Cyg 20h1'2" 42d21'1"
Cara Alyson Brown
Gem 6h29'0" 26d9'34"
Cara and Zak
Psc 0h44'34" 14d31'44"
Cara Ann
Ari 2h12'50" 13d24'26"
Cara Ann
Vir 12h47'0" -10d19'31"
Cara Ann Benjamin
Lib 15h15'23" -16d47'24"
Cara Anna
And 23h5'19" 40d52'19"
Cara Anne
Lyn 9h13'2" 39d34'13"
Cara Anne Holloway
Cas 0h6'8" 56d45'55"
Cara Ashley Gorham's Star
Gem 6h27'22" 27d5'29"
Cara Bella
Sgr 17h53'43" -21d43'5"
Cara Beth
Vir 13h51'40" -6d6'6"
Cara Blue
Cnc 7h56'22" 16d52'33"
Cara Bowman
Ari 2h48'57" 12d58'9"
Cara Boyes
Cas 23h27'2" 55d48'58"
Cara Carmines Howard
Vir 13h47'43" 3d24'41"
Cara Danielle Nachtrab
Aql 19h21'51" 3d57'11"
Cara E Waters
Sco 16h50'4" -41d11'45"
Cara Elizabeth's Star
Gem 7h38'12" 31d44'56"
Cara Elyse Sefcik
Cnc 8h4'38" 26d48'11"
Cara Gigstad
Ari 3h12'20" 22d25'37"
Cara Goodman
Lmi 10h46'22" 27d19'54"
Cara Harrington
Lmi 9h32'39" 37d3'57"
Cara Heather Richardson
Cas 0h21'58" 62d47'25"
Cara Henderson Reichert
Tau 5h47'26" 22d26'8"
Cara J. Luechtefeld
Leo 9h45'50" 18d34'32"
Cara Jacqueline
And 1h42'55" 45d20'55"
Cara Leigh Holcomb
Aqr 22h33'30" -0d47'7"
Cara Lisa Lechtner
And 0h20'40" 32d15'7"
Cara Lucy Lazzerini -
Cara's Star
And 1h14'14" 45d11'35"
Cara Lux
And 0h17'27" 28d46'1"
Cara Lynn
Ori 5h50'36" 6d43'48"
Cara Lynn Morelli
And 1h20'51" 45d11'22"
CARA - LYNNE
Umi 14h39'52" 68d1'2"
Cara Marie
Lib 15h50'56" -3d53'4"
Cara Marie Kieu Nga
Charles
And 23h30'30" 37d24'9"
Cara Marie King
Sco 16h51'33" -32d44'1"
Cara Marie Silvestro
Uma 12h5'21" 56d18'9"
Cara May Perrotti
And 1h43'14" 45d36'13"
Cara Mia
Lyn 8h51'23" 45d1'32"
Cara & Michael Thompson
Ori 6h11'18" 6d3'59"
Cara Michele Bailie
Sgr 18h55'47" -19d13'49"
cara mo anam
Cyg 21h10'8" 36d53'19"
Cara My Baby Girl
Lyn 8h31'6" 33d11'15"
Cara Noell
And 0h40'40" 38d22'0"
Cara Olson
Lyn 6h59'34" 50d28'4"
Cara Rene Sullivan
And 1h24'1" 36d35'44"
Cara Rose Cox
And 23h0'57" 40d5'27"
Cara Rose Martin
And 2h26'51" 39d0'29"
Cara Rosina
Leo 11h40'41" 24d41'0"
Cara Sims
Mon 6h46'11" -0d31'58"
Cara Sorella
And 0h48'9" 38d6'9"
Cara Wisniewski
Cam 7h57'13" 70d36'20"
Cara-11
Lep 4h58'15" -22d44'29"
cara811
Dra 19h29'56" 64d39'58"
Cara-Beth Lillback
Lib 14h38'5" -9d6'5"
CaraBrooke
Gem 6h44'27" 23d50'25"
Carah
And 23h56'28" 35d31'33"
Carah Louise
Ari 2h33'56" 27d39'43"
Caraher
Gem 7h17'43" 19d50'50"
Caralee, the light of my life
Tau 5h47'3" 22d14'8"
Caralen Jude
Lib 15h39'32" -28d34'9"
Caraline
Del 20h32'32" 7d3'2"
Cara-Lynn Jae Neault
Leo 10h32'33" 23d19'29"
Caramae
Uma 12h49'38" 61d58'21"
Caramello
Gem 7h29'42" 29d50'32"
CaraMia
Crb 16h10'38" 35d39'1"
Caramia Elizabeth
McLaughlin
Lib 15h6'52" -3d51'53"
Caras Little Piece Of
Heaven
Per 3h8'36" 56d14'44"
Cara's Star
Lmi 10h10'37" 39d24'28"
Cara's Sunshine
Leo 10h51'37" 17d42'5"
Cara's Vita Bella
And 23h0'27" 52d0'12"
CaRay's Moon
Crb 15h35'9" 37d52'59"
CarBert
(JP&LT)01/22/1996
Aqr 22h34'33" 1d4'39"
Carburos Metálicos
Gem 7h24'59" 31d24'36"
Cardinal Song
Ari 2h9'12" 12d36'51"
Care
Ori 6h2'51" 13d5'57"
Care Bear
Cnc 8h15'42" 8d10'17"
Care Bear
Ori 6h24'34" 10d55'13"
Care Bear
Leo 9h57'30" 21d46'46"
Care Bear
Gem 8h4'8" 28d41'6"
Care Bear
Uma 11h45'55" 52d3'3"
Care Bear KJW
Cyg 21h19'27" 40d55'52"
Care Bear Roslyn
Cap 21h8'18" -19d44'2"
Care Care
Ori 5h14'3" 7d43'48"
Care for the Planet First &
F.G.W.
Eri 3h42'32" -7d46'17"
Carebear
Cap 21h9'47" -16d7'16"
Carebear
Uma 8h45'7" 67d17'21"
Carebear
Sgr 19h53'40" -25d14'7"
Carebear
And 0h24'21" 45d30'23"
CareBear 1
Umi 16h9'33" 71d54'48"
Carebear and Butthead
Uma 11h29'37" 45d21'53"
Carebearicus Prime
Cas 0h31'11" 54d42'43"
Caree 2-22-71
Psc 22h59'58" 7d53'54"

Caren
Lmi 10h39'53" 33d1'0"
Caren (Bird) Schultz (10/68-12/01)
Uma 12h3'41" 54d31'51"
caren cotton
Ari 2h48'33" 12d49'50"
Caren Cramer
Tau 4h11'43" 27d28'10"
Caren Heacock
Lmi 10h29'8" 37d6'9"
Caren Marre a.k.a. Bling - Bling
Cyg 20h34'13" 48d39'21"
Carena F. Petrello
Crb 16h3'32" 26d14'11"
Carena van Riper
Ori 4h59'22" -3d40'12"
Carenza
Gem 6h26'0" 24d23'52"
Caressa
Uma 9h23'28" 57d18'13"
Caresse Kelly
Psc 0h11'22" 8d8'17"
Caresse Lynnette Wesley
And 0h40'44" 39d36'35"
Carey
Cas 0h23'59" 59d8'4"
Carey
Psc 1h36'40" 11d58'16"
Carey
Ori 6h19'10" 14d3'3"
Carey
Sco 17h12'51" -42d41'55"
Carey and Kimberely Dvorak
Cnc 8h51'28" 25d27'54"
Carey Ann Wulz
Lyr 18h58'22" 26d33'18"
Carey Clay Gunter
Ori 5h10'51" 6d18'53"
Carey Dale
Uma 9h37'15" 45d24'0"
Carey Eugene Ore
Cep 22h28'58" 86d25'24"
Carey Koehnen
Lib 15h22'13" -7d20'54"
Carey McNeal
Ori 4h52'52" -0d1'54"
Carey Randi 04/12
Ari 2h53'15" 26d40'58"
Carey Steinberg Baron
Gem 7h28'41" 20d33'7"
Carey Thomas Boyle~12/05/2002
Aql 19h50'15" 6d38'59"
Carey Thomas Cavaliere
Aql 19h53'47" -0d19'3"
Carey Thompson-Lutz
And 0h26'53" 34d4'55"
CAREY'S ANGEL STAR 21769GRP22498
Uma 11h46'21" 36d40'10"
Carey's Love
Uma 13h16'6" 62d12'43"
cargretly
Vir 12h42'38" 0d24'42"
Cari
Aql 19h43'25" 15d53'13"
Cari
Peg 21h32'18" 21d44'20"
Cari
Cyg 20h59'54" 43d52'44"
Cari
Uma 10h38'40" 55d5'35"
Cari Ann Goin
Tau 5h8'3" 25d24'11"
Cari Ann Thomson
Leo 11h12'34" 12d17'35"
Cari Anne
Uma 10h49'27" 58d6'42"
Cari Beeman
Ari 2h15'35" 25d30'10"
Cari Card
Psc 1h22'20" 6d26'23"
Cari & Chicharin
Pup 6h41'18" -48d51'51"
Cari Corbran
Aqr 22h33'8" 1d46'33"
Cari Jean Honeyman
And 23h12'3" 51d29'48"
Cari Jill Cummins-Baskin
Leo 9h46'10" 27d50'18"
Cari Lynn Cook
Cap 20h41'11" -24d17'50"
Cari Lynn Evans
Leo 9h49'46" 13d3'35"
Cari Lynn Jewell
Dra 9h55'33" 76d39'4"
Cari Schjolin-Rutland
Uma 9h47'37" 42d57'40"
Caria Elodie
Umi 14h16'48" 66d23'3"
Cariad
Cyg 21h45'22" 54d22'25"
Cariad
Per 3h6'30" 54d43'25"
Cariad
Cyg 21h15'3" 43d4'17"
Cariad-Munchkin
Lyn 8h28'47" 38d18'58"
Carianne
Vir 12h44'8" 4d2'9"
Carib Blue Apts.
Ori 6h8'29" 20d40'22"
Caribbean Dream
Cyg 20h7'52" 30d44'46"
Cariboo
Del 21h8'10" 18d45'27"
Caricatland
Lyn 8h36'13" 39d1'58"
Carico
Aqr 21h56'27" -0d12'55"
Caridad Garcia Ruiz
Ari 2h7'51" 11d8'35"
Carie
Leo 10h15'47" 11d33'54"
Carie Jurcak
Gem 6h58'8" 35d2'30"
Carie Kay Lahey
Uma 8h52'16" 47d43'24"
Carie Renee
Cyg 21h14'42" 42d17'29"
Cari-Lyn-Shane
And 0h18'44" 44d55'37"
Carin
Crb 15h31'35" 27d8'38"
Carin
Aql 19h18'46" -0d53'57"
Carin Elizabeth Andrews
Cas 0h13'49" 56d2'39"
Carin Erickson
Cas 0h42'32" 57d50'29"
Carin Smirman
Uma 9h22'54" 52d27'41"
Carin the Loving Mother
Cas 1h36'13" 66d35'42"
Carin van den Nieuwenhoff
Ari 2h16'46" 22d13'43"
Carina
Vir 12h46'48" 4d0'28"
Carina
Gem 7h25'6" 20d39'30"
Carina
Cas 0h28'23" 54d8'50"
Carina
Cyg 21h56'43" 45d7'49"
Carina
And 0h47'55" 41d56'1"
Carina
Crb 15h51'0" 34d10'3"
Carina
Uma 12h27'52" 60d17'30"
Carina
Uma 8h24'23" 59d50'40"
Carina<3
Lib 14h23'45" -11d37'56"
Carina Alisha Fröhlich
Lyr 18h42'0" 41d6'4"
Carina Angela Tomasso
Lib 14h51'43" -8d38'38"
Carina Anne Neal
And 0h22'56" 42d43'36"
Carina B.
And 23h56'9" 37d33'21"
Carina (Beanie)
Cyg 20h39'53" 46d36'40"
Carina Bergauer
And 23h30'54" 42d4'7"
CARINA; follow your heart
Ori 6h7'34" 9d29'41"
Carina Frances
Car 10h6'36" -62d57'39"
Carina Hope
Uma 11h21'16" 50d25'58"
Carina Lagache
Tau 4h41'34" 23d27'34"
Carina Lau
Cas 0h51'0" 61d0'28"
Carina Lin's Birthday Star
Car 10h6'29" -63d1'35"
Carina & Lydia
Uma 11h34'11" 32d2'33"
Carina Mary Walter
Car 7h44'38" -51d20'46"
Carina Molly, Dear Little One
Uma 8h50'38" 56d52'43"
Carina Noble
Cru 12h18'14" -60d0'22"
Carina Olsen
Peg 22h38'57" 23d12'33"
Carina Saarela
Cas 23h7'1" 58d9'20"
Carina Stewart
Aqr 22h39'19" -2d46'19"
Carina und Mathias
Ori 5h56'39" 11d49'13"
Carina Wüst
Lyn 7h36'17" 36d53'18"
Carina, 04.12.1975
Cas 1h16'12" 54d2'32"
Carina's Wish
Cap 20h20'36" -11d3'57"
Carine
Aqr 20h42'11" -7d20'48"
Carine
Per 2h46'21" 50d4'7"
Carine
Tau 5h46'16" 18d7'54"
Carine Brousse
Lib 14h59'23" -2d44'12"
Carine Bruun Johansen
And 2h34'36" 49d47'34"
Carine Chami
And 23h35'1" 42d36'10"
Carine Lefin 23.09.1970
Lib 15h22'18" -15d24'4"
Carine Leriche
Uma 9h6'38" 59d0'14"
Carine M. Lteif
And 2h13'54" 49d54'3"
Carine Maso 14/11 /1965
Sco 17h51'4" -41d19'32"
Carine Reynette épouse Parisy
Sco 17h58'6" -31d22'19"
Carino
Uma 9h22'50" 52d2'4"
Carino
Ori 6h9'55" 9d41'14"
Carisa Lee
Dra 18h49'19" 52d21'50"
Carisa Wright
Cnc 8h53'43" 27d58'49"
Carisia Soltis Francisco
Crb 15h46'22" 32d1'0"
Carissa
Her 17h6'5" 32d23'8"
Carissa
Cnc 8h45'52" 27d5'52"
Carissa
Ori 5h46'9" -2d44'42"
Carissa and Hayden
Ori 5h7'22" 7d34'21"
Carissa Ann
Aqr 22h21'44" 0d29'38"
Carissa Ann Schreck
Psc 1h22'19" 32d25'59"
Carissa Anne
Cap 20h32'12" -17d12'23"
Carissa Clair Leto
Cas 0h17'27" 51d51'19"
Carissa DeWeese
And 2h17'54" 38d16'14"
Carissa Duncan
Peg 23h55'11" 24d38'31"
Carissa G Mineo
Psc 0h29'52" 8d41'51"
Carissa K Lee
Lyn 8h25'55" 56d9'35"
Carissa Kay
Lib 14h50'47" -1d18'21"
Carissa "Kryptonite" Trigo
Vir 14h50'3" 4d16'57"
Carissa Lynn
And 1h15'1" 34d12'21"
Carissa Marie Strimer
Aqr 22h11'47" 0d13'48"
Carissa Nicole Marsh
Cas 0h29'49" 58d8'58"
Carissa Renee Happy 24th Birthday
Ori 5h32'46" -0d54'47"
Carissa Rysanek
Sco 17h46'53" -42d20'35"
Carissa's "Slice Of Heaven"
Ari 2h34'49" 29d28'16"
Carissia M. Bailey
Gem 7h23'37" 14d46'22"
Carista Schiermeyer
Cnc 8h26'45" 22d30'57"
Carita Jill Boshoff
Leo 9h47'35" 8d33'54"
Carito
Cas 0h10'36" 54d57'40"
Carito
Uma 11h38'12" 53d12'6"
Carjen Gartko
Sgr 19h12'58" -15d4'6"
CarKel
Leo 10h42'40" 14d36'34"
Carkenord
Uma 11h25'31" 49d18'43"
Carl
Per 3h49'43" 49d16'3"
Carl
Per 3h31'41" 44d53'1"
Carl A. Arzillo
Cma 6h39'6" -18d11'56"
Carl A. Blechner
Per 3h4'52" 41d58'14"
Carl A. DiFranco
Per 3h37'36" 48d27'27"
Carl A. Nadvornik
Ori 6h1'58" 17d45'50"
Carl A. Stolberg, M.D.
Cep 23h22'58" 72d23'13"
Carl Agostinacchio
Sgr 20h23'59" -29d3'21"
Carl Albert MacDonald
Cru 12h17'12" -59d15'52"
Carl and Audrey Sedacca
Gem 7h3'38" 35d4'24"
Carl and Katherine Pietrzyk
Cnc 9h21'25" 18d11'12"
Carl and Kathy
Cyg 21h15'22" 47d14'48"
Carl and Laura
And 23h50'38" 40d33'43"
Carl and Lorraine Hummelsheim
Cyg 20h11'38" 48d44'59"
Carl and Margaret Ferris
Her 16h48'16" 41d43'33"
Carl and Melissa
Sge 19h25'16" 18d18'23"
Carl and Norma's Golden
Cnc 8h36'9" 24d38'59"
Carl and Rita
And 23h6'38" 42d3'46"
Carl and Sian, Forever In Love
Cyg 20h19'0" 53d23'13"
Carl & Anita
Eri 3h43'41" -0d27'14"
Carl Anthony Booth
Per 3h1'34" 44d4'57"
Carl Anthony Harris
Uma 9h44'47" 58d22'38"
Carl Anthony Paladino III
Tau 5h50'49" 17d8'40"
Carl Arthur McIntyre
Ori 6h20'45" 2d38'59"
Carl Binsky
Uma 10h8'25" 61d1'17"
Carl Brewster
Uma 11h57'28" 52d51'31"
Carl "Bubby" Butzin
Peg 0h3'9" 26d0'52"
Carl C. and Sandra J. Hook
Cyg 19h42'44" 53d49'14"
Carl C. Lester
Cap 21h18'25" -15d45'18"
Carl Cassista - Collette Lapointe
Uma 8h41'16" 64d8'53"
Carl & Charlotte Miles
Umi 14h1'10" 87d13'0"
Carl Chester Horsley, Jr.
Uma 8h51'15" 53d17'6"
Carl Christopher Kwaku Ford
Leo 10h55'24" 8d16'56"
Carl Clausen
Uma 11h57'33" 36d10'33"
Carl Corso
Her 16h28'8" 48d55'8"
Carl D Hill
Vir 13h19'45" -5d12'3"
Carl Danley
Cnc 8h17'58" 10d17'40"
Carl David Randall
Uma 12h3'7" 47d43'49"
Carl Dean Wakefield
Ori 5h57'10" 18d11'27"
Carl & Denise
Uma 10h27'14" 59d3'32"
Carl & Derna Lucero Family Star
Peg 23h15'29" 33d11'19"
Carl Duane Brunn aka Weiner
Uma 11h26'59" 34d22'34"
Carl E. Meitzen
Leo 11h42'59" 14d21'21"
Carl E. Waters
Her 17h20'29" 31d49'51"
Carl Edward Fabian
Her 16h47'22" 28d49'21"
Carl Edwin Hughes
Uma 13h49'58" 56d10'5"
Carl & Elaine Jamele
Cyg 20h12'57" 35d44'8"
Carl Ernest Dahl
Sco 16h5'41" -29d28'57"
Carl Everett Pierce, Sr.
Ori 5h35'27" -2d20'49"
Carl Everett Siddens
Uma 11h25'3" 29d1'49"
Carl F. Boes
Aql 19h48'22" 11d9'53"
Carl F. Miller, Jr.
Aqr 20h48'2" 0d15'43"
Carl F. Savino, Sr.
Lib 14h49'27" -12d18'48"
Carl Fred Roach
Aql 19h7'37" 0d46'32"
Carl "Funzy" Boe
Uma 11h48'12" 61d59'15"
Carl G. Berg
Lib 15h34'13" -4d12'3"
Carl G. Monteith
Vir 12h6'35" -1d42'9"
Carl Gilbert Hayes IV
Uma 11h10'7" 29d58'7"
Carl & Gina D'Alonzo McLaughlin
Psc 0h57'26" 10d43'54"
Carl & Gloria Selvidge
Cyg 21h13'27" 42d21'36"
Carl Gorychka
Cyg 20h16'46" 46d15'27"
Carl Gray my guiding star always C
Per 3h3'33" 46d46'46"
Carl Gunther
Ori 5h52'3" 3d22'31"
Carl Harper Bishop
Sco 17h2'59" -43d33'53"
Carl Herbert Ekstrom
Aql 19h46'11" -0d18'24"
Carl Higgins
Her 17h20'37" 27d45'58"
Carl "Honey Buck"
Per 4h10'26" 33d28'0"
Carl Ip
Uma 13h28'32" 60d14'0"
Carl J. Caminske, III
Ari 3h9'45" 27d57'51"
Carl J Hein
Gem 6h43'59" 26d35'24"
Carl J. Johnson
Cap 21h28'50" -14d33'57"
Carl J. O'Brien Sr.
Uma 11h20'36" 67d31'54"
Carl J. Robinson
Per 2h43'12" 54d5'58"
Carl J. Scicchitano
Psc 23h32'26" 6d34'32"
Carl J. Urbanas, Sr.
Leo 10h19'28" 9d39'14"
Carl Jacob Dupper
Uma 9h5'35" 47d44'24"
Carl James
Crb 16h17'26" 35d8'9"
Carl James Maier
Leo 10h12'23" 9d58'47"
Carl James Sibley
Vir 12h42'58" 6d31'56"
Carl John Radcliff Sr.
Cep 23h53'57" 73d49'41"
Carl Jonah
Cnc 9h15'2" 12d25'2"
Carl Joseph Elliott
Vir 11h54'43" -0d24'21"
Carl & Kelly Watson
Cyg 19h52'9" 33d5'28"
Carl Krueger - The Big Wolf
Lib 15h52'40" -18d49'9"
Carl L. Cox
Sco 17h18'48" -32d40'43"
Carl L Davis
Cnv 12h46'56" 39d3'19"
Carl Lebman
Tau 4h13'44" 27d45'3"
Carl Leroy Fleming
Ori 4h55'17" 10d3'46"
Carl & Lois Allen
Lyr 18h48'4" 35d13'26"
Carl Louis Barker
Uma 9h23'57" 44d34'11"
Carl Lucarelli - Shine Bright
Pho 1h16'51" -43d4'4"
Carl Lundberg
Lib 15h5'11" -0d29'50"
Carl Lyndon Geis
Per 3h15'14" 51d46'22"
Carl M. Isaacs
Per 3h11'39" 51d22'34"
Carl & Mae Rohne
Cyg 20h49'34" 45d24'14"
Carl Magno
Lyr 18h40'54" 35d35'22"
Carl & Marie Hagen
Cas 23h43'13" 54d16'38"
Carl Mark III
Sgr 19h45'50" -12d52'51"
Carl Martin
Per 2h43'24" 50d12'52"
Carl Martin Larson
Uma 13h48'21" 53d30'12"
Carl McCormack
Cam 4h31'11" 53d58'33"
Carl Mendolia
Leo 11h17'21" -2d5'5"
Carl Meurig Broadis
And 0h51'49" 44d40'55"
Carl Meyette
Boo 14h45'49" 18d12'54"
Carl Michael Yastrzemski Jr.
Leo 10h17'6" 16d4'55"
Carl Muller
Cyg 20h5'32" 31d27'58"
Carl Murray VanAntwerp
Aqr 22h53'49" -4d39'2"
Carl n' Kitten
Ori 5h8'10" 12d53'2"
Carl Nevoso
Per 3h0'34" 55d14'49"
Carl Odell Watkins
Uma 10h18'29" 59d14'19"
Carl Osterlof
Aql 19h20'22" 5d50'29"
Carl Otis Ash Sr.
Aur 5h39'10" 44d22'20"
Carl P. Gianino
Dra 17h42'51" 56d26'50"
Carl Pemberton
Cep 22h55'14" 72d13'4"
Carl Percival
Uma 10h12'3" 71d25'18"
Carl Pfeil
Uma 11h59'43" 46d43'21"
Carl (Pop Pop) Shmidt
Dra 14h22'32" 63d18'8"
Carl Priemon
Ori 6h4'57" 4d41'40"
Carl R. Jantzen
Crb 15h39'15" 37d32'54"
Carl R. Schiebler
Lyn 7h39'28" 42d1'15"
Carl R. Siler
Ari 2h11'24" 25d46'52"
Carl Raymond Ernst
Scl 0h51'25" -33d14'47"
Carl Raymond Vege, Sr.
Sgr 18h24'38" -27d33'34"
Carl Read
Uma 10h53'57" 42d17'22"
Carl Reisman
Per 3h27'40" 34d49'30"
Carl Rhodes Staedler
Uma 8h14'50" 70d54'5"
Carl Richard Wilson
Tau 4h58'29" 18d51'6"
Carl Robert Roehling
Lyn 7h9'52" 49d16'53"
Carl Robert Woods II
Ori 5h33'23" 14d11'25"
Carl Rosenblum
Leo 10h21'18" 20d8'57"
Carl Sanders
Cyg 20h22'0" 37d55'37"
Carl Scavello
Dra 18h36'36" 51d15'25"
Carl Scott Benedict
Per 4h7'58" 42d6'34"
Carl Sylvester Scavello
Tau 5h9'24" 24d30'36"
Carl Thomas Appel
Per 3h6'33" 51d58'20"
Carl & Tina Dennis
Sgr 19h46'45" -23d43'35"
Carl Together Forever Jules
Cyg 21h55'27" 37d29'42"
Carl "Tony" Vail
Psc 1h14'29" 22d23'41"
Carl Tozzi
Vir 14h33'54" 1d46'0"
Carl W. Phelps - Father's Day 2005
Ori 6h8'42" 18d26'25"
Carl W. Rhodes
Ori 5h57'12" 6d38'9"
Carl Waldeck
Psc 0h10'21" 3d8'31"
Carl Walter Pollard
Tau 3h30'34" 12d54'27"
Carl Webb
Cnv 13h52'11" 30d5'42"
Carla
Lmi 10h43'57" 31d13'17"
Carla
And 23h12'59" 36d0'1"
Carla
And 1h14'25" 41d46'28"
Carla
Cas 1h10'23" 51d16'7"
Carla
And 23h39'57" 43d29'45"
Carla
And 23h30'18" 46d32'6"
Carla
Aqr 21h23'21" 2d7'45"
Carla
Vir 13h32'49" 1d18'23"
Carla
Psc 0h21'19" 15d37'40"
Carla
Ari 2h52'25" 19d22'21"
Carla
Gem 6h58'12" 15d20'47"
Carla
Gem 7h24'57" 15d11'32"
Carla
Boo 14h57'39" 16d4'14"
Carla
Tau 5h50'0" 26d18'59"
Carla
Vul 19h47'30" 27d53'25"
Carla
Leo 9h52'6" 22d54'28"
Carla
Lib 15h11'30" -29d2'1"
Carla
Umi 14h20'37" 72d53'43"
Carla
Lib 15h5'15" -5d49'26"
Carla
Vir 13h26'1" -21d52'22"
Carla A Smith
Com 12h52'33" 26d39'51"
Carla and Dale
Dra 19h39'16" 71d44'19"
Carla Anderson
Lyn 6h36'57" 58d44'2"
Carla Andreu Rosell
Psc 1h42'22" 12d43'6"
Carla Ann Cash
Cas 0h6'6" 57d8'6"
Carla Ann Lippy
Aqr 21h51'28" 0d15'55"
Carla Arnold
Cas 1h56'19" 67d22'7"
Carla Ayubi
Cap 21h31'9" -16d31'43"
CARLA BLANCETT
Tau 4h32'59" 28d0'19"
Carla Blanche
Cas 1h1'53" 62d24'27"
Carla Blazer
Leo 11h57'39" 20d13'32"
Carla Buerge Harvey
Mon 6h39'56" 7d22'51"
Carla Caldwell
Ori 5h35'9" 6d56'20"
Carla Chambers
Lib 15h18'47" -21d12'38"
Carla Cigoy Thacker
Vir 12h45'34" 12d43'54"
Carla C.R.
Ori 6h18'21" 14d8'31"
Carla Davison
Tri 2h7'43" 29d57'18"
Carla Drezner
Psc 1h25'14" 16d21'8"
Carla Duncan cdd-kdh
Cyg 19h43'37" 29d18'14"
Carla e Marco
Peg 23h54'29" 26d40'59"
Carla Elaine Romo
Cas 1h35'15" 63d21'1"
Carla Elizabeth Englund
Crb 16h12'0" 34d17'21"
Carla & Emarie
Leo 11h34'56" 17d53'23"
Carla & Fares
Cyg 20h11'55" 51d37'1"
Carla Fontenot Turner
Aqr 22h40'48" -1d29'1"
Carla Francesca Kingsman
And 23h44'45" 45d43'46"
Carla Gail Young
Leo 11h24'48" 10d9'44"
Carla Gaye Barnes
Ari 2h50'29" 18d40'36"
Carla Hayes Your Wings Are Waiting
Lyr 18h51'47" 32d23'57"
Carla Horton
Mon 6h55'37" -0d59'33"
Carla Isler
Per 4h47'9" 40d18'22"
Carla J. Hilse
Lyn 8h8'22" 34d11'5"
Carla J. Schoppelrey
Lyn 8h12'45" 50d5'27"
Carla J Werner "My Shining Star"
Sgr 19h33'24" -38d15'58"
Carla Jan Mitchell
Gem 7h38'19" 24d6'59"
Carla Jane Pullen
And 1h16'11" 48d52'19"
Carla Jean
And 23h9'11" 47d16'20"
Carla Jean
Sgr 18h25'46" -16d46'21"
Carla Jean
Lib 16h1'43" -9d48'15"
Carla Jean Bzik
Cap 20h21'46" -19d3'22"
Carla Jean Cragun
Cnc 9h10'26" 32d11'48"
Carla Judith Rovello
Aqr 22h32'52" 1d53'24"
CARLA JULANE PAYNE LOVE OF LIFETIME
Per 3h16'14" 41d35'25"
Carla K. Colton
Cam 4h54'18" 59d55'51"
Carla & Klaus-Peter
Uma 9h53'49" 62d55'16"
Carla L. Tranberg
Gem 6h55'46" 16d29'17"
Carla Landry
And 0h6'6" 45d32'46"
Carla Louise Gugenberger
Cru 12h28'33" -61d57'34"
Carla Louise Johnson
Uma 10h22'34" 65d26'2"
Carla Louise Van Ligten
Sco 16h8'44" -19d41'19"
Carla Marie
Cyg 19h51'51" 37d22'46"
Carla Marie
Cyg 19h43'41" 27d57'46"
Carla Marie 15.03.2006
Uma 10h56'13" 43d2'48"
Carla Marie Rabon
Uma 9h31'40" 50d48'52"
Carla Mariela Fiszman
Cra 18h12'22" -41d3'35"
Carla McAllister
And 23h50'10" 40d19'47"
Carla McNutt
And 0h35'27" 39d22'12"
Carla Megan
Sgr 19h55'40" -12d38'59"
Carla Michele Parrish
And 2h35'41" 44d27'29"
Carla Monk - Adamo, Aeternus
And 2h38'1" 37d40'5"
Carla Nicole Schneider
Lib 15h7'3" -16d5'13"
Carla Pejakovich
And 1h33'59" 46d50'59"
Carla Pendexter
Mon 7h19'12" -6d27'40"
Carla Pinotti
Ori 5h53'16" -8d18'14"
Carla R. Melton
Srp 18h21'57" -0d26'20"
Carla Rae Menasco-Dossman
Uma 9h47'57" 44d45'13"
Carla Rebimbas Benson
Aqr 21h50'28" -7d55'24"
Carla Renee
Gem 6h45'25" 17d43'39"
Carla Rigoni
And 2h6'48" 39d55'13"
Carla Roig Fernandez
Sco 16h56'48" -36d16'49"
Carla Roxann Ringhofer
Lyn 8h41'38" 45d59'8"
Carla S. Adams
And 0h33'0" 29d39'50"
Carla Savat Loves Kalea Lagrand
And 1h9'19" 35d39'31"
Carla Schiavone
Uma 10h19'10" 50d37'8"
Carla Schwan
Vir 13h19'48" -7d39'37"

Carla Suzanne
Crb 16h21'6" 27d51'21"
Carla Terri Patterson
Cru 12h26'30" -61d14'23"
Carla Thompson
Cap 21h31'51" -13d47'7"
Carla Tschümperlin
Crb 16h22'1" 39d16'4"
Carla V X IL
Cam 4h36'40" 58d14'15"
Carla Wood
Crb 15h50'1" 37d29'49"
Carla Workman
Lyr 18h22'4" 32d3'6"
Carla Zwart Brockman
Uma 9h32'49" 65d46'24"
Carlaaron
Uma 9h7'46" 72d38'49"
Carla-Maria
Tau 5h46'0" 16d52'41"
Carlanmuse
And 1h58'49" 41d47'38"
Carla-Poo Child
Cyg 21h12'14" 38d19'13"
Carlaries
Ari 3h8'54" 19d41'21"
Carla's Always and Forever
Umi 16h7'18" 70d9'33"
Carla's Candle Of Love
Psc 1h21'3" 15d58'2"
Carla's Joy
Ari 3h6'17" 29d48'52"
Carla's Mornin' Star
Gem 7h30'35" 19d13'46"
Carla...tu mi completi!!!
Cam 7h32'55" 77d38'14"
Carlea Lynn Campagna
Ari 2h4'8" 22d3'38"
Carlee
Cap 20h27'44" -26d45'40"
Carlee Alyse
Uma 10h50'28" 57d28'20"
Carlee Horan
Cas 1h5'44" 61d11'39"
Carlee Jean
Sco 17h29'35" -40d48'36"
Carlee L. Murphy
Umi 18h32'45" 88d59'10"
Carleen
Dra 19h16'40" 60d21'4"
Carleen Ann Connor
Uma 9h12'23" 50d23'11"
Carleen & Dave Star Of Lasting Love
Uma 13h59'58" 53d19'16"
Carleen Robertson
Aqr 23h48'32" -15d9'2"
Carleen Yuh
Equ 21h6'21" 8d30'29"
Carlee's Diamond in the Sky
Uma 9h50'43" 71d15'32"
Carlee's Star
Gem 6h56'43" 13d23'0"
Carleigh Amber McGurk
And 1h32'27" 41d29'2"
Carleigh Christina
Cnc 9h7'2" 32d30'37"
Carleigh Francesca Paruta
And 2h21'23" 44d2'59"
Carleigh Jean
And 1h1'11" 37d52'10"
Carleigh Michelle Nestor
Gem 7h1'20" 26d26'21"
Carlen
Aql 19h10'26" 4d10'23"
Carlena Johnson
Cas 1h53'9" 60d32'13"
Carlene
Psc 0h31'33" 14d49'47"
Carlene Cookie Johnson
Peg 23h3'30" 9d50'39"
Carlene Isola
And 0h29'28" 45d25'2"
Carlene Martin Best Nana Ever
Cas 0h51'47" 58d5'7"
Carles Pitarque Duran
Uma 10h40'23" 39d46'48"
Carlette Biggs
Aql 19h48'3" -0d39'4"
Carlette Duncan
And 23h25'3" 41d27'51"
Carley and Dustin
And 2h33'17" 50d35'59"
Carley Bevevino
Sgr 18h22'24" -32d57'5"
Carley Billington
And 0h16'9" 28d55'42"
Carley Dawn Nestor
And 1h8'47" 40d6'13"
Carley Kercheval
Cnc 8h36'25" 19d35'47"
Carley Larie
Lib 15h3'57" -17d6'51"
Carley Larissa Felix
And 0h28'10" 32d59'5"
Carley Maria Jackson
Cap 20h36'19" -17d51'57"
Carley Marie Fenton
And 0h49'16" 21d53'40"
Carley Nichole Burtt
Sco 16h54'52" -41d8'40"
Carley Paige Donnell
Vir 12h46'48" 11d35'16"
Carley's Celestial Spirit
Gem 6h29'49" 22d11'21"
Carley's Heart
Cyg 21h33'29" 41d4'52"
Carli Anne Reinecke
Cyg 20h51'11" 36d34'4"
Carli Arielle
Tau 4h9'55" 8d49'54"
Carli E. Turner
Aqr 20h52'25" 0d43'15"
Carli Faith Valent
Vir 13h59'25" -16d8'47"
Carli Lynn
Lyn 9h20'42" 41d15'50"
Carli Margot Castillon
Cnc 8h50'38" 13d59'26"
Carli Nicole Holland
Ori 5h52'51" 18d27'57"
Carli Nicole Morgan
Leo 9h42'17" 28d30'3"
Carli Shea Sharp
Peg 22h38'51" 33d16'11"
Carli the Beautiful Woodspryte
Uma 10h24'22" 53d57'41"
Carlie Ann Bedford
Pho 0h11'4" -44d35'45"
Carlie Cornell
Ari 2h51'45" 19d57'9"
Carlie Elise Miller
And 0h18'3" 22d27'50"
Carlie Gaquerel
Lyr 18h45'22" 31d49'23"
Carlie Joann Lemiska
Leo 9h43'29" 29d27'15"
CARLIE LOUISE JUPP
Sco 17h19'3" -37d34'25"
Carlie Madison Smith
Tau 4h38'39" 18d1'59"
Carlie Maree
Aqr 23h12'36" -21d2'2"
Carlie Marie
Uma 13h15'35" 62d29'34"
Carlie Martinelli xx
Sco 17h2'2" -39d41'11"
Carlie M.J. Jachimski
Lib 14h56'35" -23d27'49"
Carlie N. Caruso
Uma 9h55'7" 69d56'22"
Carlie Spaziani
Cap 21h33'28" -14d49'36"
Carlin Happy Sweet 16
Aqr 21h39'18" -1d33'0"
Carlin Priscilla Meditz
Gem 7h5'53" 34d41'15"
Carlina
Leo 10h44'46" 16d29'31"
Carlina Kuy
Sgr 17h58'31" -16d5'2"
Carline's Star
Sgr 18h9'58" -24d45'13"
Carlis
Cyg 19h38'21" 34d44'31"
Carlis Lee Wilson
Her 16h48'32" 47d13'13"
Carlita
Leo 11h6'19" 4d32'23"
Carlito & Brooke
Lmi 9h47'33" 38d21'40"
Carlito Miguel Otero Wheeler
Per 3h33'44" 42d41'39"
Carlla Jean McCullough
Leo 11h20'35" 11d34'28"
Carlo
And 1h1'9" 48d34'50"
Carlo
Cep 21h35'39" 67d21'56"
Carlo A. A. De Piano
Gem 6h36'20" 23d12'7"
Carlo Alberto
Crb 15h58'54" 38d34'49"
Carlo Anima Gemella Mia
Sco 16h9'34" -10d13'17"
Carlo&Costanza
Ori 6h12'44" 20d45'51"
Carlo Derek Coi
Cyg 21h44'33" 44d26'13"
Carlo e Lorenzo
Per 3h18'20" 39d29'38"
Carlo Ferreira
Per 2h41'21" 54d8'22"
Carlo Gallmann von Stern
Boo 14h44'57" 30d13'15"
Carlo Gonnella
Uma 9h40'42" 42d4'44"
Carlo Importa
Ser 18h21'37" -11d54'2"
Carlo & Janice Duregon
Cnc 8h18'54" 7d32'47"
Carlo Michael Boccia
Aql 19h49'25" -0d29'50"
Carlo Nicoletta's Special Star
Vir 14h24'35" 1d23'1"
Carlo Orlandini
Cyg 20h41'55" 38d59'51"
Carlo Riccucci
Ori 6h9'23" 14d58'37"
Carlo Ronnow
Cap 20h21'49" -11d26'11"
Carlo Scandura
Cma 6h43'8" -22d20'59"
Carlo & Shirley Borgi
Cyg 21h46'13" 41d17'47"
Carlo & Soheila
Vir 14h25'53" -0d52'41"
Carlo Teuber
Psc 1h9'25" 8d20'54"
Carlo Thian 7/2/1977
Ori 6h7'54" 21d3'43"
Carlo "Tommy" Togni
Umi 13h48'7" 72d49'13"
Carlo Voigt
Cas 23h28'42" 55d57'23"
Carlo82
Tau 5h47'29" 15d34'33"
Carlon R. Hutton
Ari 3h28'23" 28d24'8"
Carlos
Per 3h12'21" 54d40'22"
Carlos
Her 17h8'45" 34d1'12"
Carlos
Aqr 22h30'26" -3d44'1"
Carlos
Cep 22h56'4" 79d14'0"
Carlos
Umi 16h29'47" 82d7'49"
Carlos A Barila
Her 17h40'31" 26d6'37"
Carlos A. Formeza
Lib 15h37'12" -21d15'31"
Carlos A. Morales, Jr.
Lib 15h59'21" -9d41'15"
Carlos Abraham Villegas
Per 4h10'10" 49d59'27"
Carlos & Agata
Cas 1h26'23" 58d27'16"
Carlos Alberto
Aur 6h53'33" 42d5'3"
Carlos Alexis Rodriguez
Cnc 8h55'36" 24d24'1"
Carlos and Kelly
Cyg 19h20'56" 28d55'19"
Carlos and Lauren Villarreal
Cyg 19h40'55" 37d55'53"
Carlos and Nancy Wilson
Cyg 19h55'31" 32d1'2"
Carlos Andres Olin
Cap 20h20'31" -11d47'44"
Carlos & Ann Acosta
Aqr 22h11'52" -2d40'54"
Carlos Asdruval Bribiescas
Vir 13h39'55" 3d14'7"
Carlos Berrelez Star
Sco 16h55'34" -42d36'21"
Carlos Burges y Pilar Ruiz de Gopegui
Cas 1h56'23" 61d20'55"
Carlos Cabral
Per 3h51'53" 52d25'27"
Carlos Cespedes
Per 3h8'28" 43d51'23"
Carlos Chao
Uma 8h38'57" 72d30'47"
Carlos "Charlie" Marmolejo, Mi Amor
Vir 13h15'37" -1d12'31"
Carlos Claudio
Lyr 19h7'35" 42d29'39"
Carlos Claudio Fernandez Arroyo
Leo 11h37'18" 26d15'52"
Carlos Colon
Her 16h18'6" 19d34'54"
Carlos (Cutie)
Her 17h55'1" 44d56'58"
Carlos de la Garza
Ori 4h52'10" 5d18'10"
Carlos DeAngelo Garvin
Psc 1h18'29" 18d0'44"
Carlos E.
Aqr 22h10'57" -8d37'16"
Carlos E. Luque
Ori 5h38'11" 0d53'12"
Carlos E. Pagán, amado padre
Hya 9h37'40" -0d9'27"
Carlos Eduardo Amaya
Psc 0h45'41" 14d38'0"
Carlos Escardo
Ari 2h11'12" 25d33'48"
Carlos & Esther
Cnc 8h11'5" 16d45'25"
Carlos Gardea
Her 16h36'38" 36d53'28"
Carlos Garrido
Per 3h13'0" 43d36'45"
Carlos Gonzalez, MD #1 Dad & Papi
Ori 5h37'0" -0d27'28"
Carlos Gustillo III
Her 16h44'12" 32d7'59"
Carlos Guzman
Tau 4h25'25" 29d41'42"
Carlos Henrique Teixeira
Ari 2h6'13" 27d20'9"
Carlos Hernandez Torres
Lmi 10h5'7" 38d47'14"
Carlos H.Villescaz, Jr.
Lib 15h23'38" -7d23'44"
Carlos Ignacio de Jesus
Scl 0h8'16" -25d54'24"
Carlos J. Pagan
Aqr 22h29'50" -1d32'26"
Carlos & Johanna
Cyg 20h20'49" 55d3'35"
Carlos Jones
Aur 6h8'39" 30d20'50"
Carlos Jose Madrid
Lib 14h59'13" -9d30'3"
Carlos Leizan
Gem 6h54'13" 18d52'46"
Carlos Lejnieks
Per 3h13'15" 41d16'46"
Carlos Lepe
Boo 14h36'29" 32d10'45"
Carlos Loves Amanda
Cyg 21h41'38" 52d38'50"
Carlos M. Baez
Vir 14h27'45" -3d45'33"
Carlos Manuel Ortiz, Jr. Star
Peg 21h41'16" 12d54'44"
Carlos Mari Varo
Per 4h10'57" 48d24'48"
Carlos Martin
Her 18h8'22" 27d48'12"
Carlos Mejia
Uma 9h55'49" 56d34'50"
Carlos Mendoza
Cap 20h27'41" -13d41'30"
Carlos Miller
Sgr 18h7'51" -27d5'54"
Carlos My Baby
Gem 7h17'45" 31d7'43"
Carlos N Jessica
Uma 11h36'27" 61d16'33"
Carlos Nahas
Cnc 8h52'25" 32d20'46"
Carlos Nunes
Cap 21h16'39" -17d11'51"
Carlos Obando
Her 17h4'32" 32d17'26"
Carlos Otiniano
Del 20h41'4" 16d39'15"
Carlos Peralta
Per 4h4'24" 49d56'46"
Carlos Raven Riot
Her 17h11'18" 29d52'48"
Carlos Rosas
Leo 11h46'55" 25d13'36"
Carlos Ruiz
Cmi 7h43'7" 5d37'2"
Carlos Salvador
Uma 8h51'52" 52d50'58"
Carlos Serrano
Dra 19h31'42" 66d58'48"
Carlos Sosa
Her 17h20'0" 20d24'44"
Carlos Summer
Per 3h49'23" 42d51'24"
Carlos Tiggs Perez
Cep 22h36'27" 58d3'6"
Carlos Tovar
Her 18h44'28" 12d22'21"
Carlos Valdez III
Leo 10h53'37" 2d11'54"
Carlos Valencia
Ori 5h34'52" 1d59'4"
Carlos Valencia
Pup 7h57'12" -33d2'8"
Carlos Ygnacio Benavides III
Vir 14h21'26" 3d22'18"
Carlosaurious Major
Uma 14h15'13" 58d32'37"
Carlota P. Cervantes
Aqr 21h34'21" 2d4'56"
Carlota Zahn Munuera
Cnc 9h13'28" 10d14'4"
Carlotta
Cyg 19h47'47" 29d30'57"
Carlotta
And 0h22'37" 32d5'1"
Carlotta
Cas 0h24'30" 52d14'33"
Carlotta
Lyr 19h2'23" 45d8'16"
Carlotta
Ori 4h49'38" 4d37'16"
Carlotta
Her 17h36'54" 21d16'10"
Carlotta Bottazzi
Peg 21h40'57" 21d16'27"
Carlotta Fissi
Cma 7h16'53" -13d24'2"
Carlotta2004
Lyn 8h54'19" 33d20'25"
Carlotta's Light
Aqr 22h19'28" 0d18'0"
Carloyn DeVore
Uma 13h38'7" 60d3'44"
Carl's Cosmic Kindness
Eri 4h0'8" -28d35'45"
Carl's Haven
Lmi 10h48'4" 26d20'26"
Carl's Sentinel
Per 2h22'20" 54d23'11"
Carlsberg Marketing Sdn Bhd
Ori 5h33'44" 2d47'28"
Carlson Kamaka Kukona III
Aqr 22h49'19" -15d8'51"
Carlson Wagonlit Travel
Aur 5h18'22" 43d55'35"
Carlton & Betty Paulus 9/25/1948
Cyg 20h7'21" 35d52'10"
Carlton Buddy Howorth III
Uma 10h2'32" 58d26'10"
Carlton Daniel Williams
Her 16h37'45" 33d16'57"
Carlton Luke
Uma 9h31'20" 62d40'48"
Carlton & Melissa Goody
Dra 12h16'42" 66d20'41"
Carlton's Star
Dra 17h52'12" 56d50'26"
Carly
Sco 16h4'28" -18d40'28"
Carly
Aqr 22h58'20" -8d59'40"
Carly
Cnc 8h8'3" 32d55'40"
Carly
And 0h33'8" 39d18'32"
Carly
And 1h35'44" 40d42'21"
Carly
Leo 10h10'56" 11d1'48"
Carly
Peg 22h36'21" 18d45'9"
Carly
And 0h8'59" 29d7'59"
Carly
Gem 7h28'16" 19d42'58"
Carly Acierno
Men 5h35'26" -74d13'10"
Carly Allison
Cma 6h42'25" -14d59'22"
Carly and Gary
Ori 5h14'33" 8d35'23"
Carly and William
Tau 3h40'6" 30d11'32"
Carly & Andy Dunne
Cyg 21h39'21" 53d4'46"
Carly Ann Thomas
And 23h10'50" 47d55'6"
Carly Annmarie Hall
Aqr 21h19'39" -14d3'20"
Carly Barbara Lyon
And 2h21'29" 43d4'8"
Carly Bess Rogers
Cnc 8h37'51" 23d4'41"
Carly Brown
And 0h55'46" 34d27'32"
Carly Cardellino
Lyn 7h35'48" 49d46'55"
Carly carlyflower Davis
Cap 21h10'58" -19d20'33"
Carly Caterbone
Mon 7h15'38" -0d39'21"
Carly Ciarletta
Uma 8h55'15" 62d50'45"
Carly Clifford
Cyg 21h13'52" 47d0'27"
Carly Collins Shelton
Mon 6h46'25" -0d28'1"
Carly Constance Saunders
Gem 6h24'35" 18d1'54"
Carly Cribley
Cam 4h51'32" 55d43'28"
Carly D'Aun Sheldon
And 0h48'30" 30d0'36"
Carly Dawn
And 1h33'55" 42d17'14"
Carly Denise - My Valentine
Cnc 9h1'43" 11d56'46"
Carly Dillard
Aqr 22h52'29" -8d8'43"
Carly Elizabeth El-Zoghby
Leo 9h22'26" 17d6'1"
Carly Emma Ward
Dra 19h35'1" 65d54'20"
Carly Fannin
Psc 1h26'14" 13d49'0"
Carly Fay Ann Krall
Cnc 8h19'37" 12d25'46"
Carly Florence Alice Chick
And 0h12'30" 26d11'44"
Carly - Flower
Uma 9h18'22" 50d23'57"
Carly Fryer
Lib 14h53'6" -1d21'8"
Carly Golleher Ursa Major
Uma 9h40'22" 62d11'1"
Carly Hayes Longenecker
Tau 3h48'23" 2d59'19"
Carly Jane Hare
Pho 0h23'57" -47d11'5"
Carly Jane Phillippie
Lib 15h4'57" -22d32'35"
Carly Jane Quinn
Ari 2h48'10" 26d59'54"
Carly Jaye Potts
Tau 5h2'52" 15d48'22"
Carly Jean
Lib 14h56'14" -18d2'47"
Carly & Jerry
And 23h44'31" 42d43'46"
Carly Jill Popofsky
Ari 2h16'33" 24d14'3"
Carly Jo
Cas 1h36'1" 63d34'35"
Carly Jo
Uma 11h20'54" 56d42'41"
Carly Jo Zuba
Ori 5h53'46" 19d14'38"
Carly Jordan Traub
Aqr 21h58'13" 0d3'32"
Carly Lee Underwood
Sgr 19h57'13" -28d49'14"
Carly Little
Umi 14h57'28" 73d6'32"
Carly Lohnberg
Crb 15h46'56" 26d51'54"
Carly Lorraine Confer
Vir 12h9'2" 12d2'42"
Carly Lynn Hughes
Sco 17h52'14" -35d47'47"
Carly Lynn Jones
And 0h29'30" 38d49'8"
Carly M Reed
Sco 16h9'20" -12d19'24"
Carly Marie
Cas 1h26'6" 61d8'4"
Carly Marie Gianinoto
Dra 19h39'4" 70d0'12"
Carly & Mark
Cyg 19h50'55" 36d16'42"
Carly May Harris
Uma 10h3'27" 44d0'1"
Carly Milan
Uma 9h18'52" 52d15'37"
Carly Nelson
Ari 3h7'46" 29d0'28"
Carly Nicole
Ari 3h14'49" 27d50'41"
Carly Paige
Cnc 8h50'11" 31d25'40"
Carly Paige Johnson
And 1h28'41" 40d37'5"
Carly Patrick & Paul Courtney
Cru 12h24'5" -58d10'54"
Carly Rachel Rabner
And 1h17'0" 41d54'26"
Carly Rachelle Maita
Leo 9h34'16" 32d36'28"
Carly Rae
Leo 10h5'11" 19d15'31"
Carly Rae Collard-Cottone
Psc 1h42'20" 12d48'5"
Carly Rae Curtis
Vir 14h50'8" 0d50'29"
Carly Raye
And 0h18'51" 32d33'21"
Carly Renee
Lib 15h32'57" -4d43'47"
Carly Rios
Crb 16h14'42" 33d21'36"
Carly Ryan
Sgr 18h6'52" -28d29'27"
Carly Ryan, "Sweet Child of Mine"
And 23h45'15" 46d25'1"
Carly Shaina
Psc 1h27'14" 27d59'38"
Carly Shaniqua McElroy
Lyn 9h20'42" 40d15'59"
Carly Sue Bain
Lyr 18h51'39" 32d21'1"
Carly Valentine
Umi 13h41'31" 75d42'17"
Carly W
And 1h28'39" 46d36'31"
Carly Walters
Cru 12h20'52" -57d15'58"
Carly White
Cas 0h2'43" 55d55'47"
Carly, and Anthony's
Cap 20h54'12" -19d18'42"
Carly-Ann Miller
And 2h26'57" 48d57'35"
Carlye Kitchens
Uma 8h48'13" 70d49'38"
Carly-Miller-Baker
Crb 15h54'54" 26d25'33"
Carlyn
Lmi 10h43'21" 38d47'5"
Carlyn C. Bryan
Umi 15h4'38" 69d56'36"
Carlynne Brooke Clayton
And 1h15'29" 41d19'59"
Carlyon Wilson
Sco 16h20'47" -12d42'41"
Carly's Star
Leo 10h42'43" 6d56'37"
Carly's Star
Tau 3h30'28" 8d44'23"
Carm Lalama
Umi 16h46'53" 79d27'46"
Carma Leigh D'Amico
Sco 17h55'29" -40d32'46"
Carma May
Cnc 8h35'2" 28d7'0"
Carma Sweeney
Vir 12h40'15" -0d29'0"
Carmalene S. Bakowski
Vir 12h41'44" 6d19'42"
Carman Louise Brown
And 1h59'55" 46d53'35"
Carmarhandyl
Psa 22h24'52" -26d19'57"
Carmaricheli
Uma 8h43'55" 54d32'3"
Carmax
Uma 10h46'10" 45d11'14"
Carme Docampo Corral
Cyg 20h50'17" 35d26'43"
Carmel
Cyg 21h40'10" 51d24'22"
Carmel
Cap 20h34'6" -9d6'24"
Carmel Apple
Ori 5h45'4" 2d43'13"
Carmel Cucinotta
Uma 10h55'26" 63d26'0"
Carmel Lea
Vir 13h10'42" 7d1'46"
Carmel Lee Anne Pennington
Aqr 22h12'38" 1d56'57"
Carmel Lewis Wood, Jr.
Aqr 22h8'58" -21d26'39"
Carmel Louise King
Aqr 23h0'2" -12d17'4"
Carmel Mae Roma
Col 5h41'37" -35d26'31"
Carmel Marie Smith
Tau 5h53'20" 24d37'19"
Carmel McGinley
Gem 6h28'32" 15d27'16"
Carmel Of Jesus, Mary and Joseph
Ori 4h48'32" -3d5'57"
Carmel T. Pizighelli Carter
Cnc 9h8'43" 32d52'20"
Carmela
Lyn 8h4'29" 44d9'3"
Carmela
Per 3h1'59" 44d1'15"
Carmela
Ori 4h59'14" -3d37'6"
Carmela Ardizzone
Cyg 19h35'20" 32d9'52"
Carmela C. Smith 6/6/35
Gem 7h29'3" 20d9'11"
Carmela Campagna
Cyg 19h38'8" 31d56'29"
Carmela Candell
Lib 15h10'21" -11d22'27"
Carmela DeFrancesco
Cas 1h30'47" 65d45'34"
Carmela G. Lacayo
Lyn 6h32'37" 57d49'53"
Carmela Gorman
Tau 4h34'14" 28d11'37"
Carmela Grasso
Cnc 8h4'18" 21d1'51"
Carmela Leyva-Grueter
Lib 14h54'16" -2d5'8"
Carmela Libro
Lyr 18h23'40" 30d35'14"
Carmela Prust
Her 16h42'12" 5d44'37"
CARMELA ROSE
Cyg 19h33'43" 28d57'18"
Carmela Toscano
Cas 1h24'55" 62d14'31"
Carmelina
Ori 5h49'34" 4d11'30"
Carmelina Durante
Per 4h9'42" 45d4'2"
Carmelina Micalef
Lyn 8h50'48" 46d3'54"
Carmelita Mae Smith
Cas 1h20'52" 62d29'23"
Carmella Gambino
Cnc 8h23'35" 22d5'51"
Carmella Gissi
And 23h30'43" 48d3'23"
Carmella Isabella
Cas 23h57'35" 53d51'18"
Carmella LaRosa
And 1h24'47" 47d36'19"
Carmella Woodburn
Cas 0h2'40" 54d19'24"
Carmella's Birthday Star
Ari 2h39'7" 29d17'29"
Carmelo
Lyr 18h31'54" 33d42'25"
Carmelo
Uma 9h3'26" 65d9'31"
Carmelo Di Cara - 6th September 1976
Cru 12h32'37" -62d11'22"
Carmelo Frank Lofaso
Cep 21h33'6" 64d46'57"
Carmelo L. Morales Morales
Ari 3h7'49" 11d16'6"
Carmelo Loves Cassandra
Tau 5h16'0" 22d37'6"
Carmelo Pascali
Lyr 19h17'52" 33d7'31"
Carmelo Restifo
Sco 17h8'8" -36d34'58"
Carmelo Vincent Monaco
Gem 6h4'27" 23d15'40"
Carmelo Vitello
Cnc 8h38'24" 11d14'13"
Carmelo's Star
Lyn 8h58'43" 36d52'20"
Carmel's Diamond
Cru 12h28'43" -60d42'16"
Carmen
Sco 16h6'57" -25d46'9"
Carmen
Uma 10h43'58" 42d49'59"
Carmen
Uma 11h9'18" 72d12'35"
Carmen
Umi 14h4'46" 70d34'30"
Carmen
Uma 9h10'41" 62d55'13"
Carmen
Vir 13h37'5" -13d19'19"
Carmen
Ori 4h51'31" -2d37'32"
Carmen
Leo 11h14'14" -3d48'24"

Carmen
Sco 16h5'57" -12d40'33"
Carmen
Uma 10h11'56" 44d43'57"
Carmen
Cas 0h13'6" 53d18'14"
Carmen
Cam 4h12'33" 57d37'48"
Carmen
Ori 4h53'22" 15d40'31"
Carmen
Ari 2h9'16" 16d47'29"
Carmen
Psc 1h9'32" 12d21'35"
Carmen
Leo 11h20'8" 2d42'30"
Carmen
And 0h16'34" 29d31'26"
Carmen
Peg 22h5'22" 16d35'57"
Carmen
Her 18h41'56" 21d35'59"
Carmen "12-26-83"
Lyn 7h38'51" 41d25'50"
Carmen 22.03.1970
Uma 8h59'43" 48d1'44"
Carmen A. Kartiganer
Com 12h37'10" 23d13'3"
Carmen A. McCoy
Cas 0h31'8" 62d51'59"
Carmen A. Petrarca
Uma 10h31'22" 65d45'50"
Carmen Acuyo-Cespedes
Hya 8h45'19" -8d24'46"
Carmen and Bryan Forever
Ori 5h30'36" 2d19'44"
Carmen Andreola
Umi 17h32'1" 88d4'51"
Carmen Angela Binder
Psc 1h22'51" 16d13'51"
Carmen Anita
Crb 16h9'6" 25d46'4"
Carmen Anita Lynch
Aqr 23h24'44" -10d6'43"
Carmen Ann Teixeira
Vir 13h0'44" 3d10'33"
Carmen Apodoca
Uma 9h13'11" 56d37'11"
Carmen Archambault-Perazelli
Umi 11h56'23" 86d28'1"
Carmen Asturias San Juan
And 2h15'16" 50d31'44"
Carmen Asturias San Juan
Cyg 19h43'51" 33d53'16"
Carmen Aurora
Ori 6h19'5" 9d22'47"
Carmen Aviles
Gem 6h49'51" 26d51'55"
Carmen Bell
And 0h19'12" 27d1'0"
Carmen Bevan
Umi 14h4'49" 76d18'45"
Carmen Cecilia Osorio Parada
Vir 14h37'50" -0d2'9"
Carmen Coyne
Crb 15h44'55" 35d50'58"
Carmen Daisy Welch
Per 4h45'8" 46d48'52"
Carmen & Dany's Glücksstern
Uma 12h45'11" 53d44'29"
Carmen Delgado
Uma 8h21'52" 65d17'39"
Carmen Duenas
Aqr 22h38'49" -0d16'43"
Carmen Duran
Uma 9h30'47" 53d37'30"
Carmen Elena Taran
And 0h36'32" 30d20'20"
Carmen Elena's 1st Christmas Star
Cyg 20h52'50" 44d36'47"
Carmen Elizabeth Ojeda
Cnc 8h17'8" 15d10'2"
Carmen Eulenfeld
Crb 15h36'19" 26d17'44"
Carmen Guadix
Ari 2h45'55" 14d49'30"
Carmen Guevara
Peg 22h10'9" 6d47'42"
Carmen Haas
Vir 12h18'59" 3d14'34"
Carmen Isabel Nazario Lopez
Cyg 20h15'23" 50d48'1"
Carmen J. Porter
Leo 11h34'36" 27d46'44"
Carmen Jessy Griffin
Gem 7h3'59" 16d26'44"
Carmen Joy Delaney
Lib 15h3'52" -6d41'37"
Carmen Krähenbühl, 07.12.1987
Boo 15h7'7" 14d44'50"
Carmen Lee Lewelling
Col 6h20'52" -34d26'13"
Carmen Lee Strome
Cyg 21h45'16" 33d13'24"
Carmen M. Sierra
Sgr 17h55'29" -28d57'19"
Carmen Mama Gomez
Uma 9h12'50" 65d57'23"
Carmen Maria Ladogana Rivera
Ori 6h11'47" 19d4'49"
Carmen Maria McAleer
And 23h17'12" 50d50'58"
Carmen Marie
Cap 21h6'35" -20d2'40"
Carmen Marie Artrip
Tau 3h27'22" -0d0'59"
Carmen Marie & Darrell Alan Doomy
Her 17h18'21" 16d14'56"
Carmen Marie "Mitita's Star"
Sco 16h6'53" -14d32'13"
Carmen Marie Sawyer Roy
Hya 9h26'37" -0d23'56"
CARMEN - MELA - ANZO
Uma 8h28'49" 63d29'23"
Carmen Melendez Olmedo
Umi 19h7'3" 88d22'25"
Carmen Meza
Dra 17h29'2" 63d37'10"
Carmen & Michael
Uma 9h5'2" 56d28'21"
Carmen Michelle
Pho 0h21'8" -48d2'39"
Carmen Michelle Jackson
Aqr 22h33'31" 1d25'38"
Carmen Michelle Taylor
Vir 12h18'3" 0d52'23"
Carmen Milagros Ramos
Tau 5h5'11" 23d17'52"
Carmen Nadine Reynolds
Cap 20h23'28" -10d24'43"
Carmen Nunzio Gazzo
Her 16h37'25" 47d52'50"
Carmen O 82363
Vir 14h18'48" -5d2'15"
Carmen of the Heavens
And 0h28'42" 25d12'55"
Carmen Oleta
Ori 6h14'12" 6d4'0"
Carmen Orlinda
Lib 15h24'23" -7d14'22"
Carmen Ortiz
Lyn 8h7'24" 39d48'59"
Carmen Patterson
Ari 1h50'12" 23d39'52"
Carmen Perez
Aql 19h10'9" 2d33'19"
Carmen Petra Heuberger
Cam 6h19'26" 72d16'51"
Carmen Purizaca
Lib 15h43'34" -19d4'7"
Carmen R.
Cnc 8h46'43" 32d54'16"
Carmen R Clemente
Sgr 18h25'31" -22d27'37"
Carmen R. Skillman
Her 17h41'23" 48d5'53"
Carmen Rene' Moraca
Ori 6h11'39" -3d10'31"
Carmen Restine
Cas 0h9'49" 57d3'8"
Carmen Reyes
Sco 16h59'7" -32d35'50"
Carmen Rita Calderon
Psc 0h54'27" 29d16'45"
Carmen "Rosie" Núñez
Sco 17h0'59" -44d24'52"
Carmen Ruiz
Gem 6h45'26" 28d35'42"
Carmen Santiago
Uma 8h43'15" 57d11'27"
Carmen Shere Robinson
Lib 14h28'34" -15d24'52"
Carmen Stansfield
Per 3h46'6" 48d50'31"
Carmen Starlight
And 1h9'48" 42d17'0"
carmen the angel
And 23h3'3" 47d21'47"
Carmen Thomas DeRosa
Tau 3h48'4" 27d44'10"
Carmen Tripodi
Cru 12h29'26" -61d50'33"
Carmen V. Gonzalez
And 0h33'26" 33d55'46"
Carmen Vallejo
Psc 0h52'2" 6d57'32"
Carmen Vasquez Fitzgerald
Com 13h29'31" 27d44'54"
Carmen Wong
Crb 16h14'47" 34d21'41"
Carmencita Tan Ting
Ari 3h11'26" 22d46'48"
Carmenred
Uma 10h59'27" 65d1'0"
Carmen's Eternal Light
Cnc 8h48'2" 29d47'27"
Carmen's Love
Uma 13h4'28" 53d23'36"
Carmen's Own Birthday Star
Aqr 21h19'54" -5d27'49"
Carmen's Star
Cas 0h59'22" 70d12'10"
Carmen's Star
Peg 23h11'3" 16d5'13"
Carmen-Stern-27-12-1992
Dra 17h57'2" 79d35'2"
Carmenza Stellato
Cam 5h46'22" 60d44'41"
Carmer
Leo 10h45'13" 19d46'12"
Carmie Rose
Boo 14h42'47" 31d28'16"
Carmie's Star
Dra 18h32'53" 58d20'44"
Carmin E. Brown
Cyg 20h47'38" 34d45'41"
Carmina
Uma 11h8'6" 60d29'20"
Carmina
Cap 21h54'42" -13d27'6"
Carmina Cecilia
Aqr 22h42'1" -1d28'59"
Carmina Gallo
And 1h6'28" 34d45'2"
Carmina Iuso
Uma 8h42'58" 68d47'20"
Carmina Winona
Cas 23h13'9" 55d18'44"
CAR-MINE
Uma 8h55'39" 71d3'4"
CARMINE
Ori 4h50'51" -0d19'9"
Carmine
Sco 16h9'41" -15d26'18"
Carmine
Leo 9h36'22" 32d43'23"
Carmine A. "Nino" Porreca
Uma 9h23'42" 64d19'22"
Carmine Aliberti
Ari 2h48'9" 28d50'23"
Carmine Andranovich
Per 3h9'26" 48d32'35"
Carmine & Annie Ferraro
Sgr 18h3'54" -25d2'58"
Carmine Anzalone
Psc 22h53'50" 7d42'36"
Carmine - Caroline - Marco Franco
Uma 11h29'43" 35d21'45"
Carmine Cicchino
Lib 15h25'13" -18d47'56"
Carmine D. DeZuzio
Uma 10h26'32" 71d26'47"
Carmine Disibio
Ori 6h10'18" 16d2'11"
Carmine & Irene Ruotolo
Uma 10h58'40" 51d0'18"
Carmine J. Prudente
Her 17h39'24" 26d4'32"
Carmine J. Rubino
Tau 4h58'31" 23d14'35"
Carmine Minervino
Uma 11h54'7" 28d56'45"
Carmine Samperi
Uma 13h58'19" 50d31'4"
Carmine Sebastiano III
Aql 20h6'50" 9d15'40"
Carmine 'Skipper' Rea, Sr.
Lib 15h34'55" -13d20'41"
Carmita
Umi 16h25'57" 82d25'48"
Carmita Laboy Llorens
Mon 7h59'52" -8d59'11"
carms
Psc 0h3'45" 5d50'47"
Carnation Yellow
Hya 8h45'39" -8d47'19"
Carnevaletti Giada
Cam 6h39'16" 84d37'27"
Carni
Tau 5h19'30" 26d13'49"
Carnot Ramanibaï Mouttou
Ari 2h12'51" 11d17'50"
Caro
Uma 9h11'43" 57d29'46"
Caro Alidar Cornelia Rottschäfer
And 0h56'17" 45d52'44"
Caro190282
Psc 1h22'34" 29d43'46"
Carocchi Giuseppe
Cas 0h27'24" 62d24'57"
Carol
Uma 13h39'7" 58d2'3"
Carol
Ori 5h33'9" -7d45'48"
Carol
And 0h37'24" 28d32'47"
Carol
Her 18h13'38" 18d34'26"
Carol
Leo 9h28'19" 28d13'3"
Carol
Aqr 22h30'32" 2d18'49"
Carol
Crb 16h19'50" 38d31'8"
Carol
Uma 10h45'26" 45d30'49"
CAROL
Uma 11h0'8" 46d30'3"
Carol
Cas 0h15'18" 52d54'18"
Carol
Cam 4h43'43" 58d43'35"
Carol
Crb 16h9'6" 34d32'3"
Carol
Per 3h5'50" 36d37'40"
Carol
Cyg 20h31'55" 35d0'57"
Carol
And 0h40'8" 41d21'4"
Carol
Lyn 7h49'32" 37d56'34"
Carol , forever my Princess
Leo 9h46'7" 30d19'27"
Carol A. 2/24/1948
Uma 9h26'21" 57d46'39"
Carol A. Coleman "12/18/1933"
Cas 2h5'10" 70d33'22"
Carol A. Delker
Sgr 20h14'42" -43d38'48"
Carol A. Hamilton
Uma 12h5'46" 65d28'0"
Carol A Harris
Cas 1h10'28" 59d57'41"
Carol A. Harty
And 2h0'23" 42d24'52"
Carol A Hayford Harvey
Peg 23h28'37" 33d14'14"
Carol A. Lambert
Tau 5h30'12" 18d44'32"
Carol A. Lotane
Crb 15h20'54" 28d34'10"
Carol A. Michelson-Beer
Sco 17h46'1" -37d50'37"
Carol A. Payne
Cap 20h22'59" -25d53'5"
Carol A. Sparks
Tau 4h18'16" 12d25'18"
Carol A. Taylor-Zimmerman
Vir 14h7'44" 6d16'35"
Carol A Walton
Cas 23h27'59" 59d26'38"
Carol & Alfred Jones
Cyg 21h25'24" 34d52'5"
Carol & Amber Powell Forever
Cyg 20h36'29" 52d49'19"
Carol and Cort LaMee
Cyg 20h5'53" 35d39'15"
Carol and Michael
Cyg 20h28'13" 31d44'10"
Carol and Odis
Crb 16h13'15" 34d2'45"
Carol Anderson
Uma 11h38'0" 32d11'56"
Carol Ann
Lmi 10h12'58" 35d17'2"
Carol Ann
Gem 6h43'26" 31d4'10"
Carol Ann
Per 3h34'38" 35d6'35"
Carol Ann
Cas 1h19'35" 59d3'31"
Carol Ann
And 23h0'54" 51d1'0"
Carol Ann
Crb 16h10'38" 39d19'54"
Carol Ann
Tau 4h17'49" 0d32'30"
~Carol Ann~
Leo 11h32'47" 12d15'29"
Carol Ann
Cnc 8h57'43" 11d37'28"
Carol Ann
Cas 0h47'29" 61d1'8"
Carol Ann
Lyn 7h10'21" 53d49'49"
Carol Ann
Cam 8h24'47" 76d48'16"
Carol Ann and Nicholas Carbone, Jr.
Cyg 20h5'2" 54d47'57"
Carol Ann Arciniega
Sgr 17h44'54" -27d55'37"
Carol Ann Bacon
Tau 4h39'27" 29d3'41"
Carol Ann Barrett
Lib 14h51'44" -1d41'50"
Carol Ann Bass
Cap 21h0'46" -20d24'10"
Carol Ann Buchek-O'Brian
And 23h44'8" 47d51'35"
Carol Ann Bump
Psc 1h11'39" 28d24'58"
Carol Ann Buss (
Aql 19h25'4" -7d31'55"
Carol Ann Chapple
Tau 5h25'44" 26d38'29"
Carol Ann Chatel
Psc 1h15'1" 17d7'57"
Carol Ann Chiaramonte
And 0h23'37" 41d51'28"
Carol Ann Cody
Leo 10h54'19" -1d34'53"
Carol Ann Comboy
And 1h15'58" 48d46'28"
Carol Ann Comeau
Cyg 20h34'41" 35d8'16"
Carol Ann Conway
Peg 22h32'41" 3d15'29"
Carol Ann Cristiano
Sgr 18h8'22" -20d4'30"
Carol Ann Cubler
Cyg 20h20'22" 55d52'39"
Carol Ann Curran
Uma 10h50'46" 62d19'39"
Carol Ann Currell
Cas 23h28'18" 57d28'57"
Carol Ann Duber
Cas 1h15'41" 55d3'56"
Carol Ann Ducey
Lib 14h42'13" -21d22'27"
Carol Ann Ferris' Corner of Heaven
Cas 1h39'16" 60d20'3"
Carol Ann Gilligan Gebhardt
Tau 4h24'43" 23d40'28"
Carol Ann Hahn
Cas 1h18'40" 64d7'16"
Carol Ann Hamel
Cnc 8h25'35" 20d43'30"
Carol Ann Helms
Ari 2h44'0" 14d34'51"
Carol Ann Helstrom
Ari 2h48'59" 25d14'13"
Carol Ann Houseman
Crb 15h53'1" 34d35'43"
Carol Ann Johnson
Uma 10h55'9" 40d15'58"
Carol Ann Johnstone - Carol's Star
And 0h48'39" 27d23'19"
Carol Ann Kohler Rodriguez
Psc 0h27'29" 9d7'7"
Carol Ann Krisnosky
Uma 9h52'56" 42d4'57"
Carol Ann Manuele
Tau 4h21'40" 29d23'18"
Carol Ann Masterson
Cap 20h19'56" -20d18'6"
Carol Ann Meyers
Aql 19h14'30" 7d49'50"
Carol Ann: Mo Cuishle
Crb 15h48'41" 32d21'53"
Carol Ann Morning Star
And 0h24'56" 40d54'41"
Carol Ann Nallin
Uma 8h50'7" 67d55'34"
Carol Ann Palmisano
Uma 12h1'54" 44d30'56"
Carol Ann Payne
Uma 13h39'38" 51d37'48"
Carol Ann Pookie Fried
Uma 9h30'13" 52d6'10"
Carol Ann Puckett Bush
Ari 2h47'25" 17d13'4"
Carol Ann Quackenbusch
Crb 15h34'34" 28d49'18"
Carol Ann Sage
Sgr 18h20'56" -34d54'28"
Carol Ann Sheber
Lyr 18h49'20" 34d52'25"
Carol Ann Soper
Ari 2h34'7" 26d2'17"
Carol Ann Taylor
Uma 9h17'38" 55d47'29"
Carol Ann VanderMeer
And 0h59'34" 41d1'43"
Carol Ann Vinecour
Gem 7h35'9" 29d30'37"
Carol Ann Willis
Lib 15h25'17" -9d39'50"
Carol Ann Woods
Psc 1h44'5" 6d56'6"
Carol Anne
Vir 11h50'47" 6d55'31"
Carol Anne
Per 1h31'33" 51d50'35"
Carol Anne Caouette
Psc 1h22'28" 7d13'20"
Carol Anne Kiesel
And 0h37'32" 36d53'56"
Carol Anne MacConnachie
Cyg 19h55'56" 46d5'28"
Carol Anne Maclary
Cas 0h31'50" 62d4'9"
Carol Anne Morton
And 23h32'33" 48d4'59"
Carol Anne Nickell
Vir 13h19'51" -14d47'55"
Carol B. Duncan
Sco 17h9'5" -41d45'58"
Carol B. Sheppard
Ori 5h53'3" 18d41'49"
Carol Barbour, Ph.D.
Lib 14h59'36" -16d34'2"
Carol Battaglia
Vir 13h51'11" -15d53'58"
Carol Becker
Sgr 18h1'42" -27d21'58"
Carol Bennett & Jason Chilson
Cyg 20h28'39" 34d44'59"
Carol Beth Moreland
Uma 11h7'59" 46d40'11"
Carol Blasie
Cyg 20h1'16" 34d55'4"
Carol Bohanna
And 0h44'42" 38d33'20"
Carol Booth
And 23h47'38" 33d27'51"
Carol Bowes
Uma 8h33'23" 61d23'1"
Carol Boyd "Our Shining Star"
Crb 15h32'39" 38d46'19"
Carol Brennecke
Cnc 8h51'36" 27d24'20"
Carol Briggs Elliott
Tau 5h27'3" 20d41'50"
Carol Broos
Lyr 18h41'26" 37d13'30"
Carol Brown
Uma 13h51'28" 51d22'32"
Carol Budnik "My Shining Star"
Sgr 20h4'28" -12d51'8"
Carol Bunner Merrill
Aqr 21h40'25" 1d26'46"
Carol C. Ericson
Cas 23h44'40" 53d16'12"
Carol C: Loving Mother and Healer
Umi 14h15'58" 75d21'6"
Carol Campbell
And 0h32'13" 40d4'50"
Carol Campbell
And 0h40'58" 42d26'23"
Carol Canes-Lightle
Cas 23h8'29" 54d41'42"
Carol & Carl for Eternity
Cap 21h35'7" -13d40'46"
Carol Catherine St. John
Cas 0h26'27" 55d40'2"
Carol Caughron Whitaker
And 0h31'54" 30d4'48"
Carol Chambliss
Lib 15h33'38" -8d22'34"
Carol Channing and Harry Kullijian
Uma 11h47'1" 36d26'34"
Carol Chartier
Crb 16h14'14" 38d56'12"
Carol Christie Genest
Sgr 18h22'59" -24d4'20"
Carol Claus
Crb 16h24'4" 27d42'24"
Carol Coleman
Lib 15h4'49" -6d40'56"
Carol Coloyan
And 2h17'43" 49d34'7"
Carol Condon
And 23h12'55" 35d45'45"
Carol Cosenza
Cnc 7h57'7" 15d53'8"
Carol Crane
Sgr 18h51'39" -27d40'53"
Carol Crites
Sgr 19h36'37" -26d7'27"
Carol Croley
Uma 11h33'6" 32d57'0"
Carol Cullen
Tau 5h7'8" 18d9'28"
Carol Cunliffe Shines On BJ Forever
Cru 12h44'21" -58d14'28"
Carol D. Baker
Lyn 7h19'47" 52d55'36"
Carol/Darcy
Lyn 8h4'12" 53d23'14"
Carol Dawn
Cen 13h31'47" -59d58'51"
Carol Dawn's Saphire
Vir 11h50'57" -4d17'47"
Carol DeJac
Ari 3h5'9" 20d17'6"
Carol Delene Knight Wren
Cyg 20h52'34" 54d15'12"
Carol DelGaldo
And 2h18'48" 48d54'51"
Carol Denton
Cas 1h32'46" 67d43'24"
Carol DeSimeo Angel Star
Psc 1h21'14" 17d22'12"
Carol Diamond
Ari 3h12'31" 22d2'42"
Carol Diane Gomez
Lib 15h30'0" -15d29'14"
Carol Dianne
Ori 5h36'24" 4d56'28"
Carol Donkerbrook Weeks
Ari 3h13'38" 22d44'10"
Carol Dufour
Uma 8h40'28" 58d57'53"
Carol E. Arnett - Billy's Love
Psc 0h6'15" 9d8'58"
Carol E. Dial
Cnc 8h43'54" 18d53'59"
Carol E. Nolte
Cas 2h41'34" 57d45'53"
Carol Eifert
And 0h42'57" 35d53'8"
Carol Elaine Childers
Sgr 19h13'30" -23d12'4"
Carol Elizabeth
Lib 15h12'38" -11d51'42"
Carol Elizabeth Stack
Tau 3h30'33" 18d0'7"
Carol Erdman Merkle
Lib 14h48'52" -4d29'12"
Carol & Eric's 30th Year Star
Lib 14h52'24" -8d35'41"
Carol Ernst
Aql 20h29'43" -0d36'15"
Carol Ezzelle
Cnv 12h33'36" 43d53'36"
Carol Farnsworth Moody
Cam 5h15'18" 61d26'59"
Carol Fay Gregor
Umi 15h49'17" 75d24'54"
Carol Ferguson
And 1h34'45" 40d39'1"
Carol "Fitzsimmons" Russell
And 1h7'4" 41d17'48"
Carol Fleck Davis
Srp 18h18'38" -0d21'20"
Carol Fox
Cnc 8h55'38" 26d39'41"
Carol Frances Tolentino Hurst
Uma 12h43'19" 54d25'19"
Carol G. Deegan
Cas 2h40'33" 57d57'58"
Carol Garcia
Cap 20h42'43" -22d10'6"
Carol Gayle Wood
Uma 10h12'22" 62d9'5"
Carol Genieve Caron
Her 17h2'25" 32d30'37"
Carol Gissing
Lyr 19h4'49" 36d3'27"
Carol Goldston
Sco 16h9'47" -17d19'46"
Carol Goodhue
Tau 5h46'57" 21d33'6"
Carol Green
Lmi 10h48'58" 28d43'46"
Carol Green Yachtis
Ari 3h15'0" 29d26'44"
Carol Hadley
And 23h14'12" 48d29'56"
Carol Hamilton
Cap 20h46'13" -15d59'21"
Carol Hartford McCormack
Ori 5h47'24" 2d10'46"
Carol Hayes
Cap 20h26'39" -10d52'11"
Carol Hemingway
Ori 6h12'58" 16d5'33"
Carol Higdon
And 0h40'48" 36d24'47"
Carol Hokaj "Sign Chi Do Seed"
Crb 15h41'50" 26d40'55"
Carol Holman
And 1h40'54" 38d45'6"
Carol Hornblower & Fred Weber
Uma 10h53'30" 68d31'43"
Carol Hughes
Cas 23h4'8" 54d16'19"
Carol I Chabot
Leo 10h10'8" 15d13'42"
Carol I Love You . Love Kevin
And 23h25'41" 47d42'51"
Carol Irene
Cyg 20h2'49" 56d47'28"
Carol Irvin's Star
Ori 5h35'1" -2d22'56"
Carol J. Brock
Her 16h53'50" 36d56'18"
Carol J. Maracle
Cap 20h47'43" -24d47'0"
Carol J. McGann
Uma 10h13'13" 41d32'0"
Carol J. Mohr
Cap 20h24'18" -22d42'59"
Carol J. Stropus
Gem 6h51'37" 32d55'47"
Carol J. Taylor A Peace of Heaven
Lyn 7h44'5" 53d47'7"
Carol Jane Harper Reed
Psc 0h24'44" 8d49'29"
Carol Jane Hinkle
Tau 3h47'36" 16d46'31"
Carol Jean
Vir 12h12'36" -0d53'23"
Carol Jean
Aqr 23h23'36" -13d14'59"
Carol Jean Anderson Duncan
Lyn 7h0'0" 51d35'48"
Carol Jean Attoe
Leo 10h10'54" 21d42'6"
Carol Jean (Bean) Greenwood Kortas
Cas 1h40'50" 62d45'24"
Carol Jean Brady
Sco 16h4'59" -13d52'33"
Carol Jean Clarke
Gem 7h44'58" 33d47'14"
Carol Jean Guelzow
And 0h12'31" 37d42'50"
Carol Jean Hodge
Leo 10h15'28" 12d27'18"
Carol Jean Schmidt
Gem 6h32'53" 16d43'25"
Carol Jean Simpson
Cyg 20h35'27" 43d10'38"
Carol Jean Stacke
Uma 10h58'27" 46d5'10"
Carol Jeanne Claypool Letteriello
Lib 15h9'55" -21d15'37"
Carol Jeanne Driscoll
And 0h43'50" 27d11'24"
Carol Jeanotilla
Vir 13h16'48" -18d21'22"
Carol & Jeff's Ruby Wedding Star
Umi 14h39'42" 69d39'46"
Carol Jenkins Barnett
Cam 5h8'27" 62d43'27"
Carol Joan Fraleigh
Sgr 18h15'11" -19d9'11"
Carol Jones
Ori 4h49'14" 11d30'5"
Carol Jordan
Umi 15h23'38" 67d36'46"

Carol Joy
Mon 7h1'27" 8d19'13"
Carol Joy 50th and BEYOND 5/10/1957
Tau 4h25'13" 23d26'59"
Carol Joyce Harden
Ari 2h39'54" 21d13'4"
Carol Joyce Loll
And 23h58'25" 45d18'7"
Carol Joyce Stern
Cnc 8h45'22" 13d15'26"
Carol Juanita Serrapica
Ari 2h12'4" 11d17'14"
Carol Julie
Cas 1h29'39" 60d58'44"
Carol Kaecher
Leo 10h39'9" 14d46'4"
Carol Karam AG
Ari 2h49'8" 12d0'32"
Carol Kay
Sgr 19h16'23" -15d57'10"
Carol Kay Strickler
Cas 0h54'37" 59d23'10"
Carol Kehring
Sgr 18h53'36" -31d21'29"
Carol Keith Frazier
Ori 5h32'54" -0d47'23"
Carol Kendall
Sgr 17h56'58" -16d53'50"
Carol Kies
Psc 0h50'21" 9d42'36"
Carol Knosp
Cnc 8h55'0" 28d1'36"
Carol Kristine Capulong Castro
Crb 16h10'25" 33d1'32"
Carol Kuhl
Cas 0h14'23" 51d24'22"
Carol L. Hise
And 0h24'27" 28d40'52"
Carol L. Krawiec
Leo 11h24'46" 5d31'33"
Carol L. Meier
Com 12h41'40" 28d43'17"
Carol L O M L Mummy
Uma 10h26'52" 44d6'45"
Carol L Reesman
And 2h17'33" 49d47'34"
Carol L. Rodriguez
Uma 10h8'24" 59d15'33"
Carol Lampe
Ari 2h37'25" 13d4'10"
Carol Laurette Kirkman
Lyn 7h16'39" 53d8'19"
Carol Lea
Cas 0h16'19" 62d34'15"
Carol Lee
And 1h14'59" 38d26'55"
Carol Lee Ortman
Vir 13h4'27" 5d22'30"
Carol Leigh Polosky
Uma 11h33'43" 59d42'57"
Carol Lenetzky
Cyg 20h31'51" 55d57'28"
Carol Lesley Boam Symonds
Cep 1h39'12" 77d44'16"
Carol - "Lesley's Shining Star"
And 2h29'20" 42d49'25"
Carol Libby
Cam 5h41'32" 58d0'54"
Carol Lim
Cap 20h13'58" -9d41'23"
Carol Lock
Sgr 19h46'50" -13d57'59"
Carol Logan
Cru 12h39'40" -58d58'48"
Carol Lorace
Vir 12h14'19" 1d45'5"
CAROL LOVES JESSE
Her 18h1'38" 22d3'41"
Carol Loves Michael Forever
Lyr 19h5'51" 42d4'55"
Carol Lynn
And 2h19'35" 47d54'45"
Carol Lynn
Ari 2h26'57" 18d2'33"
Carol Lynn
Tau 4h10'53" 18d3'7"
Carol Lynn
Sgr 18h46'20" -19d53'35"
Carol Lynn Messley
Lyr 18h36'42" 31d30'48"
Carol Lynn Moauro
Cas 23h29'44" 56d35'49"
Carol Lynn Nesdill
And 0h0'33" 34d45'2"
Carol Lynn Robejsek
Ori 5h28'46" 1d56'51"
Carol Lynn Wilder-Enge
Sgr 18h29'31" -16d31'26"
Carol "Lynnie" Hunn
Sco 16h59'33" -39d30'10"
Carol M. Combs
And 0h7'53" 44d17'39"
Carol M. Smith
Crb 15h26'2" 25d41'55"
Carol M. Van Hoesen's Star
Ori 5h11'48" 6d43'8"
Carol M. Volckmann
Ari 2h34'39" 24d35'15"
Carol M. Weinberg
Gem 7h19'54" 34d10'6"
Carol Maddalena
Umi 15h23'20" 73d13'24"
Carol Marie
Aqr 23h19'47" -9d50'4"
Carol Marie
Tau 4h11'32" 2d36'3"
Carol Marie Barwick
Lib 14h54'38" -4d6'14"
Carol Marie Indri
Uma 11h0'56" 45d57'48"
Carol Marshall
And 0h46'4" 37d58'7"
Carol Martin
Crb 15h29'8" 27d31'37"
Carol Mary Meador, My Mother
Cnc 8h46'33" 31d32'55"
Carol Mary Snyder
Lib 15h21'34" -19d23'45"
Carol McMahon & Tony Gerhing
Tau 5h35'37" 25d5'12"
Carol Medina
Umi 13h48'15" 73d32'26"
Carol Melissa Jovanelly-Star Mommy
Cas 1h42'29" 61d4'34"
Carol Merryman-Rees
Dra 16h59'31" 52d9'1"
Carol Michell
Cnc 8h57'44" 8d30'11"
Carol Middleton - Myron Cleaver
Cyg 20h36'52" 36d40'43"
Carol & Mike Triano
Srp 18h29'23" -0d14'34"
Carol Miller-Baker
Leo 11h34'24" 14d58'17"
Carol Mittelsdorf
Sgr 19h39'41" -12d21'30"
Carol Murphy
Cap 21h5'14" -22d18'57"
Carol Nelson Sagan
Cas 0h29'38" 62d3'11"
Carol Nicholls
Cas 1h45'55" 60d23'39"
Carol of the Stars Love from Wendy
Uma 9h49'38" 49d46'19"
Carol! Our Bright and Shining Star
Psc 1h45'25" 18d45'20"
Carol - Our Shining Star
Ori 5h30'23" 0d28'37"
Carol Oyama
Lib 15h38'3" -7d18'34"
Carol Pacelli
Gem 6h33'37" 25d13'34"
Carol Paton
Tau 4h17'52" 23d20'12"
Carol Pellegrino
Cas 0h7'28" 53d56'43"
Carol Pennachio
And 0h38'59" 33d40'3"
Carol Peterson
Tau 4h43'28" 29d4'16"
Carol Polanish
Vir 13h16'25" 11d15'49"
Carol Priestley
Uma 12h0'59" 62d37'0"
Carol Princess of Gelli
And 0h25'36" 38d44'52"
Carol R. Garber
Lmi 10h6'40" 38d26'25"
Carol R Payne
Uma 10h23'18" 54d35'22"
Carol R. Rosenberg
Psc 0h37'13" 17d53'48"
Carol Rae Again Hunter
Cas 0h12'3" 57d6'32"
Carol Raman
Lyn 7h51'0" 57d24'50"
Carol Reese Walker
Cnc 8h40'54" 16d13'30"
Carol Richards
Uma 10h44'45" 46d20'47"
Carol Rinehart
Dra 17h37'57" 63d35'9"
Carol & Robert Bennett 50th Wedding
Gem 7h30'6" 32d57'2"
Carol Rosalie Patty
Cap 20h22'51" -19d46'46"
carol rose
Vir 12h30'21" -2d4'30"
Carol Rose
Uma 11h56'48" 30d17'24"
Carol Rose
Leo 11h52'40" 27d42'25"
Carol Rose Minnitti
Uma 9h47'20" 44d40'27"
Carol Ruberto
Cap 20h18'59" -12d36'32"
Carol Russell
Cas 0h33'49" 67d12'24"
Carol Russo-Pennachio
Cas 0h16'39" 64d34'34"
Carol Ruth Lucke
Per 3h36'41" 31d44'28"
Carol Ruth Mason
Psc 0h10'52" 6d15'6"
Carol S. Fleisher
Cas 23h26'38" 51d33'42"
Carol Saint of Lists
Pho 0h16'57" -40d19'3"
Carol Sanderson Waters
Lmi 10h15'29" 36d24'53"
Carol Sawyer
Aqr 21h55'53" -0d5'7"
Carol Schuerlein
Sgr 18h6'48" -26d45'15"
Carol Seaward
Uma 11h55'17" 64d27'34"
Carol Seddon
Uma 11h30'49" 47d19'0"
Carol Seelman
Uma 9h59'33" 48d0'24"
Carol Sheldon Hylton
Aqr 23h27'6" -19d35'23"
Carol & Shelly Miller
Sge 20h2'57" 18d21'1"
Carol & Shirley
Ari 2h10'36" 24d32'59"
Carol Shoff
Lib 14h51'7" -2d17'56"
Carol Shore
Tau 5h42'47" 18d34'1"
Carol Simowitz
Uma 10h46'23" 55d54'36"
Carol Siosa
Cyg 20h2'32" 32d8'23"
Carol Slee
And 0h18'31" 32d8'31"
Carol Sodman-Elliott (Acceptance)
Aqr 22h29'52" -0d37'11"
Carol Soloveiko *Angel In Heaven*
Sgr 18h34'1" -22d48'3"
Carol Spadafora
Lyr 18h33'6" 36d16'56"
Carol Sparkle Warhanik
And 0h34'14" 45d13'32"
Carol Stall
Cap 20h13'38" -27d24'2"
Carol & Steve's Heavenly Swan
Cyg 20h24'17" 51d44'0"
Carol Stola
Aqr 23h3'42" -9d21'55"
Carol & Stuart
Uma 10h5'5" 58d18'49"
Carol Stumme
Uma 9h54'40" 67d21'41"
Carol Sue
And 1h48'23" 41d20'4"
Carol Sue Anderson
And 23h34'27" 48d26'41"
Carol "Sue" Bass
Crb 15h51'21" 29d23'3"
Carol Sue Henson
Ori 6h7'24" 3d46'42"
Carol Sue Lotz
Sco 17h22'39" -44d15'20"
Carol Sue Mohamed
Vir 14h29'11" -5d31'29"
Carol Sue Studnicka
Gem 6h46'39" 31d34'28"
Carol Swancutt
And 0h46'31" 40d23'25"
Carol Swords
Tau 5h35'56" 26d19'46"
Carol Sylvester
Sco 17h28'48" -39d30'33"
Carol Szewczyk
Aql 19h39'21" 14d33'59"
Carol T. Buchner
Uma 11h9'57" 35d40'2"
Carol Tenedine
Psc 1h21'49" 21d20'3"
Carol Thomas
Uma 11h56'0" 33d37'2"
Carol Toland
Cas 0h23'18" 56d59'14"
Carol "Toodie" Ann
Leo 9h35'1" 6d39'46"
Carol V. Hidell
Cyg 19h43'7" 51d48'42"
Carol Vezeau Pinedo
And 0h36'37" 35d52'46"
Carol W. Stewart
Ari 2h13'5" 17d23'25"
Carol Ward Pearce
Cas 1h50'27" 61d36'40"
Carol Wassmer
Uma 11h34'1" 63d45'31"
Carol Watson
Sgr 18h30'40" -16d46'45"
Carol Webster
Cam 6h14'3" 58d19'10"
Carol Weiner - Min Aengel uf Erde
And 0h14'12" 38d53'57"
Carol White
Vir 12h43'25" 3d41'43"
Carol Wicklund-McNally
Cnc 8h41'59" 14d40'17"
Carol Wildman
Cas 1h37'17" 60d25'18"
Carol Wolfer
Cap 20h22'44" -26d21'57"
Carol y Xavi
Cap 21h52'39" -17d50'45"
Carol YaYa Camolli
Cas 0h16'46" 62d46'34"
Carol Younger
Lyn 7h42'1" 42d1'5"
Carol Yvonne Mittag
Sco 17h12'58" -35d53'34"
Carol Zug Krott
Lib 15h0'39" -21d43'4"
Carol-0625
Cnc 8h48'16" 18d13'5"
Carola Vallarino Gancia
Ori 6h3'59" 5d51'28"
CarolaJano60
Uma 11h47'21" 64d1'37"
Caroland
Umi 15h3'28" 72d55'37"
CarolAnn
And 1h33'45" 46d16'6"
Carol-Ann Lauzon
Leo 10h36'59" 20d44'19"
Carol-Ann & Lino - Shining Forever
Cyg 21h14'10" 42d28'8"
Carolann - The Brightest Star
And 22h58'49" 39d29'30"
Carolanne
Cap 21h49'56" -9d1'43"
Carolanne
Sco 17h31'40" -30d32'56"
Carolanne I & II
Vir 11h42'29" 3d2'14"
Carole
Psc 1h39'39" 13d12'19"
Carole
Ori 5h5'23" 13d38'50"
Carole
Psc 1h0'5" 28d14'33"
Carole
Crb 16h7'0" 26d1'56"
Carole
Uma 8h45'7" 47d5'13"
Carole
Uma 11h27'32" 51d27'55"
Carole
Gem 7h8'51" 30d30'55"
Carole
Sgr 18h29'19" -27d21'29"
Carole 40
Uma 8h45'42" 61d35'4"
Carole A. Hanzyk
Vir 11h53'16" 6d5'49"
Carole A. Harrison
Cas 23h29'53" 51d50'57"
Carole A. Origer
And 0h11'52" 37d38'47"
Carole A. Ryan
Leo 11h10'44" 7d56'33"
Carole and Bob
Cap 21h2'46" -26d42'1"
Carole and Nadim
Cyg 19h56'23" 40d11'54"
Carole and Phil
Eri 3h43'20" -0d35'30"
CAROLE AND SANDRO
Cas 1h42'54" 58d37'3"
Carole and Wissam For Ever
Uma 9h3'26" 58d39'26"
Carole Anita Mary
And 1h34'41" 50d6'26"
Carole Ann
Uma 11h14'55" 29d24'17"
Carole Ann and Dave's forever star
Umi 15h38'37" 74d25'28"
Carole Ann DeLuca
Leo 10h31'6" 9d26'38"
Carole Ann Friedberg
Umi 9h44'40" 88d55'0"
Carole Ann Tsoukalas
Sco 16h13'11" -10d35'37"
Carole Ann Valle
Sco 16h5'38" -13d3'51"
Carole Anne Mann
Ari 2h40'53" 25d1'33"
Carole "Bellissima" Eitingon
Cnc 8h22'7" 16d51'40"
Carole Benson
Peg 22h53'35" 32d23'16"
Carole CAP Gaspard
Crb 16h21'3" 31d44'6"
Carole Cerka O'Leary
Uma 8h24'6" 66d23'56"
Carole Collar
Uma 9h4'8" 62d24'16"
Carole Dawn
Cyg 20h17'47" 35d43'36"
Carole de la lune mon ami Sultan
Cru 12h24'2" -57d21'49"
Carole Decker
And 0h33'34" 40d31'12"
Carole Deidre Apsley
Cas 0h55'50" 51d59'55"
Carole DeRuiter
Cas 1h45'40" 64d15'59"
Carole Dulaine Carey
Psc 1h15'38" 4d11'44"
Carole E. Baker
And 2h37'3" 45d47'26"
Carole Elaine Wycik
Lib 14h46'1" -8d41'0"
Carole et Loïc
Umi 15h44'1" 88d15'51"
Carole Farley
Sgr 18h51'0" -17d45'56"
Carole Ferrazzini
Cam 7h55'27" 72d16'1"
Carole Gdula Kulwicki
And 23h35'49" 38d25'6"
Carole Harmon
Aqr 21h0'24" -12d32'56"
Carole J. Branda
Oph 17h24'43" -0d36'42"
Carole J. Davis
Uma 11h13'59" 53d49'2"
Carole J. Washburn
Cap 20h14'28" -16d38'55"
Carole Jean Caskin Preston
Umi 15h6'1" 74d9'21"
Carole Jean Pickle
Cas 0h12'42" 53d2'2"
Carole Karen Dobeck
Sgr 19h58'56" -30d26'15"
Carole & Kenneth Jaggers' Star
Lyn 8h16'23" 58d11'25"
Carole Kroon
And 0h54'35" 40d15'6"
Carole Lee
And 0h42'53" 40d52'16"
Carole Leigh Thomas Nordella
Cyg 19h41'40" 37d17'20"
Carole "Lulu" Hanson-Holt
And 1h41'34" 45d53'28"
Carole Patricia Moore - A Special Mum
Cas 1h21'40" 68d44'37"
Carole Philistin
Cap 21h23'35" -16d3'24"
Carole Pie
Cap 20h10'54" -18d25'25"
Carole: Princesse de l'univers
Uma 8h23'44" 62d8'33"
Carole Provenzano
Aqr 21h40'22" -6d14'17"
Carole Pulliam
Ori 5h45'11" 4d41'54"
Carole Reinhard
And 2h24'46" 38d23'41"
Carole & Robert Evans
Cyg 20h52'12" 46d6'58"
Carole Roni Rosenberg
Uma 8h48'45" 65d20'14"
Carole Rowland
Tau 4h36'30" 5d8'23"
Carole Sanfelippo
And 1h40'56" 42d39'58"
Carole Schwabland Kurtz
Aqr 22h35'51" -0d22'37"
Carole Sheklin
Uma 10h16'26" 64d42'31"
Carole Sorchinski
Lyn 7h13'49" 50d39'8"
Carole Stanislawiak
Uma 13h47'42" 60d21'25"
Carole Sue Doscher
Cap 21h29'42" -22d56'7"
Carole Sue Oiler/Bray
Uma 9h57'30" 45d28'32"
Carole T. Dolan
Cyg 20h56'29" 45d27'52"
Carole Tobias Nannes
Crb 16h14'22" 34d55'55"
Carole Verpillon
Uma 9h34'46" 41d51'42"
Carole Walderman
Psc 1h27'42" 26d28'50"
Carole Weinberg
Leo 9h38'39" 29d34'42"
Carole Wilder
Cyg 19h59'6" 34d6'4"
Carole Wright
Cam 6h31'9" 62d7'18"
Carole Young
Psc 0h17'15" 7d24'37"
Carolea Scott Hassard
Sgr 19h28'18" -41d59'32"
Carolee
Tau 4h27'48" 15d47'42"
Carolee Del Re
Lyr 19h8'49" 42d18'30"
Carolee Robinette
Uma 11h57'26" 35d40'40"
Carolee Solt
Dra 13h26'36" 68d20'31"
Carolee Waddell
And 23h40'20" 38d11'30"
CaroleFaith
Leo 11h4'56" 16d33'35"
Caroleo Star In Honor Of Christina
And 1h4'40" 35d31'22"
Carole's Laugh
Aql 19h17'22" 15d26'10"
Carole's Light
Crb 15h39'57" 26d21'30"
CaroleUrsa
Uma 11h23'45" 54d27'20"
Carolgene Evadney
And 1h52'18" 45d3'56"
Carolhoney
Uma 11h19'45" 60d34'28"
Carolin
Ori 5h17'54" 1d38'58"
Carolin Barbara-Anne 24.07.1987
Uma 9h55'24" 51d12'2"
Carolin Jenik
Cas 0h52'22" 60d35'14"
Carolin Maria Schneider
Cas 0h50'16" 73d19'26"
Carolin Waldschmidt
And 23h9'57" 43d9'31"
Carolina
Cnc 8h26'59" 14d26'3"
Carolina
Crb 15h48'28" 27d34'59"
Carolina
Gem 7h48'3" 17d59'47"
Carolina
Ori 6h8'45" 20d40'25"
Carolina
Uma 13h15'1" 62d22'18"
Carolina
Psc 0h24'31" -2d57'50"
Carolina Allison
Uma 9h3'37" 57d38'3"
Carolina Andrea
Ari 2h5'44" 24d41'30"
Carolina Angela Ferrier
Cap 21h28'51" -23d54'59"
Carolina Bauer
And 23h51'39" 44d35'21"
Carolina Behm
And 0h25'44" 38d6'39"
Carolina Beltran
And 0h23'31" 38d16'46"
Carolina Boehly
And 0h19'56" 27d57'6"
CAROLINA C. M. B. DASILVA
Cen 13h35'35" -51d46'49"
Carolina Campodonico
And 1h39'54" 48d59'40"
Carolina Cassani
Cam 4h13'12" 59d19'24"
Carolina Castano's Shining Isabel
Aqr 22h8'54" 2d7'48"
Carolina Cedano
Lyn 6h41'33" 52d49'36"
Carolina Coelho Star Of Freedom
Uma 10h30'15" 65d48'21"
Carolina Cordioli
Boo 14h45'10" 22d52'52"
Carolina E. Reyes
And 23h4'47" 48d26'37"
Carolina Falcone
Aqr 22h27'46" 1d51'1"
Carolina Freitas
Psc 0h55'49" 27d40'48"
Carolina Grace Frisoli
Cyg 19h25'2" 53d8'6"
Carolina Maria Arderia
Sco 16h13'19" -15d59'30"
Carolina Marie
Cyg 19h42'57" 40d11'3"
CAROLINA MENDOZA
Uma 11h10'21" 68d51'7"
Carolina Michael George
And 2h35'18" 49d5'19"
Carolina & Molly Girl
And 0h39'18" 26d11'56"
Carolina Plaza
Lyn 8h52'27" 36d21'53"
Carolina Rodriguez
Ori 5h58'18" 21d58'44"
Carolina Uruena
Psc 23h38'13" 4d42'50"
Carolina Valle Bayly
Peg 22h41'4" 29d11'18"
Carolina Werneck
Cen 14h22'3" -32d59'21"
Carolina Wheeler
Lyn 6h19'54" 58d1'29"
Carolina Wilckens Jimenez
Cru 12h48'56" -58d0'34"
Carolina's Light
Cas 1h30'46" 65d34'22"
Carolinas Neuro-Surgical ICU Star
Ori 5h32'58" 6d23'1"
Carolinda
Ari 3h15'5" 19d31'12"
Carolinda 23rd February 1946
Cas 23h29'19" 51d4'5"
Caroline
Cas 0h14'2" 59d24'16"
Caroline
Cas 1h24'14" 53d11'4"
Caroline
Lyn 8h26'49" 50d23'4"
Caroline
Uma 11h19'29" 36d31'49"
Caroline
Lmi 10h12'50" 30d35'0"
Caroline
Lyr 19h4'13" 32d34'54"
Caroline
Per 4h49'56" 39d23'49"
Caroline
Peg 22h17'7" 22d56'46"
Caroline
Com 12h31'36" 27d38'0"
Caroline
Cnc 8h12'55" 15d24'49"
Caroline
Leo 10h9'25" 21d37'2"
Caroline
Cas 0h42'30" 66d4'0"
Caroline
Uma 10h53'29" 58d23'6"
Caroline
Umi 16h3'21" 77d12'25"
Caroline 06/28/04
Cnc 8h37'35" 8d41'50"
Caroline 20.08.1981
Leo 10h33'21" 6d28'27"
Caroline 21: laetabilis pectus
And 22h58'5" 40d29'40"
Caroline 2285
Cas 23h16'30" 58d12'26"
Caroline 518
Cep 21h28'13" 64d11'22"
Caroline A. Spang
And 23h44'19" 42d25'35"
Caroline Abdel Kader
Dra 17h52'58" 53d17'35"
Caroline Adele McCullough
Cap 20h31'46" -11d48'23"
Caroline Alberta Webbe
Lyn 7h17'34" 45d48'37"
Caroline Alexander
Cap 21h37'25" -18d41'20"
Caroline Alexandra
Uma 11h4'24" 41d46'55"
Caroline Allison Brooks
Cyg 20h9'44" 35d57'50"
Caroline Alves Da Sylva
Sco 16h0'3" -22d56'39"
Caroline Amanda
And 1h11'51" 35d8'32"
Caroline and Andrew's Nuptials
Del 20h18'30" 9d43'39"
Caroline and Dan Forever
Cyg 21h14'7" 46d33'9"
Caroline Andrychowski
Leo 10h47'44" 16d3'6"
Caroline Angel Scibana
Uma 10h29'15" 47d43'21"
Caroline Ann
Crb 16h20'46" 33d34'36"
Caroline Ann Carter
Sgr 18h7'15" -27d19'8"
Caroline Ann Costello 4/13/1994
Ari 3h15'7" 29d11'42"
Caroline Ann Louise Sallee
Leo 11h2'46" 11d20'1"
Caroline Anna Maria
And 1h38'20" 45d53'21"
Caroline Austin Cubberley
Cnc 8h53'56" 14d32'49"
Caroline B Carrie
Sco 16h31'37" -32d30'5"
Caroline Beaudoin-Morin
Aqr 22h34'38" -22d24'33"
Caroline Becknell
Lib 15h8'24" -5d15'33"
Caroline Bell
Cnv 12h35'11" 40d31'4"
Caroline Bethany
Cap 20h10'8" -8d48'27"
Caroline Betsy
Cnc 8h18'11" 19d19'3"
Caroline Birnie - Family & Friends
Hya 8h41'50" -8d38'42"
Caroline Boulware Blue
Cas 0h25'6" 48d25'55"
Caroline Brady 'Our Superstar Mum'
Cyg 19h57'21" 32d41'27"
Caroline Brein
And 23h12'22" 35d48'30"
Caroline Bright Reeder
Lib 15h25'58" -4d54'31"
Caroline C. Bachand
And 23h14'9" 47d11'47"
Caroline C. Ittig
Cyg 20h21'30" 43d31'2"
Caroline Calee Fish
And 1h57'45" 41d28'24"
Caroline *Caro*
Crb 16h10'26" 37d1'3"
Caroline Carter Payne
And 0h17'50" 44d15'9"
Caroline Cibelli Genereux
Aur 6h57'3" 41d30'46"
Caroline Clare Favetti
Aqr 22h23'12" 0d16'49"
Caroline Clemens
Lyn 7h50'46" 40d18'16"
Caroline Clerkin
Cnc 8h58'19" 19d9'46"
Caroline Coleman
Cas 23h5'15" 56d19'56"
Caroline Colletta Trent
Lib 14h37'26" -12d29'8"
Caroline Constance Duckworth
And 23h4'17" 36d5'12"
Caroline Copensky
Lyr 19h7'49" 42d10'57"
Caroline Counts
Vir 11h44'45" 4d57'54"
Caroline Davies
Uma 11h24'43" 59d16'55"
Caroline Davis
Ori 5h35'25" 4d10'28"
Caroline Davis
And 1h33'10" 49d6'0"
Caroline Dawn Oskam
Uma 9h3'2" 51d7'37"

Caroline Dean Doikos Cnc 8h54'50" 13d27'30"
Caroline Doran Peg 21h9'29" 19d24'1"
Caroline Dubanik Lib 15h7'54" -12d58'56"
Caroline E. Bond Uma 9h24'33" 49d58'50"
Caroline Elisabeth Stanton Leo 10h9'26" 20d36'48"
Caroline Elizabeth Chapman And 1h19'4" 34d37'42"
Caroline Elizabeth Harber And 2h21'34" 48d40'20"
Caroline Elizabeth Keating And 2h22'46" 46d23'19"
Caroline et Jérome Psc 1h24'3" 2d50'25"
Caroline et Patrick 5 décembre 2003 Cyg 21h42'49" 44d17'19"
Caroline Eve Locanto And 1h15'55" 45d4'30"
Caroline Eve Ransden Vir 12h29'55" 11d14'4"
Caroline Favre Lyn 7h55'32" 34d6'29"
Caroline Forsyth 9/7/1978 And 23h38'28" 42d54'56"
Caroline Frances Mary Parker Per 4h48'27" 46d48'51"
Caroline G. Grady Crb 16h3'36" 36d42'22"
Caroline G. Weidig Leo 10h40'32" 13d45'9"
Caroline Gallivan And 23h43'18" 37d29'57"
Caroline Gamelin Cap 20h36'14" -25d6'9"
Caroline Gardner And 2h32'58" 48d0'59"
Caroline Gernay Tau 5h47'55" 13d3'21"
Caroline Gochna Psc 1h22'7" 32d21'28"
Caroline Goulette Cap 20h22'43" -11d37'31"
Caroline Grace Aqr 20h44'18" -2d41'9"
Caroline Grace Cunningham Cap 21h43'26" -11d15'27"
Caroline Grace Tinsley And 23h21'33" 39d15'0"
Caroline Grossman Gem 7h33'49" 16d38'37"
Caroline Guerrero Cap 21h36'36" -13d39'37"
Caroline Hannah Rose And 0h24'41" 33d18'26"
Caroline Harkins Cnv 12h50'34" 38d26'39"
Caroline Haurena Uma 9h34'44" 60d34'14"
Caroline Holly Tau 3h39'20" 16d34'56"
Caroline Holsinger Cas 0h44'20" 53d23'27"
Caroline Hopley - Kieran's Love Sco 17h49'28" -32d27'52"
Caroline Hoppe And 0h12'47" 45d6'35"
Caroline Hughes Cap 20h28'39" -11d2'41"
Caroline J. Vines Aqr 22h4'11" -14d51'45"
Caroline & Janet Cnc 8h55'55" 12d30'2"
Caroline Jasmine Kovacs Ari 2h35'58" 30d15'44"
Caroline Jean Rose And 1h31'21" 46d42'14"
Caroline Jean Zyjewski Cas 23h59'54" 57d9'45"
Caroline Josephine Michael Cam 5h19'25" 59d35'29"
Caroline Jungo Dra 20h10'16" 71d42'20"
Caroline Katherine Hill Uma 10h5'41" 65d15'2"
Caroline Kay Leo 10h6'39" 24d46'7"
Caroline Kay Marcum Cnc 9h12'26" 20d58'6"
Caroline Knoblock Cnv 12h39'3" 39d57'14"
Caroline L. Jackson Cam 7h48'49" 66d45'22"
Caroline Ledel Vir 12h33'46" -9d25'8"
Caroline Lehner & Dominik Herzog Lyr 18h52'41" 26d41'1"
Caroline Leigh Holloway Crb 15h46'9" 25d44'42"
Caroline Leigh Senn Aqr 23h0'16" -8d10'44"
Caroline Lewis Cap 21h37'51" -13d28'59"
Caroline Louise Coborn Gem 7h16'17" 34d27'1"
Caroline Louise Martin Gem 6h52'1" 19d23'20"
Caroline Lufkin Cas 1h54'33" 60d2'11"
Caroline M. Hereford Ari 2h17'42" 26d28'23"
Caroline Marcoux Uma 9h36'55" 48d37'32"
Caroline Marie Sgr 18h14'29" -19d11'4"
Caroline Marie Alsop Lyr 19h18'38" 29d49'49"
Caroline & Mark McGann Cyg 21h40'8" 46d6'1"
Caroline & Martin Davies Lyn 9h8'16" 38d47'11"
Caroline & Matthew Lyr 18h28'52" 28d8'41"
Caroline Minh Bui Cnc 8h41'25" 24d54'35"
Caroline Nell Cas 2h47'59" 58d4'11"
Caroline Nicole Pauls Lyr 18h48'15" 34d7'17"
Caroline Niggl Umi 14h28'53" 84d13'23"
Caroline Noble Hoagland Cyg 20h36'35" 54d46'56"
Caroline Noel Kedeshian Eri 4h29'36" -7d37'35"
Caroline Nutter Uma 11h4'38" 51d6'53"
Caroline P James Cnc 8h7'30" 17d1'29"
Caroline Paige Kuduk And 0h13'13" 27d36'58"
Caroline & Patrick Derrick Ara 16h37'21" -58d13'55"
Caroline Perez Lyn 8h0'23" 57d53'28"
Caroline Pia Walsh Tau 5h47'18" 17d11'3"
Caroline Platek Uma 10h51'50" 39d56'10"
caroline poul Peg 22h34'26" 28d48'47"
Caroline pour l'éternité Uma 10h21'35" 58d1'14"
Caroline Queen Vege Sgr 18h27'58" -25d56'29"
Caroline Raney, my Coco Ari 2h40'12" 31d5'15"
Caroline Raviscioni Uma 12h22'53" 53d0'43"
Caroline Rebecca Martin Gem 7h46'4" 26d51'51"
Caroline Reilly Cap 21h24'51" -19d9'51"
Caroline Roberts And 1h14'34" 36d3'45"
Caroline Rosanne Williams Leo 11h23'56" 4d53'51"
Caroline Rose Callaway Sgr 19h53'31" -12d57'14"
Caroline Rose Frey Tau 3h54'17" 25d35'53"
Caroline Rose Nunez Lib 14h49'14" -2d36'8"
Caroline Rose Watson Uma 10h23'39" 47d3'30"
Caroline Rostkowski Lyr 19h2'58" 47d11'41"
Caroline Sarah Gillette Sgr 18h50'36" -16d40'17"
Caroline Sauvan Sco 16h7'41" -23d43'32"
Caroline Schlanger Lib 15h6'16" -6d45'4"
Caroline Schmitt Cnc 8h37'45" 15d42'10"
Caroline Seger ~07/19/1941 Cnc 8h45'57" 31d23'21"
Caroline Seibt Tau 5h45'31" 13d33'33"
Caroline Souter Lib 14h30'18" -12d16'33"
Caroline Squires Cyg 21h32'10" 34d28'55"
Caroline & Stéphane And 23h36'7" 47d46'38"
Caroline & Steve Haddon Cyg 21h32'55" 50d2'50"
Caroline Sue Lindsey Uma 9h46'31" 47d2'48"
Caroline Susan Lasser Sgr 18h21'50" -32d4'2"
Caroline Suzanne Gem 6h33'37" 20d21'1"
Caroline Tamai Dos Reis Jaqua Del 20h52'22" 14d54'6"
Caroline Tew Mon 6h32'41" 7d45'28"
Caroline The World's Brightest Star Lyr 19h7'52" 26d34'10"
Caroline Thomas Gem 6h21'59" 19d45'50"
Caroline Turchi Cas 1h13'15" 65d4'2"
Caroline W Willits And 0h37'50" 43d30'23"
Caroline Williams Gray Mon 6h38'31" 11d22'5"
Caroline Zarza Mourzou Lyn 7h17'26" 57d26'58"
Carolinea Luminosus Cas 0h51'3" 60d49'6"
Caroline-Anne Cyg 20h49'44" 33d8'59"
CAROLINE-MARTINE Ori 5h53'19" 12d23'8"
Carolines Ausblick Uma 12h18'22" 58d23'56"
Caroline's Beauty And 0h17'29" 42d19'28"
Caroline's Destiny Love Star And 2h38'28" 42d26'30"
Caroline's Evening Glory Cas 23h11'59" 58d18'25"
Caroline's Eyes Leo 9h47'56" 29d2'53"
Caroline's Green Eyed Night Cnc 8h31'4" 28d36'33"
Caroline's Most Beautiful Star Cma 6h48'8" -23d1'50"
Caroline's Serendipity Gem 6h54'0" 28d14'33"
Caroline's Smile Cap 20h36'43" -14d25'13"
Caroline's Star Lmi 9h52'4" 33d48'19"
Caroline's Star, Love you Princess x And 23h56'6" 36d2'42"
CAROLINKA Cyg 20h58'34" 30d26'37"
Carolin's Haven - 26 February 1966 Cru 12h31'36" -60d4'53"
Carolita Crb 15h25'56" 27d54'48"
Carolmae Aqr 21h39'26" 1d53'17"
CarolNedStar Cyg 20h7'17" 44d1'25"
Carol's 65th Birthday Star Dra 16h50'45" 54d39'31"
Carol's Celestial Chaos Lyn 9h7'6" 34d24'54"
Carol's Diamond in the Sky Sco 16h11'26" -28d52'40"
Carol's Dream Aqr 23h7'1" -17d34'40"
Carol's Heart And 1h14'51" 40d53'19"
Carol's Light Cap 20h24'18" -13d28'9"
Carol's Shining Light Cas 0h50'56" 57d39'28"
Carol's Shining Star And 0h40'40" 32d21'29"
Carol's Shinning Star Psc 0h2'50" 4d39'9"
Carol's Sparkle Leo 11h17'49" 16d5'38"
Carol's Star Ori 6h7'27" 17d32'26"
Carol's Star Cyg 20h58'53" 31d10'31"
Carol's Star And 23h40'36" 37d1'40"
Carol's Star Cap 20h20'54" -10d21'0"
Carol's Star Aqr 22h49'38" -21d15'7"
Carol's Star "I Love You" Umi 13h30'54" 74d7'39"
Caroly R Helmuth Uma 9h7'16" 60d43'1"
Carolyn Uma 10h44'6" 70d46'42"
Carolyn Uma 11h31'49" 57d50'2"
Carolyn Cap 21h3'49" -22d0'22"
Carolyn Lib 15h28'1" -12d6'19"
Carolyn Aql 19h47'51" -0d16'10"
Carolyn Sco 16h40'11" -30d59'46"
Carolyn And 1h43'22" 42d20'15"
Carolyn And 2h6'53" 43d28'12"
Carolyn Cnv 12h43'20" 36d47'24"
Carolyn Crb 15h53'49" 39d7'10"
Carolyn Lyn 7h15'57" 50d38'7"
Carolyn Cnc 8h8'10" 16d45'4"
CAROLYN Ari 1h53'39" 24d15'58"
Carolyn Leo 11h36'31" 26d7'17"
Carolyn Tau 4h41'9" 23d58'52"
Carolyn Aqr 22h16'43" 1d25'21"
Carolyn Vir 14h15'59" 3d44'45"
Carolyn A. (Irish) Cook Umi 15h15'23" 67d33'17"
Carolyn A. Rojas Sgr 18h51'47" -28d26'5"
Carolyn A. Winkelman Ari 2h48'26" 11d7'50"
Carolyn Abney Uma 10h17'25" 51d45'19"
Carolyn Adams Short Leo 10h6'0" 9d42'18"
Carolyn Adin-Owens Copstead Leo 11h4'9" 20d16'49"
Carolyn Agosta & Bill Rauschert Cyg 21h19'20" 32d37'18"
Carolyn Alice Anderson Com 12h40'51" 28d11'8"
Carolyn Alyssa Tri 2h12'40" 34d41'16"
Carolyn Amelia Coffey Psc 0h57'40" 32d47'9"
Carolyn and Daniel Beehler Aql 20h0'35" 8d56'36"
Carolyn and Jason And 1h47'7" 49d35'25"
Carolyn Anderson Cas 0h53'40" 54d21'54"
Carolyn Ann And 1h38'18" 47d10'27"
Carolyn Ann Crb 15h32'26" 28d49'7"
Carolyn Ann Becton Lee Uma 9h41'50" 70d45'10"
Carolyn Ann Geraci And 0h41'41" 32d55'54"
Carolyn Ann Hanson Oeth Aqr 21h3'23" -10d22'34"
Carolyn Ann Kack Vesecky Peg 21h33'0" 17d23'45"
Carolyn Ann McCanna Psc 1h33'4" 21d11'52"
Carolyn Anna Calvert Grimes Crb 15h46'37" 30d59'30"
Carolyn Anna Creager Uma 9h37'19" 42d10'7"
Carolyn Anne And 23h36'31" 49d32'57"
Carolyn Anne Psc 1h32'30" 10d39'53"
Carolyn Anne Uma 8h56'0" 68d32'16"
Carolyn Anne McKell Sgr 18h2'8" -32d7'11"
Carolyn Anne Mendez-Luck Gem 6h33'55" 26d34'2"
Carolyn Anne Vernon Uma 11h58'49" 40d1'21"
Carolyn B Vir 13h18'1" -22d0'53"
Carolyn B. Glumack Ori 5h9'47" 15d15'25"
Carolyn B. Jackson Gem 6h41'11" 24d12'32"
Carolyn B. Kiel Uma 10h39'27" 57d5'56"
Carolyn "Babie" Angeles Gem 7h14'50" 19d33'14"
Carolyn Babygirl Kiskinis Sco 17h23'37" -37d45'28"
Carolyn Bayliss RDS225 Ari 2h52'15" 14d35'25"
Carolyn Benfield Carter Psc 0h59'31" 15d45'52"
Carolyn Blau AKA Star Lyn 7h47'49" 41d44'17"
Carolyn Borrelli Ari 3h25'3" 29d50'0"
Carolyn & Breck 022102 Psc 1h17'27" 6d52'20"
Carolyn Bridget Rhody And 1h48'15" 42d6'14"
Carolyn Brook Cas 23h41'57" 51d41'39"
Carolyn Brown Lib 15h39'7" -6d27'15"
Carolyn Burkhimer Star of Hope Leo 11h35'21" 27d9'10"
Carolyn C. Holley Cam 5h33'49" 62d40'2"
Carolyn C. Vaught Cam 3h34'46" 65d26'20"
Carolyn C3 DeSanto Lyn 7h10'3" 58d10'18"
Carolyn Campbell Cas 2h23'47" 59d24'9"
Carolyn "Care" Goodman Psc 1h9'54" 4d38'30"
Carolyn "CareBear" Welch Vir 13h4'49" -18d28'24"
Carolyn Carlson Uma 9h44'46" 44d0'3"
Carolyn Carney Uma 9h42'58" 55d7'11"
Carolyn & Carolyn - Friends Forever Cru 12h39'29" -59d41'11"
Carolyn & Christian Aqr 22h30'36" -0d46'24"
Carolyn Clarke Cru 12h18'31" -58d55'27"
Carolyn Cole Dunlap Cas 0h29'51" 58d29'1"
Carolyn (Cookie) Guerra Tau 5h48'5" 17d8'7"
Carolyn Corbet Uma 12h0'28" 38d11'13"
Carolyn D. Wheeler Ari 3h27'57" 28d55'33"
Carolyn Dennis Willingham And 0h26'48" 32d34'8"
Carolyn Diann Gralewski Cma 6h33'2" -16d54'1"
Carolyn Dilg Crb 16h10'58" 34d49'17"
Carolyn Dornstauder Ari 3h21'13" 12d17'25"
Carolyn Dover And 2h16'51" 49d37'46"
Carolyn E. Adey Smiling Bright Sgr 17h59'11" -17d1'17"
Carolyn E Doerner Del 20h41'49" 15d19'40"
Carolyn E Egers Umi 15h56'39" 74d6'56"
Carolyn E. Kelly Sgr 19h43'43" -15d21'47"
Carolyn Eisenschmid Lee North Sco 16h12'40" -11d9'59"
Carolyn Elaine Wachtstetter Sco 16h17'35" -15d34'10"
Carolyn Elizabeth Imel Star Aqr 21h33'38" 1d48'37"
Carolyn Elizebeth Harty Tau 4h7'27" 27d10'10"
Carolyn Eziemefe And 0h27'11" 33d1'34"
Carolyn F. Worthington Aqr 22h56'33" -8d43'4"
Carolyn Fredette Crb 16h15'42" 39d19'20"
Carolyn G. Cnc 9h13'57" 25d6'50"
Carolyn G Rahder Cam 4h11'8" 70d14'41"
Carolyn Gail Psc 1h27'20" 17d18'20"
Carolyn Gail McNish Vir 12h28'18" -5d31'47"
Carolyn Grace Kahn And 1h48'14" 38d17'14"
Carolyn Green Leo 10h32'30" 13d41'15"
Carolyn Green Crb 16h0'44" 26d34'17"
Carolyn Hedinger Cnc 9h18'4" 10d14'29"
Carolyn Hedrick Uma 9h34'34" 69d12'5"
Carolyn Helen Umi 14h7'38" 67d59'4"
CAROLYN HONG Lmi 10h25'7" 35d32'16"
Carolyn Hope Ellis And 23h32'10" 35d58'18"
Carolyn Hopson Sonic 1 And 0h31'15" 38d2'13"
Carolyn Hunt Psc 1h21'8" 30d57'29"
Carolyn J. Harris I love you Mom Uma 11h5'46" 62d36'52"
Carolyn J. Jackson Cas 23h45'42" 53d1'42"
Carolyn J. Kennedy Leo 11h19'48" 11d21'45"
Carolyn J. Pickett Cnc 8h42'16" 18d37'5"
Carolyn J. Sweeney Cas 0h17'38" 59d2'17"
Carolyn J. Wooton Lyn 7h34'41" 55d20'12"
Carolyn Jane Grundy Ari 1h49'7" 19d23'50"
Carolyn Jane Shea *CJ* And 1h51'6" 45d47'45"
Carolyn Jane Woodley Cas 1h16'47" 55d5'38"
Carolyn Jean Chapel Cap 20h33'53" -15d27'56"
Carolyn Jewell Dix Uma 9h40'42" 43d7'12"
Carolyn Jill Duff Leo 10h58'45" 14d7'56"
Carolyn Joan Towle Gem 7h29'47" 26d45'5"
Carolyn Johannes Cas 1h20'19" 56d26'8"
Carolyn Joyce Granger Whitlock Cas 23h12'1" 58d51'2"
Carolyn Joyce Hall Walker Com 12h4'41" 19d2'43"
Carolyn Joyce Way Savage Del 20h30'39" 18d18'56"
Carolyn June Jackson And 23h39'53" 37d4'56"
Carolyn Katz Lyn 8h10'14" 37d51'50"
Carolyn Kay Ramsey Sgr 18h26'38" -17d54'45"
Carolyn Keen And 0h15'44" 27d24'58"
Carolyn Kinnison Cnc 9h5'24" 17d14'10"
Carolyn Kollar Ori 5h38'10" -0d55'59"
Carolyn L. Giddinge Leo 11h37'30" 27d44'33"
Carolyn L. Sanders Cas 2h41'46" 57d37'17"
Carolyn L. Wade-Jones Crb 15h36'19" 26d9'34"
Carolyn Laura Cas 0h49'7" 48d31'51"
Carolyn Lee And 1h9'8" 44d8'33"
Carolyn Lee Hyde Cyg 21h11'56" 47d40'15"
Carolyn Leigh - Lovies Forever Cyg 19h48'30" 42d37'2"
Carolyn Lock And 0h7'18" 45d21'54"
Carolyn Louise McKay Barbitta Lib 14h54'47" -16d44'13"
Carolyn Louise Theriot Rose Uma 8h53'26" 62d59'5"
Carolyn Luck Mon 7h16'31" -9d5'14"
Carolyn/Lynn Komar Aqr 23h38'31" -7d54'51"
Carolyn M. Callahan Uma 11h55'41" 54d21'20"
Carolyn M Perry Vir 14h15'44" 1d4'18"
Carolyn M. Stiegman Aqr 22h17'28" 0d24'29"
Carolyn "Mama" Keel Cap 21h34'47" -20d25'10"
Carolyn "Mama's Magic" Ari 2h47'30" 21d33'22"
Carolyn Mann Cap 21h30'39" -9d2'14"
Carolyn Maree Gelme Cru 11h56'46" -60d20'59"
Carolyn Margaret Wilcox Reid Crb 15h43'55" 37d47'2"
Carolyn Marie Cas 0h58'32" 50d22'59"
Carolyn Marie Jeffers Cam 5h37'38" 58d39'7"
Carolyn Marie Leskowsky Gray And 23h14'16" 37d12'57"
Carolyn Marie Rossetti Uma 8h13'49" 63d29'55"
Carolyn Marshall Uma 9h39'58" 47d33'31"
Carolyn Mary Stone Mon 5h59'20" -6d25'6"
Carolyn Matthews Sco 17h41'23" -35d41'13"
Carolyn Merz-Mothernature Crb 15h40'11" 26d11'46"
Carolyn Miyoko Shinsato Cas 1h14'0" 56d31'21"
Carolyn Montgomery Cas 1h33'7" 66d36'54"
Carolyn Moore Cam 7h21'23" 64d17'56"
Carolyn Morrison Cyg 20h29'27" 52d53'41"
Carolyn N. Sink Gem 6h17'48" 26d49'16"
Carolyn & Nestor And 1h12'28" 42d18'32"
Carolyn Nicole Aql 18h55'18" -0d43'4"
Carolyn Patricia Hill Sgr 18h17'15" -22d36'44"
Carolyn Patricia Wetterholt Cnc 8h52'22" 31d46'20"
Carolyn Perrine Leo 11h35'14" 9d12'5"
Carolyn Pesterfield Cnc 8h14'28" 30d32'1"
Carolyn Pluta Aqr 22h8'49" -0d48'23"
Carolyn Protz Lib 14h35'42" -12d26'59"
Carolyn R. Burkart's Lucky Star Uma 11h4'0" 54d15'30"
Carolyn R. Dehorty Leo 10h57'58" 5d28'12"
Carolyn R. Hall Vir 12h44'18" 2d54'2"
Carolyn Ramsey Cyg 20h36'48" 48d54'28"
Carolyn Randolph Hammond And 1h22'41" 44d39'18"
Carolyn Regina And 0h38'50" 38d32'20"
Carolyn Renee Smith Vir 12h5'48" 7d3'15"
Carolyn Richards Shining Star Cap 20h47'24" -20d45'53"
Carolyn Rodman Lib 15h3'34" -29d6'56"
Carolyn Rose Farren Boo 14h50'5" 35d20'42"
Carolyn S. Myers Aqr 21h39'2" 0d51'13"
Carolyn S Rogers And 0h31'25" 28d26'10"
Carolyn 's star And 23h51'23" 44d40'4"
Carolyn Schwarz Tisdale Lyn 8h17'19" 46d55'45"
Carolyn Shearer Per 3h3'55" 48d21'20"
Carolyn Shiflett Cap 20h22'49" -22d52'50"
Carolyn Smith Mon 8h6'17" -0d37'39"
Carolyn Snyder Kline Sgr 18h17'27" -24d9'46"
Carolyn Spears Detrick Aqr 22h23'35" -5d36'59"
Carolyn St.Dennis Root Cas 1h25'21" 58d1'5"
Carolyn Stouffer Cas 1h27'56" 57d58'36"
Carolyn Sturgis Ori 6h8'4" 20d52'34"
Carolyn Sue And 1h57'33" 45d40'11"
Carolyn Sue Ori 5h55'42" -0d41'41"
Carolyn Sue Sgr 18h30'38" -18d42'29"
Carolyn Sue 1 Sco 17h14'34" -31d14'1"
Carolyn Sue Mason Crb 15h35'47" 28d39'17"
Carolyn Sue Nerdrum Cas 0h42'41" 53d37'35"
Carolyn Sue Pate Boner Aqr 22h28'39" -21d20'49"
Carolyn "Sue" Sidman Cyg 19h56'56" 54d21'23"
Carolyn T. Dessain Uma 11h3'14" 35d20'5"
Carolyn Tannehill Ari 2h10'29" 24d10'56"
Carolyn Theresa Becher Hunt Tau 4h43'11" 19d21'58"
Carolyn Thomas Christy Cap 20h7'44" -20d53'52"
Carolyn Thomas Julian Sgr 18h13'17" -17d47'10"
Carolyn Tolley ASI Health Services And 2h11'58" 38d25'8"
Carolyn Tucker Vir 13h42'27" 2d34'11"
Carolyn Velchik And 23h10'32" 48d12'1"
Carolyn Wilkinson's Love Star Cas 1h26'47" 62d5'38"
Carolyn Wolf Jelliffe's Star Umi 15h10'56" 73d31'22"
Carolyn Yendral Harman Mon 6h53'10" 1d28'29"
Carolyn, My Star Ori 5h50'19" 12d19'57"
CarolynE Zimmer ScottJ Cody11.14.04 Lmi 10h30'20" 38d45'58"
Carolyne-A Kimahaba Vir 12h46'48" 8d10'48"
CarolynMichelle Lyn 6h22'21" 61d15'15"
Carolynn Ann Blair Cas 0h10'0" 56d19'12"
Carolynn Ann Marchena 2/9/1954 Aqr 22h36'5" 2d9'35"
Carolynn Forever Ori 6h21'33" 15d42'27"
Carolynn Harris Cas 0h32'25" 50d19'15"
Carolynn J. Cerul Gem 7h40'20" 23d8'46"
Carolynn Starr Francavilla Mic 20h55'57" -31d53'5"
Carolynne & Garrett Griffin Per 3h17'55" 31d33'8"
Carolynnus Potteria Vir 13h27'36" 12d10'43"
Carolyn's Celebritas Juventus And 23h12'51" 42d47'25"
Carolyn's Essence Uma 10h28'33" 47d45'44"
Carolyn's Jewel Psc 23h33'16" 6d48'16"
Carolyn's Love Uma 12h0'9" 47d23'0"
Carolyn's Nebula Cru 12h48'41" -58d7'3"
Carolyn's Scott Tau 4h11'54" 10d2'2"
Carolyn's Star And 0h39'34" 31d10'31"
Carolyn's Star And 1h0'27" 40d36'48"
Carolyn's Twinkling Star Mon 6h54'55" -0d15'51"
Carolyn's Wish Cap 21h6'10" -24d0'37"
Carolyn's Wishing Star Psc 0h55'53" 9d10'22"
Caro-Matth for ever Cap 21h5'51" -22d50'13"
Caron Campbell Cnc 7h57'25" 12d34'57"

Caron Kay Hausinger
Lmi 10h43'37" 37d55'28"
Caron's Star
Peg 23h57'54" 27d58'53"
Carosella
Equ 21h2'7" 4d25'52"
Carpachito
Crv 12h5'38" -14d15'53"
Carpe Diem
Psc 23h50'51" -1d9'23"
Carpe Diem
Sco 16h12'29" -12d35'9"
Carpe Diem
Cas 0h57'43" 66d43'15"
Carpe Diem
Ari 2h48'16" 26d52'57"
Carpe Diem
Ori 5h56'30" 18d40'42"
Carpe Diem 4/23/67
Cam 4h32'48" 67d27'43"
Carpe Diem Philippe
Richebois
Cas 3h14'25" 66d52'22"
Carpe Diem, Always
Cet 1h6'24" -0d41'12"
Carpe Mecum Sempiterne
Noctem
Ari 2h9'59" 12d45'50"
CARPEDIEM
Per 4h26'5" 42d3'5"
Carpenter
Lyn 7h20'56" 54d16'25"
Carra
Vir 12h56'28" 0d21'24"
Carrah is Awesome
Gem 6h40'17" 20d43'11"
Carrandon
Mon 6h31'7" 8d44'57"
Carrapicho
And 1h10'9" 46d44'16"
Carre de Malberg
Cas 23h36'10" 52d47'15"
"Carrejo"
Ori 5h30'15" 9d28'38"
Carrel
Cyg 20h11'24" 41d42'23"
Carren Caraan Henderson
Leo 10h12'32" 15d42'58"
Carren Ward
Vir 12h9'15" 9d48'40"
Carren's Guiding Light
Uma 13h50'24" 52d13'38"
Carri Ann
And 1h48'44" 42d45'16"
Carri Ann Cheever
Her 17h53'48" 24d3'17"
Carrie
Tau 5h35'39" 24d13'57"
Carrie
Ari 2h32'27" 25d53'11"
Carrie
Ari 2h28'35" 24d14'47"
Carrie
Del 20h43'22" 16d36'8"
Carrie
Ori 6h17'42" 14d49'14"
Carrie
Aqr 21h32'39" 1d0'45"
Carrie
Vir 12h8'8" 7d23'18"
Carrie
Ori 6h8'53" 2d17'32"
Carrie
Cnc 8h52'55" 30d28'50"
Carrie
Uma 11h24'56" 53d51'45"
Carrie
Aqr 22h39'12" -0d57'21"
Carrie
Cap 21h56'44" -14d52'14"
Carrie
Lib 14h48'59" -18d5'19"
Carrie 21
And 0h21'26" 26d31'54"
Carrie - A Shining Star
Forever!
Leo 9h44'20" 21d44'33"
Carrie A. Steen
And 1h16'5" 45d38'12"
Carrie Abrams
Cyg 19h35'52" 29d9'4"
Carrie Ana 72
Ori 6h8'58" -1d24'41"
Carrie and Clint
Quisenberry
Mon 6h41'5" 7d9'28"
Carrie and Jim
And 0h7'35" 45d19'39"
Carrie Andrews
Lyn 8h2'31" 59d5'14"
Carrie Angel
Cyg 21h36'7" 46d26'41"
Carrie Ann
Cas 0h46'13" 52d59'15"
Carrie Ann
And 0h54'40" 37d34'24"
Carrie Ann
Vir 12h43'30" 7d13'19"
Carrie Ann
Uma 10h36'7" 54d38'1"
Carrie Ann
Cas 1h30'23" 68d31'30"
Carrie Ann Bayer
Per 3h9'15" 51d49'16"
Carrie Ann Cabell
Vir 12h58'11" -17d29'25"
Carrie Ann Dennison
Cnc 8h52'8" 30d23'23"
Carrie Ann Hall
Vir 13h45'28" -5d14'1"
Carrie Ann Hinojosa
Ari 2h27'34" 19d59'52"
Carrie Ann Hurley
Vir 12h28'51" 5d5'42"
Carrie Ann Jasiak
Gem 6h45'48" 15d52'5"
Carrie Ann Powell
Cnc 9h7'29" 27d59'14"
Carrie Ann Rodgers, A Star
Mommy
Uma 10h6'35" 54d1'54"
Carrie Anne
Cap 21h31'53" -14d57'16"
Carrie Anne Stribling
Uma 11h44'38" 29d46'26"
Carrie Atkinson
And 1h3'27" 38d27'15"
Carrie B. Warren
Psc 0h30'59" 8d46'9"
Carrie (Baby)
Lyr 19h18'9" 28d43'27"
Carrie Banty
Cnc 9h8'43" 14d53'29"
Carrie Bartoletti
Ori 6h10'13" -0d17'39"
Carrie Becky Milner
Robertson
Tau 4h21'34" 20d16'49"
Carrie Bedingfield
And 0h30'40" 23d25'57"
Carrie Beth Heffel
Vir 12h3'59" 6d4'19"
Carrie Bobek
Sco 16h15'23" -12d22'37"
Carrie Breann
And 23h24'2" 47d40'11"
Carrie Brougher's Guiding
Light
Cap 20h24'50" -16d27'54"
Carrie "Carebear" Lynch
Sgr 19h5'26" -23d59'17"
Carrie Carruth
Lmi 9h46'24" 40d33'39"
Carrie Celeste Uy
And 1h32'28" 40d49'50"
Carrie Christine Wolski
BBA
Uma 9h50'52" 57d14'36"
Carrie Conlen Manz
Vir 12h31'6" 10d9'31"
Carrie Crye
Uma 11h6'9" 58d27'21"
Carrie & Dan Sparrow
Col 5h59'22" -29d46'39"
Carrie Danielle Richardson
And 23h49'49" 38d18'21"
Carrie+Dave
Tau 5h31'57" 27d21'19"
Carrie Dawn Parker
Psc 0h45'19" 16d59'0"
Carrie Delane Price Knight
Mon 6h27'21" -1d8'59"
Carrie Denning
Vir 13h51'52" -1d31'48"
Carrie Diven
Cyg 21h54'17" 42d38'58"
Carrie Doege
Crb 16h17'18" 30d38'49"
Carrie E. Imlah
Cap 21h56'22" -14d8'18"
Carrie E. Kindrix
Ori 4h55'48" -0d38'29"
Carrie Elaine Emory
Vir 12h47'42" 8d13'1"
Carrie Elizabeth
Cyg 20h56'15" 45d35'47"
Carrie Elizabeth Dudley
And 0h5'11" 43d16'46"
Carrie Eylander
Ori 5h37'18" 3d12'57"
Carrie G. Del Vecchio
Sgr 18h17'39" -23d48'7"
Carrie George Johnston
Vir 11h48'9" -3d6'58"
Carrie Grace
And 0h14'33" 32d32'24"
Carrie Hatcher "The Carrie"
Ari 3h26'13" 20d22'51"
Carrie Hepburn
Sgr 18h6'46" -22d22'20"
Carrie J Moniz
Sgr 18h31'17" -15d58'23"
Carrie J., With love from
your sons
Leo 11h43'0" 26d1'52"
Carrie Jack
And 2h38'26" 43d43'32"
Carrie Jackson
Gem 6h44'54" 19d58'32"
Carrie & Jeff
Cyg 20h52'3" 35d5'36"
Carrie Jo
Lmi 10h52'24" 28d43'20"
Carrie Jo
Psc 22h51'27" -0d1'43"
Carrie Jo Bockbrader A.k.a
Sexy
Tau 4h30'20" 18d47'38"
Carrie Joy Creedle
And 0h46'55" 34d33'3"
Carrie Julia Ryall
Leo 11h36'5" 26d34'5"
Carrie Kablean
Cra 18h7'35" -37d12'53"
Carrie Kate Shanafelt
Ari 2h38'16" 13d41'11"
Carrie Kay
Vir 13h14'42" -16d53'45"
Carrie Kinsler
Cas 1h32'51" 64d7'55"
Carrie Knific
Uma 11h47'10" 39d33'30"
Carrie Koch
Gem 6h49'37" 18d9'14"
Carrie L Lagg
Cap 20h57'49" -19d28'15"
Carrie L. Reichert
Sco 17h4'4" -37d38'25"
Carrie Lane ~ Letzter Engel
Ari 3h7'43" 15d48'25"
Carrie Lanier Willingham
Cnc 8h56'45" 11d11'18"
Carrie Lee
Ori 5h27'51" 7d27'56"
Carrie Lee
Psc 1h45'48" 24d44'36"
Carrie Lee
Lmi 10h31'26" 31d38'33"
Carrie Lee
Lib 14h59'19" -12d38'53"
Carrie Lee 2-23-03
Cnc 8h48'13" 7d9'29"
Carrie Lee "Matakia"
Cyg 21h11'49" 51d48'37"
Carrie Lee Paddeck
Lyn 7h29'46" 45d37'29"
Carrie Lee Pleasant
Cnc 8h11'59" 9d2'21"
Carrie Lee Whitaker
Uma 12h1'53" 46d55'35"
Carrie Leigh Kraemer
Apu 14h26'13" -78d53'34"
Carrie Lienhard
Sgr 18h17'26" -29d35'28"
Carrie Lightizer
Leo 11h10'53" 8d36'46"
Carrie Linton
Ori 6h3'30" 17d50'0"
Carrie Louise Dziejma
And 2h23'16" 47d57'18"
Carrie Louise Taylor
And 23h34'50" 51d49'6"
Carrie Lyn Hansen
Vir 15h6'3" 4d6'51"
Carrie Lynn
Mon 6h43'16" 7d13'11"
Carrie Lynn
Ari 1h56'50" 21d19'45"
Carrie Lynn
Uma 9h44'11" 70d41'28"
Carrie Lynn Bleigh
Lib 15h52'40" -19d56'2"
Carrie Lynn Diehl
Lmi 10h34'27" 38d20'3"
Carrie Lynn Holley
Sgr 18h48'25" -26d57'48"
Carrie Lynn Middleton
Tau 5h45'24" 18d0'25"
Carrie Lynn Tripolone
Vir 13h58'48" -15d15'6"
Carrie Lynn Uhl
Cnc 9h17'15" 10d14'47"
Carrie Lynn Wyatt
Lyr 18h29'48" 36d26'58"
Carrie M Bell
Vir 11h57'38" 2d39'17"
Carrie M. Kopplin
Cnc 9h9'47" 31d38'49"
Carrie Mae Sass
Aqr 21h12'59" -3d16'48"
Carrie Margret Eckenrode
Cas 0h44'15" 62d22'5"
Carrie Marie
Cap 21h47'34" -18d19'34"
Carrie McCoy
Umi 19h15'52" 87d5'6"
Carrie Michelle Brinkley
Aqr 22h9'33" -24d2'38"
Carrie Nicole Weber
Sco 17h57'16" -38d44'12"
Carrie Noel Medici
And 1h43'58" 49d8'5"
Carrie Purvis
Cas 2h19'40" 68d53'9"
Carrie R. Fielder
Cas 0h36'29" 48d27'18"
Carrie Rae
Vir 13h16'18" 11d16'19"
Carrie Ray Smith...Daniel
12:3
Uma 11h8'50" 50d30'32"
Carrie Rhodes
Aqr 21h54'16" 1d54'50"
Carrie Rochelle Bain
Cas 0h47'39" 54d13'3"
Carrie Simin
Cap 21h35'8" -14d10'56"
Carrie Stern
Cyg 20h17'57" 45d45'59"
Carrie Sue
Aqr 21h9'28" 1d44'39"
Carrie Sue
Lib 15h12'49" -15d58'36"
Carrie Taylor
And 0h2'16" 48d4'11"
Carrie Taylor ~Loves~
Timothy Ware
Cyg 21h21'29" 44d53'20"
Carrie Tomovich
And 0h32'13" 42d40'48"
Carrie Totashnick
Cyg 20h28'6" 47d56'53"
Carrie Triola
Leo 11h24'27" 8d24'12"
Carrie Tyler Robertson
Cyg 21h41'27" 41d35'52"
Carrie Van Brunt - Ursa
Lyn 7h48'45" 38d39'34"
Carrie Weigand Kuester
Klumb
And 1h36'4" 43d44'18"
Carrie Weir
Lib 15h45'26" -19d34'2"
Carrie Wright
Sgr 19h45'0" -13d5'32"
Carrie Zigler
Sgr 18h4'26" -27d1'15"
Carrie, Brad, Keagan, and
'Benny'
Lib 15h25'13" -6d46'45"
Carrieanna-Boo
Cas 0h32'30" 67d18'48"
Carrie-Anne Baker
Cyg 19h31'8" 30d5'0"
Carrie-Anne Williams
And 1h26'56" 34d30'6"
Carrie-Beth
Cnc 9h10'35" 22d0'43"
CarrieCarmenHaleyMary
Umi 15h57'2" 86d15'30"
CarrieDavid
Uma 11h53'15" 35d15'47"
Carriedog
Cma 6h51'47" -20d43'17"
Carrientommy Forever
And 0h56'47" 34d41'24"
Carriers Dream
Col 5h48'39" -34d56'12"
Carries Comet
Cyg 21h33'43" 46d36'51"
Carrie's Diamond
And 23h29'48" 43d1'58"
Carrie's Fate
Psa 22h22'46" -31d40'32"
Carries "My childhood
dreams"
Sge 20h6'46" 16d45'19"
Carrie's Star
Leo 9h38'13" 30d27'5"
Carrie's Star
Gem 7h17'49" 32d46'35"
Carrie's Wink
Aqr 21h40'33" 0d17'27"
Carrigan
And 23h25'27" 48d0'50"
Carrigan Leigh Sherlock
Lyn 7h56'7" 33d57'3"
Carrigan's Faith
Vir 12h24'11" 12d29'49"
Carrington Madison Mead
Tri 2h24'53" 37d2'50"
Carrington's Corbon
Aqr 21h42'54" -0d42'58"
Carrin's Star
And 0h28'6" 37d50'10"
Carrisa
Cnc 9h18'57" 24d31'18"
Carrissa
Cap 21h13'20" -14d55'47"
Carrissa Larsen
Tau 4h9'58" 28d23'56"
Carr-Moreithi
Uma 8h41'11" 63d58'50"
Carroll & Charlotte
Cyg 19h41'59" 28d30'2"
Carroll Elliott Buras
Uma 12h31'13" 57d47'11"
Carroll F. Costner
Uma 9h18'24" 52d29'41"
Carroll Family
Uma 8h14'4" 64d18'17"
Carroll Homer Ruinon
Uma 14h12'34" 55d39'55"
Carroll Houston
Cas 2h46'41" 65d26'40"
Carroll & Justin McCarthy
Aqr 22h7'14" -1d14'0"
Carroll L. Spangler
Cas 1h35'16" 64d16'23"
Carroll Leslie Gross IV
Uma 9h11'22" 51d27'2"
Carroll star 72
Uma 14h27'46" 56d30'29"
Carroll-Colarelli
Crb 15h31'44" 30d23'56"
Carroney
Uma 11h47'50" 43d28'42"
carrot
Vir 14h29'31" 4d2'13"
Carrots =)
Vir 13h4'48" 11d47'6"
Carrubin
Ori 6h10'56" 20d48'42"
Carryadd Lesley
Cas 0h28'14" 67d20'29"
Car's Corner Of The Sky
Cap 20h23'12" -14d35'37"
CarSandra
Uma 9h30'39" 47d40'19"
Carsen John Wiley
Dra 17h37'42" 67d30'12"
Carsim
Sco 16h41'40" -25d5'39"
Carson
Per 3h35'41" 50d33'14"
Carson
Tau 4h29'2" 28d19'33"
Carson Alexander White
Psc 0h49'40" 17d17'40"
Carson Alton Floyd
Umi 16h49'57" 79d7'0"
Carson and Claire Jones
Uma 11h22'35" 28d44'52"
Carson Cade
Umi 14h4'41" 71d41'30"
Carson Catherine
And 1h22'47" 45d1'51"
Carson Chandler Roy
Uma 10h5'36" 65d20'55"
Carson David Gaspar
Umi 15h16'54" 69d3'40"
Carson Diarro Culter
Sco 17h31'26" -34d37'4"
Carson Donley Spindler
Ori 6h4'56" 21d17'48"
Carson Evan Bloom
Aur 5h41'3" 41d41'24"
Carson Glenn Smoak
Leo 11h58'17" 12d27'40"
Carson Hacker
Her 16h34'54" 36d42'19"
Carson Hayes
Uma 9h59'14" 48d56'5"
Carson Hicks
Umi 16h54'33" 78d9'18"
Carson Jo Silverthorn
Aqr 23h2'53" -24d47'29"
Carson Joseph Potter
Uma 11h15'18" 29d4'30"
Carson King Diamond
Tau 5h44'58" 25d49'50"
Carson Lee Mann
Cnv 12h48'14" 42d35'54"
Carson Loren Kinsch
Ori 6h24'53" 15d46'8"
Carson Matthew
Mazurkiewicz
Aql 20h3'13" 11d51'23"
Carson McClain VanWinkle
Ari 1h47'35" 24d26'10"
Carson McDonald
Tau 5h5'36" 16d40'37"
Carson Michael Daman
Aqr 21h6'32" 1d30'1"
Carson Micheal Janes
Umi 16h47'27" 81d10'18"
Carson Michele Stainback
Cas 0h36'49" 51d32'11"
Carson Myers
Per 3h10'46" 49d32'24"
Carson Patrick Sauter
Vir 13h16'55" 6d43'53"
Carson Randolph Wilkie
Dra 18h56'38" 64d55'20"
Carson Riley Hollimon
Cnc 9h14'52" 23d17'47"
Carson Riley Salyers
Cap 20h12'11" -18d27'24"
Carson Ryan Carrow
Lib 15h26'43" -24d37'24"
Carson Schroeder
Lyn 7h27'29" 44d37'41"
Carson Walter Kotula
Tau 3h46'59" 24d5'21"
Carson Wellman
Leo 10h42'41" 17d56'41"
Carson William Kubczak
Cnc 8h5'44" 11d27'37"
CarsonJ513
Tau 4h10'54" 5d5'40"
Carsonolie
Umi 14h47'59" 75d3'11"
Carson's Christmas Star
Uma 11h32'10" 53d43'17"
Carson's Star
Per 3h45'33" 37d19'19"
Carsons514
Tau 5h58'12" 25d44'7"
Carsta
Uma 9h55'1" 51d23'25"
Carsten
Uma 11h33'43" 29d6'7"
Carsten Brandt
Uma 8h16'16" 62d12'55"
Carsten & Iris
Uma 10h1'46" 61d34'20"
Carsten mein Engel
Uma 11h34'8" 32d52'12"
Carsten & Nina
Uma 9h55'43" 57d13'16"
Carsten Winter
Uma 11h2'42" 46d15'57"
Carstensen, Michael
Uma 9h58'54" 62d21'58"
Carsyn Ainsley Fink-
7/24/05
Equ 21h17'27" 8d45'10"
Carter
Cnc 8h21'5" 10d59'4"
Carter
Her 17h10'19" 49d4'34"
Carter
Aur 5h54'0" 43d34'12"
Carter
Uma 8h58'55" 54d20'37"
Carter
Sgr 18h9'21" -19d17'34"
Carter
Pav 21h6'32" -63d14'32"
Carter Aiden Peligri
Ori 5h46'56" 12d11'49"
Carter Alexandra Dutton
Cnv 12h38'19" 32d34'16"
Carter and Caleb Eskridge
Cnc 8h41'22" 25d22'34"
Carter Bay Audino
Cnc 9h14'2" 14d43'51"
Carter Bean
Ori 5h24'36" 9d9'25"
Carter Blake
Vir 14h12'27" 4d59'0"
Carter Douglas Barber
Umi 15h31'40" 74d37'4"
Carter Edward Duncan
Cap 20h33'38" -12d19'53"
Carter Evan Lynn
Aql 19h23'38" 8d38'49"
Carter Evan Mckie
Ser 18h35'46" -0d26'9"
Carter Evan Moore
Aql 20h14'34" 3d0'50"
Carter Everet Smith
Her 18h22'6" 21d23'59"
Carter Gene Tyson
Aur 5h3'53" 34d10'54"
Carter Harrison Wilson
Gem 7h28'54" 19d2'42"
Carter Heyward Morris
Her 16h36'28" 5d53'3"
Carter J. Davis
Cnc 8h59'24" 10d39'13"
Carter Jackson Martin
Ori 5h30'58" 6d31'33"
Carter James
Leo 10h50'39" 14d10'54"
Carter James Brown
Cnc 9h7'23" 31d30'19"
Carter James Forsyth
Tau 3h57'42" 17d5'0"
Carter James Sasser
Ori 5h50'32" 6d15'12"
Carter Joe
Sco 16h5'44" -13d58'27"
Carter John Anderson
Her 17h42'42" 36d53'6"
Carter John Lowery
Ori 6h3'44" 7d15'6"
Carter Joseph Rubin
Lib 15h30'0" -16d22'11"
Carter Joseph Willey
Vir 12h33'52" 11d11'41"
Carter Lane Bevins
Uma 10h59'2" 47d50'22"
Carter Lee Gunnells
Uma 11h42'54" 39d6'31"
Carter Lee Wentworth
Lanman
Sgr 18h54'35" -19d10'57"
Carter MacNaughton
Simons
Her 16h46'57" 7d37'49"
Carter Martin
Her 16h8'4" 24d4'32"
Carter Matthew LeBlanc
Her 16h47'30" 13d4'31"
Carter McClure
Ari 2h17'41" 24d42'48"
Carter Michael Rolka
Psc 1h54'58" 8d12'21"
Carter Moore
Umi 16h16'13" 74d30'36"
Carter Nash
Ari 2h8'47" 23d5'10"
Carter Noah
Lyn 7h25'27" 48d52'50"
Carter Owens
Aqr 22h11'37" -7d31'49"
Carter Paul Douglas
Hinderliter
Ori 5h34'52" 3d4'27"
Carter Pierce
Uma 9h41'7" 58d17'22"
Carter R. Belair
Cep 22h59'48" 76d22'42"
Carter Reid Landis
Per 4h1'27" 36d24'53"
Carter Richard Klein
Umi 14h31'13" 73d39'44"
Carter Rodeen's Radiant
Lyn 6h22'40" 57d2'9"
Carter Ryan
Sco 16h17'36" -13d23'11"
Carter Rylan Smith
Lyn 9h10'58" 34d53'44"
Carter Smith
Uma 10h59'7" 54d28'45"
Carter Star Floyd and
Wanna Forever
Lyr 18h43'4" 39d57'0"
Carter Thomas
Sacharewitz
Cyg 19h39'28" 55d33'0"
Carter Wayne Cameron
Uma 9h18'17" 57d18'36"
Carter Yoon
And 2h37'11" 40d22'15"
Carter's Bit of Heaven
Uma 11h55'3" 51d55'18"
Carter's Dream
Ari 2h29'7" 17d54'58"
Carter's Star - Heather's
Godson
Cnc 8h50'13" 29d41'19"
Carter-William's Gift
Uma 10h28'34" 39d28'36"
Cartier
Gem 7h22'7" 33d41'27"
Cartier La-Falaise Hebert
Psc 1h39'12" 10d49'9"
Carudah
Ori 5h54'43" 18d34'58"
Carus
Crb 16h8'33" 35d11'26"
Carus Ken Fabbre
Leo 9h27'1" 22d47'59"
Caruso MD
Per 3h8'41" 52d47'4"
CarvEliSydEmAmanJuliAnn
aJPreK2007
Uma 10h15'2" 46d14'57"
Carver Elementary
Uma 11h7'44" 53d18'59"
Carver Martin
Aqr 22h2'53" 1d21'48"
Carverkarsar
Peg 21h47'17" 7d19'14"
Cary
Cnc 8h44'36" 18d34'9"
Cary
Sco 17h53'25" -36d32'21"
Cary Alan and Michele
Lynn Bakker
Lyr 18h46'58" 33d55'8"
Cary and Marissa
Sgr 17h53'57" -29d45'49"
Cary Collado
Lyn 7h15'52" 59d8'56"
Cary Grant Fais
Tau 4h25'21" 23d40'38"
Cary Ingram Clark
Aqr 22h27'6" -12d54'55"
Cary Jack Bowen D.M.C
Umi 16h53'52" 78d52'44"
Cary Johnson
Aql 19h56'27" -0d28'24"
Cary Kitty Powers
Ari 2h34'7" 31d11'11"
Cary Marie Robinson
Cap 21h31'18" -16d29'27"
Cary McCutchen
Vul 19h30'51" 26d11'0"
Cary Philip Gavan
Psc 1h22'25" 15d23'40"
Cary S. Washburn
Uma 11h29'54" 46d16'43"
CarYen
Mon 6h40'19" -0d6'7"
Caryl Payne Bray
Uma 10h24'24" 47d36'50"
Caryl Wallace
Ori 6h10'46" 8d34'7"
Caryl Weinstein
Cam 3h25'56" 65d50'58"
Caryllinda
Per 3h27'9" 44d12'4"
Caryn
Vir 13h21'59" 12d51'19"
Caryn
Del 20h31'38" 16d38'53"
Caryn 40
Leo 11h10'59" 23d35'52"
Caryn Antos
Tau 4h8'2" 15d31'45"
Caryn Beth Johnson's Star
Leo 11h22'39" 0d5'24"
Caryn Crane
And 0h47'10" 30d51'49"
Caryn Daughter of The
Promise
Lib 14h50'50" -8d7'16"
Caryn Denise Howell
Cyg 20h59'20" 33d15'44"
Caryn E. Bishop
Psc 1h20'21" 28d51'40"
Caryn Elizabeth Wilson
Aqr 22h33'28" -8d29'32"
Caryn Lillo
Peg 0h3'38" 28d6'0"
Caryn M. Lakeman (1971 -
2007)
Gem 7h8'17" 33d55'12"
Caryn Rebecca's Passion
Leo 9h25'5" 15d47'17"
Caryn-Brian
Per 3h39'43" 40d52'56"
CARYNS' STAR
Leo 11h45'54" 21d39'2"
Carys
And 23h0'48" 48d21'12"
Carys
Dra 18h32'54" 65d41'42"
Carys
Uma 12h42'19" 56d53'28"
Carys Morgan Francis
Cyg 20h31'38" 38d17'23"
Carys Noelani Storm
Ori 5h31'1" 1d16'43"
Carys Star
And 1h24'51" 49d8'58"
Carys' Star
And 2h30'52" 43d42'13"

Carys Violet Louise Byford
And 2h30'24" 49d51'40"
Carys-Grace
And 23h46'7" 37d43'2"
CAS
Vir 14h9'55" -16d40'19"
Casa de Plaza
Uma 11h43'56" 36d36'30"
Casa DeVita
Per 4h37'29" 40d6'50"
Casa Rondota
Uma 11h6'26" 53d0'26"
Casa Savageau
Cas 0h12'25" 53d30'29"
Casablanca
And 22h58'42" 50d40'56"
Casalena-Knight
Cyg 19h40'0" 51d29'53"
Casalino
Uma 9h23'5" 66d43'21"
Casandra
And 23h9'38" 47d35'18"
CaSandra
Tau 5h8'11" 25d23'48"
Casandra Ann Friesner
Lmi 10h29'55" 31d39'16"
Casandra Anne Richards
Ari 3h27'52" 23d12'53"
Casandra D'Nae
Mon 7h20'21" -4d16'54"
Casandra James
Aqr 21h2'20" 1d10'46"
Casandra Lynn Burnham
Lib 15h4'32" -2d17'22"
Casandra Maria Sandoval
Lyn 7h2'25" 51d10'43"
Casanna's Light
Psc 1h13'46" 27d8'8"
Cascapuce
Uma 11h37'21" 30d36'22"
Caschew
Ori 5h28'39" 2d55'44"
Cascio
Crb 15h52'29" 38d12'46"
CASE
Sgr 19h48'54" -35d18'1"
Case Star
Uma 11h11'42" 32d2'29"
Casella
Sgr 19h4'27" -14d57'18"
Caselli Simone
Her 17h17'49" 48d15'41"
Caseman
Lyn 8h20'50" 37d27'3"
Casexy Casto
Leo 11h47'58" 19d14'49"
Casey
Ari 2h25'57" 24d53'6"
Casey
Her 17h39'52" 15d17'49"
Casey
Lmi 10h37'8" 29d18'51"
Casey
Leo 10h28'36" 12d52'58"
Casey
Lyn 7h30'55" 35d55'34"
Casey
Lyn 8h37'6" 39d14'16"
Casey
Uma 11h14'34" 47d24'2"
Casey
Cas 1h23'55" 57d35'11"
Casey
Cyg 21h33'35" 39d10'16"
Casey
Lib 14h41'19" -14d50'16"
CASEY
Tau 3h29'57" -0d54'15"
Casey
Dra 18h45'44" 63d2'52"
Casey
Uma 11h6'8" 70d33'10"
Casey
Uma 10h41'40" 67d34'17"
casey
Uma 13h0'47" 53d51'7"
Casey
Uma 13h13'30" 53d14'2"
Casey
Pup 7h6'25" -41d27'37"
Casey
Cap 21h21'58" -27d4'2"
Casey#1
Cnc 9h16'40" 10d4'36"
Casey A. Kermond
Lilsleepybunny
Ori 6h19'5" 8d40'9"
Casey Abramson- Always a
star
Aql 19h26'39" 12d15'19"
Casey Alexander
Kretzschmar
Dra 16h4'49" 68d15'24"
Casey Alexandria Howard
Tau 3h44'4" 26d1'45"
Casey and Jacob Forever
Mon 6h47'8" -2d58'16"
Casey and Jen
Cam 4h20'33" 65d42'51"
Casey and Josh Dollahan
Umi 14h22'22" 73d57'13"
casey and scott
Uma 10h9'16" 57d34'23"
Casey Ashton Granata
Lib 15h48'32" -19d40'31"
Casey Austin
Lyn 6h45'55" 53d4'45"
Casey (Beautiful)
And 0h53'33" 44d24'53"
Casey Brainerd
Cap 20h27'20" -12d34'25"
Casey Brianne Mashburn
Lib 14h52'16" -18d31'12"
Casey Brooke Splinter
Dra 18h45'20" 53d16'41"
Casey - Brown Eyes
Cyg 19h35'2" 31d38'43"
Casey Butler
Uma 8h24'42" 69d13'16"
Casey Carma Owens
Sgr 18h52'22" -19d10'7"
Casey Casavant
Cnc 8h8'34" 14d31'17"
Casey Chojnacki
And 0h42'31" 38d15'0"
Casey Coleen Hewson
Leo 11h56'55" 25d26'21"
Casey Craig Conner
Uma 10h59'22" 63d33'14"
Casey Craugh
Mon 6h36'46" 8d8'21"
Casey Cronin
Sco 16h17'45" -34d39'19"
Casey Dayan
Mon 8h5'6" -0d34'57"
Casey Deanne Thompson
Cru 12h57'6" -58d28'0"
Casey DeWitt
Sgr 18h47'16" -27d13'28"
Casey ~ Duwatt
Uma 11h57'48" 46d47'41"
Casey Dwayne Gibson
Sco 17h52'2" -43d2'15"
Casey East Trest
Uma 11h23'29" 38d9'55"
Casey Elizabeth
Cap 20h46'5" -15d50'21"
Casey Elizabeth Bennett
And 0h10'17" 43d21'57"
Casey Elizabeth Legeyt
Cyg 19h47'55" 35d6'1"
Casey Elizabeth Wig
Tau 3h43'55" 27d34'17"
Casey Erin Noll
Cyg 19h37'37" 51d41'35"
Casey Eugene Griffith
Equ 21h1'18" 9d26'50"
Casey Evan Payne
Her 17h18'20" 36d50'35"
Casey Francis
And 23h8'19" 35d22'9"
Casey Gene
Aql 20h36'46" -0d54'49"
Casey Gene Naylor
Leo 11h45'30" 19d41'1"
Casey George Anderson
Psc 2h3'4" 10d7'58"
Casey Gerard Sweeney
Vir 14h41'30" -2d19'12"
Casey Gillen
Uma 14h21'49" 56d24'10"
CASEY H. BENNETT
Vir 13h39'41" -15d32'16"
Casey Haynes
Dra 16h35'13" 66d16'51"
Casey Hedden
Cap 20h13'47" -10d31'25"
Casey Hoffman Carlesimo
Gem 7h13'26" 16d18'27"
Casey Holland
Cmi 8h2'20" 0d12'36"
Casey Hynes
Gem 7h13'24" 33d41'35"
Casey Is The Greatest
Umi 13h47'54" 78d40'34"
Casey J
Psc 0h35'55" 20d2'5"
Casey James Potters
Ari 3h22'52" 22d32'42"
Casey Jane Harmon
Crb 15h41'50" 36d30'43"
Casey & Janet
Uma 14h16'4" 62d12'59"
Casey Jay Bloyd
Gem 7h34'19" 27d23'15"
Casey Jean
And 23h12'33" 37d49'35"
Casey Jewell
Sgr 18h52'12" -36d30'54"
Casey Jo Ralph
Leo 9h28'41" 27d11'43"
Casey Johnson
Lyn 7h59'38" 42d34'7"
Casey Jones
Umi 14h59'29" 85d16'8"
Casey Jordan
Per 2h52'27" 51d39'8"
Casey Justin Howard Smith
Tau 4h37'21" 18d38'19"
Casey Kalmenson
Tau 5h30'9" 25d6'59"
Casey & Kristine
Sgr 17h52'3" -17d34'30"
Casey Kuhns
Sco 17h51'54" -35d56'28"
Casey I Costello
Her 17h16'56" 19d45'44"
Casey Lane Moore
Gem 6h51'28" 27d8'5"
Casey Lanier Meeks
And 23h0'21" 48d42'12"
Casey Lawson
Uma 12h39'0" 59d3'44"
Casey Lee
Uma 11h1'30" 61d35'10"
Casey Lee Heggtveit
Leo 10h5'10" 25d13'46"
Casey Lee Micajah Mitchell
Her 16h57'40" 27d51'11"
Casey Lee Robinson
Sgr 19h50'18" -13d2'7"
Casey Leigh Ashly
Roucher
Sgr 19h12'5" -21d36'21"
Casey Lemura
Sgr 18h25'1" -16d7'8"
Casey Loughman
Vir 12h54'54" 2d12'56"
Casey Lumia
Uma 9h12'16" 62d53'27"
Casey Lyn Casper
Mon 6h49'52" 8d32'49"
Casey Lynde
Umi 16h18'46" 76d30'10"
Casey Lynn
Uma 10h51'50" 68d10'3"
Casey Lynn
Sgr 18h3'55" -26d40'18"
Casey Lynn Bellew
Leo 10h10'44" 21d49'0"
Casey Lynn Conklin
Cas 23h42'29" 52d57'10"
Casey Lynn Costa
Peg 23h15'11" 17d37'6"
Casey Lynn Elstad
Gem 6h13'20" 26d41'58"
Casey Lynn Leggitt-Malloy
Vir 13h21'9" 8d24'41"
Casey Lynn Miller
Uma 10h18'46" 49d34'50"
Casey Lynne
Lib 15h7'42" -1d24'57"
Casey M. Rogina
Leo 10h26'5" 21d53'51"
Casey Mae Mooch
Gem 7h5'7" 23d19'33"
Casey Marie
Sgr 19h11'49" -32d59'13"
Casey Marie Marcus
Cyg 21h46'38" 40d12'19"
Casey Marie Parker
Mon 8h7'30" -0d32'23"
Casey Marie Poore
Cap 21h47'53" -16d9'51"
Casey Mattson
Cnc 8h25'27" 13d13'42"
Casey Michael Cohen
Sgr 18h10'50" -28d29'44"
Casey Michael Moen
Gem 6h53'15" 22d41'0"
Casey Michael Sullivan
Ori 6h1'10" 6d21'53"
Casey Michelle Herger
Cap 20h11'7" -25d28'38"
Casey Michelle Lindsey
Uma 9h15'20" 68d9'17"
Casey 'Munkie' Word
031804
Uma 8h32'11" 61d56'50"
Casey Nicole Walker
Ori 5h41'33" 8d4'34"
Casey Noel
Lib 15h28'18" -24d26'51"
Casey Nolan
Aqr 22h43'36" -9d1'49"
Casey O'brien
Sco 16h7'33" -27d54'52"
Casey Oliver Smith
Dra 17h32'6" 68d1'9"
Casey O'Neill
Cnc 7h59'28" 12d22'26"
Casey Ott
Umi 17h10'1" 75d15'43"
Casey Padilla
Sco 16h6'20" -9d25'6"
Casey Patrice Willette
Leo 10h25'59" 26d50'50"
Casey Patrick Hynes
Uma 8h37'16" 46d52'53"
Casey Paul
Tau 4h9'23" 18d23'37"
Casey Peach
Lib 15h4'9" -3d17'11"
Casey Pirnik
Aql 19h52'18" -0d35'41"
Casey Ray McCoy
Uma 10h52'1" 67d14'23"
Casey Renee' Loftin
Lyn 6h18'56" 58d44'0"
Casey Richardson
Crb 15h39'39" 32d30'0"
Casey Schultz
Uma 9h58'50" 55d49'32"
Casey - Seasquared777 - 7
- 1760
Uma 11h35'40" 46d57'27"
Casey Sherman Rolfes
Leo 10h59'14" 5d43'20"
Casey Shifflett
Ori 6h14'34" 15d0'15"
Casey Shira Tsabari
Tau 4h21'31" 16d59'32"
Casey Stevens
And 23h10'41" 43d15'53"
Casey & Stuart
Umi 16h36'10" 79d18'6"
Casey Sue Scanlon
Psc 0h23'25" -1d14'48"
Casey Sullivan
Gem 7h11'38" 32d14'42"
Casey Tayler Pearce
Gem 7h45'55" 32d52'52"
Casey Taylor
Ori 6h16'23" 13d32'24"
Casey The Love of My Life
Cyg 21h36'19" 31d35'35"
Casey Therese McKay
Cnc 8h41'8" 17d50'1"
Casey Tyler Duckworth
Uma 11h49'58" 29d43'1"
Casey Uldricks Crescendo
Lyn 7h4'24" 52d5'3"
Casey Walden
Tri 2h33'12" 33d31'3"
Casey Ward
Uma 8h53'2" 64d38'2"
Casey Watkins
Sco 16h10'7" -12d0'4"
Casey Wesolowski
Uma 11h37'45" 50d22'3"
Casey White
Aql 19h0'49" 9d27'38"
Casey William Ference
Cnc 8h49'59" 32d42'58"
Casey Wujkowski
Cyg 21h16'37" 44d47'1"
Casey-Lyn Bird
Aql 19h49'46" 5d33'48"
Casey-Lyn, Greg, and
Dakota
Vir 11h47'37" -6d28'7"
Casey's
Oph 17h8'27" -0d7'28"
Casey's
Crb 15h36'43" 27d17'7"
Casey's Light
Per 3h17'54" 42d30'10"
Casey's Light
Lib 15h46'45" -24d30'53"
Casey's Place-Catch a
Falling Star
And 2h8'20" 43d42'38"
Casey's Rock Star
Dra 18h56'16" 61d50'40"
Casey's Star
Uma 10h19'31" 63d41'21"
Casey's Star
Cnc 8h50'30" 9d39'57"
Casey's Star 3 and 8
Cnc 8h39'24" 15d16'30"
Casey's Wish
Uma 12h45'37" 57d20'10"
Casey's Wishing Star
Leo 11h11'10" 16d26'46"
Cash Czyszczon
Aqr 23h13'28" -13d9'58"
Cash Emerson Palmateer
Umi 14h32'24" 85d34'26"
Cash Jackson Durao
Vir 13h33'13" -8d40'23"
Cash & Sofia & Maryann's
Star
Umi 15h31'19" 77d52'1"
Cash Star
Umi 14h53'30" 76d31'21"
Cash the Whitcomb Moore
Vel 8h34'10" -44d17'7"
Cash Warrick Walsh
Pochynok
Cas 2h41'47" 68d18'23"
Cashawn Kiki Kirbas
Vir 13h50'34" -8d12'43"
Cashly
Vel 10h41'36" -56d40'58"
Cashman-Lento
Cnc 9h19'13" 14d13'55"
Cashmere Joy Walker
Cas 1h12'41" 67d20'25"
Casi
Cmi 7h32'3" -0d9'12"
Casi
Aql 19h17'31" -10d48'24"
Casi & Mimi Platzer Hotel
Victoria Ritter
Cas 0h22'0" 55d27'45"
Casia
And 0h14'27" 32d5'42"
Casie
Vir 13h45'58" 2d12'51"
Casie & Andy Wodzien
Lib 14h56'59" -24d12'46"
Casie L Gendron
And 0h55'36" 41d6'1"
Casie Lynn
Uma 9h26'21" 50d6'8"
Casie Lynn
Per 3h22'33" 45d0'21"
Casie Milam "2-28-93"
Cas 1h39'21" 61d16'33"
Casie Nadine
Leo 9h42'45" 9d16'17"
Casimir G. Oksas
Gem 7h7'39" 24d6'41"
Casimir Joseph Wolnowski
Sco 17h55'50" -34d39'10"
Casimir, Günther
Uma 13h4'28" 52d50'16"
Casimira F. Walker
Vir 11h56'34" -0d10'28"
Casimiro y Socorro
Psc 1h8'25" 20d19'47"
Casini Marco
Peg 22h41'11" 24d0'41"
Cason
Ari 2h46'45" 22d7'16"
Cason
Uma 11h15'1" 59d24'36"
Caspar Neamand
Her 17h27'6" 46d25'35"
Casper - Capper Too too
Cep 20h40'51" 61d39'20"
Casper Jerome
Cnc 8h42'21" 31d16'38"
Casper Yanker
Uma 10h34'32" 69d10'44"
Cass
Umi 14h47'40" 68d5'16"
Cass A Hillstrand
And 0h42'46" 39d0'34"
Cass and Janet DAmato's
Hope Star
Cas 1h26'43" 66d6'56"
Cass919
Vir 14h44'20" -0d10'54"
Cassandra
Lib 14h54'50" -6d3'28"
Cassandra
Aqr 21h49'30" -0d19'55"
Cassandra
Sco 16h12'24" -14d1'31"
Cassandra
Aqr 22h50'59" -7d49'22"
Cassandra
Dra 16h30'10" 57d10'49"
Cassandra
And 0h59'29" 39d57'38"
Cassandra
Uma 10h56'12" 48d40'26"
Cassandra
Cas 0h27'12" 56d44'35"
Cassandra
Cas 1h21'54" 52d22'53"
Cassandra
Psc 0h29'32" 15d49'48"
Cassandra
Psc 23h7'20" 7d30'54"
Cassandra
Aqr 22h39'19" 0d51'27"
Cassandra
Gem 7h37'54" 24d20'59"
Cassandra
Ari 3h15'25" 29d45'35"
Cassandra
Cnc 8h42'33" 18d24'12"
Cassandra
Cnc 9h19'11" 18d53'57"
Cassandra Ahlers
Cas 23h42'6" 59d45'37"
Cassandra AKA DordieBob
Ori 6h3'31" -0d46'44"
Cassandra Alesha
Huizenga
Psc 1h42'16" 18d35'0"
Cassandra Alice Stuart
And 0h26'32" 38d11'37"
Cassandra and Hendrik
Lammers
Dra 17h59'55" 59d11'57"
Cassandra Anderson
Cas 0h2'10" 53d35'0"
Cassandra & Andrew
Cru 12h47'2" -56d33'37"
Cassandra Ann
Ari 2h39'22" 14d25'19"
Cassandra Ann
Ari 2h53'27" 26d50'19"
Cassandra Anne Huber
Gem 7h20'34" 14d11'22"
Cassandra Anne Scotto
Lavina
Sgr 19h9'37" -18d2'4"
Cassandra & Anne-Louise
Roll
Uma 9h51'44" 49d40'29"
Cassandra Aragona
Cas 0h23'20" 51d0'54"
Cassandra Armstrong
Tau 4h33'18" 21d47'43"
Cassandra Bonita Lazenby
Cnc 8h56'30" 13d15'33"
Cassandra Brielle
Lib 14h36'58" -13d54'15"
Cassandra "Cassie"
Perpich
Sgr 20h25'34" -37d49'15"
Cassandra "Cassie" Sund
Uma 9h20'50" 71d0'53"
Cassandra "CC" Kinnaman
Ari 2h41'2" 25d23'27"
Cassandra Charles
Cas 0h9'14" 57d10'18"
Cassandra Cole
Lib 15h24'6" -8d11'40"
Cassandra Colleen Powell
Psc 0h56'20" 10d16'39"
Cassandra Conway
Cyg 19h32'59" 30d12'28"
Cassandra Daniella
Bodden
Cas 23h55'25" 56d27'18"
Cassandra Denise Floyd
Vir 12h4'36" 7d21'10"
Cassandra Elizabeth
Gordon
Cyg 21h9'54" 46d3'25"
Cassandra Fielder
Cyg 19h50'24" 50d40'19"
Cassandra Frances
Lib 15h24'17" -5d23'46"
Cassandra G
Cap 21h31'15" -13d49'33"
Cassandra Gillespie
Ari 2h16'32" 23d2'1"
Cassandra Gonzales
Psc 23h18'29" 7d5'50"
Cassandra Grace Noah
Ori 6h13'52" 6d31'52"
Cassandra H Contreras
1/03/2003
Cas 1h26'32" 52d52'1"
Cassandra Hawkins
Aqr 22h48'22" -21d42'39"
Cassandra Irene Elizabeth
Thurow
And 2h19'18" 45d28'24"
Cassandra Jane
Gem 7h36'10" 33d13'41"
Cassandra Jane Hardy
Ori 5h20'59" -6d42'17"
Cassandra Joelle Lakatos
Lyn 7h1'47" 45d34'23"
Cassandra Kay Berg
Vir 12h0'42" 4d30'21"
Cassandra LaDawn
Uma 11h16'45" 68d9'21"
Cassandra Lanay
Tau 4h41'48" 16d18'52"
Cassandra Ledessa 77
Sco 16h49'26" -31d38'25"
Cassandra Lee
Tau 5h6'22" 20d58'22"
Cassandra Lee Pridemore
Per 3h17'33" 45d36'55"
Cassandra Levin
Lyn 7h40'46" 56d14'38"
Cassandra (lilymarie) Lewis
Ari 2h1'27" 17d37'55"
Cassandra "Love Bunny"
Favero
Ori 5h44'24" 12d25'26"
Cassandra L.P. Hammett
Cas 2h22'38" 72d36'55"
Cassandra Lucy Carson
Lyr 18h41'13" 30d8'18"
Cassandra Lyn
Tau 5h58'43" 25d20'19"
Cassandra Lynn
Leo 10h34'17" 13d12'55"
Cassandra Lynn
Cas 0h57'47" 62d59'1"
Cassandra Lynn Breffitt
Uma 9h21'48" 55d46'40"
Cassandra Lynn Remaley
Aqr 22h50'5" -22d15'21"
Cassandra Lynne Delli
Carpini
Sco 17h57'24" -43d29'43"
Cassandra M.
Uma 11h25'50" 45d53'16"
Cassandra M Connell
Psc 1h9'40" 3d57'53"
Cassandra M Pardee
Leo 10h15'18" 8d29'41"
Cassandra Mariani
Cnc 8h25'23" 21d8'5"
Cassandra Marie
Ori 5h21'37" 6d48'8"
Cassandra Marie Cohen
And 1h38'37" 41d25'18"
Cassandra Marie Fortune
Tau 4h26'34" 17d23'46"
Cassandra Marie Garcia
Ari 2h9'52" 12d56'8"
Cassandra Marie Johnson
Sco 16h4'53" -13d25'56"
Cassandra Marie Nichols
Sco 17h6'51" -38d12'11"
Cassandra Marie Roio
And 1h32'28" 42d29'3"
Cassandra Marie Switzer
Cnc 9h10'13" 15d43'31"
Cassandra Marie Tamez
And 1h10'57" 46d16'43"
Cassandra Marie Yvette
Lambros
Psc 1h57'3" 8d7'27"
Cassandra Marion Thoburn
Lyn 8h3'4" 57d59'45"
Cassandra McCandliss
Cas 1h36'48" 65d19'59"
Cassandra McMillen
Cas 0h19'46" 57d29'19"
Cassandra Michelle
Cas 1h9'52" 62d2'3"
Cassandra Morgan
Ari 2h59'5" 31d6'13"
Cassandra Murchison
Aqr 22h54'1" -8d17'43"
Cassandra My Wife
Lyr 18h50'0" 39d8'53"
Cassandra Nichols
Cas 0h29'7" 51d14'15"
Cassandra Nicole
Cas 0h3'1" 50d10'24"
Cassandra Nicole
Choquette
Cet 1h7'10" -0d22'45"
Cassandra Olivia
And 0h34'51" 33d36'30"
Cassandra Olivos
Sgr 18h13'51" -30d57'29"
Cassandra (Our Baby)
And 0h56'11" 24d19'32"
Cassandra Paige
Uma 11h39'39" 64d17'14"
Cassandra Pardee
Tau 4h24'35" 13d5'14"
Cassandra Peddy
Psc 1h16'1" 6d22'51"
Cassandra Pough
Ori 5h53'37" 19d1'44"
Cassandra Rae
Vir 12h40'6" -9d8'39"
Cassandra Rae
Deatherage
Aqr 22h59'53" -11d22'53"
Cassandra Ray
And 23h29'52" 36d30'59"
Cassandra Raylene
Ari 2h37'59" 19d9'56"
Cassandra René Depew
Vir 13h40'53" -3d20'29"
Cassandra Renee
Psc 1h19'0" 15d55'17"
Cassandra Renee Stough
Cas 1h41'55" 63d20'15"
Cassandra Roach
Cru 12h0'20" -63d8'10"
Cassandra Robertson
Lyn 7h3'23" 50d52'37"
Cassandra Rose Kendall
And 23h11'40" 35d24'11"
Cassandra Roy
And 2h24'7" 48d7'23"
Cassandra "Sassy" Conder
Cas 0h46'14" 50d13'41"
Cassandra Starr Ballard
Leo 9h36'5" 10d21'46"
Cassandra Sue Katzer
Lib 15h4'35" -0d30'44"
Cassandra Sue Levers
Cas 0h17'56" 59d33'7"
Cassandra The Great
Aqr 22h36'8" -0d29'52"
Cassandra Viater
Ari 2h36'6" 13d36'22"
Cassandra Victor
Aqr 22h14'34" 1d18'32"
Cassandra Your Guiding
Light - Lucky
Cap 21h44'8" -16d37'0"
Cassandra Zografos
And 0h31'44" 37d23'32"
Cassandra-angel
Lib 15h32'38" -13d37'26"
Cassandra's Star
Leo 9h56'22" 14d42'9"
Cassandre
Mon 6h50'2" -0d11'6"
Cassandre
Dra 17h29'13" 66d37'18"
Cassandria Laliana
Cnc 8h55'39" 11d14'12"
Cassarah Consuelo Marie
Farson
Lyr 18h30'44" 36d47'54"
Cassaundra Rose Prillhart
Ari 2h37'39" 25d10'48"
Cassedy
Uma 10h37'4" 63d58'36"
Cassee Stem
Psc 1h24'24" 21d26'38"
Cassell Star
Umi 14h51'21" 75d37'11"
Cassels
Cyg 19h36'5" 34d27'33"
Cassey
Psc 0h40'25" 15d21'39"
Cassey Lynn Watkins
Leo 9h23'13" 32d50'29"
Cassi
Aqr 23h4'43" -9d33'46"
Cassi Alise
Cas 0h34'41" 55d58'25"
Cassi Honey
Uma 8h57'29" 60d25'40"
Cassi the Goober
Cam 6h24'30" 62d54'22"
Cassia Hills
Leo 11h38'59" 26d0'16"
Cassiana - estrela
brasileira
Uma 9h51'22" 44d9'1"
Cassidi Grace Summers
Leo 11h34'46" 16d27'45"
Cassidia
Cas 23h2'19" 56d7'44"
Cassidy
Gem 7h15'20" 15d43'31"
Cassidy
Vir 12h6'40" 8d12'55"
Cassidy
Lyn 8h36'11" 37d25'13"
Cassidy
Cas 0h59'7" 55d7'16"
Cassidy Amber Klier
Gem 6h36'42" 15d6'14"
Cassidy and Nichole Moore
Peg 21h13'25" 13d8'21"
Cassidy Anne Brennan
Uma 8h40'19" 55d13'4"

Cassidy Autumn Schmidt
Lyn 7h57'48" 42d10'28"
Cassidy Belle Maher - Bright Star
Cru 11h58'41" -55d46'42"
Cassidy Bonk
Cma 6h36'52" -12d15'29"
Cassidy Edwards
And 1h28'58" 42d30'14"
Cassidy Elaine Cochlan
Cas 1h57'36" 61d6'57"
Cassidy Elaine Sheahan
Gem 7h18'11" 33d4'20"
Cassidy Elizabeth
Gem 7h29'51" 32d0'18"
Cassidy Elyse Apeles
Ari 2h16'52" 13d44'55"
Cassidy Emma Bruckschen
Dra 19h38'27" 70d39'27"
Cassidy Emma Heaney
Ori 5h58'21" 6d47'38"
Cassidy Grace Ralph
And 0h30'25" 41d32'38"
Cassidy Jayne
Cap 21h27'35" -11d7'7"
Cassidy Jean
Peg 0h3'43" 17d20'15"
Cassidy Jean McGinnis
Vir 14h31'17" -0d6'27"
Cassidy Kay Ward
Gem 6h49'10" 26d42'53"
Cassidy Lear
Leo 9h25'41" 15d34'47"
Cassidy Logan LoCascio
Cas 1h29'1" 69d11'32"
Cassidy Lynn Sullivan
And 1h18'36" 37d4'52"
Cassidy Maia Takacs, Great Niece
Cnc 8h52'23" 12d30'56"
Cassidy Mariah Kaneski
And 1h33'55" 46d19'33"
Cassidy Marie
Psc 0h35'3" 10d47'53"
Cassidy Marie LaPointe
Ari 3h19'50" 27d54'5"
Cassidy Melanie Atkins
Cap 20h52'18" -25d13'16"
Cassidy Nicole Toler
Ari 2h24'36" 12d29'44"
Cassidy Ogden
Peg 21h55'42" 14d25'0"
Cassidy Pedram
Cas 0h55'20" 70d13'41"
Cassidy Quinn Louise Hersey
Gem 7h35'41" 24d24'14"
Cassidy Rachele Hyatt
And 0h19'29" 30d24'32"
Cassidy Rose
And 0h29'26" 36d18'3"
Cassidy Wesolowski
Uma 11h4'52" 53d37'25"
Cassidy's Girl
Sgr 19h29'7" -18d9'33"
Cassidy's Star
Cnc 9h3'40" 19d1'0"
Cassidy's White Rose
Lib 15h10'21" -9d1'41"
cassidysummers
Lyn 6h29'28" 55d59'35"
Cassidythesongofthesouth windsfuture
Uma 13h51'3" 52d42'20"
Cassie
Uma 13h58'1" 57d3'51"
Cassie
Cas 23h21'51" 57d46'31"
Cassie
Cas 1h31'39" 68d7'21"
Cassie
Lib 15h21'11" -9d43'6"
Cassie
Vir 13h23'5" -6d10'43"
Cassie
Vir 14h0'24" -11d44'14"
Cassie
Mon 7h9'20" -0d23'3"
Cassie
Vol 8h52'35" -71d39'7"
Cassie
Psc 1h43'37" 24d36'6"
Cassie
Cnc 9h20'2" 13d35'2"
Cassie
Lyn 7h51'13" 39d32'26"
Cassie
Cas 0h42'13" 52d32'32"
Cassie
Cas 0h4'43" 53d38'50"
Cassie & Alex Almos
Cyg 21h27'14" 52d44'37"
Cassie Anais
And 23h54'52" 48d36'24"
Cassie and Chuck Carlson
Mon 6h52'23" -0d54'48"
Cassie And Nick
Uma 11h32'45" 58d38'55"
Cassie Anne Monbleau
And 23h57'16" 47d50'18"
Cassie Beauchamp
Cas 0h32'20" 55d27'14"
Cassie Boese
Cas 0h25'59" 59d17'37"
Cassie Breshock
Cas 0h29'23" 59d41'8"
Cassie Cassanova
Leo 11h9'13" -4d36'24"
Cassie Delayne
Crb 15h37'31" 38d21'41"
Cassie Disch
Aql 20h5'13" -0d51'53"
Cassie Dixon
Cas 1h14'2" 51d13'50"
Cassie Eaglefeather
Aql 19h35'55" 13d22'15"
Cassie Elizabeth Cain
Cas 2h7'18" 62d55'44"
Cassie Elizabeth Wright
Sco 17h6'39" -31d57'44"
Cassie & Garrett
Cas 1h23'52" 63d58'39"
Cassie "Jo Jo Girl" Boone
Lyn 7h12'39" 57d16'9"
Cassie Jo Sanderlin
Cas 0h28'7" 53d41'11"
Cassie Jones
And 0h50'2" 44d3'58"
Cassie Jordan
And 0h59'52" 40d31'35"
Cassie Kohlhagen
Sco 16h13'53" -8d30'26"
Cassie Lauren Jones
Cas 1h53'39" 64d0'3"
Cassie Lee
Cas 1h26'51" 61d8'33"
Cassie Lee
And 0h25'43" 31d7'15"
Cassie Lee Bonvissuto
Gem 6h28'13" 22d46'32"
Cassie Lewis
Aqr 20h39'14" -13d15'36"
Cassie Lynn
Lyn 6h20'48" 56d31'6"
Cassie Lynn
Cas 1h32'9" 69d44'33"
Cassie Lynn Hockaday
Uma 11h52'55" 52d30'54"
Cassie Lynn Leary
Cnc 8h11'52" 10d11'53"
Cassie Lynn Spence
Peg 22h20'35" 33d2'41"
Cassie Lynn Taylor
Cap 21h46'23" -17d43'29"
Cassie M. Davie
Boo 14h45'33" 23d33'51"
Cassie Mabbitt
Sco 17h50'30" -41d59'40"
Cassie MacInnes
Uma 11h48'56" 41d6'52"
Cassie Manwaring
Tau 5h39'32" 26d38'42"
Cassie Marie
Sgr 18h41'59" -30d16'37"
Cassie Marie Pawlak
And 23h36'58" 47d33'34"
Cassie Mckee "Our Shining Star"
And 2h24'25" 49d39'10"
Cassie McNichol
Cyg 20h39'2" 45d5'15"
Cassie & Ming Zalucky
Lyn 7h30'23" 57d58'45"
Cassie Mishell Waldrop
Cas 1h14'52" 59d30'28"
Cassie Morley
Vir 14h7'53" -17d23'25"
Cassie Munzing
Cas 0h43'58" 61d14'23"
Cassie Nelson
Sgr 18h10'25" -16d42'4"
Cassie Nicole
Aqr 22h19'21" 0d24'44"
Cassie Picha
Lib 14h50'44" -5d20'1"
Cassie Rachel Tamblyn
Cap 21h9'48" -16d10'43"
Cassie Reimer
Uma 9h43'16" 47d15'25"
Cassie Robinson 15/09/1989
Vir 11h59'35" 7d54'8"
Cassie Schembari
Cas 0h26'36" 53d21'29"
Cassie Smith
Gem 6h48'16" 32d8'38"
Cassie Stevens
Ori 6h8'12" 13d53'27"
Cassie Swanson
Cra 19h5'36" -37d55'37"
Cassie the Passionate
Cma 7h22'22" -15d35'38"
Cassie Yang Fan
Vir 11h38'21" -0d13'2"
Cassie, Brianna & Isabelle
And 1h19'44" 44d8'48"
Cassie, diligo of meus vita
Aur 5h37'27" 33d24'7"
Cassie42
Cas 23h25'0" 56d28'44"
Cassie-ann Egan
Pho 2h14'58" -43d0'23"
Cassiel
Cap 20h9'0" -13d29'39"
Cassiem
Sgr 19h31'17" -17d33'51"
Cassie's Comet
Umi 15h6'11" 70d25'57"
Cassie's sassy star
Peg 22h45'48" 25d7'15"
Cassie's Star
Tau 5h11'1" 21d40'17"
Cassie's Star
Cas 0h40'35" 51d33'34"
Cassie's Star
Cas 1h25'54" 62d28'25"
Cassiopeia
Cas 23h5'58" 56d30'19"
Cassiopeia
Cep 20h57'37" 59d58'0"
Cassiopia Star Tierney
Cas 1h12'51" 54d17'37"
Cassius Luvious
Ari 1h53'0" 17d25'32"
Cassius William Robertson
Umi 15h11'13" 79d31'45"
Cassius's Star
Peg 21h59'1" 14d16'24"
Cassondra
Cap 20h33'49" -22d18'57"
Cassondra D. Ground
Vir 12h36'11" 6d10'56"
Cassondra Elizabeth
Lib 15h13'48" -10d12'8"
Cassondra Jean
Her 17h22'40" 29d28'41"
Cassy Ford
Vir 14h1'46" -13d34'20"
Cassy & Gudni
Sge 19h9'17" 18d29'52"
Cassy R.
And 2h13'24" 38d1'4"
CASTAPELLA
Ari 1h58'5" 14d33'22"
Castellano
Per 2h38'24" 55d13'48"
Castelow's Constellation
Peg 22h18'33" 7d39'42"
Castilho-Kelley
Umi 15h33'43" 74d55'52"
Castillo
And 2h1'37" 38d37'1"
Castin Griffaw
Crb 15h48'46" 34d51'18"
Castine Meghan Stowe
And 2h25'1" 50d37'9"
Castle In The Sky
Umi 15h52'26" 71d6'21"
Castle Wright Phelps
Tau 5h6'19" 21d19'54"
Castorsita
Vir 13h9'8" -11d43'11"
Castula Amador Orbillo
Ari 3h7'23" 29d5'32"
Casuarina Beach Club
Lyr 18h40'43" 46d57'30"
Casukes
Umi 15h57'46" 72d33'56"
Casydi Renee Dotts
Gem 6h54'37" 22d19'35"
Cat
Lyn 6h47'27" 57d50'55"
CAT
Aqr 21h45'4" -7d13'51"
Cat
Cyg 21h34'55" 38d8'59"
Cat
Leo 9h48'0" 24d58'1"
C.A.T.
Psc 1h17'34" 17d20'58"
CAT
Ari 2h47'16" 14d34'54"
CAT 2/18/1983
And 1h44'20" 50d9'16"
CAT 5
Lib 15h11'56" -6d47'36"
Cat Callahan
Lyn 7h26'44" 46d19'29"
Cat Chavez
And 23h1'38" 48d13'30"
Cat Cheese Francisco
Leo 10h51'40" 16d45'50"
Cat Eye 12940
Lyn 8h42'52" 45d4'50"
CAT (Katrina Louise Sillars) 3.25pm
Cnc 8h21'53" 15d14'49"
Cat Man Rod
Aqr 22h52'48" -23d0'14"
Cat Marnell
Mon 7h21'37" -4d37'40"
Cat Purple
Ori 5h15'11" 6d13'1"
Cat & Sima
Uma 10h42'26" 52d58'25"
Catalin Marcel Mocanu
Umi 14h18'49" 65d36'17"
Catalina
Aqr 23h12'49" -11d2'33"
Catalina
Peg 22h53'42" 10d14'39"
Catalina
Leo 10h57'37" 16d16'10"
Catalina
Cas 0h27'55" 47d36'40"
Catalina Falla
Tau 4h56'47" 22d31'37"
Catalina & Jared together always
Peg 22h4'37" 6d53'48"
Catalina Pequeno
Vir 11h46'3" -2d44'38"
Catalina Ponor - Cata
Her 17h36'47" 36d17'36"
Catalina Renee
And 0h37'49" 37d31'17"
Catalino S. Millonida
Uma 8h25'53" 67d49'50"
Catalyst
Lyn 7h45'56" 47d23'4"
Catarina
And 0h41'8" 27d9'58"
Catarina Lopes
Cam 7h28'42" 79d37'28"
CatarinaYeung
Gem 6h35'15" 17d15'51"
Catch me cause Im falling for you!
Uma 9h33'47" 44d57'9"
Catcher Bissell Hyde
Psc 1h12'40" 22d32'20"
Catchmeinyourdreams
Lyn 8h11'34" 55d47'18"
(Catchpen) Patti
Cas 1h30'56" 60d27'1"
Cate
Cas 0h26'47" 62d20'50"
Cate
Umi 14h49'49" 73d33'31"
Cate and Frank
Sco 17h0'4" -32d23'12"
Cate and Steve's star of love
Cyg 20h7'4" 53d51'56"
Cate Bailey Weinberger
Umi 16h26'38" 77d7'23"
Cate Knuth
Vir 13h17'36" 6d34'33"
Cate Loves Daddy - Randy Janey
Leo 10h18'15" 22d3'8"
Cate Marie Anne Connell Paxton
Uma 11h48'59" 39d27'50"
Cate Mastrianni
Uma 13h45'20" 56d5'33"
Catelyn Elizabeth Allard
Psc 0h13'29" 5d30'42"
Catelyn's and Andrew's Star
Ori 5h54'8" 20d52'3"
Cateina
Vir 13h14'6" 4d32'22"
Caterina
Cas 23h27'51" 57d42'7"
Caterina
Uma 13h28'27" 53d39'35"
CATERINA 12.02.1974
Umi 16h55'16" 76d19'56"
Caterina Aloi
And 1h40'39" 46d49'9"
Caterina Elisa Gabriella Bòrea
Her 16h21'10" 4d52'30"
Caterina Masia
And 0h52'14" 44d6'40"
Caterina Neri Burgos
Tau 4h16'5" 19d49'15"
Cate's Star
And 2h30'54" 50d33'12"
Catfish Troy
Aqr 22h10'45" -20d30'53"
Cath
Cyg 20h27'26" 37d29'53"
Cath Case, Martin's Darling Wife
Cyg 21h11'5" 45d46'11"
Cath "Honey Bun Midget" Mathitak
Lib 15h4'56" -6d12'3"
Cathal John Coogan
Tau 4h43'58" 28d3'46"
Cathal Uttley
Dra 17h50'28" 63d12'51"
Cathalyne Restivo
Cyg 21h52'9" 49d12'59"
Catharina "Nana" Oerlemans
Cas 2h12'6" 67d31'15"
Catharine Elizabeth Hosseinzadeh
And 0h42'19" 36d28'24"
Catharine (Kitsy) Roemer
Vir 13h10'38" -4d5'37"
Cathe and Andrew
Uma 12h54'35" 56d39'1"
Cathe and Andrew
Uma 11h40'51" 45d19'2"
Cathedral School
Uma 12h35'10" 57d44'51"
Catherina Capua
Sgr 19h48'51" -12d53'50"
Catherine
Cap 21h39'3" -19d3'13"
Catherine
Lib 15h14'58" -5d31'8"
Catherine
Lib 15h8'44" -6d51'4"
Catherine
Vir 13h46'18" -13d29'1"
Catherine
Lyn 8h34'45" 56d42'7"
Catherine
Cas 1h55'19" 63d2'31"
Catherine
Uma 14h2'40" 54d41'48"
Catherine
Cas 3h17'39" 70d17'53"
Catherine
Uma 8h45'36" 70d38'15"
Catherine
Uma 10h48'35" 69d11'38"
Catherine
Dra 19h22'33" 61d6'23"
Catherine
Uma 11h21'52" 66d21'20"
Catherine
Cas 2h49'54" 63d51'25"
Catherine
Cam 5h14'44" 63d25'57"
Catherine
Uma 9h45'50" 63d34'55"
Catherine
Equ 21h8'32" 7d9'42"
Catherine
Cas 0h50'23" 48d11'49"
Catherine
Lyn 9h7'50" 39d56'4"
Catherine
Psc 1h11'54" 16d30'35"
Catherine
Vir 13h18'38" 10d39'34"
Catherine
Psc 1h3'21" 7d38'59"
Catherine
Psc 0h22'42" 11d3'36"
Catherine A. Benzin
Cas 1h38'9" 63d51'50"
Catherine A. de Lyra
Tau 5h46'48" 26d35'5"
Catherine A. FlordeLiza
Ari 2h19'10" 21d18'25"
Catherine A. Howard
Aql 19h33'32" -10d31'38"
Catherine A. MacVeagh
Equ 21h12'48" 10d54'50"
Catherine A. Wicinski "1911-2006"
Cas 0h34'7" 60d36'47"
Catherine A. Ziegler
Sgr 18h59'28" -17d37'59"
Catherine Alexa & Elizabeth Ayana
Sgr 18h43'18" -30d17'25"
Catherine Allen
Col 5h45'52" -36d8'52"
Catherine Alvarado
Ori 6h20'31" 6d58'28"
Catherine Amanda Schmidt
Lyn 9h41'45" 41d17'33"
Catherine Amelie Herard
Lyn 7h45'26" 42d55'51"
Catherine Amsterdam
Dra 19h46'1" 66d43'37"
Catherine and Allen Hay
Lib 15h5'49" -23d20'19"
Catherine and Bill Kent
Sge 20h7'1" 17d12'37"
Catherine and Keith Vendola
Cyg 21h16'50" 46d15'32"
Catherine and Victor Murgatroyd
Cnc 8h54'4" 27d1'41"
Catherine Ann
And 0h19'20" 27d48'5"
Catherine Ann Beatrice Flynn Price
Cas 0h38'20" 64d24'35"
Catherine Ann Hill
Gem 6h28'13" 24d38'39"
Catherine Ann Kaderavek
Leo 11h34'32" 24d53'41"
Catherine Ann Kursar
And 0h6'49" 35d21'31"
Catherine Ann Latham
Psc 0h16'38" 19d14'4"
Catherine Ann Mckinnon
And 0h20'47" 27d15'58"
Catherine Ann Mepham
Sgr 19h18'10" -30d51'31"
Catherine Ann Nicholson
And 23h14'8" 41d46'48"
Catherine Ann Norris
Per 4h30'37" 31d46'11"
Catherine Ann Roe
Uma 9h43'3" 58d43'45"
Catherine Ann Schulz-Mikula M.S.N.
Crb 15h48'51" 36d23'7"
Catherine Ann Spears - Catherine's Star
And 2h30'48" 39d8'46"
Catherine Ann Voss
Aqr 22h31'10" -1d48'52"
Catherine Anna Roundy LM 21005
Aqr 22h30'40" -24d42'55"
Catherine Anne
And 23h48'10" 34d52'59"
Catherine Anne
And 1h12'40" 40d12'26"
Catherine Anne Dalton
Sco 16h15'37" -16d39'10"
Catherine Anne Diamond
Gem 7h38'27" 21d35'46"
Catherine Anne Halfon
Vir 11h57'59" -11d29'28"
Catherine Anne Halzle
Psc 1h3'14" 24d48'4"
Catherine Anne Schell
And 0h51'57" 38d32'55"
Catherine Anne Schwaab
Cap 21h19'16" -25d7'19"
Catherine Anne Sharpley
Cap 21h28'50" -14d40'2"
Catherine Anne Tiedt
Lyn 8h33'47" 45d41'40"
Catherine Antoinette Rauccio
Lyn 6h43'22" 60d35'37"
Catherine Arnold
Cas 1h28'40" 54d34'41"
Catherine Attfield
Cas 23h46'38" 50d12'1"
Catherine Avery Duncan
Cap 20h19'30" -10d58'59"
Catherine "Baby Girl" Hobbs
Sco 16h59'40" -37d43'55"
Catherine Bailey
Cnc 8h16'11" 32d49'32"
Catherine Baird
Gem 7h19'18" 24d0'37"
Catherine Baird Escalante
Lyr 18h46'31" 30d54'35"
Catherine Barbara Yacker Winter
Cas 23h3'18" 59d27'40"
Catherine Baxter
Aqr 22h37'16" -3d22'55"
Catherine Beatrice Rose
Uma 9h11'9" 48d44'36"
Catherine "Bee" Mary
And 0h32'12" 28d58'14"
Catherine Benge
Cas 0h19'43" 52d8'52"
Catherine Bilay-Oum
Dra 19h0'22" 53d6'31"
Catherine Blair-Wilhelm
Lib 15h7'42" -26d47'40"
Catherine 'Blue Eyes' Sue Riley
Leo 11h33'11" 17d15'34"
Catherine Bojko
Umi 15h34'9" 72d14'3"
Catherine Boss
Cnc 8h35'10" 31d31'53"
Catherine Bowen Orndorff
Tau 4h19'46" 13d6'27"
Catherine Bradford
Cas 1h5'51" 49d14'19"
Catherine Brooke Oles
Tau 4h21'54" 12d58'35"
Catherine Brown
Crb 15h54'2" 27d33'54"
Catherine Campo
Cas 23h10'38" 58d18'50"
Catherine & Carl Yochum
Psc 1h31'23" 10d55'39"
Catherine Carla Nemeth
Cnc 7h56'16" 14d32'19"
Catherine & Carlin Elizabeth Healey
Gem 7h35'56" 29d21'26"
Catherine Carter
Ori 5h28'12" 2d38'33"
Catherine Carter Allen
Vir 11h56'42" -2d39'50"
Catherine Catarax
Lyn 8h35'50" 45d14'47"
Catherine (Cathy) R.M. Stibbard
Crb 15h40'23" 26d48'30"
Catherine (Cathy) R.M. Stibbard
Crb 15h47'36" 26d3'50"
Catherine Celeste Stolworthy
Tau 5h48'40" 27d15'36"
Catherine Chant
Uma 8h51'24" 67d44'43"
Catherine Chen
Vul 20h39'39" 25d47'56"
Catherine Chenevey
And 1h48'55" 43d18'52"
Catherine Cheung ah Seung
Uma 8h45'40" 69d11'34"
Catherine Claire Olson
Sgr 19h57'20" -28d0'3"
Catherine Cooper
Gem 6h26'18" 27d3'4"
Catherine Crew White
Lib 16h1'58" -15d19'48"
Catherine Crowte
Sco 16h9'27" -13d34'56"
Catherine Cushing Colmery
Cas 0h10'25" 57d9'59"
Catherine D. Riddle
Psc 1h20'58" 18d11'37"
Catherine D. Rosa
Cam 4h8'29" 58d16'9"
Catherine Dang
Cap 21h7'18" -16d34'26"
Catherine Diane Rusher
Sco 16h8'34" -19d11'56"
Catherine DiFazio
Gem 6h58'13" 19d37'6"
Catherine Donald
Gem 6h50'23" 33d34'15"
Catherine Doris Koleniak
Cyg 20h12'21" 34d39'33"
Catherine Douglas Devlin
Cas 23h21'16" 58d22'11"
Catherine Doyle
Peg 21h56'27" 16d49'4"
Catherine Dubus
And 2h22'41" 48d54'32"
Catherine Durakis
Cap 20h13'6" -17d37'18"
Catherine E. Good
Cas 0h50'3" 57d49'26"
Catherine Eileen Crolius Jones
And 23h50'55" 47d26'46"
Catherine Elizabeth
Cyg 21h4'4" 48d13'51"
Catherine Elizabeth Blower
Cam 4h4'46" 56d25'14"
Catherine Elizabeth Campbell
And 0h48'40" 26d46'58"
Catherine Elizabeth Carter
Aqr 21h58'46" -8d17'46"
Catherine Elizabeth Kinard
And 1h6'50" 42d42'25"
Catherine Elizabeth Martin
And 2h37'25" 44d3'16"
Catherine Elizabeth Mc Cormick
Peg 21h57'37" 29d49'31"
Catherine Elizabeth McCarthy
Cap 20h56'16" -25d1'7"
Catherine Elizabeth Taylor
And 0h25'37" 35d27'14"
Catherine Elizabeth Thompson
Lyn 6h21'2" 60d13'57"
Catherine Elizabeth Trombly
And 0h54'39" 40d15'26"
Catherine Elizabeth Wolff
Sco 17h58'11" -35d44'27"
Catherine Ellerbee Lipton
Cas 1h11'57" 62d41'55"
Catherine Emily Cator's Star
Lyn 7h30'32" 35d37'42"
Catherine Evelyn Rohr
Mon 7h27'1" -6d17'44"
Catherine Fay Holliday
Cas 2h22'3" 67d57'17"
Catherine Forero
Cnc 8h45'14" 17d43'26"
Catherine Forgrave Hicks
Lib 15h49'37" -20d7'1"
Catherine Fox
Psc 0h37'28" 9d1'28"
Catherine Frances Maxon
Cnc 8h41'56" 28d9'50"
Catherine Frank
Com 13h5'29" 23d11'31"
Catherine Fredericksen
Uma 10h44'52" 62d5'55"
Catherine G. Keyton
Lib 15h17'9" -8d49'2"
Catherine Gallo
Vir 13h0'48" 11d1'9"
Catherine Gibbons
And 1h16'54" 49d8'6"
Catherine Gismondi Veres
Cas 0h12'12" 53d14'18"
Catherine Glapa
Uma 8h52'13" 65d51'21"
Catherine Gosnell
Ari 2h59'28" 25d10'34"
Catherine Gould
Umi 15h14'29" 84d19'19"
Catherine Grace Ayers
And 2h17'24" 47d29'29"
Catherine Grace Chardo
Ari 3h6'46" 20d40'59"
Catherine Grace Hermanson
Tau 5h18'38" 18d10'56"
Catherine Guzman
And 1h37'7" 47d29'23"
Catherine Hart Curtis
Dra 17h18'49" 61d47'4"
Catherine Hartwell
Sgr 19h18'45" -16d21'36"
Catherine Heggestad 70th Birthday
Uma 9h26'37" 43d14'40"
Catherine Helen Collins
And 1h15'56" 42d57'20"
Catherine Helen Ingleton
And 1h46'25" 45d48'8"
Catherine Hijazi
Uma 8h43'10" 60d38'17"
Catherine Hilda Bowman
Ori 5h37'2" -0d41'11"
Catherine Hill Millard
Sgr 18h14'13" -20d26'26"
Catherine Holley Venables
Gem 7h16'46" 31d47'5"
Catherine Hollyhead
Cas 0h2'12" 56d8'18"
Catherine Holthaus
Ari 2h6'36" 19d55'18"
Catherine Hope Jackson
Ori 5h21'14" 0d5'56"
Catherine Hudson
Cas 0h51'7" 58d1'36"
Catherine Hwa Lewis
Lyn 6h47'40" 50d39'15"

Catherine In The Sky With Diamonds Uma 10h45'50" 60d54'49"
Catherine Ingledew Lee Lyr 18h43'15" 39d37'55"
Catherine J Carr And 1h57'20" 46d20'48"
Catherine J. Wasson Leo 10h14'27" 22d11'58"
Catherine Jane Blackbee Aqr 23h19'53" -11d54'10"
Catherine Jane Horgan Hyde Cas 1h48'42" 61d7'15"
Catherine Janiaud Tau 3h30'47" 7d31'56"
Catherine Jennings Farnan Uma 10h18'39" 40d20'24"
Catherine Jo Sutherland (sunshine) Ari 2h9'33" 23d34'49"
Catherine Joy Gem 7h29'49" 20d34'41"
Catherine Judith Campbell Cap 21h49'33" -17d12'33"
Catherine Juliet Scaife And 23h26'9" 36d39'49"
Catherine Justine Leo 10h43'58" 14d34'21"
Catherine K L Chow Cam 3h26'4" 62d44'30"
Catherine K. Lieberman And 1h14'22" 48d56'21"
Catherine "Kay" Bastardi Cas 1h44'28" 65d6'58"
Catherine Kehoe Cas 23h3'33" 53d22'19"
Catherine Kelly Uma 11h18'15" 66d54'19"
Catherine Kerns Cas 0h36'32" 52d19'12"
Catherine Kim Cas 0h0'37" 56d39'21"
Catherine "Kitty" Alvis Knaub And 0h6'24" 45d18'8"
Catherine "Kitty" Culbertson Cas 0h44'58" 58d33'27"
Catherine "Kitty" Gannon Uma 9h51'51" 46d35'25"
Catherine Kjerste Lyr 19h9'39" 31d49'32"
Catherine & Kristian Stenberg Dra 17h42'20" 60d52'42"
Catherine Kroll Cyg 19h58'11" 34d33'41"
Catherine Krystyna Kofin Psc 0h9'14" -4d15'53"
Catherine L. Bifareti And 23h35'29" 47d28'26"
Catherine L Brodie - Exteriors by CLB Cnc 8h36'22" 8d30'8"
Catherine L Clark Ori 5h50'11" 7d21'40"
Catherine L. Hanna Aqr 21h38'54" 1d19'55"
Catherine L. Shento Uma 9h0'31" 52d25'18"
Catherine L. Stewart Sco 16h15'34" -14d40'56"
Catherine L. Vietro-Corrado And 23h4'52" 50d49'53"
Catherine Lachlan Fisher And 0h24'2" 34d32'9"
Catherine Lamar Magnussen Wilson Cnc 8h16'53" 23d10'2"
Catherine Landt Currie Williams Umi 14h0'11" 72d35'7"
Catherine Larotonda Lyn 7h14'0" 49d4'2"
Catherine Laurette Uma 10h10'35" 65d27'31"
Catherine Laverdure Vir 12h57'40" 9d55'37"
Catherine Lee Vir 13h19'3" 8d3'27"
Catherine Lee Cunningham Cas 0h39'13" 59d2'38"
Catherine Lee O'Neill Lyn 7h48'50" 38d22'34"
Catherine Lim Gregorio And 1h4'38" 37d40'24"
Catherine Lobjois Uma 10h31'50" 63d31'42"
Catherine Lorraine Knight Cam 6h9'9" 68d29'3"
Catherine Louise And 23h48'49" 33d47'14"
Catherine Louise Howells And 23h17'28" 51d44'37"
Catherine Louise Jones Del 20h46'43" 13d16'27"
Catherine Louise Mason Crb 15h34'24" 27d28'37"
Catherine Louise Stidham Del 20h26'0" 16d39'52"
Catherine Lucia Sgr 19h34'2" -13d4'26"
Catherine Lucille Aqr 22h15'39" -5d30'21"
Catherine Lynne Thrift Peg 22h46'46" 26d43'8"
Catherine M. Bryant Gem 7h11'6" 34d24'35"
Catherine M. Dotto Vir 12h25'6" 3d39'4"
Catherine M. Graham Sgr 18h11'55" -20d6'39"
Catherine Mae Bussey Tau 4h5'50" 6d5'57"
Catherine Magee Psc 0h25'15" -1d6'42"
Catherine Malina Tau 4h51'21" 26d12'6"
Catherine Malkemus Crooks Sgr 19h13'21" -14d15'8"
Catherine Maragni Cyg 20h25'22" 39d57'11"
Catherine Margaret Cas 0h37'16" 60d50'47"
Catherine Maria Mon 6h47'54" -0d3'42"
Catherine Maria Cote And 23h48'6" 47d47'59"
Catherine Marie Cas 0h31'33" 52d29'29"
Catherine Marie Brinley Sgr 17h54'21" -29d37'39"
Catherine Marie Howard Sco 16h3'21" -12d7'58"
Catherine Marie Hudek Lyn 8h36'45" 33d41'41"
Catherine Marie Johnson Cnc 9h18'21" 9d45'46"
Catherine Marie Lavers Ori 5h41'55" -1d40'58"
Catherine Marie McCauley Tau 5h48'13" 15d2'19"
Catherine Marie Spencer Tau 4h28'10" 16d43'24"
Catherine Marie Thompson Cas 1h10'31" 55d11'28"
Catherine Marie Underwood And 0h50'10" 41d59'27"
Catherine Markham Way Sco 16h18'33" -27d57'28"
Catherine Marlowe Cas 0h19'28" 61d13'58"
Catherine Marra Cas 0h2'59" 57d14'45"
Catherine Mary MacQueen And 23h7'58" 51d5'35"
Catherine McIlwee One Tau 5h8'12" 17d52'50"
Catherine McPherson Moir Cas 23h25'33" 57d4'16"
Catherine (MDOT) Kiefer Psc 0h32'28" 9d27'24"
Catherine Melissa And 2h33'29" 43d47'26"
Catherine Melissa 1966 Leo 11h8'24" 20d3'56"
Catherine Michelle Gem 7h49'51" 30d15'51"
Catherine Morgan Cas 0h57'12" 57d7'52"
Catherine Murphy Lyn 6h43'50" 50d29'51"
Catherine Murphy Sco 16h35'31" -28d7'57"
Catherine (My Cat) Kewin Lyn 7h18'26" 49d26'10"
Catherine My-Hanh Ngo Ari 2h53'13" 25d34'34"
Catherine Myrick Tau 3h55'9" 27d36'36"
Catherine Nance, Star of Hope Psc 1h27'28" 26d41'52"
Catherine Nasso Cnc 8h50'51" 27d16'26"
Catherine & Neil Soulmates 4 Ever And 0h1'37" 38d47'10"
Catherine Newton Lib 15h9'29" -5d53'32"
Catherine Nicole West Ari 2h6'30" 23d22'23"
Catherine Nicoloff Loboda Ori 5h6'41" 5d30'37"
Catherine Norvell Cyg 19h56'12" 36d30'38"
Catherine of Dublin Cnc 8h7'28" 18d22'11"
Catherine Olivia Cedillos Sco 17h34'8" -34d16'34"
Catherine Ollin Saad And 2h5'44" 43d58'0"
Catherine Orlando Reynolds Vir 12h32'52" 11d24'16"
Catherine Panizza And 0h42'24" 38d45'59"
Catherine Patricia Gem 6h37'1" 14d16'15"
Catherine Patricia Kelly Cas 23h3'51" 58d4'59"
Catherine Pettibone Hochman Aqr 21h15'9" 1d54'31"
Catherine Peyronnard Umi 10h55'59" 87d12'42"
Catherine Phillips Lyn 8h23'59" 55d3'48"
Catherine Piercy Lyn 6h46'13" 54d4'10"
Catherine "Poopey" Johnstone Cyg 20h2'15" 55d58'38"
Catherine Powell Uma 9h46'48" 50d51'25"
Catherine Prentice Deutsch Cas 1h29'0" 62d56'41"
Catherine R. Mahardy Cyg 21h41'28" 46d7'22"
Catherine Rachael Hinojosa Ori 5h8'0" 8d36'19"
Catherine Ramirez And 2h16'43" 50d42'25"
Catherine Rand - "Cat" Lyn 7h33'42" 38d8'43"
Catherine Ranieri Lib 15h20'36" -11d29'37"
Catherine Rebecca Kennedy Leo 11h21'0" 15d51'30"
Catherine Rebuldela Psa 22h50'39" -28d8'58"
Catherine Reininger Gem 7h52'17" 30d41'33"
Catherine Renee Delgado Uma 8h44'15" 67d43'36"
Catherine Rivera (BPRQ) Ari 2h50'42" 24d24'29"
Catherine Rochelle Hendricks Baker And 0h49'0" 43d7'30"
Catherine Rodriguez Ari 3h16'10" 29d7'37"
Catherine Rosaura Santos Cap 20h21'34" -12d13'11"
Catherine Rose Tau 3h50'32" 27d40'55"
Catherine Rose Woodall Cochran Lyn 8h27'11" 45d7'6"
Catherine Russo Lyn 8h52'11" 34d23'49"
Catherine Ruth Allen-Calegari And 1h28'15" 47d1'26"
Catherine S. Moreland Uma 12h57'42" 60d32'17"
Catherine S. Plowcha Cas 0h1'16" 57d8'37"
Catherine Sarah White And 0h57'36" 45d38'54"
Catherine Scarcelli Peg 22h30'32" 9d21'41"
Catherine Schär Cmi 7h24'53" 1d59'32"
Catherine SCHMITT Crb 15h53'58" 35d26'42"
Catherine Schobert Cas 1h30'12" 63d14'37"
Catherine Shoff Cnc 9h13'55" 25d48'35"
Catherine Stephenson And 23h58'17" 36d28'51"
Catherine Sue Hillary Vir 13h34'54" -16d27'55"
Catherine Sullivan Uma 12h36'29" 61d4'22"
Catherine- The Angel Above Psc 0h36'1" 8d41'49"
Catherine Tsakalis Ari 2h58'25" 25d16'34"
Catherine Tully And 0h34'22" 22d30'33"
Catherine Turner Cam 6h4'14" 62d13'2"
Catherine Van Scyoc Vir 12h9'32" 10d55'13"
Catherine Vanhandsaeme Mon 6h51'53" 8d33'0"
Catherine Vera Cussigh "Katie" And 2h34'58" 49d58'45"
Catherine Victoria Hoyem Del 20h37'8" 15d0'21"
Catherine Victoria Mejia Crb 16h22'23" 27d5'19"
Catherine Victoria West And 23h36'14" 48d15'48"
Catherine Viviers Cas 1h59'57" 61d55'45"
Catherine W. Haglund Uma 13h27'1" 57d59'48"
Catherine Wheeler Gem 6h53'26" 32d9'23"
Catherine Williams And 2h26'42" 47d54'31"
Catherine Wing Cnc 8h20'46" 27d19'47"
Catherine Wong Sgr 19h10'16" -15d45'55"
Catherine Young Lyn 8h13'17" 58d48'50"
Catherine Zhang Sgr 18h30'40" -34d42'30"
Catherine Zuppardo Mon 6h52'39" 7d27'32"
Catherine's beauty Cnc 7h57'55" 13d39'38"
Catherine's Calelamar Crb 16h12'34" 29d59'45"
Catherine's Garden Lib 15h29'32" -20d25'49"
Catherine's Lullaby Aqr 22h3'48" -2d22'48"
Catherine's Own Wish Upon A Star Cnc 9h7'54" 30d42'23"
Catherine's Star And 23h45'33" 42d25'53"
Catherine's Star Cas 0h51'48" 56d32'17"
Catheryn Mancini - VAVA's Cathy Umi 14h39'17" 72d50'51"
Catheryn Virginia King Gem 7h23'32" 30d47'30"
Cathi Psc 1h40'56" 26d55'17"
Cathi Psc 1h19'23" 16d10'40"
Cathi Ori 5h44'48" 2d30'4"
CatHi Aqr 21h32'22" 0d46'39"
Cathi Lynn Nicholson Ryan Cas 1h8'55" 56d9'22"
Cathialine Andria Cas 22h59'29" 55d11'7"
Cathie L 09 05 1979 Cas 0h12'4" 54d55'42"
Cathie LaMacchia Sco 16h8'30" -14d52'31"
Cathie Layman Leete Uma 8h25'18" 62d44'55"
Cathie Leete Cas 1h29'24" 62d34'17"
Cathie Patrick Pierotich Sgr 18h49'29" -24d47'21"
Cathie Purushothaman Uma 10h25'22" 68d39'44"
Cathie Rouillec Psc 2h4'24" 5d50'16"
Cathie Sue Fisher Sco 17h45'19" -39d5'26"
Cathie V. Reinold Peg 21h41'22" 27d29'14"
Cathleen Admirand Lyn 7h0'4" 61d0'14"
Cathleen and Dan Greenberg Cyg 19h37'11" 37d35'35"
Cathleen Conley Uma 10h45'3" 51d51'47"
Cathleen Conley Sco 17h53'55" -36d10'57"
Cathleen Dougherty Vir 13h15'4" 5d21'10"
Cathleen Frances L. S. And 0h56'50" 38d9'42"
Cathleen Gail Com 12h15'1" 32d32'48"
Cathleen Hellen Turner Ari 2h43'45" 22d28'5"
Cathleen Johanns Lyn 9h2'19" 37d28'28"
Cathleen Lily Per 4h40'52" 45d23'14"
Cathleen M. Potter Cyg 21h31'23" 48d20'40"
Cathleen Marie Bannigan Ori 5h36'12" -1d10'41"
Cathleen Marie Neumann Lib 15h43'26" -24d57'47"
Cathleen Maura Leach Lib 14h50'41" -16d24'46"
Cathleen Mow Mow Leo 9h49'15" 13d7'48"
Cathleen Rose Barron Sgr 18h3'25" -27d52'6"
Cathleen Theresa Cishek Leo 9h45'23" 27d37'36"
Catholic Charties, AOH Aqr 22h34'2" -22d17'55"
CatholicKitty Crb 16h10'53" 39d17'33"
cathpat Tri 2h20'27" 33d12'56"
Cathrine Crb 16h15'28" 32d58'54"
Cathrine Dalene Weale Leo 10h25'37" 12d36'51"
Cathrine Maliepaard And 0h30'56" 32d35'18"
CATHRON GLENN TREPAGNIER Gem 7h12'19" 27d27'2"
Cathryn Lyn 9h12'36" 34d55'56"
Cathryn Sgr 19h39'57" -15d9'42"
Cathryn Ann McClelland Psc 1h8'48" 17d45'52"
Cathryn Connolly Blair Vir 13h45'3" -7d28'51"
Cathryn Elizabeth Haskins Crb 16h10'36" 37d44'23"
Cathryn Elizabeth Lacy Crb 16h17'50" 37d41'54"
Cathryn Esposito Magliocco Uma 12h31'41" 52d40'26"
Cathryn Marie Bologna Cas 23h11'52" 55d50'41"
Cathryn Mitchell Cap 21h37'6" -16d23'3"
Cathryn Vloch (The Princess) Cas 0h8'52" 54d47'41"
Cathryne G. Rose Cyg 20h11'41" 47d48'18"
Cathryne Viktoria Millsap Tau 5h1'9" 23d56'29"
Cathus Chau Maximus Sco 17h20'19" -42d48'40"
Cathy Aqr 22h6'50" -24d0'2"
Cathy Lib 15h5'8" -3d8'22"
Cathy Umi 17h11'50" 80d52'43"
Cathy Uma 14h7'49" 57d45'13"
cathy Cyg 20h36'38" 53d27'26"
Cathy And 0h32'35" 32d15'14"
Cathy And 0h38'50" 26d0'50"
Cathy And 0h57'30" 41d31'40"
Cathy 1983 And 23h55'5" 39d54'45"
Cathy 2211 JRP Cas 0h1'17" 54d48'5"
Cathy A. Brown, Star Mom Per 2h20'15" 55d40'56"
Cathy A Nowak Lib 14h43'50" -10d2'3"
Cathy A. Wuchter Sgr 19h6'53" -24d39'36"
Cathy a.k.a. Munch Per 3h24'32" 49d22'34"
Cathy and Mark Stamm Family Cyg 20h8'37" 31d33'21"
Cathy Ann Brumbaugh Anderson And 23h10'30" 42d10'48"
Cathy Ann Crapo Tau 3h44'44" 25d29'58"
Cathy Ann Dann Uma 10h3'21" 45d34'34"
Cathy Ann Nace Lib 14h51'18" -4d0'52"
Cathy Ann Zadnik Cep 20h44'16" 62d3'8"
Cathy Anna Lucarz Uma 8h43'5" 70d1'11"
Cathy Anne Psc 1h46'25" 20d37'28"
Cathy Athanson Uma 11h17'32" 48d51'54"
Cathy Bardascino "Guiding Light" Cas 0h58'41" 54d5'19"
Cathy & Bob Travers Ori 4h57'25" -0d33'55"
Cathy Booth Lib 15h30'17" -9d30'26"
Cathy Brown Cyg 21h52'40" 47d35'50"
Cathy Bryn Ori 5h8'55" 7d31'4"
Cathy "Cat" Corcilius Cap 20h26'22" -14d20'27"
Cathy "CoCo" Beal Ari 2h16'50" 26d10'53"
Cathy Copeland Lyn 6h41'58" 56d56'42"
Cathy Czudek Cas 0h49'5" 49d36'6"
Cathy Danette Swygert Ari 2h35'20" 22d8'4"
Cathy DeGraw A Star is Born 121268 Ori 5h32'22" -0d17'41"
Cathy DelPozo Taylor - GaMPI Star Lyr 18h47'10" 41d20'31"
Cathy Diane Pure Godess of the Moon Uma 9h41'54" 68d34'7"
Cathy Domanski Leo 10h8'22" 22d2'44"
Cathy Donaghy And 23h0'27" 48d17'39"
Cathy Drew Leo 9h38'55" 29d31'38"
Cathy Duckett Hya 9h32'51" -0d3'33"
Cathy Eddy Tau 4h23'2" 15d14'56"
Cathy Elaine Walquist Psc 1h16'52" 31d32'46"
Cathy Elizabeth Callari Sco 16h33'52" -27d20'20"
Cathy Felton Cas 0h41'54" 51d51'10"
Cathy Fernandez Leo 9h33'57" 32d13'31"
Cathy Fitzsimmons Cyg 19h51'12" 31d4'7"
Cathy Gaddy Cam 6h21'19" 62d11'59"
Cathy & Gary Sherrer Cyg 20h35'17" 55d1'36"
Cathy Gaughenbaugh Cam 4h17'32" 57d34'29"
Cathy Gennaro Aql 19h14'39" 6d7'9"
Cathy Greenwood Boyer Crb 15h40'57" 37d10'53"
Cathy Gruszie Lyr 18h26'2" 33d55'52"
Cathy Harmon Cnc 8h59'51" 10d59'0"
Cathy Hartford Cas 23h1'9" 59d15'53"
Cathy Haynes Uma 10h49'5" 43d58'20"
Cathy Hugghins - Love of My Life Mon 6h49'30" -0d5'37"
Cathy J. Robinson Leo 9h28'6" 10d28'54"
Cathy Jo Ant 10h53'44" -35d45'57"
Cathy Jo Sco 17h37'9" -33d15'7"
Cathy & John Leo 9h46'56" 26d6'42"
Cathy K. Hanlon Cas 1h20'7" 65d23'25"
Cathy K Tabor (Super Mom) Psc 1h26'40" 16d1'43"
Cathy Karrenbrock Tau 5h27'39" 26d39'30"
Cathy Lawson Crb 15h26'25" 26d57'55"
Cathy LeCompte Uma 9h43'42" 50d59'55"
Cathy Lee Drenning Cap 21h15'3" -25d14'31"
Cathy LeMieux Per 3h6'48" 50d51'55"
Cathy LeMieux Per 3h7'14" 50d46'50"
Cathy Leung Cam 4h5'49" 66d34'24"
Cathy Loomer Cap 21h21'20" -16d11'16"
Cathy Lou Karwales O'Hara Psc 1h26'21" 23d7'40"
Cathy Louise Tarrel Lib 15h0'10" -9d22'14"
Cathy Lund Dra 19h33'33" 66d29'55"
Cathy Lynn Abend, always our star And 1h44'33" 43d55'11"
Cathy Lynn Campagna Cas 23h0'49" 55d50'26"
Cathy M. Martell Crb 15h36'56" 26d46'44"
Cathy Maria Cassell Tafaoa Gem 7h14'18" 28d16'12"
Cathy Marie Col 5h57'3" -38d9'3"
Cathy Marie Jensen-Warnecke Gem 7h32'30" 30d33'16"
Cathy Marie Sanders Sgr 18h54'50" -17d50'58"
Cathy & Mark Carden Crb 16h9'53" 30d34'50"
Cathy McDonald Leo 9h42'38" 6d32'22"
Cathy Michele Risinger Crb 16h20'32" 33d3'11"
Cathy Mimuchacha Crb 16h12'27" 30d36'32"
Cathy Min-May Alcudia Psc 1h27'6" 25d46'56"
Cathy Montgomery And 0h25'26" 43d23'29"
Cathy MO-TF1 Uma 12h11'25" 55d18'8"
Cathy my angel Lib 15h37'37" -2d19'49"
Cathy - my sister, my friend Lib 15h37'16" -6d39'18"
Cathy & Peter Wilson-Fry's Flame Cru 12h43'22" -59d55'44"
Cathy Powers Poeggel Cap 21h37'11" -22d48'9"
Cathy R McLaughlin Leo 9h23'49" 26d23'46"
Cathy R. O'Dell-Our Guiding Light Lyn 8h7'53" 34d51'49"
Cathy Rae Cas 1h23'42" 60d52'45"
Cathy Reinard Vir 12h34'17" 11d43'20"
Cathy Rick Lyn 9h8'55" 37d58'47"
Cathy Robin Threlkeld Ori 6h4'3" 6d29'4"
Cathy Ruiz Vir 13h17'53" 4d40'21"
Cathy Sae Wong Aqr 22h30'42" -2d56'51"
Cathy Scibelli Sgr 18h28'44" -16d32'18"
Cathy Smiley Scarborough Vir 13h23'3" -18d52'29"
Cathy Smith Uma 10h58'29" 35d34'22"
Cathy Spadafora Smith Lyr 18h29'13" 32d18'52"
Cathy & Steve 1st Anniversary Star Leo 9h41'6" 29d24'26"
Cathy Struck - Superstar Teacher Cas 0h32'14" 63d0'42"
Cathy Sue Cyg 21h2'52" 48d20'42"
Cathy "Sunshine" Massoni Lyn 8h0'7" 50d35'1"
Cathy T. Steffens Psc 1h23'2" 20d51'39"
Cathy The Eternal Flame Cas 1h6'44" 64d1'22"
Cathy Towles(mother) Her 16h33'48" 14d3'57"
Cathy Turner And 1h10'35" 42d6'55"
Cathy V Rojas Gem 7h5'20" 33d0'40"
Cathy Wery Vir 11h56'48" -7d4'15"
Cathy White Gem 7h2'40" 18d12'22"
Cathy Wills Leo 11h30'50" 26d16'24"
Cathy Wood Psc 1h13'23" 28d2'10"
Cathy Worthington And 1h1'6" 43d50'33"
Cathy Yokley Per 1h49'1" 51d22'13"
Cathy Zdrojewski Ari 2h6'3" 12d27'50"
Cathy Zecchin Cam 4h25'10" 69d32'29"
Cathy, Loving Mom Sent From Heaven! Cas 1h13'10" 57d37'5"
Cathy, Soul Mate to Roy Crb 16h7'13" 26d50'24"
Cathy, the love of my life Vir 12h4'18" -3d28'36"
CATHYDON Leo 9h38'28" 14d59'43"
Cathye Bjerum Sco 16h6'7" -16d56'29"
CathyJoe Vir 11h59'14" 7d52'36"
Cathy's 31st Anniversary Star Ari 2h56'14" 25d22'50"
Cathy's Christmas Star Ori 5h34'40" 4d57'56"
Cathy's driving light Sgr 17h59'54" -23d58'35"
Cathy's Forever Cnc 9h6'42" 12d24'25"
Cathy's Shining Star Cnc 9h0'53" 28d1'7"
Cathy's Star Tau 4h16'20" 29d58'18"
Cathy's Star Peg 21h29'49" 11d53'21"
Cathy's Star Cas 23h40'53" 58d5'24"
Catia Ori 6h13'28" 9d13'13"
Catia Sinigaglia Aur 6h2'51" 50d24'57"
Catie Sgr 18h55'0" -20d26'28"
Catie Alicia Downey Cas 23h46'2" 54d38'36"
Catie Baker And 23h10'16" 52d6'13"
Catie Eller And 1h19'21" 49d30'24"
Catie Glass Psc 1h3'31" 5d4'46"
Catie Good 1995 Cap 20h8'43" -19d55'28"
Catie Grace Pettibone Her 16h41'24" 47d46'52"
Catie Lady Aqr 22h17'49" -0d15'22"
Catie Leahy Star And 23h36'38" 48d26'44"
Catilin Andrews Psc 1h13'38" 24d23'22"
Catimus Lyr 19h19'32" 27d3'35"
Catinita Ori 5h57'43" 5d49'3"
Catisme Leo 11h18'57" 18d54'9"
Catlin Mary Hickerson Vir 12h19'55" 12d27'29"
CatMan Lyn 7h12'24" 58d34'34"
Cato Cra 18h52'59" -42d16'44"
Catortha Tau 3h51'13" 23d30'54"
Catreena Shalimar Aqr 21h51'55" -0d58'21"
Catricia Leo 11h42'20" 25d38'34"
Catrin Ari 2h15'17" 24d21'30"
Catrin Walters And 0h45'21" 31d33'2"

Catrina
Aqr 22h25'13" -4d16'11"
Catrina
Psc 0h38'55" 6d28'6"
Catrina Michelle Wyss
Gem 6h35'32" 17d25'24"
Catrina Rehak
And 2h24'42" 40d16'7"
Catrini, Claudio
Uma 9h8'26" 70d21'27"
Catriona
Ari 3h25'29" 29d1'10"
Catriona and Peter
Cyg 20h1'1" 30d9'43"
Catriona Anne Ley
And 23h36'35" 47d30'25"
Catriona Blyth Gillies
Cas 23h0'27" 58d56'47"
Catriona Bo
And 23h54'20" 35d24'51"
Catriona Heather Reid
Lyn 7h25'41" 57d10'47"
Catriona Irene
Cap 21h9'38" -19d31'44"
Catriona Mairi
And 23h41'5" 39d0'35"
Catriona & Marcus
Devaney
Cyg 19h36'49" 37d13'4"
Catriona Mhairi Robinson
Lmi 10h27'47" 39d4'5"
Cat's Eye
Tau 4h20'56" 22d31'16"
Cat's Paw Print
Cnc 9h5'47" 15d0'51"
Cat's Place in Heaven
Lyn 7h48'41" 37d30'13"
Cat's star
Ari 2h3'50" 22d57'17"
Cattlena May Changpriroa
Lyn 7h53'43" 48d21'17"
Cattus Caerulus
Sco 17h51'42" -41d10'48"
Catty P (Mrs Catherine
Whalley)
Cyg 21h30'21" 33d24'47"
Catty: Star of All Things
Magical
Psc 0h2'48" -3d49'25"
Catuntoochins
Uma 10h28'43" 64d25'41"
Catwalkers
Lac 22h56'27" 44d42'2"
Catweasel - Sheesh
Aqr 22h55'27" -11d5'52"
Caty Eidemiller
Aqr 23h18'30" -19d35'57"
Caty Rocca EmG
Uma 8h42'19" 66d6'55"
Catya Tranchida
And 22h58'42" 50d52'46"
Caulen Marcus Finch
Cas 0h54'39" 53d1'48"
Cavalier 1
Ori 5h42'32" 0d9'3"
Cavan David Jack
Umi 14h35'30" 73d12'50"
Cavan Maurice Gable
Sgr 18h52'43" -20d22'11"
C.A.Vega & I.Rogovoy
Amor Navsegda
Psc 1h22'8" 31d4'23"
Cavemin
Aqr 23h20'19" -18d41'23"
Cavendish Castle
Leo 11h42'16" 24d11'38"
Cavi
Her 18h43'45" 19d11'55"
Cavion Holloway
Gem 6h28'37" 26d51'18"
C.A.W
Leo 11h15'49" 14d20'2"
CAW
Uma 8h56'43" 57d46'35"
Cawsey-Clark
Uma 8h51'16" 67d0'0"
Caxton William Acker
Krohnfeldt
Per 4h36'57" 44d53'51"
cây phong lan ngoc bich
nang chang
Ori 6h13'38" 18d58'47"
Cay Savino
Leo 9h44'22" 29d16'11"
Cayce Leavell Ward
Cas 0h16'49" 57d39'1"
Cayce's Angel (Lauren's
Star)
Cyg 20h38'56" 36d4'25"
Caydance E Cudd
Cam 3h56'15" 69d20'54"
Caydance Leeann Kinard
And 1h58'18" 38d6'54"
Cayden Burke
Uma 10h16'22" 71d2'30"
Cayden Hupe
Vir 12h3'22" 4d11'16"
Cayden Joseph Baldwin
Tau 4h16'12" 13d33'0"
Cayden Lee Edward Arnold
Umi 14h20'38" 77d47'9"
Cayden "Mama's Baby
Doll" Boogher
Cyg 20h58'53" 47d50'25"
Cayden "Monkey1"
Dra 15h55'8" 63d13'30"
Cayden Thomas Puppe-
Wotipka
Sco 17h20'2" -32d32'46"
Caydence
Sco 16h27'32" -41d40'8"
Cayenne
Peg 23h15'32" 32d22'33"
Cayenne Bacardi Isaksen
Sgr 18h29'40" -22d48'47"
Cayenne Carter
Uma 9h1'29" 47d9'44"
Cayford-Afton
Umi 16h13'38" 71d46'4"
Cayla
Peg 22h30'26" 11d21'5"
Cayla Elizabeth
Umi 10h3'16" 89d24'3"
Cayla Fern Crilly
And 1h15'36" 38d23'34"
Cayla Madison Bernstein
Lib 15h30'48" -11d43'21"
Cayla Marie
Gem 7h7'57" 25d51'52"
Cayla Marie Barton
Leo 10h12'11" 15d9'21"
Cayla Marie Forewright
Sco 17h58'12" -30d53'12"
Cayla Mireille Pagano
Uma 13h44'15" 48d16'43"
Cayla Nicole
Lyn 6h35'42" 54d13'9"
Cayla Rae
Vir 13h5'40" 12d5'42"
Cayla Rose
And 0h4'20" 45d40'11"
Cayla Townes
Aqr 22h36'19" -2d3'15"
Cayla's Dancing Star
Lyr 19h9'43" 39d47'7"
Caylee Louise
Cas 1h34'19" 67d19'54"
Caylee Marie
Vir 12h48'15" 1d39'55"
Caylee Marie Herren
Vir 12h59'14" -7d45'12"
Caylee Marie Kmiec
Per 3h34'21" 45d35'46"
Cayleigh Beth Schwartz
Mon 7h18'20" -3d24'36"
Cayli Lyn
Ori 6h5'16" 10d3'27"
Caylie
Sco 16h14'14" -9d51'48"
Caylie Elizabeth Abbett
Ari 2h41'22" 26d2'19"
Caylum James McMahon
Conlon
Umi 17h46'10" 81d18'58"
Cayman
Lyn 7h13'28" 44d55'6"
Cayman Sinjin
Uma 10h34'30" 66d58'41"
Caz and Will - Love
Without End!!
Cyg 19h42'36" 33d5'6"
Caz Star
Umi 14h55'29" 69d27'16"
Cazstar
Uma 11h11'39" 31d0'16"
Cazz Clifford
Gem 6h42'8" 28d30'29"
Cazza
And 22h58'58" 40d19'9"
CB
Cap 20h18'34" -14d34'8"
CB CB FOREVER
Tau 3h47'23" 27d33'53"
C.B. III
Aqr 22h38'6" -2d6'8"
CB Oreo
Uma 13h12'5" 53d37'53"
CB4
Uma 11h18'22" 29d59'11"
CB-5
Sgr 19h21'53" -29d44'18"
cBd*FoReVeR*pNh
Sco 17h52'58" -34d56'56"
Cbella52805
Gem 7h23'28" 27d33'14"
CBJRH75
Uma 9h22'18" 69d11'28"
Cbrown691982
Aur 6h27'24" 51d43'16"
CC
And 23h15'57" 39d59'14"
CC
Cas 0h26'50" 61d43'38"
CC '07
Cyg 20h15'11" 52d29'59"
CC Coppola Star
Dra 17h24'3" 57d49'3"
C.C. Godwin
Leo 10h59'48" 6d40'43"
C.C. McGurr
Uma 9h13'17" 67d36'27"
CC Sparkles
Lib 15h11'32" -13d48'51"
CC West
Aqr 21h53'15" 1d10'5"
CC25
Cru 12h27'52" -60d49'42"
CCamileH
Cyg 20h46'15" 32d44'48"
C.C.C. Loves L.M.S.
Vir 14h49'23" 5d35'55"
CCE's Shining
And 0h55'52" 38d41'20"
CCM
Gem 6h47'58" 33d25'29"
CCM
Her 18h6'20" 18d48'7"
C.C.McConnell
Sco 17h36'13" -42d16'0"
CCOsborne #13
Leo 11h57'35" 25d15'14"
C-Cross
Her 16h36'38" 35d4'29"
CC's Flyer
Ori 5h39'36" 6d28'51"
CC's Graceland
And 0h31'18" 39d34'36"
CCT32666
Ari 3h13'29" 28d17'44"
CCU Stars
Ari 2h49'58" 18d21'53"
CD Balbach 60TH
Cap 20h56'6" -17d16'29"
cd270505
Uma 9h15'44" 58d17'32"
CDB'S Shining Star
Ori 5h36'20" -0d29'54"
CDC 10182004
Per 4h29'54" 52d32'27"
CDM-Hylephila
Aqr 22h2'47" -24d52'5"
CDM-Skippers
Sgr 18h7'31" -16d41'46"
CDP17666
And 1h19'32" 49d7'42"
Cdr. Francis L. Cundari,
USN
Aql 19h40'28" 3d5'50"
CDS XOXO KDS
Lyn 7h11'16" 58d49'37"
Ce Ce, Bekah and Zane
Tri 2h12'24" 33d35'31"
C.E. "Chief" Arthofer
Srp 18h24'38" -0d7'33"
C.E. Winslow Intergalactic
Realty
Leo 11h8'19" 16d40'16"
CE523
Aql 19h29'58" 3d58'44"
Cea
Her 18h51'22" 23d11'40"
CEA
Mon 6h44'48" -0d15'53"
Ceadda
Cnc 8h27'44" 9d42'49"
Céadsearc
Ori 5h58'58" -0d41'52"
Ceangal
Umi 16h39'53" 82d37'52"
Ceara
And 2h19'4" 41d17'39"
Ceasar
Uma 10h49'49" 71d43'12"
Cebum
Aql 19h37'41" 10d46'52"
Cecco Ciampini 2x3
Peg 21h29'41" 17d45'33"
CeCe
And 2h9'28" 42d28'37"
Cece
Cap 20h23'52" -25d13'4"
Cece Bloom
Leo 9h53'32" 16d2'6"
Cécélia
Uma 9h28'10" 42d44'53"
Cecelia
And 0h45'34" 41d26'40"
Cecelia
Cas 1h33'55" 66d37'52"
Cecelia
Lyn 6h49'31" 55d14'30"
Cecelia Allmann
And 0h20'12" 35d37'43"
Cecelia Ana
And 1h50'59" 37d53'41"
Cecelia and Heidi
Boo 15h19'2" 49d36'13"
Cecelia B. Nee
Cas 0h36'28" 60d36'8"
CeCelia Carroll
Cap 21h7'25" -25d24'46"
Cecelia Christy Shively
Cas 23h37'45" 56d7'40"
Cecelia Collier
Cas 1h41'23" 62d51'27"
Cecelia Curcio
Cap 20h26'49" -18d16'48"
Cecelia Dove
Mon 7h32'0" -0d51'11"
Cecelia Gallagher
Cas 1h38'50" 64d26'54"
Cecelia George
Tau 5h45'39" 25d30'32"
CeCelia Irene Baughn
Hya 9h14'54" -0d56'42"
Cecelia Lou Cubra
Aqr 22h17'31" 2d36'9"
Cecelia Marcella Pinel
Cnc 9h7'31" 11d54'44"
Cecelia Mutze
Psc 1h9'29" 31d43'20"
Cecelia Nonna DiBurro
And 2h20'3" 48d26'16"
CECELIA "PURE LOVE"
Cas 1h47'27" 63d28'28"
Cecelia Royster Lynch
Uma 10h54'21" 49d23'57"
Cecelia Stewart Travers
Peg 21h19'11" 16d15'1"
Cecelia Themis
Psc 1h43'7" 24d35'57"
Cecelia Therese Kora
Tau 4h3'33" 5d11'21"
Cecelia Yvette
Aqr 20h49'31" -12d33'21"
Cecelia, "God's counselor
of Love"
Cap 20h44'33" -25d37'19"
Ceceliastar
Del 20h55'30" 15d46'55"
Cecere 50
Uma 11h24'1" 43d30'25"
CeCe's Celestial Star
Cnc 8h54'11" 26d6'34"
CeCe's First Birthday
Crb 15h46'39" 26d35'15"
Cecil
Cap 21h35'54" -15d50'28"
Cecil
Umi 13h29'10" 72d14'7"
Cecil and Silvey Robinson
Cyg 21h41'8" 45d49'0"
Cecil E. Booth
Cam 3h53'45" 67d50'10"
Cecil Eugene Newberry
Lib 15h18'9" -5d59'16"
Cecil Glennon Camick
Uma 11h26'29" 60d49'44"
Cecil H. Snyder
Cyg 19h58'59" 44d3'0"
Cecil Lee Walker
Ori 5h42'20" 8d17'16"
Cecil M Lopes III
Umi 16h9'39" 75d36'37"
Cecil M. & Stella O.
Elledge
Cap 20h28'39" -12d51'47"
Cecil Milton
Ari 2h15'34" 14d31'54"
Cecil Orji
Per 4h43'31" 41d43'46"
Cecil Vernon Clagg
Uma 8h35'31" 48d29'4"
Cécile
Boo 15h32'35" 49d0'50"
Cecile
Lyr 18h36'41" 35d52'47"
Cécile
Uma 11h54'34" 34d21'59"
Cécile
Psc 1h6'59" 10d27'45"
Cécile
Cap 21h51'2" -10d58'4"
Cécile et Raymond
Dalebroux
Per 2h23'51" 51d6'31"
Cecile Faubert
Umi 17h6'47" 80d23'55"
Cecile Floramarie
Longobardi
Cap 21h28'31" -15d46'41"
Cecile K. Delaney Collum
Cnc 8h18'58" 9d48'1"
Cécile Koza
Crb 16h1'25" 28d0'53"
Cécile La Monica
Ari 2h48'1" 11d51'37"
Cécile & Lena
Cas 23h3'55" 57d56'57"
Cecile Marie Musselman
Uma 11h32'1" 58d4'27"
Cécile Michel
Uma 12h9'1" 59d37'45"
Cécile Revol
Vir 14h46'3" -1d11'33"
Cecile, 14.10.1989
Aur 6h39'28" 38d48'55"
Ceciledean
Sge 19h15'56" 19d35'5"
Cecile's Star From now till
Forever
Psc 0h38'0" 12d23'43"
Cecilia
Ari 3h17'10" 15d38'33"
Cecilia
Gem 6h54'44" 28d57'10"
Cecilia
Crb 15h58'31" 37d15'43"
Cecilia
Crb 16h12'17" 34d58'40"
Cecilia
Leo 9h37'27" 32d23'16"
Cecilia
Per 3h21'40" 47d3'36"
Cecilia
Uma 10h48'1" 49d50'0"
Cecilia
Cas 0h57'15" 54d9'21"
Cecilia
Lib 14h31'43" -9d41'49"
Cecilia
Uma 10h34'6" 71d39'17"
Cecilia
Cep 1h54'19" 82d23'6"
Cécilia
Sco 17h32'34" -32d57'27"
Cecilia Anne
Sco 16h9'23" -15d45'50"
Cecilia Barcelona
Sgr 19h41'12" -14d12'25"
Cecilia Bell
Cas 1h47'17" 63d37'48"
Cecilia Belmont
And 0h41'52" 46d7'38"
Cecilia Bernardo
Cam 3h59'34" 58d3'7"
Cecilia C. Cano
Aql 19h53'34" -0d21'24"
Cecilia Calderon
Cap 21h34'48" -18d55'45"
Cecilia Calgaro
And 23h57'29" 40d26'59"
Cecilia Cantos
Ari 2h8'35" 15d58'5"
Cecilia Carolina Yudin
Cyg 20h11'56" 40d22'54"
Cecilia Catanese Happy
Birthday 50
Cas 0h42'0" 60d4'31"
Cecilia Catherine Georgitis
And 23h20'51" 48d6'44"
Cecilia Ceballos
Dra 15h32'45" 55d9'56"
Cecilia Clare Jackson
Crb 15h44'25" 25d41'36"
Cecilia Collier Brown
Lib 14h48'0" -1d58'20"
Cecilia Cruz 18th B-Day
Cap 20h14'44" -12d32'53"
Cecilia Duque
Aqr 22h38'58" 1d1'28"
Cécilia E. M. Fridlender
Cas 1h22'30" 61d34'50"
Cecilia Eliane Rosa
Paasche
Uma 11h1'53" 64d16'33"
Cecilia Erin Rogers
Vir 12h30'4" -1d23'30"
Cecilia F. Scheffler
Ari 2h52'36" 24d49'11"
Cecilia Faith Runge, Rising
Star
Aqr 22h46'22" -8d36'9"
Cecilia Faricy Hood
Crb 15h35'36" 28d9'19"
Cecilia Faure
Vir 12h33'15" -5d27'48"
Cecilia Flores
Cas 0h8'56" 54d34'33"
Cecilia Gomez
Ser 18h17'36" -15d34'31"
Cecilia HB 1
And 1h2'4" 34d47'17"
Cecilia Islas
And 23h17'46" 42d31'41"
Cecilia J. McLaughlin
And 0h0'57" 39d40'36"
Cecilia J Molis
Lyn 6h17'54" 60d59'11"
Cecilia Lina Bolduc
And 23h34'49" 41d25'42"
Cecilia Lorraine Bueso
Matheu
Crb 15h51'23" 27d27'1"
Cecilia M. Perez
Uma 12h42'57" 53d26'1"
Cécilia Maï
Uma 12h45'8" 53d31'53"
Cecilia Maria Vaughan
Leo 11h16'33" 23d15'20"
Cecilia Martinez
And 0h28'14" 41d30'8"
Cecilia Mary Grace
Ari 3h17'45" 22d18'42"
Cecilia Montano
And 23h45'15" 47d3'46"
Cecilia - My Beautiful
Darling
Crb 15h36'22" 28d17'18"
Cecilia Neuman
Ori 6h15'51" 9d27'7"
Cecilia Nicole
Crb 15h37'10" 30d19'59"
Cecilia Pasiciel
Sco 16h6'15" -11d9'14"
Cecilia & Peter Laurits
Srp 18h16'34" -0d5'27"
Cecilia Prado Salisbury
Brown
Sgr 19h1'21" -28d54'40"
Cecilia Rose Likos
And 1h4'17" 39d41'28"
Cecilia (Silly)
Cnc 8h53'42" 14d59'33"
cecilia super-star 02.07.76
Ori 6h4'9" 9d58'30"
Cecilia's Everlasting Angel
Vir 13h19'25" 5d25'7"
Cecilia's Love
Cyg 20h9'43" 35d0'9"
Cecilia's Silver Star
Cyg 19h42'38" 29d28'21"
Cecilie Toudal Pedersen
Leo 11h7'28" -3d26'39"
Cecilius Maximus
Fortitudinus
Uma 11h27'50" 32d32'21"
Cecil's Star
Cnc 8h3'28" 10d5'13"
Cecily
Cnc 8h3'16" 7d14'33"
Cecily
Peg 21h43'8" 23d24'48"
Cecily
Lib 15h1'51" -0d41'43"
Cecily
Uma 10h13'14" 63d34'10"
Cecily Ann
Sco 16h47'23" -30d41'25"
Cecily Brielle Herchek
Psc 1h15'5" 16d55'46"
Cecily Celmer-Shubin
Ari 2h8'50" 24d9'0"
Cecily Clark
Lyn 6h59'18" 45d3'8"
Cecily Dalilah
Vir 12h17'59" 11d27'27"
Cecily Kolos
Cas 0h40'17" 50d46'46"
Cecily Rose Corpina
And 1h9'31" 42d39'52"
Cecily Worley
Lyn 7h38'12" 56d35'44"
Cedia
Gem 6h49'25" 34d5'0"
Cedarhurst Gardens - T&J
Jandro
Psc 0h21'34" -1d54'26"
Cedelle
Cru 12h6'25" -61d53'42"
Cedie Lynn Mock
Uma 9h44'3" 61d57'42"
Cédric
Cas 1h46'36" 61d45'25"
Cedric
Men 5h32'58" -75d3'0"
Cedric
Per 4h47'49" 46d41'45"
Cedric
Ori 6h18'24" 14d41'57"
Cédric
Ori 6h0'24" 5d6'52"
Cédric
Com 12h0'32" 28d2'1"
Cedric Adrian Schoenberg
Dra 20h19'27" 64d45'3"
Cedric and Nichole
Cyg 20h0'1" 37d32'47"
Cedric Brooks
Per 3h35'38" 32d32'55"
Cédric Forever
Umi 17h2'6" 75d52'13"
Cedric Gene
Leo 11h11'56" 21d58'32"
Cédric Pagès
Her 18h49'37" 22d21'38"
Cédric Rochereau
Del 20h39'54" 14d56'36"
Cedric Stephanie Thiebaud
Uma 11h13'0" 32d32'10"
Cedric Waßmer
Lac 22h47'1" 52d54'17"
CEE
Peg 22h9'51" 6d21'10"
CEEBEE
Leo 11h4'32" 9d38'33"
CeeCee
Psc 1h26'34" 15d53'7"
Ceegoid
Uma 9h28'44" 65d9'4"
CeeJay Larino
Cep 22h14'7" 56d43'0"
CEELESE'S DREAM
Gem 7h23'12" 23d39'9"
Cees Kikstra
Psc 1h6'9" 24d33'14"
Cef & Naisha
Sge 19h49'10" 18d51'22"
CEG II
Psc 0h44'41" 13d20'12"
Ceglarek, Volker
Uma 11h3'19" 56d19'20"
Ceije
Tau 4h43'11" 26d55'12"
Ceil & John Valeri
Gem 7h44'13" 29d52'28"
Ceili
Gem 6h8'40" 22d59'23"
Ceilia Rose Cantarella
Uma 9h58'1" 60d24'25"
Ceilidh Anna Giove-Little
Dragonfly
Uma 9h57'27" 71d1'40"
Ceiridwen Davies
Cru 12h13'55" -57d59'56"
Ce-Jo
Uma 10h14'31" 62d26'49"
Cel 1967
Uma 12h34'23" 57d53'3"
Celais
Lib 15h47'47" -17d31'21"
CELAL
Sgr 19h39'46" -26d7'16"
Celarcio
Leo 11h23'0" 5d52'58"
Celebrate-50-Years-of-BILL
BRAUN!
Crb 16h1'19" 39d17'55"
Celebrating Al Johnson's
60th Year!
Cep 22h1'20" 61d3'52"
Celebrating Mike's Good
Health
Cep 21h34'8" 58d55'39"
Celebration
Lib 15h8'20" -12d23'7"
Celebration Of Love
Cyg 21h21'44" 37d25'20"
Celebritas Aureus Julia et
Militis
Gem 6h40'8" 17d15'1"
Celecassdereen
Uma 13h45'27" 53d1'34"
Celece & Mark Forever In
Love
Cyg 20h3'12" 59d17'24"
Celena
Uma 11h47'22" 57d25'32"
Celena "Ceny" Careen
Schirmer
Uma 11h33'1" 56d50'36"
Celena's Wish
Leo 9h22'58" 12d42'20"
Celene Elisabeth
Haberkost
And 0h56'55" 35d5'20"
Celenia
Tau 4h40'22" 5d30'25"
celeripes
Mon 7h10'25" -0d0'59"
celest
Uma 9h0'14" 49d9'55"
Celesta Our Christmas
Angel
Uma 9h56'45" 66d44'5"
Celeste
Cma 6h45'22" -14d39'12"
Celeste
Sgr 18h35'42" -24d53'51"
Celeste
Uma 11h50'14" 46d33'7"
Celeste
And 0h43'57" 42d47'38"
Celeste
Mon 6h37'21" 4d6'43"
Celeste
Vir 12h48'49" 5d21'18"
Celeste
Ari 2h33'32" 21d10'34"
Celeste
Gem 6h23'33" 19d0'34"
Celeste Beatrix MoLaw
Menders Huang
Cam 4h51'55" 66d30'30"
Celeste Cecilia Ruiz
Lib 15h15'50" -7d13'32"
Celesté Chavkin-Moreno
Sco 16h7'3" -17d37'32"
Celeste D. Hamler
And 2h31'7" 39d26'27"
Celeste Denise Ochoa
Her 17h51'33" 42d45'39"
Celeste Deshaies
Cap 20h54'16" -24d22'29"
Celeste Dominique
Cas 0h1'27" 53d49'50"
Celeste Donnelly Happy
Five-Oh!!!
Gem 7h11'22" 34d11'41"
Celeste Duggan
Cet 1h12'57" -0d20'43"
Celeste F. Bradway
Cyg 20h18'41" 40d49'0"
Celeste Holsheimer-Genet
Ari 1h49'9" 25d13'37"
Celeste Jeanne
And 22h59'23" 47d6'59"
Celeste June Hendrickson
Gem 6h52'28" 15d0'24"
Celeste Katherine Petro
Cnc 9h2'56" 22d18'12"
Celeste Kimberly Perry-
Keefe
Ari 3h24'13" 29d48'17"
Celeste Kovein
Del 20h42'40" 15d33'27"
Celeste Leeann
Cap 21h12'18" -18d52'37"
Celeste M. Gibson
Sgr 19h7'59" -13d36'21"
Celeste Mares
Peg 22h54'46" 15d30'10"
Celeste Marie Duvall
Leo 11h39'22" 21d54'50"
Celeste Marie Gibson
And 23h45'41" 44d46'55"
Celeste McGrogan
Vir 12h52'36" 6d41'10"
Celeste Noel 21
Cam 4h49'9" 56d14'32"
Celeste Noelle
Cap 20h10'6" -14d22'25"
Celeste Nora
Tau 5h33'19" 24d0'32"
Celeste Paula Flosman
Ari 3h3'7" 19d27'10"
Celeste Pino
And 0h45'45" 45d26'56"
Celeste Rau
Uma 8h28'13" 60d20'52"
Celeste Shotts
And 23h2'5" 49d56'19"
Celeste Virginia Hayo
And 0h19'5" 28d5'9"
Celeste Viviana Rodriguez
And 0h38'20" 33d48'14"
Celeste Vuturo
Cas 0h54'29" 53d31'18"
Celeste Whatley
Crb 16h3'42" 31d17'31"
CelesteAnne
Ari 2h20'28" 17d36'18"

Celeste's Heavenly Light And 1h21'30" 48d0'0"
Celestia Charity Rosali Rohling Vul 19h1'15" 22d36'55"
Celestia Janelle Lyn 7h48'53" 52d51'4"
Celestial Amaranthine Cap 21h5'55" -26d56'5"
Celestial Angel Sgr 18h45'33" -16d59'31"
Celestial Angler Ed Monk Sgr 18h58'37" -29d14'19"
Celestial Beauty Cnc 8h53'23" 31d38'19"
"Celestial Beauty" Valerie M Kim And 0h37'29" 45d50'42"
Celestial Belle Lyn 9h38'10" 41d16'22"
Celestial Cherrys Oph 17h52'34" -0d0'47"
Celestial Gift And 1h13'46" 43d2'19"
Celestial Luciana Lib 15h37'55" -17d25'3"
Celestial Morgan Brooke Vir 12h49'27" 6d10'59"
Celestial Mother of Ty Lib 15h4'23" -18d4'22"
Celestial Music for Jeff and Karen Lyr 18h51'1" 34d31'29"
Celestial Queen Cas 1h22'46" 62d37'12"
Celestial Rae Mills Vir 13h5'24" -5d12'45"
Celestial Smile Col 6h4'41" -28d30'8"
Celestial Spark21 Dra 17h13'19" 53d59'34"
Celestial Strings Lib 14h51'54" -2d36'24"
Celestial Suzanne Ari 2h23'49" 11d16'1"
Celestially Bound Umi 15h48'55" 70d43'54"
Celestina Umi 14h12'35" 78d49'49"
Celestina Acela Sco 17h57'45" -37d55'57"
Celestine Short Tau 4h32'1" 13d9'38"
Celestino Ruiz IV Dra 19h15'29" 60d34'12"
Celestuna Cyg 20h54'11" 30d15'52"
Celewen Cyg 20h45'45" 47d26'48"
CELI Cam 4h2'58" 65d30'14"
Celia Cas 23h50'10" 53d36'38"
Celia Sgr 18h17'50" -21d49'15"
Celia And 23h26'14" 48d13'40"
Celia Tau 5h42'49" 24d1'17"
Celia and John- star designers Her 17h40'6" 16d2'33"
Celia Ann Williams Cyg 21h31'42" 48d50'37"
Celia Braga Lib 14h27'32" -15d53'38"
Celia Claire Hopkins Young Car 9h50'48" -61d2'42"
Celia Ellenberg Cyg 20h9'11" 41d19'13"
Celia Elyse Geraghty Cyg 19h50'27" 56d45'54"
Celia Estelle Shields And 2h15'51" 45d35'36"
Celia Ford Lmi 10h25'53" 32d0'43"
Celia Forester McGrath Lyn 7h39'31" 49d45'30"
Celia Gaillard Cap 20h46'38" -14d57'8"
Celia - Ich Och - Cedric Lib 15h10'52" -1d40'47"
Celia J. Allen Sgr 18h15'42" -19d47'17"
Celia June Uma 10h14'59" 60d59'18"
Celia M Iovinella Sgr 18h59'56" -17d57'13"
Celia M. Maysonet Ortiz Aqr 21h11'24" -12d47'42"
Celia Mattei Cnc 9h17'47" 11d38'59"
Celia & Ramon Uma 12h41'55" 56d15'0"
Celia Ramos Cyg 21h53'12" 41d43'38"
Celia Rose Elk Costa Uma 10h2'11" 70d15'52"
Celia Rose Fitzgerald Her 17h4'31" 29d19'56"
Celia Rose Millett Lyr 18h47'59" 42d52'51"
Celia Storey, Queen of the Universe Ori 4h50'20" 12d3'11"
Celia Switzer Leo 10h15'35" 22d2'51"
Celia Sy Dra 18h48'25" 62d59'8"
Celia Torres, Lelé Mon 7h1'43" 3d48'33"
Celia Valentina Buckner Cnc 9h11'26" 26d38'42"
Celia Villanueva Cotto Cas 1h15'39" 62d30'29"
Celia-Jade And 2h7'42" 46d40'52"
Celia's Dragon Eye Cap 20h28'10" -12d55'20"
Celias Favorite Star Aqr 22h4'28" 0d1'11"
Celia's star Tau 4h16'15" 2d48'19"
Celia's Star Cam 4h6'26" 55d17'46"
Celica Tau 4h5'3" 5d8'50"
Celie Carnley Cyg 20h30'54" 35d43'41"
Celie Kate Cowart And 0h33'23" 32d14'16"
Celina Ori 6h8'0" -0d21'33"
Celina Uma 13h53'51" 53d19'54"
Celina Cam 5h26'18" 64d31'55"
Celina 525 Cyg 19h54'11" 33d13'24"
Celina "Angel" Uong Cap 21h37'29" -17d35'25"
Celina Brescia Crb 15h57'27" 27d6'34"
Celina Jo Boyles Lyn 8h56'41" 37d54'2"
Celina m. e. Tuttolomondo Ori 6h7'9" 19d26'8"
Celina Miyo Ori 5h42'59" 0d47'32"
Celina Ruiz De Miranda Cas 2h35'27" 63d56'18"
Celina's Fantasy Leo 10h12'46" 24d40'16"
Celinda Per 3h22'56" 46d51'3"
Céline Cas 0h19'15" 53d33'22"
Celine And 1h34'32" 39d54'27"
Céline Peg 22h6'32" 27d45'55"
Céline Peg 22h42'58" 23d13'18"
Céline Uma 8h51'40" 55d15'36"
Celine Umi 16h17'28" 81d47'25"
Celine Arianna Cap 21h18'56" -24d37'56"
Céline Beuchillot Tau 5h48'0" 17d27'25"
Celine Butland And 0h10'30" 46d13'24"
Celine Cunha Sgr 19h21'7" -31d23'50"
Céline et Jean-D Gem 6h38'44" 25d0'41"
Céline Fiona Cyg 21h10'27" 31d42'9"
Céline Forfait De Ski Ori 6h20'37" 13d52'11"
Céline Genier Mon 6h23'23" 8d18'19"
Céline Haederli And 23h14'34" 41d14'40"
Céline Jelk Lib 14h35'34" -11d44'8"
Céline Lamongesse Crb 15h57'24" 37d41'41"
Céline Lescureux Cam 7h29'58" 71d15'0"
Céline Marclay Cas 23h49'44" 57d3'23"
Celine Martino Cas 1h34'54" 67d38'15"
Céline Olivieri Sgr 18h7'59" -17d4'55"
Céline Rollée Peg 21h54'14" 7d53'17"
Céline Touren de Mèze (34) Sgr 18h15'1" -22d32'44"
Céline Volant And 1h13'6" 41d31'57"
Céline, Tao, Tep & Compagnie Cen 14h30'26" -61d52'44"
CELINECZKA Umi 16h17'51" 75d46'10"
Céline's eyes Lib 15h44'17" -11d22'41"
Celinia Ori 5h40'1" 12d38'8"
Celio Chapa, Jr Gem 6h8'9" 25d42'32"
Celipsimas III Psc 0h21'16" -5d21'40"
Celita Uma 11h22'48" 70d22'35"
Cellar Door Herr England Peg 22h7'20" 6d31'42"
Cellbrot, Ingrid Uma 8h57'53" 55d10'47"
Cello Man Uma 11h59'33" 29d52'54"
Cello & Waschi Cep 22h52'59" 69d41'41"
Cellot Boo 15h39'30" 48d1'3"
Cellsal Ari 2h34'25" 18d9'35"
Celt Ori 6h24'15" 17d4'51"
Celtic Caity Ari 2h39'11" 29d37'3"
Celtic Dreams Tau 4h43'15" 18d39'31"
Celtic Mama Tau 5h44'47" 17d26'4"
Celtus Mobius 2000 Umi 15h23'44" 72d9'16"
Cem Boyner Lib 15h10'15" -20d5'40"
Cem Siyahi Psc 0h50'1" 28d7'39"
Cemal Noyan Sgr 17h52'53" -29d25'54"
Cemalettin SARAR Umi 14h38'58" 68d12'58"
Cemiac Lac 22h27'55" 43d20'23"
Cemma, David's Angel Cma 7h6'38" -29d2'45"
Cendrillon, 01.10.2004 Umi 14h7'27" 78d26'17"
Cenestin Cyg 22h1'49" 45d25'45"
Cengilay Dra 18h49'57" 55d43'8"
Cenie Harris Lyr 19h6'46" 33d14'21"
Cent'Anni Jimmy & Jenni Cyg 19h51'20" 50d25'20"
Central Collegiate "Class of 2006" Uma 11h28'54" 59d21'37"
Central Elementary Teachers 2006 Uma 11h52'19" 59d15'9"
Cephas Sylvester Snider Sgr 18h14'28" -18d47'20"
Cepheus (the king) Alec Aqr 22h18'25" 0d38'48"
Cera Cyg 21h50'1" 37d20'39"
cercle de la vie de l'alma Cnc 8h43'54" 25d29'12"
Cerel Reverie Uma 13h39'22" 55d26'43"
Ceren Alikaya Lyn 7h59'21" 58d27'48"
Ceres Gem 7h28'46" 26d18'21"
Ceres Leo 10h37'42" 18d30'19"
Ceres Christine Ori 5h31'32" 6d51'26"
Ceri Ann Hale "Caru" To Love Cas 0h56'6" 47d41'28"
Ceri Joanne Roscoe Ori 5h36'37" -2d42'51"
Ceridwen Cleaver Uma 10h37'37" 46d12'12"
CERIII Her 18h6'12" 23d18'25"
Cerina Cas 23h7'1" 59d27'44"
CerinaDamionSean..... Uma 10h56'21" 48d13'58"
Cerino Ro'onui Tihoni Cyg 21h14'57" 46d43'8"
Cerisa Meunier Tau 4h21'35" 17d39'28"
Cerissa Lib 15h15'54" -9d29'36"
Cerlette Bridgette Roberts Uma 13h2'17" 55d46'13"
Ceron Dra 18h27'1" 55d20'51"
Ceronda Marlo Eri 3h12'59" -21d12'46"
Cerridwen Leo 9h47'19" 20d31'58"
Cerrie's Star And 1h11'9" 45d54'4"
Certeza Uma 10h34'56" 64d43'51"
Cerys Coombes And 1h10'10" 48d17'21"
Cerys Eleri Wakeling Umi 14h55'12" 72d30'50"
Cesaire Cyg 20h37'22" 41d7'47"
Cesar Her 18h27'43" 17d42'38"
César Sco 16h52'43" -27d3'56"
Cesar Sgr 18h10'49" -27d29'18"
Cesar A. Capo Cnc 8h12'1" 7d11'33"
Cesar&Ambra Ori 5h46'58" 6d2'56"
Cesar B. Garcia - "AMOR ETERNO" Ari 3h0'10" 13d36'2"
Cesar C Jovillar, Jr Ori 5h42'0" 8d2'23"
Cesar F. Silva Uma 8h15'0" 69d11'59"
Cesar Gallardo Ori 5h33'49" 11d4'26"
Cesar Gonzalez Dra 19h32'28" 73d51'23"
Cesar Guixe Cap 21h54'17" -10d43'30"
Cesar M. Figueroa Lyn 6h51'0" 58d58'19"
Cesar Padilla Uma 11h34'1" 64d41'17"
Cesar (Vanessa's heart) Gem 7h33'0" 28d14'43"
Cesarano Psc 1h1'45" 31d22'15"
Cesarano Cherubina Umi 15h43'59" 75d28'19"
Cesare Boldrini Cap 20h28'17" -10d4'13"
Cesare Buttarelli 2 gennaio 1982 Lac 22h57'38" 44d8'49"
Cesare Cella Lyr 19h16'34" 43d26'44"
Cesare&Manuela Gem 6h49'33" 16d39'1"
Cesare Stefano Bernardinelli Crb 16h8'34" 30d15'44"
Cesare Zagarolo Uma 8h16'4" 61d53'36"
Cesca Lang Cas 3h12'21" 61d34'5"
"Ceschia", You're a Beautiful Woman And 23h49'11" 42d44'3"
CESI Schriber Sge 19h59'18" 16d57'35"
Cesia Cruz Cam 7h14'18" 70d10'7"
Cesis-800 Uma 9h31'9" 59d15'6"
c'est nous - 63 Cap 21h12'8" -19d19'44"
C'est tout pour vous Cyg 20h11'11" 53d18'0"
Cesy Sgr 19h10'32" -24d9'37"
Cete, Alpan Uma 10h32'46" 72d46'4"
CETUS Uma 11h25'53" 54d41'21"
Ceurvorst-Digges Cnc 8h9'26" 26d12'45"
Ceutaiche Hayden Elyse Psc 1h10'28" 21d3'32"
Cevi Wülflingen Cas 23h36'39" 51d35'15"
cey moon Tau 3h27'1" 6d23'52"
Ceyda Parmaksiz Uma 13h20'33" 57d56'5"
Ceylan Arbel Leo 11h6'32" 20d22'56"
Ceylene Lib 15h19'44" -21d50'7"
C.F. Hornecker III Family Star Lyn 7h45'21" 45d12'27"
CF N RL 115's star Sge 19h9'40" 19d4'1"
cFNg Cam 3h57'17" 70d50'12"
CFW II Uma 10h31'19" 51d35'27"
CG Ari 2h59'25" 30d35'17"
C.G. Amoroso Her 17h51'11" 21d0'12"
cGm Apu 16h9'47" -76d37'33"
CH. Briarbrook's Pard'n My Dust Boo 14h41'31" 36d55'50"
CH Chefhaus Taragon V Himmlisch Ori 6h3'26" 17d49'48"
CH Chestnut Hill's Stone E's Tug Uma 12h29'45" 58d48'43"
Ch<\@>d<\@> Cap 20h58'55" -25d42'51"
CH. Darkover Wind off the Skagit Cnv 13h18'17" 47d36'23"
CH DeLeao's Unchained Melody CD Umi 16h47'27" 81d3'41"
CH Deschutes Sir Zechariah JH RN Cnv 13h48'46" 34d49'27"
CH. Fantasy's DDA Thunderstruck Cma 7h9'36" -27d58'41"
Ch. Kanabec's Vision of Black-Gold Uma 11h20'14" 48d14'48"
C.H. Kenneth Knisely Cnv 13h50'45" 34d46'31"
CH Lace's Princess Penelope Rose Psc 1h55'42" 5d48'38"
CH Lin-el's Blaze'n Uno Bleu JH Cnv 12h39'14" 41d50'4"
Ch. Meade's Buster Brown, U.D. Umi 14h3'39" 71d51'51"
CH MIKELL'S RANAH OF LEATHERNECK Uma 11h45'57" 29d2'56"
Ch. Misty Mtn's & Little Mermaid Cma 6h52'12" -13d22'29"
Ch. Misty Mtn's & Ltd Edition Eddie Cma 6h40'25" -13d8'23"
Ch. Starr's Outside Chance Aql 19h36'36" 0d9'43"
Ch Sunset Ducat Chippewa Cosmos Uma 11h0'36" 64d31'57"
Ch Tag Along's Lani Makana Faith Uma 11h18'43" 63d43'27"
CH TOP HAT POLO Cyg 20h5'57" 35d45'21"
Ch. XTC Macado MacDiarmid Uma 10h5'16" 57d39'31"
Ch, BrigGen Alphonse J. Stephenson Cep 22h34'11" 74d17'43"
Cha Umi 15h17'34" 73d15'6"
Cha Cha Dra 19h29'37" 64d18'24"
Cha Cha Girl And 1h35'55" 41d27'31"
cHa iT's ChRis Uma 12h44'44" 57d28'40"
Cha&Klavs Her 17h49'1" 34d33'12"
Cha Stead Peg 23h13'36" 27d12'42"
C.H.A. Superstars Per 2h59'47" 42d20'31"
Chabmer Uma 8h47'39" 58d51'5"
Chace Benjamin John Brogan Umi 7h10'59" 88d40'4"
Cha-cha-cha Cap 20h35'39" -14d18'32"
Chachi 1 Uma 8h39'1" 58d4'37"
Chachouna Alexandra Fernandez Uma 12h14'17" 62d50'42"
Chacon Uma 9h13'35" 62d4'56"
Chad Uma 10h37'50" 47d1'40"
Chad Leo 10h31'23" 11d50'16"
Chad A. Prince Umi 14h28'24" 68d4'5"
CHAD ALEIN ARM-BRUSTER Sgr 17h57'56" -17d17'29"
Chad & Ali Warren and Family Vir 14h45'12" 5d40'45"
Chad Allyn and Nikki Ann Coverston Cyg 21h37'6" 51d12'11"
Chad and Alayna Lyr 19h6'15" 31d58'55"
Chad and Alice Gryder Tau 3h39'10" 2d22'49"
Chad and Andrea Matheny Tau 4h46'14" 26d38'21"
Chad and Brittney's Cyg 21h19'36" 41d19'7"
Chad and Katelyn Ostrander Lyr 19h16'20" 28d37'8"
Chad and Tara Gerstmeyer Cyg 20h28'4" 34d52'53"
Chad Andre Vir 14h30'6" 3d44'26"
Chad Andrew DiBiase Lib 15h12'43" -20d11'41"
Chad Andrew Pilatsky Sco 17h22'45" -42d26'22"
Chad Anthony Pierce Ori 5h53'52" 6d11'40"
Chad Aric Cap 20h15'8" -15d48'11"
Chad Ariel Gamez Ori 5h48'33" 2d31'2"
Chad Bailey Cep 21h46'32" 70d38'15"
Chad "Big Daddy" Brandel Her 17h7'35" 31d17'7"
Chad Boda Gem 7h5'0" 15d9'38"
Chad Bryant Sale Her 17h16'2" 35d8'34"
Chad Burks Tau 4h5'56" 6d11'0"
Chad Christianson Cyg 21h40'19" 47d40'6"
Chad Christopher Probst Ori 5h36'0" 1d52'12"
Chad Crow Ori 6h6'48" 17d52'25"
Chad Daniel Whisenant Cyg 19h58'34" 55d47'20"
Chad David Rencher Cap 20h50'53" -18d49'9"
Chad Eatherly Psc 0h12'8" 7d10'5"
Chad Eric Buckland Lib 15h10'18" -13d19'4"
Chad Eric Daniel Boo 13h45'54" 17d28'46"
Chad Ernest Gates Uma 9h33'0" 68d0'34"
Chad Everett Compton Uma 9h7'2" 47d9'46"
Chad Everett Johnston Dor 5h16'24" -68d31'58"
Chad F. Curry Ori 5h50'30" 3d23'40"
Chad F. Sharrow Cnc 8h52'23" 28d59'21"
Chad Fekany Uma 11h23'28" 37d53'39"
Chad Frazier Uma 10h14'17" 59d55'45"
Chad Fugere Ori 5h33'50" 8d35'33"
Chad G. Caporuscio Psc 1h0'32" 11d21'55"
Chad G. Wood Her 17h15'20" 25d29'12"
Chad Gerstel Her 18h6'41" 17d53'48"
Chad Gould Lyn 7h59'36" 50d35'2"
Chad Jandrositz Uma 13h41'35" 58d28'29"
Chad & Jenna Williams Cyg 20h28'36" 45d27'32"
Chad John Hunder Cep 22h33'29" 82d50'54"
Chad Johnson Family Star Tau 5h24'56" 26d43'34"
Chad loves Kelly Psc 23h57'16" 6d58'14"
Chad Lowe Uma 8h42'9" 70d23'33"
Chad & Marion Sweethearts Over TX Uma 11h47'4" 34d31'18"
Chad Marshall Sabrsula Aql 19h29'13" -3d10'8"
Chad Martin 34 Uma 9h49'23" 51d8'8"
Chad Matthew Sinclair Vir 13h42'14" 3d3'20"
Chad Michael Evans Per 3h22'3" 51d45'39"
Chad Michael Halstead Per 4h28'53" 34d26'54"
Chad Michael Harrington Cap 21h11'1" -19d25'33"
Chad Michael Herrick Her 18h3'55" 20d35'48"
Chad Michael Krantz Leo 9h49'41" 15d30'55"
Chad Michael Shafer Boo 14h27'7" 24d32'37"
Chad Micheal And 1h16'37" 41d18'14"
Chad Miller Ser 16h15'13" -0d2'11"
Chad Mock Per 3h14'36" 52d20'9"
chad -n- nichole Ari 2h26'41" 19d17'50"
Chad Nelson Per 3h33'4" 51d32'22"
Chad Oliver Ruesch Lib 15h6'26" -3d39'41"
Chad Perry Berger Gem 7h19'28" 19d6'7"
Chad Punkie Wheeler Ari 2h48'16" 29d20'41"
Chad R. Dack Per 3h12'48" 50d54'43"
Chad R Harding 'leben und lachen' Cra 18h49'17" -39d42'38"
Chad Reisner Aqr 22h49'17" -10d51'55"
Chad Robert Long Uma 10h57'55" 42d19'4"
Chad&RobinFletcher Uma 11h11'6" 60d37'38"
Chad Robinson Cap 21h10'49" -22d11'28"
Chad Roger Johnson Aqr 20h41'25" -12d18'18"
Chad Roy Geiger Ori 5h16'21" 7d42'19"
Chad Rozowski Her 18h38'58" 15d58'9"
Chad Ryan Engstrom Leo 11h44'11" 24d46'46"
Chad Scanlon Umi 14h17'33" 78d42'39"
Chad Schnuelle Lyn 6h28'9" 57d34'37"
Chad Shannon Boyette Tau 4h28'34" 24d50'0"
Chad Simmons Cas 0h34'25" 54d2'23"
Chad & Susan Sgr 18h26'45" -16d20'33"
Chad Thomas Lasher Aqr 22h41'24" -5d53'18"
Chad Urnikis Ori 5h31'22" 9d57'30"
Chad Valitchka Uma 11h45'6" 39d50'28"
Chad Waleed Vir 14h5'57" -12d11'35"
Chad Walmer Lib 14h47'31" -7d43'28"
Chad Wesley Her 16h37'21" 28d46'55"
Chad Westall Talbott Umi 15h11'53" 71d33'22"
Chad Weston Gardner Uma 9h4'5" 54d34'3"
Chad "Wicky" Harjer Sco 17h55'13" -39d50'25"
Chadd Hambidge Per 3h11'28" 31d43'42"
Chadd & Kari's Eternity Cra 19h2'16" -38d28'12"
Chadd Leland Huddon Vir 13h17'24" 8d20'36"
Chaddasah Sunshine Cyg 21h45'12" 38d33'36"
Chadder "The Fanch" Boo 14h43'34" 17d53'12"
Chaddwick William Wooster Sco 16h14'45" -14d35'10"
Chaddy Cap 21h42'6" -24d6'57"
Chaden Sugihara Mon 7h57'21" -4d32'5"
Chadopolis Tau 5h20'53" 20d58'54"
Chadralin Uma 9h21'3" 49d52'45"
Chad's Angel Her 17h35'26" 33d32'58"
"Chad's big one that got away" Uma 9h32'9" 69d40'3"
Chad's Star Sgr 17h55'56" -27d34'53"
Chad's Star Uma 9h57'14" 41d52'52"
Chadwick Alexander Brooks Aql 20h7'3" 6d45'33"
Chadwick David Brown Gem 7h40'12" 21d16'35"
Chadwick David Vansky Lib 14h50'37" -3d48'4"
Chadwick J. Bunch Lmi 10h37'2" 26d50'3"
Chadwick Russell Sgr 18h6'9" -27d34'54"
Chadwick-Davis Sge 20h3'16" 17d4'19"
Chadwyn Boo 14h32'31" 29d58'54"
Chae Her 17h59'46" 42d40'14"
Chaelene Bransdorf Psc 1h38'11" 19d52'34"
Chaelis Mendoza And 2h23'7" 46d21'3"
Chaem Tau 4h3'25" 17d24'26"
Chäfer Equ 21h5'40" 5d7'14"
Chäfer Cep 23h10'16" 70d8'2"
"Chäfer" Reni Cep 2h29'3" 84d17'28"
"Chäferli" Cep 3h33'40" 82d53'27"
"Chäferli" Christian Umi 13h25'1" 71d26'32"
Chagee Aqr 22h55'14" -7d22'31"
Chahana Tau 5h37'54" 25d39'16"
Chahnaze Somrani "Zampetta" Uma 11h9'7" 42d30'10"
Chai Mehra And 23h5'3" 48d17'52"
Chaia Sara y Mordecai Sco 15h54'2" -27d20'40"
Chaila Marie Cochran Umi 14h0'11" 70d0'46"
Chaille Patrick Stovall Cap 21h7'34" -17d22'20"
Chairath Panicharoensakul Umi 16h39'18" 76d53'0"
Chakapachaka Umi 14h49'37" 79d34'9"
Chakoonia Crt 11h36'54" -8d22'36"
Chakri Cap 21h5'25" -16d15'51"
Chalan Harper, my favorite trouble Ori 6h4'20" 20d47'0"
Chalarose And 0h44'31" 43d54'37"
Chalice Powell Gem 7h36'50" 18d2'2"

Chalie & Tina
Ari 3h15'23" 28d33'20"
Chalifour in the Sky
Aqr 23h1'26" -6d21'9"
Chalise Wells
And 1h49'22" 38d5'43"
Challis Star
Pho 0h41'23" -44d2'7"
Chalmer & Helen
Cyg 20h51'41" 48d42'44"
Chalmer Leland Zabrok
Per 2h30'27" 55d42'42"
Chalonda
Sgr 18h41'4" -33d7'44"
Chalupa Marie 0304
Cnc 8h54'32" 14d56'37"
ChaLyce Stancil
Com 12h30'28" 22d14'34"
Chalyse Angelina
Cas 0h58'27" 60d53'27"
Cham Ocondi
Ari 2h11'40" 12d0'5"
Chamaine
Ori 5h34'36" 10d25'35"
Chamandiak
Sco 16h15'48" -10d7'16"
Chambers' Star
Aql 19h38'9" -0d56'25"
Chameleon Dancer
Tau 3h39'58" 16d48'7"
Chami
Cap 20h51'12" -16d15'50"
Chammie
Cma 6h44'34" -11d59'49"
Chamois
Cma 6h48'47" -13d5'21"
Chamois McGill
Cma 6h23'45" -16d28'57"
CHAMP
Sgr 19h38'30" -21d28'3"
Champ
Uma 11h0'8" 63d43'3"
Champ
Cyg 20h17'8" 47d24'14"
Champ
Her 16h43'8" 46d57'38"
Champ Tara and Straton Covington
Lyr 18h40'54" 38d38'8"
Champ The Beloved Beagle
Uma 10h3'48" 54d29'35"
Champagne
Tau 5h50'38" 26d3'39"
Champagne Supernova
Uma 10h10'38" 49d56'12"
Champion Rkarth
Aql 19h42'37" 13d42'23"
Champion Sumersong Night Tracks
Ori 6h2'23" 11d52'23"
Champ....superstar dad
Uma 8h31'45" 72d28'42"
Champurradita (Kitty)
Lyn 7h16'44" 53d45'37"
Champus
Umi 15h36'51" 80d2'16"
Chamroeun Phou
Per 1h34'30" 54d33'53"
Chamron
Sco 17h21'53" -42d30'25"
Chan
Tel 18h25'27" -49d10'29"
Chan
Cam 4h53'12" 64d6'38"
Chan
Uma 9h15'24" 47d16'17"
Chan Cherri Ngo Kang
Sco 16h4'40" -13d35'56"
Chan Ching Man & Your Family
Sgr 18h6'59" -27d0'35"
Chan Gosselin
Uma 9h53'14" 69d51'43"
Chan Nheb
Lib 14h52'55" -1d27'53"
Chana
Gem 7h25'47" 14d19'54"
Chanae
Cap 20h14'6" -23d19'51"
Chanar
Peg 23h40'53" 16d44'26"
Chance
Leo 11h36'41" 18d28'9"
Chance
Cmi 7h26'8" 3d42'43"
Chance
Tau 4h5'59" 5d39'25"
Chance
Sgr 18h19'52" -32d39'15"
Chance
Cam 4h14'48" 67d54'59"
Chance
Umi 14h22'46" 72d31'33"
Chance
Uma 9h27'53" 60d39'21"
Chance
Uma 11h25'39" 67d28'11"
Chance
Uma 14h15'37" 61d24'59"
Chance Arnold Oubre
Cru 12h42'37" -56d52'51"
Chance Barnett
Per 4h18'20" 42d44'30"
Chance & Candace Jackowski
Cyg 20h40'17" 42d20'7"
Chance Chihanik
Mon 6h45'1" 11d23'50"
Chance Cole Walker
Umi 15h28'31" 72d5'3"
Chance Dakota Moore
Umi 14h10'19" 71d30'26"
Chance Dakota Moore
Uma 11h24'7" 32d53'38"
Chance Encounter
Aqr 22h31'27" -1d55'29"
Chance Jamerson Aguirre
Aql 19h50'42" 11d21'54"
Chance Llandon Livingston
Aql 19h33'52" 9d59'9"
Chance Mason Miller
Lyn 8h39'17" 45d1'15"
Chance McKinley Esposito
Peg 23h44'47" 27d39'8"
Chance Murphy
Gem 7h18'59" 21d19'53"
Chance Peterson
Uma 11h38'38" 53d16'40"
Chance Robert Bowne
Dra 17h43'27" 53d20'20"
Chance Slade Bennett
Ori 5h34'5" 13d51'15"
Chance, Sr.
Tau 3h52'18" 14d9'40"
Chances Are
Uma 11h40'33" 63d10'6"
Chances Are
Lib 15h24'26" -7d45'32"
Chancey Lee Newcum
Sgr 18h42'56" -28d18'10"
Chanda Renee Jacobi
Uma 14h1'2" 48d44'54"
Chanda, phenomenal mother & my wife
Cas 1h44'56" 65d49'56"
Chandan Rekha Sharma
Cas 23h25'36" 57d41'20"
Chandanith Douk
Tau 3h30'19" 0d59'22"
Chandelle Bodziony
Cyg 19h51'29" 36d51'49"
Chandi Leigh
Cas 0h53'36" 48d40'51"
Chandler
Her 17h10'46" 25d19'35"
Chandler
Aql 19h55'39" -0d48'36"
Chandler
Vir 11h45'43" -3d23'36"
Chandler Bailey Friedrich
Aqr 21h33'45" 2d29'57"
Chandler D. Kezele
Cap 20h37'46" -10d23'17"
Chandler Dale Newton
Uma 10h45'4" 70d14'33"
Chandler Elizabeth Fleming
And 0h26'38" 27d23'44"
Chandler Elizabeth Valentine Spivey
Lyn 7h56'9" 57d49'58"
Chandler Ellen Curran
Sgr 19h3'42" -12d42'33"
Chandler Jan Bradley
Sgr 19h12'2" -23d55'5"
Chandler Levar Glover
Psc 1h3'33" 31d57'57"
Chandler May
Psc 23h20'44" 3d56'27"
Chandler Neal McCauley
Boo 13h48'22" 24d58'40"
Chandler Renee Kardaras
Ari 2h44'23" 17d3'44"
Chandler Thomas McCraight
Ari 2h19'34" 25d27'54"
Chandler Viking AR Star
Her 17h52'26" 22d28'12"
Chandler Wade
Aqr 22h21'21" -3d47'36"
Chandler's ChaCha Star
Psc 0h16'30" 19d25'11"
Chandlers Eternity
Lyr 18h33'37" 40d14'10"
Chandni
Uma 11h7'14" 35d14'33"
Chandon
Vir 13h11'9" -7d27'39"
Chandora
Psc 23h56'50" 9d34'30"
Chandpal & Harjit
Cyg 19h55'51" 55d56'17"
Chandra
Vir 13h28'40" -18d26'58"
Chandra
Sco 17h49'21" -40d9'32"
Chandra
Aql 20h1'26" 3d24'29"
Chandra
Lyn 8h22'34" 44d8'49"
Chandra Boli
Psc 1h6'22" 31d37'37"
Chandra Dawn
Cnc 8h48'18" 13d34'58"
Chandra Hayes
And 0h24'55" 38d12'51"
Chandra Hillenbrand
Lib 15h19'44" -7d4'8"
Chandra Lee Stannard
Leo 11h2'56" 10d18'44"
Chandra Niedzwiecki
Cam 6h32'28" 68d9'31"
Chandra Sam 3
Del 20h38'14" 14d37'30"
Chandra Shrum
Crb 16h9'55" 38d7'59"
Chandra White *The Brightest Star*
And 23h7'59" 42d50'24"
Chandrew
Her 17h41'22" 33d42'59"
Chandrima
Cas 2h34'43" 68d15'1"
Chanel
Cas 23h27'1" 57d53'18"
Chanel
Cmi 7h29'58" 11d9'16"
Chanel Amour
Aql 19h47'12" -0d35'34"
Chanel Cartier
Lyn 7h36'4" 48d59'4"
Chanel Catherine Little
And 23h49'47" 34d56'46"
Chanel Dominique Holbrook
Cam 4h12'43" 67d42'55"
Chanel I Love You Yours Truly Jack
Uma 11h25'50" 54d16'12"
Chanel Le Jones
Cnc 8h31'11" 25d22'0"
Chanel Peel
Ori 4h47'34" 4d8'18"
Chanel Ruiz-Bricco
Boo 14h50'56" 50d13'15"
Chanell Amelia Timm
Aql 19h4'55" -7d0'35"
ChanelRene2
Sge 20h6'18" 17d44'10"
chanel-shawn
Uma 9h3'12" 59d42'55"
Chané.......Precious Angel
Gem 6h8'50" 25d18'51"
Chaney Carolyn King
Uma 14h1'31" 50d54'1"
Chaney Sinovich
Leo 10h3'21" 7d28'0"
Chang Chen
Eri 3h41'47" -41d10'56"
Chang Ning Tzyy
Sco 16h38'38" -25d8'10"
Chang Ruiting
Umi 14h4'17" 72d32'52"
Changa
Agr 23h16'35" -14d32'35"
CHANGAL
Sco 17h32'34" -38d32'37"
CHANGE BY STRAUSS
Lib 15h8'52" -27d35'11"
Changement
Gem 6h38'25" 20d37'17"
Changthy
Umi 14h13'12" 76d40'37"
Chani
Lyn 7h20'16" 44d53'13"
Chani Rachel Wheeler
Crb 15h50'30" 37d40'24"
Chanida Matheis
Peg 23h5'17" 26d3'59"
Channa Deneal Haines
Vir 12h44'17" -0d27'33"
Channa Lynne Ortiz Speckman
Ori 4h50'35" 4d18'26"
Channasa, l'amour de Hemang
Cas 2h49'33" 61d15'33"
Channel
Aqr 22h30'43" -1d47'48"
Channel ONE
Vir 12h43'19" 5d4'44"
Channing
Lyn 7h23'37" 49d55'38"
Channing Barnes
And 1h43'37" 46d5'30"
Channing Humphries
Lyn 7h34'35" 35d37'49"
Channing kehoe
Lyn 7h28'23" 45d16'18"
Channtah Marie
Cas 23h28'30" 58d46'47"
Chans Kong
Aqr 20h56'54" -12d46'9"
Chanse Austin Miller
Uma 11h22'54" 39d20'21"
Chantal
Peg 21h28'16" 15d35'37"
Chantal
Peg 22h48'36" 18d27'44"
Chantal
Lib 15h42'53" -11d24'24"
Chantal
Uma 9h39'22" 65d8'57"
Chantal
Sgr 18h34'29" -29d29'46"
Chantal Bernadette Gervais
Uma 11h22'50" 59d7'0"
Chantal Cyr
Aqr 23h28'22" -21d56'43"
Chantal Daniele
Crb 15h47'37" 29d8'55"
Chantal Depraeter
Crb 16h13'35" 27d49'46"
Chantal Dubois
Uma 9h10'14" 63d42'41"
Chantal et Jean-Claude
And 1h15'11" 46d0'43"
Chantal Giroux
Uma 11h31'16" 57d43'17"
Chantal Gomez
Vir 13h38'45" 3d53'29"
Chantal Gregory
Per 4h11'59" 34d13'2"
Chantal Holcombe's Gateway
Uma 11h17'51" 43d37'3"
Chantal J "Snowdrop" Graves
Cnv 13h34'36" 38d50'21"
Chantal Lapierre
Umi 14h6'14" 67d32'1"
Chantal Lardier
Lib 15h23'31" -3d53'45"
Chantal Leduc Michel Charbonneau
Crb 15h58'23" 39d21'55"
Chantal & Luke's Forever Star
Cyg 19h51'20" 38d55'27"
Chantal Marie Burke
Vir 14h48'52" -0d19'37"
Chantal Marie Moser
Crb 15h44'50" 31d21'44"
Chantal Mercy Shrauger
Aqr 21h52'7" -7d49'43"
Chantal Mousis
Cnc 8h34'23" 17d45'59"
Chantal Porfido
Lyr 19h3'42" 42d32'48"
Chantal Rousel
Ori 6h0'25" 22d18'58"
Chantal Santos
Aqr 22h1'48" -9d44'23"
Chantal Thompson
Uma 11h40'7" 29d4'47"
Chantal van't Hof Aujla
Cyg 19h44'53" 30d3'54"
Chantal Weeland
Cap 21h9'33" -22d8'56"
chantal westerhoff
Cnv 12h53'26" 41d37'55"
Chantale
Sge 19h52'17" 18d43'51"
ChantaLise
Sgr 18h22'6" -23d0'17"
Chantal's Star
And 2h18'21" 49d50'9"
Chantana Bishop
Psc 1h4'59" 28d12'28"
Chanté Danielle Dacourt
And 0h21'42" 46d0'37"
Chantel
Lib 14h49'1" -1d46'35"
Chantel Anne Kent
Lyr 18h53'1" 36d14'21"
Chantel Danae Armstrong
Lyn 7h38'0" 48d7'33"
Chantel Gomes
Cas 1h15'31" 60d23'27"
Chantel L. Bonner
Cap 21h11'6" -16d50'57"
Chantel Licata
Cap 20h18'17" -25d1'42"
Chantel M. Randell
Tau 5h46'58" 22d1'8"
"Chantel" Maine
Cet 0h43'50" -1d31'21"
Chantel "Shawny" Gale
Uma 8h45'20" 49d33'40"
Chantell
Cas 1h41'22" 62d45'14"
Chantell and Eric's star
Cyg 21h30'17" 34d10'3"
Chantell Marie Garvey
Cas 0h59'0" 67d23'28"
Chantell Tracey
Her 17h0'41" 26d30'7"
Chantelle
Aql 19h37'7" 7d18'1"
Chantelle
And 22h59'6" 52d3'16"
Chantelle Allise Lawrence
Psc 0h55'7" 32d12'34"
Chantelle - Always Amazing
And 23h53'47" 45d11'7"
Chantelle and Michael
Cap 21h50'26" -10d28'11"
Chantelle Anderson
Vir 13h0'13" 5d44'12"
Chantelle Groene
Uma 11h48'14" 54d46'23"
Chantelle My Love
Cru 12h52'51" -63d3'21"
Chantelle Robinson
Cru 12h49'8" -64d39'31"
Chantelle Sonya Brown
And 1h16'2" 42d10'34"
Chantell's Angel Star
Gem 7h5'34" 25d6'29"
Chantell's Guiding Light
Uma 10h55'55" 65d21'50"
Chanti Claire
And 1h57'54" 42d59'26"
Chantilly
Gem 7h13'44" 31d3'12"
Chantilly
Ari 2h51'21" 24d37'12"
Chantra's Star
Vir 13h55'9" -11d21'25"
Chanyn Marie Deshong
Her 17h16'5" 16d48'56"
Chaos
Aql 20h11'42" 13d52'52"
Chaos
Lmi 10h6'2" 34d31'51"
Chaos
Aur 5h27'39" 40d20'15"
Chaos
Cam 4h47'49" 69d22'51"
Chaos & Barek Willey
Cma 6h47'43" -20d1'56"
Chaos McKimmie
Ori 5h25'26" -5d10'31"
ChaosCK
Vir 14h6'42" -11d16'21"
Chaotic Collection of Curiosities
Ori 5h34'40" -5d26'42"
Chaotic Serenity
Sct 18h34'2" -7d27'13"
Chapman Northcutt Jones
Umi 14h38'41" 73d41'3"
Chappell Star
Lyn 7h49'31" 57d8'38"
Chappet Sabrina
Gem 6h50'55" 20d41'5"
Chappo's star
Lmi 10h45'54" 23d20'27"
Chapril
Cam 5h31'31" 57d7'2"
Chaquin
Leo 9h22'49" 27d15'4"
Chaquita
Tau 4h12'34" 27d28'35"
Char
Leo 9h56'58" 10d5'50"
Char
Gem 6h51'28" 31d59'37"
Char Ahroni
Cap 21h20'7" -22d52'5"
Char and John
Cyg 20h1'25" 52d9'12"
Char Gascon
Ari 3h9'57" 12d54'42"
Char Star
Psc 1h32'42" 11d16'13"
Char Star
Uma 9h20'48" 46d33'2"
Char Star 060111
Sgr 18h21'50" -17d47'3"
Char the Star
Cnc 8h43'54" 12d58'23"
Char the Sunshine Star
And 23h34'1" 37d21'16"
chara
And 0h37'48" 41d18'37"
Charanne Denise Cook-Wilson
And 2h19'16" 48d16'16"
Charbel K. Hatem
Lib 15h39'49" -9d32'24"
Char-Char Binks
And 1h44'33" 42d22'47"
Chardonnay Andrews-Butta
Lyn 8h52'12" 46d11'37"
Charee's Smile
Lib 15h26'10" -17d2'31"
Charelle08
Cnc 9h18'4" 17d21'15"
-Charemon-Charile—Kiana-Jeff-John-
Psc 1h12'25" 26d6'28"
Charey Del Rosario
Gem 6h51'19" 18d58'3"
Chari
And 0h46'50" 40d20'35"
Charice Roberts
Uma 10h47'33" 44d37'27"
Charilaos Patazos
Ori 5h55'42" -0d21'18"
Charine Taoatao Clores
Leo 9h42'47" 24d53'2"
Charis
Her 18h0'14" 25d4'53"
Charis
Ori 5h35'7" -2d36'6"
Charis
Umi 15h24'51" 74d38'53"
Charis Alexa
Cru 12h28'45" -55d56'52"
Charis Dell Mayton
Ori 5h56'50" 2d31'38"
Charis Entwisle
Uma 8h38'54" 59d39'9"
Charis Olivia Holmes
And 23h59'42" 40d5'35"
Charise Mericle Harper
And 1h17'9" 49d45'9"
Charise Rose
And 1h8'15" 34d4'15"
charish
And 0h21'7" 22d3'37"
Charisma & Yanne
Sgr 18h3'25" -28d26'10"
charissa
Ari 2h52'10" 24d43'28"
Charissa Ann Jarrett
Pho 23h54'12" -45d59'28"
Charissa (Babe-A's Star)
Mon 7h34'29" -0d20'52"
Charissa D. Olson
Ori 6h16'40" 19d1'13"
Charissa Hom
Cnv 12h48'45" 39d52'31"
Charissa Leaha
Ori 5h58'1" 3d37'5"
Charissa Marie Gonzalez-Carmona
Crb 15h27'47" 31d35'47"
Charissa Sherray Mallory
Psc 1h3'33" 5d40'3"
Charissa Vercillo
Cas 23h52'30" 57d47'42"
Charisse
Umi 16h21'40" 71d56'5"
Charisse
Ari 1h52'39" 18d4'46"
Charisse Ann Pinzone
Lyn 8h21'47" 35d38'29"
Charisse Bolden
Peg 21h56'53" 13d13'24"
Charisse Dudkowski
Uma 10h29'49" 40d19'57"
Charisse Joyel Ellis
Cap 20h46'35" -21d15'38"
Charisse Monique Segee
Leo 11h1'8" 23d22'18"
Charisse My Angel
Leo 11h1'11" 9d29'51"
Charisse Praileau-Dunn
Aqr 21h53'2" -0d11'57"
Chariti Rae Small
And 2h19'51" 47d20'3"
Charity
Lyr 18h46'7" 32d25'43"
Charity
Vir 12h24'40" 9d29'43"
CHARITY
Tau 3h46'11" 18d8'14"
Charity
Crb 15h26'53" 26d53'42"
Charity
Vir 13h40'3" -6d51'31"
Charity
Cap 21h45'36" -14d1'50"
Charity
Dra 18h24'32" 67d24'52"
Charity
Cas 22h57'19" 57d9'59"
Charity
Sgr 18h19'9" -22d51'37"
Charity Burkholder
Oph 17h24'1" -0d28'25"
Charity Dawn
Aqr 22h13'59" -0d12'27"
Charity Harvey
Her 16h31'48" 24d32'10"
Charity Kay Lynn Renee Scevers
Sgr 18h35'29" -28d13'45"
Charity Lea Kofahl
Cyg 21h44'39" 41d14'58"
Charity Leah Washam
Cap 21h33'49" -9d10'58"
Charity "Lilred" Claire Boilesen
And 0h44'56" 37d34'1"
Charity Lopes Kelton
Lyn 8h34'22" 33d17'30"
Charity Lyn Nash
Ari 2h36'42" 25d36'58"
Charity Nichole Dugas
Vir 12h34'8" -0d49'32"
Charity Richardson
Lib 14h57'21" -16d30'49"
Charity "Unbreakable" Lansall
Aqr 20h53'18" -13d44'39"
Charl Viljoen
Cru 12h42'32" -62d44'55"
Charla
Uma 9h25'58" 53d40'15"
Charla
Uma 11h25'6" 29d56'12"
Charla 1976
Tau 4h36'43" 27d16'46"
Charla A. Dufour
Aqr 20h40'4" -2d0'25"
Charla Allison Lobb
Tau 5h11'52" 21d55'32"
Charla Ann
Crb 15h46'25" 28d36'5"
Charla Ann Brown
And 23h35'49" 43d26'23"
Charla Marie
Leo 11h46'26" 23d56'17"
Charleane
Gem 7h21'55" 25d37'31"
Charlee
Uma 12h38'50" 60d32'23"
Charlee Ann Laycock
And 0h2'6" 37d54'38"
Charlee Hawkins
Psc 0h41'13" 6d19'2"
Charlee Marie
Cnc 8h59'40" 15d45'31"
Charleen Alese
Ari 3h2'33" 15d28'14"
Charleen B. Rosebery
Uma 11h19'21" 30d51'52"
Charleen Bell
Cyg 19h35'4" 35d25'59"
Charleen Brianne Sims
Leo 11h30'16" 10d18'20"
charleen guy
Leo 10h45'7" 17d19'0"
Charleen Young
Cyg 21h9'48" 37d1'41"
Charleene Clift
Cas 1h25'46" 51d54'0"
Charleigh Hope Bechand
Sco 17h17'53" -33d45'23"
Charleigh Maria Morrison
Cnc 8h51'51" 32d38'8"
Charlena
Peg 22h1'11" 9d26'20"
Charlene
Vir 12h24'51" 8d4'23"
Charlene
Cnc 8h39'8" 8d42'27"
Charlene
Mon 6h43'48" 8d28'49"
Charlene
Aqr 21h10'24" 1d36'22"
Charlene
Gem 7h19'12" 20d19'42"
Charlene
Ari 2h42'37" 27d48'36"
Charlene
Cnc 9h9'16" 27d19'2"
Charlene
Leo 11h20'1" 24d47'51"
Charlene
Lyn 8h45'16" 45d1'51"
Charlene
Psc 1h5'31" 4d15'9"
Charlene
Dra 20h11'31" 62d49'5"
Charlene
Uma 10h14'47" 56d13'15"
Charlene
Cas 23h25'22" 55d54'42"
Charlene
Cas 23h27'11" 56d14'31"
Charlene
Cas 0h50'38" 64d12'42"
Charlene
Lib 15h45'43" -13d52'14"
Charlene 30
Tau 4h41'25" 24d39'29"
Charlene 420-143-111804
Psc 23h44'28" 6d41'52"
Charlene A. Knope
Cas 1h55'22" 64d16'41"
charlene and frank
Tau 4h30'53" 17d8'32"
Charlene Ann Jones
And 23h8'17" 47d31'53"
Charlene Ann Lawson 09/23/1963
Umi 13h59'33" 71d39'42"
Charlene Ann Middleton
Psc 0h59'23" 6d59'51"
Charlene Anne
Ari 2h13'15" 21d41'51"
Charlene Belloso
Tau 4h3'57" 23d58'41"
Charlene & Bob ~ The Loving Star
Cyg 20h46'16" 43d55'26"
Charlene Cascino
Lib 15h4'1" -15d32'34"
Charlene Castillo
Ari 2h38'40" 21d9'50"
Charlene Conway
Lyn 6h26'38" 56d32'30"
Charlene Cooper
Psc 1h13'59" 14d52'11"
Charlene Cunningham
Gem 6h35'59" 19d39'52"
Charlene Dennis
Pav 19h55'34" -68d6'11"
Charlene Dotschkalof Pennsylvania
Cas 1h21'11" 63d38'13"
Charlene Elgin
And 23h23'20" 38d55'47"
Charlene Elizabeth Asbjorn
Gem 6h37'59" 15d33'36"
Charlene Elizabeth Cub
Sco 17h38'0" -39d29'35"
Charlene Francis Ellis
Sgr 18h23'28" -29d52'23"
Charlene Gibb
Ari 2h54'35" 20d8'21"
Charlene Gibson
And 23h16'39" 51d28'38"
Charlene Grillo
Aqr 22h9'18" 0d31'7"
Charlene Hatch Pearson aka Charlie
Equ 21h18'41" 7d35'59"
Charlene Higgs
Cas 1h17'57" 58d29'6"
Charlene Hollingshead
Ori 5h2'44" -0d58'32"
Charlène Jessie Boucher
Cas 23h38'40" 55d11'33"
Charlene K. Swetman
And 1h9'39" 39d43'25"
Charlene Kay Strader
Cyg 20h13'48" 35d10'7"
Charlene Keihl "Our Shining Star"
And 0h57'30" 43d59'1"
Charlene Kemmerer
Vir 12h23'33" 2d55'33"
Charlene King
Psc 1h47'10" 5d23'56"
Charlene M. Kleen
Uma 13h42'22" 50d51'52"

Charlene M Stipo
"Mommie's Star"
Leo 11h26'27" 14d56'54"
Charlene Mae Latimer
Sco 16h33'20" -26d59'44"
Charlene Marie Earls
And 23h51'4" 40d47'39"
Charlene Marie Weathers
Cas 0h24'50" 61d14'26"
Charlene Merket
Her 17h14'8" 33d34'51"
Charlene Miller
Aqr 22h5'22" -23d28'23"
Charlene Mullin
Cas 0h31'41" 61d59'39"
Charlene Neuert
Lib 15h49'30" -11d7'0"
Charlene Nicole
Ori 5h56'45" 16d59'11"
Charlene Osborn
Tau 4h41'39" 5d4'52"
Charlene Rose Vargas
Cnc 9h7'49" 12d31'57"
Charlene (Runyon) Toler
And 0h57'52" 38d56'58"
Charlene Ruth
Slomski/Morgan
Psc 1h47'0" 7d51'30"
Charlene Sher "60
Sparkling Years"
Cas 1h25'42" 62d15'47"
Charlene Williams Eastern
Star
Cas 2h51'3" 71d54'53"
Charlene Wilson
Cas 0h59'24" 60d35'35"
Charlene Yague
Ari 2h26'11" 25d41'17"
Charlene Yvette's Star
Sgr 18h37'0" -27d40'18"
Charlene-Beautiful Angel
Tau 4h30'2" 3d2'39"
Charlene's Star
Mon 6h30'31" 4d31'39"
Charlene's Star
Uma 12h0'42" 43d47'3"
Charles
And 1h42'29" 42d33'9"
Charles
Her 17h13'23" 32d36'52"
Charles
Aql 19h42'27" 0d24'23"
Charles
Aql 19h40'39" 12d35'54"
Charles
Umi 13h23'31" 73d17'37"
Charles A. and Samantha
J. Sosinski
Cyg 21h27'58" 53d58'39"
Charles A. Arceneaux
Leo 11h27'17" 16d12'9"
Charles A. "Chick"
Mazzella
Cam 3h28'44" 67d8'45"
Charles A. Johnson Sr.
Aur 6h24'8" 41d10'3"
Charles A. Lascola
Uma 12h49'16" 52d55'44"
Charles A. McDowell
Lib 14h45'49" -16d54'33"
Charles A. Miess
Leo 10h41'21" 16d6'55"
Charles A. Miller III
Sco 16h13'25" -12d23'35"
Charles A. Ogden
07/30/1934
Per 3h15'37" 46d9'49"
Charles A. Wasserman, Jr.
Her 18h3'26" 23d10'54"
Charles Adam Langston
Ori 6h13'32" 16d22'37"
Charles & Agnes Saladino
Cas 0h36'2" 52d43'38"
Charles Aidan Walker
Wright
Tau 5h47'33" 22d21'26"
Charles Albert Faul
Tau 3h36'55" 2d55'10"
Charles Alden Schuessler
(Al)
Per 3h13'39" 50d26'24"
Charles Alexander
McIlveen
Vir 13h4'34" -21d52'33"
Charles Aliahmad
Per 3h8'24" 53d50'45"
Charles Allan Christensen
9-23-67
Per 3h22'12" 43d1'8"
Charles Allen Close
Psc 23h19'41" -2d50'57"
Charles Allen Gaudino
Leo 11h49'54" 25d40'9"
Charles Allgood
Ori 5h26'58" 6d50'47"
Charles & Ally
Tau 5h57'23" 25d23'6"
Charles Alugbuo & Leslie
Williams
Pho 0h42'57" -39d28'27"
Charles and Alice
Cyg 19h40'3" 35d18'59"
Charles and Angela
Astudillo
Cyg 19h35'19" 33d57'29"
Charles and Becky
Manning
Aql 19h24'44" -0d11'43"
Charles and Carmina
Carlson
Cyg 21h14'59" 45d57'23"
Charles and Carolena
Rosenberger
Lyr 18h53'0" 34d57'3"
Charles and Crystal
Vir 12h14'10" 5d42'58"
Charles and Eleanor
Lenling
Cap 21h18'36" -26d22'56"
Charles and Grace
Leo 11h8'23" 16d28'0"
Charles and Janice Luisi
Tau 5h59'26" 28d13'17"
Charles and Jean
Crb 15h20'0" 30d28'59"
Charles and Joseph
DeCapua
Uma 10h34'10" 63d58'44"
Charles and Karen Eaton's
Love
Cru 12h13'29" -61d34'16"
Charles and Marie Sears
Vir 12h23'31" -9d56'37"
Charles and Marion
Kornegay
Cyg 20h34'0" 50d35'13"
Charles and Ontha Hull
Ari 2h17'58" 26d0'46"
Charles and Sandra
Chrietzberg
Cyg 20h59'15" 37d20'50"
Charles and Sylvia Erhart
Cyg 20h30'33" 40d44'3"
Charles and Toni Hornsey
Uma 8h20'49" 63d15'31"
Charles Andrew
Calderwood
Aql 19h13'12" 14d51'47"
Charles Andrew Campbell
Leask
Umi 16h13'34" 84d26'3"
Charles Andrew Freuchtel
Per 3h17'45" 45d39'27"
Charles Andrew Jones
Ori 5h4'7" 15d5'14"
Charles Andrew Levada
Tannous
Tau 5h21'2" 27d49'37"
Charles Andrew Nahumyk
Lyn 7h47'37" 39d53'39"
Charles Anthony Messina
Tau 3h28'55" 3d3'20"
Charles Anthony Rendle
Per 2h55'44" 56d6'48"
Charles Anthony Rizzo
Lyn 8h22'38" 34d14'13"
Charles Anthony
Sepulveda
Her 17h18'12" 32d10'39"
Charles Anthony Skowron
III
Cep 21h36'14" 57d21'32"
Charles Anthony Wilhelm
Cnv 12h40'54" 44d37'17"
Charles Aquarion
Aqr 23h5'55" -6d18'32"
Charles Arnold Ray Pickles
Per 3h22'22" 53d29'40"
Charles Arthur Meyer
Ori 6h12'49" 20d0'47"
Charles Arthur Schmidt
Cam 9h0'40" 77d22'27"
Charles Arthur Spencer
Her 18h38'37" 16d14'47"
Charles Arthur Turner
Sco 16h47'41" -35d17'9"
Charles Ashton Moseley
Dra 16h58'40" 53d7'24"
Charles & Atriece Murdock
Cyg 20h22'59" 46d7'21"
Charles Auerbach/Jeffrey
Leith
Uma 10h42'26" 56d20'59"
Charles Austin Hussey
Aur 5h41'36" 50d10'14"
Charles B. Cuono II
Lib 15h47'6" -9d17'48"
Charles B Graham
Per 4h30'40" 32d34'13"
Charles B. Lipsen
Per 2h45'58" 52d7'49"
Charles Barnes
Aur 7h30'26" 43d17'1"
Charles Beaudet
Per 3h10'26" 55d11'39"
Charles Bellinger
Aqr 23h2'21" -7d18'41"
Charles Benjamin Justice
Uma 12h11'31" 53d48'40"
Charles & Bernice Hoffman
12/27/04
Cyg 19h38'11" 38d39'43"
Charles Berry & Jenise
Quiles
Crb 15h43'26" 27d6'42"
Charles Best
Cep 22h36'10" 57d57'16"
Charles Board Ashley Jr.
Psc 0h37'23" 13d30'7"
Charles Bona
Ori 5h46'24" 6d6'21"
Charles & Brandi Farch
06/16/2001
Cyg 21h38'20" 46d10'2"
Charles Brandon Osborn
Uma 10h44'6" 45d20'21"
Charles Brandon Phipps
Aql 19h50'57" -0d13'31"
Charles Brian
Per 4h37'54" 37d32'0"
Charles Brosius
Ori 6h9'57" -0d40'3"
Charles Browder
Peg 22h15'42" 33d12'8"
Charles Bryan Bolling
Psc 1h35'38" 15d41'34"
Charles Bryce Deneen
Psc 1h11'21" 29d3'13"
Charles "Bud" Christner
Cep 23h41'30" 74d50'28"
Charles Bundrant Hundley
Tau 5h42'24" 24d31'37"
Charles Burgese, Jr.
Crb 16h23'9" 35d31'10"
Charles Burgess Nansteel
Cnv 13h10'32" 38d52'59"
Charles Burton Foster
Aqr 23h17'49" -13d41'19"
CHARLES C CLARK
Vir 12h15'17" 13d1'26"
Charles C Ulfig III
Sco 17h51'13" -36d17'48"
Charles Cameron Hilton
Cep 22h31'27" 73d56'28"
Charles Cannon
Boo 14h52'47" 50d27'0"
Charles Castro
Cep 21h36'37" 69d22'44"
Charles Cecil Smith
Uma 9h26'58" 47d4'50"
Charles *Charlie* Beckert
Cep 5h24'23" 85d30'43"
Charles & Charlotte Hunt
Sge 19h59'45" 19d17'56"
Charles "Chase" Marcone
III
Tau 3h31'17" 7d30'54"
Charles "Chickie" Goodwin
Psc 1h18'57" 32d58'27"
Charles Christopher Bucher
Leo 9h47'31" 7d39'31"
Charles Christopher
Jennings
Sco 17h22'34" -32d56'41"
Charles "Chuck" Breo
Boo 15h48'17" 52d18'38"
Charles "Chuck" Wayne
Martin
Uma 9h45'19" 41d43'14"
Charles "Chuckie" Ostberg
Cep 20h40'6" 59d39'59"
Charles (Chuckie) R.
Benton V
Lib 15h2'25" -1d3'9"
Charles Chumley
Aql 19h47'49" 16d10'0"
Charles Clayton Randolph
III
Uma 12h0'37" 58d0'55"
Charles Clough
Uma 11h11'5" 35d48'18"
Charles & Connie
Cyg 21h24'52" 46d45'32"
Charles Cook
Psc 23h50'33" 0d47'11"
Charles Cornelius
Livingston III
Her 17h2'31" 30d12'4"
Charles Coughtry Stryker
Uma 10h4'48" 65d2'54"
Charles & Courtney East
Sgr 19h23'17" -11d49'4"
Charles Cuneo
Cnc 9h15'39" 10d37'26"
Charles D Krugman &
Ruben Donovan
Peg 23h14'30" 26d17'54"
Charles D. Ortiz
05/27/1938
Gem 7h15'31" 25d27'43"
Charles Dallas Horner
Ari 2h3'33" 18d20'13"
Charles Damon Pilon
Boo 14h33'21" 14d38'14"
Charles Daniel Granat
Leo 10h47'57" 18d56'58"
Charles Daniel Kudrle Jr.
Cap 20h34'13" -15d11'24"
Charles Daniel Thompson
III
Umi 15h18'46" 68d14'43"
Charles David Robinson
Her 16h29'50" 24d40'34"
Charles David Tontz
Cap 21h4'36" -21d47'44"
Charles Davis Belcher III
Her 17h13'56" 49d12'10"
Charles de Scorraille
Leo 11h53'41" 23d22'56"
Charles Denman's Eastern
Star
Her 18h9'31" 20d11'36"
Charles DeSocio
Uma 9h48'26" 53d42'42"
Charles Devers
Her 17h25'6" 34d35'20"
Charles Devose (Poppy)
Uma 11h57'5" 65d10'52"
Charles DeWitt Cantrell
Ori 5h21'59" 4d56'20"
Charles Dexter Scovill
Boo 14h44'52" 27d57'18"
Charles Dixon Dove
Cap 20h55'39" -23d5'52"
Charles Do LeBold
Lib 15h3'14" -0d43'53"
Charles Dodd
Ori 6h8'57" 16d5'9"
Charles Donald Gunther
Uma 9h58'54" 67d1'26"
Charles "Duke" Rajaniemi
Cep 21h42'26" 61d27'17"
Charles E. Anders
Cap 20h28'25" -12d24'31"
Charles E. Blondis 1983 till
2003
Mon 7h21'26" -0d19'39"
Charles E. "Bud" Whitney
Cyg 19h51'20" 58d20'14"
Charles E. Burt
Crb 15h42'55" 27d35'37"
Charles E. Bussey
Leo 11h25'28" 11d7'38"
Charles E. Carriere III
Vir 13h24'40" 13d7'48"
Charles E. Colson
Dra 17h43'45" 63d36'53"
Charles E. Garrity, Sr.
Lib 15h40'35" -17d7'24"
Charles E. Goodwill - 70th
Birthday
Tau 3h45'29" 24d14'43"
Charles E. Hamrell, M.D.
Ari 2h40'13" 16d32'10"
Charles E Harris
Uma 8h59'56" 65d58'46"
Charles E. Hoffa
Gem 6h34'22" 24d3'3"
Charles E. Hyde II
And 1h2'9" 38d50'27"
Charles E Jackson III
Lib 15h3'39" -1d2'0"
Charles E. Jackson Jr.
Ori 5h55'39" 11d22'36"
Charles E. Lane Jr.
Ori 5h59'53" 22d44'36"
Charles E. Lanning
Ari 3h24'54" 26d3'35"
Charles E. Mayberry
Crb 15h51'39" 33d52'46"
Charles E. Roberts III
Mon 6h51'2" -0d6'52"
Charles E. Smith
Lib 15h29'46" -10d19'34"
Charles E Snyder
Ari 2h27'34" 24d12'52"
Charles E. Taylor III "Trey"
Ori 6h19'56" 13d21'47"
Charles E. Williamson IV
Sgr 18h12'51" -16d8'38"
Charles E. Zibell
Her 16h41'39" 27d8'13"
Charles Earl Hallum
Ori 5h50'32" 5d39'3"
Charles Earl Weichselbaum
Her 17h32'45" 27d27'2"
Charles Eaton
Ori 6h0'36" 10d49'3"
Charles & Edith Zipperlen
Cyg 20h32'46" 31d15'48"
Charles Edward Cordell, Jr
Aqr 21h6'19" 1d20'13"
Charles Edward Falcon
Psc 0h38'29" 8d21'19"
Charles Edward Farris
Ori 5h25'51" 1d33'28"
Charles Edward Francis
Per 4h20'31" 41d27'49"
Charles Edward Hawk
Jefferson
And 2h18'5" 45d11'45"
Charles Edward Heintz
Sco 16h11'40" -15d28'54"
Charles Edward Jones III
Aur 6h3'7" 50d18'14"
Charles Edward McWhirter
Uma 10h14'8" 45d45'0"
Charles Edward
Ohennesian
Cep 23h41'46" 79d8'47"
Charles Edward Williams
Sr.
Cyg 21h20'12" 36d54'48"
Charles Edwards Collins-
builder
Cyg 21h36'27" 50d4'31"
Charles Edwin Beck
Per 2h25'50" 55d30'40"
Charles Edwin Peavy 1V 2-
20-90
Ori 6h2'24" -0d1'19"
Charles & Elizabeth
Cnc 8h42'3" 13d14'42"
Charles & Ella Ann Spargo
Sge 20h7'18" 17d10'0"
Charles Elmo West
Gem 7h24'44" 16d4'10"
Charles Elton Fuller
Aql 19h32'7" -10d47'22"
Charles Emil Ruckstuhl
Her 17h16'35" 35d42'58"
Charles Enriquez
Psc 1h40'58" 7d53'14"
Charles Ernst
Uma 10h37'44" 41d38'37"
Charles Eugene Adams
Uma 10h27'24" 46d54'34"
Charles Eugene Carson
Uma 9h16'52" 47d31'0"
Charles Eugene Martin
Ori 6h14'30" 16d2'22"
Charles Eugene Price
Tau 5h1'20" 24d56'21"
Charles F. and Mildred G.
Hardman
Ori 6h15'25" 6d40'24"
Charles F. Cosgrove, Jr.
Ari 2h14'50" 14d1'42"
Charles F. Dietz Jr.
Vir 12h4'27" -1d11'22"
Charles F. Hughes, III
Tau 4h41'47" 26d40'44"
Charles F. Miller, Jr.
"Woodchuck"
Ori 5h35'28" 8d37'23"
Charles F. Reilly
Gem 7h45'53" 31d54'4"
Charles F. Thomas, Jr.
Vir 13h17'20" -3d33'42"
Charles F. Tupper Jr.
Her 18h16'25" 21d56'28"
Charles Feldman
Sgr 19h4'4" -14d23'12"
Charles Feldman
Lyn 7h9'41" 56d58'5"
Charles FF Beauregard
Per 2h51'52" 47d29'7"
Charles Fleckenstein
Her 18h53'23" 22d32'45"
Charles Foster Schaffer
3.12.04
Psc 1h25'0" 25d56'47"
Charles & Frances
Monanian
Tau 4h34'9" 28d5'48"
Charles Francis Page 7-12-
2005
Cru 12h19'7" -59d13'59"
Charles Francis Stout
Uma 10h6'25" 70d43'27"
Charles Frank Ventura
Psc 1h12'45" 27d43'47"
Charles Franklin Schiedel
Psc 23h2'33" 7d35'37"
Charles Frederick Cartner
Cep 22h13'46" 60d50'5"
Charles Frederick Kline
Lib 15h1'51" -3d34'24"
Charles Frederick Marryatt
Per 3h9'15" 53d57'44"
Charles G. Lowry, Attorney
at Law
Aql 19h21'48" 7d38'56"
Charles G. McLaughlin
Cep 21h36'20" 62d55'20"
Charles G. Rutledge
Leo 10h13'37" 12d12'14"
Charles G. Wagner
Cam 4h17'44" 66d14'22"
Charles George Banyas
Vir 11h38'35" 4d43'22"
Charles Gerard Herbella
Uma 12h8'31" 54d16'26"
Charles Githler
Uma 13h29'32" 52d30'0"
Charles Glenn Hollins
Cet 0h56'42" -0d25'33"
Charles Glenn Roseneau
III 01/08/02
Per 4h14'51" 41d25'49"
Charles Goers, Sr.
Lib 16h0'4" -8d52'28"
Charles Griffin Cale
Ori 5h29'59" 4d6'7"
Charles Guy Moore
Cep 21h49'31" 69d58'1"
Charles H. Bensel
Her 17h28'49" 38d37'41"
Charles H. Crist, King
Cam 4h14'34" 69d27'47"
Charles H. Diamond
Her 17h8'11" 13d16'29"
Charles H. Grace
Psc 0h59'16" 16d20'0"
Charles H. Hilliard
Her 17h23'5" 35d55'20"
Charles H. Koski
Courageous Heart
Psc 1h9'34" 23d5'45"
Charles H. Peck
Leo 10h17'32" 24d58'37"
Charles H. Prescott
Cep 23h19'59" 75d48'35"
Charles H. Savage
Gem 6h43'14" 29d13'45"
Charles H. Simms
Lyr 19h7'14" 41d59'34"
Charles H. Snyder
Tau 4h7'15" 27d57'11"
Charles H. Turner
Cyg 20h47'38" 39d41'15"
Charles Hackney
Sgr 17h50'57" -17d30'4"
Charles Hal Dayhuff III
And 0h48'22" 37d11'42"
Charles Hal Dayhuff IV
And 0h37'56" 34d36'32"
Charles Hal Dayhuff V
And 0h53'26" 37d20'29"
Charles Hamilton Allen
Ori 5h40'16" 2d44'55"
Charles Hamilton Small Jr.
Lib 15h28'44" -23d45'49"
Charles Harriman
Ari 2h18'55" 24d33'15"
Charles Harris Grant
Her 16h48'2" 28d3'46"
Charles Hawkinson
Lyn 6h30'13" 61d0'28"
Charles & Helen
Westerfield
Cyg 21h43'10" 43d52'42"
Charles Henery Nelson
Her 17h18'19" 32d14'39"
Charles Henri
Cet 1h19'57" -11d37'45"
Charles Henry Benton's
Star
Sco 17h16'4" -34d1'25"
Charles Henry Mercer
Sgr 18h7'24" -26d41'37"
Charles Henry Seitter VI
Her 17h41'58" 20d18'15"
Charles Henry Spicer, Jr.
Ari 2h10'20" 15d54'16"
Charles Herbert Zilch
Per 3h15'24" 46d46'30"
Charles Herschel Martin
Srp 18h29'17" -0d44'41"
Charles Highfill
Her 18h4'32" 36d58'57"
Charles Hirschler
Gem 7h11'22" 25d29'5"
Charles Hiscock
Ori 6h14'21" 8d18'15"
Charles Hogue
Per 4h3'56" 33d36'4"
Charles Howard
Uma 10h28'27" 48d14'22"
Charles Howse Partee
Gem 7h45'8" 14d40'58"
Charles Hufstedler
Crb 15h54'57" 29d28'43"
Charles Hunter Olson
Cnc 8h8'37" 24d56'32"
Charles Hutchinson
"Chuck"
Uma 11h29'26" 53d24'57"
Charles I Engelhart
Cep 3h50'52" 82d37'39"
Charles Insinga
Aur 5h42'17" 47d59'27"
Charles Isaac Moris
Pho 1h7'17" -42d8'34"
Charles J. Cialdella
Charlie
Per 4h22'57" 52d43'31"
Charles J. Hellier, III
Aur 7h14'15" 42d35'44"
Charles J. Hubbard
Leo 10h24'20" 11d57'18"
Charles J. Kunkle
Uma 8h36'38" 62d25'3"
Charles J Mastandrea
Uma 9h37'51" 53d7'34"
Charles J. Norris
Aqr 22h58'55" -9d30'40"
Charles J. & Rita Emerling
Cyg 20h59'48" 37d43'27"
Charles "Jack" Persinger
Cyg 19h58'40" 34d59'36"
Charles Jacob Kirkland
Sgr 19h10'22" -31d46'9"
Charles James Arnold
Dra 16h38'3" 59d40'31"
Charles James III
Cnc 8h27'29" 32d1'30"
Charles James Murray IV
Vir 13h40'3" -0d2'18"
Charles & Jane Phillips
Cyg 21h11'59" 31d35'59"
Charles Jaros
Cep 3h50'46" 81d12'49"
Charles & Jeannette
Hokanson
Uma 10h26'14" 51d17'16"
Charles Jeffrey Litteral
Ori 5h56'3" 12d30'57"
Charles Jeffrey Payne
Her 17h56'30" 24d22'48"
Charles John Anderson
Uma 8h47'33" 49d26'36"
Charles John Hawes
Sgr 19h34'6" -30d46'5"
Charles John Poulos
Vir 13h20'20" 5d45'7"
Charles John Vescovo
Ori 6h11'10" -3d56'57"
Charles John Walter
Cicarella
Aqr 21h7'3" -8d55'51"
Charles Jones
Lyn 9h12'13" 34d21'54"
Charles Joseph Bartels
Her 16h50'7" 11d4'31"
Charles Joseph Bartels
Cep 21h29'3" 61d27'3"
Charles Joseph Buttaci
Cnc 8h44'52" 31d13'28"
Charles Joseph Camp II
Uma 11h19'5" 56d21'27"
Charles Joseph Conners
Her 16h58'56" 31d10'47"
Charles Joseph Dollwet
Boo 15h20'36" 41d15'4"
Charles Joseph Englehart
Her 17h30'49" 46d11'51"
Charles Joseph Internicola
Aqr 22h42'55" 0d39'12"
Charles Joseph Murray
Uma 11h9'17" 33d30'52"
Charles Joseph Olson
Her 18h3'42" 29d14'11"
Charles Joseph Schwartz
Ori 5h15'47" 8d9'53"
charles josephson
Lmi 10h18'50" 39d14'27"
CHARLES JR
Uma 9h24'24" 66d8'18"
Charles Kelly Harrell
Per 4h48'49" 46d24'22"
Charles Kelly Hicks Jr.
Tau 5h12'6" 16d38'15"
Charles Kent Cheek
Ori 5h48'46" 5d58'45"
Charles Kerwin Pollock
Uma 14h0'50" 54d23'50"
Charles Kilgus
Aqr 22h24'26" -7d5'42"
Charles Kim
Lib 14h50'29" -8d12'24"
Charles Kinnard
Per 4h23'16" 46d3'42"
Charles Kipp Astrologer of
the Aeon
Ori 5h34'3" 7d12'36"
Charles Kramer
Gem 6h37'43" 27d41'34"
Charles L. Apisdorf
Lib 15h29'9" -20d20'42"
Charles L. Back and
Joshua W. Back
Uma 13h45'25" 47d53'29"
Charles L Brown
And 1h32'10" 43d0'41"
Charles L. De Re
Lyn 7h45'55" 39d17'6"
Charles L. Huff
Per 3h45'59" 51d5'32"
Charles L. & Marian K.
Wicker Jr.
Eri 3h49'51" -0d38'4"
Charles L. Mullen
Vir 14h1'40" -16d0'54"
Charles L. "Pappy" Swain
Cyg 19h59'43" 37d13'11"
Charles L Smith Sr.
Vir 14h29'45" 6d2'41"
Charles L Wingfield
Cap 20h31'19" -10d35'15"
Charles Lanska
Psc 0h28'13" 6d3'7"
Charles Largado Garrison
II
Ori 5h52'37" 11d46'20"
Charles & LaVerne
Gem 6h30'56" 23d51'14"
Charles Lawrence
Umi 14h18'31" 73d48'51"
Charles Lawrence Williams
Ori 5h14'40" 16d4'40"
Charles Lee
Her 18h25'4" 14d23'3"
Charles Lee Browning, Sr.
Leo 11h24'39" 15d28'53"
Charles Lee Moore Jr.
Per 2h15'30" 56d34'53"
Charles Lee Porter
Sgr 18h6'18" -27d34'9"
Charles Lee Rose "Charlie"
Uma 9h19'28" 58d52'37"
Charles Lee Wilkinson
Her 17h38'46" 46d42'49"
Charles Legare
Leo 10h13'37" 11d23'13"
Charles Lehr Weidner
Uma 10h56'21" 58d9'17"
Charles Leigh
Uma 8h15'12" 69d0'27"
Charles Lewis Hess III
Uma 8h35'13" 52d4'2"
Charles Livingstone Mace
Tau 5h33'46" 25d27'41"
Charles Lloyd Rogers
Cnc 9h14'7" 6d58'45"
Charles & Lorraine Powers
Uma 10h36'15" 55d6'43"
Charles Louis Kane
Ori 5h33'53" 13d53'38"
Charles Louis Lawson
Gem 6h40'54" 20d42'24"
Charles Louis (Pete)
Blanchetti
Psc 1h34'20" 13d32'35"
Charles Lucas Mann
Gem 7h11'32" 22d58'58"
Charles Lyons, Sr.
Aqr 23h53'23" -4d15'27"
Charles M B McIntire III
Sco 16h5'16" -10d38'48"

Charles M. "Chuck" Williams
Umi 15h43'11" 73d10'24"
Charles M. Layman, Jr.
Pho 0h16'40" -48d1'49"
Charles M. Mercer
Lib 15h3'6" -0d48'26"
Charles Malloy
Sco 16h29'25" -32d6'51"
Charles Marco Ferreira
Cep 21h59'54" 62d3'40"
Charles & Marion Gross
Lyn 9h31'4" 40d22'19"
Charles Marion Tanner
Tau 4h1'10" 27d0'4"
Charles Martin Rowland
Eri 3h45'44" -38d1'9"
Charles Martin Staples, II
Umi 15h21'9" 73d19'8"
Charles Mason Kleintaurus
Tau 3h44'5" 21d12'58"
Charles Matays
Vir 14h56'27" 6d28'17"
Charles Matsler
Per 3h33'26" 35d35'25"
Charles Matthew Mandala
Gem 7h6'54" 23d23'8"
Charles Matz
Psc 0h20'49" 19d36'51"
Charles McCleary
Uma 10h49'1" 71d47'16"
Charles Mcrae Wharton
Sco 17h33'42" -45d21'5"
Charles & Melissa Hing
Gem 7h43'0" 21d55'12"
Charles Michael Gantz (Lucky Charm)
Psc 1h24'57" 7d1'25"
Charles Michael Raible
Aql 20h36'27" -0d35'7"
Charles Michael Seabolt
Aur 6h31'5" 51d54'9"
Charles Michael Seergy
Tau 4h12'12" 28d39'18"
Charles Michael Williams
Uma 11h52'20" 39d9'15"
Charles & Michele Love Eternal
Uma 10h26'15" 40d17'3"
Charles Mila
Her 17h57'10" 14d52'7"
Charles Miles Leigh
Uma 11h3'38" 69d9'27"
Charles - "Mister"
Ari 2h46'26" 21d36'23"
Charles Mitchell Collins
Aql 18h52'1" -0d44'32"
Charles Moffitt Lineberry
Cyg 20h53'59" 47d9'37"
Charles Mongrandi
Cap 21h36'45" -14d4'41"
Charles Monnone
Vir 13h19'39" 12d56'39"
Charles Moses Rogers
Lib 15h23'4" -7d10'43"
Charles N. West
Cep 20h12'30" 60d1'25"
Charles Nageotte
Uma 9h55'26" 58d5'57"
Charles Nathan Johnson
Uma 10h51'58" 50d17'40"
Charles Nethersole
Sco 16h21'20" -26d51'19"
Charles Noble Butler III
Lyr 19h1'6" 30d38'15"
Charles Norman Barrow
Sco 17h33'27" -41d37'25"
Charles O. Green
Psc 1h18'3" 15d5'56"
Charles P. Heinrich
Vir 12h45'21" 2d37'15"
Charles P. & Lois Brown Nash
Aql 19h27'43" -0d21'36"
Charles P. Miska
Per 3h53'31" 45d1'53"
Charles P. Sangee
Aur 6h58'25" 37d39'26"
Charles "Pappy" Wilder
Per 3h1'58" 55d56'52"
Charles Patrick Oles, III
Ari 1h53'11" 17d43'12"
Charles Patrick Smith
Cnc 9h10'51" 32d15'55"
Charles Paul Kyncl
Cap 20h36'0" -23d25'3"
Charles Paul "Niff" Allen
Boo 14h48'45" 16d15'49"
Charles Pavlick
Her 18h47'41" 24d30'50"
Charles "Paw Paw" Standley
Sgr 19h59'23" -22d24'33"
Charles Perkins
Cap 20h40'32" -17d3'28"
Charles Peter John Ireland
Her 17h40'32" 16d42'17"
Charles Peter Lanza
Psc 1h16'35" 28d21'6"
Charles "Phil" Van Camp
Uma 11h7'29" 29d50'19"
Charles Phillip Armour
Vir 13h17'7" 4d21'7"
Charles "Po" Tartaglione, Sr.
Uma 12h48'57" 56d21'59"
Charles (Poy Poy) Moore
Vir 14h8'52" -18d34'38"
Charles Preston McElheney III
Umi 14h18'47" 68d1'25"
Charles Price Boyd
Sco 16h12'5" -15d33'31"
Charles R. Breaux
Her 17h20'5" 37d5'17"
Charles R. Church
Cap 21h48'46" -10d12'20"
Charles R. R. R. Hammersley Bullock
Ori 6h3'22" 10d33'37"
Charles Rabinowitz,MD
Oph 17h35'19" -0d7'41"
Charles Ramon (C.J.) Merritt
Cep 22h44'22" 66d4'10"
Charles Ray Bales-Guardian Angel
Lmi 10h39'1" 26d40'8"
Charles Ray Bradford
Ari 2h24'50" 26d9'19"
Charles Raymond Horton
Aql 19h49'2" -0d58'24"
Charles (Red) Jackson
Cep 21h30'17" 58d13'3"
Charles Richard Cooper, Jr.
Aqr 23h51'22" -16d25'28"
Charles Richard Green
Lib 15h56'19" -18d21'37"
Charles Richard Long
Ari 2h14'8" 27d23'41"
Charles Riviere
Gem 6h48'7" 28d21'47"
Charles Robert Austin
Per 3h2'33" 45d0'38"
Charles Robert McConaghy
Lyn 8h31'55" 45d35'7"
Charles Robert McTiernan
Per 3h12'15" 52d1'22"
Charles Robert Sherman
Aql 19h6'3" 15d12'59"
Charles Robert Usina Sr.
Ori 5h36'44" 0d27'35"
Charles Roderick Herron
Ori 6h3'24" 17d15'49"
Charles Roger Arbogast
Her 16h42'27" 8d14'58"
Charles Roger Michaelson
Leo 11h13'5" 0d0'13"
Charles Roland Zelinski III
Lmi 9h56'23" 33d9'52"
Charles Ross
Lib 15h47'36" -20d16'30"
Charles Roughley
Cep 22h39'30" 66d57'42"
Charles Ryan 7/31/97
Leo 11h14'36" -4d43'39"
Charles S. Gucciardo
Gem 6h48'55" 33d55'32"
Charles S & Laura A Haynes Forever
Uma 11h40'23" 57d57'9"
Charles S. Loa
Per 3h45'1" 43d48'49"
Charles S Schiel
Sco 16h14'13" -15d23'20"
Charles S. Soper
Ari 2h52'51" 29d8'25"
Charles Samuel Deneen, III
Cnc 8h18'16" 10d36'4"
Charles Seitter (Daddy)
Dor 5h12'6" -59d13'44"
Charles Semrad Quinn
Leo 11h33'58" -6d22'32"
Charles Sferrazza
Uma 12h13'14" 53d35'48"
Charles Shane Diggins "Charlie"
Uma 11h11'1" 60d56'36"
Charles Shannon
Cep 3h34'59" 82d51'16"
Charles Shapiro
Lib 14h53'59" -1d24'11"
Charles Shelton
Cep 22h58'9" 62d56'28"
Charles & Shirley Terrill
Cyg 20h8'29" 54d23'52"
Charles Simpson Porter Jr
Cnc 8h46'22" 14d27'33"
Charles Smerglia
Vir 13h46'29" -6d55'3"
Charles Smith (Dad)
Her 16h36'17" 45d11'53"
Charles Soo McCandless
Ori 5h19'11" 7d3'14"
Charles Spurlock
Leo 10h41'35" 6d54'9"
Charles Spurlock
Leo 11h19'13" 18d54'23"
Charles Stanley DeLozier
Her 17h52'1" 28d6'52"
Charles Steven
Lib 15h13'42" -7d26'46"
Charles Steven Knowles
Lmi 10h22'30" 29d47'46"
Charles & Sue
Sco 17h29'19" -39d13'55"
Charles Sullivan
Cep 22h35'56" 65d1'37"
Charles & Susan Ford
Uma 11h13'20" 51d9'23"
Charles T. Jones
Lib 14h31'0" -10d19'15"
Charles T. Pheeney
Tau 5h47'47" 23d14'56"
Charles T. Smith
Per 3h37'17" 51d38'13"
Charles Taieb Gasperment
Cet 1h58'14" -1d57'42"
Charles Taylor
Cyg 19h34'3" 29d21'15"
Charles Thaddeus Trullinger
Cas 23h50'59" 55d3'5"
Charles Thayer Montague
Ori 5h38'20" 2d20'36"
Charles "The Wookie" Wucherer
Gem 7h39'24" 14d21'37"
Charles Thomas Brooks
Psc 0h43'42" 6d37'16"
Charles Thomas Eagan
Cep 22h40'52" 73d46'5"
Charles Thomas Giles
Vir 11h56'36" -3d18'51"
Charles Thomas Helm
Uma 10h59'33" 35d18'25"
Charles Thomas Jacob
Vir 13h44'34" -7d7'54"
Charles Thomas Murray
Uma 11h26'57" 63d25'31"
Charles Thomas "Tom" Subock
Ori 5h49'47" 1d24'35"
Charles Thompson
And 2h34'12" 50d4'52"
Charles Tidwell
Uma 9h6'8" 49d53'38"
Charles Timmons Chong
Psc 23h28'2" 1d24'45"
Charles Tron
Uma 9h43'54" 52d59'33"
Charles Ventrano 12041938
Sgr 18h4'58" -27d44'13"
Charles Vincent Mahoney Jr.
Her 17h55'30" 18d40'55"
Charles von Flotow
Cap 21h28'38" -17d59'50"
Charles Von Scoyk
Per 3h4'0" 52d38'12"
Charles W. (Bill) Wilson
Cen 13h28'41" -33d43'25"
Charles W. Davis
Leo 10h27'23" 16d23'55"
Charles W. Dillon
Lib 15h43'14" -28d55'2"
Charles W. DuVal
Aqr 20h54'39" -10d43'55"
Charles W. Gamm, III
Uma 9h32'39" 47d30'37"
Charles W. McCarthy
Lib 14h52'20" -9d57'24"
Charles W. Miller
Tau 5h47'0" 26d11'21"
Charles W. Rockett
Uma 11h19'15" 39d7'19"
Charles W. Stevens
Her 18h17'38" 22d29'44"
Charles Walker
Cnc 8h51'0" 11d54'32"
Charles Wayne Kempske
Her 16h26'41" 43d53'45"
Charles Weaver Baughn
Lib 15h5'12" -27d2'38"
Charles Wesley Hicks
Ori 4h58'50" -0d21'42"
Charles Whitman, Jr.
Her 17h5'17" 36d58'2"
Charles Whitney
Per 3h21'54" 53d36'17"
Charles Wigington
Sgr 17h58'44" -26d19'10"
Charles William Conner
Tau 5h54'10" 25d58'27"
Charles William Kepler
Cap 20h29'11" -23d44'13"
Charles William Petrosky
Mon 8h4'59" -0d56'52"
Charles William Raitt
Her 16h34'9" 20d38'47"
Charles William Smith
Her 17h12'32" 35d12'20"
Charles William Turner
Gem 6h33'15" 23d29'47"
Charles Wilmer Dixon
Vir 14h12'43" 3d9'16"
Charles Wilson Dare
Cap 20h31'15" -22d1'46"
Charles Wilson Thompson
Lyn 7h38'30" 43d15'50"
Charles - You're the Best!
Tau 4h41'43" 23d36'1"
Charles-Antoine
Umi 13h39'51" 70d35'31"
Charles-Antoine Lavoie
Uma 9h51'55" 70d30'56"
Charles-Arthur Mayaud
Her 17h54'36" 45d35'10"
Charles.E.Oliver
Vir 13h25'33" 11d58'59"
Charles-Robert William Jenkins
Her 16h17'35" 48d53'44"
CharlesRT24
Leo 11h36'48" 14d42'30"
Charles's Star
Her 17h36'58" 35d2'37"
Charlette
Ari 3h18'10" 19d21'7"
Charlette Desiree Harvell
Tau 3h53'37" 9d55'6"
Charley
Lac 22h29'45" 50d20'20"
Charley 92
Umi 15h53'22" 79d0'53"
Charley Anne Behr
Vir 12h26'49" 12d32'59"
Charley Bie
Psc 1h22'41" 6d36'47"
Charley Bubbles
Cru 12h42'23" -58d36'47"
Charley Charmelo
Aqr 22h20'49" -18d42'57"
Charley Elizabeth Faucett
Cyg 20h3'43" 38d57'15"
Charley Lynn
Lyn 6h33'18" 58d2'12"
Charley Marie
Vir 13h39'55" 2d52'23"
Charley Rachel
Ori 5h36'12" -2d19'31"
Charley Ray & Martha Jean Harris
Ori 5h18'20" 6d54'53"
Charley & Renee Ford
Uma 11h51'0" 61d42'57"
Charley SeRae Hieb
And 2h37'58" 45d22'24"
Charley Thomas Mariott
Vir 14h40'14" 3d27'42"
Charley Tiger Helen Matulewicz
Eri 3h29'41" -22d9'32"
Charley W Coffland
Cyg 20h24'44" 35d23'28"
Charleygus
Uma 13h18'19" 58d59'54"
Charleyne
Col 6h33'15" -42d36'0"
Charley's Angel
Cnc 8h41'34" 15d21'6"
Charley's Star
Per 3h36'46" 46d46'46"
Charli
Lyn 7h47'39" 52d26'38"
Charli
Cru 12h43'25" -60d39'56"
Charli
Lib 15h34'38" -7d39'15"
Charli Dae
Umi 14h3'30" 69d2'25"
Charli Girl
Gem 7h11'48" 24d8'46"
Charli Hope - 04/02/2006
Cru 12h7'40" -61d51'26"
Charlianna
Gem 7h16'2" 24d3'56"
Charlie
Cnc 9h15'47" 25d9'30"
Charlie
Ari 3h8'43" 29d35'25"
Charlie
Her 17h18'21" 28d42'28"
Charlie
Leo 9h59'4" 15d11'1"
Charlie
Gem 7h1'31" 22d14'13"
Charlie
Ori 5h15'30" 7d27'42"
Charlie
Ori 5h25'9" 1d45'19"
Charlie
Ori 5h26'32" 1d59'10"
Charlie
Ori 4h52'24" 10d43'10"
Charlie
Her 16h18'37" 14d14'19"
Charlie
Ari 2h32'52" 20d31'44"
Charlie
Peg 22h17'52" 7d30'26"
Charlie
Lyn 8h1'43" 49d59'47"
Charlie
Aur 5h54'29" 52d14'24"
Charlie
Per 4h45'50" 45d32'26"
Charlie
Per 2h54'16" 49d27'34"
Charlie
Cyg 20h8'54" 40d8'44"
Charlie
Per 2h22'16" 56d22'42"
Charlie
Per 2h18'50" 55d28'32"
Charlie
Her 17h48'48" 47d11'19"
Charlie
Cyg 21h33'48" 30d38'24"
Charlie
Lyn 9h21'14" 33d14'50"
Charlie
Per 4h0'54" 34d34'52"
Charlie
Sgr 18h48'43" -32d40'4"
Charlie
Umi 13h16'23" 73d16'26"
Charlie
Uma 9h47'56" 62d39'46"
Charlie
Uma 8h40'57" 63d39'43"
Charlie
Uma 12h25'0" 56d2'12"
Charlie
Uma 11h24'3" 58d7'36"
Charlie
Uma 10h43'7" 56d1'11"
Charlie
Uma 9h0'19" 55d24'28"
Charlie
Dra 16h49'39" 53d10'39"
Charlie
Lib 14h34'13" -10d27'10"
Charlie
Aqr 20h51'3" -11d52'12"
Charlie
Cma 7h23'41" -20d5'9"
Charlie
Umi 15h41'58" 80d10'40"
Charlie
Umi 15h14'16" 79d58'46"
Charlie
Cep 23h2'54" 83d14'7"
Charlie
Aqr 22h15'2" -4d30'45"
Charlie 7
Uma 12h6'0" 64d34'15"
Charlie Albert Miller
Her 18h28'57" 25d11'29"
Charlie Alexander
Umi 14h44'39" 78d11'58"
Charlie and Barbara's bright star!
Leo 11h30'17" 15d52'34"
Charlie and Freddy Boy
Gem 6h50'19" 26d28'13"
Charlie and Jennifer
Ori 6h15'14" 15d20'26"
Charlie and Jim
Uma 9h31'49" 61d15'55"
Charlie and Joanna
And 1h16'24" 36d24'6"
Charlie and Marilyn Fields
Cyg 20h6'34" 38d41'57"
Charlie Anderson
Cep 22h48'37" 66d17'11"
Charlie Arnott
Lib 14h58'52" -11d29'29"
Charlie Ashton Burns
Uma 8h29'47" 63d46'6"
Charlie Athena
Dra 17h24'11" 52d7'43"
Charlie Atkinson
Dra 17h23'28" 54d46'29"
Charlie Azzue Zelen
Gem 7h11'21" 34d31'2"
Charlie B
Gem 7h25'54" 33d8'50"
Charlie Bachmann
Gem 7h56'3" 32d27'10"
Charlie Bean
Leo 10h57'13" 0d45'3"
Charlie Bedgood
Boo 14h36'19" 40d52'13"
Charlie Betta-Lam
Uma 9h34'38" 47d2'39"
Charlie Binns
Gem 6h8'55" 26d32'35"
Charlie Boutwell
Ori 6h3'11" 17d10'18"
Charlie Boy
Crb 15h28'7" 31d26'30"
Charlie Boy Lanza
Umi 13h15'23" 72d12'30"
Charlie Brown
Uma 10h56'46" 62d42'44"
Charlie Brown
Her 17h20'58" 36d19'31"
Charlie Brown
Ori 6h0'59" 18d59'14"
Charlie Brown
Ori 5h34'10" 1d22'21"
Charlie Carter
Dra 20h22'2" 64d41'46"
Charlie Charl
Hya 13h40'1" -27d20'23"
Charlie Chubb's Star
Tau 5h52'22" 23d38'19"
Charlie Clifford
Uma 10h46'46" 47d22'16"
Charlie Clifton Cosby
Ari 2h48'6" 27d58'8"
Charlie Coffman
Tau 5h31'25" 21d51'7"
Charlie Daniel Bloomer
Lmi 9h59'26" 37d2'19"
Charlie Dann
Cap 20h31'1" -12d22'25"
Charlie David Hobbs
Sco 16h50'8" -27d19'4"
Charlie Davison
Per 4h16'28" 31d53'3"
Charlie Denson
Umi 14h21'10" 67d35'49"
Charlie Des George Amass
Uma 13h46'33" 48d51'0"
Charlie Diegel
Uma 9h36'26" 51d53'11"
Charlie & Edna Rossi
Cyg 19h27'3" 32d27'12"
Charlie Emby
Aur 5h47'54" 45d56'29"
Charlie & Eve Callas
Cyg 20h12'55" 35d1'37"
Charlie & Evelyn Together Forever
Cyg 21h2'7" 36d58'44"
Charlie Ford
Aqr 22h6'36" -0d48'45"
Charlie Fox
Cam 4h4'45" 73d5'44"
Charlie G.
Ori 6h11'23" 13d44'39"
Charlie Gordon Weiss
Leo 11h20'26" 1d57'48"
Charlie Harris
Com 12h7'59" 21d23'48"
Charlie Harvey
Cep 0h16'35" 70d32'22"
Charlie Hazell
Crb 16h1'28" 32d2'27"
Charlie Henry Middleton
Umi 15h49'59" 75d21'0"
Charlie Himes
Lmi 9h24'29" 38d32'32"
Charlie Imbraglio
Sco 16h13'12" -10d39'38"
Charlie J Johnson
Tau 4h44'23" 22d24'19"
Charlie J Previti
Lib 14h52'48" -2d41'18"
Charlie James Morton
Cnc 9h17'48" 15d44'5"
Charlie James Rendall
Del 20h24'47" 6d20'49"
Charlie James Savarese
Uma 9h57'32" 48d32'22"
Charlie James Savarese
Cap 20h9'38" -19d12'1"
Charlie & Jana...Laska Nebeska...
Cap 21h49'31" -9d51'45"
Charlie & Jennifer
Psc 0h19'49" 4d19'17"
Charlie Ji
Leo 10h28'31" 20d1'0"
Charlie Jo
Del 20h41'34" 9d45'23"
Charlie Joe Robinson
Cep 3h48'54" 80d59'26"
Charlie John
Umi 15h34'2" 83d4'36"
Charlie John Patrick - Charlie's Star
Ori 6h12'23" 21d23'18"
Charlie John Pike
Cep 0h38'36" 77d48'29"
Charlie Johnson
Umi 15h19'40" 80d26'46"
Charlie Johnson's Star
Uma 8h54'47" 72d3'33"
Charlie Jones
Umi 15h57'51" 76d22'28"
Charlie Joseph Foster
Uma 9h28'7" 66d43'11"
Charlie & Judy Eberhardt
Lib 14h52'29" -10d4'52"
Charlie Kadle Holder
Ori 4h56'30" -0d59'16"
Charlie Kai Beshoff
Per 2h50'51" 51d3'54"
Charlie & Laura Waller
Aqr 21h43'10" -1d14'15"
Charlie Le Brun
Uma 8h10'18" 61d44'2"
Charlie Leighton James
Dra 18h39'43" 52d36'51"
Charlie Long
Cyg 20h1'26" 44d6'15"
Charlie loves Sofia
Cap 20h28'16" -9d46'25"
Charlie Martin Linn
Cap 21h25'1" -27d11'14"
Charlie Matthew Straw
Tau 5h43'44" 21d17'2"
Charlie Maupin
Uma 11h35'12" 28d51'11"
Charlie Max Robin
Gem 7h30'44" 26d14'15"
Charlie McAuliffe
Lib 15h10'20" -9d32'53"
Charlie McKenna
Psc 0h30'26" 12d19'34"
Charlie Milan Hallam
Her 18h12'2" 20d18'17"
Charlie Monson
Cep 22h52'12" 71d50'51"
Charlie Morgan
Lib 15h5'43" -7d26'46"
Charlie Mottram
Umi 16h11'51" 79d34'55"
Charlie O' Hanley
Sgr 18h18'23" -32d55'29"
Charlie Oades
Umi 6h0'58" 88d21'18"
Charlie Orlando Garcia
Gem 6h13'14" 23d27'47"
Charlie Parker Brethes
Tau 3h39'13" 1d9'32"
Charlie & Peggy Lyman
Cyg 21h15'30" 50d48'46"
Charlie Peper
Aur 5h12'32" 42d57'15"
Charlie Peter Johnny Hayes
Her 18h22'57" 25d36'41"
Charlie Peter Szabo - Charlie's Star
Ori 6h19'16" -1d49'51"
Charlie Philip Clarke
Lmi 10h8'13" 33d21'9"
Charlie Quinn Kolbrener
Aqr 21h42'50" 1d39'58"
Charlie R Thomas
Uma 9h56'47" 58d31'34"
CHARLIE RADAU
Cnv 12h27'16" 33d6'19"
Charlie Reeves
Vir 12h0'55" 3d28'0"
Charlie Robert Pringle Scott
Cep 0h19'14" 73d53'43"
Charlie Robinson
Cep 22h9'6" 53d28'7"
Charlie Rock Gumbrell
Per 4h19'21" 41d42'28"
Charlie Rose
Cap 20h57'25" -19d37'33"
Charlie Rose
Pho 0h10'46" -49d56'45"
Charlie Samenus
Cam 5h39'14" 78d41'13"
Charlie Schmidt
Cep 2h27'5" 82d45'40"
Charlie & Sharon Soul Mates Forever
Uma 11h57'22" 38d31'18"
Charlie & Shelli Puhl
Aqr 20h43'43" 1d34'28"
Charlie Slater
Uma 10h20'19" 40d17'55"
Charlie Snell
And 0h1'21" 43d10'24"
Charlie Star 71
Ari 2h15'57" 26d28'18"
Charlie "Star Of Love"
Uma 9h3'24" 71d31'45"
Charlie Steinbauer
Uma 9h32'15" 63d42'4"
Charlie Stevens - Charlie's Star
Umi 15h48'29" 79d52'4"
Charlie Stockley
Ori 5h29'22" 1d10'43"
Charlie T. Gilardi, Sr.
Her 17h43'24" 33d12'41"
Charlie Taylor
Umi 15h9'43" 77d1'23"
Charlie The Kid
Psc 1h47'36" 5d23'3"
Charlie Thierauf
Cap 20h30'34" -20d24'58"
Charlie Thomas Lambert
Per 3h19'0" 43d12'44"
Charlie Thomas - My Little Girl.
And 22h58'54" 51d11'36"
Charlie Tyrrell's Star
Ori 5h5'6" 6d43'44"
Charlie Utting
Her 17h12'4" 28d20'7"
Charlie V. Harris
Sco 17h24'37" -40d8'40"
Charlie Vega
Gem 6h55'37" 19d23'7"
Charlie Victor Napoleon Högfeldt
Ari 2h53'2" 25d11'39"
Charlie Warner
Psc 0h49'41" 15d17'34"
Charlie Warren
Ori 6h22'23" 10d5'42"
Charlie Watts
Cnc 9h5'48" 27d50'56"
Charlie Wolfe
Uma 10h20'45" 68d12'1"
Charlie Yates - "Papa"
Uma 9h18'38" 68d54'37"
Charlie Zak Firmage
Cep 22h42'31" 63d55'14"
CharlieB
Cas 23h36'29" 59d24'45"
CharlieJohn
Ori 6h19'39" 5d50'23"
Charlielou
Per 2h31'0" 54d47'8"
Charlie's Angel
Cep 0h54'54" 86d49'30"
Charlie's Angel
Ori 5h41'7" -2d42'20"
Charlie's Chopper
Gem 6h56'47" 34d33'24"
Charlie's Christening Star
Umi 15h24'44" 78d12'43"
Charlie's Christmas Star
Uma 11h11'4" 33d43'50"
Charlie's Destiny
Lib 15h26'32" -10d53'9"
Charlie's Guiding Star
Her 17h29'17" 20d35'24"
Charlies heart
Cyg 21h44'47" 50d23'14"
Charlie's Piece of Heaven
Ori 5h44'30" 1d59'49"
Charlie's Shining Nemo
Dor 5h56'37" -68d51'54"
Charlie's Smile
Ori 5h30'44" -1d31'0"

Charlie's Star
Umi 15h32'12" 75d25'48"
Charlie's Star
Cep 22h56'4" 64d58'26"
Charlie's Star
Uma 9h38'52" 63d57'28"
Charlie's Star
Ori 5h56'50" 6d59'12"
Charlie's Star
Uma 13h52'46" 50d10'10"
Charlie's Star
Her 17h49'22" 41d24'9"
Charlie's Star
Per 4h6'5" 31d53'26"
Charlie's Star
Per 4h2'48" 38d36'41"
Charlie's Starway to Heaven
Leo 11h26'40" 10d51'42"
Charlies'Ange
Gem 7h21'5" 13d56'1"
Charline
Sco 17h17'57" -31d36'29"
Charline Claes
Ori 6h16'14" 8d59'44"
Charline G. White
Cas 0h34'49" 48d2'4"
Charlize
Leo 9h41'3" 11d10'33"
Charlize Bailey Haislip
Aur 5h7'19" 34d2'2"
Charlize Butzin
Cru 12h26'0" -61d8'37"
Charlize {Charlie Girl} Liane Reith
Psc 1h10'54" 25d49'26"
Charlize Nichole Solek
And 23h25'41" 48d5'1"
Charlize Veronica Clark-Osborne
And 23h0'16" 36d11'25"
Charllie
Uma 11h17'18" 32d41'19"
Charlot
And 0h36'34" 41d16'53"
Charlott
Lac 22h40'57" 54d17'48"
Charlott Bluemellott
Tau 5h24'54" 19d46'14"
Charlotte
Del 20h37'54" 13d29'23"
Charlotte
And 0h45'39" 24d26'38"
Charlotte
Tau 5h8'38" 24d41'13"
Charlotte
And 0h39'17" 38d11'22"
charlotte
And 0h13'26" 40d56'35"
Charlotte
And 23h55'24" 34d53'35"
Charlotte
And 23h35'4" 47d37'5"
Charlotte
And 2h36'11" 50d23'34"
Charlotte
And 2h14'3" 49d2'57"
Charlotte
And 1h54'29" 46d28'0"
Charlotte
Cas 0h58'56" 46d59'6"
Charlotte
Uma 11h19'44" 61d38'35"
Charlotte
Uma 14h20'19" 60d30'47"
Charlotte
Cep 21h30'30" 62d1'9"
Charlotte
Sgr 18h45'38" -16d14'15"
Charlotte A. Norfleet
Tau 4h37'57" 22d38'56"
Charlotte A. Tingler
And 23h45'57" 43d28'40"
Charlotte Adele Williamson
And 23h52'50" 47d25'9"
Charlotte Ahrnsbrak Miller
Ori 6h3'31" 3d50'32"
Charlotte Alexander
Peg 22h29'19" 34d0'58"
Charlotte Allyn
Cam 5h16'21" 62d36'14"
Charlotte - Always Shining Over Us
Leo 9h22'40" 15d44'32"
Charlotte Amanda Susan Simpson
Uma 12h10'19" 57d40'42"
Charlotte and Adam
Umi 14h56'0" 70d25'21"
Charlotte and Andy Horwath
Cyg 20h58'12" 35d6'13"
Charlotte and Gracie Lardner
And 0h47'22" 27d53'2"
Charlotte and Jamie
Tau 5h26'11" 19d24'59"
Charlotte and Joe's Star
Cap 20h25'26" -12d39'47"
Charlotte and John
Cyg 19h44'12" 47d53'54"
Charlotte and Richard Carpenter
Leo 10h22'4" 11d45'52"
Charlotte and Sotiris
Leo 9h39'55" 21d11'37"
Charlotte Anderson Fougere
Uma 12h5'8" 52d5'28"
Charlotte & Andrew
Cyg 19h40'33" 39d8'12"
Charlotte Anita Stucky
Crb 16h17'16" 31d0'46"
Charlotte Ann
Sgr 17h58'16" -17d32'30"
Charlotte Ann
Cas 0h15'50" 61d5'34"
Charlotte Ann Cornell
And 0h53'53" 34d39'4"
Charlotte Ann Ruffin Vaughn Crummie
Psc 1h16'37" 13d59'16"
Charlotte Ann Spencer
Aqr 22h45'54" 2d32'34"
Charlotte Anne Brower
Psc 1h10'18" 31d45'53"
Charlotte Anne Butler
Lib 15h4'18" -11d10'51"
Charlotte Anne Haslam
And 0h16'20" 29d55'27"
Charlotte Anne Roller
Vir 13h30'19" -17d1'2"
Charlotte Augle 2/13/27 to 6/9/04
Cas 23h30'2" 51d48'28"
Charlotte Aurelia Townsend
Aqr 22h29'27" 0d16'41"
Charlotte Bambridge
And 23h12'42" 43d54'2"
Charlotte Barkas
And 0h24'33" 35d57'39"
Charlotte Belle 970218
Aqr 21h17'19" -8d8'45"
Charlotte Bellis
Cas 23h32'51" 51d41'46"
Charlotte Bellotte-Yeater
And 23h14'47" 46d36'31"
Charlotte Benette Hodges
Uma 11h10'36" 29d23'44"
Charlotte & Bo
Crb 15h54'23" 28d7'23"
Charlotte Bruce
And 2h16'37" 42d10'14"
Charlotte Burgess
Ari 2h18'32" 26d42'51"
Charlotte Burgess Ussery
Lyn 6h44'52" 51d47'15"
Charlotte Burton
Peg 21h43'32" 25d16'0"
Charlotte C. Kepler
Psc 1h31'5" 11d27'20"
Charlotte Cain
Ari 2h21'55" 22d12'28"
Charlotte Caroline
Ori 5h56'55" 21d58'23"
Charlotte Charlie (Elvis) Lawson
Cru 12h29'39" -60d20'32"
Charlotte Chavira
Cas 1h20'50" 67d46'52"
Charlotte Chung-Sook Kim
Ari 1h56'13" 19d11'2"
Charlotte Cicero
And 1h50'34" 37d53'10"
Charlotte Coaton
Aqr 22h38'29" 0d39'37"
Charlotte Croci-Maspoli
Her 17h4'5" 13d18'22"
Charlotte Daisy Hill
Cas 1h14'2" 55d11'54"
Charlotte Daniels
Cas 1h26'0" 60d35'8"
Charlotte Day's Birthday Star
Vir 13h50'19" 6d26'14"
Charlotte Dean and Stephen Ross
Cyg 20h50'45" 35d54'26"
Charlotte Diane Crawley
Sco 16h43'43" -31d57'33"
Charlotte Dutch
Uma 10h44'7" 49d9'19"
Charlotte E.
Lyr 18h29'9" 36d37'45"
Charlotte Elaine Buth
And 23h51'32" 45d30'39"
Charlotte Elizabeth
And 2h38'41" 48d54'39"
Charlotte Elizabeth
And 0h47'8" 45d19'42"
Charlotte Elizabeth
Tau 4h25'11" 17d15'43"
Charlotte Elizabeth Cohen
And 1h26'14" 43d7'38"
Charlotte Elizabeth Curry
Lyr 19h21'29" 31d23'42"
Charlotte Elizabeth Switzer
Lyr 18h46'47" 37d24'3"
Charlotte Ella Hocknell
And 0h46'27" 26d54'3"
Charlotte Emily Cestnik - 28/11/03
Cru 12h46'25" -63d56'4"
Charlotte Emily Cresswell
Cas 1h8'4" 63d18'15"
Charlotte Emily Gee
And 23h15'44" 50d44'48"
Charlotte Evans
And 2h24'53" 48d54'1"
Charlotte Fay Murphy Sterett
Her 17h6'35" 40d0'18"
Charlotte Faye Handel
And 23h11'42" 41d50'51"
Charlotte + Florian
Umi 14h6'44" 76d33'46"
Charlotte Fox - Mummy's Star
Cas 0h55'44" 60d24'51"
Charlotte Fumeau
Cyg 20h26'19" 50d10'54"
Charlotte Gail Johnson Rogers
Lib 15h28'27" -4d47'36"
Charlotte Grace
Cru 12h28'19" -61d17'28"
Charlotte Grace
And 23h25'18" 50d48'41"
Charlotte Grace Chandler
Vir 12h57'5" -7d46'10"
Charlotte Grace Freeman
Leo 10h13'43" 21d51'39"
Charlotte Grace Minton
And 0h48'10" 31d48'45"
Charlotte Grace Mougey Porter
Vir 12h13'1" -0d56'21"
Charlotte Grace Payton
And 1h12'17" 46d8'46"
Charlotte Grace Peacock
And 1h31'15" 47d0'57"
Charlotte Grace Rutter
Cyg 21h33'41" 47d3'25"
Charlotte Gräfe
Cap 20h28'43" -11d4'20"
Charlotte Hampshire
Cyg 21h21'41" 33d38'2"
Charlotte Hannah
Peg 22h24'59" 10d31'58"
Charlotte Hansen Brooks
Vir 12h27'20" -8d32'30"
Charlotte Hildebrand
Cas 23h34'31" 56d10'37"
Charlotte Howseman
Crb 16h4'52" 30d46'39"
Charlotte Hua-Ying Wei
Leo 9h33'40" 11d20'23"
Charlotte I. Carnes
Uma 14h14'49" 59d50'21"
Charlotte Irene Neighbours
Cyg 21h18'31" 45d3'31"
Charlotte Isabella Slovick
Umi 13h29'14" 72d6'35"
Charlotte Jacques
Cyg 19h32'51" 28d1'6"
Charlotte Jane Curley
And 0h23'14" 26d32'20"
Charlotte Jane Moore
Cas 0h13'28" 54d44'18"
Charlotte Jane Mottram
Cru 12h44'41" -57d45'36"
Charlotte Jayne
And 1h54'46" 38d19'35"
Charlotte Jean Brickey
Cas 23h52'47" 58d37'14"
Charlotte Jean Copping
And 23h43'30" 41d46'6"
Charlotte Jenkins
Gem 7h16'5" 20d30'3"
Charlotte Joy Momma Bear
Vir 13h40'40" 1d2'11"
Charlotte Joyner Johnson
Sgr 19h21'4" -31d49'20"
Charlotte Kane
Lib 15h9'53" -9d23'1"
Charlotte Kay Sheen
Cnc 8h20'12" 22d23'24"
Charlotte Kelly
Cyg 19h56'41" 30d16'18"
Charlotte King
Aqr 21h13'11" -10d33'15"
Charlotte Klostergaard
Ori 5h30'54" 8d28'55"
Charlotte L. Clayton
Uma 12h2'58" 65d7'22"
Charlotte L. Davis
Cap 21h39'17" -14d15'25"
Charlotte L. Squires
Uma 11h22'59" 55d45'19"
Charlotte Laffoucrière
Ori 4h46'25" 5d4'42"
Charlotte Lang BA (Hons)
And 23h30'17" 43d4'9"
Charlotte Lauren Smith
Ari 3h10'37" 19d11'2"
Charlotte Lavelle
Lyr 18h47'53" 44d20'19"
Charlotte Liberty Blaker-Hemsley
And 0h13'53" 33d27'31"
Charlotte Lilly Broomhead
And 23h18'11" 41d27'20"
Charlotte Louise
And 23h25'28" 39d10'47"
Charlotte Louise
Gem 6h59'6" 17d9'19"
Charlotte Louise
Ori 5h20'8" 6d54'3"
Charlotte Louise Francesca Lewis
And 23h12'40" 42d23'13"
Charlotte Louise Holland
And 1h6'14" 46d41'8"
Charlotte Lucy Cijffers
Aqr 23h8'28" -21d25'32"
Charlotte Lütolf
Cas 1h58'30" 60d21'36"
Charlotte Lynn Rivers
Psc 0h20'54" 8d49'30"
Charlotte M. Matthews
Her 17h14'24" 24d16'29"
Charlotte M Sargent
Cnc 8h42'26" 24d6'14"
Charlotte Mae Byrne
Sgr 18h13'56" -19d59'41"
Charlotte Mae Steele
Cap 21h48'13" -18d59'50"
Charlotte Mae Toon
Ori 5h27'34" 7d31'14"
Charlotte Marie 2006
Ori 6h13'48" 7d44'48"
Charlotte Marie Blumers
Lib 14h47'52" -1d21'5"
Charlotte Marie Leonard
Ori 6h9'8" 18d57'6"
Charlotte Marie Oscar
Ari 2h27'9" 20d29'36"
Charlotte Marie Pickell
Lib 14h52'10" -1d45'41"
Charlotte Mary- Our Mother Our Star
Leo 10h2'31" 19d36'18"
Charlotte Mary Rigney
Aqr 23h23'32" -12d32'39"
Charlotte Massie Joseph
Umi 14h38'37" 72d21'38"
Charlotte Masterson
Cnc 8h45'58" 27d13'5"
Charlotte Maxine
Vir 12h57'7" -19d8'48"
Charlotte May
Cas 0h50'3" 61d33'20"
Charlotte McCafferty
And 2h24'41" 46d34'10"
Charlotte McClain
Gem 6h41'40" 28d6'38"
Charlotte McElroy
Cas 23h53'18" 53d51'32"
Charlotte Meden
Uma 10h8'57" 70d27'0"
Charlotte Michael Connelly 11/6/02
Per 3h26'56" 43d58'15"
Charlotte Moore
Crb 16h7'0" 38d53'56"
Charlotte Mordin
Cap 21h18'5" -19d19'1"
Charlotte Nestor a.k.a. The Duchess
And 0h25'30" 36d33'8"
Charlotte Neve
Pho 0h43'53" -47d3'31"
Charlotte Nora Magon
Cas 0h30'54" 62d56'43"
Charlotte Olivia Pollard
And 23h14'50" 41d2'54"
Charlotte Pace
And 23h26'48" 50d40'56"
Charlotte Palczewski
Cas 1h56'3" 65d1'50"
Charlotte Parker
Sgr 18h33'6" -26d45'58"
Charlotte Parlett
Sgr 19h19'49" -37d44'11"
Charlotte Patricia Halk
Cnc 8h40'20" 28d34'38"
Charlotte Pearce
Peg 22h23'38" 32d59'39"
Charlotte Piersa
Cas 1h26'52" 66d47'38"
Charlotte Priscilla
Cas 23h4'11" 59d44'56"
Charlotte R. DiMaggio
Uma 12h5'17" 60d14'18"
Charlotte R. Gordon
Lyn 8h29'43" 39d3'29"
Charlotte Rae Franc
Cas 1h45'17" 61d31'7"
Charlotte Reeves
Cas 0h36'29" 53d20'18"
Charlotte Riopel
And 0h8'42" 29d10'58"
Charlotte Rizzi
Cnc 8h55'57" 13d48'12"
Charlotte Rose
Leo 9h31'28" 14d18'55"
Charlotte Rose Hunter
Ori 5h53'32" 17d0'39"
Charlotte Rose Rogerson
And 23h0'4" 52d21'37"
Charlotte Rose Trent
Cam 3h21'46" 59d51'14"
Charlotte Rosenthal
Cas 0h17'6" 51d11'8"
Charlotte Roxanna
Cap 20h59'20" -16d54'29"
Charlotte Schiff
Uma 12h59'32" 63d7'33"
Charlotte "Scotti" White
Sgr 19h39'38" -13d41'35"
Charlotte & Simon Semel
Ari 3h19'10" 16d33'30"
Charlotte Simpson Connor
And 0h29'17" 41d0'59"
Charlotte Sky
Cas 0h59'15" 58d33'12"
Charlotte Sophia Demos
Lib 15h36'14" -16d55'55"
Charlotte (Steffi) Hunter Brown
Aqr 22h20'2" -9d52'34"
Charlotte Taylor
Ori 5h28'41" 14d34'43"
Charlotte Teresa Maya
Aqr 23h7'36" -6d36'36"
Charlotte Texas Burnette
Cas 23h40'33" 53d17'24"
Charlotte Thackray
Ori 5h52'49" 18d3'49"
Charlotte Thorne
And 23h24'6" 42d35'22"
Charlotte Van Ness
And 23h29'23" 41d28'8"
Charlotte Vanderhurst
Lib 15h57'49" -12d14'21"
Charlotte W. Myatt
Aqr 22h45'36" -23d14'50"
Charlotte - We Love You Mimi
And 23h19'57" 41d38'14"
Charlotte Whitcomb
Psc 1h9'56" 11d17'51"
Charlotte Xiong
Peg 21h44'59" 25d44'1"
Charlotte Yvonne McNally
Cam 4h54'45" 74d15'4"
Charlotte, Sophie, Pascal
Lyn 7h56'40" 35d5'57"
Charlotte-Best Mom in the Universe
Lib 15h29'11" -18d53'32"
Charlotte—Carlos 10-22-2004
Sge 19h44'19" 18d7'34"
Charlotte-Hannah Cabrera Pepino
Vir 14h39'47" -3d51'3"
Charlotte's Angel
Leo 10h21'31" 16d59'25"
Charlotte's Celeste Astrea
And 0h19'50" 22d59'0"
Charlotte's First Christmas
And 0h24'36" 46d0'28"
Charlotte's Light
Psc 0h30'9" 6d41'5"
Charlotte's Smile
And 23h3'42" 52d3'30"
Charlotte's Star
Cas 1h21'38" 57d42'57"
Charlotte's Star
And 0h11'28" 45d8'49"
Charlotte's Star
And 2h30'50" 50d1'12"
Charlotte's Star
And 23h48'34" 37d24'38"
Charlotte's Star
Cru 12h41'1" -64d37'39"
Charlotte's Star
Pho 23h31'45" -43d54'47"
Charlotte's Star
Lib 14h52'39" -16d27'3"
Charlotte's star of my eye, sweetie
Tau 4h41'59" 9d6'32"
Charlotte's Wolfeyes
Leo 10h26'25" 14d34'35"
Charlotte-Sophie Stephanie
Uma 10h38'44" 59d53'34"
Charlsey Martha Cornejo
Gem 7h0'25" 17d39'36"
Charlsie Christopher
Cas 1h49'20" 63d48'52"
Charlton Douglas Taylor
Her 16h27'29" 10d42'0"
Charlton Michael Balcombe
Cyg 21h56'56" 54d15'7"
Charly
Tri 2h25'4" 34d3'51"
Charly Cordelet
And 23h8'11" 52d43'1"
Charly Katelyn Cooper
And 23h27'30" 41d18'32"
Charly Rose Singleton
Ari 2h50'44" 11d47'8"
Charly, my love, my Sam
Ari 3h16'56" 28d24'14"
Charlyanna Elizabeth
And 23h13'39" 52d15'7"
Charly-Bär No 1
Her 16h12'41" 5d57'5"
Charlye Mae & Janice Edwards
Aqr 22h36'41" -1d41'27"
Charlyn Yang Siew Hui
Aqr 21h7'29" -9d24'22"
Charm Angel Eyes Garofalo
Cmi 7h28'58" 7d6'39"
Charm: Charon and Matthew Berry
Uma 8h42'47" 47d8'25"
Charm Conley
Uma 11h30'32" 47d17'31"
Charmaigne J. Saba
Mon 6h31'43" 10d49'11"
Charmaine
Sgr 18h38'11" -35d7'6"
Charmaine Denise Geer
Vir 12h56'27" 12d9'45"
Charmaine Felicia
Psc 0h13'53" 8d10'52"
Charmaine Garcia Soriano
Lmi 10h34'47" 32d37'4"
Charmaine Kay Woods Styles
Cas 2h36'43" 66d7'12"
Charmaine Smith
Uma 9h27'34" 41d36'48"
Charmalee
Sgr 18h52'47" -18d49'35"
Charmania Night Star
Gem 7h3'7" 20d6'34"
Charmant Réveil
Sge 20h16'23" 17d34'59"
Charmel
Leo 11h57'45" 20d9'7"
Charming and Sensuous Dr. Anne
Psc 0h33'43" 8d1'9"
Charms the Fisherman
Aqr 22h29'49" -0d5'32"
Charneill
Lyn 6h25'46" 57d49'26"
Charnell & Erica
Cyg 19h58'59" 55d30'28"
Charol Marshall
Uma 11h12'36" 60d27'48"
Charolene Denise Smith
Del 20h41'14" 18d24'49"
Charolette Belle Tanner-Lacher
Uma 10h35'22" 49d24'46"
Charowl
Uma 10h40'27" 46d59'50"
Charrisa McCloud
Ari 2h38'41" 28d11'34"
"Charron"
Uma 9h46'56" 63d33'30"
Charronkathleen Graham
Aqr 23h19'22" -19d30'10"
Charron-Smith
Cyg 21h28'26" 46d14'11"
Char's Star
And 0h53'50" 39d15'37"
Char's Star
Com 13h14'26" 18d30'39"
Char's Star
Vir 14h42'34" 3d15'41"
Char's Star
Aqr 21h50'28" -6d33'46"
Charsie
And 0h44'37" 42d39'17"
Charyan
Gem 7h24'15" 32d20'33"
Chas
Uma 13h33'2" 58d27'5"
Chas Carter
Leo 11h47'24" 21d40'52"
Chas Laspata
Aql 19h55'43" 4d46'44"
Chase
Psc 0h53'15" 8d43'19"
Chase
Gem 6h34'40" 13d43'57"
Chase
Psc 0h24'17" 15d16'53"
Chase
Lyn 9h13'38" 34d20'21"
Chase
Uma 9h29'47" 53d18'5"
Chase
Dra 14h29'29" 60d1'44"
Chase a saurus
Cma 6h54'33" -21d25'28"
Chase Aaron
Lyn 9h8'54" 44d43'51"
Chase Alexander
Psc 23h47'45" 2d27'43"
Chase Alexander Nealson
Umi 14h20'54" 74d21'24"
Chase Allen Neville
Ori 5h17'14" 7d0'47"
CHASE AND AMANDA
Cyg 19h53'46" 31d52'33"
Chase and Cindy
Uma 9h38'41" 68d41'33"
Chase and Jess Forever
Cyg 20h39'41" 53d43'28"
Chase and Jessica
Sge 20h4'31" 17d46'11"
"Chase Austin" ; Love Forever ; Mommy
Aqr 22h43'36" -0d59'12"
Chase B. Alward
Psc 0h30'15" 16d19'10"
Chase Ballard
Cmi 8h0'30" -0d4'33"
Chase Bartlett
Cep 22h3'6" 69d37'44"
Chase Blatherwick
Cha 10h37'58" -76d31'23"
Chase Blue Koehler
Uma 14h4'56" 58d0'55"
Chase Boswell Newsome
Sgr 19h35'11" -15d14'39"
Chase Clark Catlin Zucker
Uma 12h3'47" 52d38'18"
Chase Coghill
Leo 9h43'57" 13d57'47"
Chase Cole
Ori 6h5'33" 13d53'54"
Chase Colgur
Uma 9h36'10" 50d17'37"
Chase & Connor McKay
Sge 19h44'24" 17d6'51"
Chase & Conor Montanye
Cam 3h23'42" 61d26'16"
Chase David White
Her 17h17'2" 36d57'50"
Chase David Wood
Uma 11h31'21" 62d57'5"
Chase DeLong Is The Love Of My Life
Her 17h17'5" 17d16'0"
Chase Edward Huggard
Sco 17h34'0" -45d7'52"
Chase Elaine Holton
Mon 7h14'24" -9d35'1"
Chase Elizabeth Roberts
Cap 21h27'59" -15d53'41"
Chase Elizabeth Shepley
And 1h13'41" 38d18'39"
Chase Eric Patterson
Equ 21h9'43" 11d25'51"
Chase Erick Radermacher
Her 17h53'39" 26d39'26"
Chase Evan Westphal
Lmi 10h59'54" 32d24'14"
Chase Faith
Sco 16h41'6" -30d41'13"
Chase Family
Uma 9h6'47" 59d40'27"
Chase Fry Bradford
Psc 0h48'9" 8d25'30"
Chase Furr
Peg 21h18'25" 14d43'28"
Chase Heger
Uma 11h37'58" 54d22'53"
Chase Jaeger
Uma 10h19'14" 42d9'46"
Chase Johnson
Vir 14h3'14" -15d2'36"
Chase Joseph
Her 17h8'30" 31d14'34"
Chase Kamikawa
Oph 17h23'19" -22d44'51"
Chase Macarthur Crandall
Del 20h38'10" 15d15'19"
Chase Martin
Boo 14h21'21" 17d36'48"
Chase Michael Leite
Ori 5h12'34" 11d15'31"
Chase Montgomery Keith Buck
Tau 4h27'16" 14d54'34"
Chase Mounts
Dra 16h43'56" 58d39'16"
Chase Nathaniel Bain - Chase's Star
Peg 21h38'41" 14d11'17"
Chase Nico DaBella
Uma 9h36'13" 66d30'36"
Chase Noah Moskowitz
Cnc 8h11'49" 32d22'57"
Chase Oliver Gosselin
Ari 3h16'33" 28d45'4"
Chase Paisley - Sparlin
Peg 22h10'31" 31d32'58"
Chase Pierce Cahill
Gem 7h13'42" 34d48'30"
Chase R. Fornengo
Umi 16h17'18" 75d52'49"
Chase & Rachel
Cam 3h48'33" 69d51'5"
Chase Rebecca Stern
Lyn 9h4'4" 38d20'10"
Chase Robert Neanen
Her 17h33'12" 36d9'53"
Chase Rockne Poglajen
Lib 15h30'50" -25d6'44"
Chase Rovero
Psc 1h19'16" 31d28'4"
Chase Ryan Caviness
Sgr 18h30'1" -25d8'59"
Chase Ryan Whitham 8-22-1982
Leo 9h33'8" 27d28'49"
Chase & Scott Duran
Cnc 8h8'33" 11d42'43"
Chase Thomas McKinney
Her 16h41'45" 47d7'15"
Chase Weston Lehocky
Sct 18h49'34" -9d30'25"
Chase William Carlson
Cas 1h8'49" 54d22'35"
Chase Wilson
Cnc 8h39'8" 24d52'22"
Chase Yeranoohie Bond
And 2h20'12" 43d57'21"
Chase Your Dreams
Leo 11h23'28" -0d39'34"
Chase1
Sco 17h27'46" -42d7'43"
Chasen & Tyler
Gem 7h14'4" 27d57'56"
Chase's Christmas Star
Uma 11h40'0" 31d59'16"
Chase's Star
Per 3h41'37" 34d18'30"
Chase's Star
Ori 5h47'29" 5d29'34"
Chasey Leigh Bailey
And 0h21'38" 46d6'56"
Chashley
Cyg 20h42'31" 42d6'44"
Chasidy Ann Cowdell
Leo 11h31'33" 6d51'28"
Chasity and Hank
Cas 1h13'14" 54d31'12"
Chasity Campbell
Psc 0h0'3" 8d7'4"

Chasity Chas Lyn 6h24'46" 56d49'3"
Chasity Lea Yeargan Aqr 22h3'22" 0d30'28"
Chasity Lovely Aqr 23h21'53" -13d35'22"
Chasity Mae Cas 0h28'0" 56d57'38"
Chason Michael Johnson Gem 7h23'18" 29d21'20"
Chastity Lynn Peck Uma 11h13'41" 34d46'13"
Chastity Neal And 0h25'55" 33d26'36"
Chastity Rose Wheeler And 23h50'18" 39d34'31"
Chata Lyn 7h31'8" 54d28'14"
Chatavilles Cap 20h35'33" -20d19'26"
Chatch Pho 23h29'35" -39d47'24"
Chatham Peg 22h45'56" 29d18'31"
Chato & Eva's Star Cyg 20h23'3" 38d42'51"
Chatti Cathy And 23h15'12" 42d2'29"
Chau Bao Lyn 8h12'28" 35d46'46"
Chau N. Nguyen Tau 3h47'42" 3d9'42"
Chau Trinh Sco 16h42'37" -38d34'5"
ChauMary Lyn 7h26'45" 53d54'18"
Chauncey Cas 23h24'32" 54d45'4"
Chauncey Cra 18h49'22" -39d15'36"
Chauncey Uma 10h12'27" 43d40'2"
Chauncey and Marie De Pew Lib 14h50'38" -1d19'0"
Chauncey Billups Lib 14h52'11" -1d51'39"
Chauncey Thompson Her 17h9'4" 31d47'56"
Chaundra Cam 7h26'21" 63d33'2"
Chaunliyah Sco 17h39'35" -39d34'34"
Chauntae Ruppe Per 3h28'39" 45d34'48"
Chauntel Gallardo Eri 3h45'26" -13d10'58"
Chauntelle Siobhan Letitia Walsh Cas 2h15'55" 72d31'44"
Chaunty Jean Heggtveit Leo 10h22'23" 26d5'28"
Chauri Mechelle Logan Crb 15h47'7" 35d27'11"
Chauvet Sandrine Uma 8h55'28" 52d56'45"
Chava Blodek Kopelman Vir 14h40'2" -3d38'26"
Chavalah Madeline Pilmaier Lib 15h36'48" -10d47'24"
Chavalah Rebekkah West Tau 3h59'46" 5d2'17"
Chavis Moultry Cnc 8h11'51" 21d36'52"
Chawndra Billie Lyn 8h25'39" 48d58'39"
Chay & Claire - Our Star Cyg 19h37'57" 29d28'27"
Chay La Cyg 20h38'20" 31d32'53"
Chay Michael Pena Cyg 19h54'28" 36d55'26"
Chay Woodward Ori 6h5'45" 6d46'3"
Chaya Ori 6h17'49" 5d58'17"
Chayce Greggory Ford Ori 6h1'59" -0d3'17"
Chayden's Dream Her 17h34'3" 27d9'22"
Chayito Beltran Cas 1h33'43" 57d58'20"
Chayla Bignold Cas 23h28'11" 57d43'52"
Chayla Eden Creighton Lyn 6h55'16" 59d52'5"
Chaylee Jaide Estabrook Lib 15h5'25" -12d55'52"
Chayne Scott Standage Uma 11h34'43" 48d23'1"
Chaynord 4/1/03 Umi 16h16'58" 73d42'27"
Chayse Sgr 18h56'8" -29d8'44"
Chayse Jewel Woehrle Cyg 21h16'19" 44d29'49"
Chayse Olivia And 0h21'17" 32d45'30"
Chayseng Saechao Uma 11h28'9" 51d50'55"
Chaz Vir 13h57'17" 2d38'59"
chaz Lib 15h37'38" -7d9'24"
Chaz & Catie Matses Psc 23h56'35" -4d10'14"
Chaz McGuire IV, Quinn's Daddy And 23h31'12" 47d21'45"
Chaz Michael Carruth Gem 6h48'9" 32d54'21"
Chazman75 Uma 10h6'55" 63d45'57"
Chazmine Rose Contreras Aqr 22h42'14" -1d53'16"
Chazwald Her 18h5'41" 18d57'51"
Chazz My Love Per 4h30'59" 48d24'51"
Che Vir 12h23'17" -11d17'26"
Che Arosemena Ari 2h50'40" 28d33'31"
Che Che Chang Uma 9h12'58" 61d55'21"
Che & Scott's 1st. Ori 5h57'49" -0d46'38"
Che' Vyfhuis Cyg 20h58'46" 47d12'42"
Chealse Nichole Lep 5h32'39" -18d50'18"
Cheanté Cas 23h38'12" 57d5'3"
Chechuchachu Sgr 18h3'9" -29d25'54"
Check- Tim DeWayne Inman Uma 8h39'44" 63d49'17"
Chedel Sco 16h12'39" -23d38'20"
Chee - Saw Cam 8h46'19" 83d24'47"
CheeBumBum Cas 1h51'49" 61d32'29"
Cheeka's Heart Lyn 8h11'46" 45d1'57"
Cheekiness of Chadd Shining Always Pho 23h56'34" -44d34'32"
Cheeko's Star Her 16h49'47" 9d0'46"
Cheeks And 1h28'24" 45d27'24"
Cheeky And 2h30'55" 39d14'3"
CheekylashNin Dra 17h57'42" 69d36'1"
CheekyMonkey Tau 5h15'15" 18d34'21"
Cheer Star Malinda Hasen Cap 20h8'49" -14d25'3"
Cheesey Poof 7 Sgr 19h22'41" -27d17'17"
Cheetony Cru 12h57'41" -63d53'23"
Chef Elaine Leo 10h3'22" 15d21'10"
Chef Gary Spitnale C.E.C. Her 18h47'14" 18d53'42"
Chef Tammy Lynn Shuttera, berta Ori 4h52'14" 11d25'4"
Chehalem Valley Middle School Uma 11h23'19" 58d12'56"
Che'HEM Sgr 18h4'48" -25d0'10"
cheifJohn08WSFD Uma 10h14'1" 67d36'3"
Chel Belle Gem 6h22'42" 22d22'8"
Chel & Ryan Ori 5h49'42" 8d26'4"
Chela Lmi 11h3'31" 28d36'2"
"Chela" Cas 1h30'46" 60d34'27"
Chela & Julio Uma 10h26'37" 52d37'6"
Chelby Uma 9h10'41" 48d32'44"
Chel-C Crb 16h12'47" 37d23'9"
Chelcee Cap 21h37'9" -11d18'0"
Chelcee Dawn And 0h36'43" 29d18'36"
Chelchy Ori 5h6'32" 11d26'13"
Chelcie Devin Dieterle Psc 1h0'53" 4d14'7"
Chelcie Gene Taylor Jr. Ari 2h3'6" 11d14'8"
Cheli Arussy Tau 4h39'54" 27d44'42"
Cheligo Del 20h35'45" 6d17'27"
Chelise And 0h30'54" 33d19'54"
Chelito Aqr 23h37'6" -13d29'4"
Chella Tau 4h45'43" 19d4'21"
Chella Uma 10h39'9" 41d6'42"
Chelle Gem 6h1'33" 23d7'40"
Chelle Belle Crb 16h12'16" 32d23'19"
Chelle Star Leo 9h23'1" 25d55'4"
Chelleigh Cas 1h39'40" 60d45'55"
Chelle's Star And 1h23'47" 48d45'25"
Chellie Plummer Leo 10h44'59" 10d2'43"
C-H-ELLIOTT Uma 10h40'16" 46d46'40"
Chelly Elizarraraz Uma 10h31'18" 52d25'57"
Chelly's star Sgr 19h9'11" -16d24'45"
Chelo Lyr 18h52'50" 39d18'7"
Chels Lib 15h6'40" -7d17'7"
Chels and Matt 27-7-02 to forever Cru 12h43'49" -58d2'44"
Chels Babe And 2h21'43" 47d38'44"
Chelsa Uma 10h57'32" 52d14'52"
Chelsa Cam 4h18'10" 70d39'58"
Chelsae Kosman Sgr 18h34'31" -23d1'35"
Chelsea Sgr 17h53'19" -28d19'21"
Chelsea Umi 15h22'24" 74d48'21"
Chelsea Cas 2h31'11" 66d56'14"
Chelsea Uma 10h6'13" 57d36'23"
Chelsea Uma 11h37'27" 55d33'36"
Chelsea Lyn 6h24'51" 56d57'5"
Chelsea Lyn 7h25'8" 55d2'33"
Chelsea Uma 10h35'21" 51d14'57"
Chelsea And 23h44'13" 47d20'43"
* ~ Chelsea ~ * And 2h21'4" 47d58'11"
Chelsea Uma 8h37'29" 50d33'57"
Chelsea And 23h54'16" 38d16'47"
Chelsea Gem 7h11'16" 32d31'11"
Chelsea Gem 7h41'58" 32d39'34"
Chelsea Uma 9h35'29" 43d55'45"
Chelsea Uma 11h22'53" 39d20'34"
Chelsea And 2h8'2" 41d8'16"
Chelsea And 2h31'20" 43d42'4"
Chelsea And 0h52'27" 39d28'28"
Chelsea Aql 19h41'23" 3d58'5"
Chelsea Aqr 21h27'5" 2d1'37"
Chelsea Cmi 8h4'41" 1d7'17"
Chelsea Tau 3h44'19" 7d0'45"
Chelsea Leo 11h39'52" 25d19'1"
Chelsea Gem 7h3'5" 26d56'38"
Chelsea Del 20h43'26" 18d14'27"
Chelsea Leo 11h0'50" 15d57'35"
Chelsea A. Krick Leo 10h6'17" 22d51'19"
Chelsea Alexandra And 0h12'1" 42d25'13"
Chelsea Alexis And 0h10'55" 39d53'52"
Chelsea Alexis Rosenbloom Sco 16h6'33" -17d14'4"
Chelsea Allen 5-1-1992 And 0h24'22" 35d49'57"
Chelsea Alyssa Tonsic Tau 5h30'20" 19d35'46"
Chelsea and Andy And 1h16'39" 42d12'31"
Chelsea and Jason Waldram Gem 7h48'0" 33d17'44"
Chelsea and Johnny Gem 7h13'30" 25d40'53"
Chelsea and Justin Vir 12h50'31" 6d56'31"
Chelsea and Justin Love Eternal Cyg 20h8'0" 41d21'51"
Chelsea Anita Roy Lyn 7h20'4" 45d43'1"
Chelsea Ankrum Her 18h0'59" 17d54'23"
Chelsea Ann Fredrikson Leo 11h3'40" 16d29'14"
Chelsea Ann Palermo Gem 7h31'29" 16d24'33"
Chelsea Ann Parker Cnc 8h46'28" 30d48'54"
Chelsea Ann Shaffer Sgr 19h18'37" -12d47'50"
Chelsea Ann Steinmetz And 2h8'7" 45d9'7"
Chelsea Anna Tallarico And 0h0'21" 46d4'24"
Chelsea Anne Lara Lyr 18h48'15" 33d46'59"
Chelsea Arrianna Ari 3h15'51" 19d37'13"
Chelsea Bell Cnc 9h17'35" 32d10'24"
Chelsea Blair Stone And 2h11'11" 45d13'37"
Chelsea Bree Crb 15h45'18" 38d56'4"
Chelsea Caitlin Moody Cnc 8h21'13" 12d2'7"
Chelsea & Caleb Sge 19h57'26" 18d59'12"
Chelsea Camp Vir 14h54'31" 6d48'10"
Chelsea Campagna Aqr 21h2'23" -12d0'56"
Chelsea Campbell Crb 16h0'54" 30d45'30"
Chelsea Carmelita Ari 2h39'59" 17d31'11"
Chelsea Carroll, a face of an angel Cas 3h8'11" 57d29'39"
Chelsea Cecilia Baird Crb 15h50'2" 26d11'55"
Chelsea Champagne Cas 23h59'9" 64d17'33"
Chelsea Charms And 23h20'40" 38d26'20"
Chelsea Chevonne Reed Dra 15h23'1" 58d46'19"
Chelsea Claeys Aql 20h3'30" 12d20'58"
Chelsea Dale Uma 11h47'59" 48d4'46"
Chelsea & Dave Kusky Cyg 20h51'0" 42d45'46"
Chelsea Denise Williams Lyn 9h15'4" 36d44'31"
Chelsea Diane Smith And 2h20'38" 50d14'3"
Chelsea Doll Ori 5h22'13" 1d31'51"
Chelsea Dorthea Isbel And 23h44'0" 47d13'30"
Chelsea Elaine Padova And 0h11'29" 35d59'58"
Chelsea Elisabeth Alderman Tau 3h47'4" 26d24'3"
Chelsea Elizabeth Biggs Psc 1h11'25" 10d56'12"
Chelsea Elizabeth Hoover Vir 14h14'52" 5d25'6"
Chelsea Elizabeth McLaren Psc 1h40'18" 10d34'24"
Chelsea Elizabeth Stamey Sco 16h28'42" -35d4'18"
Chelsea Erin Aqr 20h55'19" 1d19'21"
Chelsea Faith Wellborn Cyg 20h28'38" 36d1'28"
Chelsea Falcone And 1h18'32" 38d48'3"
Chelsea Glenette Gahagan And 0h22'54" 25d30'15"
Chelsea Grace Parsons' C P Star Cap 21h56'56" -18d34'55"
Chelsea Gray Tau 4h0'48" 30d32'51"
Chelsea Greer Leo 11h56'41" 23d35'34"
Chelsea Groves Uma 8h36'48" 48d16'27"
Chelsea Hartle Uma 9h31'22" 68d10'50"
Chelsea Hegge Psc 0h2'43" 1d3'56"
Chelsea Hope Uma 11h39'46" 44d22'41"
Chelsea Hope Boyles Uma 10h26'3" 57d49'50"
Chelsea Hughes And 23h20'39" 48d7'38"
Chelsea Jane Tau 5h11'30" 25d8'52"
Chelsea Janelle Zenner Uma 8h39'45" 70d58'39"
Chelsea Jeffers Ori 5h32'37" -0d13'40"
Chelsea Jordan Hylton Sgr 18h40'40" -29d28'2"
Chelsea Juanita Frost Ori 5h50'16" 6d35'15"
Chelsea Jye Cas 1h0'40" 48d58'51"
Chelsea Kathleen Prevost Cas 1h41'31" 64d52'2"
Chelsea Kathleen Sullivan Mon 6h51'51" -0d55'58"
Chelsea Kyal Gill Sco 17h7'2" -37d23'38"
Chelsea L. C. Strand Lmi 10h38'27" 32d13'14"
Chelsea Laine Peterson "Petey" Uma 10h49'58" 71d57'17"
Chelsea Lane Tau 4h41'19" 23d14'10"
Chelsea Lauren Wood Cas 23h9'48" 55d48'4"
Chelsea Lea Keller And 0h55'20" 38d56'9"
Chelsea Lee Sco 17h40'29" -42d22'32"
Chelsea Lee Kersting Cas 1h25'54" 63d29'7"
Chelsea Leialoha Pabo Psc 1h40'40" 10d54'43"
Chelsea Louise Miller Psc 1h5'43" 23d9'55"
Chelsea Lynn Uma 10h32'4" 56d53'38"
Chelsea Lynn Heaton Psc 1h35'16" 3d41'58"
Chelsea Lynn/ My Love Psc 0h31'21" 3d14'38"
Chelsea Lynn Thompson Tau 4h8'16" 8d33'32"
Chelsea Madison Gaines Tau 5h52'3" 24d49'21"
Chelsea Mae And 0h18'15" 25d46'26"
Chelsea Maitland McQuarrie Aqr 22h28'22" -2d3'52"
Chelsea Mandell Uma 11h50'11" 34d20'16"
Chelsea Margaret Leo 11h53'17" 22d12'15"
Chelsea Marie Vir 13h3'1" 6d55'41"
Chelsea Marie Cmi 7h25'3" 1d59'25"
Chelsea Marie And 1h39'12" 41d58'21"
Chelsea Marie Cap 20h7'37" -20d25'52"
Chelsea Marie Lib 14h45'29" -17d24'14"
Chelsea Marie Sco 17h46'33" -43d6'33"
Chelsea Marie Sco 16h53'40" -38d38'19"
Chelsea Marie Brown Ori 6h17'33" 2d30'18"
Chelsea Marie Dahms Ari 2h4'49" 19d23'55"
Chelsea Marie Hall Ari 2h54'39" 22d11'37"
Chelsea Marie Rainbolt Uma 9h23'56" 69d7'5"
Chelsea Marie Sims Ant 9h48'50" -32d45'52"
Chelsea Mercedes Garcia Dra 17h45'17" 51d13'6"
Chelsea Michelle Aqr 22h25'2" -1d49'27"
Chelsea Moran Sgr 19h24'4" -12d14'30"
Chelsea Moreira Ari 2h31'17" 17d38'32"
Chelsea Morgan Ashcraft Ori 6h9'58" 9d12'17"
Chelsea Morgan Hoffmann Lyn 8h2'4" 58d35'19"
Chelsea Morgan Rosnick Sgr 19h7'15" -19d8'54"
Chelsea Morgan Thiboutot And 0h13'26" 43d9'13"
Chelsea My Love Sco 16h5'37" -11d54'19"
Chelsea My One And Only True Love And 0h14'37" 42d46'39"
Chelsea Nicole Brown Leo 10h10'5" 24d40'3"
Chelsea Nicole Crouch Cap 21h34'19" -8d48'29"
Chelsea Nicole Hoover Uma 10h21'53" 57d52'51"
Chelsea Nicole Stocking Vir 13h43'48" 3d26'6"
Chelsea Nicole Weller Gem 7h4'1" 29d14'36"
Chelsea *Our Shining Star* And 1h43'54" 49d1'21"
Chelsea P. Ori 5h33'31" 6d13'58"
Chelsea Page Atwell Ori 5h54'4" 7d17'48"
Chelsea Paris Bennett And 23h20'47" 41d24'9"
Chelsea Poo And 23h40'54" 48d10'48"
Chelsea Potter And 0h40'2" 26d32'56"
Chelsea R. Hibbitt And 0h51'32" 36d58'44"
Chelsea Rae Evans Ari 2h15'34" 24d30'34"
Chelsea Rene Gem 6h49'47" 26d31'21"
Chelsea Renee Umi 15h25'40" 70d52'43"
Chelsea Renee Carrier Psc 0h38'35" 9d18'58"
Chelsea Renee Pennick Crb 15h43'49" 26d59'52"
Chelsea Resnick Uma 13h37'2" 55d57'19"
Chelsea Rheaume Lib 15h1'55" -2d42'46"
Chelsea Rodgers Cas 1h39'29" 65d27'48"
Chelsea Rose Vir 13h54'13" -0d35'3"
Chelsea Rose Tau 5h37'39" 25d52'57"
Chelsea Rose Cnc 8h17'34" 20d19'26"
Chelsea Rose Donovan Cas 0h36'7" 52d24'36"
Chelsea Rose Geiger Psc 1h57'29" 7d44'50"
Chelsea Rose Hall Cru 12h21'38" -61d39'36"
Chelsea Rose McCann Weaver Crb 16h23'7" 30d23'32"
Chelsea Rose McDonnell Lib 15h49'55" -6d25'33"
Chelsea Rose Snyder Lib 15h8'40" -4d59'4"
Chelsea Rose Trubell And 0h27'28" 40d4'55"
Chelsea Sharisse Martinez And 2h27'28" 49d1'9"
Chelsea Stapleton Leo 10h20'39" 10d48'13"
Chelsea Stargaret Acunis Graham Lyn 6h51'35" 50d16'0"
Chelsea Stockmaster Cyg 20h28'0" 36d37'8"
Chelsea "Sugar Lump" Wellman Uma 11h14'30" 35d20'32"
Chelsea Summer Gray Tau 5h39'30" 18d32'11"
Chelsea Susan Shapiro Aqr 21h35'6" 0d47'39"
Chelsea Tait And 0h10'55" 24d58'56"
CHELSEA TAYLOR MINIX Leo 10h40'32" 7d31'40"
Chelsea Victoria Vela Vir 12h59'39" -19d29'39"
Chelsea Webb Tau 3h25'5" 19d11'33"
Chelsea Welden And 0h17'28" 28d35'43"
Chelsea Wilson And 0h21'52" 43d17'27"
Chelsea Woods Gem 7h45'50" 18d28'48"
Chelsea Zora O'Barr Cap 20h32'39" -11d48'47"
Chelsea Zuspan Aqr 22h10'4" -3d7'25"
Chelseabelle Elizabeth Leiter Ori 5h38'14" 1d52'4"
Chelsea's Daddy Cnc 8h52'45" 27d31'21"
Chelsea's Star And 0h27'46" 42d37'11"
Chelsea's Star Lyn 8h11'29" 55d23'26"
Chelsea's Star Sgr 19h6'17" -31d7'15"
Chelsea's Wish Aqr 23h22'37" -10d49'38"
ChelseaStar Vir 12h33'29" 12d30'37"
Chelsee Lynn Sawai Mon 6h10'1" -7d37'8"
Chelsee Waters Gem 7h43'25" 24d35'32"
Chelsey Ari 3h6'53" 12d26'41"
Chelsey And 2h38'26" 46d44'35"
Chelsey And 1h43'29" 45d10'47"
Chelsey Aqr 22h38'54" -3d3'57"
Chelsey Cas 2h50'27" 61d44'17"
Chelsey and Cole's Star Lyn 7h22'35" 44d28'17"
Chelsey Ann Marie Sco 17h51'11" -39d5'9"
Chelsey Anne Horsley Gem 6h1'48" 21d36'5"
Chelsey Blair And 0h33'23" 36d5'26"
Chelsey Elizabeth Cnc 8h56'58" 12d49'7"
Chelsey Jane Mendelsohn Psc 1h25'35" 27d51'17"
Chelsey Lauren Smith And 0h53'52" 46d8'55"
Chelsey Leonard Per 3h21'19" 41d54'4"
Chelsey Lynn Lib 14h27'5" -10d36'27"
Chelsey Marie Williams Sgr 18h54'5" -31d34'21"
Chelsey Nicole Howser Lyn 8h39'31" 37d11'5"
Chelsey Panchot Uma 13h24'17" 55d28'35"
Chelsey Peterson Uma 11h12'53" 55d8'58"
Chelsey Rhyan Youse Cnc 8h47'19" 12d15'5"
Chelsey Teeter Cas 0h13'27" 55d30'1"
ChelseyBlue Lyn 8h56'21" 35d41'49"
Chelsey's Star Sco 17h35'12" -39d19'10"
Chelseywig Cap 21h42'59" -10d10'12"
Chelsi Fowler And 2h3'23" 42d25'20"
Chelsi Leann Robinson Cyg 20h2'36" 35d16'58"
Chelsie And 1h32'21" 42d22'21"
Chelsie Lyn 8h24'39" 38d52'50"
Chelsie Lyr 18h56'20" 34d46'41"
Chelsie Breana Tau 4h35'2" 8d39'31"
Chelsie Collera Gem 6h26'19" 16d33'19"
Chelsie Itsell Uma 12h0'8" 62d3'53"
Chelsie Jane Little And 0h27'51" 46d34'55"
Chelsie Lauren Reese And 0h10'44" 39d29'34"
Chelsie Lynn St. Peter Lib 15h11'26" -23d38'8"
Chelsie Marie Scansen Leo 10h16'16" 12d28'27"
Chelsie Reneé Anthony And 0h51'28" 44d51'41"
Chelsie's Heart Tau 3h41'38" 9d58'59"
Chelsy L. Petersen Ori 5h46'41" 7d21'16"
Chelsye DeBoor Her 16h17'59" 7d3'17"
Chelu Arenas Ari 2h59'39" 27d42'53"
Chen Chen Aqr 21h29'52" -0d37'53"
Chen Dandan Sgr 19h36'28" -13d13'46"
Chen Lihong Peg 23h51'20" 19d24'20"
Chen & Nhu Her 17h37'38" 32d4'2"
Chen Qi Uma 9h40'51" 54d8'21"
Chenellebra Umi 13h4'45" 73d36'43"
Chenenceau Uma 8h59'24" 61d19'45"
Cheney Davidson Aur 5h39'30" 33d28'13"
Cheney Micol Her 17h56'19" 21d12'24"
Cheney-Havatone Sco 16h12'8" -15d53'53"
Cheng Vue Ari 3h23'33" 24d3'56"
ChEng William Christopher Corey Lmi 10h27'23" 34d47'15"
"chenko" Louis G. Montoya 9/5/1960 Vir 12h0'6" 7d14'46"
Chennie Ari 2h52'13" 29d27'36"
Chenoa Hope Kimana Cas 3h27'36" 70d7'11"
Chenoah's Beauty And 1h16'36" 38d52'27"
Cheo Mi Amor Lyn 6h58'34" 45d58'26"
Chepa Ser 16h10'59" -0d15'32"
Chepe Phelps Leo 11h35'2" 19d35'25"
Chepina Ori 5h40'43" 11d25'8"
Cher Uma 11h4'42" 35d37'1"
Cher Crb 15h41'5" 35d22'26"
Cher Chirico "Zeva" Sco 16h14'21" -9d6'38"
Cher Walters Cam 3h37'32" 59d42'38"
Cheral's Night Light Uma 11h24'9" 57d39'43"
Cheraylyn A. Willyard Psc 1h51'59" 10d7'38"
Cherbear (Cher Riplett) Cas 0h16'10" 56d11'21"
Cherelle Barbara Ori 5h59'25" 20d37'57"
Cherelle xXx And 0h55'39" 22d23'23"
Cheri Cnc 7h58'6" 10d23'29"
Cheri Vir 12h52'12" -0d9'15"
Cheri Adams And 23h41'42" 46d59'47"

Cheri and Tricia
And 23h19'17" 48d44'41"
Cheri Ann
Vir 13h1'29" 12d3'55"
Cheri Arnold
Crb 16h8'23" 37d26'29"
Cheri Becker
And 2h27'50" 40d26'27"
Cheri Bootz
Gem 7h49'58" 16d19'52"
Cheri Collin Our Shining Star
Crb 16h2'26" 36d17'55"
Cheri Dale Haislip
Crb 15h38'46" 27d18'44"
Cheri & Doug Howlett
Aql 19h8'21" -0d18'28"
Cheri Elizabeth Shannon
Aql 19h5'28" -0d53'53"
Cheri Ellen
Uma 9h27'42" 71d6'37"
Cheri Flansburg
Vir 13h51'58" -17d45'8"
Cheri Gail
Cas 1h12'13" 58d3'45"
Cheri Ganong
Lyn 8h42'15" 43d52'55"
Cheri Hightree
Leo 11h30'44" 21d38'19"
Cheri Jean Wittbold Roush
Crb 16h6'34" 35d52'22"
Cheri K Morrison
Ari 2h21'51" 22d18'12"
Cheri Kurata-Kimura
Umi 15h13'31" 67d46'58"
CHERI L. ADAMS
And 2h13'40" 37d36'55"
Cheri Lamore
Leo 10h7'6" 22d42'7"
Cheri Lavonne Harden
Lib 15h34'28" -12d3'38"
Cheri Lyn Simonenko
Leo 9h38'35" 27d14'1"
Cheri Lynn
Mon 6h51'30" 8d0'21"
Cheri Lynn Dijak
Cnc 8h49'3" 28d30'19"
Cheri Lynn Sanchez
Sgr 19h40'31" -16d53'1"
Cheri Lynn Sheeks
Vir 13h58'54" 4d13'19"
cheri & manu
Cam 7h53'25" 63d40'11"
Cheri & Mike's Wedding Star
Cyg 19h44'5" 32d5'19"
Cheri My Love
Ari 2h7'51" 22d53'25"
Cheri Poo
Uma 10h6'27" 48d13'0"
Cheri Renee'
Cas 1h52'5" 65d22'18"
Cheri & Rudy
Leo 9h22'49" 16d55'12"
Cheri T
Cnv 13h45'1" 30d15'51"
Cheri Yvonne Sawicki Marks Tronrud
Umi 15h30'34" 87d53'33"
Cherida
Cyg 20h10'28" 38d55'35"
Cherie
Crb 15h46'4" 33d3'37"
Cherie
And 0h46'10" 38d33'46"
Cherie
Uma 11h27'49" 29d2'57"
Cherie & Andy
Cyg 20h51'19" 46d35'19"
Cherié Cara-Mae Caviness
Cas 1h35'14" 68d30'46"
Cherie Cox
Aqr 22h20'0" 0d3'6"
Cherie & Craig
Cyg 21h21'52" 52d12'3"
Cherie D. Kmetz
Crb 16h2'29" 29d35'35"
Cherie Haney
Tau 4h39'49" 21d10'40"
Cherie & Jade Forero
Cyg 21h22'6" 34d22'10"
Cherie Kim Moore
Sco 16h45'33" -36d0'6"
Cherie La Rosa
Psc 1h40'4" 10d25'47"
Cherie Lee Waterstraat
Uma 13h34'30" 57d43'6"
Cherie LeopardStar
Aur 5h44'34" 45d42'41"
Cherie Lynn Dersch
Aqr 22h26'13" -23d43'55"
Cherie Melissa Yturralde
Gem 6h20'15" 25d7'45"
Cherie My Love
Lyn 7h47'12" 42d42'8"
Cherie Nicole
Ari 3h26'28" 23d4'52"
Cherie & Paul
Uma 14h16'12" 56d40'31"
Cherie Perkins
Cap 20h35'44" -17d58'55"
Cherie R. Williams
Ari 2h56'34" 24d26'56"
Cherie S. Dick
Leo 9h37'4" 28d43'56"

Cherie SAS Seeley
Uma 11h44'17" 63d12'50"
Cherie Scalzi
Vir 12h17'8" 6d6'1"
Cherie Sigala
Sco 17h55'50" -38d31'18"
Cherie & Simon - 5 November 2004
Cru 12h40'55" -56d13'37"
Cherie Trahan
Ari 2h0'39" 21d6'19"
Cherilyn Schuff and John Shao <3
Cyg 20h13'55" 53d34'12"
Cherin Suzanne Rossbach
And 1h49'53" 46d35'24"
CHERINA
Her 17h35'7" 34d48'1"
Cherina Sparks & David Pfaff
Psc 1h26'48" 29d0'51"
Cherine K Hatem
Lyr 18h51'59" 36d21'23"
Cheri's Diamond In The Sky
Uma 11h54'53" 59d47'51"
Cheri's Firefly
Uma 9h16'59" 65d26'19"
Cherisa Michelle
And 0h20'40" 32d9'31"
Cherise
And 0h49'56" 36d41'55"
Cherise E. Gonzales
Peg 22h38'42" 12d6'32"
Cherise Leanne Dormer
Vir 12h17'2" 1d58'44"
Cherise Lynne
Uma 11h6'27" 62d33'40"
Cherish
Aqr 22h33'34" 1d20'56"
Cherish
Cyg 21h31'55" 32d1'6"
Cherish
Lyn 7h35'45" 49d43'24"
Cherish
Cyg 20h30'27" 46d8'44"
Cherish
Cyg 20h15'3" 47d47'53"
Cherish
Cyg 21h14'3" 46d48'54"
Cherish
Uma 10h25'29" 49d2'16"
Cherish Cassella
Gem 7h32'13" 34d3'34"
Cherish Faith
Aqr 23h34'2" -22d25'13"
Cherish Hall Tucker
Vir 13h20'21" 11d51'26"
Cherish Hinton
Lyn 6h45'21" 60d51'56"
Cherish Mariah Bird
Uma 11h46'44" 54d45'0"
Cherish "Miracle" Bostwick
Uma 10h1'28" 65d35'51"
Cherish Till Eternity
Sgr 19h39'4" -21d9'25"
Cherished Champion Sonny Hwang
Com 13h0'38" 21d5'22"
Cherished Friend Samantha McLeish
Peg 22h49'14" 28d13'48"
Cherished One
Cas 0h42'21" 66d53'52"
Cherisse Jones
Lib 14h48'39" -11d22'26"
Cherista
Cnc 8h18'51" 31d16'31"
Cherl A. Stegemeyer
Vir 13h54'32" -17d23'29"
Cherlece Laduca and Herve Paul
Cnc 8h51'22" 28d43'48"
Cherlene
Cap 20h23'31" -12d47'8"
Cherokee Mom 326
Cas 0h9'5" 62d33'9"
Cherokee Strickland
Peg 22h18'15" 15d29'35"
CHERON "KIT-KAT"
Aqr 21h30'19" 0d16'39"
Cherrelle's Song
Cas 0h22'35" 59d54'24"
Cherri Canady
Uma 9h37'52" 55d57'55"
Cherri Carbonara
Vir 12h29'49" -5d8'9"
Cherri L. Labadie
Lib 15h26'2" -6d14'46"
Cherri Lynn, Justin & Cacee
Tri 2h10'56" 34d9'41"
Cherri Rae
And 0h51'45" 43d32'14"
Cherrie loves Joshua 4ever & always
Cyg 19h55'38" 59d37'56"
Cherrish Sharron Jo Lee
Leo 11h8'7" 2d5'24"
Cherry
And 0h16'8" 31d48'37"
Cherry
Crb 15h35'20" 30d7'28"
Cherry (Beth) Smith
And 1h33'50" 50d7'56"

Cherry Cristina Barbier
Per 3h13'13" 51d45'24"
Cherry Flamex
Psc 1h5'58" 8d31'11"
Cherry Lynelle Boyle
Sgr 18h27'41" -32d49'46"
Cherry Minga Gleason
Cas 23h24'23" 56d45'37"
Cherry Mosesz & Nasser Goedar
Cyg 20h17'49" 48d2'33"
Cherry Queen
Cas 1h35'11" 61d44'30"
Cherry R. Childers
Aqr 22h25'38" -1d38'58"
Cherry Street
Her 16h44'44" 7d16'49"
Cherryl Lynn Baethge
Vir 14h16'58" -13d28'59"
Cherry's Star
Eri 3h14'8" -39d8'57"
Cherub L. Gardner
Aql 19h46'13" -0d18'7"
Cheryboom
Cas 2h11'13" 70d45'27"
Cheryl
Lib 15h44'13" -6d11'8"
Cheryl
Vir 14h42'37" -2d56'50"
Cheryl
Cra 19h0'49" -38d8'0"
Cheryl
Vir 11h41'47" 2d44'21"
Cheryl
Crb 15h58'47" 28d8'56"
Cheryl
And 23h8'4" 48d8'50"
Cheryl
Lyn 7h21'41" 45d46'22"
Cheryl
And 23h35'48" 42d42'50"
Cheryl
And 23h23'23" 38d57'29"
Cheryl
Lyn 7h34'2" 36d35'25"
Cheryl
Cnc 8h41'29" 30d56'33"
Cheryl
And 1h15'16" 42d7'15"
Cheryl 140
Aqr 22h24'38" -2d33'17"
Cheryl 29
Sgr 17h44'21" -20d23'43"
Cheryl A. Bellingeri
Tau 4h22'55" 20d15'31"
Cheryl A. Carlton
Aqr 20h51'0" 1d54'12"
Cheryl A. Leisenring
Leo 10h58'0" 5d59'33"
Cheryl A. Pennock
Ari 3h15'13" 22d6'37"
Cheryl A. Sisti
Lyn 7h48'20" 42d46'30"
Cheryl A. Vincent Is Ed's Great Joy
Lyr 18h27'23" 33d40'34"
Cheryl A. Wade
Her 17h29'26" 27d21'22"
Cheryl Abramowitz
And 0h15'52" 26d10'24"
Cheryl and Frank
Uma 11h16'54" 50d28'26"
Cheryl and Haley
Umi 14h20'59" 76d18'1"
Cheryl and John
Pup 7h45'29" -22d26'29"
Cheryl and Karl's Everlasting Love
Cyg 19h45'34" 53d25'56"
Cheryl and Kevin
Uma 11h44'3" 32d18'48"
Cheryl Ann
Mon 6h45'53" 6d54'6"
Cheryl Ann
Ori 6h1'32" 9d32'53"
Cheryl Ann
Cap 20h7'27" -10d42'48"
Cheryl Ann
Sgr 18h51'7" -33d29'13"
Cheryl Ann
Sgr 19h16'38" -22d51'9"
Cheryl Ann Balint-Tarrant
Uma 11h15'24" 28d24'41"
Cheryl Ann Bendle
Cas 23h58'26" 54d0'44"
Cheryl Ann Bozzo
Sgr 20h4'17" -30d11'46"
Cheryl Ann Crane
Cnc 8h30'21" 21d28'18"
Cheryl Ann Davis
Ari 2h33'16" 17d45'48"
Cheryl Ann Garr
Ari 3h5'2" 14d41'22"
Cheryl Ann Howard
Gem 7h12'32" 15d58'54"
Cheryl Ann Keuhlen
Ori 5h59'3" 20d51'29"
Cheryl Ann Myers
Psc 2h1'33" 6d38'37"
Cheryl Ann Neibarger
Cyg 20h34'42" 58d44'37"
Cheryl Ann Neikam
Aqr 23h28'18" -18d34'17"
Cheryl Ann Ogletree
Lib 15h6'51" -5d41'47"

Cheryl Ann Rompf Lewis
Cnc 8h54'29" 11d30'42"
Cheryl Ann Schenkel
Cas 1h24'9" 62d1'27"
Cheryl Ann Tierney
And 1h43'18" 39d56'20"
Cheryl Ann Whitney
Tau 3h53'30" 25d12'18"
Cheryl Anne Findlay
Cap 20h24'41" -23d32'18"
Cheryl Anne Rich
Lyr 19h17'35" 29d34'2"
Cheryl Anne Truax
Cnc 8h40'18" 8d23'21"
Cheryl Anne White
Cnc 7h58'8" 18d15'52"
Cheryl Ann's Heartbeat
Lib 15h46'37" -19d44'21"
Cheryl Baker
And 1h8'34" 45d6'53"
Cheryl Barrel
Her 15h51'57" 44d33'24"
Cheryl Beckwith
And 0h33'59" 36d57'13"
Cheryl Bergquist
Cas 1h24'56" 51d5'12"
Cheryl - Best Star By Far
Cru 12h35'46" -62d30'11"
Cheryl Bonfiglio
Uma 12h29'42" 59d24'3"
Cheryl Boswell
Cas 0h32'49" 55d48'18"
Cheryl Brown
Aql 19h10'34" -9d35'26"
Cheryl Bruder
And 1h16'5" 50d37'9"
Cheryl Byquist
Aqr 22h46'52" -8d11'28"
Cheryl C. Hardy
Lmi 10h50'8" 26d29'21"
Cheryl Catherine
Cas 23h22'13" 57d58'23"
Cheryl Catherine LeVior
Cas 1h28'8" 58d52'26"
Cheryl Chapin
Sgr 19h14'30" -23d37'32"
Cheryl Chaplin
Lib 14h57'22" -16d31'28"
Cheryl Clinton
Cas 23h31'32" 57d30'24"
Cheryl Condes Andrade (mahal ko)
Sgr 19h39'16" -12d41'13"
Cheryl Davis, RN
Aqr 22h27'18" -1d26'25"
Cheryl Denise Harwell
Leo 11h6'20" 21d40'17"
Cheryl Denise Shepherd
Gem 7h26'26" 27d20'0"
Cheryl Dinihanian
And 0h38'13" 39d49'48"
Cheryl Doepping-Ruggio
Gem 7h25'48" 17d27'15"
Cheryl E. MacArthur
Peg 22h36'36" 9d26'20"
Cheryl & Edward A. Sowul Jr.
Cyg 20h22'43" 55d46'28"
Cheryl Elaine "Binky" Rockwell
Cam 4h26'23" 64d40'21"
Cheryl Escoe
Ari 3h13'14" 16d23'40"
Cheryl Faithe Jordan
Sgr 18h33'22" -29d46'19"
Cheryl Feder
Sgr 18h2'12" -28d36'21"
Cheryl Fitzsimmons
Cnc 8h39'36" 31d50'5"
Cheryl Folkner
Crb 16h2'41" 34d29'33"
Cheryl Fritsch
And 0h55'54" 37d27'15"
Cheryl Furlong (Chicago)
Per 3h21'12" 45d10'21"
Cheryl Galford Harris
And 0h11'23" 42d25'41"
Cheryl Gettman
Tau 5h33'38" 17d45'59"
Cheryl & Gordon Ramsbottom 02/29/1992
Eri 2h58'13" -14d23'2"
Cheryl Halibrand
Lyn 7h43'7" 36d38'50"
Cheryl Hans
Gem 7h18'47" 14d16'46"
Cheryl Harper
Tau 4h59'9" 22d49'26"
Cheryl Harris
Lyn 8h16'31" 49d48'22"
Cheryl Herman
And 0h19'3" 40d57'40"
Cheryl Hersey-Anastasia
Sgr 19h23'10" -14d51'23"
Cheryl Houser
Aqr 23h11'3" -5d48'5"
Cheryl Houston Bradburn
Crb 16h18'55" 35d9'15"
Cheryl H.Z. "the Brightest Star"
And 1h47'52" 46d12'42"
Cheryl I Love You More
Lyn 8h22'40" 50d45'18"

Cheryl Ingersoll & Howard Halligan
Cyg 20h26'0" 49d49'48"
Cheryl Irene Jenovino - MA
Crb 15h31'7" 31d45'37"
Cheryl Jean Dunn Gordon
Cas 1h50'5" 60d17'14"
Cheryl Jean Willis-Hazen
Lac 22h19'20" 53d13'3"
Cheryl Jo
Uma 12h30'40" 58d44'37"
Cheryl & Joel's Wedding Star
Cyg 20h19'19" 46d45'28"
Cheryl Josserme
Lib 15h19'7" -18d30'49"
Cheryl June Forth
Gem 6h44'19" 26d20'36"
Cheryl & Justin Gibbs
Lib 14h34'55" -9d59'47"
Cheryl Kay 143
Sco 17h34'35" -40d11'25"
Cheryl Kay Barton
Cyg 20h15'18" 34d54'11"
Cheryl Kay Hudson
Cyg 19h36'23" 42d9'42"
Cheryl Kay Sabino Shewmake, Allstar
Leo 10h22'53" 18d24'31"
Cheryl Kennett
Cap 21h10'33" -18d50'17"
Cheryl Kishbaugh
Gem 6h43'28" 20d7'31"
Cheryl Kramer
Mon 6h39'7" 4d47'45"
Cheryl Kramer
Cam 4h43'23" 55d48'28"
Cheryl L. Ainsworth
Cam 4h13'19" 53d45'51"
Cheryl L. Hewson
Cyg 19h45'3" 33d31'15"
Cheryl L. Shultz
Sgr 18h3'56" -27d42'41"
Cheryl L. Zegle-Garcia
Her 17h12'0" 18d50'23"
Cheryl Laird-Graham
Cas 23h38'36" 51d46'26"
Cheryl Lau
And 2h19'53" 47d23'57"
Cheryl Lavertue Love Always, Steve
Aqr 23h19'5" -18d50'20"
Cheryl Lee Carson
Lyn 8h0'49" 40d47'29"
Cheryl Lee Javer
And 0h58'43" 43d13'36"
Cheryl Leigh Morales
Cma 7h14'21" -30d13'48"
Cheryl Lewis
Lep 5h37'46" -14d27'44"
Cheryl Lindbloom
And 1h30'38" 45d26'9"
Cheryl Louise
Del 20h38'37" 18d9'18"
Cheryl Louise Bogdanowitsch
And 2h0'10" 47d13'39"
Cheryl Louise Stenton
And 1h0'45" 44d58'56"
Cheryl Lyn Brown
Tau 4h25'1" 1d19'16"
Cheryl Lynette Braxton
Vir 14h39'12" 5d0'18"
Cheryl Lynn Bowers
Psc 0h0'34" 10d37'49"
Cheryl Lynn Brandt
Cap 20h12'16" -10d49'34"
Cheryl Lynn Davis
Vir 12h35'57" 1d29'38"
Cheryl Lynn Duckworth
Cnc 9h8'22" 25d15'36"
Cheryl Lynn Nelson
And 2h0'9" 44d28'2"
Cheryl Lynn Patmon
Leo 10h15'47" 13d52'14"
Cheryl Lynn Pickell Birthday Star
Lib 15h22'44" -10d26'41"
Cheryl Lynn Pitchford
Leo 9h28'11" 26d28'41"
Cheryl Lynn Svoronos
Ari 2h36'2" 27d27'53"
Cheryl Lynne Duff
Ari 3h16'26" 26d52'17"
Cheryl M. Sullivan
Cas 0h57'13" 56d1'13"
Cheryl M. Yokoyama
Cyg 21h47'35" 39d42'42"
Cheryl Mabey Doherty
Sgr 18h54'31" -36d30'33"
Cheryl Maciver
And 23h46'8" 36d1'21"
Cheryl MacKenzie
Mon 7h55'17" -0d27'6"
Cheryl Marcia Hennessy
Cmi 7h29'31" -0d4'26"
Cheryl Marie Lemons
Ori 5h11'15" 6d21'5"
Cheryl Marie Sticha
Cyg 19h42'14" 33d27'28"
Cheryl Marie Vernel
And 1h30'27" 40d35'7"
Cheryl Mason
Cas 0h56'30" 52d47'52"

Cheryl Mathair Chrionna
Cyg 20h49'19" 39d23'14"
Cheryl McCarrick Seamster
And 23h11'14" 45d7'28"
Cheryl McCarron
Cap 21h34'35" -22d56'55"
Cheryl & Mike Schubmehl
Cyg 21h26'36" 45d0'10"
Cheryl Mimi Gort
Vir 13h8'15" 11d38'23"
Cheryl Moore
Cru 12h6'37" -63d50'22"
Cheryl N Eric
Cyg 19h33'7" 28d54'24"
Cheryl "Paintbrush" Matthews Diggs
Sgr 19h11'20" -36d25'42"
Cheryl Peacock ElHamahmy
Uma 13h47'54" 52d6'28"
Cheryl Pekarek
Sgr 20h3'47" -37d41'46"
Cheryl Pellecchia
Tau 5h16'49" 16d21'3"
Cheryl Perez
Aql 19h44'12" -0d2'36"
Cheryl Piche
Lib 14h51'7" -1d54'24"
Cheryl R. McGuire
And 0h46'17" 46d35'32"
Cheryl Radloff
And 23h17'23" 48d30'4"
Cheryl "Riverhoney" Robinson
Cap 21h40'18" -10d31'4"
Cheryl S. Allyn
Cas 1h14'26" 64d57'43"
Cheryl Sanchez
Tau 5h47'31" 18d38'23"
Cheryl Sanders
Aqr 23h8'21" -9d48'9"
Cheryl Scribner
Cas 23h33'5" 59d45'10"
Cheryl Shatford
Per 4h33'44" 43d57'42"
Cheryl. Shearer - Savy
Cnc 8h13'25" 12d20'23"
Cheryl Staubs
Del 20h48'10" 5d45'10"
Cheryl Stump's Bright Light
Her 16h10'55" 22d53'47"
Cheryl Sue Milford - April 29, 1946
Tau 4h19'38" 22d51'14"
Cheryl Thompson
Lmi 10h41'20" 31d28'32"
Cheryl "Tootie"
Cas 0h26'19" 48d32'24"
cheryl trowbridge
Cas 2h12'50" 65d20'30"
Cheryl V. Williams
Del 20h41'49" 10d6'14"
Cheryl Vanderwende
Vir 13h17'9" 6d29'19"
Cheryl & Wade
Ari 2h48'5" 27d53'16"
Cheryl Ward
And 0h19'17" 32d37'15"
Cheryl Watts
Cap 21h14'59" -17d12'23"
Cheryl Whalen
Uma 11h15'10" 68d12'11"
Cheryl Wolff
Cap 20h30'47" -10d55'28"
Cheryl Wright
Uma 11h19'12" 70d52'44"
Cheryl Yvette Talbot { Lovebug }
Lyn 8h19'47" 56d50'46"
Cheryl, Emily & Ashley
Ori 6h15'58" 10d21'55"
Cheryl-1
Ori 5h51'2" 18d44'34"
Cherylanne Marshall
Cas 0h3'30" 57d28'42"
Cheryl-Bebe-Mommy
And 1h5'36" 41d10'11"
Cheryle A. Beuoy Richardson
Ari 2h28'56" 20d31'13"
Cheryle and Michael Ross
Lyn 7h45'20" 53d3'7"
Cheryle Anne Hungle
Leo 10h58'58" 17d25'35"
Cheryle Isreal
And 23h35'25" 44d17'35"
Cheryle Jean Ferinac
And 1h51'22" 38d41'18"
Cheryle Rae
Ari 2h0'44" 13d39'4"
Cheryle, My Shining Star Forever
Sco 16h42'38" -28d55'38"
Cheryll Lynn Stracick
Cas 2h24'5" 66d2'41"
Cherylle Rae Reed
Aql 19h49'48" -0d30'36"
Cheryll's Diamond in the Sky
Uma 9h32'6" 49d11'30"
Cheryl's Angel
Cyg 19h57'39" 38d29'15"
Cheryl's Eternal Hope
Aur 5h53'40" 53d48'37"

Cheryl's fire
Crb 16h8'15" 29d57'45"
Cheryl's Lucky Star
Tau 4h44'23" 20d17'17"
Cheryl's Lucy
Psc 1h48'55" 6d53'17"
Cheryl's Place in Heaven
Cam 6h4'51" 60d19'40"
Cheryl's Quest
And 0h44'15" 31d33'39"
Cheryls star
Psc 1h49'6" 8d18'22"
Cheryls Star
Sco 15h53'36" -20d54'3"
Cheryl's Traveler
Equ 21h13'26" 4d49'18"
Cheryl's Way
Uma 11h12'24" 51d42'46"
Cheryl's Yellow Rose
Uma 12h22'12" 62d12'5"
CherylSaraHelenElijah
Dra 15h43'23" 58d58'44"
Cherylstar - Angel In The Night Sky
Aql 18h44'29" -0d13'22"
CherylZane
Ari 3h11'9" 18d35'43"
Cheryn
Ari 2h49'53" 14d56'12"
Cheryol Rae Kramer
Uma 12h39'3" 61d58'52"
Chery's Star
Cap 20h32'57" -9d1'56"
Cheshire Bate
Psc 1h7'45" 28d53'19"
Chesley Bonesdale The Boy Cat
Lyn 8h0'41" 41d36'3"
Chesna Christene Lee
Sgr 18h3'33" -29d5'20"
Chesnee
And 0h25'57" 42d57'17"
Chesney Ryan 4.2.00
Ari 1h54'20" 22d55'40"
Chesney William Noel Kennedy
Ori 6h23'58" 10d32'3"
Chesnie Mary McCann
And 1h51'56" 40d50'13"
Chessa Shea Quenzer
Ori 5h26'15" 13d48'47"
Chessie
Uma 10h36'36" 66d35'58"
Chessie Ann Garcia
Gem 7h13'7" 26d32'23"
Chessmaster Rogers
Sco 17h39'44" -39d9'32"
Chester Adam Goehring
Aur 4h58'5" 35d44'0"
Chester F. Golimowski
Cep 22h0'35" 57d1'57"
Chester Gordon Leonard
Uma 11h50'57" 37d36'23"
Chester Gronostalski
Uma 11h40'34" 56d7'39"
Chester H. Glant
Uma 9h31'9" 45d43'42"
Chester Lindsay Jones III
Uma 11h45'54" 64d3'9"
Chester Montgomery
Cap 20h31'43" -12d17'57"
Chester "Papa" Glass
Uma 9h56'23" 47d36'13"
Chester Prince
Tau 4h10'23" 26d54'24"
Chester Ray Flanagan
Boo 14h22'54" 34d22'29"
Chester Sellers
Uma 12h3'46" 49d51'44"
Chester T. Vogel
Cap 21h42'22" -12d1'55"
Chester V. Ault
Dra 16h59'3" 66d33'53"
Chesterbomax Levandoski
Uma 11h52'55" 28d37'5"
Chester's Beacon
Aur 5h37'22" 41d40'13"
Chestnut Hill-Fayette Co. TN
Uma 9h21'50" 48d58'21"
Cheston & Abigail Hipskind
Uma 13h2'51" 62d45'16"
Chet
Ori 5h1'32" 9d42'6"
Chet Badowski
Her 18h6'56" 32d2'34"
Chet Cole Barrett
Boo 15h39'31" 43d33'44"
Chet Dee Harris
Uma 12h35'12" 60d37'37"
Chet Fassio
Uma 10h5'7" 59d50'30"
Chet Ray Hamel
Her 17h46'3" 32d51'33"
Chetna
Crb 15h33'41" 30d45'27"
Chett Oneill Lyncker
Per 2h54'46" 42d45'41"
Chetter James
Uma 10h55'16" 48d37'7"
Cheung Man Chiu the Chi Err
Cam 5h30'1" 65d49'44"
Cheung Nok Yan, Phoebe
And 0h4'16" 32d24'48"

Cheung Suk Yee Mandy
Leo 11h2'42" 8d21'18"
Cheve
Ori 5h29'49" 1d25'44"
Chevela
Cap 21h46'2" -14d2'48"
Chevelle
Umi 15h39'3" 69d50'39"
Chevelle
Umi 16h6'6" 74d7'15"
Chevey
Lyn 7h21'38" 53d2'10"
Chevron
Ori 5h20'20" -1d51'52"
Chevy
Cma 6h51'23" -14d27'4"
Chevy
Peg 22h41'34" 14d41'37"
Chevy Boy
Ari 3h13'12" 29d26'40"
Chew Kian Seng
Ari 2h36'18" 12d30'54"
Chew Yen Li
Cnc 8h13'34" 27d26'54"
Chewbacaraca
Cnc 8h44'0" 17d36'19"
Chewbacca
Cnc 8h31'55" 27d50'37"
Chewey
Tau 4h37'32" 7d33'13"
Chewy Pickles
Peg 23h13'45" 22d24'5"
Cheyanna Sierra Angeline Byrd Murat
Peg 22h19'12" 6d50'26"
Cheyanne Alexandra
Psc 0h46'41" 11d59'53"
Cheyanne Estrella
Dra 19h45'35" 74d53'5"
Cheyanne Jewel Evans
Leo 11h17'54" -0d37'22"
Cheyanne Katherine Woods
And 0h38'18" 31d16'1"
Cheyanne Noel
Cap 21h7'46" -23d52'15"
Cheyanne Noel Taylor
Cap 20h32'19" -17d3'31"
Cheyenne
Uma 8h32'18" 62d31'37"
Cheyenne
Uma 12h18'21" 54d43'5"
Cheyenne
Uma 12h39'1" 52d42'14"
Cheyenne
Sgr 19h1'59" -35d43'42"
Cheyenne
Cnv 12h32'24" 44d29'57"
Cheyenne
Cyg 20h56'52" 46d22'56"
Cheyenne
Peg 22h19'42" 6d49'7"
Cheyenne
Tau 5h46'22" 26d14'54"
Cheyenne Alissa Vallette
Ari 3h12'11" 29d23'4"
Cheyenne and Logan Dorans star
Cyg 20h35'33" 47d46'13"
Cheyenne Anderson
And 2h37'36" 43d29'3"
Cheyenne Cier Zwicker
Sgr 19h40'46" -14d10'34"
Cheyenne Cimmaron
Cyg 21h59'54" 54d16'54"
Cheyenne Cimmaron
Cyg 19h39'59" 29d27'13"
Cheyenne & David Dykstra
Cyg 19h47'35" 54d12'47"
Cheyenne Dunlop
Lyn 7h48'32" 57d36'42"
Cheyenne Elizabeth
Umi 14h31'55" 75d13'31"
Cheyenne Elizabeth Braden
And 0h50'31" 37d35'39"
Cheyenne Faith
Dra 19h7'48" 64d31'27"
Cheyenne Heevonehe
Cas 0h19'1" 50d31'11"
Cheyenne Jackson
Her 17h59'28" 23d25'29"
Cheyenne Jeran-James Henson
Leo 11h5'8" 6d37'10"
Cheyenne Kit Barron
Uma 9h43'51" 52d39'42"
Cheyenne LaShann
Tau 5h54'55" 24d47'34"
Cheyenne Logue
Tau 5h53'22" 25d8'1"
Cheyenne Luckey
Ori 5h57'4" 17d24'15"
Cheyenne Makayla Marie Berkmyre
And 2h30'35" 49d51'36"
Cheyenne Marie Murray
Uma 10h26'56" 54d17'56"
"Cheyenne" — (my lil boo-boo)
Sgr 17h52'9" -28d22'4"
Cheyenne Perez
Gem 7h33'49" 17d25'9"
Cheyenne Rebelo
Tau 5h51'13" 25d58'8"

Cheyenne Rene Key
Uma 8h38'48" 46d56'20"
Cheyenne S. Walker
Uma 11h8'19" 57d23'58"
Cheyenne Silver Pratt
Uma 10h16'48" 65d16'41"
Cheyenne U Shine
Her 18h47'41" 18d45'10"
Cheyenne Weh
Gem 6h46'11" 18d7'36"
Cheyenne's Dreams
Tau 6h0'36" 24d33'55"
Cheyenne's Legacy
And 0h47'13" 35d37'40"
Cheyenne's love for Daddy
Ori 5h53'55" 18d17'46"
Cheyenne's Smile
Leo 11h36'54" 26d36'1"
Cheyne Michael Jetton
Ari 2h57'34" 26d12'24"
Cheythan Rao Winter
Umi 14h42'33" 81d14'59"
Chez Ronnie and Laurence
Ori 5h36'40" 4d34'13"
Chez Tinou et La Schtroumpfette
Sgr 18h36'35" -23d40'47"
Chi Chi
Ori 5h36'28" 5d56'57"
Chi Chi
Cyg 19h35'28" 31d53'58"
Chi Chi la Bean
Sgr 18h48'36" -30d49'49"
Chi Chi Mon
Tau 3h35'51" 20d13'17"
Chi Mindy Phung
Gem 7h29'59" 26d46'53"
Chi trova un amico trova un tesoro
Per 4h21'2" 34d58'5"
Chia Banus
Mon 6h32'33" 0d47'40"
Chia Butchie Frimid I
Cma 7h5'38" -26d2'22"
Chiacam
Crb 15h36'52" 26d55'46"
Chia-hui Baird
Uma 9h5'17" 61d38'22"
Chiaki
Uma 11h33'44" 53d45'8"
Chia-Ming Wu
Crb 15h43'20" 34d42'40"
Chiana Claire Tubbs
Sgr 19h34'40" -14d46'12"
Chiara
Vir 12h41'39" -7d25'41"
Chiara
Uma 13h28'29" 54d50'37"
Chiara
Uma 13h20'0" 57d38'50"
Chiara
Umi 16h53'48" 83d41'17"
Chiara
Crb 16h23'29" 32d37'40"
Chiara
And 1h15'58" 42d2'12"
Chiara
And 0h2'26" 37d33'0"
Chiara
Cyg 20h14'36" 36d8'39"
Chiara
Per 3h24'2" 41d33'24"
Chiara
Per 3h26'55" 41d32'49"
Chiara
Per 3h26'16" 40d0'21"
Chiara
Per 3h32'10" 49d8'30"
Chiara
And 23h36'20" 37d44'7"
Chiara
Boo 14h58'59" 52d28'51"
Chiara
Per 2h22'23" 54d18'32"
Chiara
Vul 19h31'27" 26d2'41"
Chiara
Peg 22h2'41" 17d16'46"
Chiara
Tau 3h48'55" 5d26'17"
Chiara Aburrà
Cyg 19h54'20" 38d6'31"
Chiara Adel Hyman
Aql 19h35'38" 0d45'12"
Chiara Arbore Mauro
Cnc 8h4'29" 27d27'20"
Chiara Baraldi
Lyr 18h19'9" 41d18'35"
Chiara Bastoni
Cam 6h35'19" 68d33'45"
Chiara Bima
Cyg 21h7'48" 30d19'35"
Chiara Brandini
Umi 17h55'29" 87d9'46"
Chiara Castello
Cam 7h19'16" 70d18'58"
Chiara Cohen
Cyg 21h34'26" 39d35'0"
Chiara Coletta
Aur 4h59'50" 37d34'58"
Chiara Colombo
Her 17h59'11" 45d59'34"
Chiara Costazza
And 0h43'15" 44d17'58"

Chiara Cuccari
Lyr 19h6'17" 26d49'16"
Chiara De Simoni
Her 18h33'45" 18d56'4"
Chiara Durando Opalka
Cyg 19h38'59" 29d35'14"
Chiara e Fabio
Boo 14h31'5" 54d31'21"
Chiara Eleonora
Cep 22h1'59" 63d28'7"
Chiara Ermito
Cap 20h17'24" -14d15'20"
Chiara Giacomelli e Michele Tacca
Her 17h17'11" 48d12'24"
Chiara & Guli
Her 18h38'46" 13d58'47"
Chiara Laura Piticchio
Ori 5h55'47" 7d14'4"
Chiara Leone
Uma 8h13'14" 61d46'32"
Chiara Lucia Grace Carbone
Tau 4h8'47" 22d44'58"
Chiara&Massy
Lyr 18h45'24" 29d13'14"
Chiara Nardi
And 1h44'20" 49d49'15"
Chiara Orsini
Cyg 21h21'10" 30d54'49"
Chiara Parenti
And 2h22'7" 38d51'32"
Chiara Perricone
Lmi 10h46'46" 24d29'3"
Chiara Possemato
Cas 0h40'17" 54d25'35"
Chiara Ralli
Aur 6h38'36" 41d25'32"
Chiara Rizzo
Cyg 19h32'29" 29d17'31"
Chiara Rocco
Lyn 8h21'11" 33d36'45"
Chiara Serra
Lib 14h36'33" -24d29'0"
Chiara Urbanetto
And 2h13'27" 37d24'42"
Chiara Vanessa Fanchini
Cas 1h17'41" 53d45'55"
Chiara Venard - - Millet
And 2h34'8" 45d29'58"
Chiara Wild
And 23h36'57" 50d3'36"
Chiara Zambetta
Uma 9h29'45" 51d11'51"
Chiara Zanoli
Cam 4h27'35" 54d9'26"
Chiaramore
Uma 11h5'12" 60d23'49"
Chibi-chan Sophia
Psc 1h10'30" 27d52'58"
Chica
Ara 17h23'51" -48d15'53"
Chica D'angel
Aql 20h4'4" -11d28'28"
Chicago
Uma 8h31'7" 62d13'33"
Chicago
Uma 13h21'17" 56d31'44"
Chicago Blonde
Cyg 21h39'20" 33d39'45"
Chicago Julius
Lib 15h37'35" -8d7'57"
Chicago Sky - Harvey Alter
Umi 14h29'4" 73d46'15"
Chicago Sky - Michael Alter
Umi 15h33'32" 72d28'31"
Chicca
Cyg 19h31'32" 31d43'28"
Chicca-Chris Amo-Re
Uma 10h45'49" 56d47'32"
Chicco e Minny
Uma 10h38'30" 72d40'32"
Chicco77
Cam 9h25'36" 82d4'19"
Chi-Chi
Cmi 8h10'44" 3d31'2"
chick
Crb 16h18'26" 34d3'17"
Chick
Cyg 21h16'49" 51d16'46"
Chick and Evelyn Belotti
Cyg 20h17'32" 38d54'21"
Chick and Ida Genova Forever
Cyg 21h21'27" 51d2'57"
Chick King
Cyg 19h46'16" 34d0'47"
Chick & Lynn Hancock
Sge 19h11'14" 18d59'32"
Chick Pick
Lyn 7h42'52" 47d30'41"
Chicken Boy & Chicken Girl
Gem 6h22'29" 18d32'52"
Chicken Butt
Vir 12h8'11" 10d17'49"
Chicken & D.J.
And 1h54'8" 41d56'16"
Chicken isn't Shoite
Cas 0h1'25" 57d20'27"
Chicken Man
Leo 9h47'36" 26d13'11"
Chicken Pimp 20
Tau 4h36'56" 19d4'4"
Chickie
Psc 1h9'13" 3d57'4"

Chickie
Uma 9h40'6" 54d18'18"
Chickie
Uma 13h40'58" 58d17'22"
Chickie - Choo
Uma 9h47'25" 51d18'25"
Chickie Marie Cline
Per 2h25'22" 55d35'50"
Chickie Yager
Tau 4h49'29" 18d3'45"
Chickius Momius
Crb 15h47'51" 26d34'50"
chicklepea
Sgr 19h3'11" -33d16'46"
Chicklet 1967
Ori 5h45'4" 2d43'59"
Chicklet Charm
And 2h25'20" 46d39'4"
Chicklette Cherub
And 0h11'29" 39d27'42"
Chickosokane
Ori 5h40'50" 2d54'56"
ChickPeaBlossom
Leo 10h16'25" 25d9'28"
Chicky
Gem 6h47'49" 32d17'19"
Chickypoo Erin Schumacher
Psc 0h31'59" 16d39'2"
CHICO Ch. Belmar's Heir Apparent
Ori 5h47'4" 6d38'1"
Chico & Jason
Sgr 18h59'23" -34d15'13"
Chie
Cap 21h29'27" -23d17'30"
Chie Ishikawa
Ori 5h20'6" -6d31'31"
Chief
Ori 5h58'28" -0d43'31"
Chief
Crb 16h21'15" 38d19'19"
Chief 1
Cep 22h49'52" 74d1'16"
Chief and Ace
Aqr 22h14'48" 1d55'4"
Chief and Denise's Anniversary Star
Cyg 20h30'8" 40d26'47"
Chief and Weezy
Cyg 20h10'59" 36d41'21"
Chief Davey-U R 4ever in my heart!
Sco 16h55'47" -38d40'51"
Chief David A. Santor S.V.P.D.
Per 3h14'19" 56d44'40"
Chief Denise "Baby" McGann Bledsoe
Uma 11h40'3" 50d50'19"
Chief Donald Fred Anthony
Per 3h26'44" 49d39'10"
Chief Dream Maker
Sco 16h5'11" -16d19'54"
Chief Golden Eagle - Marge Taylor
Cnc 8h52'59" 20d12'6"
Chief James F. Bale
Tau 4h24'3" 17d20'34"
Chief Kevin Sowers
Lmi 9h55'20" 38d44'43"
Chief Maier
Ori 5h27'50" 6d20'40"
Chief Mrs. Celina Maiki
Crb 15h32'48" 31d14'2"
Chief Roger J. Aikin
Dra 19h9'4" 64d5'7"
Chief William V. Walbourne
Uma 9h18'42" 65d30'3"
Chiefy
Boo 14h6'36" 38d8'17"
chieko
Cnc 8h50'24" 16d13'28"
Chieko I. Melcher
Aql 19h49'56" -0d46'52"
Chiemi
Psc 0h3'46" -1d19'47"
Chiesa
Uma 10h25'41" 69d0'30"
Chiggar
Psc 1h26'15" 27d50'17"
Chiggy
Umi 17h29'27" 81d7'51"
Chiharu
Lmi 10h0'59" 27d56'15"
Chihlin
Gem 6h36'38" 23d58'53"
Chika Yoshizaki
Dra 17h52'5" 64d37'16"
CHIKE
Uma 9h23'12" 61d2'26"
Chikis
Dra 15h57'6" 57d6'21"
Chiko
Uma 9h24'29" 71d45'45"
Child Devel. Practice of Dr. Huffer
Cyg 21h54'7" 53d7'35"
Childhooddreams.m.servin. 62407
Cnc 9h15'57" 27d30'55"
Childlicious
Cam 4h8'32" 62d59'35"
Children of Belgrade
Uma 9h53'47" 47d24'58"

Chili Peppers
Uma 11h29'55" 44d15'5"
Chilli Pepper and Suena's Star
Uma 11h40'24" 61d39'17"
Chillie
Tau 4h28'59" 25d6'58"
Chillson Family
Uma 13h41'19" 50d25'48"
"Chilly" Our Mom Wilmetta Thompson
Cas 1h26'21" 68d1'59"
ChiLLyBerryBab' - HUKILAvvvvvvm
Lac 22h56'31" 50d2'36"
Chim
Psc 23h46'46" 7d18'31"
Chim Star
Uma 12h3'49" 38d22'50"
Chimène
Peg 22h21'5" 21d2'13"
Chimp Paw
Uma 10h53'15" 68d23'40"
Chimu
Ori 6h13'48" 7d42'20"
Chin On Ni
Peg 21h50'6" 25d44'17"
China Cat Sunflower
Crb 16h7'51" 38d49'32"
China Gordis Aguirre
Sco 17h20'1" -42d52'53"
China Lee Bowyer
Uma 11h46'42" 63d50'12"
China Lynn
Gem 6h16'32" 23d45'7"
China Man
Sco 17h38'16" -36d34'17"
China Star
Gem 7h14'42" 20d24'34"
China Tourism Management Institute
Ori 6h20'7" 20d12'2"
Chin-Chila
Per 3h6'36" 54d36'44"
Chinchilla
Leo 11h19'13" 15d9'39"
chindan
Lyn 7h18'59" 45d26'51"
Chindian and Tempest
Pho 2h24'28" -46d2'42"
Ching Yee Tse
Lib 15h7'3" -11d53'28"
Chinni and Jiju
Cap 20h7'10" -10d30'2"
Chinook
Gem 7h24'23" 14d24'24"
Chinook "Houdini" Dreyer
Uma 9h20'12" 46d1'58"
Chintamani
And 23h31'5" 43d1'24"
Chinue Vang
Ari 1h48'23" 18d4'45"
Chinupi
Gem 7h39'2" 32d20'50"
...chio rivera...
Lyn 6h49'33" 55d42'32"
Chiodi Valeria
Ori 6h24'31" 13d51'27"
Chioma Ukaegbu
Aql 19h34'33" -10d38'12"
Chiori
Crb 15h47'0" 31d18'38"
Chip
Per 4h6'41" 43d51'32"
Chip
Vir 14h37'35" 5d46'59"
CHIP
Pup 8h12'29" -39d34'25"
Chip & Ainsley
Peg 22h39'16" 24d46'53"
Chip and Cher
Ari 3h14'5" 19d37'37"
Chip Brown
Gem 6h32'1" 14d59'18"
Chip Chuprinko
Uma 10h15'32" 66d12'21"
Chip Davis
Uma 9h51'6" 66d5'59"
Chip Hackler
Uma 10h34'8" 55d4'24"
Chip Johnston
Vir 13h19'25" -20d45'5"
Chip LeMar & Joyce Sun
Cyg 21h32'31" 47d48'44"
Chip Loves Lisa
Psc 1h12'4" 26d10'28"
Chip O' Bude
Psc 1h2'46" 4d50'15"
Chip Richbourg
Uma 11h2'13" 50d14'17"
Chip Stenberg: Mogul One
Cap 21h22'3" -24d32'44"
Chip Stinnett
Ori 6h12'56" 17d38'22"
Chip The Wine Star
Lac 22h24'20" 53d5'21"
Chip Wagner
Ori 5h27'50" 13d58'1"
Chip628
Uma 11h56'33" 36d3'48"
chipmunk
Leo 10h59'19" 8d55'48"
Chipmunk
Cnc 8h38'34" 15d0'20"

chipper
Sco 16h11'40" -13d35'2"
Chipper
Cma 7h5'29" -23d43'56"
Chippy & Hayley Bear
Tau 4h26'22" 21d1'16"
Chipy Chip Chipmunk (Jessica)
Uma 11h34'45" 35d1'34"
Chipyis Poodlearis
And 0h47'19" 39d46'20"
chiquita
Cru 12h2'51" -61d57'38"
Chiquita Linda
Crb 15h46'21" 32d36'39"
Chiquita Mama Lorena
Aqr 22h47'58" -14d32'6"
Chiquitica
Peg 23h10'28" 17d40'19"
Chiquitita Bella
Sgr 17h51'47" -28d51'37"
Chiquitta Y. Fultz "Quitta's Star"
Cam 4h22'49" 66d9'6"
Chira Star
Oph 17h14'40" -0d10'43"
Chirag
Vir 12h22'19" 11d59'32"
Chirag Patla
Uma 9h36'37" 48d28'28"
Chiravuri
Cep 21h33'8" 62d45'45"
Chireen and Scott's Eternal Flame
Cyg 19h40'44" 38d10'11"
Chireland
Leo 10h34'39" 8d34'32"
Chiroclark
Uma 10h46'16" 47d8'59"
Chirp
Dra 18h47'49" 51d38'55"
Chirp Chirp, Penguin, Lemon
Uma 9h27'54" 65d51'55"
Chirstopher Van Gilder
Uma 9h49'14" 70d52'19"
Chisa Irene Pinion
Gem 7h5'18" 16d5'43"
Chisato Seipp
Umi 14h18'50" 65d48'51"
Chis-N-Midy
Peg 21h27'22" 11d46'0"
Chispa
Psc 1h9'50" 12d21'16"
Chispa
And 1h4'52" 37d55'59"
Chispita
Umi 15h5'27" 77d13'10"
Chissy, my heaven
Vir 12h20'28" -5d47'53"
Chistina
Sgr 18h8'43" -17d12'37"
Chistius ac Gregius Addo ab Sidius
Uma 14h3'43" 59d5'2"
Chistopher Bence Hall
Col 5h49'4" -31d45'55"
Chistopher Orr
Aql 19h40'13" 7d38'37"
Chisum
Per 4h49'17" 50d25'2"
Chita
Psc 1h47'30" 5d46'20"
Chitalino
Leo 11h50'49" 23d21'27"
Chitty's 40
Uma 11h19'56" 56d38'2"
CHIU Wai Ching & MA Yee Man
Dra 9h38'36" 79d18'42"
Chiu-Chuw Wu
Aql 19h46'3" 13d14'27"
Chivas
Cma 7h2'21" -23d45'29"
Chivi Linda
Ori 6h10'14" 19d33'30"
Chivi y Pili
Uma 14h1'19" 49d52'5"
Chi-yo no Tomo
Uma 11h55'42" 44d50'40"
Chizzo
Aqr 21h9'45" -14d7'3"
Chlesea Kruse Marlowe
Aqr 22h33'37" 0d36'34"
"chline Tiger"
Cep 0h26'20" 82d2'28"
Chloé
Uma 11h11'4" 60d15'4"
Chloé
Uma 10h6'54" 60d39'39"
Chloe
Uma 9h1'52" 61d17'13"
Chloe
Lyn 6h46'0" 59d16'1"
Chloë
Uma 8h37'6" 56d34'16"
Chloé
Sgr 19h28'42" -18d28'47"
Chloé
Cap 21h48'35" -18d33'50"
Chloe
Vir 11h58'33" -6d7'19"
Chloé
Ser 15h58'3" 4d45'14"

Chloe
And 1h19'6" 38d57'51"
Chloe
And 0h39'22" 44d2'39"
Chlöe
And 0h28'19" 39d49'40"
CHLOE
Gem 7h37'44" 32d11'20"
Chloe
And 2h29'43" 46d44'48"
Chloe
And 0h33'40" 45d53'32"
Chloé 13 février 2006
Cas 23h52'15" 50d0'55"
Chloe Abigail Patch
Gem 6h7'35" 25d29'35"
Chloe Adele Glay
And 2h30'25" 48d20'10"
Chloe Aimee Gray
And 2h29'23" 42d52'15"
chloe alanna
Lib 14h49'24" -1d42'35"
Chloe Alexandra Prayon
Del 20h23'8" 19d9'41"
Chloe Alice Daybell
And 1h17'5" 46d14'15"
Chloe Alisha Bower
Uma 9h24'58" 65d14'38"
Chloe Alyssa
And 1h2'41" 47d0'30"
Chloe Amelia Cawley - Our Tiny Angel
Cru 12h26'37" -61d20'37"
Chloe Anais Moreno
Cas 1h46'30" 65d42'41"
Chloe Angel
Cap 20h59'34" -17d47'45"
Chloe Ann
And 2h19'44" 45d34'25"
Chloe Ann
Vir 12h49'8" 3d19'28"
Chloe Annabelle Leah - Chloe's Star
And 23h23'40" 39d0'0"
Chloe Anne
Crb 15h31'32" 27d45'24"
Chloe Antoinette Santilli
Lib 15h21'59" -11d14'29"
Chloe Apolline Fischer
Psc 1h34'59" 19d24'53"
Chloé Bella
Dra 18h50'49" 67d16'24"
Chloe Bella Lombardo
Crb 16h10'17" 36d20'16"
Chloe & Benjamin Robison
Umi 15h27'49" 68d41'20"
Chloe Bernardini
Ori 6h6'16" 10d38'37"
Chloe Bethany Bourne - Chloe's Star
And 1h17'30" 49d48'5"
Chloe Betty Neubarth
And 1h2'17" 34d53'45"
Chloe Cannella
Leo 9h24'49" 26d51'49"
Chloé Cardinal
Ori 6h6'20" 7d2'57"
Chloe Charlotte Kuhar
Gem 7h20'35" 32d19'53"
Chloe Cheryl Garrick
Lib 15h24'41" -16d26'56"
Chloe Christine Thompson - LUV U 4EVA
Pho 0h24'27" -56d51'26"
CHLOE COCO BLAIR
And 23h52'21" 45d0'6"
Chloe Cohen
Umi 13h47'15" 75d25'59"
Chloe Copeland
And 2h22'48" 38d29'22"
Chloe Cynthia Lewis
Cas 1h13'1" 57d53'14"
Chloe Danielle Lyon
Lyn 7h53'13" 55d24'14"
Chloe Darlene Walters
Ori 5h25'15" 5d50'49"
Chloe De Buyl
And 22h57'45" 51d57'42"
Chloe Denice Phelan
Cap 21h35'19" -20d45'56"
Chloe Depamphilis
And 2h22'4" 49d26'46"
Chloé Dernoncour 05/08/2004
Cas 0h54'50" 56d43'41"
Chloe Eavan Barnes
Oph 17h2'37" -0d23'29"
Chloe Elena Mike
And 1h46'34" 43d54'37"
Chloe & Eli Notz *DADDY*
Per 2h38'35" 51d14'21"
Chloe Elijah
Aqr 21h20'24" 1d56'55"
Chloe Elisabeth Burdette
Leo 10h10'57" 18d3'3"
Chloé Elise
Per 2h46'29" 39d51'9"
Chloe Elise Casey
Tau 3h37'13" 18d35'49"
Chloe Elizabeth
Ori 5h53'18" 17d6'48"
Chloe Elizabeth
Gem 7h35'35" 19d58'51"

Chloe Elizabeth James Marie Luzza Cnv 13h9'18" 41d21'55"
Chloë Elizabeth Lee Aqr 22h24'19" -0d58'4"
Chloe Elizabeth Smith And 0h24'51" 31d20'33"
Chloe Elizabeth Wareham Psc 0h22'8" 9d28'39"
Chloe Elsa Cnc 8h39'6" 22d47'36"
Chloé et Laurie Moueix Dra 9h41'56" 77d11'37"
Chloe Frank Cnc 9h1'46" 32d29'49"
Chloe Geraldine Schofield Col 5h49'8" -29d53'59"
Chloe Gibson Lib 15h6'50" -3d6'8"
Chloe Grace Psc 0h0'23" -3d40'55"
Chloe Grace Psc 23h12'55" 0d3'18"
Chloe Hajduk Crb 16h6'51" 39d23'14"
Chloe Hammock Psc 0h0'25" 4d47'38"
Chloe Harper Findlay And 23h38'10" 42d2'1"
Chloe' Hattermann Sco 16h9'56" -13d39'17"
Chloe Helena Blish And 23h49'37" 48d32'0"
Chloe Helena Bluth Cnc 9h9'47" 31d25'15"
Chloe Hope Levine And 1h39'31" 44d27'45"
Chloe Houghton's Birth Star And 23h36'56" 47d42'39"
Chloe Hunter Morris And 1h50'17" 35d39'30"
Chloe Jade And 23h38'24" 37d21'17"
Chloe Jade And 23h18'41" 48d52'24"
Chloe Jade Cap 20h49'21" -22d7'41"
Chloe Jane Small Gem 6h1'8" 24d12'11"
Chloe Jean-Reilly Moschberger Cnc 9h11'28" 13d20'21"
Chloe Joy Light of my Life And 1h25'35" 35d37'10"
Chloe Karla Vogt And 23h27'7" 47d15'10"
Chloe Kate Bowyer Umi 16h21'50" 70d27'25"
Chloe Kathryn Coleman Cap 21h39'27" -21d51'36"
Chloe Korsmo Sgr 18h54'32" -32d4'53"
Chloe Kresge Nord And 23h16'11" 48d27'19"
Chloé la plus belle étoile Ori 6h8'54" 12d0'54"
Chloe Laverty And 1h21'5" 40d51'35"
Chloe Liston And 2h34'35" 42d54'46"
Chloe Lorin Smith Psc 1h21'23" 17d30'12"
Chloe Louise And 1h48'12" 38d8'17"
Chloé Louise And 0h2'40" 48d5'21"
Chloë Louise Barker Psc 23h10'2" -0d34'35"
Chloe Louise Posterino - 9.11.2005 Cru 12h24'10" -62d48'43"
Chloe Lugg Lib 15h50'30" -11d59'34"
Chloe Lyn Juntunen Vir 13h35'24" -4d52'59"
Chloe Lynn Arceneaux Ari 2h45'28" 14d35'10"
Chloe Lynn Burley And 23h22'35" 48d38'6"
Chloe Lynn Melby-Beloved Vizsla Cma 6h44'25" -15d57'24"
Chloe Lynn Santos Uma 10h45'20" 66d32'51"
Chloe Mackenna Sellers And 23h21'18" 48d5'14"
Chloe Madison Coker And 1h30'39" 44d21'18"
Chloe Mae Zastrow Cas 0h21'40" 57d5'50"
Chloe Mai Peggy Lennon And 0h0'43" 43d12'58"
Chloe Maree Chetcuti 20.11.05 Sco 17h42'34" -37d3'17"
Chloe Maren Johnson Crb 16h16'20" 31d12'13"
Chloe Marie Sgr 18h11'1" -17d57'50"
Chloe Marie Barnum 3:33 Lib 15h19'46" -23d40'48"
Chloe Marie Copeland Cam 5h12'58" 60d49'21"
Chloe Marie Dooley Boo 15h7'6" 47d58'0"
Chloe Marie Gibson And 2h22'32" 49d14'9"
Chloe Marie Martin Sgr 19h14'40" -18d53'45"
Chloë Markevics And 23h19'31" 51d13'59"
Chloe May And 1h29'34" 49d59'58"
Chloe May White And 0h12'40" 33d21'17"
Chloe Mesaford Her 16h34'17" 46d1'53"
Chloe Michele Sgr 17h58'12" -16d53'28"
Chloe Michelle Collar And 1h29'21" 35d31'31"
Chloe Michelle DeRuyter Cnc 7h58'37" 11d20'45"
Chloe Michelle Lochart Sgr 19h10'43" -13d59'29"
Chloe Myers Umi 16h41'58" 84d19'15"
Chloe Nichole Ari 3h5'38" 17d34'6"
Chloe Nicole Abosch Lib 14h53'38" -0d33'5"
Chloe Nicole Fitchett Cas 0h36'19" 62d48'25"
Chloe Noelle Vir 14h20'30" 5d15'11"
Chloë Philemina Leo 9h28'17" 24d36'58"
Chloë Philippart And 1h57'55" 41d42'27"
Chloe Price Uma 9h9'30" 61d2'38"
Chloe Rae Amon Ari 2h8'16" 11d5'4"
Chloe Renee Lyr 18h45'57" 38d8'38"
Chloe Robinette Tau 5h31'41" 18d57'56"
Chloe Rodengen Aql 19h26'32" 8d6'18"
Chloe Rose And 1h3'18" 45d51'21"
Chloe Rose Gem 7h35'1" 34d27'54"
Chloe Rosenblatt Sgr 18h18'17" -27d6'44"
Chloe Samantha Sco 16h8'44" -16d16'24"
Chloe Scott Price And 1h34'57" 40d54'37"
Chloe Sheila Settle Umi 16h9'26" 79d33'36"
Chloe Simone Bowles Cyg 21h28'27" 54d8'43"
Chloe Skye Bishop And 0h33'7" 25d36'41"
Chloé Star Com 13h28'32" 15d47'24"
Chloe St.Clair Eri 4h15'53" -5d0'47"
Chloe Taryn Hermonat And 0h10'57" 45d24'20"
Chloe Tennessee Gem 6h56'7" 17d45'21"
Chloe Turner And 0h13'36" 28d6'50"
Chloe Turner Gem 6h45'32" 31d51'39"
Chloe Twiggs 21st Birthday Star Cru 12h8'21" -62d10'38"
Chloe Victoria Poirot Lyr 18h48'55" 29d24'25"
Chloe Violet And 2h38'23" 45d54'43"
Chloe Wakeham And 1h46'5" 50d28'15"
Chloe Ward And 23h51'23" 36d21'26"
Chloe Young And 1h18'3" 34d10'23"
Chloe's Beginning... Eri 3h51'11" -29d16'6"
Chloe's Christening Star And 23h24'57" 51d15'40"
Chloe's Light Ari 3h21'19" 28d44'58"
Chloe's Star Lib 15h38'41" -4d55'4"
Chloe's Wishing Star Lib 15h50'20" -17d49'17"
Chloey Noel Babson Leo 9h50'4" 12d55'31"
Chlopek, Friedhelm Vir 12h52'24" 1d35'0"
Chlopkowiak Parents Sons Daughters Uma 12h39'35" 58d24'34"
CHNIGAL Sco 17h31'55" -40d15'27"
Cho Jeong Ah Tau 5h9'4" 21d15'53"
Cho Young Youn Peg 21h11'17" 19d11'21"
Chocalate Ed & Crystal Eri 3h53'53" -0d32'55"
Choc-e-Claire Louise Morris Ser 18h30'49" 1d4'50"
Choch Leo 9h30'50" 26d51'46"
Chocolate Leo 11h9'40" 4d19'58"
Chocolate Bar Cam 4h10'42" 65d12'28"
Chocolate Girl Lyn 8h36'0" 43d30'5"
Chocolate Kisses Aqr 22h35'24" -3d31'18"
Choeun & Rattana Tau 5h47'18" 22d22'3"
Choey Aur 6h53'49" 37d57'47"
Choi Ji Woo Gem 7h44'18" 21d35'30"
Choice - INZAMD Ori 5h24'46" 7d30'49"
Chole Barker Umi 13h46'47" 75d12'39"
Chole Rebecca Vardi And 23h11'46" 49d15'30"
Chollie Uma 9h8'31" 55d3'47"
CHOLON - ...eldenn1 And 2h18'0" 40d18'22"
Chomsavanh Latsavong Boody Lyn 6h54'22" 52d16'45"
Chon & Diem Cyg 20h52'29" 46d55'29"
Chona Star Gem 6h43'56" 13d18'24"
CHONES Boo 14h31'47" 19d10'4"
Chong Lok Nam Aqr 22h38'1" -2d56'23"
Chong Min Kim Vir 11h51'15" -6d6'23"
Chong Suk Chong Lyn 7h8'47" 47d1'25"
Chong Yun Kim Lyn 7h14'4" 56d18'8"
Chonsie Joy Uma 8h57'10" 67d12'22"
Choo Choo Rocket Vir 12h38'1" -7d1'4"
Chooch And 0h18'31" 25d12'16"
Chooch Gem 7h18'15" 23d39'54"
chooch Ari 2h50'41" 18d42'13"
Choon-Yi and Myong Cyg 19h51'59" 33d12'25"
Chooshaa Uma 11h14'42" 54d49'46"
Chopa Cam 7h23'58" 67d42'37"
Chop-Chop Ori 6h1'11" 19d38'4"
Chopper Sgr 17h49'8" -24d24'45"
Chopper Jim Her 16h43'55" 28d2'49"
Choppy D Dra 15h19'28" 59d27'47"
Chops & Kitters Forever Crb 15h42'13" 26d49'3"
Chora Luz Carleton Lib 15h8'18" -28d44'52"
Choraria Uma 11h50'9" 45d13'39"
Chorley 25th Mon 6h49'58" -0d21'44"
Chornarith Psc 0h13'28" 8d15'45"
chorni Lyr 18h33'10" 37d1'23"
CHORO Y CHORI Eri 4h24'22" -31d11'27"
Chouchou And 1h43'28" 42d52'34"
Chouchou Her 18h10'59" 16d2'44"
Chouchou alias Benoît Melato Aqr 20h40'9" -9d24'40"
Chouchou Loulou Forever Uma 8h32'25" 69d27'29"
Choupette Crb 16h20'27" 26d43'51"
Choupette Uma 9h30'4" 44d23'23"
Choupipi de Choupinette Jolie Umi 16h22'57" 71d6'18"
Chovys Estrella Lib 15h29'53" -22d35'5"
Chowilawu Cap 20h25'51" -14d49'50"
Choyster's Choyce Ori 6h8'4" 8d8'36"
CHP Officer Todd Laraway Aql 19h50'25" -0d7'47"
CHP1 Gem 7h46'7" 30d43'20"
Ch.Perlybab Goldcoast Rhett TDIAOV Lib 14h47'17" -7d46'5"
CHRABELI IG LIABE DI Lyr 18h33'13" 36d18'47"
Chraura Lierben Leo 10h4'21" 15d31'45"
Chregle And 1h12'43" 47d45'2"
Chri & Dan Pour La Vie Umi 15h28'58" 77d27'30"
Chrigi Umi 15h26'35" 83d37'46"
Chrigi Cyg 20h2'8" 42d50'52"
Chrigu Lyr 19h27'17" 37d41'2"
Chrindsay Vir 13h18'41" 5d50'44"
Chris Gem 6h25'31" 21d3'36"
Chris Leo 9h52'40" 24d36'37"
~Chris~ Uma 11h25'5" 47d35'18"
Chris Gem 7h34'40" 34d0'7"
Chris Cnc 8h20'14" 31d29'50"
Chris Umi 4h35'9" 89d7'8"
Chris Umi 17h5'36" 81d1'35"
Chris Lib 14h53'38" -2d16'23"
Chris Lib 16h1'28" -12d58'51"
Chris Aqr 23h1'26" -8d2'45"
Chris Cap 20h35'14" -21d8'5"
Chris Umi 14h44'44" 72d5'36"
Chris Uma 12h45'35" 60d22'2"
Chris Sgr 18h24'45" -32d7'31"
Chris Sco 16h54'33" -42d6'18"
Chris_100263 Uma 13h46'25" 49d35'1"
Chris Adam's Corner of the Sky Lmi 10h27'23" 36d44'6"
Chris & Alexa Gem 7h10'57" 17d21'52"
Chris & Allyson Leo 9h57'33" 7d1'58"
Chris Althouse Uma 11h9'59" 31d29'48"
Chris Alvarez Cnc 8h22'5" 11d21'46"
Chris & Amy Sge 19h35'3" 17d36'56"
Chris and Amanda Sge 19h19'13" 17d29'46"
Chris and April Forsythe Forever Mon 8h4'58" -0d21'35"
Chris and Ashley's undying love Sge 19h24'29" 18d22'42"
Chris and Beth Marshall Cyg 20h23'7" 58d14'17"
Chris and Bev Serwy Ori 6h11'0" 2d46'22"
Chris and Brenda Fitzgerald Cap 21h6'47" -16d36'49"
Chris and Catherine Uma 10h57'42" 37d39'16"
Chris and Chloe~Always and Forever Per 3h49'37" 33d44'49"
Chris and Chrissy Sge 19h50'48" 18d27'42"
Chris and Cindy Cyg 21h33'55" 40d25'5"
Chris and Crystal Cas 2h15'50" 63d50'33"
Chris and Crystal "PoohTater" Lib 15h7'43" -11d25'25"
Chris and Dani forever Dra 19h31'1" 70d12'2"
Chris and Debbie <3 Cyg 21h30'23" 45d56'25"
Chris and Dorothy Uma 9h36'44" 61d24'57"
Chris and Edwina Roybal—11/11/00 Cyg 20h1'27" 59d41'19"
Chris and Ele Tau 5h12'41" 17d11'2"
Chris and Elizabeth Forever Sge 20h2'26" 18d19'44"
Chris and Erin's Love Star Cyg 20h9'15" 45d48'13"
Chris and Helen Lindsay Lyr 18h46'36" 38d52'19"
Chris and Jackie Terbrock Aql 19h44'9" -0d33'22"
Chris and Jeanne Kalemba Vir 12h24'27" 0d9'9"
Chris and Jo Eternal Cyg 21h21'2" 53d53'11"
Chris and Julie Van Tassel Lyr 18h48'21" 37d18'28"
Chris and Kathy Cnc 8h11'47" 24d24'5"
Chris and Katie Wukitch Cyg 21h21'20" 51d3'1"
Chris and Lauren's Infinite Love And 1h34'34" 46d26'55"
Chris and Lisa Metternich Uma 11h59'58" 32d22'50"
Chris and Meagan Tau 4h10'53" 23d24'4"
Chris and Mike Lyr 18h34'42" 35d38'22"
Chris and Peggy's 1 Year Star Tau 3h58'16" 28d0'54"
Chris and Rachel Tau 3h50'43" 7d3'22"
Chris and Reagan Price Lib 15h9'36" -12d49'11"
Chris and Sam 4 E & E & E Per 3h25'2" 34d11'7"
Chris and Sibyl together again Cyg 19h43'41" 33d56'30"
Chris and Teri Jankowitz Tau 3h31'54" 21d22'31"
Chris and Theresa Floren Peg 22h27'36" 29d0'29"
Chris and Tina Silva Cyg 20h6'53" 43d17'15"
Chris and Vanessa Cruz Cas 23h17'15" 58d7'55"
Chris and Vess Sgr 19h15'17" -11d59'14"
Chris - Andever Psc 0h54'3" 11d36'28"
Chris Andrews Gem 6h44'1" 15d21'41"
Chris & Andy Together Forever. Cyg 19h32'11" 43d7'35"
Chris Ann Macritchie Crb 16h13'0" 39d6'59"
Chris Ann Smith Cas 1h26'53" 63d41'22"
Chris Anne Aqr 22h5'7" -14d39'33"
Chris Anthony Bores Ari 1h50'12" 18d16'5"
Chris Arlen Fitzsimmons Per 3h18'37" 52d39'43"
Chris Arnold Ori 6h19'59" 10d30'37"
Chris' Asfar Star And 0h55'23" 22d4'46"
Chris B Dorang Best Dad Ever! Star Cnc 8h15'55" 31d27'23"
Chris Badowski Ori 6h12'38" 15d45'38"
Chris & Barbara Grealish Cyg 20h1'24" 48d55'29"
Chris Barrere Vir 14h36'10" 7d12'33"
Chris Bennett - All My Love - Jill Leo 10h22'14" 7d4'34"
Chris Best Dad Ever Hutcherson Her 17h25'53" 37d33'56"
Chris & Betty Golden Reign Mon 6h30'26" 9d21'47"
Chris Biederman Psc 23h29'49" -3d8'29"
Chris Bonneville And 0h54'19" 36d38'22"
Chris Bourne Scl 0h7'20" -26d59'39"
Chris Bozzone Umi 16h46'12" 76d14'20"
Chris Breest Lyr 18h15'37" 39d32'28"
Chris Brion forever 1958 Uma 11h44'14" 39d36'51"
Chris Brookins Aql 19h21'54" 6d29'39"
Chris Bruhns Uma 9h30'32" 65d37'20"
Chris Budion Uma 11h29'24" 51d52'55"
Chris Burn Uma 12h23'45" 56d0'29"
Chris Burson Pho 23h57'38" -57d20'59"
Chris Caples Uma 9h13'5" 52d22'27"
Chris Cappiella Umi 14h41'29" 76d27'17"
Chris Carella & Rebecca Adkins Star Sgr 18h37'24" -35d18'20"
Chris & Carrie Peeples Cyg 19h18'53" 48d29'52"
Chris Carroll Ari 2h7'32" 23d5'8"
Chris Carter - Laelia's Papa Lyn 8h24'46" 43d48'17"
Chris Cat 1 Leo 9h47'59" 26d10'56"
Chris Cattelain idem For Ever Ari 2h8'32" 11d4'45"
Chris Chappelow's Star Cru 12h27'38" -63d7'57"
Chris & Chuck Gray Cyg 21h12'15" 46d36'47"
Chris Cigainero Lib 15h37'14" -7d59'6"
Chris Combis Family Star Uma 13h51'38" 61d31'2"
Chris & Cricket Eternally Entwined Dra 15h42'39" 59d7'39"
Chris Cross Uma 11h42'44" 51d7'17"
Chris & Cyndi Phillips Star Ser 15h21'2" -0d9'47"
Chris D Beautyman 'Big Bro' Uma 10h3'16" 48d16'7"
Chris & Danny Edwards Uma 10h10'55" 48d6'30"
Chris & Dawn Smith And 2h18'43" 42d58'53"
Chris DeBonaventura Ori 5h56'30" 22d47'42"
Chris & Dick 19 Sge 19h41'48" 17d59'40"
Chris Dickerson Her 18h34'32" 19d55'16"
Chris Dillon - I Love You Eternally Gem 7h16'23" 20d31'58"
Chris Dortch Her 17h12'19" 15d54'15"
Chris Downer Tau 3h34'58" 4d2'0"
Chris Drozdowski And 0h41'18" 32d41'28"
Chris Echerivel Leo 11h28'23" 14d41'21"
Chris Eckl Gem 6h47'15" 19d40'10"
Chris & Emily Seybert Lyr 18h48'44" 34d8'37"
Chris Everett Beisel Tau 5h7'49" 17d59'3"
Chris Faherty Ori 5h53'55" 21d54'59"
Chris Fara1 Vir 12h34'42" -6d9'55"
Chris Florio Aql 19h9'20" 1d12'6"
Chris - Forever My Star - Torie Ori 6h9'56" 11d21'38"
Chris Franz-Joseph Dobner Uma 8h36'8" 72d57'50"
Chris Funai Aql 19h58'5" -7d18'19"
Chris G. Doms Cyg 21h35'17" 33d8'44"
Chris Gaunt Ori 6h9'50" 7d25'2"
Chris & Gayle - Together Forever Cru 12h29'5" -59d15'52"
Chris Giannone Per 3h42'37" 44d46'28"
Chris Gillen Umi 17h18'13" 84d21'6"
Chris Goeken and Elaine Grimes Uma 11h41'27" 37d24'39"
Chris Goessling 6/1/86 - 8/14/05 Cyg 20h50'0" 36d31'27"
Chris Golias Pyx 8h49'58" -30d0'38"
Chris Gray Umi 15h20'2" 72d52'49"
Chris Greasley's Dream Psc 23h56'51" 5d49'28"
Chris Greb Per 3h7'12" 54d50'42"
Chris Grevengood Cyg 19h49'28" 35d58'16"
Chris Gwinn Umi 14h9'46" 78d11'6"
Chris "Hamburger" Hough Per 4h26'5" 40d11'32"
Chris Hampton Sgr 18h4'6" -35d57'5"
Chris Harmon Ori 5h55'8" 6d13'36"
Chris Hawkins Boo 14h47'10" 51d34'30"
Chris Haywood Pho 0h36'38" -50d26'41"
Chris Hazilias Peg 22h23'19" 28d9'48"
Chris & Heather McCarty Cyg 21h51'37" 52d26'36"
Chris & Helena: Love Eternal Cyg 19h40'16" 30d18'25"
Chris Hensley Uma 9h10'13" 49d1'21"
Chris Hoffman Her 18h51'57" 24d22'14"
Chris & Holly Tri 2h24'51" 30d54'31"
Chris Hoskins Cep 23h58'15" 74d20'18"
Chris House Per 3h48'46" 41d5'36"
Chris Israel Buckmaster Per 3h9'7" 53d17'22"
Chris Jacob Butler Cyg 21h35'17" 51d6'27"
Chris & Jaime Tau 4h2'4" 13d20'7"
Chris James Her 18h31'52" 24d51'56"
Chris & Jane's Ruby Wedding Star Cyg 20h59'49" 48d26'16"
Chris & Janet Cyg 19h54'29" 33d13'51"
Chris & Jeni Milam Forever & A Day Umi 14h45'33" 76d33'33"
Chris & Jenna 12/12/04 <3 Cyg 20h19'25" 36d20'24"
Chris & Jenna's Star Leo 10h35'17" 27d4'16"
Chris & Jerry Star Her 16h10'24" 24d51'30"
Chris & Jessie Mommens Ari 2h41'3" 26d53'45"
Chris Jones Uma 8h16'49" 71d35'51"
Chris & Judy Dra 18h5'34" 71d35'53"
Chris&Julie Shine As One Eternally Cyg 20h19'35" 44d46'16"
Chris K. Lilman Vir 14h56'7" 1d29'56"
Chris & Kate: first 5 of forever Ori 6h7'26" 19d13'0"
Chris & Katie DeDobbelaere Cyg 21h28'58" 34d3'10"
Chris & Kaye Hunter "Always" Peg 21h35'34" 22d8'6"
Chris & Kellie Cyg 20h18'52" 54d36'53"
Chris & Kelsey's Happily Ever After Cnc 8h36'23" 24d14'1"
Chris Kennedy Uma 9h2'5" 53d35'10"
Chris Kennedy Dra 17h0'0" 67d6'58"
Chris & Kim Sge 20h20'18" 19d37'57"
Chris & Kim Ades Cyg 21h16'48" 43d15'22"
Chris Klinger Gem 6h13'16" 26d36'17"
Chris Knowlton 1981 Cap 21h3'51" -16d16'12"
Chris Kokta Leo 9h23'49" 15d46'31"
Chris Kotsis, my munchkin forever Cru 12h7'43" -62d9'39"
Chris "Kringle" Dougherty Aql 20h14'3" 2d32'26"
Chris L Uma 9h44'12" 57d18'7"
Chris & Lacie Valletta and Baby And 2h32'3" 45d45'56"
Chris LaCivita, My Forever Love Lyr 18h51'24" 42d21'31"
Chris Landry Aqr 22h52'48" -5d26'12"
Chris & Lani Uma 12h6'38" 61d32'54"
Chris Lanning Uma 11h42'49" 53d58'6"
Chris Lapidus Uma 13h55'9" 60d6'43"
Chris & Laura's Memory Star Cyg 19h38'54" 30d0'13"
Chris & Lauren Lord And 2h31'53" 44d31'9"
Chris Le Mon 7h56'57" -0d43'20"
Chris Lehman Her 18h6'42" 27d35'29"
Chris & Lindsay Arentz Tau 5h27'6" 27d6'42"
Chris Lockwood ESQ. Aqr 22h5'31" -22d52'2"
Chris Loewen Per 3h19'48" 44d23'1"
Chris Lord Cas 23h49'44" 56d56'5"
Chris Lundberg Leo 11h42'34" 24d49'39"
Chris Lynn Ferguson Vir 13h10'35" 11d35'9"
Chris M. Cobb Vir 12h4'4" -10d52'46"
Chris & Mackenzie Sgr 18h55'57" -33d59'31"
Chris & Maggie Mon 7h46'52" -0d34'26"
Chris Manino Jr Psc 0h5'53" 7d44'34"
Chris Manton Uma 9h25'15" 59d7'0"
Chris Marcum Aql 20h7'45" 3d28'28"
Chris Marden Sco 17h58'49" -38d18'49"
Chris Marion Uma 10h27'58" 60d28'14"

Chris Martucci
Ari 3h19'50" 27d44'22"
Chris & Maryann "MLC"
Eri 4h30'25" -2d19'38"
Chris McBride
Her 18h10'54" 15d12'27"
Chris McCarthy
Sco 16h10'39" -10d30'10"
Chris McCowan
Per 3h12'19" 42d46'21"
Chris McDonnell Schriver
Vir 14h8'15" -10d35'11"
Chris McKeown
Per 3h18'14" 44d46'8"
Chris McKown
Uma 12h2'31" 36d32'56"
Chris Michaels
Cnc 8h38'36" 22d8'48"
Chris Molly Khriss Paige Aden
Uma 8h38'56" 51d48'51"
Chris Morata
Pic 5h58'7" -55d24'16"
Chris Moyers PhD
Her 17h11'24" 38d48'23"
Chris Must Lose
Umi 14h55'27" 76d4'20"
Chris My Love
Sco 16h16'27" -17d14'54"
Chris Myers
Ori 6h9'42" 11d45'22"
Chris N. Christie
Aqr 22h16'44" -2d31'0"
Chris&Natalie In The Name Of Love
Vir 12h41'50" 2d58'58"
Chris & Nicole September 19, 2005
Ori 5h25'15" -7d31'58"
Chris & Nicole, Always & Forever
Ori 6h6'6" 0d29'41"
Chris Noble
Aqr 23h4'45" -8d34'30"
Chris Ockwell
Per 3h24'16" 51d15'26"
Chris Oliveira
Ari 2h20'6" 26d57'23"
Chris Olsen
Umi 15h10'24" 70d56'55"
Chris Parker's shining moment
Aqr 22h10'5" 0d30'46"
Chris Parton
Cep 22h1'39" 61d21'42"
Chris Pedroley
Cas 1h25'6" 69d18'59"
Chris Peper
Uma 13h27'18" 58d5'59"
Chris Perez
Uma 12h54'50" 59d29'51"
Chris Peterson's Light in the Sky
Sgr 18h55'59" -35d19'32"
Chris' Place in Heaven
Uma 12h38'34" 56d18'51"
Chris "Pooh" Berk
Uma 10h22'30" 60d52'1"
Chris Prentiss
Crb 15h59'6" 38d42'52"
Chris Pringle
Aqr 22h28'57" -2d15'22"
Chris 'Pumpkin' Rudd
Ari 2h34'27" 17d48'28"
Chris R. Anthony
Dra 15h39'11" 57d24'56"
Chris Reid
Aqr 23h32'18" -23d52'56"
Chris & Rich Wyrwas
Ori 5h35'27" -5d13'14"
Chris & Robin's Everlasting Star
Ori 5h51'58" 10d27'49"
Chris Robinson
Uma 11h35'51" 33d50'7"
Chris Robinson I Love You
Cyg 19h50'59" 51d50'31"
Chris Ryan
Aqr 22h37'30" 0d26'35"
Chris S. Cancelleri
Uma 8h16'21" 65d8'17"
Chris & Sarah Howell
Uma 11h57'45" 29d34'21"
Chris Scala "25"
Her 18h52'27" 22d53'12"
Chris Sevilla
Per 2h54'33" 33d21'1"
Chris & Shannon Hrubec
Cyg 20h28'52" 56d54'53"
Chris Sherlock
Uma 9h40'22" 64d29'41"
Chris & Shirley Miller
Cyg 19h51'4" 36d36'53"
Chris & Shnoa Clements
Dra 17h37'55" 53d14'41"
Chris Simone
Her 17h19'59" 19d4'7"
Chris Smaragdakis
Uma 9h25'19" 48d35'34"
Chris' Star
Her 16h15'15" 43d11'59"
Chris' Star
Cmi 7h27'3" 3d27'24"
Chris & Steph forever
Col 6h9'58" -39d10'7"

Chris & Stephanie
Dra 19h17'19" 60d15'36"
Chris Stone - 2006 Teacher of Year
Uma 11h21'33" 63d16'10"
Chris Storr Stevens
Ori 5h42'52" 6d55'28"
Chris & Tahcia Rudder
Pho 0h25'22" -52d47'5"
Chris Tate & Leith Johanson 22/06
Pho 2h1'59" -42d5'18"
Chris "The #1 Dad" Barkes 12/12/66
Uma 12h49'12" 59d28'16"
Chris & Tiffany - Forever & Always
Psc 0h20'31" 15d43'34"
Chris Tocatlian
Uma 13h25'2" 62d44'24"
Chris Tygy One Neil
Per 4h6'28" 37d30'8"
Chris & Uschi
Uma 9h38'21" 72d39'35"
Chris & Vaso
Cyg 20h46'15" 41d48'52"
Chris Villella
Her 16h55'57" 40d28'19"
Chris Vinson Loves Jessica Galvan
And 23h3'18" 41d3'26"
Chris Wagner
Sco 17h2'30" -40d3'19"
Chris Walsh
Uma 9h23'12" 48d54'15"
Chris Walsh
Cnc 8h53'6" 7d12'33"
Chris Watson - 'Light of Yarrow'
Per 4h14'6" 48d13'6"
Chris Weston
Ori 6h11'20" 21d6'51"
Chris Wheeler
Per 3h43'19" 42d29'54"
Chris Williams
Dra 18h43'12" 50d37'3"
Chris' Wishing Star
Gem 6h59'52" 25d24'24"
Chris Wollin
Uma 11h59'23" 37d25'51"
Chris Woodring - Shining Star
Lac 22h22'28" 48d51'13"
Chris "Word Girl" Henson-Spencer
Leo 11h31'54" 22d12'19"
Chris Wright
Aql 19h59'22" 2d41'31"
Chris - you're my star!
Uma 13h33'45" 54d16'11"
Chris, An Answer to Prayer
Vir 12h22'18" 0d23'43"
Chris, my love and shining star
Her 17h49'46" 44d43'16"
Chris, Susan and Amy Butcher
Uma 9h3'10" 48d16'6"
Chris, Wendi, Madison Arno
And 1h3'54" 35d20'47"
Chris04-15-06Lea
Cyg 20h16'38" 52d4'47"
Chris45683968
Uma 9h26'26" 57d4'41"
Chris-Alex-Brielle-Mohn
Uma 13h14'9" 56d53'10"
Chris-Alex-Brielle-Mohn
Uma 13h46'30" 54d59'36"
Chrisanda
Uma 11h10'13" 61d24'49"
Chrisandra Nevette Allen
Vir 12h16'19" -11d24'24"
Chrisann
Cyg 20h12'31" 34d30'24"
ChrisAnne
Gem 7h11'48" 33d12'30"
ChrisAnne a star which we all love
Cas 1h17'41" 58d44'22"
Chrisbea
Cep 3h55'32" 81d16'19"
Chris-Bob
Umi 17h44'0" 82d57'22"
Chriscia'
Vir 13h21'6" 11d40'26"
Chrisclaigiambell
Uma 10h45'16" 42d15'36"
ChrisErika Destined Forever
Sge 20h3'35" 17d21'8"
chrisherri
Lyr 18h58'16" 35d50'24"
ChrisHimes
Uma 9h32'4" 51d54'2"
CHRISJEN
Cyg 21h59'49" 47d49'33"
chrisjohnbaynes
Ori 5h18'31" -7d9'16"
Chriskelly
Boo 14h34'39" 19d22'40"
Chrislynn
Uma 9h50'55" 56d45'15"
chrismelissa4ever
Cyg 21h54'57" 47d12'9"

Chrismomgramma
Ori 5h30'49" 6d47'48"
Chris-N-Jen
Uma 9h34'49" 50d22'52"
Chris-N-Rick2000
Psc 0h24'29" -4d59'0"
Chriso
Cyg 21h32'2" 38d26'17"
Chrisophiabella
Uma 11h18'57" 62d59'43"
Chrisoula
Uma 10h19'15" 65d39'56"
Chris's Dancing Star In The Sky
Tau 5h8'14" 17d55'50"
Chris's Heart
Cnv 12h35'49" 44d13'37"
Chris's Princess
Crb 16h18'24" 35d10'27"
Chris's Shining Star
Cnc 8h36'47" 19d45'44"
Chris's Star
Ari 2h11'54" 20d45'10"
Chris's Star
Per 3h19'52" 46d14'36"
Chris's Stelar
Aqr 21h51'59" -0d18'27"
Chrisse 4/2/74
Lyr 19h20'18" 28d23'16"
Chrisselda "Snuffalupagus" Leal
Cmi 7h30'9" 9d12'30"
Chrissi Lynn Presser
Cyg 19h32'35" 32d15'40"
Chrissi Puk
And 2h14'4" 37d56'8"
Chrissie
And 1h44'38" 48d49'53"
Chrissie
Del 20h34'52" 16d48'13"
Chrissie "Chris-Dawg" Ullman-Schaeffer
Uma 11h29'51" 60d46'16"
Chrissie Cucinotta
Cyg 21h51'39" 47d24'23"
Chrissie Leslie Anne Nadeau
And 0h22'5" 25d44'28"
Chrissi's Light
Cyg 20h59'54" 47d7'35"
Chrissi's nightlight
Umi 16h20'2" 70d15'31"
Chrissy
Uma 10h35'46" 71d26'45"
Chrissy
Lyn 6h47'21" 55d43'56"
Chrissy
Cap 21h35'46" -9d47'4"
Chrissy
Sgr 19h3'59" -33d42'11"
Chrissy
Sgr 18h5'3" -27d40'39"
Chrissy
Uma 9h50'13" 46d31'10"
Chrissy
Uma 9h8'46" 48d4'55"
Chrissy
And 23h18'29" 50d43'14"
Chrissy
And 23h19'14" 42d50'39"
Chrissy
Boo 14h37'35" 34d5'44"
Chrissy
Gem 7h25'44" 34d38'25"
Chrissy
Psc 1h6'25" 31d44'47"
Chrissy
Leo 11h4'43" 15d6'31"
Chrissy
Gem 7h19'25" 19d54'16"
chrissy
Ari 3h7'57" 29d31'17"
Chrissy
Tau 5h54'14" 25d14'28"
Chrissy
Aqr 22h51'10" 1d42'38"
Chrissy 1-25-91
And 0h28'55" 45d58'57"
Chrissy and Tom Ward
Sco 16h13'7" -14d7'40"
Chrissy Baker & Andy Coysh
Cyg 19h43'22" 39d18'59"
Chrissy Baker Loves Gregg Pepper
Uma 11h52'55" 60d20'34"
Chrissy "Christmas" Miller
Uma 11h25'18" 35d11'59"
Chrissy Closson My Shining Star
Uma 11h8'5" 29d20'14"
Chrissy Conrad
And 2h13'38" 45d25'28"
Chrissy Critter Gagliardi
Srp 18h19'42" -0d3'49"
Chrissy Eriksson
Gem 7h2'41" 22d47'22"
Chrissy Ford
Cyg 22h2'14" 50d56'2"
Chrissy Herman
Umi 15h48'9" 82d44'23"
Chrissy Johnson
Lyn 8h12'13" 33d22'38"
Chrissy (Kaliki)
Lyn 6h34'53" 56d44'21"

Chrissy Klosowski
And 1h43'51" 43d49'36"
Chrissy Magnone
Gem 7h0'50" 14d20'7"
Chrissy Mixx
Crb 16h23'22" 38d56'21"
Chrissy Nix
Ari 2h51'28" 27d54'46"
Chrissy of Inspiration
Leo 9h58'46" 16d26'34"
Chrissy Palazzolo " Hotness "
Cap 21h46'29" -19d17'7"
Chrissy Pettman
Psc 23h20'0" -1d0'24"
Chrissy Poo
Ori 6h15'12" 8d57'10"
Chrissy Savitri Singh
And 2h27'59" 48d3'33"
Chrissy Stalnaker
And 0h23'20" 25d16'24"
Chrissy You're A Super Star
Cru 12h26'57" -60d5'44"
Chrissy-0410
Ari 2h38'26" 14d21'32"
Chrissy's Blue Light
Ari 2h48'7" 13d19'37"
Chrissy's Dream
Psc 1h19'29" 24d53'24"
Chrissy's Gumption
Tau 5h58'57" 28d17'49"
Chrissy's Star (12/25/1968)
Cap 20h21'27" -13d17'32"
Chrissy's Wishing Star
Vir 13h13'5" -19d58'54"
Christ Star
Gem 6h49'21" 33d36'25"
Christa
Uma 8h40'54" 50d57'9"
Christa
Cas 0h0'38" 57d7'30"
Christa
Ori 4h51'35" 13d8'30"
Christa
Lib 15h11'40" -17d41'1"
Christa
Cap 20h12'1" -14d25'34"
Christa
Aqr 22h29'48" -0d35'26"
Christa
Lib 15h6'14" -26d18'23"
Christa A. Nigro
Vir 13h26'3" -21d52'42"
Christa Ammann
Uma 11h22'59" 40d45'24"
Christa Brunner/Zollinger
Crb 15h50'8" 27d27'45"
Christa Byrd
Sgr 19h43'8" -24d47'0"
Christa Camille Beavers
Lyn 7h22'2" 44d39'37"
Christa Clifford
Cyg 20h1'7" 47d32'10"
Christa Cooman
Tau 4h35'44" 28d18'44"
Christa Diane Els
Uma 13h37'8" 50d1'2"
Christa Elisabeth Sessler
Lmi 10h28'24" 34d49'37"
Christa F. Sporer
Cap 21h43'50" -24d54'47"
Christa Fanelli
Cas 1h32'16" 68d54'43"
Christa Flynn
And 23h20'47" 49d52'46"
Christa Gaumann
Mon 6h46'19" -0d32'11"
Christa Geisler-Franz
Uma 9h27'29" 53d1'18"
Christa Hansen
Leo 11h10'5" 23d50'45"
Christa J. Burleson
Cap 21h57'20" -18d30'30"
Christa Jean
Ari 2h35'29" 25d16'6"
Christa Jean
Cnc 9h2'54" 31d49'44"
Christa Joanna Lee
Gem 7h18'31" 20d11'52"
Christa Joy
Lyn 7h24'27" 51d30'43"
Christa Lattyak
Ori 6h7'6" 1d23'14"
Christa Lee Brown
Cnc 8h14'20" 15d22'29"
Christa Leigh Goshen
Tau 4h30'6" 8d51'58"
Christa Lynn Dunne
Cnc 8h21'48" 18d40'44"
Christa M. Finer - High School Grad
Sco 16h4'56" -20d45'19"
Christa Marie
Leo 10h32'35" 20d10'46"
Christa Marie McConaughy
Aqr 20h45'26" -8d49'21"
Christa Marlene Geuppert 11.01.1956
Uma 14h12'4" 60d25'16"
Christa McGrew
Cas 0h37'10" 54d2'11"
Christa Nagel
Uma 14h4'20" 58d19'21"

Christa Penelope
Leo 11h24'49" 5d14'59"
Christa Prestifilippo
And 0h57'33" 22d54'34"
Christa ~ The Star of Love
And 1h1'54" 37d43'31"
Christa Wagner
Ori 4h56'16" 10d20'10"
Christa Wais
Lac 22h2'4" 39d56'12"
Christa Watermann
Uma 12h8'53" 48d20'2"
Christabel
And 0h20'25" 32d22'36"
Christabell
Sco 17h53'40" -35d51'51"
Christabelle
Sco 17h13'37" -31d9'22"
Christain Douglas Bruno
Aqr 21h51'11" 2d8'4"
Christain Matthew Castle
Ori 5h6'35" 2d36'2"
Christal
And 0h55'38" 37d18'15"
Christal
Cap 21h15'59" -25d15'30"
Christal Leigh Bonsaint
Per 3h13'37" 48d16'22"
Christal Light
Aqr 21h5'12" -2d10'6"
Christal Lyn
Peg 22h38'0" 14d38'35"
Christal Sharma
Uma 11h29'16" 60d21'45"
Christal Sheppard
And 23h23'15" 42d7'43"
Christalicious
Per 2h22'48" 53d53'9"
Christalina's Light
Cru 12h32'18" -55d42'30"
Christal's Star
Mon 7h20'28" -0d36'26"
Christan Marie Van Mater
Uma 11h51'4" 60d7'23"
CHRISTAN MASTERS
Cas 0h31'41" 52d28'54"
Christanie Lynn Dickison
Uma 8h59'13" 48d46'56"
Christan's Glow
Tau 5h8'15" 21d35'58"
Christan's Glow
Tau 5h8'16" 18d23'9"
Christany
And 23h5'37" 38d35'59"
Christar
Umi 14h44'26" 74d54'59"
Chri-star fantastic
Cru 11h58'24" -59d24'39"
Christar4 Lewis USA
And 0h35'48" 40d31'34"
Christe, Sassy and Jake
And 1h45'47" 48d12'23"
Christeen
Cyg 20h19'21" 44d7'52"
Christeen
Vir 14h27'24" -4d30'26"
Christeenia Cargill
Ori 5h59'55" 22d35'12"
Christel
Cnc 8h34'22" 17d10'2"
Christel
Cas 0h34'3" 62d4'30"
Christel
Lib 15h38'28" -22d35'20"
Christel House Academy
Uma 11h29'23" 55d19'6"
Christel Ilse Irene 04091935
Uma 12h39'44" 52d46'15"
Christel Jo Zissimos
Lib 15h30'49" -25d58'53"
Christel & Josef Shopp - 24.12.1944
Ori 6h20'41" 10d47'53"
Christel Marie Welch
Psc 1h26'50" 19d58'21"
Christel Vinson Loves Liam
Cap 21h28'34" -20d44'15"
Christel Vinson Sheridan
Vir 13h51'35" -5d47'9"
Christella
Vir 12h57'2" -17d46'9"
Christelle
Ori 4h53'51" -2d56'43"
CHRISTELLE
Leo 11h39'43" 21d39'29"
Christelle Balcon
Uma 11h54'57" 33d19'52"
Christelle Conti
Peg 21h41'33" 16d38'29"
Christelle et Alain Hoarau
Cas 0h0'21" 59d9'9"
Christelle et Richard
Peg 22h27'46" 6d40'57"
Christelle Jadra
Uma 9h29'32" 56d0'9"
Christelle Joli coeur
Leo 11h57'14" 23d45'46"
Christelle l'étoile de ma vie
Gem 7h53'58" 30d27'44"
Christelle Maton Mon tit Chat
Cnc 8h1'47" 18d16'12"
Christemini 6447
Crb 16h22'16" 37d7'5"

Christen
Uma 10h24'20" 45d29'21"
Christen
Vir 14h27'30" 1d1'1"
Christen
Sco 16h10'42" -8d40'18"
Christen
Lib 14h42'45" -10d50'40"
Christen "Babies" Veguilla
Cnc 8h47'32" 21d20'27"
Christen Haley
Crb 15h45'53" 31d14'0"
Christen Jung
Cnc 8h4'48" 26d31'37"
Christen Marie Koets
Lyn 7h57'22" 44d28'7"
Christen Marie McElroy
Lyn 8h32'9" 43d18'0"
Christen Marie Murray
Aur 5h35'11" 41d11'24"
Christen Morgan Amundsen
Dra 19h24'23" 69d36'19"
Christen My Love
Sco 16h51'37" -38d28'25"
Christen, Klaus
Ori 5h2'13" 5d56'42"
Christena Marie Alvarez
Cnc 8h40'38" 7d2'14"
Christene Beth Hamilton
Leo 11h41'1" 11d17'44"
Christen's Star
Ari 2h53'0" 20d45'42"
Christer 54
Cnc 8h51'23" 13d44'23"
Christeve
Gem 6h36'52" 26d19'38"
Christi
And 0h29'32" 43d16'42"
Christi
Gem 7h12'34" 31d53'49"
Christi
And 1h24'43" 35d24'32"
Christi
Psc 1h18'1" 32d21'20"
Christi
Cas 0h40'10" 57d25'16"
Christi
Cap 20h58'18" -16d22'25"
Christi and Emma Lee
Uma 11h14'32" 57d3'16"
Christi Ann Taylor
Lib 15h30'58" -6d55'34"
Christi Anna Brandle Kujanek
And 1h41'51" 43d48'46"
Christi Annette Pieper
Cyg 21h55'30" 50d15'29"
Christi Brownlow
Col 5h33'9" -40d53'34"
christi cook
Lib 15h35'51" -6d10'44"
Christi "D"
Sco 17h58'43" -39d45'52"
Christi Jo Meier-Bowman
Leo 11h25'33" 9d7'28"
Christi Joy French
And 23h23'32" 52d25'40"
Christi June
Mon 6h30'52" -0d56'18"
Christi L. Morse Edwards Verstoep
Crb 15h56'13" 36d51'28"
Christi L. Underwood
Cap 21h33'44" -14d17'39"
Christi Lynn
And 23h25'45" 47d24'57"
Christi Lynne O'Neill
Psc 1h29'30" 23d25'12"
Christi Stroud
Uma 9h48'21" 48d30'29"
Christiaan Paul Van Deur Jr.
Cyg 20h28'3" 30d49'25"
Christian
Uma 11h20'1" 48d30'31"
Christian
Lyn 8h15'31" 51d2'10"
Christian
Per 4h20'41" 45d9'22"
Christian
Psc 1h7'2" 25d36'35"
Christian
Cnc 8h43'58" 20d25'45"
Christian
Leo 11h7'39" 21d47'44"
Christian
Gem 7h12'48" 17d38'46"
Christian
Gem 6h30'59" 24d43'31"
Christian
Tau 5h19'22" 23d31'57"
Christian
Her 18h49'19" 12d39'14"
Christian
Ori 6h18'30" 10d35'30"
Christian
Ori 5h37'30" 1d56'36"
Christian
Aql 19h40'8" 3d49'36"
Christian
Aqr 20h52'39" -11d10'32"
Christian
Umi 15h32'26" 75d15'43"

Christian
Aqr 21h13'4" -0d49'28"
Christian
Ori 5h52'8" -8d23'44"
Christian
Uma 9h47'28" 56d5'28"
Christian
Uma 12h30'36" 58d0'24"
Christian
Uma 10h3'18" 54d31'24"
Christian
Cep 20h42'25" 59d58'47"
Christian
Uma 11h52'29" 61d45'34"
Christian
Cru 12h5'38" -58d31'10"
Christian
Dor 5h0'39" -65d49'33"
Christian
Sco 17h57'45" -31d37'18"
Christian & Aidan McGurr
Umi 15h9'17" 78d40'28"
Christian Aiden Mostert
Her 17h15'53" 29d35'34"
Christian Albert
Psc 0h56'38" 26d14'19"
Christian Alessandro Sciolla
Cap 20h15'56" -10d47'47"
Christian Alexander Rosse
Ori 5h37'11" -3d10'31"
Christian Alexander Smith
Ori 5h31'52" 5d11'33"
Christian - Alissa
Cnc 8h16'15" 16d6'47"
Christian Amagua
Per 3h6'13" 36d28'39"
Christian A.M.Bussman-Cortez
Lyn 7h57'17" 57d10'39"
Christian and Rebekah
Per 3h20'14" 35d32'58"
Christian Andrew Graves
Aqr 22h29'58" -0d33'2"
Christian Anitelea-Tsioussis Our Baby
Sgr 19h51'20" -44d29'53"
Christian Anthony
Lyn 6h53'31" 53d45'21"
Christian Anthony Kurzyna
Sgr 18h51'38" -31d17'51"
Christian Anthony Sjogren
Per 3h14'10" 48d41'23"
Christian Anton Matthew Babcock
Sgr 18h52'20" -19d48'25"
Christian Artus
Del 20h23'43" 15d10'53"
Christian B.
Lib 15h34'14" -22d46'54"
Christian B. Enser
Ori 5h25'22" 2d5'29"
Christian B. Parker-Hayes
Dra 17h14'48" 62d29'48"
Christian Bamm Budziszewski
Her 16h21'1" 3d58'26"
Christian Beck
Per 2h57'53" 54d34'5"
Christian Beels
Cha 12h57'16" -79d10'22"
Christian & Betty Hansen
Lib 15h11'19" -29d5'51"
Christian Bischof
Uma 8h26'12" 69d53'33"
Christian Blake Mueller-Vogel
Cru 12h15'23" -57d54'11"
Christian Brett Allen-Bell
Lib 15h13'58" -15d6'48"
Christian Brian Gamecho
Dra 11h16'15" 78d2'57"
Christian & Brooke Gimenez Forever
Ori 6h15'18" 15d25'44"
Christian Buban
Cap 21h0'10" -15d48'48"
Christian Buono
Lyn 8h37'17" 38d18'38"
Christian Bush
Uma 8h44'47" 53d50'55"
Christian Busquets-Mayol
Psc 0h27'33" 19d11'30"
Christian C. Fields
Lyn 6h40'37" 61d34'5"
Christian C. Koutsoumpas
Uma 11h39'23" 64d8'20"
Christian Camacho
Psc 0h58'18" 9d11'36"
Christian Cardenas
Lyn 8h39'33" 35d2'25"
Christian Carl Merrick
Leo 11h6'34" 23d24'37"
Christian Castillo
Uma 9h20'56" 45d13'11"
Christian Charles Sagona
Ori 5h31'36" -2d29'33"
Christian Connolly
And 23h23'43" 47d41'22"
Christian Connor Vivian Peterson
Uma 12h46'37" 59d30'7"
Christian Dames
Tau 4h31'8" 25d21'33"

Christian Dante's B-day star Psc 1h17'56" 21d43'22"
Christian David Krieg Cyg 21h52'38" 52d28'0"
Christian David Wiggins Ori 6h11'13" 5d42'55"
Christian Deflorin Her 17h8'42" 20d28'19"
Christian Douglas Bourbon Ori 6h1'20" 18d2'30"
Christian Edward Guster Aqr 20h57'1" -2d43'5"
Christian Edward Stiner Per 4h30'25" 42d50'28"
Christian Ellis Lynch 12/08/2005 Cru 11h58'11" -58d59'17"
Christian et Peggy Cas 23h45'45" 56d57'1"
Christian Faircloth Ari 2h52'38" 30d14'35"
Christian Faith Foster Ori 6h19'17" 13d51'14"
Christian Faith Sassi Crb 16h7'56" 38d9'46"
Christian Faith Thomas Tau 3h52'20" 8d28'45"
Christian Filomeno Crb 16h22'44" 35d46'36"
Christian Foster Aqr 22h11'53" -2d50'19"
Christian Foster Pfeiffer Aubrecht Vir 11h55'7" -4d27'9"
Christian (Fresh) Schwalm Uma 13h55'59" 54d42'15"
Christian Friedrich Ori 4h44'26" 0d21'18"
Christian Fuess Uma 11h21'41" 59d10'3"
Christian G. Villarreal Cyg 19h34'17" 29d56'58"
Christian George Stewart Sco 16h57'26" -43d15'42"
Christian Gerard Ewals Crb 15h38'6" 28d24'12"
Christian Gravier Aql 19h4'23" -7d33'16"
Christian Grimes And 0h33'2" 25d42'28"
Christian Groß Ori 6h12'40" 8d9'36"
Christian Hans Rauch Her 17h10'8" 28d33'17"
Christian Hans Shaw Cnc 8h28'59" 13d46'52"
Christian Hart Fritts Her 16h26'13" 22d23'46"
Christian Heath Kitchen Ori 5h39'32" -0d38'56"
Christian Heer Her 16h41'50" 7d27'42"
Christian Hertsens 6469 Uma 11h24'50" 41d2'24"
Christian Holton Ori 6h7'2" 9d53'52"
Christian Ian Morris Uma 10h17'15" 44d41'34"
Christian Iannotta Her 16h49'39" 41d41'55"
Christian J. Brandan Ari 2h32'7" 25d51'33"
Christian Jacobsen Sco 17h9'29" -34d39'3"
Christian James Bucher Ari 1h49'24" 17d39'45"
Christian James Dorsey Ari 2h33'50" 27d54'37"
Christian James O'Canna Uma 9h43'1" 58d36'38"
Christian James Parmer Ori 5h55'14" 6d28'24"
Christian Jens Rimoldi Psc 1h36'17" 9d37'32"
Christian Jess Novay Tau 5h46'11" 23d59'52"
Christian Jesus Jimenez Cap 20h36'36" -15d19'2"
Christian John Desjardins Lib 15h4'1" -0d38'53"
Christian John Gladd Leo 11h57'52" 25d16'12"
Christian John Zech Her 18h43'4" 15d29'14"
Christian Jorge Ayala Leo 11h19'56" 17d44'5"
Christian Joseph Kushon Dra 16h5'24" 60d30'12"
Christian Joseph Morata Lib 15h10'44" -7d43'35"
Christian Justice Gabriel Sco 16h19'52" -18d13'36"
Christian K. Weidenauer 25.02.05 Her 18h53'45" 23d17'57"
Christian Karnasch Uma 10h12'38" 48d36'19"
Christian & Kim Free Pho 0h20'37" -39d34'32"
Christian Kody Forsyth Umi 14h29'46" 67d39'13"
Christian Lea (Peaches) Hamlin Lyr 18h53'43" 37d23'36"
Christian Lee-Alan Buckman Cap 20h15'30" -25d58'35"
Christian Lee-Allen Buckman Uma 9h12'53" 53d16'55"
Christian Lopez Lib 14h26'22" -20d39'19"
Christian Loves Reesa Cummings Cnc 8h35'47" 16d50'59"
Christian Luis Bernal Dra 15h40'17" 58d9'20"
Christian M. Garcia - Shining Star Uma 10h24'34" 48d23'26"
Christian Mack Aql 19h41'36" -7d23'35"
Christian Macpherson Ori 5h55'13" 21d25'19"
Christian Mamprin Per 3h20'33" 40d11'37"
Christian & Manuel Pereira's Star Cyg 20h44'14" 31d42'49"
Christian & Margaret Buen Cru 11h57'42" -56d8'22"
Christian Mark Tuite Umi 16h7'8" 82d41'47"
Christian Martin Lara Sco 17h50'29" -35d15'33"
Christian Matthaeus Zehntner Leo 10h3'7" 15d5'51"
Christian Matthew Colin Badcoe Umi 15h0'1" 67d37'48"
Christian Matthew Corneau Leo 9h37'52" 13d23'0"
Christian Maupin Lyn 7h40'7" 42d9'47"
Christian Mbassi Col 6h7'55" -28d16'17"
Christian Melchers Uma 10h46'32" 71d50'50"
Christian Michael Altebrando Crb 16h16'34" 29d5'49"
Christian Michael Gagnon Cyg 20h16'37" 57d14'26"
Christian Michael Gravener, Jr. Ori 6h10'36" 15d2'56"
Christian Michael Homan Her 17h44'24" 17d4'45"
Christian Michael (Kiss) Per 3h51'39" 31d14'39"
Christian Michael Nigro Aql 19h30'2" 4d20'44"
Christian Miguel Esparza Lmi 10h38'29" 36d54'52"
Christian Miguel Taylor Boo 14h36'28" 29d1'23"
Christian Mirelle Dra 15h33'51" 60d21'52"
Christian (Muxica) of Cassiopeia Cas 0h46'5" 56d43'58"
Christian N. Glaze Cnc 8h1'58" 12d14'15"
Christian Neil Stinner Dra 16h42'31" 62d27'33"
Christian Nelson Psc 0h45'3" 5d29'22"
Christian Nick Dedvukaj Uma 13h55'11" 51d33'43"
Christian~Nicole Uma 11h5'35" 68d2'31"
Christian Ocaranza Sgr 17h58'5" -24d7'20"
Christian Oertel Lmi 10h10'35" 40d53'4"
Christian & Olivia Payne Pho 23h39'23" -45d4'49"
Christian Orozco Her 17h28'36" 39d15'11"
Christian Owens Uma 11h45'10" 58d29'53"
Christian P. Bauch Psc 1h19'31" 18d49'59"
Christian Padilla Aqr 20h50'59" -10d21'0"
Christian Patrick G. Darbois Uma 11h6'37" 60d12'41"
Christian Patrick Hesse Cas 1h20'29" 55d39'53"
Christian Paul Hanko Uma 9h56'41" 42d44'34"
Christian Peter Lac 22h41'21" 50d39'19"
Christian Pfeiffer-Bellt Uma 9h55'1" 41d38'32"
Christian Pfuhl Per 3h44'35" 51d52'7"
Christian Philip Clarke-Smerk Crb 15h47'4" 36d23'35"
Christian Powell Per 3h17'25" 54d1'43"
Christian & Priska Cam 6h39'16" 65d47'1"
Christian Quidu Cyg 20h8'30" 46d53'57"
Christian R. Cayer Per 4h24'55" 52d33'42"
Christian Rae King And 0h54'47" 37d45'33"
Christian Raghias Crb 16h2'5" 35d37'14"
Christian Rahn Ari 2h40'20" 29d58'28"
Christian Rankin Per 3h46'18" 36d14'53"
Christian Reiner Meixner "Frosch" Uma 14h18'1" 58d12'39"
Christian Rekkedal Aqr 21h14'32" -7d44'54"
Christian Rene Lopez Boo 14h28'7" 17d38'35"
Christian Renee Uma 10h32'59" 46d16'47"
Christian Renk Uma 11h11'1" 71d37'12"
Christian R.G Dra 18h21'17" 54d8'54"
Christian Rodas Vir 14h18'39" 1d54'0"
Christian Roy Erdmann Her 18h55'39" 24d2'13"
Christian & Sabine Ori 6h14'49" -1d46'47"
Christian Sawyer Uma 9h38'32" 64d41'18"
Christian Schouwey Ori 6h12'37" 9d21'57"
Christian Sean Plouffe Tau 4h46'8" 21d11'57"
Christian Seth Leeds Lyn 7h41'18" 42d27'13"
Christian Shane Randle Dra 17h20'50" 54d19'28"
Christian Sobol Ori 6h13'1" 12d2'29"
Christian Spencer Umi 14h15'39" 78d2'27"
Christian Star Uma 11h49'13" 59d31'28"
Christian Stern Uma 9h24'30" 49d9'38"
Christian (Super Feet) Rubio Lib 14h23'57" -24d11'9"
Christian Sylbert Ari 3h11'17" 28d50'9"
Christian Tabernigg Cas 23h31'40" 59d8'6"
Christian Thieblot Cap 20h15'41" -21d29'2"
Christian Thomas Deiches Per 4h22'44" 43d32'9"
Christian Thüring Umi 15h21'7" 77d0'23"
Christian Tretta And 23h17'16" 47d52'39"
Christian tu m'illumini d'immenso Aql 19h30'56" 5d7'41"
Christian Tyler Gilmore Boo 14h36'10" 31d42'42"
Christian Tyler Headrick Cap 20h53'29" -16d2'27"
Christian und Carolin Uma 9h41'31" 42d36'4"
christian und jennifer gollubits Uma 11h36'51" 49d22'17"
Christian van Boeckholt Uma 9h58'56" 63d25'45"
Christian Victor Gemora Psc 0h47'23" 9d16'16"
Christian Vincent Vir 12h50'0" -2d0'25"
Christian Vincent Compolongo Lib 15h13'2" -19d54'24"
Christian W. Parrott Sr. Uma 11h52'47" 35d17'18"
Christian Wardell Armstrong Boo 14h24'17" 32d33'40"
Christian Wayne Beck Ori 5h59'35" 22d36'58"
CHRISTIAN WEBER Lac 22h35'12" 38d8'58"
Christian Weiß Uma 11h0'55" 58d29'34"
Christian William Quandt Umi 15h45'25" 74d56'56"
Christian Wolter Ori 5h58'39" 17d9'15"
Christian Worsch 50 Ori 6h10'23" 6d2'44"
Christian Young-Melcher Cap 21h34'9" -19d4'47"
Christian Zogolla Uma 12h59'29" 56d33'49"
Christian Zurlinden Dra 18h53'4" 55d42'37"
Christiana And 23h39'11" 47d14'59"
Christiana Cam 5h50'7" 56d6'57"
Christiana Lyn 8h46'47" 40d54'34"
Christiana + boyfriend Tommy Ori 5h33'43" 13d24'17"
Christiana Dahl-Hansson Sco 16h4'51" -20d6'5"
Christiana E. Drescher Gem 7h25'52" 34d17'55"
Christiana Grace Cas 23h53'51" 58d10'17"
Christiana Hilmer Cnc 8h12'6" 15d38'1"
Christiana Montgomery Cyg 19h41'25" 33d13'23"
Christiana Nicolaou Ioannidou And 2h35'0" 39d43'52"
Christiana Paloma Negron Lyr 19h22'56" 38d29'12"
Christiana Panagi Leo 11h2'17" -3d9'4"
Christiana's Star Ori 5h40'54" -2d3'2"
Christiane Lib 14h37'59" -17d53'57"
Christiane Uma 13h9'54" 54d2'56"
Christiané Cru 12h28'57" -59d55'7"
Christiane Umi 17h32'3" 84d35'51"
Christiane Cyg 20h55'19" 39d43'50"
Christiane Uma 9h22'48" 49d21'5"
Christiane Daigre Bastide Leo 9h41'8" 11d30'7"
Christiane et Serge Moxel Uma 9h16'1" 61d43'51"
Christiane Meugnier Ori 6h3'37" 0d49'53"
Christiane Nakonz Uma 9h54'26" 69d58'54"
Christiane Pepitone's 21st Birthday Crb 16h2'18" 34d24'14"
Christiane "Spätzle" Kopp Vir 13h39'13" -6d55'31"
Christiane Valladon Robillard Com 13h33'29" 27d37'21"
Christiane Wolgast Uma 9h0'8" 51d45'33"
ChristianJessJenJasonPeg Aql 19h25'49" 7d18'54"
Christianna Cas 2h37'39" 67d20'57"
Christianna Marie Cooper Tri 1h59'16" 34d22'29"
Chris-Tianne Sgr 19h27'43" -32d50'56"
Christianne, my shining star. Beck Cap 20h23'54" -14d24'49"
Christian's 18th Dra 18h58'16" 58d41'46"
Christian's Corazon Umi 15h27'59" 67d47'20"
Christian's Eye Tau 5h7'55" 23d9'45"
Christian's heart, Victoria Lyn 8h49'8" 41d12'13"
Christiansen Family Ori 6h12'54" 6d23'48"
Christi-Chiwa Ori 4h57'6" 11d14'7"
CHRISTIE Leo 11h51'1" 18d51'8"
Christie And 22h58'0" 37d35'45"
Christie And 23h23'37" 38d30'59"
Christie Cam 5h58'30" 64d30'43"
Christie Sgr 18h30'31" -22d30'35"
Christie_0626 Cap 21h33'41" -15d4'44"
Christie Ann Tau 4h13'22" 7d43'0"
Christie Ann Ferguson And 23h15'32" 47d12'18"
Christie Anne Uma 11h7'51" 65d45'14"
Christie Applewhite Lmi 10h10'5" 33d54'10"
Christie Bargfrede Lib 15h8'16" -3d51'19"
Christie Blake Brown Crb 15h47'34" 27d34'44"
Christie Braun Cam 6h6'10" 82d45'8"
Christie Callie Bridges And 1h29'44" 43d7'51"
Christie Cazad And 0h43'23" 39d36'7"
Christie Chiu Ori 5h17'22" -5d26'16"
Christie Cornia Cyg 20h44'29" 48d52'24"
Christie Davisson Cyg 19h53'33" 37d47'20"
Christie DiLorenzo Aqr 20h53'2" -13d56'29"
Christie Emeree Uma 8h48'10" 52d48'52"
Christie Fabrigas Burch Umi 15h52'18" 71d24'30"
Christie Gray And 1h9'26" 40d7'21"
Christie Hetrick Cnc 8h22'1" 29d27'49"
Christie Hewitt Cas 0h11'53" 53d18'4"
Christie Jager Ori 5h13'41" 7d34'16"
Christie Jones Leo 10h18'34" 22d37'41"
Christie K. Hajela Peg 21h55'8" 9d1'55"
Christie Katherine Simmel Ferri Cas 23h5'9" 59d44'52"
Christie Kay Mann 9/17/1977 Uma 12h29'56" 57d22'49"
Christie L. Heckler Sgr 18h20'34" -29d44'30"
Christie & Larry Cyg 21h31'32" 37d4'56"
Christie loves Marcel Ori 6h15'49" 8d24'36"
Christie Lynn Ari 3h15'10" 23d59'53"
Christie Lynn Bethune-Koehler And 0h30'39" 34d4'3"
Christie Lynn Radcliff Leo 10h56'39" 6d24'45"
Christie Marie Crb 15h47'35" 31d40'32"
CHRISTIE MARY WELCH Cnc 8h34'19" 31d14'34"
Christie Michelle Hayes Gem 6h26'21" 21d54'21"
Christie Morrissey Uma 9h35'12" 42d10'37"
Christie Nicole Marina Gem 7h6'51" 29d6'41"
Christie Seyfert Gem 6h54'31" 27d40'1"
Christie Star Sco 16h9'8" -14d50'11"
Christie Swehla Cnc 8h35'0" 24d43'9"
Christie T. Rembetsy Aqr 22h18'16" -15d48'41"
Christie the Super Star! Ari 2h32'19" 13d11'32"
Christie Vanessa Clovis Lyn 7h23'25" 44d47'57"
Christie Wilson Uma 9h22'39" 43d27'49"
Christie Wo Dra 15h17'25" 58d12'48"
Christie, Jessica, Caitlin & David Cyg 20h46'42" 37d28'26"
ChristieBear Sco 17h6'7" -37d30'32"
ChristieEidson Cas 22h59'46" 53d46'39"
Christie's Ori 6h16'4" 15d16'15"
Christie's Charm Ari 2h19'21" 18d18'43"
Christie's Kisses to Dan Cnc 8h40'44" 10d33'54"
Christie's Shining Star Del 21h7'20" 19d20'35"
Christie's Sparkle Sgr 18h53'59" -17d35'34"
christin Uma 12h18'23" 52d51'26"
Christin Sgr 18h35'42" -22d33'24"
Christin And 0h13'20" 29d4'5"
Christin Ori 6h19'40" 18d53'43"
Christin Anne Chiles Del 20h49'16" 6d26'45"
Christin & Betty We Love You Sharon Cyg 20h22'28" 41d17'20"
Christin Blistin And 0h18'3" 39d23'37"
Christin Dawn Tripp Lib 14h49'41" -3d21'24"
Christin Deana Campbell Uma 13h19'1" 52d47'16"
Christin F Johnson Dra 19h1'33" 74d34'26"
Christin Gallion Sgr 17h55'38" -17d59'29"
Christin Howell Umi 14h17'26" 86d18'54"
Christin M. Darling (Daddy's Girl) And 0h23'39" 42d23'24"
Christin M. Gnad Tau 5h25'38" 19d35'18"
Christin Marie Tau 3h38'56" 3d27'53"
Christin Marie Donnelly Aqr 23h8'47" -9d48'15"
Christin Michelle Yonley Leo 11h14'44" 14d4'0"
Christin Oldewurtel Dra 19h11'38" 69d8'14"
Christin T James Tau 3h51'47" 19d49'46"
Christin & Torsten Uma 13h33'6" 55d30'9"
Christina Uma 13h29'31" 59d47'17"
Christina Cas 1h22'56" 62d27'59"
Christina Cas 1h26'46" 63d9'56"
Christina Uma 10h1'28" 55d14'36"
Christina Lyn 7h6'37" 59d47'22"
Christina Uma 9h31'44" 58d45'33"
Christina Cas 2h16'52" 71d37'52"
Christina Uma 14h5'10" 62d7'54"
Christina Cas 1h34'34" 63d48'34"
Christina Cap 20h28'51" -10d12'58"
Christina Sgr 19h54'43" -12d55'32"
Christina Lib 14h42'17" -13d35'14"
Christina Ori 5h41'37" -3d40'15"
Christina Ori 6h14'47" -0d41'41"
Christina Psc 23h10'40" -0d11'29"
Christina Sgr 18h20'51" -34d23'7"
Christina Cap 20h27'26" -27d12'49"
Christina Cap 20h31'55" -26d59'40"
Christina Cap 21h21'51" -24d26'18"
Chris'Tina Ori 6h17'3" 15d6'43"
Christina Vir 12h4'53" 8d48'58"
Christina Vir 13h16'24" 11d23'44"
Christina Mon 6h37'37" 8d14'18"
Christina Gem 7h1'0" 14d16'38"
Christina Ori 5h47'2" 11d32'34"
Christina Ari 2h35'5" 25d37'34"
Christina Gem 7h30'38" 29d29'28"
Christina Cnc 8h37'46" 23d9'14"
Christina Gem 6h47'29" 23d22'15"
Christina Tau 5h13'5" 27d54'46"
Christina Ari 3h5'54" 27d27'37"
Christina And 0h39'16" 38d10'11"
Christina And 1h3'51" 37d58'41"
Christina And 1h5'47" 38d14'12"
Christina And 1h43'58" 43d28'27"
Christina And 1h55'21" 39d5'21"
Christina And 0h58'23" 43d23'27"
Christina Uma 9h35'39" 44d49'29"
Christina Lyn 8h1'53" 41d39'36"
Christina Uma 11h32'17" 38d10'0"
Christina Lyn 7h39'35" 36d57'53"
Christina And 1h6'0" 33d45'22"
Christina Cyg 21h6'22" 42d58'9"
Christina And 23h59'27" 42d12'35"
Christina And 1h11'59" 45d7'28"
Christina And 23h17'38" 48d44'13"
Christina Cas 1h10'36" 59d49'59"
Christina Uma 13h55'15" 52d8'39"
Christina Uma 8h51'11" 51d8'50"
Christina 090405 Ori 5h9'27" 6d37'54"
Christina 13 03 1979 Baobab*** Psc 1h18'10" 12d11'51"
Christina 13 - 07 Uma 11h46'6" 49d43'35"
Christina 13-05 Uma 10h48'25" 51d17'11"
Christina 22592 Lyn 7h32'7" 55d39'45"
Christina 381724 And 1h23'37" 50d10'42"
Christina 9/30 Lib 15h49'59" -9d48'6"
Christina A. Gulotta Cas 23h11'57" 54d31'12"
Christina A. Owens Leo 11h12'39" 16d18'49"
Christina A. Wasson Cnc 8h50'26" 31d50'0"
Christina Abu Dayeh Gem 6h9'50" 24d32'52"
Christina Aceves Mon 6h45'4" 9d36'59"
Christina Aline Cannon Lyn 8h36'36" 34d42'23"
Christina Althouse Peg 23h47'58" 28d33'8"
Christina Anastasia Gem 7h13'25" 19d25'55"
Christina and Christopher Fetko Cyg 20h16'9" 52d28'40"
Christina and Jeffrey Tau 4h37'38" 17d7'57"
Christina and Jeff's Star Aql 20h17'49" 5d13'43"
Christina and John Mills Cyg 21h21'51" 31d4'10"
Christina and Lance Maiden Cyg 19h53'24" 31d1'11"
Christina and Travis Brass in Love Cyg 20h17'50" 46d29'19"
Christina Angela Jensen Psc 1h20'23" 33d18'3"
Christina Ann And 1h5'8" 45d39'43"
Christina Ann And 23h12'19" 42d42'4"
Christina Ann Cnc 9h7'40" 24d10'32"
Christina Ann Uma 10h45'51" 53d1'5"
Christina Ann Bolinger Cnc 9h11'31" 15d17'0"
Christina Ann Bowser Leo 11h38'22" 17d54'14"
Christina Ann Conley Tau 4h9'16" 26d18'15"
Christina Ann Jasso Aqr 22h39'18" -1d42'2"
Christina Ann Koegel Uma 11h48'18" 57d26'1"
Christina Ann Locke Uma 10h41'5" 68d47'50"
Christina Ann Marie Wanger And 23h2'24" 50d43'50"
Christina Ann Neddoff Her 17h40'53" 32d14'29"
Christina Ann Robotham Peg 22h55'53" 25d25'8"
Christina Ann Sorrels Ori 5h32'40" 1d44'53"
Christina Anne And 0h4'15" 47d1'22"
Christina Anne Foley Psc 22h53'9" 7d24'59"
Christina Anne Fuhrman Lyn 9h32'43" 41d17'46"
Christina Anne Lowery Cet 1h26'50" -0d58'30"
Christina Annmarie Dye-Cheung And 23h13'33" 51d33'40"
Christina Antonelli And 23h29'54" 47d23'58"
Christina Apicella Cas 1h22'37" 57d44'10"
CHRISTINA ARGIROKAS-TRITI Cas 0h31'54" 50d11'49"
Christina Ashley Cobb Lyn 7h49'0" 57d28'58"
Christina Athena Strates Psc 1h20'32" 31d4'30"
Christina B. And 2h33'55" 38d59'36"
Christina Baker And 0h19'11" 24d14'17"
Christina Barone Cas 1h29'4" 60d50'19"
Christina Barton Stringer Cnc 8h0'3" 12d54'23"
Christina Bates Psc 0h26'39" 17d44'13"
Christina Bear 1 Umi 14h49'5" 73d23'39"
Christina Beauffet Gray Cyg 20h54'30" 38d0'2"
Christina Beltran And 1h0'11" 36d46'46"
Christina Berghoff Uma 10h44'4" 44d56'38"
Christina Beth Smith Per 4h46'19" 38d16'5"
Christina Beverly Majoris Gem 6h41'17" 19d27'39"
Christina Bittner Tau 4h28'48" 26d24'37"
Christina Brabata Crb 15h25'10" 27d31'28"
Christina Brandt Andersen Uma 13h37'47" 56d39'37"
Christina Broccoli Lyn 6h21'42" 57d13'34"

Christina Brock
Crb 15h29'59" 28d25'24"
Christina Brownlow
Uma 9h53'39" 50d16'9"
Christina Brücker
Uma 8h52'33" 68d18'10"
Christina Bui
Leo 9h22'12" 10d32'14"
Christina Bush
Cyg 20h29'51" 59d33'7"
Christina Byrd
Lyn 7h45'59" 53d44'56"
Christina C
Umi 16h46'23" 81d48'56"
Christina C. Menke
Cnc 8h21'9" 10d32'7"
Christina C. Parish
And 23h34'6" 41d35'0"
Christina Cafero
Ari 2h54'21" 25d14'7"
Christina Campana
Vir 13h14'38" -1d20'36"
Christina Canady
Sgr 19h34'26" -11d55'43"
Christina Capullo
Cas 23h27'19" 55d28'32"
Christina Carissima
Cas 23h0'18" 59d6'5"
Christina Carla Burmeister
Mon 7h42'51" -0d28'40"
Christina Carlson Viotti
Uma 11h20'27" 69d49'37"
Christina Carmelita Ramirez
Uma 13h50'28" 49d25'25"
Christina Carrion
Uma 10h11'34" 70d57'35"
Christina Catherine Ferrugiaro
Lyn 8h31'5" 52d33'40"
Christina Catherine-Gina Montecalvo
Aqr 22h9'41" 2d7'48"
Christina Cathern McCray Halsey
Mon 6h50'44" 3d12'54"
Christina Champagne Sakran
Ari 2h35'43" 13d45'9"
Christina Chardell Carey
Gem 7h19'57" 30d30'3"
Christina Chimics
Cas 2h37'33" 64d56'36"
Christina Choueifaty
And 2h37'17" 45d38'25"
Christina Cimini
Cyg 20h23'57" 52d31'33"
Christina Coler
And 23h28'0" 41d23'9"
Christina Concepcion
Cas 2h15'55" 62d11'37"
Christina Connolly
Leo 10h45'38" 18d33'12"
Christina Crabtree
Uma 9h20'25" 55d47'8"
Christina Curiel
Leo 10h11'45" 17d25'5"
Christina Curry - Rising Star
And 2h26'39" 39d13'22"
Christina D. Sierra
Lyn 6h51'58" 61d17'12"
Christina Danielle Fischer
Cas 0h37'24" 54d30'22"
Christina De Maria
Leo 9h41'9" 28d54'53"
Christina Defrancesco
Cap 20h28'14" -20d22'51"
Christina Denise Contreras
And 0h35'17" 38d40'47"
Christina Denise Holzberger
Psc 1h19'5" 3d31'41"
Christina Denton
Psc 1h11'26" 29d32'21"
Christina & Derek
Cyg 22h1'34" 51d16'36"
Christina Devald
Crb 16h15'32" 38d35'5"
CHRISTINA DIANA DIAZ
Aqr 22h21'56" -24d46'44"
Christina Diane Ellis
Aqr 23h5'12" -13d57'16"
Christina Dopp
Mon 7h59'23" -0d35'42"
Christina Dost
Dra 15h53'55" 54d11'57"
Christina E. Durham
Uma 11h27'48" 33d42'32"
Christina E. Frazen
Aqr 21h24'12" -0d39'42"
Christina Eayrs
Cnc 8h46'49" 19d51'22"
Christina Elaine Neal
Cma 6h42'56" -13d35'21"
Christina Elizabeth Partlow
Gem 7h34'51" 33d59'20"
Christina Ellen Avalos
Cas 1h25'20" 53d55'30"
Christina Erin Garofalo
And 23h43'1" 42d50'23"
Christina Evelyn Bentley Casey
Aqr 22h52'49" -21d28'46"
Christina Falsone
Umi 9h57'45" 88d52'39"
Christina Faye Ellis
Cnc 8h37'27" 24d0'23"
Christina Fernandes Pimentel
Ari 3h10'6" 21d26'24"
Christina Figueroa (Kippy)
Lyn 9h10'40" 34d39'52"
Christina Florence
Lyr 18h53'55" 34d59'57"
Christina Forever 0210
Cas 1h20'32" 54d35'4"
Christina Fountain
Sco 16h34'47" -40d33'26"
Christina G Dixon
Mon 8h6'7" -1d18'38"
Christina Gabriele Humbert
Psc 0h15'0" 15d1'46"
Christina Gabriella Tibbals
Lmi 10h30'38" 34d41'13"
Christina Gayle Welsh
And 23h48'25" 38d14'35"
Christina Giacobbe
Ori 6h24'8" 17d2'33"
Christina Gilpin
Cas 0h19'42" 60d35'35"
Christina Graziano
Lyn 8h3'4" 40d28'38"
Christina Grillo
Cas 23h57'4" 53d7'50"
Christina Guardado
Lyr 19h12'28" 35d9'38"
Christina Guilmette
Ari 3h19'52" 29d5'16"
Christina Han
Crb 16h12'19" 33d37'49"
Christina Harney
Vir 12h44'40" 7d15'51"
Christina Hart
Uma 11h57'48" 46d13'29"
Christina Haskew
Lyn 8h25'54" 50d49'58"
Christina Heisler
Lyn 7h39'20" 36d45'13"
Christina Helen
Pav 19h24'59" -62d19'30"
Christina Henning
Uma 8h10'9" 61d51'48"
Christina Hinrichs
Aqr 21h59'49" -17d7'10"
Christina Horan Williams
Cas 23h43'34" 54d43'13"
Christina I Graterole
Gem 6h48'53" 12d51'32"
Christina I Love You
Ori 5h23'51" 3d5'44"
Christina Iwaniec
And 23h24'51" 43d34'53"
Christina J.
Tau 4h6'38" 6d29'40"
Christina J. Hubeli
Umi 15h0'40" 84d25'40"
Christina Jane
Ari 3h21'2" 19d9'11"
Christina Jane
Gem 6h46'50" 29d8'5"
Christina Jay
Cap 21h6'49" -17d13'39"
Christina Jean
Sco 16h55'24" -40d28'22"
Christina Jeanne'
Dra 19h8'11" 75d48'53"
Christina & Jeff
Cyg 19h26'12" 54d44'15"
Christina Jessica Gibson
Gem 6h45'43" 31d54'45"
Christina Jo 2-2-89
Cas 1h20'26" 60d18'55"
Christina & Joe
Cnc 8h43'30" 31d44'12"
Christina Joesten
Cap 21h53'35" -9d19'30"
Christina Jordan
And 2h33'33" 48d53'21"
Christina Joy Mueller
Ari 2h0'31" 22d41'26"
Christina Joyce Bell
Lyr 19h11'9" 26d31'58"
Christina Julia Ferguson
Vir 12h41'47" 11d39'11"
Christina Julie
And 2h25'4" 46d32'9"
Christina & Justin
Vir 13h8'10" -12d50'21"
Christina K Cook
Uma 8h51'14" 57d18'26"
Christina K. Gunther
And 1h6'33" 42d25'54"
Christina Kastanaki Liberatos
Lib 15h49'18" -19d20'9"
Christina Kendzor
Tau 3h43'27" 24d59'40"
Christina Kennedy
Ori 5h45'26" 12d7'37"
Christina Kiyoko de Groot
Sgr 18h19'43" -30d47'46"
Christina Kronenberg
Uma 11h11'2" 43d6'42"
Christina L. Barndt
And 0h53'26" 38d43'22"
Christina L. Bieranowski
Lmi 10h3'57" 37d31'52"
Christina L. Cameron
Lib 15h36'38" -19d41'32"
Christina L. Choma
Cap 20h20'54" -20d27'33"
Christina L. Trentham
Uma 11h53'0" 34d0'27"
Christina L. Whisman
Ari 2h52'17" 28d19'54"
Christina La Bonita
Umi 13h5'25" 74d31'17"
Christina Lamertz
Ori 5h58'36" -2d46'22"
Christina Lancaster
And 1h29'40" 42d49'16"
Christina Lauren
Sgr 19h31'29" -12d15'49"
Christina Leah Berger
And 23h9'40" 45d15'21"
Christina Leandra
Tau 4h9'21" 16d29'59"
Christina Lee
Lyn 7h53'19" 39d48'18"
Christina Lee Evers
Cyg 20h35'29" 35d9'28"
Christina Lee Link
Aqr 22h8'51" -0d46'42"
Christina Lee McPherson
Uma 9h4'32" 62d46'49"
Christina Lee Weltmer
Ori 5h41'9" 11d48'22"
Christina Lee KW
Gem 7h38'52" 33d4'20"
Christina Leigh Carter
Cas 23h9'32" 55d0'42"
Christina Leigh Green
Lmi 10h8'36" 39d29'3"
Christina Lemke
Leo 11h16'55" 10d11'9"
Christina Leva
Cyg 20h30'21" 42d35'49"
Christina Lewallen
Aqr 22h38'36" -1d33'21"
Christina Lindell Nichols
Ari 1h48'14" 23d7'27"
Christina LoBiondo
And 1h27'49" 48d54'57"
Christina Lohrum
Boo 15h9'58" 13d19'13"
Christina Lopez
And 23h3'55" 51d20'34"
Christina Lopez / Fly
Lyn 8h25'22" 55d4'20"
Christina Lopez Grieger
Aqr 22h33'28" -9d1'22"
Christina Louise Alcala
Tau 5h45'46" 18d15'33"
Christina Louise Higgs
Lyn 7h55'46" 41d40'19"
Christina Loves Christopher
Uma 9h16'42" 63d49'46"
Christina Lunde
Sgr 19h6'28" -13d30'4"
Christina Lynda
Cas 23h57'20" 53d59'22"
Christina Lynn
Leo 10h23'13" 21d49'43"
Christina Lynn Bohner
And 0h35'13" 29d30'30"
Christina Lynn Harley's Star
Lib 15h16'0" -16d25'6"
Christina Lynn Kolodziej
Ari 2h53'31" 22d18'23"
Christina Lynn Lucas
Uma 12h19'46" 56d8'38"
Christina Lynn McDaniel
Sgr 18h22'4" -18d21'38"
Christina Lynn Root
Ari 3h28'52" 26d47'2"
Christina Lynne Champ
And 0h31'49" 40d19'59"
Christina M. Ackerman
Ori 5h31'17" 14d21'15"
Christina M. Bradley
Cyg 19h54'11" 31d4'13"
Christina M Byrd
Cam 3h57'42" 65d16'53"
Christina M. Camillone
Cyg 20h0'58" 45d2'51"
Christina M. Crofoot
Vir 11h39'34" -4d53'15"
Christina M. Havrilla
Aqr 20h44'28" -9d43'55"
Christina M. Kochan # 1 Mom
Lyn 7h48'38" 59d28'54"
Christina M Parapon
Gem 7h12'33" 16d21'32"
Christina M Riedel
Cnc 8h7'42" 17d0'23"
Christina M. W. Beilenson
Tau 4h14'12" 2d42'36"
Christina M. Wagner
Tau 4h28'3" 15d8'17"
Christina M. Woodward
Ari 2h52'2" 25d34'45"
Christina Mae Anderson
And 0h42'55" 40d33'10"
Christina Maiorano
Lyr 18h40'10" 40d51'14"
Christina Major
Ori 5h36'58" 4d10'17"
Christina Maloomian "02/09/1992"
Aqr 20h44'28" -10d41'5"
Christina "Manxy" Corkhill
Cam 4h7'28" 53d47'4"
Christina Margaret Bone
Cyg 21h43'26" 44d34'49"
Christina Maria
Ori 5h37'14" 11d18'11"
Christina Maria
Del 20h16'34" 14d30'57"
Christina Maria Derer
Vir 12h22'46" 13d2'52"
Christina Maria LeGrand
Ari 1h47'16" 19d3'56"
Christina Maria Lippi
Cyg 19h57'16" 39d7'44"
Christina Maria Pasquale
Psc 23h13'24" -0d50'37"
Christina Maria Rudnytzky
Cas 23h10'34" 58d53'52"
Christina Maria Swomley
Cas 1h2'35" 56d59'54"
Christina Marie
And 0h58'1" 44d14'31"
Christina Marie
Lyr 18h53'14" 34d16'34"
Christina Marie
Psc 1h12'55" 20d28'44"
Christina Marie
Ori 5h39'37" 4d17'56"
Christina Marie
Ari 2h22'27" 24d26'28"
Christina Marie
And 0h28'11" 28d20'46"
Christina Marie
Ori 5h57'3" 18d42'1"
Christina Marie
Cas 1h25'20" 62d2'23"
Christina Marie
Cas 0h46'46" 62d56'0"
Christina Marie
Cyg 20h29'28" 59d7'46"
Christina Marie
Lyn 6h50'12" 61d44'25"
Christina Marie
Ori 5h30'28" -7d49'2"
Christina Marie
Ori 5h38'35" -2d32'52"
Christina Marie
Eri 3h46'26" -1d8'27"
Christina Marie
Cap 20h29'0" -20d22'20"
Christina Marie Bilton
Aqr 21h42'12" -2d4'42"
Christina Marie Book
Cas 0h13'41" 59d12'32"
Christina Marie Boydston
Lib 14h45'6" -19d55'37"
Christina Marie Campbell
Cas 0h21'24" 64d1'15"
Christina Marie Cavallaro
Cas 1h43'1" 64d2'1"
Christina Marie Cecelia Vermiglio
Cnc 8h41'21" 17d45'14"
Christina Marie Daber
Psc 1h6'50" 10d44'24"
Christina Marie Devingo
And 1h35'30" 46d21'38"
Christina Marie Fernandez
Aqr 22h46'4" -24d20'11"
Christina Marie Fitzpatrick
Lib 14h24'48" -10d41'44"
Christina Marie Frenette
And 0h17'23" 37d59'38"
Christina Marie Gibson
Uma 12h54'7" 62d25'3"
Christina Marie Glover
Cap 21h35'32" -14d28'11"
Christina Marie Harris
Lyn 8h11'32" 54d5'48"
Christina Marie Harris
Tau 3h45'23" 28d35'40"
Christina Marie Hertel
Mon 6h42'54" -0d26'46"
Christina Marie Honner
Cyg 19h59'48" 37d49'11"
Christina Marie Hovanec
And 1h44'32" 46d50'35"
Christina Marie Howard
Tau 4h59'0" 25d52'34"
Christina Marie Kent
Ari 3h6'59" 29d30'48"
Christina Marie Krancic
And 2h31'7" 39d35'6"
Christina Marie Legendre
Cas 2h19'43" 73d7'42"
Christina Marie Lepe (Chrissy Bear)
Psc 1h13'11" 13d33'3"
Christina Marie Love
Aqr 21h58'10" 1d37'15"
Christina Marie Miners
And 0h23'52" 39d46'11"
Christina Marie Monteagudo
Cap 21h31'25" -11d1'25"
Christina Marie Olga Manolis
Lyr 18h57'50" 35d41'18"
Christina Marie Poggioli
Psc 1h21'34" 31d17'7"
Christina Marie Prewitt
Cnc 8h17'4" 23d33'4"
Christina Marie Santos
Lib 15h7'22" -7d38'33"
Christina Marie Scullari
Cas 23h25'2" 53d34'51"
Christina Marie Shock
Psc 23h34'56" 2d12'42"
Christina Marie Short
Gem 7h20'27" 31d20'12"
Christina Marie Skidmore
And 2h8'49" 44d16'15"
Christina Marie Trimeloni
Gem 7h28'41" 20d47'35"
Christina Marie Vanholt
Tau 4h26'24" 28d10'20"
Christina Marie Williams
And 0h35'34" 42d48'46"
Christina Marie Wright
And 1h43'21" 41d55'52"
Christina Marie Yarbrough
Aqr 22h27'47" 1d43'5"
Christina Martell
Equ 21h5'49" 9d33'51"
Christina Martinez
Cyg 21h31'6" 37d44'38"
Christina Mary Marciano
Sgr 18h44'29" -29d11'8"
Christina Massey
Lyn 9h18'59" 37d54'0"
Christina Mayer
Cnc 8h52'43" 9d40'14"
Christina Medlin
Per 4h36'46" 48d52'2"
Christina Merl
And 1h7'5" 46d40'11"
Christina Michele Krecthun
Psc 1h4'21" 21d57'10"
Christina Michella
And 0h37'32" 45d55'24"
Christina Michelle
Ori 5h1'53" 15d0'29"
Christina Michelle Corea
Leo 11h55'27" 24d28'37"
Christina Michelle Cox
Cam 3h59'8" 59d34'49"
Christina Miller 1986
Vir 14h27'31" -0d55'22"
Christina mit Dirk und Vanessa
Uma 9h35'14" 69d16'45"
Christina Montgomery
Ari 2h55'35" 17d52'22"
Christina Muller
Leo 9h29'44" 11d8'31"
Christina "My Love for Life"
Psc 1h24'31" 27d49'1"
Christina - My Star In The Sky
Lib 14h28'38" -17d27'27"
Christina N Kamm-Carpenter AGD
Vir 13h0'50" 5d38'56"
Christina Nelson
And 23h14'15" 47d24'45"
Christina Nelson
Mon 6h46'24" -0d3'24"
Christina & Nicholas Walker-Potts
Crb 16h11'6" 35d0'5"
CHRISTINA NICOLE
Tau 5h12'54" 27d58'49"
Christina Nicole Besheer
Psc 1h21'19" 23d27'27"
Christina Nicole Herrera
Psc 1h37'18" 11d3'30"
Christina Nicole Smith
Sgr 18h40'34" -22d5'3"
Christina Nicole Uvalle
Sgr 19h28'21" -14d10'50"
Christina Nicole Warner
Gem 6h52'29" 16d19'22"
Christina Nicole Witte
Psc 1h40'9" 5d43'35"
Christina Nilson
Ori 5h37'17" -2d10'25"
Christina Niston
Uma 11h33'8" 54d1'7"
Christina Noelle
Cas 1h36'20" 68d16'35"
Christina Ongoma
Sco 17h41'49" -38d37'56"
Christina Opimo Sharit
Sgr 19h7'47" -18d3'26"
Christina Palmieri
Cyg 21h16'4" 52d53'59"
Christina Pappas
Crb 16h12'16" 37d16'37"
Christina Pearston
Gem 7h47'43" 31d33'53"
Christina Peichler
Umi 16h16'18" 79d43'43"
Christina Perez SB
And 1h54'31" 38d29'0"
Christina Perry
Lib 15h7'56" -22d16'40"
Christina & Peter Hadgimallis
Gem 7h39'24" 24d1'6"
Christina Peterson
Lib 15h18'32" -4d26'31"
Christina Petrovits
Cnc 8h53'34" 30d37'1"
Christina Piotrowski
And 2h15'0" 50d10'27"
Christina Pitcher
Crb 15h30'52" 28d48'35"
Christina Porto
And 0h31'29" 42d11'15"
Christina Prate
Gem 7h3'49" 32d42'29"
Christina Prime
Tau 4h10'14" 4d13'9"
Christina R. O'Neill-Redmond
And 1h43'26" 41d50'0"
Christina R. Peterek
Com 12h48'44" 28d4'34"
Christina Rae Follis
Psc 1h37'34" 15d43'14"
Christina Ramirez
Ori 4h50'23" 13d19'36"
Christina Rancke
Vir 14h37'41" -0d19'48"
Christina Raye
Aqr 22h34'5" -1d59'23"
Christina Refugia Rubio
Crb 15h42'35" 38d36'40"
Christina Reiss
Sgr 19h32'28" -13d52'0"
Christina Rene Shaffer
Ori 5h37'48" -6d5'39"
Christina Renee'
And 1h17'30" 38d51'25"
Christina Renee Michaud
Aqr 23h13'58" -13d23'5"
Christina Ress Spyratos
Cam 3h29'52" 60d59'43"
Christina & Rey
Gem 6h47'40" 32d18'34"
Christina Reyes
Gem 7h5'25" 22d38'51"
Christina Robedeau
And 23h53'12" 45d35'9"
Christina Rosales
Vir 12h6'14" -11d8'1"
Christina Rose
Cap 21h21'52" -19d53'43"
Christina Rose
Her 17h35'52" 48d56'40"
Christina Rose
Lyn 8h27'36" 35d52'50"
Christina Rose
And 0h28'12" 43d0'46"
Christina Rose
Lyn 7h33'14" 39d10'30"
Christina Rose Conlon
Cap 21h19'4" -16d42'52"
Christina Roseann Cameron
And 0h35'17" 27d20'35"
Christina Rudolf
Cas 1h19'15" 67d53'48"
Christina Rueck
Psc 1h43'20" 24d33'40"
Christina Sachs
Uma 8h42'49" 56d44'8"
Christina Saman
And 2h16'8" 46d25'2"
Christina Schneider
Gem 7h4'2" 35d0'41"
Christina Schoenknecht
Ori 5h55'37" 18d21'7"
Christina Schreiner
Sco 16h15'24" -9d56'8"
Christina Sellitto
Cam 4h29'38" 65d58'20"
Christina Silbermann (Gissi)
Tri 1h46'17" 31d47'8"
Christina Simpkin - Angel of Love
And 1h28'29" 33d57'45"
Christina Skye Wilson
Aqr 23h1'21" -22d21'7"
Christina Snyder
Sgr 18h15'48" -34d56'1"
Christina & Sophia Horn Eternally
Ari 2h41'44" 26d26'59"
Christina Stavrou My Love
And 2h15'42" 47d47'14"
Christina Suzanna
Cnc 7h55'51" 10d59'12"
Christina Sykora's Birthday Star
Leo 9h47'58" 26d22'16"
Christina Sylvia
Cam 7h50'26" 66d42'49"
Christina T. Bates
Aqr 23h31'3" -22d38'6"
Christina Tesoro
Cap 20h25'45" -9d27'33"
Christina The Beautiful
Leo 10h30'37" 22d41'25"
Christina Theresa Picciano
Uma 9h25'0" 71d2'26"
Christina Tilghman
Aqr 20h48'26" -8d54'16"
Christina Treglia
Sco 17h52'6" -36d35'30"
Christina Tresca
Leo 9h30'24" 19d16'1"
Christina Trupia
Her 18h35'54" 15d35'29"
Christina VanMoorsel
Cap 20h52'29" -25d38'54"
Christina Viciconte
Lib 16h0'52" -6d55'40"
Christina Victoria Lee
Cas 0h37'40" 62d26'31"
Christina Victoria Patane
Del 20h41'36" 4d14'47"
Christina Vo
Her 16h42'49" 48d15'28"
Christina Von Marie Schneider
Her 17h6'12" 43d54'32"
Christina Vosbikian
Psc 1h32'45" 5d25'6"
Christina Voß
Uma 9h18'21" 64d57'33"
Christina Waszak Love Of My Life
Leo 11h44'11" 25d14'7"
Christina Waynette Paxton
Psc 0h45'51" 14d43'28"
Christina Weeks
And 0h45'5" 39d41'58"
Christina Weston Wise
Cam 4h34'42" 63d57'40"
Christina White Passariello
Gem 6h26'49" 27d14'12"
Christina Whitson
Umi 16h56'42" 85d23'58"
Christina Willey
Uma 10h19'55" 49d39'59"
Christina Williams
Cap 20h34'19" -27d15'23"
Christina Willis
Psc 1h11'40" 24d38'14"
Christina Winget
Sgr 18h10'8" -19d41'20"
Christina Wolfe
Ori 5h29'31" 7d16'8"
Christina Yanushewski
Sco 17h45'5" -37d14'3"
Christina Zedeker
And 1h33'13" 39d24'25"
Christina, Happy Birthday,Love Matt
Crb 16h2'53" 36d19'26"
Christina122468
Lyn 8h21'31" 48d42'32"
ChristinaCarbonaro
Uma 11h5'6" 42d4'27"
Christina-Marie
Uma 9h48'2" 43d18'6"
Christina-McKenzie-Casey
And 2h4'29" 37d13'15"
Christin-Andreas
Ori 6h21'55" 16d17'20"
ChristinaRobyn
Leo 10h29'32" 17d56'41"
Christina's Aspirations
And 1h19'24" 42d57'57"
Christina's Bear
Lib 15h11'35" -4d53'32"
Christina's Beauty
Sco 16h8'5" -19d7'40"
Christina's Bridge
Aqr 22h6'44" 2d27'39"
Christina's Dream
Tau 4h5'43" 5d2'36"
Christina's Enduring Love
Uma 11h18'55" 61d16'57"
Christina's First Mother's Day
Sgr 18h33'31" -23d15'10"
Christina's Forever Wish
Aqr 23h41'4" -8d30'6"
Christina's Gone with the Wind
Crb 15h46'52" 25d52'1"
Christina's Grace
Aqr 21h49'21" -1d23'58"
Christina's Heaven
Uma 11h49'51" 56d18'54"
Christina's Kiss
Cam 3h39'16" 67d7'24"
Christina's Shining Star
Vir 13h7'50" 12d1'41"
Christina's Smile
Gem 6h46'17" 14d43'2"
Christina's Star
Cnc 8h53'48" 13d59'58"
Christina's Star
Gem 6h21'20" 17d41'12"
Christina's Star
And 0h48'49" 25d24'39"
Christina's Star
Ari 2h40'19" 27d8'50"
Christina's Star
And 1h7'58" 44d8'35"
Christina's Star
Cas 1h39'8" 62d40'10"
Christina's Star
Aqr 21h45'24" -1d44'58"
Christina's Star
Sco 16h52'46" -37d48'41"
Christina's Star of Hope
Uma 9h9'26" 55d9'37"
Christina's Sunshine
Uma 11h52'7" 38d12'33"
Christina's Super Star
Uma 14h27'35" 57d26'45"
Christina's wishing star
Sco 16h36'13" -37d10'12"
Christine
Cap 20h54'26" -27d2'26"
Christine
Sgr 18h9'21" -22d44'0"
Christine
Col 5h49'19" -27d34'15"
Christine
Sco 17h24'8" -41d48'59"
Christine
Dra 16h43'26" 54d55'5"

Christine
Cyg 20h31'21" 58d12'41"
Christine
Cas 22h59'38" 54d13'40"
Christine
Cas 23h51'51" 58d49'19"
Christine
Cas 23h7'14" 56d49'38"
Christine
Uma 9h1'26" 54d18'10"
Christine
Lyn 8h21'47" 59d16'52"
Christine
Uma 8h44'36" 59d50'42"
Christine
Uma 10h33'47" 59d10'18"
Christine
Uma 10h16'53" 56d53'38"
Christine
Cas 1h26'42" 66d54'58"
Christine
Cam 7h59'30" 73d0'8"
Christine
Cas 1h29'17" 72d42'44"
Christine
Umi 15h20'56" 70d15'14"
Christine
Dra 19h30'12" 68d26'19"
Christine
Lib 14h51'58" -3d32'51"
Christine
Lib 14h59'47" -0d58'49"
Christine
Umi 13h37'50" 76d24'42"
Christine
Aqr 21h14'1" -8d50'53"
Christine
Cap 20h26'5" -13d36'7"
Christine
Lib 15h12'37" -9d36'35"
Christine
Sco 16h6'0" -21d33'25"
Christine
Lib 15h6'30" -21d37'13"
Christine
Cap 21h12'31" -15d7'40"
Christine
Cap 21h10'38" -18d58'42"
Christine
Cap 21h0'44" -19d32'0"
Christine
Cap 21h45'40" -17d51'42"
Christine
Sgr 19h2'57" -18d1'8"
Christine
Uma 11h17'17" 42d30'33"
Christine
Uma 10h18'6" 44d53'46"
Christine
And 1h41'24" 44d25'43"
Christine
And 0h51'45" 40d41'19"
Christine
And 0h43'30" 38d39'55"
Christine
And 0h28'41" 41d46'6"
Christine
And 0h36'40" 39d47'49"
Christine
Cyg 21h23'45" 33d6'7"
Christine
Gem 6h53'33" 32d17'14"
Christine
Cnc 8h53'44" 31d46'28"
Christine
Crb 15h47'32" 32d3'52"
Christine
Lyn 8h32'54" 50d14'35"
Christine
Lyn 7h44'18" 45d23'7"
Christine
Lyn 7h14'29" 46d53'30"
Christine
Cyg 20h31'45" 42d19'50"
Christine
Cnv 12h42'18" 48d6'34"
Christine
And 23h18'41" 46d5'41"
Christine
Cam 5h56'4" 56d20'12"
Christine
Psc 1h39'36" 27d52'38"
Christine
Gem 6h29'33" 15d18'5"
Christine
Gem 6h31'42" 18d29'0"
Christine
Leo 11h7'21" 15d50'51"
Christine
Leo 11h9'48" 16d47'45"
Christine
Com 12h48'19" 21d29'42"
Christine
Leo 9h58'21" 19d1'52"
Christine
Cnc 8h40'42" 16d59'57"
Christine
Crb 15h29'45" 27d45'41"
Christine
Leo 9h32'8" 27d37'30"
Christine
Leo 9h42'39" 26d30'53"
Christine
Gem 7h13'21" 25d26'51"
Christine
Tau 3h45'52" 27d51'11"
Christine
Ori 6h4'25" 16d19'5"
Christine
Tau 3h34'56" 18d29'54"
Christine
Psc 22h51'57" 7d46'27"
Christine
Peg 21h27'44" 3d27'37"
Christine
Aqr 21h13'13" 1d34'25"
Christine
Ori 5h9'37" 8d6'16"
Christine A. DeBacco
And 1h12'6" 34d32'16"
Christine A. Henault
Lyn 7h53'15" 46d49'44"
Christine A. Nagy
Uma 8h58'44" 53d19'50"
Christine Abrams
Cnc 8h10'31" 10d29'23"
Christine Adams
Tau 4h18'52" 17d23'17"
CHRISTINE aka KINA
Lib 14h29'47" -12d40'29"
Christine Alinen
Sco 16h53'13" -42d20'21"
Christine Amber Brokate
Per 3h11'2" 49d52'8"
Christine Amelia Magdalina
Heen
Del 20h49'44" 14d46'32"
Christine and David
Col 6h0'58" -35d22'21"
Christine and Glenn
Ritzenthaler
Sgr 19h23'15" -22d21'54"
Christine and Moua
Cyg 21h8'27" 47d1'58"
Christine and Nick's Star
Psc 1h0'3" 16d42'6"
Christine Anderson
"Sunshine"
Lyn 9h41'6" 41d17'49"
Christine Andrews Morey
Lib 15h31'0" -18d47'59"
Christine Angela
DeGiacomo-Kimball
Equ 21h7'38" 11d48'31"
Christine (Angelic and
Beautiful)
Cyg 19h59'45" 30d18'55"
Christine Angelos
Sco 17h44'47" -36d1'59"
Christine Ann
Peg 22h39'57" 14d10'48"
Christine Ann Ayres
Uma 8h41'3" 50d15'6"
Christine Ann Bozzone
Uma 9h24'46" 43d49'23"
Christine Ann Capelouto
Lib 15h30'17" -16d15'19"
Christine Ann Edwards
And 23h25'19" 52d29'44"
Christine Ann Giglio
And 23h13'51" 43d27'14"
Christine Ann Larsen
Cas 2h12'12" 64d11'55"
Christine Ann Makuh
Cas 1h59'41" 63d41'26"
Christine Ann Roomet
Aqr 23h8'42" -17d29'51"
Christine Ann Shepherd
And 0h14'49" 44d30'23"
Christine Ann West
Leo 11h8'17" 22d10'19"
Christine Anne Lowe
Vir 13h32'9" -10d50'38"
Christine Anne Seeburger
Cap 21h56'27" -23d4'49"
Christine Anthony
Cam 5h15'21" 66d58'1"
Christine Ariel Pochiluk
Aqr 21h12'55" 0d54'7"
Christine Asbach
Uma 11h6'53" 42d6'44"
Christine Ashley
Sco 16h35'17" -38d7'53"
Christine Aspinall Taylor
And 2h10'33" 38d4'34"
Christine Bailey
Crb 16h12'28" 37d52'40"
Christine Barreto
Cas 0h23'44" 51d18'19"
Christine Baur
Aqr 22h11'33" -19d31'56"
Christine "Bean" Abram
Cyg 21h43'45" 55d14'59"
Christine Belko
Cap 20h20'1" -13d29'39"
Christine Berglind
And 2h12'7" 38d29'45"
Christine Bernadette
Thayer
Vir 13h30'53" 10d56'28"
Christine Beth Hamilton
Smith
Mon 6h51'42" -0d49'17"
Christine Biederer
Lyn 7h40'6" 37d11'26"
Christine "Bivy" Kaczynski
Lmi 10h13'6" 34d38'59"
Christine Blalock
Gem 7h23'26" 20d2'45"
Christine Blanco
Lmi 11h4'41" 31d2'8"
Christine Bokel
Uma 9h25'41" 48d23'39"
Christine Boustani
Lib 15h58'22" -8d16'56"
Christine Boyle Mitchell
Ari 2h2'34" 14d15'59"
Christine Braund
Psc 1h15'45" 16d19'48"
Christine & Brian's Star
Psc 0h56'39" 10d39'7"
Christine Brown
And 0h47'25" 43d5'48"
Christine Brunner Foley
And 23h9'31" 43d15'46"
Christine Bühner 250965
Uma 14h17'22" 59d44'46"
Christine Burns
Cyg 19h40'40" 32d0'25"
Christine C. Miller
Lep 5h13'26" -12d39'5"
Christine Caiafa Guarino
Uma 9h27'21" 46d19'43"
Christine Cantatore
Sco 16h52'52" -28d47'54"
Christine Capote
And 0h14'15" 29d12'4"
Christine Carlson Kenny
Tau 5h55'41" 25d22'44"
Christine Carney
Crb 16h1'9" 29d12'6"
Christine Caroline Mandjik
Cap 20h41'0" -18d25'12"
Christine Carr
Com 12h49'44" 26d17'39"
Christine Caruso-Italian
Batallion
Tau 4h16'2" 2d35'45"
Christine Carvara
Cap 20h9'59" -22d16'47"
Christine Cassolino
Uma 10h52'58" 63d51'49"
Christine Cattie
Cap 20h23'37" -27d18'56"
Christine & Cesar Forever
Lyr 18h49'50" 37d1'32"
Christine Chambers
Cas 1h21'53" 57d10'43"
Christine Cheevers
Tau 4h40'22" 24d50'11"
Christine Childress
Gem 6h55'56" 19d36'54"
Christine Clark
Cas 0h1'1" 53d45'52"
Christine Clark
And 1h44'26" 39d8'29"
Christine "Clever Crow"
Geisler
Uma 10h29'19" 67d15'44"
Christine Clugston
Sco 17h27'38" -41d27'59"
Christine Colarusso
And 23h3'57" 47d44'35"
Christine Collier
Ori 5h57'54" 18d11'37"
Christine Compton
Cas 0h23'11" 52d24'13"
Christine Confalone
Aqr 22h54'16" -22d17'19"
Christine Cowburn
And 1h15'38" 34d24'0"
Christine Cretin Gogniat
Her 16h44'43" 41d35'38"
Christine Crombie
Lib 14h35'38" -13d42'1"
Christine Cute Face
And 23h6'17" 45d7'20"
Christine D. Alexy
Lib 14h50'46" -14d0'25"
Christine D Price
Cap 21h52'12" -13d21'58"
Christine Denise Breaux
Vir 12h35'30" -2d19'37"
Christine Denise Harris
Cap 21h23'45" -20d17'27"
Christine Denise Souza
Fidalgo
And 0h33'19" 42d52'57"
Christine Denker
Cas 1h43'30" 64d27'21"
Christine Dianne Ellis ,
Angel
Aqr 21h39'41" 0d40'16"
Christine DiLauro
Cyg 20h19'38" 59d22'0"
Christine D'Lea Shanks
Cnc 8h13'53" 10d45'25"
Christine & Domenic
Sge 20h18'11" 18d51'57"
Christine & Donald
Sco 16h8'49" -14d38'45"
Christine Donohue
Lyn 8h15'9" 35d33'54"
Christine Dorico
And 0h40'25" 28d0'2"
Christine Doyle
Ori 5h11'48" 7d18'58"
Christine Drapkin
Lyn 6h55'24" 51d43'29"
Christine Dreier
Lyn 8h5'53" 55d35'49"
Christine Duffy
Cyg 20h30'25" 30d11'41"
Christine E. Alexander
Cmi 7h22'40" -0d2'20"
Christine E. Brown
Cap 21h53'1" -18d21'27"
Christine E. Davistar
And 0h46'33" 39d16'13"
Christine E Milton
Gem 6h16'30" 25d16'51"
Christine Edel
Cas 0h44'17" 61d50'44"
Christine Eidolon Stanly
042904-NOW
Lyr 18h43'33" 40d55'11"
Christine Elaine
Psc 0h39'7" 18d9'19"
Christine Elaine Thomas
Cyg 20h9'44" 42d52'26"
Christine Elektra
Lyn 7h33'49" 50d52'20"
Christine Elise
Uma 10h2'15" 68d29'10"
Christine Elise Rodriguez
Cyg 19h47'13" 34d17'49"
Christine Elizabeth
Aqr 22h39'10" 0d51'50"
Christine Elizabeth
Aqr 21h47'4" -2d20'7"
Christine Elizabeth
Crawford
Peg 21h58'1" 34d55'8"
Christine Elizabeth Hayes
Cas 2h33'50" 67d28'43"
Christine Elizabeth Holmes
Uma 13h21'28" 55d35'52"
Christine Elizabeth Hudnall
Sco 16h5'25" -17d5'19"
Christine Elizabeth Janek
Psc 0h14'15" 9d15'41"
Christine Ellen Sorrell
Crb 15h35'46" 36d18'35"
Christine Ellis Ames
Cas 2h26'27" 63d42'1"
Christine Emily Mueller
And 23h15'4" 40d35'18"
Christine Erin
Psc 23h50'22" 7d10'28"
Christine et Laurent
Cas 1h0'17" 67d17'21"
Christine Farkas
Cas 23h36'37" 54d28'20"
Christine Fauneil
Cas 1h31'2" 67d44'46"
Christine Ford
Cyg 20h26'25" 48d47'46"
Christine Foronda
Cyg 19h42'57" 30d18'11"
Christine Fortin
Per 3h29'41" 36d36'36"
Christine Fox
Lyn 6h19'7" 60d4'38"
Christine Freije
Cap 21h12'4" -18d57'48"
Christine G. Meyer
Aqr 22h48'44" -8d8'28"
Christine Gagliardi
And 0h41'16" 44d12'30"
Christine Gainsborough
And 2h36'8" 46d51'56"
Christine Gallagher
Cyg 20h14'32" 31d56'27"
Christine Geneva Starkey
And 0h32'40" 46d8'42"
Christine Gerchow
Umi 15h15'42" 73d11'37"
Christine Grosso
Aqr 22h37'51" 0d54'47"
Christine Guarino
Cas 0h26'19" 59d50'8"
Christine Guisinger's
Princess Star
And 0h29'59" 32d22'24"
Christine H. Patten
And 2h24'49" 50d18'31"
Christine Hagaman
Sco 17h24'26" -38d22'56"
Christine Hamms
And 1h11'3" 45d37'10"
Christine Helen Mcgaffic
Crb 16h6'0" 32d27'24"
Christine Hierl
Vul 20h39'43" 25d49'26"
Christine Honestus
Cnc 8h49'45" 30d47'9"
Christine Hübner
Ori 6h13'0" 7d8'9"
Christine Huda
And 0h24'7" 45d36'51"
Christine I. Rivera
Tau 5h42'17" 22d51'2"
Christine "Ickle Princess"
Lawrence
And 23h20'23" 51d56'53"
Christine & Jeff Wiebe
Love 4ever
Cyg 21h37'1" 44d44'29"
Christine & Jeffrey
Cnc 8h12'31" 25d3'57"
Christine Jester
Lyn 7h43'8" 40d52'16"
Christine Joan Glowacka
And 23h22'59" 48d14'26"
CHRISTINE JOANN
Psc 0h31'21" 8d32'2"
Christine Joanne Williams
Cyg 20h34'23" 46d6'42"
Christine & John Quirk
Umi 16h28'53" 76d45'39"
Christine June Pipke
Cas 2h47'4" 58d7'11"
Christine K. Deloney
And 1h44'39" 46d8'57"
Christine Kangas
Gem 7h38'49" 31d49'11"
Christine Karen Stepanian
10-1-84
Lib 14h48'32" -12d29'59"
Christine Kathleen Doren
And 2h10'13" 43d33'53"
Christine Kay Evans
And 0h19'16" 28d11'26"
Christine Kennedy
Cyg 19h41'11" 34d4'29"
Christine Kennedy
Uma 10h19'40" 50d21'44"
Christine Keyaerts
14/02/1957
Aqr 20h54'56" -10d33'8"
Christine King
And 0h48'33" 35d24'24"
Christine Kourik Maloney
Cyg 19h57'28" 47d31'24"
Christine Kuczinski
Leo 9h59'7" 7d35'54"
Christine L. Alfonso
And 2h20'46" 48d59'11"
Christine L. Gaglione
Mon 8h8'4" -0d55'36"
Christine L. Hooten
Mon 6h46'44" -0d18'49"
Christine L. Houser
Vir 12h12'30" 2d51'46"
Christine L. Steben
Cam 4h44'29" 58d23'38"
Christine Lambert
Orfanedes
Ori 5h54'19" 10d48'15"
Christine Laurie
Ori 6h3'21" 18d37'41"
Christine LaVerne Hooper-
Davis
Psc 0h37'19" 12d52'26"
Christine Le Pottier
Gem 6h42'46" 23d13'13"
Christine Lee Bell
Ser 18h31'13" -0d51'25"
Christine Lee Smith
Leo 9h48'41" 24d35'51"
Christine Lemire
Sgr 18h42'59" -27d58'36"
Christine Letendre
Lib 15h15'24" -20d12'36"
Christine Louisa Skippen
And 2h23'31" 47d52'2"
Christine Louise Cram
LaBarre
Aqr 23h2'30" -9d46'33"
Christine Louise Lemonds
Aqr 22h6'12" -8d54'59"
Christine 'Luna' Ellis
Cnc 9h3'2" 27d21'41"
Christine Luther
Her 17h54'21" 27d41'14"
Christine Lynn
Gem 7h38'12" 34d18'41"
Christine Lynn Carpenter
Gem 6h49'33" 21d12'57"
Christine Lynn Moyer
Cet 3h17'58" -0d44'37"
Christine Lyons
Leo 11h24'46" 9d44'28"
Christine M. Leyva
Vir 12h37'6" 0d2'8"
Christine M. Manarite
Psc 1h4'29" 32d59'7"
Christine M. Osborn
Cnc 8h23'46" 23d30'42"
Christine M. Pepe
Aqr 21h2'7" -7d20'43"
Christine M. Rielly
Cap 21h2'21" -19d5'14"
Christine M. Rutherford
Ari 2h54'20" 15d0'0"
Christine Magee
Cas 1h28'46" 53d27'45"
Christine Manuel
Uma 9h29'25" 54d20'38"
Christine Marckese
Cnc 8h33'41" 8d59'58"
Christine Margaret
Schramm
Mon 6h58'34" 9d48'9"
Christine Marie
Ori 6h16'51" 10d27'34"
Christine Marie
And 0h13'35" 30d12'49"
Christine Marie
Uma 11h56'24" 49d36'28"
Christine Marie
And 2h20'1" 46d37'22"
Christine Marie
And 2h38'9" 50d11'5"
Christine Marie
And 23h24'49" 41d36'57"
Christine Marie
And 0h33'37" 41d5'54"
Christine Marie 143
Cap 21h36'1" -13d39'9"
Christine Marie (Angelini)
Wright
Psc 1h11'44" 27d26'28"
Christine Marie Benskin
And 0h46'42" 40d2'2"
Christine Marie Fellows
Emmons
Uma 13h52'26" 50d11'7"
Christine Marie Gagnon
Leo 9h42'43" 26d17'12"
Christine Marie Giang
Sgr 18h5'39" -30d36'12"
Christine Marie Grell
Lyn 8h35'31" 46d0'11"
Christine Marie Hanna
"FyrFlwr"
Sco 17h17'39" -32d12'10"
Christine Marie
Hollendorfer
And 23h33'22" 48d5'6"
Christine Marie Joyner
Psc 0h21'8" 7d35'40"
Christine Marie Kathleen
Johnstone
Gem 7h16'50" 20d15'29"
Christine Marie Koch Malik
Uma 11h21'54" 63d39'46"
Christine Marie Lefevre-
Blair
Uma 9h47'33" 50d2'42"
Christine Marie Marobella
Uma 9h27'29" 44d12'6"
Christine Marie McAdams
And 23h24'10" 47d15'30"
Christine Marie Miller
Sco 17h26'19" -40d54'26"
Christine Marie Nunez
Ari 3h4'3" 26d7'52"
Christine Marie
O'Shaughnessy
Cnc 8h45'34" 28d1'20"
Christine Marie Rexroad
Tau 4h20'25" 13d12'36"
Christine Marie Rice
Aqr 20h44'14" -8d57'55"
Christine Marie Rodriguez
Leo 11h38'35" 26d53'18"
Christine Marie Rose
Chevalier
Umi 16h14'2" 81d15'5"
Christine Marie Schmidlin
Sgr 19h28'23" -18d29'20"
Christine Marie Schwander
Crb 15h36'20" 31d2'16"
Christine Marie Serson
Aqr 22h54'9" -15d14'14"
Christine Marie Shannon
Crb 15h49'3" 28d12'52"
Christine Marie Stultz
Uma 12h18'41" 54d0'27"
Christine Marie Swain
And 24h0'0" 36d29'32"
Christine Marie Valdez
Uma 11h10'42" 36d14'27"
Christine Marie Zurvalec
And 23h7'58" 45d58'38"
Christine Marks
Tau 3h53'12" 20d12'4"
Christine Mary
Cap 20h39'4" -20d4'21"
Christine Mary Geiser
Cnc 8h15'25" 6d48'51"
Christine Mary Hill
Vir 12h35'50" -0d23'52"
Christine Mary Pistillo
And 0h45'42" 40d2'54"
Christine Mary Simon
Cas 1h22'55" 57d2'19"
Christine Mary Stokes
Uma 11h3'56" 50d34'25"
Christine Masko
And 1h52'40" 36d10'17"
Christine Matteson
Psc 23h8'14" 7d41'48"
Christine Mc 10 28 51
And 23h20'9" 51d57'23"
Christine McCafferty M P
Her 17h56'43" 21d27'50"
Christine McCaughey
Watkins
Uma 12h24'14" 56d51'37"
Christine McDonald
Cas 1h20'5" 69d37'3"
Christine McDonald
And 0h19'1" 29d3'39"
Christine McGrory
Cas 1h18'57" 51d58'2"
Christine Meredith
Gallagher
Crb 16h6'41" 39d17'46"
Christine & Michael
Tau 4h4'14" 6d30'42"
Christine Michele Jawor
Lib 15h22'40" -27d13'53"
Christine Michelle Nicoletti
Lyn 7h42'9" 57d25'57"
Christine Misencik Bunn
Lyr 18h20'42" 36d28'31"
Christine Moore
Psc 0h5'50" -0d9'11"
Christine Mounayar
Lyn 7h48'36" 52d38'43"
Christine Muller
Cyg 20h44'59" 41d49'37"
Christine "MY LOVE"
Moeller
Gem 6h27'28" 17d33'5"
Christine - My Soulmate
Cru 12h7'51" -63d42'20"
Christine Nakazawa
Cyg 21h29'25" 31d13'29"
Christine Nicole
Crb 15h39'20" 27d2'32"
Christine Nicole Boland
Lmi 10h50'49" 29d21'50"
Christine Nicole Edwards
And 0h59'35" 39d33'35"
Christine Nicole Garner
And 0h4'23" 47d31'55"
Christine Nicole Poor
Leo 11h40'14" 21d38'29"
Christine Nicole Thomas
McClure
Cnv 12h49'12" 42d50'23"
Christine Ninkovich
Ari 1h58'33" 20d28'49"
Christine Noel Kick
Uma 11h8'15" 37d44'42"
Christine Noella
Lyr 18h49'51" 35d43'43"
Christine Norah Terraciano
And 23h2'58" 50d54'35"
Christine P. King
Leo 9h52'7" 22d53'12"
Christine P. McNutt
And 2h6'49" 42d33'34"
Christine Paska
Aqr 22h11'56" 1d49'22"
Christine Petchonka
Uma 8h51'16" 71d6'22"
Christine & Pete Herting
Umi 15h19'46" 73d30'15"
Christine & Peter Lewis'
Star
And 0h27'57" 45d28'9"
Christine Pham
Leo 10h4'44" 15d19'30"
Christine Philippeles MD
96/2006
Uma 9h14'13" 59d15'44"
Christine Pichler
Uma 10h16'44" 72d41'19"
Christine Pierquin
Uma 9h1'49" 48d59'10"
Christine Pillo, Paralegal
Lyn 8h32'15" 41d21'40"
Christine Polonko
Vir 11h47'19" -3d18'37"
Christine Proenza Whitner
And 0h39'45" 38d10'40"
Christine Puhl
Sgr 19h19'22" -23d36'9"
Christine- Queen of Baugh
Lib 15h28'1" -4d39'36"
Christine Quinn Zuendt
Dra 19h42'54" 70d8'16"
Christine Raboni
Cas 0h44'13" 64d22'48"
Christine & Rafael Until
Time Ends!
Lyr 18h45'20" 32d1'43"
Christine Randall
And 23h21'28" 39d14'45"
Christine Rath
Gem 7h19'2" 28d40'18"
Christine Ray Nievaard
Tau 5h39'8" 26d52'3"
Christine Renae Henson
Tau 4h38'26" 23d53'43"
Christine Rene
Psc 0h48'56" 11d28'0"
Christine Renee
Tau 4h40'58" 13d46'10"
Christine Renee Pitera
"Chrissy"
Ari 3h9'25" 28d25'59"
Christine Renee Teeters
Pho 1h9'16" -55d50'41"
Christine Renee Tyson
Tau 3h32'17" 22d52'29"
Christine Reuther
Ari 2h44'36" 20d32'2"
Christine Rey
And 23h10'3" 42d35'22"
Christine Rita Giles
Gem 6h45'19" 15d58'11"
Christine Robin Fairchild
Vir 14h7'23" -16d41'3"
Christine Robin Vaughn
Tau 5h23'40" 18d45'12"
Christine Robinson Belle
Rubisca
Tau 5h50'12" 17d0'29"
Christine Rodgers
Cas 1h24'3" 62d0'10"
Christine Romano
Lyn 6h31'52" 61d2'39"
Christine Rose Craig
Lib 14h33'6" -19d33'2"
Christine Ruch-Hugi
Cyg 19h34'44" 31d35'44"
Christine Salama
Vir 13h3'21" -3d51'41"
Christine Sandoval Gomez
Sgr 18h38'42" -31d7'33"
Christine Sangster
Cap 21h39'24" -15d15'5"
Christine Schalcher
Lac 22h48'35" 52d51'24"
Christine Scheil
Her 16h19'15" 25d30'22"

Christine Schultz God's Angel #1 Cnc 8h34'16" 8d28'51"
Christine Scullin Cas 0h10'55" 53d37'49"
Christine Seiler Cam 5h43'8" 62d51'54"
Christine Shayna Thomas (Christi) Uma 8h26'50" 70d39'59"
Christine Shirtcliff Tau 4h19'51" 27d20'30"
Christine Shroka Uma 11h53'44" 35d45'12"
Christine Sikorski Lib 15h6'40" -22d33'23"
Christine Snoopy Giovanelli Aqr 23h17'34" -17d42'47"
Christine Spichiger Cyg 19h30'59" 30d50'47"
Christine & Spyros Ari 2h50'49" 28d4'38"
Christine Stocco Koeb Gem 7h2'23" 26d40'14"
Christine Strathie Uma 11h29'31" 39d33'12"
Christine Stripey Angel Uma 9h22'10" 46d31'18"
Christine "StrongIsland" DeLota Com 12h33'21" 24d25'47"
Christine Suero Aqr 21h6'36" -9d44'4"
Christine Suh Psc 1h10'42" 29d36'10"
Christine "Sunshine" Marie Poole Crb 15h49'14" 28d7'8"
Christine T Barbadillo Ari 2h19'24" 25d11'49"
Christine T. Gieckel Uma 11h4'46" 55d41'11"
Christine the Alluring Lyn 8h20'37" 36d18'14"
Christine The Love Of My Life Aqr 23h39'58" -22d47'14"
Christine Theresa Hayes Aqr 20h49'47" 1d51'37"
Christine Theresa Votta Her 16h15'39" 24d52'16"
Christine Therese Voss Sgr 17h51'41" -16d28'50"
Christine (Towally) Hodge Aqr 23h11'18" -21d38'26"
Christine Tran Her 17h58'56" 49d51'16"
Christine Tuomey Cas 1h15'21" 54d36'48"
Christine Valarie Vir 14h21'44" 1d21'16"
Christine Vanwart Tau 3h39'49" 3d1'15"
Christine Vavallo Vir 12h50'5" -3d13'21"
Christine Vu-Tran Huynh And 1h0'34" 41d33'4"
Christine W Sgr 19h51'19" -23d8'20"
Christine Wagner And 23h28'17" 38d34'10"
Christine Walsh And 0h39'31" 30d48'29"
Christine Warmbein Ari 2h46'27" 20d39'34"
Christine Wasko Psc 1h34'51" 13d42'44"
Christine Wetherington Leo 9h30'27" 23d55'40"
Christine Wigginton Sco 16h39'16" -35d54'6"
Christine Winbury And 2h28'26" 43d2'57"
Christine Wit Wu Gem 7h51'55" 29d8'14"
Christine Witte Uma 8h35'14" 47d57'43"
Christine Woodbine Leo 11h47'22" 25d41'8"
Christine Wright Cas 23h16'5" 54d25'9"
Christine Y. Pezzulo Gem 6h2'36" 25d40'7"
Christine Yanson Psc 1h27'17" 19d40'23"
Christine Yoakum Tau 3h49'42" 15d34'45"
Christine ~ Yu jah Lyr 18h57'54" 26d6'10"
Christine Yuree Cha Cyg 20h33'41" 36d2'52"
Christine Zack Tau 3h38'3" 25d2'28"
Christine Zurcher Cep 22h56'8" 69d48'41"
Christine Nicole And 23h43'31" 46d52'49"
Christine16 And 23h28'27" 47d34'30"
Christine1968 Crb 16h3'58" 28d9'10"
Christine-Amore Ori 5h52'50" 21d13'55"
Christinemarie Dra 18h27'8" 58d50'58"
Christine's 60th Birthday Cma 6h49'16" -21d51'45"
Christine's and Hector's Love Cyg 20h57'22" 46d40'47"
Christine's Brilliance And 0h27'59" 39d2'50"
Christine's Corner Ari 3h2'56" 25d56'15"
Christine's Glow And 1h5'7" 41d11'1"
Christine's Kaleidoscope Gem 6h50'17" 18d41'29"
Christine's Krümel Uma 11h37'24" 31d40'35"
Christine's Pelican Cove Ori 6h11'32" 15d45'17"
Christine's Shining Star And 0h57'17" 43d52'42"
Christine`s shinnng star Leo 9h45'48" 28d22'33"
Christine's Star And 0h30'13" 37d9'5"
Christine's Star Cyg 20h52'54" 46d50'49"
christine's star of God Mon 7h13'44" -9d5'12"
Christine's Triple "J" Supernova Cnc 8h10'50" 9d54'35"
Christine's Wishing Star Cyg 21h51'33" 40d20'20"
ChristinJuelSmith Christ bearer Uma 12h56'49" 55d40'38"
Christi's Babygirl Vir 13h19'35" 8d24'16"
Christle Dawn Aqr 21h1'15" 0d22'15"
Christmann, Günter Ori 6h16'35" 15d11'16"
Christmas Ori 5h31'56" 5d49'29"
Christmas Cam 3h58'46" 57d58'39"
Christmas Bride Aur 4h50'47" 36d2'24"
Christmas Carol Cap 21h49'45" -13d28'59"
Christmas Carol Miller Cap 20h14'43" -25d55'37"
Christmas Eve Lib 15h53'22" -11d16'27"
Christmas Spirit Lyn 7h57'30" 49d17'28"
Christmas Star Uma 10h41'33" 40d18'33"
Christo Umi 13h4'26" 73d14'2"
Christo Uma 11h36'30" 62d45'33"
Christo Sgr 18h24'2" -26d23'55"
Christo the Magnifico Psc 1h33'19" 21d51'49"
Christof Sgr 17h49'52" -25d44'22"
Christof Blessing Uma 11h48'53" 57d48'40"
Christoff Cep 21h50'1" 62d23'14"
Christoffel Tau 4h17'48" 13d39'43"
Christoffel Johannes Boer Lib 14h25'21" -17d7'23"
Christoffer Vir 13h44'25" 2d9'37"
Christoffer Reiss Psc 1h17'51" 24d56'33"
Christoher Michael Outland Sco 17h27'15" -45d31'50"
Christoleighpher And 2h26'52" 47d11'29"
Christoper Ori 6h25'4" 17d12'44"
Christoph And 2h21'48" 42d55'14"
Christoph Com 12h44'15" 30d58'25"
CHRISTOPH Lyr 18h35'45" 36d35'5"
Christoph & Amelie Uma 11h22'15" 66d10'2"
Christoph Gottfried Uma 11h1'11" 42d51'14"
Christoph Nicolas Sigg Cep 22h46'12" 66d50'14"
Christoph - Peter Cyg 21h37'1" 53d45'30"
Christoph Peter Breitling Uma 10h0'24" 43d13'39"
Christoph & Petra Brunner Uma 12h21'40" 57d42'40"
Christoph & Roxanne Lyr 18h16'54" 33d19'40"
Christoph Sager Cas 0h37'57" 61d18'47"
Christoph Spengler Uma 9h9'0" 72d39'0"
Christoph Wilger Uma 11h5'27" 71d6'49"
Christoph Zangerl Cas 0h29'22" 60d8'49"
Christophe Sgr 18h34'46" -24d11'17"
Christophe Casa papillon Crb 16h22'51" 35d50'0"
Christophe Chemin Cnc 8h21'47" 14d18'13"
Christophe Cheminaud Uma 9h36'44" 61d48'3"
Christophe Decadi petit homme Umi 16h57'31" 80d56'23"
Christophe Depierre 25/08/1981 Vir 13h6'7" -12d22'10"
Christophe Epy Gem 6h41'4" 32d8'47"
Christophe et Stéphanie Crb 16h16'33" 28d10'8"
Christophe Gonçalves Lib 14h34'36" -11d47'34"
Christophe Henry Sco 16h16'39" -20d35'44"
Christophe Lemaitre Crb 16h24'4" 30d50'55"
Christophe mon Kiko chéri Uma 8h55'1" 56d11'19"
Christophe Moussaoui Ori 5h16'51" -8d36'38"
Christophe Palierse Uma 11h15'38" 32d2'13"
Christophe Thomas Leo 10h31'30" 24d52'10"
ChristopheNataliia Ori 6h19'32" -0d37'38"
Christopher Vir 14h3'5" -17d40'16"
Christopher Lib 15h49'50" -9d11'37"
Christopher Lib 15h41'39" -10d42'11"
Christopher Uma 9h58'21" 53d41'22"
Christopher Uma 8h33'30" 63d41'44"
Christopher Umi 15h14'33" 70d26'46"
Christopher Col 5h10'56" -36d41'53"
Christopher Sco 17h26'0" -36d7'10"
Christopher Boo 14h13'44" 26d8'21"
Christopher Tau 4h44'16" 25d24'0"
Christopher Gem 7h1'40" 28d21'50"
Christopher Gem 6h57'7" 19d45'48"
Christopher Boo 14h19'33" 17d29'29"
Christopher Ari 2h40'0" 29d28'17"
Christopher Ari 2h35'22" 26d19'37"
Christopher Her 17h55'52" 19d18'36"
Christopher Peg 21h25'12" 17d23'53"
Christopher Ori 6h20'51" 10d54'14"
Christopher Ori 6h18'15" 10d50'59"
Christopher Del 20h34'45" 9d15'57"
Christopher Ori 6h21'6" 16d42'22"
Christopher Ori 5h5'16" 15d11'12"
Christopher Vir 15h6'6" 5d46'27"
Christopher Vir 12h45'42" 1d6'29"
Christopher Ori 6h1'41" 10d45'7"
Christopher Gem 7h24'51" 32d31'20"
Christopher Per 4h8'12" 44d57'34"
Christopher Per 4h23'44" 40d34'20"
Christopher And 0h43'36" 38d56'40"
Christopher Per 3h23'44" 45d7'9"
Christopher Per 4h42'29" 45d31'14"
Christopher Cyg 20h24'7" 38d52'25"
Christopher Cam 5h46'8" 58d22'44"
Christopher Cas 0h40'0" 52d59'51"
Christopher A. Fuse Vir 13h57'20" -20d42'3"
Christopher A. Lee Umi 16h7'5" 80d23'25"
Christopher A. Traettino Aql 19h29'55" 9d26'53"
Chris"Topher" Abner Toves Gem 6h28'10" 20d38'53"
Christopher Abram Jr "Sonshine" Cyg 20h37'26" 39d52'25"
Christopher Adam Bridges Ori 6h10'41" 15d9'54"
Christopher Admirabilis Col 6h8'4" -36d54'7"
Christopher A.J. Hogg Gem 6h36'47" 16d35'52"
Christopher Alan Ori 6h8'40" 20d45'4"
Christopher Alan Per 3h49'4" 50d49'8"
Christopher Alan Cameron And 1h20'44" 48d54'57"
Christopher Alan Erickson Tau 3h54'42" 7d32'55"
Christopher Alan Gnehm Cyg 19h53'4" 39d7'51"
Christopher Alan Morton Cru 12h38'59" -57d12'47"
Christopher Alan Spong Cam 4h40'10" 62d36'10"
Christopher Alan Taylor Cnv 12h51'45" 49d45'28"
Christopher Alcala Leo 9h38'10" 32d26'53"
Christopher Alexander Dallo Uma 11h3'39" 36d57'58"
Christopher Alexander Santora Vir 12h16'7" -0d58'51"
Christopher & Alissa Cyg 20h58'17" 47d44'1"
Christopher Allan Hale Her 17h32'10" 40d55'2"
Christopher Allan Thompson Uma 11h54'58" 61d8'37"
Christopher Allen Beasley Aql 19h18'18" -0d36'6"
Christopher Allen Harris Sco 17h43'23" -39d37'17"
Christopher Allen Myers Dra 17h45'53" 53d27'21"
Christopher Ambrosini Tau 4h38'38" 24d53'47"
Christopher Ancheta Psc 1h12'20" 27d24'40"
Christopher and Christina Psc 1h21'32" 24d34'19"
Christopher and Elizabeth Snipe Cyg 21h27'59" 48d40'49"
Christopher and Jamie Cyg 21h44'30" 52d38'5"
Christopher and Jennifer Peg 21h36'16" 10d53'20"
Christopher and Jordan Uma 11h46'14" 63d34'32"
Christopher and Kristi Neyman Peg 23h53'59" 22d12'23"
Christopher and Mary Wurster Cyg 20h43'17" 47d41'59"
Christopher and Pamela Hankins Ori 5h24'49" 2d11'0"
Christopher and Sierra Derrick Umi 15h24'16" 74d52'8"
Christopher Andrew Lib 14h41'13" -17d50'22"
Christopher Andrew Signore Lib 15h15'23" -15d37'50"
Christopher Andrew Turner Her 17h56'7" 22d11'32"
Christopher Angel Lyn 7h38'40" 56d52'36"
Christopher Anthony Berotti Uma 10h38'39" 65d36'48"
Christopher Anthony Brown Leo 11h23'28" 10d6'17"
Christopher Anthony Dahmen Cnc 7h59'10" 11d19'50"
Christopher Anthony D'Onofrio Her 17h9'18" 25d59'40"
Christopher Anthony Gomez Aqr 21h51'52" 0d26'33"
Christopher Anthony Grieve Per 3h25'1" 54d14'37"
Christopher Anthony Sciotti Uma 8h47'29" 47d39'1"
Christopher Antony Ready Uma 12h13'26" 52d44'33"
Christopher Arthur Young Psc 0h49'52" 11d43'39"
Christopher Ashley Boudreau Cyg 20h34'20" 49d49'9"
Christopher Ashley Jones Vir 12h30'49" 9d36'23"
Christopher & Ashley Plantone Cyg 21h33'7" 41d30'43"
Christopher Ashley Stokes Uma 11h32'44" 60d46'52"
Christopher Aubrey Braham Her 18h43'3" 19d43'43"
Christopher Austin Merrick Aqr 22h32'56" -3d20'33"
Christopher Avallone Sgr 18h34'33" -31d31'59"
Christopher Avery Aur 5h43'2" 45d40'44"
Christopher B Benson Ori 5h24'26" 2d56'53"
Christopher B. Kennedy Cyg 19h41'11" 35d5'47"
Chris-topher B. Rich Cap 21h9'58" -20d45'52"
Christopher Baker Lyn 9h11'46" 39d30'1"
Christopher Barden Lewis Lyn 8h21'1" 56d9'8"
Christopher Barron Quilisch Psc 1h23'48" 25d8'22"
Christopher Barry Chelko Uma 10h27'31" 46d49'0"
Christopher Beck Uma 9h19'27" 72d22'8"
Christopher Bensinger Tau 4h29'55" 22d26'25"
Christopher Bland-Ward Cep 23h55'42" 85d36'44"
Christopher Bogdan Vir 13h24'19" 10d48'8"
Christopher Boriss Aql 20h2'12" 10d16'1"
Christopher Bowman Wiedbusch Her 18h14'55" 15d5'7"
Christopher Boyce Cep 21h29'22" 57d3'53"
Christopher Braden Jones Uma 10h16'52" 60d13'9"
Christopher Bradley Edwin Newton Cep 22h35'51" 57d51'33"
Christopher & Brayden Gack Tau 4h38'17" 28d21'51"
Christopher Brian Cas 0h31'39" 52d15'12"
Christopher Brian Hingley Vir 12h45'18" 11d39'18"
Christopher Brockel Sgr 19h9'16" -22d54'16"
Christopher Brooke Rierson Ari 2h4'53" 14d7'43"
Christopher Broughton Psc 1h43'45" 14d38'47"
Christopher Brown Vir 12h41'25" 12d12'46"
Christopher Bruce Highley Ari 2h34'37" 17d58'48"
Christopher Brune Her 17h15'46" 49d12'34"
Christopher Bryan Miller Per 3h9'46" 47d17'52"
Christopher Bryant Ingram Ari 2h22'18" 26d2'37"
Christopher Bryen Ori 5h35'50" -1d46'47"
Christopher Buchanan Aql 19h3'43" -0d3'20"
Christopher Buttimer's Star Cnc 8h50'19" 30d41'22"
Christopher Byron Doell Per 3h17'24" 44d11'52"
Christopher C. Pettit Cyg 21h37'40" 39d31'50"
Christopher Calvin Lockhart Sgr 19h3'51" -14d50'14"
Christopher Campbell & Alice Hoover Leo 9h24'38" 6d31'53"
Christopher CancioBello Cap 20h32'48" -18d42'44"
Christopher Carl Cox Aql 19h35'45" 5d41'52"
Christopher Carl Crisp Psc 0h42'31" 7d2'48"
Christopher Carl Rydell Cep 23h6'5" 70d18'57"
Christopher Carter Ori 5h12'7" 15d55'6"
Christopher Carter, Star and Angel Leo 11h24'56" 17d53'41"
Christopher Caturano Her 16h48'1" 24d44'31"
Christopher Cayman Perkins Pic 5h11'22" -51d18'23"
Christopher Chapman Gem 6h45'40" 24d33'0"
Christopher Charles Hubnik Ori 5h51'58" 7d26'2"
Christopher Charles Hund Boo 14h35'35" 33d22'52"
Christopher Charles Johnson Psc 22h55'42" 4d45'51"
Christopher Charles Mosby Uma 8h34'1" 48d31'5"
Christopher Charles Salvatore Cnc 7h59'2" 13d24'45"
Christopher Charles Vukovich Her 17h11'39" 39d49'54"
Christopher Charoensook Gem 7h26'34" 21d31'59"
Christopher Cheshire Uma 9h34'53" 60d51'38"
Christopher Chrisman Uma 8h29'5" 63d21'35"
Christopher Cimino Conrad Aur 5h53'25" 46d26'6"
Christopher "CJ" Alexander Ducker Her 18h46'26" 21d17'4"
Christopher Cline 10-03-1981 Her 17h8'32" 20d59'56"
Christopher "Cochi" Young Ori 6h5'19" 11d3'39"
Christopher Codie Harris Her 16h43'53" 29d7'22"
Christopher Cole Seger Sco 17h20'11" -32d5'48"
Christopher Colton Butler Dra 19h49'36" 61d8'19"
Christopher Columbus Cas 0h47'44" 60d9'57"
Christopher Connell Cap 20h36'12" -9d23'44"
Christopher Conway Her 17h10'12" 35d26'55"
Christopher Conwell Dra 20h24'54" 74d50'15"
Christopher Coogan Uma 10h8'52" 59d59'1"
Christopher Cooper Tau 3h54'43" 24d30'17"
Christopher & Cordova Forever Cap 20h36'18" -26d50'18"
Christopher Costroff Her 18h40'5" 17d35'21"
Christopher "Cream-sickle" Higdon Cap 21h11'47" -15d42'53"
Christopher Cummings Aqr 23h1'27" -12d1'11"
Christopher Curry Her 16h52'28" 36d52'19"
Christopher Cy Leister Sgr 18h59'46" -21d44'5"
Christopher D. Eckstorm Aqr 21h16'43" -13d12'56"
Christopher D. Hale Vir 13h21'43" 5d35'54"
Christopher D Harner, MD Cap 20h8'37" -26d12'10"
Christopher D. Jester Her 17h45'22" 23d6'20"
Christopher D. Johnston Cep 20h35'9" 62d33'16"
Christopher D. Sweet Uma 12h4'4" 62d36'4"
Christopher Dahle Cep 20h14'0" 61d10'41"
Christopher Dale Magoon Aql 19h32'9" 4d23'1"
Christopher Daniel Brown Lyn 8h57'16" 41d51'1"
Christopher Daniel Mailloux Vir 12h16'18" 11d23'52"
Christopher Daniel Saiz Leo 10h12'50" 10d52'5"
Christopher David Clemens Cnc 9h3'18" 21d9'53"
Christopher David Ebab Her 17h36'37" 26d41'52"
Christopher David Jasper Tau 4h24'18" 10d52'16"
Christopher David Moore 72604 Leo 9h42'34" 28d37'34"
Christopher David Mullendore Ari 2h57'18" 22d1'42"
Christopher David Norton Ori 5h18'31" 0d33'16"
Christopher David Rivera SR. Lac 22h51'13" 52d14'56"
Christopher David Sandgren Cep 20h54'19" 69d5'53"
Christopher David Smith Cep 22h46'9" 78d35'57"
Christopher De Vita Psc 0h12'8" 7d52'8"
Christopher Deacon Dra 18h21'1" 53d24'39"
Christopher Dean Loran Inmon Aql 19h42'4" -0d6'49"
Christopher Dean Mitchell Umi 14h57'46" 76d7'43"
Christopher Dean Mitchell Per 3h6'9" 47d30'18"
Christopher DeHaven Cap 20h16'13" -9d25'23"
Christopher Denman Cam 4h24'13" 66d46'37"
Christopher Dennis Poteet Ari 3h18'21" 20d42'55"
Christopher Devlin Beaton Aqr 21h54'50" 1d36'16"
Christopher Dickson Psc 1h2'12" 4d39'25"
Christopher Dolberg Tau 4h19'0" 9d42'46"
Christopher Domenic Sweet Aur 5h45'13" 38d32'50"
Christopher Dominic Boucher Uma 10h34'3" 46d8'27"
Christopher Dominic King Sgr 18h19'0" -27d35'11"
Christopher Dominic Spencer Aqr 22h49'29" 2d32'33"
Christopher Don Sgr 18h2'41" -27d33'39"
Christopher Donald Reilly Vir 12h7'43" -0d23'47"
Christopher Dumas Cep 23h19'16" 79d49'59"
Christopher "Dupa" Yitts Aqr 23h15'17" -21d21'59"
Christopher Dylan Block Cnc 8h23'30" 14d27'51"
Christopher E. Cartagena Per 3h13'20" 55d32'8"
Christopher E. Stewart Ori 4h51'24" -0d25'35"
Christopher E. Vehmas Ori 6h7'12" 14d12'11"
Christopher Edward Earl Sco 16h56'49" -44d2'13"
Christopher Edward Fahrman Leo 11h40'24" 26d27'33"
Christopher Edward Reeder Psc 0h11'50" 11d27'29"
Christopher Edward Roxbury Aur 5h24'17" 40d43'2"
Christopher Edward Vernon Johnson Uma 11h12'5" 35d38'27"
Christopher Edward Wade Jonson Per 3h13'55" 53d47'13"
Christopher Edwin Gilson Per 3h7'5" 42d0'48"
Christopher Edwin James Boo 14h29'9" 34d52'38"
Christopher Elias Allsup Lib 15h7'32" -9d44'39"
Christopher Elliott Hammett Uma 10h46'51" 52d58'43"
Christopher Emery Zolan Per 4h37'22" 41d37'48"
Christopher Eric Gandeza Aqr 22h6'25" -3d50'50"
Christopher Eric Montana Uma 9h42'57" 54d26'38"
Christopher Erich Feige Tau 4h32'0" 17d14'44"
Christopher Erin Dann Cas 23h49'14" 50d32'35"
Christopher Ernest Schlachter, Jr Cnc 8h28'12" 17d9'32"
Christopher Ethan Her 17h13'27" 32d49'6"
Christopher Euthemius Sampson Gem 7h8'15" 15d9'5"
Christopher Evan Breedlove Lyn 6h23'57" 60d53'44"
Christopher F. Wrenn Ori 6h3'29" 16d59'31"
Christopher Faherty Ori 6h2'19" 18d21'9"
Christopher Fiduccia Cnc 9h7'6" 17d10'47"
Christopher Forster Per 3h17'13" 52d52'35"
Christopher Francis Roy Ari 2h29'14" 26d32'21"
Christopher Francis Scibilia Pho 0h59'50" -46d26'24"
Christopher Francis Wren Dra 18h32'1" 59d34'47"
Christopher Frank Pero Aqr 22h47'57" -16d1'38"
Christopher Freisem Her 17h56'35" 29d0'45"
Christopher French Tau 5h37'26" 25d20'47"
Christopher Fritz Petermann Sgr 17h52'28" -29d38'28"
Christopher G. Uma 11h16'9" 72d13'18"
Christopher G. Fierce Equ 21h12'0" 8d44'30"
Christopher G. Gore Uma 9h31'36" 53d24'53"
Christopher G. Hayes Cap 20h11'25" -17d28'56"
Christopher Gaddes Her 17h43'45" 19d15'36"
Christopher Gene Borden Tau 4h30'45" 15d28'16"
Christopher George Cnc 8h7'59" 25d20'55"
Christopher George Hay Per 4h9'31" 45d59'34"
Christopher George Sarris Aqr 23h27'40" -8d45'2"
Christopher Gerard Frase Leo 11h47'15" 20d57'17"
Christopher Gettings 21st Bday Star Per 3h9'14" 54d11'17"
Christopher Gianelloni Aqr 21h14'49" -13d30'24"
Christopher Gibson Sco 16h22'56" -29d26'31"

Christopher Gilberto
Lyn 7h57'20" 46d4'44"
Christopher Girard
04/16/60
Ori 5h33'14" -3d45'50"
Christopher Glen Arnold
Lib 15h0'17" -13d59'43"
Christopher Glen Ford
Psc 0h32'32" 12d31'38"
Christopher Gordon
Erskine 6-26-38
Uma 11h21'14" 60d8'36"
Christopher Grant Johnson
Uma 11h26'16" 37d7'28"
Christopher Gulledge
Per 2h17'35" 51d38'5"
Christopher Gutierrez
Leo 10h51'26" -3d0'43"
Christopher H. Chapin
Boo 14h22'19" 14d19'23"
Christopher Haar
Cyg 20h14'58" 59d23'23"
Christopher Haber
Boo 13h50'44" 22d47'6"
Christopher Haines
Leo 11h9'41" 22d11'41"
Christopher Harold
Alexander III
Ari 1h54'4" 18d41'39"
Christopher Harriman
Dann, Jr.
Per 3h17'52" 47d0'52"
Christopher Harris Langley
Tau 5h57'16" 27d28'23"
Christopher Harvey
Cru 12h4'23" -62d51'52"
Christopher Hayes Roney
Cra 18h17'43" -39d33'55"
Christopher Henry Lucko
Aqr 22h57'3" -21d39'34"
Christopher Himes
Cam 4h11'49" 67d14'12"
Christopher Hinde
Cap 20h24'1" -17d22'16"
Christopher Horning
Per 2h7'46" 55d45'38"
Christopher Howland
Cyg 19h40'7" 28d5'13"
Christopher Huddleston
Her 17h26'46" 24d36'45"
Christopher Huff
Ari 2h32'25" 20d46'59"
Christopher Ian Slovick
Boo 15h35'31" 45d51'44"
Christopher Irwin McKinnon
Cen 14h6'14" -43d30'11"
Christopher Iwanski
Ori 5h26'19" 8d40'29"
Christopher J. Anderson
Sco 17h35'34" -31d16'58"
Christopher J. Begley
Boo 15h32'50" 41d54'59"
Christopher J. Bradford
Cap 20h26'46" -12d41'18"
Christopher J. Brower
Lyn 8h1'5" 42d15'31"
Christopher J. Carenza
Ori 5h56'10" -0d29'16"
Christopher J. Chomo
Her 16h39'51" 23d54'11"
Christopher J. Cook
Ser 15h59'5" -0d56'58"
Christopher J. Durso
Aql 19h2'22" 3d8'53"
Christopher J. Grullon
Sgr 18h32'15" -25d11'37"
Christopher J. Maines
Aqr 22h24'34" -23d24'36"
Christopher J. Mega
Sco 17h40'8" -42d26'35"
Christopher J. Mundy
Uma 12h1'21" 60d36'37"
Christopher J. Nordengren
Her 17h38'22" 36d21'18"
Christopher J. Sharpe
Vir 13h9'30" -2d25'7"
Christopher J. Toland
Ser 15h26'53" -0d53'12"
Christopher Ja Bau Chang
Del 20h37'24" 16d10'56"
Christopher & Jackie
Maher
Vir 13h18'19" 12d36'6"
Christopher Jacob
Drummond
Ori 5h16'45" 6d45'26"
Christopher Jake Sauls
Her 17h37'56" 27d21'24"
Christopher James
And 22h57'30" 50d51'10"
Christopher James
Cep 22h50'16" 65d21'3"
Christopher James Baluta
Her 16h42'35" 48d4'59"
Christopher James Brunn
Tau 4h38'20" 29d3'21"
Christopher James Burgess
Cyg 19h43'52" 31d31'16"
Christopher James Busch
Dra 18h26'34" 72d25'6"
Christopher James
Capasso
And 0h39'0" 42d24'40"
Christopher James Cogdill
Ori 5h29'45" 1d10'57"
Christopher James
Conneally
Ari 2h23'59" 12d15'4"
Christopher James Costa
Cap 20h26'59" -13d48'33"
Christopher James Fatta
Vir 13h11'41" 11d43'22"
Christopher James
Goatham
Uma 9h22'46" 50d1'15"
Christopher James Halabi
Col 6h23'44" -34d46'11"
Christopher James Huber
Cnc 8h41'21" 32d6'48"
Christopher James
Johnston
Cnc 8h0'43" 21d27'57"
Christopher James Nealon
Sco 16h15'16" -8d22'13"
Christopher James Ruper
Vir 13h44'32" -6d47'6"
Christopher James the
Wonderful
Gem 6h24'38" 18d18'51"
Christopher James
Vazoulas
Lmi 10h44'55" 34d15'26"
Christopher Jason Witt
Lib 15h7'54" -19d47'33"
Christopher John
Gem 6h50'27" 34d51'20"
Christopher John
Uma 11h38'11" 47d42'8"
Christopher John
Her 18h52'12" 23d23'59"
Christopher John Arvid
Anderson
Cnc 9h15'9" 26d18'6"
Christopher John Booth
Uma 10h18'56" 45d51'43"
Christopher John Graham
Lyr 18h49'49" 44d12'42"
Christopher John Hilding
Lib 15h7'17" -6d33'49"
Christopher John Kefalos
Per 3h20'48" 41d16'17"
Christopher John
McCubbins
Pho 1h13'32" -47d32'18"
Christopher John
McLafferty
Her 17h8'18" 39d8'56"
Christopher John Moore
Vir 12h40'34" 0d7'47"
Christopher John O'Kula
Vir 13h14'21" -10d5'53"
Christopher John Piazza
Lib 15h21'52" -26d4'55"
Christopher John Portman
Uma 10h52'48" 42d25'26"
Christopher John Simpson
Tau 3h48'38" 27d43'58"
Christopher John
Terwilliger
Per 2h20'14" 57d6'33"
Christopher John Thomas
King
Cep 22h23'53" 72d52'39"
Christopher Johns
Cnc 8h39'2" 17d21'57"
Christopher Joseph
Sgr 18h59'39" -32d0'27"
Christopher Joseph Bell
Sco 17h17'33" -32d37'47"
Christopher Joseph Leake
Her 17h24'40" 15d40'21"
Christopher Joseph Lewis
Her 18h38'0" 15d15'29"
Christopher Joseph Lewis
Vir 13h20'40" 4d16'15"
Christopher Joseph
McDonald
Sco 17h46'7" -39d24'56"
Christopher Joseph
Michael Hill
Uma 10h47'3" 55d44'19"
Christopher Joseph
Migliozzi
Tau 5h42'33" 22d59'43"
Christopher Joseph
Schippers
Tau 3h52'13" 5d25'24"
Christopher Joseph Wilkin
Psc 1h35'35" 11d32'26"
Christopher Joson Yard
Uma 9h16'5" 48d2'39"
Christopher Jude O'Connor
Leo 9h25'19" 14d20'7"
Christopher Juhl Wilsnack
Cnc 8h21'46" 18d22'55"
Christopher Junda
Tau 4h40'32" 19d34'52"
Christopher K. Cummins
Ari 3h4'17" 13d4'26"
Christopher K Wilson
Her 16h54'36" 17d37'26"
Christopher & Kara
Westerfield
Lyr 18h52'6" 36d3'28"
Christopher Keeks Williams
Cep 22h2'44" 63d49'26"
Christopher Kelley
Per 3h33'40" 47d0'32"
Christopher Kendall
Dra 18h48'39" 52d58'23"
Christopher Kent Bingaman
Boo 14h30'34" 41d16'15"
Christopher Kerr
Uma 11h27'6" 41d28'31"
Christopher Kevin Ferris
Sgr 19h27'52" -12d43'22"
Christopher Kevin Williams
Lib 15h35'34" -10d49'24"
Christopher King
Cep 21h29'15" 58d4'29"
Christopher King Seglem
Crb 15h40'53" 26d7'4"
Christopher Kjentvet
Lib 15h2'52" -25d13'41"
Christopher Knoch
Lmi 10h1'23" 32d6'15"
Christopher & Kristin
Umi 14h33'19" 85d42'8"
Christopher Krogen
Uma 11h13'43" 42d50'23"
Christopher Kyle 17
Lib 15h16'37" -25d56'2"
Christopher L. & Jennifer J.
Skiles
Lib 14h59'39" -2d0'25"
Christopher Lanza
Charles
Vir 13h18'25" -18d3'30"
Christopher Larcen George
Psc 1h33'55" 27d47'54"
Christopher Lawrence
Marchese
Cet 2h14'50" -0d25'10"
Christopher Lee
Cap 21h10'10" -14d53'54"
Christopher Lee
Ari 3h9'22" 18d30'42"
Christopher Lee Anderson
Uma 10h18'24" 40d55'38"
Christopher Lee Bishop
Lib 15h38'38" -17d30'16"
Christopher Lee Bowling
Sgr 18h51'29" -29d5'7"
Christopher Lee Boyton
Her 17h13'18" 32d22'41"
Christopher Lee Hibbard
Uma 9h15'36" 55d58'47"
Christopher Lee McLain
Sco 16h13'40" -14d37'26"
Christopher Lee Rhoads
Sco 17h49'53" -36d18'51"
Christopher Lee Rivers
Equ 21h16'17" 7d38'27"
Christopher Lee Roth
Lib 15h5'58" -13d53'59"
Christopher Lee Shepherd
Ori 5h1'29" 15d30'17"
Christopher Lee Shortell
Tau 4h37'37" 21d59'17"
Christopher Leon
Livingston
Sco 16h37'8" -26d41'58"
Christopher Lewis Treece
Crb 16h18'6" 27d11'11"
Christopher & Lindsey
Lep 5h14'48" -11d29'18"
Christopher & Lisa Salinas
Cyg 20h43'1" 35d48'0"
Christopher Logan Woods
Uma 11h27'21" 59d10'16"
Christopher Louis
Her 16h27'46" 20d12'27"
Christopher Louis
Her 16h17'56" 19d28'22"
Christopher Loves Sara
Forever
And 23h15'13" 52d14'32"
Christopher Loy
McCandless
Per 3h17'38" 42d41'10"
Christopher Luca
Uma 9h51'11" 67d54'40"
Christopher Lucas Hamilton
Ori 5h41'30" 2d40'40"
Christopher Lucchesi
Cap 21h44'55" -23d39'28"
Christopher Luis Cirino
Ari 2h30'49" 17d45'18"
Christopher Luke Tonich
Cru 12h33'13" -61d4'29"
Christopher Lynn Cathey
Uma 10h29'20" 42d15'5"
Christopher Lynn Shepherd
Lyn 8h33'8" 55d21'28"
Christopher M.
Lyn 7h59'26" 36d42'22"
Christopher M. Black
Uma 9h21'35" 67d16'59"
Christopher M. Blosser
Her 16h24'21" 19d9'52"
Christopher M. Domenick
Lib 14h44'49" -11d30'21"
Christopher M. Earley "Big
Guy"
Psc 0h27'23" 17d38'23"
Christopher M. Gross
Per 3h17'48" 52d21'43"
Christopher M. Merle
Cep 21h39'2" 73d47'44"
Christopher M. Schatz
Cep 21h38'34" 61d37'7"
Christopher M Simo-Kinzer
Cas 23h14'27" 59d23'34"
Christopher M. Totora
And 1h41'45" 39d13'3"
Christopher M. Ward
Per 2h31'20" 56d52'39"
Christopher M. Wilson
Her 16h22'18" 19d3'49"
Christopher M. Witkowski
Uma 9h3'59" 61d13'46"
Christopher Mahlon
Her 17h31'54" 41d13'3"
Christopher Maier
Cnc 7h57'9" 17d19'44"
Christopher Marie
Cep 0h14'22" 77d16'13"
Christopher & Marisa
Bienek
Cyg 19h59'51" 30d17'29"
Christopher Mark
Gem 7h48'45" 27d30'17"
Christopher Mark Webb
Ser 16h1'39" -0d29'45"
Christopher Mark Wood
Her 17h18'27" 34d57'11"
Christopher Martin
Mon 6h51'44" 8d2'35"
Christopher Martin Olsen
Vir 12h56'52" 5d30'16"
Christopher Martin Zapata
Boo 14h32'53" 15d32'16"
Christopher Martinez
Leo 10h4'9" 13d2'2"
Christopher Marvin
Huffaker 1982
Aqr 22h57'13" -8d36'16"
Christopher Maslbas
Leo 9h33'10" 25d45'25"
Christopher Mason
Dickerson
Ari 2h0'6" 12d46'3"
Christopher Mathison
Cap 21h13'35" -21d49'6"
Christopher Matthew Cox
Psc 0h26'3" 0d54'45"
Christopher Matthew
Kavanaugh
Cyg 19h49'44" 43d8'40"
Christopher Matthew Smith,
Jr.
Psc 1h8'1" 32d2'55"
Christopher Matthew Vail
Gem 7h6'22" 20d35'14"
Christopher Matthew's
Comet
Per 3h45'37" 41d46'33"
Christopher Mawdsley
Leo 10h33'29" 8d46'39"
Christopher Maxwell
Michael Bowen
Lib 15h33'7" -5d20'13"
Christopher McCarty
Uma 8h22'0" 61d33'28"
Christopher McCoy
Uma 11h27'1" 47d58'30"
Christopher Meilstrup
Cas 0h57'52" 61d26'27"
Christopher Merritt
Ori 5h59'23" 17d18'50"
Christopher Metivier-
Daniels
Aqr 22h50'54" -7d12'56"
Christopher Michael
Lib 15h29'31" -16d7'35"
Christopher Michael
Her 17h23'53" 16d30'55"
Christopher Michael
Ori 6h3'59" 13d31'55"
Christopher Michael
Per 3h8'10" 53d59'1"
Christopher Michael
Per 2h54'36" 55d49'51"
Christopher Michael
Her 17h45'3" 41d29'44"
Christopher Michael
Anthony
Crb 15h44'10" 34d59'44"
Christopher Michael
Bowling
Gem 6h9'34" 23d3'55"
Christopher Michael
Bressler
Uma 13h52'50" 54d17'6"
Christopher Michael Burtt
Sco 17h37'16" -35d55'2"
Christopher Michael "BZ"
3/30/73
Tau 4h52'22" 23d14'47"
Christopher Michael
Campbell
Boo 15h27'19" 41d30'12"
Christopher Michael
Constable
Lib 15h33'51" -8d12'14"
Christopher Michael
Desmond
Psc 1h6'3" 25d24'58"
Christopher Michael
Doherty
Cyg 19h41'22" 30d56'44"
Christopher Michael Fazzie
Leo 11h28'34" 20d54'53"
Christopher Michael Frey
Her 17h24'13" 18d14'27"
Christopher Michael
Hauser
Lmi 10h47'50" 39d3'49"
Christopher Michael Joiner
Aql 18h52'32" -0d59'58"
Christopher Michael
Kaufman
Lyn 7h49'27" 55d5'28"
Christopher Michael & Kelly
Marie
Sco 16h14'13" -15d11'15"
Christopher Michael
Lipscomb
Lib 15h9'36" -6d53'13"
Christopher Michael
Melrose
Ari 2h17'31" 12d39'9"
Christopher Michael
Mulligan
Uma 9h2'0" 63d12'18"
Christopher Michael Nagorr
Her 17h1'0" 27d25'47"
Christopher Michael Pelz
Uma 11h36'38" 49d53'54"
Christopher Michael
Perkins
Her 18h4'44" 23d33'11"
Christopher Michael
Ramirez
Ori 6h16'33" 9d47'22"
Christopher Michael Saras
Sgr 17h50'15" -24d37'52"
Christopher Michael
Scherer, Jr.
Umi 14h41'33" 73d18'5"
Christopher Michael Sibley
Vir 14h16'3" 2d57'12"
Christopher Michael Steele
Aql 19h45'22" 7d9'5"
Christopher Michael
Stewart, Junior
Ari 3h11'7" 29d12'28"
Christopher Michael Stout
Ari 2h50'58" 28d50'0"
Christopher Michael
Tessmer
Uma 10h27'17" 61d19'23"
Christopher Michael Thorn
Her 16h28'6" 43d38'38"
Christopher Michel
Boo 15h14'14" 41d21'47"
Christopher & Michelle
Haigood
Peg 22h17'17" 6d12'57"
Christopher Miles
Ori 5h35'30" -0d41'49"
Christopher Minyone Smith
Cep 22h17'42" 55d55'47"
Christopher "Misiu" Ciolino
Cnc 9h16'21" 11d4'5"
Christopher Morrison
Cap 20h26'55" -18d31'21"
Christopher "My Squid"
Knipp
Sco 17h29'49" -40d32'40"
Christopher N Antonella
Her 18h4'28" 26d37'5"
Christopher Nasta
Boo 15h5'19" 34d13'46"
Christopher Neal
Cnc 8h41'32" 18d0'48"
Christopher Neal Salerno
Gem 6h49'45" 34d38'13"
Christopher Neil Loeffler
Gem 6h32'36" 26d44'27"
Christopher Neil Young
Uma 8h31'14" 71d33'32"
Christopher Nelson
Umi 16h26'40" 76d31'18"
Christopher Nelson Barlow
Cnc 8h4'57" 11d54'29"
Christopher Neundlinger
Cas 1h53'47" 64d33'26"
Christopher Niklaus Gilgen
Per 3h17'24" 42d10'46"
Christopher Noland
Aur 5h40'54" 43d8'25"
Christopher Otha Fink
"Petey"
Per 3h21'16" 52d7'56"
Christopher P. Hinman
Her 16h7'57" 44d54'27"
Christopher P. Westfield
Uma 9h0'6" 71d7'43"
Christopher Pacal
Psc 0h35'29" 17d33'39"
Christopher Pajonk
Aur 5h46'31" 29d59'24"
Christopher Patrick
Sex 10h13'32" -0d6'32"
Christopher Patrick Farina
Uma 11h46'48" 55d13'5"
Christopher Patrick Strand
Cnc 9h8'52" 30d26'17"
Christopher Paul
Tau 4h27'46" 27d49'33"
Christopher Paul
Her 17h18'41" 21d7'35"
Christopher Paul Athey
Ori 6h10'52" 7d46'34"
Christopher Paul Beattie
Her 17h45'2" 45d42'8"
Christopher Paul Boutwell
Per 3h39'3" 46d23'29"
Christopher Paul Crenshaw
Cru 12h49'56" -56d14'28"
Christopher Paul Curtain
Ori 5h6'37" 5d30'45"
Christopher Paul
Goncalves
Gem 7h50'46" 32d50'45"
Christopher Paul Irizarry
Ori 5h32'27" -0d30'25"
Christopher Paul Maguire
Gem 7h33'28" 28d15'57"
Christopher Paul Pupa
Aqr 23h8'51" -8d42'2"
Christopher Paul Raymond
Stannix
Gem 6h34'27" 22d21'11"
Christopher Paul Thomas
Ori 5h50'20" 3d34'3"
Christopher Pendergraft
Per 4h32'12" 31d4'34"
Christopher Peter Herget
Cyg 20h16'4" 39d28'19"
Christopher Peter Puchalla
Jr.
Uma 9h26'6" 50d49'19"
Christopher Peter Smith
Cnv 12h29'29" 47d2'24"
Christopher Pfeiffer
Aql 19h8'3" 6d32'0"
Christopher Pleasant
Uma 9h41'33" 56d17'2"
Christopher Price
Uma 12h25'36" 55d3'32"
Christopher Quigley
Her 18h20'43" 26d19'14"
Christopher R. Delorme
Ori 6h20'18" 16d50'58"
Christopher R. Dimon
Boo 15h9'11" 13d47'8"
Christopher R. Gates
Per 3h43'33" 37d10'37"
Christopher R. Gick
Vir 13h8'16" 10d15'22"
Christopher R. Karamitsos
Lyn 9h15'32" 45d32'32"
Christopher R. Rose
Sgr 18h14'16" -29d42'24"
Christopher Rajlal
Cep 22h18'1" 59d9'54"
Christopher Randall
Bomhardt
Sgr 18h47'33" -16d4'10"
Christopher Ranita
Cep 23h36'19" 78d25'54"
Christopher Ray Patton
"Sugar Man"
Tau 3h54'59" 23d8'58"
Christopher Reed Blevins
Uma 8h36'46" 50d55'14"
Christopher Reed's Dream
Cap 21h5'42" -19d38'59"
Christopher Reed's Star
Ori 5h27'0" 4d44'48"
Christopher Reeve
Uma 11h36'31" 54d30'1"
Christopher Regina
Uma 12h9'17" 52d47'7"
Christopher Reidy
Her 16h35'14" 20d48'20"
Christopher Reiff
Ori 5h4'41" 15d38'21"
Christopher Rene Hinojosa
Her 17h0'23" 16d30'35"
Christopher Ricard
Tau 5h4'1" 21d21'59"
Christopher Richard Arnold
Lyn 8h49'39" 45d56'28"
Christopher Richard
Bradley
Uma 10h28'14" 71d36'1"
Christopher Richard
Eastburn
Per 3h21'58" 41d28'38"
Christopher Richard
Holland
Uma 9h30'2" 66d40'11"
Christopher Richard
Stringini
Uma 12h4'56" 43d34'18"
Christopher Robert Andrew
Higdon
Her 17h42'44" 20d26'4"
Christopher Robert Ankeny
Leo 10h20'21" 20d17'51"
Christopher Robert Brun
Psc 1h32'45" 21d14'18"
Christopher Robert Carreiro
Lib 15h38'58" -4d11'16"
Christopher Robert Ertmer
Her 17h36'31" 37d18'31"
Christopher Robert
Nicholson
Lib 15h12'22" -7d14'54"
Christopher Robert Prior
Waddell - Topher
Cep 0h53'41" 85d1'26"
Christopher Robert
Schwarz
Aql 19h46'36" 12d35'38"
Christopher Robert
Stephens
Aql 18h57'32" -0d44'43"
Christopher Robert Weaver
Ari 2h43'11" 16d34'21"
Christopher Robert Williges
Uma 10h39'1" 46d16'51"
Christopher Robin
Ori 5h16'56" 7d26'49"
Christopher Robin
Cnc 8h6'17" 15d33'21"
Christopher Robin Brundige
Uma 10h24'21" 65d5'38"
Christopher Robin Dove
Lmi 10h18'51" 32d48'5"
Christopher Robin - Eternal
Sparkle
Cru 12h24'1" -62d46'54"
Christopher Rody
Gem 7h1'17" 15d14'13"
Christopher Ronald Horn
Uma 8h35'3" 61d49'34"
Christopher Ross Ashcraft
Her 16h46'8" 18d54'54"
Christopher Ross Mason
Per 2h48'1" 50d28'39"
Christopher Rowan (Floyd)
And 0h22'44" 34d2'37"
Christopher Rushton
And 1h5'58" 42d50'33"
Christopher Ryan Burnham
Ori 6h17'48" 7d34'46"
Christopher Ryan Ellis
Aqr 22h12'17" -2d16'22"
Christopher Ryan
Hendricks
Her 17h37'53" 33d0'33"
Christopher Ryan Saddler
Ori 6h14'3" 6d7'9"
Christopher Ryan Simpson
Psc 1h12'18" 21d20'59"
Christopher Ryan Skwirut
Ori 5h48'18" 5d49'10"
Christopher S. Amato
Uma 12h55'52" 55d9'58"
Christopher S. Floyd
Cyg 21h23'30" 32d26'34"
Christopher S. Hensley
Vir 14h4'20" -14d59'36"
Christopher S. Sapienza
Aql 19h44'43" 13d49'18"
Christopher S. Saunders
Ari 2h15'50" 27d27'32"
Christopher S Tringali
Sgr 19h13'17" -15d49'19"
Christopher S. Trosky
Umi 16h19'17" 77d25'9"
Christopher Saitta
Ari 3h7'48" 18d7'23"
Christopher Sam Gerakelis
Her 17h59'57" 27d34'32"
Christopher Sasser
Ori 5h27'35" 9d21'45"
Christopher Schafer
Per 3h49'24" 51d6'32"
Christopher Scott
Boo 14h44'54" 34d30'17"
Christopher Scott Blanken
Sco 17h43'8" -41d18'26"
Christopher Scott Blanton
Dra 16h37'11" 63d45'38"
Christopher Scott Casella
Aqr 21h53'38" -7d3'18"
Christopher Scott Fedele
Leo 11h41'51" 25d7'45"
Christopher Scott Mazurek
Cap 20h35'47" -26d12'29"
Christopher Scott Ranum
Per 4h44'32" 45d38'43"
Christopher
ScottVanBuskirk Tenicki
Tau 5h11'35" 17d35'2"
Christopher Sean O'Neal
Aqr 22h49'7" 0d15'1"
Christopher Shawn
Apu 16h0'20" -78d44'32"
Christopher Sidor
Her 16h41'16" 31d56'34"
Christopher Silvestro
Spangher
Boo 14h59'10" 36d34'56"
Christopher Simon Ridley
Gem 6h40'41" 15d39'44"
Christopher Skone-Roberts
Umi 13h45'5" 77d58'36"
Christopher Slocomb
Aql 19h5'12" 9d49'3"
Christopher Smith
Uma 10h17'24" 56d47'19"
Christopher Smudge
Wuelling
Her 17h15'18" 32d45'10"
Christopher Spain
Uma 8h26'0" 72d6'5"
Christopher St. Pierre
Per 3h48'4" 42d24'22"
Christopher St. Pierre
Gem 7h15'22" 19d20'23"
Christopher Stark
Ori 6h4'4" 18d59'56"
Christopher Stephen
Lawrence
Sgr 19h40'43" -12d39'51"
Christopher Stephen
Reuter
Psc 1h36'48" 4d8'57"
Christopher Stephen Vega
Ari 3h24'46" 29d10'59"
Christopher Stewart
Lmi 10h39'10" 32d46'10"
Christopher Stewart
Jakubowski
Per 3h5'54" 33d7'40"

Christopher Stewart Leicester
Aql 18h51'57" -0d57'49"
Christopher Stronstad
Lyn 7h7'57" 60d19'47"
Christopher & Susan
Umi 15h35'25" 73d33'51"
Christopher & Susan
Pho 23h56'52" -48d30'11"
Christopher & Susan
Per 3h12'7" 48d14'5"
Christopher T. Gentile
Per 3h8'25" 54d38'14"
Christopher T. Gerrity
Ori 5h52'58" 22d41'34"
Christopher T. Miceli
Psc 0h5'9" 0d25'48"
Christopher Talarico
Vir 12h53'2" 12d9'45"
Christopher Tarantino
Cen 13h29'47" -31d56'49"
Christopher Taylor Trianosky
Ari 3h14'2" 29d19'13"
Christopher The Cub
Gem 6h52'18" 32d37'40"
Christopher... The Love of My Life
Lyn 7h33'7" 38d2'59"
Christopher Therkelsen
Vir 14h15'23" -11d40'20"
Christopher Thomas Carbone
Psc 2h3'37" 9d24'57"
Christopher Thomas Flynn
Vir 15h10'17" 1d56'3"
Christopher Thomas Kesterson
Psc 1h18'19" 25d23'14"
Christopher Thomas MacPherson
Tau 4h39'58" 11d53'54"
Christopher Thomas Musser
Ari 2h34'31" 24d35'53"
Christopher Thomas Scott Salcido
Ori 5h30'14" 1d29'31"
Christopher Thomas Sheppard
Vir 12h12'37" -4d41'54"
Christopher Thripp
Tau 4h37'43" 27d40'4"
Christopher Timothy
Umi 13h46'40" 71d39'58"
Christopher Todd Gronsman
Uma 9h29'6" 65d55'45"
Christopher Tooley
Gem 7h48'48" 31d55'30"
Christopher Turnbull
Her 16h32'29" 7d58'51"
Christopher Ufford
Uma 9h26'2" 45d21'8"
Christopher V Shoester Jr.
Her 18h34'46" 20d2'6"
Christopher Veazey
Cas 23h52'57" 53d50'1"
Christopher Vega: We love you.
Vir 12h4'41" -6d7'39"
Christopher Verdugo
Mon 6h40'52" 8d37'57"
Christopher Vick
Uma 9h11'56" 61d14'9"
Christopher Victor Nubelo Jr
Aqr 22h53'39" -7d49'58"
Christopher Vincent Glianna, Sr.
And 0h23'18" 33d43'46"
Christopher W. Jones
Per 2h34'21" 54d19'25"
Christopher Wade Clodfelter
Dra 17h39'43" 60d12'28"
Christopher Wallace McGilvray
Per 3h21'49" 47d12'51"
Christopher Wallschlaeger
Uma 10h54'31" 39d10'15"
Christopher Walter Gainer
Per 3h34'3" 47d43'0"
Christopher Walter Hargis
Cap 20h28'20" -13d32'3"
Christopher Walter Jorgensen
Lyn 8h33'20" 56d3'25"
Christopher Wayde Shepherd
Ori 5h41'19" 0d27'51"
Christopher Wayne George Roller
Cnc 8h19'33" 20d24'8"
Christopher Webb
Per 3h35'54" 33d39'45"
Christopher Wesley Bitney
Umi 15h43'55" 79d9'8"
Christopher Whitmore: 15.03.1984
Ori 6h11'21" 16d36'50"
Christopher Wiackley
Uma 13h18'51" 56d31'45"
Christopher Wilbur
Lyn 8h36'30" 42d16'25"
Christopher William
Her 16h35'5" 33d43'44"
Christopher William Cascone, Jr.
Cap 20h7'45" -14d44'54"
Christopher William Hoff II
Ori 5h20'35" 4d8'21"
Christopher William Mertes Star
Psc 23h52'31" 7d54'37"
Christopher William Pinkus
Her 17h41'12" 15d30'53"
Christopher Wilson
Lib 15h5'34" -21d41'46"
Christopher Wilson
Psc 0h43'3" 6d11'4"
Christopher Wolfling Lotz
Gem 6h20'48" 21d33'53"
Christopher Worden Sr.
Per 2h52'6" 53d49'45"
Christopher Wright
Lyn 7h25'35" 56d35'14"
Christopher, Erica, Alexa Marie
Cap 21h35'2" -14d5'7"
Christopher-Dallas Joseph Prince
Ori 6h17'19" 9d15'56"
ChristopherJoelDewberry
Cnc 8h25'53" 12d18'11"
ChristopherKareck
Ori 6h2'32" 17d6'13"
Christopher-Michael
Umi 14h12'2" 67d6'13"
Christopher's Comets
Cnc 9h3'58" 30d53'48"
Christophers Connection
Tau 4h9'55" 27d42'15"
Christopher's Dream
Uma 10h57'3" 39d16'12"
Christopher's Light
Gem 7h21'8" 16d28'49"
Christopher's Radiance
Her 16h54'10" 17d36'12"
Christopher's Shining Rainbow Star
Ara 17h11'5" -54d16'8"
Christopher's Star
Ori 5h24'5" 2d28'17"
Christopher's Wish
Ori 5h39'33" -0d24'25"
Christophers's Light
Tau 3h41'15" 23d20'24"
christophertoo&kim
Cyg 20h22'32" 43d39'32"
Christophia TerrySmith
Uma 12h32'12" 54d27'17"
Christos Anthony Gazis
Aqr 22h5'11" 0d40'58"
Christos Arvanitakis
Uma 9h31'19" 46d13'56"
Christos & Eleana
Eri 4h30'50" -23d8'45"
Christos Kusulas
Boo 14h45'6" 53d9'52"
Christos Lambrakis
Psc 0h14'43" 6d12'19"
Christos Lanis
Dra 18h33'53" 57d13'3"
Christos Manolas
Sco 16h9'7" -21d20'57"
Christos Theodore Serdenes
Aql 19h56'27" 6d52'12"
Christos Volonakis
Sco 16h12'46" -13d20'28"
Christos Wouralis, Jr.
Psc 1h10'50" 31d5'25"
Christpher Azmeh
Dra 9h23'2" 73d0'44"
Christy
Cam 3h18'56" 61d17'27"
Christy
Vir 14h11'41" -12d42'50"
Christy
Psc 23h59'45" -3d31'6"
Christy
Cnc 8h46'28" 31d36'46"
Christy
Per 3h8'44" 54d30'30"
Christy
Ari 3h16'23" 20d41'25"
Christy 8605
Lyn 8h20'44" 49d28'26"
Christy A. Baldomino
Psc 0h53'24" 15d11'55"
Christy Ann Cammarata
Lib 14h47'26" -11d5'22"
Christy Ann Chevallier Tolley
Uma 11h24'35" 32d13'38"
Christy Ann Leathers
Sgr 18h5'16" -21d27'24"
Christy Ann Spence 12/30/1973
Uma 11h43'10" 36d10'31"
Christy Anthony
Crb 16h7'55" 33d20'23"
Christy Anthony
Vir 13h37'14" 4d3'36"
Christy Brown
Psc 1h49'40" 7d1'42"
Christy Capricorn
And 23h3'43" 41d52'14"
Christy Case
And 0h14'13" 30d10'13"
Christy Cooper
And 0h41'10" 37d8'14"
Christy Corder
Cas 1h33'9" 58d15'47"
Christy Couvillion Cooper
Psc 2h1'20" 8d28'23"
Christy D. Sica
Leo 10h55'32" 16d30'57"
Christy Eppich
Ori 5h39'34" 2d34'40"
Christy Fasekas
And 23h7'44" 51d6'24"
Christy Francis
Cyg 19h33'55" 28d39'2"
Christy Fregia
Sex 9h43'10" -0d24'48"
Christy Fust & Mark Linkiewicz
Cyg 21h17'11" 53d16'57"
Christy Fust & Mark Linkiewicz
Lyr 18h58'34" 26d57'13"
Christy Gabriel
Her 16h43'13" 32d42'10"
Christy Hausgen
Com 12h38'31" 21d43'36"
Christy Hemme
Sco 16h13'49" -9d10'0"
Christy Henry "The Rock In My Life"
Psc 0h44'3" 6d14'9"
Christy Hicks Dodel
Cas 1h24'41" 62d42'25"
Christy Hofman
Psc 0h55'22" 7d10'11"
Christy Joseph Jordan
Umi 14h48'57" 70d28'23"
Christy Kareokowsky
And 0h43'52" 41d12'12"
Christy Kepins
Cas 1h45'52" 60d38'36"
Christy L. Gunn
And 0h17'27" 44d34'9"
Christy L. Smith
Vir 14h27'9" 1d34'32"
Christy Lee Bowman
Cam 5h17'56" 58d36'37"
Christy Lee McConnell
Gem 7h42'19" 20d35'29"
Christy Lee Ulrich
Sgr 19h14'29" -33d6'5"
Christy Lou Tilton
And 2h18'34" 42d29'8"
Christy Lynn Barga
Uma 8h51'7" 61d32'25"
Christy Lynn Galzen
Leo 10h31'45" 13d56'58"
Christy Lynn Kuklinski
Crb 15h41'14" 38d39'36"
Christy Lynn Pursell
And 23h51'13" 34d9'34"
Christy Lynn Sweet
Tau 3h46'41" 29d38'0"
Christy M. Adamson
And 23h24'44" 41d18'11"
Christy M. Naughton
Tau 3h52'44" 17d57'25"
Christy Marics
And 23h0'25" 48d27'45"
Christy Marie
Aqr 22h0'59" -10d17'45"
Christy Marie
Aqr 22h33'53" -1d14'17"
Christy Marie Cockrill
Sgr 19h18'48" -15d7'56"
Christy Marie Kosary
Lib 14h50'42" -1d46'37"
Christy Marie Loretto 3/18/77
Psc 23h26'32" 3d16'7"
Christy Maxey
Mon 7h4'43" -5d46'59"
Christy McCann
Cnc 8h29'24" 11d36'2"
Christy Michele & Sue Loretto
And 0h27'18" 36d16'1"
Christy Miedzinski
Uma 8h41'26" 51d14'58"
Christy Morgan 26.AUG.1976
Vir 12h48'4" 11d4'36"
Christy N. Frost
Tau 4h19'21" 2d35'50"
Christy Neill
And 1h29'38" 46d29'51"
Christy Nichole
Cap 20h56'7" -14d39'24"
Christy Noel
And 1h1'36" 37d51'9"
Christy Pacini
Uma 9h1'54" 56d2'28"
Christy Patrick, Will You Marry Me?
Aqr 22h20'4" 1d5'57"
CHRISTY PATTERSON
Psc 1h45'46" 5d35'7"
Christy Pooky Davis
Cap 20h30'36" -15d0'8"
Christy Roomie Schultz
Uma 13h57'55" 59d51'16"
Christy Rose
Crb 15h23'30" 30d5'30"
Christy S. Callahan
Gem 6h5'45" 24d10'19"
Christy & Scott's Wedding
Cyg 20h44'58" 32d18'49"
Christy stix sweeter then pixy stix
And 0h43'40" 41d57'56"
Christy Stoklosa
Cam 3h30'53" 62d42'52"
Christy Tran
Lyn 7h53'22" 38d6'46"
Christy Wilkie
And 2h28'2" 45d32'14"
Christy Winkler 143
Mon 6h47'37" -0d46'58"
Christy, Brett, Jared and Haylee
Aql 19h42'56" 3d34'29"
Christy, Will You Marry Me?
Peg 0h2'25" 27d25'29"
Christy-Anne
And 2h37'59" 37d24'29"
ChristyJoy822
Leo 11h1'49" -3d3'5"
Christylynn Ku'uipo Nachor
Cra 19h7'52" -38d18'30"
Christy-My Shining Star!
Ori 5h12'21" 1d33'26"
Christyn Ann
And 1h32'49" 48d59'9"
Christyna
Cnc 8h13'24" 15d24'11"
Christyna Shay Stewart
Lib 14h48'26" -17d11'50"
Christy's & Anthony's Star
Umi 18h58'56" 89d7'41"
Christy's Ocean
Del 20h36'58" 15d7'9"
Christy's Star
And 0h55'47" 35d48'56"
Christy's Star
Uma 11h24'1" 59d55'40"
Christystar
Mon 7h56'40" -0d49'49"
Chrisula
Lyn 7h1'58" 53d53'46"
ChrisUrs
Umi 15h7'26" 67d44'50"
ChrisWard
Cap 20h34'47" -15d8'29"
chriswatson297
Dra 17h48'38" 51d9'26"
Chrisy Williams
Cnc 9h5'4" 12d7'2"
Chrögeli my love
Cep 22h37'48" 67d10'52"
Chronicle
Cnc 8h41'7" 8d58'3"
Chrysalis
Sco 17h51'50" -36d7'12"
Chrysalis Andrea Bennett
Umi 15h48'59" 75d39'52"
Chrysalis Montessori
Uma 9h40'11" 51d43'13"
Chrysanthi
Uma 9h24'49" 56d32'35"
Chrysi Kavalidis 13.05.04
Tau 3h30'58" 2d4'39"
Chrysi "Mummy" Katsamoundis
Ori 5h41'38" -1d54'37"
Chryslin
Lyn 7h34'45" 48d0'9"
Chrysos's Shepherd
Lyn 9h1'34" 38d8'49"
Chrysoula
Uma 11h34'46" 61d48'44"
Chrysoula & Yiannaki I.X.
Cas 2h19'29" 74d41'55"
Chryssanthy and her Joey Truffles
Lyr 18h46'7" 40d30'37"
Chrystabel
Leo 11h0'53" 15d49'50"
Chrystal
Ori 5h14'23" 7d26'32"
Chrystal
Ori 5h52'50" 7d4'19"
Chrystal
And 0h31'29" 38d56'44"
Chrystal
Cap 21h45'51" -13d10'42"
Chrystal Bowles
Ori 6h15'34" 10d22'29"
Chrystal Brilliance
Leo 11h53'58" 25d41'6"
Chrystal Clear
Cyg 20h30'25" 56d51'32"
Chrystal Huson
Sgr 18h38'42" -18d40'17"
Chrystal Jean
Uma 11h50'51" 44d18'59"
Chrystal Lee Watts
Ari 3h14'33" 20d11'36"
Chrystal Renae Retzlaff
Lyn 7h27'53" 54d32'25"
Chrystal Rose
Sco 17h20'59" -42d36'44"
Chrystal Tideman
And 0h56'56" 23d36'29"
Chrysteen Lenore Buford
And 23h31'56" 48d26'27"
Chrysti Lyn Stecker
And 1h50'25" 36d13'43"
Chrystina
Vir 14h4'14" -7d12'32"
Chrystine Major
Uma 8h48'35" 56d38'54"
CHT
Sgr 17h59'15" -27d6'58"
Chtistopher Edward Hawkins
Lib 15h35'15" -8d31'28"
Chuanchom
Sgr 18h16'59" -29d48'59"
Chub Snuggle Wub
Cyg 19h28'33" 52d52'37"
Chubb
Psc 1h8'40" 19d30'18"
Chubby Face
Umi 16h18'43" 76d24'33"
CHUBI
Pic 5h37'19" -46d10'25"
- Chubs -
Cru 12h37'59" -59d42'12"
Chuchi
Uma 10h27'14" 56d55'11"
Chuchi
Leo 11h9'46" 9d18'9"
Chuchi
Vir 14h18'19" 0d15'34"
Chuck
Cnc 8h38'58" 19d41'59"
Chuck
Aur 5h23'17" 30d52'23"
Chuck
Her 17h36'4" 32d4'46"
Chuck
Cyg 19h37'31" 52d28'19"
Chuck
Per 3h8'35" 47d16'35"
Chuck
Uma 11h10'14" 60d16'12"
"Chuck"
Sco 16h14'32" -10d28'57"
Chuck 75
Leo 9h55'47" 29d23'3"
Chuck AKA Herman
Lib 15h33'0" -19d2'48"
Chuck and Amy Star
Lyr 19h17'49" 29d42'27"
Chuck and Barbie - Forever
Cyg 20h34'44" 41d29'4"
Chuck and Carol's Celestial Ruby
Sge 19h41'51" 18d51'36"
Chuck and Christi Neff
Uma 10h31'37" 58d36'44"
Chuck and Christie Kozlik
Cyg 20h33'28" 43d29'26"
Chuck and Elyse's True Love Star
Leo 10h31'15" 13d53'2"
Chuck and Heather Forever
Cyg 21h53'56" 50d6'3"
Chuck and Holly Hradisky
Cas 1h36'12" 64d13'8"
Chuck and Kylene's Dream
Psc 1h8'56" 14d52'24"
Chuck and Laura P. Schmutzer III
Per 4h33'25" 41d7'44"
Chuck and Mary Divin
Cyg 20h54'40" 47d1'19"
Chuck and Tara
Ori 6h14'58" 15d40'54"
Chuck Andrews
Sco 16h52'23" -33d8'5"
Chuck & Andy Cassity Mr.& Mrs Santa
Cyg 20h4'56" 54d17'1"
Chuck & Audrey Maciunas
Cyg 19h17'21" 53d26'35"
Chuck Bishop
Uma 11h23'11" 56d53'30"
Chuck Brans
Lyn 8h8'10" 53d35'16"
Chuck Bryan
Sco 16h47'35" -27d21'1"
Chuck Cahoe
Aur 5h17'40" 41d57'4"
Chuck Chavez
Sco 17h34'1" -41d34'53"
Chuck Curtis
Cep 21h22'54" 76d46'36"
Chuck & Debbie Hill
Sco 17h22'12" -38d51'22"
Chuck & Diane DeVaughn Stokes
Cyg 20h43'57" 43d41'15"
Chuck Doyel
Lib 15h54'10" -9d44'50"
Chuck Ferrell
Per 2h15'17" 56d29'17"
Chuck Fouser
Tau 5h33'36" 18d15'10"
Chuck Fryer 747 Captain
Aql 19h15'6" 2d54'15"
Chuck Gorby
Tau 3h29'27" 24d57'24"
Chuck H. Powell IV
Aqr 22h14'2" -1d31'21"
Chuck Hayes
Uma 11h10'17" 48d37'59"
Chuck Hershberger
Her 17h37'27" 35d50'17"
Chuck Huffman
Lyn 8h22'22" 43d6'33"
Chuck Jackson Superstar
Cyg 20h30'23" 30d42'3"
Chuck & Joan "Together Forever"
Sge 20h3'45" 17d6'47"
Chuck & Julie Forever
Cyg 21h49'29" 38d45'44"
Chuck & Kris Pratt - Forever
Sgr 18h10'5" -16d23'52"
Chuck LaRue
Leo 11h44'10" 19d27'6"
Chuck Marshall
Per 3h30'29" 47d0'39"
Chuck Mascari
Tau 4h44'49" 20d34'0"
Chuck -n- Laurey
Cyg 20h59'0" 35d16'29"
Chuck & Nancy
And 0h16'41" 27d25'53"
Chuck Norris
Pho 0h33'58" -41d22'10"
Chuck 'ol Boy
Ari 2h10'51" 25d35'0"
Chuck O'Star
Her 18h10'38" 18d15'21"
Chuck Rainey
Sgr 18h7'20" -27d43'1"
Chuck Shuptrine
And 2h24'31" 44d50'0"
Chuck Siefert
Sco 16h12'52" -11d12'3"
Chuck Slaton
Ori 5h59'46" 19d4'17"
Chuck & Tammy Scott
Lyr 18h46'50" 40d52'52"
Chuck the Fisherman
Uma 10h51'16" 56d20'19"
Chuck & Tina
Cyg 19h49'18" 33d20'11"
Chuck Townsend
Lib 14h31'0" -21d21'36"
Chuck Tringale
Lib 14h53'58" -3d9'34"
Chuck-Bud 30
Ori 5h30'41" 6d36'33"
Chuckie
Aur 5h38'10" 40d14'34"
Chuckie and Ollie
Uma 8h34'51" 63d18'33"
Chuckie Beyer
Uma 14h3'23" 49d52'47"
Chuckie's Patty-cakes :-)
Sge 19h51'11" 18d58'16"
Chuckles
Her 16h12'30" 45d23'42"
ChucknSteve's Anniversary
Vir 12h38'10" 6d51'1"
Chuck's Bad-To-The-Bone Birdhouse
Ari 2h7'36" 16d58'21"
Chuck's Belt
Ori 5h28'21" 3d35'37"
Chuckstar
Uma 12h9'32" 54d47'20"
ChuckVirgieShillingford-11
Lyn 8h25'41" 53d36'13"
Chucky
Tau 4h26'28" 22d22'2"
Chudika Mahadevan
Cap 20h18'2" -15d23'18"
Chuen Li Bell
Cap 21h51'24" -23d33'2"
Chuffy
And 23h52'24" 46d8'5"
Chuggy
Cyg 19h36'59" 52d7'24"
Chui lee
Cap 21h19'26" -15d25'21"
Chuk Hillman
Sco 16h26'8" -18d38'9"
Chula Agacia
Aqr 21h39'27" 0d32'8"
Chulack 13.03.1976
Uma 9h30'58" 62d34'12"
chullie
Del 20h34'14" 13d45'7"
Chumbo
Ori 6h10'34" 6d48'19"
CHUMKI CHATTORAJ
Aql 19h46'27" -0d42'35"
Chumley
Ari 3h1'51" 24d29'55"
Chun Cun
Aqr 20h52'25" 0d31'6"
Chun & Yee With Good Luck 11-12-70
Lyn 7h26'32" 45d17'46"
ChunChira
Cnc 8h43'19" 15d8'2"
Chung-Hsuan Lin
Aql 19h18'48" 16d1'45"
Chung's Star
Leo 10h4'55" 17d16'52"
Chunkbutt
And 0h53'50" 38d28'28"
Chunky
Leo 9h22'17" 20d19'38"
Chunky
Ori 6h14'55" 15d5'16"
chunky monkey
Aqr 21h44'40" 1d58'47"
Chunky Monkey
Uma 10h4'57" 64d23'46"
Chunky Monkey
Uma 12h32'37" 57d2'22"
Chunnah ben Yitzchak v'Malka
Per 3h52'55" 48d41'0"
Chus
Ori 6h1'4" 20d47'39"
Chusetts
Uma 8h53'27" 63d21'19"
Chu-Sheng Yang
Sco 16h16'30" -10d42'25"
Chusina
Ori 5h49'47" 12d21'30"
Chutney
Cap 21h3'59" -26d43'5"
Chuto
Gem 7h33'25" 33d49'17"
Chyanne
Sco 16h9'40" -21d31'40"
Chyi Chyi
Leo 11h26'28" 11d52'35"
Chyi Lyn
Leo 9h53'2" 23d56'57"
Chylander Martin
Cyg 21h25'8" 33d53'32"
Chyna Rose
Peg 22h28'34" 4d25'31"
Chynna Paige Smyth
Tau 3h43'42" 3d33'22"
Chynna Shipp
Leo 9h55'31" 6d26'34"
Chyren
Leo 10h18'50" 18d53'6"
Ciadellys
Crb 15h37'30" 32d53'52"
Ciamo
Cas 23h9'31" 58d8'42"
Cian Annan Lynch
Uma 10h34'49" 48d1'28"
Cian Daniel
Peg 21h35'33" 16d21'59"
Cian Joseph Crowley
Psc 1h25'46" 22d54'33"
Cian Patrick Marks
Umi 16h26'27" 72d57'45"
Cian Sullivan
And 0h16'15" 28d2'47"
Cian Whelan
Per 3h19'12" 43d28'56"
Ciana Marie
Com 13h12'17" 17d43'50"
Cianci Jr.
Leo 10h52'45" 1d57'42"
Cianciola
Lib 15h46'42" -25d15'20"
Cianfrani
Ori 5h1'2" -0d36'40"
Ciann
Leo 10h55'5" 19d7'54"
Cianna's Wish
Umi 14h46'34" 68d30'25"
Ciara
Cma 6h24'49" -12d15'44"
Ciara 040902
Peg 22h18'42" 17d35'47"
Ciara A. Naylor
Peg 21h45'49" 15d8'26"
Ciara Anastasia
Psc 0h20'53" 6d19'33"
Ciara and Mark McCrohan
Aql 20h22'19" -8d10'22"
Ciara (Baby Girl)
Gem 7h16'3" 22d37'28"
Ciara Byrne
Peg 21h29'40" 18d12'54"
Ciara de los Reyes
Uma 8h53'47" 51d26'55"
Ciara Elizabeth Hart
Cas 1h1'33" 65d30'43"
Ciara Farley
Vir 11h43'26" -0d56'43"
Ciara Farr
Cas 1h5'17" 63d19'38"
Ciara Grace
And 0h21'32" 38d54'5"
Ciara Jordan Pimental
Ari 2h28'28" 26d53'21"
Ciara Kimberly Willis 8:58 a.m.
Psc 1h55'45" 6d18'13"
Ciara Louise
And 1h33'22" 41d13'56"
Ciara Margaret Hughes
And 23h14'23" 52d52'8"
Ciara Marie Ellis
Aqr 22h40'23" -2d45'40"
Ciara Marie Palillero
Tau 4h31'11" 17d19'58"
Ciara Neave Freya Woolf
And 2h29'47" 48d54'2"
Ciara Nichole Miller
Aqr 22h8'22" -0d46'57"
Ciara Nicole Titus
Ori 5h28'39" 9d10'45"
Ciara Renee Steele
Lib 14h37'34" -9d57'51"
Ciara Rose O'Neill-Absher
And 23h22'26" 44d57'26"
Ciara Sosnowski
And 2h13'9" 45d28'49"
Ciara Tranaé
Crb 15h46'43" 29d8'37"
Ciara Victoria Stachnik
Ari 2h59'35" 25d51'19"

Ciara, Will you marry Ari?
And 0h16'1" 27d21'44"
Ciaral Lynn Anika
Crb 15h47'46" 27d24'32"
Ciaran
Boo 14h41'25" 21d35'35"
Ciaran George Kettles
Umi 10h45'54" 88d36'35"
Ciarán Noel Lally
Cep 21h40'49" 66d8'29"
Ciaran O Tuama
Aur 7h16'31" 42d48'1"
Ciaran Patrick Murphy
Dra 15h8'38" 59d26'44"
Ciara's Dream Star
Cap 21h32'50" -8d40'15"
Ciara's Star
And 2h34'50" 39d3'28"
Ciara's Star
And 2h3'41" 37d21'45"
Ciarra Arye Brown
Vir 13h34'56" 3d48'52"
Ciarra Nicole Sagona
Umi 14h8'40" 74d49'23"
Cibola Vista
Umi 11h10'27" 89d31'48"
Ciccio Salemme
And 1h58'31" 39d9'0"
Cicciolotto
Ori 5h10'28" 8d44'3"
Cicely
Cma 7h6'1" -29d37'4"
Cici De Lucio, The Princess Star
Sco 17h49'12" -43d0'32"
Cicunak Szerelemmel 2005
Sgr 17h44'10" -23d20'42"
Cid, Gertrude, and Arthur
Tau 4h21'17" 25d51'11"
Cielo Brillit 13032003
Lyr 18h42'26" 37d33'39"
Cielo Enviastes
Leo 11h48'49" 22d44'45"
Cielo Kimberly Carrier
Tau 4h11'39" 13d21'44"
Cienna Jadine Laird
And 23h34'32" 47d16'20"
Cienna Marie aka Bonnie 2003
Sco 17h18'34" -44d40'48"
Cienna Pyke
Ori 5h6'23" 10d35'55"
Ciepiel
Peg 22h17'46" 12d51'38"
Ciera Aspen Aguirre
Sgr 19h46'36" -12d25'28"
Ciera Damon
Umi 15h56'11" 80d20'11"
Ciera Rowley
And 0h34'39" 42d44'26"
Ciera's Light
Del 20h45'7" 7d5'7"
Cierra Ann Villegas
Aqr 22h14'35" -0d2'51"
Cierra Baker
Her 18h53'56" 23d40'43"
Cierra Dawn Naef
And 1h23'21" 47d23'18"
cierra jade
Ori 6h3'42" 9d58'43"
Cierra Josephine Bae Roach
Psc 1h9'4" 21d42'5"
Cierra Loretta Mosbrucker
Psc 1h8'1" 5d42'55"
Cierra M. Shannon
Cam 3h42'50" 59d55'30"
Cierra Rayne Burke
Aqr 21h13'8" -10d49'36"
Cierra Sather "My Guiding Light"
Ori 5h54'1" 21d52'52"
Ciesiel
Uma 12h42'46" 58d23'40"
Cignasty
Cnc 8h30'29" 11d4'52"
Cila
Uma 8h56'10" 62d21'0"
Cilantro"Cil" Rogers
Uma 10h3'7" 44d36'40"
Cilecia T
Aqr 21h47'41" -3d12'30"
Cileya
Ori 5h35'51" 11d0'30"
Cilia Ormesher
Lib 15h10'9" -5d52'59"
Cilissa Marie Leroux
Psc 1h8'45" 26d43'12"
Cilla
And 0h43'37" 32d3'24"
Cilla B from Berea
Psc 0h53'17" 15d47'40"
Cillian Mark
Aur 6h32'30" 38d2'27"
Cillian Niall Byrne
Umi 15h16'18" 80d11'15"
Cillian Shea
Vir 13h35'20" 5d20'15"
Cillian's Lucky Star
Uma 11h35'17" 37d32'5"
Cilou-151092
Uma 8h57'51" 51d35'22"
Cim Lee Ware
Lac 22h24'35" 37d14'36"
Cimpurcino
Aql 20h11'46" -0d31'38"
Cina Wong
Lyn 7h38'1" 36d1'20"
CINBA
Ari 2h38'51" 26d44'34"
Cinci
Uma 13h57'11" 54d25'51"
Cincy
And 1h52'49" 35d59'57"
Cinda June
Cyg 20h52'3" 37d6'26"
CindaWayne
Crb 16h8'58" 36d36'37"
Cindee
Cas 0h13'14" 59d22'35"
Cinder Road
Pho 0h41'55" -40d19'16"
Cinder & Smiley
Cmi 7h20'43" 8d44'7"
Cinderalla 26
Sgr 19h39'32" -13d38'16"
Cinderayla
And 0h12'56" 43d10'46"
Cinderbug 10963
And 0h42'28" 36d23'25"
Cinderella
And 0h18'12" 36d46'53"
Cinderella
Uma 8h53'32" 51d0'33"
Cinderella
Vir 13h18'41" 5d23'13"
Cinderella
Aqr 21h53'55" 1d9'5"
Cinderella
Tau 4h35'33" 28d22'56"
Cinderella
Cam 7h33'8" 77d55'46"
Cinderella (aka Kara)
Ari 2h43'59" 24d55'13"
Cinderella & Jonathan
Cyg 20h46'4" 47d0'46"
Cinderella's Castle
Tau 4h1'2" 21d29'1"
CindeReno
Dra 17h25'16" 58d51'57"
CinderKelly
Lib 14h50'36" -1d9'30"
Cindi
Ori 5h49'7" -2d41'0"
Cindi
Cas 23h43'33" 56d7'59"
Cindi
Lyn 6h41'8" 61d5'55"
Cindi
Psc 1h53'11" 8d38'33"
Cindi
Lyr 19h24'28" 37d50'22"
Cindi
Lmi 9h41'51" 38d37'49"
Cindi Champlin
Lib 14h57'43" -6d22'35"
Cindi Diane Baker
Lmi 10h52'58" 28d27'28"
Cindi Farr
And 1h50'20" 45d8'54"
Cindi Haynes Store Manager MNO 007
Cas 1h25'18" 55d2'51"
Cindi Moore
Ori 5h23'51" 13d50'11"
Cindi & Ron's Star
Crb 16h19'54" 26d44'21"
Cindi Summers
Peg 22h44'26" 14d49'5"
Cindi Yeager
Cma 6h25'12" -15d32'17"
CindiJ40
Uma 11h23'28" 35d35'50"
Cindolph
Tau 3h54'47" 29d6'37"
CinDrew
Lyn 7h50'25" 42d6'21"
Cindrick
Tau 4h11'57" 9d36'52"
Cindy
Ari 2h45'56" 13d33'36"
Cindy
Vir 13h2'6" 2d53'0"
Cindy
Del 20h51'54" 12d11'19"
Cindy
Tau 4h15'21" 16d47'34"
Cindy
Tau 5h33'41" 23d28'23"
Cindy
Cnc 9h18'49" 25d51'8"
Cindy
Ari 2h22'39" 26d49'54"
Cindy
And 0h23'18" 26d27'51"
Cindy
And 0h26'10" 25d11'28"
Cindy
Her 16h56'34" 32d28'40"
Cindy
Gem 6h49'2" 30d24'22"
Cindy
Cyg 21h12'15" 45d14'48"
Cindy
Cyg 21h17'18" 45d3'1"
Cindy
Cyg 20h4'45" 52d17'14"
Cindy
Lib 15h23'27" -10d14'1"
Cindy
Lib 15h54'31" -17d9'46"
Cindy
Lib 14h53'56" -2d59'37"
Cindy
Cas 23h47'24" 61d30'26"
Cindy
Uma 9h21'41" 71d22'9"
Cindy
Lyn 7h17'16" 54d57'9"
Cindy
Cap 21h26'19" -25d37'34"
Cindy and Bob Forever Love
Lmi 10h50'29" 34d10'54"
Cindy -Angel Eyes- Urena
And 0h20'46" 26d35'43"
Cindy Ann
And 0h20'22" 38d0'39"
Cindy Ann
Sgr 18h46'51" -26d18'37"
Cindy Ann LaVine
Ori 5h19'45" -3d22'36"
Cindy Ann Stellato
Aqr 22h41'59" -0d59'33"
Cindy Ann's Light
Ari 2h45'52" 21d14'20"
Cindy Arthur
Uma 10h31'8" 59d57'54"
Cindy Au
Cap 21h5'24" -19d37'14"
Cindy Barnette
And 0h45'24" 36d36'30"
Cindy Bennett
Lyr 18h31'35" 36d29'7"
Cindy (Bindy) Boehringer 9/13/2004
Uma 11h32'27" 35d42'6"
Cindy Brenner
Gem 6h53'37" 13d44'13"
Cindy Briney
Lib 15h47'51" -9d47'32"
Cindy Bures
Psc 2h3'44" 6d39'59"
Cindy Burley
Cap 21h1'12" -15d59'28"
Cindy C
Ari 3h5'1" 24d56'47"
Cindy C. Alvarez
Vir 13h42'27" -3d1'29"
Cindy Capps
Uma 11h15'30" 43d14'1"
Cindy - CARA of the universe
Aqr 22h1'18" -14d30'24"
Cindy Caughman Corona
Crb 16h15'59" 32d29'1"
Cindy Chambless
And 0h20'57" 37d42'7"
Cindy Chong
Uma 8h48'59" 51d19'5"
Cindy Cline
Cyg 20h11'0" 41d35'23"
Cindy Clore
Uma 8h26'10" 70d58'48"
Cindy Cobb
Ori 6h6'58" 21d16'25"
Cindy Couturier
Ori 5h34'24" 10d33'55"
Cindy Cusano
Sco 17h36'7" -44d28'50"
Cindy D. Besel
Vir 13h26'37" 7d14'26"
Cindy Dindy
Cyg 20h16'19" 58d43'48"
Cindy & Dino
Lmi 9h24'39" 34d12'38"
Cindy DiPadova
Aqr 21h52'35" -1d33'37"
Cindy Dixon
Sco 17h14'46" -44d30'55"
Cindy Doo Vasquez
Crt 11h2'32" -18d25'22"
Cindy & Doug, Soulmates Forever
Cyg 21h34'32" 46d19'32"
Cindy Douglas Super Nova
Mon 7h59'33" -7d20'5"
Cindy Eastman
Crb 15h55'41" 38d58'23"
Cindy Falcone
Lac 22h20'5" 50d4'46"
Cindy Ford
Sco 17h9'51" -40d54'40"
Cindy Fortuno
And 23h21'0" 42d54'40"
Cindy Frayer World's Greatest Nurse
Cas 0h50'43" 56d37'20"
Cindy G. Morrison
Cyg 19h49'8" 32d42'43"
Cindy Godwin
Cra 18h6'51" -37d28'9"
Cindy Goss
Vir 12h31'17" -3d32'7"
Cindy Granata
Cyg 20h13'36" 47d23'17"
Cindy Grant Sprewell, You Are Loved
Ari 3h9'5" 28d15'56"
CINDY GREEN
Sco 17h32'38" -42d47'19"
Cindy Hayek
Cyg 19h44'47" 33d50'51"
Cindy Hopkins
Leo 11h4'5" -0d8'50"
Cindy Huffman
Ari 2h3'57" 13d4'1"
Cindy Jane Kilmer
Lyn 9h27'37" 40d24'6"
Cindy Janeth Rosales
Leo 11h0'50" -5d17'53"
Cindy & Jeff Ellis
Umi 15h31'43" 74d13'53"
Cindy & Jim's Place
Cyg 19h47'0" 51d31'47"
Cindy Jo Keck
Lib 15h44'25" -18d36'22"
Cindy Johnson
Sgr 19h42'14" -13d37'10"
Cindy Joseph
Cyg 21h26'19" 36d26'3"
Cindy Kay
Uma 9h16'51" 71d23'6"
Cindy Koch
Uma 10h11'33" 59d40'45"
Cindy Kramme
Aqr 20h55'34" 0d7'42"
Cindy L. Campbell
Lmi 10h40'25" 30d6'0"
Cindy L Mcneice
Cnc 8h48'50" 21d25'47"
Cindy L. Webster (Angel Eyes)
Uma 12h48'4" 62d10'7"
Cindy Lee Perryman
And 1h33'39" 39d40'17"
Cindy Lee Wingate
Sgr 19h7'55" -13d46'58"
Cindy Leppell
Cap 20h56'54" -19d31'15"
Cindy Liu
Aql 19h0'8" 8d20'57"
Cindy Lomanto
Aur 5h3'22" 40d27'20"
Cindy Long
And 23h17'1" 35d56'31"
Cindy Loo
And 0h48'51" 43d35'58"
Cindy Lou
And 0h16'44" 44d59'31"
Cindy Lou
Uma 11h14'44" 51d8'37"
Cindy Lou
Psc 1h23'46" 7d29'29"
Cindy Lou Dixon
Lyn 7h42'51" 47d30'52"
Cindy Lou OH
Mon 6h47'58" -0d3'9"
Cindy Lou + Two ~ 5/23/57
Gem 6h42'25" 24d25'15"
Cindy loves Jenna
Ari 1h59'36" 23d40'39"
Cindy Lowe
Aqr 21h11'28" -3d5'35"
Cindy Lynn Aune
Cas 1h47'11" 63d2'3"
Cindy M. Franklin
Uma 11h11'16" 53d47'23"
Cindy M. Hansen
Sco 16h13'41" -16d18'56"
Cindy Mairena
Tau 4h19'57" 18d37'24"
Cindy Makepeace
Cas 1h2'13" 61d12'5"
Cindy Mallett
Vul 19h24'51" 24d22'55"
Cindy Marie
Lib 15h58'49" -18d40'43"
Cindy Marie Domingue-Hendrickson
Ari 1h58'41" 23d36'31"
Cindy Marie Ellis
Cyg 20h27'26" 40d54'53"
Cindy Marie Hayhurst
Cas 0h41'2" 61d59'0"
Cindy Marie Kindle
Cas 1h40'57" 62d38'24"
Cindy Marie Koenigsknecht
Cas 1h23'18" 64d9'41"
Cindy Marie Stepp
Psc 23h40'0" 7d6'51"
Cindy & Mathéo Paul
Uma 11h41'6" 37d3'36"
Cindy Mayfield
And 0h43'53" 40d28'51"
Cindy Mulder
Uma 8h36'2" 49d43'22"
Cindy My Love
Uma 9h4'48" 49d22'29"
Cindy N. King
Vir 13h23'45" -19d43'58"
Cindy Neal
Lyn 8h13'45" 44d53'1"
Cindy Newville
And 1h15'2" 38d32'16"
Cindy P.
Per 2h10'40" 57d0'47"
Cindy Pham
Leo 11h11'35" 9d32'37"
Cindy Pindy 2665
And 1h54'43" 46d50'12"
Cindy Pratt's Star
Gem 7h52'0" 26d14'5"
Cindy Rachell Muñoz-Filson
Sgr 18h40'35" -34d57'3"
Cindy Rae Stinner Tomai
Cnc 8h44'57" 15d59'19"
Cindy Reilley
Cyg 21h37'57" 37d48'57"
Cindy Renee "Scooby"
Dra 10h51'4" 73d54'47"
Cindy Reyna
Crb 16h8'14" 38d23'10"
Cindy Rossiter
Cyg 19h37'20" 33d47'24"
CINDY {RYNNE} VENDAL
Vir 12h44'32" -8d58'14"
Cindy Sabino
And 0h57'21" 45d51'36"
Cindy Saucedo
Vir 12h36'10" 6d8'4"
Cindy Schafer
Crb 16h11'5" 38d10'54"
Cindy Schleuss
And 1h26'31" 36d31'2"
Cindy Schrimpf
Aqr 20h50'34" -13d25'28"
Cindy Sider
Cas 23h29'30" 51d6'13"
Cindy & Steve Marks
And 1h55'4" 38d59'29"
Cindy Stowers
Crb 15h25'36" 27d3'59"
Cindy Sue Embrey
Uma 9h53'41" 43d29'4"
Cindy Sue Korman Hodkin
Lib 14h50'35" -0d37'24"
Cindy Sue's Huckleberry
And 23h23'32" 42d55'3"
Cindy Sweetheart
And 23h21'37" 48d20'33"
Cindy Thompson
Uma 10h43'37" 52d2'8"
Cindy Torna
Gem 6h46'44" 23d47'20"
Cindy Turbow
Aqr 21h21'34" -8d32'32"
Cindy u Jens
Ori 5h26'35" 3d7'41"
Cindy Vanderaa
And 0h29'50" 25d57'56"
Cindy & Victor
And 2h3'9" 40d44'47"
Cindy Vu
And 0h52'55" 35d56'42"
Cindy Walczak
Uma 9h48'36" 46d8'4"
Cindy Walls
Cnc 7h58'13" 12d29'41"
Cindy Walsh
Cyg 19h47'38" 33d20'41"
Cindy & Walt
Cyg 20h40'43" 41d52'59"
Cindy Warren
Vir 13h7'11" 5d54'56"
Cindy Wilhelmsen Love of My Life
Gem 7h11'59" 21d33'54"
Cindy Williams
Crb 15h32'56" 26d59'33"
Cindy Yukie Morikawa
Gem 7h42'2" 14d5'46"
Cindy Zaso
Cnc 8h41'12" 7d15'21"
Cindy Zibierski
Vul 21h3'29" 21d46'58"
Cindy, a friend for life!
Cyg 21h1'22" 47d2'24"
Cindy, Forever Grateful
Cnc 8h13'30" 15d21'0"
Cindy, Louie & Ryan
Cas 0h57'48" 60d40'5"
Cindy, My Love
Uma 10h16'1" 64d51'39"
cindyacerno
Mon 7h20'34" -10d19'24"
CindyB
Lyn 8h0'53" 37d44'25"
Cindyku Sayang ^o^
Cap 20h11'6" -9d40'29"
Cindylamb
And 2h36'48" 45d47'27"
Cindy.M.Plasencia
Lyn 9h9'18" 39d56'7"
Cindy's "Ditty Dipper"
Lmi 11h5'29" 25d50'0"
Cindy's Heavenly Body
Sgr 18h31'2" -16d3'33"
Cindy's Light
Cyg 20h41'53" 36d27'54"
Cindy's Star
Lyn 8h46'40" 41d40'1"
Cindy's Star
Vir 13h35'37" 2d20'39"
Cindy's Star
Ori 6h24'28" 13d10'11"
Cindy's Star
Sco 16h14'25" -17d0'0"
Cindy's star
Lib 15h10'32" -5d20'5"
Cindy's Star
Mon 7h32'55" -0d57'47"
Cindy's star
Dra 19h12'58" 69d30'9"
Cindy's Star "Her Giving Heart"
And 0h27'45" 21d54'0"
Cindy's Twinkie
Gem 7h19'51" 32d5'55"
Cindy's Twinkling Smile :)
And 1h4'1" 33d43'55"
Cindy's Valentine Star, From Chris
And 0h12'38" 43d7'3"
Cindystar
Lmi 9h56'40" 35d55'32"
Cindy-Sue
Her 17h14'3" 28d36'54"
Cindysue
Sgr 19h3'43" -31d12'5"
Cinematic
Psc 1h16'46" 11d4'14"
Cinja
Uma 12h2'30" 49d28'42"
Cinjerdondoog
Pup 8h8'14" -21d56'2"
Cinnamon
Vir 12h9'51" -0d53'15"
Cinnamon
Vir 13h37'45" -11d26'40"
Cinnamon
Cap 20h9'22" -23d37'18"
Cinnamon
And 0h27'39" 36d40'44"
Cinnamon
Psc 1h17'39" 12d13'15"
Cinnamon Dulcy Bennett
Umi 15h14'2" 73d14'57"
Cinnamon Marie Thompson
Ari 2h36'49" 27d2'16"
Cinnamon Sunshine Mirigian
Uma 11h20'37" 54d48'57"
Cinque Sorelle
And 1h18'34" 46d33'45"
Cinta
Sco 17h53'37" -30d7'8"
Cinthea Thomsley Coleman
And 2h16'12" 39d53'2"
Cinthia C. Arduengo
Uma 9h6'16" 68d16'37"
Cinthia Claros
Cas 1h47'43" 60d32'24"
Cinthia L. CXLIII Fleur O'Neill
Ori 5h33'0" 5d16'37"
Cinthia Lynn Irvine
Ori 6h17'41" 15d15'1"
Cinthya
Her 16h42'2" 28d59'37"
Cinthya and Colton
Leo 11h43'17" 24d24'43"
Cinthya Islas
Sco 17h35'38" -30d44'58"
Cinthya Yesenia Newburn
Tau 5h10'2" 18d7'26"
Cintia Bravo
Gem 6h45'52" 14d42'10"
Cintia Pla Garcia
Tau 5h47'37" 21d30'40"
Cintilacao
Cma 6h58'7" -15d21'5"
Cinzia
Cma 6h47'1" -11d41'14"
Cinzia
Umi 13h41'4" 77d21'42"
Cinzia
Ori 6h9'49" 5d40'19"
Cinzia
Aur 6h1'53" 52d49'13"
Cinzia Bertolin
Boo 15h6'48" 24d46'46"
Cinzia Bianchi
Peg 22h38'24" 18d46'22"
Cinzia Biondi Giunta
Her 17h17'11" 48d12'24"
CINZIA BRILLY
Cas 1h20'27" 65d52'25"
Cinzia Cattaneo
Her 17h18'28" 16d9'11"
Cinzia Cazzaniga
Per 3h39'20" 40d53'16"
Cinzia e Federico
Cas 23h31'25" 53d7'45"
Cinzia e Luca
Peg 23h11'17" 31d39'51"
Cinzia la top
Cam 9h25'30" 82d2'23"
Cinzia Lenzi
Aur 5h17'49" 40d41'25"
Cinzia Orsi
Umi 17h32'28" 80d11'38"
Cinzia's Shining Star
Cam 4h38'17" 68d35'52"
Ciocia Gene
Uma 10h22'48" 57d56'49"
Ciocia Leg
Tri 2h23'29" 32d20'2"
Ciocie Kathy
Sco 17h56'19" -30d8'39"
Ciocie Suzy
Sgr 17h55'44" -29d44'24"
Cioli
Cam 5h34'23" 71d30'27"
Cion
Uma 10h46'13" 39d21'44"
Cion Aeon
Cnc 9h8'41" 32d14'7"
Ciota D
Lyr 18h15'33" 31d49'22"
Cipollina
Uma 9h50'51" 62d3'51"
Cipriano Garza
Ori 5h26'32" 5d38'53"
Circinus Gareth Enfys
Cir 14h49'13" -64d34'14"
Circle
Uma 10h6'27" 69d27'11"
Circle of Friends
Ori 6h9'54" 19d22'16"
Circle W
Leo 11h33'58" 19d32'9"
Cirena Faye Gioia
Cas 1h28'49" 62d39'35"
Ciri Falcone
And 0h34'42" 23d18'25"
Cirino "Charles" Lanzafame
Uma 8h36'24" 57d54'10"
Ciro Bello De Mi Alma
Vir 13h13'17" 8d13'35"
Ciro Gatta
Ori 6h6'11" 11d38'14"
Cisco
Uma 10h46'32" 71d24'58"
Cisco and Megan del Valle
Uma 12h46'54" 57d51'8"
Cisco " The One" and Nataly
Cnc 7h58'32" 17d18'36"
CiscoAndJess
Lyr 19h10'20" 37d58'6"
Ciss
Ori 6h7'13" 10d20'15"
Cissy
Mon 6h38'33" 1d11'48"
Cissy Tang
Uma 12h15'45" 60d14'45"
Citlali
Cyg 20h43'9" 38d26'24"
Citlali9
Psc 1h5'4" 32d6'47"
citmaDon
Uma 11h26'8" 71d5'8"
City Bowee and Country Girl
Ori 5h34'36" 10d46'2"
City of Angels
Uma 11h31'4" 41d28'7"
Ciumpy
Cyg 20h9'5" 36d40'33"
Civita Papagni
Lyn 7h39'50" 58d29'21"
Cizzle Gail Caliva
Crb 15h44'3" 31d5'23"
CJ
Per 4h25'42" 44d27'32"
CJ
Ori 5h4'58" 10d41'51"
CJ
Ori 4h50'26" 11d49'57"
CJ
Psc 0h34'36" 9d28'28"
CJ
Tau 5h46'12" 22d26'27"
CJ
Ari 2h27'11" 19d44'46"
CJ
Her 16h19'52" 19d5'9"
CJ
Peg 23h17'51" 16d7'7"
CJ
Aql 18h42'54" -0d32'50"
CJ
Aql 19h29'36" -10d42'57"
CJ
Sgr 19h41'27" -16d38'35"
CJ
Sgr 20h15'28" -44d8'57"
CJ
Cap 20h29'50" -26d50'17"
CJ and ALLISON.... 3/14 In love
Cyg 20h11'12" 55d27'7"
CJ and Bea's Wishing Star
Cyg 20h35'40" 43d10'6"
CJ and Chuck Franek
Gem 6h36'32" 13d21'8"
CJ and Jill Angersbach
Uma 8h41'7" 65d53'14"
CJ Blaine
Gem 7h43'13" 31d20'27"
C.J. Boling
Ori 5h28'57" 14d40'22"
CJ Brooks
Ori 5h30'47" -0d33'11"
C.J. Buckmaster's Light
Uma 10h33'5" 58d30'34"
CJ Cluster
Uma 10h16'32" 45d24'38"
C.J. Cuddy
Lyn 7h13'33" 54d28'42"
C.J. FORBES
Leo 10h10'57" 21d46'32"
C.J. Forever
Cas 0h32'57" 61d39'17"
C.J. II
Leo 10h53'3" -2d28'36"
CJ Mamamghani "50"
Umi 15h3'27" 78d13'34"
CJ MASTERS
Her 18h3'17" 15d4'55"
CJ Mcauley
Leo 11h16'12" 13d7'5"
C.J. McCann
Aql 19h19'42" -10d23'30"
C.J. Mitchell
Crb 15h40'55" 25d37'3"
CJ Neuman, Destined for Greatness
Uma 10h31'57" 65d56'11"

C.J. Rembowski
Aql 19h59'21" 14d49'57"
CJ Rosado
Uma 9h44'3" 55d59'59"
CJ/TATA Star
Lyr 18h44'34" 39d53'26"
CJ. Vagnone's Star
Cyg 19h59'29" 38d38'54"
CJ Watkins
Cap 21h36'50" -17d16'3"
C.J. Wink - My One And Only
Uma 11h57'19" 41d25'5"
CJ Zimmer
Ori 5h8'41" 11d38'21"
CJ1066
Cyg 21h15'55" 44d25'20"
CJ-A3
Uma 11h20'39" 47d28'43"
cjc123162
Uma 11h53'57" 40d51'26"
cjc123162
Aql 19h3'14" -11d10'54"
CJD
Dra 18h53'3" 53d28'25"
cjdm2006
Cyg 20h8'10" 52d13'16"
CJH073177most cherished one
Uma 11h10'20" 60d33'20"
CJK 218
Aqr 21h34'16" -0d5'37"
CJM
Cam 4h23'48" 66d32'46"
CJM 50
Aqr 22h27'13" -23d0'40"
CJM1 (CallieJospehineMcGillis)
Lmi 10h28'44" 36d56'38"
CJN'S Lucky Star
Lyn 8h39'41" 46d7'31"
CJoe DiMatteo
Ari 2h54'55" 27d27'32"
cjs
Gem 6h34'43" 24d12'59"
CJ's Duzzie Star
Sco 16h47'41" -37d43'13"
CJ's Guiding Light
Pho 23h59'34" -46d5'42"
CJS + JCW Forever
Pho 1h29'37" -47d39'56"
CJ's love will always shine on us!
Cap 20h26'4" -12d34'8"
CJ's Shining Heart
Lib 14h41'48" -12d23'7"
Cj's Star
Gem 6h58'30" 22d18'30"
CJSmith2
Lyn 7h40'58" 42d40'57"
cjuni
Psc 23h46'7" 5d27'47"
CJVinz27
Aqr 23h3'49" -5d55'1"
CJWJRZ4EVA
Uma 11h40'43" 46d27'35"
C.K. & L.J. 5/28/05
Cyg 19h35'12" 28d42'38"
C.K. McDonald
Crb 16h11'3" 28d29'4"
CK Yagel
Ari 3h5'55" 15d21'11"
CK22
Cet 1h21'10" -0d32'32"
C-K-L-ove
Gem 6h40'19" 18d34'5"
CKO
Aqr 22h38'20" -6d40'38"
Ckrisarm
Crb 16h1'31" 36d35'22"
CKSF4EVER+SOME
Tau 4h45'12" 20d52'35"
C-L; C-S-V-L-J; FLOYD FAMILY
And 0h19'14" 25d46'17"
Cl. IV Nikki Basnight
Cam 4h1'28" 67d38'15"
Claddagh
Leo 9h50'10" 21d9'17"
Claes Paul Hegraeus
Cnc 8h24'21" 30d3'50"
Clair Boy
Cas 0h44'57" 57d4'20"
Clair Dean/Mamie Jo Whitehead Star
Lyr 19h20'10" 29d50'29"
Clair Drew Gorton
Ari 2h30'25" 18d45'43"
Clair Elizabeth Williamson
Uma 9h34'49" 60d43'16"
Clair Gould
Cas 23h17'24" 56d4'5"
Clair Katherine Cullen-Ottaway
Cas 2h17'56" 68d47'28"
Clair Louise
And 1h57'12" 37d41'58"
Clair Rooney
Cyg 21h32'14" 34d56'19"
Clair S. Sheaffer
Lyn 7h47'46" 35d58'26"
Clair T. Steggall
Aql 19h34'20" 3d21'30"
Claira De Jersey
Eri 3h55'53" -23d17'42"

Claire
Sgr 18h35'27" -22d31'3"
Claire
Sco 17h34'0" -45d27'35"
Claire
Uma 13h55'9" 59d10'56"
Claire
Aqr 20h41'8" -11d39'39"
Claire
Psc 1h43'4" 13d21'31"
Claire
Ori 6h20'32" 13d45'23"
Claire
And 2h23'12" 42d44'7"
Claire
Uma 11h12'56" 37d50'40"
Claire
Lyn 9h7'58" 39d39'7"
Claire
And 2h18'49" 46d38'18"
Claire
And 23h18'38" 42d57'29"
Claire
And 0h18'22" 45d46'48"
Claire 1
Psc 0h26'19" 8d43'44"
Claire 21
Cas 0h45'48" 51d51'0"
Claire A. Grinde
Cnc 9h11'17" 17d32'9"
Claire Aberasturi
Ori 5h19'2" -4d14'51"
Claire Adele-My Star
Cap 20h17'10" -10d15'55"
Claire Aine Traynor
Lyn 7h59'29" 45d34'55"
Claire Alexandra Anderson
Lib 15h1'25" -7d8'45"
Claire Aline Hogan
Gem 7h5'39" 16d57'16"
Claire Alpert
Cas 0h48'45" 57d54'44"
Claire Alyse
Sgr 20h19'4" -34d16'31"
Claire Alyson
Uma 9h51'42" 53d15'37"
Claire Amigo
Del 20h47'40" 14d18'39"
Claire and Bryan For Ever
Aqr 21h36'18" -0d53'57"
Claire and Dave
Ori 5h52'19" 9d11'15"
Claire and Emanuel G. Rosenblatt
Per 3h40'44" 34d2'33"
Claire and Kevin
And 23h37'33" 48d11'12"
Claire and Lucy Pollack McGovern
And 2h13'48" 45d28'47"
Claire and Michael
Ori 6h15'19" 2d34'26"
Claire and Robert Stokes
Cyg 20h38'51" 40d57'37"
Claire Angela Bellone
Cas 0h18'25" 47d30'50"
Claire Ann Havercamp
Cap 20h7'49" -21d40'10"
Claire Anne
And 0h23'41" 33d28'48"
Claire Anne Conlon
Psc 1h32'20" 14d4'43"
Claire Anne Nicole Bromley
Ori 6h4'24" 14d0'55"
Claire Ashely Wingerd
And 1h10'22" 42d54'39"
Claire Bailey Johnston
Crb 16h19'33" 32d42'25"
Claire Bear
Gem 6h42'13" 23d25'3"
Claire Bear
Cnc 8h35'38" 23d15'29"
Claire Bear
Sgr 18h3'57" -18d43'11"
Claire Bear
Uma 11h38'12" 63d36'44"
Claire Bear
Dra 18h53'32" 60d22'24"
Claire Bear Bowie
And 23h13'49" 43d0'1"
Claire Bear Douglass
Leo 10h57'36" 7d42'39"
Claire Bear's Bright Little Star
Ori 5h22'53" 3d12'28"
Claire Bears Star
Aqr 23h16'57" -10d17'34"
Claire Bernadine
Sco 16h52'6" -26d58'38"
Claire Binnie
And 2h21'32" 45d31'39"
Claire Bodrian Keefe
Gem 7h11'1" 32d42'46"
Claire Britt Dunn
Uma 8h26'23" 67d16'46"
Claire Brown
Psc 1h36'52" 26d28'15"
Claire Bruni Reed
Lyn 7h21'24" 49d24'36"
Claire Bunney
Uma 10h54'54" 41d55'12"
Claire C. Hindes
And 23h46'9" 41d27'4"
Claire Cabral
Lib 16h0'43" -7d16'30"

Claire Campbell
And 1h9'21" 43d4'14"
Claire Carroll Takes
Cas 0h18'16" 64d20'34"
Claire Catherine McDonough
Cnc 7h57'57" 15d18'35"
Claire Chenoa Arbour
Leo 10h10'48" 13d33'31"
Claire Cheyenne
Aql 20h24'39" 2d13'13"
Claire Colrein
And 0h52'14" 44d19'52"
Claire Connolly
And 2h23'18" 46d2'14"
Claire Cootes
And 1h11'15" 45d34'23"
Claire Cordeux Crowley
Aqr 22h57'49" -7d52'35"
Claire Cucchiari-Loring
Her 16h18'41" 47d44'6"
Claire Curristin Robinson
Umi 15h52'21" 81d19'56"
Claire Da Loon
Cas 1h32'58" 61d31'25"
Claire & Dave
Aql 19h15'35" 7d34'42"
Claire Davis Warthen
Lyr 18h49'42" 38d27'9"
Claire Destiny Sonnichsen
Tau 3h33'25" 17d1'9"
Claire Dianne
And 1h10'29" 37d59'29"
Claire Drake Fielding
And 1h27'43" 44d28'10"
Claire E. Lewis
Cnc 8h50'38" 30d48'16"
Claire E O'Neill
And 0h19'31" 25d51'13"
Claire Elise
Gem 6h47'39" 22d26'7"
Claire Elise
Peg 21h55'29" 12d58'33"
Claire Elise Cunningham
Sgr 18h21'16" -22d35'29"
Claire Elizabeth
Cnc 8h18'40" 22d44'41"
Claire Elizabeth
Cas 0h55'22" 47d51'23"
Claire Elizabeth Colyer
Uma 8h35'50" 52d35'13"
Claire Elizabeth Duguid
And 2h18'27" 39d11'33"
Claire Elizabeth Elliott
Lib 14h45'39" -18d38'17"
Claire Elizabeth Manley
Cas 0h34'7" 51d15'53"
Claire Elizabeth Shaughnessy
Umi 16h22'53" 72d25'57"
Claire Elizabeth Vandenbergh
Leo 11h55'24" 21d14'42"
Claire Elizabeth Weikle
Leo 9h26'10" 8d41'57"
Claire Emma Bentley
Cap 21h11'4" -19d52'0"
Claire Emma O'Neil
Cas 0h44'46" 63d11'37"
Claire Erin
Lib 15h16'57" -5d55'16"
Claire Faller
Cas 23h51'13" 53d43'35"
Claire Frances Brown
Psc 0h48'13" 15d1'24"
Claire Grubbs
Ari 2h1'31" 13d10'26"
Claire Guffroy
And 1h54'38" 45d56'6"
Claire Guignart
Uma 10h40'44" 44d52'53"
Claire Haley Masters
And 0h8'48" 43d13'35"
Claire Haven Hooper
Mon 7h8'57" -0d42'54"
Claire Helen Mackay - Claire's Star
Uma 10h9'5" 44d2'8"
Claire Helena Williams
Sco 16h41'28" -40d55'28"
Claire & Herb Falk 50 Years of Love
Cyg 20h19'31" 47d40'57"
Claire Hotchkiss Gervais
Leo 10h36'37" 14d25'19"
Claire Ingram
And 0h39'1" 38d52'50"
Claire Isabel Webb
Cyg 20h4'1" 46d35'58"
Claire Isabella Morano
And 2h16'55" 47d48'26"
Claire Isabella Rusoff O'Neill
And 23h50'45" 40d35'17"
Claire Jorja
Umi 13h7'23" 69d32'50"
Claire Julianna McQuade
And 0h17'2" 46d23'5"
Claire Kaido
Lyn 8h28'57" 34d49'45"
Claire Karrer
Crb 16h0'13" 36d42'43"
Claire Kathryn Thomas
And 2h28'45" 41d16'7"

Claire Kaufman
Tau 5h27'47" 17d12'37"
Claire Kelly
And 23h23'53" 42d1'27"
Claire Kosasky
Srp 18h23'27" -0d34'16"
Claire L. Bordeau
Tau 3h57'46" 8d31'50"
Claire L. Nein
Vir 13h17'46" -3d1'43"
Claire Labriola
Uma 10h56'2" 39d25'52"
Claire Lauren Cousineau
Cap 20h23'46" -15d8'36"
Claire Lee Campassi
Psc 0h37'13" 6d55'53"
Claire & Lee Jones
Cyg 21h27'44" 34d16'41"
Claire Leona
And 1h31'31" 47d11'47"
Claire Lofgren
Cap 20h28'54" -27d31'49"
Claire Lopizzo
And 23h16'5" 43d56'26"
Claire Louie
Lib 15h25'10" -24d51'41"
Claire Louise
And 23h12'40" 47d44'24"
Claire Louise
And 1h56'8" 38d19'41"
Claire Louise Gillespie
Peg 22h14'52" 8d38'28"
Claire Louise Hoeffel
Cyg 21h20'56" 32d29'32"
Claire Louise Norman
Lyn 6h38'19" 55d13'28"
Claire Louise Smythe
Mon 6h31'46" 0d12'56"
Claire Love
Cyg 19h21'28" 53d43'1"
Claire Lynn Bichalski
Uma 9h49'30" 62d2'8"
Claire Lynn Graham
Aqr 22h25'6" -24d35'16"
Claire Lynne Wilke
Leo 11h6'37" 21d30'44"
Claire Madison Moores
Lyn 8h15'56" 58d14'48"
Claire "Mama Bear" Schramel
Cnc 8h13'26" 12d26'59"
Claire Maria Flanagan
And 1h9'42" 46d45'12"
Claire Marie
Cap 21h40'8" -14d16'51"
Claire Marie
Sgr 18h56'3" -34d4'39"
Claire Marie Ann Krause
Cas 0h18'0" 62d44'52"
Claire Marie Huzar
Cas 0h1'51" 59d42'35"
Claire Marie Sigler
Mon 6h53'45" -0d52'46"
Claire Marie Wilkins
Vul 19h1'16" 24d11'48"
Claire Marsh I Will Always Love You
Cyg 21h10'51" 47d26'2"
Claire Martinet
Uma 11h18'38" 44d24'34"
Claire McKenzie Cabral
Cap 21h36'59" -11d56'23"
Claire Middleton
And 23h1'12" 47d50'30"
Claire Mossiat
Ari 3h14'32" 27d45'44"
Claire Nagy Anderson
Cas 1h7'33" 64d21'21"
Claire Nakata
Lep 5h51'6" -12d54'35"
Claire Natalie Verzello
Cas 1h26'50" 65d58'5"
Claire Noel
And 23h22'21" 48d16'42"
Claire Noelle Hirsch
Leo 11h17'25" 22d12'21"
Claire Nolli
Ser 18h17'52" -12d8'28"
Claire Olivia Holliday
Tau 4h29'34" 19d47'37"
Claire Opal Clements
Tau 5h11'7" 24d24'36"
Claire P. Rogus
Cap 21h42'54" -13d20'53"
Claire Parker's Star
Sge 20h9'27" 18d15'26"
Claire Pelster Brown
Cap 20h19'11" -8d37'21"
Claire Ranko Doyle
Crb 16h7'48" 37d34'31"
Claire Rebecca Heath
Sco 17h40'17" -32d3'8"
Claire & Ron Rebele
Cyg 19h49'24" 31d30'19"
Claire Sharood
Lib 14h50'33" -10d27'49"
Claire Soucy
Uma 11h35'21" 40d7'38"
Claire Stuvland
Tau 4h38'22" 20d1'9"
Claire Suzanne Ferretti
Cas 2h11'2" 59d10'19"
Claire Swift
And 23h44'1" 39d17'34"

Claire Tannenbaum
Psc 0h51'6" 16d39'15"
Claire Therese Dionne
Cas 0h58'10" 61d0'38"
Claire Thoron Pyle
Peg 22h23'53" 16d20'25"
Claire Troth - Godmother
Claire's Star
And 1h31'44" 39d30'26"
Claire "Twinkle" Mawson
And 1h37'37" 39d48'39"
Claire Vacherot
Umi 16h15'33" 72d13'42"
Claire Victoria Schekeloff 26.7.90
Cru 12h37'49" -58d49'29"
Claire Vinson
Tau 4h5'39" 30d48'0"
Claire Walton
Lyn 7h54'35" 56d31'37"
Claire Wilson McCloskey
Uma 10h46'0" 53d52'16"
Claire Zimlinghaus
Tau 5h35'32" 25d12'56"
Claire626
Cnc 8h15'22" 15d42'16"
Claire-Agathe
Cmi 7h38'13" 2d42'31"
ClaireAndJosh
Cyg 21h47'11" 47d25'50"
Clairebear Twomey
Tau 4h16'10" 24d29'11"
ClaireColemeisterColeyRab bitColeman
Dra 15h54'13" 53d46'51"
Claire's Beauty
Ori 6h5'53" 21d12'16"
Claire's Love
And 23h57'10" 45d55'10"
Claire's Star
And 0h7'17" 45d27'11"
Claire's Star
Sco 17h29'38" -42d47'40"
Claire's Twinkling Star
And 1h24'59" 33d56'13"
Clair-Marie Risdon
And 0h19'28" 26d55'19"
Clairon
Uma 10h32'45" 44d25'35"
Clam Hold
Crb 15h31'59" 32d0'0"
Clamaransa 1980
Cyg 19h38'39" 31d59'6"
Clamdigger
Lyn 9h6'28" 41d44'48"
CLAN Deux
Cma 7h12'12" -27d47'33"
Clan McDevitt
Cyg 19h53'49" 31d26'2"
Clan Rudd
Ori 6h15'8" 8d8'19"
Clancy
Tau 4h50'59" 21d51'4"
Clancy Anne
Gem 6h28'56" 26d6'26"
Clancy Carter Kinney
Gem 7h29'17" 34d12'35"
Clancy John Schwartz
Sco 16h14'50" -23d40'26"
Clancy's Nite Lite
Cnc 8h49'44" 30d44'7"
Clancy's Star
Lib 15h4'32" -26d51'16"
Clanhaney Clark's Cosmic Comet
And 2h35'9" 48d1'18"
Clannad
Ori 5h59'23" -0d43'37"
Clara
Ori 4h49'13" -2d38'25"
Clara
Aqr 23h13'50" -8d47'53"
Clara
Cas 23h42'14" 53d56'30"
Clara
Dra 19h6'57" 69d54'38"
Clara
Cma 7h26'20" -32d55'27"
Clara
Lyn 7h27'25" 52d9'42"
Clara
And 23h53'26" 45d52'43"
Clara
Cas 0h5'18" 53d35'51"
CLARA
Lmi 10h30'54" 31d48'16"
Clara
Psc 1h15'43" 28d12'0"
Clara
Cnc 8h17'16" 19d56'50"
Clara
Ori 5h33'56" 4d47'9"
Clara Ann
And 0h42'8" 36d19'29"
Clara Ann Cyr
Cyg 21h17'19" 40d9'2"
Clara B. Mason Peavy 1-27-36
Mon 7h48'29" -0d46'46"
Clara Baigneres
Lyn 9h3'8" 37d42'43"
Clara Belle Cates
Tau 3h49'7" 18d58'26"
Clara Bokrossy
And 2h36'56" 43d36'59"

Clara Brenna
Com 12h28'38" 26d27'24"
Clara Caldwell
Psc 0h10'9" 3d5'2"
Clara Campbell
Ari 1h59'49" 22d32'28"
Clara Catherine Sue
Crb 16h6'18" 36d2'50"
Clara Claeys
Lib 14h53'12" -4d5'4"
Clara D. Cardillo
Cas 1h15'23" 57d56'59"
Clara Danielle
Vir 13h19'40" -13d41'46"
Clara Detta Boyer-Williams
Per 2h16'21" 55d25'36"
Clara Dillon
Uma 10h30'7" 47d24'12"
Clara Elizabeth Berry 80
Mon 6h51'36" 7d35'44"
Clara Elizabeth Corkill
And 2h27'48" 39d18'14"
Clara Elizabeth Hollingsworth
Del 20h31'40" 19d13'26"
Clara Elizabeth Lake
Cyg 19h28'39" 57d50'24"
Clara Faye Doyle
And 0h14'35" 44d35'6"
Clara Forslund
Lyr 18h53'11" 41d5'31"
Clara Freed
Lyr 18h42'41" 33d59'35"
Clara Haia
Psc 0h14'34" 6d18'20"
Clara Hope
Leo 10h29'37" 13d15'47"
Clara Irena Payne
And 2h37'10" 50d1'3"
Clara Jane
And 0h24'13" 31d23'26"
Clara Jayne
And 2h22'19" 41d1'37"
Clara Jean (CJ) Hurley
Vir 12h47'36" 5d7'13"
Clara Jean Noland McGill
Cap 21h53'7" -19d27'29"
Clara & John Allison, Larry Lewis
Tau 4h29'4" 3d31'44"
Clara K
Cyg 21h10'15" 42d27'58"
Clara L. Parker
Peg 21h51'17" 12d37'36"
Clara Leigh Anne Villmer
Cas 0h20'15" 62d6'49"
Clara Lou
Ori 6h3'21" 3d58'10"
Clara Louise Gay
Boo 14h42'46" 28d55'49"
Clara Louise Herring
Lyr 19h12'36" 39d43'14"
Clara Luz
Del 20h56'52" 15d34'23"
Clara Maria Paez
Lep 5h38'40" -22d16'55"
Clara Maria Thompson
Cas 0h54'37" 70d37'29"
Clara Marie
Dra 18h28'55" 59d36'30"
Clara Marie
Vir 12h37'58" -7d29'17"
Clara Marie Pranger
Tri 1h48'9" 28d1'56"
Clara Mormile
Psc 1h16'57" 28d41'14"
Clara Morrison Reimers
Aqr 20h40'29" -12d55'5"
Clara My Angel
Cap 20h31'37" -9d30'56"
Clara Nell Estep
Cam 5h55'56" 60d35'54"
Clara "Pupka" Puzan
Uma 11h5'13" 48d25'24"
Clara S. Harding
Crb 15h37'35" 32d34'43"
Clara Timke
Cas 0h28'40" 56d59'48"
Clara Tomb
And 23h53'58" 48d34'21"
Clara Vande Lanoitte
Cep 5h9'2" 84d12'41"
Clara Vasiles-LeBlanc
Cyg 20h7'48" 41d58'57"
Clarabelle
Cma 7h11'7" -31d18'46"
Claralina
Cnc 9h4'48" 31d10'16"
Clara's Light
Aqr 23h20'36" -20d3'10"
Clare
Cam 4h55'58" 54d59'6"
Clare
Cnc 9h0'23" 17d36'20"
Clare
Psc 0h19'42" 19d26'54"
Clare 17
Sgr 18h2'26" -27d16'35"
Clare and Lilly Superstar
Peg 22h45'15" 4d16'28"
Clare & Anthony Williamson
Cyg 20h7'9" 44d20'17"
Clare Bretz Main
Aqr 22h53'13" -7d17'58"

Clare Burke Deangelis
Cas 1h57'13" 65d4'16"
Clare C. Duncan
Ari 3h13'42" 16d28'34"
Clare Carey Willard
Sgr 18h33'14" -16d11'18"
Clare Catrin Smith - James
And 1h56'55" 38d4'45"
Clare Christina
And 23h47'53" 47d21'4"
Clare Dockins
Lyn 7h51'54" 41d15'9"
Clare Eastment
Sco 17h58'58" -37d27'13"
Clare Faye Goodwin
And 2h12'5" 46d10'59"
Clare Flossie Dude
Psa 22h0'16" -29d27'19"
Clare Foster-Sample
And 0h33'18" 46d7'2"
Clare Francis Leary
Cyg 20h33'49" 40d29'39"
Clare Frogley Lang
Cmi 7h30'14" 4d46'16"
Clare Gibson
Psc 1h15'30" 22d16'51"
Clare Grantham
Tau 4h32'1" 18d52'41"
Clare Greenwell
Ori 5h24'16" 14d54'29"
Clare Jennifer Clarke
And 1h17'12" 34d12'44"
Clare. K
And 23h10'50" 35d27'18"
Clare L. Berger
Per 3h25'22" 45d39'38"
Clare Leask
And 2h34'32" 38d47'30"
Clare Louise
And 2h6'32" 46d14'58"
Clare Louise Bradshaw
And 1h58'40" 37d9'14"
Clare Maclean
Crb 16h23'47" 38d30'29"
Clare Mahoney
Tau 3h43'22" 28d30'37"
Clare Margaret
Uma 10h5'18" 45d32'5"
Clare Marie Carroll
Cas 1h27'0" 64d12'46"
Clare Marie Mahood
Sco 17h41'2" -38d5'0"
Clare Marie Roberts
Cyg 20h50'32" 46d18'17"
Clare MOMO
Gem 7h15'45" 24d10'7"
Clare Monckton
And 1h59'6" 39d41'22"
Clare Moult
Uma 13h51'26" 48d24'55"
Clare My Shining Babe
Cas 1h37'33" 61d10'37"
Clare Raimondi Gavin
Tau 5h26'25" 21d54'12"
Clare "Schmoops" Templin
Leo 11h26'6" 22d53'4"
Clare Upton
Umi 16h13'59" 79d17'22"
Clare Van Houwelingen
Gem 6h46'30" 13d46'42"
Clare Vesta Velez
Lyn 8h4'35" 54d12'48"
Clare Weisenfluh
Aqr 21h40'59" -4d46'4"
Clare, Bryan, Mark, Emma
Tri 2h15'31" 34d40'3"
Clarebear
Uma 12h15'58" 55d32'26"
Clarebears Star
Uma 10h47'43" 59d24'40"
Clarefiona 18
Lmi 10h47'47" 39d4'20"
Claremorris
Uma 10h28'43" 56d30'50"
Clarence
Cap 21h29'44" -18d25'12"
Clarence A. Anderson
Crb 16h18'15" 31d13'39"
Clarence and Charlene
Uma 13h22'17" 57d54'40"
Clarence and Winnie Zeches
Cyg 20h20'25" 36d35'41"
Clarence Arens
Sgr 18h8'52" -32d19'17"
Clarence Arthur Clayborn
Aqr 23h50'51" -10d30'26"
Clarence Boop Wattum
Peg 23h7'22" 21d38'44"
Clarence Cantargis
Aqr 22h36'22" -11d16'19"
Clarence "Chuck" H. Boudreau
Cep 22h33'46" 73d31'6"
Clarence Cunningham Beloved Daddy
Ari 1h57'48" 14d18'8"
Clarence Frances Leach
Ori 5h44'0" 11d19'55"
Clarence G. Poligratis
Uma 9h40'56" 46d32'43"
Clarence Herbert Thompson "Pot Licker"
Uma 9h52'31" 71d48'24"

Clarence J. Dickinson
Uma 8h32'30" 63d14'36"
Clarence Jim Arches
Her 16h55'12" 17d24'46"
Clarence Kiessling
Crb 15h26'46" 27d5'22"
Clarence Lorain Clark
Uma 9h46'33" 72d5'46"
Clarence & Mayme Wendl
Cyg 20h50'16" 46d58'39"
Clarence McClanahan 11-2-1920
Uma 9h30'34" 46d44'16"
Clarence "Mike" & Helen Vank
Cyg 21h16'57" 45d4'31"
Clarence P. Martin
Aur 5h31'50" 52d53'31"
Clarence Paul Kroll
Uma 10h14'30" 49d18'59"
Clarence R. Pentico 9/16/29 3/4/07
Per 2h3'11" 50d6'11"
Clarence T.
Aql 20h1'43" -0d53'28"
Clarence Victor Kunkel Jr.
Sco 16h10'30" -29d3'1"
Clarence Wilcox
Uma 11h26'42" 55d14'58"
Clarence Williams
Cap 21h37'13" -9d33'5"
Clareo
Equ 21h2'57" 11d33'48"
Clare's a Star!
Sgr 20h4'37" -12d17'14"
Clare's Light
Cas 1h5'12" 60d18'50"
Clare's light
Mus 13h10'46" -72d31'47"
Clare's Shining Star
Leo 11h34'58" 25d28'9"
Clare's Star
Cas 2h15'45" 70d57'58"
Clareyopolis
Peg 22h59'12" 9d40'17"
Clari
Lib 15h24'24" -24d48'47"
Claribel Ayala
Tau 4h10'58" 2d42'0"
Claribel Natal
Tau 3h38'7" 28d30'58"
Clarice
Ari 3h16'15" 29d17'12"
CLARICE ANN
Lib 14h46'14" -10d44'50"
Clarice Ann Kennelly
Leo 11h47'5" 17d16'56"
Clarice Anna
Leo 10h13'54" 26d17'17"
Clarice Dannels
Lyn 7h18'48" 58d24'48"
Clarice Jeane Richard 1916-2004
Lyr 19h14'15" 27d4'6"
Clarice Wrightson
Cas 23h10'31" 55d48'42"
Clarin
Lyn 7h54'25" 54d45'16"
Clarina Zolezzi Ebensberger
Vel 9h31'38" -53d48'12"
Clarion
Lib 15h31'25" -14d22'51"
Clarion Virgina Cunningham
Per 3h49'35" 41d34'18"
Clarisa Michelle Sanchez
Ari 2h53'44" 28d58'3"
Clarise Webb
Peg 23h32'28" 18d24'19"
Clarissa
Ori 6h19'5" 3d7'17"
Clarissa
Cas 0h1'27" 54d13'24"
Clarissa
Sco 17h53'43" -32d45'35"
Clarissa 13
And 0h37'23" 39d45'32"
Clarissa A. Davis
Cas 2h28'1" 67d14'52"
Clarissa Anne Thompson
Leo 10h25'19" 22d35'49"
Clarissa Botello
Uma 11h13'26" 46d14'14"
Clarissa Castro, BFA
Cap 21h44'46" -14d8'1"
Clarissa Caye Cross
Lmi 10h22'39" 33d29'17"
Clarissa Elizabeth Painter
Gem 7h39'23" 26d47'23"
Clarissa -Guide Me Home- 10/07/1996
Cru 12h45'2" -56d10'37"
Clarissa Javanaud - Clarissa's Star
Oph 16h29'43" -23d19'16"
Clarissa Jean Rossiter
Tau 5h25'35" 21d30'30"
Clarissa Kahealani
Vir 13h25'56" -12d15'50"
Clarissa M Sarver
And 0h31'5" 27d45'32"
Clarissa Marie
Del 20h36'34" 13d35'42"
Clarissa Marquez
And 1h56'54" 47d26'25"
Clarissa Peixoto (KK)
Psc 23h31'2" 4d39'56"
Clarissa Rahne
Psc 0h20'6" 4d1'48"
Clarissa Rose
And 23h8'35" 45d17'5"
Clarissa Tiedke
Aql 20h6'13" 15d32'25"
Clarissa Walls
Uma 13h53'48" 54d11'33"
Clarissa97
Ori 5h44'7" 9d14'28"
Clarissa's Heavenly Star
Tau 5h51'57" 24d58'32"
Clarisse
Lyr 18h43'40" 38d29'52"
Clarisse Ang
Lib 15h21'2" -11d26'22"
Clarisse Marcelo-Tanner
Ari 3h8'49" 12d44'54"
Clarita
Ori 5h29'31" 1d27'21"
Clarita
Lyn 8h18'39" 49d7'18"
Clarita Espelita Siapco, RN
Ori 6h2'15" 0d22'43"
Clark
Her 17h55'31" 20d11'41"
Clark
Uma 11h39'26" 41d55'4"
Clark
Uma 9h4'32" 55d22'54"
Clark
Uma 10h26'24" 61d7'27"
Clark A Bridges
Uma 10h30'48" 40d36'57"
Clark and Andra
Tau 5h48'20" 19d34'34"
Clark Benjamin Zimmerman IV
Ari 2h16'28" 23d43'38"
Clark Curtis Keller III
Ari 1h59'43" 14d19'10"
Clark D. Heath
Her 17h43'36" 48d35'16"
Clark & DeeAnn Robinson
Cyg 19h45'40" 43d0'50"
Clark Edward & Barbara Sue Kelley
Cyg 21h40'50" 34d16'35"
Clark F. Walden
Umi 16h26'10" 82d42'50"
Clark Grimm
Uma 10h46'32" 53d38'48"
Clark J. Jones
Ari 2h36'42" 28d57'30"
Clark Lee McBride
Uma 11h54'8" 62d42'10"
Clark Leslie
Aqr 22h26'20" 0d29'40"
Clark & Mary Ellen
Vir 13h29'43" -21d51'21"
Clark Nathaniel Gueringer
Lib 14h57'35" -10d58'37"
Clark O. Thorton
Ori 5h48'1" 9d49'56"
Clark Pickle
Uma 8h55'34" 64d47'4"
Clark Robert Mallder
Cep 23h5'52" 78d17'10"
Clark Sparky Ausloos
Ori 6h10'27" 17d17'7"
Clark Stanley Wille
Her 18h34'57" 20d24'20"
Clark Stossel
Boo 14h37'10" 24d37'34"
Clark Strangward Merrill
Cep 22h14'18" 60d4'45"
Clark Thomas Colbert
Her 17h33'38" 37d7'44"
Clark "us"
Vir 12h48'8" 5d59'43"
Clark Van Dresser Swift
Ari 3h15'6" 21d36'43"
Clark-acercer
Gem 6h18'39" 25d42'50"
Clarkestar
Tri 2h41'23" 34d20'46"
ClarksatClovelly 5th
Cnc 8h4'22" 10d54'24"
Clarkson010164
Cap 20h28'22" -23d7'9"
Clarus Optos Pepperalus
Psc 22h53'2" 3d26'29"
Clary "Ducky" Nigels
Umi 14h26'3" 83d11'28"
Claryce E. Erickson
Aqr 22h2'58" -3d26'9"
Clas-Eirik Strand
Aur 6h30'5" 39d4'59"
CLASHEIGH
Ari 2h44'4" 25d4'47"
Class of 1940, Woburn H.S. (Mass.)
Umi 12h40'30" 89d49'46"
Class of 2005
Cam 4h26'3" 78d59'30"
Class of 2005 Builders of The Dream
Uma 11h51'53" 57d8'47"
Class of 2006, St. Joseph Erie, MI
Uma 11h14'48" 59d39'31"
Class of 2007 St. Joseph Erie, MI
Uma 11h41'57" 56d33'47"
Classy Brian Williams
Uma 11h52'39" 38d56'52"
CLAU
Uma 9h2'37" 68d39'38"
clau_ruiz te amo munequita
Lyn 8h11'22" 39d35'28"
Clauday Gabriel Jayy
Sgr 18h28'55" -26d23'12"
Claude
Lmi 10h32'14" 37d17'43"
Claude
Gem 6h19'14" 23d24'39"
Claude "Best Dad Ever" Johnston
Per 2h40'51" 53d19'5"
Claude Butler
Cyg 19h35'37" 35d37'17"
Claude & Cathy Pelourson
Umi 15h4'17" 75d33'42"
Claude d'Anna
Cmi 7h22'59" 5d18'0"
Claude DiGenova
Per 3h13'8" 54d4'53"
Claude Dubois
Uma 8h51'22" 59d33'23"
Claude E. Cray, Sr.
Lib 15h29'40" -7d58'51"
Claude Gaudy
Her 16h47'11" 4d42'13"
Claude Höliner
Lac 22h6'21" 40d34'46"
Claude Hubert LeCarpentier
Cep 22h30'17" 62d33'37"
Claude la perle des pères
Vir 11h45'22" 8d39'19"
Claude Mancel
Sco 16h10'4" -28d39'43"
Claude Marion Warren, Jr.
Per 4h45'28" 44d12'57"
Claude Merrill Barnard, Jr. 8/22/25
Leo 11h10'32" 8d19'23"
Claude my true Love
Boo 14h28'50" 19d4'44"
Claude Pascal Fischer
Uma 9h34'24" 43d9'59"
Claude Richard "Dickie" Bartell
Uma 12h0'7" 29d10'35"
Claude Tèdou
Ori 6h14'57" 13d13'5"
Claude - Tristan - Alexandra
Aqr 23h31'2" -23d27'38"
Claude Turcotte
Cep 22h21'4" 55d49'28"
Claude W. White Jr. "Love Star"
Ori 6h19'49" 16d16'4"
Claude Watkins
Lib 15h11'19" -3d42'4"
Claudecampagne
Cas 1h30'33" 67d34'18"
Claude-Gabrielle Masse
Ori 5h39'16" -2d29'57"
Claudell Moncure 1928
And 1h35'19" 49d28'0"
Claudene Winingham Corbin
Uma 8h33'8" 67d52'34"
Claudete Gereaue
Uma 13h52'11" 48d38'58"
Claudette
Cas 0h34'10" 57d12'25"
Claudette Arlene Mitchel-Weismantel
Sgr 18h17'40" -28d46'28"
Claudette Benson
Umi 14h55'43" 68d48'57"
Claudette Bhola
And 0h47'20" 37d54'1"
Claudette C. Simard
Her 16h8'39" 48d55'58"
Claudette Corcoran
Cyg 21h38'32" 41d35'45"
Claudette Dzieciuch
Leo 10h47'39" 7d46'28"
Claudette Foisy
And 2h11'58" 43d47'30"
Claudette Lillian Gregory
Leo 11h27'23" 25d4'38"
Claudette Marie Spalding
And 0h21'34" 39d55'2"
Claudi
Gem 6h29'27" 20d26'50"
Claudi Stärnestaub Angeli
Lac 22h54'25" 37d31'30"
Claudia
And 23h16'5" 43d55'42"
Claudia
And 1h34'44" 46d49'6"
Claudia
Lyn 7h3'56" 46d36'11"
Claudia
Her 17h38'18" 49d31'31"
Claudia
Uma 10h10'40" 48d16'35"
Claudia
Uma 10h8'22" 47d58'31"
Claudia
Uma 10h2'59" 47d36'18"
Claudia
Cas 0h37'25" 57d46'12"
Claudia
And 23h17'7" 52d0'0"
Claudia
Uma 10h49'29" 44d57'58"
Claudia
Crb 16h20'33" 34d23'38"
Claudia
Lmi 10h19'13" 36d22'55"
Claudia
Lmi 11h4'15" 32d55'56"
Claudia
Gem 7h30'54" 19d9'56"
Claudia
Peg 22h55'30" 15d47'9"
Claudia
Leo 9h41'31" 28d28'30"
Claudia
Vir 13h33'3" 8d34'0"
Claudia
Mon 6h43'38" 8d37'43"
Claudia
Peg 23h55'13" 10d33'18"
Claudia
Vir 13h33'7" 5d16'23"
Claudia
Vir 14h5'2" 3d25'30"
Claudia
Ori 5h59'47" 6d9'50"
Claudia
Ori 5h27'34" 2d51'42"
Claudia
Peg 22h27'12" 5d18'51"
Claudia
Umi 15h9'11" 69d8'59"
Claudia
Uma 8h11'26" 67d55'12"
Claudia
Uma 8h17'49" 69d17'59"
Claudia
Cas 23h59'7" 60d55'7"
Claudia
Cas 1h49'57" 64d58'53"
Claudia
Dra 19h19'15" 66d29'28"
CLAUDIA
Dra 16h45'38" 64d49'2"
Claudia
Uma 10h37'24" 55d36'44"
Claudia
Leo 10h51'39" -1d39'45"
Claudia
Aqr 22h52'41" -12d9'49"
Claudia 02-16-62 & Ricardo 03-25-48
Eri 3h54'49" -0d54'14"
Claudia 1976
Cas 0h32'35" 52d32'45"
Claudia 21
Aqr 22h39'20" -2d21'44"
Claudia 24.1.1968
Uma 8h52'47" 63d17'36"
Claudia 55
Her 17h15'53" 25d8'17"
Claudia 62
Ori 6h9'52" 6d1'9"
Claudia 629
Cnc 8h17'6" 19d36'35"
Claudia 9450
Ori 5h24'6" 0d30'17"
Claudia A. Johannsen Scott
Leo 10h7'14" 26d4'10"
Claudia Aisha Braam
Aql 19h31'4" 13d15'28"
Claudia ali di farfalla
Lib 15h22'59" -5d40'47"
Claudia Almaraz
Uma 10h15'52" 63d42'33"
Claudia Alt
Umi 15h17'1" 71d8'9"
Claudia Amrein 1971
And 23h13'50" 40d38'33"
Claudia Ariel
Vir 14h8'9" 2d44'40"
Claudia Arteaga
Lyn 7h22'30" 53d43'53"
Claudia Ayala
And 1h25'48" 47d14'17"
Claudia Balduzzi
And 0h42'9" 30d18'42"
Claudia Bazley
Cyg 20h37'51" 48d10'48"
Claudia Bogana
Cas 23h36'25" 58d39'5"
Claudia Brunelli
Ori 5h46'41" 6d49'41"
Claudia C Navarro
Lyn 7h14'32" 55d29'34"
Claudia Cabrera Ventura
Cap 21h55'41" -17d5'56"
Claudia Cacciabue
Per 3h44'37" 45d18'0"
Claudia & Cameron's Shining Star
Col 5h41'21" -29d50'19"
Claudia Carletti
Crb 15h46'8" 29d26'9"
Claudia Carlos Orteza
Uma 11h20'34" 31d13'22"
Claudia Cifuentes
Vir 11h40'11" -3d57'52"
Claudia C.L.A.F - I <3 You Forever
Cas 1h2'17" 63d2'10"
Claudia Clas-Becerril
Sgr 18h26'2" -32d21'3"
Claudia Contreras
And 0h25'28" 42d32'33"
Claudia De Gennaro
Oph 17h13'20" 8d49'20"
Claudia de la Guardia
Gem 7h19'54" 16d3'36"
Claudia Dergin
Sco 16h16'14" -11d56'11"
Claudia Di Crescenzo
Cam 4h3'20" 55d5'47"
Claudia e Massimiliano
Peg 22h22'26" 26d33'14"
Claudia E. Sheffer Lohnes
Aqr 21h13'44" 1d12'39"
Claudia Ebinger
Aql 19h34'33" 5d58'29"
Claudia Emily Macchiarelli
Sgr 18h7'49" -25d27'9"
Claudia et Louis Giraud
Gem 7h46'44" 21d28'30"
Claudia & Farhad Toutouni
Ara 17h29'41" -47d25'30"
Claudia Ferrel
Sgr 20h7'21" -42d8'21"
Claudia Gabriela Gonzaga Jauregui
And 0h8'36" 35d22'38"
Claudia Gaxine Bores
Umi 16h22'6" 84d26'48"
Claudia Germain
Uma 8h41'39" 51d37'57"
Claudia Giselle Pertuz Fina
Cap 20h9'35" -13d12'26"
Claudia Gonzalez
And 1h12'16" 42d28'3"
Claudia Gonzalez
Cnc 8h19'15" 9d2'57"
Claudia Grace Bridges
Leo 11h9'2" 21d12'17"
Claudia Guth
And 0h51'48" 39d10'10"
Claudia Gutknecht
Ori 5h49'12" -4d18'57"
Claudia Hagens
Her 18h22'10" 16d30'29"
Claudia & Hanno
Uma 10h9'54" 47d38'55"
Claudia Hansen
And 23h7'47" 48d43'45"
Claudia Haunreiter
Crb 15h38'47" 36d35'0"
CLAUDIA HELDT
Mon 6h37'14" 5d38'11"
Claudia Huezo
Psc 1h50'19" 6d46'21"
Claudia Hutzli
Peg 21h29'10" 21d45'59"
Claudia Jannette Montoya
Sgr 18h7'28" -33d53'3"
Claudia Jasmine
Gem 7h13'15" 15d3'29"
Claudia Jean
Peg 23h8'55" 19d58'47"
Claudia Jean
Ari 3h7'11" 23d6'29"
Claudia Jean Drimmel
Aqr 23h2'21" -9d11'44"
Claudia & Jean-Marc Buschini
Lyr 19h24'45" 37d56'43"
Claudia KLA' Gaetani
Cam 7h2'59" 71d42'52"
claudia "knuddelbär"
Cam 6h35'29" 78d7'29"
Claudia Koeppe
Uma 13h37'6" 55d31'57"
Claudia & Lewis Gentile
Cyg 19h55'1" 33d18'15"
Claudia Lillian Bush - 11/11/1996
Sco 16h11'11" -13d29'8"
Claudia Lizeth De Los Rios
Dra 16h5'10" 52d42'36"
Claudia LoNardo
Sgr 18h26'24" -31d55'46"
Claudia Lowden Nahorodny
Lib 14h49'25" -6d49'50"
Claudia Lucia Zesiger-Ortiz
Lmi 10h35'13" 31d39'28"
Claudia Madani
Gem 6h49'32" 14d54'34"
Claudia Mae Chisum
And 0h27'22" 28d36'26"
Claudia & Manfred
Ori 6h11'26" 2d10'37"
CLAUDIA & MARC FRANK *1981*1980*
Uma 9h34'54" 57d56'48"
Claudia Maria
Vir 12h20'8" 12d13'45"
Claudia Maria Macias
Sco 17h30'49" -39d20'8"
Claudia Marie Heard
Lib 15h16'18" -15d32'23"
Claudia Marie Svendsen
Ari 2h54'50" 25d56'29"
~* Claudia Marina *~
Cap 20h23'54" -13d23'18"
Claudia Marquez-Benavidez
And 0h54'40" 35d55'2"
Claudia Marti
Uma 9h51'54" 49d26'24"
Claudia~Mary~Atherton
Lib 14h55'10" -20d9'54"
Claudia Mastrodomenico
Umi 14h49'29" 72d4'27"
Claudia - Mauri
Sct 18h53'37" -12d45'6"
Claudia "Me Me" Shoemake
Lyn 6h22'17" 61d16'59"
Claudia Menjivar
And 0h4'9" 44d12'51"
Claudia Messmer
Umi 14h46'59" 75d45'49"
Claudia Montoya Escobar
Psc 0h47'4" 10d22'30"
Claudia Mullen & Shadows
Cnc 8h29'20" 10d32'56"
Claudia Orozco
Psc 0h20'19" 11d5'37"
Claudia & Pascal
Umi 15h49'14" 83d33'7"
Claudia Peltz
Psc 1h52'12" 9d3'25"
Claudia Pereira
Vir 13h45'8" -9d49'25"
Claudia Peter
Aqr 22h47'54" -13d2'20"
Claudia Plaz "Pläzli"
Boo 15h2'14" 13d15'41"
Claudia "Precious" Robles
Lyn 7h56'56" 36d39'21"
Claudia Quiñonero Gómez
Umi 13h28'10" 76d0'5"
Claudia Rismonda Petrone
Ari 2h5'45" 21d30'21"
Claudia Roca
Sco 17h27'58" -35d10'11"
Claudia Rodriguez
Aqr 20h55'39" -11d59'33"
Claudia Rogosch
Psc 23h7'17" 2d0'58"
Claudia Rubi
Psc 1h28'37" 18d51'0"
Claudia ` s Gründel Stern
Uma 10h22'52" 45d1'29"
Claudia' s & Nico' s Hochzeitsstern
Ori 6h13'36" 5d49'44"
Claudia Sajtar
Mon 6h57'10" -6d38'13"
Claudia Saner
Peg 22h2'0" 14d22'31"
Claudia & Sascha
Vul 20h20'11" 24d0'47"
Claudia Shamara
Sgr 18h35'48" -23d47'0"
Claudia Siliven
Lyn 8h46'37" 42d2'1"
Claudia Stell
Gem 6h19'23" 26d5'4"
Claudia Suter
Ori 6h18'27" 15d35'25"
Claudia T Simeone
Sco 16h8'17" -11d47'15"
Claudia Teresa Matrinez Chacon
Tau 5h57'15" 25d3'18"
Claudia - The Brightest Star
Cru 12h49'30" -64d15'1"
Claudia ti amo
Her 17h17'49" 48d15'41"
Claudia V. Deegan
Psc 0h10'12" 7d39'29"
Claudia Valdez
Gem 7h14'53" 23d21'11"
Claudia Vianino
Aql 19h33'16" -0d3'31"
Claudia von Arx
Cas 23h37'13" 51d10'17"
Claudia Wiederkehr (Clödi)
Cyg 21h22'8" 32d52'56"
Claudia, stella per un Angelo
Lac 22h49'11" 53d21'2"
ClaudiaAngel1980
Leo 10h36'10" 8d3'38"
ClaudiaKendrick
Vir 12h13'14" -0d54'27"
Claudia's Star
Lyn 8h4'34" 55d54'6"
Claudias Stern
Boo 13h56'9" 22d57'47"
Claudina
Sgr 18h57'28" -29d49'13"
Claudine
Her 18h56'24" 15d38'19"
Claudine
And 1h15'26" 38d12'10"
Claudine
Lyn 7h16'1" 45d44'18"
Claudine 19011959
Col 5h12'14" -31d42'2"
Claudine Allison
Cam 7h57'50" 78d4'40"
Claudine Calacsan
Sco 16h29'27" -28d20'13"
Claudine Lienhard
Leo 11h47'29" 15d41'53"
Claudine Liggins
Tau 4h33'16" 18d33'39"
Claudine the Great!
Ori 6h4'9" 10d16'59"
Claudine Thibault Maxime Harvey Love
Cyg 20h40'50" 36d13'15"
Claudine Winkel
Tau 3h45'37" 23d3'4"
Claudine's Lights
Cra 18h57'29" -39d25'0"
Claudio
Sgr 18h32'4" -28d5'6"
Claudio
Cam 6h59'7" 68d59'43"
Claudio
Eri 3h57'41" -0d13'0"
Claudio
Cas 23h41'37" 58d11'20"
claudio
Umi 14h25'22" 73d9'7"
Claudio
Aur 5h22'25" 39d27'53"
Claudio
Per 4h29'18" 32d58'24"
Claudio
Lyr 19h26'20" 43d29'40"
Claudio Alberti...nemo
Leo 10h7'42" 11d0'41"
Claudio and Linda Castaneda
Pyx 8h36'8" -32d43'34"
Claudio & Anny
Her 16h59'21" 31d17'39"
Claudio Cavini
Uma 10h22'25" 63d17'7"
Claudio Comiti
Lyn 8h27'18" 43d25'54"
Claudio D'Angelo
Cas 23h12'38" 55d5'18"
Claudio e Nicole
Per 3h24'56" 50d1'52"
Claudio Espinoza Benites
Aql 19h27'53" 5d19'37"
Claudio Garcia
Her 18h47'51" 20d20'21"
Claudio Giacomelli
Uma 13h37'51" 54d0'15"
Claudio Maglie
Ori 5h12'38" 0d45'22"
Claudio Martinelli 15 dicembre 1957
And 23h0'20" 36d11'34"
Claudio Maso 13/11/1981
Sco 17h16'8" -31d20'57"
Claudio Monteleone
Lyn 7h23'28" 57d52'0"
Claudio Monteleone
Mon 7h7'7" 5d16'3"
CLAUDIO PER SEMPRE
Umi 18h27'16" 86d51'22"
Claudio's star
Psc 1h6'52" 14d30'54"
Claudis & Michis Stern
Uma 9h41'23" 68d23'32"
Claudishis Pepper Link
Lib 15h29'6" -26d48'50"
Claudiu Rusu
Leo 11h38'25" 14d16'49"
Claudius Augustus "Saga" Francis
Leo 11h15'12" -4d35'23"
Claudius Ollivierum
Her 18h47'13" 22d10'30"
Claudy
Cyg 20h1'52" 35d42'28"
Clauray
Ori 5h45'7" 10d8'14"
Claus Christin (Domenica)
Dra 19h46'13" 61d29'45"
Claus Jørgen Christensen
Uma 12h23'14" 53d26'8"
Claus Swierzy
Uma 9h16'9" 64d14'40"
Clausi
Ori 5h58'32" 8d57'20"
Clavien
Lib 15h3'9" -29d54'23"
Clay
Ori 5h44'22" 7d4'58"
Clay and Anna Zobell
Gem 7h28'1" 33d31'50"
Clay Boston-My Love, Best Friend
Ori 6h20'12" 14d48'48"
Clay Bryson Jeffries
Ori 6h12'5" 19d44'54"
Clay Burch #7
Ori 5h37'16" 2d17'44"
Clay Davis Redding
Cyg 19h39'45" 39d9'7"
Clay & Diane Edwards - Wedding Star
Tri 2h25'37" 31d26'49"
Clay Durr
Ori 5h20'35" -7d19'1"
Clay Face
Psc 1h27'56" 13d1'49"
Clay Hamilton Haywood
Cyg 21h46'53" 38d43'0"
Clay Hayward
Sco 17h51'8" -33d14'13"
Clay Lilley
Per 4h46'24" 45d42'7"

Clay Lowe
Ori 5h1'58" 5d42'29"
Clay Price
Her 16h17'45" 19d27'44"
Clay & Samyra Fulton
Uma 8h57'22" 60d5'57"
Clay Tatum
Boo 14h37'16" 41d17'44"
Clay73
Cru 12h36'39" -60d50'6"
Claybasket's
Cnc 8h35'14" 11d5'14"
Clayberly
Uma 9h9'45" 47d59'2"
Clayco 143
Per 2h44'11" 56d16'19"
Clayden Leybovich
Boo 14h46'32" 27d13'54"
Clayford Gene Armstrong
Vir 13h17'8" 6d50'12"
Clay's Star
Leo 10h40'32" 14d34'58"
Claytam
Aql 19h9'12" 5d38'58"
Clayton
Cap 20h25'42" -10d47'14"
Clayton Adams
Lib 15h13'17" -23d38'21"
Clayton and Amanda Oxford
Aqr 21h55'44" -0d0'4"
Clayton and Cassie
Tau 3h46'37" 29d58'44"
Clayton and Sarah O'Neal
Umi 15h54'22" 79d45'12"
Clayton Bell Browne
Per 3h10'23" 50d49'53"
Clayton Charles Kruse
Aqr 22h48'12" 1d32'7"
Clayton Charles Van de Bogert
Aql 19h55'3" 13d15'58"
Clayton D. Kramer
Leo 10h20'59" 10d54'48"
Clayton Daniel Yell May 14, 1982
Cep 22h4'37" 60d58'47"
Clayton Douglass
Ori 6h17'58" 11d7'16"
Clayton Earnest Bush
Peg 22h44'36" 31d34'47"
Clayton Frazer
Vir 13h20'56" 12d31'20"
Clayton G. Whillackers
Sgr 19h49'2" -13d34'37"
Clayton Hemingway
Leo 11h45'48" 19d58'46"
Clayton James Garland
Uma 13h33'8" 55d35'16"
Clayton James Wallen
Her 16h35'16" 5d23'3"
Clayton James Witman, "Daddy"
Aql 19h23'13" -8d3'31"
Clayton Joseph Whitworth
Psc 0h51'23" 8d52'4"
Clayton Laje & Darleen Cabalitasan
Uma 10h8'48" 42d56'49"
Clayton Lee Wallace
Her 18h21'12" 28d20'48"
Clayton M. Hubler
Uma 13h16'13" 56d34'9"
Clayton Miles
Lac 22h21'5" 54d13'53"
Clayton Miller
Leo 11h25'5" 25d38'15"
Clayton Mulkey
Her 17h23'36" 46d20'44"
Clayton *My Love My Life My Star* Ryan
Psc 1h23'52" 30d45'17"
Clayton Nevell
Cep 22h31'41" 72d58'38"
Clayton & Pauline Valley
Tau 5h34'10" 23d33'53"
Clayton R. Farwell (Pa)
Uma 8h56'47" 62d0'19"
Clayton R. McKenney
Aur 5h55'49" 52d27'42"
Clayton Reeves
Boo 14h29'5" 24d27'20"
Clayton Reinhart
Uma 12h28'40" 53d44'13"
Clayton Robert Nuce
Per 2h36'0" 51d54'36"
Clayton Robert Nuce
Cyg 19h56'56" 42d51'55"
Clayton S. Wylie
Cru 12h26'19" -60d11'52"
Clayton Shane Schauer
Her 17h48'5" 24d50'57"
Clayton (Superman) Cannon
Per 3h32'54" 48d47'9"
CLAYTON TAYLOR KOONCE
Gem 7h26'18" 31d27'30"
Clayton the Crusher
Her 18h44'16" 14d0'29"
Clayton Thomas Green
Per 4h23'16" 50d28'29"
Clayton William Loudermilk
Peg 21h38'54" 24d20'31"
Clayton's Barrett
Lyn 7h51'16" 49d53'49"
Clayton's Star From Pamela
Sgr 18h23'29" -32d34'26"
CLC III
Ori 5h24'29" 2d11'32"
CLD Mac 4/22/52
Cam 3h54'8" 68d19'42"
Clea
And 23h6'42" 51d52'49"
Clean Tom
Lib 15h45'21" -9d39'42"
Clear Channel University
Per 4h9'21" 33d43'43"
Cleary
Cen 13h38'18" -41d29'32"
Cleburne
Vir 14h36'7" 5d0'34"
Cleeve P. Morrison
Cnc 9h13'2" 8d12'55"
Clelia
Uma 8h37'20" 70d25'8"
Clellie
And 23h57'5" 33d7'22"
Clem
Cas 0h32'42" 66d11'10"
Clem
Mon 6h44'14" -0d1'17"
Clem Watts
Uma 10h43'36" 51d25'5"
Clem, Inc.
Uma 10h32'46" 56d57'6"
Clémence
Cas 0h57'4" 54d1'30"
Clémence Delmont
Tau 3h55'11" 2d59'50"
Clémence Delmont
Umi 16h13'22" 79d44'36"
Clémence Minguely
Mon 6h21'26" -5d9'10"
Clemencia
Com 12h59'37" 20d14'15"
Clemens and Joyce
Per 3h23'1" 45d4'57"
Clemens Carl
Uma 8h26'23" 64d9'25"
Clemens Gämperle
Dra 18h59'38" 52d25'51"
Clemens, Bernhard
Ori 5h1'14" 15d28'9"
Clément
Uma 11h38'15" 41d40'52"
Clément Amandine Maxime
Umi 15h28'35" 77d29'36"
Clement Bonenfant
Uma 11h26'53" 60d26'19"
Clément Carbonneau
Umi 15h50'33" 73d19'25"
Clément Coque
Sco 16h8'52" -29d10'44"
Clement Cyprien Musumba Wekullo
Eri 4h26'33" -31d56'18"
Clement George Moutou
Cru 12h26'56" -60d9'9"
Clément Ruf
Tau 5h40'34" 24d24'32"
Clément-DMSM2005
Dra 19h45'6" 68d43'55"
Clemente-Mike, Kate & Kaitlynn
Lyn 6h30'32" 61d9'27"
Clementina Rosa
Psc 1h23'23" 24d45'17"
Clémentine
Cam 4h15'21" 55d38'56"
Clementine
Uma 11h1'36" 54d57'57"
Clementine
Cap 20h37'9" -21d37'0"
Clementine
Pho 1h15'50" -41d34'30"
Clémentine
Lib 15h22'20" -23d29'13"
Clementine Lewis
Cas 1h9'4" 69d43'0"
Clementine Michael Sapp
Lyn 6h23'0" 60d3'16"
Clementine Treadwell
Tau 4h30'30" 8d30'31"
Clémentine V
Cnc 8h17'21" 23d38'24"
Clemenzer Robertson
Aql 19h18'45" 2d39'8"
Clemson Princess
Vir 13h18'7" -2d8'12"
Clemy's Rose
Mon 6h27'6" -10d52'33"
Cleney Taymil
Sco 17h29'28" -44d3'10"
Cleo
Cap 21h26'55" -13d42'55"
Cleo
Uma 12h19'24" 52d58'42"
Cleo
Cas 1h1'14" 60d39'17"
Cleo
Gem 7h33'18" 22d39'36"
Cleo 66
Peg 21h41'38" 9d29'45"
cleo + aladdin = infinity
Cas 0h17'10" 51d11'19"
CLEO Beloved cat of Lisa McCrum
Lyn 8h52'54" 40d13'5"
Cleo Darcia
Tau 3h49'10" 23d52'5"
Cleo Esposito
Cap 21h36'54" -20d20'37"
Cleo & Jay Meldrum
Lyn 7h24'41" 55d40'20"
Cleo La Reine de Mon Coeur
Cap 20h29'11" -18d58'52"
Cleo M. Elkinton
Tau 5h48'34" 24d20'56"
Cleo & Tom Bamford
Cyg 20h17'7" 38d7'38"
Cleodon
Lib 15h51'4" -6d24'50"
Cleone
Uma 10h2'1" 63d11'36"
Cleonice
Crb 15h31'57" 26d50'59"
Cleopatra
Uma 12h53'28" 53d49'20"
Cleopatra's Star
Cas 0h6'46" 56d47'27"
Cleta Elizabeth Colbert
Cas 1h25'12" 62d15'29"
Cleta Marie
Gem 6h55'59" 26d54'25"
Cleta Marie Henchman
Mon 6h50'26" -0d9'19"
Cleta May Price
Uma 11h39'20" 46d30'14"
Clete DiNero
And 2h17'40" 45d23'45"
Cleto Marcelino Limon
Tau 4h36'25" 19d19'22"
Cletsa
Psc 1h3'34" 6d52'59"
Cletus Dean Zerr
Ari 2h36'45" 24d58'35"
Cletus E. Ireland
Cas 2h50'3" 60d52'31"
Cleusa Antonio
Cru 12h16'11" -56d36'13"
Cleveland
Gem 7h43'31" 24d51'20"
Clevs Evelyn Star
Cas 1h7'11" 69d58'58"
clh & bmw, Forever and Always
Lmi 10h9'28" 38d49'57"
CLH25
Lyn 7h20'59" 46d56'39"
Click Click
Uma 10h20'32" 43d1'18"
Clickner Star of Strength
Leo 9h37'5" 26d29'4"
Clif Carey,Devin Carey,Avery Carey
Tri 2h11'48" 33d55'37"
Clif F. Gillette
Crb 15h47'48" 26d32'8"
"Cliff"
Cnc 8h37'10" 23d28'1"
Cliff and Betty Shaw
Crb 15h54'10" 29d34'32"
Cliff and Kath Critchley
Cyg 20h2'26" 34d49'52"
Cliff and Kay Paul
Cam 4h19'36" 66d56'47"
Cliff and Kim Corbet
Lib 15h44'9" -18d5'23"
Cliff Bateman
Gem 6h59'44" 33d11'5"
Cliff Beattie
Uma 13h53'5" 49d20'29"
Cliff Charles Fisk
And 0h10'59" 32d3'46"
Cliff Cranston
Cep 21h4'50" 57d3'48"
Cliff Fawcett
Cyg 19h43'13" 37d45'23"
Cliff Jackson
Her 17h41'52" 20d8'51"
Cliff Joncich
Aql 19h54'39" -0d28'2"
Cliff King-Cherished Loved One
Per 4h48'7" 43d44'27"
Cliff Marshall
Vir 14h2'16" -11d11'27"
Cliff Nelson
Uma 13h25'3" 54d18'50"
Cliff Netten
Cep 20h52'6" 62d5'17"
Cliff Patterson's Star
Aql 19h14'0" 7d20'37"
Cliff Radziewicz
And 1h55'0" 37d23'11"
Cliff & Toni Frank
Ori 6h12'32" 2d34'17"
Cliff Yakelewicz
Sgr 18h50'42" -28d53'5"
Clifford
Vir 14h18'40" -21d4'20"
Clifford Adams
Cep 23h21'28" 77d15'30"
Clifford and Noella Paradis
Cyg 20h25'51" 53d9'34"
Clifford Andrew Perucci 8/15/1995
Leo 10h13'40" 20d33'1"
Clifford Brooks
Ori 5h46'36" 2d36'34"
Clifford Clark, Jr.
Ari 2h25'10" 24d27'13"
Clifford & Fatima Bredenberg
Pup 6h56'33" -39d3'34"
Clifford Franklin Patrick Donnelley
Her 16h43'16" 28d53'16"
Clifford French
Cep 20h49'9" 59d21'52"
Clifford Gorman
Her 17h15'52" 17d29'31"
Clifford Gundle
Uma 11h47'26" 54d22'36"
Clifford J. "Tish" Drenttel
Cep 20h39'14" 63d49'52"
Clifford James Crockford, Sr.
Her 17h46'46" 15d31'46"
Clifford James Phillips
Uma 10h20'2" 59d37'32"
Clifford & Kaylyn Welch Golden Star
Cyg 21h51'17" 52d37'29"
Clifford Leo Collier
Leo 10h22'30" 25d37'55"
Clifford Manley
Her 18h42'50" 20d16'21"
Clifford Murry Sanders Sr.
Per 3h18'1" 51d1'53"
Clifford & Opal Sieh
Cyg 19h35'34" 39d10'27"
Clifford R. Brown
Cep 22h13'27" 60d11'33"
Clifford Ray Ramsey
Her 17h55'36" 23d28'34"
Clifford Reed
Cap 20h17'29" -10d57'38"
Clifford Thomas Mooney
Aqr 22h12'10" 0d20'23"
Clifford "Uncle Cliff" Ammons
Her 17h49'11" 45d33'56"
Clifford V. Jones
Leo 11h16'3" -3d37'58"
CliffordsGoldenKat
Per 4h25'50" 48d32'3"
Cliffy
Aqr 22h42'37" 0d23'4"
Clifton J. Watkins
Ori 6h10'41" -0d14'15"
Clifton Lane Vasquez
Aur 5h18'48" 31d41'17"
Clifton O. Clark
Leo 9h50'41" 12d36'28"
Clifton Rattenbury
Lyr 18h50'5" 41d25'15"
Clifton Warren Watson
Dra 20h12'15" 72d43'23"
C-Lil's New World
Cyg 19h41'59" 30d10'11"
Climaco Pedraza
Cap 20h21'21" -10d51'10"
Climberdusk
Crb 15h39'5" 33d32'17"
ClinT
Cas 0h19'25" 51d23'39"
Clint
Her 16h13'52" 46d22'2"
Clint Alicia Connell Quayle
Gem 7h15'58" 27d26'29"
Clint & Amy Priest
Aql 19h37'35" 5d26'26"
clint and bec's piece of heaven
Cru 12h27'45" -60d56'48"
Clint and Jen for Love and Eternity
Umi 15h18'35" 74d13'44"
Clint and Tiffany Brown (Pookie!)
Cyg 19h49'49" 36d34'0"
Clint and Vanessa (Forever in love)
Uma 10h41'6" 65d35'55"
Clint Faddis
Leo 11h5'9" 15d33'43"
Clint Floyd
Uma 8h42'42" 62d21'37"
Clint Lee Kuntz
Tau 4h12'50" 13d46'43"
Clint M. Puller
Uma 11h55'59" 34d6'50"
Clint Michel
Boo 14h34'10" 32d32'24"
Clint N Suez
Cyg 20h6'7" 40d3'39"
Clint Reed Nokes
Cap 21h53'14" -13d15'53"
Clint Robert Hendricks
Uma 9h16'55" 57d0'6"
Clint Satorre
Tau 5h48'9" 21d9'52"
Clint the Cat
Ori 5h33'3" 2d32'54"
Clint & Tina Ordone
Cyg 20h28'32" 34d13'53"
Clint Wallin
Her 16h4'18" 18d4'47"
Clinton
Uma 9h57'30" 67d24'2"
Clinton A Lacarte Born Oct 15/1965
Uma 11h31'51" 56d55'13"
Clinton Alexander
Lyn 8h16'41" 51d28'9"
Clinton "Butch" Kehr
Uma 11h21'17" 57d34'38"
Clinton Cameron Huscher Johnston
Ari 3h11'52" 28d55'30"
Clinton Curtis Peery
Lib 15h49'37" -16d39'38"
Clinton Dennis Hutt
Lib 15h21'24" -19d44'52"
Clinton E. Ward
Aql 19h37'48" -0d38'58"
Clinton G. Major
Lyn 7h3'3" 50d58'27"
Clinton Gene Whitman
Uma 13h25'48" 60d15'20"
Clinton James Dunham
Vir 13h15'40" -22d24'55"
Clinton Michael Welsh
Psc 23h2'12" -0d1'52"
Clinton Moore
Cas 0h23'57" 50d5'38"
Clinton Mosley
Per 4h4'44" 36d52'7"
Clinton & Rebecca
Uma 9h2'35" 60d45'29"
Clinton Rossi
Uma 11h32'32" 63d36'23"
Clinton & Sharan Peck
Cyg 19h57'15" 33d31'35"
Clinton Zachary Womack
Cyg 19h49'59" 38d5'51"
Clinton's Gem
Dra 19h25'34" 65d12'8"
Clint's Ecstacy
Her 16h34'12" 46d36'44"
Clint's Star
Boo 13h48'54" 10d0'31"
CLINTY WINTY
Aql 20h20'23" 7d57'4"
CLIO
Uma 13h41'14" 47d52'50"
Clio Bodie
Tau 4h41'51" 22d37'9"
Cliodhna
Vir 14h36'17" 5d31'43"
Clionadh
And 2h33'32" 41d23'32"
Clive A Pearce
Aur 6h1'56" 39d55'1"
Clive Allison 60 A Star Is Born
Cma 6h39'51" -16d57'49"
Clive and Ann Cummis
Cyg 21h35'20" 50d46'12"
Clive Downing's Retirement Star
Cep 21h19'1" 58d40'27"
Clive Hackett
Tau 3h52'21" 3d27'56"
Clive Hinton
Equ 21h7'1" 5d4'37"
Clive of Flitton
Crb 16h7'46" 30d21'28"
Clive Richard Higgins - 12.04.1955
Cru 12h7'47" -62d56'23"
CLKMKR
And 1h4'42" 42d27'29"
CLMJRB041203
Uma 10h43'16" 44d9'19"
CLNilesIII-143
Lyn 8h7'1" 57d8'15"
Clodagh A.M.
Tau 5h55'23" 26d27'36"
Clodagh Gray
Cyg 20h14'27" 36d24'10"
Clodagh Marie
Aqr 22h18'52" -8d54'37"
Clodagh Power
Col 5h24'8" -35d29'7"
Clodinoro Balestri
Lib 15h18'32" -6d25'36"
Clödle, 12.01.1988
Umi 15h26'3" 69d35'58"
Cloé Butterfly
Ari 2h10'22" 10d38'13"
Cloe Grace
Uma 8h14'57" 70d56'53"
Cloe Kiev Klein
Vir 12h45'35" 3d47'1"
Cloe Silva
Leo 11h35'17" 25d13'18"
CloeAlexander
Cyg 19h58'20" 38d16'25"
cLoEbear
Lmi 10h29'15" 35d33'35"
Cloenina
Cnc 8h23'29" 14d19'12"
Cloey
Lyr 18h54'29" 33d47'59"
Cloffa Donald Vidrine
Ori 5h28'42" 1d32'35"
Clohn
Del 20h33'59" 15d2'6"
CloneAALIYAH.com
Mic 20h28'21" -32d47'56"
Clonestar (Jemma Ritto)
Cnc 8h36'53" 31d19'37"
Clora Bigley
Lyr 18h54'56" 36d49'32"
Clorinda Elizabeth Rabon
And 0h34'3" 42d25'37"
Close to You
Cep 23h7'13" 71d9'7"
Clotee
Umi 16h51'42" 76d47'0"
Clotine Surratt Davoren
Ori 5h10'45" 1d10'46"
Clôture et Début
Pho 1h30'5" -46d58'14"
Cloud 9 - Paul and Cheryl
Ori 5h1'48" 6d44'24"
Cloud Dancing Angel Baby
Pup 8h8'33" -22d11'1"
Cloud Nine
Cyg 19h23'34" 51d37'57"
Cloud Rainbow
Tau 3h37'19" 8d34'28"
Cloude
Cap 20h28'58" -25d56'12"
Cloud-Pixie829
Lep 5h37'7" -25d55'46"
Clover
Umi 14h47'28" 71d47'51"
Clover
Cnc 8h42'24" 30d48'0"
Clover Procopion
Leo 10h21'56" 10d30'32"
Clovis Theiler
Cas 0h56'26" 60d22'56"
Clownfish
Ori 5h40'1" 7d12'14"
Cloyd&ShirleyStaffords BlazeToGlory
Cyg 19h55'30" 33d43'16"
Cloyds 4 Stars
Aqr 22h31'13" -18d21'34"
CLT34
Cas 0h16'35" 51d46'2"
Club Electric Avenue
Uma 8h29'10" 64d26'12"
Club Rotes Kliff
Uma 9h53'59" 63d33'59"
Clums
Lib 15h33'13" -6d5'56"
ClustivikTR
Ori 5h31'35" -0d15'35"
Clute Barrow Nelson
Ori 4h51'52" 11d27'30"
Clyde
Ori 5h50'43" 2d9'56"
Clyde Allen Davis 39
Per 3h39'5" 44d7'39"
Clyde and Mildred West
Crb 15h44'19" 29d19'28"
Clyde and Ruth Ann Smith
Cap 21h34'16" -9d10'29"
Clyde & Annie Mae Helton
Cas 0h43'7" 47d36'26"
Clyde & Betty Karnes
Cyg 19h36'7" 31d52'49"
Clyde Clark Prestridge
Ari 2h50'43" 19d35'1"
Clyde Damron
Cnc 8h52'59" 28d21'48"
Clyde David Hice
Ori 5h30'36" -0d13'59"
Clyde E. Hollie
Boo 14h46'13" 33d0'8"
Clyde Henry Raper
Ori 6h11'31" 21d2'13"
CLYDE HIGNITE
Ari 2h2'40" 14d40'30"
Clyde Huffman
Ori 4h53'51" 11d0'25"
Clyde Kingston
Cap 20h26'38" -11d10'29"
Clyde L. Selvidge
Dra 17h31'57" 58d22'42"
Clyde Leon Loves Claudia Flores
Uma 10h48'14" 40d24'26"
Clyde Max Lawrence
Cnc 8h19'33" 31d39'3"
Clyde Mcdonald 1957
Vir 14h12'37" 3d47'52"
Clyde Merritt Johnson
Her 17h25'4" 38d20'54"
Clyde N. Lattimer
Uma 10h9'26" 55d55'16"
Clyde Neidigh
Dra 19h44'22" 61d48'27"
Clyde R. Culross
Her 17h19'55" 26d52'23"
Clyde R. Spears Jr. 12-19-1946
Sgr 18h53'33" -27d33'49"
Clyde Rathbone
Pho 0h47'48" -44d52'15"
Clyde Rojas
Cma 6h22'38" -14d35'0"
Clyde & Virginia Burch
Cyg 19h42'38" 30d55'26"
Clyde W. Hanks Jr.
Boo 14h17'2" 17d58'53"
Clyde W. Sedgeman
Her 17h3'43" 27d45'43"
CLYDER
Crb 15h54'20" 25d47'44"
Clyde's Corner
Her 17h4'33" 45d11'11"
Clyde's Dale
Lib 14h32'27" -10d55'58"
Clyde's Frog Star
Her 18h6'43" 26d40'42"
Clyde's Gleam for Clyde Lansdowne
Uma 14h24'24" 61d6'53"
Clynne T. Jones From Idabel,OK
Uma 11h25'19" 35d18'48"
ClyRan
Crb 15h43'20" 29d2'4"
Clytie
Lib 15h10'59" -6d21'34"
C.M Linda D
Lib 15h17'38" -4d58'8"
CM Stellaris
Dra 18h25'54" 50d38'21"
cmam2004
Ori 6h4'52" 10d34'40"
C.M.B
Sgr 19h35'48" -15d34'26"
CMB & MJB
Ori 6h2'14" 18d55'4"
Cmdr. Paul G. Bryant USNRet
Tau 5h48'0" 14d39'59"
CMo
Boo 13h50'26" 16d23'44"
C-Monster
Cyg 19h41'24" 36d43'28"
cmoorestars
Lyr 19h13'2" 37d53'3"
CMR LOVE
Cyg 19h42'55" 48d55'22"
CMSgt James A. Davidson
Sco 16h5'46" -25d52'7"
CMSgt Robert E. Dunn II
Aql 19h53'26" 13d40'0"
CMT
Vir 14h31'54" 3d7'22"
CMTZ
Leo 11h28'47" 27d17'26"
Cnedra My Angel
Pho 1h59'34" -44d14'23"
CNH&ACAR
Ori 6h11'48" 16d3'24"
CNimsk (aka-Charles "Chuck" Nimsk)
Sgr 18h27'13" -25d27'19"
C-Note
Lyr 18h47'25" 43d48'30"
C.N.S. Caseybells06*
Leo 9h33'58" 29d7'33"
Coach
Cnc 8h4'12" 23d44'7"
(Coach)
Aur 6h4'23" 46d59'30"
Coach A
Uma 11h44'8" 40d53'59"
Coach Bill Fanning
Ori 6h3'45" 14d41'8"
Coach Bob "Flea Flicker" Pritchard
Her 16h48'35" 37d19'36"
Coach Carole W. McArthur
Vir 12h31'32" 2d45'41"
Coach Clair Altemus
Ari 2h3'49" 24d0'35"
Coach Doug McKay
Uma 8h17'48" 62d34'24"
Coach Falc 54
Aqr 23h13'44" -13d22'9"
Coach Gary Pittillo
Pho 0h16'17" -54d48'11"
Coach Jake
Her 16h50'26" 13d42'36"
Coach Jerry Oyler
Her 17h22'17" 37d27'6"
Coach Jimmy (Daddy)
Per 3h12'59" 44d34'23"
Coach Jon Adams
Her 18h29'13" 21d55'37"
Coach Kathy
Vir 13h17'36" 8d27'14"
Coach Lin Havron
Cyg 20h29'25" 36d52'50"
Coach Maureen Trapp
Cas 1h37'35" 59d26'20"
Coach Mike Star
Aur 6h35'39" 36d1'48"
Coach Paula Cunningham
Lyn 7h38'41" 58d21'38"
Coach Petsch
Sgr 19h9'50" -15d29'12"
Coach Ron Wilhelm
Her 17h35'19" 36d32'35"
Coach Stowell- Best In The Universe
Aql 19h11'36" 4d52'55"
Coach Teri Rowe
Leo 11h22'9" 15d40'44"
Coach Terry Henry
Per 3h41'0" 41d21'28"
Coach Tom Curtiss
Her 17h21'38" 38d57'22"
Coach Wally Furtado
Uma 11h20'28" 49d44'31"
Coagulata
Dra 17h20'51" 65d31'20"
COAL
Tau 5h28'28" 19d54'59"
Coan
Lib 15h38'56" -25d56'11"
Coanabo Arnold De La Rosa
Aqr 21h59'30" -9d15'46"
Coattree Sorenson
Cnc 8h51'4" 21d42'19"

Coban's Light
Cnc 7h58'29" 14d50'27"
Cobbie
Lib 14h52'7" -2d49'50"
Cobbiebeline
Ari 2h42'32" 28d12'21"
Cobblers Cove
Lmi 10h6'16" 40d54'26"
Cobby White
Cnv 12h35'44" 39d10'42"
Cobecost IV
Cnv 13h50'7" 36d18'18"
Cobol Lipshitz
Cyg 20h7'5" 40d14'27"
Cobra Jim Ramasun Nemec
Sco 17h54'11" -33d50'59"
Coby
Sgr 19h36'18" -13d32'9"
Coby Bingham
Umi 14h46'19" 69d47'46"
Coby "Swigga" Hinkle
Her 18h4'54" 16d21'43"
Coby Wade
Uma 11h27'48" 33d25'43"
Coca
Cas 1h18'39" 53d29'48"
Coca
Tau 3h38'31" 16d25'47"
Cocamo-Brock
Aql 19h41'57" 2d36'16"
Cocca
Dra 19h30'50" 73d57'1"
CocElena03940795
Ari 3h28'52" 27d24'24"
Cochav Moshe
Leo 9h34'45" 29d21'59"
Cocheze
Cap 21h48'48" -16d54'55"
COCHICONG
Leo 10h8'12" 9d58'37"
Cochiliztli Xochitl & Huei Tochtli
Leo 11h7'34" 25d16'5"
Cochinillo Jabalin
Per 4h49'28" 46d26'6"
Cochran's Joy
Tau 4h54'9" 20d9'28"
Coco
Cmi 7h27'20" 11d22'50"
Coco
Psc 23h28'36" 6d31'37"
Coco
Tau 3h52'25" 24d9'4"
Coco
Lyr 18h43'31" 38d25'31"
Coco
Cas 0h33'33" 54d3'59"
CoCo
Lyr 18h51'23" 31d58'40"
Co-Co
Psc 1h10'14" 30d40'37"
Coco
Sco 16h3'25" -15d55'29"
Coco
Aqr 21h37'51" -0d4'49"
Coco
Umi 15h57'39" 74d28'56"
CoCo
Uma 10h27'57" 60d39'14"
Coco and Gramps
Sge 19h51'48" 18d44'37"
CoCo Baby Casillas
Umi 14h3'27" 72d22'13"
CoCo Coe Wiley
Cnc 8h57'15" 22d43'38"
Coco Love Nick
Cnc 8h44'3" 14d37'38"
Coco Loves Kyky
Lyn 8h46'0" 33d14'55"
Coco Miranda McErlain
And 0h7'53" 40d15'43"
Coco Puff
Vir 12h48'10" 5d35'48"
Coco Thaler
Cap 20h27'34" -27d8'7"
Cocoa
Uma 10h33'54" 72d2'48"
Cocoa Bean
Lyr 18h30'39" 43d57'11"
CoCoa Chippy
Umi 15h39'13" 77d53'55"
Cocoa Moo
Uma 8h30'16" 62d2'33"
CocoaNutts
Cyg 19h44'53" 48d5'14"
COCOJEN
Ori 5h31'55" 6d45'13"
Coconut
Umi 16h36'8" 77d1'55"
Coconut Court Beach Resort
Lac 22h27'45" 50d27'34"
Coconut Creek Hotel
Dra 18h31'39" 67d53'4"
Coco's Creative Star
Uma 12h41'42" 62d0'21"
COCOTYME (Cory, Coby, & Tyler Metz)
Tri 2h8'56" 33d34'41"
Cocozza Valentina
Cam 5h57'4" 60d21'31"
Cocuyo
Lib 15h5'19" -8d15'38"
CODA
Gem 7h36'17" 24d1'51"
Code 4
Cyg 21h34'46" 39d15'58"
Codee Michelle Scott
Cyg 21h31'5" 47d26'52"
Codeigh Blaire Thompson
Sco 17h6'17" -34d6'29"
Codeman
Umi 15h10'27" 78d43'0"
Codey
Sco 16h14'30" -10d13'39"
Codey Jane Rykert
Ari 3h15'56" 29d9'28"
Codey Jarie
Ari 2h56'57" 18d48'33"
Codi Mae Rogers
Lib 14h59'30" -4d1'38"
Codi Sarles
Uma 10h14'11" 58d11'16"
Codi Sugihara
Cmi 7h29'5" 2d30'16"
Codie Lynn 07021988
Cnc 9h4'48" 13d43'54"
Codine Assaad
Uma 11h46'35" 37d31'15"
Codi-Rose
Ari 3h29'17" 23d47'11"
Codruta
Uma 10h11'12" 43d44'53"
Cody
Aur 5h32'23" 39d35'14"
Cody
Aur 5h49'2" 32d24'19"
Cody
Her 16h54'26" 30d18'35"
Cody
Uma 8h36'19" 60d17'23"
Cody
Umi 14h31'18" 74d41'14"
Cody
Vir 14h1'54" -1d55'47"
Cody
Lup 15h40'49" -32d45'26"
Cody Aaron Zink
Psc 0h3'31" 2d18'58"
Cody Alexander
And 23h21'19" 50d59'14"
Cody Allan Wilson
Uma 9h12'3" 46d33'22"
Cody Allen Stormer
Gem 6h50'27" 30d37'31"
Cody and Amber together forever
Cyg 19h42'51" 39d14'13"
Cody and Bellas Goodnight Star
Cas 0h56'51" 62d1'15"
Cody and Casi's Star
Ara 16h46'6" -58d14'56"
Cody and Jessica
Gem 7h14'20" 33d48'0"
Cody Armand Thibeault
Her 17h46'49" 38d54'25"
CODY BEAR
Leo 11h2'11" 20d35'17"
Cody Bear Matthews
Uma 11h21'18" 34d52'39"
Cody BH
Uma 9h26'59" 66d56'24"
Cody "Big Diesel" Hansel
Aqr 22h9'43" -0d7'16"
Cody Brian Moore
Uma 13h21'20" 61d14'25"
Cody Bug
Per 2h49'4" 53d38'59"
Cody Clark
Sco 17h23'56" -37d12'46"
Cody Colton
Lib 16h0'6" -18d25'0"
Cody D. McKinney
Uma 11h27'49" 48d14'40"
Cody Dan LivingGood
Cep 23h17'26" 75d28'38"
Cody David Hagen "Forever"
Boo 15h17'30" 40d28'21"
Cody "DJ Trumac" Busch
Tau 4h27'47" 12d24'14"
Cody Donovan Hanna 12/24/1994
Cap 21h45'45" -12d57'13"
Cody Douglas McCoy
Per 3h27'1" 51d15'43"
Cody Evan
Gem 7h57'6" 30d47'7"
Cody Frazier Coon
Uma 11h42'2" 43d22'4"
Cody "Freddy" Shane Huff
Her 17h12'37" 19d16'42"
Cody G.
Vir 12h4'48" -0d13'36"
Cody Garrett Bruno "CODEMAN"
Cap 20h39'37" -19d23'16"
Cody Haynes
Dra 17h54'26" 63d19'53"
Cody Honey
Cap 21h47'16" -11d11'11"
Cody Huddleston
Equ 21h23'24" 3d58'51"
Cody J. Harrett
Uma 9h9'22" 62d55'4"
Cody James
Dra 20h25'19" 62d28'50"
Cody James Kegley
Aur 5h43'26" 45d4'25"
Cody James Robbins
Umi 15h50'55" 75d4'43"
Cody James Saugestad
Per 3h7'14" 52d19'27"
Cody James Smith
Cru 12h46'16" -64d30'12"
Cody James Wright
Cnc 9h4'48" 31d27'48"
Cody Jeffers
Cyg 20h5'23" 58d26'15"
Cody Jefferson Christopher
Tau 4h1'41" 16d26'23"
Cody Jett Villalon
Aur 5h38'55" 48d35'14"
Cody Joe
And 23h21'57" 43d37'53"
Cody John Glad-21!-John 15:13
Cas 23h45'14" 56d29'44"
Cody Joseph Otts
Her 17h3'46" 22d33'47"
Cody Katie Chelsey Pepper
Cnv 13h52'41" 38d58'37"
Cody & Katie Dollinger
Cyg 21h34'22" 40d44'53"
Cody Kevin Steele
Per 3h28'22" 41d6'20"
Cody Lee Lockley
Cap 21h36'51" -9d11'9"
Cody Lee McGullion
Ari 3h6'59" 18d20'44"
Cody Leisner
Uma 10h47'22" 49d24'17"
Cody Leon, My Love
Per 3h20'5" 32d47'17"
Cody Luke Morris
Her 17h2'26" 27d47'53"
Cody Lynne
Ori 5h39'17" 4d39'44"
Cody Marlowe
Her 18h21'28" 15d27'8"
Cody Martin Tederick
Psc 1h26'53" 17d19'48"
Cody Matheson
Uma 9h31'48" 52d46'6"
Cody Matthew
Cru 12h15'12" -63d36'49"
Cody Matthew Thomas
Leo 9h25'44" 15d49'50"
Cody Michael Moy
Psc 0h55'14" 27d8'6"
Cody Michael Pado
Ori 5h31'20" 14d25'57"
Cody Nolan
Her 17h22'14" 18d12'23"
Cody Perel
Uma 9h51'3" 52d30'7"
Cody Potter A Shining Star
And 0h27'27" 26d49'16"
Cody R. Petersen
Ori 5h37'50" 8d24'3"
Cody Reid
Per 2h28'0" 53d32'14"
Cody Robert Jones
Crb 16h15'40" 37d5'43"
Cody Robert Taborda
Cyg 19h43'21" 30d55'32"
Cody Robert Yohn
Cap 21h16'27" -14d59'8"
Cody Ryan
Ori 5h24'12" 3d18'29"
Cody Ryan Baker
Cap 20h7'54" -20d38'56"
Cody Ryan King
Lib 14h32'24" -15d29'1"
Cody Schoonover
Uma 12h40'14" 59d24'13"
Cody Scott
Sgr 18h15'6" -27d39'9"
Cody Shayne
Lup 15h42'14" -30d56'58"
Cody Smink, Eternal Smile
Her 18h39'50" 20d30'9"
Cody Smith
Umi 14h37'37" 74d5'28"
Cody Smith Mattioli
Cma 6h32'30" -17d32'17"
Cody Sneed
Aur 6h27'11" 34d55'43"
Cody Steven Schmurr
Per 2h55'24" 42d45'6"
Cody "Sweet Puppy" Marcinkus
Cap 20h38'16" -13d3'49"
Cody T
Her 17h24'44" 25d10'28"
Cody Thacker Davis
Sgr 19h5'43" -27d59'14"
Cody Thomas Bailey
Gem 7h14'32" 24d25'39"
Cody Thomas Knippen
Aql 19h27'3" 6d47'42"
Cody Toohey
Ari 3h13'21" 29d8'52"
Cody Tyler Darling
Lyn 7h55'55" 43d5'19"
Cody Vincent Gottberg
Tau 4h41'52" 20d19'52"
Cody W. Hillman
Cnc 8h43'16" 31d16'3"
Cody Wenning
Aur 6h18'26" 47d33'44"
Cody Whalen
Vir 11h44'57" 4d20'31"
Cody Wilbanks
Vir 11h52'44" 6d11'47"
Cody William Clarke
Leo 11h16'38" 18d30'22"
Cody Wyatt Smith
Tau 5h12'4" 16d58'3"
Cody Young
Umi 13h45'58" 74d14'13"
Codye Eddings-Hamm
Psc 23h41'26" 6d2'17"
Cody-Leigh P Mullin (Clem) Sweet 16
Umi 15h56'47" 82d53'22"
Cody's Courage
Per 3h19'16" 51d52'49"
Cody's Komet
Cep 22h59'0" 66d4'58"
Cody's Star
Uma 12h36'52" 55d36'3"
Cody's Star
Cru 12h25'5" -57d8'46"
Cody's Star
Gem 7h17'19" 20d8'43"
Coel Logan Bryant
Col 5h36'39" -34d22'31"
Coen James Chaney
Uma 12h5'8" 43d22'47"
Coen Ray
Sco 17h17'22" -32d34'8"
Coenraad Fourie Stassen
Her 17h13'37" 48d48'25"
Coeta Evelyn Russell Lewis
Umi 17h30'40" 80d5'32"
coeur de croissant
Peg 21h23'41" 19d25'12"
Coeur indigo
Cnc 8h45'38" 9d16'55"
Coffaa's Family Star
Ori 6h10'48" 20d54'37"
Coffee*
Cnc 9h13'2" 25d12'37"
Coggie
Dra 10h35'6" 73d47'0"
Cognationis spiritus
Aqr 22h8'19" -1d59'44"
Cohen Jay Bonus
Per 4h27'57" 49d59'30"
Cohen Jeremiah Davidson
Ori 6h14'40" 5d51'21"
Cohen Michael Clark
Aur 6h36'1" 31d48'26"
Coinneach MacCoinnich Grieve
Uma 11h55'5" 35d8'35"
Coke & Oi
Aqr 22h38'36" -18d37'46"
Cokes
Cru 12h25'55" -60d58'45"
Coko Toomey's Star
Umi 16h13'57" 79d27'23"
Coky
Vir 14h14'57" -14d32'20"
Col
Uma 11h39'20" 37d7'34"
Col & Cate
Cru 12h53'33" -59d21'29"
Col. Elmer A. Horne
Per 3h21'16" 42d35'37"
Col Gaylen Roberts
Ori 5h38'18" -0d9'7"
Col. George Franklin Blalock
Uma 11h39'55" 60d10'8"
Col. James Henry Hayes
Uma 13h35'0" 54d1'8"
Col. James Wade Stanley, USAF (ret)
Aql 19h43'34" 7d31'10"
Col Joseph F. De Rienzo, USAF
Ori 6h13'22" 9d8'18"
Col. Lyle Maritzen
Aql 19h7'47" -0d18'24"
Col. (ret.) Joseph A. Kuhn
Aql 19h9'51" -0d29'1"
Cola
Uma 11h10'5" 53d31'54"
CoLa
Psc 1h31'40" 19d34'10"
Cola
Com 13h7'24" 28d11'48"
ColandMar
Gem 6h50'14" 24d16'30"
Colavita-Simone
Cyg 20h0'57" 39d23'3"
Colbi
Uma 9h20'58" 48d16'35"
Colburty K
Ori 5h9'7" 3d42'28"
Colby
Ori 5h36'55" 9d35'44"
Colby
Cap 21h9'22" -18d48'24"
Colby Aiden Welborn
Mon 6h27'3" -9d55'45"
Colby Allen Whitacre
Cap 21h37'42" -16d54'55"
Colby Andrew Egkan
Tau 5h53'32" 23d16'52"
Colby Bennett McCaskill
Aqr 21h54'27" 0d49'36"
Colby Boroff
Uma 10h11'21" 54d50'13"
Colby Bruin Curran
Her 17h13'54" 25d49'17"
Colby Camalier
Umi 16h20'30" 76d52'53"
Colby Cochran
And 0h28'5" 29d36'26"
Colby Foster Beckett
Dra 18h46'31" 54d17'26"
Colby Garrett Beck
Ari 1h58'55" 11d38'41"
Colby H. Smith
Uma 12h6'6" 48d9'21"
Colby & Jacqueline Ramos
Cyg 21h30'32" 45d19'2"
Colby James King
Ori 5h28'58" 4d34'43"
Colby James Martin
Aqr 22h13'22" -3d16'36"
Colby Jay
Gem 7h9'47" 33d24'0"
Colby Jayne Lima
Cnc 9h6'52" 30d22'36"
Colby JR & SR
Lyn 6h56'49" 54d4'8"
Colby Kenneth Brent
Lib 15h20'10" -27d35'3"
Colby Lee Ingo
Umi 19h16'42" 88d52'14"
Colby Lin
Lib 15h23'41" -5d16'38"
Colby Loomis
And 2h33'37" 45d38'22"
Colby Noel Munoz
Cen 13h12'7" -45d42'11"
Colby Parker
Aqr 22h25'1" -2d4'19"
Colby Reasor
Boo 14h19'38" 46d7'56"
Colby Robert Schlingmann
Uma 10h55'39" 64d47'33"
Colby Royce Keusch
Ori 5h56'10" 18d23'32"
Colby Scott Villard
Sgr 18h8'3" -18d13'35"
Colby Toshimitsu Lee
Uma 13h13'32" 60d30'35"
Colby William Hurtubise "Alligate"
Aqr 21h2'59" 0d50'48"
Colby William Shewbridge
Dra 16h21'46" 52d24'54"
Colby's Dreams "11-11-86"
Uma 11h23'12" 59d45'50"
Colbys' Light
Ari 3h29'13" 21d23'36"
Cole
Leo 10h34'54" 13d53'43"
Cole
Uma 9h42'48" 46d8'0"
Cole
Uma 11h30'24" 47d39'0"
Cole
Uma 10h57'49" 35d40'9"
Cole
And 1h14'30" 38d31'28"
Cole
Sgr 19h46'31" -22d12'32"
Cole
Sco 16h13'16" -10d13'10"
Cole
Umi 14h31'9" 76d19'3"
Cole Alan Minnick
Cnc 8h17'21" 16d52'56"
Cole Alexander Dawkins
Leo 9h41'23" 27d40'31"
Cole Alexander Morehead
Her 17h36'29" 46d23'36"
Cole and Drew Vianello
Sgr 19h8'50" -16d56'15"
Cole and Wade Wichard
Umi 15h15'14" 74d0'18"
Cole Anthony Drouin
Sco 16h37'2" -30d57'52"
Cole Anthony Turner
Per 3h6'39" 47d1'6"
Cole Armon Cutchin
Peg 23h32'17" 8d51'22"
Cole Avery Bastarache
Vir 13h59'44" 3d24'56"
Cole Avery Delarso
Aql 19h47'52" 2d42'4"
Cole B. Knox
Cep 22h57'17" 59d51'32"
Cole Benjamin Knox
Her 16h27'42" 41d54'6"
Cole Benson
Aqr 22h32'48" -19d11'6"
Cole Berkowitz
Aqr 21h23'39" 1d54'33"
Cole Bernard Jaeger
Col 5h47'32" -30d25'3"
Cole "Bucket" Williams
Leo 10h21'12" 26d33'34"
Cole Butler
Tau 4h19'19" 29d0'18"
Cole Calkins
Her 17h17'41" 36d8'15"
Cole Chandler
Her 17h52'24" 23d46'5"
Cole Charles Parmley
Peg 21h12'13" 15d50'24"
Cole Christine
Cas 0h9'53" 56d3'13"
Cole Daniel Harrison Smallwood
Sco 17h52'38" -36d16'17"
Cole David Rogers
Sgr 18h15'50" -17d43'7"
Cole Davis Kennedy
Umi 14h25'44" 77d40'41"
Cole Dennis Mumm
Uma 10h37'41" 47d51'42"
Cole Dinan
Ori 6h6'47" -1d58'59"
Cole Dominic Hudson
Ori 5h27'55" 6d12'3"
Cole Donald Sonnichsen
Lib 15h6'49" -9d27'24"
Cole Douglas Hathaway
Per 2h47'32" 41d9'53"
Cole Duane Boushee
Ari 2h53'25" 28d19'44"
Cole Duane Boushee
Uma 12h13'9" 58d1'52"
Cole Dubois
Aur 5h10'16" 41d50'30"
Cole Eaton
Lyn 8h7'32" 59d16'5"
Cole Edward
Her 17h5'49" 31d59'56"
Cole Edward Firlie
Ori 5h1'29" 15d22'10"
Cole Edward Hamilton
Umi 14h26'46" 75d22'47"
Cole Edward Wagner
Per 4h8'49" 45d3'59"
Cole Engin Arslanbas
Cnc 8h14'32" 26d0'9"
Cole Ethan Edwards
Her 17h45'51" 46d51'38"
Cole Family 70
Cyg 20h33'34" 58d32'17"
Cole Ferry
Tau 4h24'17" 24d38'9"
Cole Higbee's Wishing Star
Gem 6h52'35" 24d44'46"
Cole Hilliard Price
Lyn 8h0'15" 44d18'21"
Cole John Evans
Cep 22h1'12" 60d4'54"
Cole Jon Richardson
Umi 14h37'45" 81d5'46"
Cole JT
Aqr 22h36'23" -0d39'55"
Cole Kareliussen
Uma 9h18'21" 50d24'40"
Cole Kennedy
Lyr 18h33'40" 36d13'20"
Cole Kenneth
Cyg 21h32'30" 53d50'39"
Cole Kevin Rogers
Ari 1h47'43" 21d38'6"
Cole L. Kistner
Cyg 20h34'1" 36d26'0"
Cole Landen McNally
Aqr 22h7'21" -3d5'26"
Cole Larsen
Tau 4h31'2" 23d2'35"
Cole Laurie
Psc 1h34'37" 14d47'1"
Cole Lawrence Sashkin
Vir 13h29'22" 1d10'28"
Cole Lewis
Cap 21h3'37" -23d31'40"
Cole Lloyd Crotty
Ori 6h5'45" 11d9'6"
Cole Matthew Leblanc
Leo 11h2'35" 9d42'1"
Cole Matthew Sisk
Uma 11h8'39" 70d9'15"
Cole McMillan Hess-Sallach
Tau 5h37'46" 26d29'30"
Cole Michael Andrews
Cnc 9h6'58" 21d48'49"
Cole Michael Bright
Her 17h52'31" 24d3'25"
Cole Michael Fiore
Sgr 19h44'55" -12d29'3"
Cole Michael Fiore
Sgr 18h5'5" -29d57'38"
Cole Milia Leighton
Uma 11h7'32" 35d8'33"
Cole "Moose" Corwin
Ari 1h53'5" 19d13'59"
Cole Myers
Her 18h15'54" 15d49'18"
Cole Newcomer
Her 18h15'16" 29d56'12"
Cole Patrick Horan
Sco 17h53'28" -36d41'46"
Cole Presley
Cma 6h50'24" -13d55'41"
Cole R Flanagan Star of Lives
Her 17h11'14" 33d45'20"
Cole "Rascole" Sleeter #343
Lib 15h24'26" -16d44'46"
Cole Rasic
Hor 2h58'18" -52d37'29"
Cole Richard
Per 3h40'11" 51d54'44"
Cole Richard Durham
Per 3h6'53" 45d21'52"
Cole Richard Jenkins
Vir 14h5'31" 7d11'19"
Cole Robert Marshall
Peg 22h43'7" 25d3'25"
Cole Robert McGee
Per 3h31'17" 45d24'39"
Cole Roby
Cyg 19h55'53" 37d0'20"
Cole Russell Planamento
Tau 3h49'19" 29d38'30"
Cole Ryan
Sco 17h57'11" -38d6'19"
Cole Ryland Sorensen
Ari 2h44'41" 17d19'15"
Cole Sandelin
Ari 2h6'14" 21d46'29"
Cole Sherlock Hersey
Gem 7h46'29" 26d38'8"
Cole Smith
Tau 4h1'39" 22d50'25"
Cole Spencer Pruitt
Lyn 8h45'23" 37d58'52"
Cole Thomas Cartalino
Her 16h33'15" 36d15'50"
Cole Thomas Stuart
Her 17h38'42" 36d52'6"
Cole Thomas Tidwell
Ori 4h54'13" 2d28'10"
Cole Vincent Charles
Ari 2h49'46" 24d27'6"
Cole W. Hamilton
Uma 10h33'15" 59d58'37"
Cole Wessel Ten Broeck
Ori 6h5'55" 18d25'12"
Cole White
Cap 20h15'48" -26d7'29"
Cole Whitten
Ori 6h9'49" 3d16'57"
Cole William Nilles
Per 2h45'52" 53d26'37"
Cole Wilson
Her 18h19'47" 15d23'23"
Cole, for your pocket. Love Ashley.
Cep 21h25'26" 66d0'11"
Colebug
Equ 21h12'58" 5d36'23"
Coledust
Ori 5h37'31" 0d36'51"
Coleen April
Cas 0h24'21" 61d36'27"
Coleen Harting
Cyg 20h17'48" 41d4'53"
Coleen J.
Cap 21h33'31" -9d4'30"
Coleen K Humphries
Cap 20h46'31" -15d24'53"
Coleen MacDonald
Leo 11h52'56" 18d47'34"
Coleen Mary Tracey
Cas 1h17'37" 59d41'3"
Coleen My Love
Gem 6h29'56" 16d27'26"
Coleen Petrello
Cas 0h46'49" 57d29'31"
Coleen RuthAnn Crane
Lyr 19h1'23" 39d29'4"
Coleen's Love
Aqr 21h14'32" -12d47'20"
Colegio Maria del Rosario
Uma 11h16'10" 51d45'14"
Coleman Alexander
Uma 8h51'54" 66d56'11"
Coleman Ann
Lib 15h46'46" -18d51'23"
Coleman Davis Laye
Cap 20h49'12" -25d42'41"
Coleman Edward Espinoza
Aql 19h52'45" 12d0'13"
Coleman Ferns.t.a.r.
Uma 9h45'22" 58d47'34"
Coleman Hartigan
Psc 0h59'51" 24d36'57"
Coleman & Jen Forever
Cyg 19h44'39" 37d23'29"
Coleman Joseph Bell
Her 17h56'54" 49d40'41"
Coleman Katz
Lyn 7h21'0" 47d42'50"
Coleman Luke Fryer
Peg 23h19'24" 32d38'28"
Coleman Max Weintraub
Psc 1h39'37" 16d29'21"
Coleman Michael Hurt
Ori 6h12'42" 15d46'51"
Coleman Vegvari "1993-2006"
Uma 10h34'10" 52d42'16"
Coleman William Jeter
Lyn 7h44'39" 53d34'46"
Coleman's Little Slice of Heaven
Cyg 19h33'57" 51d59'40"
Colene Denise Doughty
Aqr 22h54'57" -6d49'23"
Colene Smith
Vir 14h1'36" -16d41'26"
Colene's Paramount
Uma 9h12'58" 55d7'51"
Col.E.R.Johnston
Leo 11h2'34" 2d36'24"
Cole's Shining Light
Sco 16h17'17" -16d50'54"
Cole's Star
Cnc 8h39'55" 17d26'59"
Coleson Lee Kiser
Umi 13h34'6" 72d21'26"

Coletan Anthony Traughber
Ari 2h6'23" 12d39'56"
Coleton Charles Elliott
Lyn 6h51'53" 52d58'16"
Coletta Jean Robertson
Vir 11h52'32" 5d27'52"
ColettaJ80
Tau 4h23'45" 8d37'28"
Colette
Ari 2h20'16" 14d8'41"
Colette
Vir 15h5'38" 4d1'27"
Colette
And 22h59'50" 48d32'52"
Colette 120447
Cas 23h28'20" 56d4'21"
Colette A Hanley
Cyg 19h34'23" 28d15'15"
Colette Anne Waid
Mon 7h34'50" -1d1'52"
Colette Dube
Lib 14h53'39" -8d27'3"
Colette et Claude Pellegrin
Mon 7h17'45" -5d32'8"
Colette Helen Vandiver
Sco 15h55'29" -22d33'24"
Colette Kimberlin Reagan
And 0h25'49" 23d25'48"
Colette Lucia Wood Sime
Cap 20h47'36" -20d42'26"
Colette Miller's Star 5-17-1983
Tau 4h21'10" 27d10'5"
Colette Rener
Cas 0h37'56" 58d46'43"
Colette & Roman
Cyg 19h40'55" 35d57'59"
ColetteMinuit
Per 3h14'41" 47d15'53"
Coletti
Cnc 8h26'53" 28d24'2"
Coley
Vir 13h8'43" 10d59'11"
Coley
Sco 16h14'43" -10d9'58"
Coley
Umi 16h17'51" 78d33'37"
Coley Cole
Lib 15h0'7" -6d16'22"
Coley Daniel Tyler
Per 3h47'31" 49d6'1"
Coley Lynne
Cas 0h7'20" 56d12'34"
Coley Star
Lib 15h6'11" -17d36'40"
Coley12096
Uma 10h18'5" 67d36'34"
ColeyRichardson*EndlessLoves'Star*
Crb 15h35'38" 27d32'54"
Coley's Star
Crb 16h7'27" 35d48'18"
Colie
And 0h33'10" 40d18'30"
Colie
Ari 2h49'52" 25d57'0"
Colie
Uma 9h15'16" 63d17'14"
Colie J!
Psc 23h59'15" 7d15'11"
Colie Jo Smigliane
Per 2h11'1" 56d9'43"
Coliemyn - Nov 24 1956 - C&L Langhout
Cyg 20h19'6" 40d37'1"
Colin
Her 17h43'56" 36d48'0"
Colin
Gem 7h35'32" 27d39'12"
Colin
Cam 7h6'47" 68d58'35"
Colin
Psc 1h52'41" 2d48'4"
Colin A. Davis
Cnc 8h54'44" 10d30'53"
Colin A Ferguson - 19 April 1917
Cru 12h28'28" -57d4'11"
Colin A Gregory
Psc 23h25'2" 3d13'28"
Colin A. McCabe
Her 17h36'29" 37d25'7"
Colin Alexander William Chisholm IV
Aqr 22h50'40" -22d24'41"
Colin and Devin Forever
Aqr 23h4'48" -9d46'44"
Colin and Kelly Larkin
Cyg 21h14'20" 44d45'43"
Colin and Lindsay 2004
Boo 15h1'47" 45d51'42"
Colin and Lindsay's Special Star
Cyg 20h8'46" 53d52'2"
Colin Anthony Hammel
Lib 15h9'20" -7d35'51"
Colin Ballantyne (UB 6)
Ori 5h26'16" 5d31'23"
Colin Bauer
Lyn 6h30'36" 61d55'10"
Colin Biedermann
Umi 14h26'19" 76d55'46"
Colin Blake Scoggins
Gem 6h51'1" 28d32'4"
Colin Bownes
Peg 22h24'32" 18d44'34"
Colin Christopher Soltis
Her 16h27'28" 48d23'50"
Colin Clark "Tigger Ruler of Zog!"
Cnc 8h33'23" 17d11'30"
Colin Coady
Peg 22h34'26" 18d1'21"
Colin Cooper
Ori 6h14'38" 15d36'22"
Colin Daniel Schrage
Gem 7h2'8" 17d30'37"
Colin David Bailey 10 January 1956
Cap 21h33'6" -9d38'52"
Colin David Scott
Uma 10h53'30" 49d13'4"
Colin DeMarlie
Dra 18h31'13" 57d15'45"
Colin Earl
Aql 19h27'47" 7d39'6"
Colin Edward Ryan
Uma 9h29'19" 51d23'17"
Colin & Elizabeth Dutton 24-12-2002
Lyn 8h11'12" 38d6'39"
Colin & Elke
Per 3h18'35" 42d21'12"
Colin Fitzpatrick
Cyg 20h3'21" 39d9'4"
Colin Fragar - My Love For Eternity
Cru 12h27'57" -60d32'50"
Colin Gawronski
Aqr 23h6'35" -4d55'32"
Colin George Millar
Leo 11h29'39" 27d46'5"
Colin Gregory Waugh
Gem 7h17'1" 28d57'40"
Colin Hayden
Cep 21h52'4" 63d4'51"
Colin J Delahunty
Per 1h42'21" 50d49'15"
Colin J Good
Ori 5h48'48" 3d5'13"
Colin Jack Dugan
Ori 5h33'18" 7d5'29"
Colin Jacob's Celestial Sparkle
Ori 6h3'18" 18d7'2"
Colin James Boyarski
And 0h27'19" 45d21'28"
Colin James Fagan
Lib 15h36'40" -24d35'45"
Colin James Harris
Ori 6h10'47" 9d11'6"
Colin James Hernandez
Psc 1h20'0" 18d0'52"
Colin James Loudermilk
Peg 21h38'55" 23d27'55"
Colin James Masco
Lmi 10h37'6" 37d44'23"
Colin James McGovern
Dra 17h14'54" 63d18'45"
Colin Jason Lee
Sco 17h55'54" -41d34'37"
Colin Jeffery Pfister(Colin's Star)
Psc 1h15'15" 32d36'2"
Colin - Jérémie 2005
Cyg 21h26'30" 34d23'29"
Colin Jian McCabe
Sco 16h17'58" -10d15'31"
Colin John Rose (C.J.)
Aqr 23h21'9" -23d26'46"
Colin Johnson
Aqr 22h47'47" -20d49'20"
Colin Josef McDonald
Dra 17h45'16" 55d15'7"
Colin Joseph
Uma 8h28'47" 66d50'12"
Colin Joseph Donnelly
Uma 11h46'59" 46d33'10"
Colin Joseph Maximus Nicassio
Cnc 8h4'10" 17d35'22"
Colin Jude Saragusa
Sco 17h53'38" -35d43'26"
Colin Kai
Cnc 9h6'4" 14d32'57"
Colin L. Essig
Her 16h28'32" 43d7'15"
Colin Leeds
Per 3h16'5" 52d25'56"
Colin Lewis
Per 2h33'4" 51d11'16"
Colin Louvain Mortimer, "Big Grandad"
Uma 10h30'13" 61d46'9"
Colin Low
Lyn 7h8'23" 53d40'33"
Colin Mackie
Per 3h36'14" 34d33'51"
Colin MacNeil
Ser 15h29'12" 24d28'22"
Colin Maddox Pape
Lac 22h23'32" 47d1'26"
Colin Masely
Lib 15h31'30" -21d6'32"
Colin Mathew
Cap 21h41'22" -10d50'40"
Colin Matthew Nguyen
Umi 14h17'54" 78d27'4"
Colin McNelley
Ori 6h15'47" 14d58'32"
Colin McPherson
Vir 14h4'36" -15d55'9"
Colin McPherson Smith
Ori 4h59'7" -2d32'10"
Colin McTaggart
Peg 21h58'31" 17d21'13"
Colin Michael Dwyer
Per 2h50'1" 43d56'6"
Colin Michael Groody
Cap 21h16'15" -15d46'35"
Colin Michael McCarthy
Sco 16h11'9" -10d43'5"
Colin Michael McNiven
Boo 14h40'36" 28d11'39"
Colin Michael O'Connor
Her 17h15'54" 32d34'9"
Colin Michael Pasquale
Per 4h45'47" 49d34'28"
Colin Michael Peters 12/12/1996
Ori 5h56'37" 12d56'41"
Colin Michael Winston
Ari 1h55'55" 23d58'54"
Colin Morgan
Her 17h16'17" 18d3'7"
Colin Nicholas Chimento
Aqr 23h15'41" -14d54'55"
Colin O'Donoghue
Uma 9h39'22" 53d25'1"
Colin Patino
Ari 2h53'26" 21d25'52"
Colin Patrick Kilgore
Uma 10h33'8" 41d43'36"
Colin Peirce O'Brien
Sco 16h48'55" -34d46'20"
Colin Peter Carey
Uma 9h5'45" 48d47'33"
Colin Robert Davis
Psc 0h35'28" 8d3'33"
Colin Scott Kirschner
Ori 6h13'15" 16d52'10"
Colin Stephen's Star
Lib 14h22'24" -23d13'51"
Colin Thomson McKay
Tau 5h12'14" 17d51'48"
Colin Timbrell. Dad's Ray of Light
Aql 20h22'57" 5d42'54"
Colin Timothy Johnson
Dra 19h47'35" 64d33'55"
Colin Vose Reagan
Cep 22h45'6" 72d58'4"
Colin Wolfe
Uma 11h14'13" 68d50'51"
Colin Y. Miyajima
Tau 3h48'28" 18d29'53"
Colin's Star
Tau 3h50'51" 23d19'45"
Colin's star-EAGLE 1
Cap 20h24'12" -11d8'11"
Colinterri
Aql 19h49'42" 11d29'9"
Coll 10-22 "You Are Stellar"
Uma 9h55'11" 42d10'11"
Collard Bovy Lionel
Umi 15h51'37" 81d54'9"
Colleen
Sco 16h14'54" -14d26'56"
Colleen
Cap 20h29'46" -20d5'1"
Colleen
Lib 15h13'54" -20d4'5"
Colleen
Cyg 19h45'47" 57d21'10"
Colleen
Sgr 18h24'10" -22d53'41"
COLLEEN
Sgr 18h35'48" -23d22'33"
Colleen
Crb 15h44'47" 36d32'4"
Colleen
Crb 15h46'48" 35d41'23"
Colleen
Lmi 10h2'34" 30d44'16"
Colleen
And 23h35'14" 50d3'29"
Colleen
And 23h20'34" 48d19'0"
Colleen
Cyg 20h53'15" 37d42'20"
Colleen
Tau 5h48'13" 26d2'59"
Colleen
Psc 1h25'26" 27d34'5"
Colleen A. Farrell
Uma 11h7'36" 30d59'59"
Colleen A. Yokel
Cas 0h14'55" 54d25'13"
Colleen Adrienne
Vir 12h46'32" 2d43'15"
Colleen and Anthony Turner
Cyg 21h18'21" 45d30'50"
Colleen and Duane
Crb 15h34'53" 28d16'45"
Colleen and Tori sisters forever
Leo 10h49'42" 19d31'45"
Colleen Andujar
Lyn 8h42'18" 37d11'5"
Colleen Ann
Crb 16h15'49" 32d9'24"
Colleen Ann Bowyer
Uma 11h34'59" 48d52'33"
Colleen Ann Brown-Choate
Her 17h16'1" 32d50'24"
Colleen Ann Condito
Lib 14h51'21" -0d33'59"
Colleen Ann Tigue
Cas 23h3'28" 55d24'39"
Colleen Ann Wall
Cyg 20h36'31" 30d22'15"
Colleen Anne Logan "Leenie"
And 1h53'44" 39d20'40"
Colleen Annette Cahill
Col 5h51'37" -35d57'1"
Colleen "Ashley" Cleary
And 23h30'54" 41d24'43"
Colleen Balzano
Lyn 7h37'15" 50d41'9"
Colleen Barkauskas
Ari 2h35'24" 23d56'14"
Colleen Bock
And 2h11'16" 46d47'15"
Colleen Brazil
Aqr 21h42'4" -2d19'36"
Colleen Bridget Baumann (Beanie)
Uma 12h3'17" 60d15'37"
Colleen Calhoun
Aql 19h33'16" 2d49'47"
Colleen Campbell
Ari 2h58'19" 25d13'49"
Colleen Chutich
Lyn 8h25'54" 39d52'12"
Colleen Claudette
Cas 0h10'7" 56d53'19"
Colleen Cluster
And 23h14'54" 50d37'27"
Colleen Colabella
Lib 14h51'55" -2d14'38"
Colleen Cowan
Mon 6h35'22" 4d50'45"
Colleen Cross
And 1h9'3" 36d6'44"
Colleen Crystal Bolander
Aqr 22h43'17" -6d23'59"
Colleen Danelle Larke
Lib 15h39'42" -12d53'35"
Colleen Danielle Barwick
Sco 16h12'16" -10d0'52"
Colleen & David Mun
Uma 11h49'1" 61d0'56"
Colleen Denise Hurst
Cra 18h37'1" -41d50'54"
Colleen Diane
Cyg 20h14'59" 34d54'52"
Colleen Donegan Gioioso
Umi 15h46'32" 84d46'52"
Colleen Duke
Leo 11h42'51" 28d3'23"
Colleen "Elephant" Varney
Cas 1h22'20" 67d29'1"
Colleen Elizabeth Flynn
Cas 23h2'38" 56d8'54"
Colleen Elizabeth O'Day
Lyn 7h18'41" 58d58'52"
Colleen Elizabeth Vidal
Cas 1h39'17" 64d4'25"
Colleen Erck
And 23h10'57" 47d9'52"
Colleen Erin Hickey
Cnc 8h38'37" 17d50'19"
Colleen F. Hunter
Lmi 10h23'5" 37d3'2"
Colleen Faith Williams
Cas 0h37'30" 64d47'50"
Colleen Gerrard My Special Love xxx
Aqr 23h6'43" -21d25'56"
Colleen Graley
Tau 5h4'56" 23d29'48"
Colleen - Hannibal Slaying Goddess
Cas 23h47'35" 52d54'42"
Colleen Hunter
Cap 21h0'58" -19d43'12"
Colleen Hyland
And 23h35'7" 36d52'5"
Colleen J. Davis
Lyr 18h29'24" 37d16'43"
COLLEEN JANA CARROW
Psc 1h41'28" 20d23'22"
Colleen Jane Hart BVAD Grad Dip Ed
Psa 22h27'54" -30d34'33"
Colleen Joy Bryan
Leo 11h40'31" 25d18'39"
Colleen Kathleen Fairs
Leo 11h40'42" 25d31'50"
Colleen Kay Cooper
Psc 1h7'52" 14d23'36"
Colleen Kelly
Vir 12h11'45" 7d32'21"
Colleen Kelly Mahaffey
Mon 7h35'22" -0d20'24"
Colleen Koba's Star
Lib 14h59'16" -9d26'2"
Colleen Lake
Mon 6h56'50" 9d9'2"
Colleen Le Page
Cas 23h12'38" 54d46'2"
Colleen Leszczynski
Uma 9h50'9" 44d49'51"
Colleen Lorraine Hallisey
Lib 15h36'13" -9d39'22"
Colleen Lynn Roberts
Lyn 7h53'24" 47d29'24"
Colleen Lysiak
Sgr 18h23'21" -31d23'4"
Colleen M Clinton
Cas 0h55'55" 65d19'52"
Colleen M. Stefhon
And 23h19'34" 49d43'4"
Colleen Macort
Ari 2h12'12" 24d48'14"
Colleen Magruder
Vir 12h0'48" 7d26'45"
Colleen Malone
Vir 12h42'36" 6d48'34"
Colleen Manni
Leo 11h11'37" -0d10'3"
Colleen Margaret Jolly
Sco 16h15'50" -13d10'24"
Colleen Marie
Vir 13h16'23" -18d50'18"
Colleen Marie
Lyn 7h48'55" 56d23'34"
Colleen Marie
Leo 9h38'14" 29d36'32"
Colleen Marie
And 1h3'21" 41d49'24"
Colleen Marie
Cnc 8h14'45" 32d49'16"
Colleen Marie (Dreamz)
Cas 1h32'35" 63d46'16"
Colleen Marie Germain
And 23h15'50" 47d42'14"
Colleen Marie Posner
Cas 1h28'8" 53d42'6"
Colleen Marie Shawda
And 1h58'28" 39d15'1"
Colleen Marie Wolf
Cas 1h34'31" 72d2'49"
Colleen Martin
Vir 13h30'57" -20d31'58"
Colleen Mary
Eri 4h33'49" -0d22'29"
Colleen Mary
Leo 10h21'45" 17d52'7"
Colleen Mary McNulty
Lyr 19h13'32" 26d55'23"
Colleen Meagan Carson
Psc 23h5'12" 2d24'59"
Colleen Michele
Lyn 6h51'5" 53d43'53"
Colleen Mohen
Peg 23h53'58" 31d37'6"
Colleen Moira Clinton'Reilly
Gem 6h44'33" 33d41'12"
Colleen Moody
Cas 2h44'14" 64d42'28"
Colleen Morris
Ari 2h59'14" 25d38'59"
Colleen Much
And 0h43'52" 32d35'41"
Colleen O'Brien
And 0h35'45" 28d50'8"
Colleen (our loving mum)
Cyg 19h42'49" 28d1'45"
COLLEEN P DONAHOE
Sco 17h0'33" -35d40'38"
Colleen Quinn Byrne
Psc 1h11'8" 27d45'7"
Colleen R. Willms-Cook DVM, DACVECC
Cas 0h37'22" 54d25'7"
Colleen Richmond
Cas 0h59'14" 59d2'51"
Colleen Riddle-Vest
Dra 19h37'32" 60d50'7"
Colleen Rose Miller
And 1h31'29" 48d3'0"
Colleen Rossiter
Uma 10h45'11" 41d59'8"
Colleen Scarlett
Peg 22h5'5" 11d40'21"
Colleen Scussa
Uma 8h14'8" 72d41'14"
Colleen Smith
Cap 20h31'51" -19d18'22"
Colleen Star
Gem 6h48'43" 12d45'59"
Colleen "Stoney" Stone
Sco 16h47'37" -45d34'18"
Colleen Sullivan
And 2h30'53" 49d56'26"
Colleen Sunshine Hedberg
Aqr 22h6'11" -16d32'4"
Colleen "Superstar" Ford
Cnc 8h23'41" 31d1'54"
Colleen Susan Looney
Sco 16h12'1" -9d7'38"
Colleen Terese McCormick
Tau 5h51'11" 14d15'32"
Colleen "Topsy" Vogel
Vir 15h6'30" 4d8'31"
Colleen Trepp
And 1h15'59" 46d27'46"
Colleen Vicki Hazel Hardina
Cap 20h27'18" -19d56'32"
Colleen Virginia Taylor
Mon 6h55'30" -0d50'0"
Colleen Wade
And 23h44'1" 35d15'55"
Colleen Wong Smith
Cas 0h29'31" 59d10'36"
Colleen Yanosh
Vir 12h4'39" 2d26'53"
Colleena
Cyg 19h35'22" 30d6'7"
Colleena
Sgr 18h31'7" -27d0'36"
Colleen's Celestial Presence
Cyg 21h16'35" 54d54'17"
Colleen's Star
Vir 14h7'29" -9d30'50"
Colleen's Star
Leo 11h23'19" 23d43'58"
Collette Joyal
Uma 12h17'22" 59d8'43"
Collette & Mandy 2 The Moon & Back
Cyg 21h28'3" 47d2'35"
Colletti-Youngs 092803
And 0h25'20" 27d29'5"
Collide
Uma 13h31'29" 56d51'2"
Collier Allan Cerny
Cep 21h51'34" 64d58'58"
Collin
Lib 15h6'57" -9d21'54"
Collin
Ari 1h55'14" 24d47'24"
Collin and Ashley Osbourn
Lib 15h33'48" -7d27'41"
Collin Andrew Buonomo
Boo 14h50'4" 19d25'10"
Collin Andrew Slaybaugh
Umi 16h12'2" 75d59'16"
Collin Christopher McGuire
Leo 9h46'14" 25d57'16"
Collin Doyle's Star of Hope
Uma 10h48'52" 41d33'20"
Collin Dustin Pillars
Psc 1h29'32" 13d25'28"
Collin Ethan Shelly
Cap 20h43'42" -24d36'29"
Collin Fatcat 11
Tau 5h47'59" 22d21'57"
Collin J. Bickel
Per 3h51'42" 47d40'47"
Collin Jake
Boo 14h31'48" 53d43'45"
Collin James Scull
Cyg 20h4'40" 54d37'58"
Collin John Jones
Her 16h14'14" 16d41'2"
Collin Johnson
Aur 5h17'18" 43d2'51"
Collin Joseph Ledet
Uma 9h54'6" 53d41'12"
Collin Kinsworthy Blye
Uma 10h43'41" 66d43'14"
Collin Michael Boettner's Star
Dra 17h44'53" 55d59'4"
Collin Michael Nollin Brady
Ori 5h27'42" -4d58'13"
Collin Michael Spader
Uma 11h19'28" 63d32'21"
Collin Nathaniel Edelman
Cap 20h20'37" -12d59'44"
Collin Place
Sco 16h32'26" -30d37'6"
Collin Ryan Franklin Pepper
Gem 6h47'29" 21d44'17"
Collin Smigelski
Lyn 8h31'23" 42d33'35"
Collin Thomas Mennie
Aql 19h7'17" 3d42'28"
Collin Thomas Utley
Gem 7h13'43" 15d53'31"
Collin Tiller
Uma 10h1'12" 42d22'50"
Collin Walters
Aql 19h13'2" 8d53'41"
Collin Wesley Riddle
Leo 9h52'37" 23d55'28"
Collins Family Star
Umi 15h50'51" 73d40'52"
Collin's Fireball
Uma 9h24'50" 43d19'0"
Collins3
Uma 11h19'53" 60d7'7"
Colleen T. Rossiter
Uma 10h19'27" 54d59'14"
Collmar's Light
Uma 12h20'10" 56d54'52"
CollovesStace
Aqr 22h49'34" -15d59'40"
Colm P. Browne
Sgr 19h14'9" -22d12'11"
Colm Richard Galvin
Leo 11h13'11" -6d35'10"
Colombina
Cas 0h31'14" 62d26'6"
Colonel and Mrs. Eric D. Garvin
Aql 20h14'53" 6d32'32"
Colonel Charles D. Maynard
Aql 19h29'32" 11d33'43"
Colonel Dean Vangundy
Leo 10h18'11" 21d53'5"
Colonel F.P.C. Ralph, B.Bus.
Psc 0h13'48" -1d23'56"
Colonel H. L. Rauch
Aql 19h16'44" -0d7'48"
Colonel Mosby
Aql 19h4'30" 14d46'26"
Colonel Rich
Sgr 19h1'41" -34d36'1"
Colonel Ronald L. Weaver
Dra 20h17'35" 65d46'9"
Colonel Thomas & Sunday Spencer
Aql 19h57'13" -0d21'32"
Colonia
Uma 9h17'18" 67d11'50"
Colonia
Uma 9h29'8" 54d30'14"
Colonia
Uma 8h42'32" 50d14'15"
Colonia
Uma 10h30'10" 49d26'50"
Colony Club Hotel
Cyg 20h25'17" 58d36'59"
Colorado Mountain Chick: Rachel
Gem 7h39'58" 31d47'20"
Colorful Marvie
Sgr 19h16'33" -34d12'44"
Colors Tubba Waubba
Cam 3h55'12" 66d56'1"
Colt J Bernhard
Cnc 8h14'3" 30d25'25"
Colt Sand
Equ 21h9'13" 11d35'26"
Colten Riley Pitsch
Her 17h7'31" 34d29'24"
Colter Haldane
Sgr 18h25'7" -32d22'3"
Colton
Per 3h34'34" 42d29'3"
Colton A. Pollock
Her 17h52'36" 17d18'37"
Colton C. Rose
Leo 10h37'35" 15d20'46"
Colton Chance Hayden
Her 18h28'21" 12d30'35"
Colton Conner MacMinn
Sgr 18h11'36" -20d1'38"
Colton Craig
Aur 5h18'58" 30d45'36"
Colton Dean Edwards
Cyg 20h29'29" 53d0'22"
Colton Dixon Notterman
Uma 11h34'13" 61d33'53"
Colton E.J. Saugestad
Aur 5h54'8" 51d53'6"
Colton G. Chism
Her 18h5'14" 32d4'48"
Colton Gerald Pierson
Cap 21h40'15" -21d59'21"
Colton Goad
Ori 5h52'18" 6d41'41"
Colton Henderson Sorce
Cyg 20h4'1" 40d0'48"
Colton Jerome Markel
Umi 13h36'9" 72d57'35"
Colton John DeVos
Her 17h7'17" 39d52'50"
Colton Joseph Beall
Lac 22h20'35" 52d52'54"
Colton Joshuah Hoffman
Psc 1h16'39" 23d19'48"
Colton Lear
Sco 17h24'43" -32d58'34"
Colton Lee
Ari 3h21'51" 15d39'44"
Colton Lineman
Dra 19h26'45" 67d51'53"
Colton Lynn Coco Taylor
Per 3h3'45" 54d37'20"
Colton Marconi
Lyn 7h55'47" 58d6'38"
Colton Michael Posanski
Umi 16h15'5" 71d49'1"
Colton Pannell
Lib 15h13'44" -18d19'33"
Colton Perdriau
Her 17h49'33" 17d17'49"
Colton Sonne
Lmi 9h49'57" 37d45'36"
Colton T. Thompson
Ori 4h52'45" 14d16'49"
Colton Thomas Knight
Aql 18h50'2" -0d12'34"
Colton Wayne Shelar
Ori 5h48'43" 1d28'2"
Colton Zoellner
Umi 15h48'35" 72d47'9"
Coltster
Cap 20h38'45" -22d34'57"
Columbia Bove
Uma 13h41'53" 50d17'8"
Columbina
Sco 17h22'44" -38d8'1"
Columbus & Allison Ayers
Lyr 18h52'36" 34d12'29"
Colvic
Cyg 20h7'16" 51d49'15"
Colvin Bruce VanDeBogart
Boo 14h49'14" 26d13'5"
colzheart12
Lyn 7h50'19" 36d31'6"
ComAnderKing
Ori 4h56'22" 15d7'10"
Comarade d'ame
Leo 10h35'36" 22d10'25"
Come Away With Me
Cyg 20h26'11" 36d48'22"
Come what may
Uma 12h34'2" 53d54'2"

Come What May
Oph 17h1'24" -0d16'20"
Come What May~R&M
Cyg 19h44'33" 43d1'34"
Comella-Fienga
Umi 15h31'52" 75d13'29"
Comerci Maria
Ori 5h51'27" -8d24'10"
Comerio Carolina
Umi 17h11'37" 80d15'17"
Comes Family Star
Cap 20h24'42" -11d57'52"
COMET
Aqr 22h7'47" 2d3'0"
Comfort
Cyg 20h37'44" 35d50'49"
Comfort & Kindness
Uma 8h25'44" 63d35'20"
Comic
Cep 21h16'4" 59d21'29"
Cominciare Destino Insieme
Lyr 18h53'2" 43d19'45"
Comini Marco
Ori 6h5'7" -2d6'54"
Comitto Amplexus Candidus 7.8.6
Uma 11h27'13" 58d54'19"
Commador
Cnc 8h50'5" 29d54'36"
Command Chief Stephen Moore
Sco 16h8'47" -11d45'9"
Commander Charles "Trip" Tucker III
Uma 11h7'3" 46d18'21"
Commander Katherine Zak
Cyg 21h48'16" 47d15'0"
Commander Ozmo
Tau 4h57'10" 26d56'41"
Commander R. Lang
Her 16h53'0" 34d58'42"
Commander Richard Walter Davis
Uma 13h10'35" 55d54'7"
Commemini Adamo
Ori 6h0'35" 13d40'2"
Commencement of a Dream
Uma 11h22'44" 35d15'55"
Commissioner F. D. Edmondson
Her 16h46'16" 33d48'48"
Commissioner Richard A. Cowen
Cyg 20h49'7" 46d11'4"
Commissioner Stan Inman
Ori 5h0'18" 4d18'23"
Commitment
Uma 9h25'8" 56d29'33"
Commitment
Cep 21h28'52" 64d9'55"
Committee
Cnc 8h46'47" 16d18'49"
Commodore R. W. Warwick
Cyg 21h51'10" 38d55'44"
Commoneo Meus Angelus Aundrea
Vir 12h46'37" 4d50'33"
Commonwealth Treason - Gemma & Royce
Ori 5h34'56" -0d48'32"
COMMUNIS - BEOGRAD
Her 18h50'11" 21d0'4"
Community Animal Hospital
Lyn 8h8'10" 34d59'52"
Community EMS
Leo 11h2'29" 13d10'53"
Community Titles
Pyx 8h52'5" -34d54'24"
Compagno di anima
Col 5h42'32" -34d13'23"
Compagnon d'amy
Ori 6h2'4" 19d52'6"
Compañerito
Ari 2h42'3" 13d35'4"
Compañeros Del Alma
Lyn 7h9'46" 57d20'47"
Compassionate Melissa
Sco 17h41'58" -41d31'1"
Compassus Heidi
Cas 1h0'4" 60d55'14"
Complete Balance Star
Peg 21h39'41" 21d32'36"
Completed Destiny
Sco 17h42'24" -38d29'35"
Completely Amazing
Cap 21h31'48" -15d58'43"
Complicité
Cas 0h58'20" 66d24'7"
Comstock
Per 4h22'12" 44d11'41"
Comstock Family Star
Uma 9h37'14" 50d4'6"
comsummatus eluceo A.A.A.
Sco 16h13'23" -13d1'30"
Con Amore Siempre
Cnc 9h9'0" 7d59'8"
Con Colovos
Cru 12h17'12" -56d56'51"
con te in eterno Yoshinori & Sachiko
Tau 5h32'59" 19d48'26"
Con Te Partiro
Peg 22h7'5" 15d15'35"
Conal Traverse
Umi 10h25'11" 88d35'17"
Conalkil
Peg 22h25'26" 4d4'44"
Conamy
Lyn 6h51'58" 52d34'38"
Conan Paul Cebek
Cep 20h35'36" 64d6'19"
Conan T
Tau 3h37'18" 0d53'22"
Concepcion
Sgr 18h32'44" -30d36'37"
Concepcion Gutierrez
Gem 7h42'48" 26d1'59"
Concepcion Zambrano
Sgr 19h23'31" -15d23'27"
CONCETTA
Ari 3h17'2" 28d5'18"
Concetta Barbara
Cas 0h15'21" 60d20'21"
Concetta Franconero 12.12.1938
Sgr 19h55'56" -29d1'37"
Concetta Gambino-Inzirillo
Uma 9h19'57" 45d6'0"
Concetta Maria Ruggeri
Cas 2h40'4" 58d2'37"
Concetta Sibilia
Cas 23h34'1" 51d36'9"
Concetta Theresa Culotta
Lyr 18h25'26" 31d4'45"
Concetta "Tina" Albanese
And 1h7'57" 39d9'12"
Concettino Dino Calcaterra
Her 18h41'12" 24d24'32"
Concetto & Carmela Zappulla
Cyg 20h53'36" 38d9'24"
Concha Morales
Cyg 19h47'30" 43d44'28"
Concha Vasquez Perez
Crb 16h12'39" 34d20'20"
Conchi
Ori 6h25'13" 16d52'46"
Conchi
Ori 5h22'11" 2d0'48"
Conchi
Aql 20h9'45" 10d48'24"
Conchita and Bill Easton - 30th Anniversary
Cyg 20h21'0" 52d44'10"
Conchita Espinosa
Pho 0h16'42" -43d2'44"
Conchita Espinosa
Crb 15h26'34" 29d3'13"
Conchita Espinosa
Crb 15h51'25" 26d52'29"
Conchita R. & Kananai K. Moore
Ari 2h47'11" 25d48'59"
Concorde
Umi 14h3'27" 68d41'14"
Concorde Marguerite Ledoux Lachance
And 2h28'36" 49d13'57"
Concrete ad vitam
Uma 9h28'15" 61d35'45"
Concrete Pete
Uma 11h39'50" 36d8'18"
Condawg
Vir 14h14'49" -15d20'19"
Condor
Cas 0h48'53" 52d44'42"
Condra Star
Her 17h5'3" 34d47'53"
Conehead + Sloth = Gooden
Cyg 20h2'38" 44d23'36"
Conejo Caprika
Cam 6h18'55" 67d20'4"
Conely
Lyn 8h42'16" 40d28'57"
Coner Voller Wodzień
Col 5h52'24" -37d32'3"
Conesus Laketime Dream
Uma 10h56'7" 38d5'46"
Confido
Psc 0h17'41" 11d31'10"
Conforti's World
Cnv 14h3'48" 35d18'58"
Congratulations Hope Marie 2007
Ari 2h40'58" 26d12'3"
Coni
Cnc 8h8'2" 6d47'38"
Coni Jo's Gyroscopic Precession
Crb 16h1'49" 35d18'55"
Conie's Star
Cap 21h33'42" -17d22'13"
Conlan A.R.
Sco 16h2'54" -18d51'27"
Conlan E Crosley
Aqr 22h1'57" -14d39'42"
Conley
Cas 0h29'14" 60d39'36"
Conley & Dustin Wolfswinkel
Sge 19h51'17" 18d7'27"
Conley Robinson
Aql 19h23'31" 1d20'28"
Conluvio
And 0h17'45" 46d6'16"
Connelley Rose Erwin
Lmi 10h49'42" 28d15'35"
Connelly
Uma 13h21'15" 57d56'8"
Conner
Psc 1h22'56" 29d57'30"
Conner
Lmi 10h37'58" 37d43'1"
Conner Aaron Chabino
Lyn 7h39'55" 58d41'56"
Conner Bruce Smith
Boo 14h25'56" 17d31'32"
Conner David Nolt
Leo 9h48'21" 27d52'3"
Conner Dean Ottinger
Uma 11h8'58" 55d26'53"
Conner Dean Paul
Umi 14h5'12" 77d27'19"
Conner Douglas Knehr
Ori 5h31'35" -0d59'43"
Conner Glenn Mueller
Cru 12h12'12" -62d14'6"
Conner Hibberd
Dra 18h20'47" 58d40'3"
Conner Ian Anderson
Psc 1h6'38" 23d31'1"
Conner Impey
Umi 17h27'33" 83d30'6"
Conner Joel Schultz
Ari 2h12'57" 24d1'46"
Conner M. Lemm
Psc 0h35'34" 7d44'59"
Conner McDermott
Uma 11h24'28" 49d51'20"
Conner Merimee
Uma 10h37'38" 46d27'43"
Conner Reese Rucks
Aql 19h30'37" 7d51'5"
Conner Thomas Maher
Per 4h46'1" 40d27'39"
Conner Thoreau Vance
Sgr 17h59'24" -17d28'1"
Conner Welch
Her 17h44'0" 19d49'57"
Conner's Way
Aqr 22h34'56" -20d39'39"
Conner's Way
Aqr 23h35'37" -9d9'26"
Conney May Richardson
Vir 13h3'54" 7d48'34"
Conni BF Logan
Aql 19h2'56" 12d34'15"
Connie
Tau 5h44'21" 17d18'32"
Connie
Peg 21h33'8" 16d28'4"
Connie
And 0h35'27" 27d8'35"
Connie
And 0h41'12" 27d22'25"
Connie
Leo 10h28'14" 27d27'59"
Connie
Ari 3h20'3" 28d28'35"
Connie
And 0h53'3" 41d14'56"
Connie
Crb 15h44'0" 32d38'4"
Connie
Lyn 8h31'46" 34d15'27"
Connie
Tri 2h24'32" 37d15'35"
Connie
And 0h25'6" 37d22'43"
Connie - Always In Our Hearts
Cas 23h39'31" 52d3'10"
Connie Ann
Psc 1h29'42" 11d56'7"
Connie B Carter
Cas 1h27'4" 66d2'20"
Connie B. Mann
Cap 21h47'3" -14d58'58"
Connie B. Mohl
Dra 17h46'49" 72d16'45"
Connie Bailey
Uma 11h30'46" 33d23'44"
Connie Bourke
Cnc 8h22'1" 13d37'4"
Connie Brown
And 0h29'5" 43d37'36"
Connie & Charlie
Cyg 19h41'25" 41d9'15"
Connie Cross
Peg 22h5'8" 12d27'19"
Connie Crowley
Sco 16h27'53" -26d23'45"
Connie Ditz Hagen
And 2h24'23" 49d39'30"
Connie Dixon
Tau 4h5'15" 6d28'59"
Connie & Ed Star of Love
Cyg 20h29'47" 47d43'50"
Connie Ellenburg
Eri 4h36'19" -0d57'10"
Connie F. Fast
Leo 11h17'49" 18d15'48"
Connie George - My Sweetheart
Vir 13h39'12" -8d42'0"
Connie Gipe
Cas 1h24'15" 66d7'13"
Connie Golightly
Uma 10h35'30" 57d2'14"
Connie Greathouse "Shining Star"
Cap 21h7'53" -19d17'36"
Connie Griffith-Anderson
And 23h0'47" 47d58'20"
Connie Guerrero
Ari 2h33'16" 13d46'27"
Connie & Harry Culver
Cyg 19h47'2" 52d17'24"
Connie Haware
Cnv 12h46'14" 41d11'32"
Connie Haydon
Aqr 22h19'19" 0d57'55"
Connie Herriott
Cas 0h16'46" 51d26'8"
Connie Hoopes Race
Cap 21h32'1" -20d12'11"
Connie J Hanson
Uma 13h9'7" 62d46'34"
Connie J Lewis
Mon 6h36'48" 9d8'13"
Connie J. O'Neill
Lyr 18h35'43" 40d19'23"
Connie & Jack Leonard
Her 18h31'39" 17d49'26"
Connie & Jack Luetkemeyer's 56th.
Uma 10h35'20" 47d41'33"
Connie Jayden Adams
Peg 21h45'11" 23d29'38"
Connie Jensen
Leo 11h26'19" 10d21'56"
Connie Jo
Tau 3h45'1" 9d2'25"
Connie Jo
Leo 10h29'22" 6d31'19"
Connie Jo Guzman
Aql 19h6'34" 8d57'11"
Connie Jo Ostrander Delfino
Leo 9h46'7" 27d14'21"
connie jo seaburg
Cyg 20h2'43" 30d8'59"
Connie Joe Harmon
And 1h10'29" 36d26'31"
Connie K. Grayling
Lib 15h42'38" -10d40'58"
Connie K. Thomas Enfield
Peg 21h39'8" 27d48'19"
Connie Kaczorowski
And 1h23'45" 49d58'45"
Connie Kay Hutchinson
Cap 20h50'34" -17d15'57"
Connie King "Au Mau"
Crb 15h49'52" 34d48'2"
Connie Klein's Forever Love
Uma 10h35'8" 44d27'34"
Connie Koob
Ori 5h39'57" 5d40'41"
Connie L. Curtsinger
Cnc 9h7'47" 24d12'33"
Connie L. Greenwood
Dra 16h28'31" 58d0'43"
Connie L Hanlin
Uma 10h23'33" 70d0'29"
Connie L. Johanknecht
And 1h3'27" 35d52'51"
Connie L. Shank
Tau 4h30'30" 21d51'31"
Connie Lanciloti
Lib 15h1'19" -16d10'50"
Connie Larue
Sge 19h54'17" 18d57'47"
Connie Lau
Lyn 7h7'8" 58d24'33"
Connie Lee Phillips Hughes
Psc 0h9'7" 5d38'28"
Connie Lloyd
Vir 13h27'7" 12d39'46"
Connie Lynn
Leo 11h3'27" 11d54'3"
Connie M. Richards
Mon 7h15'2" -0d38'44"
Connie M Wagner
Lyr 18h41'17" 46d56'52"
Connie Mae Sayre
Leo 11h24'47" 14d58'30"
Connie Marie Norton
Sco 17h51'7" -38d47'26"
Connie Marker
Oph 17h46'15" -0d17'2"
Connie McNamara
Lyr 18h51'48" 43d16'23"
Connie McReynolds
Sgr 18h51'10" -18d43'22"
Connie (Megan's Mom)
Uma 11h28'53" 59d54'18"
Connie Million
Cas 23h45'47" 52d57'54"
Connie Moscardini
Lyn 6h31'20" 57d51'46"
Connie Moser
Leo 11h14'52" 16d34'15"
Connie (Mother G-G) Pennington
Tau 3h51'38" 15d23'7"
Connie Nay
Cyg 19h50'21" 59d24'31"
Connie Orsini
Leo 9h34'41" 7d14'17"
Connie Park
Srp 18h35'14" -0d32'10"
Connie Parkin
Cas 0h37'17" 62d49'49"
Connie Pinto
Umi 15h18'12" 69d11'56"
Connie Riemen
Aqr 23h0'34" -21d24'24"
Connie Riley
Vir 12h44'14" 1d58'0"
Connie Roberson My Shining Star
Cyg 19h39'35" 30d16'56"
Connie Rogozinski
Uma 8h24'5" 64d6'49"
Connie Rojas
Gru 21h45'30" -37d13'12"
Connie S. Chriscoe
Tau 4h44'41" 25d34'15"
Connie S Covington
Mon 6h43'8" 6d25'28"
Connie S. Kennedy
Lyr 18h49'12" 37d45'40"
Connie Shanta Carmony
Cet 3h8'56" 7d1'23"
Connie Short 0401
Cas 1h9'39" 66d1'52"
Connie Shreffler
Lyr 18h28'42" 28d11'8"
Connie Stabile
And 23h30'6" 42d58'20"
Connie Sue
Gem 6h51'54" 13d45'16"
Connie Sue Braswell
And 0h55'16" 40d46'13"
Connie Sue Cullop
Dra 17h32'52" 64d25'8"
Connie Sue Curran
Sco 16h23'20" -32d27'11"
Connie Sue Galietti
Leo 11h27'9" 2d17'23"
Connie Sue Johnson
Tau 3h34'37" 12d11'3"
Connie Sue Kennedy
Leo 11h20'47" 17d55'39"
Connie Sunnergren
Ori 6h13'53" 11d43'0"
Connie Trappeniers ik hou van u
And 23h0'33" 49d50'29"
Connie Violet Evans
And 1h21'33" 43d23'32"
Connie We Love you Mom
And 0h51'49" 40d51'30"
Connie Westerfer
Her 17h13'33" 29d34'16"
Connie Wilson soul-sisters
Cas 0h44'30" 64d8'29"
Connie, Jamie, & Jacob
Psc 1h8'31" 9d36'33"
ConnieCarm
Vir 11h54'39" 8d51'51"
Connie-Jo Kennedy
Cnc 9h2'17" 15d28'2"
Connie-Linda
Sco 17h57'1" -41d27'30"
ConnieMissico.bestmom.int heuniverse
Aqr 21h38'47" 0d49'37"
Connie's Angel Star
And 23h32'35" 45d27'16"
Connie's Light
And 0h27'17" 40d43'58"
Connie's Lucky Star
Crb 15h37'16" 27d30'26"
Connie's Piece of Heaven
Cas 1h32'43" 65d45'39"
Connies Spark
Psc 0h18'29" 6d2'46"
Connie's Star
And 0h49'34" 39d58'28"
Connolly - Stojack
Lac 22h27'42" 48d32'33"
Connon's Star
Cyg 19h46'4" 34d16'13"
Connor
Ari 2h59'6" 28d23'59"
Connor
Gem 7h27'11" 17d29'49"
Connor
Vir 12h51'23" 7d30'51"
Connor
Umi 14h16'31" 70d12'28"
Connor
Lib 15h45'58" -4d36'9"
Connor Alan Murphy
Sgr 18h28'23" -24d24'35"
Connor Alan's Beam Of Light
Gem 6h42'44" 23d48'52"
Connor Alexander Tardif
Lib 14h54'11" -1d38'45"
Connor Allan Dyer
Ari 1h52'26" 18d8'54"
Connor and Laura's Star
Ori 6h19'26" 19d36'11"
Connor Andrew
Aqr 22h4'36" -8d37'40"
Connor Andrew Franklin Taylor
Her 18h33'10" 22d32'24"
Connor Andrew Walk
Cep 22h38'8" 62d35'40"
Connor Anthony
Per 3h0'25" 42d46'54"
Connor Anthony Banks
Sco 16h9'45" -13d41'28"
Connor Aragorn McDowell
Uma 11h12'51" 53d6'45"
Connor Aron Rush
Ori 5h54'21" 16d54'55"
Connor Avery Howell
Cep 21h21'25" 64d24'23"
Connor Beckham Gentry
Vir 12h7'25" -4d7'21"
Connor Bedgood
Ori 5h23'29" 9d59'47"
Connor Blaine Finch
Sgr 17h48'12" -28d0'1"
Connor Blore
Umi 17h23'54" 86d5'23"
Connor Blye Brown
Cam 4h24'54" 71d26'26"
Connor Bradley Breneman
Lyn 8h9'42" 38d5'14"
Connor Brannan Schwaab
Cap 20h29'40" -24d29'10"
Connor Brian Dunlop "12/22/01"
Lyn 8h47'18" 34d25'31"
Connor Bryce
Uma 10h32'14" 52d27'13"
Connor & Caden
Umi 15h9'22" 70d10'2"
Connor Ciesielski
Aql 19h53'5" 2d5'42"
Connor Coen
Lib 14h55'28" -16d38'21"
Connor Craig
Boo 14h39'5" 25d14'37"
Connor David
Her 16h6'57" 47d29'29"
Connor Edward Currie June 10, 2002
And 23h25'37" 45d26'42"
Connor Edward Ledesma
Cnc 8h51'5" 21d42'5"
Connor Fauteux
Ori 5h30'2" 4d43'43"
Connor Fennell
Cep 22h56'15" 67d18'25"
Connor Fidoe's Star
Her 18h16'30" 28d29'45"
Connor & Gavin Black
Uma 11h40'22" 64d32'28"
Connor Glen Jones, Jr.
Cep 22h22'39" 62d4'24"
Connor Gordon Baine
Uma 9h29'44" 72d52'2"
Connor Grant Smith
Ori 5h32'38" 14d17'49"
Connor Gray
Per 3h39'28" 39d0'10"
Connor Havison
Uma 8h52'12" 50d3'7"
Connor Hughes
Cas 0h36'3" 60d14'9"
Connor Ishihara
Lyn 7h22'32" 53d31'41"
Connor J. Moesel
Psc 0h45'38" 17d23'10"
Connor J. Walters
Uma 11h32'45" 53d26'0"
Connor Jack Snell
Leo 10h6'34" 9d53'43"
Connor Jackson
Uma 11h14'0" 54d17'38"
Connor Jacob Guerin
Gem 6h50'28" 31d53'10"
Connor Jakob Wille
Sco 16h13'10" -12d46'18"
Connor James Green
Aqr 22h12'14" -12d43'55"
Connor James Quigley
Cnc 8h21'20" 20d10'29"
Connor James Thomas Sheerin
Leo 9h46'59" 29d25'15"
Connor James Winiarczyk
Ari 2h52'8" 28d23'15"
Connor Jay McKillop
Ori 5h44'23" 7d17'26"
Connor Jay Robinson
Cru 12h15'2" -57d7'2"
Connor John McAuley
Cma 6h46'5" -13d42'48"
Connor John McGee
Tau 4h45'53" 23d4'33"
Connor John West
Umi 15h28'8" 73d55'23"
Connor Jon
Cap 20h36'9" -23d26'58"
Connor & Jordan - My Stars My Angels
Cru 12h17'40" -56d49'35"
Connor Joseph Dolgos
Gem 7h42'59" 32d26'35"
Connor Lachemann
Lib 15h25'18" -19d30'7"
Connor Lance Charles Normand
Dra 9h26'18" 73d15'57"
Connor Lange Oatman
Aql 18h51'20" -0d3'58"
Connor Lee
Uma 9h13'33" 55d52'21"
Connor Lee
Peg 23h45'53" 24d35'4"
Connor - Levi
Ori 5h54'13" 17d5'13"
Connor Lewis Engelhardt
Psc 1h8'52" 27d28'33"
Connor Lewis Tuttle
Cnc 9h14'42" 29d2'57"
Connor Loren Spaulding
Lmi 10h46'57" 26d54'25"
Connor Mason Hutcherson
Ori 5h20'28" 7d8'11"
Connor Mathew Peters 12/26/1998
Ori 5h29'9" 9d12'58"
Connor Matthew Edo
Tau 3h50'11" 28d15'3"
Connor Matthew Fields
Lyn 8h8'35" 58d1'56"
Connor Matthew McHugh
Gem 6h45'33" 32d30'44"
Connor Michael Murphy
Ori 6h12'49" 9d5'30"
Connor Michael Olson
Uma 9h17'53" 61d26'37"
Connor N. Larson
Sco 16h12'29" -15d47'30"
Connor Nelson
Lib 15h8'2" -5d28'57"
Connor Newton
Cam 3h31'22" 62d13'44"
Connor Noel Bjorland
Crb 16h4'24" 26d28'6"
Connor Nolan Clifford
Tau 5h32'45" 21d15'26"
Connor O'Leary
Uma 9h35'23" 46d41'9"
Connor Owen
Psc 23h45'27" 5d31'35"
Connor Parker Reid
Boo 14h9'16" 19d2'52"
Connor Patrick O'Mahoney
Aur 5h27'12" 41d1'58"
Connor Patrick Poglajen
Gem 7h12'2" 18d54'16"
Connor Patrick Quintard
Sgr 19h11'32" -18d42'34"
Connor Patrick's Star
Leo 11h18'40" 21d54'46"
Connor Paul Tolley
Uma 8h35'40" 57d45'4"
Connor Philip O'Brien
Uma 9h50'5" 52d43'22"
Connor Ralph
Per 3h28'24" 47d2'58"
Connor Reece Makai
Cen 13h36'56" -36d33'0"
Connor Richard Giles
Sgr 20h4'54" -39d10'54"
Connor Riley Anderson
Gem 6h47'13" 33d31'22"
Connor Robert James Purdy
Gem 6h51'2" 25d21'25"
Connor Ryan Masters
Leo 11h17'31" 14d21'58"
Connor Ryan McLaughlin
Ori 5h27'46" 2d19'28"
Connor Rye Gretz
Umi 16h16'52" 76d2'10"
Connor Scatena
Peg 22h53'28" 27d27'32"
Connor Scott Troost
Aqr 22h36'46" -2d46'48"
Connor "Squeaky" Vaughn
Uma 10h59'38" 49d45'2"
Connor Stallings
Sco 16h41'31" -36d41'39"
Connor Stephens
Sco 17h57'58" -38d57'51"
Connor Stewart McGlashan
Cru 12h2'28" -61d53'4"
Connor Theodore William
Leo 9h23'51" 11d57'55"
Connor Thomas Hughes
Leo 11h8'32" 20d50'27"
Connor Thomas McInerney
Pho 0h52'25" -48d22'5"
Connor Walsh
Her 18h48'52" 16d8'14"
Connor Wei Jun
Pho 0h19'4" -43d32'26"
Connor William Alonso
Dra 17h25'56" 58d1'31"
Connor Witman
Gem 7h10'16" 33d5'3"
Connor Wood
Umi 14h17'47" 66d27'29"
Connor......hope & inspiration
Vir 14h18'44" 3d38'6"
Connoriseros
Dra 16h51'20" 69d49'2"
Connor's Christmas Star
Lyn 7h31'29" 35d15'59"
Connor's glory
Her 17h51'18" 47d39'57"
Connor's own bright star up there
Tuc 0h16'52" -71d38'14"
Connor's PAX2U
Lyn 9h11'25" 45d8'26"
Connor's Special Star
Aql 20h20'42" -0d48'58"
Connor's Star
Umi 14h28'13" 68d57'10"
Connor's Star
Cas 1h37'22" 61d13'48"
Connor's Star, Forever Shining
Del 20h26'24" 20d32'38"
Conny
Lyn 7h52'39" 45d6'38"

Conny Baume-Grütter
And 0h4'20" 39d44'28"
Conny & Christian
Her 17h50'37" 21d30'25"
Conny Eggler
Crb 15h59'47" 28d20'10"
Conny Gustafsson
Cru 11h57'35" -64d8'5"
Conny Lange
Uma 9h2'52" 63d26'54"
Conny M. Knecht
Per 4h19'44" 44d2'47"
CONNY MULLER
Uma 9h46'26" 48d59'52"
Cono Anthony Sciamanna
Cap 20h33'7" -14d46'35"
Conor 21
Lyr 18h29'8" 37d47'21"
Conor Boo
Boo 15h4'51" 48d31'17"
Conor Bradley Montalbano
Psc 23h12'28" 5d13'56"
Conor Brian
Her 18h8'21" 48d32'24"
Conor Bryce McLeod Daly
Aur 5h51'44" 36d56'41"
Conor + Celeste
Cnc 8h22'27" 28d13'14"
Conor Clark Galligan
Her 16h49'48" 28d9'9"
Conor David Hemming
Umi 10h16'4" 86d15'37"
Conor Delaney Farrell
Crb 15h48'18" 26d5'27"
Conor Hynes
Lib 15h1'43" -0d54'50"
Conor J Keegan
Uma 11h33'13" 59d7'46"
Conor Jeremy McAuliffe
Cru 12h10'23" -61d32'43"
Conor Joseph
Umi 14h39'5" 76d4'48"
Conor Joseph Murphy
Ari 2h22'35" 22d13'55"
Conor Kane Quinn
Cam 5h24'53" 64d43'49"
Conor Lawrence Bayer
Umi 14h36'8" 74d9'3"
Conor Lynch Hicks
Ori 6h2'23" -0d20'58"
Conor Man
Ari 2h41'50" 27d53'38"
Conor Mark Harris
Umi 17h26'17" 83d44'23"
Conor Michael Hart
Her 18h10'16" 15d34'20"
Conor Morano
Crb 15h36'13" 29d24'47"
Conor Niall Callan
Cyg 20h9'17" 37d14'25"
Conor O'Callaghan
Vir 13h41'12" -5d19'39"
Conor O'Neill
Umi 14h28'3" 74d6'53"
Conor & Shane - Star
Twins
Gem 7h1'29" 26d54'34"
Conor Sid Campbell
Uma 9h8'47" 60d5'38"
Conor Thomas
Ari 2h5'6" 23d57'10"
Conor - Zac - Odin
Cru 12h38'45" -64d8'2"
Conor's Star
Umi 15h11'27" 67d40'14"
Conowitz 2004
Lmi 10h3'35" 36d34'56"
Conqueror
Crb 16h22'0" 29d5'4"
Conrad
Cap 20h27'42" -27d33'24"
Conrad & Amy's Deka-Star
Uma 11h25'39" 51d45'55"
Conrad Anderson
Cas 0h9'8" 53d53'49"
Conrad Celinski
Ari 2h44'18" 27d48'33"
Conrad De La Rosa-
Anderson
Tau 3h40'3" 19d47'47"
Conrad Elton Smith II
Cyg 21h54'28" 44d18'53"
Conrad H. Francis
Aur 5h54'19" 39d48'47"
Conrad Hall
Uma 11h59'36" 60d24'47"
Conrad Ingersoll Dube
Leo 9h25'8" 17d38'1"
Conrad Joseph Paglia
Leo 9h31'28" 27d44'51"
Conrad Kannengiesser
Ori 4h48'14" 4d26'18"
Conrad Kupferman
Tau 5h51'6" 27d36'49"
Conrad Loomis Hall
Aur 7h15'43" 42d35'25"
Conrad Peter Polys
Cyg 21h56'15" 48d53'46"
Conrada Major
Uma 9h45'51" 52d47'45"
Conrado Bokoles
And 0h18'46" 41d38'39"
Conrado Indart -
September 2, 1937
Vir 14h36'6" 2d14'53"

Conri Harry Robinson
Per 3h49'57" 36d4'47"
Conright 4/9/62
Uma 9h21'32" 72d34'49"
Conry
Cra 19h0'48" -42d12'49"
ConSam
Dra 20h33'17" 67d43'33"
Consdad 50
Leo 9h55'52" 30d28'40"
Consepcion "Mama
Concha" Gonzalez
Uma 12h0'3" 57d18'45"
Consha
Per 3h22'30" 44d34'33"
Consiglia Niro
Per 3h15'1" 48d20'38"
Consigliere D'Ambrosio
Pho 1h11'31" -50d44'14"
Consorcia
And 0h11'16" 45d0'14"
Constance
And 23h42'3" 37d58'36"
Constance
And 2h19'25" 47d56'36"
Constance
And 23h22'34" 51d2'55"
Constance
And 0h5'39" 44d36'29"
Constance
Cyg 19h55'17" 33d30'43"
Constance
Vir 12h25'44" 1d21'50"
Constance
Ari 3h9'45" 28d7'33"
Constance
Tau 4h42'24" 23d49'5"
Constance
Sgr 19h13'56" -17d52'15"
Constance
Sgr 18h8'24" -17d54'54"
Constance A. Cross
And 1h2'37" 44d34'21"
Constance Berneking
Uma 11h28'40" 47d14'48"
Constance Brayton
Cas 0h21'21" 51d22'28"
Constance Bush Rauscher
Aqr 22h50'12" -18d15'15"
Constance Caliope
Costaris
Vir 12h23'18" 8d38'36"
Constance Catherine
ThomasWhitworth
Aql 19h52'31" 8d12'43"
Constance Cauble
Dra 19h39'48" 60d55'55"
Constance Clara Gionis
Pereire
Vir 12h28'7" 2d44'1"
Constance Clendenin
Lyn 6h22'7" 54d33'54"
Constance "Connie"
Horwitz
Cas 23h25'57" 59d49'58"
Constance (Connie) Jones
Uma 10h7'5" 47d47'39"
Constance Dorothea
Scannell - Connie's Star
Lib 15h11'9" -22d47'30"
Constance Elaine
Lyn 8h49'57" 35d42'39"
Constance Eve King
Cep 21h59'26" 61d34'29"
Constance F. Carper
Sco 16h45'12" -42d26'59"
Constance Guy
Cap 21h29'35" -15d44'27"
Constance J. Pilkington
Lyn 8h28'56" 38d28'45"
Constance Jeanne Swan
And 23h30'13" 48d22'17"
Constance Judith
And 22h59'41" 37d14'27"
Constance Kinnamon Fava
Sco 16h3'42" -10d33'49"
Constance L. "Connie" Edie
Cas 1h32'56" 60d49'11"
Constance L. Lacey
Crb 16h13'38" 37d39'35"
Constance L. Thome
And 2h32'41" 38d55'16"
Constance Leanore Lewis
Cas 1h37'1" 63d50'58"
Constance Lewis
Tau 4h12'56" 12d14'16"
Constance Lou Wall
Lyn 7h5'9" 61d28'35"
Constance Louise Albair
Cas 1h7'56" 60d31'55"
Constance Lucia
Cas 0h24'23" 62d6'5"
Constance Lynn
Sgr 20h0'53" -24d49'31"
Constance Lynn Heck
Psc 1h46'41" 22d41'15"
Constance M Efstration
And 23h15'31" 45d11'52"
Constance M. Jensynn
Cap 20h9'55" -21d43'7"
Constance M. Sullivan
Per 3h17'32" 45d16'6"
Constance Maher
Lyn 7h44'49" 43d47'42"

Constance Mahoudeau
Lib 15h0'53" -1d23'55"
Constance Marie
Sco 16h6'18" -13d6'13"
Constance Marie
And 1h49'11" 38d41'23"
Constance Marie Angrabe
Sco 16h6'26" -8d51'41"
Constance Marie Herbst
Vir 12h46'44" 5d41'3"
Constance Melrose Martin
And 23h12'17" 44d4'48"
Constance Morris
Ori 6h1'13" 20d6'18"
Constance Odenheimer
Cap 21h40'2" -19d20'49"
Constance Ramkissoon
Cap 21h0'25" -23d37'36"
Constance Renee
Cas 0h39'37" 61d8'18"
Constance Rochelle
McKenzie
Aqr 22h39'31" -0d57'13"
Constance Stasia Mosseau
Cas 1h6'39" 53d34'21"
Constance Susan Esters
Cas 2h19'25" 69d41'8"
Constance Urantian
And 1h22'49" 50d4'33"
Constance Winifred Jones
Cas 0h1'0" 56d44'41"
Constance Xanthe
Pappadis Lewis
Cas 0h45'30" 56d30'0"
Constance Zeimba
And 0h41'37" 38d47'25"
Constance's Delfino Nella
Notte
Del 20h36'16" 15d3'9"
Constant Euphoria
Aqr 22h10'44" -14d1'16"
Constant Love
Cnc 9h11'20" 16d19'54"
Constantia
Cen 13h13'46" -49d27'17"
Constantina Gregoriadis
Cas 0h37'24" 58d22'31"
Constantina Kostka
Uma 13h10'12" 55d21'4"
Constantine
Lac 22h47'17" 36d1'50"
Constantine
Cnc 8h48'7" 32d38'34"
Constantine
Tau 5h8'43" 24d29'53"
Constantine C. Petropoulos
Per 2h17'43" 57d9'16"
Constantine George
Avdalas
Lib 14h55'5" -16d36'33"
Constantine George
Pergantis
Per 3h5'16" 53d4'15"
Constantine James
Maroulis
Vir 14h5'15" 3d40'13"
Constantine Kekeris
Her 18h49'4" 25d22'46"
Constantine Maroulis
Ori 5h22'54" 10d48'10"
Constantine Maroulis
Lyr 18h46'44" 39d59'58"
Constantine S Efstathis
(Ticka)
Cra 19h3'31" -42d0'26"
Constantine Tsatsos
Aql 19h28'46" -8d32'57"
Constantine, the brightest
star. x
Lib 14h58'34" -9d31'10"
Constantino Michael
Ceriello
Leo 9h40'52" 27d59'19"
Constantinos
Cra 18h5'5" -40d22'58"
Constantinos
Egkolfopoulos
Ori 5h50'13" 11d49'56"
Constantino's Mooring
Tau 4h22'44" 14d33'30"
Constantly
Oph 16h57'2" 0d34'3"
Constanza Duque
Crb 15h19'9" 26d32'34"
Constanze + Arne
Uma 11h25'23" 60d47'5"
Constanze Gerstmayr
Cas 23h54'5" 57d59'27"
Constanzia
Lib 14h47'54" -20d5'52"
constellation
Cyg 20h41'40" 49d25'31"
Constellation Cocca Twins
Cnc 7h57'45" 12d39'40"
Constellation Kaeding
Vir 13h11'20" 12d45'10"
Constellation MaD
Dra 18h20'10" 53d54'44"
Constellation Marian
Torolski
And 1h33'55" 44d51'32"
Constellation of Dreams
Her 18h30'16" 14d31'30"
Constellation Yummy
Cas 0h25'1" 56d21'8"

Consuelo
Ari 3h23'34" 28d40'28"
CONSUELO AMANTE
Cap 20h24'14" -27d1'37"
Consuelo Arelis Rodriguez
Cyg 20h10'14" 33d41'2"
Consuelo Portes Presnillo
Umi 14h5'45" 75d9'8"
Consuelo (Suello)
Uma 9h11'24" 54d48'58"
Consuelo V. Santos
Tau 5h49'20" 23d23'58"
Consuelo's Star
Cnc 8h22'35" 16d11'50"
Consuelo's Star
Uma 9h36'8" 49d17'0"
Consummo Adamo
Lmi 10h36'1" 36d7'3"
Contact One
Cap 20h37'49" -27d6'26"
Contellucia Major
Cyg 19h38'21" 28d32'34"
contenda
Lyn 6h23'55" 61d11'40"
Contessa Diane Elizabeth
Burbank
Lib 15h56'58" -6d6'22"
Contessa Rodina
And 2h33'13" 38d27'27"
Conti Antonella
Per 3h25'16" 41d34'26"
Contzen, Horst Josef
Uma 8h21'37" 63d27'36"
Conundrum
Mon 6h42'28" 9d23'7"
Convallaria Venusta
Octobris
Uma 10h28'55" 47d8'58"
Convallis Quercus
Umi 14h49'31" 85d37'12"
Conway Figueroa
Uma 10h6'16" 43d12'25"
Conway Figueroa
Her 16h34'36" 28d36'18"
Coochis
Mon 6h44'1" 6d54'0"
Coogg
Vir 14h18'22" -0d51'2"
Coogley's Piece of Heaven
Cas 1h10'47" 54d1'33"
Cook-Devereux Etoile
Col 5h48'54" -30d23'34"
Cooke's Nook
Cap 20h46'55" -18d43'3"
Cookie
Cma 6h43'41" -14d32'50"
Cookie
Cam 4h11'41" 66d40'25"
Cookie
Cam 5h4'56" 57d50'29"
cookie
Crb 15h34'40" 32d32'1"
Cookie
Lyr 18h52'57" 35d45'17"
Cookie
Cnc 9h9'4" 31d48'40"
Cookie
Vir 13h33'21" 3d22'36"
Cookie
Tau 3h41'31" 27d27'56"
Cookie
Ari 3h8'55" 27d7'20"
Cookie
Leo 11h55'10" 20d4'15"
Cookie 13
Uma 10h44'57" 43d42'12"
Cookie and Pretzel
Ari 1h54'28" 18d28'54"
Cookie Bear
Umi 13h21'15" 71d48'45"
Cookie Cutter Man
Aqr 23h15'34" -9d20'34"
Cookie Girl
And 23h25'59" 46d19'23"
Cookie Jar Star
Uma 11h2'38" 67d21'55"
Cookie Milo
Lyn 7h23'36" 58d29'19"
Cookie Murphey
Lib 14h52'8" -11d38'49"
"Cookie" Niehuser
Sgr 19h36'40" -13d26'12"
Cookie & Noel Goodman
Cyg 21h50'4" 49d46'47"
COOKIE STAR
Aqr 22h41'42" 1d29'53"
Cookie (Thelma) Bailey
Lib 14h52'36" -0d47'3"
Cookie121302
And 2h22'5" 48d3'17"
cookiepie
Vir 14h16'22" 4d47'40"
Cookies & Babies
Ori 6h8'33" 20d0'49"
Cookie's Freedom Star
Aqr 21h46'25" -1d36'52"
"Cookie's Star"
Uma 12h43'51" 57d18'18"
Cookie's star
Cas 0h48'18" 61d47'6"
Cookie's Star
Uma 9h52'7" 62d42'49"
Cookie's Star
Sco 17h13'9" -38d20'11"

Cookie's Star
And 0h42'10" 39d31'21"
Cookie's star 05142006
Leo 9h25'48" 32d38'52"
Cookie-Sweetie
Aur 5h10'59" 46d59'30"
Cookoops
Uma 8h55'25" 69d16'4"
Cook-Sowder
Ori 5h18'44" -4d17'34"
Cool Breeze & Barby
Parrish, always
Uma 12h32'2" 55d28'29"
Cool Bug-A-Boo
Sco 17h34'2" -35d10'29"
Cool Cat Carissa
Ari 3h19'11" 24d41'29"
Cool Justin
Per 3h35'12" 32d8'26"
Cool man Jared
Lib 15h20'44" -24d10'58"
COOL2401
Leo 10h10'28" 27d47'57"
CoolDogJoe
Cma 6h40'20" -12d40'25"
Coolpopa
Uma 14h24'54" 57d46'52"
coondog &ms. dog &pups
Cma 6h47'26" -15d35'18"
Cooney
Ari 2h42'10" 27d39'57"
Coop
Gem 6h30'14" 26d30'47"
Coop PV
Umi 17h3'22" 79d41'6"
Cooper
Umi 15h6'9" 67d34'8"
Cooper
Ari 3h20'25" 27d11'39"
COOPER
Ari 2h36'54" 27d18'23"
Cooper
Ori 6h19'36" 16d11'21"
Cooper Atticus Silverstein
Cam 4h22'15" 66d47'48"
Cooper Chapman
Cap 21h28'55" -19d40'21"
Cooper Chase Griebel
Crb 15h32'8" 29d59'20"
Cooper DeFord
Weyerhaeuser Driscoll
Vir 12h10'46" -0d23'45"
Cooper Douglas Hett
Uma 12h3'36" 59d55'40"
Cooper Hafiz 30-12-04
Cap 20h48'52" -22d39'13"
Cooper Halford
Aql 19h51'16" -0d12'40"
Cooper Hollis
Her 17h21'39" 15d5'53"
Cooper Jack
Cma 6h40'31" -16d17'51"
Cooper Jackson Witek
Peg 22h43'18" 9d31'48"
Cooper Jacob Morrow
Lib 15h17'34" -5d0'23"
Cooper Jake Smith (11-19-
2003)
Sco 16h12'32" -15d25'31"
Cooper James Lavey
Cap 21h41'4" -8d38'17"
Cooper James Law
Cap 21h41'18" -15d2'39"
Cooper James Ward
Cap 20h53'6" -25d14'12"
Cooper Janice
Leo 11h3'16" 7d15'4"
Cooper John Bay -
3/3/2003
Umi 17h5'25" 75d24'26"
Cooper Jye Smith
Ari 1h50'34" 18d18'8"
Cooper Lane Larkin
Sgr 19h2'10" -33d48'2"
Cooper Lawrence Evans
Uma 9h51'19" 58d38'49"
Cooper Lee 6-20-2002
Gem 7h35'45" 20d40'55"
Cooper Lee Cleaves
Cap 20h25'40" -9d57'25"
Cooper Leif Younger-
Hughes
Sco 17h42'36" -37d13'16"
Cooper Michael Davis
Her 16h52'41" 34d10'56"
Cooper Morgan Wilson
Boo 14h59'0" 54d50'2"
Cooper Nathan Jenkins
28.11.2006
Cru 12h25'7" -60d58'3"
Cooper Prindle
Uma 9h20'57" 66d18'45"
Cooper R. Civera
Uma 10h59'41" 50d1'44"
Cooper Ray Villereal
Aqr 22h52'10" -15d16'1"
Cooper Ryley Helmers
Cru 12h0'41" -60d7'18"
Cooper Sherman
Uma 12h44'48" 57d17'15"
Cooper Singleton
Sgr 19h21'49" -16d44'53"
Cooper Thanos Hudson
Cra 18h47'15" -38d57'42"

Cooper the Inventor
Uma 10h38'0" 66d42'4"
Cooper Theo
Cen 13h49'43" -59d53'47"
Cooper Theodore
Mortensen
Tau 4h17'19" 23d11'36"
Cooper Thomas
Pennington
Lib 15h7'2" -20d29'49"
Cooper Tiffen
Cru 12h47'39" -57d22'12"
Cooper Westin Lew
Tau 5h49'16" 21d17'5"
Cooper White
Aqr 21h11'37" -10d48'9"
Cooper William Welk
Pho 1h19'34" -41d6'58"
Cooper Z. I
Cap 21h36'29" -23d11'30"
cooper-man
Vir 13h6'50" 13d53'6"
Cooper-Reed
Dra 18h45'59" 60d15'2"
Cooper's Destiny
Cnc 8h38'48" 29d14'26"
Cooper's Place
Ori 4h56'20" 12d41'48"
Cooper's Star
Per 3h25'19" 47d35'14"
CoopHeat
Per 4h50'10" 46d36'19"
Coos in the sky ...
Mon 6h43'53" 6d1'19"
Copack Pirillo
Uma 12h43'3" 59d14'26"
Copareliz
Cep 21h24'36" 61d34'32"
Coppelia Sotelo "March
13, 1962"
Psc 1h15'54" 2d42'41"
Copper
Lyn 9h6'25" 42d14'7"
Copper Pierce
Lib 14h49'29" -1d46'33"
Coppersmith Star
Leo 10h32'0" 12d39'17"
Copson-Oakes 10th
Anniversary
Cyg 21h53'16" 47d26'50"
Coquetita D'Amico
Uma 10h10'26" 51d44'29"
Coquillette
Peg 22h6'51" 6d43'12"
Cor Willi
Lyn 7h29'38" 50d39'45"
Cora
Leo 9h47'56" 15d24'24"
Cora
Sco 16h4'9" -19d59'48"
Cora "Angel Star"
Tau 5h49'31" 16d11'54"
Cora Ann Lawson Duval
Psc 1h15'11" 20d26'51"
Cora Anne Stratuliak
Aqr 23h50'28" -9d44'54"
Cora Bell Apger 1921-2005
Cep 22h17'1" 58d20'39"
Cora Belle
Gem 6h59'26" 15d3'33"
Cora Dee Jenkins
Sco 16h44'41" -39d5'1"
Cora Elizabeth
Aqr 23h5'41" -9d6'6"
Cora Emeline Burns
Cas 0h46'21" 62d21'18"
Cora Fuller
Mon 6h48'50" -0d5'4"
Cora Glanman - Friends
Cas 0h33'52" 55d10'5"
Cora Hope Winberg
Psc 1h30'48" 27d37'23"
Cora Lea
Tau 5h27'27" 25d36'32"
Cora Lee
And 0h17'38" 25d52'4"
Cora & Lon Forevermore
Cyg 20h53'32" 54d39'51"
Cora Lynn Sealfon's
Shining Star
Leo 9h46'33" 25d52'38"
Cora Miller
Aqr 22h15'15" 1d14'27"
Cora Russell Muschinsky
Umi 14h33'5" 74d41'23"
Cora Sprinkle
And 2h12'37" 46d49'15"
Cora Thorndike
Cyg 20h38'45" 56d49'38"
Coraggiosa Bella Buona
Dorina
Ari 2h23'57" 21d47'0"
Coral
And 2h35'51" 49d23'52"
Coral Ashlee Brown
Cas 2h24'18" 73d7'54"
Coral Calina Linzey
Lyr 18h49'35" 32d11'15"
Coral Elizabeth Nigolian
Vir 12h33'59" 1d13'1"
Coral Grace Bennett-Arnold
Del 20h36'31" 18d29'59"
Coral Harris
And 0h45'31" 37d19'18"

Coral Jade 10/29/96
And 0h31'26" 37d48'27"
Coral Lee Brodak
Vir 13h37'53" -4d5'24"
Coral Lee Link
Sgr 18h47'38" -21d52'57"
Coral Mist Beach Hotel
Per 2h8'18" 57d24'27"
Coral Reef Club
Boo 15h39'12" 51d4'51"
Coral Renee Igyarto
Cam 3h59'47" 59d33'9"
Coral Rose
Ari 3h11'11" 29d26'57"
Coral Sands Beach Resort
Her 16h25'18" 48d26'43"
Coral Stoudt
Lyn 6h48'1" 56d20'14"
Coral Veronica Mae
Mermaid
Gem 7h36'24" 17d56'46"
Coralee
Cyg 19h35'37" 51d45'6"
CoraLee Siders
Leo 10h19'0" 27d4'52"
Coralia
Uma 8h22'23" 61d59'54"
Coralie
Cam 4h33'20" 58d10'51"
Coralie Bessette
Peg 21h55'20" 11d50'20"
Coralie Elizabeth
Cap 21h2'44" -18d41'51"
Coralie Grandjean
Cyg 20h44'15" 45d56'36"
Coralie Sophie
Ori 4h44'34" 9d52'25"
Coralifidanza
Cas 23h54'24" 58d29'26"
Coral's Star
Uma 11h51'0" 58d5'24"
Coratt
Cyg 20h34'6" 44d3'40"
Corazon
Uma 10h38'52" 45d40'33"
Corazon
Leo 11h27'3" 14d32'3"
Corazon Bello
Ari 2h22'48" 24d23'45"
Corban
Cyg 21h56'58" 52d43'2"
Corban Joseph Bennett
Shea
Gem 6h52'44" 14d32'4"
Corbin
Equ 21h24'25" 12d37'16"
Corbin
Ori 4h52'38" 6d5'54"
Corbin
Uma 9h18'52" 68d8'13"
Corbin and Nieve Galey
Sge 20h3'20" 16d48'48"
Corbin Gillis-Vigil
Uma 11h51'54" 59d14'37"
Corbin Handsome-
Donovan Clay
Gem 7h0'32" 35d6'18"
Corbin J. McCrary
Lib 14h49'33" -4d20'24"
Corbin Jovan
Lyn 6h58'1" 52d26'55"
Corbin Micah Ebersbach
Her 17h25'49" 36d29'42"
Corbin O'Neill Tierney
Sgr 19h19'28" -17d8'38"
Corbin Robert Rust
Tau 4h47'29" 21d43'17"
Corbin William Richard
Jensen
Uma 8h12'10" 64d29'10"
Corbrea
Her 17h19'10" 15d24'18"
Corbstar
Ari 3h24'14" 29d0'54"
Cord & Karen 9/19/05
Cyg 20h21'11" 52d17'12"
Corda Johnson
Eri 3h47'59" -11d3'36"
Cordale 1959!
Umi 14h34'50" 78d20'1"
Cordelia
Uma 10h16'21" 59d8'7"
CORDELIA
Peg 21h28'18" 15d28'21"
Cordelia
And 0h22'44" 25d13'28"
Cordelia Katherine Keeney
Vir 13h28'15" -19d11'29"
Cordell Anthony Szot
Cep 22h32'11" 77d56'15"
Corder's Cosmos
Uma 9h47'9" 69d46'0"
Cordomi Chloé
Cas 0h14'1" 54d55'51"
Cordovano
Per 3h13'52" 41d50'52"
Cordula Uleman
Aqr 23h12'27" -19d57'50"
Corduroy
Umi 14h12'22" 75d20'4"
Core 4
Cas 1h22'38" 59d41'23"
Coree Heighway
Leo 9h26'7" 10d8'28"

Coreen
Peg 22h29'46" 17d56'40"

Coreen #1
Leo 9h52'1" 20d7'44"

Coreen Bernstein 60
Cap 20h42'23" -26d0'33"

Coreena
Cyg 19h50'57" 29d48'37"

Coreen's Dream
Sco 17h51'47" -36d49'12"

Corena Bisby
Lib 15h39'45" -12d3'32"

Corena Bonnie Hinman
Sgr 18h5'39" -27d18'27"

Corentin Augris
Ori 6h16'17" 15d46'2"

Corentin Merle
Cas 0h46'15" 65d6'24"

Corentin Merle
Cas 1h35'50" 62d32'31"

Corentin Petit Poussin
Uma 10h18'10" 59d38'18"

Coreo
Uma 11h36'24" 48d57'7"

Corestates Capital
Uma 11h7'37" 28d53'13"

Corey
Cnc 8h41'37" 15d43'2"

Corey
Her 17h20'21" 13d55'17"

Corey Allan
Her 17h41'5" 37d59'11"

Corey Allan Hill
Leo 11h27'34" 20d7'27"

Corey Allen Vanhorn
Her 16h39'56" 36d11'16"

Corey Allen Werfel <3 I love you!
Uma 8h28'25" 65d44'57"

Corey and Marko's Forever Love!
Leo 9h55'49" 19d20'15"

Corey and Mollie
Gem 7h28'2" 33d33'51"

Corey Anne
Lyn 8h14'10" 35d29'32"

Corey Anthony Spencer
Uma 13h10'35" 52d26'59"

Corey Antonio Glyenn
Uma 11h49'54" 43d7'40"

Corey Bennette Gunn
Dra 11h44'30" 69d43'48"

Corey Brendan
Uma 11h32'38" 62d26'56"

Corey Brooke Fields
And 0h37'28" 41d50'4"

Corey Christian Couch
Boo 15h29'59" 44d29'44"

Corey David Cole
Tau 5h44'8" 25d55'24"

Corey David Phelan
Lib 14h31'2" -20d47'5"

Corey Dean Facemire
Ori 5h49'47" 2d18'17"

Corey + Dioneshia Lang 4-Ever
Cyg 19h51'58" 38d7'8"

Corey Don Goldrick
Uma 11h2'17" 50d33'34"

Corey Douglas Pierce
Umi 16h23'52" 77d52'59"

Corey Elizabeth Chavis
Sgr 17h54'53" -29d53'52"

Corey Everett Belk
Ori 5h6'7" 14d19'28"

Corey Gibbs
Umi 14h31'21" 71d54'11"

Corey Grace
Vir 14h15'54" 3d9'40"

Corey & Hillary
Tri 2h5'11" 33d56'25"

Corey J. Dumais
Ari 3h11'15" 29d9'20"

Corey James Albert Sarin
Her 18h45'34" 18d8'59"

Corey James Pike
Lib 16h0'35" -15d25'17"

Corey James Rayment
Dra 16h16'58" 68d13'23"

Corey Jay Angwin
Umi 15h54'23" 73d14'9"

Corey Kearns
Cnc 9h0'22" 15d16'10"

Corey King 'Daddy's Mate' 19-12-95
Dor 4h59'57" -68d37'6"

Corey L. Russ
Lib 15h8'16" -12d33'57"

Corey & Landrine
Cyg 21h21'33" 50d57'27"

Corey Lincoln McFadgen
Gem 6h41'54" 21d42'44"

Corey Loren Schwartz
Cnc 8h48'36" 12d23'55"

Corey M. Fling
Uma 8h49'52" 57d26'11"

Corey M. Turner
Tau 3h40'47" 25d18'46"

Corey Martin Dill
Leo 11h43'57" 17d47'0"

Corey Matthew Ross
Cyg 21h37'57" 37d37'34"

Corey Michael Carpentier
Uma 10h20'46" 48d30'33"

Corey Michael Gittinger
Lib 14h36'9" -24d40'4"

Corey Michael Masters
Sco 16h12'15" -16d15'52"

Corey Michael Yaklin
Lyn 7h43'0" 46d49'5"

Corey Michael Zak Mohney
Umi 16h3'2" 78d7'18"

Corey Michelle
Aqr 23h45'19" -10d37'45"

Corey Michelle Hastings
Aqr 23h12'35" -19d51'35"

Corey Miller
Her 16h34'47" 48d43'44"

Corey Morris
Her 17h13'31" 36d47'40"

Corey Patrick
Cnc 9h11'5" 9d6'55"

Corey Patrick Mason
Her 18h39'31" 17d10'16"

Corey R. Walker
Cnc 8h35'22" 27d36'12"

Corey Rose Roberts
Vir 11h44'37" 8d16'40"

Corey Rudolph Griffis
Cap 20h18'36" -26d1'58"

Corey Scott Murphy
Gem 6h30'1" 22d7'8"

Corey Shannon Ritchie
Uma 8h15'16" 59d44'57"

Corey & Shannon's Star
Ori 5h11'48" 12d44'14"

Corey Siegmund
Ori 4h49'39" -0d20'19"

Corey Solis
Cam 7h35'18" 60d8'0"

Corey Steven
Ari 2h39'5" 31d0'27"

Corey Thomas DeCarlo
Cnc 8h22'2" 23d53'30"

Corey Thomas Sabol
Her 16h39'28" 47d33'47"

Corey Watson
Per 4h11'59" 51d16'58"

Corey Wenrick
Psc 0h47'25" 20d41'15"

Corey Whitney
Uma 12h5'46" 34d0'49"

corey wilenken
Uma 10h7'32" 55d23'12"

Corey William Walker Swan
Psc 0h56'52" 9d29'45"

Corey Wojcik
Uma 12h39'40" 58d58'1"

CoreyBoyHartman
Aql 19h28'3" 6d22'26"

Corey's
Lmi 10h8'12" 39d25'7"

Corey's love & passion
Ori 6h5'4" 18d35'57"

Corey's Star
Ori 6h21'9" 13d37'51"

Cori
Leo 10h50'51" 6d27'8"

Cori
Sco 16h11'20" -17d29'30"

Cori
Sco 16h42'8" -36d33'24"

Cori and Tim Dustan
Cyg 19h58'1" 49d22'3"

Cori Doll
Cas 1h37'9" 64d30'29"

Cori Eldridge
Aur 5h28'24" 45d56'23"

Cori Gregory
And 0h26'30" 32d51'55"

Cori K. Valois-Stewart
Umi 13h5'42" 72d16'7"

Cori Lee Musgrove
Ari 2h49'26" 22d3'24"

Cori Lenckus
Cap 20h47'10" -14d44'31"

Cori 'Licensing Super Star' Hartje
Sgr 19h35'14" -30d55'54"

Cori Lynn
Her 16h25'55" 45d46'39"

Cori Lynn Schaufelberger
Cnc 9h21'31" 11d53'7"

Cori Lynn Sieger
Gem 6h34'31" 12d7'30"

Cori Marie Sutherland
Cet 0h54'18" -0d31'30"

Cori McCormick Engel
Lyr 19h18'59" 34d20'47"

Cori Michelle Mooney
Tau 5h0'25" 20d26'8"

Cori Renai' Cavazos
Lyn 8h15'7" 42d23'44"

Cori, the Dumb Girl
And 23h57'4" 42d49'34"

Corian Jacob Halvorson
Cnc 9h9'1" 32d17'0"

Corianda Dimes
Uma 10h22'58" 61d24'5"

Corianna_0815
Mon 7h17'4" -0d54'59"

Coribeth Sawders
Per 3h29'3" 49d50'36"

Corie
Psc 1h26'28" 30d46'9"

Corie Ann Lynch
Ari 2h25'12" 26d44'18"

Corie Jacobsen
Leo 9h42'41" 28d12'22"

Corie L.V. Scotto
Tau 5h57'38" 24d59'47"

Corie Lynn Duckett
Uma 9h38'59" 59d36'46"

Corie Marie Sims
Her 17h36'28" 32d2'44"

corie33
Aqr 20h51'45" 1d41'39"

Corielle Kathleen
Ari 2h28'40" 18d49'55"

Corilyn Kelly Oliva
And 1h41'51" 39d30'53"

Corin
Equ 21h3'10" 4d41'39"

Corin "Corky" Nemec
Ori 5h59'32" 18d34'10"

Corin DeLancey
Sco 16h3'48" -19d31'13"

Corin Garland
Aqr 21h7'54" -12d50'37"

Corin & Justin
Psc 0h19'56" 3d45'36"

Corina
Cap 21h41'48" -10d36'59"

Corina
Lib 15h3'58" -8d13'46"

Corina
Gem 7h42'3" 19d53'19"

Corina
Cnc 8h45'39" 17d37'35"

Corina
Cnc 8h6'10" 9d16'7"

Corina
Leo 9h45'46" 10d41'41"

Corina Ann Dawson
Tau 4h19'21" 3d46'20"

Corina Ann Ryland
Lyr 19h5'31" 42d3'23"

Corina Cano...My Goober...
Ari 3h2'26" 13d40'7"

Corina Ching
Aqr 22h50'45" -3d50'7"

Corina & Dale
Cnc 8h22'1" 15d20'41"

Corina Marie Krieser
Lyr 19h22'38" 38d16'47"

Corina Neufeld
Lib 14h53'17" -2d51'7"

Corina Raisigl
Uma 9h23'20" 47d12'18"

Corina Rena Clem
Cas 1h21'17" 57d53'20"

Corina Rose Joyce Cornish
Lyn 7h34'31" 41d39'35"

Corina Schärer
Cas 1h33'32" 72d5'46"

Corina Taylor
Cnc 8h51'32" 11d50'4"

Corine
Aur 5h15'26" 30d2'34"

Corine
Uma 9h48'49" 52d1'2"

Corine D Gandulla
Lib 15h16'9" -3d43'23"

Corine Hobbs
Mon 8h3'13" -0d54'20"

Corine Isabel Unruh
Cyg 19h35'19" 37d54'40"

Corine Jamil Saab
Uma 11h21'49" 30d18'14"

Corine Parker
Vir 13h24'38" 11d41'45"

Corine V. Liverpool
Sgr 19h38'13" -13d17'4"

Corinia 6-6-40
Lyr 19h16'58" 43d11'33"

Corinn Cannelli
Cas 0h43'59" 51d34'9"

Corinn Elizabeth Walters
Cap 20h14'5" -21d25'50"

Corinna
Lyr 18h44'13" 38d56'55"

Corinna
Peg 21h30'38" 15d13'7"

Corinna Clara Breitling
Uma 14h27'18" 61d11'6"

Corinna Hilpisch
Uma 12h11'13" 53d50'20"

Corinna Hyatt
Cyg 19h47'2" 35d32'41"

Corinna Kimi Guits Gutierrez's Star
Crb 15h58'54" 38d35'38"

Corinna Löser
Uma 9h52'10" 49d51'12"

Corinna Lynn Swanson
Cas 0h25'32" 62d59'8"

Corinna Marquerite Wild
Lyn 7h11'2" 58d50'48"

Corinna Pasley
Leo 11h45'13" 20d8'49"

Corinna's Star
Crb 15h46'8" 28d28'35"

corinne
Peg 21h25'13" 15d25'49"

CORINNE
Sge 19h38'45" 17d26'50"

Corinne
Uma 9h56'36" 46d31'2"

Corinne
Uma 9h36'55" 46d9'3"

Corinne
Lyr 19h20'4" 38d14'39"

CORINNE
Cas 0h13'11" 52d27'52"

Corinne
And 23h40'17" 41d6'43"

Corinne
And 2h37'38" 49d19'24"

Corinne
Cyg 19h33'47" 30d4'11"

Corinne
Gem 7h11'36" 33d40'24"

Corinne
And 1h31'19" 42d2'34"

Corinne
Aqr 22h6'9" -13d5'13"

Corinne 11
Vir 13h17'10" 7d7'18"

corinne 3
Com 13h15'9" 14d37'29"

Corinne 8
Her 18h22'28" 17d11'28"

Corinne A Feehan
Cas 0h20'26" 61d58'17"

Corinne Abaterusso
And 1h4'3" 46d32'15"

Corinne and Rob Ulrich
Per 2h54'0" 51d44'56"

Corinne (Angel)
Lib 14h56'38" -4d27'58"

Corinne Autino
Uma 11h43'16" 48d15'18"

Corinne Bielmann
Cep 22h20'20" 65d51'58"

Corinne & Bob Bennett
Ori 6h6'24" 17d23'45"

Corinne C. McClane
Vir 12h47'45" 3d33'45"

Corinne Case
And 0h22'34" 29d6'49"

Corinne Celice Wilson
Lib 15h47'53" -18d42'58"

Corinne Chebeon Mary Allen
Gem 6h55'34" 21d37'36"

Corinne Cherigny
Cas 23h57'33" 53d37'2"

Corinne Daily Moran
Uma 8h28'37" 65d43'50"

Corinne E. Whidden
And 0h16'46" 45d18'28"

Corinne Elise
Tau 3h46'3" 23d44'15"

Corinne Elizabeth
Ori 6h4'16" 19d2'49"

Corinne Elizabeth 1988
Umi 13h47'8" 78d44'49"

Corinne Elliott Brooks
Sgr 18h11'7" -28d21'3"

Corinne est amoureuse de Ryan
Vir 12h9'0" 6d35'17"

Corinne et Stéphane
Del 21h6'56" 14d2'45"

Corinne Fisk
Cas 0h36'42" 64d22'31"

Corinne Hasler-Kunz
Cep 22h27'47" 80d26'47"

Corinne Haslinger
Her 17h33'12" 15d47'31"

Corinne Hernet
Psc 1h20'39" 19d7'48"

Corinne Huet
Tau 5h51'38" 23d32'25"

Corinne Isabelle Scott
Sgr 18h23'36" -16d35'20"

Corinne J Bohling
Ari 3h4'27" 25d44'22"

Corinne L. Oravits
Cap 20h17'2" -15d10'49"

Corinne L. Portner
Srp 18h18'38" -0d3'58"

Corinne Lois Karpinski
Lib 14h59'23" -3d8'53"

Corinne M. Greenberg
Lyn 7h56'24" 51d18'24"

Corinne Mae
And 0h3'33" 44d4'30"

Corinne Major
Uma 11h21'27" 34d19'42"

Corinne Margaret Erickson
Cas 2h25'25" 72d35'48"

Corinne Marie Coleman
Leo 11h6'44" 7d34'2"

Corinne Marie Hayward
Aqr 22h40'23" 1d45'12"

Corinne Meier
Lyn 7h37'46" 56d49'11"

Corinne Meuwly - Lindinger
Tri 2h10'53" 32d21'57"

Corinne Michelle Cundiff
Lib 14h34'9" -11d9'36"

Corinne Mitchell
Cas 23h43'29" 55d28'58"

Corinne Moretti
Oph 17h38'21" -0d36'5"

Corinne Muff
Cyg 20h33'1" 37d33'45"

Corinne Nicole
Vir 12h46'38" 7d25'26"

Corinne Rachel
Uma 11h5'37" 50d56'32"

Corinne Rita
Uma 11h37'19" 37d32'7"

Corinne Rose Witt
Uma 8h59'47" 49d56'9"

Corinne Sergent
Sgr 19h4'1" -36d28'4"

Corinne Stebler
Dra 15h5'1" 56d59'25"

Corinne Stewart
Uma 10h18'12" 46d40'12"

Corinne und Michel Köpfer
Peg 21h30'36" 15d29'49"

Corinne Victoria November 25, 2004
Sgr 18h57'26" -30d54'42"

Corinne Walker
Cap 20h24'27" -11d16'48"

Corinne, I Love You. Robert
Leo 11h16'46" -5d43'53"

Corinne's Corona
Sco 16h6'55" -16d28'17"

CorinneStar Roller
And 0h53'54" 35d59'6"

coriponimus77
Lib 14h37'9" -9d26'4"

Cori's Star
Uma 9h41'0" 68d6'23"

Corisa
And 23h25'7" 52d25'59"

Corissa
Gem 6h36'47" 16d29'47"

Corissa Anne Manikhi
And 1h55'26" 46d11'55"

Corissa "Polly" Roberts
Gem 6h37'18" 26d31'21"

Corky Blalock
Tau 4h20'46" 24d42'59"

Corky Trelstad
Cnc 9h8'3" 27d58'36"

Corky Tuttle
Cyg 20h57'33" 47d37'57"

Corky Underwood
Leo 11h9'49" -1d20'43"

Corley James Hammond
Cru 12h38'26" -58d35'1"

Corlisikle
Ari 3h4'44" 23d13'54"

Cormac and Laura
Lyr 18h26'57" 39d26'7"

Cormac Myles Merriman
Uma 8h32'43" 69d28'43"

Cormac Seán Gaynor
Lmi 11h0'29" 31d41'4"

Cormac William Hayden
Cru 12h55'52" -63d32'33"

Corman Lee Bennett
Vir 14h58'6" 4d13'30"

Cormier
Gem 7h24'57" 17d54'30"

Cormiki
Gem 6h50'0" 34d32'45"

Corndog & Piglet
Aqr 22h53'43" -15d58'34"

Corneillius Love
Cep 20h50'4" 60d24'5"

Cornel Füglistaller
Peg 22h40'38" 3d25'18"

Cornel Keller
Crb 15h53'15" 26d58'57"

Cornel Leuthard-my love
Vul 20h23'3" 26d38'56"

Cornelia
Her 17h36'43" 21d16'30"

Cornelia
Gem 7h7'37" 32d2'19"

Cornelia
Tri 1h51'56" 34d29'24"

Cornelia
Uma 11h13'16" 41d59'4"

Cornelia
Uma 11h35'54" 61d36'14"

Cornelia
Cap 20h25'30" -8d43'51"

Cornelia A. Kirkpatrick
Tau 5h49'11" 26d16'41"

Cornelia Eugster
Cnv 12h45'46" 43d0'6"

Cornelia Hellmann 50
Cas 0h56'29" 58d23'27"

Cornelia Hinze
Uma 8h16'39" 68d57'59"

Cornelia Kull, 07.08.1955
Umi 14h7'18" 67d48'45"

Cornelia Löffler
Uma 14h20'28" 56d55'22"

Cornelia Maria Schuh
Tau 4h18'19" 17d29'1"

Cornelia Muther-Rutz
Cyg 20h55'24" 47d1'55"

Cornelia Rhodes McPherson
Tau 4h28'34" 29d55'17"

Cornelia Ross
Tau 4h32'50" 25d0'16"

Cornelia VanWinkle
Lib 14h51'0" -0d38'1"

Cornelia, Corey, Brett Valencia
Cap 21h29'19" -10d3'55"

Cornelio Taroliki
Sco 17h29'50" -34d8'44"

Cornelius E Mara, Jr, Super Genius
Lib 15h53'25" -17d54'45"

Cornelius Grocholl
Uma 12h2'22" 37d6'52"

Cornelius James Ryan VI
Lib 15h4'42" -11d20'3"

Cornell And Sammy Evans
Cnc 8h7'3" 6d57'45"

Cornell LeVonne Waddell
Sco 16h46'0" -29d19'37"

Cornell Murray
Cyg 21h15'59" 54d49'47"

Cornellia
Uma 13h38'16" 57d22'13"

Cornelus Marinus Langeveld
Uma 11h52'22" 63d1'11"

Corni Astrum
Eri 3h35'36" -41d32'9"

Cornucopia
Ori 5h14'23" -1d19'41"

Corona de Michael
Crb 16h12'49" 28d49'5"

Corona Mariscal
Lyn 8h28'11" 41d59'27"

Coronado
Lyr 18h37'58" 39d26'1"

Coronarosey
Ori 5h42'33" 7d16'13"

Corporal Jeremiah Wayne Robinson
Uma 9h54'22" 51d26'58"

Corporal Kyle Wesley Powell, USMC
Per 4h1'52" 36d43'3"

Corporate-Third Party Accounting
Her 17h3'51" 46d59'33"

Corpus Christi Parish
Ori 5h44'7" 4d13'48"

Corrado
And 23h18'45" 44d22'57"

Corrado D'Antoni
Cam 5h37'8" 59d58'43"

Corral Orbet - Robert & Carol Shurey
Ori 5h42'30" -2d55'53"

Correale
Cas 0h39'32" 62d34'11"

Correa-Sorich
Her 17h15'56" 27d14'51"

Corri
Vir 13h10'35" 11d8'17"

Corrie
Vir 11h50'56" -0d15'40"

Corrie 131
And 0h27'54" 33d19'35"

Corrie Duffee
And 1h5'27" 40d2'56"

Corrie James Flegg, My One And Only
Sco 16h3'11" -25d57'14"

Corrie Jill Dean
Hya 9h22'19" -0d44'22"

Corrie Leone Anderson Primous
Cap 21h26'4" -14d47'29"

Corrie Marie - October 18th
Cnc 8h54'34" 29d47'3"

Corrie Waynoka Hawkins
Leo 10h56'49" 11d21'24"

Corriena
Ori 5h43'28" 7d46'40"

Corrin
Umi 15h14'56" 72d35'13"

Corrin Coelho
Umi 16h17'40" 76d18'14"

Corrina 1
Lib 15h17'23" -5d55'3"

Corrina "CNC" Conley
Lib 15h56'55" -5d41'0"

Corrina Colquhoun
Cas 0h28'11" 50d28'21"

Corrina De Shazo
Sgr 19h8'37" -17d8'29"

Corrina Marie
Cap 21h33'32" -10d47'8"

Corrina Munoz
Crb 16h8'47" 26d30'10"

Corrine
And 23h20'41" 48d36'42"

Corrine
Psc 0h26'24" 3d37'19"

Corrine Anne Petrushonis
Del 20h41'57" 17d2'44"

Corrine Atkinson
Cas 0h53'45" 58d10'55"

Corrine (BaBa) Younger
Ari 2h54'7" 19d58'32"

Corrine Carr
Sco 17h6'26" -45d30'2"

Corrine Driver
Cas 1h16'31" 69d51'46"

Corrine Kimber Heck 8/25/1961
Vir 13h52'4" -2d7'35"

Corrine Lee
Crb 15h46'44" 26d19'18"

Corrine Owolabi's Star
Sge 20h9'45" 18d42'56"

Corrine Rose
Cas 0h38'44" 58d6'38"

Corrine Vicinenzo
And 1h9'49" 41d12'53"

Corrinne
Umi 13h35'5" 70d56'11"

Corrinne Nicole McQuade
Leo 9h30'35" 28d16'10"

Corris R. Boyd
Gem 7h36'44" 27d14'44"

Corris R. Boyd
Gem 7h23'59" 28d36'15"

Corry Coogan
Sgr 18h8'50" -20d37'2"

Corry's ModZIckAMaDiklYiYiWhatI Do
Cyg 21h13'42" 42d28'1"

Corso-Turner
Sco 16h27'58" -40d38'19"

Cort Townsend Scheel
Ori 5h35'39" -1d57'2"

Corte Weatherall
Lmi 10h6'43" 28d44'47"

Cortez Emmanuel Stallings
Dra 10h14'28" 73d7'5"

Cortez Ford
Gem 7h31'36" 19d42'40"

Cortez Kennedy
Vir 13h33'55" -9d23'10"

Cortnea Regina DeMaio
Sgr 19h41'26" -12d33'43"

Cortnee Alyse White
And 1h11'45" 43d2'34"

Cortnee Lee Dalton
Uma 8h36'58" 56d34'33"

Cortnee Louise Raymond
Cnc 8h34'7" 23d31'33"

Cortney
Aqr 21h47'24" -2d0'48"

Cortney
Aqr 20h39'43" -7d20'23"

Cortney and John
Lyr 18h48'15" 34d48'30"

Cortney Ann Elliott
Aqr 22h18'36" -3d12'11"

Cortney Ann Holloway
Aqr 21h45'9" -8d0'20"

Cortney Ann Kelly
Cap 20h29'38" -8d55'18"

Cortney Bauer
Uma 11h13'37" 70d44'4"

Cortney Cadby
Psc 1h26'5" 10d20'8"

Cortney Caramanico
Sgr 19h36'22" -14d52'17"

Cortney Cruttenden
And 2h18'2" 41d24'44"

Cortney F. Watterson
Umi 14h17'32" 70d23'34"

Cortney Foti
Aqr 22h55'49" -8d41'14"

Cortney Henry
Lyr 19h8'36" 42d6'37"

Cortney J'Ann Malone
And 23h15'43" 46d41'19"

Cortney J'Ann Malone
Vir 14h18'50" -6d26'35"

Cortney Johnson's star
And 23h14'25" 52d9'53"

Cortney Leigh Cochran
Ari 2h45'30" 21d42'23"

Cortney M. Brown
Mon 6h53'51" -0d6'49"

Cortney M. Urbanek
Per 4h43'4" 46d52'28"

Cortney Nichole Shewell
Cap 21h22'31" -27d10'47"

Cortney Severyn
Sco 16h7'22" -13d20'43"

Cortney Suzanne Thomas
Sgr 19h31'21" -14d3'47"

Cortney Swenson
Cap 21h15'16" -27d22'28"

Cortney Swift
Per 3h45'1" 43d46'40"

Cortney's
Lyn 7h15'9" 48d31'44"

Cortney's Eternal Love
Sgr 18h0'44" -17d36'37"

Cortneys World
Uma 11h8'47" 52d35'27"

Cortni Elise
Lyr 19h8'29" 26d59'4"

Coruscant
Lyn 9h10'24" 38d24'0"

Coruscating Cassie
Cas 1h37'53" 55d11'41"

Corwin Matthew Cummins
Tau 4h33'47" 9d4'57"

Corwon Finley
Sco 17h52'26" -32d55'5"

Cory
Uma 14h24'48" 58d44'4"

Cory
Sgr 18h13'50" -16d19'56"

Cory
Aqr 23h16'15" -15d7'59"

Cory
Vir 12h29'20" -5d26'45"

Cory
Aqr 22h36'7" -1d4'12"

Cory
Ari 3h5'10" 29d24'31"

Cory
Ari 2h53'17" 28d24'49"

Cory 7/18/82-7/13/06
Cnc 8h12'44" 9d20'46"

Cory A. Slome
Ari 2h46'59" 25d14'0"

Cory Allen Strassburger
Gem 6h44'9" 14d56'3"

Cory and Brie Carvalho
Vir 12h46'38" 5d28'54"

Cory and Laura 12/25/03
Cyg 20h43'45" 38d32'50"
Cory and Nikki's Someday
Uma 11h47'30" 46d30'6"
Cory Anne Koerselman
Cas 23h13'19" 56d11'49"
Cory Ashley Mukai
Ori 5h34'23" -3d46'3"
Cory Bast
Tau 5h43'33" 26d24'57"
Cory Benjamin Tackett
Per 3h50'14" 42d6'40"
Cory Britton
Dra 10h42'25" 74d41'23"
Cory Brooke
Cap 20h59'41" -23d47'4"
Cory Cassell
Lib 15h35'56" -18d3'14"
Cory Charles Hare
Cru 12h20'6" -56d33'48"
Cory Christian Peach
Cru 12h38'30" -60d46'44"
Cory Christopher Harding
Per 3h9'40" 55d10'42"
Cory Cissell
Aqr 22h24'11" -13d22'55"
Cory Dale Antonakos
Ori 5h36'37" 1d8'22"
Cory David Merz
Ari 2h45'41" 10d41'47"
Cory Dillon
Vir 13h20'34" 8d25'43"
Cory & Donielle Davies
Cyg 21h34'33" 53d21'48"
Cory Edward Varga
Umi 14h35'47" 76d59'24"
Cory Einhorn
Cap 21h21'6" -22d50'44"
Cory Elizabeth Hauswirth
Uma 9h51'48" 49d18'12"
Cory Green
Ori 5h35'27" 11d7'11"
Cory James McBride
Sgr 19h6'49" -18d35'10"
Cory Jerald Shepard
Aur 5h49'12" 32d49'42"
Cory & Jessica's Marriage
Cyg 20h29'20" 48d33'47"
Cory Jon Sytsma
Leo 10h6'7" 12d23'0"
Cory Kemble
Ori 5h15'48" -0d37'36"
Cory Kendall
Uma 11h30'49" 56d5'59"
Cory Leon Vincent "A Great Son"
And 0h39'15" 37d18'9"
Cory Lyle
Uma 11h21'50" 35d55'54"
Cory Mathews
Gem 7h46'19" 34d7'26"
Cory Matthew Frederick
Cnc 8h34'11" 32d9'38"
Cory Natofsky
Aql 19h6'19" 3d15'54"
Cory Quinn Wilson
Gem 7h26'31" 31d53'53"
Cory & Rebekkah
Cyg 20h15'36" 44d40'6"
Cory - Robert
Umi 15h32'50" 78d13'37"
Cory Robert Snyder
Uma 10h57'39" 56d32'23"
Cory Rowlands
Cep 22h27'17" 57d23'41"
Cory Rudolph
Boo 14h52'16" 22d37'7"
Cory Sean Martin
Uma 9h40'24" 58d27'40"
Cory Sklareski
Lac 22h26'36" 46d12'55"
Cory & Stephanie~Always n Forever
Sge 20h5'50" 18d49'30"
Cory Vaske
Per 3h15'51" 46d54'21"
corybeth
And 2h7'55" 41d16'36"
Corylus
Cas 0h56'38" 57d43'5"
Corymazi
Lyr 18h53'0" 31d48'41"
Coryn Noelle Sevigny
Lib 15h8'23" -4d17'44"
Corynn Hannah Boresi
And 0h47'31" 43d46'25"
Corynne Schnaars
Cas 0h25'46" 54d31'41"
Cory's Light
Umi 16h23'38" 79d37'13"
Cory's Star
Dra 16h51'54" 61d29'57"
Cory's Star
Cru 12h38'44" -60d11'12"
CoryWery2005
Uma 9h8'44" 68d7'40"
Cosa
Cyg 19h58'36" 30d24'12"
cosak
Per 4h8'30" 33d54'27"
Coshis "Te Amo"
Cap 20h21'15" -11d10'6"
Coshley
Her 17h19'49" 15d35'9"
Cosima
Uma 10h53'10" 46d24'0"
Cosima
Cyg 21h14'11" 44d20'28"
Cosi's Own
Uma 14h15'57" 61d50'6"
Cosita
Psc 23h51'40" 6d19'52"
Cosita - Gemma
Gem 6h52'45" 17d20'26"
Cosmic Craft of the Constellations
Ori 5h29'9" 9d59'14"
"Cosmic Crimmit"
Cep 20h39'56" 61d30'47"
Cosmic Dancer
Cep 21h28'8" 63d47'59"
Cosmic Interlude
Ori 6h14'48" 15d38'54"
Cosmic Kaz
And 1h58'36" 38d2'41"
Cosmic Kramer
Uma 11h27'46" 45d44'52"
Cosmic Liz
Cas 1h31'46" 61d9'11"
Cosmic Ray
Uma 11h14'28" 70d22'1"
"Cosmic Super Nay" Nadia 24/5/1977
Gem 7h19'19" 24d17'15"
Cosmicsky
Uma 13h45'15" 55d6'57"
Cosmic-Turtle
Umi 15h41'48" 74d18'23"
Cosmo
Uma 11h29'19" 63d58'50"
Cosmo
Aqr 20h42'52" -8d50'18"
Cosmo
Uma 11h22'46" 43d44'45"
Cosmo Anthony George Martin
Dra 18h34'6" 58d15'26"
Cosmo James Esposito
Peg 23h47'7" 30d17'12"
Cosmo Rocket Nugen
Uma 10h19'43" 49d51'26"
Cosmo Vincent Ingenito, Sr.
Umi 15h25'14" 81d30'28"
Cosmopolitan
Aqr 21h39'20" 0d3'52"
Cosplay Neko
And 23h25'58" 47d30'6"
Cossette Arias (Blondie)
Sco 17h22'10" -37d48'42"
Cossey Cosmos
Cyg 20h0'52" 56d26'55"
COSTA DINO
Psc 1h20'15" 32d35'3"
Costanza
Cyg 20h3'18" 51d26'4"
Costanza Homes- star craftsmen
Her 17h44'8" 17d28'44"
Costanzo and Ilde Montecalvo
Cyg 20h22'11" 43d59'55"
Costanzo Cosentino
Peg 21h41'1" 26d59'28"
Costas Gianacopolos' Bright Star
Cyg 19h39'49" 44d12'30"
Costas Tantis
Her 18h29'57" 14d59'19"
Costello-Ryan
Tau 5h25'34" 24d59'8"
Costello's Embrace
Lyr 19h8'34" 42d50'19"
COTC Forensic Criminalitics I, 2005
Lyn 8h47'52" 38d49'57"
COTC FORENSIC SCI-ENCE 2005
Her 16h50'28" 34d56'51"
Cotra Enterprises Sdn Bhd
Ori 5h31'15" 2d48'19"
Cotton David Mendenhall
Lib 14h59'15" -5d16'32"
Cotton Doc
Cep 21h42'6" 56d28'48"
Cotton & Hod ~ Forever in Love
Cyg 21h36'2" 50d30'21"
Cotton & Tha Kid
Cyg 20h11'50" 42d6'6"
Coty
Umi 14h30'37" 74d49'6"
Cotye, Fred's Loyal Companion
Uma 11h24'17" 56d53'1"
Cotyes Lorene
Lyn 7h47'31" 53d47'4"
Cougar
Psc 0h52'22" 27d39'20"
Cougar Schnitzel
Leo 9h42'52" 25d8'24"
Coughlan's Twinkle
Uma 12h40'59" 58d49'2"
Couldn't Have Asked For More (Gail)
Cap 21h42'43" -13d44'23"
Coulter Family Star
Tau 4h38'44" 3d36'57"
Coulter Lee Mariott
And 0h41'37" 36d4'31"
Coumbe-Forte
Ori 5h28'30" -0d32'38"
Count Fetid Rabbit Boodoir De Stink
Car 7h30'33" -51d46'14"
Count Richard
Boo 14h59'14" 19d30'22"
Count Scotula
Vir 13h24'32" 11d50'44"
Countess Duskie
And 2h35'54" 44d38'29"
Countess of the Midnight Sky
Lib 15h32'25" -11d44'6"
Countess Sarah R. Pence
Ari 3h15'38" 24d21'36"
Country Boy
Uma 12h16'39" 61d50'17"
Country Huckelberry
Psc 1h24'0" 18d56'59"
Country's Western Star
Aql 19h59'10" 11d19'36"
Coupines VB
Her 17h59'12" 49d45'9"
Courage of Helen Kidd
Peg 22h32'35" 14d56'40"
Courage, Strength, and Happiness
Uma 11h32'42" 48d52'0"
Courageous Love
Cnc 8h26'36" 6d56'35"
Coureur des Cieux
And 1h28'6" 49d33'28"
Cournoyer-Azze
And 2h20'32" 45d39'39"
Courson
Per 3h27'16" 45d6'55"
COURT 143
Cyg 21h12'18" 47d0'10"
Court Bonis
Aql 19h40'17" 4d58'48"
Court Harward
Psc 1h20'12" 28d6'58"
Courtania
Cam 4h9'37" 69d30'7"
Courtenay
Psc 1h21'26" 25d52'51"
Courtenay Elizabeth Lanyon
Psc 1h16'16" 22d22'31"
Courtlandt Stasiewicz
And 2h29'35" 41d28'6"
Courtley
Ori 5h57'29" 17d25'34"
Courtlyn
Lyn 8h2'34" 33d50'4"
Courtnee Jean Hoff
And 0h45'30" 30d5'48"
Courtney
And 0h39'12" 35d46'51"
Courtney
Uma 10h57'55" 36d49'0"
Courtney
Lyn 8h35'20" 44d11'54"
Courtney
And 0h25'6" 40d57'52"
Courtney
And 0h49'36" 40d55'46"
Courtney
And 22h58'25" 48d0'49"
Courtney
And 23h7'19" 46d1'9"
Courtney
Ori 5h56'50" 21d3'9"
Courtney
Leo 9h43'49" 19d24'15"
Courtney
And 0h32'10" 26d44'55"
Courtney
Tau 5h26'15" 17d2'21"
Courtney
Leo 11h18'2" 7d43'26"
Courtney
Leo 10h48'20" 10d43'13"
Courtney
Vir 14h40'25" 6d9'55"
Courtney
Uma 8h21'25" 66d6'50"
Courtney
Cas 1h41'15" 64d6'35"
Courtney
Cas 1h32'9" 65d48'12"
Courtney
Uma 8h37'27" 61d21'26"
Courtney
Lyn 6h27'54" 57d11'14"
Courtney
Cas 1h29'15" 64d11'11"
Courtney
Cap 21h32'14" -11d35'16"
Courtney
Cap 20h33'4" -10d39'38"
Courtney
Vir 13h12'12" -0d53'48"
Courtney 2005
Aqr 22h37'27" -17d9'48"
Courtney 2007
Ori 5h50'51" 9d33'22"
Courtney 42604
Uma 11h6'35" 49d34'27"
Courtney A. Johnsen
Lib 15h1'39" -18d44'35"
Courtney A. Lace
Gem 7h3'27" 27d0'5"
Courtney & Adam
Umi 13h17'7" 70d23'40"
Courtney Alexis Chavis
Leo 10h32'49" 15d3'11"
Courtney Alicia Walsh
Gem 6h34'22" 22d5'44"
Courtney Allison Mabee
Ori 5h33'59" 4d16'11"
Courtney and Chris
Tau 4h13'51" 20d0'37"
Courtney and Sean
Cyg 19h47'36" 54d24'19"
courtney ann christian
And 0h43'2" 37d10'13"
Courtney Ann Good
Cas 1h33'26" 61d34'18"
Courtney Ann Gross
Cet 3h18'11" -0d45'7"
Courtney Ann Jallo
Lyn 6h57'0" 58d47'14"
Courtney Ann Korn
Aql 19h35'3" 7d32'23"
Courtney Ann Lawson
Lyn 7h37'49" 38d23'27"
Courtney Ann Philbin
Lyn 6h20'32" 61d15'22"
Courtney Ann Schaudel
Crb 16h7'8" 37d54'40"
Courtney Ann Shelstad 5/18/82
And 0h36'9" 26d25'5"
Courtney Anne Lewis
And 23h41'54" 43d50'3"
Courtney Anne Parent
Psc 23h30'17" 6d16'45"
Courtney Anne Whitman
Tau 5h45'50" 22d28'33"
Courtney Armas
And 1h12'26" 34d53'25"
Courtney Ashlan Miller
Tau 4h2'22" 25d9'31"
Courtney Aston
Vir 12h52'35" 2d14'35"
Courtney Austin (Sweetheart)
Gem 6h29'39" 27d38'46"
Courtney Autumn Gold
Uma 13h40'47" 54d14'2"
Courtney Bailey
Per 2h21'32" 55d11'3"
Courtney Baker the Costa Surf Queen
Tau 5h36'11" 26d2'35"
Courtney Bay
Leo 9h26'48" 15d38'2"
Courtney Bennett Kessler
Sgr 18h29'10" -23d23'43"
Courtney Beth
Aqr 22h40'34" 2d34'20"
Courtney Blaire Stephens
Cnc 8h51'47" 15d10'35"
Courtney Brooke
Cas 1h42'27" 67d34'3"
Courtney Burke Gillis
Uma 10h2'7" 51d17'10"
Courtney Camille Wade
Gem 6h7'6" 22d25'5"
Courtney Cantrella
And 2h15'46" 50d33'5"
Courtney Capra
Vir 13h4'4" -21d33'23"
Courtney Case
And 0h27'47" 29d52'25"
Courtney Christina
And 1h11'57" 36d55'52"
Courtney Crin Stookey
Cyg 20h59'32" 37d6'59"
Courtney Crocco
Sco 17h21'13" -37d44'25"
Courtney & Curtis Manning
Cyg 21h11'0" 44d16'34"
Courtney Danielle
Sge 19h45'39" 16d55'59"
Courtney Davidson
Cas 1h25'36" 59d1'5"
Courtney Dawn
Leo 10h15'31" 22d17'16"
Courtney Dawn Howard
Cyg 21h21'0" 54d24'1"
Courtney Dawn Mask
Uma 8h37'13" 68d46'26"
Courtney Day XVII
Tau 5h27'56" 20d48'23"
Courtney DeArros
And 1h39'14" 38d3'43"
Courtney DeCiuceis
Sgr 19h34'58" -15d47'50"
Courtney Demuth
Psc 1h31'15" 19d9'30"
Courtney Denise
Cap 21h35'50" -14d19'45"
Courtney Denzlinger
Ori 5h41'42" 7d47'45"
Courtney Diana Bond
Sco 17h56'38" -31d22'40"
Courtney Diane Eads
Dra 15h36'28" 56d8'23"
Courtney DiGuardi
Cas 0h28'23" 62d59'11"
Courtney Dina
Leo 9h53'56" 15d20'38"
Courtney Dumont
Uma 9h21'33" 46d19'47"
Courtney Dunlop
Vul 21h18'49" 25d5'39"
Courtney E. Rollins Gold
Ari 3h9'17" 26d17'16"
Courtney Elise
And 2h18'31" 48d46'24"
Courtney Elizabeth
And 0h28'30" 43d5'6"
Courtney Elizabeth
Sgr 18h9'45" -24d59'0"
Courtney Elizabeth Baker - Schroat
Aqr 22h39'14" 0d26'16"
Courtney Elizabeth Cullop
And 0h22'41" 42d8'15"
Courtney Elizabeth Murphy
Tau 5h40'4" 26d10'12"
Courtney Elizabeth Small
And 1h7'20" 41d38'2"
Courtney Elizabeth-Belle
Lyn 7h6'57" 59d37'1"
Courtney Ellison
Srp 18h20'15" -0d21'8"
Courtney Ellyse Young
Uma 11h27'30" 68d58'45"
Courtney Erin Brown
And 22h59'15" 50d54'44"
Courtney Eska Palmer
Aur 5h3'45" 35d46'2"
Courtney Eubanks
Dra 18h39'22" 77d21'13"
Courtney F. Lozowski
Uma 11h16'55" 55d22'0"
Courtney Faith Evans
Del 20h54'37" 6d23'39"
Courtney Fitzgerald
Vir 13h5'26" 6d50'51"
Courtney Furlong
Uma 8h43'22" 61d49'46"
Courtney Galanis
Vul 20h25'28" 23d12'43"
Courtney Gerrison
Cap 21h9'26" -19d23'36"
Courtney Gibson
Del 20h34'29" 7d15'49"
Courtney Giles
And 23h13'45" 48d52'21"
Courtney Gloria Smith
Lyn 7h36'48" 35d14'20"
Courtney Grace
Cas 1h23'20" 63d9'15"
Courtney Grace Meade
Sco 17h29'48" -42d18'52"
Courtney Gratchic
Cnc 8h10'52" 31d57'19"
Courtney Gurwell
Crb 16h8'27" 36d8'32"
Courtney Heather Snyder
Tau 3h58'56" 22d59'7"
courtney hennes
Lyr 18h15'50" 32d18'38"
Courtney Howels "Star on the Beach"
Sgr 18h46'16" -30d24'58"
Courtney Isabella
Cas 2h17'7" 67d14'59"
Courtney Iversen
Lyn 7h22'37" 48d33'42"
Courtney J. Henderson
Sco 16h4'59" -17d0'39"
Courtney Jane Olson
Cnc 7h57'39" 15d26'53"
Courtney Jo Kelleher
Ori 5h54'3" 21d38'0"
Courtney Jo Rangel
Psc 1h25'44" 24d46'3"
Courtney Joyce Simpson
Lib 15h48'12" -18d15'14"
Courtney & JuanCarlos
Cyg 20h26'3" 39d18'53"
Courtney Juba
Gem 6h57'10" 24d18'31"
Courtney Kanaby
Uma 10h10'12" 61d45'36"
Courtney L.. Hild
Uma 8h35'9" 62d23'39"
Courtney Lachae Edwards
Aqr 22h38'44" -2d5'17"
Courtney Laine Smith
Sgr 18h50'55" -28d53'8"
Courtney Lake
Aqr 22h58'33" -8d24'26"
Courtney Lane Hundley
Leo 9h29'8" 12d39'52"
Courtney Larren Parker
Uma 8h37'28" 51d35'14"
Courtney Lauren Cox
Ari 2h34'3" 21d29'32"
Courtney Lauren Hill
Lyn 7h47'10" 45d26'10"
Courtney Lauren O'Rourke
Vir 13h21'9" -12d27'27"
Courtney Lee
Vir 14h49'23" 1d55'41"
Courtney Lee Thomas
Lyn 6h27'8" 56d22'58"
Courtney Leigh
And 0h55'25" 40d18'42"
Courtney Leigh Baker
Aqr 22h40'3" 1d58'38"
Courtney Leigh Cooter
Lib 14h59'11" -4d19'1"
Courtney Leigh Johnson (AppleCore)
And 0h31'58" 45d51'45"
Courtney Leigh McGaughy
Aqr 22h25'17" -3d12'22"
Courtney Leigh Merriam
Leo 11h33'44" 19d20'59"
Courtney Leilani Soileau
Lmi 10h36'25" 32d7'20"
Courtney Letson
Ari 2h42'57" 29d43'59"
Courtney Lindahl
Vir 13h47'5" -11d50'26"
Courtney Louise Leith - Courtney's Star
And 23h59'38" 45d58'16"
Courtney loves Corey
Uma 13h50'32" 59d22'15"
Courtney Lyn
Lyn 9h11'32" 39d12'6"
Courtney Lyn Fontenot
Cas 0h30'27" 65d37'7"
Courtney Lyn Gould
Cas 0h7'54" 56d30'41"
Courtney Lynn
And 2h30'23" 50d15'35"
Courtney Lynn
And 0h48'32" 26d31'0"
Courtney Lynn
Vir 12h19'37" 4d9'41"
Courtney Lynn Burns
Ari 2h38'25" 21d34'22"
Courtney Lynn Cahill
Uma 12h40'4" 52d20'37"
Courtney Lynn Cooper
Cyg 19h43'1" 29d2'42"
Courtney Lynn Crain
And 23h45'59" 46d6'59"
Courtney Lynn Crozier
Cap 20h34'45" -20d59'12"
Courtney Lynn Edger
And 1h41'12" 43d20'6"
Courtney Lynn James
Uma 10h12'37" 58d32'59"
Courtney Mailhot
Per 3h22'43" 39d32'26"
Courtney Margaret Train
Cru 12h32'20" -62d21'58"
Courtney Mariah
Aqr 21h45'0" -1d0'1"
Courtney Marie
Lyn 7h27'12" 53d2'42"
Courtney Marie
Tau 4h30'32" 30d3'16"
Courtney Marie
Psc 1h39'22" 11d47'53"
Courtney Marie Forsman
Uma 11h28'5" 50d0'41"
Courtney Marie Kreller
Ari 2h40'3" 30d6'18"
Courtney Marie Lyons
And 0h27'48" 42d13'26"
Courtney May
Leo 10h10'46" 13d35'3"
Courtney Mayo
Aqr 22h7'48" -0d42'45"
Courtney McClure
Psc 0h30'56" 6d29'37"
Courtney McHale Paddick
Dra 18h22'17" 76d46'28"
Courtney McIver
And 0h24'49" 26d1'27"
Courtney Meier
Cnc 8h35'28" 26d21'20"
Courtney Michael Garner
Tau 4h21'9" 6d6'13"
Courtney Michelle
Lyr 19h17'19" 29d56'23"
Courtney Michelle Day
Vir 12h26'57" 7d40'50"
Courtney Michelle Lysykanycz
Ari 3h24'26" 27d25'3"
Courtney Michelle McMullen
Cyg 20h35'30" 41d59'56"
Courtney Michelle Sessions
And 23h45'36" 42d48'24"
Courtney Michelle Spear
And 23h54'41" 40d27'31"
Courtney Michelle Tessmer
Uma 10h48'13" 63d8'22"
Courtney Michelle Westdyke
And 1h53'1" 41d36'6"
Courtney Moltion
Sco 17h6'42" -40d56'50"
Courtney (my baby)
And 0h26'36" 42d59'38"
Courtney My Love
And 23h10'44" 47d6'51"
Courtney My True Love
Sco 16h15'47" -10d19'35"
Courtney ~N~ Justin
Crb 15h51'29" 33d56'4"
Courtney Newsome
Ari 2h44'53" 25d11'51"
Courtney Nicole
Cnc 8h35'30" 17d51'52"
Courtney Nicole
Sco 17h27'58" -44d41'31"
Courtney Nicole Carter
And 23h16'9" 43d11'39"
Courtney Nicole Grzyb-Cuzick
And 1h21'55" 39d23'12"
Courtney Nicole Hauer
Vir 15h6'32" 3d47'17"
Courtney Nicole Hernandez
Uma 9h57'50" 50d12'41"
Courtney Nicole West
Sco 16h5'58" -14d53'12"
Courtney Noel Mason
Cap 21h51'12" -9d10'22"
Courtney Noelle Redmond
And 1h46'21" 37d47'34"
Courtney O'Grady
Psc 2h3'2" 3d2'15"
Courtney Paige Fee
Sgr 19h44'48" -11d46'3"
Courtney Payne
Gem 6h10'55" 24d8'2"
Courtney Pettit - My True Love
Uma 11h23'31" 42d18'17"
Courtney Quintana
And 1h8'50" 44d15'59"
Courtney R. Toole
Cnc 8h32'9" 12d11'51"
Courtney Rae
Gem 7h32'41" 26d54'4"
Courtney Raeann Peck
Uma 10h58'9" 38d6'37"
Courtney Renee
Tau 3h44'59" 29d30'10"
Courtney Renee
Tau 5h27'53" 24d58'3"
Courtney Renee
Ari 2h38'22" 18d19'13"
Courtney Renee Gideon
And 0h16'3" 40d53'36"
Courtney Robberson
Cap 20h58'34" -19d48'55"
Courtney Roberts
Uma 11h24'20" 54d24'20"
Courtney Rose Halverson
Her 16h55'28" 25d36'54"
Courtney Rummell
Uma 8h34'28" 64d44'19"
Courtney Ruth Kaneski
Cas 1h48'56" 63d52'13"
Courtney Ruth Roerden
Aqr 22h23'16" 1d40'12"
Courtney Ryan
Tau 4h8'18" 20d19'1"
Courtney S. Worth
Uma 13h57'25" 54d27'43"
Courtney Sands
And 23h25'33" 51d33'22"
Courtney Sarah Gardner
Sgr 18h39'38" -27d53'57"
Courtney Sarah Gardner
Sgr 18h25'41" -26d26'25"
Courtney Semkewyc
And 23h42'32" 45d47'46"
Courtney Shea
Vir 14h6'11" -14d23'12"
Courtney Sheree Shaw
Cas 23h2'26" 57d52'33"
Courtney Stevens
Crb 15h51'57" 28d3'3"
Courtney Steward
And 0h45'51" 42d45'20"
Courtney Stewart Legg
Lib 15h27'14" -8d3'39"
Courtney Struthers-Honors Star
And 2h38'10" 50d3'43"
Courtney Sue Wright
Sgr 19h51'38" -13d5'23"
Courtney "Sweet Pea" Covington
Uma 9h36'53" 66d47'19"
Courtney Taylor Kemp
Uma 10h51'2" 57d3'43"
Courtney The Beautiful Star
Cap 20h8'45" -22d21'15"
Courtney the Star
Uma 10h14'31" 70d21'24"
Courtney Thomas
And 0h3'38" 44d23'42"
Courtney & Tori Dixon
Sge 19h54'37" 19d8'33"
Courtney Villani
Lib 15h42'12" -4d6'27"
Courtney Waldron
Cyg 21h11'33" 45d10'38"
Courtney Walker
And 0h37'1" 40d18'41"
Courtney Walker
Vir 14h32'17" 3d42'12"
Courtney Ward
Gem 7h0'56" 27d39'34"
Courtney Whelan
Cap 20h47'52" -21d3'51"
Courtney Wilkes #24
Gem 7h18'10" 33d21'35"
Courtney Wolfgang
Aqr 21h31'12" 0d32'26"
Courtney Worley
Uma 8h42'57" 62d10'53"
Courtney Wynn Rollins
Tau 5h19'5" 18d9'34"
Courtney Zgraggen
Tau 4h9'22" 16d31'2"
Courtney, My Lioness
Leo 11h42'36" 25d19'10"
Courtney, My Love
Lmi 10h29'46" 36d37'12"
courtneyhyde
Sco 17h1'15" -36d57'56"

CourtneyKGregory
Gem 7h20'48" 33d14'52"
Courtney-Lee
Tau 4h35'24" 13d36'15"
Courtney's Bisou
Cas 0h32'44" 57d26'23"
Courtney's First True Star
Cyg 20h53'4" 47d25'14"
Courtney's Galaxy of Stars
Ari 2h24'54" 10d47'2"
Courtney's Grace
And 23h50'17" 45d58'3"
Courtney's Light
Aql 20h19'51" 0d21'50"
Courtney's love
Her 16h34'48" 14d14'35"
Courtney's Mystery and Wonder
And 23h23'1" 48d38'7"
Courtney's Radiance
Uma 11h27'35" 36d19'14"
Courtney's Salve
Uma 9h33'10" 45d9'47"
Courtney's Star
Tau 5h10'40" 18d7'39"
Courtney's Star
Tau 4h8'31" 7d59'46"
Courtney's Star
Sgr 18h10'43" -18d44'55"
Courtney's Sun
Per 4h24'27" 52d45'23"
Courtni Stryker
And 0h17'2" 29d25'52"
Courtnick
Del 20h39'11" 12d10'12"
Courtnie
Cen 12h9'12" -51d38'14"
Courtniff
Ori 5h32'5" 7d12'43"
Courtsmyhero
Her 18h18'50" 22d4'18"
Courtster
Cap 20h39'2" -20d42'46"
Courynn Syl'vil'naius
Vir 13h44'17" -10d57'4"
Cous Cous
Vir 14h19'4" 1d42'1"
Cousin Edie #1
Crb 15h32'58" 32d6'22"
Cousin Jeffrey
Gem 7h47'55" 18d32'46"
Cousin Marion's Night Diamond
Umi 15h25'13" 71d18'13"
Cousin Rodney
Her 16h25'14" 18d9'19"
Cousin Steven David DiPietro
Per 3h16'43" 53d45'1"
Cousin TK Maximus
Ari 2h52'49" 10d51'22"
Couver Richard Smith
Leo 11h5'20" 6d38'21"
Cove Spring House
Uma 9h15'32" 65d51'55"
Covella Corter
Ari 2h14'21" 21d49'47"
Covin Bon Hassler
Umi 17h29'14" 83d23'38"
Cowboy
Equ 21h6'37" 10d1'36"
"Cowboy"
Boo 14h35'11" 51d8'1"
cowboy and monkey
Uma 9h38'41" 65d22'47"
Cowboy "Big Daddy" Mitch Wilson
Boo 14h32'47" 36d0'10"
Cowboy Bill
Lyn 7h36'59" 57d8'26"
Cowboy Bradley
Her 16h41'42" 6d49'48"
Cowboy Chris Constellation
Cep 23h53'4" 87d37'0"
Cowboy Pete
Cru 12h48'25" -62d17'35"
Cowboy Todd
Ari 2h43'57" 28d12'19"
Cowgirl Jeannie
Vir 12h43'36" 12d8'18"
Cowherd/May
Her 16h50'29" 31d13'46"
Cowpea
Ori 6h6'42" 17d21'23"
"CowTow"
Uma 11h50'25" 32d48'11"
Cox Digital Diamond
Uma 11h28'48" 33d23'28"
COX070723
Cap 20h31'41" -19d51'54"
Coy Lynn Richardson
Lib 14h56'46" -16d15'43"
Coy Wayne Brooks
Aql 19h46'44" 12d5'45"
Coye Lynch
And 23h16'0" 50d25'17"
Coyle's Star, the Love of My Life
Sco 16h7'12" -15d24'57"
Coyote Woman
Cma 6h40'6" -15d58'32"
Coy's Sunshine
Tau 4h22'59" 18d8'39"
Coz
Vir 12h42'21" 8d35'10"
Cozy
Gem 7h10'44" 26d46'34"
Cozy Derzic McNeal Jr.
Ori 5h3'38" 3d37'43"
CozyBear Errington
Sco 17h25'6" -37d42'57"
Cozzens
Lyn 8h25'54" 34d8'8"
Cp
And 0h58'17" 36d18'25"
CP+AO Forever
Psc 1h24'27" 30d56'52"
C.P. Beh-Beh Victoria
Ori 5h54'19" 20d58'45"
CP Nebula
Leo 10h36'2" 14d53'31"
CP Stationery Sdn Bhd
Ori 5h33'23" 2d49'20"
CP Susie
And 1h15'0" 42d4'31"
CP&TS7182006
Vir 14h37'1" 4d0'33"
CPA Realty - Robin Hirsch, Owner
Sco 16h51'4" -44d46'13"
CPETME100703
Ori 5h8'50" 15d21'14"
C-Pickens
Uma 11h33'42" 45d42'43"
Cpl. Brian C. Zimmerline
Uma 12h59'8" 57d36'55"
Cpl Christopher E Esckelson USMC
Per 4h20'58" 35d55'43"
CPL Gordie Lewis
Psc 23h53'32" 5d24'31"
Cpl. Joseph Heredia
Uma 11h21'2" 55d45'5"
Cpl. Kenneth John Voss
Ori 5h30'0" 7d17'39"
Cpl Kyle J Renehan
Per 3h45'47" 42d26'51"
Cpl. Nathan A. Schubert
Aqr 22h39'57" -9d4'39"
Cpl. Nick Ziolkowski
Uma 11h8'0" 36d52'19"
Cpl. Stephen Patrick Johnson
And 0h49'10" 40d3'25"
Cpl. Timothy Gibson
Ori 5h38'8" 2d0'34"
(CPSport) Your Chosen Starname Here !
Per 2h48'2" 50d22'32"
Cpt. Dorothy A Bartoletti R.N.,N.P.
Ari 2h6'21" 24d15'7"
Crabby Cakes Nebula
Cnc 8h4'44" 12d37'26"
Crabshell
Lyn 8h0'36" 40d51'40"
??!???& ??!??? & ??!Cracker
Uma 8h59'37" 57d53'16"
Crae Allen Justice
Sco 16h7'34" -17d0'59"
Craig
Crt 11h41'50" -17d58'33"
Craig
Tri 2h19'13" 35d54'21"
Craig
Aur 5h28'38" 46d34'23"
Craig Alan Hill
Aql 20h8'32" -0d44'58"
Craig Allen Lane
Ori 5h33'34" 10d5'9"
Craig Allen Stevens
Uma 13h32'11" 57d26'13"
Craig & Anah
Sgr 18h13'3" -19d16'45"
Craig and Ariel
Sgr 18h41'4" -17d13'31"
Craig and Deborah
Leo 11h26'27" 17d5'38"
Craig and Jen's Starlight Love
Cas 0h9'24" 55d16'0"
Craig and Kristeen
Leo 11h35'58" 26d19'44"
Craig and Leanne
Cyg 20h47'56" 47d8'55"
Craig and Maryanne Sibley
Gem 7h12'36" 32d18'56"
Craig and Rita
Leo 9h58'37" 17d20'42"
Craig and Sandra Forever Love Star
Per 3h49'25" 34d26'29"
Craig & Andrea's Valentines Star
Cru 12h51'24" -62d55'17"
Craig Andrew Chromoga
Uma 11h7'32" 60d2'52"
Craig Anthony Livingston Joyce
Tau 5h36'13" 23d9'9"
Craig Anthony Rees
Uma 11h39'30" 31d50'21"
Craig Arthur LeValley
Lib 15h17'43" -23d29'54"
Craig Austin Caldwell
Cap 21h38'13" -10d11'50"
Craig B. Clark, DTE
Gem 6h58'55" 25d8'22"
Craig B. Pettit
Uma 12h55'4" 63d16'17"
Craig B. Sargent
Psc 1h15'2" 25d41'33"
Craig Belisle
Her 18h47'54" 19d29'24"
Craig Benjamin Feser
Lib 14h42'10" -12d33'7"
Craig Bodie
Lib 15h27'19" -27d20'14"
Craig Borschmann
Cra 18h7'4" -37d29'31"
Craig Bray - The Star of the Cornish Pasty
Umi 13h23'34" 71d14'51"
Craig Butcher
Cep 22h59'54" 79d51'57"
Craig C. Keller
Lyn 8h0'5" 37d40'55"
Craig Cadogan
Psc 1h7'50" 27d3'27"
Craig Case
Her 17h28'41" 38d50'7"
Craig Cherry
Cyg 20h38'33" 34d46'56"
Craig & Clare Lee
Uma 9h58'12" 45d21'13"
Craig Cooney
Sco 17h14'14" -41d30'13"
Craig Cooper
Uma 9h39'54" 51d57'5"
Craig Corcoran
Per 3h8'51" 47d7'2"
Craig D. Bates - My Prince
Sgr 18h15'40" -19d12'35"
Craig David, Forever
Uma 8h32'30" 60d19'35"
Craig Dawson
Cep 20h49'37" 56d39'32"
Craig Dean Zimmerman II
Uma 11h35'0" 35d24'53"
Craig Ducote
Sco 17h8'12" -33d57'47"
Craig Dyrack
Aur 5h44'8" 41d12'57"
Craig Dysert
Crb 15h47'47" 36d17'12"
Craig E Trout
Her 16h37'33" 42d25'18"
Craig Edward Blanchard
Her 18h14'42" 26d59'19"
Craig Emerson
Ari 2h20'41" 24d48'45"
Craig Evan Pawling
Uma 10h31'42" 41d22'20"
Craig Evans - My Buddha Forever
Cru 12h40'33" -60d32'17"
Craig Family
Uma 11h53'36" 55d21'59"
Craig & Fay Forever xX
Cyg 20h24'23" 54d4'49"
Craig Forte
Cyg 19h16'38" 50d47'24"
Craig Fox
Sco 16h16'25" -18d5'25"
Craig Fredrick Bauer
Sgr 18h42'5" -28d10'35"
Craig Gillett
Peg 22h34'9" 17d11'44"
Craig H. Kelch
Lyn 9h3'38" 33d19'1"
Craig H. Robinson
Sgr 18h7'2" -25d38'36"
Craig Herbert Lundberg
Aql 19h15'0" 9d31'50"
Craig Hunter
Ori 5h37'51" 8d11'32"
Craig Iain Watt
Ari 2h41'43" 25d15'26"
Craig James
Cap 21h3'52" -15d23'4"
Craig James Cochrane
Uma 11h26'8" 58d34'42"
Craig James Holliman
Umi 16h17'20" 80d53'41"
Craig James Just
Per 3h14'36" 55d19'46"
Craig & Jen
Sco 16h9'29" -10d36'45"
Craig Joel Deuchie Lootis Delfino
Aqr 20h47'38" -9d44'27"
Craig John Campbell
Aqr 21h9'31" -10d56'44"
Craig Joseph Arledge
Per 4h5'52" 43d47'39"
Craig Joseph Laudicina
Her 17h37'46" 31d45'46"
Craig Joseph Maurice
Her 16h35'8" 38d34'23"
Craig Joseph Poland
Aqr 23h7'15" -13d48'11"
Craig & Kat's
Tau 3h59'44" 17d18'32"
Craig "Kevorkian" Rife 1973-2007
Uma 11h33'6" 57d25'36"
Craig Kissel
Aur 5h24'59" 41d14'59"
Craig Kleist
Aur 5h27'10" 47d21'50"
Craig Kleist
Uma 11h56'35" 52d7'52"
Craig & Kristina
Cyg 20h34'16" 37d42'10"
Craig & Larissa - 23rd March 2007
Cru 12h28'46" -60d49'9"
Craig & Lee - A Symbol Of Our Love
Tau 4h24'27" 10d47'49"
Craig Lee McCrystal
Cnc 8h49'33" 11d44'56"
Craig LeFoe
Uma 12h6'40" 62d59'1"
Craig & Lisa
Cyg 19h33'21" 44d0'23"
Craig Little
Umi 15h12'14" 72d28'59"
Craig Louis Blaise Cainkar
Lib 15h54'56" -7d0'39"
Craig Luis Woehr
Gem 6h33'48" 12d35'48"
Craig & Lyndsay "Eternal Love Star"
Ori 5h39'1" -2d9'33"
Craig & Lyndsey
Cyg 21h24'47" 32d39'38"
Craig Lynn Carey III Always Loved
Umi 15h26'45" 73d8'42"
Craig MacDonald
Per 3h23'54" 49d40'5"
Craig MacKenna
Dra 18h37'32" 57d57'0"
Craig Mackenzie Chisholm
Per 3h52'23" 35d12'35"
Craig & Marina Ginn
Uma 10h27'33" 58d38'11"
Craig Marshall
Cru 12h27'0" -61d31'16"
Craig Martin Hapanovich
Uma 13h55'4" 58d27'59"
Craig Matthew DeSimone
Aur 5h44'25" 42d52'8"
Craig Matthew Schulz
Uma 8h37'50" 62d28'13"
Craig Merrick Willuweit
Uma 8h26'17" 63d42'35"
Craig Michael Carignan
Per 3h14'24" 49d16'17"
Craig & Michele Morrissey
Dra 17h50'51" 63d37'25"
Craig Morton
Aql 19h31'37" 11d19'34"
Craig Nathan
Cru 12h30'32" -61d21'57"
Craig Nation's Class of 2005
Dra 19h27'31" 67d48'27"
Craig Nicolas
Lib 15h37'31" -20d7'10"
Craig Noe
Uma 11h33'15" 30d45'53"
Craig Nottingham I Love You Forever
Lyr 19h17'0" 28d37'32"
Craig P. Christ
Cap 20h27'3" -18d16'32"
Craig P. Cochran
Sgr 18h36'30" -22d34'16"
Craig P. Rieders
Leo 11h6'49" 20d29'48"
Craig Paul Henchey
Cma 7h15'47" -16d37'24"
Craig Phillip Forster
Dra 17h52'34" 54d0'28"
Craig Queen
Cep 3h54'3" 84d27'27"
Craig Quigley
Uma 9h7'39" 69d8'18"
Craig R. Stauffer
Cap 21h54'47" -14d22'47"
Craig Robert Sincock II
Ori 6h11'39" 6d4'37"
Craig Roger Grovic
Aur 5h39'40" 43d43'48"
Craig Rogers Indyke Jr
Tau 4h19'33" 26d42'0"
Craig Ross Wilson "My Bright Star"
Her 18h28'58" 12d22'42"
Craig Royston Dicken
Uma 9h59'7" 66d13'29"
Craig Salazar
Sco 16h14'26" -16d27'23"
Craig Shannon Burden (Paw Paw)
Uma 9h26'15" 45d30'48"
Craig Standish Higdon
Ori 5h35'55" -3d33'48"
Craig Steven Bandalin
Psc 23h54'39" -0d17'17"
Craig Steven DX
Sco 16h8'57" -28d58'51"
Craig Steven Schmitz
Cap 20h58'9" -18d12'9"
Craig Stoker
Cyg 20h0'30" 41d30'8"
Craig Summers
Aql 19h8'2" 13d1'11"
Craig & Tammy
Cyg 19h59'22" 39d39'33"
Craig & Teresa Scott 09-18-2004
Cyg 21h20'41" 33d37'21"
Craig Thomas Joseph White
Ori 5h10'51" 8d51'19"
Craig Turner
Cen 13h47'23" -43d51'25"
Craig Van Baal
Sco 16h44'40" -29d28'5"
Craig Waterman
Gem 7h45'54" 34d26'57"
Craig Wayne George
Uma 9h18'20" 69d59'24"
Craig William
Her 16h36'48" 33d50'51"
Craig William Baldwin
Cnc 9h17'33" 14d11'59"
Craig William Bauer
Her 17h46'47" 48d34'36"
Craig William Takahama
Her 17h39'43" 27d49'12"
Craig y Dalia Para Siempre
And 0h47'30" 40d24'24"
Craigalie
Ori 5h36'55" -0d36'3"
CraigandTracy
Sgr 18h15'15" -22d8'55"
Craige Micelle Jephsonius
Ori 6h2'46" 18d28'17"
Craiger
Ari 2h47'56" 14d50'12"
Craiger
Lac 22h45'16" 48d49'20"
Craiger-24
Sgr 17h58'2" -29d57'36"
Craigle-McDaigle
Cnc 8h32'19" 16d34'24"
CraigLovesAmber
Lib 14h29'26" -17d7'41"
Craig's Obsession
Cas 1h42'1" 60d55'49"
Craig's Star
Vir 14h42'40" 4d9'36"
Craig's Star
And 23h1'13" 47d2'27"
CraigSandraDamonJustinRoanValentine
Cru 11h59'9" -63d4'10"
Craig-The Shining Light In My Night
Ori 5h28'33" -4d6'23"
Cram Cram
Her 17h43'27" 19d4'37"
Crämer, Josef
Uma 10h39'16" 47d11'13"
Crane Beach Hotel
Umi 15h30'57" 67d56'25"
Cranson Franças Johnson
Uma 12h20'21" 58d5'35"
Crapino
Uma 11h10'46" 46d42'50"
Craquo (Hugues Bourgeault)
Cas 1h57'4" 65d27'21"
Craraland
Tau 4h28'49" 21d14'45"
CRASH
Per 2h24'20" 52d27'3"
Crash
Dra 19h4'26" 79d18'38"
Crash
Aqr 22h12'36" -1d6'51"
Crash
Lib 15h12'59" -15d39'7"
Crash - Star of Strength & Love
Uma 9h0'42" 50d58'37"
Crash Trick 280682
Cnc 8h58'31" 24d13'57"
Crashaundra M. Goins
Lyr 19h18'21" 29d31'55"
Crater Ketubah
Crt 11h36'47" -7d23'36"
Craton
Dra 19h51'46" 75d31'37"
Craven
Dra 15h58'40" 59d40'19"
Crawford Vincent Kier
Uma 9h53'12" 51d28'53"
Crayton & Nikki
Tau 5h37'54" 22d40'32"
Crazee Man
Her 17h56'3" 14d31'44"
Crazy
Cnc 8h12'45" 12d42'10"
"Crazy"
Cap 20h57'32" -20d6'41"
Crazy Beautiful
Lyn 6h17'53" 58d52'6"
Crazy Beautiful
Vir 12h48'27" 3d51'13"
Crazy Cake
Uma 11h3'57" 51d55'51"
Crazy Dee
Tau 4h18'7" 4d56'58"
Crazy Diamond
Crb 16h3'56" 26d49'20"
Crazy Frank
Aqr 22h49'38" -21d36'24"
Crazy Horse
Sgr 19h1'38" -21d21'12"
Crazy Horse Mac
Peg 21h57'51" 23d57'51"
Crazy Laur
Leo 11h24'29" 9d52'33"
Crazy Lina
Cap 21h20'15" -19d20'50"
"Crazy Love"
Crb 15h40'12" 37d37'26"
Crazy Pam Ayers
Psc 1h28'23" 25d50'36"
Crazy Papa
Boo 14h23'20" 22d46'16"
Crazy Queso
Lyn 7h55'56" 34d2'51"
"Crazy" Steve Dalton
Cnc 8h19'5" 8d57'1"
CrazyDaizy86
Tri 2h21'6" 33d17'5"
crazydiamond
Gem 6h30'59" 24d0'11"
Crazyeyes Erin Smith
Aqr 23h2'14" -17d22'27"
Crazynun
Sgr 18h29'37" -16d48'9"
crazyrj
Cyg 21h36'43" 44d33'34"
C.R.C. + A.J.J. 8-23-2003
Lyr 18h49'17" 43d19'55"
Cre8eve Brenda
Ari 2h11'43" 24d42'43"
Cream Puff
Cyg 21h59'12" 49d42'47"
Creata Promotion
Ori 5h29'0" -4d40'26"
Creative Gymnastics Center
Uma 8h32'26" 61d36'39"
Creative&PerformingArtsHighSchoolMR
Peg 23h26'41" 33d48'33"
Credence Jacob Philip Shaw
Aql 20h10'44" 14d6'50"
Credere
Aqr 22h35'21" 2d26'15"
Creed "Lucky" Littlefield
Tau 4h32'9" 12d39'57"
CreedOstler2005
Per 3h3'52" 31d36'34"
Creekside at Shallowford
Cyg 21h13'57" 34d36'23"
Creeky
Uma 11h10'6" 44d58'1"
Crépin Christian Bayonne
Cep 23h39'32" 74d18'56"
Crescendo
Aql 19h49'58" -0d14'31"
Crescent DeMieri
Aqr 21h31'40" 2d4'11"
CresStar
And 2h30'46" 49d18'25"
Crestwood Challengers
Cas 0h50'47" 60d26'31"
Crestwood Scarlet Guard
Cas 1h24'0" 61d21'23"
Creswell Stanley Joynt
Cru 12h21'28" -59d27'28"
Crew 3, 120 Sqn, RAF Kinloss
Ori 5h53'54" 17d40'24"
Crew Elijah Payne
Uma 9h42'41" 55d45'2"
Crew Lanning Hensley
Cam 9h18'56" 73d53'34"
Crew Lanning Hensley
Per 3h29'12" 51d34'56"
Cri 41 kg
Cyg 21h27'15" 37d2'8"
Cri 64
Lyr 18h57'45" 27d54'54"
Cri & Lela
Uma 9h59'13" 57d57'9"
Criag Robert Elston
Leo 10h56'59" -0d51'13"
Crias
Cnc 8h0'40" 24d1'0"
Cribees
Umi 14h42'1" 73d52'44"
Cric
Sco 17h24'29" -42d51'50"
Cricca
Uma 9h8'21" 55d19'54"
Criccy Nugent
Peg 23h44'11" 12d15'1"
Cricia
Gem 6h42'59" 34d32'4"
Cricket
And 0h48'46" 42d10'12"
Cricket
Lyn 7h21'27" 48d56'9"
Cricket
Cyg 21h14'5" 45d32'26"
Cricket
Ari 2h3'5" 18d31'20"
Cricket
Tau 4h37'11" 18d13'34"
CRICKET
Sgr 18h25'32" -32d5'59"
Cricket
Cam 4h16'16" 69d43'16"
Cricket
Cam 3h17'56" 60d51'16"
Cricket
Uma 10h10'7" 64d12'21"
Cricket
Uma 9h15'33" 61d33'34"
Cricket
Cyg 20h30'0" 52d44'48"
Cricket Aldrich
Lmi 10h28'21" 36d0'42"
Cricket Reed
Ori 6h16'46" 15d42'23"
cricket stacy
Peg 22h27'33" 33d4'46"
Cricy
Cyg 21h6'0" 38d54'42"
Crikett's Shining Star
Sge 19h50'22" 18d28'46"
Crikey Crocs rule Steve Irwin
Cru 12h28'15" -59d2'14"
Crimson Collector, TJM
Eri 3h51'47" -0d22'45"
Crimson Marie Napier
Lyr 18h41'55" 35d30'3"
Crimson Tide
Her 18h3'20" 14d55'46"
Cripple Creek
Leo 9h24'5" 24d37'16"
Cris
Gem 6h45'6" 21d20'44"
Cris Alan Randall
Cnc 8h16'18" 22d26'58"
Cris Bent
Cnc 8h37'51" 28d29'7"
Cris & Carrie Breneiser-Castellanos
Cnc 7h56'41" 17d6'32"
Cris -M the Q- LaCalle
Psc 0h41'29" 18d11'20"
Cris&Raquel14
Cyg 21h17'26" 43d50'15"
Cris Saponara
And 0h42'15" 27d3'13"
Cris Van Gorp
Psc 1h43'24" 5d27'7"
crisangthamum
Leo 10h9'50" 25d42'9"
Crisanna
Uma 9h19'21" 44d47'23"
Crisbel Valero Gracia
Lyn 7h53'16" 34d12'15"
Criselda Miller
Mon 6h28'51" -9d13'46"
Crisha
Cyg 21h31'12" 31d0'58"
Crisha Marie
Leo 10h42'58" 7d46'16"
Criska
Gem 6h55'44" 14d39'32"
Crispin Nicholas Vick
Her 17h24'20" 35d22'28"
Crispus Al's Three Vir Plaga Domus
Per 3h12'12" 52d40'42"
Crissa's Night Light
Cnc 8h9'7" 26d32'27"
Crissie Trigger
Cas 0h49'29" 57d29'39"
Crissy
Ori 4h51'58" 7d5'34"
Crissy
Cap 21h35'57" -9d6'39"
Crissy
Sco 16h53'28" -36d26'30"
Crissy and Jason
Lyn 8h11'15" 45d34'19"
Crissy C. Delatte
Cap 21h45'24" -13d6'42"
Crissy Sheets
Per 3h28'22" 33d52'5"
Crissy's Firefly
Peg 21h40'19" 27d29'55"
Crista
Uma 9h12'14" 56d40'25"
Crista Ann Holt
Aqr 20h52'16" -13d36'20"
Crista DeRoma
Cas 0h20'17" 62d59'48"
Crista Noel
Peg 22h29'15" 4d10'53"
Cristabel
Cnc 8h50'10" 25d31'54"
Cristal
Dra 15h28'23" 56d40'17"
Cristal A. Perkins
Vir 13h29'10" 11d10'40"
Cristal Dawn
Ori 5h14'6" 1d19'32"
Cristal Luna
Psc 1h31'10" 20d12'51"
Cristal Nichole
Ari 2h42'6" 16d43'6"
Cristal Palmatier
And 0h11'17" 45d50'28"
Cristales
Cam 5h40'31" 73d30'45"
Cristel Raynae Rivera
Aqr 22h5'46" -7d44'26"
Cristen And Brandon Forever
Psc 0h26'25" 3d37'44"
Cristen Barlow
And 1h1'16" 39d48'22"
Cristen Danielle Scott
Uma 11h25'19" 39d1'16"
Cristen Evangelista
Gem 7h8'11" 34d39'0"
Cristen Leigh
Cas 0h19'37" 59d34'39"
Cristen Marie Wolfe's star
Lmi 10h36'58" 34d11'26"
Cristen Puhl
Crb 15h49'58" 36d59'29"

Cristeta Cortez
Uma 11h56'35" 64d57'1"
Cristi
Uma 8h55'18" 60d1'51"
Cristi
Ori 6h19'50" 16d3'26"
Cristi J. Carlini
Sco 16h9'17" -26d10'20"
Cristi Janes
Com 12h35'28" 16d59'23"
Cristi Ralleo
Sco 17h36'1" -33d4'37"
Cristian
Vir 14h23'20" 3d17'58"
Cristian Dario Zeppieri
Sgr 18h30'46" -22d42'24"
Cristian e Chiara
Cam 4h10'50" 53d33'56"
Cristian e Federica
Lyr 18h42'18" 31d29'13"
Cristian Fernandez Captain Petrell
Vir 13h38'35" -15d37'18"
Cristian Scott
Sgr 19h10'25" -30d2'13"
Cristiana
Dra 10h17'18" 78d12'49"
Cristiana Damiano
Sco 17h21'7" -32d20'39"
Cristiana DiNardo
Vir 13h5'23" 14d2'0"
Cristiana Gabriella
Aqr 21h41'32" -0d30'23"
Cristiana Meroni
Ori 6h2'6" 11d31'33"
Cristiana Numico
Peg 22h3'30" 19d39'14"
Cristiana&Renato
Cam 5h34'23" 71d30'27"
Cristiana Vici
Cas 0h31'13" 62d20'24"
Cristiane Fortunato Pereira
Sco 16h10'36" -13d20'16"
Cristiani Byrne
Uma 9h15'30" 49d49'54"
Cristianne Trucks
Lyn 8h6'27" 34d16'25"
Cristiano Sarmento
Cnc 8h10'48" 10d49'48"
Cristie A.
Ari 3h2'41" 14d8'32"
Cristie Anne McCumber
Ori 6h5'39" 17d10'59"
Cristie Pateman's Heavenly Body
Cru 12h49'52" -57d35'33"
Cristin
Uma 8h53'19" 54d11'39"
Cristin Anne Gadue
Vir 13h15'6" 6d56'21"
Cristin Cornett
Ari 2h21'41" 20d31'20"
Cristin L. Cooley
Psc 0h41'32" 8d4'3"
Cristin Stafford
Cap 20h28'50" -14d20'7"
Cristin Whetten
And 2h22'41" 49d19'23"
Cristina
Per 2h57'31" 48d39'37"
Cristina
And 1h14'14" 45d53'54"
Cristina
And 23h42'52" 47d47'18"
Cristina
Tau 4h34'44" 30d12'51"
Cristina
Cyg 19h55'12" 33d6'0"
Cristina
Lyn 9h21'34" 38d34'26"
Cristina
Aur 6h51'5" 44d45'29"
Cristina
Cnc 9h3'15" 27d56'3"
Cristina
Com 12h34'55" 26d50'21"
Cristina
Sco 16h8'46" -13d23'55"
Cristina
Lib 15h45'5" -17d1'8"
"Cristina"
Cap 20h26'58" -19d29'38"
Cristina
Cam 5h53'7" 59d24'37"
Cristina
Cas 1h38'13" 62d48'28"
Cristina
Psa 22h31'57" -31d48'53"
Cristina 1976
Lac 22h17'15" 54d44'8"
Cristina 23/10/2003
Ori 5h58'46" 6d30'58"
Cristina A Brox
Uma 10h21'59" 45d38'57"
Cristina A. Donastorg
Lyn 8h53'44" 45d45'58"
Cristina Alexandra Runnman LaMar
And 1h17'25" 36d35'12"
Cristina and Julien Forever in Love
Sgr 19h31'12" -12d24'54"
Cristina Ashley Canty
Gem 6h44'30" 13d11'8"
Cristina Baby Brown Eyes
Mon 7h57'4" -4d28'56"
Cristina (Bacelis-Bush)
Sco 16h13'36" -8d38'49"
Cristina Barbara Ornelas
Sco 17h32'3" -39d52'38"
Cristina Canales
Cap 21h27'7" -9d46'20"
Cristina Carlomagno
Her 18h37'47" 18d19'19"
Cristina Carmona Botía
Ori 6h15'15" 2d2'26"
Cristina Chozas Agudo
Lib 15h39'43" -17d4'29"
Cristina Colombo
Peg 22h53'4" 26d12'42"
Cristina Coltelli
Peg 23h10'53" 31d5'57"
Cristina Dall' Olmo
Uma 10h40'48" 71d34'56"
Cristina Daniela Howard
Uma 14h3'28" 61d6'51"
Cristina Davoli
Cam 4h22'2" 56d49'11"
Cristina de la Portilla
Vir 11h53'48" 3d19'59"
Cristina del Rey Velasco
Aqr 23h6'8" -9d7'10"
Cristina Diaz
Cap 21h41'28" -24d2'21"
Cristina Di'Meglio
Gem 7h39'59" 15d49'27"
Cristina E. Glendenning
And 0h53'43" 40d45'1"
Cristina E. Martinez
And 1h35'37" 50d32'7"
Cristina F. and Philippe C.
Ari 2h21'10" 24d13'18"
Cristina Gavasheli
And 0h12'3" 34d30'34"
Cristina Gaxiola Angulo
Sco 17h52'13" -42d59'27"
Cristina Guida
Aqr 22h52'0" -9d11'54"
Cristina&Hardy
Aql 19h33'20" 10d3'57"
Cristina Isabel Martin Collazo
Aqr 22h59'49" -13d57'9"
Cristina James
Lib 14h52'13" -3d16'48"
Cristina & Joey
Sge 19h50'32" 18d5'41"
Cristina Krukar
Vir 14h10'28" -14d52'30"
Cristina Louise Roney
And 1h14'52" 42d15'38"
Cristina Lujan
Sgr 19h47'13" -32d47'14"
Cristina M. Clowers
Lib 15h43'15" -29d19'7"
Cristina Margarita Santiago Bravo
Psc 23h35'16" 2d24'11"
Cristina Maria Benitez Roman
Uma 9h46'30" 67d39'44"
Cristina Maria Henry
Gem 6h33'45" 13d52'0"
Cristina Maria Turcu
Sco 17h49'7" -33d28'53"
Cristina Marie Setzer
Gem 7h31'34" 29d18'20"
Cristina Martin Ludeña
Cas 1h58'35" 60d16'20"
Cristina Martinez
Per 3h27'42" 34d35'33"
CRISTINA MATA
Sgr 19h21'44" -16d42'28"
Cristina McCloud
Sgr 18h43'4" -33d48'39"
Cristina Meijide (Beautiful Angel)
Del 20h37'16" 4d27'43"
Cristina Moscaliuc
And 0h38'32" 28d9'1"
Cristina Mueller
Cam 7h55'23" 66d8'16"
Cristina Navarro Brown
Psc 1h19'39" 23d26'45"
Cristina Noel
Leo 10h38'22" 22d38'50"
Cristina Ojeda Martin
Cas 0h46'17" 64d50'35"
Cristina Olivieri
Cyg 19h42'35" 35d16'2"
Cristina Orona's Star
Psc 1h1'36" 3d24'15"
Cristina P.
Cas 0h48'7" 56d10'5"
Cristina Paganoni
Cam 3h30'50" 63d29'22"
Cristina Pallarol
Tau 4h13'55" 6d39'26"
Cristina Patricia Nunez
Ari 2h24'11" 25d23'45"
Cristina Perez
Sco 16h2'22" -23d40'1"
Cristina Perrone
Cas 23h2'3" 54d59'40"
Cristina (P-nut)
Cam 3h59'18" 69d4'54"
Cristina Ramirez
Ori 5h37'21" -0d45'46"
Cristina Rodriguez
Cyg 20h41'4" 36d41'13"
Cristina Ryzyi
And 2h13'17" 46d4'54"
Cristina Saenz
And 1h2'40" 42d23'54"
Cristina Saez
Vir 14h30'45" -2d28'28"
Cristina Sanchez
Lib 15h30'46" -6d28'11"
Cristina Sigillo
Uma 9h9'42" 51d33'45"
Cristina Stanfill
Ari 3h20'37" 23d58'38"
Cristina Suárez
Cas 1h47'49" 60d19'5"
Cristina Tamez Briseño
Cas 1h17'35" 65d2'25"
Cristina Valedes
Ori 6h1'19" -0d26'36"
Cristina VT
Cap 20h14'14" -8d43'47"
Cristine
Ari 3h29'7" 23d59'55"
Cristine
Lyn 9h29'48" 40d34'42"
Cristine Bada
Tau 5h44'52" 22d14'5"
Cristine Durieux
And 23h34'2" 47d20'1"
Cristine Pastore
Lib 14h53'56" -4d54'2"
Cristinita
Aql 19h41'10" -0d4'54"
Cristi's star of love
Uma 10h39'11" 71d0'46"
Cristo
Uma 13h57'2" 48d32'56"
cristo4o
Umi 13h57'53" 71d27'35"
Cristobal Rodriguez Melguizo
Peg 21h40'53" 23d19'10"
Cristofer
Cep 0h17'10" 71d55'56"
Cristoforino GILARDINO
Lib 15h19'14" -14d19'37"
Cristopher Cooper
Tau 4h32'33" 21d11'48"
CRISTOPHER JR.
Her 18h4'13" 25d26'52"
cristy allison
Sco 17h0'38" -35d51'0"
Cristy Becker
And 0h20'56" 42d6'4"
Cristy "Dreams Come True"
Uma 13h54'36" 61d46'10"
Cristy Gutierrez
Sco 16h26'56" -18d41'32"
CRISTY & IGOR
Vir 13h12'50" 11d45'15"
Cristy Lee
And 0h41'24" 26d49'11"
Cristy Maldonado
Lib 15h29'33" -8d46'52"
Cristy Marie Calvert My Love
Leo 9h42'33" 24d10'0"
Cristy Marie Fernandez
Lup 15h38'39" -40d35'46"
Cristy White
Cyg 19h21'5" 29d33'58"
Cristy Wright
Oph 17h27'28" -0d30'41"
Cristyn
Cam 4h17'30" 73d29'56"
Criswell Townsend Yantis
Aql 19h41'46" -0d0'47"
Crisy Cris "CyCi"
Col 5h44'56" -30d12'29"
Criticion
Gem 7h24'56" 28d9'57"
Critter
Cnv 12h36'19" 33d56'44"
Critter Herkert
Cap 21h45'25" -10d15'56"
Crittle Bug
Uma 9h27'11" 60d0'15"
CRK 82784
Vir 13h5'9" 0d3'22"
Croatian Woman 32
Lyn 7h53'23" 33d55'1"
Crockett
Uma 11h14'28" 42d21'28"
Crocketts Dream
Ori 5h24'51" 5d55'13"
Croia Dunlea Treanor
Peg 21h34'11" 10d53'35"
Croire
Lyn 9h1'8" 37d44'5"
Cronos
Dra 15h35'32" 57d16'29"
Crosby
Tau 5h6'10" 26d38'45"
Crosby Lee Strange
Uma 13h40'45" 54d7'54"
Crossan Family
Aql 19h47'58" -0d12'8"
Crosty2004
And 23h52'47" 40d41'12"
CRP Silver Storm
Sgr 18h17'10" -18d33'57"
Cruella
Lyr 18h51'51" 44d34'55"
Cruellen
Sco 16h35'16" -31d18'10"
Cruetopia
Gem 7h26'25" 26d30'10"
Cruise High —Rollie & Denise—
Ori 6h7'39" 21d0'6"
Crunchysiegus
Cap 21h15'33" -23d30'33"
crunker
Sgr 18h22'9" -32d33'8"
Crusawacky
Leo 11h25'17" 8d58'43"
Crush
Aur 6h59'52" 41d26'5"
Crush
Uma 10h47'4" 55d56'43"
Crusty
Uma 9h41'1" 72d52'22"
Cruz 905
Aur 5h53'14" 44d14'37"
Cruz De Isabel
Uma 9h15'38" 56d27'46"
Cruz Stahl
Tau 5h45'41" 19d20'16"
Cruz Terrazas Martinez
Uma 11h56'33" 63d34'25"
Cruz,Evelyn Margaret Cruz (Lorian)
Uma 10h7'44" 46d9'22"
Cry's Luck
Uma 11h3'47" 54d41'25"
Crysicca Lee
And 2h10'8" 45d13'8"
Cryspy
Cap 20h32'14" -13d57'3"
Crysta Marie White
Uma 11h8'47" 50d25'10"
Crystal
Uma 11h14'1" 50d21'35"
Crystal
Uma 10h30'24" 51d47'1"
Crystal
Lyn 9h14'53" 45d49'15"
Crystal
Cyg 20h7'48" 52d19'46"
Crystal
And 23h8'29" 51d54'20"
Crystal
Cam 4h30'1" 54d9'48"
Crystal
And 1h19'5" 47d38'54"
Crystal
Per 3h10'29" 47d58'54"
Crystal
And 23h49'23" 37d21'27"
Crystal
And 0h55'27" 42d7'44"
Crystal
Cyg 19h59'59" 31d49'44"
Crystal
Leo 11h28'34" 10d29'37"
Crystal
Tau 3h40'42" 2d33'11"
Crystal
Vir 13h6'56" 5d19'59"
Crystal
Ari 2h21'15" 10d37'26"
Crystal
Ori 4h50'2" 11d25'29"
Crystal
Ari 2h53'8" 27d25'35"
Crystal
Ari 2h22'46" 23d47'28"
Crystal
Peg 22h51'8" 15d28'53"
Crystal
Cnc 8h5'28" 24d1'12"
Crystal
Leo 11h38'14" 25d58'42"
Crystal
Crb 15h28'30" 28d5'42"
Crystal
Sgr 18h21'12" -18d29'4"
Crystal
Cap 21h3'40" -15d30'13"
Crystal
Aqr 23h44'56" -3d37'41"
Crystal
Aqr 23h1'25" -6d21'25"
Crystal
Lyn 8h23'45" 56d38'5"
Crystal
Lyn 7h18'59" 53d39'50"
Crystal
Cas 1h40'41" 62d59'19"
Crystal
Uma 11h48'46" 60d11'9"
Crystal
Cam 6h21'34" 63d19'28"
Crystal
Uma 9h2'33" 71d44'30"
CRYSTAL
Sco 16h31'53" -26d36'17"
Crystal
Sgr 18h53'43" -28d54'29"
Crystal A. Espinoza
Cas 0h59'37" 50d9'17"
Crystal A Wright
Lib 15h13'46" -6d4'9"
Crystal & Adrian Santoyo
Cyg 21h34'38" 44d34'1"
Crystal Agnew
Cap 21h53'15" -9d54'6"
Crystal and Henry's Engagement
Cyg 20h2'54" 52d1'55"
Crystal and Joshua Griffin
Ari 3h7'2" 12d59'33"
Crystal and Mathew
And 0h51'31" 35d49'16"
Crystal and Matt
Her 17h56'25" 21d24'18"
Crystal and Scott's Piece of Heaven
Umi 15h53'30" 79d35'17"
Crystal and Steve Speigl's Star
Cyg 19h43'19" 37d11'2"
Crystal and Wissam
Dra 11h38'15" 68d48'24"
Crystal Angelica Borrero
Crb 16h11'29" 37d6'43"
Crystal Ann
Sco 16h15'58" -14d36'42"
Crystal Ann
Sco 16h57'5" -42d1'47"
Crystal Ann Chambers
Ari 3h7'36" 13d28'26"
Crystal Ann Cooper
Leo 10h41'28" 14d17'46"
Crystal Ann & Don
Cyg 21h52'42" 48d18'20"
Crystal Ann Irwin
Cas 0h56'39" 54d15'36"
Crystal Ann Lutz Barnes
Cap 21h56'7" -24d30'4"
Crystal Ann Marie
Cnc 8h34'50" 22d14'3"
Crystal Anne
Ari 2h15'8" 24d57'53"
Crystal Anne
Psc 1h18'15" 29d5'3"
Crystal Anne
Cas 1h6'30" 54d18'45"
Crystal Anne
Crb 16h8'25" 35d27'23"
Crystal Anne
And 2h2'18" 42d49'5"
Crystal Anne
Lib 15h48'27" -12d22'29"
Crystal Anne Slezak
Psc 0h33'41" 4d19'10"
Crystal & Anthony 2006
Ari 2h10'28" 14d17'45"
Crystal Aukett
Ari 2h35'25" 31d2'2"
Crystal B. Ennis
Cap 21h29'20" -15d24'39"
Crystal "BabyBoo" Nixon
Sgr 19h33'13" -16d57'1"
Crystal Baker
Aql 19h48'11" -0d6'30"
Crystal Banasiak
Aqr 22h41'26" -0d57'35"
Crystal Bell
Cnc 8h5'42" 7d18'36"
Crystal Bissonette
Lyr 18h34'3" 33d29'45"
Crystal Booth
And 23h13'11" 41d13'32"
Crystal Brandgard
Uma 9h26'3" 50d2'19"
Crystal C. Daugherty
Crb 15h50'34" 34d11'25"
Crystal Carey
Lmi 10h53'47" 36d31'42"
Crystal Carr - The Angel Star
Gem 6h51'56" 13d56'44"
Crystal Case
And 1h56'22" 37d9'35"
Crystal Champtal
Cap 21h39'38" -12d23'9"
Crystal CJRV Rivera
And 23h49'21" 37d21'47"
Crystal Clear
Gem 6h38'12" 27d13'18"
Crystal Collins
Cnc 8h8'8" 13d38'30"
Crystal Cotell
And 0h24'25" 42d38'25"
Crystal Cove Hotel
Lmi 9h54'57" 41d0'43"
Crystal Crain
Lyr 18h58'19" 26d16'15"
Crystal D Bulla
Ori 4h51'53" 7d50'9"
Crystal D. Moore
Cas 0h34'42" 54d23'50"
Crystal D. Quebbeman
Dra 15h57'0" 59d51'8"
Crystal D. Ribail
And 0h39'47" 25d55'2"
Crystal Dabbs
Crb 15h25'57" 30d30'22"
Crystal Dale McNamara
Lyn 9h3'13" 36d59'52"
Crystal Danielle Barnosky
Tau 4h8'3" 6d49'4"
Crystal Dawn
Ori 5h40'45" 3d8'18"
Crystal Dawn
Peg 21h56'57" 13d49'16"
Crystal Dawn
Cap 20h47'18" -20d41'0"
Crystal Dawn Crees
And 1h35'53" 39d8'45"
Crystal de L'Amour
Cam 7h19'58" 65d3'58"
Crystal Dechene
Sco 16h12'28" -22d24'51"
Crystal Deeana
Ari 2h26'38" 24d14'57"
Crystal Dentis
Lyn 9h6'25" 37d25'8"
Crystal Diamond Rose
Cyg 19h35'18" 30d59'47"
Crystal Divorne
Per 3h44'44" 45d18'2"
Crystal Elyse Meske
Uma 9h52'55" 52d28'32"
Crystal Ewa Stanislawek
And 0h15'11" 29d35'33"
Crystal F Long
Psc 0h36'30" 5d9'35"
Crystal Faith
Gem 6h32'41" 24d20'29"
Crystal Faye Curp
Cnc 8h38'56" 9d22'32"
Crystal Fleury
Vir 14h6'48" 2d40'33"
Crystal Follis
Vir 13h29'51" -6d4'56"
Crystal G Blakeney
Lyn 7h43'1" 37d51'50"
Crystal Gale DeSmet
Cap 20h27'40" -12d5'5"
Crystal Galello
Psc 1h18'8" 25d51'33"
Crystal Genevieve
And 0h6'58" 40d13'22"
Crystal Gilbert
Cnc 8h21'24" 31d21'12"
Crystal Gold
And 23h29'28" 43d51'40"
Crystal Gonzalez
And 23h3'27" 47d2'44"
Crystal & Greg Together at Last
Cnc 8h45'24" 21d46'10"
Crystal Grimm
Ari 2h4'34" 23d51'13"
Crystal Hall,tsiml.J.R.
Cnc 8h26'32" 26d17'32"
Crystal Hayes
Ori 5h33'5" 6d33'26"
Crystal Heffernan
Vir 11h48'3" 8d52'36"
Crystal & Hunter Adams
Ori 5h52'20" 8d57'54"
Crystal Isela Tran
Mon 7h17'43" -0d19'11"
Crystal & Jarrick
Lib 15h39'30" -28d11'16"
Crystal & Jayson
Cyg 20h24'11" 43d19'11"
Crystal Jean
Dra 19h38'40" 61d27'43"
Crystal Jiron
Vir 13h14'51" -18d58'27"
Crystal JND
Aql 18h50'22" -1d30'15"
Crystal Joy Shire
Gem 7h9'22" 32d36'16"
Crystal June Longoria
Gem 7h40'57" 21d24'21"
Crystal June Storey
Tau 5h50'13" 23d16'7"
Crystal & Karl
Cyg 21h25'46" 53d5'16"
Crystal Kasper
And 2h6'31" 43d13'6"
Crystal "Kitten" Jester
Vir 12h11'50" -2d34'57"
Crystal Kochka
And 22h59'7" 52d12'55"
Crystal Kwon
Vir 14h16'27" -16d13'56"
Crystal L. Ashton
Sco 17h49'2" -42d23'1"
Crystal L. Chinn
And 1h18'44" 36d20'33"
Crystal L Thomas
Cap 20h36'13" -10d5'3"
Crystal Lee AnnBradford
Sco 17h43'8" -33d24'8"
Crystal Lee Turner
Lyn 7h17'24" 52d41'11"
Crystal Lee Wright
Cnc 8h31'9" 23d7'8"
Crystal Lillian
Uma 9h35'28" 71d27'16"
Crystal Lina Lopez
Vir 12h13'21" -3d33'53"
Crystal Linette Johnson
Cap 20h27'29" -9d56'15"
Crystal Loo
Crb 15h25'13" 31d45'18"
Crystal Love Walker
Ori 6h12'51" 15d5'35"
Crystal Loves Jason
Cyg 19h44'42" 43d55'54"
Crystal Lynn
Cas 1h13'28" 59d20'30"
Crystal Lynn
Lyr 18h51'55" 31d48'20"
Crystal Lynn
Psc 0h13'30" 10d27'57"
Crystal Lynn
Gem 7h48'48" 21d48'8"
Crystal Lynn
Sgr 18h16'16" -20d15'22"
Crystal Lynn
Cam 4h36'29" 66d4'13"
Crystal Lynn
Uma 11h14'25" 53d11'5"
Crystal Lynn Burke
Uma 10h5'42" 44d11'36"
Crystal Lynn Hunt
Psc 1h30'38" 15d0'28"
Crystal Lynn Ivey
Cap 20h26'31" -11d29'14"
Crystal Lynn Kokenos
And 2h23'3" 50d10'31"
Crystal Lynn Landis
Tau 4h3'3" 25d4'41"
Crystal Lynn Littell
Cas 1h32'45" 62d21'40"
Crystal Lynn Pryka
Vir 13h37'1" 3d54'35"
Crystal Lynn Shannon
Cas 1h39'24" 64d24'26"
Crystal Lynn Walters
Uma 13h50'18" 49d34'56"
Crystal M. Hardeman
Psc 1h11'7" 7d15'29"
Crystal M. Infante
Del 20h41'41" 15d54'30"
Crystal M. Nashawaty
Cnc 8h48'14" 31d59'55"
Crystal M. Walker
Dra 20h22'20" 63d11'32"
Crystal Mae Benson
And 2h22'14" 46d6'4"
Crystal Mahina Conlee
Vir 14h43'42" -0d47'2"
Crystal Maranto
Lib 15h12'45" -22d8'2"
Crystal Marguerite Rose Anderton
And 2h21'19" 50d17'59"
Crystal Marie
And 1h15'50" 50d36'44"
Crystal Marie
Leo 11h1'49" 3d28'32"
Crystal Marie
Psc 0h30'53" 9d21'29"
Crystal Marie
Sgr 18h59'1" -17d51'54"
Crystal Marie
Cas 0h47'57" 64d56'23"
Crystal Marie Bergman
Cap 20h7'54" -20d24'17"
Crystal Marie Comer
Cap 20h28'7" -11d56'28"
Crystal Marie Dancel Dejesus
Eri 3h37'48" -13d25'37"
Crystal Marie is Loved by M,N,C
Tau 4h3'51" 17d46'37"
Crystal Marie Johnson
Tau 3h40'29" 25d38'44"
Crystal Marie Peña
Leo 9h44'20" 32d29'44"
Crystal Mauldine Terry
Tau 5h23'8" 21d8'52"
Crystal May Massimo
Sco 16h8'30" -9d39'15"
Crystal Meacham
Her 16h30'40" 20d18'10"
Crystal*-Mi Angelito Forever*
Tau 4h11'39" 22d42'39"
Crystal Michele Stafford
And 23h22'15" 52d8'9"
Crystal Michelle Davis
Lib 15h58'58" -9d59'41"
Crystal Miller
Cnc 8h38'38" 24d12'55"
Crystal Millette
Vir 12h5'10" -5d36'58"
Crystal my u-s-di
Gem 6h41'8" 34d6'9"
Crystal N Cheung
Cnc 8h50'47" 31d54'29"
Crystal N. Taylor
Cnc 8h4'8" 21d46'41"
Crystal Nadine
Ari 2h34'54" 26d45'21"
Crystal Necole
Her 18h43'33" 20d28'23"
Crystal Nicole Woolsey
Gem 6h39'22" 13d45'37"
Crystal Nowak
Cyg 19h46'9" 44d6'54"
Crystal Otero Sandoval
Psc 1h17'13" 17d43'22"
Crystal Paulette Meyer
And 2h11'22" 46d17'27"
Crystal Phillips
Dra 16h36'0" 60d26'42"
Crystal Pierce
Cap 20h14'36" -14d16'52"
Crystal "Pocochito" Gomez
Cas 23h9'58" 54d48'8"
Crystal Princess
And 2h32'43" 46d33'33"
Crystal Pure
And 0h43'47" 29d25'34"
Crystal R. Cook
Aqr 22h7'52" -13d41'23"
Crystal Rangel
Vir 12h40'25" 3d56'59"
Crystal Ray
Uma 9h47'13" 50d32'30"

Crystal Ray
Lyr 18h44'43" 31d37'23"
Crystal Rebecca Jones
And 2h19'34" 46d44'24"
Crystal Renee
Uma 10h58'30" 58d24'17"
Crystal Renee Manochehri
Dra 18h49'54" 51d16'51"
Crystal Retiz
Peg 23h49'10" 21d30'19"
Crystal Rhea
Aqr 21h50'2" -3d37'39"
Crystal Rickard
Cas 22h58'43" 55d44'52"
Crystal Robles
And 0h44'37" 30d45'33"
Crystal Rochelle
Crb 16h3'21" 35d8'38"
Crystal Rohr
Cnv 12h40'20" 37d30'40"
Crystal Romero
Lep 5h51'19" -12d29'19"
Crystal Rose
Leo 11h30'32" 27d29'50"
Crystal Rose Bentz
Tau 4h33'48" 27d44'0"
Crystal Rose Garza
Cyg 19h36'19" 28d58'49"
Crystal Ruby Espinosa
Leo 11h28'50" 17d29'48"
Crystal S Winters
Crb 15h50'6" 37d8'13"
Crystal Salinas
Cap 20h30'44" -11d47'55"
Crystal Shontaè "Ladybug"
Lib 14h45'53" -24d59'2"
Crystal Skula
Lyn 8h2'32" 50d42'1"
Crystal Smith
Lyn 8h30'53" 56d29'0"
Crystal Star
Uma 8h34'55" 48d15'47"
Crystal Star
Crb 15h39'44" 27d50'58"
Crystal Star Knighton
And 23h22'44" 48d13'6"
Crystal Stone
Uma 14h3'23" 50d17'28"
Crystal Symone Myers
Lyn 7h13'49" 58d22'53"
Crystal Tanguay
Ari 2h36'31" 17d43'8"
Crystal Terra
Cas 0h41'48" 55d19'50"
Crystal Thomas
Vir 14h24'59" -3d4'0"
Crystal Thomas
Sgr 18h50'9" -35d34'50"
Crystal Thuy Nguyen
Sgr 18h1'19" -22d15'49"
Crystal Tibbets
Lyn 8h0'49" 35d11'22"
Crystal & Tim
Cyg 20h14'37" 33d20'24"
Crystal Time...
Vul 20h33'57" 24d13'36"
Crystal Torres
Ari 3h14'49" 19d8'4"
Crystal Valdez
Uma 11h54'1" 56d10'24"
Crystal Villa
And 0h15'30" 32d46'27"
Crystal Vizcarra
Cam 6h26'31" 62d25'37"
Crystal Watson
And 0h0'53" 39d56'1"
Crystal Weber
Peg 23h47'46" 15d24'8"
Crystal Wheeler
Sgr 19h16'54" -17d7'46"
Crystal Williams
Uma 11h47'20" 29d16'31"
Crystal Willis
Gem 6h47'39" 15d29'13"
Crystal Willoughby
Leo 10h58'2" 6d49'22"
Crystal Yvette Larios
And 0h43'16" 35d6'48"
Crystal Zion
Lib 15h28'34" -23d29'51"
Crystal,tsiml!J.R.
Cnc 9h20'59" 6d59'4"
Crystalack'Z
Cen 11h55'59" -61d54'48"
Crystali
Lyn 9h15'19" 33d19'24"
Crystalina Garcia
Uma 11h38'48" 59d34'4"
Crystal-Rae
Leo 11h26'46" -1d35'56"
Crystal's Angel
Lyr 18h53'53" 31d50'16"
Crystal's Beautiful Smile
And 2h22'56" 43d13'46"
Crystal's Blaze
Her 17h6'48" 33d50'27"
Crystal's Connie Gail
Uma 11h44'48" 51d29'13"
Crystal's Constellation
Cyg 21h15'38" 47d38'42"
Crystal's Cosmic Cricket
Leo 11h27'59" 2d2'26"
Crystal's Destiny
Aqr 21h4'34" -8d36'42"
Crystal's Everlasting
Candle
Tau 3h44'8" 14d0'8"
Crystal's hearts on fire
Ari 2h3'3" 14d27'29"
Crystal's Heaven Star
Lyr 18h43'55" 38d51'12"
Crystal's Home
Uma 11h9'22" 32d49'26"
Crystals Hopes and
Dreams
Sco 17h29'41" -42d20'11"
Crystal's Jewel
Tri 2h32'11" 31d41'56"
Crystals Love
Lib 15h41'58" -5d31'24"
Crystal's Star
Aqr 22h25'3" -1d29'21"
Crystal's Star
Lib 15h20'20" -15d14'32"
Crystal's Star
Uma 8h45'8" 64d14'17"
Crystal's Star
Aqr 23h31'47" -21d11'57"
Crystal's Superstar
Gem 7h36'59" 17d35'51"
Crystal's Touch
Cap 21h14'49" -16d52'10"
Crystalyn Marie Kendall
And 2h9'1" 43d2'35"
Crystel Breuker
Umi 15h48'57" 79d58'13"
CrystelleDavid
Cyg 19h55'15" 33d16'33"
Crysti and Aaron's
Cam 3h54'8" 71d11'17"
Crystle Marie
And 1h33'38" 44d42'59"
Crystyn Anne Merrill
Gem 6h20'50" 18d5'37"
CS
Umi 15h3'43" 80d30'29"
CS Blumberg 61680
Uma 11h17'51" 31d46'36"
CS McDermott
Tau 5h48'44" 18d46'26"
CS & RN
Lyn 6h58'13" 48d44'30"
Csaba Csati
Cap 21h16'29" -24d29'40"
C.Sandra Iaquinta
"Ipanema"
And 0h42'15" 26d48'1"
CSCI - JoLinda
Cyg 19h38'23" 28d25'10"
Cseh Tivadar
Uma 8h46'17" 62d25'26"
Csepregi Éva
Cas 3h26'42" 77d10'14"
CsiCsillag
Leo 10h25'34" 13d10'51"
Csicsvári Ilona
Uma 10h30'18" 60d27'4"
Csíkos Lilla
Uma 10h33'6" 64d49'28"
Csikós Róbert
Uma 10h9'10" 49d37'49"
Csilla & Csaba Csillaga
Cas 0h17'33" 58d23'0"
Csilla Zsigmond
Leo 11h22'33" 25d0'40"
Csillagaink, Gergo és Berta
Uma 11h32'28" 62d30'27"
CsillagBogár
Sco 17h15'53" -43d28'29"
CsillagMara és CsillagJocó
Cas 2h50'17" 63d51'45"
Csillagom Zoli
Ari 2h31'21" 24d35'59"
Csillagom, Andi
Leo 10h42'52" 10d24'8"
Csillagom, Ernoke
Cap 20h17'25" -12d13'15"
Csillagom, Gyöngyi
Lib 14h49'24" -12d33'16"
Csillagom, Náday Ildikó
Uma 10h48'43" 63d26'57"
Csillagsziv, Dr. Merkely
Béla
Cnc 8h4'58" 14d22'25"
Csillagunk Miklós Judit
Tau 3h44'8" 10d0'22"
CsillagVirág Szabolcs &
Virág Attila
Sco 17h41'37" -32d33'54"
Csincsi János
Uma 9h31'30" 60d16'28"
Csizmadia Balázs
Aqr 21h8'6" -2d40'10"
CSM Joe
Aql 19h28'3" 1d19'58"
CSM Johnny Ramirez
Cep 20h45'41" 64d27'12"
CSM (Retired) James
Curtis Nave
Psc 1h24'10" 17d24'40"
CSM Robert Edward
Woods
Her 16h46'26" 30d35'18"
Csodas Fohn
And 23h14'49" 44d0'43"
CSR
Crb 16h6'16" 31d58'17"
Csurilla Dániel
Uma 8h19'6" 70d17'57"
Csúzi Szilvi
Uma 10h59'9" 37d47'34"
CSW 101
Ori 5h23'5" -9d20'59"
CT
Peg 22h21'26" 22d1'39"
CT Nimitz
Psc 1h10'53" 5d31'43"
CTC Alpha
Leo 9h41'52" 27d34'41"
CTC92008THC
Umi 15h20'2" 75d52'18"
C.T.H. Perfect
Vir 14h7'11" -13d2'46"
CTRLXV
Lib 15h9'18" -5d43'33"
CTSTAR
Uma 11h42'42" 44d58'44"
CTwamh20
Tra 16h14'35" -61d28'31"
CU
Uma 12h37'12" 58d12'44"
Cuatro Fluker iv
And 0h1'38" 44d14'44"
Cuauhtémoc Balam
Sámano-Chávez
Per 3h10'47" 55d29'40"
Cub Nowakowski
Ori 6h17'21" 8d46'46"
Cub Sabo
Aqr 22h48'29" -1d33'19"
Cub Scout Pack 467, New
Milford CT.
And 2h3'56" 42d32'43"
Cub Scout Pack 509
Uma 11h18'37" 52d50'26"
Cubby
Uma 12h47'36" 58d8'27"
Cubby
Umi 14h40'20" 75d8'53"
Cubby
Tau 5h48'36" 21d18'2"
Cubmaster Andrew
Abraham
And 1h31'28" 48d40'18"
Cucci
Peg 22h42'37" 5d26'59"
cucciolino
Peg 22h26'44" 2d50'55"
Cucciolo
Uma 10h53'53" 59d3'39"
Cucciolo y Gattina
Umi 17h35'3" 84d39'50"
Cucciolotta
Uma 10h12'34" 48d9'5"
"Cucinotta's Star-Gazer"
Uma 10h12'57" 41d46'32"
Cuddle Bear
Uma 11h45'24" 33d26'31"
Cuddle Bug
Peg 0h5'48" 19d0'37"
Cuddle Tuba
Ari 3h6'6" 18d38'49"
Cuddlebug
Ori 6h9'18" 8d54'15"
CUDDLEBUG
Her 16h22'30" 48d47'5"
Cuddlebugz
Ori 5h21'25" -7d26'33"
Cuddles
Uma 8h51'29" 63d12'26"
Cuddles
Lyn 9h8'38" 46d6'26"
Cuddles
Lyn 8h16'3" 45d7'36"
Cuddles Will Love Emo
Kitty Forever
Lyn 8h39'6" 43d39'35"
Cuetei Lo-Gt Meochan La
Umi 15h30'15" 86d34'48"
Cuevas De La Cruz
4/16/84
Del 20h35'11" 11d33'33"
Cui Cui
Peg 22h41'8" 23d54'58"
Cui Zhong San
Lyn 6h28'29" 57d57'54"
"Cuirle mo Croide" Karen
Rice -mjr
Cas 1h19'46" 57d8'30"
Cujo
Cma 6h48'30" -15d57'24"
Culleen
Lib 15h5'10" -7d22'26"
Cullen
Uma 11h7'12" 66d50'35"
Cullen McClay Cooper
Her 17h47'42" 17d37'16"
Cullen Thomas Clarke
Ori 5h3'59" 6d56'51"
Cullen Whitelock
Umi 15h12'12" 69d34'33"
Cullen's Kindle
Aqr 22h41'43" -21d38'43"
Cullom-Walter 2004
Ori 6h10'57" 15d45'32"
Culvahouse
And 0h16'57" 43d15'2"
Culver Carter Craddock
Ori 5h30'28" 2d17'12"
CulverJr.'s Family
Umi 15h49'7" 73d6'31"
Cuntez-Vous
Uma 11h28'22" 38d43'7"
Cuore Di Uno Zingaro
Ori 5h31'38" 0d48'6"
Cuore d'oro
Cnc 8h19'28" 10d51'23"
Cuore Mia
Cyg 19h21'9" 51d37'11"
Cuori Appassionati
Del 20h39'17" 13d6'58"
Cuoricino
Per 3h26'26" 41d31'45"
Cup Cake
Per 3h19'3" 48d11'0"
Cup Half Full
Ori 5h42'45" -2d56'2"
Cupcake
Umi 14h6'52" 77d49'36"
Cupcake
Cap 21h23'26" -17d2'50"
Cupcake
Sco 16h10'26" -13d34'29"
Cupcake
Sco 16h10'17" -13d43'10"
Cupcake
Uma 12h33'36" 55d51'34"
Cupcake
Sgr 17h55'32" -29d25'16"
Cupcake
And 23h47'16" 35d44'8"
Cupcake
Peg 22h26'50" 9d13'20"
Cupcake
Ori 5h59'56" 3d13'34"
Cupcake
Psc 0h56'50" 14d31'15"
Cupcake Dingdong
Cas 0h40'28" 60d46'1"
Cupcake - Eternally Angel-
Eyes
Ori 5h41'0" -1d35'31"
Cupcake & Snuggles AAF
Cyg 19h27'36" 53d20'59"
Cupid
Sco 16h13'48" -12d2'54"
Cupid De Locke
Cyg 20h19'23" 46d2'8"
Cupid: My Valentine Star
for Brad
Lyn 9h23'21" 39d41'50"
cupio omnia quae vis,
Danielle
Cap 21h25'54" -24d50'44"
Cuque & Gaston
Eri 4h40'12" -0d2'31"
Curet & Family: Our
Mothers Star
Leo 10h19'6" 22d42'6"
Curious George
Uma 11h9'37" 38d46'7"
Curious George
Aur 5h27'20" 36d1'52"
CuriousCuda
Del 20h37'2" 18d32'35"
Curl
And 2h37'37" 46d30'46"
Curlee Blonde 2
And 0h46'17" 36d24'4"
Curley
Uma 10h8'18" 57d7'16"
Curley Robertson
Uma 11h51'6" 60d7'47"
Curl's Girl
Gem 7h34'29" 25d38'11"
Curly
Cas 1h11'40" 64d7'21"
Curly Bean
Crb 16h10'51" 37d27'16"
Curly "No Shoes" Jr.
Cnc 8h8'55" 14d4'24"
Currie
Ori 4h50'59" 2d47'41"
Curry
Lyr 19h14'40" 27d21'47"
Curt Alan Christian
Dra 18h28'46" 52d18'49"
Curt Dunn
Uma 9h6'36" 57d51'49"
Curt E Hathaway
Cep 21h36'44" 61d41'40"
Curt Granger
Tau 5h10'12" 18d48'35"
Curt Kaneshiro
Lyn 8h16'31" 50d23'44"
Curt Kendal Hunnewell
Cyg 19h36'26" 30d12'54"
Curt La Croix's Superfluous
Star
Ori 5h54'0" -0d34'54"
Curt McCray
Per 2h19'59" 52d9'20"
Curt R. "Kermit" Santos
Cyg 21h35'33" 49d4'35"
Curt Vance Haworth
Ori 4h53'23" 2d49'1"
Curtains Up
Cas 1h24'47" 52d5'29"
CurtiousKitten
Leo 10h52'2" 5d40'18"
CURTIS
Her 17h9'32" 32d15'48"
Curtis and Charlotte's
Eternal Fire
Cyg 20h19'37" 52d26'6"
Curtis and Nina Cooper
Ori 5h20'8" -6d43'42"
Curtis and Sarah Aho
Leo 11h55'48" 25d21'35"
Curtis Andrews Edwards
Uma 10h34'29" 68d14'22"
Curtis B.
Tau 3h51'17" 18d0'32"
Curtis Barnes
Aql 19h21'46" 3d15'29"
Curtis "Bert" Andrews
Edwards
Lib 15h3'30" -1d29'24"
Curtis Blaine Bartel
Sco 16h6'58" -17d9'7"
Curtis Buisson
Tau 3h29'46" 1d10'36"
Curtis Clampet (Mr. Wales)
Boo 14h51'28" 31d29'58"
Curtis Clarence Lockwood
"C.C. INC"
Cnv 12h45'45" 37d33'38"
Curtis Clemmer
Cnc 8h52'55" 14d46'18"
Curtis Daniel Tolson Feb-
July 1981
Aqr 22h34'31" -0d6'29"
Curtis E Moen "Dog"
And 23h21'48" 52d58'24"
Curtis F. Oman
Uma 8h48'41" 53d5'39"
Curtis Garth
Tau 3h47'27" 29d24'56"
Curtis Gingras
Cnc 8h38'49" 16d9'18"
Curtis Grayson Voisine
Tau 5h33'38" 18d26'6"
Curtis Hack
Uma 10h29'1" 70d17'5"
Curtis Isaiah Newman
Schupp
Her 17h20'56" 32d6'32"
Curtis James Thames, Jr.,
CMP
Tau 4h29'33" 14d42'18"
Curtis Jamie Brazier
Per 4h9'20" 41d23'35"
Curtis Katz
Sgr 17h52'46" -29d33'17"
Curtis Laakea Paleka
Cnc 9h11'17" 7d24'34"
Curtis Laine Winger
Ori 5h26'40" 2d24'58"
Curtis Lee Ramsey
Ori 6h8'28" 20d48'32"
Curtis Leek's Eternal Flame
Uma 12h2'52" 43d31'40"
Curtis Mason
Her 16h52'48" 17d55'51"
Curtis Oliver Stone
Uma 10h32'38" 58d29'59"
Curtis Owen Moulton
Lyn 6h35'43" 55d56'52"
Curtis & Patrice
Lyn 8h0'45" 54d3'56"
Curtis Peterson
Aur 5h39'30" 50d6'40"
CURTIS R.A. -THE
ACTOR- FLEECS
Aqr 21h53'11" 1d6'23"
Curtis Ratzlaff
Vir 13h32'25" -20d20'26"
Curtis RJ Elizabeth
Pho 1h57'32" -42d44'19"
Curtis Robert Dekker
Her 17h48'44" 45d34'7"
Curtis Spencer
Del 20h38'2" 17d25'3"
Curtis T Hiser - King of the
Sky
Aql 19h38'40" 11d55'48"
Curtis Tussey Jr
Apu 17h21'36" -70d0'33"
Curtis Williams III
Her 16h59'18" 26d5'19"
Curtiss Danniel Sullivan
Lyn 7h40'44" 47d42'23"
Curtiss Ryan Hoeft
Tau 4h29'3" 18d57'13"
Curtiss Smith
Her 17h35'22" 36d26'21"
Curtis-Warhola
Lmi 10h28'38" 28d35'58"
Curt's David
Uma 13h20'36" 62d29'23"
Curt's Star
Aqr 21h10'15" 1d4'36"
Curt's Star of Infinate Love
Tri 2h7'37" 34d6'0"
Curú
Boo 15h19'10" 42d40'37"
Curva
Cnc 9h6'36" 32d1'32"
CUS
Her 16h42'36" 7d23'57"
Cussi
Ori 5h15'42" 1d33'11"
Custom Tiling Solutions
Umi 15h6'5" 73d24'16"
Customer Service Lugano
Uma 11h7'9" 41d2'48"
Cut (Monchici)
Gem 7h11'41" 34d48'36"
Cute Boy 65
Lib 14h59'28" -13d4'22"
Cute Lady Yoshiko &
Kimiko
Aqr 23h25'42" -15d15'21"
Cute Patute & Sexymlexy
Psc 0h27'20" 8d34'8"
Cute Prince
Cen 12h51'13" -46d32'24"
Cute Stuff
Peg 23h19'22" 16d11'18"
Cutey McSleepy Britches
Leo 11h36'12" 22d4'25"
Cuthbert
Uma 8h36'16" 71d59'35"
Cutie
Cap 20h28'36" -26d44'17"
Cutie
Per 3h2'26" 41d1'1"
Cutie
Uma 10h5'58" 45d5'2"
Cutie
Uma 10h23'3" 47d25'47"
CUTIE
Cam 5h40'40" 59d22'38"
Cutie Bear
Ori 5h35'36" 2d27'21"
Cutie Booty
Uma 11h33'31" 62d6'5"
Cutie Pie
Ori 5h56'45" 6d31'35"
Cutie Pie Josie
And 22h59'54" 40d46'10"
Cutie Wahlquist
And 23h22'3" 47d35'55"
Cutie with a Bootie!! I love
you!
Lib 15h25'47" -13d11'31"
Cutiepie
Cam 3h44'23" 56d39'1"
Cutler Polk
Aql 19h37'51" 10d34'26"
Cutrina Marie Wornell
Crb 16h13'42" 25d41'51"
Cutter Kanoa Libby
Lyn 8h2'39" 37d31'8"
Cutter Scott Dittman
Per 3h8'55" 43d26'11"
Cuttino's
Vir 13h45'10" -0d43'59"
Cuxhavener Küstenstern
Uma 14h5'50" 62d8'13"
CUZ
Tau 4h23'49" 22d17'7"
Cuzn
Gem 7h27'11" 34d54'13"
Cuzzi Co.
And 0h52'46" 38d17'47"
C.V. Starr & Co.
Cam 4h30'13" 66d38'55"
CVMS Mustangs Science
2005
Ori 6h16'32" 16d18'30"
CW
Ori 6h17'19" 10d11'43"
C.W. Churchill 1962
Uma 11h25'48" 59d21'22"
C.W. Lieb, Jr.
Psc 1h8'31" 23d20'3"
C.W. Neff Nighthawks
"Shine On"
Uma 12h5'27" 35d12'34"
C.W. Sills Jr
Aqr 23h36'33" -18d15'19"
CWDI
Her 17h37'15" 26d48'10"
CWF
Lyn 8h8'54" 33d52'17"
CWF Special
Uma 12h0'1" 38d35'23"
CWO4 Donald R Slessler
Lyn 7h57'55" 42d58'46"
CWR143
Uma 9h29'59" 59d21'58"
Cy Anthony
Leo 11h38'6" 26d29'40"
CY Boston 2004-05
CityServe Team
Ori 6h3'55" 18d47'31"
Cy Curtis
Ori 5h14'45" 2d8'57"
Cy Hayden Mincey
Ori 5h58'36" 9d40'44"
CY he is star
Cyg 19h56'54" 44d19'23"
Cy "Pookie" Unruh
Sgr 18h47'50" -30d33'13"
Cyan
Uma 12h39'55" 53d51'35"
Cyann Janelle Cox
Umi 9h26'8" 86d13'57"
Cyanne
Lac 22h42'35" 52d5'7"
Cybelle
Lyn 6h53'15" 55d50'39"
cyberangel1967
Uma 11h5'19" 34d19'26"
Cybermom/Nana
Uma 12h25'55" 53d12'23"
Cybernetic
Her 18h56'10" 16d33'9"
Cybertech
Uma 9h4'9" 57d21'20"
Cybil Annette
Peg 21h42'32" 24d2'15"
Cyd & Faith
Cyg 20h38'49" 55d32'30"
Cydbeunos
Cyg 20h1'12" 34d50'38"
Cydnee
Vir 13h33'1" -6d40'36"
Cydney
Tau 4h24'50" 26d28'56"
Cydney Alexa-Sonny
Kiedaisch
Cyg 20h19'3" 39d33'49"
Cydney Camryn Dunn
Ari 2h55'27" 26d13'10"
Cydney LeZanne Keller
Psc 1h5'10" 29d45'24"
Cydney Linn Casper
Leo 10h11'36" 14d19'27"
Cydonia
Tau 4h5'21" 13d51'40"
Cyera Rayne O'Neal
Peg 0h7'29" 16d23'20"
Cygnetz
Her 17h53'54" 26d12'15"
Cygnus Princess Grace
Peanut Motion
Cyg 19h54'56" 32d6'58"
Cygnus Sophia
Ori 5h0'58" 13d14'54"
Cyle Stephen Murdoch
Cnc 8h44'20" 23d52'59"
Cymba Tenali
Tau 4h16'13" 27d2'27"
Cymbalmagic
Cnc 8h23'41" 13d36'15"
Cymvelene
Uma 11h7'54" 40d20'49"
Cyn
Lib 15h9'36" -3d40'52"
Cyn Dee Lee
Umi 14h49'37" 67d35'37"
Cynaminny!!!
Lib 15h46'4" -18d22'48"
Cynara
Cas 0h59'43" 59d49'18"
CynCyr
Cas 1h10'54" 58d16'39"
Cyndal Ann Martell
Leo 10h28'13" 21d16'1"
Cynde Meyer
Mon 8h10'2" -0d45'24"
Cyndea Turner Wendell
Lyn 8h16'15" 36d14'27"
Cyndee Johnson
And 0h21'34" 24d8'6"
Cynder Lou
Lyn 8h1'11" 44d34'17"
Cyndi A. Eschardies
Cnc 8h20'58" 26d6'58"
Cyndi Brillante
Cas 23h51'54" 56d27'13"
Cyndi Butler
Mon 6h50'20" 6d49'22"
Cyndi Chuzie-McDowell
Vir 12h53'26" 12d13'34"
Cyndi Cooper
Cnc 7h59'19" 14d54'53"
Cyndi Diderrich
Ari 2h52'52" 18d12'26"
Cyndi Elaine Aid
Lyn 9h15'17" 40d51'42"
Cyndi Emrie Caton
Uma 13h59'36" 54d25'16"
Cyndi Gale McCauley
And 0h38'37" 41d4'11"
Cyndi Gomez
Cas 1h35'41" 61d23'9"
Cyndi K. Dunn
Vir 14h3'10" -0d51'49"
Cyndi Karimi
Crb 15h34'17" 31d0'53"
Cyndi Lee Feener
Her 17h1'29" 13d33'39"
Cyndi M. Robertson
Uma 11h17'14" 53d44'10"
Cyndi " Pegasus "
Christensen
Peg 22h51'50" 13d42'47"
Cyndi Rikert
Cap 20h41'43" -21d12'20"
Cyndi Schweiger
Vir 13h12'48" 4d37'20"
Cyndi Sue Anderegg
Brohaska Miner
Pho 0h43'48" -43d6'1"
Cyndi4089
Vir 12h37'14" 3d32'3"
CynDic-60
Cyg 20h19'47" 50d20'24"
CYNDIE
And 23h38'53" 33d34'14"
Cyndilou's 40
Lyn 7h40'0" 36d4'5"
Cyndi-Pook-9-23
And 2h12'40" 41d35'8"
Cyndy Angelique
And 0h21'20" 45d7'53"
Cyndy Ann Little
Cnc 8h20'28" 19d48'25"
Cyndy Brown Carlson
Cap 21h12'16" -15d4'25"
Cyndy Pirrera
Uma 13h1'7" 59d22'24"
Cyndy's Charismatic
Cosmopolitan
Tau 4h8'21" 16d44'50"
Cyndystar
Dra 16h17'5" 58d20'59"

Cynibyte T.M.S.
Cnc 9h17'34" 11d36'12"
Cynosure
Sgr 19h44'20" -12d57'45"
Cynt-Cynt
Psc 0h19'43" 15d41'50"
Cynthea D. Degler
Sco 17h15'57" -44d59'9"
Cynthia
Cap 20h37'48" -21d24'58"
Cynthia
Lib 15h4'9" -21d13'0"
Cynthia
Uma 10h3'42" 57d10'11"
Cynthia
Lyn 8h21'50" 55d34'38"
Cynthia
Umi 15h0'26" 71d48'44"
Cynthia
Psc 0h30'51" 18d48'34"
Cynthia
Tau 4h7'52" 17d13'8"
Cynthia
Tau 5h52'6" 15d52'0"
Cynthia
Vir 13h0'46" 9d36'1"
Cynthia
Psc 23h43'35" 5d32'13"
Cynthia
Leo 9h49'10" 17d51'37"
Cynthia
Leo 11h11'31" 17d22'5"
Cynthia
Gem 7h15'34" 26d57'51"
Cynthia
Vul 20h22'33" 27d48'59"
Cynthia
And 0h27'0" 33d43'39"
Cynthia
Leo 11h38'42" 26d42'22"
Cynthia
And 1h42'48" 46d39'19"
Cynthia
Uma 10h9'24" 49d16'4"
Cynthia
Uma 10h38'30" 48d31'47"
Cynthia
Cyg 21h47'34" 46d40'10"
Cynthia
Lmi 9h43'21" 40d21'21"
Cynthia
And 23h47'2" 34d28'59"
Cynthia
And 0h45'0" 38d43'55"
Cynthia
And 1h38'5" 41d31'30"
Cynthia
Lyn 8h14'30" 35d30'53"
Cynthia
Crb 15h37'24" 37d12'21"
Cynthia 1962
Aqr 21h9'44" -3d13'11"
Cynthia 5-7-79
And 0h51'11" 43d22'4"
Cynthia A. Guzman
Cas 23h27'27" 53d52'30"
Cynthia A. Heffington-Weeks
Lyn 8h43'3" 41d8'42"
Cynthia A. White
Mon 6h53'44" -0d48'35"
Cynthia Abiodun Egerton-Shyngle
And 23h38'19" 33d43'28"
Cynthia Abraham De Meo 8/18/68
Leo 11h36'17" -0d20'20"
Cynthia Aguero
Gem 7h22'16" 31d58'35"
Cynthia Alicia Marquez
Uma 10h39'17" 46d34'56"
Cynthia Amedeo
Psc 1h4'21" 3d7'26"
Cynthia and Chris
Gem 6h42'16" 28d44'39"
Cynthia and Clayton
Uma 10h22'28" 51d7'27"
Cynthia Andreea
Ari 2h8'47" 24d50'56"
Cynthia Ann
Tau 3h34'37" 24d38'24"
Cynthia Ann
And 0h53'6" 37d5'53"
Cynthia Ann
Sgr 19h2'24" -13d33'49"
Cynthia Ann Barnes 0114
Lep 5h9'0" -10d53'31"
Cynthia Ann Beck McCandlish
Cas 1h40'34" 61d9'43"
Cynthia Ann Black Angelilli
Cap 21h26'16" -15d5'58"
Cynthia Ann De Fusco
And 23h18'28" 43d4'16"
Cynthia Ann DeHart
Gem 7h15'5" 26d28'17"
Cynthia Ann Froneberger
Uma 13h51'12" 55d19'58"
Cynthia Ann Gills
Srp 17h56'52" -0d43'23"
Cynthia Ann Gunther
Uma 8h36'0" 49d59'12"
Cynthia Ann Hopkins
Uma 8h40'18" 71d34'5"
Cynthia Ann Huntington
Cyg 20h29'46" 58d8'26"
Cynthia Ann Opsuth
Leo 10h26'37" 17d48'45"
Cynthia Ann Robb
Cyg 20h27'54" 46d13'38"
Cynthia Ann Sargent Scheuer
Gem 7h16'40" 13d33'20"
Cynthia Ann Stackwell
Cyg 21h11'32" 35d54'33"
Cynthia Ann Taylor
Cas 0h49'50" 66d49'40"
Cynthia Ann Warnken
Cnc 8h43'52" 11d19'40"
Cynthia Anne
Uma 8h48'6" 60d8'9"
Cynthia Anne Brock
Mon 6h46'6" -0d22'49"
Cynthia Anne Lockhart
Sgr 18h14'58" -20d3'41"
Cynthia Annette Hamilton
Lib 15h5'36" -7d0'59"
Cynthia Arnone
Eri 3h46'3" -2d17'43"
Cynthia Asher Silverstein
Vir 12h49'35" -11d23'3"
Cynthia Avalos
Psc 1h18'32" 12d5'11"
Cynthia Badger
Per 3h25'32" 46d29'48"
Cynthia BLUEEYES Milgrim
Lyr 19h8'38" 27d12'5"
Cynthia Bream
Vir 13h25'1" -0d16'43"
Cynthia Brown
Psc 0h31'26" 19d33'19"
Cynthia C. Ott
Lyn 8h12'29" 42d2'43"
Cynthia Cain Hendrix
Gem 7h48'50" 34d29'22"
Cynthia Calcote
Cap 21h34'5" -12d48'47"
Cynthia Campbell Burke
Leo 11h22'33" 22d15'26"
Cynthia Carol Scarborough
Ori 5h55'33" 17d19'13"
Cynthia Caroline Giammarino
Leo 9h52'11" 13d38'52"
Cynthia Christians
Leo 10h11'6" 24d40'9"
Cynthia (Cindy) Pafiolis Bragdon
Tau 4h0'17" 17d17'47"
Cynthia Clark
Leo 10h42'51" 17d0'23"
Cynthia Colleen Blevens
Cas 0h42'18" 60d55'53"
Cynthia Collins Shain
Psc 23h6'25" 0d59'56"
cynthia cray
Cas 0h23'19" 65d37'7"
Cynthia Cruz Clemente "Pitu"
Leo 11h34'49" 6d58'50"
Cynthia "Cyndi" Maria Gomez
Ari 2h53'14" 26d9'57"
Cynthia "Cyndie" Price-Robinson
Cnc 8h29'36" 26d25'28"
Cynthia D. Hunt
Del 20h33'20" 16d8'56"
Cynthia D. Munoz
And 1h38'13" 41d27'50"
Cynthia D. Reader
Cap 20h38'56" -19d43'22"
Cynthia Dagley 1016
And 2h33'22" 43d16'11"
Cynthia Danell
Sgr 18h10'35" -28d1'43"
Cynthia & David Gutshall
Lyr 18h48'27" 34d18'27"
Cynthia Dawn
Leo 11h35'13" 24d42'32"
Cynthia Dawn
Aqr 22h6'10" -1d43'30"
Cynthia Demas
Peg 22h10'47" 26d9'23"
Cynthia Denise Hobson
Cnc 9h8'28" 6d58'57"
Cynthia Diamond
Vir 13h23'10" -5d31'43"
Cynthia Dianne
Lib 15h21'43" -20d33'58"
Cynthia Dianne Eason
Peg 21h49'49" 8d23'38"
Cynthia Dixon Maronet
Sco 16h13'57" -20d19'38"
Cynthia E. Bomenka
Sco 16h40'45" -36d39'11"
Cynthia E Wagner
Ori 5h29'57" 2d41'7"
Cynthia Elaine Collins
Lyn 7h31'51" 45d23'29"
Cynthia Elizabeth
And 23h41'38" 48d25'2"
Cynthia Elizabeth
Vir 13h45'39" -17d45'4"
Cynthia Elizabeth Lee-Fernandez
Aqr 21h39'16" -6d35'33"
Cynthia Elizabeth Marie Zimmerman
Cnc 9h3'18" 23d43'30"
Cynthia Ezell, The love of my life
Leo 10h9'48" 24d45'45"
Cynthia Faith Thomas
Col 5h53'7" -34d25'48"
Cynthia Farrell
Lyn 7h6'11" 44d40'59"
Cynthia Fay
Cap 20h50'58" -23d50'1"
Cynthia Feldman
Psc 1h28'4" 23d13'13"
Cynthia Fiore-Brooks
Uma 9h38'48" 41d50'40"
Cynthia Foster
Oph 17h42'50" -0d58'18"
Cynthia Frances Bacon
Tau 4h46'59" 27d21'7"
Cynthia Francis Pagano
Lmi 9h30'7" 34d31'55"
Cynthia Gastelum
Lib 15h7'8" -16d17'49"
Cynthia & Gerald Hayes
Lyr 18h23'36" 38d56'52"
Cynthia Graf
And 2h25'49" 41d34'29"
Cynthia Guerra Boucek
Uma 10h25'1" 45d23'5"
Cynthia Hackel
And 23h11'6" 51d49'58"
Cynthia Hammon
Hya 8h26'19" -11d17'15"
Cynthia Hernandez
Com 13h6'41" 28d22'45"
Cynthia Hessberg's Star
Lyn 7h25'29" 49d28'54"
Cynthia Hobart Patten
Oph 17h35'7" -0d58'57"
Cynthia J Balbo
Psc 1h13'13" 11d28'42"
Cynthia Jane
Cas 23h44'10" 57d46'23"
Cynthia Jane Cort
Sgr 19h34'15" -15d31'10"
Cynthia Jazmin Toscano
Tau 4h16'58" 14d5'43"
Cynthia Jean
Cap 21h43'0" -17d14'52"
Cynthia Jeanine
Cap 20h48'38" -16d4'27"
Cynthia & Jeremy Quintana
Cyg 21h10'59" 45d18'18"
Cynthia Joan Burns
Uma 9h16'9" 51d12'7"
Cynthia JoAnn Rucker
Cnc 8h35'27" 7d37'32"
Cynthia Johnson
Cas 23h50'48" 52d37'43"
Cynthia Johnson
Lyn 7h30'37" 53d16'6"
Cynthia Johnson
Uma 8h31'35" 68d38'50"
Cynthia Jones
Lyn 6h36'22" 57d59'50"
Cynthia Jones
Peg 22h37'37" 18d33'31"
Cynthia Joyce Couto
Lib 14h53'25" -4d4'12"
Cynthia K. Henley - Clair
Lyn 7h27'19" 53d27'53"
Cynthia K Hill
Cap 21h26'48" -22d38'16"
Cynthia K Miller
Aqr 23h10'44" -22d20'34"
Cynthia Karen Acquavella
Gem 7h19'8" 15d2'23"
Cynthia Katherine See
Cma 6h35'27" -16d17'35"
Cynthia Kay
And 0h7'16" 43d7'22"
Cynthia Kay Harrell
Peg 23h26'43" 17d39'24"
Cynthia Kay Owen
Sgr 18h49'9" -31d0'8"
Cynthia Keyes
And 1h27'31" 44d30'48"
Cynthia Knapp
Uma 8h36'37" 62d34'31"
Cynthia L. Cline
Lyn 7h42'46" 44d55'33"
Cynthia L. Coles
Cam 5h58'8" 63d46'53"
Cynthia L. Heffran
Aur 7h17'8" 38d46'23"
Cynthia L. Schikschneit
Lib 14h54'12" -3d30'7"
Cynthia Lattimer
Cas 0h23'48" 62d50'14"
Cynthia Layton
Sgr 18h31'25" -30d40'23"
Cynthia Lee Anderson
Cas 1h36'19" 64d46'20"
Cynthia Lee Davis
Lib 15h12'47" -21d38'50"
Cynthia Lee, the Moongoddess
And 1h40'20" 42d54'41"
Cynthia Leigh Bohling
Ari 3h16'59" 23d53'47"
Cynthia Leigh Mottern
Gem 7h46'25" 24d38'49"
Cynthia Liegh
Uma 9h4'24" 57d1'10"
Cynthia Little Betty Klene Makey
Sco 16h39'41" -25d53'32"
Cynthia Lou
Cas 0h48'15" 65d25'57"
Cynthia Lou
Cas 0h55'41" 56d55'51"
Cynthia Louise Brown
Lyn 6h17'39" 60d38'22"
Cynthia Louise Corder Pelham
And 2h34'17" 43d30'31"
Cynthia Louise Dollahan-Cannon
Psc 1h9'28" 27d37'6"
Cynthia Louise Dupre
Cas 1h17'32" 59d48'19"
Cynthia Louise Long
Lyn 7h47'36" 38d57'45"
Cynthia Louise Palmieri
Cas 0h43'26" 57d6'49"
Cynthia Love of My Life
Vir 14h13'38" -15d52'57"
Cynthia "Lucky Charm" Salguero
And 1h15'43" 38d3'49"
Cynthia Luvs Emy And Vix Evermore
Ori 5h32'37" -3d2'20"
Cynthia Lux Lucius
Ori 5h6'13" 10d2'53"
Cynthia Lynch-Ehrmann
And 1h7'59" 37d48'58"
Cynthia Lynn
Lyn 8h1'21" 43d52'20"
Cynthia Lynn
And 0h28'45" 34d13'2"
Cynthia Lynn (Cindy) Darling
And 0h25'5" 27d17'30"
Cynthia Lynn Factor
Uma 9h29'55" 44d44'36"
Cynthia Lynn - Heidi Arnolds Mother
Psc 1h10'54" 7d8'27"
Cynthia Lynn Hodges
Uma 8h37'27" 50d8'18"
Cynthia Lynn Leidy
Cyg 19h52'4" 39d11'0"
Cynthia Lynn Richeson
Psc 0h15'48" 15d6'30"
Cynthia Lynn Sipe
And 2h23'6" 42d2'46"
Cynthia Lynn Stillerman
Cas 1h24'24" 58d4'40"
Cynthia M. Houck
Ari 2h51'30" 26d21'31"
Cynthia M. Lorch
Lib 14h49'33" -2d13'51"
Cynthia M Lovrovich
Leo 9h23'0" 27d15'26"
Cynthia M. Sharp
Cyg 19h57'38" 49d2'58"
Cynthia M. Sloan Ewers "Lou"
Aqr 22h7'15" -17d20'41"
Cynthia M. Terlecki
And 23h9'57" 42d21'45"
Cynthia Mallett
Crb 16h9'41" 35d51'40"
Cynthia Maria
Psc 1h15'52" 16d35'42"
Cynthia Marie
Cas 23h26'38" 55d26'45"
Cynthia Marie Hayward
Gem 7h29'14" 19d45'28"
Cynthia Marie (Kelly) Curran
Cas 2h14'31" 70d26'32"
Cynthia Marie Kraatz
Vir 12h17'38" 4d23'22"
Cynthia Marie Rodriguez
Cap 20h15'42" -9d32'51"
Cynthia Marie Satter
Cap 20h34'27" -13d54'26"
Cynthia Marie Wadley
Psc 1h22'10" 15d20'15"
Cynthia Marissa
Sgr 18h7'42" -16d51'9"
Cynthia Mason
Lib 15h31'5" -24d36'30"
Cynthia May
Lyr 19h15'31" 34d45'44"
Cynthia McCallion - Mas
Ori 5h26'11" 14d1'22"
Cynthia Michelle Ronca
And 2h26'47" 41d42'41"
Cynthia Miles-Gray
Cas 1h22'46" 59d35'42"
Cynthia & Mitchel Together Forever
And 1h28'6" 43d58'54"
Cynthia Najjar & Tarek Abou Deeb
Uma 11h26'10" 39d30'53"
Cynthia Nowlin
Lib 15h3'13" -15d7'36"
Cynthia Olsen
Sco 17h45'50" -35d51'40"
Cynthia Owen
Vul 19h43'36" 25d53'42"
Cynthia R. Otwell
And 0h26'54" 44d25'8"
Cynthia Rae
And 1h37'25" 44d27'19"
Cynthia Rae Macintyre
Umi 14h8'32" 77d45'52"
Cynthia Rauld
Cap 21h46'5" -19d43'37"
Cynthia & Rebecca Forever Love
Cyg 19h29'18" 47d8'35"
Cynthia Rene
Ari 2h21'0" 19d51'15"
Cynthia Renee Sherrod Birthday Star
Sco 16h30'46" -25d39'29"
Cynthia Rocha
Cnc 8h4'54" 20d28'53"
Cynthia Rose Terry
Cam 7h48'55" 70d12'42"
Cynthia Rose Westendorf
Cas 1h41'0" 62d58'47"
Cynthia Rowles
Cnc 8h47'10" 12d53'42"
Cynthia Salvador
And 2h23'11" 41d22'5"
Cynthia Schuster
Aqr 21h51'46" -0d37'49"
Cynthia Scott
Vir 13h36'5" -5d50'36"
Cynthia <\@> Serenity
Aqr 20h52'10" -8d13'1"
Cynthia Shaw
Cam 7h32'53" 60d11'45"
Cynthia Sheperd Jaskwhich
Uma 8h21'19" 59d40'27"
Cynthia Shurte
Tau 3h49'12" 24d14'9"
Cynthia Skunza Macioce "Mrs.M"
Sgr 19h22'45" -33d38'11"
Cynthia Smith
Boo 15h25'12" 46d26'20"
Cynthia Soowal
Cyg 21h9'15" 46d12'20"
Cynthia Sorensen
Uma 9h29'59" 69d9'5"
Cynthia SPD Faust
Psc 1h19'7" 16d19'52"
Cynthia Terrell
Cyg 20h14'16" 37d57'45"
Cynthia & Todd Forever
Del 20h29'25" 18d41'25"
Cynthia V. Eisenkerch "My #1 Star"
Sgr 19h23'55" -33d43'17"
Cynthia Van Epps
Cnc 8h59'17" 10d4'42"
Cynthia Victoria Miller
Cyg 20h18'25" 44d12'12"
Cynthia Vivona
Vir 12h19'42" 2d31'26"
Cynthia Wright "The Princess Star"
Uma 13h8'41" 54d50'6"
Cynthia Y. Palermo
And 1h4'2" 36d0'42"
Cynthia Yost
Uma 10h49'19" 56d55'59"
Cynthia Zei Piechocki
Gem 6h54'2" 28d4'26"
Cynthia19
Vir 13h16'28" 6d17'23"
Cynthialis Deni-Hayes
Leo 10h30'34" 18d59'13"
CYNTHIAMONCHER
Sgr 18h36'17" -23d48'29"
Cynthia's Forever Diamond
Sgr 19h12'16" -16d20'4"
Cynthia's Good JuJu
Col 5h48'41" -35d47'37"
Cynthia's Shining Star
Psc 0h37'43" 14d36'7"
Cynthya Nichole Acuña
Cas 0h28'50" 52d0'47"
CYPLIK DonEde
Uma 12h15'50" 53d23'59"
Cyprianna Rolle, Tupperware
Cas 22h59'43" 57d13'15"
Cyrene S. Lorenz
Cap 20h53'36" -25d13'26"
Cyrielle
Ori 5h4'58" 10d51'34"
Cyril
Lib 15h24'59" -8d24'51"
Cyril C. & Viola M. Houghton
Ori 5h33'53" 0d51'31"
Cyril & Doreen Pearce
Per 3h16'37" 40d27'58"
Cyril & Karine
Cap 20h59'25" -27d14'50"
Cyril Orcel
Tau 4h16'37" 7d24'43"
Cyril Plichon
Uma 12h1'12" 53d33'53"
Cyril Wick
Sgr 19h9'13" -17d3'42"
Cyrileah
Mon 6h51'55" -6d20'32"
Cyrill
Umi 15h0'16" 73d43'13"
Cyrill Enea
Vul 19h53'2" 22d35'18"
Cyrille Champagne
Ori 4h58'56" 15d14'16"
Cyrille Ravet
Crb 16h14'50" 38d14'0"
Cyrilou
Ari 2h8'26" 22d38'46"
Cyrogeba
Aql 19h34'56" 10d49'36"
Cyrous Matin
Peg 21h34'54" 21d7'28"
Cyrsta Star Carless
Ori 5h35'48" 0d5'31"
Cyruh
Lmi 10h1'18" 41d23'48"
Cyrus
Per 2h38'14" 54d28'53"
Cyrus David Hardin
Ari 3h25'44" 29d10'4"
Cyrus E.
Per 3h18'22" 46d59'47"
Cyrus Esmaili
Uma 8h26'30" 64d8'51"
Cyrus H Amato, Jr
Sgr 20h6'13" -32d9'7"
Cyrus Kelly Pepper
Ari 2h1'57" 22d28'56"
Cyrus Yarpezeshkan
Her 18h9'7" 18d24'42"
Cyta's Heart
Gem 7h38'55" 27d48'40"
Cyver
Lyr 18h32'35" 31d58'49"
Czech, Dietmar
Uma 8h39'34" 72d44'39"
Czegledi Istvan
Tau 3h24'3" 11d32'42"
Czubók Éva
Vir 13h36'18" -11d53'37"
СИΛВА
Aql 19h44'43" 8d22'36"
D
Ari 2h45'56" 27d45'43"
D A M - Dennis Alexander Mac Millan
Cyg 20h13'45" 51d4'25"
D. A. (Skip) Christopherson
Hya 9h39'29" -0d23'15"
D. Aldon Ferrara
Umi 14h9'11" 69d50'17"
D and J Lamb
Cyg 20h50'59" 36d24'48"
D' Ann
Leo 11h7'34" -4d11'33"
D. C. Filkins
Per 3h26'59" 45d16'0"
D&C Hough
Cru 12h5'12" -64d38'8"
D. D. B. A. - 1: The Journey
Aqr 22h51'13" -15d19'55"
D. D. Lasiter
Lyn 7h32'52" 38d12'26"
D D n J
Vir 13h58'51" 3d18'18"
D. D. Titenup Buttercup
Her 18h46'17" 20d24'36"
D. Donaldson
Aqr 21h53'6" -0d23'34"
D Dowling
Uma 8h10'52" 68d7'58"
D. Duminski
And 23h26'18" 48d37'59"
D E N N I S
Boo 14h42'37" 12d19'54"
D & G Tritt
Sge 20h12'41" 18d52'6"
D. Gantkessy
Uma 8h50'34" 65d51'33"
D. H. B. D. Tapia
Umi 15h15'48" 68d49'49"
D. H. Grass
Ori 6h17'10" -1d7'26"
D Honey
Per 3h13'58" 41d22'53"
D. Irene
Uma 9h28'28" 72d1'44"
D&J
Aqr 23h3'54" -8d51'1"
D. J. Boutot
Aql 19h39'18" 7d2'52"
D J Campbell
Uma 12h1'13" 39d50'41"
D. J. Fling
Aql 19h43'47" 14d26'23"
D&J Jackson
Vir 13h19'1" 6d42'58"
D Jean OBrien
Peg 21h30'44" 17d6'42"
D. Jeffery Curry MLJ4EVER
Gem 7h0'23" 15d20'27"
D&JW
Vir 12h57'47" 11d57'41"
D&K Dugan
Lyn 7h31'0" 56d16'5"
"D. Kathleen O'Sullivan"
Vir 12h14'49" -8d6'33"
D. Keith Kellum
Per 3h17'11" 43d44'19"
D. Kenneth Morgan
Her 17h0'35" 34d53'57"
D & K's First Christmas Star
Oph 17h53'8" -0d18'27"
D L Brault
Sco 17h15'56" -34d27'42"
D & L Forever
Tau 4h13'25" 23d40'54"
D. L. Kicklighter
And 23h55'14" 48d20'51"
D. L. Lewis Star of Eternal Hope
Sgr 18h34'13" -17d19'22"
D L Ward
Sgr 18h5'32" -27d51'20"
D&L Williams
Sge 19h45'50" 17d31'58"
D. Laird-Florida's Visioneer Star
Her 16h25'14" 17d11'19"
D LASTELLA
Lib 14h55'28" -2d28'52"
D+M
Pho 23h48'1" -49d56'48"
D M Foster
Ari 2h51'47" 19d48'7"
D. M. Lawrence
Aqr 22h14'34" -16d10'43"
D. Muskrat
Uma 9h45'59" 43d46'5"
D~n~A
Gem 6h57'57" 20d51'7"
D. N. E. 138
And 0h30'48" 36d1'15"
D N R 04.08.07
Ori 5h19'23" 5d23'18"
D. Patrick Dillon
Vir 13h48'59" -6d35'5"
D&R Bennett
Ori 5h32'40" 1d45'8"
D * R * E * A * M
Per 4h45'7" 48d3'53"
D R Thoresen
Cra 18h4'28" -37d29'51"
D R W B 3.14
Gem 6h25'32" 25d48'9"
D. Rory Barr
Ori 5h27'28" -1d34'4"
D&S
Cyg 21h38'14" 52d45'43"
D & S
Gem 6h32'50" 18d34'43"
D S
Cyg 21h49'40" 42d50'13"
D&S 01-11-01
Lyr 18h46'15" 34d39'16"
D_S My Lady
And 0h19'21" 25d40'36"
D S R D K F S M W D D
Uma 8h39'52" 66d57'57"
D & S Slate
Ori 6h0'22" 21d44'13"
D. Sammy Rangel
Ori 5h8'22" 9d15'3"
D. Scott Ridley
Lib 14h49'6" -11d28'48"
D. Sullivan 721LU4E
And 0h15'20" 32d52'24"
D und D & Kollegas
Aur 6h27'15" 34d29'43"
D & W Grice
Aql 19h58'29" -0d31'42"
D. Wayne Erickson
Uma 8h33'46" 63d21'1"
D. Wayne Page
Sgr 17h51'57" -17d27'33"
D. West the Rockstar
Tau 5h19'29" 18d8'22"
D. William Provance
Vir 11h46'51" -0d57'15"
D. Worboys 2007
Sgr 18h17'18" -16d51'30"
D Young's 25th infinity & beyond
Cen 13h45'3" -33d47'28"
D2
Psc 0h52'59" 33d11'16"
"D^2" Donald & Donna Morel
Cyg 20h14'30" 48d58'5"
D2 Gentili
Leo 11h12'13" 9d3'2"
Da
Uma 13h51'37" 62d0'4"
Dä chli Prinz-Sean Jordan Sturzenegger
Uma 8h17'57" 60d22'12"
Da Da
Per 3h9'6" 47d58'45"
Da Dawg
Leo 11h21'31" 17d54'58"
Da Di Di Da Da
And 0h28'6" 44d40'6"
Da Honey
Uma 8h53'51" 60d50'56"
Da Joyce
Ori 6h20'17" 14d21'34"
Da n Ca Evans
Uma 10h46'15" 68d55'0"
Da Ya's Blessings
Aqr 22h39'56" -6d22'3"
DaBarbarvidaRidDangecely y1966-2006
Cyg 19h30'2" 27d53'42"
DABE2002
Ori 6h19'25" 15d17'22"
Dabekah 9/10 and 9/17, 1983
Vir 11h50'35" -3d9'33"
Dabin
Cyg 21h30'6" 39d16'14"
DaBlaze
Cap 21h24'34" -16d11'11"

Dabobna
Mon 6h27'41" -4d19'44"
Dac
Cnc 8h48'15" 12d45'28"
DacCac40
Her 16h10'58" 48d1'31"
Dace Ward Dadio
Aqr 22h28'22" 1d21'22"
Dacia Hartford
Cyg 20h28'0" 52d46'32"
Dacian
Cap 21h37'49" -17d26'20"
Dackiewicz, Darius
Cap 21h0'47" -24d58'30"
Dacotah Wren Clarke
Aql 19h11'45" -7d39'54"
DACS
Scl 0h7'21" -28d7'7"
DAD
Cap 21h10'19" -16d5'14"
Dad
Cyg 19h43'35" 56d16'40"
DAD
Uma 14h4'56" 51d12'48"
Dad
Cnv 13h19'26" 47d0'3"
Dad and Laura
Crb 16h19'11" 29d10'34"
Dad and me
Uma 11h25'33" 58d55'56"
Dad and Mom Bates
Per 3h6'3" 53d0'1"
DAD and MOM Forever
Psc 1h11'57" 26d52'39"
"Dad" - Bill Black
Cep 22h34'52" 73d0'19"
Dad - Dave Zimpfer
Leo 11h21'36" 15d33'29"
Dad (Eric Hero Seem)
Aqr 21h11'19" 1d16'59"
"Dad" Ernest T. House
Her 18h14'7" 23d15'1"
Dad Hilgert
Pup 7h39'52" -13d59'39"
Dad & Jan 04'
And 1h8'22" 37d42'3"
DAD Jim Jones
Aqr 22h4'20" 0d40'29"
Dad Keithly
Uma 13h41'8" 56d13'27"
Dad Leibovitz
And 0h24'6" 32d30'9"
Dad Long
Aql 19h42'17" -0d0'11"
Dad Loves Scotty & Alexis Fraser
Lyn 7h3'50" 54d19'22"
Dad loves you Logan forever
Ori 6h1'3" 6d13'54"
Dad & Meme Arenz Forever
Lyr 18h34'5" 35d55'1"
Dad Our Shining Star
Cep 21h24'56" 65d1'49"
Dad & Rachel (the bubble)
And 23h9'28" 36d59'36"
Dad Robert Paul Krueger (Bob/Robin)
Sgr 18h35'23" -18d17'55"
Dad (Roger Martin) Full
Leo 10h13'39" 27d15'34"
Dad - Stephen John Aldred
Aur 6h0'58" 53d34'53"
Dad Storck
Vir 12h8'18" -0d6'22"
Dad T
Uma 13h51'28" 55d22'5"
Dad Tiggerman
Ari 2h18'58" 25d43'29"
Dad & Tom
Ori 6h8'55" 9d18'46"
~Dad~ Tony Francia
Uma 10h3'9" 55d42'22"
"DAD" Wallace Drews
Cep 21h40'18" 55d29'31"
Dad - Warren Tolmie
Lyn 7h39'58" 37d37'27"
Dad, Grandpa & Great Grandpa
Mon 7h13'20" -3d20'37"
Dad, Laurie, and Maggie's Star
Tri 2h20'13" 35d16'21"
Dad, my entire universe
Aqr 22h37'7" 1d39'46"
Dad,Mentor,Best Friend,Inspiration
Per 4h26'43" 44d1'58"
Dada
Per 3h1'15" 45d51'4"
D.a.d.a
Uma 13h33'7" 57d46'25"
Dada - Chris Gilmour
Per 3h15'36" 40d49'11"
dada mattonova
And 2h18'3" 38d31'53"
Dada Nari Lakhaney
Uma 10h30'8" 71d49'55"
Dada84
Cnc 9h7'44" 24d58'43"
DaDaDeDe
Del 20h30'24" 6d44'14"
Dada's star
Per 4h35'34" 47d47'26"
Dadda's Little One
Col 5h47'22" -35d49'28"
Daddie Hammer Wray
Per 4h7'12" 44d17'58"
Daddie's "Beloved" One
Uma 10h56'22" 52d5'49"
Daddie's Jewels
Crb 15h37'37" 30d54'7"
Daddio
Her 16h58'4" 30d56'59"
Daddy
Cyg 19h42'17" 32d32'48"
Daddy
Per 4h49'43" 42d52'1"
Daddy
Aqr 22h11'0" 1d47'25"
Daddy
Gem 7h26'46" 26d28'57"
Daddy
Ari 3h22'49" 29d7'43"
Daddy
Her 17h24'45" 17d39'53"
Daddy
Sgr 19h13'1" -25d24'16"
Daddy
Umi 15h28'52" 71d2'14"
Daddy
Lac 22h49'23" 54d18'20"
Daddy
Cep 21h18'51" 58d23'10"
Daddy
Cep 20h51'16" 59d22'10"
Daddy
Uma 11h47'8" 56d33'43"
Daddy
Leo 11h19'53" -0d16'41"
Daddy
Cma 6h45'29" -20d54'50"
Daddy 24 ~ #1 DAD
Uma 9h1'49" 69d15'25"
Daddy 6/27/02
Ori 5h4'45" 5d58'49"
Daddy and Divot's Star
Lmi 9h25'37" 33d32'57"
Daddy and Emma
And 1h52'45" 41d32'15"
Daddy and His Little Girl
Cyg 19h31'28" 29d27'52"
Daddy and Nathan's Star
Sgr 18h53'16" -17d53'50"
Daddy and Serena
Aqr 21h11'42" 1d18'13"
Daddy and Tootsie Chigoy
And 0h57'3" 38d59'17"
Daddy- Anthony Settimo, Jr.
Sco 16h8'20" -12d38'25"
Daddy B
Cnc 8h38'55" 9d40'28"
Daddy Batts
Sgr 20h3'54" -29d19'37"
Daddy Bear
Aqr 22h7'51" 1d34'27"
Daddy Bear Charette
Cap 21h25'53" -20d52'13"
Daddy Bear W2MQB
Ori 5h27'56" 7d13'39"
Daddy Bill
Cep 23h24'13" 76d53'41"
Daddy Brad & Mara Star
Cnc 8h51'49" 28d27'53"
Daddy Bruce McConnell
Cep 22h5'24" 67d22'43"
Daddy Cal
Per 4h49'44" 46d39'3"
Daddy Coobs
Cnc 8h40'35" 31d39'47"
Daddy cool McMeechan
Cep 20h32'56" 60d40'5"
Daddy Cool - The Best Star in the Sky.
Her 18h12'0" 24d3'51"
"Daddy" Craig Stanley Coleman
Lib 15h5'53" -7d2'13"
Daddy & Cyndi Bear's Special Star
Uma 13h10'46" 55d24'22"
Daddy DeMarino
Cep 22h46'24" 75d48'27"
Daddy Dennis
Lyn 9h26'54" 40d12'6"
Daddy Diver
Lib 15h8'23" -7d31'29"
Daddy Don
Cap 20h10'36" -16d55'18"
Daddy & Dopey
Lyn 7h4'28" 52d20'6"
Daddy Face Morley
Peg 21h36'9" 8d19'54"
Daddy Fishy Baby
Eri 3h47'10" -11d53'38"
Daddy Flynn II
Tau 4h21'34" 24d35'25"
Daddy Franks
Aqr 22h6'10" -0d58'13"
Daddy George Valenti
Boo 14h40'27" 19d1'43"
Daddy Gidi
And 1h8'47" 45d47'4"
"Daddy" Harold Dube
Per 2h21'16" 56d23'44"
Daddy Here We Are
Brenda David Nick
Uma 12h13'45" 58d32'30"
Daddy Hughes
Cep 21h32'7" 61d41'25"
Daddy In Our Hearts
Aqr 23h5'32" -9d41'59"
Daddy is a Super Star
Cyg 21h38'42" 41d0'47"
Daddy is my shining star
Cep 21h31'52" 60d37'29"
Daddy Jeff's House of Shaman
Gem 7h33'51" 31d15'4"
Daddy Jim
Vir 12h30'59" 9d11'15"
Daddy - Jim Patterson
Uma 11h40'0" 62d10'6"
Daddy Joe
Cep 22h16'7" 61d31'41"
Daddy K.C.
Aur 5h46'51" 47d27'38"
Daddy Ken
Uma 9h45'33" 49d7'9"
Daddy Kuehl
Her 16h36'36" 47d36'6"
Daddy Loves Cheyenne Nicole
Tau 4h15'29" 15d14'33"
Daddy loves Cody, Olivia & Russell
Umi 15h34'43" 75d1'17"
Daddy Mann
Ori 5h12'7" -5d44'57"
Daddy Mark and Baby Sarah Gholson
Leo 11h6'49" 9d0'35"
"Daddy" Michael Jarvis
Uma 13h20'55" 58d22'10"
Daddy My Hero
Cep 21h41'44" 59d34'11"
DADDY MY HERO
And 23h39'32" 45d42'39"
Daddy My Shining Star Hero
Gem 6h41'4" 12d21'16"
Daddy O
Cep 21h57'14" 63d19'6"
Daddy - Paul Joseph Daulby
Cep 23h6'50" 70d2'25"
Daddy Ray Hoffmaster-Our Teddy Bear
Ari 3h7'50" 24d34'20"
Daddy Richard Pomerville
Uma 8h44'50" 59d8'18"
Daddy Rodriguez
Cnc 9h5'28" 13d1'5"
Daddy Rolfe
Lib 15h15'57" -19d38'42"
Daddy Ryan
Uma 11h50'16" 56d8'13"
Daddy Schneids
Cnc 8h33'20" 24d13'44"
Daddy Simon Star
Boo 15h27'57" 43d54'7"
Daddy Soldier
Aqr 22h2'55" -2d30'6"
"Daddy Spears DjkoolT"
Cep 22h10'12" 58d30'35"
Daddy Star
Her 17h5'5" 32d37'28"
Daddy Star
Vir 11h55'35" 6d16'51"
Daddy Tom
Cru 11h59'40" -61d9'36"
Daddy Tony
Gem 7h38'35" 33d6'1"
Daddy War-Bucks
Boo 15h3'22" 41d31'33"
Daddy/Will
Her 16h40'21" 45d19'0"
Daddy & Zachary Lichtie
Cap 20h23'34" -13d12'57"
Daddy, Our Hero - (Korey Hatch)
Per 3h25'24" 47d43'22"
Daddy30 Louis J. DeGasperis
Uma 11h7'48" 35d26'1"
DaddyBearDan
Uma 13h6'12" 60d44'31"
Daddykins
Gem 7h19'27" 18d51'23"
DaddyNige
Per 3h26'29" 36d4'0"
Daddy-O
Cep 21h23'54" 64d38'30"
Daddy-O Higgins "9-26-76"
Uma 9h22'51" 49d57'12"
Daddy-O's Candy-O
Uma 11h31'2" 45d35'53"
Daddy-Pops-G-Paw-George Jackson Sr.
Aqr 22h7'3" -8d7'50"
Daddy's 40th
Uma 11h35'42" 62d38'52"
Daddy's Angel Dana Suzanne
Cas 23h46'38" 59d52'27"
Daddy's Angel Erika Michelle
And 0h23'41" 26d20'57"
Daddys Angels
Peg 22h39'8" 25d24'24"
Daddy's Baby Donna
Cnc 9h7'45" 17d5'20"
Daddys Babygirl Amiee
And 23h36'2" 47d58'34"
Daddy's Belle Star
Psc 1h34'31" 10d55'34"
Daddy's Blessing, Lucas McAbee
Ori 5h34'18" 11d1'50"
Daddy's Boy Too, Evan David
Psc 23h10'25" 1d57'57"
Daddy's Christmas Star
Her 16h18'51" 19d22'32"
Daddy's Dat Star
Uma 9h42'16" 64d4'53"
Daddy's Dreamcatcher
Aqr 22h54'34" -24d14'2"
Daddy's Eliza Butterfly
Umi 15h46'57" 70d47'26"
Daddy's Eyes
Ari 2h45'55" 28d4'1"
Daddys Eyes-The Marty Martel Star
Her 16h36'33" 33d46'2"
Daddy's Girl
Cap 21h42'59" -13d46'56"
Daddy's Girl
Cru 12h42'50" -60d11'43"
Daddy's Girl Caitlin
And 0h20'4" 42d43'49"
Daddy's Girl Delaney
Gem 6h27'20" 22d33'32"
daddy's girls
Peg 22h35'10" 30d41'25"
Daddy's Girls
Uma 11h25'6" 59d37'14"
Daddys good girl Kira
And 2h28'31" 44d36'47"
Daddy's Infinity Dot
Cap 20h35'18" -8d53'32"
Daddy's Joonie Moon
Sco 17h53'46" -35d46'53"
Daddy's Little Angel
Sco 17h20'6" -37d22'0"
Daddy's Little Angel
Leo 10h15'4" 18d23'48"
Daddy's Little Angel - Nikita Louise
And 2h31'36" 49d25'59"
Daddy's Little Angel, Deanna Ochi
And 0h48'9" 35d29'21"
Daddy's Little Angel, Ruby Rianna
Lib 15h32'5" -19d44'51"
Daddy's Little Emma
Vir 12h12'50" 8d15'21"
Daddy's Little Girl
Cnc 9h9'12" 8d33'24"
Daddy's Little Girl
Aqr 21h59'22" 0d19'45"
Daddy's Little Girl
Vir 12h11'55" 6d52'40"
Daddy's Little Girl
Uma 10h18'38" 43d36'19"
Daddy's Little Girl
And 0h22'5" 43d2'55"
Daddy's Little Girl
Per 3h1'38" 55d12'31"
Daddy's Little Girl
Uma 11h57'35" 52d44'10"
Daddy's Little Girl Nicole
Vir 12h13'54" -0d27'33"
Daddy's Little Girls Forever
Col 5h41'41" -31d39'7"
Daddy's Little Heaven
Cyg 20h16'14" 32d1'18"
Daddys Little Man
Cyg 20h3'13" 45d43'17"
Daddy's Little Princess Evelyn
And 0h58'59" 39d28'20"
Daddy's Little Star
Per 4h47'57" 46d45'19"
Daddy's Little Star Alexander
Pho 0h31'21" -43d23'59"
Daddy's Love
And 23h36'26" 47d58'24"
Daddy's Love for Rosie
Uma 8h16'3" 61d20'8"
* Daddy's Love Forever *
Ori 5h41'28" -2d19'41"
Daddy's Nightlight
Cap 20h24'53" -10d13'3"
Daddy's Shining Star
Umi 14h46'49" 72d55'37"
Daddy's Shining Star "L.H.L."
Uma 10h0'8" 57d27'49"
Daddy's Soulmate/EFD
Leo 9h47'26" 27d23'39"
Daddy's Spot in the Sky
Cap 21h49'4" -13d18'41"
Daddy's Star
Uma 11h31'19" 58d23'14"
Daddy's Star
Cep 21h27'19" 64d4'27"
Daddy's Star
Uma 9h45'8" 51d22'18"
Daddy's Star
Her 18h4'20" 38d41'57"
Daddy's Star
Peg 22h38'56" 33d31'51"
Daddy's Star
Per 3h34'19" 41d40'53"
Daddy's Star
Her 16h38'57" 35d57'56"
Daddy's Star
Cyg 19h55'57" 33d22'43"
Daddy's Star 23
Lyr 18h36'47" 27d8'16"
Daddy's Star - Brendan Corner
Uma 11h30'7" 56d16'18"
Daddy's Star for KJL
Her 17h24'54" 38d45'47"
Daddy's Star For Taylor - Tom Wernig
Sgr 17h52'37" -17d23'40"
Daddy's star forever, Billie-Jade
Peg 22h23'24" 20d41'34"
Daddy's Star in Heaven (AJJ)
Uma 11h9'34" 68d53'31"
Daddy's Star (LUM With Love)
Cep 22h4'51" 62d10'56"
Daddy's Star - Stuart Richard Newland
Uma 11h0'36" 38d43'18"
Daddy's Star-Jeffrey Alan Abel
Sco 17h47'11" -42d58'21"
Daddy's Sunshine
Umi 15h24'24" 75d33'32"
Dade
Cyg 21h29'29" 36d57'46"
Dad-e-o: *Hope*Faith*Love*Family*
Uma 8h28'47" 70d43'55"
Dádi 80
Cas 1h35'7" 68d22'11"
Dadi & Tato
Uma 12h59'35" 57d18'38"
Dad...my love, hero & shining star!
Her 17h17'44" 17d36'26"
Dado
Peg 22h16'7" 17d11'37"
Dadogram
Uma 11h37'28" 44d16'22"
Dads Dream, Moms Heart Cain
Ari 3h13'2" 19d58'25"
Dads Dream, Moms Soul Chase
Ari 2h36'1" 24d27'45"
Dad's Eternal Light
Ori 5h59'10" 11d24'36"
Dad's Halo-in Memory of Dick Hupe
Uma 13h23'43" 60d28'18"
Dad's little bit of heaven
Lyn 7h7'20" 60d2'50"
Dad's Little Ratbag
Cru 12h18'7" -56d28'39"
Dad's Rumple Star
Tau 4h2'42" 11d57'21"
Dad's special star
Uma 8h28'39" 72d39'20"
Dad's Special Star
Cep 22h44'54" 79d41'53"
Dad's Star
Sco 17h43'34" -42d2'38"
Dad's Star
Ori 5h47'46" 6d37'8"
Dad's Star
Her 18h35'31" 17d41'10"
Dad's Star
Crb 15h46'51" 35d51'38"
Dad's Star
Uma 8h40'2" 50d15'54"
Dad's Star
Per 3h8'16" 45d15'0"
Dadulko
Cas 1h31'21" 68d2'3"
Dae Joon Yoo
Ori 6h25'6" 10d58'20"
Daemon's Destiny
Tau 3h55'16" 5d1'46"
Daena
Gem 6h38'57" 23d34'12"
Daena Schweiger
Lyn 7h31'46" 37d14'55"
Daena Switzer
Gem 7h37'18" 21d56'22"
Daffodil For Amber
And 1h38'2" 48d43'47"
Dafi Man
Boo 15h37'41" 45d31'24"
Dafni
Col 6h17'6" -34d37'51"
Dag Jorgen
Lyr 19h1'40" 31d19'3"
Dagan's Star
Ori 5h30'35" -7d44'41"
Dageför, Gudrun
Uma 12h53'49" 53d52'25"
Daggett
Uma 11h16'7" 50d15'39"
Dagmar
Uma 9h16'11" 51d47'36"
Dagmar Jäger
Umi 15h9'9" 68d43'8"
Dagmar Keller
Cra 18h16'1" -39d55'47"
Dagmar Pereira
Sgr 18h7'7" -33d35'8"
Dagmara
And 23h34'58" 44d42'0"
Dagmara Burzynska
Cas 0h31'58" 60d9'33"
Dagmara P. Baranska Czarnowieska
Cnc 9h16'21" 23d27'49"
Dagmar's Nova Zembla
Srp 18h5'34" -0d2'29"
dagmarV
Peg 22h3'49" 27d19'47"
Dagny Allison Moore
Aql 20h6'33" 3d48'3"
Dagny Deidre Clarabal
Lib 15h38'21" -26d57'5"
Dagny & Sigge
Vir 13h54'19" 4d50'33"
Dagny Sorensen
And 23h49'19" 37d49'50"
DagnyWoo92860
And 0h17'20" 21d53'50"
Dagoberto's Star of Hope and Love
Sgr 19h44'33" -18d13'41"
DaGovoni
Uma 11h36'2" 39d41'51"
Daguana
Lac 22h20'3" 46d49'35"
Dahiana W Paredes
Aqr 21h5'2" -14d27'18"
Dahlheimer, Walter
Uma 12h48'49" 54d12'47"
Dahlia
Uma 11h41'6" 62d36'10"
Dahlia
Lib 14h53'45" -3d37'34"
Dahlia
Gem 6h24'24" 23d1'56"
Dahlia Balladares
Sco 16h39'13" -36d40'39"
Dahlia Tova Segal
And 23h18'25" 48d59'44"
Dahlia Yadira Rodriguez
Uma 12h3'25" 58d54'45"
Dahlke, Dietmar
Ori 5h7'36" 15d59'38"
Dahl's Star
Aqr 21h56'52" 1d39'56"
Dahman-Leiseca
Lyr 18h40'47" 37d53'28"
Dahna Sue Arensberg Proctor
Cap 20h42'16" -18d26'26"
Daht in the Sky
Crb 16h8'1" 29d32'39"
Dahyanni
Cap 20h54'46" -18d58'16"
Dai Bach Rebecca
Ori 5h53'29" 6d16'36"
Daia Nolo
Lmi 11h2'28" 33d16'43"
Daianne
Lib 15h27'45" -22d43'35"
Daicali
Cet 1h18'31" -0d16'21"
Daidein's Reannag
Aqr 23h37'44" -14d43'39"
Daija King
Leo 10h16'24" 24d12'59"
Daijai
Cam 5h34'20" 58d50'34"
Daila
And 1h16'22" 42d43'39"
Dailene Jenae
Lyn 8h46'32" 39d3'26"
Dailey Shelly
And 1h55'58" 42d24'21"
Daily
Sgr 18h45'6" -30d4'6"
Dailyn Ray White
Aql 20h4'27" 11d33'22"
Daimones nassmah Aura
Uma 9h51'26" 62d28'17"
Dain Bishop Yoder
Cnc 9h7'36" 23d24'20"
Dain Lee Martin
Ori 5h29'50" 6d50'35"
Dain Robert Susman
Ori 6h16'59" 14d43'6"
Daina Ivie
Cyg 19h35'30" 30d12'18"
Daine Crockford
Uma 8h44'24" 69d32'35"
Dainin Michael Christopher Zawislak
Gem 7h21'32" 32d19'10"
Daira Rae Speer
Vir 13h42'13" -14d40'6"
Daire Michael Gavin
Umi 15h49'58" 76d24'44"
Daisaku & Kaneko Ikeda
Per 4h30'42" 41d0'55"
Daisey Mae
Sgr 18h7'47" -30d49'55"
DaiseyJuneDenson
Cnc 8h45'32" 16d56'58"
Daisie Mae Prince
And 23h32'58" 42d26'43"
Daisuke
Cas 1h57'56" 62d0'16"
Daisy
Lyn 7h41'27" 56d53'29"
Daisy
Uma 9h25'14" 64d24'49"
Daisy
Lib 15h29'3" -11d14'58"
Daisy
Cma 6h21'46" -27d37'50"
Daisy
Cas 1h5'55" 54d5'39"
Daisy
Uma 10h10'50" 50d49'33"
Daisy
Uma 11h7'48" 48d7'23"
Daisy
Lyn 8h23'17" 39d51'40"
Daisy
And 0h42'51" 38d2'3"
Daisy
And 0h36'54" 26d43'57"
Daisy
Gem 6h37'38" 25d48'19"
Daisy
Ari 3h21'33" 29d56'36"
Daisy
Lyr 18h32'28" 28d35'58"
Daisy
Psc 0h47'25" 16d13'48"
Daisy
Ori 5h17'21" 15d39'59"
Daisy "9-16-1990"
Crb 15h46'10" 34d15'18"
Daisy Alexandra McLean
Ari 3h21'46" 25d8'8"
Daisy Alice Barr
And 23h53'4" 41d24'25"
Daisy Amerie Borba
Sgr 19h42'26" -12d5'31"
Daisy and Lily Dustan
Gem 6h36'52" 19d58'33"
Daisy and Rosie Joseph
Umi 15h22'50" 71d34'37"
Daisy B. Worley
Lep 5h41'37" -20d0'11"
Daisy Beatrice Volker
Uma 9h36'54" 47d39'10"
Daisy Belle / Benson Bear
Umi 14h41'10" 70d25'19"
Daisy Berry
Aql 19h28'1" 10d41'43"
Daisy Braith Seren Jones
And 23h39'20" 37d31'54"
Daisy Brazil Carcia
Vir 14h44'9" 0d28'36"
Daisy & Chris
Vir 13h27'7" 13d36'57"
Daisy Dixon-Amphlett
And 23h41'45" 42d55'30"
Daisy Elaine Spencer
Uma 10h43'3" 71d13'31"
Daisy Emily Mahon
And 23h27'26" 41d39'49"
Daisy Emma Eaton
And 0h48'32" 25d29'55"
Daisy Emma Lord
Lyr 19h7'9" 27d14'7"
Daisy Florence Bartlett Young
And 0h9'40" 46d9'29"
Daisy Grace
Cam 11h9'47" 79d22'41"
Daisy Grace Naylor
And 0h41'24" 22d23'13"
Daisy Hang
Aqr 22h33'19" -1d32'51"
Daisy Honey Bryant
And 2h29'37" 39d1'14"
Daisy in the Sun
Cap 20h19'15" -20d3'13"
Daisy Isabel Taylor - Daisy's Star
And 0h17'46" 35d54'37"
Daisy Isabella O'Halloran
And 2h30'37" 39d16'14"
Daisy Isabelle Fawcett
Umi 17h22'58" 77d51'44"
Daisy June Bonner - Daisy's Star
And 22h58'55" 44d29'29"
Daisy Kate
And 0h11'57" 27d42'25"
Daisy Lambert
Vir 12h28'0" -0d24'32"
Daisy Laurel Solomon
Uma 10h23'26" 64d40'13"
Daisy Lavern Sumner
Sco 17h30'25" -42d34'21"
Daisy Lucchesi
And 0h57'28" 38d46'44"
Daisy Lynetta Oates
Mon 7h43'30" -5d14'25"
Daisy M. Leverett
Leo 10h19'3" 16d43'9"
Daisy Madeleine
Uma 9h29'16" 62d40'53"
Daisy Mae
Psc 1h15'12" 19d0'35"
Daisy Mae
Ari 3h11'49" 18d49'47"
Daisy Mae
And 23h57'54" 41d15'23"
Daisy Mae Belcher
And 0h2'2" 44d50'49"
Daisy Mae Lambert
Gem 6h39'57" 18d38'43"
Daisy Mae Langley Jenkins
Del 20h40'30" 19d54'27"
Daisy Mae Sutcliffe
Umi 14h44'46" 71d44'10"

Daisy Mae Walker
And 23h29'44" 38d3'6"
Daisy Mary Catherine Church
And 23h33'43" 41d32'30"
Daisy May
And 1h19'32" 34d36'55"
Daisy May
Cas 1h36'26" 63d43'49"
Daisy May Kate Warren
Gem 7h46'45" 16d4'30"
Daisy May Singleton
And 2h16'58" 42d38'19"
Daisy & Micky
Uma 10h35'39" 47d21'22"
Daisy Oakey
And 0h5'44" 40d49'22"
Daisy Pooks Bear
Uma 10h39'22" 43d25'52"
Daisy Princess
And 23h35'56" 43d4'43"
Daisy Ray
Cnc 7h58'34" 10d56'20"
Daisy Roberts (Dudley, West Mids)
Umi 16h7'3" 72d7'59"
Daisy Rose Summerfield
And 1h2'17" 42d43'18"
Daisy Ryan Hutnik
Ori 6h14'22" 9d19'44"
Daisy Saylor
And 23h7'49" 49d3'35"
Daisy Silva
Vir 13h6'5" -8d40'28"
Daisy Summer Michelle Skinner
And 0h17'33" 39d9'44"
Daisy Swartz
Cma 6h20'56" -30d44'17"
daisy23
And 0h55'5" 35d41'49"
daisy's Rainbow Brite
Cnc 9h20'54" 22d32'5"
Daisy's Star
Dra 19h3'0" 52d25'12"
Daisy's Star
And 23h17'50" 41d1'55"
Daithi Leith Bigadmaya Wade
Gem 6h49'38" 13d46'46"
Daivid Pontzer
Lib 15h1'59" -15d37'19"
Daizle
Uma 14h7'40" 62d12'55"
Daizy Kay Schmid
Her 16h54'36" 33d2'28"
DaizyR111
Cap 20h31'35" -10d7'14"
Dajana
Cyg 20h46'9" 33d37'29"
DAJO
Gem 6h49'2" 25d20'48"
DAJ's Imagine
Ari 2h31'6" 12d21'52"
DaJsha LaShae Peaches Miller
Cnc 8h6'49" 15d46'16"
Da-Ka-Ti
Uma 10h55'25" 48d18'32"
Dakato James Knorr
Lyn 8h29'16" 41d26'55"
Dake
Ori 5h24'16" 2d7'41"
Daker Star
Aqr 23h33'39" -14d14'6"
Dakhari
Uma 10h13'38" 47d27'30"
Daki
Lyn 6h53'56" 52d15'49"
Dakkota Rain MacFarlane
Ori 5h33'18" 3d38'46"
Dakota
Ari 3h26'27" 20d48'49"
Dakota
Peg 21h43'33" 9d37'37"
Dakota
Her 17h2'25" 29d45'54"
Dakota
Leo 9h38'16" 32d0'27"
Dakota
Cma 6h50'49" -15d21'58"
Dakota
Aqr 22h31'42" -0d8'8"
Dakota
Umi 16h5'48" 79d50'53"
Dakota
Uma 8h40'11" 55d59'25"
Dakota
Uma 12h20'44" 55d32'55"
Dakota Ann Barrineau
Lyr 18h36'25" 36d9'39"
Dakota Anne 4-16-1998
Ari 2h3'7" 20d45'57"
Dakota Bad Ears
Lib 15h29'28" -7d27'6"
Dakota Benton Burns
Lyn 6h42'29" 57d14'57"
Dakota Blaine Orlando
Vir 13h37'16" 5d4'41"
Dakota Christian Lee Berkmyre
Her 16h19'40" 17d9'32"
Dakota Cree
Lyn 7h48'18" 42d47'52"
Dakota Cunneen
Uma 11h15'21" 63d26'30"
Dakota Everett Schmidt
Uma 11h41'19" 43d2'36"
Dakota Horn
Cma 7h12'30" -14d16'25"
Dakota Jade
And 0h4'52" 37d27'8"
Dakota James
Uma 9h34'30" 63d4'38"
Dakota James MacDonald
Uma 9h56'53" 47d28'39"
Dakota Jayne Butchart-Adams
Pup 6h36'2" -48d52'16"
Dakota Joelle DeAnthony
Peg 23h11'56" 17d13'55"
Dakota Johnson
Umi 14h40'1" 75d45'2"
Dakota Joy Kuczynski
Psc 0h34'18" 20d10'30"
Dakota Kade Gripp
Tau 3h58'25" 29d3'58"
Dakota Kai Brown
Uma 12h5'21" 36d2'18"
Dakota Lee
Gem 7h37'26" 33d21'12"
Dakota Lee
Psc 0h45'42" 11d19'7"
Dakota Lee Spivey
Cnc 9h12'19" 8d12'42"
Dakota Linn Hitchcock
Aqr 22h57'51" -5d26'12"
Dakota Michael " Little Man's Star"
Cep 22h50'38" 77d44'7"
Dakota Ostrenger
Tau 4h11'10" 14d50'39"
Dakota Ray-Ann Werenka
Uma 11h31'48" 58d9'6"
Dakota Rayne Sloop
Vir 14h20'9" 1d12'47"
Dakota Reeann Willis
Lib 15h59'7" -12d20'53"
Dakota Reise Facinelli
Leo 11h54'11" 19d8'55"
Dakota Ruby Pequeno
Uma 8h52'10" 52d58'17"
DAKOTA RYAN ROW-LAND
Her 16h44'7" 22d3'22"
Dakota Sand
Lmi 10h30'1" 37d20'36"
Dakota "Scarecrow" Zacharias
Uma 10h22'15" 55d28'53"
Dakota Skye Clifford
Peg 23h11'37" 17d48'56"
Dakota Skyler Bergstrom
Umi 15h33'8" 68d20'5"
Dakota Sunset
Lyn 7h2'47" 46d44'6"
Dakota Taylor Gross
Gem 6h46'17" 27d34'44"
Dakota Warner
Ori 5h33'27" -1d15'34"
Dakota Woodward
Dra 19h45'11" 80d49'47"
Dakota, The Fireman
Her 17h52'3" 31d43'46"
Dakotah
Cam 4h43'39" 58d33'26"
DAKOTAH
Tau 4h14'48" 15d33'54"
Dakotah David Schumacher
Aqr 22h28'42" 0d3'31"
Dakota-Kid VAZQUEZ
Ori 6h17'20" 10d28'12"
Dakota-Kid VAZQUEZ
Ori 6h18'33" 10d40'53"
Dakota-Kid VAZQUEZ
Ori 6h19'11" 10d41'26"
Dakota's Rose
Crb 15h38'37" 34d53'29"
Dakota's Star
Uma 9h30'18" 56d44'10"
Dakota's Star-Cosmically Loving You
Uma 10h35'41" 66d50'7"
Dak's Star
And 2h37'31" 40d15'7"
Dakshina Murthy
Tau 5h26'6" 26d14'17"
Dal 'n Ris
Cam 4h45'45" 55d40'24"
Dal - Strom
Leo 9h43'46" 32d18'50"
Dalal Abdullatef Al-Ghanim
And 1h26'19" 47d10'4"
Dalal Al Ghunaim
Peg 22h50'52" 21d40'28"
Dalaney Vilamaa
Gem 6h21'12" 26d29'53"
Dalbey Laine
Cap 21h29'55" -8d59'59"
Dale
Sex 9h52'11" -0d38'26"
Dale
Vir 14h15'48" -11d38'13"
Dale
Uma 8h37'56" 62d10'53"
Dale
Uma 11h58'17" 54d51'53"
Dale
Sco 16h24'44" -26d25'17"
Dale
Ari 2h7'8" 26d31'14"
Dale A. Lumbra
Per 3h2'2" 55d38'50"
Dale A. Rice
Uma 12h34'33" 56d48'0"
Dale A. Smith, Sr.
Tau 3h57'20" 23d4'37"
Dale A. Tice
Her 18h52'29" 23d53'3"
Dale Alan Marrinan
Uma 11h25'57" 49d3'52"
Dale & Alexandra
Aqr 22h36'24" -2d23'1"
Dale Allen Knight
Her 17h39'5" 27d36'16"
Dale Allison Edmands
Psc 1h22'14" 14d20'21"
Dale and Barbara Price
Cyg 20h27'32" 57d6'7"
Dale and Carmen Goodwin
Crb 16h1'32" 33d45'24"
Dale and Carol Inman 12-20-69
Sgr 19h4'8" -30d28'33"
Dale and David
Psc 23h45'11" 7d2'23"
Dale and Doris' Mercury Star
Cyg 20h44'58" 40d16'32"
Dale and Erika's Star
Sgr 20h14'16" -40d42'26"
Dale and Sharron Wullner
Aqr 21h42'33" -0d4'35"
Dale and Shirley Holly
Cnc 8h45'14" 15d47'31"
Dale and Wendy's Eternal Love Star
Vir 13h16'13" -14d43'51"
Dale Andrew Davis
Boo 15h30'41" 43d29'36"
Dale Ball
Uma 11h22'24" 63d45'39"
Dale & Barb Smith
Pho 0h15'38" -44d45'25"
Dale Baughman, The Upholstery Man
Psc 23h35'44" 2d59'32"
Dale Blackburn
Cas 1h0'35" 62d33'57"
Dale & Charlotte Killian
Cyg 20h39'5" 39d10'19"
Dale Clark
Sgr 19h42'54" -43d47'31"
Dale Cronin
Sco 16h16'1" -32d22'13"
Dale Dahl
Lmi 11h2'7" 25d59'28"
Dale David Dale
Aqr 22h35'17" -1d36'12"
Dale Dean
Ori 5h40'46" -0d51'37"
Dale DiDomenico
Ari 3h14'49" 19d27'48"
Dale Dobrovolskiene
Aqr 20h45'52" -13d59'16"
Dale & Dolores
Lyr 18h17'59" 33d29'11"
Dale Dudley
Cas 1h39'17" 63d27'39"
Dale Duncan
Her 17h13'19" 36d2'54"
Dale E. Craft
Her 18h27'47" 12d9'31"
Dale E. Fridell
Uma 9h32'48" 68d31'12"
Dale E. Hill
Dra 12h39'3" 72d12'15"
Dale Fitzpatrick Brenner
Com 11h58'46" 27d27'10"
Dale Fleetwood
Leo 11h21'29" 7d24'55"
Dale & Frank
Uma 10h36'27" 59d52'20"
Dale Gibson
Vir 12h25'10" 10d55'29"
Dale Halaway
Per 3h12'1" 44d44'22"
Dale Howard Henderson
Cep 21h40'22" 61d1'25"
Dale Hugh Creach
Aur 6h30'18" 34d30'34"
Dale Jacobs Is The Best Mom Ever!
Ari 2h9'39" 20d20'9"
Dale John Fisher
Per 3h15'24" 52d27'5"
Dale & Judy
Com 12h42'21" 30d51'51"
Dale Keith Mast
Uma 11h15'41" 31d24'11"
Dale Kent Henry
And 0h38'53" 43d45'20"
Dale Leslie Goodell
Leo 11h35'30" 18d4'10"
Dale Loves Shannon (Twila)
Cyg 21h16'15" 42d37'55"
Dale Michael White
Her 17h34'35" 35d53'30"
Dale Morris Brazil "Dbravonna"
Aur 5h26'13" 39d40'34"
Dale Mossy
Dra 18h53'29" 55d13'24"
Dale My Soulmate
Sco 17h26'56" -40d48'31"
Dale Nelson
Lib 14h59'29" -18d43'11"
Dale Nunnally's Family Star
Sct 18h46'32" -5d11'42"
Dale P. Herrman
Psc 23h57'28" 2d16'42"
Dale Pershyn
Ari 3h10'8" 23d27'54"
Dale R. Garbush 2-17-1920
Aqr 22h48'53" -10d3'55"
Dale R. Parsze
Uma 10h45'35" 61d52'15"
Dale Racine
Aql 19h33'38" 8d36'2"
Dale Ritter My Love
Crb 15h21'55" 26d40'14"
Dale Robert Hall
Vir 14h16'43" -1d38'22"
Dale Robert Smeltz
Ori 5h35'54" 13d32'35"
Dale Ryan Hebert
Aqr 21h12'41" -13d15'33"
Dale & Sherry Mires
Sgr 18h24'47" -16d10'1"
Dale & Sheryl 25 Years
Lmi 9h51'36" 33d39'4"
Dale "Sitting Duck" Denning
Uma 11h50'56" 50d38'0"
Dale Spierenburg - our little star
Col 5h28'12" -39d23'17"
Dale Steven Martin
Uma 12h1'21" 64d24'15"
Dale T. Johnson
Ori 5h53'56" -0d18'33"
Dale The Great
Sgr 19h37'34" -23d19'37"
Dale & Theresa Hoover
Aql 19h9'58" -7d34'15"
Dale Towell (Mum)
Aqr 22h53'29" -11d39'47"
Dale Vern Hibner
Boo 15h36'4" 39d49'53"
Dale Ward
Cnc 8h49'10" 30d45'46"
Dale Warrick
Uma 11h29'33" 42d54'27"
Dale Wayne Thompson
Her 17h12'10" 28d20'43"
Dale Wesley Mendenhall
Vir 14h19'7" -2d45'31"
Dale Westmark
Dra 20h23'2" 67d25'36"
Dale032096
Umi 15h30'48" 72d37'10"
Daleiden, Heike
Ori 6h15'4" -2d13'11"
DALEN
Vir 13h54'3" -3d18'27"
Dalen Andrew Thierry
Peg 22h34'17" 29d39'55"
Dalen Richard Durnan
Uma 10h39'4" 41d43'24"
Dalene G. Cox
Cas 1h36'46" 66d6'11"
DalenRich
And 1h37'19" 49d33'45"
Dale's Challenger
Uma 11h47'3" 62d55'6"
Dale's Falcon Star
And 1h40'10" 46d5'24"
Dale's Little Love
Ori 5h37'34" 4d12'1"
D'Alessio Angelo Franco
Umi 17h8'4" 79d44'31"
Daleuski "12-1-69"
Umi 16h16'45" 76d16'1"
Dalex
Leo 11h8'39" 23d49'17"
Dalexisclau 1 IWALY
Lyn 7h20'4" 53d18'7"
Daley's Comet
Uma 9h31'38" 72d8'57"
Dali Chachashvili
Cas 23h40'38" 52d37'41"
Dali Mari Rivera
Vir 13h33'54" 2d12'30"
Dalia
Lmi 10h42'23" 30d39'11"
Dalia
Uma 12h16'13" 55d27'49"
Dalia
Cam 5h46'40" 60d7'45"
Dalia Akel
Cnc 8h51'54" 29d23'59"
Dalia Amira Charif
Mon 6h42'41" -0d17'32"
Dalia Aurora Orozco
Aql 19h26'12" -0d27'20"
Dalia Darwish
Uma 9h14'28" 71d29'45"
Dalia Jaffal
Uma 9h38'36" 71d59'59"
Dalia Pardo
And 1h35'42" 49d12'2"
Dalia Pursley
And 23h41'11" 34d12'18"
Dalia & Russell
Cyg 21h41'16" 52d38'36"
DALIAH
Cyg 19h38'17" 32d23'41"
Daliah Brooke
Tau 4h26'55" 21d43'31"
Dalibor i Snezana zauvek
Her 18h55'43" 19d29'38"
Dalice
Uma 11h25'49" 62d8'48"
Dalicetean
Vir 13h9'33" 13d27'37"
Dalila
Leo 10h36'5" 13d29'25"
Dalila
Psc 0h34'28" 8d28'17"
Dalila
Cnc 9h12'19" 22d33'6"
Dalila
And 0h44'30" 41d57'29"
Dalila Villareal
Aqr 22h35'21" -4d36'5"
Dalilah Elizabeth Petersen
Crb 15h37'4" 31d57'18"
Dalilah Monae
Lyn 7h30'35" 37d32'58"
Dallary Cornelious
And 0h25'32" 33d0'1"
Dallas
Gem 6h29'41" 20d47'59"
Dallas
Ori 6h12'34" 19d32'13"
Dallas
Ori 5h44'19" 12d8'28"
DALLAS
Cap 20h24'8" -11d32'49"
Dallas
Cas 23h26'17" 56d30'56"
Dallas Andrew Palumbo
Aql 19h22'42" 4d29'59"
dallas bethany boedeker
Tau 3h38'6" 16d15'1"
Dallas Danielle Thompson
Cas 0h51'25" 48d49'9"
Dallas Ealy
Ori 5h26'26" 10d0'6"
Dallas Fisher
Uma 9h5'19" 72d9'37"
Dallas Gerodi Harrelson
Ari 2h57'13" 28d8'38"
Dallas Greggory McFadden
Psc 1h3'9" 13d34'57"
Dallas James Taylor
Cru 12h35'4" -60d19'0"
Dallas Junior Jarnigan
Cap 21h5'2" -14d54'56"
Dallas Munday
And 0h44'22" 37d58'39"
Dallas Paige
Tau 3h53'40" 12d24'59"
Dallas "Red" Felver
Lyn 8h27'47" 39d35'29"
Dallas Sanders
Her 17h11'47" 35d16'53"
Dallas Tommy West
Sco 16h4'29" -17d1'19"
Dallas Walter Albert Koehler
Per 2h21'44" 57d8'53"
Dalles H. Westling
Psc 1h3'17" 17d36'27"
Dallin
Umi 15h6'1" 71d0'53"
Dallin and Destiny's Star
Cyg 19h57'42" 40d0'50"
Dallin Richard Zook
Uma 12h1'9" 55d30'11"
Dally Dixie Belle
Ari 2h14'57" 11d23'9"
Dalodeal
Uma 9h19'23" 62d21'40"
daloveyou-peia
Lyn 6h41'57" 56d44'23"
Daltiebug
Com 12h59'42" 22d45'40"
Dalton 5
Cap 20h26'35" -25d0'10"
Dalton Alexandra
And 2h18'48" 49d59'21"
Dalton Alford
Aur 5h32'59" 45d53'9"
Dalton Anthony Rosenfarb
Aur 5h40'54" 53d10'6"
Dalton Austin Rios
Dra 18h59'17" 57d0'3"
Dalton Blaine
Cyg 20h38'41" 50d6'50"
Dalton Blakley
Her 17h19'0" 36d44'2"
Dalton Burchette
Cyg 19h35'58" 28d8'11"
Dalton Edward Atkinson
Tau 5h10'54" 19d23'43"
Dalton Ford Sunday
Ori 5h22'33" 9d54'45"
Dalton James Mann 4/21/03
Uma 11h42'40" 50d52'20"
Dalton James Van Aalst
Ori 5h23'57" 9d4'35"
Dalton Keith Walkup
Vir 12h45'30" -10d21'11"
Dalton Lamonte
Per 3h11'17" 42d54'23"
Dalton Max Holt
Uma 12h24'53" 58d39'41"
Dalton Sellers, Jr
Uma 10h43'47" 68d43'49"
Dalton Skye
Gem 6h51'21" 26d27'43"
Dalton Sterling
Boo 14h7'41" 18d6'21"
Dalton Strickland
Ori 5h24'26" 2d34'8"
Dalton T. Spillman
Per 3h42'35" 41d5'35"
Dalton Tucker
Vir 14h8'58" -2d58'27"
Dalton Tyler Herb
Gem 7h30'44" 31d39'34"
Dalton Young
Cap 20h34'32" -15d24'7"
Dalton's Blaze
Lyn 7h53'53" 36d44'33"
Dalton's Star
Aqr 21h37'39" 0d42'26"
Daltyn James Quick
Leo 9h52'27" 32d11'40"
Dalusev
Dra 19h0'28" 68d19'21"
Dalvinder "Meenu" Sangha
Cap 20h28'12" -20d18'52"
Dalvinder Singh Badial
Leo 9h29'59" 19d35'50"
Daly
Tau 4h11'39" 15d33'51"
Daly Whitney Marquard
And 1h56'0" 41d29'49"
Dalya & Abdallah
Per 3h3'55" 43d53'20"
Dalyn Rose
Mon 6h46'45" -0d7'6"
DaLynn Eve
Sco 16h19'34" -18d44'57"
Dalys Baum
Ori 5h30'11" 0d16'56"
Dalystar
Cas 23h30'55" 57d6'21"
Dalyte
Dra 20h15'21" 65d1'18"
DalzellJR1958
Uma 8h24'12" 68d35'48"
Dalzel's Grace
Cyg 20h50'23" 47d0'17"
Dalzire Ragusa
Uma 10h32'41" 64d28'10"
D.A.M. Donna Ardie Mary
Tau 3h47'5" 20d10'26"
Dama
Dra 15h30'0" 62d7'27"
Dama Diana
Crb 15h48'6" 34d19'48"
Daman Ryan Curtis
Per 2h35'30" 53d55'24"
Damares
Cnc 9h11'51" 32d9'31"
Dámaris
Gem 6h28'46" 19d30'41"
Damaris
Ari 3h10'11" 23d12'10"
Damaris Lopez
Lyn 8h28'25" 38d54'43"
Damaris M. Berrios
Sco 16h14'38" -18d7'54"
Damaris Reyes
Uma 8h39'3" 58d6'26"
Damaris Yusuf
Uma 9h31'48" 47d50'25"
Damary's Passion
Sgr 18h1'2" -32d40'2"
Damaya Mowbray
Sco 16h11'49" -12d26'1"
D'Ambrisi
Psc 1h10'9" 22d39'22"
D'Ambrosio
Psc 0h42'5" 16d16'12"
Damenica
Vir 12h0'6" -4d59'14"
Dameon and Cara Huether
Cyg 19h47'11" 57d50'43"
Damian
Uma 12h7'21" 53d38'9"
Damian
Uma 8h35'9" 51d31'17"
Damian A. Tearney - Amor Bis Morte
Per 2h11'0" 57d23'37"
Damian Acosta and Nikki Cady
Sge 19h52'59" 18d47'34"
Damian Alesbury
Sco 17h17'57" -38d35'7"
Damian Ansel Harry
Tau 3h53'23" 7d40'11"
Damian Anthony Islas
Cnc 8h22'51" 9d18'47"
Damian Boyle
Her 16h20'41" 10d31'4"
Damian Carrera
Ari 2h5'34" 17d34'48"
Damian Claudio
Vir 14h41'47" 3d38'14"
Damian F. Tola
Cap 20h35'18" -22d20'44"
Damian Gorse
Ori 5h54'51" 17d42'30"
Damian Grob
Crb 16h18'34" 36d58'33"
Damian Hunter Hamilton
Uma 11h20'36" 55d6'7"
Damian J. Charron
Vir 13h48'58" 7d16'49"
Damian Jacob Duncan
Per 3h29'46" 51d20'22"
Damian Jahn Emmertz
Sgr 18h4'2" -21d19'32"
Damian James Terrizzi
Psc 0h3'50" 10d32'3"
Damian Joseph Martino
Ari 3h16'38" 22d39'11"
Damian Kempin und Jeannine Rayroux
Aur 5h57'37" 41d36'55"
Damian Laframboise
Uma 11h24'29" 58d42'20"
Damian Narine
Ari 2h14'42" 25d58'57"
Damian Nathaniel Chasteen
Lyn 8h18'5" 41d59'50"
Damian Scott Prickett
Cap 20h52'24" -17d52'50"
Damian te iubeste Anastasia
Cnc 8h4'21" 21d8'12"
Damian Tobias
Sco 17h51'48" -39d14'33"
Damian Waller
Her 17h41'42" 36d3'53"
Damiana
Umi 14h12'32" 76d55'32"
Damiana Huber
Psc 1h28'43" 26d14'35"
Damiano
Psc 1h9'53" 29d45'20"
Damiano Arciello
Lyr 18h34'49" 32d20'26"
Damiano Roberti
Lyn 7h3'12" 59d20'43"
Damidace
Cyg 19h28'25" 51d43'23"
Damiem Marcus Young
Leo 11h25'40" 13d0'51"
Damien
Ari 2h53'54" 25d32'6"
Damien
Cnv 12h36'35" 43d22'30"
Damien
Lib 14h39'53" -10d43'24"
Damien
Aqr 23h5'1" -9d36'3"
Damien
And 2h19'48" 48d21'59"
Damien
Sgr 18h0'41" -28d35'55"
Damien and Claire's eternal love star.
Cyg 19h26'48" 30d2'59"
Damien & Brown Player
Lyr 18h44'43" 38d48'9"
Damien & Charley's Eternity Star
Cyg 20h58'5" 46d40'3"
Damien Fisher
Aur 5h42'51" 39d3'41"
Damien Fisk
Lib 15h17'23" -4d0'42"
Damien G.
Cnc 9h21'11" 17d10'32"
Damien Jens Steel-Baker
Cen 13h24'38" -35d59'2"
Damien Louis 18 02 2003
Aqr 20h57'58" -9d54'19"
Damien Mathew Troue
Gem 6h48'2" 33d13'39"
Damien Michael Jordan
Cyg 21h14'45" 40d54'32"
Damien & Nicky - lit up by our love
Dor 6h1'32" -66d8'16"
Damien Olsovsky
Aqr 20h55'14" 1d38'49"
Damien Phillip
Boo 15h3'21" 40d0'14"
Damien Pollock
Sgr 18h48'33" -16d9'12"
Damien Thomas Fitzgerald
Vir 14h47'45" 1d40'24"
Damien Turner
Cnc 9h0'52" 18d34'37"
Damien V. Cross
Lib 14h48'5" -2d22'36"
daMillen
Aql 19h18'46" -11d8'58"
Damilyn Joy
Cnc 8h31'39" 32d22'0"
Damion A. Richards
Aur 5h50'50" 53d53'27"
Damion Anthony Desantis
Uma 11h56'41" 53d10'55"
Damion Caleb Quinn
Ari 2h11'51" 23d47'27"
Damion Santiago
Aqr 22h34'49" -2d31'12"
Damir & Kristina
Cas 23h47'40" 59d5'45"
Dami's Sweet Caroline
Cnc 8h4'44" 21d6'24"
Damjan Djokic
Tau 5h54'34" 26d27'19"
Dämmig, Thomas
Ori 5h35'31" 10d40'7"
Damo
Pho 0h51'42" -40d7'13"

Damon
Psc 0h7'18" 10d11'16"
Damon and Natasha Jefferson
Lib 15h5'25" -2d3'33"
Damon Antonio Wright
Her 16h53'53" 35d46'43"
Damon C. Sands (my SET)
Ori 5h35'12" 10d36'32"
Damon & Cynthia
Leo 10h34'47" 18d29'15"
Damon Darryl Hill
Psc 1h10'15" 15d48'43"
Damon Friedman
Aqr 22h54'44" -5d26'36"
Damon G. Laffey
Cnc 9h9'20" 30d53'27"
Damon & Gina Forever
Cyg 20h26'57" 53d18'45"
Damon Lambert
And 2h34'39" 49d6'29"
Damon & Meredith
Sge 19h40'33" 19d5'26"
Damon Michael Jones
Uma 12h20'17" 62d28'4"
Damon Michael Seeber
Per 2h53'37" 55d55'20"
Damon Paul Gardner
Psc 0h45'36" 15d39'21"
Damon Peter Brazenor Schneider
Cep 21h49'47" 61d25'55"
Damon, Ginger, Bronwynn 2-14-2006
Psc 0h35'7" 17d47'19"
Damond McElhaney
Ari 2h16'29" 24d13'12"
Damon's Star
Ori 5h34'39" -1d42'15"
D'amore
Gem 6h45'59" 19d31'19"
Damorea Michael Faloon
Uma 10h13'13" 58d3'38"
Damoym
Sco 17h15'14" -31d19'19"
D'AMY STEWARD
Cnc 8h54'31" 28d43'40"
Dan
Tau 3h34'5" 24d26'40"
Dan
Leo 10h25'24" 13d37'34"
Dan
Lib 14h33'24" -24d59'1"
Dan
Lib 14h58'54" -18d11'9"
Dan & Aldora
Umi 15h17'39" 73d4'18"
Dan and Allison's Star
And 23h19'44" 51d41'10"
Dan And Amy's Anniversary Star
Uma 11h14'44" 43d21'40"
Dan and Celia
Gem 6h47'23" 19d50'9"
Dan and Claire
Cyg 20h35'13" 47d32'39"
Dan and Danielle
Cyg 20h17'50" 54d30'54"
Dan and Ellie
Lyn 7h0'1" 60d9'6"
Dan and Emily
Cyg 20h56'39" 48d38'33"
Dan and Hannah's Star
And 23h29'23" 39d4'39"
Dan and Jen - Forever in Love
Cru 12h42'49" -57d27'40"
Dan and Joan Walkowiak
Dra 12h43'37" 71d11'24"
Dan and Kara
Cyg 19h57'10" 56d0'9"
Dan and Karen Bobo Baumbach
Lib 15h2'46" -1d2'16"
Dan and Kate Always and Forever
Cyg 20h3'9" 30d24'58"
Dan and Kelly Bevan
Ori 5h56'44" 20d54'45"
Dan and Lesa West
Uma 11h11'57" 52d16'26"
Dan and Lynn's Star
Lib 15h17'20" -5d14'15"
Dan and Mary LaMont
Cyg 20h1'25" 38d17'18"
Dan and Mary's Star
Cyg 21h37'34" 49d20'28"
Dan and Nicole Forever
Cnv 13h33'30" 50d30'27"
Dan and Rach
Uma 10h36'39" 64d29'20"
Dan and Zita Chubbuck
Ori 5h44'2" 11d49'29"
Dan & Ang - Amor Ad Infinitum
Gem 6h53'27" 23d21'21"
Dan Anh - Worthy of being loved
Dor 4h50'11" -68d59'55"
Dan & Ash
Lyn 8h58'22" 39d59'18"
Dan & Bev Eldredge 4 ever
Sge 19h48'52" 18d27'9"
Dan & BG Lyman
Sge 19h53'1" 17d31'33"
Dan Blasy
Sgr 18h37'4" -34d5'45"
Dan Brewer
Her 18h47'5" 15d54'35"
Dan Calvin Crum
Ori 5h21'37" 5d38'51"
Dan Cannon
Ori 5h10'9" 7d25'27"
Dan Castilla
Cep 22h36'22" 67d29'13"
Dan Clancey, ATOS Star
Per 4h25'5" 43d24'34"
Dan Creamer
Gem 6h38'20" 14d53'21"
Dan Curtis Stout
Tau 5h19'10" 24d22'22"
Dan Dan 12 9
Sgr 18h50'6" -18d48'38"
Dan Danaila
Tau 5h8'2" 17d22'14"
Dan Daniell
Cyg 20h57'10" 34d53'5"
Dan (Danny) Eltzroth
Uma 9h48'40" 69d21'44"
Dan Davenport
Ari 2h1'28" 14d33'28"
Dan Davies
Ori 4h58'37" 15d8'50"
Dan DeBroux
Sgr 18h19'58" -34d36'57"
Dan DeCesare
Leo 11h17'9" 21d30'26"
Dan&Denise Pedroza
Cyg 20h24'48" 38d8'26"
Dan Derk
Uma 10h54'18" 42d0'27"
Dan Devine
Ari 2h27'42" 21d26'59"
Dan Ding
Lib 15h44'15" -23d18'28"
Dan E. Blanco
Lyn 6h29'27" 57d45'24"
Dan E. King
Cep 22h38'30" 74d48'20"
Dan Folk
Her 18h11'35" 29d55'51"
Dan Ford
Aql 20h8'27" 10d48'4"
Dan G. Breeton
Cnc 8h7'47" 13d49'32"
Dan G., Our Hero and Friend
Uma 11h52'23" 39d24'25"
Dan & Giota
Pho 0h15'50" -41d7'42"
Dan Glover
Per 2h53'24" 53d8'22"
Dan Gomba
Cam 5h46'27" 58d21'51"
Dan Grauman
Leo 10h16'39" 26d3'2"
Dan Guthridge
Per 3h9'24" 53d32'2"
Dan Hall
Her 18h53'4" 25d27'2"
Dan & Helen's
Sco 16h28'39" -35d38'25"
Dan Helms
Sco 17h57'18" -39d21'57"
Dan Herring, Jr.
Uma 8h40'16" 57d58'39"
Dan & Hien
Vol 8h40'43" -72d30'50"
Dan Hill "My Beautiful Kathy I Love You"
Uma 11h54'19" 31d10'16"
Dan Hopp
Uma 9h39'51" 65d43'59"
Dan Ira Shotz
Leo 9h34'27" 23d47'20"
DAN IS THE ONE AND ONLY STAR
Her 18h47'6" 18d58'37"
Dan J. Lautenbach
Ori 5h25'11" 4d58'25"
Dan & Jade - Eternity of Love
Ara 17h11'40" -55d51'5"
Dan & Jan Pratt
Cyg 20h49'10" 50d55'3"
Dan & Jen
Cyg 20h12'18" 33d14'57"
Dan & Joan Williams - 50 Years
Cyg 19h36'20" 31d19'6"
Dan Johnson
Cap 21h41'55" -14d28'24"
Dan & Julie Mallam
Cyg 20h16'39" 44d17'0"
Dan Kelley
Sgr 19h8'54" -20d38'13"
Dan Kirkwood
Uma 11h59'26" 56d17'51"
Dan Knott
Dra 17h2'49" 58d10'30"
Dan Knudson
Dra 19h6'40" 64d10'40"
Dan&Kristin Souza Est.Aug 6th 2005
Mon 6h28'30" 9d20'22"
Dan Krupin
Aur 4h51'4" 31d53'50"
Dan Lavelle
Cet 1h43'0" -5d43'4"
Dan/Linda O'Tooles' Moon River Star
Lyr 18h40'50" 41d4'33"
Dan & Lindsay Minton
Ori 6h0'16" 10d6'34"
Dan Lynch
Sco 17h52'23" -35d49'43"
Dan MacDonald
Aql 19h26'16" 6d32'16"
Dan Magnane
Per 4h50'5" 48d53'0"
Dan & Margaret Slagle
Sge 20h3'54" 18d0'59"
Dan & Marilyn's Star
Lyr 18h34'59" 36d6'23"
Dan & Marion
Cyg 19h48'57" 36d7'34"
Dan & Marta
And 1h56'53" 36d39'59"
Dan Mayid
Cep 23h19'36" 77d57'19"
Dan Mc Donald
Ari 2h12'52" 25d23'30"
Dan McDowell
Ser 16h1'26" -0d14'11"
Dan McNesby
Aqr 22h39'42" -17d10'33"
Dan Michael Opacich
And 2h25'20" 42d44'5"
Dan & Michelle's Destiny
Cyg 20h40'57" 53d34'15"
Dan Minchin
Vir 13h15'28" 6d28'57"
Dan Monsonego
Uma 11h57'46" 58d24'28"
Dan Morris
Per 3h12'44" 51d19'4"
Dan Morrison
Ari 2h1'56" 23d24'51"
Dan Narlock
Her 17h7'57" 50d11'2"
Dan & Natalie Beach
Vir 13h19'1" 6d18'49"
Dan & Nataliya Inselman
Cyg 21h26'42" 53d28'11"
Dan Nugent
Cep 1h29'51" 80d22'7"
Dan Ollis
Dra 19h18'42" 61d17'28"
Dan Pace
Vir 12h23'59" 6d45'18"
Dan Parker
Uma 11h11'45" 29d43'33"
Dan & Paula Nagel
Eri 4h24'45" -0d1'56"
Dan Perez
Uma 13h43'7" 56d42'2"
Dan Piette
And 23h4'51" 41d4'12"
Dan Pingbone Ellis
Uma 13h24'29" 54d14'54"
Dan Pittis
Lac 22h50'58" 54d16'50"
Dan Poresky
Cep 22h46'49" 71d57'47"
Dan Reed Moreillon
Ori 4h51'8" 0d24'7"
Dan Ritt
Cet 0h24'35" -12d54'33"
Dan Robert Radke, Jr.
Her 17h39'21" 20d54'25"
Dan Romero
Her 17h33'14" 26d47'39"
Dan Rosner
Vir 14h1'31" -14d39'33"
Dan & Sara Wolfgram
Cyg 21h25'26" 45d38'2"
Dan Scharch
Per 3h49'1" 51d30'26"
Dan Schworer
Gem 6h44'23" 29d32'59"
Dan Smith
Uma 11h35'27" 47d16'4"
Dan Smith McCain
Per 4h48'33" 46d52'29"
Dan & Stacie
Cyg 21h37'1" 50d22'3"
Dan & Stefan Times
Sco 16h18'27" -9d39'23"
Dan - Sugar Lips - Moore
And 0h31'47" 26d31'50"
"Dan The Man"
Ori 5h45'27" 6d50'53"
Dan the Man
Aql 19h25'17" 3d44'57"
Dan the Man
Ari 3h2'57" 11d14'56"
Dan the Star Man - Daniel M. Burns
Uma 8h40'7" 52d22'10"
Dan "Tough Guy" Kovacs
Boo 14h43'12" 27d29'38"
Dan Towbin
Sco 17h53'11" -36d7'6"
Dan V. Mericle
Aur 5h42'52" 50d13'53"
Dan & Vinnie Farina
Gem 6h4'18" 25d8'53"
Dan & Vivian Bonitata
Gem 7h3'43" 10d51'31"
Dan Wadell
Dra 18h20'23" 57d43'17"
Dan "Wangstar" Slepak
Cap 21h43'50" -11d49'20"
Dan Way
Aql 19h32'52" 12d24'15"
Dan Wetherbee
Cep 22h48'0" 76d1'2"
Dan Whitaker
Dra 19h10'25" 70d49'37"
Dan Woods
Ari 2h36'3" 20d22'14"
Dan & Yvonne Mielke
Lyn 7h4'46" 48d32'15"
Dan Zipay
Uma 9h39'43" 62d33'58"
Dan, Kathi and Lian
Uma 11h38'59" 47d45'17"
Dan, Rebekah, Ryan
And 23h10'58" 48d5'58"
Dan, Tammy and Daniel Modl
Cam 4h14'15" 70d32'19"
Dana
Uma 12h33'21" 62d15'45"
DANA
Cas 23h37'17" 56d30'37"
Dana
Lyn 6h26'3" 59d32'44"
Dana
Sco 16h9'27" -11d46'7"
Dana
Sgr 18h25'59" -18d33'38"
Dana
Psc 0h49'5" 6d37'3"
Dana
Sco 17h55'5" -39d15'8"
Dana
Lyn 7h31'44" 38d22'59"
Dana
And 0h40'44" 43d55'51"
Dana
And 1h21'7" 39d15'19"
Dana
Gem 6h48'26" 32d41'3"
Dana
Boo 14h26'58" 35d34'41"
Dana
Peg 23h10'47" 16d24'59"
Dana
Peg 21h39'41" 21d13'43"
Dana
Leo 11h7'14" 21d0'5"
Dana
Cnc 9h6'51" 19d4'28"
Dana
Gem 6h49'1" 16d56'39"
Dana 06
Psc 1h10'11" 13d58'24"
Dana A. Driscoll
Ori 5h57'14" 21d1'1"
Dana A. Hibri
Uma 11h51'32" 38d0'18"
Dana A. Riley
Srp 18h4'25" -0d29'16"
Dana Africa
Psc 1h3'0" 11d37'4"
Dana Always Remember
Psc 1h36'36" 27d10'44"
Dana Amick
Oph 16h39'11" -0d52'18"
Dana and Ben Forever
Cyg 20h27'0" 44d47'15"
Dana and Charlie
Pho 0h45'11" -42d8'55"
Dana and Chelsea's Star
Cyg 21h35'22" 45d53'22"
Dana and Dave Amato
Psc 23h41'40" 7d0'56"
Dana and Don Always
Cyg 20h2'39" 50d39'14"
Dana and Gary
Lyn 6h20'53" 56d34'53"
Dana Ann
Aqr 22h0'50" -15d21'58"
Dana Ann Fugler
Uma 10h29'30" 42d59'18"
Dana Annette
Cap 20h38'53" -9d4'48"
Dana Baby
Pup 7h34'53" -29d7'6"
Dana Baby
Ori 5h31'32" 1d10'4"
Dana Bernhard
Uma 11h12'19" 54d26'30"
Dana & Brian Nielsen
Cnc 9h9'11" 21d5'57"
Dana Britting Wilton
Aur 6h3'17" 47d39'0"
Dana Brittni Escue
And 1h54'24" 38d8'33"
Dana Brooke Duncan
Lib 15h27'50" -25d42'54"
Dana Brooke Sobe Sean Alex Jordan
Lib 15h47'10" -19d44'25"
Dana Bryant Duell
Ari 2h35'47" 14d18'10"
Dana Carlson
And 0h53'12" 42d26'53"
Dana Carrington
Leo 11h27'42" 19d18'39"
Dana Charlotte Jones Lettkeman
Aqr 21h57'2" 1d39'52"
Dana Chirico
Ori 5h27'41" 4d29'7"
Dana Christina Morgan
Mon 6h46'51" 10d29'10"
Dana Corelli
Aqr 22h2'29" -17d40'19"
Dana Craaybeek, Beloved Father
Aur 5h15'39" 43d10'36"
Dana D Arco
Leo 10h21'30" 26d47'19"
Dana D. Cable
Psc 1h15'27" 7d43'46"
Dana Daniels
Lyr 18h51'34" 45d39'14"
Dana Darnell Tate
Cnc 9h21'20" 32d18'25"
Dana Dawson-My Brightest Star
Cnc 8h38'46" 20d34'36"
Dana & Diana
Lyr 19h20'52" 34d18'3"
Dana Dominic
Uma 11h31'29" 32d15'33"
Dana Dorries
Leo 10h41'33" 14d34'3"
Dana Dougherty
Aqr 22h46'52" -2d41'40"
Dana Dynazel Terrill
Ari 3h26'55" 23d2'3"
Dana Efron Flier
And 23h12'18" 49d29'3"
Dana Eileen O'Toole
Aqr 21h57'48" 1d11'46"
Dana Elizabeth Ashouri
And 0h11'8" 43d28'51"
Dana Elizabeth Manning
Cas 0h37'44" 63d48'15"
Dana Elizabeth Passaro
Gem 7h7'16" 34d51'5"
Dana Elizabeth-Ellen
And 23h20'0" 49d4'13"
Dana Elliott Wilson
Tau 4h3'5" 16d58'18"
Dana Evans Farrar
Uma 9h38'37" 71d56'46"
Dana Faith
Cas 1h26'7" 69d37'58"
Dana Fenelon
Vir 13h11'39" 6d17'40"
Dana Ferrera Burns
Psc 0h20'39" 7d48'20"
Dana Francoeur
Ari 2h52'55" 27d26'2"
Dana Franz
Cam 6h57'53" 76d38'7"
Dana Froio
Uma 13h12'28" 56d7'55"
Dana & Gary
Mon 8h2'6" -0d54'3"
Dana Gauthier
Uma 14h14'44" 59d46'17"
Dana Grace
Dra 17h49'8" 51d43'39"
Dana Haase
Cam 4h28'16" 53d31'46"
Dana Hazel
Mon 6h31'50" 8d10'38"
Dana Hengl
Vir 13h39'31" 3d47'38"
Dana Hightower Angel 83
Sgr 18h50'53" -18d39'50"
Dana Howard
Ari 3h8'0" 13d20'13"
Dana Huszar
Cas 0h7'24" 56d28'53"
Dana J. Sciandra
Sgr 20h2'31" -30d41'52"
Dana & Jeff, Love Forevermore
Uma 10h48'49" 56d45'16"
Dana Jennifer Park
Uma 8h15'39" 71d34'33"
Dana Jo
Sgr 19h4'20" -32d36'0"
Dana Jo Morse
Sco 17h54'58" -43d0'42"
Dana Johnson
Lyn 7h59'49" 56d17'52"
Dana Jones
Per 3h22'39" 48d23'3"
Dana Joy & Aytan Todd Gabai
Dra 16h27'23" 53d2'17"
Dana Joy Dacey
Lib 14h29'28" -13d24'17"
Dana K. Riffle
Leo 9h22'37" 6d28'34"
Dana Kathleen Hawley
Psc 0h59'23" 27d13'2"
Dana Kathryn O'Gara
Cas 23h27'59" 56d4'10"
Dana Kaufman
Uma 12h38'13" 58d35'57"
Dana Kay Daniels
And 1h6'11" 38d6'15"
Dana Kay Newman
Uma 10h16'49" 67d32'48"
Dana Kaye
Uma 13h48'12" 58d20'58"
Dana Kuglin
Sco 16h10'14" -14d3'56"
Dana L. Moss
Ari 2h7'37" 23d29'13"
Dana L Nisley
Peg 22h27'59" 4d25'22"
Dana L. Richardson 8-30-1956
Vir 13h5'35" 11d59'19"
Dana Lacy Odom
Gem 7h20'59" 26d45'26"
Dana Lanae Nelson
Cnc 8h26'47" 12d30'4"
Dana Landreth
Cnc 9h10'14" 29d32'44"
Dana Lauro
Uma 11h31'21" 31d25'32"
Dana Ledbetter
Sco 16h46'3" -29d17'26"
Dana Lee DuGay
Vir 13h38'56" -0d0'54"
Dana Lee Omer
Ori 6h9'16" 19d43'4"
Dana Leigh
Cnc 9h0'30" 7d48'22"
Dana Leigh
Srp 18h18'43" -0d15'30"
Dana Leigh
Aqr 22h24'48" -13d0'2"
Dana Leigh Canter
Psc 0h23'5" 8d37'31"
Dana Leigh Wilson
Cas 3h15'54" 61d43'55"
Dana Lewis
Leo 10h31'41" 7d41'2"
Dana & Lindsay Sheetz
Tau 5h34'56" 17d15'40"
Dana Linhart
Psc 23h18'2" 6d34'36"
Dana Lites
Psc 0h31'48" 14d8'27"
Dana Lockyear
Ori 5h59'14" 3d32'31"
Dana Logan
Cap 20h22'45" -11d31'40"
Dana Long
Sco 16h14'34" -12d57'50"
Dana Lou
Vir 11h51'24" 5d43'46"
Dana Louise Pearson
Uma 11h13'9" 63d3'14"
Dana Loves Samantha 1206
Lib 14h51'42" -2d28'48"
Dana Luv
And 2h20'3" 39d39'43"
Dana Lyman
Lib 15h44'53" -6d28'52"
Dana Lynn Ferguson
Sco 16h55'22" -39d7'32"
Dana Lynn Field
Lib 15h54'30" -11d2'51"
Dana Lynn Hogancamp
Cam 4h20'58" 57d34'49"
Dana Lynn Novack
And 1h17'42" 43d33'39"
Dana Lynn Richards
And 23h11'36" 35d31'7"
Dana Lynn Sanger
Crb 16h7'53" 36d30'18"
Dana M Baker
Ari 2h46'15" 27d46'30"
Dana M. Dickens
Psc 0h8'0" 8d11'16"
Dana M. Scott
Dra 19h21'28" 70d54'41"
Dana Maishoua Vue
And 23h57'48" 40d8'59"
Dana Mama Wheezer Squeezer P. N. L.
Uma 12h36'58" 59d54'37"
Dana Mann Veazey
And 23h28'1" 42d45'7"
Dana Marie
And 23h20'29" 42d22'5"
Dana Marie
Ori 6h8'44" 6d20'35"
Dana Marie
Uma 12h15'12" 57d10'7"
Dana Marie
Cas 23h43'1" 57d47'26"
Dana Marie 1991
And 0h49'40" 37d17'30"
Dana Marie Ambruzs
Gem 6h54'42" 14d15'59"
Dana Marie DeMarcello
Tau 3h49'55" 29d50'43"
Dana Marie Galloway
Gem 7h47'30" 17d6'22"
Dana Marie Guerino
Gem 7h19'22" 16d33'6"
Dana Marie Kanoena
Lyr 18h52'32" 35d45'32"
Dana Marie Ku'ulei Nepo
Ari 3h16'3" 29d24'40"
Dana Marie Sekelsky
Tau 5h53'34" 24d13'4"
Dana Mary DeNiro
Tau 4h16'44" 27d26'19"
Dana May
Sco 16h10'3" -14d2'4"
Dana Michelle Burks
Lib 15h21'31" -20d31'26"
Dana Michelle Holland
Cas 0h39'47" 55d33'30"
Dana Monsees
Aql 20h1'38" 14d52'50"
Dana - My Little Angel - Giufre
Cru 12h19'10" -61d28'13"
Dana my Love
Uma 11h34'41" 53d14'27"
Dana My Love
And 23h45'20" 38d29'32"
Dana my one and only
Tau 4h10'47" 8d11'53"
Dana n Derek
And 0h32'20" 25d11'29"
Dana Nafziger
Cra 18h13'14" -37d9'29"
Dana Nicole Kovaric
Uma 9h25'15" 59d28'51"
Dana Nicole Walker
Uma 12h4'8" 38d22'14"
Dana Noel Whitaker
And 2h21'30" 49d8'44"
Dana Osborne
Peg 23h10'28" 13d10'57"
Dana "Pal" Keller
Sco 17h54'27" -36d43'12"
Dana Palmer 5-7-02
Lyr 19h18'13" 29d7'44"
Dana Payne
Uma 8h53'53" 60d19'11"
Dana Pershyn
Per 3h7'14" 53d45'3"
Dana Plume
Lyr 19h7'21" 33d9'4"
Dana Pondt the Girly Girl star
Vir 13h11'40" -4d59'43"
Dana Poyer
Aur 5h44'52" 29d57'59"
Dana Prophett Johnson
Sex 10h30'17" -0d29'44"
Dana Rachel Sara
Ori 6h6'35" 20d53'24"
Dana Rae
And 2h29'45" 44d44'45"
Dana Rae Smith
Leo 11h1'10" 2d36'47"
Dana Raley
Umi 14h24'9" 72d53'1"
Dana Rasmussen Lindamood
Cap 21h7'49" -18d1'0"
Dana Ray Dickson Moon Child X 4
Cnc 8h40'26" 7d28'57"
Dana Renae Johnson
Ari 2h16'36" 26d27'53"
Dana Renee Emge
Crb 15h47'50" 37d43'51"
Dana Renee French
Mon 7h8'24" -0d38'15"
Dana Rexford
Cas 1h33'0" 66d4'35"
Dana Rose Villari
Vir 12h23'44" 11d33'6"
Dana Rusnak
And 23h19'27" 47d56'31"
Dana S Virgin
Cap 20h24'57" -13d7'16"
Dana Samaha
Cyg 21h33'0" 35d25'49"
Dana Samuelson
Cyg 21h32'12" 34d58'8"
Dana Sanders
Ari 3h3'47" 27d16'58"
Dana Savino
Uma 10h27'58" 58d27'56"
Dana & Scott
Ori 5h27'11" 4d23'40"
Dana Seraina Gruber
Cyg 19h35'4" 30d14'15"
Dana Stumbers
Ori 5h53'49" 18d26'57"
Dana Suzanne Tebbe
Ori 5h42'31" 8d47'17"
Dana Tanner - GaMPI Shining Star1
Lyn 8h11'17" 41d51'41"
Dana Thompson
Lyn 7h47'17" 36d13'6"
Dana und Dennis
Uma 9h5'34" 54d8'25"
Dana Victoria Martin
Uma 11h29'34" 39d18'18"
Dana Virginia Statton
Tau 5h10'8" 18d12'38"
Dana Wagner
Lyn 7h47'19" 38d53'33"
Dana Waters Mook
Vir 12h27'39" 12d19'33"
Dana Will You Marry Me
Psc 1h26'13" 17d4'52"
Dana Wood
Lyn 8h24'47" 36d33'24"
Dana Wyatt
Cas 23h50'46" 53d58'20"
Dana Zimmerman
Lyn 6h41'10" 55d12'1"
Dana, Jessica, and Joshua
Uma 11h40'34" 46d17'58"
Dana, Mother of Love, For all Times
Cas 1h41'38" 68d45'36"
Dana, Ms. Pooh Bear
Cnc 8h23'25" 22d29'0"
Dana, The Brightest Star in my Sky
Ori 5h52'26" 7d23'17"
DANACAKES
Lyn 6h59'43" 47d30'11"
Danae K. Fegan
Lib 14h51'46" -3d41'53"
Danae' Nae' Nae' Kinnett
Crb 16h9'58" 27d57'43"

Danae Welty
Cap 20h38'5" -17d11'12"
Danaea Marlo
Gem 7h25'8" 33d2'24"
DANA-GER
Psc 1h32'15" 15d30'50"
Danah Al-Fayez
Leo 11h19'26" 4d4'59"
Danah My Queen
Cas 23h32'55" 53d7'35"
Dana-Hercules Goldstein
Tau 5h59'47" 25d56'24"
Danai 21
Ori 6h18'41" 11d4'13"
Danalmax
Uma 10h13'29" 66d21'19"
Dana-Marley
Sgr 19h48'34" -14d29'33"
Danan Earls
Ori 6h15'20" 14d19'10"
Danana
Uma 13h51'8" 54d48'22"
DANANAD
Umi 16h37'6" 84d22'52"
Dana's 19th Frumingle
Sco 17h47'19" -42d58'39"
Dana's and Rayden's Star
Aqr 23h6'48" -9d41'23"
Dana's Biggest Fan
Cnc 8h57'24" 27d38'13"
Danas Heart
Ari 2h42'31" 30d41'42"
Dana's Mocha
Boo 14h52'41" 31d53'56"
Dana's Star
Per 3h10'23" 50d31'56"
Dana's Star
Tau 5h55'14" 26d1'46"
Danasia Monet
Cyg 20h47'5" 35d30'5"
Danasty
Sco 17h53'23" -36d20'25"
Danay
Sgr 17h56'13" -25d29'12"
Dance Doc 5678
Cas 23h36'57" 53d0'36"
Dance Stars
Eri 4h30'1" -0d40'53"
Dance USA - Mark
Varchulik-Marino
Cyg 20h57'25" 36d38'39"
DANCENELIA
Cap 21h47'27" -19d16'54"
Dancer
Lyn 7h33'15" 38d2'0"
Dancer
Lyr 18h49'9" 37d16'4"
Dancer
Gem 6h57'22" 12d26'54"
Dancer In The Sky
Cmi 7h36'58" 6d32'16"
Dances On The
Wind~10/18/1944
Peg 22h45'3" 2d56'27"
Dances-Show-Tunes
Ori 5h33'46" 0d50'1"
Dancing Bob Greer
Psc 0h8'50" 2d0'5"
Dancing D
Lib 14h32'53" -11d28'35"
Dancing Danielle
And 0h30'40" 33d12'35"
Dancing Hillary
Tau 4h21'26" 18d45'0"
Dancing Larissa
Lyr 18h32'58" 32d25'48"
Dancing Nora
Uma 8h35'58" 61d10'59"
Dancing on the tops of
buildings
Ari 2h6'34" 19d49'45"
Dancing Sally
Cam 3h54'32" 67d12'12"
Dancing Staci
Cyg 19h59'57" 46d7'34"
Dancing Stars forever Alley
& Cliff
Uma 11h15'59" 62d42'43"
Dancing Water Woman
And 23h6'46" 46d58'18"
Dancing with Harry
Cap 21h46'35" -16d10'36"
DanCon 2-14-97
Uma 12h2'32" 29d29'25"
Dancypants
Tau 4h48'24" 28d51'4"
Dandi Renee Mickler
Leo 11h12'47" 12d9'1"
Dandona
Uma 10h48'49" 56d14'4"
Dando's Star
Cen 13h51'41" -36d2'47"
Dandoun-28
Cyg 21h51'34" 38d32'51"
Dandrea
Cnc 9h18'5" 24d49'2"
Dandrea1
Her 17h10'20" 23d46'9"
D'ANDRIA
Cnc 8h54'10" 23d56'54"
Dandy Reed Chex
Ori 6h0'5" -0d16'53"
Dane Andrea Donabedian
Gem 7h49'42" 31d8'9"
Dane Christian Pinelli
Cnc 8h51'53" 32d13'15"
Dane Christian Thomsen
Psc 0h39'31" 21d17'12"
Dane Cupitt
Cap 20h11'58" -19d12'12"
Dane Eagle
Aql 19h7'26" -0d14'9"
Dane Edwards
Uma 12h46'47" 57d42'12"
Dane Heady
Uma 10h7'56" 55d46'12"
Dane Holmes
Dra 9h26'57" 77d28'9"
Dane Howard Floyd
Ori 5h22'46" 2d13'15"
Dane i Ljilja
Ori 5h52'27" 20d50'20"
Dane i Ljilja
Peg 21h53'55" 17d43'54"
Dane Landers
Per 2h16'18" 56d38'15"
Dane & Leonie's Bright
Wedding Star
Uma 9h39'59" 56d45'31"
Dane Loves Alyssa
Psc 0h17'7" 6d38'39"
Dane Michael
Umi 16h38'5" 77d45'22"
Dane Michael Larsen
(Rock Star)
Per 3h0'51" 54d44'1"
Dane Richard Eubanks
Tau 4h14'51" 17d55'51"
Dane Robert McGuckian
Leo 10h42'1" 6d48'58"
Dane Sprince Smith
Sco 16h47'57" -27d35'11"
Dane & Tegan Thomas
Col 5h49'33" -35d21'51"
Dane Upton
Psc 1h53'13" 7d23'47"
Dane Upton
Cam 6h6'20" 69d24'6"
Dane William Esposito
Tau 5h46'32" 24d10'3"
Danecis Baczanakis :-D
Tri 2h10'36" 35d7'45"
Daneen Lamb
Lyr 18h51'14" 39d0'15"
Daneisha Latrice
Cas 23h5'45" 55d35'35"
Daneiva Doris Lovett
Cam 5h18'18" 63d37'57"
DaneK
Psc 1h42'37" 24d40'47"
Danelia
Sco 17h51'36" -36d19'35"
Danell
Lyn 6h28'41" 58d8'52"
Danella Swann
Aqr 23h3'39" -7d43'25"
Danelle Barr
Uma 12h52'17" 59d27'47"
Danelle Lynn Pampinella
Tau 5h34'26" 18d17'44"
Danelle Rose Novotny
Lep 5h38'56" -11d53'34"
Daner
Sgr 18h0'50" -26d12'23"
Danerys Billaroman
Pangelinan
Lyn 7h7'28" 59d47'1"
Danesha Mcleod
Cas 0h13'46" 63d2'44"
Danessa's Light
Uma 11h35'23" 41d43'39"
Danette N. Casserino
Sco 16h4'30" -16d9'2"
Danford Scott Ferrazano
Lib 15h1'30" -17d8'2"
Danforth
Lyn 8h49'7" 33d28'1"
D'Angelo
Cyg 21h33'26" 34d20'34"
Dangerboy Thomas
Cyg 20h10'37" 43d13'14"
dangermouse
Cnc 8h15'46" 16d38'8"
Dangerous Fantasy
LP102304
Aqr 23h8'8" -8d16'20"
~*'Danger'saurus Rex*~
Cap 20h8'50" -26d5'11"
Dangie
And 23h1'2" 40d7'56"
Danglee's Star 2
Uma 11h28'4" 36d33'43"
Danglee's Star 3
Uma 11h48'50" 35d2'29"
Dangles
Tel 20h28'1" -48d46'28"
dani
Sco 16h53'33" -38d49'35"
Dani
Sgr 18h59'13" -31d14'58"
Dani
Lib 15h19'16" -13d9'51"
Dani
Umi 14h52'51" 80d19'38"
Dani
Uma 8h49'12" 57d19'53"
Dani
Uma 12h5'38" 61d39'0"
Dani
Umi 15h52'18" 73d38'37"
Dani
Uma 11h12'15" 70d8'13"
Dani
Lmi 10h47'34" 37d18'52"
Dani
Uma 11h12'24" 42d37'53"
Dani
Per 4h22'23" 43d28'59"
dani
Crb 15h48'12" 27d51'46"
Dani
Ori 5h28'49" 2d40'1"
Dani
Psc 0h5'47" 7d43'53"
Dani 30. házassági évfor-
dulóra
Uma 11h44'50" 34d50'45"
DANI A
Cap 20h22'17" -10d58'45"
Dani boy and Aubee girl
Umi 16h16'47" 78d7'12"
Dani & Brigitte Oetterli
Per 2h15'47" 51d22'36"
(Dani) Danielle Lee
Burigsay
And 2h6'28" 45d4'38"
Dani Elissa
Peg 23h59'47" 20d14'51"
Dani Elstins
Vir 13h29'57" -7d33'16"
Dani & Gillian
Cas 23h40'10" 56d10'40"
Dani Girl
Vir 12h7'5" -0d19'33"
Dani Girl
Lyn 8h6'4" 38d40'25"
Dani & Godi
And 23h54'53" 48d7'27"
Dani Gonzalez Gonzalez
Uma 9h3'32" 63d8'23"
Dani Lyn (Bam Bam)
Eri 3h46'34" -0d25'42"
Dani Lynne O'Byrne
Mon 7h17'49" -0d6'52"
DANI mein SONNEN-
SCHEIN
Ori 5h23'23" 3d9'35"
Dani Robinson
Sgr 19h2'18" -22d4'34"
Dani "The Reyning Star"
Rosenberg
Aqr 22h13'58" 2d5'28"
Dani Uetz
Umi 15h59'29" 77d24'45"
Dani und Christian
Uma 12h39'6" 58d2'18"
Dani und Markus
Uma 10h44'9" 43d49'1"
Dani143
Psc 1h10'40" 32d0'4"
Dania
Tau 4h9'46" 6d44'52"
Dania
Cnc 9h19'23" 24d49'3"
Dania
Uma 12h20'45" 61d40'15"
Dania
Psc 1h50'23" 6d28'4"
Dania & Joel
Eri 4h22'44" -22d21'20"
Dania Weaver
And 23h8'0" 51d31'49"
Danial Umbrell
Lib 15h25'25" -19d34'7"
Danica
Uma 13h46'2" 61d43'11"
Danica
Uma 8h12'53" 65d9'24"
Danica
Uma 11h51'22" 53d9'58"
Danica
And 23h11'38" 49d36'44"
Danica
And 0h43'34" 37d38'31"
Danica
Vir 15h6'6" 7d4'34"
Danica Ann Bailey
Ori 5h28'48" 0d9'22"
Danica Diane Nielsen
Sco 16h18'44" -15d9'53"
Danica Humphries
Uma 11h8'17" 62d15'2"
Danica Jade "Forever
Fifteen"
Tau 5h32'17" 24d41'17"
Danica Jane Carpenter
Sgr 19h57'18" -32d28'40"
Danica Karic
Boo 15h32'49" 45d13'2"
Danica Marullo
Cas 1h38'30" 66d30'52"
Danica Rae
And 0h40'39" 40d15'2"
Danica Rae & Brianna Lee
Sgr 17h59'21" -27d59'21"
Danichelle
Sge 19h27'38" 18d17'16"
Danicia Lynn Lohmier
Lib 15h34'17" -12d42'0"
Danie Lalaenja Williams
And 0h40'35" 22d26'32"
Danie Lee
Ori 5h12'30" 15d13'17"
Danie-Gurl
Sco 16h13'27" -11d23'58"
Daniel
Lib 15h30'45" -9d35'52"
Daniel
Lib 14h51'17" -11d9'43"
Daniel
Cap 21h46'40" -13d17'37"
Daniel
Sgr 18h48'51" -17d38'21"
Daniel
Ori 6h8'5" -1d35'55"
Daniel
Ori 5h29'24" -1d23'29"
Daniel
Uma 8h18'28" 63d52'14"
Daniel
Uma 12h10'49" 63d9'47"
Daniel
Dra 18h10'33" 72d13'47"
Daniel
Cep 21h20'13" 65d17'58"
Daniel
Sco 17h22'13" -40d46'28"
Daniel
Psc 1h10'21" 20d29'55"
Daniel
Aql 18h58'49" 7d51'51"
Daniel
Ori 5h31'13" 5d2'6"
Daniel
Psc 0h44'31" 11d59'51"
Daniel
Ori 4h51'9" 10d32'56"
Daniel
Psc 1h6'9" 28d22'13"
Daniel
Ori 6h1'11" 21d5'30"
Daniel
Gem 6h54'37" 15d25'54"
Daniel
Ari 3h12'43" 27d50'51"
Daniel
Ari 3h21'39" 23d46'33"
Daniel
Her 17h35'11" 26d23'24"
Daniel
Vul 20h26'47" 26d34'4"
Daniel
Per 3h28'46" 33d56'43"
Daniel
Uma 11h25'51" 36d32'57"
Daniel
Her 17h10'51" 30d16'49"
Daniel
Uma 13h41'28" 51d39'58"
Daniel
Uma 14h4'5" 50d57'45"
Daniel
Boo 14h22'59" 47d10'32"
Daniel
Cyg 21h47'43" 48d7'23"
Daniel
Per 3h19'0" 53d39'55"
Daniel
Aur 5h55'42" 52d39'34"
Daniel
Cas 0h58'16" 54d2'27"
Daniel
Per 3h42'22" 47d39'54"
Daniel
Aur 5h41'24" 45d10'17"
Daniel 1590
Aur 5h43'38" 41d8'34"
Daniel 26
Cap 20h59'36" -24d39'30"
Daniel 30
Leo 9h31'15" 8d44'3"
Daniel A. Brownstein, Sr.
Gem 6h54'6" 14d35'38"
Daniel A. Langston
Her 17h7'46" 23d7'50"
Daniel A. Letteriello
Lmi 10h55'47" 25d53'48"
Daniel A. Miller
Tau 5h51'30" 15d38'5"
Daniel A. Richard
Uma 8h17'21" 70d38'9"
Daniel A. Saltzman
Uma 11h59'50" 46d28'57"
Daniel A. Westphal 8/9/51-
2/13/03
Ari 1h55'14" 19d5'2"
Daniel Aaron Farley
Uma 8h40'34" 53d21'13"
Daniel Aaron Fox
Per 3h44'45" 31d8'45"
Daniel Abraham Benjamin
Sgr 18h24'39" -32d35'33"
Daniel Abraham "Fish"
Ori 5h44'38" -3d42'40"
Daniel Abraham
Zelenovskiy
Vir 14h6'17" -13d29'54"
Daniel Adam Janicki
Ari 3h17'49" 27d57'43"
Daniel Alan Beach Sr.
Umi 14h18'24" 67d12'44"
Daniel Alan Beach, Jr.
Umi 15h26'53" 67d53'51"
Daniel Alan Hoke
Dra 15h53'51" 54d59'0"
Daniel Alberto Escamilla
Lib 14h38'42" -12d16'47"
Daniel Alejandro
Rodriguez
Vir 12h52'12" 0d45'9"
Daniel Alexander Buckley
Ori 4h50'21" -0d40'59"
Daniel Alexander Traficante
Uma 12h2'29" 52d35'31"
Daniel Alfred Hoffler
Lib 15h24'40" -7d18'39"
Daniel Allen Hovis
Gem 6h36'48" 20d55'56"
Daniel Allen Rhodes
Uma 11h58'59" 40d46'14"
Daniel Allen Shipley
Uma 8h41'57" 66d48'37"
Daniel Allen Spencer
Tau 4h8'12" 6d35'13"
Daniel & Amber Forever
And 0h25'44" 33d39'11"
Daniel Ancel Seecharan
Uma 10h7'32" 58d23'58"
Daniel and Amelia
Uma 11h54'54" 62d37'24"
Daniel and Connor Lee
Bartz
Lmi 10h31'54" 37d14'32"
Daniel and Diane Arnold
Vir 12h25'11" -5d48'18"
Daniel and Elizabeth Lloyd
Forever
Vir 12h49'37" 3d53'37"
Daniel and Emily
And 0h32'40" 40d11'37"
Daniel and Erika Fore's
love
Tau 4h30'8" 16d5'45"
Daniel and Hali's Love Star
Cnc 7h58'16" 11d44'55"
Daniel and Jackson Year
Ori 5h39'21" -0d20'3"
Daniel and Joyann Brake
Ori 5h53'0" 7d11'0"
Daniel and Judy Estrada
Ari 2h23'4" 11d8'19"
Daniel and Judy Newlin
Cyg 21h16'25" 32d19'47"
Daniel and Lee Ann Key
Pyx 8h29'7" -36d23'16"
Daniel and Lidia Gubanski
Cyg 20h19'37" 44d31'26"
Daniel and Mary
Per 3h43'18" 46d45'1"
Daniel and Nicole
Ori 4h46'33" -2d23'27"
Daniel and Sophie Dawson
Gem 7h45'37" 20d20'59"
Daniel and the Three
Youths
Umi 13h10'19" 74d45'26"
Daniel André Martin
Dra 17h27'57" 56d56'47"
Daniel Andresen
Gem 6h33'18" 13d59'54"
Daniel Andrew Brandon
Peg 21h37'57" 16d49'15"
Daniel Andrew + Janine
Pho 1h23'29" -41d2'14"
Daniel Andrew Shettle - 11
June 2006
Umi 11h1'35" 87d47'49"
Daniel Angel
Per 4h22'43" 52d31'10"
Daniel Angel Luciano
Cnc 8h52'21" 11d50'42"
Daniel Angeron
Uma 8h40'5" 62d50'1"
Daniel Antener
Uma 9h57'5" 57d10'44"
Daniel Anthony Cooney
Sgr 18h34'2" -16d54'53"
Daniel Anthony Piloseno
Sr.
Ori 4h47'0" -0d22'26"
Daniel Aparicio
Sgr 19h27'31" -17d42'14"
Daniel Archer
Ori 5h32'55" 4d12'24"
Daniel Aron Ohligschlager
Aql 20h10'26" 12d54'19"
Daniel Arthur Kazarian
Leo 9h38'34" 23d31'33"
Daniel Arthur Pogue
Her 17h44'47" 45d34'20"
Daniel Arthur Scholze
Vir 13h34'27" -4d21'55"
Daniel Asad Haddad, Jr.
Cyg 20h43'18" 44d32'8"
Daniel & Ashley Greene
Ori 6h21'31" 10d15'23"
Daniel Atkinson
Per 3h12'35" 55d33'36"
Daniel Austen Ferretti
Cyg 20h2'27" 47d3'23"
Daniel B. Langford
Her 17h56'8" 26d15'35"
Daniel B. McVeigh
Sgr 19h1'7" -13d50'39"
Daniel B. O'Connell "DOC"
Cnc 8h44'55" 9d38'10"
Daniel B. Small
Cyg 20h32'36" 34d47'6"
Daniel "Ba" Woo
Cnc 8h47'43" 30d16'20"
Daniel Bandy Miller
Cnc 8h35'25" 13d25'10"
Daniel Bartlett
Sgr 18h13'11" -29d44'59"
Daniel Bartz
Cap 20h24'47" -14d46'40"
Daniel Bastidas
Ori 5h11'2" 3d45'15"
Daniel Baumgartner
And 1h17'42" 39d41'44"
Daniel Bazan, Jr.
Uma 12h2'52" 38d6'31"
Daniel Beckwith&Catharina
Cranford
Cnc 8h41'38" 22d42'39"
Daniel Bednarski
Cep 23h28'50" 77d10'2"
Daniel Bena "Captain Dan"
Cep 21h38'23" 60d50'38"
Daniel Benjamin George
McKeating
Cep 22h32'57" 61d45'49"
Daniel Benjamin H. Miller
Homer
Lib 15h3'16" -5d6'19"
Daniel Benjamin Hirtle
Umi 17h6'42" 82d3'0"
Daniel Bernstein
Uma 9h19'35" 51d3'26"
Daniel & Beth
Tau 4h32'11" 6d24'57"
Daniel "Big Dan" Flynn
Vir 15h5'26" 0d18'51"
Daniel Blank
Ari 2h54'53" 29d37'27"
Daniel "Blue Eyes"
Benjamin Beech
Lib 15h30'50" -17d10'53"
Daniel Boczek
Per 2h46'1" 48d35'35"
Daniel Boehle
Uma 8h49'31" 50d24'34"
Daniel Boone Area School
District
Uma 8h41'0" 65d39'31"
Daniel Bottass
Uma 9h17'16" 48d27'33"
Daniel Bottass
Per 3h17'1" 54d3'15"
Daniel Boyce Foster
Pho 23h29'0" -52d14'12"
Daniel Bram Jiho Seo
Stack
Uma 14h11'34" 58d32'35"
Daniel Brandon's Valentine
Uma 9h3'3" 55d56'7"
Daniel Brann - Rising Star
Lmi 10h20'16" 31d53'46"
Daniel Brian
Her 18h23'50" 23d20'37"
Daniel & Briana
Cyg 20h35'6" 37d29'12"
Daniel Brown
Vir 14h9'41" -15d33'52"
Daniel Bruce Vantrease
Uma 8h11'44" 70d5'28"
Daniel Bryant
Per 3h11'15" 44d23'12"
Daniel Bryniczka
Sgr 19h29'40" -22d44'9"
Daniel Bühler
Tau 4h24'53" 10d48'37"
Daniel C. Adams
Com 12h11'14" 18d49'40"
Daniel C Leronowich
Her 18h25'33" 27d17'38"
Daniel C. Macklin
Lyn 7h51'51" 36d21'50"
Daniel C. Rylee
Aql 19h26'23" -0d2'29"
Daniel C Tough, Sr
And 0h47'56" 39d17'57"
Daniel Cane
Per 4h21'53" 40d45'28"
Daniel Cardoza
Uma 8h45'25" 56d12'49"
Daniel Chance Hernandez
Lup 15h49'3" -31d2'52"
Daniel Charles
Cru 12h32'47" -56d12'33"
Daniel Chavez
Dra 15h56'46" 63d35'35"
Daniel Cherbuin
Uma 14h23'15" 56d59'44"
Daniel Chmara Collins
Umi 14h39'38" 79d47'24"
Daniel Christian Suppiger
Ori 6h13'10" -1d23'15"
Daniel Christiansen
Uma 11h36'13" 34d53'5"
Daniel Christopher
Lmi 9h39'31" 35d54'51"
Daniel Christopher
Dra 17h21'35" 53d33'26"
Daniel Christopher Boylan
Lyr 18h38'55" 46d56'54"
Daniel & Christopher's
Shining Star
Cep 21h33'32" 62d29'40"
Daniel Ciampa's Shining
Star
Cru 12h28'26" -63d6'33"
Daniel Clive Wheldon
Aql 19h3'55" 0d51'22"
Daniel Clyde Ewing
Ori 5h54'0" 20d46'41"
Daniel Cobb
Uma 9h22'6" 53d22'44"
Daniel Cole
Her 16h39'40" 21d18'16"
Daniel Conroy
Psc 1h23'0" 31d38'47"
Daniel Cordova
Ori 4h54'37" 5d19'28"
Daniel Cox
Psc 1h45'58" 9d11'55"
Daniel Crowe
Gem 7h14'22" 25d9'10"
Dániel csillaga
Leo 10h11'10" 11d57'31"
Daniel D. Pronti
Ari 2h0'44" 20d26'37"
Daniel D. Siders
Leo 10h38'23" 17d42'23"
Daniel D. Velasco
Cap 20h22'41" -12d44'32"
Daniel D. Wagnon
Lib 15h8'26" -8d51'35"
Daniel D. Wheeler
Her 18h43'38" 21d24'52"
Daniel Daggett
Ori 5h36'24" 10d44'44"
Daniel Dallin Floyd
Per 4h22'54" 50d47'59"
Daniel & Daniela
21.10.1994
Ori 5h42'58" -2d3'50"
Daniel Danny
Lib 15h33'2" -15d30'54"
Daniel "Danny" Abbott
Gem 7h52'43" 33d1'43"
Daniel "Danny" Cowhey
Ari 2h17'7" 25d49'11"
Daniel "Danny" Hocevar
Vir 12h12'23" 2d6'3"
Daniel 'Danny' Smith
Lmi 10h43'19" 28d55'48"
Daniel David
Her 18h17'29" 23d2'36"
Daniel David
Aqr 20h48'25" -2d57'36"
Daniel David John Gibbons
Aql 19h18'35" 0d28'20"
Daniel Day II
Gem 7h32'13" 26d50'3"
Daniel "Dazur5" Buval
Ori 6h4'51" 20d41'37"
Daniel & Denise Ewing
Cyg 21h34'0" 34d48'47"
Daniel Desmond McCarthy,
Jr.
Lib 14h52'23" -15d15'28"
Daniel Devers Newsome
Lib 15h47'15" -28d55'2"
Daniel DeWarrenne
Rogers-deCastillo
Her 18h29'11" 13d9'20"
Daniel Dodd
Sco 16h30'53" -29d10'16"
Daniel Dominic Manetti
Vir 12h11'32" 10d17'4"
Daniel Douglas Moore
Cep 1h9'36" 80d2'3"
Daniel Dresbach
Per 3h30'18" 40d47'16"
Daniel Drucker
Umi 15h18'40" 73d7'50"
Daniel Duckett
Her 17h24'43" 22d14'58"
Daniel Dwain Burnett
Sgr 19h8'14" -23d1'58"
Daniel E. Bailey
Gem 6h37'27" 15d15'30"
Daniel E Campbell
Ori 5h47'46" 7d11'40"
Daniel E. Gargel
Cap 20h50'1" -17d36'36"
Daniel E. Levin
Aqr 22h41'9" -0d14'35"
Daniel E Ryder
Uma 9h56'50" 65d33'38"
Daniel E. Vaupel
Her 17h13'41" 15d31'36"
Daniel E. Wolaniuk
Sco 17h23'42" -33d14'10"
Daniel Eaton Curtis
Umi 14h52'47" 70d29'25"
Daniel "Eddy Boy" Masuga
Uma 11h32'32" 38d21'8"
Daniel Edwin Paschal
Psc 23h42'45" 5d6'5"
Daniel Elizabeth
And 0h20'38" 27d16'32"
Daniel & Elizabeth Brigman
Wedding
And 1h42'39" 36d7'47"
Daniel Elliott Barry
Lib 15h20'14" -18d25'40"
Daniel Erbsmehl
And 1h6'19" 41d32'24"
Daniel Erdos
Aql 19h6'53" 15d10'25"
Daniel Ernest Hawkins
Uma 11h41'2" 52d37'26"
Daniel Estrella
Sco 16h6'29" -14d26'43"
Daniel Eugene Kahnke
Ori 5h52'14" 18d27'41"
Daniel Eugene Little
Cnc 8h28'45" 22d27'7"

Daniel Eugene Robert Hedge Sr.
Tau 5h39'40" 21d9'46"
Daniel Evan Hoffman
Dra 18h21'12" 56d50'58"
Daniel Evans: Faith, Hope, Love
Per 3h32'19" 32d37'19"
Daniel Everest Brown
Ori 6h18'52" 14d49'50"
Daniel F Owen
Tau 4h34'17" 6d7'21"
Daniel Fanchini
Uma 13h58'2" 56d2'59"
Daniel Faull
Ori 5h42'51" -9d6'46"
Daniel Felber
Aur 5h28'5" 47d19'20"
Daniel Ferguson
Uma 12h32'44" 53d18'10"
Daniel Ferrazzano, Jr.
Lib 15h8'8" -0d46'22"
Daniel Fischer
And 23h59'16" 42d53'13"
Daniel Fitzsimons
Umi 15h45'35" 76d54'27"
Daniel Floren
Aql 18h51'18" -0d4'37"
Daniel Foley
Uma 11h28'24" 36d28'53"
Daniel Forbes
Umi 13h28'39" 71d33'54"
Daniel Forcey
Uma 11h30'8" 60d24'46"
Daniel Francis Patrick Leahy
Uma 11h52'12" 44d59'40"
Daniel Frank Colverson
Cru 12h18'38" -62d26'38"
Daniel Fredrick Baenziger
Del 20h39'33" 13d40'7"
Daniel Frydrych
Ori 5h28'50" 3d5'55"
Daniel Füglistaler
Dra 19h37'47" 71d5'31"
Daniel G. and Callum R. Thompson
Aql 20h16'8" -2d5'38"
Daniel G. Harrison
Gem 6h43'49" 12d55'54"
Daniel G. Kuy M.D.
Psc 23h44'2" 5d26'16"
Daniel G. Schultze
Leo 9h26'3" 10d55'8"
Daniel Galea
Gem 6h42'0" 23d2'19"
Daniel Gard Shearer
Gem 7h13'48" 19d59'3"
Daniel Gardner McCollum
Gem 7h43'24" 25d24'56"
Daniel Gary Clark
Mon 8h5'50" -0d43'59"
Daniel Gene Rigdon
Uma 11h36'14" 30d21'23"
Daniel Gennaro
Per 3h6'27" 44d2'25"
Daniel George Clark
Psc 1h1'24" 14d10'34"
Daniel George Donnelly
Vir 12h24'49" -7d17'7"
Daniel George Mauger
Her 16h49'4" 27d6'37"
Daniel George Moxon
Uma 12h59'32" 55d35'10"
Daniel Gettings 21st Birthday Star
Per 3h16'4" 52d5'14"
Daniel Giancarlo Zanelli
Per 3h49'37" 32d27'24"
Daniel Gibb
Cep 20h44'16" 61d3'23"
Daniel Gilbert Mojica
Oph 16h38'1" -0d52'9"
Daniel Gionet
Umi 14h39'7" 75d46'1"
Daniel Gladwin
Boo 14h47'55" 26d20'48"
Daniel Glo
Ori 5h16'37" 5d41'49"
Daniel Goedken
Lyn 7h16'10" 47d44'25"
Daniel Goldman & Rachel Wallman
Uma 12h31'47" 62d41'39"
Daniel Gomilar
Her 18h40'53" 20d55'30"
Daniel Grabarkiewicz
Aqr 22h50'1" -8d15'48"
Daniel Grande
Cas 0h18'32" 53d38'21"
Daniel Greene
Per 2h44'2" 53d43'31"
Daniel Gregory Farry
Tau 3h43'16" 20d26'12"
Daniel Gregory Phelps
Aqr 21h48'35" -1d47'5"
Daniel Gregory Thrasher
Uma 12h0'7" 65d36'49"
Daniel Gregory Ziegenfuss
Uma 11h2'41" 69d13'21"
Daniel Guido Huertas
Cmi 7h30'45" 4d37'41"
Daniel H. Barnak Star
Uma 10h41'23" 66d40'48"
Daniel H. Coe, Jr.
Sco 16h16'0" -12d42'39"
Daniel H. Deutsch
Lmi 10h12'31" 29d34'9"
Daniel H. Fenn
Cap 20h33'5" -19d8'2"
Daniel H. Schulte
Uma 11h53'31" 29d9'26"
Daniel H. Stockwell
Dra 17h4'56" 56d21'1"
Daniel H. Zirlott
Cnc 8h32'23" 13d6'35"
Daniel Haghighi
Sgr 18h1'54" -31d40'39"
Daniel Hallock
Umi 16h1'3" 79d57'37"
Daniel Hamby
Uma 9h46'21" 48d51'7"
Daniel Hammer
Lib 14h43'57" -11d52'27"
Daniel Handalian
Uma 9h41'9" 44d7'49"
Daniel Hansen
Her 17h58'51" 49d40'43"
Daniel Hanson
Uma 12h58'6" 54d33'26"
Daniel Hartz
Uma 13h28'58" 61d41'35"
Daniel Hartz
Her 16h41'18" 6d11'57"
Daniel Hayes Morgan
Ori 5h18'40" 6d47'52"
Daniel Heinz Tietz - Braveheart
Cru 12h55'9" -62d56'38"
Daniel "Helmutfork" Gabel
Lib 15h17'26" -15d2'12"
Daniel Henri Huguenin-Virchaux
Ori 6h15'41" 10d9'52"
Daniel Henson
Her 18h35'40" 23d37'18"
Daniel H.J. Pronti
Sgr 18h57'55" -18d40'33"
Daniel Hobbs
Umi 16h40'56" 75d12'31"
Daniel Hohrath 12/23/1970
Uma 11h2'28" 35d5'43"
Daniel Holt Sr.
Cap 20h23'17" -12d0'42"
Daniel Horton
Psc 1h13'30" 17d48'26"
Daniel Hudgins Cassell
Ori 5h28'38" 4d12'8"
Daniel Ian Campbell "1969-2006"
Sco 16h9'41" -17d22'41"
Daniel Ian Tieger
Aqr 21h36'56" 0d49'28"
Daniel Ihle
Srp 18h27'28" -0d25'42"
Daniel & Illary
Cam 7h25'48" 66d38'39"
Daniel Isaac Katz
Her 17h54'35" 34d17'25"
Daniel & Isabella=Soulmates Forever
Lyr 19h9'19" 42d28'49"
Daniel Isiah Feliciano
Sgr 20h10'27" -38d53'18"
Daniel J. Bevans Jr.
Psc 1h7'28" 28d8'42"
Daniel J. Bredesen
Per 4h29'52" 49d59'16"
Daniel J. Cella / Scenic View
Ari 2h4'27" 11d18'35"
Daniel J Dlugozima
Tau 4h28'8" 18d11'43"
Daniel J. Flick
Aqr 22h32'53" -2d9'2"
Daniel J Frey
Gem 6h46'21" 33d20'0"
Daniel J Hafrey
Eri 3h47'11" -0d36'52"
Daniel J. Honovich
Col 5h59'28" -27d47'25"
Daniel J. Kojack, Sr.
Cnc 8h20'36" 16d49'9"
Daniel J. Leary, Jr.
Psc 1h4'16" 28d59'13"
Daniel J. Leone
Ari 1h59'49" 24d23'26"
Daniel J Marten
And 1h19'30" 46d23'37"
Daniel J. McDuffee
Cep 22h59'1" 63d30'30"
Daniel J. Mendelsohn
Ori 6h19'2" 9d26'59"
Daniel J. Monahan
Uma 11h11'33" 51d36'34"
Daniel J O'Connell, Jr.
Uma 11h54'57" 63d9'21"
Daniel J. Oshea "Danny Boy"
Per 4h50'10" 46d26'48"
Daniel J. Politi
Uma 11h59'50" 31d27'18"
Daniel J. Popp, M.B.A.
Equ 21h12'27" 10d5'44"
Daniel J. Quinn Jr.
Lib 14h46'3" -19d10'42"
Daniel J. Sapienza
Lib 14h57'31" -24d20'54"
Daniel J. Szelag
Cap 21h38'55" -12d46'50"
Daniel J. Tuzinowski
Psc 1h11'2" 26d20'9"
Daniel J. Weber, Jr.
Cnc 8h55'36" 26d43'24"
Daniel Jack
Sgr 19h9'32" -16d42'57"
Daniel Jack Passey
Cyg 21h36'18" 35d46'41"
Daniel Jacob
Leo 11h45'26" 25d20'59"
Daniel Jacob
Cnc 8h42'21" 10d55'10"
Daniel Jacob Hotra
Sco 17h46'16" -37d6'7"
Daniel Jacob Radcliffe
Leo 10h16'54" 24d53'50"
Daniel Jacob Winchester
Sgr 19h39'8" -12d54'15"
Daniel Jacquemai
Cas 23h59'58" 53d1'14"
Daniel Jäggi te amo para siempre
Lyr 18h43'19" 35d24'23"
Daniel Jake Hoskins
Cap 21h44'0" -11d0'8"
Daniel James Adams
Her 18h48'26" 24d39'33"
Daniel James Bredehorn
Leo 10h21'24" 15d21'23"
Daniel James Brigham
Psc 0h32'32" 8d55'47"
Daniel James Bruso
Ori 5h30'13" 0d43'13"
Daniel James Chambers
Cep 1h25'53" 87d28'39"
Daniel James Compton
Cnc 8h23'51" 12d8'1"
Daniel James Cornell
Sco 17h26'28" -32d0'20"
Daniel James Daggett
Ori 5h13'47" 1d45'9"
Daniel James Dodson
Cap 21h30'20" -17d41'1"
Daniel James Dotro
Sgr 19h12'0" -30d22'27"
Daniel James Edward Kelly
Lup 15h30'47" -40d16'30"
Daniel James Faulkner
Cyg 19h36'54" 31d53'34"
Daniel James Henderson
Umi 14h50'5" 73d27'10"
Daniel James Marks
Ori 5h50'3" 18d56'20"
Daniel James Packwood
Per 2h19'48" 54d25'41"
Daniel James Parker
Uma 13h23'57" 58d30'31"
Daniel James Skjegstad
Sco 16h12'2" -16d27'40"
Daniel James Taylor
Uma 10h43'41" 50d53'24"
Daniel Janis
Cap 20h14'49" -10d53'59"
Daniel Janquart
Uma 11h34'45" 54d53'16"
Daniel Jarvis
Cru 11h57'31" -62d41'4"
Daniel Jason Greenhalgh
Per 3h26'37" 43d54'58"
Daniel Jason Haas
Cap 20h24'33" -24d45'39"
Daniel Jay Mancini
Sco 17h50'5" -38d31'3"
Daniel Jay Stuart
Uma 10h1'9" 48d47'58"
Daniel & Jennifer Nolt
Cyg 20h18'19" 45d6'21"
Daniel & Jenny
Psc 0h4'40" 6d2'33"
Daniel Jimmy Schmitz "1961-2005"
Tau 4h14'11" 14d43'38"
Daniel (Joey) Guth III
Cap 20h42'18" -15d11'38"
Daniel John
Sco 16h7'18" -15d27'25"
Daniel John
Per 2h43'40" 53d21'35"
Daniel John Cicciaro, Jr.
Per 4h47'26" 41d24'34"
Daniel John Coleman
Psc 1h1'14" 24d59'16"
Daniel John Dreves
Leo 11h34'24" -0d20'6"
Daniel John Foster
Aql 19h58'46" -0d27'17"
Daniel John Frishkorn
Sco 16h53'31" -42d22'26"
Daniel John Gillard
Per 2h55'58" 40d48'8"
Daniel John Greenwaldt
Sco 17h48'22" -36d44'42"
Daniel John Hazelgrave
And 0h2'58" 39d50'28"
Daniel John Jopek III
Tau 4h16'42" 10d57'26"
Daniel John Klementovich
Tau 4h28'41" 16d51'10"
Daniel John Martin
Uma 10h20'19" 48d26'46"
Daniel John Posbrig
Lyn 7h24'16" 53d53'28"
Daniel John Randall - Danny's Star
Her 18h13'41" 19d3'7"
Daniel John Rennie
Per 2h50'29" 45d17'37"
Daniel John Schweitzer
Per 3h26'25" 48d20'8"
Daniel John Stogner
Aqr 22h35'29" -2d54'21"
Daniel John Van Haften
Ari 3h4'44" 26d49'54"
Daniel Joseph Alleman
Cyg 20h14'55" 43d31'8"
Daniel Joseph Bayer
Cap 21h35'31" -8d38'17"
Daniel Joseph Bruno, MD, Ph.D.
And 0h37'37" 31d10'53"
Daniel Joseph Fazzino
Leo 10h29'53" 14d35'5"
Daniel Joseph Grahn
Lmi 9h50'54" 38d20'25"
Daniel Joseph Heaney
Ori 5h3'19" 4d57'3"
Daniel Joseph Hodder
Cen 13h43'22" -40d40'36"
Daniel Joseph Marcocci
Lib 15h0'23" -11d59'28"
Daniel Joseph McGrath
Lyr 18h32'35" 33d53'5"
Daniel Joseph McKenna
Vir 12h42'29" 12d3'23"
Daniel Joseph Muro/Trophy
Uma 12h24'30" 54d3'23"
Daniel Joseph Paradis
Leo 11h44'8" 19d17'11"
Daniel Joseph Ricciardi
Cap 20h22'39" -19d38'49"
Daniel Joseph Rizzo
Per 3h12'54" 53d39'49"
Daniel Joseph Rumbolo
Her 17h15'51" 46d0'4"
Daniel Joseph Sloma
Cnc 8h5'44" 20d19'58"
Daniel Joseph Sullivan
Ari 3h16'58" 29d43'27"
Daniel Joseph Van Dyke
Psc 2h4'5" 6d41'19"
Daniel Juba
Psc 1h2'44" 6d47'13"
Daniel Judge
Uma 8h11'0" 65d50'7"
Daniel K Case
Ori 5h59'57" 22d34'5"
Daniel Katherine Scott
Ari 2h8'36" 17d54'59"
Daniel Keith Elledge
Leo 11h15'27" 15d12'40"
Daniel Keith Franks 2/05/1991
Ori 5h40'16" 2d7'44"
Daniel Keith Watts
Umi 16h46'11" 83d53'43"
DANIEL KELLER
Boo 14h8'24" 7d59'17"
Daniel Kennedy
Ant 10h31'11" -35d34'12"
Daniel Kenneth
Gem 6h33'24" 24d43'43"
Daniel Kenneth Mullen
Sco 16h13'32" -12d14'45"
Daniel Klarnet
Sco 16h4'43" -10d18'30"
Daniel Konstantin Faust
Vir 13h14'33" 11d8'41"
Daniel Korica
Cnc 8h39'7" 15d29'11"
Daniel "Kudanrt" Solis
Cam 3h33'20" 67d19'52"
Daniel Kuebler
Sco 16h42'17" -26d7'18"
Daniel Kyriakos Riskas
Col 5h30'27" -35d26'25"
Daniel L. Bingham
Cyg 20h54'30" 30d7'5"
Daniel L Hines
Leo 9h37'2" 32d16'51"
Daniel L. Losey
Her 17h52'40" 15d32'4"
Daniel L Myers
Ara 17h22'19" -51d56'19"
Daniel L. Owen
Lib 15h5'57" -14d9'9"
Daniel L. Pletcher
Cyg 20h12'19" 40d53'34"
Daniel L. Shannon
Per 4h20'39" 51d58'6"
Daniel L. Wood
Cyg 19h42'23" 38d36'12"
Daniel Lacey
Cep 23h11'38" 70d29'21"
Daniel Lane Van Cleave
Tau 3h50'39" 24d11'45"
Daniel Latka
Ari 2h11'24" 21d40'7"
Daniel Lawrence Salters
Her 16h49'48" 46d15'55"
Daniel Lee Boyd
Her 16h57'50" 40d28'37"
Daniel Lee Kehrer II
Lyr 18h47'42" 35d1'57"
Daniel Lee Kubesh
Aql 19h51'17" 12d11'0"
Daniel Lee Otteson
Sgr 19h31'25" -14d49'34"
Daniel Lee Root 1978-2004
Tau 5h49'31" 17d15'33"
Daniel Lee Tillman IV (ELS)
Per 4h26'29" 44d37'58"
Daniel Lee Wilson
Ori 6h15'59" 15d22'51"
Daniel Leighton Flather
Psc 23h2'20" -0d58'26"
Daniel Leland Diamond
Sco 17h39'31" -32d1'32"
Daniel Lemp In ewiger Liebe Nadine
And 1h41'11" 43d16'9"
Daniel Levi
Cap 20h46'15" -14d35'32"
Daniel Lewis Brodsky
Her 17h17'5" 19d20'41"
Daniel Lewis Loves Sarah Lingard
Dra 18h46'22" 53d45'39"
Daniel Liam Maistelman
Umi 16h23'4" 76d43'23"
Daniel Limouzin
Gem 6h21'59" 17d58'2"
Daniel Lindsey
Per 2h51'30" 54d4'19"
Daniel Loewen
Per 4h16'15" 43d51'44"
Daniel Louis Hoffman
Lib 15h20'17" -26d47'47"
Daniel Loving's Star
Her 18h28'21" 13d9'39"
Daniel & Lydia Zumwalt
Cyg 19h45'24" 31d15'10"
Daniel M. Sheehan Jr.
Uma 8h24'6" 70d29'6"
Daniel MacDonald
Psc 0h16'23" 19d18'14"
Daniel Malawy
Ari 3h20'16" 28d24'31"
Daniel Manfre 02.10.1984
Ori 5h50'50" 19d38'49"
Daniel Mani Mega Star
Aql 18h58'30" -0d29'26"
Daniel Manix Kindred
Cap 20h54'27" -25d19'49"
Daniel Marbach
Cyg 21h7'5" 36d49'41"
Daniel & Marianne
Cyg 19h40'49" 38d28'36"
Daniel & Marie Richard
Uma 13h41'0" 56d26'41"
Daniel Marino and Ronalee Francioni
Sco 17h54'24" -35d58'3"
Daniel Mark DenBraber
Uma 11h10'50" 58d7'35"
Daniel Mark Holdaway
Cen 13h22'48" -54d56'55"
Daniel Mark Johnson
Lyn 7h9'1" 53d7'6"
Daniel Martin
Gem 7h7'1" 33d57'6"
Daniel Martin Figueroa
Per 2h35'30" 52d1'52"
Daniel Martin Hunt
Tau 4h30'19" 16d42'33"
Daniel Matthew Ferrandi
Psc 0h37'58" 8d38'6"
Daniel Matthew Maine
Her 18h33'33" 16d15'55"
Daniel Matthew Robert Mayhew
Uma 9h56'40" 49d47'27"
Daniel Max
Uma 10h30'57" 60d23'48"
Daniel - Max 10.02.82
Tri 2h6'42" 33d23'25"
Daniel May
Vir 13h53'31" -1d3'2"
Daniel McCafferty
Umi 13h10'9" 75d13'43"
Daniel McDeavitt
Lyn 7h35'57" 36d56'2"
Daniel McKay Maher 11 17 45
Sco 17h16'35" -41d4'18"
Daniel McKenna
Aql 19h26'43" 3d36'41"
Daniel McMillan
Cra 17h59'49" -37d22'26"
Daniel & Meghan Baldridge
Cyg 20h55'13" 47d26'26"
Daniel & Melissa
Leo 9h42'58" 27d59'35"
Daniel & Melissa
Aqr 21h46'36" -0d14'25"
Daniel Mengel
Her 16h13'32" 47d31'21"
Daniel Mettraux
Uma 8h12'24" 72d23'55"
Daniel Michael Bressler
Uma 12h20'0" 60d30'13"
Daniel Michael Giannone
Lib 14h54'47" -1d15'18"
Daniel Michael Kuzian
Sgr 18h49'34" -26d48'3"
Daniel Michael Perry
Psc 1h12'4" 24d59'17"
Daniel Miles Preston
Vir 13h39'55" 1d25'51"
Daniel Miller
Psc 0h35'57" 18d3'20"
Daniel Miller
Umi 16h38'39" 77d49'52"
Daniel Milton
Lmi 10h16'5" 32d0'14"
Daniel Mirabile
Aql 19h15'18" 6d29'14"
Daniel Mitchell's Star
Leo 9h57'48" 14d9'0"
Daniel Moe
Lib 14h49'20" -1d35'48"
Daniel Morin
Cep 20h35'25" 65d14'30"
Daniel Murphy
Lib 16h1'1" -16d54'25"
Daniel My Brother
Uma 11h58'15" 34d48'33"
Daniel N. Delenela
Boo 13h50'36" 17d4'49"
Daniel Narden
Uma 10h12'7" 41d46'2"
Daniel Neil Stevenson
Uma 9h42'43" 71d49'13"
Daniel Noel
Aql 20h28'33" -0d30'13"
Daniel Nolt
Ari 2h32'8" 23d37'36"
Daniel Norman Fitts
Her 17h56'30" 14d57'16"
Daniel Normile
Cap 21h2'52" -26d43'7"
Daniel Note
Ari 2h16'27" 24d49'19"
Daniel Nowak
Cyg 20h21'13" 53d13'17"
Daniel O. A. Lundqvist Fernholm
Uma 8h28'25" 66d40'56"
Daniel O. Wagster 11-27-1927
Sgr 19h1'58" -17d33'57"
Daniel O'Brien Fulcoly
Ori 6h13'24" -3d47'26"
Daniel Olande Jones
Per 3h1'14" 44d58'53"
Daniel O'Neil
Cyg 21h23'47" 34d47'36"
Daniel Osguthorpe
Uma 9h10'33" 65d29'42"
Daniel O'Sullivan
Leo 11h5'21" 8d25'56"
Daniel P. Kempka
Umi 16h9'2" 76d17'28"
Daniel P. Maher
Ari 2h19'22" 24d38'23"
Daniel P. Miner
Her 17h51'3" 18d16'11"
Daniel P. O'Brien, Jr
Tau 5h31'2" 21d53'8"
Daniel P Pilachowski
Tau 4h33'12" 28d24'48"
Daniel P. Rodriguez
Her 17h6'50" 34d28'3"
Daniel P. Wilson, Jr.
Uma 12h32'11" 59d31'6"
Daniel Pagan
Per 3h9'54" 56d29'32"
Daniel Palmer
Sgr 18h23'48" -26d27'21"
Daniel Patrick Fitzgerald
Aqr 22h35'55" -20d33'35"
Daniel Patrick Leavitt
Leo 9h28'50" 6d28'29"
Daniel Patrick Moran
Cep 21h25'53" 64d27'43"
Daniel Patrick Murtagh
Aql 20h14'36" 4d22'29"
Daniel Patrick Reed Phillips
Lyn 8h16'36" 50d23'42"
Daniel Patrick Reno
Leo 9h29'56" 27d0'18"
Daniel Patrick Roach
Cnc 8h34'36" 30d53'32"
Daniel Patrick Thompson
Ori 5h56'13" -2d16'26"
Daniel Paul
Her 17h26'7" 34d46'35"
Daniel Paul DeLancey
Uma 10h37'32" 48d21'54"
Daniel Paul Higgins
Leo 9h36'40" 26d57'21"
Daniel Paul Magnuson
Aql 19h38'11" 7d53'14"
Daniel Paul Silvestri
Aql 19h58'11" 9d12'58"
Daniel Perry
Her 17h27'47" 36d17'11"
Daniel Perry
Uma 13h46'49" 58d37'16"
Daniel Peter - 12.11.2004
Cru 12h50'47" -64d21'10"
Daniel Peter 28
Aur 5h43'28" 40d6'16"
Daniel Peter Donahue
Sco 16h8'42" -12d12'30"
Daniel Peter Eisner
Sco 15h56'15" -21d0'40"
Daniel Peter Ingersoll
Ori 6h6'28" 20d12'23"
Daniel Peter Kirton Oakes
Lyn 6h46'34" 51d2'17"
Daniel Peter Murphy
Cep 22h33'51" 85d33'20"
Daniel Peter Robert Armitstead
Tau 4h26'32" 21d0'28"
Daniel Peter Spillane
Cru 12h15'50" -59d48'58"
Daniel Philip Brown - Dan's Star
Uma 11h57'10" 34d12'33"
Daniel Philip Quinn
Cap 20h23'15" -14d47'59"
Daniel Phillip Bar Miller
Sco 16h11'19" -34d44'27"
Daniel Phillip Lopez
Leo 11h24'59" -3d44'35"
Daniel Phillip Savona
Per 4h39'25" 49d34'51"
Daniel Pires
Uma 11h20'2" 56d0'35"
Daniel Pohle
Uma 8h51'54" 63d1'6"
Daniel Pomerleau
Uma 13h9'35" 60d54'38"
Daniel Pooler
Sgr 18h8'41" -17d0'41"
Daniel POP Thompson
Uma 11h15'9" 51d56'18"
Daniel pour toujours
Cnv 12h50'52" 48d2'6"
Daniel Prokes
Cnc 9h7'33" 32d17'16"
Daniel Pulliam
Ori 5h49'38" 5d16'47"
Daniel Q. Leon
Gem 6h48'47" 16d11'14"
Daniel Quintana
Cmi 7h22'56" -0d1'13"
Daniel R. Earl
Uma 11h53'4" 56d11'40"
Daniel R. Mazza-Manser
Aqr 23h45'37" -11d11'23"
Daniel R. Smith
Uma 10h3'56" 68d9'17"
Daniel R. Solomon
Cap 21h9'25" -16d40'49"
Daniel R. Watson
Aqr 22h3'46" -6d48'11"
Daniel Ralph Hirby
Cnc 8h50'36" 19d13'58"
Daniel Rand
Uma 11h39'57" 59d45'43"
Daniel Rascon & Patricia Munoz
And 1h57'21" 45d59'19"
Daniel Ray Orlando
Sco 16h18'47" -8d59'18"
Daniel Raymond Beauchaine
Lib 15h7'34" -2d51'32"
Daniel Reinhart
Tau 3h51'36" 26d28'28"
Daniel Rennhard
And 1h1'7" 48d39'29"
Daniel Repetti
Cnc 8h2'3" 10d50'55"
Daniel Rescher
Boo 13h49'29" 24d45'26"
Daniel Rhein
Aql 19h15'32" 13d3'23"
Daniel Ricardo Yanez
Her 16h26'12" 17d4'50"
Daniel Richard Montez
Uma 9h53'35" 61d0'31"
Daniel Richard Stickles
Leo 9h37'0" 28d37'3"
Daniel "Rigger Dan" Beard
Ori 6h5'28" 12d56'17"
Daniel Rivera
Vir 13h12'38" 4d35'20"
Daniel Robert Abrams
Aql 19h46'54" 5d31'14"
Daniel Robert Crawford
Leo 10h12'7" 10d58'33"
Daniel Robert Eyrle Rohlsen
Aur 6h8'45" 29d39'33"
Daniel Robert Mondragon
Gem 7h33'4" 26d33'0"
Daniel Robert Pizzarelli
Her 16h18'25" 48d43'36"
Daniel Robert Squance
Cep 20h16'12" 61d41'0"
Daniel Robin Kennedy
Lyn 8h1'42" 54d36'53"
Daniel Rodrigues
Aur 4h52'37" 37d48'58"
Daniel Rolland
Lib 14h31'9" -10d42'49"
Daniel Rooney
Aql 19h45'7" -0d33'59"
Daniel Russo
Leo 10h48'5" 17d51'8"
Daniel Ryan Duffy
Dra 18h31'14" 64d30'39"
Daniel S Côté
Cep 0h18'25" 80d16'58"
Daniel S. Doupe
Vir 12h30'58" 12d28'20"
Daniel S. Eisenklam
Cep 22h7'53" 72d52'58"
Daniel S. Espinoza
Lyn 6h53'3" 52d56'8"
Daniel S. Feeley
Aur 6h26'3" 41d16'55"
Daniel S. Foster "1927-2006"
Leo 11h30'1" -0d58'45"
Daniel S. Lass
Ari 1h49'39" 17d40'54"

Daniel S. Marroquin
Gem 7h32'48" 27d45'34"
Daniel S. Russell "my heart keeper"
Ari 2h7'2" 17d52'46"
Daniel Sabino Jude Yodice
Uma 9h0'15" 56d56'2"
Daniel Sacknoff
Lib 15h55'32" -10d44'51"
Daniel & Samantha
Cir 14h50'7" -64d13'2"
Daniel Samela
Cep 20h50'17" 69d32'20"
Daniel Sammon
Per 3h39'29" 34d23'40"
Daniel San Martin-Love of My Life
Vir 14h49'9" 6d37'48"
Daniel Sanchez Soria
Cap 20h43'52" -17d32'32"
Daniel Satoshi Yokomizo
Ori 5h8'54" 12d40'26"
Daniel Saunders
Her 18h17'42" 28d19'18"
Daniel Sawyer Baxter
Lib 15h31'1" -24d23'58"
Daniel Scalise's Death Star
Leo 10h51'20" 15d15'31"
Daniel Schindler
Leo 9h55'40" 23d25'44"
Daniel Scott
Cyg 21h38'30" 52d17'17"
Daniel Scott
Lib 15h36'4" -16d33'7"
Daniel Scott Anthony Beaulieu
Per 2h14'54" 56d26'41"
Daniel Scott Colangelo
Hya 9h0'42" -0d34'55"
Daniel Scott Fruchtman
Cep 21h46'57" 55d42'36"
Daniel Seidl
Uma 11h34'36" 56d54'11"
Daniel Shawbitz
Ori 5h12'38" 15d17'50"
Daniel Shelley Ashworth
Ori 6h22'35" 14d9'15"
Daniel Shiveley
Sco 17h15'14" -32d46'38"
Daniel Silva
Uma 9h56'46" 52d55'10"
Daniel Simmons
Per 4h12'9" 50d59'27"
Daniel Skyphantom
And 0h42'35" 36d52'51"
Daniel Smith
Aqr 22h56'5" -4d40'51"
Daniel Song 8:7
Cep 20h25'43" 60d55'33"
Daniel Sorum
Pho 1h7'17" -41d53'27"
Daniel Spradling
Cyg 20h7'26" 35d30'32"
Daniel "Star Colon" Clark
Her 18h1'22" 22d46'55"
Daniel Starker
Gem 7h22'37" 25d52'31"
Daniel Stavin
Crb 16h11'1" 28d4'11"
Daniel Steenerson
Cep 21h54'36" 63d30'23"
Daniel Stephen
Umi 14h52'36" 69d3'11"
Daniel Stephen Gdula
Aqr 22h31'3" -0d51'17"
Daniel Stephen Schafers
Tau 4h4'20" 12d4'59"
Daniel Steven Barton
Vir 14h29'7" 0d5'29"
Daniel Steven Sigala
Vir 12h41'6" 3d9'35"
Daniel Straus
Uma 11h59'20" 33d40'59"
Daniel "Superman" Uster
Uma 9h8'15" 52d1'26"
Daniel Swift
Umi 15h28'1" 80d1'30"
Daniel Swith (My Love)
Uma 13h52'21" 61d59'33"
Daniel T. Altersitz
Per 3h31'31" 34d53'15"
Daniel T. Murphy
Ari 2h11'42" 23d51'17"
Daniel + Tanja
Uma 13h2'31" 53d53'25"
Daniel Tarondo
Peg 23h45'5" 10d16'58"
Daniel Tarvin's Shining Star
Sco 17h11'18" -31d52'56"
Daniel The Great
Gem 7h51'18" 31d12'19"
Daniel Thomas
Leo 9h25'30" 17d1'23"
Daniel Thomas Algie, Jr.
Gem 6h43'15" 34d59'7"
Daniel Thomas Christenson
Her 16h27'19" 20d52'9"
Daniel Thomas Douglas Hercules
Ori 5h9'40" 8d34'48"
Daniel Thomas Edward Evans
Per 4h8'5" 44d25'48"
Daniel Thomas Engle
Cma 6h27'10" -29d6'37"
Daniel Thomas Jackson
Her 17h25'9" 38d31'52"
Daniel Thomas Osborne
Her 17h48'27" 26d36'19"
DANIEL THOMAS SPELL-MAN JR
Lib 14h27'47" -9d35'21"
Daniel Thomas Stamper
Crb 15h56'3" 27d48'0"
Daniel Thompson
Uma 9h52'54" 42d50'53"
Daniel Todd Luckey
Her 17h23'9" 37d57'39"
Daniel Tomos Springer
Dra 16h42'27" 65d54'44"
Daniel Tonery, Sr.
Ari 2h43'15" 28d49'21"
Daniel Tosh
Her 17h55'42" 21d26'6"
Daniel Townsend
Her 16h19'36" 46d37'51"
Daniel Tscheschlog
Aur 6h12'34" 31d43'16"
Daniel Tucker
Her 18h10'4" 41d25'55"
Daniel Turner
Her 18h29'37" 22d3'6"
Daniel Tyler White
Vir 15h9'23" 3d32'57"
Daniel Urbina
Uma 9h7'6" 51d53'48"
Daniel Urbina
Sgr 18h58'48" -21d19'3"
Daniel & Valarie Milam
Uma 11h31'34" 34d9'19"
Daniel Valdes
Ori 6h16'18" 10d28'7"
Daniel Vincent Bradley
Umi 14h5'19" 70d4'36"
Daniel Vincent Cioffi
Umi 12h22'44" 86d6'14"
DANIEL VINCENT GUER-RINO
Cap 21h0'54" -18d52'41"
Daniel Vincent Long
Sco 17h50'0" -38d34'36"
Daniel & Virgina Cahill 9-1-56
Uma 13h34'32" 53d22'45"
Daniel W. Bowman
Psc 1h9'38" 9d51'39"
Daniel W Livingston
Sco 17h51'59" -35d50'55"
Daniel Walsh Haskett
Gem 6h24'3" 22d24'33"
Daniel Ward
Vir 14h53'55" 1d9'20"
Daniel Ward
Cyg 20h16'38" 38d30'8"
Daniel Warren (Danny Boy)
Ori 5h57'33" -0d26'43"
Daniel Wassmer
Aur 5h17'26" 30d3'11"
Daniel Wayne Booth
Aql 19h49'21" -0d57'39"
Daniel Wayne Zimmerman
Gem 6h57'10" 13d42'38"
Daniel - we U as much as chocolate
Uma 11h23'42" 57d36'39"
Daniel Westwell
Aur 6h31'49" 35d35'9"
Daniel Wil Kinnunen
Lyn 7h5'7" 47d8'6"
Daniel William
Per 3h2'12" 42d12'14"
Daniel William Alexander McKenzie
Her 16h37'44" 8d37'21"
Daniel William Morgan
Sgr 18h53'39" -29d4'2"
Daniel William Tirendi
Crb 15h37'7" 38d55'9"
Daniel Wilson LoPresti
Uma 10h40'25" 69d44'12"
Daniel Windsor Jackson
Her 16h18'36" 44d45'35"
Daniel Woo Sr.,MD
Oph 17h23'0" -24d11'49"
Daniel Wright
Umi 13h24'12" 70d57'16"
Daniel Wultz
Per 3h23'19" 38d57'43"
Daniel y Elisa
Cyg 19h39'54" 28d8'56"
Daniel Y. Melcher
Aql 19h44'43" -0d54'33"
Daniel Yee Chak Fung
Tau 4h20'19" 22d47'36"
Daniel Zachary McMullen
Her 17h3'25" 34d0'57"
Daniel Zamichieli
Uma 11h9'31" 45d33'45"
Daniel Zemanek this love is forever
Tau 4h28'52" 22d51'50"
Daniel Zimmermann
Ori 6h17'31" 7d29'15"
Daniel, Katie & Laura Hawes
Ori 5h56'10" 21d12'59"
Daniel033185
Ori 5h53'47" 20d40'22"
Daniela
Gem 6h37'39" 25d52'42"
Daniela
Ori 6h20'40" 6d16'8"
Daniela
Ari 2h54'47" 11d14'18"
Daniela
Cam 3h16'37" 59d8'33"
Daniela
Cas 1h1'47" 56d50'53"
Daniela
Lac 22h22'42" 48d29'22"
Daniela
Lac 22h40'47" 50d13'12"
Daniela
And 1h40'26" 47d49'5"
Daniela
Cap 21h10'43" -15d35'24"
Daniela
Sgr 18h17'29" -18d13'25"
Daniela
Cam 5h31'26" 71d28'49"
Daniela
Uma 8h33'58" 59d57'37"
Daniela
Uma 14h4'28" 55d48'13"
Daniela
Psc 0h46'47" 7d28'0"
Daniela & Adrian
And 0h46'57" 42d30'43"
Daniela Alejandra Amaya
Tau 5h7'38" 17d51'28"
Daniela Alica
Uma 13h47'29" 52d52'13"
daniela amato
Uma 10h57'1" 59d49'13"
Daniela Berardi
And 23h11'44" 50d40'54"
Daniela Better
And 23h30'36" 41d59'18"
Daniela Bojorquez
Leo 11h18'52" 13d3'35"
Daniela Candinas
Cas 23h32'18" 56d14'34"
Daniela Castro
And 1h23'38" 35d56'7"
Daniela Catanese
Vir 13h9'14" -10d56'2"
Daniela Colandrea
Uma 10h8'20" 56d27'2"
Daniela Costa Diniz
Sgr 19h29'48" -28d26'59"
Daniela & Daniel XOX D2D
Pho 0h40'40" -40d40'6"
Daniela dLiu
Uma 11h26'38" 49d46'40"
Daniela & Dominik
Ori 6h6'58" 18d59'30"
Daniela e ....
Cyg 21h55'58" 53d47'9"
Daniela E. Britt
Mon 6h48'28" -0d7'26"
Daniela Elena Cecilia Martinengo
And 0h56'39" 39d2'14"
Daniela Elizabeth Cabrera
Cap 21h36'21" -12d55'45"
Daniela Ferrari - 20/07/1951
And 2h13'41" 42d15'23"
Daniela Francesca Tomasiello
Ari 3h13'34" 15d37'17"
Daniela Geinitz
Leo 9h27'1" 10d21'31"
Daniela&Gianniforever
Cyg 20h44'9" 43d22'38"
Daniela & Jason
Uma 9h21'3" 43d43'8"
Daniela Kancler
Umi 16h54'5" 75d25'22"
Daniela Knizova
Aqr 23h3'57" -6d13'19"
Daniela Knizova
Aqr 21h10'18" 0d25'16"
Daniela Künzel
Lac 22h35'46" 38d9'35"
Daniela La Rocca
Leo 9h28'4" 32d13'43"
Daniela Landini
And 23h52'0" 34d2'1"
Daniela Leo
Lac 22h22'12" 41d28'12"
Daniela Magdalena
Sgr 20h26'44" -33d22'58"
Daniela & Manuel
Boo 15h10'10" 48d25'57"
Daniela Maria Puliti
Sco 16h15'10" -11d58'16"
Daniela Maria Ticchio
Aur 5h11'34" 32d19'34"
Daniela Marino
Tau 4h8'25" 27d22'22"
Daniela Martinez
Col 5h39'28" -32d15'11"
Daniela Minacapilli
Lac 22h33'8" 42d38'38"
Daniela Nell Matz
Uma 11h56'7" 30d50'48"
Daniela Nicoleta Sofronie-"Dana"
Her 17h21'2" 39d54'50"
Daniela Onorati
Per 3h1'35" 51d40'47"
Daniela Pergjoka
And 2h24'32" 43d1'22"
Daniela Polo
Ori 5h56'27" 17d36'47"
Daniela Roccatello
Cas 0h28'18" 62d22'3"
Daniela Scuncio
Cap 20h24'9" -9d41'24"
Daniela Sima
Aqr 20h41'23" -9d36'58"
Daniela Simic Geistwite
Leo 9h51'5" 15d7'41"
Daniela Sofia Granaroli
Mon 6h26'0" -0d46'21"
Daniela Stanickova
Cas 0h38'9" 60d26'8"
Daniela Susana Solis Rodriguez
And 2h21'21" 45d13'46"
Daniela SV
Vir 12h23'6" 0d13'11"
Daniela Turconi Polpetta Di Mostro
Uma 9h18'36" 53d14'16"
Daniela Twinkle
Psc 0h8'30" 6d16'3"
Daniela und Christian
Uma 9h59'48" 72d46'38"
Daniela und Georg Eigner
Uma 8h25'33" 63d48'21"
Daniela und Nathalie
Uma 8h45'35" 50d32'43"
Daniela und Philipp's Hochzitsschtärn
Uma 8h24'31" 65d30'35"
Daniela Wenger
Ori 5h9'10" 6d10'30"
Daniela's Diamond In The Sky
Lib 14h44'44" -13d45'9"
Daniela's Stern
Ori 5h20'56" 5d21'21"
Danielaura
Uma 10h35'37" 68d2'59"
Daniele
Umi 15h30'0" 68d37'19"
Daniele
Ser 18h19'45" -11d4'21"
Danièle
Tau 4h29'53" 21d23'11"
Daniele
Ori 6h18'6" 16d0'17"
Daniele
Aur 5h53'59" 32d44'56"
Daniele Bocci
Aur 6h32'9" 33d56'4"
Daniele C. Hughes
Crb 15h22'53" 30d59'7"
Danièle et Daniel Prost
Peg 21h51'54" 12d14'37"
Daniele Haywood
Leo 10h24'51" 9d40'48"
Daniele Iafolla
Ori 5h9'58" -0d10'6"
Daniele & Paola
Cyg 21h29'15" 36d58'20"
Daniele Rocchi
Cam 7h21'5" 74d21'56"
Daniele Salvatelli
Lyr 19h2'7" 26d14'4"
Danielele's Wish
And 23h12'51" 48d14'1"
Danielhero
Her 16h49'28" 32d14'4"
Daniel-ke4nok
Psa 22h43'13" -25d49'34"
Daniella
Tau 4h40'36" 23d10'24"
Daniella
Ori 6h1'59" 5d34'5"
Daniella and Diana Rosales
Cyg 21h39'20" 44d47'56"
Daniella Cooke
Leo 10h54'32" 16d1'42"
Daniella Froio
Uma 13h34'23" 56d22'37"
Daniella Giorgio
Ari 3h10'51" 18d40'22"
Daniella J. Wilcher
Psc 0h13'47" 4d18'26"
Daniella JNN Mitchell
Cnc 9h12'48" 15d58'14"
Daniella Joyce
And 0h44'50" 28d52'55"
Daniella Krayzelburg
And 1h23'55" 37d15'40"
Daniella Lopez
Del 20h34'38" 19d17'55"
Daniella Marie Michelle Martinez
Com 12h18'44" 27d40'5"
Daniella Mattioli Knight
Gem 7h43'10" 31d21'39"
Daniella Nicole Peer
Lib 15h2'6" -2d34'45"
Daniella Pantoja
Lmi 10h45'12" 36d39'21"
Daniella Paula
Aqr 23h5'49" -7d14'59"
Daniella Ruby
Leo 11h53'2" 19d19'54"
Daniella S. Oleyek
Crb 15h21'30" 28d30'20"
Daniella Schaffner Jr.
Crb 16h5'55" 28d14'2"
Danielle
Lmi 10h53'22" 25d4'59"
Danielle
Tau 4h9'3" 28d37'19"
Danielle
Ori 5h52'20" 22d49'39"
Danielle
Tau 5h58'20" 28d8'15"
Danielle
Tau 6h0'32" 26d59'31"
Danielle
And 0h14'11" 29d36'27"
Danielle
Ari 2h52'38" 27d57'9"
Danielle
Ari 2h53'44" 18d32'22"
Danielle
Ori 6h19'52" 16d23'32"
Danielle
Aql 20h2'33" 11d49'56"
Danielle
Ari 2h1'8" 17d42'11"
Danielle
Ari 2h16'35" 21d54'10"
Danielle
And 0h48'4" 22d2'49"
Danielle
Gem 6h55'19" 13d27'0"
Danielle
Ori 6h16'59" 10d1'56"
Danielle
Cnc 8h49'5" 13d24'36"
Danielle
Leo 9h28'14" 7d19'55"
Danielle
Vir 13h29'41" 6d29'54"
Danielle
Psc 23h24'33" 3d40'58"
Danielle
Ori 5h5'20" 10d12'40"
Danielle
Ori 5h49'34" 11d42'14"
Danielle
Her 16h45'39" 30d50'46"
Danielle
Cnc 9h3'9" 32d33'52"
Danielle
And 0h41'39" 35d14'22"
Danielle
And 0h11'42" 40d20'57"
Danielle
And 0h55'8" 39d40'6"
Danielle
And 2h19'13" 37d50'43"
Danielle
And 1h49'1" 38d6'58"
Danielle
Lyn 7h54'46" 38d24'38"
Danielle
Lyn 8h16'32" 42d59'28"
Danielle
And 23h41'36" 38d34'20"
Danielle
And 23h27'8" 42d17'9"
Danielle
And 0h37'58" 45d41'22"
Danielle
Lyn 7h24'53" 51d17'58"
Danielle
And 2h22'25" 50d24'46"
Danielle
And 22h59'16" 52d19'40"
Danielle
And 23h33'14" 48d23'12"
Danielle
Cam 4h11'9" 56d17'7"
Danielle
Lib 15h4'48" -5d45'54"
Danielle
Aql 19h40'50" -0d1'13"
Danielle
Lib 15h48'46" -6d30'52"
~Danielle~
Sco 16h4'25" -10d43'34"
Danielle
Cap 20h9'33" -10d59'34"
Danielle
Aqr 22h1'53" -12d59'52"
Danielle
Aqr 23h42'6" -10d26'37"
Danielle
Cap 21h14'54" -19d2'34"
Danielle
Cap 21h58'3" -19d44'13"
Danielle
Aqr 22h36'4" -15d10'55"
Danielle
Cap 20h13'55" -20d38'15"
Danielle
Lyn 7h8'12" 55d47'10"
Danielle
Dra 18h25'38" 72d41'22"
Danielle
Psc 0h25'46" 4d54'4"
Danielle
Cra 19h1'13" -37d13'37"
Danielle
Sco 17h52'37" -35d49'51"
Danielle
Lib 15h45'8" -26d11'15"
Danielle 0720
And 0h48'3" 39d47'16"
Danielle 18
Gem 7h30'20" 26d21'36"
DANIELLE/1965
Sgr 18h14'35" -18d13'5"
Danielle 2004
Uma 9h48'7" 58d45'52"
Danielle <3 Bimonte
Sge 20h19'10" 18d20'0"
Danielle - 5583
Tau 4h41'10" 7d30'15"
Danielle A. Gagnon
Cyg 19h40'42" 33d31'26"
Danielle A. KJ
Gem 7h39'37" 24d36'48"
Danielle A Meier
And 2h14'46" 46d23'34"
Danielle A. Shoup
Aqr 22h26'11" -2d57'20"
Danielle Abbott
Uma 11h32'5" 46d42'40"
Danielle Abril Harrison
Sgr 19h32'44" -13d49'32"
Danielle Aleta Engle
And 0h12'41" 45d1'4"
Danielle Alexandra Joy Rothenberg
Equ 21h13'17" 10d20'57"
Danielle Alexis
Dra 18h49'49" 52d36'37"
Danielle Alicia Cross, Special Star
Uma 12h12'20" 56d32'45"
Danielle Amanda
Sgr 18h31'4" -28d5'1"
Danielle and Adam
Aqr 23h26'38" -21d27'17"
Danielle and Baby Bump
And 23h27'44" 35d41'40"
Danielle and Eric 1:01
Lib 14h38'52" -19d32'19"
Danielle and Joe forever
Uma 11h17'14" 71d26'28"
Danielle and John
Lyn 6h45'58" 52d38'28"
Danielle and Steve's Star 21
Dra 12h26'47" 72d58'37"
Danielle Anderson
And 0h21'4" 29d39'1"
Danielle Anderson
Psc 1h13'34" 28d14'42"
Danielle Angela Lora
Cap 20h49'41" -20d17'33"
Danielle Angelina Brockman
Gem 7h46'48" 32d46'22"
Danielle Anita
Cas 23h37'10" 52d2'46"
Danielle Ann
And 23h42'57" 43d59'53"
Danielle Ann Della Valle
Cas 1h18'31" 59d9'13"
Danielle Ann Reed
Mon 7h33'6" -0d22'54"
Danielle Ann Scott
Psc 1h42'30" 19d33'52"
Danielle Anne Ruggieri
Lib 15h24'44" -20d45'36"
Danielle Annis
Dra 16h27'13" 65d56'50"
Danielle Arcate
Tau 5h19'10" 28d32'54"
Danielle Ashley Ulmer
Cap 21h28'44" -14d44'7"
Danielle Austin Salomon
Tau 3h37'1" 7d54'31"
Danielle & Ava Kuczarski
Cyg 20h8'49" 56d39'7"
Danielle Ayleen Kukuljan
Leo 11h36'7" 23d44'8"
Danielle Ayles
Cru 12h4'58" -61d22'26"
Danielle B. Dancey
Sgr 18h58'33" -30d45'44"
Danielle "Baby Penny" Sullivan
And 0h42'40" 42d38'43"
Danielle Bachar
Cyg 19h55'20" 33d16'57"
Danielle Barber
Sco 16h52'6" -25d4'4"
Danielle Benedetto
Per 3h36'7" 45d53'33"
Danielle "Betty" Mattix
Dra 18h35'14" 79d43'57"
Danielle Boeck
And 1h46'1" 46d37'51"
Danielle Borsh
Cyg 21h46'9" 47d24'13"
Danielle Boyce & James Buckingham
And 22h58'30" 43d0'17"
Danielle Bradley
Crb 16h11'8" 34d25'4"
Danielle Bree Nelson
Uma 10h31'14" 50d59'22"
Danielle Broadbent
Ari 3h19'42" 27d28'31"
Danielle Brooke
Lyn 7h46'5" 43d43'9"
****Danielle & Bryan****
Cyg 20h9'30" 31d44'28"
Danielle Burrows
Cas 1h39'54" 67d54'14"
Danielle C. Heidmann
And 1h10'53" 41d54'1"
Danielle C. Spataccino
Boo 14h21'48" 14d18'35"
Danielle C Star
Uma 8h39'42" 47d46'41"
Danielle C. Strickland
Cnc 9h16'22" 20d22'8"
Danielle Calhoun
Cyg 20h5'40" 51d19'6"
Danielle Campagna
Crb 15h41'48" 38d24'9"
Danielle Campoli's Shining Star
Aur 5h44'17" 43d36'20"
Danielle Catherine-Nicole Anderson
And 22h58'57" 41d39'42"
Danielle Christina Capriotti
Ari 3h17'17" 29d8'59"
Danielle Christine
Sgr 18h30'52" -17d2'19"
Danielle Christine Benson
Sgr 19h42'48" -13d55'31"
Danielle Christine Holladay
Psc 2h4'6" 3d54'8"
Danielle Christine Schmidt
Cap 21h37'34" -13d45'6"
Danielle Clardy
Sco 17h28'7" -30d29'35"
Danielle Connie Riley
Vir 13h4'46" 11d55'34"
Danielle Crawford McLinden
Cas 23h40'40" 51d41'46"
Danielle & Curtis Mustapich
Uma 10h27'45" 46d29'7"
Danielle D. D'Amato
And 1h25'14" 47d3'38"
Danielle *Dani* Dexter
Uma 13h50'53" 48d45'20"
Danielle Dekofski, Baby Nellie
Vir 13h26'39" 5d9'45"
Danielle DeLallo
And 0h29'22" 28d58'14"
Danielle Delisio
Leo 9h31'28" 12d29'16"
Danielle Denny-Leach
And 0h1'19" 42d5'14"
Danielle DeSandis
And 0h49'15" 40d15'22"
Danielle DeToy
Cas 1h40'18" 67d41'41"
Danielle DeVore
Peg 23h6'7" 29d46'25"
Danielle Dewhurst
Del 21h8'38" 14d36'24"
Danielle Digiovanni
Ari 2h38'26" 30d4'13"
Danielle Dinten
Sco 17h53'41" -34d36'25"
Danielle DNA
Cam 4h32'32" 56d38'20"
Danielle Dobry
Aqr 22h44'11" -20d23'37"
Danielle Dolente
Cyg 21h52'52" 42d10'38"
Danielle Donahue
Aqr 22h38'2" -6d31'27"
Danielle Duce
And 23h1'55" 40d41'52"
Danielle Dumas
Lmi 9h59'49" 36d8'6"
Danielle Duncan's Star
Ori 6h1'32" 21d10'39"
Danielle E. Romano
And 2h35'22" 40d5'47"
Danielle Eason
Sco 17h55'4" -43d53'24"
Danielle Eileen Scott
And 0h24'12" 42d8'42"
Danielle Elaine Stone
Uma 9h0'48" 55d49'40"
Danielle Elena Dick
Uma 10h13'38" 52d57'22"
Danielle Elena Weliczko
Cam 7h24'24" 74d5'15"
Danielle Elizabeth
Uma 10h57'40" 59d59'58"
Danielle Elizabeth
Cas 23h2'58" 55d33'58"
Danielle Elizabeth
Cap 20h54'35" -18d37'35"
Danielle Elizabeth
And 23h25'50" 51d17'45"
Danielle Elizabeth Kempsell
Her 16h27'58" 6d46'55"
Danielle Elizabeth Manahan
Vir 13h26'0" 11d47'33"
Danielle Elizabeth Marcuccio
And 23h16'2" 44d51'31"
Danielle Elizabeth Medina
Lac 22h52'36" 52d39'42"
Danielle Elizabeth Muise
Uma 10h44'44" 48d42'45"
Danielle Elizabeth Struzziery
And 2h17'20" 44d7'44"
Danielle "Ellie" Schilling
And 0h43'9" 37d14'50"

Danielle Elyse And 23h9'1" 41d1'42"
Danielle Ericka Franklin Uma 9h10'15" 72d55'53"
Danielle Espinoza Vul 21h22'24" 27d23'16"
Danielle Esposito Aql 19h0'58" 3d49'44"
Danielle Esther Tamir And 0h37'39" 34d20'16"
Danielle Eva Gibson Cyg 21h10'24" 42d28'51"
DANIELLE EVE MACKENZIE Sco 16h20'9" -34d12'37"
Danielle F. Basil Peg 23h2'53" 13d21'52"
Danielle F. Glendenning Vir 12h17'47" 11d20'16"
Danielle Faircloth Lac 22h42'47" 42d44'26"
Danielle Faith Wynne And 0h23'6" 31d46'25"
Danielle Falorni Cnc 8h50'28" 9d19'53"
Danielle Fleming And 0h54'21" 45d48'39"
Danielle Franks And 23h52'48" 32d2'58"
Danielle Gail Kulp And 1h30'39" 44d10'15"
Danielle Gallock is Amazing And 0h34'36" 33d36'29"
Danielle Garcia Ari 2h50'53" 28d52'25"
Danielle & Geoff Miller Gem 6h42'38" 33d2'49"
Danielle Girardin Cnc 9h2'22" 21d51'16"
Danielle Giudici Wallis Psc 0h55'48" 9d0'51"
Danielle Goldsworthy And 1h45'35" 40d50'3"
Danielle Grace Lib 15h41'21" -17d55'54"
Danielle Guerin And 2h19'33" 37d58'13"
Danielle Guidi Cas 0h14'20" 62d49'9"
Danielle Guy ILYFE Cyg 20h1'14" 51d35'38"
Danielle Harrison My Love Lyn 8h27'28" 57d35'53"
Danielle Hayden I love you And 1h35'18" 46d56'46"
Danielle Helliar And 2h22'52" 49d28'24"
Danielle Hess Sgr 18h6'11" -16d57'32"
Danielle Hoover Ori 6h13'35" 3d10'7"
Danielle Hope Griswold-Wheeler And 22h57'42" 51d14'34"
Danielle Hope Jared Uma 9h22'18" 59d5'1"
Danielle Hopkins Cyg 19h56'36" 40d6'16"
Danielle Horak And 0h42'4" 39d35'0"
Danielle Howard Aqr 23h20'41" -18d55'14"
Danielle Hughes And 0h39'54" 40d29'13"
Danielle Hulsey And 0h15'22" 28d40'23"
Danielle Hunter And 0h0'51" 48d19'33"
Danielle Hutchinson And 0h32'24" 25d12'46"
Danielle I Love You And 23h16'41" 44d0'29"
Danielle I Love You With All My <3 Her 16h43'8" 31d50'3"
Danielle Irene Saavedra And 2h10'5" 39d3'28"
Danielle J. Griffis Leo 10h48'4" 16d13'59"
Danielle J Guell Ari 3h14'42" 16d22'31"
Danielle J. Lancour And 23h5'45" 41d19'22"
Danielle Jade Myers Umi 15h21'57" 69d3'28"
Danielle Jae Biswell Lib 15h31'57" -24d42'25"
Danielle Janine Kelsey Aqr 22h53'20" -20d22'11"
Danielle Jean Callahan And 23h21'47" 42d13'32"
Danielle&Jeramy Ori 6h11'40" 15d43'48"
Danielle & Jimmy Uma 10h5'5" 47d17'40"
Danielle Johnson Vir 12h32'48" -6d47'7"
Danielle K. K. DeMatta Vir 12h16'28" 7d18'1"
Danielle K. Ungaro Lyn 7h2'18" 45d8'57"
Danielle Karen Romero Gem 7h37'17" 26d39'10"
Danielle Kass Cnc 8h45'16" 26d20'41"
Danielle Kay Zehnder And 2h37'14" 50d36'53"
Danielle Kinley And 1h17'42" 45d9'50"
Danielle Kirsten Pankhurst Sco 16h57'30" -39d29'27"
Danielle Kohut Cnc 8h50'23" 32d12'7"
Danielle Korin McCormack Lyn 8h57'1" 45d15'22"
Danielle Kristine Bell Uma 11h38'21" 50d27'18"
Danielle "Kristine" Cleary Cyg 20h8'24" 36d33'0"
Danielle L. McCallum Aql 19h15'13" -7d39'43"
Danielle L Meador Gem 6h45'9" 27d11'32"
Danielle LaChance Lyn 8h52'41" 34d41'52"
Danielle Ladd Cnc 8h15'38" 16d20'15"
Danielle Ladd Ori 6h23'54" 14d6'10"
Danielle Lair Lmi 10h34'58" 33d14'4"
Danielle Lamorie Lyn 7h48'17" 40d30'54"
Danielle Langelier Vul 19h52'33" 28d25'35"
Danielle Lauren Griffiths Psc 0h44'6" 9d59'14"
Danielle Lauren Watkins Vir 13h35'27" -20d44'37"
Danielle Lee Cyg 20h33'19" 52d8'27"
Danielle Lee Beauregard Crb 15h44'53" 26d50'20"
Danielle Lee Hewitt - 12.01.1984 Cru 12h26'22" -62d0'39"
Danielle Lee Jones Ari 2h22'23" 26d4'47"
Danielle Leeann Frank And 23h20'2" 42d4'0"
Danielle Leigh Cyg 20h58'4" 34d40'9"
Danielle Leinassar Uma 14h21'58" 62d6'20"
Danielle Lenee And 0h50'12" 36d31'8"
Danielle Lindley Cyg 19h53'27" 36d30'58"
Danielle Logiudice Aqr 22h2'5" -14d46'13"
Danielle Lomax And 0h3'39" 44d4'22"
Danielle Lorraine Follette Ari 2h6'8" 11d41'58"
Danielle Louise Born Sco 17h52'44" -37d33'53"
Danielle Louise McMinn - Bik's Star And 23h4'22" 46d47'34"
Danielle Louise Williams And 2h7'11" 36d23'25"
Danielle Luisi Cyg 19h55'56" 32d50'16"
Danielle Luree Faircloth Tau 4h33'12" 18d15'55"
Danielle Lyne Cas 23h4'46" 54d48'51"
Danielle Lynn Vir 12h32'40" -2d4'11"
Danielle Lynn Bolte Tau 4h20'58" 23d40'47"
Danielle Lynn Deck And 23h46'14" 37d49'45"
Danielle Lynn Hernandez Ari 2h47'14" 20d11'17"
Danielle Lynn Perrigan Tau 4h2'48" 22d14'16"
Danielle Lynn Reed Uma 9h32'14" 41d49'31"
Danielle Lynn Santana Lyn 7h3'21" 52d47'22"
Danielle Lynn Scheuer Ori 5h56'37" 12d29'50"
Danielle Lynn Wentworth Cap 20h46'59" -21d11'38"
Danielle M. Bennetzen And 0h41'50" 24d43'1"
Danielle M. Cantelli Leo 9h59'11" 19d30'1"
Danielle M. Karrat Cyg 20h30'39" 34d43'24"
Danielle M. Shones Cnc 9h7'20" 31d46'53"
Danielle MacKenzie And 2h23'48" 43d19'7"
Danielle Margaret Fata Tau 4h34'34" 19d39'38"
Danielle Marie Tau 4h44'37" 26d37'18"
Danielle Marie Leo 11h32'25" 26d13'56"
Danielle Marie And 1h29'23" 44d22'49"
Danielle Marie Crb 15h42'37" 33d33'42"
Danielle Marie And 1h35'52" 46d16'3"
Danielle Marie And 23h27'17" 48d5'22"
Danielle Marie Vir 13h0'19" -21d6'40"
Danielle Marie Uma 9h37'45" 62d34'40"
Danielle Marie Alexis Aql 19h57'25" 11d10'8"
Danielle Marie Amato Leo 9h26'18" 18d11'35"
Danielle Marie Brown And 23h17'11" 49d24'29"
Danielle Marie Clark And 0h38'51" 35d20'35"
Danielle Marie Clites Sco 16h13'35" -9d56'10"
Danielle Marie Discepoli Cnc 9h17'15" 25d51'21"
Danielle Marie Duva Lyn 7h40'31" 41d18'59"
Danielle Marie Galati Cru 12h0'39" -62d0'22"
Danielle Marie Gorectke Uma 10h53'18" 57d18'40"
Danielle Marie Greco Ari 3h21'48" 27d18'42"
Danielle Marie Gurtler Dra 18h50'20" 67d2'32"
Danielle Marie Hall Tau 4h10'17" 15d42'58"
Danielle Marie Hockenberger Vir 12h32'25" -9d58'49"
Danielle Marie Hoke Leo 10h11'33" 12d21'5"
Danielle Marie Hope LeMay Lyr 18h39'40" 30d12'56"
Danielle Marie Huston-Brown Tau 5h18'32" 18d44'35"
Danielle Marie Lynn Jolly Tau 4h29'0" 24d1'29"
Danielle Marie Mannix Crb 16h6'5" 27d0'55"
Danielle Marie Olenik Cen 13h47'43" -33d11'51"
Danielle Marie O'Neill Lib 14h36'23" -10d29'37"
Danielle Marie Opyt Cyg 19h49'16" 32d24'18"
Danielle Marie Piccione Cap 21h47'20" -11d9'53"
Danielle Marie Steinberg Mon 6h53'15" -0d19'29"
Danielle Marie Ventura Sgr 18h20'58" -25d16'4"
Danielle Marie Zemko Sgr 19h7'57" -21d42'42"
Danielle Marissa Rodin Cnc 8h50'33" 32d30'52"
Danielle Marta - Marie Austin Lyn 8h6'44" 42d17'57"
Danielle Martin And 23h47'47" 37d35'48"
Danielle Mary LaBianca Sco 16h5'27" -11d48'35"
Danielle Maryke Andersen Lyn 8h1'3" 56d15'6"
Danielle Mathouser Cyg 21h36'56" 39d50'24"
Danielle & Matt Sge 19h51'8" 19d4'12"
Danielle Mauk Tau 3h33'14" 23d0'24"
Danielle Mause Vir 11h51'23" 3d23'34"
Danielle May And 22h59'34" 48d7'27"
Danielle M.C.A-73 Gem 7h3'44" 18d4'37"
Danielle McCauley Cyg 19h56'19" 32d19'7"
Danielle McHaney Lyr 19h12'25" 46d17'40"
Danielle McKenna Knick Leo 11h16'22" 21d49'4"
Danielle McKenzie And 0h27'17" 27d41'6"
Danielle Menard Cas 0h11'28" 61d28'36"
Danielle Meyerowitz Umi 14h6'31" 70d33'4"
Danielle Michelle Ireland Lyn 7h35'6" 36d51'33"
Danielle Michelle Singer Vir 12h42'51" 3d21'57"
Danielle Minet Merrick Vir 13h14'16" 8d51'6"
Danielle Mitchell And 2h4'37" 39d15'31"
Danielle Mitchell And 23h34'28" 44d42'39"
Danielle Monsibais And 1h44'20" 38d24'23"
Danielle Moose McDonald Lyn 7h55'57" 58d4'43"
Danielle "My Angel" Cas 2h22'25" 63d40'32"
Danielle My Belle Umi 14h5'32" 77d7'38"
Danielle (My Boo) Cyg 21h53'53" 39d18'56"
Danielle MY LOVE Cam 3h52'18" 61d56'16"
Danielle "My Love My Life" Cap 20h25'5" -15d16'19"
Danielle N. Ferrazzano Gem 7h19'49" 33d2'33"
Danielle N. Healey Ori 5h51'30" 6d49'46"
Danielle N. Lucas Gem 6h51'21" 28d16'54"
Danielle N. Puryea Uma 8h9'52" 63d30'44"
Danielle N Seklejian Per 3h18'49" 40d53'5"
Danielle Naselsker And 0h29'40" 31d17'28"
Danielle Nichole Sgr 18h36'24" -31d55'44"
Danielle Nickcol Preuss Sgr 18h16'43" -16d49'7"
Danielle*Nicole Ari 2h31'5" 11d53'25"
Danielle Nicole Benner Del 20h33'38" 13d34'39"
Danielle Nicole Choy And 1h58'5" 42d51'30"
Danielle Nicole Halsey Lyr 19h21'56" 37d38'1"
Danielle Nicole Howard-The Best One Lyn 6h57'44" 53d14'13"
Danielle Nicole Johnson Lib 15h10'53" -12d47'9"
Danielle Nicole Leggett Uma 9h11'33" 49d58'52"
Danielle Nicole McLean Uma 11h25'51" 62d19'13"
Danielle Nicole Ricke And 2h5'18" 37d34'15"
Danielle Nicole Rycyk Aqr 22h20'40" 2d12'47"
Danielle Nicole Wood Uma 10h45'48" 72d23'32"
Danielle O'Connor And 1h7'46" 34d12'22"
Danielle Opolka and Josh Czech Uma 13h40'2" 58d11'11"
Danielle Osborne Sgr 18h40'34" -21d27'40"
Danielle - Our Divine Intervention Cnc 8h17'53" 30d28'49"
Danielle Our Precious Diamond Per 2h57'21" 48d49'53"
Danielle P Eastwood And 1h24'35" 49d39'13"
Danielle P. Rayos Cam 4h8'52" 57d27'34"
Danielle Pacheco Gem 6h19'29" 21d54'56"
Danielle Palminteri Lyn 9h22'7" 37d46'48"
Danielle Papaleo Cru 12h39'28" -58d38'40"
Danielle Patricia Smith Gem 6h28'58" 26d0'16"
Danielle Pauline Koch Vir 11h50'21" -2d54'28"
Danielle Pennock And 2h27'55" 50d4'57"
Danielle Penrod And 23h15'40" 49d25'18"
Danielle Peterson Umi 16h46'1" 76d40'33"
Danielle -Princess Of My Universe- And 0h16'0" 31d47'25"
Danielle R. Bargovic And 1h50'55" 36d30'43"
Danielle R. Korman Ari 3h1'2" 20d15'4"
Danielle Raabe Cnc 9h3'26" 31d15'25"
Danielle Rachel Heft Cnc 8h31'38" 6d47'50"
Danielle Rachel Rosenblum Gem 7h44'13" 32d39'33"
Danielle Rae Long And 23h46'38" 48d17'9"
Danielle Ray Modlin Sco 17h19'48" -31d17'5"
Danielle Raye Uma 12h17'17" 61d22'2"
Danielle Rebecca Britt And 0h33'30" 45d29'21"
Danielle Reese Heckler Mon 7h41'51" -2d46'22"
Danielle Renea Gillmen Aqr 22h38'29" 0d55'9"
Danielle Renee Craft Sco 16h5'58" -18d20'47"
Danielle Renée Starr Gem 6h28'49" 16d41'2"
Danielle Renee Westerlund Ari 2h4'45" 20d29'10"
Danielle Rhodes Ari 2h52'39" 19d54'58"
Danielle Richard Uma 13h57'1" 51d20'38"
Danielle Richardson And 23h54'20" 46d48'26"
Danielle Rigo Her 16h47'33" 48d31'54"
Danielle Rinehart Per 3h45'38" 43d56'7"
Danielle Robb Pho 0h29'41" -41d50'48"
Danielle Roberts Sgr 19h15'26" -16d57'30"
Danielle Roberts Gem 6h50'54" 19d15'29"
Danielle Rose Psc 1h21'59" 23d58'5"
Danielle Rose Aqr 21h31'1" 1d32'20"
Danielle Rose Cyg 20h10'51" 41d38'21"
Danielle Rose Dix Vir 12h17'52" -0d54'28"
Danielle Rose Estep Cyg 19h34'43" 28d19'18"
Danielle Rose Gabor Ari 2h12'34" 24d57'14"
Danielle Rules! Gem 6h51'2" 26d6'16"
Danielle Sabrina Stranc Lib 15h10'31" -6d24'59"
Danielle Sallou Uma 8h37'15" 56d36'15"
Danielle Salquist Gem 7h14'30" 23d52'25"
Danielle Sanders Uma 10h27'13" 41d56'48"
Danielle Sapiens And 2h30'42" 41d49'8"
Danielle Savino Uma 11h10'2" 48d32'38"
Danielle Scagliola Cyg 19h57'55" 45d40'42"
Danielle Scheunemann And 0h59'52" 45d26'33"
Danielle Schneeloch Scharlach Cap 20h23'3" -17d17'9"
Danielle Silva Gem 7h2'50" 34d12'1"
Danielle Simone Novak Ari 3h20'26" 15d37'56"
Danielle Simone Vionnet Umi 13h58'7" 71d47'17"
Danielle Singleton Uma 11h41'53" 54d21'33"
Danielle Smith Com 12h17'4" 32d9'50"
Danielle - Smitty's Guiding Light And 0h33'23" 25d25'23"
Danielle Soltis And 1h0'42" 36d46'0"
Danielle Spangler Sgr 19h50'47" -21d42'21"
Danielle Spark Bier Cap 20h24'18" -19d0'2"
Danielle Sprague Cas 23h30'47" 51d26'57"
Danielle St. Peak Cap 20h24'23" -10d46'45"
Danielle Stephanie Pope Cnc 8h47'8" 27d36'59"
Danielle Stone Ori 5h34'52" 3d40'11"
Danielle Storhoff And 23h23'59" 51d47'32"
Danielle Summer Cap 20h15'15" -21d56'55"
Danielle " Sunshine " Houston Uma 13h46'11" 58d29'0"
Danielle Susan Cohen And 2h29'47" 44d46'16"
Danielle "Sweets" Snapp Vir 11h41'3" 10d14'21"
Danielle Sylvia Rose Frisina Cyg 20h53'18" 32d9'18"
Danielle T. K. D'Alessio Uma 8h54'17" 55d48'57"
Danielle Theberge Herard Peg 23h40'49" 25d10'43"
Danielle Theresa Sgr 18h48'22" -16d54'8"
Danielle Theresa Riley Psc 23h52'14" 7d23'51"
Danielle Thomas Lib 15h5'45" -2d48'24"
Danielle Valdez Del 20h25'13" 9d56'51"
Danielle Veronica Baurle Ori 5h12'27" 3d32'15"
Danielle Veronica, My Sunshine Star And 23h37'32" 47d49'53"
Danielle Vick Forever Loved Peg 22h41'58" 18d11'4"
Danielle Victoria Lyn 6h40'31" 55d3'48"
Danielle Villanueva Lyn 7h10'7" 48d11'26"
Danielle Warbois And 0h28'48" 33d1'0"
Danielle Williams Cas 2h12'39" 64d31'58"
Danielle Wood Tau 3h33'3" 3d48'5"
Danielle Works Uma 11h21'35" 38d8'1"
Danielle Xylena Saydeh Cap 21h15'26" -15d3'30"
Danielle Yliece Womack Ori 5h8'44" 0d20'47"
Danielle Yuckert Mon 6h57'18" -0d59'31"
Danielle Zubek Uma 11h32'50" 36d30'54"
Danielle, Catherine+ Mark Stutzmann Cam 3h47'56" 62d3'16"
Danielle041303 Umi 13h31'8" 73d8'48"
Danielle-Forever my lucky star Gem 7h31'37" 32d55'28"
Danielle-Louise And 1h1'53" 43d54'23"
DanielleNicole Vir 12h48'2" 5d47'10"
DanielleNicole Vir 13h38'17" -11d24'3"
Danielle's Diamond Cas 1h7'29" 49d27'28"
Danielle's Kiss And 1h30'34" 46d50'3"
Danielle's Light Ari 2h44'58" 23d27'25"
Danielle's Little Star Peg 22h25'26" 4d26'39"
Danielle's Little Way Vir 13h49'46" -2d36'27"
Danielle's Love Uma 11h33'32" 36d34'27"
Danielle's Playground And 23h35'23" 41d18'39"
Danielle's Shining Star Sco 17h17'25" -36d52'8"
Danielle's Star And 1h27'13" 34d0'26"
Danielle's Star Uma 11h35'12" 44d46'42"
Danielle's Star Vir 13h19'0" 2d33'36"
Danielle's Star And 0h32'4" 26d48'21"
Danielle's Star Leo 11h44'11" 24d41'12"
Danielle's Starr Ori 5h19'24" -8d53'52"
Danielle's Sunshine Lib 14h48'51" -2d33'47"
Danielle's Sweet Sixteen Lib 15h22'2" -24d39'5"
Danielle's Very Own Mars Cap 20h9'53" -18d27'27"
DanielleV Uma 10h59'15" 47d0'15"
Daniel's Ajax Umi 16h22'37" 76d28'59"
Daniel's Apology Uma 11h39'8" 34d31'54"
Daniel's Baseball Her 16h56'50" 23d45'52"
Daniel's Birvalmas Cap 20h55'58" -22d7'58"
Daniel's Light Cru 12h23'11" -57d42'53"
Daniel's Light Lyn 8h33'41" 39d34'38"
Daniel's Love Uma 10h17'15" 43d45'42"
Daniel's Ray Lib 15h32'55" -18d40'0"
Daniel's Star Umi 16h21'36" 82d34'40"
Daniel's Star Lib 14h53'12" -1d16'58"
Daniel's Star Aqr 23h2'43" -6d55'13"
Daniel's Star Leo 9h31'12" 15d35'4"
Daniel's Star Psc 1h6'48" 10d7'45"
Daniel's Star Psc 1h34'5" 15d24'42"
"Daniel's Star" forever Leanne Ara 17h10'10" -51d0'16"
Daniel's Star Of Time Tri 2h13'0" 33d50'0"
Daniel's Torch Gem 7h27'44" 29d18'52"
Daniel's van Gogh Lyn 7h32'16" 36d13'40"
Daniel-Sweety-Chouchou-Swiedzi Sco 16h40'35" -37d56'37"
DaniePatin Dra 15h56'11" 52d14'3"
Danie's worlds Ari 2h52'23" 27d44'20"
DaniJohnx3 Lyn 7h57'19" 41d44'28"
Danik Semenchyuk Umi 14h48'54" 72d53'55"
Danika Uma 10h27'12" 56d36'16"
Danika Uma 9h42'20" 50d12'4"
Danika Gray Lindsay Lyn 6h56'14" 60d33'25"
Danika Marie Vir 13h11'6" 11d21'10"
Danika Potgieter Pho 0h15'51" -42d32'32"
Danika Rae Anderson "Morning Star" And 23h36'7" 36d36'20"
Danilo Cra 18h56'16" -39d30'53"
Danilo Uma 9h9'17" 55d47'11"
Danilo Curioso Jr. Leo 9h23'7" 28d57'6"
Danilo Gaitan Uma 11h59'19" 31d41'17"
Danilo Galvagno Lyr 18h29'13" 32d30'2"
Danilo Lesnjak Dra 17h17'58" 56d9'8"
Danilo Zelenovic Ari 2h48'36" 28d50'27"
Danimal Uma 9h29'43" 41d57'10"
Danimal Per 3h11'40" 53d20'31"
Danimal Uma 10h38'34" 58d38'28"
Danis Dancing in Light Beautiful Gem 7h18'6" 23d2'55"
Dani's Destiny Sco 17h53'55" -36d49'13"
Dani's Diamond Dra 16h36'56" 63d2'20"
Dani's Dreaming Star Sco 17h28'39" -40d57'5"
Dani's Dukte Gem 6h40'52" 23d14'3"
Dani's Light of Hope Gem 7h21'40" 25d46'27"
Dani's Star Ori 5h32'13" 8d0'48"
Dani's Star And 23h53'6" 46d24'11"
Dani's Star Umi 14h19'41" 75d48'24"
Dani's Star Sgr 19h50'37" -15d3'17"
Danisa Mon 6h29'7" -3d53'23"
Dani-Schatz Cnc 8h57'38" 15d48'0"
Danish Dreamboat Gem 6h28'19" 25d50'28"
Danish Shariat Siddiqui Sco 17h36'55" -40d15'34"
Danisha & Wynn Written in the Stars Cyg 20h45'7" 35d24'3"
Danit Tyler & Sequoia Christensen Her 18h38'42" 19d40'43"
Danita Uma 10h36'43" 71d35'17"
Danita Sgr 18h25'47" -21d38'58"
DANITA Lib 15h47'52" -11d29'17"
Danita A. Day Gem 6h50'41" 27d47'10"
Danith Boo 14h12'55" 20d20'39"
Danitza & Lazaro for Eternity Gem 7h19'53" 17d1'13"
Daniva Ori 5h50'51" 9d20'44"
DaNiVeneCK Uma 9h32'4" 66d43'25"
Danivia Col 5h48'22" -34d59'22"
Daniyella Psc 1h8'25" 10d57'42"
Danja Cep 22h30'43" 66d52'13"
Danjewel* Her 17h8'48" 14d30'15"
DanJoe Her 16h12'20" 17d24'59"
Danka Prilepkova - Gale Lib 15h6'0" -1d36'9"
Danko Stojnic Ori 5h59'0" 20d45'36"
Dann Saladin Sco 16h49'55" -32d23'50"
Dann W Carson Sgr 19h21'29" -23d22'38"
Danna Sgr 18h0'10" -30d10'42"
Danna Byers Cnc 8h18'37" 21d33'44"
Danna Marie Oldham Ari 3h22'6" 22d50'3"
Danna Morris Uma 8h13'31" 60d34'18"
Danna Valenti Uma 8h35'29" 63d58'33"
D'Annabell Aqr 22h5'36" -13d27'39"
DannaHarry And 0h34'17" 42d58'20"
Dannelys Raposo-Isabel Oph 17h21'25" 3d35'44"
Danner Dra 18h34'24" 69d29'32"

Dannette D. Olivarez
Lyn 7h1'23" 53d51'56"
Dannette Dee Lynch
Del 20h36'56" 15d13'33"
Dannette Lynn Bowers
Cap 20h17'40" -25d10'44"
Danni
Dra 18h28'39" 59d59'27"
Danni Holman's 25th
Birthday Star
Leo 9h37'36" 13d7'37"
Danni Kat Moore
Cas 1h25'38" 51d0'5"
Danni - pour toujours Vôtre
Vir 13h58'35" 6d45'59"
Danni Ranae
Ori 6h7'1" 20d31'5"
Danni Sloan
Cyg 19h26'32" 36d14'43"
Danni & Tom Forever
Cyg 19h50'3" 31d22'12"
DANnibroc
Lyn 7h39'52" 58d4'19"
Dannica
Vir 12h11'56" 6d35'33"
Dannie & Basti
Uma 9h10'7" 65d23'9"
Dannie Campbell
Psc 0h52'12" 28d21'59"
Dannie West
Ori 5h37'22" 2d46'28"
Dannie Zapata
Cap 21h37'27" -9d58'4"
Dannielle
Crb 16h20'45" 27d34'32"
Dannielle and Kyle
Uma 13h38'41" 48d9'23"
Dannielle Ayre, the star.
And 0h14'46" 44d15'38"
Dannielle R. "Pickle"
Ireland
Gem 7h5'3" 34d16'44"
Dannielle Spivak
Lyn 8h6'3" 41d49'43"
Dannielle xxx My Princess
And 1h20'17" 47d39'24"
Dannii's Sanasaa Star
Peg 22h45'21" 27d33'26"
Danni-Lynn
Tau 4h42'4" 6d48'25"
Danni's Star
And 2h30'29" 48d22'54"
Danno...
Sgr 18h2'11" -17d12'50"
Dannon
Uma 9h10'59" 50d27'33"
Danny
Uma 9h35'29" 47d58'3"
Danny
Uma 9h56'39" 49d25'55"
Danny
Uma 10h7'33" 49d32'22"
Danny
Cyg 20h48'23" 49d14'3"
Danny
Per 2h53'48" 46d15'22"
Danny
Per 3h24'25" 48d22'33"
Danny
Aur 7h15'39" 42d43'23"
Danny
Her 18h50'58" 25d35'30"
Danny
Cnc 8h51'36" 25d45'7"
Danny
Her 18h38'7" 16d13'7"
Danny
Lib 15h42'58" -6d54'35"
Danny
Aqr 22h27'24" -1d24'30"
Danny
Aqr 22h37'44" -0d53'31"
Danny
Dra 16h46'56" 74d36'40"
Danny
Car 7h21'38" -51d17'31"
Danny ,Roberta, and
Chiquita
Lyr 19h2'8" 42d7'41"
Danny Alan Eplin
Cyg 20h6'35" 34d37'7"
Danny Allen Hoskins, Sr.
(Grandpa)
Uma 9h48'31" 57d58'15"
Danny Altersitz
Per 3h24'6" 51d2'40"
Danny and Alison
Uma 13h52'8" 55d17'49"
Danny and Christine's Star
Uma 9h20'40" 41d50'52"
Danny and Dawn
Sge 19h53'14" 16d36'7"
Danny and Marlena
Cyg 19h23'2" 52d53'0"
Danny and Megan
Ori 5h33'12" 5d53'22"
Danny Anglin
Aur 5h42'18" 46d19'24"
Danny Attubato
Uma 14h20'49" 57d23'56"
Danny Ayers
Dra 19h22'6" 60d3'8"
Danny B. Flanery Star #1
Equ 21h1'6" 4d21'32"
Danny Baby Boy Worley
Lyn 7h26'37" 59d31'18"
Danny Boo Boo Greenlee
Aql 18h52'21" -0d24'20"
Danny Boy
Lyn 7h54'53" 57d4'10"
Danny Boy
Uma 10h34'3" 52d54'50"
Danny Boy
Sco 17h4'48" -33d21'20"
Danny Boy
Del 20h36'5" 16d0'9"
Danny C. Adams
Her 17h38'17" 27d3'30"
Danny C. Robinson
Vel 9h30'50" -41d35'55"
Danny Christopher Inglis
Her 18h41'46" 12d18'39"
Danny D. Grigsby
Boo 15h0'56" 49d10'5"
Danny Deangelo Davila
Aql 20h2'51" 4d54'50"
Danny Deckert
Uma 12h52'0" 62d23'39"
Danny Do
Uma 9h47'35" 54d59'3"
Danny "Double D" Dugan
June 5-1975
Gem 7h47'49" 22d47'59"
Danny Douglas Pitcher
Tau 4h44'57" 28d53'22"
Danny E. Cuff
Sco 16h18'40" -11d54'52"
Danny E. Van De Hey
Uma 9h30'2" 52d25'47"
Danny Eugene
Leo 9h26'36" 10d22'29"
Danny Farnsworth
Uma 11h28'26" 43d34'44"
Danny Ferreira
Cnc 9h9'1" 26d1'40"
Danny Franks, Jr.
Lmi 10h27'25" 38d10'40"
Danny Frey Great Star in
the Galaxy
Uma 13h41'50" 56d3'42"
Danny Gagliano
Cap 20h21'1" -10d54'18"
Danny Glover
Her 18h32'49" 20d20'44"
Danny Gniazdowski
Uma 11h25'30" 71d45'1"
Danny Griffin
Her 16h39'34" 27d18'50"
Danny Guillermo
Sgr 18h29'12" -28d40'3"
Danny H. Donovan Star
Per 3h27'47" 44d57'34"
Danny Hamlin
Her 18h38'13" 24d15'5"
Danny Hawkins
Per 3h1'46" 46d20'28"
Danny&Heather
Her 16h42'10" 33d2'43"
Danny heroicus angelicus
cum Deus
Uma 10h30'27" 49d56'4"
Danny Holly
Vir 12h38'40" -0d2'5"
Danny Honeycutt
Sco 16h56'43" -38d18'3"
Danny J. Scaduto
Vir 12h36'13" 8d13'44"
Danny & Jane Breen
Lyr 18h45'17" 32d8'45"
Danny Johnson
Gem 7h47'45" 28d17'15"
Danny Kay
Ori 4h54'10" 3d37'30"
Danny Kost
Ori 5h20'35" -1d23'54"
Danny L. Cone
Crb 16h12'45" 36d57'51"
Danny L Cooper II
Lyn 8h10'1" 54d14'41"
Danny&Laura Butterflyz
Forever
Cyg 20h1'43" 32d56'15"
Danny Lee
Dra 20h18'24" 69d40'56"
Danny Lee Cornelius Sr
Uma 11h24'57" 48d43'54"
Danny Lee Mayo
Cep 21h29'0" 58d59'32"
Danny Lees
Sco 16h12'14" -25d31'32"
Danny Leon
Lib 15h4'38" -6d45'56"
Danny Liberatoscioli
Sgr 19h2'25" -28d59'13"
Danny "LionHeart" Diieso
Leo 9h53'9" 18d54'46"
Danny & Lissa Moore
Cyg 20h23'44" 32d24'21"
Danny Lynch
Cep 2h15'10" 80d58'8"
Danny Lynn Sherrod
Uma 11h8'35" 53d12'3"
Danny MacClellan Darby
Her 18h28'32" 12d9'25"
Danny Mae
Cam 7h41'7" 77d58'11"
Danny Mancuso
Ori 5h41'9" 0d48'57"
Danny Michael
Per 4h43'1" 47d10'29"
Danny Miller
Uma 10h37'16" 48d21'14"
Danny My Circinus, Carol
my Perseus
Cir 15h20'19" -56d11'2"
Danny Myers
Leo 10h10'18" 20d53'38"
Danny Navarro
Cap 21h21'46" -22d43'14"
Danny Neuman
Lyr 18h38'33" 31d39'38"
Danny & Nicole
Her 17h25'17" 38d2'50"
Danny Nogo
Leo 9h27'54" 9d58'42"
Danny Nykyforuk
Uma 9h53'30" 70d14'49"
Danny O
Ori 6h19'7" 12d33'27"
Danny Oleksa
Cnc 9h4'32" 28d34'35"
Danny P
Aql 19h32'51" 15d8'24"
Danny P. Edwards
And 1h40'40" 41d22'20"
Danny Parker
Uma 9h17'13" 46d59'2"
Danny Quinn
Uma 9h43'31" 52d23'31"
Danny R. Etue
Aqr 22h34'4" 1d0'49"
Danny R. K. Williams
Her 17h28'15" 37d54'59"
Danny R. Kirstine
Psc 0h30'12" 5d54'22"
Danny R. Rainey Sr.
Leo 10h40'22" 13d17'8"
Danny R. Willard
Tau 5h45'59" 22d5'38"
Danny Ray
Uma 10h1'29" 45d10'15"
Danny Ray Brown
Lib 15h17'18" -21d43'14"
Danny Ray Collins
Dra 18h18'59" 73d46'7"
Danny Ray Tryan
Aur 5h51'29" 45d42'47"
Danny Reyes
Ari 2h33'25" 22d6'32"
Danny Robertson
Sco 16h10'32" -15d40'25"
Danny "Rogue" Haas
Her 18h2'30" 21d58'5"
Danny Ropitzky
Uma 11h3'49" 50d10'43"
Danny Roselle
Lib 15h13'40" -19d1'7"
Danny Schmidt
Cnc 8h52'35" 27d21'59"
Danny Shields
Boo 13h51'37" 10d43'55"
Danny Star
Per 4h46'40" 49d19'4"
Danny Sullivan
Cep 22h22'53" 61d58'21"
Danny & Susan - Soul
Mates Forever
Cyg 21h36'47" 48d53'41"
Danny & Susan wedding
star
Cyg 21h43'21" 52d44'10"
Danny Tarasiewicz
Lmi 10h36'2" 37d25'39"
Danny Terlecki
Cnc 8h21'53" 13d56'25"
Danny Todd Haworth
Aql 18h52'33" -0d42'10"
Danny Turzi
Cep 3h30'18" 83d47'2"
Danny Velez
Sco 16h8'45" -10d40'58"
Danny Wendell
Lib 15h48'49" -16d54'55"
Danny Whaley
Aur 5h8'59" 34d3'11"
Danny William Chatham
Umi 14h13'6" 67d2'34"
Danny Winkler
Vir 12h41'1" -6d39'48"
Danny Wussow
Uma 10h52'23" 52d31'33"
Danny Z.
Cap 20h29'17" -13d58'47"
Danny Zumbrun
Sco 16h18'50" -18d6'12"
DannyBoy
Cmi 7h22'48" 8d37'12"
Dannyboy
Her 18h47'40" 12d50'48"
Dannyelle
Uma 11h12'39" 32d6'52"
DannynJoni#9
Tau 4h36'18" 29d31'20"
dannyrand
Mon 6h48'8" -0d17'52"
Danny's Heaven
Cap 21h37'51" -10d21'7"
Danny's Little Sister
Aql 19h25'44" 0d28'5"
Danny's Shining Star
Uma 9h33'33" 66d34'57"
Danny's Smile
Uma 12h6'44" 60d8'55"
Danny's Smile
Per 3h30'59" 47d10'33"
Danny's smile=God's
LOVE
Aur 5h51'53" 50d53'55"
Danny's Star
Her 17h32'50" 30d13'13"
Danny's Star
Cen 13h9'22" -44d58'10"
Dano
Her 16h22'32" 9d35'19"
Danobbi
Cnc 8h21'58" 12d21'53"
Dano's Shining Star
Gem 7h45'48" 33d59'6"
Danquan
Uma 14h11'7" 58d21'52"
Dan's Country Girl
Cap 21h58'13" -18d20'11"
Dan's Dream
Psc 0h11'16" 3d48'31"
'Dan's Dream Star'
Uma 12h3'25" 48d25'42"
Dan's little Duder
Sco 17h18'37" -32d30'33"
Dan's Lucky Star
Gem 6h7'27" 22d3'4"
Dan's Peace and
Happiness Star
Uma 8h48'1" 52d23'7"
Dan's Place
Boo 14h47'26" 54d7'4"
Dan's Star
Leo 11h52'34" 21d52'45"
Dan's Star of Faith
Tau 4h21'48" 17d57'23"
Dan-Sandy's Love Star
Cap 20h54'55" -17d39'42"
DanSar
Cyg 20h28'5" 45d38'14"
Danse pour Françoise
Lowinsky
Umi 16h54'19" 75d9'41"
Danseur Canadien, Etoile
de l'amour
Lyr 18h38'40" 31d6'54"
DansJesseBetsySandyZac
arielerene H.
Uma 10h33'1" 65d48'16"
Danskerstjerne halvtreds
årsdag
Sge 19h53'53" 18d46'47"
Dantanalee
Dra 18h47'14" 51d53'53"
Dante
Umi 15h33'26" 72d33'21"
Dante Aubert
Crb 15h50'16" 33d43'45"
Dante Biagio Salvatore
Uma 11h23'13" 43d38'23"
Dante DePinto
Uma 11h18'19" 48d31'59"
Dante Duri
Boo 14h37'46" 38d0'34"
Dante Fantos
Dra 19h14'58" 68d8'57"
Dantè & Jennifer Rasicci
Sco 16h9'20" -16d52'26"
Dante Lanzi
Per 2h17'16" 57d19'18"
Dante Lazaro
Uma 9h10'33" 57d42'50"
Dante Lucio Aceto
Sco 17h29'30" -42d47'32"
Dante Marcus
Sgr 17h57'25" -24d42'28"
Dante Michael Masucci
"6/8/2004"
Umi 14h33'33" 73d27'14"
Dante Reed Frisbie
Uma 8h42'50" 69d17'7"
Dante Sebastian Ciabattari
Uma 11h48'10" 50d35'58"
Dante van Wrinkle
Uma 9h10'48" 68d39'53"
Dante Xavier Rosa
Psc 0h8'56" 3d5'23"
Dantina
Cas 1h12'26" 59d12'15"
Dänu Wüthrich
Peg 21h33'39" 20d43'35"
Danuta
Ori 5h55'48" 17d51'0"
Danuta K. Smith
Cam 4h28'16" 76d18'56"
Danuta Szatynska-
Raszkowska
Cnc 8h52'10" 28d10'38"
Dany
Boo 14h43'27" 19d20'21"
Dany
Aqr 22h48'42" -22d45'20"
Dany
Cam 4h31'53" 58d12'29"
Dany 07.02.1976
Uma 11h56'47" 62d53'45"
Dany Bédar
Cnc 8h3'34" 14d19'3"
Dany Donnen
Com 12h45'35" 17d43'0"
Dany Fdz Duque y Ramiro
Gzz Urquijo
And 0h42'50" 27d1'18"
Dany J. Elyoussef
Umi 17h7'45" 77d41'4"
Dany Musset
Vir 13h14'23" -17d42'58"
Dany & Patty
Uma 10h32'43" 53d41'31"
Dany Watson 07/09/1961
Fr 62
Cas 0h12'24" 53d51'15"
Dany, bella come un ange-
lo
Cyg 20h6'10" 34d58'52"
Danya Marie
Ari 3h20'50" 19d23'9"
Danya Waggoner
Gem 6h17'0" 25d57'1"
Danyce M. Henslick
Radiance
Aqr 20h55'45" -9d41'31"
Dany-Céline
Cam 5h11'16" 66d28'34"
Danyel My Baby
Crb 15h59'9" 27d38'51"
Danyel Nicole Parmer
Peg 21h58'12" 25d59'10"
Danyell 613
And 1h38'18" 37d48'28"
Danyell & Jenn! Forever &
Always!
Cyg 19h18'59" 51d15'24"
Danyell Kristine Chavez
Vir 11h52'35" 7d15'41"
Danyelle
Ori 5h40'58" 2d17'4"
Danyelle Clabin
Uma 10h34'46" 66d41'52"
Danyelle Knight
And 0h59'3" 45d29'24"
Danyellebug
Gem 7h27'52" 14d37'9"
Danyle
Psc 23h27'45" 8d6'43"
Danyle's Shining Star
Cap 21h41'35" -24d3'9"
Danyn Rose
Dra 20h17'37" 65d12'1"
Danysue
Sco 17h30'23" -31d34'32"
Danziger Family Star
AARS
Ari 2h55'7" 26d52'44"
DAPA 1020
Her 18h4'50" 48d26'1"
Daphna M. Haynes WGM
of IL 2004 OES
Crb 15h43'37" 29d21'38"
Daphne
And 23h33'21" 46d43'48"
Daphne
Cyg 20h56'0" 30d50'7"
Daphne
Vir 13h17'43" -18d14'28"
Daphne A.
Leo 10h59'26" -1d53'33"
Daphne Angelique
Truelove Ellis
Vir 13h21'26" -14d5'48"
Daphne Arabella Sturm
Cyg 19h44'33" 34d5'50"
Daphne Dawn
Cas 0h55'11" 54d19'19"
Daphne Isabel
Cas 0h40'49" 61d52'57"
DAPHNE LAURIE
And 23h19'33" 46d58'18"
Daphne Newcomer
Sco 17h49'0" -37d20'33"
Daphne Rahel Schwann-
Doegah
Gem 6h53'58" 32d56'25"
DAPHNE ROSE
Cam 6h14'36" 60d24'28"
Daphne Ruth West
Lyn 6h53'24" 58d58'37"
Daphne & Terry's Ruby
Wedding Star
Cyg 20h34'25" 47d40'42"
Daphne Victoria Woodham
Cru 11h59'41" -58d47'31"
Daphne Williams Liedy
And 0h18'43" 30d10'40"
Daphnée
Cas 0h58'1" 54d15'9"
Daphne's Star of Eternal
Love
And 1h13'52" 43d6'36"
Daphnyy Zyanya
And 1h50'46" 38d3'11"
Daquan Wimbush
Peg 23h14'14" 25d9'48"
Daquiri
Gem 6h37'29" 22d49'58"
Dar
Cru 12h25'9" -61d30'55"
Dar & Gal: The Star of Kian
& Yian
Lib 15h40'2" -28d31'47"
Dar Johnson
Psc 0h38'20" 18d54'1"
Dara Alison Caggiano
And 1h36'52" 44d45'54"
Dara Cruz
Sgr 18h2'14" -27d34'51"
Dara forever
Tau 4h5'36" 6d18'49"
Dara Ida Schlachter
Cyg 19h44'5" 36d0'49"
Dara June De John
Her 16h55'43" 34d21'14"
Dara Lynn Saputo
Ari 3h15'10" 27d14'11"
Dara Najera
Uma 10h48'26" 49d56'40"
Dara Renee Cotler
And 0h2'2" 46d31'14"
Dara Riccardi
Umi 15h56'4" 73d34'37"
Dara Rose
Aqr 23h39'7" -15d47'3"
Dara Ryder
Aur 5h12'13" 35d3'26"
Dara5384
Lyn 8h4'15" 42d46'20"
DaRaFF & Littlemoon
4ever
Cas 23h56'32" 50d27'12"
Daragh Sheridan
Cap 20h19'42" -13d3'4"
Daralynn Fleischer
Goodman
Psc 0h50'41" 27d30'20"
Dararat (Pui)
And 1h34'32" 46d53'0"
DarayTala
Sgr 20h3'5" -41d17'56"
Darbie
Psc 0h37'37" 10d43'50"
Darby
Sgr 20h1'39" -34d50'29"
Darby
Lyn 6h24'9" 60d46'16"
Darby Coleman Dryden
God's Angel 19
Psc 0h5'39" 6d19'39"
Darby Heather Jill
Her 18h36'29" 15d35'31"
Darby Leann
And 23h37'9" 48d8'32"
Darby Marie and Hooli
Dra 17h19'40" 65d25'19"
Darby Moeller
Mon 6h54'26" -0d49'37"
Darby Shea Grimm
Lyn 9h0'59" 44d1'3"
Darby's
Dra 17h37'14" 56d18'42"
Darby's Star
Eri 4h4'59" -11d55'17"
Darc Vu
And 0h15'51" 24d23'9"
Darcee Weber
Psc 0h32'42" 4d44'47"
Darcey
Sgr 19h33'40" -17d11'50"
Darcey 24.09.2004- Ruby
04.04.2005
Vir 13h41'50" -19d25'3"
Darcey Antonia Banks
And 2h10'52" 39d41'24"
Darcey Elizabeth Clark
And 0h10'51" 24d21'14"
Darcey Hiatt
Cyg 20h21'26" 52d36'9"
Darcey Maria Eckel
Cas 0h14'25" 53d59'58"
Darcey Scarlett Jones
Cas 0h19'21" 54d57'16"
Darcey Woodbury
Cyg 19h42'25" 27d50'55"
Darci
Mon 6h51'16" 6d51'4"
Darci
Lib 14h52'27" -1d49'17"
Darci Alice
Lyn 8h23'13" 39d31'13"
Darci Lynn
Lyn 6h30'1" 57d17'50"
Darci Lynn Rawhouser
Psc 1h4'52" 18d27'5"
Darci M. Graves
Sgr 19h17'51" -36d31'38"
Darci Michelle Smith
Psc 0h40'42" 7d43'23"
Darcie A. Goodenough
And 23h20'38" 41d15'38"
Darcie Bigelow
Cas 1h37'2" 65d16'6"
Darcie Crifase
Per 3h23'13" 37d54'5"
Darcie Erin Lortie
Cap 21h0'39" -27d16'24"
Darcie Hutchinson
Aql 19h53'31" -0d39'25"
Darcie Jorgensen
And 23h0'6" 48d16'48"
Darcie Lea
And 2h26'24" 50d23'56"
Darcie Lily
And 23h39'58" 47d0'2"
Darcie Lucisano
Uma 9h23'56" 59d28'32"
Darcie Marie
And 1h18'44" 48d55'36"
Darcie Nicole
Lib 15h52'55" -17d31'21"
Darcie Reeve Smith
And 1h39'28" 50d0'7"
Darcie & Rupert in
Stellante Caelo
Leo 11h48'56" 12d4'7"
darcieolarry
Ori 5h46'31" -2d2'3"
Darcie's Gaze
Leo 11h53'18" 22d56'38"
Darcie's Star
Psc 1h24'13" 25d23'55"
Darcy
Gem 7h39'52" 33d3'14"
Darcy and Rich 10 Year
Anniversary
Cyg 19h36'36" 41d32'19"
Darcy and Sophie
Gem 6h53'8" 24d44'1"
Darcy Bliss of Leo Minor
Lmi 10h38'46" 28d47'56"
Darcy Cinderella Cervantes
Cnc 8h40'34" 7d25'58"
Darcy Cinnamon Gliddon
Peg 22h13'37" 11d0'23"
Darcy Curnow Jones
Per 3h28'43" 54d24'30"
Darcy & Deiana
Lib 15h35'39" -17d34'55"
Darcy e Jose Francisco-
Amor Eterno
Cru 12h25'55" -58d22'23"
Darcy Gilmore
Crb 15h41'5" 32d31'48"
Darcy Green's Star
Umi 17h39'30" 81d43'41"
Darcy J Lehr & Sarah N
Pusc
Aur 6h9'32" 54d0'57"
Darcy James Patrick
Charles
Cru 12h9'39" -62d3'26"
Darcy Jennings
Vir 15h7'43" 2d37'39"
D'Arcy Kautz
Ari 2h47'10" 27d33'16"
Darcy Linne Erickson
Powers
Cas 1h51'11" 61d41'10"
Darcy Lynn Williams
Sgr 17h52'23" -25d4'26"
Darcy Lynne
Uma 11h2'22" 55d13'42"
Darcy M. Smith
Vir 13h9'57" 6d23'8"
Darcy Mae Woodward
Peg 21h49'28" 16d49'1"
Darcy Marie
Aqr 21h30'16" -1d45'40"
Darcy Messner
Tau 5h48'25" 14d28'47"
Darcy P. Holland
Cap 21h19'57" -14d31'53"
Darcy Sidgwick
And 23h22'43" 52d21'58"
Darcy Vinson
Lyn 7h24'21" 46d2'45"
Dare E. Kelley
Ori 5h5'0" 15d32'43"
Daredevil Dalton "2-8-99"
Lyn 7h52'57" 49d30'19"
Darée's Guiding Light
Lib 15h0'6" -10d21'58"
Darel Troy Jones
Uma 10h7'22" 43d24'5"
DAREMAMA
Sco 17h19'54" -43d42'19"
Daren
Ori 5h26'27" 3d27'30"
Daren Dillon
Psc 1h14'23" 29d41'30"
Daren Fech
Per 3h27'4" 44d47'1"
Daren Henley 1973 -
(Dish's Jewel)
Ari 3h2'14" 21d25'27"
Daren James Karol
Ori 5h52'2" 20d49'34"
Daren Mark John Wilson
Cyg 19h36'42" 31d57'55"
Daren Miles LaFayette
Per 4h15'36" 51d15'32"
Daren Mongno
Cnc 8h41'59" 32d8'49"
Daren & Paula Moore
(United-as-one)
Cyg 19h47'33" 33d9'19"
Daren Reid (Star of
Gosnells)
Cru 12h38'35" -59d22'26"
Daren Schleicher
Per 2h43'31" 51d27'59"
Daren W. Koch
Aql 19h46'24" -0d11'59"
Daria
Cas 1h16'17" 57d46'27"
DARIA
Per 4h7'32" 34d9'17"
Daria Ashley Jones
Cyg 20h26'18" 39d31'18"
Daria Paolucci
Peg 22h44'10" 4d47'32"
Daria ~ Princess of the
Stars
Aqr 22h33'3" -11d9'6"
Daria Rapp
And 2h24'23" 39d36'50"
Darian
Ori 5h23'50" 7d41'48"
Darian A. Epherson
Dra 20h22'49" 63d27'38"
Darian Brooke Scott
Lyn 8h38'38" 33d46'23"

Darian Simpson
Leo 9h24'56" 9d38'52"
Darian "Wedgy" Baker
Vir 12h17'11" 10d44'6"
Darianne J.S. Perez
Cnc 8h16'19" 30d45'14"
DARIAUS IR JURATES
Lyn 6h32'30" 58d0'32"
Dariel Watts
Aqr 23h22'41" -17d33'40"
Darien & DeAngelo
Psc 1h24'21" 15d44'3"
Darien Lamar Fripps
Uma 9h46'0" 69d7'54"
Dariena
Cyg 21h15'38" 46d54'46"
Darienlee Jennifer Castorina
Lyn 7h32'43" 57d16'22"
Darija9000
Uma 10h6'46" 43d6'57"
Darijo Baresic
Boo 14h47'51" 15d45'44"
Darijus Spakauskas
Umi 14h49'16" 74d44'27"
Darikim
Lyr 18h25'29" 39d29'39"
Darima
Lib 15h38'42" -10d47'33"
Darin Gloe
Aql 19h59'19" 14d11'15"
Darin Hudec
Psc 1h17'57" 10d56'41"
Darin Justin Phillips
Aql 19h43'10" -0d1'43"
Darin L. Septon
Sco 17h28'38" -44d13'41"
Darin Lee Schumacher
Cnc 8h30'23" 10d23'22"
Darin Michael Topel
Sgr 18h25'56" -16d28'52"
Darin Patrick Lewandowski
Psc 1h10'12" 11d0'21"
Darin Pinchuk
Leo 9h49'2" 27d59'23"
Darin St George
Lib 14h47'37" -16d34'5"
Darin Timothy McPhee
Cyg 20h3'12" 45d9'54"
Darin William
Uma 9h9'32" 53d41'56"
Darin Yoder
Per 3h30'36" 34d14'3"
Darinsk
Uma 11h51'35" 51d48'58"
Dario
Her 18h43'4" 20d10'35"
Dario Angelo Balzan
Boo 14h12'26" 52d21'15"
Dario Berte
Ori 6h15'46" 10d33'15"
Dario Doron
Ori 6h13'50" 10d10'20"
Dario Jeisy
Com 13h7'51" 18d34'47"
Dario Luciano Grassini
Ori 5h32'54" 3d26'10"
Dario & Marina
Cam 3h17'32" 65d0'2"
Dario Morante
Cep 22h12'31" 73d43'48"
Dario's Stern
Her 17h54'14" 14d38'34"
Darius
Her 17h49'39" 38d42'11"
Darius
Uma 9h13'27" 66d21'14"
Darius
Lib 15h43'44" -20d11'47"
Darius & Cheryl Bishop
Cyg 20h24'33" 55d35'5"
Darius Coleman
Uma 8h26'51" 69d19'5"
Darius Isaiah Tanksley
Dra 18h36'16" 59d37'13"
Darius & Janis
Lib 15h58'52" -10d11'33"
(Darius) Light of Phyllis Ozoux
Cru 12h1'12" -63d2'28"
Darius Romas
Psc 1h20'24" 10d52'58"
Darius Thomas
Uma 8h38'58" 52d53'53"
Dark Blue
Sgr 19h35'23" -13d26'48"
Dark Eyes
Lyn 9h13'44" 34d12'58"
Dark Knight (I.M.S.D-K)
Per 4h33'42" 31d3'49"
Dark Mavin
Sco 16h6'14" -10d29'11"
dark star
Gem 7h5'16" 19d4'17"
Dark Star
Leo 9h42'14" 24d18'6"
Dark Vador
Per 4h23'38" 40d21'24"
DarkAngel CamelaNicHollin
And 22h58'3" 50d58'33"
Darkhorse
Sco 16h6'52" -10d12'3"
Darklos
Men 4h51'28" -71d39'55"
Darkly Bier
Aql 19h16'42" 4d28'56"
Darkon
Gem 6h46'8" 32d1'50"
DarkRivers (Todd)
Cyg 20h8'44" 38d45'46"
Darkstar13 Diego Winkens
Ari 3h10'24" 18d56'43"
Darky
Com 12h45'9" 15d34'42"
Darky
Uma 10h10'47" 42d53'20"
Darla
Cnc 8h53'1" 31d44'19"
Darla
Uma 11h33'20" 35d0'33"
Darla
And 0h23'12" 29d3'3"
Darla
Uma 9h36'5" 65d21'39"
Darla , love of my life
Sgr 18h5'21" -27d37'13"
Darla Bodfield
Cyg 21h31'59" 35d55'4"
Darla Castleberry
Aqr 20h45'37" -11d13'28"
Darla Eden
Vir 12h45'18" -1d27'48"
Darla Gail Peck
Sco 15h54'0" -20d53'47"
Darla J. Buttles
Psc 0h27'15" 7d8'36"
Darla J. Farrell
Vir 11h39'29" -0d37'32"
Darla Jean
And 2h15'25" 50d1'56"
Darla Jeanne
Psc 1h42'10" 12d36'6"
Darla Kay
Psc 1h2'38" 29d16'15"
Darla Kennon Hill
Crb 15h45'18" 33d16'47"
Darla L. Miller
Ori 5h31'35" 3d8'40"
Darla Lynn Gordon
Cap 20h23'25" -26d51'39"
Darla Mae
Psc 2h3'9" 4d3'40"
Darla Marie
Ori 6h15'57" 9d47'12"
Darla Rachelle
Sgr 19h34'16" -39d46'29"
Darla Twoey
Aqr 22h41'36" 0d16'50"
Darla Yohner "Mom's Star"
Lib 15h0'41" -21d40'28"
Darlana Wolf
Cnc 9h8'17" 21d26'37"
Darleen Carnell Horton
Ari 2h13'30" 23d29'50"
Darleen Lundgren
And 0h35'31" 26d50'40"
Dar-Lene
Crb 15h52'5" 26d20'30"
Darlene
Ori 5h17'5" 7d53'51"
Darlene
And 1h48'17" 36d6'25"
Darlene
And 2h16'9" 49d45'28"
Darlene
And 23h10'30" 51d41'37"
Darlene
Per 2h26'49" 57d26'8"
Darlene
Uma 8h48'2" 50d2'58"
Darlene
Sgr 20h4'57" -12d25'45"
Darlene
Uma 9h19'2" 63d15'55"
Darlene
Uma 8h33'31" 72d40'10"
Darlene
Sco 16h9'15" -23d28'22"
Darlene 0606
And 0h51'13" 36d17'2"
Darlene A. Bennett
Cas 0h14'46" 63d13'10"
Darlene Adeline
And 1h8'21" 42d51'6"
Darlene Alberts
Lyn 7h7'10" 47d42'28"
Darlene Allee
Aqr 22h31'11" -2d6'26"
Darlene Ann Sikorski
Sco 17h23'54" -31d28'50"
Darlene Annette Gildersleeve
Cas 23h46'0" 52d33'48"
Darlene & Austin White
Cyg 20h22'2" 41d30'0"
Darlene C. Jordan
Lyn 6h17'4" 59d15'41"
Darlene Cancino 143
Lyn 7h50'55" 48d3'42"
Darlene Cerise
Crb 15h25'33" 26d32'52"
Darlene Colesworthy
And 0h10'22" 43d57'27"
Darlene Donaldson
Tau 4h35'39" 28d29'51"
Darlene D'Ottavio L.P.
Lyr 18h48'49" 34d5'39"
Darlene Dowsey
Leo 11h6'44" 22d26'34"
Darlene E. Rolen (Thweety)
Lib 15h16'54" -6d25'2"
Darlene Fulks " Star of Love"
Ari 2h50'2" 27d4'22"
Darlene Hermony
Lyr 19h11'41" 26d31'48"
Darlene Hildenbrand
And 0h43'19" 40d3'45"
Darlene J. Nowocien
Sgr 18h7'17" -19d23'23"
Darlene K. Huxtable
And 0h46'40" 39d48'14"
Darlene Kay Foster 9-16-1951
Aqr 21h32'43" 1d58'15"
Darlene Keyoite
Umi 16h3'57" 80d26'43"
Darlene King
Vir 12h48'40" 4d34'49"
Darlene L. Beveridge
Crb 15h47'3" 27d38'45"
Darlene lady love
And 0h36'29" 33d9'50"
Darlene Lalime
Cas 1h29'14" 60d39'53"
Darlene Lawson
And 23h9'39" 48d24'4"
Darlene Lee Lewis
Mon 7h32'1" -0d51'24"
Darlene Louise Evenson
Cas 2h36'55" 67d3'12"
Darlene M Lampi
Cnc 8h9'38" 13d10'28"
Darlene M Smith
Vir 12h55'22" 4d16'22"
Darlene Marie Confair April 16 1954
Ari 3h4'10" 18d22'5"
Darlene Marie Elizabeth Andro
Psc 23h34'21" 0d55'0"
Darlene Marie McDonnell
Dra 17h24'30" 54d5'10"
Darlene Marie Skinner
Psc 0h21'0" 7d39'22"
Darlene Marie Tomasiello
Aqr 20h51'54" 0d6'17"
Darlene Marie Wright
And 23h50'25" 39d52'43"
Darlene & Mark's Wedding Star
Uma 13h20'5" 57d36'24"
Darlene McRae
And 0h12'59" 29d45'18"
Darlene Michelle Wulff
Cyg 20h1'17" 32d18'18"
Darlene Mignon Scheyer
Sgr 18h57'5" -28d18'52"
Darlene Milan
Sco 16h11'16" -12d25'14"
Darlene Montalvo
Sco 17h10'19" -34d0'42"
Darlene Ness
Lyn 8h17'12" 48d3'30"
Darlene Newsome Kahler
Vir 14h11'2" 2d53'37"
Darlene Oglevee
Peg 22h19'51" 12d28'20"
Darlene Patricia Lucia Caird *MOM*
Lib 15h25'6" -27d25'18"
Darlene Pennie Stafford
Tau 4h11'34" 19d20'26"
Darlene Rathbone
Aqr 21h56'37" 0d24'36"
Darlene Rockhill
Cyg 19h43'4" 54d53'51"
Darlene & Rowan
Uma 9h54'54" 48d27'54"
Darlene Rowell Contreras
Gem 7h27'20" 15d39'26"
Darlene S Burt
Lac 22h48'4" 50d4'59"
Darlene Schimmelpfening
Cam 3h57'49" 59d25'34"
Darlene spark in my heart
Del 20h36'47" 7d21'45"
Darlene Star Schweickert
Cas 0h55'15" 63d38'34"
Darlene "Stella" McGrath
Tau 5h52'13" 27d5'11"
Darlene "White Chocolate"
Cnc 8h50'58" 29d33'36"
DarleneKayHaysQueenMother
Sgr 18h49'53" -20d29'38"
Darlene's Dream Star
Vir 13h8'28" 10d49'37"
Darlene's second star to the right
And 1h15'55" 46d6'8"
Darlene's Shining Star
And 23h45'29" 42d57'29"
Darlene's Star
Tau 3h37'46" 26d9'13"
Darlene's Star
Umi 8h56'46" 86d9'11"
Darley Buns
Ori 5h25'4" 8d24'17"
Darlin
Psc 0h32'35" 7d29'28"
Darlin' Debra Mauser
Per 2h48'40" 37d34'22"
Darlin Nikkie
Lyr 18h44'46" 35d13'41"
Darling Andrew
Cnc 8h5'51" 8d50'9"
Darling Anthony
Sgr 19h3'47" -32d5'5"
Darling Belleza
Cap 21h32'41" -14d11'21"
Darling Britlyn
Tau 4h8'17" 16d59'36"
Darling Darlene
Cyg 20h22'46" 46d7'28"
Darling Dear
Cyg 20h25'29" 45d10'36"
Darling Ella
And 0h42'8" 36d16'20"
Darling Gayle
Cyg 21h44'34" 48d27'53"
Darling George
Ori 5h49'20" -0d9'0"
Darling Husband Jon
Cyg 20h53'50" 39d17'25"
Darling Jane
Uma 11h36'17" 35d25'36"
Darling Janet Marie Caddick
Sgr 19h47'34" -13d47'16"
Darling Jen Riley
Peg 21h45'39" 17d50'17"
Darling Jilly, I love you
Uma 10h22'17" 64d58'47"
Darling Maria
And 23h43'8" 34d21'5"
Darling & Mi Amor Praguin
Tau 3h57'27" 15d17'53"
Darling Miranda
Psc 1h36'58" 7d48'26"
Darling Morgan
And 0h44'11" 31d0'25"
Darling Niki
Ari 2h3'6" 23d52'29"
Darling Star
Uma 9h14'12" 65d59'28"
Darling Sylvia Marconi
Lib 14h45'51" -24d23'32"
Darling—Angela
Leo 10h23'21" 13d28'19"
DarlingYuki I Love You
Vir 13h12'3" -13d51'33"
Darla's Star
Cyg 19h35'20" 32d4'23"
DarlMario
Vir 14h53'18" 4d11'59"
Dar-Lou Star
Cam 3h34'58" 65d15'54"
Darlyn Charlyn
Cnc 8h40'12" 24d35'51"
Darmstädter, Helga
Uma 8h56'14" 49d4'34"
Darnell Levy
Gem 6h29'8" 24d39'55"
Darnell Perella
Umi 16h19'39" 71d9'6"
Darol F. Evans
Vir 12h39'30" 2d57'19"
D'Arps "Superstar"
Uma 11h39'37" 58d54'34"
Därr, Manfred
Uma 8h55'23" 48d53'44"
Darra Deanne
Lib 15h4'53" -4d16'9"
Darragh
Her 17h45'48" 15d58'23"
Darragh's Love
Lyn 7h24'35" 44d25'22"
Darragh's Star
Uma 10h31'24" 60d53'47"
Darrah
Uma 10h37'41" 49d45'9"
Darrah Elizabeth
And 23h23'44" 41d22'10"
Darral and Anita Hendricks
Cyg 20h36'54" 30d7'46"
Darrel & Sheryl McDaniel
Uma 10h9'48" 44d38'47"
Darrel Wilburs
Aql 19h8'43" 5d13'34"
Darrell and Sherri Bowman
Lyr 18h34'21" 29d28'53"
Darrell Andrew Bates
Ari 2h32'26" 23d43'2"
Darrell Christenson
Gem 7h22'29" 30d52'4"
Darrell Conkle
Uma 14h17'35" 58d57'37"
Darrell Crews
Leo 11h43'55" 21d42'10"
Darrell Dwaine Norris
Sgr 18h6'36" -27d16'10"
Darrell Dwayne Elser
Aqr 22h21'57" -3d52'25"
Darrell E. Massingille, Jr.
Her 16h32'13" 37d41'56"
Darrell Edward Zepp
Vir 13h21'12" -0d1'22"
Darrell G. Whitehead
Psc 1h14'37" 18d32'20"
Darrell James Becker
Cyg 21h30'26" 42d4'3"
Darrell James Russell
Ori 6h19'27" 14d25'54"
Darrell Keith McAlexander 5/18/1953
Tau 5h13'15" 27d1'4"
Darrell King
Her 17h19'30" 27d12'3"
Darrell L. Holloway
Cap 20h39'43" -22d22'1"
Darrell L. Jones
Per 4h26'14" 40d43'41"
Darrell Lawrence Smith
Psc 1h23'5" 17d34'29"
Darrell Layne Marley
Ori 5h55'3" 10d23'19"
Darrell Leon Hayes 1954.12.14
Tri 2h23'13" 33d10'27"
Darrell Lomain Bolyard
Sgr 18h35'1" -25d10'14"
Darrell M. Hunt
Ori 5h39'38" -5d14'13"
Darrell Morrison
Per 2h13'40" 56d58'56"
Darrell & Norma Ibach "50 Years"
Cyg 21h30'50" 36d54'16"
Darrell Pope
Cep 0h0'13" 73d5'16"
Darrell R. DeGraff
Uma 11h38'16" 31d42'36"
Darrell Ray Gentry
Sgr 19h22'27" -26d20'56"
Darrell Scott Heen
Her 16h22'14" 20d35'3"
Darrell Steffes
Umi 11h33'18" 88d25'12"
Darrell (Tex) Huling
Per 4h17'8" 36d57'40"
Darrell Thomas
Leo 10h9'26" 24d40'20"
Darrell W. Cook & Shawna M. Briggs
Cnc 8h29'51" 24d51'30"
Darrell Warren Kirnbauer
Psc 23h56'46" 2d14'14"
Darrell Wayne Bunger, Sr.
Sgr 19h49'6" -13d33'12"
Darrell Weatherford
Ari 2h31'49" 27d2'54"
Darrell West
Aql 19h11'1" 10d48'38"
Darrell Winn
Uma 13h45'16" 57d50'42"
Darrell, Joe & Meredith Marshall
Uma 11h41'42" 52d11'3"
DarrellCaralineNathanaelMaggie
Peg 23h12'21" 14d36'33"
Darrell's 40th `a new car not star'
Cnv 12h35'30" 38d16'25"
Darrell's Guiding Light
Gem 7h16'45" 13d31'22"
Darrem Charles
Her 16h55'12" 28d10'57"
Darren
Cap 20h46'5" -14d42'9"
Darren A. Luckhardt
Per 2h21'51" 57d15'51"
Darren & Abby Forever
Cyg 20h45'36" 48d58'19"
Darren Alan Bridges
Aql 19h30'33" 7d51'38"
Darren and Brooke Andrews
Cru 11h58'2" -59d21'8"
Darren and Janine
And 1h42'14" 42d29'42"
Darren & Anne
Cyg 21h30'36" 44d41'28"
Darren Anthony Bryant
Tau 3h26'45" 15d37'36"
Darren Anthony Dries
Aql 19h24'50" -8d4'15"
Darren Appel - 31st March 1971
Ori 6h6'29" 14d17'52"
Darren & Aprile Vereen's Wedding
Cyg 19h35'40" 32d31'57"
Darren Bassett
Dra 19h13'12" 62d20'27"
Darren Blythe Busbee
Cnc 8h34'48" 20d14'45"
Darren Boyd
Ori 6h14'36" 11d29'41"
Darren Bragg
Per 4h36'52" 43d59'3"
Darren Bridgman
Lac 22h35'12" 41d34'4"
Darren Brown
Per 2h56'42" 53d36'59"
Darren Carico Brandt
Ori 5h36'13" 4d50'15"
Darren Christopher Travis Eller
Vir 12h18'37" 7d22'54"
Darren David Shepherd
Ori 6h21'25" 16d35'57"
Darren "Daz" Durham
Dra 16h47'5" 67d23'52"
Darren Duane
Dra 17h29'36" 64d14'35"
Darren Emory Bragg
Uma 8h36'38" 65d4'37"
Darren Ensunsa Star
Ari 2h8'25" 11d11'25"
Darren Fortier
Aql 19h14'54" -10d26'59"
Darren Foster
Ori 5h17'37" 8d35'29"
Darren & Gloria
Leo 10h20'54" 10d30'13"
Darren "Gummybear" Fritzinger
Vir 12h44'34" 7d22'51"
Darren Hart
Cru 11h57'40" -63d53'28"
Darren Heath McPhail
Cru 12h0'52" -60d30'4"
Darren Ian Hopkins
Cep 3h43'16" 78d25'54"
Darren James Newmark
Cep 20h53'37" 69d1'14"
Darren James Terrado
Dra 18h44'10" 51d53'41"
Darren & Janine Buckley 11.9.99
Cyg 20h16'30" 40d3'25"
Darren & Keri Leigh Anamacha Cairde
Cyg 19h39'32" 31d37'5"
Darren & Kristi Dondero
Cyg 20h17'46" 55d15'48"
Darren Lawson
Uma 8h46'57" 57d40'28"
Darren Lee Parks
Per 3h4'11" 52d31'42"
Darren Longo
Ari 3h13'23" 28d46'33"
Darren Lowe
Dra 18h21'49" 75d18'46"
Darren M. Parsons
Per 3h6'4" 46d10'1"
Darren Martyn Marks
Gem 6h33'47" 24d24'49"
Darren & Mhari Forever
Cyg 19h39'10" 55d50'46"
Darren Michael
Per 3h37'29" 44d24'3"
Darren Moore
Her 18h50'32" 22d29'30"
Darren & Nicola
Cyg 20h9'52" 46d15'47"
Darren Park
Umi 14h25'49" 78d23'1"
Darren Patrick Dooley
Cyg 20h37'32" 38d17'53"
Darren Paul Laurie
Per 3h13'10" 40d47'28"
Darren Picone
Per 3h9'38" 54d6'22"
Darren Richard Floyd
Uma 11h2'29" 50d35'20"
Darren Robert Havermale
Uma 8h11'38" 61d25'2"
Darren Ross Jones
Lmi 9h56'18" 39d18'50"
Darren Selwyn Bryce Cleaven
Psa 22h4'50" -33d7'54"
Darren Shinya Shirai
Per 3h32'6" 48d31'30"
Darren T Campo
Tau 3h55'56" 11d22'35"
Darren W. McDaniel
Aur 5h38'37" 55d28'43"
Darren Wayne Bonner
Per 2h14'29" 56d45'20"
Darren Wells
Per 3h44'51" 42d28'39"
Darren William Gietz
Ari 2h38'57" 21d42'30"
Darren Wylam
Oph 16h58'59" 3d23'25"
Darren & Yvette
Cra 18h5'46" -37d4'16"
Darren, Angela, Matt & Megan
Uma 11h3'32" 67d23'4"
Darren,Jennifer, Andrew&Olivia Lamb
Peg 22h25'18" 11d2'14"
Darren-Ron-Bruce-Vick
Lib 15h10'1" -5d51'17"
Darren's Princess, Katie Tomko
Vir 12h51'39" 11d13'49"
Darren's Spirit
Cnc 8h25'16" 29d30'7"
Darren's Star
Leo 10h40'16" 14d51'33"
Darrent Williams
Lib 15h27'40" -28d23'14"
Darrian Elizabeth Mouille
Psc 1h43'53" 6d51'23"
Darrian & Suzy for Eternity
Cru 12h39'1" -58d37'57"
Darrick & Jessica Kouns
Umi 14h35'9" 68d26'56"
Darrin & Berinda Futter
Cyg 19h42'48" 52d47'9"
Darrin Bezilla
Uma 10h58'12" 47d55'39"
Darrin Robert Jay
Lib 15h19'34" -6d35'20"
Darrin Stokes
Vir 13h23'12" 3d45'48"
Darrin Yarbrough
Lib 15h6'15" -5d2'49"
Darrion Marcus McNamara Sardo
Per 3h27'36" 50d42'50"
Darroll
Aur 6h14'47" 36d58'43"
Darron & Merav
Cyg 20h6'44" 39d7'49"
Darron's Star
Cnv 12h43'15" 40d26'23"
Darrrlinnnng
Cyg 19h15'13" 44d57'32"
Darryl A. Thompson
Psc 1h28'11" 8d6'59"
Darryl Andre Nicholas
Her 16h36'28" 36d56'11"
Darryl Atkinson
Peg 21h55'23" 17d34'35"
Darryl D. Bunting
Aur 5h58'6" 29d23'34"
Darryl & Dorothy McMicking
Cyg 21h13'8" 43d4'35"
Darryl Edmund Keens
Aur 6h23'51" 42d19'2"
Darryl Glen Purswell
Aql 19h16'1" -0d5'25"
Darryl & Holly Sorensen
Col 5h47'47" -32d26'58"
Darryl L Superczynski Jr
Tau 4h35'38" 9d42'28"
Darryl Lamont Adams
Gem 7h15'55" 34d48'54"
Darryl Michael Almarza
Cyg 19h57'56" 39d45'46"
Darryl R. L. Smith a Virtuous Man
Lib 15h2'10" -15d58'11"
Darryl S. Tungate
Cap 21h43'13" -12d42'55"
Darryl Sue
Crb 15h36'24" 35d5'49"
Darryll A.Pico
Cmi 7h59'40" 0d3'10"
Darryl's Love
Gem 6h46'3" 31d25'57"
Darryl's Star
Uma 10h39'23" 56d28'37"
Darryl's Star of Hope
Psc 1h36'45" 20d19'5"
Darryn Scott Potter
And 1h12'42" 41d47'3"
Dar's Dream
Ori 5h34'12" -1d20'55"
Dar's Star
Cap 21h29'55" -15d7'14"
Darsan Letrice Glover
Del 20h36'47" 18d3'16"
Darsej
Tau 4h0'29" 15d38'37"
Darshar
Cyg 19h48'33" 31d54'33"
Dart Champion
Cas 0h27'2" 55d52'57"
DaRuji
Uma 11h23'44" 57d46'35"
Darvas László
Uma 9h52'49" 52d24'34"
Darville 2004
Her 17h58'28" 49d2'23"
Darwin Ellsworth Rogers
Aqr 21h41'0" -6d27'2"
Darwin G Broughton
Cep 21h10'40" 67d6'20"
Darwin H. Wright
Ari 2h31'59" 26d5'45"
Darwin Idio Llamido
Cap 21h24'3" -23d40'48"
Darwin M Dockum III
Uma 10h2'12" 49d16'56"
Darwin 'Pa' Sweetalla
Cyg 20h49'7" 35d47'48"
Darwin Prestesater
Lyr 18h33'8" 38d22'1"
Darwin Scott Sanders
Ari 3h27'14" 22d16'57"
Darwin's Star
Cru 12h40'18" -60d22'44"
Darya
Psc 0h46'59" 15d24'10"
Darya
Psc 0h41'12" 14d13'49"
Darya
Aqr 21h56'28" 1d23'56"
Darya Tkach
Crb 15h54'0" 35d27'4"
Daryen
Psc 1h8'48" 27d5'49"
Daryian Rae Vernon
Tau 5h34'3" 27d48'15"
Daryl
Vir 13h26'17" 3d12'46"
Daryl
Cas 23h59'50" 54d0'29"
Daryl and Janice Olson
Lmi 10h41'40" 24d50'18"
Daryl and Mayra's Love Star
Pho 0h20'0" -42d38'39"
Daryl Anthony West
Uma 10h42'40" 45d4'47"
Daryl Boris
Dra 18h50'44" 59d47'43"
Daryl Cow Conlu
Per 3h9'40" 38d41'46"

Daryl Laux
And 0h54'44" 38d59'10"

Daryl Levi Pierce
Sco 17h53'9" -35d48'4"

Daryl Lewis-Hendrickson
Psc 1h22'4" 25d13'9"

Daryl "Mark Wills" Williams
Leo 11h25'12" 11d15'12"

Daryl Moody And Ginny Moody
Cyg 20h27'41" 46d45'17"

Daryl Paul Phipps
Sco 16h56'35" -30d44'30"

Daryl Riley
Uma 10h51'17" 56d26'38"

Daryl Robert
Ori 4h54'25" -0d26'7"

Daryl S. Brown
Her 18h19'28" 17d23'45"

Daryl "The Rose" Rothman
Uma 11h32'13" 61d21'42"

Daryl Wayne Ressler
Cep 23h14'18" 76d49'47"

Daryl Wendell Durrett
Vir 14h14'15" -10d31'48"

Daryn Jay Lotz
Cyg 20h31'40" 48d26'20"

Daryn Smith
Vir 13h52'17" 5d13'3"

Daryn's Compass
Uma 8h50'47" 58d24'36"

Daryon Shariati
Gem 6h33'39" 27d1'54"

Darzin Shadowleaf
Peg 22h50'14" 29d5'34"

Das Pooh
Cnc 8h11'21" 11d59'29"

Da's Star
Per 3h17'55" 52d1'16"

Das Star 151
Cnc 8h38'37" 22d16'30"

Dasa
Lyr 19h21'8" 37d30'41"

Dasan Johnson
Vir 12h44'2" 5d31'3"

Dasha
Aqr 23h17'30" -19d17'59"

Dasha and Lesha Fursov
Tau 4h9'36" 27d11'3"

Dasha Bezuglaya
Lyn 6h18'1" 56d17'44"

Dasha Svitinskaya
Lib 14h48'41" -4d37'20"

Dasha & Vanya Sukman
Cyg 20h36'28" 52d55'9"

Dashevskiy, Alexey
Ori 5h6'59" 5d37'33"

Da'Shira
And 2h36'52" 45d11'44"

Dasia Anahi Diaz
Leo 10h21'5" 22d59'26"

DaSilva / Gordon
Cyg 19h59'59" 46d14'41"

Dassler, Rudolf
Ori 5h12'15" 6d0'25"

dasso39
Mon 6h48'25" -0d20'18"

DaStrul
Uma 9h15'19" 47d56'5"

DATA (Dana Gerolimatos)
Aqr 22h17'35" 1d36'2"

D-Atherlyopia
And 2h29'27" 48d45'55"

DaTho
Cyg 19h40'14" 34d9'6"

Datwan & Hannah
Psc 1h28'1" 26d31'39"

Da'ud
Leo 11h1'50" 23d51'41"

Daughter and Son sparkle as One
Cru 12h2'3" -62d54'12"

Daughter Lisa L. Piche
Uma 9h16'50" 57d46'25"

Daum, Klaus-Wilhelm
Ori 6h18'57" 15d52'55"

Dauntay Phelese Glaze
Gem 6h48'44" 33d33'28"

Daunte Thompson
Cnc 9h18'3" 29d6'35"

Daura
Ori 5h21'50" -5d20'32"

Dauwn Myshell
Lep 5h28'31" -12d46'1"

D'Avanzo Manuela
Umi 17h43'3" 81d15'8"

Dave
Cap 21h58'9" -15d28'46"

Dave
Lib 14h35'54" -19d39'26"

Dave
Gem 6h43'12" 26d59'36"

Dave
Her 18h35'23" 18d49'9"

Dave
Tau 3h41'50" 12d55'29"

Dave
Cyg 20h56'3" 33d20'38"

Dave A Longo A Light Forever Bright
Uma 11h30'22" 59d12'6"

Dave Ainoa
Cet 2h7'31" -20d53'9"

Dave and Ali
Cyg 21h44'57" 32d12'37"

Dave and Deb Humbracht
Gem 6h51'43" 23d46'13"

Dave And Doreen Forever
Cam 4h18'16" 63d12'19"

Dave and Fay
Cap 21h10'52" -14d53'25"

Dave and Heather - Love Forever
Cyg 20h25'33" 40d29'13"

Dave and Holly Thornell
Tau 4h32'51" 15d32'25"

Dave and Jan
Lmi 10h4'25" 36d19'17"

Dave and Janet
Lac 22h30'32" 41d47'15"

Dave and Jennifer
Cyg 19h51'40" 35d56'7"

Dave and Joelle's World
Boo 14h59'47" 32d54'23"

Dave and Jules Forever
Tau 3h56'16" 8d8'5"

Dave and Julia Pohlman's Star
Cyg 20h22'28" 47d11'54"

Dave and Kate Chesters
Cyg 21h35'25" 39d15'49"

Dave and Laura Kaiser
Cyg 21h27'42" 55d6'32"

Dave and Laurens star
Cyg 19h37'23" 30d40'33"

Dave and Leigh's Star
Tau 3h55'20" 23d16'11"

Dave and Lesley Demos
Cet 3h1'32" -0d35'24"

Dave and Mercedes Star
Gem 6h22'41" 21d23'46"

Dave and Michele
Lmi 10h49'57" 35d51'17"

Dave and Sarah's Star
Cyg 20h47'3" 47d23'52"

Dave and Sue Martin
And 1h47'37" 49d38'17"

Dave and Tereza
Per 4h16'59" 45d17'13"

Dave and Terri
Crb 16h16'52" 39d21'6"

Dave & Angel Stiffler's 25th
Gem 6h37'55" 16d37'20"

Dave & Avalon Winn
Uma 10h58'9" 47d5'39"

Dave Ball
Gem 6h50'43" 33d51'45"

Dave & Barbara Abell
Aur 5h25'23" 32d20'30"

Dave Bélisle
Her 17h21'57" 48d9'39"

Dave Bosi
Cam 3h57'32" 68d19'27"

Dave Burke
Dra 17h8'1" 55d21'51"

Dave Carey
Per 2h14'11" 56d9'5"

Dave Carlson
Lyn 8h17'50" 56d52'45"

Dave & Cathy Hoffman
Umi 16h21'52" 73d46'15"

Dave & Chris Silver Anniversary
Uma 8h45'28" 57d36'7"

Dave & Christine Begin
Lyr 19h18'58" 28d26'23"

Dave Claunch
Lyn 7h33'33" 59d26'26"

Dave Cohen
Per 3h48'47" 47d5'57"

Dave Cozby
Ari 2h23'13" 20d22'5"

Dave Croft
Boo 14h50'17" 32d46'54"

"Dave" Dads Star Watching Over Us
Umi 15h41'53" 69d38'24"

Dave "Dassie" Bradbrook
Tau 5h7'43" 24d12'37"

Dave & Dawn Messerla
Sgr 19h29'11" -13d49'28"

Dave & Deb Afterall
Uma 14h9'16" 57d47'20"

Dave & Dee Jewell
Cyg 21h12'47" 45d6'8"

Dave Denham
Uma 11h39'58" 44d59'34"

Dave DeWitt
Lyn 9h29'26" 40d50'22"

Dave & Dotti Dean
Vir 13h58'29" -16d14'35"

Dave & Dottie Serpas
Lyr 19h19'1" 29d51'29"

Dave Dougherty
Cnc 8h14'22" 28d59'32"

Dave E Weekly
Uma 9h22'19" 43d25'28"

Dave Edwards luvs Le Thi Kieu Trang
Cru 12h12'59" -62d19'25"

Dave Farm
Her 16h20'8" 18d5'33"

Dave Garets
Gem 7h27'41" 33d41'11"

Dave Gary, Jr.
Gem 6h10'42" 25d14'28"

Dave Grande
Her 18h31'49" 17d50'19"

Dave Hall
Cap 20h8'19" -9d25'50"

Dave Hancock
Tau 5h31'24" 22d15'15"

Dave Hayes
Cep 21h33'55" 63d57'14"

Dave & Hellie - little Star of Love
Cyg 20h51'32" 30d30'50"

Dave Huizenga
Aur 7h25'25" 39d52'12"

Dave & Jean
Aqr 23h5'30" -8d34'47"

Dave & Jenn Forever
Cyg 21h37'3" 50d50'33"

Dave & Jo Jolley
Cyg 21h46'7" 44d17'28"

Dave & Joann Honigman
Sgr 18h58'7" -17d47'54"

Dave & Jodie
Crb 16h19'36" 36d8'46"

Dave Kachele
Per 4h39'37" 39d15'8"

Dave Kahn
Per 2h58'35" 31d54'18"

Dave & Karen
Cyg 21h45'22" 48d6'31"

Dave & Katrina Couture
Cyg 21h30'5" 46d47'39"

Dave Kelley's SteelinAway
Her 17h58'7" 28d55'28"

Dave Kelly
Tau 4h47'6" 18d49'53"

Dave Kelly
Cep 23h42'42" 75d52'56"

Dave Kong
Her 16h35'5" 38d41'12"

Dave & Laura *Love Thy Neighbor*
Cyg 19h55'57" 44d43'48"

Dave & Leanne
Cyg 21h41'41" 48d38'23"

Dave Levasseur - My Light
Ori 6h10'41" 15d19'17"

Dave & Lin Cairns
Cyg 19h55'1" 30d26'25"

Dave & Lisa
Cap 20h53'14" -19d32'58"

Dave&Lorraines 25thAnniversary Star
Cyg 20h15'51" 38d59'45"

Dave Loves Mikki
Ori 6h1'11" -0d57'16"

Dave & Lynn
Her 17h17'32" 46d51'37"

Dave & Maggie's Love & Inspiration
Gem 7h25'25" 29d20'34"

Dave Marcus
Uma 11h50'24" 30d37'35"

Dave & Margaret Lumia
Uma 8h54'16" 48d47'19"

Dave Martin
Leo 10h26'38" 14d20'26"

Dave Merlin Morris
Cma 6h54'24" -15d56'5"

Dave Mesirow
Aql 19h3'40" -0d5'22"

Dave Minetti's Special Star
Uma 10h47'52" 63d17'10"

Dave Miranda & Alex Dunning
Cyg 21h40'23" 42d29'5"

Dave Mitchell
Vir 13h24'38" 12d1'56"

Dave my one true love
Sgr 18h29'40" -27d46'53"

Dave: My One, My Only, My Love.
Her 17h47'25" 44d59'21"

Dave N Yee
Her 16h44'35" 44d24'8"

Dave Narehood 10/30/1980
Uma 10h41'5" 58d22'10"

Dave Nobile
Cyg 20h24'59" 36d4'8"

Dave Notcher
Umi 14h57'43" 73d5'39"

Dave Odeggard
Dra 15h56'23" 53d14'57"

Dave Painchaud
Cyg 20h50'19" 45d16'2"

Dave Palais
Ari 2h48'42" 22d24'14"

Dave Patton
Crb 16h4'27" 28d25'10"

Dave & Patty Eastham
Cap 20h19'55" -14d19'16"

Dave Pepper Brooks
Uma 10h0'13" 63d42'50"

Dave R. Anderson
Gem 6h26'34" 19d23'59"

Dave Rapp
Per 2h10'11" 56d22'26"

Dave & Rebecca
Lmi 10h49'45" 33d5'29"

Dave Richards Family Star
Per 3h13'3" 46d57'3"

Dave Richmond
Cas 1h17'43" 62d24'39"

Dave Robertson-DMS
Per 4h32'47" 40d28'42"

Dave~Ronda~Greg Doble
Uma 14h2'39" 52d20'24"

Dave & Ryan
Ori 5h18'39" 7d2'51"

Dave & Sam's Sweet Serenade
Uma 11h55'19" 37d54'16"

Dave & Sandy Wetzel
Uma 8h24'17" 68d24'6"

Dave Savino
Sgr 18h14'23" -19d33'47"

Dave Scelba
Per 3h16'34" 54d10'32"

Dave Shackelford
Aql 20h14'7" 13d24'39"

Dave & Sherolyn Forever
Dra 18h37'9" 54d11'57"

Dave & Sherry Gold
Cyg 21h43'55" 30d8'1"

Dave Siembieda
Cas 0h29'41" 54d50'41"

Dave Sitton
Lyn 7h22'49" 52d18'10"

Dave & Skippy
Sgr 18h50'38" -32d48'40"

Dave Smellie
Uma 8h30'40" 65d43'31"

Dave Snyder ~ A Perfect 10
Cep 21h36'33" 67d7'22"

Dave Soehnlen
Lyn 8h45'22" 34d44'54"

Dave & Sonia
Cyg 19h55'2" 33d16'38"

Dave Sorenson
Tau 4h30'46" 22d8'50"

Dave Spangenberg
Boo 14h42'16" 18d24'46"

Dave Spangenberg
Lac 22h44'20" 54d8'5"

Dave Statton
Gem 6h32'33" 26d5'43"

Dave Stroud the "Cool Man"
Sgr 18h28'52" -18d24'2"

Dave & Sue Mason
Cyg 20h9'3" 53d23'41"

Dave "Super Nova" Becker
Sgr 18h15'22" -23d34'11"

Dave T. Rock
Psc 1h37'20" 10d8'7"

Dave Taboada
Sco 16h15'13" -15d32'27"

Dave & Tammy
Per 3h50'13" 33d24'38"

Dave Tautkus Jr.
Cap 20h14'35" -10d7'17"

Dave - The Brightest Star
Cap 20h33'15" -8d57'46"

Dave the Digger from Turtle Point
Cyg 20h7'4" 35d25'54"

Dave & Toni
Lib 14h43'31" -11d9'9"

Dave "Tra Il Sole E Il Cielo 2006"
Vir 12h22'41" -3d39'53"

Dave Ulmer
Uma 10h48'44" 47d31'7"

Dave W. Cook
Ori 6h15'37" 15d15'32"

Dave Webster
Gem 7h9'20" 17d14'31"

Dave Williams
Uma 9h32'41" 42d58'54"

Dave Winkel
Uma 13h26'55" 57d57'33"

Dave Yakutchik
Cep 21h0'32" 68d34'22"

Dave Young - Blitzemon
Tau 4h8'37" 8d21'43"

Dave Zawolkow
Aql 19h32'47" 6d8'58"

Dave, star of love
Her 17h14'8" 15d25'10"

Daveanand James Ramnarace
Umi 14h33'53" 76d48'23"

Davear
Cap 20h39'23" -21d59'48"

Dave-Boy
Peg 22h17'23" 34d31'3"

daveed
Tau 4h18'17" 27d13'2"

Daveen LaRay Ferguson
And 2h13'26" 49d41'41"

DavEmma Futurus - Together For Life
Cru 12h17'32" -57d40'17"

Daven Nataupsky
Uma 10h47'1" 61d54'26"

Daven'Kar
Ori 5h37'22" 15d22'46"

DaveO
Ari 2h20'46" 19d36'50"

Davero
And 1h27'46" 34d36'32"

Dave's big one
Gem 7h25'21" 32d37'2"

Dave's Cruising to Happiness Star
Ori 5h9'1" 15d34'4"

Dave's K.C.
Tau 3h48'10" 16d24'15"

Dave's Plaice
Uma 12h4'43" 31d10'38"

Dave's Shining Light - Nerada
Ari 1h56'1" 18d55'57"

Dave's Star
Ori 6h2'38" 6d45'27"

Dave's Star
Ari 2h31'12" 24d55'17"

Dave's Star
Per 3h21'56" 42d20'47"

Dave's Star
Lyn 7h15'6" 55d31'43"

Dave's Star
Sgr 17h57'36" -27d0'57"

Dave's Starberry Field
Aql 19h16'16" 0d22'19"

Dave's Westernstar Ironbark Aurora
Cru 12h44'30" -58d27'59"

DavesgirlJane
Peg 22h32'3" 11d23'27"

Davey
Sge 20h4'28" 18d19'14"

Davey
Her 16h36'1" 27d29'39"

Davey
Tri 2h34'9" 33d51'26"

Davey's Star
Cru 12h29'17" -59d46'11"

Davey's Star
Uma 11h37'3" 56d0'37"

Davi and Amie Star of Dreams
Umi 16h16'8" 79d35'59"

Davian Connor Blankenship
Her 16h45'34" 16d27'38"

David
Her 18h43'30" 21d3'2"

David
Psc 1h40'33" 27d56'38"

David
Ari 2h30'31" 27d45'6"

David
Her 17h26'53" 26d57'55"

David
Her 17h34'53" 28d26'15"

David
Peg 22h59'50" 26d5'59"

David
Leo 9h44'6" 28d30'47"

David
Leo 10h25'5" 22d58'49"

David
Boo 14h39'48" 23d42'5"

David
Cnc 8h1'11" 27d35'40"

David
Vir 13h6'36" 13d14'41"

David
Cnc 7h59'23" 12d14'20"

David
Ori 6h15'55" 13d22'7"

David
Leo 11h35'20" 7d29'30"

David
Tau 4h30'38" 4d46'34"

David
Del 20h53'58" 5d33'30"

David
Ori 5h3'9" 10d0'47"

David
Psc 1h24'40" 31d2'29"

David
Lyn 9h18'58" 36d38'18"

David
Boo 15h16'48" 37d10'22"

David
Aur 7h23'40" 40d30'44"

David
Lyn 7h53'53" 42d9'18"

David
Per 3h9'25" 45d10'9"

David
Lyn 7h40'51" 47d33'16"

David
Boo 14h33'53" 42d5'52"

David
Uma 11h13'6" 49d43'6"

David
Aur 5h52'14" 53d26'17"

David
Dra 18h17'50" 76d32'13"

David
Cmi 7h19'29" -0d10'46"

David
Ori 5h49'42" -5d23'43"

David
Ori 5h3'26" -0d24'4"

David
Lib 14h49'2" -1d37'45"

David
Cap 20h30'21" -16d24'5"

David
Lib 15h56'9" -17d14'16"

David
Sco 16h8'36" -11d28'37"

DAVID
Sco 16h17'18" -13d56'10"

David
Aqr 20h53'27" -14d7'11"

David
Aqr 23h4'22" -9d12'41"

David
Uma 12h38'9" 53d21'39"

David
Uma 8h37'47" 54d9'52"

David
Uma 10h21'35" 67d16'41"

David
Cep 20h43'43" 69d15'0"

David
Cep 20h38'11" 61d52'29"

David
Sgr 17h55'36" -29d11'33"

David
Cap 20h25'0" -23d56'39"

David
Aqr 23h43'51" -19d25'26"

David 1984
Ori 5h30'17" 3d3'35"

David -22
Vir 11h42'21" 3d22'13"

David 30/05/2004
Sco 17h23'6" -32d45'4"

David 65
Sgr 18h32'2" -36d1'9"

David A. Bolton
Vir 13h3'47" -9d16'55"

David A Cianci is loved eternally
Lyn 7h13'59" 49d29'42"

David A Consigli
Cnc 8h15'13" 11d40'39"

David A. Cott
Cep 5h3'2" 84d5'3"

David A. D'Angelo
Cep 21h21'29" 66d54'28"

David A. D'Angelo 1-27-1958
Cep 22h21'32" 72d19'47"

David A. Dross Will You Marry Me
Ori 5h34'24" -0d22'34"

David A. Epstein
Sgr 18h12'36" -19d49'56"

David A. Fee III
Lmi 10h28'42" 30d43'49"

David A. Hannah, Sr.
Leo 11h28'42" 24d5'14"

David A. Johnson: My Enigma
Mon 6h23'9" -2d54'1"

David A. Joviak
Uma 9h52'42" 63d10'47"

David A. Kain
Cnc 8h27'9" 15d47'9"

David A. Kanson
Psc 0h37'4" 18d5'44"

David A. Kelso
Her 17h38'58" 28d30'50"

David A. Kenney, M.D.
Uma 10h27'37" 57d11'54"

David A. Kiernan
Tri 2h30'18" 29d29'48"

David A Launius
Ari 2h24'15" 22d0'39"

David A Logue
Uma 10h16'58" 51d28'1"

David A. Moore
Uma 9h18'12" 54d28'39"

David A. Rosenthal
Cet 1h12'9" -0d24'37"

David A Silverman
Psc 23h59'28" 5d20'31"

David A. Smith
Leo 9h53'49" 16d21'6"

David A. Thompson 2/20/54
Psc 23h18'11" 1d2'26"

David A. Thompson 2-20-54
Psc 1h45'44" 7d35'13"

David A. Toland
Uma 10h2'17" 53d57'4"

David A. Tomassi Jr.
Leo 10h59'54" 6d40'39"

David A. Tomer Noel Major
Her 16h18'25" 25d40'48"

David A. Winjum OPD 322
Uma 9h57'18" 56d59'26"

David Aaron Breault Jr.
Cyg 21h57'42" 44d32'24"

David Abdoo
Ori 5h30'31" 6d27'9"

David Abilez
Vul 19h30'36" 23d53'2"

David Abraham Simpson
Leo 10h54'18" 7d54'41"

David Adams
Ori 6h17'24" 9d28'26"

David Adams
Her 17h4'43" 15d3'38"

David Addisons' Lucky Star
Dra 17h32'41" 56d46'28"

David Aelion
Cap 20h7'53" -20d9'7"

David Aguilar Antonio
Mon 6h59'20" -0d8'48"

David Aguirre
Her 17h44'18" 31d24'25"

David Aidan Polite
Vir 13h24'56" 12d45'44"

David Aiello
Cnc 8h45'35" 32d39'51"

David aka Boobala
Sgr 18h2'31" -28d0'3"

David Akbarali Nagji Darling
Psc 1h14'3" 10d23'40"

David Alan
Uma 9h15'49" 61d12'34"

David Alan Crowne 12/20/57
Sgr 18h34'10" -23d42'56"

David Alan Freeman
Ari 1h56'0" 19d3'9"

David Alan Fuller
Cap 20h34'59" -21d28'34"

David Alan Kazeck
Cep 3h57'55" 81d2'52"

David Alan Kezerle
Vir 12h20'38" 4d41'6"

David Alan O'Hare
Ori 6h5'17" 1d44'33"

David Alan Post
Her 17h25'19" 14d14'36"

David Alan Reece
Cas 0h55'56" 60d15'6"

David Alan Reed
Her 17h49'47" 49d33'21"

David Alan Stinson
Aql 18h46'13" -0d9'11"

David Alan Tweedale
Ori 5h8'42" 9d59'40"

David Alan Walls
Ori 5h25'45" 2d47'52"

David Alan Wilson
Leo 9h33'18" 29d57'38"

David Alan Yadon
Ori 4h50'18" 11d23'17"

David Albershardt
Nor 15h57'8" -53d42'11"

David Albert
Uma 10h4'43" 67d22'13"

david albert herman
Her 16h51'34" 23d23'35"

David Alejandro Cruz Gonzalez
Cru 12h28'24" -59d25'50"

David Alexander
Cnc 9h3'12" 21d43'27"

David Alexander
Per 3h23'51" 44d11'46"

David Alexander Allweiss
Sgr 18h51'43" -36d12'44"

David Alexander Gaines
Her 17h27'55" 45d16'13"

David Alexander McGavin
Per 3h36'35" 41d1'1"

David Alexander Nance
Pic 5h19'0" -55d15'12"

David Alexander Pinto
Her 17h47'31" 36d7'35"

David Alexander Stadler
Uma 10h54'21" 71d38'33"

David Alexander Wilson
Cyg 21h40'20" 53d22'7"

David Alfonso Miller Levine
Sco 17h30'28" -39d14'38"

David Alister McKie
Ori 6h24'15" 10d24'58"

David Allan
Lib 14h54'20" -1d36'24"

David Allan Elliott
Cep 20h49'41" 59d56'16"

David Allan Gammon II
Cep 4h33'7" 86d21'46"

David Allan Hoover 02-17-1952
Her 17h16'15" 26d20'32"

David Allan Lantz
Uma 9h53'30" 72d31'6"

David Allen Bailey
Cep 22h55'57" 59d35'38"

David Allen Bradley
Psc 1h4'26" 26d59'40"

David Allen Fischer
Vir 13h17'52" -10d56'53"

David Allen Gregory
Aql 19h48'9" -0d6'39"

David Allen Hempel
Ari 2h39'40" 20d19'6"

David Allen Nebrig
Cnc 8h31'43" 24d44'31"

David Allen Osmon
Psc 1h17'11" 21d18'27"

David Allen Patten, Jr.
Ari 2h40'23" 15d39'38"

David Allen Sievers
Uma 10h13'40" 49d2'46"

David Allen Tarver
Her 18h32'24" 15d41'55"

David Allen Walker AKA Festus
Cnv 14h2'33" 35d11'34"

David Allen Young
Uma 11h21'54" 31d26'15"

David Allyn Hitchings
Ari 2h27'36" 23d49'13"

David Always & Forever
Uma 10h14'0" 43d23'56"

David & Amy
Lyr 18h51'14" 33d6'54"

David Anchell
Sgr 19h24'2" -20d20'59"

David and Amy Farrell
Cep 20h47'20" 60d38'3"

David and Anna
Aqr 23h40'33" -4d21'30"

David and Anna Forever!
Sgr 18h39'22" -24d52'19"

David And Ashlee (The Puppy Star)
Sco 17h8'13" -41d5'40"

David and Belinda Deacon
Cyg 20h6'11" 51d51'9"

David and Brittany's Lucky Star Cyg 20h25'14" 32d39'23"
David and Carol 1973-Eternity Umi 15h29'34" 73d49'40"
David and Cheryl - Ten Years Cyg 21h59'12" 47d29'25"
David and Cindy Matson Cyg 20h20'50" 55d6'2"
David and Clare Lomas Uma 11h55'3" 30d23'2"
David and Connie 10th Anniversary And 1h17'37" 45d34'14"
David And Dania And 23h15'4" 44d39'42"
David and Destiny Lee Sgr 19h10'18" -17d43'27"
David and Diane Eternity And 23h33'50" 49d39'21"
David and Elane Aqr 22h35'54" 1d38'22"
David and Fran Bloom Uma 10h21'7" 66d22'27"
David and Hermine Vir 14h4'6" -19d2'52"
David and Isobel Baker Cyg 21h22'19" 39d57'44"
David and Jennifer Cep 0h53'12" 82d48'40"
David and Jennifer Uma 10h45'20" 54d57'47"
David and Jessica DeBerry Cap 21h3'51" -27d11'11"
David and Jodi Kidd Cyg 21h21'3" 50d28'41"
David and Juanita Cyg 21h14'54" 47d17'55"
David and Kaitlin Huber And 0h17'4" 33d12'31"
David And Katherine Umi 15h31'49" 74d35'29"
David and Kathleen Eustace Crb 16h1'10" 30d58'17"
David and Kelly Rumler Umi 14h15'2" 74d21'48"
David and Linda Burriss Uma 11h38'19" 47d11'19"
David and Lindsey Allen Cyg 19h51'33" 47d15'40"
David and Lyna Valdez Dra 12h36'3" 73d40'28"
David and Margaret Her 16h50'5" 23d36'36"
David and Michael Briggs Sge 19h13'15" 18d17'43"
David and Morgan to the Moon Uma 13h45'28" 61d39'16"
David and Nancy Schuck Cyg 21h12'32" 47d1'21"
David and Natalie And 0h12'5" 46d4'50"
David and Nicole Cyg 21h58'25" 49d8'35"
David and Paulina Zwang Aqr 21h56'21" 0d48'19"
David and Peggy Sgr 18h0'51" -31d59'40"
David and Penny Doman Lyn 7h56'43" 54d8'24"
David and Priya Cyg 20h45'26" 43d55'52"
David and Rachel Sferra Sco 17h22'56" -42d33'19"
David and Sandra Pappenfus Uma 11h6'26" 60d24'16"
David and Sharon *Two Become One* Cas 23h59'0" 60d29'12"
David and Stephanie's Star Uma 13h16'24" 57d36'1"
David and Susan Brooks Cyg 20h39'49" 44d48'25"
David and Susan Graham Cyg 19h57'55" 33d11'27"
David and Suzanne Thistle Lib 14h50'22" -2d58'14"
David and Terri Frolich Cyg 21h24'59" 52d2'42"
David and Victoria Wish Col 5h44'55" -33d17'28"
David and Virginia Seiter Cyg 20h11'23" 30d22'49"
David and Yvonne Parker Cyg 21h46'40" 35d6'56"
David and Zella Wilkins Uma 13h6'44" 57d16'12"
David Anderson Cas 0h59'26" 54d21'54"
David & Andrea Foote Cyg 20h18'10" 54d39'50"
David Andresen, PHD Uma 10h22'6" 42d41'7"
David Andrew Uma 11h19'20" 52d51'7"
David Andrew Fogleman Her 17h48'16" 39d26'7"
David Andrew Read Aql 19h45'5" 11d49'29"
David Andrew Richard Willis Per 4h0'44" 43d47'12"
David Angelo Antonio Fata Her 16h21'1" 5d51'0"
David Anthony Ayo Her 16h49'18" 22d49'44"
David Anthony DiFabio Sgr 19h13'13" -12d52'8"
David Anthony Doherty Cap 21h36'41" -19d35'53"
David Anthony Doherty Col 6h16'41" -35d25'27"
David Anthony Fischer Uma 12h28'52" 59d6'19"
David Anthony Gedville Her 16h41'39" 30d36'42"
David Anthony Nobles Uma 11h8'40" 34d12'56"
David Anthony Vasquez J.r. Ari 3h18'31" 23d33'16"
David Antony Gonzales Aqr 22h30'30" -2d8'44"
David Apodaca Leo 11h42'42" 12d42'35"
David Arnold Kish Leo 9h41'4" 10d46'14"
David Arron Chavis And 23h37'29" 48d39'29"
David Arron Chavis And 23h27'40" 47d42'17"
David Arthur Duncan Cyg 20h8'1" 32d7'35"
David Arthur Henry Starr Uma 14h14'56" 58d23'40"
David Assenzio Lib 15h24'1" -29d13'52"
David Atwood Bliss Aql 19h29'29" 7d52'52"
David Auchinleck Regan Lib 15h37'17" -18d55'48"
David August Julian Lawergren Jr. Uma 11h2'2" 44d41'2"
David Austin Naylor Lib 15h39'35" -28d51'49"
David Austin Sky Aur 5h41'9" 40d34'27"
David B. and Helen A. Garner Lyr 19h15'27" 28d58'18"
David B Avison Lyn 9h33'8" 40d17'13"
David B. Cochrane Her 17h0'12" 30d2'31"
David B. Cook Vir 13h10'37" -4d29'5"
David B Davis Family Star 2/25/1984 Ari 3h3'11" 12d52'56"
David B Donnelly Cnc 8h2'26" 20d31'1"
David B. Hargis Lyn 7h49'46" 41d40'5"
David B. Kerwien Vir 13h51'49" 3d38'48"
David B. Mayer Leo 10h27'1" 9d13'11"
David B. Port Per 3h26'29" 36d11'24"
David B. Snyder Her 17h19'56" 36d41'22"
David Babbit Cep 22h23'49" 63d11'36"
David Babiarz Uma 9h37'23" 41d50'47"
David Bahr - November 20, 1964 Sco 16h8'43" -17d23'3"
David Balsamo Cap 21h0'40" -17d33'12"
David Barlow Uma 10h20'31" 70d6'50"
David Barnes Vir 15h4'22" 6d29'59"
David Barrett John Cap 21h58'39" -20d1'33"
David Barrow Dick Her 16h15'42" 5d47'34"
David Barry Connors Aur 5h50'22" 32d48'35"
David Barton Hazelden Aur 5h38'54" 47d25'4"
David Bartrum Cep 21h53'37" 66d54'24"
David Beamertron Uma 9h15'55" 47d29'14"
David Behring An Angel From Heaven Per 2h50'30" 45d54'57"
David Bell Lyn 7h24'46" 49d32'17"
David Bell Uma 9h54'59" 63d49'10"
David Belles Her 16h7'52" 46d27'56"
David Bennage Vir 13h14'23" 5d26'11"
David Bennett Ari 2h33'16" 26d25'53"
David Berman Per 3h21'38" 50d51'41"
David Bernal, El Sol de Amor Leo 11h50'18" 25d5'49"
David Berni Uma 9h52'0" 48d48'24"
David Bisby Her 17h13'1" 19d23'19"
David Black Cap 21h36'40" -20d9'25"
David Blagg Cyg 19h30'41" 30d16'7"
David Blair Cru 12h20'34" -57d32'46"
David Blake Uma 10h7'49" 47d13'8"
David Blazek Dra 16h28'46" 54d7'47"
David Bonyak's Wish Cep 21h41'33" 57d33'10"
David Booth Hirshlag Ori 5h20'41" 6d12'26"
David Borgert's Shining Star Per 2h55'22" 45d11'5"
David Boyd Kness Peg 22h44'49" 33d0'6"
David Boyes Lyr 19h9'57" 26d30'6"
David Bradford Carten Leo 11h45'39" 23d59'7"
David Bradshaw Gem 7h48'55" 15d33'40"
David Brandon Chanin Aqr 22h30'15" -1d29'29"
David Brantly Rudisill Tau 5h44'16" 22d26'33"
David & Brenda Coveney Together Forever Cyg 20h52'48" 51d42'54"
David & Brenda Deighton Leo 11h57'50" 20d48'54"
David Brenner Ori 6h8'46" -0d28'21"
David Brett Rood Uma 11h30'18" 54d4'11"
David Brian Crompton Cap 20h29'40" -19d5'15"
David Brian Hughes Dra 18h31'10" 53d57'53"
David Brian McCarty Her 17h13'26" 15d36'13"
David Brian McDermott Uma 13h9'50" 53d34'19"
David Brian McKinley Lib 14h28'15" -24d28'47"
David Brice Her 18h45'38" 16d46'40"
David Brown Gem 7h33'19" 20d3'29"
David Brown Vir 12h30'51" 3d19'19"
David Brown Uma 10h53'53" 42d34'26"
David Brownlee & Melanie Maholick Cyg 20h28'27" 51d47'39"
David Bruce Ash Vir 13h20'1" 3d39'55"
David Bruce Wilner Hya 9h23'51" 3d37'14"
David Brumfield Her 16h28'55" 32d45'25"
David Bruno Wendt Ari 2h28'25" 18d57'20"
David Bryant Elton Watts Dra 19h31'59" 69d12'20"
David Bryant Schneck Vir 14h32'28" -1d58'2"
David Buck Per 3h49'54" 33d14'7"
David Burns Beaver Aqr 23h38'20" -4d29'24"
David Byron Engstrom Her 17h17'59" 26d33'20"
David Byron Lamb 04-09-54 Peg 21h59'43" 19d14'42"
David C. Cary Gem 7h29'37" 20d12'36"
David C. Eason Aur 5h38'40" 50d11'29"
David C Fisher wish upon your star Col 5h27'53" -28d50'10"
David C. Flaherty Cap 20h48'26" -24d33'52"
David C. Hanson Lyn 8h20'31" 37d39'36"
David C. Migdal Aqr 21h6'10" -8d36'8"
David C Stevens Her 18h52'54" 23d50'2"
David Cafferty Lyr 18h29'24" 28d15'17"
David Caldwell Lyn 8h42'1" 33d33'16"
David Caplan Sgr 18h33'51" -22d51'34"
David 'Captain' Campbell Cra 18h57'6" -39d52'12"
David Carey Dra 14h37'47" 56d52'52"
David Carl Wilson Per 3h42'43" 39d6'59"
David Carl Yarborough Uma 14h25'43" 61d45'34"
David & Carlene Cyg 20h18'48" 56d11'32"
David & Carmen Cap 20h7'43" -26d10'43"
David & Carol Louise Hand Cyg 19h53'16" 32d3'14"
David + Carole Ti amo Cyg 21h56'0" 46d57'32"
David Carrier / Ethan Carrier Ori 5h32'3" -0d35'18"
David Carroll Bryan Cap 20h25'59" -14d53'34"
David & Cassi Sorensen Cyg 21h24'4" 52d10'1"
David & Cassie Brooks Cyg 20h45'48" 54d22'27"
David Castillo - Rising Star Ori 6h19'58" 10d43'52"
David Castor Uma 8h44'0" 62d54'9"
David Castro Leo 11h58'1" 22d41'53"
David Cavallaro Uma 14h25'55" 61d38'14"
David Chaim Steinbrecher Ori 5h43'56" 1d27'43"
David Charba & Brittney Reyes Aqr 22h40'2" -2d6'48"
David Charles Aker Leo 11h17'13" 14d6'7"
David Charles Earl Leo 9h30'45" 22d46'37"
David Charles Gravely Aur 5h36'29" 38d22'45"
David Charles Hall Cep 21h17'35" 62d27'11"
David Charles Hauser II Vir 13h15'17" -3d28'59"
David Charles Madoux Aql 19h34'34" -11d11'8"
David Charles Newbould Ori 5h21'25" -5d27'10"
David Charles Officer Uma 11h47'37" 44d58'58"
David Charles Solvig Ari 2h18'20" 27d24'25"
David Charles Tucker Lac 22h35'5" 54d43'51"
David & Charlotte 4ever Her 16h48'22" 35d58'30"
David & Charlotte Steinbach Ori 6h8'27" 21d0'58"
David Chaves Aql 19h51'57" 12d12'44"
David Chester Per 3h10'32" 47d12'10"
David Chester Bryant Ori 5h35'35" -2d20'47"
David Chester Smith Ori 5h31'45" 10d42'51"
David & Chrissy The Foxy Johnson's Aqr 23h16'51" -13d5'50"
David Christian Dickinson Psc 0h57'32" 29d14'46"
David Christiansen Uma 11h41'22" 35d38'16"
David & Christine Pfenninger Per 2h25'56" 51d9'19"
David Christopher Brockman Gem 7h31'55" 34d26'43"
David Christopher Flores Her 16h37'57" 35d19'12"
David Christopher McDonald Lac 22h22'26" 53d21'11"
David Christopher Monk Per 4h17'45" 38d44'9"
David Christopher Volpe Leo 9h34'20" 28d47'47"
(David Christopher's Star) Hope Cas 1h52'54" 63d30'32"
David Cicchesi Cam 4h51'9" 62d46'42"
David Ciesluk Psc 1h20'37" 31d30'24"
David Clarence Boo 14h37'50" 33d45'21"
David Clark Uma 10h18'58" 69d46'30"
David Clark Hicks Uma 11h57'17" 28d30'54"
David Cleal Rinker Ari 3h16'28" 28d51'21"
David Clinton Whitley Per 3h31'12" 50d40'19"
David Colin Birch Lib 15h42'0" -15d15'35"
David Collier Aqr 20h40'27" 0d50'22"
David & Colyn Cullum Uma 13h15'23" 58d26'19"
David Contant Her 16h23'57" 44d51'2"
David Cooper Tri 2h35'47" 31d13'46"
David Cotter Cep 21h28'48" 55d31'58"
David Craig Brown Tau 3h49'5" 19d21'56"
David Craig Doerfler Lyn 7h57'5" 43d0'56"
David Craig Dreyfuss Cep 23h15'6" 79d17'2"
David Craig Hamilton Uma 9h27'3" 52d13'28"
David Crest Paper Anniversary Star Cyg 20h2'25" 35d18'26"
David Cruz Ori 5h56'47" 6d29'51"
David Curtis Psc 1h10'0" 4d14'52"
David & Cynthia Stanchak Eri 4h11'59" -0d56'25"
David Cyril Wheatcroft Uma 9h56'4" 70d13'20"
David D. Johnson Uma 11h50'58" 48d12'15"
David D. Mason Cep 22h39'44" 71d53'10"
David D Stucker Her 17h16'18" 32d43'42"
David D. Velez Ori 5h49'29" -0d40'22"
David Dale Arnesen Cep 21h29'36" 65d10'54"
David Dale Smith - immortalis amor Cap 20h54'58" -25d59'3"
David D'Alessandro Uma 9h34'10" 68d40'7"
David Daniel Drebsky Uma 8h37'41" 52d11'16"
David Daniel Edwards Per 3h9'1" 56d25'36"
David Daniel Gallegos Per 4h37'51" 44d53'44"
David Daniel Krus Gem 7h7'23" 24d3'32"
David & Danny's Star Uma 13h53'3" 52d46'57"
David "Dave" L. Fast Sco 17h34'25" -36d24'54"
David Davis Her 17h44'34" 17d49'55"
David & Dawn Cyg 20h13'33" 38d6'27"
David & Dawn Together Forever Gem 7h1'38" 23d12'51"
David Dawson And 0h57'15" 37d21'13"
David D.B.M.G. Wright-Spaner Crb 16h6'5" 36d27'11"
David de la Cruz Garrido Lyn 8h8'47" 34d36'19"
David Dean Johnson Peg 22h16'49" 23d15'26"
David Dean McGill Tau 3h57'43" 4d52'59"
David & Deborah Cyg 21h46'53" 49d38'28"
David DeHerrera Uma 10h17'19" 43d20'8"
David & DeLea Poljak Umi 15h23'19" 73d49'29"
David Dellorso Aur 5h13'0" 47d10'12"
David Desteno Cnc 8h48'5" 18d10'33"
David (Devil's Fan) Fisher Dra 16h39'32" 63d41'3"
David Diamond & Jessica Galati Her 18h9'15" 18d24'30"
David & Diane Saia Cyg 20h53'1" 34d9'30"
David- DiddyKong-Grayson Leo 11h16'21" 19d25'58"
David Diggs Aur 6h59'47" 38d17'10"
David Dipaolo Gem 7h42'42" 24d30'3"
David DiSalvo Uma 11h41'11" 29d18'47"
David Disbrow Aur 5h45'29" 48d52'53"
David Dixon Clark Ari 2h54'28" 24d28'48"
David - DLH #1 Uma 11h33'19" 60d45'19"
David Dodge Per 3h19'22" 51d53'19"
David Dolman - "D60 DOL" Aur 7h26'45" 39d26'16"
David & Dolores Robison Uma 12h7'34" 50d31'1"
David & Doreen Forest Cnc 9h6'14" 30d59'46"
David Dossett Dra 18h38'45" 70d12'27"
David Douglas Henderson And 23h8'25" 52d6'33"
David Douglas Trosin Gem 6h52'19" 32d0'10"
David Douglas Zahn Sgr 18h28'36" -18d24'23"
David Driscoll Leo 11h43'33" 24d47'32"
David "Duke" Marlowe Cnc 9h5'57" 21d2'24"
David Dunkis My Diamond in the Sky Per 3h20'15" 39d38'8"
David "Dusty" Rhodes Per 4h47'24" 45d19'53"
David Duzenski Ori 5h24'22" 2d1'43"
David E. Bressler Uma 11h8'28" 42d41'57"
David E Brodrick Tau 5h48'56" 16d58'54"
David E. Caywood Per 4h43'37" 46d26'5"
David E. Daugherty Per 2h30'28" 54d19'49"
David E. Fox Vul 19h20'10" 21d42'22"
David E. Gallagher: God's Warrior Vir 13h15'29" 6d26'4"
David E. Morrow Pho 1h10'27" -49d48'9"
David E. Sand - A Guiding Light Her 17h21'22" 35d48'42"
David E. Severt Her 17h8'14" 34d1'23"
David E. Walchak Per 3h11'46" 55d26'48"
David E Worster Vir 12h59'23" -0d27'53"
David Earl Gregory Sgr 19h8'51" -26d8'41"
David Earl Smith Sco 16h22'36" -8d26'9"
David Earp Uma 11h32'30" 59d42'50"
David Easley & Jack Forrest Aql 19h46'38" 5d53'29"
David Eaton McQueen Her 17h59'23" 49d38'58"
David Edgar Breault Her 18h44'42" 21d25'44"
David Edward & Elizabeth Stroughair Cyg 20h30'31" 49d35'51"
David Edward Sammarco Aqr 22h30'37" -10d8'24"
David Edward Wilczynski Sr. Cap 21h4'36" -19d24'48"
David Edward Will Cep 23h14'48" 75d8'50"
David Edward Wingate Ori 5h58'0" -0d43'25"
David Eichler Her 17h34'51" 36d17'30"
David & Elaine Lyr 19h23'17" 37d46'55"
David Eldon Hammond Her 16h25'28" 7d1'12"
David & Elizabeth Metraux Cyg 20h5'16" 34d26'34"
David & Elizabeth Pellett Cyg 21h9'26" 30d57'52"
David Ellen Ori 6h17'4" 1d49'48"
David Ellis Rea Vir 13h55'2" -21d57'42"
David Elwin Bee And 23h24'34" 43d14'36"
David Emerson Cap 21h49'51" -13d39'0"
David Emilio Cornejo Ari 3h8'27" 18d21'47"
David Emmet Shepard Ari 2h10'50" 12d31'34"
David Emmett Newton Lib 14h55'45" -7d7'49"
David Emmett Villamarin Uma 10h37'49" 61d36'40"
David Eric Higgins Sco 16h10'27" -12d2'58"
David Eric Trick Cep 22h40'53" 64d54'24"
David Erickson Ori 5h33'22" 4d42'34"
David Eschmann Per 2h54'40" 33d6'24"
David Escobar Her 16h12'3" 48d11'16"
David Eskra Cep 22h42'22" 88d31'26"
David et Coralie Peg 22h21'1" 20d54'15"
David et Sandy Cnc 8h17'14" 19d54'23"
David et Stéphanie Cnc 8h22'27" 22d29'36"
David Etson Ori 5h38'42" 2d19'23"
David Eugene Kline Ori 5h57'44" 11d17'27"
David Eusebio Guillen Aqr 22h30'51" -1d22'57"
David Evan Boo 14h36'36" 51d49'44"
David Evans - "Pop's Star" Uma 12h24'30" 58d25'52"
David Ezra Flatau Aqr 23h5'59" -7d36'8"
David F. Burg Cas 1h6'43" 53d15'33"
David F. Drew II Psc 1h38'50" 7d40'33"
David F. Hoyle Her 17h39'12" 45d6'13"
David F. Irwin Jr. Psc 1h38'25" 24d36'13"
David F. M. Todd Aqr 22h24'12" -1d11'8"
David F. Nelson Hya 9h33'40" -9d45'0"
David Faust Uma 9h58'45" 56d18'41"
David Fawcett Per 3h14'10" 52d54'37"
David Feckley Tau 4h37'29" 27d22'36"
David Fedorko, Akela And 2h38'42" 44d39'0"
David Fernando De la Torre Cap 20h46'38" -27d24'29"
David Fiegl Cep 20h38'36" 64d12'58"
David Fischer Uma 11h17'30" 38d25'3"
David & Flavia Leo 11h26'58" 19d50'10"
David Fleener (eyes of heaven) Lib 15h0'49" -7d23'20"
David Fleetham - 8th June 1973 Umi 16h21'41" 70d39'53"
David Flores Lib 15h32'49" -7d15'49"
David "Forever To Go" Uma 10h46'54" 58d7'1"
David Francis Babcock Vir 13h17'30" 4d54'52"
David Francis Canfield Per 3h10'2" 46d20'47"
David Franco Ori 5h30'56" 7d35'49"
David Franklin McWhorter Uma 10h5'26" 53d26'23"
David Frederick Bahr Sgr 19h15'37" -16d12'11"
David Frederick Sheinker Leo 9h28'3" 28d37'20"
David Fredericksen Uma 10h22'36" 62d25'57"
David Fresne, Owner of the Universe Ari 2h40'13" 28d33'46"
David G Naugle Psc 1h29'46" 14d3'33"
David G. Paulus Aqr 22h9'29" 2d8'40"
David G. Sterken Ori 6h10'41" -1d4'33"
David G. Willis Aqr 22h36'3" -0d30'37"
David Gabriel Hoffman Lib 15h25'51" -10d3'35"
David Gabriel Pell Cyg 20h53'52" 46d19'30"
David Gage Allen Aql 19h26'36" 4d58'40"
David & Gail Anderson Cas 0h20'35" 53d56'9"
David Galasso Uma 9h1'1" 72d17'45"
David Gardner Uma 9h11'37" 58d16'46"
David Garfield Yontz Her 17h46'29" 43d57'7"
David Garman Uma 14h28'9" 56d29'8"
David Gates Peg 22h58'6" 26d12'40"
David Geoffrey Eisner Cnc 9h7'6" 32d20'40"
David George Crb 15h31'40" 37d32'55"
David George Hof M.D. Cep 21h25'15" 56d55'15"
David George Williams Dra 16h35'41" 58d58'26"
David Gerald Taylor Uma 8h28'30" 61d34'32"
David Gerard Uma 10h23'29" 58d29'46"
David Gerard Matthews Per 4h11'20" 49d35'37"
David Getsfrid Uma 10h51'45" 64d19'33"
David Gettig's Star Ori 5h30'56" 1d57'21"
David Giammarco Boo 14h37'5" 18d58'6"
David Giannino Vir 13h2'28" 11d4'20"
David Gibbs Loges Leo 11h12'21" 16d4'58"
David Gibilie Per 3h18'50" 51d54'16"
David Gibson Perry Psc 23h57'6" -0d8'49"
David & Gidget Cyg 20h43'52" 31d6'13"

David Gilbey
Ori 6h15'12" 5d37'40"
David Gilhooley's Star
Cep 20h7'51" 60d55'2"
David Gingerich
Per 2h42'52" 51d22'59"
David Giuliano
Boo 14h34'46" 24d6'7"
David Glen Smithson
Tau 4h10'55" 5d35'55"
David Glenn Cerven
Lyn 7h57'58" 37d59'18"
David Glowacki
Gem 6h54'42" 22d2'53"
David Godwin
Aur 5h40'27" 53d14'19"
David Gold
Per 2h46'43" 50d31'47"
David Goonewardene
Umi 14h32'34" 71d42'55"
David Gow Scott
Cep 20h43'3" 59d2'6"
David Grave
Lib 15h36'35" -11d34'56"
David Gray
Uma 14h12'27" 56d32'21"
David Gray Wilhelm
Cyg 19h48'25" 31d13'35"
David Grayson Hansard II
Aur 5h48'35" 49d45'43"
David Greenwald
Cam 4h38'54" 67d42'39"
David Grigoryan
Aur 7h25'54" 40d8'39"
David Grosser
Uma 8h48'28" 51d39'6"
David Gubernick
Per 4h32'27" 39d57'13"
David Guest
Umi 15h56'14" 71d54'48"
David Gustin
Vir 12h42'21" -0d33'9"
David Gutman
Ari 2h57'44" 18d21'55"
David Guy Lawrence Terenzio 5-15-37
Tau 5h25'17" 27d16'0"
David H. Berg
Umi 14h5'49" 75d17'56"
David H. Britt
Ori 6h16'19" 15d17'58"
David H. Kirk, Jr.
Ari 3h9'47" 24d25'3"
David H Mastbrook
Uma 10h49'59" 68d26'24"
David H. Solano
Cyg 20h47'57" 31d51'7"
David H. Williams
Tau 4h38'16" 6d31'4"
David Haddaway
Lib 15h12'6" -4d37'14"
David Halloran
Tau 3h53'56" 21d48'48"
David Halsey-Wells
Cyg 20h7'1" 30d49'21"
David Hamilton
Uma 12h33'32" 60d36'40"
David Harris DeMerritt
Uma 9h28'59" 63d58'22"
David Harrison Hice
Ori 4h55'14" 11d8'59"
David Harvey Mattice
Dra 16h56'57" 52d56'11"
David Hawkes
And 23h4'41" 50d59'53"
David Hay - 19 June 1967
Cru 12h29'49" -60d8'4"
David Hayes
Ori 5h14'11" 12d6'6"
David Headly
And 0h45'42" 23d52'5"
David Heath Jr.
Her 17h35'15" 34d3'37"
David & Heather Evertsen
Cas 0h32'5" 53d39'59"
David & Heather Forever
Cyg 20h12'56" 34d13'15"
David & Heidi
Cyg 20h31'55" 33d52'6"
David Heim
Aur 5h54'13" 52d15'4"
David Henry
Boo 15h5'8" 40d30'19"
David Herman Spilfogel
Per 2h13'1" 53d14'54"
David Hester
Cyg 20h21'22" 48d44'54"
David Hickson
Gem 6h46'34" 31d45'50"
David Hill
Leo 10h20'41" 25d51'33"
David Hill May 23rd, 1963
Uma 10h58'31" 52d51'46"
David Hobbs & Kelly Martin 09/19/03
And 22h59'52" 48d3'5"
David Hodge, Jr.
Vir 13h19'56" 13d21'47"
David Hodgkinson
Per 4h4'0" 33d52'28"
David Hoffman Aeschlimann
Ari 2h16'27" 14d15'12"
David Hohimer Jr.
Aql 19h17'57" -3d50'9"
David Holan
Peg 22h28'39" 17d41'23"
David Horne's Musical Star
Aql 19h46'33" -0d43'6"
David Horton
Uma 11h14'25" 60d35'4"
David "Hoss" Matthew Gray Fielder
Uma 8h20'24" 70d50'42"
David Hostetler
Her 16h43'47" 43d45'54"
David Howard Youse
Cap 20h8'40" -18d12'42"
David Howell
Pho 0h44'16" -40d50'39"
David "Hubba" Haltom
Vir 11h58'31" 6d2'34"
David Hugendubler
Mon 7h50'4" -0d52'57"
David "Huggy Bear" Renick
Uma 11h46'15" 61d24'37"
David Hunt
Her 17h42'41" 16d14'3"
David Hussell Hole in One
Ari 3h24'32" 26d37'58"
David I. Kamhi
Umi 16h23'35" 77d55'22"
David "I Love You Baby"
Lyn 9h15'58" 46d21'44"
David I. Savage
Vul 19h47'35" 27d12'7"
David I. Williams
Cnc 8h46'27" 17d31'0"
David Inman
And 0h36'24" 22d4'3"
David Italia
Uma 10h18'59" 50d41'54"
David Ivan James Palmer
Per 3h52'33" 33d48'41"
David J. Camardella
Per 2h23'22" 57d24'11"
David J. Collins
Aql 20h0'7" 9d3'22"
David J. Conklin
Aqr 21h30'34" -0d53'11"
David J. Coon
Boo 14h50'39" 29d2'32"
David J. Crispin 040175
Ari 2h3'20" 24d6'24"
David J. Denino's 30 years <\@> SCSU
Per 2h51'2" 54d5'57"
David J. DeZuzio, Sr.
And 1h16'45" 39d19'8"
David J. (Freddie) Duarte, Jr.
Per 3h43'14" 52d32'38"
David J. Hanarty
Her 17h49'47" 44d56'14"
David J. Herzig
Sgr 18h1'6" -33d36'33"
David J. Knote
Umi 14h26'56" 69d17'29"
David J. Maness
And 23h26'2" 46d0'57"
David J. Murray
Vir 14h42'39" 3d35'58"
David J. Neff
Cnc 8h56'31" 8d9'1"
David J Ordonez
Her 17h51'42" 47d58'31"
David J. Perez Jr.
Her 16h46'10" 38d47'21"
David J. Picker, Esq.
Ari 2h16'34" 24d3'40"
David J. Ronske & Jessica K. Kovacs
Uma 8h12'38" 64d11'56"
David J. Stafford, Jr.
Ori 5h25'25" 6d5'51"
David J. Stern
Aur 6h5'1" 48d17'9"
David J. Williams
Sgr 18h3'2" -17d34'27"
David J. Woodward
Gem 6h52'53" 30d18'34"
David Jace Medina
Per 3h26'48" 47d52'25"
David Jackson
Per 2h10'42" 54d13'44"
David Jackson Carmichael Bonnen
Tau 3h36'45" 5d56'42"
David Jacob Glick
Dra 18h23'2" 57d33'7"
David Jacob Wrobel
Vir 14h0'31" -13d5'19"
David Jacob Wrobel
Per 3h26'49" 47d6'15"
David Jacobs
Lib 14h53'0" -1d36'59"
David Jacot
Cas 0h37'32" 67d4'34"
David Jakubek
Boo 14h45'29" 20d31'21"
David James
Psc 0h37'15" 14d38'47"
David James
Per 2h54'43" 54d13'28"
David James Becker, Jr.
Her 17h52'47" 32d36'21"
David James Bowick
Sge 20h20'9" 18d46'22"
David James Colalillo
Cma 6h30'31" -17d42'39"
David James Derusha
Gem 6h45'38" 33d34'13"
David James Doneker, Jr.
Uma 13h48'5" 52d35'1"
David James Etherage
Ori 5h16'45" -4d22'2"
David James Ewing
Ari 2h29'55" 21d1'17"
David James Hardie
Uma 9h50'41" 42d9'34"
David James Hess
Per 3h29'47" 43d34'40"
David James LeRose
Psc 1h4'42" 14d34'33"
David James McCarthy
Her 17h29'3" 22d36'20"
David James Milligan
Uma 12h1'28" 49d1'37"
David James Murrison
Sgr 18h29'16" -21d54'19"
David James Pugh
Cep 22h36'12" 57d47'35"
David James Roth
Aql 19h46'1" -0d30'40"
David James Russell
Uma 10h47'56" 40d27'37"
David James Spearly
Aqr 22h27'21" -0d34'31"
David James the Optimist
Cap 20h33'35" -21d25'19"
David James Warner - E2 DJW
Cep 21h54'52" 63d22'42"
David James Zajchowski
Sgr 18h29'48" -22d14'43"
David Jameson
Sgr 19h50'7" -13d15'17"
David & Jamie
Cyg 20h12'33" 54d20'12"
David & Jara Chambers 10.2.04
Uma 11h12'13" 30d19'0"
David Jason Mercieca
Cap 20h52'32" -24d58'27"
David Jason Molash
Boo 15h30'43" 40d21'40"
David Jay Budbill
Cep 21h34'31" 58d52'4"
David Jay Kerr
Cap 21h22'46" -14d32'22"
David Jay Slaughter
Uma 8h57'59" 49d25'9"
David & Jean White 50th Anniversary
Cyg 20h25'9" 46d16'57"
David Jeffers
Ori 5h8'32" 5d34'28"
David Jeffrey Ray
Uma 10h36'41" 55d31'45"
David & Jenny Ann Walker
Cyg 19h45'52" 32d26'33"
David Jerome Lavach
Cnc 8h52'31" 26d15'16"
David Jerome Metheny
Sgr 18h52'14" -18d13'22"
David Jerry Collins
Tau 3h42'28" 12d53'24"
David Jimenez
Uma 13h45'15" 59d40'14"
David John Brickwood 13-12-1982
Cru 12h36'59" -63d53'56"
David John Cowham
Uma 9h12'10" 49d10'44"
David John Dinubilo
Cep 20h41'2" 60d33'38"
David John Edwin Capps
Uma 10h7'19" 45d1'54"
David John Gerhard
Tau 4h27'20" 18d26'52"
David John Gulliver
Uma 11h17'27" 64d23'56"
David John Guth
Her 17h14'15" 35d11'56"
David John Hickmott - Puppy Eyes
Cyg 21h33'29" 40d59'4"
David John Holmes
Uma 11h3'14" 71d20'35"
David John Horton Badger
Umi 16h32'25" 77d21'19"
David John Phillips Our Daddy G2
Ari 2h35'0" 18d22'13"
David John Roberts
Uma 9h58'29" 55d2'24"
David John Savord
Psc 1h36'54" 12d2'35"
David John Scanlon
Per 4h6'16" 34d5'41"
David John Schlabach
Uma 11h53'58" 45d4'18"
David John Starr
Gem 6h30'55" 25d55'56"
David John Wright
Psc 1h12'3" 11d54'52"
David Johnson
Her 16h15'9" 6d6'5"
David Jolly
Ori 6h2'41" 21d1'51"
David Jon Shepard
Her 18h7'28" 31d37'22"
David Jonathan Kainath
Uma 10h16'48" 47d33'4"
David Jonathan Spray
Uma 12h29'51" 63d9'49"
David Jones
Aqr 22h18'13" -8d55'43"
David Jones
Uma 9h7'24" 52d0'48"
David Jones
Gem 6h31'41" 25d3'56"
David Jones Millennium Falcon
Per 2h46'56" 39d54'21"
David Jones, My Soulmate
Cyg 19h46'27" 33d22'30"
David Joseph
Dra 18h41'4" 58d42'22"
David Joseph "D J"
Her 17h41'39" 20d7'46"
David Joseph Getto
Cen 14h32'34" -30d5'24"
David Joseph Jaquint
Uma 12h27'49" 53d12'54"
David Joseph Kashkin
Uma 8h40'9" 52d27'22"
David Joseph Lane
Uma 12h8'58" 52d29'25"
David Joseph Patrick Barry
Per 4h9'24" 35d45'40"
David Joseph Presley
Tau 4h6'15" 9d10'40"
David Joseph Schaefer
Her 17h46'22" 19d50'13"
David Joseph Staab II
Lup 15h48'10" -33d53'52"
David Joseph Takacs
Sgr 19h11'54" -29d46'43"
David Joseph Verrico
Lib 15h15'7" -23d29'36"
David Joseph Villanueva
Sco 16h8'11" -15d58'37"
David "JR" Bacon
Uma 11h31'13" 57d57'55"
David Julian
Cnc 8h21'24" 28d52'16"
David & Julie
Cyg 21h55'58" 46d38'23"
David Junka
Lyn 7h41'57" 45d9'3"
David Justason Rennert
Vir 12h27'41" 12d18'57"
David Justin Cameron
Lyn 7h12'59" 47d5'9"
David Justin Piegaro
Umi 16h30'44" 76d56'38"
David K. Berke
Uma 9h47'22" 43d3'23"
David K. Frahm
Cep 21h10'39" 66d20'13"
David K. O'Brien III
Sgr 20h17'17" -35d26'4"
David K. Reinstein
Cep 22h39'4" 72d0'7"
David K. Tumbarello
Lib 15h26'34" -23d44'33"
David K. Wilkinson
Per 3h37'39" 51d49'18"
David Kahn
Ari 3h14'3" 29d14'56"
David & Kamisha
Cap 21h7'56" -25d36'37"
David & Karen Gladman
Cyg 20h31'49" 59d19'54"
David Karl Elsrod
Cyg 20h21'11" 40d46'43"
David Kasler
Aur 5h44'22" 38d1'31"
David & Katrina Elliott - 19th May 07
Cen 13h51'34" -60d11'24"
David Kaufman
Ori 5h37'18" -0d41'41"
David Keith Bottom
Cyg 19h38'37" 34d51'42"
David Keith Hodges
Leo 10h9'59" 21d45'3"
David & Kelly's Wedding Star
Leo 10h55'9" 19d38'8"
David Keltch
Per 3h13'59" 45d33'10"
David & Kendra
Cyg 20h49'51" 54d18'50"
David Kendrick #7
Her 17h36'10" 33d10'49"
David Kenneth Baumbach
Her 17h11'27" 39d26'18"
David Kenneth Lochner
Cnc 9h7'26" 28d38'31"
David Kent Graves
Uma 12h54'38" 57d21'38"
David Kerwin - Star
Crb 16h17'31" 30d22'45"
David Key Loves Beth Hutcheson
Sco 16h16'1" -30d11'47"
David Kim
Uma 8h31'50" 70d58'26"
David Kimberling
Cnc 8h46'48" 22d30'26"
David Kinman
Her 17h9'3" 30d37'34"
David Kirchen
Gem 7h2'17" 16d37'27"
David Konig
Lac 22h52'20" 50d25'37"
David Kosoglow
Psc 1h8'22" 7d59'42"
David Kozubal
Aql 19h49'26" -0d8'29"
David Kranker
Gem 7h56'24" 29d36'17"
David & Krystle
Sge 20h5'33" 17d8'29"
David Kuchinsky
Sco 16h41'29" -39d45'37"
David Kulczycki
Uma 11h13'14" 65d30'28"
David Kurrle 8/4/84
Leo 11h9'37" 23d20'7"
David Kurrus Buzan
Ori 5h29'24" 0d12'31"
David "Kyle" Clarkson
Her 17h7'6" 34d7'27"
David L. Bibey
Uma 10h10'39" 54d6'17"
David L. Bobinger "Pa Pa"
Her 18h47'35" 25d28'27"
David L. Flippin Sr. - "Lovee"
Aqr 21h47'16" -3d31'35"
David L. Griffith & Susan H. Krol
Cyg 20h37'22" 30d29'58"
David L. Heckt
Cep 22h44'12" 64d20'13"
David L Henders
Aqr 23h42'28" -4d43'40"
David L. Herzog
Gem 7h19'21" 25d39'2"
David L. Hyman
Crb 15h49'23" 39d21'21"
David L. Kurtz
Cyg 20h17'0" 47d14'20"
David L. Landers
Aql 20h8'17" -0d4'54"
David L. Linke, Jr.
Uma 11h7'4" 62d42'28"
David L. Miller
Psc 1h53'24" 6d30'37"
David L. Moor
Cnc 9h8'25" 28d29'6"
David L. Neely
Dra 17h53'20" 63d27'56"
David L. Norton
Ori 6h7'48" 7d18'15"
David L. Perry, Jr
Per 3h14'37" 44d58'55"
David L. Roach
Her 16h19'14" 26d30'30"
David L. Rydquist
Cyg 19h35'45" 28d3'55"
David L. Sindel
Sgr 19h22'17" -26d56'20"
David L. T. Ito
Equ 21h17'29" 8d42'18"
David L. Turner
Aql 19h15'46" -1d31'34"
David L. Vennard "The Myth"
Ori 5h16'27" 5d40'7"
David Lachapelle
And 2h11'38" 50d22'25"
David LaFuze
Aql 19h52'38" 12d5'11"
David Laman, MD
Crb 15h50'47" 31d7'19"
David Lanciano, Jr.
Gem 7h46'53" 31d32'4"
David Languerre forever xxx
Uma 12h30'57" 59d59'43"
David Large
And 22h59'25" 39d31'47"
David Larkin
Del 20h43'19" 9d43'43"
David Lawrence Gaither
Gem 7h13'11" 32d25'59"
David Lawrence Proud
Aql 19h22'3" 15d59'57"
David Lawrence William Fodor
Ari 3h12'12" 27d17'47"
David Le Meur
Psa 21h57'49" -33d57'16"
David & Leanne Adkins' Star of Love
Cep 22h14'31" 57d58'5"
David Lee Broughton
Lib 15h5'13" -27d31'19"
David Lee Chaney, Jr. (D.C.)
Her 17h34'22" 33d57'28"
David Lee Cornelison
Aql 19h39'44" 3d41'17"
David Lee DeScalzo
Aql 19h22'28" -0d9'8"
David Lee Frevert
Vir 13h23'36" -7d10'33"
David Lee Heatwole
Boo 14h44'33" 23d17'24"
David Lee Herrera
Uma 11h57'0" 52d51'50"
David Lee Holt
Uma 11h46'48" 43d35'27"
David Lee & Jennifer Zee
Cyg 19h48'17" 38d28'10"
David Lee Loves Julie Lee IDST
Cam 5h3'22" 60d51'58"
David Lee Matthews
Sgr 18h6'20" -27d35'34"
David Lee Page
Lyr 18h48'29" 45d34'7"
David Lee Ralph III
Vir 13h39'17" 3d12'58"
David Lee Raulerson
Cap 20h29'43" -11d29'52"
David Lee Renahan
Leo 10h42'7" 13d2'24"
David Lee Reynolds
Aqr 22h33'5" 1d50'35"
David Lee Shumate
Dra 17h37'47" 62d44'29"
David Lee Watkins
Uma 10h12'46" 51d52'52"
David Lee Wortman
Sco 17h51'19" -36d22'34"
David Lee-Wei Liu
Aql 19h44'25" -0d6'26"
David Leigh
Cas 1h17'38" 55d8'43"
David Leigh Peterson
Per 4h6'54" 51d33'32"
David Lennox
Cyg 21h40'55" 32d20'6"
David Leo
Ori 5h44'49" 0d4'8"
David Leo Benner
Per 2h48'30" 53d23'44"
David & Lesley Cowles
Cyg 21h20'45" 44d34'15"
David Lester Jenkins
Gem 7h47'36" 25d52'20"
David Leuw
Uma 11h26'1" 68d50'19"
David Li Vuillemin
Her 17h4'42" 13d29'16"
David Lincoln
Cep 23h19'3" 77d8'5"
David & Linda Tanner
Sge 19h55'45" 17d26'43"
David Lord Sutter
Ori 6h4'7" 17d22'57"
David Lorenc
Uma 8h12'2" 71d13'40"
David Lorenzo Heaton Guardian Star
Per 4h42'14" 45d35'9"
David Louis Ross
Per 3h16'45" 46d16'25"
David Louis, Shirley's Shining Star
Umi 19h9'53" 88d1'11"
David Love
Her 17h49'11" 41d26'23"
David Loves Jennie Forever
Umi 14h21'3" 67d35'17"
David Loves Lauren
Tau 4h42'41" 21d50'55"
David Lowell Maxwell
Tau 4h55'50" 17d58'15"
David Lowenstern 3/14/56
Psc 1h8'1" 33d19'59"
David Loy Rogers
And 23h10'9" 42d28'4"
David Loya, Jr.
Vir 12h43'18" -2d27'36"
David Loyall Morgan
Lac 22h21'42" 47d18'55"
David "LPD 973" Hernandez
Tau 3h26'28" 4d37'44"
David Lucas
Uma 10h7'41" 50d5'10"
David Lukas Rhodes
And 0h43'21" 34d0'30"
David Luke
Lib 15h52'9" -12d32'56"
David Lydick #1 Dad
Lyn 7h47'13" 38d1'38"
David Lynn and Barbara Baur
Pho 1h12'50" -40d20'42"
David Lynn Watkins
Aqr 22h35'35" -2d18'45"
David Lyon
Uma 10h4'23" 56d47'45"
David M. Birdwell~My Love & My Life
Umi 15h2'24" 73d55'23"
David M. Brodhead
Leo 11h6'43" -4d44'37"
David M Gizynski
Her 18h45'36" 23d57'29"
David M. Haigh
Lyn 9h9'50" 38d47'18"
David M. J. Wilkens
Tau 4h27'11" 20d45'34"
David M. Meyler
Sgr 18h5'54" -27d24'39"
David M. Owens & Wendy L. Herrick
Cyg 20h41'25" 36d45'12"
David M. Prior
Sgr 18h6'14" -27d11'48"
David M Ratz
Del 20h38'15" 14d42'6"
David M. Stanley
Gem 6h17'26" 23d26'17"
David M. Strohl Jr.
Cnc 8h17'25" 22d31'45"
David M. Tenenbaum
Cyg 20h26'8" 46d51'56"
David M. Tolli
Ari 2h43'57" 20d51'17"
David M. Tyler, Jr.
Her 16h25'44" 12d11'53"
David Mac., Brown, soul-mate 2 PCB
Uma 11h43'9" 64d54'1"
David Mack
Aqr 22h11'17" -21d11'45"
David Mackintosh
Her 17h22'2" 15d42'30"
David Maddox
Cep 0h4'54" 73d20'45"
David & Madeline's Eternal Star
Uma 11h49'32" 57d5'57"
David Malcolm Allman
Cyg 20h32'42" 35d3'28"
David Malick
Tau 3h32'33" 8d23'42"
David Mangion
Cru 11h57'42" -62d22'17"
David Manzini
Sco 17h53'44" -33d28'58"
David & Marge Biesiada Forever
Lyn 8h24'37" 55d57'21"
David & Marianne Bates
Umi 16h16'22" 71d27'8"
David Marine
Cep 22h18'57" 60d20'29"
David & Marissa 6.6.02 till Forever
Cyg 20h21'12" 47d39'16"
David Mark Curran
Ori 6h16'8" 14d42'12"
David Mark Eblen
Gem 6h40'15" 28d30'1"
David Mark Parrillo
And 23h15'4" 47d15'21"
David Mark Tolley
Vir 12h16'48" 2d33'21"
David Marks
Uma 10h4'11" 64d29'27"
David Marshall Wallace III
Uma 11h26'12" 33d30'32"
David Martin
Cnc 8h44'12" 19d1'22"
David Martinez, love of my life
Her 17h13'41" 19d3'44"
David & Mary Hager
Vir 13h25'31" 8d11'56"
David Mataya
Cnc 8h47'34" 28d32'47"
David Mathew Lawrie
Cap 21h6'2" -16d23'25"
David Matthew
Ari 2h12'9" 24d20'19"
David Matthew Cummings
Sgr 18h8'34" -17d56'26"
david matthew eisel
Cyg 20h47'48" 39d45'23"
David Matthew Koble
Her 18h51'50" 24d18'30"
David May
Boo 15h34'27" 40d36'48"
David Maynard
Lib 15h10'8" -27d55'28"
David Mc Caleb
Uma 9h27'4" 47d57'37"
David McCarthy
Peg 21h57'25" 11d29'14"
David McCloughan, Jr.
Vir 13h19'57" -17d56'55"
David McGuckin's Star
Peg 22h16'46" 11d34'13"
David & Melinda Mahosky ~TRUE LOVE~
Sco 16h55'59" -44d49'40"
David Mendez
Per 2h52'42" 49d56'7"
David & Meredith Ward 11/30/00
Per 3h8'15" 55d26'23"
David Michael
Leo 11h44'45" 26d16'55"
David Michael
Peg 22h46'53" 17d53'28"
David Michael Anderson
Lyn 8h14'34" 35d19'15"
David Michael Bates
Ari 3h17'24" 25d49'38"
David Michael Buhidar
Ari 3h3'0" 15d54'6"
David Michael Carter B.C.F.D. 2005
Per 3h27'23" 51d26'36"
David Michael Eplin
Cyg 20h7'58" 34d23'38"
David Michael Flannery
Per 3h26'7" 48d44'22"
David Michael Hendrix
Her 17h48'57" 18d24'11"
David Michael Hernandez
Ari 2h36'23" 18d16'42"
David Michael Hutchinson
Vir 13h14'47" -4d16'36"
David Michael Jennings II
Tau 3h45'25" 15d49'24"
David Michael Jezek
Cyg 19h49'40" 35d55'19"
David Michael Kipphorn
Aql 19h53'39" 13d10'18"

David Michael Long
Umi 13h0'39" 88d57'26"
David Michael Malishchak Jr.
Ori 5h39'36" 3d5'55"
David Michael McLean
Ori 5h23'41" -4d36'27"
David Michael Myers
Aqr 22h55'24" -16d54'49"
David Michael Paskell, Jr.
Sgr 18h54'7" -28d4'20"
David Michael Pugh
Vir 13h16'40" 9d6'46"
David Michael Riter
Ori 6h0'9" 22d31'1"
David Michael Schum
Uma 14h0'44" 54d3'49"
David Michael Shindle
Cap 21h34'17" -16d5'33"
David Michael Sommers April 3 1956
Cep 22h59'28" 78d51'14"
David Michael Thomas - Alex's Daddy
Uma 11h39'50" 57d4'30"
David Michael Vidales
Her 17h13'56" 34d40'56"
David Michael Watts 04.01.1976
Cru 12h47'5" -56d53'54"
David Michael Whitlock
Cep 21h44'23" 64d12'55"
David Michael Wooten
Uma 11h39'43" 42d0'55"
David Michaels
Her 16h17'11" 44d44'54"
David Micheal Hobson
Lib 15h3'28" -3d23'17"
David Migliore
Ori 5h36'2" -0d37'46"
David & Mildred Vail Family Star
Cyg 19h49'14" 33d31'23"
David Miles Whitaker
Ori 5h40'59" -0d0'33"
David Miller
Gem 6h51'34" 19d28'38"
David Million
Ori 5h33'13" 5d57'18"
David Mills "Hunklet" Hay
Sco 17h54'9" -31d30'37"
David & Mindy Moya
Cyg 21h52'33" 44d42'36"
david mon coeur
And 23h43'45" 36d7'22"
David Monfils
Lib 15h5'43" -3d50'33"
David Moniz 28.04.2000
Lac 22h35'8" 55d50'11"
David "Monkey" Goldsmith
Sco 16h17'42" -9d22'0"
David Moran
Gem 6h47'49" 24d11'54"
David Moreno y Lizett De Leon
Per 4h34'0" 41d52'55"
David Morgan Ermer
Cep 21h14'7" 65d41'39"
David Morley
And 2h20'36" 47d52'23"
David Mount
Aqr 22h32'49" 2d0'59"
David M.R Hughes - Il Mio Oceano
Tau 4h14'9" 21d14'9"
David Mulhall loves Megan Lackey
Cra 18h39'22" -38d51'2"
David Muradyan
Boo 14h52'24" 53d38'35"
David Murdock Jr.
Cma 7h3'47" -16d49'58"
David Murphy
Per 4h40'32" 40d14'10"
David Mykel Royce Oliva
Lyr 18h48'22" 34d30'23"
David N Andrews - Our Three Wise Men
Cru 11h58'50" -63d20'21"
David N. DeGreef
Her 17h29'51" 42d34'39"
David N. Sillars, Ph.D.
Umi 15h21'59" 73d17'13"
David Nadasi
Ori 6h4'45" 19d55'28"
David & Nancy Nunke
Equ 21h15'24" 10d59'44"
David Nash Payne III
Cap 20h26'42" -9d32'23"
David Nathan Isadore
Her 16h13'2" 16d18'20"
David Nathaniel Brown
Gem 7h6'18" 23d48'47"
David Nathaniel Sumwalt
Umi 14h50'3" 72d22'29"
David Navarro/Most Generous
Uma 10h14'21" 58d12'30"
David Neal Geubelle
Per 3h34'51" 32d57'22"
David Neel
Per 3h21'39" 43d54'40"
David Negron
Vir 14h18'35" -4d27'35"
David Neil Adams
Her 17h2'14" 30d25'28"
David Neiman
Aur 5h52'13" 37d28'57"
David Nesbitt Matthews
Cas 23h45'43" 57d47'34"
David Nestlebush - ECVHS 2006
And 1h42'7" 42d33'38"
David Nicholas
Tau 4h11'10" 4d12'58"
David Nicholas
Cep 3h39'35" 78d36'8"
David Nisky
Psc 0h58'57" 14d55'51"
David Nolan Davis
Psc 1h33'57" 23d6'33"
David Norman Morrow
Cyg 20h29'42" 43d26'43"
David Nuno-Murguia
Psc 1h8'54" 29d35'20"
David O. Brink
Dra 17h58'57" 54d58'2"
David O McConnaughey
Vir 13h18'39" 1d0'0"
David Oakes' Own G-D Star
Ori 5h20'22" 5d7'17"
David O'Connor
Aql 20h12'50" 7d16'59"
David Olivier
Ori 5h45'6" 10d17'51"
David Ollivier
And 2h35'25" 39d0'15"
David O'Loughlin
Sgr 17h52'33" -28d52'43"
David Olsen
Gem 6h48'33" 17d12'56"
David Oscar Becker
Her 18h26'54" 18d43'10"
David Otto
Cyg 21h44'40" 44d56'49"
David Overton
Lyn 7h37'33" 35d51'29"
David P. Bryson
Uma 11h3'22" 55d50'9"
David P. Calabro
Aur 5h50'43" 41d43'1"
David P. Conley
Uma 10h39'50" 65d11'21"
David P Hart
Aql 18h43'5" -0d37'59"
David P. Jasso
Cnv 12h41'30" 36d20'57"
David P. Koch
Umi 14h9'22" 74d9'19"
David P Milward
Leo 11h51'28" 22d1'33"
David P. Ramsey 111771
Sco 16h12'39" -13d49'29"
David P. Silva, Jr.
Cnc 8h17'2" 22d13'17"
David P. Tenzel The Star Surgeon
Her 16h49'47" 17d24'32"
David P. Waite
Peg 22h33'24" 21d56'43"
David "Pa" Bryant
Cnc 8h54'50" 30d40'6"
David Palmer Kindt
Her 17h32'6" 34d18'7"
David & Pamela Rozanc
Uma 14h3'43" 51d51'29"
David & Pam's lucky star
Ori 5h27'14" 5d0'12"
David "Papa" Cynowa
Psc 1h17'39" 31d28'16"
David Pardee
Uma 11h34'11" 31d21'9"
David & Patricia Golden Anniversary
Lyr 18h54'47" 31d37'5"
David Patrick Adams Taylor
Aur 5h53'10" 38d54'10"
David Patrick Foley
Leo 10h8'0" 24d9'31"
David Paul Audoma January 22, 1940
Crb 15h39'17" 33d3'41"
David Paul Carlson
Aur 6h18'51" 52d14'24"
David Paul Diehl My Lucky Star
Cyg 19h47'42" 30d47'12"
David Paul Mann - David's Star
Uma 11h20'42" 40d49'48"
David Paul Oliver
Sco 17h46'54" -37d17'33"
David Paul Reed Sr.
Vir 14h43'17" 2d30'24"
David Paul Wilson
Uma 10h24'13" 42d58'11"
David & Paula Hamm
Cap 20h11'2" -8d34'14"
David & Pauline's light to eternity
Ari 2h42'20" 26d13'39"
David Pavliscak
Uma 10h34'18" 68d0'32"
David Pavlow
Lib 14h49'10" -2d20'49"
David Payne
Cnc 8h41'35" 9d30'6"
David Pedley - "Star of David"
Cep 23h52'51" 74d39'23"
David Penman
Per 4h17'56" 43d40'0"
David Perry I Love You
Cyg 21h18'45" 53d7'15"
David Peter
Cyg 20h13'32" 41d16'58"
DAVID PETER GRAF
Peg 22h5'6" 26d28'15"
David Peter Martinelli, Sr.
Per 3h57'10" 32d58'21"
David Peter Walsh
Cyg 19h40'44" 28d22'17"
David Philip Kirson
Per 3h42'14" 45d32'41"
David Phillip Talarico,July 8, 1983
Cnc 7h59'53" 12d12'7"
David Pickett
Peg 22h45'58" 27d39'18"
David Pilcher
Vir 12h18'8" 3d58'36"
David Pitt
Sgr 19h32'44" -28d50'33"
David Polnerow
Cnc 8h22'13" 32d32'42"
David Prager
Cnc 8h50'27" 27d33'9"
David Prather - Honoring 56 Years
Aql 19h3'39" -0d7'13"
David Preston Weltz
Lib 14h52'7" -3d40'18"
David Price
Her 17h22'5" 31d54'48"
David (Punky) Drake Duncan
Her 16h35'29" 5d45'58"
David Quinn Greene, Jr.
Her 16h41'2" 26d50'36"
David R. Dembinski
Aql 19h12'38" -0d4'33"
David R. Engles
Her 16h29'7" 17d5'37"
David R. Fletcher
Ari 2h37'48" 17d55'27"
David R. Kelly
Tau 5h59'23" 26d22'24"
David R. King
Lyn 8h13'16" 40d56'32"
David R. Lenton
Ori 6h22'58" 16d31'17"
David R. Masse
Sco 16h16'44" -11d53'35"
David R. McWhorter
Her 16h38'6" 32d52'41"
David R. Pasick
Ori 6h15'31" 10d44'0"
David R. Schreier
Per 2h11'22" 57d9'22"
David R. Smith
Sco 16h33'1" -28d55'40"
David R. Swope 1437
Aur 6h3'54" 47d23'6"
David R. Wetzel, Ph. D.
Leo 11h30'9" 9d6'58"
David Rakowsky
Her 18h36'40" 19d43'44"
David Ramos
Vir 12h26'3" 0d47'37"
David Randall
Boo 14h53'29" 31d5'58"
David & Randi
Lyr 18h25'1" 39d46'14"
David Raulins
Uma 11h1'54" 59d58'21"
David Ray Cave
Cap 20h14'54" -10d42'7"
David Ray of Hope and Brilliance
Per 4h2'26" 35d23'58"
David Ray Ritcheson
Uma 12h46'27" 62d24'12"
David Ray Vosberg
Uma 12h15'56" 55d12'51"
David Reagle
Lib 14h52'43" -4d46'2"
David Reddick
Her 16h36'39" 38d16'4"
David Reich "Father Time"
Ori 5h58'11" 17d18'25"
David Renzi
Lib 15h37'2" -9d40'25"
David Reser's Birthday Star 2007
Gem 7h14'30" 22d54'28"
David Resnick
Cyg 21h38'49" 37d42'45"
David Reyes
Ara 17h30'46" -62d43'50"
David Reynolds
Aur 6h4'50" 47d52'11"
David Reynolds Hyne
Tau 4h33'52" 4d20'24"
David Richard Ostrom
Ari 2h48'8" 22d22'12"
David Richard Rouza
Cap 21h35'45" -10d25'56"
David Riley Coffee
Her 18h43'46" 17d25'44"
David Risdon Crosley
Tri 1h46'32" 33d28'57"
David Robert Anderson, Sr.
Leo 10h2'34" 16d2'34"
David Robert Bittner
Ori 5h11'6" 1d35'2"
David Robert DuBois
Cnc 8h15'59" 18d16'52"
David Robert Elkin
Gem 6h35'17" 14d50'8"
David Robert MacRae
Uma 13h18'58" 60d20'58"
David Robert Marquis
Psc 23h9'49" 1d2'16"
David Robert O'Connor
Cru 12h36'10" -58d41'56"
David Robert Pollock
Aur 5h35'17" 33d26'32"
David Robert Ross
Vir 13h0'18" 11d57'13"
David Robin Powers
Her 17h40'44" 35d59'51"
David Roger Charles Smith
Ari 2h25'13" 24d34'50"
David Rogers — David's Star
Per 3h20'33" 32d18'49"
David Rose
Uma 8h57'30" 59d30'58"
David Rose
Lib 14h27'47" -19d15'45"
David Rosen
Cap 21h4'11" -23d27'41"
David Rosenberg
Uma 14h10'55" 62d10'48"
David Rosenbluth
Aur 5h29'9" 40d14'28"
David Ross Caplan
Pho 1h8'49" -44d44'13"
David Ross Dorward
Lmi 10h51'17" 38d58'34"
David Ross Nadel
Vir 11h48'0" -3d55'47"
David Rowland
Leo 11h50'46" 25d59'34"
David Roy
Cep 22h14'35" 61d25'40"
David Roy Eaton
Cyg 21h30'12" 41d14'45"
David Roy Hunt
Aqr 20h42'26" 0d38'56"
David Roy Teich
Dra 14h37'36" 63d31'50"
David Royal Pierpont
Psc 0h57'56" 28d37'18"
David Royce Bell
Cap 21h54'45" -10d25'30"
David Royce Doran
Vir 13h29'4" -6d18'54"
David Rudman
Cep 22h38'55" 60d20'45"
David Rudolph Glad
Cep 21h36'7" 61d44'30"
David Russell Krichbaum
Aur 6h31'34" 35d23'50"
David Russell Payne
Dra 17h13'58" 65d41'13"
David Russell Petrak
Her 16h57'48" 29d22'46"
David Ryan
Per 4h11'2" 47d21'13"
David Ryan Hedrick
Cap 21h26'37" -17d48'49"
David Ryan Keach
Her 16h38'34" 4d26'2"
David Ryan St. Louis
Her 18h46'23" 14d49'27"
David Ryan Staley
Dra 17h21'40" 52d24'27"
David Ryan Weinraub
Vir 12h58'14" -16d31'49"
David S. Asbill, Jr. M.D.
Uma 9h46'11" 47d58'35"
David S. Goodman
Uma 8h47'45" 63d3'6"
David S. Miller II
Uma 10h8'13" 63d49'13"
David S. Morrison
Uma 8h36'48" 47d22'59"
David S Parker
Psc 0h17'11" -4d11'9"
David S. Rayman
Cep 0h36'46" 82d5'7"
David S. Straus
Umi 14h39'41" 74d49'50"
David S. Wiggin
Aur 5h54'26" 42d35'39"
David Sainato Super Star
Dra 17h34'1" 50d53'29"
David & Sally Boesel
Uma 11h25'27" 61d14'30"
David Samuel Owens
Cnv 13h45'57" 30d39'45"
David Samuel Rosenbaum
Uma 11h53'41" 29d40'48"
David Sanchez Jr.
Aqr 22h19'6" 0d38'47"
David Sandler, CEO
Per 3h25'33" 52d25'0"
David Santiago
Ori 5h39'2" -9d13'1"
David & Sarah Sterling Family Star
Ari 2h44'28" 27d33'8"
David Sawyer Thompson Jr.
Uma 9h27'40" 43d37'52"
David Schall, III
Uma 11h8'21" 34d15'6"
David Schechter
Uma 8h35'28" 68d39'28"
David Scholes
Cep 21h48'10" 80d16'0"
David Schultz
Ori 6h19'43" 2d29'55"
David Schwanke
Gem 7h43'41" 33d11'27"
David Scott
Uma 13h25'53" 61d47'39"
David Scott Bible
Aur 6h6'53" 54d24'47"
David Scott Brannon
Umi 16h7'7" 77d21'21"
David Scott Calverley
Hya 9h33'50" -0d29'22"
David Scott Chenier
Per 3h49'26" 40d50'24"
David Scott Dixon
Cnv 12h43'28" 43d16'8"
David Scott Kuss
Cas 0h8'52" 53d15'27"
David Scott Maxwell
Tau 5h15'11" 19d45'4"
David Scott Taylor
Her 17h17'56" 27d3'10"
David Scott Thompson
Ari 2h12'0" 23d15'45"
David Scott Wiseman
Aql 20h7'47" 10d34'16"
David Searfoss
Umi 14h50'42" 73d35'35"
David Sears
Ori 5h8'14" 15d0'43"
David Shamoun
Psc 1h8'59" 4d34'52"
David Shane Ostrom
Tau 3h30'41" 4d56'3"
David "Shawn" Patton "Honey Man"
Vir 13h12'9" -13d13'11"
David Sheffield
Sgr 17h56'53" -18d49'19"
David Shen
Psc 1h25'46" 15d9'56"
David Sherman Devault
Ori 5h33'29" -2d30'37"
David & Sherry
Cyg 20h51'45" 32d37'33"
David & Sherry Forever on 12-31-76
Cap 21h28'55" -8d47'7"
David Sidney Smith
Her 18h25'5" 13d23'58"
David Siegel
Her 17h44'23" 47d2'37"
David Sierra
Boo 14h29'10" 51d23'31"
David Simpson
Boo 15h42'18" 45d46'44"
David Sinfield - The Wise One
Cep 21h55'36" 63d34'5"
David Siroty
Her 18h45'31" 25d54'56"
David Sisteck
Uma 11h6'56" 49d36'21"
David Skidmore
Lib 15h59'30" -8d47'30"
David Sloan
Sco 17h57'12" -40d15'15"
David Smith "Father of the Bride"
Boo 14h10'5" 52d2'55"
David "Snookie" Brown
Uma 14h16'57" 58d29'44"
David Sofia
Boo 13h37'31" 21d41'1"
David & Sonya's Eternal Light
Col 5h14'34" -35d48'49"
David Spellman
Uma 9h41'29" 44d56'14"
David Spence
Dra 19h43'11" 61d8'4"
David Spencer Orr
Her 18h15'31" 15d50'8"
David Spierling
And 0h52'24" 38d2'49"
David Spinks
Psc 0h56'51" 31d9'51"
David Squicquero
Per 3h11'59" 49d8'32"
David Stacy
Lyn 7h47'49" 39d8'42"
David Stanford
Per 2h51'27" 37d45'6"
David Stebbing
Cep 0h15'30" 73d29'34"
David Stehl
Cnc 7h56'4" 15d53'21"
David Steiner
Aql 19h27'56" 8d54'57"
David Stendeback
Cma 6h36'13" -23d22'58"
David Sterling 9/30/1956
Uma 11h11'42" 57d38'39"
David Steven Benfield
Ori 6h10'13" 19d27'4"
David Stewart
Per 4h17'27" 39d32'25"
David Stuart Duhon 07281960
Leo 9h38'39" 14d45'16"
David Suberman
Cas 0h34'45" 53d55'31"
David Sulteanu
Lac 22h41'55" 51d23'35"
David Sumner
Cam 3h49'46" 64d29'37"
David Suppan
Her 16h42'12" 47d8'13"
David & Susen Latchman
Lyr 18h48'45" 33d54'48"
David Sven Nelson - My Angel Above
Cru 12h44'11" -56d43'32"
David Sweet Velvet Love Hentz
Crb 15h46'3" 34d40'1"
David Sylvester
And 0h51'27" 23d39'24"
David & Sylvia Laird
Cyg 20h55'8" 47d11'17"
David & Sylvie
Tau 3h24'23" 13d57'50"
David T. Anderson 09 Sept 46 Star
Ori 5h40'14" -2d41'21"
David T. Davis
Lyn 7h58'48" 35d29'9"
David T. Meyer's Star
Aqr 22h30'0" 0d35'33"
David & Tammy's Star
Umi 15h21'12" 74d6'40"
David Tenho Toyryla
Psc 0h22'13" 5d32'51"
David Tennant
Her 18h4'21" 29d22'40"
David Terry Lawson Shainberg
Peg 23h18'44" 26d0'36"
David Thomas
Tau 3h48'37" 7d11'21"
David Thomas Cline
Cap 20h26'40" -11d53'25"
David Thomas Collier
Her 16h43'14" 41d56'6"
David Thomas Fiala
Aur 5h33'3" 37d53'18"
David Thomas Johnson Jr.
Psc 23h10'42" 0d32'26"
David Thomas Meade
Ori 6h18'10" -1d19'0"
David Thomas Williamson
Her 18h30'49" 13d0'59"
David Tiffin
Her 17h1'57" 32d10'0"
David Townes
Lyn 9h40'46" 40d8'38"
David Truong Luu
Leo 9h48'41" 9d6'19"
David Tweeton
Ant 10h48'48" -35d55'53"
David Ujenski
Aqr 21h55'44" 0d40'33"
David und Marina
Ori 5h56'14" 7d13'43"
David V. Foley II
Per 3h45'54" 51d37'6"
David Van Dyke
Dra 15h58'58" 53d54'0"
David VanMiddlesworth
Ori 6h7'14" 6d4'18"
David Vannoy
Her 18h3'33" 29d57'39"
David Varney
Tau 4h17'6" 30d25'24"
David Vartanian
Leo 10h50'32" 9d55'4"
David Vaughan
Uma 9h55'2" 50d21'42"
David Vaughn
Leo 10h24'39" 25d22'27"
David Vela
Umi 16h45'41" 76d56'28"
David Veytsman 12/12/1937-3/21/2001
Uma 9h50'8" 49d22'26"
David & Vicki
Umi 15h26'57" 73d31'23"
David Vieyra
Lib 15h3'8" -15d9'42"
David Villeneuve "Funky D"
Her 15h52'23" 39d50'55"
David Vincent Ferguson
Psc 1h30'14" 24d24'31"
David Vinnedge
Eri 3h52'37" -0d47'31"
David Vought Squires
Cyg 21h5'43" 42d16'59"
David W Boyt
Cap 20h22'10" -12d38'59"
David W. Burnis
Uma 11h12'49" 51d5'21"
David W. Carr
Her 17h49'52" 33d57'59"
David W. Chesser
Psc 1h43'56" 24d46'2"
David W. Ellershaw
Cep 21h36'54" 55d41'51"
David W. Fegley
Aqr 20h57'0" -9d23'8"
David W. Fiedrich
Aur 5h16'44" 42d58'11"
David W. Haub
Aur 5h54'54" 36d34'24"
David W. Henson, Jr.
Her 17h42'37" 16d1'57"
David W. Kauffman
Uma 11h24'4" 36d39'4"
David W. Kay
Ori 6h14'42" 8d19'5"
David W Keely
Her 17h7'50" 30d34'10"
David W. Lemerand Family
Cep 21h40'46" 61d27'2"
David W. McPherson
Lib 15h23'23" -5d35'31"
David W. Miller
Cap 20h9'40" -11d19'36"
David W. Miller
Tau 4h12'28" 27d42'8"
David W. Mosley
Cap 21h45'15" -9d4'42"
David W. Paquettes' Happy 49th
Aqr 22h19'34" 0d19'17"
David W. Sellars
Aql 19h51'27" -0d33'8"
David W Thornton
Her 16h40'13" 33d43'16"
David Wagner
Cep 22h41'1" 66d31'11"
David Wagner
Uma 9h3'37" 68d26'58"
David Wagner
Uma 9h48'20" 60d32'15"
David Walder Rose AKA Honey Roseman
Sgr 18h9'11" -26d29'56"
David Walker Gardner
Cep 22h5'0" 57d18'25"
David Wallace Mack
Ori 5h5'48" 11d45'34"
David Wallace Trapp
Lib 15h46'59" -18d22'6"
David Wanamaker
Aur 5h39'33" 45d13'21"
David Ward Brown
Cyg 20h58'18" 34d51'14"
David Waters
Aql 20h21'38" -0d34'5"
David Watts
Tau 3h44'38" 27d32'58"
David Wayne Brown, Jr.
Lib 14h39'23" -22d20'6"
David Wayne Buol
Aur 7h21'45" 39d31'14"
David Wayne Collins
Dra 17h31'22" 64d55'46"
David Wayne Crow
Crv 12h14'36" -21d0'23"
David Wayne Donnellan
Cyg 19h39'59" 38d50'11"
DAVID WAYNE HALE
Ori 4h46'44" 12d20'0"
David Wayne Morrison
Lyn 6h22'50" 59d23'59"
David Wayne*Nicole Lynn
Vir 13h19'24" -14d33'49"
David Wayne Powers
Tau 4h40'38" 1d0'17"
David Wayne Soley
Uma 11h34'50" 55d11'53"
David Wayne Stanley
Gem 7h19'48" 32d0'45"
David Weaver
Tau 5h13'55" 19d58'49"
David Weaver
Leo 10h4'32" 23d18'25"
David Weller
Sex 10h19'58" -9d49'34"
David Wells
Per 3h9'47" 46d16'37"
David & Wendy
Cyg 20h55'22" 46d12'9"
David Wesler
Aur 5h19'48" 37d36'59"
David Wheatley
Lib 15h5'39" -3d44'5"
David Wheaton Berg
Peg 22h13'2" 33d7'35"
David Wheeler's Eastern Star
Uma 11h25'27" 35d18'45"
David White
Lyn 6h33'52" 61d0'17"
David White Sr.
Her 17h2'52" 30d35'21"
David Whitman
Cap 20h57'57" -22d1'36"
David Wilds
Lib 15h40'41" -11d57'31"
David William Dysinger, Senior
Her 16h32'16" 7d49'27"
David William Hyman
Cap 20h37'15" -21d29'12"
David William Krajanowski
Cyg 19h30'20" 28d39'13"
David William Mathies
Uma 8h31'31" 61d53'8"
David William McDonner, Jr.
Her 16h36'44" 37d17'47"
David William Ockenden
Sco 17h21'58" -38d0'7"
David William Pettigrew
Per 3h9'35" 56d19'47"

David William Reiter
Uma 10h28'7" 62d21'21"
David William Schweikhardt
Cep 23h46'47" 82d36'43"
David William Sikorski
Leo 10h15'13" 25d52'1"
David William Stanton
Tau 4h33'54" 0d18'38"
David William Taylor
Ori 6h18'13" 15d33'37"
David William Thein
Uma 11h31'9" 58d19'0"
David William Ward
Vir 13h50'33" -18d24'55"
David William Waters
Leo 10h8'19" 20d41'59"
David William Wright
Lyn 6h32'23" 61d9'5"
David Wilson
Sgr 20h0'44" -39d33'56"
David Wilson
Uma 10h37'17" 47d53'55"
David Wilson
Her 17h5'45" 33d0'33"
David Windsor Buck
Tau 4h35'52" 27d51'19"
David Winston Kiser
Uma 9h16'52" 70d30'20"
David Winston Leff
Uma 11h6'42" 54d34'20"
David Wojtkowiak
Umi 15h21'28" 72d11'54"
David Woods and Tammy Burnside
Lib 14h59'0" -16d52'44"
David Wormstone
Dra 18h37'57" 57d19'8"
David Wright
Peg 23h45'36" 25d12'39"
David Wright
Ori 5h59'36" 6d11'32"
David Wyatt Wittlin
Sco 16h46'19" -26d34'49"
David Xianghe Ma
Cyg 21h30'0" 31d34'38"
David Young
Vir 12h45'59" 5d33'20"
David Z. Kolen
Cas 23h49'48" 53d59'18"
David Zeltner Chesnoff
Per 3h41'28" 44d49'58"
David Zfania
Per 4h16'43" 52d21'43"
David & Zoe - Just the beginning
Cru 12h49'28" -56d40'26"
David, Matt, Jay & Jotina Willett
Her 18h52'34" 24d3'7"
David, Patrick Stoner's Grandson
Her 17h41'30" 44d37'5"
David, you weirdo!
Ori 6h2'17" 10d46'23"
David, you'll always have my heart
Vir 13h23'34" -19d32'39"
David30Dodo
Uma 8h39'27" 59d37'1"
davidandheather
Gem 7h39'16" 15d28'52"
DavidAPappagalloPerAngustaAdAugusta
Uma 11h17'27" 45d43'7"
davidbrown
Lyn 8h24'38" 36d13'6"
Davide
And 1h17'49" 42d20'7"
Davide
Cam 3h37'10" 55d13'21"
Davide
Ari 3h11'16" 24d47'35"
Davide
Peg 22h11'0" 10d18'13"
Davide
Cep 21h31'27" 69d34'51"
Davide
Uma 8h53'51" 67d59'32"
Davide Barzaghi
Her 15h54'11" 44d47'12"
Davide Carboni
Oph 17h10'48" 1d38'27"
Davide Ferrarese
Uma 9h52'11" 67d3'59"
Davide & Flavia
Lyr 18h54'51" 28d2'22"
Davide Gelisio
Cam 7h27'31" 66d31'4"
Davide Lasi
Aur 5h28'27" 35d46'6"
Davide Leith Rizzo
Cen 13h58'34" -61d3'54"
Davide Marzullo
Cyg 20h4'31" 32d45'7"
Davide Paladino
Sco 16h5'16" -8d47'15"
Davide Riccardo Comi
Umi 13h37'3" 76d8'41"
Davide Samuel Eibo
Ori 5h56'37" 2d9'26"
Davide Veluti
Ori 5h56'35" 17d41'49"
Davide Vincenzo Cuomo
Her 17h38'42" 41d39'24"
Davide...Dado
Uma 9h23'39" 41d40'56"
David-Erickson2002
Del 20h42'15" 11d6'17"
Davidet Guilhem Gauthier
Cmi 7h29'28" 2d31'18"
David-HDS&S
Lyn 8h7'40" 36d30'20"
DavidJacobs1
Boo 14h40'40" 36d44'20"
Davidji 05051957
Ser 18h19'59" -14d29'58"
DavidMitchellMantz23
Sco 17h42'20" -40d2'30"
David-Paul Cook
Dra 19h0'26" 52d24'45"
David's Birthday Star
Sgr 17h55'38" -22d19'13"
David's Celestial Star
Lac 22h47'39" 52d28'36"
David's Cookie
Uma 11h20'15" 49d55'2"
David's Diamond
Ori 5h36'30" 10d12'9"
David's Diamond
Cnc 8h55'2" 8d26'13"
David's D-Light for Joan
Per 3h11'49" 52d16'30"
David's Eternal Light
Aql 19h14'8" 2d12'46"
David's Eternal Light
Psc 1h12'53" 28d43'0"
David's Lamp
Ari 2h20'56" 22d45'42"
David's Love for Becky is this Star
Cyg 19h33'3" 27d48'1"
David's lucky star
Leo 11h48'39" 19d0'59"
David's Lucky VI
Per 4h1'58" 33d3'48"
David's Point of Light
Per 2h48'1" 54d12'8"
David's Reflection
Sgr 18h52'27" -20d15'34"
David's Star
Sco 16h19'57" -19d9'59"
David's Star
Uma 11h27'7" 60d9'49"
David's Star
Uma 14h13'4" 54d58'31"
David's Star
Uma 13h51'43" 58d19'19"
David's Star
Aqr 22h52'31" -22d50'45"
David's Star
Cyg 20h25'24" 49d1'23"
David's Star
Per 3h0'4" 46d11'23"
David's Star
Lyn 9h7'2" 33d16'58"
David's Star
Uma 11h8'9" 33d17'42"
David's Star
Her 17h51'11" 29d46'13"
David's Star
Cnc 8h56'39" 6d39'57"
David's Victory!
Tau 3h42'32" 15d43'19"
davids2424
Sco 17h38'15" -33d30'46"
David-S4
Uma 11h40'9" 52d43'37"
DavidSasha
Ari 2h44'39" 27d19'5"
DavidSiri
Uma 12h8'57" 45d25'32"
Davidson Andrew James Ungurait
Vir 12h53'12" -10d53'57"
DavidTracey
Umi 14h4'35" 73d53'53"
Davidushka
Cep 23h1'42" 80d31'51"
davidwardadamslight
Lib 15h49'3" -18d8'44"
Davie
Umi 14h12'36" 77d0'39"
Davie Campbell
Umi 12h28'11" 85d58'27"
Davie Joy Oriolo
Cap 20h12'35" -22d18'22"
Davie My Love
Lib 15h5'55" -5d26'50"
Davies Adastral Musicallis
Sgr 18h7'22" -27d54'27"
Davila
Ori 5h12'4" 6d11'42"
Davin Ted Purtle
Ori 6h22'25" 14d28'34"
Davin Thierry
Lyr 19h27'22" 40d31'18"
Davina
Ari 1h55'32" 19d40'44"
Davina Bedsole
Vir 13h40'54" -4d34'22"
Davina & Ian McClung
And 0h26'34" 22d12'14"
Davina Quan-yin Tong
Psc 0h35'29" 14d7'18"
Davine & Sanford Levy
Cyg 19h35'52" 34d24'10"
davinia
Cap 20h39'34" -15d7'52"
Davinora
Cyg 20h19'12" 51d12'53"
Davin's Heart
Aqr 23h7'11" -15d40'26"
Davion Manuel Benson
Uma 12h47'21" 52d48'18"
Davis
Cra 18h33'32" -42d45'38"
Davis 60th Anniversery Star 1945-05
Uma 10h30'50" 64d4'13"
Davis A. Graff (Beautiful Boy)
Her 17h16'20" 26d41'50"
Davis Alexander Fryhoff
Ori 4h53'51" -0d25'51"
Davis Alexander Saputa "9:55a.m."
Psc 1h29'50" 6d51'21"
Davis Brian Rose
Dra 16h31'17" 57d33'50"
Davis Camalier
Umi 16h11'34" 75d18'42"
Davis Christopher Cox
Psc 1h21'55" 24d48'54"
Davis Crosby
Her 16h48'14" 34d11'48"
Davis Gould
Uma 9h42'33" 54d49'1"
Davis Hardin Royall
Ari 3h9'1" 21d35'43"
Davis Lee Troutman
Mon 7h45'42" -0d20'45"
Davis Murray
Psc 1h6'7" 19d44'56"
Davis My Love
Tau 5h44'3" 26d57'34"
Davis (the Pooh) Sanders
Cma 6h38'40" -14d35'59"
Davis Wayne Ferguson
Ari 2h36'53" 24d52'6"
DavisClan Deuce
Dra 18h33'55" 55d38'6"
Davis-Edmonds
Aqr 23h52'42" -24d20'54"
Davis-O'Dell
Cet 2h10'28" -13d18'38"
Davita Bryant
Uma 13h42'41" 59d24'27"
Davon Alexander
Cap 20h20'44" -14d3'1"
Davone E. Madison
Uma 11h38'9" 63d38'59"
DavonMarcel Horne
Uma 11h20'5" 46d26'16"
Davonna Helen Carson
And 1h10'2" 46d38'45"
Davor Doma
Cnv 12h41'57" 42d48'7"
Davphencer
Dra 18h51'43" 63d41'58"
DAW
Uma 12h32'50" 57d19'12"
daw&clc...'Till The Stars Burn Out
Ori 6h10'27" 12d3'34"
Dawg
Cnc 9h4'46" 32d10'24"
DawKeeKei
Lyn 8h8'13" 54d12'48"
Dawn
Dra 18h31'39" 58d41'56"
Dawn
Cas 23h12'45" 59d17'28"
Dawn
Cas 2h21'23" 68d53'19"
Dawn
Sco 16h14'41" -13d34'6"
Dawn
Sco 16h7'13" -12d38'45"
Dawn
Lib 15h51'42" -9d56'46"
Dawn
Cap 20h50'33" -17d5'42"
Dawn
Sgr 18h52'23" -19d26'32"
Dawn
Sco 16h13'50" -16d15'0"
Dawn
Sco 16h47'47" -32d47'52"
Dawn
Sco 17h37'31" -35d2'6"
Dawn
Sgr 18h0'13" -29d39'12"
Dawn
Gem 6h41'13" 34d45'39"
Dawn
And 1h12'39" 36d26'4"
Dawn
Crb 16h5'44" 33d18'7"
Dawn
Lyn 8h7'53" 39d28'39"
Dawn
And 2h25'32" 42d52'0"
Dawn
Peg 22h24'0" 35d11'26"
Dawn
And 0h38'29" 39d14'29"
Dawn
And 23h9'32" 49d46'37"
Dawn
Tau 5h35'20" 26d20'19"
Dawn
Tau 5h56'0" 23d6'11"
Dawn
Ari 3h17'0" 27d17'18"
Dawn
Leo 10h19'40" 24d30'1"
Dawn - 3204
Crb 16h2'4" 28d3'24"
Dawn 65
Cnc 8h52'10" 22d49'45"
Dawn A. McKenna
Gem 6h2'4" 26d6'39"
Dawn Adair and Tom Wride
Cyg 21h32'40" 31d52'23"
Dawn Alene Walsh
Sco 16h6'12" -12d3'15"
Dawn Alexander
Lyr 18h51'28" 35d18'15"
Dawn Alicia Ohanian 2/20/63-1/10/06
Psc 1h18'37" 17d58'33"
Dawn Alo
Psc 0h55'38" 9d36'44"
Dawn Amber Schuler
Uma 10h47'5" 52d57'35"
Dawn Amber Trash Powell
Psc 1h17'26" 10d7'26"
Dawn and Aaron
Uma 11h26'14" 33d9'23"
Dawn and Steve
Lmi 10h34'44" 29d43'29"
Dawn Anderson - Mummy's Star
Cas 23h56'46" 54d9'24"
Dawn Angel
Cas 1h57'18" 64d45'17"
Dawn Angelique
Crb 15h38'29" 38d16'24"
Dawn Ann
Sgr 19h4'25" -25d58'54"
Dawn Arlene
Lyr 18h38'23" 38d16'55"
Dawn Ausborn
Cyg 20h0'13" 30d45'16"
Dawn Badger
Uma 8h38'51" 54d42'51"
Dawn Bass
Lyn 9h14'49" 40d0'5"
Dawn Bennett
Gem 6h16'41" 22d16'43"
Dawn Breden
Vir 12h7'16" 5d53'36"
Dawn Camara Andersen
Uma 11h25'57" 39d56'16"
Dawn Campbell
And 22h57'41" 50d49'43"
Dawn Camponelli
Uma 13h23'45" 54d38'44"
Dawn Capolongo
Cep 22h24'17" 67d31'46"
Dawn Carella
Cas 0h39'37" 64d37'13"
Dawn Carey The Light Of My Life
Cas 0h57'34" 63d37'49"
Dawn Casale
Uma 9h52'42" 57d40'12"
Dawn Casale
Lyn 7h53'30" 38d1'58"
Dawn Celeste Young-Pavasco
Tau 4h50'16" 17d42'33"
Dawn Cernak
Mon 6h26'34" 9d10'41"
Dawn Chere Gipson
Lyn 7h53'27" 34d39'58"
Dawn Choiniere
Gem 7h33'6" 34d3'50"
Dawn Christa
Tau 3h39'59" 4d20'10"
Dawn Christian
And 0h47'11" 30d45'19"
Dawn Christine Nichols
Aqr 22h13'37" -13d35'47"
Dawn Colleen Bonagofsky
Uma 11h24'10" 60d26'44"
Dawn D
Cnc 9h9'59" 23d11'16"
Dawn & Daniel Cox TogetherForever
Cap 20h45'23" -20d59'45"
Dawn DaSilva
Leo 11h22'59" -5d28'9"
Dawn Daylene
Com 13h6'48" 18d34'45"
Dawn DeEtte Costa
Lyn 9h4'0" 44d30'22"
Dawn Denee
Sgr 18h6'36" -27d0'43"
Dawn Devoe
Cas 23h22'46" 57d58'18"
Dawn Donner
Leo 10h41'1" 15d26'43"
Dawn E. Garofalo
Cmi 7h24'37" 5d28'21"
Dawn. E. Guthrie
And 2h33'48" 40d29'5"
Dawn Elaine Zuhlke
Vir 13h49'25" -7d17'22"
Dawn Elizabeth
And 23h9'59" 41d26'4"
Dawn Elizabeth Minardi
Ori 5h12'19" 2d33'24"
Dawn Elizabeth Parlapiano
And 0h50'10" 40d27'39"
Dawn Engelbrecht
Umi 15h37'43" 78d36'5"
Dawn Eternal
Uma 13h37'27" 56d41'51"
Dawn Evonne - In Loving Memory
Cru 12h44'41" -58d0'19"
Dawn Ford
Lyn 6h24'15" 61d3'11"
Dawn Foster
Mon 6h37'44" 6d54'2"
Dawn Fuoco
Vir 13h23'17" -12d50'10"
Dawn G Darling
Aqr 23h9'15" -10d22'6"
Dawn Gallegos
Crb 15h35'40" 31d35'10"
Dawn Gentile
And 23h25'37" 49d34'28"
Dawn Grulick
Lib 15h20'29" -4d4'56"
Dawn Hasty
Uma 10h38'9" 42d58'16"
Dawn Heffernan
Ari 3h12'53" 29d11'16"
Dawn Hudson
And 0h32'23" 32d11'8"
Dawn Ionascu
Cas 23h45'49" 58d12'9"
Dawn J. Westland
Cas 0h49'34" 50d36'30"
Dawn Jacques
Cas 23h38'4" 53d42'13"
Dawn Jillson
Uma 14h17'53" 57d23'19"
Dawn John Star
Cas 23h1'45" 56d36'8"
Dawn Jordan
Ari 2h25'5" 19d51'43"
Dawn & Jude's Star
Cyg 20h31'37" 47d48'49"
Dawn Kanselaar-The Love of my Life
Cap 21h21'10" -23d4'25"
Dawn Kathlyn Radford
Aqr 21h49'53" -7d32'33"
Dawn Kennett
Psc 0h51'4" 16d48'26"
Dawn Kenniston
Lyr 19h14'59" 34d24'38"
DAWN KUBIK
Lyn 6h55'51" 52d41'22"
Dawn L. Bibby's 30th Birthday Star
And 1h45'16" 42d41'45"
Dawn L. DeWerdt
Cas 0h45'41" 60d9'53"
Dawn L. Johnson
Gem 7h42'18" 32d38'46"
Dawn L. Watkins
And 0h53'0" 38d32'16"
Dawn L Witmer
Uma 8h45'34" 60d32'39"
Dawn "La Bella Donna"
Cap 20h10'24" -15d25'52"
Dawn Lapore
And 1h53'14" 44d40'6"
Dawn Lee Bales
Cam 3h30'3" 55d59'0"
Dawn Lehner
Ori 5h41'24" 7d53'36"
Dawn & Liam
Cnc 9h6'41" 14d50'9"
Dawn Louise
Pho 0h43'20" -41d42'24"
Dawn Louise O'Brien, My Star x x
Cas 0h58'19" 65d37'19"
Dawn Loves Megan
Sge 19h50'17" 17d32'5"
Dawn M. Bisbee
And 0h50'42" 42d39'36"
Dawn M Davis
Cyg 19h57'20" 47d7'51"
Dawn M. Halferty
Cap 20h34'38" -14d56'15"
Dawn M. Kirgan
Uma 9h36'14" 52d53'16"
Dawn M. Lo Presti
Gem 6h46'17" 27d36'15"
Dawn M. March 22, 1961
Cyg 19h59'43" 40d46'1"
Dawn M. Rodriguez
And 1h10'37" 42d54'58"
Dawn Macdonald
Cas 1h25'55" 52d5'41"
Dawn Marie
Cam 3h21'19" 57d33'23"
Dawn Marie
Cas 0h18'14" 54d11'37"
Dawn Marie
Gem 7h45'17" 21d14'25"
Dawn Marie
Vir 12h55'41" 5d5'9"
Dawn Marie
Cas 3h15'6" 67d31'58"
Dawn Marie
Cap 20h23'20" -13d41'36"
Dawn Marie
Lib 15h13'59" -14d4'52"
Dawn Marie
Lib 15h6'18" -2d18'56"
Dawn Marie
Psc 1h6'52" 3d34'41"
Dawn Marie
Sco 16h48'9" -36d12'12"
Dawn Marie
Lib 15h12'9" -24d35'41"
Dawn Marie Baxter
Cnc 9h10'28" 32d4'4"
Dawn Marie Bonakowski
Umi 14h21'5" 84d4'20"
Dawn Marie Brannon
Tau 3h50'20" 7d38'10"
Dawn Marie Cahill
Cnc 9h10'7" 15d6'42"
Dawn Marie Chamberlin
Uma 10h8'49" 68d38'57"
Dawn Marie Connell
Sco 16h19'55" -9d59'14"
Dawn Marie D'Arcy
Sgr 19h27'43" -28d45'2"
Dawn Marie Dean
Aqr 23h16'21" -8d59'26"
Dawn Marie J. Rivera
Psc 0h48'55" 8d33'46"
Dawn Marie Kriedeman
Tau 4h37'33" 24d36'33"
Dawn Marie MacCollum
Tau 4h31'34" 21d26'9"
Dawn Marie Marcusse
Lib 15h27'58" -26d17'11"
Dawn Marie Martin
Lmi 10h26'23" 28d52'23"
Dawn Marie McCune
Sco 17h56'33" -40d40'31"
Dawn Marie Morales
Uma 11h1'4" 39d42'40"
Dawn Marie Newton
Aqr 23h56'4" -10d48'46"
Dawn Marie Ozment
Cap 20h38'7" -18d58'10"
Dawn Marie Paulick
And 23h25'29" 52d18'52"
Dawn Marie Pietroforte
Aql 19h15'11" 3d8'32"
Dawn Marie Simone
Aqr 21h59'43" 1d4'51"
Dawn Marie Tilton
And 2h32'40" 49d50'15"
Dawn Marie Tuhowski
Cas 23h10'9" 54d26'12"
Dawn Marie Vandover
Tau 3h30'29" 26d37'3"
Dawn Marie Virgilio
Aql 18h47'55" 7d43'10"
Dawn Marie (Wright) Hoobler
Uma 11h1'3" 37d28'51"
Dawn Marousek
Uma 13h47'46" 57d37'13"
Dawn Mason
Peg 21h59'9" 10d36'39"
Dawn McGowan
Aql 19h47'29" -0d16'54"
Dawn McGrew - Love of my life
Aql 19h12'49" 14d57'21"
Dawn McVey
Leo 10h42'9" 17d50'41"
Dawn Michelle Powers
And 23h36'14" 47d19'23"
Dawn Millicent
Cyg 19h59'33" 34d1'16"
Dawn Mitchell
And 0h17'1" 27d59'34"
Dawn Monique
Cap 21h37'48" -8d48'19"
Dawn My Shining Star
Cas 1h27'57" 63d54'27"
Dawn Myers
Aql 19h17'56" 12d59'34"
Dawn Nethercott
Vir 14h13'33" 0d21'5"
Dawn Nicole
And 23h15'50" 39d39'26"
Dawn Nixon
Cas 23h34'44" 55d17'57"
Dawn of a New Day
And 0h55'23" 42d42'51"
Dawn of Amber
And 22h58'56" 52d14'31"
Dawn Of Kelly
And 2h23'59" 45d16'39"
Dawn of the Universe
And 2h38'20" 44d53'0"
Dawn O'Neil
Peg 21h40'50" 15d19'30"
Dawn Orion Rowland
Ori 5h37'43" 3d14'34"
Dawn Ortega
Psc 0h59'26" 18d26'4"
Dawn Ozanne
And 1h37'1" 42d35'32"
Dawn P. Constantino
Gem 7h14'7" 14d34'35"
Dawn P. Malone
Gem 7h26'6" 17d38'6"
Dawn Pehrson
Aqr 20h40'40" -2d56'44"
Dawn Phillips
Gem 6h56'1" 19d22'12"
Dawn Pisturino
Gem 6h11'54" 27d41'12"
Dawn R Fojtik
Tau 3h52'39" 28d27'24"
Dawn R. Radford
Cnc 8h12'25" 32d12'37"
Dawn Ragucci
Aqr 20h54'32" 0d2'31"
Dawn Ramnath
Aql 18h59'42" -7d32'14"
Dawn Renae Harvey
Cnc 8h56'6" 6d49'18"
Dawn Renate
Crb 15h52'16" 38d12'58"
Dawn Rene Jackson
And 0h53'17" 42d27'50"
Dawn Rene Stewart
Cyg 21h37'49" 46d48'20"
Dawn Renee Hedrick
Vir 12h6'58" -7d46'59"
Dawn Renee Lowery
And 2h16'19" 38d38'54"
Dawn Renee Wiederstein
Vir 13h38'51" 3d9'7"
Dawn Rider
Tau 4h25'22" 28d21'17"
DAWN & ROLAND CORKERN
Cyg 21h55'22" 51d37'41"
Dawn S.
Sco 16h27'57" -26d18'16"
Dawn S. Moody
Cas 2h39'41" 66d44'4"
Dawn Sandal Windsor
Gem 7h26'19" 22d28'38"
Dawn Sapp
Vir 13h54'31" -0d28'41"
Dawn Shahan
Cas 0h49'8" 61d18'22"
Dawn Sharon
Sco 16h12'23" -15d17'44"
Dawn Spinner
Mon 7h52'54" -2d39'42"
Dawn Star
Lyn 6h53'11" 52d13'20"
Dawn Star Patton-Valentine
Lib 15h8'19" -6d39'15"
Dawn Starr Wyman
Leo 9h31'17" 10d58'29"
Dawn Steel
Umi 14h48'18" 68d38'41"
Dawn Steelman
Uma 9h30'32" 48d57'43"
Dawn Susan
Psc 1h42'38" 18d37'46"
Dawn Taylor
Cyg 19h56'30" 41d17'6"
Dawn (the Baptist) Quenzer
Uma 13h48'12" 56d18'52"
Dawn Thorine Bladet
Aqr 21h17'56" -11d59'58"
Dawn Training Centre
Uma 9h25'31" 63d40'46"
Dawn Treader
Ori 5h2'5" 1d55'37"
Dawn Vollmer
Lyn 7h18'56" 45d28'6"
Dawn Webb
Cap 21h8'54" -24d37'50"
Dawn Wendy Godwin
Sco 17h56'3" -42d47'49"
Dawn Willams My Princess
And 23h24'27" 42d10'31"
Dawn Youdan STAR Learner Angel
Sgr 17h45'18" -29d28'38"
Dawn, Mark & Mark Main
Umi 15h43'39" 82d30'24"
Dawn, Shigeo and Maddie
Sco 16h3'47" -12d25'41"
Dawn, The Star of My Heart
Sco 17h9'42" -40d44'4"
Dawna
And 2h2'51" 42d17'47"
Dawna and Jacob Bridges Forever
Cyg 19h49'32" 43d36'0"
Dawna Comeaux
Cyg 20h21'29" 39d0'46"
Dawna Jean
Cnc 8h22'27" 24d49'5"
Dawna Ree
Leo 11h25'27" 5d29'45"
Dawna Vir
Vir 13h30'35" -2d8'1"
Dawnabelle
Gem 7h25'40" 18d37'21"
Dawnar
Peg 23h21'13" 17d44'56"
Dawn-Dawn
And 1h44'0" 43d45'28"
Dawne and Craig Barnes
Cyg 21h14'0" 52d53'26"
Dawnelle J. Henretty
Aqr 22h40'43" 1d3'11"
Dawnelle Thome's Heavenly Light
Psc 1h57'8" 8d15'36"
Dawnielle
Ori 5h59'10" 21d33'48"
Dawnielle
Mon 6h48'6" -0d6'35"
Dawnielle Kristen Conger
Cap 20h32'46" -20d32'27"
Dawnielle Nichole Cox
Crb 15h47'48" 37d9'32"
Dawning Star
Gem 7h8'6" 22d43'32"
dawn-mommy's star
Psc 1h19'18" 28d3'38"

Dawn's Angels: Frisky, CJ & Charlie
Cyg 21h46'51" 47d54'30"
Dawn's Chance
Cnc 8h35'15" 24d34'39"
Dawn's Diamond
Cnc 9h11'52" 26d22'5"
Dawns Heavenly Light
And 23h44'38" 47d46'15"
Dawn's Light
Lyn 8h16'51" 40d33'51"
Dawn's Light
And 0h40'18" 24d43'57"
Dawn's Light
Cap 20h41'50" -21d23'9"
Dawn's Light
Aqr 20h41'17" -7d13'47"
Dawn's MorningStar
Tau 4h19'15" 7d47'44"
Dawn's Sapphire
Cas 1h4'44" 63d43'27"
DAWNS STAR
Ori 5h51'59" 5d42'1"
Dawny Prawny
Aur 7h28'16" 39d46'18"
DawnyBB
Cap 20h21'14" -9d31'33"
Dawson
Psc 1h23'59" 32d8'38"
Dawson Harold Swindell
Aqr 20h56'0" -2d38'0"
Dawson Michael Quick
Leo 9h30'0" 31d8'50"
Dawson Mitchell Talley
Uma 11h1'17" 59d31'32"
Dawson "Saucy" Vaughn
Uma 9h59'37" 49d23'33"
Dawson Steckel
Uma 10h24'32" 45d35'42"
Dax Gregory Nguyen
Aqr 22h11'34" 1d3'44"
Dax Michael Dundon
Lib 15h36'7" -14d41'57"
Dax Mitchell Corey
Leo 11h28'2" 6d24'43"
Daxter
Mon 7h3'4" -1d7'47"
Daxton Casey DeRyan Shipley
Sco 16h32'26" -31d48'34"
Day
Uma 9h46'7" 52d51'34"
D.A.Y. '40' (06.08.1967)
Lmi 10h12'48" 29d11'58"
Day and Jay
Cnc 8h26'24" 17d32'57"
Day Dream
Dra 18h50'37" 52d29'32"
Dayalan Prakash Vignarajah 12.11.80
Ori 5h59'33" -1d48'28"
Dayana
Leo 11h19'47" -1d45'13"
Dayana Fracarro
Tau 4h24'58" 22d28'1"
Daycha
Gem 6h45'26" 17d17'16"
Daye Leigh Long
Sgr 18h5'34" -27d16'17"
Dayla Sabate
Lyn 7h47'36" 37d17'19"
Daylan Alejandro Medrano
Her 17h49'17" 28d24'26"
Dayle Childers
Psc 0h22'23" 7d59'18"
Dayle Lee Baker
Uma 8h23'13" 64d34'19"
DayleMR
Aqr 22h13'28" 0d43'23"
Daylen Makenzie Wilks
Cap 21h18'42" -25d23'12"
Daylight
Leo 9h46'25" 27d16'22"
Daymein Kekoa
Cap 20h53'36" -24d26'45"
Daymian
Lib 15h20'20" -26d6'51"
Dayn Conrad
Ari 3h17'42" 28d32'34"
DAYNA
Tau 5h57'18" 25d1'29"
Dayna
And 1h9'12" 35d26'55"
Dayna
And 0h50'15" 43d1'11"
Dayna
And 0h48'21" 40d13'37"
Dayna
Lib 15h13'57" -24d7'29"
Dayna Ann Verhey
And 1h9'18" 37d52'37"
Dayna & Armando's Star
Her 16h54'11" 17d3'57"
Dayna "Batbabe" Christie
Uma 14h27'4" 60d31'0"
Dayna Beth Boyles
And 23h13'30" 47d2'40"
Dayna Blattman
Oph 16h43'47" -0d31'3"
Dayna Boswell
Crb 16h3'1" 29d31'11"
Dayna Deneshie
Aqr 22h40'25" 0d58'56"
Dayna Dixon
And 1h37'55" 36d30'16"
Dayna Dos
Cap 21h46'55" -14d42'48"
Dayna L. Emerson
Vir 14h24'44" 1d34'59"
Dayna L. Giordano
Lib 15h10'54" -1d49'42"
Dayna Lynn
Cyg 19h31'41" 30d8'22"
Dayna Lynn Ahern
Vir 13h24'7" -21d19'38"
Dayna Lynn Dubin
Uma 11h59'45" 64d56'4"
Dayna Mae Patrick
Uma 13h6'21" 59d59'30"
Dayna Michelle Kenyon
And 2h27'53" 41d12'22"
Dayna Monique
Sgr 19h46'31" -39d1'15"
Dayna Roxanne Mitchell
And 2h3'8" 45d28'47"
Dayna Star
And 0h23'27" 43d2'51"
DaynaMarie79
And 2h36'15" 40d42'51"
Dayne Boyk
Leo 10h11'41" 24d20'9"
Dayne Kamau Preston 12:03 P.M.
Aqr 21h40'37" 1d15'56"
Dayne Smoter
Her 18h1'52" 24d45'10"
Daysha
Pho 0h45'16" -41d43'8"
Daysha Jahns
Sgr 18h15'48" -32d31'15"
Daysi
Tau 4h37'14" 11d22'54"
Daysi D. Rivera
Pho 0h28'29" -43d24'54"
Daysi Rivera
Cap 21h56'8" -8d55'18"
Daysia Monae Walker
Gem 7h6'49" 27d54'8"
Daysia-Dana-David
Uma 12h43'57" 61d39'33"
Dayson Trent Hawkins
Uma 11h59'41" 47d1'31"
Dayton Amey
Uma 13h25'1" 59d28'46"
Dayton Bradley Eller-Olsen
Lyn 6h42'26" 54d10'39"
Dayton C. McLaughlin
Umi 14h47'9" 79d7'9"
Dayton Dave Sloas
Peg 21h14'40" 18d50'59"
Dayton Hobson
Aql 20h21'45" 8d1'40"
Daytrana
Umi 18h57'18" 88d40'40"
daywalkerdaz 050166
Cap 21h43'10" -19d34'30"
Dayzi El-Ghafari
Cnc 8h39'50" 24d20'45"
Daza 187
Umi 9h22'7" 87d55'19"
Dazaviri
Sgr 17h52'55" -29d40'5"
Dazen012-Cheddah25
Cyg 21h27'59" 45d46'12"
DAZ—SJZ(1)
Uma 9h9'22" 57d14'40"
Dazz
Cru 12h15'32" -60d54'16"
Dazz A Roni 11/17/99
Uma 9h14'18" 55d40'9"
Dazzeling Dr. D. 2005
Aqr 22h49'4" -16d55'26"
Dazzle of Destiny
Psc 0h41'59" 12d37'34"
Dazzler
Lyn 7h3'6" 53d22'18"
Dazzling and Lovely Cynthia Lea
Uma 10h2'59" 44d3'30"
Dazzling Dali
Cnc 8h55'44" 19d40'18"
Dazzling Dawn
Cas 0h19'31" 57d5'47"
Dazzling D.C. III
Uma 11h10'43" 46d51'47"
Dazzling Decade
Per 2h48'14" 42d53'37"
Dazzling ~En~
Ori 6h3'42" 16d18'20"
Dazzling Jas
Psc 0h41'33" 21d21'33"
Dazzling Skyler
Cnc 9h13'31" 13d3'0"
Dazzling Stargazer Lily
Cas 0h19'17" 53d23'8"
Dazzling Tana
Leo 10h41'7" 22d31'24"
DB Eternity
Crb 16h20'35" 32d53'15"
DB Goggans
Ori 6h19'34" 9d37'40"
D.B. Naik
Crb 15h43'26" 29d23'30"
DB512ST
Uma 12h34'29" 54d51'33"
D.B.A
Lib 15h41'48" -28d36'28"
Dbear
Uma 10h31'55" 64d9'59"
D-Bebe
Aqr 23h51'51" -4d56'16"
Dbl N & Girly Girl of Decatur Blue
Cyg 21h13'13" 44d26'9"
D.B.N.J.
Leo 10h53'13" 16d11'41"
DBoswellTHawkins
Cet 2h44'35" -0d25'48"
DBoyz19
Her 17h49'35" 43d18'17"
D.B's Wish
Uma 11h31'3" 60d50'56"
DBYS Cowboy
Aqr 21h41'21" 2d18'9"
DC 4 Eternity
Her 18h33'46" 19d58'30"
DC 974
Uma 11h43'11" 38d23'13"
DC Bacchus
Gem 7h36'3" 26d44'4"
D.C. Kelley
Vir 13h17'48" 5d57'32"
DC Winter Guard
Pic 5h51'28" -54d54'25"
DC Young
Cap 20h11'42" -10d58'58"
DCGS Financial Team
Her 16h39'57" 9d22'6"
DCJ-2004-12-12
Cyg 21h34'56" 33d33'43"
D.Class + S.Dogg
Ari 2h41'32" 27d53'8"
DCOwen80
And 1h34'28" 42d17'30"
DCRS
Cyg 21h30'3" 34d36'44"
DCS
Ari 2h20'27" 27d21'44"
dcwmonkey
Umi 15h33'37" 70d32'32"
DD Hornby 25
Uma 9h28'47" 55d46'26"
DD Lerner
Cap 21h46'52" -14d2'31"
D.D. Maloney
Leo 10h13'35" 14d16'7"
DD Star
Psc 1h13'9" 15d57'30"
DD555
Eri 4h13'40" -0d11'2"
DDA 50
Cap 20h15'46" -14d55'48"
DDC
And 22h57'58" 46d18'23"
DDPH
Lyn 7h32'10" 41d15'27"
DD's Star
Cyg 20h59'47" 31d29'57"
DD's Star
Aqr 20h49'30" -8d46'41"
De Allen
Per 4h48'19" 50d27'55"
DE BEAR
Uma 9h18'43" 55d3'32"
De Ciechi Cesare
Cam 4h8'10" 56d55'22"
De corde totaliter Et ex mente tota
Dra 9h25'28" 76d29'49"
De Grandis Antonio
Uma 10h33'40" 70d46'23"
De Joncker Monique
Tau 3h50'19" 8d24'41"
De Laun and Elizabeth Martin
Cyg 20h9'31" 47d10'50"
de Luca, Mario
Uma 9h27'24" 47d32'1"
De Moura Aurélie
Uma 8h22'8" 71d57'43"
De Novo
Sge 19h51'7" 17d20'32"
De Petris Federica
And 23h36'44" 33d8'4"
De Renee
Pho 0h40'40" -40d46'24"
De Santis Melania
Cyg 21h21'33" 38d40'41"
de Vivies
Uma 11h23'40" 29d51'33"
De Yarelle Wolf
Ari 2h3'41" 20d2'31"
Dea Lynn Petkus
And 1h4'59" 46d28'36"
Dea Roma
Aqr 21h26'34" 2d29'20"
DEA TSUTSKIRIDZE
Cyg 20h53'29" 37d12'37"
Deacon
Per 3h50'42" 37d26'21"
Deacon Byron Champagne
Cyg 19h57'46" 37d32'33"
Deacon Joseph Pollock
Cyg 19h58'33" 33d45'45"
Deacon Kai George Bickley
Ori 4h45'15" 4d8'41"
Deacon Minus Hall Sublett, Sr.
Uma 9h41'28" 61d56'15"
Deacon Peter James DeLuca
Cyg 21h10'51" 31d47'1"
Deacon Richard Bilella
Psc 0h10'22" 9d6'25"
Deacon Wells
Sgr 18h0'16" -29d55'9"
Deadis
Lib 15h41'51" -26d7'9"
Dean
Cnc 9h3'26" 28d10'33"
Dean
Her 17h43'15" 15d59'1"
Dean
Lyn 7h10'18" 49d29'29"
Dean A. Bagley
Ari 3h2'54" 10d41'14"
Dean A. Frescura
Ori 5h51'18" 6d40'21"
Dean Allen DeBoer
Vir 13h45'9" -6d18'24"
Dean and Gwen Edwards
Crb 16h21'11" 34d46'28"
Dean and Karen Roussel
Uma 9h38'7" 54d52'44"
Dean and Kyra
Psc 23h51'39" 7d38'17"
Dean And LeAnn Forever In The Stars
Uma 9h51'55" 47d37'35"
Dean and Lucy Sargeant
Cyg 19h41'27" 34d44'57"
Dean and Paige
Cap 20h22'15" -11d38'33"
Dean Anthony Bard
Uma 11h44'30" 40d17'30"
Dean Anthony Peters
Cap 20h31'12" -19d55'50"
Dean Arrington
Gem 7h19'54" 13d53'54"
Dean Borgmeyer
Her 17h50'10" 24d13'23"
Dean & Brenda
Cyg 20h4'28" 42d50'49"
Dean Brett Sitler
Cnc 8h27'51" 30d38'34"
Dean Brian Lomas
Uma 11h44'9" 43d27'29"
Dean & Chantelle - April 7, 2006
Cru 12h28'12" -56d11'14"
Dean Cornell
Peg 21h45'3" 22d53'0"
Dean (DAD) Durfey
Vir 13h18'27" 5d6'53"
Dean Davidson
Lyr 19h14'6" 30d1'24"
Dean Dodds Ramsey
Cnc 8h56'30" 18d56'27"
Dean Edward Dagermangy
Eri 3h28'43" -9d19'52"
Dean Elliott Mills
Dra 17h6'41" 57d38'51"
Dean Emmanuel Kalani Lake Jr.
Ari 3h21'45" 27d22'23"
Dean Eyler- Gray Plant Mooty Star
Umi 17h27'18" 83d26'38"
Dean - Forever Ours, Forever Us - Lucy
Ori 5h20'36" -2d38'22"
Dean Gaetano Zazzara
Aqr 21h31'53" 1d10'13"
Dean (G.G.) Clark
Lyn 6h58'7" 48d31'35"
Dean Griffin
Aur 5h17'23" 41d57'50"
Dean Hitsos
Ori 5h50'49" -0d33'35"
Dean Holschbach
Per 3h16'49" 55d29'57"
Dean Izzo
Boo 15h37'20" 49d19'29"
Dean J. Dahlem
Gem 7h39'18" 32d15'55"
Dean J. Stoker
Boo 14h45'59" 22d54'44"
Dean J Wessel
Her 17h38'32" 16d30'7"
Dean James Poxon
Cru 12h51'26" -60d39'6"
Dean Jason Meintjes
Ari 2h21'46" 12d36'50"
Dean Johnson (Bear)
Her 18h51'57" 23d1'59"
Dean Joseph Andretta
Her 18h8'2" 28d2'40"
Dean & Kellie #5
Ori 4h55'43" -2d56'46"
Dean Kelly
Lyr 18h34'59" 36d4'14"
Dean Kittinger
Uma 13h48'49" 54d17'33"
Dean Kleist
Uma 9h29'39" 41d29'15"
Dean Lawrence Sweeney
Vir 14h4'6" -1d56'17"
Dean Lizzi
Uma 9h31'18" 63d32'55"
Dean Louis
Aqr 22h1'39" 1d4'29"
Dean Lowell Lauritzen
Gem 6h43'21" 26d17'33"
Dean Luca Roth
Cyg 20h14'34" 49d32'11"
Dean M. Binkley
Her 17h14'52" 49d0'38"
Dean Mark Lortie
Her 17h52'17" 16d53'49"
Dean Marquis
Boo 14h39'21" 51d17'2"
Dean Marshall Stewart
Per 3h16'35" 54d25'23"
Dean Matthew Hughes
Boo 14h26'42" 47d37'51"
Dean Michael Hasik
Cap 20h52'25" -19d10'41"
Dean Michael Scorpios Borealis
Cnc 7h56'9" 10d12'40"
Dean Michael Thomas
Uma 9h23'1" 60d49'12"
Dean Miller
Her 16h15'40" 45d29'17"
Dean Mitchell Blevins
Umi 20h1'50" 88d49'5"
Dean Munkeby
Lmi 9h41'40" 35d18'9"
Dean Murray
Aur 5h40'47" 33d0'29"
Dean & Natalie Gerow
Cyg 19h53'46" 30d30'24"
Dean P. Cernek
Tau 3h45'16" 19d45'58"
Dean Paterson
Cru 12h28'24" -61d1'12"
Dean Patrick Marano
Per 4h49'45" 41d53'39"
Dean Perry
Cep 0h32'15" 82d4'26"
Dean R Halton
Her 18h3'15" 36d57'17"
Dean R. Wagnon
Vir 14h35'11" 2d42'34"
Dean Ray Adams
Her 17h18'32" 25d32'11"
Dean Richard Redmon
Her 16h39'3" 37d22'10"
Dean Robert George Simrick
Cep 21h45'3" 56d35'46"
Dean Roy Kregger
Leo 10h14'19" 15d47'56"
Dean S.
Psc 22h51'29" -0d10'27"
Dean Scardino
Gem 7h43'19" 33d19'48"
Dean Scott Dansky
Her 18h7'8" 31d57'16"
Dean Scroggins
Vir 14h34'43" 3d2'59"
Dean & Sharon Peace Love & Harmony
Cyg 22h0'35" 45d13'50"
Dean Soots
Uma 12h18'45" 60d21'37"
Dean Stanley Farmer
Boo 14h55'6" 17d46'41"
Dean & Susie Gushikuma
Ori 6h20'54" 9d19'12"
Dean Sutton
Cyg 20h22'49" 47d46'33"
Dean Taylor
Uma 11h49'6" 60d21'17"
Dean Taylor
Cru 12h6'33" -63d25'56"
Dean & Teara
Cyg 19h25'53" 48d38'9"
Dean Thomas Hunter
Her 17h58'18" 23d34'39"
Dean Thomas McLean Cochran
Col 5h37'27" -35d32'19"
Dean Thomas Ralston
Aql 19h2'27" 14d58'1"
Dean Travis Kraft 6-15-1980
Ori 5h35'23" -1d11'25"
Dean Trevor Draper - 1968
Ori 5h10'29" 15d25'10"
Dean Winston Franke II
Leo 11h28'38" 26d41'19"
Deana
Lmi 10h43'37" 35d33'7"
Deana
Lyn 8h10'41" 41d26'8"
Deana
Aqr 22h12'28" -18d57'13"
Deana Allington
And 0h50'46" 36d4'0"
Deana Ayala
Cnc 8h10'52" 26d26'25"
Deana Ballerina
And 23h52'38" 47d10'36"
Deana Christine Chizinski
Cam 3h17'52" 59d50'46"
Deana Dale Bedor
Psc 0h54'47" 30d59'55"
Deana Hunter
Uma 11h3'57" 49d35'19"
Deana Jean
Vir 12h7'20" -4d19'22"
Deana Lynn Moss
Lyn 8h27'21" 53d11'50"
Deana M. Ligda
And 2h25'6" 49d54'22"
Deana Mae Heselton
Vir 13h53'51" 2d38'12"
Deana Maria Darnell
And 1h15'11" 41d18'58"
Deana Marie
Lyn 8h42'47" 41d50'1"
Deana Marie Nazzarese
Lib 15h35'50" -10d34'5"
Deana Michelle
Vir 12h8'48" 5d52'2"
Deana Michelle " DEE " Siroki
Sco 16h15'43" -11d50'50"
Deana My Love For Ever and Always
Dra 16h15'2" 52d10'41"
Deana Prock
Mon 7h48'36" -0d32'25"
Deana Tart & Tomasz Obara Forever
Leo 11h28'38" 13d31'36"
Deana Yi Kane
Per 3h34'2" 46d35'42"
Dean-Bean and Juna-Willa
Umi 17h37'3" 82d5'54"
Deandra
Cam 4h2'10" 56d22'41"
DeAndrea
Crb 16h8'25" 37d19'27"
DeAndrea T. Thompson
And 1h26'17" 47d30'58"
Deane Finnegan "Class of XXII"
And 23h33'4" 41d26'9"
Deane & Kate Forever
Aqr 22h11'32" -14d5'25"
Deangelo
Psc 0h35'25" 17d53'58"
Deanie Piner
Psc 0h48'20" 15d42'2"
Dean-L
Cen 13h29'53" -38d4'56"
DeAnn
Cas 23h39'3" 53d55'26"
Deann Guerra
Lib 15h46'58" -15d16'16"
DeAnn of the Faeries
Vir 12h12'6" 5d58'56"
Deanna
Leo 10h43'20" 8d0'56"
Deanna
Cnc 8h32'37" 28d12'55"
Deanna
Tau 3h45'20" 29d7'20"
Deanna
Leo 10h28'10" 23d41'2"
Deanna
Ari 2h38'36" 27d25'23"
Deanna
And 0h41'52" 26d10'44"
Deanna
And 0h43'12" 25d56'53"
Deanna
Cas 0h48'10" 48d49'35"
Deanna
Lyn 7h56'35" 49d20'55"
Deanna
And 23h29'49" 45d46'28"
Deanna
And 23h1'27" 47d46'32"
Deanna
Boo 14h36'26" 52d4'15"
Deanna
Lib 14h45'54" -9d37'20"
Deanna
Lib 15h4'8" -13d53'43"
Deanna
Lib 14h54'54" -1d15'27"
Deanna
Sgr 20h24'41" -42d51'28"
Deanna
Cap 21h6'40" -25d50'36"
Deanna 21st
Cru 12h48'41" -63d3'38"
Deanna Angarola
And 23h8'45" 43d36'47"
Deanna Arble
And 0h52'23" 40d26'15"
Deanna Ariel Nathan
Gem 6h44'23" 19d30'16"
Deanna Barrett
And 1h20'51" 46d25'10"
Deanna Casey
Psc 0h54'59" 28d59'27"
Dean-na Castaldo
Lyn 8h22'55" 53d36'28"
Deanna Celsi "Little Dee"
Umi 14h7'29" 72d11'17"
Deanna Cervantes
Crb 15h45'15" 28d27'47"
Deanna Cross
Hya 9h16'18" -0d28'37"
Deanna Dawn
Gem 6h30'30" 23d33'31"
Deanna "Dee" Nehmeh
Aur 5h50'36" 51d38'37"
DeAnna Douglass
Cyg 20h26'9" 30d38'39"
Deanna Edwards
Cyg 21h25'59" 34d9'31"
Deanna Faith Chesser
Tau 4h9'11" 26d51'32"
Deanna Favata
And 1h11'12" 34d26'44"
Deanna & Gerald Hagerman
Cyg 21h26'58" 32d46'0"
Deanna Hauser
And 0h45'25" 39d21'20"
Deanna Jean
Dra 18h53'28" 51d29'28"
Deanna Jo
Tau 4h58'50" 19d26'16"
Deanna K Newman
Lyn 6h58'20" 52d39'27"
Deanna Kate Pyper
Gem 6h56'4" 19d16'43"
Deanna Kay Malseed "Wave Dancer"
Cap 20h12'23" -15d54'41"
Deanna Kelley
Ori 5h19'54" -8d27'6"
Deanna Kochman
Cas 0h22'1" 56d35'44"
Deanna Lauren Manniello
Leo 9h40'28" 32d41'1"
Deanna Lee Kelner
And 1h30'0" 37d48'21"
Deanna Leigh Sagedy
Cas 1h33'16" 60d39'7"
Deanna Lieu
And 0h17'50" 33d12'20"
Deanna Lynn
Lyn 8h25'8" 39d55'46"
Deanna Lynn Spivey
And 23h9'54" 38d32'46"
Deanna Lynn Todd
Lyn 8h9'22" 46d49'39"
Deanna Lynne
Ori 6h6'7" 0d15'22"
Deanna Mae Bartron
Aqr 22h30'13" -3d54'12"
Deanna Mae Bartron
Sco 16h8'29" -10d24'35"
Deanna Maria Cravens
Crb 16h18'3" 32d54'18"
Deanna Marie
And 2h35'16" 45d46'12"
Deanna Marie
Lib 15h42'51" -14d58'54"
Deanna Marie Casey
Peg 22h34'36" 32d46'27"
Deanna Marie Gosser
Ari 3h16'22" 29d52'33"
Deanna Marie McLaughlin
Crb 15h42'3" 26d56'24"
Deanna May & Christopher John Grace
Cyg 21h17'29" 40d3'44"
Deanna Michelle Ray
And 0h42'19" 40d38'56"
Deanna Mohamed Orra
Lib 15h15'43" -4d5'8"
Deanna "Muse 6" Diaz
And 23h52'23" 42d13'36"
Deanna My Star
Lyn 7h22'14" 44d38'50"
Deanna Mychalishyn June 15, 1992
Gem 6h35'54" 19d40'56"
Deanna (NANA) Ratcliffe
Col 5h32'57" -33d9'47"
Deanna Nicole
Cas 0h40'35" 63d50'10"
Deanna Nicole Davis
Umi 15h20'27" 68d19'2"
Deanna Orfanelli
Sgr 18h12'49" -21d36'2"
Deanna Our Angel In The Sky
Her 17h4'10" 33d3'2"
Deanna Rae
And 1h27'41" 48d41'52"
Deanna Rae Strand
Uma 8h53'29" 53d56'41"
Deanna René Honea
Uma 10h21'33" 46d22'43"
Deanna Renee Jessika Brown/ Cortez
Cas 0h57'38" 59d32'23"
Deanna Rizzo
Gem 6h37'23" 21d15'38"
DeAnna Rose Sellards
And 23h7'54" 48d20'50"
Deanna Ruggiero
And 23h32'29" 42d41'33"
Deanna Ruth Malone
Leo 11h3'5" 0d19'37"
Deanna Senna
Aql 19h59'55" -7d34'43"
DeAnna & Stephanie-Fayleen's Angels
And 2h10'50" 43d26'35"
Deanna Sue Davis
Sco 16h9'34" -10d55'30"
Deanna The Spirit
Per 4h5'29" 45d14'4"
Deanna Therese Westmark
Aqr 22h41'30" -5d6'12"
Deanna Vallejo
And 23h39'37" 47d0'10"
Deanna Yard
And 2h3'29" 46d25'8"
Deanna, My Love
Sgr 18h54'1" -20d27'2"
Deanna, My Specious Love
And 2h1'1" 41d32'37"
DeannaMarie
Sgr 19h28'6" -35d58'54"
Deanna's Moon
Gem 6h43'21" 18d20'51"
Deanna's Radiance
Ari 3h2'39" 18d39'29"
Deanne
Gem 7h44'5" 21d56'56"
Deanne
Sgr 19h27'50" -16d46'11"

Deanne Alpha 1
Sgr 18h10'45" -19d35'58"
Deanne and David's Dream
Cru 12h41'19" -62d38'27"
Deanne and Pat
Cap 21h32'21" -20d58'7"
Deanne Christine Reeve
Leo 11h19'54" 1d22'46"
Deanne "Dee" Ray Swift
Vir 13h7'37" 7d12'1"
Deanne F. Peters
Cap 20h11'8" -9d2'33"
Deanne McFadzien
Cen 13h9'15" -61d19'26"
Deanne Owens DiRado
Uma 11h55'47" 57d0'34"
Deanne & Robert—Eternal Soul Mates
Cas 2h0'32" 70d28'38"
Deanne, Craig, Riley & MacKenzie Lum
Cru 12h38'15" -59d37'6"
Deano
Per 3h21'58" 32d20'35"
DeanoStar
Cnc 8h39'23" 8d36'38"
Dean's Boy
Lib 14h59'9" -13d59'8"
Dean's Star
Umi 16h24'27" 71d23'44"
DeAnthony Salazar
Lib 14h46'4" -9d25'11"
Deany Fondren
Her 17h28'1" 36d20'46"
Dear Jens- My one in a trillion
Uma 9h37'38" 67d25'42"
Dear Shunsuke
Vir 13h11'48" -8d15'41"
Dearest Carol
Cap 21h9'29" -22d3'59"
Dearest Chassity
Sgr 19h33'42" -16d28'56"
Dearest Elena
Gem 7h21'59" 33d29'43"
Dearest Ever
Cyg 21h40'31" 51d54'11"
Dearest Kathleen
Ori 5h55'55" 11d29'38"
Dearest Meredith
Lib 15h25'28" -8d23'23"
Dearest Minna
Vir 13h21'39" 5d18'38"
Dearest Mother Filomena
Cas 23h0'45" 53d53'25"
Dearest Mother Linda Van Raden
Lyn 8h10'34" 37d7'21"
Dearly Beloved Iris
Cru 12h48'3" -60d49'7"
Deary Weary
Uma 8h35'56" 47d12'21"
Deasia
And 1h14'10" 46d6'58"
Deathless Affinity Healing the Star
Mon 7h16'27" -0d51'58"
Deatra
Tau 3h30'20" 24d35'12"
DeAuna Noel Marks
And 0h42'17" 35d18'9"
Deb
And 0h14'17" 43d46'33"
Deb 2
Dra 18h26'9" 73d26'7"
Deb 2626
Cyg 20h31'50" 37d39'21"
Deb and Marilyn
Uma 10h8'38" 49d44'22"
Deb Brighton
Lyr 19h8'33" 26d31'13"
DEB CANTON
Cyg 21h19'51" 53d17'8"
Deb Collins
And 2h28'26" 47d53'0"
Deb Daehlin
Cas 23h30'12" 58d1'49"
Deb Delaney
Ari 2h41'58" 12d51'35"
Deb & Dwight Derry
Sco 16h8'37" -24d35'10"
Deb Edack
Cnc 8h55'39" 23d51'24"
Deb Gabor
Aqr 22h26'10" -4d7'46"
Deb Gallagher
Lyr 19h10'29" 26d27'7"
Deb Goldblatt (ox y)
Uma 9h24'55" 46d29'47"
Deb Niemeyer
Cam 3h59'11" 55d37'34"
Deb Odell
Lyn 9h4'32" 36d32'44"
Deb or Deborah
Sco 16h23'8" -27d50'41"
Deb Russell-2005 Technology Allstar
Crb 15h36'24" 27d47'9"
Deb Saari
Ori 5h41'10" 0d47'51"
Deb Saylor
Aqr 23h0'45" -8d53'6"
Deb Solomon and Linda Stovall 1991
Cyg 20h52'38" 31d28'29"
Deb Star
Crb 16h7'5" 26d26'40"
Deb U Lon
Cas 1h19'46" 57d39'29"
Deb11561
Ari 3h9'10" 18d7'30"
Debanjali
Cap 21h34'25" -9d53'11"
Debanoff
And 2h17'42" 47d17'26"
Debasish Ghosh
Ori 5h54'13" 7d6'25"
Debb. B
Aqr 22h15'58" -1d14'25"
Debbe Cole
Leo 11h31'14" 26d18'31"
Debbi
Tau 3h38'26" 26d6'48"
Debbi Donlin
And 0h50'15" 39d21'55"
Debbi J Moffett
Lib 15h37'59" -3d55'47"
Debbi Jo
Crb 15h41'32" 38d22'26"
Debbi Kethley
Leo 11h36'13" 26d0'49"
Debbi Lynn, Duchess of Oak Ridge
Sco 16h10'55" -16d51'53"
Debbi Strand
Cas 23h24'34" 55d35'48"
Debbi Strand
Cas 0h12'11" 55d25'4"
Debbi-do
Cnc 8h30'58" 15d28'29"
Debbie
Cnc 9h13'45" 16d15'29"
Debbie
Ari 2h28'20" 26d43'12"
Debbie
Leo 10h12'55" 26d23'40"
Debbie
Peg 22h48'8" 27d29'19"
Debbie
Vir 12h49'38" 5d52'6"
Debbie
Cnc 8h41'47" 7d27'13"
Debbie
Ari 3h26'47" 20d30'57"
Debbie
Mon 6h27'54" 8d19'50"
Debbie
Vir 13h11'46" 10d23'5"
Debbie
And 23h23'4" 42d33'32"
DEBBIE
Cyg 20h53'31" 35d47'18"
Debbie
Lyr 18h36'48" 35d54'14"
Debbie
Cas 0h23'16" 61d57'39"
Debbie
Lyn 6h25'41" 59d42'32"
Debbie
Lib 15h27'47" -22d17'2"
Debbie
Cap 21h3'11" -22d22'19"
Debbie
Lib 15h46'35" -12d45'46"
Debbie #1 Friend
Uma 10h57'1" 38d30'49"
Debbie A.
Uma 12h46'45" 61d36'0"
Debbie A. Burk
Uma 13h42'16" 60d46'36"
Debbie A Golanka 052859 DARKPH 04
Gem 7h17'3" 27d32'25"
Debbie A. Mandell
Cas 1h32'39" 64d45'57"
Debbie A. Tomb
Cma 7h3'5" -22d57'46"
Debbie Aaberg
Cnc 8h26'42" 25d52'13"
Debbie Akeroyd "My Shining Star"
Uma 8h9'55" 62d5'22"
Debbie Alexander & Harley
Uma 8h40'50" 48d31'2"
Debbie and Bob Frostic
Cyg 19h36'36" 34d14'17"
Debbie and Frank's Friday.
Cyg 21h6'0" 42d15'30"
Debbie and Jesse Sharpes
Leo 9h38'43" 31d12'31"
Debbie and Pat
Sge 19h51'1" 18d21'56"
Debbie and Russell
Crb 15h45'23" 28d44'44"
Debbie and Vishal
Srp 18h40'38" -0d24'33"
Debbie Anderson
Leo 9h42'43" 13d50'57"
Debbie Ann
Lib 15h3'47" -5d38'33"
Debbie Ann Price
Uma 9h21'50" 42d12'18"
Debbie Ann Sargent
Lib 15h12'24" -7d7'22"
Debbie Ann Williamson
Uma 11h43'20" 31d54'6"
Debbie Annabelle Gonzales
And 0h31'9" 27d10'0"
Debbie Anne Gordon
Sco 16h52'36" -40d2'17"
Debbie Arnold
And 0h14'57" 39d51'21"
Debbie Ashley's Dreamcatcher
Uma 10h7'41" 62d11'7"
Debbie Barakat
Lyn 7h49'1" 57d11'58"
Debbie Bates
Per 4h21'1" 35d17'33"
Debbie Beaudry
Leo 10h18'40" 22d31'54"
Debbie 'Bella' Bennett
Leo 11h15'35" 7d2'38"
Debbie Bieksha
Her 16h27'45" 17d37'49"
Debbie Blinder
Crb 15h41'46" 29d40'13"
Debbie Blinder
And 2h1'17" 38d9'19"
Debbie Bullington
Cyg 19h32'4" 31d33'33"
Debbie Cooke
And 23h33'7" 38d20'22"
Debbie D. Holloway
Cas 1h40'13" 62d57'49"
Debbie D. Wood
Lmi 10h45'44" 28d2'30"
Debbie & Dan Stearman
And 0h38'17" 28d33'53"
Debbie Darmofal - # 1 Mom
Cnc 8h25'33" 23d34'28"
Debbie D.Daniel
And 2h22'15" 47d16'30"
Debbie Dean
Dra 17h51'15" 63d6'42"
Debbie 'Deb' Lovato
Cap 21h27'8" -18d33'22"
Debbie DeLuna
Sco 16h9'17" -12d26'44"
DEBBIE DEMOGENES THE GREAT
Cnc 9h19'17" 7d5'4"
Debbie Denise Thomas 50th
Cnc 9h2'18" 23d18'40"
Debbie Digate
Ari 2h42'25" 15d54'40"
Debbie -Do Lang
And 2h0'4" 42d37'6"
Debbie Doll
Leo 11h48'33" 21d25'45"
Debbie Doll
Cap 20h59'31" -22d13'1"
Debbie Doo
Cyg 19h55'12" 42d58'18"
Debbie Doots
Uma 11h36'11" 38d19'40"
Debbie Egglestone
And 23h30'44" 50d32'37"
Debbie Fecsko
Cap 20h23'56" -27d7'57"
Debbie Fitch
Cas 0h29'17" 58d13'7"
Debbie "Flower" Bullen
Sgr 18h20'23" -22d57'46"
Debbie G. Skinner
Her 18h13'17" 28d29'22"
Debbie Geis
Crb 15h50'44" 27d0'45"
Debbie & Glenn Mangham
Sge 19h46'23" 17d28'48"
Debbie Gonzales
And 2h28'54" 38d37'47"
Debbie Grady
And 2h34'52" 49d50'17"
Debbie Gregory Bermudez
Leo 10h55'5" -4d46'17"
Debbie Griffin
Cas 0h36'25" 57d59'14"
Debbie Grigg-Morikawa
Crb 15h44'55" 34d43'31"
Debbie Hacon
Leo 11h56'22" 19d42'36"
Debbie Hancox
Uma 10h13'12" 43d52'6"
Debbie Handley 'Pure Elegance'
Cyg 20h1'20" 54d7'54"
Debbie Harbinson "Sign Chi Do Seed"
Crb 15h49'17" 27d25'17"
Debbie Hennessy
Vir 11h46'14" 8d33'58"
Debbie Henson
Lib 14h50'27" -19d14'29"
Debbie Houston
Vir 14h16'37" 1d28'52"
Debbie Hoysa
Cnc 8h12'46" 25d31'49"
DEBBIE IS MY DENSITY, MY LIFE
Vir 12h58'10" 12d11'9"
Debbie J. Lovejoy
Lib 15h47'55" -16d56'33"
Debbie J. Martin
Sgr 19h49'14" -14d5'47"
Debbie & Jerry
Cyg 21h40'50" 51d6'15"
Debbie Jo Branham
And 1h9'9" 34d42'57"
Debbie Jo McMillian
Tau 5h1'54" 25d11'48"
Debbie & John Webber
Umi 16h17'49" 80d21'37"
Debbie Jones
Cas 23h9'7" 56d24'38"
Debbie Kauerauf
Umi 15h13'26" 75d37'33"
Debbie Kaufman
Leo 9h34'11" 17d25'27"
Debbie Kies 3-28-1969
Mon 6h59'2" -2d14'57"
Debbie Klipstein
Cnc 8h32'21" 17d44'21"
Debbie L. Bell
Cas 1h48'9" 65d17'34"
Debbie Lamb
Cnv 12h47'36" 38d5'20"
Debbie Laney
Lyr 18h33'49" 36d12'1"
Debbie Lash
Uma 9h32'39" 53d20'35"
Debbie Lee
And 0h18'17" 37d18'50"
Debbie Lee
Leo 11h47'36" 24d28'57"
Debbie Lee Pray
Crb 15h23'34" 29d42'23"
Debbie Louise
And 2h27'17" 38d31'12"
Debbie Lumia
Uma 12h50'6" 53d53'4"
Debbie Luszcz
Cap 21h9'52" -15d25'7"
Debbie Lynch
And 2h0'10" 38d50'51"
Debbie Lynn
Psc 1h37'52" 13d19'11"
Debbie Lynn
Cap 21h12'20" -18d58'7"
Debbie Lynn Bailey
Tau 5h19'35" 25d37'54"
Debbie Lynne Sneddon
And 23h59'21" 48d2'7"
Debbie M.
Lyn 7h53'26" 52d59'7"
Debbie M. LaBella
Sco 16h11'27" -13d40'13"
Debbie MacGillivray
Lyn 7h3'56" 47d27'8"
Debbie Mandichek
Aqr 22h0'18" -16d53'18"
Debbie Manocchio
Leo 9h29'59" 9d47'0"
Debbie Marchiafava
Uma 10h16'26" 44d19'55"
Debbie Martinez
Cyg 21h29'28" 51d30'28"
Debbie Matthews
Lib 15h10'59" -7d48'9"
Debbie May
Uma 12h55'15" 63d1'26"
Debbie McCann
And 0h42'34" 40d43'17"
Debbie McDonald
Cas 23h34'19" 55d13'43"
Debbie Miller & Armen Mzrakian
Umi 13h24'52" 72d16'28"
Debbie Min Ju Glencross
Ori 5h36'55" -1d57'46"
Debbie Murtha
Cam 3h40'54" 58d16'6"
Debbie My Love
Cas 1h31'24" 63d36'42"
Debbie N.G. Beautiful & Brilliant
Cas 23h31'51" 53d5'55"
Debbie Norrgard
Cyg 20h32'32" 41d50'33"
debbie nowa
Cas 23h43'10" 56d33'36"
Debbie Olivia Morris
Psc 1h0'10" 16d36'26"
Debbie Osteen
Cas 1h31'11" 58d43'10"
Debbie "Our Shining Star"
Vir 13h20'20" 4d22'6"
Debbie Parker
Cyg 19h59'15" 30d55'54"
Debbie & Paul
Uma 10h36'0" 55d17'45"
Debbie & Paul - Silver Anniversary
Cru 12h38'38" -62d22'13"
Debbie Pendlebury
Lep 6h2'13" -25d53'39"
Debbie Perdue
Psc 23h53'47" -3d6'12"
Debbie Phillips
Cyg 20h21'41" 38d16'31"
DEBBIE PIERCE
Sco 16h10'47" -17d0'50"
Debbie Renney
Cap 20h10'17" -10d21'38"
Debbie Rew
Vir 14h13'23" -6d21'55"
Debbie Robinson
And 1h45'34" 50d14'53"
Debbie Russell
Cap 21h12'26" -25d47'36"
Debbie Ryan
And 23h24'25" 45d37'45"
DEBBIE S. CLAY
Sco 16h36'54" -32d29'57"
Debbie S. Janeczko-DOE #1
Lyn 7h9'41" 51d1'2"
Debbie Sabella
Lib 15h13'27" -3d54'18"
Debbie Scammell
Aqr 22h34'1" -1d47'24"
Debbie Selikman
Col 5h59'24" -35d8'38"
Debbie Shilling
Sgr 18h4'47" -28d29'38"
Debbie Sibley "Angel To All Dogs"
Cmi 7h26'16" 8d13'27"
Debbie Smart LaFever
Psc 1h9'55" 7d14'6"
Debbie Snelson
Cru 12h0'37" -63d11'11"
Debbie Spunkybum
Aqr 22h52'36" -19d20'21"
Debbie "star gazer" Mangano
Uma 11h43'59" 60d15'45"
Debbie Stubbington - Web Queen
Cas 0h20'47" 50d40'37"
Debbie Sue Holtz AKA Bubbles
Crb 15h42'25" 27d7'14"
Debbie Sue Tullock
Cyg 21h35'29" 53d52'8"
Debbie Tanaka
And 0h14'59" 46d35'3"
Debbie Tennick
Psc 23h47'50" 4d49'48"
Debbie the Great
Sgr 19h31'54" -18d40'39"
Debbie Thurston
Leo 11h8'17" 25d28'31"
Debbie Tribble
Cnc 7h58'31" 15d22'16"
Debbie Tweedel
Cas 0h55'39" 61d30'59"
Debbie Vega
Lyr 18h50'12" 33d37'23"
Debbie W. Crotty -Twin - 3-3-1955
Psc 1h50'23" 6d2'13"
Debbie Wareham
Sco 17h52'34" -36d50'44"
Debbie & Wayne Gilman
Aql 19h33'51" -10d40'34"
Debbie Windham
Uma 9h49'14" 47d17'1"
Debbie Wright
Cas 1h6'39" 63d35'52"
Debbie Yameen
Tau 4h32'24" 28d49'20"
Debbie, Gary's Love
Cnc 8h38'26" 27d47'40"
Debbie, Honey
Uma 10h31'53" 66d14'53"
Debbie, Kathy's Guardian Angel
And 0h8'40" 45d37'57"
Debbie-Doo
Uma 11h13'0" 62d53'3"
Debbie-Lee Enders
Uma 13h34'51" 61d6'50"
DebbieLinc
Cas 0h37'38" 51d13'46"
Debbie-Rachel Connection
Uma 10h35'46" 43d24'14"
Debbie's Delight
Tau 5h48'53" 19d25'17"
Debbie's Diamond In The Sky
Cas 0h39'0" 59d27'49"
Debbie's Dream
Leo 11h31'27" 12d16'46"
Debbie's Dream Diamond
Sgr 18h40'0" -30d27'51"
Debbie's Heart
Cas 0h52'47" 61d26'44"
Debbie's Infectious Smile
Lib 14h49'41" -2d11'20"
Debbie's Irish Smile
Gem 6h19'40" 21d56'32"
Debbie's Light
Vul 19h24'31" 21d53'21"
Debbie's Light
Crb 15h41'17" 26d35'35"
Debbie's Petsirius
Cas 23h29'18" 55d47'39"
Debbie's Spirit
Ari 2h15'3" 23d58'23"
Debbie's Star
Crb 16h3'47" 33d57'58"
Debbie's Star
Vir 14h40'19" -1d54'54"
Debbie's Star
Sgr 18h23'58" -24d12'59"
Debbie's Star of Hope
Aql 19h3'28" 3d4'36"
Debbie's Swan Super-Nova
Crb 15h51'47" 38d12'30"
Debbie-SUNSHINE-Puckett
Aqr 20h48'55" -7d7'3"
Debbilicious
Vir 14h49'12" 3d58'39"
Debbi's Rock
Lac 22h47'22" 53d48'7"
Debbi's Smile
Cas 0h27'43" 54d19'22"
Debbi's Star
Uma 11h42'56" 53d20'8"
Debborah J. Duval
Uma 9h40'52" 60d42'55"
Deb-Bra
Uma 11h27'58" 47d43'30"
Debbric
Aqr 22h14'5" 1d6'22"
Debby
Gem 7h42'34" 16d16'21"
DEBBY AND BOB
Lib 15h6'22" -1d45'8"
Debby Ann Tevere
And 23h21'31" 43d19'17"
Debby Billingsley
Sco 16h10'46" -13d32'27"
Debby Coshal
Ari 2h44'9" 25d37'24"
Debby " DJ " Hassett
Cru 12h17'41" -60d42'52"
Debby Frazer
Leo 10h24'43" 27d24'34"
Debby Lynn Stinson
Cap 21h29'15" -13d34'33"
Debby Roberts
Mon 6h45'6" -0d2'48"
Debby Seckel
Cas 23h38'2" 59d2'14"
Debby Sue Boswell
Sgr 18h45'10" -31d34'36"
Debby von Dettmanns Clan
Uma 9h17'44" 50d55'40"
Debby Woodruff
Mon 6h45'6" 7d22'32"
Debbye Kinney "Chicago Super Star"
Uma 12h0'48" 40d32'8"
Debe Christnacht
Uma 10h41'8" 65d36'5"
Debella
Ori 6h15'25" 10d28'10"
DeBello Nebula
Uma 11h1'22" 69d9'38"
Deben's Star
Aqr 22h14'48" -12d45'39"
Debera Hope
Uma 13h47'13" 61d40'27"
Debera Schuman Kettenhofen
Uma 11h6'38" 63d35'16"
Debhora
And 2h27'11" 42d21'59"
Debi
Cnc 8h43'16" 25d50'13"
Debi Boo Bear My Love
Cas 0h42'29" 58d49'4"
Debi Boobie Kins
Cap 20h26'25" -22d2'1"
Debi Carroll
Cap 20h32'15" -27d30'26"
Debi Gutierrez
And 1h45'13" 43d24'1"
Debi Mattingly
Vir 11h42'55" -3d17'28"
Debi & Nati
Uma 10h2'49" 68d39'12"
DEBI of C.B.
Cnc 8h30'51" 29d1'15"
Debi Wilmeth
Uma 11h42'15" 41d15'32"
"Debi", My Pretty Lady
Cas 23h36'5" 54d5'1"
DEBI'S HEART LIVES ON
Her 17h58'28" 14d54'44"
Debita-ELF
Tau 5h3'29" 25d34'28"
DebLaRon717
Cyg 21h34'17" 41d2'0"
Deblasio
Aqr 20h52'50" -13d24'17"
DeBlois - RDTJC
Umi 15h39'51" 74d31'29"
Debora
Dra 17h4'50" 53d19'41"
Debora
And 1h13'59" 43d14'58"
Debora
Leo 11h26'0" 3d53'7"
Debora A. Gatto
Lyr 19h11'52" 46d14'15"
Debora Gaskins
Cap 20h39'13" -21d10'23"
Debora Hawkins
Vir 12h33'52" 8d10'29"
Debora Kaye Marsh
Cas 23h31'7" 52d1'31"
Debora M. Boyer
Uma 11h21'40" 69d21'15"
Debora Malouf
And 0h38'25" 40d30'24"
Debora Sue Eichelberger
Vir 13h34'54" -3d49'14"
Debora Wimmer
And 2h21'27" 50d19'44"
Deborah
Cyg 21h47'55" 39d50'23"
Deborah
And 23h8'32" 41d28'13"
DEBORAH
And 0h7'31" 46d43'41"
Deborah
Cnv 13h15'24" 45d24'44"
Deborah
Uma 13h52'45" 49d38'48"
Deborah
Uma 9h39'43" 46d0'29"
Deborah
Per 3h24'50" 41d32'12"
Deborah
Cnc 8h24'27" 14d31'8"
Deborah
Ari 3h8'0" 22d9'32"
Deborah
Psc 0h26'26" 16d55'6"
Deborah
Vir 12h41'24" 4d11'33"
Deborah
Psc 0h26'30" 7d37'33"
Deborah
Crb 15h49'47" 28d49'37"
Deborah
Ari 2h35'48" 24d9'30"
Deborah
And 0h41'36" 23d47'17"
Deborah
Umi 14h23'13" 75d52'11"
Deborah
Cra 18h48'37" -39d41'13"
Deborah A. Burns
Lib 14h59'52" -1d55'9"
Deborah A. Cummins
Psc 0h2'5" 2d30'4"
Deborah A. Dilley
Ori 5h52'23" 21d26'57"
Deborah A. Dugan Beaulieu 1952-2005
Uma 11h39'20" 34d58'43"
Deborah A. Eigel
Gem 7h9'10" 33d39'37"
Deborah A Fisher
Del 20h35'56" 13d35'50"
Deborah A. Smeets 3/6/1960
And 22h59'36" 48d11'21"
Deborah A Tucker
Crb 15h39'53" 28d39'31"
Deborah Alexander
Gem 6h56'41" 20d23'30"
Deborah and Caron
Lyr 19h17'18" 28d49'29"
Deborah and Eric Probst
Cyg 21h31'14" 46d58'40"
Deborah Anderson
Cnc 8h11'15" 21d17'37"
Deborah Ann
Ori 6h24'16" 16d59'49"
Deborah Ann
And 2h12'26" 38d21'33"
Deborah Ann
Lyn 6h28'8" 55d5'47"
Deborah Ann Aucoin
Lib 16h0'23" -12d29'9"
Deborah Ann Barnett Brown
Hya 8h47'53" 0d5'20"
Deborah Ann Bender
Sgr 18h58'16" -35d4'13"
Deborah Ann Burden (Mam Maw)
Uma 9h22'27" 43d34'35"
Deborah Ann Carlson
Cnc 8h55'57" 8d23'53"
Deborah Ann Criscitiello
Sgr 18h26'26" -32d16'57"
Deborah Ann Ferguson
And 0h30'43" 38d29'5"
Deborah Ann Fraser
Uma 9h25'40" 51d41'26"
DEBORAH ANN JENKINS
Cnc 9h3'57" 7d51'46"
Deborah Ann Lini
Cnc 8h5'44" 18d46'59"
Deborah Ann Logan
Lyn 9h0'32" 38d2'42"
Deborah Ann Meucci Stark
Cnc 8h41'0" 14d42'34"
Deborah Ann - November 25, 1953
Uma 9h35'39" 67d40'40"
Deborah Ann Orr
Lyn 7h15'33" 54d28'25"
Deborah Ann Prell
Gem 7h9'15" 20d49'32"
Deborah Ann Reid
Cnc 7h59'26" 16d55'26"
Deborah Ann Robinson
Tau 4h57'5" 23d21'27"
Deborah Ann Rubin
Lib 15h17'8" -21d14'35"
Deborah Ann Small
Vul 19h26'1" 25d18'34"
Deborah Ann Southerly
Tau 3h27'7" 8d28'20"
Deborah Ann Story
Cnc 8h45'1" 15d3'51"
Deborah Ann Wade
Tau 5h9'6" 18d34'51"
Deborah Anne
Uma 8h29'7" 65d44'27"
Deborah Anne Cummings
And 23h10'1" 36d18'51"
Deborah Anne Dickson
Lyr 18h33'56" 33d9'42"
Deborah Anne Leonard
Psc 1h24'34" 27d31'17"
Deborah Ayers
Cas 1h11'27" 63d42'0"

Deborah Ayn of Christmas Lake
Tau 5h38'10" 25d45'20"
Deborah Belinda
Cas 0h25'25" 50d14'32"
Deborah Bezdikian Star Assistant
Cas 0h24'13" 55d49'3"
Deborah Bridges
Cas 0h16'16" 51d22'51"
Deborah Bush
Ari 2h58'17" 21d41'5"
Deborah Campbell
And 1h7'55" 44d39'45"
Deborah & Chris Ivey
Mon 7h15'22" -0d22'15"
Deborah Clark
Sgr 17h59'8" -17d33'33"
Deborah Cooke
And 1h4'25" 40d28'55"
Deborah Cothern
Uma 11h8'39" 50d34'52"
Deborah Cruthirds
Cyg 20h49'2" 33d13'20"
Deborah D. Hickok
Aqr 22h3'38" -14d4'40"
Deborah D Hopkins
And 0h25'1" 42d30'18"
Deborah Daugherty
Leo 11h28'21" 3d0'11"
Deborah "Debbie"
And 2h37'39" 43d48'39"
Deborah "Debbie" Ann Diogostine
And 23h6'43" 48d53'32"
Deborah Dee Ellis
Cas 0h52'8" 50d12'58"
Deborah Dee Walling Brechler
Vir 13h42'40" 2d48'20"
Deborah Denise Lee
Psc 1h29'10" 25d27'14"
Deborah DeRespinis
And 0h15'31" 27d51'0"
Deborah Di Giovanni
Ari 3h18'27" 28d44'5"
Deborah Diane
And 1h15'42" 45d27'21"
Deborah Dietrich Lynas
Aqr 21h46'22" -3d29'18"
Deborah Dimino (Mom) Star
Cas 0h18'59" 54d37'53"
Deborah & Donald Parker
Uma 9h29'54" 50d53'31"
Deborah Dooley
Lyn 7h25'44" 57d56'49"
Deborah Dowling
Sco 16h50'15" -31d59'36"
Deborah Dulabaum
Cnc 8h51'36" 11d46'34"
Deborah E. Campbell Johnson
Uma 10h48'28" 55d31'19"
Deborah E. McDonald nee Rohn
Umi 16h57'58" 77d10'26"
Deborah E. Sherwin
Uma 9h11'3" 56d33'30"
Deborah E [The Twin(B)] Frederiksen
Umi 15h31'49" 75d57'14"
Deborah Elayne Hodgson
Vir 14h59'46" 0d18'30"
Deborah Elgynia
Cnc 8h46'0" 26d46'42"
Deborah Elizabeth Barry Marchand
Cnc 8h18'42" 22d16'26"
Deborah Ellen Rock
Ori 5h59'42" 18d13'18"
Deborah Esther Aronow
Ari 2h56'53" 13d37'27"
Déborah et Laurent
Dra 19h46'58" 61d55'7"
Deborah Farnham
And 1h16'30" 45d15'33"
Deborah Five O
Lmi 9h23'47" 34d10'8"
Deborah Frank
Gem 7h35'40" 30d40'0"
Deborah Franks Normand
Uma 8h39'46" 51d44'51"
Deborah Fultz Wright
Lyr 18h49'42" 32d30'38"
Deborah Gaume
Ser 15h41'10" 23d4'4"
Deborah Gay - Bubbles
Cru 12h44'34" -62d33'25"
Deborah Gaylord
Leo 9h32'54" 27d28'50"
Deborah Gill Hilzinger PhD
Ari 2h15'18" 17d15'3"
Deborah Glick
And 0h21'36" 35d12'25"
Deborah Graff
Vir 11h55'39" -0d4'40"
Deborah Gwen Fiss
Ari 2h57'8" 27d10'21"
Deborah H. Gonzalez
Leo 11h54'39" 25d4'3"
Deborah Hope
Cap 21h0'35" -16d5'36"
Deborah Houlihan
Ori 5h38'5" 2d54'44"
Deborah Hradek
Lyn 7h7'1" 48d31'12"
Deborah Hunt Cook Burke
Cas 1h26'47" 57d49'6"
Deborah Ivory
Sgr 19h0'11" -16d9'4"
Deborah J Bivens
Uma 10h44'31" 51d5'19"
Deborah J. Fidler and David E. Most
Men 5h5'21" -70d44'31"
Deborah J Garrity
And 1h30'32" 41d25'32"
Deborah J. Gleason
Cas 1h27'3" 68d48'56"
Deborah J. Hart
And 1h20'24" 47d32'0"
Deborah J. Townsend
Uma 9h10'9" 56d47'20"
Deborah J. Ukish
Cas 0h35'13" 61d39'13"
Deborah J. Williamson
Leo 9h42'24" 6d28'37"
Deborah Jackson
Cam 5h9'8" 60d19'18"
Deborah Jan Ritchie
And 2h37'40" 38d24'44"
Deborah Jane
Aql 20h4'47" -11d36'39"
Deborah Jane
Lib 15h30'8" -6d10'57"
Deborah Jane Johnson
Cap 21h34'40" -9d28'32"
Deborah Jane Mitchell
Psa 22h5'52" -31d41'4"
Deborah Janet Peacock's Star
And 1h1'44" 37d4'20"
Deborah & Jay
Uma 12h2'48" 57d41'10"
Deborah Jean
Leo 10h22'46" 21d7'33"
Deborah Jean Constantineau
Leo 10h55'32" 16d59'0"
Deborah Jean Kirk
Cru 11h58'18" -58d42'46"
Deborah Jean Macik
Leo 9h53'59" 20d42'35"
Deborah Jean Parsons
Cas 0h23'38" 60d57'34"
Deborah Jean Sugarbaker
And 22h58'24" 51d31'3"
Deborah Jean Turoff
Sco 16h12'57" -11d16'38"
Deborah Jean Wodicker
Cas 1h1'47" 53d50'4"
Deborah Jeanne Rice
Gem 6h40'13" 29d37'40"
Deborah & Jim Van Valkenburgh
Cyg 20h54'29" 47d55'41"
Deborah Jo Saathoff
Cyg 20h12'36" 35d46'57"
Deborah Jolly Hirschhorn
Cas 1h19'28" 59d3'22"
Deborah Justine Leimbach
And 23h22'1" 47d43'47"
Deborah Ju-ting Yen
Sco 17h12'56" -32d23'12"
Deborah K. Doherty
Ari 1h51'43" 19d19'39"
Deborah K Gegenheimer
Ari 2h54'24" 17d32'32"
Deborah K. Petty
And 23h41'16" 43d1'48"
Deborah Karen Fowler
Srp 18h6'5" -0d42'57"
Deborah Katherine - 02.02.65
Cyg 20h38'7" 46d47'50"
Deborah Kay
And 2h22'20" 45d55'41"
Deborah Kay
Uma 11h50'54" 35d44'15"
Deborah Kay
Vir 14h51'39" 5d14'54"
Deborah Kay
Sgr 17h54'19" -29d45'56"
Deborah Kay
Sgr 18h52'25" -32d11'57"
Deborah Kay Alexander
Crb 15h50'10" 28d50'0"
Deborah Kay Hopkins
Ari 2h34'10" 25d55'44"
Deborah Kay Skorr
Ari 2h40'5" 21d4'26"
Deborah Kay Swinney Sims
Lib 15h49'49" -17d20'26"
Deborah Kaye Liseo
Ari 2h17'5" 13d50'27"
Deborah Kelly
Psc 1h15'31" 12d23'37"
Deborah Kopald
Psc 1h27'16" 11d5'3"
Deborah Krywinski
Lyr 18h47'47" 43d20'31"
Deborah L. Auciello
And 0h24'51" 41d27'1"
Deborah L. Dulaney
And 1h19'10" 45d25'8"
Deborah L. Goldman
Lyn 7h47'50" 54d16'50"
Deborah L. Holloway
Ori 5h5'51" 14d16'30"
Deborah L. Summa
Uma 10h38'53" 48d31'16"
Deborah L Tibbs
And 0h37'21" 42d49'42"
Deborah L Whitney
And 23h59'7" 39d40'5"
Deborah Lee
Cas 0h13'31" 62d47'12"
Deborah Lee Michael
Vul 20h18'6" 23d53'55"
Deborah Lee Smith
And 23h36'24" 47d56'38"
Deborah Lee's Diamond
Nor 16h16'40" -43d18'22"
Deborah Lichon
Uma 11h57'13" 57d20'9"
Deborah Linette
Cnc 8h36'37" 31d32'31"
Deborah Longwith Wasik
Cas 23h33'2" 58d19'25"
Deborah Lou 1954
Cas 23h31'6" 59d56'31"
Deborah Louise - Forever Shining
And 2h21'7" 41d12'8"
Deborah Louise McKie
Cas 23h46'31" 55d58'9"
Deborah Louise Todd
Cru 12h25'6" -59d2'19"
Deborah Luther
Vir 13h53'15" -2d9'12"
Deborah Lyn Ramey
Sgr 17h54'45" -29d34'8"
Deborah Lynn
Sgr 18h11'45" -32d29'23"
Deborah Lynn
Cyg 19h51'46" 33d28'44"
Deborah Lynn
And 23h16'35" 47d12'31"
Deborah Lynn
Lyn 7h15'30" 49d57'33"
Deborah Lynn
Tau 4h4'23" 27d9'59"
Deborah Lynn Byers
Umi 15h37'9" 81d9'14"
Deborah Lynn DeBiew
And 0h7'52" 44d4'42"
Deborah Lynn Escobedo
Tau 4h26'53" 5d34'1"
Deborah Lynn Hutzler McGowan
Lib 15h2'21" -10d15'8"
Deborah Lynn Mai
Sgr 18h35'28" -24d11'18"
Deborah Lynn Menefee
Per 2h48'14" 48d8'2"
Deborah Lynn Merz
Cnc 8h14'34" 23d34'47"
Deborah Lynn Murphy
Uma 9h52'5" 49d20'19"
Deborah Lynn Powell
And 1h28'7" 49d52'34"
Deborah Lynn Ramey
Sgr 19h2'5" -30d29'43"
Deborah Lynn Shields
Lyr 19h19'42" 29d59'13"
Deborah Lynn Swope
Gem 7h17'11" 32d31'16"
Deborah Lynn Willis
Leo 9h44'27" 29d57'3"
Deborah Lynne Herron
Leo 11h15'10" 17d30'54"
Deborah Lynne, The Elegant
Umi 15h56'10" 79d20'11"
Deborah M Doyle nee Nemeth
Gem 6h44'35" 18d47'17"
Deborah M Galvin
Psc 1h12'39" 12d6'17"
Deborah M. Karpf
Psc 0h52'43" 6d20'32"
Deborah M. Mohr
Lib 15h7'19" -3d42'38"
Deborah Mae Walag
Lib 15h14'12" -10d30'36"
Deborah Malakoff
Lyn 8h7'24" 47d49'18"
Deborah Maribito
And 2h10'37" 41d36'28"
Deborah Marie
And 1h55'34" 37d46'39"
Deborah Marie
And 23h44'16" 48d6'8"
Deborah Marie Eller
Lib 15h18'26" -7d33'25"
Deborah Marie Harlan
Lib 14h53'40" -19d36'8"
Deborah Mary
Cas 0h56'11" 63d11'4"
Deborah Mary Malouf
Cru 12h48'6" -56d19'49"
Deborah McChesney
Uma 9h8'12" 50d55'46"
Deborah McGuffey
Cas 1h43'25" 65d13'16"
Deborah & Michael
Sgr 19h37'50" -37d43'18"
Deborah & Micheal
Lyr 19h11'3" 45d4'1"
Deborah Miller Trent
Cas 0h31'13" 61d36'59"
Deborah Mills
Lyr 18h33'34" 37d8'18"
Deborah Monson
Cap 21h3'55" -21d56'43"
Deborah Monzo "2/18/1955"
Lib 14h55'20" -17d5'9"
Deborah Moore 1939-2007
Uma 8h44'1" 56d35'38"
Deborah Mullen Baker
Crb 15h27'24" 27d1'54"
Deborah Mumford
Vir 12h43'45" 9d32'56"
Deborah - My Princess
And 23h23'27" 48d7'48"
Deborah Nevarez
Ari 2h48'56" 21d42'57"
Deborah Norton
Uma 9h35'27" 42d26'14"
Deborah Oeschger
Mon 6h28'51" 8d55'5"
Deborah Ortiz
Aqr 22h20'21" -1d14'14"
Deborah P. (Sutherland) Geasley
Psc 0h49'48" 9d37'28"
Deborah Pariseau
Cas 0h40'7" 61d15'7"
Deborah Parker
Cas 0h51'29" 56d18'37"
Deborah (Pat) Hoffmann
Psc 1h11'50" 11d48'50"
Deborah & Peter Sullivan
Cyg 20h59'10" 31d5'2"
Deborah Pinnell
Aql 20h10'10" 2d3'55"
Deborah Pitre
Cap 20h19'13" -9d16'57"
Deborah Porcaro
And 0h42'1" 38d54'57"
Deborah Pruitt
Vir 12h50'20" 7d40'42"
Deborah R. Davis
Vir 12h30'29" -9d44'38"
Deborah Radosevich
Ari 2h46'19" 30d46'56"
Deborah Raye Shalosky and Family
And 0h13'19" 35d25'26"
Deborah Regnery
Tau 4h9'32" 16d32'18"
Deborah Robbins Duke
Tau 3h37'20" 20d52'13"
Deborah Rueschman Carter
Her 17h5'43" 24d6'5"
Deborah Russo
Vir 12h31'40" 12d16'44"
Deborah S. Condon
Sco 17h54'12" -38d52'47"
Deborah S Hayward - My Snow Queen
Lib 15h49'4" -14d9'12"
Deborah Scammell
Col 5h19'30" -37d33'48"
Deborah Scott Morgan
Cra 18h36'52" -39d26'27"
Deborah Sharleen - Shar's Star
Ori 5h57'10" 18d24'52"
Deborah Sidelnik
Cyg 20h13'23" 38d18'28"
Deborah Sofia
Lib 15h20'36" -5d59'54"
Deborah Sparks
Vir 14h31'12" -4d29'49"
Deborah Stankwytch Norton
Lyn 8h31'20" 34d17'38"
Deborah "Star" Venner
Sco 17h47'12" -35d19'38"
Deborah Stow Reeder
Cas 1h10'2" 62d7'33"
Deborah Sue Horn
Gem 6h48'49" 17d23'33"
Deborah Sue Jarrett
Vir 12h29'1" 7d58'39"
Deborah Supowit
Ori 5h0'33" 4d5'32"
Deborah Suzanne Gauthier
Ori 5h38'28" -1d27'38"
Deborah Taylor-Sole
And 1h47'12" 42d46'54"
Deborah Townsend Carlisle
Psc 1h8'48" 22d58'19"
Deborah Upham Pifer "Grammy"
Cas 23h48'57" 58d18'47"
Deborah Waldman
Cas 1h40'41" 63d18'26"
Deborah Ward The light of my life
And 1h16'3" 42d53'55"
Deborah Watson
Ari 1h59'26" 17d54'6"
Deborah White McEniry
Her 16h41'34" 7d11'39"
Deborah Williams
Sex 9h47'17" -0d32'43"
Deborah Wood
Uma 9h51'37" 43d57'1"
Deborah Woolsey-Heim
Cas 0h3'26" 56d29'39"
Deborah Wooten
And 23h43'31" 36d57'14"
Deborah York
Tau 4h22'54" 27d11'11"
Deborah Yvonne Pinch
Sgr 18h16'44" -17d10'47"
DeborahElizabeth
Vir 12h36'46" -1d8'13"
Deborah-Lee Bodkin
Cas 1h47'31" 61d23'14"
DeborahLouden
Cnc 8h13'14" 25d45'45"
Deborah's Andromeda
And 1h50'26" 38d34'2"
Deborah's Destiny
And 0h55'1" 38d30'25"
Deborah's Home
And 1h3'31" 48d24'33"
Deborah's SHEERO
Cas 1h57'38" 64d39'0"
Deborah's Twinkling Star
Psc 1h39'3" 17d56'18"
Deborha Brush
Cap 21h16'30" -19d47'22"
Deborialis
Leo 11h25'45" 12d25'8"
Deborist
Cas 1h18'36" 71d46'46"
Deboruth
And 0h54'1" 40d36'52"
Debra
Cyg 20h8'0" 44d7'50"
Debra
And 23h16'16" 47d46'31"
Debra
Psc 1h45'43" 12d52'19"
Debra
Tau 4h23'25" 14d20'19"
Debra
Ori 5h55'42" 8d0'8"
Debra
Boo 15h3'58" 23d36'21"
Debra 381
Uma 8h49'26" 52d4'35"
Debra A. Anderson
Aqr 23h1'35" -9d13'57"
Debra A. Burlingham Johnson
Ari 3h13'58" 27d11'47"
Debra A. Cuccinello
Gem 6h22'35" 20d6'11"
Debra A. Hamilton
Cas 0h10'9" 53d54'9"
Debra A. Mendelsohn
Cas 0h11'6" 57d8'47"
Debra A. Wolf
Her 18h18'7" 15d18'45"
Debra Allmer
And 23h27'49" 41d28'6"
Debra and Gary Marks
Uma 12h16'51" 62d33'32"
DEBRA AND TIM
Cyg 20h14'33" 40d12'25"
Debra Anderson
Cas 0h12'45" 53d11'36"
Debra Ann
Uma 11h30'36" 51d37'35"
Debra Ann
And 0h5'32" 46d29'33"
Debra Ann
And 0h56'59" 38d17'43"
Debra Ann
Cnc 9h10'18" 32d14'46"
Debra Ann
Cap 20h38'0" -24d19'11"
Debra Ann Buttacavoli
Vir 13h5'58" 8d6'3"
Debra Ann Caselli
Lib 15h53'7" -17d19'39"
Debra Ann Dawson
Aql 20h20'56" -7d38'7"
Debra Ann Fargo
Vir 15h4'37" 3d8'26"
Debra Ann Lane
Psc 1h23'30" 20d13'31"
Debra Ann Margison
Cas 1h44'39" 61d23'43"
Debra Ann Marie
Ori 5h11'59" 6d19'25"
Debra Ann Oswald -angel child
Uma 9h49'45" 54d17'19"
Debra Ann Pitchford
Cyg 20h8'1" 37d13'7"
Debra Ann Turney
Ari 2h6'48" 14d59'16"
Debra Ann Viti and Heath Alejandro
Cyg 21h35'51" 51d50'46"
Debra Anne Blackstock
Cam 4h54'7" 54d20'4"
Debra Annette My Shining Star
Cnc 8h52'6" 31d45'51"
Debra Barrie
Cas 23h25'55" 58d41'16"
Debra Beautifalis
Crb 15h51'17" 39d35'51"
Debra & Bill Moran
Cyg 20h8'55" 31d26'35"
Debra Breton
Ari 2h44'5" 28d38'25"
Debra & Brian
Cyg 21h26'7" 37d9'27"
Debra Brower, RN
Lmi 10h40'16" 33d50'15"
Debra Carr
And 1h45'49" 44d58'7"
Debra Carrick
Uma 11h56'15" 32d14'33"
Debra Caviness Best Mother Ever
Crb 15h39'0" 25d48'58"
Debra Chambers
Aqr 22h10'49" 0d8'5"
Debra Christine Schwingle
And 1h14'14" 40d55'20"
Debra Chumley
And 2h23'38" 50d34'6"
Debra Cler Mikell
Hya 9h32'53" -0d54'52"
Debra D. Siebert
And 1h31'4" 50d26'7"
Debra D. Taylor
Uma 9h41'32" 49d41'46"
Debra Dawn Cogan
Cyg 20h11'41" 37d1'31"
Debra Deacon
Cru 12h36'29" -60d44'22"
Debra Diane
Dra 16h41'51" 58d7'46"
Debra "dobboo" Harper
Leo 10h23'4" 22d0'13"
Debra Doris Johnson
Ser 18h35'6" -0d11'57"
Debra & Ed Davalos
Vir 13h14'58" 7d8'55"
Debra Elaine Yarbrough
Umi 15h32'52" 73d52'25"
Debra Espinoza
And 0h8'24" 40d9'2"
Debra Faye Disharoon
And 0h52'33" 36d4'7"
Debra *Freya* Thibault-Cross
Tau 3h48'27" 19d14'43"
Debra G. Cruz
And 23h2'46" 52d35'9"
Debra Garrison
Tau 5h31'9" 26d54'55"
Debra Glerum
Cyg 19h44'23" 28d28'33"
Debra Guastella
Psc 0h43'11" 11d25'36"
Debra Haase
Ori 6h10'55" 19d16'4"
Debra Hamilton
Lyr 19h20'15" 28d53'50"
Debra Helene Ploss
Cap 20h48'14" -25d59'57"
Debra Hood Koch
Vir 13h15'51" 11d10'28"
Debra I40
Cas 1h15'44" 55d48'55"
Debra Iachini-Lux
Cyg 19h52'53" 39d23'40"
Debra J. Brumbaugh
And 23h23'32" 51d37'58"
Debra J Pacheco
Leo 11h24'25" 11d32'31"
debra j poole
Psc 23h5'11" 6d22'34"
Debra J. Zidich Gibbons
Sgr 18h26'43" -32d25'18"
Debra Jane Foglietta
Cas 23h6'6" 57d6'50"
Debra Jane Thomas
Cnc 8h24'34" 14d13'47"
Debra Jean
Cyg 20h32'21" 39d37'29"
Debra Jean
Cas 23h10'14" 56d14'12"
Debra Jean
Sco 16h5'13" -14d58'20"
Debra Jean Creese
Gem 7h21'51" 32d9'39"
Debra Jean Ellstrom
Leo 11h16'17" 14d57'50"
Debra Jean Franklin
And 0h17'43" 31d49'32"
Debra Jean Ibbetson
Sgr 18h46'36" -31d7'3"
Debra Jean Mesick
And 0h43'46" 39d34'52"
Debra Jean Pinkard
Aqr 22h9'33" 1d59'33"
Debra Jensen
Tau 4h36'23" 22d52'18"
Debra Joy Ryan
Sgr 17h52'47" -25d55'24"
Debra June Tewell-Hunt
Cnc 9h16'10" 17d42'44"
Debra K Connelly
Cas 0h39'48" 54d19'39"
Debra K. Scribner
Tau 4h23'9" 26d19'59"
Debra K Shifflett
Vir 13h11'10" -20d38'9"
Debra Kathleen Gibbs
Tau 5h45'9" 17d7'2"
Debra Kay
Aqr 21h49'48" -8d11'38"
Debra Kay Moody
Mon 6h46'45" 7d20'21"
DEBRA KAYE
Ari 1h49'17" 20d12'6"
Debra Kenney
Ori 6h4'13" 17d25'47"
Debra L. Andersen
Cap 21h48'32" -14d42'42"
Debra L. Bass
Cas 0h34'55" 57d40'20"
Debra L. Fox
Gem 7h20'11" 16d14'1"
Debra L. Harrison
Crb 15h44'39" 38d24'13"
Debra L. Kaplan
Psc 1h28'4" 5d55'40"
Debra L Keller
Tau 5h46'43" 26d48'29"
Debra L. Linke
Gem 7h6'41" 33d9'0"
Debra L. Musbach
Vir 13h44'54" -14d26'59"
Debra L Smith
And 2h27'24" 48d0'0"
Debra L. Wright / My Debbie
Leo 9h43'4" 27d45'27"
Debra Lea
Vir 14h46'7" 5d31'45"
Debra Lee Scala-Krill
Ari 2h39'20" 30d22'24"
Debra Lee Wilber
Sco 16h25'25" -26d3'54"
Debra Leigh Faust
Gem 7h46'54" 23d28'19"
Debra Longo
Vir 12h23'22" 2d21'39"
Debra Loria
Ari 2h10'53" 13d12'39"
Debra Lorraine Fountain
And 23h12'1" 35d33'21"
Debra Lou
Psc 0h54'58" 27d59'49"
Debra Lou Morris
Sco 17h53'54" -36d38'49"
Debra Louderback
Crb 15h34'19" 39d3'38"
Debra Lyn Swinford Best Mommy Ever
Leo 11h14'52" 14d24'27"
Debra Lynn
Aql 19h39'11" 11d24'32"
Debra Lynn
Tau 4h27'55" 19d36'49"
Debra Lynn
Lyr 18h29'1" 28d28'4"
Debra Lynn
And 23h19'40" 47d56'21"
Debra Lynn
Uma 9h35'1" 45d2'25"
Debra Lynn
Cap 20h42'37" -26d42'48"
Debra Lynn Brennan
Psc 1h20'22" 20d55'57"
Debra Lynn Haney
Sgr 19h23'40" -27d36'46"
Debra Lynn Steiner
Cnc 8h59'27" 10d10'4"
Debra Lynn Walden
Aqr 23h25'18" -6d22'18"
Debra Lynn Watson
Sco 17h46'49" -32d20'36"
Debra Lynne Corradino
Gem 6h20'48" 26d58'57"
Debra Lynne Rich
Aur 5h50'41" 52d13'38"
Debra M. Donohue
Lyn 6h43'19" 60d24'49"
Debra M Smithey
Cap 20h32'0" -17d19'17"
Debra M. Tamez
Cap 20h22'28" -24d1'31"
Debra Marie Remington
Lyn 7h24'6" 47d3'8"
Debra Marrocco
Uma 11h13'19" 43d9'20"
Debra Martsching
Cas 1h19'3" 66d22'9"
Debra Masters
Col 5h56'13" -35d31'33"
Debra McDonald
Cyg 19h42'45" 30d46'27"
Debra Mullens*** Luv of My Life
Sco 17h48'6" -33d33'41"
Debra - Ned 11-1-1996
Lyn 8h3'7" 42d27'51"
Debra P.
Cas 0h37'46" 58d1'57"
Debra Pauline
Tau 4h30'8" 25d33'48"
Debra Rae
Cap 20h48'9" -24d37'31"
Debra Ray
Leo 11h58'6" 21d50'32"
Debra Reinisch
And 0h39'20" 42d32'9"
Debra Renae
Ari 3h2'17" 25d29'23"
Debra Renee Bellamy
Umi 16h34'30" 77d29'0"
Debra Robyn Tischler
Ari 2h36'39" 24d54'38"
Debra Schneider
Cam 4h1'6" 71d28'14"
Debra Sellers
Cas 0h23'9" 60d22'50"
Debra Sims
And 1h50'5" 44d1'24"
Debra Star Cigal
Lib 14h55'27" -3d27'58"
Debra Stern
Lyn 7h11'13" 57d22'7"

Debra Sue
Pho 0h51'52" -52d37'54"
Debra Sue
And 0h23'14" 36d25'5"
Debra Sue Purtle
Lib 15h6'0" -29d12'23"
Debra Sue Webster
Cas 23h0'36" 54d4'22"
Debra Susan, The greatest MOM ever
Lyn 7h27'23" 53d9'49"
Debra Susser
And 2h23'29" 50d20'38"
Debra Suzanne Diamantos
Leo 9h34'48" 31d8'35"
Debra T. Kuznicki
Lib 15h35'7" -21d8'36"
Debra Termine "Deb's Light"
Cas 1h12'1" 62d2'29"
Debra Towles
Cap 20h20'24" -9d36'5"
Debra V. Pasquale
Psc 1h14'59" 16d52'21"
De'Bra W. Jacobs
Leo 9h22'22" 17d50'16"
Debra Waltz
Uma 11h28'4" 49d9'10"
Debra Ward IK
Cam 3h28'37" 67d12'43"
Debra Warren
Cam 6h47'39" 68d3'38"
Debra Way
Aqr 22h57'23" -9d5'54"
Debra Woodward Battle
Leo 11h33'34" 21d49'37"
Debra Wydra
And 0h56'48" 36d23'17"
Debra Yacino
Sgr 18h9'58" -27d28'24"
Debra Yacino
Sco 17h2'51" -32d19'14"
Debra1954
Cap 20h59'14" -19d31'33"
Debraalex
Ori 5h27'15" 2d51'17"
Debrah Lynn Lytle; AKA (Mom)
Lyn 7h32'34" 42d0'56"
DebraKay
Lmi 10h5'5" 37d11'22"
DebraKay
Sco 16h57'39" -45d2'33"
DebRan Stoudt Hope Star
Uma 11h43'1" 40d58'18"
Debra's Angel
Gem 6h44'25" 23d3'26"
Debra's Eternal Light
Ari 2h46'2" 27d41'56"
Debra's Space Debris
Ori 6h15'47" 13d32'48"
Debra's Star
Uma 11h37'6" 62d43'49"
DebRay
And 0h13'25" 26d26'29"
Debrett's Kefitzat ha-Derekh
Cru 12h55'33" -60d7'27"
DeBrian724
Lmi 10h28'15" 36d20'49"
deBruza - forever one
Cru 12h36'59" -58d11'21"
Debs
Cas 1h38'29" 61d10'57"
Deb's Destiny
And 23h23'9" 47d27'13"
Deb's Eye
Cap 20h38'33" -10d54'41"
Deb's Light
Cas 2h4'52" 62d56'4"
Deb's Shining Star
Leo 11h26'43" 25d14'19"
Deb's Star
Ori 6h18'1" 14d33'58"
Debs you really are a star!
And 2h29'41" 49d56'19"
DebsTAR Polaris-Magni
Aqr 22h34'32" -21d52'6"
Debula
Ari 2h24'59" 11d58'47"
Debvora
Com 12h30'0" 27d53'36"
Debz
And 23h3'7" 47d53'51"
Dec
Aur 5h57'37" 37d4'1"
DECA Oompa
Tau 4h33'34" 15d28'4"
Decan
Lmi 10h27'33" 29d39'9"
Decano Gavilanes
And 23h37'25" 33d36'25"
Deceased Members St Mary's Parish
Umi 8h49'18" 88d20'59"
December
Ari 3h3'6" 26d56'24"
December
Ori 6h1'52" 18d9'39"
December 17, 2005
Cyg 20h21'58" 46d47'59"
Decenna Cadiz
Cap 21h3'35" -19d4'23"
Dechanel Irving
Her 17h25'2" 17d43'55"
Deck 143 labh
Psc 23h47'31" 6d18'2"
Decker Ashton Wold
Umi 16h49'58" 81d44'53"
Deckie
And 1h9'55" 45d56'59"
Declan Chase
Sgr 18h24'12" -16d3'58"
Declan Dooley
Per 3h28'24" 42d11'31"
Declan Gould
Uma 9h38'58" 49d32'51"
Declan Gregory Cuozzo
Her 17h15'28" 29d15'41"
Declan Griffiths
Umi 13h44'36" 86d22'12"
Declan James Barry
Uma 10h5'11" 50d55'1"
Declan John Richter
Uma 10h28'42" 54d11'21"
Declan Jolowski
Sco 16h8'8" -17d48'37"
Declan Joseph Geoghegan
Sgr 18h46'4" -29d32'3"
DECLAN KUNKEL
Ori 5h26'8" 2d2'42"
Declan Lindsay Alan Garvey
Cru 12h13'28" -62d9'35"
Declan Magee
Uma 11h50'35" 41d17'44"
Declan Michael Grady
Umi 14h48'50" 67d42'9"
Declan Michael McLaren
Umi 13h8'53" 69d21'8"
Declan Michael Sinnott
Cnc 8h52'46" 12d47'46"
Declan Michael Swift
Umi 15h1'8" 72d54'55"
Declan Smith
Uma 13h48'17" 53d55'55"
Declan Starfighter
Tau 5h50'28" 27d27'55"
Declan T. Reinert
Uma 10h26'36" 45d31'24"
Declan's Star
Mon 7h20'37" -4d32'8"
Decob
Uma 13h21'49" 57d19'43"
Decorus Angelus
And 2h35'39" 39d9'3"
DECORUS ARDOR
Tau 4h29'13" 20d14'37"
Decorus Caltha
Psc 1h20'48" 32d58'51"
Decorus Laureatus
Uma 11h11'35" 52d39'57"
Decorus Puella
Lyn 7h15'32" 58d52'40"
DeDa
Cyg 20h19'1" 36d50'46"
Deda
Gem 6h44'21" 26d9'23"
Deda Igor Alexandrovitch Ilin
Dra 10h51'43" 78d57'35"
DeDaJaBrAnLoTy-My Blessings
Lyn 7h41'13" 42d11'19"
DeDe
Tau 5h58'29" 25d31'54"
Dede Dewey's out of this worldwheel
Lyn 8h26'51" 53d1'55"
Dede "Doodles" Renick
Uma 9h25'6" 59d56'2"
Dede's 50th Shines For Eternity
Ari 3h18'3" 28d46'44"
DeDe's Sunshine
Cnc 9h16'42" 30d54'56"
Dedi 535
Lib 14h53'13" -11d48'36"
Dedicated to Aaron Thomas Thacker
Her 18h46'30" 22d32'16"
Dedra Lynette Ricks
Cap 20h29'30" -23d7'13"
Dedrick
Her 17h40'33" 20d30'37"
Dedrick Demartin Washington
Psc 0h6'7" -4d30'59"
Dee
Sgr 18h29'7" -19d7'1"
Dee
Lib 15h8'56" -20d40'15"
Dee
Uma 9h33'5" 58d14'58"
Dee
Sgr 18h1'3" -29d2'24"
Dee
Cru 12h22'14" -58d33'58"
Dee
Leo 9h44'10" 28d24'44"
Dee
Psc 0h21'22" 10d0'33"
Dee
And 1h44'42" 42d22'38"
Dee
Cyg 20h36'4" 47d40'30"
Dee
And 23h6'15" 40d16'23"
Dee 1978-2007
Cru 12h32'11" -58d0'52"
Dee A. Munoz
Uma 9h23'47" 45d22'48"
Dee and Julia
Gem 7h7'22" 24d23'21"
Dee and Terry Saddler
Uma 11h17'47" 62d14'31"
Dee Ann Roshong
And 0h43'3" 34d41'33"
Dee Anne
Cnc 8h27'54" 32d35'32"
Dee Anthony
Aqr 22h42'24" -0d47'30"
Dee Bailey
Her 18h36'32" 20d4'49"
Dee & Bill Barker
Lyn 9h4'50" 36d4'2"
Dee & Billy
Lyr 18h42'15" 31d7'11"
Dee Chesko
Aqr 23h48'55" -10d45'40"
Dee Damico "Our Shining Star"
And 0h40'44" 42d2'32"
Dee Dee
Aqr 23h23'10" -14d41'39"
Dee Dee
Vir 11h56'22" -0d10'9"
Dee Dee
Cam 3h39'29" 67d28'23"
Dee Dee
Cap 20h25'21" -27d24'42"
Dee Dee and Jmo 4ever in Love
Per 4h41'30" 49d28'22"
Dee Dee Ann Jones
Lyn 6h24'44" 58d25'16"
Dee Dee Forever - 26th January 1956
Cen 13h47'54" -45d46'4"
Dee Dee Roney
Lib 15h38'3" -5d42'0"
Dee Dee Stopar
Ari 3h12'26" 16d10'44"
Dee Dee Thorburn
Uma 9h21'23" 65d16'40"
Dee Dee's Shining Star
Ori 6h0'29" 10d45'18"
Dee Dee's Star
Cas 0h5'2" 56d43'28"
Dee Dee's Star
Aqr 23h33'18" -20d0'53"
Dee Elizabeth Globetti
Cas 1h26'45" 52d6'47"
Dee Foster
Lyn 6h43'18" 52d37'58"
Dee Hackney Streeter
Lyn 8h46'20" 41d28'14"
Dee J Iman
Uma 10h38'55" 47d24'39"
DEE & JEE
Cap 21h1'0" -21d52'17"
Dee Katz
Sco 16h13'55" -14d50'7"
Dee Mahl
And 23h29'41" 49d20'9"
Dee Marie Engler
Uma 9h6'10" 71d39'47"
Dee Marie Ericksonmoen
Tau 4h3'51" 6d15'37"
Dee Miller
Lyn 7h34'46" 53d8'59"
Dee Morgison
Sco 16h50'50" -32d40'52"
Dee Morris
Leo 11h54'8" 22d59'43"
Dee Murphy
Cap 21h46'17" -9d43'12"
Dee Parham
Tau 4h12'47" 17d44'18"
Dee Ryan
Psc 23h36'1" 5d50'38"
Dee Schuiteboer
Lyn 8h11'32" 33d55'56"
Dee & Sully's Shining Star
Aqr 22h16'53" 0d55'28"
Dee & Tim Krey
Cyg 20h44'51" 39d33'4"
Dee Tirocchi
Tau 3h36'40" 25d26'49"
Dee Wood
Cyg 20h31'0" 34d29'1"
Dee1993
Aqr 21h10'39" -14d12'9"
DeeAdra
Ari 2h34'12" 27d55'55"
DeeAnn CLaire Marie Lenard
Vir 12h16'21" -3d41'42"
DeeAnne Elizabeth Hendrickson
Cas 23h4'48" 56d37'2"
Deeberlilaz
Uma 11h53'19" 30d47'17"
Deebo's Star
And 23h44'6" 34d20'57"
Deeda
Cma 6h57'2" -12d43'37"
DeeDee
Cas 2h36'58" 68d43'35"
DeeDee
Uma 11h19'0" 35d26'40"
DeeDee
Uma 11h21'55" 52d29'57"
DeeDee
Psc 0h50'57" 9d40'12"
DeeDee Baby
Lib 15h19'45" -21d25'20"
DeeDee Cecilia Pavone
Cyg 21h24'52" 39d51'24"
DeeDee Girl 725
Leo 11h4'43" 17d3'37"
DeeDee My Love. Our light, our star
And 2h16'53" 50d12'4"
deedeemarie
Cam 5h25'55" 63d25'58"
deedeldeedee
Uma 11h23'45" 32d55'37"
Deedy
Cyg 20h25'12" 47d15'3"
Deegan22
Ori 5h32'54" 10d22'9"
deegee
Sco 16h10'16" -8d40'15"
Deejitz
Ari 2h23'58" 24d25'56"
Deeken, Horst
Gem 7h17'45" 20d25'8"
Deeksha H. Dave
Tau 4h23'14" 23d40'7"
DeeLaine Millard
Uma 13h13'34" 52d47'18"
DeeLight
Ari 3h20'25" 28d51'43"
Dee.M.G
Aqr 23h15'19" -15d6'49"
Deena
And 2h17'1" 50d22'49"
DEENA AND NICK
Cyg 21h16'16" 42d55'5"
Deena Black
And 1h2'15" 38d51'24"
Deena Delight Gallagher
Lmi 10h9'36" 28d35'58"
Deena Lynn Norfleet
Lyn 8h32'6" 48d46'54"
Deena Marie
Cyg 19h27'47" 53d18'0"
Deena Marie Tonning
Ari 2h10'49" 25d20'5"
Deena Musharbash
Cas 0h27'25" 54d21'20"
Deena Peters
Cas 23h47'45" 53d11'32"
Deena & Todd 35th Anniversary
Gem 7h28'13" 20d19'56"
Deena's star
Lib 14h23'58" -13d11'32"
Deenna Lynette McKenzie
Uma 13h9'23" 53d34'40"
Deep & Bindi
Cyg 20h59'5" 35d42'27"
Deep Sea Treasure Hunter
Aqr 22h7'56" -4d8'52"
Deep Sky Dale
Psc 1h7'59" 13d40'11"
Deep Space Hallie
Psc 1h13'21" 16d0'54"
Deep Space Hiker
Ori 5h50'15" -0d5'41"
DEEPA
Sgr 19h34'25" -14d29'4"
Deepa Bansal
Lib 15h21'1" -16d33'44"
Deepa (Gagoon)
Umi 16h25'8" 72d14'17"
Deepak Gupta
Cep 23h56'29" 68d59'24"
Deepak's Star
Cep 1h56'35" 81d55'49"
Deepanwita Prusty
Ari 3h4'12" 18d51'26"
Dee's Fahey Muzzey
Lyn 6h20'39" 56d57'29"
Dee's Light
Uma 9h59'43" 59d4'15"
Dee's Star
Sge 20h7'33" 16d6'55"
DeeSegrest
Psc 1h17'8" 23d27'34"
Deester
Leo 9h30'15" 7d19'33"
Deeta's Star
And 0h39'49" 40d6'6"
Deeter
Psc 1h24'2" 20d35'40"
Deeter
Uma 10h23'21" 68d10'27"
D.E.F.
Aql 19h40'28" 6d41'57"
DEFEKT!!!
Ori 5h20'21" 1d11'30"
DEFEKT!!!
Ori 5h55'30" 3d41'50"
DEFEKT!!!
Uma 11h47'24" 39d26'17"
DEFEKT!!!
Uma 11h42'1" 40d23'8"
DEFEKT
Uma 11h36'3" 33d41'5"
DEFEKT
Uma 13h52'46" 50d29'20"
DEFEKT
Uma 9h39'23" 47d7'37"
DEFEKT
Uma 9h42'31" 46d48'27"
DEFEKT!!!
Uma 8h40'7" 50d39'2"
DEFEKT!!!
Uma 11h11'6" 69d50'7"
DEFEKT!!!
Uma 10h57'0" 68d21'46"
DEFEKT!!!
Uma 8h38'58" 69d42'36"
DEFEKT
Uma 11h24'25" 70d10'5"
DEFEKT!!!
Uma 11h22'25" 70d48'40"
DEFEKT
Uma 9h48'33" 57d19'23"
DEFEKT
Uma 12h59'19" 55d38'5"
DEFEKT!!!
Uma 13h9'42" 56d16'17"
DEFEKT!!!
Uma 12h53'46" 56d23'0"
DEFEKT!!!
Uma 12h59'54" 56d49'54"
DEFEKT
Uma 12h59'54" 58d11'40"
DEFEKT
Uma 13h8'13" 59d20'35"
DEFEKT!!!
Uma 14h14'29" 54d59'38"
DeFeo Twin Baby Boys
Umi 14h41'33" 77d26'11"
Defiant-Ronald Wayne Bonesz
Ori 4h59'53" -0d5'13"
Defining Grey
Uma 9h8'39" 66d12'44"
Defjiplr
Cyg 21h10'27" 44d15'28"
Defne Beyza Zeybek
Psc 0h55'18" 25d55'6"
Defne Rana Kursunoglu
And 2h26'10" 40d23'6"
Deforrest Noriss Hilligus
Cas 1h16'38" 61d54'38"
Defrance and Alyssa
Lib 15h20'36" -7d16'37"
Deftig Mark
Sco 17h18'57" -32d25'51"
DeFy + Riaki
Aql 19h51'50" 12d18'20"
DEG
Ori 5h59'57" 6d30'6"
Degas Ballerina Nina
Psc 1h9'36" 14d50'15"
Degener, Gerhard
Psc 1h14'42" 2d54'6"
Dehami
Uma 10h26'35" 53d55'33"
Deia
Ori 5h56'10" 20d59'45"
Deidle Moonwatyr
Lyn 6h41'51" 58d59'51"
Deidra
And 0h16'4" 38d8'11"
Deidra
And 2h10'7" 38d53'52"
Deidra Ann DeHarts Star
Cap 20h35'39" -16d26'29"
Deidra Elizabeth Marsh
Uma 9h44'19" 49d13'45"
Deidra Fabian
Lib 15h44'3" -18d43'20"
Deidra Lynae
Cap 21h36'18" -10d1'49"
Deidra Nicole Boardwine
And 23h26'8" 48d39'58"
Deidre A. Kelly
Tau 3h54'56" 26d7'21"
Deidre And George Forever
Vir 13h21'38" -0d38'32"
Deidre Ann Hassell
Leo 10h41'6" 14d11'34"
Deidre Ann Meade
And 2h15'33" 50d11'40"
Deidre Anne
Cru 12h8'46" -62d26'32"
Deidre Anne Urso
Ari 3h14'58" 24d25'7"
Deidre Duvel
Cas 1h20'15" 56d1'58"
Deidre Lea Walker
Sgr 17h52'46" -29d31'30"
Deidre Lea Walker
Sgr 17h52'46" -28d54'46"
Deidre Leigh Seidel
Leo 9h32'32" 28d47'33"
Deidre M. Rhea
Crb 15h44'24" 26d39'14"
Deidre Michelle Long
Ori 6h14'7" 13d28'38"
Deidre Muth
Uma 10h17'8" 70d35'10"
Deidre Nicole Duncan
Dra 18h59'20" 54d29'46"
Deidre Pearce Weilano
Sgr 18h27'54" -27d19'58"
Deidre Smith Martons Eternal Star
Sco 17h39'29" -36d12'0"
Deidre's Twinkle
Cas 1h20'15" 61d53'47"
D-Eillot-9
Cnc 8h44'52" 12d18'30"
Deimian Jeffri
Her 16h31'41" 37d56'8"
Deion James Quirk, Our Angel
Cru 12h21'20" -62d38'28"
deiraust
Umi 14h58'29" 77d59'10"
Deirdre
Lib 15h8'6" -27d55'54"
Deirdre A. Mooney
Gem 7h42'32" 32d52'51"
Deirdre and Artie's Star
Vir 12h28'32" 1d36'49"
Deirdre Ann- #1 Sister In The World
Her 17h0'8" 32d17'21"
Deirdre Ann Pfeffer
Uma 11h52'29" 48d27'32"
Deirdre B. Dunham
Gem 7h3'23" 14d20'30"
Deirdre baby
Ari 3h25'58" 27d12'13"
Deirdre Balfe
And 0h41'50" 31d28'4"
Deirdre Berk 10.9.57
Uma 12h34'56" 62d40'36"
Deirdre Byrne
Crb 15h42'49" 27d18'26"
Deirdre Clifford
Uma 10h0'52" 67d31'51"
Deirdre DKWR
Psc 1h11'46" 22d33'32"
Deirdre Donaldson
Ori 5h30'34" -0d17'55"
Deirdre - love you always - Nicole
Cru 12h22'31" -62d24'0"
Deirdre Mary Kennedy
Aqr 20h39'35" -11d33'27"
Deirdre O'Sullivan
Sgr 19h46'55" -12d24'8"
Deirdre Star
Vir 12h46'56" -9d29'21"
Deirdremac
And 2h33'36" 39d16'33"
Deirdre's Sunflower
Vir 12h42'42" 12d2'57"
Deisi
Psc 23h36'26" 4d7'4"
Deisy Karina
Cas 23h28'22" 51d26'6"
Deity
Cam 3h58'27" 65d33'36"
Deiwis Dos Santos
Leo 11h39'15" 26d18'56"
Deizy Yamile
Ant 10h21'29" -30d37'25"
Dej DeJohn
Gem 6h45'58" 17d40'25"
Deja Lynn Wimbish
Uma 11h20'1" 64d46'46"
Deja Rene Sinclair Chavez
Uma 8h57'27" 52d8'41"
Deja Vu Marie Ryder
Gem 6h36'14" 15d14'15"
Dejah
Lib 15h12'36" -7d39'6"
Dejah Anderson October 21, 1991
Equ 21h23'54" 5d44'16"
Dejan & Nangsa
Uma 10h4'3" 68d13'39"
Dejan Nikolic
Lib 15h4'28" -1d3'16"
Dejera Marie Conrad
Uma 10h1'4" 53d10'43"
Dejeun
Cas 0h35'17" 57d53'29"
deji
Vir 13h12'33" -3d53'37"
Deka (Swiss) Privatbank AG
And 1h13'36" 45d11'31"
dekfalk
Ori 6h16'19" 16d41'50"
Dekishimenainoni
Gem 6h46'27" 26d34'41"
DeKlyen Delight
Lyn 9h35'6" 41d6'44"
Del
Tau 5h53'3" 28d12'3"
Del
Aql 20h6'18" 9d2'3"
Del Alan Ryan
Ori 5h28'5" 4d20'3"
DEL Chikara Yoko
Sco 17h49'5" -43d1'52"
Del Diablo
Ori 6h12'14" 18d2'46"
Del e Aziz
Cep 0h17'26" 78d34'51"
Del & Fred Preziosi
Cyg 20h33'49" 39d55'10"
Del Groene
Uma 13h22'11" 53d34'1"
Del Jessen
Uma 11h28'37" 62d1'7"
Del & Margaret Stephenson
Ori 5h52'27" -0d21'45"
Del Sordo
Gem 6h55'26" 21d39'34"
Del&Ta 2005
Uma 10h57'25" 34d55'53"
Dela Cruz Family
Uma 11h41'53" 51d3'14"
Dela + Derrick = Love
Sgr 20h5'25" -38d37'14"
Delaina Renee Angeline Champlin
Mon 6h45'46" -0d43'37"
Delainah's Dream Catcher
And 1h25'38" 39d53'3"
Delaine Betty
Psc 0h17'51" 17d50'47"
Delaine Leggett
And 0h11'22" 44d3'54"
Delainey-Hart
Dra 19h3'41" 65d47'6"
Delana Marie
Aqr 22h37'25" -0d50'40"
Delane
Sgr 20h3'14" -29d26'54"
DeLane Wright
Lyn 8h20'56" 43d6'37"
DeLanea Herrick
Cnc 8h39'33" 23d1'29"
Delanee Rachel Helene Barron
Sgr 18h0'5" -29d2'58"
Delanee Raechyl Johnson
And 23h28'29" 37d10'20"
Delaney Alexis Carthen
Cnc 9h18'27" 26d36'50"
Delaney Amber Pierce
Lib 15h38'1" -17d17'37"
Delaney Ann Blackwell
Lmi 10h20'38" 32d50'29"
Delaney Ann Louise
Cas 1h36'19" 57d47'46"
Delaney Christine Renneker
Aqr 22h40'29" -21d47'34"
Delaney Ella Graef
Tau 5h55'12" 24d46'48"
Delaney Elliott
Uma 11h49'23" 40d12'7"
Delaney Grace Nash
And 2h16'34" 48d10'41"
Delaney Jane
Ari 2h2'47" 14d13'30"
Delaney Jane
Cas 1h27'46" 61d57'48"
Delaney Jay Gray
Leo 11h5'14" 22d16'24"
Delaney Judith Skinner
Aql 18h59'21" 16d30'35"
Delaney Kalin Pulido
Psc 1h0'37" 16d55'38"
Delaney - Keogh
Lyr 18h40'19" 39d49'59"
Delaney Kristine
Vir 13h19'54" -7d56'16"
Delaney Lauren Newman
And 1h10'59" 38d0'49"
Delaney Leigh Franklin "Laney-Bug"
And 1h42'14" 50d7'53"
Delaney Mae
Uma 12h42'20" 56d53'21"
Delaney Maranda
Leo 11h43'51" 25d48'9"
Delaney McKenna Cox
Sco 17h58'21" -38d26'46"
Delaney Paige Wehn
And 2h29'35" 42d41'46"
Delaney Rae Wall
And 0h8'2" 40d59'20"
Delaney Rose
Leo 9h22'43" 30d28'10"
Delaney Rose
Tau 5h49'22" 17d58'57"
Delaney Wismans
Uma 11h30'8" 63d38'12"
delaneyrae
Aql 20h15'58" 5d19'20"
DelaneyRyanStewart
Tau 4h12'0" 22d34'59"
Delani Rose
Vir 13h16'11" 6d51'20"
Delanie Anne Drabek
Mon 6h45'40" -6d33'16"
Delanie Lynn Mullens
Tri 2h31'27" 30d51'25"
DeLanna Beth Robertson
Cyg 19h40'31" 35d48'56"
Delano Cooper Williams
Del 20h40'57" 13d59'23"
Delano G. Olson
Gem 7h20'27" 25d39'53"
Delany Colman
Umi 14h59'35" 73d46'18"
Delao
Umi 16h36'12" 76d17'27"
DelaPierreJoy4Ever
Cyg 20h47'56" 55d2'38"
Delara Yass Shahossini
Lib 14h29'42" -17d50'53"
Delayna Marie Pagan
Sgr 20h13'50" -38d51'26"
Delaynne Ferguson
Lib 15h4'49" -27d50'6"
D'Elbee
Mon 6h53'51" -6d26'16"
Delbert Dale Smith
Ori 5h24'29" -4d48'22"
Delbert Fentress
Ori 5h34'8" -0d41'38"
Delbert J Brown
Cnc 9h8'4" 30d33'44"

Delbert M. Alt
Ori 4h51'28" 11d41'6"
Delbert (Paga) Green
Uma 11h31'19" 43d58'36"
Delbert W. Atkins
Cam 3h49'7" 71d57'38"
Delcarlo "Lil' T.C." Chapman
Sgr 18h32'39" -17d3'28"
Dele & Amber For a Lifetime of Love
Lyn 8h26'28" 45d20'37"
De'Lee
Cas 0h50'41" 57d24'21"
Deleen
Uma 12h37'55" 56d45'22"
Deleila Destiny
And 0h17'45" 40d42'58"
Delena
Uma 11h43'44" 37d47'15"
Delene Gilmore
Cyg 20h54'42" 31d47'27"
DeLeon voor altijd
Sco 16h44'15" -30d41'38"
Deletta W. Boutwell
And 0h21'50" 33d38'30"
DELFIC AMPARO CINTRON LOPEZ
Sco 16h4'8" -21d9'3"
Delfin y Fina
Cas 1h46'10" 55d39'1"
Delfina
Del 20h34'41" 16d37'28"
Delfina D Doran
Vir 13h7'44" 11d59'26"
Delfine Ryder Bauer
Psc 1h2'3" 27d12'0"
Delgado
Lyr 18h29'36" 35d56'12"
DelGallo
Cap 20h28'47" -13d42'18"
Deli
Cet 1h54'36" -8d39'42"
Delia
Aql 20h11'46" -0d27'53"
Delia
Lib 15h38'16" -6d46'21"
Delia
Umi 15h13'50" 76d45'24"
Delia
Cap 21h5'48" -26d59'38"
Delia
Lyn 6h47'57" 50d48'23"
Delia
And 0h20'1" 45d27'59"
Delia
Tau 4h4'18" 7d7'56"
Delia Cecilia
Crb 15h28'41" 27d15'12"
Delia DiLoreto
Cap 21h40'20" -11d4'40"
Delia Elizabeth Nevins
And 0h23'5" 41d45'51"
Delia & Gianluigi
Ori 5h54'24" 22d43'3"
Delia Gonzalez
Uma 11h27'11" 37d24'13"
Delia J. Chapa
Aqr 23h6'21" -8d42'28"
Delia James Hartley
And 1h39'23" 46d15'42"
Delia M. & Caridad J. Loqui-Lopez
Cnc 9h5'40" 32d42'38"
Delia Mae Browne
Cas 0h44'43" 64d3'20"
Delia Magdalena Loqui-Lopez
Cnc 8h27'50" 26d5'41"
Delia Margaret McGrath Garner
Cas 2h18'45" 72d11'8"
Delia Mota
Crb 16h18'24" 39d12'16"
Delia Navarro Lozano
Ori 5h21'26" 12d57'25"
Delia Niedermann
Vul 19h25'54" 24d44'49"
Delia Sables
And 2h34'24" 39d36'41"
Delia Sonya Lorigo
And 23h14'43" 41d47'53"
Delia Stirling
Cas 23h10'44" 55d4'40"
Delia the Navajo Goddess
Gem 7h11'14" 30d35'5"
Deliana Willemijntje Keus
Aqr 22h43'4" 0d34'3"
Deliargyris, Peter
Uma 10h37'38" 65d40'6"
Deliberate Creation
Cyg 21h52'59" 50d49'15"
Delicacy
Oph 17h21'37" 4d25'54"
"Delicate Flower"
Aqr 21h40'24" 0d46'28"
Delicate Rose
Tau 5h28'0" 26d45'43"
Delicia Cotto
Cam 3h56'25" 69d2'28"
Delicia Pennington
Psc 1h17'59" 5d55'29"
Deliciae Valerie
Cnc 8h24'44" 18d51'37"
Délicieuse
Uma 10h8'31" 59d56'8"
Delicious Dee
Lyn 7h7'8" 58d14'39"
Delicious Hugh
Her 17h9'9" 32d50'49"
Delightful Dee
Cyg 20h1'21" 35d0'3"
Delila Rae Our Mother
Per 2h57'36" 52d28'30"
Delilah
Gem 7h1'25" 17d25'45"
Delilah
Vir 14h51'21" 5d33'31"
Delilah
Sgr 19h5'37" -17d52'21"
Delilah
Umi 14h25'4" 77d24'50"
Delilah
Lib 15h6'56" -29d1'6"
Delilah Cooper
Vul 21h17'27" 22d23'2"
Delilah & Rich
Cyg 19h41'47" 54d26'9"
Delilah Standridge Mathis
Aql 18h47'46" 7d57'24"
Delila's Fire
Sgr 18h29'40" -18d33'39"
Delina
Uma 10h15'10" 43d56'1"
Delina Kaitlynn DaSilva
Crb 15h54'47" 36d53'48"
Delinda Haley
Sco 16h26'49" -30d38'14"
Delipenguin
Vir 12h29'27" -4d42'33"
Delise Janay Jones
Uma 9h10'18" 58d29'19"
Delissa Bernadette Crowe
Psc 1h22'53" 17d28'30"
Delissia Nicole Boyd
Cap 21h3'44" -26d24'23"
Dell LaVerne Rabas
Uma 8h45'7" 53d10'45"
Della
Cam 4h31'17" 65d12'30"
Della Ash Ladwig
And 1h46'53" 39d10'6"
Della Beatrice Holley
Uma 14h14'56" 60d28'11"
Della Bella
Leo 10h17'48" 25d10'45"
Della Cohen
Eri 3h17'44" -3d12'8"
Della Doss
Uma 12h16'52" 54d27'59"
Della Faye Burks
Tau 5h24'3" 19d5'18"
Della Maria DiLeo
Vir 13h28'54" 10d38'51"
Della Marie Hodges
Lyr 19h19'11" 28d14'35"
Della Marie Millard 6/04/53
Cas 1h5'19" 63d40'54"
Dellando
Sco 17h27'56" -41d33'24"
Dellannie Phillips
Aql 20h1'56" 15d30'20"
Della's Special Star
Uma 14h10'6" 60d54'51"
Delle
Vir 13h27'47" -13d42'47"
Delle Dunnagan
Uma 10h47'17" 72d44'26"
Dellene Sandra
Vul 20h40'20" 24d57'36"
Delleo
Aqr 22h39'45" -3d52'52"
Dellibub
And 0h8'30" 45d35'41"
Dellie Cotton
Cep 20h39'46" 56d2'57"
Delling
Boo 14h44'5" 53d22'0"
Delma Jack
Umi 16h25'38" 79d40'57"
Delmar Harry Howe
Per 4h20'11" 51d10'13"
Delmar Jarrell
Uma 12h45'39" 56d25'19"
Delmar Keith Griffith
Cep 22h40'22" 65d46'12"
Delmar Leon Simmons, Jr.
Leo 9h33'6" 11d2'35"
Delmar Lewis Hardin
Srp 18h39'44" -0d1'26"
Delmis
Cap 21h16'16" -19d14'9"
Delmy Fragaso
Umi 15h26'29" 71d24'45"
Delmy Guerra
Leo 9h24'3" 11d35'30"
Delmy Payne
Gem 7h7'40" 24d12'29"
Delois Ecklund Price
Sgr 17h50'52" -24d48'19"
Delois Holloway
Del 20h49'5" 7d41'16"
DeLonna
Crb 15h43'38" 38d33'20"
Delook
Lyn 8h22'43" 42d6'36"
Deloras Ann
Psc 1h17'6" 23d6'45"
Deloras "Aunt Tootsie" Ward
Gem 7h47'6" 25d18'8"
Delores
Gem 6h46'30" 26d22'41"
Delores
Ari 2h15'56" 26d24'19"
Delores
Psc 1h12'19" 16d25'52"
Delores
Ori 4h53'9" 14d35'34"
Delores Adele Croak
And 2h36'50" 43d56'49"
Delores Ann Doll Krebs
Uma 11h32'38" 58d29'4"
Delores B. Ahlin
Aqr 22h21'7" -2d51'5"
Delores Davis
Uma 13h4'35" 54d22'27"
Delores Dawn Davis
Cnc 8h46'48" 24d46'40"
Delores "Dee" Maddox
Leo 11h11'46" 11d55'8"
Delores Eileen Humphreys
Cnc 7h56'24" 11d12'36"
Delores Haddad
Sco 16h12'27" -13d25'42"
Delores Helen Calvert
And 0h11'42" 41d13'58"
Delores Imogene Bowen
Sge 19h55'34" 18d54'50"
Delores Lister
And 0h33'8" 28d3'14"
Delores M. Berkelo Doyle
And 0h19'59" 44d40'26"
Delores M. Hanke
Lyn 6h24'3" 59d15'56"
Delores M. Lenon
Ari 2h40'31" 29d10'38"
Delores Naomi
Col 5h21'38" -38d34'59"
Delores O'Bryant Star-Gazer
Cap 20h8'36" -17d1'31"
Delores Shelley Lewis
Cas 1h18'14" 58d49'35"
Delores Sotelo-Gaulrapp
And 1h15'56" 40d54'46"
Delores Summers
Crb 16h10'43" 34d17'7"
Delores Ware
Cas 0h36'3" 56d14'57"
DeloresGordon
Uma 10h53'51" 64d56'17"
Delos Clift
Per 3h26'39" 33d11'17"
Delos W. Kerr
Per 4h38'2" 41d10'33"
Delossantos
Lib 14h56'37" -4d26'22"
DeLou
Gem 6h38'24" 22d36'53"
Delphin
Dra 17h14'24" 68d27'13"
Delphine
Cas 1h45'59" 69d10'18"
Delphine
Uma 11h42'17" 56d0'17"
Delphine
Del 20h34'37" 16d42'12"
Delphine
Ari 3h13'33" 16d20'48"
Delphine 23/05/1958
Del 20h35'8" 19d53'33"
Delphine Bensoussan
Sgr 18h50'56" -19d6'42"
Delphine Bichoffe
Cap 20h20'12" -23d31'53"
Delphine et Frank Webb
Del 20h21'6" 9d42'6"
Delphine Garrido Kowalski
Cap 20h56'39" -19d48'50"
Delphine Gregory S. Mickens
Vir 12h31'4" 1d27'33"
Delphine King
Tau 3h42'59" 23d37'43"
Delphine Lamande Frearson
Del 20h30'24" 13d3'50"
Delphine l'Ange Naissant
Cas 2h49'39" 60d40'16"
Delphine Trinquet
Del 20h34'6" 7d27'29"
Delphinus Teddy
Del 20h29'53" 11d33'56"
Delsy Gaetano
Del 20h47'19" 16d26'39"
Delta D. Combs
Cyg 20h41'47" 34d20'58"
Delta Danielle
Uma 9h14'37" 46d48'46"
Delta Gamma
Uma 9h46'26" 56d51'28"
Delta Gamma - Alpha Phi Chapter
Cas 0h16'50" 55d5'1"
Delta Gamma-Gamma Iota Chapter
Uma 11h22'8" 48d26'5"
Delta Mu
Lac 22h38'17" 53d11'29"
Delta Theta Tau Sorority, Inc.
And 1h3'48" 40d51'19"
Delta Zeta Sorority-Theta Mu
Cas 1h18'32" 57d0'53"
Delta.Star.Love.
And 0h44'54" 41d32'53"
DelTour
Sco 17h43'8" -41d25'39"
Deltrina-Niala
Vir 12h19'10" 3d0'23"
Deluca
Ari 3h15'26" 25d7'50"
DeLude
Ori 6h6'1" 10d27'38"
Delukey
Gem 7h50'34" 32d20'50"
Delverlon
Sgr 17h52'6" -29d4'10"
Delvin Dale Williams
Aql 19h40'55" 4d50'13"
Delvin Nielsen Family Star
Boo 14h37'19" 19d44'43"
Delvin-Hahn
Sco 17h31'39" -36d51'57"
Delvin's Turtle
Cnc 8h44'56" 12d27'41"
Delvis Santiago Rivera
Cam 4h57'44" 64d14'24"
Delvon
Uma 9h49'4" 51d12'25"
Delwyn Leilani Hifo
Uma 10h39'31" 45d21'32"
deLylas IV
Vul 19h50'50" 25d33'8"
DeLynne
Lyr 18h23'47" 39d56'52"
Delzanno Jean Marc
Sgr 18h3'42" -30d5'32"
Dem <3 Bwr
Lyn 7h59'18" 34d6'6"
Demaceo
Lyn 6h30'7" 60d35'10"
Demari & Desaun
Cyg 19h52'27" 39d2'56"
DeMarius Terrell Travis-Bell
Dra 17h3'58" 69d59'28"
Demea Brinsley
Gem 7h19'36" 33d27'6"
Demeka Paramore
And 0h36'51" 30d13'37"
Demeter, Manfred
Ori 5h14'51" -8d13'17"
Demetra
Crb 16h23'34" 27d14'8"
Demetra's Odyssey
Vir 14h24'14" -0d45'37"
Demetri John Karambelas
Cep 22h32'50" 65d3'21"
Demetri Nicholas Strates
Aqr 22h13'12" 1d58'1"
Demetria Lajoyce Haliburton
Cyg 20h54'26" 43d12'47"
Demetrio Otero
Cru 12h6'27" -62d39'2"
Demetrios Ioannis Hannan
Her 18h45'24" 21d11'3"
Demetrius
Tau 4h41'42" 28d46'18"
Demetrius Alexander
Cep 20h10'6" 61d10'40"
Demetrius Battle
Uma 8h44'10" 66d10'2"
Demetrius R. Black Bonnett
Sco 17h44'7" -39d41'51"
Demetrius T Fortson
Aqr 20h47'59" -8d50'55"
Demi
Col 5h38'48" -35d14'42"
Demi
Leo 10h9'26" 21d52'31"
Demi
Leo 9h27'52" 11d46'0"
Demi 11268
And 2h34'24" 44d21'16"
Demi Bourne
And 23h10'1" 39d58'41"
Demi Elisa
Her 17h48'37" 39d5'55"
Demi Leen
Cnc 8h37'48" 29d22'21"
Demibeaucathy
Umi 14h44'55" 71d21'37"
Demitrique Ashton Butchert
Cet 3h7'43" -0d9'11"
Demoiselle Belle Bisous
Tau 4h23'15" 25d57'2"
Demon
Uma 13h18'12" 58d57'33"
DeMond Michael Grant
Sco 16h13'37" -36d22'21"
Demont C. Blue
Aqr 21h23'50" 1d33'8"
DeMori-Spurney J's Shining Stars
Ori 5h39'33" 4d14'40"
Dempsey
Cra 19h3'23" -42d7'28"
Dempsey and Allister Memorial Star
Cru 11h56'51" -56d6'29"
Dempsii Reyanna Rosales
Com 12h43'26" 14d26'45"
Den
Per 2h43'59" 41d3'12"
Den Art
Per 3h43'19" 45d31'9"
Den Den
Ori 5h19'33" 5d40'54"
Den Engenieur
Dra 18h57'52" 51d51'51"
Den Na Lee Fewson
Uma 10h14'45" 53d53'34"
Den - The Light of my Life
Aqr 22h36'16" -22d16'32"
Den wunderbarsten Eltern
Uma 11h48'0" 29d46'15"
DENA
Uma 11h30'5" 48d25'12"
Dena
Sgr 18h20'32" -16d21'59"
Dena
Sco 16h19'15" -37d41'32"
Dena 40
Sgr 17h58'39" -29d54'10"
Dena Berg
Cap 21h24'6" -15d52'40"
Dena Honu
And 23h21'25" 48d17'39"
Dena K. Dionne
Lib 15h20'32" -26d37'9"
Dena Lynn Wendling
And 0h34'20" 33d14'48"
Dena Mae Crisler Higgerson
Lyr 19h3'26" 42d20'0"
Dena Marie George Obney
Vir 12h30'16" -1d49'0"
Dena Marie's Star
And 23h20'39" 41d52'6"
Dena Michelle Scott
Uma 11h20'56" 35d16'9"
Dena Mitchell
Psc 1h47'45" 8d59'46"
Dena Renee
Uma 11h48'22" 31d52'58"
Dena Rose
Cap 20h43'2" -20d44'31"
Dena Smith
Leo 10h10'53" 16d51'59"
Dena Stacy Maria Zakaib-Kunelaki
Her 16h20'38" 17d9'9"
Dena0830
And 0h41'15" 33d14'3"
Denae Allaire
Aqr 20h48'44" 0d52'18"
Denae Doyle
Umi 16h58'59" 85d31'17"
Denaki
Gem 7h31'47" 26d39'57"
Denali
Leo 11h30'14" 23d51'50"
Denali
Per 3h9'35" 56d12'16"
Dena's Gem
And 1h52'16" 45d59'7"
Dena's Star
And 23h24'13" 47d0'19"
Den-Bob
Tau 5h1'17" 19d22'21"
Denbo's Luvbug
Cnc 8h1'17" 25d5'2"
dendav
Sco 17h52'56" -35d56'53"
Dene Bryant Keely
Cas 23h48'52" 58d11'38"
Dene Truax
Uma 11h55'17" 35d7'15"
Denee
And 23h15'1" 48d39'28"
Denee'
Lib 15h9'24" -5d55'42"
Deneen
Ori 5h54'16" -0d44'19"
Deneen
Tau 3h35'43" 8d47'33"
Denemily
Uma 10h36'11" 53d51'56"
Denerica Lynn
Cas 0h45'28" 61d12'9"
Denes, Lia-Smaranda
Uma 13h48'54" 56d23'35"
Denese R. Dale
Cas 0h46'29" 56d56'19"
Deneshia's Star
Sco 16h12'52" -12d32'45"
DENG Xiao Ping China
Umi 17h36'39" 84d50'15"
Deni
Cyg 21h34'53" 50d41'48"
Deni
Uma 11h24'54" 47d48'7"
Denia
Psc 0h19'29" -0d5'43"
Denial Dee Norris
Uma 11h41'25" 31d28'28"
Denica
Umi 14h45'26" 73d59'37"
Denice E. Rissetto
Gem 7h48'0" 25d20'37"
Denice Elizabeth Gaunt
And 0h41'59" 32d56'48"
Denice Mesuro
Mon 6h49'58" -0d6'21"
Denice Wong (Bchu Bchu)
And 23h20'3" 48d18'24"
Denice Zayas
Peg 21h41'56" 7d58'36"
Deniece Alvarado
Uma 13h35'13" 55d0'24"
Denielle
And 0h41'45" 36d8'44"
Deniese Reding
Sco 17h29'24" -42d27'22"
Denika-May
Gem 7h19'18" 24d27'0"
Denim
Ori 5h24'35" 14d33'28"
Denim Rose
Lmi 10h1'27" 37d19'24"
DenineAnnRenz
Lyn 8h14'21" 37d37'46"
Denis
Cnc 8h5'19" 25d31'21"
Denis 07191994
And 0h53'17" 45d15'5"
Denis and Peggy Stillwell-O'Connor
Uma 8h49'34" 71d5'26"
Denis and Sara
Uma 10h16'47" 57d1'30"
Denis Big-D Messemer
Cap 21h28'42" -8d41'33"
Denis Brosnan
Sgr 17h52'56" -17d16'17"
Denis Campbell
Cas 0h14'59" 57d21'42"
Denis Cregan, Golden Co.Tipperary
Uma 8h58'43" 50d54'43"
Denis D. Dean
Ori 5h45'29" 4d34'50"
Denis Flynn
Lyn 7h40'5" 37d46'7"
Denis Heath
Boo 14h44'43" 37d25'39"
Denis Henry Pomroy
Sco 17h17'21" -32d47'31"
Denis Hubert Franklin
Lib 15h26'53" -5d43'59"
Denis & Jacqui - September 2nd 1978
Crb 16h6'9" 36d48'32"
Denis James Paquette
Aql 19h59'46" 14d49'44"
Denis Johnson
Lyn 8h5'32" 48d16'56"
Denis Leclair
Uma 10h28'53" 60d20'44"
Denis Pourcelet
Uma 9h54'59" 42d59'3"
Denis Q
Cma 6h34'59" -17d51'14"
Denis Quaife
Tau 5h32'29" 27d59'54"
Denis Reckner's Memorial Star
Gem 7h35'22" 33d38'25"
Denis S. Lattanzi
Cam 3h54'43" 69d0'43"
Deni's Star
Uma 11h49'24" 62d37'35"
Denis' Star - Wonderful & Wise
Cep 1h53'15" 85d13'42"
Denis & Susan King 10th Anniversary
Cep 22h17'35" 63d53'44"
Denis V. DiGrazia
Per 3h20'22" 45d45'19"
Denis Wilfred Collins
Cru 12h48'16" -61d1'27"
Denis Withington
Uma 11h32'29" 53d36'1"
Denisa 02 15
Aqr 22h6'4" 0d19'22"
Denisa Krivosikova
Psc 0h50'28" 28d53'57"
Denise
Her 17h14'14" 16d21'15"
Denise
Tau 4h54'12" 28d40'2"
denise
Gem 6h46'23" 23d11'38"
Denise
Uma 11h15'54" 28d38'20"
Denise
Uma 11h27'54" 29d21'39"
Denise
Tau 3h35'11" 0d5'52"
Denise
Leo 9h44'12" 7d11'36"
Denise
And 23h13'31" 47d34'42"
Denise
Gem 7h12'5" 32d42'11"
Denise
And 0h56'46" 34d38'56"
Denise
Aur 4h49'32" 35d42'19"
Denise
And 1h26'32" 41d5'41"
Denise
Uma 12h43'0" 55d20'17"
Denise
Cas 23h59'28" 58d40'42"
Denise
Cas 0h42'55" 64d32'33"
Denise
Umi 13h56'6" 74d18'28"
Denise
Uma 8h16'44" 61d29'36"
Denise
Psc 0h24'25" -1d26'48"
Denise
Lib 14h48'47" -3d53'39"
Denise
Vir 14h18'24" -9d57'18"
Denise
Aqr 22h8'7" -16d34'17"
Denise
Uma 10h18'59" 50d41'54"
Denise 40
And 2h5'24" 37d2'7"
Denise A Ohawk
Sco 17h21'51" -41d26'56"
Denise A.C.
Lyr 18h40'10" 38d26'19"
Denise Alida
And 1h4'25" 41d0'50"
Denise Allen
Sgr 19h1'5" -14d33'49"
Denise Allison Kane
Lib 15h34'25" -25d18'13"
Denise and Bryan's Serenity
Lyr 18h28'2" 27d4'37"
Denise and Dave Forever
Gem 7h11'48" 21d8'32"
Denise and Don
Cyg 21h14'12" 43d18'7"
Denise and Eddy's Star
Her 17h53'27" 24d37'54"
Denise and Eric Moody
Uma 9h45'46" 59d42'26"
Denise and Michael
And 0h41'10" 43d27'19"
Denise and Ryan's love star
Vir 13h16'46" 4d46'28"
Denise & Andy
Cyg 20h20'47" 56d46'33"
Denise Angel Tye
And 1h23'54" 41d20'53"
Denise Ann
And 0h36'19" 42d23'0"
Denise Ann Bainbridge
Uma 11h19'55" 60d39'4"
Denise Ann Beyer
Vir 13h26'10" -13d26'10"
Denise Ann Marie
Cnc 9h10'24" 32d24'20"
Denise Ann Norberg
Aqr 22h53'48" -7d2'23"
Denise Ann Pegg Shea
And 0h34'35" 37d41'38"
Denise Ann Vaccarino Meuser
Cas 0h18'38" 55d53'36"
Denise Anne Donaldson
Psc 1h26'37" 24d50'4"
Denise Annette Millard
Boo 14h59'53" 34d18'18"
Denise Armas Harris
Cnc 8h47'10" 23d33'23"
Denise Bagwell
Cap 21h39'2" -18d31'23"
Denise Baird - Princess
And 23h53'36" 39d22'52"
Denise Banta Mother's Day 2006
Cas 23h34'16" 55d6'59"
Denise Barker
Cyg 19h58'11" 34d41'23"
Denise Bawden
Cas 1h47'45" 65d29'8"
Denise Bell
Leo 11h2'31" -1d33'7"
Denise Bentley
Cas 0h38'53" 60d39'50"
Denise Berti & Gary Dennis
And 1h26'24" 48d44'34"
Denise Borland
Crb 15h32'21" 34d23'29"
Denise Bosnack
Cas 0h12'35" 63d16'4"
Denise Bosshardt
Ori 6h15'1" 13d52'54"
Denise Bowers
Cas 23h40'50" 54d15'32"
Denise Boyer
Cap 20h28'45" -11d2'6"
Denise Bradley
Cas 23h40'52" 58d26'45"
Denise Buisman
Uma 14h18'11" 56d39'41"
Denise C. Henderson
And 0h10'55" 31d57'8"
Denise C. Youngblood
Cap 20h33'52" -23d45'44"
Denise Cameron
Psc 23h59'0" 9d23'17"
Denise Capazzi
Gem 7h6'47" 21d16'12"
Denise Cariss
Cas 23h58'42" 53d22'33"
Denise Carol Padula
Cas 1h36'50" 64d11'3"
Denise Carol Slayko
Sco 16h41'45" -29d14'30"
Denise Carrie Frazier
Cam 5h44'37" 61d21'18"
Denise & Cheryl Ann's Guiding Light
Uma 12h17'57" 62d45'6"
Denise & Christian
Uma 11h49'25" 52d48'20"

Denise Clareen Schneider
Angel Star
Cnc 9h16'55" 25d53'10"
Denise Cleary
Cas 1h40'10" 71d46'1"
Denise Corbit "Little Dee"
And 1h58'38" 45d54'44"
Denise Coulson
Cap 20h25'43" -23d40'44"
Denise Cox
Vir 13h15'23" 5d35'6"
Denise Crawford
Gem 6h37'5" 16d38'34"
Denise Cross
Sco 17h51'5" -36d21'1"
Denise Davis
Boo 14h46'2" 33d8'3"
Denise Dawn Linkous
Lib 14h29'28" -20d42'19"
Denise(Deedee baby)
Oswald
Uma 9h0'59" 58d57'12"
Denise DeVita & Julian
Gedroye
Peg 21h41'59" 21d56'35"
Denise Dowd Welch
Lib 15h33'0" -25d22'26"
Denise Doyle
Gem 6h51'16" 32d44'44"
Denise Dufour
Crb 15h50'48" 27d41'44"
Denise Dunlop
And 2h12'5" 45d1'11"
Denise E. Dennis
Uma 11h52'14" 58d4'35"
Denise E. Glatfelter
Cas 23h51'21" 57d23'32"
Denise E. Holiway
Leo 10h43'37" 18d12'3"
Denise E. Sandlin My
Shining Star
Cam 3h29'19" 59d45'58"
Denise E. Taylor
Sco 17h23'19" -37d58'58"
Denise Ellis
And 0h15'10" 39d40'32"
Denise Etzkorn
Gem 7h30'21" 16d47'49"
Denise F. Campbell
Per 3h51'45" 32d59'38"
Denise Famularo
Crb 15h21'53" 29d34'53"
Denise Faulkner-Edwards
Sco 16h42'59" -31d31'47"
Denise & Florian
Mon 7h3'47" 5d23'24"
Denise Ford Sullivan
Leo 11h14'38" 17d11'51"
Denise Franklin
Vir 11h43'45" 4d1'26"
Denise Gaymer
Eri 4h33'15" -21d28'0"
Denise Genine Southwick
Cas 1h39'46" 62d42'9"
Denise Glessner (I love you
most!)
Leo 10h21'40" 17d11'32"
Denise Grace Minaberry
Aql 19h55'39" 15d30'51"
Denise Hagoriles Ruiz
Lib 14h56'31" -14d58'40"
Denise Harper
Lib 15h35'28" -27d8'25"
Denise Hays
Psc 1h33'34" 27d41'47"
Denise Heffner
Psc 0h6'15" 5d54'33"
Denise Heflin
And 0h44'19" 44d5'40"
Denise Herring
Lib 14h47'33" -11d48'51"
Denise Huerta
Vir 12h55'35" 5d41'33"
Denise Hutson
Cam 6h48'10" 67d10'27"
Denise & Ignazio
Sge 19h55'48" 19d9'1"
Denise & Ivan Watts
Cyg 21h44'13" 33d12'52"
Denise J. Lom....Stardom
at Last!
Lib 14h54'53" -3d43'44"
Denise J. Mazzapica
Uma 10h53'35" 49d54'59"
Denise J. Ure
Aqr 22h10'0" -13d33'23"
Denise Jacque DeNofrio
Psc 1h13'26" 23d55'11"
Denise Jakob
Cas 1h17'12" 56d59'59"
Denise James
Sgr 19h11'1" -15d17'58"
Denise Jeanne Golden
Ari 2h24'17" 21d5'5"
Denise & Joseph Marino
Cyg 20h21'58" 52d28'56"
Denise Josephine Kissane
And 0h51'16" 44d17'16"
Denise Joy
Ori 5h5'41" 4d38'36"
Denise Joy Wolfgang
Cooper
Lib 15h14'42" -15d59'8"
Denise K Eleftheratos
Leo 10h7'1" 18d10'47"
Denise Karen Worrell
Crb 16h18'46" 39d20'4"
Denise Käsler
Ori 6h19'49" 7d9'21"
Denise Kehoe
Aqr 21h9'0" -3d39'19"
Denise Kranich
Tau 4h10'3" 27d32'50"
Denise L. Airola
Mon 6h44'57" 9d37'14"
Denise L. Bizjack
Gem 6h30'31" 26d14'5"
Denise Lemitte Koerber
Uma 11h5'6" 64d25'45"
Denise Lerma
Cnc 8h38'53" 13d14'50"
Denise Lewis Dempsey
Peg 21h44'55" 8d21'11"
Denise Lowrey
Uma 13h13'28" 62d50'58"
Denise Lydecker
Tau 4h47'0" 20d49'29"
Denise Lynn
And 2h30'24" 44d20'47"
Denise Lynn Delane
Ori 6h2'23" -3d47'0"
Denise Lynn Kokis
Cap 20h20'11" -12d32'56"
Denise Lynn Rothe
Vir 13h21'54" -22d30'5"
Denise Lynn Weber
Cnc 8h50'26" 7d27'6"
Denise Lynne Mager
Uma 11h9'30" 69d53'0"
Denise M. Amirto
And 23h13'1" 47d31'34"
Denise M. Bittenbender
Lyn 9h11'53" 45d54'9"
Denise Mack
Uma 10h33'54" 49d24'51"
Denise Makofske Luhrs
Tau 4h13'45" 21d27'27"
Denise Mallin
Aql 19h52'43" 11d24'41"
Denise Mancini
And 0h22'29" 45d46'54"
Denise Manczyk
Uma 9h14'45" 51d46'50"
Denise Marie
Uma 10h27'31" 63d4'43"
Denise Marie
Uma 9h34'4" 64d5'21"
Denise Marie Bates
Tau 5h58'21" 25d59'26"
Denise Marie Berger
"McClure"
Ori 5h10'37" 7d38'50"
Denise Marie Brockmyer
Uma 10h32'9" 54d37'43"
Denise Marie Cordovano
Cas 1h20'51" 60d52'5"
Denise Marie Drum
Gem 7h15'10" 33d10'58"
Denise Marie Egan
Sgr 19h5'17" -15d17'22"
Denise Marie Perez
Cnc 8h20'28" 10d55'1"
Denise Marie Shaft
Aqr 21h52'45" -1d33'20"
Denise Marie Viau
Per 4h43'45" 46d45'19"
Denise Marie Washington
Ari 2h10'23" 25d58'19"
Denise Maru
And 2h20'54" 47d7'23"
Denise Mascaro
And 1h10'36" 33d52'50"
Denise & Matthias
25.11.2001
Lac 22h50'19" 38d43'50"
Denise McIvor
Lib 14h34'32" -20d30'2"
Denise Mellins
Cas 0h20'27" 51d47'49"
Denise Mesa
Cyg 19h56'46" 30d41'59"
Denise Mi Amor
Cap 20h56'2" -15d6'48"
Denise Michelle Kelsey
Lib 15h1'35" -10d9'10"
Denise Mills
And 1h45'7" 42d48'42"
Denise Miriam
Umi 15h43'48" 79d40'31"
Denise Mitchell
And 23h40'9" 36d17'59"
Denise Moran
Crb 16h8'8" 37d16'40"
Denise Morquecho
Vir 13h7'2" -12d10'9"
Denise Murphy
Psc 1h9'56" 27d23'3"
Denise My Taylor Girl
Cas 23h31'9" 57d42'18"
Denise My Sweetness and
Light
Cap 21h32'7" -17d42'12"
Denise Nana Jacobi
Aqr 21h42'45" 0d48'45"
Denise Nardozza
Uma 9h55'5" 48d1'54"
Denise Nebula
Cyg 20h25'18" 43d14'5"
Denise Nelson
Lyr 19h3'40" 25d42'1"
Denise Nelson Schneider
Ari 2h16'36" 12d59'18"
Denise Nicole Filipone
Sgr 18h51'40" -30d2'17"
Denise Nicole Stevens
Cas 1h44'6" 62d55'2"
Denise Nip
Cet 3h14'41" -0d42'38"
Denise "Nise" Monica Scott
Cmi 7h22'47" 3d16'12"
Denise O'Donnell
Uma 10h5'22" 48d46'12"
Denise O'Donoghue
Cyg 19h51'19" 38d1'3"
Denise Owens
Lyn 7h27'35" 56d36'54"
Denise Padilla
Cam 7h47'20" 63d53'10"
Denise Patricia Sorrentino
Ari 1h47'36" 17d34'48"
Denise Paula
And 1h42'57" 45d25'3"
Denise Pearson
Sco 17h50'27" -32d4'41"
Denise Penrod
Cas 23h36'57" 56d12'2"
Denise Perfetto
Cnc 9h9'23" 27d41'55"
Denise Petrina Jordan
Cyg 20h5'0" 57d1'7"
Denise & Pfinster Baby
Lep 6h5'18" -25d10'18"
Denise Pilcher
Cnc 8h5'42" 14d12'41"
Denise Pilkington
Cap 20h16'4" -16d11'0"
Denise Pitzen
Lyr 18h33'2" 37d2'53"
Denise Pocol
Cap 21h49'39" -18d19'50"
Denise Prainito
Del 20h36'47" 15d22'50"
Denise Przybylski
Gem 7h44'13" 33d33'22"
Denise Purzycki
Tau 5h25'25" 17d59'18"
Denise Rae
Sco 17h13'30" -32d24'45"
Denise Rene Tate
Cyg 19h39'8" 35d22'42"
Denise Renee
Sco 16h10'7" -9d38'51"
Denise Renee Bakaitis
Cap 21h54'15" -10d34'57"
Denise Renee Coyle
Psc 1h36'16" 18d53'44"
Denise Renee Heim-
Pazdan
And 2h23'16" 48d2'50"
Denise Renee Persico
And 0h50'33" 41d18'15"
DENISE RENEE
QUINONES
Cnc 8h42'37" 31d8'22"
Denise Rigney
And 23h17'44" 43d10'56"
Denise Rinaldo
Gem 6h5'54" 22d26'35"
Denise Rininger
Lyn 8h1'43" 38d30'59"
Denise Romeo Lyon
Tau 4h40'43" 19d3'50"
Denise Rose
Cyg 21h47'54" 53d21'12"
Denise & Ryan
Lib 15h36'58" -15d23'28"
Denise S. Bardong
Uma 8h34'21" 72d13'8"
Denise Seljen
And 23h56'57" 43d37'7"
Denise Sevean
Cnc 8h46'54" 15d59'29"
Denise Silverman
Cas 1h9'53" 55d2'47"
Denise Simmons
Psc 0h38'36" 8d45'30"
Denise Soble
Uma 8h59'58" 62d9'12"
Denise & Stefan *Ewigi
Liebi*
Boo 15h34'51" 45d56'54"
Denise Stephanie Callahan
Lyn 9h19'22" 38d3'32"
Denise Strebel
Boo 15h4'57" 13d47'51"
Denise Sullivan Lynch
"Gracie"
Sgr 19h50'40" -12d37'50"
Denise Super Fine Sugar
And 23h23'14" 41d18'33"
Denise Swasty
And 0h14'3" 45d35'26"
Denise Taylor
Uma 10h33'12" 68d53'30"
Denise The Shining Star
Holstein
Tau 4h20'55" 26d2'19"
Denise Thomas
Cyg 20h50'23" 38d16'34"
Denise (Tish) Triola
Cap 21h35'48" -10d23'30"
Denise Trainor
Vir 11h41'2" -2d58'59"
Denise Tri Sierra Franks
And 2h27'41" 45d40'51"
Denise Turcotte
Sgr 18h0'26" -30d50'59"
Denise Vaccaro
And 0h27'54" 42d8'55"
Denise Vaniadis
Cas 0h31'20" 52d34'44"
Denise Vera
Sco 16h13'55" -16d33'29"
Denise Wagner
Per 4h29'12" 31d37'30"
Denise Watkins
Gem 6h37'6" 22d53'44"
Denise Widmer
Lac 22h57'17" 43d9'48"
Denise Yvonne Gannon
Sco 17h58'30" -38d47'52"
Denise, The Star of
Heaven's Gate
Sgr 19h5'38" -17d12'41"
Denise-1
Cnc 9h10'42" 32d26'27"
Denise1965
Cas 1h12'37" 62d49'14"
DenisEFrunzPerle
Uma 12h35'21" 58d38'6"
Denise's Angel
Vir 13h17'28" -14d4'47"
Denise's Birthday Star
Gem 7h9'23" 27d13'44"
Denise's Heavenly
Diamond
Ari 2h20'59" 12d48'22"
Denise's Joy
Gem 6h29'36" 21d25'57"
Denise's star
Cnc 8h40'56" 15d58'18"
Denise's Star
Srp 18h19'41" -0d13'56"
Denise's Star
Uma 8h35'34" 60d22'17"
Denisse Aljure
Gem 6h32'22" 14d46'11"
Denisse & Kevin Eternally
Ori 6h16'45" 10d5'57"
Denisse Rivera
Lib 15h12'29" -28d10'19"
DeNita's Little Jonah
Lyn 8h17'49" 41d56'23"
denitsa
Uma 11h41'27" 52d4'35"
Deniz
Ari 3h15'5" 28d21'31"
deniz
Cam 4h41'21" 62d35'55"
DENIZ
Uma 9h6'47" 61d18'59"
Deniz Arkan
Cyg 20h55'0" 32d39'8"
Deniz Töreli
Del 20h48'42" 3d3'17"
DenJoan
Cyg 19h57'7" 31d33'33"
Denman Family, Suwanee
GA
Crb 15h32'27" 29d29'14"
Denna Anne Ybarra
Leo 9h33'38" 10d37'45"
Dennig 25
Cyg 21h22'9" 46d36'12"
Dennis
Her 17h35'26" 37d1'7"
Dennis
Aqr 22h44'1" 1d47'41"
Dennis
Leo 11h46'54" 24d6'13"
Dennis above us shining
down
Cen 11h35'42" -48d35'26"
Dennis Alan Bierbaum II
Lib 14h57'40" -4d36'59"
Dennis Anatolievich
Seregin
Aqr 23h12'22" -9d15'13"
Dennis and Connie
Constellation
Sge 19h50'44" 17d40'49"
Dennis and Denise Guidera
Lyn 8h42'43" 34d3'18"
Dennis and Dianne Dugan
Lyr 18h45'35" 41d5'51"
Dennis and Elaine LAM
Cap 20h42'55" -15d17'50"
Dennis and Elizabeth
Allison
Cyg 19h40'57" 39d5'1"
Dennis and Kathi
Wasserman
Cyg 20h54'41" 47d29'31"
Dennis and Pamela Collard
11.1.80
Sco 16h12'39" -15d3'8"
Dennis and Penny Johnson
Cyg 19h59'20" 56d7'19"
Dennis & Andrea Gray
Cyg 21h52'31" 49d59'55"
Dennis Andrew Beris
Cru 12h40'16" -58d10'33"
Dennis Ann Blake
Uma 11h22'56" 54d29'38"
Dennis Ash
Per 2h41'32" 57d2'10"
Dennis Ashley Shackelford
Srp 18h38'32" -0d55'43"
Dennis' Astellas Star
Gem 7h9'6" 34d21'23"
Dennis Augusto DaSilva
Lyn 9h0'3" 46d1'28"
Dennis Bailey
Lib 15h9'18" -5d50'24"
Dennis & Barbie Jones
Cyg 19h47'34" 29d42'33"
Dennis Bartholomew
Her 16h45'19" 31d9'10"
DENNIS BOND
Uma 9h42'14" 70d34'41"
Dennis Bower Connell
Uma 11h50'26" 37d22'28"
Dennis Brady
Her 18h17'20" 17d34'33"
Dennis Brown
Uma 13h19'59" 53d10'26"
Dennis C. Gebhardt
Her 18h44'46" 21d35'12"
Dennis Calligan
Lyr 19h10'15" 37d27'51"
Dennis & Cass G3 Forever!
Sgr 18h35'33" -26d46'32"
Dennis Castro a friend and
star
Sgr 18h39'20" -17d2'19"
Dennis Connors
Uma 12h45'14" 62d21'1"
Dennis & Corrine's Silver
Star
Cyg 19h39'9" 33d49'56"
Dennis - Courage of the
Wolf
Lup 15h44'21" -32d4'40"
Dennis Cunningham
Uma 12h27'59" 53d16'53"
Dennis D Bergmann
Gem 6h51'25" 21d40'32"
Dennis Dale Gaddy
Sco 17h51'7" -32d38'25"
Dennis Dale Green
Per 3h45'42" 45d10'50"
Dennis Delp
Crb 16h14'49" 37d36'44"
Dennis (Denny) Fisher, Jr.
Per 3h11'21" 55d1'44"
Dennis "Denny" Woodward
Ori 5h31'17" 0d28'20"
Dennis Devon Harris Logan
Lib 14h50'17" -1d11'34"
Dennis & Dianne Star
Uma 11h12'8" 57d33'55"
Dennis Dirks
Her 16h40'23" 34d21'52"
Dennis E. Ewing
Crb 16h7'25" 38d56'2"
Dennis E. Frenier Sr.
Aqr 21h15'14" 2d1'56"
Dennis E. Nelson
Cyg 20h20'58" 43d20'16"
Dennis Earl Keefer
Cap 20h53'23" -21d52'26"
Dennis Edward Bauer
11/17/2005
Uma 9h56'21" 43d46'15"
Dennis Elroy Smith
Vir 12h57'41" 0d1'47"
Dennis & Elvire
Cyg 20h47'13" 45d24'50"
Dennis Evans 2004
Leo 10h10'53" 10d37'58"
Dennis F. Gillum
Boo 14h24'29" 25d47'44"
Dennis Farr
Ori 5h40'42" 3d15'50"
Dennis Fedorov
Sgr 19h23'34" -14d19'54"
Dennis Ferranti
Ori 5h37'20" -0d56'51"
Dennis Fisher
Her 17h3'14" 47d0'21"
Dennis Fitzgerald
Lyn 7h32'10" 37d16'6"
Dennis Flores
Ori 5h28'53" 2d22'26"
Dennis Francis Vanbergen
Per 4h25'8" 32d19'31"
Dennis Frederick George
Beard
Per 3h59'53" 38d34'52"
Dennis G. Carr Star
Lmi 10h39'13" 30d54'12"
Dennis G. DeClark
Ari 3h0'9" 19d10'0"
Dennis Garboden
Tau 4h11'33" 11d15'37"
Dennis Gary Jordan, Jr.
Gem 6h45'44" 33d34'17"
Dennis George Harris
Dra 18h13'8" 73d34'41"
Dennis Giovannelli
Leo 10h34'23" 21d0'17"
Dennis "Gramps" Parbst
Ari 2h22'32" 24d54'19"
Dennis Green
Uma 13h30'19" 58d50'17"
Dennis Gregory
Cap 20h22'32" -22d43'3"
Dennis H. Acer
Cyg 20h7'26" 34d5'2"
Dennis H. Clatterbuck
Per 3h18'0" 43d57'30"
Dennis H. Husband
Ari 3h4'20" 18d28'34"
Dennis Hanson
Aql 19h14'37" 15d36'52"
Dennis Harrell Smith
Aql 19h32'37" 1d37'34"
Dennis Hernandez
Sco 16h12'52" -20d58'43"
Dennis Hosey
Leo 11h39'46" 25d11'56"
Dennis House 020547
Per 3h33'8" 49d48'34"
Dennis J. Blanchett
Equ 21h18'3" 10d7'30"
Dennis J Broit
Cru 12h24'2" -57d8'45"
Dennis J. LeClert KA6JLJ
Psc 1h18'10" 16d34'1"
Dennis J. Meyer
Lib 14h52'41" -1d50'21"
Dennis J. Oliva
Psc 0h56'52" 6d53'31"
Dennis J. Sosinski
Uma 11h53'23" 44d0'38"
Dennis J. Walsh Jr.
Uma 10h55'11" 50d0'23"
Dennis James Wathen
Cap 20h18'13" -14d19'28"
Dennis Jarrett Florek
Leo 9h27'17" 25d5'2"
Dennis Jeffrey Klar
Sco 16h45'55" -31d5'0"
Dennis Joe Brown
Ori 5h48'53" 3d17'15"
Dennis Jon Alexander
Sco 17h46'48" -32d15'34"
Dennis Joseph Campbell
Umi 16h15'14" 77d16'9"
Dennis Joseph Trchka
Dra 18h25'4" 51d18'27"
Dennis Jr., Tyler and
Sydney Dunne
Per 4h29'10" 49d25'25"
Dennis & Julie33
Cyg 19h36'56" 38d5'16"
Dennis Junka
Lyn 7h39'54" 50d7'32"
Dennis Karl Jones
Cap 20h27'52" -12d39'56"
Dennis & Kathy
Cyg 20h23'45" 43d35'11"
Dennis Keith Fuller
Her 17h22'20" 32d2'59"
Dennis Kelly
Umi 9h53'29" 89d17'31"
Dennis & Kelly Brandow for
Eternity
Cyg 19h49'26" 33d39'17"
Dennis King
Sgr 18h25'36" -32d13'21"
Dennis & Kirsty's Mile High
Club
Cyg 20h25'28" 51d11'52"
Dennis Koutoulogenis
Psc 1h17'19" 31d25'2"
Dennis Kozior
Uma 11h5'16" 56d29'14"
Dennis L. DeVotie
Per 1h45'29" 51d18'38"
Dennis L. Kogod
Her 17h47'56" 15d28'7"
Dennis L. Rick Jr.
Cap 20h20'17" -13d7'37"
Dennis Lawrence Knight
Her 17h56'20" 18d22'3"
Dennis Lawrence Roux
Psc 0h51'23" 8d23'11"
Dennis Lee McMahon
Cnc 8h17'32" 11d22'17"
Dennis Leo Kowalak
Uma 9h36'38" 66d22'55"
Dennis Lewandowski
Sgr 19h25'5" -24d30'42"
Dennis Lewis
Per 3h49'35" 32d28'12"
Dennis Littlefield
Aqr 21h34'27" -0d38'32"
Dennis & Lori Dryman
Ara 17h21'13" -52d4'21"
Dennis - LOVES - Manny
Cyg 20h56'25" 38d40'57"
Dennis Lynn Kyle
Psc 1h42'31" 13d45'29"
Dennis M Devlin, Jr.
Aur 5h50'43" 45d52'21"
Dennis M. Nestved
Per 3h15'23" 53d22'38"
Dennis M. Olmstead
Leo 9h38'6" 27d40'33"
Dennis M. Quinn
Uma 13h43'44" 54d7'14"
Dennis M. Sullivan
Per 3h10'53" 43d58'47"
Dennis M. Tutewohl Jr.
Aqr 23h23'32" -7d15'21"
Dennis M. Walker
Sco 17h24'6" -31d56'11"
Dennis Makoto Yamachika
Jr.
Lib 15h39'40" -28d22'52"
Dennis Malone, I love you
Dad, Matt
Cep 21h25'33" 61d2'56"
Dennis Maloney
Uma 10h17'57" 46d24'48"
Dennis McDaniel
Umi 14h12'10" 68d25'49"
Dennis McKay
Crb 15h33'12" 25d47'8"
Dennis Meehan
Uma 10h49'17" 66d1'49"
Dennis Merchant
Cru 12h32'16" -58d12'56"
Dennis Michael Dreher
Leo 11h42'26" 26d38'2"
Dennis Michael Foy
"Hondo"
Per 4h4'39" 35d1'10"
Dennis Michael Kolodge
Tau 3h44'53" 28d45'21"
Dennis Michael Patrick
Per 3h12'23" 55d13'52"
Dennis Michael
Zamiatowski
Aql 19h32'26" 10d26'43"
Dennis Miles
Per 3h22'43" 48d36'17"
Dennis Morrow
Aql 19h31'31" 7d58'44"
Dennis Neal Beeler
Lib 14h50'26" -3d23'20"
Dennis Noah Coletta
Dra 18h58'17" 66d32'54"
Dennis Norman
Cap 21h37'58" -20d10'8"
Dennis P. Circo
Tau 4h3'54" 8d48'49"
Dennis P. Hill
Cyg 19h59'52" 32d13'20"
Dennis P McMahon
Cap 20h26'58" -14d31'11"
Dennis P. Wilkins
Aql 19h9'39" 13d49'0"
Dennis Patrick Doyle
Ori 5h46'24" 8d0'41"
Dennis Patrick Roe
Dra 16h42'48" 55d58'59"
Dennis & Paula - Now and
Forever
Aql 19h25'57" -0d3'22"
Dennis "Pop" McDonald
Per 3h6'18" 42d51'38"
Dennis Proia
Vir 13h45'16" -5d13'44"
Dennis R. Hattaway
Ori 6h8'13" 2d27'5"
Dennis R. Ingram
Lib 15h21'17" -26d52'1"
Dennis R. Perrin
Aur 5h45'52" 51d58'23"
Dennis R. Robinson
Tau 5h53'37" 25d24'1"
Dennis Ray Bridgeforth
Vir 12h28'26" 5d40'10"
Dennis Ray Mundine
Aql 19h22'23" -0d8'3"
Dennis Ray Super Grandpa
Aqr 20h56'4" -11d48'26"
Dennis Raymond Cliff
Uma 9h26'19" 47d53'6"
Dennis Raymond Young
Cep 22h53'7" 64d23'22"
Dennis Rega
Lib 15h22'29" -8d9'19"
Dennis Richardson
Uma 9h54'23" 45d26'39"
Dennis Rogers 40th
Birthday Star
Per 4h44'20" 46d41'6"
Dennis Ross Martin
Leo 10h1'24" 21d8'36"
Dennis S. Lopez
Uma 8h47'25" 62d26'7"
Dennis & Sandy's Special
25th Star
Cyg 20h17'17" 42d51'18"
Dennis & Sharon Maher
Cyg 20h11'23" 46d39'16"
Dennis & Sheila Morrison-
Wesley
Lib 15h15'30" -5d16'49"
Dennis "Slick" Morris
Uma 12h5'51" 52d52'38"
Dennis "Sparky" Wise
And 22h57'52" 47d30'2"
Dennis Stanfield
Her 18h52'59" 23d46'35"
Dennis Stanislaw
Uma 8h47'46" 51d58'15"
Dennis Stinner
Vir 12h34'46" -6d15'1"
Dennis Stockemer
Cyg 20h43'27" 45d50'7"
Dennis Stokes
Per 3h14'42" 52d3'20"
Dennis T. Madrid
Ori 5h3'33" -0d9'54"
Dennis & Tamara
Pho 0h53'49" -48d45'58"
Dennis "The Professor"
Strasburg
Per 4h48'53" 49d44'49"
Dennis The Right 911
Vir 13h15'36" 6d11'49"
Dennis Upton McRann
Per 4h36'32" 44d34'9"
Dennis W. Krebs
Ori 5h28'29" -0d31'23"
Dennis Walter Lang
Uma 10h46'8" 45d58'28"
Dennis Wayne Bowes
Vir 12h56'43" 10d41'16"
Dennis Weiss
Her 18h32'12" 21d17'13"

Dennis Wicker
Cnc 8h32'49" 21d49'45"
Dennis, Becca, Breezin, Roc
Peg 21h45'52" 13d29'56"
Dennis, My Eternal Love (F.A.T.S.)
Ori 5h58'10" 16d56'10"
Dennise Diaz
Psc 1h59'39" 8d31'44"
Dennise Patterson
Cyg 20h26'6" 39d26'8"
Dennise Roach
And 2h32'39" 41d7'12"
Dennisse Mio Amore
Cas 3h15'13" 61d33'36"
Denniston-Llewellyn
Cnc 8h21'57" 21d20'48"
Denny
Sgr 19h22'21" -13d51'56"
Denny et Erin Continu
Pho 23h44'14" -41d31'35"
Denny Förster
Ori 5h59'35" 17d35'53"
Denny Fried
Her 16h45'4" 27d25'20"
Denny Guyer
Sco 16h17'50" -12d0'14"
Denny Kerper #23 Forever
Sco 16h8'16" -13d40'59"
Denny Lenoir's Wishing Star
Crb 16h17'41" 33d10'51"
Denny M
Lyn 6h50'45" 52d32'25"
Denny Mool
Uma 13h51'26" 58d25'28"
Denny 'River' Bolton
Ori 5h23'58" 4d22'36"
Denny Singer
Cas 0h31'15" 52d11'36"
Denny Sue Nicholas Elyse Chipollini
Her 17h12'51" 20d32'55"
Denny & Sue on a star made for 2
Ori 6h19'44" 5d37'47"
Denny-Ashley Strange
Lyr 18h48'32" 31d33'5"
Dennys Brite
Psc 1h45'0" 20d59'30"
Denny's Cine Star
Uma 12h2'1" 57d18'28"
Dennys Menz
Ori 6h7'52" 8d11'36"
Denny's Unconditional Love
Psc 0h40'52" 9d47'16"
Denock
Cyg 19h27'21" 36d30'22"
Denoel Philippe
Umi 16h18'31" 70d23'53"
DenPat11
Leo 9h47'49" 27d34'21"
Densey
Lib 15h22'4" -20d13'0"
"Density" for Phil & Ruth Akin
Uma 9h22'32" 54d17'23"
Denton M. Messick, Jr.
Her 17h26'12" 14d54'4"
Denver
Uma 11h44'1" 52d6'2"
Denver
Lib 14h24'42" -18d58'44"
Denver Fentress
Uma 11h30'17" 47d51'6"
Denver James Saubert
Uma 9h4'10" 55d1'53"
Denver Richardson
Sgr 19h16'17" -15d49'11"
Denver Russell Seville
Her 16h17'28" 44d38'18"
Deny Me and Be Doomed
Vir 13h6'8" -11d47'22"
Denys Noyles, my pride and joy!
Leo 11h10'16" 0d45'54"
Denyse
Her 16h43'51" 48d42'31"
Denyse D. Aivalotis
Uma 8h36'0" 64d4'46"
Denzell "Dink" Clay
Cnc 8h49'55" 28d59'12"
Deoge Dog Gurley
Per 3h34'38" 44d18'35"
Deok-Im Jean
Cas 1h25'29" 58d49'44"
Deon Costalez
Tau 4h3'49" 14d37'56"
Deon Maxine Wynn (Goose or Big"D")
Psc 0h34'51" 4d33'47"
Deo's 50th Paramount
Psc 1h17'24" 25d2'34"
Deplancke-30
Lyn 7h42'20" 37d51'27"
Depusior
Lib 14h49'13" -0d50'45"
Deputy Donald Wass
Boo 13h48'2" 17d40'9"
Deputy Galen J. Herren
Gem 6h50'53" 34d27'51"
Deputy Michael Jason Rodriguez, BSO
Ori 6h13'59" 8d49'51"
Deputy Ramon Gamez Jr.
Cmi 7h20'23" 9d29'26"
Deputy Red Gordon
Cma 6h13'19" -21d46'12"
Deputy Rick Waits
Aqr 22h46'33" -4d16'5"
DeQuan Tomlinson
Uma 11h36'56" 53d38'5"
Der Hans Bauer Stern
Ari 3h21'27" 20d35'19"
der Hochzeitstern von Anne & Frank
Ori 6h16'15" 9d46'19"
Der Hüter des Sonnenscheines
Ori 6h19'18" 19d33'57"
der immer leuchtende
Leo 9h59'4" 18d5'22"
Der Robin & Kathi Stern 23.07.2004
Psc 1h12'44" 28d49'8"
Der Stern der Liebe
Gem 6h53'50" 32d31'11"
Der strahlende Stern Ruph
Ori 6h16'44" 2d57'14"
Der Topfer
Sge 19h45'17" 17d3'13"
DeRailius Krewius
Mon 6h28'28" 1d14'19"
Derba
And 23h21'20" 52d25'5"
Derby Berlin - Rising Star
Lyn 6h59'29" 47d4'0"
Dercio Herculano Ramos
Umi 16h11'32" 82d11'26"
Dereck F. Hayes
Lib 14h50'35" -1d33'6"
Derek
Col 5h40'41" -31d35'23"
Derek
Cyg 19h52'21" 36d47'32"
Derek
Ori 5h49'1" 5d41'30"
Derek
Vir 14h32'53" 3d36'27"
Derek
Boo 14h40'55" 17d54'35"
Derek Aldaco
Lib 14h30'50" -19d17'33"
Derek Allen Park
Aur 6h38'3" 38d25'9"
Derek Allen Wholihan
Cnc 8h52'37" 25d26'27"
Derek and Brigette Barker
Cyg 19h47'31" 39d10'38"
Derek and Kam van Pelt 3-11-2007
Lyn 9h4'29" 41d24'40"
Derek and Lynn Espinoza
Cyg 20h31'21" 46d7'5"
Derek and Melissa Andrjeski
Lyr 19h18'23" 29d47'44"
Derek and Sheena Blake
Cyg 19h22'13" 52d58'26"
Derek Anthony Gilmore
Cyg 19h26'40" 53d50'56"
Derek Ashley Franks
Her 17h30'13" 34d1'14"
Derek Attwood
Dra 11h23'57" 77d47'35"
Derek Avery Clay
Cep 21h29'41" 59d59'2"
Derek Baldwin loves Pamela Richie
Ori 6h0'8" 21d25'29"
DEREK "BLUE" LINGREN
Sco 16h12'4" -10d26'51"
Derek C. W. Baine
Cyg 19h44'40" 38d49'1"
Derek Cameron 26/10/1989
Uma 9h11'22" 50d53'33"
Derek Carlson
Gem 7h9'14" 24d12'16"
Derek Charles Conway
Uma 14h24'2" 58d27'19"
Derek Christopher Stone
Cet 0h53'57" -0d43'24"
Derek DeMott
Aql 19h12'20" 0d16'4"
Derek Doane Deems: Anjia's Cowboy
Cnc 9h11'19" 30d36'54"
Derek Edward Miller
Aqr 22h33'9" -1d24'44"
Derek F. Dale
Her 16h21'32" 23d5'52"
Derek F. Menchan
Del 20h37'28" 13d50'46"
Derek & Felecia Eby-Forever
Cap 21h5'4" -16d38'5"
Derek Feltham
Cep 22h52'50" 69d57'45"
Derek Finneran
Cep 21h26'33" 81d11'38"
Derek Fuller
Cep 21h13'8" 56d0'15"
Derek G. Teele #2
Per 3h23'52" 51d22'25"
Derek H Janiak
Gem 7h6'29" 18d41'13"
Derek Hal Parrish
Psc 23h58'18" 1d19'52"
Derek Hamilton Morgan
Tau 5h45'42" 20d57'52"
Derek Hellender
Psc 0h19'15" 15d29'11"
Derek Hercule
Sco 16h7'40" -13d47'49"
Derek J. Gisburne
Per 3h17'29" 32d51'38"
Derek & Jacqueline Tazzar
Cyg 21h50'6" 47d52'17"
Derek James Pukash
Sgr 18h8'0" -26d40'36"
Derek James Simmons
Psc 23h19'15" 6d38'43"
Derek & Jeana
Psc 1h58'55" 7d39'56"
Derek Jesse Poli
Leo 9h59'14" 14d11'12"
Derek John Hart
Per 3h26'0" 37d34'21"
Derek John O'Connor
Cep 4h11'57" 80d39'9"
Derek Judd 1951
Dra 16h42'24" 56d51'56"
Derek Kenneth Heckman
Cnc 8h44'47" 16d57'48"
Derek Kieper
Uma 13h1'11" 56d57'36"
Derek Knight
Aur 5h46'14" 45d59'28"
Derek Kyle Snell Our Angel
Ori 5h21'16" -3d54'32"
Derek Lee Eisnor
Del 20h46'1" 15d45'4"
Derek Lee LeClaire
Lyn 7h45'1" 56d56'51"
Derek & Lin
Cru 12h19'2" -56d17'32"
Derek Liscoumb
Gem 6h42'9" 21d18'22"
Derek Lucien
Her 17h4'47" 29d45'11"
Derek & Maeve's Star
Cnc 8h1'59" 18d3'49"
Derek & Margaret
Cyg 19h55'47" 38d34'45"
Derek Mark Martinez
Aqr 22h2'31" -1d5'7"
Derek Mayeaux
Umi 14h31'14" 72d51'24"
Derek Means
Uma 13h56'21" 53d47'32"
Derek Michael Griffin
Lyn 8h3'27" 40d17'4"
Derek Neal
Uma 10h59'44" 56d35'15"
Derek P. Robertson
Vir 15h4'6" 5d21'19"
Derek Pershyn
Lib 14h46'59" -3d32'12"
Derek Pimentel
Her 17h16'46" 43d27'25"
Derek (Pokey) Rigman
Ari 2h9'7" 24d54'54"
Derek R. Drake
Lib 15h59'23" -12d22'25"
Derek Raymond
Vir 12h11'15" 11d37'23"
Derek & Rick
Her 17h43'30" 15d31'11"
Derek Ring
Per 2h19'16" 56d39'9"
Derek Rollan Dutton
Leo 9h40'45" 15d3'10"
Derek Rosendahl
Ari 3h10'6" 17d50'2"
Derek Schocken
Uma 11h29'58" 34d18'27"
Derek Senft
Lyr 19h16'20" 39d35'18"
Derek & Sheila Alderton 8-9-1956
Cru 12h37'1" -58d18'45"
Derek Son of Helen
Sco 16h45'2" -32d14'52"
Derek T. Markham
Per 2h43'52" 53d50'32"
Derek T. "Xtreme" Snellings
Ori 5h5'14" -0d21'17"
Derek & Tammy 2005
Cyg 19h50'40" 32d30'11"
Derek Taylor
Uma 8h19'54" 60d5'5"
Derek Thomas, Promising You Forever
Psc 0h4'42" 3d30'1"
Derek Tyler Neff & Dakota Lee Neff
Uma 11h26'41" 58d29'44"
Derek & Vanessa Strong 7-7-2007
Cyg 20h19'35" 52d43'57"
Derek Walker
Aql 19h5'28" -0d56'44"
Derek 'Wes' Snipe-Eternity & Beyond
Sgr 18h24'52" -22d46'12"
Derek William Davison - Bear's Light
Uma 12h30'1" 58d37'33"
Derek William Dressler
Ari 3h18'57" 25d0'20"
Derek William Thompson
Psc 1h2'37" 33d25'48"
Derek Wilson-Forkin
Cma 7h1'8" -29d17'33"
Derek, Lisa, Johnathan Anderson
Cyg 20h59'27" 46d38'36"
DerekandCaitrionaNoonan
Uma 12h45'19" 55d27'14"
DerekandLisa4EVER
Cnc 8h17'10" 16d26'51"
Derek's Shining Down on You
Cap 21h7'36" -19d5'30"
Derek's Star
Dra 18h39'6" 76d40'35"
Derek's Star
Dra 18h51'17" 57d13'42"
Derek's Star
Ori 6h12'7" 2d36'10"
Dereksaurus Rex
Lib 16h1'16" -18d13'31"
Derek-You are my shining star!
Cnc 9h13'52" 16d29'21"
Derelie Laidler
Pav 18h46'27" -67d5'34"
Deresha
Psc 1h12'53" 14d51'52"
DERFSEYAM
Aql 20h9'55" 4d58'35"
Deri Armagost
Leo 11h12'5" 3d35'23"
Derick Andrew Christie
Dra 18h9'0" 75d18'30"
Derick J. Daniels
Per 3h29'48" 49d13'42"
Derick James Warren
Leo 11h46'17" 27d19'10"
Derick's Ann
Per 3h20'44" 31d35'12"
Derik Miller
Ari 2h10'26" 23d53'38"
Derin Edward
Sgr 18h53'14" -21d42'49"
Derk Bass
Boo 14h48'52" 22d6'44"
Dermot Hamilton
Peg 23h47'25" 19d22'36"
Dermotallis Allenius
Cru 12h44'38" -57d25'27"
Deronie Tan
Cnc 9h12'25" 24d57'47"
Derr -N- Alyhead for ever
Pho 0h16'54" -39d36'44"
Derra
Uma 13h48'4" 59d45'8"
Derri Little Angel
And 1h24'50" 34d33'2"
Derrick
Aur 5h52'58" 38d33'16"
Derrick
Her 17h9'50" 16d42'1"
Derrick & Amanda Rigbye (desami)
Cnc 8h30'3" 22d15'48"
Derrick Amato
Uma 10h11'21" 64d3'16"
Derrick and Jessica
Cap 20h20'33" -12d1'25"
Derrick and Shaunte Clark
Umi 15h27'57" 75d32'6"
Derrick and Val - SSTT Forever
Uma 9h40'21" 72d30'10"
Derrick J. Sorweide, D.O.
Cep 20h15'23" 60d18'45"
Derrick Kern
Her 17h15'53" 46d38'28"
Derrick L. Wilhoite
Ari 2h28'32" 12d27'15"
Derrick Lee Hamilton
Ari 2h49'35" 30d37'25"
Derrick Michael Davis
Her 16h33'48" 33d26'6"
Derrick Morgan
Per 3h9'6" 31d18'32"
Derrick Reed Mi Amore
Psc 1h9'2" 11d30'46"
Derrick & Sotheavy
Gem 6h53'32" 24d45'8"
Derrick & Tracy McIntosh
Lyr 18h23'57" 38d32'25"
Derrick Tyrone Truss
Ori 5h20'48" 6d0'29"
Derrick Williams
Psc 0h42'52" 17d38'51"
Derrick Withers
Her 18h4'31" 15d28'30"
Derrick Worth
Cep 21h45'48" 58d9'17"
Derrick87
Leo 10h4'16" 20d24'10"
Derrick-n-Lynda
Uma 12h47'57" 54d58'59"
Derrick's Star
Cmi 7h22'10" 1d6'29"
Derrik's Star
Gem 7h30'39" 28d26'46"
Derrin's 8th Lovelight
Uma 10h36'21" 45d10'17"
Derri's Twinkle
Cas 0h36'21" 58d2'17"
Derry O'Malley
Uma 11h27'53" 59d17'32"
Derya Ilknur Dalprà-Uenlütürk
Crb 15h26'50" 25d50'36"
Derya Karakose
Uma 10h15'24" 42d36'55"
Deryck DeMar Toles
Ori 6h15'28" 14d31'43"
Deryck Ray
Umi 17h5'58" 82d53'23"
Deryle
Gem 6h57'8" 16d13'28"
Des Biquets
And 2h15'32" 48d43'53"
Des & Ged's Heavenly Union
Cyg 19h36'49" 29d54'2"
Des Klonk
Uma 12h23'28" 59d21'56"
Des Koalas
Per 4h45'26" 49d19'24"
DeSagana Diop
Aqr 21h4'17" -10d41'20"
Desaix Alexandra Ringger
Gem 6h35'46" 23d35'10"
Desarae Burks-Moses
Cyg 21h55'1" 37d36'30"
Desarae Nichole McLaughlin
Aqr 20h48'14" 2d23'33"
Desarae Schenk
Ori 6h3'53" 18d37'1"
Deschler Poss, Marion
Ori 5h12'53" -8d36'6"
Desdemona
Sco 17h17'53" -32d21'18"
Desdemona Pippen-Coralis
Tau 4h28'38" 21d10'11"
Dese A. Cirelli
Uma 12h38'13" 55d27'18"
DeSelms
Leo 11h39'30" 12d33'45"
Deseo
Ori 6h9'16" 14d29'46"
Deseo Contigo
Aql 19h20'30" 7d30'11"
Deserae Wykoff
Lib 15h10'36" -16d21'37"
Deseree Contreras
Lib 15h51'57" -10d42'18"
Deserie Comfort
And 23h17'59" 48d5'19"
Desert Rose
And 23h2'14" 45d47'28"
Desert Rose
Ari 2h16'17" 25d22'15"
Desert Star
Leo 11h21'20" 18d6'26"
Desert Suzie
Crb 15h25'12" 27d34'13"
Desha Bene
Aqr 21h56'28" -7d38'9"
DeShae- Your Shining Star
Uma 12h42'56" 56d34'39"
Deshante Yvonne Davis
And 0h11'16" 25d36'43"
DeShauna
Leo 11h15'39" 16d36'40"
DeShawn Michelle Straube
Lib 14h55'26" -2d35'4"
Deshea Leah Neusetzer
Cam 3h23'44" 65d33'22"
Desi
Ori 5h47'39" 2d52'45"
Desi Gallegos
Cam 3h51'14" 54d36'13"
Desiderata
Cas 0h28'21" 57d29'43"
Desideria
Her 18h48'24" 13d29'58"
Desideria
Aql 20h9'48" -0d54'54"
Desideria Anna-Maria
Cas 1h1'52" 56d35'57"
Desiderio Cubas
Psc 1h20'58" 32d41'37"
Desi-Lucy-Cleo
Cmi 7h29'1" 6d39'52"
Desir Te Epouser Moi
Cyg 19h39'30" 52d13'55"
Desirae
Cas 0h18'21" 54d27'33"
Desirae
Aqr 23h33'53" -6d31'13"
Desirae Andrea
Lyn 8h18'3" 34d47'53"
Desirae Denise Quintana
Vir 14h12'26" -4d21'17"
Desirae Leigh
Mon 6h59'30" -1d14'12"
Desirae N. Hawks
Lyn 7h37'46" 56d11'22"
Desire
Lyr 19h20'12" 38d29'27"
Desire
Psc 1h26'37" 18d5'46"
Desireah
Ari 2h8'53" 16d21'30"
Desiree
Psc 1h20'54" 16d13'36"
Desirée
Peg 23h51'51" 9d19'27"
Desiree
Aql 19h13'10" 2d12'48"
Desiree
Ori 5h14'22" 12d7'10"
Desireé
Boo 14h49'8" 23d23'51"
Desiree
Cas 0h59'6" 47d58'18"
Desiree
Cyg 21h45'34" 47d50'36"
Desiree
And 23h25'34" 51d55'1"
Desiree
And 0h44'16" 34d55'44"
Desiree
And 0h38'3" 31d15'0"
Desiree
Crb 15h37'1" 37d5'13"
Desirée
Cnv 13h34'51" 30d33'10"
Desiree
Cyg 20h45'27" 34d38'22"
Desiree
Vir 11h44'56" -3d27'4"
Desiree
Aqr 22h26'18" -9d40'53"
Desiree
Sco 16h16'56" -9d40'12"
Desiree
Sgr 18h31'25" -15d56'38"
Desiree
Cap 20h11'55" -22d8'18"
Desiree 2005
And 0h42'58" 41d46'6"
Desiree Ann Giordano
Cmi 7h28'23" 10d55'22"
Desiree Cabral
Cap 21h48'58" -19d52'45"
Desireé Cheri Anderson
Cap 20h24'5" -10d31'8"
Desiree D.
Cap 20h18'4" -13d57'5"
Desiree De Molina
Cyg 21h38'38" 53d25'52"
Desiree Delaossa
Gem 7h11'6" 32d28'53"
Desiree Delores Cougill
Psc 23h57'10" -4d24'23"
Desiree DeSi Dunaway
Tau 5h12'27" 23d34'48"
Desiree Destiney Cantu
Her 17h37'34" 17d22'34"
Desiree Diaz
Crb 15h43'31" 32d48'14"
Desiree & Eduardo 143-637
Uma 9h15'46" 53d55'23"
Desiree Garcia
Ori 5h34'49" 11d4'47"
Desiree Glover
Cnc 8h38'20" 22d43'54"
Désirée Hängärtner (Grämi)
Ori 6h15'56" 20d51'42"
Desiree Hennessy
Gem 7h19'37" 14d33'51"
Desiree Hoflock
Ari 2h19'32" 11d37'52"
Desireé Hoo
Lib 15h35'2" -10d34'55"
Desiree Jane Goodwin
And 23h11'37" 51d40'14"
Desiree Jean Bryant
Cas 1h20'59" 65d49'33"
Desiree Josephine Delhommer
And 2h34'11" 50d42'59"
Desiree Kawa's Little Night Light
Cap 20h8'41" -21d26'29"
Desire'e Kay Lauderbaugh
Aqr 22h10'11" -23d4'18"
Desiree L. Allison
And 0h34'11" 36d32'19"
Desiree Lea
Cnc 8h38'55" 8d9'26"
Desiree Lee Valentine
Tau 4h37'37" 28d40'36"
Desiree Marie Martino
Aqr 23h53'1" -14d55'41"
Desiree Marie O'Donnell
Sgr 18h9'6" -25d40'54"
Desirée Marie Ripo LaJustica Bonita
And 23h18'56" 47d34'43"
Desiree Michelle Deshields
Psc 0h59'9" 25d16'20"
Desiree Nicole Anderson Phillips
Vir 13h6'50" 6d35'20"
Desirée Noel Worstall
Sco 16h7'42" -14d26'33"
Desiree Renee Hatter Figg
Uma 8h45'51" 50d51'53"
Desiree Rita Vaccaro
Leo 11h28'41" 17d45'32"
Desiree Rodriguez
Cam 3h42'13" 58d22'20"
Desiree Russo
Cap 21h48'17" -14d20'44"
Desiree Salerno
Ari 2h4'44" 21d21'19"
Desiree Sarah
Tau 5h42'22" 25d35'4"
Desiree Stampone
Ori 6h12'54" 17d3'17"
Desiree1
Cnc 8h49'46" 11d50'35"
Désirée-Franziska Weibel
Cas 1h59'18" 60d59'59"
Desiree's Perfect Place
Aqr 21h50'48" -2d44'9"
Desi's Forever
And 2h20'13" 48d56'30"
Desi's Nightlight
Umi 14h25'14" 77d17'53"
Desislava Dolamova
And 0h21'43" 40d44'1"
Desistar
Cep 20h25'10" 60d22'20"
D.E.S.K.
Uma 9h33'16" 51d19'16"
Desmerea
Cru 12h15'4" -60d50'4"
Desmond and Gertrude Fahey
Cnc 8h37'51" 32d37'59"
Desmond Cronin
Her 18h8'43" 28d59'9"
Desmond Curzon Lacey
Cen 13h40'14" -41d44'9"
Desmond Jamal Leger
Vir 12h18'34" -10d59'25"
Desmond Mong Seng Tan
Ari 3h0'56" 17d34'28"
despina
Psc 1h40'10" 24d20'12"
Despina
Per 4h0'37" 45d3'42"
Despina
Sct 18h54'27" -12d27'11"
Despina Aivazidou
Sco 16h9'2" -26d5'55"
Despina Manvelyan
Sgr 18h53'18" -20d13'27"
Despoina
And 2h36'13" 40d47'50"
Desprez Justine
Tri 1h52'2" 30d38'7"
Dessa Marie Merritt
Lib 15h18'1" -12d12'43"
Dessa Shanahan
Cas 2h16'46" 73d4'11"
Dessel Family
Lyn 7h30'46" 52d35'19"
Dessi Lou
Umi 15h33'28" 72d38'54"
Dessi Nintcheva
Uma 8h39'37" 67d9'59"
Dessie Lynn Warwick
Gem 6h44'6" 28d45'7"
Dessiree Murphree *God's Angel 14*
Leo 10h43'21" 19d7'58"
Desta James
Leo 9h22'57" 27d15'38"
Desta Jean Shoemaker Donnell
Crb 15h31'35" 30d58'27"
Destaney Marie Bigham
Dra 18h23'27" 70d26'37"
Destany Kay
Vir 13h25'26" 8d52'25"
Destany Violet Carle
And 2h19'18" 50d16'13"
Destany Violet Carle
And 2h16'26" 45d13'53"
Destany's Dreams
Aqr 20h52'26" -9d3'17"
desteni
Cyg 20h58'35" 44d46'38"
Desti
Psc 0h51'5" 9d31'1"
Destin Council Walls
Aql 19h27'11" -0d36'9"
Destin n Chris no. 1
Sco 16h48'57" -26d47'35"
Destined for Molly
Peg 22h33'52" 5d56'29"
Destined Love
And 23h9'56" 40d42'30"
Destinee
And 23h54'59" 43d9'14"
Destinée
Aql 19h33'0" 5d58'21"
Destinee Lee Dockhorn
Aqr 22h39'32" 1d24'56"
Destinee Shining
Her 17h34'25" 31d31'47"
Destiney Lee Bolton
Tau 4h25'52" 11d13'3"
"Destiney-Dawn"
Psc 1h12'8" 24d21'25"
Destini Tyne Caldwell
And 0h29'39" 34d43'47"
Destinicky
Uma 11h30'48" 61d39'55"
Destinie Rebozo
Vir 12h36'28" -4d9'25"
destino
Lib 14h58'26" -0d32'55"
Destino
Uma 8h45'18" 71d13'53"
Destino
Lyn 8h6'40" 33d12'43"
Destino
Uma 8h54'39" 50d47'30"
Destinus Truest 11/25/01
Lyn 7h21'42" 45d28'40"
Destiny
Uma 8h40'26" 51d23'22"
Destiny
And 0h24'40" 45d12'59"

Destiny
Cyg 20h5'18" 43d54'56"
Destiny
Crb 16h13'45" 38d45'39"
Destiny
Cas 0h22'8" 57d5'42"
Destiny
Cas 1h24'15" 55d15'58"
Destiny
And 23h10'23" 52d4'48"
Destiny
Lyn 8h37'42" 33d38'33"
Destiny
Lyn 7h49'10" 36d8'44"
Destiny
Lyn 8h48'26" 34d5'43"
Destiny
Tri 2h26'37" 36d17'28"
Destiny
Aur 5h11'53" 32d55'35"
Destiny
Her 16h57'14" 32d39'43"
Destiny
Lyr 19h7'0" 32d33'40"
Destiny
Lmi 10h34'14" 36d45'11"
Destiny
Cyg 19h49'14" 36d57'32"
Destiny
And 0h54'23" 40d43'0"
Destiny
And 0h43'26" 39d10'48"
Destiny
Peg 21h18'38" 15d57'18"
Destiny
Psc 0h26'12" 8d21'14"
Destiny
Cnc 8h35'25" 14d17'39"
Destiny
Cnc 8h19'46" 13d57'9"
Destiny
Cnc 8h56'36" 12d18'18"
DESTINY
Tau 5h5'11" 21d34'55"
Destiny
Umi 13h15'56" 73d0'2"
Destiny
Uma 12h33'22" 61d59'42"
Destiny
Uma 11h38'5" 63d52'30"
Destiny
Uma 8h53'3" 64d4'51"
Destiny
Lyn 6h46'2" 56d52'17"
Destiny
Lyn 8h12'42" 57d36'32"
Destiny
Uma 9h8'44" 58d28'40"
Destiny
Uma 12h37'7" 56d16'48"
Destiny
Vir 12h42'19" -7d35'36"
Destiny
Eri 4h29'45" -11d20'6"
Destiny
Aql 19h47'30" -0d32'58"
Destiny
Umi 15h25'49" 77d10'37"
Destiny
Umi 16h24'3" 81d0'32"
Destiny
Mon 7h3'42" -7d2'17"
Destiny
Ori 4h57'38" -3d24'43"
Destiny
Aqr 21h23'22" -9d22'21"
DESTINY
Cap 20h44'45" -14d44'51"
Destiny
Lib 14h40'46" -11d1'1"
Destiny
Sgr 19h9'47" -15d27'39"
Destiny
Cap 20h10'8" -21d14'3"
Destiny
Cap 21h8'4" -17d57'48"
Destiny
Sgr 18h11'7" -17d14'7"
Destiny
Sco 17h32'5" -45d29'21"
Destiny
Cru 12h39'2" -60d31'51"
Destiny Alexa
And 2h30'54" 45d44'56"
Destiny Ann Story
Lib 15h1'52" -14d10'53"
Destiny Aquila Desmond
Aql 19h42'5" 7d2'16"
Destiny at 'Z'
Col 6h1'39" -30d12'56"
Destiny Bell
Cap 20h33'17" -17d30'47"
Destiny Christina Gonzales
Cap 21h34'58" -14d58'2"
Destiny Christine
Gem 7h9'10" 34d33'59"
Destiny Daniel
Cep 22h38'9" 72d11'7"
Destiny Faith Love
Sco 17h23'57" -33d1'35"
Destiny Faith Ryana Bone
Cnc 8h53'56" 22d59'12"
Destiny Farm
Cyg 20h27'18" 47d43'43"
Destiny for you and me.
And 0h28'58" 39d58'49"
destiny happy one year KAT
Cyg 19h36'57" 28d29'8"
Destiny Hope
Eri 4h15'38" -9d18'20"
Destiny Island
Cyg 21h32'0" 34d38'4"
Destiny Jade Berg
Uma 10h45'47" 71d24'4"
Destiny Jon L. & Brit W. 9-4-03
Cyg 21h31'32" 45d2'48"
Destiny Kazuma Stella of Maho
Psc 1h31'31" 12d33'32"
Destiny Lane DOD 5-19-05
Uma 11h31'29" 47d10'58"
Destiny Lee
And 23h25'16" 47d0'47"
Destiny Marie
And 23h37'38" 41d18'53"
Destiny Marie
Lyn 7h56'8" 35d22'50"
Destiny Moore
Leo 10h42'34" 12d47'14"
Destiny Reedy
Crb 15h46'8" 35d22'54"
Destiny Renaissance
And 0h28'28" 26d6'55"
Destiny Shalom
Uma 11h57'53" 52d14'6"
Destiny S.L.J
Cas 1h28'59" 57d57'54"
Destiny Stegall
Peg 23h41'49" 29d2'50"
Destiny T&D
And 0h43'24" 43d48'56"
Destiny Tello
And 23h35'29" 38d49'23"
Destiny Terry and Susan 31 May 04
Lib 15h56'20" -5d32'48"
Destiny Torres
Sgr 19h9'1" -21d50'43"
"Destiny" - Victor & Alexa
Sge 20h6'43" 17d49'49"
Destiny, Baylee, Izaiah & Rockoe GA
Uma 9h41'47" 62d5'17"
Destiny, Brad and Melissa Acker
Tri 2h11'25" 34d38'48"
Destiny1227
Uma 10h6'23" 64d41'48"
Destiny34
Her 17h15'53" 20d16'3"
DestinyJadeHerron'sLilPiec eofHeaven
Tri 2h5'58" 33d37'57"
Destiny's Star
Cyg 19h50'11" 36d23'40"
Destrie Johnson
And 0h31'44" 40d46'34"
Destro Walter
Cyg 19h40'16" 46d37'54"
Destructobaby
Sgr 18h0'56" -18d22'36"
Destry Brannock
Gem 6h39'29" 19d2'42"
Destry James Hansen
Ori 5h54'4" -0d2'30"
DESTYN
Cam 4h14'0" 65d58'47"
Detavius Rasjahn Anthony Niblack
Umi 14h48'35" 72d32'51"
Detective Stephen J. Strehl
Ari 1h57'41" 21d41'25"
Determination
Uma 11h16'44" 68d41'49"
DeThomas together for eternity
Sco 16h9'55" -10d14'21"
Detlef
Uma 12h55'4" 61d52'49"
Detlef Bornefeld
Uma 8h47'3" 62d43'47"
Detlef Happel
Ori 6h19'42" -1d50'9"
Detlef Kahle
Leo 9h48'29" 22d50'2"
Detlef Lentge
Uma 10h20'7" 48d59'49"
Detlef Scheel
Uma 8h37'55" 50d59'1"
Detlef Schumacher
Ori 5h46'23" -3d31'10"
Detlef Steinberg
Ori 6h3'37" 9d47'37"
DETLEF50
Aqr 22h10'19" -1d26'22"
Detoura Neal
Uma 9h53'32" 57d30'35"
Detra-Claudette 1976
Dra 17h46'10" 52d43'13"
Detrix
Sct 18h56'16" -4d0'31"
Detta Hanson
Cnc 8h50'49" 32d47'24"
D'etta Henderson
Cap 21h36'47" -24d1'35"
Dettmann, Lutz
Ori 5h52'42" 21d40'41"
Dettmar, Heinz
Uma 10h31'14" 40d24'58"
Deuce
Umi 14h42'2" 86d34'31"
Deuce Is Loved (Thomas E. Webb)
Tau 4h9'31" 5d5'11"
Deus Rumor Pluvia Nox Noctis
Ari 3h23'36" 28d12'59"
"DeuSys", Góz Péter Csillaga
Uma 12h3'23" 30d56'27"
deux âmes pour toujours
Leo 9h50'55" 6d44'40"
Deux etions et n'avions qu'un coeur
Leo 10h9'46" 21d48'36"
Deux inconnus qui s'aiment...
Cas 0h21'8" 57d56'23"
Dev
Lyr 18h52'56" 44d54'7"
DeVal'l Anthony Banks
Aur 5h45'25" 29d29'9"
DeVan
Cnc 9h2'45" 26d36'6"
Devan
Lyn 8h2'36" 47d57'20"
Devan Brian Shortridge
Lib 15h34'14" -26d40'42"
Devan DiLauro
Cyg 19h43'9" 30d42'54"
Devan In memory of pappy Vaughn
Uma 10h46'45" 62d10'56"
Devan Lawrence Curtis
Per 4h14'51" 36d33'22"
Devan & Lindsay Muir
Cyg 20h58'8" 41d8'55"
Devan Marie Bartels
Vir 12h25'44" 10d30'17"
Devan Michael
Per 4h20'6" 50d35'32"
Devan Michelle Williams
Uma 8h51'32" 60d11'14"
Devan Sanders
Cap 20h9'29" -8d44'0"
Devan Taylor Pontius
Her 16h23'24" 11d32'8"
Devang & Shangruti Desai
Aqr 22h8'28" -19d10'45"
DEVANJALI
Col 5h54'31" -36d5'32"
Devannie
Cyg 19h36'5" 37d55'33"
Devanous Majora
Leo 11h40'41" 19d32'51"
DeVastar
Tri 1h49'50" 34d9'59"
Deven Elizabeth White
Vir 12h20'53" 9d26'46"
Deven Howard River Trotman
Eri 4h23'56" -9d29'56"
Deven Jordan Asselin
Leo 9h57'42" 18d20'1"
Deven Tribble
Leo 9h36'15" 32d13'49"
Devencenzi
Dra 17h19'33" 53d29'50"
Deventria
Cam 4h32'54" 69d59'42"
DeVera 22
Ori 5h51'30" 8d35'16"
Devereaux Shawn Wilson
Per 1h33'21" 54d23'5"
Deveta Alisa
Uma 9h38'19" 60d32'56"
Devhan / Megin
Gem 7h19'34" 31d18'2"
Devi Kosala Ekanayake
Umi 15h44'39" 81d3'4"
Devid Lee Gilmore
Cep 20h37'48" 64d1'38"
Devil Woman
Umi 14h37'26" 70d29'8"
Devin
Cap 20h15'47" -16d52'49"
Devin
Sco 16h18'17" -11d31'25"
Devin
Aql 20h19'30" 2d9'54"
Devin
Vir 15h4'24" 0d29'11"
Devin
Gem 7h18'28" 15d38'45"
Devin 4-4-1985
Ari 2h18'53" 23d19'18"
Devin Adele Lindstrom
Peg 21h51'10" 15d46'53"
Devin Andrew Downs
Gem 6h54'15" 27d19'8"
Devin Andrew Gdula
Ari 2h37'28" 27d29'38"
Devin Ashley Maurer
Cnc 8h49'26" 27d38'57"
Devin Bean
Umi 14h19'12" 68d49'16"
Devin Bernard Briggs
Her 16h51'25" 30d30'18"
Devin Brandes
Ori 6h11'10" 13d49'47"
Devin Brian Shaw
Peg 23h50'33" 16d12'37"
Devin Charles Turner
Leo 10h17'39" 16d18'57"
Devin Clark
Umi 17h10'9" 82d28'26"
Devin Conrad Woodard
Cap 21h8'23" -15d33'44"
Devin Cox
Sco 17h33'20" -39d1'6"
Devin D/R
Aql 19h14'9" -0d22'20"
Devin Daniel Duckstad
Lyn 7h29'49" 44d24'0"
Devin Deep Gabriel
Tau 3h29'26" 8d25'1"
Devin Derosia
Uma 9h54'32" 57d0'54"
Devin Devost My Love
Vir 12h4'56" 3d56'43"
Devin Dick
Leo 11h21'11" 22d44'52"
Devin Elijah
Lyn 6h30'14" 56d48'54"
Devin Francil
Leo 11h2'5" 16d43'41"
devin gale
Leo 10h31'3" 17d50'8"
Devin Gene Brennan
Leo 10h56'37" 11d57'41"
Devin Glenn Bryant
Uma 10h10'44" 67d28'14"
Devin Glunt's Dynasty
Uma 9h10'17" 54d17'59"
Devin Hildebrand
Aqr 22h57'22" -7d31'15"
Devin Jacob Hoover
Sco 16h7'22" -12d27'34"
Devin James Addington
Cnc 8h34'31" 8d27'9"
Devin James Seay
Uma 8h59'35" 48d20'8"
Devin Jensen
Her 18h34'16" 19d55'14"
Devin John Henderson
Uma 11h56'31" 32d19'45"
Devin John Henderson
Sco 16h4'24" -13d7'42"
Devin Jordan
Aqr 22h13'14" -12d39'1"
Devin Lee James
Cap 20h19'55" -10d13'24"
Devin Leigh Lake
Gem 7h31'48" 28d35'12"
Devin Logan Griffey
Uma 9h24'15" 43d9'40"
Devin Lorraine Campbell
Uma 9h42'34" 46d8'8"
Devin M. Hamilton
Leo 9h41'50" 28d5'47"
Devin M Loeffler
Vir 12h57'51" 12d31'28"
Devin Mathew
Cap 20h34'2" -16d5'39"
Devin Matthew
Her 17h8'41" 38d19'33"
Devin Matthew Smith
Her 18h46'48" 15d35'36"
Devin McKenna
And 2h34'54" 44d42'32"
Devin Michael Gasser
Lib 14h54'4" -2d15'23"
Devin Michael Rigolino
Dra 19h5'52" 64d3'19"
Devin Mitchell Carson
Gem 6h29'55" 21d31'47"
Devin Munro
Psc 23h38'55" 0d18'54"
Devin "My Heart"
Ori 6h17'31" 2d38'34"
Devin Nicole Ammann-Jones
Uma 10h48'33" 63d22'35"
Devin Nicole Blair
Uma 12h6'22" 53d49'27"
Devin Noelle Wilson
Cnc 8h39'8" 20d53'45"
Devin O'Leary
Uma 14h10'9" 58d22'22"
Devin Oliver Perry
Leo 10h12'22" 13d55'9"
Devin Olivia Hunter Dalrymple
Aqr 23h54'25" -10d14'2"
Devin Papp
Per 3h13'22" 53d6'23"
Devin Patrick Fleming
Lib 14h53'37" -3d25'34"
Devin Patrick Murphy McQuade
Dra 20h22'22" 67d28'10"
Devin(punky)
Uma 12h24'12" 60d41'7"
Devin Rogozinski
Cyg 21h41'30" 52d51'38"
Devin Rose
Boo 14h4'51" 20d16'50"
Devin Smoter
Her 18h2'10" 24d48'46"
Devin son of Dennis+Nicole Petrino
Psc 1h11'19" 24d26'47"
Devin Steelman Jarvis
Crb 15h46'42" 33d33'25"
Devin Sullivan Murria
Boo 14h37'14" 33d45'43"
Devin Thomas Perlman
Her 18h56'26" 15d48'9"
Devin Trame
Peg 22h56'20" 27d15'6"
Devin Walker
Ari 2h19'39" 21d38'7"
Devin Wayne Hutteball
Tau 4h39'41" 19d18'11"
Devin Wideman-Levin
Uma 13h33'39" 55d41'28"
Devin William Ream
Leo 10h10'14" 24d38'48"
Devina "Isis" Davis
Sgr 18h16'34" -31d51'18"
Devina Jasmine Deo
Leo 11h36'35" 17d13'44"
Devinder S. Mangat M.D.
Vir 13h25'3" -3d32'47"
Devine
Ari 2h26'34" 24d14'52"
Devinne, Brightest Star In My Sky
Peg 21h36'28" 24d56'40"
Devin's Delight
Uma 10h29'0" 69d10'6"
Devin's Little Heaven
Uma 13h19'4" 57d55'14"
Devin's Poem
Cnc 7h56'37" 16d47'56"
Devin's Star
Dra 18h30'13" 52d20'33"
devk-12
Aql 19h4'7" -10d20'13"
Devlin Berry
Uma 12h8'26" 60d26'51"
Devo
And 1h3'3" 41d28'18"
Devon
Cyg 20h36'46" 30d48'42"
Devon
Psc 0h8'22" 8d38'42"
Devon
Uma 10h44'5" 66d37'18"
Devon
Cyg 20h5'39" 56d10'19"
Devon
Vir 14h13'5" -13d17'26"
Devon Ainsley Smolca
Leo 9h50'29" 7d5'52"
Devon Amber Woodward
Cas 23h30'44" 56d56'55"
Devon and Tim Goldrick FOREVER
Ari 2h21'17" 24d19'55"
Devon Anne Lugano
Leo 11h57'31" 19d57'27"
Devon Anthony Arizon
Umi 14h31'9" 72d21'10"
Devon Ariel Mulrine
Tau 3h51'45" 23d11'57"
Devon Carrington
Uma 9h39'59" 49d20'4"
Devon Chance Taylor
Lib 14h56'12" -20d27'46"
Devon Clare
Lyn 6h34'55" 56d56'13"
Devon Daniel Paul
Cap 21h39'19" -11d14'57"
Devon Dante' Swink
Lyn 8h51'18" 34d24'28"
Devon Dares
Psc 0h12'52" 4d42'30"
Devon DeMaria
Per 3h7'49" 35d51'30"
Devon E. & Tiffany G. (Are Cool)
Per 2h23'49" 55d16'35"
Devon Elizabeth
Lib 15h4'32" -21d14'32"
Devon Elizabeth Dent
Lib 14h54'25" -7d11'48"
Devon Florea
Umi 14h57'56" 76d6'30"
Devon George Mah Lee
Psc 2h2'13" 5d20'53"
Devon Gillis DeKorver
Ari 1h51'28" 18d8'41"
Devon Gregory Fritzgerald Cajuste
Boo 15h27'37" 43d46'20"
Devon Holden Harper
Aql 19h20'53" -1d47'51"
Devon J. Wildermuth
Umi 16h54'10" 79d44'5"
Devon Jackson
Dra 16h33'43" 57d55'50"
Devon Jay Flem
Uma 11h44'43" 64d23'41"
Devon Joe Piper
Ari 2h8'57" 24d24'16"
Devon Kathryn Haberski
Cas 0h21'52" 50d26'15"
Devon L. Carroll
Umi 16h13'39" 82d25'49"
Devon Leanne
Uma 9h9'27" 68d49'25"
Devon Lee Ward
Tri 1h55'7" 29d54'45"
Devon Looney
Vir 14h49'0" -0d14'9"
Devon Louise
Sgr 18h3'58" -34d4'50"
Devon Louise Yacka
Vir 15h3'7" 4d20'53"
Devon M. Cowell
Per 1h44'23" 50d51'32"
Devon Michael Robbins
Lib 14h49'11" -6d25'6"
Devon Michael Snider
Lyn 7h1'5" 51d33'57"
Devon Myers
Aql 19h11'43" 12d26'44"
Devon Newton
Gem 6h22'26" 21d14'47"
Devon Rogers
Lib 15h56'25" -17d38'51"
Devon Rose
Cnc 9h9'33" 31d1'54"
Devon S
Lib 14h28'2" -11d48'44"
Devon Saunders
Ori 5h50'12" 19d10'19"
Devon Scott Readinger
Cas 1h3'6" 53d2'34"
Devon "Sparky" Garner
Cap 21h4'55" -24d56'39"
Devon Walker Ovall
Uma 9h19'10" 61d39'12"
Devon Wrue Flores
And 0h28'15" 25d52'24"
Devon11Susan
Hya 9h27'42" 2d21'35"
DeVona J. McElroy
Lyn 6h50'3" 52d30'22"
Devonia
Cyg 20h0'31" 30d43'11"
Devora Cespedes
Aqr 22h17'45" 2d36'50"
Devorah Giselle Reyna
And 23h40'12" 46d53'22"
Devorahhency
Uma 8h34'20" 71d19'42"
DeVore
Cam 4h14'52" 68d0'4"
Devotion
Ori 5h44'52" 8d20'59"
Devotion: Nick&Tiffany Tocco 9.3.05
Cyg 20h2'41" 35d50'59"
Devotion P. Fata Hip Hop Soulchild
Peg 23h9'56" 27d51'23"
Devotion to Carol
Cyg 20h14'21" 34d15'7"
Devoyon Bruno
Ori 4h50'19" -2d34'48"
Devozione
Uma 11h39'54" 64d31'31"
Devra Marie
Uma 11h6'54" 40d32'25"
Devra's Rock Star - Love, Jay
Cnc 8h44'21" 15d3'13"
DevResh
Eri 3h48'42" 0d1'16"
Devry Bell
Uma 11h52'35" 46d26'2"
Devyn
Lib 14h47'20" -8d36'54"
Devyn
Aqr 22h15'1" -9d35'13"
Devyn Alexis "Boo Bear"
Uma 12h34'38" 57d5'49"
Devyn Danielle Pollock
Per 4h35'45" 41d53'27"
Devyn Dean Brown
Sgr 18h59'14" -25d41'1"
Devyn & Gaby Always & Forever
Cyg 21h24'43" 39d19'10"
Devyn Jude Murphy
Uma 13h50'18" 60d41'20"
Devyn & Kaitlyn Tracy
And 0h27'52" 31d54'18"
Devyn Krystine
Vir 13h15'23" 12d26'6"
Devyn Leigh Drusjack
Gem 7h33'45" 32d45'37"
Devyn Mitchell Watson
Lyn 7h55'13" 34d50'37"
Devyn Rangiah
Cru 12h33'15" -59d32'58"
Devyn Robert Harvick
Lyn 8h52'40" 40d54'50"
Devyo Leigh
Umi 15h19'25" 69d39'27"
Dew Dropius
Tau 3h53'18" 7d4'44"
Dewars
Psc 1h10'27" 29d3'53"
DeWayne
Uma 11h29'25" 55d33'6"
Dewayne Bolt
Cru 11h58'26" -56d37'5"
DeWayne J Fellows
Her 16h41'8" 35d47'35"
Dewayne Scott Rooks
Uma 11h5'52" 35d3'6"
Dewayne's Divine Light
Uma 12h5'39" 55d36'47"
Dewey
Psc 1h17'5" 26d33'36"
Dewey
Tau 4h35'20" 28d35'57"
Dewey and Lillie Mae Holst
Lyr 18h36'6" 31d50'39"
Dewey and Linda
Lyn 9h36'7" 41d18'31"
Dewey Leclair
Per 4h46'37" 39d4'48"
Dewey Lynch Pritt
Cnc 8h45'47" 21d4'44"
Dewey Robin Driggars
Sgr 19h28'59" -11d59'9"
Dewey W. Clark III
Leo 10h52'3" 16d4'42"
Dewey Wilson
Del 20h17'11" 12d2'28"
DeweyandJewey
Lyr 18h40'13" 35d21'41"
Dewi Fonna
Uma 8h26'39" 72d2'41"
Dewitt Talmadge Purdue Jr. TX Gent
Leo 10h56'8" -1d3'10"
Dewon "Infinite" Wilson
Aqr 22h2'40" -19d34'35"
Dexheimer
Her 16h41'29" 8d34'39"
Dexter
Cet 1h2'11" -0d27'28"
Dexter
Psc 0h18'39" -2d34'35"
Dexter
Umi 15h27'54" 69d4'15"
Dexter
Sco 17h19'30" -42d42'52"
Dexter Cullum Calhoun
Psc 1h21'35" 28d39'2"
Dexter Dee
Lmi 10h39'32" 25d27'4"
Dexter "Earth Angel" Tisby I
Psc 0h58'0" 9d40'42"
Dexter & Francis Greenfield
Peg 22h5'54" 26d21'42"
Dexter LaMont
Psc 0h21'9" 14d6'13"
Dexter Mayer 42
Cnv 12h52'53" 48d34'43"
Deyala Al Najjar
Umi 13h52'27" 70d23'36"
Deyanira Zujeith Michener
Umi 14h25'3" 75d9'33"
Deyes Place
Sco 16h9'53" -15d11'16"
Dez&FiFi
Aqr 20h40'3" -12d44'41"
Dezaray LeAnna Kuemmerle
Mon 6h35'4" 1d3'52"
Dezel Telage
Cma 6h47'21" -15d30'10"
Dezerland Prime
Cet 2h48'23" -0d30'12"
Dezirae Devoe
Tau 3h56'47" 17d59'43"
Deziree
And 23h16'38" 47d27'28"
Deziree Jill Leyba
Cas 23h2'15" 54d4'15"
Deziree Rachael
Ori 6h5'12" 19d7'25"
DezroK
Vir 13h25'3" -9d45'32"
Dezso Paul Bajusz
Cap 20h49'9" -19d24'56"
DFAS Orlando Vendor Pay Team
Uma 9h20'37" 46d19'34"
DFVOS1/Dillon Festags Very Own Star
Aur 5h53'35" 38d31'29"
DG:URSOMLM
Ori 6h12'54" 8d9'12"
DGF
Cap 21h44'35" -12d57'19"
DGiglio529
Ori 6h10'56" 13d50'3"
DGJ
Psc 1h41'50" 20d32'48"
DGP RLT Amour vrai
Gem 6h28'38" 22d15'8"
Dhaani
Uma 10h41'44" 47d25'46"
Dhanmatte Indira Dattoo
Cyg 20h9'51" 33d37'17"
Dhanya Renata Zimmermann
Dra 18h41'6" 74d56'4"
Dhara
Crb 15h35'42" 35d30'7"
Dhara Patel
Boo 14h30'27" 53d42'38"
DharamJet Banjaradil Captain-Saab
Uma 9h21'22" 54d9'44"
DhaubhadelSzerbakowski-CuppieCakes
Lib 15h41'44" -14d49'22"
Dhaval Patel
Cep 22h3'8" 53d47'46"
Dhebora Piva
Ori 6h15'38" 16d17'18"
Dheer K Patel
Leo 9h59'14" 9d46'39"
DHM10428
Lib 15h6'30" -12d49'16"
DHO 12-7-44
Sgr 18h28'11" -16d8'54"
Dhodho
Cap 20h57'21" -18d45'40"

Dhruv Jhamb
Cyg 20h11'20" 40d31'13"
Dhruva
Lmi 10h35'41" 35d33'34"
Dhwani's Star
Aql 19h37'34" 3d24'1"
Di
Uma 9h8'40" 58d22'1"
Di and Clem Ruby Anniversary
Cyg 21h46'40" 44d47'40"
Di Domenico Simona
Ori 4h54'47" -1d57'4"
Di & Don's Inspiration Shines On
Tau 4h39'30" 27d39'18"
di Jeannene
Cap 21h4'5" -20d45'18"
Di Luna Forever
Ori 6h10'32" 16d14'18"
Di Mellis
Cas 1h0'18" 61d21'26"
Di Menza Julie 21.09.2006
Uma 11h12'32" 72d31'10"
Di Nardo Chiara
Boo 14h45'10" 53d58'58"
Di & Nebs
Cyg 21h17'14" 42d39'36"
Di Prima Laura
Cam 6h51'50" 69d35'33"
DI RUI MIN
Ari 2h12'43" 24d10'24"
di Siro, Luca
Uma 10h12'27" 68d11'0"
Di Terlizzi Eleonora
And 0h46'6" 45d49'39"
Dia Duit
And 0h4'14" 43d18'4"
Dia En Ozi
Tau 3h54'42" 21d13'8"
DiabloTTin
Del 20h38'18" 9d16'23"
Diabolique Fantastique Rachelique
Leo 11h49'57" 20d2'33"
Diafed
Cyg 19h34'10" 30d14'36"
DiAg
Lyr 19h3'57" 29d18'3"
Diahn Micheal Rainey
Psc 0h3'31" 8d2'41"
Diahna Rivera
Lib 15h45'24" -29d5'12"
Diala 78
Cyg 21h52'43" 41d34'30"
Diala and Hassan
Ori 6h19'33" 14d1'21"
Diala S. Itani
Uma 13h4'48" 59d33'25"
DIALHOTZ
Cap 21h37'39" -14d52'40"
Diamant de mon coeur
Umi 15h12'29" 85d56'52"
Diamantenpaar Inge & Willi Bottor
Uma 9h5'45" 60d43'59"
Diamantides' Dream
Her 18h16'2" 18d37'47"
Diamantis Masoutis
Uma 10h44'4" 40d16'31"
Diamarie Vargas Rivera
Sgr 18h15'18" -35d34'55"
Diamond
Uma 11h3'2" 69d9'51"
Diamond
Uma 11h2'24" 44d45'6"
Diamond
Cnc 9h20'37" 30d49'29"
Diamond 171101
Crb 15h50'4" 26d50'2"
Diamond Anniversary
Cyg 20h21'39" 42d19'2"
Diamond Anniversary Star
Cyg 19h22'59" 46d56'55"
Diamond Billington
Tau 5h24'51" 25d40'59"
Diamond Braces
Uma 14h20'57" 58d24'40"
Diamond Bum
Leo 11h4'15" 20d27'17"
Diamond Dan
Uma 8h59'21" 47d36'40"
Diamond Dave Cochran
Aql 19h9'42" -0d20'34"
Diamond Eyed Angel Baby Carolyn
And 22h59'14" 37d47'31"
Diamond G
Psc 1h9'41" 27d45'1"
Diamond Girl
And 0h22'7" 25d55'14"
Diamond In The Rough
Cnc 8h52'1" 27d13'28"
Diamond In The Sky
Ori 5h40'36" -0d18'5"
Diamond Stardust
Boo 14h46'55" 24d17'45"
Diamond Star-Mark Ness
Her 17h51'14" 22d13'13"
Diamond Star-Mark Neveu
Boo 14h51'17" 26d25'6"
Diamond Star-Melissa Paulson
Cas 23h56'40" 57d24'47"
Diamond Star-Mike Neuschwander
Per 2h58'2" 46d48'31"
Diamond Star-Pat Pevan
Uma 12h30'28" 53d42'2"
Diamond Star-Troy Lightfield
Umi 16h57'31" 77d26'27"
DiamondBrandi
Uma 9h28'58" 65d55'6"
Diamond's Smile
Leo 9h50'14" 29d8'59"
Diamonds Star
Lib 15h40'30" -6d41'2"
Dian
Mon 8h10'32" -0d47'44"
Dian
Uma 11h34'44" 35d35'20"
Dian 1
Ari 2h45'42" 27d41'42"
Dian#2
Ari 2h10'48" 26d54'8"
Dian and Pat Disibio
Lyr 18h45'54" 39d56'31"
Dian Martha Evans Olsen
Mon 6h52'24" -0d42'31"
Dian Stai
Uma 11h29'40" 32d38'30"
Diana
Cnv 13h55'33" 34d1'28"
Diana
Lyr 19h14'33" 34d32'12"
Diana
And 1h9'2" 34d34'18"
Diana
And 2h34'7" 39d51'30"
Diana
And 1h26'4" 44d58'12"
Diana
And 0h17'12" 42d54'25"
Diana
And 2h19'12" 46d52'25"
DIANA
Lyn 7h45'15" 49d32'50"
Diana
Uma 11h0'25" 45d25'22"
Diana
And 23h59'13" 48d7'50"
Diana
And 23h5'14" 47d2'32"
Diana
Cyg 21h49'39" 50d4'21"
Diana
Ari 2h53'7" 25d29'26"
Diana
Leo 10h12'15" 16d22'0"
Diana
Leo 9h50'34" 22d25'47"
Diana
Lmi 10h51'29" 26d35'7"
Diana
Cyg 19h43'38" 29d26'58"
Diana
Tau 4h44'34" 23d19'10"
Diana
Ari 3h18'46" 27d53'13"
Diana
Gem 6h39'56" 13d34'43"
Diana
Cnc 8h23'5" 11d55'49"
Diana
Tau 4h32'8" 1d52'28"
...diana...
Umi 15h17'51" 80d36'8"
Diana
Vir 12h32'21" -1d53'34"
Diana
Cap 20h25'16" -14d8'37"
Diana
Lib 15h56'26" -11d40'32"
Diana
Cap 21h40'44" -18d26'3"
Diana
Uma 10h49'22" 68d3'18"
Diana
Cam 4h10'51" 70d42'25"
Diana
Uma 9h38'31" 71d40'52"
Diana
Uma 9h58'13" 58d36'49"
Diana
Cas 0h37'40" 62d52'1"
Diana
Cas 1h51'50" 60d29'55"
Diana
Sgr 18h28'13" -31d15'39"
Diana
Ant 10h21'12" -30d50'13"
Diana
Sgr 19h20'25" -26d16'49"
Diana
Sgr 18h5'19" -27d34'10"
Diana A. Buitrago
And 0h34'33" 42d44'26"
Diana "A Star To Guide You"
Ori 6h19'1" 18d41'18"
Diana & Al Spruill ~ August 5, 1967
Cyg 21h35'15" 46d40'53"
Diana Alcaraz
Cap 20h25'25" -13d56'33"
Diana Alvear
Cap 21h47'23" -18d11'36"
Diana and Eduardo Morgan
Cyg 21h25'53" 53d53'51"
Diana and Ross
Car 9h50'13" -60d48'20"
Diana and Savi Lividini
Cyg 21h22'37" 50d30'1"
Diana Anh Nguyen
Lib 15h6'19" -21d43'39"
Diana Ann Mokry
Tau 3h43'38" 17d9'55"
Diana "Annie" Woods
Cas 0h0'13" 55d39'20"
Diana Anselmi
Cam 4h27'17" 53d46'29"
Diana Apostolovska
Vir 13h16'47" -22d16'36"
Diana Arena
Vir 13h24'35" 10d58'49"
Diana Arencibia
Cas 0h41'12" 64d27'58"
Diana Ariya Chanthavisouk
Uma 8h36'24" 67d48'50"
Diana Arutyunova
Uma 11h40'16" 48d12'20"
Diana Ayala-Carrillo
Cnc 7h57'13" 16d41'38"
Diana B. Boggess
Ori 6h2'11" 1d53'48"
Diana B. Oaks—Our Shining Star
Uma 9h2'33" 53d58'44"
Diana Babychou
Leo 10h22'22" 6d42'30"
Diana Barnett
Cnc 9h8'54" 32d15'12"
Diana Baumann
Cas 22h58'53" 56d51'45"
Diana Berry Bostick
Sco 16h54'5" -34d57'16"
Diana Bonfil
And 0h27'24" 35d15'44"
Diana Bonney
Lyn 8h29'54" 39d18'27"
Diana Cailteux
Uma 10h55'58" 40d1'4"
Diana Carol Neuman
Cas 1h3'17" 49d18'34"
Diana Cedrone
And 1h29'20" 48d50'33"
Diana Comin
Aqr 23h12'8" -8d8'2"
Diana Cormier
Mon 7h0'0" -0d48'15"
Diana Cristina Moisescu
Lib 14h58'33" -9d34'48"
Diana D.
Lyn 6h59'3" 45d46'12"
Diana D. Lyles
Cas 0h38'12" 53d16'22"
Diana & Daniel
Cas 1h2'10" 56d44'2"
Diana Daniels
Cam 6h28'19" 68d0'51"
Diana Darrell
Cyg 19h55'47" 33d35'19"
Diana Dell
And 23h35'34" 36d24'56"
Diana Denese Drum aka Pooh
Aqr 23h53'8" -3d29'33"
Diana Dill, My Angel in the Sky
Cas 0h46'23" 53d23'6"
Diana Duarte
Vir 12h8'54" 7d14'35"
Diana Dunn
Aqr 22h40'53" -3d23'14"
Diana Dupre
And 1h13'12" 46d31'32"
Diana Edwards
Cnc 8h20'14" 9d32'45"
Diana Elena Sfrijan
Com 13h5'56" 29d14'18"
Diana Elizabeth Keenan
Sgr 20h4'0" -44d44'25"
Diana Elizabeth Mack
Vir 12h17'13" -0d59'50"
Diana Emery
Leo 9h24'25" 23d12'53"
Diana Evans
Sgr 18h22'5" -33d3'48"
Diana Fox
Dra 19h0'9" 70d17'54"
DIANA GAIL BLACKBURN
Sgr 20h22'59" -35d12'15"
Diana Gallardo's Fire
Mon 7h15'12" -0d38'17"
Diana Garcia
Ori 4h52'59" -0d14'41"
Diana Garcia
Lib 15h2'34" -1d26'15"
Diana Gayle
Tau 3h49'35" 27d13'12"
Diana Gissel
Tau 5h17'38" 17d8'35"
Diana Gorbunova
Sgr 18h55'50" -20d34'58"
Diana Grace Allegra
Gem 7h34'48" 22d50'30"
Diana Graves
Peg 22h33'15" 16d53'30"
Diana Grossman
Lib 14h49'53" -1d51'11"
Diana Guenther Tunigold
And 1h42'37" 38d1'1"
Diana Haigney
And 23h29'57" 47d39'43"
Diana Harrison
Tau 5h20'50" 17d9'9"
Diana Hatfield
Gem 6h33'30" 19d16'46"
Diana Hinckley
Crb 15h25'45" 29d54'49"
Diana Hofer Loves Paul Frenette
Umi 16h36'31" 82d59'33"
Diana Holder
And 23h1'32" 48d10'8"
Diana Irene LaPasha
Cas 1h28'49" 57d56'24"
Diana J. Cardoza
And 0h24'42" 44d59'30"
Diana J. East
Uma 10h53'42" 46d14'9"
Diana Jackson Weber Palms 11181940
Cyg 19h30'29" 28d28'44"
Diana Jade Skipper
And 0h33'1" 40d52'28"
Diana Jane Avery Weaver
And 23h23'30" 35d32'42"
Diana Jasmine Arredondo
Lmi 10h0'17" 41d4'1"
Diana Jayn
Lib 14h32'26" -24d52'1"
Diana Jean
Sgr 19h18'0" -22d38'50"
Diana Jean
Sgr 18h0'42" -17d32'17"
Diana Jean Casler
Sgr 18h41'44" -18d20'35"
Diana Jean Fowler Duck
Lib 15h20'48" -6d26'6"
Diana Jean Kenson
And 0h18'41" 31d40'44"
Diana Jean Lawrie
And 0h46'21" 39d55'23"
Diana Jeanne
Ari 2h12'0" 26d17'37"
Diana Jim and Adrian 05
Psc 1h18'27" 14d36'7"
Diana Joanne Kell
Aqr 21h47'33" -0d32'9"
Diana Joy
Sco 16h12'12" -15d34'25"
Diana Joy Bundoc
Sco 16h53'51" -45d18'20"
Diana Joyce
Boo 13h53'1" 24d47'27"
Diana Julianna
Cnc 8h35'10" 18d47'18"
Diana K. Bishop
And 2h18'50" 46d47'13"
Diana Kay
Psc 1h3'45" 3d2'2"
Diana Kay Hoganson
Tau 3h31'22" 24d30'30"
Diana & Kenneth Marx Oct.16,1965
Cyg 19h55'57" 31d18'49"
Diana Koehne Family
And 0h26'23" 44d48'55"
Diana Kramer & Thomas Vig
Cyg 20h4'55" 33d43'47"
Diana Krepich
Sgr 18h55'52" -36d4'8"
Diana L Nevins
Ari 3h9'18" 29d42'0"
Diana Lake Reidy
Aqr 21h51'7" -6d19'41"
Diana Lang
Lmi 10h26'56" 32d33'45"
Diana Lautenberger
And 1h17'56" 38d36'55"
Diana Lee Falkowski
Sgr 18h38'56" -27d8'33"
Diana Lee West
Gem 6h55'49" 35d14'23"
Diana Lee Wetmore
Psc 1h22'7" 15d41'14"
Diana Leigh Bocek
Uma 9h51'2" 72d14'24"
Diana Leigh Rigozzi
Psc 0h29'49" 4d8'37"
Diana Leon Taylor
Ori 5h52'27" 18d8'29"
Diana Lichtenwalter
Cas 1h26'38" 60d58'36"
Diana Littrell
Com 13h2'43" 18d8'46"
Diana Louise Wakefield
Crb 15h51'51" 36d16'2"
Diana Luther
Cap 20h17'11" -20d34'14"
Diana Luvsbaskets
And 0h44'35" 39d48'48"
Diana Lynn
Cyg 21h14'15" 38d23'9"
Diana Lynn
Cnc 8h22'45" 21d31'25"
Diana Lynn
Vir 12h51'39" -1d26'22"
Diana Lynn Anastasia
Cas 0h9'47" 57d5'27"
Diana Lynn Ballew
Gem 6h49'50" 34d13'37"
Diana Lynn Cerny
Sco 16h54'0" -38d25'44"
Diana Lynn Davis
Lyn 8h36'22" 38d56'44"
Diana Lynn Millard
Boo 15h18'33" 36d13'52"
Diana Lynn Werner Niemann
Sco 16h9'52" -14d22'3"
Diana M. Goodwin, CMP-GaMPI Star
Lyr 18h52'52" 34d51'51"
Diana M. Stephens
Mon 6h47'40" -0d20'17"
Diana Mae
Mon 6h44'2" -0d20'44"
Diana Malke
Cnc 8h34'25" 8d9'55"
Diana Malouly
Ori 5h28'21" 3d20'55"
Diana Marie
Aqr 20h52'25" -13d8'30"
Diana Marie Bray
Cnc 8h14'56" 16d55'48"
Diana Marie Cox
Cas 23h21'50" 58d2'58"
Diana Marie Curfman
And 23h35'55" 47d35'32"
Diana Marie Di Mola
Cnc 8h53'0" 28d59'1"
Diana Marie Driscoll Schoner
Uma 9h30'43" 57d14'54"
Diana Marie Escobedo
Tau 4h37'45" 7d19'30"
Diana Marie Migliore
And 2h20'17" 39d21'17"
Diana Marie Santillan
Lyn 8h4'32" 42d22'1"
Diana & Matt
Tau 5h38'42" 23d57'4"
Diana & Matthew
Psa 22h33'51" -27d17'5"
Diana Mayhall
Lyn 7h7'36" 57d6'46"
Diana Michelle Lacey
Leo 9h39'43" 28d8'56"
Diana Mirella Cam Diaz
Gem 7h20'56" 24d38'52"
Diana (Moon Goddess)
Uma 9h56'53" 52d24'30"
Diana Morales
And 1h8'15" 45d58'25"
Diana Moreau
And 23h19'47" 52d54'15"
Diana Munzlinger
Cas 0h26'33" 55d37'4"
Diana Murchison
Uma 10h39'10" 46d19'53"
Diana My Shining Star
Ari 2h52'27" 26d36'55"
Diana Mykris
And 2h6'50" 37d29'9"
Diana Ngai
Tau 5h40'49" 25d30'33"
Diana Offler
Cyg 21h22'6" 34d8'0"
Diana Olivares
Psc 1h1'59" 19d27'41"
Diana Osseyran
Cas 0h26'5" 50d54'36"
Diana Ott
Ori 5h8'33" 12d25'9"
Diana P
Crb 15h21'31" 30d45'42"
Diana P. Sanderson
Ari 1h51'5" 23d43'42"
Diana Perez
Cap 20h24'28" -11d11'33"
Diana Perrett
Crb 16h14'38" 38d42'33"
Diana Pink Huber
Cnc 8h40'14" 32d21'6"
Diana Pou
Lyn 7h17'10" 47d59'1"
Diana Rachael
Cam 4h41'54" 54d50'36"
Diana Rae Waldrop
And 23h30'9" 42d38'59"
Diana Razen
Uma 11h8'27" 33d53'48"
Diana Rebecca Traynis
Vir 12h40'43" 5d38'56"
Diana Renee Carr
Lib 15h34'46" -14d37'33"
Diana Richelle
Cas 0h39'46" 61d43'34"
Diana + Robin Städler
Ori 6h5'26" 20d54'1"
Diana Rodrigues
And 2h8'2" 37d22'55"
Diana Rose Street
Uma 10h56'8" 48d28'59"
Diana S. Archer
Cas 1h24'32" 61d26'20"
Diana Saad Shalabi
And 23h3'38" 52d22'21"
Diana Salazar
Aqr 20h55'18" -8d38'55"
Diana Sanchez
Peg 22h1'8" 33d54'19"
Diana (Saurus) Mulligan
Tau 4h56'15" 28d26'20"
Diana Schumann
Uma 11h44'19" 35d23'50"
Diana Silva
Cas 23h5'53" 53d44'14"
diana sky
Sco 17h8'38" -36d57'2"
Diana (Skygirl) Castro
Oph 17h1'50" -16d23'54"
Diana "Smoopy" Philip
Sco 16h36'47" -38d11'31"
Diana: star of my soul 4-ever
Tau 4h57'59" 16d43'3"
Diana Stormy Barbara
Umi 17h36'8" 82d54'26"
Diana Stout
Uma 8h22'48" 60d24'39"
Diana Sue Chamblin
Psc 1h35'36" 10d11'57"
Diana & Tateshi's STAR!!!
Cyg 21h24'42" 45d54'32"
Diana Terry
And 1h16'36" 37d6'22"
Diana the Little Beamer
Cnc 9h11'59" 17d46'31"
Diana the Sweet Pea
Lib 15h30'13" -17d58'44"
Diana & Tom Fink
Peg 22h54'49" 14d3'6"
Diana Tomchick
Mon 6h51'29" -0d6'5"
Diana Travelute
Uma 11h30'40" 32d43'0"
Diana Ung
Cnc 8h29'17" 28d3'16"
Diana Villanueva
Lib 15h31'4" -4d51'27"
Diana Vogt O'Brien
Cap 21h26'18" -17d8'57"
Diana Welfel
Vir 11h42'54" 7d40'50"
Diana Wilkotz
Cam 6h27'5" 62d34'19"
Diana Zalewski
Leo 9h44'27" 30d42'43"
Diana Zeberlein
Uma 11h56'1" 33d53'59"
Diana Zeylikman
Tau 3h45'35" 27d20'37"
Diana, Precious Earth Day Daughter
Tau 5h27'43" 21d47'23"
Diana2036
Lyn 7h47'6" 52d52'5"
Diana74
Crb 15h50'15" 30d20'43"
Diana-Elena Ciornei
Sco 17h52'28" -35d57'54"
Dianaesthesia
Her 17h5'24" 34d15'5"
Dianah Patricia Espinoza
Ari 3h13'8" 22d3'37"
DianaLou Princess Raymundo Bernardo
Gem 6h16'39" 22d2'48"
Diana-My Falling Star-Trevor9/10/05
Cru 12h37'56" -59d45'43"
Diana's Dream
Sco 16h6'8" -14d39'1"
Diana's Dream
Cam 3h39'34" 57d54'46"
Diana's Eyes
Ori 5h33'30" -8d0'2"
Diana's fairy
Ori 5h32'22" 3d55'50"
Diana's Happiness
Ari 2h59'20" 25d44'14"
Diana's Love
Psc 0h52'52" 17d33'47"
Diana's Lucky Star
Cap 20h46'19" -25d16'51"
Diana's Star
Sco 16h50'36" -28d32'37"
Diana's star
Gem 6h38'20" 19d52'35"
Diana's Star
Crb 15h45'45" 29d25'16"
Diana's Wish
Psc 1h26'28" 19d13'21"
Diana-Scott
Per 3h20'48" 38d52'4"
Dianastar
Per 3h33'49" 43d54'21"
Diandra
Gem 6h43'40" 17d43'59"
Diandra
Aqr 20h46'59" -7d13'18"
Diane
Aqr 22h34'39" -6d33'30"
Diane
Cam 7h26'41" 78d28'7"
Diane
Sco 16h12'50" -10d38'0"
Diane
Cap 20h37'33" -16d47'19"
Diane
Lib 15h4'42" -17d25'9"
Diane
Dra 15h15'59" 56d52'42"
Diane
Cas 0h54'21" 66d43'2"
Diane
Cas 0h19'17" 61d22'46"
Diane
Uma 11h7'15" 60d35'5"
Diane
Sco 17h17'14" -32d54'25"
Diane
Leo 10h3'20" 20d59'58"
Diane
Peg 22h33'39" 15d49'1"
Diane
And 0h34'53" 31d26'32"
Diane
Leo 10h10'9" 25d57'41"
Diane
Cnc 8h53'59" 25d49'41"
Diane
Gem 7h32'17" 25d41'8"
Diane
Tau 4h8'4" 6d44'28"
Diane
Psc 0h42'16" 14d4'53"
Diane
And 2h5'12" 40d36'14"
Diane
Peg 22h11'13" 33d28'33"
Diane
Cyg 19h48'53" 36d21'57"
Diane
And 0h17'15" 34d50'42"
Diane
And 1h10'13" 33d56'9"
Diane
Cas 2h53'11" 58d0'20"
Diane
Uma 9h0'46" 51d1'12"
Diane
Cas 0h13'41" 51d23'46"
Diane
And 23h10'2" 43d8'37"
Diane A. Rittenhouse
And 0h15'28" 27d22'11"
Diane A. Satterthwaite
Cas 0h16'49" 53d46'44"
Diane A. Satterthwaite
Cas 0h16'25" 53d5'36"
Diane Aguinaldo
Uma 12h2'15" 60d35'38"
Diane Alderton Canham
Cru 12h2'26" -59d43'36"
Diane Alessi Arnold
Gem 7h31'56" 15d0'11"
Diane and Andrzej 30th Anniversary
And 0h9'47" 45d42'55"
Diane and Bob Garett
Dra 16h37'14" 53d12'4"
Diane and Charles Ewan
Lyr 18h50'42" 31d29'57"
Diane and Cole Tvrdik
Uma 10h11'26" 71d38'44"
Diane and George Sheahin Star
Crb 16h0'1" 30d2'56"
Diane and Greg George L/P Cuz
Cyg 19h57'35" 30d53'31"
Diane and Jim Yarrito
Cyg 19h35'44" 34d20'27"
Diane and Maynard Stratton
Uma 10h1'11" 45d51'7"
Diane Andre Hubisz
Sgr 18h45'5" -28d8'47"
Diane Ardith
Tau 5h45'2" 23d36'47"
Diane Azevedo
Vir 12h33'18" -0d39'0"
Diane B. Ambrose
Psc 1h24'55" 32d58'38"
Diane B. Dudley
Leo 11h25'46" 6d0'46"
Diane B. Molnar
Tau 4h24'58" 19d21'13"
Diane Baker
Tau 5h50'0" 17d7'41"
Diane Beaulieu
Aqr 21h45'22" 1d7'32"
Diane Benn
Tau 4h13'58" 26d46'49"
Diane Beth Jacobitti 3/18/64
Psc 0h48'14" 9d40'30"
Diane Blanshard - Mum's Star
And 1h34'28" 36d13'58"
Diane Bowdoin
Vir 13h57'42" -2d6'35"
Diane Boyle
Gem 6h33'3" 18d41'30"
Diane Brenner
Cas 0h37'53" 63d54'30"
Diane Breton
Ari 3h11'49" 27d59'21"
Diane Bringham
Tau 4h8'22" 5d54'5"
Diane Brown
Cas 1h26'42" 57d53'54"
Diane Bruce
Cas 0h16'38" 47d56'16"
Diane C. Biedermann
Sgr 19h16'56" -34d18'9"
Diane C. Miller 4-4-1941
And 23h15'37" 48d2'2"
Diane Cach
Aqr 23h56'8" -19d46'59"
Diane Caldwell
Vir 13h14'23" 6d11'34"
Diane Campbell
Cnc 7h57'57" 19d12'38"

Diane Carley Mon 7h37'55" -0d30'28"
Diane Carol Divine Sge 20h3'20" 18d54'21"
Diane Carol Dulaney And 23h36'45" 44d55'30"
Diane Carol Thompson Del 20h45'38" 2d41'57"
Diane Catherine Giannascoli Cas 23h19'0" 60d56'28"
Diane Catino Keweshan Uma 11h13'29" 62d52'50"
Diane Cautero Sgr 18h20'25" -31d27'7"
Diane Cheetham Cas 23h35'22" 54d17'29"
Diane Choplosky Cas 0h41'4" 61d31'32"
Diane Coffman Prince Mon 6h35'33" 8d42'38"
Diane Czaszewicz Cnc 9h6'30" 16d52'28"
Diane Damen Cas 1h28'36" 63d33'39"
Diane & Dan - D² Friends & Lovers Lyr 19h5'7" 42d22'31"
Diane D'Angelo Mon 6h27'54" 9d4'16"
Diane "DeDe" Tetreault Cas 1h4'40" 69d56'46"
Diane (Dee Dee) Her 17h56'35" 50d2'44"
Diane Delikat Cap 20h44'55" -27d8'46"
Diane Dillon-Walley Cnc 8h48'50" 10d13'40"
Diane DiMella Uma 10h3'44" 58d35'52"
Diane DiPalma Cnc 8h31'23" 25d46'9"
Diane Dowlan Psc 1h2'39" 24d26'8"
Diane E. Quarry Leo 11h42'6" 24d47'27"
Diane E2Z Cnc 8h8'13" 16d31'9"
Diane Edmonds And 23h37'40" 48d20'21"
Diane Edna Aqr 22h41'50" 0d43'10"
Diane Elaine And 1h13'45" 39d7'40"
Diane Elaine Stamp Gem 7h14'4" 18d59'47"
Diane Elizabeth Psc 1h18'47" 3d4'51"
Diane Elizabeth Fluit And 0h23'41" 38d1'40"
Diane Elizabeth Stone Lib 14h50'45" -5d33'1"
Diane Elsie Cas 23h16'29" 54d53'40"
Diane Emily Kassner Cas 1h20'14" 59d34'12"
Diane Engels-Velten And 0h36'36" 42d11'12"
Diane & Erv Cas 23h48'20" 50d35'34"
Diane Estella Redmond Psc 1h27'33" 19d30'33"
Diane Eulalia Edney Uma 9h33'0" 53d36'50"
Diane Farlay Per 3h56'29" 33d40'27"
Diane Forden Umi 15h44'21" 75d35'9"
Diane Galigher Oph 18h13'27" 6d30'41"
Diane Gardner Cas 0h22'35" 56d26'17"
Diane Geraldine Malone Gem 6h42'5" 15d31'54"
Diane Giannone Uma 10h42'43" 65d41'17"
Diane & Giles' Shining Star Lyr 18h25'14" 30d32'57"
Diane Gilly Crb 15h53'28" 25d49'48"
Diane Glass Ari 2h50'24" 29d28'32"
Diane Goldstein Her 18h17'41" 22d14'17"
Diane "Grandma" Anita Sahagian Sgr 18h47'30" -25d42'59"
Diane Guadara Sgr 18h43'12" -27d56'35"
Diane H. Canary Cnc 9h0'43" 14d45'15"
Diane H. McCue Lyn 9h16'12" 34d52'38"
Diane Hajdinak Wisniewski And 23h47'47" 39d10'14"
Diane Hammond Ali Cap 20h27'20" -10d53'0"
Diane & Hans Paulsen 1959 Lyn 8h44'0" 33d52'22"
Diane Harriman Leo 9h58'0" 17d34'42"
Diane Harris Lib 14h23'54" -10d50'23"
Diane Harrison Umi 14h23'7" 73d35'39"
Diane Hayashi Peg 21h53'7" 12d41'24"
Diane Hochstettler Her 16h42'56" 32d56'10"
Diane Homiak Mon 6h49'25" 8d13'22"
Diane Honey "Panda" 271066 Sco 17h42'59" -39d49'4"
Diane Hudson Cnc 8h38'17" 24d30'44"
Diane Hunt Jones Lyr 18h42'9" 34d28'35"
Diane Hurley Tau 4h42'3" 20d37'43"
Diane Huth Lib 14h54'42" -17d11'37"
Diane Ingram Uma 11h30'42" 57d7'1"
Diane Javaid Sco 16h8'21" -16d57'12"
Diane Jean Sgr 18h3'35" -17d34'28"
Diane Jessie Lyn 8h11'48" 53d11'54"
Diane Joan Cas 2h22'5" 72d27'0"
Diane & Joe Pascale Cyg 19h47'15" 30d53'13"
Diane & Joschua Stern Uma 10h55'30" 64d48'49"
Diane K. Bolger Cnc 9h9'8" 31d1'11"
Diane K. Brennan Tau 5h47'49" 22d17'54"
Diane K Queen Sgr 18h41'39" -35d26'58"
Diane Katherine DeGuzman Sco 16h54'44" -36d22'56"
Diane Käthner Boo 14h32'16" 19d38'22"
Diane Kay Johnni Leo 9h46'47" 8d12'17"
Diane Kay Rodgers Psc 0h44'59" 11d24'28"
Diane Kelley Vir 13h48'20" -3d59'46"
Diane Kidd And 0h12'30" 45d16'45"
Diane Kim Nguyen Aqr 23h14'23" -7d38'48"
Diane Kohler Cas 1h39'10" 64d19'57"
Diane Krebbs Aql 19h9'32" 3d18'14"
Diane L. Battaglia And 0h51'10" 39d54'2"
Diane L. Copeland And 0h5'31" 44d5'47"
Diane L Doyle And 2h13'24" 44d27'58"
Diane L. Hamblin Umi 13h24'34" 76d19'54"
Diane L Maturani Umi 15h18'14" 75d26'25"
Diane L. Morgan And 0h14'58" 28d0'35"
Diane L. Wilder Cyg 19h35'27" 37d49'48"
Diane "Lady Di" Lolich And 23h24'49" 50d55'53"
Diane LaMarche, Mademoiselle Artist Cas 1h24'6" 61d57'3"
Diane Lee Durdle Sgr 17h51'4" -28d33'49"
Diane Letko Ari 2h56'25" 28d0'52"
DIANE - L'etoile de ma vie! Uma 9h58'27" 61d21'47"
Diane Leydon Aql 20h0'37" -0d30'57"
Diane (Lieblings-Muckel von Hase) Uma 9h43'47" 54d48'8"
Diane "Lil D" Tan Lyn 6h43'35" 55d33'25"
Diane Loiselle Cap 20h10'13" -15d14'7"
Diane Lori Avila Leo 9h29'33" 10d57'14"
Diane Louise Ari 1h59'34" 18d6'32"
Diane Louise Harri Ori 5h34'36" -6d5'35"
Diane Love Cas 1h10'54" 65d39'39"
Diane Loves Etienne Forever Ori 5h37'58" 8d45'41"
Diane Lucky Uma 14h5'2" 56d56'17"
Diane Lusk Aqr 20h47'40" -11d10'20"
Diane Lynn Cas 2h19'17" 72d29'45"
Diane Lynn Butler Cap 21h5'36" -26d25'49"
Diane Lynn Cholewa-Betts Cas 23h53'52" 54d18'22"
Diane Lynn Cross (Princess) Cap 20h18'22" -13d30'13"
Diane Lynn Frederick Ori 6h2'52" 18d45'34"
Diane M. Jacob Cas 1h17'31" 65d51'17"
Diane M. Miller Ori 6h19'37" 7d50'51"
Diane M O'Brien And 2h30'5" 43d21'59"
Diane M. Park Leo 10h53'12" 6d42'7"
Diane M. Rochford Com 12h15'6" 19d50'55"
Diane M. Taslov Cas 1h37'3" 64d20'9"
Diane M. Wilson - Loved Sister Aqr 23h17'25" -19d13'43"
Diane M. Zanetti Tau 3h59'31" 1d18'7"
Diane Mackey Lib 15h26'36" -15d6'10"
Diane Macom Mullen Cas 0h51'24" 57d37'19"
Diane Major Cyg 21h58'53" 46d2'41"
Diane Manns Vir 13h4'46" 0d27'33"
Diane Maree Cru 12h52'26" -57d55'5"
Diane Margaret Wright Ari 3h13'17" 23d16'42"
Diane Marie Gem 7h30'46" 26d16'16"
Diane Marie Psc 1h18'56" 15d53'30"
Diane Marie And 0h44'37" 22d7'54"
Diane Marie And 0h56'3" 45d30'3"
Diane Marie Sco 16h11'59" -13d35'52"
Diane Marie Cyg 20h25'42" 58d17'56"
Diane Marie Uma 10h4'2" 57d5'48"
Diane Marie Brooks Uma 11h0'56" 56d27'59"
Diane Marie Ford And 23h14'41" 49d15'46"
Diane Marie Johnson Gruber And 0h35'8" 37d28'56"
Diane Marie Kezerle Gem 7h20'38" 32d45'19"
Diane Marie Lehuta-Rogers And 23h18'3" 51d43'55"
Diane Marie Markham Cap 20h49'19" -19d52'18"
Diane Marie Straight Uma 11h49'13" 54d12'54"
Diane Marie Taylor Cas 1h39'37" 68d5'41"
Diane Marie Tessmer Uma 10h18'38" 63d29'42"
Diane Marie's Devotion Crb 15h50'12" 36d51'43"
Diane Marshall Leo 10h21'49" 18d53'2"
Diane Mary Gem 6h49'31" 34d9'22"
Diane Mary Mother Vir 14h39'17" 1d4'11"
Diane Maureen Hilton Cas 23h24'37" 54d58'36"
Diane Mavis Psc 0h53'2" 9d49'19"
Diane McClure Vacca And 2h38'5" 41d0'29"
Diane Melinda Gonzalez Tau 4h22'43" 18d35'24"
Diane Mendenhall Cas 0h42'38" 66d50'2"
Diane Menna Per 3h26'32" 47d11'33"
Diane Michele Felicissimo And 23h58'29" 43d42'25"
Diane Michele Gonzalez Dra 17h57'28" 60d18'18"
Diane Miller Tau 5h14'59" 19d27'16"
Diane Millhouse Cas 1h25'12" 52d4'58"
Diane Mizelle's Goodness Cas 0h19'49" 57d28'50"
Diane Mizelle's Light Cas 0h35'9" 56d18'1"
Diane Mizelle's Love Cas 0h8'20" 58d31'28"
Diane Mizelle's Patience Cas 0h27'56" 57d39'37"
Diane Morgan Dunn Tau 4h11'34" 7d40'33"
Diane Mularoni-Is The Best!-M,P,&M Vir 13h25'41" 6d17'57"
Diane Murphy Aqr 20h50'1" 0d37'32"
Diane Murphy And 1h33'55" 49d6'29"
Diane Nancy Griggs Crb 16h15'26" 33d52'9"
Diane Nicole Psc 1h23'19" 4d5'49"
Diane ODell Ari 3h15'13" 26d43'12"
Diane of Great Heart Cyg 20h7'59" 30d15'30"
Diane of Wovenstar Her 16h48'35" 41d3'38"
Diane P. Uma 11h51'37" 39d24'7"
Diane P. Palmer Cas 2h17'36" 63d28'59"
Diane Palmer Cap 20h32'24" -26d26'18"
Diane Paris Kiesnowski Cas 1h7'49" 63d58'48"
Diane Patricia Peg 21h47'6" 21d23'12"
Diane Pendgraft Witte Crb 15h44'43" 35d0'54"
Diane Porter Leo 11h37'20" 27d17'56"
Diane Priestman And 2h11'22" 48d46'0"
Diane Rae Lawson And 23h13'14" 42d57'44"
Diane Renee Leininger Cas 0h55'56" 54d55'1"
Diane Reposa Vir 13h23'33" -16d39'57"
Diane Richard And 23h2'11" 41d5'42"
Diane & Richard Morris Cyg 21h38'13" 49d25'4"
Diane Rivaud Cas 1h16'40" 71d57'45"
Diane Rizk Cyg 19h36'18" 30d13'44"
Diane Ronconi And 23h16'4" 39d0'18"
Diane Rooney Lmi 10h37'2" 28d43'58"
Diane Rose Hockenberry Turner Cap 20h52'58" -18d32'7"
Diane Rosen Gem 6h46'30" 30d9'22"
Diane Rossier Lib 15h10'38" -8d7'27"
Diane Roy Per 3h15'58" 41d13'9"
Diane Ruth James-Sharpless Gem 6h59'32" 14d39'37"
Diane Ryan Ori 5h40'43" 7d2'17"
Diane Sauve Cas 0h39'46" 56d3'32"
Diane Self Vir 13h0'15" -5d40'27"
DIANE SEVEN Ari 1h58'14" 21d21'1"
Diane Severance Vir 12h41'47" 12d24'58"
Diane Shelton Cam 4h25'32" 56d12'50"
Diane:Shining Star Of Our Lives Cap 20h34'23" -20d15'22"
Diane Speer Leo 10h0'9" 20d18'3"
Diane St. Germain-Dewey Cnc 8h21'46" 20d23'2"
Diane(Stellar Mom)Jaracz Psc 0h8'13" 4d17'50"
Diane Suchon My Goddess
Dianna Tau 5h4'58" 20d37'48"
Diane Susan Baumann Vir 11h59'11" 6d18'17"
Diane Suzie d Nancy Theodore Cas 1h13'41" 65d44'44"
Diane Tankersly Lyn 7h17'35" 52d43'11"
Diane Tanski Sgr 18h23'49" -23d21'21"
Diane Tatterson Cas 2h43'55" 58d5'16"
Diane Therese Helen Tau 4h17'4" 29d44'40"
Diane Tina Sgr 19h3'10" -31d44'30"
Diane Vivien Guillaume Ari 3h13'7" 15d10'44"
Diane Walsh Psc 1h39'27" 27d56'0"
Diane Warren Cam 6h48'40" 65d36'58"
Diane Warren "Light Of My Life" Leo 9h27'21" 26d3'38"
Diane Weiss Psc 1h41'14" 27d25'21"
Diane Wiley Crb 15h36'34" 31d12'0"
Diane (Wollypog) Thomas Aql 19h43'47" 6d56'59"
Diane Wyer Lyr 19h10'56" 42d58'59"
Diane & Yvan Uma 11h38'48" 34d58'41"
Diane Yvonne Linder Cam 7h58'59" 60d37'27"
Diane Zambory Uma 8h29'53" 70d40'7"
Diane Zivkovic And 2h35'24" 49d42'44"
Diane, for ALWAYS in ALL ways,Ron And 0h38'48" 32d49'51"
Diane, Mesha, & Jeff Uma 11h48'14" 36d57'41"
Diane-05-08-05 Uma 9h13'56" 57d6'29"
Diane07 And 0h22'12" 41d9'2"
Diane120304 Uma 11h1'24" 45d11'50"
Diane130405 Uma 12h4'28" 33d46'42"
Diane-1963-Princess And 23h58'51" 42d11'59"
Dianelisse Ayala Pico Cmi 8h7'55" -0d0'25"
DianelovesDustin Tau 5h35'32" 24d27'37"
Diane's candidus lumen Psa 22h8'51" -29d40'41"
Diane's Grain of Sand Ari 2h37'27" 24d48'50"
Diane's Little Star Del 20h27'44" 20d36'11"
Diane's Love Sgr 18h55'36" -29d18'52"
Diane's Love Uma 11h36'51" 65d38'49"
Diane's Loving Light Ori 5h32'8" 9d41'24"
Diane's Prince Leo SLW IV - 2005 Tau 4h30'38" 22d54'29"
Diane's Shinning Star And 1h56'5" 45d11'32"
Diane's Star Cas 1h10'31" 60d34'26"
Diane's Star 4/11/53 Uma 10h31'39" 45d23'52"
Diane's Twinkle And 0h46'54" 37d12'36"
Diane's-60th Birthday Lib 15h43'44" -11d26'2"
DianeTod Cnc 8h5'24" 7d1'46"
DianeXOXO Pho 1h32'35" -43d54'19"
Dianick Bochef Uma 9h37'34" 71d39'0"
Dianmlie Crb 15h38'2" 36d3'17"
Diann Davis Cyg 19h27'49" 36d31'11"
Diann Edwards Uma 11h38'25" 30d24'42"
Diann Elaine Milford Ori 5h57'32" 12d12'59"
Diann Hamant Leo 9h30'46" 10d37'33"
Dianna Ari 2h38'51" 18d51'2"
Dianna Aql 19h11'15" 2d27'26"
Dianna Ari 3h5'42" 29d49'52"
Dianna Cyg 20h27'12" 35d20'35"
Dianna and Bob Crowder Sge 20h5'12" 18d50'39"
Dianna Bethany Cam 7h48'1" 69d37'32"
Dianna CD Ari 2h38'30" 12d59'4"
Dianna E Schoenrock Tau 5h5'49" 21d45'52"
Dianna Kim Scott Lyr 19h14'4" 26d23'34"
Dianna L. Minger Cas 1h28'31" 65d34'29"
Dianna Lee Olsen Black Uma 10h31'30" 42d44'10"
Dianna Lou Lyn 7h42'54" 49d17'42"
Dianna Louise Quatraro Leo 11h14'27" 18d19'31"
DiAnna Loy And 0h29'50" 25d47'7"
Dianna Lynn Watson Lib 14h53'54" -4d42'57"
Dianna Lynne Potter And 2h5'23" 45d56'34"
Dianna M. Wheat Lib 15h22'51" -16d27'53"
Dianna Malett Kauder Cnc 8h9'12" 18d4'32"
Dianna Marie Cam 4h51'22" 59d56'31"
Dianna Mary Dunbar And 23h21'55" 52d21'53"
Dianna Michelle "Ryan's Baby 4ever" Leo 10h19'32" 25d44'31"
Dianna Monraz Crb 15h39'45" 37d24'21"
Dianna Romero Lyn 7h13'34" 59d22'26"
dianna s. green Sco 17h17'56" -32d53'36"
Dianna Samantha And 0h6'41" 28d49'44"
Dianna Siderea Vir 14h8'36" -14d31'56"
Dianna Speranza And 2h20'21" 46d6'7"
Dianna "Starshine" Reeves And 0h26'17" 42d34'52"
Dianna Stockwell Cam 3h57'31" 55d58'32"
DiAnna Suzanne Livingston Cnc 8h13'26" 10d21'19"
Dianna The Hustler And 0h54'28" 37d43'23"
Dianna's When you wish upon a star! And 0h50'5" 42d14'17"
Dianne Uma 11h10'3" 29d35'47"
Dianne Lib 15h12'49" -5d40'0"
Dianne Mon 6h48'56" -0d26'17"
Dianne Cam 11h50'26" 77d17'53"
Dianne Sgr 19h7'7" -20d57'16"
Dianne Bassham Aql 20h2'47" 0d50'28"
Dianne Benjamin Vir 13h59'31" -13d41'48"
Dianne Caputo Gem 7h10'7" 21d35'28"
Dianne Duhaime Ori 6h14'12" 1d59'41"
Dianne Elderedge Cyg 20h31'45" 38d22'41"
Dianne Faith Fogarty - 3.1.1947 Cap 20h26'9" -15d3'50"
Dianne forever and a day And 2h27'57" 43d50'6"
Dianne Gadzuk "What a Star" Uma 13h46'48" 54d4'3"
Dianne Greller Cyg 21h39'9" 50d48'48"
Dianne Hart Crb 16h6'0" 36d31'13"
Dianne Jalalaty Leo 9h22'19" 12d47'51"
Dianne Joy Hauger Cam 7h17'1" 80d46'17"
Dianne Kaylene Cardwell Aqr 23h19'31" -22d29'29"
Dianne Kennedy Vir 13h17'52" 0d7'1"
Dianne Loftin Uma 10h18'4" 44d9'33"
Dianne M Ballino Cas 0h18'29" 58d6'10"
Dianne M Poole And 23h41'54" 36d32'17"
Dianne M Tharp Gem 6h33'2" 21d25'26"
Dianne Marie Keller Lib 15h0'42" -23d53'46"
Dianne Marie Leidy Cnc 8h53'58" 14d54'11"
Dianne Martin and Sam Houston V Psc 0h56'9" 25d6'27"
Dianne Martino Del 20h52'12" 13d15'6"
Dianne McAlister Davis Uma 11h42'34" 40d28'3"
Dianne Michelle Aql 19h40'12" 11d55'12"
Dianne Ouellette Psc 0h0'25" -0d30'7"
Dianne Pagano Sco 17h55'16" -42d38'8"
Dianne Perry Vanderlip Tau 3h43'44" 27d48'46"
Dianne Renee Com 13h13'44" 17d34'57"
Dianne RJDD Hayhurst-Pigg 811 And 0h59'30" 40d10'39"
Dianne S. Dalenberg Cas 1h30'23" 57d6'9"
Dianne Sloan (Lady Di) Cra 19h1'57" -37d59'3"
Dianne Sullivan Aqr 22h27'3" -9d7'3"
Dianne Terrell Cam 3h57'14" 74d53'5"
Dianne Walsh Ari 3h5'1" 23d53'33"
Dianne York Psc 1h10'24" 27d19'15"
Dianne, Samantha & Natalie's Star Cru 12h38'1" -60d16'20"
Dianne's Star Crb 15h57'2" 35d4'50"
Dianoosh Abdolrazagh Lib 15h30'53" -21d53'29"
Diaron Uma 12h0'44" 51d21'22"
Diaundra Jackson Lib 14h44'7" -18d18'30"
Diazen Joshua Gru 22h33'53" -37d58'32"
Dibee's Huggie Bear Uma 11h42'7" 61d53'6"
DiBello Gem 7h43'22" 16d0'56"
Dicataldo Sco 17h24'13" -42d23'36"
Dicey The One Pure Heart In The Sky And 22h59'50" 47d30'24"
Dick Boo 14h41'41" 33d16'1"
Dick Lyn 8h27'50" 42d54'33"
Dick Alen Uma 8h30'2" 71d55'43"
Dick and Addie Young Her 18h4'59" 37d7'57"
Dick and Betsy Cyg 20h31'24" 52d10'29"
Dick and Jane 22 Cyg 20h13'30" 31d26'15"
Dick and Katie Davis Aql 20h11'44" 15d29'59"
Dick and Mae Ungerman~sweethearts Uma 9h7'36" 49d54'41"
Dick and Marcie Ziegler Lyr 19h8'8" 27d20'3"
Dick and Marie Staffin Cyg 21h11'11" 45d38'10"
Dick Baughn Cap 21h0'4" -17d44'47"
Dick Beidler Her 17h47'3" 47d27'8"
Dick Cartelli Crb 16h8'0" 37d9'51"
Dick Chamberlain Leo 11h11'57" 6d39'42"
Dick Clardy Srp 18h10'51" -0d5'15"
Dick Clausen Psc 1h17'24" 10d21'7"
Dick Erhardt Per 3h14'39" 51d33'32"
Dick Foucart Her 17h15'1" 46d35'2"
Dick & Gerry Walz Uma 9h32'12" 48d40'0"
Dick Gorsuch Ori 6h15'56" 10d0'33"
Dick Grimmer Per 2h44'7" 54d30'30"
Dick Hartzell Sgr 19h39'47" -14d49'55"
Dick Hitchcock Uma 11h27'25" 45d55'57"
Dick Larson ~ My Lucky Star Forever Aur 5h29'59" 45d40'8"
Dick Linklater Eri 1h57'6" -50d31'2"
Dick & Lu's Star Ari 2h6'10" 10d46'34"
Dick Mader Lyn 8h53'15" 45d8'7"
Dick Maguire Her 16h35'39" 37d53'11"
Dick - n- Jane And 23h50'4" 40d29'56"
Dick Overmann Cep 22h15'15" 66d2'57"
Dick "PC" Parenti Uma 10h32'50" 65d36'25"
Dick "Peanut" Sturm Per 3h37'21" 48d46'22"
Dick Pickell Birthday Star Aqr 22h36'37" -1d50'36"
Dick Roeder Jr Cep 23h59'55" 74d35'19"
Dick Sessing Sgr 19h43'34" -12d15'47"
Dick Simon Ori 5h17'47" 5d52'22"
Dick Stresser Ori 5h33'28" -4d30'21"
Dick Thomas Ori 5h21'36" 1d32'41"
Dick Vittitow's Rebirth Dra 16h31'28" 57d57'19"
Dick Wilcox Boo 15h25'31" 44d54'14"
Dick Wiley Psc 1h11'16" 15d38'42"
Dick Wilson Uma 9h50'41" 59d14'11"
Dick Wining Her 18h2'30" 16d25'4"
Dickens Cma 6h45'58" -14d43'19"
Dicker Crance Cap 21h11'58" -18d45'47"
Dickerson 40 Cyg 20h23'45" 46d50'31"
Dickerson Roberts Watkins Uma 13h21'17" 52d42'5"
Dickerson Twins Cnc 7h56'24" 13d59'41"
Dickey Loy Uma 10h9'42" 42d36'37"
Dickie & Angela Ari 3h24'47" 27d30'45"
Dickie Hinton And 2h6'56" 37d52'13"
Dickie of FoxCroft Cmi 7h33'14" 4d5'22"
Dickinson Ori 5h47'17" -0d55'50"

Dickran & Geraldine
Tau 4h29'44" 25d28'11"
Dickran Gulesserian, M.D.
Aur 7h13'11" 42d2'29"
DICKY
Sgr 19h18'8" -23d39'20"
Dicky Bird's Fancy Dancy
Tau 4h2'5" 20d55'16"
DiCor
Agr 21h48'37" -2d32'54"
DiDa
Vir 14h13'23" 3d14'13"
Dida
And 23h56'50" 48d40'23"
Didako
Cru 12h16'20" -63d12'8"
Didderz
Lyr 18h45'18" 38d25'28"
Diddly
Ori 5h12'52" -1d27'49"
Diddo
Aql 18h46'0" 8d16'59"
Diddy
Lyn 8h37'47" 37d54'47"
"Diddy"
Sco 17h35'27" -36d9'10"
Didem
Ari 2h11'10" 13d28'2"
Di-De-Mi Love
Psc 1h28'3" 13d42'44"
DiDi -50
Cep 20h26'22" 60d6'28"
Didi Gluck
Mon 8h9'56" -2d2'28"
Didi Vivier
Cyg 19h57'45" 38d37'45"
Didier
Uma 9h24'25" 52d44'29"
Didier Bedel
Tau 5h51'27" 17d15'41"
Didier Chaulier
Lyn 7h5'44" 49d17'27"
Didier et Laurence
Crb 16h14'41" 33d31'12"
Didier Jeanmart
Sco 17h2'24" -38d21'9"
Didier Laurent
Uma 9h56'22" 42d57'31"
Didier Uzureau - Exaltation
Ari 2h34'32" 31d9'55"
Didirocks2005-"Alcoelevja" Pheasant
Umi 16h22'39" 77d55'18"
Didi's Glücksstern
Ori 6h23'56" 11d6'8"
Didi's Star
Lmi 10h21'56" 37d28'37"
Dido
Sco 16h47'23" -37d19'55"
DiDonato's Love
Uma 10h48'2" 45d51'13"
Die=Ana
Vir 13h18'21" -14d14'28"
die leuchtende kraft unser-er liebe
Cas 23h28'37" 51d28'18"
"Die Magie des Lebens"
Uma 9h52'22" 47d37'42"
Die Prinzessin
Agr 21h9'39" -2d13'0"
Die Wonderlike Luzan
Lib 15h16'23" -21d23'47"
Diede Hagens
Cep 22h19'44" 86d20'12"
Diedra Ann Healy
Crb 15h57'40" 34d21'6"
Diedrich, Paul-Heinz
Uma 9h13'34" 63d26'24"
Diegito
Agr 20h59'9" 1d47'28"
Diego
Tau 3h44'18" 28d3'28"
Diego
Her 17h40'8" 21d12'35"
Diego
Lyr 18h31'6" 32d24'9"
Diego
Uma 12h19'36" 56d39'34"
Diego
Ori 5h34'11" -4d1'58"
Diego
Cas 0h31'3" 62d25'24"
Diego Alvarez
Per 3h43'35" 44d13'32"
Diego Calle
Uma 9h58'49" 67d58'58"
Diego Corcobado
Lmi 11h6'27" 28d38'25"
Diego e Elisa
Her 16h35'14" 17d32'3"
Diego Echeverria
Tau 3h29'55" 8d47'0"
Diego Fabbri
Ori 5h24'49" -4d34'47"
Diego Foglini & Giulia Masiero
Aur 6h12'11" 28d36'41"
Diego Guiza Perezarizti
Agr 23h20'33" -20d19'7"
Diego Joseph Ramirez
Lac 22h42'17" 49d43'30"
Diego Le Metayer
Leo 11h58'10" 26d11'45"
Diego Lee Hargis
Vir 13h40'14" 3d33'11"

Diego Luna
Cam 4h32'45" 72d12'57"
Diego Martini
Cnv 12h48'48" 37d53'19"
Diego Mateo Perez Vick
Ori 4h52'15" 3d17'27"
Diego Minto
Cas 0h28'18" 62d22'3"
Diego Mondragon Primer Dia
Col 6h39'22" -33d13'41"
Diego Moresi
Cyg 20h12'26" 35d58'5"
Diego PPG
Cyg 20h8'20" 42d28'40"
Diego Prader
Cyg 21h7'27" 47d4'56"
Diego Retamoza
Sco 16h23'30" -31d47'14"
Diego Siragusa (sexy tesorino)
Vir 15h2'18" 4d19'39"
Diego Valdivia
Sgr 18h45'49" -25d47'24"
Diego Villasenor
Lib 15h12'10" -19d58'53"
Diego y Mariann Para Siempre
Her 17h45'12" 48d56'47"
Diegos estrellan de sus princesas
Uma 10h28'4" 69d18'2"
Diehl, Wolfgang
Uma 10h26'55" 65d4'31"
Diekmann, Sönke
Gem 6h25'16" 20d24'17"
DiEIE, la nostra buona stel-la
Leo 9h29'37" 18d13'49"
Diem Pham
Lib 15h29'22" -20d33'35"
Diemer Hervé Bapascha
Her 17h58'54" 17d36'25"
Diemos
Lib 15h7'24" -3d45'53"
Diena and Kim
Cyg 19h42'2" 42d45'41"
Diendorfer, Christian James
Ori 4h55'51" 11d5'17"
Dienstag
Lyn 7h40'14" 52d8'44"
Dienutza
Sgr 18h0'4" -30d4'12"
Dierdre Rosiejka
And 23h10'13" 47d54'46"
Dierk& Svenja
Uma 9h19'21" 66d48'12"
Dierker
Umi 13h30'49" 75d26'51"
Diesel
Sgr 18h52'53" -16d18'10"
diesel
Psc 1h11'2" 23d51'11"
Diesel Burgan
Uma 13h37'43" 55d14'24"
Diesel Hutcherson
Aql 20h4'54" 7d46'56"
Dieter 24.04.1947
Uma 12h28'32" 53d35'46"
Dieter 24.05.1970
And 23h23'56" 41d40'31"
Dieter Adolf Wilhelm Lowe
Her 18h30'2" 19d21'24"
Dieter Bacher
Per 4h20'5" 44d18'38"
Dieter Biehl
Uma 9h55'24" 56d57'29"
Dieter Blocher
Uma 10h36'51" 42d9'50"
Dieter Bolten
Ori 5h58'9" 7d16'41"
Dieter Czerner
Uma 11h4'17" 57d32'31"
Dieter (Dadi) unser Licht
Sco 17h53'1" -40d59'37"
Dieter Dallmeier
Cap 21h30'22" -23d43'45"
Dieter Ernst Eckert
Oph 16h42'48" -0d54'9"
Dieter Fischer
Uma 8h23'20" 64d55'21"
Dieter Fuxius
Uma 11h5'34" 64d43'13"
Dieter Gaglin
Lyn 7h39'19" 42d29'47"
Dieter Georg Meixner "Grosser"
Uma 10h5'19" 51d50'31"
Dieter Götzl
Uma 10h1'6" 41d29'50"
Dieter Hallek
Uma 9h17'33" 67d48'56"
Dieter Hentschel
Uma 8h35'46" 67d38'31"
Dieter Hofmann
Uma 8h55'45" 61d6'13"
Dieter Kobrock
Uma 13h8'43" 53d18'47"
Dieter Kruber
Uma 9h48'8" 50d8'27"
Dieter Kubli
Cep 20h44'39" 65d35'4"
Dieter & Lynette
Cru 12h30'16" -58d17'2"

Dieter Röhlen
Ori 6h7'51" 7d42'46"
Dieter Schmidt
Uma 13h45'42" 52d3'27"
Dieter Walz
Uma 8h18'25" 61d57'59"
Dieter Weiß
Uma 11h47'52" 51d15'48"
Dieter Wild
Uma 10h7'54" 64d28'58"
Diethelm (M) Sdn Bhd
Ori 5h35'31" 2d48'48"
Dietmann, Gerd
Uma 8h37'40" 59d48'47"
Dietmar Breintner
Uma 13h54'25" 57d32'25"
Dietmar Haubold
Uma 8h38'53" 62d56'38"
Dietmar Hofmann
Uma 11h0'37" 63d39'54"
Dietmar Struch
Uma 13h29'20" 58d50'6"
Dietmar Werner Jaeckle
Lyr 19h16'36" 34d48'20"
Dietra J.
Cnc 8h47'19" 14d33'21"
Dietrich Busing
Uma 12h12'19" 62d18'25"
Dietrich Claassen
Ori 5h22'56" 8d18'17"
Dietrich Reimann
Uma 11h8'30" 28d57'58"
Dietrich Stobbe
Uma 13h40'7" 59d47'25"
Dietrich-Jürgen Poeplau
Ori 6h18'7" 8d55'17"
DIETY594
Tau 3h48'1" 23d38'36"
Dietze, Clara
Uma 10h34'56" 69d18'3"
Dievole
Vir 12h44'53" 3d13'22"
DiFranco
Uma 9h18'41" 48d3'36"
Diga
Aql 19h50'1" 11d42'45"
Digger
Uma 11h43'44" 41d25'17"
Digger
Cru 12h10'9" -57d33'48"
Diggity Dion
Lib 15h38'13" -20d44'49"
Diggitybear
Ori 4h56'44" -2d43'17"
Digits
And 1h26'1" 48d4'15"
Digna
Lyr 18h31'28" 36d47'38"
Digna A. Calderon
Ari 2h36'33" 30d17'10"
Digna D Estevez
Mon 6h42'23" -0d19'26"
Dignelis Taymi Jimenez
Sco 17h38'30" -42d30'4"
Dignity
Ari 3h11'42" 22d55'14"
Dignory Reina
Cam 4h17'26" 67d33'54"
Dijana Andrasic
Peg 22h23'4" 20d27'21"
Dijana's Glücksstern
Uma 9h59'44" 42d8'8"
Dil' Boo
And 23h21'42" 48d12'3"
Dilbagh Singh Virdee
Cep 21h50'33" 67d41'6"
Dildine (BooBoo) LyBarger
Ari 1h47'6" 13d41'18"
Dileas
Gem 6h42'56" 16d6'3"
Dilek Tunc 1984 - 2005
Cru 12h22'16" -56d32'14"
Dilemma
Umi 13h27'28" 76d1'0"
Diletta Giustozzi
Ori 6h7'18" 11d48'35"
Dilia S .T . A . R .
Uma 9h25'1" 43d41'5"
Diligo of Meus Vita Chloe Nicole
Crb 15h37'12" 26d41'45"
Diligo quod Amicitia
Lib 15h53'2" -4d13'30"
Dilini (Baba)
Cas 0h32'29" 51d26'43"
Dilip Keshu
Cyg 20h52'13" 33d0'5"
Diller, Andrea
Uma 11h37'30" 38d32'42"
Diller, Günter
Ori 6h15'11" 15d59'11"
Dilleta
Sco 17h55'34" -34d37'17"
Dillion (Boogie) Craig
Lac 22h47'40" 52d22'36"
Dillon
Ori 6h7'54" 20d48'18"
Dillon Alexander Buff
Ori 4h54'7" 13d3'16"
Dillon Ashton
Ori 5h32'54" 0d22'46"
Dillon Dermot King
Cep 21h42'21" 64d33'4"
Dillon Doucette
Sgr 19h42'16" -14d49'2"

Dillon Douctte
Per 3h11'15" 50d35'44"
Dillon Douglas Gresham
Her 17h0'42" 15d45'22"
Dillon G. Haugh
Ari 2h11'55" 23d2'16"
Dillon Hilderbrand "Sparkly"
Sgr 17h58'36" -29d5'43"
Dillon James Blankenship
Lyn 7h55'55" 43d0'24"
Dillon James Dorr
Ari 2h43'31" 14d55'54"
Dillon John Hudson
Psc 1h10'30" 25d27'8"
Dillon Joseph Phillips
Her 17h34'59" 16d27'6"
Dillon K. Wise
Uma 13h24'5" 58d44'57"
Dillon Lee ONail
Her 17h40'8" 39d4'36"
Dillon Michael Mondino
Cep 21h47'55" 69d55'27"
Dillon Moody
Gem 6h47'44" 19d34'55"
Dillon Nicholas Carney
Agr 23h18'39" -18d21'25"
Dillon Ostrenger
Tau 4h15'44" 18d57'45"
Dillon Ray
Lib 15h5'16" -5d29'26"
Dillon Rhys Mantei
Sgr 18h34'5" -32d25'21"
Dillon Thomas Laffitte
Tau 4h10'46" 19d57'59"
Dillon Tichy Thomas
Umi 14h26'4" 74d55'45"
Dilwel57
Cru 12h27'49" -61d2'50"
Dilyara (Chudo)
Uma 10h32'7" 45d21'18"
Dim 9
Tau 5h34'22" 22d5'55"
Dima
Uma 10h17'18" 44d1'36"
Dima Merkulov
Sgr 19h6'14" -35d55'26"
DimAnna
Ari 2h21'41" 26d27'44"
DiMarcus L. Mayes
Agr 21h50'32" -4d59'35"
Dimas de la Cruz
Uma 11h37'58" 58d44'20"
Dimbat, Günther
Uma 8h49'21" 58d15'54"
Dime - a - Dance Dan
Uma 11h5'26" 33d27'0"
Dimi
Umi 13h27'4" 72d7'50"
Diminishing Marginal Utility
Ari 3h8'22" 24d8'6"
Dimitra Kalafati
Vir 12h17'30" -0d3'46"
Dimitra Tzoumas
Uma 11h34'46" 45d33'13"
Dimitri Chaliotis
Agr 23h36'33" -22d57'7"
Dimitri Chernyak
Umi 13h54'53" 69d36'31"
Dimitri Drakopoulos
Dra 17h53'12" 54d12'51"
Dimitri George
Her 16h5'53" 46d45'15"
Dimitri "Jimmy" Djordjevic Coukos
Uma 12h3'30" 57d57'45"
Dimitri Kotsalis
Cru 12h28'30" -59d41'21"
Dimitri Maslennikov
Pho 0h38'53" -42d3'29"
Dimitrios Alexiou
Cep 20h43'17" 58d7'37"
Dimitrios Klonaras
Ari 2h41'11" 22d5'36"
Dimitrios Tzallas
Uma 11h29'25" 46d33'36"
Dimitrios Zanikos
Umi 17h10'56" 82d6'9"
Dimitris
Psc 0h38'29" 17d53'42"
Dimitris
Peg 21h41'19" 7d27'29"
Dimitris Bardanis
Per 3h0'19" 52d52'25"
Dimitris Kontominas
Uma 10h24'38" 72d23'51"
Dimitris M. Stavrakakis
Cep 20h40'14" 62d45'52"
Dimitris Markopoulos
Uma 10h47'25" 39d35'59"
Dimitris Mi Masas
Cap 20h43'7" -15d11'21"
Dimitris Thomopoulos
Uma 11h45'39" 32d49'53"
Dimity
Tau 5h55'51" 25d8'26"
Dimivat
Cru 12h36'7" -63d18'42"
Dimka And Tilik's Dreams
Lyr 18h27'32" 32d31'23"
Dimochka
Sgr 18h55'35" -32d29'6"
Dimo's Dream
Aur 5h35'13" 42d2'52"
Dimple
Uma 10h53'3" 49d5'28"

Dimple
Uma 11h7'24" 59d17'11"
Dimple "Momo" Drysdale
Uma 11h51'11" 63d16'32"
Dimples
Leo 9h58'54" 18d42'49"
Dimples Montes
Psc 0h49'53" 16d18'9"
Dimples Stacey Lynn Duncan
Uma 11h35'50" 48d1'49"
Dimus Vinograd
Ari 2h53'38" 28d25'15"
Din Din
Lyn 8h8'1" 34d32'21"
Din Goldschatz
Lmi 10h42'0" 34d16'56"
Dina
Lyn 7h44'14" 36d2'49"
Dina
And 0h28'7" 39d45'0"
Dina
Uma 14h2'17" 47d53'44"
Dina
Psc 2h1'19" 6d43'21"
Dina
Umi 15h30'59" 71d5'17"
Dina 1989
Uma 10h36'47" 45d50'18"
Dina Abramova
Vir 12h13'31" 11d32'6"
Dina Agha
Cas 2h15'49" 70d35'37"
Dina Ann Ditmar
Lib 15h6'56" -0d44'27"
Dina Ann Turner
Sco 17h28'29" -42d47'55"
Dina Arnold-Rossi
And 1h57'8" 40d51'2"
Dina As Our Two Souls Meet
Cap 21h20'11" -17d27'21"
Dina B.
Lib 14h29'44" -19d51'53"
Dina Beans
And 0h55'6" 40d31'49"
Dina Cidade
Gem 7h18'34" 21d9'49"
Dina Diruscio
Dra 15h3'39" 58d33'58"
Dina Eller
And 0h32'2" 34d0'3"
Dina Excelsior
Tau 5h52'37" 13d59'22"
Dina Figarola
Peg 23h55'45" 24d28'30"
Dina Firter
Uma 11h17'21" 42d41'4"
Dina Garcia
Lyn 8h21'31" 52d11'3"
Dina Gildeh
Cyg 19h37'20" 30d6'13"
Dina H. Poulin
And 2h30'42" 44d48'21"
Dina L. Lloyd
Vir 15h2'55" 2d16'51"
Dina Lepow Spector
Sco 17h54'17" -32d38'13"
Dina M. DiGiacomo
Vir 14h1'12" -9d50'51"
Dina Marie
Agr 23h7'15" -8d18'26"
Dina Marie
And 0h16'57" 27d39'52"
Dina Meyer
Uma 9h34'44" 59d32'44"
Dina Mostafa
Leo 9h56'33" 7d21'38"
Dina Perratto
And 1h35'49" 48d5'29"
Dina_Rob
And 1h35'53" 45d46'7"
Dina Simoes
Aql 19h40'53" 3d33'33"
Dina V. Pina
Leo 11h55'47" 22d5'15"
Dinah
Gem 6h48'29" 27d35'22"
Dinah
Cap 20h42'7" -26d59'29"
Dinah and Dolores Yorkin
Gem 7h28'36" 24d13'40"
Dinah and Gary's Eternal Love
Sco 17h10'34" -31d8'30"
Dinah Emma
Ori 5h53'22" 7d15'26"
Dinah Estelle Wharton
Mon 7h2'53" 5d37'6"
Dinah Lynn Rollison
Cam 7h50'4" 61d45'12"
Dinah Nasty
Cnc 8h41'42" 30d52'14"
Dinah Tapper
Umi 0h35'4" 88d57'29"
Dinah's Pride
Dra 16h44'25" 56d38'8"
Dinah's wish
Dra 16h7'48" 59d14'54"
DinaKa
Sgr 19h44'27" -16d31'42"
Dina-Marie and Daniel
Sco 16h7'59" -12d23'53"
DinAntoniettaSturAnse
Aur 5h17'34" 49d37'45"

Dina's Rome
Lib 15h4'30" -0d51'7"
Dina's Star
Cyg 20h55'20" 35d14'47"
Dindo Delicano Vivar (Norman)
Cap 21h13'27" -23d6'58"
Dine & Don
Cyg 21h10'7" 38d16'53"
Dine Yazhi
Aql 19h21'11" 15d5'17"
Dinee
Uma 12h42'0" 57d7'6"
Dineen Michelle Talerico
Gem 6h26'32" 17d50'21"
Dinesh Gokal
Cep 22h41'48" 76d17'24"
Dinesh Matthew
Ari 2h52'54" 28d49'51"
Dinesh Ravishanker
Tau 5h32'24" 18d44'50"
Ding Family Star
And 2h35'31" 43d23'5"
Ding Fang
Sco 16h13'42" -14d20'43"
Ding-Ding
Sco 17h9'43" -31d11'52"
Dingo's Star
Uma 11h35'51" 50d52'21"
Dini
Ori 4h47'15" 5d17'18"
Dini
Ori 6h0'39" 10d27'27"
Diniah
Cas 1h27'29" 64d31'34"
Dinina
Psc 1h41'45" 22d31'17"
Dink
Gem 6h28'4" 24d37'5"
Dinka
And 1h9'7" 42d27'8"
Dinka Marin
Lib 15h31'40" -27d59'46"
Dinkel
Ori 5h54'20" 17d17'56"
Dinkel, Ralf
Uma 9h31'53" 42d5'35"
"Dinker May"
Tau 4h13'19" 21d48'23"
Dinkle Major
Lyn 8h11'59" 47d3'25"
Dinkum
Lyr 18h56'14" 33d56'36"
Dinkum
Agr 22h33'27" -0d34'57"
Dinky Barvinchak
Per 2h22'16" 53d20'1"
Dinky Monkey
Ori 5h11'59" 6d37'24"
Dinky Mouse
Umi 15h16'31" 69d34'6"
Dinky Welbury
Lyn 6h38'47" 57d9'19"
Dinno + Noelle
Cyg 20h19'45" 58d57'20"
Dinny L.
Cnc 9h2'15" 12d40'18"
Dino
Her 16h39'51" 6d14'23"
Dino
Psc 0h36'45" 8d53'37"
Dino
Cnc 9h10'6" 25d40'48"
DINO
Tau 5h49'19" 24d16'33"
Dino Alessandro Palatiello
And 23h41'37" 33d23'34"
Dino D'Argenzio
Lib 15h4'14" -7d8'11"
Dino Deane
Cnc 7h57'21" 12d50'28"
Dino Latino
Dra 14h53'54" 55d37'9"
Dino Mavroudis
Uma 12h29'24" 56d19'10"
Dino Michael Rudolph
Uma 9h46'46" 56d12'21"
Dino & Tiger's Playground
Lib 14h53'41" -3d1'29"
Dino Valentino Raponi
Uma 11h23'23" 57d37'58"
Dinokitty
Umi 14h42'2" 75d51'49"
Dinolino
Uma 8h54'25" 63d8'32"
Dinorah Garcia
Sgr 18h27'20" -27d18'25"
Dino's Star Of Bella
Aql 19h57'25" 9d25'19"
Dinosaur
Per 3h1'35" 40d20'57"
Dinullini
Agr 20h39'9" -7d21'7"
Diodato Francesca
Lyn 8h22'27" 37d47'51"
Dion Ahmetaj
Her 18h40'42" 25d39'4"
Dion (foggy) Romanowski
Uma 11h10'58" 51d5'32"
Dion J. Burmaz
Lib 15h14'39" -6d38'35"
Dion V. Raudenbush
Ori 6h8'39" -0d5'24"
Dion Wamsley
Sco 16h41'38" -31d51'41"

Dion Webster
Tau 3h56'18" 12d0'25"
Diona Daniele Jones
Sgr 19h44'21" -25d27'38"
Diona Macaraeg
Sgr 18h38'28" -22d16'11"
Dionde
Crb 15h45'41" 26d16'20"
Dione R. Adams
Cyg 19h37'55" 38d2'58"
Dione Scheltus
Cru 12h4'24" -62d34'0"
Dione Vaios Alexon
Gem 6h40'25" 14d58'10"
Dionisi
Lib 14h50'31" -2d19'28"
Dionisia Tedesco
Umi 14h3'40" 72d25'43"
Dionna Evelyn
Ari 3h18'2" 27d29'57"
Dionna McDonald
Cap 21h47'22" -10d38'21"
Dionna Trane`
Uma 13h35'23" 58d5'9"
Dionne
And 0h45'36" 21d56'58"
Dionne
Psc 1h17'35" 14d19'22"
Dionne Hamlin
Gem 7h42'14" 34d35'56"
Dionne's Corona
Uma 11h4'51" 40d23'20"
Dionysia Sweeney Ruht
Cap 20h58'32" -19d15'11"
Dionysis
Per 2h54'33" 49d55'18"
DionysusJ
Agr 22h43'9" -9d24'8"
Dior Chinai Strong
Uma 8h31'10" 67d59'11"
Diorio's light
Uma 10h2'13" 71d36'50"
Dios es siempre cerca
Sco 17h46'22" -42d30'52"
Dios Mandó
Her 18h15'31" 22d6'42"
Dios me ha bendecido con Stephanie
Cru 12h19'55" -57d54'14"
Diosa
Sco 16h56'51" -31d33'36"
Diosa Em
Cas 0h52'39" 48d10'22"
Dioscaris
And 23h24'45" 47d34'14"
Dioscuri
Gem 6h35'40" 21d43'4"
Diossy Gonzalez
Cap 20h24'12" -16d42'34"
Dipak Bagchi
Per 3h53'5" 32d35'26"
DIPI
Cyg 19h46'47" 34d0'25"
Dipiglas
Gem 6h31'6" 13d53'8"
Dipl.- Phys. Peter Weinrich
Vir 11h50'28" 3d40'28"
Diplodocus Maximus
And 23h55'3" 37d59'48"
"dips"
Peg 22h35'40" 2d52'56"
Dirce D'Agnolo Cassini
Cas 22h58'21" 53d48'7"
Direktor Martin Povodon
Umi 14h34'4" 72d24'17"
Dirge - My Love
Ari 2h4'19" 13d18'56"
DiRienzo's Passion
Uma 10h51'28" 69d52'11"
Dirk
Lac 22h10'19" 52d51'55"
Dirk
Lyr 18h44'58" 41d23'2"
Dirk
Uma 11h22'48" 48d47'25"
Dirk (Debi's angel)
Agr 23h42'41" -11d9'28"
Dirk + Eri Love forevermore
Uma 8h13'30" 60d14'43"
Dirk Evan's Bright Light
Ori 5h41'11" 12d18'30"
Dirk Genkens
Her 17h59'11" 23d47'59"
Dirk Gliese
Uma 8h20'43" 66d39'54"
Dirk Jan Eric van Eeuwen
Psc 1h10'8" 19d31'58"
Dirk Jansen
Uma 14h12'16" 57d53'56"
Dirk Jope
Uma 10h36'17" 68d14'23"
Dirk Lorenz
Ori 6h23'4" 15d4'8"
Dirk März
Ori 5h21'54" 9d13'57"
Dirk Mönig
Uma 10h41'48" 70d27'31"
Dirk Peter Meier
Uma 13h26'13" 61d28'18"
Dirk Ploegmakers "Dirks' Plow"
Per 3h38'41" 46d27'10"
Dirk Vermeulen
And 1h17'1" 42d21'37"

Dirk Wiese
Uma 9h59'24" 63d30'22"
Dirk, auf ewig unvergessen !!!
Uma 9h6'8" 50d24'31"
DIRT:STAR
Lib 15h7'51" -11d2'4"
Dirty Faccia Lorde
Aur 5h9'3" 40d49'14"
Dirty LiL Secret
Gem 6h47'42" 12d40'38"
Dirty M
Uma 11h14'37" 53d41'55"
Dis
Aur 7h19'38" 41d36'56"
Di's Angel
Sgr 19h9'47" -19d21'49"
dis mami
Ori 5h39'28" -10d28'30"
dis schätzli
Her 17h25'36" 37d23'41"
Disc
Uma 12h12'36" 55d26'3"
Disco
Umi 14h26'58" 66d35'21"
Disco Dad R M H's Daddy Star
Aql 18h55'47" -0d44'20"
Disco Jonathan Beckley
Crb 15h31'58" 26d59'20"
DISCO & JOY
Uma 10h54'19" 48d28'16"
Discovery Bay Hotel
Lyr 19h5'9" 47d24'4"
Dislay
Gem 6h54'22" 18d56'7"
Disney Dawn VanHaitsma
And 1h39'31" 48d13'57"
Disney Girl AKA: Lisamaeous
Lib 14h50'33" -1d29'8"
Dissa Patinka
Dra 20h20'16" 67d7'13"
Dissauer 1964
Uma 8h12'58" 67d29'59"
Distant Reflection
Lyn 7h51'4" 51d28'2"
Disten
Psc 23h23'21" 2d33'5"
DISTROYER
Ori 6h2'7" 13d36'39"
Ditda
Aur 5h5'12" 47d23'50"
Ditria Wilburn
Tau 4h15'10" 26d28'48"
Dittmann, Karl-Michael
Uma 8h20'29" 63d36'37"
Dittmann, Marko
Uma 11h49'4" 37d0'25"
Dittmeyer, Brigitte
Uma 9h10'35" 51d12'15"
Ditto
Cyg 20h15'50" 51d26'51"
Ditto
Cyg 21h57'5" 48d3'42"
Ditto
Tau 5h49'8" 26d47'12"
Ditto
Ari 2h44'23" 29d23'2"
Ditto
Dor 5h59'8" -68d9'5"
Ditto 41
Tau 3h40'52" 29d12'20"
Ditto Day
Uma 9h31'50" 57d36'6"
Ditto Man
Peg 22h45'18" 25d45'7"
Ditto Marchetti
Ori 5h45'8" 5d58'5"
Ditto More
And 23h36'9" 47d44'41"
Dittrich-Wenzel
Uma 12h50'36" 62d27'32"
Div
Leo 9h30'44" 26d29'55"
Diva
Uma 9h39'4" 52d19'12"
Diva
Cas 23h50'49" 52d52'38"
* Diva Dezy *
Her 18h10'59" 47d39'14"
Diva Laurie Kingston
And 2h25'55" 46d49'6"
Diva Margaret Jeanne
Aqr 22h39'45" 2d10'57"
Divachelle
Cap 21h12'47" -15d13'27"
DivaKri
Crb 15h47'7" 33d21'49"
Divalysscious Moms
Cas 1h6'18" 61d25'4"
Divani P. Singh
Vir 13h15'53" -2d40'41"
Diventare
Cas 2h19'3" 74d38'6"
Divi Southwinds
Lyr 18h18'29" 44d0'10"
Divina Eve
Peg 22h38'28" 16d48'48"
Divina Gracia
Leo 11h17'57" 6d35'1"
Divina Marlene Florice
Lyr 18h36'31" 37d23'20"
Divine and Lovely Emily Ivy
Ori 5h45'14" 5d44'12"
Divine Guidance
Cra 18h4'23" -37d10'16"
Divine Intervention
Uma 11h45'46" 42d28'8"
Divine Justine
Cas 0h36'19" 58d14'54"
Divine Star of Olivia
Sgr 18h28'12" -18d9'6"
Divinely Sweet Eudora Beckley
And 2h34'44" 45d13'21"
Diving For Dreams
Ari 2h22'29" 19d27'33"
Divinius
Mon 6h40'42" 9d33'38"
Divino
Cas 0h40'39" 61d16'58"
Divinus
Ori 6h5'1" -0d39'55"
Divorah Muravsky - "Fluffy"
Cnc 9h16'50" 9d46'35"
divum est terminus
Aql 19h4'34" 15d49'57"
Divya
Psc 0h47'54" 16d47'15"
Divya Ajitkumar Patel
Tau 5h48'19" 22d17'36"
Divya Bahl
Sco 17h5'25" -43d27'29"
Dixcee Loo
Vir 13h41'43" -16d36'28"
Dixie
Lyn 7h34'18" 56d23'8"
Dixie
Gem 6h8'7" 26d36'14"
Dixie
Lyn 8h15'59" 40d33'1"
Dixie Arnold
Lyn 9h19'6" 38d40'42"
Dixie Casford
Uma 11h52'41" 37d40'49"
Dixie Doodle
Uma 11h10'28" 58d44'19"
Dixie Elizabeth Lincks
Gem 7h27'13" 26d39'13"
Dixie Helen David
Cap 21h4'26" -19d0'30"
Dixie & Henry Posert
Peg 22h4'33" 6d29'40"
Dixie J. Saalwaechter
Uma 11h14'55" 59d19'35"
Dixie Joyce Sawicki Fritts
Umi 15h17'47" 86d27'18"
Dixie Marie
Aqr 22h9'46" -8d1'22"
Dixie Martin
Leo 11h21'47" 12d24'40"
Dixie Ophelia Connolly
Cap 21h5'23" -19d5'9"
Dixie Rose Callinicos
And 23h49'18" 34d58'46"
DixieDew
And 1h36'25" 41d48'51"
Dixieme Anniversaire Etoile
Crb 15h30'44" 30d45'29"
Dixie's Diamond
Lmi 10h26'48" 38d30'42"
Dixie's Star
And 0h35'24" 30d16'4"
Dixon-Balaoing
Uma 10h57'8" 34d7'5"
Dixons A.G.L.O.
Lmi 11h0'16" 27d41'28"
Dixon's Annabelle Lee
Cnc 9h9'57" 25d17'53"
Dixxie
Leo 11h12'51" 22d46'31"
Diya Patel
Ari 2h36'55" 19d13'54"
DIYA - Tetyana Kondratyuk
Cnc 8h17'27" 32d52'14"
Diz Best The brightest star of all
Sco 16h18'2" -42d3'14"
Dizzy Lizzy
Uma 8h39'28" 70d35'34"
DJ
Lib 14h28'9" -19d41'22"
D.J.
Her 18h10'28" 41d15'19"
D.J.
Uma 9h3'26" 46d51'57"
D.J.
Her 16h52'23" 17d18'8"
DJ 2-11
Aqr 21h52'9" -3d51'28"
D.J. Boyle - Chain
Aur 5h32'40" 33d34'59"
DJ Collins
Cnc 8h26'56" 26d34'58"
D.J. Crater
Psc 1h19'7" 31d18'5"
D.J. Daum (1/2/82-1/30/05)
Cap 20h26'56" -12d31'13"
D.J. DeZuzio, Jr.
And 1h4'31" 37d22'25"
DJ Eternal Darkness
Ori 6h9'54" 9d7'22"
D.J. Fobert
Sgr 18h22'8" -22d57'20"
DJ & Kimmy: Always and Forever
Uma 12h41'54" 61d3'30"
DJ Larsen
Dra 12h32'31" 72d41'4"
DJ Madigan
Uma 11h44'12" 63d37'32"
D.J. Maurer "My Everything"
Aqr 22h1'39" -19d1'39"
DJ McKellar "starring" as Peter Pan
Uma 9h22'3" 65d3'28"
DJ Mike B
Per 4h34'28" 44d47'59"
D.J. Monteiro
Uma 10h0'8" 55d21'18"
DJ Murf
Lac 22h38'16" 37d51'50"
D.J. Nufer
Hya 9h33'26" -0d30'8"
DJ OTZI
Uma 10h42'4" 48d43'16"
"DJ PETE" MUNOZ
Boo 14h8'28" 17d51'6"
DJ PJ
Uma 10h26'43" 68d13'11"
DJ & Rachel Footh
Lmi 9h29'49" 37d1'30"
DJ R-Fresh
Gem 6h17'41" 25d35'17"
DJ Rockin Russ
Psc 0h58'56" 25d22'34"
DJ Scholl
Uma 10h58'31" 68d16'35"
DJ Signo 60
Ori 6h18'15" -0d28'53"
DJ Space RLD
Cas 2h40'38" 57d47'20"
D.J. Stewart
Leo 11h8'50" 27d45'12"
DJ "Sweet Dreams" Underwood
Cnc 8h25'57" 11d56'16"
DJ - The Girl Of My Dreams
Leo 9h28'39" 12d7'42"
DJ White Gravey
Ori 5h37'42" 4d56'7"
DJ WhiteGravey
Per 3h1'26" 54d52'10"
DJ922
Mon 6h39'23" -6d1'25"
D'Jamiel & Jason
Leo 11h44'31" 20d4'36"
Djamila
Tau 4h32'51" 22d1'40"
Djamilla Anna
Cep 22h53'14" 80d20'4"
DJANGO
Cra 18h45'36" -38d15'57"
Djavad Mowafaghian
Uma 11h23'40" 57d54'16"
DJay Euphoria
Lyr 18h45'48" 34d46'29"
DJC Forever My Dad Love Sam
Per 4h32'23" 40d19'35"
D.J.Crawford
Vir 14h36'19" 4d8'33"
Djeiar
Tri 2h5'59" 33d17'9"
Djenaver - Black Swan
Cru 12h13'43" -63d35'10"
Djidji
Uma 9h26'25" 55d23'24"
Djillali
Leo 11h7'11" 8d27'34"
Djoest
Crb 15h31'53" 38d19'24"
DJong "ditto" 4EVA
Pho 1h10'36" -44d57'59"
~*Dj-Q*~
Her 16h30'43" 8d23'54"
DJRVC Ryder
Aqr 22h5'54" -15d23'40"
D.J.'s Narnia
Ori 5h18'26" 11d21'39"
DJ's Star
Uma 9h48'43" 59d22'46"
DJ's Wishing Star
Leo 11h27'28" 12d5'54"
dk
Cyg 21h24'0" 52d36'0"
DK
Uma 9h53'36" 62d16'23"
D.K. Gall
Cyg 19h16'46" 52d48'6"
DK Grill
Sco 17h26'40" -42d52'34"
DKH-1
Uma 13h8'56" 58d42'11"
DKJA Students Past Present & Future
Per 3h51'37" 33d53'16"
DKS_CNN
Gem 7h14'16" 27d28'57"
DKSH Marketing Services Sdn Bhd
Ori 5h36'12" 2d54'5"
DKT1225
And 23h19'31" 48d3'22"
DL Erwin
Uma 9h34'28" 42d17'27"
DL Humphreys
Cyg 21h18'51" 52d19'39"
D.L. Young
Cap 21h7'5" -18d14'0"
dla mamy ELI SZYMCZAK od dzieci AGT
Ari 2h48'22" 24d51'50"
Dla Mojego Meza Krzyska
Cyg 20h52'7" 46d58'45"
D-Lanz
And 0h50'33" 42d18'3"
DLCotton67
Aqr 22h8'56" 1d35'19"
DLDHDryden
Aql 19h38'35" -7d27'23"
D-Lee 01
And 1h13'1" 42d48'8"
DLG Baquero
And 2h11'27" 38d30'5"
DLGN89
Lib 15h12'9" -6d3'32"
Dlineman
And 0h18'45" 26d42'1"
DLKG - 911
Her 17h19'33" 45d42'23"
dlmx2
Dra 19h30'28" 72d50'9"
DLS 10/02/04
Sge 20h10'33" 18d4'29"
DLS SRL
Ori 6h5'12" 19d51'12"
DLT True Patriot Hero
Per 3h29'51" 50d28'50"
Dlúth Chara Gcónai, Jenny-Girl
Cnc 8h20'33" 9d27'4"
Dlymiester
Cas 1h19'59" 51d45'36"
D'Lyn
Umi 16h20'2" 79d43'28"
DMA/HAS
Vir 13h11'0" 13d20'53"
"D-Mac" Dana McAfee
Aql 19h17'36" 7d3'57"
DMaki2
Lib 15h41'41" -12d52'51"
D-Man Deardorff
Tau 4h9'48" 7d42'47"
DMAN Hero 01/25/01 - 07/17/05
Per 4h21'47" 45d1'45"
dman1954
Aql 19h4'51" 9d3'0"
D.Martin's girls
Uma 10h59'56" 47d34'18"
DMazzulo
Tau 4h36'9" 14d21'51"
DMBIRCH2005IAS
Uma 13h2'38" 52d31'10"
DMDSAT
Lyn 7h46'33" 44d43'47"
DMG-PBT102106
Cyg 20h15'47" 51d44'42"
DMH30
Her 17h12'14" 18d51'24"
Dmitri Petrov
Dra 17h16'50" 70d37'10"
Dmitri & Vlade
And 1h51'37" 41d28'41"
Dmitrij Weiß
Uma 9h26'37" 49d23'58"
Dmitrius Mahon-Haft
Lib 15h39'7" -19d41'35"
Dmitry Baskakov
Her 18h1'57" 21d37'1"
Dmitry Gursky
Uma 13h48'27" 48d52'20"
Dmitry Remy
Psc 1h14'30" 27d7'52"
DMK & EML (Bratnie Dusze)
Cyg 19h29'21" 54d10'56"
DMN
Cyg 20h48'21" 41d9'51"
D-Mo
Uma 11h34'33" 47d53'7"
DMOC
Uma 11h34'51" 62d26'48"
DMS 4/14/2001
Aqr 21h49'31" -5d17'18"
DMS-40
Uma 9h14'56" 53d27'59"
DNA
Lib 14h53'47" -7d54'5"
D-N-A
Ori 6h13'58" 8d33'47"
DNA Sarmiento Moore
Ari 2h33'57" 26d51'22"
dnablue
Umi 15h55'2" 76d51'53"
DNANVSSS*
Vir 12h39'55" 1d37'14"
DnCfor70
Uma 11h58'17" 40d16'49"
DNCNXTC
Cam 3h52'11" 67d22'33"
D'Nessa LeShea
And 0h21'27" 35d14'50"
DNR
Ori 5h1'50" -0d59'10"
Do
Lyn 7h3'37" 59d16'31"
Do -Dah
Sco 17h20'43" -41d14'0"
Do Dieu Linh
Leo 11h53'2" 18d45'18"
Doan Quynh Mai - clcyckl!
Cas 0h21'33" 50d0'39"
Doane Star
Leo 11h56'40" 22d20'48"
Dobbie
Uma 10h27'54" 53d45'33"
Dobbin Santa Ana
Cap 21h43'30" -18d32'33"
Dobbins
Aql 20h5'14" 5d16'44"
Dobi
Gem 7h9'15" 28d5'34"
Dobie Wayne Carver
Psc 1h14'3" 11d43'48"
Dobriyin*
Vir 12h50'20" 3d11'23"
Dobronz
Lib 15h38'12" -6d45'34"
Doc
Aqr 22h59'28" -6d49'47"
Doc
Uma 14h9'46" 59d18'27"
Doc
Dra 19h12'53" 60d27'52"
Doc
Leo 9h49'24" 32d11'8"
Doc
Lyn 9h27'4" 41d9'56"
Doc Baird
Cap 21h56'31" -17d38'31"
Doc & Bel forever
Ara 16h48'12" -62d41'58"
Doc Haimes
Oph 17h41'9" -0d37'30"
doc. Ing. Hana Brezinová, Csc.
Lac 22h1'6" 38d5'38"
Doc Leigh Cognitive
Uma 11h37'16" 48d58'25"
"Doc Neal"
Sco 17h57'14" -31d2'35"
Doc Paice
Per 3h10'29" 49d49'17"
Doc (Poppy)
Lib 15h56'40" -14d16'7"
Doc Rainer MBL
Uma 13h52'40" 57d2'14"
Doc Wendy
Peg 23h9'18" 17d25'3"
DOC WILHELM
Her 16h8'14" 24d22'0"
Docherty Ferocity
Her 17h51'51" 23d56'33"
Doci Kaniuk "Sloneczko"
Leo 9h28'10" 19d57'55"
docinboca
Del 20h36'46" 15d20'11"
Docqueline
Vir 13h24'38" -0d33'44"
Doc's Classical Theatrical Red Baron
Tau 5h8'38" 27d43'0"
Doc's little bit of heaven
Uma 12h23'5" 60d44'45"
Doc's Star
Cyg 21h54'8" 38d5'13"
DocStar082051
Leo 11h17'48" 0d42'25"
Doctor Alan and Linda Levene
Umi 13h43'48" 71d58'37"
Doctor Alla Maslova
And 1h15'0" 48d49'37"
Doctor B
Her 17h12'31" 20d31'57"
Doctor Dani
Umi 15h5'40" 73d19'54"
Doctor Francois Sun
Ari 2h36'2" 19d50'45"
Doctor G
Aql 19h30'2" 1d33'37"
Doctor Kirk H. Packo
Uma 10h11'38" 53d41'14"
Doctor Mike
Per 3h42'0" 46d41'22"
DOCTOR MUSIC
Lyr 18h49'4" 33d58'32"
Doctor Patricia Owens
Ari 2h19'41" 23d44'2"
Doctor Paula King
Oph 17h20'17" 7d43'2"
Doctor Phelps
Per 2h54'1" 47d26'48"
Doctor Roman Zabarskiy
And 1h13'17" 47d25'14"
DOCTOR SAMI EL KHAS-SAWNEH
Ori 5h55'49" 12d50'6"
"Doctor Silens" - Silent Teachers
Uma 9h11'58" 52d58'34"
Doctor Stephen Campbell
Per 3h26'18" 49d2'6"
Doctor Wilma Westensee
Dra 14h31'56" 56d5'35"
Doctors Beth and Jeff Shuster
Cyg 20h58'5" 40d41'7"
Dodarlee
Cru 11h56'56" -60d34'11"
Dodd Adam Pfeffer
Ori 6h6'12" 16d19'2"
Doddius Mobius
Leo 9h41'54" 30d25'58"
Doddsy The Rock
Uma 9h24'2" 45d51'8"
Doddy
Mon 6h49'2" -0d25'58"
Dodecathesis
Gem 7h7'54" 31d50'17"
Doderer, Peter
Uma 13h14'10" 52d45'44"
Dodger
Ori 5h54'27" 17d26'53"
Dodgey!
Leo 10h33'2" 13d20'10"
Dodi Carl Star
Ari 2h15'23" 23d19'22"
Dodi Merrill
Lyn 7h8'7" 54d51'1"
Dodie
Uma 10h18'38" 55d23'23"
Dodie -N- Al
Her 17h36'40" 39d33'52"
Dodie Nelke
Sgr 18h59'35" -16d27'22"
Dodie Roxanne Asbury
Mon 6h44'43" 9d0'15"
Dodie Verlanic
Ari 2h52'56" 13d55'11"
Dodie's Light
Tau 3h51'32" 26d14'27"
Dodie's Star
Vir 12h19'44" 12d13'32"
Dodo
Uma 8h35'59" 52d14'13"
Dodo
Cnv 13h32'45" 30d26'26"
Dodo
Uma 11h32'10" 43d7'42"
Dodo
Umi 15h16'24" 80d32'58"
Dodo Land
Per 2h38'59" 42d33'45"
Dodojon
Cas 23h15'2" 55d20'49"
Dodo's Lucky Star
Per 4h30'15" 48d45'40"
Dodo's Star
Aqr 21h48'7" 1d57'29"
Dody
Cam 3h58'50" 72d34'18"
Dody
Vir 13h47'36" -0d21'17"
Doe
Peg 22h30'31" 6d19'46"
Doe In Breeze
Psc 0h7'37" 7d47'33"
Dog God
Uma 10h42'34" 60d21'1"
dog nose
Sco 16h43'25" -24d55'35"
Doga Egin
Psc 0h38'45" 21d25'30"
dogcodseye
Boo 14h14'56" 28d18'36"
Doggie Heaven
Uma 12h59'25" 54d1'32"
Doggles
Cnv 13h39'45" 37d6'25"
Doggy
Psc 0h16'28" 19d8'28"
DogManX-MILKMAN
Tau 4h27'54" 13d27'1"
Dogskin
Gem 7h18'37" 32d17'38"
Doherty's Star
Ori 5h32'15" 9d54'34"
Doig's Law
Uma 10h33'5" 46d37'29"
Doina
Ari 2h54'36" 18d51'4"
Doina Aniela Corneanu
Lib 14h50'29" -7d53'54"
Dokaste
Cap 20h9'44" -18d0'32"
Dola
Crb 15h22'59" 29d3'32"
Dolan
Uma 8h38'4" 60d9'56"
Dolan Thomas Wynne
Dra 18h51'26" 65d30'7"
Dolç Estel
Gem 6h43'7" 23d7'1"
dolce adorabile Dora
Umi 13h59'45" 88d20'34"
Dolce Amore
Cas 1h24'26" 59d43'26"
Dolce cara amica mia Alessandra Dei
And 23h39'36" 48d5'8"
Dolce Claudia
Her 17h50'54" 21d6'28"
Dolce Ellen
Crb 16h17'36" 35d44'13"
Dolce Mamy
Cas 0h29'3" 62d22'14"
Dolcezza Mia Grandma's Giada
Aql 20h5'1" 7d30'24"
Dolci Grace Humphreys
And 0h44'2" 30d40'5"
Dolci Labbra
Vir 14h0'14" -15d1'25"
Dolcissima Patrizia
Cas 1h1'11" 60d26'13"
Dole II
Tau 3h44'42" 26d35'26"
Doliana
Her 16h24'19" 17d10'34"
Doll
Lib 15h24'33" -22d39'48"
Doll Face
Vir 11h51'19" -5d39'44"
Dolla Miriam
Vir 12h46'7" 12d7'47"
Dollar
Dra 11h5'32" 77d41'18"
dollbaby
Lib 14h22'3" -15d39'39"
Dollface
Crb 15h34'30" 28d17'12"
Dolliana, Herbert
Uma 9h56'4" 47d29'12"
Dollie-o-o 's
Sco 16h39'41" -35d48'4"
Dollstar
Ari 3h7'23" 25d55'58"
Dolly
Gem 7h47'22" 22d58'48"
Dolly
Uma 11h19'45" 49d55'9"
Dolly
And 23h6'45" 41d29'54"
Dolly
Lyr 18h40'18" 35d49'37"
Dolly
Lyn 8h3'22" 38d4'42"
Dolly
Umi 13h35'52" 71d40'37"
Dolly
Uma 11h20'34" 69d10'11"
Dolly
Uma 13h38'14" 59d9'53"
Dolly A. Chapman
Cap 20h38'43" -19d1'15"
Dolly Buser
Aql 20h25'12" 2d3'53"
Dolly Chapman
Cas 23h9'2" 56d3'58"
Dolly Chung
Mon 6h41'28" 1d41'18"
"Dolly" Hartmann
Tau 3h43'35" 24d35'19"
Dolly I. Shaw
Uma 8h50'23" 55d22'43"
Dolly J. Mohr
Sgr 19h19'51" -23d32'32"
Dolly Kathryn Stiles
Psc 1h38'34" 20d48'59"
Dolly (Linda Pariello)
And 0h37'8" 26d10'27"
Dolly Lynn Dumas
Lib 15h18'7" -12d46'41"
Dolly M Turner
Leo 11h18'4" 2d8'17"
Dolly Mastrofillippo
Ari 3h21'43" 21d3'44"
Dolly Mayenzet
Mon 6h47'39" -0d1'50"
Dolly R. Skiles
Aqr 22h36'6" 0d56'42"
Dolly Rivera
Cap 20h41'2" -22d4'47"
Dolly Sands
Lyr 18h49'28" 35d44'49"
Dolly Semel
Ari 3h20'20" 16d25'5"
Dolly the Pisces Queen
Psc 1h5'30" 29d53'55"
Dolly, Sister of Lisa
And 0h37'18" 32d57'33"
Dolly-Rok Star
And 0h14'22" 40d30'19"
Dolma
Cyg 20h58'22" 48d24'10"
Dolo Jlenya
Peg 21h26'52" 21d29'53"
Dolores
Psc 1h8'46" 23d20'39"
Dolores
And 0h25'23" 42d19'12"
Dolores
Sgr 18h2'39" -21d44'51"
Dolores A. Delgado
Cas 1h27'55" 61d2'11"
Dolores A (Grondahl) Stredwick
Crb 16h11'14" 29d48'57"
Dolores A. Smith
And 1h44'33" 43d1'37"
Dolores A. Vezzetti
Cas 0h29'20" 61d20'36"
Dolores Alexander
Lyr 19h8'28" 28d3'50"
Dolores Alvarado Rios
Mon 6h28'58" 10d9'35"
Dolores and Casey Padilla-Stanton
Psc 0h40'42" 8d43'28"
Dolores and Harold Wrege
Lyr 18h52'47" 31d57'36"
Dolores Ann
Cnc 8h34'46" 27d47'45"
Dolores Ann Mazzocchi
Psc 1h16'48" 24d35'43"
Dolores Anne Buonopane
Cyg 19h56'29" 32d31'48"
Dolores Benante
Lyn 8h35'11" 46d7'15"
Dolores Bracale
Cap 20h36'33" -11d3'28"
Dolores C. Manning
Per 4h30'51" 43d2'25"

Dolores C Ziccardi
Vir 12h55'47" 11d51'32"
Dolores Camilleri
Vir 11h51'31" -5d33'36"
Dolores Chavez
Oph 17h33'13" 6d39'2"
Dolores Delgado
Lyr 19h14'22" 26d58'30"
Dolores Didio
Cnc 8h27'24" 32d7'11"
Dolores DiLegge
Aqr 22h0'22" -1d54'4"
Dolores DiOrio
Cas 1h39'22" 68d49'10"
Dolores DiPonti
Sge 19h45'16" 18d52'12"
Dolores "Do" Horn
Cyg 21h3'47" 54d58'53"
Dolores E. Horn
Cyg 20h15'49" 37d20'35"
Dolores E Kehr
Cas 0h44'0" 64d55'22"
Dolores F. Colombo The Best Mother
Cas 1h38'53" 63d16'14"
Dolores Francis Toole
Cas 1h37'8" 72d57'29"
Dolores Gardner
Cap 20h23'44" -10d1'27"
Dolores Gokie Plotner
Boo 14h51'3" 29d56'5"
Dolores Gutierrez
Her 17h52'40" 49d51'53"
Dolores J. "Dolly" Hamill
Cas 23h47'21" 52d8'35"
Dolores J Reilly
Leo 10h21'58" 13d31'5"
Dolores J Reilly
Leo 10h1'37" 13d14'42"
Dolores J Reilly
Leo 11h13'16" 9d53'50"
Dolores Jane Davis
Lyn 8h4'55" 52d52'55"
Dolores Jane Leary
Gem 6h28'44" 26d14'9"
Dolores Jean
Gem 6h49'14" 28d34'18"
Dolores & John R. Hurst
Sge 19h15'22" 18d0'42"
Dolores Johnson
And 0h26'50" 38d46'22"
Dolores L. Garbush
Tau 4h35'17" 18d51'21"
Dolores L Marshall
And 1h26'20" 46d17'1"
Dolores Lorhsa QueenSilk
Lib 14h27'56" -9d43'48"
Dolores Louise (Bunny) Smiley
Cnc 8h12'34" 15d16'38"
Dolores M. Rice
Gem 6h26'5" 22d19'25"
Dolores M. Walsh
Cas 0h25'44" 58d56'54"
Dolores Mae Shoot
Lib 15h20'3" -4d3'30"
Dolores Maria Cruz
Cas 1h11'19" 50d53'0"
Dolores Marie Beller
Lep 5h16'1" -11d24'22"
Dolores Marie Everitt
Crb 15h31'33" 29d26'37"
DOLORES MARSHALL
Cas 0h26'12" 56d47'8"
Dolores May Mark
Tau 4h31'48" 17d18'33"
Dolores McKenna
And 23h17'58" 48d49'4"
Dolores McNamara
Cas 0h17'25" 62d58'10"
Dolores Micili
Her 18h32'12" 16d34'29"
Dolores Morency
Ari 2h10'28" 23d27'1"
Dolores Murdock
Cas 0h40'42" 60d44'29"
Dolores Neal Montague
Sco 16h45'36" -42d43'5"
Dolores Nevin Barnett
Cnc 8h34'26" 32d36'2"
Dolores Rainsbury ILG's 1st 10 star
Cas 1h22'53" 54d52'40"
Dolores Saliba The Glitterin Star
And 0h43'14" 36d39'16"
Dolores Shawe
Sge 20h10'49" 18d12'8"
Dolores Shea
Uma 9h23'59" 51d36'3"
Dolores Sintich
Cas 0h11'30" 60d51'34"
Dolores Sour Scott
Ori 5h34'17" -1d5'37"
Dolores Stanley
And 23h13'18" 47d32'9"
Dolores Stöckli
Tri 1h47'18" 31d9'34"
Dolores The Poet
Sgr 18h50'3" -29d47'24"
Dolores The Shining Star
And 23h22'44" 47d22'18"
Dolores Theresa
Cas 1h42'31" 62d9'59"
Dolores Thomas
Lib 15h8'11" -22d38'17"
Dolores Vaughn
Cyg 20h14'9" 48d53'31"
Dolores Waldschmidt
And 1h17'25" 34d31'10"
Dolores y Viola y Penn
Cap 20h23'36" -12d51'8"
Dolores Yaneza Aguilar
Crb 15h27'54" 31d9'58"
Dolorita La Prad
Tau 3h34'57" 11d45'34"
Dolphin Rob Christie
Psc 0h19'33" 15d13'43"
Dolphin Sister
Leo 10h12'24" 26d36'10"
Dolphin yuko
Gem 7h26'39" 23d33'30"
Dolphinia
Uma 11h25'23" 64d45'39"
Dom and Buddy
Cnc 8h35'52" 7d38'35"
Dom and Cherry
Cyg 20h9'32" 48d20'45"
Dom Montalbano
Uma 13h37'26" 52d16'19"
Domanik Ethan Valdovinos
Uma 8h48'53" 64d4'30"
Domanyi
Sco 16h40'37" -38d41'7"
Domar I
Mon 6h43'41" 9d2'9"
Domar II
Peg 22h22'40" 6d45'5"
Domè
Uma 11h4'11" 60d6'6"
Dome, der Glücksritter
Vir 13h25'32" 2d37'39"
Domenic
Crb 15h34'5" 26d20'48"
Domenic
Lib 15h28'56" -4d8'40"
Domenic and Giovanni
Boo 14h38'32" 37d19'37"
Domenic Carrocce
And 23h18'27" 51d13'55"
Domenic D'Amore
Lyr 18h29'52" 34d1'1"
Domenic Michael
Aur 5h28'14" 35d37'51"
Domenic Nunziante Micieli
Ari 3h10'15" 28d41'51"
Domenic S. Ranocchia
Ari 2h22'22" 11d29'8"
Domenic Virgilio
Umi 4h53'54" 89d47'9"
Domenica
Cap 20h54'18" -21d11'49"
Domenica
Uma 11h3'26" 35d30'47"
Domenica Vitale 19 settempre 1959
Ori 6h23'3" 13d10'54"
Domenici Federica
Per 3h5'1" 51d40'6"
Domenick A. Sanseverino
Per 2h19'18" 56d6'28"
Domenick Jordan Swan
Lib 14h44'53" -24d36'55"
Domenick Liberatore
Sco 17h51'32" -42d23'21"
Domenick Pucillo
Lac 22h34'47" 41d28'17"
Domenico Antonio Agostino
Lyr 19h10'51" 38d45'30"
Domenico Christopher Masi
Her 18h46'26" 22d14'55"
Domenico Creta
Lyr 18h39'26" 31d19'15"
Domenico Maiuri
Uma 11h18'32" 50d16'35"
Domenico Mascagni
Ori 5h51'27" -8d24'10"
Domenico Postiglione
Per 4h6'3" 47d1'7"
Domenico Roger Monaco
Per 3h9'20" 57d0'10"
Domenic's Guiding Light
Cru 12h33'49" -60d24'36"
Domenique Minervini
Aql 19h34'1" -10d58'19"
Domer Doo
Vir 11h37'57" -3d28'38"
DOMI Stern - MORANO
Uma 11h34'0" 35d32'2"
Domiano Colarutolo
Her 17h17'2" 35d46'23"
Domienik
Uma 10h12'8" 57d26'58"
Domin
Lib 14h31'6" -15d42'15"
Domina
Umi 16h23'23" 73d20'21"
Dominator TKS
Uma 9h6'56" 47d41'10"
Domingo A. Rodriguez (Chacho)
Ari 2h3'24" 21d26'54"
Domingo Pagan
Gem 6h49'41" 23d14'20"
Dominic
Psc 1h16'46" 14d20'2"
Dominic
Tau 3h47'42" 14d13'47"
Dominic
Uma 11h26'8" 49d38'57"
Dominic
Uma 11h37'25" 57d47'30"
Dominic
Sgr 17h54'38" -29d42'44"
Dominic A. Lachimia
Gem 7h46'26" 16d2'21"
Dominic A. Trocchia/Poppy
Gem 6h29'10" 19d33'21"
Dominic Alcide Fedrizzi
Aqr 22h42'4" -1d23'24"
Dominic Alexander Guidry
Tau 4h41'22" 19d24'52"
Dominic Alexander Veloso
Lmi 9h48'13" 37d1'54"
Dominic and Angela "Molocules"
Cep 21h50'47" 58d44'21"
Dominic and Maryann Carlino #13
Sgr 18h26'12" -36d26'52"
Dominic Andrew Benassi
Per 3h3'10" 33d3'38"
Dominic Andrew Lewis
Her 17h39'41" 38d16'12"
Dominic Anthony Abucejo
Uma 10h57'42" 47d56'34"
Dominic Anthony Suzik
Dra 20h19'45" 67d6'0"
Dominic Bernard terWeeme
Her 18h1'6" 37d24'55"
Dominic bonhomme Masse
And 23h8'50" 51d50'47"
Dominic Calla
Cep 22h41'29" 61d58'17"
Dominic Carl Hibbs
Uma 8h36'23" 50d32'17"
Dominic Ceravone
Per 3h8'52" 53d55'43"
Dominic Chin
Her 16h44'54" 42d44'50"
Dominic Christa
Sgr 18h42'34" -27d47'0"
Dominic Craig DiSilvestro
Umi 15h19'20" 67d51'10"
Dominic DeGiosa
Uma 11h54'20" 31d13'8"
Dominic Delucchi
Her 17h31'12" 32d29'46"
Dominic Deola
Tau 5h52'42" 12d44'6"
Dominic Dimauro
Cru 12h56'41" -60d26'17"
Dominic Dupree Emanuel
Uma 8h41'43" 60d44'10"
Dominic Ficarrotta Jr
Vir 14h19'5" 3d17'39"
Dominic Ford
Cep 3h18'57" 83d30'48"
Dominic Francis Rotondi
Psc 1h10'7" 28d44'27"
Dominic Gentile
Umi 14h12'46" 74d54'51"
Dominic Giangregorio
Lyr 18h53'54" 39d12'30"
Dominic Griffin Soreco
Dra 18h31'4" 67d4'31"
Dominic J. Gherardini
Uma 13h46'6" 55d50'8"
Dominic J. Romano
Tau 4h43'42" 21d15'37"
Dominic James Bedford
Per 4h14'17" 35d41'43"
Dominic James Brinkley
Leo 11h49'50" 20d11'16"
Dominic James DeMatteo
Boo 14h42'25" 22d45'48"
Dominic John Simon
Ori 5h8'40" -0d47'28"
Dominic Johnson CF
Ori 6h17'54" 18d52'12"
Dominic Jordan Ricigliano
Leo 10h15'22" 26d19'35"
Dominic Joseph Cerullo
Sco 17h53'21" -35d25'19"
Dominic Joseph Georgiana
Ari 3h17'18" 27d38'25"
Dominic Joseph Granata
Vir 12h58'15" -12d49'4"
Dominic Joseph Lumia
Uma 9h30'9" 48d24'41"
Dominic & Josie
Cru 12h32'19" -57d34'50"
Dominic Logan Johnson
Boo 14h39'44" 33d54'35"
Dominic Louis
Cep 22h31'4" 65d20'0"
Dominic & Luke's Lucky Star
Ori 5h32'31" -4d21'36"
Dominic MacNeil
Umi 15h31'24" 71d58'53"
Dominic Malegni
Per 4h22'7" 37d52'17"
Dominic Martinez
Pho 0h26'2" -50d9'51"
Dominic Maxwell Garcia
Cyg 19h54'57" 31d35'8"
Dominic Michael Amore
Her 16h27'49" 43d37'56"
Dominic Michael Castro
Vir 12h12'39" 11d54'12"
Dominic Michael DelMonte
Tau 4h35'3" 9d19'47"
Dominic Michael Piazza
Aqr 22h41'38" -20d7'55"
Dominic Nicolo SuperStarLoverFriend
Sco 16h13'52" -16d24'43"
Dominic Noah
Leo 9h30'23" 19d40'45"
Dominic Paul
Her 17h54'2" 29d9'46"
Dominic Plüss
Ori 5h11'18" -1d31'31"
Dominic Riesen
Per 3h5'1" 41d53'11"
Dominic Robert
Vir 13h8'43" 8d39'45"
Dominic Salvatore DeSimone
Lib 15h16'22" -11d52'50"
Dominic Santangelo
Uma 8h29'17" 67d7'21"
Dominic Sblendorio
Ori 6h10'22" 18d10'48"
Dominic Scott Wuertz
Her 17h12'14" 34d9'21"
Dominic Stites
Ori 5h42'50" 5d37'47"
Dominic T. Burgese
Per 4h35'20" 40d33'18"
Dominic Thomas
Lib 15h29'56" -7d14'12"
Dominic Thomas Oliveira
Vir 11h38'31" -4d23'54"
Dominic von Gunten
Sco 17h42'7" -36d3'14"
Dominic William O'Neill
Her 17h45'8" 15d43'13"
Dominic, Son of Laura and Noel Rice
Lib 15h13'10" -4d14'41"
Dominic.C. & Emma.W. Forever as One
Cas 0h33'21" 50d30'18"
Dominick
Gem 7h0'53" 17d3'35"
Dominick
Cnc 9h3'10" 15d4'31"
Dominick
Cap 20h37'24" -17d38'45"
Dominick and Antoinette Sblendorio
Tau 4h23'17" 17d40'10"
Dominick Anthony Conetta
Ari 3h21'6" 27d23'19"
Dominick Anthony Daniele
Her 17h20'32" 23d30'45"
Dominick Barrie Thomas Love
Ori 5h53'13" 7d11'35"
Dominick Brandon 1992
And 0h17'34" 27d8'28"
Dominick Bulat
Leo 11h18'6" 10d48'57"
Dominick Carl Ragusa, Jr.
Per 3h27'9" 52d5'9"
Dominick F. Scoccimarro
Dra 18h59'31" 67d12'27"
Dominick Fielder
Cap 21h0'40" -16d41'26"
Dominick J. Mortellito
Umi 16h26'13" 77d21'4"
Dominick J. Mumolo
Ari 2h3'47" 19d51'16"
Dominick & Josephine Previte
Uma 13h41'48" 56d1'1"
Dominick Lawrence Trippiedi
Cam 7h18'42" 60d5'35"
Dominick Leonard Simms
Her 18h28'15" 14d43'59"
Dominick Lieze
Cnc 8h26'31" 10d10'58"
Dominick Lipscomb
Her 17h48'20" 23d7'7"
Dominick Lucas Edwards
Ari 2h47'30" 28d51'0"
Dominick Mercadante
Umi 15h29'45" 71d51'17"
Dominick Nathaniel Powers
Vir 13h0'47" 5d37'53"
Dominick P. Ciola
Lib 15h21'1" -8d39'28"
Dominick R. Virgilio
Per 2h44'3" 56d17'42"
Dominick Rocco Picicco
Gem 7h0'39" 17d59'32"
Dominick Sean Victor
Cnc 8h14'59" 11d8'40"
Dominick V. Amato, Jr., 1985-2007
Umi 14h18'11" 78d20'23"
Dominicka's Galaxy
Uma 13h56'39" 58d22'38"
Dominic's In Love
And 0h20'55" 30d59'9"
Dominic's Passion
Gem 7h23'50" 26d15'5"
Dominie
Psc 0h34'22" 7d35'30"
Dominik
Ori 5h12'36" 15d24'16"
Dominik
Cyg 20h44'41" 45d50'27"
Dominik
Uma 8h46'18" 72d47'52"
Dominik Günter Meixner "Mollmopf"
Uma 8h19'17" 72d0'24"
Dominik Stutz
Cas 23h48'7" 53d56'22"
Dominik Woll & Sarah Kuhn
Ori 6h21'16" 13d20'9"
Dominik Wyss
Vul 20h25'36" 26d50'33"
Dominika
Cyg 21h30'45" 43d1'50"
Dominika
And 1h12'54" 38d3'1"
Dominika
Uma 10h13'50" 55d28'46"
Dominique
Cas 0h59'51" 64d35'37"
Dominique
Umi 13h8'35" 72d37'54"
Dominique
Sgr 19h42'46" -13d32'24"
Dominique
Cap 21h39'38" -17d51'59"
Dominique
Sgr 18h14'42" -20d2'7"
Dominique
Sco 16h10'6" -30d8'8"
Dominique
And 0h44'48" 37d55'19"
Dominique
Lmi 10h44'6" 32d55'15"
Dominique
Lyn 8h51'38" 33d26'10"
Dominique
Leo 9h37'19" 28d21'29"
Dominique
Gem 6h21'56" 19d25'8"
Dominique
Cnc 9h1'21" 15d40'46"
Dominique
Her 17h12'6" 16d25'57"
Dominique Alexandria
And 0h33'14" 36d29'43"
Dominique Anne Trepiccione
And 23h33'42" 43d12'56"
Dominique Bastide
Leo 9h22'20" 7d23'0"
Dominique Bertallo
Per 3h28'55" 50d32'48"
Dominique Buccieri
Uma 11h53'46" 29d38'46"
Dominique Candice Gustard
Del 20h38'21" 15d7'11"
Dominique Cheree Clinton
And 2h21'6" 45d18'52"
Dominique Cheyenne Briddell
Ori 4h49'41" 3d40'12"
Dominique Cindy Rouhana
Cap 21h29'43" -20d1'47"
Dominique Daleiden Yokosawa
And 23h38'11" 45d17'9"
Dominique Dion
Cep 20h47'19" 64d53'57"
Dominique Donzel
Uma 11h9'0" 44d36'40"
Dominique Fidanza
Cyg 19h39'15" 28d29'22"
Dominique Florence
Tri 1h53'44" 26d1'49"
Dominique Galuscio
Aqr 21h56'28" 0d21'59"
Dominique Grund
Psc 1h3'49" 33d11'33"
dominique il mio angelo
Dra 16h28'59" 52d21'54"
Dominique J. Connell
Lyn 7h39'53" 57d0'10"
Dominique J. Smith
Gem 7h35'49" 28d8'48"
Dominique Johner
Sgr 17h59'17" -21d7'15"
Dominique Kayleene Reh 7-3-1998
Lyn 8h27'49" 42d55'0"
Dominique L&S TFITH
Vir 11h47'19" -3d24'46"
Dominique Lalonde
Cas 0h42'58" 53d50'53"
Dominique Lamer
Cep 21h44'55" 72d11'34"
Dominique Lendi
Her 17h54'49" 23d4'48"
Dominique LoBiondo
And 23h0'54" 41d36'34"
Dominique Madu
Lib 14h26'0" -19d50'9"
Dominique Marie
Cas 0h37'30" 57d21'4"
DOMINIQUE MARSHALL
Mon 6h52'37" -0d14'32"
Dominique Mellin 25/03/1955
Ori 5h17'20" 15d55'51"
Dominique Michael Cook
Sco 17h35'14" -35d13'11"
Dominique Mikaits
Uma 10h20'38" 49d17'47"
Dominique Millas
Tau 4h21'58" 25d37'45"
Dominique Monaco's Lucky Star
Lib 14h54'44" -3d46'21"
Dominique My Shining Star
Cyg 19h36'38" 32d41'47"
Dominique "Nikki" Bragg
Cnc 8h18'16" 15d23'31"
Dominique Peleriaux Dream
Leo 9h57'32" 11d31'20"
Dominique S. Sanders
Cyg 21h13'28" 46d57'38"
Dominique Samei
And 0h46'35" 25d8'45"
Dominique Sitton
Psc 0h57'15" 18d46'11"
Dominique Vogel
Cas 1h15'1" 50d46'24"
Dominique's Shining Star
Leo 11h39'8" 24d20'4"
Dominique's Star
Sgr 18h52'14" -16d0'19"
Dominko
Ori 6h12'50" 7d20'9"
Domino
And 23h5'40" 48d27'26"
Domino "Dominian" Burgan
Uma 13h55'10" 52d55'12"
Domino Dominique Monet
Cyg 21h21'20" 34d32'54"
Domino Harvey
Leo 11h7'30" 22d42'13"
Domino La Farr
Her 17h7'20" 34d5'37"
Domino - Mummy's Spunky Boy
Cru 12h52'13" -57d32'59"
Domino O'Driscoll
Cnc 7h59'14" 15d7'27"
Domino's Spot
Cma 6h42'45" -28d45'39"
Dominus Domino " Master, Lord "
Uma 10h18'37" 64d5'39"
Domitille
Umi 16h17'15" 75d25'21"
Domiziana
Boo 14h40'1" 23d20'41"
Dom-Lucy Mercurio
Uma 8h40'36" 53d11'28"
Dommie
Gem 7h35'58" 26d27'56"
Dommy & Dana - True & Eternal Love
And 23h9'45" 51d45'43"
Domokos Sándor
Cap 21h58'42" -18d47'38"
Domonique and Wayne
Lib 15h5'48" -28d43'11"
Domonique - Drew
Cet 1h28'57" -0d50'40"
Doms
Gem 7h27'58" 15d30'36"
Domty
Uma 10h16'17" 71d25'54"
Domus Luminarium
Pyx 8h41'46" -35d49'23"
Don
Uma 11h23'6" 71d20'52"
Don
Cep 2h23'48" 81d5'56"
Don
Boo 14h53'6" 14d55'42"
Don
Per 2h20'12" 51d14'42"
DON
Per 4h24'0" 50d41'15"
Don
Her 16h50'48" 40d49'24"
Don 4CA
Ari 3h3'59" 13d15'14"
Don Alderman
Uma 11h18'54" 51d16'9"
Don Alfonso
Per 2h22'37" 51d44'7"
Don Alfredo
Her 17h47'40" 42d40'35"
Don & Amishka Hughes
Cnc 9h0'3" 10d14'47"
Don and Alice Moore
Uma 11h14'31" 37d34'35"
Don and Elaine Dennehey
Cyg 20h54'54" 41d35'8"
Don and Janie Shelden
Ori 5h40'32" -8d40'53"
Don and Kathleen Wilkinson
Sgr 18h34'33" -27d32'54"
Don and Lucille Forever
Cyg 21h34'34" 35d52'52"
Don and Marjorie Bane
Cyg 20h21'1" 55d18'18"
Don and Mel Whitrow
Uma 10h43'12" 44d54'36"
Don and Neil
Cyg 20h35'28" 34d37'31"
Don and Olive Neterer
Sge 19h50'42" 17d24'11"
Don and Rosie Pattison
Lyr 18h48'56" 34d21'31"
Don and Wanda, Always Together
Leo 10h18'15" 27d4'46"
Don and Wendy
Cnc 8h11'26" 29d20'17"
Don Anderson
Aql 20h7'52" 2d45'59"
Don Angelo Lipari My Eternal Star
Her 18h8'49" 33d33'5"
Don Arterburn
Sco 17h57'39" -30d5'32"
Don Baldini
Uma 9h21'19" 56d59'58"
Don Benoist
Peg 21h50'19" 12d51'24"
Don Biederman
Vir 12h39'26" -9d36'45"
Don Blackburn I love you forever
Leo 9h54'30" 13d7'50"
Don Blankenship
Cas 23h42'43" 53d9'0"
Don Bohdan Wynnyczok
Cap 21h16'54" -27d18'0"
Don Bohrer
Ori 5h6'37" -1d16'35"
Don Boulanger
Cep 20h43'58" 62d30'28"
Don Bowers
Ori 5h50'39" 6d39'24"
Don Bradley
Her 17h33'25" 23d15'50"
Don & Bridget
Cyg 20h43'27" 43d8'52"
Don Bulk
Gem 7h46'23" 32d45'35"
Don Burke, Sr.
Her 17h48'53" 46d30'47"
Don Butler 24/09/24 Our Star
Lib 15h16'8" -9d9'43"
Don "Buzz Lightyear" Huffner
Lmi 9h55'39" 36d1'50"
Don C. Quast, M.D.
Leo 10h28'45" 19d0'2"
Don Cain
Uma 12h6'25" 58d51'54"
Don Cave, Light of Our Lives
Cep 22h58'32" 71d8'0"
Don Chizek
Aur 5h55'52" 32d27'41"
Don Chizek
Per 2h14'15" 54d39'29"
Don Christian Willie
Uma 12h15'35" 56d54'57"
Don Clary
Aql 20h2'29" 10d48'33"
Don Collins - Soul Mate
Tau 5h52'18" 25d24'41"
Don Conejo
Vir 13h0'59" 2d20'57"
Don Conry 1927
Gem 6h53'23" 33d3'59"
Don Crook "8-22-30"
Per 3h38'57" 46d29'8"
Don Croteau-Hanson
Uma 11h19'4" 72d4'58"
Don Crouse
Her 17h11'6" 34d49'21"
Don "Daddy" Harrison
Cep 22h39'17" 84d47'31"
Don & Debby Fonda
Cyg 19h43'38" 43d25'7"
Don DeBock
Her 17h58'21" 29d9'16"
Don Dixon
Tau 4h28'9" 1d40'40"
Don Doty
Ari 3h15'17" 22d49'43"
Don Edgerton
Boo 14h12'27" 28d5'55"
Don Edwards
Aql 19h2'23" 1d11'34"
Don & Elaine Murray
Cap 20h36'18" -21d20'27"
Don Enrico P.
Per 3h3'16" 51d41'51"
Don Ermold
Ori 6h22'36" 13d36'46"
Don & Esther Leiby's Star
Leo 9h32'25" 32d4'34"
Don & Evelyn Smith
Cyg 20h23'59" 39d49'20"
Don & Evelyne Rogers
Vir 13h18'43" 3d58'18"
Don Fawlk
Sco 16h46'21" -41d31'10"
Don Field
Per 4h6'9" 33d49'36"
Don & Gerry Searle
Cyg 21h10'21" 30d10'19"
Don Gould Jr. "1950-2006"
Cep 0h15'41" 82d17'26"
Don Gracey
Ori 6h13'7" 16d20'59"
Don Griffith
Cep 0h19'40" 76d47'59"
Don H. Morrison
Per 3h6'49" 57d16'36"
Don Hassletine
Lac 22h17'31" 54d16'4"
Don & Heidi 20th Anniversary
Cas 1h32'31" 61d43'3"
Don & Helen Amsden
Uma 11h18'19" 43d38'10"

Don & Helen Marlow
Uma 13h35'15" 61d29'46"
Don Hilton
Uma 13h29'16" 53d4'55"
Don Holmes
Aqr 21h51'11" -0d1'36"
Don Horton
Lyn 6h47'20" 53d26'52"
Don Hugh Fleming
Ori 5h26'57" 1d46'48"
Don Imus
Hya 9h8'57" -0d46'11"
Don IU2
Aur 4h51'8" 39d12'21"
Don J. Tice
Her 15h59'31" 44d32'40"
Don Jackson
Leo 10h57'57" 14d22'40"
Don & Jen Warrant Forever
Pup 8h2'19" -17d23'9"
Don Jones
Vir 12h43'3" 1d46'19"
Don Jones
Ari 2h7'31" 23d18'31"
Don & Joyce's Love Eternal
Cyg 22h2'11" 51d21'32"
Don Juan Polit-Klein
Cma 6h24'38" -14d59'53"
Don Karcz
Cma 6h55'49" -13d4'37"
Don Kauffman
Sgr 17h56'36" -27d12'29"
Don Kirby
Ara 17h35'16" -46d8'11"
Don Lau 7/29/48
Leo 11h7'28" 10d19'50"
Don Lee Loxley
Per 2h21'49" 56d42'31"
Don & Linda
Cyg 21h27'4" 39d0'6"
Don & Lois Tharp
Uma 10h32'15" 56d19'49"
Don Lova
Peg 23h37'59" 13d59'24"
Don & Marge Kenyon
Ori 5h55'19" 11d38'48"
Don & Mary Cashion
August 3, 1955
Cyg 21h13'31" 37d5'42"
Don & Mary Friedman
Uma 10h27'31" 44d26'45"
Don Matthew Servidio
Leo 11h22'57" 12d40'54"
Don Medal
Tri 2h12'23" 34d24'43"
Don N. Peska
Per 4h13'13" 46d49'18"
Don & Nancy Maushart
Cyg 19h29'28" 31d15'29"
Don Newroth
Cap 20h10'7" -12d53'10"
Don of Dewars
Boo 14h8'46" 29d18'46"
Don Ojibway
Uma 11h46'0" 60d1'22"
Don Osborn
Uma 12h59'50" 55d4'16"
Don Palmer
Uma 11h17'41" 47d36'29"
Don Parish - My Shining Star
Her 16h55'13" 16d55'23"
don Pedro
Ori 5h32'47" 0d46'1"
Don Pellin
Tau 5h51'11" 13d33'38"
Don Pepper
And 0h26'49" 43d24'51"
Don Pierpaolo
Her 17h49'31" 36d46'3"
Don Polizzi
Dra 17h4'12" 62d48'27"
Don Potts
Uma 10h34'1" 68d57'59"
Don Radig's Special Star
Sgr 18h20'1" -32d7'17"
Don Ray
Cyg 19h29'4" 48d23'33"
Don Richards
Uma 11h21'42" 57d44'17"
Don Richardson
Aql 19h9'55" 14d55'3"
Don Rigsby
Her 17h19'26" 27d28'18"
Don (Rolltide) Milner M.D.
Vir 12h33'45" 10d59'45"
Don Rox Smith
Per 4h20'1" 51d11'53"
Don Rynard
Crb 16h16'43" 28d17'6"
DON' S
Psc 23h44'23" 0d35'55"
Don S. Williams
Her 16h26'30" 11d56'35"
Don & Sally Sutterfield
Sge 19h13'34" 18d55'1"
Don Seubert
Cep 23h20'0" 79d28'33"
Don Shinn
Uma 9h58'7" 49d31'35"
Don Shortreed
Ori 6h1'35" 19d47'52"
Don Smith
Her 16h49'1" 40d2'43"
Don Smith TGO "2-12-24"
Dra 10h36'29" 74d11'2"
Don Sommers
Boo 14h47'14" 25d43'56"
Don T. Matthews
Uma 11h37'32" 51d33'20"
Don Thomsen
Per 2h13'10" 57d17'24"
Don "Trey" Capps
Per 3h55'38" 32d55'59"
Don Tribble
Ori 5h28'59" 3d34'30"
Don VanZandt
Tau 4h41'11" 27d43'41"
Don Vecchione
Gem 7h26'8" 30d57'21"
Don W. Cooper
Ori 6h18'1" 11d9'41"
Don W. Mann
Sgr 18h24'31" -16d15'51"
Don W. Vaughn
Aql 18h46'18" 8d51'41"
Don W Walker
Cnc 8h27'58" 12d2'14"
Don & Wanda
Cyg 20h49'40" 33d15'16"
Don White
Ori 6h7'11" 17d24'46"
Don_Whitesel
Cap 20h37'49" -8d52'49"
Don Williams
Aqr 21h42'22" -0d48'18"
Don Wilson II
Lib 15h10'36" -5d33'52"
Don "Wow Factor" Blanton
Uma 10h34'58" 54d24'58"
Don Yielding
Cep 22h44'34" 64d4'15"
Dona
Cam 3h21'25" 58d7'36"
Dona
Cyg 21h12'44" 45d32'50"
Dona Angelina Argiriadou
Cap 21h19'38" -22d42'27"
Dona Delia & her Daughter
Cnc 8h37'15" 32d19'26"
Dona Dierling
Cyg 20h35'9" 42d19'49"
Dona & Donny Dotson
Cyg 19h47'0" 36d54'20"
Dona Elena E
Uma 8h33'32" 71d51'45"
Dona L Marion
Uma 11h49'55" 51d37'17"
Dona - la piccola stella
Mon 7h0'43" 7d29'15"
Donahue
Uma 11h28'31" 39d42'23"
Donakeli
Tau 4h32'51" 5d21'7"
Donal
Lyn 6h45'39" 53d56'55"
Donal Mel O'Brien
Aqr 23h40'38" -15d0'31"
Donald A. Brandt
Cyg 20h4'55" 33d48'48"
Donald A. Chew
Vir 11h44'49" -3d32'11"
Donald A. Glaser
Lyn 8h11'41" 44d21'59"
Donald A. Kurtyak
Dra 16h56'14" 69d29'28"
Donald A. & Margaret E. Rankin
Ori 5h26'36" 14d47'34"
Donald A Murphy
Uma 10h48'42" 46d55'38"
Donald A. Musick
Her 16h16'10" 13d14'22"
Donald A Wurden
Lyn 7h19'51" 46d42'19"
Donald Aeternalis
Sco 16h57'23" -33d30'17"
Donald Alexander Jones
Sgr 19h22'36" -16d11'16"
Donald Alexander McBride
Cam 3h36'2" 55d24'27"
Donald Allen Duke
Her 16h20'27" 45d53'12"
Donald Allen Tam
Ori 5h45'43" 7d13'25"
Donald and Danielle Schwier
Cyg 20h15'55" 52d34'20"
Donald and Dionne Caldwell
Eri 3h59'33" -8d33'26"
Donald and Joyce McClellan
Cyg 19h52'22" 33d43'39"
Donald and Kathleen Kenneally
Uma 13h30'47" 57d28'43"
Donald and Margaret Black
Cyg 20h20'43" 54d51'58"
Donald and Muriel Noonan
Cas 2h7'21" 64d32'16"
Donald and Sandra Neice
And 23h10'42" 44d4'27"
Donald and Steven: Cosmic & Forever
Cyg 20h22'31" 58d22'11"
Donald Arthur
Peg 22h31'44" 5d54'0"
Donald Arthur Smith
And 1h29'48" 42d22'32"
Donald Arthur Woodside
Her 17h44'32" 36d14'0"
Donald Ayotte
Uma 11h54'43" 40d28'3"
Donald B. & Jane H. Liddell
Sgr 18h55'38" -16d28'48"
Donald B. Macon
Uma 11h17'50" 65d43'9"
Donald Bass
Dra 16h5'42" 67d40'32"
Donald Beverley
Her 17h24'54" 34d54'42"
Donald Black
Lyr 18h50'24" 37d41'0"
Donald Bond Callaway
Aql 19h11'17" -0d43'10"
Donald Bowden
Ori 6h20'49" 14d20'2"
Donald Brayton 11-23-04
Ori 5h42'11" 4d37'24"
Donald Brent Schumacher, 11
Aql 19h46'8" -0d21'43"
Donald Brescia, M.D.
Cyg 20h36'1" 56d17'34"
Donald Brown
Aql 19h24'14" 4d5'12"
Donald Bryan Burgess
Aqr 23h0'21" -14d28'4"
Donald Burd
Tri 2h16'27" 35d16'13"
Donald C. Gras IV
Uma 13h38'17" 49d31'45"
Donald C. Mills
Lyn 7h5'41" 52d27'41"
Donald C. Sterling
Psc 1h17'32" 18d53'8"
Donald C. Warner
Per 3h9'32" 54d6'37"
Donald Cameron
Uma 11h36'13" 32d24'17"
Donald Campbell
Per 3h25'23" 52d55'11"
Donald Carl Taylor
Lib 15h48'56" -11d40'27"
Donald Chambers - Hello Again
Per 3h24'12" 50d28'36"
Donald Charles Sieber
Leo 9h44'23" 29d8'46"
Donald Chester Ernest
Ori 6h5'10" 15d51'15"
Donald Clabaugh
Her 16h33'55" 45d37'33"
Donald Clark Leibbrand
Uma 8h44'19" 56d29'36"
Donald Clifford Taylor
Cap 20h11'25" -16d52'33"
Donald Collett, Heavenly Father
Uma 9h44'12" 49d56'38"
Donald Colquitt Hartsfield
Gem 6h47'14" 20d58'33"
Donald Constantine Grzelak-Zehar
Umi 14h21'3" 74d37'2"
Donald Craig Noonan
Cap 20h45'51" -26d40'16"
Donald Cunningham
Uma 11h21'45" 37d24'19"
Donald Curran Olson
Per 3h30'48" 41d57'56"
Donald D. Crumrine
Uma 12h39'12" 59d5'22"
Donald D. Dow
Per 3h47'47" 42d47'14"
Donald D. Mckinney
Cap 21h37'28" -9d7'49"
Donald (Dan) Forman
Ori 5h14'58" 8d27'2"
Donald David Burnett
Sgr 17h58'50" -20d21'26"
Donald David Dixon Ronald O'Connor
Vir 13h33'10" 2d22'1"
Donald David Ferguson
Ori 5h34'47" 3d26'33"
Donald Day
Her 16h49'47" 36d51'37"
Donald Dean Goolsby
Her 16h49'7" 47d40'6"
Donald Dean Wise
Aur 5h22'16" 29d23'9"
Donald Decker
Ori 5h44'17" 6d56'52"
Donald DeGrazia
Vir 13h37'36" -1d4'12"
Donald DiAmico
Ari 2h49'39" 29d2'26"
Donald Diamond
Tau 5h11'37" 18d15'57"
Donald Dillon
Aqr 21h2'25" -11d37'48"
Donald "Donny" Hint
Uma 9h45'35" 49d47'36"
Donald Dunow
Per 4h4'49" 47d7'15"
Donald E. Conway
Uma 9h55'58" 43d7'45"
Donald E Fisher II
Uma 9h34'39" 72d5'6"
Donald E. LaGuardia
Cep 22h13'9" 61d20'38"
Donald E. McCardle
Ori 6h4'45" 13d13'44"
Donald E. Stephens
Cap 21h21'10" -19d59'1"
Donald E. Wahlberg
Leo 11h16'7" -1d16'56"
Donald Edward and Diane Olson Lange
Aqr 23h8'53" -8d10'5"
Donald Edward Baustian
Her 17h16'1" 34d22'56"
Donald Edward Fraker
Her 17h53'4" 21d16'56"
Donald Edward Kilian, Senior
Her 18h22'39" 18d33'49"
Donald Elbert Mason
Gem 6h46'45" 33d32'31"
Donald Emerich
Lac 22h20'14" 43d9'59"
Donald Emond
Her 16h49'6" 22d22'10"
Donald Erven Feickert
Ari 1h59'46" 22d11'9"
Donald Erwin Vanek
Uma 10h19'4" 49d29'2"
Donald Esselborn
Hya 9h28'43" 4d28'58"
Donald Eugene Mason
Psc 0h50'11" 15d46'8"
Donald & Evelene Ratliff
Cap 21h52'53" -17d49'5"
Donald F Delpha II
Per 3h10'3" 52d28'26"
Donald F. Doucette Jr.
Tau 4h12'27" 29d7'37"
Donald F Halladay
Per 2h54'1" 53d50'47"
Donald F. Houston
Psc 1h21'40" 21d36'31"
Donald F. & Laura Lee Anthony
Ari 2h18'14" 16d15'24"
Donald F. O'Brien
Uma 9h37'18" 54d11'27"
Donald F. Schermetzler, Jr.
Vir 12h41'56" 6d19'59"
Donald F. Thurston
Aur 5h48'1" 39d55'21"
Donald Fargo Beebee Ehlert
Uma 10h32'56" 49d2'15"
Donald "Farmer" Howard
Boo 14h37'39" 28d45'3"
Donald Floyd
Ori 5h8'47" 9d1'13"
Donald Francis Lamb
Gem 6h46'13" 13d6'3"
Donald Frederick Walker, Jr.
Cyg 20h32'11" 50d14'55"
Donald G. Pike
Gem 7h2'59" 21d19'29"
Donald G. Ranieri Sr.
Psc 1h8'34" 3d44'21"
Donald G. Robinson
Psc 0h40'1" 16d9'7"
Donald G. Sword
Aur 5h5'16" 34d43'59"
DONALD GARY SCHOEF-FLING
Cep 22h48'22" 68d28'3"
Donald Gene Bigler
Per 2h41'27" 40d50'57"
Donald Gene Hillyard
Lib 15h4'38" -7d48'37"
Donald Gene Slone
Her 16h45'37" 48d0'55"
Donald Gerard Belfiore
Psc 0h45'1" 4d0'38"
Donald Gerard Beyer
Ser 15h54'21" -0d34'39"
Donald Graham IV
Per 3h15'20" 33d2'37"
Donald H. Dippner Sr. (Pappy)
Uma 11h22'25" 48d57'38"
Donald H. Sparklin Sr.
Leo 11h42'33" 25d44'1"
Donald Hallahan
Lyr 19h8'53" 39d35'6"
Donald Hansen
Uma 10h53'27" 49d28'5"
Donald Heisler
Uma 12h4'58" 56d14'31"
Donald J. Ambrose
Psc 1h6'29" 30d37'23"
Donald J. and Bernadine E. LaLonde
Uma 9h54'49" 71d0'58"
Donald J. Bolling
Ori 5h32'18" 2d20'37"
Donald J. Coppola
Aqr 22h57'58" -9d48'52"
Donald J. Dallaire
Cep 21h36'57" 69d26'1"
Donald J Dudek
Cnc 8h37'34" 32d11'10"
Donald J. Kalina, Jr.
Vir 13h7'21" 5d28'57"
Donald J. Meyer
Sco 16h38'57" -29d59'54"
Donald J. Robinson
Cma 6h40'43" -22d20'41"
Donald J. Warren
Cep 21h24'9" 64d26'3"
Donald Jack Bennett
Per 2h16'48" 56d17'21"
Donald James
Ari 2h7'24" 23d0'22"
Donald James Lambert
Cnc 8h52'55" 16d25'31"
Donald James Lulu
Per 2h57'5" 32d31'49"
Donald James Otto Payne
Per 2h10'8" 54d38'7"
Donald James Yavorcik
Boo 14h15'1" 28d26'16"
Donald & Joan Anderson
Uma 11h52'36" 49d15'2"
Donald Joe Aguilera
Aql 19h38'51" 0d37'4"
Donald John Lee 6/18/1937
Gem 6h50'42" 34d13'12"
Donald Joseph Allebach Jr.
Lup 15h51'23" -32d36'13"
Donald Joseph Dugan
Her 17h56'25" 45d55'34"
Donald Joseph Hadley
Uma 11h53'55" 46d27'8"
Donald Joseph Raimo
Gem 7h42'23" 33d52'25"
Donald Joseph Sr.
Cyg 20h21'7" 47d36'56"
Donald Keith Gilbert
Uma 12h9'0" 60d31'28"
Donald Kenneth Gardner
Leo 10h12'17" 26d20'54"
Donald Kenneth Harris
Leo 9h44'59" 30d0'2"
Donald Kobaly, Jr
Leo 11h18'9" -1d30'14"
Donald L. Anderson
Ari 2h48'32" 18d32'11"
Donald L. Bennett
Aur 5h46'21" 49d36'33"
Donald L. Courtney
Per 3h12'21" 54d30'41"
Donald L. Dunham
Psc 0h46'19" 20d0'21"
Donald L. Huyett
Sco 17h18'35" -43d58'8"
Donald L. Janis
Boo 14h33'1" 52d27'40"
Donald L. Jeffries
Uma 12h23'54" 56d9'20"
Donald L. Keplinger
Ori 5h42'44" 5d56'35"
Donald L. Kintz
Ori 6h20'10" 9d29'52"
Donald L. Metcalf
Aqr 22h14'8" 0d38'12"
Donald L. Mutton
Her 16h54'33" 16d58'59"
Donald L Swartz
Ori 6h18'7" 10d43'27"
Donald L. Wunderler
Aql 19h45'45" 16d20'53"
Donald Lambert
Cnc 8h18'7" 13d13'34"
Donald Laru Mitchell
Aql 19h48'20" -0d16'11"
Donald Le Pard Artist Son Brother Friend
Uma 8h39'51" 61d15'58"
Donald Lee Butz
Psc 1h8'0" 32d56'18"
Donald Lee McCombs
Her 16h35'5" 28d31'2"
Donald Lee Smith
Aur 6h28'9" 52d0'28"
Donald Leland Smith
Cep 21h15'40" 62d13'50"
Donald Leo Baker II
Aqr 23h44'8" -23d15'41"
Donald Leonard Walson
Cap 21h20'22" -21d19'38"
Donald Leslie Foss
Uma 11h27'57" 59d2'37"
Donald Lewis
Lyr 18h46'52" 38d41'18"
Donald Lynn Sperry
Lib 15h40'27" -13d49'37"
Donald M. Fleming
Her 17h11'7" 27d36'3"
Donald M. McNair Sr.
Per 3h22'9" 41d53'47"
Donald M. Saer
Sco 16h37'44" -34d24'15"
Donald Maberry
Gem 7h8'47" 17d2'18"
Donald MacMillan Mayberg
Sco 16h12'38" -12d44'51"
Donald Magnotta ~ "Our Champion"
Her 16h33'19" 48d42'55"
Donald & Margot Pardee
Uma 11h32'11" 30d59'10"
Donald & Marian Greif 75th Birthday
Ori 5h57'24" 17d31'47"
Donald Mark Achey
Her 16h30'50" 27d18'55"
Donald Martin Alstadt "7-29-21"
Per 2h22'54" 57d11'45"
Donald Matthew Briggs
Boo 14h48'33" 35d43'13"
Donald McNaught Jackson
Per 4h11'17" 52d2'34"
Donald Merkley Campbell
Uma 11h10'38" 57d4'33"
Donald Meyers
Uma 11h32'7" 34d56'57"
Donald Mill GaMPI Shining Star
Aql 19h43'44" -0d0'47"
Donald Million
Ori 5h46'42" 6d43'47"
Donald Nelson
Per 4h36'57" 40d13'41"
Donald Nelson Paulson
Gem 6h59'14" 31d29'28"
Donald Nestor
Lyn 6h20'49" 58d55'2"
Donald Norman Hunsinger
Psc 23h32'22" 3d29'50"
Donald Orville LeMar Sr.
Per 3h37'1" 47d25'22"
Donald Otis Emerson
Cep 21h33'12" 60d38'32"
Donald P. Buchanan Jr. BJ
Cnc 8h14'55" 6d54'13"
Donald Pace
Leo 10h16'53" 18d6'39"
Donald "Papa" Beaudry
Gem 6h39'6" 15d15'42"
Donald "Papa" Boelhauf
Gem 6h52'9" 15d10'51"
Donald "Papa Red" Rogers
Cnc 8h39'3" 28d18'30"
Donald "Papa" Stepro
Uma 8h51'50" 52d41'21"
Donald Patrick Kenealy Sr.
Lmi 10h24'41" 38d12'9"
Donald Paul Beane
Cnc 8h7'49" 13d7'40"
Donald Paul Parker
Cap 20h31'29" -26d50'36"
Donald Paul Roenigk
Tau 5h8'56" 17d50'20"
Donald Paul Trapp
Per 3h7'3" 56d1'55"
Donald Peckinpaugh
Uma 11h24'46" 45d16'30"
Donald Pile
Cep 20h58'8" 68d2'1"
Donald Piper
Leo 11h15'46" 12d36'37"
Donald Pittavino
Aql 19h27'52" 9d11'33"
Donald Poe, 06-07 NJHS Sponsor
Uma 12h4'1" 58d26'6"
Donald Port
Lyr 18h40'29" 34d12'5"
Donald Pray Markham
Per 2h55'56" 53d39'21"
Donald R. Burden, Sr.
Per 4h44'42" 41d33'55"
Donald R. Camirand
Her 17h19'29" 37d6'2"
Donald R Midkiff
Uma 9h29'37" 53d5'18"
Donald R. Wolfe
Ari 2h57'20" 26d43'11"
Donald Radican
Sco 16h38'48" -29d40'15"
Donald Randolph Steinmetz
Uma 9h56'55" 59d32'42"
Donald Ray Ashworth Jr.
Uma 11h22'4" 28d58'28"
Donald Ray Bellew
Per 3h21'44" 47d10'56"
Donald Ray Fisher III
Psc 1h38'0" 18d43'59"
Donald Ray Gebhardt
Cyg 21h21'11" 37d58'14"
Donald Ray Smith
Tau 4h14'10" 27d49'15"
Donald Ray Wasserman
Her 17h13'21" 20d33'45"
Donald Raymond Goodman
Oph 17h41'3" -0d11'17"
Donald Rechler
Sct 18h57'7" -4d7'22"
Donald Reed Thomas
Her 16h28'54" 13d33'57"
Donald Richard Anderson, Jr.
Aql 19h30'54" 3d13'6"
Donald Richard Rhoades
Leo 9h45'4" 8d13'1"
Donald Richard Willey
Her 18h46'51" 20d7'11"
Donald Robert Blood, Jr
Ori 6h18'2" -0d12'24"
Donald Robert Bowen
Cap 20h25'46" -27d10'55"
Donald Robert Kracke
Cnc 8h40'50" 23d59'9"
Donald Robert MacMillan
Boo 14h35'40" 22d19'56"
Donald & Rosemary Dummer
Leo 10h48'56" 13d39'58"
Donald Ross Wyman
Ori 6h2'45" 12d20'44"
Donald & Ruth McGraw's 50th 2-29-52
Lyr 19h7'46" 32d5'2"
Donald S. Randall
Ari 2h8'5" 25d31'36"
Donald S. Welch
Cma 6h46'43" -20d30'46"
Donald Sal Vinci
Her 16h54'37" 17d57'48"
Donald Santopietro
Uma 9h42'0" 41d48'16"
Donald Schmitt
Cep 22h1'15" 58d17'10"
Donald Sciościa aka Scios
Cas 1h43'2" 63d23'10"
Donald Scott Williams
Her 17h40'55" 34d13'25"
Donald Sherman
Her 17h49'28" 44d58'11"
Donald Singer (The Don)
Cep 0h2'31" 73d58'42"
Donald Snookiepuss Gagne
Psc 0h20'55" 17d19'24"
Donald "StarDon" Gagné
Umi 16h21'4" 72d32'50"
Donald Stokes Macneir "Our Star"
Ori 5h39'18" 1d5'28"
Donald Stommel's Shining Star
Uma 13h44'21" 48d47'55"
Donald Stuart Kramer
Lmi 10h0'48" 29d34'47"
Donald Sutkus
Cap 20h25'6" -9d57'20"
Donald T. Chivas
Cep 23h31'3" 74d17'34"
Donald T. Robertson
Per 2h50'27" 45d25'31"
Donald Thomas Walters
Ori 5h27'35" 1d58'13"
Donald & Tracy Cummings 12/20/03
Uma 8h26'12" 60d19'22"
Donald Van Teyens - Band Director
Lyr 18h31'4" 31d46'55"
Donald Vincent Stevenson
Oph 16h53'55" -0d32'46"
Donald & Virginia Lord
Sco 16h50'22" -38d22'32"
Donald W. Jones
Ari 2h6'38" 23d50'57"
Donald W. Spellman
Sco 16h16'15" -16d24'40"
Donald W. Yutes
Tau 5h35'40" 26d21'3"
Donald Wagner
Per 4h11'10" 50d25'13"
Donald Ward Suman
Sco 16h48'2" -25d41'9"
Donald Wayne Brooks
Aql 19h39'16" 11d24'27"
Donald Wayne Brown
Cep 21h43'14" 63d32'42"
Donald William Blosser Jr.
Ari 2h39'15" 23d26'45"
Donald William Cone
Psc 0h26'4" 16d45'38"
Donald William Keith Stubbs, Sr.
Per 4h25'57" 34d52'43"
Donald William Mauck
Sgr 18h3'11" -28d34'48"
Donald William Moss "Papa"
Per 2h19'29" 57d9'48"
Donald William Pipe
Tau 3h52'24" 29d28'30"
Donald William Weston Sherk
Ori 6h18'36" 9d40'15"
DonaldB
Lmi 10h6'12" 36d36'59"
Donald-Clary
Lyn 6h48'22" 52d58'57"
DonaldHaynesLovesPatriciaDeMartin
Cyg 21h37'2" 47d24'29"
DonaldLewis12/28/1931Didier75th
Cep 22h46'3" 61d43'57"
Donaldo bird
Aql 19h37'32" 16d20'20"
Donalds view
Lib 15h34'12" -4d59'30"
Donaldson A. Miele
Ari 2h21'1" 24d4'59"
DonAnn Sally Campbell Love
Uma 10h34'26" 55d57'50"
Donara Tanatsky Minor
Umi 15h29'41" 73d26'27"
Dona's Astro Believe
Cyg 20h45'15" 46d24'27"
Donata Agnes Randall
Ori 6h11'18" 9d16'53"
Donata Peletta
Ari 3h9'16" 12d30'50"
Donatella
Uma 9h45'45" 46d20'55"
Donatella Cacace
And 23h56'0" 35d55'7"
Donatella Ciaschini
Tau 5h51'56" 14d54'37"
Donatella Grossi
Psc 23h46'26" 6d35'33"

Donatella Malizia Per 2h50'1" 47d51'43"
Donatella Niccolai Per 4h17'59" 45d13'43"
Donatella Ortolani And 23h51'33" 34d1'20"
Donatella Scupola Dra 18h9'46" 72d39'55"
Donato Aur 6h32'30" 39d5'7"
Donato Edmund Basso Lib 14h50'55" -3d40'59"
Donavan James Per 3h23'16" 49d28'47"
Donavan72 Sco 16h14'24" -15d16'46"
Donavan's Star Aur 5h38'57" 41d26'57"
Donbar Uma 10h33'49" 46d15'41"
Dönci bácsi csillaga Lib 15h47'42" -17d51'40"
Dondi Per 3h37'51" 41d4'23"
DONDO Uma 12h17'37" 55d18'24"
Dondy's Beyond Umi 14h23'31" 68d36'30"
Done Right Engine Cnc 8h38'36" 17d20'0"
Donell McGill Vir 12h33'36" -0d0'42"
Donelle Ari 2h13'56" 25d40'37"
Donerz Psc 0h1'57" 4d46'39"
Donette Miranda Vir 11h51'11" 6d28'46"
Donette Sue Phillips-Espinoza Lyn 9h8'36" 39d22'11"
Dong Ho Cho Per 2h23'9" 56d20'25"
Dong Jing Lyn 7h37'43" 46d41'27"
Dong Yom And 0h29'1" 41d27'22"
Donger Psc 1h20'51" 16d29'9"
DongYi Cap 20h9'26" -25d41'56"
Doni Uma 11h34'50" 40d15'45"
Donia Cas 23h28'14" 58d26'41"
Donica's Star And 1h1'50" 35d51'18"
Donicia Lib 15h14'36" -20d32'29"
Donika Mrijaj Lib 15h21'18" -19d47'22"
Donis Putnam Sparks "Mama Putt" Umi 13h6'8" 72d20'51"
Donise Hardy Lyr 18h32'7" 35d59'58"
Doniya Docherty Sco 17h52'13" -36d18'0"
Donkasarus Hill Gem 6h7'30" 27d34'54"
Donley C. & Kelly P. Cyg 20h19'10" 55d23'57"
Donley O. Niskanen Psc 1h14'2" 13d11'25"
DonLo Cep 20h46'37" 60d24'48"
Don'Lyn Feather Aqr 21h5'49" -8d45'46"
Don...My Fate, My Destiny Sco 16h7'4" -25d52'19"
Donn E. Kinzle Cas 23h0'36" 57d49'41"
Donn Emory Coval Barone Sgr 18h0'11" -25d14'4"
Donn Wilkerson Cap 20h11'0" -23d49'51"
Donna Lib 14h22'55" -23d16'41"
Donna Sgr 18h2'16" -34d23'15"
Donna Psc 1h10'53" 3d18'3"
Donna Vel 9h30'38" -40d49'0"
Donna Sco 17h4'9" -42d23'58"
Donna Lyn 7h7'53" 56d55'21"
Donna Umi 14h23'3" 71d23'41"
Donna Uma 9h38'20" 60d1'11"
Donna Cap 21h57'31" -8d48'43"
Donna Sgr 18h10'19" -17d16'24"
Donna Sgr 18h27'3" -17d38'31"
Donna Cap 21h26'58" -15d47'6"
Donna Cap 20h36'19" -15d49'29"
Donna Mon 6h52'49" -0d6'33"
Donna Vir 13h52'40" -11d31'55"
Donna Ori 5h56'58" 18d5'58"
Donna And 23h24'22" 36d56'45"
Donna And 1h34'53" 42d56'35"
Donna And 0h27'31" 43d13'32"
Donna And 1h5'47" 40d34'2"
Donna Cyg 21h13'18" 42d30'11"
Donna And 23h34'1" 41d43'55"
Donna Cas 0h14'41" 51d47'30"
Donna 30X2 Leo 9h42'48" 28d43'39"
Donna 72565 Cnv 13h41'33" 33d45'59"
Donna * 812 Lib 15h13'41" -4d41'48"
Donna A Basile Lyn 6h43'53" 60d5'51"
Donna A. Corrao Lyr 18h46'29" 37d59'54"
Donna A. Nivens-Aragon Lib 15h37'49" -11d25'26"
Donna and Aurora Mochrie Psc 1h29'44" 17d36'9"
Donna and Bo Her 17h2'57" 34d28'31"
Donna and Joe July 4, 2002 Uma 8h15'54" 66d36'51"
Donna and Joseph Eaton Uma 8h42'54" 72d34'0"
Donna and Raven Mcalpin Star Her 17h36'30" 35d52'53"
Donna and Robert Lester Dra 19h4'39" 64d41'44"
Donna and Roger Cagle Uma 9h43'22" 72d11'53"
Donna And Rosemary Lib 14h48'10" -0d41'15"
Donna Angelicata Tau 4h15'38" 13d46'36"
Donna Ann D'Agostino Sco 16h7'17" -9d24'48"
Donna B. Fawcett Psc 23h4'51" 1d12'17"
Donna Babe Aqr 21h59'48" -14d4'0"
Donna Bailey Sgr 18h38'38" -32d57'47"
Donna Bayern Lyn 6h32'12" 58d2'30"
Donna Beck-Kirkham Uma 13h45'28" 62d8'32"
Donna Belle Warcholak Cas 1h30'16" 64d9'43"
Donna Berger-Steinman Leo 9h26'12" 26d49'35"
Donna Blough Vir 13h0'7" 3d5'55"
Donna "Blue Daisy" Cnc 9h15'47" 20d4'41"
Donna Blundell Cru 12h10'44" -63d52'47"
Donna & Bob Ward And 0h12'45" 27d4'22"
Donna Bonita And 0h49'41" 37d43'13"
Donna Boots Van Horn And 2h15'47" 38d16'33"
Donna Bortree Cas 0h24'39" 60d54'40"
Donna Brady Lyn 7h41'8" 42d8'53"
Donna Brandsema And 0h18'0" 44d37'57"
Donna Brown Cam 3h43'7" 52d59'16"
Donna Burchell Lyr 18h37'43" 28d21'2"
Donna Butt Pres. WMD ALA '06 - '07 Cas 0h19'19" 57d21'21"
Donna C. Evans And 0h37'45" 28d9'18"
Donna & Calvin Borrowman 50th Uma 9h43'53" 42d8'17"
Donna Carrie Lyn 7h2'20" 47d37'14"
Donna Chadock Jacobs Ori 5h8'33" -0d1'10"
Donna Chapman Pate Uma 9h45'2" 56d49'12"
Donna Christine D'Angelo Sgr 18h57'41" -18d46'55"
Donna Christine Hebreard Del 20h18'15" 15d56'6"
Donna Claire Psc 0h54'24" 26d48'34"
Donna Cole-Matthews Cra 18h10'52" -40d57'16"
Donna Conley Pierce "5-8-34" Lmi 10h37'50" 31d45'19"
Donna Couch And 0h39'19" 35d43'0"
Donna Crawford Sco 16h52'23" -27d24'28"
Donna Czak Ori 5h15'14" 15d52'40"
Donna D. Blasko Cam 5h44'22" 59d3'26"
Donna Daley Cas 0h23'54" 60d58'23"
Donna Daniels Ari 2h58'58" 28d13'19"
Donna D'Annunzio Lyn 6h45'25" 52d12'0"
Donna Darlene Patri Crb 15h16'39" 26d28'48"
Donna Davis Cyg 20h48'41" 45d40'17"
Donna De Sco 16h44'12" -32d0'36"
Donna Deckard Sco 17h13'5" -34d45'10"
Donna Delgoffe Uma 11h38'30" 50d53'30"
Donna DeLorenzo And 2h24'14" 48d45'35"
Donna Denison Cnc 8h53'16" 26d25'25"
Donna Denoon Cnc 8h34'44" 14d54'17"
Donna Disney Boo 15h38'38" 42d20'20"
Donna E. Boone Aqr 21h4'55" -8d54'6"
Donna Elaine Streeter Cyg 20h47'14" 40d41'11"
Donna Elizabeth Durnin Umi 16h25'4" 71d26'35"
Donna Ellen Ballard Cnc 8h47'21" 17d11'47"
Donna Emily Serrano Com 12h59'6" 17d49'39"
Donna Engard Lyn 8h18'0" 55d14'27"
Donna Esposito Leo 10h59'11" 14d41'6"
Donna Estrela Martins Lyn 7h4'22" 59d29'31"
Donna F. Miller Lib 14h36'8" -23d6'37"
Donna Farkas "My True Love" And 23h23'35" 38d4'50"
Donna Faye Cam 4h44'57" 70d42'20"
Donna Ferrino Umi 16h31'56" 81d13'4"
Donna Field Cas 0h15'42" 56d34'29"
Donna Fleming Cyg 20h54'27" 39d26'32"
Donna Fleming Cas 0h56'38" 64d9'49"
Donna Fox Uma 9h32'32" 46d47'15"
Donna Frieden Gem 6h40'39" 20d2'57"
Donna G. Klein Aqr 22h32'3" -8d49'11"
Donna Gail Catanzaro Aqr 22h7'41" 2d3'25"
Donna Gail Haines Peg 23h29'59" 27d38'33"
Donna & George Cyg 20h56'49" 31d7'11"
Donna & Gil Bryan Cyg 20h4'40" 39d47'40"
Donna Gillissie-Cross Cas 0h54'50" 54d10'6"
Donna & Gordon Bickley Mon 8h5'7" -0d41'9"
Donna Grace Vir 12h21'51" 10d44'29"
Donna Gramma Jelly Her 16h49'37" 32d51'47"
Donna Hamblin And 1h39'1" 45d45'7"
Donna & Harold Crb 15h22'2" 30d56'50"
Donna Harshaw And 22h58'10" 40d1'24"
Donna Heebner Ari 2h46'18" 25d43'43"
Donna Hiller Per 2h48'43" 56d19'0"
Donna Hoffman Ari 3h20'23" 29d20'5"
Donna & Howard Kohlbacher Per 3h46'26" 34d8'19"
Donna Hudson Uma 11h0'11" 72d24'37"
Donna Hughes And 1h38'11" 45d21'29"
Donna Hurley Aql 19h43'54" 3d2'23"
Donna Hurlock Lib 15h49'11" -6d20'57"
Donna Ieraci Uma 10h46'43" 51d37'42"
Donna Irene Rossi Cap 21h47'49" -19d57'14"
Donna J. Clapsaddle Vir 14h2'39" 5d51'22"
Donna J. McGrew Gem 7h46'29" 16d5'59"
Donna & Jason Always and Forever And 23h21'35" 47d50'13"
Donna & Jayna Marie's Star Cyg 20h1'59" 52d21'34"
Donna Jayne Williams Cnc 8h53'35" 30d39'12"
Donna Jean And 1h0'52" 41d43'35"
Donna Jean And 23h19'20" 47d50'20"
Donna Jean Cam 3h24'41" 55d29'52"
Donna Jean Ori 5h51'22" 3d22'24"
Donna Jean Sgr 18h13'36" -16d32'49"
Donna Jean Ebner Uma 14h14'17" 57d43'14"
Donna Jean Guyer Psc 1h23'50" 27d55'4"
Donna Jean Hamblin Ari 3h20'57" 26d0'59"
Donna Jean Kerekes Psc 1h17'12" 32d46'49"
Donna Jean LPN Cas 1h26'9" 60d5'9"
Donna Jean McComas And 0h29'7" 28d45'22"
Donna Jean Phillips Cotten Aqr 21h48'55" -0d32'50"
Donna Jean Rosenking Ori 6h16'33" 10d41'54"
Donna Jean Schulze Per 3h16'52" 54d13'30"
Donna Jean Wood Wilkins And 1h41'30" 43d4'28"
Donna Jean Zitelman Vir 13h24'9" -21d6'7"
Donna Jeanne Johnson, Jan. 29, 1947 Aqr 22h42'17" -1d17'8"
Donna Jeanne Neilson-Zimmerman Lyn 7h50'36" 56d34'17"
Donna/Jenna/TicTac Family Star Uma 8h44'13" 63d13'34"
Donna Jo Harrison Uma 9h52'10" 70d30'5"
Donna Jo Roysdon And 23h11'29" 43d42'19"
Donna Jo Simons "DJ" Cap 21h24'48" -15d27'9"
Donna Jo Treat Lmi 10h0'33" 37d49'56"
Donna Joy Momb Cnc 8h25'18" 23d4'39"
Donna Joy Wandling 4-22-2000 Umi 17h40'30" 83d41'55"
Donna Joyce Anderson Vir 11h53'30" -0d57'9"
Donna June Deti Cyg 21h33'2" 37d45'38"
Donna JVM Templeton Sco 16h14'56" -25d18'59"
Donna K Watkins Lib 15h40'43" -18d56'19"
Donna Kanowitz Cas 1h38'59" 64d14'16"
Donna Karr Peg 22h58'39" 23d54'16"
Donna Kathleen Luther Leo 11h50'6" 19d57'23"
Donna Kay Aqr 22h41'15" -3d18'51"
Donna Kay Sgr 19h1'10" -35d56'35"
Donna Kay Lovell Cap 21h42'7" -19d4'40"
Donna Kay McComas Vir 12h2'37" -0d8'43"
Donna Kay Rehm Frazier And 2h15'30" 49d18'47"
Donna Kaye Cato Per 3h45'7" 43d47'15"
Donna Kaye Grady Sgr 20h24'6" -41d32'13"
Donna Kaye Humphrey And 23h9'46" 46d23'13"
Donna "Keebab" Chamberlain And 2h31'13" 39d35'53"
Donna Kinard Cas 0h32'59" 51d17'29"
Donna Kristina Dra 9h46'30" 79d15'37"
Donna L. Bialozor Cas 2h2'39" 58d14'28"
Donna L. Black Cyg 20h48'19" 47d8'34"
Donna L. Cavanaugh Uma 8h43'53" 64d58'13"
Donna L. Dolliver 2-14-1946-8-14-04 Aqr 22h37'27" -2d5'2"
Donna L. 'Nipponnut' Jordan Lyn 9h40'48" 39d49'32"
Donna L Ragatz Uma 11h38'48" 64d35'27"
Donna L Stinziano Lib 15h54'49" -5d36'5"
Donna Lane Uma 11h14'3" 48d12'32"
Donna Laura, Little Bug, My Love Cnc 9h4'32" 7d18'33"
Donna Lea Cyg 20h36'16" 56d51'20"
Donna Lea Chavez Gem 7h21'23" 13d37'37"
Donna Lea Falvey - Trapp..."Purd" And 0h30'4" 37d35'30"
Donna Leadbetter Brown's Star Aqr 22h27'9" -2d30'40"
Donna Lease Uma 10h48'33" 60d39'26"
Donna LeBlanc Per 1h44'35" 52d26'58"
Donna Lee And 1h24'0" 33d54'30"
Donna Lee Leo 10h11'32" 14d19'7"
Donna Lee Uma 11h21'57" 52d57'46"
Donna Lee Cra 18h37'33" -41d2'10"
Donna Lee Failor Aqr 22h42'29" 0d39'50"
Donna Lee Green Lac 22h25'19" 51d0'24"
Donna Lee Irwin Crb 16h8'32" 38d55'31"
Donna Lee Meng Cyg 21h32'38" 35d52'57"
Donna Lee Schafer Uma 10h0'49" 66d54'4"
Donna Lee Staab Walker Vanderwall Cas 0h41'1" 47d37'10"
Donna Lee Thomas Child Aql 19h46'5" 6d19'9"
Donna Lee Viola And 0h24'32" 42d27'13"
Donna Lee's Point of Light Crb 15h35'13" 25d47'3"
Donna Leigh Ori 5h32'46" 7d14'12"
Donna Leigh Brown Crb 16h2'2" 33d14'41"
Donna & Lenny's Shining Star Cyg 20h15'37" 55d28'30"
Donna Lisa Phillips Cap 20h59'7" -20d20'20"
Donna Litkenhaus Cas 1h17'36" 62d25'8"
Donna Loomis Rubin Lib 15h17'51" -5d18'48"
Donna Lorraine Watt Vir 12h19'12" -4d57'46"
Donna Louise Cas 23h29'15" 52d58'0"
Donna Louise Clarke Allen And 0h1'36" 45d22'44"
Donna Louise Robinson Cas 0h51'48" 61d41'58"
Donna Louise Russell And 23h28'45" 50d23'7"
Donna Lyn Huggins Ori 5h8'7" 5d51'19"
Donna Lynn Vir 13h35'14" 10d8'1"
Donna Lynn Vir 13h34'37" -7d39'1"
Donna Lynn Carr Lib 15h1'50" -17d23'54"
Donna Lynn Iarusso Lyr 18h33'29" 36d16'8"
Donna Lynn Laughter Lib 15h54'2" -10d21'45"
Donna Lynn Lucas Lib 14h39'12" -12d16'42"
Donna Lynn Mares (Nee) Michalek Her 17h8'18" 32d8'15"
Donna Lynn Olsen Cnc 8h46'36" 15d25'48"
Donna Lynn & Paul Joseph Caruso Cyg 20h8'46" 32d12'2"
Donna Lynn Rogers Tau 4h11'11" 10d8'59"
Donna Lynn Rotunno Psc 22h54'20" 3d31'21"
Donna Lynn Sellnow Cas 23h22'39" 57d13'26"
Donna Lynn Shafar Cas 0h25'53" 63d40'56"
Donna Lynn Sink Uma 8h57'58" 49d58'14"
Donna Lynn Stewart Uma 11h16'2" 46d58'10"
Donna Lynne Saul Cam 4h30'53" 54d1'35"
Donna M. B. Lyn 9h16'57" 42d37'36"
Donna M. Bigley Psc 1h23'0" 28d21'2"
Donna M. Condron Cas 1h27'9" 52d4'18"
Donna M. Goehring And 1h51'12" 39d21'36"
Donna M. Meloun Aur 5h23'6" 48d57'46"
Donna M. Moore Uma 9h16'13" 57d21'33"
Donna M. Regan Psc 1h20'46" 27d22'16"
Donna M. Wiezycki Lib 15h23'39" -26d57'44"
Donna MacCarthy aka DM Boo 14h27'44" 14d35'36"
Donna Mackey Lmi 10h45'29" 33d40'49"
Donna Mae Ari 2h51'54" 22d27'32"
Donna Mae Cnc 8h3'44" 20d40'19"
Donna Mae Andree-Parrill Ori 6h16'47" 14d16'7"
Donna Mae Heidbreder Steeples Ari 2h14'26" 26d52'58"
Donna Mae Kunder Dra 18h12'4" 57d39'5"
Donna Mardell Dunn Umi 13h57'44" 71d52'10"
Donna Maria Ori 6h16'5" 6d13'32"
Donna Marie Ori 6h2'40" 6d6'21"
Donna Marie Peg 21h40'28" 13d39'39"
Donna Marie Ari 2h32'48" 25d41'9"
Donna Marie Psc 1h8'22" 22d59'56"
Donna Marie And 0h10'29" 28d29'36"
Donna Marie Crb 15h37'21" 33d49'25"
Donna Marie Lyn 8h2'46" 40d8'3"
Donna Marie And 1h59'41" 46d3'10"
Donna Marie And 23h14'45" 44d20'29"
Donna Marie Cap 20h30'2" -18d13'41"
*** Donna Marie *** Lib 14h59'21" -3d48'56"
Donna Marie Vir 12h36'8" -1d4'21"
Donna Marie - Always and Forever Aqr 23h4'33" -11d12'5"
Donna Marie Andrews Angel Star Cas 1h16'13" 53d51'49"
Donna Marie Buckley Cas 0h0'35" 57d1'9"
Donna Marie Corry Cap 21h49'22" -22d38'21"
Donna Marie Crumit Ari 2h13'23" 24d51'42"
Donna Marie Cummans Gem 6h51'45" 24d36'47"
Donna Marie (DMTC) Her 16h47'21" 33d0'51"
Donna Marie Donzi Per 3h38'40" 38d37'23"
Donna Marie Evans Sco 17h25'46" -38d41'49"
Donna Marie Fazzini Aqr 23h10'6" -5d48'49"
Donna Marie Fitzhugh Lyn 7h36'8" 37d39'19"
Donna Marie Galier Lyn 7h57'50" 40d22'2"
Donna Marie Hopkins Ari 2h27'9" 26d36'14"
Donna Marie Imhoff Psc 1h10'49" 32d7'34"
Donna Marie Iris Aqr 20h41'39" -9d37'9"
Donna Marie Kennedy And 1h56'59" 36d28'4"
Donna Marie Marinucci Cap 20h29'9" -24d38'5"
Donna Marie Miller, Sr. Sco 16h2'56" -22d31'40"
Donna Marie O'Brien And 2h22'16" 50d12'29"
Donna Marie Rock Tau 4h48'0" 16d11'5"
Donna Marie Russell Barriga Lyn 7h18'22" 57d21'48"
Donna Marie Squires Gem 6h52'50" 21d35'47"
Donna Marie T Cam 5h5'50" 60d20'58"
Donna Marie Walsh Sco 17h55'28" -32d6'24"
Donna Marie West Cas 0h32'45" 54d58'0"
Donna Marie"Mariskils" DeRusso Ari 2h46'10" 14d45'4"
Donna Marine Kelley Lib 15h35'33" -24d59'30"
Donna Marlene Crispell Umi 17h1'20" 79d30'10"
Donna Martin Crb 16h9'39" 38d37'13"
Donna Martin Uma 11h35'54" 32d56'27"
Donna Mary & Donald Wayne Westbrook Sco 16h42'20" -37d11'28"
Donna Mary Eva Martin Leber Aqr 23h2'19" -11d28'39"
Donna Mary Louise Monger at 21 And 1h48'29" 46d28'25"
Donna May LaBelle Tau 3h49'13" 19d55'24"
Donna May Marshall Cas 23h25'24" 58d42'0"
Donna McCarthy Ori 5h16'23" 0d47'49"
Donna McCormick Pav 18h19'32" -59d40'54"
Donna McCue Cas 0h7'15" 57d22'54"
Donna McDonald And 23h54'55" 44d46'50"
Donna McHale Aqr 21h10'18" -11d22'9"
Donna McPheters Leo 11h20'48" -4d5'0"
Donna Mellon Cas 0h52'0" 62d41'44"
Donna Meyers-Krzyzewski Ori 5h33'0" -1d48'30"
Donna & Michael Goodfred Ori 5h32'58" -0d17'41"
Donna Michaels And 1h44'6" 50d12'45"
Donna Michaud Cam 4h8'9" 53d55'40"
Donna Mock And 1h26'14" 43d32'51"
Donna Mullen Crb 15h40'6" 35d52'58"
Donna - My Little One Cap 21h35'52" -15d58'28"
donna - my shinning light Ori 5h51'10" 7d21'53"
Donna My Sister Psc 23h6'56" 1d5'39"
Donna N. Cas 0h0'58" 58d44'4"
Donna N. Cas 0h0'39" 59d15'42"
Donna "Nanny" Cas 0h36'14" 53d48'38"
Donna Nilsson Uma 8h54'24" 47d1'53"
Donna O'Shaughnessy Cas 23h23'13" 55d0'40"
Donna Ouellet Cnc 9h3'0" 9d39'21"
Donna Our Shining Star Cas 0h12'2" 54d47'26"
Donna P. Holliday Sco 16h46'45" -34d58'58"
Donna Pacurai And 23h40'40" 47d15'30"
Donna Patrice Nichols Her 18h2'4" 36d21'8"
Donna Patterson Cas 0h20'22" 53d35'26"
Donna & Paul McElyea Cyg 20h32'30" 30d26'20"
Donna Peach Raychel And 1h8'14" 46d47'44"
Donna Peck And 1h40'8" 43d55'6"
Donna Pittman Peg 21h18'1" 19d29'34"
Donna Powers Cas 1h39'44" 61d18'17"
Donna R Cercel Leo 11h36'37" 0d49'1"
Donna Rae Psc 0h41'10" 13d32'36"
Donna Rae Cas 1h20'5" 57d35'17"
Donna Rae Cap 20h26'21" -26d52'28"
Donna Rae Johns And 22h58'3" 37d26'5"
Donna Rae Richardt Leo 10h25'6" 9d53'25"
Donna Rae Schneider Cyg 21h48'17" 47d17'55"
Donna & Ralph-till the end of time Cyg 19h17'53" 54d43'11"
Donna Raman Lyn 7h56'20" 42d29'19"
Donna Rappich Carr Psc 1h49'25" 6d3'29"
Donna Rauch Uma 10h41'36" 68d4'22"
Donna Ray Ari 2h26'44" 12d3'23"
Donna Ray Repass Sco 17h53'36" -36d29'13"
Donna Rayburn And 1h33'4" 37d58'13"
Donna Raye Richards Cyg 21h14'37" 47d7'6"
Donna Renae Sanderson Leo 10h7'40" 18d24'12"
Donna Russell Lyn 7h7'34" 61d16'22"
Donna Ruth Hermann Crb 16h17'20" 31d47'42"

Donna' s Heart
Sgr 20h3'36" -30d37'17"
Donna Samolyk
Cam 3h22'38" 59d0'4"
Donna Savelle
Aqr 21h29'23" -2d56'43"
Donna Schaeffer
Equ 21h15'53" 9d1'24"
Donna Serviolo
Cas 1h11'42" 64d6'10"
Donna Shining Brightly Forever
Aql 19h55'11" -0d33'33"
Donna Shining Star
Sgr 19h15'23" -14d22'28"
Donna Small
Sco 17h37'12" -35d43'39"
Donna Sue
Sgr 18h9'50" -27d9'59"
Donna Sue
Vir 13h54'36" -1d21'1"
Donna Sue Presley
Gem 7h16'39" 28d16'0"
Donna Susan Mayne
Peg 21h54'27" 16d20'12"
Donna "Sweetness" McCallum
Cnc 8h39'28" 23d30'55"
Donna (The Mom) Vandergrift
Cas 1h18'51" 62d40'34"
Donna Ucciferri
Aqr 22h1'16" 1d30'21"
Donna Urbanowicz
Cas 1h31'50" 65d29'18"
Donna Utley
And 23h27'24" 38d47'28"
Donna V. WMAJ
Leo 10h44'0" 14d44'22"
Donna Vagedes
And 0h17'12" 25d46'23"
Donna Vekni 24
Vir 12h59'48" 4d24'38"
Donna Venezia
Lyn 9h12'52" 34d59'46"
Donna Walther
Per 3h12'58" 44d6'25"
Donna Webster Drew
And 0h34'46" 34d3'16"
Donna Wee
Lib 15h43'31" -17d13'50"
Donna Weinstein
Crb 16h17'10" 30d53'14"
Donna Wilson
Lib 15h6'18" -3d44'35"
Donna Wright
Vir 11h50'11" -5d50'59"
Donna, my brightest star
Cma 6h40'53" -14d41'33"
Donna, My Heavenly Star
Uma 12h57'42" 53d37'50"
Donna, the star of my heart
Sgr 18h0'41" -18d7'45"
Donna-BabyDoll
And 1h23'46" 39d50'6"
Donna-Beloved Mother of Three
Uma 12h0'16" 41d12'30"
DonnaCollegius
Cas 23h34'32" 54d32'17"
Donnajean Pierson
And 1h7'43" 43d33'35"
Donnajean Young
Sco 16h12'4" -15d30'51"
Donna-Jo Banik Haffke
Sgr 17h55'48" -27d21'45"
DonnaJoy Osorio
Aql 19h33'31" 11d4'59"
Donna-Lee
Cas 23h9'9" 54d40'59"
Donnalee "Hoy" Vaughn (God's Light)
Tau 4h24'50" 22d27'30"
Donnamarie
Vir 12h47'34" 4d30'36"
Donna-Marie
Cas 0h44'45" 51d51'52"
DonnaNova
Crb 16h12'3" 32d58'6"
Donna's Angel
And 2h33'5" 50d12'2"
Donnas Angel Angelina
Vir 12h49'41" -11d5'42"
Donna's Diamond
Sco 16h52'43" -35d33'17"
Donna's Dream
Ori 5h35'12" -1d36'9"
Donna's Dream Maker
Cyg 20h37'5" 38d19'2"
Donna's Light of Hope
Tau 5h16'24" 23d10'59"
Donna's Miracle
Uma 9h30'8" 41d36'59"
Donna's Portrait
Lib 15h47'8" -5d57'55"
Donna's Shining Star
Cap 20h51'4" -21d59'3"
Donna's Silver Dream
Sgr 19h13'33" -21d41'41"
Donna's Spirit Shines On
Sco 16h47'31" -31d35'45"
Donna's Star
Uma 8h57'37" 50d13'5"
Donna's Star
Leo 10h19'37" 23d58'29"
Donna's Star
Cnc 8h35'26" 21d13'40"
Donna's Star
Gem 7h18'50" 18d54'0"
Donna's Star. "You are" My love XXX
Cru 12h35'43" -60d34'40"
Donna's Wish
Cnc 8h16'4" 9d22'12"
DonnaTerry
Lac 22h56'34" 38d46'56"
Donncha
Tau 4h33'36" 6d20'8"
Donne Bernstein
Dra 18h55'58" 50d26'43"
Donne Holden, M.D.
Lib 14h45'15" -11d51'28"
Donnell Allen King
Ori 5h1'54" -0d24'17"
Donnelley
Tau 5h49'2" 20d38'13"
Donnelly
Uma 10h34'26" 65d9'5"
Donner
Psc 0h28'13" 7d38'35"
Donner, Harald
Ori 5h14'14" -8d5'15"
Donnesha Yolande Sandidge
Psc 23h13'56" 5d49'18"
Donni-Brooke Briar
Crb 16h11'5" 31d8'21"
Donnie and Brooke Musick
Cyg 19h58'58" 33d21'28"
Donnie and Nicky
Cap 21h19'8" -15d32'53"
Donnie Anthony Toups, Sr.
Umi 15h4'29" 82d23'59"
Donnie Bowen's Angel
Umi 14h58'3" 76d7'37"
Donnie C. Kehler
Aqr 22h6'27" -3d5'38"
Donnie D
Cir 15h19'16" -58d43'40"
Donnie Gene Cogsdill
Her 16h32'34" 37d6'44"
Donnie Gomes
Leo 11h38'6" 26d41'17"
Donnie Hathaway
Her 17h20'56" 26d55'27"
Donnie Jerina
Umi 16h28'23" 77d36'32"
Donnie L. Weaver
Her 17h44'30" 38d18'43"
Donnie Lee Henton
Ori 5h58'37" 20d48'33"
Donnie Lee Triplett
Ari 2h3'29" 22d41'16"
Donnie Lynn Holland
Tau 5h14'35" 19d58'26"
Donnie - My Precious Angel
Uma 11h34'45" 46d44'47"
Donnie Myles Crocker
Uma 11h21'10" 58d55'24"
Donnie Pearl Stevens
Lyr 18h39'2" 32d0'39"
Donnie Rae Dreampierre1982
Uma 9h1'28" 55d2'40"
Donnie Ray Johnson
Gem 6h4'38" 26d0'47"
Donnie Ray Jones-Gilliland
Psc 23h47'41" 1d50'36"
Donnie Reed
Lyn 7h36'38" 48d37'6"
Donnie Saxon
Dra 18h46'34" 59d25'12"
Donnie Vaughan
Uma 11h16'52" 67d18'25"
DonnieGayle
Aql 19h41'27" 14d38'8"
Donnie-Star Gazer
Umi 13h8'34" 72d39'19"
Donnis Forever
Uma 11h21'22" 63d56'5"
DonnisGary
Per 4h6'35" 41d11'15"
Donnnnnt!
Her 17h36'39" 44d24'1"
DonNora
Cas 0h53'54" 64d36'28"
Donny
Cru 12h38'52" -59d1'27"
Donny
Cnc 8h48'35" 15d19'18"
Donny and Jacki's Bay
Gem 6h23'2" 21d35'20"
Donny And Jenn 11-19-04
Ori 5h59'54" 20d38'44"
Donny and Jenn's Star
Lyr 18h28'23" 39d46'39"
Donny Cartwright
Boo 15h19'59" 47d38'17"
Donny Cervantez
Sgr 18h51'9" -32d4'32"
Donny Montross, My Buddy
Aur 5h16'46" 29d24'2"
Donny Mowlds
Cma 6h19'29" -28d52'58"
Donny My Love
Sco 16h39'45" -36d38'2"
Donny Ray Wilson
Nor 16h7'28" -48d18'27"
Donny Rosen
Lib 15h51'13" -13d21'22"
Donny Shope
Leo 9h32'39" 29d9'3"
Donny & Veronica
Cyg 19h48'53" 47d16'6"
Donoren
Pho 0h31'8" -40d58'20"
Donovan & Amy Butler
Cyg 20h2'33" 32d13'13"
Donovan Bodie Hall
Aqr 22h35'30" -3d30'43"
Donovan D Paull
Her 18h48'39" 16d24'49"
Donovan Earl Recore
Cep 22h44'4" 69d11'21"
Donovan family
Vir 12h42'1" 12d18'56"
Donovan Gary Elliott
Per 3h22'41" 49d21'30"
Donovan Huge James McGraw
Vir 13h30'35" 4d27'57"
Donovan Hunter Scott Mackay
And 23h58'28" 45d34'46"
Donovan James Quinn
Sgr 18h49'16" -20d17'16"
Donovan Joseph
Sgr 19h34'35" -14d15'29"
Donovan Loss
Leo 10h15'33" 25d44'0"
Donovan Sherard Tennimon
Psc 0h17'49" 9d53'24"
Donovan Thelander
Ori 5h46'30" 2d7'49"
Donovan's Grá
Leo 10h59'0" 18d23'16"
Donovan's Shining Star
Leo 10h2'9" 18d45'28"
Donovan's Stars 2004-2006
Gem 6h31'8" 15d50'27"
Donresa
Lmi 10h0'23" 28d45'57"
Don's Dream in the Heavens
Lib 15h8'51" -6d45'23"
Don's Love
Sgr 20h19'44" -41d27'6"
don't let go
Uma 13h49'10" 50d21'8"
"Don't Stop Believing"
Uma 11h35'29" 57d59'47"
Donta Richardson
Psc 0h57'12" 32d51'9"
Dontae
Tau 5h48'59" 17d32'56"
Donté Louis Wilder
Gem 7h37'7" 14d39'42"
Donte Payne
Lyn 7h14'18" 59d19'46"
Donte' Shay Hamilton
Vir 13h24'55" 12d49'28"
DonToine
Uma 10h39'50" 46d0'28"
Donum di Diligo
Gem 6h44'47" 12d18'43"
donum mater matris
Crb 15h25'43" 27d55'51"
Donvidaelle
Lyn 8h14'43" 44d42'25"
Donya
Ori 6h4'40" 7d17'51"
Donya Bentley
Psc 1h7'26" 12d23'27"
Donya Christine
Lyn 7h49'35" 35d44'49"
Donya Denise Parnell
Lyn 8h5'28" 40d6'52"
Donzner
And 2h23'22" 41d44'24"
Doo
Per 3h24'50" 48d52'19"
Doo Doo
Umi 15h7'27" 79d10'13"
Doodah
Cnc 8h38'16" 8d40'44"
Doodah
Ari 2h22'28" 17d44'52"
Doodethzkie
Lyn 8h20'59" 58d40'50"
Doodle
Cas 1h18'29" 62d41'12"
Doodle
Lib 15h49'24" -19d30'57"
Doodle
Psc 0h46'52" 8d20'51"
Doodle
Del 20h45'49" 16d0'13"
Doodle Bug
Uma 9h45'19" 42d10'11"
Doodle & Ethel Tobert
Ori 6h1'51" 20d48'50"
Doodle Lou
Tau 4h16'25" 26d26'37"
doodlebug
And 0h20'20" 37d31'50"
DOODLES
Gem 7h35'44" 32d53'56"
Doodles
Umi 16h14'3" 71d39'53"
Doodles. Always and Forever
Leo 9h31'37" 25d15'5"
Doodles Destiny
And 2h24'38" 50d36'9"
Doodlidoo
And 1h33'7" 47d5'43"
Doody Star
Uma 8h58'14" 58d30'7"
Doody1
Vir 13h14'48" -22d9'8"
Doofy
Vir 12h29'20" -2d13'54"
Doogie
Lmi 10h26'33" 39d20'33"
Doogle
Leo 10h52'2" 14d33'9"
DoogleDog - Douglas Maynard
Uma 11h19'25" 47d50'59"
DOOLA-MARIE
And 0h57'8" 35d59'7"
Dooley
Cam 3h48'34" 55d37'57"
Dooley
Ori 5h25'1" 10d16'39"
Dooley's Dragon
Cnc 8h51'11" 32d35'17"
Doozer
And 0h59'4" 45d40'49"
Doozie Star
Her 17h5'9" 33d40'38"
Dora
Cyg 19h40'24" 46d40'29"
Dora 4-22-29 & Joe 10-31-26 Marker
Umi 15h21'31" 72d7'55"
Dora A Guiding Light
Crb 16h8'31" 33d10'27"
Dora Ann Smith
Cas 23h50'29" 57d13'46"
Dora Bowie
Psc 1h28'35" 12d10'34"
Dora Bowman
And 1h43'46" 45d36'16"
Dora Dayton
Uma 11h37'50" 52d9'18"
Dora DeMaglie
Ari 2h12'48" 23d35'46"
Dora E. Bright
Cas 0h49'18" 56d32'41"
Dora Francisca Marin
Lib 15h30'18" -27d22'32"
Dora Girardi
Uma 14h4'23" 57d0'15"
Dora & Jim
Cyg 20h28'30" 53d12'54"
Dora & Joseph Huser
Ori 5h59'8" 13d52'16"
Dora Kayla
Crb 15h26'27" 27d36'48"
Dora Kidd
Lyr 18h34'19" 27d1'45"
Dora Lee Mendoza (Doe')
Umi 11h26'6" 88d22'12"
Dora Loves Siegfried forever
Uma 10h59'53" 47d17'3"
Dora Maldonado
Sco 17h9'48" -38d49'0"
Dora Rodriguez Livi
Aqr 22h30'53" -1d14'21"
Dora Root
Uma 9h18'33" 59d58'43"
Dora the Explorer
Leo 11h24'32" 17d8'15"
Dora Vasilatos
Leo 9h54'32" 30d30'27"
Dora Villalobos
Del 20h40'32" 5d32'40"
Dora Wallace
Sgr 18h29'33" -34d32'53"
Dora, Tou Hou
Leo 11h5'37" 15d3'25"
Doraine Jodlowski
Vir 12h23'29" 4d6'21"
Doralisa (Charito) Vasquez Santana
Gem 6h56'21" 16d3'48"
Dorange
Crb 16h18'26" 26d43'9"
Dorathea "DOTTIE" Chase
Uma 8h26'42" 65d30'51"
Dorbrian
Dra 16h52'17" 62d16'2"
Dorcas and Lewis Faulkner
Her 18h40'31" 17d36'25"
Dor-Dor's Star
Cyg 20h55'25" 36d36'22"
Dordy
Lib 14h51'42" -3d48'40"
Doreatha & Thomas Woody
Uma 8h21'56" 69d51'46"
Doreen
Uma 8h51'25" 69d51'14"
Doreen A. Stevenson 08/28/1965
And 23h23'12" 41d39'0"
Doreen and Dennis O'Connor
Cyg 20h35'41" 59d30'23"
Doreen and Eric Leeds
Lyr 18h27'34" 38d0'50"
Doreen and Terry Scott
Cyg 20h46'4" 38d38'41"
Doreen Ann Crow
Sco 16h31'54" -27d17'55"
Doreen Anthony
Cnc 8h26'12" 11d36'15"
Doreen Babcock
Cas 23h43'5" 56d40'7"
Doreen Berne
Ori 5h30'46" 14d29'20"
Doreen Blanche Roach
Hya 10h16'15" -17d11'13"
Doreen Brown Welsh
Aqr 22h4'48" -10d57'7"
Doreen Chapman Martin
Uma 11h12'23" 62d59'24"
Doreen Davidson
And 2h30'23" 37d34'27"
Doreen Donnelly
Aqr 22h44'50" 1d42'26"
Doreen Douglas
Cas 2h10'18" 66d45'43"
Doreen Dyce 02/03/1927
Psa 22h27'48" -29d6'51"
Doreen Edwina Williams
Uma 10h26'20" 46d52'34"
Doreen Elizabeth French
Aqr 23h11'27" -6d42'13"
Doreen Elizabeth Zekas
Cnc 8h4'43" 26d33'34"
Doreen Ellen Gaborow
Uma 8h46'32" 53d49'54"
Doreen Gauthier - Librarian
Cas 0h40'39" 60d45'28"
Doreen Grima
Cru 11h58'22" -60d9'54"
Doreen J Eyre
Cas 23h42'56" 58d22'12"
Doreen & Joseph Mondelli
Cyg 20h20'11" 51d20'57"
Doreen Kinley
Cyg 20h58'59" 48d42'25"
Doreen Louise Dawson
Cnc 8h36'15" 21d16'45"
Doreen Lyn Reese
Gem 7h30'15" 27d16'21"
Doreen May Howes
Cas 0h56'35" 63d15'19"
Doreen Navarro/Most Precious
Uma 10h8'8" 67d16'24"
Doreen Nelson
Ari 2h21'45" 14d49'49"
Doreen Nobile
Cyg 20h35'1" 34d50'49"
Doreen Polden (Our Mum & Nan)
Cas 2h0'33" 70d28'51"
Doreen Sheppard
Cas 0h26'58" 63d1'43"
Doreen Spencer
Cas 0h21'23" 55d41'19"
Doreen & Tom's Golden Wedding Star
Cyg 20h50'59" 48d44'57"
Doreen Tucker
Cas 0h42'16" 54d0'30"
DoreenDeHaven01032004
Ari 2h10'9" 24d12'22"
Doreene Grace Fish
Aqr 20h57'37" -10d24'49"
Doreen's Delight, Shine4Ever!
Cyg 19h27'26" 44d52'17"
Dorelle Beck
And 23h0'55" 41d45'58"
Dorene Bethel
Dra 15h40'7" 57d17'20"
Dorene Kaplan
Lyn 7h41'22" 38d42'20"
Dorene Leann Malloy
Lib 14h58'30" -13d26'18"
Dorene Marie
Ari 2h22'1" 19d11'45"
Dorene (Mom)
Cas 1h21'22" 64d27'46"
Dorene Priolo Barker
Cas 0h46'50" 47d34'52"
Dorene Sziede
Tri 1h51'13" 29d38'3"
Doretta E. Johnson
Cyg 21h20'58" 54d46'48"
Dori
Dra 15h50'31" 52d54'33"
Dori
Per 3h24'50" 41d32'12"
Dori
Lac 22h38'58" 41d20'42"
Dori Ann Timko Boucher 2Infinity...
Ari 2h21'10" 20d43'48"
Dori Márquez García
Her 18h39'54" 21d17'5"
Doria
Lib 15h11'44" -12d14'28"
Doria Moyun Xu
Sco 17h0'27" -36d18'3"
Dorian Alfredo van der Laan
Psc 23h50'1" 6d42'4"
Dorian Bon
Per 3h7'26" 39d30'37"
Dorian Crist
Lib 14h38'52" -17d45'52"
"Dorian Drake" Maine
Del 20h46'54" 18d44'35"
Dorian Jacquelyn Franco
Cas 0h30'0" 59d4'51"
Dorian Tamara Paulette Bernier
Tau 5h21'56" 17d55'10"
Doriana Caiffa
Lyn 8h33'41" 42d57'15"
Doriana Croce
Vul 20h13'50" 22d2'28"
Dorianne
Sgr 17h51'11" -28d8'14"
Dorian's Bella
Uma 9h46'45" 68d9'9"
Dorian's B.T. Star
Uma 8h29'34" 62d13'36"
Dorie Erin Dugan Stites
Aql 19h4'35" 2d33'28"
Dorie L²
And 0h30'21" 33d32'29"
Doriel May Bertram (Mudley)
Leo 11h10'17" -1d30'58"
Dorie's Forever Star
Lyn 7h26'32" 44d58'2"
Dorinda
Peg 22h16'50" 5d45'0"
Dorinda
Aqr 22h41'56" -16d16'2"
Dorine Morgan
Ari 2h46'36" 16d3'25"
Doris
Tau 5h4'5" 20d39'49"
Doris
Ori 6h0'25" 13d36'8"
Doris
Psc 23h31'18" 4d55'23"
Doris
Gem 6h29'32" 23d6'37"
Doris
Tri 2h27'14" 29d45'22"
Doris
Cyg 19h41'7" 31d40'37"
Doris
And 23h56'48" 42d42'32"
Doris
Her 16h49'31" 41d38'18"
Doris
Cap 21h32'35" -18d47'0"
Doris
Sgr 18h53'41" -26d55'29"
Doris A. Dragich
Uma 8h15'14" 68d21'25"
Doris A Halpin
Lyn 7h48'52" 37d49'51"
Doris Alicia Delgado
Ari 2h40'35" 26d39'42"
Doris and Ernie
Cyg 21h54'29" 49d56'20"
Doris and Francis McDermid
Uma 11h41'20" 46d59'3"
Doris Ann Thomas
Mon 6h52'37" -0d19'21"
Doris Anne Brennen
And 1h56'8" 42d55'31"
Doris Antonieta Smith
Tau 4h22'51" 24d37'1"
Doris Archuleta
Lib 15h41'45" -16d59'45"
Doris B. Riley
Cnv 12h39'19" 40d15'12"
Doris Barbara Palmer Robbins
Gem 6h50'54" 30d8'12"
Doris & Bernard
Tri 1h54'8" 25d41'0"
Doris Brierley
Cas 1h24'37" 67d32'40"
Doris Brown Ware
Lib 15h18'24" -4d16'18"
Doris C. Boyer
Cas 23h28'15" 53d25'41"
Doris Clews, My Nanna
Cas 0h7'57" 54d18'22"
Doris Correia
Lyr 19h3'13" 41d31'52"
Doris Crowely
Uma 9h0'26" 51d1'51"
Doris Dauphinais
And 1h6'12" 40d46'36"
Doris & David
Cyg 20h49'8" 36d56'35"
Doris DeLyte Baer
Ori 5h45'28" 6d11'54"
Doris E. Abramson
Cas 1h12'26" 58d7'57"
Doris E. Hildebrand
Cnc 8h51'44" 31d29'21"
Doris Elaine Slade
Leo 11h15'7" 0d58'24"
Doris Elisabeth Theysohn
Ori 6h9'37" 10d33'33"
Doris Ellen Carlson Blank
Mon 7h38'2" -8d12'5"
Doris Ellen Curran
Cas 23h33'5" 58d16'21"
Doris Ethel Wolgamott Rice
Ori 6h19'15" 6d8'47"
Doris Eva Almeida
Lib 15h26'2" -18d57'49"
Doris Evelyn Beck
Cnc 8h29'8" 11d18'34"
Doris Fausch
Her 17h16'4" 16d0'17"
Doris Fielding
Cas 1h24'8" 63d27'49"
Doris Frank
Cap 21h6'51" -18d42'11"
Doris G. & John F. Coppola; 7/23/55
Uma 10h24'18" 71d37'40"
Doris Goodall
Cnc 8h59'12" 7d48'2"
Doris Hall
Cas 2h34'48" 65d57'32"
Doris+Hans Helds "Wedding Star"
And 1h41'13" 46d19'35"
Doris Hanson
Leo 10h20'11" 18d52'10"
Doris Hart Colbert
Vir 12h43'37" 1d16'34"
Doris Hazel Thompson
And 2h37'24" 46d16'2"
Doris Hazlett Mark
Crb 16h16'33" 30d56'59"
Doris Heard
Aqr 22h39'1" -15d33'37"
Doris Higdon
Uma 10h21'9" 48d5'59"
Doris Hoffarth (Mom)
Uma 10h20'56" 53d14'24"
Doris Irene Gorski
Cas 0h46'46" 57d45'26"
Doris Isobel
Lyn 8h1'31" 45d6'52"
Doris J. Wolf
Uma 8h58'28" 64d40'30"
Doris Jean Bassett
Leo 9h47'57" 8d33'43"
Doris Jean Eberhart
Ari 2h48'51" 27d55'38"
Doris Jean Gregory
Uma 10h18'50" 50d59'10"
Doris Jean Rabik
Sgr 19h21'48" -27d52'47"
Doris Jean Sampson Owen
Ori 5h5'5" 3d50'48"
Doris Jeanette Alexander
Crb 15h31'4" 27d4'38"
Doris Jeanette Cooke
Vir 12h23'17" -5d17'24"
Doris Jenkins Giuliano
Sgr 19h38'24" -14d54'37"
Doris Jewel
Tau 4h33'41" 23d39'48"
Doris & John Baldwin
Tau 5h52'13" 23d48'35"
Doris Johnson
And 1h13'28" 38d58'22"
Doris Jordan
Uma 13h20'28" 57d41'56"
Doris Kathleen Sherrard Roherty
Cap 21h2'36" -24d36'53"
Doris Kristek
And 1h13'47" 42d23'50"
Doris L. Brown
Uma 11h6'27" 66d55'2"
Doris L. Powell
Psc 1h25'0" 29d8'54"
Doris L. Scharffenberger
And 23h28'14" 48d12'42"
Doris LaFave-Miltz
Sgr 18h25'16" -24d49'2"
Doris Lawrence
Lyn 7h31'52" 43d11'28"
Doris Levenduski
Tau 4h26'56" 18d36'45"
Doris Lillian Williamson
Ari 1h55'1" 18d24'46"
Doris M. Brown Aug. 31, 1922
Vir 13h57'0" -0d47'52"
Doris M. Frank
And 1h29'25" 40d25'2"
Doris M. Kurtze
Cas 1h44'38" 66d13'26"
Doris M. Pye
Cas 0h47'10" 62d11'8"
Doris Mae
And 0h16'7" 33d24'30"
Doris Mae Menz
Aql 19h33'12" 1d27'3"
Doris Mae Turner
Crb 15h30'7" 28d38'28"
Doris Mae Varella Welch
Cap 20h59'7" -18d9'22"
Doris Major
Aqr 23h25'0" -7d43'8"
Doris Malloy
Lyr 18h51'33" 43d17'32"
Doris Marie
Sgr 19h21'14" -15d57'15"
Doris Marie Akerman - 27/12/1933
Cap 21h16'1" -23d28'15"
Doris Marie Drost Bennett
Ori 6h0'53" 1d10'8"
Doris Marie Sanders
Lyn 6h41'12" 56d55'1"
Doris Marie Wells
Cyg 20h33'28" 50d29'49"
Doris Mary Hess
Ori 6h12'7" 5d40'16"
Doris Mathlin
Crb 15h24'23" 30d33'38"

Doris Mauersberger
Aqr 22h24'41" -2d52'38"
Doris May
And 1h49'59" 38d34'53"
Doris May Stumpf
Cnc 8h24'40" 18d9'44"
Doris Miller Osterhout
Her 17h44'10" 22d42'2"
Doris Morgan Richards
Sgr 18h54'36" -16d14'36"
DORIS MORRIS
Cnc 8h53'2" 32d0'3"
Doris Nanny Abston
Lyn 8h47'52" 36d31'6"
Doris Neill
Cam 6h58'28" 67d15'25"
Doris "Nikki" Marie Summers
Dra 17h44'32" 52d19'42"
Doris Osthoff
Peg 22h21'29" 21d30'54"
Doris Patricia Sibley
Her 16h36'53" 38d14'7"
Doris Picardi
Uma 11h33'21" 42d44'42"
Doris Pile
Cas 23h36'53" 54d59'26"
Doris Plum-Eßmann
Uma 9h25'0" 48d38'49"
Doris Podrasky
Uma 10h41'49" 66d26'26"
Doris Post
Aqr 21h3'58" -8d45'3"
Doris Pratt Ballsun
Leo 11h43'55" 24d10'0"
Doris Price
Cap 21h26'51" -23d30'44"
Doris Reisdorf
Uma 9h16'56" 64d11'10"
Doris & Richard Zimmerman
Uma 8h41'30" 70d14'7"
Doris S. King
Lyn 8h20'18" 41d1'36"
Doris S. Robinson
Uma 11h21'54" 55d13'35"
Doris Sanford
And 2h19'33" 46d44'11"
Doris Santanello
Ari 2h48'40" 29d43'33"
Doris Schiemann Boeff
Ari 2h49'59" 28d16'42"
Doris Schmitz
Sco 17h34'9" -41d9'28"
Doris Schor
Cas 0h38'6" 50d45'55"
Doris Shuey
Cas 1h14'0" 59d46'2"
DORIS SINDLINGER
Dra 18h53'16" 50d1'40"
Doris Slatkin
Uma 12h29'1" 57d33'0"
Doris' Star
Crb 15h42'15" 31d48'8"
Doris Stary-Treichler
Uma 9h33'6" 43d12'38"
Doris & Stefan Krummenacher
Dra 16h39'10" 56d58'59"
Doris Stradcutter
Lyr 18h25'12" 33d56'44"
Doris Stresser
Ori 5h36'46" -5d8'16"
Doris Strobel
Uma 11h51'19" 30d13'13"
Doris Stump
Cas 1h11'43" 55d3'17"
Doris Taylor
Cas 23h38'41" 56d23'20"
Doris & Thomas
Boo 14h35'56" 16d3'59"
Doris Topham
Sco 16h15'48" -12d21'52"
Doris und Peter
Cyg 21h36'45" 54d53'19"
Doris Virginia Colligan
Ari 2h7'26" 24d36'37"
Doris Vivian Hendrickson Shelton
Psc 1h15'39" 26d26'48"
Doris Wells
Cam 4h51'53" 58d31'19"
Doris & William McAdams, Sr
Cyg 19h37'51" 32d10'24"
Doris Winger Walstead
Cas 0h0'3" 58d21'39"
Doris Wood Kerbs 7/23/54
Leo 11h29'7" 25d31'54"
Doris Y. Gavazzi
Ori 5h35'57" -0d43'48"
Dorisann
Lib 15h6'1" -5d9'3"
Dorisela Ortiz-R.
And 0h54'19" 37d18'36"
doris-marianne
Boo 15h7'51" 47d31'17"
DorisMejia LEMAA
Uma 9h5'46" 64d9'44"
dorisnoeliacavallo
Lib 15h36'45" -6d28'24"
DoriZ
Sgr 18h54'47" -19d20'21"
Dorjan and Jill's
Aqr 21h5'22" -13d32'53"
Dork
Cnc 8h1'57" 21d15'4"
DORK
Ori 5h20'56" 5d48'41"
Dorkstyle
And 0h31'57" 29d39'51"
Dorli
Ori 6h9'45" 8d36'49"
DorListar
Gem 6h43'41" 16d43'15"
Dorlyn
Ori 4h58'29" 15d17'28"
Dorman Bermingham
Leo 9h49'45" 28d3'11"
Dorn, Juliane
Ari 3h13'39" 18d52'3"
Dorn, Melanie
Uma 12h13'16" 61d8'22"
Dornell
Uma 11h17'34" 30d55'31"
Dorner
Uma 9h42'12" 66d56'18"
dornique
Uma 9h39'7" 56d39'13"
Doron + Brittany
Lyn 7h51'40" 41d17'32"
Doron Kotkowski
Uma 11h10'41" 35d18'40"
Doros Konstantinou
Uma 8h35'31" 70d34'3"
Dorota
Uma 8h41'32" 54d35'22"
Dorota
Leo 9h47'57" 28d57'20"
Dorota
Ari 2h50'37" 25d10'5"
Dorota Adamek
Cas 23h5'25" 58d44'9"
Dorota Adamska
Umi 15h34'2" 71d32'11"
Dorota Muria Tarnowski
Gem 6h25'59" 20d27'44"
Dorota Pradziad-hyc
Cam 4h14'12" 66d48'7"
Dorotha Schreiner Smith
Cyg 21h23'4" 50d32'50"
Dorothea
Uma 11h33'39" 44d25'30"
Dorothea
Cas 1h56'8" 68d54'52"
Dorothea
Cap 20h23'39" -13d35'3"
Dorothea
Sco 16h10'52" -8d46'36"
Dorothea Faye Lippincott
Uma 12h0'19" 30d35'56"
Dorothea Grommes
Uma 9h37'38" 52d2'30"
Dorothea Hohmuth
Uma 12h30'33" 54d41'55"
Dorothea Holdener
Cas 1h45'49" 66d50'59"
Dorothea Martin
Uma 8h31'22" 61d3'16"
Dorothea McAnulty Olsen
Cru 12h46'46" -56d29'47"
Dorothea McRae
Cas 1h28'45" 58d50'22"
Dorothea Montgomery
Uma 9h0'23" 50d35'32"
Dorothea Vine
Cap 21h35'56" -13d34'17"
Dorothea Welch
Uma 12h20'55" 60d42'32"
Dorothea Wöhler
Uma 8h54'39" 67d58'53"
Dorothea's Dot
Uma 9h45'10" 55d34'51"
Dorothee Faith Golemme
Cas 0h39'0" 64d1'59"
Dorothy
Cas 23h26'0" 53d1'30"
Dorothy
Lyn 7h54'45" 52d50'14"
Dorothy
Sgr 19h37'44" -15d21'21"
Dorothy
Aqr 22h41'48" -0d27'52"
Dorothy
Psc 0h17'58" -3d45'7"
Dorothy
Umi 4h29'8" 88d26'0"
Dorothy
Sco 17h2'19" -31d44'57"
Dorothy
Cas 0h32'2" 55d40'33"
Dorothy
Lyn 8h30'22" 46d42'21"
Dorothy
Aur 5h28'35" 46d19'58"
Dorothy
Crb 15h49'34" 27d56'23"
Dorothy
Cnc 8h45'35" 7d26'17"
Dorothy A. Arbogast
Ari 3h12'55" 29d17'21"
Dorothy A. Everett Fenty
Uma 14h25'41" 62d12'58"
Dorothy A. Hallman
Uma 9h6'47" 56d51'54"
Dorothy A. Johnson
Cas 1h27'11" 62d55'22"
Dorothy A. Martin
Cnc 8h56'11" 10d9'23"
Dorothy Ada
Uma 8h36'27" 49d15'14"
Dorothy Alessi
Lyr 18h32'26" 37d8'55"
Dorothy Amy Zavala Burke
Lyr 18h56'55" 33d40'23"
Dorothy and Arthur Roth
Lib 15h34'25" -6d41'42"
Dorothy and Frank
Cyg 19h41'49" 55d5'53"
Dorothy and Hayward Love Star
Mon 7h30'6" -0d23'53"
Dorothy and Hyman Chudnofsky
Cyg 19h42'13" 50d49'12"
Dorothy and Jack Mylott
Crb 16h19'17" 27d58'46"
Dorothy and Joe
Cyg 21h33'14" 51d21'26"
Dorothy and Virgil Criscola
Cyg 21h27'58" 46d14'55"
Dorothy Anderson
Cas 1h56'7" 60d52'56"
Dorothy Anita Deer
Crb 15h34'57" 27d59'36"
Dorothy Ann and Gene Sadler
Lac 22h49'59" 49d43'6"
Dorothy Ann Bainton
Gem 7h5'55" 29d42'21"
Dorothy Ann Culpepper
Aqr 22h38'34" -1d20'34"
Dorothy Ann Cunningham
Aqr 22h39'31" 2d2'26"
Dorothy Ann Daane
Cap 20h8'42" -14d1'0"
Dorothy Ann Forte
Her 16h34'48" 45d36'14"
Dorothy Ann King "my Cinderella"
Leo 11h35'33" -2d21'27"
Dorothy Ann Sweet
Dra 18h34'58" 61d38'52"
Dorothy Ann Toaldo (Row)
Cnc 8h38'12" 16d45'3"
Dorothy Anna Denzine
Psc 1h4'37" 21d0'11"
Dorothy Annette Heydorn
Sge 20h14'24" 18d6'1"
Dorothy Arbogast
Uma 11h30'50" 52d21'26"
Dorothy Arlene
Cam 6h23'21" 68d43'51"
Dorothy Armstrong
Ari 2h11'55" 24d59'19"
Dorothy B. Termine
Vir 13h29'22" 11d44'7"
Dorothy "Baba" Error
Lib 15h45'53" -25d45'33"
Dorothy Barker
Uma 10h49'43" 50d30'36"
Dorothy Barker "Our Dot"
Crb 16h1'14" 33d47'17"
Dorothy Bassemir
Cas 0h44'17" 60d51'34"
Dorothy Becvar
Ori 6h20'2" 14d48'33"
Dorothy Bernice Matricciani
Gem 7h23'37" 28d14'11"
Dorothy Bernice Sanders
And 0h44'16" 41d11'4"
Dorothy Best Taurus
And 2h35'16" 43d52'19"
Dorothy Bladt
Sco 16h6'54" -11d11'40"
Dorothy "Bobby" Lamerson
Uma 9h55'8" 48d39'35"
Dorothy Bonanno
Cnc 9h3'48" 31d47'48"
Dorothy (Boogy) Strom
Vir 12h13'8" -8d51'15"
Dorothy Brady
Leo 10h20'56" 9d48'18"
Dorothy Bridges Poet Star
Vir 13h44'23" 4d43'9"
Dorothy Brown
Peg 22h13'18" 26d0'6"
Dorothy Brown
Uma 9h10'6" 62d17'43"
Dorothy Buckley
Cap 21h11'27" -21d51'25"
Dorothy Buffington
Cas 1h41'41" 63d33'34"
Dorothy C. Bordeau
Vir 12h28'6" 11d38'37"
Dorothy C. Moyer
Cas 0h5'28" 55d9'49"
Dorothy Callahan McDonald
Umi 14h21'4" 85d32'48"
Dorothy Christy Bryant
Cep 22h19'2" 70d21'46"
Dorothy Clarke Perpich
Cas 1h30'8" 57d49'0"
Dorothy Clay Sims
Cyg 21h36'46" 50d10'27"
Dorothy Clem Garner
Uma 11h2'25" 50d6'43"
Dorothy Clinch
Leo 11h33'50" 23d21'59"
Dorothy Cohen Hasson
Dotti Mom Nana
Per 3h9'40" 40d36'55"
Dorothy Collins
Cam 6h3'55" 69d1'14"
Dorothy Collison
Cap 20h59'26" -20d19'24"
Dorothy Crystal Scaggs
And 23h23'23" 46d15'44"
Dorothy D
Crb 16h0'46" 33d56'28"
Dorothy Dale & Lloyd Lester Kretz
Ari 2h27'27" 12d37'36"
Dorothy "Dar" Lee Collison
And 23h44'16" 46d49'7"
Dorothy Dedmon
Lyr 18h47'57" 38d5'23"
Dorothy Deyne
Cas 0h53'31" 63d5'12"
Dorothy "Dot" Pullen
Vir 12h39'42" 8d42'23"
Dorothy (Dottie) Ann Marshall
Lib 14h53'51" -3d9'26"
Dorothy Durkin
Mon 7h0'20" 4d8'36"
Dorothy E. Deering
Per 3h32'52" 50d4'47"
Dorothy E. G. Simmons
Aqr 21h14'15" -11d39'38"
Dorothy E. Robillard
Cas 0h31'20" 63d51'50"
Dorothy E. Spurgin-Hicks
Crb 16h15'46" 35d4'33"
Dorothy E. Young
Cas 1h0'12" 53d37'9"
Dorothy Eaton Loving Mother
Uma 10h53'32" 49d43'24"
Dorothy Edana
Ari 3h21'17" 29d59'44"
Dorothy Edna Kenny
Cyg 19h37'35" 31d26'14"
Dorothy Elberfeld
Gem 7h23'31" 14d50'31"
Dorothy Elizabeth Bryant
And 1h17'39" 38d53'52"
Dorothy Elizabeth Cerajewski-George
Uma 10h24'27" 45d17'32"
Dorothy Elizabeth Eaton
Uma 8h58'24" 67d37'32"
Dorothy Ellis
Cyg 19h49'23" 31d16'18"
Dorothy & Eric Whitaker
Cyg 19h41'56" 30d41'20"
Dorothy Erna Beetle-Pillmore
Uma 11h21'29" 53d19'49"
Dorothy F Holloway-MOM
Lib 15h28'17" -6d6'13"
Dorothy F. Mueller
Cas 0h20'14" 56d44'24"
Dorothy F. Murata
Cas 0h52'55" 52d21'31"
Dorothy Fleury
Cas 23h5'14" 55d56'23"
Dorothy Fowler Williams
Cas 1h44'22" 67d2'13"
Dorothy Friedman Klein
Lib 15h3'9" -24d3'59"
Dorothy Frye
And 0h27'29" 42d57'30"
Dorothy G. Ruse
Cas 1h19'28" 71d59'14"
Dorothy Gee
Peg 21h56'59" 8d18'38"
Dorothy Glover
Cas 0h24'35" 54d36'15"
Dorothy Goldman
And 23h9'17" 52d3'1"
Dorothy Goscinski
Uma 12h40'46" 58d5'0"
Dorothy Grace Clarke
Crb 16h22'42" 30d12'27"
Dorothy "Grammy" Gouveia
Leo 9h56'45" 23d1'47"
Dorothy Gray
Crb 15h49'20" 27d6'1"
Dorothy H Brown
Cnc 8h52'39" 7d19'3"
Dorothy H. Marsh
Cap 21h13'6" -15d4'35"
Dorothy Hagberg
Mon 6h48'32" -0d9'58"
Dorothy Hames
Psc 1h35'41" 20d17'55"
Dorothy Harris
Sco 16h18'0" -12d36'5"
Dorothy Harter
And 23h41'56" 40d40'55"
Dorothy Harter Pierce
Vir 12h43'58" 11d1'32"
Dorothy Hawthornthwaite
Cas 1h9'43" 49d3'40"
Dorothy Hedgpeth
Gem 6h58'23" 33d32'13"
Dorothy Helen DeVitis
Lyn 8h34'19" 34d16'51"
Dorothy Hembree
Cam 7h57'0" 70d35'49"
Dorothy & Herb Goldsmith Forever
Uma 9h19'27" 66d2'12"
Dorothy Hilda Crocco
Sgr 20h11'58" -44d32'38"
Dorothy Howell
Vir 14h17'14" 1d44'13"
Dorothy Hoyer Linholm
Lyn 8h53'54" 33d7'36"
Dorothy Hudson
Vir 13h59'26" -10d21'19"
Dorothy I. Mieskowski
Cas 0h29'46" 65d4'9"
Dorothy Ilene Davison
Uma 11h28'59" 36d48'39"
Dorothy Irene Johnson
And 0h46'39" 45d45'56"
Dorothy Irene Parker-Hull
Uma 11h32'30" 52d51'2"
Dorothy J. And William A. Elder
Pho 23h44'24" -40d42'39"
Dorothy J. Hartsock
Cas 1h15'32" 55d1'37"
Dorothy J. Kramka-Miller
Cnc 8h49'41" 14d22'52"
Dorothy J O'Neill
Vir 14h39'9" 3d17'43"
Dorothy Jane King
And 0h47'5" 37d48'33"
Dorothy Jane King
Lib 15h28'27" -10d7'46"
Dorothy Jean
Mon 6h42'7" -1d17'41"
Dorothy Jean
Lyn 8h15'19" 59d32'36"
Dorothy Jean BeBe Arant
Sgr 19h46'57" -12d26'47"
Dorothy Jean Pruner
Tau 4h29'16" 25d41'1"
Dorothy Jean Watson Federspiel
Uma 11h5'57" 37d9'51"
Dorothy Jean Welch
And 0h15'12" 25d14'9"
Dorothy Joan
Sgr 19h20'51" -15d38'34"
Dorothy Joan Ballantyne
Cas 23h15'31" 59d28'23"
Dorothy Jones
Peg 22h10'14" 30d46'26"
Dorothy Joy Enslen
Uma 11h13'21" 38d41'47"
Dorothy Juanita Helton
Dra 18h40'21" 70d29'42"
Dorothy (Jupiter Mom) Persell
Cma 7h3'35" -30d11'43"
Dorothy K
Aqr 22h27'51" -0d59'54"
Dorothy K. Stix
Sgr 19h42'4" -16d11'38"
Dorothy Kelly Bosch
Cas 0h39'6" 61d41'25"
Dorothy L Robinson
Psc 23h52'41" 6d4'44"
Dorothy L. Towle
Cas 0h32'16" 58d1'4"
Dorothy Lamb
And 1h20'18" 47d37'44"
Dorothy Lamb Hansen
Ari 2h6'56" 26d14'40"
Dorothy LaRue Lafoon
Cas 1h38'7" 63d16'15"
Dorothy Lee
Nor 16h9'30" -44d49'44"
Dorothy Lee Cook
Ori 6h1'31" 9d59'38"
Dorothy Lee Maginnis Doremus
Ari 2h48'11" 26d24'2"
Dorothy Lewis
Cas 23h42'52" 52d42'15"
Dorothy Long
Cyg 20h27'12" 37d14'16"
Dorothy Lou Nance Stewart
Gem 6h47'56" 21d35'6"
Dorothy Louise Chasteen
And 0h20'25" 32d27'4"
Dorothy Louise Howard
Lyr 18h42'4" 35d28'18"
Dorothy Louise Jacobsen
Cas 0h37'34" 54d41'56"
Dorothy Louise McAdoo
Ari 2h25'45" 11d41'50"
Dorothy Louise Novy Wilson
Crb 15h43'4" 28d20'56"
Dorothy Loyd
Cas 0h54'48" 57d58'15"
Dorothy & Lyle Martin's Star
Crb 15h52'58" 25d44'51"
Dorothy M. and Francis H. Custance
Crb 15h32'25" 28d3'19"
Dorothy M and John V Vaughn
Uma 10h52'3" 63d26'22"
Dorothy M. Dewing's Eastern Star
Mon 6h43'22" 11d44'40"
Dorothy Mae
Psc 0h39'38" 8d2'6"
Dorothy Mae
Uma 9h42'22" 69d3'7"
Dorothy Mae
Cas 1h39'1" 62d6'45"
Dorothy Mae Clark Ragatz
Ori 5h30'26" 5d4'20"
Dorothy Mae Hillstrom
Cas 0h17'37" 47d25'18"
Dorothy Mae Neill
Ari 2h57'43" 22d25'40"
Dorothy Mae Page
Uma 11h18'26" 41d18'14"
Dorothy Mae Preiss
And 0h44'33" 39d59'7"
Dorothy "Ma-Grandma" T. Monko
Crb 15h42'59" 26d29'42"
Dorothy Majka
Lib 14h53'55" -13d9'24"
Dorothy Margaret
Lib 15h51'59" -13d51'43"
Dorothy Margaret Devlin
Cap 21h4'34" -23d30'22"
Dorothy Marie Cowen
Cas 1h12'24" 57d50'57"
Dorothy Marie Dillon - God's Gift
Lib 15h41'23" -19d10'18"
Dorothy Marie Kujawa
Uma 11h51'5" 57d24'45"
Dorothy Marie Newhall
Cas 1h25'34" 56d11'52"
Dorothy Marie Smalley
Crb 15h37'40" 33d46'56"
Dorothy Marie (Smith) Preston
Uma 11h45'40" 40d0'42"
Dorothy Mary Lauer McDevitt
Uma 13h30'4" 53d42'5"
Dorothy Mastrolia
And 23h28'35" 46d55'34"
Dorothy May
Cam 5h3'27" 65d49'33"
dorothy may roberts
Aqr 22h54'14" -8d56'37"
Dorothy McAtee
Uma 9h58'33" 44d47'26"
Dorothy McDonald
Cyg 20h2'30" 30d25'20"
Dorothy "Mommy" Viehman, from Mikey
Uma 9h35'29" 58d15'27"
Dorothy Moriarty
Uma 8h35'52" 61d37'41"
Dorothy Morrison Dunigan
Crb 16h16'15" 37d53'2"
Dorothy Moyer
Peg 22h6'34" 31d35'32"
Dorothy "Nana" Hutton
Cas 0h31'2" 58d33'10"
Dorothy "Nana" Rush
Ori 6h10'11" 20d52'27"
Dorothy "Nanny" Coates
Cas 0h3'29" 56d39'36"
Dorothy Naylor
Uma 11h3'47" 53d52'9"
Dorothy Niles Wilson
Cas 1h10'23" 69d23'15"
Dorothy "Nutch" Sebastian
Crb 16h8'36" 34d12'6"
Dorothy O'Brien
Ari 2h38'1" 25d22'1"
Dorothy Okun
Cas 23h3'3" 54d7'7"
Dorothy Pai-San Lee
Psc 0h2'45" 2d28'41"
Dorothy Pammenton
Cyg 21h45'39" 46d37'55"
Dorothy Pekrul
Sgr 17h52'22" -27d13'35"
Dorothy Plavin
Cam 7h43'33" 67d14'13"
Dorothy - Queen of the Mae
Cas 0h1'59" 55d37'8"
Dorothy R. Washington
Dra 19h50'21" 59d51'45"
Dorothy Rae Jackson Reynolds
Tau 4h29'1" 25d51'5"
Dorothy Reed
Lib 15h14'58" -22d15'9"
Dorothy & Richard Berry
Uma 11h6'47" 45d38'41"
Dorothy Richards
Crb 15h50'28" 37d22'54"
Dorothy Roberson
And 1h10'46" 44d58'36"
Dorothy Roberson
Cas 1h39'40" 65d48'45"
Dorothy Rodham
And 0h22'55" 30d50'1"
Dorothy & Roger
Cyg 20h1'32" 37d12'47"
Dorothy ~ Roger ~ Casey
Cas 0h13'43" 55d1'35"
Dorothy Romanies
Uma 10h33'19" 56d26'33"
Dorothy Rose
Psc 23h38'56" 1d40'14"
Dorothy & Roy - Together Forever
Vir 14h14'22" -4d0'6"
Dorothy Ruth
Crb 15h54'9" 27d20'17"
Dorothy Ruth Copp
Cap 20h35'18" -16d4'48"
Dorothy Ruth Harris
Aqr 22h23'30" -13d25'22"
Dorothy Ryan
Aqr 22h46'20" -0d56'52"
Dorothy Sack-Johnson
Cap 20h16'28" -13d29'17"
Dorothy Scarvace
Lep 5h32'19" -13d48'28"
Dorothy Schnabel
Uma 10h2'48" 65d31'49"
Dorothy Schwarz
Aqr 21h6'43" -9d40'29"
Dorothy Senninger
Uma 11h49'46" 50d47'9"
Dorothy Shaneberger
Cas 1h38'47" 60d15'57"
Dorothy Sheffield
Uma 12h39'14" 52d42'33"
Dorothy Sherry
Lyr 19h3'55" 37d25'22"
Dorothy Silverman
Cas 0h31'26" 60d57'55"
Dorothy Simonelli
Lmi 10h52'47" 28d53'27"
Dorothy Simpson
Gem 7h33'54" 30d41'43"
Dorothy Smalley
Cas 2h43'52" 63d53'5"
Dorothy Sonia Jacobs
And 2h36'59" 48d59'12"
Dorothy St Aubin
Cnc 8h30'0" 20d28'19"
Dorothy Sutherland Brown
Ori 6h8'16" 8d50'10"
Dorothy Tarabocchia
Uma 10h17'29" 50d32'8"
Dorothy Taylor Brooks
Crb 15h28'33" 27d53'34"
Dorothy Tirabassi Aliano
Cap 21h48'54" -9d47'0"
Dorothy Vassallo
Cyg 21h33'25" 45d41'39"
Dorothy Vaughn Palmer
Uma 13h13'16" 55d18'43"
Dorothy Voltz
Vir 14h49'50" 5d15'57"
Dorothy Weidel
And 0h21'23" 30d43'42"
Dorothy Williams Smith
Psc 1h23'12" 20d1'35"
Dorothy Winterbottom
Cas 22h58'45" 55d21'49"
Dorothy Wyatt Chandler
And 23h48'43" 47d14'58"
DorothyLeonard
Uma 8h56'44" 55d3'9"
Dorothy-Pete
Psc 1h21'23" 22d3'19"
Dorothy's Dot
Cnc 8h35'0" 23d45'47"
Dorothy's Piano Star
Lyr 18h46'0" 37d9'51"
Dorothy's Star - Susan's Mommy
Aqr 22h25'51" 0d59'33"
Dorotka B Zdunek
Leo 10h26'17" 23d15'22"
Dorr Buckley Begnal
Gem 7h12'21" 29d46'15"
Dörr, Felix Benedikt
Sgr 18h24'34" -22d53'41"
Dorrace Moses
Leo 9h26'48" 27d43'24"
Dorri
Uma 10h1'16" 42d13'6"
Dorrie Williamson
Per 2h30'39" 56d11'56"
Dorris L. Dalrymple
Vir 14h40'38" -0d5'25"
Dorsai
Ori 6h13'29" 15d55'33"
Dortechrisang
Crb 15h49'57" 26d28'17"
Dortha Helen & Merl Dean Schultz
Cyg 19h50'4" 50d56'21"
Dorthe & Claus
Tau 3h47'43" 24d16'56"
Dorthea Evans
Cnc 7h59'48" 11d15'45"
Dorthy and Chuck Amie
Cyg 21h40'5" 34d59'15"
Dorthy and Newell Carman
Uma 11h35'55" 58d53'33"
Dorthy Keck
Vir 13h18'5" 12d6'49"
Dorthy Klassen
Cas 1h26'4" 55d35'5"
Dory
Umi 15h24'33" 72d11'30"
Dory
Lyn 8h7'58" 44d6'50"
Dory Santiago
Vul 20h32'12" 27d36'13"
Doryan Jacob Ward
Her 16h27'1" 47d54'7"
Doryan-Auriana
Cnc 8h12'12" 18d37'27"
Doryn
Pho 0h11'54" -39d33'19"
Dosch 43
Cyg 21h30'54" 41d36'57"
Doshin Jr.
Cyg 20h12'1" 44d47'0"
Dosi & Tom
Crb 16h0'34" 32d59'13"

Dosiehn, Norbert
Ori 6h17'18" 10d8'34"
Dot
Uma 10h37'4" 54d32'20"
Dot
Cap 21h45'0" -10d22'24"
Dot 9 10
Lyn 7h27'50" 54d19'0"
Dot Cobb
Leo 10h31'17" 8d17'52"
Dot Devlin
Aql 19h3'33" 2d12'24"
Dot & Ed Slade - "Far Out" Friends
Gem 7h43'35" 34d20'3"
Dot Incorvati
Cas 2h25'52" 60d46'18"
"Dot" Our Sparkling Star In The Sky
Uma 11h21'40" 54d9'59"
D.O.T. Pholmann
Lib 15h5'42" -3d57'13"
Dot Robinson
Vir 13h35'57" 2d5'57"
Dot Salter's Little Angel
Cas 23h2'9" 58d8'8"
Dot Star
Sgr 19h2'8" -32d49'52"
Dot Walker
Lyr 18h42'20" 34d15'25"
Dot Wilson
Sco 17h44'6" -39d15'14"
DOT11291925
Sgr 18h40'37" -16d57'13"
DOT75
Aql 20h20'27" 4d4'47"
Doted
Lyr 19h18'50" 28d37'27"
Dots Centrifugal Motion
Uma 10h27'15" 66d49'19"
Dot's Dazzling Dot
Sgr 19h17'58" -16d7'0"
Dot's Star
Cru 12h17'57" -60d15'16"
Dot's Twinkle
Cap 20h34'53" -22d55'51"
Dötsch, Manfred
Ori 6h10'41" 16d33'38"
Dotsy
Crb 15h43'54" 26d4'18"
Dott. Michele Campanelli
Cas 23h0'46" 58d28'28"
Dott. Silva
Uma 12h8'21" 44d46'52"
Dotti Crisp
Crb 15h37'30" 30d43'47"
Dottie
Cas 23h10'8" 55d50'50"
Dottie
Sco 16h33'20" -41d30'29"
Dottie and Bill Welde
Vir 13h12'44" -2d54'15"
Dottie B
Vir 13h44'56" -8d9'10"
Dottie "B-A-L" Berry
Leo 11h48'44" 20d28'49"
Dottie Bonney
And 0h56'50" 40d19'43"
Dottie Boswell & Karen Boswell
Crb 16h12'4" 32d12'38"
Dottie Bozek
Gem 6h48'4" 33d54'23"
Dottie Dockery
Lyr 18h51'0" 31d6'23"
Dottie Lacell
Cap 20h18'31" -10d29'53"
Dottie Lovell Vazzana
Lyr 19h8'10" 46d2'19"
Dottie Lu
Cnc 8h30'1" 15d17'46"
Dottie Mc
And 23h19'58" 47d25'44"
Dottie Miles
Cas 0h13'29" 55d39'35"
Dottie Murdock
And 2h37'43" 43d51'56"
Dottie R. Ganter
And 2h17'37" 50d18'34"
Dottie Robinson
Psc 1h23'55" 25d42'53"
Dottie-Jean
Cam 5h19'29" 73d7'43"
DottieNall
Leo 10h29'36" 12d45'10"
Dottie's Star
Lyn 7h25'42" 44d49'53"
Dottie-Sue
Aqr 22h42'1" -15d41'8"
Dottor Enrico Tozzi 40
Ori 5h58'18" 6d32'11"
Dott.ssa Agnese Cacciola (NEJA)
Cam 4h20'10" 56d53'27"
Dott.ssa Anca Modrogeanu
Mon 6h20'14" -4d48'21"
Dotty Brancato
Cas 0h1'25" 53d38'3"
Dotty Lee Bice
Her 18h38'25" 18d58'37"
Dotty Porter
Cas 23h57'44" 56d39'43"
Dotty's Shining Star
Lyn 7h42'6" 50d0'0"
Dotty's Star
Cas 0h23'34" 53d53'21"
Doty Jordan Kempf
Tau 4h32'3" 21d46'40"
Dotzel, Kerstin
Tau 4h28'55" 29d39'10"
Dou Dou Ray
Psc 0h21'38" 15d40'7"
Double C
Cyg 21h56'7" 52d7'50"
Double D
Her 17h17'44" 46d40'9"
Double D
Per 3h33'13" 45d6'36"
Double Dimple
Uma 11h49'17" 56d25'42"
Double Ell
Psa 22h6'13" -27d57'1"
Double "G"
Uma 11h34'19" 64d19'27"
Double J
Uma 10h44'19" 56d42'15"
Double R's Chigger Boo
Uma 9h1'6" 49d3'1"
Double Stars Dancing Eternally
Pho 1h47'6" -43d25'18"
Double Trouble Always
Del 20h38'57" 15d13'48"
DoubleDee
Lib 15h15'7" -5d1'31"
DoubleR
Lib 14h53'45" -2d48'1"
Double-star Irma-Gene
Cyg 19h42'14" 38d26'24"
Douce Aurine
Uma 10h36'19" 58d12'49"
Douce Magali
Uma 11h22'57" 70d5'49"
Douceur Brigitte Olivier
And 23h11'56" 43d43'50"
Doudou et Fatiha pour la vie
Ari 2h23'12" 11d30'26"
Doudouine Cora
Cas 22h57'20" 57d59'28"
Doug
Vir 14h51'6" 3d0'31"
Doug
Tri 2h35'28" 35d4'43"
Doug
Per 3h51'29" 34d48'58"
Doug and Cara's Star
Ari 2h11'20" 24d18'8"
Doug and Chris Raarup
Cyg 21h11'12" 44d9'6"
Doug and Elizabeth Drummond
Cyg 19h49'40" 56d30'46"
Doug and Jamie Marsh
Cyg 21h39'17" 46d13'33"
Doug and Julie Presley
Aur 6h5'22" 30d50'51"
Doug and Linda Tussing
Leo 9h27'17" 11d16'57"
Doug Anderson
Cra 18h59'36" -39d6'58"
Doug B. Huntley
Cnc 8h46'37" 25d31'51"
Doug Bahrey
Gem 7h45'2" 33d55'25"
Doug Barnett 1946
Umi 4h55'43" 89d18'10"
Doug Beach
Per 4h20'33" 50d50'23"
Doug Bojack
And 22h59'36" 45d35'26"
Doug & Bonnie Forever
Cyg 20h53'5" 31d55'36"
Doug Brendel
Lib 14h54'24" -7d20'14"
Doug Bui
Psc 1h16'17" 18d39'48"
Doug & Cara Hartinger
Uma 9h27'34" 67d28'37"
Doug & Cindy Davidson
Cyg 21h41'5" 38d13'56"
Doug Cottingham
Her 16h43'23" 24d57'32"
Doug + Crystal Forbes
Sge 20h4'55" 16d45'30"
Doug DeCluitt
Aql 19h32'26" 10d1'34"
Doug "Doogal" Tipper our star always
Cru 12h24'59" -64d40'34"
Doug Draper
Cep 20h39'7" 64d22'43"
Doug Ellis
Boo 14h27'57" 43d12'53"
Doug Feltus Lost and Loaded
Uma 10h42'17" 52d21'42"
Doug Guy's Premier Power
Pho 0h34'52" -42d22'48"
Doug Harrington
Her 18h10'4" 28d31'17"
Doug Higgins
Per 3h24'8" 43d28'12"
Doug "Hopper" Gamble
Uma 13h10'55" 58d3'3"
Doug Ibach
Lyn 7h49'54" 57d49'32"
Doug J. Terry
Uma 8h34'37" 70d43'40"
Doug James
Cep 23h39'35" 74d11'53"
Doug Johnson
Cnc 8h50'24" 18d12'24"
Doug Kantrowitz
Her 18h28'33" 25d6'31"
Doug & Karen
Cyg 19h32'50" 53d22'54"
Doug LaTant loving father & friend
Per 3h21'1" 45d23'50"
Doug & Lee Mooney
Cas 0h23'15" 54d37'12"
Doug Lemire V.M.D. - Leo
Psc 0h10'10" 12d27'58"
Doug Lenhoff
Cep 23h55'53" 88d30'48"
Doug & Linda Walt
Sge 19h51'9" 17d34'37"
Doug Logan
Her 16h22'41" 44d50'39"
Doug & Louise Johnson's 50th Anniv.
Uma 9h51'20" 57d56'27"
Doug Love-Dad Forever In Our Hearts
Cru 12h36'21" -59d21'10"
Doug Lunning
Ari 2h35'39" 22d10'44"
Doug "Maverick" Schwab
Cyg 19h44'27" 28d8'29"
Doug McQuirter
Ori 5h11'25" 15d46'18"
Doug Merrill
Uma 9h26'24" 63d4'15"
Doug Morris
Aur 5h44'26" 43d45'22"
Doug P Shell
Uma 10h1'15" 70d59'8"
Doug Robins # 55 (Grandpa)
Uma 10h14'3" 45d23'48"
Doug (Rolltide) Milner
Cap 21h37'44" -19d38'23"
Doug Rouen
Ori 5h36'23" 14d1'21"
Doug S. Vale
Uma 9h21'1" 60d41'6"
Doug Schmitt B.F.L.
Her 18h34'20" 25d38'58"
Doug Simonton
Vir 14h33'38" -2d31'51"
Doug Stone
Cnc 8h49'3" 24d8'4"
Doug Supler
Ori 5h54'32" 22d34'9"
Doug & Susan McCormick
Umi 16h4'15" 75d46'26"
Doug Tabb
Ara 17h19'46" -52d7'6"
Doug & Tracy
Uma 11h36'6" 49d30'17"
Doug & Trisha
Tau 5h40'32" 24d48'58"
Doug & Vicki Hill
Cyg 19h45'51" 50d27'55"
Doug Wilson
Her 17h49'9" 21d53'37"
Doug Wright
Lyn 8h0'19" 45d3'8"
Doug Young
Uma 10h0'7" 67d51'46"
Doughboy
Sco 16h36'52" -26d44'59"
Doughnut
Crb 16h20'59" 31d58'43"
Dougie
Tau 5h59'15" 24d46'38"
Dougie Joe Rick
Uma 8h35'49" 56d15'28"
Dougie Murante
Uma 10h41'57" 72d25'17"
Dougie Rillo
Umi 16h35'20" 78d54'37"
DOUGIE STAR
Cap 20h50'49" -25d30'3"
Douglas
Cru 12h27'13" -61d6'37"
Douglas
Cap 20h39'30" -21d26'23"
Douglas
Lib 15h10'37" -13d56'7"
Douglas
Uma 11h30'53" 56d17'40"
Douglas
Ori 5h55'17" 8d25'21"
Douglas
Tau 4h38'47" 10d8'32"
Douglas
Aur 5h46'55" 42d46'43"
Douglas A. Chini
Uma 10h56'43" 51d20'44"
Douglas A. Fonner
Her 18h42'42" 13d43'52"
Douglas A O'Handley
Uma 10h54'58" 54d53'44"
Douglas Aaron Dubrowski
Uma 10h9'31" 59d55'15"
Douglas Alan Hathaway
Cep 23h1'11" 70d20'5"
Douglas & Alana Stone
Cyg 21h39'33" 36d15'26"
Douglas Allan Orr
Aqr 22h42'49" -0d43'18"
Douglas Allen Coggeshall
Aql 20h19'21" -4d7'53"
Douglas Allington
Vir 11h39'36" -3d53'46"
Douglas and Betty Terry
Cap 20h47'24" -18d7'48"
Douglas and Kimberly
Cyg 20h19'57" 55d33'45"
Douglas and Margaret Le Santo
Cyg 20h36'13" 30d41'53"
Douglas and Monica Keene
Cyg 20h26'5" 37d18'16"
Douglas Andrew Stroh
Vir 12h14'43" -10d35'53"
Douglas Andrew Welsh
Per 2h51'3" 48d33'4"
Douglas Andrew White
Per 2h40'23" 56d28'42"
Douglas Andrews, Jr.
Uma 9h46'7" 47d29'27"
Douglas Anthony Grindel
And 0h44'46" 37d14'56"
Douglas B. Goodall
Ori 6h14'35" 16d18'50"
Douglas Ballew
Lmi 10h32'37" 37d4'53"
Douglas Barton "D.B." Stevens
Per 2h55'27" 41d48'22"
Douglas Bean
Her 17h37'51" 16d46'36"
Douglas Boyce
Vir 13h49'14" -7d26'36"
Douglas Brian Beach
Leo 10h24'43" 21d36'39"
Douglas Buehler
Lib 15h51'32" -19d9'58"
Douglas Burton McColm
Vir 14h8'12" 7d3'25"
Douglas Bush
Aur 5h46'6" 47d41'30"
Douglas C. Hofmann
Psc 0h33'38" 6d54'56"
Douglas C. Johnson
Uma 11h0'38" 54d33'46"
Douglas C. Nicholson
Cap 20h18'48" -14d58'11"
Douglas Charles Tritton (Australia)
Pyx 8h34'59" -23d43'9"
Douglas Charles Workman
Uma 13h14'3" 57d14'4"
Douglas Citro
Peg 23h29'25" 21d32'42"
Douglas Codere Fleming
Per 3h13'17" 51d10'51"
Douglas Cory Todd Pearce
Umi 16h37'16" 82d46'5"
Douglas Craig Frimmet
Vir 11h53'43" -0d4'9"
Douglas David Stogner
Leo 11h33'13" 5d25'18"
Douglas Descher Revel
And 23h21'2" 48d38'11"
Douglas & Dianne Baulch
Cyg 20h48'56" 52d3'4"
Douglas Dolansky
Her 18h24'58" 21d32'52"
Douglas Donald Herbst
Uma 11h57'44" 47d6'41"
Douglas Dunbar
Per 4h11'26" 36d0'50"
Douglas E Hemenway
Aur 6h5'36" 48d10'15"
Douglas E. Void
Cyg 20h2'35" 39d46'31"
Douglas E. Wolan
Uma 9h8'8" 51d50'24"
Douglas Earl Mittag
Cnc 8h0'47" 22d35'44"
Douglas Earl Page
Sco 17h36'7" -39d10'55"
Douglas E.Barden
Sco 16h42'14" -34d25'20"
Douglas Edward Hodge
Cyg 20h23'38" 36d10'47"
Douglas Edward Meyerkord
Lyn 7h59'7" 57d41'37"
Douglas Eugene
Uma 10h20'40" 65d15'16"
Douglas Eugene Armstrong
Ari 2h30'41" 24d51'16"
Douglas G Allen
Cnc 9h17'41" 10d38'34"
Douglas G Hall
Aqr 22h23'40" -21d51'40"
Douglas G. Plonski
Sco 16h19'26" -18d41'27"
Douglas G. Riley
Crb 15h42'28" 25d36'56"
Douglas George Jupp
Cep 22h3'4" 56d31'32"
Douglas Glen Spahr Jr.
Uma 11h25'43" 54d10'22"
Douglas Graham
Her 17h33'27" 32d30'50"
Douglas Guzman
Cnc 9h6'45" 31d0'29"
Douglas H. Lewis
Ori 5h24'27" 1d8'1"
Douglas Haldeman
Vir 12h50'26" 5d6'41"
Douglas Hall
Uma 8h59'36" 53d43'54"
Douglas Hamilton Linn
Per 3h28'19" 45d47'50"
Douglas Hancock
Umi 14h5'15" 66d2'47"
Douglas Haydin LeGrand
Sco 17h0'17" -38d51'58"
Douglas & Helen Duke
Uma 9h41'33" 67d38'3"
Douglas Hofman
Per 4h34'47" 34d37'19"
Douglas Huntington Olson
Ori 5h49'59" 3d30'1"
Douglas J. DeRusha
Leo 11h24'7" 25d45'10"
Douglas J. Pliska
Peg 22h13'45" 12d7'28"
Douglas J Zaruba - Dreamgate
Cma 6h24'42" -30d54'50"
Douglas "Jabo" Jablonski
Uma 9h22'33" 72d7'43"
Douglas James Beck
Uma 11h50'34" 49d20'41"
Douglas James Law
Cas 3h26'18" 75d59'38"
Douglas James McCoy
Ori 5h16'22" 15d58'40"
Douglas Jerry Troutman
Per 3h3'49" 52d27'0"
Douglas John Bottume
Cnc 9h16'20" 23d4'7"
Douglas John Richard Ward
Cep 21h26'2" 68d48'41"
Douglas Jon Vogt
Uma 14h0'31" 49d2'29"
Douglas Joseph Ales "DJ"
Aql 19h29'26" 0d31'46"
Douglas Joseph Evans - 11/17/79
Sco 16h13'36" -12d13'36"
Douglas Julian Gross
Lib 15h9'22" -7d40'23"
Douglas Kane Bolton
Her 17h27'9" 44d12'9"
Douglas & Kathy Siglow
Cyg 20h24'12" 34d56'58"
Douglas Kent Ivey
Tau 5h30'13" 20d38'48"
Douglas Kent Weber
Aqr 21h23'19" -9d56'40"
Douglas & Kimberly Wallick
Dra 15h24'16" 62d37'30"
Douglas Kirkmeyer
Uma 10h59'9" 58d1'40"
Douglas Krechnyak
Tri 2h40'3" 33d52'49"
Douglas L. Bradshaw
Uma 11h12'47" 72d31'44"
Douglas L. Reeder
Tau 5h21'5" 26d40'9"
Douglas L. Schultz
Sco 17h52'27" -35d52'29"
Douglas Lee Hargis
And 0h32'3" 39d35'42"
Douglas Lee Stratton
Per 3h48'39" 48d8'1"
Douglas Lewis Wanke
Ori 5h9'54" 11d23'36"
Douglas Lloyd Barber
Umi 4h57'27" 88d30'14"
Douglas M. Chumley
Ori 5h25'59" 1d15'13"
Douglas M. Mosher and Lisa A. Smith
Per 2h58'51" 54d48'50"
Douglas M. Strawderman
Uma 10h35'10" 41d12'14"
Douglas M. Sutton, Jr.
Uma 11h17'30" 41d56'52"
Douglas M. Wray
Aqr 21h4'58" -9d22'20"
Douglas & Marilyn
Cyg 20h36'55" 52d51'39"
Douglas Matthew Fowler
Uma 11h19'38" 55d58'53"
Douglas Meacham
Cyg 19h57'34" 30d3'19"
Douglas Merola Stisi
Ori 5h7'4" -1d5'32"
Douglas Michael Carroll
Her 18h27'32" 13d40'31"
Douglas Michael McDonald
Her 17h5'42" 32d12'9"
Douglas Muscheid
Lmi 9h33'10" 38d2'10"
Douglas N. Stanton, Sr.
Gem 6h46'43" 17d35'44"
Douglas Owen Fleming
Dra 18h50'5" 60d28'1"
Douglas Patrick Brown
Sgr 17h54'43" -17d59'12"
Douglas Paul Johnson, II
Her 17h26'19" 21d13'18"
Douglas Paul McDougall
Uma 9h26'36" 50d12'47"
Douglas Paul Nielsen
Ori 6h2'43" 18d15'22"
Douglas Pete Meehan
Ori 5h38'27" 2d8'30"
Douglas Possenreide
Per 3h11'35" 55d48'4"
Douglas Poston
Per 2h26'3" 56d27'50"
Douglas Quigley
Umi 13h34'3" 70d48'54"
Douglas R. Mathie
Sco 16h47'38" -43d16'39"
Douglas Randall Wilkerson, II
Cap 21h23'24" -23d27'22"
Douglas Ray Howell, Jr.
Umi 15h47'1" 77d14'10"
Douglas Roark III
Sgr 18h51'4" -31d32'30"
Douglas Robert Gavin
Ori 4h51'47" 6d3'11"
Douglas Robert James Okey
Ari 2h36'41" 21d25'4"
Douglas Roy Lee Cook
Vir 14h1'31" -15d28'56"
Douglas Ryan Kinard
Cnc 9h5'12" 27d1'52"
Douglas S. and Vicki-Marie DeLuca
Uma 13h55'39" 50d26'35"
Douglas S. Reiter
Ari 2h44'8" 15d55'0"
Douglas S. Rosensky
Ori 5h55'50" 22d32'29"
Douglas Sanford Gorton
Cap 20h37'59" -13d33'35"
Douglas Scott Mason
Vir 13h25'33" -11d6'44"
Douglas Scott Neale
Vir 13h51'51" -1d44'36"
Douglas Sidney McCall
Cep 22h52'6" 71d43'21"
Douglas Silcock
Her 18h50'28" 23d39'9"
Douglas Stephen Smith
Per 4h48'7" 49d29'25"
Douglas Testa
Gem 6h47'0" 19d5'50"
Douglas (The Hooligan) Morgan
Sco 16h5'22" -11d42'11"
Douglas & Tracy's Star
Uma 9h49'39" 45d29'0"
Douglas W. Doane
Psc 1h51'2" 6d30'4"
Douglas W. Gailey
Umi 15h56'39" 70d29'45"
Douglas W. Macke
Lyr 19h7'50" 26d40'16"
Douglas Ward Aston
Cep 23h39'9" 79d41'13"
Douglas Watson Patrick
Cnc 8h41'38" 31d11'16"
Douglas Wayne Monahan
Aql 18h58'18" -0d10'42"
Douglas White Kremer
Psc 1h8'29" 29d15'33"
Douglas Whittington
Cnv 12h43'25" 40d58'5"
Douglas William Good
Per 3h19'28" 52d29'46"
Douglas William Reiser
Cep 22h21'29" 58d39'6"
Douglas, Kristin and Stephen DeLuca
Uma 13h26'12" 58d23'35"
Douglas8Michele Newman
Psc 0h6'24" -2d45'29"
Douglaslee Haley
Ori 5h24'46" 1d2'55"
Douglass & Annabelle
Ari 3h8'20" 25d11'45"
Douglass Wishing Star
Gem 6h11'18" 25d16'13"
DougLexlan
Lyn 8h39'8" 33d43'49"
DougnKristenEcklund
Lib 15h15'20" -4d25'7"
DougnRachel B.
Lyr 18h49'50" 32d30'57"
Doug's Dream
Uma 8h44'3" 60d37'53"
Dougs' Love Shines Down On Us
Uma 11h36'55" 57d33'18"
Doug's Star
Leo 11h24'40" 26d12'19"
Doug's Star
Ori 6h10'38" 8d13'30"
Dougy
Per 2h56'18" 53d46'4"
Dougy
Uma 9h10'14" 61d22'37"
Dougy Virr - 06.09.1933
Vir 13h21'35" -0d28'3"
Douie Jarman
Her 17h26'6" 35d30'0"
douleur, plaisir, amour..perfection
And 23h0'22" 47d15'1"
Douma 2003
Tau 5h29'43" 20d3'38"
DOURS
Lmi 10h1'20" 36d54'35"
Douthit Family
Uma 13h40'50" 54d11'12"
Dove
Oph 16h57'31" -8d36'19"
dove
Uma 11h32'26" 40d7'31"
Dove Star-Walter R Kiernan, Jr.
Lib 15h31'33" -15d7'42"
Dover Beach Hotel
Lyn 8h26'28" 57d53'36"
Dover Castle
Peg 23h40'0" 26d35'42"
Dovie Lee Catchen
And 2h25'38" 46d41'10"
Dovile M Zduoba
Cru 12h41'2" -56d42'12"
Dowd's Pride
Lib 15h5'9" -29d3'22"
Downrose
Dra 10h18'54" 74d18'59"
Doyal W. Armitage, Jr., Kels love
Aql 18h59'46" -4d54'13"
Doyle Bryant
Aur 5h14'54" 28d39'47"
Doyle Ellis Shining Star
Her 17h43'53" 31d31'47"
Doyle Owen Gray - Giant
Vir 13h8'19" -0d54'2"
Doyle Ward
Per 4h46'33" 47d2'11"
Doyle, Robert, Joyce, Timothy
Leo 11h24'48" 23d52'16"
Doyle-Maher Star
Vir 14h15'40" 1d52'29"
Doylene
And 2h12'50" 41d13'27"
Doyle's Pride of Cuchulain
Sco 16h49'58" -37d35'1"
Doys Brown
Ori 6h14'52" -0d32'13"
Dozer
Uma 9h49'51" 62d53'31"
Dozier Jones
Lyn 7h57'6" 33d33'10"
Dozna
Uma 14h3'0" 61d13'48"
Dozzie's Dazzler
Aql 19h16'22" 1d37'19"
dp
Uma 9h53'0" 44d5'53"
DP
Cep 22h18'51" 61d19'6"
DP + AB
Pho 0h40'46" -50d37'3"
DP & EP Forever
Leo 11h5'29" 22d31'9"
D.P. Streed
Psc 1h13'53" 23d32'35"
DPeacock
Lib 14h52'36" -2d47'18"
DPJL Billings Star
Her 18h53'25" 23d42'37"
DPLHAF1
Uma 9h33'27" 58d21'11"
DPT Olivia Webster
And 1h35'30" 38d6'57"
Dr A S Abul Kalaam Azad
Sco 17h53'15" -37d29'19"
Dr. Aamna
And 23h32'39" 42d54'0"
Dr. Aileen Mendoza
Tau 3h29'22" 8d27'56"
Dr. Alan Delman
Cap 20h19'16" -9d52'11"
Dr. Alan Edward Zloto
Cyg 19h58'32" 30d19'59"
Dr. Alan John Hamilton
Per 3h20'58" 41d53'38"
Dr. Albert Raymond Klinski
Lib 14h55'18" -4d28'25"
Dr. Alfonso Velasco
Per 3h9'2" 43d40'31"
Dr. Alison Leanne Smith
Vir 14h23'34" -0d59'12"
Dr. Allan John Hamilton
Cep 22h27'23" 63d14'50"
Dr. Allan Metzger
Dra 18h48'40" 56d1'49"
Dr. Allister McQuoid
Uma 9h38'18" 56d55'57"
Dr. Andrew Lawrence
Cyg 20h7'10" 56d10'40"
Dr. Angela Ruiz
Ari 2h7'48" 21d9'4"
Dr. Angie Nazaretian
Uma 10h36'44" 54d27'58"
Dr. Ann Durshaw
Per 2h10'37" 57d28'38"
Dr. Ann Frasier Ph. D.
Tau 4h25'33" 27d5'45"
Dr. Anthony Atalla
Boo 14h18'28" 21d17'41"
Dr. Anthony C. "Tony" DiMaio
Lib 14h52'38" -2d16'37"
Dr. Anthony Michael D'Agostino
Aql 19h3'42" -0d4'5"
Dr. Anthony Moretti
Her 18h34'8" 19d5'7"
Dr. Anthony R. Zembrodt
Uma 14h13'20" 62d9'13"
Dr. Anton Krotky
Uma 13h34'17" 55d25'3"
Dr. Armen Tashchian
Her 16h36'26" 32d52'22"
Dr. Arne Possing
Uma 9h19'10" 45d1'18"

Dr. Arnold "Arnie" Chanin
Leo 11h28'57" 12d47'51"
Dr. Augustus Saled
Boo 15h39'51" 41d58'53"
Dr. Axel Hoffmann
Uma 9h32'25" 49d28'27"
Dr. Ayhan Uzuner
Her 16h53'57" 14d49'49"
Dr. Balla Zoltán
Tau 4h8'17" 6d35'46"
Dr. Barbara J. Carter
Uma 11h39'12" 64d43'55"
Dr. Barry and Michelle Cromer
Lyr 19h18'6" 29d9'26"
Dr. Becky Barnum
And 0h39'17" 24d34'5"
Dr. Benjamin Gephart
Cnc 8h44'46" 18d25'47"
Dr. Benny Lee
Aqr 22h33'45" 2d28'54"
Dr. Bernard "Bernie" Katzen
Leo 9h58'2" 15d51'19"
Dr Bernard Edward Weller
Uma 10h11'41" 59d10'55"
Dr. Bernard Goffe
Uma 13h32'47" 58d12'1"
Dr. Bernd Matschulat
Uma 8h47'50" 68d41'45"
Dr. Betty Hathaway
Lyn 6h41'36" 52d12'22"
Dr. Bilby Barnett
Cnc 8h44'1" 30d20'44"
Dr. Bill Descovich
Her 16h51'29" 29d14'29"
Dr. Bill Harrison's Caring
Per 3h27'5" 32d53'19"
Dr. Blake
Uma 13h58'49" 51d45'39"
Dr. BlueSky
Aql 19h51'3" 3d6'50"
Dr. Bob
Aqr 21h34'9" 1d5'39"
Dr. Bob
Per 1h45'33" 50d52'33"
Dr. Bob
Uma 8h16'43" 61d29'54"
Dr. Bob
Cep 23h13'39" 82d39'33"
Dr. Bob & Kathy Dornbach
Uma 10h34'22" 39d21'28"
Dr. Bob Monsour
Ori 6h15'0" 7d24'10"
Dr. Bobbie Reker-Dickason
Sco 17h51'20" -30d58'41"
Dr Bonnie-Marie Doughty-Jenkins EdD
Cas 1h34'19" 66d10'28"
Dr. Bradley Hyman
Lib 14h54'33" -2d36'43"
Dr. Brian A. Mason
Uma 11h43'13" 57d40'49"
Dr. Briana
Cyg 21h26'49" 32d33'27"
Dr. Bridget
Cas 0h22'53" 53d12'46"
Dr. Bridget Jones
Tau 5h24'35" 25d42'14"
Dr. Brieanna Nation
Mic 20h38'37" -36d6'2"
Dr. Brigitte Theiner
Uma 11h33'48" 29d11'17"
Dr. Bruce H. Feldman
Aqr 21h13'18" -7d16'23"
Dr. Bruce Hart Esq
Tau 4h57'37" 25d13'41"
Dr. Calvin D. Jamison, Sr.
Oph 17h18'55" -22d59'21"
Dr. Carl E. Wick Sr
Her 18h36'42" 20d21'14"
Dr. Carl G. Zweig
Her 17h15'10" 33d43'21"
Dr. Carolyn J. DeForte
Uma 11h21'37" 55d8'35"
Dr. Charles Borden
Dra 15h15'16" 64d52'2"
Dr. Charles E. Poletti
Cep 21h29'34" 66d11'28"
Dr. Charles Graves Kingston Jr.
Uma 11h41'59" 34d23'39"
Dr. Charles J. Blackstock
Pho 0h15'46" -43d36'3"
Dr. Charles Leroy Anderson
Pyx 8h45'8" -29d7'59"
Dr. Chia Hui Lee
Tau 3h32'34" 7d33'57"
Dr. Christopher Andrew Babbage
Mon 6h53'31" -0d58'40"
Dr. Chuck Olds
Aqr 22h12'2" 0d3'47"
D.R. Clarks-P.
Leo 9h57'51" 16d4'56"
Dr. Claud Pitts III
Uma 8h48'54" 64d12'9"
Dr. Cliff Hudis - "The Wizard"
Cnc 8h20'47" 17d24'14"
Dr. Colin Abrams
Aur 6h30'27" 41d54'31"
Dr. Colin Thomas
Cas 23h39'34" 52d59'39"
Dr. Colleen M. Schwartz
Cas 23h57'55" 56d56'17"
Dr Cosmo Esquire
Gem 6h57'9" 24d32'24"
Dr. Curry
Cnc 8h35'13" 21d51'12"
Dr. D
Boo 15h9'57" 34d30'2"
Dr. D. Thomas Curry
Dra 15h21'12" 61d47'49"
Dr. Dad
Psc 1h33'59" 17d12'43"
Dr. Daddy Reynolds
Ori 5h45'37" 7d52'2"
Dr. Daniel Carter Holsenbeck
Ori 5h30'6" 1d24'28"
Dr. Daniel G. Marsalek
Aur 5h37'19" 44d54'9"
Dr. Daniel Lawhon
Lyr 18h30'24" 30d51'59"
Dr. Danielle Desjardins
Cnc 8h32'59" 16d23'46"
Dr. Darius Ameri
Her 16h25'35" 41d41'28"
Dr. Dave
Uma 11h55'33" 43d34'30"
Dr. David A. Finley
Tau 5h2'4" 25d4'53"
Dr David A. Latter
Cyg 20h33'37" 51d38'31"
Dr. David A. Loftus D.D.S.
Uma 9h36'43" 67d36'10"
Dr. David Augie Pommerenck, Jr.
Her 16h41'36" 34d29'35"
Dr. David C. Bors
And 1h35'45" 49d7'29"
Dr. David Goodman Simons
Oph 17h15'33" 2d24'32"
Dr David Jung
Psc 23h18'21" 1d23'4"
Dr. David Kingsley, Ph.D.
Cam 5h6'40" 66d47'18"
Dr. David L. McCarthy Sr.
Sco 17h57'52" -30d27'36"
Dr. David Pence
Ori 6h19'23" 13d36'30"
Dr. David Rutlen
Per 4h7'48" 43d21'13"
Dr. Dawn R. Brown
Uma 11h26'32" 67d47'10"
Dr. Debra Reichert Hoge
Per 2h50'2" 42d18'34"
Dr. Dennis' Dynamic Dream
Cnc 8h13'47" 22d19'16"
Dr. Dennis Michael Goretsky
Her 18h50'53" 25d4'41"
Dr Dennise Mary Broderick
Cap 20h22'8" -20d17'47"
Dr. Derek George Fraser
Cep 21h43'30" 55d53'58"
Dr. Dieter Schön
Uma 12h2'48" 30d7'12"
Dr Dog
Vir 13h9'22" 11d52'55"
Dr. Donald C. Alfano
Gem 6h26'36" 24d26'19"
Dr. Donald E. Simmons
Gem 6h59'41" 14d34'19"
Dr. Donald L. Holladay, III
Lyn 7h55'0" 39d12'38"
Dr. Donald L. Sweet
Cyg 20h21'33" 59d48'17"
Dr. Donald Murray Martyn
Boo 14h59'39" 40d23'23"
Dr. Donald Van Den Berghe
Per 3h43'40" 48d50'40"
Dr. Donna Smith
Cap 21h43'37" -18d36'37"
Dr. Dorothy Cowser Yancy
And 22h57'41" 49d58'25"
Dr. Doug De Vore
Psc 0h11'47" -1d33'43"
Dr. Douglas Galuk Family Star
Uma 13h16'50" 56d55'18"
Dr. Duke, Stephen, Randy Calabria
Sgr 18h10'21" -21d0'24"
Dr. E. Joy Jones Lynch
And 0h38'12" 37d31'19"
Dr. E. L. Pautler
Leo 11h49'37" 15d29'13"
Dr. E Michael Fox
Her 17h47'50" 46d24'44"
Dr. E.C. Smith
Boo 14h43'29" 21d30'26"
Dr. Edith J. White
Cam 4h27'7" 56d41'3"
Dr. Ed's Star
Tau 4h18'34" 0d39'21"
Dr. Edward Ira Gould
Lib 14h22'30" -9d3'16"
Dr. Edward M Ruscitti
Ari 3h8'7" 28d40'4"
Dr. Elaine German
Ari 2h52'5" 28d6'25"
Dr. Elias Naguib Bedewi - The Great
Gem 6h50'4" 33d9'29"
Dr. Elischer Zoltán 60
Cas 2h48'8" 65d25'53"
Dr. Ellis Strick
Her 16h59'37" 41d28'28"
Dr. Elona G. Marcy
Umi 15h10'15" 67d50'51"
Dr. Elvis Arterbury
Oph 17h36'51" 6d38'47"
Dr. Ephraim Suhir 1937
Uma 8h37'40" 64d55'3"
Dr. Erin Lynne Silsby.O.D.
And 0h49'48" 41d47'24"
Dr. Ernst Moerk
Psc 0h46'15" 15d56'48"
Dr. Eugene Swella & Dr. Jeff Swella
Uma 12h15'17" 57d43'43"
Dr. Euro Lovecrush
Lib 15h10'21" -7d56'57"
Dr. F. Stanley Hoffmeister
Cep 23h19'27" 76d12'17"
Dr. Fayza Al Khrafi
Cyg 20h36'2" 60d59'48"
Dr. Fekete Tamás
Gem 6h40'26" 24d47'14"
Dr. Floyd Lee Covey
Leo 9h39'19" 13d43'37"
Dr. Francis Leonard Robertaccio
Cep 20h21'31" 61d23'59"
Dr. Francisco Garcia Quiroz
Tau 4h5'40" 17d17'18"
Dr. Frank Feldman
Ari 1h59'21" 12d49'7"
Dr. Frank T. Patrick
Oph 17h41'34" -0d33'43"
Dr. Franklin P. Green
Her 17h41'47" 34d29'43"
Dr. Franklyn & Mrs. Alfleda Jenifer
Sge 19h50'9" 18d16'2"
Dr. Franz Gernot Köhler
Uma 8h22'34" 65d20'11"
Dr. G
Aqr 21h46'11" 2d8'44"
Dr. Gabriel Santiago
Ori 5h28'40" 14d51'9"
Dr. Garfield Fizzard
Sco 17h16'14" -41d44'7"
Dr. Gary De Sesa
Per 3h32'11" 45d26'31"
Dr. Gary Graham
Cas 23h50'6" 54d1'59"
Dr. Gary W. Nickel
Uma 13h32'59" 57d29'52"
Dr. Gedeon Perneczky
Uma 12h18'27" 59d30'5"
Dr. Gemma & His Crew
Uma 13h33'10" 57d22'55"
Dr. George A Cosper
Uma 10h57'41" 34d41'45"
Dr. George Salvator Fidone
Gem 6h32'34" 20d48'4"
Dr. Gerald Klaz *70th*
Lib 15h24'50" -21d21'37"
Dr. Glen Cooley
Uma 11h0'44" 48d2'43"
Dr. Glenn D. Gates "GEE"
Her 17h1'43" 36d34'0"
Dr. Glenn Scott Fuoco
Ori 5h31'7" 1d51'46"
Dr. Grant Houghton
Cnc 8h55'51" 11d12'51"
Dr. Gregory Kubik
Vir 13h5'6" -2d59'3"
Dr. Gregory Rinaldi
Aqr 22h45'37" -0d58'15"
Dr. Guido A. Merkens
Cyg 21h43'33" 39d41'35"
Dr. Handley
Oph 17h30'50" 7d59'51"
Dr. Hansa Jayakumar
Cyg 19h37'19" 35d55'17"
Dr. Hans-Günther Berg
Uma 10h52'32" 48d15'35"
Dr. Harald Biesold
Uma 10h0'27" 53d55'29"
Dr. Harold Book
Crb 15h52'25" 33d30'25"
Dr. Harold David Katz-Dad's Star
Cap 21h24'7" -23d48'17"
Dr. Harold 'Skip' Crisp Cox, Jr.
Gem 6h53'13" 33d40'47"
Dr. Harris Alan Lappin
Vir 12h15'23" 4d24'23"
Dr. Harry A. Nahorney
Uma 10h57'28" 53d39'22"
Dr. Harry Dunscombe, cellist
Cas 0h36'53" 57d3'5"
Dr. Hayden Kho, Jr.
Her 17h18'1" 24d55'59"
Dr. Hazel Carney England
Uma 10h10'53" 50d12'5"
Dr. Herbert I. McCOY
Oph 17h27'13" 6d45'21"
Dr. Herbert J. Dunn
Leo 9h57'52" 20d27'22"
Dr. Herbert Wogatzki
Uma 9h25'41" 59d41'4"
Dr. Holly Knor
Cyg 19h47'53" 31d28'5"
Dr. Horváth Levente 1976
Uma 9h48'34" 61d5'10"
Dr. Hosea Turk (Son of Thunder)
Uma 11h29'52" 33d58'54"
Dr. Howard Mandel
Aql 19h18'19" 6d40'34"
Dr. Howard Willner
Per 3h29'51" 50d9'47"
Dr. Hugh M. Bowen PhD
Cyg 20h57'57" 33d20'3"
Dr. Indira Parthasarathy
Cnc 8h37'50" 9d25'2"
Dr. Inglewood Quaid Jenkins,Esquire
Tau 3h35'19" 5d43'31"
Dr. Ira P. Monka
Leo 11h42'47" 25d20'58"
Dr. J. David Cassel
Uma 8h59'13" 63d39'30"
Dr. Jack Wright
Uma 13h36'38" 54d14'11"
Dr. Jaime Mullin
Mic 20h31'20" -33d7'15"
Dr. James E. Parsons
Sco 17h23'12" -37d41'38"
Dr. James Hagen
Ari 3h12'50" 27d53'40"
Dr. James L. McGaugh
Uma 11h11'2" 51d22'24"
Dr. James Lloyd
Dra 17h5'36" 59d1'8"
Dr. James Naplacic "Blue Star"
Ori 5h36'11" 3d23'11"
Dr. James W. R. Thomas
Del 20h51'0" 6d32'38"
Dr. Jan Richards
Mon 6h45'21" -0d16'31"
Dr Jane Whittaker
Aql 19h1'34" 15d58'43"
Dr. Janza Frigyes 60
Uma 8h38'42" 68d43'13"
Dr. Jay Scott Neale
Lup 15h46'47" -31d29'49"
Dr. Jean-Marc Sansot
Leo 10h23'5" 13d19'42"
Dr. Jeffrey Howard Sturza
Leo 9h38'44" 31d16'4"
Dr. Jerome Benson
Ori 5h35'26" -0d16'37"
Dr. Jerry L. Harvey
Cru 12h12'4" -62d13'58"
Dr. Jill Quirin
Men 4h52'31" -71d5'17"
Dr. Jill's Star
Cas 0h22'20" 51d24'25"
Dr. Jim Flugrath
Lib 15h28'12" -6d24'53"
Dr. Jim Pace
Gem 6h40'30" 14d2'36"
Dr. Jo Amburgey
Lib 15h44'31" -19d59'27"
Dr. Joan Welker
Cas 0h56'47" 63d36'28"
Dr. Joanne Christopherson
Cas 0h37'22" 53d42'12"
Dr. Joe Henry Stout
Aql 19h50'56" -0d34'48"
Dr. Joe "Star" Gromada
Cap 20h19'22" -19d28'49"
Dr. John
Ori 6h1'2" -0d44'50"
Dr. John
Uma 13h41'30" 56d37'44"
Dr. John and Willa Woods
Cyg 19h42'8" 39d22'29"
Dr. John Charles Adams
Cap 20h13'52" -14d51'30"
Dr. John Douglas Bramwell
Per 3h9'31" 51d12'37"
Dr. John E. Occhuizzo
Cep 7h27'46" 86d51'57"
Dr. John G. McCall
Ori 5h44'46" -5d25'34"
Dr John Galvin Shining Once Again
Cru 12h39'16" -62d29'4"
Dr. John Henry Kauffman III
Tau 4h6'0" 22d43'9"
Dr. John Huber Wasilik
Aur 5h15'57" 29d18'58"
Dr. John J. Smith
Leo 10h27'23" 21d5'51"
Dr. John (Jack) Nash
Per 4h24'52" 48d42'1"
Dr. John Joseph Owens
Ori 5h41'30" -0d59'37"
Dr. John Joseph Walsh
And 0h18'4" 39d49'14"
Dr. John Lamb Reynolds
Tri 2h16'36" 32d4'48"
Dr. John Lu M.D.
Boo 14h31'8" 17d30'9"
Dr. John Swanda
Cnc 8h37'16" 24d55'42"
Dr. John William Thomas
Aql 19h39'45" 9d26'30"
Dr. Johnny Lawton
Lib 14h51'10" -6d32'41"
Dr. Jon Mastrobattista
Per 3h4'36" 54d50'7"
Dr. Joseph F. Dooley
Per 3h11'57" 53d9'22"
Dr. Joseph "Fred" King
Cep 21h23'9" 63d21'1"
Dr. Joseph J. Zaydon
Uma 12h14'28" 53d5'57"
Dr. Joseph Martin Utay
Cnc 9h11'28" 31d43'7"
Dr. Joseph Rowland Hedgpeth
Psc 0h50'32" 21d15'45"
Dr. Joseph Wade Schumer
Aqr 22h45'22" -0d59'6"
Dr. Joseph Zito
Uma 14h1'23" 52d38'14"
Dr. Jozenia Colorado
Lyn 6h35'6" 56d33'52"
Dr. Judith G. Miranti
Tau 4h19'47" 14d20'50"
Dr. Judith Kim Eckerle Kang
Leo 10h17'38" 26d50'44"
Dr June Selim 04/06/1945
Cru 12h6'39" -63d0'45"
Dr. K
Her 17h23'34" 27d20'12"
Dr. Karen B. Burgess
Lyn 9h1'30" 33d24'25"
Dr. Karen Seiter
Sco 17h48'25" -37d55'9"
Dr. Kari Beth Law
And 0h43'43" 43d56'24"
Dr. Katherine Charlap
Uma 10h1'27" 71d55'35"
Dr. Kathleen M. Bonsick
Cyg 21h3'45" 36d16'47"
Dr. Kathleen Walsh
And 0h23'0" 41d6'51"
Dr. Kathleen Wells DDS and Staff
Cas 0h49'47" 48d17'15"
Dr. Kathryn A. Agard
Psc 0h19'58" 9d3'21"
Dr. Kathryn Payne
Lib 15h12'12" -7d16'26"
Dr. Kathy J.
Gem 6h6'0" 25d18'37"
Dr. Kathy McNally
Lyn 8h57'3" 40d53'38"
Dr. Kaye S. Andrews
Uma 11h47'18" 35d7'20"
Dr. Keith Brian Annapolen
Per 3h10'35" 51d46'12"
Dr. Keith Niesenbaum, VMD, healer
Umi 14h43'0" 77d30'57"
Dr. Ken A. Pettine
Ari 2h33'53" 21d2'23"
Dr. Ken: Life Saver of the Big Bend
Her 16h28'27" 8d51'10"
Dr Kenn Stevenson
Gem 6h40'26" 27d46'36"
Dr. Kenneth David Mack
Cyg 19h57'10" 38d22'12"
Dr. Kenneth H. Cooper, MD
Equ 21h0'43" 3d49'55"
Dr. Kenneth Murphy
Ori 5h33'9" -0d52'31"
Dr. Kian Kooros
Ori 5h49'41" 11d24'58"
Dr. Kimberly Edgel
Sgr 18h39'34" -28d18'23"
Dr. Klaus J. Porzig
Aur 6h24'35" 32d30'21"
Dr. Koncz István csillaga
Cap 20h13'54" -21d4'16"
Dr. Kovács Judit Katalin
Vir 13h51'4" -14d54'54"
Dr. Kristi Compton
Ori 6h16'13" 1d52'58"
Dr. Kristie A. Spangler
Cyg 20h55'9" 35d38'20"
Dr. Kristina Obom
Psc 1h11'57" 23d46'1"
dr. Kuksi
Cas 0h5'12" 52d17'4"
Dr. Laura Bridges
Uma 9h26'31" 55d3'52"
Dr Laura Corrigan
Cas 23h37'23" 55d38'20"
Dr. Lauren Ferrara Md.
Cas 1h25'28" 62d21'16"
Dr. Laura's Star
Her 17h38'12" 19d49'42"
Dr. Lawrence Goodman
Uma 10h36'32" 42d28'1"
Dr. Lefkos Aftonomos
Leo 11h11'8" 9d14'20"
Dr. Leo McCormick
Lyn 8h48'40" 43d11'56"
Dr. Leonard Lecks
Per 3h22'58" 52d1'40"
Dr. Lesha Dawn Hickman
Cnc 9h7'35" 20d11'32"
Dr. Lester W. Blair
Cep 22h29'47" 57d58'3"
Dr. Liebert - Anywhere From Here
Uma 13h50'57" 50d57'30"
Dr. Lillian Buchanan
Dra 19h41'12" 65d55'33"
Dr. Linda M. Ziegler
Sgr 19h2'38" -16d39'12"
Dr. Linda Seger - 25th Anniversary
Oph 17h29'54" 11d1'47"
Dr. Louis E. Barbosa
And 23h16'17" 43d11'24"
Dr. Maan Anbari
Vir 13h10'20" 6d42'30"
Dr. Maggie Mamen
Lyn 8h8'15" 35d23'47"
Dr. Magic
Cap 21h51'27" -12d51'48"
Dr. Makeen Yacoub
Aur 5h40'9" 33d30'24"
Dr Marcin Wieloch
Lib 14h38'23" -23d2'8"
Dr. Marcy Black-Alamodin
Tau 3h33'37" 2d24'46"
Dr. Maria
Cap 20h36'21" -17d10'59"
Dr. Marisa S. Briones
Cnc 8h49'18" 23d16'3"
Dr. Mark A. Welch
Aql 19h36'34" 10d42'12"
Dr. Mark Andrew Lambert
Sco 16h46'15" -39d59'27"
Dr. Mark R. Jones Memorial Star
Her 16h55'13" 32d59'9"
Dr. Marla Wendy
Uma 13h15'2" 56d56'20"
Dr. Martin Weiner
Cep 0h4'13" 79d59'59"
Dr. Mary
Cas 0h30'2" 53d25'36"
Dr. Mary Frame
Cnc 8h11'52" 16d44'9"
Dr. Mary Jane Stamm
Lac 22h35'17" 41d56'1"
Dr. Mary Jo Parker
Her 16h54'34" 41d55'41"
Dr. Mary Macedonio
And 23h18'27" 42d20'47"
Dr. Mary W. Loveland
Umi 13h24'47" 71d35'6"
Dr. Massoud Tehrani
Per 3h27'36" 47d1'17"
Dr. Maxson's Miracle
Uma 11h30'28" 58d9'4"
Dr. May Livingstone
Tau 4h17'34" 18d40'36"
Dr. Meg Hoffmeyer
And 2h38'36" 41d50'19"
Dr. Meredith H. Duncan
Uma 13h13'30" 56d3'41"
Dr. Michael Alessandri
Ori 6h0'27" 21d49'58"
Dr. Michael Duzy
Leo 10h43'28" 10d13'35"
Dr. Michael Endres
Uma 9h25'1" 44d46'41"
Dr. Michael J. Noetzel
Ari 3h3'23" 29d17'48"
Dr. Michael Joseph Brosnahan
Uma 9h44'13" 56d30'58"
Dr. Michael Raska
Apu 14h29'25" -80d37'19"
Dr. Michael Rotberg
Vir 12h9'7" 9d3'21"
Dr. Michael S. Kline
Gem 7h31'21" 29d28'39"
Dr. Michael T. Beachem
Uma 12h49'5" 60d27'0"
Dr. Michael Walsh
Pho 23h42'46" -41d2'15"
Dr. Michelle Gargagliano
Cnc 8h21'40" 23d55'58"
Dr. Miguel Mora
Aur 5h52'33" 46d48'3"
Dr. Mikki Meadows-Oliver
Crb 15h33'6" 26d20'52"
Dr. Millie Schaefer
Psc 1h22'24" 32d41'35"
Dr. Milton Beck
Del 20h53'38" 4d29'20"
Dr. Milton Wilner
Aur 5h55'49" 42d26'11"
Dr. Moss
Sgr 18h7'27" -29d40'48"
Dr. & Mrs. Jack Tabaska
Cyg 20h56'40" 50d24'5"
Dr. & Mrs. J.C. Wade, Jr.
Cyg 19h42'55" 33d17'5"
Dr. Music Master
Cnc 7h58'2" 15d4'49"
Dr. Myrna E. Garcia
Uma 11h33'51" 61d5'38"
Dr. Myrna Thomas-MacArthur
Crb 16h9'26" 31d2'8"
Dr. Myron Brin
Cnc 8h29'46" 27d35'53"
Dr. Naomi Louise Boness
And 1h51'23" 38d52'40"
Dr. Nelson Macy Walker
Cas 23h48'12" 56d34'30"
Dr Nick
Per 3h20'25" 42d32'49"
Dr. Nick Andriacchi
Sco 16h30'56" -41d44'51"
Dr. Norman W. Crisp, Jr.
Cyg 21h14'28" 51d24'52"
Dr. Numb
Cap 21h46'27" -10d58'59"
Dr. O.A. Jacobson
Uma 10h24'21" 66d3'12"
Dr. Odell Nickelberry
Uma 11h19'55" 58d4'3"
Dr Olli ihre chlii Stärn
Her 18h29'9" 12d52'21"
Dr. Oran Richard White
Lib 15h45'44" -5d43'21"
Dr. Oscar Nepomuceno
Psc 1h39'35" 22d8'13"
Dr. P
Uma 8h36'17" 64d13'56"
Dr. P. Crowley Life's Shining Star
Cas 1h16'55" 51d25'53"
Dr. Papp István csillaga
Uma 9h36'46" 44d17'7"
Dr. Pat Gianaris Booth
Oph 17h46'14" 4d44'49"
Dr. Patrick Louis Gianopoulos
Ori 5h39'38" -0d47'51"
Dr. Patrick Ross
And 23h37'11" 42d29'59"
Dr. Patti Gault
Lmi 10h14'2" 38d45'17"
Dr. Paul and Suzie Hersey
Uma 8h48'33" 66d48'54"
Dr. Paul David Monsour
Psc 1h26'7" 13d32'39"
Dr. Paul F. Murphy, D.M.D.
Ori 5h46'17" 7d10'58"
Dr. Paul F. Warms
Aur 5h48'55" 29d49'23"
Dr. Paul Martin Weinhold
Per 3h22'5" 47d3'34"
Dr. Pepper
Lyn 8h3'8" 41d9'13"
Dr. Pepper
Cmi 7h36'37" 5d59'25"
Dr. Pepper & Mr.Pibb's Special Star
Cnc 9h7'18" 25d1'40"
Dr. Pete Zafirides
Lyn 7h55'15" 56d47'33"
Dr. Peter Garner
Uma 10h32'12" 67d8'41"
Dr Peter Holman
Cru 12h17'19" -56d28'32"
Dr. Peter James McDonald Tenicki
Lib 16h1'23" -12d21'26"
Dr. Peter Ulbrich
Aqr 23h35'22" -22d50'19"
Dr. phil. nat. Anselm Erlich Oberholzer
And 23h39'17" 41d35'58"
Dr. Philip J. Cilio
Ori 5h3'50" 13d29'45"
Dr. Philip Maurice Steen
Uma 13h35'42" 51d13'3"
Dr. Philip Tippin
Per 3h36'28" 34d53'56"
Dr. Philip W.Cooke
Cap 20h33'50" -24d38'15"
Dr. Phillip Gill
Ori 5h50'25" 6d49'46"
Dr. Phlox
Per 4h16'37" 40d32'8"
Dr. Potts
Psc 1h38'38" 23d31'43"
Dr. Prasad G. Kilaru
Per 4h0'26" 39d56'59"
Dr. Pual Solomon
Per 3h43'40" 44d51'9"
Dr. Quentin Giorgio
Uma 10h20'43" 52d19'57"
Dr. Radoje Ilic
Cas 23h34'56" 53d55'34"
Dr Ralph Chapman
Cru 12h36'42" -60d51'46"
Dr. Randi Darling - 01/01/1965
Gem 7h7'20" 26d32'40"
Dr. Randy Craig Moze
Ari 2h10'41" 21d39'45"
Dr. Raviv Balfour
Ori 5h27'10" 5d45'56"
Dr. Raymond E. Grizzle
Ori 5h20'6" -0d31'34"
Dr. REK
Crb 15h18'17" 26d8'20"
Dr. Renee Exelbert, my Shining Star
Cap 21h47'6" -14d57'15"
Dr. Rich Shapiro Our Shining Star
Uma 11h42'5" 53d13'3"
Dr. Richard A. Shlofmitz
Per 3h8'23" 56d12'4"
Dr. Richard Benoit
Ori 6h1'26" 19d41'0"
Dr. Richard Cyburt
Aql 19h56'54" 7d4'7"
Dr. Richard J. Clement
Cma 6h40'2" -14d20'17"
Dr. Richard Marchand
Lmi 10h4'4" 34d46'42"
Dr. Richard Rieger
Umi 13h21'30" 72d4'2"
Dr. Ricky Steven Gutierrez
Oph 16h40'7" -0d53'9"
Dr. Riley
Aql 19h48'30" 8d0'49"
Dr. Ripul R. Panchal,DO
Oph 17h29'36" 10d31'37"
Dr. Robert Falk
Per 4h44'13" 40d1'11"

Dr. Robert Graziano
Per 3h23'17" 52d15'30"
Dr. Robert H. Latter
Lib 15h15'23" -18d52'57"
Dr. Robert J. Kramer
Her 16h27'59" 47d1'25"
Dr. Robert John Wojnar
Aqr 22h16'10" 2d14'54"
Dr. Robert L. Slavens
Cam 5h56'30" 66d24'48"
Dr. Robert Lane
Cep 22h17'57" 58d1'55"
Dr. Robert Steinberger
Tau 5h9'48" 23d24'13"
Dr. Robin L. Beaumont
Tau 3h42'33" 16d30'1"
Dr. Robin L. Chambers
Ari 3h20'50" 27d12'4"
Dr. Rod McLeod
Per 2h13'39" 52d1'5"
Dr. Rod Paige
Aql 19h49'55" 6d43'20"
Dr. Ron Nagel 1956
Umi 16h42'43" 88d37'49"
Dr. Ronald Peter Mack
Tau 5h7'44" 26d52'55"
Dr. Ronald Waxman
Cep 22h27'48" 68d21'30"
Dr. Rosanna Santini
Uma 10h5'38" 57d6'32"
Dr. Rosemarie Rizzo Parse
And 0h21'53" 44d45'10"
"Dr." Ross Cortez
Per 4h0'38" 44d0'45"
Dr. Roy's Shining Star
Her 17h30'26" 48d3'8"
Dr. Russell L. Sturzebecker
Aur 5h54'20" 46d37'50"
Dr. Salomon Goldberg aka Salamoosh
Gem 7h31'20" 32d45'36"
Dr. Samuel David Thomas
Aql 19h49'38" -0d0'11"
Dr. Samuel "Poppy" Anderson
Psc 1h35'27" 12d52'44"
Dr. Sandra R. Brown
Crb 15h37'0" 37d14'7"
Dr. Sandra Tirado
Cas 0h36'45" 48d37'31"
Dr. Seeger
Cep 22h12'34" 62d2'33"
Dr. Seymour J. Rieger
Uma 9h18'26" 42d45'27"
Dr. Sharon J. Kaczmarek
Cas 1h25'56" 59d34'58"
Dr. Sheri Lynn Gostomelsky
Her 16h41'40" 18d7'1"
Dr. Sherwin & Mary Lou Miller
Psc 1h4'10" 21d26'38"
Dr. Shiama Sanmugaratnam
Cru 12h11'26" -60d34'34"
Dr. Shirley Nuss
Cas 1h19'40" 58d13'13"
Dr. Shirley Rohl Daniel
Sco 16h10'53" -14d37'0"
Dr. Sidney H. Sobel-70
Boo 15h19'2" 48d52'12"
Dr. Silvester Carl Henderson
Per 3h11'41" 40d14'52"
Dr. Sipka Péter Máté
Lib 15h49'42" -17d51'33"
Dr. Sipka Rózsa
Cnc 9h17'5" 10d7'38"
Dr. Sipka Sándor
Sgr 18h30'10" -16d47'4"
Dr. Sipka Sándor Mihály
Cnc 9h18'30" 10d42'14"
Dr. Sipkáné Serfozo Rózsa
Aqr 21h42'3" -0d24'14"
Dr. Stanley Furman
Leo 11h2'44" 21d4'56"
Dr. Steinmetz Barnabás
Lib 15h50'47" -17d50'39"
Dr. Stephen Charles Gleason
Her 16h34'48" 18d44'31"
Dr. Stephen J. Clark
Per 3h57'55" 33d31'45"
Dr. Stergios Stergiopoulos
Psc 22h52'29" 3d0'37"
Dr. Steve, IOG
Cnc 8h22'51" 12d39'5"
Dr. Steven Klein
Cep 22h2'19" 69d45'48"
Dr. Steven M. Burns, PhD
Uma 9h54'9" 44d18'13"
Dr. Suchindran 'Chat' S. Chatterjee
Gem 6h49'42" 17d25'22"
Dr. Sudhir I. Patel
Lyr 18h54'45" 33d54'54"
Dr. Susan L. Whiting A Florida Star
Cnc 8h41'35" 22d44'43"
Dr. Suzanne Goldberg
Oph 17h37'26" 2d13'8"
Dr. Suzanne V. Ernst
Ari 2h48'27" 31d9'21"
Dr Sydney Tatsuno
Psc 23h53'40" 5d56'34"

Dr. Sylven Beck
Lyn 7h54'55" 58d25'31"
Dr. Szigeti Zoltán
Vir 15h0'7" 5d41'11"
Dr. Sztrókay Ottó csillaga
Cnc 8h8'59" 31d2'16"
Dr. Tammy
Lyn 8h33'39" 35d52'15"
Dr. Tammy
Tau 3h45'39" 23d45'16"
Dr. Tara Devine
And 0h30'37" 42d58'31"
Dr. Ted Varas
Aql 19h51'54" 16d20'40"
Dr. Terrance Dengler
Cyg 20h24'59" 33d31'1"
Dr. Terry
Sgr 17h44'49" -20d59'27"
Dr. Theodore Slutsky
Cep 22h5'9" 62d51'29"
Dr. Thomas J. Fahey
Uma 10h11'52" 45d52'14"
Dr. Thomas J. Norris
Gem 6h32'57" 26d13'25"
Dr. Thomas Kerns
Cep 21h34'13" 66d35'8"
Dr. Tim Blessing
Vir 12h41'42" 11d28'3"
Dr. Tim Vierheller
Her 18h56'30" 15d21'28"
Dr. Timothy J. Whelan, DC "Doc Tim"
Uma 11h13'4" 28d38'14"
Dr. Tnadticdai
Uma 10h46'51" 69d12'6"
Dr. Tnadticdai
Lyn 7h19'33" 58d57'9"
Dr. Todd "Bonezy" Bannen's Star
Lyr 19h14'38" 36d15'33"
Dr. Tom J. McDaniel President OCU
Her 17h14'26" 17d35'45"
Dr. Tracy Kemble
Uma 10h27'8" 44d33'20"
Dr. Trav & Margie Hindman
Oph 17h43'59" -0d48'31"
Dr. Valya E Visser
Lup 15h50'47" -33d52'5"
Dr. Vámos Tibor
Aur 5h46'4" 28d46'58"
Dr. Varga Aladár ás családja
Uma 9h58'13" 50d11'22"
Dr. Victoria Franchetti Haynes
Lib 14h30'57" -18d30'55"
Dr. Vincent Quagliarello
Per 2h14'36" 55d54'25"
Dr. Vivian Smith
Cam 4h3'10" 69d41'37"
Dr. W Star Of Radiology
Her 17h0'49" 29d58'12"
Dr. Walter J. Rogers, Jr.
Mic 21h9'40" -31d49'0"
Dr. Walter L. Nickells
Cyg 21h38'46" 54d58'37"
Dr. Walter Wilson Gale
Dra 19h50'18" 76d19'55"
Dr. Warren A. Stewart
Leo 9h56'43" 6d58'16"
Dr Wesley N Shellen
Ori 5h57'3" -0d2'38"
dr. Wieslaw Wiznerowicz
Dra 18h2'16" 50d42'22"
Dr. William Albert Seng, DC
Sco 17h43'7" -43d30'53"
Dr. William "Bud" Calley
Aql 19h55'9" 2d18'35"
Dr. William David Henderson
Per 3h7'54" 54d13'58"
Dr. William Michael McCormack, Jr.
Leo 10h8'53" 25d11'21"
Dr. Winnie Lam
Oph 16h39'10" -0d51'53"
Dr. Wonderful Rink
Tau 4h18'32" 26d33'26"
Dr. Yieu Chen Wu
Psc 1h11'28" 30d1'53"
Dr. Yvone B. Flynn
Oph 17h21'46" 7d13'13"
Dr. Zella Elizabeth Moore
Cas 23h43'55" 54d32'38"
Dr. Ziad Magued Mamish
Cyg 19h28'12" 35d49'8"
Dr. Zoidberg
Dra 17h39'19" 52d58'18"
Dr. Zoran Trifunovic
Cas 0h35'13" 65d5'34"
Draelos
Lyn 8h4'28" 51d12'42"
Drága én Csillagom
Leo 9h48'28" 11d18'44"
Dragan & Sandra
Uma 10h4'31" 55d5'41"
Dragan Vujic - 06.11.1956.
Her 17h41'17" 30d21'45"
Dragana Kustura
Ari 2h13'44" 23d51'6"
Dragana Markovic
Tau 4h30'53" 29d37'45"

Draghetto 74
Ori 6h11'33" 15d54'21"
DRAGI 78
Umi 13h18'39" 70d5'33"
Dragica Bulic
Lmi 10h16'19" 38d34'35"
Dragica Katic
Dra 19h31'2" 65d2'57"
"Drago" Alexander J. Hall
Dra 17h44'28" 54d54'34"
Drago354
Dra 15h50'43" 54d8'29"
Dragon
Vir 12h6'49" 5d42'13"
Dragon Alexander
Cap 20h22'52" -10d13'20"
Dragon Fly
Leo 11h18'26" 16d27'38"
Dragon Lady
Dra 17h32'51" 61d12'35"
Dragon Light
Dra 16h25'49" 52d22'49"
dragon_master
Ori 5h58'31" 10d35'38"
Dragon Og
Dra 15h49'30" 62d9'8"
Dragon Star
Aqr 22h29'45" -7d6'16"
Dragon Star
Leo 11h9'30" 10d36'18"
dragonfly
Ari 2h55'38" 28d4'56"
Dragonfly
Aqr 22h22'14" -0d39'43"
Dragonfly
Dra 17h44'11" 53d58'36"
Dragonfly
Lib 15h31'26" -28d1'55"
Dragonfly Lesli
Tau 5h50'15" 23d4'24"
Dragonfly - Lynn & Steve Carr 17/8/07
Cyg 20h14'15" 39d29'44"
Dragonius
Dra 14h44'32" 55d21'35"
Dragon's Shadow
Aqr 22h17'11" 1d5'58"
Dragonwigga
Psc 0h38'31" 17d38'46"
Dragovich
Aqr 22h8'37" -13d12'44"
Drahcirnitsua III VII MCMLVIII
Psc 1h19'58" 33d12'11"
Drahonovsky, Renate
Ori 6h3'38" 19d51'16"
Draimen Maris
Uma 8h38'45" 67d12'29"
Drak Goad
Peg 22h19'5" 24d9'47"
Drake Anthon Stenberg
Leo 11h19'32" -6d25'54"
Drake Landon Buchanan
Uma 13h12'57" 62d54'30"
Drake Loves Natalia Yi Always
Sco 17h29'57" -31d41'44"
Drake O. Williams
Cap 21h54'56" -23d57'3"
Drake Sebastian
Umi 15h0'43" 69d55'46"
Drake "The Snake" Cain
Cep 20h37'13" 62d31'3"
Drake Venson Burgeis
Uma 10h50'18" 39d35'56"
Drakestar
Ori 6h6'51" 6d20'32"
Drakon
Tau 5h46'7" 22d28'50"
Drama Queen
Cas 0h55'46" 55d43'25"
Dr.Arie Maman
Oph 18h22'43" 11d25'11"
Drashti Gandhi
Tau 4h20'31" 6d14'59"
Draude, Karl
Aqr 21h16'5" -2d1'57"
DRAVEN
Leo 10h41'50" 16d28'44"
Draven Galbraith 7-1-03
Lyn 8h10'4" 40d56'13"
Draven Staffelbach
Sco 17h27'41" -45d21'25"
Drawde
Cnc 8h21'55" 23d26'28"
D-Ray
Her 17h18'59" 32d36'58"
Draya Monster
Dra 15h37'24" 58d28'21"
Drayton C. Parker
Psc 1h1'46" 28d39'28"
Drazen Lukic
Cru 12h31'35" -61d19'0"
Dr.Bob
Gem 6h56'23" 17d48'51"
DRE
Lyn 9h5'3" 37d27'51"
Dr.E & Lee Whitfield The BigTripper
Umi 15h32'31" 71d46'57"
Drea
Cap 20h17'13" -13d24'15"
Drea
Aur 5h17'38" 31d52'12"

Drea
Ari 3h5'14" 28d37'36"
Drea Jane
Cas 1h45'56" 65d57'28"
Drea Rome Cusano
Umi 15h30'31" 81d3'18"
Drea21
Tau 5h23'56" 28d37'8"
DreaLovely
Lib 14h54'17" -2d19'56"
Dream
Umi 16h14'4" 85d5'16"
Dream
Cyg 19h27'13" 54d7'35"
Dream
Ari 2h33'44" 27d31'56"
Dream
Cyg 20h14'9" 35d2'2"
Dream
Cyg 20h20'57" 51d54'18"
Dream & Believe
Uma 11h22'1" 57d59'14"
Dream Big 32
Uma 13h3'44" 58d6'6"
dream big and biggest it shall be
Uma 11h0'34" 50d49'7"
Dream Catcher
Cyg 21h54'7" 50d17'19"
Dream Catchers
Psc 1h12'37" 3d19'14"
Dream Chaser
Peg 0h4'56" 16d46'52"
Dream Crush
Mon 7h16'3" -0d34'41"
Dream Don Spry Forever
Pyx 8h53'51" -34d29'24"
Dream Girl...Mandy
And 0h19'45" 30d33'41"
Dream Jackie Dream, You shine on us
Lyn 8h2'18" 43d42'55"
Dream Land
Tau 4h31'14" 20d53'45"
dream lover
Gem 7h19'56" 25d15'45"
Dream Lover
Cap 20h39'7" -18d59'42"
Dream Star
And 0h37'0" 39d17'55"
Dream Sweetly Lauren
Peg 21h52'55" 17d21'7"
Dream Weaver
Cet 2h31'19" 6d58'42"
Dream Weaver
Cyg 20h46'26" 30d35'37"
Dreama
Leo 9h49'9" 21d40'0"
Dreambeam
Aqr 21h15'14" -13d18'37"
Dreamclo
Vir 12h35'30" -9d52'0"
Dreamer
Cnc 9h16'1" 29d1'13"
Dreaming of You
Lib 15h13'36" -5d14'33"
Dreamline Rabert
Cyg 20h11'31" 45d47'40"
Dreams Really Do Come True
Cep 21h45'9" 61d50'0"
Dreamshine
Uma 9h9'39" 52d41'17"
Dreamteam Angeline Meyer & Martin Wälti
Tri 1h44'21" 31d25'33"
Dreamward-246
Cyg 19h37'40" 35d37'31"
Dreamway
Cyg 20h30'53" 43d35'50"
Dreamy Butter Cup RX Boy
Aqr 22h15'49" 0d56'21"
Dreamy Dave
Per 2h22'10" 57d3'3"
Dreamy Dhunji
Uma 12h42'34" 56d51'5"
dreamz611
Sgr 18h47'23" -27d40'43"
Dreanna Michelle Laughlin
Sco 17h54'21" -35d51'1"
Dreap
Aql 19h46'46" 7d25'22"
DREEMBEEKEN
Cam 4h31'42" 63d34'17"
Dreena Ponder
Sgr 18h36'39" -23d47'31"
Drees, Andreas
Uma 9h16'12" 70d44'51"
D.Reid
Vir 12h39'25" -6d57'19"
Dreïlalexanmure
Sgr 18h23'8" -22d42'41"
Drema Darlene Swader
Oph 16h48'29" -0d24'8"
Drema Keller 23
Gem 6h5'7" 24d12'58"
Drennen Michael Duel
Umi 15h38'38" 73d2'7"
Dresser Family
Uma 10h59'35" 58d52'15"
Dreux
Leo 9h34'49" 28d47'5"
Drew
Uma 11h34'21" 29d31'40"

Drew
Her 16h39'17" 23d21'20"
Drew
Gem 6h39'17" 15d54'14"
Drew
Ari 2h14'49" 25d12'22"
Drew
Crb 16h23'3" 31d24'27"
Drew
Ori 5h37'41" -0d25'28"
Drew
Psc 0h2'20" 3d27'3"
Drew A. Moesel
Vir 14h42'10" 2d44'57"
Drew A. Pate
Aqr 22h16'27" -1d27'26"
Drew Abraham Slane
Uma 10h34'6" 63d50'27"
Drew Alesander Palcek
Uma 11h55'59" 60d59'4"
Drew and Brittany
Aqr 22h51'1" -9d38'18"
Drew and Claire Bedingfield
Lac 22h6'32" 45d1'28"
Drew and Emily's Love Star
Uma 13h56'28" 60d47'51"
Drew and Melissa's Love
Cnc 8h51'27" 11d53'26"
Drew and Roxy
Vel 9h27'51" -49d32'22"
Drew Anders Hegner
Vir 12h53'20" -6d13'2"
Drew Anderson
Aql 18h59'38" -10d34'24"
Drew Bedard
Tau 3h46'41" 28d37'1"
Drew Benjamin Fisher
Lyn 8h38'53" 33d20'11"
Drew Burbank
Lyn 7h26'33" 45d18'52"
Drew Cahn
Aqr 22h25'3" -1d38'17"
Drew & Cate Forever
Umi 14h42'42" 69d48'31"
Drew Corbusier
Lup 15h49'54" -31d59'8"
Drew Curtis Bunn
Ari 2h26'16" 11d34'43"
Drew Dario
Her 17h13'19" 48d17'2"
Drew & Debra, 2 Beautiful Years!
Cyg 19h36'45" 51d11'11"
Drew Deneher
Sco 16h53'1" -38d16'35"
Drew "D-Giddy" Gardner
Uma 11h21'54" 32d32'13"
Drew Edward Gordon
Dra 16h29'14" 63d55'10"
Drew Edward Perry
Cap 21h5'26" -15d54'39"
Drew Eggleton
Her 17h31'21" 31d19'9"
Drew Elise Chattic
Aqr 23h20'24" -17d42'27"
Drew Flanagan
Ori 6h19'13" 21d23'18"
Drew Hopay
Per 4h26'32" 35d12'32"
Drew Johnson's Christmas Star
Aql 19h51'13" 8d51'27"
Drew Johnston 08-04-00
Leo 11h15'25" -1d50'0"
Drew Joseph Sheils
Ori 6h18'9" 8d57'18"
Drew Kief Goodin
Per 3h9'0" 46d43'49"
Drew Kitchen
Her 17h10'15" 28d5'59"
Drew Lintker
Sco 16h9'43" -12d33'28"
Drew & Lorraine
Umi 15h18'50" 73d44'16"
DREW LOVES KAYLA
Lib 14h56'18" -4d14'27"
Drew Meinhardt
Lyn 6h20'34" 56d15'40"
Drew & Mel True Love Always Forever
Cru 12h47'13" -57d41'24"
Drew Michael Wash
Tau 4h45'22" 21d11'24"
Drew Miller Loves Pauli Rousseau
Lib 14h29'33" -17d46'48"
Drew Model School Staff 2005-2006
Dra 16h44'44" 60d34'0"
Drew Monsta
Cru 12h18'41" -56d18'39"
Drew Parisette
Vir 13h16'32" 4d17'40"
Drew Parkison
Sgr 18h12'53" -27d54'39"
Drew Parks Warman
Umi 15h17'19" 69d7'38"
Drew (Richard AnDREW Panettieri)
Uma 12h3'48" 29d1'35"
Drew Robert Johnson
Lib 15h48'42" -7d43'57"

Drew Rosen - CCIE 4365, CCSI 22045
Sgr 18h22'30" -26d7'1"
Drew Scatena
Peg 22h42'40" 25d14'52"
Drew Sweetman
Per 3h46'31" 48d14'46"
Drew Trimmer So Smart Gone Too Soon
Ori 5h56'26" 3d41'11"
Drew William Charles Hignite
Boo 14h46'51" 27d3'24"
Drew William Dolgos
Vir 14h47'52" 0d51'32"
Drew William Duval
Cap 20h23'1" -19d18'4"
Drew William McGarrity
Sco 17h28'13" -33d39'45"
Drew & Zel
Gem 7h8'39" 21d47'2"
Drew, Shining Star, Brow
Aur 5h37'5" 41d33'58"
Drewanda Gailia
Uma 11h27'11" 30d33'49"
Drewbrent
Lyn 7h0'26" 51d36'16"
Drewdick
Psc 1h20'18" 17d28'15"
DrewJetta Sagittarius 1986
Uma 11h27'21" 66d58'30"
Drew's Star
Uma 10h26'43" 66d33'6"
Drewsilla 6
Umi 16h32'53" 79d45'26"
Drewsita
Lyn 6h29'30" 58d50'54"
Drewy
Sco 16h40'24" -32d14'59"
Drewy
Uma 10h27'39" 43d57'19"
Drexetta Perez
And 1h23'16" 39d49'58"
Drey & Karen
Tau 5h55'14" 25d18'29"
Dreyfuss Lewis
Lyn 9h8'48" 37d30'5"
Dri Francis Star
Leo 11h7'26" 21d27'33"
Drianna Jaden Brown
Uma 11h25'30" 71d49'36"
Drifter
Eri 3h40'10" -10d22'23"
Drifter's Dream
Uma 8h46'46" 70d29'0"
Drifting
Eri 4h36'39" -13d38'33"
Drika's Star
Aqr 22h37'34" -21d6'44"
Driko & Buttercup's Star
Sco 16h11'9" -11d5'14"
Drill Sergeant K. Richie
Tau 5h44'16" 17d17'5"
Drin Haxhijakupi
And 23h39'8" 39d7'44"
Drinette
Dra 14h37'43" 60d8'34"
Driscoll- King Star
Lyn 9h38'27" 40d48'22"
Driskoll
Lyr 18h54'31" 32d23'16"
Driss
Cam 3h44'21" 55d28'53"
Drita Angel Selca
Cnc 8h11'28" 22d25'13"
Dritan & Susanne Luca
Uma 9h38'27" 64d51'43"
Drive-In Fisherman
Cep 23h1'11" 72d16'56"
Dr.J.Jayalalitha
Psc 0h34'4" 12d9'39"
Dr.John Metropoulos A Gift From God
Aql 19h20'53" 15d1'2"
Dr.Laurie Joseph
Her 15h59'30" 44d37'15"
Dross, Herbert
Ori 6h16'34" 18d53'14"
Drowning Pool
Ari 2h37'0" 17d50'47"
Droz
Ori 5h56'57" 19d34'23"
Drs. George and June Unger
Cyg 20h24'47" 32d5'55"
Drs. Shira and Andrew
Vir 12h49'22" 6d20'14"
Drs. Thoms & Melhuish
Uma 13h16'22" 56d50'7"
Drs. Tom and Leslie Tworoger
Uma 8h45'42" 60d43'53"
DRU
Her 18h7'47" 17d52'34"
Dru
Uma 8h36'17" 46d54'23"
Dru Bear
Gem 7h2'13" 34d7'27"
Dru Daniel Pittsley
Psc 1h4'0" 7d28'48"
Dru Our Wonderful Mom
Sco 16h5'49" -9d19'22"
Dru Rhodes
Lyr 18h26'28" 46d35'13"

Druce Gavin
Umi 13h26'31" 74d45'0"
Drucie Baby
Aql 19h56'27" 8d24'59"
Drucilla
Sgr 18h19'48" -23d21'40"
Drue Ellen Uman Sable
Crb 16h5'50" 31d41'34"
Drue Noelle
Sgr 18h57'5" -34d27'14"
Druid Lords' Higher Testosterone
Cmi 7h41'50" 5d7'59"
Drum Master
Lib 15h9'33" -19d37'37"
Drunky Munky
Ari 2h5'10" 18d20'18"
Druscilla
Aqr 22h34'53" 1d34'47"
Drusilia Gutierrez
And 2h19'42" 47d25'17"
Drusilla Christine Taylor
Psc 0h28'36" 16d44'33"
Druskoczi Ilona Szerelmem
Uma 8h17'35" 64d16'20"
DrWKMackeyNASAstaratLincoln&Cheyney
Uma 12h35'50" 59d50'24"
Drzewiecki Family
Uma 10h30'25" 66d1'50"
D.S.
Pho 0h33'23" -48d34'33"
ds 5-5
Uma 13h13'27" 57d53'51"
D's Dazzle
Ori 5h31'20" 3d41'36"
D.S. Love
Lyr 18h25'53" 34d1'40"
D's Star
Tau 4h37'34" 5d30'20"
ds85 Denise Jane
Leo 11h19'0" 24d7'46"
Dselestial
Sge 19h48'3" 17d3'24"
Dshinga
And 0h48'16" 35d6'6"
DSJ
Sgr 19h21'47" -22d52'11"
DSLN4
Umi 15h12'24" 81d2'51"
D.S.McLellan
Crb 16h12'29" 37d30'1"
D-Stat 33
Aqr 22h38'17" -3d31'26"
DSwan's Galactic Super Nova 4-25-46
Cyg 21h9'43" 34d16'21"
DSXC Superstar
Uma 10h14'59" 47d34'3"
D.T.
Mon 7h42'55" -0d59'37"
DTA Dream Catcher
Aql 19h31'31" 7d36'11"
DTina
Ori 5h23'46" 13d34'57"
DTMW
And 0h13'10" 25d11'15"
DTOII
Aqr 22h7'28" -1d5'35"
du Berger
Her 18h41'20" 22d23'7"
Du bist mein Stern der Liebe
Uma 9h5'56" 53d32'26"
Du & Loan
Gem 6h45'0" 24d19'4"
Du Papounet
Cas 0h12'43" 57d11'6"
Duain Darrell Bennett
Her 18h6'13" 17d49'44"
Duan Lian
Aur 6h30'48" 48d44'51"
Duane Allen
Cnc 8h29'51" 13d20'35"
Duane Allen Mathews Sept. 24, 1969
Uma 12h37'20" 57d40'1"
Duane and Cobi Forever
Sge 19h41'2" 17d55'36"
Duane "Army Boy" Eddy, Jr.
Uma 13h10'32" 58d30'6"
Duane E. Voorheis III
Uma 8h45'54" 69d3'46"
Duane Edward
Her 17h8'47" 30d47'16"
Duane Edward Taylor
Her 16h39'56" 25d20'32"
Duane Edwin Holt
Tau 4h23'23" 25d16'54"
Duane Emmett
Psc 0h22'58" 15d50'38"
Duane John Schnee
Her 16h33'43" 31d42'37"
Duane Lee
Tau 4h13'2" 5d6'5"
Duane Lorenzo Plano
Ari 3h15'20" 24d1'28"
Duane McDaniel
Ari 3h20'33" 19d10'32"
Duane "Osito" Hargis
Ori 4h51'44" 3d5'23"
Duane Pennock
Cnc 8h10'54" 16d8'42"

Duane & Sandy Zeltwanger Mon 6h53'24" -0d21'33"
Duane T Hall Uma 12h12'5" 62d17'51"
Duane Thomas Dow Keenan Cen 13h25'47" -37d43'20"
Duane Thomas Hewitt Cep 21h38'24" 55d27'54"
Duane Tubach Gem 6h51'0" 18d42'40"
Duane & Vera Pearce Lyr 18h48'56" 32d7'26"
Duane Wigham Uma 10h16'49" 50d1'11"
Duane William Dingman Cnc 8h45'51" 32d38'44"
Duane's Star Cyg 19h39'54" 32d12'33"
Duann Kier Sywanyk 50 Crb 15h57'57" 32d53'24"
Duanne Melissa Leo 11h5'14" 5d18'51"
DUANYS MILAY Cnc 8h29'26" 17d0'4"
Dub Porter Uma 13h29'5" 58d1'30"
Dubble Break 5 Uma 11h6'11" 40d5'35"
Dubi Cas 1h29'32" 72d41'31"
Dubie Psc 1h8'21" 4d37'23"
Dubrez Nicolas Ori 5h23'20" -4d13'0"
Duc and Trin Lyn 7h37'13" 53d41'57"
Ducasse Lib 14h50'53" -11d37'20"
DUCE 22 Gem 6h53'59" 31d47'49"
Duchess Caralyn Ann Uma 10h54'48" 35d1'26"
Duchess Chelsea Ori 5h38'59" -2d45'8"
Duchess Emch Cas 23h56'8" 59d24'33"
Duchess Nike Of Forest Grove Leo 11h49'55" 22d5'28"
Duchess of Kent Uma 10h25'43" 42d22'42"
Duchess of Sanspants Ari 2h10'18" 26d30'55"
Duchollet Valérie Vir 14h14'36" -9d38'5"
Duckarfish Mon 6h44'46" -0d7'15"
Duckie Dra 20h20'58" 75d15'58"
Duck's Dragonfly Leo 9h32'46" 6d49'43"
Ducky Umi 16h16'41" 78d28'2"
Ducky Srp 18h41'23" -0d9'32"
Ducky Sgr 18h18'9" -32d40'11"
Ducky & FuFu Uma 10h45'18" 46d43'12"
Ducky & Sweetpea Forever Cyg 19h56'10" 37d38'16"
Ducrocq Céline Vir 13h52'55" -19d43'32"
Duda Cru 12h28'31" -57d9'38"
Duddy Dunn Uma 11h23'18" 35d15'10"
Dude Uma 13h25'35" 55d32'24"
Dude Uma 11h51'48" 54d14'32"
Dude Cep 22h20'50" 68d35'6"
Dude Le'Dawasch'73 Ori 5h33'44" 2d34'41"
DUDE! Where's My Star? Cyg 20h0'48" 35d49'40"
Dudeman Uma 10h13'47" 45d54'27"
DUDER Per 3h14'36" 47d32'35"
Duder Dra 16h28'9" 56d33'29"
Dudgeon5 Aqr 23h25'29" -14d55'24"
DUDIE-SUE Umi 13h13'18" 72d11'14"
Dudley Ari 2h6'50" 24d27'27"
Dudley Bentivegna Lmi 10h34'37" 38d23'38"
Dudley (Chilsham Shropshire Lad) Her 18h6'49" 21d47'16"
Dudley & Peggy Crawford Aqr 22h26'4" -1d32'46"
Dudley Sr. Gem 7h56'21" 19d51'21"
Dudzy Uma 10h52'3" 71d52'59"
Due Cuori Un Amore Her 16h12'8" 17d32'34"
Duene Frances Cyg 21h51'0" 46d0'1"
Duffey Col 5h39'35" -34d27'3"
Duffy Aqr 22h34'28" -0d48'34"
Duffy Ori 5h14'46" -0d49'36"
Duffy Gem 6h37'9" 19d44'50"
Duffy Ann Williams Aqr 21h54'13" -8d12'7"
Duffy Jackson Aql 19h45'25" -0d36'35"
DuffyCeithir JohnMaryCatherineBrian Uma 9h32'15" 61d37'18"
Dug Pomeroy: A Star Dad Lib 15h31'8" -11d53'17"
Dugan Leo 10h13'46" 21d22'33"
Dugan and Jovi Cnv 13h22'18" 29d20'29"
Duggan's Star Per 3h22'33" 41d29'2"
Dugu Ari 2h43'56" 26d20'15"
DUHBUL Sgr 19h33'28" -16d47'2"
Duke Uma 9h21'29" 61d31'33"
Duke Cnv 13h48'33" 38d27'26"
Duke Clark Gem 7h23'45" 32d8'50"
Duke Cloran Per 2h41'59" 57d1'52"
Duke Estherson Cam 4h1'41" 70d26'15"
Duke Haven James Cherpeski Ori 5h55'18" 20d55'22"
Duke JC Rood Leo 11h58'12" 10d48'42"
Duke LaBerge Uma 10h53'22" 43d23'53"
Duke Riley And 1h43'0" 50d15'31"
Duke T. Stemerick Vir 13h26'22" -17d38'19"
Duke Twins Her 16h21'31" 19d16'29"
Duke, 29/09/04 Peg 21h30'23" 23d26'55"
Dukey Umi 14h44'45" 75d18'47"
Dula Bug Uma 10h51'18" 57d12'42"
Dulani Umi 13h36'10" 75d6'12"
Dulce Ori 5h15'54" -8d42'28"
Dulce Uma 8h43'20" 68d44'9"
Dulce Lmi 10h29'49" 37d13'45"
Dulce de Leon Lyr 18h43'44" 31d41'58"
dulce et utile Lib 14h25'54" -13d57'9"
DuLce Maria Ca$tro And 0h17'5" 28d7'32"
Dulce Maria de las Mercedes Uma 9h29'12" 49d11'56"
Dulce Mia Lib 15h50'52" -12d43'50"
Dulce Paez Uma 12h50'46" 60d14'56"
Dulcey Marie Degilio Crb 16h3'4" 35d39'49"
Dulche Del 20h37'50" 16d7'8"
Dulcia_Serinitas Umi 15h59'19" 74d47'55"
Dulcie and Bill - The Lovers Cru 12h39'5" -59d16'8"
Dulcie Marie Bartsch ~ 24 Feb 1914 Cru 12h46'9" -61d3'32"
Dulcinea Cas 23h39'31" 57d48'27"
Dulcinea Uma 11h23'25" 34d2'21"
Dulcius ex asperis Peg 22h29'51" 12d42'28"
Dülge, Bernd Uma 8h11'33" 63d51'6"
Dumas-Muckerman Cyg 20h30'24" 35d55'58"
Dumb Dumb Lyn 8h36'6" 34d4'20"
Dumdog Slept On Cowch Cma 7h24'16" -32d50'18"
Dunatix Gem 6h46'23" 35d4'21"
Dunbar 2007 Lyn 7h51'39" 40d9'52"
Dunbar Wright Bostwick Uma 13h46'55" 55d1'6"
Duncan Sgr 19h20'48" -12d51'11"
Duncan Per 2h48'3" 53d41'48"
Duncan Ori 4h52'39" 10d34'1"
Duncan Alexander McMurrey Tau 5h41'30" 21d57'16"
Duncan and Renee Harrison Cyg 20h47'53" 37d21'19"
Duncan Billing Umi 15h27'31" 86d56'31"
Duncan bringer of Joy Umi 13h32'17" 74d13'16"
Duncan Campbell Anderson, Sr. Sgr 18h42'36" -23d38'47"
Duncan Ford Brackin Ari 2h2'54" 21d31'22"
Duncan J Stuart Cas 0h18'57" 50d41'26"
Duncan James Ari 1h50'43" 25d14'17"
Duncan James Umi 14h1'54" 72d33'33"
Duncan James "D.J." Dickenson Sex 9h44'47" 3d20'32"
Duncan James Nowling Ori 5h37'1" 4d45'59"
Duncan & Marianne Webb Cyg 21h52'33" 53d15'20"
Duncan McRice Leo 11h31'16" 17d9'14"
Duncan R. Seguin Her 16h38'34" 45d44'2"
Duncan Senghaas Lyn 7h1'32" 60d50'47"
Duncan & Sharon First Anniversary Cyg 21h46'13" 41d30'31"
Duncan Weeks Aql 19h42'38" 5d4'15"
Duncan's Eternal Light Gem 6h58'22" 19d34'59"
Duncanus Ivonas Gem 7h40'57" 33d7'38"
Dünckmann, Regina Uma 8h13'20" 68d28'14"
Dundee Abby Darby And 23h25'49" 47d10'52"
Dundika Crb 15h30'42" 30d17'44"
Dune Per 3h32'31" 48d34'37"
Dune Dolly Cas 23h21'25" 57d41'16"
Duner Uma 10h2'9" 51d53'25"
Duni Zenaye Ari 2h22'44" 23d50'11"
Dunja Kankaraš Ori 6h19'6" 16d34'9"
Dunja Kihr Her 17h25'10" 19d35'7"
DunJen Cyg 20h15'12" 47d52'32"
Dunks Psc 1h8'41" 4d23'19"
Dunkster Psc 0h20'14" 16d56'7"
Dunlap Forbes Cyg 19h13'18" 53d50'15"
Dunn Leo 10h10'19" 22d18'50"
Dunn Family Uma 8h37'37" 57d45'2"
Dunne Richardson Cyg 21h14'53" 42d29'15"
Dunstan Pasterfield Her 18h49'5" 15d35'32"
dunx Cra 19h5'11" -39d23'40"
Duong/Kamine Uma 10h7'28" 47d46'57"
DuPage SAC Teams Her 16h56'45" 28d30'44"
Duphorn, Alexander Ori 5h27'31" 14d36'27"
Dupla KáVé Cas 22h58'52" 58d44'12"
Dupuis Ford Lincoln Ori 5h1'55" 15d30'46"
Durango-Roca Lep 6h5'0" -26d32'22"
Durco Petronella Cru 12h56'51" -59d51'24"
DuRene Marie Cnc 8h42'48" 17d32'20"
Durham Castle - Gill, Callum, Caitlin And 2h10'38" 42d19'13"
Durham Regent Uma 11h44'46" 42d55'56"
Durjaya Neysmith Eri 4h13'45" -12d55'2"
Durka Durka Dra 20h16'42" 64d5'36"
Dusan Milosav Gojkovich Uma 13h20'17" 55d25'40"
Dusha Umi 15h30'46" 81d34'41"
Dusica Glisic Cas 1h26'12" 51d12'35"
Dusseau's ShiningStar Cam 4h30'30" 66d22'35"
Dussier Magali Del 20h36'15" 17d12'35"
Dustan Wells Costine Oph 17h43'24" -0d32'40"
Dustee Nikole Pho 0h15'23" -39d31'44"
Duster Uma 10h43'11" 57d13'55"
Dusti Leigh Taylor And 0h12'12" 27d59'33"
Dusti Marie Scruggs Vir 13h39'35" 2d19'23"
Dusti, Sam & Avery Ukeiley Uma 11h59'4" 60d41'31"
Dustie Katchmarik Tau 4h12'2" 17d52'31"
Dustin Boo 14h40'35" 36d1'6"
Dustin Lyn 7h13'58" 59d36'55"
Dustin Sgr 19h18'39" -12d31'54"
Dustin Cap 21h39'46" -12d41'7"
Dustin A Forsman Tau 4h29'42" 12d9'40"
Dustin A Harkness Psc 23h27'12" 2d23'31"
Dustin Aamodt Tau 4h28'7" 15d14'18"
Dustin Alan Splawn Ori 5h37'58" 0d21'41"
Dustin Alexander Padilla Lib 14h35'12" -17d45'35"
Dustin & Alicia Cox Mon 8h7'20" -0d48'56"
Dustin & Alyssa Uma 14h19'24" 59d34'7"
Dustin & Amanda Uma 9h35'32" 60d13'35"
Dustin and Danielle's Star Ari 2h9'20" 24d35'48"
Dustin and Holly Sge 20h19'43" 17d58'53"
Dustin and Jaime Mon 6h52'27" -0d54'55"
Dustin and Julie Ann - Endless Love And 1h15'57" 36d51'23"
Dustin and Julie Forever Cyg 21h44'49" 46d44'32"
Dustin and Sam's Eternal Love Sge 19h27'56" 18d30'51"
Dustin and Shannon Cyg 20h33'53" 50d9'35"
Dustin Boughton aka Midnight Cobra Cas 23h32'44" 55d8'15"
Dustin Carl Steele Her 16h22'38" 19d20'27"
Dustin "Charlie 8" Uma 10h23'50" 44d36'3"
Dustin Christopher Agner Uma 11h32'30" 32d8'35"
Dustin David Duncan Snadden Per 3h42'44" 40d58'49"
Dustin Don Riley Uma 9h9'4" 50d15'50"
Dustin "Dust" Sams 1982-2004 Leo 9h52'39" 19d33'30"
Dustin Edward Henry Leo 9h59'57" 16d5'18"
Dustin Edwin Wampler Aqr 21h49'10" -1d42'53"
Dustin Elliot Chrissley Ori 5h21'16" -4d35'42"
Dustin Eric Reinhart And 2h22'32" 47d41'26"
Dustin Flugrath Sco 17h21'25" -33d5'47"
Dustin Hamblin Per 4h7'47" 46d12'9"
Dustin J. Friesner Aqr 20h40'41" -2d35'13"
Dustin James Staska Ori 6h15'55" 14d54'57"
Dustin James Welch Cyg 20h10'29" 59d28'13"
Dustin Jennings Uma 9h6'14" 65d28'7"
Dustin Jimmie Provost Ori 5h32'14" -0d55'17"
Dustin Joseph Meraz Lmi 10h13'47" 33d20'16"
Dustin Keeton Ross Ori 5h29'10" 5d12'45"
Dustin Keith Loehr Her 18h1'0" 26d59'43"
Dustin Kell Sco 17h56'33" -31d10'1"
Dustin & Kelly... Inseparable Ori 6h2'49" 9d33'26"
Dustin Khoa Minh Ari 2h0'43" 22d43'36"
Dustin L. Stuppy Psc 0h57'16" 13d8'31"
Dustin Lane Lib 15h37'7" -12d41'0"
Dustin Leon Gravelle Uma 11h52'52" 43d53'0"
Dustin "Little-D" McCullough Psc 1h7'13" 3d35'57"
Dustin littles Schrantz Per 2h46'45" 53d47'1"
Dustin loves Torrey Lyr 19h8'30" 26d36'20"
Dustin M. Silvas Cap 20h56'38" -15d32'52"
Dustin M. Wood Boo 14h32'59" 26d59'59"
Dustin Michael Wing Leo 11h5'8" 20d9'27"
Dustin Moon Psc 23h34'4" 4d27'52"
Dustin N. Rutledge Ori 5h56'10" 20d53'28"
Dustin Olson Her 17h49'9" 45d20'24"
Dustin Parent Lib 15h0'24" -1d35'12"
Dustin Paul Husch's Heavenly Star Her 17h42'2" 15d6'25"
Dustin Paul, Beautiful Son Sco 17h5'32" -38d23'13"
Dustin Perry Aqr 22h7'43" -8d25'35"
Dustin Ray Brandt Gem 7h32'47" 29d18'49"
Dustin Rey Colgan Cnc 9h1'20" 18d49'32"
Dustin Roman Ward Mon 6h53'43" -1d27'38"
Dustin Schell Aqr 22h12'32" -3d21'52"
Dustin Scott Reece Birthday Star Sgr 18h26'48" -23d9'3"
Dustin Tyrone West Gem 7h23'35" 13d56'18"
Dustin Vail Ori 5h10'36" 3d11'7"
Dustin "VGN" Engel Lib 14h57'54" -17d42'49"
Dustin William and Erin Lee Forever Sco 16h51'11" -38d56'6"
Dustin William Eichler Her 17h29'51" 38d20'37"
Dustin Wilson Psc 1h11'21" 27d6'22"
Dustin's Brianna Vir 13h4'56" -21d21'51"
Dustopolis Uma 8h39'59" 49d3'40"
Dustora Lyn 7h40'58" 54d23'59"
Dusty Umi 16h32'31" 77d6'35"
Dusty Per 3h22'12" 42d8'8"
Dusty Ori 5h32'2" 4d43'22"
Dusty and Daisy Taylor Lyn 8h12'36" 50d12'50"
Dusty Braun Cam 7h19'57" 82d6'9"
Dusty Cheyenne Autumn Sco 17h45'19" -33d55'36"
Dusty Cooper Ori 5h29'54" 12d1'52"
Dusty Edwards Vir 12h17'48" 10d22'16"
Dusty Hall Umi 14h30'49" 75d39'9"
Dusty James Gleason Cap 21h45'17" -12d47'0"
Dusty L. Hendrickson Cap 20h58'25" -25d3'0"
Dusty Lashley Boo 14h47'36" 29d32'1"
Dusty Lee Her 18h14'12" 14d24'7"
Dusty Lovick Uma 8h34'20" 70d13'41"
Dusty Marie Creech Uma 12h1'48" 61d24'13"
Dusty Raye Uma 10h22'28" 56d46'3"
Dusty Reeves Sco 17h56'45" -38d1'53"
Dusty Rose Ori 4h51'19" 11d39'12"
Dusty Rose Lyn 7h11'22" 44d46'51"
Dusty Rose Lynn Lmi 10h24'35" 36d46'28"
Dusty Wall The Woman of my Heart Lib 15h34'32" -27d17'56"
Dustymyson 2 Cyg 20h51'38" 33d25'54"
Dustyn Ramey Uma 11h58'48" 31d54'6"
Dustyn's Glow Sgr 19h11'36" -20d54'7"
Dusty's Star Uma 11h51'42" 37d6'40"
Dutch Lady Milk Industries Berhad Ori 5h35'24" 2d34'23"
Dutchess Bolander Fawber Cnv 13h53'27" 34d20'24"
Dutchess Erica Godwin Cas 23h44'12" 55d50'38"
DutchessM27 Lib 16h1'43" -10d4'28"
Dutchman Umi 15h36'47" 81d8'4"
Duvessa Cyg 21h8'17" 30d20'45"
DuWayne A. Warner Ori 5h21'36" 1d11'50"
Düwel, Peter Uma 10h25'6" 67d18'32"
Duy yeu Quynh Anh Lib 16h1'8" -18d41'10"
Duyane and Inez Dorsey Uma 10h14'34" 68d41'55"
Duzie Cyg 21h23'23" 44d30'31"
DV Norfolk Ori 6h2'6" 6d28'38"
DVAC Sewol Lyn 6h31'34" 56d38'22"
Dve Andjele And 2h12'28" 47d56'21"
Dvora Aqr 22h38'47" -21d21'3"
Dvora Fields Leo 9h27'6" 28d40'31"
DVParisotto Col 5h43'49" -33d0'57"
DW 2112 Uma 10h18'33" 72d37'49"
DW Boyd Uma 12h45'19" 52d23'32"
Dwain P. Gambrell Leo 9h26'29" 15d27'1"
Dwan & Bill Twyford Uma 11h18'7" 57d54'22"
D'wanda Aur 6h3'5" 47d39'20"
Dwarf Lord The Forbidden Zone Dra 17h36'41" 65d50'24"
Dwayna Sue McCombs Cap 21h35'42" -19d50'7"
Dwayne Foster Cnc 8h23'35" 25d42'57"
Dwayne Lewis Bennett Her 17h35'32" 47d35'48"
Dwayne Lorenzo Valentine Murray Uma 8h51'53" 53d51'37"
Dwayne M. Hungerford Per 3h27'3" 44d19'6"
Dwayne Michael Kauffman Ari 3h18'54" 18d48'2"
Dwayne Scott Kurfirst Cyg 21h6'46" 46d44'3"
Dwayne "The Pilot" Bell Aqr 22h17'8" 2d31'33"
DWC 6956 Uma 12h26'59" 62d15'16"
D-Web Lyn 8h11'54" 34d20'5"
Dwels Uma 11h12'20" 34d8'49"
D.Whitehead43 Uma 10h18'1" 61d13'23"
Dwight Cyg 19h54'18" 38d33'50"
Dwight A. Rouse Her 18h13'8" 15d34'48"
Dwight (Abe) Myers Per 3h9'14" 47d12'13"
Dwight Argil Bartholomew "Daddy" Gem 6h34'4" 19d44'16"
Dwight David Cep 21h42'43" 67d10'39"
Dwight Dean Dunmoyer Her 17h26'28" 35d24'40"
Dwight Dow Psc 1h5'7" 31d58'27"
Dwight E. Jones Her 16h56'46" 35d27'55"
Dwight Edward Buck Menefee Vir 12h18'0" 7d25'31"
Dwight Harrah Tau 4h35'26" 14d37'47"
Dwight J. Guynn Boo 14h37'1" 35d29'25"
Dwight James Elwood Gem 6h49'47" 28d47'38"
Dwight L Tucker Ori 5h22'1" 4d49'52"
Dwight & Liz Umi 15h20'15" 75d6'20"
Dwight Loines Cas 0h25'55" 61d28'19"
Dwight Martin Cap 20h9'20" -25d27'36"
Dwight Raymond Yager Cnc 8h27'2" 12d23'54"
Dwight & Rosaria Ellis Cyg 19h39'14" 48d41'38"
Dwight's Peace Aql 20h0'47" 9d52'40"
DWKING EXCALIBUR Cep 21h54'30" 62d31'11"
DY+KS Together Forever in the Stars Per 4h24'12" 38d1'53"
Dyan Juliett Ari 3h15'1" 26d59'8"
Dyan Stott (McCrary) Umi 15h29'49" 69d38'23"
Dyan Young Umi 14h32'50" 75d33'3"
Dyana A. Silva Minor Cnc 9h11'55" 18d51'5"
Dyana Brown Cas 1h10'47" 57d58'31"
Dyani Anais Colon Sgr 18h28'59" -17d0'12"
Dyania Leo 9h37'0" 32d27'59"
Dyann Vir 13h13'51" -3d36'22"
Dyann Langella Crb 15h34'59" 27d18'16"
Dyann Roberts Tri 2h18'19" 33d37'30"
Dyki, Agata Leo 10h22'12" 18d28'2"
Dylan Del 20h39'3" 17d33'45"
Dylan Lmi 10h52'32" 33d39'48"
Dylan Lmi 10h23'20" 31d28'38"
Dylan Cas 1h48'57" 59d24'58"
Dylan Lib 15h24'36" -7d13'3"
Dylan Aqr 23h49'34" -11d50'58"
Dylan Umi 15h19'34" 73d9'17"
Dylan Sgr 18h56'57" -30d5'48"
Dylan A. Hill's Birthday Star Uma 11h15'51" 65d48'33"
Dylan Aaron Gaissert Uma 9h28'41" 69d5'57"
Dylan Alexander Smith Ori 5h27'42" 2d22'26"
Dylan Alexander Zammit Cru 12h36'15" -58d25'46"
Dylan & Amara Adams Aql 19h16'10" 5d34'16"
Dylan and Barrett Talbert Uma 9h48'51" 56d26'32"
Dylan Andrew Stephen LaBrake Aql 19h26'6" 6d49'5"
Dylan Anthony Wickowski Uma 10h15'20" 51d40'25"
Dylan Bailey Dra 14h38'41" 55d16'35"
Dylan Baker's Oliptesa And 23h21'38" 44d51'23"
Dylan Benjamin McCarthy Uma 9h24'1" 69d56'8"
Dylan Bowne And 0h57'38" 23d23'13"
Dylan Bradden Mace Ari 3h9'59" 17d30'36"
Dylan Bruce Wilkins Ori 6h5'19" 17d18'47"
Dylan Burns Uma 10h30'31" 69d27'16"
Dylan C. Cooper Ori 5h22'6" 1d8'27"
Dylan Catledge Vir 14h29'52" 0d41'29"
Dylan Charles Grady "Pickle" Aur 5h18'44" 41d49'29"
Dylan Charles Lynn Aql 19h17'32" 8d51'31"
Dylan Christopher Agro Cnc 9h20'36" 29d8'41"
Dylan Christopher Egan Higgins Boo 14h54'34" 24d35'32"
Dylan Christopher Markle Lib 15h44'54" -8d38'45"
Dylan Christopher Phoenix Hinson Lyn 7h19'9" 56d47'37"
Dylan Curtis Edwards Lib 15h24'13" -9d1'29"
Dylan Cy Addington Ori 5h44'7" 0d33'45"
Dylan Dakota Gane Cra 19h1'37" -39d33'54"
Dylan Daniel Chowka Umi 13h15'2" 73d33'11"
Dylan Darlyn Victor Tau 3h58'57" 27d52'52"
Dylan Dennis Feller Ari 2h13'46" 25d27'58"
Dylan Douglas Dyer Sco 16h25'0" -19d29'12"
Dylan Dwyer Boo 14h50'25" 22d4'54"
Dylan Edward Dennis Arnold Per 4h10'27" 44d21'37"
Dylan Elika Roach Cmi 7h26'42" 1d59'38"
Dylan Elizabeth Kirk Vir 13h24'55" 12d37'37"
Dylan Elyse Sampert Dra 18h49'50" 74d18'52"
Dylan Emery Ori 6h6'8" 14d23'58"
Dylan Enodoc Rowles Dra 17h32'40" 57d16'29"
Dylan Erich Cannon Per 2h52'31" 56d45'16"
Dylan F. Rothschild Cnc 9h4'16" 15d7'20"

Dylan Fazio Winters
Cnc 8h45'45" 25d26'21"
Dylan Frederick King
Cep 22h27'34" 63d47'52"
Dylan Gage Mills 8-12-1994
Leo 10h24'10" 9d43'44"
Dylan Gregory Bruckner
Sgr 19h5'32" -12d42'42"
Dylan Griggs
Sco 17h56'24" -31d6'47"
Dylan Haynes Davis
Gem 6h44'39" 14d23'14"
Dylan Hays Agron
Sco 16h5'28" -12d46'39"
Dylan Henry David Wilder
Per 2h20'49" 53d51'50"
Dylan Hinda Pollack
Lyn 7h20'44" 59d29'10"
Dylan J
Dra 20h18'47" 64d18'25"
Dylan J. Burgin
Pho 0h15'34" -40d40'12"
Dylan J. Hunt
Ori 6h11'18" 9d14'51"
Dylan J Lyonnais
Uma 11h35'2" 64d7'21"
Dylan Jahanbigloo
Her 16h33'37" 34d31'0"
Dylan Jake Stilwell
Ori 6h17'4" 5d52'46"
Dylan James
Sco 16h52'8" -38d59'56"
Dylan James Adkison
Uma 11h58'43" 37d32'52"
Dylan James Austin
Her 18h31'36" 17d55'57"
Dylan James Baer
Uma 10h45'34" 60d11'52"
Dylan James Coatney
Her 17h45'31" 48d16'1"
Dylan James Hayden
Her 17h32'18" 44d25'48"
Dylan James Manning
Uma 12h15'0" 54d0'11"
Dylan James Mayotte 3/10/94-3/05/04
Sgr 18h35'32" -23d6'7"
Dylan James Pelkey
Tau 5h40'18" 22d19'31"
Dylan Jean Eli Platt
Cyg 19h38'17" 38d6'5"
Dylan Jean Younie
Uma 10h13'40" 49d28'35"
Dylan & Joelle's Love Star
Uma 9h25'48" 65d28'49"
Dylan John
Uma 13h55'43" 49d11'55"
Dylan John
Leo 11h45'1" 26d4'40"
Dylan Joseph
Ari 3h6'9" 27d53'41"
Dylan Joseph Baldasare
Umi 14h30'7" 77d33'21"
Dylan Joseph Borges
Sgr 19h4'24" -26d23'12"
Dylan Joseph Fontenot
Sco 16h53'37" -42d25'8"
Dylan Joseph Harvatt
Leo 11h57'31" 22d32'10"
Dylan Joseph Kellenberger
Umi 14h13'7" 67d1'21"
Dylan Joseph Ljiljanich
Ari 2h5'55" 21d30'25"
Dylan Joseph Patton
Aur 5h28'28" 47d50'13"
Dylan Joseph Stewart
Her 17h54'7" 34d15'58"
Dylan Justice Donovan
Aqr 22h12'8" -0d56'0"
Dylan Keith Chesser
Sco 16h29'58" -27d41'40"
Dylan L. Quinerly
Her 17h39'50" 19d57'17"
Dylan Larsen
Cap 21h38'58" -14d40'22"
Dylan LaVone Bowser
Cyg 21h32'33" 34d27'6"
Dylan Lee
Umi 15h31'39" 68d29'11"
Dylan Lee Anthony Quinerly
Cep 1h42'18" 78d9'24"
Dylan Lee Redd
Lib 15h24'47" -21d58'26"
Dylan Leidig
Cnc 8h15'5" 32d42'22"
Dylan Liam Crackower
Her 18h29'58" 14d41'58"
Dylan Luca Melillo
Lib 15h9'45" -6d53'57"
Dylan Mae Ann Johnson
Lib 15h2'27" -11d21'59"
Dylan Marley Steinberg
Her 16h31'14" 35d43'32"
Dylan Marquis
Lyn 8h20'36" 34d2'45"
Dylan Matthew Fitzpatrick
Umi 16h25'34" 76d10'11"
Dylan Matthew John Drake
Ori 5h55'30" 12d20'35"
Dylan Matthew Tavares 2000-2002
Umi 16h39'0" 77d21'9"
Dylan Matthew Tuch
Per 3h30'4" 34d34'51"
Dylan Matthew Vaughn
Ori 4h51'30" -0d38'0"
Dylan Matthew Yuknavage
Ari 2h11'4" 15d53'26"
Dylan Maya Tyler Fischer
Uma 11h50'46" 31d10'10"
Dylan McCoy Wesley
Ari 2h4'45" 27d32'42"
Dylan McDonie
Tau 3h48'33" 18d21'46"
Dylan McGovern
Ari 1h54'21" 21d59'38"
Dylan Michael
Umi 16h28'11" 78d6'5"
Dylan Michael Clarke
Leo 11h37'25" 15d26'18"
Dylan Michael Frise
Sgr 19h35'20" -17d23'17"
Dylan Michael Lee St. Louis
Lib 14h49'43" -1d35'46"
Dylan Michael Warner
Cma 7h8'40" -28d37'8"
Dylan Michaela Weber
Umi 16h11'56" 73d5'42"
Dylan Miles Dorman
Uma 10h34'23" 43d51'23"
Dylan Miller
Uma 11h36'7" 62d14'58"
Dylan Montoya
Lmi 10h23'1" 36d10'27"
Dylan Nicholas Gordon
Lmi 9h27'15" 33d43'51"
Dylan Nicholas Gordon
Umi 13h53'39" 86d20'44"
Dylan Palmer, Gentleman & Scholar
Pho 0h43'34" -47d7'45"
Dylan Park
Umi 14h40'38" 73d25'6"
Dylan Patrick Clyne
Her 18h28'9" 22d3'39"
Dylan Patrick Janssen
Psc 0h6'56" 7d43'58"
Dylan Patrick Jones
Lib 15h17'56" -4d36'8"
Dylan Patrick Steele
Uma 9h54'20" 68d19'31"
Dylan Paul Ludbey
Cru 12h44'46" -55d49'39"
Dylan Paul MacArthur
Lib 14h24'7" -19d51'0"
Dylan Piquininho
Dra 10h49'4" 78d19'19"
Dylan R. Richardson
Tau 5h3'48" 19d23'16"
Dylan Rae Dick
Umi 16h12'39" 73d55'26"
Dylan Rae Silverthorn
Sgr 17h52'59" -17d53'40"
Dylan Rathburn Perry
Lyn 7h34'14" 59d37'42"
Dylan Robert
Leo 10h15'24" 9d27'16"
Dylan Robert Brown
Per 3h22'21" 42d27'43"
Dylan Robert Pirylis
Her 16h32'10" 14d22'35"
Dylan Robert Woods
Leo 10h15'40" 13d59'12"
Dylan Ross
Cnc 9h18'53" 17d49'59"
Dylan Roy Jones
Cam 3h55'52" 66d12'57"
Dylan Sage Beasley
Psc 1h9'18" 8d45'2"
Dylan Sage Frusciante
Cnc 9h16'51" 24d45'19"
Dylan Sage Sachs
Psc 1h8'30" 28d13'44"
Dylan Skye Panganiban
Psc 2h0'59" 7d14'56"
Dylan Stanbridge
Leo 10h9'39" 20d37'51"
Dylan Starr McNair
Peg 22h7'8" 7d28'15"
Dylan Steven Pavlesich
Cnc 8h17'49" 15d2'20"
Dylan T. Kezele
Aqr 22h18'59" -15d45'42"
Dylan Thai Phan
Sco 17h29'17" -38d50'32"
Dylan Thomas Benjamin
Her 16h18'55" 10d7'34"
Dylan Thomas Cavanagh
Psc 1h6'44" 31d16'46"
Dylan Thomas Hocevar
Tau 5h38'58" 27d8'14"
Dylan Thomas Merz
Gem 7h33'3" 19d44'50"
Dylan Thomas Scripture
Psc 1h33'10" 16d1'5"
Dylan Thomas Timothy Schaefer
Umi 13h56'5" 73d49'52"
Dylan Thomas Vigil
Cnc 8h45'59" 18d6'49"
Dylan Timothy Kai Riley
Scl 0h11'36" -26d48'15"
Dylan Tobias Jenkinson
Lmi 10h13'23" 31d11'43"
Dylan Torres
And 23h26'31" 48d34'31"
Dylan Veronica Leary
Lib 15h46'54" -23d21'11"
Dylan Wade Heins
Cep 21h31'54" 63d40'23"
Dylan Watson
Per 3h9'30" 45d57'45"
Dylan Wells
Uma 13h55'48" 51d0'16"
Dylan Williams
Gem 7h51'1" 28d54'25"
Dylan Wommack
Her 17h49'18" 42d20'55"
Dylan Wyrick
Uma 11h24'2" 43d46'53"
Dylan Young
Umi 16h21'42" 77d39'59"
Dylanbear
Umi 13h51'14" 78d28'10"
Dylanie's Star
Umi 16h34'48" 75d36'24"
DylanJohnHaile
Lyn 7h10'0" 57d57'37"
Dylanova
Ori 5h53'26" 20d48'0"
Dylan's Divine Destiny
Cnc 8h45'35" 12d18'1"
Dylan's Dream
Ari 2h32'18" 10d47'28"
Dylan's Dynamic NOVA
Tau 4h41'10" 4d11'29"
Dylan's Shining Star
Vir 12h28'30" 11d56'33"
Dylan's Sky Treasure
Uma 9h26'17" 66d19'20"
Dylan's Star
Vir 12h34'57" 1d37'11"
Dylan's Star
Cnc 8h40'7" 31d40'10"
Dylanstar
Aql 19h16'1" 4d39'8"
Dylcia Daniela Ramos
Cnc 8h23'53" 9d37'21"
Dyllan
Lmi 10h27'37" 29d42'7"
Dyllan Elizabeth Hersey
Ari 3h18'50" 18d58'25"
Dyllan Joseph
Cyg 20h37'0" 51d12'25"
Dyllis Martin
Sgr 18h1'56" -32d50'22"
DYLNNM
Dra 17h59'44" 57d19'58"
Dylon Cargill
Cen 12h34'1" -35d8'43"
Dylon Jacob Robinson
Lmi 10h4'30" 40d6'31"
Dyl's Star
Per 4h28'56" 44d25'25"
Dynamic Wellness
Cma 6h55'22" -30d27'6"
Dynamic-Dylan
Dra 20h36'23" 79d31'34"
Dynargie
Uma 8h51'5" 46d33'43"
Dynell Sarasvati Yeshe Webber
Lib 14h53'8" -3d45'38"
Dyrilious
Uma 8h15'47" 62d27'35"
Dyroff 1
Dra 18h18'37" 77d6'17"
Dyson - Rotherham
Cyg 20h13'20" 47d43'59"
Dzeneta My Love
Aqr 20h39'23" -8d59'24"
Dzentutu Mawuli Kpoh
Vir 12h35'10" 12d17'6"
dziadzi
Uma 12h29'14" 60d45'49"
Dziadziu and Haley
Lyr 19h20'1" 37d50'6"
Dzidzia
Aur 5h46'43" 53d58'48"
Dzióbka
And 1h20'13" 49d46'37"
Dzsenifer, a mosolygós csillag
Aqr 21h17'22" -11d10'34"
Dzulija Plusnina
Cnc 8h12'17" 27d59'7"
E
Gem 6h46'3" 20d47'16"
~8~
Vir 12h42'31" 13d6'14"
"E"
Aql 19h52'17" 5d6'25"
e
Cam 4h30'31" 65d44'49"
8120181918
Lyr 18h46'6" 31d34'50"
8/21/79 Giovanni Favis 11/8/04
Ori 6h3'42" 10d45'59"
823-Ro
Lyn 8h13'30" 51d49'31"
82Sapphire0930 Marzel Lea Williams
Lib 15h41'56" -19d9'41"
831 - AAJ
Uma 10h40'6" 54d8'40"
831 ~ Jimmy and Christa Always
Tau 5h48'36" 17d21'7"
84 Bob Bjaranson, Valentine Forever
Aqr 23h39'41" -24d29'52"
8/5/06-KC Erika'Lee 18-20
Cap 20h46'4" -14d43'41"
8-5-1946-Patience-60
Leo 11h4'44" 10d13'26"
8700 Orion Place
Ori 6h12'39" 8d30'44"
8704
Per 3h19'40" 41d35'45"
88 Wraith
Ori 5h35'42" -2d4'33"
88MELISSA88
Her 17h37'58" 32d11'45"
88TAMMY88
Lmi 10h23'33" 33d38'51"
E & A Moose
Cyg 19h32'17" 52d16'4"
E. A. "Pete" Peterson
Per 1h40'11" 54d29'40"
E Allan Farnsworth JR
Per 2h53'39" 52d41'19"
E. Angel
Cnc 8h47'26" 31d24'12"
E. Anthony Rodriguez
Aur 6h3'37" 47d39'44"
E Bunny aka Crystal Coleman's Star
Lep 5h15'56" -12d58'54"
E. Burns McLindon - Stellar Father
Psc 0h18'5" 8d4'31"
E. Burton Wallace
Cyg 21h58'41" 50d55'40"
E. C. Cheechie
Cru 12h48'10" -60d42'48"
E Chelsea Campbell
Uma 9h39'34" 49d46'22"
E. Courtney Pearson
Cnc 9h14'27" 7d54'17"
E. D. J. R. V. A.
Sgr 18h33'44" -26d49'1"
E. E. S. Z. Zarbun
Uma 12h37'33" 52d36'47"
E. Emory Davis
Ari 2h23'41" 24d33'33"
8 Eshghe-Javdaneh 8
Ori 5h57'19" 13d11'11"
E. Florence Morseth
And 2h34'7" 39d37'22"
E. Fred Luebke
Sgr 19h23'40" -17d12'45"
E. Gayle McDaniel Long
Col 5h46'44" -36d17'16"
E+H Always and Forever
Pho 0h39'50" -42d30'37"
E. Henry Mellusi
Boo 14h37'24" 27d32'17"
E. Idolina Dickman
Lyn 8h0'42" 54d28'1"
E. J. Downton
Cyg 20h5'51" 38d14'0"
E J V 6 5 6
Tau 3h47'15" 13d43'33"
E. Jane Whetzel
Ari 3h13'50" 18d48'7"
E. John and Barbara Themios
Her 17h9'27" 32d1'51"
E. John DeYot
Ori 5h12'24" 7d41'14"
E. Kaitrin Kuchera
Leo 9h36'45" 27d38'30"
E. L. A.
Cap 21h9'12" -15d12'4"
e Leapheart
Cnc 8h31'40" 7d8'11"
E. Leroy Austin Jr.
Uma 9h43'46" 64d51'35"
E*Louks
Lib 14h35'12" -12d22'49"
E M F
Uma 12h41'44" 56d55'4"
E. M. G. I
Leo 9h28'34" 18d50'49"
E M P
Lib 15h2'47" -2d11'41"
E&M Storer
Uma 9h54'29" 46d12'16"
" E. Marie "
Lib 15h9'0" -11d59'49"
E. Martin von Kanel
Sgr 18h13'17" -22d34'39"
E. Michelle Ruddick
Tau 5h23'48" 24d17'59"
E/Mo
Uma 12h2'42" 62d14'43"
8 Months of Heavenly Bliss
Cap 20h21'28" -11d3'56"
E R & S L Gaylor
Gem 7h2'16" 16d55'46"
E. Rena
Mon 6h42'58" -0d21'50"
E. Robert Gamble
Ori 6h1'42" 20d41'27"
8 Sideways
Dra 17h29'24" 51d52'52"
E Stella
Crb 16h8'31" 36d23'11"
E. T. Masters
Uma 10h13'6" 57d30'44"
E. Thomas Tsai. Our eternal light
Sgr 17h53'2" -29d53'12"
E. Victor and Betty S. Wilson
Col 5h42'16" -34d40'54"
E. W. Johnson 9-12-21
Aqr 21h10'53" -1d58'58"
E. Wayne Christopher
Uma 14h23'15" 59d40'56"
E. West Hornor, Jr.
Ori 6h15'48" 10d25'58"
e^2 = dtr
Uma 12h48'44" 60d22'41"
EA
Peg 22h39'23" 20d4'6"
Ea
Cam 3h52'31" 56d57'51"
EA MEL OU
Peg 21h38'25" 23d40'10"
Eabha
Her 17h56'33" 28d30'47"
Eabha
And 23h12'8" 47d20'21"
EAC
Sco 16h8'22" -23d29'21"
Éadaoin O'Raw
And 2h27'35" 46d20'41"
Eadin
Lyn 8h18'10" 42d1'19"
Eagan
Uma 10h0'29" 45d30'48"
Eagle 1 Class of 1956 RT66Joplin,MO
Aql 19h43'22" 7d54'53"
Eagle 101 Wayne "Sarge" Randall
Tau 3h31'13" 17d46'10"
Eagle 6 Mark-David
Aql 19h9'56" -0d22'27"
Eagle Eye 76
Vir 14h6'24" -15d47'59"
Eagle Scout Brandon Ryan Poli
Cnc 8h52'24" 31d17'37"
Eagle Scout Christopher DeGirolamo
Aqr 22h9'52" 1d41'56"
Eagle Scout James T. Fernandes III
Cap 21h42'1" -12d11'56"
Eagle Scout Mark S. Gymiski
Cnc 9h7'35" 31d41'50"
Eagle Scout Nicholas Joseph Poli
Aqr 22h37'5" 1d11'0"
eaglepoint1993
Sco 16h6'20" -11d9'5"
Eagle's Pride
Aql 19h4'35" 14d59'23"
Eakdom The Blumenstein Family
Uma 10h35'45" 46d46'20"
Ealasaid
Cma 7h18'50" -18d8'15"
Eamon Buehning
Psc 0h58'42" 15d48'0"
Eamon Day Nathan McHugh
Lib 15h51'38" -6d56'17"
Eamon Fitzgerald
Cyg 19h32'57" 28d55'34"
Eamonn & Amanda
Cyg 20h3'6" 46d50'5"
Eamonn Branchaud
Cap 20h37'52" -22d47'34"
Eamonn Corcoran
Her 16h30'58" 20d35'39"
Eamonn Owenie Ned
Aqr 21h43'44" -0d11'50"
Eamonn Roche
Sex 10h21'38" -0d37'18"
Ean James Smith
Ori 5h8'25" 16d5'11"
Ean M. Davis
Cap 21h53'42" -24d28'43"
Ean Phillips
Umi 16h22'46" 72d42'30"
Ean Scott McGuinness
Ori 5h34'14" 14d6'55"
Eáránë Itarillë Tiwele
Aql 19h34'32" 4d14'57"
Eärendil
Crb 15h43'33" 30d43'41"
Earl and Alice Benson
Sge 19h40'32" 17d23'38"
Earl and Alice Kalchik
Sge 19h45'34" 17d55'59"
Earl and Gloria McCormick
Lib 15h35'19" -10d53'35"
Earl and Mary Fryer
Aql 19h42'12" 7d13'58"
Earl B. Chappell III
Gem 6h8'43" 25d34'18"
Earl Bakken
Eri 3h58'32" -0d22'1"
Earl Berry
Tau 3h40'22" 26d4'20"
Earl Brasfield's Star
Ori 5h25'57" 0d59'33"
Earl (Bud) Nelson
Cep 23h5'4" 75d24'5"
Earl C. Woodall
Cyg 21h12'26" 52d25'42"
Earl & Cathryne Crocker
Uma 11h7'51" 42d50'38"
Earl Cleveland Bryant II
Uma 9h38'9" 47d38'27"
Earl Cranston Scurlock, Jr.
Aql 19h43'36" -0d7'36"
Earl Creger
Cep 3h47'9" 80d54'13"
Earl & Cyndi Rediske
Tau 4h16'4" 3d38'29"
Earl D. Simmons
Sgr 18h15'56" -26d23'9"
Earl E. Kuecken
Her 16h59'24" 18d24'28"
Earl Franklin Staley
Her 16h46'7" 36d50'19"
Earl Gosvener
Per 2h12'4" 51d58'55"
Earl Gronniger
Boo 15h32'14" 47d44'44"
Earl Howell 12-3-1988
Her 17h8'21" 31d5'1"
Earl Hulsey
Nor 16h7'33" -48d1'57"
Earl James Abner
Per 2h44'59" 42d44'1"
Earl James Folmer
Cap 21h53'53" -17d37'41"
Earl Laks
Cep 21h54'55" 61d2'28"
Earl Lee Jones
Mon 7h6'7" -0d2'25"
Earl & Lela Avery
Cyg 20h24'16" 35d8'9"
Earl & Mabellynn, EverMore 29.06.05
Cru 12h35'41" -58d54'33"
Earl & Mary Nickerson Gold Star
Uma 8h18'44" 69d33'1"
Earl Mayeresky's Lucky Dawg
Cnv 12h36'45" 44d49'50"
Earl Miller Jr. Dad My Shining Star
Lib 15h13'45" -6d12'28"
Earl & Nancy Stewart *1-16-2003*
Uma 8h55'36" 60d17'14"
Earl Orson Greene
Cep 20h19'2" 60d20'16"
Earl Palmer Glasscock, Jr.
Uma 11h57'40" 30d50'43"
Earl & Patricia Johnson-We love you
Uma 10h15'48" 54d28'44"
Earl Peter Wondra
Uma 12h33'12" 56d47'35"
Earl Peterson
Crb 16h19'39" 33d20'38"
Earl S. Holtsclaw
Aql 19h16'44" -0d22'6"
Earl & Shirley Groff
Cyg 20h14'6" 32d44'15"
Earl & Sue Lane
Lyr 19h15'41" 28d24'50"
'Earl The Pearl'
Cep 22h6'33" 53d37'28"
Earl Timmerman
Her 16h33'25" 35d17'59"
Earl Vogelsong Lindsay, Jr.
Aql 19h19'21" 14d28'42"
Earl Walters
Gem 6h53'3" 12d7'18"
Earl Wayne McMillan
Uma 11h18'24" 35d57'7"
Earl Wilson
Cma 6h40'49" -17d12'24"
Earle 04 28 1926
Uma 13h36'1" 49d32'14"
Earle and Mary Litzenberger
Leo 9h22'52" 15d45'5"
earle and rachel mccants
Uma 11h53'3" 38d52'9"
Earle Martin Runnion
Aqr 22h21'42" 2d19'6"
Earle Morrow Cassidy
Leo 11h36'49" 10d59'30"
Earle O. and Alba G. Fromm
Cyg 20h28'8" 46d59'12"
Earle P. Stover, Jr.
Her 17h28'10" 14d43'3"
Earle & Valti
Cyg 20h56'44" 34d58'44"
Earle Vernon
Gem 6h48'20" 16d57'29"
Earleen Kelley Miller
Lib 15h46'50" -17d31'56"
Earlene
Lib 14h54'2" -6d59'1"
Earlene Virginia Lee Houston
Aqr 22h31'45" -22d43'47"
Earlene-Jessie
Gem 7h2'27" 22d51'5"
Earlene-Precious 4/05/44
Uma 10h27'29" 41d46'16"
Earlexia Montoya Norwood
Cas 23h32'22" 56d21'12"
Earley Family
Lyn 8h23'51" 35d22'54"
Earli May
Ari 3h6'15" 29d26'33"
Earl's Girl
Cas 0h37'38" 48d35'33"
Earl's Pearl
Sgr 18h12'51" -18d7'39"
Earl's Star
Uma 11h14'46" 46d54'43"
Earlyn C. Pike
Crb 15h53'13" 27d43'29"
Earnestine May Gwin
Gem 6h34'10" 22d50'1"
EaRnldKyA
Cas 23h20'10" 55d10'52"
Earnie Swartz
Psc 23h2'33" 5d43'7"
Earth
Lyn 7h21'7" 51d19'56"
Earth 2
Aqr 22h50'32" -7d51'16"
Earth Angel
And 1h4'4" 39d52'56"
Earth Angel Laura
And 23h48'27" 40d8'38"
Earth Mother Annyce
Sco 16h18'25" -11d58'49"
Earth to Ori
Ori 5h38'50" -0d56'0"
Earthator
Sco 16h12'43" -18d1'18"
Earth's Angel E.L.L.
Aqr 20h45'6" -7d18'23"
Earthworms Marketing PTE LTD
Umi 14h17'24" 75d41'49"
EASLA
And 0h32'8" 42d9'50"
East Bridgewater Class of 2007
Uma 10h30'13" 56d14'21"
East Coast Firestar
And 23h20'25" 42d18'21"
East Side Center Spirit Center Star
Uma 12h17'11" 54d36'45"
East, Nicholas
Uma 11h14'6" 34d35'41"
easter
Sge 19h11'20" 19d0'3"
Easter Baby Ella Jayne
Ari 2h19'45" 17d22'53"
Easter "Esther" Rufino
Uma 11h42'24" 43d12'33"
Eastern Decorator Sdn Bhd
Ori 5h36'58" 2d26'0"
Easton Patrick Greco
Umi 14h35'3" 77d27'20"
Easy Odie - der kleine Bär
Tau 5h7'56" 25d36'25"
E.A.T. Konstantine
Sco 16h5'8" -13d54'13"
Eat, Drink, and be Merry
Uma 9h32'2" 45d52'37"
Eavan Ty Johnson
Dra 12h57'53" 70d31'16"
Eavy Noelle Christensen
Lib 14h35'26" -12d17'15"
ebab
Lib 15h28'38" -3d58'25"
Ebba
Uma 11h30'47" 42d8'34"
Ebbie & Dinky
Lib 15h19'21" -6d16'36"
EBeau
Umi 15h21'26" 83d2'11"
Ebeling, Klaus-Konrad
Ori 6h19'29" 14d23'52"
Eben Lassen
Per 3h17'33" 51d29'39"
Eben Matthew Ohime
Cap 20h20'48" -26d18'59"
Eben Moore
Uma 13h52'4" 49d30'33"
EBEO
And 2h21'0" 50d23'6"
Eberhard Grobbecker
Uma 11h9'5" 58d40'24"
Eberhard Klaas
Uma 9h58'23" 57d5'47"
Eberhard Koch
Uma 11h3'1" 58d11'39"
Eberhardt (aka) "MICKEY" 1938
Leo 9h46'9" 25d48'4"
Ebert, Hans Jürgen
Uma 10h31'15" 68d27'51"
Ebo Kobina Jeketi Addae
Umi 15h38'58" 84d33'40"
Eboladoc
Gem 7h5'36" 14d57'58"
Ebonie Nicole Irvine
And 23h59'10" 45d12'44"
Ebonnae
Cas 1h12'52" 60d23'33"
Ebony
Tau 4h12'32" 23d50'14"
Ebony Christine Evans
And 0h24'17" 37d22'34"
Ebony Denise Dabney
Leo 10h19'43" 18d28'50"
Ebony Eyes
Psc 1h0'55" 8d3'8"
Ebony Gary
Leo 11h23'53" 18d23'11"
Ebony Haigh
And 1h25'44" 43d53'40"
Ebony Harris
Cap 21h14'44" -27d10'50"

Ebony & Ivory
Sco 16h37'2" -41d18'55"
Ebony Jade
Cra 18h15'3" -41d10'21"
Ebony Kelly
Cru 11h57'26" -64d4'21"
Ebony Mollie Allen
Lib 15h19'7" -22d29'45"
Ebony Rose Guard
Lib 14h33'18" -12d47'51"
Ebony's Cosmic Butterfly Star
Crb 15h53'51" 27d10'26"
EBSM 7.8.06
Aql 19h35'57" 12d24'3"
Eburn Lara Grace
Cyg 20h31'32" 33d3'14"
Eby Jane Anthony
Leo 11h11'41" -2d27'20"
Ebzabah
Uma 12h15'48" 54d20'39"
ECB91047
Vir 13h1'41" -19d29'2"
Ecccko 5
Ori 5h19'51" 0d13'37"
Ece
Cap 21h22'39" -19d57'56"
ECERIN03
Aqr 23h56'20" -10d3'39"
ECH Galaxy
Uma 12h57'54" 57d19'3"
Echo
Lyn 7h34'18" 53d48'40"
Echo
Leo 11h55'51" 22d9'53"
Echo Dawn
Lib 14h52'38" -2d30'19"
echo eleven
Sgr 19h4'13" -19d56'11"
Echo Kiddo
Leo 9h35'31" 28d40'48"
Echo Phelps
Lyn 7h16'55" 48d3'39"
Echo Vanderpool
Per 4h13'32" 32d0'48"
Echo Volfson
And 1h34'30" 42d49'5"
Echo Von Schnee
Uma 9h35'6" 43d11'13"
Echo Wood
Uma 10h31'23" 51d27'16"
Echobe
Mon 6h52'59" -0d24'42"
Echoe Marie Averill
Lib 14h26'30" -17d30'39"
Ecila Antunes
Tau 3h51'26" 28d10'20"
Ecilop
Uma 11h26'24" 44d8'55"
Ecin
Eri 4h45'24" -27d7'22"
Eck, Dietrich
Uma 8h37'52" 59d25'25"
Eckard Lammel
Uma 10h35'59" 41d10'13"
Eckardt, Lutz
Uma 13h57'42" 56d7'14"
Eckerslyke
Uma 13h50'7" 50d48'22"
Eckert
Lmi 10h20'46" 34d17'42"
Eckert
Uma 9h58'56" 53d15'21"
Eckert, Anka
Sgr 18h7'30" -27d12'33"
Eckert, Lucienne
Uma 8h47'23" 49d2'52"
Eckhard Fick
Uma 11h43'50" 38d29'59"
Eckhard Rudolf Watson
Cep 22h13'17" 60d54'55"
Eckhardt
Uma 9h35'39" 42d32'47"
Eckles/Rohlff
Per 2h17'37" 51d43'51"
Ecky Dee
Lmi 10h10'59" 28d44'46"
Eclipsed
Cnc 9h1'5" 15d11'20"
École Saint-Patrice
Uma 11h32'12" 56d2'53"
ECP's Star
And 2h12'23" 49d5'7"
Ecstasy
Psc 0h18'14" 7d14'9"
Ecstasy 1210
Leo 11h2'11" -2d49'8"
Ecy
Ori 6h17'54" 7d57'2"
Ed
Cep 23h41'55" 74d14'20"
Ed Abbott
Aur 7h14'38" 43d6'14"
Ed Alsafi
Her 18h19'22" 21d12'36"
Ed and Brandy, Buddies Forever
Lyr 18h46'44" 31d8'20"
Ed and Diane
Ori 5h22'31" 7d55'17"
Ed and EllenJeanne's 50th Star
Lyr 18h43'42" 38d39'58"
Ed and Judy Manley
Vir 12h19'17" 5d0'13"
Ed and Kaye Wicander
Cap 20h17'50" -14d9'40"
Ed and Linda Maina
Cyg 19h36'35" 27d57'23"
Ed and Marg
Cyg 19h35'21" 30d44'32"
Ed And Patty Papesh *LOVE FOREVER*
Cyg 20h43'27" 45d44'59"
Ed and Petzel Klugman
Boo 14h51'9" 27d18'33"
Ed and Ruth Fairbanks
Uma 12h23'29" 53d32'46"
Ed and Tina Zielomski
Crb 16h19'56" 35d4'30"
Ed & Ann
Gem 6h30'10" 23d54'42"
Ed & Anne Roelli
Umi 15h30'2" 74d41'8"
Ed & Audrey
Cyg 20h38'11" 31d12'26"
Ed Babylove Oates
Cma 6h55'32" -17d5'54"
Ed Baehr
Ari 2h32'33" 22d0'49"
Ed "Bear" Seward
Mon 7h21'59" -2d56'31"
Ed Biezenthal
Lib 15h29'9" -20d20'7"
Ed Bitondo
Per 3h22'45" 48d36'5"
Ed Boen
And 1h44'4" 45d58'6"
Ed Boggs
Her 17h20'36" 17d3'0"
Ed Bostetter
Sgr 18h26'36" -32d1'18"
Ed Brown
Per 3h3'25" 46d17'21"
Ed Brown Jr.
Sgr 17h57'50" -28d53'14"
Ed "Bud" Wilke
And 1h33'10" 47d12'1"
Ed Bullock
Uma 11h31'48" 60d12'52"
Ed & Carol McGuire 50th Anniversary
Uma 8h40'26" 71d23'32"
Ed Cassidy
Dra 16h38'20" 66d40'23"
Ed Cefola
Boo 14h39'55" 19d40'26"
Ed & Celina Haines
Sgr 18h36'3" -28d46'48"
Ed & Cindy Bauer
Cyg 20h24'55" 54d25'57"
Ed & Clerisa
Crb 15h37'39" 29d22'36"
Ed Cox T.H.E. Robotics Team 2005
Ori 5h28'4" 5d4'50"
Ed D. Vaniman
And 1h31'5" 34d55'28"
Ed Daller
Tau 4h2'22" 12d22'8"
Ed & Debbie Mac
Aur 5h45'46" 53d42'44"
Ed Dilworth's Star
Uma 10h29'24" 40d5'18"
Ed Domine
Uma 9h14'22" 63d5'29"
Ed & Donelda Gallagly
Lyr 18h47'57" 39d59'43"
Ed & Dot
Lyn 7h33'22" 43d23'26"
Ed Dugan "Our Special Star"
Cep 22h40'2" 86d28'0"
Ed Dutchess
Her 17h37'7" 26d34'24"
Ed Eliasson
Per 3h46'0" 51d38'56"
Ed Erbe
Cnc 9h0'50" 12d13'3"
Ed Fasulo
Per 4h16'44" 45d5'22"
Ed & Gay's 20th Anniversary Star
Gru 21h48'56" -36d49'37"
Ed H. Perales
Sgr 18h5'8" -24d15'33"
Ed Harrison
Cap 20h22'20" -13d40'5"
Ed Heaberlin
Per 3h7'40" 39d6'38"
Ed Herider
Uma 8h13'50" 64d51'40"
Ed Hubbard
Boo 14h44'13" 30d9'35"
Ed Hyman
Cep 3h23'57" 80d56'41"
Ed & Irene Morris-Together Forever
Eri 3h49'37" -0d3'46"
Ed is very handsome
Ari 3h5'57" 29d24'58"
Ed J McCormick-my shining star
Lyn 7h54'43" 55d40'51"
Ed Jacobs
Cam 3h59'59" 69d0'22"
Ed Jen Nick And Chris Ochoa Family
Uma 10h27'41" 56d26'34"
Ed & June Miller
Cyg 21h15'23" 32d26'7"
Ed K. Ford
Her 17h38'55" 32d28'13"
Ed & Karen Until The 12th Of Never
Uma 8h54'11" 55d13'48"
Ed Keeler
Her 17h51'17" 44d41'41"
Ed Krafczyk
Cyg 19h35'18" 29d11'24"
Ed Kueffer
Mon 7h35'0" -0d35'43"
ed l stankeys star
Per 3h12'18" 41d21'28"
Ed Large
Her 16h53'19" 40d36'4"
Ed London
Her 18h14'46" 15d33'21"
Ed & Lori
Per 3h14'25" 41d22'52"
Ed & Madge Hanly - 54 years of Memories
Lyr 18h49'12" 35d0'53"
Ed Maggi
Uma 11h38'3" 54d48'22"
Ed Marinucci
Her 16h52'23" 34d18'47"
Ed "MC squared" Santilli
Sgr 19h16'25" -14d3'30"
Ed McVey Family Tree Star
Tau 3h33'6" 18d9'58"
Ed & Meg Together Forever
Sge 19h48'35" 18d19'12"
Ed 'MSL' Bouchard
Aqr 22h32'55" 1d22'32"
Ed n Susan's love light
Crb 16h23'23" 38d27'55"
Ed Nessel
Crb 15h40'10" 35d29'8"
Ed Oquendo
Ori 6h4'22" 20d42'18"
Ed & Pat Street's 50th. Anniversary
Umi 15h26'25" 70d56'32"
Ed & Peggy Sumner
Tau 3h58'58" 11d38'13"
Ed Quek
Cep 21h46'19" 55d53'10"
Ed & Renee's Star
Cyg 21h9'52" 35d56'14"
Ed Robert's Star
Uma 8h47'35" 69d29'21"
Ed Ross
Cep 21h29'55" 60d52'48"
Ed Ross
Her 17h7'36" 31d22'51"
Ed Ryan
And 23h12'51" 49d51'2"
Ed S. Grant & Consuelo Ortiz
Sge 19h53'48" 18d48'3"
Ed Sanders "The Bear Hunter"
Ori 6h17'39" -0d22'22"
Ed Satterfield
And 0h16'6" 31d58'33"
Ed Schneider
Uma 10h23'56" 52d20'3"
Ed Schulman Vishnu
Uma 11h34'36" 50d47'10"
Ed Sherman - Legacy of a Teacher
Her 17h21'29" 38d12'45"
Ed Shines On
Cas 23h43'15" 55d44'41"
Ed Squires
Her 17h11'39" 31d26'40"
Ed Starink
Ori 6h1'38" 13d22'35"
Ed & Stella Rickwood
Cma 7h19'31" -30d22'31"
Ed St.Germaine
Lib 14h49'39" -5d18'57"
Ed + Terry Dillehay
Cyg 19h32'16" 30d28'40"
Ed Tubel
Lib 15h23'19" -8d2'26"
Ed Tuller
Uma 13h39'6" 58d4'26"
Ed & Val
Dra 19h2'56" 70d56'39"
Ed & Vickie Nunn
Cyg 21h50'25" 49d56'20"
Ed Washington
Uma 10h7'8" 61d19'22"
Ed Zoladz
Col 5h21'37" -31d50'23"
Ed, Alice & Peachy
Cyg 20h11'33" 43d19'18"
Ed, Donna and Red Nolte
Cma 6h27'52" -29d7'21"
ed154
Gem 6h46'56" 23d14'9"
Eda Duran
Umi 15h43'41" 82d50'11"
Edaliz
Peg 23h14'42" 18d23'32"
edaney
Uma 11h23'33" 35d53'30"
Edda
Cas 0h6'46" 58d45'24"
Edda
Cma 7h19'22" -13d42'53"
Edda
Cap 21h58'35" -19d48'59"
Edda 12244
Uma 12h37'54" 58d16'38"
Edda The Little Goat Girl
And 23h24'20" 38d58'53"
Eddeb Eternal Love
Cnc 8h16'57" 18d39'10"
Eddie
And 0h28'29" 29d31'4"
Eddie
Ari 3h26'53" 22d16'23"
Eddie
Psc 23h28'0" 4d10'30"
Eddie
Per 2h45'44" 56d31'54"
Eddie
Her 17h40'28" 32d47'57"
Eddie
Uma 13h38'42" 55d59'10"
Eddie
Uma 10h8'7" 61d28'43"
Eddie
Lib 15h50'20" -13d20'14"
Eddie
Vir 13h31'50" -9d49'17"
Eddie
Sgr 18h9'37" -28d39'21"
Eddie Alton Wilson
Uma 11h17'32" 58d9'40"
Eddie and Gracie
Uma 10h13'35" 53d45'25"
Eddie and Vals Shining Star
Gem 7h12'49" 20d30'24"
Eddie B Sterling
Ori 5h38'5" 6d43'33"
Eddie Budd
Per 2h26'0" 55d46'39"
Eddie C Millar
Cyg 20h36'16" 53d50'32"
Eddie Cappiello
Per 3h23'30" 51d58'9"
Eddie Cerise, Jr. light of my life
Her 17h6'30" 15d38'22"
Eddie & Christina Hall
Cyg 20h8'35" 44d37'52"
Eddie Ciampa
Uma 11h43'59" 49d48'37"
Eddie Clark
And 1h29'49" 45d38'4"
Eddie Cramer
Lib 15h21'50" -4d10'2"
Eddie & Doreen Wheeler
Ari 2h36'49" 29d26'6"
Eddie Dougan
Cep 22h4'45" 56d52'31"
Eddie Fernandez 1979-2002
Sco 16h11'28" -14d6'39"
Eddie Finn
Lib 15h11'46" -26d59'57"
Eddie Fisher
Ori 6h12'25" 12d48'0"
Eddie Fiskaa
Uma 10h0'27" 65d38'4"
Eddie Franco
Her 16h59'10" 34d6'21"
Eddie Freeman
Ori 5h6'0" 5d0'44"
Eddie Fuentes
Cap 21h5'26" -16d3'56"
Eddie Gertken
Her 17h12'7" 19d18'15"
Eddie Graves
Tau 4h9'42" 5d50'24"
Eddie Hall, just passing!
Per 3h47'17" 48d32'8"
Eddie Hernandez
Aqr 22h36'50" -1d5'14"
Eddie Hrebic
Per 3h27'8" 48d20'3"
Eddie Huggins
Lyn 7h15'33" 53d12'34"
Eddie & Jacqueline
Eri 4h23'57" -21d44'27"
Eddie Jean Norris
Cnc 8h50'47" 26d11'23"
Eddie John Miller
Tau 4h19'41" 3d46'8"
Eddie Jr.
Leo 9h32'20" 27d55'55"
Eddie Keher
Psc 0h51'55" 16d10'22"
Eddie King Jr.
Aql 19h21'24" -5d23'45"
Eddie Kirsch
Aqr 22h53'18" -9d38'8"
Eddie L
Leo 11h11'17" -1d57'34"
Eddie Lamont Frazier
Cam 4h29'1" 68d18'42"
Eddie Lawrenson
Cep 22h39'34" 60d12'29"
Eddie Lazarski
Her 17h24'53" 23d32'56"
Eddie Lee
Leo 11h10'51" 15d42'9"
Eddie Lemoine's Longest Drive
Aql 20h2'59" 15d25'4"
Eddie Lewis - Daddy's Star
Per 2h25'12" 56d17'23"
Eddie & Lisa
Vir 11h43'49" 3d20'34"
Eddie & Lisa Berg's 1st Anniversary
Cyg 19h47'11" 35d32'14"
Eddie Loves Momy
Psc 1h19'11" 26d2'31"
Eddie Luchansky
Sgr 17h50'16" -25d38'45"
Eddie M. McNally
Per 4h25'37" 39d56'13"
Eddie "Mac" McVerry
Uma 10h5'30" 57d9'33"
Eddie Monaco
Sgr 18h20'16" -32d8'8"
Eddie Monroe's Star
Cap 21h24'40" -17d18'11"
Eddie Morse
Lib 15h21'55" -17d42'14"
Eddie Mosqueda
Uma 14h22'3" 60d55'12"
Eddie "Mr. G." Greenwald
Tau 4h11'0" 2d56'11"
Eddie Muir
Ari 2h0'50" 22d0'5"
Eddie My Love
Psc 23h1'10" -0d16'46"
Eddie My Lovey
Psc 1h22'2" 22d0'42"
Eddie My Star
Uma 11h23'41" 54d3'33"
Eddie "O"
Per 2h41'14" 57d12'21"
Eddie Ortega - June 18, 1997
Ori 5h25'57" 5d40'30"
Eddie "Pa" Toler
Cyg 19h44'53" 36d18'22"
Eddie Pietro Johnson
Umi 13h21'40" 71d43'37"
Eddie Pozos
Ser 15h53'28" -0d34'44"
Eddie Rios
Lyn 8h10'15" 56d52'58"
Eddie Sakhleh
Ari 2h29'18" 11d7'36"
Eddie Sanchez
Cnc 8h54'38" 28d6'22"
Eddie Settle-Smith
Uma 10h23'17" 46d37'18"
Eddie Smith
Cep 21h50'32" 62d28'27"
Eddie spaghetti
Umi 17h3'11" 76d7'5"
Eddie Sr.
Aur 5h54'57" 36d47'37"
Eddie T. Potts Jr.
Boo 14h49'27" 24d47'28"
Eddie Turner
Her 18h49'43" 24d51'41"
Eddie Walsh
Uma 11h40'56" 31d3'26"
Eddie Whitmore
Ori 5h53'32" 12d46'17"
Eddie Wolly Pog
Per 3h26'12" 44d51'38"
Eddiekins
Per 4h47'38" 46d25'43"
Eddie's First Birthday
Uma 11h11'51" 56d14'50"
Eddie's Starr
Per 2h15'34" 54d4'34"
Eddison Alleen Theberge-LeDuc
Cap 20h15'16" -11d23'22"
Eddleblutanus
Cnc 8h19'26" 8d32'41"
Eddo Plowman
Crv 12h24'40" -14d33'13"
Eddy
Cma 6h54'24" -17d36'54"
Eddy Agueros
Uma 9h13'23" 47d36'17"
Eddy Dal Santo B.A.A.
Lib 15h55'31" -14d35'0"
Eddy Delsignore
Per 3h43'0" 37d42'56"
Eddy Juan Park
Ori 5h5'44" 3d47'54"
Eddy Meneses
Tau 3h58'52" 7d20'9"
Eddy Puckett, my Special Ed
Her 17h58'6" 49d36'8"
Eddy Welch
Boo 15h10'55" 42d21'50"
Eddy Winata Spectacular
Ari 2h42'14" 11d29'21"
Eddy's BooBoo
Per 4h22'28" 44d23'56"
Edee Cathryn Clark
Psc 0h8'3" 12d22'24"
Edelina Lanzon Vitkus
Lyr 19h12'13" 26d17'13"
Edelmira
Cnc 8h33'39" 10d49'53"
Edeltraud McMullan (Australia)
Psc 0h56'46" 9d51'52"
Eden
Leo 11h27'52" 13d23'26"
Eden
Crb 15h53'24" 36d35'58"
Eden
Leo 9h37'28" 31d52'28"
Eden
And 23h25'32" 41d34'57"
Eden
And 23h16'35" 51d41'45"
Eden
Mon 6h49'5" -0d49'14"
Eden
Cyg 21h22'49" 53d43'42"
Eden Callaway
Tau 4h20'9" 25d49'50"
Eden Cara
Lyn 7h7'10" 59d34'31"
Eden Delaine Newman
Lyr 19h1'12" 33d15'46"
Eden Elizabeth
Cnc 8h9'51" 9d4'7"
Eden Ellis
Cnc 8h12'46" 10d22'36"
Eden Espinosa
Uma 11h41'11" 43d18'59"
Eden Grace
Lmi 9h30'20" 33d38'4"
Eden Grace
And 23h43'45" 41d50'47"
Eden Hana
Sco 16h6'53" -10d11'52"
Eden Hatle
Tau 4h28'34" 3d55'2"
Eden Joi Harris
Sgr 19h29'28" -24d21'56"
Eden Joseph Roe
Uma 8h37'38" 53d16'34"
Eden Joseph Roe
Uma 9h56'3" 56d53'16"
Eden Lily Hargreaves
Aur 5h53'13" 51d41'57"
Eden Malone Wettstone Swails
Psc 0h53'7" 11d9'16"
Eden Marguerite Lorentzen
Lyn 8h8'30" 53d43'16"
Eden Marie Blumberg
And 0h14'9" 43d44'56"
Eden Mark Bailey
Umi 13h23'47" 74d14'1"
Eden Mok
Lib 14h33'6" -24d54'22"
Eden Patrick Gillispie
Lyn 8h26'25" 42d5'19"
Eden Rebecca Shaw
Vul 21h2'6" 23d3'3"
Eden Rose
And 1h25'37" 44d50'35"
Eden Rose
Umi 13h38'47" 74d40'32"
Eden & Sydney
Cyg 19h46'40" 36d46'51"
Edes feleségem, Farkas Gabriella
Ari 3h28'51" 29d57'17"
Édes Kincsem, Joci Csillaga
Uma 8h35'33" 58d26'5"
Édes Panni Mami és Béla Papi
Cas 1h43'24" 77d35'3"
Édesanyám Csillag
Sgr 18h19'1" -25d43'8"
Édesapám
Uma 13h47'35" 51d35'29"
Edgar
Uma 8h36'29" 65d26'18"
Edgar
Leo 11h19'1" -0d29'2"
Edgar A. Franz
Ori 5h58'35" -0d42'8"
Edgar Amrein
Leo 11h54'1" 25d3'55"
Edgar Anahit Kasumian
Sgr 19h9'46" -23d52'2"
Edgar Bölt
Ori 5h55'47" 11d54'3"
Edgar (Boysie) Pike Shine on
Her 16h48'13" 41d37'35"
Edgar Butler Jr & Homie Mae Smith
Uma 10h50'38" 68d7'17"
Edgar C Kolb
Tau 5h3'43" 28d2'42"
Edgar Crum "The Big bud"
Ori 6h18'25" 8d37'41"
Edgar Diggs
Ori 4h51'59" 12d49'17"
Edgar Donte Tapia
Cnc 8h39'56" 15d51'12"
Edgar Gross
Her 18h35'42" 17d9'12"
Edgar & Henrietta Micallef
Mon 7h30'35" -0d15'51"
Edgar I. Wilson
Cap 20h29'35" -10d40'37"
Edgar Isaac Friedmann
Aur 5h36'55" 47d34'43"
Edgar Isaac Tyson Jr.
Leo 10h56'31" -4d44'28"
Edgar Jimenez.The Beautiful.wbs
Uma 10h42'18" 59d1'53"
Edgar John L'Heureux, Jr.
Tau 4h18'23" 19d0'33"
Edgar Kurz
Uma 10h14'53" 47d52'51"
Edgar Lawrence Cross
Per 3h49'4" 38d8'59"
Edgar Leonardo Yrusta
And 1h13'26" 47d14'18"
Edgar Manucharyan
Uma 11h29'12" 60d47'6"
Edgar Mathieu-Henri Chaix
Sgr 19h10'13" -24d58'17"
Edgar Miceli y Melissa Trevino
Uma 11h34'30" 32d48'34"
Edgar P. Cambere
Uma 13h19'27" 59d17'33"
Edgar R. Starkey, Sr.
Umi 14h31'26" 78d51'15"
Edgar Rousseau
Cas 23h33'12" 55d24'1"
Edgar Shlomo Bondy
Boo 14h42'41" 54d27'29"
Edgar - The Star of Pegasus
Peg 0h10'57" 16d40'23"
Edgar Thomas Padgett
Her 16h35'42" 34d30'1"
Edgar Ulmi, 19.03.1973
Sge 19h43'57" 18d16'19"
Edgard C. Barreto
Uma 10h31'38" 62d33'35"
Edgardo Silo 8 agosto 1957
And 23h6'49" 51d34'9"
EdHead
Lib 15h39'40" -17d0'37"
Edi Stähli
And 0h29'1" 46d22'14"
Edie
Aql 19h31'36" 11d57'32"
Edie
Cas 1h35'43" 61d17'16"
Edie
Sco 17h14'4" -33d49'48"
Edie Akin
Sgr 18h52'59" -24d31'49"
Edie Blakely
Ori 5h16'8" -7d10'0"
Edie Mae Carruthers
And 1h24'54" 50d19'26"
Edie May
And 22h59'14" 50d46'58"
Edie of the Stars
Cas 0h23'18" 62d55'39"
Edie Rodgers
Uma 9h40'2" 67d33'38"
EdieMirman
Eri 3h44'10" -0d9'2"
Edie's Daddy Hits 50!
Vir 12h58'38" 11d7'53"
Edie's Love
Cma 6h53'24" -19d45'27"
Édike és Pampó
Uma 9h58'8" 45d38'35"
Edil William Roldan Rivera
Aqr 23h16'50" -20d54'29"
Edin Hotilovac
Per 3h9'53" 41d33'22"
Edina
And 2h35'35" 40d27'59"
Edina
Ori 5h55'8" 2d30'5"
Edina
Cap 20h17'28" -11d57'11"
edina
Cas 23h58'26" 56d42'25"
Edina et Doda
Leo 10h17'12" 13d4'44"
Edina S. Fields
Gem 7h44'4" 33d17'5"
Edina, Child of the Sun
Uma 8h28'4" 60d6'54"
Edio Forchielli
Umi 15h2'43" 73d19'51"
Edison Fan
Sco 17h58'27" -40d5'30"
Edit Urban
Vir 12h54'20" 12d59'58"
Edita
Psc 1h14'33" 28d8'54"
Edith
Ari 3h0'11" 29d5'33"
Edith
Per 3h22'5" 47d6'34"
Edith
Uma 8h47'21" 62d17'42"
Edith
Uma 11h32'44" 64d56'6"
Edith 263
Dra 18h49'20" 59d27'27"
Edith Alicia Viloria
Del 20h38'35" 15d59'12"
Edith Anne Ackenhusen Haskett
Aqr 22h42'56" -17d5'12"
Edith Arnold
Lmi 10h16'47" 31d28'51"
Edith Barnett Grandy
Cas 2h27'19" 65d12'47"
Edith & Bernie Shoor
Crb 16h12'8" 34d51'37"
Edith Buffington
Crb 15h44'59" 35d47'11"
Edith C. Caricofe
Lyn 7h44'33" 39d20'24"
Edith C. Kern
Cap 20h28'50" -9d26'28"
Edith Challis Erickson
Lib 15h33'20" -6d45'20"

Edith Chauvel
Lyr 18h23'52" 32d47'11"
Edith Denis
Umi 14h35'24" 80d5'6"
Edith Domke
Lyr 18h46'56" 34d46'26"
Edith Ferrett Duncan
Cas 1h8'40" 51d25'38"
Edith Fischer
Cas 23h50'54" 57d36'26"
Edith Harvey
Peg 22h36'26" 23d15'37"
Edith Helen Pettus
Cyg 21h7'10" 46d44'35"
Edith Herman
Cas 1h8'20" 62d9'39"
Edith Hughes
Lib 15h25'6" -8d29'36"
Edith Ibarra
Gem 7h1'53" 20d47'10"
Edith J. Boettcher
Cas 0h11'7" 56d15'42"
EDITH JAMES
And 23h3'4" 47d6'56"
Edith Joy Williams
Cam 4h15'50" 73d39'28"
Edith Keresi
Cas 1h15'16" 57d21'29"
Edith Klee
Uma 11h0'32" 50d38'11"
Edith L. Pacillo
Vir 11h57'1" -0d26'1"
Edith M. Veloz
Sgr 18h4'36" -17d13'15"
Edith Mae Buanjug - Shining Star
Lep 5h18'0" -15d27'52"
Edith Mae Tanner
Cnc 8h20'26" 15d3'12"
Edith Marie
Lyn 6h51'10" 57d51'4"
Edith Marie Gore
Tri 2h30'17" 31d23'46"
Edith Mary Crank
Cas 0h19'27" 51d25'49"
Edith May
Cas 1h28'29" 55d42'15"
Edith May
Uma 10h0'17" 44d42'15"
Edith May Pickett
Cas 23h16'59" 54d59'15"
Edith Nan Page Stevens
Cam 6h26'37" 64d43'8"
Edith Neal Montague
Aqr 21h13'10" 0d54'34"
Edith P. Hurst
Leo 11h23'22" 17d44'35"
Edith Paetzold Novak
Peg 22h27'23" 7d26'24"
Edith Parenti
Lyr 18h46'46" 41d20'51"
Edith Robertson
Uma 9h53'41" 47d40'14"
Edith Robinett
Uma 8h37'5" 51d4'4"
Edith S. Sands
Aqr 23h17'33" -17d9'6"
Edith Schultheiss
Boo 14h19'7" 13d59'22"
Edith Shaw
Uma 13h39'35" 58d40'1"
Edith Smuin Starke
Cap 20h22'25" -23d2'38"
Edith Stuart Phillips—ESP
Aql 19h49'11" 1d48'33"
Edith Sutphin
Com 13h6'10" 16d33'10"
Edith Szolna
Lyn 7h10'12" 58d22'29"
edith + thomas
Ori 6h3'32" 21d22'30"
Edith Thyra Lind Ell
Leo 10h37'48" 18d29'53"
Edith Van Riper-Haase
Per 3h19'29" 47d56'14"
Edith Velasco & German Padilla Star
Umi 14h23'20" 79d21'38"
Edith Wander
Com 13h7'31" 26d51'26"
Edith Yanson "Stellar Teacher"
Cas 0h39'18" 58d27'46"
Edith Zeller
And 1h7'31" 46d53'4"
Edith-Ann Faoro
Tau 4h13'24" 5d12'27"
Edith-R
Mon 7h37'19" -1d6'43"
Edith's Star
Sco 16h41'57" -42d2'54"
Editke
Cnc 9h7'42" 8d58'35"
EdjamuhkaytedQT-DMSJ
Her 17h32'59" 26d55'58"
Edje
Lyr 19h6'41" 35d8'8"
Edker L. Tinnerman
Tau 4h28'30" 20d10'17"
EDL
Sco 16h47'26" -35d20'19"
Edlin Bradley
Leo 10h38'5" 22d21'6"
Edman
Psc 1h15'39" 16d13'54"
Edmea Costa
Cam 5h35'19" 71d27'16"
Edmée Vanier
Uma 10h59'29" 46d17'31"
Edmond Charles Gregorian
Her 17h17'57" 31d39'14"
Edmond Francis Rondepierre
Vel 10h23'16" -41d27'58"
Edmond M. Layton
Uma 11h14'19" 35d20'20"
Edmond Malka
Uma 11h53'49" 62d26'31"
Edmond Naughton
Uma 8h45'40" 46d53'29"
Edmond Sullivan
Aql 19h31'29" 5d51'12"
Edmondo
Lyr 19h1'36" 42d42'17"
Edmondo Faienza
Uma 12h40'53" 55d53'43"
Edmund A. Ryniec, "Mooner"
Sgr 18h27'43" -16d28'59"
Edmund A. Vail
Sco 17h45'42" -40d50'52"
Edmund Albert Santella
Uma 9h16'15" 62d38'43"
Edmund and Susan Kaminski
And 23h14'40" 51d9'52"
Edmund Creelman
Per 2h15'2" 56d44'52"
Edmund Eugene Crowell
Aur 5h14'40" 29d20'23"
Edmund F. Picano
Leo 11h32'46" 2d14'43"
Edmund Francis Vogel III
Psc 1h3'7" 6d36'15"
Edmund James McLelland
Ori 5h30'6" 1d54'28"
Edmund John Hill
Pic 4h44'36" -46d40'26"
Edmund Joseph Doherty Jr.
Uma 11h29'0" 40d4'27"
Edmund - Katie's Light & Life
Sgr 18h6'47" -28d4'19"
Edmund Kraemer
Cnc 8h25'23" 22d56'33"
Edmund L. Gritton Jr.
Sco 16h10'48" -13d39'49"
Edmund L. Whitehead
Cnc 8h53'30" 7d37'11"
Edmund Lisecki
And 23h13'45" 44d9'2"
Edmund Mallamo
Uma 8h43'45" 56d10'27"
Edmund Martins
Per 3h27'30" 52d21'38"
Edmund Reynolds Wood
Gem 7h30'14" 28d20'48"
Edmund Skrocki II
Vir 13h27'58" -8d9'8"
Edmund Yee
Leo 11h11'23" 22d31'28"
Edmundo M.
Crb 16h21'45" 35d0'49"
Edmund's Light
Umi 15h6'14" 72d20'36"
Edna
Sgr 18h32'28" -16d57'28"
Edna
Pyx 8h35'41" -31d2'10"
Edna
Cas 0h29'11" 55d28'37"
Edna
Com 13h14'37" 17d46'37"
Edna
Ari 2h33'32" 25d37'37"
Edna & Allie Shorette
Cyg 20h48'17" 46d54'26"
Edna Amador
Vir 12h49'36" 7d10'1"
Edna Barz
Lyr 18h50'16" 41d48'21"
Edna Bell Strickland Smith
And 23h14'26" 47d48'13"
Edna Belle Ball Cornish
Cnc 9h7'40" 30d26'40"
Edna Briand
Lyr 18h39'7" 36d48'13"
Edna Carolan
Uma 12h27'12" 61d25'15"
Edna Catherine Arnold
Uma 10h36'6" 44d28'7"
Edna Comp
Uma 13h35'25" 58d41'18"
Edna Cordova
Lyr 18h58'46" 33d33'41"
Edna Darlene
And 23h12'43" 42d27'42"
Edna Diane Vossler
Uma 10h15'46" 58d52'47"
Edna Donnelly
Umi 14h31'21" 69d0'57"
Edna Downey's Guiding Light
Cet 1h23'44" -0d24'47"
Edna Faubus Sparks
And 22h59'44" 47d12'31"
Edna Foy
Her 16h20'7" 5d51'18"
Edna - Granny
Sco 17h24'13" -42d11'3"
Edna Haslam
Cyg 21h5'10" 47d5'24"
Edna Hertz
Cas 1h40'19" 58d1'43"
Edna Hinckley 1919
Cas 0h54'23" 70d2'58"
Edna Irene Saylor Thatcher
Ori 5h38'30" 5d12'52"
Edna j
Lyn 9h27'5" 40d43'54"
Edna Joe Soulmates
Uma 10h2'7" 67d20'8"
Edna J.R. Farrell
Uma 12h41'45" 60d21'49"
Edna Katherine Goff
Aql 19h42'3" -0d2'4"
Edna Knight
Cas 1h29'53" 57d28'13"
Edna Kukulski
Lib 15h55'34" -9d34'39"
Edna Lois Upchurch Jeffreys
Her 18h50'14" 25d28'35"
Edna Louise Langwell
Cap 20h19'34" -12d42'17"
Edna M. Lom
Ori 6h4'10" 12d1'27"
Edna Mae
Sgr 19h58'27" -30d25'4"
Edna Mae Mondello
Cas 0h36'54" 51d27'46"
Edna Marie Hanes
Lyr 18h48'41" 43d29'15"
Edna Mary Bishop
Cas 23h39'50" 55d52'56"
Edna Matsue
Lyn 7h21'11" 56d15'32"
Edna May
Cas 2h30'15" 65d50'51"
Edna May
Sgr 19h30'10" -41d5'30"
Edna May Oubridge
Cas 0h52'16" 73d16'24"
Edna Mitchell Grant
Ari 2h43'49" 14d39'11"
Edna Omalley Clinton Pope
Uma 12h31'5" 61d7'36"
Edna Pickens
Uma 10h1'47" 56d7'44"
Edna Q. Adams
Crb 15h48'45" 27d13'21"
Edna Robles Brutus
Lib 15h6'41" -10d31'22"
Edna Rosa Cruz
And 23h2'3" 50d46'35"
Edna Ruth Woofter
Lyn 7h40'7" 45d0'19"
Edna Sanderson
Cru 12h38'37" -60d28'47"
Edna Schwartzman
Cas 2h17'2" 64d1'2"
Edna Shultis
And 23h12'52" 51d41'0"
Edna - "Sid's Shining Star" & Grandchildren
Cas 1h18'29" 69d41'15"
Edna Soderquist
Tau 4h20'15" 25d56'41"
Edna Tenney
Cas 1h17'33" 57d19'22"
Edna Weatherbee
Psc 0h50'6" 9d37'34"
Edna West Ryals
Cas 1h48'37" 61d32'12"
Edna Wilson
Aur 5h10'18" 28d42'35"
Edna90
Cnc 8h25'22" 14d4'22"
Edna's Light
Cyg 20h14'32" 33d56'55"
EdNay
Cam 4h54'57" 70d55'32"
Edoardo
Umi 14h50'39" 70d58'47"
Edoardo Martinelli
Lyr 19h15'11" 29d1'50"
Edoardo Vittorio
Peg 22h52'19" 26d50'24"
Edouard Belanger
Peg 21h36'43" 19d16'3"
Edouard Edilber
Sgr 18h38'2" -28d17'26"
Edouard Fornas
Uma 12h54'7" 54d41'50"
Edouard Sworowski né de Gayant
Sco 17h42'3" -40d41'12"
Edra Napier
Cam 4h33'54" 67d32'21"
Edrian Oliver
Cam 5h1'40" 74d37'40"
Edrick
Lib 15h43'36" -6d47'35"
EDRICK1293 Edward Patrick Marshall
Cap 20h45'54" -17d36'28"
Edris Kapchan
Sco 16h9'44" -10d43'42"
Edris Lenora
Cnv 12h37'40" 39d44'42"
Ed's 40
Per 3h4'55" 55d24'34"
Ed's Forever Perservering Spirit
Aqr 22h36'5" 0d16'49"
Ed's Own Star
Sco 16h10'38" -13d12'38"
Ed's Place
Aql 19h31'45" 16d10'20"
Ed's Soaring Spirit
Sco 16h45'35" -33d56'3"
Ed's Special Star
Her 17h16'23" 19d55'36"
Edson Ariza
Umi 15h46'27" 78d3'43"
Edstar
Dra 19h2'0" 74d31'3"
Edsue Woodruff
Uma 11h43'8" 43d2'7"
Eduard & Erika Glavinskas
Cyg 20h23'23" 48d29'41"
Eduard Gschrei
Uma 10h56'28" 69d32'26"
Eduard Theodore Beringer
Cnc 9h2'49" 20d43'36"
Eduarda Graber
Cas 0h48'44" 69d26'24"
Eduardito Gonzalez
Umi 15h8'38" 76d8'0"
Eduardo
Boo 14h39'13" 28d2'27"
Eduardo
Boo 14h36'46" 34d16'11"
Eduardo A Agudelo
Dra 9h57'9" 74d36'33"
Eduardo A Chavez
Tau 5h26'33" 18d55'19"
Eduardo Diosdado Dillon
Her 17h57'55" 29d6'14"
Eduardo Echagarruga
Dra 19h59'28" 78d16'56"
Eduardo & Elisalex
Cra 18h48'27" -42d28'14"
Eduardo Feitosa
Aur 6h30'53" 35d9'49"
Eduardo Gonzalez
Pho 23h51'11" -51d28'52"
Eduardo Guilherme
Per 2h31'15" 52d20'26"
Eduardo Luis Saavedra
Ori 5h56'15" -0d16'36"
Eduardo Moses Guzman
Boo 14h20'11" 33d52'49"
Eduardo Sansone
Her 18h28'18" 25d43'29"
Eduardo Seijo
Uma 11h22'56" 52d11'9"
Eduneo
Sgr 17h52'2" -29d18'21"
Edvin & Selin
Sco 17h6'8" -39d58'45"
EDW
Cnv 12h22'2" 33d40'10"
Edw. Duda, Jr~Bunnie & Edw. Duda,Sr
Cyg 19h13'6" 50d0'14"
Edward
Cyg 20h11'40" 33d15'14"
Edward
Leo 11h55'23" 21d56'21"
Edward
Gem 6h49'16" 20d40'59"
Edward
Cep 20h42'22" 65d26'5"
Edward A. Angelinas
Per 2h41'11" 50d7'28"
EDWARD A. CARNEY
Per 3h55'56" 33d49'47"
Edward A. Trzcinski
Per 3h12'5" 55d25'5"
Edward A. Welsch
Gem 6h49'31" 19d47'50"
Edward Aaron Fields
Lyn 7h57'49" 57d9'35"
Edward Alan Struttman
Cep 21h12'27" 81d21'42"
Edward Alexander Bienz
Cep 21h25'49" 59d47'20"
Edward Alfred Burnham
Per 2h11'32" 51d44'29"
Edward Allen Demonbreun
Tau 3h30'4" 7d10'54"
Edward Allen Rupert
Tau 3h38'23" 1d43'7"
Edward and Addie Guinn
Lyr 18h58'12" 33d38'59"
Edward and Anna Zalar
Ser 15h34'22" 17d46'14"
Edward and Gulsina
And 23h39'52" 47d45'2"
Edward and Josephine Peterschick
Uma 10h2'10" 58d25'25"
Edward and Rita West
Lyr 19h16'35" 27d15'5"
Edward and Sue Brayboy
Cyg 19h58'57" 51d2'9"
Edward and Teresa Thomas
Lyn 6h45'1" 61d37'20"
Edward Andrew Fiskaa
Cep 22h30'8" 63d55'49"
Edward Anthony Renaltner
Ori 5h57'42" 17d29'44"
Edward Anthony Thorpe - Eddie's Star
Her 16h14'32" 43d12'6"
Edward Ashley Wagner
Lib 14h40'44" -17d34'49"
Edward A.Silva
Per 3h28'49" 50d58'44"
Edward B. Finnerty
Uma 11h20'1" 57d19'21"
Edward B. Huff
Oph 17h25'32" -2d18'29"
Edward "Ba" Krizek
Her 17h33'0" 34d32'32"
Edward & Barbara Hughes
Uma 10h2'0" 57d50'35"
Edward Benedick Morley Cottrell
Umi 14h44'47" 78d3'10"
Edward Benjamin Hutchinson
Per 3h12'1" 53d13'2"
Edward Benjamin Slominski
Uma 10h6'21" 51d28'15"
Edward Big Dog Galis
Gem 6h56'1" 15d50'34"
Edward Britt Adams, Jr.
Sco 16h8'32" -10d11'16"
Edward Brown
Cep 22h43'59" 74d31'35"
Edward Bruce Williams Jr.
Mon 7h44'19" -0d40'34"
Edward Brunetti
Lyr 18h55'44" 32d9'45"
Edward Bryant
Lib 15h32'18" -7d22'6"
Edward C. Kaps
Uma 10h32'11" 63d5'52"
Edward C. Kerber
Cep 21h30'41" 59d36'42"
Edward C. O'Connor
Per 3h16'8" 54d38'50"
Edward C. Portner
Vir 14h11'52" 4d0'43"
Edward C. Vogel, Sr.
Uma 9h52'44" 45d14'59"
Edward Callow
Cep 0h6'29" 77d24'12"
Edward & Carolyn
Lmi 10h38'2" 32d0'12"
Edward Carter Johnston
Per 4h26'2" 43d39'4"
Edward Charles DeSimpelaere
Ori 5h50'22" 11d30'30"
Edward Charles Kruse
Sgr 19h16'43" -15d14'58"
Edward Charles Perry
Leo 10h12'58" 22d17'59"
Edward Cohen
Aur 5h35'41" 38d59'18"
Edward Conlon
Per 4h28'49" 34d30'17"
Edward Cranshaw
Cru 12h36'56" -58d23'28"
Edward Crew
Sco 16h12'55" -10d47'0"
Edward D. & Brenda Amelia Delafield
Vir 13h16'37" -13d11'38"
Edward Dale Ellis
Leo 9h50'30" 29d0'51"
Edward Daum
Ori 5h14'8" -9d19'38"
Edward David Benjamin Jr.
Per 3h5'35" 54d24'51"
Edward David Dahl
Leo 10h21'32" 25d2'21"
Edward David Ehrlich
Aqr 21h10'57" -7d1'16"
Edward David Jenkinson
Cyg 20h17'44" 45d9'12"
Edward David Leah - 19 May 2006
Pho 23h32'18" -44d17'9"
Edward David Llewelyn James BSc.
Gem 6h56'23" 17d37'12"
Edward Dolman
Umi 15h12'24" 68d42'58"
Edward Dunne
Lyn 7h57'18" 53d11'50"
Edward Dwight Sly
Cep 21h3'57" 65d28'35"
Edward E. Kogutkiewicz
Per 3h9'28" 53d56'50"
Edward E. Miranda
Cep 22h8'20" 64d13'3"
Edward E. Riley
Cep 22h36'16" 57d2'9"
Edward E. Riley
Cep 22h37'22" 56d55'26"
Edward E. Scully Jr.
Cyg 19h38'35" 29d8'44"
Edward E. Zorn, Jr.
Cyg 19h46'49" 31d38'20"
Edward "Ed" Orekar
Aur 5h53'20" 36d33'4"
Edward "Eddiee" Gutknecht
Cep 22h51'53" 70d14'18"
Edward Ehlert
Uma 11h40'30" 39d14'7"
Edward Emerson Dougherty, Jr.
Cnc 9h10'18" 18d22'45"
Edward & Emily Hart's Star
Cyg 21h33'23" 46d41'43"
Edward Eugene Smyth, Jr.
Uma 10h16'13" 61d58'38"
Edward Eugene Stone "Buzz"
Gem 6h53'15" 29d58'18"
Edward F. Betsch
Uma 13h46'19" 55d59'28"
Edward F. Hughes
Cep 22h30'53" 63d3'9"
Edward F. Peters
Per 4h21'54" 44d45'52"
Edward F. Walsh
Aql 20h16'12" 1d48'36"
Edward F. Warner
Uma 12h52'18" 56d31'11"
Edward Falco IV
Umi 15h13'53" 70d37'31"
Edward Feest
Uma 14h18'35" 57d33'15"
Edward Foy Kimball
Leo 10h37'10" 22d7'58"
Edward Frances Disser
Aur 5h38'48" 41d42'9"
Edward Francis Coleman
Cnc 8h4'30" 25d14'59"
Edward Francis Drake
Psc 0h40'0" 8d27'5"
Edward Francis Nelson VI
Boo 15h27'2" 49d11'49"
Edward Francis Norris
Aur 5h23'5" 31d0'20"
Edward Francis Xavier Scott III
Uma 9h28'7" 41d34'26"
Edward Franklin Possessky
Ari 2h31'0" 24d40'39"
Edward G. Haase
Her 16h58'42" 33d19'31"
Edward G. Tronca
Per 3h6'41" 55d49'19"
Edward George Curington
Cep 22h53'26" 64d12'57"
Edward Giannantonio
Aqr 22h45'29" -24d25'13"
Edward Gibson Camolilla
Aqr 22h10'40" 0d47'48"
Edward Grant Hall
Per 3h33'31" 48d32'49"
Edward Grayson Fox
Umi 14h54'53" 72d38'22"
Edward H. Fogelman
Sgr 17h51'15" -26d23'33"
Edward H. Hegarty
Cep 22h36'23" 65d39'28"
Edward H. Luning
Lib 15h44'35" -23d21'8"
Edward H. Schwannecke II
Per 3h13'56" 37d8'6"
Edward Hardin Mills
Uma 12h1'12" 53d29'9"
Edward Harry Cohen
Lyn 7h52'38" 49d56'28"
Edward Hawkins
Gem 6h59'7" 26d45'7"
Edward Hayes Benson
Ori 5h38'27" -0d0'40"
Edward Henderson Corddry
Psc 1h6'2" 8d36'50"
Edward Howe
Aql 18h45'28" -0d56'18"
Edward Hugh Delfino, Esq
Lyn 7h14'20" 59d2'17"
Edward Hughes Flora
Crb 15h28'15" 28d46'46"
Edward & Irene Szczepaniak
Cap 21h5'8" -19d46'18"
Edward J Baker
Per 2h43'45" 53d47'2"
Edward J. Cody
Lib 14h48'49" -1d2'18"
Edward J. Conklin
Aur 5h52'25" 51d55'29"
Edward J. Cush, Sr.
Uma 10h15'49" 64d0'55"
Edward J. (Eddie) Brophy
Cyg 20h22'3" 44d20'37"
Edward J. Jerin, Sr.
Cma 6h44'57" -13d17'20"
Edward J. & Joan E. Daly
Uma 9h58'54" 59d23'24"
Edward J Jones
Ari 2h16'28" 24d46'35"
Edward J. & Loretta M. Walsh
Gem 6h52'19" 19d39'41"
Edward J O'Neill
Aur 5h50'25" 52d2'53"
Edward J. Piersa Sr.
Ari 2h32'58" 10d46'23"
Edward J. (Pop) Gardner
Uma 10h32'35" 66d23'50"
Edward J Reilly IV
Aql 18h50'39" -0d18'41"
Edward J. Rogacki
Aqr 23h15'7" -23d35'59"
Edward J. Sacre
Sco 17h33'36" -43d16'20"
Edward J. Schirg
Tau 4h55'9" 22d39'50"
Edward Jackson's Precious
Her 16h50'32" 36d3'56"
Edward James Bobbett
Vir 13h49'51" -19d15'32"
Edward James Brown
Uma 14h6'30" 61d57'4"
Edward James Christie
Ari 2h5'29" 21d38'40"
Edward James Goode
Her 17h39'16" 20d36'35"
Edward James Howell
Her 17h0'45" 32d23'3"
Edward James Stearns IV
Uma 14h1'23" 51d34'26"
Edward James Sweeney, Jr.
Uma 11h1'57" 59d59'17"
Edward Jay Rankin
Her 18h52'26" 22d9'51"
Edward Jerome Riggie IV
Uma 10h49'23" 63d38'47"
Edward Joel Fox
Leo 10h52'46" 14d24'12"
Edward John 10-12-14
Ari 2h1'42" 21d47'10"
Edward John Chataway
Per 4h38'41" 47d12'9"
Edward John Hottin
Cep 21h16'25" 66d50'0"
Edward John King
Cep 22h0'39" 61d26'57"
Edward John Lichtenhahn
Ari 2h50'19" 29d3'18"
Edward John Olivero
Boo 14h53'44" 17d47'58"
Edward John Reeves
Uma 11h33'2" 63d10'35"
Edward John Sherstone
Uma 9h13'52" 64d25'22"
Edward John Snyder, Jr.
Cep 22h18'40" 60d21'49"
Edward John Wishnoff
Leo 10h13'5" 15d39'33"
Edward Jordak
Per 3h40'28" 45d55'16"
Edward Joseph Dreyling
Ori 5h30'59" 2d53'40"
Edward Joseph McGurn
Uma 13h6'28" 56d20'0"
Edward Joseph Mello Jr.
Sco 16h14'3" -14d18'39"
Edward Joseph Steinbeigle
Her 16h29'37" 43d4'44"
Edward Kamp
Aur 5h51'28" 53d30'37"
Edward Koseck
Per 4h34'59" 33d52'45"
Edward L. Bold (Pop)
Uma 11h21'33" 63d35'15"
Edward L. & Carolyn J. Krause
Cyg 19h46'10" 33d16'28"
Edward L. Myers, III 's Star
Her 18h43'51" 20d1'8"
Edward L. Saunders
Cnc 8h21'8" 24d6'56"
Edward L. Thomas
Aur 5h49'24" 52d19'41"
Edward Lee Webb
Uma 10h43'42" 47d52'2"
Edward Leon Ford
Lib 16h1'17" -10d27'22"
Edward Leonard Dive
Her 16h34'3" 14d14'9"
Edward Lewis McCourt Jr.
Sgr 19h26'51" -23d22'40"
Edward Lilley #835 Beta Xi
Aqr 22h33'48" -9d36'52"
Edward Locke Anderson
Del 20h34'32" 13d15'55"
Edward Louis Lee "Ed's Star"
Tau 4h39'16" 11d51'4"
Edward Lynn
Psc 1h31'58" 22d33'17"
Edward M. Bodie Jr.
Sco 16h6'52" -25d24'28"
Edward M. Lukash
Lyn 7h1'39" 58d22'16"
Edward M. McGowan
Cap 21h21'22" -23d29'33"
Edward M. Norton
Aur 6h25'25" 41d33'31"
Edward Madden
Uma 13h8'12" 59d37'15"
Edward Mahoney
Lyr 18h28'24" 41d39'56"
Edward Martirosyan
Per 3h26'39" 44d4'42"
Edward Matthew Zatwarnicki
Cnc 9h13'59" 26d23'47"
Edward Maunsell Van Rensselaer
Ori 4h51'29" 12d23'48"
Edward Maurice Ross
Uma 8h32'44" 60d47'0"
Edward Maurice Weinheimer Sr.
Ori 5h37'39" 1d18'33"
Edward McCants
Ori 5h23'19" 13d39'39"
Edward Melvin Katz
Her 16h55'36" 28d33'6"
Edward Michael Clark
Cap 20h9'11" -15d48'26"
Edward Michael Corley
Sgr 18h12'36" -20d3'58"

Edward Michael & Laurie Ann Morse
Per 2h13'49" 51d9'24"
Edward Michael Scullion III
Umi 15h22'39" 70d49'24"
Edward Miles
Cep 0h0'49" 73d12'27"
Edward Miljon Weglarz
Ori 5h32'40" -6d30'18"
Edward Moran
Hya 9h35'16" -0d16'45"
Edward & Myrna Macugay
Per 3h46'34" 42d24'1"
Edward N. Pasqual
Uma 8h46'27" 63d55'28"
Edward N. Schwalje
Per 2h15'41" 56d38'22"
Edward "nak-ed" Keljik
Tau 4h6'4" 19d26'10"
Edward Newhamn Miller
Cap 20h47'35" -14d50'46"
Edward Newman
Cmi 7h23'53" 3d3'47"
Edward O'Halloran
Leo 11h34'38" 25d53'49"
Edward Omar Zahed
Leo 10h22'17" 9d31'2"
Edward Osborn, my dad
Leo 9h29'59" 12d4'3"
Edward Oscar Ramos
Boo 14h47'52" 20d53'0"
Edward P. Brundrett, Jr.
Uma 9h32'55" 42d40'24"
Edward P. Donofrio
Per 3h25'49" 36d50'48"
Edward P. Dumala
Cep 21h15'6" 65d47'3"
Edward P. Evans
Umi 16h2'25" 74d31'55"
Edward P. Guindon, Jr.
Leo 11h8'22" 24d45'22"
Edward P. House
Cep 22h14'21" 63d21'46"
Edward P. Mansfield Jr.
Sco 16h12'49" -14d19'55"
Edward P. Tenn, Jr.
Cap 20h38'26" -14d45'12"
Edward Peter Promuto
Vir 14h45'25" -1d36'24"
Edward Petruskevich
Cap 20h23'38" -11d6'30"
edward philip luzzi
Tri 2h33'49" 34d21'39"
Edward Phillips
Boo 14h34'23" 40d50'41"
Edward Pichs
Cnc 9h3'52" 15d12'12"
Edward R. Davis
Uma 11h39'37" 53d6'14"
Edward R. O'Brien
Uma 12h39'20" 52d59'9"
Edward R. Starsine II
Cnc 8h4'43" 25d51'24"
Edward R Torresen
Nor 16h15'32" -48d12'15"
Edward Reardon
Cyg 20h27'9" 54d17'8"
Edward & Rebecca Lowicki
Cyg 20h47'42" 38d12'43"
Edward Regalado Santos III
Ori 5h52'3" 6d13'27"
Edward Reid Caldwell
Sgr 17h54'6" -22d50'11"
Edward Reid Luckey
Vul 19h50'3" 22d33'37"
Edward Reid Luckey
Vul 19h30'2" 22d15'1"
Edward Ricca
Tau 4h56'38" 22d46'58"
Edward Richard Dagermangy
Eri 3h0'57" -9d0'3"
Edward Robicheaux
Ori 5h42'0" 0d35'27"
Edward Romo
Cnc 9h20'16" 13d53'38"
Edward Ross Lara
Cnc 9h17'35" 7d26'53"
Edward Rouzie Baird IV
Vir 14h11'55" -11d56'41"
Edward Roy Webb
Aqr 22h20'10" -12d46'21"
Edward S. Baldacci
Sgr 19h26'50" -12d23'9"
Edward S. Bevis
Aql 19h56'39" 13d15'30"
Edward S. Carr Jr.
Uma 12h58'2" 57d7'25"
Edward S. Harriman, Jr.
Sct 18h42'5" -5d0'50"
Edward S. Taft
Cep 22h9'12" 60d8'27"
Edward Sawicki
Ari 3h21'37" 29d0'33"
Edward Schulze
Cap 20h24'26" -20d21'36"
Edward Scovel
Per 2h42'32" 54d3'15"
Edward Sebastian Barrington Coleman
Uma 9h53'13" 43d37'29"
Edward Seeberger
Ori 5h48'33" -0d22'50"
Edward Shroyer "Our Protector"
Per 3h11'23" 46d0'6"
Edward Silva
Sco 17h21'40" -37d57'37"
Edward Sokolosky
Cma 6h49'7" -16d19'14"
Edward & Stephanie Celkis
Umi 15h43'1" 81d37'10"
Edward Stephen Thomas Chiffriller
Tau 3h59'1" 22d23'53"
Edward Stephens
Her 18h48'17" 16d5'45"
Edward Stofko
Ori 6h25'29" 16d59'15"
Edward Stuever
Tau 4h16'17" 2d52'27"
Edward Sullivan
Cyg 20h41'44" 51d37'20"
Edward T Cahill
Leo 11h7'24" 10d14'57"
Edward T. Strasler
Psc 23h42'13" 1d46'2"
Edward Tabela
Leo 11h25'31" 23d58'51"
Edward Talbott
Per 3h15'33" 52d23'35"
Edward Tanner
Crb 15h37'8" 27d47'27"
Edward Testa
Ari 3h17'32" 28d14'18"
Edward Theodore Kemen
Cep 22h0'57" 57d10'18"
Edward Thomas
Cnc 8h4'45" 13d5'41"
Edward Thompson III
Cap 20h7'44" -18d49'49"
Edward Thompson Webb
Ori 5h23'34" 2d25'25"
Edward Thurber Haines
Gem 6h37'37" 21d1'8"
Edward Ufford
Per 4h18'41" 50d51'33"
Edward Urena
Aql 19h14'49" 6d27'9"
Edward V. Ciavolino
Per 2h47'19" 53d36'32"
Edward W. De Domenico II
Her 17h48'41" 17d58'51"
Edward W. Robbins, Sr.
Cyg 20h8'17" 40d22'32"
Edward Wagner
Aqr 22h25'50" -24d50'59"
Edward Wainwright
Her 16h18'51" 21d45'48"
Edward & Wallace's Star
Lyr 18h47'3" 38d55'48"
Edward Warren Miller
Her 17h40'19" 38d5'3"
Edward Wayne Hall
Per 2h40'17" 55d18'23"
Edward Werner
Umi 15h38'33" 69d42'49"
Edward Wesly Sheets
Ori 6h17'0" 14d47'19"
Edward Wight
And 1h22'34" 43d32'57"
Edward William Davies
Lib 15h55'48" -5d6'19"
Edward William Eells 111
Vir 11h59'22" -0d9'51"
Edward William McAfee "Dub"
Her 16h38'46" 41d1'49"
Edward William Patrick O'Loughlin
Cep 21h53'52" 58d26'18"
Edward Wujcik & Judy Flaherty
Cyg 21h30'5" 53d22'21"
Edward York - Edd's Star
Per 4h17'12" 44d27'37"
Edward Zerbo
Umi 16h20'22" 70d29'56"
Edward Zoltan Semanate
Cnc 8h4'37" 20d20'38"
Edward-Lawrence D. Garcia-Garnica
Ori 5h14'17" 0d20'17"
Edwardo
Cra 18h47'40" -39d58'42"
EDWARDS
Uma 8h28'13" 64d41'38"
Edwards Red Devil
Per 3h12'1" 50d28'54"
Edward's Star
Uma 10h17'56" 63d57'26"
Edward's Superstar
Umi 14h9'8" 69d46'48"
Edward's Utopia
Cnc 8h42'33" 25d11'2"
Edwige 25.03.2006
Ari 2h43'29" 21d53'14"
Edwin A. Brown
Umi 11h54'20" 89d10'51"
Edwin A. Forrest
Dra 20h18'57" 67d4'56"
Edwin Albert Haugh, Jr.
Vir 14h37'48" 1d2'43"
Edwin Allan Snyder
Lmi 9h34'51" 34d40'49"
Edwin Anthony
Per 2h26'0" 55d15'21"
Edwin Arvelo III
Psc 1h37'50" 27d55'56"
Edwin B.
Her 16h47'43" 20d48'19"
Edwin B. Bolz
Sgr 18h5'56" -26d25'47"
Edwin B Dunbar
Uma 8h48'27" 67d51'7"
Edwin Belzer & Evelyn Hartshorne
Sco 16h20'11" -15d39'42"
Edwin David Beaty Marlowe
Lib 15h13'44" -16d8'8"
Edwin Dean Hampton
Aqr 23h38'39" -17d20'28"
Edwin Dolin
Per 2h55'13" 46d39'13"
Edwin Earl Moore
Ori 6h1'7" -0d3'35"
Edwin Elden Schroeder
Gem 7h43'11" 24d9'14"
Edwin Emerson Brooksbank
Lac 22h25'58" 44d25'33"
Edwin Eugene Beatty
Uma 11h0'11" 45d44'16"
Edwin Eugene Nothnagel
Ori 5h44'54" 0d23'31"
Edwin G. Drumheller, II
Cep 21h29'29" 65d14'22"
Edwin Gillett
Her 16h13'50" 47d58'50"
Edwin Hanks
Leo 11h36'6" 15d40'45"
Edwin & Hilda Peck Sr.
Uma 11h16'13" 35d10'34"
Edwin Hochstettler
Her 16h59'13" 33d25'36"
Edwin J. Fitch
Cep 0h8'4" 66d57'7"
Edwin J. Hudgins
Sco 17h54'34" -33d18'44"
Edwin James
Ori 5h9'0" -1d5'41"
Edwin James Saunders
Aql 19h22'40" 9d36'47"
Edwin Jennings NeVille
Lib 15h3'28" -22d50'39"
Edwin Joseph Mackowiak
Per 3h0'43" 37d15'59"
Edwin Justiniano, Jr.
Aur 5h28'7" 46d9'36"
Edwin K. Mahlo
Aur 5h35'43" 45d17'18"
Edwin L Boardman
Ari 2h42'48" 30d44'57"
Edwin L. Rosenberger Jr.
Per 3h19'2" 49d11'43"
Edwin Leon
Sco 16h40'52" -31d53'28"
Edwin & Lisa Friskey
Lyn 8h29'27" 55d7'29"
Edwin M. Antonucci
Aqr 23h9'59" -20d59'25"
Edwin M. Miller
Cnc 8h42'57" 18d48'13"
Edwin Manuel Diaz
Cep 21h26'37" 64d39'4"
Edwin McCumber - Rising Star
Boo 15h12'5" 47d53'20"
Edwin Mendez Jr.
Ori 5h52'28" 22d41'0"
Edwin Merino "My True Love"
Cyg 20h13'13" 39d37'3"
Edwin Miller Bush
Cep 23h45'41" 74d5'56"
Edwin Minzenmayer
Cnc 8h29'33" 14d38'13"
Edwin O. Robertson
Per 3h3'55" 56d31'54"
Edwin Omar Salazar
Sgr 19h17'54" -14d31'3"
Edwin Pagan III
Umi 15h40'30" 78d43'38"
Edwin Pardee
Uma 11h32'59" 31d31'8"
Edwin Pearce
Peg 22h42'38" 33d25'35"
Edwin Pearce
Lyr 19h13'6" 38d22'22"
EdwiN RiVeRa
Psc 1h34'14" 18d38'15"
Edwin Slennagiel
Gem 6h49'58" 23d20'26"
Edwin Stanley Davis
Her 17h26'36" 34d49'18"
Edwin Velez Villar
Tau 3h26'38" 8d56'12"
Edwin Wentworth MacNeill
Uma 11h11'22" 65d51'59"
Edwin Wolf
Dra 19h48'47" 76d36'4"
Edwina Ings-Chambers
Com 12h22'45" 21d9'48"
Edwina Leo
Uma 12h23'54" 57d11'31"
Edwina Mary Vardey
Cas 1h17'50" 50d46'32"
Edwina Pirotte
Uma 9h46'49" 54d32'40"
Edwordine J. Vecchiolla
Psc 0h57'28" 26d59'22"
Edy
Aur 5h57'42" 33d8'13"
Edy Always and Forever
Her 17h13'14" 35d8'54"
Edy Ambrose
Vir 13h55'46" -6d19'29"
Edy J. Guerrero
Boo 15h18'58" 37d11'22"
ed.yanyan
Tau 4h10'40" 5d42'12"
Edye
Uma 10h3'29" 44d14'47"
Edyn Jaye Diaco
Ari 2h22'42" 11d52'55"
Edyta
Vir 12h44'0" -6d25'43"
Edyta Rhodes
Ari 2h54'15" 18d46'20"
Edythe Irene Weichselbaum
And 23h18'27" 44d10'10"
Edythe Mae Kaneiwa
And 1h28'45" 49d24'26"
Edytka
And 23h30'26" 41d27'45"
Edzar
Aql 20h11'21" 11d45'21"
EDZIA
Psc 1h27'44" 18d33'30"
Ee-a-Kee
Aql 19h8'14" 0d57'37"
Eeedy und Marscherite
Uma 8h51'8" 54d20'59"
eefje buisman
Uma 8h41'15" 69d53'29"
EEM
Lib 15h54'32" -17d42'36"
Eemeli Lintukorpi
Per 2h51'19" 50d51'25"
Eepa
Cyg 21h23'52" 46d36'59"
EES & JMB
Umi 14h56'48" 69d21'43"
Eeuwige Liefde
Ari 2h49'11" 28d17'6"
Eeva
Ori 5h6'17" 10d58'26"
Eeyore
Cyg 19h49'16" 38d24'29"
eeyorerose
Tau 5h4'48" 22d5'39"
Eeyore's Serendipity
Peg 23h41'56" 23d49'54"
Efe Siyahi
Ari 2h47'8" 12d19'15"
Effie
Lyn 7h35'11" 49d23'50"
EFFIE
Lmi 10h7'22" 32d38'5"
Effie Bastas's Star of Compassion
Ori 5h47'42" 6d45'30"
Effie Florence Weaver
Uma 11h48'30" 36d29'21"
Effie Lee Taylor Saunders
Gem 7h3'58" 34d11'24"
Effie May Upton
Psc 0h35'0" 6d15'34"
Effie Nelson
Uma 10h35'38" 58d15'49"
Effie Nelson
Lyr 18h53'32" 39d13'30"
Effie (Princess) Marosouli
Ari 2h49'50" 27d54'5"
Effie T
Uma 12h22'45" 58d58'55"
Effigy
Uma 13h31'29" 57d10'0"
Effortless
Gem 6h42'2" 18d52'44"
EFG
Her 16h24'39" 48d32'26"
Efi Karakitsou
Uma 9h10'47" 61d18'31"
Efimia
Cyg 20h59'3" 43d14'21"
EFKAR Prime
Dra 18h47'11" 53d19'38"
Efrain A. Marrero
Aql 19h49'12" 6d53'59"
Efrain D. Rodriguez
Dra 20h2'40" 71d12'22"
Efrain Garfias
Boo 14h36'30" 34d6'33"
Efrain Ibarra
Ori 5h32'0" 3d7'22"
Efrain V
Gem 6h9'32" 24d19'57"
Efrain Velazquez-Rodriguez
Cap 21h30'38" -9d7'11"
Efrati
Tau 5h37'39" 26d7'1"
Efren Perez-Collazo
Cep 20h14'22" 60d2'59"
Efren Sandoval
Uma 12h35'59" 52d35'3"
Efstathios Drivas
Uma 11h28'36" 43d19'0"
E.G. and Cleo L. Lewis
Cyg 20h7'54" 47d2'33"
E.G. "Big Papa" McBride
Ori 6h14'59" 2d47'33"
E.G. Lewis
Uma 11h29'32" 52d47'33"
Egads! It's All That Glitters
And 2h36'15" 44d44'24"
Egan - Rafferty
Cyg 21h25'1" 37d11'58"
EGC 1924
Uma 10h39'22" 60d54'56"
Egerton and Norah Taylor's Star
Peg 21h40'27" 20d45'51"
Eggert, Heiko
Uma 10h19'10" 66d53'27"
Eggertsen
Uma 8h25'8" 64d37'37"
EGH *My One True Love*
Lib 15h36'36" -17d37'36"
Egidijus Kasaciunas (Tobulas)
Leo 11h26'12" 2d48'34"
EGIDIO
Uma 10h44'33" 59d39'23"
Egle Matkeviciute
Umi 14h50'4" 76d33'51"
EGO
Leo 11h29'34" 4d8'21"
Ego Amor Tu
And 23h23'36" 43d49'27"
Ego Diligo Vos
Lmi 10h10'37" 31d13'1"
Ego Philo Su
And 23h19'40" 47d1'35"
Egon Krause
Ori 6h17'14" -1d27'23"
Egon Lüdtke
Ori 6h9'47" 7d53'2"
Egon Mansfeld
Ori 5h56'27" 12d10'34"
Egon Ross
Uma 9h0'25" 50d40'18"
Egon WOLF
Cam 7h43'46" 62d38'2"
EGOR
Lyn 7h21'40" 50d12'15"
Egrabine
Gem 6h42'55" 21d58'9"
Egymásért
Uma 9h0'36" 62d26'27"
Egypt
Cma 6h50'47" -17d8'3"
Egyptian Camel
Gem 7h39'50" 32d28'35"
EH&MG4EVA
Uma 9h22'30" 44d44'13"
*EHA *Jr. Midgets* The Brats*
Cha 10h36'45" -75d48'18"
EHC Visp
Her 17h12'44" 16d21'43"
Ehfar
Gem 7h47'51" 14d21'36"
Ehli
Cyg 20h22'58" 41d43'23"
Ehmad Barnieh
Cyg 21h14'7" 38d46'2"
Ehmry
Gem 7h0'18" 16d54'53"
Ehren Kellogg
Cnv 13h55'51" 37d57'0"
Ehrin Michael Davis
Leo 11h50'52" 28d4'41"
Ehrler, Roland
Uma 13h49'54" 56d58'49"
Ehsan
Aqr 22h23'23" -1d38'8"
Eiblmeier, Helmut
Uma 8h54'47" 50d12'6"
Eichholz, Liv Stella
Uma 10h14'2" 62d29'55"
Eickemeyer, Jörg
Uma 10h41'23" 66d20'23"
Eickhoff, Bettina
Uma 13h27'57" 52d23'48"
Eickhoff's 50th Wedding Anniversary
Peg 22h44'28" 16d38'52"
Eidas dna Netsirhc
Sgr 18h51'35" -19d18'1"
Eifiona
Ori 5h8'29" 15d15'53"
Eigee
Uma 10h51'6" 67d40'19"
Eila
Cas 0h39'21" 60d11'1"
Eila Wan - Nimmy Jox
Ori 5h35'1" -6d15'17"
Eilea Ruth
Lyr 18h48'29" 45d7'54"
Eilean L Myer
Aqr 21h51'9" -1d6'19"
Eileen
Lib 14h26'20" -17d58'49"
Eileen
Cas 0h18'21" 61d39'2"
Eileen
Cam 5h20'51" 56d10'18"
Eileen
Lyn 7h34'45" 37d46'3"
Eileen
And 0h45'9" 39d14'59"
Eileen
Vir 12h52'21" 12d38'51"
Eileen
Leo 9h26'1" 14d20'9"
Eileen
Crb 16h21'23" 28d29'38"
Eileen A. Brien - Guiding Light
Sco 16h50'53" -27d48'15"
Eileen A. Chavez
Leo 9h43'30" 7d17'32"
Eileen Albinson
Cep 22h1'32" 68d25'49"
Eileen Alexander
Ori 5h23'2" 9d4'23"
Eileen Amedeo
Uma 11h23'28" 51d45'35"
Eileen and Allan
Cyg 21h36'20" 38d10'43"
Eileen Anne Johnston
And 0h28'28" 35d59'41"
Eileen Anne Sheila McAuliffe
Ari 2h47'22" 25d8'44"
Eileen Annette Bowden
Ori 6h7'59" 9d12'44"
Eileen Barbara Rieder
Lib 14h49'44" -0d58'58"
Eileen Barker
And 2h38'16" 47d20'52"
Eileen Bates
Ari 2h42'23" 21d37'45"
Eileen Bengtson
Per 3h10'4" 55d36'23"
Eileen Bergbreiter
Cas 1h32'4" 62d21'11"
Eileen Berris Glazer
Sco 16h9'16" -12d59'43"
Eileen Blaney
Sgr 18h22'54" -33d2'21"
Eileen Blum
Uma 8h57'9" 62d16'28"
Eileen Buttonow
Sco 17h17'44" -44d31'20"
Eileen Capitoli
Cas 1h55'44" 63d53'21"
Eileen " Cara Mia"
Cyg 21h38'18" 42d55'5"
Eileen Catherine Walsh
Sco 16h13'54" -15d24'44"
Eileen Clem
Leo 11h10'15" 3d25'33"
Eileen Cochran
Uma 9h14'24" 58d44'35"
Eileen Cosgrove
Cam 4h8'15" 70d21'44"
Eileen Daly
Cas 2h50'14" 61d48'58"
Eileen DaRold
Sco 17h50'4" -39d12'16"
Eileen Elizabeth Anne Rio Hall
Vir 14h6'47" -13d40'36"
Eileen Elizabeth Cabrera Ward
Vir 11h43'35" -0d51'30"
Eileen Eller
Tau 3h49'57" 7d57'50"
Eileen Esther - October 11, 1934
Lib 15h18'31" -23d17'54"
Eileen Franco
Umi 15h39'9" 74d44'38"
Eileen G. Demblowski
Crb 16h7'7" 37d27'52"
Eileen Georgina
Cas 0h19'36" 51d14'55"
Eileen Glass
Sco 16h2'29" -10d40'44"
Eileen Greene
Cnc 8h39'44" 23d1'5"
Eileen Guidice
Sco 17h58'19" -39d52'30"
Eileen Helick, a Woman of Grace
Cnc 9h11'35" 23d16'24"
Eileen Hoelter
Uma 9h15'53" 58d57'59"
Eileen Holroyd Brunner
Tau 4h19'33" 29d53'55"
Eileen J. Brooks' Star
Uma 8h46'31" 62d46'2"
Eileen J. Chase
Mon 6h39'21" -1d45'38"
Eileen & Jim Castaldo
Cyg 20h32'53" 43d31'33"
Eileen Joy
Del 20h47'27" 17d37'15"
Eileen Joyce Barratt Wellard
Vir 11h54'0" -0d31'47"
Eileen Kelleher
Cas 1h30'3" 60d58'24"
Eileen Kelly
Tau 5h38'16" 25d53'0"
Eileen Kelly
Her 16h18'33" 12d28'27"
Eileen Kennedy
Tau 4h42'55" 27d18'4"
Eileen Kowler
Psc 0h56'55" 33d6'31"
Eileen Krieger
Cap 20h39'55" -18d50'32"
Eileen Kuszak
And 23h37'46" 47d28'43"
Eileen L. Cinquemane
Tau 3h45'28" 21d16'39"
Eileen Lee 40
Psc 1h12'18" 24d30'55"
Eileen M. Page
Cnc 8h58'13" 10d37'31"
Eileen Madlmayr
Vir 12h50'18" 3d29'32"
Eileen Maguire
Cas 0h21'26" 50d56'50"
Eileen Maraziti
Cas 1h29'39" 61d46'2"
Eileen Margaret
Cas 0h20'22" 55d39'44"
Eileen Margaret Daly
Sgr 17h55'14" -17d35'51"
Eileen Marie Heraty "Kitten"
Lyn 7h42'35" 38d40'19"
Eileen Marron Keating
Vir 13h22'20" 11d56'7"
Eileen Mary Casbon
Cas 1h28'47" 52d14'56"
Eileen Mary Garry
Lyn 6h41'7" 50d49'37"
Eileen Mary Regan
Ari 3h15'8" 27d52'50"
Eileen McLaughlin
And 0h41'59" 38d47'15"
Eileen McWilliams
Cas 1h35'49" 63d26'23"
Eileen & Melisa Gspandl
Uma 13h21'39" 61d45'10"
Eileen Michele Giglio
And 2h23'8" 43d6'16"
Eileen Molloy-Starr
Psc 23h15'19" 1d44'21"
Eileen Morell
Sgr 19h20'4" -41d37'31"
Eileen Mortenson
Aqr 22h47'45" -14d56'33"
Eileen Mottram
And 23h34'6" 47d22'9"
Eileen Murphy
Psc 1h11'27" 28d12'51"
Eileen (Nell) Aherne
Cas 2h11'58" 62d24'36"
Eileen O'Brien-Meade
Lib 15h22'10" -6d35'11"
Eileen O'Connor
Crb 16h11'49" 37d48'42"
Eileen Okurowski
Uma 14h5'55" 49d32'18"
Eileen Patti
Per 2h52'26" 54d9'43"
Eileen Perdue
Cas 1h28'34" 56d52'5"
Eileen Powell
And 1h7'23" 36d43'3"
Eileen "Puddin' Pop" Clark
Lyn 7h23'54" 44d41'3"
Eileen R. Gray
Uma 8h45'11" 72d46'4"
Eileen R. & Richard W. Flinch
Lyr 18h47'21" 40d44'34"
Eileen Riviello Giardino, RN, PhD,
And 23h20'17" 42d20'23"
Eileen Robertson
Crb 15h44'54" 35d22'3"
Eileen Roche-Clark
Cyg 20h14'6" 46d54'32"
Eileen Rose Puppos
Ori 5h8'37" 15d42'57"
Eileen S. King
Cap 20h14'40" -9d43'0"
Eileen Sever
Peg 22h7'57" 6d47'54"
Eileen & Shane McClung
Uma 9h31'29" 45d25'8"
Eileen Speidel
Cap 20h35'26" -9d37'6"
Eileen Stanley
And 0h13'54" 36d20'50"
Eileen & Steve Kokulak
Cyg 20h8'18" 33d15'4"
Eileen Walsh
Cyg 21h56'13" 48d46'43"
Eileen Wrennall
Sgr 19h51'58" -12d23'36"
Eileen060980
Gem 7h27'6" 33d2'55"
Eileena Amor
Mon 6h25'26" 8d9'53"
Eileen's Angel
Leo 11h9'52" 16d13'43"
Eileen's illumination
Leo 9h43'6" 28d17'41"
Eileen's In the Garden Star
Lyr 18h50'25" 36d59'25"
Eileen's Light in the Sky
Aqr 20h51'47" -10d42'39"
Eileen's Phantom Star
Cap 20h14'15" -25d24'55"
Eileen's Star
Lib 15h7'37" -2d53'26"
Eilen M. Zimmerman
Cnc 8h3'52" 22d32'43"
Eilert Voge
Psc 0h56'21" 15d17'56"
Eilidh Marris
Cyg 19h51'6" 39d44'23"
Eilidh's Guardian Angel
And 2h31'55" 39d25'6"
Eilin
Cap 21h49'18" -21d45'49"
Eili's star of Hasting's
Eri 4h21'37" -31d50'9"
Eimear Mulhern
Uma 11h9'8" 31d50'31"

Eimear Olivia
Per 4h48'17" 42d3'48"
Eimear O'Toole
Leo 9h40'17" 10d15'38"
Eimear Rose Traynor
And 0h48'40" 30d13'12"
Eimeir
Per 3h14'53" 47d15'53"
Eimhin-Sean
Cnc 9h2'0" 7d36'39"
Ein Blinseln Weg (One Blink Away)
Cyg 21h55'46" 54d50'49"
ein neues Leben
Sco 17h53'52" -42d29'15"
Eina Cherrine
Ari 2h52'37" 26d51'16"
Einar Ingenting DMB 1955 SC, IA
Oph 17h29'21" 8d5'27"
Einhäuser, Eberhard
Uma 8h17'7" 63d49'34"
EINS
Uma 12h6'14" 60d14'41"
Einspender, Burkhard
Uma 11h22'42" 37d37'58"
Einstein
Boo 14h34'2" 40d20'31"
Einstein
Her 17h5'9" 46d3'23"
Einstein Machado & Vicky Holguin
Psc 0h49'10" 19d16'57"
Eipo
Cep 22h55'17" 57d56'42"
Eir Bear
And 0h43'22" 25d55'44"
Eiram and Revilo Notsnhoj
Sge 20h12'35" 20d29'52"
Eire
Cma 6h27'12" -15d39'56"
Eireann Nicole Carter
Aqr 21h42'20" -7d45'56"
Eireen
Aqr 21h6'36" -11d1'31"
Eiren O'Hara
Cas 1h41'6" 64d37'56"
Eirin Berg
Sco 17h57'33" -30d39'8"
Eirye
Lyn 6h28'29" 54d19'13"
Eisai to asteri mou!
Ari 3h5'24" 18d42'54"
Eisblume
Cnv 13h52'38" 33d54'8"
Eisha Sharma
Aqr 23h24'5" -14d0'47"
Eishayion Luxshreail Kennedy
Vir 14h2'19" 5d1'51"
Eisler, Manfred
Ori 6h18'48" 10d51'46"
Eisley
Uma 13h11'20" 53d55'13"
Eitan
And 0h56'47" 40d36'52"
Eitan's Zaidy
Per 4h48'13" 50d33'36"
Eivind Hiis Hauge
Ori 5h17'0" 9d7'59"
E.Ivory
Ori 5h38'26" -0d19'54"
EJ
Vir 14h10'5" -16d15'56"
EJ
Aql 19h9'14" 0d28'48"
EJ 2004
Gem 7h3'47" 24d14'15"
EJ Batchelder
Tau 4h7'10" 15d43'2"
EJ Bun-E
Ari 2h59'32" 24d59'52"
E.J. Conroy
Psc 1h20'21" 29d57'30"
EJ Duffy
Uma 8h17'24" 70d58'12"
E.J. Winks
Lib 14h42'18" -18d48'24"
EJami (EJ Wells and Samantha Brady)
Pho 1h32'40" -42d5'6"
Ejan Peterris III
Cnc 8h28'19" 11d28'47"
EJB
Leo 10h34'7" 13d49'19"
EJB1
And 0h6'50" 45d18'54"
EJJBAS
Uma 12h5'15" 31d13'20"
EJMc Poppa
Psc 0h15'16" 6d24'50"
Ejon Prelvukaj
Gem 6h7'2" 25d27'24"
E.J.'s Love
Cyg 19h46'18" 34d53'28"
EJ's Star
Psc 0h16'8" 6d15'3"
EK&SW
Vir 14h33'26" 2d26'35"
Eka Wiryastuti - Forever Shining
Lep 6h10'53" -23d2'54"
Ekalynn
Gem 6h45'28" 18d42'56"
Ekaterina
Vir 13h11'24" 10d35'48"
Ekaterina & Aleksej Tükanko
Uma 13h36'9" 48d5'8"
EKATERINA KRYCHTALE-VA
Cnc 8h30'5" 14d39'29"
Ekaterina Marinina
Lac 22h48'37" 54d15'49"
Ekaterina Pankova
Cnc 8h55'28" 11d4'32"
Ekaterina Popova
Leo 9h30'29" 10d25'30"
Ekaterina Slesar
Gem 7h7'45" 27d3'51"
Ekaterina Zhuk
Cap 21h44'41" -20d7'1"
EKD2000
Leo 11h43'27" 24d24'32"
Ekin-Fuat
Umi 17h47'43" 81d56'39"
Ekkehardt
Agr 21h57'51" 0d10'38"
Ekky's Sparkle
Vir 12h29'43" 8d22'27"
Ekos
Sco 16h39'48" -27d31'39"
El Alvarito
Sco 16h39'29" -37d29'12"
El amor de mi vida
Cas 1h0'42" 60d31'9"
El Amor De Yadira
Psc 1h46'3" 8d29'53"
El Bambino y la Flaca
Cap 21h5'14" -20d17'11"
EL BLANQUITO
Psc 0h19'3" 19d24'41"
El Camino Wade
Psc 0h22'10" -4d20'48"
El Cazador Grande
Umi 15h0'56" 67d39'52"
El Corazon de los Amantes
Col 5h10'40" -34d27'34"
El Espacio Atras De La Nuve 9
Leo 10h11'15" 24d44'17"
El Faro
Umi 15h11'31" 69d58'34"
El Farts
Aqr 21h12'12" -9d22'10"
el gallo y la monita
Lib 15h24'58" -13d9'49"
El Josco
Uma 8h27'13" 68d35'34"
El Kada
Tau 4h30'17" 28d15'39"
El Mar de Juan De Haro Martin
Ari 3h17'12" 16d7'49"
El Mariachi
Ori 4h51'21" 10d36'22"
El Miranda Star of Wonder & Delight
Aqr 22h36'35" -23d7'54"
El Nina
Ori 5h39'10" -1d10'29"
El ojo
Per 2h57'31" 54d17'4"
el otro media de mi naranja
Lyn 8h21'6" 34d47'22"
El Pablo
Ari 2h36'36" 14d33'4"
El Papa Oso Entrecano
Uma 10h1'33" 67d48'20"
El Poder De Los Sueños
Her 18h38'20" 23d48'54"
el polluelo caliente
Ari 1h59'21" 14d20'42"
El Roberto
Gem 7h8'50" 25d50'47"
el shaddai
Uma 9h42'51" 57d32'33"
El Sol de Jack and Irene Framke
Uma 10h15'18" 53d51'30"
'El Teds' McGrath
Cyg 19h58'33" 30d32'36"
El Tigre
Lyn 6h34'56" 58d3'17"
El Tigre
Sgr 18h12'59" -18d59'31"
'El Torro Fuerte'
Ari 2h11'49" 22d53'55"
eL143eN
And 0h27'53" 41d41'21"
Ela
Ori 5h59'42" 21d48'10"
Ela
Sco 16h18'31" -10d53'57"
Ela
Uma 9h38'21" 63d16'3"
ELA 1963
Equ 21h2'51" 4d5'0"
Ela 21.10
Uma 9h12'49" 58d25'58"
Ela Moana Oberli
Ori 5h7'9" 11d20'52"
Ela Saari
And 23h55'56" 37d39'53"
Elaben Nazashca
Gem 6h57'56" 26d34'53"
Eladio Aviles
Gem 7h34'42" 16d53'33"
Elae
Psc 1h18'15" 16d58'50"
Elaheh Azodi
Cap 21h44'5" -19d38'2"
Elaia Delbosc-Iron
Cam 3h50'6" 53d29'57"
Elaimar
Lyn 8h5'40" 40d3'19"
Elain Greene
Sco 16h13'7" -10d4'12"
Elaina
Uma 9h19'54" 53d49'40"
Elaina Aquino
Tau 4h34'44" 17d54'24"
Elaina Bellone
Gem 6h46'41" 15d56'43"
Elaina Cipcic
And 23h12'35" 46d54'26"
Elaina Claire
Sco 17h24'24" -40d8'6"
Elaina Danielle Hines
Lib 14h26'32" -17d28'51"
Elaina Eaton
Cyg 20h55'13" 30d43'41"
Elaina K. Clayton
Vir 12h10'53" -11d18'46"
Elaina Marie Eckardt Adams
Cam 7h18'52" 62d20'44"
Elaina Marie Martinez
Tau 4h6'1" 7d56'20"
Elaina November 2, 2004
And 23h2'47" 39d56'40"
Elaina Tae
Cnc 8h25'59" 20d20'57"
Elaina Terraza Bartolo
And 23h48'11" 46d6'40"
Elaine
Cas 1h2'41" 54d3'50"
Elaine
Uma 9h45'16" 49d51'15"
Elaine
Per 3h22'2" 47d12'33"
Elaine
Lyr 18h33'47" 36d54'36"
Elaine
Crb 15h40'22" 30d25'19"
Elaine
Lmi 10h19'22" 32d37'29"
Elaine
Gem 6h41'4" 27d44'42"
Elaine
Tau 4h8'23" 25d45'24"
Elaine
Crb 15h26'43" 27d46'53"
Elaine
Leo 10h50'28" 13d28'18"
Elaine
Uma 11h59'10" 55d45'17"
ELAINE
Sgr 18h55'52" -31d5'12"
ELAINE
Sco 17h51'50" -35d39'59"
Elaine
Sgr 18h24'42" -23d7'33"
Elaine 220504
Cas 1h31'35" 63d16'42"
Elaine A Rodney - Mum, Oma, Star!
Cru 12h19'19" -57d41'26"
Elaine & Alan
Cyg 20h9'25" 55d3'42"
Elaine Alma Booth
Cyg 21h32'36" 49d10'21"
Elaine Amanda
Ori 6h16'14" 10d18'43"
Elaine Amy Chou
Leo 10h2'41" 17d17'37"
Elaine and JR Wakasugi
Cyg 19h42'24" 28d2'40"
Elaine and Merlin Schwambach
Ori 6h3'54" 18d41'54"
Elaine and Paul Truncellito
Uma 10h10'18" 58d56'21"
Elaine and Steven
Cyg 19h46'2" 55d31'18"
Elaine Ann Schneibolk
And 0h43'7" 37d44'7"
Elaine Annette O'Canas
Uma 12h28'53" 60d14'56"
Elaine Antonia Kastrenos
Tau 3h23'27" -0d14'54"
Elaine Balitz
Umi 16h59'9" 78d40'54"
Elaine Barbara Hotopp
Aqr 22h39'56" -1d49'35"
Elaine Barnes = My Angel Mom
Uma 10h48'17" 58d27'44"
Elaine Batt
Cas 0h48'1" 65d25'35"
Elaine Bavington
Eri 3h54'52" 0d5'22"
Elaine Berridge
Gem 6h45'40" 13d21'58"
Elaine Beussink
Aqr 22h9'33" 2d9'39"
Elaine Bonanno
Gem 6h21'20" 19d58'55"
Elaine Brandl's Eastern Star
Lyn 7h25'54" 48d8'5"
Elaine Bricknell - brightest star
Ari 1h57'9" 17d10'31"
Elaine Brooks
And 2h28'55" 44d52'44"
Elaine Caras Couchman
And 2h22'29" 49d16'2"
Elaine Chiu
Uma 12h43'33" 54d24'34"
Elaine Christiani
Vir 13h26'3" 13d24'35"
Elaine Christie Yap Rafol
Cap 20h20'36" -18d30'4"
Elaine Cindy Muchamel
Cap 21h42'50" -18d28'50"
Elaine "Cookie" Cormier
Tau 4h1'37" 21d45'44"
Elaine D'Allura
And 23h47'18" 34d46'32"
Elaine Davis
Cyg 19h58'7" 58d8'52"
Elaine DeAnni
Uma 11h12'4" 42d34'3"
Elaine D'Farley
And 2h18'17" 50d40'42"
Elaine D'Farley
Cam 7h23'25" 77d31'54"
Elaine (Elena) Ann Miles
Cnc 8h45'33" 10d5'34"
Elaine Ella Mattinson-Lorimer
Tau 3h47'38" 28d36'4"
Elaine Evan Elizabeth = E3
Uma 10h49'6" 48d57'45"
Elaine Faith "Bunny" Gordon
Cnc 8h25'44" 13d23'21"
Elaine Fox Whitney
Ori 5h44'49" 4d1'15"
Elaine Francis Hosier
Lyn 8h54'7" 35d24'9"
Elaine Garceau
Uma 9h58'1" 51d2'3"
Elaine & George Maskin
Uma 10h6'23" 44d17'31"
Elaine Goggin Stieritz
Sco 17h57'2" -32d10'9"
Elaine Harris
And 2h25'44" 49d30'8"
Elaine Hernandez
Lyn 8h13'38" 46d48'18"
Elaine Hernandez
Vir 13h18'0" 6d42'25"
Elaine HGTV 8/3/58
Leo 10h43'36" 22d37'14"
Elaine Holmgren
Crb 16h23'49" 30d54'0"
Elaine Howell's Eastern Star
Lyn 7h1'37" 45d1'26"
Elaine Irene Perry Bass
Aql 19h6'52" 14d24'43"
Elaine J. Burrow Mother/Grandmother
Uma 10h17'27" 41d47'21"
Elaine Jaffe
Cas 0h25'21" 56d10'17"
Elaine Jane Montanini Roche
Lyn 8h2'16" 54d25'26"
Elaine Josephson
Cas 1h35'51" 60d53'24"
Elaine Joy
Lyn 7h52'28" 44d52'48"
Elaine Joyce Hampshire
Psc 2h2'44" 10d27'24"
Elaine Joyce Singer
Lib 15h9'42" -7d19'51"
Elaine K Hughes
And 23h25'47" 41d43'30"
Elaine Katherine de la Mata
Crb 15h38'27" 33d26'4"
Elaine Kee
Lyn 7h25'27" 57d3'21"
Elaine Koblinski
Uma 8h41'30" 59d26'45"
Elaine Leah Rehm
And 1h53'9" 46d59'45"
Elaine Lee
Peg 21h14'34" 15d17'27"
Elaine Leite
Vir 12h42'25" 6d11'19"
Elaine Lester
And 1h45'26" 41d18'55"
Elaine Levey
Gem 6h42'44" 13d16'11"
Elaine Lidia Martinez
Vir 12h42'29" 12d0'11"
Elaine Liles
Lyr 19h4'14" 31d45'9"
Elaine Lipsun
Lyn 9h21'45" 34d5'2"
Elaine "Lucy"
Cas 1h10'4" 51d9'32"
Elaine M. Bosilevac
Cas 0h32'5" 54d25'13"
Elaine M Lieser
Aqr 22h23'52" -12d49'1"
Elaine M. Oprysko
Cas 0h46'2" 60d27'28"
Elaine M. Sweeney
Uma 9h18'50" 69d7'54"
Elaine Maley
Cmi 7h54'43" 12d32'48"
Elaine Mangham
Equ 21h14'38" 11d2'41"
Elaine Marguerite Tomasiello
Cap 20h14'45" -22d20'49"
Elaine Marie
Mon 7h30'13" -0d48'55"
Elaine Marie
Lyn 7h0'0" 55d5'5"
Elaine Marie
Tau 4h51'4" 20d27'0"
Elaine Marie Bellistri
Cas 22h57'21" 55d28'4"
Elaine Marie Campione
Cnc 8h24'55" 14d11'22"
Elaine Marie Entler
Gem 7h9'0" 33d51'15"
Elaine Marie Jacobi
Uma 8h21'5" 60d21'12"
Elaine Marie Louviere
Lmi 10h20'40" 29d13'51"
Elaine Marie Nolan
Uma 9h30'58" 48d12'48"
Elaine Martir
Sco 17h57'12" -37d33'13"
Elaine Mary Duffy
Ori 5h52'30" 18d29'12"
Elaine McDermott Carey
And 23h23'10" 49d26'1"
Elaine McGahey
Leo 10h18'17" 26d6'22"
Elaine McKenzie
Ori 5h32'7" 9d53'58"
Elaine Meredith Bordner
Ari 3h17'52" 23d52'42"
Elaine Michelle Hutchen
Dra 10h11'20" 78d29'10"
Elaine Murphy
Uma 11h27'42" 58d44'49"
Elaine - My Best Friend
Per 3h3'40" 55d16'4"
Elaine Nahyan Kang
Sco 17h17'56" -32d13'15"
"Elaine Okoro"
Cas 23h4'25" 53d33'29"
Elaine Olin
Ari 3h12'23" 28d5'3"
Elaine Osborne & John Bicknell
Aur 6h28'33" 42d49'17"
Elaine Papis
Cnc 8h50'53" 26d16'4"
Elaine Pauline Condos
Psc 0h47'37" 8d33'45"
Elaine Qualtiere
Cnc 8h46'12" 30d14'57"
Elaine Raschke (Mom)
Crb 16h6'21" 36d20'20"
Elaine Reeves
Gem 7h48'41" 26d27'48"
Elaine Rempala
And 0h16'37" 41d14'52"
Elaine Robinson
Leo 9h44'14" 31d13'30"
Elaine Romano
Lyn 7h29'49" 45d9'1"
Elaine Rossiter - Lainey's Rock Star
Lyn 7h22'45" 54d13'48"
Elaine S. Edelman
Tau 4h6'26" 11d8'21"
Elaine Scott Hoffman
Del 20h34'22" 15d19'51"
Elaine Scott Stoll
Aqr 22h9'24" -2d38'49"
Elaine Sherman
Uma 10h16'9" 56d29'27"
Elaine Smyth
Umi 14h3'23" 71d31'32"
Elaine Sneed
And 0h7'42" 36d58'33"
Elaine Speller
Cas 0h36'58" 62d40'58"
Elaine & Stephen
Vir 15h10'16" 1d56'7"
Elaine Sutherland International
Cas 23h10'54" 59d40'24"
Elaine Tanner
Cas 23h34'24" 54d0'44"
Elaine Tate
Lyn 7h47'41" 38d37'40"
Elaine Taube
Ori 6h5'31" 11d19'48"
Elaine the Omega
Cas 0h42'45" 66d34'22"
Elaine Therese Parker
Cas 0h19'11" 50d7'9"
Elaine Valliere-Wittig
Sco 16h9'10" -8d53'56"
Elaine Varriano
Cas 1h48'1" 60d34'23"
Elaine Violyn Luzine
Aqr 20h46'22" 2d1'29"
Elaine Virginia Parker
Vir 14h5'21" -9d9'38"
Elaine "Wish Bear" Lipsun
Umi 15h32'25" 70d55'52"
Elaine Wong
Cnc 8h50'19" 19d2'20"
Elaine Wood
Cyg 19h37'4" 36d29'33"
Elaine Wood Swearingen
Gem 6h39'57" 26d17'53"
Elaine Woodard
Uma 8h23'8" 59d57'49"
Elaine090397
Sco 16h14'58" -16d46'55"
Elaine1005
Crb 15h23'30" 30d13'7"
ElaineJoyce
And 0h51'39" 37d53'46"
Elaine's Eyes
Cyg 20h10'10" 59d29'30"
Elaines' Guiding Star of Hope
Cnc 9h15'44" 23d2'36"
Elaine's Lucky Star
Vir 12h59'31" -20d49'38"
Elaine's Point
Ori 5h53'32" -0d47'55"
Elaine's Prayer
Tau 3h47'21" 26d32'16"
Elaine's Sixteenth
Cap 21h22'23" -15d14'27"
Elaine's Star
Lyn 6h59'44" 53d24'29"
Elaine's Star
Cas 1h18'38" 69d1'15"
Elaine's Star
Cam 3h24'15" 65d58'27"
ELAM
Ser 15h30'29" 19d6'50"
Elam Boila
Aql 19h21'17" -10d31'33"
ELAN
Lyr 18h50'8" 43d7'38"
Elan Malach Sigler
Umi 15h40'41" 71d27'57"
Elana Dadd
Cru 12h17'47" -56d49'28"
Elana Faleck
And 2h28'40" 50d7'36"
Elana Michelle Blatt
And 1h21'41" 47d46'53"
Elana Nicole
Leo 10h13'54" 26d42'46"
Elana Winter
Leo 10h52'52" 9d57'21"
Elandra Grace Christopher
Cas 1h15'9" 63d28'34"
Elane C Tohmc
Vir 13h26'18" 12d34'36"
Elango
Leo 9h36'28" 32d22'2"
Elanor
Cas 23h33'39" 54d50'43"
Elastico 22-01-2005
Lyr 18h32'39" 36d28'41"
Elayna Beth Kroona
Gem 7h15'19" 31d18'23"
Elayna Kyle Robins
Cas 23h36'43" 54d36'13"
Elaysha Brooke Stipanovich
Vir 12h22'38" 11d3'8"
Elba Abundez
And 1h53'10" 41d22'35"
Elba Barbara Linares
Cap 20h51'57" -16d40'48"
Elba Durieux
Uma 9h35'1" 68d13'55"
Elba Julia Munoz Doerner
Tau 5h13'23" 17d21'38"
Elba Lorena Paredes Guerrero 102103
Uma 13h50'37" 61d38'51"
Elba Rosemary Gonzalez-Morera
Lib 15h5'7" -4d55'53"
Elbereth
Vir 13h50'58" -17d55'41"
Elbert D Brooks
Her 17h56'13" 29d54'59"
Elbert M. & Marilyn T Jones Pittman
Cyg 20h46'12" 42d10'2"
Elbert, Kristin
Uma 10h40'11" 46d11'58"
Elberta
Cnv 13h13'23" 43d55'33"
Elbiris Adymargarite Brown
Tau 3h39'18" 18d11'2"
El-Bow
Uma 11h7'25" 60d6'58"
Elbunit
Ari 3h23'31" 16d32'13"
Elby Gonzalez
Psc 23h51'40" 5d50'30"
Elda
Cas 23h1'28" 53d51'30"
Elda & Ernie Bengert June 24 1950
Aql 19h33'43" 11d11'36"
Eldar Bagirov
Uma 10h11'13" 64d35'1"
Eldar Kadymoff
Lyn 7h24'23" 51d3'20"
Eldarwen Lúinwë
Cnc 7h57'11" 11d26'41"
Elden & Laurette Thompson
Uma 10h27'10" 43d9'22"
ELDER
Cyg 19h56'51" 42d50'59"
Elder "Blue Eyes" Walsh
Her 16h25'19" 18d44'44"
Elder "Brown Eyes" Roberts
Her 17h14'53" 13d32'43"
Elder Nathan Kevin Roberts
Aql 19h32'5" 13d32'18"
Eldin Smith
Her 18h8'47" 39d9'47"
Eldon and Mildred's Heaven
Ari 3h28'34" 27d32'57"
Eldon J. Stoutenburg
Uma 8h15'20" 66d35'39"
Eldon L. Sheffer
Uma 13h27'40" 56d1'8"
Eldon W. Jenkins "Love You More"
Ori 6h15'43" 14d19'7"
Eldon W. Wands
Aql 18h55'20" -0d21'31"
Eldon's 4 Hearts Forever
Uma 10h48'49" 60d39'30"
Eldora Jakie Star
Gem 6h37'2" 15d3'25"
Eldora Starr
And 2h34'41" 47d19'37"
Eldrid Grindvold
Psc 2h2'36" 7d44'33"
Ele
And 2h14'28" 42d13'0"
ELE 11.10.06
Ori 4h49'17" 13d23'12"
Ele la mia più grande amica
Crb 15h59'49" 36d50'28"
Elea Raiswell
Cnc 8h5'25" 13d5'27"
Elea1012004
Cyg 19h26'30" 35d57'2"
Eleana Marie Mauceri
And 23h14'4" 51d44'8"
Eleanna
Uma 11h40'17" 54d6'31"
Eleanne Soto
Ari 2h51'8" 19d10'19"
Eleanor
Peg 21h42'14" 14d40'17"
Eleanor
Vir 13h12'46" 4d57'40"
Eleanor !!!
Gem 6h48'47" 15d3'34"
Eleanor
And 1h13'53" 45d19'14"
Eleanor
Uma 8h48'32" 48d22'12"
Eleanor
Lyn 8h18'1" 46d41'54"
Eleanor
And 23h34'1" 43d46'50"
Eleanor
And 1h40'11" 39d3'50"
Eleanor
Uma 9h18'7" 54d49'27"
Eleanor
Cas 0h49'8" 61d50'51"
Eleanor
Uma 9h8'47" 65d19'20"
Eleanor
Cma 6h50'6" -21d23'42"
Eleanor A Torresen
Del 20h38'7" 15d49'57"
Eleanor and Claire's Star
Uma 9h37'48" 57d47'13"
Eleanor and Ernie Fleury
Cyg 19h18'10" 52d48'25"
Eleanor and Sam Carr
Cyg 20h24'8" 33d29'37"
Eleanor Ann
And 1h54'45" 46d7'28"
Eleanor Avery
Vir 13h9'25" -20d25'35"
Eleanor Barrett
Lyr 18h29'52" 31d40'2"
Eleanor Bechler
Uma 8h40'9" 49d48'2"
Eleanor Belen
Tau 4h25'4" 3d5'6"
Eleanor Betz
Cas 1h45'32" 64d40'41"
Eleanor Blanche Brooks Membrino
Uma 11h1'55" 38d16'56"
Eleanor Blythe Armstrong
Leo 9h37'25" 24d23'13"
Eleanor Boulanger
Cas 1h22'35" 60d33'3"
Eleanor Brady Schlegel
Sco 16h13'25" -14d26'15"
Eleanor Brass Guenther
Leo 9h34'10" 17d7'42"
Eleanor Brown Goldberg
Psc 1h38'37" 22d33'30"
Eleanor Bryn Power
Leo 11h25'3" 7d29'30"
Eleanor Burg
Peg 22h38'3" 23d10'17"
Eleanor Catherine Weglein 10/4/1914
Cas 1h19'47" 62d27'30"
Eleanor Christine
Aqr 23h7'59" -7d24'15"
Eleanor Christine Salgado
Cap 21h14'8" -23d56'34"
Eleanor Constance
And 23h16'11" 41d42'7"

Eleanor D. Kelley
Uma 9h32'58" 65d33'11"
Eleanor Decaro
Cas 23h5'40" 54d35'47"
Eleanor DeNoble
And 23h25'1" 47d29'21"
Eleanor Dufner
And 23h24'8" 48d1'52"
Eleanor E. DeBelles
Umi 14h28'14" 72d21'13"
Eleanor & Edward Steinmetz
Cyg 21h14'9" 45d59'45"
Eleanor Elizabeth Ann Collins
Cas 0h49'56" 63d52'49"
Eleanor Elizabeth Cassidy
Mon 7h31'30" -0d19'12"
Eleanor Elizabeth Hall
Peg 23h18'49" 23d58'0"
Eleanor Elizabeth Kelly
Uma 10h50'36" 50d59'57"
Eleanor Evans
Leo 11h0'35" 8d0'40"
Eleanor Ewing
Cas 1h1'18" 63d44'48"
Eleanor Felicia
Lib 14h50'15" -1d0'4"
Eleanor Flynn Cartland
Psc 23h35'8" 4d36'2"
Eleanor Foley
Ari 3h14'9" 23d28'44"
Eleanor Fookes
Vir 12h56'20" 4d12'53"
Eleanor G. Allen
Cnc 9h17'7" 13d58'44"
Eleanor Gamicchia
Cas 23h34'25" 52d59'24"
Eleanor Grace Downs
And 0h32'40" 42d41'36"
Eleanor Grace Hazell
And 1h14'45" 45d45'43"
Eleanor Grace Sheridan
And 1h54'51" 36d35'0"
Eleanor Grube Lovret 12/25/01
Cas 2h29'1" 66d5'21"
Eleanor Hailpern
Ret 4h11'0" -61d32'59"
Eleanor Hamblin
Cyg 20h1'42" 37d23'32"
Eleanor Harrington
And 0h43'22" 40d24'20"
Eleanor Holberg
Psc 0h36'12" 12d59'17"
Eleanor Hunt
Ori 5h40'33" 4d46'40"
Eleanor Israel Ponder
Her 16h39'46" 37d18'45"
Eleanor Jane "Miss Ellie"
Cas 23h38'16" 57d16'29"
Eleanor Jenner
Vir 12h33'42" -7d20'23"
Eleanor Jennings
Ori 4h43'37" 0d46'48"
Eleanor Joy December 1
Lyn 7h53'17" 56d34'47"
Eleanor K. Hastie
Gem 6h44'46" 33d19'29"
Eleanor K. Shoemaker
Lyn 8h18'13" 33d28'45"
Eleanor Katherine Cowan
Cyg 21h47'10" 54d22'21"
Eleanor Kathleen Cawthon
Cnc 8h27'59" 14d54'26"
Eleanor Kaye Benson
Lib 15h46'17" -18d33'42"
Eleanor L. Fontes
Uma 13h53'27" 57d59'31"
Eleanor L Phail
Lmi 10h53'39" 25d48'6"
Eleanor Lampo
Cas 1h24'17" 61d58'3"
Eleanor Langone "Happy 90th"
Gem 7h13'37" 26d23'57"
Eleanor Langston
Uma 10h55'46" 50d19'0"
Eleanor Leifer
And 23h1'33" 47d17'56"
Eleanor Licata
And 0h20'9" 28d33'16"
Eleanor Loretta Newman Mattingly
Cas 1h27'25" 57d38'10"
Eleanor & Louis Golden Anniversary
Uma 11h20'54" 62d32'29"
Eleanor Louise Kuntz
Sgr 17h57'57" -23d50'37"
Eleanor Louise Pattenden
And 1h15'58" 42d17'55"
Eleanor Lynette Dunn
Cru 12h23'42" -57d11'22"
Eleanor Lynnette Malcho
Lyn 6h33'59" 59d12'56"
Eleanor M. Christman
Uma 11h31'45" 53d5'18"
Eleanor M. Gavron
Tau 3h47'11" 16d28'0"
Eleanor M. Kinney
And 22h58'6" 51d44'5"
Eleanor Magdalena McGovern
Aqr 22h15'26" 1d18'18"

Eleanor Mallamo
Uma 8h47'24" 55d35'35"
Eleanor Margaret O'Hara
Peg 22h57'13" 8d26'8"
Eleanor Mary Feuster
Sco 16h3'0" -20d36'34"
Eleanor Mary McNicholas
Cas 0h49'20" 64d11'51"
Eleanor Mary McQuillan
Cas 0h38'27" 51d4'32"
Eleanor Masciotra
Cas 0h27'39" 57d8'7"
Eleanor May Owen
And 23h54'45" 38d35'54"
Eleanor (McGinley) Torrens
Peg 22h47'27" 25d0'50"
Eleanor Mitchell Givens
Lyn 8h54'30" 39d43'11"
Eleanor Mlott
Uma 10h12'18" 63d12'28"
Eleanor Nelson Jones
Cnc 9h13'43" 10d6'57"
Eleanor O'Leary
Ori 5h32'5" -1d44'30"
Eleanor P. O'Connor
Cas 23h32'9" 55d58'36"
Eleanor Philomena Rakonitz
Sgr 17h58'50" -25d15'15"
Eleanor Pizzino
Ori 6h18'21" 10d25'55"
Eleanor Plummer
Umi 16h31'1" 76d5'50"
Eleanor Poopy
And 2h20'34" 38d57'11"
Eleanor Psichos
And 23h51'25" 42d21'49"
Eleanor (Put) Randle Atkins Burris
Lyn 7h7'23" 51d32'54"
Eleanor Quill Barry
Lyr 19h13'28" 39d1'8"
Eleanor R. Znack
Cap 20h14'19" -18d21'22"
Eleanor Rachel McNamara
Ori 5h32'29" -2d14'45"
Eleanor Rachel McNamara
Uma 12h6'15" 45d30'59"
Eleanor Rachel Parry
Ori 6h16'34" 6d22'17"
Eleanor Rae Wambach
Lib 14h49'27" -24d2'29"
Eleanor Regina Speece
And 2h35'50" 42d6'4"
Eleanor Reyes Moreno
Cnc 8h34'45" 26d2'24"
Eleanor Rhyne Coates
Tau 4h19'37" 2d36'15"
Eleanor Riesenberg Graham
Per 4h48'7" 45d38'0"
Eleanor Rosaire Rader-Caso
Lib 15h16'49" -15d41'9"
Eleanor Rose Pearson
And 1h44'37" 45d38'14"
Eleanor Rose Poore
Gem 6h59'6" 34d28'45"
Eleanor Rospierski
Cap 20h15'26" -14d49'23"
Eleanor Sanchez
Cap 21h46'17" -9d13'42"
Eleanor Shields
Lyr 18h46'56" 39d14'10"
Eleanor Slaten Wald
And 23h40'55" 48d19'39"
Eleanor Smith
Gem 7h1'2" 26d57'6"
Eleanor Stokes
And 2h35'2" 49d5'15"
Eleanor Ventimiglia
Cas 1h43'32" 63d33'12"
Eleanora Joachimsen
Uma 9h9'7" 69d8'25"
Eleanore and Bill Cummings Forever
Cyg 20h21'13" 39d48'1"
Eleanore Catherine Cullen
Uma 10h25'0" 47d49'21"
Eleanore Dombrowski Antonowicz
Uma 10h34'57" 41d21'48"
Eleanore Jaeger
Uma 11h32'6" 51d30'48"
Eleanore Ryder
Cyg 20h30'29" 51d12'19"
Eleanor's 75th Birthday Star
Cas 0h4'34" 56d19'54"
Eleanor's Heart
Tau 4h17'18" 26d35'29"
Eleasha
Psc 23h21'48" 4d51'14"
Eleasha
Psc 1h4'26" 32d32'32"
Eleazer Graham
Uma 9h0'20" 59d20'47"
Electra
Sgr 17h49'10" -26d11'3"
Electra Flash
Lyn 8h4'10" 48d18'36"
Electra kai Olympios (Pipinos)
Cas 23h35'33" 52d41'0"

Electra Sky
Aqr 22h19'19" 0d23'28"
Electra Vasilopoulos
Uma 10h30'55" 68d49'35"
Electra Webb Bostwick
Uma 13h36'25" 55d2'27"
Eleen
Tau 3h58'8" 27d2'42"
Eleesha *030388*
Psa 22h27'39" -34d9'11"
Elef Melek Ozder
Cyg 20h5'30" 33d16'31"
Eleftheria Stefanou
And 23h20'41" 47d47'46"
Eleftherios Vlazakis
Uma 11h12'28" 29d39'51"
Elegant Hotels Group
Aur 6h11'15" 52d35'32"
Elegantly Exquisite Erinn Marie
Cap 21h28'41" -10d3'31"
Elek John Paris
Ari 2h21'47" 22d15'57"
Elektra Luxx- Something-er of Noun.
Aqr 22h41'14" 0d47'23"
Elelia Grace
Lmi 9h28'38" 36d40'28"
Elemér & Nadia Szerelemcsillaga
Cas 23h6'35" 55d45'35"
Elena
Cam 6h21'20" 68d37'35"
Elena
Uma 10h47'26" 64d40'53"
Eléna
Cap 20h28'20" -12d54'27"
Elena
Ori 5h38'21" -2d5'48"
Elena
Dra 10h10'41" 79d50'19"
Eléna
Psc 1h25'35" 3d24'1"
Elena
Ser 18h18'48" -14d54'5"
Elena
Per 3h7'5" 51d38'43"
ELENA
Gem 7h6'59" 34d39'58"
Elena
And 23h41'55" 36d46'17"
Elena
Per 2h56'48" 44d17'14"
Elena
Cam 3h55'7" 59d15'56"
Elena
Cam 4h32'58" 56d35'12"
Elena
And 23h56'5" 47d48'11"
Elena
Cyg 21h20'32" 46d49'2"
Eléna
Boo 15h16'38" 46d38'35"
Elena
Uma 10h38'46" 48d35'36"
Elena
Lyn 6h49'20" 52d27'31"
Elena
Lyr 18h45'0" 42d36'11"
Elena
Her 16h25'3" 44d24'21"
Elena
Psc 23h6'26" 0d22'1"
Elena
Ori 5h18'29" 0d8'42"
Elena
Vir 12h10'26" 5d41'16"
Elena
Mon 7h3'28" 6d5'23"
Elena
Mon 7h11'12" 0d9'12"
Elena
Mon 6h48'56" 9d24'30"
Elena
Ari 3h25'24" 26d44'50"
ELENA
Leo 11h23'38" 15d21'54"
Elena 21/10/2004
Ori 6h9'45" 2d27'56"
Elena 64
Per 3h7'10" 51d38'12"
Elena A. Sachenko
Crb 16h16'27" 30d9'1"
Elena - A Wish Upon a Star
Psc 1h11'10" 12d15'29"
Elena Alexandra Ali
Crb 15h35'50" 26d39'37"
Elena Aliece Douget
Uma 8h30'15" 61d30'25"
Elena Amel Sophie
Per 4h10'39" 34d50'52"
Elena Andersen
Cas 1h5'0" 49d42'5"
Elena Angel
Tau 5h46'3" 17d43'16"
Elena Ann Schipani
Crb 16h11'59" 34d43'32"
Elena Ariani
Peg 23h53'16" 27d36'25"
Eléna Ava Esther Tran
Sgr 19h26'34" -21d5'47"
Elena Beautiful Star
And 23h40'58" 48d0'55"
Elena Bethany
Aqr 22h53'15" -21d45'14"

Elena Blinova
Psc 1h14'48" 27d24'38"
Elena Botti
Cyg 19h47'8" 46d35'38"
Elena Bou Ortiz
Cap 20h41'5" -25d47'26"
Elena C. Tipton
Cas 23h2'3" 54d50'16"
Elena Carasso
Leo 11h29'25" 26d23'16"
Elena Carina
Cas 0h31'31" 60d13'52"
Elena Concettina Busson
And 1h19'14" 42d32'34"
Elena Contreras
Cnc 9h0'2" 28d49'17"
Elena Cronin
Cap 21h50'33" -17d0'11"
Elena Devitskaya
Uma 11h9'13" 34d15'54"
Elena ed Andrea
Cyg 19h44'44" 32d20'59"
Elena Elizondo
Leo 11h29'8" 24d53'16"
Elena & Evan Wies
Gem 7h7'29" 32d28'46"
Elena Faith Lewis Mohan
Sgr 17h59'25" -17d46'56"
Elena Falzoni
Sgr 20h17'32" -35d14'15"
Elena Francesca Buckley
Ori 5h33'27" -1d17'13"
Elena G.
Crb 15h27'1" 29d57'29"
Elena Gallotta
And 2h27'26" 42d16'9"
Elena Garduno
Leo 11h51'18" 23d45'37"
Elena Gautschi
Uma 9h49'36" 48d8'21"
Elena + George = L.F.E.
And 0h6'17" 46d41'36"
Elena Geuna
Leo 9h48'48" 28d11'40"
Elena Ilieva - Liony
Sco 17h57'38" -39d28'51"
Elena Ivashevich
Leo 11h6'24" 16d46'25"
Elena Jeannette Witt
Sco 17h43'5" -32d41'29"
Elena Jordan
Crb 15h47'2" 35d17'2"
Elena Joy Gustman
And 23h45'8" 35d51'16"
Elena Karic
Vir 13h15'30" 12d11'58"
Elena Karmen
Cas 23h0'26" 58d15'6"
Elena Klinduhova
Ari 2h6'14" 21d52'43"
Elena Kostka
Uma 12h42'6" 56d26'37"
Elena Kuzmoff
And 22h59'34" 46d43'53"
Elena + Laszlo
Ori 5h5'19" -0d17'27"
Elena Laurenne Marlovits
Cyg 21h41'1" 41d7'12"
Elena Laurenne Marlovits
Cyg 20h45'11" 34d37'17"
Elena Lee
Lib 15h2'35" -1d25'8"
Elena Lee Jaworski
Sgr 19h30'49" -16d27'19"
Elena Lynn Parry
And 0h15'9" 30d9'38"
Elena M Marmo February 18, 1974
Uma 10h17'48" 72d46'0"
Elena Maccione
Psc 0h26'20" 19d31'36"
Elena Maggi
Per 4h48'35" 41d18'46"
Elena Margarita Mira Coddington
Psc 1h18'10" 19d43'51"
Elena María López-Camacho
Gem 7h13'50" 33d41'3"
Elena Marie
Uma 10h53'1" 44d35'18"
Elena Marie Baltazar
Cap 20h19'51" -11d49'40"
Elena Mariotta
Cyg 20h3'25" 46d14'10"
Elena Martha Varela
Lib 14h40'55" -18d2'57"
Elena Maso 01/03/1987
Psc 1h18'18" 23d20'36"
Elena Miranda
Lyn 7h21'41" 44d52'17"
Elena Moosmann
Com 13h33'49" 21d27'17"
Elena Mostes
Her 18h54'50" 18d22'13"
Elena Ogneva
And 2h2'28" 45d54'9"
Elena Oteiza
Ari 3h10'32" 18d2'16"
Elena Pagnamenta
Lyr 18h47'5" 35d49'14"
Elena Popova
And 23h25'49" 49d6'55"
Elena Rigatuso
Tau 4h13'57" 5d44'14"

Elena Rivera
Cas 2h20'51" 62d50'31"
Elena Rose Lowther
And 2h32'41" 45d59'31"
Elena Sandrone
Lyr 18h45'52" 30d50'14"
Elena Shevereva
Dra 18h44'28" 58d59'48"
Elena Staar
Her 16h21'2" 16d12'40"
Elena 'Starfish' Catherine Blake
And 2h26'30" 39d23'7"
Elena Sturman
Ari 3h20'2" 27d44'18"
Elena Suarez
Lyn 8h23'4" 35d57'46"
Elena Suglia
And 2h2'26" 46d46'17"
Elena "Sunshine" Parsley
Lyn 7h50'10" 40d58'1"
Elena Travkin
Gem 6h54'56" 26d26'48"
Elena Troc
Tau 4h10'26" 27d50'55"
Elena Vartanian
And 23h46'46" 42d9'50"
Elena Wren
Aqr 21h43'39" -0d42'17"
Elena Zueva
Sco 16h4'42" -20d0'48"
Elena, My Filipina
Lmi 11h0'22" 29d12'18"
ElenaH
Cap 20h43'56" -14d45'15"
Elena-"Happy 40th Birthday!"
Sco 17h35'0" -32d45'28"
Elena"My Angel" DeLuca
Tau 5h49'16" 18d57'24"
Elena's Paul Tarr
Lyn 7h4'23" 55d34'6"
Elena's Star "Stretch"
Peg 21h43'48" 23d43'36"
Elenda Selina
Uma 9h17'8" 55d27'29"
Eleneth
Gem 7h20'14" 33d7'15"
Elengus
Her 18h13'54" 17d53'13"
Eleni
Gem 6h42'42" 35d11'28"
Eleni
Lyn 9h5'22" 44d35'4"
Eleni
Uma 10h25'39" 63d20'42"
ELENI 28/5/2005
Sco 16h10'20" -17d21'6"
Eleni Farmakis
Cyg 21h30'53" 46d17'36"
Eleni - March 10, 1916
Psc 1h15'9" 23d54'38"
Eleni Mihailidou
Mon 6h39'23" 5d20'7"
Eleni Papapostolou
Uma 8h48'11" 56d44'21"
Eleni Vakalopoulos
Vir 13h5'50" -0d2'22"
Elenita
Cam 3h34'40" 53d18'15"
Elenitsa
And 0h31'11" 29d53'33"
Elenore Eichman
Uma 12h22'40" 62d49'54"
Elent Vorontsova
Uma 8h54'30" 63d34'9"
Eleonora
Cas 23h31'59" 56d31'25"
Eleonora
Sgr 19h43'31" -13d46'20"
Eleonora
Cam 4h35'29" 58d13'55"
Eleonora
Peg 21h30'35" 23d30'56"
Eleonora
Cyg 19h43'46" 29d31'41"
Eleonora
Cyg 19h45'7" 46d34'47"
Eleonora Belyaeva
Uma 11h9'27" 41d36'23"
Eleonora Cardamone
Lyn 7h53'35" 52d37'39"
Eleonora Del Fattore
Her 17h32'21" 12d26'26"
Eleonora Francesca Orena
Ori 5h42'1" 7d26'10"
Eleonora Gemini
Umi 16h57'7" 85d21'45"
Eleonora Maria Diletta Scaggion
Uma 10h47'21" 58d59'42"
Eleonora Maria Seminara
Peg 21h51'5" 16d35'38"
Eleonora Sanguinetti
Cam 6h51'41" 69d33'24"
Eleonora Scacchetti
Crb 16h22'15" 30d46'5"
Eleonora Verducci
Lyr 18h15'58" 39d29'58"
Eléonore
Cap 21h56'50" -8d28'49"
Eléonore Lauga
Uma 13h45'8" 50d33'31"
Eleos
Lyr 18h32'21" 27d30'35"

Elephant Juice
And 23h26'13" 41d33'8"
Elephant Sanctuary Tennessee
Uma 9h27'58" 61d25'40"
Elephant Shoe
Psc 1h11'54" 26d0'41"
Elephant Shoes!
Leo 10h56'52" 7d22'16"
Elephant Shoes
Umi 18h51'45" 88d7'41"
Ele's Star
Sgr 18h17'14" -18d51'2"
Elesa Fahlen
Cyg 20h5'9" 48d32'22"
Elesha Emily Kulavere
Sgr 19h30'11" -41d55'2"
Elesia - World's Best MOM
Lyr 18h51'20" 39d35'5"
Elessar Aragorn Myne
Ori 5h2'29" 5d1'38"
Életem értelme, Frisch Gábor (dr.)
Lib 15h25'53" -19d51'46"
Életem Fénypontja
Uma 9h48'21" 51d26'33"
Életem Szerelme Kukk
Uma 12h1'46" 60d27'39"
Eleuthero Asteri
Pho 0h35'1" -51d4'22"
Elevation
Aql 19h8'14" 11d22'56"
Eleven Steps
Lyr 18h52'7" 39d24'43"
Eleven TLR
Uma 10h42'36" 62d59'31"
Eleven-Eleven
And 23h21'39" 51d8'17"
Elexia Belle
Cyg 19h44'7" 30d32'31"
elf
Cam 6h49'41" 66d54'48"
ELFASM
Uma 10h26'44" 61d36'15"
Elfentraum
Oph 17h11'5" -23d9'11"
Elfers McGreal
Tau 5h26'38" 17d16'56"
Elfi
Leo 9h59'30" 19d11'31"
Elfi Kostelanik (Mama)
Leo 11h1'47" 23d16'47"
Elfie Kotzer
Uma 12h24'2" 52d54'35"
Elfie Zeidler
Gem 6h23'5" 22d26'28"
elfiran1943
And 23h1'32" 46d44'2"
Elfrerieda Leisalotte Palmer
Leo 10h22'2" 27d29'41"
Elfrida Garcia
And 23h23'1" 40d45'45"
Elfriede & Johann
Uma 10h54'27" 72d24'40"
Elfriede M.A. Stadlwieser
Tau 5h1'45" 24d52'32"
Elfriede Petrovitch
Uma 11h38'43" 28d31'13"
Elgie 50
Cep 22h2'38" 57d6'46"
Elgilmelthil
Crb 16h10'59" 28d29'49"
Elgin M Wells
Uma 9h47'46" 49d26'38"
Elham Kaveh
Cyg 20h51'4" 48d21'35"
Elham Salamé
Boo 13h50'33" 11d18'44"
Elhamisin
Uma 9h8'24" 64d45'33"
Eli
Sco 16h9'41" -13d4'17"
Eli
Sco 16h26'24" -40d39'54"
Eli
Uma 10h39'35" 71d33'38"
Eli
Psc 1h39'59" 7d35'39"
ELI
Tau 4h33'4" 8d20'28"
Eli
Her 17h14'50" 27d11'21"
Eli
Cnc 7h56'5" 16d15'59"
Eli
Per 2h43'37" 54d9'59"
ELI_1967
Lyr 18h48'1" 33d8'50"
Eli A. Rosenberg
Lib 15h15'42" -3d58'5"
Eli Amadae Bradford
Umi 15h6'49" 80d27'16"
Eli Borovoy's special star
Her 18h41'58" 20d6'15"
Eli Cruise Zimmerman
Umi 8h54'3" 89d7'52"
Eli Franklin Ramer
Uma 8h37'30" 67d23'18"
Eli Fuchsman
Her 17h34'48" 43d39'32"
Eli Gonsalves
Uma 13h55'30" 53d59'30"
Eli Harris Ashby
Aql 19h58'3" 7d51'29"

Eli Hudson
Aqr 22h40'30" 0d36'45"
Eli Hunter Bardin
Oph 16h33'37" -0d11'13"
Eli Jacob Schroeder
Aql 18h49'31" -0d3'1"
Eli James Jepsen
Uma 9h11'25" 58d59'11"
Eli Jesse
Ari 2h35'58" 21d2'17"
Eli Ko
Ari 3h24'5" 23d29'45"
Eli Lee
Uma 12h35'31" 61d23'48"
Eli Locke Crystal
Sgr 18h24'40" -33d2'11"
Eli Lopez
And 0h58'4" 34d37'41"
Eli Maxwell Tedrow
Sco 17h57'9" -38d19'10"
Eli My Bella
Cas 2h20'24" 71d52'12"
ELI = PALADIN=
Her 17h37'39" 20d22'39"
Eli Pierce & Kelly & Brad
Cyg 20h11'56" 52d21'57"
Eli Ravi's Light
Umi 16h48'2" 82d8'24"
Eli Salomon Contreras
Uma 9h25'0" 49d22'37"
Eli Sandy
Umi 17h9'47" 76d33'32"
Eli Sweigart
Umi 14h32'39" 74d15'2"
Eli & Tito
Tau 3h55'25" 17d40'8"
Eli Victor
Her 16h15'15" 7d34'33"
Eli Warren Gaines
Cap 20h33'13" -9d15'26"
Eli William Larsen
Vir 14h6'39" -2d19'21"
Eli Yazdjerd "Eli's Star"
Cap 20h32'55" -24d21'54"
Elia
Cap 20h9'3" -25d47'7"
Elia
Vir 13h33'22" -11d33'26"
Elia
Ori 5h34'34" -3d4'42"
Elia Amezquita Diaz
And 23h22'22" 37d40'33"
Elia and Chris 12-28
Cap 20h36'14" -9d27'40"
Elia e Franco Azzani
Lyr 18h42'2" 29d34'12"
Elia Francesca
Vir 13h0'27" -0d30'56"
Elia Gian Lorin
Umi 16h22'2" 71d59'18"
Elia Magliulo
Lyr 19h7'21" 36d18'34"
Elia - 'The Bebu'
And 1h21'56" 39d59'28"
Elia Valdez
Cap 21h39'57" -14d29'5"
Elia Yaritza Berbaum Barrios
Crb 15h34'16" 39d5'16"
Eliad Adam Becker
Cap 21h6'1" -15d48'57"
Eliana
Uma 10h5'4" 72d25'55"
Eliana Grace Anglada
And 0h39'18" 31d21'43"
Eliana Grace Sander
Lib 14h35'58" -24d40'21"
Eliana Guadalupe Montalvo
Cam 6h59'33" 64d30'11"
Eliana Kurzum
Gem 7h49'48" 21d14'10"
Eliana Marie Gomez
Lyr 18h51'27" 41d53'30"
Eliana Mirabilis
Cas 0h17'58" 51d24'40"
Eliana Pezzilli
Cas 23h14'39" 55d24'41"
Eliana Villalta
Leo 9h42'31" 26d36'30"
Eliana's Star
Aqr 22h13'55" -17d43'57"
Eliane
Cas 2h19'21" 72d59'13"
Eliane
Uma 10h30'58" 72d48'33"
Eliane
Vir 12h45'45" 5d49'33"
Eliane
Per 2h23'36" 51d24'17"
Eliane Duhayon
Cap 20h14'43" -9d50'0"
Eliane & Harry
Per 2h17'42" 51d16'45"
Eliane Joumard
Lib 15h49'45" -20d3'25"
Eliane Marcelin
Ari 1h58'39" 23d3'1"
Eliane Rossi
Leo 10h7'36" 24d6'29"
Elianna Alberotanza
Aur 5h21'19" 49d48'18"
Elianna my Bobolina
Leo 11h43'54" 14d7'48"
Elianna Rayne Politte
Vir 14h50'2" 4d7'32"

Elianna Sheva
Lmi 10h18'36" 36d1'22"
Elias
Cnc 8h46'50" 30d38'55"
Elias
Peg 22h54'30" 30d54'43"
Elias
Dra 18h53'22" 50d46'13"
Elias and Alem
Umi 15h19'45" 71d40'51"
Elias Argyropoulos
Uma 10h49'6" 39d58'46"
Elias Baumgartner
Leo 9h46'44" 28d18'9"
Elias Dionysios Levada
Tannous
Tau 4h51'54" 24d5'14"
Elias Evjen
Cam 5h40'53" 69d44'16"
Elias Johann
Cas 0h11'33" 65d31'15"
Elias John Castillo
Ari 2h8'15" 15d59'12"
Elias Khyar Hussan
Uma 12h4'57" 48d43'11"
Elias Luis Ivone Chajet
Ari 3h22'53" 24d31'2"
Elias Mario Rocha II
Tau 5h41'56" 26d29'20"
Elias N. Fasolas
Sco 16h13'55" -9d56'32"
Elias Papachristos
Uma 9h13'28" 46d37'52"
Elias Papaelias
Cap 20h29'54" -10d2'54"
Elias Plastiras
Cru 12h17'3" -57d9'24"
Elias Quinn Stephens
Cyg 21h30'16" 41d42'33"
Elias Scott Mohr
Umi 16h9'1" 70d23'11"
Elia's Star
And 0h15'24" 26d33'50"
Elias Tomas Miller
Per 4h22'20" 32d49'1"
Elicia
Cnc 8h54'59" 17d38'42"
Elicia and Wayne Kessler
Lac 22h30'58" 51d11'5"
Elicia Anneke Lopez
Cnc 8h45'26" 26d26'40"
Elicia Bonamy-Price
And 23h42'29" 47d20'34"
Elicia Cloutier
Uma 14h23'29" 60d37'48"
Elicia Hope Izaldo
And 23h11'7" 52d18'21"
Elicia Martinez - Rising Star
Crb 16h10'10" 38d58'30"
Elida
Cnc 8h1'49" 21d26'6"
Elida
Tau 5h42'28" 19d22'32"
Elida Elizondo Guerra
Ori 5h36'27" -1d21'0"
Elida M. Cruet
Aql 20h1'1" 10d17'38"
Elidia
Ori 6h6'56" 17d26'5"
Elidia P. Andrade
And 1h14'36" 43d0'14"
Elie Akel
Cyg 19h58'46" 30d46'30"
Elie Cabanel
Uma 10h6'1" 44d42'0"
Elie Matta
Crb 15h24'3" 32d11'42"
Elie Rabih Hammouche
Uma 10h52'25" 69d29'34"
Elie Rahme
Cyg 20h6'4" 59d36'37"
Eliezer
Cep 22h22'43" 73d47'55"
Elija - R1
Tau 4h27'48" 25d40'55"
Elijah
Her 17h39'45" 27d50'15"
Elijah
Cnc 9h21'36" 10d29'14"
Elijah
Umi 13h43'43" 75d58'0"
Elijah
Sgr 18h9'39" -16d16'20"
Elijah Alon
Her 17h25'8" 32d20'22"
Elijah and Rebekah Frank
Per 3h18'27" 47d21'16"
Elijah Asher Zimmerman
Umi 15h46'36" 77d29'21"
Elijah Barrett Thompson
Ari 2h3'20" 23d38'46"
Elijah Beck Rowe
Lib 14h33'28" -18d42'44"
Elijah Benjamin Flores
Ori 4h55'45" 2d34'24"
Elijah Benjamin Snead
Tau 5h22'54" 23d19'59"
Elijah & Brooke
Cyg 19h50'14" 39d41'12"
Elijah Carver Owens
Gem 7h6'43" 24d15'37"
Elijah Christian Ford
Lyn 9h5'0" 41d6'53"
Elijah Christopher Sambar
Ari 2h12'58" 22d57'19"
Elijah Dafydd Watkins
Gem 6h40'12" 20d3'23"
Elijah Douglas
Her 16h33'18" 32d19'34"
Elijah "Eli" Frailey
Leo 9h43'59" 27d57'26"
Elijah Enrique Reyna
Cap 21h38'52" -14d38'27"
Elijah Frankle
Uma 9h19'19" 49d55'3"
Elijah Jacob Winkler
Sgr 18h4'16" -20d40'31"
Elijah James
Sco 16h42'22" -43d47'7"
Elijah James Douglas
Psc 0h28'17" 7d38'23"
Elijah James Pozzi
Ari 3h24'43" 28d2'39"
Elijah Ja'quez
Per 3h3'33" 45d3'24"
Elijah Jared John - Our
Little Star
Cru 12h33'43" -56d3'4"
Elijah Jay Hoff
Her 17h36'56" 26d42'11"
Elijah Jeffrey Hale
Per 3h52'28" 32d13'4"
Elijah John Atherton
Per 3h5'39" 44d6'0"
Elijah Joseph Kozak
Tau 4h41'59" 7d54'28"
Elijah Joseph Von
Gnatensky
Gem 7h48'47" 32d0'7"
Elijah Jovani
Sgr 17h56'50" -17d27'24"
Elijah Laconna Harris
Umi 15h45'38" 72d32'57"
Elijah Luke
Her 17h56'13" 23d36'43"
Elijah Michael Haddad
Ori 5h28'20" 4d56'5"
Elijah Murray Freeman
Her 17h29'23" 36d58'23"
Elijah Patrick Kenneth
Knowles
Cep 22h49'53" 77d0'1"
Elijah Praise Nickerson
Uma 11h32'42" 64d5'0"
Elijah Quinn Hanson
Aqr 22h37'3" 1d16'1"
Elijah Ramsey Hollander
Psc 1h14'49" 25d19'4"
Elijah Samuel Steven
Norbert
Per 3h10'59" 54d23'26"
Elijah Santana Grant Lopez
Per 2h54'35" 36d49'58"
Elijah Shihad
Uma 10h34'4" 44d20'28"
Elijah T. McWilliams
Ari 2h9'33" 23d15'42"
Elijah Water Cook
Per 4h36'56" 31d32'51"
Elijah, 28.01.1981
Lyr 18h45'51" 33d9'2"
Elijah's Blue Giant
Ori 6h9'8" 20d48'28"
Elijah's Light
Ari 3h2'43" 27d53'53"
Elika
Uma 11h31'18" 31d59'47"
Eli-Max-Hartman
Peg 23h11'19" 17d46'7"
ELiN
Vir 13h29'41" 2d21'55"
Elin
Cam 3h53'27" 58d53'23"
Elin Abad
Ari 2h3'28" 17d39'19"
Elin Christine Wright
Cet 2h45'58" -0d26'57"
Elin Hammarbäck
Ori 6h16'57" 14d37'44"
Elin Marlene Thuner
Gem 7h35'36" 34d35'40"
Elin Marloes Hoekstra
Sct 18h30'3" -15d18'17"
Elin Shim Rydstrom
Vir 13h6'22" 6d29'31"
Elin & Tiger
Ori 5h59'44" 19d49'19"
Elina
Leo 10h44'39" 10d41'52"
Elina Cara Vogel
Ori 6h22'54" 10d7'54"
Elina Eganova
Uma 12h29'21" 55d3'2"
Elina Vayzman
Uma 11h49'40" 55d3'55"
Elina Vilenskiy
Vir 13h6'47" 2d11'16"
Elinaki Mou
Tau 3h56'0" 26d51'49"
Eline
Cnc 9h7'26" 31d22'49"
Eline
Vir 14h6'56" -15d58'42"
Elinor AKA The Angel Otto
Lyn 6h30'43" 55d49'10"
Elinor Diane Seth
Cas 1h39'32" 62d26'35"
Elinor Foltz
Umi 15h28'58" 72d48'49"
Elinor Geraldine (Lamb)
Kenneally
Peg 21h47'45" 27d15'25"
Elinor R. Bernholz
Gem 7h14'30" 28d57'29"
Elinor Roth Catsman
Cnc 8h37'48" 7d0'20"
Elinor S. Elliott
Uma 8h37'17" 72d21'35"
Elinore Petty Praet
Cas 2h38'13" 66d40'9"
Elio
Cas 0h33'14" 61d10'47"
Elio Bruno
Uma 10h50'17" 61d41'1"
Elio Rodriguez
Aqr 22h9'18" -13d19'55"
Elio S. Carra
Aql 19h59'32" -0d29'9"
Eliot
Cap 20h38'9" -9d24'17"
Eliot
Sgr 18h26'35" -17d58'44"
Eliot G. Clemons III
Cen 13h36'22" -33d7'8"
Eliot & Lilith Rockett
Lyr 18h34'23" 28d5'59"
Eliot Marchioni
Dra 17h23'59" 58d33'57"
Eliot Middleton
Lyn 8h26'40" 56d25'19"
Eliot Pierre Philippe
Gantaume
Cas 23h42'12" 53d27'52"
Eliot Stephen Roosa
Lyr 18h41'12" 37d52'22"
Elioth & Cie
Dra 17h35'50" 68d18'41"
Eliot's Hockey Star
Uma 10h51'14" 63d0'55"
Eliott and Bettina Krems
Per 3h17'45" 48d16'5"
Eliott Knight Browne
Per 3h42'34" 43d22'11"
Eliott Le Moël
Psc 0h38'47" 8d31'13"
Eliris
Tri 2h14'58" 33d9'2"
Elis Helena Oliveira
Leo 10h15'14" 25d43'12"
Eli's Light
Per 4h11'20" 43d48'16"
Elis Reed Lorenzo
Dra 15h32'6" 56d13'54"
Elisa
Uma 10h7'45" 66d14'51"
Elisa
Vir 14h10'37" -17d57'37"
Elisa
Lib 15h35'17" -8d25'27"
Elisa
Cyg 21h1'59" 45d1'3"
Elisa
Lyn 7h42'27" 38d1'48"
Elisa
And 23h12'17" 36d34'56"
Elisa
Cyg 20h2'3" 31d6'11"
Elisa
Uma 8h54'26" 47d6'48"
Elisa
And 1h46'39" 46d27'31"
Elisa
Cyg 21h1'59" 45d1'3"
Elisa
Leo 10h57'46" 5d19'14"
Elisa
Ori 6h20'45" 4d22'35"
Elisa
Ori 5h40'42" 1d50'20"
Elisa Annaloro
Her 18h23'52" 22d11'9"
Elisa Arrizola
Cyg 19h47'28" 32d37'12"
Elisa Aven
Ari 3h9'52" 19d57'53"
Elisa B. Cohen
Cam 5h2'5" 63d20'31"
Elisa Bannon
Lyr 18h27'23" 32d12'0"
Elisa Baroni - Magenta
Uma 9h41'55" 68d35'4"
Elisa Chiminello-Lelli
Cas 23h24'28" 58d6'39"
Elisa Choochie Ortega
Tau 5h52'14" 28d27'30"
Elisa Corradini
Aqr 22h47'13" -12d48'49"
Elisa Costantino
And 1h12'6" 45d54'2"
Elisa Demi Janes
Aqr 20h41'19" -2d11'38"
Elisa e Daniele
Cyg 20h1'30" 33d25'48"
Elisa Feroldi di Anita
Uma 9h29'13" 56d38'42"
Elisa GEMMS Star
Aqr 21h40'3" -3d50'7"
Elisa Giuranno
Cap 20h25'51" -15d12'55"
Elisa Grollo
Psc 0h32'39" 7d49'22"
Elisa Hansen
And 23h23'47" 47d8'9"
Elisa Hult
Uma 10h41'0" 42d26'55"
Elisa Iannacone Velasco
Cas 0h59'9" 64d38'9"
Elisa Jean
Crb 15h46'29" 38d21'31"
Elisa Juliana
Tau 5h6'51" 25d24'34"
Elisa K Stroh
Sgr 19h29'16" -16d15'4"
Elisa Kay
Ori 4h59'52" 5d33'46"
Elisa Kay Lambert
Sco 16h57'45" -39d47'48"
Elisa la mia stella
Lyr 18h33'16" 28d40'30"
Elisa Licata Iacovacci
Cnc 8h52'47" 31d23'39"
Elisa Loaiza
Vir 13h41'56" -11d54'11"
Elisa luce
Lyn 8h6'23" 44d6'26"
Elisa Mae
Dra 16h49'47" 56d29'5"
Elisa Mary
Ari 1h57'43" 19d41'16"
Elisa Miller
Peg 22h48'36" 21d34'42"
Elisa Morelli
And 23h14'40" 44d33'58"
Elisa Nanni
Ari 2h40'7" 17d8'12"
Elisa Nicole
Cas 0h40'52" 60d5'8"
Elisa "Nonna" Scali
Lib 15h11'20" -22d21'21"
Elisa & Ocal
Tau 3h54'22" 22d23'31"
Elisa Orsini
Boo 14h43'12" 54d42'39"
Elisa Pontecorvi
Cyg 21h28'30" 35d11'38"
Elisa Puma
Ori 5h52'21" 6d42'20"
Elisa Sandoval
Tau 4h23'57" 24d8'22"
Elisa Schwarz
Sgr 18h3'48" -33d4'32"
Elisa Virginia Palmier
Lyn 7h27'42" 45d38'40"
Elisa Waldi Alvarez
Mon 6h51'47" 3d15'24"
Elisa Zuniga
Cap 20h52'23" -16d7'35"
Elisa, my love
Uma 9h27'22" 57d45'52"
Elisabel
Umi 14h41'56" 79d8'13"
Elisabete Cathrine Bradley
Aql 20h5'50" 2d16'34"
Elisabete Nunes
Uma 10h59'18" 48d36'25"
Elisabeth
Per 4h8'59" 34d19'20"
Elisabeth
Gem 7h16'55" 26d12'24"
Elisabeth
Gem 7h18'12" 27d46'56"
Elisabeth
Psc 1h9'31" 29d27'48"
Elisabeth
Umi 15h4'55" 81d44'19"
Elisabeth
Lyn 8h28'24" 53d48'38"
Elisabeth
Cep 1h31'33" 78d10'22"
Elisabeth 27.04.1945
Uma 8h11'10" 60d8'59"
Elisabeth Achter
Uma 9h54'7" 45d41'32"
Elisabeth Alexandra Collins
And 0h14'43" 28d28'11"
Elisabeth Ann Eisenhart-
Alexander
Uma 8h53'41" 71d24'5"
Elisabeth Bättig
And 0h19'37" 46d17'26"
Elisabeth Bendel Ryan
Tau 4h39'5" 11d44'33"
Elisabeth Betsy
Cas 1h42'45" 61d20'18"
Elisabeth Boxleitner
Uma 8h31'54" 63d56'5"
Elisabeth Brooks
Aql 19h14'23" 2d20'46"
Elisabeth Cebrián Scheurer
Cap 20h8'33" -17d55'3"
Elisabeth D 05111964
Sco 17h17'3" -32d28'32"
Elisabeth D. Smith
Sex 10h25'37" -2d44'41"
Elisabeth Dröser
Ori 4h50'48" -1d55'14"
Elisabeth Forrest
Tau 4h43'3" 6d19'12"
Elisabeth & Fritz Schärer
Cyg 20h49'13" 47d32'1"
Elisabeth Gassmann
Cas 22h58'26" 58d8'7"
Elisabeth Grace
And 23h36'48" 41d32'30"
Elisabeth Grace
And 0h59'21" 40d17'21"
Elisabeth H. Dickson
Cyg 20h6'51" 35d56'23"
Elisabeth H. Fish
Uma 11h15'54" 32d14'31"
Elisabeth Haag-Lidl
Lyr 19h23'16" 37d52'39"
Elisabeth Jane
And 1h17'59" 44d46'16"
Elisabeth K McKeon
Gem 6h20'50" 26d20'51"
Elisabeth & Lac Léman
Leo 11h54'4" 26d13'57"
Elisabeth Lehman
Lmi 9h53'22" 35d2'27"
Elisabeth M. Portman
Cyg 20h12'50" 58d58'24"
Elisabeth Mary Jovanis
And 2h15'26" 37d37'44"
Elisabeth Meddins
And 23h33'48" 41d20'52"
Elisabeth Melle
Col 6h22'43" -34d24'31"
Elisabeth Mergl
Uma 8h15'20" 65d25'37"
Elisabeth Michelle Strouse
Cas 1h40'9" 64d37'40"
Elisabeth M.L. Milstein
Peg 22h26'20" 28d56'3"
Elisabeth Peschka
Dra 17h7'55" 52d57'56"
Elisabeth Prikler
Psc 0h49'40" 21d14'12"
Elisabeth Rebecca
Psc 0h21'59" 6d19'22"
Elisabeth Rüegger-Meier
Umi 16h7'27" 71d46'35"
Elisabeth & Scott
TwoHeartsOneLove
Ara 17h34'32" -47d45'27"
Elisabeth the good-hearted
Lyr 19h7'27" 34d44'4"
Elisabeth Torgler
Uma 8h14'23" 62d24'12"
Elisabeth Tortora
Ari 2h39'5" 21d26'24"
Elisabeth Towne Wright
Ari 2h26'44" 25d38'47"
Elisabeth Trott Hills
Psc 0h17'15" 2d8'42"
Elisabeth Willett
Lib 15h46'36" -13d4'2"
Elisabeth, Forever Holds
My Heart
Uma 11h31'34" 55d12'12"
Elisabeth-Ann Grottke -
9.12.21
Aqr 22h19'51" -22d9'32"
ElisabethsLight
Sgr 17h52'30" -29d13'37"
Elisabetta
Cma 7h9'57" -12d40'22"
Elisabetta
Peg 22h15'27" 23d52'25"
Elisabetta
Mon 7h4'12" 6d23'26"
Elisabetta
Her 17h12'48" 49d27'14"
Elisabetta Alba
Her 18h39'29" 19d7'13"
Elisabetta Fiore
Equ 21h15'53" 10d39'38"
Elisabetta Giannotti
Crb 16h19'56" 31d28'21"
Elisabetta Monterosso
And 2h2'40" 46d41'40"
Elisabetta Nava
Cas 1h51'46" 73d47'10"
Elisabetta Re(Ely)
Uma 14h21'9" 55d46'2"
Elisabetta Soders
Lac 22h22'33" 51d41'18"
Elisange
Aqr 21h40'36" -0d21'9"
Elisaveta
Cyg 20h57'16" 31d45'16"
Elise
And 1h9'10" 37d4'6"
Elise
And 2h28'57" 49d21'24"
Elise
Uma 8h39'11" 51d52'59"
Elise
Vir 13h11'42" 11d6'28"
Elise
Ori 5h48'44" 9d4'38"
Elise
Leo 10h33'54" 24d57'6"
Elise
Sco 16h11'1" -10d12'35"
Elise
Sco 16h5'52" -11d36'1"
Elise
Aqr 22h53'2" -7d50'59"
Elise
Lib 15h34'29" -27d15'14"
Elise 1404
Lib 14h55'10" -16d24'41"
Elise Alene Case
Gem 6h36'53" 22d48'30"
Elise Anne
Cam 3h26'49" 64d23'50"
Elise Anne Jennings
Cap 21h32'7" -18d58'31"
Elise April Chong
And 0h42'45" 37d33'48"
Elise Avera
Cap 20h33'45" -10d40'44"
Elise Bradt
Ori 6h17'6" 14d49'8"
Elise Carol Coleman
And 23h40'54" 46d50'49"
Elise Carolyn D'Adamo
Psc 0h34'54" 5d49'3"
Elise Claire David
Ori 4h49'9" 11d11'2"
Elise Collins
And 1h56'31" 41d41'52"
Elise D. Grimm
Col 5h43'23" -33d30'33"
Elise Danielle
Uma 8h30'32" 66d17'47"
Elise & David Forever
Pho 23h35'49" -40d26'50"
Elise DiBattista
Leo 9h45'44" 28d2'20"
Elise & Dick Martino
Cyg 19h34'1" 33d10'12"
Elise Dray
Cmi 7h59'17" 6d18'15"
Elise Estelle Parker
Ari 2h53'13" 14d35'54"
Elise et Lucie Mores
Umi 16h26'37" 72d10'1"
Elise F Rosenblum
Cyg 20h54'59" 35d9'43"
Elise Flack
Vir 12h33'47" 2d20'59"
Elise Georgina Severson
Cnc 8h44'11" 15d12'8"
Elise Gombos
Peg 22h33'18" 18d30'32"
Elise Hurley
And 0h23'47" 36d29'20"
Elise James
Cnc 8h12'2" 16d41'48"
Elise Jean Camp
Mon 7h15'0" -10d17'10"
Elise & Jonathan
Cap 20h29'4" -15d46'29"
Elise Kassie
Leo 11h18'18" 22d39'21"
Elise Kimberly Manzke
Psc 23h45'27" 2d0'39"
Elise Lalley Hellmann
Cnc 8h43'14" 20d13'51"
Elise Louise Kyle
Cas 0h58'46" 57d2'31"
Elise Mahaut
Leo 11h55'19" 22d21'59"
Elise Major
Ori 6h23'37" 9d58'18"
Elise Margaret Mungovan
And 0h42'43" 37d20'36"
Elise Marie
Umi 16h27'29" 81d47'49"
Elise Maxwell
Lyn 7h20'0" 51d49'55"
Elise May Deighan
Lib 15h22'28" -26d2'57"
Elise May Prunty Shining
Star
Cap 21h50'23" -15d47'35"
Elise Mona Lydia
Dra 19h10'1" 56d29'22"
Elise my Sweetheart
Leo 9h59'26" 16d13'52"
Elise Nicole Brandt
Vir 12h46'51" -10d49'13"
Elise Olivia Fernihough
And 23h23'34" 47d51'10"
Elise Overgaard
Leo 11h53'2" 22d14'14"
Elise Quirindongo
Uma 9h54'33" 70d44'36"
Elise Regenwether
Leo 11h24'34" 23d55'59"
Elise Renee Veach
Cnc 8h4'20" 7d58'2"
Elise Russo
Aqr 22h37'16" -2d46'47"
Elise Sauer "Queen Of
Sales"
Cas 0h47'6" 56d22'2"
Elise Tordella
Cas 1h6'50" 62d22'53"
Elise Victoria Bass
Psc 0h18'0" 5d8'40"
Elise White
Lib 15h25'7" -4d22'58"
Elise102305
Sco 17h25'1" -44d11'45"
Eliseo Corazo
Umi 16h25'11" 71d21'36"
Eliseo Leon
Her 16h38'17" 22d0'31"
Elise's Special Grandad
Mark
Uma 13h31'57" 56d49'28"
ELISHA
Uma 11h33'51" 32d28'12"
Elisha Cole Henderson
Crb 15h38'9" 27d16'44"
Elisha Dawn
Cnc 8h10'2" 29d37'7"
Elisha Dawn Peralta
Cas 23h54'8" 56d3'19"
Elisha M. Worthington
Cnc 8h28'18" 32d59'56"
ELisha Macking
Per 2h16'20" 51d48'46"
Elisha Marie Hanna
And 0h39'4" 32d1'9"
Elisha Marie Lopez
And 1h1'38" 36d11'38"
Elisha Marie Torento
Crb 16h22'13" 38d38'10"
Elisha May
And 23h12'42" 43d54'27"
Elisha Pemper
And 23h5'56" 36d45'3"
Elisha Rhea Singer
Sco 16h11'4" -10d30'37"
Elisha Shipe
Uma 8h13'42" 62d55'44"
Elisha1281
Lac 22h24'24" 48d25'28"
Elishama Joy Spicer
Leo 9h39'30" 20d17'51"
Elishia
And 23h53'24" 39d25'42"
Elisiana Zahara Hill
Leo 9h35'1" 31d49'27"
Eliska
Vir 14h52'15" 3d34'30"
Elislane Ferreira
Sco 16h18'52" -11d21'50"
Elislane Ferriera
Sco 16h54'0" -38d38'14"
Elison Michael Bishop
Uma 8h40'45" 71d33'35"
Elissa
Aqr 23h48'19" -4d12'29"
Elissa
Ari 3h7'32" 19d31'47"
Elissa
Ari 2h51'42" 26d57'0"
Elissa
Lmi 9h49'10" 35d16'44"
ELISSA
And 2h36'50" 44d3'49"
Elissa A. Schrock
Cap 20h29'29" -14d50'59"
Elissa A. Wilson
Lyn 6h19'17" 60d30'33"
Elissa Bassima Freiha
And 23h33'7" 41d32'59"
Elissa Carol Fetty
Dra 20h26'29" 67d28'10"
Elissa Catherine Girratano
Ari 3h16'13" 12d15'42"
Elissa J Schlumpf
Ari 2h41'40" 26d20'22"
Elissa Jean Southon
Cap 21h48'19" -8d43'30"
Elissa Lee
Lib 15h16'35" -23d55'20"
Elissa Maria
Uma 11h45'50" 48d7'26"
Elissa Marie
Sco 16h9'4" -15d18'27"
Elissa Marie the Brightest
Star
Aqr 21h59'44" -15d21'52"
Elissa Marion Gene Jordan
Lib 14h31'27" -24d52'1"
Elissa Mills
And 2h38'20" 49d47'41"
Elissa Virginia Cazan
Valenzuela
Lyr 19h8'45" 27d48'19"
Elissa's Star
Psc 23h27'24" 4d52'38"
Elisse Noelle Kangas
Cnc 8h49'50" 13d54'54"
Elissia
And 2h36'45" 49d45'21"
Elita Chanel Tubbs
Umi 17h30'2" 80d33'42"
Elix
Vir 14h6'28" -12d56'32"
Elix / Berni
Ori 5h19'6" 6d11'27"
Eliya Berger
Uma 10h23'24" 60d52'7"
Eliz
Uma 9h7'36" 56d25'57"
Eliz
Psc 1h28'8" 16d32'41"
Eliza
Ari 3h23'11" 27d27'15"
ELIZA
Per 4h39'9" 48d51'20"
Eliza
And 0h1'17" 42d28'11"
Eliza
And 0h52'30" 40d4'29"
ELIZA
Aqr 22h35'54" -1d2'0"
Eliza
Aqr 22h6'58" -14d8'12"
Eliza
Cma 6h54'22" -28d11'12"
Eliza 1998
Uma 11h40'56" 41d12'45"
Eliza *802*
Psc 0h22'12" 2d29'52"
Eliza Arnold
Psc 1h16'6" 28d36'37"
Eliza Battles
And 1h0'53" 44d52'14"
Eliza Crowley Jackson
Aqr 21h59'6" -7d0'41"
Eliza D. Lehrman
Cap 21h46'54" -11d19'13"
Eliza Elixer Lischin
Sgr 18h8'34" -19d23'16"

Eliza Grace
Pav 19h38'13" -59d44'1"
Eliza Grace
And 0h32'46" 27d46'29"
Eliza Hayley Babb
Lib 15h43'4" -4d35'38"
Eliza Isabella Silva
Aqr 21h52'3" -1d44'19"
Eliza Jane Canter
Uma 10h56'17" 72d1'26"
Eliza Jane Scovill
And 1h35'31" 42d29'12"
Eliza Jo Morgan
Sgr 17h54'58" -29d59'49"
Eliza Jo Oppenheimer
Cas 1h27'46" 61d29'33"
Eliza & Justin's Dream Star
Ori 6h19'36" 14d24'32"
Eliza Kate Douglas
Uma 14h5'46" 57d14'21"
Eliza Lee Cherry
Cap 20h36'12" -18d24'6"
Eliza Loch Lowman
Cnc 8h49'15" 31d52'15"
Eliza Lynne
Ari 2h32'51" 25d0'48"
Eliza Marie McElfresh
Sco 17h50'42" -42d5'36"
Eliza Peyton Donnell
Gem 7h2'7" 17d10'59"
Eliza Raye
Uma 9h55'36" 42d23'44"
Eliza Reynolds
Ari 2h21'17" 12d32'4"
Eliza Rose
Lyn 7h59'1" 39d0'15"
Eliza Timbol
Crb 16h14'12" 33d14'12"
Eliza True
Leo 11h10'8" 4d19'47"
Eliza1
Sgr 18h49'32" -15d57'56"
Elizabet Jane Paxton "Ellie"
Ari 2h43'48" 25d20'50"
Elizabet & Rhonda
Cyg 19h46'59" 42d5'52"
ELIZABETA DURCIK
Per 4h40'48" 41d10'55"
Elizabeth
Uma 11h13'37" 43d56'2"
Elizabeth
And 1h36'22" 40d47'49"
Elizabeth
And 0h48'24" 39d24'10"
Elizabeth
And 0h53'19" 37d44'4"
Elizabeth
Cyg 21h10'25" 30d15'6"
Elizabeth
Lyr 18h23'55" 36d5'3"
Elizabeth
Lyr 19h14'29" 34d35'54"
Elizabeth
Crb 16h5'28" 35d21'44"
Elizabeth
Crb 15h36'13" 35d43'16"
Elizabeth
And 1h19'46" 34d46'38"
Elizabeth
And 0h48'53" 36d38'44"
Elizabeth
And 0h47'1" 33d30'47"
Elizabeth
Cyg 19h54'48" 38d6'13"
Elizabeth
Lyr 18h34'5" 39d36'9"
Elizabeth
And 0h36'54" 45d30'33"
Elizabeth
And 1h36'48" 46d6'40"
Elizabeth
And 23h6'40" 51d48'15"
Elizabeth
And 23h21'36" 48d23'4"
Elizabeth
Cas 23h29'20" 51d38'24"
Elizabeth
Cam 4h58'56" 53d46'5"
Elizabeth
Aur 5h54'36" 53d59'7"
Elizabeth
Cas 1h26'12" 58d51'49"
Elizabeth
Ari 2h9'53" 22d42'35"
Elizabeth
Gem 7h27'14" 22d11'50"
Elizabeth
Gem 6h29'3" 16d48'57"
Elizabeth
Ari 3h12'4" 29d53'7"
Elizabeth
Tau 4h16'20" 27d29'56"
Elizabeth
Cnc 8h52'51" 25d37'45"
Elizabeth
Crb 15h54'4" 27d36'46"
Elizabeth
Leo 11h15'9" 24d27'39"
Elizabeth
Leo 10h17'20" 24d9'15"
Elizabeth
Tau 4h37'31" 5d32'18"
Elizabeth
Ori 5h44'7" 4d24'13"
Elizabeth
Ori 5h36'31" 4d12'41"
Elizabeth
Ori 5h53'46" 8d49'45"
Elizabeth
Aqr 21h31'9" 0d25'3"
Elizabeth
Ori 6h18'56" 10d50'17"
Elizabeth
Vir 12h17'16" 12d57'38"
Elizabeth
Psc 1h20'28" 16d47'46"
Elizabeth
Tau 4h4'56" 17d2'25"
Elizabeth
Tau 5h23'19" 19d50'53"
Elizabeth
Tau 5h45'1" 22d23'53"
Elizabeth
Cap 21h20'8" -18d9'40"
Elizabeth
Aqr 22h57'54" -8d13'12"
Elizabeth
Cap 21h50'24" -8d38'52"
Elizabeth
Aqr 22h16'58" -2d25'40"
Elizabeth
Lib 15h33'39" -7d10'11"
Elizabeth
Leo 11h8'23" -0d29'34"
Elizabeth
Vir 11h49'26" -3d30'44"
Elizabeth
Mon 6h46'53" -6d46'25"
Elizabeth
Mon 6h54'59" -0d36'44"
Elizabeth
Cas 23h24'0" 56d31'17"
Elizabeth
Uma 13h3'55" 52d35'20"
Elizabeth
Uma 12h37'5" 52d44'28"
Elizabeth
Uma 8h28'59" 72d59'7"
Elizabeth
Uma 8h33'39" 65d45'38"
Elizabeth
Uma 10h9'13" 61d59'50"
Elizabeth
Cra 19h1'50" -38d13'5"
Elizabeth
Sco 17h13'39" -38d44'31"
Elizabeth
Cap 21h11'21" -23d10'25"
Elizabeth 2004
Aqr 23h2'23" -5d30'35"
Elizabeth 21
And 23h54'40" 38d41'38"
Elizabeth 247
Cas 0h50'30" 54d16'23"
Elizabeth 308
Cam 7h9'40" 79d25'51"
Elizabeth A. Alexander
Mon 6h53'17" -0d1'18"
Elizabeth A. Anthony
Cam 5h20'29" 64d48'2"
Elizabeth A. Avalos
Cas 0h2'48" 56d15'9"
Elizabeth A. Berger
Cnc 8h4'30" 21d54'41"
Elizabeth A. Bernhardt
Ari 3h22'30" 29d39'29"
Elizabeth A. Brown
Uma 12h10'13" 60d23'51"
Elizabeth A Cromer
Sco 17h56'9" -38d42'37"
Elizabeth A. Cruz
And 1h32'24" 49d38'1"
Elizabeth A Dyke
Vir 13h28'54" 8d46'10"
Elizabeth A. Gatto
Cyg 19h36'31" 32d24'47"
Elizabeth A. Hearn
Gem 6h48'42" 27d44'43"
Elizabeth A Janiak
Lib 15h29'21" -26d34'19"
Elizabeth A. LaFreniere
Gem 7h3'44" 26d59'21"
Elizabeth A Nace
Cyg 21h38'9" 47d15'4"
Elizabeth A. Sanna
Uma 8h44'7" 63d23'26"
Elizabeth A. Scavarda
Ori 5h17'20" 7d39'41"
Elizabeth A. Schiever
Her 16h39'3" 7d22'47"
Elizabeth A. Shadle
Aqr 20h44'46" -2d54'30"
Elizabeth A. Smith
Tau 5h7'21" 17d49'14"
Elizabeth A. Taylor,
PharmD
Cap 21h57'57" -8d29'23"
Elizabeth A. Wade
Mon 6h42'40" -0d55'16"
Elizabeth A. Zane
Vir 13h27'38" -8d45'26"
Elizabeth Ada Valenza
Lib 15h51'2" -7d6'7"
Elizabeth Agothy Colby
Cap 21h42'54" -21d51'51"
Elizabeth Akiko Hamaguchi
Souza
Lib 15h1'58" -2d39'39"
Elizabeth Alexander Kocher
Cas 0h35'40" 51d42'53"
Elizabeth Allen
Cyg 20h19'53" 47d27'14"
ELIZABETH ALLEN
Car 10h55'28" -59d36'34"
Elizabeth Allison
Gem 6h44'54" 19d24'37"
Elizabeth Allyn Lotz
And 0h9'34" 45d33'58"
Elizabeth Alvizo
Leo 9h49'3" 23d58'52"
Elizabeth Amber
Cap 21h30'47" -23d3'20"
Elizabeth Amy Rada
And 23h12'22" 48d14'12"
Elizabeth and Lindsay
Carter
And 0h56'52" 46d18'22"
Elizabeth and Michael
Brindley
Dra 20h11'33" 62d20'42"
Elizabeth and Ryan
Mon 7h6'25" 4d6'42"
Elizabeth and Steven
Hornyak
Cyg 19h36'49" 31d21'0"
Elizabeth Ann
And 2h22'16" 49d30'45"
Elizabeth Ann
Tau 4h24'41" 21d48'28"
Elizabeth Ann
Gem 6h34'46" 13d49'11"
Elizabeth Ann
Leo 10h40'40" 17d9'30"
Elizabeth Ann
Leo 11h25'2" 15d39'4"
Elizabeth Ann
Mon 6h56'46" -0d49'22"
Elizabeth Ann
Lib 15h16'7" -28d37'31"
Elizabeth Ann - 20 - Keep
On Shining
And 2h19'49" 45d45'46"
Elizabeth Ann Anderson
Cnc 9h17'39" 23d29'42"
Elizabeth Ann Archer
And 0h43'23" 46d27'6"
Elizabeth Ann Barry
Gem 6h55'25" 13d28'26"
Elizabeth Ann Basham
Vul 19h50'14" 23d44'45"
Elizabeth Ann Bashian
Ori 5h56'55" 5d45'21"
Elizabeth Ann Bettencourt
Crb 15h41'28" 35d38'20"
Elizabeth Ann Biehn
Ori 6h16'0" 15d56'58"
Elizabeth Ann Brooks
Vir 12h47'8" -5d30'14"
Elizabeth Ann Clement
Uma 9h9'34" 67d46'47"
Elizabeth Ann Combs
Lib 14h51'43" -5d50'8"
Elizabeth Ann Crosby
And 1h12'53" 36d9'32"
Elizabeth Ann Esposito
Leo 9h40'34" 13d12'11"
Elizabeth Ann Evans
Psc 0h4'14" 1d53'53"
Elizabeth Ann Fortune
Cas 23h58'48" 63d47'14"
Elizabeth Ann Fritz
Cam 3h17'5" 60d47'45"
Elizabeth Ann Galloway
Sco 16h38'4" -43d5'47"
Elizabeth Ann Harman
And 0h38'55" 41d59'21"
Elizabeth Ann Hendrix-
Sutton
Uma 11h46'36" 52d39'22"
Elizabeth Ann Hoffman
Boonie
Cyg 20h14'59" 49d7'57"
Elizabeth Ann Hower
And 2h26'34" 46d3'10"
Elizabeth Ann Jackson
Vir 13h53'25" -17d48'25"
Elizabeth Ann Keslo
Lyn 6h17'11" 58d44'21"
Elizabeth Ann Krinkey
Vivian
Vir 13h20'11" -3d42'23"
Elizabeth Ann Lainchbury
Bootle
And 0h5'56" 41d55'19"
Elizabeth Ann Lake
Lib 15h14'44" -20d14'34"
Elizabeth Ann Laney
Cap 21h19'30" -15d22'28"
Elizabeth Ann Lommerin
And 1h54'41" 41d20'10"
Elizabeth Ann Macleod
Cyg 22h1'9" 50d19'41"
Elizabeth Ann McCall
Sco 17h41'35" -39d57'51"
Elizabeth Ann McMurry
Cap 20h25'56" -27d28'44"
Elizabeth Ann Melampy
And 23h0'29" 47d19'41"
Elizabeth Ann Nasby
Psc 0h17'57" 19d9'40"
Elizabeth Ann Rief
Gem 7h6'0" 10d35'34"
Elizabeth Ann Roddy
Cap 20h11'29" -18d36'10"
Elizabeth Ann Rogers
Cas 23h55'51" 56d13'59"
Elizabeth Ann Russ
Crb 15h36'13" 26d2'56"
Elizabeth Ann Shier
Ari 3h11'41" 28d41'45"
Elizabeth Ann Smith
Uma 8h44'43" 62d46'1"
Elizabeth Ann Spaeth
Uma 11h22'9" 36d37'10"
Elizabeth Ann Sullivan
Isenhour
Gem 7h7'4" 18d34'4"
Elizabeth Ann Tate
Lib 15h36'0" -6d48'44"
Elizabeth Ann Tripplehorn
Psc 1h38'2" 10d37'28"
Elizabeth Ann Trupiano
And 1h15'20" 49d8'45"
Elizabeth Anna
Psc 23h11'43" 4d23'15"
Elizabeth Anna
Sco 16h14'8" -22d28'10"
Elizabeth Anna Christensen
9-23-67
And 0h3'51" 46d9'47"
Elizabeth Anna Steckler
Ori 5h38'9" 1d37'2"
Elizabeth Anne
Gem 6h49'23" 14d47'40"
Elizabeth Anne
Leo 9h34'3" 10d6'31"
Elizabeth Anne
And 0h16'26" 27d41'30"
Elizabeth Anne
And 1h12'35" 45d16'13"
Elizabeth Anne
Lmi 10h16'52" 36d54'10"
Elizabeth Anne
And 2h25'35" 38d29'31"
Elizabeth Anne
Lib 15h0'5" -7d18'9"
Elizabeth Anne Alexander
And 23h20'15" 52d11'32"
Elizabeth Anne Brosius
Aqr 22h16'6" -17d20'25"
Elizabeth Anne Duckworth
Cas 1h5'47" 49d54'14"
Elizabeth Anne Englund
Gem 7h9'50" 32d42'19"
Elizabeth Anne French
Ori 4h48'33" -3d12'14"
Elizabeth Anne Holton
Purcell
Cap 20h24'2" -14d56'16"
Elizabeth Anne Howard
Leo 10h30'20" 19d35'11"
Elizabeth Anne Johnson
Lib 15h42'11" -5d41'40"
Elizabeth Anne Klovstad
And 2h10'53" 42d29'41"
Elizabeth Anne Longo
Lib 14h50'38" -5d28'23"
Elizabeth Anne Lupori
Vir 12h3'51" 6d59'3"
Elizabeth Anne Mimnaugh
Sco 17h10'5" -44d38'42"
Elizabeth Anne Schaub
Sgr 18h43'52" -27d33'51"
Elizabeth Anne Smith
Leo 9h38'29" 23d21'32"
Elizabeth Anne Stark
Ari 2h21'57" 22d19'20"
Elizabeth Anne Sullivan
Sco 17h54'5" -36d24'18"
Elizabeth Anne Tabb
Ori 5h34'45" 3d42'35"
Elizabeth Anne Teracino
And 1h32'55" 48d30'54"
Elizabeth Anne Yeager
And 23h26'50" 35d39'12"
Elizabeth&AnthonyFlynn
LeCheileStar
Ori 6h12'57" 15d27'36"
Elizabeth Aphrodite 221951
Cyg 21h35'51" 34d12'21"
Elizabeth Apple
Leo 11h7'59" 22d3'47"
Elizabeth Ariel Thilenius
Psc 1h3'36" 27d7'35"
Elizabeth Arkwright
And 23h35'11" 38d41'36"
Elizabeth Armstrong
"Grammie"
Vir 13h8'50" 12d20'48"
Elizabeth Arredondo of
Harmonia
And 0h48'13" 26d15'36"
Elizabeth Ashley
Ari 3h17'42" 28d30'49"
Elizabeth Ashley Babiuk
Leo 11h25'36" -4d59'18"
Elizabeth Ashley " Boo"
Bisland
Umi 14h50'15" 86d9'45"
Elizabeth Ashley Manser
And 0h12'16" 48d0'9"
Elizabeth Ashley Rine
Lyr 18h31'57" 33d41'14"
Elizabeth Audrey
And 1h49'50" 43d13'19"
Elizabeth Avery
Cas 1h4'45" 54d19'57"
Elizabeth Awerbuch
Sco 17h17'53" -44d51'55"
Elizabeth B. Ameling
Uma 8h36'44" 59d49'27"
Elizabeth B. Hauser
Ori 4h54'43" 12d11'1"
Elizabeth "B" Morse
Cas 0h2'22" 55d59'14"
Elizabeth Babicz
Ari 2h49'18" 20d41'7"
Elizabeth (babygirl) Coley
Cnc 9h8'45" 15d48'34"
Elizabeth Baker Henderson
Col 5h54'28" -37d2'37"
Elizabeth Banon Keys
Vir 15h2'20" 5d13'19"
Elizabeth Barnard
Cas 1h42'37" 61d53'39"
Elizabeth Barnett
And 2h22'31" 43d3'55"
Elizabeth Beanie Wampler
Vir 13h56'35" -16d14'38"
Elizabeth Beaulieu
Psc 0h59'47" 25d2'46"
Elizabeth Behrens
Ari 3h9'6" 27d52'56"
Elizabeth Bell Aldridge
And 1h16'41" 48d48'35"
Elizabeth "Belle" Williams
Leo 10h23'55" 20d17'45"
Elizabeth Benge
Aqr 21h46'38" -1d29'31"
Elizabeth Bergen Adams
Vir 13h20'17" 3d49'19"
Elizabeth Beserra
And 23h2'16" 42d7'25"
Elizabeth (Beth) Roddick
Hardy
Tau 5h18'48" 22d24'30"
Elizabeth "Bets" Dibiase
Ori 6h4'35" 17d36'1"
Elizabeth (Betsy)
Uma 11h53'39" 35d13'36"
Elizabeth 'Betty' Jane
Munson
And 1h46'30" 45d15'15"
Elizabeth Betty Jean Arold
Cnc 8h41'56" 16d17'22"
Elizabeth "Betty" Koziol
Uma 10h33'17" 56d28'30"
Elizabeth "Betty" Skretch
Uma 14h19'23" 56d19'34"
Elizabeth Bi Huang Smith
Ori 5h45'52" 3d42'58"
Elizabeth Bibeau "Our
Lucky Star"
Leo 10h20'28" 7d53'20"
Elizabeth Biggs
Leo 10h24'21" 25d45'2"
Elizabeth Black Summerlin
Lib 15h41'26" -19d22'41"
Elizabeth Blazuk
Aqr 22h32'24" -0d11'53"
Elizabeth Bonney &
Brenton Garrett
Cyg 20h35'22" 56d59'42"
Elizabeth Borts Forever
Young
Vir 13h17'16" 5d46'12"
Elizabeth Brantley McKain
Gem 6h22'45" 21d34'51"
Elizabeth Brenneman
And 0h39'5" 25d44'55"
Elizabeth Breckenridge
Armstrong
Gem 6h40'42" 35d5'39"
Elizabeth Breen's Family
Star
Ori 5h3'19" 14d38'19"
Elizabeth Breshears
Uma 13h46'57" 58d39'0"
Elizabeth Brewer
Ori 5h30'46" -0d12'26"
Elizabeth Brewer
Lyn 8h6'44" 33d19'30"
Elizabeth Brianna Anisah
Marissa
And 0h54'18" 37d12'45"
Elizabeth Briggs Nisbet
Lyr 18h26'23" 32d21'54"
Elizabeth Brooke Farmer
Vir 13h13'2" 3d18'38"
Elizabeth Burg
Mon 6h45'41" 11d37'20"
Elizabeth Burt Lukather
Peg 22h25'6" 34d12'59"
Elizabeth Buxton
Sgr 19h22'47" -15d7'12"
Elizabeth C
Cyg 20h53'15" 31d26'22"
Elizabeth C. Gaiser
Vir 12h24'15" 9d57'38"
Elizabeth C. Harrison
Lyr 18h34'28" 27d20'39"
Elizabeth C. Kane
Cas 0h36'28" 61d24'47"
Elizabeth C. Metcalf
And 23h25'41" 38d48'37"
Elizabeth Carmany's
Saphire Star
Cas 1h30'43" 63d43'16"
Elizabeth Carnall
Cap 21h38'43" -9d56'11"
Elizabeth Carol Bernard
Cyg 20h20'14" 51d46'20"
Elizabeth Cash and Brent
Ecenbarger
Cyg 21h48'19" 47d52'45"
Elizabeth Cassandra
Ruggiero
Cnc 8h59'40" 15d4'0"
Elizabeth Catalina Starr
Lib 15h11'23" -21d46'30"
Elizabeth Catherine Hunt,
My Love
Aqr 22h27'56" -1d13'19"
Elizabeth Catherine White
Ori 5h8'30" 7d24'49"
Elizabeth Catriona
Johnston
And 2h0'3" 42d35'13"
Elizabeth Chandler
Crb 16h7'24" 37d53'10"
Elizabeth Cherkasov
Cap 20h29'21" -13d57'13"
Elizabeth Christine
Lyn 7h55'38" 58d35'0"
Elizabeth Christine
Alexanian
Tau 3h40'7" 3d29'13"
Elizabeth Christine
Jackman
Lib 15h34'49" -4d59'55"
Elizabeth Christine Long
Lyn 8h12'47" 55d10'57"
Elizabeth Christine
Zimmerman
And 2h28'0" 48d2'43"
Elizabeth Claire
Cnc 8h0'33" 21d59'52"
Elizabeth Claire Heffernan
And 1h9'37" 46d41'49"
Elizabeth Claire Negola
And 2h17'28" 45d10'5"
Elizabeth Clara Cook
Dra 17h35'19" 67d58'51"
Elizabeth Clarrisa Ann
Jardine
Vir 13h0'47" -3d55'55"
Elizabeth Clayton Garner
Dra 12h55'44" 70d5'15"
Elizabeth Clifford
Cas 0h24'0" 51d51'4"
Elizabeth Cole O'Hearn
Crb 15h42'51" 32d46'0"
Elizabeth Colvard
Lyn 6h53'48" 57d2'50"
Elizabeth Conley
Psc 0h32'14" 18d58'51"
Elizabeth Conrad
Cyg 20h19'56" 40d10'19"
Elizabeth Constance
Lemke
Sco 17h25'7" -41d34'54"
Elizabeth Contreras
Cap 21h14'22" -22d56'22"
Elizabeth Corrao
And 0h23'27" 30d54'44"
Elizabeth Courtney Minor
Peg 23h29'27" 24d3'20"
Elizabeth Creature
Cas 1h39'10" 61d54'16"
Elizabeth Cristy Montano
Cap 21h15'15" -19d20'21"
Elizabeth Crockett
Mon 6h29'10" 9d9'27"
Elizabeth Cuneo
Cru 12h50'18" -59d34'50"
Elizabeth Curia Rizzo
Psc 0h55'34" 26d33'27"
Elizabeth Cyr
Cap 21h6'21" -17d25'37"
Elizabeth D. Evans
Cas 1h19'44" 55d57'49"
Elizabeth D. Smoyer
And 0h0'33" 34d37'59"
Elizabeth Dana
Cap 21h5'33" -21d2'57"
Elizabeth Dancer
McClanahan
Lmi 10h10'8" 38d17'42"
Elizabeth Danielle Nagy
And 1h28'32" 39d8'31"
Elizabeth & Danny
Tau 5h24'37" 20d6'59"
Elizabeth Dara
Aqr 20h43'14" 1d25'52"
Elizabeth David
And 2h33'21" 44d25'43"
Elizabeth Dawn
Cas 1h15'5" 53d27'12"
Elizabeth Dawn
And 2h24'59" 49d8'11"
Elizabeth De Camp
And 1h40'17" 48d30'45"
Elizabeth DeAnn Mumma
Lyn 6h24'26" 55d27'19"
Elizabeth DeMello
Cas 1h24'31" 62d22'29"
Elizabeth Demuth
Psc 0h21'6" -2d48'54"
Elizabeth Denise Batson
And 0h18'30" 29d26'54"
Elizabeth DeYoung
Lyn 8h52'27" 35d52'8"
Elizabeth Diossi Coscia
Lib 15h33'45" -23d58'53"
Elizabeth Dorothy "Best
Mom Ever"
Lyn 7h25'35" 56d35'14"
Elizabeth Drysdale
Uma 8h41'51" 61d42'50"
Elizabeth Duffy Johnson
Aqr 22h41'50" 1d12'43"
Elizabeth Dumin
Aqr 23h21'56" -20d59'42"
Elizabeth E. Fry
Uma 11h4'19" 67d0'17"
Elizabeth E. Hardin
And 23h19'36" 47d42'27"
Elizabeth E. Meder
Lac 22h23'4" 53d6'12"
Elizabeth E. Murrell
Cap 20h31'38" -13d17'22"
Elizabeth E. Watson
Gem 6h34'15" 21d51'45"
Elizabeth Early
And 1h33'37" 49d2'24"
Elizabeth Ebinger
Uma 12h37'37" 54d3'15"
Elizabeth Eckerman
Cyg 20h2'9" 58d30'8"
Elizabeth Edelman
And 23h10'39" 48d7'22"
Elizabeth Edgecomb
Cnc 8h11'31" 25d47'46"
Elizabeth Edmonds
Mon 6h48'27" 7d39'1"
Elizabeth Elease
Cap 21h56'17" -20d45'22"
Elizabeth "Eli" Greenman
Aqr 21h40'30" 1d31'14"
Elizabeth Ellen Jasina
Ari 2h11'57" 24d51'9"
Elizabeth Ellen Lang
Psc 1h16'5" 28d2'34"
Elizabeth Ellen McGibbon
Uma 14h23'30" 59d42'3"
Elizabeth Ellen Minahan
Cra 18h47'4" -40d29'33"
Elizabeth Ellen Rastede
Uma 10h14'0" 41d26'33"
Elizabeth Engkjer
Cyg 20h23'8" 55d51'14"
Elizabeth Erin Himes
Cam 4h9'48" 57d33'59"
Elizabeth Erin Murray
Vir 13h49'45" -20d2'8"
Elizabeth F. Montgomery
Sgr 17h49'38" -25d5'32"
Elizabeth Faith
Cap 20h23'30" -10d38'6"
Elizabeth Faith Garza
And 0h40'2" 26d1'37"
Elizabeth Falzon
Uma 10h2'51" 49d53'36"
Elizabeth Fay Vorel
Cas 0h7'17" 62d37'49"
Elizabeth Faye
Leo 10h22'17" 25d10'18"
Elizabeth Flannelly
Ari 3h18'19" 27d52'13"
Elizabeth Flink
Uma 8h38'59" 51d53'44"
Elizabeth Flowers
Cas 1h9'53" 65d49'7"
Elizabeth Fookes
And 2h35'30" 41d5'25"
Elizabeth Frances
Cnc 9h10'17" 27d48'30"
Elizabeth & Frank Ihrig
Cyg 19h57'4" 34d53'41"
Elizabeth Freeman
Sco 16h45'37" -39d11'31"
Elizabeth Fuller
Lyr 18h36'41" 31d40'39"
Elizabeth G. Medina
Cap 21h26'14" -18d57'24"
Elizabeth Gadbois
And 1h3'24" 41d50'17"
Elizabeth Gail's Flaming
Star
And 23h27'42" 37d31'37"
Elizabeth Gallagher My
Addiction
Vir 12h43'48" 10d51'15"
Elizabeth Gallivan Rosa
Psc 1h16'20" 32d31'1"
Elizabeth Galope
Boo 14h41'9" 47d15'26"
Elizabeth Garcia
Uma 9h1'53" 56d58'27"
Elizabeth Gatter's Star
Lyn 8h9'14" 37d16'3"
Elizabeth Gaudino
Tau 5h43'33" 26d40'36"
Elizabeth Geddes
Aqr 22h36'56" -0d1'52"
Elizabeth Giasemedis -
VAVA's
Umi 14h55'19" 73d8'1"
Elizabeth Gift From God
Gem 6h29'26" 18d40'34"
Elizabeth Gonzalez
Aqr 22h42'23" -15d47'30"
Elizabeth Gordon LaRoche
Vir 13h8'22" 6d15'45"
Elizabeth Grace
And 23h25'13" 47d14'49"
Elizabeth Grace
And 23h37'25" 44d13'32"
Elizabeth Grace
Umi 14h54'53" 74d9'15"
Elizabeth Grace
Dra 16h37'0" 63d0'50"

Elizabeth Grace Chadbourne
Cap 20h32'33" -21d22'17"
Elizabeth Grace Collins
Cas 23h13'27" 56d6'17"
Elizabeth Grace McGarry
Lmi 10h14'4" 31d44'54"
Elizabeth Grace McMullen
Vir 13h6'31" 4d39'7"
Elizabeth Grace Seiffert
Cas 0h39'27" 60d46'22"
Elizabeth Grace Seward
Ari 2h29'0" 24d51'21"
Elizabeth Grace Smith
Cap 21h39'14" -9d54'1"
Elizabeth Grace Strong
And 1h2'46" 42d52'14"
Elizabeth Grace Tice
Cap 21h10'8" -25d49'47"
Elizabeth Grace Trotter
Lyn 8h46'20" 46d27'23"
Elizabeth Grace Vieira Côté
Umi 4h14'21" 88d42'29"
Elizabeth Graeber Mears
And 0h40'57" 25d2'39"
Elizabeth "GrandMa" McLean
Leo 11h14'2" 16d54'35"
Elizabeth Green
Gem 7h6'49" 22d53'48"
Elizabeth Grissinger
And 0h21'25" 29d37'43"
Elizabeth Groh
Cyg 20h19'56" 39d10'15"
Elizabeth H. Soglin's Special Star
Umi 14h46'47" 73d56'2"
Elizabeth Haass
Her 18h44'10" 20d22'52"
Elizabeth Haleigh Schultz
Ori 5h46'49" 8d11'31"
Elizabeth Haley Miller
Uma 9h4'38" 60d16'16"
Elizabeth Hannah Jenkin Fairlie
Cas 1h31'57" 68d17'15"
Elizabeth Harper
And 23h20'23" 42d27'21"
Elizabeth Harper Howard
And 1h41'2" 37d42'57"
Elizabeth (Harrigan) Youngberg
Cas 1h37'55" 58d26'52"
Elizabeth Harrison Cunnington
Col 5h59'42" -35d44'51"
Elizabeth Harriss
Cas 1h16'50" 58d7'54"
Elizabeth Hartgrove
Cas 1h42'22" 63d24'28"
Elizabeth Haskins
Uma 11h17'35" 29d45'15"
Elizabeth Hayes Lozada
Tau 5h10'31" 16d43'33"
Elizabeth Hazan
Ari 3h14'33" 27d40'41"
Elizabeth Healey
Gem 6h33'22" 24d34'0"
Elizabeth Heidinger
Uma 10h23'33" 44d47'24"
Elizabeth Hemmerling
Aqr 22h37'1" 2d17'34"
Elizabeth Henry
Cet 2h25'23" 8d22'15"
Elizabeth Hensley
Vir 13h7'36" 11d40'27"
Elizabeth Herrera
Lep 5h54'28" -12d23'12"
Elizabeth Hill
Vir 14h33'33" 2d4'16"
Elizabeth Hoffman Aeschlimann
Leo 9h39'39" 11d51'31"
Elizabeth Holiday Ficek
Uma 11h51'52" 56d7'43"
Elizabeth Hope
Vir 13h1'2" -18d53'8"
Elizabeth Hope Newman
Gem 6h49'3" 17d2'37"
Elizabeth Hope Weber
Psc 0h14'1" 13d8'12"
Elizabeth Horton Goheen
And 1h5'4" 46d21'27"
Elizabeth Hoyt Eberle
Mon 7h35'53" -0d56'4"
Elizabeth Hubbard Jennings Pettus
And 1h12'53" 45d50'35"
Elizabeth Hugo Mohr
Crb 16h7'23" 37d51'48"
Elizabeth Hunt Walker
Lyn 8h3'7" 38d17'27"
Elizabeth Hutchinson Scattergood
Crb 15h41'29" 39d18'57"
Elizabeth I. Gallup
Cyg 20h1'55" 45d35'34"
Elizabeth Iconomopulos
Mon 6h30'18" 0d4'51"
Elizabeth Ida
Crb 15h52'25" 34d24'29"
Elizabeth Ida Elledge
And 23h43'8" 39d35'7"
Elizabeth Ida Marie Cronin
Cyg 20h30'40" 44d34'39"
Elizabeth Isley Riley
Vir 14h28'42" 1d34'43"
Elizabeth "Itty" Mohr
Uma 9h2'52" 50d47'37"
Elizabeth Ivette Ramirez
Ari 3h2'3" 11d43'57"
Elizabeth Ivy
Vir 13h21'44" 11d39'25"
Elizabeth Ivy Kenyon
Tau 4h12'38" 27d46'20"
Elizabeth J Bengert
Gem 7h15'15" 22d43'9"
Elizabeth J. Jones
Leo 11h6'6" 12d5'54"
Elizabeth J. Olechowski
Uma 8h49'18" 50d38'21"
Elizabeth J. Shroyer
Aqr 22h2'19" -8d6'46"
Elizabeth J. Springer
Scl 23h29'27" -29d30'21"
Elizabeth J. Waldeland
Gem 7h48'44" 31d40'21"
Elizabeth Jademan Chu
And 23h25'14" 51d27'2"
Elizabeth & Jamel
Cyg 21h24'1" 34d6'8"
Elizabeth Jami Schader
Peg 21h56'15" 13d25'46"
Elizabeth Jane
Cas 0h26'41" 58d4'58"
Elizabeth Jane Fitzsimmons
Vir 13h27'38" 13d37'44"
Elizabeth Jane Forrest
And 23h31'9" 42d15'12"
Elizabeth Jane Furgiuele
Uma 10h41'13" 45d5'48"
Elizabeth Jane Hoefsloot
And 1h34'26" 39d9'8"
Elizabeth Jane Kearney
Cas 1h30'53" 62d21'11"
Elizabeth Jane Kurkowski
Lac 22h31'35" 51d33'7"
Elizabeth Jane Linklater
And 1h21'18" 39d52'49"
Elizabeth Jane Mulligan
Cas 1h27'18" 63d23'36"
Elizabeth Jane Sinner
Cyg 19h52'50" 42d12'4"
Elizabeth Jane Webb
Per 4h21'51" 35d51'39"
Elizabeth Jane-Foster Robey
Sgr 19h14'53" -21d41'53"
Elizabeth Janes
And 2h15'33" 45d5'0"
Elizabeth Jay Hageman
Ari 2h4'45" 22d15'15"
Elizabeth Jayne Ackerman
Cam 5h39'28" 69d10'38"
Elizabeth Jayne Adkins
Cap 20h44'50" -17d16'30"
Elizabeth Jayne Arber
And 0h38'20" 22d7'35"
Elizabeth Jean McMenamon
Lyr 18h28'4" 39d48'24"
Elizabeth Jean Stephens
Tau 4h12'30" 12d1'13"
Elizabeth Jean Weaver
And 2h21'47" 42d10'30"
Elizabeth Jenee Mitchell
Cam 4h2'25" 57d28'23"
Elizabeth Jenkins
Peg 22h59'28" 25d3'52"
Elizabeth Jennings Cole
Sco 15h51'56" -21d22'45"
Elizabeth Jo DeBolt
Vir 14h10'40" 2d55'36"
Elizabeth Joan Loughran Robinson
Col 5h10'42" -37d10'9"
Elizabeth Joan Scatamacchia
And 0h36'30" 41d51'15"
Elizabeth Johnson
Cas 0h14'14" 53d26'28"
Elizabeth Joy Hall Friends Forever
Ari 3h12'35" 27d47'57"
Elizabeth Julia Hug
Lib 15h29'26" -12d16'39"
Elizabeth June
Aqr 22h10'37" 2d22'48"
Elizabeth June Bennett
Cnc 7h57'15" 19d21'43"
Elizabeth K
Tau 4h14'46" 3d0'40"
Elizabeth K. Belcher
Cas 0h32'2" 65d41'53"
Elizabeth K Grisham
Cap 20h31'6" -11d21'38"
Elizabeth K. Johnson
Vir 12h46'27" 1d44'54"
Elizabeth Kaitlin Ann
And 23h20'56" 51d11'52"
Elizabeth Kale Whaley
Aqr 22h30'9" -2d21'15"
Elizabeth Kate Fury
Cyg 20h3'30" 58d24'3"
Elizabeth Katerina
Lyn 7h59'42" 48d54'25"
Elizabeth Katharine Roberts
Umi 14h27'41" 72d22'50"
Elizabeth Katherine
Aqr 21h21'47" -10d15'31"
Elizabeth Katherine
And 2h21'39" 47d0'59"
Elizabeth Kathleen Garrett
And 0h50'6" 35d39'9"
Elizabeth Kathleen Murray
Lyn 6h20'30" 61d8'40"
Elizabeth Kathleen Remshaw
Mon 6h51'57" -0d10'53"
Elizabeth Kathryn Raczkowski
Crb 16h21'26" 30d41'34"
Elizabeth Katona
Mon 7h20'5" -10d23'59"
Elizabeth Kavanagh
Cas 23h49'23" 59d20'3"
Elizabeth Kay
Sgr 18h59'17" -26d54'27"
Elizabeth Keeven
Aql 19h25'50" 6d30'31"
Elizabeth Kelley Childress
Cas 0h3'55" 53d38'15"
Elizabeth Kelly
Cas 1h38'0" 60d36'12"
Elizabeth Kenderes
Leo 11h2'24" 3d34'31"
Elizabeth Kibbe
And 1h14'46" 37d52'36"
Elizabeth Koch
Uma 10h14'13" 55d24'51"
Elizabeth Koehler Johnson
Gem 6h50'37" 26d12'41"
Elizabeth Kowalski
And 2h31'29" 45d36'49"
Elizabeth Kristine Wright
Lyr 18h43'59" 38d42'34"
Elizabeth Kroll
Cas 0h24'54" 51d9'12"
Elizabeth L. Dickinson
Cas 23h10'13" 58d40'14"
Elizabeth L. Roach
Ori 6h0'25" 20d52'11"
Elizabeth L. Todd
And 23h19'27" 47d19'5"
Elizabeth Lahree Harris
Per 4h6'6" 44d13'33"
Elizabeth Lane Tyo
Leo 10h53'23" 11d30'36"
Elizabeth Langston Hester
Lyn 7h1'17" 58d48'35"
Elizabeth Lara Jenkins
And 1h31'16" 48d7'11"
Elizabeth Lauren
And 23h13'32" 43d11'41"
Elizabeth Lavino Zona 26 marzo 1991
Ori 5h54'48" 6d17'54"
Elizabeth LeAnn Gillette
Aqr 23h2'46" -6d59'39"
Elizabeth Lee
Peg 22h10'28" 31d18'40"
Elizabeth Lee Courtney Dolan
Cir 14h39'35" -64d56'35"
elizabeth lei wilson 4/9/2005
Leo 11h19'56" 14d22'11"
Elizabeth Leigh Hammer
Cap 20h21'18" -23d8'44"
Elizabeth Leigh Singh
Cas 1h36'52" 65d29'41"
Elizabeth Lenore
Lyn 7h25'24" 49d6'8"
Elizabeth Leona Haak
Col 5h30'32" -32d57'14"
Elizabeth 'Libby' Jean
And 1h32'22" 49d54'25"
Elizabeth Lillia Bogey
And 2h19'49" 47d2'15"
Elizabeth Limberg
Lib 14h49'12" -17d57'0"
Elizabeth Lindsay Watt
Uma 13h44'58" 52d1'16"
Elizabeth Livingston Wilder
And 0h34'23" 45d29'6"
Elizabeth "Liz" Hayes Conway
Cas 0h42'46" 61d50'51"
Elizabeth "Liz" Jean Rabenberg
Cas 0h39'2" 57d30'50"
Elizabeth "Liz" Lividini
Cam 4h26'57" 58d22'57"
Elizabeth "Liz" Pulaski
Mon 6h53'39" -0d16'31"
Elizabeth "Liz" Williams
Hya 9h24'18" 4d33'32"
Elizabeth "Lizzie" Morales
Uma 10h15'42" 56d52'18"
Elizabeth Locke Burkett
And 1h1'14" 45d22'43"
Elizabeth Lopez
Cyg 20h29'31" 33d7'45"
Elizabeth Lopez
Com 13h13'14" 27d9'20"
Elizabeth Loraine Leffman
And 2h21'25" 49d44'55"
Elizabeth Lori
Aqr 22h29'58" -18d41'5"
Elizabeth Loring Gaines
Uma 11h57'8" 49d25'59"
Elizabeth Lorraine Johnson
Uma 8h47'51" 69d45'16"
Elizabeth Lou Latham
Uma 8h57'47" 51d32'57"
Elizabeth Lou Maginn Cilano
Cnc 8h18'32" 7d8'37"
Elizabeth Louise Appel
And 2h37'8" 49d56'34"
Elizabeth Louise Cianflone
Crb 15h49'53" 26d34'55"
Elizabeth Louise Giordano
Sgr 18h46'58" -28d23'23"
Elizabeth Louise Kershaw
And 2h35'22" 50d1'38"
Elizabeth Louise Kinerk
And 1h45'25" 49d37'55"
Elizabeth Louise Millar
Lac 22h47'21" 53d35'7"
Elizabeth Louise Palmer
Tau 5h18'16" 27d13'8"
Elizabeth Louise Watkins
Psc 1h10'45" 29d24'3"
Elizabeth Louise Wheeler
Cap 20h39'40" -22d1'57"
Elizabeth Lulu
Sgr 20h2'25" -25d51'48"
Elizabeth Lund Covish
Cam 5h17'33" 65d38'46"
Elizabeth Lynn Bennett
Leo 11h21'21" 18d57'31"
Elizabeth Lynn Ford
Aql 19h44'37" 4d11'0"
Elizabeth Lynn Nulman
And 0h22'50" 45d59'37"
Elizabeth Lynn Proal
Ari 3h1'20" 11d32'30"
Elizabeth M DeLuca, Miss Elizabeth
Vul 18h59'8" 21d52'19"
Elizabeth M. Fernstrom
Lmi 10h38'16" 37d46'7"
Elizabeth M. Fick
Cam 5h52'18" 60d3'44"
Elizabeth M. Getty
Cas 1h20'31" 72d44'55"
Elizabeth M. Lane
Psc 1h21'39" 29d42'33"
Elizabeth M. Lundquist
Cas 23h30'6" 53d48'39"
Elizabeth M. Metz
Aqr 20h43'4" -8d8'17"
Elizabeth M. Patterson
Sco 16h10'30" -14d13'39"
Elizabeth M. Patterson
And 1h5'20" 42d11'26"
Elizabeth M. (Peggy) Ketterle
Umi 14h20'38" 77d12'30"
Elizabeth Macedonia Adame
Tau 4h29'48" 18d41'10"
Elizabeth Madeleine Chandler
And 1h16'18" 46d42'17"
Elizabeth Madeline Holman
Cnc 8h46'9" 30d56'45"
Elizabeth Mae Patete
Sgr 17h57'9" -17d41'6"
Elizabeth Mae Yoder
And 0h50'52" 42d45'59"
Elizabeth Mahla
Tau 4h11'18" 14d55'38"
Elizabeth Mair
Umi 14h58'12" 71d52'48"
Elizabeth Manning Schumm
Cas 1h58'37" 60d8'19"
Elizabeth Marie Aibner
Lib 15h16'58" -22d23'11"
Elizabeth Marie Allen
Ori 6h17'6" 6d22'21"
Elizabeth Marie Barrientos FDJ
Uma 8h24'30" 64d39'44"
Elizabeth Marie Elliott
And 2h16'34" 49d54'2"
Elizabeth Marie Gallant
Gem 7h37'19" 31d22'21"
Elizabeth Marie Hammerly
Cyg 21h58'9" 54d42'49"
Elizabeth Marie Ingraham
Gem 6h49'5" 33d6'28"
Elizabeth Marie Jackson
Tau 3h56'25" 3d50'24"
Elizabeth Marie Judy
Sco 16h12'20" -10d28'10"
Elizabeth Marie Manino
Lib 15h31'1" -7d44'44"
Elizabeth Marie Mills
And 2h23'51" 46d44'23"
Elizabeth Marie Musgrave
And 0h13'1" 29d45'49"
Elizabeth Marie Rezac
Umi 15h30'37" 71d23'12"
Elizabeth Marie Sova
Sco 17h19'49" -35d10'40"
Elizabeth Marie Sproul
Sgr 17h57'41" -22d10'9"
Elizabeth Marie Wallis
Cap 20h22'54" -14d53'59"
Elizabeth Marie Weakley
Srp 18h8'7" -0d8'17"
Elizabeth & Mark Meenan
Sgr 18h8'36" -31d9'30"
Elizabeth Markward
Gem 7h27'37" 33d2'6"
Elizabeth Marrone
Uma 12h25'24" 58d6'23"
Elizabeth Marry Chetfield (Clawson)
And 1h29'59" 39d39'27"
Elizabeth Marsiglia
Cas 1h36'27" 63d27'51"
Elizabeth Martin
Uma 10h22'42" 63d27'54"
Elizabeth Martin
And 1h31'35" 38d52'28"
Elizabeth Martin
Cyg 19h59'38" 32d22'10"
Elizabeth Martin
Tau 4h49'27" 21d44'59"
ELIZABETH MARTIN PEREZ
Lyn 6h24'0" 60d14'32"
Elizabeth Mary Beckert
Ori 4h58'49" 11d2'3"
Elizabeth Mary Fadel
Her 16h31'46" 48d39'24"
Elizabeth Mary Matkins
Vir 13h25'40" -13d32'39"
Elizabeth Mauer
And 23h16'37" 41d17'1"
Elizabeth May
Cyg 21h47'3" 37d29'42"
Elizabeth May Wood
Uma 9h28'9" 66d22'23"
Elizabeth Mc Cord
Lib 14h26'42" -21d30'50"
ELIZABETH MCCANN
Gem 6h29'34" 16d57'3"
Elizabeth McCarthy Kirwan (Lizzie)
Cap 21h34'51" -24d40'0"
Elizabeth McCauley Gray
And 22h58'0" 47d55'36"
Elizabeth McGee Miller
And 1h39'5" 49d30'57"
Elizabeth McGuire
Cas 0h10'56" 53d42'43"
Elizabeth McKain
And 23h12'58" 52d18'28"
Elizabeth McKaughan King
Her 17h52'56" 49d41'39"
Elizabeth McKean
Crb 16h20'32" 28d41'24"
Elizabeth McKeehan Black
Sco 17h46'15" -32d6'5"
Elizabeth McKenzie Oxendine
Leo 11h35'55" 6d36'21"
Elizabeth McLaughlin
Psc 0h33'8" 20d34'46"
Elizabeth McManus
Cas 0h38'35" 54d34'4"
Elizabeth Mead
Psc 1h44'47" 12d50'45"
Elizabeth Meeks
Lmi 10h50'7" 28d42'45"
Elizabeth Megan
And 2h25'13" 41d45'15"
Elizabeth Mehaffey
And 23h37'23" 43d6'43"
Elizabeth Mencke
Lyn 8h10'22" 34d28'6"
Elizabeth Merkh
Lmi 10h26'31" 34d57'32"
Elizabeth Metcalf
Cas 0h22'7" 53d37'58"
Elizabeth & Michael
Uma 9h34'33" 52d25'10"
Elizabeth / Michael 9-27-91
Lib 15h22'18" -14d27'51"
Elizabeth Miller
Cas 0h29'5" 60d26'30"
Elizabeth Miller
Gem 7h39'49" 24d17'32"
Elizabeth Milo Coop
Lyn 6h57'1" 59d16'47"
Elizabeth Misamore
Ori 6h16'46" 10d8'12"
Elizabeth Mitchell
Lmi 10h3'47" 35d45'37"
Elizabeth Moffat
Cyg 19h30'19" 31d1'56"
Elizabeth Monroe Boggs
Uma 13h53'8" 48d27'41"
Elizabeth Morgan
And 1h39'44" 48d36'11"
Elizabeth Morgan Fleetwood Price
Leo 10h58'16" 15d3'48"
Elizabeth Morgan McCandless
Cnc 8h45'15" 32d27'7"
Elizabeth Morgan Miller
Cap 20h55'42" -25d45'52"
Elizabeth Moyer
Cnc 9h2'48" 31d37'2"
Elizabeth Mulligan
Gem 6h42'57" 22d59'10"
Elizabeth Munoz
Cnc 9h18'43" 10d36'55"
Elizabeth Murnin Carter Boyle 80th
Cas 0h5'41" 54d22'37"
Elizabeth - My Fairy
Uma 10h29'56" 61d31'15"
Elizabeth N. Clark
And 0h51'57" 40d34'48"
Elizabeth Nakata
Crt 11h23'36" -13d51'57"
Elizabeth / Nathan 2004
Crb 15h48'39" 39d17'9"
Elizabeth Nelson
Uma 10h21'54" 45d42'46"
Elizabeth Neri Haberski
Uma 11h14'49" 62d29'41"
Elizabeth Nesa
Leo 9h58'57" 8d39'42"
Elizabeth Nester
Ori 6h24'25" 13d56'24"
Elizabeth Neuman
Cyg 20h43'11" 46d54'39"
Elizabeth Neumann
Gem 7h15'1" 24d19'3"
Elizabeth Newell
Sco 16h50'19" -27d9'22"
Elizabeth Nichole Tuttle
Aqr 22h13'40" -2d1'47"
Elizabeth Nichole Wicks
Cap 20h19'37" -8d35'24"
Elizabeth Nicole
Tau 5h33'59" 27d11'32"
Elizabeth Nicole
Ari 2h50'49" 29d23'31"
Elizabeth Nicole Curd
And 0h39'45" 27d44'45"
Elizabeth Nicole Hatfeild
Leo 10h40'59" 14d17'40"
Elizabeth Nicole Riddle
Lyn 7h41'18" 35d38'56"
Elizabeth Nissen
And 2h36'32" 44d27'25"
Elizabeth Noelle Luna
Sco 16h13'54" -29d5'19"
Elizabeth Nolan
Cas 0h41'21" 66d53'37"
ELIZABETH NOLEN
Mon 6h29'58" 8d27'24"
Elizabeth Nye
Ari 2h46'5" 27d17'44"
Elizabeth Ochoa Steward
Cyg 21h18'52" 47d16'48"
Elizabeth O'Halloran
Cap 21h37'52" -14d17'8"
Elizabeth Orion Michelson
Ori 5h54'6" 21d44'1"
Elizabeth Orndorff Westermark
Aqr 22h32'10" 0d21'50"
Elizabeth Page Duncan
Lyn 6h19'17" 59d3'34"
Elizabeth Paige Bloch
Psc 1h44'32" 6d56'8"
Elizabeth Paige Wilson
Lib 15h5'13" -5d44'17"
Elizabeth Parker
Uma 9h35'22" 46d58'3"
Elizabeth Patino
Leo 9h31'1" 13d13'57"
Elizabeth Patricia Kline
Tau 4h37'41" 28d23'0"
Elizabeth Patton Pirone
And 23h31'4" 37d9'37"
Elizabeth Paul
Lib 14h50'46" -0d36'3"
Elizabeth Pauline Sandford Frampton
Cas 0h35'11" 53d41'11"
Elizabeth Payne
Aqr 21h50'13" -1d26'42"
Elizabeth Peall
Cnc 9h9'36" 32d0'55"
Elizabeth Perez
Sgr 18h47'0" -23d3'9"
Elizabeth Perri
Uma 10h20'40" 48d59'10"
Elizabeth Peryer
Ori 6h18'59" 7d18'54"
Elizabeth Pesch Stafford
Cam 4h21'12" 56d8'38"
Elizabeth Pottenburgh
Uma 11h0'5" 45d34'47"
Elizabeth Poynor
Cas 1h19'13" 55d39'47"
Elizabeth Prather
Ori 5h28'58" 6d41'37"
Elizabeth Premo "Grandma Bluehouse"
Uma 9h52'20" 51d57'32"
Elizabeth Primrose Duffield
Leo 11h37'4" 24d5'46"
ELIZABETH PRINCESS LEEDY
Uma 11h8'7" 54d34'55"
Elizabeth Protheroe
Cas 23h46'54" 50d7'38"
Elizabeth R. Fell
And 0h44'52" 45d46'49"
Elizabeth R. Gallendo
Psc 0h46'41" 8d47'10"
Elizabeth R Riddell
Aqr 22h15'53" -0d4'55"
Elizabeth Rae Sexton
Tau 4h26'20" 13d9'54"
Elizabeth Rae Smith
And 23h52'56" 47d6'0"
Elizabeth Raelee
Leo 11h10'59" 10d49'49"
Elizabeth Raven
Sco 17h53'25" -43d18'55"
Elizabeth Reade Morris
Leo 9h30'55" 28d21'40"
Elizabeth Renee
Leo 10h49'57" 19d42'20"
Elizabeth Renee
And 2h22'24" 50d17'23"
Elizabeth Rennolds
Psc 22h56'14" -0d5'2"
Elizabeth Renwick Johnston Ferguson
Tau 4h48'33" 20d38'50"
Elizabeth Robert
And 0h43'10" 44d24'11"
Elizabeth & Rocco Lemongello
Cap 20h22'19" -10d49'34"
Elizabeth Roddick Hardy
Gem 7h23'21" 20d42'16"
Elizabeth Rodriguez
And 1h13'21" 37d44'51"
Elizabeth Roman
And 23h16'29" 52d25'28"
Elizabeth Romanaux
Lib 15h20'35" -22d59'9"
Elizabeth Ronn
Ori 4h47'32" 4d51'23"
Elizabeth Rooney
Sco 16h4'44" -13d27'42"
Elizabeth Rose
Lyn 7h19'26" 54d19'13"
Elizabeth Rose
Uma 13h44'54" 53d51'59"
Elizabeth Rose
Cnc 8h22'14" 12d30'50"
Elizabeth Rose
Cas 0h16'52" 55d54'21"
Elizabeth Rose
Gem 7h37'12" 33d15'48"
Elizabeth Rose Barrett Walters
Sco 16h12'52" -15d58'40"
Elizabeth Rose Cohn
Ari 2h51'33" 20d51'18"
Elizabeth Rose Ennis
And 0h54'5" 38d50'29"
Elizabeth Rose Fleming
Psc 0h21'36" 0d29'6"
Elizabeth Rose Gosselin
Lib 15h33'34" -10d47'47"
Elizabeth Rose Lee Ward
Lyn 7h46'14" 39d36'44"
Elizabeth Rose Profita
Uma 9h10'44" 64d9'18"
Elizabeth Rose Seymour
Uma 10h41'11" 53d7'28"
Elizabeth Rose Wilson
Umi 15h50'32" 80d40'35"
Elizabeth Rose Young
And 2h3'24" 37d43'10"
Elizabeth Rösener
Umi 15h59'22" 75d46'13"
Elizabeth Rostankovska
Cra 19h9'39" -42d59'51"
Elizabeth Ruth
Sco 17h22'55" -30d24'29"
Elizabeth Ruth Dick
Uma 8h56'26" 52d42'56"
Elizabeth Ruth Hood
Cap 20h32'36" -20d55'14"
Elizabeth Ruth Keyes
Cas 0h6'38" 53d21'6"
Elizabeth - Rylee
Tau 5h33'6" 23d0'39"
Elizabeth S. Densmore
And 2h29'0" 43d12'32"
Elizabeth S. Mirabito
Lmi 9h35'37" 34d56'41"
Elizabeth Salazar
Cnc 8h41'38" 22d20'34"
Elizabeth Salerno 143
Leo 10h43'50" 10d32'27"
Elizabeth Sally McAllister
Cas 0h25'41" 47d37'25"
Elizabeth Sanders Stuckey
Aqr 21h16'55" -9d9'48"
Elizabeth Sanderson
Lyn 8h3'30" 40d37'48"
Elizabeth Sarah
Uma 10h19'18" 48d54'6"
Elizabeth Sarah McDowell
Lyn 7h43'17" 39d45'8"
Elizabeth Schaeffer
Psc 1h33'48" 12d28'16"
Elizabeth Scheidig Knapp
And 0h5'36" 38d43'54"
Elizabeth Scheu
Cas 2h1'11" 58d21'47"
Elizabeth Schreiber
And 22h58'46" 47d47'40"
Elizabeth Scott
Gem 7h41'32" 24d24'1"
Elizabeth Scott Crosby
Aqr 21h33'25" 0d36'52"
Elizabeth Sencion
Per 3h27'33" 45d32'39"
Elizabeth Sharlene Logsdon
Ari 2h45'46" 27d45'39"
Elizabeth Sharon Francisco
Cap 20h14'42" -15d20'13"
Elizabeth Shaw
Psc 23h35'51" 0d21'53"
Elizabeth Shaw Ward
Col 5h35'52" -27d33'9"

Elizabeth Shehee
Sco 17h51'31" -41d20'23"

Elizabeth Sherrill
And 0h56'25" 24d23'30"

Elizabeth Shine Nace
Tau 5h37'28" 19d18'21"

Elizabeth Short
Aqr 23h9'16" -20d48'32"

Elizabeth Simon
And 23h39'17" 36d45'55"

Elizabeth Skelton
Lyn 7h42'16" 38d13'30"

Elizabeth Smith
Cas 23h32'38" 53d24'41"

Elizabeth Sparkles
Cap 20h27'47" -25d56'3"

Elizabeth Spears
And 1h43'14" 45d36'28"

Elizabeth Stamatakos
Aqr 22h1'23" -13d41'7"

Elizabeth Star
Vir 13h50'23" 0d13'36"

Elizabeth Starr
Cas 0h16'38" 52d12'0"

Elizabeth Steele
Crb 15h45'16" 27d31'44"

Elizabeth Sterling Oles
Sgr 18h46'21" -25d21'1"

Elizabeth "Storyteller" Eorgan
Sco 17h38'22" -39d34'30"

Elizabeth Sue Donahue
Tau 5h4'19" 18d52'37"

Elizabeth Sue Mock
Lib 14h50'1" -1d33'59"

Elizabeth Suraci D'Amico
Ari 3h12'40" 27d29'15"

Elizabeth Suzanne Gettum
Vir 11h47'28" 2d32'35"

Elizabeth T. "Libby" Roach
Ari 2h27'9" 18d57'21"

Elizabeth T. Roberts
Cap 20h27'44" -22d22'30"

Elizabeth T Spinella
Cnc 7h57'39" 19d4'51"

Elizabeth Tamez
Psc 23h53'46" 6d28'31"

Elizabeth (Tani) Harrison
Uma 10h9'44" 64d40'4"

Elizabeth Tanner Azrak
Uma 9h32'40" 67d51'5"

Elizabeth Taylor
Lib 15h20'40" -23d12'41"

Elizabeth Thayer
Ori 5h18'49" 1d38'29"

Elizabeth the Beautiful
Uma 11h9'51" 29d47'14"

Elizabeth the Beautiful
Cas 2h18'36" 72d3'53"

Elizabeth "The Diva" Nelson
And 0h29'9" 33d36'54"

Elizabeth Theresa Grace Hebb
Ori 6h15'11" 15d26'38"

Elizabeth Therese Huddleston
Aqr 22h39'34" -8d53'37"

Elizabeth Thompson
Vir 12h41'36" 5d52'9"

Elizabeth Thors
Gem 6h21'55" 27d26'24"

Elizabeth Titus
Cas 1h23'47" 63d2'59"

Elizabeth Torres
Uma 9h8'51" 67d42'12"

Elizabeth Townsend
Leo 10h14'9" 15d50'49"

Elizabeth Traci Curtis
And 2h10'37" 41d19'42"

Elizabeth Truesdale
Uma 11h59'11" 44d4'15"

Elizabeth Tufo Forte
Vir 13h22'49" -5d27'47"

Elizabeth Tullo
Uma 10h19'40" 51d58'29"

Elizabeth Turner
Uma 9h25'4" 58d24'19"

Elizabeth Tyus, Seth Loves You!!
Ori 5h35'13" 1d13'20"

Elizabeth Uihlein
Gem 7h36'1" 16d35'42"

Elizabeth Underwood
Cas 2h47'50" 72d8'42"

Elizabeth Van Horn Keenhold
Cam 7h37'48" 76d10'20"

Elizabeth Vanderklay
Sco 16h17'50" -9d59'36"

Elizabeth Vasquez
Lib 15h25'22" -22d7'37"

Elizabeth Vaughan
Psc 0h17'10" 18d57'27"

Elizabeth Vazquez Bagnole
Cas 23h45'11" 61d37'50"

Elizabeth Victoria De Kleine
Cyg 20h32'24" 45d0'6"

Elizabeth Victoria Nosser McGaughey
Tau 4h7'0" 5d51'29"

Elizabeth Vigelius Maxwell
Cas 0h50'29" 56d55'17"

Elizabeth Virginia Szklany
Lmi 10h31'8" 37d10'14"

Elizabeth Vulin
Psc 0h53'24" 32d0'29"

Elizabeth Wathen Tyler
Uma 9h46'3" 43d4'19"

Elizabeth Weeks
And 1h46'42" 49d0'26"

Elizabeth Wessman McGee
Lib 15h55'56" -15d14'42"

Elizabeth Whittaker 21st Star
Tau 4h19'18" 6d23'46"

Elizabeth Wilcox
Psc 1h16'3" 26d22'42"

Elizabeth Williamson
And 1h0'12" 39d36'38"

Elizabeth Wood
And 2h31'8" 44d10'30"

Elizabeth Wright
Psc 0h57'23" 32d38'14"

Elizabeth Yarmola Manzione
And 23h28'15" 43d5'20"

Elizabeth Yinger
Aqr 23h4'18" -21d19'13"

Elizabeth Yula Cowan
Uma 8h55'1" 67d16'18"

Elizabeth Zangara
Ori 5h55'8" 21d12'41"

Elizabeth Zychowka
Ori 5h45'9" 6d52'40"

Elizabeth, James, & Amelia Howell
Umi 17h5'32" 77d25'34"

Elizabeth, My Light
And 0h29'9" 35d17'5"

Elizabeth102x3
Lyn 6h56'28" 53d9'11"

Elizabeth-226
Ori 4h48'46" 11d6'41"

ElizabethAMaillis
Gem 6h44'22" 23d3'38"

Elizabeth-anne Paonessa
Cnc 8h46'2" 17d7'34"

ElizabethCaroline
Cas 1h14'3" 55d13'7"

ElizabethD2005
And 0h57'26" 38d53'41"

Elizabethe Anne Cavanah Crowson
Leo 11h56'3" 21d6'37"

ElizabethGratar999
Vir 13h26'53" 2d51'57"

ElizabethGuy/TeamDraper Nelson
Cap 20h26'58" -10d30'12"

Elizabeth-Jeanette Tomlinson
Cap 20h9'12" -10d12'36"

ElizabethnDaniel4ever
Uma 9h36'55" 68d41'34"

Elizabeth's Beacon
Lmi 10h32'7" 32d5'55"

Elizabeth's Birthday Star
And 0h18'29" 44d56'50"

Elizabeth's Courage
Lib 15h58'30" -15d35'26"

Elizabeth's Daddy
Dra 17h10'27" 56d2'28"

Elizabeth's Inspiration
Her 17h52'34" 21d3'25"

Elizabeth's Jem
Uma 12h5'0" 33d48'14"

Elizabeth's Joy
And 0h11'11" 40d21'8"

Elizabeth's Love
Lyn 7h39'34" 47d23'19"

Elizabeth's love
Umi 17h38'7" 80d27'34"

Elizabeth's Place For Dreams
Aql 20h12'0" 15d51'43"

Elizabeth's Star
Ari 2h50'4" 26d39'23"

Elizabeth's Star
Vir 13h30'34" 5d27'19"

Elizabeth's Star
And 1h48'17" 41d40'26"

Elizabeth's Star
Boo 15h5'31" 32d57'33"

Elizabeth's Star
Psa 21h48'23" -34d37'41"

Elizabeth's Starlight
Uma 9h25'0" 49d22'58"

Elizabeth's Starr
Ori 5h22'59" 3d23'18"

Elizabob
Uma 8h14'6" 71d12'42"

Elizadeanna (the aries queen)
Cas 0h18'31" 61d26'17"

Elizebeth
And 1h20'29" 35d24'53"

Elizebeth Marie Jensen
Uma 13h33'23" 52d38'30"

Elizebeth Williams
Peg 21h58'39" 18d10'41"

Elizet Perez
Dra 18h46'41" 61d11'0"

Elizie's Irish Star
Cap 20h58'56" -20d57'38"

Elizita y Moritta
Cap 21h27'1" -15d38'3"

Elizma
Cnc 8h27'14" 27d38'51"

Eljean
Col 5h54'13" -34d56'21"

Éljenek az Anyák és a Nagymamák!
Uma 9h47'27" 46d23'25"

Eljona Jahaj
Cap 20h23'23" -11d21'47"

Elk'chen
Uma 13h29'31" 55d16'6"

Elke
Cas 0h16'18" 51d49'33"

Elke
Gem 7h24'3" 29d18'23"

Elke
Gem 7h12'28" 27d23'24"

Elke
Leo 11h41'29" 18d17'58"

Elke
Leo 11h4'55" 15d12'44"

Elke
Ori 6h11'43" 8d51'50"

Elke Anne Karola Krieg
Vir 13h29'11" -18d3'22"

Elke Binder
Cap 20h49'0" -20d38'14"

Elke Eva Claes
Gem 7h16'46" 22d42'2"

Elke Marina
Uma 13h38'2" 48d39'19"

Elke - Phil
Uma 12h4'56" 54d16'34"

Elke Wiegerich
Aur 5h10'16" 42d23'8"

Elken
Dra 17h18'42" 60d2'47"

Elken Gold - Together Forever
Cyg 20h36'7" 36d15'30"

Elke's LiLi Lil 3003
Ari 2h50'13" 12d11'28"

Ell Family
Lyr 18h50'0" 33d24'57"

Ella
Her 17h4'26" 30d29'5"

Ella
Uma 9h27'12" 41d33'8"

Ella
Cas 0h56'34" 55d16'28"

Ella
And 23h37'28" 42d6'41"

Ella
Uma 8h37'23" 52d24'45"

ELLA
Aqr 22h33'53" 0d48'47"

Ella
Psc 1h31'35" 18d6'47"

Ella
Uma 8h30'38" 65d3'18"

Ella*16
Sco 16h15'47" -41d57'44"

Ella and John
Sgr 18h10'18" -21d36'4"

Ella and Sheyvakh
Cyg 20h36'57" 59d58'34"

Ella Angliss DeSane
Ari 3h22'36" 28d4'57"

Ella Anne
And 0h10'51" 27d31'9"

Ella Anne Rorabaugh
Crb 15h39'8" 35d30'24"

Ella Annelore Bauer
Uma 8h14'30" 62d59'56"

Ella Arbuckle
And 0h21'49" 31d43'44"

Ella Arwen Nicoloff
Uma 11h28'19" 32d45'7"

Ella bean
And 23h20'38" 46d59'40"

Ella Bella
And 23h52'46" 45d33'24"

Ella Bella
Tau 5h34'50" 22d35'5"

Ella Bella
Uma 11h25'4" 58d47'35"

Ella Bella
Sgr 18h27'51" -30d2'44"

Ella Bella's Star
Sco 16h47'52" -41d31'16"

Ella Belles
And 23h13'12" 47d35'49"

Ella Birgitta Scherer
And 2h22'50" 37d33'8"

Ella Bleu
Aqr 20h58'48" -11d15'23"

Ella Borello
And 0h4'43" 34d20'41"

Ella Bridge McCann
And 2h17'25" 37d22'20"

Ella Brooke
Ari 2h41'2" 16d28'2"

Ella C Booth
Crb 15h45'2" 38d15'58"

Ella C Plumstead
Leo 11h54'46" 19d20'16"

Ella Carla
Cnc 8h19'8" 31d38'2"

Ella Casperson Shope
Cnc 9h6'26" 27d20'13"

Ella Catherine Budelis
Uma 11h29'36" 30d9'12"

Ella Catherine Hart
Cnc 8h48'14" 30d16'9"

Ella Catherine McLean
And 0h22'0" 36d48'28"

Ella Catherine Nunnally
Sct 18h47'14" -11d50'20"

Ella Cavell Bigg
Uma 10h57'38" 43d4'26"

Ella Christine Young
Sgr 18h22'18" -23d39'17"

Ella Clare Russo
Cap 20h36'23" -10d32'17"

Ella - con lei sempre
Cru 12h12'2" -63d41'39"

Ella Concetta
Psc 1h49'34" 6d22'1"

Ella Day
Lmi 10h46'54" 32d47'41"

Ella Dorothy Johnson
And 23h11'10" 42d7'54"

Ella E. Hood
Uma 9h33'45" 41d59'22"

Ella Elizabeth Balding
Cas 0h24'9" 56d51'23"

Ella Elizabeth Tucker
Uma 9h27'10" 42d5'14"

Ella Emmons Bennett
Aqr 22h17'53" 1d46'14"

Ella "Ezzy" Moree
Lyr 18h45'25" 39d3'46"

Ella F Cameron "Pixie"
Cyg 19h21'59" 48d29'56"

Ella Fante
Cep 21h43'37" 61d54'57"

Ella Gamble
Cyg 19h51'37" 39d28'47"

Ella Gilmore
Cep 21h34'58" 76d16'23"

Ella Gogel
Cas 0h23'54" 54d18'36"

Ella Grace
And 23h17'56" 48d5'6"

Ella Grace
And 1h29'35" 50d32'44"

Ella Grace
And 0h48'6" 40d56'0"

Ella Grace
Lmi 10h41'12" 31d12'14"

Ella Grace
Sgr 18h18'58" -23d45'19"

Ella Grace Amiet
Tau 3h37'48" 25d11'2"

Ella Grace Bailey
Cru 12h30'8" -63d0'41"

Ella Grace Blackerby
Mon 6h29'57" 5d34'12"

Ella Grace Bradley
Lyn 8h43'33" 35d25'59"

Ella Grace Collins
And 1h26'34" 49d35'19"

Ella Grace Macdonald
Vir 12h22'55" -5d22'23"

Ella Grace McCarthy 15th April 2006
And 2h29'38" 49d34'54"

Ella Grace Schoenberg
Cnc 8h27'3" 23d51'13"

Ella Grace Tibbals
Tau 4h7'18" 20d56'0"

Ella Grier Avant
Lyr 18h39'23" 35d1'55"

Ella Hazel
Lib 15h47'1" -28d38'16"

Ella Hope's Eternal Light
Sco 17h51'40" -37d32'14"

Ella Ines Berry
And 23h4'23" 51d13'46"

Ella Jane
And 0h45'5" 43d35'17"

Ella Jane
Umi 14h49'25" 70d34'17"

Ella Jane B
Ari 2h47'1" 30d41'28"

Ella Jane Betts
Cas 23h34'59" 53d16'9"

Ella Jane Kofke
Gem 7h45'26" 23d44'17"

Ella Jane Moriarity
Cap 21h51'2" -17d30'26"

Ella Jay
Umi 16h3'20" 82d9'19"

Ella Jean
Per 3h29'11" 55d2'55"

Ella Jo Parcewski
Aqr 22h9'46" -0d45'11"

Ella Jo Whiston
And 23h31'41" 42d49'55"

Ella Joan DuBorg
Leo 10h51'31" 4d34'28"

Ella Karaoglanova
Sco 16h11'4" -25d8'58"

Ella Kate
And 0h31'59" 46d13'57"

Ella Kate Hageman
Lyr 18h34'19" 27d14'2"

Ella Kathleen Davenport
And 0h48'14" 43d21'51"

Ella Kaye Thompson
Tau 4h2'26" 21d33'42"

Ella Lauren
Dra 18h46'25" 69d16'29"

Ella Louise
And 2h6'17" 40d4'35"

Ella Louise Algajer Needham
Umi 15h44'51" 74d49'2"

Ella Louise Butherus Munson
Mon 6h44'56" -0d26'16"

Ella Louise Payne
And 0h57'10" 46d7'41"

Ella Louise's Christening Star
Uma 12h53'29" 55d57'44"

Ella Lucia Mashura
Peg 22h38'13" 29d11'1"

Ella Madeline Schwartz
Mon 6h28'54" 4d53'6"

Ella Mae
Tau 5h46'28" 26d40'50"

Ella Mae Haeberle
Tau 4h30'42" 3d58'4"

Ella Mae Hesse
Leo 10h9'29" 20d58'6"

Ella Mae Kent
Com 12h10'29" 20d3'36"

Ella Mae Knapman xxx "Ella Bella"
Umi 13h55'39" 73d33'59"

Ella Mae Louise Michaud
Cas 1h23'24" 63d36'3"

Ella Mae Robertson
Uma 8h33'20" 59d47'6"

Ella Mae Zingery
Uma 11h8'14" 67d18'50"

Ella & Manny Krauss
Crb 16h13'39" 37d6'12"

Ella Marat
Cyg 20h21'41" 51d18'36"

Ella Marianne Klokow
Uma 13h3'1" 58d41'34"

Ella Marie
Uma 9h25'18" 60d21'48"

Ella Marie
And 2h26'11" 48d59'29"

Ella Marie Aller
Cnc 8h33'52" 32d17'14"

Ella Marie Armstrong
Crb 15h42'29" 31d57'5"

Ella Marie Lombardo
Sco 16h4'56" -12d42'53"

Ella Marie Malee
Sgr 19h38'27" -11d59'37"

Ella Marie Moore
Dra 18h50'56" 54d43'36"

Ella Marie Valenzuela
Aqr 22h55'40" -9d15'6"

Ella Mary Evans
Cnc 8h47'23" 14d54'53"

Ella Mary Hales
And 0h22'44" 40d16'41"

Ella May Donaldson
And 23h38'24" 48d15'59"

Ella May Floren
Cap 20h22'37" -9d59'6"

Ella May Veglio Thomas
Lmi 10h59'45" 28d33'10"

Ella McFaull
And 23h15'57" 51d6'52"

Ella Michelle Everett - 21 Nov 2001
Sco 16h19'0" -41d54'18"

Ella Morgan
Uma 11h31'21" 56d34'21"

Ella Mosley
Cas 1h8'42" 63d20'9"

Ella My "Bella"
Lib 15h6'22" -25d30'56"

Ella Myfanwy Sole
Cam 13h34'27" 76d28'49"

Ella Myka Aleszczyk - Our Angel
Leo 10h37'44" 21d29'39"

Ella Ng
Ori 4h48'7" 4d38'27"

Ella Nicole
Ori 5h53'33" 6d56'29"

Ella Nicole
Ari 3h4'43" 21d0'9"

Ella O'Brien
Gem 7h46'52" 23d48'1"

Ella Orion Neal
Ori 6h7'44" 18d26'50"

Ella Parker Hill
And 1h50'1" 45d11'40"

Ella Pauline Smith
Cas 23h31'17" 53d19'4"

Ella R. Cottrell
Sco 16h15'54" -11d16'20"

Ella Rae Lesh
And 23h36'16" 47d50'5"

Ella Ray
Gem 6h31'8" 22d41'45"

Ella Reece Wohlers
And 0h36'0" 36d10'46"

Ella Reece Wohlers
Sgr 18h2'50" -27d50'18"

Ella Rita & Kate Hanorah Moroney
Aqr 22h12'5" -4d50'17"

Ella Rose
And 1h18'18" 45d49'42"

Ella Rose
And 23h28'34" 44d45'44"

Ella Rose
Cyg 21h45'3" 43d53'15"

Ella Rose
Leo 9h52'23" 24d33'39"

Ella Rose
Cnc 8h21'22" 16d21'47"

Ella Rose Bedinotti
Cnc 9h11'53" 15d24'37"

Ella Rose Cost
Peg 0h11'9" 17d26'19"

Ella Rose DeDecker
Ari 2h45'46" 22d6'48"

Ella Rose Everall
Uma 10h49'15" 43d20'23"

Ella Rose Farrington
Cyg 20h16'13" 43d20'51"

Ella Rose Mannick
And 1h41'33" 36d47'24"

Ella Rose Tommer
Lyn 8h11'45" 41d20'19"

Ella Rose Whittington
Gem 6h49'55" 29d50'29"

Ella Rose Yearwood
And 23h58'39" 45d35'29"

Ella Ruby
Ori 5h41'32" -0d15'13"

Ella Ruby Harper Ball
Ari 3h12'28" 11d20'51"

Ella Ruby Mahoney
Cap 21h49'14" -15d50'54"

Ella S.
Uma 9h47'14" 42d21'7"

Ella Sarah Harpur
Gem 6h44'45" 32d37'57"

Ella Schreiner Delgado
Umi 15h40'14" 76d55'30"

Ella Somers
Aqr 22h2'53" -18d48'39"

Ella Sophia Watson
Cam 3h33'21" 60d58'40"

Ella / Sunny / Bunny
Aqr 22h53'21" -4d58'9"

Ella Teresa Condon
Cas 1h14'46" 54d0'3"

Ella Thomas
Cas 0h58'57" 57d15'56"

Ella Underwood
Cas 1h12'26" 57d2'55"

Ella Victoria Buckley
And 2h31'25" 46d31'29"

Ella Whitcomb Morris
Cyg 20h34'12" 57d43'4"

Ella-Boo
And 23h29'17" 40d43'30"

Ellah Bear Smith
Uma 9h10'57" 50d31'57"

Ellah Donnelly
Cyg 21h58'58" 47d47'6"

EllaHannahLouLou
Uma 11h40'16" 38d35'2"

Ella-Mae Drabble-Middlebrook
Cyg 20h0'1" 33d53'39"

Ellamaurice
Uma 10h10'56" 64d44'29"

Ella-May
Cas 1h0'12" 57d22'14"

Ellan Fitzgerald
Cam 3h55'26" 63d59'1"

Ellard Ladon Thomas
Lib 15h37'3" -23d50'1"

Ella-Rose Olivia Hirsch Angus
And 2h29'30" 40d52'16"

Ella's Eternal Flame
Per 3h48'17" 49d24'51"

Ella's Light
Tau 3h29'37" 25d20'28"

Ella's Nightlight
And 1h0'33" 45d53'5"

Ella's Star
And 23h1'39" 51d17'50"

Elle
And 0h13'54" 44d54'47"

Elle
Uma 13h57'56" 53d29'18"

Elle 11.11
Uma 11h23'16" 64d21'9"

Elle - "Angel on my Shoulder"
Cyg 21h52'17" 47d58'15"

Elle B. Izzo
Aqr 23h50'57" -11d24'8"

Elle Belle
Mon 6h48'29" 9d31'20"

Elle e Marco
And 2h13'21" 44d6'53"

Elle Holcomb - My Love - My Life
Per 3h16'56" 32d43'41"

Elle & Lily
And 1h39'27" 49d40'37"

Elle Munder
And 0h32'47" 25d58'6"

Elle Munder
Psc 0h53'52" 24d41'17"

Elle Ramsay's Star
Sgr 19h16'59" -14d6'55"

Elle Renee Bywaters
Uma 11h18'2" 35d49'26"

Elle St. John
Gem 7h17'47" 13d46'25"

Elle & Vinny
Aql 19h5'11" -6d19'0"

Elleanor E. Balliet
Cyg 21h46'50" 46d11'37"

Ellee Nicole
Col 5h53'34" -40d37'12"

Elleen Marzolino
Umi 17h5'58" 80d5'27"

Ellegra Gabriel Davis
Cap 21h28'2" -16d40'13"

Ellehcim
Sco 17h0'38" -34d53'36"

Ellen
Sco 17h25'16" -38d51'12"

Ellen
Cap 20h11'18" -14d16'3"

Ellen
Uma 9h57'48" 68d35'16"

Ellen
Cas 0h42'45" 64d54'56"

Ellen
Cas 0h23'41" 56d32'53"

Ellen
Lyr 18h51'57" 42d20'24"

Ellen
Lyn 7h47'25" 37d19'13"

Ellen
Uma 10h58'15" 44d5'49"

Ellen
And 0h51'25" 40d20'42"

Ellen
Leo 11h26'38" 8d21'22"

Ellen
Ori 5h55'9" 6d29'3"

Ellen
Cnc 8h47'36" 24d49'57"

Ellen
Crb 15h42'30" 28d27'38"

Ellen A.
Crb 16h18'40" 35d33'27"

Ellen & Aaron
Gem 7h10'13" 17d46'42"

Ellen and David Jaffe
Cyg 19h34'48" 28d56'2"

Ellen and Jeanne - 2 CCCW
Aur 6h30'22" 38d11'55"

Ellen and Jerry golden
Gem 7h21'19" 24d59'50"

Ellen and Jim
Vir 12h13'18" 11d11'5"

Ellen Ann
Leo 9h30'36" 9d25'9"

Ellen Ann Seeback
And 0h40'51" 41d12'42"

Ellen Ashworth Abbott
Uma 12h26'31" 55d32'27"

Ellen Aurora
Ari 2h44'51" 30d26'50"

Ellen B. Lockamy
Tau 3h41'35" 24d51'53"

Ellen (babydoll) Macioce
And 0h17'0" 29d47'42"

Ellen Beach
And 0h20'11" 23d51'7"

Ellen Bee Coffman Phillips
And 23h17'47" 45d36'35"

Ellen Benton
And 23h15'50" 51d26'30"

Ellen Berg - Glen's Shining Star
Gem 6h55'50" 29d7'55"

Ellen Bertram
Uma 12h25'29" 56d8'14"

Ellen Bielefeld ~ August 16th 1973
Leo 11h33'55" 25d51'0"

Ellen Boyle-20 Years HVB BDB
Cas 0h37'42" 62d39'0"

Ellen Butler
Cyg 20h47'27" 38d55'44"

Ellen Callaway Cleaveland Koebler
Cap 21h29'34" -14d59'29"

Ellen Cecelia Hudson Trexler
Ori 5h3'0" 9d29'40"

Ellen Celeste
Leo 11h18'46" -1d51'49"

Ellen Charlesworth Atkinson
Cas 0h38'46" 56d51'30"

Ellen Cullen 2007 Visitation Star
And 1h39'17" 43d36'15"

Ellen Daniels
Lac 22h12'7" 51d51'26"

Ellen Darleen Blackwell
Psc 23h9'40" 0d59'12"

Ellen Davenport
Uma 8h32'57" 61d3'22"

Ellen DellaValle Milito
Cap 20h23'19" -11d3'52"

Ellen E. Weaver
Uma 11h59'33" 36d10'39"

Ellen E Wheeler
Sgr 18h10'23" -16d41'20"

Ellen Eastman
And 1h22'39" 39d28'29"

Ellen Elaine Fernandez
Cyg 19h46'42" 36d41'6"

Ellen Elise
Sco 17h40'31" -33d14'19"

Ellen Elizabeth
And 2h10'59" 39d8'41"

Ellen Elizabeth deBernière Given
Cam 7h45'18" 71d8'20"

Ellen Elizabeth Levin
Lmi 10h11'33" 35d51'33"

Ellen Fadian McHugh
Cyg 21h27'32" 50d24'12"

Ellen Faith Young
Cyg 19h50'22" 36d47'26"
Ellen Filer
Cas 0h0'45" 52d47'21"
Ellen Foster
Ori 5h14'53" 15d45'40"
Ellen Friesen Schmidt
Crb 15h48'14" 35d57'30"
Ellen G. and A. Paul Victor
Cyg 20h42'8" 51d50'1"
Ellen Gries Loves Billy Randall STG
Uma 11h17'43" 64d23'25"
Ellen Heatley
And 23h16'33" 39d42'27"
Ellen Hernon
Uma 11h13'3" 61d10'59"
Ellen & Heyward Brown
Uma 10h13'44" 47d4'45"
Ellen J. Hayes
Cas 0h28'19" 64d54'47"
Ellen J. Unruh
Uma 11h21'55" 45d36'8"
Ellen Jacobs
Ori 5h38'48" -1d44'52"
Ellen Jane
And 0h55'45" 43d15'42"
Ellen & Jimmy Stephens
Cyg 19h33'48" 29d5'46"
Ellen Johanne 11-26-1971
Sgr 18h14'23" -19d10'5"
Ellen Johnson Liek
Umi 13h10'8" 69d27'52"
Ellen Joyce
Leo 10h19'54" 24d36'8"
Ellen Kasprzyk
Cas 1h5'41" 53d6'51"
Ellen Kathleen
Psc 1h37'4" 19d28'47"
Ellen Kay Mitchell
Ari 2h21'27" 25d25'56"
Ellen Kristin
And 0h10'35" 28d50'11"
Ellen Laura
Vir 13h17'17" 1d26'8"
Ellen Lee Chamberlain
Ari 2h37'22" 20d7'6"
Ellen Lee Williams
Psc 1h18'56" 31d45'19"
Ellen Leigh Hart
Vir 13h16'8" 6d15'37"
Ellen Lent
Ari 2h34'28" 25d27'32"
Ellen Lily
And 0h14'48" 33d25'51"
Ellen (Little One)
Cyg 21h39'18" 51d15'27"
Ellen Louise Cox
Aqr 22h33'1" -1d17'8"
Ellen Louise McSpedon
Tau 5h29'49" 26d39'8"
Ellen Louise Warren
Cas 1h6'25" 63d43'8"
Ellen loves Ken till the stars fall
Her 16h39'52" 25d44'32"
Ellen M Dubin
Mon 6h52'26" -0d28'40"
Ellen M. Sullivan
Psc 23h24'20" 1d41'30"
Ellen Mac Donald
Vir 14h30'25" 4d25'18"
Ellen Maggie
And 23h42'55" 35d41'2"
Ellen Margaret
Uma 11h9'54" 52d58'22"
Ellen Marie
Cyg 20h28'56" 36d42'51"
Ellen Marie
And 0h39'37" 36d10'52"
Ellen Marie 1944
Per 3h18'12" 46d52'48"
Ellen Marie Ahdemar
Umi 12h26'6" 86d11'55"
Ellen Marie Caldwell
And 1h7'48" 45d57'49"
Ellen Marie Cecilia Croke
Uma 13h53'10" 60d2'21"
Ellen Marie Kaysen
Uma 10h48'55" 56d25'2"
Ellen Marie Roach
Leo 10h56'49" 14d23'40"
Ellen Marie Sykes: Rare Fortitude
Per 3h34'51" 36d19'12"
Ellen Mary Heighton
Crb 15h35'35" 37d43'2"
Ellen Mary Shindle
Cap 21h42'19" -15d29'47"
Ellen McGrane
Uma 13h26'23" 54d32'20"
Ellen Merritt Feierabend
Sgr 18h54'58" -31d36'31"
Ellen Moisan
Tau 4h30'49" 22d13'5"
Ellen Molly Nicholls
And 23h59'17" 40d33'54"
Ellen Murphy Jernigan
Psc 1h9'36" 4d11'24"
Ellen Nagai
And 2h11'39" 38d49'23"
Ellen Nickles
Cnc 9h9'59" 8d42'41"
Ellen P. Mitchell
Uma 12h58'42" 60d28'44"

Ellen P. Wayne
Leo 11h6'44" 7d53'44"
Ellen Parker
Sco 16h14'44" -9d22'47"
Ellen Patricia De-Leston
Leo 10h17'21" 10d37'38"
Ellen Pavia
Uma 8h42'20" 61d50'52"
Ellen Perez
Vir 13h18'43" 12d26'56"
Ellen (Princess) Hagen 09041984
Ari 3h29'17" 27d1'31"
Ellen Priscilla
And 0h31'0" 45d2'11"
Ellen Ray
And 2h24'38" 45d23'33"
Ellen Rebecca Dreyer
Lib 14h50'20" -5d13'28"
Ellen Robertson Martinez, Bembi
Tau 5h37'44" 26d45'54"
Ellen Rumler
Umi 14h39'11" 73d37'19"
Ellen Samuels
Per 4h2'52" 33d45'49"
Ellen Sara Gitomer
Aqr 20h59'34" -7d4'59"
Ellen Sorenson Watkins
Cas 0h40'50" 51d47'34"
Ellen Sue
Cnc 9h16'22" 26d6'52"
Ellen Sue
Tau 4h40'34" 4d42'4"
Ellen Tangry
Lyn 7h30'23" 53d16'35"
Ellen Therese Porter
Hya 9h26'11" -0d12'6"
Ellen Thurmond
Crb 15h43'42" 31d12'22"
Ellen Tien
Cyg 19h50'24" 32d52'16"
Ellen Timney
Uma 11h1'46" 43d0'25"
Ellen V. Posker
Psc 23h26'17" 5d58'43"
Ellen Vaughn and Baby Boy Faircloth
Tau 3h41'54" 29d42'55"
Ellen Waltman
And 1h19'5" 43d54'12"
Ellen Williams
And 1h23'40" 37d51'31"
Ellen Winifred Marett
Lmi 9h57'33" 38d17'49"
Ellen Young Gunn
Psc 0h55'30" 17d40'29"
Ellen Yvette Gifford
Sco 16h13'31" -10d44'18"
Ellen, Sunshine
Lmi 10h31'21" 37d3'37"
Ellena
Aur 6h35'56" 39d3'35"
Ellen-Mika and Richard Brown
Cyg 19h44'1" 42d24'16"
ELLeN-m-n
Cam 7h12'1" 80d55'10"
Ellen's Legacy
Mon 6h54'22" -0d28'27"
Ellen's Stella Brillante
Cru 12h32'29" -59d48'39"
Ellen's World
Uma 10h27'15" 43d16'32"
Elleoz
And 1h10'6" 41d39'14"
Ellerie Fuller
Uma 10h17'21" 49d56'11"
Ellery Fugui Gaffin
Del 20h33'13" 13d38'5"
Ellery John & Shannon Lee
Cyg 19h22'39" 52d5'58"
Ellery Reese Schmidtke
Cnc 8h24'38" 16d50'58"
Ellery Ruben
Her 17h7'50" 43d28'23"
Ellery Victoria Long
Tau 4h38'27" 18d31'45"
Ellery Wagner Faller
Uma 12h15'26" 53d6'18"
Ellevicfairbir
Mon 6h47'7" 6d53'25"
Elley
Umi 16h40'4" 83d27'48"
Elley 1
Tau 3h28'7" 1d21'52"
Elli
Ori 6h4'15" 6d32'59"
Elli
Uma 11h29'1" 36d54'18"
Elli
Uma 11h31'26" 61d43'50"
Elli Arlene Granfors Bellegie
Crb 15h43'14" 27d55'0"
Elli Beth
And 1h2'24" 47d26'49"
Elli Schoen
Sgr 19h46'16" -43d36'57"
Elliana
Uma 13h51'12" 57d39'6"
Ellice "Leecie" Chase
And 0h36'36" 37d17'36"
Ellicia
Ori 4h51'29" 13d2'6"

Ellie
Tau 4h45'52" 19d7'26"
Ellie
Leo 10h33'46" 23d19'7"
Ellie
Ari 3h18'7" 29d16'29"
Ellie
Cnc 8h27'9" 19d52'23"
Ellie
Ari 2h32'30" 25d40'28"
Ellie
And 0h9'9" 28d52'25"
Ellie
And 0h24'34" 34d55'18"
Ellie
Crb 16h17'54" 37d14'17"
Ellie
Uma 10h34'56" 43d57'32"
Ellie
And 2h18'51" 41d57'39"
Ellie
And 23h0'12" 41d0'58"
Ellie
Cas 0h6'7" 56d33'8"
Ellie
Uma 13h11'32" 58d20'41"
Ellie
Uma 10h31'56" 66d53'53"
Ellie
Cam 3h34'24" 66d17'19"
Ellie
Uma 9h53'21" 71d3'38"
Ellie
Lib 14h40'2" -14d37'14"
Ellie
Aqr 23h54'7" -10d39'3"
ELLIE
Cap 20h55'4" -22d20'58"
Ellie
Sgr 18h33'41" -23d19'29"
Ellie Ager
Peg 21h52'59" 22d55'17"
Ellie Allyn Miller
Tau 4h22'26" 28d3'33"
Ellie and Jerry Hirschberg
Pho 23h36'29" -55d53'6"
Ellie & Anna Gustavsen
Cyg 20h34'1" 47d45'59"
Ellie Anne Dow
And 22h59'24" 42d42'38"
Ellie Barbara Johnston
And 1h22'18" 40d12'55"
Ellie Beth Campbell
And 1h33'42" 40d20'27"
Ellie Boisen Depuydt
Uma 9h30'10" 58d18'20"
Ellie Brooke Sytsma
Leo 10h15'51" 22d57'7"
Ellie Caitlin Turton
And 1h17'30" 42d9'30"
Ellie Cookie Jace Cassidy
And 0h20'4" 45d42'15"
Ellie Daun Stitzer
Umi 15h3'59" 81d32'35"
Ellie De Wit - Our Mother
Cnc 9h13'4" 26d34'15"
Ellie Dee Danaceau
Aqr 23h37'39" -21d59'14"
Ellie Donnelly
Umi 16h49'44" 78d56'25"
Ellie Eileen Sebastiano
And 0h36'51" 35d57'12"
Ellie Fasso
Tau 4h4'17" 16d57'41"
Ellie Georgia
And 23h16'58" 41d31'26"
Ellie Grace
Cap 21h55'14" -20d23'17"
Ellie Grace Doggen
Cru 12h2'8" -61d39'14"
Ellie Grace One
Cas 23h35'22" 53d56'24"
Ellie Hinds
Peg 21h38'24" 22d43'51"
Ellie Irena
Gem 6h46'27" 33d46'44"
Ellie J Y McNamara
Umi 16h10'4" 74d15'1"
Ellie Jakubowski
Vir 12h29'48" 1d9'58"
Ellie Janet Naylor
And 0h32'10" 33d38'26"
Ellie Jay
And 1h48'22" 45d47'17"
Ellie Jayne
And 0h13'21" 25d59'25"
Ellie Jayne Heath
And 1h37'56" 40d41'55"
Ellie Kane
Uma 10h57'7" 49d49'3"
Ellie Lauren Beach
Mon 7h44'26" -1d59'29"
Ellie Louisa Pugh - "DitDot's Star"
Cas 1h57'10" 64d46'3"
Ellie Louise
And 1h0'2" 45d41'5"
Ellie Louise Mills
And 0h8'19" 41d12'45"
Ellie Louise Williams
And 23h37'6" 48d32'48"
Ellie Mae Carradus
Psc 0h0'7" -6d10'32"

Ellie Mae Heard - Ellie's Star
And 2h16'38" 50d28'45"
Ellie Marie
Lyn 7h41'13" 41d18'2"
Ellie Marie
Uma 11h34'52" 31d7'5"
Ellie Marie
Vir 13h52'3" -7d34'32"
Ellie Marie Sanders
Gem 6h31'47" 16d25'45"
Ellie Martin Sternat
Cma 7h7'44" -24d37'4"
Ellie Mayginnes
And 0h16'35" 42d29'17"
Ellie Mocabee ~09/05/1927
Cas 0h16'0" 55d54'45"
Ellie Mullen
And 1h43'33" 49d19'14"
Ellie n Wayne's guidin' Star xxx
Cas 0h35'55" 57d6'30"
Ellie Olivia Caggegi
Cae 4h52'8" -35d23'57"
Ellie Rebecca Jakubowski
Uma 8h23'2" 64d43'19"
Ellie Rose
Sgr 18h29'37" -27d53'9"
Ellie Rose
Ori 6h3'28" 11d2'2"
Ellie Rose Gullickson
Mon 7h39'27" -0d43'35"
Ellie Rose Knobel
Gem 7h9'38" 21d1'16"
Ellie Rose May
And 0h11'3" 45d52'50"
Ellie Sherwood
Agr 21h34'12" 2d14'41"
Ellie Somers
Crb 16h16'6" 36d39'53"
Ellie Suzanne
Lib 14h45'2" -10d41'6"
Ellie T. Henwood
Cas 0h53'57" 53d17'34"
Ellie Teresa Carroll - Ellie's Star
And 23h7'39" 50d56'37"
Ellie Wood
And 1h16'30" 34d42'41"
Ellie, a bright little star
And 23h41'36" 37d29'41"
EllieAnn Johnson
And 0h57'29" 41d30'34"
Ellie-Ann Tait
And 23h5'35" 51d24'58"
Ellie-Belly
Agr 21h26'33" 0d45'36"
Ellie-K
Lmi 10h13'59" 35d56'20"
Ellie-Mae
Umi 14h3'17" 70d40'9"
Ellie-May
Lib 14h31'35" -9d53'12"
Ellie-Rose
And 1h20'5" 41d0'24"
Ellie's Belly
Cas 0h3'23" 57d31'57"
Ellie's Destiny
Lib 15h45'1" -5d43'53"
Ellie's Light
Pyx 8h39'45" -22d11'34"
Ellie's Star
Lyn 7h22'15" 53d10'49"
Ellie's Star
And 23h39'35" 42d49'21"
Ellie's Star
And 1h15'38" 42d59'6"
Ellie's Star
Ari 2h30'28" 22d16'19"
Ellifford
Agr 23h37'54" -9d54'42"
Ellika & Raymond
Cyg 19h36'4" 32d32'55"
Ellin Quinn Graduation Day - My Star
Cru 12h48'44" -60d33'19"
Ellina - "Curator of Fidelis Amor"
Ari 3h13'57" 22d39'0"
Ellinger Family Star
Ori 6h9'23" 15d16'33"
Elliot
Leo 9h26'53" 14d37'38"
Elliot
Vir 14h18'7" -19d22'17"
Elliot Alexander Frueh
Lib 15h25'46" -15d21'44"
Elliot Burnell Jordan Gartan
Umi 14h9'57" 78d22'34"
Elliot Forrest
Aql 19h47'14" 14d1'19"
Elliot Forrest
And 1h51'59" 38d6'51"
Elliot Jay Thompson
Lib 14h53'35" -7d20'20"
Elliot Joel Stern
Uma 11h33'5" 59d51'45"
Elliot John Busby - Elliot's Star
Per 4h45'53" 41d45'22"
Elliot John Tiede
Leo 10h53'56" 6d25'4"
Elliot Kenneth Zimmer
Leo 9h42'12" 31d14'48"

Elliot Kurt Latzsch
Vir 14h7'12" -15d12'48"
Elliot Ly
Lyn 8h4'40" 34d20'52"
Elliot Mark Smallshaw
Psc 1h7'39" 32d3'17"
Elliot Michael Trento 2:32 P.M.
Vir 13h17'3" 11d47'32"
Elliot Otto Abt
Uma 10h26'31" 66d10'53"
Elliot Our Gentle Soul <\@> The Bridge
Lyn 7h8'56" 58d49'27"
Elliot Paniagua
Aur 6h54'2" 41d29'34"
Elliot Paul Day
Per 4h10'21" 33d27'59"
Elliot Pindall Nordstrom Griffiths
Per 3h29'7" 51d46'2"
Elliot S. Taynor
Per 4h30'49" 40d51'12"
Elliot the Innovator
Per 3h41'43" 46d48'16"
Elliot Thomas Moss
Her 16h15'5" 6d50'5"
Elliot Wendell Wright
Cap 20h29'37" -23d56'26"
Elliot William Esposito
Cnc 8h12'38" 30d39'21"
Elliot William Hedley
Per 4h24'29" 32d6'40"
Elliot Wynn Davison
Cap 20h19'16" -12d25'21"
Elliot, Jack & Sophie Rumble
Uma 9h30'55" 49d28'4"
Elliot's Star
Tau 3h57'26" 28d13'11"
Elliot's Wishing Star
Ari 3h16'41" 23d43'55"
Elliott
Ari 3h29'20" 24d11'28"
Elliott
And 1h30'14" 40d6'18"
"Elliott"
Dra 18h37'17" 78d59'21"
Elliott
Cep 22h12'5" 72d50'13"
Elliott Alasdair Davies - Elliott's Star
Peg 21h56'20" 17d35'38"
Elliott Alfie Steven Byers-Hulme
Uma 13h29'39" 58d51'35"
Elliott Bensinger Thompson
Ori 5h47'18" -3d42'13"
Elliott Blaufuss
Tau 5h9'5" 24d21'18"
Elliott Cameron
And 1h24'43" 49d28'47"
Elliott Edward Fitzgibbons
Sco 16h7'6" -14d24'49"
ELLIOTT G
Lup 15h22'11" -41d1'48"
Elliott Gasche
Lyr 18h53'34" 38d14'5"
Elliott Glass
Gem 7h43'38" 31d11'50"
Elliott Gordon Gray
Ori 5h51'8" 19d56'8"
Elliott James
Cnc 7h55'49" 17d28'7"
Elliott Kurt Klein
Ari 3h9'41" 23d13'30"
Elliott Langley Dunham Smith
Lib 15h56'29" -18d35'24"
Elliott Manley
Per 4h36'5" 42d57'25"
Elliott Millet
Vir 12h47'17" -10d53'55"
Elliott Pacaux
Psc 23h3'5" 7d0'4"
Elliott Richard Hearn
Cep 23h26'44" 72d11'6"
Elliott Sarah Calhoun
Sgr 19h1'18" -25d3'21"
Elliott Straite
Her 16h34'10" 34d51'23"
Elliott Taylor Marburg
Del 20h42'38" 16d38'1"
Elliott Teipen Parrott
Vir 13h53'23" -20d7'1"
Elliott Thomas Stallwood
Umi 17h21'41" 85d47'22"
Elliott Zakk
Umi 15h9'56" 72d12'0"
Elliott's Eternal Love/Light
Uma 9h33'12" 56d0'40"
Elliott's Star
Peg 21h45'5" 20d59'26"
Ellis
And 23h53'3" 45d19'24"
Ellis
Umi 14h0'41" 70d15'58"
Ellis
Uma 11h20'40" 72d4'34"
Ellis and Johnny
Psc 1h12'56" 27d6'53"
Ellis' Angel Lyn
Boo 14h42'51" 45d32'47"
Ellis Brownridge
Tri 1h57'31" 32d54'40"

Ellis Clay
Vir 14h35'31" 3d48'32"
Ellis Franklin Daniels
Uma 9h40'10" 46d33'46"
Ellis Igor Kay
Cas 1h30'11" 62d31'46"
Ellis Jay Hurndall
Cep 23h42'47" 86d30'11"
Ellis Judson Rich
Cyg 20h2'53" 46d3'45"
Ellis Lee Houghton
Ori 6h22'2" 15d5'43"
Ellis Lloyd Jones
Uma 8h39'10" 68d19'50"
Ellis MacKenzie Kemper
And 0h55'30" 45d20'16"
Ellis Martin
Uma 12h29'43" 55d55'57"
Ellis & Michael
Uma 9h14'20" 50d54'49"
Ellis Robinson
Lib 15h22'58" -24d46'56"
Ellisa Emma Mackey
Lyn 7h31'29" 38d58'22"
Elliya Chloe Farr
Uma 9h21'50" 54d49'52"
Ellizabeth Skyler Madison
Aqr 21h37'38" 2d46'45"
Elllie Elizabeth
Aqr 21h42'18" 1d48'3"
Ello Lovely
Her 16h59'25" 13d42'19"
Ellodie Reynolds
Cyg 20h56'28" 35d41'14"
Ellorah Ann Scalf
And 1h36'35" 39d31'33"
Ellory Blythe Dickerson
Ari 2h26'47" 24d27'6"
Ellsworth August Rolfs
Aqr 22h50'29" -10d47'20"
Ellweese Linden
Aql 20h17'26" 5d20'39"
Ellwood Werry
Uma 10h53'30" 65d29'50"
Elly Antons
Lyr 18h34'13" 28d9'48"
Elly B B
Cas 1h25'53" 56d22'5"
Elly B. Waibel
Vir 14h48'16" 0d33'52"
Elly Culucundis
Sco 17h37'44" -39d33'53"
Elly G
Lyr 19h17'56" 28d42'25"
Elly Harvey
And 23h40'14" 48d16'17"
Elly Moayer
Uma 10h33'24" 46d24'8"
Elly Rae
Gem 6h58'55" 33d33'57"
Ellyce Aubrey Klida
Psc 0h52'7" 28d20'3"
Ellychka
Tau 4h30'31" 19d38'14"
Ellyn
Leo 9h48'21" 6d37'47"
Ellynn McLean
Lyn 8h21'1" 40d30'25"
Ellyson Grace
Cas 0h43'5" 66d33'21"
Elma delos Reyes
Cyg 20h35'58" 34d39'41"
Elma Ledoux
Sgr 17h52'35" -28d6'55"
Elma Lee Walters
Uma 13h0'16" 54d27'18"
Elma "Meme" Looney
Peg 23h49'31" 24d1'57"
Elmar Brähler
Uma 12h45'48" 55d53'38"
Elmarr Stevenson
Uma 9h3'2" 58d46'8"
Elma's Clifford
Sgr 18h14'1" -19d29'54"
Elmer
Uma 10h24'44" 40d27'43"
Elmer Broomhall "Superman"
Gem 7h0'30" 31d42'52"
Elmer Burrill
Cap 20h58'32" -19d56'25"
Elmer Cornelius
Lyr 18h41'54" 38d50'17"
Elmer E. Eychaner III
Umi 14h17'29" 68d36'46"
Elmer E. Scott Jr.
Boo 14h22'42" 13d52'53"
Elmer E. Walcker
Psc 0h51'40" 12d4'21"
Elmer Eli
Aqr 23h51'1" -12d15'33"
Elmer & Evelyn Zeckser
Uma 12h2'31" 40d25'9"
Elmer F. Miller, Jr.
Cyg 19h51'5" 36d13'36"
Elmer J Worden
Uma 12h21'40" 53d18'0"
Elmer Johnson
Her 16h10'50" 47d10'30"
Elmer Joseph Kish
Boo 14h35'52" 27d7'20"
Elmer Koob *100th*
Psc 23h12'13" 1d10'59"
Elmer & Margie Forever
Cyg 20h12'18" 35d52'52"

Elmer S. Wald
And 1h18'14" 49d18'7"
Elmer Shattuck Patterson
Aql 19h42'22" -0d3'21"
Elmer Warner Cromie
Leo 10h48'53" 16d49'51"
Elmira Josephine Broughman Waddell
Cas 23h57'23" 53d45'23"
Elmira Safina
Lyn 7h27'11" 44d25'2"
Elmo
Lib 14h49'12" -2d39'58"
ELMO
Cap 21h55'46" -23d4'26"
Elna M. Scheinfeld
And 2h4'9" 37d15'52"
ELNO
Uma 10h4'59" 46d52'55"
Elo
Leo 9h42'12" 20d36'57"
Eloah
Uma 11h29'40" 35d42'31"
ELOCIN
Uma 11h33'50" 50d42'48"
Elodia
Cas 23h41'24" 59d31'10"
Elodie
Cyg 19h12'5" 52d36'5"
Elodie
Dra 10h28'42" 73d26'0"
Elodie
Col 6h6'34" -28d15'12"
Elodie
Cnv 13h49'45" 37d37'50"
Élodie
Vir 11h58'8" 6d2'23"
Elodie Colin
Uma 9h45'29" 60d6'26"
Elodie Colleen Soustelle
Ari 2h5'2" 20d53'48"
Elodie Constantin
Uma 13h6'40" 54d56'0"
Elodie Cuhat
Uma 9h36'34" 44d7'46"
Elodie Curado
Cas 0h38'14" 58d42'0"
Élodie & Frédéric
Cas 0h1'36" 59d50'29"
Elodie Gayraud
Sco 16h13'26" -29d10'2"
Elodie Gernay
Sco 17h49'5" -33d37'56"
Elodie Lafitte
Leo 11h16'2" 0d5'38"
Elodie ma Princesse
Ori 6h22'4" 10d44'50"
Elodie Soleil + Cédric Maillot
And 23h47'14" 40d20'34"
Elodie-Anne
Ari 2h47'35" 11d7'31"
Éloïk Dicaire
Ori 6h0'20" 10d18'8"
Eloisa
Cap 20h36'12" -11d8'40"
Eloisa Cortez
Cap 21h29'19" -13d29'31"
Eloisa Del Carmen Ortez-Henriquez
And 23h22'20" 48d47'6"
Eloisa Laurente
Sgr 18h36'49" -30d44'27"
Eloise
Sgr 19h43'0" -20d9'30"
Eloise
Eri 3h46'54" -13d51'27"
Eloise
Umi 16h27'38" 81d0'33"
Eloise
Lyn 7h37'4" 57d3'12"
Eloise
Uma 8h27'30" 60d18'56"
Eloise
And 23h24'47" 52d24'35"
Eloise Aliski
And 2h11'36" 38d16'31"
Eloise Amy Morris - "Ellie's Star".
And 2h32'0" 46d50'31"
Eloise and Murice Herrick Memorial
Cyg 20h56'22" 48d20'6"
Eloise Ann Cobain's Christening Star
And 1h14'13" 44d14'1"
Eloise Elizabeth Olden
Crb 15h44'1" 26d44'19"
Eloise Joan "Bubbles" De Santis
Aqr 21h10'22" -12d32'8"
Eloise Lyons
Uma 9h56'23" 64d59'2"
Eloise Marlowe Andrews
Aqr 20h48'29" 0d1'17"
Eloise * Mom
Umi 14h58'0" 72d56'50"
Eloise Runyan
Lyr 19h7'0" 36d2'52"
Eloiza Laguer
Tau 3h39'28" 24d14'26"
Elona Bright
Gem 6h19'46" 26d49'15"
Elora
Crb 16h11'38" 33d1'21"

Elora Anne
Cyg 21h11'55" 31d43'43"
Elora Geier
Uma 10h24'4" 64d42'37"
Elora Jade
Uma 11h38'58" 51d31'52"
ELORAC 51
Uma 10h19'59" 56d5'45"
Elora's Shining Star
Uma 8h49'33" 61d40'48"
Elores Faye Arnold
Crb 15h44'58" 27d29'55"
Eloria
Sco 17h55'58" -31d58'38"
Elouisa Rebecca
Cap 20h35'18" -21d32'4"
Elouise Mesnard
And 23h10'52" 41d45'55"
Eloy
Aur 6h33'18" 39d3'51"
Eloya - Nabil Ibrahim Atallah
Mon 7h6'0" -6d47'56"
Elpiniki
Lib 15h5'57" -20d35'50"
Elrond
Pho 0h44'22" -43d27'51"
El-Roooo7(Saud & Fatma forever06)
Mon 8h10'50" -3d20'17"
Elroy Blinky
Ori 4h52'12" 4d33'43"
Elroy of Larry and John
Cma 7h27'26" -18d30'32"
Elsa
Sgr 19h15'4" -16d37'28"
Elsa
Cam 5h27'0" 78d55'2"
Elsa
Pho 0h34'4" -46d37'38"
Elsa
Aqr 23h15'43" -20d42'17"
Elsa Anna Olsen
Uma 11h1'27" 43d52'4"
Elsa Aurora
Cas 1h19'45" 55d33'14"
Elsa Chen
Sco 17h0'49" -30d30'2"
Elsa Cormier
Psc 1h4'1" 6d59'31"
Elsa Diane Carman Wright
Cnc 8h46'27" 26d47'21"
Elsa Dorca
Cap 21h34'23" -19d51'24"
Elsa Doris
Leo 11h16'53" 17d31'16"
Elsa et Hélène
Aqr 22h24'46" -19d20'11"
Elsa Guadalupe Canales
Cam 6h18'33" 64d10'58"
Elsa Hajjar
And 2h32'14" 45d41'57"
Elsa Jean
Lyn 7h12'8" 53d17'14"
Elsa Kyriakopoulou
Lib 14h24'34" -11d12'34"
Elsa Mae Bohlender May 1988
Tau 4h42'28" 20d25'35"
Elsa Maria
Cas 1h26'32" 60d39'47"
Elsa McGee
Gem 6h42'13" 12d34'48"
Elsa My Love
Cyg 22h0'21" 53d16'26"
Elsa Patricia Alvizar Sandoval
Ori 5h49'10" -5d29'0"
Elsa Rae Sather
Crb 15h29'0" 31d30'59"
Elsa Teixeira (tite cuillère)
Dra 16h56'59" 57d1'37"
elsadie28
Aqr 21h27'6" -0d33'52"
Elsanico
Cas 0h54'51" 56d39'59"
Elsbeth-RuMaDa Stern
Cas 1h21'55" 71d39'14"
ELSE
Uma 9h15'5" 58d10'5"
Else Christensen
Pho 0h18'14" -41d41'37"
Else Gausel Hyman
Aql 19h58'28" -0d27'39"
Else Stähli
And 1h3'1" 46d24'5"
Elsi Rivas
And 23h15'16" 36d9'43"
Elsie
Aqr 23h3'21" -6d30'17"
Elsie
Umi 15h42'10" 73d14'25"
Elsie
Sco 17h51'33" -36d28'16"
ELSIE
Sgr 17h54'15" -28d47'43"
Elsie Alice Payne
Crb 16h1'3" 38d45'30"
Elsie Anne
And 1h6'17" 45d11'10"
Elsie Anne Jack
And 2h25'19" 41d2'4"
Elsie Blumer
Lyn 7h39'14" 38d15'32"

Elsie Bonner
Cas 23h16'1" 54d30'4"
Elsie Brownfield Caton
Aqr 22h27'4" -2d22'51"
Elsie Cole
Umi 15h0'18" 76d56'26"
Elsie Fang & Lawrence Tai
Tau 5h46'47" 25d43'36"
Elsie Frances Yawa Ampofo
And 2h36'59" 38d57'7"
Elsie Garwood Lyons
Leo 9h26'32" 25d7'51"
Elsie Irene Lam
Aqr 21h58'10" -2d46'6"
Elsie J Young
Cas 0h40'48" 61d4'43"
Elsie Jean Bartel-Clayton
Uma 8h47'24" 72d58'41"
Elsie June Olmstead
Cas 2h14'30" 71d3'39"
Elsie K Bryant
Sgr 18h36'14" -26d52'56"
Elsie L. Hylt
Uma 10h24'24" 56d35'41"
Elsie LaSalle
Uma 12h58'34" 58d1'6"
Elsie & Linda BFFE
Cam 4h47'42" 55d53'52"
Elsie Lowe Henderson
Psc 2h1'24" 6d53'58"
Elsie Marie Cheramie Horton
Lyr 18h40'31" 33d53'22"
Elsie Maud Roberts
Cas 23h2'56" 58d29'24"
Elsie Mercedes
Aqr 23h6'49" -7d6'8"
Elsie Muriel Newton
Cas 0h56'20" 48d14'16"
Elsie Nancy Avedon
Gem 6h49'20" 21d47'31"
Elsie Patterson
Leo 10h29'5" 27d9'16"
Elsie Pieroni
Uma 11h36'32" 34d36'14"
Elsie Rhind
Cas 23h32'18" 55d12'38"
Elsie Schwarz
Lyn 8h1'50" 41d45'10"
Elsie Tamsin Taguchi-Medd
Umi 15h58'34" 77d19'31"
Elsie Tipton & Kim Fuchs forever
Gem 7h48'11" 28d15'4"
Elsie Torres
Cyg 21h52'3" 47d57'39"
Elsie's Halo
Crb 15h37'7" 28d7'8"
Elsie's Lighthouse
Cas 22h57'33" 53d14'2"
ElsieSara
Cnc 8h34'28" 28d48'12"
Elsita
Cnc 9h5'35" 21d8'53"
Elsker Stjerne
Lyr 18h49'22" 37d16'31"
Elsmar 40
Cyg 20h53'50" 48d32'23"
Elspeth Mary
Cas 0h47'53" 64d1'55"
Elspeth Zara Cameron
Lyn 6h51'26" 53d43'32"
Elspeth's Star
Ori 5h47'50" -2d5'26"
ELSRON
Dra 12h2'50" 71d28'40"
Elster
Her 16h49'6" 12d11'35"
ELSY
Crb 16h12'54" 35d2'7"
Elsy Abi Gerges
Tau 5h3'45" 16d8'46"
Elsy Arriaga
Tau 5h8'35" 23d10'14"
Eltern von Franziska Pastuschka
Uma 9h38'41" 54d48'28"
Elton H. J. LaBell
And 23h15'40" 47d32'35"
Elton R. & Geraldine E. Williams
Sge 19h28'11" 18d59'33"
Elton R. Kerr
Cnc 8h32'6" 13d42'55"
Elton R. Sullivan
Per 3h49'44" 51d13'34"
Elum James VanEst
Lib 15h0'5" -10d34'7"
Elushka Cudovnaya
Sco 16h38'22" -41d52'35"
Elva Alliene
Cnc 8h22'45" 24d29'35"
Elva Cruikshank
Cma 6h41'50" -18d58'41"
Elva Del Porto
Cap 20h9'46" -21d43'0"
Elva Juarez
Uma 11h31'41" 46d30'46"
Elva L. lid
Cas 0h26'20" 55d34'33"
Elva Riley
Uma 11h32'19" 56d15'18"

Elva Root
Uma 13h48'50" 58d42'18"
Elva Rosa
And 0h25'17" 31d17'33"
Elvera Hetterman Herdrich
Lyr 18h49'1" 31d10'3"
Elvera Leuch
Lyn 8h0'19" 51d30'48"
Elverage Allen
Gem 7h36'10" 27d10'54"
Elverlena A. Cormier
And 23h11'56" 48d5'10"
Elveta
Cap 21h42'51" -12d24'44"
Elvia
Sco 16h17'36" -12d36'20"
Elvia
Ori 4h54'9" 1d36'27"
Elvia and Tom
Cyg 20h44'29" 35d47'52"
Elvia Isabel Hernandez Pacheco
Com 13h16'17" 18d59'51"
ElviaEduardo
Sgr 18h49'0" -28d13'14"
Elvie Charles Austin "Ace
Uma 12h2'30" 65d13'22"
Elvin
Uma 11h59'39" 36d51'56"
Elvin Fleetwood Frost
Cnc 9h3'35" 28d49'17"
Elvin Fowler
Ori 5h32'24" 3d6'58"
Elvina Anne Wilson
Cas 1h27'53" 68d47'36"
Elvina Samantha
And 1h24'56" 44d17'38"
Elvina's Shooting Star
Cnc 8h29'28" 28d17'15"
ElvinOceanusKirbyDemetri usPucket
Ori 5h37'41" 11d56'48"
ELVINRAY
Ori 6h11'48" 8d3'5"
"ELVIO" Il mio AMORE
Boo 14h31'40" 42d8'6"
Elvira
And 2h9'35" 45d44'18"
Elvira
Ori 6h15'47" 11d0'38"
Elvira
Lmi 10h56'15" 29d9'11"
Elvira
Sct 18h52'38" -15d30'31"
Elvira Coulson
Aql 19h48'49" -0d46'17"
Elvira Elizabeth Sprieck Heil
Cap 21h9'51" -15d38'55"
Elvira Galgano
And 23h40'55" 42d11'0"
Elvira Grace
Vir 13h17'55" 4d25'31"
Elvira Jakob-Eggenberger
Umi 13h28'2" 71d33'17"
Elvira Lehmann - Feusi
Uma 10h56'12" 33d58'11"
Elvira M. Robinson
Ser 15h54'24" -0d17'7"
Elvira Maria
Aqr 22h20'36" -17d35'33"
Elvira Martinez De Cardenas
Vir 12h29'44" -2d49'19"
Elvira Quadri De Stefani
Crb 15h27'3" 26d17'48"
Elvira Rashidovna Kharisova
Umi 14h46'31" 84d22'3"
Elvira Taylor
Aql 19h7'39" 7d37'49"
Elvira the Everlasting Beauty
Vir 13h7'36" 13d52'53"
Elvira's Light
Vel 9h20'58" -42d17'58"
Elvis
Cep 22h27'35" 83d16'24"
Elvis
Sgr 19h15'28" -16d45'53"
Elvis
Eri 4h35'1" -15d2'45"
Elvis Walton
Ari 3h17'36" 29d48'30"
Elwanda
Del 20h48'12" 15d9'17"
Elwood Albert Dance
Gem 7h15'59" 20d31'43"
Elwood Campbell
Leo 9h29'35" 15d38'33"
Elwood Lefever
Aqr 22h13'4" 2d11'25"
Elwood & Rose Seasholtz
Cyg 21h13'27" 51d21'37"
Ely
Lyr 18h43'54" 30d5'47"
Ely
Uma 9h1'42" 57d34'1"
Ely & Javi
Sge 19h50'8" 18d50'58"
ELY1774
Cap 21h47'41" -15d55'36"
Elya
Tau 4h59'19" 19d34'18"

Élyane Deslauriers
Del 20h22'40" 20d3'43"
Elyanne Breton
Uma 9h40'0" 50d15'38"
Elyce 122
And 0h19'46" 32d44'37"
Elyce Louise Pearson
And 1h23'35" 39d44'19"
Elyce Michele Franks
Aqr 22h1'32" -16d53'13"
ELYHJA
Leo 11h39'31" 25d21'4"
Elyn Beth Lawhorn
Cap 21h39'1" -13d51'47"
Elynne Chaplik Aleskow
Gem 6h43'28" 15d10'45"
Elys Grace Walker
Cyg 20h36'15" 54d10'27"
Elysce Ontiveros
Psc 23h15'26" 7d2'23"
Elyse
And 0h19'32" 43d42'52"
Elyse
And 0h21'31" 37d4'27"
Elyse
Uma 11h54'19" 47d43'8"
Elyse
Crb 15h42'21" 38d25'32"
Elyse
Cap 21h39'30" -11d57'8"
Elyse
Umi 16h59'23" 78d52'39"
Elyse
Lib 14h48'46" -1d14'44"
Elyse Ackerly
Psc 1h14'31" 27d26'32"
Elyse and Billy
Vir 13h49'1" -3d10'40"
Elyse Arline Luciano
And 0h20'4" 42d10'54"
Elyse Camarra
Aqr 22h17'22" -22d43'20"
Elyse Edward's Father
Per 2h18'30" 53d20'19"
Elyse Erin Terry
Leo 11h26'29" 14d44'38"
Elyse et Mickaël
Tau 5h40'31" 23d52'42"
Elyse/ Lenny
Lyn 6h29'12" 56d49'23"
Elyse Mae
Cap 20h34'48" -21d59'1"
Elyse Morschauser
Aqr 20h45'55" 2d12'24"
Elyse Nicole Skowronek
Psc 1h8'30" 33d32'38"
Elyse Piwonka
Sgr 18h51'8" -31d42'7"
Elyse Price
Sco 17h17'8" -44d17'59"
Elyse R. Kaplan
Uma 8h37'10" 49d21'1"
Elyse Rose Keenan, Beautiful & Wise
Cyg 20h46'50" 52d20'56"
Elyse Sophia
Mon 6h50'37" -0d14'26"
Elyse The Beautiful
And 23h10'31" 37d51'48"
Elysia
Psc 1h28'31" 28d3'27"
Elysia 220803
And 1h6'34" 41d37'56"
Elysia Joy Winebrenner
And 0h7'9" 46d15'57"
Elysia Pooh Bear
Sgr 17h57'35" -16d54'16"
Elysian
Mon 6h27'44" 8d34'32"
Elysium for Lucy M. Bean
Cnc 8h23'58" 20d44'57"
Elyssa
Mon 6h34'55" 9d29'21"
Elyssa Alters
Aqr 22h35'8" -7d50'46"
Elyssa Jill Caplan
Cap 20h19'12" -24d4'54"
Elyssa Sharon Greenberg
Psc 1h19'35" 7d25'10"
Elyza Cole
Cyg 19h35'32" 32d24'26"
Elz
Lyr 18h49'56" 35d5'48"
ELZA
Aql 19h34'20" 10d20'44"
Elzbieta Bakula
Cnc 9h9'29" 31d9'44"
Elzbieta Starzyk
Ari 2h28'12" 22d59'4"
Elzbieta Uliasz - kubus
Del 20h34'9" 14d47'28"
Elzbieta Zakrzewska
Ari 2h48'59" 25d5'14"
Elzina
Peg 22h23'9" 3d42'32"
Elzo B. Walden
Sco 17h8'12" -31d43'49"
Em
Psc 0h1'31" 7d6'18"
Em
Sgr 19h28'5" -18d4'56"
Em
Tau 4h38'7" 27d42'46"
Em
And 1h39'43" 40d8'12"

Em
Boo 15h23'43" 44d24'3"
em and rach
Leo 11h38'23" 20d4'16"
Em Blaze
Sco 16h10'0" -29d50'55"
Em & Si - A Love Etched on the Stars
Pho 23h55'11" -49d58'23"
Em The Champ
Cru 12h40'2" -62d44'30"
Em Vision
Ori 6h20'38" 9d41'26"
e-mac
And 1h20'34" 44d42'13"
Emad
Cap 20h17'22" -12d38'44"
Emad Abdrabou
Uma 10h48'27" 46d49'54"
Emaien Dahn Nabrohnia
Leo 10h46'24" 15d25'50"
Ema-Leigh Jo Hamilton
And 1h8'27" 35d17'22"
Emalese Rose Fuelling
Psc 0h19'44" 19d2'9"
Emalia Rose Collins
And 1h46'33" 39d4'37"
Emalyn & Joe
Per 2h46'27" 40d58'25"
Eman Anis
Vir 13h32'10" -12d7'31"
EMANJULA
Sco 17h21'7" -38d25'8"
Emanuel
Aur 5h23'58" 40d54'39"
Emanuel
Lac 22h47'24" 51d22'41"
Emanuel
Her 17h11'21" 24d47'43"
Emanuel A Paris IV
Psc 0h29'10" 19d35'16"
Emanuel Defty - Rachel Wyatt Patton
Umi 16h2'32" 77d34'54"
Emanuel Klein
Her 17h34'7" 47d53'34"
Emanuel N. Catelanos
Boo 14h30'54" 28d7'3"
Emanuel R Galosson
Ori 6h23'47" 14d37'21"
Emanuel Rodriquez
Cet 0h24'33" -14d21'6"
Emanuel T Conomos
Cru 12h37'50" -61d2'36"
Emanuela
And 2h24'49" 43d54'53"
Emanuela
Ori 5h52'8" -8d23'44"
Emanuela
Ori 6h10'41" -2d6'41"
Emanuela
Umi 17h6'22" 80d34'7"
Emanuela
Her 18h50'29" 20d55'32"
Emanuela
Crb 15h58'53" 39d10'58"
Emanuela
And 2h20'42" 44d9'5"
Emanuela Arrigoni 40
Mon 7h4'17" 5d45'19"
Emanuela DeNicola
Cnc 8h19'45" 11d27'33"
Emanuela e Roberto
And 1h40'39" 46d49'9"
Emanuela Markova
Psc 0h50'37" 12d55'51"
Emanuela Mazzarelli & Andrea Carvelli
Lyr 18h44'44" 31d38'55"
Emanuela Monzino
Cyg 20h39'47" 47d44'12"
Emanuela Nifoudis
Ori 6h17'17" 15d18'14"
Emanuela Pagin
Lyn 7h46'29" 46d0'50"
Emanuela Storchi
Aql 20h16'24" -0d40'54"
Emanuela Ti voglio tanto tanto bene
Her 16h35'7" 13d36'56"
Emanuela, la stella più splendente
Uma 10h11'1" 66d14'59"
Emanuele
Per 3h48'54" 33d3'27"
Emanuele
Ori 5h40'6" 13d3'9"
Emanuele
Ori 6h0'16" 6d15'14"
Emanuele
Lyr 19h13'5" 38d57'15"
Emanuele
Dra 16h21'12" 52d18'22"
Emanuele Amaru
Sgr 18h27'39" -16d3'29"
Emanuele Anthony Accetta
Her 17h29'26" 36d10'5"
Emanuele Barzacca
Dra 15h28'52" 63d33'55"
Emanuele Biamonti
Uma 11h24'50" 62d25'14"
Emanuele Iacopelli
Ori 5h12'35" -1d32'25"
Emanuele Innaco
Umi 16h1'13" 79d43'10"

Emanuele & Nicole
Cas 23h32'39" 57d2'34"
Emanuele Scoletta
Lyn 6h20'37" 58d34'4"
Emanuelio
Her 18h46'31" 12d40'5"
EmanuelP1
Leo 10h35'24" 13d8'41"
Emanuel's Star
Per 3h21'7" 43d7'48"
Emarie
Lib 15h29'14" -10d2'9"
Emarr Anthony Castillo
Boo 14h29'12" 24d49'4"
Ematum
Uma 11h30'49" 48d3'56"
EmBa
And 1h19'4" 45d15'11"
Embacher, Hildegard
Uma 9h50'1" 43d10'53"
Embassy (MBC)
Umi 14h51'1" 68d0'29"
Ember Anna
Lib 15h8'30" -4d0'8"
Ember Lucia Regaspi
Sgr 20h24'28" -33d27'33"
Ember Lynn
And 0h22'10" 25d40'17"
Ember Pearson
Cyg 20h31'15" 40d0'44"
Ember Rose
Cas 1h51'28" 63d27'8"
Emberlee
Aql 19h8'44" -7d7'50"
Emberly Doherty
Cyg 21h40'30" 35d26'43"
Ember's Heavenly Light
Umi 15h23'22" 74d53'49"
Ember's Red Flame
Crb 15h36'16" 32d15'2"
Embry
Ori 6h4'2" 10d18'41"
"EMBUT" - Emma Howell
Cyg 21h44'49" 44d54'54"
Emdall 7/4
Peg 21h45'54" 22d58'29"
EmDobs
Sgr 19h49'41" -16d49'9"
Emdon Lucky
Uma 11h49'9" 62d33'41"
EMEJ
Tau 5h44'6" 13d51'53"
Emeka
Gem 7h24'11" 31d18'33"
Emel and Tuncay = Eternity
Cru 12h15'1" -63d44'33"
Emel Joanne Sensoy
Umi 14h51'37" 84d30'28"
Emel Topal
Lib 15h2'45" -9d29'32"
Emelelowe
Cnc 8h43'20" 17d13'8"
Emelia Grace Te Paa
Cru 12h19'56" -62d30'24"
Emelia Janthina
Gem 6h44'57" 23d22'54"
Emelia Nicole Phelan - "My Star"
Cnc 8h3'58" 14d41'21"
Emelia Rose
Psc 1h8'52" 31d32'30"
Emelie Gustafsson
Cas 0h29'23" 61d33'21"
Emelie M. McDermott
Lib 15h8'46" -10d36'27"
Emelin E. Allen
Aur 5h23'14" 44d37'24"
Emeline
Crb 15h59'58" 38d10'59"
Emeline
Her 17h53'16" 23d45'27"
Emeline
Sco 17h14'18" -33d7'40"
Emeline et Marie
Uma 9h29'2" 41d41'8"
Emeline Marseillès
Her 17h6'26" 46d31'11"
Emelini Lowe
And 2h4'8" 37d43'22"
Emelita Terrado
Umi 17h38'47" 81d25'45"
Emely & Anthony
Ari 2h50'52" 15d49'29"
Emely Hernandez
Leo 10h44'20" 15d38'16"
Emelye
Cru 12h50'5" -62d26'28"
Emeola 0805
Cnv 13h48'6" 32d55'50"
Emer
Mon 6h54'39" -1d8'16"
Emer and Brad
Aql 19h35'6" 12d28'10"
Emerald Beatrice Albrecht
Cas 0h37'7" 58d14'29"
Emerald Diamond
Psc 1h32'45" 19d17'1"
Emerald Diana Munoz
Aqr 22h39'27" 1d31'51"
Emerald Eyes
Cas 0h31'7" 53d1'28"
Emerald Eyes (Jessica Compton)
Cas 2h40'5" 72d28'57"

Emerald Molly's Dream
And 0h36'36" 44d22'27"
Emerald Rose
Crb 16h3'58" 39d24'5"
Emerald Star of Jeannette
Cyg 21h37'42" 43d1'57"
Emerald Travis
Crb 15h37'49" 26d26'51"
Emeri
Leo 10h40'3" 15d35'16"
Emeric
Her 16h37'41" 11d49'50"
Emerie Pimental
Sco 17h28'26" -44d57'59"
Emerik Leclair
Uma 11h26'45" 59d1'31"
Emerson Anthony
Per 3h17'20" 42d56'11"
Emerson (Dad's Little Star)
Sgr 19h20'49" -16d55'30"
Emerson David Gatchel Jr. 05-07-49
Aql 19h53'12" 15d51'59"
Emerson Grace
Uma 12h35'58" 54d9'27"
Emerson & Mitchell
Uma 11h24'33" 58d2'51"
Emerson Paige Burgett
Cap 21h26'22" -16d12'57"
Emerson Paige Burgett
Cap 21h17'16" -16d9'0"
Emerson Rose Lemisch
Cam 5h44'20" 67d35'22"
Emerson Rose Medina
Sgr 18h50'26" -18d45'21"
Emerson Rose Peters
Uma 11h45'28" 50d5'54"
Emerson Runyan
Cap 20h57'39" -15d50'34"
Emerson Samfilippo
Ari 3h15'0" 29d10'50"
Emerson Sei Wardally
Uma 11h4'27" 66d33'59"
Emerson Way
Cas 23h4'43" 59d33'7"
Emerson Weinman
Sgr 18h3'6" -25d16'15"
Emerson's Christened Star
Psc 0h6'15" 4d17'2"
Emery
Dra 18h32'46" 75d8'52"
emery
Leo 10h31'45" 14d46'25"
Emery Hewlett 100th Birthday
Leo 11h26'26" 16d34'47"
Emes
Lac 22h46'23" 54d8'31"
EMF
Gem 7h45'24" 14d22'20"
Emi
Uma 9h55'20" 41d54'4"
Emi and Yoshinobu Kitahara
Lyn 9h7'49" 33d59'50"
Emi Girl
Tau 5h43'48" 23d19'49"
Emi Loomis
Cam 5h22'11" 60d45'6"
Emi Mitsuno Takeo Gagnon
And 23h17'10" 51d28'1"
Emi, Feri/Philemon, Bauchis/Mama, Papa
Uma 10h29'29" 64d17'37"
Emichan
Cam 3h46'32" 53d54'32"
Emidio
Cyg 21h32'33" 33d53'37"
Emidio B. Vera, Servant of God
Aur 4h47'16" 32d58'39"
Emidon
Uma 10h7'57" 56d15'15"
Emiko
Psc 1h38'12" 17d44'55"
Emiko "Obachan's Star"
Sco 17h18'23" -30d34'9"
Emiko Ono
Gem 6h3'8" 23d28'28"
Emil
Umi 14h46'57" 72d9'8"
Emil and Peggy's Point of Light
Psc 23h52'13" 4d53'20"
Emil Bucceroni
Psc 1h11'20" 28d38'21"
Emil Canestrino
Per 3h26'43" 51d53'46"
Emil Chizek
Gem 7h45'25" 17d19'46"
Emil & Gayana
Umi 13h46'16" 77d27'23"
Emil J. Johnson
Ori 6h8'3" 2d5'44"
Emil Jecmen Jr.
Ori 4h50'18" 14d25'46"
Emil Jones, Jr.
Her 17h2'25" 30d34'25"
Emil Karol-Chik
Lib 14h54'35" -3d59'18"
Emil Klinkow
Uma 10h10'51" 64d28'47"
Emil Pollansky
Per 2h51'35" 54d22'4"

Emil Quaglino
Uma 12h0'44" 61d24'53"
Emil R. Ritter, M.D.
Leo 9h22'43" 12d14'8"
Emil Sebastian Tomescu
And 0h48'16" 45d59'18"
Emilce
Gem 7h46'38" 33d26'15"
Emile & Anne Marie Bergeron, Jr.
Ara 16h54'42" -49d13'51"
Emile Bartholomew Lewis
Psc 22h52'5" 5d28'47"
Emile Mahieu
Leo 11h43'33" 19d45'42"
Emilee
Ori 5h31'27" 0d44'38"
Emilee
Psc 0h56'33" 5d29'13"
EMILEE 111303
Sco 16h43'44" -31d7'35"
Emilee Anne Allison
And 0h54'26" 36d49'30"
Emilee Burch
Crb 15h42'44" 26d3'10"
Emilee Catherine Skillman
Lyn 7h27'58" 51d34'17"
Emilee Elizabeth Carnall
And 1h53'12" 45d29'27"
Emilee Jeane Puckett
Lmi 10h28'50" 33d30'44"
Emilee Lynn Fisher
Leo 11h49'33" 16d41'42"
Emilee Nadine Jackson
Cnc 8h22'40" 15d54'2"
Emilee Noelle Indivero
And 2h17'51" 46d2'13"
Emilee Rebekah Andrade
Aqr 22h8'59" 2d3'2"
Emilee Sue Bullock
And 2h13'52" 41d15'25"
Emilee Super
Cas 0h48'56" 60d40'42"
Emilee- The Star of Beauty
Cnc 9h19'41" 10d25'40"
Emilee's Bright and Shining Star
And 0h24'48" 45d48'5"
Emilee's Inspiration
And 23h14'11" 47d9'37"
Emilee's Star
Aqr 21h37'13" 2d32'30"
Emile-L
Cen 13h46'55" -38d25'14"
Emileo
Cam 5h22'24" 60d5'38"
Emili and David <3 1.5.07
Cyg 21h35'18" 52d20'45"
Emili Horner Combs and Kaden Combs
Aqr 21h13'20" -8d8'59"
Emilia
Aqr 22h37'43" -3d1'9"
Emilia
Uma 8h40'34" 60d44'42"
Emilia
Uma 10h6'28" 51d24'28"
Emilia
Ori 5h10'20" 6d10'27"
Emilia Aine Grace McMonigall
Mon 6h24'15" 8d26'0"
Emilia Ann Elizabeth Ochoa
Psc 1h13'30" 25d0'2"
Emilia Anne Dellinger
Ori 6h18'8" 8d31'1"
Emilia Benedek
Gem 6h51'18" 20d40'24"
Emilia Courtney
And 1h59'49" 37d13'46"
Emilia Deakova
Cas 1h22'51" 69d45'23"
Emilia Grace Oliver
Cru 12h28'56" -58d41'16"
Emilia Jean Barksdale
Tri 2h23'36" 32d8'49"
Emilia Rose Ponce
Aqr 22h40'22" -22d45'13"
Emilia Rosenfeld
Psc 0h56'40" 28d30'18"
Emilia Sue
And 23h30'49" 41d21'44"
Emilia Summer Darkins Rutsch
Gem 6h48'30" 17d24'48"
Emilia Teresa Michalak
Cru 12h11'3" -61d59'38"
Emilia Vazquez Mendez
Cnc 8h40'15" 27d16'12"
Emilia,Tony,Alexander&Ariana Badia
Uma 12h1'40" 29d53'51"
EMILIAN
Sco 17h43'30" -32d5'25"
Emiliana
Ori 5h58'17" 9d21'25"
Emiliana 79
Cyg 20h1'55" 39d44'39"
Emiliano
Cas 2h41'7" 57d44'59"
Emiliano Gizzi
Cas 23h35'20" 53d1'33"
Emiliano Rodriguez Melguizo
Mon 6h45'55" 10d58'56"
Emilia's Geburtsstern
Uma 10h5'0" 64d38'30"
Emilie
Dra 10h20'47" 73d20'34"
Emilie
Cas 1h59'30" 61d39'27"
Emilie
Lib 15h50'44" -11d44'49"
Emilie
Sgr 19h28'7" -34d29'22"
Emilie
Ari 2h58'37" 27d20'4"
Emilie
Per 2h52'51" 45d27'32"
Emilie
And 0h53'12" 35d38'42"
Emilie A. Meyer
Vir 13h18'50" 6d11'12"
Emilie Alexandra
Her 17h47'13" 21d21'16"
Emilie and Josh's Star
Gem 7h15'29" 23d38'44"
Emilie Bullins
Lyn 8h32'23" 36d29'3"
Emilie Caitlin Garratt
Cyg 21h55'10" 50d55'20"
Emilie Charlotte Rose Nicolson
Ari 2h28'27" 21d2'11"
Emilie Crochet
Uma 11h26'53" 51d46'22"
Emilie Dakota Jones
Gem 6h51'46" 15d33'44"
Emilie Dupuis
Uma 10h21'5" 68d53'55"
Emilie Easter Desrosier
Uma 10h11'9" 68d27'9"
Emilie Elizabeth Morris
Sgr 19h35'16" -39d19'52"
Emilie Genevieve Vinyard
Sgr 19h10'29" -17d1'33"
Emilie Grace Theado
Ari 3h2'55" 19d3'30"
Emilie Guerin
Vir 13h5'15" 12d31'39"
Emilie Inca Maxime
Uma 8h26'51" 61d24'36"
Emilie Jo Pedro
Gem 6h29'10" 20d34'26"
Emilie Katherine Carney
Cas 0h22'31" 54d55'18"
Emilie Lafontaine
Cap 20h19'3" -8d43'35"
Emilie Martin
Ari 2h41'1" 21d34'15"
Emilie Miller
And 1h56'7" 46d2'55"
Emilie mon Ange
Cap 20h41'21" -22d0'20"
Emilie Müller
Cas 3h13'44" 71d26'2"
Emilie Rose
Leo 10h9'14" 18d12'18"
Emilie Simone
Lyn 8h16'45" 33d15'20"
Emilie-anne Elethe Nichols
Lac 22h26'2" 49d47'21"
Emiliechérie
Leo 9h28'6" 10d24'14"
Emilien
Del 20h24'53" 4d3'50"
Emilie's Fear
Cnc 9h5'8" 31d1'26"
Emilija
Uma 10h38'56" 57d55'55"
Emilijan Cucek
Vir 13h20'27" 4d2'32"
Emilio Alexander Pedriani
Vir 13h21'1" 11d48'0"
Emilio Alistair
Uma 11h7'41" 38d41'5"
Emilio and Jennifer Lobato
Pup 8h9'52" -22d45'48"
Emilio and Maria
Lyr 18h47'8" 39d18'54"
Emilio Francisco
Sco 16h2'58" -19d4'4"
Emilio Garavaglia
Her 16h47'31" 27d43'41"
Emilio "Vinny" Iannone
Lib 14h54'25" -0d54'37"
Emilio Y Gladista
And 0h15'17" 29d19'41"
Emilios George Leris
Cnc 8h39'8" 17d28'44"
Emilita
Cet 3h18'1" -0d44'31"
Emilka
Uma 9h42'1" 54d51'4"
Emilli Louise Blair
And 2h6'50" 36d30'43"
Emillia Georgina Edwarda Hedge
And 23h38'28" 48d14'11"
Emillie Ramos
Psc 23h3'27" 6d14'6"
Emillin
Cas 23h32'25" 57d19'2"
Emilly K Elliott
Cap 20h53'26" -20d33'51"
Emily
Sgr 18h32'1" -18d7'44"
Emily
Vir 14h14'49" -16d10'0"
Emily
Aql 19h35'36" -10d53'47"
Emily
Sco 16h14'33" -9d58'17"
Emily
Vir 11h38'23" -3d35'56"
Emily
Lib 14h56'2" -5d52'37"
Emily
Lib 14h53'58" -5d50'49"
Emily
Aqr 22h30'42" -2d56'47"
Emily
Aqr 21h53'50" -0d21'1"
Emily
Lep 5h48'17" -11d8'11"
Emily
Lyn 7h46'26" 59d0'15"
Emily
Uma 9h52'1" 58d34'15"
Emily
Uma 8h38'47" 61d23'1"
Emily
Uma 12h32'26" 62d0'47"
Emily
Cap 20h8'24" -24d51'51"
Emily
Ori 5h43'58" 7d36'19"
Emily
Ari 3h4'40" 13d9'32"
Emily
Vir 13h17'59" 4d24'12"
Emily
Vir 12h45'1" 6d43'38"
Emily
Leo 11h21'19" 3d57'16"
Emily
Cnc 8h42'26" 14d4'43"
Emily
Del 21h7'6" 14d45'52"
Emily
Gem 7h25'51" 18d34'45"
Emily
And 0h28'43" 28d19'30"
Emily
And 0h45'6" 29d58'2"
Emily
Her 18h39'55" 15d38'20"
Emily
Gem 6h27'38" 23d50'23"
Emily
Tau 4h6'34" 25d45'56"
Emily
Tau 5h43'1" 26d10'1"
Emily
Tau 4h18'8" 27d1'14"
Emily
And 23h47'0" 47d15'31"
Emily
And 23h35'46" 49d44'20"
Emily
Cas 23h45'45" 51d45'23"
Emily
Cas 0h45'50" 57d55'57"
Emily
Uma 11h3'0" 50d26'59"
Emily
Cyg 19h43'37" 43d14'59"
Emily
And 23h16'46" 43d51'49"
Emily
And 1h13'33" 46d3'46"
Emily
And 2h31'21" 50d36'9"
EMILY
Lyn 7h3'0" 46d28'17"
Emily
Uma 8h53'5" 49d4'26"
Emily
And 0h46'46" 37d8'26"
Emily
Psc 1h18'34" 32d33'24"
Emily
And 0h39'38" 35d2'55"
Emily
And 0h36'13" 30d21'19"
Emily
Lyn 8h38'46" 33d54'35"
Emily
Lmi 10h5'33" 34d10'2"
Emily
Uma 11h50'16" 33d57'24"
Emily
Crb 15h39'44" 35d18'24"
Emily
Uma 10h9'57" 44d0'31"
Emily
Lyn 9h2'29" 37d41'39"
Emily
And 1h38'10" 42d56'32"
Emily
And 1h34'21" 38d36'14"
Emily
And 1h22'20" 39d9'51"
Emily , Michelle Habrylo.
Cyg 20h9'36" 31d5'2"
Emily 10-21-03
Cyg 20h47'55" 47d20'2"
Emily 21
And 0h16'51" 44d18'37"
Emily A. Binkowski
Psc 1h20'10" 32d44'14"
Emily A. Dunn
Uma 10h44'32" 45d7'43"
Emily A. Hill Laughing
And 2h35'50" 39d33'17"
Emily A Quintana
Lyn 7h11'15" 48d10'13"
Emily A. Thornton
Vir 12h34'35" 1d43'19"
Emily Abrielle
Vir 12h47'46" 13d0'13"
Emily Addington Milheim
Aqr 22h0'49" -13d57'57"
Emily A.G. Levy
Gem 6h32'31" 13d42'27"
Emily Agnes Etelka Hetrick
Cnc 9h6'3" 28d12'31"
Emily Alexandra
Crb 16h15'2" 38d51'36"
Emily Alexis Bailey
Sco 16h38'10" -28d9'58"
Emily Alice Akimoto
Ari 2h23'28" 23d54'44"
Emily Alice Barrett
And 23h20'18" 52d24'5"
Emily Alice Rolph-Dickinson
Umi 15h51'21" 82d40'20"
Emily Allison
Aqr 22h22'26" -1d2'53"
Emily Amanda Boyd
Vir 12h46'50" -3d14'15"
Emily Amore Mio
Ari 2h23'39" 23d58'14"
Emily and Ady Wyatt
Cyg 19h48'45" 31d56'30"
Emily and Chris
Cyg 19h32'28" 52d57'40"
Emily and Colt Forever
Uma 11h5'28" 35d38'39"
Emily and David Anmuth
Cyg 19h35'2" 28d17'43"
Emily and Kyle
Leo 11h27'22" 11d6'39"
Emily and Mark
Sge 20h16'43" 18d42'27"
Emily and Mark's Star
Uma 8h12'47" 66d34'30"
Emily and Molly Hingle
And 1h46'53" 45d24'16"
Emily and Paul
Cyg 21h29'2" 52d59'59"
Emily Andrews Petraitis
Lyr 19h13'35" 32d53'5"
Emily Ann
Cyg 19h48'36" 33d10'2"
Emily Ann
And 23h18'35" 38d22'59"
Emily Ann
Ari 1h55'9" 23d20'26"
Emily Ann
And 0h34'15" 31d33'10"
Emily Ann
Gem 6h47'51" 12d18'19"
Emily Ann Ballard
Gem 7h55'14" 31d20'7"
Emily Ann Bock
Vir 13h24'36" 13d52'11"
Emily Ann Bowles
Leo 10h14'53" 25d59'23"
Emily Ann Boyle
And 0h25'27" 42d11'8"
Emily Ann Brown
Ari 2h10'8" 23d43'55"
Emily Ann Bynum
Ari 3h3'56" 26d23'22"
Emily Ann Clarke - Embleton
Tau 5h46'29" 27d34'38"
Emily Ann Combs
Leo 11h13'44" 16d54'4"
Emily Ann Cummings
Sgr 18h19'8" -31d53'16"
Emily Ann Dillard
Uma 11h35'10" 65d38'47"
Emily Ann Gencarelli
And 0h25'38" 45d6'37"
Emily Ann Gilchrist
Uma 8h25'13" 68d17'22"
Emily Ann Hokanson
Cas 23h43'28" 59d48'17"
Emily Ann Margaret Sturt
And 23h50'6" 36d3'25"
Emily Ann Morgan
Cam 6h15'6" 59d49'31"
Emily Ann Quigley
Sco 16h32'5" -28d26'33"
Emily Ann Schlenk
Lyn 7h43'23" 56d44'12"
Emily Ann Siegal
And 0h40'11" 40d6'56"
Emily Ann Spatafore
And 0h58'17" 39d31'21"
Emily Ann Taylor
Ori 6h16'59" 14d41'46"
Emily Ann Terech
Psc 1h5'1" 6d21'19"
Emily Ann Wade
Cap 21h51'52" -9d34'11"
Emily Ann Webster
Cyg 20h23'52" 48d57'19"
Emily Anna Osbourn
Cas 23h13'11" 54d55'43"
Emily Anne
Sco 17h57'26" -40d41'22"
Emily Anne
And 23h34'8" 47d27'39"
Emily Anne
And 0h37'7" 35d1'27"
Emily Anne Criste
Uma 9h42'59" 63d43'22"
Emily Anne Demenkow
Cap 20h35'6" -10d48'58"
Emily Anne Edgerley
And 0h33'9" 41d21'24"
Emily Anne Gunstream
Aqr 23h51'33" -11d1'38"
Emily Anne Zell
Cru 12h39'13" -60d37'20"
Emily Ann's Hat Cleaning Service
And 23h27'51" 42d3'54"
Emily Augustine Mancourt
And 0h47'49" 35d51'21"
Emily B. Martin
Tau 4h32'28" 21d32'9"
Emily Bailey
And 0h4'51" 48d8'23"
Emily Banks Clippinger
Vir 12h30'0" -11d15'0"
Emily Barclay
Pho 2h16'47" -45d37'23"
Emily Barratt
And 23h38'50" 41d22'19"
Emily Bealessio
Com 13h2'6" 18d24'50"
Emily Beer
Cap 21h40'3" -13d56'27"
Emily Behn's Shining Star
And 23h1'51" 50d49'26"
Emily Belle James
And 23h8'50" 35d28'11"
Emily Bellush
Cas 1h25'16" 63d45'56"
Emily & Ben together 4 eva
Cap 20h19'23" -25d58'46"
Emily Beth
Lib 15h28'47" -6d18'38"
Emily Beth McPherson
Lmi 10h31'11" 37d33'12"
Emily Beth Morrris
And 23h34'6" 50d33'1"
Emily Bethan Jones
Lmi 9h52'19" 38d35'21"
Emily Billings
Ori 5h55'6" 18d34'21"
Emily Billingsley
Sgr 18h2'27" -24d9'1"
Emily - "Blessed Angel"
And 1h41'59" 41d37'17"
Emily Blythe
Cnc 8h51'14" 17d45'30"
Emily Boggs
And 0h32'27" 32d40'49"
Emily Boyd
Sco 16h59'6" -38d29'9"
Emily & Brandy
Lyr 18h52'31" 32d11'51"
Emily Bridget Morley
Lib 14h54'36" -2d55'42"
Emily Bristow
And 23h42'23" 37d13'55"
Emily Brooke
Leo 10h5'38" 26d37'37"
Emily Brooke
Lyn 6h22'32" 58d19'21"
Emily Brooke Pilla
Leo 9h36'56" 7d29'12"
Emily Brooke Shields
Cas 1h41'41" 67d20'32"
Emily Browne
Sgr 18h44'19" -32d6'3"
Emily C. Hammond God's Angel #10
Ari 3h16'49" 20d13'11"
Emily C. Hise
And 0h39'37" 28d24'3"
Emily C Inge
Vir 13h45'4" 4d22'58"
Emily Campbell - Emily's Star
And 2h14'59" 42d24'46"
Emily Cancer R.M.W.C.W.S.M.C.
Cnc 8h29'48" 27d35'48"
Emily Canterbury
Uma 10h33'36" 51d43'56"
Emily Capper
Sgr 19h53'10" -16d6'58"
Emily Carol Ransom
And 0h24'32" 30d26'47"
Emily Caroline Abdelsalam
And 0h13'34" 37d5'43"
Emily Carolyn Joyce Tokson
Ori 6h5'48" 18d7'39"
Emily Catherine
Leo 11h36'50" 14d14'55"
Emily Catherine Bohannon
Lyr 18h49'39" 37d21'5"
Emily Catherine Bostick
Aqr 23h7'19" -8d8'43"
Emily Catherine Hakeman
Uma 12h4'54" 43d2'12"
Emily Catherine Manzella
Gem 6h55'18" 16d4'19"
Emily Catherine Michelle Wall
Lyn 7h25'42" 48d31'4"
Emily Catherine Watts
And 23h11'51" 42d23'56"
Emily Cecile Leonard
Lac 22h23'7" 52d26'25"
Emily Chapman
Uma 10h31'6" 49d1'5"
Emily Charlotte
Lyr 18h45'42" 38d16'47"
Emily Charlotte
Sgr 17h53'36" -16d53'31"
Emily Cheung
And 23h6'26" 45d16'6"
Emily Cho
Cam 6h5'34" 60d36'44"
Emily Christina Demalis
And 2h28'19" 49d34'15"
Emily Christina Girard
Umi 14h54'28" 73d46'38"
Emily Christina Skowronek
Sgr 19h39'25" -12d15'10"
Emily Christine
Lyn 7h57'36" 35d31'23"
Emily Christine Coole
Umi 15h21'29" 69d59'4"
Emily Christine Hanson
Lyn 8h43'39" 40d34'36"
Emily Christine Marsh
And 23h25'19" 36d19'52"
Emily Christine Morgan
And 23h28'39" 49d59'14"
Emily Christine Schmehil
Ori 5h25'0" -7d32'4"
Emily Claire
Sgr 19h9'26" -29d49'39"
Emily Claire Anderson
Leo 11h32'45" 22d26'0"
Emily Claire Cole
Ari 2h47'0" 27d38'25"
Emily Claire Macdonald
Gem 7h34'47" 34d27'53"
Emily Clare
And 2h20'4" 42d38'22"
Emily Clare Proudlock
And 23h2'53" 52d28'10"
Emily Claveau
Uma 9h26'11" 43d22'11"
Emily Coates & Matt Terhune
Sco 16h44'9" -33d58'18"
Emily & Cody's Love Child
Umi 13h28'3" 73d3'9"
Emily Cole
Cru 12h55'45" -60d48'51"
Emily Coombs
And 1h50'35" 39d16'36"
Emily Coral Mary Winfield
Ari 3h8'21" 11d34'29"
Emily Crandall
And 0h30'17" 32d49'53"
Emily Cubellis (Emma)
And 1h38'43" 37d52'49"
Emily Daniel
Lyn 9h27'8" 40d9'36"
Emily Danielle Leonard
Tau 5h34'52" 21d19'28"
Emily Danielle Stout
Uma 10h50'34" 43d59'9"
Emily Darker
Tau 3h50'34" 23d5'27"
Emily Dawn
Lib 15h5'54" -22d50'1"
Emily Dee Isabella Roe
Cas 1h20'15" 56d27'4"
Emily Deere Reed
Lmi 10h44'26" 32d5'11"
Emily Diane Rose
Del 20h36'30" 13d17'42"
Emily Dietz
Ari 2h34'19" 25d51'49"
Emily DiPietro
And 0h25'35" 44d46'52"
Emily Dlugosh
Ari 2h6'48" 25d35'43"
Emily Dodge
And 23h21'12" 42d58'55"
Emily Dolvin-Schultz
Aqr 22h12'48" -21d41'3"
Emily Dorkface
Leo 10h24'7" 25d43'13"
Emily Dorsey Ford
And 0h39'46" 36d17'34"
Emily Dougherty
Com 12h47'11" 18d19'8"
Emily Dougherty
Com 13h26'9" 15d9'32"
Emily Dow
Sco 16h40'1" -28d44'31"
Emily E. Geibe's "Pooh"
And 2h20'17" 40d55'25"
Emily E Monger
Uma 9h12'46" 49d34'54"
Emily Efron Flier
And 23h5'39" 47d11'48"
Emily Ehret
Tau 5h17'45" 18d57'55"
Emily Elaina Allor
Lib 15h38'28" -6d12'45"
Emily Elaine Brite
Psc 1h15'9" 30d50'33"
Emily Elana
Vir 12h43'59" -7d43'24"
Emily Elayne
Uma 11h2'33" 53d51'6"
Emily Elizabeth
Lyn 7h41'51" 42d16'1"
Emily Elizabeth
Tau 4h11'5" 8d15'52"
Emily Elizabeth
Gem 7h10'7" 17d25'17"
Emily Elizabeth
Ari 2h19'36" 23d58'50"
Emily Elizabeth
And 0h31'11" 31d42'3"
Emily Elizabeth
Gem 7h15'22" 24d50'30"
Emily Elizabeth Bakker
Gem 6h49'40" 33d56'18"
Emily Elizabeth Betts
And 23h20'9" 48d10'22"
Emily Elizabeth Butcher
Tau 4h20'36" 27d53'55"
Emily Elizabeth de la Mare
Cas 23h55'11" 53d34'27"
Emily Elizabeth DeCost
Tau 4h1'12" 19d37'14"
Emily Elizabeth English
Tau 4h19'1" 24d3'52"
Emily Elizabeth Garvey
Cnc 8h52'42" 30d41'13"
Emily Elizabeth Gibbons
Cap 20h57'35" -15d20'17"
Emily Elizabeth Haberberger
And 2h34'9" 49d45'8"
Emily Elizabeth Howe - My Sunshine
Cnc 8h32'7" 21d16'44"
Emily Elizabeth Kaatze
Cas 1h10'30" 65d48'44"
Emily Elizabeth Stoughton
Aqr 23h24'44" -12d5'23"
Emily Elizabeth Ware
Dra 17h33'36" 69d24'40"
Emily Elizabeth Wills
And 1h25'57" 39d29'7"
Emily Elizabeth Wilson
Lyr 18h58'16" 33d34'37"
Emily Elizabeth Winchell
Lib 15h16'4" -15d1'39"
Emily Ellen's Star
Cas 23h27'24" 57d4'23"
Emily Ells Addesa
Sco 16h9'27" -9d23'53"
Emily "Ema" Anne Hodgson-Soule
Cas 2h18'19" 66d40'38"
Emily Erin Day
Lyn 6h41'44" 54d8'32"
Emily Errico
Ari 2h18'22" 23d31'46"
Emily Eugenia Read
Ari 2h16'25" 10d35'53"
Emily F. Barr
Ori 5h32'5" -4d21'3"
Emily Faith Collins
Uma 8h47'48" 63d40'24"
Emily Farsakian
Cyg 20h22'51" 47d52'13"
Emily Fay
Cap 20h23'27" -11d22'0"
Emily Faye
Ari 2h36'31" 20d22'10"
Emily Ferstle
Cyg 20h53'6" 45d3'56"
Emily Fields
Umi 15h15'4" 79d52'21"
Emily Fincher
Lyn 7h16'21" 51d33'13"
Emily Fitz
Psc 1h33'12" 9d27'24"
Emily Flames Duncan
And 1h46'18" 41d20'6"
Emily Fletcher
Cru 12h4'45" -62d50'46"
Emily Fleur Gilmour
And 0h22'17" 25d51'35"
Emily Fontaine
Vir 11h41'0" 4d33'11"
Emily Foster
Lyn 9h22'10" 41d23'39"
Emily Frances Bushnell on her 16th
Gem 7h29'21" 15d3'34"
Emily Frances Doreste
Crb 16h7'46" 33d27'35"
Emily Frances Tomko
And 0h41'44" 42d43'55"
Emily Francis
Vir 13h43'28" -15d26'6"
Emily Francis
Lyn 6h33'15" 56d36'56"
Emily Funkhouser/Preston
Gem 7h33'45" 22d39'51"
Emily G. Mandy
Leo 10h30'59" 12d7'59"
Emily Garber
Ori 5h15'50" -4d45'13"
Emily Gavidia
Leo 10h13'33" 25d9'3"
Emily George
Sgr 19h44'47" -14d55'29"
Emily Goline
And 0h10'52" 31d54'0"
Emily Goodlin
Cas 0h11'25" 54d1'45"
Emily Gordy Dolvin
Lib 15h9'1" -0d36'29"
Emily Grace
Lib 14h53'11" -6d27'4"

Emily Grace
Lyn 6h32'44" 61d14'4"
Emily Grace
Psc 0h10'2" 2d1'2"
Emily Grace
And 2h31'56" 38d24'58"
Emily Grace
Leo 11h15'48" 0d19'14"
Emily Grace
Vir 13h13'26" 5d3'24"
Emily Grace
Tau 4h6'6" 10d45'20"
Emily Grace Crago
Tau 5h47'54" 17d58'32"
Emily Grace Driscoll
Leo 10h32'52" 6d21'41"
Emily Grace Elding
Cas 23h28'25" 52d53'23"
Emily Grace Fagge
And 23h50'22" 42d33'49"
Emily Grace Federico
Leo 11h29'10" 25d40'35"
Emily Grace & Flopsey Kastner
Vir 14h6'51" -15d51'16"
Emily Grace Kuczynski
And 0h19'47" 44d53'4"
Emily Grace Manson
And 0h19'47" 32d44'9"
Emily Grace Purcell
Mon 6h49'5" -0d8'54"
Emily Grace Tuin
Leo 9h29'43" 19d10'49"
Emily Griffin
Ori 5h54'37" 19d25'19"
Emily H. Huffman
Leo 9h34'27" 28d45'36"
Emily Haines
And 2h18'3" 43d3'12"
Emily Ham
Cap 20h51'2" -26d1'22"
Emily Hamer
Ari 3h8'42" 25d33'47"
Emily Hannah Lennox
And 0h48'10" 25d33'44"
Emily Harris
And 23h3'30" 42d28'45"
Emily Hart
Vir 13h20'9" -14d23'55"
Emily Hart
Lyn 6h47'49" 55d59'21"
Emily Heim
Cnc 8h44'23" 17d21'55"
Emily Helena
Aqr 23h0'48" -5d56'26"
Emily Helene Ketchley
Cas 1h21'4" 55d21'18"
Emily Hernandez Alegria
Psc 0h44'59" 11d28'44"
Emily Hogan
Ari 2h52'33" 28d19'31"
Emily Hollerbach
Vul 21h19'37" 24d45'5"
Emily Hollon 1993
Lib 14h48'10" -19d27'13"
Emily Hope Code
Sgr 19h36'28" -14d51'30"
Emily Horsfield
Tau 5h45'26" 27d4'34"
Emily Howell
Mon 6h47'29" -0d6'48"
Emily HuaJian Hendrickson
Lyn 9h17'22" 34d42'59"
Emily Hull
And 23h26'46" 47d10'58"
Emily Hutchins
Crb 16h21'57" 27d46'55"
Emily Iannuzzelli
Vul 20h49'13" 28d1'21"
Emily Imogen Charlton
Ari 2h56'54" 21d25'52"
Emily Imogen Mortimer
And 23h38'6" 42d56'24"
Emily is a Bright Star
And 1h16'1" 41d18'41"
Emily Isabella Klenk
Sco 16h38'10" -28d29'11"
Emily Ivonne
Gem 6h32'34" 22d52'13"
Emily Ivy Salkeld
And 23h14'15" 37d36'11"
Emily J. Martin
Peg 23h53'57" 26d58'3"
Emily Jackson
And 1h47'22" 37d56'54"
Emily & James
Cyg 21h57'21" 55d20'2"
Emily James Maddox
Leo 10h31'39" 27d13'46"
Emily Jane
Lib 15h24'24" -22d8'22"
Emily Jane - 10/15/62
Cas 1h12'36" 59d31'40"
Emily Jane Beynon - Em
Lyn 7h34'28" 36d56'52"
Emily Jane Bruss
Lyn 7h22'26" 54d19'20"
Emily Jane Kellogg
And 0h29'30" 37d13'24"
Emily Jane Kerr
Cas 1h24'9" 53d17'51"
Emily Jane Paris
Cas 2h3'44" 62d14'19"
Emily Jane Satterthwaite
Umi 16h50'43" 84d1'49"

Emily Janicz
Leo 9h35'12" 31d24'13"
Emily Jansen Smith Van Beek
Leo 11h31'30" 5d1'22"
Emily Jasmine
Cas 1h6'2" 63d36'15"
Emily Jasmine Ranyak 4-2-05
Psc 23h51'26" 2d24'56"
Emily Jay Balden
And 23h15'53" 38d17'13"
Emily Jay Roberts
And 2h12'53" 43d25'55"
Emily Jayne Smith
Cam 4h30'23" 57d26'36"
Emily Jean
Lyn 8h30'51" 33d8'9"
Emily Jean Faulk-Maxwell
Gem 6h26'56" 23d36'56"
Emily Jean Leonard
Gem 7h44'15" 32d36'7"
Emily Jean Nichols
Gem 6h13'17" 25d4'41"
Emily Jean Phillips
Leo 10h52'43" 18d37'49"
Emily Jean Theno
Aqr 21h44'35" 1d4'14"
Emily Jeanne
Leo 11h4'1" 21d25'32"
Emily Jeanne
Cas 2h21'57" 71d16'22"
Emily Jessica Rosslyn
Gem 6h56'8" 22d2'16"
Emily Jie-Ou Ma
Cyg 21h16'55" 31d40'33"
Emily Jo
And 23h25'37" 38d5'24"
Emily Jo
Aqr 22h43'6" -19d58'30"
Emily Johnstone's Star
And 23h54'29" 32d7'4"
Emily Josephine
Cap 20h52'16" -14d33'8"
Emily & Josh Pearcey
Per 3h53'26" 51d1'45"
Emily Joy
Crb 15h33'31" 38d7'32"
Emily Joy Peterson
Cap 21h13'38" -16d54'26"
Emily Joy Plimpton
And 23h3'24" 45d47'39"
Emily Juchem
Lyn 7h42'9" 54d5'37"
Emily Jumelle Gill
And 23h33'55" 41d37'17"
EMILY JUNE McCANN
Crb 15h31'29" 28d27'27"
Emily Junz
Cas 1h35'16" 64d37'35"
Emily K. Felling
Psc 2h5'37" 3d59'32"
Emily K. Nordengren
And 0h17'7" 41d25'57"
Emily K Schlimm
Leo 9h31'6" 10d36'31"
Emily K. Schuster
And 23h42'55" 47d54'33"
Emily Kaiti Kelley
Leo 10h20'18" 13d12'34"
Emily Kang
Ori 6h9'29" -0d38'3"
Emily Karvecky
Tau 3h45'2" 27d58'58"
Emily Kate
And 0h20'24" 42d58'17"
Emily Kate
Uma 9h3'41" 54d5'20"
Emily Kate
Cen 11h15'33" -45d36'58"
Emily Kate Anglin
Cap 21h45'14" -18d46'50"
Emily Kate Maltby-Smith
And 1h12'59" 35d21'24"
Emily & Katherine
Mon 6h53'31" -0d31'47"
Emily Katherine Morris
Aql 19h22'11" -8d25'43"
Emily Katherine O'Leary
Uma 13h21'58" 56d4'9"
Emily Katherine Snow
Ari 2h23'18" 23d51'39"
Emily Kathleen
Tau 4h26'2" 0d59'55"
Emily Kathleen
And 1h35'22" 38d41'0"
Emily Kathleen Fruge
Aqr 22h2'3" -17d33'55"
Emily Kathleen Rhodes
And 0h38'50" 34d3'17"
Emily Kathleen Wood
And 2h19'56" 39d21'28"
Emily Kathryn
And 2h26'59" 47d38'43"
Emily Kathryn Wood
Sco 16h48'2" -32d45'46"
Emily Kay
Her 16h51'12" 40d27'13"
Emily Kay Buckle
Cnc 8h56'44" 12d47'31"
Emily Kay Patterson
Oph 17h6'21" -0d33'2"
Emily Keith
Cyg 20h16'33" 33d5'38"

Emily Kelly Koffel
Dra 17h1'55" 52d33'55"
Emily Kendall Murdock
Cnc 8h22'8" 10d55'59"
Emily Kendall Pollock
Leo 9h52'17" 28d11'42"
Emily Kenna Linden
Cyg 20h8'40" 43d47'19"
Emily Kessler
And 0h37'56" 27d12'19"
Emily Kikue 010
Gem 7h45'15" 25d13'30"
Emily Klabough
Leo 10h54'22" -2d15'52"
Emily Knox
Gem 7h13'48" 27d46'32"
Emily Koreen Rodenbough
And 1h1'17" 36d45'15"
Emily Krasnow
And 23h14'49" 46d58'16"
Emily Krist
Cas 0h25'9" 52d18'57"
Emily Kunselman
Gem 6h25'38" 26d21'7"
Emily L. Chapman 1985
Ari 3h12'13" 28d48'26"
Emily L. Gerdes
Lyr 18h53'11" 42d57'49"
Emily L. Michlewski
Cnc 9h9'0" 30d41'33"
Emily Lane
Lyn 7h45'9" 40d15'37"
Emily Lassetter
Mon 8h0'33" -6d33'8"
Emily Lauder Wilson-Luke Loves You
Uma 8h48'56" 61d53'32"
Emily Laura Ryan
Gem 7h18'47" 31d58'47"
Emily Lauren Augello
Sco 17h20'34" -32d36'52"
Emily Leanne
Peg 21h53'52" 20d41'41"
Emily Leanne & Charles Adam Forever
Cyg 20h46'38" 37d36'49"
Emily Lenore Wahl
Lmi 9h46'18" 40d10'52"
Emily Lenore Walters
Uma 11h28'59" 55d20'11"
Emily Libby
And 2h23'4" 49d5'8"
Emily Long
And 1h12'3" 47d45'39"
Emily Lookabaugh
And 1h8'47" 33d48'6"
Emily Loren Cooke
Crb 15h19'10" 29d27'23"
Emily Lorna Brooks
And 22h58'57" 51d55'1"
Emily Lou Coulter-awe-some godmother
Cyg 21h57'55" 48d52'10"
emily louise
Leo 10h52'28" 17d6'45"
Emily Louise
Cas 1h55'25" 61d13'2"
Emily Louise Brooke
Uma 12h10'51" 58d8'7"
Emily Louise Green
And 2h34'29" 40d36'42"
Emily Louise Hanbury
Peg 22h24'57" 15d15'11"
Emily Louise Hartzell
Uma 11h44'53" 39d32'39"
Emily Louise Maddock
And 23h14'59" 42d3'21"
Emily Louise Peterson
Sco 17h48'26" -34d9'32"
Emily Louise Smith
Cap 20h45'10" -25d7'13"
Emily Louise Smith
Cam 7h49'32" 76d50'24"
Emily Louise Spina
Cyg 20h41'54" 45d58'1"
Emily Louise Street
Crb 15h35'54" 33d22'35"
Emily Louise Webb
And 1h30'40" 47d44'10"
Emily Love
And 0h14'49" 46d27'34"
Emily Lowe
Psc 1h6'40" 31d45'9"
Emily Luise Skeens
And 23h17'46" 42d34'46"
Emily Lynn
And 23h8'5" 41d32'56"
Emily Lynn Bartlett
Crb 15h41'8" 31d38'30"
Emily Lynn Blake
Gem 7h15'38" 20d3'0"
Emily Lynn Shaw
And 0h57'13" 35d42'21"
Emily Lynn Smyth
And 1h16'55" 34d58'18"
Emily Lynne Wenrick
Cep 23h15'8" 73d8'22"
Emily M. Barresi (EM819)
Leo 10h11'48" 27d26'52"
Emily M. Chow
Uma 10h14'41" 60d8'33"
Emily M. DesForges
Dra 18h46'55" 54d58'3"
Emily M Motrini
Lyn 7h24'35" 53d12'47"

Emily Macenko
Cnc 8h43'9" 13d40'49"
Emily Machones
Leo 10h14'19" 7d58'44"
Emily MacKinnon
Cap 21h47'43" -11d5'17"
Emily Madison Stodghill Love Star
Cas 0h16'34" 59d15'51"
Emily Mae Mauck
Gem 6h53'16" 24d44'19"
Emily Mae Sears
Gem 6h14'42" 27d24'15"
Emily Mae Vail
Uma 10h33'31" 68d3'48"
Emily Maki
Sco 17h16'30" -34d3'18"
Emily Margaret
Psc 0h48'31" 10d17'59"
Emily Margaret Kinney
Aqr 23h22'4" -18d37'9"
Emily Margaret McLaughlin
Ari 2h7'51" 24d5'15"
Emily Margaret Weber
Sco 17h35'13" -39d49'24"
Emily Marie
Mon 6h44'3" -0d5'58"
Emily Marie 02/25/1984
Psc 0h24'49" 6d31'31"
Emily Marie Aten
Lyn 6h38'2" 57d15'38"
Emily Marie Insley
And 23h45'13" 47d27'46"
Emily Marie Jane Castillo
And 1h15'42" 45d13'7"
Emily Marie Jordan
Vir 11h39'44" 4d28'34"
Emily Marie Kyllo
Ari 3h24'57" 28d40'13"
Emily Marie Myers
And 1h20'27" 44d39'33"
Emily Marie Reiman
And 22h58'31" 48d37'59"
Emily Marie Reynolds
Ori 5h35'24" -0d9'13"
Emily Marie Rivera
Cap 21h23'34" -25d50'30"
Emily Marie Sappa
Tau 5h46'32" 26d7'14"
Emily Marie Trotter
Aqr 21h58'42" -2d53'7"
Emily Marie Vincent
Psc 0h52'26" 10d33'2"
Emily Marshall
Lyn 7h40'35" 56d50'13"
Emily Marten's Big Ball of Gas
Uma 9h0'43" 49d43'19"
Emily Mary
Sco 16h12'27" -26d4'17"
Emily Mary Dadswell
Uma 11h29'0" 39d48'51"
Emily Mary English
Cet 2h44'24" -0d27'58"
Emily Mary Kearney
Ari 3h26'15" 27d46'51"
Emily Massey Prince
And 1h25'46" 38d47'36"
Emily Masterson
Tau 4h37'38" 14d26'7"
Emily & Matt
Uma 8h35'55" 58d12'12"
Emily & Matt Forever
Cyg 19h47'25" 33d18'55"
Emily May Forbes - Emily's Star
Sco 17h3'6" -44d39'3"
Emily May King
Psc 1h10'43" 27d44'38"
Emily May Taylor
And 23h1'49" 42d52'44"
Emily McCleary
And 23h24'25" 45d43'53"
Emily McConnaughey
Lyr 19h17'27" 29d51'38"
Emily McMaster
And 23h15'14" 42d29'34"
Emily Mei Qin
Cnc 8h21'47" 22d14'41"
Emily Merin
Cyg 20h4'15" 58d5'25"
Emily Michaela Holmes
Lyr 18h41'45" 35d19'38"
Emily Michale
Aqr 21h59'15" 1d23'35"
Emily Michelle
Aqr 21h42'26" 1d45'32"
Emily Michelle Gist
Vir 14h41'9" 3d42'46"
Emily Michelle Mike
And 1h51'38" 43d29'30"
Emily & Mike
Tau 4h8'59" 27d35'9"
Emily Miller
Uma 10h13'2" 48d40'46"
Emily Miller
Uma 9h3'31" 64d48'43"
Emily Ming Johnson
Uma 9h39'6" 70d57'56"
Emily Missino
Sgr 17h50'4" -17d1'35"
Emily Monique Shelley Taconet
And 0h42'31" 39d27'14"

Emily Morano
Leo 10h40'38" 15d33'31"
Emily Morgan Hoskins
Ari 3h20'39" 29d0'0"
Emily Morgan Hughes
Uma 8h56'4" 57d40'1"
Emily Morgan Scott
Uma 8h57'21" 61d58'44"
Emily Morggan Gwyn
Cap 21h41'39" -9d48'32"
Emily Mroz
Leo 11h28'39" -3d19'15"
Emily Murray
And 2h16'9" 41d59'13"
Emily "My Little Miracle"
Psc 1h46'31" 12d43'36"
EMILY: my starmate and soulmate
Dra 16h5'52" 65d45'41"
Emily Nation
Leo 11h12'44" 15d45'28"
Emily Niamh Haine
Cma 7h5'0" -16d58'27"
Emily Nichole Kellerman
Cap 20h32'54" -22d56'1"
Emily Nicole
Gem 7h21'35" 33d15'0"
Emily Nicole BFF :)
Cnc 8h49'56" 26d32'3"
Emily Nicole Freeman
Cap 21h27'26" -14d53'17"
Emily Nicole Gunderson
Sco 17h39'53" -42d36'17"
Emily Nicole Jones
And 0h16'45" 28d45'23"
Emily Nicole Lewis - My Soul Mate
Gem 7h1'49" 28d40'54"
Emily Nicole Paller
Gem 6h55'21" 15d29'53"
Emily Nicole Puls
Vul 21h19'41" 22d27'6"
Emily Nicole Shukers
Ari 2h38'30" 24d43'34"
Emily Nicole Triolo's Star
Gem 6h33'56" 25d0'3"
Emily Nicole White
Lib 14h44'34" -9d9'22"
Emily Noel
Tau 3h48'41" 21d27'55"
Emily Noyes
And 0h37'51" 25d50'56"
Emily O
Crb 15h27'23" 31d51'13"
Emily Oleyer
Psc 1h14'24" 13d1'33"
Emily P Wilhelm
And 0h51'17" 39d50'38"
Emily Pacula
Sco 16h30'56" -33d41'38"
Emily Paige Keim
And 1h31'25" 49d20'51"
Emily Paige
Lmi 10h27'39" 36d41'27"
Emily Paige Craven
Ori 5h45'20" 4d56'56"
Emily Paige Runkle
And 0h34'50" 35d22'27"
Emily Paige Sands
Lyn 7h34'7" 38d13'36"
Emily Paige, Angel of the Stars
Cru 12h22'15" -58d5'33"
Emily Palmquist
Peg 22h49'47" 18d54'31"
Emily Paluba
And 23h21'1" 41d52'48"
Emily Pamela
Lyn 8h20'55" 33d59'58"
Emily Patricia Bohatch
And 0h25'57" 33d57'52"
Emily Patricia Prusnek's Star
Tau 4h34'27" 19d55'2"
Emily & Paul
Cyg 20h49'58" 34d31'18"
Emily Pearl Spine
Per 2h17'26" 57d6'42"
Emily Pearson
And 0h41'20" 32d46'28"
Emily "Peep"
Gem 7h44'20" 34d11'17"
Emily Perez
Uma 10h0'32" 59d26'28"
Emily Perretti
Uma 9h22'44" 44d58'8"
Emily Phillips
Cru 11h56'40" -60d3'45"
Emily Poe
Leo 10h14'19" 23d9'26"
Emily Polydoros
Cyg 20h2'55" 35d2'47"
Emily Pope
Aur 6h38'18" 38d51'16"
Emily Praveena
Cas 0h4'11" 61d37'42"
Emily R. Carta
And 2h37'27" 45d35'52"
Emily R. Gradeless
Uma 9h11'10" 70d52'17"
Emily R. Jervis
Dra 18h48'37" 72d57'8"
Emily R. Kerns
Uma 11h16'54" 45d1'52"

Emily R. Krueger
Lib 15h49'51" -18d9'39"
Emily Raab Vacha
Lib 14h23'14" -19d45'19"
Emily Rachel Ferman
And 1h6'41" 46d10'29"
Emily Rae
Uma 11h39'47" 36d2'54"
Emily Randall
And 23h22'10" 42d16'47"
Emily Rayne Johnson
Uma 11h11'31" 38d29'0"
Emily Rebecca Helsel
Uma 9h18'5" 46d54'15"
Emily Rebecca Resnik Bat Mitzvah
Uma 8h47'39" 59d18'7"
Emily Rendell
Uma 13h4'24" 52d20'50"
Emily Renee - D
Cap 20h14'40" -19d5'15"
Emily Renee Kell
And 1h32'21" 41d25'24"
Emily Richard
Uma 11h16'44" 52d11'11"
Emily Riley
Cas 0h19'30" 53d54'4"
Emily & Rob Cincotta
Sge 19h50'50" 17d35'4"
Emily & Robbie Forever
Cru 12h43'23" -58d25'42"
Emily (Robin) Kushnick
And 23h12'35" 47d28'56"
Emily Romine
Lib 15h39'33" -11d4'8"
Emily Rooks
Cnc 8h55'38" 26d58'20"
Emily Rosa Lopez
And 0h24'31" 21d44'25"
Emily Rose
Ori 5h29'0" 1d29'54"
Emily Rose
Psc 23h43'35" 0d32'19"
Emily Rose
Ari 3h11'22" 28d46'22"
Emily Rose
Tau 3h37'45" 27d46'33"
Emily Rose
Her 17h41'8" 18d58'27"
Emily Rose
Cyg 21h19'30" 50d48'17"
Emily Rose
And 23h44'2" 45d22'15"
Emily Rose
And 23h7'49" 42d45'28"
Emily Rose
Cyg 20h30'32" 39d48'57"
Emily Rose
Per 3h16'45" 49d32'28"
Emily Rose
Uma 9h2'37" 57d3'22"
Emily Rose
Uma 12h59'51" 57d8'2"
Emily Rose
Sco 16h43'4" -32d21'39"
Emily Rose
Sco 17h40'15" -35d54'24"
Emily Rose Atkins
And 2h19'29" 41d32'46"
Emily Rose Blum
Ari 2h51'50" 29d29'16"
Emily Rose Byway - "Emily's Star"
And 23h45'0" 48d33'38"
Emily Rose Erdman
Aqr 21h56'9" -8d14'13"
Emily Rose Fisher
And 23h10'46" 48d37'28"
Emily Rose Kirk
Cas 23h35'39" 53d41'55"
Emily Rose LaHood
Tau 3h43'52" 6d24'42"
Emily Rose Suplinskas
Per 3h47'26" 38d52'57"
Emily Rose Townsend
Cyg 19h39'25" 30d14'16"
Emily Rose Van Alphen
Gem 7h17'50" 23d24'9"
Emily Rose Walburn 3-18-2004
Psc 1h18'8" 31d13'27"
Emily Ross
Sco 16h14'18" -9d36'48"
Emily Ross Peterson
Per 4h16'42" 42d54'46"
Emily Ruth
And 1h45'27" 41d20'8"
Emily Ruth
Mon 6h45'2" 9d5'57"
Emily Ruth Clarke
And 23h8'44" 44d17'59"
Emily Ruth Kendall
And 0h38'30" 25d19'51"
Emily Ryan Cole
And 1h33'50" 40d28'3"
Emily Ryan Garza
And 0h18'12" 29d43'40"
Emily Sage Cochrane
And 2h19'42" 42d23'4"
Emily Santiago Giles
Cap 21h35'4" -19d59'36"
Emily Sara Murphree Brannon
Sco 16h9'45" -15d38'7"

Emily Sara Pearlstein
And 1h3'17" 42d36'27"
Emily Sarah Goldstein
And 0h57'11" 23d51'3"
Emily Savannah Jedzinak
Uma 8h17'36" 69d11'9"
Emily Savarino
Aqr 22h53'17" -3d47'41"
Emily Schuler Anderson
And 2h10'8" 46d38'14"
Emily Scott Fawcett
Psc 1h4'56" 15d35'51"
Emily & Sean
Cyg 20h20'43" 47d33'19"
Emily Seratch
And 2h14'29" 45d20'24"
Emily Shay Duenas
Tau 4h10'18" 30d46'58"
Emily Shin "4-13-03"
Lyn 7h50'35" 38d19'36"
Emily Siobhan Potter
Cru 12h52'46" -57d53'30"
Emily Smith
And 1h18'36" 38d8'8"
Emily Sousa
Lmi 10h22'18" 35d19'13"
Emily Stacey Evanow
Gem 6h48'52" 33d13'29"
Emily Starr
Uma 8h40'2" 67d6'13"
Emily Stevenson
Cap 21h46'37" -13d13'4"
Emily Strobino
Cnc 8h45'23" 15d33'49"
Emily Strong
Col 5h59'25" -35d14'47"
Emily Sue
Vir 12h27'57" -9d17'3"
Emily Sue
Tau 5h38'14" 20d57'1"
Emily Sue Regnier-Franz
Cyg 20h19'37" 50d36'58"
Emily Sue Stolzenberg
And 1h0'51" 42d3'47"
Emily Susan Kloeppel
Com 13h11'55" 18d5'39"
Emily Suzanne
And 0h16'59" 37d50'30"
Emily Suzanne Irons
Tau 4h39'33" 1d34'45"
Emily Suzanne Prior
Ori 5h9'42" -0d43'51"
Emily Sweet Pea Rose
Lib 15h57'42" -19d15'25"
Emily Tarpey
Ori 6h3'14" 20d19'26"
Emily Taylor
Leo 11h36'22" -5d2'9"
Emily Tedesco
Uma 10h51'52" 58d50'31"
Emily (The Eternal Lover)
Vir 12h56'6" -7d5'53"
Emily - The Light of my Life
Cra 18h52'58" -39d51'27"
Emily Theresa
Dra 16h34'36" 54d27'23"
Emily Theresia Ribitch
Gem 6h39'46" 14d7'50"
Emily / Timothy
Gem 6h35'39" 15d49'56"
Emily To Ryan: Happy 30th my love
Leo 11h37'19" 25d44'35"
Emily Tozzo
Uma 12h5'22" 37d8'7"
Emily Tuckerman Allen (Tuck)
Per 3h9'9" 48d5'34"
Emily Ueno-Dewhirst
Cru 12h51'11" -60d35'43"
EMILY V
Cyg 20h12'39" 35d7'2"
Emily V. Brotherton
And 0h52'40" 40d3'1"
Emily V. Trinidad & Sal Giuffrida
Peg 22h25'3" 15d18'8"
Emily Van Cleave
Sco 17h34'25" -32d27'47"
Emily Victoria Fenway Belmont
And 23h33'53" 41d38'19"
Emily Victoria McBurney 25.09.03
And 1h26'45" 44d9'7"
Emily Virginia
Ari 2h11'58" 26d31'17"
Emily Virginia Fade
Per 3h15'43" 41d17'48"
Emily Vitemb
Cas 23h5'9" 53d12'39"
Emily W Horton
Lib 15h45'39" -17d1'52"
Emily Walker 27
And 0h43'28" 26d52'13"
Emily Walsh
And 0h14'41" 44d45'4"
Emily Warren
Peg 21h37'32" 26d5'3"
Emily Watson
And 2h17'55" 45d46'35"
Emily Webster
Umi 15h46'53" 75d5'40"
Emily Wei - Shorty
Sgr 18h49'14" -33d7'56"

Emily Westermann
And 0h53'13" 38d13'12"
Emily Westermann
Vir 13h19'49" 12d46'14"
Emily White
Ari 2h54'8" 27d43'27"
Emily Wooten
Uma 10h48'58" 55d4'37"
Emily y Juan
Lyr 18h54'51" 45d7'18"
Emily Yott
Uma 14h25'0" 61d53'49"
Emily Yvonne Mercier
And 23h17'59" 48d7'26"
Emily Zhang Xiu-Ju
Tau 3h41'27" 28d24'49"
Emily, I loveyou! Happy Valentines!
And 0h20'44" 28d7'20"
Emily, LOVE GODDESS of the Hanisits
Leo 9h27'14" 13d40'50"
Emily, "The Feely"
Lyn 6h20'47" 56d20'36"
Emilyanne and Clint's 2 year star!!
Cnc 8h28'42" 15d55'55"
EmilyBarney
Col 5h34'20" -32d28'44"
Emily-Claire
Del 20h24'32" 9d41'19"
EmilyiusAstrobrightStargazerToo
Crb 15h44'35" 26d44'28"
EmilyJasmine143
Lib 15h34'31" -27d24'10"
Emily-Jean
Gem 7h35'47" 32d59'0"
Emily-Jean Reynolds
Cas 23h24'10" 62d21'28"
Emily-Joy 23041999
Ari 1h50'7" 17d46'49"
EmilyMaddison
Psc 1h26'26" 23d40'40"
emilyrod
Sco 16h39'14" -41d59'25"
Emilyrose
Umi 17h24'24" 78d57'36"
Emily's
Cap 20h27'7" -13d2'47"
Emily's 13th
Lyn 7h5'50" 45d15'24"
Emily's Angel
Ari 2h8'2" 24d7'9"
Emily's Diamond
Ori 5h57'41" 18d20'15"
Emily's Guardian
Ori 5h40'23" -2d43'7"
Emily's Guardian Star
Uma 11h47'53" 28d29'25"
Emily's Haven
Leo 10h7'54" 23d50'47"
Emily's Hope
Tri 2h20'2" 33d1'2"
Emily's Light
Gem 6h37'20" 14d37'20"
Emily's Little Piece of Heaven
Aqr 21h17'19" -8d5'19"
Emily's Marbles
Crb 15h35'46" 26d6'30"
Emily's Space
Vir 13h20'12" 6d51'40"
Emily's Sponsored Star
Umi 15h50'34" 80d20'58"
Emily's Star
Cas 1h16'58" 62d9'38"
Emily's Star
And 1h50'46" 35d46'24"
Emily's Star
And 2h1'22" 41d32'13"
Emily's Star 10/15/99
Mon 6h53'0" -0d4'35"
Emily's Star "Forever"
Uma 8h47'3" 69d41'5"
Emily's Star of Faith
And 1h8'19" 37d3'32"
Emily's Twinkling Star
Crb 16h4'5" 37d2'42"
Emina
Ori 6h2'49" 13d26'3"
Eminemily's Star
And 1h50'8" 43d17'4"
Eminger, Gerhard
Uma 11h49'14" 54d13'35"
Emir Ali Yazgan
Per 3h23'38" 33d37'17"
Emir und Dorothee für immer
Cap 21h20'6" -17d33'29"
Emi's Star
Psc 0h9'22" 12d6'45"
EMIWalter
Tri 2h10'3" 34d3'2"
EmJay
Cyg 20h21'6" 59d9'59"
EmJoMat Clarke Trinity
Ori 5h57'26" 11d33'8"
Emlékül férjemnek, Nagy Lacinak
Ari 2h35'7" 23d5'44"
Emlien
Psc 1h22'25" 11d25'42"
Emlyn Buck
Vul 19h29'3" 26d14'19"
Emlyn Hughes
Per 4h9'10" 35d46'5"
Emlyn's Star
Cyg 20h1'36" 38d17'43"
Emma
And 0h23'37" 45d14'59"
Emma
And 2h20'21" 46d3'11"
Emma
And 2h24'17" 46d38'25"
Emma
And 1h33'27" 49d8'48"
Emma
And 23h17'2" 50d1'15"
Emma
And 23h13'52" 49d27'16"
Emma
Cas 0h13'58" 58d34'39"
Emma
Cyg 21h2'23" 46d7'56"
Emma
And 1h32'28" 37d27'37"
Emma
And 0h50'40" 37d11'16"
Emma
And 0h39'39" 37d13'0"
Emma
Crb 16h1'39" 30d2'46"
Emma
And 0h11'28" 43d50'44"
Emma
And 0h27'56" 43d26'10"
Emma
And 0h34'8" 43d34'38"
Emma
And 2h19'44" 41d10'4"
Emma
And 0h35'34" 33d8'38"
Emma
Crb 16h10'26" 28d56'18"
Emma
Leo 11h37'54" 24d17'46"
Emma
Ori 6h7'15" 19d56'13"
Emma
Leo 11h35'20" 19d7'24"
Emma
Cnc 9h7'9" 18d12'51"
EMMA
Ori 5h53'58" 12d31'33"
Emma
Ori 5h2'56" 10d22'52"
Emma
Tau 4h10'59" 1d46'58"
Emma
Uma 13h29'38" 57d5'3"
Emma
Umi 13h24'57" 73d4'46"
Emma
Cap 21h18'24" -19d55'21"
Emma
Sgr 19h34'37" -15d3'0"
Emma
Lib 15h53'43" -18d20'23"
Emma
Aqr 20h46'8" -11d28'19"
Emma
Cap 20h18'57" -9d17'45"
Emma
Ori 6h12'22" -1d23'39"
Emma
Cma 6h42'43" -11d39'26"
Emma - A Beautiful Star
Lib 15h12'5" -23d1'34"
Emma A. Manning
Aqr 22h24'12" -23d39'29"
Emma & Abby
Lyr 19h27'58" 38d39'59"
Emma Abigail Donnelly
And 0h22'12" 29d59'56"
EMMA ADAMS
Lib 14h27'9" -16d56'29"
Emma Ainsley Hinkleman
And 0h44'9" 35d9'6"
Emma Alexandra Lister
And 2h34'54" 43d40'35"
Emma Alice Cainey
And 2h33'35" 49d33'49"
Emma - All Embracing
Cru 12h19'36" -57d40'4"
Emma - Amaranta
Cyg 20h49'51" 35d22'28"
Emma and Anne
Cyg 20h23'41" 46d2'57"
Emma and Brian's star
Ori 6h18'58" 16d1'11"
Emma and Michael Spataro
Cnc 9h18'20" 19d4'43"
Emma and Victor Harrington
Cep 0h6'19" 75d33'27"
Emma Ann
Vir 13h26'20" 12d13'20"
Emma Ann Brown
Lib 15h4'12" -5d21'2"
Emma Ann Palmer
And 1h44'51" 45d38'11"
Emma Ann Sally Jassenoff
And 0h22'46" 32d57'0"
Emma April Heydon
Lmi 10h6'19" 34d56'41"
Emma Aurora Fiore di Pesco
Crb 16h2'27" 26d6'49"
Emma Bambridge
And 2h37'4" 44d7'1"
Emma & Beth Roskell
And 23h47'8" 42d19'40"
Emma Bettson
Lib 15h44'13" -29d9'54"
Emma Blasucci
Cap 20h18'52" -11d41'31"
Emma Brodie
Ari 2h23'40" 11d57'36"
Emma Brookley
Gem 6h49'46" 20d31'14"
Emma Buck Anderson
Lmi 10h42'0" 27d42'29"
Emma Caitlin
Ori 5h36'15" -3d17'14"
Emma & Caitlin Ballantyne (UB7 & 8)
Ori 5h20'36" -7d29'21"
Emma Caitlyn Kirkwood
Col 5h7'44" -30d6'9"
Emma Callen
Cru 11h57'59" -62d45'34"
Emma Cardarelli
And 0h51'58" 39d9'35"
Emma Carmella Cingari
Cap 20h23'49" -19d59'51"
Emma Carol Boam Symonds
Cep 21h8'36" 81d59'57"
Emma Cassidy
Tau 4h23'29" 18d0'26"
Emma Catherine
Vir 13h22'21" 5d53'11"
Emma Catherine Champion
Uma 11h1'2" 52d54'50"
Emma Catherine Ibex
Sgr 17h52'20" -28d29'7"
Emma Catherine Kennealey
Leo 9h28'12" 24d34'24"
Emma Catherine Ringelberg
And 2h18'43" 48d21'11"
Emma Charlotte Brown
Col 5h59'14" -34d58'4"
Emma Cherie Fields
Lmi 10h43'22" 28d3'39"
Emma Chichester Clark
Lib 15h47'33" -10d3'33"
Emma Choi
Lmi 10h24'21" 29d28'32"
Emma Christian
Cru 12h17'17" -58d41'46"
Emma Christine Nys
Aqr 23h9'53" -9d33'44"
Emma Christine Phillips
Sgr 19h44'42" -16d21'1"
Emma Clair Akers-Douglas
And 1h58'30" 39d20'1"
Emma Claire
And 1h11'35" 36d41'33"
Emma Claire
Leo 10h51'42" 13d20'5"
Emma Claire Hollingsworth
Lib 15h43'31" -6d27'40"
Emma Clara Rule Lucas, Beloved
And 0h16'30" 30d26'14"
Emma Clare Harral-Ball
And 23h0'37" 43d0'23"
Emma Colleen
Ori 5h25'28" 11d57'52"
Emma Corienne Garcia
Uma 12h3'27" 49d15'19"
Emma Cu, I Love You.
Cyg 21h34'31" 37d34'24"
Emma & Dave's Guiding Light
Lib 14h29'44" -13d14'19"
Emma Dean
Sco 17h53'22" -41d37'55"
Emma Delaney Wunderlin
And 0h22'10" 44d17'4"
Emma Diane Skilman
Cnc 9h0'29" 31d39'4"
Emma Doncaster
And 1h24'56" 37d22'31"
Emma Duckie Wright
Pho 0h19'53" -44d40'47"
Emma DuPre Vaughn
Gem 7h14'35" 19d48'5"
Emma E. Norris
Gem 6h23'41" 22d26'38"
Emma Eaves
And 0h54'18" 46d12'18"
Emma Elisabet Wikman
Sgr 19h14'36" -22d13'12"
Emma Elisabeth
And 0h37'17" 35d25'56"
Emma Elisabeth Casey
And 1h18'11" 46d9'14"
Emma Elizabeth
Cas 23h54'14" 52d57'54"
Emma Elizabeth Brown
Sco 16h10'44" -20d16'5"
Emma Elizabeth Budreau
Lib 15h40'3" -19d16'29"
Emma Elizabeth Flynn
Psc 0h53'1" 28d39'48"
Emma Elizabeth Guare
And 2h27'41" 48d45'9"
Emma Elizabeth Hilley
Vir 13h14'44" -22d30'18"
Emma Elizabeth Kovats
Aqr 20h49'33" 0d10'37"
Emma Elizabeth Krieg
And 23h19'11" 47d14'43"
Emma Elizabeth Lee
Sco 17h57'31" -38d9'22"
Emma Elizabeth Lemmink
Tau 4h35'17" 23d24'40"
Emma Elizabeth Moir
Uma 11h29'45" 32d37'2"
Emma Elizabeth O'Leary
Ari 3h21'56" 27d51'3"
Emma Elizabeth Schultheis
Aqr 23h31'42" -7d3'39"
Emma Elizabeth Steggles
Cra 18h45'8" -40d24'35"
Emma Elizabeth Vickers
Cap 21h48'43" -9d50'2"
Emma Elizabeth Wilson
Aqr 22h35'6" -11d18'33"
Emma Estelle Ryan
Dra 16h19'47" 57d45'45"
Emma Faith Henley
And 0h31'19" 30d22'12"
Emma Faith Shuster
And 0h7'11" 45d51'40"
Emma Fielding
And 1h18'14" 47d17'19"
Emma Fields
Cas 0h22'9" 62d50'37"
Emma Finley-Gillis
Uma 11h38'59" 44d22'30"
Emma Frances Haas
Cap 21h8'24" -15d14'36"
Emma Frances Tellepsen
Uma 11h49'13" 56d49'46"
Emma Gath
Cyg 19h37'16" 32d14'43"
Emma Gaughran
Cas 0h15'18" 53d46'49"
Emma Giselle Moore
Lyn 7h49'50" 47d38'42"
Emma Glennalee
And 0h38'52" 33d59'18"
Emma Gonzales-Rodriguez
Psc 23h7'37" 1d55'54"
Emma Grace
Tau 4h29'23" 28d6'46"
Emma Grace
Dra 19h59'0" 70d38'25"
Emma Grace
Vir 12h15'27" -3d23'50"
Emma Grace Belcher
Psc 1h20'7" 32d32'16"
Emma Grace Fitzgerald
Vir 13h11'30" 7d25'9"
Emma Grace Hittinger
Cas 0h19'7" 51d41'34"
Emma Grace Hixson
Umi 13h9'56" 70d48'49"
Emma Grace Marie
And 23h13'46" 35d35'20"
Emma Grace Mount
And 1h23'0" 50d3'39"
Emma Grace Patterson
Gem 6h35'12" 16d11'40"
Emma Grace Schulz
Sco 17h22'16" -39d51'52"
Emma Grace Wolters
Psc 0h50'47" 16d41'17"
Emma Grace Zewatsky
Tau 5h11'22" 21d40'58"
Emma Graham
And 1h2'54" 45d57'21"
Emma Grande
Lib 15h10'47" -10d55'10"
Emma Gunavahara
Peg 22h35'48" 17d41'41"
Emma Hailey Gabrys
And 1h14'49" 37d53'26"
Emma Halse's Grandad - Les Jones
Sco 17h11'8" -32d31'43"
Emma Harper
Lyn 7h11'5" 59d29'19"
Emma Havana Keedy
Del 20h20'7" 10d39'7"
Emma Heidi
And 2h28'7" 42d6'51"
Emma Hickey
Vir 12h4'54" -0d43'25"
Emma Hiscocks
Crb 15h16'44" 29d23'23"
Emma 'Honey' Wardle
And 1h55'41" 35d53'54"
Emma Humphrey and Paul Shine
Cyg 19h57'9" 39d6'30"
Emma Illuminated
Cap 21h51'38" -9d39'47"
Emma Irene Gragg
Uma 10h5'45" 50d17'48"
Emma Irene Lang
Aqr 22h27'43" 0d17'17"
Emma is a Safety Superstar
Psc 1h26'17" 15d31'28"
Emma Isabelle
Cas 0h21'29" 50d43'4"
Emma J R
And 2h34'57" 50d25'8"
Emma Jade
Tau 4h41'14" 27d13'54"
Emma & Jadon
Gem 7h27'54" 32d10'20"
Emma Jane
And 1h29'42" 34d41'59"
Emma Jane
And 2h21'22" 42d15'24"
EMMA JANE
Gem 7h37'16" 26d19'52"
Emma Jane
Uma 11h20'9" 53d53'39"
Emma Jane Beasant's Star
Cnc 8h52'0" 26d47'17"
Emma Jane Brown - Emma's Star
Uma 11h6'21" 39d13'1"
Emma Jane Browne
Aqr 22h4'17" -14d17'53"
Emma Jane Newell
Leo 9h36'29" 28d54'13"
Emma Jane Orr
And 2h38'15" 49d5'41"
Emma Jane Roby
Dra 19h59'29" 75d27'43"
Emma Jane Stubbs
Cas 2h18'21" 73d35'42"
Emma Janicki
Cyg 20h41'43" 51d58'42"
Emma Jayne Bradt
Leo 11h5'16" -4d26'49"
Emma Jayne Tunstall - Emma's Star
And 0h9'26" 36d54'48"
Emma Jean Ginty
Tau 6h0'27" 25d19'48"
Emma Jean Romine
And 1h2'30" 43d31'2"
Emma Jeanette Starr
Sco 17h51'57" -42d55'47"
Emma Jessica Casey
And 23h40'33" 38d16'38"
Emma Jessmin
Ari 2h35'56" 20d54'15"
Emma Jobson
And 1h51'25" 38d34'20"
Emma JoHanna
Leo 10h16'47" 25d41'8"
Emma Jones
Aqr 22h11'25" -23d18'22"
Emma Jordyn Martelock
And 23h22'12" 46d42'50"
Emma Joy Carmel Brice
Psc 1h28'48" 15d36'56"
Emma Joy Handschuh
Umi 15h18'30" 68d7'51"
Emma Kaitlyn
And 0h31'7" 46d23'34"
Emma Kate Kelly
And 0h17'43" 45d20'21"
Emma Kate Sarnat (NEO)
Sgr 19h43'37" -21d42'44"
Emma Katharine O'Donnell Ginther
Mon 7h22'25" -0d22'45"
Emma Katherine
Gem 7h4'41" 15d29'56"
Emma Katherine Anne Rehfeld
Sgr 18h8'34" -29d25'15"
Emma Katherine Elkins
And 0h18'33" 30d4'20"
Emma Katherine Smiddy
Cnc 9h11'40" 8d59'39"
Emma Katherine Squires
Lyn 8h34'28" 37d32'33"
Emma Kathleen Boucher
Cas 23h47'46" 57d20'10"
Emma Kathleen Grandas
Cap 21h40'23" -11d34'8"
Emma Kathryn Albright
Psc 1h15'10" 3d8'44"
Emma Kathryn Williams
And 1h9'49" 44d7'8"
Emma Kaye
And 0h55'43" 44d17'32"
Emma Kehoe
And 23h40'38" 38d38'47"
Emma Kennedy
Lyr 18h42'31" 30d34'55"
Emma Knightime
Aql 19h10'56" 5d9'4"
Emma Koscielny
And 23h39'49" 46d20'12"
Emma Kovalcik
Lib 14h57'53" -17d33'23"
Emma Kristy Bell
Lep 5h27'44" -20d59'32"
Emma K-Tracey Burden
Crb 16h5'30" 37d2'22"
Emma Kylie Ruzic
Ari 3h5'23" 28d43'50"
Emma Labedis
And 0h14'58" 28d58'11"
Emma Laughlin
Lyr 18h34'21" 32d39'21"
Emma Lee
Per 3h19'38" 43d54'37"
Emma Lee
Aqr 20h51'53" 0d5'32"
Emma Lee
Cyg 21h39'51" 53d57'8"
Emma Lee
Sgr 18h0'54" -22d49'21"
Emma Lee Aisthorpe
Sgr 17h54'35" -28d39'33"
Emma Lee Gist
Uma 9h33'26" 47d30'4"
Emma Lee Guth
Tau 5h22'54" 20d2'20"
Emma Leigh
And 1h17'31" 47d18'22"
Emma Leigh Ergott
Ori 6h4'46" 17d46'0"
Emma Leigh Smith
Peg 23h13'43" 15d32'15"
Emma Lily Leamy AKA Granny
Psc 0h50'9" 15d45'10"
Emma Lily Sterne
Leo 11h37'40" 13d57'33"
Emma Logutenkova
Leo 9h39'34" 13d30'15"
Emma Lois
Lyn 8h24'44" 48d33'44"
Emma Loo Bunny Hunny
Uma 10h37'25" 48d0'11"
Emma Lorene Hamilton
And 23h39'13" 37d57'2"
Emma Lou
Ori 6h8'5" 15d25'45"
Emma Lou
Cas 1h11'3" 61d57'1"
Emma Louisa Boreham
Peg 22h47'18" 4d28'39"
Emma Louise Brindle
Peg 22h37'6" 15d5'10"
EMMA LOUISE CORRADI
Lib 15h47'8" -23d8'18"
Emma Louise Demers
Cas 1h38'56" 61d29'41"
Emma Louise Filkohazi
Lib 14h50'57" -2d18'58"
Emma Louise Foster
Uma 8h48'50" 61d56'34"
Emma Louise Gavin
And 1h22'20" 43d41'54"
Emma Louise Henderson
And 23h3'42" 35d40'18"
Emma Louise - "Little Princess"
And 23h1'2" 45d20'3"
Emma Louise Paton
Cas 0h32'43" 52d15'39"
Emma Louise Skelton
Cyg 21h24'20" 35d54'38"
Emma Louise Whitbeck
Cyg 19h52'47" 33d15'50"
Emma Louise's Wish Star
Vir 11h50'37" 5d22'42"
Emma Love
And 23h17'30" 52d15'4"
Emma Luciana Engle
Cap 21h39'1" -13d49'30"
Emma Lucille Harden
Vir 14h6'40" -11d37'35"
Emma Lydia Groves
Gem 7h8'28" 23d33'59"
Emma Lyn Kindl
Tau 3h29'0" 8d49'25"
Emma Lyn Trembone
And 0h21'33" 26d59'57"
Emma Lynn
And 2h15'13" 41d32'28"
Emma Lynn Coble
Ari 3h6'23" 26d4'32"
Emma Lynn Cochran
And 23h49'55" 34d31'57"
Emma Lynn Danowski
Ari 2h42'51" 28d14'10"
Emma Lynn Marie Massie
Uma 8h29'12" 66d55'40"
Emma Lynn Oppedisano 7lb. 11oz.
And 1h57'41" 46d15'36"
Emma Lynn Pells
Uma 8h26'33" 60d46'8"
Emma Lynn Pritchard
Umi 9h44'12" 87d0'24"
Emma Lynn Sajewski
And 1h42'32" 41d34'14"
Emma Lynn Sajewski
Ari 3h6'3" 28d44'40"
Emma Lynn Thurston
And 2h27'43" 46d9'5"
Emma Lynn Wotzka
Peg 22h4'27" 12d33'51"
Emma Lynne Milton
Tau 5h28'14" 24d33'47"
Emma Lynne Thurston
And 2h16'4" 47d34'23"
Emma Lynn's Star
Aqr 23h32'38" -22d34'38"
Emma M Hall - My Ray of Sunshine
Cyg 19h34'47" 29d26'50"
Emma Madison Leduc
And 1h12'4" 38d22'16"
Emma Mae
Cnc 8h17'1" 15d1'31"
Emma Mae Simons
And 23h44'1" 46d10'19"
Emma Makaela
And 23h14'35" 41d16'27"
Emma Maltman
Lib 15h7'8" -6d16'49"
Emma Marconcini
Uma 9h6'25" 50d42'44"
Emma Marette ~ A Star is Born!!
And 0h19'13" 33d1'12"
Emma Margaret O'Brien
Ari 3h7'28" 23d11'22"
Emma Marguerite Hagan
And 23h12'7" 52d16'26"
Emma Maria
Cnc 9h2'43" 17d57'47"
Emma Maria Bancroft
Cas 23h14'2" 55d36'5"
Emma Marie
Cas 0h21'4" 64d12'51"
Emma Marie
Uma 9h14'1" 62d51'6"
Emma Marie Halvorson
Umi 15h41'52" 75d56'2"
Emma Marie Loparco
Cet 1h8'11" -0d41'27"
Emma Marie McMillin
And 0h17'47" 37d47'53"
Emma Marie Stagg
Lyn 7h15'29" 57d26'43"
Emma Marie Weishaar
Ari 2h31'3" 25d24'49"
Emma Marin Domingo
Psc 1h2'36" 4d33'56"
Emma Markosyan
Cas 23h7'33" 56d21'36"
Emma Mary
Aqr 20h41'52" -8d47'58"
Emma Mary Boxall
Cas 23h45'37" 57d23'16"
Emma Mary Festoff
Uma 9h35'36" 68d43'2"
Emma Mary Festoff
And 0h20'8" 29d58'5"
Emma Mary Kazanchyan
And 0h48'32" 35d53'42"
Emma Maryellen Brotz
And 1h11'1" 37d55'11"
Emma Maxine Pfeiffer
Uma 13h14'48" 62d22'18"
Emma May Featherstone
Umi 14h1'52" 69d33'23"
Emma May Narotzky
Gem 7h34'22" 33d17'24"
Emma May Radley
Lmi 9h34'11" 33d49'36"
Emma May Rose
Col 5h31'35" -37d46'0"
Emma McCosh
Cap 21h38'13" -12d27'49"
Emma Mercedes
And 0h55'20" 40d9'54"
Emma Mitsel
And 1h14'43" 46d27'11"
Emma Mosher
Dra 18h49'11" 61d15'50"
Emma Nadine Donnelly
Umi 13h50'57" 86d44'21"
Emma Nanialoha
Cnc 8h1'39" 24d30'9"
Emma Nicole
And 23h37'9" 39d1'6"
Emma Nicole
And 0h41'19" 40d52'0"
Emma Nicole
Uma 10h57'7" 55d6'42"
Emma Nicole Boshnick
Gem 7h41'40" 31d47'23"
Emma Nicole Frizzo
Aqr 22h24'12" 0d39'43"
Emma Nicole Schmiedigen
Psc 0h15'28" 5d59'43"
Emma Nicole Timblin
And 1h46'2" 45d18'31"
Emma Noel Gibbons
Cnc 8h52'45" 31d41'1"
Emma Noelle
Crb 15h43'47" 35d20'42"
Emma O. Rotchford
Cas 1h13'56" 60d20'23"
Emma Olivia Youngs
Cyg 21h13'23" 42d3'57"
Emma Paige
Ari 3h22'48" 28d8'41"
Emma Paige Corrigan
Lmi 10h34'54" 38d1'57"
Emma Parker
Psc 1h19'29" 16d39'18"
Emma Parry-Jones
And 0h53'30" 43d28'0"
Emma Patricia Holland
Her 18h14'55" 16d56'33"
Emma Pearl
Boo 14h31'19" 44d54'15"
Emma Pearman
Ari 2h26'17" 18d19'22"
Emma Peetje
Psc 1h22'6" 17d46'38"
Emma Pflugfelder
Ari 2h33'1" 24d39'1"
Emma Plasner Mijanovic
Uma 14h1'45" 48d23'53"
Emma Quade
Psc 1h17'51" 16d45'18"
Emma Quattrochi
Lyn 8h36'35" 39d16'6"
Emma Rae
And 1h37'33" 47d2'27"
Emma Ray
Lyr 18h42'5" 30d52'50"
Emma Ray
Lyn 6h30'9" 55d49'46"
Emma Rebecca - A Star For Eternity
Cru 12h39'2" -59d0'24"
Emma Reda
And 1h22'10" 38d14'24"

Emma Reese Nophut
Ari 1h57'4" 14d14'4"
Emma Regan Landry
Cap 21h32'23" -8d29'57"
Emma Remondelli
And 0h10'6" 46d44'37"
Emma Renee
Cap 21h41'47" -16d21'55"
Emma Renee Gross
Lyr 18h51'29" 31d54'4"
Emma Riley Poulin
Uma 11h59'24" 51d31'21"
Emma Ritter
And 1h31'4" 38d55'52"
Emma Robin Bliss
And 1h36'3" 44d10'35"
Emma Roby
Cyg 19h34'4" 28d58'9"
Emma Rosato
Cap 20h13'1" -10d6'46"
Emma Rose
Uma 11h30'58" 56d16'42"
Emma Rose
Gem 6h29'3" 17d55'6"
Emma Rose
Cnc 8h50'14" 11d19'0"
Emma Rose & Abigail Grace Dunn
And 2h7'57" 42d40'31"
Emma Rose Lane
Ori 6h6'32" 10d56'6"
Emma Rose Martin
And 2h35'42" 50d9'46"
Emma Rose Mc Daid
Cas 1h28'14" 55d27'58"
Emma Rose O'Flanagan
Psc 23h12'22" -1d32'1"
Emma Rose Santora
Lyr 18h51'57" 33d11'15"
Emma Rose Sciscione
Ari 2h44'59" 27d56'34"
Emma Rose Verbeke
Cap 20h13'50" -11d2'48"
Emma Ruth
Aqr 21h58'54" 1d1'27"
Emma S. Studt
And 0h41'1" 44d20'9"
Emma Samantha Grunin
Cap 20h27'14" -9d29'43"
Emma Sarah Levinson
Cas 0h20'38" 50d4'46"
Emma Schield
Ari 3h0'51" 12d27'40"
Emma Shae Filonow
And 1h26'33" 40d0'48"
Emma Shaw
Uma 13h55'18" 49d34'53"
Emma Singh
Lyn 7h6'11" 60d23'33"
Emma Sofia Allsopp
And 23h31'39" 41d22'34"
Emma "Sparky" Priest
Ori 5h40'23" -1d47'56"
Emma Starr
Cap 21h28'0" -9d32'42"
Emma Sue
Peg 22h32'23" 5d24'47"
Emma Swain
Uma 11h35'38" 50d3'27"
Emma Sweetman
And 2h2'16" 37d16'31"
Emma T. Gogel
Cnc 9h2'38" 15d52'22"
Emma Takagi 'Babe' Trujillo
Peg 22h45'29" 12d22'21"
Emma Trissy
Ori 5h33'7" 3d25'35"
Emma Turner
Aqr 21h46'42" -2d46'6"
Emma Tuttleman
And 0h15'10" 27d28'13"
Emma Utting
Ari 1h54'44" 13d58'37"
Emma V. Hook born 30.12.1983
And 0h29'41" 36d42'31"
Emma Victoria
Cnc 8h32'16" 10d34'56"
Emma Victoria Henderson
Tau 4h37'12" 2d24'48"
Emma Victoria Hubbard
And 0h26'36" 26d42'29"
Emma Virginia Kerezsi
Sco 16h51'46" -34d57'34"
Emma Vivian Bleyzer
Cas 1h23'30" 60d44'44"
Emma Vivian Bleyzer
Tau 4h45'11" 24d1'26"
Emma Walker
Aqr 22h9'29" 1d3'43"
Emma Wallis
And 23h27'26" 42d43'46"
Emma Webb
Psc 0h40'18" 21d14'46"
Emma Weiler
And 1h10'48" 45d51'51"
Emma Wendt
Cru 11h56'55" -61d22'39"
Emma WenXue Kershaw
And 1h30'48" 44d12'50"
Emma Wheeler
Cas 1h27'37" 59d15'58"
Emma Woodland's Star
And 23h35'45" 49d17'46"
Emma Ying Eiden
Peg 23h12'47" 27d16'37"
Emma, My Emma - Emma's Star
Cas 23h51'23" 56d18'37"
Emma.21
And 0h46'20" 26d25'26"
EmmaCaroline
And 1h53'13" 42d46'39"
Emma-Elizabeth Gwinn
Her 16h35'56" 7d24'14"
EmmaEllen Pugsley Brown's Butterfly
Lyn 7h6'37" 60d59'4"
Emma-Jalin-Oldenburg
Vir 14h25'45" 0d40'29"
EmmaJames Trott
Gem 7h28'4" 26d9'26"
Emma-Jane's Star
And 2h32'15" 49d31'29"
Emmajean
Psc 1h7'50" 24d52'42"
Emmalaya Jaleel Owen
Umi 14h20'49" 75d54'4"
Emmalee
Cap 21h26'59" -22d54'5"
Emmalee Annmarie
And 23h39'11" 36d31'26"
Emmalee Margaret
Cyg 21h22'12" 38d44'27"
Emmaleigh Jane Ashcraft
And 0h12'50" 40d12'8"
Emmalemagus
Ari 2h34'46" 24d44'26"
Emma-Lily Broom
And 1h15'55" 34d47'24"
Emmalin Taylor Kettell
Uma 11h56'10" 44d14'21"
Emmaline Anna Marie McKenzie
Uma 10h31'5" 42d20'9"
Emmaline Claire Goldberg
Psc 0h55'51" 3d19'57"
Emma-Louise Cattell
And 23h18'50" 41d14'49"
Emmalyn Claire
Per 2h24'13" 56d8'6"
Emmalyn Everett
And 0h19'59" 43d58'45"
Emmalynne Olson
And 0h21'27" 32d4'43"
Emma-Mai Heathcote - Emma's Star
And 23h42'30" 35d44'33"
Emmanoella
And 23h42'57" 47d49'20"
Emmanuel Alvarado
Ori 5h32'54" -2d54'55"
Emmanuel Cruz
Uma 8h41'3" 48d10'14"
Emmanuel Cucchiarelli
Peg 23h45'53" 25d57'58"
Emmanuel & Dylan Fafalios
Her 17h38'39" 24d36'37"
Emmanuel Etienne
Umi 14h48'54" 77d12'24"
Emmanuel Moire
Dra 18h35'53" 79d22'33"
Emmanuel N. Dessypris
Her 17h23'41" 37d32'27"
Emmanuel Reyna AKA Daddy
Uma 11h29'19" 44d33'9"
Emmanuel Soubrouillard
Sgr 19h45'53" -24d20'59"
Emmanuel Tapia Jr.
Cep 22h31'28" 62d42'10"
Emmanuel, May our lips touch always
Vel 10h28'18" -42d46'43"
Emmanuelle
Uma 14h4'57" 56d37'15"
Emmanuelle
Tau 4h42'7" 17d10'34"
Emmanuelle Bracops
Sco 17h23'45" -42d41'44"
Emmanuelle Di Grazia
Uma 9h56'22" 43d28'35"
Emmanuelle Doize
Lyn 7h43'32" 51d13'53"
Emmanuelle Obeuf
Cas 1h28'58" 51d54'53"
Emmanuelle Soum
Umi 14h43'44" 75d34'55"
Emmanuelle07fr
Ari 2h23'59" 17d56'9"
Emmarie Elizabeth Cook
Sgr 18h31'30" -16d21'40"
EmmaRuth
And 0h47'35" 35d55'45"
Emma's Gaze
And 0h17'12" 36d14'4"
Emma's Love
Aqr 20h51'42" -12d12'24"
Emma's Lucky Star
And 0h2'5" 47d41'46"
Emma's Smile
And 0h55'39" 36d37'40"
Emma's Star
Cyg 20h29'6" 37d29'57"
Emma's Star
Cas 0h12'18" 55d18'58"
Emma's Star
Vir 12h21'36" 0d6'12"
Emma's star
Ori 5h1'30" 4d57'40"
EmmaStar 07-11-2001
And 1h27'57" 45d37'29"
Emmat
Crb 16h16'56" 28d7'58"
Emmaus
Uma 10h16'49" 41d35'54"
Emme
Aur 5h25'8" 37d40'15"
Emme Corbet
Cas 23h45'32" 51d10'53"
Emme Katherine
Lib 14h51'42" -2d21'26"
Emmelia Joan Beckinsale
Cru 12h35'29" -58d33'31"
Emmeline Carlotta Steiner-D'Angelo
Cap 21h45'6" -13d30'55"
Emmeline Morris
Tau 3h36'44" 21d21'45"
Emmeline T. Wilson
Sgr 18h36'46" -21d40'55"
Emmerich, Siegfried
Uma 10h46'46" 41d25'36"
Emmermacher, Richard
Uma 10h34'35" 68d22'14"
Emmerson Samuel Snair
Her 16h49'51" 25d40'46"
Emmet
Per 3h1'18" 43d11'5"
Emmet
Cap 21h11'51" -19d2'29"
Emmet Douglas Anderson, Sr.
Lib 14h50'25" -6d41'57"
Emmett Alex
Umi 15h31'50" 75d23'8"
Emmett Andrew Fraley
Vir 11h50'27" -0d16'11"
Emmett Doherty
Umi 13h43'1" 72d9'47"
Emmett Ernest Banks
And 0h29'42" 28d18'35"
Emmett Henry Kallmeyer
Cnc 8h53'0" 25d53'21"
Emmett Layne Anderson
Umi 14h20'28" 68d18'10"
Emmett Ostrander
Leo 9h35'38" 10d19'3"
Emmett R. Schmitz
Sco 17h11'2" -43d22'7"
Emmi
Sco 17h29'12" -42d13'8"
Emmi Reece McLain
Aqr 21h34'0" -7d38'56"
Emmi Schraner-Keller
Her 17h10'54" 19d43'37"
Emmie Faith Frederick
And 2h29'36" 44d34'8"
Emmie J.
Uma 12h37'50" 55d34'0"
EmmieMatt
Lyn 9h19'6" 38d45'52"
Emmiline Doris Phin
Boo 14h31'53" 18d17'2"
Emmiline Elizabeth Fish
And 23h28'2" 48d35'6"
Emmily Crazy
Leo 11h13'40" 20d25'30"
Emmily Louise Morris
Cap 20h27'26" -19d26'49"
Emmisen Lynae
Lib 15h28'42" -5d48'39"
Emmita
Cas 1h16'0" 67d20'9"
Emmitt
Lyn 6h19'8" 56d45'15"
Emmitt James Crenshaw III
Per 3h16'56" 48d0'6"
Emmy
And 23h2'0" 51d24'59"
Emmy
And 0h1'6" 33d55'55"
Emmy
Lmi 10h34'11" 32d53'20"
Emmy
Leo 10h35'56" 21d28'50"
Emmy
Ori 5h52'18" 18d3'16"
Emmy
Psc 0h42'31" 14d43'21"
Emmy
Cam 7h31'20" 61d26'0"
Emmy
Mon 6h51'29" -0d3'57"
Emmy
Cap 21h43'41" -10d56'29"
Emmy
Lib 15h3'34" -29d43'55"
Emmy Bear
And 1h34'26" 42d21'55"
Emmy Bright Eyes Bridges
And 1h34'17" 49d22'10"
Emmy & Daniel Ryan
Mon 7h16'57" -10d54'26"
Emmy & Henri Lichtenberg
Cyg 20h38'45" 50d59'32"
Emmy Lane
Psc 1h21'0" 20d10'33"
Emmy Loves Jason
Oph 17h31'15" 6d11'29"
Emmy Luna
Cas 0h36'18" 57d28'30"
Emmy Maria
And 2h21'46" 48d49'40"
Emmy Nichelle Allee
Ori 5h37'38" -3d5'17"
Emmy O'Toole
Uma 11h31'57" 59d29'39"
Emmy Rose
Psc 2h0'10" 9d10'37"
Emmy-G 171005
Lib 15h43'53" -24d26'57"
Emmy-Rose St-Pierre
Uma 11h31'58" 57d58'39"
Emmy's Light.......May 13, 2006
Leo 10h19'52" 24d23'21"
Emmy's Pale Blue Eyes
And 23h2'13" 49d54'18"
Emo
Ori 5h43'56" 5d23'22"
Emo Buddy <3
Aqr 22h26'29" -2d29'34"
Emo és Zoli csillaga
Uma 8h57'52" 71d34'15"
E-Mo the shining star <\@> Purdue Univ
Uma 8h23'3" 68d19'45"
EMOD
Peg 21h50'59" 21d29'25"
Emodire
Leo 11h8'42" 16d35'20"
Emogene Mae
Lyn 8h57'17" 44d15'49"
EMONA
Gem 7h28'40" 19d56'52"
Emoral
Ori 5h38'28" 0d10'34"
Emorina
Lyr 18h50'15" 43d21'16"
Emory Chase
Leo 11h18'49" 15d17'58"
Emory Elaine Verret
Psc 23h8'11" 0d49'36"
Emory Jordon Bailes IV
Ori 5h41'5" 2d40'22"
Emory Katherine
Lyn 6h29'9" 56d58'18"
Emory Logan Keys
Her 16h34'57" 18d39'7"
Emory Yun-Hsi Broek
Aqr 21h12'11" -8d53'8"
Emperoress Stella
Lyn 7h31'2" 38d0'14"
Empheria
Tau 3h30'23" 20d48'43"
Empiress Katherine Mussoline
Aql 19h5'33" 3d8'10"
Empourious Carlious The Great
Cep 22h5'26" 61d16'57"
Empress Aislynne de Chartier
Cas 23h43'55" 56d7'1"
Empress Kiki
Tau 5h47'5" 16d55'19"
Empress Meghan Holland
Tau 3h25'34" -0d27'43"
Empress Nancy
Cnc 8h20'3" 24d36'56"
Empress of English
Cas 1h39'1" 63d59'45"
Empress Vanessa Vuong
Psc 1h9'19" 11d4'38"
Empyrean 3
Leo 9h38'4" 20d30'56"
EmRae
Uma 11h57'54" 28d54'51"
Emrecan Pirinccioglu
Cnc 9h0'10" 16d32'56"
Emren
Uma 8h32'49" 63d10'57"
EmreYCarlos(FFE)NYC'07
Gem 6h46'49" 34d2'53"
Emrick Maheu
Del 20h49'13" 13d46'53"
EMS 1974
Lyr 18h35'55" 37d1'26"
EMS Boys
Ori 5h29'11" 3d7'31"
Em's Little Corner of Heaven
Cnc 9h9'11" 12d22'10"
Em's star
And 23h14'35" 41d37'15"
Em's Star
Cma 6h23'16" -21d37'58"
EmShady
Cnc 8h38'51" 23d48'20"
Emski
And 23h57'26" 41d56'22"
Emstar
Uma 11h52'37" 39d27'37"
EmStar
Cas 23h28'57" 57d56'41"
emtogana
Dra 20h10'42" 73d7'7"
EMW BWK Cool
Per 3h28'10" 44d20'24"
EMY
Gem 7h38'31" 24d8'14"
Emy
Peg 21h34'51" 16d48'44"
Emy
Psc 0h28'13" 8d42'26"
"Emy"
Uma 8h36'43" 57d22'11"
Emy Aime Jocelyn
Ori 6h7'20" 6d4'10"
Emy Center
Crb 16h17'50" 33d44'28"
Emy Eidson
And 23h49'19" 45d23'4"
Emy66
Dra 18h28'30" 79d25'33"
Emy-Jeanne
Umi 13h34'23" 69d46'45"
Emylee Grace Whitten
And 0h11'4" 34d44'33"
Emylie & Louis... Now and Forever
Cas 23h21'24" 57d57'49"
Emyllee Anne Jensen
Lyr 18h42'40" 35d14'7"
Em-Zack
Cyg 19h49'4" 33d44'50"
Emzlite
Cyg 20h14'55" 41d13'0"
En101
Aqr 22h27'39" -12d38'3"
Ena
Umi 14h15'47" 68d24'25"
Ena Egli
Her 18h39'27" 19d3'18"
Ena McDonald
Sgr 19h20'52" -22d12'38"
Ena Spahic - 24.12.1988.
Cyg 20h57'6" 31d10'52"
Enaysa Tishelle Payne
And 1h49'41" 37d45'1"
Encanación Fernández Candel
Umi 14h56'4" 77d59'1"
Enchant
Cyg 19h34'58" 48d39'31"
Enchanted
Lib 15h4'46" -8d14'3"
Enchanted Starlight
Crb 15h44'14" 27d19'30"
Enchanted Wizard 6/29/01
Lyr 18h47'43" 43d26'15"
Enchanters Lady Lilly Rollins
Cas 23h37'29" 54d21'53"
Enchanting Angel
Cyg 20h2'25" 46d4'23"
Enchanting & Eternal, Eva Louise Noble
And 1h23'56" 49d29'43"
Enchantin's Star
Tau 5h40'28" 25d58'40"
Enchantment
Tau 4h44'43" 27d3'19"
Enchiridion
Leo 11h58'9" 20d47'18"
Encounter
Uma 12h46'59" 54d41'6"
Encsillagom
Vir 14h41'29" -4d40'5"
ENCYPHER
Ari 2h3'5" 24d19'41"
Enda
Cnc 8h40'1" 32d9'43"
Endeavour
Dra 10h48'48" 77d14'46"
Endellion King & Nick Carter's Star
Cep 23h25'12" 85d51'43"
Endicoot
Tau 4h56'0" 23d9'18"
Endless
Cyg 20h44'3" 46d44'54"
Endless Devotion
Ari 3h14'29" 28d21'32"
ENDLESS LOVE
Sge 19h24'6" 17d8'53"
Endless Love
Cyg 19h42'53" 47d59'55"
Endless Love
Cyg 20h11'2" 50d15'49"
Endless Love
Per 3h12'14" 31d42'48"
Endless Love
Aur 5h27'14" 40d41'39"
Endless Love
Lyn 8h44'1" 40d58'16"
Endless Love
Uma 11h28'49" 58d24'46"
Endless Love
Uma 9h23'42" 52d44'40"
Endless Love
Cyg 20h21'43" 52d52'35"
Endless Love
Cyg 19h51'21" 53d16'35"
Endless Love
Sco 16h36'18" -41d48'46"
Endless Love for Melanie & Sascha
Ori 6h18'5" 16d43'10"
Endless Love For Nate G From MOe
Ori 6h13'12" 8d56'21"
Endless Love: Rachel & Jordan
Ori 6h5'6" 10d11'12"
Endlose Liebe- 2 Jahre, viele mehr!
Cyg 21h26'53" 46d56'28"
Endor
Per 3h9'6" 47d44'1"
Endor Enternity
And 23h16'14" 51d21'45"
Endora
Ari 2h6'18" 22d46'49"
Endriß, Julius
Ori 5h7'30" 7d37'14"
Enduring Devotion
Uma 11h7'24" 52d37'17"
Endymion
Uma 13h45'15" 52d5'7"
Endzone
Vir 12h15'17" 2d27'54"
Enea Jan
Her 16h50'36" 11d35'52"
Enedina Josephine
Aqr 22h28'44" 0d32'39"
Eneida Masdeu
Dra 18h27'40" 55d1'54"
Energee!
Uma 10h39'52" 50d59'49"
Enevy
Cam 4h6'18" 71d32'12"
Engel
Uma 8h16'53" 66d17'20"
Engel
Uma 9h25'0" 63d14'12"
ENGEL
Com 12h8'45" 14d48'3"
Engel - Bengel
Uma 10h42'12" 42d38'55"
Engel Elsa Marie
Her 16h59'49" 13d14'16"
Engel Sonja Werner
Aqr 22h28'3" 0d51'18"
Engel Tanja
Uma 10h42'24" 41d44'10"
Engelchen
Ori 6h19'7" 19d10'53"
Engelhardt, Doris
Ori 6h19'53" 19d51'22"
Engelyna
Tau 3h58'10" 7d42'43"
Engin Celebi
Ari 2h12'21" 26d42'25"
Eni
Leo 9h30'41" 11d51'45"
Enid
Aql 19h26'32" 9d13'40"
Enid and Frank Spillett
Cyg 19h26'49" 54d17'45"
Enid and Len Boxer
Umi 16h15'0" 75d39'25"
Enid B
Cyg 19h51'35" 38d6'26"
Enid Bekker
Crb 16h17'20" 30d25'55"
Enid Claire Whittenberg Tweeton
Cas 0h20'48" 56d26'13"
Enid Goodwin
Crb 16h4'32" 38d22'29"
Enid McMurdo
Cnv 12h43'47" 39d32'58"
Eniko B.
Cyg 21h54'52" 50d12'43"
Eniomel
Del 20h25'1" 16d8'17"
Enise Koch
Cru 12h26'15" -59d7'11"
Enki
Uma 10h6'37" 56d22'39"
ENL
Lib 15h37'16" -6d42'13"
Enlightus
Cap 21h45'18" -14d45'51"
Enmity
Per 3h47'15" 47d31'15"
Enna Mink Dishaw
And 23h58'41" 45d23'20"
Ennis Elizabeth Mitchell
Cap 20h44'3" -18d36'2"
Ennis Muharemi
Gem 6h41'16" 17d28'31"
Enno Van Dam Jr.
Uma 12h42'36" 52d33'44"
ENOLA
Lac 22h4'33" 52d1'38"
Enormous Joy
Cru 12h26'53" -63d48'12"
Enravishka
Cap 21h23'8" -22d0'44"
Enric Amigo
Lyr 18h53'23" 26d29'11"
Enrica
Ori 5h48'27" 4d7'50"
Enrica
Boo 14h41'41" 53d59'54"
Enrica Però
Cas 0h25'0" 62d22'42"
Enrica Segond Romeo
Leo 10h12'26" 15d14'38"
Enrico
Com 11h59'13" 28d18'3"
Enrico Baccioni
Ori 5h42'25" 7d12'53"
Enrico Mc.Giver
Her 16h41'45" 24d3'17"
Enrico Morante
Cep 22h4'48" 73d36'56"
Enrico Paone, Jr.
Cap 20h26'42" -10d6'26"
Enrico Rinaldi
Ari 2h2'56" 22d14'26"
Enrique
Psc 1h16'33" 22d31'17"
Enrique
Cap 21h57'43" -22d20'30"
Enrique and Eleni's Star
Umi 15h29'26" 75d44'7"
Enrique Carvajal My Favorite Person
Lmi 10h48'9" 38d51'43"
Enrique Francisco Heredero
Sco 16h56'56" -44d48'22"
Enrique Guadalupe
Aur 5h4'49" 36d9'18"
Enrique Ignacio Tapia Balladares
Cep 21h26'30" 67d27'16"
Enrique Martinez Gustavo Boquin
Cnc 8h21'9" 12d39'19"
Enrique Moran Marin
Sco 16h32'19" -35d38'15"
Enrique & Rafael Enrique Paris
Ori 5h34'55" -1d35'45"
Enrique Rams
Aql 19h5'14" -0d5'7"
Enrique Roberto Beltra
Aqr 21h14'51" -11d14'25"
Enrique Rubio Rivera
Dra 15h44'45" 59d23'52"
Enriqueta Aguilon
Lyr 19h2'51" 33d33'58"
Enriqueta Elizabeth Saucedo
Uma 10h33'31" 59d29'42"
Ensaneh Kaamel
And 0h9'35" 43d13'58"
Ensign Hoshi Sato
Aql 19h9'6" 15d2'56"
Ensign Travis Mayweather
Aql 19h18'31" 14d56'6"
Enterprise
Per 2h50'20" 42d36'18"
Enterprise NX-01
Uma 11h40'20" 47d23'26"
Entwined Souls
Uma 11h40'22" 52d15'40"
ENV
Crb 15h32'32" 33d15'46"
EnV - AnF
Uma 11h44'31" 37d11'4"
Enver IV
Cap 21h33'34" -17d28'0"
Enymig
Sco 16h11'11" -41d27'24"
Enza e Lillo
Uma 8h41'32" 61d9'44"
Enza Litzler
Ori 6h9'50" 11d43'3"
Enzo
Dra 17h32'3" 62d35'45"
ENZO
Umi 14h40'56" 73d40'31"
Enzo
Cnc 8h27'23" 6d36'53"
Enzo Anthony Brun III
Gem 7h9'0" 34d36'6"
Enzo der Glückliche
Uma 10h9'39" 48d44'52"
Enzo Fedir Scatola
Per 4h47'59" 42d25'17"
Enzo Gioia
Cyg 20h32'6" 34d23'9"
Enzo Gioioso
Lyr 18h49'42" 33d4'12"
Enzo Giuseppe Molina
Cap 21h51'26" -13d10'35"
Enzo Joseph
Tau 4h8'43" 27d1'58"
Enzo Masani
Gem 7h32'0" 16d24'35"
Enzo Renato Miraglia
Col 5h58'5" -31d14'44"
Enzo Savary
Cap 21h25'55" -27d25'14"
Enzo Tettamanti
Gem 6h54'50" 28d3'51"
EO
Crb 16h20'10" 39d11'2"
Eoghan Neil Doheny
Boo 15h0'23" 42d23'20"
Eoghan O'Callaghan
Umi 15h33'8" 71d51'54"
Eoghan Peadar Keogh
Uma 9h0'19" 51d1'19"
Eoghan Thomas Kelly
Her 16h49'43" 27d54'38"
Eòin Caòlan Vallely
Umi 15h15'25" 68d42'21"
Eoin Christopher Kenny
Her 17h27'32" 23d10'59"
Eoin Fox Reilly
Lmi 9h48'34" 40d45'55"
Eoin Haughey
Uma 12h37'53" 55d37'7"
Eoin James McKernan
Per 4h27'21" 32d35'46"
Eoin McGroddy
Per 4h9'37" 33d12'45"
Eoin Redmond
Per 4h15'15" 39d21'16"
Eon Andre Dyal
Uma 11h26'8" 60d33'6"
EONS GIRL
Aqr 22h41'10" -3d24'42"

Éowyn Josie Ruzzon Falla
And 1h15'53" 49d4'28"
Epcot
Lib 15h25'30" -18d45'35"
Ephemera
Ori 6h3'26" 17d55'51"
Epi und Opa
Sgr 18h7'35" -24d2'48"
Epifania Toribio
Peg 23h26'54" 10d45'10"
Epifanio Figueroa
Uma 11h27'56" 63d52'23"
Episkopos of Harrison
Uma 9h8'53" 63d32'34"
E.P.K.21
Dra 18h59'11" 54d14'7"
EPM
Aql 18h59'18" -0d45'30"
epoc
Lib 14h56'44" -2d2'17"
Eppy - Stefano Pasqualetti
Ori 6h1'25" 12d11'10"
Epsilon Delta
Peg 22h35'40" 25d15'44"
Epsilon Farris 5
Uma 10h37'55" 58d18'29"
Epy
Psc 1h21'33" 15d37'39"
Equidae
Aqr 22h24'56" -17d40'49"
ER FOREVER
Lyr 19h6'26" 36d18'57"
ER MICKEY
Cnc 8h33'7" 27d38'15"
Era Dell (Gamma)
McWhorter
Uma 8h42'34" 59d49'57"
ERA The Masiello Group
Umi 14h11'51" 75d46'3"
eraina
Uma 9h2'33" 47d23'40"
Eraina Verchiens
Uma 11h36'26" 53d45'26"
Eraj Torian
Umi 14h27'14" 86d27'55"
ERALGAL
Sco 16h9'53" -18d16'43"
Eramathea
Lib 15h25'11" -26d45'36"
Eran Raven Feigenbaum
Per 3h25'56" 42d30'31"
Erasmo Trionfo
Cep 22h42'20" 66d54'34"
Erawan
Cyg 19h50'43" 32d37'38"
ErBear
Ori 5h45'36" 8d15'51"
Erbey
Umi 17h0'5" 87d14'40"
Erbolita
Eri 3h20'4" -3d18'18"
Erdelt Pulsar
Uma 10h58'26" 65d26'38"
Erdész Katalin
Leo 11h24'28" 6d23'33"
Erdman 60
Ara 16h39'40" -59d34'8"
Erdogan
Lyn 8h3'13" 39d30'33"
Erdos Zoltán 30, az
Egyetlen
Uma 12h16'57" 62d58'55"
Eredità Del Alfred
Gem 7h16'14" 33d43'56"
Eremeyeva Katerina
Nikolaevna
Lib 15h12'56" -6d45'21"
Erén
Ari 2h16'58" 15d27'43"
Eren Burmaci
And 0h17'40" 46d20'15"
Eren Gondogdu
Cap 21h40'0" -14d0'24"
Erendira
Vir 14h7'43" -7d53'58"
Erendira Barraza Cardenas
Crb 16h5'52" 35d10'50"
Eres Mi Estrella Brillante
Aqr 22h41'59" 0d9'15"
ERG22321204
Ori 6h17'16" 16d48'20"
Ergick 8-18-82
Uma 11h8'56" 44d11'24"
e.rhapsody
Cyg 20h12'21" 52d5'1"
Erhard Bruns
Uma 8h17'0" 60d4'30"
Erhard Kaupa
Uma 9h53'16" 44d17'44"
Erhard Lothar Schietzel
Uma 14h10'4" 58d2'22"
Erhardt Wunderlich
Oph 17h31'18" -0d36'44"
Eri
Eri 3h46'36" -13d35'16"
Eric
Cas 0h46'42" 60d31'7"
Eric
Uma 12h15'26" 58d1'41"
Eric
Uma 10h22'53" 52d31'47"
Eric
Dra 17h48'20" 62d1'34"
Eric
Cra 18h6'24" -44d31'50"
Eric
Sco 16h37'5" -29d36'21"
Eric
Aur 6h32'46" 42d59'23"
Eric
Cyg 21h26'46" 36d20'3"
Eric
Her 16h36'44" 37d20'10"
Eric
Cnc 8h14'48" 32d40'10"
Eric
Her 16h21'40" 48d24'3"
Eric
Aqr 22h2'9" 1d41'31"
Eric
Gem 7h31'49" 28d16'1"
Eric A Feinstein
Lib 15h4'42" -16d42'23"
Eric A. Hozubin
Per 4h16'32" 32d47'1"
Eric Aaron Blum
Uma 14h27'43" 55d29'9"
Eric Aberman
Uma 9h27'10" 72d41'56"
Eric Abernathy
Uma 14h23'45" 57d40'29"
Eric Alan Meier
Ori 5h34'6" 10d51'8"
Eric Albert
Dra 19h5'20" 61d35'5"
Eric Albert Houghton
Lyr 19h12'48" 27d21'30"
Eric Allen
Ari 2h2'35" 11d30'33"
Eric Allen Armstrong
Psc 1h13'59" 27d13'41"
Eric Allen Hoover
Uma 9h21'50" 62d31'59"
Eric Allen Pelton
Uma 11h16'33" 64d44'20"
Eric & Anabelle's Star
Lyn 7h57'23" 33d18'25"
Eric and Amber
Cyg 21h36'58" 53d11'3"
Eric and Anthony "10-07-
1982"
Uma 10h22'58" 67d54'6"
Eric and Ashley Yokoyama
Col 5h52'58" -36d14'1"
Eric and Charity Forever
Her 17h3'25" 34d8'25"
Eric and Danielle
And 23h21'16" 47d33'19"
Eric and Deanna Floyd
Pav 19h6'48" -58d20'32"
Eric and Emily Horn
Cyg 20h37'24" 42d0'2"
Eric and Erica's Silver Light
Lyn 7h46'28" 49d43'49"
Eric and Erin Louttit
Cap 21h58'9" -18d38'24"
Eric And Katy
Cas 1h0'31" 61d33'10"
Eric and Kelli
Lyn 7h31'4" 48d53'13"
Eric and Lauren
Cyg 20h22'17" 47d11'15"
Eric and Louise Reeves
Uma 10h13'35" 64d0'10"
Eric and Marcy
Lib 14h52'59" -13d9'29"
Eric and Martina Olson
Psc 23h27'7" 4d52'41"
Eric and Meredith
Uma 11h19'38" 50d30'33"
Eric and Nicky Khan
Cyg 20h31'59" 48d32'31"
Eric and Sage
Cyg 19h18'6" 53d7'54"
Eric & Andrea
And 1h26'31" 50d33'22"
Eric Andrew Neilsen
Aqr 22h29'47" -3d3'44"
Eric Andrus
Sex 10h13'39" -0d15'13"
Eric & Ann's Paradise
Uma 8h37'17" 66d19'15"
Eric Anthony Fatigate
Her 18h30'51" 13d19'56"
Eric Anthony & Kelley Anne
Centeno
Cyg 19h35'18" 37d55'40"
Eric Anthony Mayfield
Aur 5h21'47" 33d1'28"
Eric Anthony & William
Howard Byron
Lib 15h34'57" -28d5'23"
Eric Arthur Kuusela
Cyg 19h59'12" 50d43'1"
Eric Ast
Leo 9h40'3" 12d37'9"
Eric B. Stallone aka. GuGu
Psc 1h32'27" 18d38'39"
Eric B. White
Sco 16h5'40" -24d55'23"
Eric Bard Kronner
Per 3h51'9" 32d19'45"
Eric Barnhart
Cmi 7h44'47" 5d20'32"
Eric Beguelin
Uma 11h18'59" 53d59'58"
Eric Bernier
Umi 15h31'35" 75d11'5"
Eric & Betsy's Star
Cyg 19h30'23" 31d58'45"
ERIC BLAKE
And 23h25'16" 47d57'56"
Eric Borgerson
Sgr 18h26'36" -32d32'56"
Eric Borges
Uma 10h26'13" 53d32'43"
Eric Brady a much loved
family man
Uma 13h59'9" 57d13'8"
ERIC BRAMLETT
Cam 4h12'58" 67d0'31"
Eric Bramwell Jackson
Per 3h16'44" 48d3'7"
Eric Brian Borth
Sco 16h46'24" -32d9'48"
Eric Brian Chavez Jr.
Cnc 9h7'23" 24d58'10"
Eric Brian Roberts
Aqr 23h53'40" -3d32'18"
Eric & Brittney
Uma 11h27'21" 40d33'32"
Eric Browne
Aql 19h7'14" -0d58'2"
Eric C. Schwochow
Cep 21h33'53" 64d17'56"
Eric Campbell
Lyr 19h3'24" 39d31'24"
Eric Castells Sanllehí
Sgr 18h14'59" -28d11'47"
Eric Cedergren
Her 17h37'21" 37d23'37"
Eric Chad Reinhard
Uma 11h40'35" 28d32'59"
Eric Champagne
Uma 9h49'36" 64d9'47"
Eric Champiney
Umi 15h2'10" 87d48'10"
Eric Charles Freeman
Leo 9h51'16" 21d56'13"
Eric Charles Hathaway
Cap 20h26'51" -12d26'58"
Eric Cholewa
Sco 17h3'50" -34d50'58"
Eric & Christine's Dreams
Cyg 20h56'39" 45d35'47"
Eric Christopher Chappano
Uma 9h54'1" 45d0'12"
Eric Christopher Lowrie
Tau 4h23'16" 21d52'38"
Eric Clinkenbeard
Per 3h11'13" 52d24'59"
Eric Clory
Aur 5h14'47" 41d47'9"
Eric Cody Risher
Umi 14h32'29" 70d3'56"
Eric Cooper Little
Cap 20h22'35" -24d44'32"
Eric Crump SVT
Psc 1h46'36" 13d33'16"
Eric D. Kingston
Tau 4h18'45" 29d7'5"
Eric D. Plourde
Dra 15h47'46" 53d41'54"
Eric D. Schurk
Her 17h14'30" 29d10'57"
Eric Dakota Ruegsegger
Cep 19h13" 86d39'15"
Eric Daniel Altman
Sco 16h21'51" -32d4'14"
Eric Danielsson
Sgr 17h53'0" -21d45'50"
Eric Daucher
Cma 6h33'35" -14d54'28"
Eric Daudelin
Sgr 18h47'22" -16d55'55"
Eric David
Per 3h7'36" 55d3'56"
Eric David Cioffi
Uma 9h27'7" 46d37'59"
Eric David Grabow
Per 3h45'59" 46d37'8"
Eric David McMillian
Oph 17h26'27" 8d0'10"
Eric David Trent
Uma 11h30'59" 40d55'51"
Eric Dawson
Leo 9h37'17" 28d42'28"
Eric Demmons
Vir 14h8'44" -16d31'56"
Eric Donavan Route
Per 3h32'58" 34d27'52"
Eric Donovan Gallardo
Aur 5h13'29" 29d37'54"
Eric Duncan Tollefson
Dra 17h8'15" 64d21'11"
Eric Enriquez
Lyn 6h17'35" 58d52'21"
Eric & Erika 7th grade til
forever
Umi 13h45'47" 76d31'21"
Eric E.S. Brown No.1 Dad
& Husband
Sco 17h21'32" -42d29'57"
Eric Escher
Aur 5h22'2" 31d34'35"
Eric Esteves
Lib 14h22'38" -9d47'1"
Eric et Caroline
Uma 11h52'18" 57d36'56"
Eric et Sylvia
Sgr 18h59'17" -17d37'41"
Eric "Evan" Carasco
Hernandez
Her 17h51'0" 15d13'6"
Eric & Evelyn
Cyg 19h53'16" 33d13'41"
Eric Everett Head
Her 16h56'52" 33d49'31"
Eric F Keita
Tau 5h48'12" 26d22'11"
Eric Feinberg
Her 17h25'6" 39d4'48"
Eric Fleming
Her 16h25'20" 12d48'33"
Eric (Fluffy) Applegate
Leo 11h43'20" 23d9'49"
Eric Forrest
Lmi 10h55'56" 26d18'11"
Eric Fournier
Cep 22h50'2" 79d19'45"
Eric Francis
Lib 14h56'51" -17d26'45"
Eric Franklin
Her 16h34'25" 11d22'26"
Eric Freeman
Sgr 18h43'19" -30d6'33"
Eric Fuller
Leo 11h28'24" 21d53'31"
Eric G. and Kathryn M.
Johnson
Uma 11h9'58" 32d54'52"
Eric G. Mounce
Per 3h51'47" 37d9'29"
Eric Gaedtke
Uma 9h9'46" 64d19'0"
Eric Gagnon & Sarah
Swanson
Cyg 19h31'59" 54d26'29"
Eric Gary Travioli
Cap 21h50'39" -18d20'30"
Eric Gates
Lib 15h19'10" -16d27'4"
Eric Gelman
Cnc 8h47'58" 18d48'46"
Eric Germain
Aqr 23h45'39" -11d28'21"
Eric Gibbons
Leo 11h14'11" -1d17'44"
Eric Goldstein
Per 2h40'1" 56d21'20"
Eric Gray
Lib 15h28'3" -26d0'2"
Eric Guilliams
Sco 17h42'38" -37d58'35"
Eric Havely
Her 16h41'5" 34d32'16"
Eric&Heather4Ever
Cyg 21h20'16" 38d10'27"
Eric Hernandez
Oph 17h32'59" -3d27'53"
Eric & Hilary - Boogums
Star
Uma 8h44'40" 61d4'23"
Eric Hirschi
Lyn 7h53'46" 34d3'11"
Eric Hoin
Lib 15h30'9" -6d23'16"
Eric I. Holtzman
Ori 6h14'1" -1d28'59"
Eric is in Love with
Stephanie Boyd
Uma 9h47'8" 51d45'41"
Eric Ivie
Cet 2h43'29" -0d25'20"
Eric J. Schmertz
Per 3h42'42" 51d11'40"
Eric J Seidel
Ori 5h6'43" 4d15'40"
Eric J. Van Blunk
Psc 0h40'21" 21d5'53"
Eric Jacob Schmid
Leo 9h36'52" 32d55'36"
Eric & Jacqueline
McGourty
Cyg 20h11'40" 51d22'48"
Eric James
Psc 1h23'20" 33d3'38"
Eric James Adair
Gem 6h46'16" 23d7'33"
Eric James Bear Trepepi
Cyg 19h55'43" 35d49'50"
Eric James Compton 1-9-
76
Cap 20h53'21" -21d1'39"
Eric James Hallagan
Sct 18h50'28" -9d41'42"
Eric James McCoy
Gem 6h27'15" 26d48'0"
Eric James Ortiz
Ori 5h20'45" 3d37'53"
Eric James Tobin
Uma 13h36'21" 56d9'59"
Eric James Warriner
Ari 3h9'24" 29d53'23"
Eric James Woloszyn
Vir 13h22'5" 11d42'57"
Eric Jamison
Her 16h31'55" 47d35'11"
Eric Jason Gonzales
Her 18h49'51" 15d56'7"
Eric Jemique Royster
Lyn 8h31'58" 53d39'26"
Eric & Jessica Ottnod
Forever
Uma 11h9'20" 47d48'29"
Eric & Jody
Ori 5h3'54" 2d35'44"
Eric John
Her 17h7'30" 32d18'52"
Eric John de Oude
Cyg 19h37'22" 44d11'42"
Eric John Edeen
Umi 14h43'6" 72d8'36"
Eric John Ellenbogen
Her 17h19'23" 36d52'21"
Eric John McDevitt
Her 17h41'32" 47d28'25"
Eric John Ramirez
Leo 10h21'37" 6d39'4"
Eric John Zielinski
Dra 16h17'21" 56d37'18"
Eric Johnson
Ori 6h5'18" 19d33'55"
Eric Jordan Chastain
Cnc 8h31'56" 14d37'15"
Eric Joseph
Sco 16h50'5" -44d14'41"
Eric Joseph Biordi
Uma 10h2'58" 52d8'37"
Eric Joseph Castillo
Cnc 8h45'5" 29d15'27"
Eric Joseph Haner
Per 3h33'4" 41d23'58"
Eric K. Sullivan
Aql 19h10'41" 8d14'26"
Eric K. Tofty, My Shining
Star
Aqr 20h51'37" -11d23'54"
Eric Kabakoff
Cnc 8h14'20" 23d5'7"
Eric Kaijankoski
Uma 11h14'28" 48d31'46"
Eric Kedzielawa
Ari 3h22'57" 19d34'34"
Eric Keith Butler
Peg 21h44'16" 27d46'10"
Eric Kelly
Gem 7h44'27" 25d3'2"
Eric Kemal Elliott
Cnc 8h42'50" 23d56'51"
Eric Kent Rekdahl
Cnc 9h3'12" 21d28'5"
Eric Knapp
Ori 5h54'34" 8d52'0"
Eric Kurt Viljanen
Ari 2h54'23" 21d57'34"
Eric L. Dubuc
Gem 7h39'26" 32d22'33"
Eric L. Hightower
Per 4h49'30" 50d30'7"
Eric L. King
Her 18h16'10" 27d33'1"
Eric L. Wruck and Amy
Psc 1h9'43" 32d18'18"
Eric Lacueil
Cnc 8h23'54" 17d56'6"
Eric Laferrière
Psc 1h53'53" 6d6'39"
Eric Lapointe
Uma 10h34'16" 68d30'23"
Eric Lawrence
Cyg 20h50'38" 42d7'15"
Eric Lee
Uma 8h44'32" 50d53'43"
Eric Leigh & Dayna Marie
O'Canna
Uma 13h52'10" 55d3'41"
Eric Leigh Marquis
Per 3h14'18" 47d12'39"
Eric Leon DuBose III
Uma 11h9'29" 45d7'12"
Eric LeRiche
Lac 22h31'24" 41d19'20"
Eric Levy
And 23h14'1" 51d26'34"
Eric Lewandowski
Lyn 7h21'9" 57d15'11"
Eric Lim
Uma 10h23'3" 44d1'58"
Eric & Lisa
Lib 14h51'18" -7d22'57"
Eric & Lisa Sorge
Tau 4h29'56" 16d54'51"
Eric & Lizzi Neighbors
Forever
And 0h8'34" 45d48'25"
Eric & Lori Fendelander
Cyg 20h37'49" 37d48'1"
Eric Louisseize (Coco)
Cep 23h27'37" 77d52'15"
ERIC LOVES LACEY
And 23h24'45" 41d53'21"
Eric M. Davis
Aql 20h5'28" 12d6'58"
Eric M. Dolan
Per 4h4'29" 32d45'47"
Eric Mantos
Vir 12h56'46" 6d1'30"
Eric & Maria Pfeiffer
Ori 5h10'34" 9d15'42"
Eric Mark Hoener
Ori 5h59'43" -0d46'4"
Eric Mark Vicario
Uma 10h11'24" 65d28'16"
Eric Martinez
Cnc 8h50'18" 14d57'11"
Eric Matthew Boyd
Her 17h19'55" 32d39'36"
Eric Matthew Weissman
Sgr 18h23'43" -23d56'49"
Eric Maust
Uma 8h46'18" 50d39'9"
Eric McLamb
Aur 5h15'40" 28d43'37"
Eric Medlen
Leo 11h37'27" 12d5'19"
Eric & Melissa's Star
Cyg 19h41'54" 37d44'36"
Eric Michael
Cap 20h28'51" -12d43'41"
Eric Michael Flint, M.D.
Vir 13h9'37" -22d25'49"
Eric Michael Gonzales
Gem 7h21'13" 15d6'29"
Eric Michael Oeth
Cap 20h22'40" -15d10'47"
Eric Michael Perrin
Uma 10h48'57" 60d42'27"
Eric Michael Sarrafian
Vir 12h33'42" 11d44'49"
Eric Michael Scavnicky
Ori 6h17'32" 11d1'2"
Eric Michael Wagner
Lep 5h21'13" -11d59'42"
Eric Michael Wong
Uma 9h33'38" 44d24'2"
Eric Michon
Cnc 8h8'34" 22d7'16"
Eric Miller
Sco 16h50'11" -38d16'41"
Eric Morson
Per 2h28'43" 54d20'54"
Eric "My Love" Geisert
Cyg 20h0'2" 35d0'39"
Eric Nehemiah Fondren
Cnc 8h34'12" 24d0'4"
Eric Nelson
Uma 11h23'14" 44d2'13"
Eric Nelson
Boucher/Faithful
Uma 10h10'45" 50d23'29"
Eric Nerison
Sco 16h27'13" -25d16'15"
Eric Nielsen
Gem 6h50'32" 34d2'57"
Eric "Owen" Kerney
Sgr 18h33'14" -17d25'24"
Eric P. Loveless
Gem 6h46'41" 12d40'19"
Eric Palmiter
Gem 6h25'25" 19d45'59"
Eric (Papa) MacDonald
Cap 21h38'31" -13d55'32"
Eric Paul Best
Leo 9h30'42" 28d43'29"
Eric Pearce Johnson
Psc 23h14'25" 6d22'20"
Eric Peter Hilstrom
Lac 22h50'4" 52d8'41"
Eric Pfeiffer
Tau 4h34'36" 24d34'30"
Eric Phan
And 0h37'5" 29d0'28"
Eric Philip Buettner
Her 16h41'32" 7d33'22"
Eric Polis
Per 3h42'47" 48d26'0"
Eric R. Blumeno
Gem 6h27'1" 27d41'24"
Eric R. Manuel
Dra 17h45'24" 53d54'40"
Eric R Ostendorf
Uma 11h32'34" 33d21'18"
Eric R. Sax
Cnc 8h32'27" 17d54'58"
Eric R. Theil "E.T." ~ Oct.
6, 1964
Lib 14h50'11" -2d5'30"
Eric Reid Heinzman
Her 16h43'37" 41d4'25"
Eric Ricardo Mapp
Uma 10h28'53" 46d10'51"
Eric Richard Levitt
Per 2h24'1" 54d8'15"
Eric Richard St.Germain
Sco 16h4'53" -14d21'46"
Eric Robert Sears
Ori 5h43'2" 8d6'15"
Eric Robert Vandezande
Gem 7h39'6" 16d34'17"
Eric Robles
Psc 1h2'27" 28d58'10"
Eric Rosenqvist
Gem 7h37'6" 18d53'41"
Eric Ryan Closson
Gem 7h21'35" 34d10'23"
Eric Ryan Otto
Sgr 18h39'45" -25d22'44"
Eric S. Holzberg - YRMW
Sgr 17h58'28" -27d41'46"
Eric S. Mundt
Vir 13h54'20" -10d47'8"
Eric Salley
Lyn 8h26'11" 35d24'58"
Eric Schatzlein
Aur 5h14'59" 42d28'49"
Eric Scott
Her 17h26'24" 47d50'12"
Eric Scott Coomer
Sco 16h1'39" -20d41'43"
Eric Scott Dunn
Per 2h45'50" 52d38'21"
Eric Scott Helgeson
Uma 10h40'4" 62d15'6"
Eric Scott Kotora
Her 18h39'32" 20d15'22"
Eric Scott Russell
Uma 10h5'34" 42d45'11"
Eric Scott Williamson
Cnc 8h14'2" 22d30'32"
Eric Scott Windle
Ari 2h37'22" 30d21'48"
Eric Scott Young
Ari 3h25'6" 27d24'18"
Eric Shannon Moulds
Cnc 8h38'56" 17d21'39"
Eric Shawe
Sge 20h4'23" 19d32'32"
Eric & Shawn
Ori 6h16'1" 14d14'50"
Eric Sligar
Lyn 8h28'42" 56d30'55"
Eric Stephen
Ori 6h3'46" 19d55'45"
Eric Steven Benda
Gem 6h53'52" 26d42'46"
Eric Steven Nielsen
Sco 16h7'22" -13d30'21"
Eric Stremmel
Sgr 19h50'3" -12d19'27"
Eric & Suzie Hernes
Cyg 19h45'20" 39d5'39"
Eric Sylvester
Cas 23h1'44" 55d8'53"
Eric Sylvestre
Per 2h54'24" 53d35'20"
Eric & Tanell Nordling
Cyg 21h26'59" 46d12'56"
Eric Tanguay & Claudine
Rivest
Cyg 19h59'59" 44d39'1"
Eric Thoman
Boo 14h52'7" 52d0'31"
Eric Thomas Johnson
Cep 22h14'35" 66d32'27"
Eric Thomas Zvaniga
Cep 23h51'29" 73d58'54"
Eric Thrash "ET"
Her 16h15'23" 12d33'39"
Eric Townsend
Sco 16h42'22" -36d57'42"
Eric Trujillo
Sco 16h9'33" -10d17'33"
Eric U Will Always Be My
RisingSTAR
Vir 13h26'7" -4d54'29"
Eric Valaika
Her 16h10'41" 47d10'48"
Eric Valdez
Sgr 19h11'54" -26d51'29"
Eric & Vanessa DeMarcus
Cyg 20h48'2" 43d13'31"
Eric VDP
Vir 14h10'58" 1d30'12"
Eric & Veronika Friedberg
Cyg 20h43'1" 34d44'40"
Eric Vestfal
Cyg 20h0'42" 43d3'2"
Eric & Vivienne
Dra 17h36'18" 68d11'55"
Eric Volpe
Per 2h45'50" 54d28'16"
Eric W. Bergmann
Gem 7h5'2" 18d11'30"
Eric W. King
Per 3h49'49" 50d47'44"
Eric W. Wallen
Dra 17h21'24" 69d9'14"
Eric W. Walters
Lyn 8h13'30" 47d4'58"
Eric Wabey
Tau 5h4'58" 23d13'22"
Eric Wan & Ann Wong
And 1h19'55" 35d21'18"
Eric Wayne Hayden
Uma 12h16'30" 60d43'7"
Eric Wayne Watts
Cap 20h36'42" -11d51'28"
Eric Wayne Watts
Her 17h35'25" 35d21'23"
Eric & Wendy
Cyg 20h16'11" 38d4'6"
Eric Westphal
Sgr 18h14'31" -19d25'28"
Eric Wetzel
Cnc 8h11'13" 15d23'47"
Eric White
Ari 2h2'5" 19d40'37"
Eric "Whoadie" Patterson
Ori 5h57'2" 6d15'19"
Eric William and Rachael
Lauren
Cyg 21h42'11" 54d25'40"
Eric William Letz
Per 3h47'31" 33d30'54"
Eric William Snyder
Vir 13h17'5" 8d51'53"
Eric William Walker
Cnc 8h12'35" 11d56'58"
Eric William Walker
Per 4h13'35" 49d52'54"
Eric Wilson
Leo 10h31'24" 26d36'17"
Eric Winholt
Lib 14h32'4" -19d9'1"
Eric Winkelman
Boo 15h34'24" 41d31'25"
Eric Wolf & The Wolf Pack
2002-2005
Cma 6h35'34" -30d38'39"
Eric Yanni
Uma 13h36'43" 59d55'3"

Eric Yule & John Netland Cyg 19h43'46" 28d30'40"
Eric, Bradley, Kendra & John Shafer Uma 11h8'22" 59d2'57"
Erica Uma 11h30'17" 59d10'18"
Erica Uma 10h12'57" 59d41'49"
Erica Uma 10h49'32" 64d19'57"
Erica Cam 4h33'23" 61d3'51"
Erica Umi 14h10'44" 68d44'54"
Erica Vir 13h54'52" -18d13'43"
Erica Sco 16h9'26" -10d7'24"
Erica Lib 15h9'43" -6d28'4"
Erica Aqr 22h13'12" -1d19'9"
Erica Umi 17h11'17" 83d13'18"
Erica Pho 23h55'25" -40d7'23"
Erica Per 3h24'2" 41d33'24"
Erica And 0h33'48" 30d58'21"
Erica Ari 3h24'27" 26d30'42"
Erica Tau 5h35'9" 26d59'10"
Erica Her 18h18'24" 15d13'0"
Erica Leo 10h30'47" 12d23'14"
Erica Vir 15h7'9" 4d31'3"
Erica Aur 5h38'58" 32d59'39"
Erica Cnc 9h2'14" 31d39'4"
Erica Cnc 8h53'29" 31d28'10"
Erica Cnc 8h44'13" 32d35'12"
Erica Her 16h55'38" 32d19'4"
Erica Lmi 10h32'6" 34d25'36"
Erica Cyg 20h52'47" 35d17'56"
Erica And 0h14'26" 42d58'20"
Erica A. Canty Leo 11h51'37" 22d52'23"
Erica A Decker & My Baby Spook Dra 18h33'34" 68d40'21"
Erica A. Kline Leo 9h42'23" 24d58'18"
Erica A Soto Tau 4h24'50" 21d14'24"
Erica A. Weitzel And 0h23'43" 40d58'41"
Erica Abney Hardy "The Red Lion" Leo 10h21'40" 11d48'43"
Erica Adam's Cnc 9h8'2" 32d19'51"
Erica Amanda Cartmill Uma 8h55'37" 53d14'45"
Erica Anaya Cyg 20h44'30" 37d49'50"
Erica and Braeden Uma 9h48'9" 56d58'4"
Erica and Grace Ari 3h5'51" 27d40'4"
Erica Ann And 0h18'16" 43d19'13"
Erica Ann Clark-Maglieri Mon 6h56'1" -0d7'47"
Erica Ann McCollam Cyg 21h1'19" 46d41'13"
Erica Ann Whitford Sco 16h12'57" -17d7'45"
Erica Ann Wyatt Lib 15h20'48" -9d48'3"
Erica Ashley Soma Cnc 8h14'3" 9d13'43"
Erica Baby Leo 10h35'43" 12d45'49"
Erica Barkshire Cyg 19h38'50" 38d1'58"
Erica Camille Sgr 17h56'21" -27d8'5"
Erica Candice Peterson Lyn 7h50'5" 37d32'25"
Erica Carol Sue Dunbar Cas 0h46'29" 56d59'27"
Erica Casarez And 2h16'22" 49d7'59"
Erica Chandris Lib 15h44'48" -19d22'21"
Erica + Chris = Forever Uma 8h45'50" 67d54'13"
Erica Cricket Carr Lib 15h15'17" -9d48'2"
Erica Cristini Gem 6h27'6" 17d34'9"
Erica D Rigopoulos And 0h10'56" 44d54'55"
Erica Danielle Cyg 19h40'19" 31d16'53"
Erica Danielle Cap 21h40'0" -17d1'0"
Erica Danielle Dra 19h48'24" 59d52'25"
Erica Danielle Pearson And 23h9'5" 42d33'1"
Erica Dawn Leo 9h27'42" 25d49'13"
Erica Deanne Phillips Lib 15h42'31" -28d42'28"
Erica Debellis Aqr 22h50'20" -6d28'2"
Erica Deutchman Lib 15h0'14" -22d46'47"
Erica Diane Kirkland Ari 2h36'19" 18d38'7"
Erica Elana Crb 15h26'43" 29d4'15"
Erica Elizabeth Thoreson Lib 14h22'25" -9d39'41"
Erica F. Austin And 0h31'33" 38d40'20"
Erica Ford Crb 15h46'14" 38d8'40"
Erica Garcia Leo 11h7'56" 5d55'39"
Erica Gardeski And 2h15'14" 47d45'42"
Erica Geoffroy Gem 6h58'6" 16d4'22"
Erica Golchin's Star Lyn 8h35'24" 35d17'17"
Erica Goodrich And 1h43'38" 46d26'32"
Erica & Greg Lyn 7h21'38" 49d22'19"
Erica Gregorio & Janzen Rivera Cyg 21h22'13" 45d17'26"
Erica & Hayden's Naming Day Star Leo 9h53'21" 13d49'48"
Erica Held Uma 10h23'57" 51d59'36"
Erica Hennessy Cnc 8h39'38" 32d38'30"
Erica Hernandez Ori 6h17'59" 13d43'46"
Erica Hyche Vir 12h6'29" 2d21'55"
Erica is the Coolest! Cnc 8h16'24" 20d50'42"
Erica J Crescenzo Lib 15h15'33" -14d12'30"
Erica Jane Bohn Lyn 7h42'53" 42d11'20"
Erica Jantos Lmi 10h33'6" 38d58'29"
Erica Jean And 0h42'29" 30d32'48"
Erica Jean Gem 6h29'48" 25d34'36"
Erica Jean Mazaika Lib 14h50'45" -1d43'52"
Erica Jean Victoria Miller Cas 0h45'7" 66d3'41"
Erica & Jereff's Star Uma 8h36'23" 55d50'19"
Erica Joelene Uma 12h36'3" 59d53'29"
Erica & John Gunning Ori 6h10'12" 18d59'41"
Erica Johnson And 0h42'27" 44d4'52"
Erica Kathryn Hess Peg 22h39'48" 32d59'5"
Erica Kathryn Scowley And 0h45'47" 43d23'12"
Erica Kim Lyn 6h54'21" 58d56'50"
Erica Korolik And 23h48'53" 46d4'56"
Erica Kozma Crb 15h48'54" 33d29'17"
Erica Kristin Alhart's Perfect Star Cma 7h9'0" -31d23'53"
Erica L. Montanez Vir 12h30'14" -9d34'7"
Erica L. Walker And 23h13'0" 48d21'48"
Erica Laird Tau 4h5'19" 16d34'7"
Erica Larrea Per 2h19'49" 54d29'48"
Erica Lauren Acuna Tau 3h38'36" 6d20'45"
Erica Lea Langhorst Cyg 20h24'21" 59d33'19"
Erica Leahy Irish Angel Gem 6h58'29" 17d46'20"
Erica Leann Showler Mon 7h31'42" -1d13'45"
Erica Lee Cas 2h25'59" 63d9'35"
Erica Lee And 1h11'39" 37d5'58"
Erica Leigh Ari 2h33'56" 18d45'48"
Erica Leigh Lib 15h24'42" -19d48'43"
Erica Leigh Hoban And 0h50'16" 39d32'41"
Erica Leigh Stewart And 23h9'51" 49d40'45"
Erica Leigh Stopper Vir 13h25'17" 12d45'57"
Erica Leigh Thibeault Aqr 20h46'18" -7d1'58"
Erica Leighty Super Star Lib 15h36'15" -9d23'39"
Erica Lindsay Pho 23h43'18" -55d7'50"
Erica Louise Lanier And 1h20'42" 39d34'45"
Erica Lyn Cahill Cas 0h33'47" 63d28'23"
Erica Lynn Uma 10h7'0" 66d27'28"
Erica Lynn Cap 20h15'19" -13d15'43"
Erica Lynn Aqr 23h3'34" -7d36'33"
Erica Lynn Anderson A.K.A. Missy And 23h24'10" 48d21'9"
Erica Lynn Bell Uma 11h28'5" 35d32'43"
Erica Lynn Terran Lib 15h10'53" -14d29'16"
Erica Lynn Woestman Cap 21h44'26" -10d59'44"
Erica M. Graham Cnc 8h54'57" 30d58'34"
Erica M. Spitulski And 0h42'49" 37d6'28"
Erica Maria Gomez Corona Crb 15h44'55" 35d20'31"
Erica Marie Anderson Vir 12h42'18" 6d42'41"
Erica Marie Garza Gem 7h37'11" 31d50'36"
Erica Marie Kling Sco 16h56'55" -32d4'48"
Erica Marie Taylor And 2h13'15" 45d54'35"
Erica Marie Thompson Cas 1h36'13" 65d27'11"
Erica Marlene Rydzewski Vir 12h44'51" 8d3'55"
Erica Marnell Ori 6h24'12" 16d54'16"
Erica Maurer Dra 19h37'7" 76d25'24"
Erica McLaughlin And 0h47'11" 41d13'29"
Erica McNees And 0h31'33" 42d26'28"
Erica Melissa Alonso And 2h5'9" 39d0'15"
Erica Metzger Lyr 19h24'44" 31d47'38"
Erica Michelle Aguilar Tau 3h58'42" 20d59'21"
Erica Moore And 23h59'28" 45d11'3"
Erica Mortensen Crb 15h57'16" 34d9'31"
Erica Mother of Linc, Tria & Kenya Cas 0h0'5" 56d37'15"
Erica Munn And 23h28'11" 38d25'24"
Erica Nicole Davis Tau 4h21'52" 17d45'33"
Erica Nicole Hay Lib 14h50'50" -11d58'43"
Erica Nicole Sara Deutch Psc 23h48'39" 1d1'5"
Erica Paulina Schindelheim Umi 14h49'1" 71d4'10"
Erica Pauline Cru 12h38'20" -60d29'59"
Erica PEG Forever Cas 2h13'7" 64d21'30"
Erica Petrie Cnc 9h13'8" 9d27'45"
Erica Pullum Cas 0h7'23" 58d44'43"
Erica R. Ayala Leo 10h8'46" 24d28'11"
Erica R. Grande And 23h23'15" 41d37'33"
Erica R. LaMantia Lib 14h51'52" -1d36'33"
Erica R. Michael Leo 10h24'29" 23d50'21"
Erica Rachael Psc 23h18'26" 1d35'30"
Erica Rae Sadler and Kevin S. Fiur Cyg 20h19'29" 45d35'34"
Erica Raelynn Nein Cas 0h39'31" 61d48'41"
Erica Raquel Shaw Ori 5h57'45" 22d5'43"
Erica Regan Sco 16h12'44" -10d8'27"
Erica Renee Mullins Crb 15h31'2" 37d29'20"
Erica Rose Coronado Psc 1h41'45" 24d45'46"
Erica Rose Martinez Sgr 19h56'13" -26d7'48"
Erica Rovetta Her 16h52'0" 27d49'8"
Erica Ruta Cyg 19h59'12" 45d44'7"
Erica Santos Santarelli And 0h55'47" 43d59'2"
Erica Sarah Swillum Tau 5h11'59" 23d51'9"
Erica Saree Chavez And 1h52'9" 41d38'26"
Erica Scarlett Steinhouse - Sweetpea Uma 11h38'43" 51d48'32"
Erica Seel Ari 2h17'23" 23d27'6"
Erica Sepeda Lmi 10h30'23" 34d53'52"
Erica Shaye Nagy's Sunshine Star Del 20h47'50" 19d26'8"
Erica Shelley Lyn 7h46'57" 57d16'46"
Erica Sillings Cyg 21h25'35" 39d25'58"
Erica Starks And 1h10'53" 42d52'49"
Erica Suzanne Backman Cam 4h28'11" 66d42'52"
Erica Suzanne Kahn Vir 13h30'40" -1d25'47"
Erica Theresa Georgeo Sco 16h58'7" -34d37'16"
Erica & Tom Lyr 18h58'17" 26d30'20"
Erica Tucker Mon 7h21'39" -0d18'26"
Erica V. Morton Aqr 20h57'43" 1d6'5"
Erica V. Shur Leo 11h35'21" 24d31'22"
Erica Vasquez Uma 11h58'15" 51d18'41"
Erica Williams And 23h0'47" 48d16'58"
Erica Wolfe Lib 15h33'40" -4d25'1"
Erica Zanotti Psc 1h22'57" 22d28'6"
Erica Zuniga Hya 9h14'58" -0d56'24"
Erica, My Love And 1h12'2" 45d48'36"
EricaGonzalez&NicholasDamian4/Ever Cnc 8h45'15" 18d49'10"
EricaJason Cyg 21h52'52" 46d51'34"
Erica...la stella nanetta Peg 0h2'54" 26d15'37"
Erica's Dreams Umi 14h49'33" 72d15'43"
Erica's Smile Cam 5h51'12" 65d28'3"
Erica's Star Sco 16h15'44" -10d42'54"
Erica's Star Lib 15h43'21" -17d27'32"
Erica's Star Lib 15h56'24" -18d18'14"
Erica's Star Vir 13h27'21" 11d59'56"
Erica's Star Psc 0h51'31" 11d19'30"
Erica's Star Tau 3h40'42" 12d56'53"
Erica's Star Lyr 19h23'54" 39d5'38"
Erica's Star Aur 5h57'22" 36d51'36"
EricCourt 9378 Vir 13h10'38" 6d17'27"
Ericdreammaker Tau 4h37'21" 28d32'52"
Erich Ori 4h59'15" 2d35'37"
Erich And 23h25'19" 48d7'46"
Erich Alfred Sge 19h52'10" 18d38'0"
Erich Conrad Cap 21h55'28" -18d12'44"
Erich Daniel Drumm Lyn 7h41'30" 41d50'59"
Erich & Hayley Neubauer - Infinity Cru 12h48'0" -63d48'46"
Erich Köster Uma 8h46'58" 63d26'57"
Erich Leumann Umi 14h54'54" 68d26'28"
Erich Preiser Ori 5h48'38" 11d51'6"
Erich Spirgi Lac 21h59'23" 39d32'59"
Erich Vincent Malone Aqr 22h28'33" -0d56'57"
Erich Wade Schacherl Cnc 9h5'53" 13d30'10"
Erich William Sneideraitis Uma 10h38'5" 44d19'48"
Erichthonius Wendler Del 20h39'15" 15d56'9"
Erick Cnc 8h40'14" 11d21'46"
Erick Anderson January 3, 1993 Lmi 10h14'52" 31d25'59"
Erick Brandon Palanker Uma 9h11'3" 64d58'12"
Erick Eduardo Lara De Leon Cap 21h6'13" -19d50'13"
Erick Frederick (Fritz) Taenzer Leo 9h34'29" 27d51'22"
Erick Grinham Her 17h49'58" 44d18'2"
Erick*I*Love*U Her 16h53'36" 47d29'23"
Erick Lee Ericksen Cnc 9h18'21" 10d28'42"
Erick Lopez Sct 18h29'26" -6d4'37"
Erick "Rainman" Romero Her 16h16'42" 17d51'54"
Erick & Tracy McDaniel Vir 13h18'36" -21d48'21"
Erick William Inman Ari 2h33'49" 25d25'15"
Erick y Ariana Sco 16h15'12" -15d4'21"
Erick Z. Westphal Per 2h57'40" 42d27'59"
Erick1112006 Sco 17h55'3" -39d55'24"
Ericka Cam 3h45'37" 57d39'14"
Ericka Beatrice Lyn 8h35'15" 44d50'49"
Ericka Blake Fehl Lyn 6h24'51" 57d27'23"
Ericka Briscoe Cnc 7h57'26" 13d32'55"
Ericka Chea Mon 6h51'0" 8d18'18"
Ericka DeLaine Gem 7h38'22" 24d2'35"
Ericka Kristeena Martin And 1h15'58" 42d16'36"
Ericka L Crocker Cnv 12h50'22" 43d52'13"
Ericka Lee Emling Cyg 20h12'13" 35d45'20"
Ericka Leigh Lyr 18h30'1" 28d56'28"
Ericka Leigh Aqr 21h41'50" -3d37'36"
Ericka Marie Gem 7h22'48" 32d5'50"
Ericka Mendoza Cas 1h41'21" 64d11'30"
Ericka Nicole Larson Lyn 6h59'4" 51d13'37"
Ericka Paz Cas 2h5'27" 64d50'53"
Ericka Perez Peg 22h54'18" 19d53'2"
Ericka Ramos Cas 1h41'33" 62d11'31"
Ericka Selene Cnc 8h31'48" 10d46'42"
Ericker Lmi 9h54'13" 36d14'58"
Ericker Camarillo Sgr 18h27'3" -26d46'56"
Eric's Angel Gem 7h12'29" 19d36'36"
Eric's Guiding Light Per 3h17'55" 51d46'47"
Eric's Place Psc 1h8'2" 21d6'50"
Eric's star Tau 4h46'46" 20d58'51"
Eric's Star Her 16h34'44" 12d51'17"
Eric's Star Wars Galaxy Lib 15h13'5" -7d29'13"
Ericus Pattonius Uma 11h23'47" 71d39'51"
EricV Cas 0h0'9" 58d18'48"
Erie Gem 7h46'26" 16d49'7"
Erie Sgr 18h43'4" -20d40'8"
Eriel Tau 4h16'4" 26d59'15"
Eriel and Shea's Jacey girl Tau 5h21'19" 25d30'5"
Erik Boo 14h29'29" 27d35'44"
Erik Ori 5h53'6" 21d35'31"
Erik Tau 4h10'59" 19d38'31"
Erik Aqr 22h41'38" 1d6'21"
Erik Sgr 18h28'39" -24d42'25"
Erik and Erica Duane Mon 6h51'21" -0d14'57"
Erik and Melissa Siekmann Ari 2h21'45" 24d56'19"
Erik Burrito-Burritz Flores Gem 6h21'41" 22d34'0"
Erik C. Rhebb Cnc 8h11'35" 32d18'30"
Erik C. Shuttleworth Uma 10h55'59" 45d21'2"
Erik & Chrissey, January 9, 2000 Aql 20h4'18" -0d24'39"
Erik Christian Krauss Sgr 18h15'11" -18d56'35"
Erik Christian Triebenbacher Per 3h37'52" 38d29'51"
Erik Christian Wirtner Lib 15h39'53" -10d0'32"
Erik Crespo Gem 6h2'28" 22d24'7"
Erik Dean Palmer Sharman Her 17h47'43" 38d1'54"
Erik Denning Aur 7h18'5" 41d17'11"
Erik DiNardo Aqr 22h38'33" 1d43'57"
Erik Duris Her 16h25'3" 9d37'23"
Erik et Laysa And 0h30'30" 31d58'35"
Erik Fleury Ori 5h42'6" -1d53'41"
Erik Francis Dornbush April 29,1993 Tau 5h21'46" 21d24'7"
Erik Frithjof Johnsen Cep 21h58'4" 84d33'0"
Erik Hansen Uma 13h43'49" 53d50'47"
Erik Helleskov 4447 Boo 14h9'22" 27d14'1"
Erik Herbster Sco 17h33'35" -36d8'46"
Erik J. Eisenhauer Uma 12h49'54" 62d52'52"
Erik J. & Inna Swenson Gem 7h46'24" 33d29'21"
Erik J Legath Aqr 20h52'4" -8d33'2"
Erik Jacob Wells Ori 5h25'44" 14d28'1"
Erik + Jenn Cyg 20h24'51" 59d29'36"
Erik K. Cote Pho 0h41'49" -41d49'42"
Erik Karl Umi 15h44'52" 71d31'34"
Erik & Katie's Wedding Star Uma 12h5'42" 52d59'25"
Erik Krogen Aql 19h37'36" -7d21'2"
Erik Ludlow Lib 14h47'7" -3d33'16"
Erik MacPherson Sgr 19h45'58" -11d46'4"
Erik Martin Currie Aql 19h14'40" -0d3'21"
Erik Michael Ori 6h4'59" 6d52'11"
Erik Micheal Cunningham Boo 14h41'59" 28d56'53"
Erik & Michele 100105 Uma 10h27'42" 40d50'59"
Erik Newhook Gem 6h47'53" 34d7'31"
Erik Parsons Plantiff Lib 15h41'35" -20d32'49"
Erik Polyniak Sco 17h53'36" -34d16'53"
Erik R Striegel Lyn 6h37'37" 56d30'47"
Erik Ragnar Spellerberg Boo 14h53'57" 35d12'37"
Erik Robertson Sco 16h42'48" -43d51'16"
Erik S. McCrae Ori 6h15'58" 11d9'40"
Erik & Sara Forever Love Lib 15h23'12" -8d34'2"
Erik Steven Bennett Mon 6h46'24" -4d55'6"
Erik Swetech Cep 22h23'49" 69d31'27"
Erik Szmania Vir 13h33'27" 6d29'26"
Erik "The Hurricane" Marshall Cep 22h18'50" 60d36'57"
Erik und Lars Hennies Tau 5h14'14" 16d50'12"
Erik Unneland Pagenhart Uma 11h22'16" 31d41'19"
Erik Vaag Her 17h0'10" 24d21'55"
Erik Videbeck And 0h23'23" 26d35'41"
Erik W. Wyche Boo 14h28'18" 22d22'39"
Erik Walker Psc 1h22'4" 23d26'30"
Erik Werner Gorman Ari 2h3'2" 18d49'31"
Erik Wilcox Per 3h14'3" 52d2'54"
Erik William Leable Uma 10h17'11" 42d24'11"
Erik Wright Uma 8h41'45" 47d15'36"
Erik Y Laura Paredes 4 ever Tau 5h2'39" 19d18'34"
Erika Peg 21h54'36" 8d47'2"
Erika Ori 6h20'48" 10d1'22"
Erika Psc 1h11'5" 11d52'56"
Erika Ari 2h43'32" 28d27'18"
Erika Peg 22h46'0" 20d14'40"
Erika Cnc 8h52'48" 18d23'51"
Erika Tau 3h45'20" 28d45'28"
Erika Lyn 8h30'25" 47d2'35"
Erika And 2h21'43" 50d42'11"
Erika Cyg 21h49'40" 42d47'16"
Erika Cam 4h23'45" 56d42'58"
Erika Cyg 21h47'5" 46d18'16"
Erika And 1h22'36" 41d0'3"
Erika Uma 11h33'7" 31d52'4"
Erika Uma 8h49'27" 65d18'48"
Erika Cas 23h46'6" 53d17'26"
Erika Lib 15h38'59" -7d14'11"
Erika Vir 13h36'7" -7d50'0"
Erika Lep 5h45'34" -15d15'13"
Erika Sgr 18h17'9" -28d44'44"
Erika 80 And 0h24'27" 27d48'31"
Erika and Regan Uma 12h31'2" 58d20'38"
Erika Andrea Cap 21h44'27" -23d30'19"
Erika Angel Cnc 8h50'41" 31d56'15"
Erika Ann Strohmayer Aql 19h42'8" -0d28'57"
Erika Ashleigh Wedum And 0h28'5" 41d1'37"
Erika Banina And 1h38'46" 49d56'42"
Erika Beauvillain de Montreuil Lyr 18h43'20" 30d15'38"
Erika Beth Apu 15h16'15" -71d25'15"
ERIKA BRETT DAYBOOK Tau 3h29'32" 24d28'3"
Erika Browne Aqr 22h15'43" -2d22'54"
Erika Cavazos Sgr 18h47'12" -30d0'51"
Erika Chemaly Uma 10h26'34" 44d52'8"
Erika Cox And 2h22'42" 48d10'57"
Erika Cummins And 0h4'4" 46d49'11"
Erika D. Day Leo 11h49'34" 10d21'55"
Erika Dawn Cas 1h31'47" 63d17'15"
Erika Dawn Weber Gem 6h44'22" 26d50'11"
Erika Desiree' Schwager And 0h46'38" 38d30'55"
Erika E. Iovacchini Uma 10h27'13" 66d29'45"
Erika Eby Tau 5h39'10" 26d37'50"
Erika Edwards Lyn 6h55'28" 54d51'24"
Erika Elisha Dunn Sco 17h37'29" -33d43'31"
Erika Fainsilber The Great And 0h24'42" 43d17'21"
Erika Frasier Uma 9h34'57" 67d42'44"
Erika Garcia Aqr 22h55'17" -6d52'50"
Erika Gonzalez Psc 1h9'11" 32d17'12"
Erika Gould Leo 10h22'15" 23d36'11"
Erika Hope Seaquist Tau 3h53'44" 15d53'10"
Erika J Machuca Cap 21h41'53" -14d29'30"
Erika J. Schwartz Aqr 23h19'19" -19d46'54"
Erika Janitz Uma 8h43'5" 68d0'48"
Erika Jean And 0h11'45" 40d31'27"
Erika Jeanette Garza Harper Ori 5h58'24" 3d15'54"
Erika Jordan A. Psc 0h27'47" 9d3'19"
Erika Joy Rasi Uma 11h11'18" 54d12'11"
Erika Kindsfather And 1h15'5" 41d34'0"
Erika Kolton Gem 7h16'15" 32d37'18"
Erika Kürti "Celestial Heather" Cas 23h13'19" 55d39'50"

Erika L. Nestler
Mon 7h31'51" -0d42'27"

Erika L. Seeling
Lyn 7h28'54" 49d8'12"

Erika Langley Bunpermkoon
Ori 4h53'9" 2d43'54"

Erika Lea
Vir 13h30'46" 3d7'13"

Erika Lee Bennett
Lib 14h54'17" -2d19'4"

Erika Longardi
Aql 20h13'36" -0d36'13"

Erika Louise Rempe
Col 5h39'7" -27d34'10"

Erika Lynn
Dra 19h38'27" 62d35'20"

Erika Lynn Auclair
Vir 15h6'56" 3d48'2"

Erika Lynn Flagg
Cap 21h31'53" -23d0'30"

Erika Lynne Stocks
Cnc 9h6'7" 30d58'5"

Erika M. Zapata
Sco 16h5'53" -18d47'48"

Erika Manieri
Cam 5h34'2" 59d23'2"

ERIKA MANRIQUE
Sgr 17h58'59" -16d54'28"

Erika Maria Leich
Tau 4h47'39" 27d19'20"

Erika Marie
Gem 6h36'9" 17d0'42"

Erika Marie
Cam 6h17'25" 69d26'54"

Erika Marie Attoma
Tau 5h5'51" 27d31'58"

Erika Marie Nelson
Ori 5h12'15" 15d26'49"

Erika Marie Steinbrenner
Dra 18h34'21" 75d30'39"

Erika Marie Windish
Crb 15h39'1" 37d51'17"

Erika Maryline Marconato
Tau 5h8'3" 17d42'20"

Erika McCants
Lyr 19h17'16" 28d30'8"

ERIKA + MICHEL
Umi 13h37'15" 88d54'1"

Erika Michele Bostwick
And 0h44'13" 32d58'59"

Erika Michele Ritter
Peg 23h10'15" 19d43'21"

Erika Minh Crowley
Aur 5h47'17" 46d17'41"

Erika Montes
Gem 7h46'0" 19d22'24"

Erika Morales
Lib 15h43'28" -28d38'22"

Erika Morgan
Psc 0h38'40" 8d27'37"

Erika Nichole
Leo 9h31'12" 15d34'40"

Erika Nicole Ford - Cowgirl Angel
Peg 21h54'10" 14d55'55"

Erika Nicole Porter
And 0h37'36" 27d59'39"

Erika Noel
Cap 20h21'2" -13d23'37"

Erika Paige Flores
Cas 24h0'0" 56d52'14"

Erika Paige Kmieciak
Psc 1h17'56" 32d19'33"

Erika Perron
Umi 15h11'52" 70d16'9"

Erika Rae Benavides
Cnc 8h49'19" 12d26'44"

Erika Rae Hansen
Uma 10h35'21" 58d56'14"

Erika Raehse
Ori 5h11'46" 15d55'31"

Erika Reimus
Uma 12h59'1" 57d23'34"

Erika Reindl
Tau 3h57'30" 19d55'54"

ERIKA RENEE
Gem 7h49'7" 27d16'21"

Erika Renee 1989
Umi 14h44'41" 68d27'59"

Erika Rogers
Vir 13h17'26" 5d18'51"

Erika S. Jannetty
Cas 1h34'23" 63d41'50"

Erika Sage Houston Johnson
Pho 0h34'31" -42d50'20"

Erika Simonvaros
Ari 2h6'36" 18d54'49"

Erika Skye McManus
Cas 23h16'11" 59d28'36"

Erika & Stephen
Tau 4h9'3" 4d47'18"

Erika Stotz 20.05.1952
Uma 10h13'27" 66d43'31"

Erika Suda Avery
Lyr 18h47'12" 44d4'40"

Erika Suzanne My Love
Lib 15h30'16" -10d10'14"

Erika Suzuki Alonso
And 0h4'15" 40d58'38"

Erika Tomlinson - Happy Anniversary
Cru 12h29'54" -59d59'40"

Erika Trimble
Lyn 8h4'24" 50d0'43"

Erika Turner
Leo 11h9'10" 23d59'51"

Erika Viezzoli
Boo 14h44'11" 29d46'53"

Erika Vorndamme
Ori 6h19'34" 10d25'51"

Erika W. Magana
And 1h0'34" 43d52'14"

Erika & Wolfgang
Uma 8h18'57" 69d2'56"

Erika Woodams & Christine Ablett
Psc 0h50'41" 19d54'45"

Erika Xochitl deHollan
Uma 11h24'56" 46d49'44"

Erika Yandira Campos Reyes "Yandy"
Gem 7h22'51" 16d25'25"

ErikaEhrhard
Cru 12h53'46" -56d36'49"

Erika-Gonzalez Decker
Cap 21h42'26" -17d12'5"

ErikaLynn
Crb 15h48'41" 36d3'30"

Erika's - First Love Never Dies
Vir 12h41'21" 5d59'27"

Erika's Shining Star
And 0h13'23" 40d54'46"

Erika's Shining Star
Uma 10h29'43" 71d26'13"

Erika's Star
Uma 11h6'6" 57d34'53"

Erika's Wish
Gem 6h21'52" 18d33'4"

Erik-Erling Sund
Tau 3h43'46" 26d12'23"

Erikinha
Leo 9h34'42" 23d1'33"

Erikita
And 2h13'14" 45d26'7"

Erikka Aidan Hayes
Cap 20h36'8" -13d39'52"

Eriknsam
Ori 5h59'35" 17d56'49"

Eriko
Sco 16h28'3" -26d37'36"

Eriko Baby
Sgr 19h14'48" -12d47'16"

Erik's Star
Uma 11h49'17" 57d10'37"

Erille
And 0h23'4" 36d31'9"

Erimax
Lyn 7h57'40" 41d25'44"

Erimel
Ori 5h41'8" 0d38'49"

ERIMMIE
Lyr 18h42'40" 37d42'19"

Erin
Lyr 18h48'2" 39d7'11"

Erin
Cyg 20h12'35" 38d23'38"

Erin
And 23h31'31" 41d41'54"

Erin
And 1h20'31" 50d18'35"

Erin
And 1h43'11" 49d28'38"

Erin
Her 17h9'16" 46d10'39"

Erin
And 23h12'13" 51d2'43"

Erin
And 1h7'17" 43d51'39"

Erin
And 0h48'37" 44d54'55"

Erin
Ori 5h34'50" 5d27'27"

Erin
Aql 20h7'39" 4d58'17"

Erin
Aql 19h40'36" 0d34'3"

Erin
Cnc 9h20'16" 11d50'9"

Erin
Vul 19h20'24" 25d10'45"

Erin
Ari 3h27'6" 28d47'56"

Erin
Ari 3h28'34" 22d36'36"

Erin
Tau 5h53'30" 26d8'53"

Erin
Cas 1h39'57" 62d26'54"

Erin
Dra 10h39'58" 74d59'59"

Erin
Umi 15h7'59" 70d28'19"

Erin
Uma 12h50'44" 61d55'51"

Erin
Sco 16h17'26" -18d0'42"

Erin
Umi 15h21'33" 77d19'14"

Erin
Aqr 22h18'9" -0d51'8"

Erin 05
Lib 14h40'10" -12d25'14"

Erin 21
Vir 13h59'52" -6d39'40"

Erin 5/5
Lyn 7h31'58" 36d13'52"

Erin Adles
Vir 12h43'15" 4d18'12"

erin alanna
Lyn 6h52'19" 60d27'39"

Erin Alexander
Aqr 22h26'28" -0d34'6"

Erin Alyse O2
Pho 1h11'50" -43d34'19"

Erin Amanda Taylor
Ari 2h58'5" 19d7'43"

Erin Amber Lamb
Cas 1h37'27" 69d18'31"

Erin Amelia Larthe
And 1h31'56" 45d47'48"

Erin and Brett Woodruff
Cyg 20h15'17" 51d38'47"

Erin and Chris
Cyg 19h38'57" 36d54'26"

Erin And Duane II
Sgr 17h52'20" -29d59'40"

Erin and Pop
Cnc 9h3'2" 21d26'32"

Erin and Sean's Star
Ari 2h30'11" 11d22'45"

Erin Ann
Cir 15h21'30" -58d7'1"

Erin Anne
Sgr 18h12'2" -22d16'37"

Erin Anne Cameron
Leo 10h36'20" 9d38'2"

Erin Anne Elizabeth Ball
Cas 1h27'58" 63d15'39"

Erin Anne Hoare
Gem 6h54'27" 14d23'22"

Erin Anne Sweeney
And 2h18'37" 46d42'40"

Erin Arnold Smith
And 23h1'35" 50d35'50"

Erin Ashley Hall
Psc 0h35'19" 6d57'34"

Erin Axline
Leo 9h32'49" 27d32'33"

Erin "Baby Girl" Hastings
Cap 20h34'24" -20d36'3"

Erin "Babydoll" Bungay
Peg 21h33'23" 5d13'20"

Erin Bailey
Ari 2h56'48" 18d26'59"

Erin Bearss' Special Star
Vir 12h47'25" 3d32'12"

Erin Bertrand
Vir 15h1'25" 4d6'14"

Erin Birmingham Salas
Tau 5h44'13" 13d27'13"

Erin Bolema Ramirez
Lyn 7h31'52" 37d54'56"

Erin Boothman
Sco 16h51'36" -44d35'49"

Erin Brayshaw
Psc 1h19'4" 33d16'9"

Erin Brooke
Gem 7h35'48" 27d25'46"

Erin Calteux
Leo 11h24'1" 18d24'30"

Erin & Carl Davis
Tau 3h32'47" 29d49'0"

Erin Carly
Cnc 8h49'25" 31d25'7"

Erin Carroll
Leo 11h10'0" 8d59'34"

Erin Catherine Hartranft
And 0h59'10" 43d18'10"

Erin Cecilia Carpentier
Aql 19h5'40" -0d1'32"

Erin Chadwick
Ori 6h25'27" 16d51'35"

Erin Champaigne
And 2h12'48" 43d55'56"

Erin Che' Wright
Aqr 22h26'16" -1d16'56"

Erin Chick
Leo 10h14'21" 22d40'56"

Erin Christina O' Malley
And 23h14'32" 46d13'50"

Erin Christine Bass
Psc 0h28'46" 5d6'50"

Erin Christine Battaglia
And 23h39'22" 47d22'26"

Erin Christine Behrends
Psc 1h25'9" 17d47'18"

Erin Christine McHugh
Ari 2h47'16" 28d52'21"

Erin Christine Rose
Cap 20h39'9" -15d39'24"

Erin Christmas
Leo 9h34'13" 29d19'18"

Erin Claire Solley
Vir 13h14'45" -3d7'57"

Erin Clark
Psc 1h22'22" 26d57'40"

Erin Colleen Tierney
Vir 13h40'8" -10d38'33"

Erin Conway
Lyn 6h36'20" 59d48'34"

Erin Corley
Vir 12h7'56" 11d41'39"

Erin Courtney Dunlap
Lyn 7h37'28" 42d54'55"

Erin Curtis
Cru 11h58'20" -59d37'51"

Erin Dae
Ori 6h23'38" 14d19'47"

Erin & Darin Our Names Rhyme 4ever
Cyg 20h16'14" 51d55'16"

Erin David Boath - 15 October 1975
Lib 15h37'17" -11d14'49"

Erin DelCour
Cnc 8h46'30" 18d10'21"

Erin Delia Moore
Uma 10h45'11" 52d12'14"

Erin Delia Moore
Vir 13h36'11" -5d30'38"

Erin Dene Burkhart
Sgr 19h51'33" -12d40'1"

Erin Diane Lindemann
Psc 23h51'56" 6d0'45"

Erin Dittmann
Ori 5h44'36" 3d46'53"

Erin Doede
Ori 5h58'7" 6d10'42"

Erin & Donald Schwartz
Lyr 18h47'30" 45d11'48"

Erin Dorval Stacey
Cap 20h54'5" -19d50'8"

Erin E. Carter
Lib 15h3'10" -2d50'24"

Erin E. Kasperek
Ari 2h6'20" 18d7'43"

Erin E. Mulcahy
Tau 4h5'50" 5d18'59"

Erin E. Murphy
Lib 15h6'19" -25d14'27"

Erin E. Smith 4-16
Ari 3h16'42" 26d32'23"

Erin E. Viviano
And 1h55'24" 39d12'41"

Erin E. Yea
And 0h30'28" 27d33'2"

Erin Eden
And 2h13'31" 44d58'27"

Erin Eileen Gendreau
Ori 5h31'2" 0d24'57"

Erin Eileen Kelly
Leo 11h3'52" 5d56'15"

Erin Eileen McGinnis
Psc 1h44'20" 22d10'45"

Erin Eileen Mulroy
Tau 4h6'37" 8d32'21"

Erin Eileen O'Neill
Lmi 10h9'46" 37d28'17"

Erin Eilish Brianna Cole
Leo 11h21'22" 15d50'24"

Erin Elise
Sgr 17h44'4" -26d44'9"

Erin Elise Stevenson
Lyn 7h18'27" 55d18'39"

Erin Elizabeth
Cas 0h39'24" 64d30'17"

Erin Elizabeth
Uma 13h42'32" 55d1'22"

Erin Elizabeth
Gem 7h3'24" 17d47'59"

Erin Elizabeth
Ari 2h20'13" 23d46'18"

Erin Elizabeth
Ari 3h28'7" 22d38'6"

Erin Elizabeth
Cnc 8h19'29" 28d34'41"

Erin Elizabeth
Crb 15h37'39" 35d50'45"

Erin Elizabeth
And 0h35'31" 41d22'9"

Erin Elizabeth
And 22h57'55" 46d47'26"

Erin Elizabeth
And 23h33'28" 41d24'1"

Erin Elizabeth Birmingham
Uma 10h21'28" 60d30'28"

Erin Elizabeth Calley
And 0h36'2" 41d28'59"

Erin Elizabeth Casey
Cas 1h22'12" 63d54'46"

Erin Elizabeth Chandler
Lyr 18h48'23" 31d22'18"

Erin Elizabeth Docker
And 1h25'42" 34d9'55"

Erin Elizabeth Duff
Uma 12h8'38" 59d27'25"

Erin Elizabeth Etter Garner
Vir 13h12'20" 11d43'19"

Erin Elizabeth Fischer
And 23h15'44" 44d27'37"

Erin Elizabeth Grasse
Aqr 23h2'33" -24d19'24"

Erin Elizabeth Husbands
Crb 16h12'43" 33d32'30"

Erin Elizabeth Knode
And 0h56'11" 38d24'55"

Erin Elizabeth Koehler
Psc 0h43'25" 13d31'34"

Erin Elizabeth Lorberter
And 23h15'30" 47d41'5"

Erin Elizabeth McDonough
Gem 6h42'58" 33d43'13"

Erin Elizabeth Mitchell
And 23h24'33" 42d43'48"

Erin Elizabeth Morrisey
Cnc 8h6'17" 21d13'50"

Erin Elizabeth Morrison
Lib 15h17'31" -27d59'20"

Erin Elizabeth Murphy
Cyg 21h13'20" 44d44'30"

Erin Elizabeth Schultz
Cnc 8h18'13" 23d56'44"

Erin Elizabeth Vassallo
And 0h4'45" 41d5'28"

Erin Ellis
Cam 4h28'7" 67d35'40"

Erin Emily Winn
Lyn 8h3'42" 39d3'11"

Erin Enchanta 11
Cap 20h53'31" -19d54'2"

Erin (Er-Bear) Pinkham
Psc 1h25'14" 10d16'1"

Erin & Eric Sun
Cas 0h8'44" 58d32'44"

Erin Eternal
Uma 8h41'10" 68d31'42"

Erin Fairlie Bishop
Cra 18h48'35" -39d28'42"

Erin Flaherty
Lyn 8h25'49" 36d36'6"

Erin Flock Tully
Gem 7h3'19" 26d58'44"

Erin Forever
Tau 4h5'57" 26d14'11"

Erin Francis
And 2h24'51" 42d58'46"

Erin Fransen
And 1h30'33" 49d42'49"

Erin & Gary Lewis - Tupperware
Sge 19h43'3" 18d9'40"

Erin Giaime
Aqr 22h26'14" -22d57'2"

Erin Gill
Lyr 18h41'47" 34d47'34"

Erin Glynn
Per 3h39'1" 34d51'37"

Erin Grace
Psc 23h41'46" 3d25'26"

Erin Greenlaw
Aqr 21h9'47" -3d1'6"

Erin Griffith
And 0h20'31" 29d8'11"

Erin Griffiths
Sgr 17h52'14" -28d53'0"

Erin Grube
Cnc 8h43'50" 18d5'53"

Erin Guerinot
Leo 10h4'6" 19d4'9"

Erin Hackett
And 1h49'44" 46d47'41"

Erin Haillie Corbeil
And 0h23'21" 44d21'45"

Erin Hall
Dra 12h51'51" 72d14'31"

Erin Hanna "Yellow"
Gem 6h6'33" 26d35'40"

Erin Harris
Cas 0h41'15" 61d35'31"

Erin Hawley
Uma 8h35'12" 57d38'33"

Erin Henry
Ari 3h22'33" 26d53'52"

Erin Herman
Cas 23h2'39" 56d56'31"

Erin Hicks
Uma 11h46'50" 52d54'11"

Erin Hill No, you're the star
Tau 4h20'19" 23d52'5"

Erin Holmes Peabody
Vir 12h24'53" -2d51'20"

Erin Hope
And 23h17'33" 47d44'57"

Erin Horne
Psc 1h17'38" 31d25'29"

Erin Hutchison
Cyg 19h36'40" 55d34'30"

Erin Ikeda Woner
And 2h7'24" 42d48'21"

Erin Is A Star Forever
Uma 12h0'9" 53d48'8"

Erin & James
Uma 8h40'31" 61d28'36"

Erin Jane
Leo 9h24'7" 11d44'54"

Erin Jane "E.J."
Cyg 20h16'30" 59d45'0"

Erin Jane Howard
Lyn 6h59'32" 51d9'26"

Erin Jane Tornquist
Aur 5h52'32" 37d5'19"

Erin Janice Tjoe
Ori 5h41'46" 4d52'56"

Erin Jayne
And 0h42'58" 30d19'31"

Erin Jayne Clarke
Leo 9h35'2" 18d34'34"

Erin Jean Perez
Cam 3h29'23" 63d39'19"

Erin Jeanne Gill
Lyn 8h11'12" 56d53'11"

Erin Jennifer Niland
Cap 20h52'51" -24d57'48"

Erin Jessica Pryor
Leo 11h42'25" 27d0'23"

Erin Jewell Byrne
Uma 11h3'0" 60d39'16"

Erin Jill Angel Balzanna
Cnc 8h36'40" 32d8'22"

Erin J'Kia
Leo 11h21'49" 22d6'44"

Erin Jo
And 1h33'44" 39d41'29"

Erin & John Clogston
And 0h24'23" 29d28'51"

Erin June Drewes
Tau 4h42'25" 4d59'17"

Erin Juneau
Psc 1h12'23" 13d49'38"

Erin K Gallagher
Gem 6h55'42" 13d31'56"

Erin Kasey OConnor
Lyn 7h57'10" 37d36'19"

Erin "Kate" Cleary
Cas 23h5'50" 57d21'16"

Erin Katherine McGrew
And 23h6'54" 50d45'27"

Erin Kathleen
Peg 22h44'33" 15d27'35"

Erin Kathleen 271285
Col 5h12'58" -31d52'7"

Erin Kathleen Forker
Ari 2h8'54" 25d59'35"

Erin Kathleen Hughes
Sco 16h11'59" -15d4'49"

Erin Kay Goldstrom
Ori 5h36'56" 3d57'39"

Erin Kaye
Gem 7h11'55" 24d15'27"

Erin Keepers
Psc 0h50'4" 8d18'2"

Erin Kelley
Dra 13h18'25" 67d49'52"

Erin Kelly Ciborowski
Cyg 20h33'4" 52d37'8"

Erin Kelly Little
Sgr 19h53'28" -24d10'39"

Erin Kennedy Mathews
Sco 17h21'7" -32d34'33"

Erin Kisby
Col 5h59'9" -35d3'55"

Erin Kiyomi Skedeleski
Cas 1h25'25" 51d52'22"

Erin Kostowicz "sweet angel"
Cas 0h16'25" 63d23'12"

Erin L.
And 23h7'22" 48d32'47"

Erin L Magee
And 1h52'15" 46d14'32"

Erin L. Moseley
Ari 1h49'11" 18d13'18"

Erin L. Smith
Uma 11h47'5" 44d30'32"

Erin Lacey Bradley
And 23h52'47" 45d55'2"

Erin Landesman
And 0h53'43" 38d57'0"

Erin Lane Duncan
Ori 6h19'4" 15d29'31"

Erin Lathem
Aqr 20h47'52" -9d15'22"

Erin & Lauren
Crb 15h56'51" 27d59'0"

Erin Lee Genevieve Tush
And 23h13'8" 43d41'49"

Erin Lee McDermott
Vir 12h53'58" 1d48'55"

Erin Leigh
Cap 20h54'6" -15d56'40"

Erin Leigh Benes
Uma 9h24'18" 58d0'56"

Erin Leigh Berger
And 1h33'50" 44d3'32"

Erin Leigh Davis
Per 4h31'39" 31d47'12"

Erin Leigh Fuller
Lyn 7h35'18" 45d55'38"

Erin Leigh Ridgeway
Sco 16h3'33" -17d25'52"

Erin Leigh Stephens
Ari 2h47'18" 14d50'27"

Erin Lemmond
Leo 11h34'54" -0d47'34"

Erin Leslie
Sgr 18h32'1" -17d23'46"

Erin Lin
Vir 12h20'43" -4d6'32"

Erin Lindsay Kane
Lib 14h52'55" -3d0'39"

Erin Lindsey Walters
Cru 12h22'12" -57d17'35"

Erin Linfield Evans
Vir 13h47'51" -10d41'42"

Erin Linsday
Ari 2h22'58" 25d46'40"

Erin Logterman
Cas 23h44'24" 54d15'47"

Erin Louise Sefton
And 0h18'22" 45d58'42"

Erin Louise's Christening Star
And 2h24'21" 43d50'35"

Erin Lynn Guevara
Tau 4h30'52" 23d42'2"

Erin Lynn McFarland
Sgr 19h11'50" -16d29'31"

Erin Lynne McLaughlin
Gem 6h37'54" 12d18'4"

Erin M. Clack E=MC18
Umi 13h29'24" 71d7'50"

Erin M. Rudolph
Uma 11h21'53" 35d4'36"

Erin M. Sahagun
Ari 3h23'17" 26d0'22"

Erin M. Treacy
Uma 10h27'52" 64d24'36"

Erin Mae Bouldin
Cap 20h15'12" -15d45'48"

Erin Mae McGorm
And 1h16'55" 50d19'5"

Erin Mahlstedt
Uma 11h26'16" 29d42'48"

Erin Malia Throop
Cap 20h50'30" -26d31'44"

Erin Margaret Adamski
Ori 5h7'46" 7d15'20"

Erin Marguerite Cassidy
Cyg 20h46'49" 32d55'34"

Erin Marie
Cnc 8h47'5" 7d17'30"

Erin Marie
Vir 12h49'15" 9d37'5"

Erin Marie
Psc 1h47'24" 5d25'33"

Erin Marie
Lyn 7h58'50" 53d40'13"

Erin Marie Babineau
Cnc 8h57'33" 24d50'41"

Erin Marie Bracht
Ori 5h44'42" 5d9'39"

Erin Marie Brock
Lib 15h57'25" -18d38'24"

Erin Marie Brown
Psc 1h8'5" 9d27'23"

Erin Marie Byrnes
And 0h3'0" 45d13'51"

Erin Marie Cavanaugh
Uma 9h9'25" 66d1'38"

Erin Marie Devlin
Cas 23h32'41" 58d39'9"

Erin Marie+Everette Wayne [Always]
Uma 11h44'14" 41d33'25"

Erin Marie Harrold
Aqr 22h5'32" 0d4'42"

Erin Marie Huggins
Umi 14h49'50" 68d32'12"

Erin Marie King Morrow Clarkson
Psc 0h35'8" 8d40'22"

Erin Marie Latham Birthday Star
Vir 13h34'28" -1d19'53"

Erin Marie Methered
Cap 20h29'41" -13d43'51"

Erin Marie Murphy
And 23h26'28" 50d25'7"

Erin Marie O'Shea
And 23h18'53" 48d52'59"

Erin Marie Patrick Theriault
And 0h11'48" 44d57'31"

Erin Marie Reilly
Lyn 7h5'45" 51d5'41"

Erin Marie Sanguiliano
Gem 7h24'7" 34d56'19"

Erin Marie Wick
Sgr 19h2'32" -33d59'38"

Erin Marie Wren
Leo 11h2'15" 24d47'26"

Erin Martin
Ari 2h38'52" 17d33'23"

Erin Mazur
Cas 1h27'33" 58d4'42"

Erin Mc Crickard
Peg 22h55'22" 18d38'26"

Erin McAllister
Uma 12h38'22" 60d30'0"

Erin McCarthy
Psc 0h53'5" 29d7'35"

Erin McCarthy
Ori 5h31'39" 9d51'13"

Erin McGough
Leo 10h24'53" 13d19'44"

Erin McGurk
Vir 13h18'54" -11d40'11"

Erin McKenzie Barnes
Vir 12h55'41" 5d8'40"

Erin McNicholas
Ori 5h41'43" 0d13'24"

Erin McQuarrie
Umi 15h16'10" 68d39'14"

Erin Merdinian
And 1h4'57" 41d3'0"

Erin Michael
Umi 16h38'39" 83d10'19"

Erin Michael Hart
Aqr 21h44'7" -0d8'40"

Erin Michele
Gem 7h20'45" 29d10'4"

Erin Michele Budzyn
Cen 11h31'36" -38d35'22"

Erin Michele Halsey
Vul 20h17'47" 24d5'5"

Erin Michelle
Vir 14h22'3" 4d3'34"

Erin Michelle
And 1h57'34" 46d38'43"

Erin Michelle Baker
Cas 0h56'38" 63d43'0"

Erin Michelle Brinton
Ori 5h28'42" 4d7'54"

Erin Michelle Crawford
Vir 12h14'35" -0d59'41"

Erin Michelle Duffy
And 23h8'30" 51d8'0"

Erin Michelle Kelleher
Uma 11h39'15" 38d14'55"

Erin Michelle Ludlow
Gem 7h16'50" 31d0'49"

Erin Michelle Merritt
Cas 0h14'1" 50d13'52"

Erin Michelle Redfield
And 0h35'35" 36d11'46"

Erin Michelle Riley
Uma 9h31'30" 52d37'21"

Erin Michelle VanKampen
Cam 3h45'32" 70d18'0"

Erin Michelle Vanta
Dra 18h59'5" 70d38'20"

Erin Michelle, My Heart, My Love Tau 5h54'42" 22d55'59"
Erin Miller Mon 7h17'2" -1d28'58"
Erin Miller's Love Star Psc 0h26'56" 3d46'21"
Erin Molloy Cnc 8h35'6" 17d50'28"
Erin Molly Aqr 21h25'20" 2d26'35"
Erin Molly Cyg 20h55'3" 30d18'2"
Erin Montana Gilbertson Cas 23h46'20" 57d16'21"
Erin Morgan Wilcox Uma 11h7'19" 61d9'52"
Erin Morris Lyr 18h24'48" 45d2'59"
Erin Mozeika "A Star For Mommy" Vir 13h24'28" 12d56'10"
Erin Munk Per 3h19'41" 45d37'39"
Erin Murphy And 23h17'38" 42d13'58"
Erin My Girl Ari 2h49'51" 14d29'14"
Erin *My Love* Aqr 22h1'8" -18d11'28"
Erin & Nate And 1h29'44" 40d46'1"
Erin Naylor-Gray And 23h32'10" 41d22'24"
Erin Nichole And 1h16'0" 41d38'50"
Erin Nicholle Cnc 7h59'9" 10d33'38"
Erin Nicole Gem 7h44'11" 16d1'4"
Erin Nicole Uma 11h18'36" 47d33'26"
Erin Nicole Aqr 22h30'43" -22d4'2"
Erin Nicole Uma 10h31'42" 61d16'17"
Erin Nicole Arnold Gem 7h12'36" 29d11'13"
Erin Nicole Davidson Peg 22h59'24" 25d7'12"
Erin Nicole Kendrick Dra 19h0'26" 71d13'0"
Erin Nicole McCord Uma 10h45'27" 71d24'47"
Erin Nicole Meckly Tau 3h37'48" 13d13'13"
Erin Nicole Neel "Lil' French Girl" And 0h53'26" 38d57'47"
Erin Nicole Rios Leo 10h26'18" 22d59'53"
Erin Nicole Wray And 0h27'35" 31d42'53"
Erin Nila Dianne Sheppard Crb 15h35'20" 27d38'58"
Erin Nixie Fauber Ari 2h48'29" 18d22'11"
Erin Noel Sgr 19h19'53" -35d0'45"
Erin Noel & William the Fifth Cyg 19h39'6" 54d7'8"
Erin Nowassa Cnc 9h4'15" 26d10'25"
Erin O'Donnell Uma 8h30'3" 64d5'41"
Erin O'Neill Crb 16h8'5" 37d19'36"
Erin Ora Pfeffer Ori 6h3'57" 17d42'33"
Erin O'Shea Vir 12h24'37" 11d4'58"
Erin Paige Siller Sco 16h54'44" -42d4'12"
Erin Patricia And 23h9'44" 51d13'19"
Erin Patricia Wagner Sco 17h44'38" -32d38'19"
Erin Paula Ruehr 08071980 Leo 10h43'20" 15d53'20"
Erin Pearle Sco 17h45'4" -42d56'31"
Erin Philley Her 16h31'53" 18d50'53"
Erin Pinky Chalmers Ari 1h59'26" 18d9'29"
Erin Piper Edmondson Psc 0h43'2" 16d10'58"
Erin Porras And 0h45'36" 39d57'51"
Erin Punkin Mahoney Oph 16h43'6" -0d39'9"
Erin Purcell & Kyle Coffrin Cyg 20h12'33" 53d18'15"
Erin R. Jones Her 18h26'44" 12d10'23"
Erin Rachel McFaull Del 20h40'6" 6d55'35"
Erin Rebecca Cannon Leo 11h48'15" 23d59'3"
Erin Rebeckha Hardy " Baby Erin " Vir 12h56'38" -11d6'3"
Erin Rebekah McKinzie Tau 3h42'14" 4d59'47"
Erin Rehberg Ori 6h2'9" -0d8'3"
Erin Renae Cam 5h17'4" 60d25'50"
Erin Renfree Cas 23h3'15" 53d36'10"
Erin Rose Cnc 8h11'15" 7d13'10"
Erin Rose Ballenger Sgr 18h27'3" -16d56'40"
Erin Rose Deslippe Uma 11h23'3" 60d45'36"
Erin Rose Floyd Cyg 21h53'56" 49d46'36"
Erin Rose Thoele And 0h33'42" 42d3'20"
Erin Rosenberg Lib 15h34'54" -15d38'51"
Erin Ryan Soper Dra 18h41'52" 54d14'42"
Erin & Ryan's 1st Anniversary Star Her 17h18'38" 27d38'17"
Erin S. Ehrhard Cnc 8h42'40" 21d59'55"
Erin Schwochow Sco 17h51'4" -36d15'11"
Erin Sebele Cas 0h20'30" 50d51'9"
Erin Shaun Doyle Tau 3h57'13" 2d50'48"
Erin Sigourney Peach And 1h8'40" 45d49'0"
Erin Siobhan Smith Lyn 6h58'56" 50d46'35"
Erin Smith Cam 9h10'35" 77d44'18"
Erin Sprague Ari 2h43'15" 26d18'36"
Erin Squared Uma 11h36'50" 49d30'43"
Erin Stimer's Hope Lib 14h54'39" -2d9'9"
Erin Sutton Cnc 8h46'27" 19d13'21"
Erin Swann Gem 7h5'11" 10d46'37"
Erin Sydney Powell Aqr 22h7'13" -19d50'44"
Erin T. Fitzgerald Vir 14h23'50" -1d42'45"
Erin Taylor Kelley Vir 13h19'19" -0d14'57"
Erin Terry & Travis Allenton Cyg 20h26'45" 43d12'28"
Erin the Kitten Lyn 8h0'39" 57d50'3"
Erin the Peaceful Ori 5h23'11" 1d13'53"
Erin Toohil Aqr 23h3'33" -7d42'8"
Erin Tranchina Vir 11h42'10" -0d48'34"
Erin Trista Hawkins Lyr 18h49'40" 35d10'15"
Erin Tumacder The Most Beautiful And 1h38'13" 44d50'0"
Erin V. Harmison And 2h24'35" 48d13'19"
Erin Valle Kellogg Cnc 8h26'52" 14d37'33"
Erin Vannoy Uma 9h46'6" 46d17'33"
Erin Virginia Tierney Sco 17h45'51" -42d57'56"
Erin Ward Ari 2h10'58" 23d4'18"
Erin Warren Vir 14h30'51" 5d7'28"
Erin Weber Crb 16h3'56" 39d20'23"
Erin Wells Uma 10h35'12" 41d2'54"
Erin&Wes:symbol of my eternal love Ori 5h33'39" 0d47'40"
Erin West Gem 7h13'26" 32d52'46"
Erin Whitaker Gem 7h38'51" 32d46'17"
Erin White Aqr 21h37'58" -6d24'45"
Erin Wiemelt Lyn 7h42'27" 47d59'15"
Erin Wilson, Apple Queen 2003 Uma 11h11'15" 30d51'45"
Erin Woodbury Aql 19h50'50" -0d29'34"
Erin Yaboo Thursby Ori 6h15'50" 10d34'18"
Erin1025 Leo 11h17'38" 27d21'43"
Erina Crb 15h42'20" 32d55'25"
Erina Arya Cra 18h13'21" -39d45'56"
Erina Sophia Cyg 20h36'42" 30d23'45"
Erina " The Best Mom In The World " Uma 9h18'22" 47d48'19"
Erin-Bear Tau 4h9'58" 10d14'38"
erinbrookewood Sgr 18h37'27" -25d35'56"
ErinDale Cap 21h41'44" -8d58'48"
Erinello Cantu Aqr 22h18'13" -3d10'20"
Erini Fragou-Laleni Uma 11h15'12" 29d42'22"
Erinia Cygnae Cyg 21h40'51" 49d53'26"
Erinlvey Faeriesuperstar Libra-1013 Lib 14h52'35" -13d22'54"
Erin-Jane Ori 4h58'10" 5d26'15"
Erin-My True Love & Dream Come True And 23h36'58" 35d50'56"
Erinn Uma 9h19'37" 52d34'19"
Erinn and Rich Laffoon Eri 4h28'45" -1d18'33"
Erinn Janine Foster Cap 20h18'48" -10d13'27"
Erinn Louise Van De Kamp Per 4h29'29" 35d13'19"
Erinn McCarthy Uma 10h43'38" 61d11'47"
Erinn McGrew Umi 16h14'33" 73d55'42"
Erinn Nigro Uma 10h27'22" 60d24'2"
Erinn "Osita" Walsh Cyg 19h42'54" 35d16'38"
Erinn Shay Tybush Sco 16h19'55" -37d17'7"
Erinn, Chris, Jackson, and Max Cyg 19h45'32" 30d43'6"
Erinne Lily Haitana Cnc 8h38'44" 11d31'27"
ErinO051783 And 0h16'0" 44d28'36"
Erin's Beautiful Shining Star And 22h59'33" 46d11'29"
Erin's Brilliance Gem 7h22'51" 16d15'45"
erin's christmas star Aql 18h59'10" 7d2'31"
Erin's Christmas Star 2004 Lyr 18h29'21" 28d40'20"
Erin's Dancing Light Ari 3h22'8" 27d57'24"
Erin's Diamond Ori 5h12'23" 15d24'34"
Erin's Dream Sco 16h13'25" -14d35'41"
Erin's dreams Ori 5h35'34" 12d1'2"
Erin's Hole To Heaven Cnc 7h58'2" 17d39'7"
Erin's Howdy Star Ori 5h25'20" -4d12'4"
Erin's Island Ori 6h18'38" 15d41'43"
Erin's Kiss And 1h9'28" 34d22'56"
Erin's Love And 0h14'2" 46d9'46"
Erin's Love Leo 9h26'27" 10d37'9"
Erin's Place in the Cosmos 15051983 Col 6h8'2" -28d14'29"
Erin's Rock Star Ori 6h2'43" 17d36'57"
Erin's Soul Gem 7h13'35" 22d17'29"
Erin's Star Leo 11h29'35" 11d39'39"
Erin's Star Ori 5h16'53" 5d48'22"
Erin's Star And 2h22'35" 49d55'4"
Erin's Star And 1h0'47" 47d2'10"
Erin's Star And 23h49'7" 48d5'9"
Erin's Star Psc 0h52'23" 32d23'15"
Erin's Star Leo 9h59'25" 30d6'9"
Erin's Star Vir 14h24'17" -3d19'21"
Erin's Star Uma 11h15'2" 72d21'17"
Erin's Star Uma 9h14'38" 53d58'38"
Erin's Star Cas 1h9'6" 64d13'47"
Erin's Sweet 16 Ari 2h54'43" 24d26'48"
ErinStar Sco 16h49'29" -35d49'39"
ERION Her 18h22'11" 16d17'20"
Erion Gjoni Ori 6h17'50" 6d46'5"
Eris Cam 4h22'2" 56d49'11"
Eris Michael Davis Ari 2h19'49" 12d32'6"
Eriva Sgr 17h59'21" -18d49'31"
E.R.Janowicz Ari 2h35'17" 26d49'53"
Erk Harris Cam 4h20'4" 67d36'30"
ERK-77 And 0h45'23" 37d47'15"
"Erka" Gulyás Erika csillaga Uma 11h36'52" 63d56'7"
Erky Gem 6h51'59" 33d13'5"
Erky Judit Gem 6h29'55" 24d41'39"
Erlangga & Indra Sgr 17h55'30" -28d27'49"
Erlean M. Reynolds Lyr 18h27'13" 37d6'7"
Erlene Hafer Cyg 20h14'54" 50d0'4"
Erlinda Mosqueda Porguez Sco 16h7'46" -17d38'53"
Erling Her 18h11'24" 17d29'40"
Erling Olaf Elverum Sco 17h53'27" -36d33'17"
Erma at 70 Ari 3h22'19" 28d1'43"
Erma Crayne Lyr 18h44'58" 38d20'29"
Erma Gwendolyn Brown Miller Uma 11h27'46" 37d43'40"
Erma Hattie Darkes Morgan Lyn 8h25'36" 41d8'44"
Erma Jane Rollins Psc 23h16'26" 3d27'37"
ERMA LOUISE NEW-COMBE Lyn 7h30'41" 49d37'37"
Erma R Koslo mother/friend/grandmom Crb 15h29'9" 28d18'57"
ErmaBob(Riebel) Rawa wed 12-22-1951 Tri 2h21'30" 31d58'47"
Ermadel Ludwick Uma 12h8'47" 45d30'45"
Ermandude Lmi 10h30'2" 39d2'42"
Ermel, Elmar Uma 9h43'39" 49d7'36"
Ermelinda Sgr 18h34'13" -22d41'2"
Ermie Rossi Cnc 8h18'20" 32d11'42"
Ermine(jack)Sargent Ari 3h2'59" 28d4'44"
Erminia Per 3h5'10" 51d43'15"
Ern and Eugenia Jones Umi 15h31'52" 73d15'27"
Erna Uma 9h9'43" 46d45'13"
Erna "Annie" Uma 9h28'27" 51d44'52"
Erna Clair Wyler Cap 20h19'3" -15d36'43"
Erna Dzeba Uma 12h35'58" 54d29'14"
Erna & Erwin Federer 09.09.77 Dra 17h10'44" 51d6'26"
Erna - Galaxy's Greatest Grandma Del 20h35'57" 19d30'44"
Erna Gottschalk Uma 14h1'13" 59d21'47"
Erna Hildegard Rosa Ori 5h3'39" 7d5'3"
Erna Krause Gem 7h39'3" 22d37'44"
Erna Michel Brown Cas 0h15'49" 51d25'26"
Ernest Per 3h12'42" 47d37'40"
Ernest Ori 5h55'28" 11d40'11"
Ernest Cnc 8h19'0" 14d13'3"
Ernest A Amende Ori 4h50'10" 4d10'20"
Ernest A. Ingenito Uma 11h31'9" 39d26'47"
Ernest A. Sneddon, MD Per 3h12'28" 57d9'53"
Ernest Arnold Ricker Leo 9h45'58" 28d47'41"
Ernest Augustus Morel Her 16h31'32" 37d56'14"
Ernest Baidoo Aql 19h10'19" -0d2'3"
Ernest Blackburn Vermont Aqr 22h12'9" -2d24'4"
Ernest Carranza 111 Ori 6h3'52" -0d13'40"
Ernest D. Lopez Lyr 18h50'4" 31d42'42"
Ernest DeMarco Vir 14h17'13" -3d51'51"
Ernest E. Charles Cas 0h27'32" 58d59'29"
Ernest Fredrick Struckman IV Dra 10h37'23" 75d39'3"
Ernest G. Carbone II Aqr 22h33'15" -8d1'16"
Ernest G McIntosh Uma 11h17'41" 46d30'21"
Ernest i Núria Ori 6h1'19" 21d3'13"
Ernest Irizarry Sco 17h41'54" -39d16'1"
Ernest J. "Butch" Ackley Tau 5h51'49" 25d42'52"
Ernest J McCluskey Boo 14h24'31" 14d53'47"
Ernest John Cibik Ori 6h17'50" 13d57'7"
Ernest Joseph Marquez - Tin Man Aql 19h45'11" -0d35'33"
Ernest Joseph Perron Cyg 20h2'33" 43d52'3"
Ernest Kenjiro Ichioka Uma 9h38'16" 63d21'32"
Ernest Kopstein Tau 4h43'45" 17d40'39"
Ernest L. Beaudrie - Pa Cep 22h12'36" 61d47'55"
Ernest L. Fitzhugh, Jr. Cma 6h41'41" -12d54'37"
Ernest L. Stephens Her 17h11'46" 13d37'53"
Ernest L. Waldin III Gem 6h6'56" 26d54'59"
Ernest Lawing Lyn 6h22'54" 61d27'24"
Ernest Leon Lutz Per 3h33'53" 46d16'11"
Ernest Leroy White Sco 17h50'2" -38d34'30"
Ernest M. Jones Ori 6h8'15" -0d16'17"
Ernest Mario, PhD Gem 7h38'20" 24d39'48"
Ernest Martsching Per 2h39'56" 55d33'1"
Ernest Moya Ori 5h51'41" 6d37'7"
Ernest Nicholas Urfer Uma 8h43'29" 46d52'29"
Ernest O. Horn Jr. Cyg 20h5'13" 37d20'14"
Ernest R. Rheaume Uma 8h57'57" 47d2'52"
Ernest R. Walls Per 3h42'20" 38d59'33"
Ernest Raphael Aur 5h45'49" 55d0'43"
Ernest Robert Gonzales Jr. Per 3h47'10" 47d27'33"
Ernest Rolland Comstock Gem 7h45'16" 20d30'30"
Ernest S. Pharo Cnc 8h18'55" 16d22'2"
Ernest (Sammy) and Cheryl Jochen Uma 11h33'13" 34d15'54"
Ernest Sr. & Carmen Rivera Cyg 20h11'47" 36d42'2"
Ernest Torocco Her 18h10'49" 18d39'15"
Ernest Wayne Gregory Her 16h54'43" 33d53'13"
Ernest Y. Kagiwada, Jr. Per 3h11'45" 52d27'12"
Ernest Young, Jr. Sgr 18h34'34" -23d42'56"
Ernesta Dra 9h35'32" 78d2'3"
Ernestina Jobu Rose Gershwind Uma 8h32'55" 72d0'15"
Ernestina Sakyi And 0h49'56" 44d4'38"
Ernestine Uma 10h37'42" 40d42'34"
Ernestine And 23h44'31" 47d6'41"
Ernestine A. Reed Cas 1h3'18" 62d51'17"
Ernestine Collie Haynes Sgr 19h14'28" -16d33'25"
Ernestine Lynn Cyg 19h42'3" 37d43'25"
Ernestine Randlett Bernstein And 23h11'27" 47d57'45"
Ernestito Mio Umi 15h22'14" 74d9'12"
Ernesto Aqr 23h0'25" -13d59'18"
Ernesto Alejandro Iglesias Uma 12h56'15" 52d25'35"
Ernesto & Ana Mary Arias Cyg 20h32'10" 57d13'55"
Ernesto and Veronica Cyg 21h13'31" 36d18'46"
Ernesto Guadalupe Mota Sgr 18h16'23" -16d39'56"
Ernesto Joaquin Regales Uma 9h0'23" 49d3'56"
Ernesto Mau Quidilla II Cen 13h34'33" -36d58'36"
Ernesto Ortiz Vir 14h40'27" -2d44'8"
Ernesto Paniccioli And 1h44'4" 46d48'32"
Ernesto Ponton Boo 14h35'11" 27d26'19"
Ernesto Radames Sierra Lib 14h55'18" -12d46'49"
Ernesto Rodriguez Leo 9h45'55" 29d36'45"
Ernesto Saladino Vir 13h5'22" -15d18'23"
Ernesto Santaniello Ori 5h53'51" -8d22'51"
Ernesto Silva Aur 5h11'50" 38d57'36"
Ernesto & Starr And 23h31'37" 42d52'20"
Ernesto.MG.Umo50 Gem 6h51'23" 24d46'22"
Ernie Leo 10h43'32" 15d59'45"
Ernie Ori 5h33'7" 10d14'0"
Ernie Her 17h28'42" 34d38'48"
Ernie and Nannette Umi 13h37'9" 73d9'53"
Ernie B. Luna Lib 15h23'57" -19d53'9"
Ernie Berardinelli Sgr 19h7'59" -16d12'45"
Ernie Charles Blake Ragiel Sco 16h52'13" -38d14'7"
Ernie & Cheryl ~ True Love Forever Cyg 19h41'10" 42d17'25"
Ernie Dingo Cru 12h24'44" -61d54'40"
Ernie Hess Uma 10h47'0" 55d51'40"
Ernie John Watson Umi 15h30'1" 78d41'9"
Ernie Juarez Boo 13h47'27" 17d18'25"
Ernie L Mixon Aqr 22h24'37" -0d19'9"
Ernie Steele Sco 16h13'29" -13d10'40"
Ernie Tamashiro Aql 19h45'21" -0d42'17"
Ernie Topolnycky Boo 14h43'31" 41d37'35"
Ernie's Star Her 17h24'49" 27d20'23"
Ernie's Star Umi 14h42'36" 69d21'4"
Ernie's Star - The Light of My Life Cap 20h44'19" -21d40'46"
ErnJoVal Crb 15h46'53" 31d0'22"
Ernst Aichner Uma 9h28'44" 47d31'41"
Ernst and Claudia deHaas Her 18h32'40" 19d26'31"
Ernst Brechbühl Per 4h32'21" 40d37'24"
Ernst Freitag Uma 10h2'55" 63d6'40"
Ernst Ghenzi Com 12h18'49" 28d45'51"
Ernst Heinrich Nisius Ori 6h20'22" 15d10'43"
Ernst Heinz Wallrapp Ori 6h15'14" 6d34'37"
Ernst Hummel Uma 10h15'22" 48d54'55"
Ernst Lothar Scheuble Ori 6h17'10" 5d55'45"
Ernst Schnellmann Cas 23h26'39" 54d15'17"
Ernst Werner Uma 10h2'51" 62d15'52"
Ernst Wilhelm Ari 2h41'35" 14d47'8"
Ernst, Günter Uma 13h18'19" 52d57'48"
Ero F. Rifelli 20 Years HVB BDB Crb 16h15'26" 28d31'13"
Eroc50 Sco 16h5'51" -20d38'34"
E-Rock Cnc 8h51'22" 25d6'43"
Erock And 0h31'3" 36d17'26"
ErockStar Leo 11h57'14" 23d52'42"
Eroe Gem 7h4'10" 10d12'18"
Eroica Uma 8h33'50" 65d31'53"
ErOkDrAh! Cet 0h59'35" -12d26'20"
Erol Esen Her 18h57'16" 15d34'48"
Erol Tabanca Psc 0h54'31" 29d22'8"
Eronberg Ori 5h38'44" -2d0'31"
Eronne Umi 14h56'21" 70d31'20"
Eros Cap 21h10'35" -25d50'1"
Eros Her 17h33'25" 32d11'26"
Eros Lyr 19h2'31" 32d55'52"
Eros Lyr 18h55'44" 36d44'36"
Eros Lyn 7h35'45" 39d59'7"
Eros Cyg 21h26'46" 51d33'11"
Eros And 23h27'12" 41d47'21"
Eros Aion Lyn 7h15'26" 58d5'10"
Eros Aiónios Cyg 20h18'11" 50d24'57"
Eros and Aphrodite* Cyg 20h59'30" 30d34'42"
Eros Camellia Blue Dra 19h18'21" 66d24'47"
Eros Gama Uma 12h25'29" 62d21'49"
Eros Passionate Love of Friendship Crb 16h9'7" 37d16'20"
Eros und Athos And 0h1'0" 44d57'8"
Eros Zanaboni Dra 19h21'7" 78d58'28"
Eros, My love for Bonnie Jean Cyg 21h16'45" 44d55'23"
Eroswen Cas 1h12'3" 63d19'56"
Eroyn Sgr 18h40'15" -32d0'26"
Erric Fruntel Grandy Ari 2h38'11" 25d19'15"
Errico Savarino Lmi 11h1'47" 25d14'49"
Errin's Star Gem 6h49'9" 24d15'45"
Erris And 23h11'16" 35d57'32"
Errol Sgr 19h53'24" -28d36'53"
Errol Duck-Chong - Serving the Lord Cru 12h0'53" -62d25'38"
Errol Schuh Cru 12h23'46" -61d52'15"
Errol Together forever Like A Star Ori 5h22'21" 14d52'55"
Errol Van Stralen Srp 18h4'46" -0d31'49"
Errolyn Evelyn Boston Lizotte Tau 3h38'39" 22d7'3"
Erryn Tam Eri 3h43'55" -0d15'44"
Ersan Baskurt Uma 13h51'20" 48d2'11"
Ersha Peg 21h45'49" 8d19'50"
Ersi Tsami Leo 9h29'35" 14d23'19"
Erskine H B Per 3h12'34" 43d19'34"
Erthtajojota Uma 8h38'40" 64d20'28"
Ertler, Nina Cnc 9h11'20" 25d23'25"
Erv Clark Uma 11h17'36" 40d13'52"
Erv Euler Aqr 21h36'25" 0d32'37"
Erv & Pam Dobson Cyg 21h23'31" 41d24'36"
Ervanna Her 17h15'53" 25d8'17"
Ervin Cango Dra 19h0'38" 74d15'38"
Ervin Williams Dra 18h37'20" 70d26'1"
Erwan Berot Uma 9h51'58" 43d2'38"
Erwan Geneste Ori 6h0'12" 13d40'23"
Erwin Uma 10h45'40" 50d0'0"
Erwin Beltle Uma 13h51'7" 58d53'47"
Erwin Cherwin Ari 2h43'22" 29d37'5"
Erwin David Solloway Sgr 19h33'6" -13d18'57"
Erwin E. Petschauer Uma 8h56'11" 50d4'15"
Erwin "Erv" Weiland Eri 4h27'16" -0d0'53"
Erwin (Larry) Starosta Jr. Her 16h56'29" 13d43'51"
Erwin Pearl Sco 17h28'3" -45d20'17"
Erwin Pfeil Uma 10h14'36" 55d42'39"
Erwin-Kennel Leo 10h33'51" 19d7'13"
Eryan Umi 13h56'11" 75d3'38"
Eryca Del 20h34'34" 19d46'21"
Erycee Cam 4h1'7" 67d42'45"

Eryka Lynn Aberle
Cas 1h37'21" 67d16'49"
Eryn
Cas 1h17'44" 62d40'13"
Eryn
Aqr 23h44'10" -8d15'41"
Eryn Courtney Atha
Uma 8h49'17" 58d43'1"
Eryn Dawn Goodall
And 23h0'3" 51d17'48"
Eryn Jade
Aqr 23h8'58" -10d37'1"
Eryn Lea Dunbar
And 2h14'33" 38d41'2"
Eryn Margaret Young
Gem 7h53'41" 14d0'42"
ERYN MARIE
Ari 3h13'47" 25d50'4"
Eryn Sturdivant
Tau 3h30'32" 28d3'50"
Eryn, The twinkle in my eye!
Dra 19h4'20" 63d50'20"
Erynn
Aur 5h44'32" 42d51'7"
Erynn Alice
And 23h16'47" 47d16'39"
Erynn Lynn
Ori 5h28'10" 4d14'59"
Erynn Patricia Konza
Lib 14h58'29" -17d55'23"
Erynne Victoria Johnson
Lib 14h27'40" -17d45'18"
Erynn's Luminosity
Ari 2h50'33" 13d57'9"
Eryns Wisher
Ori 5h8'36" 15d20'14"
ERZSEBET ENIKO MOL-NAR
Her 18h8'38" 18d43'36"
Erzsebet Szabo-Backer
Cyg 21h21'56" 34d0'10"
Erzsi
Mon 6h54'55" -0d18'30"
Erzsike-csillag
Gem 7h36'10" 19d8'39"
Esa Rustholkarhu
Uma 9h36'20" 47d41'38"
Esau Alarcon
Vir 12h0'48" 8d39'38"
Escalante's
Uma 9h0'50" 47d24'4"
Eschenhorst, Ursula
Uma 8h18'13" 65d56'32"
Eser Yildiz
Leo 11h19'12" 7d23'14"
Eshan
Sgr 18h13'17" -22d29'48"
Esildut
Dra 17h16'44" 68d12'54"
Esilei Danden
Lyn 7h31'52" 38d0'48"
Esio Trot
Uma 11h32'8" 37d24'15"
Eskimo & Butterfly: J.T.S. & L.C.H.
Cnc 9h4'1" 31d30'59"
Eskimo Kisses
Ori 6h1'11" -0d0'12"
Esli Medrano
Sgr 18h38'7" -17d45'47"
Esma Etan
Dra 19h24'20" 76d36'7"
Esme' 1926
Aqr 22h49'46" -16d29'46"
Esme Calder Callaghan
And 23h48'50" 45d53'24"
Esme Judith & Xanthe Maria Banyard
Gem 6h30'18" 24d36'14"
Esmé Louisa Stevens
Umi 14h10'46" 69d18'22"
Esme Louise Froning
And 0h41'53" 45d36'53"
Esme Louise Pattinson
Uma 9h22'19" 66d14'42"
Esme Marie Peschke
Leo 9h37'14" 29d31'0"
Esmeralda
Tau 4h24'36" 14d9'41"
Esmeralda
Uma 8h53'23" 47d25'50"
Esmeralda
Cas 1h51'21" 60d2'46"
Esmeralda
Cap 21h22'36" -14d59'48"
Esmeralda
Sco 17h57'42" -30d7'57"
Esmeralda Aiello Gomez
Vir 12h44'47" 0d9'20"
Esmeralda Dacosta
Uma 11h25'16" 39d27'52"
Esmeralda "Estrella" Maria Rosales
Vir 14h12'16" -4d41'14"
Esmeralda Herrera
Cap 20h24'36" -24d53'47"
Esmeralda Isabel
And 1h14'13" 35d35'56"
Esmeralda Lopez
Cam 5h17'18" 57d30'47"
Esmeralda Q
Her 18h52'2" 25d18'31"
Esmeralda y Alicia
Leo 10h50'53" 6d48'43"
Esmeralda's Radiant Emerald
Tau 3h52'11" 20d57'47"
Esminia 51791
Per 3h22'7" 41d46'21"
Esmy Villarreal
Ori 5h30'48" 1d59'4"
Esoterica
Ori 5h48'21" 6d8'16"
Esoterica "3-20-91"
Uma 11h43'16" 50d51'22"
E.S.P.
Cap 21h3'20" -19d48'12"
Esperanca
Cyg 20h25'26" 33d41'8"
Esperance
Dra 18h52'21" 64d35'26"
Esperanza
Uma 11h10'49" 66d13'36"
Esperanza
Lib 15h17'46" -21d48'29"
Esperanza
Mon 7h30'1" -0d59'1"
Esperanza
Cyg 19h35'20" 34d35'31"
Esperanza
Cyg 21h16'40" 47d17'49"
Esperanza A. Anderson
And 0h23'26" 43d42'35"
Esperanza Angelica Aparicio
Vir 13h42'52" -11d23'38"
Esperanza Baires
Cap 21h8'20" -25d39'43"
Esperanza Barranon Gurrola
Cas 0h34'7" 55d10'51"
Esperanza D. Hansen
Cam 7h50'59" 70d16'37"
Esperanza Garcia
Dra 18h55'5" 51d49'16"
Esperanza Shafer
Sgr 19h37'36" -16d56'8"
Esperer
Cnc 8h2'55" 23d50'35"
Esperetem
Aqr 20h54'10" -11d2'43"
Espinola-Burke
Cap 21h39'12" -9d36'19"
Espoir
Col 6h33'34" -36d35'43"
Esposa Hermosa
Ori 5h8'11" 9d4'30"
Esposa mas atractiva ymadre amorosa
Cnv 12h43'6" 41d7'43"
Esposito Tania
Uma 11h4'21" 45d28'40"
Espree Zolkowski
Uma 9h13'17" 69d8'47"
Esprit de Kona
And 0h59'49" 45d23'7"
Espronceda
Peg 22h29'42" 7d38'18"
Esquimau
Cas 0h27'7" 54d30'44"
Esra Jah
Del 20h48'27" 18d43'13"
Esra Turk
Mon 7h51'4" -0d49'45"
Essau
Uma 10h52'43" 57d55'10"
Essence
Uma 10h20'6" 61d21'30"
Essence Michelle Mueller
Com 13h2'40" 17d55'6"
Essence of Christine
Sco 17h18'28" -32d33'53"
Essence of Heart Always Loving Rose
And 1h48'0" 46d37'0"
Essence of Lori Newswanger
And 0h8'3" 31d28'14"
Essence (Reed & Jill's Star Planet)
Cyg 21h43'35" 48d6'48"
Essential Stellaris Nikos
Ori 5h11'52" -9d32'27"
Essenza
Umi 17h36'37" 84d40'31"
Essie Burns
Psc 1h18'41" 11d56'44"
Essie Marguerite Boles
Psc 23h33'42" 3d20'11"
EssieWarner
Cap 20h32'28" -11d41'26"
Esso Arnold Spencer Askew
Her 16h34'3" 35d23'36"
Esta
Sgr 18h26'0" -27d51'52"
Esta Meyer
Sgr 20h15'7" -45d3'15"
Esta of Diandra Amerie
Peg 21h37'59" 22d18'46"
Esta Shin
Lyn 8h12'20" 34d9'55"
Estar
Ari 2h14'33" 23d28'50"
Estar
Mon 7h31'48" -0d23'46"
Estasia
Lmi 10h33'40" 35d38'16"
Este Louise
Crb 15h46'55" 32d37'2"
Este Rodriguez Water Moccasin, Jr.
Cas 23h43'55" 53d13'47"
Esteban
Dra 16h44'3" 54d28'4"
Esteban
Uma 9h58'47" 46d7'33"
Esteban Andrea's Maldonado
Umi 14h28'20" 72d12'0"
Esteban Daniel Rivero
Leo 9h45'14" 27d10'21"
Esteban Ponce
And 2h19'7" 41d5'22"
Esteban Sevilla
Her 17h27'27" 31d51'39"
Esteban y Mariajose
Sge 19h51'7" 18d53'48"
ESTEFANIA 31-05-1992
Gem 6h17'55" 23d39'19"
Estefania Haro Aparicio
Leo 11h14'7" -3d23'44"
Estefania Macchiavello-
Lyn 8h8'6" 50d25'44"
Estefania Rivera
Lib 15h42'57" -17d11'1"
Estefany Ines
Cap 20h27'9" -13d5'55"
Esteffy
Uma 11h13'26" 50d28'34"
Estela
Sco 16h8'4" -12d22'18"
Estela Falcon Guajardo
Aur 5h54'34" 49d0'59"
Estela Figueroa Contreras
Aqr 22h27'30" -6d4'59"
Estela Pena Leyva
Aqr 22h29'41" 0d35'2"
Estela Treto
And 0h2'28" 35d29'35"
Estela Vigoa
Ori 6h25'14" 16d59'14"
Estela Zeik Hinojosa
And 0h23'23" 41d43'40"
Estelaluminous 040187
Cep 23h37'40" 80d2'28"
Estelita & Beatriz Cardoso Perez
Her 17h12'42" 18d53'6"
Estella
Sgr 18h42'18" -29d9'59"
Estella A. Boman
Uma 9h24'45" 45d48'52"
Estella Arias
Cas 3h14'5" 58d4'49"
Estella Josephine
Vir 12h55'46" -10d5'54"
Estella Louise Gross
Uma 9h45'25" 44d56'12"
Estella Stigler
Cas 1h25'54" 57d8'28"
Estelle
Aur 6h36'16" 48d32'3"
Estelle
Uma 11h37'15" 38d46'52"
Estelle
Aql 18h54'22" 7d51'13"
Estelle
Uma 9h55'10" 66d29'45"
Estelle
Uma 9h55'36" 66d3'32"
Estelle
Cas 1h19'45" 66d15'1"
Estelle
Cas 0h45'24" 61d46'34"
Estelle; A Star Named For A Star
Ari 2h58'19" 30d13'24"
Estelle Andrzejewski
Ari 2h55'51" 28d52'55"
Estelle Bianca Amaitis
And 2h37'56" 45d56'40"
Estelle Brosset
Vir 13h34'14" 6d15'13"
Estelle Chabot
Cap 20h14'35" -8d57'47"
Estelle Cohen-Solal
Uma 9h54'13" 43d33'11"
Estelle Cole
Crb 15h30'13" 28d30'28"
Estelle Dawn Zorman
Tau 3h59'2" 28d8'49"
Estelle Drew
And 23h59'24" 46d34'8"
Estelle et David
And 2h33'34" 45d59'43"
Estelle Farbstein
Ari 2h21'59" 23d51'41"
Estelle Flicher Leo Sun 30
Cas 23h53'11" 50d20'21"
Estelle Golby
And 0h54'47" 40d20'5"
Estelle Hill's 100th Birthday Star
Sco 16h44'32" -39d12'34"
Estelle Hoffman
Com 13h8'5" 26d29'5"
Estelle " I did it my way"
Aqr 20h47'52" -11d39'41"
Estelle Jane Abraham
Cas 0h59'51" 50d32'21"
Estelle LeBenger
Vir 11h40'55" -0d4'32"
Estelle M. Flett
Lyn 7h39'1" 52d39'9"
Estelle M. Hines
And 23h18'54" 43d58'49"
Estelle M. Jones
And 0h55'40" 38d24'28"
Estelle Marie
Psc 1h44'31" 6d19'6"
Estelle Nuage Rêveur
Psc 1h48'36" 7d27'37"
Estelle Rae
Cnc 8h5'10" 18d43'58"
Estelle Ray Bonner
Gem 6h33'40" 24d27'7"
Estelle Raymundo
Tau 4h14'15" 27d18'47"
Estelle Reale
Umi 16h45'11" 76d58'47"
Estelle Robichaux
And 23h14'50" 47d55'10"
Estelle & Ron Davis
Leo 11h32'18" 13d18'52"
Estelle Stern
Cyg 19h37'28" 32d43'38"
Estelle & Tony
Psc 0h46'35" 15d21'26"
Estelle's Star
Tau 4h24'14" 17d17'38"
EStephania
And 1h34'31" 49d51'11"
Ester
Cnc 8h36'6" 8d0'33"
Ester
Col 5h42'39" -33d18'30"
Ester de Araújo Cypriano - Tequinha
Cru 12h32'19" -58d12'47"
Ester Esteban TAPS
Cap 20h35'52" -25d1'36"
Ester Figueras Puig
Cep 21h32'38" 57d0'56"
Ester J. Perrotti
Cas 0h9'38" 56d22'40"
Ester Meldgaard Andersson
And 23h24'23" 50d37'52"
Ester Sandtrø
Lmi 10h41'42" 31d25'31"
Ester Sleutelberg
And 0h2'15" 37d29'15"
Ester Suh
Umi 15h22'46" 68d2'1"
Ester & Toni Bover
Cas 0h45'16" 64d52'41"
Ester-03071971
Vul 19h50'57" 21d42'15"
EsteraP1
Leo 11h8'27" 6d18'32"
ESTERMANCEBO
Sco 17h55'51" -42d43'41"
Estes Oil
Umi 14h16'2" 87d55'3"
Esther
Mon 6h24'13" -5d1'15"
Esther
Cep 21h43'8" 56d18'33"
Esther
Cae 4h53'22" -27d56'46"
Esther
Ari 2h35'0" 14d6'16"
Esther
Tau 4h54'17" 21d40'1"
Esther
Her 18h9'46" 19d42'53"
Esther
And 0h28'25" 32d26'56"
Esther
Gem 6h52'44" 33d58'46"
Esther
And 1h19'9" 42d3'47"
Esther
Cas 0h55'38" 52d23'32"
Esther
Lyn 7h51'4" 49d59'48"
Esther #1 Mom Wong
Umi 16h19'14" 85d12'4"
Esther & Anaida (Forever Love)
Vir 13h14'15" 11d31'17"
Esther Anna
Psc 1h45'8" 13d10'39"
Esther Beasley
Cas 1h18'35" 60d33'55"
Esther Bouman-Hulst
Cyg 21h52'19" 49d45'19"
Esther Bubbie Frailer
Uma 10h24'21" 67d21'4"
Esther C. Levin
And 0h30'41" 31d16'39"
Esther Carigiet
Per 4h11'26" 35d34'51"
Esther Chitwood
Crb 16h10'44" 36d36'41"
Esther E. Gross
Cyg 21h12'16" 41d20'56"
Esther Erickson Wojcik
Sgr 19h11'53" -12d24'24"
Esther Flieg
Crb 16h22'6" 30d43'21"
Esther G. Blevin
Cas 0h36'22" 55d50'4"
Esther Gonzales
Psc 23h54'34" 0d8'1"
Esther Grace
Uma 10h55'11" 55d46'21"
Esther Hernandez
Psc 1h16'31" 15d19'33"
Esther Hui Ai HO
Vir 12h53'56" -9d7'8"
Esther & Ira Kay
Pho 23h46'12" -43d10'48"
Esther J. McMahon
Crb 15h38'10" 31d41'29"
Esther K. Howard
Crb 15h39'39" 28d49'54"
Esther K. Nakao
Crb 15h32'50" 32d55'52"
Esther Kantrowitz
Uma 10h16'8" 46d33'48"
Esther Kaufman
Psc 23h26'52" 5d43'48"
Esther Kim
Leo 9h24'27" 11d34'52"
Esther Kim
Uma 13h39'26" 54d51'33"
Esther L. Patterson
Uma 9h53'2" 69d54'27"
Esther Leigh
And 2h20'40" 39d2'26"
Esther Louisa
And 1h44'4" 45d15'0"
Esther M. Furst
Tau 3h25'34" -0d47'53"
Esther M. Saldivar
Sgr 18h7'29" -27d7'7"
Esther Maria Ayala
And 0h37'55" 37d1'4"
Esther Marie
Cas 1h29'22" 59d8'4"
Esther Mary
Tau 3h43'21" 28d40'38"
Esther & Matias
Gem 6h37'59" 26d14'31"
Esther Matz
Cap 21h41'0" -9d6'23"
Esther May Black
And 23h35'32" 47d24'36"
Esther Mercedes Garcia
Lyr 19h27'27" 42d46'32"
Esther Merki
Mon 6h32'6" 9d9'34"
Esther Mojica
Vir 13h13'34" -4d4'21"
Esther Morse
Lyn 7h11'39" 45d17'25"
Esther Nava
Tau 4h56'58" 16d22'16"
Esther Nawrocki
Ari 3h6'25" 10d29'31"
Esther Nicholas
Lyn 7h43'34" 41d13'5"
Esther Olivia Foggin
Aql 19h5'47" -6d17'0"
Esther P See
Sco 16h19'26" -10d8'54"
Esther Pfammatter
Peg 22h49'34" 27d51'52"
Esther Preston
Crb 15h47'10" 31d19'35"
Esther & Rolf
Cas 23h51'57" 53d30'33"
Esther San Millan
Aql 19h2'52" -8d27'5"
Esther Slachter
Uma 12h31'55" 56d45'4"
Esther Sloshberg
Psc 1h14'21" 28d31'38"
Esther Snyderman MD
Sco 17h31'22" -45d28'34"
Esther Speicher
Sco 16h17'12" -10d0'19"
Esther Susanne
Cas 23h41'36" 55d8'21"
Esther T. Berger (E.T.)
Lib 15h45'33" -10d47'2"
Esther Travis
Umi 13h15'15" 71d47'14"
Esther Virginia (Nana)
Cas 22h59'23" 54d35'30"
Esther Wanerman
Lyn 7h49'51" 35d36'45"
Esther y Abraham
Her 18h29'58" 15d52'21"
Esther Yendukor Lare
Ari 3h4'45" 22d4'47"
Esther & Yves
Cas 1h3'26" 60d18'7"
Estherella
Cnc 8h18'27" 18d1'29"
Esther-Fretz
Uma 11h21'51" 43d37'30"
Esther-Im
And 0h4'41" 44d37'12"
Esthermay "30 Years"
Uma 8h48'0" 55d18'34"
Esther's Glückstern
Peg 22h38'9" 26d30'52"
Esther's "GOLD"en Star
Gem 7h17'22" 31d17'35"
Esther's star
Leo 9h37'58" 30d40'18"
Esther's star
Cyg 19h45'55" 56d18'34"
Esther's Star "Everybody's Mom"
Lyr 19h17'30" 30d6'39"
Esthi + Wädi
Uma 11h28'16" 31d38'11"
Estie Adams
Sco 16h10'56" -11d23'30"
Estil A Vance Jr.
Per 2h43'56" 55d51'16"
Esto perpetua
Peg 21h42'19" 16d16'7"
ESTO PERPETUA
Del 20h40'58" 16d32'37"
Esto Perpetua
Psc 0h33'52" 8d22'7"
Esto Perpetue - Gary Barnet
Sco 17h29'4" -32d10'17"
Estoban's Elysium
Her 18h19'30" 21d26'4"
estralla de sylvia
Ori 6h3'8" -0d4'56"
Estrela
Pho 1h3'49" -48d32'4"
Estrela da Sara Bertram
Crb 16h2'15" 35d39'23"
Estrela de Graziella
Leo 9h57'58" 15d51'33"
Estrela de Nair Giupponi Franca
Leo 10h17'5" 13d0'11"
Estrela Para Elsa
Psc 23h58'56" -0d10'36"
Estrela Zuda
Leo 9h29'52" 29d7'42"
Estrelita Little Star
Lyn 7h27'23" 52d19'38"
Estrella
And 0h12'28" 39d1'54"
Estrella
Com 12h25'6" 17d39'52"
Estrella
Leo 11h30'28" 10d17'1"
Estrella
Dra 10h14'11" 73d6'41"
Estrella
Sgr 18h54'47" -30d49'37"
Estrella Alvarado
Ari 2h14'19" 24d44'53"
Estrella Amalia Vargas
Psc 0h32'38" 18d6'29"
Estrella Anabela
Sgr 18h50'25" -25d37'51"
Estrella Arellano
Tau 5h41'42" 25d39'36"
Estrella Blossom
Sgr 18h45'6" -27d9'42"
Estrella Bonita
Sgr 18h40'40" -17d0'3"
Estrella de Alicia
Mon 7h21'12" -0d16'32"
Estrella de Amaysin
Leo 11h5'31" 5d20'18"
Estrella de Arturo Arriaga
Per 4h41'24" 43d58'56"
Estrella de Blume
And 23h25'6" 47d43'35"
Estrella de Carlos Davant
And 1h14'50" 35d9'25"
Estrella de Cesar Astorga
Ori 5h55'16" -0d38'3"
Estrella de Conner Kids
Uma 9h39'34" 54d45'35"
Estrella de Denise
Vir 11h54'25" 2d56'7"
Estrella de Esperanza
Uma 13h50'11" 49d40'12"
estrella de gordis
Lyr 18h39'54" 35d52'49"
Estrella de Hernandez
Aur 5h8'32" 33d52'10"
Estrella de Huebner
And 0h50'20" 23d12'8"
Estrella de Italina
Lyn 6h42'30" 61d7'53"
Estrella de Jami
And 1h27'53" 42d17'41"
Estrella de Jennifer
Psc 0h38'24" 21d26'59"
Estrella de Karina
Lib 14h48'47" -10d39'45"
Estrella de la Esperanza
Tri 2h3'1" 32d17'4"
Estrella De La Perla
Tau 4h32'1" 21d11'29"
estrella de la suerte
Ori 6h18'36" 11d1'47"
Estrella de Lucas
Cnc 8h32'58" 32d52'41"
Estrella de Marcita
Com 12h8'0" 18d51'8"
Estrella de Marine
And 0h46'21" 40d21'11"
Estrella De Mary
Per 2h58'13" 55d34'26"
Estrella de Melbourne
Leo 11h2'15" 0d28'1"
Estrella de mi amor para Tara
Cyg 20h0'25" 33d14'12"
ESTRELLA DE MI VIDA
Aqr 23h31'0" -7d19'46"
Estrella de Negrón y Silman
Lib 15h24'54" -7d54'45"
Estrella de Phaedron Mirabella
Sco 17h12'37" -34d1'0"
Estrella de Sandra
Lyn 7h34'34" 40d59'5"
Estrella del Amor
Peg 23h10'8" 24d43'53"
Estrella del Carmen Polo
Uma 8h43'58" 48d49'41"
Estrella del Padre Número Uno
Lib 15h44'59" -18d59'12"
Estrella Dunst
Boo 14h13'55" 29d38'14"
Estrella Eterna de CBR X4M
Per 2h15'51" 51d41'39"
Estrella Eterna Del Amor
Gem 6h33'23" 24d13'32"
Estrella Italombiana
Peg 22h9'48" 34d4'8"
Estrella Jennifer Staples
Cap 21h35'56" -8d38'48"
Estrella Joan Valborg Anderson
And 0h16'23" 41d35'32"
Estrella linda
And 1h23'27" 34d33'23"
Estrella Maeda
Cas 1h20'9" 55d34'46"
Estrella Marcela
Sgr 18h23'41" -30d44'59"
Estrella Marina Pons
Crb 15h37'47" 32d3'27"
Estrella Molina Lopez
Aql 20h13'25" -0d1'13"
Estrella Paez
Cam 9h38'30" 82d20'49"
estrella peralez
Leo 10h39'19" 19d6'47"
Estrella Renard
Uma 10h57'18" 35d19'25"
Estrella Rochelle O'Connor
Uma 12h0'35" 36d14'33"
Estrella Secreta de Stephanie
Cap 20h25'23" -11d19'3"
Estrella Serna
Lyn 7h5'57" 50d37'33"
Estrella Sylvia
Tau 4h27'59" 1d20'47"
Estrelle
Uma 8h30'33" 72d53'30"
Estrellita
Cap 20h26'40" -12d33'9"
eSTRELLiTA
Vir 13h15'8" 12d49'15"
Estrellita
Tau 4h40'34" 22d40'20"
Estrellita Drago Urso
Dra 17h45'32" 54d5'55"
Esveidet
And 1h30'4" 42d41'33"
Eszter & Viktor esküvoi csillaga
Uma 9h6'37" 72d4'47"
EszterCsillag
Lib 14h49'8" -5d56'21"
Eszti és Domi Csillaga
Vir 14h15'7" -17d9'6"
Esztike
Cas 0h10'32" 51d12'58"
E.T.
Cyg 19h51'46" 52d18'40"
E.T.
And 0h57'28" 43d12'23"
E.T.
Ori 5h55'42" 17d6'18"
Et Jaime Tu
Uma 10h42'48" 57d38'56"
E.T. Wenzel
Dra 19h15'58" 65d19'13"
Eta 581
And 0h20'45" 38d20'50"
Eta17 Carinae Tauri Tonti
Uma 10h42'3" 49d7'25"
Etai Eshel
Crb 16h7'49" 33d49'0"
Étain Mary McCarthy
Leo 11h19'16" 18d28'31"
Etaks
And 2h6'5" 41d41'52"
etamas
Gem 7h30'41" 16d59'47"
Etana
Aqr 21h13'26" 0d31'29"
Etat
Aqr 22h31'21" -5d13'57"
ETEMPA
Ari 2h6'16" 20d12'18"
Eterna Amor Hasta My Hija Nicole
Aqr 23h3'55" -13d21'16"
Eternal
Cru 12h10'48" -61d56'57"
Eternal
Gem 7h21'36" 28d12'36"
Eternal
Her 17h57'51" 23d29'24"
Eternal
Crb 16h15'55" 30d16'19"
eternal
Cnv 13h19'17" 48d10'47"
Eternal
Uma 13h38'35" 50d20'7"
Eternal Aquarius Love
Cyg 20h31'43" 36d51'8"
Eternal Bliss
Sco 17h17'12" -34d8'3"
Eternal Blissfulness
And 0h57'43" 36d25'6"

Eternal Bond
Her 17h58'8" 23d28'47"

Eternal Bryony
Crb 15h48'6" 34d3'3"

Eternal Butterfly
Psc 1h23'18" 30d45'16"

Eternal Champion
Ori 5h32'47" 14d7'23"

Eternal Cruising
Cen 12h29'35" -50d7'52"

Eternal Daddy
Cru 12h35'35" -60d34'43"

Eternal Destiny
Tau 4h43'1" 23d13'34"

Eternal Devotion
Cyg 20h22'55" 40d16'1"

Eternal Faith
Vir 13h22'40" 11d57'46"

Eternal Fire
Psc 0h50'21" 16d35'22"

Eternal Fire, Eternal Love
Uma 13h21'35" 57d37'0"

Eternal Flame
Cen 14h41'57" -41d30'27"

Eternal Flame
Cma 7h3'12" -24d14'11"

Eternal Flame
Cnc 8h8'28" 22d16'54"

Eternal Flame Belinda
Pav 20h47'52" -60d12'58"

Eternal Flame Of Rachel and Craig
Cyg 20h13'2" 49d40'4"

Eternal Flaming Heart
Cap 20h24'16" -15d32'10"

Eternal Friendship
Lmi 9h56'11" 36d31'37"

Eternal Happiness
Cyg 21h23'4" 51d35'28"

Eternal Holly
And 2h32'56" 48d46'47"

Eternal Hope
Umi 14h56'22" 72d4'6"

Eternal Interlacement
Cyg 20h36'33" 40d33'35"

Eternal Ione
Psc 0h15'8" -3d30'44"

Eternal - Jas & Baj
Cyg 20h29'9" 30d29'49"

Eternal Jeffrey Simon Cohen 5-30-76
Lyr 18h50'32" 36d8'27"

Eternal Joy
And 1h2'26" 35d16'32"

Eternal Joy
Sco 17h54'33" -36d0'27"

Eternal Kevin J. Hill
Cnc 8h15'51" 25d44'54"

Eternal Light
Cru 12h26'9" -61d34'11"

Eternal Light
Pho 0h37'8" -51d2'0"

Eternal Light
Cap 20h28'15" -9d47'42"

Eternal Light for Vivian
Uma 10h10'17" 46d16'5"

Eternal Love
Uma 11h45'41" 47d32'47"

Eternal Love
Cyg 20h39'45" 48d46'38"

Eternal Love
Cyg 21h45'27" 47d1'32"

Eternal Love
Lyr 18h36'26" 34d4'7"

ETERNAL LOVE
Cyg 19h41'59" 32d5'54"

Eternal Love
Cyg 20h51'10" 35d6'22"

Eternal Love
Cyg 21h21'21" 30d55'20"

Eternal Love
Her 17h4'41" 29d58'36"

Eternal Love
And 0h17'45" 33d20'23"

Eternal Love
Ori 5h52'44" 21d25'53"

Eternal Love
Ori 6h8'44" 21d6'56"

eternal love
Ari 2h14'59" 26d7'6"

Eternal Love
Sge 19h41'3" 17d28'10"

Eternal Love
Her 17h52'50" 21d52'50"

Eternal Love
Sge 19h49'4" 18d36'21"

Eternal Love
Del 20h25'28" 9d39'16"

Eternal Love
Leo 11h34'55" 10d9'42"

Eternal Love
Leo 10h25'8" 12d34'34"

Eternal Love
Leo 9h32'17" 13d50'12"

Eternal Love
Cma 6h34'31" -14d50'4"

Eternal love
Aql 19h51'43" -0d16'4"

Eternal Love
Aqr 22h27'54" -1d16'56"

Eternal Love
Umi 15h16'56" 74d33'12"

Eternal Love
Cyg 19h43'51" 54d10'48"

Eternal Love
Cyg 19h46'6" 52d31'35"

Eternal Love
Lyn 7h20'12" 55d6'33"

Eternal Love
Sco 17h48'34" -39d19'49"

Eternal Love
Psc 1h17'33" 3d27'44"

Eternal Love
Sgr 18h7'36" -23d18'40"

Eternal Love ···Love is Here!
Lyn 8h29'18" 58d30'53"

Eternal Love Blessed By Angels
Aqr 22h52'16" -23d48'27"

Eternal Love & Ecstasy !
Sco 16h9'37" -11d38'28"

"Eternal Love" Elizabeth S & Doug C
Vir 12h57'40" 13d35'6"

Eternal Love For Barbara
Cyg 20h16'25" 38d20'21"

Eternal Love for Carin and Roman
Cas 0h10'54" 59d21'6"

Eternal Love for Lauren St. Pierre
Cyg 21h26'26" 30d24'27"

Eternal love - Gregory A. Williams
Psc 1h45'46" 10d36'16"

Eternal Love J & H
Sge 20h5'35" 17d9'8"

Eternal Love Joseph And Amanda Tuck
Cyg 19h35'28" 30d11'22"

Eternal love of Chris & Steeve
Sco 17h49'12" -39d5'6"

Eternal Love of Jennifer and Eric
Uma 9h38'52" 47d48'12"

Eternal Love - Ron and Karol
Hor 4h8'31" -42d19'34"

eternal love Stephanie&Olivier
Sgr 17h53'29" -20d18'20"

Eternal Love wtih Saori & Hiroyuki
Ari 2h31'38" 13d35'19"

Eternal Love, Sonya
Psc 1h29'0" 31d11'15"

Eternal Love-Barrett & Katie
Ari 3h29'3" 23d1'28"

Eternal LovErotica
Umi 14h57'28" 85d28'48"

Eternal Lovers
Cyg 20h18'59" 37d54'50"

Eternal Presence
Cyg 21h31'46" 45d59'52"

Eternal Rapture
Cyg 20h0'7" 33d27'46"

Eternal Reflection
Sgr 18h32'20" -24d52'5"

Eternal Rose
Peg 21h42'30" 18d6'44"

Eternal Sarah
Ori 6h19'53" 2d3'16"

Eternal Son
Gem 7h52'44" 27d43'47"

Eternal Sparkle
Ari 2h48'24" 27d43'44"

Eternal Star
Uma 10h33'51" 59d28'22"

Eternal Star
Cas 23h29'7" 53d42'22"

Eternal Star of Takamasa & Yulia
Cnc 8h32'57" 9d2'43"

Eternal Star of Yuka Kosaka
Cnc 8h4'24" 17d55'32"

Eternal Starfire
Cyg 20h15'55" 55d50'2"

Eternal Starshine Gina Calvert
Ari 2h16'5" 25d44'34"

Eternal Sunshine
Pho 1h28'48" -41d1'30"

Eternal Teddy Bear
Ori 6h7'38" 10d40'52"

Eternal Treasure Yuko & Nori
Gem 7h56'8" 20d55'39"

Eternal Valentine For My Beloved
Cyg 19h45'34" 34d7'53"

Eternal-Lee
Psa 21h58'30" -29d24'47"

Eternally Bain's
Uma 13h50'3" 50d11'45"

Eternally Beautiful April
Leo 11h26'46" 14d19'50"

Eternally Casey
Ori 6h18'33" 10d20'4"

Eternally CK's boys
Lyn 7h21'48" 54d19'58"

Eternally Connected
Cep 22h9'29" 65d47'12"

Eternally ENJ
Cyg 20h22'15" 46d38'47"

Eternally I-Chieh
Uma 11h55'41" 57d33'12"

Eternally Jesse
Umi 13h29'43" 72d32'32"

Eternally Lianne
Gem 6h48'40" 28d51'6"

Eternally Loved Tiffany Lynn Wieand
Cnc 9h4'41" 31d49'40"

Eternally Loving Peter Bosson
Her 16h45'14" 18d19'43"

Eternally Madeline
Sgr 18h16'44" -28d12'39"

Eternally Nancy
Psc 1h18'0" 23d9'23"

Eternally Stacey's Star
Cyg 22h0'11" 50d4'5"

Eternally TaRee
And 2h29'52" 44d12'23"

Eternally Twinkling Brandi
Uma 9h48'49" 53d45'10"

Eternally Yours
Sco 16h51'55" -41d34'18"

Eternally Yours
Tau 5h48'17" 22d44'36"

Eternamente bellissimo
Uma 9h42'17" 68d12'45"

Eternamente Geoff e Hilary
Aqr 23h11'22" -11d12'30"

Eternellement Amour
Sgr 19h39'20" -40d13'39"

Eterni Arameyne
Psc 23h51'25" 5d8'33"

Eternidad
Boo 14h18'28" 20d28'51"

Eternité
Her 17h12'57" 13d38'43"

Eternité
Sgr 18h34'49" -24d9'24"

Eternité
Uma 10h22'32" 63d49'10"

Eterniti
Cyg 20h11'55" 42d26'13"

Eternity
Lyr 19h21'40" 38d37'25"

Eternity
And 1h13'46" 46d15'17"

Eternity
Cyg 21h57'52" 46d35'36"

Eternity
Cyg 21h57'46" 47d21'49"

Eternity
Per 3h3'37" 54d35'5"

Eternity
Cyg 19h58'48" 51d42'43"

Eternity
Uma 10h31'57" 50d47'32"

Eternity
Lyn 9h6'13" 39d52'39"

Eternity
Cyg 20h25'36" 36d50'28"

Eternity
Aur 6h1'26" 36d34'44"

ETERNITY
Psc 1h16'56" 32d4'3"

Eternity
Cyg 19h51'54" 31d49'0"

Eternity
Vir 13h16'45" 12d52'37"

Eternity
Ori 6h12'37" 8d29'47"

Eternity
Tau 5h18'41" 18d5'38"

Eternity
Psc 0h38'19" 8d43'0"

eternity
Ori 5h44'7" 5d0'28"

Eternity
Cnc 8h41'44" 7d19'0"

Eternity
Gem 6h41'27" 18d26'34"

Eternity
Psc 1h11'57" 24d33'29"

Eternity
Her 18h15'37" 21d35'45"

Eternity
Sge 19h45'17" 17d28'43"

Eternity
Gem 6h38'37" 22d53'25"

Eternity
Leo 10h7'13" 27d17'33"

Eternity
Uma 8h57'4" 65d27'46"

Eternity
Uma 10h39'32" 68d30'6"

Eternity
Umi 15h18'1" 71d41'31"

Eternity
Umi 15h16'4" 74d22'1"

Eternity
Uma 8h55'56" 55d58'29"

Eternity
Uma 8h42'22" 53d56'17"

Eternity
Uma 11h57'0" 54d19'58"

eternity
Uma 11h54'57" 55d54'15"

Eternity
Uma 9h58'44" 59d6'56"

Eternity
Uma 10h0'16" 55d26'43"

Eternity
Cyg 20h38'49" 53d50'46"

Eternity
Uma 14h13'39" 55d18'3"

Eternity
Aqr 22h15'23" -8d3'24"

Eternity
Sco 16h9'12" -18d0'35"

Eternity
Cam 4h35'40" 75d10'17"

Eternity
Eri 4h15'37" -0d11'4"

Eternity
Pho 23h42'13" -42d53'15"

Eternity
Col 6h22'44" -39d24'58"

Eternity
Dor 6h19'53" -67d13'58"

Eternity
Cru 12h28'15" -59d40'4"

Eternity 11041989
Lyn 7h42'25" 45d58'25"

Eternity 66
Leo 9h39'47" 30d15'30"

~Eternity~ Alex & Shelly forever!
Lib 15h11'58" -7d27'37"

Eternity Aurélia-Jean-Philippe
Uma 11h6'36" 60d15'22"

Eternity "Craig & Nikki"
Cyg 19h52'53" 38d6'37"

Eternity Moment
Ari 2h6'36" 21d45'56"

Eternity of Dearest Hidei
Sco 16h47'33" -41d23'29"

Eternity of Love
Tau 4h4'27" 26d2'7"

Eternity Of Love, N.Morris & T.Bull
Cru 12h6'47" -58d38'34"

Eterno Amare
Uma 11h42'2" 41d55'2"

Eterno amore
Cam 3h28'12" 61d13'31"

Eternus Amicitia
Sgr 19h7'16" -18d25'47"

eternus amicitia
Psc 0h45'48" 19d23'51"

Eternus Amor
Cyg 20h46'45" 39d42'54"

Eternus Amor
Sgr 19h43'57" -16d12'20"

Eternus Amor
Lyn 6h23'7" 54d26'39"

Eternus Amor STJW NJTW
Cyg 20h8'11" 36d23'18"

Eternus Diligo
Lib 14h49'48" -0d33'22"

Eternus Diligo
Vir 12h27'49" -1d3'50"

Eternus diligo: Amado et Cookie
Uma 11h33'52" 31d22'38"

Eternus Impetus
Sco 17h33'39" -44d35'39"

Ethal Marie Shiffer
Cap 20h28'50" -12d34'28"

Ethan
Cam 4h19'50" 76d12'13"

Ethan
Eri 4h7'1" -11d28'56"

Ethan
Her 17h20'37" 46d40'14"

Ethan
Vir 13h9'56" 3d18'47"

Ethan
Aqr 22h19'25" 1d53'49"

Ethan
Leo 11h45'40" 24d10'0"

Ethan
Her 17h57'27" 29d32'32"

Ethan
Boo 14h54'54" 17d36'7"

Ethan
Leo 11h2'36" 22d16'9"

Ethan Abel Lloyd
Sgr 18h23'22" -23d17'6"

Ethan Alexander Conlin
Ori 5h35'59" 2d0'43"

Ethan Allan Conklin Simas
Cnc 8h59'18" 8d22'53"

Ethan and Allison Renda
Gem 7h25'48" 31d3'37"

Ethan Andrew Morning
Tau 3h46'40" 23d29'53"

Ethan Anthony Bennett
Her 16h31'49" 37d4'37"

Ethan Anthony McHugh
Ori 5h0'52" 5d12'0"

Ethan Antonio Floyd
Cyg 21h18'11" 31d31'3"

Ethan Aron Duff
Umi 14h9'54" 66d31'20"

Ethan Asher Frank
Aqr 21h4'42" 0d45'25"

Ethan. B.
Vol 6h37'48" -66d0'30"

Ethan Bakker
Aql 19h49'22" 5d58'27"

Ethan Banaag
Per 3h26'52" 43d31'5"

Ethan Beyreis-Heim
Psc 1h8'57" 26d55'58"

Ethan Biggers
Her 16h32'17" 38d8'17"

Ethan Blumenthal
Aql 20h1'6" 7d51'14"

Ethan Bowen Dymond Burchell
Uma 13h4'6" 53d6'24"

Ethan Boyd Everett - 18 Nov 2003
Sco 16h19'5" -42d3'19"

Ethan Bradly Wagner
Sco 16h14'56" -12d48'30"

Ethan Brent Johnston
Boo 15h19'28" 32d42'50"

Ethan Brock
Ori 4h58'4" -2d25'33"

Ethan Brodie
Cnc 8h34'45" 32d25'27"

Ethan "Bug-a-Boo" James
Aqr 22h18'48" -0d2'22"

Ethan Claude Naidoo
Her 16h57'13" 17d22'0"

Ethan Clinton
Leo 10h55'37" -0d6'10"

Ethan Cole Torpey
Cam 4h18'22" 73d51'26"

Ethan Cook
Her 17h37'38" 24d25'52"

Ethan Cox
Psc 1h8'4" 29d34'7"

Ethan Curtis
Per 3h32'2" 45d14'48"

Ethan D Grantis
Umi 16h22'45" 73d34'11"

Ethan Daniel McDermott
Crb 16h14'12" 26d44'49"

Ethan David
Uma 13h39'12" 57d6'24"

Ethan David Armour
Her 18h10'15" 18d58'34"

Ethan David Schwartzbaum
Dra 18h55'44" 50d19'23"

Ethan David's Christening Star
Per 3h10'16" 46d30'15"

Ethan Devon 10498
Lib 15h46'38" -13d0'31"

Ethan E. Vernick
Gem 7h7'9" 34d27'25"

Ethan Edvard Rysanek
Cnc 9h4'50" 32d37'13"

Ethan Edward
Umi 17h29'51" 82d24'27"

Ethan Edward Shapiro
Cap 20h58'16" -21d36'9"

Ethan Edward Sprague
Cam 3h47'51" 66d48'23"

Ethan Edward Walder
Aql 18h59'28" -10d23'56"

Ethan & Elise
Peg 22h43'10" 24d24'27"

Ethan Foreman
Ori 5h47'17" 7d53'10"

Ethan - Forever My Star
Aqr 22h23'10" -7d14'1"

Ethan Frederick
Sco 17h19'32" -39d29'25"

Ethan G. Bohm
Lac 22h27'12" 50d23'8"

Ethan Gahan
Lib 15h24'42" -15d54'11"

Ethan H. Naftalin
Her 18h43'48" 21d59'37"

Ethan Henry Brandt
Sgr 19h17'43" -18d55'29"

Ethan Henry Hirsch
Cnc 8h24'4" 14d0'5"

Ethan Henry Leopoldus
Her 18h50'31" 22d51'47"

Ethan Henry Pfeiffer
Cnc 8h42'46" 24d15'57"

Ethan Henson Kaplan
Del 20h40'6" 16d44'26"

Ethan H.M. Yoshioka 9-19-1999
Vir 13h5'57" -10d3'18"

Ethan & J. Lyn
Lyr 19h15'28" 29d34'8"

Ethan Jackson
Uma 12h2'52" 33d46'17"

Ethan Jackson Shines Forever
Umi 14h15'10" 78d32'56"

Ethan Jacob Rolo
Ori 4h53'41" 8d20'30"

Ethan Jacob Trend
Per 2h58'16" 42d53'0"

Ethan Jake Stockham
Umi 14h33'19" 68d7'5"

Ethan James
Dra 17h38'42" 55d24'22"

Ethan James
And 2h19'56" 46d32'27"

Ethan James - A star for eternity
Cru 12h20'26" -57d51'44"

Ethan James Dush
Cnc 8h31'52" 19d14'13"

Ethan James Evans
Ori 5h18'56" -1d18'9"

Ethan James Forbis
Sgr 19h40'35" -16d56'39"

Ethan James Gledhill
Umi 13h41'15" 85d58'28"

Ethan James Hoggan
Dra 19h24'19" 64d1'25"

Ethan James Jauch
Vir 12h21'57" 11d23'31"

Ethan James Krieger
Per 4h10'9" 34d54'15"

Ethan James Landes
Cnc 8h53'36" 19d20'9"

Ethan James Putnam
Ori 6h15'57" 14d42'13"

Ethan James Randall Jagoe
Cyg 19h37'17" 41d46'34"

Ethan James Traynor
Lyn 7h47'41" 41d16'35"

Ethan James Williams
Uma 9h35'19" 60d3'49"

Ethan James Williams
Vir 13h40'25" -21d7'2"

Ethan James Woods
Umi 13h59'57" 72d30'7"

Ethan Jeffery Stillwell
Vir 12h39'46" 5d41'1"

Ethan Jett - 9 June 2006
Cru 12h28'49" -60d23'15"

Ethan Jett Kalem
Cap 21h32'57" -13d50'30"

Ethan Joe
Per 4h15'45" 45d33'30"

Ethan John
Aqr 21h19'3" 1d12'50"

Ethan John
Lib 15h19'44" -23d15'50"

Ethan John Benjamin
Sgr 19h0'41" -30d54'28"

Ethan John Brown
Umi 15h10'27" 76d17'4"

Ethan John Dewhirst - E J's Star
Ori 5h42'24" 13d4'8"

Ethan John James Ritter
Umi 14h19'3" 76d17'4"

Ethan John Smith
Umi 14h29'48" 68d0'55"

Ethan Joseph Burger
Gem 6h3'29" 25d41'47"

Ethan Joseph Cloud
Ari 2h28'8" 25d26'12"

Ethan Joseph DeGuevara
Tau 4h0'26" 27d6'23"

Ethan Joseph Leone
Cyg 20h34'41" 37d37'36"

Ethan Joseph Mehlick
Ori 5h31'11" 0d23'24"

Ethan Joseph Trolinger
Uma 10h3'23" 48d21'57"

Ethan Jozsef Derbowka
Umi 15h30'59" 75d15'7"

Ethan Koepke
Sco 17h6'34" -34d48'28"

Ethan Kyler Tinsley
Ori 6h12'48" 15d57'19"

Ethan Lars Murphy
Ari 3h9'28" 26d48'59"

Ethan Laurance
Umi 13h47'49" 72d46'46"

Ethan Lawrence Diaz
Vir 12h29'13" -2d53'49"

Ethan Lawrence Diaz
Vir 12h7'53" -8d5'34"

Ethan Lawrence Diaz
Uma 13h58'48" 49d3'38"

Ethan Lawson Beier
Vir 14h16'20" -13d52'25"

Ethan Layne Berg
Cap 21h41'53" -23d14'7"

Ethan le Petit Prince de nos coeurs
Ori 6h8'35" 12d24'15"

Ethan Lee Van Ornum
Gem 7h3'2" 15d21'32"

Ethan Lewis Thomas
Per 2h15'24" 53d59'11"

Ethan Link Griggs
Cnc 8h10'47" 17d3'44"

Ethan Lloyd Gereke
Umi 15h14'38" 72d32'19"

Ethan Lucas
Her 17h57'28" 22d29'47"

Ethan Lukas Marijosius
Uma 11h59'51" 55d1'31"

Ethan Luke Arrowsmith O'Connor
Per 3h12'41" 44d41'31"

Ethan M. Clough
Umi 15h36'0" 78d29'54"

Ethan Magid
Cyg 19h41'35" 34d16'13"

Ethan MaKyle Fritz
Uma 10h6'57" 43d13'5"

Ethan Martin Alan Cooley
Lmi 10h37'16" 29d17'17"

Ethan Martin Berg
Cap 21h33'48" -13d52'21"

Ethan Matthew Fulmer
Sgr 18h3'20" -20d58'3"

Ethan Max Levin
Cap 20h24'8" -11d9'25"

Ethan Michael
Lib 15h31'4" -11d41'19"

Ethan Michael Bohi
Cnc 8h14'32" 25d57'57"

Ethan Michael Coulter: Ethan's Star
Her 18h55'10" 16d30'53"

Ethan Michael Ehnert
Cap 21h2'32" -18d13'37"

Ethan Michael Horrighs
Aqr 23h7'23" -19d13'6"

Ethan Michael Maus
Leo 9h45'43" 29d36'19"

Ethan Michael Palmer
Aql 19h34'36" 10d42'8"

Ethan Michael Wunar
Gem 6h29'22" 16d29'56"

Ethan Michot Audain
Uma 8h33'37" 66d13'6"

Ethan "Monkey" Todd
Cap 20h40'17" -20d53'5"

Ethan Montgomery Fann
Vir 13h2'6" 12d22'23"

Ethan Morris
Aql 19h27'43" 0d18'54"

Ethan Neville Holt
Umi 14h48'51" 67d42'13"

Ethan Nicholas
Aqr 23h0'17" -8d40'32"

Ethan Noah Paul Martin
Vir 13h10'14" 14d3'41"

Ethan Oliver
Ori 5h31'13" 1d59'50"

Ethan Pannizzo Couto
Cas 0h55'17" 62d0'37"

Ethan Paul Todd - Ethan's Star
Ori 6h19'40" 8d11'4"

Ethan Peter Duborg
Leo 10h55'51" 6d23'30"

Ethan Phillip Brasher
Peg 23h44'35" 23d21'47"

Ethan Quinn
Aql 19h57'10" 15d13'19"

Ethan Raymond
Ori 6h21'1" 11d1'12"

Ethan Richard Phillip Fersch
Ori 6h17'5" 11d7'2"

Ethan Richard Wood
Per 4h22'47" 35d37'41"

Ethan Riley Gould
Her 16h43'4" 48d25'48"

Ethan Robert Ellis
Her 16h37'37" 46d39'15"

Ethan Robert Gallo
Cap 20h20'39" -10d39'31"

Ethan Roseman
Boo 14h26'13" 14d43'33"

Ethan Rothkrug Morris
Aql 18h58'11" -0d45'21"

Ethan Ryan Gersh
Her 17h55'32" 24d48'28"

Ethan Ryan Kay
Per 3h19'31" 44d43'32"

Ethan S. Altshuler
Her 16h29'46" 11d36'56"

Ethan Saheb Dudley Ziggy
Sgr 18h20'1" -32d15'4"

Ethan Samuel
Cam 4h0'23" 53d45'27"

Ethan Scott Huey
Tau 3h40'29" 9d0'9"

Ethan Scott Price
Cyg 21h23'27" 37d15'18"

Ethan Shane Brandwene
Her 17h41'54" 48d41'59"

Ethan Snow
Lyn 7h10'54" 52d32'8"

Ethan Solomon Bender
Boo 14h38'3" 50d6'25"

Ethan Stanbridge
Leo 10h20'47" 7d39'52"

Ethan Sutton Austria
Vir 14h8'24" -9d38'38"

Ethan Swan
Cap 20h44'9" -14d53'17"

Ethan Tejeda
Per 2h13'7" 54d49'53"

Ethan Thomas
Leo 9h41'5" 27d8'31"

Ethan Thomas Berde
Her 16h11'16" 48d2'58"

Ethan Thomas Custard
Lyn 7h1'25" 46d41'44"

Ethan Thomas Eads
And 1h19'39" 42d33'13"

Ethan Thomas Gamester
Per 4h15'3" 47d58'13"

Ethan Thomas Grant Morris
Umi 16h55'30" 77d24'3"

Ethan Thomas Kinsman
Vir 11h56'37" -0d49'22"

Ethan Thomas Lewis
Cep 0h16'40" 75d25'38"

Ethan Thomas Pascoe
Her 18h57'2" 26d5'14"

Ethan Thomas Peaslee Gustafson
Gem 6h7'12" 25d39'11"

Ethan Tyler
Aql 19h11'36" -7d33'18"

Ethan Tyler Culp
Tau 4h48'5" 18d36'20"

Ethan Vincent Attard
Leo 10h59'52" 24d0'33"

Ethan Vincent Martin
Leo 9h34'29" 29d1'16"

Ethan Vinnicombe
Cep 20h51'8" 60d11'24"

Ethan Wayne Diener
Vir 14h30'46" 3d53'29"

Ethan Weiss
Cap 21h3'27" -22d1'35"
Ethan Wells Bowman
Her 16h15'21" 6d15'37"
Ethan William Andrews
Ori 5h47'1" 8d6'33"
Ethan William Goldman
Sgr 18h50'38" -28d46'53"
Ethan William Layton
Uma 10h19'14" 53d5'44"
Ethan Zane Van der Berg
Cru 11h58'31" -63d50'6"
Ethan Zhi Hui Wong
Pho 0h36'0" -50d22'10"
EthanH
Ori 6h17'45" 13d13'36"
Ethanoid
Gem 6h20'40" 26d40'5"
"Ethan's Charlie Brown"
Lyn 8h35'9" 37d4'46"
Ethan's Enduring Light
Uma 11h44'23" 51d1'55"
Ethan's Nightlight
Uma 13h53'30" 52d42'4"
Ethan's Rising Star
Lib 15h59'47" -5d31'26"
Ethan's Star
Sco 17h25'19" -38d17'25"
Ethan's Star
Her 17h53'21" 17d39'26"
Ethan's Star
Gem 6h41'30" 15d19'37"
Ethan's Star - Ethan Thomas O'Neill
Ori 6h24'55" 14d45'33"
Ethan's World
And 23h38'33" 44d19'21"
Ethel
Sco 17h39'12" -39d39'26"
Ethel A. Balduf
Cas 0h26'54" 58d19'59"
Ethel Edith Love
Cyg 20h3'30" 38d20'17"
Ethel Einna
And 0h24'31" 25d3'37"
Ethel Elizabeth Eckert
Sco 17h34'12" -45d18'47"
Ethel Elva Stotz
Cas 0h48'33" 57d11'57"
Ethel Falls Tice
Cap 21h35'5" -13d15'44"
Ethel Frieder
Cas 1h39'37" 63d57'46"
Ethel G. Lashbrook
And 1h53'50" 45d4'27"
Ethel -Granny- Donohew
Uma 12h8'10" 48d13'31"
Ethel/Junior Goldstein
Lyn 6h47'37" 60d13'44"
Ethel Kaminski
Uma 10h15'57" 54d49'2"
Ethel Koocher
Psc 1h22'38" 4d46'33"
Ethel Lafreniere
Leo 11h24'18" 8d54'58"
Ethel Lashley, Beloved Wife and Mom
Cas 2h17'4" 73d14'30"
Ethel Looney
Dra 17h48'20" 66d45'19"
Ethel Lorene Lee
Psc 23h46'0" 1d57'9"
Ethel Loretta Anderson
Uma 10h21'5" 40d37'39"
Ethel Mae
Cas 1h30'11" 66d33'43"
Ethel Mae Kelley
Cas 3h15'26" 60d13'5"
Ethel Mae Webster
Cas 1h25'30" 62d1'16"
Ethel Margaret Chatlos May 15, 1921
Tau 5h35'25" 19d37'47"
Ethel May Broaddus Burtnett
Boo 14h31'18" 16d42'33"
Ethel May Jache
Lib 15h44'41" -17d7'17"
Ethel May Pittenger
Lyr 18h48'1" 37d20'18"
Ethel Poplovski, Bushie's Star Kids
Gem 7h15'32" 27d11'57"
Ethel Scott
Uma 10h17'25" 43d42'26"
Ethel Sherman
Gem 6h37'10" 15d25'21"
Ethel Wesby
Lmi 10h8'27" 34d23'40"
Ethel Williams Allan
Uma 10h42'56" 48d55'10"
Ethel Zanan
Lyn 7h55'34" 35d17'1"
Ethel95
Crb 15h46'53" 26d41'6"
Ethel-earth to heaven eternal light
Cas 23h4'26" 53d19'11"
Ethelmae
Gem 6h40'37" 14d51'33"
Ethel's Light
Cas 1h38'59" 64d17'53"
Ethelyn Marie McCauley Barnett
Uma 11h43'35" 33d19'29"

Ethem Sancak
Tau 4h33'22" 29d10'40"
Ethereal Natasha
Tau 5h38'49" 25d10'1"
Etherene May Cline
Tau 5h44'52" 17d32'30"
Ethermancer
Sco 16h16'32" -9d30'3"
Ethernell Murray Saxon
Uma 9h57'5" 41d40'27"
e-thoran
Pyx 8h36'28" -33d36'13"
Ethylis Cornholious Goodridancius
Boo 14h25'21" 18d26'38"
ETI
And 0h24'59" 46d28'51"
Etienne
Ori 6h16'31" 14d11'8"
Etienne & Antoinette
Dor 4h53'31" -69d17'50"
Etienne Barrett Super Star
Her 16h47'0" 34d36'10"
Etienne Paquette
Her 17h49'35" 18d1'56"
E.T.Kane
Tau 5h34'20" 25d56'55"
Etoile
Cas 23h4'21" 58d30'3"
Etoile Alban Ducruet
Crb 15h54'41" 29d17'40"
Étoile Belle
Umi 16h51'18" 81d58'33"
etoile binaire des ammons
Uma 13h26'43" 58d20'51"
Étoile D' Karla
Ori 5h40'58" 3d37'48"
étoile d'amants
Uma 12h6'15" 64d59'11"
Etoile d'Amour
Uma 12h6'38" 44d38'55"
Etoile d'Anniversaire de L'Arizona
Pho 1h44'46" -42d49'3"
étoile de amour
Uma 10h13'13" 55d43'24"
Etoile de Jim Lakota
Lib 14h29'28" -11d45'27"
E'toile de Kilgore
Cyg 20h34'56" 33d51'18"
Etoile de Maitre
Ari 3h18'1" 28d22'34"
Etoile de Marie
And 0h48'15" 37d2'54"
E'toile De Mer
Cyg 21h52'10" 50d16'5"
Etoile de Michael
Del 20h24'30" 16d45'44"
etoile de mon Ambre aime
Aqr 22h17'43" 2d9'25"
Etoile de mon Amour
Ori 6h1'29" 21d20'10"
Etoile de Nous
Cap 20h8'11" -16d37'4"
Etoile de Renee T. Hallinan
And 2h30'15" 45d30'37"
Etoile De Robertson
Uma 11h22'24" 45d7'32"
étoile de Tobi
Tau 3h44'56" 30d22'24"
Etoile de un Ange
Uma 10h56'6" 49d8'32"
Etoile des Etoiles du Sport
Col 5h25'22" -31d32'36"
Etoile des reves
And 23h36'22" 46d59'1"
Etoile du coeur
Vir 15h5'7" 2d53'0"
etoile jbr
Psc 23h23'10" 6d22'14"
Etoile Lolita Foussat, d'Amour
Uma 11h55'44" 47d12'38"
Étoile Sylvie et Roger 20.12.1980
Dra 20h2'52" 71d53'24"
Étoile Treber au firmament
Ori 6h5'23" 6d32'29"
étoile735
And 1h54'26" 38d0'33"
Etre Incroyable
Her 17h36'34" 29d41'16"
Etsuko "Echo" Stephens
Uma 13h43'30" 57d26'32"
Etsuko Nagatani
Lyn 8h20'26" 56d2'17"
Etta
Peg 21h13'37" 15d45'42"
Etta
Uma 13h50'0" 52d1'46"
Etta Marie
Lib 14h58'45" -0d38'8"
Etta Marie Pittala
Cas 23h9'5" 59d7'10"
Etta Pateta - A Star in Her own Right
Cas 1h26'46" 56d14'51"
Etta Winokur b. 5/18/11
Ori 5h59'27" 12d40'13"
Ettenger
Psc 1h20'17" 32d35'4"
Ettore Bianco Fiore
Cas 0h21'22" 52d5'10"
Ettore-Antonio-Quaglia
Cas 0h26'49" 50d21'52"

Etzold, Peter
Uma 9h11'36" 72d4'41"
Euan Anthony
Per 2h50'50" 48d13'49"
Euan Fettes
Per 3h14'29" 47d17'20"
Euan James Sadler
Ori 6h12'43" 8d11'8"
Euan John Milne
Ori 5h42'18" 12d55'29"
Euber
Ori 5h34'47" -1d14'30"
Euchre Babe
Sgr 18h18'22" -21d26'38"
Eucke Warren
Ori 6h3'36" -0d33'17"
Eufemia
And 1h43'46" 49d7'21"
Eufrazia DePaula
Vir 14h29'42" -4d17'9"
Eugen & Lilo Ziegler
Cnc 8h14'13" 22d16'38"
Eugene
Her 18h36'56" 18d45'28"
Eugene
Aur 6h31'8" 48d4'13"
Eugene
Her 16h31'40" 40d52'15"
Eugene
Aqr 22h23'57" -0d54'6"
Eugene
Cru 12h14'14" -56d8'7"
Eugene A. Hawley
Ori 5h33'55" 3d46'34"
Eugene A. Rotterman, Jr.
Aur 5h17'8" 42d0'27"
Eugene Alexander Gulledge
Ori 6h17'39" 14d20'15"
Eugene and Marian Erickson
Uma 9h14'15" 54d28'21"
Eugene Anthony Bertsch
And 2h37'32" 50d33'20"
Eugene Arnold
Cyg 21h12'58" 46d25'28"
Eugene Benedict
Ari 3h19'30" 27d47'10"
Eugene Broadus II
Uma 13h15'28" 61d48'34"
Eugene- BROTHER GENE- Williams
Crb 16h9'31" 31d42'12"
Eugene Butler
Aql 19h8'18" 9d3'25"
Eugene C. Ceccarelli
Tau 4h55'3" 21d6'13"
Eugene Callow
Oph 17h42'35" -0d35'35"
Eugene & Carmella Mirandola
Umi 15h33'25" 73d34'58"
Eugene & Christina Sheldrick 9/3/00
Umi 15h29'49" 73d48'35"
Eugene (Dad & Grandpa) Orefice
Uma 13h8'41" 61d40'23"
Eugene Dennis Davis Sr
Cnc 8h27'57" 11d52'27"
Eugene E Kahler
Lib 14h38'8" -9d4'58"
Eugene F. Thompson Sr.
Aql 19h56'34" 1d20'38"
Eugene Feis
Lib 15h28'6" -23d15'10"
Eugene Fernandes
Col 5h42'7" -27d52'52"
Eugene Fowler
Per 3h38'45" 51d35'4"
Eugene Francis Ford
Leo 10h9'20" 20d48'11"
Eugene Frederick Kinder
Per 3h7'41" 46d44'0"
Eugene Gedgaudas
Lib 15h43'8" -19d57'33"
Eugene H. Sorensen
Uma 8h9'40" 60d24'37"
Eugene Jack Christie
Cnc 8h49'11" 12d33'17"
Eugene James Schultz
Ari 2h36'31" 29d6'24"
Eugene Joseph Chiaramonte Jr.
And 0h49'12" 43d27'3"
Eugene Joseph Hayman
Sco 17h27'30" -37d41'47"
Eugene K Emmart
Per 3h29'3" 39d25'46"
Eugene Karl Belden
Aqr 21h42'2" -7d46'21"
Eugene & Katherine Rudloff 2/5/1966
Aqr 20h50'43" -7d21'54"
Eugene L Anderson
Gem 7h24'16" 21d32'59"
Eugene L Pederson
Sgr 18h48'38" -28d6'34"
Eugene Lamb, Jr.
And 0h12'4" 32d33'41"
Eugene Lampone
Leo 9h58'35" 23d28'40"
Eugene Louis Folgo Jr.
Per 4h3'38" 35d35'37"

Eugene M. Hope, Jr.
Ari 3h1'13" 19d1'29"
Eugene Moglovkin
Cas 1h54'35" 64d37'30"
Eugene Neuman
Vir 14h5'3" -16d37'30"
Eugene O. Kiesling's Star
Ori 5h52'11" 6d28'41"
Eugene Phillip Juel
Vir 13h10'41" -16d36'9"
Eugene Politsch
Uma 13h48'57" 51d45'54"
Eugene Reese
Sco 17h54'20" -40d49'6"
Eugene Richard and Bobbie Lou Wells
Leo 10h43'5" 14d25'46"
Eugene Rivera-As You Wish
Ari 3h18'37" 27d52'39"
Eugene Roller
Per 4h0'7" 35d4'53"
Eugene Rowden
Lyr 18h53'16" 29d41'9"
Eugene Runion
Uma 11h8'36" 37d47'9"
Eugene S. Libby
Cap 20h56'27" -24d26'22"
Eugene S. Marquis II
Gem 6h31'41" 17d57'45"
Eugene Schwartz
Per 3h46'46" 50d49'11"
Eugene Sebastien Fortier
Per 4h11'58" 45d19'37"
Eugene Singer
Sgr 19h53'55" -35d50'50"
Eugene Steeb
Leo 11h11'55" -0d56'4"
Eugene Sudassy
Her 17h10'17" 19d8'44"
Eugene Sullivan
Ori 5h41'14" -0d11'32"
Eugene Taylor Siddell V
Aur 5h25'55" 41d10'33"
Eugene Thomas Guerin
Aur 4h59'56" 32d12'18"
Eugene V. Herrmann
Uma 12h46'55" 53d27'44"
Eugene Walsh
And 23h38'44" 41d44'21"
Eugene Wolfarth
Boo 14h43'49" 26d54'52"
Eugene Yevgeny Barash
Per 3h50'40" 41d40'7"
Eugene Yevgeny Donner
Psc 1h27'52" 24d30'30"
Eugenia
Psc 23h0'16" 7d28'11"
Eugenia
Leo 9h41'36" 30d46'46"
Eugenia
Crb 15h54'42" 38d11'52"
Eugenia
Lyn 7h52'18" 53d34'53"
Eugenia
Uma 14h22'28" 61d57'12"
Eugenia
Cap 20h23'43" -10d57'24"
Eugenia A. Skovmand
And 0h45'40" 41d13'20"
Eugenia A Trojan
Cam 3h57'0" 68d38'35"
Eugenia Boki
Lyn 7h46'44" 56d48'30"
Eugenia Guerrero I
Sco 16h33'16" -37d3'47"
Eugenia Nan
Lyn 8h16'50" 49d40'48"
Eugenia Sun
Sgr 19h59'35" -30d24'40"
Eugenia Turov
Cnc 8h36'19" 7d50'20"
Eugenie Grady
Lyr 18h42'59" 35d41'43"
Eugenie Magoon
Sco 17h22'38" -42d53'54"
Eugenija
Aqr 22h3'26" -16d7'0"
Eugenio Abay
Her 18h40'7" 17d12'11"
Eugenio Capuzzi
Sgr 18h16'36" -34d13'28"
EULA
Her 17h52'57" 38d46'20"
Eula and Bob Adkins
Uma 9h58'40" 58d19'18"
Eula Anderson
Crb 15h41'30" 28d46'57"
Eula Imogene Woodruff Brindley
Uma 11h18'16" 60d59'42"
Eula Pauline West
Vir 13h33'28" 3d3'2"
Eulalia
Gem 7h43'50" 33d47'41"
Eulalia Han Ni
Sgr 18h8'0" -25d55'19"
Eulalie Crommelin Draper Given
Aqr 21h30'38" 2d27'30"
Eulalie E. Gibbs Best Mom
Ori 6h15'56" 13d14'56"
Eully Risi
Tau 5h3'46" 18d56'16"

Eumenides
Leo 11h25'26" 21d24'49"
Eumi
Cap 21h9'13" -16d4'45"
Eun Hye Kwak
Vir 12h35'23" -1d28'49"
Eun Ju Kim
Ori 5h44'41" 11d16'33"
Eun Ryung
Lmi 10h23'5" 38d42'22"
Eunae
Cyg 20h26'14" 47d35'7"
Eunah Sarah Hyun
And 23h26'40" 42d46'14"
Eunice
And 0h36'21" 37d12'11"
Eunice 06 29 1925
Uma 11h3'0" 69d50'56"
Eunice Amelia
Cas 1h29'46" 66d24'15"
Eunice Atkins Lynn
Ori 5h45'47" 7d47'55"
Eunice Barnes-Nonie
Scl 0h35'7" -28d16'18"
Eunice Chan Nga Yan
Tau 4h7'58" 7d30'32"
Eunice Cobb Todd
Crb 15h22'18" 30d43'13"
Eunice Diaz a.k.a "NICKY"
Leo 9h35'1" 10d41'32"
Eunice G Syckes
Psc 1h1'49" 4d58'3"
Eunice G. Tillis
Cas 1h8'22" 61d34'36"
Eunice Grace Godden
And 1h31'42" 49d41'29"
Eunice Harris
Sco 16h40'27" -33d47'33"
Eunice Healey
Col 5h52'32" -38d57'58"
Eunice Irene Morris
Lib 15h33'26" -9d29'19"
Eunice Koh - Shining Star
Lmi 10h47'18" 33d50'48"
Eunice MacRitchie
Aql 20h0'46" 15d59'48"
Eunice Mae Lee
And 1h59'22" 43d8'53"
Eunice "Oma" Kimmitt
Ari 3h8'28" 27d32'29"
Eunice Remy Yomada
Psc 23h27'22" 2d34'21"
Eunice Taylor
Lib 15h23'14" -19d14'35"
Eunice Thomason
Uma 10h10'16" 55d34'58"
EunJoo
Uma 11h29'7" 62d2'26"
EUNYOUNG
And 0h46'2" 35d21'37"
Euphoria
Aql 19h32'2" 8d39'41"
Euphoria with only a Kiss
Sco 16h11'42" -31d30'17"
Eureka
Her 17h49'16" 42d26'12"
Euretta
Leo 11h25'58" 9d58'14"
Euria
Per 3h29'56" 47d42'21"
Euripia
Aqr 23h33'7" -17d12'45"
Euro Poletti
Her 18h38'29" 21d5'51"
Eurobank EFG Stedionica
Peg 21h38'35" 16d25'39"
Eurobank filijale
Peg 22h35'55" 24d24'6"
Eurobank proizvodi
Peg 21h42'27" 16d24'59"
Eurobank vrednosti
Peg 21h36'20" 16d9'28"
Eurobank zaposleni
Peg 21h56'51" 18d11'51"
eurokangas
Leo 11h12'55" 7d33'35"
Europa Mara Matzat
Cas 0h41'6" 61d36'4"
Europa Mercedes Koenig
Vir 13h25'0" 1d41'34"
Europe
Gem 6h20'36" 17d59'14"
Eurusher Jackson
Cyg 19h48'12" 33d39'6"
eusen Stern Rebi und Roli
Aur 6h34'19" 39d3'10"
Eusen, Thekla
Uma 8h36'57" 58d20'10"
Eustaccia Chanel Lanet Biggs
Aqr 21h8'39" 1d33'58"
Eva
Psc 1h29'16" 12d32'17"
eva
Ori 5h22'1" 9d4'23"
Eva
Ori 5h35'53" 1d5'25"
Eva
Vir 13h12'56" 11d27'31"
Eva
Gem 7h15'2" 26d32'30"
Eva
Lyn 7h6'45" 46d12'53"
Eva
Cas 0h4'20" 57d22'46"

Eva
Cam 4h20'10" 56d53'27"
Eva
Uma 12h38'58" 57d59'28"
Eva
Cas 2h23'13" 65d40'36"
Eva
Lib 15h1'19" -17d33'43"
EVA
Cap 20h9'4" -21d28'26"
Eva
Cap 21h52'21" -16d35'9"
Eva
Aqr 22h31'54" -18d39'53"
Eva
Cep 3h3'41" 82d45'9"
Eva
Vir 14h39'59" -0d46'13"
Eva Alessandra
Ori 5h23'15" 2d1'7"
Eva Almeida Plotts
Mon 7h18'57" -0d33'36"
Eva Anastasia
And 1h47'59" 41d46'9"
Eva and Harold Holdsworth
Lmi 10h2'45" 33d1'55"
Eva Angelic
Cnc 8h55'41" 13d22'15"
Eva Arnold Edmonds
Sgr 18h33'31" -31d56'22"
EVA_B
Gem 7h47'20" 18d21'49"
Eva Caprice Smith
Gem 6h41'21" 33d54'25"
Eva Carmella Hagaman
Sgr 19h46'26" -13d4'32"
Eva Chen
Cas 23h45'8" 53d57'22"
Eva Chen
Cyg 21h38'17" 46d33'7"
Eva Claire
And 1h1'20" 41d10'48"
Eva Darias Esteban
Dra 19h4'17" 52d23'9"
Eva Darias Esteban
Leo 10h55'31" 6d52'7"
Eva Dunn
Crb 15h55'30" 36d43'44"
Eva Eleanor Pederson Ellis
Ori 6h8'42" 19d27'9"
Eva Elizabeth
Cas 1h23'39" 61d25'14"
Eva Eternal
Tau 4h13'9" 27d26'1"
Eva Flik Calcutt
Lib 15h16'52" -6d18'18"
Eva Frederick, Mom and G
Vir 12h21'47" 12d25'42"
Eva Garfat
Ori 5h58'49" -0d44'53"
Eva Grace
And 0h5'0" 45d34'15"
Eva Grace Harris
And 23h47'6" 48d32'18"
Eva Haines - 22.08.2005
Cru 12h29'58" -64d16'29"
Eva Hannan
Aqr 23h18'36" -9d29'24"
Eva Hintereder
Uma 9h58'20" 43d20'37"
Eva Horn (Granny)
Aqr 21h22'43" -13d32'35"
Eva Hung
Uma 11h31'33" 57d55'27"
Eva Indiana Christoffel
Lib 15h27'34" -6d20'19"
Eva Ingeborg Brenner
And 0h23'44" 44d51'7"
Eva Isabel Weiner
And 0h22'28" 31d31'48"
Eva James Lupis
Leo 11h29'40" 17d30'55"
Eva Jean Wistner Nelson
Leo 10h50'31" 19d36'55"
Eva K.
Uma 13h2'11" 55d33'19"
Eva Katherine
Sco 16h54'24" -44d9'48"
Eva Kathryn
Cyg 20h15'59" 46d51'46"
Eva Kleman
And 0h17'27" 31d42'6"
Eva Kostalova
And 1h6'23" 44d55'28"
Eva Lee Shepherd
Cyg 19h51'25" 39d30'59"
Eva Leigh Schreppel
And 0h43'5" 42d20'45"
Eva Leona
Lyn 8h24'15" 56d49'59"
Eva Leone Barnes
Umi 15h6'19" 72d41'57"
Eva Louise
Lyn 7h30'35" 49d44'15"
Eva Lucero
Cas 1h18'47" 56d40'1"
Eva Lydia Margaret Barton
Crb 16h10'57" 37d35'15"
Eva Lynn
And 0h33'35" 27d26'20"
Eva M. Willett
Psc 0h13'21" 6d53'58"
Eva Madeline Samantha Hannah Mikula
Sgr 19h25'31" -12d5'13"

Eva Mae Conner
Lyn 7h33'10" 54d36'36"
Eva Mae Smith 4-14-1924
Ari 3h12'47" 28d29'27"
Eva Margarita
Gem 6h10'43" 23d0'11"
Eva Margit Joan Boros Kiss
Tau 4h26'32" 10d7'11"
Eva Mari Anna Mae
Tau 4h17'51" 23d19'21"
Eva Maria
Uma 13h56'38" 59d48'9"
Eva Maria Burda Trella
Cas 1h14'25" 59d24'54"
Eva Maria Gastelum Olivas
Leo 10h13'2" 16d45'45"
Eva Maria Harris
Uma 13h9'11" 59d48'53"
Eva Maria Wyszatycka-Moore
And 1h14'30" 46d48'56"
Eva Marie
And 0h49'22" 35d45'56"
Eva Marie Bliss
Crb 15h26'54" 32d16'18"
Eva Marie Burnham Golden Birthday
And 23h54'51" 40d2'47"
Eva Marie Cosgrove
And 23h26'0" 48d37'54"
Eva Marie Nichols
Cap 21h26'32" -19d19'9"
Eva Marie Olson
Cap 21h20'14" -17d10'33"
Eva Marie Pearson
Sco 17h29'50" -33d20'22"
Eva Marie Prappas
Com 12h0'12" 21d32'1"
Eva Marie Ryland
Aqr 22h14'28" 0d46'16"
Eva Maryn Silzer
Cas 1h5'45" 69d49'0"
Eva Massai
Ori 5h35'52" -5d8'9"
Eva Maxine Parker
Apu 15h30'52" -71d32'19"
Eva May
And 0h6'35" 28d54'8"
Eva May Brock
And 1h7'54" 45d58'22"
Eva Melanie Schwartz
Vir 13h3'59" 9d4'0"
Eva Mighetto
Psc 0h23'46" 7d37'5"
Eva Mº San Andres Fernandez
Cnc 9h10'53" 23d9'3"
Eva Morgan Ritchie
Cas 0h4'18" 57d9'39"
Eva Muñeca Ferri
Her 18h47'46" 20d3'46"
Eva Muresan, Teacher and Friend
Crb 15h54'23" 37d15'19"
Eva N. Jeirles
And 1h4'49" 35d37'4"
Eva Nadine Featherman 10/27/04
Sco 16h12'31" -29d58'43"
Eva Nancy
Umi 14h43'48" 83d40'50"
Eva Nevado
Sco 16h10'5" -41d49'6"
Eva Nordby
Cas 0h2'27" 53d22'40"
Eva Palango
Uma 8h35'2" 67d20'29"
Eva Pauline
Cap 20h27'21" -11d47'11"
Eva Pop
Sco 16h10'26" -13d22'54"
EVA Proksch
Uma 8h36'47" 48d11'1"
Eva Rae
Ari 2h6'9" 23d14'26"
Eva Reifler (mjpfsa)
Ori 5h59'24" 20d38'26"
Eva Romo
Uma 11h4'32" 65d21'45"
Eva Rose
Cen 11h37'45" -59d31'32"
Eva Rose
And 23h20'45" 42d13'25"
Eva Rose
And 1h14'22" 34d5'47"
Eva Rose Leach
And 23h22'20" 42d3'0"
Eva Ruth Smith
Vir 13h59'40" -21d31'47"
Eva Ruth Trammell "MONGA"
Umi 16h8'50" 77d38'7"
Eva Savic
Cru 12h25'17" -60d46'57"
Eva Scalese
And 0h16'53" 25d21'4"
Eva Seiller-Batten
Cas 1h22'55" 52d18'56"
Eva Sepulveda
Lyr 18h57'25" 37d7'18"
Eva Simone
Cnc 8h40'16" 30d59'33"
Eva Snell
Umi 15h21'11" 69d54'45"

Eva Solé Bonay
And 0h23'24" 30d3'29"
Eva Stamboulis
Cas 0h9'43" 62d31'9"
Eva Star
Umi 14h23'27" 69d19'15"
Eva Star Stratton
And 23h1'25" 48d52'8"
Eva - Stathis
Lyn 7h16'58" 51d25'18"
Eva Stolzenburg
Ori 5h36'0" -0d5'42"
Eva Tallulah Bourdet Gemmell
And 0h19'45" 29d35'50"
Eva Tekauerova
Ori 6h1'53" 4d26'42"
Eva und Reinhard Furrer
Cas 23h5'10" 57d21'22"
Eva & Vaclav
Cyg 21h5'19" 47d22'22"
Eva Victoria Lewis
And 1h14'49" 50d34'48"
Eva Yong Leung
And 1h15'58" 35d14'33"
Eva Yorton
Lyr 19h14'59" 38d33'17"
Evabean
Cru 12h46'7" -58d42'59"
Evácska
Cas 0h47'37" 50d13'27"
Evadne Marsteller McKinley
Psc 1h46'57" 5d49'57"
Evaggelia Fertis
Ori 6h13'44" 5d39'31"
Evalene
Dra 15h33'21" 61d21'28"
Evalina Lurye
Her 18h53'6" 23d26'0"
Evalina Zanoni
Lib 15h4'46" -0d35'48"
Evalove
And 2h16'48" 48d14'1"
Evalyn Beebe Precious Granddaughter
Ori 6h6'46" 10d26'35"
Evalyn Hatfield
Cyg 19h36'1" 34d23'37"
Evalyn R France
Sco 17h52'51" -38d51'18"
Eva-Maria Blank 27.01.1964
Uma 9h15'41" 47d40'24"
Eva-Maria&Tilman-U.Wagner22.04.1989
Uma 13h2'23" 55d47'50"
Evan
Uma 11h8'10" 58d59'3"
Evan
Umi 16h7'37" 78d20'31"
Evan
Lib 15h51'4" -17d59'19"
Evan
Col 5h10'12" -33d41'17"
Evan
Uma 9h42'22" 51d35'11"
Evan
Lyn 7h56'6" 34d0'3"
Evan
Vir 13h17'35" 8d42'34"
Evan
Cnc 8h48'41" 10d9'26"
Evan
Ori 5h54'3" 20d57'55"
Evan aeternum fortunoare
Ori 5h40'21" -2d40'25"
Evan Aidan Ridenbaugh
Gem 7h29'36" 28d36'48"
Evan Alanson Nessel
Vir 13h10'36" 11d17'27"
Evan Alexander Suty
Per 2h52'10" 49d4'4"
Evan and Barry Epoch
Per 3h17'43" 47d18'20"
Evan and Janet Lomelino 40 Years
Cyg 20h17'28" 46d57'41"
Evan and Jen
Com 12h17'45" 31d39'13"
Evan and Kari: Goobers Forever
Lib 15h9'35" -7d41'14"
Evan and Max Hechtman
Uma 11h45'43" 62d53'34"
Evan and Shawna
Her 16h58'41" 33d53'15"
Evan and Torie Forever
Crb 16h23'59" 35d50'40"
Evan Andre' Prazeres Jackson
Gem 6h53'55" 26d3'33"
Evan Andrew Caldwell
Peg 22h20'50" 32d9'41"
Evan Andrew Klane
Cam 6h15'45" 62d57'55"
Evan Anthony Gravino
Gem 6h51'4" 17d10'20"
Evan Arlen Neal Caudell
Uma 9h3'20" 70d1'3"
Evan Ayoroa
Cap 21h27'34" -9d0'35"
Evan Bailey
Lyn 7h36'31" 37d52'47"
Evan Bailey Reed
Uma 9h31'59" 47d32'31"
Evan Barrett O'Connor
Her 17h35'20" 36d44'54"
Evan Bookstaver
Umi 19h21'41" 88d10'46"
Evan Bressoud
Cap 21h27'13" -8d41'41"
Evan C Lauber
Her 17h20'52" 27d20'12"
Evan C. Nordengren
Per 3h56'42" 39d8'30"
Evan Cahill
Peg 21h34'50" 13d19'17"
Evan Christopher
Sco 16h5'13" -12d31'54"
Evan Christopher Baker
Per 2h51'13" 51d58'30"
Evan Christopher Ehrenberg
Per 3h24'17" 49d39'1"
Evan Conner
Uma 10h6'37" 48d15'25"
Evan D. Hunter
Her 17h42'5" 19d18'50"
Evan Dallmann
Cnc 8h42'41" 10d37'43"
Evan Daniel Siegel
Cap 20h29'56" -10d24'11"
Evan David Debling
Ori 6h23'9" 13d48'53"
Evan David Peck
Gem 7h32'46" 28d57'16"
Evan Doyle Lotz
Lib 15h53'24" -5d12'10"
Evan Duff
Umi 14h48'8" 80d46'26"
Evan Elise Esraelian
Lib 15h46'19" -29d25'37"
Evan Emerson Spencer
Ari 3h22'23" 27d22'35"
Evan Gregory Hewitt
Her 16h30'31" 42d21'3"
Evan Gregory Linett
Psc 1h13'34" 17d37'38"
Evan H Bergwall II
Umi 15h13'21" 71d28'22"
Evan Hall
Cep 0h9'23" 70d37'25"
Evan Hanes Pelletier
Ari 2h59'57" 20d47'52"
Evan Hangley
Uma 11h37'10" 52d54'4"
Evan Hendershot
Sco 16h25'31" -32d17'48"
Evan Heneghan
Sgr 18h25'27" -35d57'20"
Evan Hunter
Ori 5h35'8" 8d50'42"
Evan Jacob Ostrander
Dra 18h56'43" 65d39'23"
Evan James Buckmeier
Ari 2h21'13" 24d46'59"
Evan James Byers
Lib 15h8'3" -3d45'25"
Evan James Crouse
Gem 6h32'22" 21d36'56"
Evan James Hahn
Uma 9h28'2" 72d55'30"
Evan James Moseley
Aur 5h43'46" 40d49'38"
Evan James Quiros
Psc 1h5'56" 32d16'35"
Evan James Rice
Cap 20h11'42" -10d9'33"
Evan Jones
Uma 11h52'4" 44d43'34"
Evan Joseph Ignatius Clark
Lmi 10h30'50" 33d9'34"
Evan Joseph Mapleson
Cnc 9h9'37" 31d43'20"
Evan Joseph Sanford
Sco 16h10'28" -10d20'57"
Evan Karl Erickson
Psc 0h18'13" 8d20'56"
Evan Kaunas Madeira
Her 16h38'2" 48d47'35"
Evan Ladd Scavnicky
Her 17h59'26" 49d30'22"
Evan Lee Overton
Her 17h46'29" 37d52'35"
Evan Lee Stark
Vir 15h6'45" 3d51'45"
Evan Levy-Goldblatt
Aql 19h19'24" 6d48'9"
Evan Lewis Wood 'A Star is Born'
Ori 6h21'37" 16d11'1"
Evan Louis Guzman
Cas 23h27'0" 51d15'35"
Evan Louis Herman
Leo 11h12'0" -1d12'50"
Evan Low
Psc 0h59'7" 16d48'14"
Evan MacIntyre
Cyg 19h14'54" 50d47'22"
Evan Martin
Ori 5h39'55" 2d0'13"
Evan Mason Murillo
Uma 13h18'6" 53d23'52"
Evan Matthew Hughes
Crb 15h37'9" 26d56'38"
Evan Maxwell Goodman
Lib 14h29'23" -8d59'33"
Evan McGovern
Per 4h26'33" 43d9'34"
Evan Michael 5/02/90
Her 17h8'10" 30d0'24"
Evan Michael Loos
Uma 9h46'38" 60d41'8"
Evan Michael Weiland
Ori 5h32'57" -0d24'23"
Evan Mitchell
Per 2h16'53" 56d18'46"
Evan Mitchell Biava
Sco 16h39'28" -24d55'18"
Evan Mitchell Farnsworth
Aqr 22h13'45" 0d4'23"
Evan Mitchell Lyons
Cap 21h30'45" -17d35'45"
Evan Morgan Wheeler
Uma 9h41'49" 55d44'8"
Evan N. Featherstone
Cap 21h49'45" -11d21'52"
Evan Nazal
Cap 20h36'22" -18d58'8"
Evan Nees
Lyn 8h7'54" 40d54'36"
Evan Nicholas
Her 18h39'51" 18d27'59"
Evan Nicholas Schmerber
Ari 2h29'30" 18d16'4"
Evan Noah Levy
Sgr 18h32'22" -23d37'53"
Evan Orion Whetsel
Ori 5h10'45" -0d23'50"
Evan P. Ostman
Per 4h16'46" 33d57'22"
Evan Pastrikakis-Gale 30/07/04 - 12:07 am
Umi 15h2'23" 85d36'40"
Evan Patrick
Uma 9h50'20" 52d42'31"
Evan Patrick Amerson
And 2h21'55" 46d41'18"
Evan Patrick Catanzaro
Leo 11h27'43" 26d22'17"
Evan Paul Biel
Vir 13h43'53" -18d53'46"
Evan Pederson
Boo 15h38'50" 39d58'12"
Evan Phillip Tabbner
Umi 15h40'48" 68d14'36"
Evan Pickard
Uma 12h26'25" 63d6'30"
Evan R. Newman
Her 16h47'37" 4d56'34"
Evan Randall Steuer
Sgr 18h28'7" -27d53'42"
Evan Ray Bownes
Per 3h5'30" 40d40'36"
Evan Ray Kopack
Peg 22h9'42" 26d11'11"
Evan Ritson
Vir 12h32'47" 2d57'58"
Evan Robert Baxter
Sco 17h17'10" -43d21'46"
Evan Robert Miller
Uma 10h17'37" 42d31'45"
Evan Robert Smith
Ori 4h51'36" -3d4'40"
Evan Ronald Painter
Cnc 9h9'22" 31d46'56"
Evan Ross Nowels
Cyg 21h57'22" 38d28'28"
Evan Samuel Greco
Per 3h5'52" 55d49'30"
Evan Schlameuss
Lib 14h47'52" -10d26'16"
Evan Sean Ramsey
Uma 11h53'43" 40d59'36"
Evan Serge Byford
Uma 10h28'48" 62d59'48"
Evan & Shari Herbert Eternity Star
Cen 13h43'57" -42d59'6"
Evan Shulman
Leo 11h24'14" 25d34'35"
Evan Snyder
Her 18h6'32" 35d14'27"
Evan Spiegel
Tau 5h41'21" 22d49'56"
Evan Sudul
Ori 5h36'40" -1d25'29"
Evan Sydenham
Ori 5h35'20" -2d3'19"
Evan T. Webb
Her 18h45'49" 13d19'23"
Evan Tamerou
Per 3h7'17" 42d57'20"
Evan Terwey
Lyn 8h11'7" 56d48'34"
Evan Thomas Berger
Vir 13h49'24" -10d20'9"
Evan Thomas Foresman
Ari 2h23'46" 25d3'40"
Evan Thomas Nitzsche
Per 4h6'44" 47d45'45"
Evan Thomas White
Psc 1h35'47" 27d51'6"
Evan Toscano
Uma 11h18'19" 55d36'55"
Evan Tyler Grimm
Uma 10h0'34" 60d23'33"
Evan Tyler Wright
Vir 15h0'37" 2d14'6"
Evan Ulrich
Tau 4h11'13" 22d48'32"
Evan Uras-Urban
Aqr 20h40'9" -0d52'16"
Evan Vincent Miles
Lyn 7h53'1" 56d35'51"
Evan Walters
Dra 18h31'14" 59d50'57"
EVAN WHITACRE
Ari 3h17'59" 16d19'9"
Evan William Berggren
Tau 5h47'43" 13d10'6"
Evan William Bhatt
Ari 3h23'51" 19d47'17"
Evan William Bruzzichesi
Per 3h53'54" 49d38'43"
Evan William Campbell
Uma 12h34'9" 58d16'33"
Evan William DeFazio
Cap 21h30'42" -16d4'7"
Evan William Duhig
Her 16h33'49" 25d40'18"
Evan William Thole
Psc 0h32'44" 20d17'36"
Evan William Thomas
Lac 22h25'23" 46d39'49"
Evan Wyatt Orgel
Cnc 8h23'52" 26d19'27"
Evan-13184
Her 17h41'27" 15d29'47"
Evana
And 23h56'37" 34d20'38"
Evaneja Brigita Jurkovic
Uma 9h3'4" 53d51'25"
Evangelia
Aqr 22h30'45" 1d1'33"
Evangelia & Giorgos
Cnc 8h47'29" 12d58'36"
Evangelia Nipper
Ari 2h33'23" 27d46'33"
Evangelia Tsokatos
Peg 22h22'4" 4d48'59"
Evangelina
Sco 17h37'0" -44d42'36"
Evangelina Casarez
Psc 1h12'42" 20d31'30"
Evangelina Veronica
Cap 21h5'42" -26d25'4"
Evangeline
Cas 23h29'43" 57d59'48"
Evangeline Baquiran
Cyg 19h41'25" 34d49'35"
Evangeline Fae Morgan
And 23h30'55" 49d51'6"
Evangeline for Kelsey Rebekah
Leo 9h47'20" 15d36'30"
Evangeline "Happy Valentine's Day"
Uma 12h19'31" 56d24'29"
Evangeline Harmony & Astrid Eleanor
Cru 12h33'35" -57d41'13"
Evangelist Isabell Webb
Cyg 20h3'2" 33d2'52"
Evangelista
Com 13h4'16" 17d30'2"
Evangelista Lili Chérie
Leo 9h51'15" 21d21'56"
Evangelos Antonios Siamantouras
Tau 5h35'35" 22d38'20"
Evangelos Moraitis
Uma 9h1'38" 59d47'38"
EvanKing
Cnc 8h10'7" 28d41'11"
EvanMark
Her 17h21'58" 38d41'54"
EvAnn
Umi 14h28'35" 75d10'8"
Evanne Hanley
Tau 3h44'39" 23d47'29"
EvanPatrick McIntosh's Star
Vir 14h2'25" -22d21'26"
Evans
Cru 12h36'33" -58d21'16"
Evans 3
Leo 10h15'5" 15d22'58"
Evan's Gift
Aqr 22h7'40" -13d30'48"
Evan's Heavenly Compass
Cnc 8h46'30" 8d3'25"
Evan's Light
Aql 19h59'58" -0d0'7"
Evan's Star
Cas 0h41'8" 57d59'9"
Evans Varelas
Ari 2h57'11" 26d42'17"
Evan's World
Her 16h48'9" 17d56'28"
EvansDreherAlain
Uma 9h32'23" 44d7'1"
Evarie
Mon 7h34'23" -1d27'25"
Evaristo Beccalossi
Lyr 18h28'5" 33d4'59"
Evas Uniqus
Dra 16h15'55" 60d5'59"
Eva's Who Cares
Peg 23h31'37" 17d41'44"
E.V.C.
Lmi 10h15'29" 29d2'52"
evchen+momo *
Per 3h18'29" 47d33'33"
Evdokeya Sturtevant
Uma 10h31'6" 69d26'30"
"EVE"
Uma 10h43'25" 54d55'59"
Eve
Aqr 20h54'45" -8d57'59"
Eve
And 2h16'4" 48d18'9"
Eve
And 23h35'15" 42d47'32"
Eve
Vir 12h46'20" 7d38'54"
E.V.E
Ori 6h20'37" 7d11'25"
Eve Angeline
Leo 10h12'34" 11d37'17"
Eve Baird
Ori 6h19'15" -1d2'8"
Eve Barbosa
And 0h52'9" 35d50'50"
Eve Cameron
And 1h50'4" 37d33'45"
Eve Chaput
Tau 5h58'10" 26d28'56"
Eve Costello
And 23h23'31" 50d43'6"
Eve - Eros - Sagittarius
Sgr 18h4'33" -34d8'0"
Eve Farmer
And 2h24'14" 45d26'1"
Eve Katie Behenna
Lmi 10h3'42" 29d33'12"
Eve Marie Ebenezer
And 23h3'40" 40d45'21"
Eve May Dove
Cam 3h33'23" 64d7'0"
Eve McGetrick - Eve's Star
And 23h52'5" 39d21'56"
Eve Mucci
Lyr 18h46'21" 37d46'35"
Eve Nicole
And 1h59'21" 38d32'32"
Eve Noel
Cas 0h37'45" 52d1'14"
Eve Reynolds
Cas 0h22'15" 51d33'28"
Eve Ryan
Uma 12h32'59" 58d21'26"
Eve Sara
Lib 15h31'26" -25d17'15"
Eve Schwartzberg
Lmi 10h42'31" 28d40'23"
Eve Smith
Cam 5h25'53" 67d12'47"
Eve Sofia Wasylik
And 23h8'0" 41d39'17"
Eve Sohmer
Crb 15h45'9" 32d42'30"
Eve Théoret
Umi 15h40'9" 72d45'35"
Eve Yorioka - Tupperware
And 0h21'43" 29d10'54"
eve, my Taiwanese princess
Ari 3h9'10" 29d11'6"
Evelia Hernandez
Cnc 9h7'12" 15d26'24"
Eveliene Barbara Daro
Cnc 8h18'4" 19d46'50"
Evelin Roos
And 2h36'3" 50d34'14"
Evelina
Cyg 21h29'19" 37d0'16"
Evelina
Psc 23h50'37" 3d11'12"
Evelina Osip
Umi 13h37'36" 74d43'37"
Evelina Perez de Marroquin
Vir 13h44'45" -8d50'23"
Evelinda
Crb 15h48'32" 29d44'29"
Evelinda Peña
Vir 13h46'14" -18d53'20"
Eveline
Ori 6h20'53" 3d8'54"
Eveline and Marcelino forever
Crb 15h46'34" 27d19'49"
Eveline Arnet (Schtaernli)
Cnc 9h8'48" 22d17'2"
Eveline Bretschneider
Uma 8h58'9" 49d16'15"
Eveline Degenfelder
Uma 9h22'21" 49d44'27"
Eveline Graf
Mon 7h40'44" -4d17'57"
Eveline Lengen
Boo 14h10'26" 28d56'1"
Eveline Ossinger
Uma 14h22'5" 55d9'30"
Eveline Paz-Roldan
And 0h42'23" 40d44'0"
Eveline Podrecca
Eri 4h23'15" -8d8'36"
Eveline Schibli
Lac 22h52'35" 53d57'25"
Eveline's Träume
Crb 15h37'47" 36d47'43"
Evelio Antonio Quevedo Fernandez
Uma 11h38'10" 61d39'5"
Evely Hints
Del 20h35'51" 3d24'35"
Evelyn
Aqr 22h38'40" 0d37'38"
Evelyn
Tau 4h47'32" 20d7'12"
Evelyn
Ori 5h51'45" 20d52'47"
Evelyn
Com 13h11'50" 25d54'17"
Evelyn
Tau 3h52'2" 27d8'34"
Evelyn
Cnc 8h20'54" 29d46'33"
Evelyn
Leo 11h12'45" 21d57'56"
Evelyn
Ari 2h41'23" 25d33'30"
Evelyn
Uma 11h24'49" 39d13'32"
Evelyn
Lyn 7h57'59" 38d36'24"
Evelyn
Cas 0h11'30" 56d49'6"
Evelyn
Cas 0h15'20" 57d43'19"
"Evelyn"
And 2h29'37" 45d20'0"
Evelyn
Aqr 22h14'43" -1d26'38"
Evelyn
Cap 21h4'40" -17d26'26"
Evelyn
Pho 1h47'57" -50d50'59"
Evelyn 1
Psc 0h26'15" 9d9'47"
Evelyn 82802
And 23h37'15" 41d22'6"
Evelyn A. Kivi
Uma 10h28'13" 55d55'46"
Evelyn Adams
Uma 8h40'13" 68d17'16"
Evelyn Adele Bowes
Cen 13h23'8" -36d59'36"
Evelyn Adriana Arce
And 1h1'27" 40d23'21"
Evelyn Ainslie
Gem 7h29'56" 28d15'20"
Evelyn Alexandra Edwards
Vir 14h27'58" 3d45'33"
Evelyn and Bob Hall
Ori 5h58'11" -2d34'28"
Evelyn and Robert
Uma 8h18'25" 59d47'37"
Evelyn Ann Comer
And 1h1'2" 45d50'57"
Evelyn Ann Jacobson
Uma 11h33'18" 58d4'54"
Evelyn Ann Snizek
Psc 23h4'32" 7d37'14"
Evelyn Anna Royer Gnau
Leo 9h52'21" 6d28'43"
Evelyn Askew Schillaci
Lib 15h8'23" -4d39'59"
Evelyn B. Scarff
Uma 9h21'47" 50d58'50"
Evelyn "Babe"
Cam 4h10'5" 57d31'47"
Evelyn Batica & Danny Willett
Uma 12h1'45" 34d49'16"
Evelyn Betlejewski
Cap 21h56'14" -18d14'19"
Evelyn Burke
Uma 11h10'45" 57d25'56"
Evelyn Burt
Ari 2h57'36" 26d6'4"
Evelyn Butts
Ari 2h9'58" 25d7'50"
Evelyn Carol Knapp
Uma 10h34'16" 68d28'48"
Evelyn Carraher
Uma 8h57'28" 63d37'31"
Evelyn Cecelia Ann Willi
Cnc 7h56'36" 13d32'10"
Evelyn Christine Martin
Crb 16h20'19" 33d8'51"
Evelyn Claire Donnell
Leo 9h43'51" 20d56'46"
Evelyn Cotto
Psc 23h8'28" 6d5'37"
Evelyn Cruz Sanchez
And 1h2'50" 39d57'43"
Evelyn Dann
And 0h11'26" 27d22'54"
Evelyn Dansker Masciale 1925
Lib 15h13'39" -22d42'0"
Evelyn Davis Lindemuth
Lyr 19h10'58" 45d53'54"
Evelyn Dee
Uma 11h21'27" 55d41'24"
Evelyn Delores Thomas
Ari 2h46'57" 27d27'13"
Evelyn Duval
Cas 0h0'51" 56d40'8"
Evelyn Eckbold
Lib 15h20'54" -10d28'27"
Evelyn Elaine Stinebring
Vir 13h4'43" 10d48'12"
evelyn elise
Cnc 8h11'47" 16d37'31"
Evelyn Elizabeth Frechette
And 0h33'38" 31d13'15"
Evelyn Esparza
Lyn 7h1'56" 57d35'32"
Evelyn Fallin
Peg 22h26'12" 4d22'38"
Evelyn Faulk
And 0h49'43" 40d7'7"
Evelyn Fay 1
And 1h50'19" 42d52'52"
Evelyn Florence Hollenbeck Sailors
Sgr 18h1'11" -31d58'5"
Evelyn Flores
Cas 0h42'52" 60d50'9"
Evelyn Francesca Alexandra Aspinall
And 2h37'21" 38d3'21"
Evelyn Francesca Hulme
Cnc 9h19'3" 20d23'5"
Evelyn Fruges
And 0h32'8" 32d17'23"
Evelyn G.
Cas 1h28'9" 62d5'57"
Evelyn G. Dakajos
Cas 0h2'32" 56d19'4"
Evelyn Gallagher
Cas 1h43'59" 62d7'56"
Evelyn Gaylord's Heavenly Star
Lyr 18h32'47" 33d26'22"
Evelyn & Gerald Lancaster
Crb 15h28'27" 28d17'58"
Evelyn Gladys Huddart
Cas 1h40'57" 61d43'33"
Evelyn Gonzales
Aqr 21h42'27" -1d53'43"
Evelyn Grace Matthews
And 23h13'39" 47d18'58"
Evelyn Grammy Fegley
Lmi 11h0'34" 27d18'11"
Evelyn Gran
Cas 22h58'0" 54d38'47"
Evelyn Gwen Mathison
Gem 6h49'11" 16d55'8"
Evelyn Haller
Sgr 18h16'5" -27d44'17"
Evelyn Hallock
Ari 3h11'4" 28d55'0"
Evelyn Hasten
Lmi 10h32'9" 37d13'49"
Evelyn Hernandez
Lyn 9h5'40" 39d58'41"
Evelyn Jane Heaven's Gift To Me
Uma 14h23'7" 55d0'57"
Evelyn Jean Callahan
Aqr 23h38'14" -13d20'31"
Evelyn & Jesse
Uma 10h38'8" 67d22'24"
Evelyn Johnson
Leo 11h26'39" 2d12'14"
EVELYN & JUANMI
Gem 7h3'14" 17d16'56"
Evelyn Kay
Lib 14h56'50" -10d19'11"
Evelyn Kraft
Uma 9h32'2" 63d20'5"
Evelyn Laura McClung
Aqr 22h31'54" -23d26'19"
Evelyn & Leonard Rokaw
Gem 7h8'42" 14d17'59"
Evelyn Lepera
Uma 9h8'9" 63d35'5"
Evelyn Levy
Cnc 9h6'55" 31d5'37"
Evelyn Lolarga
Lyn 7h57'13" 48d57'32"
Evelyn Lombard
Cas 1h10'19" 59d51'42"
Evelyn Lopez
Cas 23h18'6" 59d57'16"
Evelyn Lorraine
Gem 7h38'8" 31d53'45"
Evelyn Louise
Lyn 7h50'48" 41d24'36"
Evelyn M. Dempsey
Cyg 21h43'14" 33d11'14"
Evelyn Marie
Vir 11h50'48" 6d32'10"
Evelyn Marie Berger
And 1h29'24" 34d32'20"
Evelyn Marie Bosko
And 1h24'56" 49d17'51"
Evelyn Marjorie Hawkins Meyer
Ari 2h12'15" 24d15'40"
Evelyn Matilda Bradshaw Gardner
Cyg 20h56'19" 35d11'34"
Evelyn Maud Higgs
Uma 9h4'42" 52d45'43"
Evelyn - Michele per sempre
Aur 6h3'27" 38d48'16"
Evelyn & Mike Forever
Lyn 7h32'47" 48d52'19"
Evelyn "Mimi" Kilmer
Uma 10h28'37" 68d8'14"
Evelyn Monahan
Uma 11h4'2" 59d1'58"
Evelyn MyGuidingStar 5/8/52-24/2/89
Cru 11h59'30" -62d29'35"
Evelyn Natalie Rose de Gruyther
And 2h36'18" 49d33'58"
Evelyn Nell Foley
Cas 0h6'42" 53d23'20"
Evelyn Niraj Pandya
Gem 6h50'25" 32d6'48"

Evelyn Norton
Uma 11h13'45" 46d16'7"
evelyn oesch
Umi 13h46'12" 70d28'59"
Evelyn Pearl Vennes
Uma 11h16'0" 56d23'57"
Evelyn Phoebe Cusato
Lib 15h12'37" -7d35'36"
Evelyn Pritchett
Sgr 19h34'37" -30d12'53"
Evelyn R. Shultz
Cap 20h18'18" -11d55'58"
Evelyn Ray
Crb 15h46'58" 26d47'53"
Evelyn Rebecca
Psc 0h19'46" 6d15'29"
Evelyn Rendle
Leo 9h38'48" 27d0'38"
Evelyn Roberts
Uma 12h10'10" 58d47'0"
Evelyn Rose
Lyn 8h6'27" 45d24'16"
Evelyn Rose
And 23h12'32" 40d21'44"
Evelyn Santiago
Cyg 19h37'25" 55d59'24"
Evelyn Seelye
Lyn 7h49'27" 41d30'38"
Evelyn & Seymour
Cyg 20h7'41" 41d45'9"
Evelyn Shaul
Lmi 10h9'47" 30d0'53"
Evelyn Shaw
Lyr 18h46'15" 38d23'43"
Evelyn & Shelby
Ori 6h15'44" 20d50'23"
Evelyn Simerick
Leo 11h18'33" 16d30'0"
Evelyn Snow Taylor
Per 4h35'37" 48d10'46"
Evelyn Solano
Uma 14h0'14" 60d12'27"
Evelyn Spanner 13.12.56
And 2h38'20" 40d16'11"
Evelyn Spiegel
Cap 21h42'48" -12d28'22"
Evelyn Swanson Barnard, 4/19/26
Leo 11h6'32" 8d13'20"
Evelyn T. Ramos
Aps 15h31'30" -71d14'2"
Evelyn T. Struthers
Psc 1h25'25" 23d58'35"
Evelyn Toner
Cas 0h21'44" 57d46'49"
Evelyn Tonti Francioli
Gem 6h52'6" 25d27'55"
Evelyn Tower Wyman
Uma 12h0'31" 41d10'6"
Evelyn Turner
Uma 13h43'29" 58d8'31"
Evelyn Viera Rivera
And 23h54'53" 45d17'22"
Evelyn Viola Begoray 1928
Cyg 21h57'39" 45d10'21"
Evelyn Virginia Quinn
Ari 2h2'7" 22d48'10"
Evelyn W. Shadd
Uma 11h30'26" 50d3'0"
Evelyn W. Ward
Cyg 21h28'30" 47d1'28"
Evelyn Yeh
Sge 19h52'28" 16d39'12"
Evelyn-46
Lib 14h55'51" -12d24'55"
Evelyne
Ari 3h27'7" 27d9'44"
Evelyne Dutil Harvey
Umi 14h15'35" 66d41'15"
Evelyne Fontaine
Umi 13h4'18" 71d56'5"
Evelyne Jacquelinet
Ori 6h19'30" 14d11'37"
Evelyne Meyer
Lyn 7h32'58" 37d1'5"
Evelyne My Babe
Cyg 19h31'21" 29d20'23"
Evelyne Rosenthal
Uma 12h1'25" 31d36'28"
Evelyne Schüpach
And 23h32'17" 40d28'1"
Evelyne Trehard
Umi 15h18'32" 82d15'6"
Evelyne Vogel
Cyg 21h5'35" 36d21'31"
EvelyneLiang
Leo 11h16'49" 15d31'59"
Evelynn Rose Pettijohn
Crb 15h53'4" 32d32'46"
Evelynne Gajardo
Cru 11h56'20" -60d27'3"
Evelyn's Light
Cap 20h57'33" -25d57'43"
Evelyn's Star
Ari 2h11'7" 19d26'55"
Evelyn's Wishing Star
Sco 17h21'8" -38d8'4"
Evenlyn Arora Serrano Ramirez
Cas 1h27'0" 71d52'30"
Ever After
Lyr 19h3'32" 33d39'12"
Ever & Betty's Love
Ori 5h52'2" 8d44'8"
ever bright star
Dra 17h9'33" 67d31'28"
Ever Lasting Love!
Cyg 21h25'23" 35d57'47"
"Everbrite" by Avicom Marketing
Uma 12h41'7" 56d19'46"
Everdina
And 0h14'39" 44d33'5"
Everett
Gem 6h30'12" 22d15'30"
Everett and Ashleigh Dreams
Cyg 21h37'27" 53d7'29"
Everett and Gladys Snowbarger
Umi 16h15'56" 76d29'43"
everett and yomaira pollard
Peg 22h47'16" 10d27'51"
Everett C. Cooper II
Sco 16h14'2" -17d51'30"
Everett C Nearburg 7/1/98 - 1/14/05
Cyg 20h14'41" 47d49'49"
Everett Charles Arden
Lib 15h4'38" -1d50'36"
Everett "Corky" Richeson
Per 3h21'15" 45d38'24"
Everett George Rhoads
Psc 1h44'42" 8d33'2"
Everett H. Roots WGP of IL 2004 OES
Ori 5h24'45" 5d41'22"
Everett Hall
Lyr 18h31'29" 33d17'30"
Everett Harrison Bauer
Tau 5h25'14" 21d57'37"
Everett L. Paluska Sr.
Gem 6h4'57" 27d21'9"
Everett Neal Carrol III
Lyn 7h52'7" 54d19'34"
Everett Philip Wesp
Psc 1h24'34" 21d3'9"
Everett Smith
Ori 5h35'22" -2d39'53"
Everett Smith
Ori 5h15'23" -5d25'56"
Everette Merdith Oswald
Sgr 19h14'54" -28d2'4"
Everette-24
Psc 0h20'10" 6d9'16"
Everhard Windscheif
Ori 6h15'0" 6d43'47"
Everlasting
Gem 7h1'42" 19d13'26"
Everlasting
Cnc 8h46'34" 19d27'30"
Everlasting
Sge 19h25'21" 17d33'34"
everlasting
Per 3h50'24" 35d16'7"
Everlasting
Uma 9h39'37" 49d53'45"
Everlasting 50th Anniversary
Uma 11h6'49" 51d13'47"
Everlasting Beauty
Gem 7h13'14" 18d51'4"
Everlasting Beauty - Vivian Chow
Cas 1h34'22" 60d7'20"
. * . Everlasting Enchantment . * .
Gem 6h33'11" 18d47'46"
Everlasting Endearment
Vir 15h1'49" 5d20'59"
Everlasting Friends Mol*Ap*Vic*Sher
Lyn 8h24'19" 38d15'14"
Everlasting Jackson
Cyg 19h41'56" 50d47'45"
Everlasting: Jacob and Anna
Vir 13h11'34" -3d18'15"
Everlasting Light of Blatchley
Per 4h7'36" 41d2'44"
Everlasting Love
And 0h42'44" 40d26'20"
Everlasting Love
And 1h24'47" 49d44'45"
Everlasting Love
Tau 4h35'2" 19d47'28"
Everlasting Love
Tau 3h36'55" 27d40'17"
Everlasting Love
Cap 21h35'25" -15d50'13"
Everlasting Love
Lib 15h6'17" -9d30'57"
Everlasting Love
Cyg 20h8'24" 54d14'37"
Everlasting Love 1966
Cyg 20h15'50" 43d18'7"
Everlasting Love: BJK & ERH
Psc 23h2'46" -1d9'7"
Everlasting Love - Calvin & Cheryl
And 2h24'50" 48d26'57"
Everlasting Love FMTM AAF
Lyn 7h54'52" 58d35'13"
Everlasting Love - Larry and Amy
Cyg 19h37'53" 31d26'43"
Everlasting Love of Pamela & Frank
Gem 7h42'58" 33d2'3"
Everlasting love - Shing and Ming
Cnc 8h31'47" 18d24'23"
Everlasting Love Stephen and Lori
Ari 2h2'42" 23d5'59"
Everlasting Sarah & Andy
Gem 7h26'57" 28d59'18"
Everlong
Vir 13h0'24" 11d10'29"
Everlong
Cas 0h19'45" 52d18'28"
Everlong
Uma 13h48'37" 56d36'2"
Everlong
Sco 16h15'19" -15d34'54"
Everlong - Chris and Hayley's Star
Cru 12h28'58" -59d39'53"
Evermore
Uma 11h14'59" 52d9'57"
EverMore's Hoss Cartwright
Cep 22h28'52" 86d44'42"
Evershining Star of Dorin and Ran
Cyg 19h37'14" 31d40'59"
Eversophia
And 1h20'39" 34d57'26"
Evert (Chasen Daniels)
Per 3h43'53" 37d19'8"
Evert (Skip) D Carter III
Her 18h24'42" 12d39'1"
Everthere
Uma 11h57'34" 31d57'57"
Every Breath You Take
Ari 1h58'31" 13d17'42"
Every night wish on our same Star
Aur 6h23'11" 41d30'32"
Everybody Conga!
Lyn 7h8'35" 59d3'3"
Everybodys Favorite
Uma 12h30'13" 61d57'18"
Everyday Love (Kelli and Nick)
Sgr 18h42'39" -16d43'30"
Everything
Aqr 23h28'56" -18d3'36"
Everything
Her 16h20'7" 19d0'26"
Everything I've Ever Dreamed Of
Umi 14h49'11" 73d24'39"
Everything You Want
Aqr 23h53'48" -11d41'15"
Everywhere & Always - The Henry's
Leo 9h25'51" 11d37'52"
Eve's Light
Crb 15h32'1" 27d49'33"
Eve's Star
Cyg 21h24'32" 30d18'57"
Evette
Aqr 22h18'30" -3d13'11"
Evette
Aql 18h43'42" -0d12'2"
Evette Elaine Graves Schluter
Lib 15h47'57" -12d59'12"
Evette Gonzalez
And 1h9'14" 40d24'37"
Evette Mendez
Gem 7h18'47" 14d39'10"
Evette Wow Murphy
Lyn 6h43'44" 57d42'42"
Evey Jolin Cloyes
Peg 22h17'47" 8d42'45"
Evey Joy
Cam 4h55'46" 53d31'58"
Eveyln Proietto
And 1h5'49" 45d38'6"
Evgeni Tcherner
Uma 11h52'30" 43d45'27"
Evgenia Listunova
And 2h27'37" 38d7'51"
Evgeny Bendersky
Vir 12h16'35" -8d32'18"
Evgeny Iosifovich Kashin
Vir 12h29'20" -6d53'45"
Evgeny Maximovich Primakov
Sco 16h11'15" -16d4'50"
Evgeny Trushin
Lyr 19h27'53" 41d46'42"
EVI 120480
Ari 2h29'1" 21d42'1"
Evi R
Cyg 21h45'57" 36d43'23"
Evi Sidonia Huffer
Leo 9h28'27" 27d39'56"
Evi und Marco
Lac 22h4'47" 45d50'27"
Evia Carettoni
Her 16h42'39" 6d34'26"
Evidence
Uma 8h17'59" 60d48'26"
Evie
Dra 20h24'58" 66d58'2"
Evie
Lib 15h59'2" -17d30'30"
Evie
Lib 15h23'39" -16d12'0"
Evie
Sco 16h48'20" -25d22'42"
Evie
Ori 6h7'41" 3d26'11"
Evie
Ori 6h12'1" 16d10'55"
Evie
Cnv 12h44'45" 39d53'9"
Evie
And 0h27'18" 46d25'9"
Evie
And 23h39'37" 42d34'52"
Evie
Cyg 21h21'29" 39d25'58"
Evie
And 1h59'21" 45d32'0"
Evie
Cyg 19h56'28" 36d37'24"
Evie
Crb 16h0'59" 33d54'53"
Evie Beatrice Goldstein
Lyr 18h37'32" 31d16'29"
Evie Belle Murray
Cas 0h29'49" 54d8'41"
Evie Blair
And 23h56'24" 34d12'51"
Evie Brown
Lib 15h59'45" -17d36'42"
Evie Cecilia
And 1h12'25" 44d14'50"
Evie Elisabeth McMullen
And 23h12'32" 43d4'17"
Evie Elisabeth Smith
And 1h49'14" 36d33'13"
Evie Elizabeth Porter
Peg 23h21'44" 18d25'39"
Evie Elizabeth Stannard
And 23h7'46" 38d36'51"
Evie & Ella's Star
Gem 7h6'17" 14d1'18"
Evie Francesca Leek
And 2h19'17" 43d53'41"
Evie Grace Christopher
And 2h30'25" 40d42'50"
Evie Jane Head
And 23h43'30" 48d31'35"
Evie Jane Read
And 0h26'54" 35d53'0"
Evie Joyce Phipps
And 23h49'30" 38d32'59"
Evie Kathleen Wortley
And 23h31'0" 49d54'5"
Evie Lauren Garmston
And 23h10'51" 42d42'37"
Evie Luv Lopez
Peg 22h4'46" 36d6'41"
Evie Lynn Altland
Gem 7h44'14" 21d5'30"
Evie Mae
And 1h18'43" 33d55'3"
Evie Mae
Ori 4h46'52" -3d19'31"
Evie Mae Larvin
Cas 0h6'4" 54d44'16"
Evie & Maye Sayers *Forever One*
Ori 4h59'14" 7d33'49"
Evie Michelle Blunden - Evie's Star
And 23h1'22" 51d3'17"
Evie Michelle Harper
And 0h11'33" 44d35'40"
Evie Piggot Richardson
Cas 1h27'6" 56d42'3"
Evie Pugliese
And 0h33'9" 42d3'12"
Evie Rose Hrydziuszko
And 2h21'18" 43d4'20"
Evie Rose I'Anson
And 1h20'10" 49d32'5"
Evie Simone Richardson
And 0h37'30" 22d1'59"
Evie Star
And 1h46'0" 42d35'13"
Evie's Eagle Superstar
Aql 19h25'43" 10d42'46"
Evil
Ser 15h30'51" 14d21'42"
Evil Fry
And 1h32'1" 44d59'58"
Evily
Tau 5h45'32" 24d7'19"
Evin and Kayla's Love
Uma 10h33'1" 67d22'12"
Evin Jacob Leite
Per 3h39'18" 46d35'14"
Evin Patrick
Ori 6h10'55" 8d9'4"
EVIN PAUL GORDON
Sco 16h13'47" -14d17'7"
Evito Domingian
Cnc 9h11'11" 17d8'8"
Eviva-Grace Kelly
And 2h38'9" 46d6'33"
EvKatanie
Sco 16h8'56" -10d46'55"
Evo J. Guerrini
Leo 10h26'50" 26d1'51"
EVO- LIZ
Uma 9h35'16" 44d9'37"
evol
Lib 15h22'19" -8d6'3"
evol
Sco 16h5'7" -16d14'38"
Evolyn Earnest Wilcox
Cas 0h38'16" 58d7'1"
Evon I. Martinez
Uma 9h57'43" 70d7'25"
Evon Seli
Aqr 22h39'44" -4d17'0"
Evonne
Aqr 20h51'13" 0d52'24"
Evonne Forelli
Aqr 22h35'25" 0d38'14"
Evonne Marie Calvillo
Psc 1h9'2" 22d59'2"
Evonne's Shining Star
Sco 17h52'20" -36d23'20"
Evrim Akyilmaz
And 0h47'28" 45d19'21"
Evryn Amelia Carlson
Peg 23h45'52" 28d37'45"
Evsei Litmanovitch
Cnc 8h57'49" 14d56'57"
Evy
Tau 3h28'22" 14d57'51"
Evy
And 0h25'35" 29d11'23"
Evy
Umi 15h44'43" 71d14'47"
Evy C. Adams
Cas 0h14'48" 55d7'53"
Evy Marie Elizabeth Hackett
Uma 8h21'19" 69d3'44"
Evy Marie Elizabeth Heins
And 1h27'4" 39d49'58"
Evy Sage Star
Psc 1h19'56" 28d22'58"
Evyan Maru
Sco 17h56'31" -30d5'24"
Evyn Lynne Humphrey
Sgr 18h13'40" -25d29'37"
Evynn Nicole Mullican - Ohnemus
Uma 12h25'30" 59d26'29"
Evyonne Iris Fountain
Cru 12h34'56" -62d37'47"
EWA
Lmi 10h55'59" 28d49'2"
EWA 1958
Uma 8h42'58" 63d20'31"
Ewa Brissing
Psc 0h33'34" 11d26'9"
Ewa J. Kosek
Ori 5h7'24" 15d28'24"
Ewa Jasiulaniec
And 23h19'12" 43d4'53"
Ewa Titz
Uma 12h53'45" 62d57'26"
Ewald Torreiter
And 23h37'33" 50d29'0"
Ewan Christopher Roe
Her 18h17'21" 28d36'25"
Ewan Harry James Stubbs
Cep 2h40'31" 86d58'43"
Ewan James Gordo Doyle
Per 2h49'5" 55d28'35"
Ewan James John
Lmi 10h50'26" 33d43'2"
Ewan Kennart V. Saunier
Aql 18h52'47" -0d12'24"
Ewan Louis Pritchard - Ewan's Star
Dra 16h35'54" 57d45'14"
Ewan Mark Smith
Her 17h29'5" 17d24'29"
Ewan-Rebecca Star of Liebe
Cap 20h53'36" -15d44'42"
ewa-sikoreczka
Cyg 20h50'13" 48d46'49"
Ewelcia
Sco 17h29'56" -40d41'56"
Ewelina Dziak
Tau 3h26'39" 4d57'27"
Ewelina Gurgul
Cap 21h22'41" -25d3'43"
Ewelina Zajac
Uma 10h8'51" 44d52'42"
Ewelinka
Aqr 21h23'9" 0d28'45"
Ewf GiWL Awyple
Leo 11h31'52" 18d23'6"
Ewig Zusammen
Uma 8h57'58" 58d3'46"
Ewige Liebe Christian, 03.09.2003
Uma 8h22'23" 59d56'11"
Ewige Liebe Juan & Andrea Cabrera-Ofner
Umi 14h25'35" 69d48'32"
Ewige Stärke
Her 17h18'30" 34d20'6"
Ewige Wiederkunft
Leo 9h30'29" 9d39'36"
Ewigi Fründschaft-Marina Steiner
Uma 9h52'21" 47d40'56"
Ewigi Liebi
And 2h36'28" 50d6'1"
Ewigi Liebi
Umi 15h8'24" 70d56'37"
Ewigi Liebi Andy & Andrea
Aur 5h59'58" 42d55'6"
"Ewigi Liebi" Daniela & Stefan
Cas 1h43'7" 68d48'5"
ewigi Liebi Patrik & Angela
Dra 18h47'7" 57d41'29"
Ewigi Liebi söu üs dä Stärn schänke!!!
Ori 6h14'13" 14d11'3"
Ewigi Liebi, 01.06.1997
And 22h58'29" 39d51'41"
Ewing "John" Stamps
Vir 13h13'51" 6d16'10"
Ewunia
Gem 6h45'58" 27d16'39"
Ewunia
Uma 10h0'33" 42d11'57"
Ewunia
Sex 9h46'36" -0d15'37"
Exardescere
Pho 1h11'3" -46d51'28"
Excalibur
Aqr 23h3'4" -7d10'16"
Excalibur Leather Network
And 1h7'21" 46d32'29"
Excellentia
Uma 12h34'50" 58d43'16"
EXCEPTIONAL Rakesh Malhotra
And 1h6'22" 45d20'0"
eximius madre
Vir 13h13'12" -16d7'50"
Exmanus
Ori 5h40'1" 1d52'4"
Exorbitant
Gem 6h58'57" 12d55'25"
Exotic MoonGoddess Sister Shireen
Cnc 8h32'49" 18d24'49"
Exotic Princess
Sgr 17h58'42" -17d59'35"
Express Financial Services
Uma 9h43'15" 49d8'38"
Express T H Sdn Bhd
Ori 5h35'56" 4d17'39"
Ext Star Symbol Of Esser Reunion
Uma 10h34'0" 56d29'51"
Extra Special Penny Jean
Psc 0h57'41" 25d17'44"
Extrorgasm
Sgr 18h7'45" -18d12'1"
Ex-Uncle Jose Ruiz
Hya 9h17'46" 3d35'52"
Exzandrite
Vir 12h19'57" 11d13'27"
Eydie Marlowe
Sgr 17h59'4" -24d42'20"
Eye of Cassandra Rachel
Cas 3h14'5" 63d33'37"
Eye of David
Uma 8h15'21" 60d23'16"
Eye of Laleh
And 0h2'10" 44d23'25"
Eye Of Melissa
Gem 6h13'0" 22d36'37"
Eye of Mirela
Ori 5h59'32" 17d4'20"
Eye Of The Angel
Vir 12h36'16" 9d2'52"
Eye of the Angel Kristi
Sco 17h7'33" -34d6'8"
Eye of Xairus
Ori 5h49'39" 6d22'54"
Eyebrow's Folly
Cyg 19h45'53" 32d32'15"
Eyes of Blue
Lyn 6h38'33" 54d6'58"
Eyes of Steele
Uma 13h36'0" 57d3'19"
Eyes of the World
Ser 15h17'54" -0d19'27"
Eyky Orellana
Sco 16h47'36" -27d46'32"
Eylem
Psa 22h26'6" -30d52'5"
Eyr Thanah
Cma 7h12'17" -11d39'27"
Eyvonne
Aqr 23h38'40" -7d49'29"
Ezekeil Connor Gilmer
Uma 11h24'7" 45d55'13"
Ezekiel Jacob Lubin
Sco 16h27'20" -26d30'7"
Ezekiel the Magnificent
Ari 2h2'28" 20d20'59"
Ezequiel Moujan
Ari 2h42'11" 20d15'48"
Ezio Capraro
And 2h38'36" 48d56'29"
EzioElena
Umi 15h56'24" 75d43'8"
Ezmeralda C
Cas 0h24'33" 61d26'54"
Ezmerelda
Cas 1h15'36" 63d42'32"
Ezperansa Gisselle Urrea Pleittes
Uma 10h28'12" 64d26'49"
Ezra Darlin'
Lib 15h43'28" -7d2'24"
Ezra Ed
Her 17h43'20" 40d43'54"
Ezra Krieg
Del 20h20'41" 15d11'43"
Ezra Levi
Umi 20h24'22" 89d10'21"
Ezra Marquez
Eri 3h50'6" -15d33'8"
Ezra Powell
Ori 6h7'20" -0d11'1"
Ezra Roati
Aqr 22h46'55" -18d30'6"
Ezra Sebastien Yewers
Cru 12h1'37" -64d37'53"
Ezz's Sparkle
Gem 6h4'59" 23d44'19"
F
Leo 11h12'44" 22d58'58"
F A M D I A Y
Aqr 20h42'33" 1d15'59"
F. Albert Herter
Gem 7h43'9" 33d39'53"
F. Allan Duncan and Sigrid Nelson
Per 3h12'29" 51d44'21"
F. Barry Keenan
Her 17h25'37" 45d1'12"
F C H
Cap 21h49'16" -10d17'45"
F & C Stargazer
Her 17h29'51" 27d20'3"
F. Chatonnet
Cas 0h7'15" 53d23'10"
F. Christopher Yandel
Ari 2h47'21" 13d1'39"
F. Darrin Perry
Ari 2h24'7" 23d43'39"
F E Bear, Jr.
Sco 17h34'17" -42d20'12"
F. E. Bus Smith
Tau 5h32'44" 24d1'1"
F F Bey Ovation "Niles"
Peg 22h53'15" 22d4'43"
F&H Forever 8/8/05
Gem 7h46'52" 32d6'48"
F. H. MILEFSKI II
Umi 14h12'55" 77d20'45"
F. Jean Conner
Tau 4h11'9" 14d11'27"
F & L Grossi 35th Anniversary Star
Gem 6h27'42" 21d10'57"
F L Martin
Aql 19h49'31" 16d15'14"
F & M Jorgenson
Uma 11h31'21" 56d22'3"
F. Marlene & C. Joseph Morris
Cyg 20h48'9" 44d14'9"
F&N Coca-Cola (Malaysia) Sdn Bhd
Ori 5h35'13" 3d44'35"
F&N Dairies (M) Sdn Bhd
Ori 5h36'45" 3d18'55"
F O R E V E R
Col 5h38'58" -31d29'14"
F. Patrick Logan
Uma 8h54'25" 66d19'50"
F. Ron Da Star
Sgr 18h26'20" -18d38'43"
F. Ronald Amos
Leo 11h7'34" 15d11'2"
F&S 10
Uma 10h49'36" 68d25'16"
F. Sherwood Rowland
Cnc 9h9'9" 31d41'51"
F. V. Oliver, Jr.
Lyn 8h17'56" 34d44'1"
F. & W. Stebbeds
Lyr 19h17'58" 29d32'52"
F. Wilson Bishop
Cnc 8h12'37" 32d12'38"
F. Wilson Jackson, III
Aql 20h6'17" 8d15'45"
F. Yayoi & Yusuke
Cyg 20h49'7" 37d10'27"
F3D
Ori 6h4'10" 11d33'22"
4 Bello Amor 29
Lyn 8h36'56" 33d49'35"
4 ever... 2 DI...
And 2h14'10" 47d57'39"
4/EVER Bill & Laura Warren 7-11-71
Cyg 20h52'20" 34d35'23"
4 Glory
Crb 16h14'15" 37d11'0"
4 L 4 B
Cam 3h54'42" 65d40'18"
4 Lizzy My Durgeon
Mon 7h17'35" -0d40'24"
4 Ms' Daddy
Her 18h6'29" 26d17'54"
4 My Baby Boo Alwayz...
Cru 12h52'29" -56d54'27"
4 My Bay Bee!
Cyg 19h45'44" 30d54'18"
4 My Best Friend Ashley, Love Ryan
Ari 2h4'48" 20d57'9"
4 "Red" In Loving Memory
Uma 10h17'38" 63d1'27"
4 Rs Mom 90
Lib 14h24'53" -19d56'29"
4007o1Life.o1Love2015 409
Ori 6h10'59" 17d36'10"

40th Anniversary: Sam & Judy Tocci
Cyg 20h10'30" 39d33'5"
40th Festiva for Margarita
Sgr 18h54'37" -28d21'32"
41 Fordman
Boo 15h15'21" 40d39'57"
4.10.04
Uma 10h52'55" 61d10'36"
41131SMSGTstarJ.E.Hiott
Cen 13h24'7" -36d6'23"
422MK2006Y5
Cyg 19h57'9" 45d59'22"
44 NANCEE
Gem 7h21'43" 14d35'38"
44622 2/8/18 "Burnetta" MKVRB 26062
Col 5h36'19" -33d15'21"
45683908
Lyr 19h20'7" 29d31'40"
458
Psc 23h2'13" -0d8'10"
*459*JWM*469*Always.....
Leo 9h26'50" 10d4'38"
4anita - A sparkling star
Cas 22h59'56" 58d40'13"
4C
Per 1h34'40" 54d31'43"
4Crawler
Cap 21h56'23" -10d31'37"
4Eva In Each Other's Hearts LA - MA
Cru 12h33'5" -62d15'29"
4Eva Soul Mates 26924
Aur 6h27'11" 36d22'55"
4ever michi
Umi 16h7'7" 72d42'39"
4evran1
Cyg 19h37'22" 52d24'54"
4Giddaboudit Mirmelstein
Uma 11h40'6" 52d22'55"
4J 2M 2K Everyone I Love My Family
Pho 2h21'42" -46d43'56"
4M4Ever
Cyg 19h40'16" 52d52'24"
4PaulandBeeEternally
Cyg 21h11'54" 47d19'2"
4-PDC-211-69-5683-MS
Peg 22h0'9" 11d10'33"
4RodnRuckus' Hartshires Sonikalemja
Sco 16h49'6" -26d22'40"
4Seasons
Ori 6h8'31" 12d42'18"
4ta1sULuvd&Lst.lLuvU<->thismuch
Leo 10h24'11" 16d53'23"
5/1/67-Stacy Darlene Fincher-5/1/05
Cas 0h8'57" 53d15'32"
#5 Brandon Michael Farmer
Eri 3h45'45" -0d50'16"
5 DWC 50
Del 20h39'36" 15d24'43"
5 Lachman Stars
Cap 21h37'45" -21d34'57"
5 PJ STRAND 5
Her 17h40'44" 32d39'24"
5 RO's Stella Brillante
Uma 8h34'56" 65d15'27"
5 Sam
Lyn 8h18'32" 45d9'31"
5 Star Academy of Cheer & Dance
Cma 6h30'7" -16d23'52"
5 Viens
Cam 4h1'42" 66d2'0"
50 Year Star ~ Cathy Watson
Vir 13h6'3" -15d42'24"
501
Uma 13h38'14" 62d11'15"
50th Anniv. Becky & Gary Locke
Ori 5h30'9" 9d54'58"
50th Anniversary Blessings
Cyg 19h39'27" 32d19'19"
50th Anniversay -Dick and Shirley
Cyg 20h45'39" 32d9'9"
50th Wedding Star Joe&Fran '54 -'04
Ori 5h36'41" -2d31'31"
50thYearAnniversaryJames &LoisOlson
Gem 7h40'26" 27d23'22"
5122 Bear Grimes
Uma 9h17'17" 60d58'43"
5-18-90 My Chop, My Love 12-8-02
Sgr 19h12'31" -16d50'1"
544 Midnight Ln, 80816 Love Eternal
Lib 14h39'50" -17d54'1"
57... ...JAWS
Uma 8h36'51" 48d18'46"
57PamMac07
Cyg 20h38'15" 34d38'31"
588*8*3673837
Sco 17h53'6" -36d17'59"
5B II (p-jaks)
Ori 5h20'12" 6d54'54"
5dd5
Ari 2h53'3" 30d19'45"
5M8W8M7
Cam 4h42'18" 65d21'20"
5th May 1951 - Whipps Family Star
Cru 12h38'33" -57d47'9"
F8.14
Leo 9h49'51" 27d45'44"
f.a. - EternalLove - w.l.
Ari 2h50'31" 20d49'42"
Fa!th
Ori 5h24'35" 10d11'26"
FAB
Psc 0h49'23" 15d25'42"
Fab 5
And 0h25'14" 32d10'42"
Fab LaMarca
Uma 10h19'16" 59d4'35"
Fab & Lio
Crb 16h4'29" 32d21'50"
Fabez
Vir 13h19'51" 6d59'21"
Fäbi
Lmi 10h40'32" 39d16'16"
Fabia
Lyr 18h48'11" 40d5'54"
Fabian
And 0h54'14" 46d12'59"
FABIAN
Uma 10h12'38" 45d37'58"
Fabian
Ori 5h58'8" 17d10'43"
Fabian Dominique Giles
Uma 10h2'26" 59d17'16"
Fabian Karl Handl
And 0h1'3" 32d42'28"
Fabian Michor
Uma 8h40'7" 53d54'16"
Fabian Nicola
Lac 22h41'6" 52d43'7"
Fabian Radtke
Cyg 19h47'2" 34d46'57"
Fabian Sciboz
Dra 16h43'59" 55d58'56"
Fabian Vasquez
Mon 6h20'58" -6d1'16"
Fabian Werner
Uma 11h21'50" 62d39'3"
Fabian Zumsteg
Dra 18h20'59" 71d54'6"
Fabiana
Aql 19h3'19" -0d18'58"
Fabiana 40th
Aqr 22h28'7" -1d5'1"
Fabiana Foresta
Lyn 6h53'26" 60d44'50"
Fabiana Freschi
Cas 23h2'57" 53d28'10"
Fabiana Ortiz
Peg 22h8'33" 8d4'15"
Fabiana Silveira
Cru 12h25'19" -57d24'10"
FABIAN-CHOUCHOU
Her 16h14'36" 42d30'51"
Fabian-Thomas Raisig
Ori 5h55'56" 7d19'39"
fabiawädi
Cas 23h42'2" 56d8'0"
Fabien Kovac
Uma 9h59'8" 46d56'39"
Fabien Ramos Camacho
Gem 6h32'45" 20d19'13"
FABIENNE
Cas 23h27'21" 54d39'2"
Fabienne
Uma 11h20'12" 68d43'57"
Fabienne
Cas 2h24'0" 70d37'46"
Fabienne
Psc 1h33'19" 3d21'14"
Fabienne&Christian
Per 4h10'14" 34d27'42"
Fabienne Désirée 3. oct. 1988
Dra 10h55'3" 73d59'21"
Fabienne/Dominik
Umi 14h39'14" 77d52'1"
Fabienne Guillaumond
Gem 6h40'58" 12d27'57"
Fabienne (Hagschi) Hager
And 0h38'42" 31d14'37"
Fabienne & Katja 12.11.1950
Uma 14h13'33" 60d1'6"
Fabienne Leone
Cas 23h53'43" 56d27'44"
Fabienne Mauchle
Umi 18h45'8" 86d41'11"
Fabienne Savrimoutou
Uma 13h5'30" 54d59'11"
Fabienne & Tanguy
Uma 10h3'11" 47d42'28"
Fabienne und Urs
Cas 1h39'51" 64d30'46"
Fabio
Cep 22h59'12" 79d48'30"
Fabio
Umi 17h40'40" 80d42'20"
Fabio
Uma 13h31'20" 60d7'59"
Fabio
Cas 23h50'11" 50d14'54"
Fabio
Lyn 7h34'39" 35d44'46"
Fabio
Her 16h39'56" 29d24'51"
Fabio - Alessia - Daniela
Ori 5h58'43" 21d18'53"
Fabio Arcangeloni
Cam 12h49'9" 77d30'13"
Fabio Brigante Colonna Angelini
Per 3h21'36" 41d26'13"
Fabio & Christine
Sco 17h54'10" -35d46'6"
Fabio Corra
Lac 22h46'33" 54d21'7"
FABIO&DANIELE
Ori 5h36'7" 15d31'42"
Fabio Di Nunzio
And 2h36'52" 45d34'29"
Fabio Gambarotto
Lac 22h56'32" 51d4'46"
Fabio Loïc
Cas 1h39'51" 69d2'40"
Fabio Meduri
Lyr 18h39'28" 30d2'11"
Fabio my love
Ori 6h9'48" 5d46'48"
Fabio Olivi
Dra 17h46'25" 69d14'25"
Fabio Paulo Emilio Sollecito
Uma 11h41'11" 45d45'15"
Fabio Rafaie Scotto Lavina
Leo 10h42'4" 18d22'20"
Fabio Ramon Hauser
Aur 4h48'22" 35d37'46"
Fabio Rececconi
Lyr 19h15'6" 32d32'12"
Fabio Scascighini
And 0h41'54" 30d9'52"
FabioElenaxsempre
Cyg 21h0'31" 40d12'31"
Fabio-Giuseppe Marucci
And 2h9'37" 42d27'59"
Fabiola
Aur 5h24'10" 41d3'14"
Fabiola
Per 4h47'32" 44d11'56"
Fabiola
Uma 10h6'11" 51d21'36"
Fabiola
Ori 6h0'7" 0d58'6"
Fabiola
Tau 5h7'11" 17d56'46"
Fabiola Angel
Uma 11h36'53" 53d36'25"
Fabiola Brignone
Uma 13h35'52" 59d22'23"
Fabiola Ceccato
Uma 8h45'11" 56d29'41"
Fabiola Cubillos
Lyn 8h47'15" 41d24'48"
Fabiola & Daniel
Peg 22h24'41" 21d16'29"
Fabiola Ivonne Napoles
Ori 5h29'30" -4d13'10"
Fabiola Robles
Lyn 7h9'35" 57d11'19"
Fabiola Villaca De Lima
Gem 6h41'57" 17d26'30"
Fable
Cam 4h52'6" 66d49'0"
Fabrezio the fabuloso
Cnc 8h25'59" 11d22'56"
Fabrice Cattin
Gem 7h23'32" 31d10'58"
Fabrice Charpin
Ori 4h50'16" -3d25'1"
Fabrice David Quinio
Uma 10h14'3" 59d39'9"
Fabrice et Lydia
Ari 2h49'17" 20d6'40"
Fabrice Lebeau
Lyr 18h40'30" 31d37'20"
Fabrice Lombard
Boo 14h32'43" 19d27'23"
Fabrice Michalet
Sgr 18h11'24" -25d11'37"
Fabrice Sophie 5.49
Uma 12h58'12" 62d7'15"
Fabrice Theodon
Cap 21h8'40" -27d26'25"
Fabris Eligio
And 1h16'57" 42d41'20"
Fabrizia
Uma 11h34'56" 35d26'38"
Fabrizio
Uma 11h12'28" 32d49'22"
Fabrizio
Aur 4h54'28" 38d32'21"
Fabrizio
Aur 6h40'0" 38d35'50"
Fabrizio
Ori 5h47'49" -3d6'11"
Fabrizio Antonio Bruto Griguoli
Cen 13h47'28" -38d18'14"
Fabrizio Balestri
Lib 15h33'26" -5d5'47"
Fabrizio Caccavale
Leo 9h44'47" 28d24'39"
Fabrizio Nicoli
Cas 0h1'30" 52d1'19"
Fabrizio Rigo de Righi
Uma 12h21'56" 60d44'29"
fabtrisha4ever
Uma 11h16'41" 46d45'12"
Fabuleuse Sylvie
Uma 11h54'47" 46d51'47"
Fabulous Forrest
Cnc 7h57'33" 11d9'34"
Fabulous Holdings Sdn Bhd
Ori 5h35'34" 3d13'56"
Fabulous Nin of Wondrous Design
Ori 5h44'50" 2d5'41"
Faccia Di Angelo
Aql 19h5'0" -0d2'35"
Facco Marco
Lyr 19h22'50" 30d56'9"
Face
Lyr 18h51'13" 35d48'59"
Faceica
Vir 13h20'57" 7d10'29"
Facelis Soto
Cas 0h29'33" 50d7'34"
Fachie
Cam 3h42'2" 58d19'10"
Facie
Aqr 22h43'14" -16d44'17"
F-A-C's Hound of Heaven
Cyg 20h47'53" 30d49'21"
Fadder Ed P. 66
Leo 11h2'54" 20d15'58"
Fade Into You Nothing, Nothing
Cas 23h48'23" 56d45'7"
Fadi Hanna
Uma 11h15'42" 28d46'52"
Fadi Najem
Mon 6h47'12" -0d10'52"
Fadi Slaieh
Sco 16h55'21" -40d25'25"
Fadia Jessica Louis Mira Junior Paul
Uma 10h7'36" 62d27'20"
Fadoua
Uma 14h1'35" 49d44'48"
Fadoua Ouallal
Del 20h37'16" 8d49'52"
Fadri Bernet
Cas 1h39'48" 77d27'54"
Fadri Francesco Fanetti
Vul 19h51'58" 22d48'44"
Fadri Noël
Umi 16h0'16" 73d44'45"
Fadrina Davaz
Cnv 13h54'5" 35d13'2"
Fady Chammas
Cyg 20h1'23" 38d36'49"
Fady Elassaad
Ori 5h56'20" 12d33'32"
Fady Elassaad
Cep 1h22'49" 80d29'15"
Fady Foud
Cas 0h9'26" 50d13'51"
Fae Daisy Bosett-Roberts - 20.08.03
Ori 5h44'46" 7d15'57"
Faedo Grazia
Boo 15h18'12" 43d8'5"
Faehner 40
Lyr 18h50'26" 35d58'49"
Faerie Divine
Sco 16h50'59" -32d53'27"
Faeryland
Mon 7h3'46" -3d42'36"
fafaoa
Aur 6h2'43" 36d23'44"
Fafi - Rocky
Uma 13h42'25" 60d56'10"
Fagan Tillman
Uma 12h4'33" 56d26'35"
Fagel Zeesa
Cnc 9h7'56" 16d26'47"
Fagel, Christian
Ori 6h16'23" 18d49'49"
Fags
Cap 20h57'38" -19d38'16"
Fahad
Dra 19h22'18" 68d17'17"
Fahad
Cyg 19h39'17" 32d36'17"
Fahad & Haifa
Lmi 10h0'41" 36d15'49"
FAHR
Uma 11h30'54" 41d10'13"
Failte
Cep 22h6'40" 60d21'48"
Faina
Aqr 21h50'48" 2d1'46"
Faion Fabrizio
Per 3h5'1" 51d40'6"
Fair of Love
Ori 5h0'19" 9d33'34"
fairchuck
Cma 6h57'18" -20d57'20"
Fairie Way
Sgr 19h9'23" -13d42'5"
Fairley-Holland Family Star
Pho 0h35'34" -50d18'17"
Fairmont 44
Cap 20h37'45" -13d27'26"
Fairmont Glitter Bay Hotel
Tri 2h32'50" 36d18'10"
Fairmont Royal Pavilion Hotel
Cnv 13h11'54" 49d51'21"
Fairn Marjorie Jones
And 1h5'18" 42d45'39"
Fairouz
Sco 17h25'29" -33d53'38"
Fairspeare Samson
Umi 14h50'44" 76d7'23"
FAIRVIEW REACHED THE STARS IN 2005
Uma 11h16'44" 61d45'43"
Fairy Angelina Ydarenia
Cap 21h51'25" -18d9'5"
Fairy Chu Ka Lan - The Cutest Baby
Lib 15h17'15" -23d28'38"
Fairy Girl
And 0h28'3" 45d34'47"
Fairy Girl Angie
Psc 1h16'37" 28d16'11"
Fairy Mindy
Lmi 9h26'59" 35d38'12"
Fairy Princess' Star
And 23h44'17" 38d59'14"
Fairytale
Peg 21h42'2" 14d5'42"
Fairytale Brownies 15th Anniversary
Uma 11h14'40" 56d46'15"
Fairytale Princess
And 23h46'34" 39d20'27"
Faisal
Sco 16h13'32" -14d5'50"
Faisal Al Ruzaihan
Cap 21h49'39" -23d4'32"
Faisal Bin Sultan
Ori 6h3'11" 14d2'32"
Faisal Emil Bou Orm
Per 3h9'55" 49d14'33"
Faisal Salam
Peg 22h24'57" 5d33'15"
Faith
Aqr 22h16'40" 1d20'56"
Faith
Ari 2h51'16" 14d23'38"
Faith
Psc 1h16'25" 12d4'51"
FAITH
Tau 3h53'44" 1d34'39"
Faith
Tau 5h48'23" 13d59'48"
FAITH
Psc 1h26'47" 16d21'17"
Faith
Tau 4h45'51" 19d16'48"
Faith
Peg 22h48'9" 20d10'26"
Faith
Ori 5h58'51" 18d1'32"
Faith
Gem 6h34'0" 15d46'53"
Faith
Gem 7h30'53" 19d25'45"
Faith
Cnc 9h10'13" 15d9'45"
Faith
Cnc 8h19'1" 17d21'10"
FAITH
Gem 7h46'14" 26d45'6"
Faith
Per 3h10'17" 47d2'34"
Faith
And 23h21'46" 41d25'16"
Faith
Cyg 19h59'8" 46d40'55"
Faith
Uma 10h27'1" 46d49'55"
Faith
Cam 4h39'41" 53d55'6"
FAITH
Cas 2h10'58" 59d41'30"
Faith
Cyg 21h41'33" 49d41'32"
Faith
Per 3h45'24" 36d56'56"
FAITH
Her 17h40'13" 37d7'27"
Faith
And 1h28'53" 38d3'0"
Faith
Lyn 8h7'24" 42d51'10"
Faith
Sgr 19h6'24" -25d32'27"
Faith
Eri 2h16'9" -53d1'49"
faith
Sco 16h8'48" -11d16'57"
Faith
Vir 13h29'31" -17d45'18"
Faith
Uma 11h30'10" 53d3'39"
Faith
Uma 11h49'26" 56d12'55"
Faith
Cas 1h6'13" 62d34'15"
Faith
Uma 14h17'2" 61d29'9"
Faith
Cas 2h18'51" 64d26'33"
Faith
Uma 10h0'14" 70d45'42"
Faith
Dra 12h35'35" 72d8'27"
Faith Abbegail
Uma 11h47'11" 49d15'45"
Faith Allie
Sgr 17h53'27" -27d26'9"
Faith Amelia
And 0h21'43" 45d44'41"
Faith and Hope
Uma 11h0'45" 53d57'3"
Faith and Jeff
Gem 7h28'41" 20d26'13"
Faith and Jeff Yeider
Cyg 20h13'45" 52d37'16"
Faith and Ken Tackett
Gem 6h3'11" 25d39'5"
Faith Angel Rhoden
And 23h14'29" 47d29'44"
Faith Ann Deguzman
Cyg 20h40'33" 45d50'0"
Faith Ann Fewtrell
Leo 9h38'57" 27d26'42"
Faith Ann Staun
Uma 13h36'48" 50d0'2"
Faith Ann Wabl
Sco 16h10'5" -21d34'36"
Faith Anne
Sco 16h17'39" -11d59'17"
Faith Anne
Col 5h43'55" -38d54'38"
Faith Antico
Uma 12h35'7" 57d37'54"
Faith Ariana McCallum
Cyg 21h25'14" 50d27'46"
Faith Barbara Larsen
Vir 13h44'15" -6d27'19"
Faith Blackford
Cyg 19h43'47" 38d38'9"
Faith Cheryl Trombly
Lyr 18h30'47" 29d56'45"
Faith Christina McGeachie
And 23h27'56" 40d38'17"
Faith Coralyn Moss Tucker
Leo 10h32'36" 13d11'21"
Faith Dow
Uma 8h56'27" 55d36'37"
Faith E. Cook
Ori 6h13'13" 15d4'6"
Faith E. Fawcett
And 2h25'39" 41d35'18"
Faith Elaine
Cas 1h21'7" 67d43'25"
Faith Eleanor
Vir 12h37'46" -10d2'47"
Faith Elizabeth
And 2h13'48" 41d38'20"
Faith Elizabeth
Lyr 18h38'39" 33d45'3"
Faith Elizabeth Andrews
And 0h33'5" 33d43'50"
Faith Elizabeth Narvaez-Tisdell
And 0h27'52" 45d28'35"
Faith Elizabeth Saathoff
Cyg 20h21'4" 39d17'51"
Faith Elizabeth Yetter
Sco 17h51'58" -41d39'50"
Faith Erin & Michael
Uma 11h28'58" 62d59'2"
Faith Etienne Daigle
Uma 10h50'57" 57d26'3"
Faith F. Fitzgerald
Col 5h45'53" -31d48'26"
Faith (Faithie) Maria Hart
Ari 2h27'14" 25d29'13"
Faith Faulk
Crb 16h16'12" 34d58'2"
Faith Freedom Forgiveness
Lib 15h13'6" -5d10'54"
Faith Greenwald
Lyn 9h37'40" 40d10'42"
Faith Gregory-Curl
Cyg 19h37'45" 35d26'28"
Faith Heald
Pho 23h41'55" -44d55'35"
Faith Hope Love
Lib 15h19'13" -8d30'28"
Faith Hope Love
Cyg 19h49'18" 31d23'9"
Faith Hope Love & Joy for Everyone
Uma 10h23'35" 57d12'47"
Faith in Love
Leo 10h25'40" 23d24'45"
Faith Iris MacIlvaine 12/25/00
Umi 17h9'25" 77d9'49"
Faith Jacobson
Lyr 18h32'36" 32d30'13"
Faith & John
Psc 1h18'1" 32d30'16"
Faith Kailani
Cnc 8h19'24" 14d8'44"
Faith Kunica
Gem 7h16'28" 28d13'0"
Faith Lane Kunkel
And 1h9'34" 40d59'27"
Faith LeeAnn Allen
Umi 13h54'48" 74d29'50"
Faith Leeila
Per 3h13'20" 40d20'52"
Faith Light Juan & Peggy Gutierrez
Per 3h3'36" 53d11'23"
Faith Lord
Ari 2h56'21" 26d18'8"
Faith Lorraine Wormsbacher
Cap 21h10'22" -19d24'30"
Faith Love & Friendship thru Trust
Ari 2h9'46" 21d23'0"
Faith Lynn Wise
Cyg 21h59'16" 54d52'55"
Faith Madison Horn
Ari 3h4'46" 28d39'18"
Faith Marie Ferguson
Cap 20h30'0" -16d47'59"
Faith Marie Fischer
Tau 4h12'26" 26d28'20"
Faith Marie Hackbarth
Psc 0h27'55" 8d4'0"
Faith Marie Hackbarth
Uma 11h37'13" 47d23'48"
Faith Marie Huang
Umi 15h25'34" 70d58'34"
Faith Marie Jones
Aqr 20h41'2" -8d16'33"
Faith Marie Maslana
And 23h21'49" 37d49'29"
Faith Marie Molino
And 0h6'34" 46d28'44"
Faith Marie Rodgers
And 0h2'35" 35d2'5"
Faith Marion Mills
And 0h17'25" 28d9'18"
Faith Mary Jane Gordon-Smith
Cnc 9h20'20" 12d24'59"
Faith Mary Johnsen
Psc 0h56'46" 32d9'13"
Faith Melanie Cupp
Sco 16h57'30" -38d47'20"
Faith - mi estrella preciosa
And 0h53'23" 36d32'27"
Faith Neryn Cary
Psc 1h13'49" 4d5'53"
Faith Nichole
And 0h33'19" 26d2'56"
Faith Nicole
Sgr 18h47'21" -29d23'19"
Faith Precilla May Woolley
And 23h12'52" 46d15'41"
Faith Rosemary
Cma 6h44'53" -13d45'13"
Faith Roy
Uma 11h37'7" 52d53'25"
Faith Scheibe Vannucci Riley
Tau 3h55'33" 9d12'7"
Faith Seserko Gaffney
Uma 11h33'38" 61d34'0"
Faith Still
Lib 15h14'39" -29d33'44"
Faith Swift
Cyg 21h41'7" 43d10'55"
Faith The Shining Star of My Life
Leo 9h26'41" 9d39'15"
Faith Victoria Pisani
Gem 7h7'24" 21d9'42"
Faith Yack
Uma 9h59'5" 57d29'2"
Faith, Hope and Love
Psc 1h20'13" 21d33'45"
Faith, Hope & Love J&G Smythe Star
Uma 14h11'1" 57d34'11"
FAITH, HOPE, AND LOVE
Ari 2h14'26" 23d10'59"
Faith-Ann
And 0h42'19" 27d44'58"
Faith-Anne
Uma 10h31'10" 60d27'37"
Faith-Csillagom csillaga BK'54.01.09
Cap 21h43'47" -19d21'33"
faithfull
Lyn 7h30'22" 57d25'51"
Faithfully
Per 1h42'30" 50d44'43"
Faith-KJT
Sco 17h21'48" -38d24'33"
FaithLoveCourage
Mon 8h5'45" -0d34'56"
Faith's Heart
Vir 13h20'41" 2d53'51"
Faiths Sparkle
And 2h12'46" 48d44'52"
Faith's Star
And 0h13'52" 45d44'37"
Faithy
Ori 5h41'37" 9d7'55"
Faithy Bob Angel 1
And 2h25'38" 39d28'42"
Faithy's Wish
Sgr 18h2'52" -21d10'31"
Faiz Gina
Leo 9h57'4" 24d18'39"
Faizah
Cap 20h23'9" -16d7'21"
Faizal
Cap 21h55'42" -10d29'34"
Fakaosi Langi & Sabrina Dastyari
Cru 12h1'54" -62d35'18"
Fakher Surji
Uma 11h16'27" 37d15'18"
Falcon
Aql 19h4'18" 14d33'45"
Falcon
Col 5h52'7" -32d55'30"
Falconi Maria Angiola
Aql 18h52'28" 8d34'28"
Falgun B. Patel
Per 3h11'16" 52d43'12"

Falguni Patel
Lib 15h38'50" -25d51'13"
faline jones
Umi 13h45'8" 78d59'35"
Faline's Fire
Sco 17h40'5" -32d51'20"
Falivene
Vir 12h39'56" -8d34'40"
Faljoe
Sgr 18h34'53" -23d8'41"
Falk
Cma 7h10'53" -25d42'9"
Falkenstein, Luca
Uma 10h18'28" 66d23'2"
Falkenstein, Tanja
Uma 11h53'40" 60d34'11"
FALKOR
Ari 3h12'56" 28d47'43"
Fallen Angel
Umi 13h37'24" 78d11'13"
Fallen Muse
Crb 16h10'35" 27d48'15"
fallin' in love again
Sge 19h35'29" 18d28'27"
Falling
Peg 21h25'25" 17d11'39"
Falling In Love With You 15 Years
Cru 12h26'3" -61d50'28"
Fallman
Uma 13h45'53" 52d41'34"
fallon
Sco 17h55'56" -37d29'1"
Fallon
Lyn 7h28'51" 52d11'27"
Fallon Ann Cosgrove
And 1h19'51" 45d17'55"
Fallon Bear
Lib 15h9'54" -6d22'42"
Fallon Bridget
Gem 7h43'23" 33d5'23"
Fallon D. West
Sgr 18h35'30" -23d34'24"
Fallon Ferguson
Uma 9h36'28" 58d40'48"
Fallon Valente
Her 17h34'46" 37d14'3"
Falon Celeste
Psc 1h19'57" 8d28'7"
Falon Dupuis
Ori 5h37'45" 7d43'48"
Falon Rae Vela
Sgr 19h50'47" -22d52'27"
Faloona
Cas 0h15'40" 55d9'24"
Faltz 4
Lyn 8h15'55" 55d8'36"
Falzon
Cru 12h17'22" -56d34'55"
Fam. Alberto Cademartori
Peg 23h41'15" 25d0'25"
Fam. Brechbühl-Weber
Psc 23h9'58" 3d21'0"
Fam. Dübendorfer
Peg 21h34'8" 21d8'59"
Fam. Luciano Morbidelli
Umi 15h43'57" 86d3'6"
FAMAY
Tri 2h21'1" 32d25'55"
Fameed Khaligue
Uma 10h38'2" 64d15'59"
Famiglia ed Amore
Uma 10h55'4" 62d3'23"
Famiglia Minello
Ori 5h54'25" 12d43'8"
Familia Arreola
Uma 10h8'15" 61d17'36"
Familia Calvo
Col 6h5'22" -42d29'30"
Familia Corcoles
Aql 19h26'4" 2d44'15"
Familia de Rivera
Uma 9h21'8" 50d52'28"
familia molina palomo
And 2h19'48" 50d10'7"
Familia Monte
Her 18h2'5" 17d37'39"
Familia Ribaudi i Compte
Ori 6h17'52" 13d49'15"
Familia Talavera
Her 18h0'25" 17d28'23"
Familia-Parentes-Filius-Astrum
Aqr 21h48'57" -2d47'19"
Familie Bauer
Per 4h46'50" 48d57'55"
Familie Blödorn-Vanheiden
Ori 6h2'59" 19d37'38"
Familie und Liebe
Umi 14h42'17" 73d29'46"
Familiestern Gearhart
Uma 11h21'19" 55d48'40"
Famille Baillargeon-Michaud
Uma 8h45'8" 64d15'17"
Famille Bonnefoy
Umi 16h21'0" 74d35'38"
Famille Claude et Véronique Germain
Aql 19h56'50" 0d26'0"
Famille Daniel Lavigne
Ori 5h53'42" 17d40'29"
Famille Paul-R. & Marie-M. Negoti
Aql 19h27'4" -7d35'29"

Family Bonds
Uma 10h29'36" 50d52'14"
Family Flude
Cyg 20h54'12" 32d42'56"
Family Forever Star
Aql 19h56'51" -6d48'8"
Family of Chris and Valerie Linhoff
Uma 10h26'40" 65d46'58"
Family of Matthew & Kirsten Demmel
Cyg 19h34'13" 28d4'9"
Family Of Togetherness
Uma 13h23'49" 60d20'17"
Fan C. Pants Lipton
Cyg 19h44'32" 36d56'55"
Fan Peiwen & Huang Yi Forever
Cyg 19h47'53" 29d35'47"
Fan Xiao Wei
And 1h3'52" 41d49'27"
Fanam'S
Cas 1h10'36" 48d47'12"
Fancis
Tau 5h43'12" 20d26'43"
Fancy Face
Lyr 19h23'55" 37d41'11"
Fancy Nancy
Peg 21h51'30" 12d51'42"
Fancy Nancy
Vir 13h27'47" 7d58'5"
Fancylyssa
Uma 11h58'0" 35d19'57"
FandE BandB
Sgr 19h35'59" -16d40'46"
Fandzi Prince de l'Univers
Peg 0h12'16" 19d42'51"
Faneeza Bibi Ramsaroop
Vir 12h17'27" 2d55'6"
Fanfan
And 2h21'12" 47d0'50"
Fanizzi
Uma 10h49'33" 62d47'30"
Fannie
Cas 1h21'19" 54d3'3"
Fannie Bailenson
Lib 14h48'3" -2d42'48"
Fannie Esther Novello Castaldo
Cas 23h56'59" 57d23'15"
Fannie Herrera
Gem 6h44'55" 22d25'40"
Fannie M. & Woodrow H. Spicer
Uma 9h58'51" 70d57'18"
Fannie Marie Tremont
Vir 14h16'27" 2d3'40"
Fannie N. Gillming
Leo 10h24'9" 25d20'47"
Fannie Rose Osran
Uma 8h10'24" 66d48'4"
Fannie Snowberg
Cas 1h7'54" 59d41'43"
Fannie, Angelic & Beautiful
Lyr 18h35'21" 37d41'17"
Fannie's Phantom
Ari 2h4'53" 24d39'47"
Fanniessa Singh
Aqr 23h56'3" -19d33'49"
Fanny
Ari 2h32'36" 25d44'53"
Fanny
Cnc 8h22'16" 19d18'7"
Fanny Adjadj
Leo 10h17'34" 20d21'27"
Fanny and Frank
Cyg 19h25'22" 48d25'33"
Fanny Baby
Tau 5h48'3" 19d58'17"
Fanny Blanchard
Leo 10h50'40" 15d11'45"
Fanny et Stéphane
Aqr 21h1'26" 2d7'10"
Fanny & Joe
Cyg 19h58'34" 39d17'35"
Fanny & Mikhail Reznikov
Sco 17h45'38" -31d9'23"
Fanny Priscilla WRIGHT
Aqr 23h46'14" -10d47'45"
Fanny Ropiteau
Cap 21h28'22" -13d12'40"
Fanny Sourisseau
Lmi 10h1'55" 31d33'51"
Fanny (ticoeur)
Cyg 19h26'23" 36d17'56"
Fanny-Alexandra-Huin-16 ans
Uma 14h14'25" 56d40'43"
Fanny-Christine-Yves
Umi 13h21'35" 73d25'21"
Fannygladhearts
Gem 7h0'50" 31d52'41"
fannynatalie
Uma 11h47'26" 36d16'17"
Fanouche & Huana per Pata viento
Cam 7h53'49" 62d42'31"
Fantasia
Mon 7h24'19" -3d42'12"
Fantasmino
Cam 7h40'23" 73d29'13"
Fantasmino Rosso CB
Lyr 18h46'13" 31d7'27"
Fantastic place
Cam 4h33'20" 58d10'51"

fantasyland
Mon 7h57'41" -4d2'31"
Fantato Paola
Umi 13h15'13" 70d41'41"
Fany
Umi 17h0'11" 76d30'18"
Fany Moulie Moroz
Sgr 17h48'12" -20d7'35"
Fara Bethany Kaner
And 1h51'10" 45d59'53"
Fara M. Resnick-Van Horn
Cnc 8h39'8" 16d15'33"
Faraaz Sadruddin
Aqr 21h5'22" -14d13'0"
Farabi Mohamed Hussain
Cnc 8h26'17" 18d40'26"
Farago Gabor
Psc 23h42'16" 2d37'17"
FARAH
Peg 21h44'11" 19d42'14"
Farah
Mon 7h18'53" -0d27'58"
Farah
Uma 11h16'35" 61d55'51"
Farah Amjadi
Lib 15h25'58" -6d12'29"
Farah Belalia
Umi 12h42'29" 86d22'36"
Farah Elakhaoui
Vir 13h40'5" 4d22'39"
Farah Leanne Matley
Ori 5h44'31" -4d17'5"
Farah Mortada
Cyg 21h29'55" 37d7'5"
Farah Nantel
Uma 11h27'23" 59d48'38"
Farah Usmani
Cyg 20h24'42" 50d4'46"
Farai
Per 3h16'15" 46d34'37"
Faranak Bolboldarashtinejad
And 0h52'2" 37d48'2"
Faraway Angel
And 2h16'29" 43d42'58"
Faraway Soldier
Ori 4h58'39" 9d32'16"
"FarberLights"
Cyg 19h44'56" 37d40'8"
FAREAVEN
Sgr 18h48'47" -29d20'14"
Faren D. Lemaster
Oph 17h8'36" -0d28'34"
Fariba Barani-Caldwell
Ori 6h8'9" -0d40'31"
Fariba Shadfar
Umi 16h39'15" 82d21'25"
Farid et Samia le 15 Mai 2005
Del 20h51'0" 10d15'20"
Farid Naffah M.D.
Psc 1h23'45" 15d59'38"
Faride DeVaio
Sco 16h7'5" -11d25'53"
Farina Love
Lyn 8h1'3" 34d5'48"
Farkas Ati 30 - az Életem
Vir 13h50'51" -14d20'26"
Farkas Harom
Vir 12h31'1" 8d5'20"
Farkas Nikita Martin
Ari 2h32'12" 23d17'25"
Farkas Timea (Picilány)
Aqr 22h6'54" -10d12'0"
Farkas Zalán Péter
Cas 22h57'40" 54d54'16"
Farley Todd Pickering
Uma 11h5'33" 38d17'55"
Farm Boy
Dra 14h36'38" 56d30'26"
Farm Chen Saetern
Peg 22h42'46" 26d57'20"
Farmboy and Buttercup
And 0h13'31" 29d41'30"
Farmer
Tau 4h59'57" 16d9'37"
Farm's Best Food Industries S/B
Ori 5h34'38" 3d20'54"
Farnaz
Uma 11h45'43" 43d35'29"
Farnaz Belleville School 2007
Per 4h13'10" 43d55'48"
Farnoush & Joel
And 2h3'44" 37d33'48"
Farnsworth Family on Stoner Lake
Cnc 8h25'59" 12d51'38"
Farout Farlee
Uma 13h21'37" 62d53'47"
Farrah
Lib 15h52'16" -18d23'52"
Farrah and Patrick
Uma 10h28'4" 64d46'31"
Farrah Ann 222
And 0h43'8" 45d39'15"
Farrah Elizabeth Wolfe
Lib 15h11'25" -7d2'35"
Farrah Gabriella
And 0h46'32" 43d0'41"
Farrah Marie Goal
Ari 2h20'17" 25d40'48"
Farrahlove's Den
Vir 13h19'54" 5d31'36"

Farrell
Lmi 10h51'48" 28d42'42"
Farrell Maria Murray
Aqr 22h20'13" 0d49'11"
Farres Sarrouh
Tau 5h11'50" 24d15'34"
Farris And Yana's Home To Be!
Uma 11h4'4" 49d42'27"
Fartblossom
Uma 10h22'10" 58d16'11"
Fartwagon
Lib 14h58'31" -3d31'32"
Faruel
Oph 17h39'13" -0d51'9"
Fary and Mark
Cyg 21h28'36" 45d37'52"
Faryal
And 23h36'48" 47d37'45"
Faryal Jafar
Leo 11h57'59" 25d18'38"
Faryn Breanne
Her 18h54'57" 23d43'18"
Faryn Floyd O'Connor
Vir 12h45'1" 8d11'43"
Farzad
Cam 4h16'51" 65d40'55"
Farzad
Cep 22h22'24" 72d5'47"
Farzana Hasan
Uma 9h28'25" 54d9'13"
Farzana Islam
Aqr 22h17'53" 2d10'16"
Fas Exigo
Boo 14h37'30" 52d4'56"
"Fa's Star"
Cep 22h33'21" 63d33'32"
Fascinant
Gem 7h10'58" 17d39'22"
Fasha Paris
Cnc 9h17'55" 23d28'5"
Fashion Princess Jessica
Peg 22h36'58" 8d52'39"
Fasser Carla
Uma 13h31'21" 60d9'8"
Fast Eddie
Boo 14h40'34" 54d17'5"
Fast Eddie
Her 18h3'7" 17d54'26"
Fat George
Lib 14h50'15" -18d33'6"
Fat Jack & Ricky
Lyn 8h29'37" 36d22'5"
Fat Pants
Umi 16h54'49" 83d50'5"
Fatalis
And 0h38'46" 41d40'21"
Fatameh
Ori 5h22'53" 13d35'34"
Fatboy
Lib 14h30'14" -16d46'9"
Fate
Cas 23h56'5" 56d59'17"
Fate
Cas 1h33'19" 60d40'48"
fate
Uma 11h17'14" 53d13'5"
fate
Umi 15h53'5" 72d37'36"
Fate
Her 16h25'59" 10d11'39"
Fate
Ari 2h29'5" 27d41'10"
Fate
Gem 7h25'29" 34d27'41"
Fate
Cyg 21h36'51" 40d43'57"
Fate Amor
Gem 6h11'19" 27d29'41"
Fate and Destiny
Cyg 20h22'57" 43d34'11"
Fate And Forever Love
Lib 15h30'27" -7d42'15"
Fate (Princess Karlie)
Tau 4h33'48" 19d39'27"
Fate, and eternal love
Sge 20h20'4" 20d12'27"
Fate'N'Faith
Sco 16h12'51" -9d44'0"
Fate's Eternal Love
Cyg 21h10'37" 44d48'19"
Fate's Fortune
Aqr 20h48'17" -11d9'47"
Fate's Kiss
Lib 15h0'32" -11d57'50"
Fate's Love - Princess Kimberly
Cnc 8h18'3" 13d11'32"
Father Cortez
Lyn 7h42'26" 39d22'39"
Father Don Golasinski
Per 1h46'50" 51d21'3"
Father Edward William McElduff
Cep 21h52'6" 61d40'1"
Father Herndon
Uma 11h40'10" 40d29'59"
Father Jim Mifsud
Aqr 21h56'25" -1d9'10"
Father John Vondras
Cra 18h27'41" -40d57'56"
Father Karl F Duerr Jr.
Cap 20h35'58" -13d57'45"
Father Keith O'Hare
Psc 1h20'34" 24d10'20"

Father Martin
Lmi 10h7'15" 38d7'22"
Father Marty
Ori 5h39'19" -0d16'1"
Father Mathew
Cep 21h27'43" 64d17'42"
Father Matthew Foley
Cep 21h32'32" 63d2'11"
Father Michael Schueller
Psc 1h46'26" 6d57'54"
Father Of Great Wisdom
Cnc 8h41'4" 31d18'47"
Father Paul Fazio
Per 3h31'0" 40d39'23"
Father "Pops" Schweigardt
Cyg 19h48'46" 54d6'37"
Father Rhoderick Araneta
Her 16h44'29" 38d44'37"
Father Richard Mullins
Tau 4h11'51" 2d37'52"
Father & Son:Larry & Allen McKinney
Uma 11h13'22" 59d51'48"
Father/Sons Pa and Papas
Uma 11h32'46" 58d28'13"
Father's Spirit Melford H Olson Sr.
Cep 20h36'51" 65d46'19"
Fathiah
Psc 1h57'5" 6d24'7"
Fathieh Mohammad
Her 16h53'1" 31d29'48"
Fathima
Leo 9h59'18" 11d58'44"
Fathollah Sadaghiani
Uma 11h26'52" 39d29'25"
Fatih Gucu
Vir 11h37'56" 4d10'19"
Fatima
Psc 1h41'10" 24d41'23"
Fatima
Cru 12h22'0" -58d33'20"
Fátima
Umi 16h49'25" 83d59'11"
Fatima Ahmed
And 0h4'16" 48d40'28"
Fatima Amin
Aqr 22h48'5" -9d12'0"
fatima lillyann diva
Leo 9h58'28" 21d59'1"
Fatima Maridj
Cap 20h38'18" -13d29'28"
Fatima Marie Cancela
Sgr 18h13'36" -34d36'48"
Fatima My Queen
Cas 0h36'56" 57d30'28"
Fatima Rhattas
Lib 15h53'52" -10d10'51"
Fatima Sotelo
And 23h22'17" 52d54'45"
Fatima Sufi Tarannum Shaikh (BOB)
Sco 17h46'34" -39d7'12"
Fatimah Hishmeh
And 2h36'6" 49d26'9"
Fatima-Zahra
Cyg 20h49'2" 48d27'1"
Fatma
Ori 6h9'28" 8d30'7"
Fatma Hasaen
Leo 9h28'48" 12d7'39"
Fatma Wahba
And 23h55'19" 40d59'33"
Fatma's Stern
Ori 6h2'1" 10d14'49"
Fatou Tall Senegalensis
Sgr 19h11'48" -28d6'23"
FATTUS
Sgr 19h29'42" -16d38'3"
Fatty I Love You Always and Forever
Tau 5h23'51" 17d45'18"
Fatullah Babadi
Cet 0h30'39" -4d24'36"
Fauck, Werner
Uma 9h56'57" 66d20'10"
Faulkner
Leo 10h22'47" 15d50'25"
Faun Clarida
Cas 1h48'48" 60d21'32"
Faunan C. Nibble-Starchips
Cas 1h28'28" 64d0'6"
Faustino DeSisto
Tau 5h37'21" 25d50'30"
Fausto Ballotti
Aqr 23h48'34" -10d21'5"
Fausto Lazzaretti
Lyr 19h12'6" 33d10'52"
Fausto Rhino Apodaca
Umi 14h12'0" 69d20'40"
Fauve Rhiannon
Cas 23h1'0" 57d22'25"
Fauxy Roxy / Suga Pwiss
Ari 2h49'17" 27d28'41"
Fauzi-Ahmed Bischr
Uma 8h56'34" 52d38'6"
Favaretto Tatiana
Umi 0h15'6" 89d31'41"
Favio
Tau 4h25'49" 18d4'53"
Favorite Aunt & Uncle
Cyg 21h47'21" 45d51'42"
Favourite Sister
Pho 1h10'29" -50d30'3"

Fawn
Psc 1h6'16" 4d47'21"
Fawn
Crb 16h12'9" 38d36'41"
Fawn
And 1h44'17" 48d59'58"
Fawn
Ari 3h0'23" 19d54'29"
Fawn
Tau 4h19'1" 2d57'0"
Fawn C Armstrong
Uma 10h38'19" 55d24'34"
Fawn Chandra
Ori 5h34'52" 1d30'19"
Fawn Garrett
Lyr 18h26'49" 44d14'31"
Fawn Howard
Lib 14h41'57" -21d0'20"
Fawn Skinner
Lib 15h4'54" -15d22'1"
Fawnell's Star
Cyg 20h26'52" 39d18'48"
Fawn-Lee Chernoff
Her 18h4'49" 37d10'3"
Fay
Cyg 21h22'29" 46d5'16"
Fay
Uma 12h49'24" 55d21'43"
Fay Audrey Hathorne Labbe
Gem 7h4'7" 21d57'37"
Fay Berg
Cas 0h37'6" 65d44'25"
Fay C. Brickman
And 1h40'13" 43d37'3"
Fay Hubbard
Per 3h19'10" 46d6'12"
Fay Lorraine Curtin
Sco 16h18'26" -9d4'44"
Fay Martin's Star
Sge 20h9'30" 16d35'22"
Fay Paula Catando
Umi 16h4'45" 70d29'32"
Fay Ray Princess Ping Pong
And 23h33'52" 35d48'11"
Fay Stone
Umi 17h3'20" 80d30'4"
Fay Weaver
And 0h13'10" 33d12'6"
Fay & Weldon Tolhurst's Eternal Star
Cru 12h38'44" -60d49'8"
Fay Zagoren
Psc 23h53'34" -2d33'56"
Fay, 23.04.1999
Per 3h19'28" 42d12'21"
Fayat
Uma 12h1'11" 35d3'19"
Faye
Psc 1h21'44" 17d10'26"
Faye
Uma 10h1'7" 69d0'25"
Faye (Antaki) Pinard
Cyg 22h0'9" 50d56'16"
Faye Arlene McEwen "1932-2005"
Cap 21h36'10" -19d36'50"
Faye B. Coates
Aqr 22h11'1" -23d14'45"
Faye Barnes Murray
Ari 2h32'17" 23d41'47"
Faye Bloom
And 0h22'58" 43d7'48"
Faye Crace
And 0h58'0" 39d0'27"
Faye Dunne
And 0h9'35" 39d50'14"
Faye E Gray
Cas 1h38'40" 63d4'14"
Faye Elizabeth Hutchinson
Umi 16h13'47" 84d26'57"
Faye Elizabeth Slater
Uma 8h46'34" 54d35'2"
Faye Emily
Lyr 19h23'44" 37d32'22"
Faye Fletcher
Crb 15h57'37" 32d21'27"
Faye Forever
Sgr 18h42'58" -27d24'32"
Faye Lorraine Rhode
Psc 1h35'35" 13d47'45"
Faye Lorraine & Sally Jeanne C.
Gem 7h43'2" 32d5'12"
Faye Lorraine Schmidt Lukens
Cyg 20h6'12" 38d19'16"
Faye Louise Allison
Cap 21h46'12" -14d4'56"
Faye Lynn
Tau 4h36'18" 22d47'13"
Faye M. Escobedo
Lib 14h50'35" -10d5'21"
Faye McKenzie
And 0h48'34" 46d14'5"
Faye McTrek McIntosh
Crb 16h5'36" 29d19'36"
Faye Mills
Srp 18h18'40" -0d10'35"
Faye Morley
And 23h26'37" 47d41'23"
Faye "Nanny" Raack
Peg 21h42'6" 27d38'12"

Faye P. Austin
Cyg 19h49'10" 33d5'55"
Faye Petromelidou
And 0h14'29" 26d49'35"
Faye Rose Paller
Cnc 9h9'39" 15d22'24"
Faye S. Lambert
And 0h29'43" 32d53'47"
Faye & Shannon - Eternal Brilliance
Pho 0h33'46" -50d59'24"
Faye Victory
Uma 14h5'50" 50d19'13"
Faye's Star
Uma 11h37'18" 60d30'13"
Fayfay
Lib 15h36'37" -7d28'48"
Faylyn
Cam 4h27'15" 59d16'37"
Fay's Garnet
Umi 15h25'29" 71d31'23"
Fayth Ann Hines
Aqr 20h41'8" -3d9'47"
Fayyaz Aslam Sheikh
Ari 2h47'14" 13d13'51"
fazee786
Aur 6h9'29" 39d0'1"
FC Thun 1898
Sgr 18h14'39" -19d34'29"
FC2
Aqr 23h31'48" -24d2'26"
FCDM Reed
Uma 13h53'57" 57d27'22"
FCHS USAD Team Class of 2005
Ori 5h33'10" 0d49'7"
Fe
Vir 13h20'54" 12d44'42"
Fe
Vir 12h6'9" -0d39'5"
Fe Flores
Cas 23h49'12" 53d4'24"
Fe Maria Pena
And 23h20'22" 38d51'9"
Feaker's Home Base in Heaven
Uma 8h37'32" 49d39'55"
Feanor
Pho 23h41'46" -51d3'34"
Fearghal Grimes
Uma 11h3'38" 53d28'55"
Fearless Leader
Her 16h54'1" 20d32'22"
Feather
Sgr 19h31'34" -22d32'18"
Feather Allison Michelle Metsch
Gem 6h20'50" 17d53'52"
Feather duster
Lib 15h58'44" -13d6'28"
Feather's Star
Lyn 7h50'59" 40d59'17"
Featherstone
Ori 4h50'10" 14d46'11"
Featuring Steve Marcus
Her 17h46'49" 43d51'6"
Feb Tullis
Ori 6h12'23" 15d8'42"
Febronee Lalagos Shear
Uma 9h6'2" 53d46'10"
Feddy
Per 3h57'33" 52d35'15"
FeDe & Andrea
Cam 6h26'30" 63d34'34"
Fede, Gioia, Amore, Pace
Psc 1h25'14" 12d35'42"
Fedele Aloè
Peg 22h0'51" 13d27'1"
Fedeli d' Amore
Crb 15h54'12" 31d52'38"
Federica
Dra 18h21'5" 49d6'26"
Federica
Her 17h48'44" 39d9'43"
Federica
And 0h1'43" 45d39'12"
Federica
Ori 5h41'23" 11d40'24"
Federica
Peg 21h23'44" 21d31'2"
Federica
Umi 14h9'51" 72d55'54"
Federica
Umi 15h54'48" 78d59'6"
Federica
Vul 20h12'13" 23d32'36"
Federica Cesarin
Lyr 18h32'47" 27d4'36"
Federica De Blasi
Oph 17h10'50" 3d15'11"
Federica De Cugni
Lyr 18h29'40" 37d20'47"
Federica Forti
Per 3h26'55" 41d32'49"
Federica Masiero
Lac 22h40'52" 44d25'45"
Federica Quintini
Uma 10h22'9" 50d51'3"
Federica Scorza
Cyg 19h33'51" 31d44'59"
Federica Spila
Cas 23h20'2" 57d45'10"
Federica Valeri
Cas 0h2'51" 55d52'41"

Federico
Lyr 18h29'16" 33d3'34"
Federico
And 2h2'12" 38d19'0"
Federico
Peg 22h29'58" 7d32'7"
Federico A. Amorini
Ari 2h23'55" 20d37'31"
Federico Amato
Uma 11h27'14" 57d30'54"
Federico Angelucci
Her 18h36'17" 17d45'35"
Federico Cercena
Cnc 8h41'44" 19d57'44"
Federico Filippucci
Ori 4h55'6" -2d53'55"
Federico Frata
Lyn 8h1'44" 53d23'46"
Federico & Marjan
Cyg 21h29'2" 30d49'21"
Federico Pasqualini
Dra 17h17'58" 51d27'55"
Federico Pietra
Uma 9h33'14" 43d20'39"
Federico Valeria 8-10-1989
And 0h46'12" 45d13'19"
Federico Zamora
Leo 11h18'16" 12d31'16"
Fediers Residence
Peg 21h28'18" 15d32'6"
Fedora Galasso
Psc 1h27'55" 7d9'48"
Fee & Jens 2005
Uma 9h37'27" 57d32'32"
Fee Sai Trajano & Leung Mei Sze
Umi 15h30'4" 72d26'39"
Fée Sandra Alpini-Sandy Night
Uma 9h35'58" 60d34'9"
Feel - Életem csillaga BK54.01.09.
Cap 21h16'18" -19d38'56"
Feels Like Home
Lib 15h11'35" -7d20'19"
Feels Like Rain
Lyn 7h55'44" 58d23'4"
Fee's Bright Birthday Star
Ori 5h12'1" 5d58'58"
Fefe
Sgr 19h51'52" -17d32'45"
Fefesa
Sco 16h58'32" -33d8'50"
Feffi
Lmi 10h1'36" 33d54'15"
FEFI...IPS
Lyr 18h36'10" 30d33'35"
Feger
Tau 4h18'28" 25d2'55"
Fehér Viktória
Mon 6h52'4" 3d7'55"
feifei
Lib 14h49'23" -16d47'43"
Feili
Sco 16h55'38" -39d1'8"
Feisty Frankie
Uma 13h50'45" 53d46'33"
Feiyan Song
And 23h53'1" 35d26'28"
Felchner, Monika
Uma 9h31'46" 49d39'2"
Feldmann, Martin
Uma 13h12'33" 52d42'0"
Felecia A. Wood
Aqr 23h17'50" -18d41'23"
Felecia and Brian's Star
Sco 17h48'1" -33d21'59"
Felecia Jenkins
Lyn 8h1'17" 40d54'3"
Felecia Lynn Pokosa
Lyn 7h43'41" 58d2'48"
Felica Haney
And 23h44'6" 48d33'17"
Felica & Robert. Always & Forever.
Tau 4h50'52" 19d50'53"
Felice
Lyn 7h38'34" 57d1'24"
Felice Edoardo
Uma 8h44'15" 56d26'37"
Felice&Maria
Dra 16h13'2" 56d38'3"
Felice Maria Vanin
Cap 20h28'4" -9d32'9"
Félice van Nunspeet
Tau 3h40'43" 21d19'46"
Felicetta Maria Mellor
Dra 19h26'5" 72d52'7"
Felici Miriam
Cma 7h19'46" -30d32'16"
Felicia
Uma 8h31'18" 70d16'20"
Felicia
Aql 19h44'37" -10d16'32"
Felicia
Aqr 21h47'26" 0d25'3"
Felicia
And 2h14'30" 42d49'16"
Felicia
And 23h56'6" 35d16'57"
Felicia
And 1h11'11" 35d19'0"
Felicia Allessandra Barger
Leo 11h22'33" 13d39'47"
Felicia Amparo I love you baby!
Cru 12h46'13" -61d58'54"
Felicia and Michael's Star
Uma 10h10'38" 56d1'16"
Felicia Annaleah Rojas Villota
Aur 5h54'53" 42d8'54"
Felicia Cruz
And 22h58'45" 40d50'44"
Felicia Danielle Haynes
Lyn 9h35'47" 41d15'38"
Felicia DeSantis & Josh Wilbur
Cyg 21h19'15" 46d17'35"
Felicia E. Stricklen
Lyn 7h40'0" 48d11'57"
Felicia Feathers
Aqr 22h0'44" -16d7'23"
Felicia Goldfine
Cap 20h48'49" -22d24'53"
Felicia I. Elliott
Per 4h10'0" 45d51'12"
Felicia Jane Gillespy
Cnc 8h10'46" 18d1'10"
Felicia Jinzo
Vir 14h38'19" 4d2'27"
Felicia Katherine
Del 20h35'35" 13d40'55"
Felicia Kim Jeffries
Uma 11h46'31" 46d11'29"
Felicia Kinder
Cas 23h28'45" 53d0'9"
Felicia Lenore Dent
And 0h43'58" 41d36'0"
Felicia Lucky Stella
Ari 3h14'41" 29d19'25"
Felicia M Torres "Amina's World"
Cas 23h1'39" 59d40'38"
Felicia Marie Thompson
And 0h42'6" 27d4'48"
Felicia Mathis
Lyr 18h31'25" 40d28'59"
Felicia Mathis (Gregory)
Cap 20h27'15" -14d5'0"
Felicia Milewicz
Lyr 18h27'13" 32d15'47"
Felicia Minarchi
And 23h7'45" 47d35'11"
Felicia Mirsky
Gem 7h45'29" 33d21'30"
Felicia Noel Thorness
Sgr 19h38'55" -17d22'46"
Felicia Pancoast
Gem 7h17'6" 22d27'18"
Felicia Pavel
Cas 1h28'42" 62d20'19"
Felicia Rachel Allen
Mon 7h32'40" -0d24'10"
Felicia Rena Guzman
Aqr 21h1'4" 0d42'14"
Felicia Renshaw
Uma 10h51'27" 53d9'11"
Felicia Rogawska Milewicz
Lyn 8h15'4" 35d40'10"
Felicia Rose
Gem 7h4'4" 20d0'54"
Felicia Rose Little
Ari 2h40'25" 27d42'12"
Felicia Smith
Cas 23h10'0" 59d33'3"
Felicia Stauffer
Lyn 7h1'18" 51d45'37"
Felicia Sue
Cas 0h51'25" 63d45'4"
Felicia T. Thomas
And 1h51'3" 38d24'17"
Felicia Turàni (szeretlek, Z.)
Cas 1h7'9" 69d35'27"
Feliciano Hogeda Trevino
Gem 7h30'31" 15d21'35"
Felicia's Eternal Love
Cap 21h40'14" -22d0'44"
Felicia's Star
Aur 7h17'21" 42d43'58"
Felicia's Star
Cas 2h6'49" 59d15'17"
Felicita Marta Gaz en Monza
Cyg 19h41'6" 40d31'45"
Felicitas
Ori 5h9'7" 14d13'9"
Felicitas
Uma 10h8'33" 55d31'59"
Felicity
Cap 21h39'13" -9d36'34"
Felicity
And 1h12'2" 37d25'3"
Felicity Barr '21' 03.01.07
And 1h44'23" 49d17'54"
Felicity Belle Redmond
And 23h16'53" 43d30'9"
Felicity Dougal
Tau 5h7'21" 27d10'59"
Felicity Farm Star
Uma 11h31'2" 62d50'58"
Felicity Irene Raymond
Crb 16h13'14" 39d1'46"
Felicity Jane Grayson Charalambides
Cru 12h1'53" -62d37'23"
Felicity Mae
And 23h31'2" 46d57'55"
Felicity Panek
Sco 16h4'57" -26d10'15"
Felicity's Star
Cas 23h51'43" 50d7'58"
Feli.G...(Gutierrez)
Cnc 8h49'13" 26d50'30"
Feliks Vinokur
Per 2h38'2" 53d5'59"
Felina
And 23h3'17" 35d10'30"
Felina, I Love You So Much
Lib 14h45'22" -19d23'39"
Felinessa
Uma 11h17'39" 40d17'42"
Felipa Garcia
Crb 15h57'2" 36d15'44"
Felipe Agustin Alvarez
Uma 12h3'53" 63d27'23"
Felipe Armenta
Her 17h12'38" 27d45'58"
Felipe Arturo García
Cas 0h51'59" 62d13'40"
Felipe Barajas
Her 16h16'42" 45d8'51"
Felipe Lindo
Leo 11h44'54" 18d47'43"
Felipe Santos Pallanes
Sco 17h51'44" -42d20'3"
Felipe Velazquez "My Tribute"
Crb 15h38'46" 34d30'51"
Felis Lignea
Lyn 7h37'16" 40d53'27"
Felisa Legaspi
Lib 15h10'28" -22d32'2"
feliscalz
Aur 4h55'55" 39d50'56"
Felisha
Lyn 8h31'26" 50d33'45"
Felisha RaNay Prosock
Umi 15h0'51" 69d11'19"
Felisha Renay
Tau 4h10'5" 1d42'5"
Felix
Boo 14h42'36" 12d17'5"
Felix
Ori 6h18'1" 13d31'2"
Felix
Uma 11h52'15" 31d5'30"
Felix
Her 17h28'29" 33d47'36"
Felix A Lacarte Born: Aug 5/1963
Uma 11h32'8" 57d30'57"
Felix Amadeo Malia
Per 4h12'32" 31d43'57"
Felix and Irene Garmendia
Cas 3h16'12" 67d39'24"
Felix Cristina
Ari 3h24'45" 26d40'35"
Felix Diaz
Ori 5h36'14" 10d55'8"
Felix Edwin Fudge III
Vir 12h14'53" 4d0'23"
Felix F. Muccioli
Cas 23h59'45" 54d14'2"
Felix (Fidscha) Schmidt
Uma 10h9'35" 49d29'34"
Felix Garcia
Aql 19h30'9" 6d54'48"
Felix Gene Gibson
Per 3h22'59" 43d17'11"
Felix Josef Wyss 31.8.1964
Peg 21h29'3" 15d30'56"
Felix Luke Edmeades
Cep 21h41'28" 66d4'42"
Felix M.H. Villars
Cam 4h0'42" 67d15'12"
Felix Michael Perry
Aql 19h43'44" -4d45'25"
Felix Morison Vere Nicoll
Ori 6h17'8" 7d17'29"
Felix Neuburger
Umi 15h30'56" 71d6'18"
Felix Niewinski
Uma 12h55'57" 54d14'39"
Felix Phaedra
Aqr 23h21'26" -18d30'29"
Felix (Pipo) Ruiz
Her 18h1'48" 22d34'45"
Felix Preston Fudge II
Lib 14h54'10" -1d2'25"
Felix R. & Sherrell M. Valdez-Loqui
Cyg 21h33'54" 39d39'39"
Felix Ricardo Loqui-Valdez 6-9-1968
Gem 7h20'35" 25d16'15"
Felix Rudolf
Uma 9h2'47" 64d58'53"
FELIX & RUTH
Her 16h27'45" 13d54'40"
Felix Sanchez
Sco 17h53'59" -41d52'13"
Felix, Norbert
Uma 11h14'28" 35d19'13"
Felixa Virginia Flores Prieto
Uma 13h36'50" 57d25'20"
Feli-X-in
Sgr 19h16'45" -25d13'17"
Felma Maydell Rowe
Uma 11h54'2" 56d57'35"
Felton & Newman Twilight Twinkle
Ori 5h37'0" -1d35'26"
Fely
Gem 7h25'30" 32d59'17"
Femi Maha
Tau 3h29'58" 9d21'43"
Femke Bink
And 23h3'13" 35d52'29"
"Femma Sutra" - Love Embraced
Ara 17h23'10" -52d4'33"
FENCOV
Cap 20h22'8" -10d50'42"
Fendissime
And 23h33'39" 39d20'49"
Fenek
Com 11h59'21" 21d56'40"
Fenella Morgan
Leo 11h54'26" 10d50'5"
Fenella's Star
Ori 5h11'11" -0d5'1"
Feng Jia
Ari 2h39'38" 25d5'52"
FengnJessie
Sco 16h55'38" -37d8'40"
Fenia Papalouka
Ari 2h8'30" 21d26'51"
Fenix
Ori 5h24'6" 2d31'52"
Fenix Clay Guthrie
Uma 11h30'8" 63d58'24"
Fenny-Shirl
Tau 3h40'31" 16d19'3"
Fenrir Drachenstein
Dra 17h42'49" 52d16'27"
Fenske, Sigrid
Uma 10h20'30" 39d52'40"
Feny
Lyr 18h59'11" 41d8'40"
Fenylo csillagom: Lengyel Laci
Sgr 19h29'20" -30d47'4"
Fenyo Miklós
Psc 23h1'43" 2d15'50"
Feola
Gem 6h30'39" 15d50'30"
Fephy
Cnc 9h0'2" 24d52'18"
FERATIL
Uma 12h6'31" 61d22'25"
ferchis
Eri 4h22'19" -24d27'21"
Ferdia Griffith Breslin
Tau 4h23'55" 24d19'56"
Ferdie
Lyn 8h2'11" 45d19'25"
Ferdie & Elsie Grobler - Ewig
Pav 17h54'51" -57d20'56"
Ferdie Miller Walker
Sco 17h51'13" -34d2'7"
Ferdinand Hofstädter
Ori 6h24'1" 14d29'43"
Ferdinand Mengerlin
Uma 11h37'54" 60d13'27"
Ferdinand Saez
Aqr 22h29'50" -7d0'30"
Ferdy Mefford
Lmi 10h35'42" 32d39'34"
FereGentz 1926
Dra 16h53'54" 61d45'7"
Ferenc Tolvaly
Uma 8h28'16" 72d36'55"
Ference
Cep 22h29'46" 86d8'17"
Ferenczy "Apu" Ivan
Uma 10h23'2" 65d48'1"
Ferg
Lyn 6h41'38" 57d31'44"
Fergel Town - AMF06
Uma 9h13'38" 49d11'1"
Fergul Mestanov
Cru 12h44'51" -57d32'11"
Fergus and Shirley Briggs
Dra 19h49'45" 67d24'21"
Fergus Anderson
Per 2h12'34" 54d14'25"
Fergus Durham
Per 2h43'32" 50d52'46"
Fergus Moore
Uma 8h9'5" 60d42'18"
Fergus Tidy
Cma 6h44'46" -20d34'17"
Ferguson
Gem 7h11'15" 15d16'32"
Ferguson-Nolan Family Star
Cyg 19h30'20" 53d49'31"
Ferguson's Heart
Umi 15h13'47" 77d48'32"
Férjecském, Szatmári István 33
Cap 20h17'2" -11d58'50"
Férjem, Bandi
Ari 2h56'7" 28d39'42"
Ferko
Del 20h35'14" 18d5'4"
Fermin & Celia Llaguno
Crb 15h41'54" 27d55'44"
Fern
Her 17h56'34" 24d4'13"
Fern
Psc 1h16'47" 7d31'44"
Fern Abbott
Ari 2h44'27" 29d9'41"
Fern Anne Panzer
And 1h10'50" 39d55'12"
Fern Hastings
And 0h16'11" 39d26'3"
Fern Marie Keefer Boyer
Uma 10h5'19" 56d35'29"
Fern Martha Hultquist
Crb 15h39'51" 26d14'0"
Fern Muller
Sgr 18h23'28" -25d18'30"
Fern & Phillip's Brilliant Christmas Star
Cyg 19h45'44" 39d7'52"
Fern Ring Elkind
Sgr 19h47'32" -12d54'51"
Fern Rose Hyry
Psc 1h19'44" 31d15'27"
Fern Sparks
Aql 19h26'4" 4d47'59"
Fern Thelma Lambert
Cyg 19h32'45" 31d54'28"
Fern Vanzwol
Uma 10h27'5" 49d48'20"
Fern Viola
Cnc 9h5'54" 21d29'4"
Fern Wasserman
Cap 21h7'7" -18d51'32"
Fernand Neuenschwander
Sco 16h57'26" -41d0'34"
Fernanda
Cma 7h18'47" -14d44'22"
Fernanda
Uma 11h17'43" 34d2'34"
Fernanda Maxey
Lyn 7h34'22" 52d36'54"
Fernanda Michalski
Mon 8h0'53" -1d31'2"
Fernande M. Piccolo
Cas 1h26'18" 52d27'50"
Fernandez24
Aql 19h30'55" 12d37'20"
Fernando
Gem 6h24'33" 21d2'30"
Fernando Alvarado
Lib 14h29'17" -23d36'57"
Fernando Anaya
Leo 10h21'9" 12d18'35"
Fernando Aquino Oyanguren
Uma 11h8'57" 46d16'38"
Fernando & Bianca
Ori 5h29'56" -0d4'20"
Fernando et Georgette Paquette
And 0h3'10" 39d12'14"
Fernando Fusco
Ori 5h43'12" -0d14'26"
Fernando Garcia loves Vanessa Bravo
Cyg 21h43'21" 42d19'46"
Fernando Joaquin Villalobos
Lmi 10h25'46" 28d16'51"
Fernando Luis Casals
Lib 15h8'37" -6d16'5"
Fernando Membrillo
Sgr 19h4'39" -31d35'44"
Fernando Pomares
Cnc 8h15'31" 29d24'25"
Fernando Ramos
Psc 1h12'32" 7d29'15"
Fernando Raul Rinaldi
Lib 14h24'9" -13d53'48"
Fernando Saldana Ramirez
Lyr 18h54'1" 35d49'51"
Fernando Xavier
Ori 5h32'16" 3d42'44"
Fernando y Tamara
Tau 5h52'11" 14d45'32"
Ferne Giles
Cas 1h32'50" 64d52'40"
Fernie Grace Tiflis
Vir 12h48'13" 11d29'2"
Fern's distant hangout
Ari 3h0'28" 19d57'18"
Fero Pytlik
Ari 3h17'26" 27d22'9"
Ferola Francesca e Ferola Roberto
Per 4h48'7" 47d2'12"
Ferqueen
Col 6h18'58" -33d37'47"
Ferrara's Light
Aqr 22h15'27" 2d30'44"
Ferrarelle
Her 17h53'31" 20d43'3"
Ferrari Carlo
Crb 16h3'39" 32d39'39"
Ferraro Family Star
Del 20h46'46" 10d45'42"
Ferri Family- Beloved Mom & Grandma
Cyg 21h32'8" 50d36'6"
Ferrier
Uma 11h58'17" 32d32'27"
FERRIS's HEART
Uma 10h58'0" 69d0'53"
Ferro
Cas 0h45'52" 69d33'43"
Ferry-Ploss
Uma 8h46'43" 54d32'35"
Fervent
Cnc 9h8'24" 31d14'23"
Feryal
Cas 3h5'35" 59d30'15"
Fess Family Star
Leo 9h41'33" 12d17'2"
Festa Cristian
Ori 6h21'43" 16d39'11"
Fethi Baddi Ouverture d'e-sprit
Lmi 10h3'36" 36d9'47"
Fethi Mehmet
Oph 18h16'5" 0d33'7"
Feuchtner, Jürgen
Uma 12h52'57" 61d26'34"
Feuerstein, Ursula
Uma 12h46'5" 61d32'4"
Fey
Dra 17h42'24" 79d24'47"
Feyrus
Dra 16h17'48" 66d16'30"
ff882
Uma 11h15'42" 47d26'39"
ffej's flamin' fireball
Lyr 19h19'3" 35d22'16"
Ffion Lowri Jones
And 1h52'23" 46d44'22"
Ffion Stokes
And 0h3'56" 45d50'41"
Ffion Vicki
And 23h49'40" 38d14'33"
FFM Marketing Sdn Bhd
Ori 5h35'1" 3d36'34"
FG Pops
Uma 11h55'38" 29d22'32"
FGF Momstar1
Cas 1h46'24" 71d15'0"
Fhelleyballs
Leo 11h36'41" 20d51'38"
F.H.T. 34
Per 4h17'6" 43d25'32"
Fi amo Fhayne del Marie del Karité
Lyr 18h46'36" 39d55'55"
Fia
Cyg 20h51'1" 47d27'31"
Fia
And 1h11'23" 42d7'11"
Fia
Leo 10h55'39" 5d8'54"
Fiachra 1.
Lyn 7h32'22" 36d26'23"
Fiachra Bohan
Ori 6h15'4" 20d52'26"
Fialas forever
Cep 22h57'18" 79d28'20"
Fiamma e Pieraldo
Gem 7h1'24" 17d30'37"
Fiamma Marchione
Umi 14h59'46" 84d55'12"
Fiamma Sdn Bhd
Ori 5h33'27" 3d25'11"
Fiamme Gemellate
Gem 6h57'53" 21d50'57"
Fiara Firebrand
And 0h34'17" 38d11'54"
ficfiric Nemanja Rnic
Per 4h19'51" 42d20'17"
Ficht, Manfred
Uma 8h56'47" 64d12'1"
ficker
Cnc 8h53'35" 17d58'24"
Fidd
Uma 9h19'14" 54d44'51"
Fiddler's Green Lt. Ron. Powell USN
Aqr 22h44'28" -17d25'21"
Fidel Amat
Her 18h43'17" 21d57'28"
Fidel Avilucea
Cep 22h23'24" 61d59'57"
Fidele
Lyn 7h45'54" 41d34'32"
Fideli
Uma 11h14'12" 66d41'10"
Fidelia Cotorcea
Sco 16h56'23" -40d37'8"
Fidelio
Tau 4h18'28" 7d38'8"
Fidelis Kathy
Sco 17h47'11" -38d2'36"
fidelis pectoris
Uma 13h40'23" 49d38'6"
Fidelity
Cyg 20h7'51" 51d9'17"
Fides
Pyx 8h41'43" -22d27'15"
Fides
Lib 15h6'11" -8d1'51"
Fides Zeiter-Kalbermatten
Her 17h10'3" 18d55'37"
Fidge
Cas 1h30'11" 65d56'43"
Fidgeon's Star
Uma 8h43'13" 69d45'29"
Fidus Achates
Sco 16h15'1" -10d10'5"
Fiebelkorn, Norbert
Uma 13h56'2" 55d16'54"
Fiery Adam
Uma 8h20'39" 62d22'44"
Fif & Gat's
Cnc 9h3'18" 16d29'23"
Fife III
Pho 0h40'32" -46d16'51"
FiFi
Uma 8h42'12" 52d46'23"
Fifi Garrick
Leo 11h13'36" -1d15'36"
FiFiTrixiBell7 #28
And 23h40'44" 41d48'8"
fifty
Uma 10h36'40" 67d10'38"
Fig
Uma 9h56'20" 65d0'51"
Fige Ballard
Cra 18h49'12" -41d43'14"
FigJean
Uma 11h34'6" 34d46'44"
Figura, H.Werner & Charlotte
Uma 14h7'33" 57d8'49"
Fijian Rose
Lyn 7h57'41" 58d55'6"
FiJohn020906
Cyg 21h42'45" 47d49'39"
Fiker
Uma 10h31'4" 42d58'18"
Fikisha L. Davidson
And 23h29'37" 42d47'49"
Fileena
Cnc 8h55'11" 29d33'42"
Filfil, Ghasan
Ori 6h8'11" 13d19'2"
Filia
And 0h22'29" 28d49'14"
Filia Deum
Cru 12h57'12" -56d1'58"
Filiolus
Leo 10h45'33" 8d13'31"
Filip Bondy
Cas 23h35'6" 53d2'9"
Filip Bondy
Cap 21h44'52" -20d37'28"
Filip deBalkany, our bright-est star
Lyn 8h3'11" 54d6'54"
Filip Noel
Dra 12h55'55" 70d48'50"
Filipa
Psc 1h8'26" 23d20'53"
Filipe De Moura
Uma 8h36'53" 50d39'13"
FilipeRuha
Per 3h42'52" 44d54'46"
Filippo
Lyr 18h26'5" 31d6'6"
Filippo
Crb 15h33'23" 38d0'26"
Filippo
Cep 22h28'1" 65d30'44"
Filippo
Cas 0h32'54" 62d24'0"
Filippo Alessandro Ercolino
Uma 8h55'14" 59d13'19"
Filippo Bellanti
Ori 5h56'30" 16d54'25"
Filippo Bozzi
Aur 5h25'16" 33d25'25"
Filippo Dell'acqua
Sgr 19h0'57" -20d8'13"
Filippo Maria Labonia
Boo 14h45'3" 50d46'6"
Filippo Monge
Uma 9h11'30" 63d32'47"
Filippo Multinu
Uma 9h57'13" 41d48'48"
Filippo Nava
Lyr 18h26'17" 41d19'58"
** Fili's Dream World **
Tau 3h53'4" 21d27'44"
Filius
Uma 8h30'40" 64d19'42"
Filiz Aktas
Peg 21h29'17" 11d47'52"
Filiz Erdoganli - 0.04.1963
Ari 1h47'52" 19d25'27"
fille inoubliable
Ori 5h0'46" 5d34'28"
Filleul
Leo 11h32'17" 25d3'58"
Filmore Noel Reynolds
And 0h44'28" 31d32'50"
Filmsniper
Dra 19h19'12" 64d34'1"
Filo Caimi
Ori 5h59'59" 21d43'8"
Filomena
And 0h22'18" 44d26'23"
Filomena Beltrani
Umi 15h43'20" 79d39'59"
Filomena - Juvenalia
Psc 0h48'12" 10d55'10"
Filomena Marino
Per 2h45'38" 53d51'5"
Filomena Mucci
Lyr 19h18'16" 28d25'38"
Filomena Salvatore 1953 - 2005
Vir 12h46'0" -3d1'55"
Filomena Skovron
Gem 6h41'58" 20d35'13"
Filson
Cru 12h34'56" -60d23'23"
Filthy-Twitches
Sco 16h53'33" -32d19'53"
Fimberley
Aql 19h55'30" 11d36'20"
FIN
Cyg 20h52'49" 47d14'6"
Fin Boy
Cep 0h3'30" 73d14'4"

Fina Gili Ramos-Ricardo
Bou Galiana
Psc 0h21'8" 16d5'6"
Finally
Ori 5h50'2" 9d20'12"
Finally
Per 4h17'27" 36d3'23"
Finan Seamus Lynch -
Finan's Star
Her 16h14'31" 3d52'13"
Finbarr & Sandra McCarthy
Vir 12h30'23" 10d21'42"
Fincely Micely
Lib 14h27'28" -9d24'54"
Find your way
Per 3h6'48" 52d27'36"
Findlay Charles Graham
Her 18h6'54" 27d27'27"
Findlay The Celtic Soldier
Ori 5h17'54" 0d15'29"
Finian Appletree
Per 2h54'15" 55d3'33"
Finkey Murray
Nor 16h32'35" -57d4'1"
Finlay
Lib 15h54'4" -13d14'37"
Finlay Alexander Ross-
Davie
Per 4h47'6" 44d39'58"
Finlay Bayard Souders
Tau 5h10'36" 23d19'54"
Finlay D D Walter
Her 18h51'27" 23d17'34"
Finlay David Page
Her 17h24'41" 22d9'25"
Finlay Eden
Sco 17h58'6" -31d9'44"
Finlay George Britton
Her 18h34'27" 12d22'25"
Finlay George High
Per 4h23'7" 41d1'43"
Finlay Hayes
Umi 15h37'59" 67d52'7"
Finlay Jack Keates
Lib 14h46'14" -24d53'9"
Finlay James Griffin
Umi 15h57'1" 83d18'48"
Finlay James Houghton
Per 4h19'57" 40d57'54"
Finlay James Wells
Dra 18h26'57" 51d24'34"
Finlay Joseph Connor
Ori 5h35'48" -2d12'42"
Finlay Josephine Cameron
Cen 13h50'22" -46d30'32"
Finlay MacDonald
Her 16h41'4" 11d34'16"
Finlay Michael
Her 18h53'14" 21d24'49"
Finlay Nils Logan McMillan
15 - 05 - 07
Per 3h10'26" 46d41'12"
Finlay Thomas Hogg
Cep 22h48'21" 63d10'31"
Finlay Walsh
Uma 9h23'21" 47d51'54"
Finlay Young Coldwell
Cep 21h2'32" 59d21'11"
Finlay's Star
Per 4h34'21" 38d53'30"
Finley
Dra 18h47'9" 51d57'59"
Finley
Umi 12h54'32" 86d26'30"
Finley Ann
Sgr 19h15'34" -22d23'54"
Finley Eden Beck
Pho 23h41'3" -46d16'20"
Finley Eric Nelson
Cep 21h32'33" 61d46'33"
Finley George Duckett
Per 4h20'38" 40d27'48"
Finley Jack Radding
Sco 17h7'46" -35d35'22"
Finley Jamae
Uma 11h56'25" 37d46'47"
Finley James Taylor
Per 3h35'8" 49d58'33"
Finley Kai Weitzen
Cnc 8h52'57" 21d59'0"
Finley Krause
Lyn 9h4'46" 40d10'19"
Finley Owen Cook
Uma 11h39'2" 47d23'22"
Finley Paget
Per 2h30'28" 52d5'42"
Finley Ralph
Umi 16h9'8" 72d54'45"
Finn
Uma 13h55'18" 57d3'8"
Finn
Psc 1h49'40" 5d36'48"
Finn
Uma 9h52'14" 50d54'26"
Finn
Gem 7h7'51" 33d28'37"
Finn Alexander Ballard
Ori 5h13'22" 15d34'14"
Finn Austen Richardson
Her 18h35'53" 13d48'31"
Finn Bryan Casebere
Gem 7h5'1" 31d31'57"
Finn de Tobon - Little
Diamond Star
Lmi 10h8'14" 33d15'41"
Finn Foght Mikkelsen
Tau 5h5'4" 26d21'5"
Finn Francesco Gallo
Gem 7h37'48" 34d39'8"
Finn Lilly Dittman
Cru 12h9'33" -62d51'1"
Finn Peter Keegans -
Flumpy
Umi 16h23'56" 70d9'1"
Finn - you little star
Col 6h12'31" -42d9'37"
Finneas James Morgan
Cru 12h39'21" -57d42'13"
Finnegan John Gray
Uma 8h53'18" 55d54'34"
Finnegan "My Bonny Lad"
Umi 16h27'27" 75d30'16"
Finnegan's Star One
Leo 9h23'22" 13d39'49"
Finnian Martin Jennings
Her 18h32'14" 23d50'49"
Finnian William Gassner
Per 4h12'18" 50d25'24"
Finnigan's Twinkle
Ori 5h27'2" 6d56'6"
Finnin Harris
Dra 19h27'52" 60d1'24"
Finnlay Jack McElrue - Inch
Sgr 17h53'38" -17d11'45"
Finnley Spence - Finn's
Star
Her 16h22'34" 8d54'2"
Finnly Rose DeFrancisco
Umi 14h41'3" 82d44'55"
Finnon
Cep 21h18'44" 55d35'9"
Finn's Albrite
Ori 5h33'51" 8d47'57"
Finn's Star
Her 16h24'5" 25d43'1"
Finn's Star
Ari 2h1'30" 23d49'6"
Finocchiaro Davide
Uma 13h54'45" 57d34'48"
Finola P A
Psc 1h3'41" 6d4'6"
Fins To Left
Cas 0h51'16" 60d32'47"
Fintan and Julliann's Star
Uma 8h27'10" 68d14'23"
Fintan's Twinkle
Cam 4h8'47" 53d24'0"
Finton Thomas Bramwell
Umi 16h52'29" 85d45'17"
Fiolka, Elfi
Ori 5h10'35" 15d53'57"
Fiona
Mon 6h51'13" 8d23'16"
Fiona
Leo 10h30'21" 23d35'2"
Fiona
And 2h18'20" 49d34'37"
Fiona
And 2h36'50" 47d11'52"
Fiona
Lyr 18h30'59" 31d35'2"
Fiona
Uma 9h46'41" 42d16'13"
Fiona
Cap 20h13'20" -10d27'10"
fiOna
Cap 20h23'27" -13d12'17"
Fiona
Cas 23h48'3" 53d0'41"
Fiona
Lyn 7h55'32" 54d34'35"
Fiona 4 Chris
Cnc 9h2'9" 21d18'45"
Fiona Abercrombie Scott
Cyg 21h48'3" 50d28'7"
Fiona Allison Cozens
And 23h4'38" 36d46'52"
Fiona & Andreas
Cyg 21h9'11" 30d2'4"
Fiona Anne Rahim
Cen 13h13'32" -48d48'21"
Fiona Arnott Price -
Mummy's Star
And 2h3'51" 45d45'24"
Fiona Barclay
Sco 17h28'2" -41d44'59"
Fiona Barclay
Cam 5h6'34" 60d1'35"
Fiona Bissada
And 1h37'24" 49d11'1"
Fiona Caterina McMurrey
Peg 21h9'59" 13d8'19"
Fiona Costa
Cru 12h5'40" -62d27'20"
Fiona & David Robbie's
Star
Cru 12h27'42" -60d38'28"
Fiona Doirean
Cnc 8h29'7" 27d57'48"
Fiona Dunton
Cru 12h9'7" -62d47'23"
Fiona Elizabeth Rutherford
And 0h4'20" 40d57'40"
Fiona Elspeth Mhairi
McBain
Ori 6h9'0" 14d36'53"
Fiona Gibson
Sco 17h52'48" -37d18'11"
Fiona Grace O'Reilly
And 2h24'56" 46d10'36"
Fiona Heather
Lmi 9h51'17" 37d21'0"
Fiona Julie Grove
Gem 6h22'6" 21d11'3"
Fiona L. Barr
Ori 5h29'33" -3d46'58"
Fiona Lillian Price
And 0h25'5" 40d2'7"
Fiona Louise Peterson
Umi 15h30'2" 70d59'13"
Fiona MacKenzie
Sge 20h9'10" 19d44'43"
Fiona Mairead Dunn
Crb 16h13'18" 37d34'51"
Fiona Margaret Welsh
2007
Umi 4h33'2" 89d28'32"
Fiona McBean
Cas 23h35'32" 57d55'34"
Fiona McColl McCarthy
Aqr 21h27'36" 2d13'16"
Fiona Mead
Lmi 10h0'48" 33d15'17"
Fiona Myra Dow
Cyg 20h24'17" 48d52'4"
Fiona & Neil's Silver
Anniversary Star
Cyg 20h32'48" 30d53'2"
Fiona no. 1
Cas 23h42'18" 53d32'45"
Fiona O'Halloran
Ari 1h58'11" 14d26'18"
Fiona O'Mahony
And 2h13'16" 39d52'39"
Fiona Parker
And 1h7'33" 41d11'0"
Fiona Patterson (My Wife)
Cyg 21h12'38" 37d43'25"
Fiona "Petal" Ronaghan
And 1h35'26" 49d3'46"
Fiona Rae
And 23h55'56" 42d17'2"
Fiona Ramshaw
Crb 16h2'56" 25d53'2"
Fiona Rosamond
Lmi 9h26'27" 37d1'36"
Fiona Ryan Walsh
Gem 6h30'39" 17d14'47"
Fiona - Sarah E Hancock
Lyr 18h42'37" 34d53'24"
Fiona Sinead McGee
And 23h5'9" 50d46'38"
Fiona Skye McMullen
And 23h5'7" 37d9'52"
Fiona Soraya
Uma 10h14'29" 62d21'16"
Fiona Thornewill MBE
Cas 0h6'55" 50d35'38"
Fiona Wallis
Cru 12h27'1" -59d34'9"
Fiona Walsh
Cas 23h43'17" 56d2'29"
Fiona Waterland
Lyr 18h35'15" 38d53'48"
Fiona Whelan
Cas 2h15'11" 73d29'37"
Fionàn O'Loingsigh
Sco 16h35'25" -27d42'14"
Fionas Infinitas
Ari 1h55'24" 19d31'22"
Fiona's Jewel
Cas 23h37'14" 54d27'37"
Fiona's Star
Cas 2h45'20" 64d10'12"
Fiona's Star
And 23h33'1" 42d51'2"
Fionn Haviland Theloneus
Leo 11h10'30" 9d21'29"
Fionnuala
Dra 9h53'48" 74d0'7"
Fionnuala Creaven
Cas 0h8'37" 50d0'28"
Fionnula Josephine
Johannsen
Leo 10h12'48" 25d59'51"
Fiore
Lac 22h31'33" 41d15'8"
Fiorega
Lyn 8h50'24" 43d15'22"
Fiorella
Uma 11h55'19" 36d44'43"
Fiorella
Gem 6h59'48" 15d38'27"
Fiorella
Peg 22h13'55" 21d35'32"
Fiorella
Cam 6h59'7" 68d59'43"
Fiorella Valdesolo
Cas 0h42'54" 50d11'20"
Fiorella Zitoli
Oph 17h10'50" 9d1'37"
Fiorello Family, Suwanee
GA
Her 16h52'5" 34d48'7"
Fiorina
Psc 0h43'41" 9d40'54"
Fiorina Barazzetta
Peg 22h29'51" 21d30'37"
Firdos Bellamod
Lyn 6h46'20" 57d9'35"
Fire Family Star
Cnc 8h54'1" 23d31'29"
Fire Fly
Lyn 6h59'45" 57d34'42"
fire of two hearts
Cap 20h17'3" -12d6'53"
Fire Station 16
Her 16h53'53" 34d14'50"
Fire Woman
Lib 15h17'22" -5d29'51"
FireBel
Ari 3h11'39" 26d57'52"
Firecracker
Cyg 19h52'52" 57d10'29"
Fire-Cracker Courtney
And 1h35'22" 38d41'35"
Firefly
Uma 11h20'53" 48d2'41"
Firefly
Tau 5h56'43" 26d33'7"
Firefly
Psc 1h29'38" 10d9'40"
Firefly
Tau 4h24'53" 17d3'43"
FireIris
Uma 9h15'51" 53d55'23"
Fireman Bob
Per 3h7'12" 51d50'31"
Fires Kiss
Eri 3h55'31" -0d10'31"
Fireside Coffee
Uma 11h7'18" 45d45'15"
Fireside Fred
Leo 10h18'1" 10d56'51"
FireStar-Tolly Burkan
Crb 15h38'10" 29d26'11"
Firetag
Ori 5h35'22" 0d11'44"
Fireworks
Cnc 9h1'30" 7d17'42"
Fireworks
Aql 19h16'39" -10d53'0"
Firnhaber's Star
Cas 1h17'12" 62d7'32"
Firouzeh E. Muhammed
Sco 16h10'47" -17d20'29"
FIRSAINDCOYRSE
Cnc 8h23'36" 20d37'45"
First Baseman Kenny
Mathias
Sco 16h18'27" -9d4'24"
First Born
Pho 23h36'19" -43d6'9"
First Discoveries
Per 3h21'7" 47d35'16"
First Independent Bank
And 23h10'48" 48d34'31"
First Kiss
Cra 17h59'50" -37d23'47"
First Kiss "Jennifer"
Cas 0h51'12" 65d0'23"
First Lieutenant Neil
Thomas
Uma 11h37'41" 56d25'14"
First Love
Sgr 18h26'3" -21d44'39"
First Love
Umi 14h34'40" 76d41'24"
First Love 3/ 18 / 07
Uma 10h1'2" 67d53'38"
~First Love~ (Chinda &
Anastasia)
Ari 2h57'48" 27d59'15"
First Love R & C
Uma 10h20'38" 40d24'43"
First of Many
Lmi 10h25'7" 35d20'30"
First Time One
Cnc 8h22'9" 20d8'42"
First "Trew" Anniversary
Cyg 19h57'34" 54d25'58"
First Unitarian Universalist
of OKC
Cam 3h58'6" 65d53'35"
Firufan
Uma 8h35'56" 69d23'25"
Fischer Martin Seymour
Gem 6h1'48" 25d24'10"
Fischer Mathieu LaFond
Psc 0h23'1" 2d42'48"
Fischer, Andreas
Uma 10h43'46" 46d23'51"
Fischer, Fritz
Uma 11h29'17" 28d50'34"
Fish
Leo 9h47'31" 27d2'12"
Fish Guy
Ori 5h51'4" 12d7'16"
Fish Tank
Gem 7h16'55" 31d37'5"
Fish4EverMore
Her 16h50'55" 16d57'19"
Fishboy
Per 4h24'48" 39d42'17"
Fishdill
Uma 8h52'43" 64d33'9"
Fisher
Dra 15h50'39" 54d59'47"
Fisher K. Lucas
Leo 11h39'45" 16d26'14"
Fisherman - Captain
Ramiro Ramirez
Psc 1h32'39" 21d31'28"
Fisher's 25th Valentine
Anniversary
Cyg 21h51'21" 43d43'41"
Fisher's 50th
Uma 11h9'45" 54d27'32"
Fishi Love
Gem 7h9'4" 22d34'12"
Fishie
Sco 16h5'5" -15d4'14"
Fisk Family
And 0h24'45" 32d15'40"
fitch
Uma 10h59'3" 34d33'26"
Fitchett
Uma 9h59'41" 41d47'38"
Fitos "Baby" Eszter
Uma 11h16'25" 28d47'49"
Fitsum Abraha Yohannes
Lyn 7h14'3" 57d44'28"
Fitz
Uma 12h40'6" 56d19'41"
Fitz
Umi 13h56'37" 74d46'3"
Fitzgerald Hunter
Cyg 20h2'59" 48d21'14"
Fitzgerald-Hill
Lib 14h26'34" -9d52'52"
Fitzgerald-McKay
Sco 16h59'58" -42d26'29"
Fitzgerald-Seavolt
Ari 2h37'50" 24d35'0"
Fitz-Lloyd George Brown
Aql 19h50'25" -0d35'51"
Fitzpatrick
Cyg 19h53'42" 36d33'35"
Fitzy5
Ari 2h53'2" 28d57'54"
"Five Angels"
Vir 13h10'15" 12d36'0"
Five Dolllar Bob Rutkowski
Cnc 9h10'10" 8d28'51"
Five Foot
Tau 4h26'53" 13d4'15"
Five Squeezes Forever
Uma 10h29'14" 43d57'43"
Five Star Fire Protection
Billy
Vir 12h40'35" 2d28'5"
Fix It
Her 17h7'59" 24d31'1"
Fizul Ally
Dra 18h59'57" 70d11'22"
Fizz
Lyn 8h7'26" 38d2'4"
Fizzix
Lyn 8h5'4" 37d7'59"
Fizzle
Umi 15h3'33" 70d32'35"
Fizzy G - 62
Vol 7h37'59" -65d29'6"
Fizzy Maltby
Leo 11h44'6" 24d3'57"
F.J. DeLaratta
Ori 5h59'34" 7d19'10"
F.J. Panebianco
Ori 5h31'28" 2d3'19"
F.James Reid
Psc 23h8'46" 2d13'12"
fjjqllqjjf
Uma 11h32'6" 56d17'27"
FJN ~ always ~ CHR
Crb 15h38'38" 38d19'0"
F.J.R.
Aur 5h12'24" 32d6'14"
Flädermuus
Uma 8h51'36" 47d33'28"
Flabersham's Light
Lyr 18h41'9" 37d41'50"
Flaggar
Lib 15h35'36" -25d49'57"
Flame of Eternity
Tau 5h21'44" 23d39'41"
Flaming Heart
Aqr 22h32'37" -1d6'54"
Flaming Star
Oph 17h27'30" 8d57'52"
Flamingo
Dra 19h50'43" 74d55'57"
Flamingo
Umi 15h39'48" 74d58'34"
Flamingo011400
Psc 1h23'43" 30d59'50"
Flamini Lukic Petra 1991
Cam 4h36'0" 58d14'6"
Flaminia Rocchi
Her 18h52'1" 21d47'28"
Flammifer
Uma 9h26'1" 42d18'23"
Flan
Dra 17h37'34" 65d59'44"
Flanney
Uma 10h22'5" 69d0'57"
Flannigan's Flame
Vir 13h1'11" 2d9'53"
Flapjack & Marbles
Per 3h55'15" 36d20'49"
Flart Coltman
Aqr 23h2'4" -14d10'14"
Flash
Per 4h40'17" 41d13'44"
Flash
Lyr 19h9'52" 39d28'44"
Flash
Ari 2h14'5" 23d34'58"
Flash
Ari 2h33'18" 27d50'18"
Flash Gordon Joseph
Bergbauer
Her 18h17'32" 15d51'49"
Flash My Shinging Star
Uma 11h43'1" 42d43'47"
Flavia
Uma 13h37'3" 50d14'39"
Flavia
Leo 11h58'19" 12d59'59"
Flavia
Umi 17h38'0" 80d0'22"
Flavia Bulza
Gem 7h12'59" 27d37'49"
Flavia Carlini
Lyr 18h45'33" 32d6'6"
Flavia Martella
Ori 6h18'12" 15d24'15"
Flavia Mauro
Uma 13h22'48" 57d28'57"
Flavia Nunes
Pho 23h52'9" -45d24'3"
Flavia Pezzano
Umi 17h46'3" 80d7'35"
Flavia Roberta Moreira
Pedrosa71880
Lyn 7h55'4" 48d54'35"
Flavia Tosti
Cap 21h36'41" -17d7'32"
Flavie - Anaïs
Ori 5h14'48" 8d31'28"
Flavie Ragonnet
Vir 11h52'46" 8d1'8"
Flavio
Lyr 18h47'35" 39d58'29"
Flavio And Jeanne
Cyg 21h13'12" 31d9'22"
Flavio Gaetano Fichera
Aql 19h41'8" 13d9'34"
Flavio Giordano
Ori 5h51'2" 4d26'49"
Flavio Iannuzzelli
Her 18h1'21" 17d39'25"
Flavio Mastroberti
Peg 22h44'20" 27d8'30"
Flaviola Garcia
Cas 0h33'2" 61d25'45"
Flavius & Lidia Lungoccia
Crb 15h41'12" 28d45'6"
Flavy Clerc
Uma 13h41'25" 55d24'20"
"FLAWLESS"
Uma 8h14'33" 62d34'51"
FleaBooger Angel
Ari 2h3'39" 24d0'48"
Fleck, Christiane & Carsten
Uma 10h37'6" 62d32'57"
"Flee"
Her 16h41'41" 23d32'15"
FLEE'
Lyn 7h16'58" 44d25'3"
Fleecy
Cep 23h51'50" 74d19'58"
Fletcher
Cep 21h53'33" 62d37'21"
Fletcher
Uma 11h18'2" 40d56'29"
Fletcher Clement Cain
Vir 12h35'18" 4d59'1"
Fletcher Harold Herrald III
Her 18h23'3" 22d4'14"
Fletcher Michael Hall
Tau 3h48'5" 24d4'19"
Fletcher Nolan Hislop
Leo 10h46'54" 17d55'54"
Fletcher Wesley Adair
Per 3h14'37" 47d15'3"
Fletcher-Bruzik 2005
Cnc 8h20'37" 10d34'21"
Fletcher's Celestial Flame
Ori 6h9'40" 12d33'48"
Fletcher's Heart
Her 17h53'1" 37d43'1"
Fleur
Ori 5h55'17" 5d50'30"
Fleur
Dra 16h41'48" 55d36'58"
Fleur
Lac 22h47'33" 53d58'24"
Fleur Claesen
Umi 15h16'42" 68d20'56"
Fleur de Linda
Cap 21h55'32" -10d30'20"
Fleur De Lys
Cru 12h38'0" -64d25'34"
Fleur McFaull
And 23h27'40" 47d44'0"
Fleur Webber
And 22h59'49" 51d20'34"
Fleurdeliz Laygo
Crb 15h39'32" 35d5'39"
FlexiMama
Lib 15h14'6" -5d59'57"
Flicka Grady
Aql 20h13'14" 13d41'2"
Flicka RSG
Cap 21h44'9" -18d25'54"
Flight of the Robbins
Aql 19h44'12" 7d5'18"
Flinkli
Per 4h36'43" 40d9'10"
Flinn William Waldron -
Flinn's Star
Her 18h19'42" 28d59'59"
Flintstar
Dra 10h50'59" 76d27'3"
Flipmode3188
Cas 1h20'2" 59d56'30"
Flipper
Cnc 8h25'57" 23d18'36"
Flipper
Leo 10h51'35" 15d22'15"
Fliss Redmond
And 23h1'43" 41d13'30"
FLL
Uma 11h54'3" 63d38'23"
Flo
Peg 22h20'40" 21d31'40"
Flo and Bill
Uma 11h30'25" 28d22'31"
Flo & Babsi
Uma 8h43'13" 49d15'6"
Flo & Jack Bernhardt
Uma 11h17'31" 57d26'50"
Flo Schenck
Uma 8h17'58" 67d44'52"
Flo + Simone
Uma 8h13'27" 61d30'12"
Flo Van Volkom
Cam 4h5'24" 58d13'45"
Flo Wilkins
Uma 10h10'36" 50d44'5"
Floater Fish
Ori 5h21'55" -7d27'43"
Flo-G
Dra 17h4'23" 56d0'42"
Flolid
Umi 14h38'28" 67d54'2"
Flonron55
Uma 9h8'10" 58d28'51"
Floody
Vir 12h28'15" 6d38'10"
Flor
And 0h23'25" 41d47'59"
Flor Erika Martinez
Lyn 9h13'49" 33d15'29"
Flor Silva y Javier Cordova
Umi 14h11'19" 73d39'46"
Flor Soriano Baladjay
Lib 14h50'47" -2d28'7"
Flor y Jorge
Cap 21h22'25" -24d28'10"
Flora
Leo 9h45'12" 28d18'12"
Flora AA61
Uma 9h27'6" 48d12'13"
Flora Albert
Cas 0h21'22" 62d48'2"
Flora B. Giffuni
Cyg 19h53'59" 31d43'34"
Flora Bluestein
Uma 10h8'19" 62d3'34"
Flora Chan
Com 12h0'32" 22d4'33"
Flora Delphine Loftin
Del 20h52'33" 8d46'40"
Flora Emily O'Rourke
Cyg 21h29'2" 36d21'54"
Flora et Giovanni
Umi 16h50'36" 75d32'9"
Flora Hattenbach
Psc 0h21'47" 0d17'0"
Flora*Jason*Jeffery*Jakup
Natali
Sgr 18h36'10" -17d19'36"
Flora Jean Fisher
Psc 23h56'31" 10d41'36"
Flora Kathleen Pleasant
Cap 20h53'7" -15d51'47"
Flora Lopez
Uma 11h44'15" 29d25'21"
Flora Loyer
Uma 11h4'9" 45d51'23"
Flora Maja Kudlinska
Crb 16h15'59" 38d33'12"
Flora Q
Lyn 7h53'23" 35d14'36"
Flora Salts
Aqr 23h6'52" -6d44'31"
Flora Sterling Solom
Ari 1h53'7" 17d28'29"
Flora Wheeler
Cam 4h12'46" 61d39'25"
Floraina
Mon 7h35'57" -0d17'40"
Floralba
Sco 16h10'32" -10d35'3"
Florale May Vencer
And 2h26'15" 48d22'54"
Florance Kathryn Snyder
Sgr 18h11'25" -19d44'9"
Florance Thomas
Mon 6h51'31" 7d52'0"
Flora's Christening Star
And 0h19'53" 45d56'30"
Flore DOREL
Uma 8h12'40" 64d27'11"
Flore Schwalm
Aqr 22h17'34" -16d39'44"
Florea Fortuna Nautica
Peg 21h55'34" 12d29'5"
Florea Fortuna Nautica
Crb 15h58'9" 27d51'26"
Florecilla Carolina
Psc 1h21'47" 16d52'30"
Florence
Leo 11h24'21" 15d41'22"
Florence
And 0h21'13" 25d1'42"
Florence
Cas 0h9'24" 57d9'19"
Florence
Cas 23h36'49" 51d12'57"

Florence
And 2h38'33" 41d10'2"
Florence
Lib 15h4'56" -8d13'37"
Florence
Umi 14h0'54" 73d30'33"
Florence
Sco 17h7'36" -35d42'17"
Florence and Irving Sokol
Leo 9h37'53" 31d43'46"
Florence Ann Kossack Bird
Tau 3h30'44" 19d27'5"
Florence Atlas
Ari 1h52'52" 17d43'52"
Florence B. Lane
Ari 2h51'22" 29d3'26"
Florence Bessom Higgs
Sgr 19h6'56" -12d54'15"
Florence Brillant
Uma 11h43'31" 49d20'24"
Florence Canter
Cnc 8h36'52" 32d23'4"
Florence Chatenet
And 0h27'27" 22d18'47"
Florence Curran
Lib 14h41'5" -11d10'3"
Florence Dah Beck
Aqr 23h6'46" -9d15'1"
Florence E. Kline
Gem 7h35'32" 27d29'27"
Florence Eileen Wells
Sgr 19h5'41" -30d22'15"
Florence et Laurent C
Uma 9h0'12" 47d11'9"
Florence Foresti
Ori 6h18'57" 6d29'32"
Florence G. Tarabek
Gem 7h47'14" 33d8'2"
Florence Galisson
Uma 9h4'3" 68d11'13"
Florence Gayot
Ori 6h17'11" 10d47'10"
Florence & Georg
Cyg 21h27'32" 49d43'29"
Florence Gerba
Cas 1h36'26" 58d1'53"
Florence & Glynn Larkin
Aqr 23h46'10" -11d27'32"
Florence Gonzalez
Cnc 8h17'53" 14d22'53"
Florence Grace Bosma
And 1h26'27" 34d36'6"
Florence Griffin
Dra 18h39'15" 54d25'37"
Florence Herman
Lyn 7h2'12" 52d40'27"
Florence High
Cas 23h37'10" 56d29'27"
Florence Hildie Rose
And 1h21'47" 35d10'54"
Florence Hoibierre
Mon 6h47'16" -4d38'54"
Florence Holbrook
Uma 9h1'10" 49d1'39"
Florence Jean
Lyn 7h49'9" 58d34'55"
Florence & Joe
Cyg 20h46'21" 36d12'6"
Florence Kalberer Farley
And 1h10'5" 41d43'7"
Florence (Kim Amanda Ellis)
Cma 6h39'41" -14d49'32"
Florence Krause
Vir 14h14'41" 4d14'39"
Florence L. Standish
And 0h31'37" 42d57'54"
Florence "Lori" Lorraine Diedrich
Cas 1h27'21" 59d11'16"
Florence Marcella Mahoney Williams
Psc 1h13'3" 6d14'10"
Florence Margaret Daniels
Cas 23h52'5" 56d25'12"
Florence Marie Greve 1918-1997
Cyg 19h19'7" 52d56'4"
Florence Marnie Ingrey
And 0h48'10" 46d9'31"
Florence Mary Quigley
Umi 16h57'18" 77d45'40"
Florence May Bright
And 1h28'0" 43d30'19"
Florence May Foster
Cas 23h17'50" 59d44'20"
Florence May Johnson
Crb 15h50'53" 35d1'46"
Florence Michau
Gem 6h35'4" 12d13'55"
Florence Mikorski
Tau 3h54'36" 8d49'3"
Florence Mildred Belling Rockswold
Cas 1h40'22" 65d12'14"
Florence Miller
Lib 15h41'43" -16d58'55"
Florence Morgenstern
Lyn 7h56'18" 57d16'3"
Florence Mui
And 0h46'30" 36d34'30"
Florence Neill
Ari 2h43'14" 28d51'54"
Florence Olsen
Lyr 18h40'6" 31d44'30"
Florence Ratcliffe
Psc 0h4'22" 3d52'46"
Florence Ratz
Cas 1h35'13" 64d1'5"
Florence Rhodes
Cnc 8h37'53" 18d48'41"
Florence Riverin
Umi 15h33'25" 68d51'13"
Florence Roth
Per 3h27'29" 50d0'10"
Florence & Roy Bartels
Aqr 22h25'23" -12d40'48"
Florence Ruble
Cnc 7h58'12" 18d45'55"
Florence Schlossberg
Crb 15h40'52" 37d42'19"
Florence Schuler
Cnc 8h35'51" 14d46'28"
Florence Sofia Gatto
Col 5h46'33" -32d34'25"
Florence Staats Johnson
Cnc 8h20'35" 24d56'17"
Florence Storz
And 0h39'52" 36d48'28"
Florence Terdeman
Cas 1h43'19" 66d44'45"
Florence Wallsh Stein
Sgr 18h56'29" -20d19'1"
Florence Wanda Marie Sullivan
Lyn 6h33'55" 57d10'17"
Florence Wensel
Cas 0h48'11" 57d11'28"
Florence Wilbrandt
Lyn 7h25'53" 51d29'40"
Florence Y. Schechtman
Crb 15h37'28" 27d33'9"
Florence, ma puce pour la vie
Ori 6h22'58" 10d22'12"
Florence-Mother of Love & Kindness
Tau 5h2'0" 16d15'48"
Florence's Brilliant Star
Lyr 19h14'8" 37d24'25"
Florence's Stardust
Leo 9h49'46" 22d49'3"
Florencia
Cru 12h24'32" -57d21'32"
Florene C. York
Uma 9h43'4" 60d7'22"
Florent
Vir 14h18'7" -6d29'21"
Florent Haché
Uma 12h12'59" 56d47'21"
Florentina Cristina Ochea
Aqr 22h32'50" 0d45'32"
Florentina's Star
Cet 1h0'4" -0d54'14"
Florentino Equilos
Uma 8h53'50" 56d26'16"
Flores
Cam 7h28'59" 63d18'43"
Flores Jean-Bernard
Leo 10h28'4" 20d49'59"
Flores Uarii Tihoni
Cyg 20h50'43" 45d17'51"
Florette Witzig Bodmer
Cyg 20h14'54" 49d7'27"
Florge
Lyr 19h7'54" 26d24'12"
Floria
Her 17h48'36" 38d44'52"
Florian
Ori 6h21'2" 10d1'6"
Florian
Uma 9h59'43" 57d24'29"
Florian
Ori 5h37'53" -2d33'35"
Florian Christoph Wegmann
Lyr 19h19'32" 29d49'19"
Florian Conus
Ori 5h53'48" 6d44'43"
Florian Garcia Bombase
Tri 2h21'30" 36d15'28"
Florian Hecht
Uma 11h45'39" 29d47'12"
Florian Henry Gebauer
Uma 11h57'43" 57d33'57"
Florian Jastrow
Uma 9h31'6" 50d41'5"
Florian Joseph Kordas, Jr.
Psc 1h26'41" 18d0'12"
Florian Müller
Mon 6h30'35" 8d31'54"
Floriana Rubino
Ori 5h26'10" 3d10'8"
Floriane-Anthony
Peg 21h40'15" 16d58'33"
Florida Seifel 7
Uma 8h46'29" 50d5'52"
Florim Ismaili
Uma 10h17'51" 54d25'25"
Florin Ghelmez
Psc 2h1'35" 6d36'55"
Florina
Lyr 19h14'4" 35d32'53"
Florina
Uma 9h47'17" 42d46'6"
Florinda
Per 3h27'11" 50d31'58"
Florinda LaFord
Ori 5h37'55" -1d14'46"
Florine
Uma 9h12'58" 50d19'36"
Florine Clements Brown
And 1h6'1" 36d53'28"
Florita
Tau 4h16'22" 29d43'30"
Flor's Artic Star
Vir 14h42'12" 4d7'16"
Flory
Cnc 8h46'2" 17d10'40"
Flo's Star
Cas 23h43'15" 55d10'22"
"Floss"
Uma 11h48'19" 50d22'13"
Flossie
Lyn 7h24'31" 45d32'12"
Flossie Mae Hobbs
Uma 10h34'47" 45d19'43"
Flossie's Star
Cyg 21h31'16" 36d45'49"
Flossy
Per 2h39'41" 57d2'51"
Flota & Jeff
Sge 20h3'3" 17d43'28"
Floux
Cap 20h23'3" -18d47'18"
Flower Lover
Vir 13h1'42" -5d16'30"
Flower of Scotland
And 0h18'15" 43d8'57"
Flower's Love
Cnc 8h25'17" 15d58'46"
Flowertje-7552
Cnc 8h38'27" 13d8'50"
Floy
Cas 0h2'35" 57d2'20"
Floy Morway
Cep 20h51'8" 58d24'8"
Floyd
Cep 23h59'37" 74d38'23"
Floyd "Bonehead" O'Neal
Sco 16h37'25" -28d31'34"
Floyd Boone
Her 16h25'9" 5d43'29"
Floyd & Doris Johnson
Uma 9h37'26" 62d8'21"
Floyd E. Duran, Sr.
Cnc 9h8'41" 30d44'41"
Floyd Ferren
Ari 1h56'32" 18d46'36"
Floyd J Ledbetter
Sco 17h12'46" -44d41'18"
Floyd Jay Vandermark III
Umi 14h9'39" 69d9'26"
Floyd & Jean
Uma 11h58'21" 56d43'33"
Floyd Johnson
Cnc 8h52'24" 31d49'10"
Floyd & Marcella
Uma 8h50'49" 68d39'22"
Floyd Marcus
Psc 1h26'20" 32d42'26"
Floyd Michael Stark Sr.
Eri 3h10'36" -5d8'17"
Floyd & Norma Nabors
Uma 11h9'57" 30d54'41"
Floyd O. Hader III
Cap 21h42'6" -24d43'23"
Floyd Overacre
Sgr 17h59'31" -17d29'55"
Floyd & Patricia Bennett
Uma 10h54'28" 72d44'56"
Floyd Roye Scarborough
Leo 10h12'16" 14d6'14"
Floyd Seiple
Cep 22h19'7" 60d9'1"
Floyd Sesler
Per 4h20'38" 52d37'4"
Floyd Venne
Leo 11h0'31" -2d50'46"
Flt Lt Steven Johnson
Ori 5h59'0" 5d58'43"
FlufferDoodle!
Gem 6h17'16" 22d12'13"
Fluffy
Ori 6h18'37" 13d59'59"
Fluffy
Uma 9h28'14" 49d41'52"
Fluffy
Uma 11h22'8" 54d5'52"
Fluffy Garrison
Ori 5h41'54" 2d15'21"
Flügelchen
Vir 13h51'26" 4d5'13"
Flugie+Sonny=50 years of Passion
Lyn 7h39'38" 38d13'3"
Flumoxed and Puddled
And 2h21'29" 44d43'40"
Flurin
Cam 7h51'49" 81d44'56"
Flurina_11_Love_Forever_Bill
Tau 4h28'16" 18d6'2"
fluteT3
Ari 2h16'57" 24d37'45"
Flutter
And 2h34'18" 43d22'26"
Flutter
Cyg 19h55'53" 38d25'3"
Flutterfly
Psc 1h8'43" 14d35'19"
Fluxus
Uma 10h54'17" 49d2'17"
Fly Boy
Per 3h15'42" 46d12'47"
Fly Bry & Julie Always
Sgr 19h19'55" -21d1'11"
Fly Me To the Moon
Umi 14h4'38" 86d1'50"
Flyboy
Aql 19h10'17" 5d41'21"
Flyboy 50
Peg 21h34'38" 22d9'55"
Flyboy's Beauty
Lib 15h6'54" -6d23'48"
FlyBug
Cap 20h12'44" -10d12'34"
FlyBy
Uma 8h22'37" 67d55'1"
Flyin' Ryan Arcia
Sgr 18h3'22" -27d48'45"
Flying Angel
Cam 7h29'9" 81d19'8"
Flying Dude dedicated from N to O
Umi 13h31'56" 75d23'43"
Flying Eagle's Little Dove
Aql 19h35'14" -9d54'5"
Flying Free & Wild
Leo 11h14'29" 15d28'27"
Flying Girl
Peg 22h3'14" 11d17'55"
"Flying" high
Vir 15h9'0" 2d45'45"
Flying High, Defying Gravity
Psc 1h13'23" 22d41'39"
Flying Solo
Aql 20h6'47" 4d54'47"
Flying Tigress Star
Cap 20h9'28" -10d0'20"
Flying Whisper
Tri 2h17'53" 37d1'39"
FlyingLady
Ant 9h47'27" -32d52'43"
Flynn Maxwell Pallot
Vel 9h55'24" -43d23'55"
Flynn Sherwood
Leo 10h21'29" 24d37'36"
Flynnie
Uma 12h35'17" 58d15'21"
Flyrageous Loves Startastic
And 23h25'2" 52d29'29"
F.M. "Buz" Busby
Uma 12h53'55" 61d4'16"
Focke, Norbert
Uma 10h16'26" 72d30'24"
Foda
Umi 16h25'44" 80d50'49"
Fodla's brown-eyes, The Star "Bre"
Cma 7h18'53" -14d53'29"
Foe
Aur 6h1'19" 47d17'31"
Foehrhaber
Per 3h51'41" 32d3'25"
Foemina
Vir 13h12'10" 6d40'10"
Fofar Truls Andren
Peg 21h35'27" 22d2'52"
FoFo
Sgr 18h23'39" -18d13'7"
Foggy Londontown
Lib 15h30'13" -23d0'24"
Foil William McLaughlin
Ori 5h16'16" 7d30'44"
Folan Love
Cyg 20h33'46" 38d33'28"
FOLAWN
Sgr 19h21'53" -27d16'14"
Folded Hills
Per 4h20'8" 31d17'49"
Földi Zsuzsi
Cap 20h13'31" -21d14'48"
Folie Amoureuse
Aqr 23h45'43" -10d38'20"
Folkert K. O.
Uma 11h47'26" 54d56'27"
Follansbee
Uma 10h37'36" 46d38'12"
Folletto
Umi 17h34'21" 81d22'17"
follow your dreams
Uma 12h43'18" 58d7'18"
Follows the Path with Heart
Cha 10h45'11" -79d39'49"
FOMF Star
Psc 1h20'39" 32d15'34"
Fonda
Leo 10h14'8" 12d10'3"
Fonda F. Fischer
Cnc 7h58'56" 19d10'33"
Fonda Kay
Aqr 21h48'49" -3d34'51"
Fonda Prue
Aqr 21h47'39" -2d55'42"
Fonesarus
Lib 15h52'9" -19d9'31"
Fongsanith
Lib 15h16'51" -15d5'30"
Fontelle Jean & Joseph Ponzoha
Lyn 6h51'41" 58d25'44"
Fontenella
Sco 16h55'31" -42d25'51"
Fonterra Brands (M) Sdn Bhd
Ori 5h33'30" 3d29'41"
Foo, may you always light my way
And 0h16'15" 42d52'4"
Food 4 Less
Per 3h49'32" 45d32'28"
Foofee
Cas 0h29'19" 58d26'25"
Fools Rush In
Cnc 8h48'58" 28d28'46"
Foosey
Dra 17h33'10" 66d50'42"
For Alicia, the only star in my sky
Cas 1h26'20" 62d35'22"
For All Eternity
Vir 12h46'25" 9d38'50"
For all Eternity Kevin and Stacy
Ori 5h30'0" 14d18'50"
For All That We Hope
Sgr 19h39'40" -15d52'54"
For all to see, The Star Kendal Lee
Cma 7h9'48" -12d48'28"
For all to see, The Star Terry
Cma 7h13'24" -14d0'11"
For Allen With All My Love Nikki
Per 4h3'57" 34d1'44"
For Always
Ari 2h25'35" 18d15'17"
For Andrew and Christina Johnson
Psc 1h27'8" 3d52'6"
For Bo
Cap 21h12'30" -23d13'57"
For Cady
Vir 12h51'29" 7d59'42"
For Cathie Broocks Honoring Me-Sue
And 0h23'8" 33d30'56"
For Deirdre and Eternity
Cyg 20h20'31" 34d14'3"
for Elizabeth from Jef version 2.0
Vir 13h16'43" -13d55'27"
For Eternity, Bill and Kim
Cyg 20h7'0" 33d43'4"
For ever
Lyn 7h46'26" 38d2'46"
FOR EVER CHRIS
Ori 6h3'52" 13d29'18"
For Ever & Ever.. Robert & Julianna
Aql 19h43'3" 8d44'26"
For ever in my heart,
Ori 6h45'52" -0d3'44"
For Ever my Love, Patrice Marie
Cyg 19h56'17" 46d28'52"
For ever Peter
Cas 2h50'43" 60d51'6"
For Ever Twisted
Gem 7h11'27" 22d44'29"
For gum in Love
Umi 15h40'30" 81d54'22"
For Hope,Wisdom,Serenity,For ever,EW
Cru 12h22'29" -61d56'37"
For Infinity - Chris & Kelly
Uma 8h21'40" 70d42'46"
For Jacqui - The brightest star
Cru 12h32'25" -60d17'51"
For Janna With Infinite Love
Mon 6h46'21" -0d32'50"
For Jeffrey Race & Kimberly Davis
Lyr 18h48'4" 32d42'15"
For Joanna
Sco 16h43'19" -30d53'21"
For John Fell with love from Laura
Lyn 7h36'53" 42d12'21"
For John, Sandy, & Laura, with love
Her 16h21'35" 5d44'5"
For Julie In Love Michi
Uma 9h30'28" 54d25'32"
For Katherine
Gem 6h47'20" 16d46'57"
For Ken Chau, on his 27th Birthday.
Leo 9h44'13" 26d26'16"
For Kylie, my lil' Brat. Love, Joey
Sgr 19h10'20" -22d30'17"
For Lawrence, Our Park In The Sky
Crb 15h48'54" 34d3'17"
For Love of Connor Jackson Smith
Sco 16h13'29" -15d4'17"
For Love's Light Shines Bright
Vir 14h17'28" -12d46'48"
For Marley "Puv" West - Love, Mom
Uma 8h56'15" 61d7'54"
For Melinda My Shining Star
Her 17h48'59" 42d1'59"
For Michael, "The Dancing Queen at 60."
Cas 0h6'5" 56d36'18"
For my Angel
Gem 7h15'2" 16d3'24"
For my ANGEL - LINDA HALL
And 1h1'45" 35d2'1"
For My Angel, Lindsay Min
Crb 15h39'26" 37d22'51"
For My Baby Robert J. P. Hinckley
Sgr 18h56'41" -25d7'42"
For my Beautiful Becky
Cas 23h34'28" 55d9'2"
For my ETERNAL love, from PL to KB
Sgr 18h59'50" -12d52'17"
For My First, for Alden
Ori 6h18'6" 10d50'6"
For my husband - Brent M. Metcalfe
Cyg 20h6'3" 37d32'10"
For my husband Mike, my hero
Per 3h18'48" 42d57'22"
For my little boy, my better half
Cap 21h21'5" -21d18'30"
For My Love Alyssa
Lib 14h36'16" -16d0'1"
For my love. Lindy Kay Verner
And 0h36'36" 29d4'26"
For My Love Michele
Cyg 21h36'6" 36d19'28"
For My Love...Jennifer Ashmore
Psc 1h29'25" 13d21'33"
For my lovely wife, Julie Ross
Cap 20h26'55" -11d36'14"
For My Mom Nancy
Cap 21h42'48" -14d0'48"
For My Mommy: Brenda Lee
Cap 20h8'19" -13d47'34"
For My One True Love Nathan Haney
Cas 23h36'36" 58d37'12"
For My Perfect Lindsay, Love Jordan
And 1h15'8" 38d17'47"
For my sweet Babboo
Ori 5h27'43" -7d54'11"
For now, forever, Scott D. Marshall
Gem 6h58'25" 16d9'27"
For Paige Elizabeth Bell
Uma 13h15'49" 63d4'31"
For Panda
Umi 14h38'44" 74d38'6"
For Past, Present, and Our Future
Tau 4h22'37" 21d15'55"
For Peace by Charles L. Kammer III
Cyg 20h54'49" 31d50'51"
For Ryan Creamer with Love
Tau 5h55'36" 25d12'4"
For The Love of Brad
Her 16h13'52" 46d0'17"
For the Love of ~Branden~
Cnc 8h53'11" 13d9'13"
For The Love of Dogs
Cnv 12h44'3" 41d11'55"
For the Love of Fre
Cap 21h38'18" -15d10'34"
For The Love of My Life Mary
Cas 1h4'57" 54d11'29"
For the love of Natosha & Vincent
Gem 7h23'43" 25d47'34"
For the love of Rob and Marlo
Lyn 6h23'6" 55d56'22"
For Us 396
Cyg 20h16'2" 55d13'26"
For you Nilay - Love always, Dave
Cas 0h23'6" 53d52'38"
For You Tami
Ori 6h10'17" 15d57'45"
For you...A star to guide you home.
Gem 7h25'26" 26d17'7"
For Your Eyes Only
Sco 16h9'21" -8d22'39"
For Your Future Dreams Joseph Snarr
Col 5h30'8" -34d30'11"
Forbade Lovers Star
Leo 10h30'40" 18d33'1"
Forbidden
Sgr 19h1'41" -24d50'25"
Forbis' Star
Ori 5h25'20" 13d48'34"
Ford Loos Bond
Dra 17h49'2" 52d31'15"
Fordazio
Uma 9h13'5" 69d6'9"
Forest Frasure
Aqr 22h43'3" -1d29'17"
Forest Galaxy 30
Gem 6h37'27" 16d11'56"
Forest y Eloisa Martin
Sge 20h6'25" 17d4'45"
Forestine King
Uma 14h3'28" 62d0'16"
Forevaus
Pyx 9h25'15" -29d33'40"
Forever
Aqr 23h48'3" -22d59'32"
Forever
Cru 12h16'0" -58d22'28"
Forever
Uma 13h10'29" 60d11'16"
Forever
Uma 9h33'25" 66d25'47"
Forever
Cam 4h31'27" 69d37'38"
Forever
Dra 19h59'43" 73d28'18"
Forever
Umi 15h11'29" 70d46'54"
Forever
Uma 13h57'12" 57d28'25"
Forever
Cyg 19h46'48" 53d52'45"
forever
Cas 0h47'39" 64d36'37"
Forever
Lyn 7h55'40" 57d18'38"
Forever
Lyn 7h57'17" 52d58'58"
Forever
Uma 12h17'12" 57d57'55"
Forever
Uma 11h7'58" 56d55'16"
Forever
Aqr 22h1'2" -3d52'51"
Forever
Eri 3h33'36" -7d52'9"
Forever
Cep 23h10'47" 76d19'16"
Forever...
Cap 20h29'1" -14d2'29"
Forever
Sge 20h5'17" 17d40'39"
Forever
Sge 20h6'51" 17d18'21"
Forever
Sge 19h50'14" 17d4'28"
forever
Peg 21h32'34" 21d12'11"
Forever
Her 18h48'54" 19d53'36"
Forever
Ori 5h57'35" 21d38'44"
Forever
Ori 5h58'20" 21d55'38"
Forever
Leo 10h44'4" 16d52'44"
Forever
Cnc 8h34'0" 15d30'59"
Forever
Leo 10h1'19" 20d23'11"
Forever
Tau 4h43'7" 27d37'39"
Forever
Ori 5h11'16" 13d49'10"
Forever
Tau 3h30'47" 13d12'8"
forever
Ori 5h22'6" 6d4'6"
Forever
Tau 5h49'6" 17d1'10"
Forever
Tau 4h17'58" 19d7'7"
Forever
Tau 3h37'0" 17d47'44"
FOREVER
Her 18h9'0" 14d44'49"
Forever
Vir 12h4'37" 8d37'16"
FOREVER
Cas 0h34'45" 58d1'25"
Forever
Cyg 20h42'15" 47d25'40"
Forever
Uma 9h20'38" 46d8'32"
Forever
Cyg 19h52'9" 37d39'47"
Forever
Per 3h7'41" 50d54'19"
Forever
Uma 8h42'14" 48d25'39"
Forever
Uma 8h37'20" 52d28'9"
Forever
Uma 9h41'36" 44d54'24"
Forever
Lyn 9h24'8" 39d49'56"
Forever
Lyn 8h28'41" 41d21'31"
Forever
Cyg 20h10'15" 36d0'13"
Forever
Cyg 19h54'25" 36d39'14"
Forever
Crb 15h50'43" 35d28'51"
Forever
Cyg 19h29'56" 30d48'12"
Forever
Lyn 9h12'47" 34d22'43"
Forever
Lyn 7h30'32" 36d56'11"

Forever 1 Robert & Donna Col 5h38'1" -31d22'28"
Forever 27 Lmi 10h51'24" 38d56'14"
Forever 8-4 Her 18h44'54" 21d35'17"
Forever a free spirit - Serendipity Peg 22h45'46" 7d37'52"
Forever a Perfect Fit Lyr 19h10'23" 26d21'27"
Forever - Adam Garofalo Lyr 18h43'51" 43d13'49"
Forever Ali Cyg 21h41'34" 43d53'34"
Forever Alice Uma 11h17'47" 34d52'53"
Forever Always Lyr 18h48'3" 36d56'43"
Forever & Always Cnc 9h5'58" 31d3'46"
Forever & Always Cnc 8h57'18" 8d26'44"
Forever & Always Ari 1h57'0" 13d26'10"
Forever & Always Cap 20h27'57" -12d49'50"
Forever & Always Bompy! (o: :o) Ari 2h20'17" 16d25'7"
Forever & Always in Love with You Cyg 20h40'48" 35d17'28"
Forever & Always Michael Lmi 10h51'55" 35d8'28"
Forever Always Remember The Stars Uma 12h6'31" 51d13'0"
Forever Amanda's Light Sco 17h53'39" -41d36'4"
Forever Amazed-Dottie 4806 Tony And 0h44'19" 28d6'13"
Forever Amy And 0h4'29" 45d13'27"
Forever Amy Conner Cnc 8h38'4" 24d44'54"
Forever an Angel ~ Dona Bathurst Uma 10h46'21" 70d7'32"
Forever and a Day Cyg 19h45'21" 52d55'7"
Forever And A Day Cap 20h46'23" -14d37'35"
Forever And A Day Sco 17h44'24" -32d24'23"
Forever and a Day Cas 0h36'35" 57d5'10"
forever and always Cyg 20h12'1" 51d33'32"
"Forever and Always" And 23h15'38" 38d59'30"
Forever and Always Cyg 20h11'46" 42d29'3"
Forever and Always Her 16h57'18" 35d21'45"
Forever and Always Uma 10h35'13" 43d7'13"
Forever and Always Her 18h53'59" 23d49'22"
Forever and Always Leo 10h20'52" 26d24'28"
Forever and Always Sge 19h41'37" 17d51'15"
Forever and Always Ori 6h15'55" 10d14'26"
Forever and Always Aqr 22h31'41" 0d55'14"
Forever and Always Aqr 22h13'51" -23d28'0"
Forever and Always Ori 5h55'33" -6d27'12"
Forever And Always Uma 9h23'16" 67d32'43"
Forever and Always Dra 19h19'50" 65d33'27"
Forever and Always Uma 8h56'18" 67d28'8"
Forever and Always Divine Pic 5h34'46" -45d5'20"
Forever And Always- Jarod & Beth Aqr 23h52'21" -22d46'40"
Forever and Always JM and KH Lib 15h1'18" -16d27'50"
Forever and Always Laurie Sge 19h19'58" 17d37'34"
Forever and Always ~ Matt and Kira Cyg 21h38'20" 46d10'2"
Forever and Always Nayara and June Leo 10h24'4" 20d5'16"
Forever and Always- Xander and Lily Cyg 19h20'4" 52d46'12"
Forever and Always Yours George Tau 5h48'53" 17d9'14"
Forever and Beyond Lib 15h37'49" -20d29'18"
Forever and ever Uma 11h55'57" 60d15'52"
Forever And Ever Umi 16h10'28" 72d10'6"
Forever And Ever Gem 6h40'27" 14d26'22"
Forever And Ever Ori 6h0'14" 22d1'22"
FOREVER AND EVER AMEN Cap 21h40'14" -12d40'52"
Forever and for always Dra 16h11'25" 62d28'24"
Forever and For Always Sge 19h43'26" 16d56'15"
Forever and For Always Cyg 19h42'14" 51d28'24"
Forever and For Always Cyg 21h47'47" 41d56'54"
Forever And For Always Laura & Josh Per 2h18'18" 56d54'47"
Forever Andreas Uma 9h13'30" 59d0'49"
Forever Angel - Shanshan Wang Tau 3h52'16" 9d50'6"
Forever Angels - Spencer & Ned Morton Cru 12h27'38" -58d30'16"
Forever Angie Tri 1h46'17" 30d21'55"
Forever Annette's Uma 8h34'17" 63d19'20"
Forever Art Gem 6h52'5" 16d5'45"
Forever ASH+GDB Cyg 21h22'52" 44d38'41"
Forever Aubrey E. Lee Lib 15h20'16" -25d20'40"
Forever Austin Her 17h53'12" 49d3'18"
Forever Ayups Per 2h12'21" 54d18'12"
Forever Bazza Barratt 24-8-1925 Cru 12h48'45" -61d6'5"
Forever BB Gem 7h44'56" 34d42'51"
Forever Beautiful Sgr 19h6'5" -15d45'7"
"Forever Beautiful" A Star For Lyn Gem 6h56'48" 18d21'53"
Forever Beautiful Amanda Lyn 7h20'3" 48d26'48"
Forever Becca Ori 6h8'39" 21d4'16"
Forever Best Friends: SAH & KMW And 2h22'49" 49d41'29"
Forever Betty Uma 8h47'50" 52d8'43"
Forever & Beyond Cas 0h31'18" 51d8'7"
Forever Bianca Leo 11h23'15" 20d12'31"
ForEver Bill Per 3h10'47" 43d45'46"
Forever Bobby Vir 12h7'16" 10d43'5"
FOREVER BOBI Aur 6h39'16" 48d42'33"
Forever Brad and Jenn Lyn 7h51'13" 47d59'14"
Forever Breana Leo 10h10'3" 26d0'58"
Forever Bright Uma 11h19'20" 64d22'33"
Forever Brigitte Per 4h45'52" 40d46'58"
Forever Brown Ori 5h53'17" 5d56'24"
Forever Bruce & Leah Cyg 22h2'36" 50d50'25"
Forever Bud Lich Leo 9h45'22" 13d49'44"
Forever Burning Umi 16h52'9" 76d8'20"
Forever Büsché Her 16h49'43" 41d47'17"
Forever By Your Side Steve&Annette Uma 9h27'31" 67d1'18"
Forever Camelot Tau 5h24'0" 23d58'41"
Forever Carin And 1h17'0" 42d9'14"
Forever Carina Mon 6h33'57" 6d16'57"
Forever Carlos Tau 4h22'40" 25d46'47"
Forever Carlos Umi 18h36'6" 87d17'22"
Forever Carol Psc 1h2'31" 29d37'53"
Forever Carter Ori 6h2'50" 10d38'30"
Forever Casie Ari 2h38'10" 26d0'58"
Forever Celeste With All My Heart Uma 8h47'53" 53d52'35"
"Forever Céline, Victoria" Cas 1h52'7" 66d27'17"
Forever Charlie and Karianne Ari 3h13'48" 15d6'23"
Forever Charlotte Umi 17h9'44" 76d27'41"
Forever Charlotte's Morn Leo 9h46'57" 30d57'23"
Forever Charly Preissel Lyr 18h47'42" 36d20'12"
Forever Chaves Cyg 20h37'7" 44d43'6"
Forever Chima Cas 0h26'51" 66d40'25"
Forever Chip Ori 4h44'30" 1d40'0"
Forever Chloe Leo 9h59'47" 7d13'55"
Forever Chloe's Leo 9h54'33" 13d1'49"
Forever Christian Uma 10h43'34" 59d40'42"
Forever Christy's Star Leo 9h44'26" 31d12'11"
Forever Clara Aqr 21h14'54" -9d32'54"
Forever Clear Skys Grandpa! And 23h33'26" 43d18'8"
Forever Connected Aqr 23h41'23" -22d51'41"
Forever Connected Aldrig Glömd Always Cherished Gem 6h19'4" 24d12'18"
Forever Connected, Aldrig Glömd Uma 13h23'36" 59d57'47"
Forever Cory Crb 15h38'4" 26d48'11"
Forever Dan Uma 10h22'33" 41d49'4"
Forever Danielle Uma 10h0'27" 70d25'17"
Forever Dave Uma 9h51'43" 41d52'35"
Forever - Dave and Cassie Uma 10h1'16" 55d15'55"
Forever David Uma 8h35'48" 62d1'34"
Forever - David & Kelly Per 3h12'25" 31d34'47"
Forever Dawn And 23h38'50" 37d17'24"
Forever Deanna Cap 20h39'56" -16d31'2"
Forever Decorus Cas 23h29'48" 55d43'26"
Forever Deric Leo 11h58'8" 20d47'12"
Forever Derrick Cep 5h7'5" 81d56'12"
Forever Dick and Bern Cnc 8h33'44" 20d28'31"
Forever DIGI Dra 17h9'32" 65d50'13"
Forever Dreanda Vir 12h28'51" -4d10'0"
Forever Duke Dra 13h29'47" 67d44'26"
Forever Ed & Karen Ari 2h42'9" 26d13'59"
Forever Emily And 0h16'8" 28d51'8"
Forever Endeavour... Ori 5h41'43" -2d36'7"
Forever Eric Uma 14h26'19" 60d43'29"
Forever Eric Ari 3h6'20" 25d43'0"
Forever Erin Gem 7h14'12" 16d34'48"
Forever Erin Lib 14h53'45" -5d50'21"
Forever & Ever Uma 14h9'55" 56d49'22"
Forever & Ever Cyg 20h14'38" 54d47'29"
Forever & Ever Sge 19h36'9" 17d2'10"
Forever & Ever Ori 5h5'53" 10d28'46"
Forever & Ever, Babe Cyg 21h42'15" 41d58'56"
Forever Everything Cyg 20h35'46" 39d30'19"
Forever Faith Cyg 20h38'8" 42d28'11"
Forever Falk Uma 9h9'5" 66d22'57"
Forever Falling For Joe Termini Uma 9h14'48" 65d53'25"
Forever Family Crb 15h52'56" 35d43'35"
Forever Felix Ori 4h54'24" 9d48'20"
Forever Fire Chief Matt Tracy Lmi 10h27'44" 36d28'55"
Forever Flashman Cyg 19h25'6" 43d49'52"
Forever & Foralways Col 5h47'47" -32d26'59"
Forever Fox Love Star Vul 19h48'25" 24d42'8"
Forever Frances73 Pho 2h22'34" -41d19'35"
Forever Frankie Uma 10h40'0" 71d1'29"
Forever Freckles & Snuggles Cru 12h34'48" -59d23'35"
Forever Friends Cru 12h26'5" -58d21'36"
Forever Friends Del 20h46'3" 15d14'36"
Forever Friends Uma 10h40'41" 52d1'49"
Forever Friends Carter & Meyers Ori 5h17'6" -0d14'6"
Forever Friends Dennis Lee Kral Jr Uma 10h38'38" 44d10'23"
Forever Friends K.J.B.C.M.R.K. 2006 Aql 19h30'29" 6d3'50"
"Forever Friends" Ross and Sarah Cas 1h16'56" 57d11'45"
Forever Garni Crb 15h37'1" 32d29'50"
Forever Gene Her 18h47'44" 20d36'59"
Forever Genka Uma 11h26'11" 39d9'33"
Forever George Boo 14h52'23" 20d57'53"
Forever George Lib 15h42'9" -16d53'3"
Forever Glenn And 0h40'23" 23d54'31"
Forever GLZ Aqr 21h42'19" -3d9'52"
Forever Gold Cyg 20h3'57" 38d50'28"
Forever Gold-E Cas 0h19'35" 58d38'22"
FOREVER GRACE'S LIGHT Sco 15h54'38" -23d33'2"
Forever Granny Cas 0h28'52" 59d56'50"
Forever Greg Vir 13h53'55" 2d28'34"
Forever Gregory's Star Vir 13h19'52" -12d49'49"
Forever Guiding Uma 11h44'13" 62d49'22"
Forever Handsome Psc 1h22'33" 15d45'41"
Forever Hans Lidforss And 1h7'26" 34d56'33"
Forever Harlow and Joan Hawes And 0h41'53" 44d1'3"
Forever Harry & Rose Marie Mills Col 5h44'54" -33d22'11"
Forever Harty ~ Arti Cas 1h38'9" 62d28'26"
Forever Here For You! Leo 9h40'19" 13d23'23"
Forever Holly Leo 9h49'52" 6d36'40"
Forever Honest Cyg 20h4'40" 57d24'16"
Forever Howard Manuel Peg 22h16'50" 6d39'56"
Forever Hugo Cyg 19h37'26" 32d24'54"
Forever I will love you Kari 2/28 Cyg 20h12'28" 50d48'18"
Forever In Dreams Uma 8h20'4" 61d55'42"
Forever in Love Dra 18h49'28" 72d58'55"
Forever in Love Umi 15h49'55" 78d42'20"
Forever In Love Ori 5h34'30" 4d48'26"
Forever In Love Her 18h4'48" 21d57'2"
Forever In Love [AL+SO] Lyr 19h20'6" 28d51'31"
Forever In Love Chris & Jessica Cyg 21h33'39" 47d42'0"
Forever In Love: Lopez & DeSisto Sco 16h9'4" -16d7'22"
Forever in Love Stephan Her 17h28'27" 37d23'6"
Forever In Love With Jane Marie Dra 12h47'20" 71d28'6"
Forever in love, Gina C and Eric V Cyg 20h29'28" 55d30'32"
Forever In My Heart Col 5h43'55" -42d23'36"
Forever In My Heart Aqr 22h38'58" 1d1'2"
Forever In Our Hearts Andrew Ori 5h39'48" 0d42'33"
Forever in our hearts-Ed Meyer, Jr Per 3h18'10" 42d47'27"
Forever Infinity: Tiffany & Justin Lib 14h49'15" -10d54'48"
Forever Isabel Lyn 6h49'23" 58d42'51"
Forever Isabelle Her 18h5'11" 21d30'28"
Forever J. B. Ari 2h41'13" 21d57'44"
Forever Jacob Aqr 23h47'37" -11d33'7"
Forever Jacqueline Fister Leicht Gem 7h28'13" 18d46'20"
Forever Jacqui Gem 6h49'4" 23d24'58"
Forever Jayne Com 13h0'8" 27d48'38"
Forever Jeanette Cma 6h49'34" -14d4'7"
Forever Jed Lib 14h52'54" -10d5'49"
Forever Jed Her 18h28'17" 13d20'15"
FOREVER JEL Ori 5h20'14" 3d56'24"
Forever Jennifer And 1h44'5" 45d22'57"
Forever Jennifer & Oileg Cyg 21h21'57" 54d10'44"
Forever Jenny Cap 21h21'24" -15d41'44"
Forever Jenny And 0h18'24" 30d3'48"
Forever Jesse's Uma 11h3'24" 33d45'30"
Forever Jessica Lac 22h26'58" 47d45'9"
Forever Jessie & Tyler Psc 1h55'58" 5d11'25"
Forever Jessy Umi 16h40'34" 75d15'35"
Forever JID Ori 6h5'40" 13d43'16"
Forever Jidiah And 23h26'49" 48d58'52"
Forever Jim Sheehey Cep 21h31'22" 68d11'11"
Forever Jizzy Lyr 19h11'52" 27d18'36"
Forever Jocelyne Col 6h25'17" -40d19'45"
FOREVER JOCHEN Ori 5h49'30" -9d37'58"
Forever Joe Cep 2h12'27" 81d13'36"
Forever John and Rena Ori 6h9'8" 16d49'27"
Forever Johnny-Cakes Aronson Per 3h3'49" 54d59'41"
Forever Jose Uma 13h45'31" 50d12'6"
Forever Josh Uma 12h35'23" 52d42'58"
Forever Joshua Kay Ori 5h26'42" -0d9'23"
Forever Joy M Tau 3h42'3" 29d0'22"
Forever Julie Uma 10h55'47" 67d16'1"
Forever Julie Ann Cnc 8h49'50" 26d9'4"
Forever Juneau Cam 7h47'39" 62d23'40"
"Forever Jürgen" Cas 2h3'35" 74d41'53"
Forever Kara and Nick Lac 22h25'56" 50d54'3"
Forever Karen Jule Ara 17h18'51" -52d14'25"
Forever Kathy Lyn 7h21'2" 44d44'22"
Forever Katy And 1h13'52" 34d40'59"
Forever Kevin & Caesar Aql 19h0'49" 15d49'43"
" FOREVER KIMBERLY " Tau 3h55'17" 21d16'14"
Forever Kimberly Cnc 8h53'24" 13d6'0"
Forever Kimuralowe Uma 12h8'52" 55d52'32"
Forever Kristin Lib 15h30'52" -20d53'55"
Forever Kristin Uma 11h32'44" 38d3'45"
Forever Kristina Lib 14h59'59" -13d7'11"
Forever Kyla Cas 23h22'31" 57d33'17"
forever lan Cyg 19h47'4" 35d17'0"
Forever Lasting Friendship Dra 15h55'58" 56d30'57"
Forever Laura Cap 20h19'24" -9d52'11"
Forever: Lauren and Greg Vir 12h33'35" 4d57'48"
Forever Laurie Lyn 7h19'37" 45d52'27"
Forever LeighAnn Cnc 8h48'15" 18d33'33"
Forever Lena Sco 16h25'31" -36d1'4"
Forever Linda Cam 6h10'58" 66d12'21"
Forever Lisa Lyn 7h20'31" 56d23'34"
Forever Lisa Aqr 21h22'40" -0d13'32"
"Forever Lisa" And 0h19'18" 30d50'26"
Forever Lisa Marie Uma 9h50'33" 51d26'53"
Forever Llano Lyn 7h49'19" 57d53'50"
Forever Lori Ann Aqr 21h9'1" -3d0'35"
Forever Louise & Chad Cyg 19h37'1" 53d9'9"
Forever Love Uma 9h21'56" 60d32'3"
Forever Love Mon 6h45'11" -0d26'42"
Forever Love Lib 15h30'7" -7d33'48"
ForEver Love Aqr 21h8'53" -9d22'24"
Forever Love Cyg 19h21'25" 46d53'38"
Forever Love Cyg 20h17'47" 51d28'27"
Forever Love Cyg 21h30'32" 44d25'24"
Forever Love Cyg 19h56'50" 30d37'45"
Forever Love Lyr 18h27'55" 31d48'33"
Forever Love Gem 7h58'37" 32d59'18"
Forever Love And 1h32'29" 43d3'16"
FOREVER LOVE Cyg 20h30'59" 35d5'5"
Forever Love Vul 19h56'57" 23d25'34"
Forever Love Tau 4h34'55" 28d7'45"
Forever Love Gem 6h45'35" 16d38'59"
Forever Love Gem 6h44'11" 17d37'29"
Forever Love Vir 13h47'18" 1d31'30"
FOREVER LOVE Tau 3h44'15" 13d14'3"
Forever Love Her 17h22'27" 14d45'22"
Forever Love - Bobbi Jo Dixon Leo 11h24'28" 10d23'31"
Forever Love Ciarra Cyg 20h38'13" 48d2'56"
Forever Love - Dawn & Tim Cummings Gem 6h50'26" 14d57'19"
Forever Love D.C.G. Cyg 20h45'3" 32d57'59"
Forever Love - Doug and Rocki Cmi 7h28'9" 8d1'33"
Forever Love of Ijen & Yoelin Cnc 8h51'25" 31d54'39"
Forever Love Patrick&Catherine Lmi 9h54'13" 36d0'59"
Forever Love Randy And Nicole Ori 6h10'14" 17d55'32"
Forever Love - Terry/Linda Hubbard Cyg 20h36'55" 34d23'9"
Forever Love Turu and Jennifer Lyr 19h7'57" 37d24'8"
Forever Love, Jesse & Alecia Ari 3h13'15" 29d28'21"
Forever Loved Burl "Jim" Bryant Her 16h49'59" 24d29'50"
Forever Loved James Jordan Miele Cep 20h47'1" 58d51'20"
Forever Loved, Nora & Albert Cannon. Ari 2h0'12" 13d3'38"
Forever Loving Shirley Vir 11h39'17" 10d13'59"
Forever Luis Aql 19h8'47" 5d39'38"
Forever Luke and Megan Cnc 9h10'7" 8d0'42"
Forever Luke Gervasi Boo 14h35'50" 18d3'5"
Forever Lynne Cas 1h33'42" 65d28'5"
Forever M&A T. Crb 16h14'33" 39d28'35"
Forever M&B Cma 6h22'12" -29d24'35"
Forever M & F Lmi 9h52'53" 38d53'58"
Forever M & M Uma 9h28'30" 55d3'8"
Forever Macy Cru 12h46'45" -55d55'52"
Forever Manuela Lyn 6h48'1" 58d51'12"
Forever Marcia And 0h9'55" 34d14'15"
Forever Marie Tau 5h34'57" 20d14'1"
Forever Mark & Sugar Vitullo And 1h6'33" 37d30'18"
Forever Martin... Her 16h51'19" 34d56'46"
Forever Mary Ori 5h55'13" 8d38'54"
"Forever" MAYA Per 3h49'39" 49d39'52"
Forever Mayra Vir 13h28'4" -5d22'12"
Forever Meemaw Cnc 8h59'50" 15d33'47"
Forever Melanie Ori 5h29'35" 3d41'45"
Forever Melissa Ori 6h2'55" 16d33'28"
Forever Melissa Umi 14h41'3" 81d31'13"
Forever Melissa Dra 20h14'30" 63d58'1"
Forever Memories Per 3h14'22" 38d25'59"
Forever Michael "Ferg" Ferguson Sgr 19h13'51" -25d10'27"
Forever Mighty Quinn Uma 12h22'30" 53d29'22"
Forever Mihaela Equ 21h23'29" 7d8'17"
Forever Mikey & Tegan Cru 12h37'51" -60d35'8"
forever mine Peg 21h35'29" 20d41'37"
Forever Molly Leo 10h16'40" 25d17'43"
Forever Mom Cap 21h40'48" -13d5'6"
Forever Mom, Ruth Minogue Houle Ori 5h36'37" 14d27'2"
Forever Mom's Sharon Del 20h25'20" 14d15'47"
Forever Monica Cas 1h22'49" 53d58'8"
Forever Moore Lyn 9h33'8" 40d19'6"
Forever & More Cra 18h47'4" -38d5'42"
Forever Mostest Her 17h10'37" 28d27'15"
Forever Mum, Eternal Nanna xx Cas 0h46'41" 61d55'46"
Forever Müsli Her 17h53'19" 38d15'17"
Forever - MWG + KRH Cyg 19h48'23" 33d42'22"
Forever My Bri Cnc 8h38'47" 18d22'40"
Forever My Cootcha Cnc 9h20'43" 9d9'37"
Forever My Juliet Aqr 22h55'13" -19d31'16"
Forever My Love Cru 11h59'29" -59d52'34"
Forever My Love Sge 19h24'10" 18d2'1"
Forever my Love Uma 12h1'53" 29d33'44"
Forever My Love Cyg 21h10'40" 47d18'9"
Forever My Love Cathy Ori 4h56'1" -0d17'57"
Forever My Love Diana Cas 0h50'55" 60d36'27"
FOREVER MY LOVE KARA D'NAE Cyg 20h15'16" 37d47'3"
Forever My Love, Paul Laskowski Her 18h51'12" 23d7'8"
Forever My Pumpkin Lib 14h40'16" -19d22'43"
Forever My Santino Col 6h27'15" -41d36'37"
Forever My Star - Tania Wittensleger Tau 5h9'24" 18d10'54"
Forever My Sweet Baby James Per 4h36'47" 49d28'39"
Forever Nam Ori 6h14'10" 20d55'50"
Forever Nancy And 0h10'9" 35d0'25"
Forever Natalie Sgr 19h48'33" -37d38'19"
Forever Nate Lyn 7h0'28" 46d11'52"
Forever Nathalie & Marco Uma 11h4'28" 59d29'40"
Forever Naylise Dra 15h39'44" 58d50'3"
Forever Nick Sco 16h38'52" -28d39'42"
Forever Nicole Her 18h5'14" 16d58'23"

Forever - Nicole Garofalo
Lyr 18h47'20" 44d9'21"
Forever Nina
Crb 16h12'33" 26d35'56"
Forever Noble Will
Ori 5h6'10" 10d3'23"
Forever Norton
Ori 4h52'3" 4d21'8"
Forever Noshi
Col 5h29'15" -29d7'22"
Forever November
Lyr 18h49'5" 32d33'55"
Forever: November 22, 2003
Per 3h36'27" 32d36'14"
Forever on my Mind
Cru 12h43'51" -62d45'15"
Forever One
Lyr 18h56'55" 44d22'32"
Forever One
Crb 16h19'25" 27d52'6"
Forever Oops
Cru 12h29'10" -60d54'14"
Forever Our Beloved Jesse
Cru 12h42'35" -57d53'39"
Forever Our Daddy
Her 16h52'21" 35d45'29"
Forever Our Love
Her 16h11'51" 25d27'45"
Forever Our Star - Tyler Dunne
Lib 14h50'6" -2d56'31"
Forever Ours
Aqr 22h33'58" -3d23'25"
Forever Ours — I Love You
And 2h32'2" 46d33'11"
Forever Pat
Cnc 8h58'9" 19d8'52"
Forever Pat & Velma Love Sanctuary
And 1h57'50" 46d4'39"
Forever Patrick
Aqr 22h4'34" -14d30'2"
Forever Paul
Her 16h48'31" 24d4'30"
Forever Peggy
Uma 9h18'11" 71d20'46"
Forever Pelle
Ori 6h7'50" -1d14'14"
forever penny - 143
Del 20h31'47" 19d19'23"
Forever Phil
Ara 16h39'20" -59d44'2"
Forever Pookies
Umi 15h18'27" 74d52'23"
Forever Poppa
Uma 10h1'27" 65d14'4"
Forever Princess Merin
And 0h33'10" 26d24'15"
Forever Priska, 17.11.1981
Lac 22h52'20" 37d42'17"
Forever Rachael
Cru 12h19'55" -62d16'51"
Forever Rachel
Cas 23h25'59" 53d2'16"
Forever Radhe
Lib 15h2'25" -25d57'35"
Forever Rae
Vir 14h11'34" -14d45'44"
Forever Rainer
Cap 20h18'50" -14d9'46"
Forever Ralph
Cam 8h45'51" 77d12'54"
Forever Remembered
Cru 12h27'41" -56d29'3"
FOREVER RETO
Umi 15h51'52" 81d13'10"
Forever Richard
Gem 7h35'22" 24d3'14"
Forever Rita
Cap 20h37'6" -14d6'0"
Forever RJBrice
Sco 17h29'35" -38d5'31"
Forever Robin
Vir 14h13'15" -17d56'47"
forever roli
Lyn 8h24'34" 39d49'21"
Forever Roman
Ori 6h14'19" 9d47'15"
Forever Ron
Sco 16h12'13" -13d58'33"
Forever Ross
Ori 6h0'19" 21d34'40"
Forever Ruedi
Uma 9h16'34" 57d37'46"
Forever Sabine and Roman
Ori 6h14'2" 11d0'17"
Forever Salome
Uma 11h44'19" 42d10'37"
Forever Sam and Lauren
Cyg 20h0'44" 53d23'22"
Forever Sandra
Cyg 21h11'59" 46d36'28"
Forever Sara
Uma 9h1'56" 60d18'45"
Forever Sarah
Uma 12h17'38" 53d40'57"
Forever Sarah
Uma 10h42'25" 48d35'16"
Forever Sarah
And 23h44'21" 39d16'43"
Forever Schnuggi
Sgr 18h5'6" -27d17'47"
Forever Scott and Nicole
Per 4h50'16" 45d26'30"
Forever Sean's
Psc 1h24'56" 27d38'34"
Forever Sharon
Lac 22h37'11" 53d56'50"
Forever Sharon
Cas 1h29'27" 63d30'18"
Forever Shin
Boo 14h27'23" 41d1'2"
Forever Shine Gregg
Aqr 21h13'31" -13d42'20"
Forever Shining Brandon's Light
Per 3h34'7" 47d36'15"
Forever Shining Drew Robert Martin
Her 18h25'11" 26d2'29"
Forever Shining Love: Mee & Vinz
And 23h1'58" 35d50'29"
Forever Shining - Willy & Ann Eder
Aqr 23h2'36" -17d0'43"
Forever Shining-Gabriel Gonzales
Uma 11h30'7" 48d47'21"
Forever Siani
And 2h38'7" 38d7'1"
Forever Skyler
Psc 0h18'53" 11d21'40"
Forever Sonja
Umi 17h6'19" 76d2'42"
Forever Soulmates
Lib 15h22'45" -19d23'30"
Forever Soulmates CK
Cyg 20h15'16" 35d19'19"
Forever Sparkling Peter Mahony
Cap 21h5'33" -24d28'24"
forever stefan
Cep 23h10'59" 79d23'35"
Forever Stefan
Cyg 19h38'38" 31d54'53"
Forever Stephanie & Peter Spezia
Lyr 19h25'38" 38d13'39"
Forever Summer Girls
Lyn 6h47'1" 61d51'2"
Forever Sunshine
Cap 20h13'58" -20d45'2"
Forever Susan
Cnc 8h26'41" 17d25'28"
Forever Sweet
Sco 17h58'13" -37d6'21"
Forever Tamara
Cap 21h43'31" -12d38'55"
Forever Tammy
Cas 23h39'26" 52d45'41"
Forever Thomas Frank
Boo 15h37'3" 45d21'19"
Forever Thuy Trieu
Ori 6h18'19" 8d17'10"
Forever Timothy
Sco 17h53'54" -39d4'46"
Forever Timothy and Denise
Psc 0h49'46" 10d59'40"
Forever Tina
Boo 15h43'30" 45d10'57"
Forever to my Tanja - Zaki
Cyg 20h37'47" 48d29'36"
Forever Together
Crb 15h27'24" 27d57'49"
Forever Together Eric and Laura
Her 18h41'25" 25d52'12"
Forever Tom & Darlene
Cyg 20h50'18" 31d47'48"
Forever Toongi
Cnc 9h10'30" 24d3'13"
Forever True Love
Ori 5h41'19" -0d19'14"
Forever Tutu & Pop
Cyg 21h28'52" 50d6'13"
Forever Us
Per 4h25'27" 44d50'58"
"FOREVER US"
Vir 12h18'38" -3d46'40"
Forever V. Quitania
Uma 13h43'14" 47d53'37"
Forever Valentines
Uma 9h44'23" 53d57'19"
Forever Victoria
Uma 11h48'36" 38d29'14"
Forever Walter Reyes
Tri 2h34'38" 35d9'36"
Forever Young
Crb 15h44'13" 32d51'2"
Forever Young
Her 17h54'9" 23d55'40"
Forever Young
Umi 15h35'16" 72d45'48"
Forever Young
Pho 23h36'7" -43d11'5"
Forever Young (Alyssa)
Psc 1h20'4" 32d45'8"
Forever your Angel Young Moggy xx
Per 3h5'53" 35d7'18"
Forever your babyy <33
Sgr 18h42'37" -33d1'52"
Forever Your North Star
Uma 11h46'36" 38d18'27"
Forever Yours
Cnc 9h2'54" 32d48'38"
Forever Your's
Uma 11h9'26" 30d44'6"
Forever Yours
Uma 11h24'25" 31d53'52"
Forever Yours
Uma 11h38'24" 31d17'6"
Forever Yours
Uma 11h53'47" 46d16'25"
Forever yours
And 0h50'35" 45d54'9"
Forever Yours
Cyg 19h42'25" 29d5'37"
forever yours
Tau 5h57'1" 27d8'28"
Forever Yours
Cnc 8h38'54" 18d58'5"
Forever Yours
Gem 7h14'43" 15d28'9"
Forever Yours
Aqr 22h41'54" 1d33'6"
Forever Yours
Ari 3h12'30" 10d51'41"
Forever Yours
Ori 6h15'55" 10d19'19"
Forever Yours
Ori 6h9'39" 13d8'59"
Forever Yours
Uma 10h47'43" 69d18'22"
Forever Yours
Dra 18h19'46" 58d9'15"
Forever yours Erin
Ori 5h7'34" 15d46'9"
Forever Yours. Forever Mine. Ours.
Ori 5h30'36" 1d1'48"
Forever Yours In Love
Cyg 21h23'55" 50d16'34"
Forever Yours Steph
Cru 12h27'28" -60d15'34"
Forever Yours. To My Angel, Kelsey
Cas 1h41'22" 68d32'27"
Forever Yours x
And 23h24'42" 47d20'29"
Forever Yours, Faithfully.
Lyn 7h59'37" 42d6'52"
Forever Yvonne
Cyg 21h3'39" 36d33'1"
Forever Zachary
Ari 3h19'59" 25d6'3"
FOREVER ZEUS
Lyn 6h40'7" 61d48'11"
Forever, Diane
Cas 0h42'37" 61d54'51"
Forever, Our Place in the Heavens
Ori 5h36'57" -0d2'7"
Forever, The Love of My Life
Per 3h33'46" 52d0'39"
Forever——Aline
Vir 13h26'30" 12d58'22"
ForeverArchie
Cma 6h51'4" -16d54'17"
Foreverlasting
Lyn 6h43'51" 56d25'18"
ForeverLove Matthew & Vanessa Hicks
Cyg 20h2'1" 55d22'42"
ForeverLynn&Jack/Dol&Jim /Em&Albert
Leo 10h15'31" 20d46'53"
Foreverness
Lyr 18h43'55" 39d33'30"
Forever's Beginning, I Love Krystal
Cyg 21h19'50" 31d43'12"
Forever's Dream
Ori 5h55'34" 11d54'13"
foreversummer
Her 16h54'53" 35d38'26"
ForeverTheThird
Gem 8h6'31" 31d32'28"
forevertinakeith
Aqr 22h55'26" -7d13'53"
ForeverYours52106
Cyg 20h2'47" 53d41'10"
ForEvroUs
Cyg 20h8'24" 58d37'51"
Forget Me Not
Lyn 7h37'32" 53d55'12"
Forget Me Not
Ori 5h34'59" -1d22'23"
Forget Me Not
Cyg 20h0'30" 30d18'33"
Forget Me Not
Per 4h34'39" 47d3'56"
Forgiveness
Cam 3h46'13" 63d47'5"
Fornasiero Manuel
Lyr 18h41'26" 41d16'25"
Fornino
Lyn 8h7'3" 37d9'38"
FORREST
Cyg 20h57'38" 46d47'33"
Forrest and Jenny Giron
Cyg 20h21'13" 44d10'35"
Forrest Anthony Welk
Aqr 23h55'46" -17d7'51"
Forrest D. Moulton
Psc 1h11'24" 2d47'30"
Forrest Dale Hudson
Uma 9h20'26" 51d43'56"
Forrest Elias Diamond
Her 17h3'19" 40d2'44"
Forrest James McDaniel
Aql 19h5'44" -0d42'0"
Forrest Karl Moe
Gem 7h1'55" 34d48'42"
Forrest Strom
Cnc 8h9'31" 25d54'36"
Forrest Wilson
Ori 5h34'30" 3d29'3"
Forrester William
Sco 16h56'49" -38d20'1"
Forster, Barbara
Sgr 19h13'5" -35d6'39"
Förster, Irene Prof. Dr.
Uma 9h3'51" 56d33'47"
Förster, Ute
Uma 9h17'17" 58d12'23"
Forsythe Sweeney Vesterholm
Crb 15h19'54" 26d57'5"
Fort Bend's Stellar Communicator
Pho 0h39'44" -43d35'33"
Fortitude Syuya Okubo
Cnc 8h4'51" 16d30'41"
Fortius quo Amorous
Sge 20h6'17" 17d45'44"
Fortress
Cnc 8h38'5" 32d25'17"
Fortuitous Bliss
Uma 11h57'42" 29d18'49"
Fortuna
Her 17h41'59" 16d32'56"
Fortuna Astrum
Cas 1h6'57" 56d42'33"
Fortunato Emilio Sanchioni
Peg 23h4'15" 28d30'21"
Fortunato Veloso
Del 20h21'45" 18d32'10"
Fortunatus Tredecim
Cru 12h37'42" -57d33'33"
Fortune
Umi 16h15'0" 77d41'50"
Fortune
Gem 6h41'0" 21d0'5"
Fortune
Lyr 18h50'2" 32d4'4"
Fortune and Steve forever 2007 xoxo
Uma 11h56'37" 57d0'49"
Forty7Twenty9
Lib 15h16'41" -6d54'6"
Fosnaught's Star
Lyn 7h48'48" 44d5'4"
Foster's Falkor
Dra 14h46'37" 56d10'51"
Foster's Star
Uma 11h43'50" 59d52'53"
Foth, Renate
Ori 5h7'24" 8d47'19"
Fotini
Cap 21h41'36" -11d21'47"
Fotis Eliodromytis
Umi 15h33'24" 78d30'14"
Fotis Kallianezos
Cnv 13h55'57" 36d31'2"
Fouad Blas
Lib 16h1'58" -20d12'51"
Fouad & Noor
Equ 21h7'38" 11d49'26"
Fouad, "Fokch"
Mon 6h37'0" 1d21'21"
Foulds
Cnc 8h33'35" 7d30'48"
Foulger Sixty
Cas 1h26'43" 55d33'54"
Found
Leo 9h36'46" 30d31'3"
Foundation
Per 3h37'59" 33d1'21"
Fountains
Uma 10h29'22" 66d39'0"
Four Aces Apts & Cottages
Crb 16h7'14" 38d50'39"
Four Starzzz Bruce
Uma 11h22'47" 35d59'57"
Fourpaws
Leo 11h44'15" 17d56'41"
Fourtees
Per 2h24'3" 55d0'3"
Fourth Grade Love
Vir 13h11'9" 11d45'39"
Fouzia
Sco 16h42'54" -41d2'51"
Fouzia
And 1h52'35" 45d32'32"
Fox
Her 17h17'27" 20d1'30"
Fox
Tau 5h45'35" 26d41'22"
Fox13
Vir 12h59'34" 11d0'33"
Fox's Guiding Light
Tau 4h27'35" 22d58'19"
FoxSkatr
Leo 11h48'52" 20d22'27"
Foxxus Parasaurolophus
Cas 0h53'1" 64d14'49"
FOXY
Cnc 8h34'33" 31d31'17"
Foxy 65
Cep 21h43'50" 58d46'52"
Foxy Rebekah of Heidelberg
Vul 19h22'34" 22d17'49"
FoxyFoxx
Psc 1h7'7" 4d28'45"
Foxy's Boy - Paul McDonald
Cap 20h51'1" -19d19'16"
Foye Johnson
Cap 20h35'21" -12d7'11"
FPM-329
Ari 3h18'32" 29d3'35"
FPX - 2450620.30208
Uma 9h33'45" 71d42'24"
Fr. Alex H. Bradshaw's Star
Cra 19h1'58" -37d54'54"
Fr. Basil Colasito
Aql 19h50'16" -10d17'23"
Fr. Bill Faiella
Aur 5h44'20" 46d51'21"
Fr. Gary McInnis Peace Be With You
Ori 5h45'44" 3d16'22"
Fr. Pedro Cartaya, S.J., Ph.D, M.A.
Ari 2h12'41" 21d57'51"
Fraccascia Monica
Ori 6h10'17" -2d11'34"
Fragolina07
Crb 16h24'50" 28d7'50"
Fraideh Taefi Torian
Crb 15h32'11" 29d17'15"
Frair
Uma 10h4'10" 47d31'44"
Frajenberg
Ari 2h4'18" 12d29'5"
Framille
Uma 8h19'7" 68d6'56"
FRAMO
Cyg 20h4'2" 48d51'4"
FRAN
Peg 22h29'0" 19d17'10"
Fran
Uma 10h58'33" 65d47'4"
Fran
Lib 15h28'24" -15d22'44"
Fran Alvey
And 1h45'30" 43d10'52"
Fran and Chris
Dra 18h49'48" 53d33'32"
Fran and Jerry Souders
Sge 19h27'18" 16d57'35"
Fran and Kelly Forever
Lyr 19h7'36" 45d9'33"
Fran and Paul Breitbach
Cyg 20h2'10" 57d54'9"
Fran Calloway
Uma 9h20'16" 57d2'57"
Fran Carr
Aql 19h1'50" 3d46'37"
Fran - Center of My Universe
Sco 17h2'39" -33d20'51"
Fran & Chap
Cas 0h19'20" 51d36'24"
Fran Diamond "Stellar Mom"
Cas 0h40'1" 53d47'27"
Fran & Fiore
Uma 9h46'54" 45d30'9"
Fran Forehand
Cam 3h55'31" 57d33'44"
Fran George Hughes
Cas 23h24'56" 55d1'38"
Fran Georgianna
And 2h15'30" 46d43'19"
Fran Harding-Williams
Uma 8h45'11" 62d10'2"
Fran Harris
And 0h35'36" 39d4'59"
Fran Kaufmann Chenoweth 8251936
Vir 12h4'10" 4d35'51"
Fran Kieselstein
Boo 15h32'48" 45d28'24"
Fran & Landon Dunaway
Cyg 19h46'18" 33d58'23"
Fran McKnight
Cap 21h8'22" -21d1'41"
Fran Rasi
Aqr 22h13'25" -1d1'49"
Fran Reich-Birthday Star
Uma 9h22'28" 67d45'26"
Fran Robbins
Cas 1h15'58" 54d25'13"
Fran Ryan Butler
Aqr 21h43'7" 1d39'49"
Fran Sadler
Ori 5h32'36" 5d29'52"
Fran Smith Jr.
Lyn 7h57'13" 56d40'36"
Fran Szymanek
Leo 9h40'12" 27d10'51"
Fran Volz
Lyn 7h50'19" 42d49'34"
Fran Willover
Uma 10h55'42" 50d1'20"
Fran X. Walls
Uma 11h58'42" 54d55'17"
Fran you are my Eternal Light...
Eri 4h36'16" -35d2'26"
Franandel 5963
Mon 6h53'10" 8d19'38"
FranBurg
Uma 14h24'36" 57d36'10"
Franca e Piero
Boo 15h19'48" 48d1'12"
Franca Forni
Umi 16h56'37" 83d4'38"
Franca Orsi
Cas 0h32'2" 54d54'6"
Franca Pugliese
Cas 1h14'50" 50d58'0"
FranCaro6155
Crb 15h29'52" 27d45'49"
France 10-01-04
Cyg 20h47'10" 41d56'50"
France Caroline Carriere
Umi 16h43'36" 83d11'54"
France Denoeud
Cnc 8h54'25" 10d34'51"
Francene Hemhauser
Cas 2h15'32" 63d51'28"
Frances
Vir 12h59'35" -21d44'47"
Frances
Sco 16h15'4" -11d50'19"
Frances
Cnc 8h33'7" 13d41'20"
Frances
Aqr 22h35'24" 0d50'59"
Frances
Ari 3h14'4" 26d17'30"
Frances
Peg 23h20'45" 16d20'40"
Frances
Gem 7h36'8" 16d56'27"
Frances
And 23h17'9" 43d11'32"
Frances
Cas 23h53'29" 50d22'44"
Frances "A Knight To Remember"
Uma 10h50'3" 43d52'57"
Frances Adelle Viera
And 0h43'49" 35d39'57"
Frances Alberta Cole
Vir 12h28'38" 4d32'47"
Frances Alberta Glaub
Leo 10h51'34" 14d10'30"
Frances Alberta Marshall
Dra 19h42'12" 63d16'24"
Frances Alexander Speed
Sco 17h51'28" -36d29'24"
Frances Alice Armstrong
Cyg 20h8'46" 36d25'22"
Frances Amendola
Sgr 18h12'46" -18d53'16"
Frances Anderson
Tau 4h23'37" 22d48'57"
Frances Ann DiMaggio
Leo 10h19'19" 20d12'2"
Frances Ann Friedman
Vir 12h26'22" 11d4'24"
Frances Ann McCormick
Lib 15h38'13" -19d49'50"
Frances Anne Augusta Donovan
Cru 12h28'21" -59d20'5"
Frances Barbara Jaskolka - Philip
Umi 13h33'3" 70d8'7"
Frances Berry Tierce
Tau 4h18'58" 3d17'52"
Frances Blower
Gem 6h36'10" 24d43'39"
Frances Brannen Hanly
Cam 4h26'27" 72d24'42"
Frances Brasch
Leo 10h10'24" 13d33'19"
Frances Bruce
Cam 7h28'26" 73d27'23"
Frances Bryant Edens
And 23h31'24" 36d31'26"
Frances C. Southworth
Uma 9h24'46" 65d7'28"
Frances Camelio
Ari 2h16'1" 25d43'57"
Frances Carolyn Pullen
Cas 1h12'56" 64d14'26"
Frances Clare
Uma 13h28'22" 52d25'9"
Frances Clark
Lib 15h40'33" -29d28'0"
Frances Cloward
Gem 7h29'0" 28d46'18"
Frances Custer
Tau 4h37'59" 29d36'25"
Frances Devitto
Aqr 20h46'37" -1d0'18"
Frances Egan
Uma 12h40'32" 56d25'3"
Frances Elaine Baugher Donovan
Aqr 22h16'51" -24d28'13"
Frances Elaine Thurston
Cam 4h20'0" 54d28'50"
Frances Elenor Perron
Lyr 18h30'40" 39d55'39"
Frances Eliane Perez
Del 20h41'20" 15d21'25"
Frances Elizabeth Crowson
Cap 20h32'33" -16d9'43"
Frances Elizabeth Howell
Cnv 12h14'57" 35d59'51"
Frances Ella Verrier
Peg 21h34'3" 19d10'12"
Frances Ellen Pennell
Lib 15h14'26" -4d36'3"
Frances Evans
Uma 9h24'43" 56d14'9"
Frances Fedroff (Mom)
Cyg 20h30'2" 34d25'2"
Frances Fee
Lyr 18h40'4" 36d57'54"
Frances Fides
Psc 0h38'40" 17d31'5"
Frances "Frankie" Oberlie
Uma 11h59'59" 42d57'49"
Frances "Frannie" R. Durre
Cas 23h8'29" 59d31'11"
Frances Freeman
Aqr 21h52'16" -2d47'51"
Frances French One in a Million
Cap 21h5'24" -15d27'39"
Frances (Fritz) Winter
Cyg 20h28'9" 52d1'29"
Frances Fromknecht
Uma 11h18'7" 56d23'51"
Frances Gabrielle
Psc 1h8'45" 13d3'46"
Frances Galbreath Locke
Cnc 8h32'2" 23d56'12"
Frances Gentile
Vir 14h8'40" 6d47'33"
Frances "G-G" Emily Randolph
Leo 9h39'55" 26d33'37"
Frances Goetz Star
Vir 12h38'7" 8d41'26"
Frances H. Muscatello
Cas 0h48'38" 60d57'7"
Frances Harris
Aqr 22h30'9" -2d21'13"
Frances Haselberger
Cyg 21h15'46" 44d33'11"
Frances Hayes
Cam 4h3'34" 67d22'53"
Frances Hoch
Uma 11h12'52" 62d55'50"
Frances Inda
Mon 6h52'2" -0d56'23"
Frances Jeanette Barnhouse
Cas 0h29'50" 61d35'6"
Frances Josephine Brocklebank
Crb 15h43'3" 26d6'0"
Frances Josephine Lewis
Lyr 18h27'46" 28d5'0"
Frances K. Taurisano
Cyg 21h57'25" 44d42'16"
Frances K. Trampy
Ari 2h27'54" 24d26'4"
Frances Kalend
Tau 4h33'7" 19d17'1"
Frances Kay Keenan
Cap 21h43'33" -22d56'36"
Frances Kaye Elliott
Sgr 19h2'20" -22d25'18"
Frances Kowalczyk Rentz
Leo 10h18'34" 26d37'22"
Frances Kusinski
Leo 11h6'15" -1d27'39"
Frances L. Anderson
Cap 20h16'48" -23d11'42"
Frances L. Samalis
Cap 21h0'18" -17d21'25"
Frances L. Steele-Knab
Per 3h36'28" 47d4'42"
Frances L. Tierney
Ari 2h27'30" 10d44'33"
Frances Langan's Star
Cyg 21h9'34" 46d14'16"
Frances Leland Christie Dancer
Psc 0h54'28" 8d8'32"
Frances Lenay Jackson
Gem 6h23'32" 22d26'51"
Frances LoPresti
Uma 12h42'14" 59d20'46"
Frances Lorraine
Tau 5h3'2" 19d23'38"
Frances Lou Black Moulin
Vir 13h26'43" -13d36'20"
Frances Louise Beazley Weaver
Tau 4h5'9" 9d50'17"
Frances Louise Shaw Stavros
Uma 10h4'51" 57d41'30"
Frances Louise Valadez
Aqr 22h13'44" -0d2'16"
Frances Lynn Vorsteg
Uma 11h53'3" 45d54'6"
Frances M. Butner
And 23h36'27" 42d8'32"
Frances M. Cavalari
Lyr 18h37'58" 31d29'22"
Frances M. Hindman Gilroy
Vir 13h11'53" -6d32'48"
Frances M. Hurley
Psc 1h12'20" 26d42'39"
Frances M. Taylor
Cam 4h8'32" 55d30'8"
Frances M. Yocum
Cyg 19h56'33" 50d52'35"
Frances Madden Jones
Tau 4h26'15" 12d54'18"
Frances Madsen
Ari 2h7'50" 20d24'24"
Frances Margaret Walsh (Mammy)
Cas 23h53'50" 53d57'56"

Frances Marguerite
Roberts Corbin
Gem 7h4'29" 25d55'26"
Frances Marie Kochian
Umi 15h14'17" 78d31'22"
Frances Marie Lewis
Cnc 8h19'44" 10d21'42"
Frances Marie Zuk
Psc 1h7'48" 27d59'44"
Frances Marshall Campbell
Uma 10h39'45" 53d26'59"
Frances Martin
Aqr 21h11'52" -7d42'35"
Frances Mary Rinaldi
Simonsen
Cap 21h9'0" -15d39'1"
Frances Mary Swatridge
Cas 23h19'5" 59d45'49"
Frances Mary Veronica
Vir 12h47'11" 1d19'14"
Frances Matthews
Sgr 19h37'23" -24d4'58"
Frances McGaha
Cyg 20h23'16" 49d59'1"
Frances McQuade ~ July
27, 1947
Leo 9h43'26" 30d46'56"
Frances Mei Liu Hardin
Uma 11h28'18" 54d59'15"
Frances Milo
Uma 10h51'45" 62d48'57"
Frances Monica Retsas
Cma 7h3'45" -16d33'9"
Frances Mountfort
Cas 23h4'14" 54d36'30"
Frances Naphegyi
Uma 11h11'15" 30d30'5"
Frances Nordstrom
Tau 5h44'53" 18d57'48"
Frances O'Connor
Uma 11h27'7" 39d42'23"
Frances O'Donnell Simeon
Lib 15h11'42" -10d16'6"
Frances Orr
Cap 21h54'42" -13d43'51"
Frances "Our Shining Star"
Uma 11h4'24" 47d25'20"
Frances P. Ficke
Sco 16h13'15" -11d59'43"
Frances Pane
Umi 16h24'34" 70d14'16"
Frances & Paul Thompson
Cyg 19h56'0" 40d20'4"
Frances Pauline
Ari 2h4'38" 13d9'38"
Frances Pecoraro
Vir 13h1'42" -10d48'2"
Frances Penelope Rumsey
Cas 0h7'33" 55d44'32"
Frances Peraino
Leo 10h14'11" 22d12'38"
Frances Poppy-Mae Slater
Cam 5h49'27" 56d33'13"
Frances Protetti LoSchiavo
Uma 9h20'45" 70d25'42"
Frances Rebecca Hart
Leo 11h26'8" 13d43'5"
Frances Rooney Cerasoli
Sgr 18h22'10" -22d45'46"
Frances Schreib
Tau 3h46'36" 29d0'23"
Frances Scott Plappert
Cyg 21h59'52" 53d55'46"
Frances Sederholm
Leo 10h2'51" 21d16'4"
Frances Skeete
And 23h19'59" 41d44'42"
Frances "Skip" Cox
Leo 9h37'59" 28d42'24"
Frances Soleil Antonio
Cnc 9h18'31" 29d14'40"
Frances Stephanie
Shininger
Ari 2h52'8" 30d28'27"
Frances Terzano-Born:
Dec. 23, 1914
Lyr 19h19'25" 29d1'49"
Frances Tsolinas
And 0h42'2" 40d33'8"
Frances Veitinger
Cyg 20h22'41" 36d32'18"
Frances Virginia Click Mise
Lib 15h23'57" -8d29'14"
Frances W. Boyd
Sco 17h13'36" -31d38'25"
Frances Walsh
Crb 16h11'28" 34d8'12"
Frances Wolowiec
Lyr 18h39'12" 30d57'20"
Frances Y. Morrow
Cmi 7h40'19" 5d28'43"
Frances Z. Hedges
Vir 13h13'34" -21d37'8"
Frances, Will You Marry
Me ?
Cyg 20h26'52" 58d12'53"
Frances-22348
Uma 9h14'21" 58d25'45"
Francesa Kathleen Weaver
Psc 0h34'48" 14d8'31"
Francesca
Mon 6h31'2" 7d7'5"
Francesca
Boo 14h52'25" 14d21'0"
Francesca
Peg 22h3'43" 27d14'25"
Francesca
Crb 16h24'57" 28d35'9"
Francesca
Ari 3h17'52" 24d11'3"
Francesca
And 0h27'35" 32d4'30"
Francesca
And 0h36'17" 41d3'11"
Francesca
And 0h27'45" 42d52'16"
Francesca
And 1h34'27" 41d20'46"
Francesca
Lyn 8h42'22" 43d24'17"
Francesca
And 23h44'36" 43d1'7"
Francesca
And 0h18'5" 45d52'40"
Francesca
Crb 15h58'1" 38d9'29"
Francesca
Cam 4h35'29" 58d13'55"
Francesca
And 23h34'14" 49d0'12"
Francesca
And 22h59'14" 50d58'12"
Francesca
Uma 9h49'49" 49d50'56"
Francesca
Cas 23h10'48" 55d29'33"
Francesca
Uma 9h55'6" 69d27'31"
Francesca
Umi 13h15'37" 73d28'16"
Francesca
Cma 7h10'27" -12d25'24"
Francesca
Umi 15h2'52" 76d51'9"
Francesca
Sgr 18h37'3" -27d5'9"
Francesca
Umi 16h58'59" 83d20'56"
Francesca Amabel
Sco 17h37'37" -35d1'43"
Francesca Amelia Evers -
my star
Lyn 7h33'36" 35d52'46"
Francesca and John
Sgr 20h17'21" -35d20'19"
Francesca and Mia
Cas 1h1'26" 50d32'46"
Francesca Andranovich
And 0h49'38" 39d50'37"
Francesca Ann Wang
And 0h54'53" 41d0'41"
Francesca Anna
Cas 1h17'4" 69d23'43"
Francesca Anna Smith
And 23h16'20" 37d45'46"
Francesca Antonelli
Lyr 18h21'44" 45d10'35"
Francesca Antonelli
Uma 11h55'1" 28d58'32"
Francesca Arcangeli
Cam 3h49'22" 58d31'5"
Francesca Arcuri
Cyg 21h7'5" 36d41'23"
Francesca Arduini
Per 3h24'10" 41d39'1"
Francesca Audrey Augelli
Uma 10h25'22" 40d4'37"
Francesca Bacco
Aur 6h31'54" 47d30'32"
Francesca Bimba Zirafa
Leo 9h38'17" 13d3'6"
Francesca Bucci
Ori 5h6'30" 11d26'36"
Francesca C. Cook
Peg 22h11'36" 7d21'4"
Francesca Carmella
Perrotti
Lyn 6h53'19" 51d12'57"
Francesca Castorina
And 0h41'28" 39d27'44"
Francesca Chiara Sbuttoni
Cyg 21h41'25" 47d29'57"
Francesca D'Albore
And 1h44'1" 42d21'49"
Francesca Dansereau
Cap 20h46'21" -21d30'43"
Francesca D'Ascoli
Uma 9h32'15" 53d51'6"
Francesca De Luca
Lyr 18h26'36" 34d6'42"
Francesca D'Elia
Leo 11h56'42" 21d44'0"
Francesca Diaz
And 2h28'58" 48d15'17"
Francesca e Andrea
And 2h20'36" 41d43'59"
Francesca e Paolo 3°
anniversario di matrimonio
Her 17h54'21" 25d6'42"
Francesca Elena
And 1h7'12" 34d21'58"
Francesca Ellie
And 2h32'52" 40d47'33"
Francesca Elyse Shirk
Sco 16h51'27" -38d56'19"
Francesca "Frenchy" M.
Rodriguez
Sgr 18h35'22" -20d14'17"
Francesca Graham
And 23h22'19" 51d18'9"
Francesca Grassi
Sgr 18h6'57" -29d7'26"
Francesca Hope Villanueva
Uma 11h19'26" 46d16'39"
Francesca Irene Deanna
White
Umi 14h31'14" 78d0'48"
Francesca Kardos
Ori 6h15'59" 15d31'46"
Francesca Leo
Uma 11h59'15" 34d34'42"
Francesca Leonardy
Mon 6h50'48" 9d47'44"
Francesca Louise
Sco 16h41'32" -41d24'21"
Francesca & Luigi
Tau 5h50'24" 14d40'52"
Francesca - Luna
Her 16h40'33" 6d51'12"
Francesca M. Arminio
Tau 3h47'53" 29d5'43"
Francesca Marfella
Cas 23h7'7" 59d16'43"
Francesca Maria Weeks
Leon - "Franceskita"
And 23h20'58" 52d6'18"
Francesca Marie
Cas 1h34'2" 67d5'15"
Francesca Marie Buzzanca
And 0h34'18" 30d21'37"
Francesca Marinella
Per 3h23'2" 49d24'37"
Francesca Marzo
Cyg 21h41'51" 45d59'2"
Francesca Mia Munday
And 1h17'38" 43d24'9"
Francesca "Mumbles"
Zappitelli
Sgr 19h46'34" -21d31'7"
Francesca Nepa
Lib 15h16'26" -5d2'25"
Francesca Nicole Lewis
Lyn 7h47'32" 38d39'18"
Francesca Noelle
Sgr 18h35'19" -28d17'59"
Francesca Norris Veitch
And 0h14'41" 44d55'26"
Francesca Olivia Louise
Versey - "Pud"
And 0h22'50" 46d26'55"
Francesca Onnis
Crb 15h25'8" 30d26'14"
Francesca Pizzuto
Peg 21h26'28" 21d27'20"
Francesca Polidetti
Cas 2h44'30" 63d40'12"
Francesca Pollara-Parsons
Psc 1h27'18" 27d10'43"
Francesca Remick
Cas 23h54'38" 53d12'35"
Francesca Romana
Per 4h14'0" 33d5'32"
Francesca Romana
Cas 1h21'20" 54d37'23"
Francesca Rose
Uma 11h27'54" 51d38'40"
Francesca Rose Bourquin
Cyg 19h36'3" 54d23'45"
Francesca Rose Trizano
Dra 16h22'2" 65d53'13"
Francesca Ryan
Cnc 8h15'20" 29d26'6"
Francesca Salczynski
And 23h18'5" 42d59'49"
Francesca. Salvino.
Aqr 21h59'8" -14d55'41"
Francesca Salvo
Cas 1h51'2" 61d30'44"
Francesca Sfrondato
Uma 11h3'0" 47d32'50"
Francesca Sibilla Alberto
Per 2h24'8" 51d40'37"
Francesca Sibra
Uma 8h59'59" 51d56'1"
Francesca Sinocruz
Leo 10h51'19" 15d48'29"
Francesca Sommerstein
Uma 8h34'0" 68d38'38"
Francesca & Tommy
Peg 22h45'3" 33d51'1"
Francesca Treccani
And 0h16'54" 40d22'5"
Francesca Trivigno
Dra 16h15'8" 61d48'23"
Francesca Ventimiglia
Aqr 21h16'34" -6d19'11"
Francesca Y. Lien
Cas 1h23'50" 60d56'41"
Francesca Ruini
Aqr 23h36'55" -18d19'51"
FrancescaRMPappagalloSi
clturAdAstra
Uma 11h26'39" 45d20'47"
Francesca's Hope
And 2h15'20" 50d22'26"
Francesca's Star
And 2h11'16" 50d17'51"
Francesca's star
Her 18h36'52" 19d16'43"
Francesco
Peg 23h42'48" 29d38'50"
Francesco
And 23h34'9" 40d40'49"
Francesco
Lac 22h30'51" 42d45'49"
Francesco
And 23h39'32" 37d24'47"
Francesco
Cyg 20h13'38" 36d10'10"
Francesco
Cyg 20h7'55" 35d50'11"
Francesco
Lac 22h45'36" 52d36'33"
Francesco
Uma 13h57'26" 55d37'56"
Francesco
Cas 1h55'47" 64d24'37"
Francesco
Uma 10h56'15" 72d35'15"
Francesco and Micheal
Pizzarello
Aqr 22h17'51" -3d21'43"
Francesco Andrea Roth
Ori 5h7'2" 7d53'56"
Francesco Biava
Aur 5h25'50" 49d22'10"
Francesco Calimazzo
"Grandpa Frank"
Leo 11h12'9" 4d1'48"
Francesco Cantatore
Cyg 20h58'56" 40d14'20"
Francesco Carli
Umi 14h11'48" 65d52'51"
Francesco d'Angelo
Her 17h30'26" 41d43'35"
Francesco De Luca
Aur 6h38'36" 31d29'57"
Francesco Di Caro
Her 18h9'30" 18d9'40"
Francesco Dorascenzi
Uma 12h25'9" 60d25'15"
Francesco e Marta
Uma 8h47'35" 71d17'52"
Francesco Elia Ciampi
Cas 23h46'12" 55d42'3"
Francesco Esposito
Umi 15h30'19" 72d54'54"
Francesco Ficicchia
Uma 10h3'48" 54d35'7"
Francesco Fusco
Ari 3h17'5" 27d36'26"
Francesco Gargaro
Peg 22h34'31" 19d36'3"
Francesco GUIDO
Peg 22h50'26" 2d51'2"
Francesco Ingenito
Cam 5h38'28" 59d12'57"
Francesco Lettieri
Lyn 9h24'46" 39d48'34"
Francesco Magni
Cep 21h2'10" 59d36'25"
FRANCESCO MANCO
Cas 23h56'39" 54d19'24"
Francesco Morabito
Cnc 8h51'57" 31d18'23"
Francesco Nardacchione
Uma 11h34'27" 39d14'9"
Francesco Oddenino
Ori 6h10'35" -2d10'46"
Francesco Pallino
Cyg 20h43'49" 32d46'56"
Francesco Sangiovanni
And 2h33'52" 39d56'6"
Francesco Tarquini
Cam 5h22'59" 60d27'47"
Francesco Visentin
Ori 4h47'45" 5d25'21"
Francesco, 20.07.1974
Cam 5h12'18" 79d22'55"
Francesco61
Ori 5h21'41" -0d46'11"
Franceska & Jr.
Her 18h3'55" 17d29'46"
Franchesca
Ari 2h54'55" 25d7'29"
Franchesca
Tau 4h49'50" 24d49'19"
Franchesca Naiomi Tapia
Sgr 19h54'8" -41d55'19"
Francheska
Dra 16h34'43" 60d25'34"
Francheska Estelle Medina
Cas 1h27'2" 63d2'23"
Francia Habib
Tau 5h17'18" 18d7'23"
Francie McDonald
And 23h13'54" 42d10'27"
Francie Silverstein
Lmi 10h42'35" 31d8'54"
Francien&Francisco
Her 16h27'31" 25d16'19"
Francienia
Mon 6h49'49" -0d10'17"
Francina
Cam 6h19'47" 63d37'21"
Francina Leong - We Love
You Mum
Col 5h47'49" -34d21'57"
Francine
Crb 15h46'4" 28d34'42"
Francine
Tau 3h35'2" 21d48'38"
Francine
Ori 6h12'50" 7d4'58"
Francine
And 0h31'3" 42d51'1"
Francine Battaglia
And 1h10'18" 41d52'25"
Francine & Charles Lask
Leo 11h22'33" 22d43'23"
Francine Chartier / Stavibel
Aqr 23h39'52" -16d57'57"
Francine Dancer
Lyr 18h49'9" 37d22'52"
Francine Denise Brunette
And 1h41'31" 49d45'18"
Francine Filippa Curran
Lib 14h52'47" -1d16'15"
Francine Gabriella
Uma 9h42'57" 62d27'15"
Francine Giglio
Aql 19h30'51" 11d28'23"
Francine Ina Mirochnik
Klein
Gem 6h21'26" 18d22'23"
Francine LaPorte
Psc 0h40'56" 6d33'12"
Francine Lizotte - 14 juillet
1947 - 2007
Uma 10h59'54" 48d32'52"
Francine Marie
Gem 7h8'45" 32d34'12"
Francine R Coplon
Uma 10h34'13" 63d20'43"
Francine Winifred Horvath
Ari 3h15'56" 29d0'3"
Francine's Celestial
Diamond
And 1h3'42" 38d57'14"
Francis
Gem 7h18'21" 16d12'43"
Francis
Sco 16h17'20" -11d8'0"
Francis A Eley
Per 3h24'30" 52d27'43"
Francis and Delrose
And 23h11'10" 43d57'3"
Francis and Lauren
And 2h17'26" 46d21'20"
Francis and Williams
Family
Her 15h51'55" 39d44'25"
Francis B. Keenan
Her 16h58'9" 17d45'19"
Francis Belen
Lyn 6h27'11" 54d8'56"
Francis Berdah
Gem 6h55'55" 16d58'12"
Francis Bouron, mon papa
Her 18h46'50" 23d30'45"
Francis Bridges - Mom's
STAR
Uma 11h25'19" 58d34'17"
Francis Campbell Brown
Uma 8h21'3" 60d8'28"
Francis Conlan - Forever
Shining
Cra 18h6'15" -45d24'55"
Francis D. Dolan
Aqr 22h50'50" -7d4'58"
Francis Daniel Donovan
Uma 9h19'9" 45d31'4"
Francis Daniel Powers
Leo 9h44'16" 31d5'15"
Francis De Los Santos
Umi 17h22'25" 84d6'25"
Francis Donadio
Vir 12h22'5" 10d59'33"
Francis Dumont
Tau 5h46'56" 27d55'22"
Francis E. O'Brien Jr
Sco 16h37'19" -27d16'18"
Francis Edward Condon,
Ph. D.
Per 4h46'38" 40d43'10"
Francis & Eileen 10292005
Cyg 20h24'49" 58d26'51"
Francis Enda Fallon
Cas 1h37'8" 59d13'2"
Francis Fales
Aqr 23h24'10" -7d33'10"
Francis Fanning
And 2h3'56" 37d57'17"
Francis "Fran"
Uma 11h8'54" 61d54'41"
Francis -Frank- A. Wolf
Sgr 19h46'44" -14d43'15"
Francis "Frank" Rocco
Santoro
Leo 9h27'3" 16d18'54"
Francis(Franny)
Lyn 7h4'31" 58d32'7"
Francis G. Gusty
Per 4h30'0" 51d7'54"
Francis George Edward
Baxter
Aqr 21h13'42" 0d7'20"
Francis Gnat
Psc 0h58'48" 32d32'14"
Francis Hammond Ball
Ari 3h18'21" 29d52'0"
Francis Hawthorne
Her 18h15'55" 23d49'42"
Francis Henry Dillon V
Leo 10h14'18" 24d36'58"
Francis Honrado Espinal
Aql 19h26'49" 9d20'34"
Francis J. Imhoff
Lib 14h54'20" -15d37'15"
Francis James King
"10/17/1978"
Lib 14h56'20" -16d21'51"
Francis John McCreith-
Shields
Umi 14h32'59" 84d31'2"
Francis Johnson
Uma 11h0'14" 42d33'53"
Francis June Cordaway
Leo 11h31'5" 5d28'5"
Francis Kamara
Cyg 21h49'45" 42d45'48"
Francis Kenneth Boner
Aqr 22h30'11" -24d47'31"
Francis L. Confer
Cep 22h33'39" 73d0'26"
Francis L. Schubkegel
Uma 11h37'36" 35d19'16"
Francis Lalanne
Leo 10h25'39" 16d21'54"
Francis M Tenney, Jr.
Her 18h31'13" 16d14'18"
Francis & Margaret Jordan
Uma 10h27'51" 66d31'34"
Francis Margaret
Warburton
Lyr 18h46'57" 30d49'55"
Francis Michael Reoch
Lib 15h30'33" -18d24'8"
Francis Mose
Cnc 8h31'24" 17d9'14"
Francis N Pellegrino
Tau 4h30'29" 22d33'17"
Francis N. Simms
Sgr 18h21'52" -17d49'38"
Francis Nicholas Cordopatri
Per 2h17'3" 54d46'26"
Francis Omar Melendez
Sanchez
Per 3h45'55" 49d20'48"
Francis P. Clark
Per 3h23'25" 44d28'5"
Francis Palladino
Uma 11h3'30" 45d1'28"
Francis Patrick Staab
Psc 1h31'39" 27d30'24"
Francis Patrick Trimble
Uma 10h5'23" 58d49'23"
Francis Paul and Ines Anna
Bottone
Cyg 20h43'56" 45d51'25"
Francis Paul Bibbo
Cnc 8h55'33" 7d37'53"
Francis Pearce Kelley
Psc 1h52'47" 9d27'52"
Francis R. Doorakian
Aql 19h34'56" 13d10'4"
Francis R. Harvey
Ari 2h21'7" 24d56'38"
Francis R. Leonard
Uma 12h41'20" 56d51'3"
Francis Raymond
Summers
Cyg 19h36'26" 29d2'6"
Francis Richard Janek IV
Vir 14h45'14" -0d20'18"
Francis Robert Burns
Uma 11h11'22" 52d53'27"
Francis Rosemarie Regan
Vir 15h6'3" 4d4'36"
Francis Santini Mancuso
Sco 17h20'31" -33d8'9"
Francis Smyth
Cep 21h38'11" 57d32'51"
Francis' Star
Dra 18h53'4" 52d44'54"
Francis Starnes
Uma 8h25'44" 65d8'31"
Francis T. Rogers, Jr.
Cyg 19h46'27" 30d1'8"
Francis V. Tursi Jr.
Psc 0h56'6" 29d36'2"
Francis Walker Conway
Mon 6h51'36" -0d3'32"
Francis William Barton
Per 4h32'10" 49d44'35"
Francis William Soderholm
3/14/1924
Psc 0h38'53" 3d3'11"
Francis Wilson
Dra 16h28'38" 66d20'39"
Francis X. Burkhardt
Ori 5h41'28" 0d49'13"
Francis X. Lee
Cep 22h43'59" 67d7'4"
Francis Xavier Miller
Uma 9h39'21" 46d22'47"
Francis Yssel
Cen 13h53'56" -46d41'57"
Francis, Cyril, John,
Connolly
Ori 6h12'26" 5d40'37"
FRANCISCA
Cnc 8h32'58" 24d50'3"
Francisca Altagracia Rivas
Crb 15h40'1" 26d31'57"
Francisca Lentisco
Cas 23h32'47" 51d42'22"
Francisca Norma Fletcher
"Pecosit"
Nor 16h5'15" -43d51'29"
Francisca Onelia Torres
Cru 12h44'55" -57d22'39"
Francisca Perfeito
Cmi 7h26'18" 1d53'59"
Francisca Silvia Magana
Sco 16h16'34" -10d41'46"
Francisco
Ori 5h38'38" 1d14'52"
Francisco
Psc 1h38'0" 14d30'9"
Francisco
Leo 11h36'51" 18d54'16"
Francisco A. Martinez 2000
Quanah
Srp 17h58'5" -0d14'25"
Francisco Carrillo
Vir 13h38'45" 5d0'36"
Francisco Castro, Jr.
Vir 13h30'50" 1d21'52"
Francisco de Assis Alves
Teixeira
Scl 0h50'59" -27d15'12"
Francisco de la red
Uma 8h57'8" 70d55'57"
Francisco & Elena
Her 17h37'51" 23d14'0"
Francisco Enriquez
Gem 6h56'44" 14d21'57"
Francisco Figueroa
Vir 14h8'9" -7d49'26"
Francisco & Holly Velez
Uma 9h40'28" 56d29'17"
Francisco Javier Morales
Cyg 19h46'56" 45d59'18"
Francisco Javier Ortiz
Godoy
Her 17h23'11" 32d18'37"
Francisco Javier Torres
Abadia
Aqr 21h48'54" -0d45'22"
Francisco Jimenez
Uma 9h33'52" 62d57'18"
Francisco L. Orozco
Cnc 8h9'32" 12d19'38"
Francisco Leggio
Cep 21h36'31" 65d49'52"
Francisco M Serrano
Uma 11h19'48" 50d39'21"
Francisco Nuno Leal Avelar
Dias
Peg 22h37'11" 24d12'51"
Francisco Oscar Perez
Leo 10h11'7" 13d53'32"
Francisco Panchito
Cisneros
Vir 13h20'27" 12d7'27"
Francisco Peniche
Pup 7h47'23" -23d39'30"
Francisco Rosario
Uma 10h46'4" 66d19'50"
Francisco Torres
Uma 8h40'3" 61d0'36"
Francisco Vitaver
Ori 5h38'12" 0d14'52"
Franciska
Gem 7h12'22" 33d16'12"
Franck Bailly
Psa 22h1'35" -26d45'25"
Franck Batta
Cyg 21h54'2" 44d33'10"
Franck Bergès
Gem 6h52'28" 25d4'51"
Franck Bourcereau
Ari 2h19'57" 12d43'4"
Franck Brun-Picard
Lyn 7h31'26" 39d12'41"
Franck et Corinne Boscato
Her 18h3'48" 24d40'56"
Franck Fiori
Uma 13h16'47" 58d18'22"
Franck Graziosi
Ori 6h17'3" 13d42'10"
Franck - Karine pour la vie
Crb 16h1'31" 32d11'20"
Franck & Katia
Per 1h49'56" 51d26'35"
Franck "Kiki" Danton
Sgr 20h2'52" -28d31'12"
Franck Mozzanini Eternel
Amour
Col 5h30'18" -35d15'54"
Francky
Cam 4h56'45" 59d33'27"
Franco
Per 2h44'6" 50d5'39"
Franco
Ori 6h18'42" 11d14'17"
Franco
And 0h2'26" 37d33'0"
Franco Acquaviva
Per 3h30'7" 52d12'1"
Franco Cavalli
Per 4h8'34" 51d32'19"
Franco Del Picchia
And 2h23'55" 43d57'24"
Franco e Rita - Mapà
Uma 12h3'25" 30d53'53"
Franco & Erin Barresi
Uma 8h46'33" 49d16'19"
Franco Fabio Immediata
Gem 6h52'0" 22d30'10"
Franco Forghieri
Uma 10h46'1" 64d25'27"
Franco Francullo
Tau 3h47'15" 28d37'45"
Franco Gabelli
Tau 3h35'30" 18d23'45"
Franco Giulio Cangelosi
Leo 11h52'31" 25d31'22"

Franco Monia Lorenzo Damiano
Lyr 18h39'39" 30d27'4"
Franco Nuschese
Dra 15h26'48" 59d12'11"
Franco Oriol "1926-2001"
Cnc 8h30'51" 12d47'38"
Franco & Phyllis Di Domenica
Ari 2h5'18" 19d47'6"
Franco r.m.
Uma 10h37'13" 62d16'17"
Franco Saglio
Cap 21h37'51" -14d53'23"
François Aurora
Cas 0h5'4" 54d4'43"
François Bernardini
Uma 9h46'39" 65d58'36"
François Boucher
Her 16h10'8" 24d55'0"
François Boudreau, amour de ma vie
Cep 23h39'25" 84d42'38"
François Cornet
Eri 4h56'49" -4d53'48"
François Dall'Asta
Cam 5h47'16" 61d13'32"
Francois DeForges
Aql 19h38'55" 15d1'23"
François Donnee
Aqr 21h1'41" -9d43'56"
Francois du Plessis
Psc 1h31'57" 13d19'50"
François et Huguette
Dra 20h10'33" 62d18'38"
François Gallays
Cep 4h58'12" 82d3'44"
François Laroche
Her 15h54'51" 40d3'4"
François Pineda
Sco 17h54'28" -36d48'18"
François Regnier
Ori 5h50'28" 11d33'32"
Francois Walker
Uma 11h53'31" 55d49'45"
Françoise
And 1h35'53" 47d44'42"
Françoise
Cyg 19h30'20" 28d55'42"
Françoise
Per 2h54'4" 46d42'17"
Françoise Anne - Marie Ponthus Gemmayel
Her 17h7'18" 13d52'29"
Françoise Burgarella
Cas 23h43'32" 57d7'59"
Françoise Coeur
Tau 3h47'33" 22d45'41"
Françoise Dähler-Martin
Sco 17h20'14" -41d23'26"
Françoise et André Rolando
Peg 22h15'2" 23d36'1"
Françoise Faye Ronchetto
Uma 10h13'21" 44d21'0"
Françoise Luna Rives
Uma 12h48'31" 60d39'23"
Françoise Mi Amas Tin
Uma 13h1'28" 57d54'59"
Françoise mon coeur
Ari 2h4'52" 22d37'49"
Francoise MyLove, MyDream Come True
Aqr 21h19'57" -9d54'52"
Françoise Zephir
And 23h53'6" 47d19'25"
Francy
Lyn 8h30'18" 57d7'57"
Francy e William
Aur 5h33'12" 32d57'11"
Francy&Taty
Lyr 19h26'58" 40d29'52"
Francymia
Cyg 20h9'12" 33d7'16"
Franette
Dra 19h23'43" 69d19'38"
Frank
Uma 10h57'14" 72d37'33"
Frank
Dra 20h25'22" 62d5'26"
Frank
Cep 21h21'4" 60d43'58"
Frank
Cap 20h49'59" -19d37'55"
Frank
Sgr 20h11'30" -40d15'35"
Frank
Per 4h21'19" 50d38'1"
Frank
Uma 9h38'2" 50d20'54"
Frank
Cnc 9h19'59" 11d19'57"
Frank A. Apesa, Jr. (Cheech)
And 1h54'19" 35d59'25"
Frank A. Catalano
Ari 3h15'46" 28d12'55"
Frank A. Ferraina
Per 4h49'10" 43d50'28"
Frank A. Fontana Sr.
Peg 22h33'26" 32d28'19"
Frank A. McGoveran's Star
Cnc 8h40'10" 10d7'49"
Frank A. Pazakis, D.D.S.
Her 18h20'13" 22d42'19"
Frank A. Williams
Uma 11h30'13" 32d32'31"
Frank A. Zaba
Aqr 23h50'29" -17d11'16"
Frank Albert Carque
Per 3h2'53" 36d17'15"
Frank Alfred Croce III
Per 2h42'35" 57d4'1"
Frank Allegresso
Cep 22h7'48" 56d24'31"
Frank Allen Richards
Uma 13h53'1" 53d45'42"
Frank Allerton Summers
Dra 18h37'0" 59d34'52"
Frank Aloisio
Uma 11h48'49" 36d20'37"
Frank Aloy LeMire
Cyg 20h21'26" 44d44'13"
Frank and Agnes Gertz
Cyg 19h45'20" 51d41'21"
Frank and Ava Forever Love
Uma 11h36'43" 52d16'47"
Frank and Bette Beeson
Cru 12h19'51" -57d47'3"
Frank and Elsie Forever
Per 3h25'41" 40d38'13"
Frank and Florence Weinhold
Cyg 20h46'52" 33d4'22"
Frank and Grace Gardner
Per 2h24'48" 51d36'1"
Frank and Joan
Dra 17h47'57" 66d15'51"
Frank and Kelly
Ari 2h51'7" 14d16'1"
Frank and Mari White
Lib 14h27'40" -18d32'31"
Frank and Maria Domé
Eri 3h47'17" -0d1'28"
Frank and Marjorie Silva
Tau 5h55'57" 24d47'35"
Frank and Martha Ratto
Cyg 20h37'24" 47d41'29"
Frank and Mary Record
Aqr 21h21'38" -9d52'45"
Frank and Sally Trovato
Lyn 9h9'57" 34d22'49"
Frank and Steff Elliott
Cyg 20h9'51" 30d57'30"
Frank and Teresa Mitzel
Cru 12h40'47" -56d57'5"
Frank and Vesta Fey
Uma 11h38'27" 58d26'34"
Frank & Angie Del Monte
Cyg 20h39'35" 44d53'54"
Frank & Anne Marie's Silver Star <3
Cyg 19h36'56" 35d11'30"
Frank Anthony Flores
Psc 23h10'43" 0d9'27"
Frank Anthony Maribito
Dra 18h47'30" 71d2'14"
Frank Antoni Zychon
Per 3h12'19" 46d53'51"
Frank Arnold
Sgr 18h26'31" -17d34'16"
Frank Arnold Poulton
Uma 8h24'53" 63d20'15"
Frank Arthur Blackmore (FAB)
Cyg 20h49'45" 47d50'59"
Frank at the Flood Zone
Lib 15h30'27" -23d37'8"
Frank Atkinson
Umi 10h44'13" 88d29'31"
Frank B. Giorgianni, Jr.
Cas 0h16'32" 53d5'5"
Frank B Jensen
Psc 0h55'34" 31d36'0"
Frank B. Mayer
And 0h52'37" 36d16'56"
Frank Baade
Lib 15h19'45" -5d34'31"
Frank Ballatore
Tau 3h33'44" 1d5'21"
Frank Barilla
Cap 21h12'19" -27d21'3"
Frank Bart Carpe Diem
Vir 13h26'7" 12d40'2"
Frank Bartel
Vir 13h4'13" -20d22'10"
Frank Bärwald
Ori 5h58'4" 11d17'23"
Frank Bastasch
Aur 5h26'13" 39d23'5"
Frank Belsar
Lib 14h36'12" -18d9'29"
Frank Benning
Ori 5h43'50" -2d22'47"
Frank Beville
Uma 11h27'35" 52d53'27"
Frank Biear
Per 2h10'59" 56d17'44"
Frank Bielec - Star of our Hearts!
Lyn 7h50'18" 45d36'19"
Frank Biribauer
Sgr 18h55'38" -19d33'35"
Frank Black
Lyn 8h3'55" 42d13'36"
Frank Blakely
Ori 5h17'56" -6d25'10"
Frank Boettcher
Per 3h25'0" 48d53'2"
Frank Bova
Psc 0h21'20" 4d10'30"
Frank Bradburn Jr.
Leo 11h40'32" 26d19'12"
Frank Brendan Leonard
Psc 1h29'34" 23d28'9"
Frank Brotz
Her 18h26'21" 13d55'24"
Frank Brundage Reynolds, III
Leo 9h26'19" 10d15'5"
Frank Brunke
Uma 8h40'9" 56d54'51"
Frank Bubba Clark
Leo 10h52'39" 5d2'5"
Frank Buck
Ori 5h42'18" -4d24'10"
Frank Budzinsky
Boo 14h39'50" 17d33'44"
Frank Burton Wright
Gem 7h5'39" 26d37'59"
Frank C. Benhatzel
Vir 14h29'33" 7d2'16"
Frank Cameron Fischer
Cep 22h22'34" 62d10'12"
Frank Candito
Uma 11h49'33" 39d47'2"
Frank & Carolyn Lumia
Uma 8h42'37" 59d28'38"
Frank Cedrone
Uma 9h22'51" 45d32'21"
Frank & Charla Malaney
Sge 20h7'13" 20d26'40"
Frank Charles Gallo
Aql 19h50'49" 8d41'27"
Frank Chauteau
Uma 11h19'29" 42d38'1"
Frank Chidsey
Vir 12h12'56" 11d23'13"
Frank Christopher Castania
And 0h46'36" 38d9'10"
Frank Ciko, Jr.
Aqr 21h50'23" -7d25'23"
Frank Cleland
Her 18h17'57" 23d23'14"
Frank Conte
Uma 10h36'3" 53d17'38"
Frank Coto III
Cap 21h44'26" -17d38'25"
Frank Coulthard
Lmi 10h57'33" 29d19'25"
Frank Crispgina III
Ari 3h18'13" 29d4'37"
Frank Crowell
Cep 22h12'23" 61d31'53"
Frank D. Lawson
Gem 6h49'5" 24d4'58"
Frank D. Lemaster II
Aql 18h56'17" -0d48'13"
Frank D'Amato
Uma 11h21'7" 52d54'16"
Frank Dambro
Pho 23h49'19" -45d40'6"
Frank Daniels
Her 17h27'48" 37d22'21"
Frank & Dawn
Cyg 21h29'4" 47d39'11"
Frank & Debbie Kessler
Uma 10h43'34" 43d3'52"
Frank DeMarco
Ari 3h5'23" 19d51'29"
Frank DeRosa
Tau 5h39'18" 25d48'17"
Frank & Dianne Are So In Love
Per 4h14'26" 37d20'29"
Frank Dominic Dulcich
Tau 3h37'12" 21d28'6"
Frank & Donna Sinistore
Umi 14h37'44" 72d34'43"
Frank Douglas Winters
Boo 14h19'34" 48d36'12"
Frank Douthit
Per 3h12'51" 42d48'5"
Frank Drabwell - Drabby's Star
Uma 12h4'33" 44d17'1"
Frank E and Doris L Cook
Cyg 20h17'42" 54d42'17"
Frank E. Flowers, Sr.
Lyn 9h11'23" 34d12'13"
Frank E Rodgers Sr
Her 16h48'17" 35d27'3"
Frank Edward Harris
Cru 12h1'21" -59d53'5"
Frank Edward Mejia
Cnc 8h6'48" 26d42'51"
Frank Edward Swansfeger III
Aqr 20h40'22" -11d44'9"
Frank & Elaine Jacobsen
Gem 7h17'16" 13d49'47"
Frank & Elaine's Star
Ori 5h27'45" 3d55'34"
Frank Esposito
And 1h17'2" 49d41'52"
Frank Eugene DeCoursey
Gem 7h3'22" 32d27'26"
Frank F. DeLise
Lib 15h52'20" -12d25'46"
Frank Falsetti
Lyn 7h37'58" 35d38'19"
Frank & Fanny
Aur 5h41'46" 52d25'1"
Frank Farr
Ori 5h7'40" 15d21'3"
Frank Fertitta III
Boo 14h35'39" 41d52'52"
Frank Fertitta, Jr.
Sco 17h58'1" -38d8'21"
Frank Fisher
Lac 22h23'50" 47d24'17"
Frank & Frederica Sackett Forever
Cyg 20h21'4" 43d13'58"
Frank Friel
Aql 19h0'21" -8d41'37"
Frank "Frumbie" Milanese
Per 4h17'51" 42d4'55"
Frank George
Uma 11h3'10" 56d57'36"
Frank Gianakos
Sgr 18h2'43" -27d52'1"
Frank Gilbert DeNauw
Vir 14h7'16" -11d14'37"
Frank Gill
Uma 11h46'8" 48d25'0"
Frank & Gill For Always
Cyg 19h31'53" 44d45'59"
Frank Giordano
Her 16h16'5" 17d36'21"
Frank "GrandPa" McLean
Leo 11h15'28" 19d14'55"
Frank "Grandpa" Sapienza
Gem 7h45'23" 17d46'12"
Frank Grundig
Her 16h34'49" 47d5'7"
Frank Guzzetta
Cep 20h29'10" 61d36'19"
Frank H. Crandall III
Uma 12h47'49" 54d47'34"
Frank H. & Doris A. McCafferty
Cnc 9h6'8" 31d22'38"
Frank H. O'Neal
Sgr 19h13'17" -36d41'56"
Frank H. Orlando
And 0h28'18" 34d43'22"
Frank Hamilton Cain
Sco 16h39'50" -37d5'55"
Frank Hartmann
Ori 6h15'28" 10d55'56"
Frank Heiner
Cnc 8h10'47" 20d46'8"
Frank & Helen Caprara
Peg 23h13'43" 32d9'37"
Frank & Helen's 50th Anniversary
Uma 11h54'49" 64d54'30"
Frank Higgins 1929-2005
Uma 12h15'16" 61d58'18"
Frank Horvath
Leo 10h50'28" 8d6'15"
Frank Host
Lyn 7h59'57" 37d26'35"
Frank Hubbard Little
Ori 5h29'56" 3d37'23"
Frank Inman Wilson III
And 1h3'6" 46d13'33"
Frank J. Casserino
Cnc 9h3'59" 28d28'10"
Frank J. Daily
Per 3h10'40" 51d38'1"
Frank J. Heger Our Angel in the Sky
Uma 11h54'30" 30d29'19"
Frank J. Jannotta Jr.
Per 2h18'13" 56d56'10"
Frank J. Keller
Psc 1h28'1" 4d4'7"
Frank J. Luton
Psc 1h8'5" 9d45'0"
Frank J. Regula
Sgr 18h18'28" -31d54'51"
Frank J. Riscili
Uma 14h8'37" 59d13'25"
Frank J. Wilson
Aur 6h54'25" 45d5'48"
Frank James Birch
Uma 10h56'40" 62d17'9"
Frank James Price II
Lib 15h2'20" -11d10'3"
Frank Janisse
Lyr 19h20'9" 29d34'13"
Frank & Jean Salantri Forever Love
Cyg 19h41'30" 54d17'37"
Frank & Jenny Quinn
Uma 11h27'3" 58d27'45"
Frank & Joanne Bednarofsky
And 2h30'23" 48d49'29"
Frank Joseph Byers
Cyg 19h55'17" 36d17'4"
Frank Joseph Celio
Her 17h53'39" 23d24'9"
Frank Joseph Delear
Ori 5h39'31" 3d4'59"
Frank Joseph John Dvorak III
Cma 7h0'36" -22d52'2"
Frank Joseph Losh
Psc 22h58'31" -0d0'34"
Frank Joseph Lyons, Jr.
Leo 9h40'29" 27d36'14"
Frank Joseph Rodriquez Jr.
Ori 5h28'28" 2d5'13"
Frank Joseph Stagnaro
Cep 21h45'59" 64d8'13"
Frank Jospeph Knaps, Jr.
Aqr 22h38'27" -1d23'10"
Frank K. Hardin
Cnc 8h43'56" 8d42'17"
Frank Kasmir
Gem 7h11'37" 16d3'21"
Frank & Katie
Vir 13h2'10" 11d4'45"
Frank & Katja
Uma 9h53'57" 45d38'14"
Frank Kellerhals - Maya Hardware
Ori 5h36'21" -0d56'58"
Frank & Kelli Stubbs
Cyg 20h36'51" 56d15'11"
Frank Kempe
Uma 8h21'54" 67d23'3"
Frank & Kirsten
Uma 10h39'30" 58d26'33"
Frank Kocur "Kokey"
Boo 14h47'5" 51d49'37"
Frank Kranich
Uma 9h12'8" 55d13'54"
Frank & Kristy
Sge 19h41'35" 19d12'42"
Frank L. Faulkner
Ori 4h51'20" 4d3'37"
Frank L. Harp
Uma 14h8'57" 54d54'35"
Frank L. Hijek
Cep 23h7'38" 87d34'10"
Frank L. Pepe
Cep 22h35'53" 74d7'31"
Frank Lang
Uma 13h3'41" 53d19'35"
Frank LaVern Bernard
Tau 3h46'29" 29d14'55"
Frank Leo
Uma 12h46'36" 57d36'35"
Frank Leonard Hunt
Gem 6h40'51" 13d8'57"
Frank Leonardis, Sr.
Ori 5h27'5" 5d8'39"
Frank Lescher
Ori 5h58'19" 12d9'23"
Frank Lewis
Uma 13h53'53" 48d14'1"
Frank & Linda Rosano
Cyg 20h17'53" 56d4'33"
Frank Lockett
Cep 22h12'45" 56d28'14"
Frank Lohne
Uma 8h39'27" 47d16'18"
Frank Long
Uma 8h42'49" 50d8'14"
Frank Lopez Garza
Ori 5h26'21" -0d3'45"
Frank & Loretta J. Miller
Uma 11h26'15" 55d11'25"
Frank Louis Lucero
Her 18h7'54" 27d41'32"
Frank Louis Vinz
Cap 21h8'6" -20d6'13"
Frank & Louise Supino:Stars Forever
Uma 11h28'36" 30d11'47"
Frank Luke Comune
Ari 3h5'44" 29d22'9"
Frank M.
And 0h46'24" 45d19'26"
Frank M. Chalaire, M.D.
Lib 15h27'18" -15d50'41"
Frank M. Oleksy
Uma 8h29'43" 71d26'38"
Frank M. Polack
Lmi 10h36'8" 30d57'25"
Frank M. Wade
Lyn 7h36'2" 38d46'42"
Frank MacMillan
Ser 16h19'47" -0d18'18"
Frank "Mankie" Scaffido
Aur 5h36'35" 33d24'17"
Frank Manu Merl NCC-1701-A
Uma 11h19'23" 67d58'22"
Frank Marion Irving III
Vir 14h49'9" 5d29'0"
Frank & Marketta
Aur 5h53'34" 42d31'31"
Frank Matra
And 2h34'19" 44d0'52"
Frank McAree
Uma 8h35'32" 55d23'12"
Frank McKinley Gray
Uma 11h3'27" 52d51'43"
Frank McLane 1931-2006
Per 2h55'30" 54d49'54"
Frank Meechan
Lyr 19h14'48" 27d4'50"
Frank Meyer
Her 18h44'41" 20d21'44"
Frank Michael Capece II
Ori 4h51'41" 9d29'9"
Frank & Michelle Russell
Cyg 21h28'46" 46d1'29"
Frank Mikan
Cyg 20h23'47" 48d10'11"
Frank Milano
Gem 6h56'58" 24d55'24"
Frank Miller
Uma 11h13'2" 56d27'40"
Frank Mizzi "Our Star"
Cru 12h40'28" -64d38'39"
Frank (Moose) Joseph Pence
Cnc 8h51'19" 14d24'7"
Frank Murphy
Uma 13h31'11" 54d40'36"
Frank n Harriet
Lyr 19h8'26" 45d58'40"
Frank N Helen Mac
Dra 15h9'2" 64d57'27"
Frank N' Shine
Aqr 22h31'20" -0d29'29"
Frank N. Stabile
Uma 9h8'47" 53d10'4"
Frank & Natalie, a year & counting
Cru 12h53'19" -64d1'24"
Frank Needell's Centurion Journey
Lib 15h15'25" -6d43'47"
Frank Norcia "10/7/1919-12/15/2000"
Uma 8h49'30" 62d19'28"
Frank Norman Clifton
Ari 1h58'4" 24d48'14"
Frank Nowak
Uma 10h35'15" 48d20'18"
Frank Otto
Ori 5h7'31" 0d57'33"
Frank P. Cote "The Cote Star"
Cap 20h7'1" -9d39'50"
Frank P. "Dutch" Culbertson
Leo 9h49'40" 27d8'18"
Frank P. Jones
Uma 8h24'53" 69d56'35"
Frank "Pa" Thomas
Cas 23h33'43" 54d33'15"
Frank Pacheco
Per 4h8'29" 52d8'33"
Frank Pancho
Dra 16h18'14" 63d56'31"
Frank "Pappy" Simmons
Cep 22h50'38" 71d53'11"
Frank Parry
Sgr 18h1'42" -27d39'12"
Frank Patricia Marrek
Uma 10h34'14" 56d57'35"
Frank Patrick Gallagher
Cap 21h6'53" -15d42'8"
Frank Paul Costa
Per 2h59'41" 55d17'31"
Frank Paul Iavernaro
Aqr 22h22'12" -6d33'3"
Frank Paul Wilcox
Uma 11h30'10" 48d11'43"
Frank Percival Goodwin
Cep 0h2'13" 80d18'21"
Frank Pessara
Ori 4h44'6" 4d29'50"
Frank Phillip Amo IV
Her 17h11'25" 28d11'49"
Frank Phillip Ferreri
Gem 7h33'48" 18d13'52"
Frank Pino Jr.
Cnc 9h2'48" 29d5'22"
Frank Pisillo
Uma 9h17'34" 53d2'16"
Frank Plawski
Lmi 10h3'16" 38d11'17"
Frank Pollera
Dra 20h33'44" 70d53'19"
Frank Powers
Cnc 8h46'18" 30d48'50"
Frank Preuß
Uma 11h14'29" 64d13'12"
Frank R. Blenman
Sgr 18h32'48" -18d43'8"
Frank R. Haines
Uma 11h0'45" 68d3'56"
Frank R. Meyers
And 0h28'39" 42d49'56"
Frank R. Musgrove Sr.
Leo 9h47'16" 17d59'52"
Frank R. Shafer, III
Dra 16h12'11" 52d52'50"
Frank R. Trizano
Uma 8h59'51" 61d53'54"
Frank R Tuttle Tut Loved Life Music
Lib 14h49'45" -2d26'12"
Frank R. Wacholz
Uma 11h57'49" 42d48'8"
Frank Racioppi
Her 17h47'49" 42d19'24"
Frank Radice
Per 4h40'3" 48d10'20"
Frank Reed
Uma 10h2'22" 59d18'48"
Frank Rene Sanfiel
Gem 6h14'53" 27d53'30"
Frank Richard Gearhart III
Peg 22h45'27" 31d20'31"
Frank Richter
Ori 6h16'57" 9d2'27"
Frank Robert Fazio
Per 3h23'21" 52d12'18"
Frank Roberts
Uma 10h37'59" 65d25'14"
Frank Rochow
Uma 10h45'26" 48d18'18"
Frank Rohde
Aqr 22h17'47" 1d50'9"
Frank & Rose Viola 60 Happy Years
Cyg 20h53'5" 36d36'20"
Frank & Rosemary Kelly
Leo 9h41'19" 20d17'49"
Frank Ruberto
Cap 20h15'43" -13d8'6"
Frank Russo
Ari 2h16'17" 23d19'56"
Frank Russo
Cnc 9h7'7" 32d44'37"
Frank Ryszczyk
Aqr 21h52'45" 0d39'35"
Frank S. Gibala
Aqr 21h43'59" -0d54'43"
Frank 's (Kiss My ...) Shining Star
Leo 9h25'36" 10d20'35"
Frank S. Nichols
Ori 5h36'17" -4d22'32"
Frank S. Sobol
Uma 8h25'48" 68d47'43"
Frank Saia's Star
Aqr 22h25'2" 0d0'1"
Frank Sakosky
Aur 5h38'25" 40d51'4"
Frank Salentin
Uma 8h40'59" 51d34'29"
Frank Schiess
Crb 15h47'57" 27d33'50"
Frank Schütz
Ori 6h19'17" 15d53'46"
Frank Scordo "Poppy"
Ori 6h10'37" 18d40'35"
Frank Simon Horowitz
Sco 16h12'34" -16d18'5"
Frank & Simone Kreuzer/Liebesstern
Uma 9h18'48" 44d10'58"
Frank "Slugger" McFadden
Psc 1h31'12" 32d21'57"
Frank Snock
Vir 12h16'29" 10d36'14"
Frank Squillace
Per 4h11'5" 46d28'57"
Frank Stanley Holman
Cep 21h20'11" 57d3'40"
Frank Steven Summers
Dra 19h4'12" 64d20'24"
Frank Stevens
Ori 5h51'45" 3d32'43"
Frank Strauss
Her 18h27'21" 21d28'42"
Frank Sulak & Stella Valoppi
Uma 10h3'58" 42d14'59"
Frank Sullivan - #1 in Universe
Leo 10h19'47" 24d48'21"
Frank T. Genova, Jr.
Uma 11h12'37" 69d56'23"
Frank T. Sorrentino
Lib 15h10'29" -13d16'12"
Frank T. Zaremba
Her 17h38'3" 48d6'9"
Frank Takemoto
Aqr 22h11'41" -1d50'20"
Frank Taurozzi
Lac 22h14'15" 48d48'38"
Frank The Fearless
Lib 15h7'46" -21d52'32"
Frank Thomas Davis
Aqr 22h44'38" 2d6'20"
Frank Thomas Lombardi
Vir 14h39'25" -2d52'6"
Frank Thomas Richo
Aqr 22h8'10" -7d50'18"
Frank Tierney
Ori 6h8'41" 8d33'37"
Frank & Torri
Uma 9h59'54" 41d57'38"
Frank Troy Bruce
Leo 9h41'35" 26d41'59"
Frank Tudela
Col 6h6'34" -27d55'39"
Frank Tull
Uma 8h34'24" 53d27'40"
Frank V. Taliercio
Cyg 19h56'57" 32d20'18"
Frank & Valerie Dunn
Lyr 19h24'30" 38d20'40"
Frank Vargas
Boo 14h46'53" 34d30'36"
Frank Vega Betancourt
Her 18h17'17" 25d37'1"
Frank Veltri
Dra 11h9'50" 77d47'42"
Frank & Vickie's Max
Uma 9h39'16" 56d14'31"
Frank Victor Dumelle
Vir 13h10'7" 6d10'47"
Frank Victory Hager
Cru 12h32'4" -56d5'38"
Frank Vidotto
Lmi 10h38'52" 26d16'30"
Frank Vincent Bonsignore
Psc 23h12'17" 7d10'43"
Frank & Virginia Andrus
Cyg 21h34'42" 44d50'43"
Frank W Gift Storyteller of Stars
Her 16h31'4" 14d15'47"
Frank W. Jelks
Sgr 18h21'27" -32d13'38"

Frank W. Lee
Leo 9h38'57" 31d3'57"
Frank Walter Card Jr
Uma 10h30'9" 67d55'10"
Frank Watson
Boo 14h17'57" 33d6'43"
Frank Wayne Vroman Jr.
Cap 20h24'33" -16d11'2"
Frank Welsh
Per 3h52'52" 52d29'41"
Frank Wesley Sims
Leo 10h24'53" 17d54'0"
Frank Wesley Stoda Jr.
Aqr 22h12'22" -1d29'38"
Frank William Samuelson
Per 3h53'58" 39d42'52"
Frank X. Desmond
Uma 9h22'14" 53d17'28"
Frank Young
Vir 13h18'29" 4d53'36"
Frank & Yvonne
Ori 6h22'0" 16d31'35"
Frank Zito
Cnc 8h33'12" 25d24'46"
Frank, Claudia
Uma 11h59'35" 36d22'26"
Frank, Leo the Great ,
Gercas
Uma 11h12'46" 34d38'29"
Frank-Catherine Hayden
Lyn 7h16'7" 44d50'19"
Franke Addison
Leo 11h50'32" 25d48'4"
Franke, Hartmut
Ori 6h19'27" 13d16'59"
Franke, Lutz
Ori 5h58'10" 21d31'45"
Franke, Michael
Uma 14h0'7" 55d18'31"
Franke, Wolfgang
Uma 11h40'53" 28d28'5"
FrankG
Aqr 22h42'5" -10d20'15"
Franki
Cas 0h38'37" 48d37'45"
Frankie
Lyn 8h1'0" 48d33'56"
Frankie
And 1h51'30" 46d47'39"
Frankie
Her 17h37'49" 32d6'8"
Frankie
Her 16h23'52" 20d14'49"
Frankie
Her 17h39'18" 18d51'49"
Frankie
Tau 5h12'48" 18d48'0"
Frankie
Psc 1h21'11" 15d7'51"
Frankie
Vir 13h16'57" -12d38'53"
Frankie
Lib 15h33'0" -7d0'26"
Frankie
Cep 20h39'22" 58d43'23"
Frankie
Lyn 7h50'10" 52d52'47"
Frankie
Dra 12h9'57" 68d41'25"
Frankie
Lyn 6h45'58" 61d0'16"
Frankie
Sgr 20h19'39" -41d48'14"
Frankie Abernathy
Uma 10h10'34" 59d32'25"
Frankie and Dee's Wish
Upon A Star
Tau 4h13'19" 9d8'50"
Frankie Angel
Cnv 12h35'42" 37d27'17"
Frankie Ashley
Aql 19h16'54" 3d37'25"
Frankie B
Sgr 18h10'8" -29d4'46"
Frankie Baldassare
Per 4h3'0" 45d6'37"
Frankie Caren
Vir 12h34'0" 6d1'23"
Frankie Ciofrone
Psc 1h26'18" 24d34'18"
Frankie Constantino
Crb 16h8'0" 31d42'31"
Frankie Constantino
"Koots"
Leo 11h22'8" 18d10'18"
Frankie D.
Ori 5h43'5" -3d44'58"
Frankie Darren
Lib 14h50'55" -21d49'30"
Frankie Davis
Cnc 9h7'17" 12d45'53"
Frankie Emberlin
Per 3h49'31" 51d10'40"
Frankie Fonseca
Uma 12h34'17" 58d37'35"
Frankie - Forever Our
Shining Star
Cnv 12h27'36" 47d13'48"
Frankie Gwen Pettley
Tau 4h23'13" 25d36'42"
Frankie John Marcell
Cap 20h9'21" -20d49'43"
Frankie Kaczmarczyk Jr.
Leo 10h31'2" 17d50'8"
Frankie Lynn
Sco 16h6'59" -16d23'19"
Frankie Marie Warren
Sgr 18h30'46" -16d5'8"
Frankie Maureen Mitchell
Ari 2h3'20" 18d51'52"
Frankie McClure
Per 3h42'32" 33d59'16"
Frankie Meeks
Uma 8h33'47" 62d23'18"
Frankie My Love
Dra 16h45'16" 65d36'28"
Frankie Noone
Uma 14h6'37" 49d10'32"
Frankie Rae Santana
And 0h31'44" 27d28'17"
Frankie Ravella
Ori 5h20'34" 6d8'8"
Frankie Reyes
Per 3h33'33" 44d17'29"
Frankie Riscile
Aqr 22h44'20" 0d6'15"
Frankie S. Bowman
Pyx 8h59'7" -32d38'5"
Frankie SilverStar
Cep 22h31'0" 69d21'19"
Frankie Sparkes
Uma 10h19'50" 40d50'56"
Frankie Sullivan
Aur 5h16'54" 49d41'25"
Frankie Vesce
Cap 21h9'20" -18d35'37"
Frankie, Jr.
Ori 5h8'4" 7d29'41"
Frankie-Boy
Gem 6h2'59" 25d27'5"
frankiemac
Cap 20h19'42" -24d37'8"
Frankie's Escape
Tau 5h20'13" 19d26'22"
Frankie's Star
Aur 7h14'22" 42d28'38"
Frankie's Star
And 0h21'33" 39d55'9"
Franklin Boone Jones
Her 16h20'52" 19d10'58"
Franklin Bunny Star
Uma 11h49'23" 39d27'9"
Franklin Chambers
Ari 2h36'36" 12d52'19"
FRANKLIN CLAIRE
Leo 11h37'33" 24d55'51"
Franklin D. Rich
Her 16h20'12" 6d36'35"
Franklin David Childs
Gem 7h47'9" 34d8'13"
Franklin Douglas Ford
Leo 11h42'19" 21d21'53"
Franklin Edwin Hunt
Aur 7h14'20" 42d1'41"
Franklin Family Star
Lyr 19h27'4" 41d35'40"
Franklin J. Marks, Jr.
Lib 15h15'33" -6d51'27"
Franklin Johnson
Gem 7h0'19" 15d34'2"
Franklin Kay
Leo 11h11'21" 9d12'52"
Franklin LaBarbara
Ori 5h15'46" 6d34'20"
Franklin Lee Barnes
Her 18h20'25" 17d4'52"
Franklin Rimalovski
Ori 5h37'28" -1d54'1"
Franklin Shannon Henson
Uma 10h2'25" 60d25'41"
Franklin Sipe-Blad
Uma 11h17'21" 45d3'33"
Franklin Stephen Pullano
Vir 12h13'39" -0d59'12"
"Franklin" The brightest star
Sco 17h1'11" -39d18'50"
Franklin Wayne
Katzenmeyer
Del 20h36'44" 14d27'37"
Franklin Woodrow Wilson
Ori 5h25'56" 3d2'10"
Franklin's Love Nest
And 0h4'0" 43d49'32"
Franklyn Harvey Goddard
Hill
Per 4h47'56" 45d38'46"
Franklyn McKean Hynes
Leo 11h20'19" 24d17'51"
Franklyn Washburn
Boo 15h36'7" 47d38'18"
Frank's 50th Anniversary
Sco 16h14'48" -11d41'56"
Frank's 85th Birthday
Vir 13h13'40" 12d59'59"
Frank's Guiding Light
Per 4h23'47" 40d15'45"
Frank's Petunia
Dra 17h51'59" 58d58'23"
Franks Twilight
Psc 0h25'3" 17d1'42"
Frankster
Vir 13h27'51" 12d21'16"
Frank-The Love Of My Life
Dra 19h48'45" 74d45'6"
Franky N Rodriguez
Aqr 22h23'33" -24d48'18"
Franky Siclari
Umi 13h29'22" 73d21'25"
Franky Yadao
Tau 4h14'52" 23d15'29"
Franky's Star
Umi 13h32'14" 76d9'42"
Frannie
Uma 9h25'10" 54d29'4"
Frannie 1
Tau 5h14'34" 21d47'53"
Frannie Jo
Crb 15h38'15" 27d6'41"
Frannie Marie
And 2h7'48" 46d29'10"
Franny
Cet 0h55'19" -0d29'14"
Franny
Sgr 18h9'15" -19d55'29"
Franny Marino
Cam 6h5'47" 73d57'26"
Frano
Vir 13h49'51" -6d0'34"
Franqui Raina Passio
Umi 14h24'42" 74d49'19"
Frans de Weers
Ori 5h26'44" 2d18'24"
Fran's Dugout
Lyr 18h33'5" 36d52'57"
Frans H. Kuipers
Cap 21h5'51" -18d35'10"
Fran's Light of Merriment
Lyr 18h41'54" 38d26'6"
Fran's Star
And 23h33'50" 48d36'51"
Fran's Star
Aql 20h5'33" 4d24'47"
Fran's Tiger Lily
Lyr 19h1'23" 32d58'26"
Franselia Hall
Ari 2h47'18" 21d37'59"
FRANSIS
Ori 6h15'23" 11d14'48"
Fransisca Maria
Sgr 19h53'47" -34d43'8"
Fransisco Eugenio Conte
(TICO)
Lib 15h8'57" -9d17'54"
Fransky
Umi 17h0'40" 75d59'1"
Frantz
Lib 15h51'3" -6d1'30"
Frantz
Cam 4h43'50" 63d47'26"
Frantz Gadbois
Dra 19h53'20" 70d11'51"
Franwood Alaminos
Ari 2h37'49" 29d22'57"
Franz Czeisler
Dra 16h14'32" 55d43'57"
Franz Dornstauder
Leo 10h49'6" 9d41'14"
Franz Geier
Uma 11h2'56" 57d33'31"
Franz & Gerlinde
Engelhardt
Cyg 21h12'40" 35d34'8"
Franz Graf
Uma 8h40'39" 57d28'53"
Franz Häusler
Uma 8h56'1" 52d30'30"
Franz Josef Berhorst
Uma 9h5'8" 61d12'19"
Franz Kröger
Uma 9h53'30" 48d17'33"
Franz - Mozart Stern der
Liebe und des Verzeihens
Uma 9h19'26" 66d5'42"
Franz Palatini
Uma 9h7'24" 59d21'17"
Franz Schaudeck
Ori 5h53'59" 3d32'42"
Franz' Star
Psc 0h41'2" 14d47'57"
Franz Tobisch
Uma 9h33'52" 59d22'35"
Franz Winkel
Ori 5h56'37" 17d45'1"
Franz Wohl
Ori 5h11'14" -6d26'22"
Franz Xaver Pichler
And 23h57'8" 46d26'54"
FRANZI
Peg 21h28'56" 22d2'58"
Fränzi Haefeli
Peg 23h43'33" 13d22'20"
Fränzi Salber
Tri 1h46'15" 30d46'58"
Franziska
Ori 6h16'45" 16d0'38"
Franziska
Uma 13h6'5" 62d23'30"
Franziska and Johan forev-
er
Cas 1h41'29" 67d30'14"
Franziska Hagius
Cas 2h31'32" 67d41'9"
Franziska und Collin
Ori 6h12'51" 14d11'47"
Franziska, mein
Glücksstern
Ori 6h0'24" 14d25'14"
Franziska's Glücksstern
Cas 1h1'1" 62d27'35"
Franzl 50
Uma 11h48'0" 52d47'35"
Frape Behr Montblanc
Leo 10h57'29" 6d5'41"
Frape Behr Zona Franca
Her 17h41'43" 15d34'33"
Frappier
Cyg 20h24'41" 39d44'52"
Frase
Lib 15h41'58" -17d0'12"
Fraser
Uma 11h26'42" 30d0'19"
Fraser
Ori 6h2'36" 21d6'56"
Fraser Archie Ames
Per 4h18'39" 35d13'9"
Fraser Henry Sherwell
Per 3h23'7" 55d4'44"
Fraser James Nicholson -
Fraser's Star
Her 17h46'59" 40d56'20"
Fraser James Sutherland
Umi 14h27'6" 69d50'23"
Fraser Klouse
Lib 15h3'14" -0d55'44"
Fraser Phillips
Her 17h53'56" 24d3'48"
Fraser Samuel Spence
Per 4h47'43" 47d22'37"
Fraser Scott
Her 16h12'29" 48d10'37"
Fraser Scott Lehmann
Cru 12h30'31" -57d53'5"
Fraser Slorach
Per 3h45'41" 38d15'29"
Fraser-Allen
Cra 18h1'54" -37d13'57"
Fratello Sorella Maggiore
Ori 5h29'13" 14d8'15"
Frater D.A.U.
Ori 5h36'26" 1d21'16"
Frater Hrumachis
Lyr 19h11'28" 26d31'50"
Frau
Cas 1h27'50" 67d54'42"
Frau Angelika
Cap 21h46'49" -19d6'25"
Frau Ruth S. Rusnock-
Greatest Mom
Gem 6h31'6" 20d19'11"
Frauke
Uma 9h53'20" 64d4'43"
Frauke Kugi
And 2h11'19" 38d11'20"
Fraulein Lichti - SB#1
Lib 15h38'21" -17d32'31"
Frawley's Star
Her 16h28'57" 45d13'50"
Frayne Kleiman
Aql 19h46'5" -0d7'28"
Frazier Golding
And 0h27'37" 42d34'34"
frazzlerazzledazzle
Aqr 22h42'44" 1d6'21"
Freakin Shweet
Uma 9h55'9" 71d44'1"
Freckle Foot
Uma 10h29'3" 44d55'25"
Freckled Angel
Aqr 22h42'44" -1d10'50"
Freckles
Uma 9h14'0" 67d27'36"
Freckles
Leo 10h25'27" 20d28'12"
Freckles
Ari 3h19'21" 29d49'31"
Freckles
Tau 5h7'42" 23d50'37"
Freckles 51292
Uma 8h30'41" 64d34'28"
Freckles Holmes
Cma 6h37'45" -16d37'18"
Freckles in the Sky
Cas 0h55'49" 54d31'50"
Freckmann, Hanns
Ori 5h8'9" 15d58'53"
Fred
Gem 6h55'6" 16d56'22"
Fred
Her 18h3'30" 18d7'35"
Fred
Her 17h14'22" 17d18'6"
Fred
Cyg 20h8'10" 37d30'33"
Fred
Aur 6h55'35" 44d7'30"
Fred
Aur 5h47'44" 36d9'54"
Fred
Tri 2h17'42" 32d40'47"
Fred
Cep 0h1'51" 70d37'50"
Fred A. Ney
Uma 10h44'8" 66d1'8"
Fred A. & Rena F. Venditti
Uma 10h13'8" 67d39'18"
Fred Aaron
Uma 9h1'2" 52d20'21"
Fred Adams
Lib 14h31'39" -24d54'14"
Fred Allen Bump 08081969
Ori 5h28'30" 4d5'49"
Fred Allen Henson III
Leo 10h45'27" 12d7'3"
Fred Allendorfer
Per 3h8'45" 51d15'14"
Fred and Bernice-A Star for
Forever
Crb 15h53'58" 33d13'56"
Fred and Ginger
Cap 20h30'43" -14d54'41"
Fred and Jean Rapp 60th
Anniversary
Cyg 19h50'33" 47d17'53"
Fred and Margaret Carroll
Forever
Sco 16h26'27" -27d56'51"
Fred and Ray Forever
Cas 1h25'48" 60d37'21"
Fred Atkin
Ari 3h11'33" 29d46'3"
Fred B. Bock, My
Inspiration
Cap 21h41'19" -9d0'41"
Fred B Marshall IV
And 1h27'37" 46d30'23"
Fred B. 'Ory-Fred' Torck
Uma 10h3'26" 52d56'8"
Fred Baker
Ari 3h12'53" 29d57'28"
Fred & Barbara Baughman
Cyg 20h1'29" 31d1'45"
Fred & Barbara Schiller
Cyg 19h43'5" 35d27'15"
Fred Bennett
Uma 9h35'28" 46d4'19"
Fred & Bernadette
Siekmann
Uma 11h47'25" 33d44'1"
Fred & Betty Means
Leo 10h13'25" 15d50'44"
Fred Brizee
Gem 7h18'8" 22d22'21"
Fred C. Houser III
Ori 6h17'40" 13d34'39"
Fred C. Houser Jr.
Cep 22h1'40" 56d49'9"
Fred & Carmen Farmer
Uma 10h51'21" 59d15'1"
Fred & Cathy McKenny
Cyg 20h52'3" 50d32'11"
Fred Chris Sorenson
Uma 11h1'29" 34d59'5"
Fred Clifford
Boo 14h36'3" 30d21'53"
Fred Corraro
Her 17h21'40" 31d47'55"
Fred D. Boutwell
And 0h8'51" 30d31'18"
Fred De Wayne Parman
Ari 3h13'21" 15d26'35"
Fred & Delia Schuster
Sge 20h5'32" 18d26'18"
Fred Deluca *Mr. Subway*
Cep 20h39'47" 56d10'38"
Fred & Diane Schuelke
Ori 5h51'40" 22d32'14"
Fred Diver
Aur 6h58'45" 38d41'53"
Fred & Dolores Martinez
Cnc 8h17'7" 17d51'33"
Fred Elias
Aqr 22h26'18" -24d26'52"
Fred Eltringham
Her 17h1'58" 13d9'23"
Fred Euliss
Uma 9h33'51" 69d2'11"
Fred Fell
Ari 3h13'48" 29d51'44"
Fred Fisher
Lib 14h34'58" -16d58'1"
Fred & Fluff
Uma 9h41'26" 44d7'49"
"Fred" Frederick Joseph
Brodka
Vir 13h13'27" -2d19'38"
Fred (Frederico) Colhard
Sco 16h41'25" -26d51'52"
Fred. G. O'Callaghan
Dra 17h44'24" 63d1'24"
Fred G. Payne
Lac 22h41'54" 49d56'44"
Fred Gaida
Uma 9h58'2" 62d35'9"
Fred & Gale
Cyg 19h55'20" 32d27'33"
Fred Germaine (Dad)
Vir 14h1'40" -22d3'34"
Fred Gilbertson
Cap 20h22'26" -24d26'27"
Fred&Glad
Cyg 19h48'41" 39d17'3"
Fred Griesbach
Dra 18h42'52" 56d39'46"
Fred "Gumpy" Silverstein
Jr.
Ori 5h4'52" 1d39'39"
Fred Hassen
Cma 6h36'59" -16d0'12"
Fred Hausman
Uma 11h0'27" 70d33'52"
Fred Heim's 80th Birthday
Star
Her 17h20'12" 37d37'30"
Fred Hicks
And 2h25'31" 49d24'49"
Fred & Irene Carr-
Anniversary Star
Uma 9h22'21" 52d48'9"
Fred J. Gambale
Sco 16h56'15" -38d38'1"
Fred J. Rieble
Cep 21h46'19" 61d36'19"
Fred James
Her 18h33'16" 15d35'2"
Fred Jaskot
Per 2h51'18" 48d4'15"
Fred Johnson Fletcher
Psc 1h21'24" 16d18'41"
Fred & Julie Kuhn
Cnc 9h16'16" 17d28'25"
Fred Kemeny
Sco 16h10'1" -10d25'47"
Fred Klauber
Lib 15h28'56" -4d34'28"
Fred Krcmar
Per 3h34'54" 47d6'3"
Fred L Cipriano
Her 17h12'28" 19d22'59"
Fred L. Marlow III
Lyn 8h4'10" 41d19'5"
Fred Lake
Ori 5h29'57" -4d11'18"
Fred Leyba
Cep 21h25'58" 60d21'33"
Fred Lopez, Jr.
Gem 7h2'44" 17d21'34"
Fred M. Nash
Uma 11h11'14" 43d56'45"
Fred M. Onorati "Dad"
Per 4h6'33" 41d29'10"
Fred & Marge Feeney's
Wedding Star
Cyg 21h10'10" 48d13'58"
Fred & Marion Kingsbury
Uma 13h51'25" 51d9'12"
Fred Meyer
Per 3h29'59" 50d31'24"
Fred Michael Battah
Peg 22h43'46" 13d55'23"
Fred Miller
Her 18h52'8" 24d18'31"
Fred N. Conklin
Per 3h35'27" 35d3'34"
Fred Orvel James
Ori 5h5'37" 3d11'26"
Fred & Pennie Gallo
Cyg 20h51'11" 43d19'45"
Fred Peter Begoray 1923
Cyg 22h1'20" 45d9'14"
Fred Petruzzi
Lyn 8h54'26" 36d44'59"
Fred Poehls
Uma 8h33'3" 69d11'45"
Fred "Pop Pop" Yohe
Umi 13h7'50" 75d18'20"
Fred "Pops" Benefield
Uma 9h6'58" 58d2'32"
Fréd & Rick
Uma 8h31'1" 61d57'10"
Fred Robinson
Sco 16h4'44" -14d45'31"
Fred Rogacki
Gem 7h25'18" 14d47'54"
Fred Rowe Jr.
Uma 11h21'22" 57d57'41"
Fred & Sandra Frost
Sco 16h20'45" -31d30'25"
Fred Schindler
Boo 14h46'45" 41d59'59"
Fred Schroeder
Psc 1h3'27" 32d20'11"
Fred Schubert
Per 3h49'21" 50d44'4"
Fred "Sea Wolf" Figgins
Uma 10h59'30" 71d54'54"
Fred Semenoff's Seventieth
Birthday
Cep 21h49'5" 63d13'36"
Fred Shectman
Ori 6h15'39" 14d4'46"
Fred Sigren
Her 18h4'8" 15d26'32"
Fred Smith, Sr
Lyn 7h22'58" 53d2'38"
Fred Spear
Cma 6h56'26" -21d3'55"
Fred Steele
Leo 9h43'35" 28d33'53"
Fred & Sue - Forever
Yours
Cyg 20h0'29" 33d22'17"
Fred & Sylvie Cohen
Cyg 20h56'36" 46d39'34"
Fred T. Holder
Aqr 22h28'49" 1d2'7"
Fred the Lovebird
Aql 19h46'5" 6d1'34"
Fred the Star
Leo 11h20'34" 15d6'33"
Fred Townsend
Lyn 7h40'9" 37d57'33"
Fred & Verna - Together
Forever
Cru 12h31'16" -60d41'37"
Fred & Vi Temple
Cyg 20h4'34" 34d8'41"
Fred Vögeli
Aur 5h33'6" 55d36'52"
Fred Wagner
Per 2h40'13" 53d25'11"
Fred Weyrick - The Star!
Uma 11h21'30" 54d25'55"
Fred Wilbanks
Uma 9h47'8" 42d55'18"
Fred William "Buddy"
Bayne
Cep 22h50'18" 61d53'13"
Fred William Evans
Umi 15h32'33" 75d43'9"
Fred Wy
Per 3h11'41" 43d17'45"
Fred Wynne
Her 17h18'37" 31d48'33"
Fred^2
Uma 10h21'8" 67d49'26"
Freda
And 2h35'29" 40d27'16"
Freda Anderson Freitas
Lyn 7h14'18" 49d48'44"
Freda B
Cyg 19h53'2" 38d56'53"
Freda B. Justan
Uma 10h32'47" 48d32'49"
Freda Baddeley
Cru 12h28'30" -60d54'46"
Freda Bell - Lady in Red
Cas 23h48'29" 57d54'40"
Freda Etchison
And 2h20'23" 45d37'52"
Freda Goulden
Uma 12h43'18" 60d19'27"
Freda Graves
Cyg 20h7'1" 39d21'27"
Freda Hofstetter
Gem 6h43'9" 33d6'40"
Freda Kerr
Cap 21h56'53" -21d23'42"
Freda May
Uma 9h34'42" 42d24'12"
Freda Richards
Uma 9h20'9" 47d37'31"
Freda Wilson Gunkler
Umi 14h19'11" 77d15'29"
FredChase100
Cep 21h54'40" 61d8'10"
Freddie
Uma 11h37'52" 63d42'51"
Freddie
Uma 8h46'4" 54d40'4"
Freddie
Her 17h22'44" 34d26'1"
Freddie Barone
Leo 10h41'38" 7d32'48"
Freddie Bloomingdale's
Star
Sco 16h45'59" -34d43'28"
Freddie Brown
Ari 2h4'56" 12d55'58"
Freddie Charles Douglas
Hale
Ori 6h17'10" 15d31'25"
Freddie Charles Herbert
Umi 14h48'26" 71d28'49"
Freddie Diaz Cruz
Aql 19h32'27" 6d4'50"
Freddie George Marshall
Peg 23h24'25" 20d30'51"
Freddie Harry Crawford
Dra 19h7'21" 60d1'23"
Freddie Hitchens
Aur 6h34'3" 34d34'59"
Freddie Jake Taylor
Her 18h38'25" 15d10'12"
Freddie James Weeks -
Freddie's Star
Her 17h35'1" 19d2'7"
Freddie & Linda Lash
Mon 7h15'24" -0d15'42"
Freddie Mahoney
Ari 3h27'30" 26d46'52"
Freddie Martin, Jr.
Cnc 8h37'29" 22d6'41"
Freddie Mckenzie Lambert
Aql 19h30'12" 7d25'3"
Freddie Michael Garrett
Umi 17h6'53" 77d12'52"
Freddie Noah Law
Her 18h29'19" 13d14'41"
Freddie Samuel's Star
Her 18h50'7" 21d37'42"
Freddie Winston Rumble
Lmi 10h20'3" 38d13'44"
Freddy
Aur 5h28'21" 42d46'52"
Freddy
Aqr 21h41'11" 2d12'4"
Freddy and Charlene
Cnc 8h58'52" 12d57'33"
Freddy and Samantha
Forever
Uma 11h57'33" 34d41'14"
Freddy Bear
Uma 10h16'51" 46d35'33"
Freddy "Bubba" Ellis, Jr.
Leo 9h56'44" 8d10'37"
Freddy Eleazar Zavaleta
Uma 11h45'50" 43d39'26"
Freddy Holloway
Per 3h42'14" 49d54'54"
Freddy Joe
Her 17h17'32" 32d18'23"
Freddy N Mary Shuey
Cnc 8h50'11" 22d21'29"
Freddy Ruiz, Jr.
Vir 12h1'12" -4d44'42"
Freddy Sanchez, best
friend
Vir 14h35'42" 4d0'38"
Freddy Vega
Per 3h11'54" 53d23'33"
Freddy Victor Young
Umi 17h6'47" 79d9'57"

Freddy, Coty, & Tistan
Cas 23h18'34" 55d47'46"

FreddyBear
Sgr 17h59'44" -17d57'53"

Fredel Bates
Cnc 9h6'24" 29d14'0"

Frédéric
Gem 7h5'8" 13d26'9"

Frederic A. "Chip" Shaw, Jr.
Aql 19h50'29" 12d52'26"

Frédéric Anton
Crb 16h20'24" 35d45'46"

Frédéric Bourdier
Gem 6h54'5" 22d5'35"

Frédéric Dafflon
And 0h20'5" 22d21'43"

Frédéric Dumas "Fern" 11-03-1983
Her 15h53'21" 39d50'4"

Frédéric Edelstein
Leo 11h0'34" 10d4'50"

Frédéric et Stéphanie LeBreton
Cam 9h15'58" 73d51'25"

Frederic Joseph Grudnowski
Gem 7h27'43" 34d32'41"

Frédéric Thivierge
Uma 11h5'2" 66d56'48"

Frédéric1993
Cas 23h59'0" 64d53'35"

Frederica Rose
Ori 6h4'25" 18d3'1"

Frederica Tannen
Ari 2h37'46" 14d5'54"

Frédérick
Cnc 8h14'10" 8d45'18"

Frederick
Tau 5h54'41" 25d6'24"

Frederick
Her 17h13'53" 48d32'42"

Frederick A Cobb (Dad)
Her 16h19'48" 8d37'47"

Frederick A. Safford
Uma 12h11'57" 54d18'4"

Frederick A. Wright Jr.
Per 3h49'7" 44d0'41"

Frederick Alan Hudson II
Per 3h18'53" 42d14'27"

Frederick Albert Scott
Peg 0h11'54" 16d13'14"

Frederick Alfred Lambo IV
Ori 6h15'44" 10d20'0"

Frederick Allen Campbell
Ori 5h39'7" 7d5'56"

Frederick Alois Fellner
Uma 10h50'37" 56d15'19"

Frederick Anthony Charles Cox
Per 3h29'1" 36d35'50"

Frederick B. Mayo
Cep 2h27'32" 80d43'39"

Frederick B Smith
Her 17h48'29" 46d51'52"

Frédérick Buell
Psc 0h19'41" 3d38'39"

Frederick C. Fritz
Per 4h27'12" 42d27'25"

Frederick C Mowery
Sgr 18h7'21" -28d2'35"

Frederick Carlyle Stebner
Ori 5h54'46" 21d17'54"

Frederick & Carmen Sapp
Aqr 23h12'4" -21d17'47"

Frederick Charles Edgington
Uma 11h26'59" 29d0'57"

Frederick Charles Palumbo, Jr.
Gem 6h48'9" 33d59'35"

Frederick Christian Kallmeyer
Aur 5h35'42" 46d10'8"

Frederick Christian Miller 3/24/24
Uma 11h27'40" 53d57'57"

Frederick Christopher McDaid
Per 4h48'18" 47d9'9"

Frederick Coffey
Cas 1h18'50" 57d20'19"

Frederick Conrad von Voigt Jr.
Crb 16h7'8" 27d18'0"

Frederick Cornelius
Leo 11h26'49" 20d18'56"

Frederick Dalein Wassenaar
Umi 14h46'52" 78d14'36"

Frederick Debruce
Ori 5h18'50" 6d18'39"

Frederick E. Calhoun
Sco 17h44'0" -32d1'4"

Frederick Earl Foster
Tau 5h18'38" 22d55'19"

Frederick Edwin Simerly
Vir 14h39'21" -3d6'46"

Frederick Eugene Swanson Jr.
Sco 16h8'29" -26d13'21"

Frederick F. Klein
Per 3h30'2" 50d29'49"

Frederick G. Beecher, Jr.
Cap 20h47'31" -16d12'22"

Frederick George Zillessen
Mon 6h39'27" 10d28'26"

Frederick Gerald Williams
Cap 21h21'49" -18d57'33"

Frederick Gilbert Kelley
Sco 16h7'2" -15d25'10"

Frederick Herbert Edwin
Her 17h42'48" 37d42'32"

Frederick Hubert Tetro
Aqr 22h27'20" 0d27'38"

Frederick Hueting's Eastern Star
Ori 5h3'49" -1d5'30"

Frederick J. Allan
Cap 20h18'44" -16d18'59"

Frederick J. Fahey
Per 3h11'54" 53d16'47"

Frederick J. Frink
Her 17h17'42" 14d34'43"

Frederick J. Jaindl
Boo 14h21'38" 36d39'18"

Frederick J. Keim
Uma 8h56'3" 68d43'57"

Frederick J. Masterman
Cap 21h2'30" -20d56'10"

Frederick J. Wallace
Uma 14h2'42" 56d54'33"

Frederick James Edmwnd Haryott
Her 16h30'29" 7d15'3"

Frederick James Waterfield
Per 4h50'21" 46d36'46"

Frederick John Ray
Uma 10h39'26" 61d41'17"

Frederick John Zunk "82"
Lyn 7h0'54" 57d56'36"

Frederick Joseph Koch
Aqr 20h48'41" -12d58'22"

Frederick Kruger
Sge 20h6'23" 18d45'20"

Frederick L. Slone, MD
Uma 11h15'46" 33d38'28"

Frederick Lawrence Young
Ori 5h27'36" 13d17'36"

Frederick Lewendon
Leo 10h53'23" 15d17'38"

Frederick Logan Lewis
Cnc 8h12'20" 10d7'8"

Frederick Marquardt
Uma 11h8'41" 37d23'40"

Frederick McEwen
Uma 8h33'18" 68d32'15"

Frederick Michael Crowley
Per 3h17'12" 46d4'34"

Frederick Morris Tiberio
Vir 12h13'41" 7d59'3"

Frederick Paul Klaburner
Her 17h19'28" 24d30'39"

Frederick Paul Kussman
Her 17h10'25" 34d23'28"

Frederick Paul Wood
Her 16h50'12" 33d7'42"

Frederick R. Ray Levier
Uma 11h40'49" 54d40'44"

Frederick R. Schuler Jr.
Uma 9h45'57" 43d6'51"

Frederick R. Wilkey, Jr.
Her 17h3'55" 29d57'45"

Frederick Rader Duck
Psc 1h12'20" 16d24'24"

Frederick Richard Bowden
Cnc 8h57'44" 7d15'21"

Frederick Schiller
Pho 0h24'57" -40d53'28"

Frédérick Simard Alias Fred
Umi 15h16'15" 78d25'14"

Frederick Smith
Lib 15h1'17" -14d38'33"

Frederick "Sonny" Gorena
Sgr 19h13'20" -28d58'8"

Frederick Stanley George Rose
Lmi 10h39'26" 25d22'24"

Frederick Stern
Gem 7h4'23" 11d52'11"

Frederick Sutherland
Ori 5h36'6" 10d42'28"

Frederick Trapanese
Cep 0h35'28" 78d23'15"

Frederick W. Wolf
Aqr 23h10'56" -7d41'9"

Frederick William Boylan
Ori 6h8'9" 7d28'36"

Frederick William Henry Dawes
Ori 6h12'55" 15d14'34"

Frederick Windsor of the Manor
Cma 7h8'17" -26d40'52"

Frederick Zupp
Tau 4h41'59" 19d26'42"

Frederick's Fury
Crt 11h29'26" -7d59'7"

Frederick's Star
Sco 17h53'45" -35d56'20"

Frederico Tarrinha
Uma 10h43'43" 61d51'9"

Frédérique
Psc 1h38'7" 20d1'24"

Frédérique
Her 18h5'13" 24d7'12"

Frédérique et Sébastien
Cyg 21h9'54" 51d46'35"

Frédérique Galtier
Tau 4h10'32" 26d10'59"

Frédérique Louichon - 23 06 73
Cnc 8h28'25" 23d29'0"

Fredes
Eri 4h29'30" -23d6'4"

Fredi
Uma 11h46'2" 28d43'26"

Fredi Budd
Uma 11h59'2" 30d10'30"

Fredi Heins
Uma 10h48'4" 72d42'15"

Fredlandia
Ori 4h59'6" 11d10'8"

fredniz
Uma 8h40'19" 47d17'38"

Fredrica
Aqr 22h21'15" -4d6'19"

Fredrick Bowers
Her 17h19'12" 35d47'30"

Fredrick C. Hammer
Tau 5h46'23" 27d48'10"

Fredrick C. Phillips
Aql 19h4'51" 3d25'20"

Fredrick G. McGavin
Psc 1h6'48" 26d5'47"

Fredrick Morris Smith
Ari 2h8'3" 14d35'40"

Fredrick Paul Shaffer
Aql 20h1'0" -0d3'21"

Fredrick R. Pullen
Nor 16h8'21" -46d40'10"

Fredrik Johansson
Her 16h28'36" 47d1'45"

Fredrik Schroer
Aqr 22h34'54" -0d18'42"

Fred's Guiding Star
Ori 5h43'6" 2d27'54"

Fred's Thelma
Uma 11h6'4" 48d17'48"

Fredy
Cyg 21h16'2" 46d51'58"

Fredy
Tri 1h48'27" 26d56'33"

Fredy
Cas 23h36'24" 55d59'41"

Fredy and Kate
Vir 13h18'33" 8d0'19"

Fredy Torres
And 23h8'18" 48d0'0"

Free
Lyn 9h11'48" 33d19'4"

Free Fallin'
Her 16h26'29" 45d7'38"

Free Inna Cage
Lyr 18h30'56" 38d42'44"

Free Love
Lib 15h20'4" -17d47'1"

Freed and Varda FOREV-ER
Cyg 20h49'25" 37d25'18"

Freedog
Peg 23h39'10" 26d2'46"

Freedom
Cnc 8h48'14" 27d21'20"

Freedom
Tau 4h15'1" 20d32'24"

Freedom
Aql 19h25'15" 4d19'3"

freedom
Cap 21h5'31" -16d8'42"

Freedom
Lib 15h4'3" -6d19'34"

Freedom 0207
Uma 12h19'28" 62d58'9"

Freedom Faye
Lyr 18h29'17" 37d1'32"

Freedom2000
Uma 10h29'10" 71d34'21"

Freedom's Light
Uma 11h22'2" 50d35'37"

FreedomStar
Lyn 8h19'56" 49d38'31"

Freedora
Uma 10h12'47" 64d56'9"

Freeman Edward Ezell
Uma 11h46'30" 57d32'3"

Freeman Family Star
And 23h28'53" 41d27'45"

Freeman J. Perreira
Sco 16h10'16" -15d3'8"

Freeman-Boucher - United March 27, 2002
Lyr 19h4'29" 33d3'43"

Freer
Lyn 7h34'50" 42d42'7"

Freeway
Leo 11h39'4" 23d38'47"

"FreezingWarm"
Cep 20h45'36" 60d33'51"

Freffer
Gem 6h4'44" 26d20'36"

Frehley Brin Jordan
Uma 11h47'43" 61d6'49"

Freija
And 2h19'56" 45d35'59"

Freily
Aqr 23h9'44" -15d31'42"

Freise Family Star
Uma 11h10'19" 60d14'1"

FrelsStar
Peg 21h42'43" 27d39'24"

FREMAR
Cyg 21h59'54" 48d51'8"

"Fremdes Land"
Uma 9h40'21" 68d52'22"

French Family Star
Lib 14h52'16" -6d43'51"

Frenchy
Uma 8h50'58" 71d5'41"

Freneau Francesca Ann
Ori 6h16'10" 9d23'52"

Frenita
Tri 2h32'27" 34d10'45"

Fre's Baby Boy Adam
Cap 21h40'25" -16d20'5"

Fre's Baby Girl Rah Rah
Cap 21h45'27" -14d41'4"

Fresh Aire
Leo 10h33'40" 9d19'49"

Frettchen Lady
Uma 11h10'42" 50d27'30"

Freund Udo
Uma 11h14'1" 69d21'31"

Freund, Elisabeth
Uma 14h5'37" 56d40'24"

Freund, Wilfried
Sgr 18h6'38" -27d37'58"

Freunde sind wie Sterne ...
Uma 11h9'25" 70d54'5"

Freundschaft
Cnv 13h55'53" 37d11'47"

Freweini Tesfaldet
Cap 20h21'9" -13d19'2"

Freya
Umi 16h31'29" 75d37'49"

Freya
Cas 23h43'15" 57d41'53"

Freya
And 0h3'23" 40d15'36"

Freya
Ari 1h55'51" 21d42'56"

Freya
Tau 4h41'25" 29d9'28"

Freya
And 0h10'55" 23d33'46"

Freya
Gem 6h24'3" 22d29'20"

Freya Banks
And 2h14'42" 39d55'45"

Freya Elizabeth Newman
And 1h57'58" 41d0'9"

Freya Elizabeth Palmer
And 23h44'26" 34d32'52"

Freya Georgia Bardill
And 23h53'22" 42d1'42"

Freya Jane Ada Herd
And 1h23'29" 35d1'15"

Freya Jasmine Campbell
Cyg 21h30'43" 41d40'8"

Freya Jayne Kikuts
And 1h24'47" 35d25'48"

Freya June Curran
Lyn 7h54'39" 35d45'24"

Freya Louise Beckham
Cnc 8h19'39" 14d3'48"

Freya Lucy
And 23h17'35" 35d34'59"

Freya Mae
And 1h9'1" 42d53'57"

Freya Marie Cook
And 1h28'18" 47d14'9"

Freya Marie Hammond
And 23h20'44" 52d9'45"

Freya May
And 0h9'24" 32d48'21"

Freya Rebecca Palmer - Freya's Star
And 0h1'9" 43d39'30"

Freya Rose Leek
And 23h44'12" 37d30'28"

Freya Rose Martin
And 2h32'41" 50d44'35"

Freya Siobhan Edmonds
And 2h35'35" 49d56'7"

Freya Susan Nelson
Cas 0h28'11" 61d56'0"

Freya-D
Dra 19h20'25" 61d1'0"

FreyaHutchx21
For 2h32'8" -31d54'15"

FRHV-FPW1
Vir 13h40'57" 2d12'13"

Fribe's Milky Way
Ori 5h34'27" 9d22'5"

Frida
Sgr 19h24'8" -13d43'12"

Frida Berggren
Peg 21h20'44" 15d49'57"

Frida Elena Urrutia
Ari 2h32'31" 26d55'32"

Friday
Uma 11h58'8" 60d46'59"

Friday
Pho 23h42'18" -42d5'28"

Fridolin Koch-Rey Büttikon, 3.1.1935
Cas 2h46'9" 57d50'59"

Friebel, Klaus
Uma 10h40'3" 45d26'48"

Frieda and Willie 65th Anniversary
Sgr 19h20'15" -34d39'41"

Frieda Franco
Lyn 7h37'34" 48d24'54"

Frieda Gendloff Number One Mother
Cap 21h53'50" -9d17'15"

Frieda my Star
Her 17h44'30" 38d0'21"

Frieden Monika
Her 18h6'49" 17d51'21"

Frieder
Leo 10h38'15" 20d13'9"

Friederike Brixler
Uma 9h13'20" 52d53'33"

Friederike Robinson - WIEN
Cap 20h54'41" -14d36'16"

Friederike Scherl
Umi 13h55'11" 86d41'53"

Friederike & Werner
Cep 23h1'23" 59d48'58"

Friedhelm
Cas 1h28'48" 68d16'12"

Friedhelm Ost
Uma 9h53'17" 44d7'8"

Friedhelm Wojeme
Uma 9h3'30" 48d56'35"

Friedman, Bärbel
Uma 11h39'45" 33d7'2"

Friedolin Luckas
Uma 8h44'19" 50d22'23"

Friedrich Baumgartner
Ori 5h59'8" 17d55'52"

Friedrich Popko
Aql 19h26'14" -11d39'1"

Friedrich Pukelsheim
Ori 6h21'3" 13d58'55"

Friedrich R. Bürkle
Uma 11h6'8" 59d57'47"

Friedrich Seibt
Ori 6h13'20" 8d10'38"

Friedrich Szwedek
Ori 5h51'3" -2d18'32"

Friedrich, Werner
Ori 6h21'4" 13d50'48"

Friel Family Star
Uma 11h35'54" 61d39'27"

Friend
Aqr 20h50'6" -2d6'37"

FRIEND
Psc 1h1'26" 16d36'21"

Friend
Cyg 21h38'21" 51d11'40"

Friend Always Kirsty Bloxsome
Cyg 19h53'9" 57d34'13"

Friend Benjamine Wilson
Cep 22h10'42" 61d23'23"

Friend,Lover,Soulmate~Shelly Graves
Cas 2h13'27" 66d24'12"

Friendly Smile
Cam 4h8'41" 70d11'29"

Friends Aimee & Missy
Cas 1h38'44" 60d34'22"

Friends by chance Sisters by choice
Psc 0h25'13" 6d24'26"

Friends Eternal
Ori 5h22'42" -0d34'3"

Friends Forever
Vir 13h13'3" 11d24'46"

Friends Forever
Tau 4h29'39" 24d55'27"

Friends of Steve Cross
Cnc 9h0'2" 9d56'29"

Friends United Across The Miles
Leo 11h26'18" 16d0'7"

Friends, Lovers, Always
Sco 17h43'39" -30d25'1"

Friendship
Aqr 22h13'16" -0d9'30"

Friendship
Lyn 7h28'18" 46d55'39"

Friendship 10
Cyg 20h26'59" 40d23'21"

Friendship Lady
And 1h49'0" 46d0'14"

Friendship Star
Lyn 6h48'14" 50d34'20"

Friendship Star
Uma 10h10'18" 47d10'32"

Friquet
Ari 2h19'10" 11d41'24"

Frisia aka F & D
And 1h55'41" 39d39'25"

Frits' star
Umi 17h31'47" 82d44'0"

Fritz
Uma 11h51'39" 43d28'55"

Fritz
Cas 0h37'19" 50d38'20"

Fritz and Cheryl Pflock
Cyg 21h56'52" 44d39'58"

Fritz A.O. Schwarz
Cep 21h31'50" 64d30'12"

Fritz Donnelly
Aqr 21h54'44" 0d50'13"

Fritz God's Shining Star
Lyr 18h26'32" 39d56'58"

Fritz Hahn
Uma 11h22'3" 40d58'33"

Fritz Heiss
Dra 20h24'57" 71d2'32"

Fritz Karl
Uma 10h21'17" 67d38'3"

Fritz Küenle
Uma 10h54'54" 69d37'17"

Fritz Spycher
Aur 5h19'44" 30d0'52"

Fritz und Manu Treichel
Per 4h13'1" 34d54'33"

Fritz Volz
Ari 2h38'5" 24d43'6"

Fritz Wilhelm Binder my star
Uma 9h31'7" 48d18'50"

Fritzi
Vir 14h24'34" 1d40'18"

FRITZIE
Cyg 21h47'34" 49d19'43"

Fritzie Mehling
Lyn 7h44'3" 38d19'5"

Fritz-Jürgen Rüger D 31228 Peine
Uma 9h5'40" 59d41'6"

Fritzner J.
Cap 21h40'12" -24d48'23"

Fritzopa
And 2h6'49" 43d6'1"

Fritzy
Cas 23h51'32" 54d26'48"

Frock
Cap 20h17'15" -16d5'28"

Frog and Toad
Uma 13h36'51" 54d29'7"

Frog Face
Ori 4h51'42" -0d34'39"

Frog Prince
Psc 1h14'55" 11d45'42"

Frogger
And 23h26'13" 47d26'56"

Froggi
Cnc 8h53'22" 13d10'35"

Froggie 143
Tau 3h46'23" 6d33'10"

Frogley
Cra 18h16'12" -39d55'32"

Fröhlich Martin 26.10.1985
Ori 5h9'16" 15d58'33"

From A to Z
Lyn 9h22'22" 40d12'1"

From Heaven Above & Stars Beneath
Leo 11h39'38" 21d50'50"

From Mer to Mom
Sgr 18h56'16" -23d40'45"

From now 'til the stars don't shine
Crb 15h26'56" 31d51'43"

From This Moment...You Are The One
Cas 0h18'31" 55d20'53"

Frömme Souris
Aqr 22h51'17" -22d45'4"

Frommherz, Rosemarie
Sgr 17h53'25" -29d23'22"

Fronie Lacey
Aqr 21h5'20" -3d1'23"

Frora Gonzalez
Mon 6h42'57" -0d0'14"

Fröschli
Tri 1h48'0" 31d4'15"

Frosted Flake
Lyn 8h3'47" 44d31'59"

Frosty
Gem 7h25'21" 28d52'12"

Frosty
Uma 9h12'19" 67d56'32"

Frosty Coyle
Leo 11h36'54" -5d48'44"

Frozen Eternity
Cnc 8h17'37" 22d32'38"

Frozen Moment
Cnc 8h42'29" 32d26'6"

Früchtenicht, Sonja Gertrud
Uma 9h34'14" 63d36'10"

Frühauf, Klaus-Dieter
Uma 10h49'20" 43d44'37"

FruitBat and Squid
Uma 9h46'46" 66d50'52"

Fruity Faerie
Leo 9h39'20" 31d38'45"

Fruma Forman
Cas 1h31'31" 59d34'27"

Frumoasa Victorita
Peg 22h31'7" 28d52'29"

Frumpkins
Uma 12h41'21" 62d40'37"

Frunzi's Ring of Fire
Cnv 12h46'49" 40d32'4"

Frye
Cnc 8h6'54" 14d10'28"

Frying Pan
Lmi 10h27'3" 36d31'33"

FTB
Uma 11h3'7" 65d6'14"

FTC "Grandpa" Star
Aql 19h51'25" 8d54'9"

F-T-L-O-M-L Joy Best
Umi 14h46'24" 69d15'6"

Fucci-Kootchie
Uma 11h13'54" 40d38'49"

Fuchs, Gernot
Ori 5h52'33" 21d47'18"

Fuchs, Harald
Uma 9h42'43" 44d24'27"

Fuchs, Horst
Uma 9h42'59" 49d13'30"

Fuchs, Petra Cornelia
Ori 6h17'9" 18d50'13"

Fuchur 18.06.1977
Cas 0h42'49" 50d6'15"

Fudge
Cas 0h49'7" 61d9'17"

Fudge
Cma 6h26'19" -24d20'58"

Fudge Weichelt
Umi 16h3'11" 76d20'53"

Fudgy
Leo 11h15'16" 3d36'35"

Fuego
Gem 7h24'19" 31d12'22"

Fuego del Faro
Cap 21h18'40" -14d39'9"

Fuentes-Nagle
Dra 12h27'33" 71d56'15"

Fuerza y corazon del Bobby
Leo 11h26'13" 13d57'3"

Fuffstern- Kamischnuppe
Ori 6h21'14" 13d38'48"

FuFu
Lyn 8h25'27" 44d57'37"

Fufy
Lib 15h29'12" -12d2'51"

Fug Bish
Ari 2h0'23" 24d34'50"

Fuji
Leo 10h42'55" 16d36'49"

Fuji Tours Mijio & Linda
Psc 1h53'49" 6d51'58"

Fujiko And Sean
Ari 2h27'28" 23d24'53"

FUKK THE STANDARD
Ori 5h37'21" -1d55'45"

Fulfill a wish of Yoshio
Cnc 8h24'39" 16d5'5"

Fulgencio Cruz
Aqr 22h43'33" 1d18'23"

Full Circle
Uma 11h28'56" 58d24'50"

Fuller
Uma 13h58'41" 54d35'26"

Füllert, Heinz
Ori 6h11'41" 16d26'32"

FullMetal Lizz
Cap 20h11'2" -14d33'40"

FulmerFXMM
Crb 16h15'57" 38d25'40"

Fulvio Patti - Babbo Speciale
Her 16h19'51" 3d52'45"

Fumi Maria Probus
Uma 10h18'50" 44d49'47"

Fumito Kazusa
Sgr 18h41'50" -19d0'43"

Fun Toby
Aql 18h54'7" -0d12'34"

Fun-2/B-Us, Rich & Kris
Cyg 20h29'44" 46d40'48"

Fungetta Y. Whitaker
Uma 10h12'6" 48d28'13"

FuNk_ReTrO_LoVe est.012806
Uma 12h33'14" 58d0'35"

Funky Munky
Per 3h7'31" 47d29'5"

"Funny Grandpa" Jim Barnes
Ori 6h8'58" 13d29'24"

Fuoco Silvia
And 2h37'11" 44d23'7"

"für Dich"
Cas 23h51'12" 52d34'47"

für Hans Peter und Irène
Umi 13h44'53" 79d4'42"

für Hansueli und Karin
Ori 5h8'32" 0d50'7"

Für immer dis Müsli
Leo 9h22'22" 12d14'54"

Für immer gebunden Jeanette und Patrick
Cyg 21h14'30" 45d57'0"

Für immer zu zweit mit Dir
Ori 4h55'59" 2d30'12"

für Jean-Marc und Irene
And 1h12'29" 34d51'5"

für Jörg und Karin
Uma 8h12'41" 64d10'12"

für Jürg und Eliane
Ori 5h9'16" 0d11'22"

für Markus und Heidi
Dra 18h41'24" 77d35'43"

Für mein SchatzJens Mahler
Uma 8h48'18" 48d28'25"

Für meinen Schatz
Uma 10h40'37" 39d46'19"

Für meinen süssen Stern Anna Kittler
Cas 1h35'39" 64d9'14"

Für Mis Müsli
Her 18h43'47" 20d8'26"

Für Roland in Liebe Sylwia
Uma 9h36'4" 51d56'28"

für Ruedi und Denise
Cas 0h18'41" 57d6'11"

Furbee - 4
Uma 12h59'43" 59d46'16"

Furbi
Uma 10h57'22" 40d11'55"

Furere aliqua
Cen 13h26'36" -36d26'55"

Furman
Uma 11h17'11" 40d25'30"

Furnelly
Cyg 19h57'7" 45d18'1"

Furry Friends
Lyn 7h22'32" 53d30'36"

Fursey
Dra 17h15'19" 62d45'48"
Furstinna
Lyn 7h16'36" 57d2'16"
Fury Bear
Aql 19h49'50" -0d13'24"
Furzee
Hya 8h15'2" 4d34'56"
Fusae
Vir 13h19'36" 5d58'49"
Fusaro's Light
Cam 4h7'55" 73d50'11"
Fusimo'omo Kalolaine Tui'neau
Gem 6h52'27" 14d46'54"
Fusion
Uma 9h21'50" 54d9'51"
Füsun
Cas 0h32'16" 61d6'28"
FUTUR LEO 2002
Lac 22h27'47" 41d27'44"
Futura
Uma 11h48'47" 47d13'11"
Futura 2006
Aur 5h24'21" 36d7'49"
Future
Ori 6h8'45" 9d11'8"
Future
Ori 6h9'23" 15d34'19"
Future
Ari 2h49'47" 17d0'34"
Future
Uma 8h42'49" 68d16'36"
Future Dream
And 23h22'13" 51d20'34"
future model
Psc 1h19'41" 31d12'5"
futureboy
Ori 6h8'24" 14d14'1"
Futures
Ori 5h55'26" 22d0'4"
Futuro di Kristina
Ori 5h14'47" -9d10'3"
Fuyo & Toyoichi Nishizawa
Cap 20h31'10" -8d34'3"
Fuzgidear
Aql 19h45'22" 13d20'15"
Fuzz
Sco 16h31'7" -26d21'28"
FuZZ
Sco 16h9'47" -14d50'17"
Fuzzball
Her 17h34'34" 49d47'31"
Fuzzy
Col 5h32'7" -29d20'38"
Fuzzy
Sgr 18h32'8" -31d2'4"
Fuzzy Bear
Uma 10h40'33" 64d4'38"
Fuzzy Lumpkins
Ori 5h25'55" 2d44'36"
Fuzzy McGee
Uma 11h11'55" 50d30'55"
Fuzzy Poof
Uma 10h19'50" 55d38'15"
Fuzzy-Ding
Cmi 7h8'49" 10d0'40"
Fuzzyfeet
Uma 12h45'31" 63d6'2"
Fuzzy's Star
Tau 4h32'53" 28d45'58"
Fuzzy's Star of Seth
Gem 6h50'47" 22d42'9"
FV15022006
Uma 12h17'15" 62d20'25"
FW and Brinn Willis, eternal love.
Uma 11h14'40" 55d32'19"
FWCDSFalconsVarsityGirls Soccer2005
Cas 1h8'15" 67d28'24"
F.X. Sterling, Jr.
Ori 5h34'49" 0d51'25"
FX-11
Her 16h50'33" 23d29'33"
Fyfe Audio - Robert Page Fyfe
Pho 23h58'52" -40d20'44"
FYHAC
Lib 14h41'32" -9d21'48"
Fyllis Hockman
Vir 13h18'31" -15d49'10"
Fynn Aidan Regan's Star
Ori 5h9'29" 8d33'34"
Fynn Lennis
Uma 14h4'3" 56d45'1"
Fynn Leopold Willy
Umi 15h44'19" 75d46'23"
Fyondo and Krystal Goma
Pyx 8h55'39" -33d41'32"
f.yu.inJ
Sgr 18h57'13" -34d21'52"
G
Ori 6h13'22" 19d38'47"
G and V Eternity
Lmi 10h52'25" 36d17'36"
G. Anthony Gelderman III
Vir 12h20'4" -0d53'36"
G. Brady Dawkins
Uma 11h44'23" 60d4'26"
G. D. Y. B. B. "Friends Forever"
Gem 6h52'58" 23d19'29"
G Daddy Dipper
Uma 11h48'23" 32d30'45"
G. David Bline
Dra 20h17'50" 62d26'28"
G Dog
Mon 6h41'18" 8d41'27"
G. Eugenia Rios Santa Cruz Polanco.
Aqr 22h26'58" -19d8'5"
G + Fi McMeekin : Union Star
Cyg 20h9'14" 45d43'24"
G. Frank (Zeke) Lindenmuth
Sgr 18h29'35" -19d9'2"
G & G
Lyr 18h47'9" 31d56'6"
G G Perkins
Vir 13h28'33" -8d32'46"
G G 'z Glow
Tau 4h33'18" 10d24'27"
G Girl
Ori 5h37'27" 4d22'3"
G G's Ginger
Peg 22h46'8" 18d37'10"
G & G's September Star
Cyg 20h28'2" 31d39'1"
G & H - Gernot & Ha Eun Ruppelt
And 1h43'49" 42d2'46"
(G)Host of Berlin & Living Museum
Mon 7h19'37" -0d31'26"
G. I. Barbie 91104
Lyr 18h42'56" 39d49'26"
G I L B E R T
Uma 14h10'52" 58d45'10"
G I N A
Cam 3h30'19" 54d10'23"
G J Bush
Uma 9h8'47" 58d32'25"
G&J Sciarabba's Guiding Light
Lmi 10h3'50" 29d14'36"
G & Jen
Sge 19h52'12" 18d46'22"
G&K 7/11
Cnc 8h52'15" 32d12'25"
G & K Guthrie
Sgr 19h21'5" -18d56'55"
G. Keith Bass
Cnc 8h49'10" 13d31'20"
G & K's Nuptial Star
Cru 12h39'11" -58d34'26"
G. L. Jones
Uma 13h41'34" 56d55'15"
G/L Richardson
Mon 6h48'34" -0d2'56"
G. Lennard Gold, MD
Cnc 8h12'53" 8d11'7"
G. Lorraine Kelly
Uma 9h20'25" 61d53'49"
G Man Clev
Lyr 19h12'30" 26d31'53"
G. Michael Bohardt
Lyn 7h36'52" 56d32'23"
G. Michael Novak
Cnc 9h4'6" 28d25'48"
G Michael Patton
Dra 16h28'30" 54d1'16"
G. Mize
Gem 7h48'20" 30d55'29"
G Money
Ari 2h49'33" 27d48'43"
G. Monte Fitzpatrick
Lyn 7h3'40" 54d22'28"
G & N Flores
Aql 19h47'45" -0d5'45"
(G)Oli
Dra 16h42'16" 57d57'15"
G P A D L
Leo 9h27'12" 12d17'11"
G. Patton Wright
Ori 6h16'29" 13d32'34"
G. Paul Keeley
Cnc 8h5'43" 8d38'36"
G&PHill
Cyg 21h32'1" 55d11'32"
G&R Always & Forever
Hya 12h20'39" -26d24'55"
G. R. Sullivan Sr.
Uma 13h36'43" 54d26'9"
G & S Jojola "Love Infinity"
Uma 11h44'10" 35d41'18"
G. Scott Dennis
Gem 6h55'10" 22d27'22"
G. Sloan
Peg 22h0'58" 14d59'4"
G. Stephen
Vir 13h1'22" -0d37'41"
G Style
Vir 12h41'42" 5d11'23"
G & T Wason - 1st Anniversary
Ori 6h18'49" 6d39'59"
G. Thomas Adams
Cnc 8h16'11" 20d12'5"
G. Troy
Cep 21h54'42" 65d13'32"
G W E
Cam 4h36'5" 57d1'34"
G1
Gem 7h12'3" 27d15'24"
G1940W2004C
Her 16h39'50" 23d6'5"
G² Forever
Cyg 20h42'12" 48d54'17"
G2 (Gidget & Gary)
Lib 14h59'7" -2d14'14"
G21434+1WYS
Ori 5h36'54" 1d44'19"
G3 - Glenda Gale Gourlay
Lyn 8h3'59" 51d26'23"
G3 My Guy
Tau 4h36'9" 19d25'51"
G4 - Jon, Mike, Ben and Matt
Lyr 18h25'7" 33d25'50"
Ga
Leo 10h13'44" 18d7'53"
Ga Ga
Sco 16h14'44" -18d20'59"
'Ga Ga' Pat McGhee
Mon 6h45'12" -0d4'50"
Ga Our shining star always! M C K
Uma 13h25'47" 56d35'28"
GA60
Vir 12h27'33" 9d57'43"
Gaajah
Uma 8h44'12" 57d54'3"
Gabarró
Her 17h40'53" 15d55'36"
Gabbert - 4
Uma 8h58'55" 49d40'28"
Gabby
And 0h58'12" 45d21'21"
Gabby
Cyg 20h54'18" 37d47'21"
Gabby
And 0h45'2" 42d20'28"
Gabby
Aqr 21h41'18" 1d47'43"
Gabby(booger face lover sauce)
Vir 13h15'14" 3d42'1"
Gabby Day
Psc 1h20'13" 28d58'38"
Gabby Gebhardt
Uma 10h49'27" 65d31'49"
Gabby Girl
And 2h36'10" 39d58'40"
Gabby Ng
Her 16h45'4" 25d7'17"
Gabby Ramirez
Psc 0h38'21" 8d2'44"
Gabby Ramirez
Psc 0h30'54" 8d23'20"
Gabby "Tiny Dancer" Rodriguez
Psc 23h50'51" 5d7'20"
Gabby's Shining Star
Uma 11h54'17" 63d20'56"
Gabcap
Sgr 18h35'34" -35d54'7"
Gabe
Lmi 10h35'15" 38d42'49"
Gabe G Green, Amazing Dad
Her 17h24'53" 16d24'34"
Gabe Lopez
Cap 21h6'8" -24d19'55"
Gabe & Mickey
Lyr 18h47'57" 37d18'42"
"Gabe" Padilla-Garcia
Tau 4h6'43" 5d59'36"
"Gabe The Babe"
Her 16h5'6" 48d8'6"
Gabe Valdez
Her 16h26'23" 25d18'36"
Gabel Lackups
Dra 19h22'11" 64d5'12"
Gabel, Norbert
Uma 12h50'4" 54d7'55"
Gäbel, Petra
Uma 12h57'1" 60d13'46"
Gabel, Uwe
Uma 13h48'59" 56d11'55"
GabeLottie 2007
And 1h48'49" 37d16'2"
Gabeluminous Simagnificat DelVictor
Sco 16h28'38" -19d18'12"
Gabers
Psc 1h21'16" 23d18'53"
GaBeS lUcKy StAr
Lmi 10h47'52" 30d42'32"
Gabey Babey
Tau 5h55'14" 25d11'3"
Gabi
Tau 4h42'34" 17d0'19"
Gabi
Lyn 8h51'53" 38d22'34"
Gabi
Uma 9h33'3" 49d33'20"
Gabi
Uma 12h44'3" 55d13'3"
Gabi
Lyn 6h41'39" 58d21'55"
Gabi Ackermann
Cas 23h53'20" 50d17'10"
Gabi AZ
Leo 11h13'51" 8d40'16"
GABI - CICA
Sge 19h53'14" 18d17'20"
Gabi da Ju
Apu 14h32'28" -79d4'50"
Gabi & Florian Divis
Vul 19h50'24" 22d31'6"
Gabi - Gerhardt - 888888
Uma 8h12'4" 64d44'17"
Gabi Keßler
Uma 11h45'46" 37d20'38"
Gabi L Munoz
Vir 14h23'48" 4d9'18"
Gabi und Dani
Cas 23h2'26" 57d17'4"
Gabi und Martin Ewigi Liebi
Lyr 18h54'0" 33d35'52"
Gabila
Cam 3h43'52" 54d47'5"
GabiniusGustatus
Col 5h50'52" -32d0'24"
Gabor Major
Leo 11h15'46" -0d17'31"
Gabrael Hansen
Uma 9h47'13" 49d17'4"
Gabriel
Uma 9h27'7" 46d23'56"
Gabriel
Uma 13h48'52" 52d13'55"
Gabriel
Lyr 19h3'2" 42d23'8"
Gabriel
Lyr 18h26'39" 39d52'49"
Gabriel
Her 16h19'0" 41d59'31"
Gabriel
Per 2h39'49" 51d57'14"
Gabriel
Her 17h17'19" 32d52'59"
Gabriel
Uma 11h54'46" 34d30'49"
Gabriel
And 23h41'57" 36d59'31"
Gabriel
Cnc 8h10'29" 17d50'44"
Gabriel
Ori 5h52'23" 18d34'50"
Gabriel
Umi 14h0'18" 77d53'51"
Gabriel
Cap 21h33'31" -9d43'53"
Gabriel
Uma 8h58'28" 62d49'47"
Gabriel
Uma 10h7'2" 62d10'46"
Gabriel
Cep 21h33'38" 63d54'6"
Gabriel
Dra 18h23'58" 72d58'59"
Gabriel
Sco 17h57'0" -44d36'2"
Gabriel
Gem 6h27'1" 27d22'20"
Gabriel A. Mendoza
Aql 20h0'18" 12d56'3"
Gabriel Abascal
Cyg 21h47'58" 44d16'3"
Gabriel Alexander Galindo
Tau 3h28'54" 8d3'31"
Gabriel Allen Wesley
Vir 14h7'11" -15d14'32"
Gabriel Alvarado Scott Parker TeAmo
Leo 11h14'44" 6d39'48"
Gabriel and Nicole
Sco 17h37'22" -36d29'40"
Gabriel Andreas Elvis Koslowski
And 0h57'12" 38d46'5"
Gabriel Angelus Tupper
Uma 10h32'30" 47d12'38"
Gabriel & Anne Gibney
Cyg 21h31'38" 32d24'50"
Gabriel Anthony
Ori 4h49'22" 13d10'3"
Gabriel Anthony
Psc 0h53'6" 16d6'31"
Gabriel Anthony Carestio
Her 17h19'3" 45d36'17"
Gabriel Anthony DiSimone
Per 3h25'7" 43d49'56"
Gabriel Anthony Garone
Tau 5h9'23" 22d2'10"
Gabriel Anthony Perciaccante
Per 2h34'57" 56d2'32"
Gabriel Antonio Cruz
Cap 20h47'14" -20d43'34"
Gabriel Arroyo
Cnc 8h13'33" 7d44'50"
Gabriel Arthur Martin
Aur 5h40'31" 46d16'8"
Gabriel & Ashley
Sco 16h13'46" -20d14'8"
Gabriel Caraballo
Pho 0h56'36" -57d25'3"
Gabriel Cavazos
Sgr 18h31'2" -17d13'51"
Gabriel Christian Jimenez Marana
Sgr 19h45'40" -16d45'31"
Gabriel & Co.
Cyg 20h48'47" 40d23'38"
Gabriel Cookie Stutzman
Uma 9h25'12" 42d51'7"
Gabriel Cravener
Umi 14h17'50" 76d49'26"
Gabriel D. Wilson "JAM"
Ari 2h49'27" 25d34'12"
Gabriel Daisy Walden-Jones
Ari 2h57'24" 18d41'17"
Gabriel Daniel
Psc 0h46'59" 13d14'7"
Gabriel David Muro/Most Precious
Uma 10h31'29" 58d47'41"
Gabriel David Peter Leeming
Umi 13h39'9" 69d49'55"
Gabriel Dworet
Psc 1h8'19" 6d19'49"
Gabriel Edward
Lib 15h3'54" -4d5'51"
Gabriel Elisha Soud
Lmi 10h48'44" 29d29'7"
Gabriel Francis McDonnell
Cap 21h40'42" -14d50'1"
Gabriel Gannon
Cyg 21h51'54" 48d35'12"
GABRIEL GARDNER
Psc 1h5'55" 3d17'15"
Gabriel Gebara
Psc 0h23'16" 6d3'0"
Gabriel Guzman
Cep 22h29'22" 65d31'21"
Gabriel H. Solano
Cyg 20h37'59" 34d17'49"
Gabriel Harry
Cnc 8h9'18" 10d24'56"
Gabriel Henri Havert
Uma 8h54'41" 59d50'31"
Gabriel Hubert Kerr
Uma 14h17'9" 60d49'50"
Gabriel Huerta
Uma 11h25'42" 62d18'44"
Gabriel Isaiah Cardenas
Lib 15h47'15" -13d4'29"
Gabriel Isaiah Jefferson
And 0h10'59" 45d25'13"
Gabriel J. Chevere
Sgr 18h43'10" -27d33'41"
Gabriel Jaime Schultz
Tau 4h12'14" 27d58'46"
Gabriel James Holden
Her 16h49'42" 22d33'50"
Gabriel James Rader
Peg 21h37'26" 22d27'37"
Gabriel James Tretter
Tau 5h3'21" 22d42'32"
Gabriel John Campo Davies
Per 4h3'38" 37d12'24"
Gabriel John Kornstad
Sgr 19h2'47" -28d46'22"
Gabriel Jorge Luis Monteiro
Lib 14h52'8" -2d35'11"
Gabriel Joseph Perkins
Vir 12h57'16" -2d14'47"
Gabriel Joseph Quinn
Sco 16h11'6" -35d23'21"
Gabriel Joseph Van Wagoner
Ori 6h5'42" 9d35'25"
Gabriel "kiki" Medina
Cyg 21h21'18" 39d53'55"
Gabriel Landon Turner
Lyn 6h30'59" 56d34'29"
Gabriel Lee Krivoshey
Per 4h15'47" 45d30'13"
Gabriel L.L.
Cnc 9h3'58" 30d36'31"
Gabriel Louis Riccio
Boo 14h44'40" 35d1'38"
Gabriel Luis Perez
Cnc 8h27'17" 25d36'17"
Gabriel Lynch
Sco 17h38'6" -39d0'57"
Gabriel Lynn Borden
Gem 6h49'56" 17d3'2"
Gabriel Marrufo
Ori 5h52'58" 12d31'19"
Gabriel Martin Lara
Aqr 22h28'14" -6d15'0"
Gabriel Martinez
Uma 11h19'19" 52d17'35"
Gabriel Maslin Baugh
Leo 9h34'31" 32d24'58"
Gabriel Massoc
Lib 15h55'1" -12d57'32"
Gabriel Medina Holmes
Per 4h27'55" 51d7'33"
Gabriel Métivet
Vir 11h52'32" 9d37'32"
Gabriel Michael Fidibus
Her 18h35'44" 19d17'23"
Gabriel Michael Schafman
Dra 18h33'24" 72d18'41"
Gabriel Mignerori-Desnoyers
Vir 11h51'8" 9d58'0"
Gabriel Ngawati Parangi
Psc 1h46'0" 22d3'3"
Gabriel Olebar - Gabey-Baby
Uma 10h43'49" 63d40'42"
Gabriel Patriarca
Cas 23h39'19" 55d24'42"
Gabriel Paul David Wilkinson
Leo 11h0'12" 14d38'30"
Gabriel Pedro Boyle Winterson 2006
Sco 17h55'35" -38d17'38"
Gabriel Pitaro
Cnc 8h55'43" 16d19'19"
Gabriel Porter Chittenden
Aqr 20h39'42" 1d8'10"
Gabriel Quinn Constant
Cep 22h25'53" 74d27'51"
Gabriel Ralph Brigantino
Aqr 22h7'26" -2d1'42"
Gabriel & Regan Castelli
Cyg 20h22'59" 52d55'27"
Gabriel Reid
Leo 11h15'4" -1d18'7"
Gabriel Richard
Lyn 8h23'48" 35d28'59"
Gabriel Robert Allsop
And 0h59'26" 40d58'54"
Gabriel Rock
Ori 6h15'59" 13d42'26"
Gabriel Roland Malek
Aql 19h12'14" 5d10'34"
Gabriel Roosevelt Shulman
Cyg 21h49'7" 45d43'39"
Gabriel Ruggiero
Boo 15h25'42" 34d18'23"
Gabriel Ryan Griffin
Vir 12h11'45" 12d58'19"
Gabriel Salemi
Cyg 19h41'59" 30d32'35"
Gabriel Saylor Prescott
Gem 7h19'4" 33d14'30"
Gabriel Shane Plunkett
Leo 9h41'40" 16d59'27"
Gabriel T. 010205-1104-806 -20
Umi 14h33'54" 69d28'28"
Gabriel Thomas Maxon
Uma 8h50'35" 63d5'50"
Gabriel Thomas Stockwell
Gem 6h51'59" 34d25'21"
Gabriel Tovar
Vir 11h41'1" 9d4'8"
Gabriel Valentine Rembisz
Ari 3h22'22" 15d35'48"
Gabriel Venot-Montrot
Dra 18h37'50" 52d15'20"
Gabriel Vincent Basile
Lac 22h37'9" 39d28'55"
Gabriel Walker Blye
Uma 9h17'32" 54d44'26"
Gabriel & Wiki Forever
Ori 5h26'57" 10d8'56"
Gabriel William Vincent
Ori 6h12'28" 15d0'30"
Gabriel Zeringue
Gem 6h38'43" 17d18'42"
Gabriela
Cyg 21h11'25" 45d38'31"
Gabriela
And 23h23'2" 51d45'5"
Gabriela
Uma 9h40'49" 54d55'16"
Gabriela
Vir 12h24'38" -3d18'26"
Gabriela
Psc 23h3'14" -0d55'50"
Gabriela
Cap 21h34'12" -18d31'21"
Gabriela
Cap 21h6'46" -19d50'12"
Gabriela
Sco 17h34'5" -36d54'54"
Gabriela
Sco 16h12'54" -23d6'20"
Gabriela Alexandra
Leo 11h36'57" 16d10'14"
Gabriela Analise Balkin
And 0h25'9" 29d38'38"
Gabriela Benešová
And 2h23'26" 42d20'52"
Gabriela Biber
Ori 6h19'31" 11d5'27"
Gabriela Bonilla
Ori 5h48'10" 3d29'22"
Gabriela Castell Vargas
Gem 6h54'28" 17d12'58"
Gabriela Cindy Lopez
Lib 15h11'10" -12d54'8"
Gabriela Cucu
Sco 16h50'26" -42d30'51"
Gabriela & Daniela Gomez-Rodriguez
And 23h24'45" 49d32'19"
Gabriela Elena Gauna
And 1h39'25" 43d1'40"
Gabriela G
Sgr 19h22'11" -16d34'7"
Gabriela Gloria Velasco Jenkins
Vir 12h7'59" 0d59'55"
Gabriela Grace Giordano "7-13-99"
Lyn 7h15'24" 58d15'38"
Gabriela Haidari
Vir 15h3'58" 0d36'47"
Gabriela Herrera Oropeza
Leo 9h27'44" 26d45'0"
Gabriela Kleger
Ori 6h8'8" 20d51'35"
Gabriela & Laurent lovestar
Boo 15h40'5" 45d34'4"
Gabriela Manosis
Tau 4h31'8" 9d33'6"
Gabriela Marie
Sco 17h59'2" -38d24'32"
Gabriela Marie Novielli
Vir 12h48'44" 5d53'45"
Gabriela Marie Rizzo-Velez
Ari 2h34'52" 30d54'8"
Gabriela "Mausi"
Per 4h32'5" 39d26'25"
Gabriela Meier
Dra 17h9'28" 65d54'51"
Gabriela Möckel
Uma 9h0'22" 68d17'38"
Gabriela my Love
Per 3h42'43" 32d23'5"
Gabriela Piccone
Lyn 9h14'18" 37d15'32"
Gabriela Ramirez Peña
Ori 5h27'54" -0d17'10"
Gabriela Salinas Bailey
Cap 20h56'17" -23d10'35"
Gabriela Sipkova
Lyn 8h52'42" 44d11'56"
Gabriela Sterki
Boo 15h32'5" 45d34'7"
Gabriela Tello
Psc 1h37'52" 11d37'30"
Gabriela Tocchio
Ari 2h12'43" 21d7'23"
Gabriela Valerio Gonzalez
Ori 6h4'34" 13d23'31"
Gabriela Valerio Gonzalez
Ori 6h6'58" 13d51'58"
Gabriela Vela
Lib 15h44'40" -22d8'0"
Gabriela Venegas
Vir 13h20'5" 1d32'58"
Gabriela Victoria Panayoti
Lib 15h25'12" -20d21'10"
Gabriela Wigger, 30.08.1974
Uma 14h22'26" 57d54'53"
Gabriela's Deseo Star
And 2h21'16" 43d56'54"
Gabriela's Gift
Psc 1h19'57" 32d2'17"
Gabriele
Crb 15h55'52" 32d56'56"
Gabriele
And 23h15'1" 43d55'30"
Gabriele
And 2h25'37" 45d8'0"
Gabriele
Cas 1h33'2" 68d56'57"
Gabriele
Aqr 22h13'49" -1d0'25"
Gabriele and Carmela Solinto
Cnv 12h43'21" 47d49'30"
Gabriele Antonello
Uma 9h42'8" 61d47'8"
Gabriele Bellini 17 luglio 1959
Cnc 8h31'33" 12d19'56"
Gabriele Bestelmeyer 02.07.1953
Uma 11h50'10" 32d14'9"
Gabriele Daniele Russo
Cep 22h31'54" 71d35'39"
Gabriele David Javier
Del 20h40'37" 15d39'2"
Gabriele Gatti
Ori 5h44'9" 12d31'27"
Gabriele Grammig Brewer
Gem 7h10'17" 26d9'57"
Gabriele H.
Vir 13h12'15" -22d8'42"
Gabriele Maria Irene Rethemeier
Uma 11h10'41" 48d5'28"
Gabriele Marruso, Jr.
Leo 9h29'27" 25d29'13"
Gabriele Neissl
Uma 14h10'32" 58d37'49"
Gabriele Rendina
Peg 21h56'23" 34d9'3"
Gabriele Ruf
Uma 11h21'0" 39d22'20"
Gabriele&Simona Cavazzini
Ori 5h57'3" 11d45'21"
Gabriele Stöckl
Uma 10h50'3" 70d39'41"
Gabriele und Gernot Neumann
Sge 20h3'51" 18d17'59"
Gabriele Verhagen
Uma 11h43'9" 52d22'18"
Gabriele Volk
Uma 10h8'40" 61d48'51"
Gabriele Westbrook
Uma 9h21'35" 72d50'45"
GabrielJoleen
Crb 16h4'18" 38d8'18"
Gabriell Lynn Runyan
Uma 10h27'38" 63d48'52"
Gabriella
Aqr 22h50'9" -9d43'36"
Gabriella
Lib 14h47'16" -6d24'50"
Gabriella
And 2h23'43" 50d15'40"
Gabriella
Cas 1h28'26" 56d3'50"
Gabriella
And 1h41'56" 44d23'9"
Gabriella
Her 18h39'36" 19d42'49"
Gabriella
Leo 10h11'53" 26d18'33"
Gabriella
Tau 5h58'2" 23d55'36"

Gabriella
Tau 4h24'14" 24d7'17"
Gabriella
Vir 14h13'1" 1d16'20"
Gabriella
Cnc 9h3'22" 12d23'21"
Gabriella
And 0h36'46" 21d42'2"
Gabriella
Tau 3h37'10" 16d8'53"
Gabriella 1711
Cas 1h18'23" 69d48'13"
Gabriella and Angelina Makary
Uma 8h26'25" 68d8'45"
Gabriella Ann DiRosse
And 23h34'33" 45d30'52"
Gabriella Anne Garcia
Lib 15h30'32" -5d25'5"
Gabriella Anne Kerop
And 1h56'41" 42d32'3"
Gabriella anything is possible
And 2h29'19" 45d59'4"
Gabriella Barbagalo
Gem 7h46'38" 32d47'25"
Gabriella Bartasek
Cas 23h23'21" 53d29'20"
Gabriella Britth
And 0h25'49" 31d31'41"
Gabriella Carlevalis
Cam 6h22'48" 67d46'32"
Gabriella Cioban
Sco 16h38'37" -34d26'5"
Gabriella Cristina Brennan
Sgr 18h57'43" -16d6'28"
Gabriella de Reynal de Saint Michel
Psc 0h19'42" -5d12'26"
Gabriella Diane Pellerite
Psc 1h36'17" 23d42'37"
Gabriella Doherty
And 23h34'46" 35d42'55"
Gabriella e Matteo
Cyg 21h42'10" 35d33'36"
Gabriella Elise Szesnat
Sgr 19h21'10" -22d35'6"
Gabriella Elyse Hubchen
Lyr 19h13'38" 38d26'17"
Gabriella Estolano Abidog
Sgr 18h15'53" -19d9'11"
Gabriella "Gabby's Star"
And 1h31'5" 46d47'6"
Gabriella Galvagni
Ori 5h11'53" 2d16'57"
Gabriella Grace
And 0h14'4" 35d11'15"
Gabriella Grace Hunter
Psc 23h48'26" 2d25'47"
Gabriella Gropo Travain
Aqr 22h20'33" 1d28'36"
Gabriella Hailey
Sco 16h9'45" -9d59'41"
Gabriella Joy
And 1h0'35" 34d47'17"
Gabriella Joy Lowhar
Umi 10h43'59" 89d44'29"
Gabriella Juliann Theresa Tanner
Uma 8h57'29" 56d57'39"
Gabriella June
And 23h24'23" 47d1'54"
Gabriella Lynn DeLucia
And 0h44'32" 38d4'18"
Gabriella Mancassola, 13.09.1963
Tri 2h14'19" 31d57'8"
Gabriella Marie
Gem 7h32'4" 31d5'17"
Gabriella Marie
And 23h17'0" 43d33'34"
Gabriella Marie Genovese
Lyn 7h40'23" 57d53'16"
Gabriella Matyas
Cam 6h3'31" 64d25'41"
Gabriella Melina Donato
And 2h31'3" 49d51'27"
Gabriella Michelle Morrotti
Cas 0h40'32" 64d17'24"
Gabriella Mindy Glickman
Lyn 8h0'8" 59d3'45"
Gabriella Myza Gaqui Laranang
Uma 8h46'8" 48d13'22"
Gabriella Nancy Schwartz
And 2h30'2" 39d35'50"
Gabriella Nickoal Adams 06/30/07
And 23h11'46" 48d14'53"
Gabriella Nicole
Mon 6h51'23" 10d29'21"
Gabriella Nicole Jeffery
And 0h33'3" 29d51'34"
Gabriella Nicole Marti
Gem 6h5'18" 26d45'0"
Gabriella Nicole Suarez
Tau 3h57'27" 15d58'7"
Gabriella Nuñez
Lib 15h6'57" -7d3'52"
Gabriella Oman
Tau 4h35'8" 28d3'37"
Gabriella Penelopea Vadon Stoffer
Cas 0h30'21" 50d20'15"

Gabriella Rebecca Stober
Tau 4h34'51" 7d56'0"
Gabriella Renee Keeler
Cap 20h9'48" -8d50'40"
Gabriella Rhian Thomas
Lyn 7h45'48" 38d33'10"
Gabriella Rose
Mon 7h21'31" -1d20'6"
Gabriella Rose Catherine Petro
Ari 2h54'4" 28d25'30"
Gabriella Rose Colonna
Sco 17h47'44" -43d33'58"
Gabriella Rose Giardella
Cas 23h43'10" 55d26'9"
Gabriella Ruth Arendt
Umi 15h18'0" 73d22'27"
Gabriella Seena Del Giorno
Gem 7h25'3" 18d10'53"
Gabriella Simone Grzan
Cap 21h42'57" -13d59'46"
Gabriella Sofia Aleman
And 1h14'50" 43d2'17"
Gabriella Sonia
Vir 14h42'13" 3d24'15"
Gabriella Uzal Teodoro
Leo 9h23'49" 18d52'9"
Gabriella Victoria Graham
Lyr 18h53'22" 36d27'0"
Gabriella Violet Cordeiro
Cas 0h55'50" 56d59'13"
Gabriella Virginia Giliberti
Psc 1h22'0" 6d25'2"
Gabriella Zumstein
Lac 22h28'50" 41d33'20"
Gabriella, My Tootsie Pop
Leo 11h42'35" 24d25'16"
Gabriella's Star
And 22h57'53" 52d11'46"
Gabrielle
Cas 0h12'56" 59d52'31"
Gabrielle
And 0h15'22" 45d39'34"
Gabrielle
Per 4h46'28" 45d41'40"
Gabrielle
And 2h23'23" 49d6'50"
Gabrielle
Psc 0h52'37" 33d11'35"
Gabrielle
Crb 15h27'22" 25d39'41"
Gabrielle
And 0h41'50" 27d23'29"
Gabrielle
Tau 4h31'15" 6d42'0"
Gabrielle
Ari 2h19'45" 14d5'28"
Gabrielle
Psc 1h10'43" 10d29'48"
Gabrielle
Leo 10h49'20" 8d52'21"
Gabrielle
And 2h17'54" 47d26'36"
Gabrielle
Sco 16h46'3" -31d38'3"
Gabrielle
Cam 7h51'57" 64d58'29"
Gabrielle
Uma 9h37'34" 63d42'57"
Gabrielle
Cas 23h0'10" 54d39'34"
Gabrielle 29 Forever
Psc 0h45'26" 18d5'24"
Gabrielle 3-28-44
Ari 3h20'15" 21d43'2"
Gabrielle A. Berlinski
Cap 21h56'14" -17d35'45"
Gabrielle Adelia Dawkins
And 23h43'47" 38d56'11"
Gabrielle Alexa Brandwene
And 2h10'17" 46d28'31"
Gabrielle Alexandra
And 1h39'25" 41d58'41"
Gabrielle Amrani
Sgr 18h45'6" -32d40'35"
Gabrielle Andrea Morales
Sgr 19h37'1" -13d8'46"
Gabrielle Anne Bowne
Cas 0h46'8" 52d18'44"
Gabrielle Anne Pred
And 23h53'33" 35d21'44"
Gabrielle Barajas
Com 12h32'55" 26d23'52"
Gabrielle Bass
And 23h20'8" 47d6'46"
Gabrielle Bélanger
Umi 14h25'51" 66d43'36"
Gabrielle Brooke Freund
Aqr 22h15'42" -15d22'44"
Gabrielle Bueno
Ari 2h25'22" 24d41'36"
Gabrielle Christina
And 23h14'17" 43d26'26"
Gabrielle Christina Dickey
Ori 5h39'42" 6d34'38"
Gabrielle Christine Aaron
Uma 10h33'39" 59d48'48"
Gabrielle Chungunco
Lyn 7h25'57" 59d35'16"
Gabrielle D. Halpin
Uma 12h17'25" 56d1'8"
Gabrielle Desgroseilliers
Vir 13h13'11" -15d10'45"
Gabrielle Ederene Smith
Per 3h14'2" 48d50'5"

Gabrielle Elizabeth
Gem 7h12'26" 28d48'58"
Gabrielle Elizabeth Fager
Sgr 18h29'9" -17d0'10"
Gabrielle Elizabeth Poulin
Uma 9h43'52" 58d3'9"
Gabrielle Esquibel
Tau 4h24'59" 17d34'5"
Gabrielle et Michel
Boo 14h40'6" 16d24'53"
Gabrielle Faith
And 0h23'38" 46d40'50"
Gabrielle Gold
Cnc 9h6'48" 31d51'35"
Gabrielle~Hadrian 2454007.42373
Ori 5h34'24" -5d1'28"
Gabrielle & Hailey Gazett
Vir 12h59'24" -11d28'52"
Gabrielle Hall
Gem 7h9'13" 16d50'18"
Gabrielle Janeen
And 1h43'7" 43d5'33"
Gabrielle & Jeffrey Altstadter
Cnc 9h7'59" 32d48'31"
Gabrielle Jolie
Ari 1h52'55" 16d18'26"
Gabrielle Jones
Aqr 22h35'22" -0d40'5"
Gabrielle Joy Loging
Ori 5h52'12" 22d4'56"
Gabrielle Leah Phillips
Ori 5h55'22" 18d22'21"
Gabrielle Leigh Miller
Crb 15h37'41" 26d49'18"
Gabrielle Lena Lamparillo
And 0h44'12" 26d53'59"
Gabrielle Lyn Hargrove
Cnc 8h18'3" 10d41'25"
Gabrielle Lynn DeVault
Tau 5h52'11" 17d46'24"
Gabrielle Lynn Minor
Tau 4h16'42" 17d20'7"
Gabrielle M. Amoretti
Ori 6h14'24" 15d29'3"
Gabrielle Maia Caminiti
Cap 20h28'24" -16d4'45"
Gabrielle Margaret Slater
And 1h18'27" 48d49'57"
Gabrielle Marie
And 1h22'57" 45d56'1"
Gabrielle Marie
And 2h21'14" 50d10'11"
Gabrielle Marie
Ori 6h19'32" 14d3'46"
Gabrielle Marie Rios
Eri 4h46'51" -26d23'20"
Gabrielle Martin
Lyn 7h50'18" 55d38'35"
Gabrielle Mary
Cnc 9h17'31" 24d9'42"
Gabrielle Mattingly
Vir 12h47'5" 4d54'52"
Gabrielle May
And 1h38'11" 44d29'26"
Gabrielle McAfee
Ori 6h16'28" 10d57'39"
Gabrielle Meli Singer
Cap 21h36'56" -15d5'53"
Gabrielle Messana Malave
Per 3h31'10" 48d46'15"
Gabrielle Molina
Ari 2h55'57" 10d27'40"
Gabrielle Mongiovi
Cyg 19h52'32" 50d48'23"
Gabrielle Montoya
Ari 2h46'26" 20d53'6"
Gabrielle Nicole Johnson
And 0h15'49" 45d32'5"
Gabrielle Nicole Stoddard
Lib 15h46'41" -11d3'11"
Gabrielle Noelle Favors
Sgr 17h46'42" -21d20'51"
Gabrielle Norma Tickner
Umi 13h50'11" 70d3'33"
Gabrielle Paige Melamed
Leo 11h19'31" 18d8'53"
Gabrielle Prévost
Uma 8h55'13" 49d54'52"
Gabrielle Record
And 2h16'1" 38d54'33"
Gabrielle & Roland St. Amand
Per 4h44'38" 45d34'47"
Gabrielle Rosaleigh Rice
Aqr 21h35'24" -0d32'0"
Gabrielle Rose
And 2h26'15" 43d20'41"
Gabrielle Rose
And 0h1'43" 36d28'49"
Gabrielle Rose Cavenas
Tau 4h15'30" 0d41'26"
Gabrielle Rose Knudtson
And 23h38'30" 44d53'11"
GABRIELLE ROSELLOE
Aqr 22h45'52" 1d6'33"
Gabrielle Rozea' Bellinger
Uma 11h51'0" 41d21'11"
Gabrielle Ryan Taylor
Lyn 7h46'34" 51d31'12"
Gabrielle Schneider
Crb 15h37'24" 37d52'43"
Gabrielle Sophia Woods
Aqr 23h37'57" -15d27'28"

Gabrielle Stewart
And 1h16'43" 43d49'49"
Gabrielle Sue Lucero
Ari 3h13'39" 11d26'51"
Gabrielle Tait 90 Years of Love
Uma 10h7'47" 64d49'39"
Gabrielle Tristen Large
Cam 4h5'35" 67d4'55"
Gabrielle Victoria Williams - Gabbie's Star
Pav 18h16'3" -59d12'23"
Gabrielle Violet Marie Roelant
And 0h45'15" 37d35'10"
Gabrielle Walters
Uma 9h52'17" 59d10'59"
Gabrielle Wire
Tau 4h20'58" 25d21'26"
gabrielle zerilli
Cnc 9h14'25" 25d34'33"
Gabrielle Zoe Peer
Ari 2h53'50" 25d36'54"
Gabrielle's Red Love
Lib 15h55'43" -17d45'21"
Gabrielle's Smile
Umi 14h18'25" 73d51'41"
Gabrielle's Star
And 0h14'20" 41d33'54"
Gabriel's Angel
Per 3h6'3" 42d45'53"
Gabriel's Gate
Per 4h35'26" 40d1'18"
Gabriel's Light
Cap 20h23'0" -23d35'50"
Gabriel's Smile
Uma 11h54'13" 63d49'52"
Gabriel's Star
Cyg 19h47'15" 37d9'0"
Gabry
Cam 6h29'35" 62d29'35"
Gabry ti amo!
Cyg 20h17'46" 52d6'19"
Gabryel E. Rosmus
Ori 5h52'35" 5d22'22"
Gabrysia
Lyn 6h35'9" 54d21'33"
Gaby
Uma 10h32'12" 71d56'14"
Gaby
Sco 16h12'23" -15d40'39"
Gaby
Umi 13h44'24" 77d29'37"
Gaby
Sco 17h50'23" -42d36'51"
Gaby
Cnc 8h55'27" 16d6'37"
Gaby
Per 4h34'59" 40d29'29"
Gaby A+G Anthony
Cyg 21h41'38" 50d0'16"
Gaby Moreno-Hensley
Lyn 7h48'43" 48d30'47"
Gaby Pohl
Ori 5h15'33" -0d41'56"
Gaby Rohner
Uma 9h35'34" 58d27'6"
Gaby & Samantha Ghostine
Psa 23h6'40" -33d44'52"
Gaby Tischler
Lac 22h24'50" 47d47'46"
Gaby und Markus
Gem 6h18'21" 21d57'2"
Gaby323
Gem 6h48'57" 20d35'40"
Gaby-34
And 1h22'33" 39d56'28"
Gabye Lee Usina
Ori 5h32'53" 1d47'17"
Gabz's Star
Cap 21h8'55" -15d5'24"
Gadget
Mon 6h46'45" 7d50'41"
Gadget
Her 17h20'16" 14d2'40"
Gadiel Gomez
Dra 19h37'57" 63d16'7"
Gae
Uma 10h46'41" 59d23'36"
Gae
And 1h19'26" 42d22'28"
Gae
Boo 14h39'1" 51d22'41"
Gae Ann
And 1h14'7" 42d19'1"
Gaea
Sgr 18h2'47" -30d38'41"
Gael
Vul 19h30'24" 26d4'15"
Gaële
Umi 13h6'35" 72d25'4"
Gaelle
Uma 9h1'19" 68d24'21"
Gaëlle
Psa 22h47'27" -34d1'7"
Gaëlle
Cnc 9h4'41" 26d7'39"
Gaëlle Derqué
Ori 6h20'14" 10d55'2"
Gaelle et Mihran
Lib 15h22'47" -28d37'22"
Gaelle Maron-Ferdinand
Gem 6h23'54" 26d1'21"

Gaëlle Poulain
Col 5h34'24" -30d15'31"
Gaétan Côté - 1955
Cas 0h54'48" 57d17'20"
Gaetan Lorenz Pettigrew
Mon 7h20'28" -0d50'53"
Gaetana Isabelle Franklin
And 1h48'2" 37d45'39"
Gaétane Hains Simard
Cas 0h55'54" 47d42'30"
Gaetano 1218
Uma 13h31'20" 57d3'45"
Gaetano DiChiara
Vir 13h19'30" 11d18'0"
Gaetano Germinario
Psc 0h50'6" 20d22'17"
Gaetano & Kelly Calandrini
Cyg 19h43'25" 42d58'41"
GAETANO MILETO
Uma 11h0'3" 60d13'27"
Gaetano Recupero, Sr.
Psc 1h21'38" 25d46'45"
Gaethe, Ingo Wolfgang
Uma 8h24'11" 64d47'9"
Gaffaar
Ari 2h50'55" 27d51'36"
GAFJWFBRVAJVZAVMN-FAMFRDF
Vir 12h25'22" 2d0'12"
Gafurr Plakolli
Her 17h48'33" 21d45'40"
GaGa
Lyn 7h34'54" 36d0'24"
Gaga
Sgr 17h54'23" -29d57'16"
GaGa Our Bright and Shining Star
Ori 6h5'24" 19d46'58"
Gagaan Kax
Uma 9h1'48" 69d34'29"
Gagan Maan
Per 4h43'43" 46d28'23"
Gage
Cyg 20h1'5" 36d3'9"
Gage
Sco 17h27'10" -39d28'50"
Gage Acha
Uma 10h24'32" 61d28'56"
Gage Anderson
Uma 11h39'50" 53d50'37"
Gage Anthony Fetters
Gem 6h54'13" 33d58'7"
Gage Arik Sears
Per 4h13'7" 32d45'28"
Gage Bousheé
Dra 15h8'56" 55d16'1"
Gage Connor Folkerth
Tau 4h48'26" 19d25'53"
Gage Daniel Inman
Cap 20h51'40" -18d56'35"
Gage Emmet Kole
Ari 2h47'12" 24d33'32"
Gage Huff
Ori 6h20'3" 9d50'2"
Gage Joshua Mann
Uma 9h52'20" 58d59'50"
Gage Justice Schirmer
Uma 11h43'13" 56d10'21"
Gage L. Kistner
Cyg 20h27'6" 35d56'3"
Gage Munoz
Aql 19h22'10" -0d58'22"
Gage Remy Pier 10/05/94 - 08/09/04
Uma 10h23'40" 71d22'33"
Gage Scott Bernard
Lmi 10h27'35" 38d42'41"
Gage Thomas Griffiths
Psc 0h41'27" 17d30'43"
Gage William Jendrzey
Cnc 8h8'56" 25d11'52"
Gage William Spreder
Her 17h3'55" 30d42'29"
Gage's Wish
Cas 0h49'56" 64d6'26"
Gahgene Gweon
Gem 7h26'14" 32d23'4"
Gahibble
Peg 22h10'21" 6d17'25"
Gaia
Per 3h28'28" 41d38'17"
Gaia
Cam 3h33'59" 55d43'27"
Gaia 17-8-2005
Her 17h55'56" 24d48'42"
Gaia Elise
Ari 1h58'17" 12d50'37"
Gaia Iaccarino
Uma 9h27'39" 44d15'3"
Gaia Vaisya
Lyr 18h28'57" 32d11'6"
Gaidi
And 0h35'0" 42d14'2"
Gaige 102406
Uma 10h45'24" 58d24'2"
Gaige Maxwell Neumann
Umi 14h25'54" 75d44'51"
Gail
Aqr 22h16'19" -9d13'15"
Gail
Dra 17h8'20" 56d28'28"
Gail
Sgr 18h6'17" -27d4'59"
Gail
Cyg 20h28'40" 31d42'39"

Gail
And 23h52'32" 43d13'39"
Gail
Vir 15h5'22" 1d53'24"
Gail
Ori 5h28'22" 3d23'29"
Gail A Bode
Lyn 8h29'39" 35d43'52"
Gail A. Shanley
Cas 2h50'15" 57d43'37"
Gail Albon
Uma 9h4'40" 52d4'24"
Gail and Dan
Uma 9h25'4" 68d27'37"
Gail Angelico Cohen
Cas 23h40'12" 56d47'32"
Gail Ann Boskovich
Lyn 8h10'19" 33d58'46"
Gail Ann Bowen
Ori 6h24'27" 17d21'48"
Gail Ann Brumett
Uma 11h10'56" 41d22'40"
Gail Ann Cawley
Aqr 20h44'12" -11d49'29"
Gail Ann Glanz
Ari 2h16'17" 11d30'28"
Gail Anne Froida
Ari 2h42'12" 27d55'51"
Gail Arlene
Crb 16h18'20" 32d36'56"
Gail B. Pendergast
And 22h59'7" 39d35'55"
Gail Barnes Worlds Greatest Mom
Sgr 18h52'39" -29d50'8"
Gail Bassin
Umi 14h29'51" 73d53'56"
Gail Blanpied Ware
Crb 15h35'8" 38d13'46"
Gail Boswell
Sco 16h53'8" -42d33'28"
Gail Boulton
Aqr 22h14'54" 1d25'50"
Gail Brazilian
Mon 8h6'36" -0d49'27"
Gail C. & Richard H. Guscott
Cyg 21h54'31" 51d14'42"
Gail Cummo
Lib 15h3'43" -2d25'5"
Gail d'Auguste
Uma 9h24'51" 41d30'4"
Gail Delgado
Cap 20h28'52" -17d50'44"
Gail Dowty
And 0h25'15" 43d58'33"
Gail Ehrler - NOD
Tau 3h41'45" 16d58'48"
Gail Eileen
Sgr 18h37'50" -23d42'41"
Gail Elaine Mayo
Cra 18h20'43" -37d14'37"
Gail Elizabeth
Lyn 6h40'21" 52d51'18"
Gail Elizabeth
And 1h49'11" 39d27'48"
Gail Ellen Bell Koslowski
Cap 20h38'16" -22d15'51"
Gail Ellen Murphy's Sun
Leo 10h18'42" 10d54'8"
Gail Esther Deverell
Ari 2h47'49" 13d50'4"
Gail Evans-Oberbeck
And 2h16'20" 49d5'56"
Gail Fernandez
Sgr 20h0'47" -24d35'15"
Gail Fletcher
Cyg 20h39'30" 44d57'20"
Gail Garrett
Cyg 21h32'29" 36d39'14"
Gail Griswold-Burkett
Sgr 17h53'59" -29d22'7"
Gail Grossman
And 0h38'27" 39d27'43"
Gail Heidtke
And 23h49'26" 43d44'31"
Gail Irene
Gem 7h22'0" 33d31'18"
Gail Irene Always A Star
And 1h48'56" 44d29'35"
Gail Isabell
Cyg 19h40'13" 29d23'52"
Gail J. Webster
Ori 6h2'32" 14d14'34"
Gail Jagarnauth
Aur 6h2'14" 39d6'53"
Gail Jo Jones
Uma 12h6'17" 65d17'30"
Gail Kadlec Meadows
Lib 15h29'49" -15d22'1"
Gail Keller Armiger
Cas 1h33'56" 68d57'17"
Gail & Kevin
Aqr 22h30'59" -4d25'17"
Gail Kingsbury
Cas 2h10'57" 65d18'7"
Gail & Kjell
Col 5h39'40" -30d43'30"
Gail L. Bond
Uma 9h31'55" 65d17'25"
Gail Langkrahr
Tau 4h0'28" 8d46'51"
Gail Lee Clarno
Tau 4h43'52" 19d45'47"

Gail Leonessa
Cnc 8h39'12" 7d31'16"
Gail Lynne Hann
Psc 1h28'7" 12d57'6"
Gail Lynne & James Richard Barnette
Crb 16h13'48" 29d5'44"
Gail M. Alford
Cyg 21h14'24" 46d25'9"
Gail M. Bulas
Aqr 21h47'57" -7d5'27"
Gail Manishor
Ari 2h21'5" 27d23'7"
Gail Maria
Cra 18h53'7" -41d30'13"
Gail Marie Iorio DeFuria
Cap 20h27'33" -9d14'11"
Gail Marie & Michael Peter
Aql 19h38'56" 7d43'50"
Gail Marie Murphy
Sgr 18h33'15" -27d36'42"
Gail Marie Payne
Cnc 8h41'55" 13d56'18"
Gail Marie Pitman
Cyg 21h30'5" 30d39'56"
Gail Marlene Jackson
Crb 15h37'16" 31d38'44"
Gail Massey Feinberg
Aqr 22h12'49" -12d59'54"
Gail Miller Hatter
Lib 15h46'33" -16d53'30"
Gail Neumann
Uma 11h24'15" 44d21'19"
Gail "Pinkies" Carrier
Vir 13h5'41" -6d44'57"
Gail Pruner
And 23h19'18" 45d12'59"
Gail Renee Badalamente
Boo 14h16'52" 34d45'46"
Gail & Richard Polin
Her 17h56'30" 49d38'20"
Gail&Richard02112006naturelovers
Sge 20h2'24" 17d19'13"
Gail Richards a super star sister
Gem 6h48'50" 32d57'25"
Gail Roach
Leo 9h45'11" 24d9'41"
Gail Rubinger
Cas 0h46'55" 64d40'26"
Gail Ryan
And 1h24'12" 34d41'13"
Gail "Shania" Sewall Feb. 8th 1965
Crb 15h42'48" 27d42'15"
Gail Shepherd
And 1h3'59" 45d1'41"
Gail Snowgrass
And 23h28'28" 50d2'27"
Gail Stewart Arlotta
Cap 20h43'5" -24d40'51"
Gail Storm
Psc 1h59'59" 5d3'39"
Gail Sturrock
Sco 17h33'27" -38d33'23"
Gail Susan Tribendis
Tau 4h10'45" 19d6'20"
Gail T. Burton
Mon 7h19'52" -0d56'55"
Gail the Hummingbird
Gem 7h4'59" 30d3'9"
Gail Uyeda
Col 5h50'39" -39d7'50"
Gail Wrightson
Cas 1h19'14" 65d0'39"
Gail Zanelotti
Cyg 21h16'3" 41d11'51"
GAIL1951
Gem 7h9'40" 21d46'19"
Gailalicious
Sco 17h31'59" -30d31'7"
Gailanalex
Gem 6h22'13" 21d39'8"
Gaillouis
Cyg 19h31'56" 29d2'16"
Gail-Maree
Aqr 22h36'59" 2d36'7"
GailMarion
And 22h59'57" 41d6'25"
Gailon Stewart
And 0h44'38" 24d58'40"
Gail's Grace
Gem 7h25'33" 25d15'30"
Gail's Place
Per 3h38'0" 42d57'14"
Gail's Wish
And 0h56'10" 44d3'38"
Gailute A. Urbonaite-Narkeviciene
Umi 15h25'55" 68d43'3"
Gailya Bonzon
Crb 15h48'20" 34d1'42"
Gaines Todd Jr.
Uma 9h14'33" 69d26'32"
gaiolovezingg
Ori 5h52'35" 11d15'4"
GAITANO
Lac 22h24'15" 47d20'2"
Gaius Augustus Peake
Ori 6h3'36" 16d55'17"
Gaius Goree Webb
Uma 8h17'9" 62d49'34"
Gajewski
Peg 21h34'16" 17d57'4"

Gál Erika 18.
Ari 2h28'0" 21d32'27"
Gala
Ari 3h10'14" 25d32'51"
Gala Garde
Psc 1h9'38" 12d11'30"
Gala Sofia
Sco 16h7'57" -18d25'46"
Galactica Galdieri
Sco 16h14'47" -15d30'57"
Galán
Cnc 8h35'7" 8d9'46"
Galarneau's Christmas Star
Her 18h4'18" 36d16'29"
Galatea
And 0h6'58" 47d36'52"
Galatia Natasha Papathomas
Lib 15h7'16" -24d12'47"
Galavejh
Lmi 9h57'39" 38d55'18"
Galaxy 1
Cyg 21h33'21" 48d21'48"
Galaxy DJ Bill Matthews
Ori 5h39'0" -1d38'0"
Galaxy's Best Mom - Carol Armstrong
Leo 10h18'47" 15d55'51"
Galbraith
Lyr 18h48'45" 46d24'28"
Galdea
Cyg 19h46'40" 33d55'47"
Gale and Steve Klayman
Lyr 18h44'54" 40d47'55"
Gale Dawn
Cyg 20h50'32" 30d40'57"
GALE DIANE POLESCHUK
Vir 13h12'49" -20d15'40"
Gale Hall C.
Cnc 8h45'4" 18d34'52"
Gale Illuminatum Spectaculorum
Sco 16h3'41" -17d53'58"
Gale L. Wells
Leo 9h44'41" 8d40'20"
Gale Morgan Hunsuck
Sgr 18h58'49" -34d50'54"
Gale Richardson
Leo 11h1'55" 24d24'37"
Gale Sales
Psc 0h18'45" 18d15'45"
Gale Susan Rosario
Cas 1h15'44" 56d35'21"
Gale Winifred O'Leary
Lyn 8h7'22" 36d57'14"
Gale Zipp
Peg 23h16'6" 15d52'58"
Galen Anthony June
Cap 20h31'36" -14d22'15"
Galen Emerson Gibbs
Uma 11h51'39" 60d8'57"
Galen Erickson
Ori 6h16'5" 10d29'25"
Galen G. Bowman
Lyn 7h54'51" 59d23'23"
Galen G. Good
Cnc 9h11'18" 32d6'44"
Galen Grant
And 2h20'32" 45d1'59"
Galen Harig-Blaine
Sco 17h37'33" -39d54'48"
Galen Jones
Leo 9h28'30" 26d20'28"
Galen&Kathleen Everlasting Est11~00
Cyg 20h19'3" 47d44'39"
Galen T. Reynolds
Psc 1h23'15" 26d32'39"
Galen Thomas Lamphere Sr.
Aur 5h15'53" 36d7'43"
Galena Belle Cox
Cyg 20h1'13" 48d59'32"
Galena Seiler
Com 13h22'29" 15d15'34"
Galen's Waterspout
Cap 20h20'37" -11d46'58"
Gale's Constellation
Leo 9h47'35" 26d45'28"
Gale's Fire-King
Tau 4h30'57" 19d38'18"
Gale's Northern Cat Star
Crb 16h17'6" 36d54'10"
Galexy Mother One
Aqr 22h8'39" -0d4'3"
Galia
Cam 7h3'29" 64d56'34"
Galic Rada - Necina i moja zvezda
Cyg 19h52'59" 30d47'29"
Galik's Consilium Ancora Imparo
Lyn 7h43'39" 35d14'43"
Galilea
Uma 9h12'47" 49d10'0"
Galina
Psc 23h35'59" 2d35'16"
Galina Anastasia Smith
Gem 6h54'30" 22d7'6"
Galina and David 10/01/05
Cyg 20h0'22" 34d6'1"
Galina Georgieva Zheleva
Cas 0h39'0" 64d41'17"
GALINA K
Cyg 19h51'35" 54d51'14"
Galina Kojuhova
Psc 1h7'32" 23d21'29"
Galina Kozlova
Uma 12h2'5" 35d59'32"
Galina Maslov
Leo 10h55'38" 5d24'29"
Galina Pastornak
Sco 17h53'58" -34d44'16"
Galina Sergey 052905
Cas 23h40'9" 54d16'25"
Galina Wolf
Uma 8h23'57" 69d1'31"
Galina, My Dream Come True
Uma 12h30'50" 59d41'24"
Galinda
Sco 17h27'32" -45d21'25"
Galine
Aql 19h37'58" 0d8'20"
Galislava
Vir 13h30'35" 12d9'23"
Galit Couture
Sco 16h13'47" -35d47'53"
Galka
And 23h10'33" 44d10'59"
Gallacher FS
Her 17h30'12" 31d49'23"
Galland-Brault Marine
Crb 16h3'30" 28d21'46"
Gallatin- Michelle L Mantz
Aqr 20h55'25" 0d42'48"
Gallitzin
Cyg 21h54'36" 41d36'26"
Galloping Gaucho George
Ori 6h9'40" 16d0'57"
Galloway
Car 10h9'38" -62d39'32"
Galloway Dreams
Ori 5h59'5" 6d57'53"
Galloway's Special Place
Oph 17h56'23" -0d53'59"
Gallygoo (Gail Reading)
Tau 5h51'30" 25d0'18"
Galo Chaves IV
Gem 7h19'18" 25d2'31"
Galochka Magidenko
And 0h41'48" 36d15'28"
Galsheimer, Ralf
Ori 6h18'45" 10d5'41"
Galway
Aql 19h31'34" 3d29'20"
Galway Rainbow
Uma 11h18'36" 61d24'17"
Galway Skylight
Aqr 22h20'42" -6d36'34"
Galzarius Engelian
Ori 5h48'42" 2d52'17"
Gama216
Lyr 18h52'25" 33d31'37"
Gamal Aziz
Uma 11h37'48" 56d47'12"
Gamble
Boo 14h42'36" 53d56'30"
Gamblemouse
Uma 12h32'38" 57d49'45"
Gamelan
Lyn 7h24'48" 45d35'13"
Gamma & Gamps
And 1h41'17" 50d2'35"
Gammapolis
Uma 9h0'50" 70d5'52"
Gammer
Psc 0h53'59" 14d47'27"
Gammie Fay
Lib 14h47'51" -1d13'29"
Gammi's Heart
Cyg 20h21'34" 50d29'13"
Gammy's Love
Crb 15h32'51" 27d46'35"
Gamper, Harald
Vir 14h38'13" -1d37'27"
Gampy's Galaxy
Cap 20h38'33" -23d51'53"
Gamsjäger, Maximilian
Uma 11h58'45" 60d51'43"
Gamze
Uma 9h35'20" 44d27'51"
Ganaa
Lmi 10h9'35" 36d16'30"
Gancia
Ori 6h2'57" 7d24'20"
GANDME4EVA
Uma 11h21'1" 34d7'19"
"Gang Gang", Our Special Angel
Pho 1h43'35" -43d44'43"
Gangar
Ari 3h1'52" 26d26'44"
Gangsta D
Ori 5h7'58" 8d22'8"
Gangy
Psc 1h3'19" 32d42'29"
Ganja Kitty
Gem 7h39'42" 22d35'0"
Ganky
Uma 10h0'30" 70d18'40"
Gannon C. Michna
Cam 3h49'39" 66d18'42"
Ganny Mac
Uma 9h31'50" 50d9'54"
Ganokorn & David Moody
Cyg 19h40'56" 34d31'39"
Gansito u will always b my #1 hero!
Per 3h30'55" 44d38'39"
ganzike
Cyg 20h11'54" 39d7'28"
Gaol mo chridhe
Peg 22h51'47" 15d14'41"
Gapa Bain
Del 20h16'57" 13d36'4"
Gapo Loves Brenda
Uma 8h47'13" 54d6'34"
Gappy the Great
Uma 9h50'40" 64d9'43"
Gar Bear
Ari 2h23'26" 23d13'28"
Gar Den
Ori 5h4'9" 6d57'0"
GARACH
And 1h1'49" 39d18'42"
Garaczi János
Cnc 8h21'34" 24d31'14"
Garas Matyi
Uma 11h12'54" 50d9'34"
Garbonzo Bean
Cyg 21h24'8" 45d4'7"
Garbus
Sgr 19h40'48" -12d26'38"
Garby
Ori 5h16'44" 1d29'7"
GARCIA #1
Uma 9h56'36" 57d30'28"
Garcia's Vision
Uma 8h37'24" 63d58'51"
Garðar Thór Cortes
Tau 4h17'22" 27d1'13"
Garden Gypsy
Lyr 18h47'32" 43d51'38"
Gardendale 2004
Uma 11h24'35" 51d16'57"
Garde'neus
Uma 10h34'19" 65d41'13"
Gardiner W Bridge
Uma 8h27'45" 69d2'51"
Gardner Morris Thompson
Crb 15h17'19" 25d51'44"
Gardner Stanton Bailey
Cep 21h0'45" 58d51'46"
Gardner's "Star" for TRE 2006
Ori 5h52'5" 12d4'22"
Gare-Bear
Leo 10h4'9" 9d43'42"
Garee'
Ari 2h35'25" 25d56'6"
Garehbaghi Agdam Monica 1988
Cas 0h26'11" 62d27'57"
Garen Hagop Yepremian "Valfigrar"
Uma 8h58'45" 51d2'29"
Garenth Lafielle
Aqr 22h55'7" -7d6'0"
Gareth
Cyg 20h14'1" 42d46'12"
Gareth
Leo 10h37'36" 7d35'52"
Gareth - Allstar Forever 10.10.2000
Gem 7h22'42" 35d7'22"
Gareth and Linda Massey
Ori 5h5'10" 5d59'4"
Gareth and Sarah's Star
Uma 8h36'25" 65d9'39"
Gareth and Zoe always.
Lyn 7h36'40" 37d52'29"
Gareth Cotter
Cap 21h10'26" -15d27'9"
Gareth Egerton
Psc 0h58'50" 25d52'48"
Gareth "GT" Best Footballer Star Number 1
Per 3h3'24" 46d45'17"
Gareth Hanks
Dra 18h54'18" 53d59'43"
Gareth Joubert's Star
Per 2h50'30" 44d10'31"
Gareth Longhurst
Cnc 8h1'55" 12d27'46"
Gareth Spencer Elliott Perry Sanderson
Cep 22h7'42" 57d1'40"
Gareth W. Hagen
Uma 9h31'30" 57d37'27"
Gareth Wheeler
Psc 22h51'40" 7d1'45"
Gareth Wingfield
Aur 5h43'40" 40d43'46"
Garett S Otterbein- Eagle Scout
Aur 5h4'21" 46d58'41"
Garfield
Ori 5h48'58" 5d52'19"
Garfield E. Greenhalgh
Cyg 21h4'0" 54d37'6"
Garfield Landry
Aqr 22h27'46" -1d36'10"
Garfield Larson
Umi 14h4'15" 77d21'47"
Garfield Meredith
Ori 5h17'26" 15d26'45"
GarfieldCrystalBigShotFire StarRoom5
Uma 11h26'56" 64d20'23"
"GarFon" Together Forever
Cyg 21h47'46" 44d38'32"
Garids Star
Uma 10h31'22" 56d16'43"
Garie Leigh Hanson
Aqr 22h55'31" -8d55'17"
Garik Robertovich Gevondyan
Dra 17h36'40" 63d57'37"
GARIMA
Cap 20h27'29" -17d2'17"
Garion Michael Baksh
Cep 23h58'14" 84d13'17"
Garjen 42164
Tau 5h46'4" 13d18'20"
Garla
Cyg 20h23'15" 43d27'51"
Garland
Cap 20h15'22" -16d13'47"
Garland Gene Davis
Uma 9h13'0" 50d44'47"
Garland L Ford Jr
Her 16h32'47" 33d44'28"
Garland Leon Gregoire
Cnc 8h20'58" 20d43'8"
Garland O'Quinn
Her 16h49'56" 33d52'2"
Garland Parental Units
Cyg 21h42'28" 46d6'21"
Garland & Patrick Fitzgerald
Uma 10h28'19" 54d11'34"
Garliz 2003
Ori 6h18'35" 10d45'4"
Garlyn
Dra 15h48'31" 60d18'41"
Garna Anetta Rohrer
Psa 22h41'20" -34d9'48"
Garner
Aql 19h29'18" 2d26'59"
Garnet Chicken Butter Chunk
Cyg 20h18'26" 41d7'9"
Garnet Marie
Lyn 7h6'34" 51d15'31"
Garnett 72
Uma 13h37'55" 51d37'44"
Garo Donabedian
Uma 9h18'6" 67d13'47"
Garofalo Raffaello
Cas 0h4'59" 56d54'54"
Garon Jayne Rose Daugherty
Sgr 19h24'4" -15d37'31"
Garould Martin
Vir 14h7'44" -15d39'36"
Garou's Star
Cas 22h59'56" 54d36'21"
Garpeg
Leo 11h52'36" 17d20'1"
Garren Joseph Dutto
Sco 16h15'34" -9d29'11"
Garren Wayne Jones-Holden
Cep 23h29'36" 80d19'57"
Garret John Cousin
Leo 11h20'23" 7d10'19"
Garret Lee Peeler
Per 3h7'11" 54d5'47"
Garret Michael
Ori 4h51'0" 13d55'3"
Garret Moore
Peg 22h21'56" 5d3'52"
Garrett
Aql 19h57'49" 1d19'5"
"Garrett"
Aur 5h58'53" 41d26'6"
Garrett
Uma 14h0'2" 55d50'27"
Garrett Addison Crouch
Psc 0h45'5" 8d49'30"
Garrett and Ashlyn
Cyg 19h46'48" 42d30'30"
Garrett Andrews "I love you!"
Umi 17h3'59" 75d1'0"
Garrett Beall Deatherage
Sco 17h52'39" -36d24'35"
Garrett Bellora
Uma 10h37'46" 66d28'41"
Garrett Brian Gorsky
Lib 15h8'49" -10d54'22"
Garrett Brown Marquis
Lib 14h25'21" -14d23'6"
Garrett Charles McLaurin
Dra 17h2'28" 63d53'31"
Garrett Colman
Her 16h47'38" 30d39'42"
Garrett Curtis Coman
Dra 18h33'32" 64d33'7"
Garrett Dean Hlavinka (Blue Star)
Ori 6h20'51" 5d40'51"
Garrett E. Kowal "10/04/1980"
Lib 15h1'40" -0d49'3"
Garrett Edward Angshed
Uma 11h23'33" 47d17'10"
Garrett Evan Cudahey
Cap 21h4'21" -20d35'57"
Garrett James
Lib 14h50'8" -2d39'45"
Garrett James Budjinski
Boo 14h43'11" 14d57'32"
Garrett James Taylor
Her 18h48'45" 20d39'13"
Garrett Jameson Weekes
Sco 16h39'3" -29d48'46"
Garrett Jamison Wimmer
Cyg 21h19'58" 45d25'35"
Garrett & Jillian
Cyg 21h57'2" 38d27'5"
Garrett Lee Burleson
Aql 19h51'25" 11d19'40"
Garrett Lee Xanders
Her 16h57'19" 19d31'0"
Garrett Martin Fogg
Her 18h19'24" 19d9'9"
Garrett Matthew Anderson
Mon 8h0'43" -0d34'51"
Garrett McKinley (G-Man) Greenlee
Her 17h40'22" 19d11'17"
Garrett Michael Fraser
Leo 10h37'38" 14d31'0"
Garrett Michael Smith
Sgr 18h38'56" -22d39'24"
Garrett Michael Woo
Uma 12h28'47" 63d9'45"
Garrett Reuther
Her 17h39'44" 29d1'24"
Garrett Robert John Linton
Tau 4h5'3" 5d15'22"
Garrett Schenk
Aur 5h48'58" 31d53'21"
Garrett & Theodora always & forever
Lyr 18h30'35" 35d52'38"
Garrett Thomas Meisinger
Tau 5h54'57" 24d12'19"
Garrett Van Davis
Aql 20h7'17" 2d8'26"
Garrett W. Blunt
Lib 14h50'7" -14d41'44"
Garrett Wang
Sgr 17h54'19" -29d51'24"
Garrett William
Ori 6h8'51" 15d27'2"
Garrett William Taintor
Uma 14h13'16" 61d57'36"
Garrett William Wyman
Uma 11h8'5" 28d21'19"
Garrett Wright Brown
Ori 5h31'8" 11d1'28"
Garrett Wynston Johnson
Lyn 9h7'17" 35d10'12"
Garrett's Glory
Lyn 8h3'25" 51d12'43"
Garrett's Star
Leo 9h58'3" 21d44'57"
Garrett-Star of Lerma
Uma 10h16'29" 48d40'57"
Garrett-Zimmerman
Uma 10h23'56" 66d46'10"
Garrick Anthony
Umi 14h40'16" 75d30'8"
Garrick Louis
Ori 6h9'4" 17d23'12"
Garrison "God's Servant" Summers
Uma 8h49'49" 66d36'2"
Garrison Wayne Fisher
Lyn 9h24'48" 40d12'51"
Garrit Brian Patrick Cerruto
Cnc 8h14'35" 31d42'24"
Garrit Dean Nixon
Uma 9h42'23" 68d58'52"
Garriy Zaslavskiy
Lyn 8h20'13" 38d3'46"
Garron - 152
Her 17h11'37" 26d46'8"
Garry
Tau 5h38'14" 27d52'20"
Garry and Theresa's Love Star
Cyg 19h58'10" 32d30'22"
Garry Bass
Ori 6h19'19" 10d35'5"
Garry Beard
Vir 13h47'50" 1d43'43"
Garry Beresford
Ori 5h20'6" 6d35'45"
Garry Butler
Uma 8h19'6" 69d11'39"
Garry Coates
Cnc 8h35'10" 9d38'19"
Garry D. Cyrus
Ori 6h7'39" -0d22'49"
Garry Eugene Wood - 'Woody'
Her 18h3'48" 36d36'21"
Garry Everett Thompsen
Per 2h44'7" 54d10'36"
Garry George Graham
Cep 23h55'4" 66d43'19"
Garry Hart
Her 17h31'27" 34d32'51"
Garry Ivan Brown
Leo 10h18'9" 18d46'1"
Garry John Clark
Sco 16h10'49" -11d39'52"
Garry John Leavy
And 0h2'47" 46d4'29"
Garry Lee
Vir 12h38'50" 8d59'43"
Garry & Linda Cucci
Cyg 20h25'4" 51d0'14"
Garry & Lois Jones 11/18/1983
Cyg 20h56'34" 46d10'5"
Garry N Hubbard
Cap 21h35'22" -19d54'0"
Garry Robles
Tau 3h27'16" 17d57'31"
Garry Shells
Ari 2h47'42" 25d39'17"
Garry Struth
Cyg 21h19'47" 32d53'27"
Garry Wayne Parrish
Cnc 8h1'13" 19d35'8"
Garry - You will never dream alone!
Lib 15h41'35" -11d7'6"
Gar's Forgiveness
Aur 5h55'55" 37d23'12"
Gar's Leading Light
Pyx 8h33'13" -23d16'59"
Garth Louis Manheim
Per 4h15'15" 51d20'37"
Garth Thomas Bull Snr
Col 6h7'57" -42d29'43"
Garthie Muggeo
Ori 6h15'53" 10d4'37"
Garth's Glorious Glow
Sgr 17h52'40" -28d47'47"
Garth's Star
Cru 12h38'59" -61d15'3"
Gartichek
Psc 1h22'57" 12d4'18"
Gartner, Elke
Ori 5h7'21" 5d52'25"
Gartner, Renate Elisabeth
Uma 8h30'27" 64d15'15"
Garvis Noyles
Vir 14h11'54" -8d40'55"
Garwood's Spirit
Lmi 9h33'28" 34d59'40"
Gary
Her 17h52'46" 47d12'46"
Gary
Cnc 9h1'20" 7d5'51"
Gary
Aql 19h35'51" 10d39'40"
Gary
Cnc 8h38'50" 19d34'49"
Gary
Uma 11h34'30" 29d14'40"
Gary
Cap 20h48'31" -21d0'6"
Gary
Cen 12h12'27" -38d23'53"
Gary
Sco 17h54'2" -35d42'47"
Gary A Miller aka RainbowRider
Ari 2h42'47" 13d36'1"
Gary A. Walker, Jr.
Tau 4h18'25" 18d16'55"
Gary A. Worsham "Pop"
Her 16h45'55" 32d44'43"
Gary Alan Golwitzer Jr.
Her 16h56'25" 36d26'40"
Gary Alan Musselman Jr.
Cyg 20h3'37" 32d24'12"
Gary Alan Ramos
Ari 2h34'53" 21d55'48"
Gary Alyn Blattenberger
Leo 11h54'0" 17d54'29"
Gary and Billy Jones
Uma 10h13'24" 50d53'22"
Gary and Bonnie Tucker
Tau 4h49'21" 22d50'6"
Gary and Glenda Carnahan
Cas 0h34'24" 60d24'17"
Gary and Hazel Conway
Cyg 20h22'22" 33d26'18"
Gary And Jamie
Cyg 19h47'42" 29d33'42"
Gary and Jenna
Vir 13h35'40" 3d3'2"
Gary and LaKiah Gregory
Gem 7h40'29" 24d6'16"
Gary and Noreen Kordosky
Gem 7h23'1" 18d12'24"
Gary and Sara 3/23/03
Cnc 9h18'55" 25d45'39"
Gary and Sheryl Starostka
Cyg 21h15'48" 47d11'35"
Gary and Suzie
Cyg 19h45'55" 53d57'5"
Gary and Tek
Her 16h21'41" 20d29'9"
Gary Anthony Sepulveda
Per 3h31'8" 51d27'59"
Gary Arthur Bryce
Ori 5h29'15" 0d49'15"
Gary Arthur Daughters
Psc 2h2'9" 3d49'46"
Gary B.
Umi 15h9'20" 71d1'30"
Gary B. Becht
Uma 13h30'17" 61d10'31"
Gary B Beckwith Jr.(The Chosen One)
Her 17h16'13" 36d25'6"
Gary Beasley
Her 16h26'29" 7d3'57"
Gary Bishop
Sct 18h22'16" -7d50'50"
Gary Blanton's Eastern Star
Aql 20h8'34" -0d1'48"
Gary Bogner
Aqr 23h28'33" -24d13'14"
Gary Bonneau
Leo 11h14'39" 17d30'33"
Gary Boyer and Michael Foy
Uma 8h14'6" 69d40'8"
Gary Brady Elliott
Cnc 8h55'53" 24d15'17"
Gary Brandt
Vir 12h7'17" -11d5'18"
Gary Brill 1
Ori 5h32'37" 3d29'32"
Gary Bruff
Psc 0h33'22" 7d56'1"
Gary Bucchino
Uma 11h9'20" 58d15'15"
Gary C. West
Her 16h58'6" 18d49'46"
Gary Carlsen
Her 17h47'30" 34d15'26"
Gary Carmen Tisone
Tuc 23h7'18" -66d40'16"
Gary Chapman - Spartan Blue Giant
Aql 19h20'23" 0d16'12"
Gary Charles Strouhal
Psc 1h10'24" 32d17'36"
Gary & Cheryl Cotten
Sge 19h34'58" 18d56'57"
Gary & Cindy Jordan
Leo 11h23'34" 3d7'16"
Gary Clothier
Sgr 19h38'2" -13d56'12"
Gary Cooper
Leo 11h27'54" 4d37'18"
Gary Copeland-Loml
Aur 6h28'11" 34d59'1"
Gary Crouch
Per 3h13'14" 52d21'39"
Gary Cummings
Lac 22h15'21" 54d9'3"
Gary D. Kerin
Cru 12h51'5" -58d34'47"
Gary D. Luster
Uma 12h3'19" 35d1'40"
Gary D. Moore
Cap 20h31'39" -19d40'34"
Gary D. Newberry ~ Flint, MI
Uma 8h57'30" 56d30'46"
Gary D Robinson
Per 4h37'21" 31d42'17"
Gary Dale Christiansen
Sco 16h50'4" -41d54'41"
Gary Dale Strater
Cen 11h40'53" -52d5'42"
Gary Dalton
Ori 5h3'22" -0d26'45"
Gary David Kent, My Shining Star
Her 18h2'56" 24d48'36"
Gary David Summers
Cru 12h26'25" -61d45'16"
Gary Dean
Gem 7h29'43" 19d47'33"
Gary Dean Whitlock
Gem 7h39'3" 25d13'35"
Gary Delos Monigold
Lyn 7h53'23" 57d16'0"
Gary DelVecchio
Ori 6h1'35" 19d29'41"
Gary Dennis
Per 2h48'52" 54d13'28"
Gary Dianne Megan Erik Sigley
Uma 10h33'40" 52d46'8"
Gary Don Abernathy
Cap 21h54'28" -18d10'2"
Gary Don Stuckey
Vir 12h13'46" 2d51'48"
Gary Donald
Peg 23h21'39" 32d45'13"
Gary Dudenhoeffer
Her 17h17'13" 37d0'59"
Gary Durand
Tau 5h12'9" 27d35'15"
Gary Durbridge
Mon 6h45'27" -0d2'25"
Gary Dwayne Smith
Aql 19h55'33" -0d25'51"
Gary E Kirkman
Sgr 17h51'37" -19d40'50"
Gary E Massoud
Gem 6h51'28" 30d59'16"
Gary E. Payton
Peg 21h40'39" 16d11'15"
Gary Edward Geers
Tau 3h50'32" 14d21'2"
Gary Edward Pollard
Lib 15h44'44" -9d50'40"
Gary Edward Vogt
Uma 9h4'0" 60d28'25"
Gary Edwards
Dra 18h28'43" 51d56'43"
Gary & Elaine Viggiano
Uma 13h49'37" 56d7'50"
Gary & Ellie's Star
Lmi 9h39'16" 37d54'18"
Gary & Emilie Peterson
Cyg 20h19'37" 53d8'45"
Gary F. Kreps
Per 4h8'14" 50d53'35"
Gary Fink
Uma 8h29'0" 71d21'34"
Gary Fisher
Ori 6h14'23" 2d46'29"

Gary - Forever In Our Hearts
Cep 22h29'9" 57d49'37"
Gary Frances Bender
Uma 10h34'40" 45d17'36"
Gary Francis Flynn
Her 17h6'35" 31d18'33"
Gary G. Gilbertson
Crb 15h38'37" 26d9'57"
Gary G. Smith
Lyn 6h57'57" 57d12'16"
Gary Gambrill
Sco 17h36'21" -35d21'21"
Gary Gayle
Sco 16h10'56" -17d45'41"
Gary Gentile
Ori 6h20'11" 10d41'36"
Gary & Geri "A Love that Shines"
Cyg 19h42'48" 41d57'8"
Gary G.G Wilhelm
Her 17h31'29" 27d29'53"
Gary Gould
Aqr 22h25'57" -23d15'18"
Gary Grant Roosa
Lib 15h33'38" -24d29'19"
Gary Graverson
Tau 3h37'46" 27d17'55"
Gary H. Santerre
Cyg 19h36'56" 54d8'49"
Gary Halliwell's Guiding Light
Uma 9h44'17" 47d34'53"
Gary Harper
Umi 15h47'26" 78d27'11"
Gary Hembd
Sco 16h15'2" -35d39'28"
Gary Hobley
Cen 13h48'25" -41d31'45"
Gary Hoffman
Vir 12h32'4" 12d39'24"
Gary "Hubble" Way
Uma 12h6'42" 46d7'3"
Gary Huffman
Lib 15h38'28" -28d39'11"
Gary Ivey
Aqr 22h26'7" -0d51'8"
Gary J. Swiden Family Star
Uma 10h32'20" 60d9'16"
Gary J. Taylor
Cnc 8h53'37" 15d13'21"
Gary Jackson Lynn
Lib 15h37'21" -11d8'28"
Gary & Janice Kubiak May 10, 1975
Cyg 20h39'35" 45d6'29"
Gary John Beamish
Ori 5h31'0" 9d27'42"
Gary John Burns
Uma 8h54'18" 51d46'41"
Gary John Gosztonyi
Sgr 19h14'18" -23d42'5"
Gary John Mason
Uma 10h42'13" 64d44'42"
Gary John Powell
Per 3h34'52" 47d21'55"
Gary & Kassandra Fitzgerald
Cyg 20h0'3" 45d29'27"
Gary Keith Swann
Cap 20h7'43" -23d10'26"
Gary Kellogg Northrop
Per 3h10'21" 53d35'0"
Gary Kerr
Per 4h1'59" 42d18'20"
Gary Kidd
Ori 6h6'14" 20d12'48"
Gary King
Cep 0h3'10" 75d23'8"
Gary Knight, Jr. & Jaime Drebit
Cma 6h31'58" -14d14'19"
Gary Koenig
Uma 11h0'31" 57d17'40"
Gary L. Eddy
Per 3h17'21" 48d25'48"
Gary L. Flack
Psc 1h22'55" 31d55'2"
Gary Lager
Per 3h11'58" 50d59'1"
Gary Landi
Her 16h30'17" 47d22'55"
Gary Landry
Ser 16h11'26" -0d17'42"
Gary Lanning
Her 18h14'21" 41d30'48"
Gary Lay
Pho 0h43'27" -48d1'15"
GARY LEE CARDUCCI
Vir 12h47'35" 12d10'41"
Gary Lee Emerson
Boo 15h38'2" 42d23'24"
Gary Lee Krebs
Uma 11h48'55" 57d15'45"
Gary Lee Maines
Her 18h47'26" 19d36'33"
Gary Lee Mathews
Her 18h18'42" 16d27'30"
Gary Lee Park
Ari 2h49'16" 19d3'50"
Gary Lee Sliefert
Her 17h48'55" 19d9'42"
Gary Lee Stephenson
Uma 9h21'13" 44d57'11"
Gary Lee Vandenberg
Cnc 8h58'6" 28d46'45"
Gary Lee Wertenberger
Cnc 8h32'4" 21d58'16"
Gary Leigh Holmes
Crb 15h46'34" 33d17'58"
Gary LeRoy Steelman
Ori 4h47'34" -0d23'22"
Gary Levar
Uma 8h35'12" 72d1'59"
Gary Livsey
Per 3h42'53" 38d49'14"
Gary Loren Bright
Her 17h19'55" 28d1'57"
Gary & Lorna
Cyg 19h31'18" 30d11'29"
Gary Louis Sparks
Tau 4h2'44" 18d36'39"
Gary Lowell Gaither
Ori 4h55'6" 10d54'11"
Gary Lun Swanson
Cyg 20h59'59" 31d12'50"
Gary Luscott
Umi 14h37'50" 83d48'43"
Gary & Lyn - 25 years of love
Cra 18h57'21" -39d23'29"
Gary Lyn Knight
Uma 10h54'29" 34d52'49"
Gary Lynn
Uma 11h16'36" 45d39'23"
Gary Lynn Green Jr.
Tau 4h4'17" 7d4'41"
Gary Lynn Stillwell Sr.
Sgr 19h34'20" -29d39'12"
Gary M. Georgeff
Ori 5h0'19" 13d18'21"
Gary M. Jackson
Uma 10h19'18" 49d13'39"
Gary M Miller
Aqr 22h22'42" -23d34'51"
Gary M Zuckerman
And 23h8'59" 47d13'28"
Gary Marousek
Uma 13h14'46" 53d49'13"
Gary Martin Furness - Gary's Star
Uma 11h8'3" 44d16'56"
Gary McLean
Cep 2h3'52" 80d11'37"
Gary Michael Gervasoni
Uma 9h11'2" 48d22'10"
Gary Michael Green
Lmi 10h27'1" 38d46'25"
Gary Michael Schafer
Lyn 7h39'18" 49d8'21"
Gary Michael Sweeden
Aqr 21h14'46" 0d57'17"
Gary & Michelle True Forever
Uma 10h29'32" 45d45'57"
Gary Midnight's Magic Star
Lyn 7h22'0" 48d56'31"
Gary Monroe
Uma 11h30'14" 54d37'45"
Gary Moore
Leo 10h22'13" 12d21'29"
Gary N. Cohen
Aur 5h11'47" 48d37'10"
Gary Nelson
Per 3h18'19" 44d26'28"
Gary Newberry
Cep 22h23'53" 62d3'13"
Gary Nicol
Vir 14h0'10" -8d17'29"
Gary P. Gauvin
Lib 15h13'20" -15d5'6"
Gary P. J. Franklin
Her 17h46'42" 22d43'16"
Gary P. Jones
Aql 19h55'20" 11d38'1"
Gary Paul Moniz
Uma 9h34'41" 45d55'14"
Gary Paul Thurman
Ari 3h6'32" 11d7'56"
Gary Paxton & Jacqueline Kouri
Mon 6h46'19" -0d41'32"
Gary Payton
Tau 3h27'16" 17d16'49"
Gary Peacock
Aql 20h2'45" 11d46'3"
Gary Peter Bychyk
Boo 15h35'29" 45d48'53"
Gary Plumley
Her 16h30'46" 32d5'13"
Gary Plunkett's Star
Umi 15h28'26" 71d6'33"
Gary R G Costin
Psc 0h57'5" 9d14'40"
Gary R. Luther
Lib 15h20'59" -6d27'13"
Gary R. Moudy
Cnc 8h30'21" 9d26'54"
Gary R Welke
Cas 23h36'11" 55d45'57"
Gary Randazzo
Lyn 7h1'26" 45d9'30"
Gary Rankin Cline
Sco 17h44'41" -41d44'55"
Gary Raymond Haas
Uma 11h55'7" 53d44'58"
Gary Raymond McDonald
Her 17h21'3" 22d43'14"
Gary & Renee Haseley
Dra 12h13'23" 73d1'3"
Gary Rhys Warnell
And 1h9'30" 37d43'12"
Gary & Richadene Childers
Vir 12h27'3" -0d38'4"
Gary Richard Kirk
Aqr 23h28'1" -19d51'37"
Gary Richard Morrison
Gem 6h6'17" 26d55'18"
Gary Richard Williams - "Lord Gary"
Cep 20h36'17" 67d20'1"
Gary Robert
Per 3h5'15" 54d18'47"
Gary Robert Molinaro
Lyn 8h25'11" 34d53'50"
Gary Robert Trock
Uma 9h27'21" 65d31'5"
Gary Ronald Darnall
Vir 14h3'8" -13d27'42"
Gary Ronald McMillan
Umi 13h7'19" 72d22'8"
Gary Ronald Vines, Sr
Aqr 21h24'18" -9d15'23"
Gary Rowland
Tau 4h41'20" 23d6'9"
Gary Runyan
Uma 9h25'10" 71d2'3"
Gary S. Addon
Vir 12h38'57" -0d3'23"
Gary S. Grosser
Lib 15h16'56" -25d11'9"
Gary S. McCumber
Tau 4h23'56" 18d32'38"
Gary S. Smith
Cep 21h11'7" 67d3'19"
Gary Salazar
Aqr 23h22'3" -11d45'0"
Gary Salt
Her 18h32'29" 24d25'16"
Gary Sardis
Aur 5h57'15" 42d46'56"
Gary Sayle Lessin
Tau 4h32'8" 29d9'50"
Gary Schaeffer
Psc 0h35'57" 5d13'57"
Gary Schlekeway
Uma 12h45'42" 56d30'57"
Gary Scott Love
Uma 10h49'15" 49d39'14"
Gary Scott Wilbur
Lep 5h33'33" -11d8'37"
Gary Sealey
Cep 21h54'37" 58d9'26"
Gary Servatius
Uma 11h15'49" 64d1'58"
Gary & Sharon Kell
Sge 19h40'9" 18d56'22"
Gary Shuckahosee
Cma 6h33'35" -15d9'45"
Gary Sikkema
Uma 11h44'41" 56d18'35"
Gary Son of Fay
Cru 12h23'34" -58d21'6"
Gary Steven Gilbert
Uma 8h47'40" 69d46'17"
Gary Stewart Dimmick
Tau 4h8'51" 7d40'58"
Gary Strzesynski
Uma 12h4'25" 37d49'46"
Gary Sullens
Her 17h16'7" 32d9'23"
Gary Sullivan
Her 17h14'27" 15d40'16"
Gary & Susan Nunn
Cyg 21h34'56" 48d21'18"
Gary & Suzi Forever
Cyg 19h32'25" 29d5'4"
Gary Svec
Aur 5h37'18" 40d58'4"
Gary T. Chant
Equ 21h11'0" 11d19'27"
Gary T. Malloy "Sweetheart Star"
Cep 21h22'48" 56d45'51"
Gary & Teresa Crowe
Cyg 19h32'46" 47d11'45"
Gary & Terrie jo Fox
Apu 15h33'32" -72d32'33"
Gary Thomas
Per 3h21'52" 45d35'2"
Gary Todd Streich
Cnc 9h3'10" 22d21'30"
Gary Tomlin
Ori 5h30'58" -2d39'5"
Gary Tretsch
Per 3h18'16" 51d18'27"
Gary Van Gieson
Cep 22h10'40" 56d36'39"
Gary Vincent Dobrynski
Tau 3h26'27" 1d34'31"
Gary W. Goodman Memorial Star
Uma 8h34'50" 71d29'2"
Gary W Masters
Ori 5h8'48" 11d27'59"
Gary W. Pederson AKA D.F.G.
Cep 22h13'53" 69d48'53"
Gary Wade Phillips
Per 3h29'9" 51d51'49"
Gary Westburg
Psc 1h13'34" 25d52'29"
Gary William Muse
Sco 16h37'30" -29d25'7"
Gary William Rush
Aqr 22h41'2" -0d36'28"
Gary William Schraufnagl
Per 3h25'11" 39d19'58"
Gary Williams Johnson
Cep 22h32'14" 62d3'30"
Gary Wm. Thompson
Cnc 8h50'56" 26d34'59"
Gary Wright
Ari 3h22'10" 29d2'32"
Gary Zimmerman
Sgr 19h22'27" -14d51'23"
Gary,Cathleen,Robert,Michael,Rachel
Uma 10h55'56" 68d12'31"
Garylso
Crb 15h47'25" 31d34'50"
garykleo
Ari 2h13'26" 25d24'38"
Garylee and Amie's Star
Aql 19h6'45" -7d23'55"
GaryMNovak011353
Cap 21h40'36" -12d23'24"
Gary-my-Gary
Ori 6h19'50" 10d13'59"
Gary's 75th Shines On Forever
Vir 13h8'22" 13d47'42"
Gary's and Stella's love star
Cyg 20h36'50" 30d26'45"
Gary's Galactic Empire
Cyg 20h44'13" 53d54'49"
Gary's Glow (G. C. Worthington)
Cru 12h47'46" -64d17'1"
Gary's Greatest Gift
Aql 19h13'14" 1d16'31"
Gary's Holy Instant
Ari 3h8'24" 23d4'28"
Gary's Kiss
Sco 16h6'40" -10d14'6"
Gary's Star
Aql 20h7'32" -0d21'10"
Gary's Star
Cep 22h48'15" 65d40'45"
Garza's Journey
Cap 21h1'33" -15d50'29"
Gas aka Rue Akros
Sco 16h9'51" -15d27'28"
Gashtoop HNM
Gem 7h23'39" 25d58'43"
Gasm Star
Cap 20h54'14" -20d41'6"
Gáspár György
Aqr 23h51'6" -15d31'20"
Gáspár Katika
Aqr 22h26'39" -21d4'52"
Gaspard et Vanessa
Uma 8h31'54" 65d30'37"
Gaspard Le Rall
And 23h3'27" 51d14'37"
Gaspare Anthony Campo
Uma 9h8'4" 54d29'18"
Gasper Hernandez
Umi 14h56'32" 76d41'56"
Gaston
Equ 21h22'34" 6d48'9"
Gaston Bérard Roeschli
Dra 15h17'9" 58d39'55"
Gaston "Louie Louie"
Lyr 19h12'43" 28d50'20"
Gaston Rimey-Maurivard
Sgr 18h53'35" -18d45'17"
Gates Hartford Councilor Jr.
Gem 6h59'34" 19d36'0"
Gates of Heaven
Ara 17h25'21" -48d8'54"
"Gates to Heaven" Stephen C. Gates
Leo 9h50'10" 8d51'39"
Gathering of Titans
Uma 11h34'9" 59d8'23"
Gatita
Cam 3h58'51" 76d33'23"
Gatita
Pho 0h21'5" -43d50'55"
Gatito & Bu!
Ori 6h20'3" 16d50'47"
Gatito Malo
Gem 7h39'52" 24d9'1"
Gatlinburg Gal
Gem 6h39'48" 15d52'35"
Gato&Coki
Cra 18h2'45" -37d9'29"
GatorBottom
Lmi 10h40'47" 26d41'35"
Gatosan
Sco 16h46'47" -30d22'7"
Gatsby's Dream
Uma 12h37'56" 58d42'56"
Gattas Star 40
Cru 12h15'2" -63d41'36"
Gattino Rich Sagitarius
Sgr 18h13'37" -19d49'19"
Gatto
Sco 17h45'33" -36d57'57"
Gattylyons, Jr.
Dra 18h39'55" 53d59'15"
Gatz, Klaus-Peter
Uma 8h54'7" 60d25'2"
GAU121750
Per 3h36'55" 45d31'29"
Gaudalupe D. Ali
Ori 6h21'35" 16d4'44"
Gaudeamus igitur 111461937
Lyn 8h6'9" 35d21'47"
Gaughan/Ommert
Lyn 8h3'41" 33d51'0"
Gauri Mohan
Sco 17h43'41" -43d5'31"
Gauri Pardesi
Per 3h32'36" 37d11'48"
Gauri Shanker Srivastava
Ari 2h13'57" 14d52'38"
Gautam Banerjea Happy 50th! 8/6/57
Leo 11h0'46" 3d41'31"
Gaven Tomos Webb
Peg 22h43'11" 27d28'12"
Gaven's Star
Dra 17h51'44" 56d29'53"
Gavin
Uma 10h35'11" 65d28'58"
Gavin
Vir 14h11'42" -16d22'3"
Gavin
Lib 15h36'13" -8d23'9"
Gavin
Ori 4h59'29" -2d56'47"
Gavin
Gem 7h33'22" 16d47'8"
Gavin
Ari 2h14'59" 25d46'12"
Gavin Alexander
Uma 11h10'52" 46d41'10"
Gavin Alexander Roberts
Sgr 20h25'38" -41d53'44"
Gavin Allen Paape
Sgr 18h48'55" -17d40'59"
Gavin Allen Vanderpool
Her 17h22'29" 35d43'40"
Gavin and Ethan x Infinity
Uma 11h26'20" 54d42'58"
Gavin Angelo Gray-Devins
Dra 18h42'52" 50d49'43"
Gavin Anthony Popick
Leo 11h44'44" 26d27'9"
Gavin Antonio Vito
Per 4h41'29" 38d4'48"
Gavin Blake Vincent
Lyn 8h34'20" 38d41'54"
Gavin Caleb Matthews
And 0h27'10" 42d20'38"
Gavin Charles Rome
Vir 13h46'22" 5d59'20"
Gavin Christopher Birkley
Lib 15h53'28" -4d9'13"
Gavin Clint Hicks
Uma 8h53'33" 48d48'35"
Gavin David
Aqr 20h42'3" -12d4'9"
Gavin Donal Hegarty
Ari 3h9'37" 17d57'29"
Gavin Dougherty Chapman
Uma 11h41'22" 42d44'10"
Gavin Easthope
And 23h10'56" 41d25'39"
Gavin Edward Pagliarini
Cap 21h11'32" -20d6'19"
Gavin Frederick Park
Cep 7h41'6" 86d21'42"
Gavin "Gavi" Mychal Walker
Leo 10h33'58" 20d16'35"
Gavin Gores
Per 4h29'37" 41d19'4"
Gavin Hannan
Aqr 22h52'12" -11d21'58"
Gavin Hanson Tubre
Aqr 23h4'41" -21d5'41"
Gavin Henry Rennie
Uma 12h29'12" 59d18'25"
Gavin Hopkins
Cyg 20h54'45" 36d25'58"
Gavin Isaiah
Uma 10h44'48" 52d12'37"
Gavin J. DeMersseman, with love
Crv 12h25'26" -17d38'36"
Gavin J. Kearns
Gem 6h47'0" 16d19'19"
Gavin James Bell
Cnc 8h54'50" 26d57'51"
Gavin James Collins
Lmi 10h4'11" 38d39'52"
Gavin James Huber
Uma 8h51'54" 60d28'34"
Gavin James Putnam
Ari 2h1'48" 14d46'54"
Gavin John Axt
Sgr 18h38'39" -21d5'52"
Gavin John Hanigan
Lmi 10h43'1" 38d8'5"
Gavin John Shirkie
Ori 6h22'53" 14d29'32"
Gavin Kenyu Kanzaki
Cyg 20h34'48" 36d4'30"
Gavin Lane Hazard Campbell
Sgr 18h5'11" -25d44'18"
Gavin Lester
Uma 8h38'5" 50d47'13"
Gavin Liam
Her 17h20'36" 28d39'45"
Gavin Luke Fredy
Per 4h7'21" 50d42'58"
Gavin M. Reiser
And 1h56'29" 38d24'33"
Gavin McPherson
Vir 14h0'1" -16d23'10"
Gavin Michael
Lac 22h28'39" 44d54'28"
Gavin Michael Filippini
Uma 9h8'28" 49d48'46"
Gavin Michael Hadlock Kitzerow
Psc 0h41'48" 15d40'14"
Gavin Michael Henry
Sco 17h48'7" -42d34'16"
Gavin Michael Jacobs
Per 3h31'0" 51d38'32"
Gavin Michael McLane
Umi 14h25'19" 76d45'58"
Gavin Michael Steven Sampey
Lyn 8h23'18" 57d16'53"
Gavin Mitchell Vestrand
Per 3h19'16" 52d44'59"
Gavin Mitchell O'Connell
Gem 6h47'4" 32d46'58"
Gavin Montgomery Finley
Ari 2h0'4" 21d11'31"
Gavin & Myrna - Anniversary Star
Cru 12h38'21" -60d49'3"
Gavin Nicholas Heitzman
Cep 21h35'59" 62d51'15"
Gavin Nicholas Ryeson
Psc 1h9'30" 27d20'52"
Gavin Patrick Curran
Leo 9h43'51" 9d39'19"
Gavin Petherick - The Brightest Star
Cru 12h29'8" -59d35'20"
Gavin Reed
Aqr 22h22'37" 1d39'56"
Gavin Richard MacGarva
Gem 7h12'45" 32d36'15"
Gavin Riley Eubanks
Lyn 7h24'29" 54d13'17"
Gavin Robert
Her 16h20'24" 8d59'10"
Gavin Robert Nys
Vir 11h44'13" 2d57'11"
Gavin Ryan
Psc 1h27'23" 17d55'38"
Gavin Ryan Hugaboom
Aqr 22h29'49" 0d48'36"
Gavin Santacross
Uma 12h13'18" 55d41'39"
Gavin Scott McCloud
Ori 5h48'11" 11d40'14"
Gavin Shaw Kennedy "Little Hawk"
Aql 19h8'51" -0d37'14"
Gavin Stephen Poteet
Ari 2h37'8" 27d4'30"
Gavin Thomas Emerson
Ari 2h22'30" 13d50'32"
Gavin Tyler Shaw
Her 18h11'43" 21d2'37"
Gavin Voisey
Aql 19h42'14" 10d54'1"
Gavin Wilkins
Per 2h47'0" 40d45'32"
Gavin Wilkins
Cep 22h5'29" 67d32'18"
Gavin William
Cma 6h48'10" -13d22'15"
Gavin, I'll love U 4ever Love, Mom
Lib 15h42'53" -10d46'28"
Gaving
Cru 12h5'17" -62d27'21"
Gavon B
Aur 5h26'34" 46d11'30"
Gavon H. Ahvah Wong
Per 4h37'51" 31d50'2"
Gavriil Kasvikis
Umi 13h37'3" 70d37'47"
Gav's Star 13
Cma 6h30'52" -15d23'58"
Gavyn Berntsen
Per 3h56'29" 43d23'25"
Gavyn Edward Bruce
Her 17h24'6" 37d45'8"
Gavyn Phillip Boyle
Gem 7h11'18" 21d11'11"
Gavyn Wommack
Ori 5h56'41" 22d30'42"
Gavynn's Starbright
Cap 21h52'7" -20d45'30"
Gawason Peace Star
Uma 9h30'44" 46d41'29"
Gay
And 23h12'47" 42d29'0"
Gay
Aur 5h4'6" 36d12'25"
Gay
Vir 13h53'47" -7d53'10"
Gay Ann
Cap 21h41'59" -15d1'10"
Gay Edgar Jones' Heaven
Lib 15h10'46" -5d8'54"
Gay Gallino
Uma 12h40'14" 56d19'34"
Gay Guthrey
Cnc 8h59'19" 14d47'39"
Gay R. Paluch
Cap 20h41'49" -21d16'17"
gayann
Cap 20h13'12" -18d22'58"
Gayathri Divya
And 0h56'18" 39d40'50"
Gayathri Samyuktha
Tau 5h19'38" 18d20'30"
Gayatri Diya Mehta
Dra 16h42'42" 53d32'40"
Gayatri GaMarsh
Gem 6h30'52" 22d41'47"
Gayatri Patel and Justin Yen
Cyg 20h5'9" 33d25'35"
Gaye Michele Goldberg
And 0h14'44" 28d37'27"
Gaye The Moose Watcher
Aqr 22h19'35" 1d6'16"
GayeLynn
Lep 5h8'40" -12d33'18"
Gayla Currie
Cru 12h40'40" -60d43'13"
Gayla Schmerbeck
Ori 5h45'34" 6d56'10"
Gaylan Annette
Lyn 7h41'29" 44d39'44"
Gayle
And 1h13'44" 45d11'41"
Gayle
Aqr 22h19'40" -0d49'26"
Gayle Anita Frey
Tau 4h24'39" 22d15'29"
Gayle Brown Fox
Vul 19h47'36" 26d17'16"
Gayle Collins
Sco 17h16'29" -33d7'40"
Gayle Ellen Pittman Walsh
Cas 0h45'46" 57d0'59"
Gayle Freeborn
Uma 12h21'33" 52d24'30"
Gayle Golden Perkins
Sco 16h50'20" -42d12'35"
Gayle Halperin Kahn
Cas 0h17'26" 47d22'7"
Gayle Kosmowski
Uma 10h59'36" 45d48'42"
Gayle "Littlest" Hooker
And 1h51'41" 37d31'32"
Gayle Lynn Bruno
Ari 3h23'25" 27d26'6"
Gayle MacFarlane
Cam 7h59'29" 60d35'43"
Gayle MacMillan
Crb 15h26'8" 29d54'10"
Gayle "MAMA" Marrocco
And 0h54'1" 41d26'9"
Gayle Mara's Star 4-15-1954 Blinky
Ari 3h18'48" 29d15'0"
Gayle Marie Boryca
Her 17h54'51" 26d56'19"
Gayle Maureen Gray-Graham
And 0h15'16" 32d58'49"
Gayle McRae
Ori 5h36'6" -1d58'37"
Gayle Moniz
Psc 0h42'50" 7d30'19"
Gayle Norton
Cas 23h59'51" 58d42'13"
Gayle Osmun
Ari 2h35'4" 22d1'20"
Gayle Palka
Leo 10h18'33" 16d28'37"
Gayle Raspberry Carmichael
Ari 2h28'46" 19d19'49"
Gayle Ruth
Sgr 18h33'32" -23d38'33"
Gayle S. Taylor
Cas 2h45'10" 57d47'8"
Gayle Schuckmann
And 1h7'49" 45d20'31"
Gayle & Terry Huntling
Lyr 18h52'25" 32d6'19"
Gayle Venice Sawyer Harrell
Cyg 21h26'20" 37d20'18"
Gayle Winston
Cas 23h55'49" 54d20'14"
Gaylee
Uma 9h27'36" 43d32'12"
Gaylene Le Bas
Pup 7h56'46" -32d2'41"
Gaylene56
Gem 6h56'34" 16d59'6"
Gayle's and Hal's Eternal Love Star
Psc 23h22'38" 7d50'37"
Gayle's Faith
Lyr 18h53'46" 35d42'25"
Gayle's Spes
And 0h31'13" 39d21'57"
"Gay-light"
Ori 6h12'15" 8d41'21"
Gaylin "Tupperware" Olson
Aur 5h47'59" 48d47'14"
Gaylon Markota
Crb 16h3'13" 29d7'24"
Gaylord & Betty Shehorn
Uma 11h45'57" 33d11'10"
Gaylord Hanes
Lyn 8h22'8" 35d35'32"

Gaylynn
Sco 17h23'7" -42d20'39"
Gay-Lynn Prescott
Ori 5h27'7" 3d11'6"
Gaynell Watson
Tau 4h5'24" 5d19'7"
Gaynelle
Cnc 8h29'43" 25d15'45"
Gaynor Lee
Lib 15h40'18" -28d26'18"
Gaynor Mapley's Star
And 22h57'59" 51d38'20"
GayStar
Her 16h49'46" 40d28'32"
Gaz
Ari 3h7'12" 19d0'41"
Gazela Marz
Psc 0h47'14" 7d34'59"
Gazer
Ori 6h18'4" 10d39'21"
Gazing Into Serenity
Sco 16h47'25" -32d33'30"
Gazman
Apu 15h16'18" -71d10'36"
Gazmend
Cyg 19h36'49" 30d6'35"
Gaz's Star
Ori 5h13'2" 0d42'46"
G-BALL 75
Uma 9h9'23" 49d39'53"
G-Bar4
Lib 14h48'32" -4d13'21"
GBD Architects, Inc.
Per 4h8'49" 46d52'37"
Gbesemete
Sco 16h46'9" -37d38'17"
GBO
Cnc 8h39'4" 30d57'37"
GBT Koelbel Family Star
And 23h13'41" 49d12'0"
GBWheeler
Eri 4h24'40" -0d54'14"
GC Taelcat Connor MacLeod
Cnc 8h1'56" 12d53'25"
GCCA Dad
Lib 15h32'55" -6d25'46"
GCD
Lyn 7h6'44" 54d37'36"
gcdebons
Uma 9h44'56" 55d34'45"
G.D.A.E.S.C. OLAFSEN
Uma 13h55'48" 52d3'7"
GDCD
Cnc 8h34'45" 30d58'18"
GDEMLVTLND
Ori 5h34'41" 5d38'50"
G-Dog
Uma 13h1'54" 58d14'30"
G-Dog
Ori 6h18'2" -0d12'2"
GDogNick
Dra 17h25'38" 67d53'32"
GDZ
Umi 14h59'11" 79d30'50"
G.E.
Crb 15h55'10" 28d17'36"
Gea
Umi 14h55'4" 80d46'24"
GeanerBeaner
Cnc 8h6'43" 10d49'45"
Geary S. Bonville
And 1h11'30" 34d48'39"
Gea's Journey
Uma 11h47'30" 41d3'11"
Géb
Uma 13h55'53" 57d31'15"
Gebetja - 60
Crb 15h48'10" 26d46'24"
Gebhard Glania
Uma 9h7'58" 59d22'4"
Gebhardt, Ralf
Lib 15h59'18" -7d21'20"
GEC111069
Sco 16h10'10" -10d36'55"
Gecky
Uma 10h28'9" 62d48'1"
Ged & Babs Dignan - 50th Anniversary Star
Per 4h12'10" 38d1'59"
Ged Ludden
Cep 23h2'37" 70d58'44"
Geddhead
Cam 4h28'50" 65d2'25"
Gedeon Róbert 50
Tau 4h17'23" 4d43'23"
GEDS
Aqr 21h59'28" -19d31'40"
Gee
Aql 19h2'48" 9d13'7"
Gee
Per 4h41'36" 42d17'42"
Gee Bee
Ori 6h17'16" -0d43'32"
Gee Dub xxx
Cyg 21h34'48" 48d23'9"
Geek 2004
Her 18h45'38" 20d10'12"
Geekfeazal
Psc 1h20'36" 18d51'53"
Geemommie
Lib 15h43'41" -15d12'4"
Geena
Cap 20h58'10" -22d6'47"
Geena (star) Anesta
Vir 14h12'18" -16d9'48"
Geena's Star
Vir 12h12'14" 9d38'16"
Gee's Star
Aql 19h31'8" 11d5'7"
Geese and Chaz are Gay!
Tau 4h52'12" 26d52'5"
Geeta
And 23h45'28" 47d21'58"
Geeta Madhuri
Ari 3h13'59" 12d18'36"
Gefen Irene Finn
Psc 2h4'0" 6d38'46"
Geffory Ian Gerber
Gem 7h13'42" 24d8'38"
Gege
Gem 7h9'25" 33d2'27"
Gegumis
Lyn 8h38'21" 36d56'6"
Gehnen, Dieter
Ori 6h19'33" 13d17'20"
Gehrig Ronan Bock
Her 18h53'17" 23d35'31"
Gehrt, Jürgen
Ori 6h11'7" -0d13'6"
Geier, Sabina
Uma 9h47'15" 44d35'29"
Geir
Uma 11h57'11" 48d33'42"
Geißler, Gunnar
Ori 5h21'23" 3d19'45"
Geiss's Golden Star
Cyg 20h46'8" 44d11'5"
Geist
Cma 6h51'4" -20d10'53"
GEKA
Cnc 9h3'28" 8d57'15"
Gel Chambers Detrick
Cas 0h31'45" 47d4'8"
Gelb's
Uma 8h25'6" 65d17'33"
Gelchsheimer, Eberhard
Uma 11h41'39" 34d17'49"
Gelder, Erich
Ori 5h2'3" 6d52'51"
Gelen
Crb 16h5'28" 26d25'4"
Geli
Aur 6h19'40" 33d30'15"
Geli & Raggs
Ori 5h22'43" 7d9'31"
Gelita
And 2h21'39" 45d37'57"
Gella
Ari 2h57'54" 26d36'37"
Geller Dora
Tri 2h17'40" 32d3'31"
Gelly
Cas 23h7'44" 55d36'11"
Gelly Kiss
Sco 17h50'38" -33d42'44"
Gelsomino Vincent Delguercio
Cep 21h10'48" 66d52'50"
Gelu
Ari 3h14'35" 22d55'53"
G.E.M.
Leo 9h55'15" 27d50'41"
Gem
Umi 14h23'2" 72d12'43"
Gem and Magical
Boo 15h8'1" 48d51'12"
GEM Keystone
Lyn 6h56'34" 50d46'56"
Gema Martin Muñoz
And 23h19'53" 48d22'15"
Gemarieleila
Sgr 17h58'40" -29d48'59"
Gema.sd
Col 6h30'43" -42d16'3"
Gemella1
Leo 10h15'5" 27d20'37"
Gemella2
Leo 10h14'28" 25d47'48"
Gemini
Leo 9h26'51" 26d41'53"
Gemini Shipley
Gem 6h31'22" 22d48'31"
Gemini Skirptunas
Gem 7h30'21" 29d53'33"
Gemini's Seredipity
Gem 6h45'55" 18d26'13"
GeminiusAstroshineHeidius
Crb 15h54'11" 26d58'30"
Gemma
And 0h33'55" 26d11'56"
Gemma
Cyg 21h52'29" 49d46'31"
Gemma
Cas 0h57'38" 55d20'50"
Gemma
And 1h21'36" 49d8'2"
Gemma
And 1h1'42" 33d45'34"
Gemma
And 1h19'10" 41d11'19"
GEMMA
Dra 17h46'8" 66d45'37"
Gemma Alison
And 2h31'14" 39d21'25"
Gemma Amaterasu
And 23h17'37" 42d45'33"
Gemma Anne Reid
Cas 2h14'23" 62d46'12"
Gemma Baines (My baba) x
Psc 1h2'5" 10d35'33"
Gemma Carroll's 21st Birthday Star
And 0h8'47" 29d3'11"
Gemma Catherine
And 0h46'18" 25d12'4"
Gemma Celeste
Vir 13h8'37" -12d0'28"
Gemma Cronk
And 2h12'22" 49d11'39"
Gemma DiFilippi
And 0h12'32" 34d38'22"
Gemma Dineen
And 0h11'26" 46d38'19"
Gemma Eleanor
And 0h44'5" 26d51'52"
Gemma Elzi
Cas 0h27'45" 62d24'25"
Gemma Goggins
Cas 23h48'5" 53d39'5"
Gemma Gundry - The Light Of My Life
Cru 12h17'51" -57d2'21"
Gemma Larson
Gem 6h51'45" 12d3'45"
Gemma Lawrence
Equ 21h7'21" 4d23'13"
Gemma Leanne Pearson
Leo 11h4'58" -3d32'8"
Gemma Llorens
Dra 18h23'32" 49d2'1"
Gemma Lord Pierpont
Leo 10h44'54" 16d59'20"
Gemma Louise
Tau 5h45'26" 25d40'39"
Gemma Louise
Del 20h36'50" 6d27'3"
Gemma Louise Brady
Cas 23h59'12" 61d24'34"
Gemma Louise de Figueiredo
Tau 3h38'3" 5d1'7"
Gemma Louise Howell - Shining Forever
Col 5h47'45" -35d28'35"
Gemma Louise Reed-Keith
Cas 0h18'33" 51d53'7"
Gemma Louise Taylor
Cyg 21h11'59" 41d19'9"
Gemma Luciana
And 2h35'2" 49d56'11"
Gemma & Matthew 29/11/03
And 22h58'59" 48d23'30"
Gemma Medrano Elias
Her 17h55'25" 28d50'44"
Gemma Mignella "Happy 75th"
Aqr 22h30'47" -10d21'59"
Gemma Paternoster
Lyn 6h59'45" 54d47'46"
Gemma Rose
Cas 1h18'5" 69d10'37"
Gemma Sky Waterson-Cochran
Crb 15h51'8" 31d22'47"
Gemma Stamp - 21st - 10.11.05
And 23h18'21" 41d31'41"
Gemma Walters & Greg Cook
Cra 18h40'7" -38d50'59"
Gemma Ward
Equ 21h5'49" 4d7'46"
Gemmarc
Dra 18h58'15" 51d56'1"
Gemma's Christmas Star
Vir 14h21'27" 1d9'26"
Gemma's Estrelya
Ori 5h40'37" 12d44'12"
Gemma's Star
Lyn 9h34'53" 41d15'33"
Gemma's Star
And 23h41'45" 37d5'13"
Gemmasrealta
Ori 6h4'53" 0d54'9"
Gemmata
Cyg 20h38'13" 50d54'59"
Gemmit's Galaxy
And 1h13'27" 41d43'6"
Gemstar
And 1h23'48" 44d39'14"
GEMSTAR
Crb 16h2'53" 26d37'10"
Gen
Lmi 9h48'13" 40d10'59"
Gen Gen
Cnv 13h3'10" 47d59'25"
Gen57
Ori 5h56'4" 7d21'27"
Gena
Cam 7h50'47" 63d16'5"
Gena & Cary
Sge 20h2'47" 17d24'22"
Gena & Danny
Cyg 20h3'17" 39d4'39"
Gena Faith
Umi 15h4'48" 71d28'51"
GENA G
And 2h28'44" 41d43'47"
Gena Leigh Silberman
Vul 21h7'24" 25d57'51"
Gena & Marina
Uma 8h31'53" 65d0'44"
Genadis Agrest
Gem 7h25'57" 14d5'28"
Genadiy
Psc 1h17'4" 24d30'3"
Genajuade Shyrelle Tade
Ori 6h2'28" 10d5'2"
Gena's Radiance
Lmi 10h18'36" 35d39'2"
Gene
Her 17h23'6" 26d49'32"
Gene 6/25/1926 ECJ
Cnc 9h19'29" 16d37'0"
Gene Addy
Lib 15h15'59" -4d28'3"
Gene and Elease's star
Cyg 19h40'34" 31d26'26"
Gene and Helen
Cyg 21h21'25" 35d17'48"
Gene and Kitty Sette
Lib 14h50'37" -2d12'37"
Gene and Yvonne Lucia
Cyg 21h16'31" 47d4'16"
Gene Bantz - My Beloved
Sgr 19h40'5" -12d39'35"
Gene Baumgartner
Ari 2h37'5" 25d5'8"
Gene Blalock
Leo 10h58'5" 17d35'22"
Gene C. Emig
Lib 15h20'39" -6d55'12"
Gene Carol Ennis Emerson my mother
Uma 12h39'29" 57d0'57"
Gene Cigw Stevenson
Tau 4h13'9" 3d0'31"
Gene Comfort
Her 18h54'50" 23d34'7"
Gene Cuvelot
Leo 10h24'38" 9d54'48"
Gene D. Shipley
Lmi 10h41'50" 25d55'30"
Gene Edward Ransom Jr.
Dra 10h17'36" 73d10'18"
Gene Fedorczyk
Boo 14h58'2" 25d50'29"
Gene Fisher - "Poppy"
Her 17h19'54" 34d21'50"
Gene Francis Alan Pitney
Uma 12h20'14" 62d43'31"
Gene "Geno" V. Bradley
Per 2h57'28" 55d56'26"
Gene Gentle
Uma 11h1'32" 62d39'24"
Gene & Gerri Britt
Lib 15h50'49" -17d54'22"
Gene Gomez
Cyg 21h35'43" 31d0'48"
Gene Grambo
Per 3h20'51" 51d35'32"
Gene "Hardrock" Downing
Ori 5h29'41" 3d48'58"
Gene Horton
Per 3h30'24" 45d21'41"
Gene Jacob Bare 2-15-1937
Aqr 22h39'1" -3d51'25"
Gene & Jill's 25th Anniversary Star
Cyg 20h20'33" 53d37'44"
Gene Kastell
Ori 6h14'1" 2d49'44"
Gene & Kathy Giggleman
Lyn 9h7'51" 37d48'27"
Gene Kato, Director Mrs. California
Per 3h24'55" 49d9'55"
Gene & Kay Naughton
Uma 11h35'35" 58d33'25"
Gene Klueg
Cep 21h46'59" 62d5'16"
Gene Kutyreff
Ari 2h45'2" 14d16'44"
Gene' L. Brinson
Tau 3h58'19" 27d21'37"
Gene Lee Holman
Sco 16h39'58" -44d38'0"
Gene L.Stang
Uma 9h15'21" 56d54'18"
Gene Lynn
Boo 14h8'39" 29d46'28"
Gene & Margie Atkerson
Cyg 19h49'18" 38d33'18"
Gene & Marj Eaton
Gem 6h19'41" 21d30'54"
Gene Martin Hooper, Jr.
Her 17h49'16" 37d14'30"
Gene & Mary Ann Williams
Lyn 7h10'51" 52d40'12"
Gene Michael Azar
Sco 16h20'16" -32d10'19"
Gene - My Ray of Sunshine
Her 17h16'44" 26d16'47"
Gene "Papa" Ford
Vir 13h7'46" -15d14'12"
Gene Pearson
Sco 17h27'24" -37d53'0"
Gene Ragsdale Lamar
Per 3h11'33" 39d13'14"
Gene Ramos
Cap 21h42'57" -10d22'24"
Gene Reynolds
Gem 7h0'48" 16d21'42"
Gene Risko
Her 17h14'48" 29d25'3"
Gene S. Bell
Sco 17h43'4" -34d8'32"
Gene Samburg
Cep 22h57'15" 80d12'34"
Gene Schiavi
Lyn 7h8'13" 44d21'41"
Gene Selfe
Uma 11h24'52" 34d56'29"
Gene Serene
Ori 5h43'6" 1d27'2"
Gene Sickles
Per 3h42'50" 40d25'47"
Gene Steinhauer
Cap 20h15'15" -10d2'7"
Gene Stephen Ammann
Uma 10h50'41" 65d25'58"
Gene Sun
Leo 11h37'19" 5d12'41"
Gene & Susan McMath
Aql 19h21'2" -0d12'45"
Gene & Terrie
Cyg 19h44'26" 48d15'42"
Gene Vaisberg's Star
Her 17h22'57" 36d11'41"
Gene William Ferris
Lib 14h47'9" -11d47'22"
Gene Wm. Singer
Per 3h15'1" 41d25'47"
Gene Worley
Cru 12h24'11" -57d0'47"
Geneffa Kainrad
Psc 0h47'50" 18d1'51"
Genel Patoloji Merkezi
Cam 3h46'8" 58d47'3"
Genellison
Cru 12h2'32" -62d4'27"
Gene-Marie Sperduti
Boo 14h17'28" 16d58'55"
Genepatiffany
Cyg 20h7'39" 44d50'31"
General Gregory S. Martin
Ori 6h13'13" 17d18'23"
General John
Her 17h42'57" 36d16'23"
General Lee
Boo 14h32'27" 13d43'48"
General Lee
Uma 13h37'20" 55d31'17"
General Olin Oedekoven
Uma 11h42'25" 32d16'52"
GENERAL William Heath Basil Bono
Her 16h19'57" 25d8'6"
generálmajor Ing. Jozef DUNAJ
Aql 19h49'27" 13d25'19"
Generosa Casas
Cas 1h33'59" 73d6'57"
Generose A. Wilson
Sgr 18h29'34" -22d35'21"
Generosity
Cep 22h5'58" 60d35'57"
Generous Sun
Peg 22h17'58" 34d52'42"
Gene's FirstLove
Uma 10h13'31" 71d35'42"
Gene's star
Cnv 13h22'54" 48d34'36"
Genesis
Tau 5h52'31" 17d46'51"
Genesis
Her 18h41'44" 19d24'42"
Genesis
Lib 15h10'46" -16d19'36"
Genesis 2:18
Pho 2h15'19" -40d42'39"
Genesis Nevaeh Miranda
Crb 16h17'35" 33d38'31"
Genesis Nizeth Guevara
Gem 7h21'8" 22d13'15"
"Genesis", Angel of my "Heart"
Lyr 18h53'1" 35d39'53"
Genesius
Per 4h12'16" 44d24'32"
Geneth Remedio
Lib 14h41'14" -9d48'48"
Genette
Lib 14h50'44" -1d16'29"
Geneva
Gem 6h41'34" 35d6'1"
Geneva
Vir 12h37'19" 11d18'21"
Geneva
Ori 5h44'19" 12d12'56"
Geneva D. Ferguson
Cas 0h27'47" 61d22'28"
Geneva Gail
Ori 4h48'18" 12d37'53"
Geneva Grace Bierce-Wilson
Aqr 22h22'26" -16d25'28"
Geneva Inez "Ginny" Cladin
Sgr 19h24'51" -40d5'31"
Geneva Jane Durham
Cap 20h8'25" -26d1'55"
Geneva M Flynn
Cap 20h34'21" -20d39'21"
Geneva Pollard
Uma 9h6'47" 52d53'59"
Geneva-N-Rick<\@>60
Uma 9h29'7" 42d59'38"
Genevie Marie
Gem 7h17'0" 14d14'29"
Genevieve
Gem 7h16'52" 14d50'48"
Genevieve
Ari 2h8'12" 17d30'51"
Genevieve
Cnc 8h27'13" 19d31'9"
Genevieve
Uma 11h57'6" 28d33'48"
Genevieve
Lyr 19h26'30" 38d18'56"
Genevieve
Lyn 8h32'23" 50d25'29"
Genevieve
Uma 12h33'12" 53d44'49"
Genevieve
Cas 1h34'6" 61d32'59"
Genevieve
Cas 1h20'3" 72d6'38"
Geneviève
Sgr 17h47'40" -22d41'21"
Genevieve
Sgr 19h15'36" -34d55'24"
Genevieve
Sco 16h35'16" -38d50'36"
Genevieve A. Abbott
Cyg 20h34'21" 48d47'26"
Genevieve and Betty Muire
Lyr 19h13'44" 27d4'1"
Genevieve Bean
Lyr 19h17'34" 29d43'5"
Geneviève Benoit
Sco 17h1'30" -37d21'22"
Genevieve Brooks
Cap 21h37'26" -10d32'11"
Genevieve Crooks Angel "Gram"
Uma 10h40'29" 46d35'29"
Genevieve Dorothy Tyler
Cnc 9h5'40" 17d22'30"
Genevieve Elizabeth Savage
Ari 2h32'9" 26d38'28"
Genevieve Elliott
Tau 3h49'24" 17d10'45"
Genevieve Emilé
Cam 3h53'44" 63d56'59"
Genevieve & Eugene
Aur 5h12'44" 35d29'6"
Geneviève Ever So Nice
Cap 20h46'32" -22d23'59"
Geneviève Filiputti
Uma 9h4'2" 68d32'8"
Genevieve Forester
Uma 10h29'7" 40d45'53"
Geneviève Garceau
Psc 0h51'18" 27d37'47"
Geneviève Hamel Mercier 1934
Cap 20h39'41" -20d32'35"
Geneviève Hébert et Benoit Gagner
Ori 5h36'42" -2d37'8"
Genevieve Kathryn Schnabel
Ari 3h19'21" 27d30'28"
Genevieve Lawrence (Mom)
Gem 7h9'23" 31d44'50"
Genevieve LEFORT
Aqr 21h49'3" -7d0'34"
Genevieve Linn
Lyn 8h53'47" 37d49'44"
Genevieve M. Studer
And 1h3'59" 39d42'38"
Genevieve Mae Claytor
Cas 1h25'13" 53d3'20"
Genevieve Marie
And 23h35'55" 48d21'2"
Genevieve Matts
And 0h53'2" 41d50'21"
Genevieve May DeSutter
Lyr 18h47'33" 38d27'33"
Genevieve May Winters
Lib 16h0'57" -7d22'8"
Genevieve Merle Mills Stone
Cas 23h43'9" 54d54'55"
Genevieve Moncreif
And 0h50'17" 41d53'25"
Genevieve Monsma
Uma 11h38'3" 39d2'48"
Genevieve Monsma
Cas 23h58'7" 64d42'12"
Geneviève Montagne
Per 3h11'3" 41d13'57"
Geneviève Moullet
Cas 23h5'22" 57d15'47"
Genevieve O'Leary
Gem 6h27'57" 18d8'13"
Genevieve P Gorewit
Cas 0h37'12" 64d34'41"
Genevieve Pasqualina Covino
Ori 6h16'51" 6d33'28"
Geneviève "Pretty Smile"
Uma 10h0'3" 47d48'34"
Geneviève Robichaud 15 novembre 1991
Cyg 21h46'56" 44d24'38"
Genevieve Schlekeway
Crb 15h46'6" 35d13'22"
Genevieve Simone Ondish
Cnc 7h58'20" 17d16'38"
Genevieve Winter
Vir 14h5'26" 4d22'16"
Genevieve's OneLove
Cap 20h27'37" -10d53'56"
Genevieve's Star
Sgr 18h42'50" -22d48'41"
Geneya & Daniel Ojinaga
Cyg 21h47'36" 41d33'13"
Geni Dieffenbacher
Lyr 19h9'21" 46d8'5"
Genie
Cnc 9h6'19" 12d23'36"
Genie
Cnc 8h55'40" 15d13'54"
Genie A
Uma 9h42'48" 66d29'30"
Genie & Marcus Rotunda
Uma 11h5'8" 50d12'0"
Genie Quy-Trung Vu
Uma 10h50'17" 39d37'37"
Geniece Brianna Francisco
Psc 0h7'37" 9d13'7"
Genii
And 0h13'46" 46d23'2"
Genise
Ori 6h18'50" 10d55'22"
Genius Jeremy
Per 3h52'18" 33d35'25"
Genius, Jörg
Ori 6h10'39" 16d49'42"
Genna Atcheson
Cyg 20h41'45" 45d54'38"
Genna Lee Lynn
Cap 21h44'34" -24d2'10"
Genna Marisa Nezowitz
Cas 1h24'36" 57d7'18"
Genna Rae
And 23h35'21" 48d17'9"
Genna Rose Gioioso
Umi 15h45'58" 84d52'33"
Genna Rose Skalski-98
Uma 10h1'29" 57d2'24"
Genna Star
Umi 15h14'27" 68d58'43"
Genna Viso
And 2h20'50" 50d22'19"
Gennadij Snytkin
Ori 6h16'47" 8d21'17"
Gennady Ayvazyan
Ari 3h10'9" 29d29'6"
GennaEve
Cnc 8h17'37" 23d30'40"
Gennaro Brascia
Lmi 10h26'40" 28d23'36"
Gennaro Buonocore
Cap 20h46'14" -15d58'59"
Gennaro Paolo
Vul 19h46'0" 28d42'18"
Gennaro Ruperto
Uma 12h12'14" 56d59'52"
Gennasis Pearl
Umi 14h29'46" 82d20'36"
Genni
Lmi 10h6'4" 27d56'34"
Genni
And 0h19'7" 45d39'12"
Gennifer Ashley Gray
Lyn 7h26'40" 56d49'29"
Gennifer Irene Robles
And 22h59'20" 46d39'36"
Gennivive Rodriguez-Abusamaha
Psc 1h6'34" 32d59'34"
Genny
Uma 10h19'37" 58d52'58"
Genny and Josh's Eternal Flame
Cyg 21h13'56" 47d7'36"
Genny Drake
Uma 10h5'16" 42d40'19"
Genny La More
And 2h28'8" 49d50'12"
Genny Lilli
Psc 0h40'46" 8d41'6"
Genny Pearl
Lyn 7h57'33" 47d48'12"
Genny Stearns
Cyg 20h1'32" 41d16'35"
Genny, David & Grace King
Aql 20h15'42" 0d52'44"
Geno
Aqr 22h36'53" 0d41'30"
Geno
Lib 14h52'21" -4d54'8"
Geno Stuart
Cep 22h8'8" 66d37'12"
Genovesi Giorgia
Lac 22h57'14" 44d11'18"
Genoveva Javier Salgado
Ari 2h55'39" 25d36'32"
Gensberger, Sonja
Uma 9h56'34" 60d14'13"
Gentil Octave
And 0h25'55" 22d19'35"
Gentile's Folly
Per 3h20'8" 42d34'57"
Gentle Giant
Per 2h57'19" 56d2'12"
Gentle Inspiration
Peg 22h38'21" 11d42'29"
Gentle Supreme Sdn Bhd
Ori 5h34'56" 3d20'3"
Gentleman Jim Hall
Gem 7h29'27" 19d1'50"

Gently Falling Rain, Saifon
Cnc 8h47'41" 12d53'29"
Gentri Jade Clark
Cyg 19h32'19" 27d54'13"
Gentry Anne's Never NeverLand
Tau 4h4'28" 5d3'41"
Gentry Slade-Mansur
Eri 3h53'12" -0d54'31"
Genure & Paige
Cyg 20h22'43" 52d14'3"
Geny
Sco 17h37'57" -40d14'4"
Genzy 1977-82
Uma 12h36'44" 55d29'20"
Geo
Cam 7h0'39" 69d3'29"
Geo
Per 3h51'43" 32d6'13"
Geo
Lyr 18h50'52" 33d14'24"
GEO
Aql 19h45'8" 2d4'28"
GEO 1-8-30
Cap 21h52'0" -22d57'46"
Geo Mozzi
Uma 14h1'50" 61d53'43"
Geo W Terembes
Sco 16h12'45" -10d33'44"
GEOCECI RUBIN
Lmi 10h31'25" 32d5'13"
Geoevwa - Grand Light in the Sky
Mon 6h57'56" -0d51'1"
Geoff
Cyg 19h46'52" 39d57'30"
Geoff and Courtney Forever
Cyg 19h52'14" 38d39'5"
Geoff and Danielle Terrett
Cyg 21h30'40" 33d25'44"
Geoff and Molly Beck's Wedding Day
Cas 0h26'33" 54d20'5"
Geoff and Sydnee Hazelwood
And 2h17'3" 46d10'5"
Geoff & Blake
Per 3h38'33" 47d44'28"
Geoff Cassidy - Rising Star
Boo 15h26'2" 43d47'27"
Geoff Chaplin
Per 2h40'18" 50d16'42"
Geoff Dannatt
Aur 6h23'15" 41d46'59"
Geoff Giles
Cep 22h11'11" 53d44'24"
Geoff & Jordan
Cyg 19h42'5" 41d1'10"
Geoff & Lacie Forever
Cap 20h37'6" -14d24'58"
Geoff & Nicole's Shining Star
Cyg 21h41'44" 45d27'54"
Geoff Philipsen
Boo 14h42'35" 37d12'11"
Geoff Porter - Eternally my Light
Ori 5h9'55" -0d4'14"
Geoff Rosenhain
Cep 22h15'24" 66d27'27"
Geoff Strickland
Cnc 8h52'23" 12d34'20"
Geoff & Tammy Ad Infinitum
Cyg 20h31'55" 46d47'51"
Geoff Wissman
Gem 6h48'16" 18d48'41"
Geofferlyn Fragmin
Crb 15h29'20" 30d25'18"
Geoffers
Ori 6h11'32" 7d57'24"
Geoffery Brown
Sco 16h45'35" -31d46'55"
Geoffrey
Aqr 22h37'42" -17d44'42"
Geoffrey *** TI AMO *** Sasha
Col 6h15'29" -42d5'30"
Geoffrey and Tara's Star
Leo 11h8'8" 16d30'0"
Geoffrey Armstrong
Ori 6h10'41" 18d13'18"
Geoffrey Avrum Prost
Cnc 9h7'6" 10d36'11"
Geoffrey & Brooke Schwartz 2006
Cap 21h36'14" -13d45'50"
Geoffrey Calvin Dunsmore
Cyg 21h35'55" 48d57'10"
Geoffrey Carduff Caruso
Lib 15h19'14" -5d51'13"
Geoffrey Cockrum
Cam 4h2'12" 57d22'27"
Geoffrey Craig Vandenberge
Umi 15h1'46" 68d32'38"
Geoffrey E. Zeigler
Dra 18h55'54" 58d0'4"
Geoffrey Edward Smith
Ori 6h19'32" 8d35'25"
Geoffrey Elliot
Per 2h56'16" 52d45'43"
Geoffrey Ethan Lai Tillisch
Her 18h9'40" 28d41'9"
Geoffrey Heanue
Her 16h23'35" 22d57'24"
Geoffrey J. Lambert
Aur 5h56'37" 53d6'4"
Geoffrey Jakeman
Cru 12h37'54" -59d2'38"
Geoffrey Kenneth Mortaley
Psc 1h24'49" 32d30'6"
Geoffrey Knoll
Lib 15h14'10" -15d18'13"
Geoffrey Kyle Alban Dunaway
Uma 9h59'33" 45d46'15"
Geoffrey Laurence Curtis
Sgr 19h32'20" -23d18'35"
Geoffrey Michael Fraser
Lmi 11h6'28" 32d3'22"
Geoffrey Owen McAllen
Ori 6h13'29" 16d29'12"
Geoffrey Rogers
Cnc 8h23'8" 16d7'13"
Geoffrey Scott
Gem 7h21'13" 20d12'19"
Geoffrey Scott Vogel
Gem 7h8'37" 24d4'30"
Geoffrey Shaw
Per 3h45'26" 36d41'57"
Geoffrey Smith
Aur 6h35'25" 34d50'52"
Geoffrey Steven Yannalfo
Uma 10h47'42" 49d54'47"
Geoffrey T. P. Poole
Cma 6h48'10" -21d27'44"
Geoffrey Tom Tuffin
Cep 0h2'27" 68d35'30"
Geoffrey Trippaers
And 1h40'22" 42d30'27"
Geoffrey Weech - Always My Danny
Eri 4h38'57" -5d59'47"
Geoffrey William Hager
Sco 16h8'11" -16d32'49"
Geoffrey William Sweetman
Per 4h15'44" 33d33'5"
Geogre Tony Miller
Her 16h22'17" 19d32'12"
geogrib gineska
Lyr 18h55'43" 33d53'51"
GEOINT by Lt Gen James R. Clapper
Umi 14h48'7" 84d58'23"
GeoKat
Tau 5h8'49" 24d55'54"
Geordie
Peg 23h51'5" 23d26'21"
Geordie Elliese Curtis
Cru 12h45'35" -60d26'33"
Geordie - Heaven Sent
Cru 12h9'14" -62d4'3"
Georg Albert Adler
Uma 9h11'53" 49d39'5"
Georg Friesenhan
Uma 11h0'56" 70d24'23"
Georg Gerhard Günther
Uma 10h18'1" 45d8'59"
Georg Janzen
Ori 5h55'52" 3d4'26"
Georg Josef Riedel
Sgr 19h9'9" -33d52'50"
Georg Köhler
Ori 5h57'17" 12d18'38"
Georg Kugler
Uma 8h20'59" 65d34'26"
Georg Lederer
Uma 11h56'33" 29d20'9"
Georg Oefelein Birthday Star
Tau 4h22'55" 24d2'12"
Georg Paul Tom Vicky Yasin Mathias
Uma 13h36'54" 50d46'2"
Georg Rumbaur
Uma 9h15'44" 47d35'20"
Georg Wilhelm Ochs
Uma 10h19'46" 42d0'29"
Georg Zak
Uma 8h27'12" 65d27'21"
Georganna Scartozzi
Cma 6h39'46" -14d44'54"
Georganna T. Sinkfield
Lyn 9h0'19" 34d29'49"
Georganne Williams
Cap 20h40'43" -17d50'12"
George
Lib 14h56'9" -16d36'52"
George
Aqr 21h2'29" -11d33'32"
George
Aqr 23h6'6" -4d27'0"
George
Cas 1h20'30" 68d41'59"
George
Her 18h7'12" 37d6'26"
George
Per 2h14'44" 56d3'39"
George A. Ball, Jr.
Aql 19h18'50" -10d2'5"
George A. Budd
Ari 2h25'16" 24d15'54"
George A. Hall
Sco 16h4'16" -12d58'34"
George A. Podlaski
Vir 12h34'52" 4d12'44"
George A. Rogers Family
Lyn 6h26'10" 60d9'24"
George A. Slebodnik
Leo 11h3'57" 15d1'22"
George A. Stedman
Pho 23h30'44" -50d56'6"
George A. Sutherland
Ori 5h10'42" 5d50'57"
George A. Tralka Jr.
Aqr 21h56'41" 1d48'4"
George Alan Lowe
Sex 9h47'28" 2d16'53"
George Albert Sinner
Cyg 19h46'19" 43d2'33"
George Alexander Clark
Aur 5h6'40" 35d48'18"
George Alexander Frederick Yeomans
Sgr 20h20'32" -42d18'11"
George Alexander Martin
Aur 5h51'59" 43d43'28"
George Alfred Arkwright, III
Uma 9h7'45" 68d36'10"
George Alfred William Royston
Umi 15h39'10" 69d37'27"
George & Alison Brokaw
Umi 15h28'18" 73d8'49"
George Allan McKellar-Angus
Uma 8h36'19" 52d19'29"
George Allen
Uma 10h37'11" 56d31'47"
George Alonso
Uma 13h42'47" 49d40'55"
George Alvin Platt's Star
Per 3h26'1" 33d41'46"
George and Abier Barakat
Cnc 8h28'58" 19d52'34"
George and Andrea's Love
Cru 12h7'52" -63d24'32"
George and Charlotte Coppens
Cyg 19h59'57" 41d39'32"
George and Cindie-Eve
Pyx 9h15'45" -31d50'5"
George and Dolores Strouse
Cyg 19h36'18" 48d6'24"
George and Doris Kauffman
Aqr 21h55'31" 0d39'38"
George and Erika Brattain
Lyn 7h7'11" 51d41'26"
George and Ernie Oliphant
Cyg 20h36'58" 51d19'58"
George and Freda Fogle
Cyg 21h51'18" 43d27'36"
George and Gail Winson
Tau 5h22'17" 17d30'31"
George and Geni Gritton
Psc 0h22'54" 8d37'10"
George and Gladys Neff
Lib 15h22'17" -6d34'36"
George and Kevin: Shining Stars
Uma 11h58'25" 41d2'14"
George and Lorraine Wadzinski
Lyr 18h54'10" 33d26'18"
George and Martha Dumas
Cyg 19h44'49" 32d1'43"
George and Minnie Mayall
Sge 19h54'10" 19d12'21"
George and Nancy Webb
Cap 20h55'34" -15d45'59"
George and Naomi Krum 25 years
Her 16h27'33" 44d37'41"
George and Rachel Azar
Ari 3h5'43" 27d26'32"
George and Shirley Cummings
Sge 20h10'27" 20d53'8"
George Andrew
Lib 14h44'46" -19d6'35"
George Andrew Jonathan Bonner
Ori 5h55'34" 17d6'36"
George Andrew Jones jr.
Peg 21h45'28" 11d6'48"
George & Anne
Cyg 20h52'11" 31d35'12"
George & Anne Covill
Col 5h11'30" -28d11'8"
George Anthony & Celeste J. Kulz
Cyg 19h40'17" 29d39'31"
George Anthony Milton-White
Ari 2h3'56" 11d12'0"
George Anthony Zavadil 10/01/40
Uma 10h22'57" 56d13'1"
George Apisak Peters
Her 18h26'22" 16d50'30"
George Archie Anderson
Uma 8h26'1" 64d8'38"
George Armadoros
Per 4h8'4" 52d9'13"
George Arthur Emery
Cep 20h43'59" 61d26'47"
George Artz Blessed Be
Per 4h14'27" 49d16'35"
George Austin Reis II
Aqr 22h12'32" 2d1'52"
George B. Diggs Esq. III
Ori 6h4'15" 13d57'34"
George "B" " For Beautiful" Keefe
Ori 4h51'13" -0d7'47"
George B. Hick, Sr.
Cnc 9h4'6" 32d21'57"
George B. Thompson, Sr.
Dra 20h5'11" 74d54'8"
George B. Weir
Her 17h9'36" 36d29'11"
George Barr Hammel III
Aqr 22h34'58" 2d21'46"
George Barron Dalton
Sgr 19h3'56" -34d58'45"
George Bauder
Umi 16h10'4" 78d33'23"
George Behrens
Cep 22h51'36" 72d30'27"
George Benham's Eastern Star
Cma 6h33'31" -15d10'51"
George Bergerman "Georgie"
Her 16h7'53" 21d37'39"
George Bernacchia
Uma 13h43'9" 49d35'48"
George Bernard Conniffe
Leo 11h11'42" 12d48'51"
George Bernard Shaw
Aqr 22h30'36" -10d25'20"
George & Blanche Graham
Psc 0h38'51" 17d9'23"
George Bolden
Boo 14h49'11" 40d2'48"
George Boppa Shelkey
Tau 3h32'15" 25d1'29"
George Bourguillon
Uma 10h59'24" 35d28'41"
George Bouri "Star Of My Life"
Uma 10h15'5" 59d31'51"
George Bourque
Uma 11h31'3" 48d46'56"
George Bradford Lesley, Jr.
Cap 20h54'37" -24d54'59"
George Bradley
Ori 5h28'17" 5d22'5"
George Brant Moseley
Sgr 19h46'53" -14d58'6"
George Bristol
Psc 0h42'55" 7d59'9"
George Brown
Leo 9h38'58" 10d44'55"
George Buckwald
Uma 11h1'20" 64d3'43"
George Burdett Haycock
Gem 6h39'23" 12d45'17"
George "Butch" Heintz
Cyg 19h45'5" 38d47'26"
George Butrus
Per 3h23'36" 50d34'17"
George Byrd
Lib 14h48'55" -18d22'4"
George C Hardin
Her 17h54'51" 49d36'11"
George C. Resh
Lyn 7h0'21" 45d31'52"
George C. Smith III
Uma 11h1'13" 54d24'49"
George C Will
Lyn 9h3'33" 41d23'48"
George Cabrera III
Cep 21h29'5" 63d51'2"
George & Carol
Gem 7h28'41" 16d7'41"
George & Carolyn Tyson Happy 52nd
Lyr 18h30'46" 36d36'21"
George Carpenter
Ori 6h0'4" -0d19'13"
George Cendo 100548 - Tamara & Lisa X
Cru 12h49'13" -57d35'45"
George Charles Campion
Her 17h14'42" 13d34'27"
George Charles Dellinger
Psc 1h1'55" 24d44'41"
George Charles Sholin
Crb 15h45'32" 32d59'20"
George & Cherie
Ori 4h54'47" 12d49'25"
George + Cheryl Mateka
Cyg 21h17'57" 31d3'12"
George & Christelle - Angel Parents
Ori 5h36'11" -1d42'47"
George & Christina Catalano
Her 17h32'25" 17d19'18"
George Christnacht
Uma 11h47'46" 59d56'21"
George Christopher Davis
Dra 17h34'4" 66d38'34"
George Christopher Terral
Uma 11h30'55" 34d10'6"
George Clark
Uma 12h13'25" 61d10'12"
George Cleverly
Cnc 9h17'17" 15d24'13"
George & Connie Dusckas
Lyr 19h10'29" 46d27'56"
George Conway
Boo 14h52'52" 35d41'13"
George Cook
Uma 11h6'48" 46d49'47"
George Cortez, Jr.
And 0h15'33" 24d3'37"
George Craig Johnson
Ori 6h7'5" 10d44'35"
George Creighton Childs, Jr.
Per 3h19'1" 39d50'33"
George D. Hunt, III
Gem 6h48'2" 33d38'0"
George "DAD" Collins
Aqr 21h14'25" -10d47'17"
George Daniel Spencer
Sco 16h12'36" -19d12'10"
George David Gogas
Cru 12h9'10" -62d0'45"
George David Town Jr.
Psc 23h26'51" 6d20'54"
George David Utz
Lyn 7h40'40" 38d10'30"
George Davis
Aqr 20h40'26" 1d16'8"
George Davis
Ori 5h19'52" 12d5'38"
George Deakin
Per 4h23'18" 49d56'24"
George DeBarros
Boo 14h39'2" 40d43'59"
George & Deidre's Love Shines 4ever
Umi 15h20'22" 75d8'48"
George DeMartino-Sullivan
Tau 4h15'37" 17d4'31"
George DeNoto IV
Vir 13h44'9" -7d14'46"
George Dew
Tri 2h10'42" 35d0'44"
George Diehl
Uma 8h36'42" 48d33'44"
George Dobbins
Uma 9h6'15" 59d57'50"
George Donato Serini
Umi 16h19'25" 74d11'23"
George Dorsey
Aqr 21h59'2" -14d28'30"
George Douglas Batchelor, Jr.
Srp 18h32'38" -0d13'8"
George Dunn
Tau 4h18'12" 17d37'23"
George E. "Buz" Jochetz III
Gem 6h25'8" 27d42'27"
George E. & Carolyn R. Schellang
Cyg 19h46'12" 35d5'13"
George E. Hawkins
Lib 15h19'48" -12d58'12"
George E. Johnson
Tau 4h12'24" 13d0'21"
George E. Kaloroumakis
Leo 10h19'46" 26d15'8"
George E. Lewis Jr.
Ori 5h6'0" 2d30'9"
George E. Neese Jr. (Your Own Star)
Ori 5h50'31" 0d44'8"
George E. Simmons, Jr.
Per 3h27'1" 33d33'3"
George E Stewart II
Leo 11h47'24" 21d45'43"
George Eatmon Sr.
Cep 20h41'22" 59d45'37"
George Edmond Bassett
Ari 1h56'51" 21d4'39"
George Edson Leonard 111
Psc 0h29'36" 4d53'56"
George Edward Martinez
Per 1h51'8" 48d11'20"
George Eldon Marchyshyn
Ori 6h15'51" 16d26'35"
George Eli White
Lmi 10h2'35" 37d14'9"
George&Ellen Wolf
Boo 14h53'18" 23d27'15"
George Elwood Anderson
Lmi 10h26'17" 36d43'0"
George Embiricos
Aql 19h32'42" 11d52'48"
George Englehardt
Aur 5h9'41" 47d1'21"
George Eugene Hilton II
Per 3h8'15" 56d31'48"
George F. Basilo
Ari 1h59'9" 18d47'42"
George F. Brand
Her 16h57'32" 30d38'44"
George F & Helen C Brown-Mom & Dad
And 2h25'23" 49d7'51"
George F Marcucci
Lib 15h5'28" -4d59'50"
George F Mitri
Cnc 8h46'19" 28d10'23"
George F Plikaitis
Ori 5h4'20" 13d32'46"
George F. Waters
Lyn 7h38'13" 53d14'53"
George F. Weisner
Uma 11h37'11" 31d19'54"
George Fagan
Ori 6h2'52" 18d23'9"
George Fehringer, Sr.
Lib 15h4'0" -5d1'40"
George Fleming
Cep 4h50'2" 82d53'35"
George & Florence Meuse
Cyg 19h48'20" 32d44'1"
George forever my star 17.4.28 - 31.7.04
Cru 12h40'55" -57d30'10"
George Forni
Uma 9h35'8" 56d43'25"
George Fox - "Duke 916"
Lyn 7h19'54" 49d55'8"
George Francis Cominsky
Cet 2h33'25" 8d16'52"
George Francis Jones
Cap 20h9'22" -18d23'0"
George Francis Urban
Cap 21h36'51" -11d3'2"
George & Frankie Langer
Cyg 21h25'49" 51d56'14"
George Franklin Mays
Cyg 20h49'29" 45d31'3"
George Frederick Daniel
Cep 21h37'37" 61d37'15"
George Frieder
Per 3h15'18" 54d20'48"
George Fusiek
Per 3h16'17" 53d25'12"
George G. Bowser
Uma 8h56'20" 48d51'34"
George G. Conlin, Jr.
Aur 5h27'10" 45d52'51"
George Garcia
Aql 19h33'5" 7d27'56"
George Gaston
Per 2h33'42" 55d40'27"
George Georgie Innes Greco
Leo 10h43'1" 14d30'35"
George Gerald Buck
Peg 21h31'17" 15d41'19"
George Gerontakis
Uma 11h18'36" 44d10'3"
George Gerontakis 11
Uma 11h11'49" 57d44'56"
George Gillyatt
Ari 3h15'19" 28d53'36"
George Gilmer "Gilly" Sale III
Uma 9h26'22" 55d3'11"
George & Gladys Barnum
Lyr 18h45'45" 32d29'20"
George & Gladys Plos
Cam 4h11'13" 68d11'32"
George Glenn Zipf 10/12/1920
Lib 15h11'38" -22d56'19"
George Green - Georgeous Star
Ori 6h20'40" -1d24'31"
George Gregory Karlin
Aql 19h44'54" 5d17'42"
George H. Clausen-drive God's train
Uma 12h4'40" 52d26'3"
George H Crane
Ori 5h57'40" 3d1'26"
George H. Jones
Uma 10h23'9" 56d1'12"
George H. Klinger Jr.
Cep 23h40'28" 86d27'52"
George H. Sudheimer
Lmi 10h9'50" 36d42'0"
George H. Talbot, Jr.
Psc 1h9'8" 10d8'13"
George Hall
Per 3h1'25" 46d31'45"
George Harlan Clawson
Uma 13h49'36" 56d40'16"
George Harper
Her 18h8'9" 32d10'29"
George & Heather Eternal Union
Cyg 19h36'35" 53d19'24"
George & Helen King
Uma 8h51'10" 52d58'44"
George Henry
Psc 1h29'2" 7d1'10"
George Henry Ainslie - "Henry's Star"
Ori 5h1'26" 5d16'56"
George Henry Lane
Ori 6h4'41" 10d34'28"
George Hertz
Pic 5h46'7" -52d8'1"
George Hijazin
Her 16h18'56" 5d41'46"
George Hilf
Vir 12h37'54" 3d34'6"
George Hollis Parker
Uma 9h44'59" 46d55'59"
George Hon
Aur 5h59'56" 42d19'6"
George Hovis
Boo 14h38'50" 25d10'44"
George Hugh Baker
Uma 11h0'6" 37d20'9"
George Hulme 90th Birthday Star
Cep 1h17'30" 84d47'50"
George Ian Felder
Her 17h51'35" 31d28'31"
George & Inez Lengvari
Cyg 19h48'7" 38d13'49"
George Ingram Jr.
Uma 8h35'8" 50d9'33"
George Inouye
Uma 11h28'52" 37d9'45"
George & Ione Graham
Cas 1h24'7" 62d28'25"
George Isaac Atterby Haynes
Her 18h20'24" 20d53'56"
George J. Baird
Cep 21h28'17" 65d0'23"
George J. Bayer
Per 3h11'58" 52d8'54"
George J. Budko
Ori 5h22'28" 13d20'2"
George J. Galli
Cnc 8h55'23" 27d10'2"
George J. Iannone, Sr.
Cep 22h21'6" 62d39'22"
George J. Kosmorsky (Dido)
Her 18h50'54" 23d58'30"
George J. Pedersen
Aql 19h50'53" -0d33'28"
George J. Rull
Boo 15h1'32" 48d19'11"
George J. Schilling, Jr.
Cas 1h22'47" 62d14'27"
George J. Trevlakis
Her 16h7'52" 47d50'7"
George J. Yapor
Aql 19h58'36" 10d24'32"
George James Felos
Ari 2h13'34" 25d37'25"
George James Gerrard Murphy
Umi 17h20'28" 85d58'27"
George & Janet Butch
Gem 7h3'23" 17d2'8"
George John Everts III
Dra 19h48'0" 58d33'34"
George JonPaul Sladek
Leo 10h21'34" 19d31'13"
George Joseph Bauman Jr.
Uma 9h37'38" 62d17'30"
George Joseph DeMers
Her 16h39'58" 10d49'55"
George Joseph Gail
Cir 14h45'57" -57d55'55"
George & Joyce's Starlight
Tau 5h34'54" 20d14'39"
George Jr. Wyszynski
Ari 2h45'17" 27d29'46"
George K. Wright Sr
Sgr 17h46'42" -18d36'20"
George Kaemen
Crb 15h46'40" 27d33'12"
George Kaercher
Her 18h2'17" 24d33'5"
George Kalmar
Her 17h26'1" 34d34'51"
George Kanellakis
Her 16h38'14" 37d49'12"
George & Kelly
Cyg 20h12'7" 35d26'47"
George Kenneth Walster
Uma 10h7'48" 55d58'0"
George Ketterer
Boo 14h26'31" 14d52'33"
George King Detzler
Sgr 18h16'18" -31d40'14"
George Klippness
Boo 15h8'30" 34d25'36"
George Kloske
Psc 1h19'1" 25d51'33"
George Knapp
Uma 11h22'1" 31d8'14"
George Korchinsky
Aqr 21h39'50" -5d54'26"
George Kouvaras
Uma 11h59'36" 42d28'25"
George Krikorian
Her 16h51'32" 22d35'41"
George & Kristy Katavic
Ara 17h14'37" -51d1'2"
George Kritikos
Uma 11h58'9" 42d40'50"
George Kulwein
Sgr 18h35'54" -23d45'30"
George Kyriakoudes
Aqr 21h41'8" 2d18'8"
George L. Erion III
Psc 0h57'6" 18d45'44"
George L. Van Valen
Cap 20h32'52" -21d2'43"
George LaFe
Uma 9h45'12" 66d9'59"
George Lawrence DeSellier
Sco 17h56'31" -33d6'18"
George Leach Stone
Uma 9h30'49" 54d51'49"
George Lee
Uma 10h25'0" 62d42'41"
George Lee Holland
Uma 11h3'48" 49d51'34"
George & Leonie Angelis - Epeteios
Col 6h11'51" -34d15'59"
George Leventen Family Star
Cnc 9h12'44" 15d6'52"
George Londors in loving memory
Cep 2h45'0" 85d22'31"
George Louda
Uma 10h0'30" 54d18'0"

George Louis Guerra Her 16h54'44" 30d23'8"
George Louis & Judith Reed Russell Gem 6h56'8" 19d30'56"
George Louis Thornton Aqr 23h35'10" -13d5'58"
George Loves Susan Cyg 19h41'35" 33d3'32"
George Luis Castillo Jr. Vir 11h39'37" 3d57'5"
George Lundberg 3/21/33, Roll Tide! And 0h29'9" 33d38'52"
George M. Ernst, III Cap 20h23'34" -23d19'8"
George M. Ferrara "Papa George" Uma 9h44'44" 64d10'41"
George M. Foreman III Cnc 8h27'37" 16d8'38"
George M. Hobel Vir 12h51'28" 6d49'9"
George M. Lusic III Ari 1h59'57" 13d50'47"
George M. North *60th* Cyg 20h5'23" 37d24'17"
George M. Ritchie Per 3h25'15" 52d16'48"
George M. Selivonchik Cep 22h31'10" 73d58'18"
George Mac Sr. Ari 3h7'18" 26d33'16"
George Maloof Aur 5h17'59" 44d33'2"
George Mark Calil Cep 21h20'16" 64d25'22"
George & Mary Umi 16h25'35" 80d10'5"
George Matejko 60 Lyn 8h52'50" 34d8'0"
George Mcconnell Aqr 22h56'2" -21d32'13"
George McGraw Aqr 22h43'4" -4d3'26"
George Mennen Cep 22h44'9" 62d7'59"
George Meredith "Dogie" Jones Cma 6h59'37" -21d57'57"
George Metrou Leo 10h38'34" 16d27'54"
George Michael Bourisk, Jr. Lib 15h46'35" -17d10'44"
George Michael Colvin Sgr 18h22'40" -32d38'9"
George Michael Gregory Psc 23h11'54" -0d28'20"
George Michael Porter Sco 17h58'21" -41d17'15"
George Michael Skove Leo 11h23'57" 15d51'44"
George Michael Weale Her 17h8'23" 31d42'7"
George Milasinovich Ori 5h11'12" -7d23'39"
George & Mildred Starr 117 Cyg 20h22'53" 54d39'52"
George Miskondra Sekerak Uma 11h2'14" 65d9'37"
George Monroe Nielsen Gem 6h54'17" 12d6'47"
George Moore Psc 1h5'43" 3d24'45"
George Mortimer Gem 6h57'49" 25d39'18"
George M.Thomas Her 18h0'29" 24d35'25"
George Musa Sahhar Gem 6h44'29" 23d25'24"
George n Tina Aql 19h49'23" 5d46'11"
George Naoum Ari 2h20'55" 14d6'8"
George Nathan-Moran Weldon Cru 12h50'49" -60d11'46"
George Nava Psc 1h10'16" 22d22'4"
George Neville Her 17h48'36" 34d39'32"
George Nicholas Pourpouras Lib 14h29'1" -9d49'12"
George Nicholas Tsaldaris Uma 9h55'37" 56d15'59"
George Norman Arnovick Leo 9h42'16" 16d42'22"
George Norman Brown Lib 15h11'45" -6d11'9"
George Novak Lyr 18h27'46" 33d40'6"
George O. Kazika Umi 16h20'5" 80d49'17"
George O. Sampson IV Her 16h19'17" 45d26'56"
George Offenburg Her 17h31'22" 39d23'12"
George Oliveira - Worlds Best Daddy Cep 22h21'26" 65d16'41"
George Owen Lib 15h17'41" -5d29'13"
George P. Fresenborg Aur 5h4'31" 35d49'8"
George P. Kalev (01/16/86-10/8/05) Cap 21h39'23" -19d30'8"
George P. Semenyak Aqr 23h3'5" -22d30'16"
George (Papa T) Tenzinger Del 20h40'2" 15d32'53"
George Papas Uma 9h15'2" 61d55'54"
George Parker Wentworth Aql 20h0'28" 6d35'53"
George & Patricia Forever *1952* Ori 5h36'44" -0d32'28"
George Patrick Fitzgerald, III Sgr 18h7'14" -26d12'23"
George Patrick John Colgan Cep 2h44'43" 82d9'27"
George Patrick Philbin Dra 17h3'48" 53d38'43"
George & Patti White Aqr 23h9'38" -7d52'5"
George Paul Bernal, II Cep 21h54'20" 62d57'31"
George & Pauline McFarland Parents Ori 5h24'7" 8d14'57"
George Pesce-Not Just A "Best Man" Uma 12h22'32" 57d30'40"
George Peter Menedis Pho 23h48'17" -45d26'11"
George Peter Stevens Ori 5h5'46" 13d45'50"
GEORGE PETROPOULAKIS Cyg 20h8'7" 33d13'5"
George Philip Oxley Her 17h37'18" 45d8'1"
George Philip VI Cep 0h16'3" 70d17'50"
George Phillips Taylor II Aur 7h19'41" 43d0'58"
George/Phyllis Cyg 21h30'33" 48d56'0"
George Pittas Uma 11h59'58" 42d30'37"
George Powell Psc 23h14'7" -2d27'50"
George Powers Ori 5h43'24" 8d13'7"
George Price Lyr 18h32'30" 36d16'6"
George Psaras Uma 9h36'38" 47d24'45"
George R. Kulka Aql 19h12'42" 6d22'0"
George R. Lewis Jr. Aqr 22h20'59" -2d27'59"
George R. Mayotte My love forever D Cep 22h58'39" 71d12'44"
George R. Miller's Magic Psc 1h15'53" 31d23'12"
George R. Moser Ori 5h57'27" -0d46'46"
George R.A. Johnson Tau 4h20'55" 27d37'0"
George Rafferty Her 18h35'53" 16d7'36"
George Ray Terral Uma 11h47'48" 34d6'54"
George Raymond Drew II Ori 6h18'25" 10d55'34"
George Retos Ori 6h1'1" 13d20'9"
George Richard Moate Ori 6h4'1" 15d23'46"
George Robert Bucchi Lib 15h1'38" -15d35'12"
George Robert Charles Leo 9h26'41" 32d0'25"
George Robert Smith Ori 5h10'30" 15d49'49"
George Rolan Thigpen Her 17h20'44" 34d22'2"
George Ronald Gilbert Sco 15h53'3" -24d30'22"
George Ronald Schowerer Ori 5h12'47" -0d12'27"
George Roppel Psc 23h50'34" 1d52'43"
George Rucker Cnc 8h41'20" 30d26'15"
George S. Aql 19h29'44" 3d49'57"
George S. Daniel Sgr 18h33'6" -33d2'41"
George S. McDermott Psc 1h4'24" 12d49'51"
George S. Moore Aqr 22h16'6" 0d37'32"
George S. Wills Psc 1h18'47" 31d11'54"
George & Sally Cyg 20h12'16" 43d53'5"
George Samuel Smile Sco 16h11'5" -13d29'4"
George Sanchez Uma 9h20'10" 42d12'39"
George & Sandra Hooper Uma 8h54'22" 58d37'26"
George Saro Avetian Cap 20h27'44" -10d58'58"
George Scanlan Per 3h54'10" 38d2'6"
George Scarlett Per 3h15'25" 49d11'43"
George Schaffer Uma 13h45'14" 58d18'5"
George Scott Mercurio Sgr 18h21'32" -33d2'35"
George (Scotty) Hamilton Cet 1h27'33" -5d42'39"
George Shelton Ori 5h56'22" 8d40'37"
George & Shirley Haas Cyg 21h24'5" 45d6'53"
George Sidoris Uma 11h9'41" 50d1'20"
George Simon Bardmesser Her 17h18'45" 18d52'27"
George Siomkos Uma 10h27'36" 47d44'23"
George Smith Pup 8h1'46" -35d9'31"
George Snow Her 17h26'42" 32d52'59"
George Snyder Uma 11h11'56" 44d1'32"
George Soyka Uma 10h3'33" 71d8'51"
George Steven Wright Cyg 20h41'22" 51d7'36"
George Stevenson Per 4h49'58" 39d18'43"
George Stewart MacGill Uma 8h10'17" 61d37'12"
George Stewart of Linlithgow Cep 20h45'23" 59d35'59"
George Stringos Lib 15h22'21" -22d44'3"
George T. Chicoine Boo 14h28'23" 41d40'20"
George T. Lacey Oph 17h18'45" -0d15'45"
George T. Wray Cep 21h42'8" 60d30'2"
George Tallman Ladd II Ori 6h1'43" 17d11'18"
George Tann Cnc 8h50'49" 31d30'32"
George Tedaldi Sgr 19h7'37" -12d43'0"
George "The Big E" Ennis Ori 6h16'31" 9d13'40"
George Thomas Coontz Sr. Lmi 10h26'5" 33d1'59"
George Thomas King Uma 8h48'3" 47d20'44"
George Thomas Roy Uma 13h28'29" 53d41'30"
George Tindall Per 4h16'39" 49d10'14"
George & TroyAnn - Together As One Cyg 20h29'23" 45d32'24"
George Tsigas Tau 5h58'2" 27d51'41"
George V. Eltgroth Tau 3h41'39" 26d30'2"
George Valerius Cep 21h34'48" 68d3'30"
George & Vera Berry Gem 7h24'50" 21d30'18"
George Vernon Brown Ori 5h1'40" 15d2'5"
George Vizcaino Cnv 13h27'0" 32d23'15"
George W. and Jeanne B. O'Day Her 16h59'47" 32d45'59"
George W. Besley, Jr. Cep 21h28'33" 64d40'46"
George W. Diersen Per 4h48'45" 43d28'55"
George W Hautanen Ori 6h18'26" 14d51'1"
George W. Johnson Per 3h15'15" 54d5'16"
George W. Moorer, Jr. Cnc 8h36'7" 31d4'30"
George W. Oliver Leo 9h52'26" 7d38'24"
George W. Rollins, Sr. Sco 17h1'7" -30d6'4"
George W. Summerford Cas 1h13'2" 66d43'16"
George W. Worthley Cma 7h12'48" -25d5'44"
George W. Wurthmann Jr. Cap 21h26'43" -24d38'21"
George Waal Psc 0h46'8" 15d13'25"
George Walker Cullen 07/01/1937 Cnc 8h19'10" 22d52'32"
George Wallace Per 3h39'58" 46d35'7"
George Wallace Bever Vir 13h6'47" 11d37'3"
George Walter Kraft, Jr. Lac 22h23'24" 48d3'14"
George Warren Booz Sco 17h0'34" -38d22'51"
George "Weirdo" Wasilenko Leo 9h33'20" 7d29'21"
George Weiss Aur 5h45'46" 50d19'42"
George Werner Knapp Cnc 8h42'27" 24d56'51"
George Wesley King Lyn 7h31'15" 43d7'21"
George Wesley Palovich Aqr 22h37'31" -3d45'19"
George Westover Gauley, Jr. Crb 16h7'7" 35d43'52"
George Whillans' Eildon Reiver Lmi 10h9'6" 38d9'30"
George Whitfield Pepper Leo 9h48'23" 7d32'46"
George Whitfield Pepper Jr. Aqr 23h45'43" -22d43'36"
George Wilhelm Her 17h11'24" 33d30'39"
George William Grosvenor Orchard Uma 8h50'23" 47d0'1"
George William Haessler, Jr Her 18h22'16" 18d10'20"
George William Hall Sr. Uma 13h35'24" 59d37'6"
George William Lucy IV Uma 11h22'46" 45d16'45"
George William Maak "Big George" Sgr 18h57'0" -33d21'9"
George William Ollinger Aur 4h58'17" 35d42'24"
George William Prescott Cap 20h16'7" -13d30'11"
George William Rabuck *08-22-1957* Uma 12h17'16" 52d29'6"
George William Ratcliffe Ori 6h15'9" 9d14'10"
George William Robert Fairbairn Umi 14h35'56" 70d0'8"
George William Starr - George's Star Her 16h41'23" 13d3'31"
George William Tucker Per 4h19'14" 40d50'44"
George Williams Fallen Leaf Angel Sco 16h36'44" -42d6'42"
George Wolverton Cap 21h41'37" -18d58'17"
George Zak Cep 22h58'32" 71d34'12"
George Zander Uma 8h30'52" 72d59'42"
George Zarglis Cru 12h45'18" -57d46'12"
George Zimmer Lyn 7h9'12" 54d48'21"
George Zimmerman Aql 19h46'53" -0d4'37"
George (Zubo's Mate) Gantt Lyn 7h11'11" 52d47'26"
George Zuras Lyr 18h38'30" 36d29'48"
George, my love my life always Nina Ori 5h54'38" 7d50'25"
GeorgeA Pho 2h16'54" -46d15'40"
Georgeanne Umi 16h14'0" 78d8'34"
Georgeanne Baker Cnc 9h20'17" 10d46'40"
Georgeanne Grove Leo 11h6'43" 15d54'50"
Georgeanne Martin Uma 10h39'17" 69d24'42"
George'n'Heathurr's Cyg 20h16'7" 50d31'55"
Georgentessa Lyn 6h35'11" 57d53'19"
Georges Dra 16h4'21" 61d55'1"
George's 70th Dra 17h54'8" 62d36'28"
Georges Basil Tarazi Aqr 21h39'30" 0d34'39"
Georges Constanty Cap 20h47'1" -25d11'14"
George's Gem Ari 3h15'7" 29d59'34"
George's J Cnc 8h45'8" 18d27'47"
Georges Laraque Lac 22h47'58" 52d7'2"
George's Love Nahgem Mon 6h52'4" -0d36'37"
George's Man On The Moon Sgr 18h41'36" -30d17'42"
Georges Moussa Ori 5h56'38" 17d51'4"
George's of Geneva Gem 6h51'13" 34d15'49"
Georges Rimondi Cam 3h24'44" 64d54'7"
George's Shining Star Uma 8h10'58" 65d12'0"
George's Star Ori 4h56'5" 10d3'9"
George's Star Her 18h5'39" 14d58'36"
George's Sydney Dra 20h17'10" 63d35'14"
George`s Viper Sct 18h31'14" -15d12'56"
Georgette Gretina & Dawn Booth Lyr 18h39'30" 38d10'34"
Georgette Lacroix Crb 16h22'22" 36d20'7"
Georgette Levandowski Sgr 18h58'36" -35d51'17"
Georgette Paulsin Mosbacher Lyn 7h9'26" 45d59'7"
Georgette Spanjich Eri 4h12'33" -0d10'26"
Georgeva Olya Umi 16h59'50" 80d12'14"
Georgi L Conlon 042595, My Lucky 7 Aql 20h8'8" 15d22'16"
Georgia Crb 15h49'48" 27d17'25"
Georgia Leo 9h37'45" 27d35'58"
Georgia And 1h4'4" 45d55'3"
Georgia Vir 12h56'53" -0d20'49"
Georgia Aqr 21h49'50" -2d46'11"
Georgia Cas 23h44'25" 55d20'8"
Georgia 15-01-2006 Vir 13h16'58" 13d30'56"
Georgia A Lea Lib 15h6'12" -5d15'25"
Georgia Ann Gebhardt Aqr 22h55'38" -16d25'11"
Georgia Ann Rolf Heckenberg CGNRA Vir 15h8'42" 5d12'33"
Georgia Ann Young Hya 8h51'47" -0d28'49"
Georgia Anne Lever And 23h15'1" 51d6'3"
Georgia Baker And 1h27'9" 46d44'32"
Georgia Bauman Cam 3h49'39" 58d57'47"
Georgia Bea Shaw Cru 12h17'24" -59d5'42"
Georgia Benson Cap 20h14'55" -19d44'5"
Georgia Brenckman Lyr 18h48'4" 36d56'14"
Georgia Burkhard Cyg 21h25'21" 53d55'29"
Georgia Calley And 0h56'1" 40d35'38"
Georgia Cameron Dunn And 1h21'27" 43d31'17"
Georgia Claire Oats Cru 12h39'31" -60d35'19"
Georgia Cornett Golden Lyr 19h13'39" 45d53'45"
Georgia Crews "All My Love" Uma 10h9'49" 49d0'6"
Georgia Crunk Leo 10h37'3" 21d9'13"
Georgia Eccles Eri 4h0'42" -21d43'49"
Georgia Ede's Star Crb 15h38'52" 27d41'26"
Georgia Emma Melvin - Georgia's Star And 1h13'5" 46d5'57"
Georgia Erin Cas 0h16'7" 55d42'25"
Georgia Florea Gem 7h10'0" 27d20'52"
Georgia Flynn Forrest Uma 8h53'15" 71d58'22"
Georgia Frances Rhodin Leo 10h42'51" 13d23'12"
Georgia Hathorne Lib 14h54'23" -2d27'33"
Georgia Jane Kim And 2h32'44" 48d54'37"
Georgia & J.B. O'Keefe Aql 20h34'58" -7d27'7"
Georgia Jenny Kane Gem 7h45'20" 34d18'19"
Georgia Jeppesen Ari 3h19'44" 28d32'31"
Georgia & Jimmy Carpenter Uma 11h8'20" 35d12'6"
Georgia Jones ~ Heavenly Angel Lyr 19h11'52" 27d8'3"
Georgia "Judd" Kline Cas 23h55'9" 56d47'27"
Georgia "Junior" Gootee Sco 16h8'54" -11d53'17"
Georgia Kanouse And 22h59'12" 48d4'30"
Georgia Kate - 16021984 Cru 12h19'21" -57d14'41"
Georgia Katlin Bolin Peg 22h5'39" 15d27'23"
Georgia Kay Sanders Tau 5h39'41" 17d42'23"
Georgia L. Burke Aqr 22h28'33" -2d4'7"
Georgia L. Shaw Leo 11h3'1" 20d14'22"
Georgia Lee Anderson Vir 12h47'40" 6d42'46"
Georgia Lee (Hoover) Stiles Aqr 22h38'53" -2d14'2"
Georgia Leigh Smith - Georgia's Star And 23h48'35" 48d33'44"
Georgia Lili Costigan Vir 13h33'8" 3d45'11"
Georgia Lou Hall Lyr 18h50'34" 43d9'1"
Georgia Louise And 1h15'39" 43d4'20"
Georgia Mae Ari 2h20'5" 12d12'47"
Georgia Mae Ori 5h33'24" -0d7'44"
Georgia Mae Psc 0h44'9" 6d12'22"
Georgia Marie Circelli And 1h11'0" 46d51'46"
Georgia Masters And 22h58'22" 38d26'27"
Georgia Mea Alice Wilson And 2h23'35" 42d23'58"
Georgia Mechelle And 1h25'41" 48d2'1"
Georgia Meletiadis Sco 17h41'14" -39d38'10"
Georgia Passas Hall Lyn 7h58'30" 36d6'31"
Georgia Peach Uma 11h59'51" 30d52'47"
Georgia "Peach" Sheehan Del 20h40'10" 13d15'47"
Georgia Perez Uma 8h59'16" 47d2'39"
Georgia Pie Turner Ori 5h24'38" 1d19'15"
Georgia Piper And 0h49'33" 26d46'1"
Georgia Rhiannon Thomson And 0h51'47" 24d0'25"
Georgia Rice Westcott-Pitt Leo 11h40'17" 25d33'8"
Georgia Rose Cru 12h39'43" -59d6'54"
Georgia Rose Psc 23h25'42" -3d2'10"
Georgia Ryan Psc 1h27'53" 11d42'28"
Georgia Stern-Super Science Teacher Uma 9h2'28" 58d48'16"
Georgia T. Young Cnc 9h1'29" 28d29'21"
Georgia Violet Lyn 7h28'39" 52d41'49"
Georgia Webb Ari 2h44'0" 15d35'9"
Georgia-Eden `Monkey' Scotchmer And 23h20'42" 42d26'14"
Georgiana Ori 5h32'54" 2d53'31"
Georgiana Vir 12h55'48" 4d26'1"
Georgiana deRopp Ducas Lac 22h52'12" 49d6'22"
Georgiana Elizabeth Huisking Gem 7h2'18" 18d55'59"
Georgiana Gaztambide HB Ori 5h5'56" 4d22'53"
Georgiana Mirabella Murray Cnc 9h3'8" 28d37'19"
Georgiana (My Love) Lyr 19h6'28" 42d10'49"
Georgiann Mon 6h35'46" 1d38'22"
Georgiann M. Fallon Aqr 23h33'53" -8d35'3"
Georgianna Jones Leo 10h19'7" 20d53'3"
Georgianna Motto And 2h36'49" 45d49'3"
Georgianna Slack Mon 6h39'8" 7d10'22"
Georgia's Star Cas 0h22'29" 56d8'33"
Georgia's Star Cru 12h43'26" -57d18'12"
Georgie Uma 12h32'39" 56d46'16"
Georgie Umi 16h7'38" 73d28'40"
Georgie Uma 12h5'9" 42d13'16"
Georgie Uma 11h37'25" 38d4'47"
Georgie Ori 5h25'22" 1d46'46"
Georgie Psc 1h18'23" 15d28'59"
Georgie Gem 6h55'4" 22d21'44"
Georgie Uma 11h58'55" 28d23'56"
Georgie Day Cas 1h18'54" 55d44'30"
Georgie Gladdys Carriage House Uma 9h29'41" 64d27'46"
Georgie Koikas Cru 11h58'43" -62d21'49"
Georgie My Special Star Aqr 23h52'47" -20d39'58"
Georgie & Richard Together Forever Sct 18h52'38" -5d7'50"
Georgieous Cap 21h43'45" -9d40'39"
Georgina Cyg 21h37'13" 40d30'29"
Georgina Amelia Mary Hyde And 22h59'18" 46d47'27"
Georgina Brito 1942-2005 Uma 11h49'34" 63d9'59"
Georgina Eileen Leo 11h8'23" -1d5'11"
Georgina Ella Catalinotto 22 Peg 23h48'27" 15d38'31"
Georgina F. Perez Leo 11h41'35" 25d25'5"
Georgina Grace Lewis-Hodgson Aqr 20h40'21" 1d45'38"
Georgina Haddock Del 20h41'28" 16d7'0"
Georgina Magina Megadu Begadu Gem 6h46'2" 25d33'29"
Georgina McWhirter Cas 23h52'38" 59d45'32"
Georgina N. Watson And 23h14'20" 41d48'5"
Georgina Pearl Smith Cas 23h43'50" 57d55'26"
Georgina Pribulsky Cas 0h26'19" 56d25'12"
Georgina Rubio Chiner Aql 19h30'54" 6d49'43"
Georgina Sanger-Train Cyg 19h42'35" 48d7'36"
Georgina & Sophie 'Percys' And 0h28'21" 39d45'33"
Georgina Wallace Smith Baird BSc Cyg 19h54'54" 33d26'29"
Georgina Ware Peg 23h43'57" 17d20'58"
Georgina's Valentine Star Vir 11h41'15" 2d44'19"
Georgonicas Nulli Secundus Pho 2h22'19" -46d2'18"
Georgous Satellite Cyg 20h26'34" 33d5'59"
Georgy Ori 6h10'41" 7d15'46"
Georgy M. Cep 20h59'11" 57d18'59"
Georibi Aql 19h33'23" 8d17'41"
Georjane Johns Gem 7h42'20" 31d35'48"
Geovani Mando Santos Uma 8h16'40" 64d50'13"
Geowiga Del 20h18'6" 9d40'24"
Geppert, Sieglinde Uma 12h31'28" 54d49'40"
Ger Boo 14h41'19" 18d1'31"
Ger Cyg 21h56'10" 40d23'37"
Ger & Abby Pup 7h50'22" -31d59'31"
Geraghty Cas 2h12'44" 59d31'45"
Gerald Per 2h33'52" 53d54'15"
Gerald Lyr 19h8'33" 31d24'22"
Gerald Tau 5h11'16" 18d44'43"
Gerald 6/28/1941 Cep 20h39'23" 63d59'22"
Gerald 60 Cep 21h54'12" 62d32'44"
Gerald A. Callan Leo 9h28'45" 26d12'54"
Gerald A. Gustus, Jr. Leo 9h34'51" 12d14'57"
Gerald and Laura Casey Sco 16h3'6" -10d42'11"
Gerald and Patsy Gant Cyg 21h58'59" 48d53'41"
Gerald Anthony Clow Cyg 21h24'49" 39d32'53"
Gerald Beykirch Uma 9h31'9" 72d51'18"

Gerald Binns
Uma 9h5'50" 56d33'1"
Gerald Bove
Umi 15h2'26" 74d12'32"
Gérald Boyer Notre Pépé d'Amour
Umi 16h48'44" 75d42'47"
Gerald Braden
Uma 9h13'24" 70d48'47"
Gerald Bredin
Her 16h5'58" 46d0'16"
Gerald C Nuffer
Ori 6h20'10" 13d20'57"
Gerald Carl Riedthaler 1966
Uma 8h38'46" 47d16'53"
Gerald Charles Oatman
Vir 13h15'2" -6d55'59"
Gerald Clarence Felts
Uma 12h49'10" 58d22'29"
Gerald Conway Neil
And 23h16'53" 47d49'9"
Gerald D.
Uma 10h22'51" 60d42'20"
Gerald David Hinkebein
Uma 9h57'5" 67d56'54"
Gerald DeCroce
Uma 9h54'17" 55d46'21"
Gerald Desaire
Sgr 19h36'31" -18d30'46"
Gerald Drake
Dra 15h46'27" 61d31'50"
Gerald E. Bond
Leo 9h39'45" 14d17'21"
Gerald Ediger
Lib 15h17'36" -5d44'38"
Gerald Edward Civis Jr
Uma 9h28'11" 58d5'23"
Gerald Emmanuel Serfass
Ari 1h55'0" 20d2'15"
Gerald F. Losey
Her 17h34'3" 36d42'17"
Gerald F. Macke
Aql 19h8'29" 7d6'43"
Gerald Filipowicz
Ari 2h45'18" 29d4'39"
Gerald G Urban
Per 4h6'46" 40d24'44"
Gerald Gorman
Sgr 19h34'7" -16d35'0"
Gerald Grady Pitts
Psc 0h36'43" 7d7'44"
Gerald Helwig
Per 3h6'25" 52d48'55"
Gerald Henry
Her 18h5'3" 23d35'42"
Gerald Hugh Wiley
Ori 6h17'45" 11d24'9"
Gerald "Ice" 11
Aqr 23h45'7" -4d17'57"
Gerald James Clark
Ori 5h31'51" 3d36'21"
Gerald Jerome Parker
Boo 14h39'34" 35d32'10"
Gerald "Jerry" Grabinski
Her 18h24'20" 12d9'0"
Gerald Jerry S. Schlosser
Cyg 19h54'24" 51d37'50"
Gerald Johnson
Aur 5h14'6" 40d20'2"
Gerald Joseph Ahern
Ori 5h32'37" 8d27'58"
Gerald King Nakashian, Jr.
Boo 14h45'20" 53d12'54"
Gerald Klemm's Stern
Ori 6h13'43" 8d11'54"
Gerald Korican
Vir 13h28'33" -5d15'33"
Gerald L. Soderstrom
Sco 16h11'9" -14d40'23"
Gerald Lawrence & Diane Jean Gay
Cyg 21h20'32" 52d45'31"
Gerald Lee Donnelly
Vir 12h52'29" 5d23'35"
Gerald Little
Uma 11h21'58" 38d45'21"
Gerald M. Moch
Dra 16h43'50" 58d36'51"
Gerald Magee McMillan
Her 18h12'46" 25d0'30"
Gerald + Marjorie Groskopf
Cyg 20h31'35" 33d15'34"
Gerald Mark Rodencal Jr
Ari 2h12'40" 24d8'19"
Gerald Martin Strauss Jr
Lib 14h49'7" -0d53'5"
Gerald Massoth Jr.
Vir 12h18'15" 4d9'34"
Gerald Maurice Blay
Per 4h16'13" 36d9'5"
Gerald McClune
Sco 16h13'22" -11d14'17"
Gerald McCullouch
Ari 3h19'23" 19d50'56"
Gerald Nethery
Ori 5h59'0" 17d12'55"
Gerald Noyles, Jr.
Tau 3h41'18" 11d36'37"
Gerald Pitzikcat Schindler
Lib 15h25'36" -25d38'58"
Gerald Pointing
Gem 7h47'46" 15d0'37"
Gerald "Pop-Pop" Bennett
Gem 6h31'13" 26d14'20"
Gérald Querido
Leo 10h3'57" 25d14'39"
Gerald Quigley
Ori 5h5'4" 3d34'4"
Gerald R Skeels
Ari 2h17'9" 14d18'10"
Gerald Raymond Elliott
Lib 14h51'42" -16d6'31"
Gerald Regis Hintz, Sr.
Aql 18h52'12" 7d19'57"
Gerald Rieflin Sr.
Uma 13h45'6" 58d9'18"
Gerald Roan
Equ 21h7'32" 4d30'43"
Gerald Robert Brown
Boo 15h11'1" 43d0'51"
Gerald Roger Knutson
Per 3h32'17" 51d47'37"
Gerald Sabin
Per 3h28'24" 36d34'30"
Gerald Scott Ader
Leo 9h46'0" 14d6'3"
Gerald Seitel
Uma 8h56'30" 52d23'18"
Gerald Silverman
Cep 21h21'27" 77d43'13"
Gerald Stanley Klein
Uma 9h23'19" 45d31'12"
Gerald Sullivan
Uma 10h50'46" 44d4'57"
Gerald T Dobishinsky
Uma 8h49'36" 55d20'0"
Gerald Terry Ovila Gelineau
Sco 17h38'53" -41d53'53"
Gerald the Giraffe
Cam 5h26'49" 56d23'6"
Gerald Theberge
Uma 8h14'3" 73d0'37"
Gerald & Theresa Seeber
Dra 19h2'21" 69d55'8"
Gerald Thomas Philip White
Per 2h58'2" 48d25'16"
Gerald Thomas Royal
Uma 14h10'30" 60d3'47"
Gerald W. Gurney
Boo 14h39'36" 24d31'40"
Gerald W. Motejunas
Dra 17h9'26" 64d15'26"
Gerald William Steven Pochynok
Cas 0h41'57" 73d34'1"
Gerald Wolf
Uma 8h55'41" 56d41'30"
Gerald X. Devereaux
Tau 4h23'44" 18d15'30"
Geraldine
Psc 0h40'35" 15d32'7"
Géraldine
Leo 9h34'11" 29d11'25"
Geraldine
Cyg 19h20'49" 29d58'45"
Geraldine
Ari 3h18'22" 27d52'23"
Geraldine
Cas 0h12'30" 57d12'59"
Geraldine
And 23h16'35" 49d21'12"
Geraldine
Cas 23h58'25" 56d51'4"
Geraldine
Cas 23h14'58" 55d56'19"
Geraldine
Cas 2h38'54" 73d5'38"
Geraldine 1953
Uma 9h59'15" 55d35'50"
Geraldine Anna Melsage
Vir 11h39'6" 4d40'57"
Geraldine Audette
Crb 16h9'9" 33d19'8"
Geraldine Bare Pope
Uma 10h3'1" 69d54'32"
Geraldine Christi DeMatteo
Uma 10h34'28" 62d41'42"
Geraldine Claire Fletcher
Vir 12h9'29" 11d1'31"
Geraldine Cole
Mon 7h1'56" 9d11'22"
Geraldine Collins
Lyn 7h36'29" 53d8'33"
Geraldine Couch
Uma 11h38'55" 47d43'41"
Geraldine & David
And 23h1'38" 52d20'33"
Geraldine DeSantis
Ari 2h22'4" 26d56'36"
Geraldine Dunraven
Cnv 12h44'41" 39d53'21"
Geraldine E Daubert
Mon 7h21'43" -0d57'52"
Geraldine Edel
Cas 1h31'46" 69d44'32"
Geraldine "Edstar" Liddy
Uma 11h59'6" 28d36'22"
Geraldine Geneva Weber
Uma 13h38'19" 53d7'23"
Géraldine Giraud
Tau 5h48'22" 17d46'26"
Geraldine Girshek
And 0h45'17" 41d44'22"
Geraldine H. LaBell
And 23h24'49" 48d30'8"
Geraldine Hill
Lyn 8h43'50" 45d30'19"
Geraldine Howey
Uma 9h53'36" 68d50'8"
Geraldine Joy Klein
And 2h20'8" 50d9'17"
Geraldine K Patten "1/10/1904"
Uma 10h43'51" 49d37'38"
Geraldine LaPaille
Aqr 22h29'16" -9d57'19"
Geraldine Lawrence
Lyn 9h12'3" 38d8'16"
Geraldine Liganor
Sco 16h12'49" -21d18'22"
Geraldine McFadden
Pyx 8h44'59" -27d19'52"
Geraldine Oldfather
Sco 16h51'10" -27d0'15"
Geraldine Patricia
Tau 4h4'30" 29d48'33"
Geraldine R. Sullivan
Cas 23h57'48" 50d4'8"
Geraldine & Roberto
Uma 10h43'36" 51d9'12"
Geraldine Romitti-Klaes
Dra 17h5'4" 52d15'7"
Geraldine Stonebank
Cas 0h29'10" 63d52'11"
Geraldine Theresa
Dra 16h1'16" 62d2'47"
Geraldine Thompson
Sgr 18h46'43" -29d54'29"
Geraldine's Love Light
Cru 12h26'39" -59d33'26"
Geraldiniana Major
Cyg 21h27'31" 32d57'27"
Gerald's Eternal Light
Cep 23h48'3" 67d9'33"
Gerald's Fancy 65
Cen 13h49'13" -60d6'58"
Gerald's Rose of Mayfield
Cyg 21h12'0" 40d26'2"
Gerald's Star
Cep 22h25'21" 84d47'2"
Geraline Mae Campbell
Per 3h31'38" 35d4'53"
Geralyn DiGeronimo
Cyg 20h15'31" 41d33'31"
Geralyn Ganesha
Cma 6h47'30" -14d15'4"
Geralyn Narkiewicz
Leo 9h58'54" 7d31'30"
Geramino
Sco 17h9'33" -33d11'24"
Gerard
Ant 9h48'31" -32d56'6"
Gérard
Cap 20h16'45" -26d8'57"
Gerard
Cru 12h51'39" -59d17'28"
Gerard
Cam 3h56'21" 66d28'54"
Gerard 343
Ori 5h35'53" -2d8'4"
Gerard Allen
Lib 15h18'26" -27d36'48"
Gerard Anthony Christopher Brierley
Cnv 12h40'58" 36d5'48"
Gerard Bredin
Cep 21h33'24" 66d6'6"
Gerard Campbell
Boo 14h49'13" 23d7'34"
Gerard Cannin(g)
Ori 5h25'43" 9d15'7"
Gerard Daniel Mulford
Her 16h39'16" 34d59'8"
Gérard Deldon
Cap 21h52'35" -10d22'21"
Gerard D'Souza
Ari 2h6'9" 20d10'0"
Gerard F. DeSantis, Jr.
Sgr 19h6'2" -21d44'27"
Gérard F. Plasse
Her 16h12'32" 47d48'51"
Gerard Forde & Sheafali Viju Patel
Lyn 7h45'20" 54d21'32"
Gérard Francis Fritz
Sgr 18h26'1" -16d24'12"
Gerard (Gerry) Bellemare
Cep 23h48'25" 73d28'30"
Gerard "GT" Erikson
Aql 19h6'3" 14d38'32"
Gerard H. Aumand
Her 16h40'42" 5d18'12"
Gerard Henry Cummings
Leo 9h23'0" 15d40'7"
Gerard Henry Lavoie
Gem 6h53'53" 30d21'40"
Gerard J Butler
Cep 23h8'8" 79d39'16"
Gerard J. DeSimone
Leo 11h43'27" 24d23'5"
Gerard J. Maloney
Cep 22h16'23" 63d46'59"
Gerard Kehoe
Cru 12h51'33" -59d9'34"
Gerard Kelly
Sgr 18h53'40" -33d24'29"
Gerard L.A. Downes
Cap 21h40'7" -21d29'50"
Gérard Lenorman
Aqr 22h4'1" -13d30'27"
Gérard Lenorman
Cmi 7h7'0" 9d34'8"
Gérard Lenzini
Leo 10h45'30" 14d52'31"
Gerard M. White
Lib 16h1'33" -10d3'28"
Gerard M. Zyla
Uma 8h39'58" 69d18'32"
Gérard Maurhofer
Peg 21h28'51" 23d32'49"
Gérard mon diamant
Uma 13h30'58" 61d13'8"
Gerard P. Pawlowski
Cep 22h39'11" 84d21'21"
Gerard Pasquale Antonellis
Cyg 20h29'54" 51d57'15"
Gérard Peter Marie Zeegers
Per 3h46'2" 33d10'46"
Gerard R Miller
Uma 10h1'36" 52d53'10"
Gerard Smeets
Gem 7h16'25" 18d52'54"
Gerard Stephen Trimboli
Gem 6h54'29" 17d3'26"
Gerard Steven Cerino
Per 4h15'9" 52d24'3"
Gerard Valentine
Lib 14h57'39" -12d17'19"
Gerard, Jr.
Leo 11h33'13" 20d3'35"
Gérard15031946
And 2h27'58" 45d12'19"
Gerarda & Bryan Culipher
Uma 9h33'23" 58d40'47"
Gerardine Elizabeth
Lyr 18h31'38" 30d53'22"
Gerardine Thornton
Peg 22h44'17" 16d7'20"
Gerardo
Ori 5h18'24" -8d35'27"
Gerardo A. Mechler (MonPère)
Cas 0h21'53" 52d55'51"
Gerardo Garcia - Shining Star
Aql 19h19'34" 16d10'50"
Gerardo G-Rock Ramos
Lyn 8h45'12" 42d20'31"
Gerardo Javier Renovales Mercado
Lyn 6h50'32" 60d9'26"
Gerardo Manuel Fundora
Cap 20h29'16" -12d1'21"
Gerardo Natividad Rico
Lib 15h12'58" -24d30'16"
Gerardus Majella Veldhuis 07111946
Sco 17h55'8" -30d38'13"
Gerarld Moran
Aqr 22h37'41" -19d59'4"
Gerasyutina Anastasiya Ivanovna
Vir 14h30'12" -2d2'36"
Gerb
Uma 8h31'52" 62d30'15"
Gerd
Uma 10h14'27" 57d46'3"
Gerd Hase
Uma 9h32'25" 51d9'24"
Gerd Lindner
Uma 9h57'54" 48d0'25"
Gerd Lundt
Her 16h46'49" 43d49'2"
Gerd Markoni
Uma 10h57'53" 57d30'41"
Gerd Neuhaus
Uma 12h3'30" 48d23'21"
Gerd Oppermann "Big-Star"
Uma 11h10'43" 38d56'8"
Gerd Petersen
Uma 11h50'24" 30d5'40"
Gerd & Ute Wegener
Cyg 21h31'18" 36d54'59"
Gerd von Poblotzki
Uma 9h30'5" 50d38'19"
Gerd Wiese
Ori 6h18'23" 7d51'18"
Gerd Wobbe
Ori 6h13'15" 20d38'38"
Gerd Zibirre
Uma 10h21'53" 56d57'49"
Gerda
Mon 8h2'34" -7d43'3"
Gerda
Lyr 19h20'20" 28d17'18"
Gerda Jablonski
Sco 16h15'22" -13d41'53"
Gerda Marie Benson
Cas 1h30'16" 60d10'58"
Gerda Marie Luise Crockett
Aqr 22h24'7" -0d46'28"
Gerda Olga Talesnik
Umi 16h16'26" 72d12'18"
Gerda "Schnütchen"
Uma 12h59'9" 62d35'49"
Gerdi's Light
Cap 20h21'49" -25d11'44"
Gere Géza
Aqr 20h45'0" -2d35'34"
Gerémy
Psc 0h42'46" 10d25'27"
Geremy George Harabedian
Per 3h16'13" 53d0'27"
Gergana Zheleva
Ari 2h0'14" 12d50'20"
Gergely
Uma 11h51'21" 41d17'6"
gergnjerg
Cnc 8h27'44" 21d45'58"
Gerhanda
Umi 14h36'10" 68d18'22"
Gerhard
Aqr 20h58'47" -0d31'5"
Gerhard Bahlinger
Uma 10h55'51" 66d3'12"
Gerhard Baier
Umi 15h36'33" 76d0'20"
Gerhard Bauer
Uma 10h53'56" 58d11'13"
Gerhard Dohnal
Umi 13h24'3" 72d15'56"
Gerhard Engert
Uma 9h27'10" 57d6'3"
Gerhard F. Wiethe
Uma 9h59'44" 63d16'47"
Gerhard Griensteidl
Uma 9h0'43" 63d59'51"
Gerhard Haufe
Ori 5h53'35" 3d12'22"
Gerhard Heering
Uma 9h15'50" 62d36'41"
Gerhard Kalweit
Ori 6h24'59" 10d4'8"
Gerhard & Lisa Kost
Cyg 21h12'22" 44d43'44"
Gerhard Mundt
Ori 6h3'24" 21d12'41"
Gerhard Preiner
Uma 8h24'11" 64d17'31"
Gerhard Rothammer
Aur 5h5'14" 37d1'16"
Gerhard Schaub
Cyg 20h7'28" 48d18'21"
Gerhard Slawik
Uma 10h44'51" 59d51'40"
Gerhard Spindler
Uma 11h56'40" 30d24'19"
Gerhard Traunecker
Ori 6h8'7" 8d33'14"
Gerhard Wanderer
Uma 10h7'8" 63d53'0"
Gerhard Werner
Ori 6h24'49" 10d16'37"
Gerhard Wonneberg
Ori 5h12'39" 8d19'37"
Gerhard's Hope
Sge 20h17'32" 18d2'45"
Gerhards, Udo
Uma 11h49'20" 54d31'39"
Geri
Ori 5h30'27" -5d1'19"
Geri
Cyg 21h24'46" 32d47'40"
Geri Ann Noll
Sgr 18h30'42" -20d44'10"
Geri Bea Kowalski
And 0h40'46" 42d26'48"
Geri C. Leon
Sgr 20h12'0" -37d33'40"
Geri J. Wallach
Aqr 22h35'10" -2d2'31"
Geri Jim Vossa Monterey Jazz Music
Lyr 18h49'31" 35d46'5"
Geri Lynn Schwendinger
And 2h38'39" 44d24'17"
Geri McGibney Star of Eternity
Aqr 22h18'35" -19d1'24"
Geri Ploskonka
Cas 0h45'36" 52d18'3"
Gerianne Fiumefreddo
Vir 13h54'51" -20d13'23"
Geriausios Drauges
Gem 6h52'30" 25d9'41"
Gericke, Brigitta
Uma 8h50'9" 60d1'34"
Geril
Uma 8h54'1" 65d44'33"
Gerilyn A. Schantz
Cnc 8h55'29" 14d29'29"
Geri's Gem
And 1h26'18" 49d4'45"
Gerison Lee Buendia
Leo 10h12'29" 24d21'45"
Gerlach, Gerald
Uma 9h54'54" 45d27'32"
Gerlinde B Goode
Sco 17h23'50" -40d54'52"
Germaine
Cas 0h15'24" 51d39'6"
Germaine "Chisi" Harris
Aqr 22h57'25" -7d17'18"
German A. Boobie Aponte
Sco 17h56'12" -30d39'19"
German Martinez Cazares
Cyg 21h49'40" 40d39'7"
German Mitschka
Uma 9h41'37" 54d32'37"
German Mutti
Vir 12h46'39" 4d33'3"
Germana
Lyn 8h26'44" 37d24'26"
Germana79
Uma 12h3'41" 50d15'54"
Germanitas
Peg 22h21'13" 7d9'8"
GERMANO MORGANTI
Cas 0h52'24" 62d56'38"
GerMar Ga De 1
Lib 15h35'3" -18d11'43"
Germer Kruy
Uma 11h45'54" 46d35'26"
Gernot & Peggy
Uma 9h46'8" 65d13'50"
Gernot Ramrath
Ori 5h53'36" -2d10'30"
Gernot Tripcke
Ori 6h22'24" 15d47'36"
Gero
Aql 19h24'41" 3d11'34"
Gerod 173
Her 17h0'1" 30d58'32"
Gerold L. Schiebler, M.D.
Dra 20h14'47" 72d17'49"
Gerolemos Chrysokhou
Per 3h43'30" 38d2'37"
Gérôme et Audrey
Vir 13h56'46" 4d41'7"
Gerome Wiesemann
Uma 12h42'22" 54d53'30"
Geronimo and Amanda Aguinaldo
Cyg 21h11'52" 43d42'9"
Geronimo Joe
Cap 21h29'39" -15d10'42"
Geronimo "MO" Mason
Boo 14h20'2" 46d29'14"
Gerrado Castaneda
Ori 6h9'23" 15d25'34"
Gerrard Jr Lock Kennedy
Umi 13h43'6" 78d11'14"
Gerrard Lee Gillespie
Ori 5h53'57" 11d39'25"
Gerrards Star
Col 5h53'41" -36d8'48"
Gerred Allen Price
Ori 4h52'58" -0d56'54"
Gerrell Piper
Cap 21h5'53" -19d11'37"
Gerrer, Lorenz
Ori 6h15'17" 9d3'2"
Gerri and Jack Callahan
Cyg 21h45'58" 31d15'9"
Gerri Ann
Crb 15h39'47" 36d13'0"
Gerri G. Collins
Ari 2h47'22" 14d10'4"
Gerri Whittredge
Vir 13h41'20" -6d43'55"
GerriAnn
Ori 4h52'27" 4d30'43"
GerriannaKeith 2005
Ori 6h3'4" 19d47'58"
Gerrie Lanning Gietema
Uma 11h11'49" 56d4'22"
Gerrie Lou Robison
Vir 14h21'18" 2d17'45"
Gerri's Star
And 0h44'30" 32d54'13"
Gerrit
Uma 12h57'5" 53d32'57"
Gerrit Haas
Uma 9h43'6" 62d45'53"
Gerrit Jacob Swiftney
Cep 21h25'19" 63d41'52"
Gerrit Van Dyk
Lmi 10h3'44" 40d20'3"
Gerritt Jacob "Rooster" Swiftney
Psc 0h54'46" 31d30'55"
Gerry
Ari 3h20'23" 29d34'4"
Gerry and Sheila Wakelin
Lmi 9h43'9" 38d35'29"
Gerry Buckingham
Her 17h12'58" 36d45'4"
Gerry Ciccarelli
Uma 10h23'0" 61d29'50"
Gerry Dodge
Cyg 21h45'46" 54d28'18"
Gerry Forever
Per 4h30'19" 39d51'34"
Gerry Haskell - Simply the Best
Umi 4h7'48" 89d44'53"
Gerry Hazel
Psc 1h31'51" 11d37'59"
Gerry Hillburn
Sco 16h7'57" -11d24'1"
Gerry Hope Milinkovich
Cyg 21h14'48" 31d14'32"
Gerry Lillo
Uma 9h25'20" 68d10'17"
Gerry M. Clarke
Cap 21h28'28" -19d14'31"
Gerry McMenemy
Gem 6h32'8" 26d13'55"
Gerry Moss
Cep 0h9'29" 72d28'9"
Gerry Nelson
Uma 9h30'40" 54d14'41"
Gerry Nicklas
Cnc 8h41'3" 29d42'8"
Gerry Porter
Cra 19h2'36" -38d47'44"
Gerry Quincey
Cep 20h17'28" 60d45'10"
Gerry Rae Jaramillo
Eri 3h44'40" -3d9'26"
Gerry Roche
Cep 3h41'43" 81d4'9"
Gerry Skyphantom
And 0h46'11" 35d59'52"
Gerry Tucker & Anne Laliberte
Psc 1h20'43" 6d34'52"
Gerry Wayne Monroe - 07.27.1918
Leo 9h34'15" 19d17'5"
Gerryann
Cap 20h30'29" -10d10'2"
Gersende Lacoste
Cap 20h7'29" -17d29'25"
Gerson
Leo 10h4'33" 20d22'17"
Gerst
Uma 9h1'21" 60d24'9"
Gert Long
Lib 15h5'39" -3d56'7"
Gert Lorenz
Ori 5h15'59" -0d10'19"
Gerta Guilbaud
Cyg 21h48'35" 45d18'41"
Gertie & Tony Salamone
Umi 18h47'37" 86d48'11"
Gertner Eva, Zanyaka
Uma 10h58'46" 69d43'32"
Gertrud Bunn
Ori 6h9'32" 8d18'11"
Gertrud Hertel-Pflugfelder
Crb 16h10'19" 37d14'17"
Gertrud Maria
Ori 5h52'53" 7d28'16"
Gertruda
Gem 6h48'52" 25d57'52"
Gertrude
Psc 1h10'28" 22d23'21"
Gertrude Downing
Cyg 19h38'57" 30d11'6"
Gertrude Ella Foster
Sco 17h5'57" -38d11'18"
Gertrude H. Dorchak
Cas 1h40'12" 63d57'52"
Gertrude Leona Darin
Uma 9h16'42" 49d37'35"
Gertrude Lorenc
Uma 8h58'35" 67d24'49"
Gertrude Louise Bourgault
Sco 17h58'26" -34d34'31"
Gertrude M. DeLorenzo
Cas 0h6'57" 53d30'30"
Gertrude Margaret Lynch Schwartz
Ari 2h36'53" 31d3'47"
Gertrude Marie
Ori 4h55'52" 1d47'45"
Gertrude May Block Valiquette
Uma 11h21'3" 47d58'7"
Gertrude Parker Johnson
Uma 11h2'14" 49d37'59"
Gertrude Sarah Thornley
Psc 1h9'22" 11d43'16"
Gertrude Smith
Cap 21h44'15" -12d57'39"
Gertrude Theresa Robert
Cyg 19h34'39" 29d58'30"
Gertrude "Trudy" Maniscalco
Aqr 21h40'24" 1d26'32"
Gertrude Vavrina Collier
Leo 11h41'58" 24d19'4"
Gertrudis Abad-Ariza
Crb 15h52'23" 27d4'16"
Gertrud-Otto
Uma 8h31'36" 59d35'28"
Gertry Comden
Cyg 21h35'6" 47d38'12"
Gertude W. Seeger
Cas 1h47'12" 60d40'17"
Gerty
Vir 11h49'52" 5d57'19"
Gervais Dionne
Ori 5h33'51" -1d21'12"
Gervaise & Simo
Vir 11h40'39" -1d48'23"
Gervase
Dra 18h25'27" 51d8'46"
Gerwyn Jones
Cep 22h1'49" 64d12'40"
Géry
Tau 4h26'46" 21d22'28"
Gesica
Psc 0h55'5" 32d26'24"
Gesierich, Marie
Cnc 8h21'34" 20d31'4"
Geslav Joseph Szczygiel
Leo 10h33'10" 18d34'48"
Gessika B. Perez
And 1h58'32" 45d16'56"
GET 381 JCQA
Crb 15h44'49" 26d41'6"
Get Gelt Director Michael Hardy
Lib 15h34'55" -25d40'34"
Gether C. Rockwell 12-20-1892
Sgr 18h3'4" -28d3'3"
Gethin James
Dra 10h42'17" 73d19'35"
Gettis Elizabeth Caruthers
Sco 16h48'27" -30d15'55"
Getty Guy 1 4 3 A & F
Psc 1h6'45" 28d3'34"
Gettysburg
Cap 21h37'26" -9d0'13"
Getzoff64
Gem 7h1'29" 22d20'14"

Geum-Jin Ka
Psc 1h23'36" 10d34'4"
Gevvy
Vir 13h4'3" -3d26'53"
gew202
Psc 0h51'15" 16d55'11"
Gewirz Anniversary Star
Cyg 19h57'40" 39d5'34"
Geyer, Wolfgang
Uma 8h15'9" 72d38'53"
Geza Pap
Uma 8h25'35" 69d22'44"
Gezegend
Leo 10h10'51" 14d19'53"
G.F. Harris
Leo 11h36'57" 17d46'0"
G.F.J.F. Holbrook Celestial Harmony
Uma 9h38'48" 68d6'36"
GG
Lyn 6h56'51" 58d19'12"
GG
Uma 11h15'6" 51d24'15"
GG
And 2h27'14" 44d38'6"
Gg and Bella
Aqr 22h51'4" -8d11'34"
GG and Minis
Cyg 19h55'26" 37d26'53"
G.G. Carmichael
Psc 1h54'2" 9d58'58"
G.G. Dorthy LoRee
Lib 15h7'55" -22d1'28"
GG Ellen Sorensen
Cas 0h12'50" 59d6'30"
G.G. Nannie & G.G. Papa
Tri 1h48'38" 34d26'39"
GG Ruth Spears
Mon 7h21'32" -0d55'28"
G.G Watching Over Us
Peg 23h56'54" 23d57'32"
G.G. Wolf's Star
Uma 8h48'59" 72d32'41"
GG070555
Uma 8h39'51" 59d51'12"
GGR & CCR, A Match Made In Heaven
Crb 15h43'54" 36d31'9"
GGreek 811955
Leo 11h48'37" 16d24'6"
G.G's Dream
Cru 12h43'30" -59d42'44"
GG's Star
Tau 5h4'40" 16d37'52"
Ghada Homad Sultan
Dra 18h23'12" 52d55'10"
Ghamz's Corner of the Cosmos
Ori 5h35'24" -2d13'48"
Ghan FayCurry
Aqr 22h50'12" -3d56'51"
Ghassan Ascha
Gem 7h6'49" 34d7'49"
Ghassan I. Shaker
Gem 7h18'54" 14d44'3"
Ghassan Taher Fadlallah
Uma 11h29'50" 35d30'34"
Ghazaleh Saidi
Ori 5h24'53" 14d58'42"
Ghazi
Ara 17h19'45" -64d32'55"
Ghazi Ali Hammoud
Gem 7h53'8" 13d28'52"
Ghazi Shaban
Sco 16h46'55" -31d1'47"
Ghe
Vir 12h12'15" 12d9'53"
Gheed Karam Qumseya
Peg 22h33'45" 18d19'53"
ghel Kalyx
Dra 12h37'12" 70d43'37"
Ghete
Uma 9h32'58" 65d30'45"
Ghetto Cowboy
Aqr 23h28'17" -17d10'22"
Ghettofabulous
Sgr 18h28'9" -26d20'43"
Ghezal
Uma 9h22'0" 67d54'47"
Ghibly Land
Uma 9h37'5" 58d41'0"
Ghila
Sex 10h2'59" 2d20'35"
Ghiloh Meaning Revealed
Lyn 7h31'41" 47d49'41"
Ghis
Uma 10h3'24" 60d22'17"
Ghiselle Dominguez
Cnc 8h46'7" 15d0'10"
Ghislaine et Marcel Auriac
Ori 5h0'23" 10d15'22"
Ghislaine Sabri
Ori 6h4'47" -2d10'5"
Gholamreza Khoshnevis
Mon 7h16'7" -0d56'30"
Ghost Kitty Lander
Lyn 7h47'23" 47d53'5"
Ghost xox Alien
Cyg 20h5'4" 44d1'39"
Ghostwolf
Sco 17h58'20" -37d53'24"
Ghouse Maqsood Hameed
Lyn 8h14'3" 56d48'48"
Ghuneim
Uma 11h22'30" 34d30'18"
Ghyslain et Isabelle
Cyg 20h52'30" 39d53'48"
Gi Gi
Psc 1h9'19" 10d24'2"
GI Puppy-Lover
Lup 15h28'33" -38d55'3"
Gia
Cap 21h39'12" -22d32'38"
Gia Bella
Sco 16h7'2" -10d23'10"
Gia C Clarke
Gem 7h49'56" 33d17'33"
Gia Colombo
Cnc 8h18'46" 12d42'58"
Già e Ceci
Per 2h31'6" 52d10'4"
Gia Luna
Dra 15h24'24" 64d10'21"
Gia Marie
Dra 15h53'59" 57d56'25"
Gia Marie Celauro
And 23h17'40" 48d28'33"
Gia Marie Rivera
Cap 21h39'45" -13d48'3"
Gia Mi Amor
Gem 7h32'27" 27d22'35"
Gia "My Sweetness"
Psc 0h16'40" 10d56'41"
Gia Panta
Lmi 10h8'50" 37d29'12"
gia panta Giota
Gem 7h12'35" 25d8'46"
Gia Peanut
Gem 7h39'26" 31d10'59"
Gia Stella Rossa
Umi 17h3'52" 75d51'18"
Gia Susanne DeStefano
Leo 11h44'8" 23d31'23"
GiaBella Mattlyn Blaser
Lyn 8h30'38" 35d6'49"
Giacinta P. Jones
Sco 16h33'13" -33d7'26"
Giacomina Paccione
Sgr 20h22'31" -33d17'14"
Giacomo
Lyn 7h49'36" 46d0'31"
Giacomo
Her 17h44'31" 31d29'46"
Giacomo
Her 18h34'36" 17d45'28"
Giacomo
Ori 5h44'38" 8d7'57"
Giacomo Bartolini
Lyr 18h51'38" 38d3'38"
Giacomo Di Mento
Cam 4h55'53" 53d53'4"
Giacomo J. "Jim" Iozzo
Per 3h7'40" 53d11'44"
Giada
And 23h17'32" 45d46'47"
Giada
Boo 14h48'48" 25d59'46"
Giada
Crb 15h35'40" 26d41'50"
Giada
Dra 18h41'16" 52d50'54"
Giada Dusina
Crb 16h2'24" 33d58'45"
Giada&Luca
Cam 5h34'37" 59d57'11"
Giada Maria
Umi 15h7'48" 84d4'55"
Giada Masino 08-10-2000
Lib 15h46'43" -27d0'57"
Giada Olivia Hudon
Cap 21h42'29" -8d37'41"
Giada Villano
And 1h47'44" 41d3'2"
Giamma
Cas 0h31'57" 61d30'22"
Giampo
Ori 5h15'30" -0d36'44"
Gian
Per 4h35'11" 40d23'10"
Gian
Leo 11h20'9" 23d20'42"
Gian
Lyr 19h12'30" 28d56'32"
Gian Andrin Lüthi
Lyr 18h57'2" 33d12'16"
Gian Julian
Umi 14h6'17" 75d29'44"
Gian Marco Beltrame
Uma 9h12'36" 57d4'36"
Gian Paolo Barzellotti
Cru 12h55'15" -58d43'3"
Giana
Cas 23h45'3" 62d16'4"
Giana Lee Swasey-Stephenson
Sco 17h41'53" -30d36'2"
Giana Marie
Lib 15h38'39" -24d45'31"
GIANANDREA
Vir 12h38'11" -9d6'8"
GIANCARLO
Aur 5h48'16" 41d14'15"
Giancarlo
Uma 10h0'45" 41d33'15"
Giancarlo
Cas 0h5'31" 55d48'35"
Giancarlo Angelo Frisina
Lyr 18h34'47" 39d39'18"
Giancarlo Chantres
Ori 6h10'7" 15d17'4"
Giancarlo Cortes
Boo 15h2'18" 42d29'30"
Giancarlo Dichiazza
Leo 10h16'22" 27d4'19"
Giancarlo Ferraresi
Tau 3h36'50" 26d37'30"
Giancarlo Genova
Peg 23h20'28" 11d33'39"
Giancarlo Giorgi
Lyn 7h46'29" 46d0'50"
Giancarlo L.
Lyn 6h20'10" 56d59'53"
Giancarlo Mi Cielo. Te Amo.
Vir 14h28'12" 3d24'1"
Giancarlo Moretto
Aql 19h27'42" -11d25'5"
Giancarlo Panerai
Peg 23h39'30" 27d48'49"
Giancarlo Romero
Aqr 21h39'37" 0d56'16"
Giancarlo Viscardi
Uma 10h41'3" 63d4'23"
Giancarlo Volpe
Sgr 20h26'44" -35d58'49"
Giandeliz Rodriguez
Sco 17h44'7" -33d24'29"
Gianfranco Albieri
Umi 13h47'41" 72d55'3"
Gianfranco Falco
Cam 7h55'7" 62d57'0"
Gianfranco Peluso
Cyg 20h39'45" 46d6'28"
Giang Nguyen
Ari 2h23'54" 22d10'35"
Gianina-Livia
Uma 12h4'28" 50d35'9"
Gianluca
Boo 14h59'58" 50d6'29"
Gianluca
Lyr 18h58'38" 45d46'2"
Gianluca
Cas 3h2'0" 57d50'6"
Gianluca
Peg 22h41'15" 3d5'50"
Gianluca Del Carlo
Cyg 19h45'46" 46d37'2"
Gianluca e Laura 17 agosto 2001
Per 3h25'13" 41d36'22"
Gianluca Enrico Bigagli
Srp 17h59'23" -0d5'33"
Gianluca Fucili
Ori 5h59'8" 9d17'49"
Gianluca Nenciarini
Cam 3h53'4" 52d54'30"
GIANMARCO BENSI
Cyg 21h28'48" 50d13'39"
Gianmaria e Patrizia
Lyr 18h57'41" 45d48'5"
Gianna
Ari 2h25'53" 10d35'21"
Gianna
Uma 9h21'49" 56d48'41"
Gianna
Uma 11h53'34" 55d21'13"
Gianna Angelopoulos-Daskalaki
Cas 23h13'50" 54d16'21"
Gianna Bella
Uma 11h50'37" 45d52'1"
Gianna Bella Yordanopoulos
Ari 2h33'40" 26d59'45"
Gianna & Carlos Rocha
Per 3h45'47" 46d21'35"
Gianna Catherine Freni
Leo 10h52'27" -2d27'47"
Gianna Cynthia
Uma 12h37'31" 62d59'8"
Gianna Dibiase
Ori 5h54'3" 19d55'15"
Gianna Elaine Louise Herb
Cap 21h37'16" -16d6'48"
Gianna Elizabeth
Crb 15h40'53" 28d45'2"
Gianna Foisy
Peg 21h35'50" 10d8'50"
Gianna Friedman
Vir 13h29'55" -21d44'2"
Gianna Gabriela Masiello
Cnc 8h24'51" 21d18'28"
Gianna Louise Gunn
Mon 6h53'9" -0d58'14"
Gianna Lucia
Ori 4h49'21" -0d5'14"
Gianna Lucia , Bright Angel
Tau 4h37'59" 18d25'21"
Gianna Lynn
And 2h9'43" 45d30'15"
Gianna Marae Lengyel
Umi 17h35'25" 83d51'35"
Gianna Marie
Sco 17h57'51" -30d13'11"
Gianna Marie
And 2h21'58" 45d2'47"
Gianna Marie
Lyn 6h48'34" 51d45'14"
Gianna Marie Branosky
Ari 3h2'49" 26d1'39"
Gianna Marie Cipolla
Tau 4h39'42" 10d13'58"
Gianna Marie Michaels
And 1h10'26" 45d42'19"
Gianna Marie Michaels
Uma 9h57'9" 64d12'7"
Gianna Marie Ruttler
Sgr 18h34'15" -32d45'31"
Gianna Marie Taliercio
Lib 14h30'6" -16d3'37"
Gianna Marie Zmigrodski
Lyn 7h57'9" 38d15'56"
Gianna Melania Aiken
And 0h28'41" 29d16'9"
Gianna Nicole Merritt
Gem 7h4'59" 15d5'14"
Gianna Nicole Rosetti
Cnc 8h53'5" 11d34'48"
Gianna Nicole Trani
And 1h34'7" 48d23'53"
Gianna Paola Gonzalez Pina
Crb 15h39'45" 27d38'52"
Gianna Patricia Boyce
Sco 15h52'55" -23d13'16"
Gianna Rose Budlong
Cyg 20h21'4" 51d33'14"
Gianna Rose Masucci "2/15/2006"
Umi 14h30'56" 74d50'26"
Gianna Savarino
Vir 13h19'12" 4d36'38"
Gianna Sofia
Aqr 21h1'45" -11d59'59"
Gianna Spatafora
Uma 8h24'38" 64d25'44"
Gianna-Carina Adelina DeRosa Lohnn
Aqr 20h54'50" -12d56'33"
Giannalberto Leni
Umi 15h3'31" 86d21'8"
GiannaMarie Elizabeth Mariano
Vir 14h10'58" -19d12'29"
Gianne Aizza Devega Videna
And 23h48'15" 46d17'54"
Gianne Cruz Udave
Sgr 18h27'35" -20d47'8"
Gianni
Sco 16h46'19" -29d37'14"
Gianni
Lyr 19h17'8" 37d31'2"
Gianni
Lyn 8h7'58" 44d6'50"
Gianni
Her 17h49'1" 30d27'9"
Gianni Bernardo Tedesco
And 0h9'25" 32d9'59"
Gianni Calia und Jasmin Bethke
Cyg 20h35'15" 38d44'47"
Gianni e Lisa Gasbarro
Uma 10h43'58" 42d49'59"
Gianni e Stefania
Uma 9h10'43" 54d49'51"
Gianni Fanelli
Uma 9h22'4" 68d59'57"
Gianni Iannone
And 0h49'52" 44d56'24"
Gianni Kochari
Cyg 21h50'54" 48d6'37"
Gianni Luigi DiSora
Gem 7h42'51" 34d0'36"
Gianni Michael Gabriel
Lyn 9h18'12" 34d49'8"
Gianni Nicola Abbruzzese
Gem 7h45'23" 34d52'56"
Gianni Rocco Maisano-Torres
Vir 13h52'20" -10d44'47"
Gianni W. DeLango
Del 20h33'46" 13d14'15"
Gianni, il mio amore per sempre
Ori 5h35'14" 10d2'57"
Giannina Alonso aka G-Dawg
Cnc 8h34'15" 16d51'32"
Giannini Dara 1991
And 2h13'19" 44d53'2"
Giannis
Sco 17h51'15" -39d43'23"
Gian-Paul Simpson
Ari 2h6'12" 13d49'26"
Gianpiero and Jen
Cyg 20h22'17" 34d5'5"
GianpiGiò
Uma 9h8'21" 55d19'54"
Giao love Tran
Cyg 20h13'49" 39d11'38"
Giavana Curcio
Boo 15h15'4" 25d34'36"
Giavana Eileen Polunci
Cnc 8h46'46" 13d28'46"
GIAVANNA CHRISTINA ACOSTA
Uma 9h5'59" 59d39'48"
Giavanna Domenica
Tau 4h21'47" 27d29'53"
Giavanna Grace Ferrucci
Ari 2h27'13" 26d33'30"
Giavanna Lena Scott
Lyn 7h26'46" 48d45'14"
Gib and Marilyn Rosenbaum
Uma 9h1'18" 57d57'53"
Gibb
Tau 4h42'7" 18d8'44"
Gibbo
Per 3h24'32" 41d44'47"
Gibbo's Star
And 1h19'36" 37d50'28"
Gibbs
Lyr 18h48'49" 38d34'10"
Gibby
Leo 11h15'24" 4d21'40"
Gibby
Tau 4h14'51" 2d29'59"
Gibby
Uma 9h30'38" 71d26'57"
Gibby Ramirez
Cnc 8h30'46" 25d31'40"
Giber
Ori 5h9'57" 9d9'53"
Gibney
Uma 9h50'42" 54d48'50"
Gibson - 4
Uma 11h57'17" 49d43'26"
Gibson Cage
Ori 5h57'12" 17d10'12"
Gibson Odes Smoak
Cnc 8h8'20" 16d16'38"
Gibson Randall-Logan
Cma 6h22'7" -21d46'38"
Gibson Todd
Uma 11h10'45" 66d45'35"
Giddo
Aql 20h5'7" 8d35'21"
Gideolyn
Tri 1h48'21" 29d14'38"
Gideon Palmer Knapp
Cep 22h15'50" 57d49'46"
Gideon Samuel
Boo 14h59'26" 52d12'39"
GiDeOnChIcK
Ari 2h51'48" 29d41'8"
Gideon's Love
Sge 19h44'49" 17d53'31"
Gidget
And 0h34'12" 32d13'20"
Gidget
Cap 21h32'14" -19d10'16"
Gidget Dale
Psc 1h25'28" 15d14'51"
Gidu
Uma 11h23'56" 57d38'35"
Gidu Shroff
Crb 15h51'57" 34d49'19"
Giebenhain Family Star
Uma 11h39'17" 61d44'36"
Giedre Dicbanyte
Lib 14h28'56" -23d57'53"
Giegerich, Esther
Uma 10h9'34" 48d30'8"
Gieling Schlösser, Gertraud
Uma 10h14'37" 60d0'23"
Gielnik, Michael
Ori 6h5'11" 11d11'23"
Gienek
Uma 8h54'47" 52d49'15"
Gierling, Neydhart
Uma 8h48'43" 58d20'4"
Giesel
Sco 16h11'46" -18d6'57"
Giff
Del 20h46'6" 16d19'35"
Gifford
Umi 14h37'9" 79d44'49"
Gifford Pierce
Psc 1h10'20" 32d11'36"
Gift from above
Sgr 17h55'1" -17d0'34"
Gift Of Moonlight
Sco 16h6'17" -17d13'26"
Gigantress
Pic 5h19'20" -49d7'47"
Giggie Pepre Paul & Maureen Lessard
Uma 13h6'28" 58d1'34"
Giggle Monster
Peg 23h12'53" 26d49'15"
Giggles
Tri 2h27'6" 36d25'9"
Giggles
And 0h57'29" 40d33'59"
Giggles
Uma 10h0'9" 53d10'12"
Giggles Bean MacGillicuddy
Cas 0h43'59" 62d8'40"
Giggle's Star
Ara 17h17'29" -55d4'7"
Gigi
Aqr 22h11'6" -21d5'2"
GiGi
Cam 7h15'8" 75d30'18"
Gigi
Psc 0h21'45" -0d41'40"
Gigi
Mon 6h52'57" -0d0'45"
GiGi
Mon 7h56'38" -5d1'53"
Gigi
Vir 13h46'31" -7d44'1"
Gigi
Aqr 22h30'42" -1d47'45"
Gigi
Aqr 21h45'41" -1d22'31"
GiGi
And 1h1'51" 42d23'46"
Gigi
Per 3h14'42" 31d8'17"
Gigi
Crb 16h12'20" 36d34'0"
Gigi
Uma 10h9'47" 48d55'37"
Gigi
And 23h9'48" 41d37'46"
Gigi
Peg 22h52'29" 19d9'18"
Gigi
Gem 7h6'34" 16d15'4"
Gigi
Leo 11h16'3" 16d56'23"
Gigi
Psc 23h25'52" 4d27'19"
Gigi
Cmi 7h17'1" 9d52'52"
GiGi
Vir 11h38'22" 8d4'37"
Gigi
Vir 13h16'27" 12d24'40"
Gigi
Vir 13h0'1" 10d6'11"
GiGi ~ Our Special Momma ~
Cas 1h34'8" 63d0'23"
Gigi~Dave's light of the universe
Tau 5h57'36" 25d2'21"
Gigi et Britney
Per 4h37'29" 39d56'25"
Gigi Gatto
Lib 15h51'11" -10d9'3"
Gigi Guet Gourmelon
Cap 21h42'27" -17d3'51"
Gigi Je t'aime
Cam 3h38'5" 58d39'28"
Gigi (Jenna) Nicole Cox
And 2h24'3" 50d57'21"
GiGi loves JB
Her 17h52'39" 23d10'11"
Gigi Nannini
Uma 9h32'51" 49d18'54"
GiGi & Phil
And 1h44'48" 50d21'15"
Gigi Robinson
Leo 11h45'49" 19d46'3"
Gigia Del Olmo
And 23h57'35" 40d30'30"
Gihovany Alcaraz
Sgr 18h16'24" -17d15'32"
Giichiro (Butch) Hayashi
Cep 21h58'54" 56d23'28"
Gijs & Marijke voor altijd samen
Lyn 6h19'5" 55d23'12"
Gil
Cam 6h18'17" 62d33'17"
Gil
Uma 10h21'26" 52d27'20"
Gil
Aur 5h53'24" 54d2'47"
Gil
Aur 5h38'38" 38d23'3"
Gil
Her 16h19'29" 25d53'8"
Gil
Aql 20h5'35" 4d22'15"
Gil Barthelemy
Vir 12h12'14" 11d2'15"
Gil & Carol the 2 brightest stars
Her 17h55'6" 14d37'52"
Gil Estel
Ori 5h31'43" 5d54'18"
Gil Fogoros
Leo 10h20'2" 24d40'44"
GIL GUILLORY
Ori 5h16'25" 10d21'58"
Gil Hollingsworth
Ari 2h30'47" 25d21'24"
Gil Howell
Gem 7h19'59" 19d56'27"
Gil & Marcia Siegert
Vir 14h23'55" 1d30'49"
Gil Osvaldo Mederos
Boo 15h32'44" 40d35'59"
Gil & Rotem
Lyr 18h38'3" 31d43'57"
Gil (Shmo) Peterson
Her 17h19'41" 38d20'38"
Gil the Great
Aql 19h2'21" -0d25'54"
Gil Torres
Gem 7h43'40" 24d32'5"
Gila Caspi
Cyg 21h43'24" 52d2'39"
Gilbert
Boo 15h31'4" 45d40'2"
Gilbert
Lyn 7h4'23" 46d32'45"
Gilbert Alan Baldwin
Boo 14h41'46" 37d31'27"
Gilbert Alan Trujillo
Dra 18h32'28" 54d43'17"
Gilbert and Dianne Mesa
Umi 15h27'28" 73d9'12"
Gilbert & Anny
Lib 15h4'59" -13d17'13"
Gilbert Beto Minjares Jr.
Uma 11h47'32" 36d58'30"
Gilbert Bittner
Uma 10h12'26" 57d40'33"
Gilbert C. Gross
Cyg 19h47'21" 38d4'23"
Gilbert Cartwright
Cnc 8h29'59" 22d26'37"
Gilbert & Clarice Griffith
Uma 10h13'12" 67d56'4"
Gilbert Curti
Leo 11h25'11" 4d12'17"
Gilbert Dedering
Cap 21h31'44" -16d1'30"
Gilbert deMalvilain
Cnc 8h56'54" 10d19'36"
Gilbert Dunbar Mead
Gem 7h26'16" 33d15'35"
Gilbert E. Martin
Psc 1h7'13" 33d19'36"
Gilbert E. Saiz
Ori 6h11'29" 17d31'1"
Gilbert Feliciano
Aqr 20h40'51" -6d29'30"
Gilbert Flores
Tau 4h36'38" 30d46'8"
Gilbert Gazzia
Ari 3h0'30" 28d28'24"
Gilbert "Gibby" Hudgens
Her 17h35'37" 36d23'20"
Gilbert Gonzalez Jr.
Aqr 20h51'3" -8d8'32"
Gilbert Ives
Lyn 8h27'31" 41d37'18"
Gilbert Jacob Nunez
Boo 14h16'0" 33d31'14"
Gilbert Joseph Poirier
Cyg 21h36'18" 48d42'36"
Gilbert Jourdain
Uma 11h22'7" 31d44'55"
Gilbert Karam - My Shining Star
Col 6h25'52" -41d39'46"
Gilbert Keith
Psc 0h19'30" 17d23'5"
Gilbert Klock
And 0h24'41" 32d48'24"
Gilbert Lieberman
Dra 16h45'20" 59d3'25"
Gilbert Loves Mandy
Sge 20h9'33" 16d42'22"
Gilbert Manning Memorial Star
Vir 12h21'2" -10d17'22"
Gilbert Mendelson
Sco 17h56'37" -30d10'20"
Gilbert Michael Rivera
Sgr 19h21'26" -16d0'54"
Gilbert n Katrine's Star of Geeks
Ori 6h13'47" 19d51'54"
Gilbert Oglesby
Aur 5h11'42" 42d14'20"
Gilbert Oswald
Cep 23h0'10" 70d13'54"
Gilbert Paul Falland
Cru 12h45'25" -63d34'58"
Gilbert R. Collins
Per 2h54'27" 47d50'54"
Gilbert Roland
Dra 16h50'10" 69d1'2"
Gilbert Sealy
Ori 5h23'29" 3d8'2"
Gilbert Westerman
Umi 21h1'16" 89d20'27"
Gilbertina
Ari 3h23'20" 16d37'26"
Gilberto
Umi 14h28'10" 68d27'50"
Gilberto Monterroso
Aqr 21h41'3" 2d10'9"
Gilberto Osorio
Ori 5h40'5" 1d14'56"
Gilby
Peg 21h48'33" 8d3'4"
Gilby's Star
Ari 2h36'9" 22d8'22"
Gilda
Cyg 20h49'39" 55d14'42"
Gilda & George
Lyr 18h30'15" 37d1'55"
Gilda Inès Laracuente Saldaña
Mon 6h50'2" -0d10'13"
Gilda Katherine Lewis
Mon 6h52'22" -0d21'36"
Gilda Lowe
Cas 0h4'2" 53d26'57"
Gilda M. Gonzalez
Aqr 21h43'24" -0d14'46"
Gilda Marie Cipriani White
Lyn 7h48'21" 53d16'8"
Gilda Natangeli
Uma 11h28'46" 64d33'37"
Gilda P. Warden
Tau 3h44'39" 26d27'27"
Gilda Salazar
Crb 15h35'42" 37d54'56"
Gilda's I Did It Star
Peg 22h37'46" 13d47'50"
Giles
Cep 20h45'2" 57d38'20"
Giles E. L. Smith
Lib 15h0'37" -21d54'50"
Giles F Stanton
Lyn 8h55'38" 36d58'44"
Giles George Howard Mowbray
Her 18h14'2" 22d11'1"
Giles Justin McCarthy
Ori 5h29'4" 1d31'13"

Giles Paul Whitton Cra 18h16'54" -39d28'10"
Giles' Star Uma 8h9'11" 60d2'14"
Gil-Estel Ori 4h50'35" 13d39'27"
Gili Claire Butler-Furlong Uma 12h44'32" 62d14'1"
Gili Swimir Sgr 18h51'51" -18d46'14"
Gillian Linda Cas 0h21'37" 55d32'6"
GiliLori's Eternal Light Cas 1h27'35" 65d21'52"
Gill & Simon Copley Cyg 19h46'1" 55d53'45"
Gilla Lyr 19h23'51" 37d53'23"
Gillen 6446 Aql 20h13'4" 3d45'26"
Gillengerten, Günther Uma 9h57'55" 45d5'28"
Gilles Alexandre Herard, Jr. Lyn 7h30'40" 37d38'33"
Gilles and Diane for Eternity! Cyg 20h32'1" 57d9'3"
Gilles Barathier Dra 14h39'30" 58d43'6"
Gilles Duhamel Cep 1h34'15" 77d53'44"
Gilles Lacroix Lib 14h41'5" -9d51'7"
Gilles Lacroix de Rognac. Uma 9h35'44" 46d42'4"
gilles laube Dra 9h57'5" 77d31'52"
Gilles Nassar Uma 12h29'9" 57d44'0"
Gilles Racicot / Stavibel Sgr 20h25'22" -29d55'58"
Gilles Stephan - Ours au miel Uma 11h1'24" 63d18'20"
Gillian Umi 15h49'38" 81d47'3"
Gillian And 1h48'34" 46d8'58"
Gillian Ari 2h6'49" 25d44'28"
Gillian Gem 7h5'41" 17d33'35"
Gillian Leo 10h4'8" 21d20'50"
Gillian and Nevaeh's Star Lyn 6h17'29" 60d51'45"
Gillian Ann Cox Leo 11h54'31" 25d29'47"
Gillian Ann Rippl And 1h25'31" 45d0'17"
Gillian Bella Yordanopoulos Ari 1h57'52" 21d50'38"
Gillian Busby Sco 17h54'11" -36d10'19"
Gillian Cate Pruteanu Cnc 7h58'33" 12d7'58"
Gillian Crawford Vir 13h52'53" -1d53'44"
Gillian Delanie Ori 6h5'17" 17d38'5"
Gillian Elizabeth Aylene Shelbourne Cas 23h44'4" 55d46'19"
Gillian Elizabeth Barrios Cyg 19h30'52" 28d20'49"
Gillian Elizabeth Fry Ori 5h22'10" 9d18'45"
Gillian Elizabeth Haycock Umi 16h20'8" 73d4'1"
Gillian Elizabeth Nelson Srp 17h56'32" -0d31'53"
Gillian Evans Sco 16h40'47" -43d6'56"
Gillian Fogg Cas 23h35'4" 58d59'28"
Gillian Grace Hanks And 0h35'22" 31d25'20"
Gillian Helen Smith Cas 1h32'22" 60d41'7"
Gillian Hope Autajay Ori 5h24'24" 7d36'24"
Gillian Hutchison Cap 20h18'46" -22d59'52"
Gillian Jump And 23h37'31" 47d4'0"
Gillian Logie Cyg 19h47'19" 38d3'37"
Gillian Louisa Phillips Cas 0h14'4" 60d58'29"
Gillian Lynne Cas 0h10'4" 53d10'37"
Gillian M. Becker Cap 20h22'34" -25d9'24"
Gillian Mary Aldam Meschino Cas 1h23'23" 52d19'25"
Gillian Mascis Cnc 8h35'55" 21d28'18"
Gillian McMaster Cas 1h19'29" 53d34'35"
Gillian Mohseni Gem 6h18'23" 26d15'25"
Gillian - My little Bundle of Perfection Cas 2h32'0" 72d10'16"
Gillian Nickens Mon 6h28'12" 0d3'7"
Gillian Paige Buterbaugh Lyr 18h54'2" 43d52'55"
Gillian Patrick Cyg 20h38'27" 41d50'34"
Gillian Rae Donato Uma 10h10'42" 47d30'23"
Gillian Ross Erickson Cas 23h49'30" 57d9'1"
Gillian Sydenham Leo 9h51'19" 7d45'22"
Gillian Taggart Ari 2h47'27" 14d53'5"
Gillian Whiddon Henderson Uma 9h19'42" 64d58'54"
Gillian's Gem And 1h43'7" 36d36'56"
Gillian's Graceful Cygnet Cyg 21h40'19" 49d35'36"
Gillian's Star Uma 8h27'36" 63d0'32"
Gillian's Star Cas 1h58'28" 61d18'26"
Gillikin Psc 1h10'15" 14d39'26"
Gillin Boo 14h56'1" 23d31'36"
Gillis & Kathy Wettermark Uma 8h50'48" 47d45'49"
Gillis Lynn Aqr 22h39'41" -0d57'36"
Gillooney Ori 5h16'9" 1d51'37"
GILLS And 0h35'25" 22d11'30"
Gill's Anniversary Ori 5h5'16" 16d4'32"
Gill's Big Four Oh! Cnc 8h52'49" 7d12'44"
Gill's Birthday Star Aqr 22h32'30" -16d36'15"
Gilly Aqr 21h43'56" -1d36'1"
Gilly Bortman Sgr 18h12'29" -16d18'33"
Gillys' Wish Per 4h26'3" 44d14'55"
Gilman (Gil) Johnson Per 3h52'18" 43d1'6"
Gilman William Page Per 3h29'15" 42d39'19"
Gilo Libetario Sgr 18h8'19" -34d32'17"
Gilou for ever And 1h0'42" 34d10'41"
Gilquin Uma 10h8'28" 56d40'20"
Gimpy's Cripple Lyn 9h3'4" 33d40'33"
Gin Lyn 6h55'55" 52d0'46"
Gin Sgr 20h8'58" -27d51'54"
Gin "Angle" Crb 15h23'39" 29d46'25"
Gina Crb 15h39'55" 26d59'55"
Gina Cnc 9h20'26" 28d51'9"
Gina Gem 7h20'18" 28d53'40"
Gina Ori 6h1'26" 0d43'52"
Gina Peg 22h44'1" 8d32'24"
Gina Leo 9h23'42" 14d20'37"
Gina Lyn 6h56'23" 52d14'15"
Gina Cas 0h40'19" 51d47'6"
Gina And 23h59'21" 45d2'19"
Gina Cam 5h54'15" 57d0'1"
Gina Cnv 12h29'26" 51d18'40"
Gina Cnc 9h5'34" 31d24'29"
Gina And 2h19'45" 44d33'55"
Gina Peg 22h1'50" 30d26'54"
Gina Pyx 8h33'42" -30d37'41"
Gina Sco 17h40'15" -39d56'37"
Gina Uma 11h15'53" 56d43'50"
Gina Cas 1h32'20" 62d10'26"
Gina Aqr 23h18'0" -18d41'39"
Gina Cap 21h1'36" -17d24'22"
Gina Aqr 22h32'22" -0d13'25"
Gina and Jacob Srp 18h8'36" -0d21'59"
Gina and Jimmy Forever! Tau 4h22'45" 26d56'57"
Gina Ann Ficarra Ari 3h14'11" 28d49'36"
Gina Ann Ingalls Per 3h20'44" 48d23'24"
Gina Anne Cam 12h58'30" 76d57'32"
Gina Arias Sgr 20h24'22" -41d40'8"
Gina Baby Princess Lover Ori 6h9'4" 9d10'58"
Gina Ballard Cap 20h28'42" -18d18'13"
Gina Belle Star Uma 10h25'17" 39d44'41"
Gina Berrong Ori 5h13'44" 15d13'31"
Gina Bisogna 22 Cas 23h21'27" 55d18'3"
Gina & Bobby-First Love is Forever Sgr 19h34'50" -28d58'48"
Gina Boche And 23h22'30" 51d35'40"
Gina "Boobie" Star Tau 4h29'42" 16d22'43"
Gina Bortolotti (Gi-Bo 20060930) Lib 14h59'21" -9d39'49"
Gina C. Ross Leo 11h26'2" 15d50'38"
Gina Capra Lib 15h12'27" -22d12'4"
Gina Cariello And 23h11'2" 51d50'14"
Gina Christina Ari 2h54'53" 26d44'51"
Gina Comitini Uma 10h38'18" 42d13'35"
Gina Dawn Vul 19h0'8" 23d11'0"
Gina Del Beccaro Sco 17h17'55" -32d0'55"
Gina Delano Ari 2h34'52" 13d27'31"
Gina DiLalla Cas 23h24'49" 57d36'1"
Gina Divinely Limitless Lyr 18h53'9" 42d8'33"
Gina Elizabeth Correa Tau 4h16'48" 8d39'42"
Gina Elizabeth Menard Leo 10h11'57" 26d18'20"
Gina Elizabeth Morelli And 1h43'1" 43d42'34"
Gina Elizabeth Saucedo Cnc 8h30'21" 12d25'53"
Gina Fares And 0h46'10" 36d31'43"
Gina Ferrer Vir 14h35'28" -0d19'42"
Gina G. Johnson Vir 14h18'9" -2d42'45"
Gina Gabrielle Sgr 19h30'24" -18d18'39"
Gina Gabrielle Caputo Lib 15h9'46" -4d44'59"
Gina Gaye Bullock Leo 11h3'50" 15d34'32"
Gina Gemma Mariella Leo 11h30'11" 16d49'16"
Gina God's Gift Vir 13h9'5" 12d13'27"
Gina Griffin and Tim Hanson Vir 13h39'53" -11d46'49"
Gina Guardino Vir 13h28'48" 11d40'50"
Gina Guglielmotti Sex 10h23'29" -0d30'15"
Gina Halbom And 23h47'38" 45d2'42"
Gina Hammett "My Shining Star" And 23h24'26" 38d14'26"
Gina Haynes And 2h20'6" 49d34'33"
Gina Hundey And 1h29'53" 42d29'6"
Gina Isabella Schmid And 0h19'56" 30d21'11"
Gina Jamal Mendello Vir 13h50'15" -6d36'20"
Gina Jennings Psc 0h10'38" 3d9'40"
Gina & Jim Sco 16h9'38" -17d49'10"
Gina K. (Sunshine) Sco 17h38'42" -40d12'1"
Gina Katherine Schwan Aqr 20h50'38" -11d49'14"
Gina Kay Acker Lib 15h36'39" -7d41'27"
Gina Kay Lile Vir 12h36'42" -8d53'0"
Gina Kelley And 1h31'24" 45d21'9"
Gina Kilpatrick Cas 0h20'55" 56d12'50"
Gina Kirby Cap 20h50'52" -14d50'14"
Gina L. Bailey Lib 15h32'27" -20d0'20"
Gina L Wilder Cap 20h30'16" -9d55'38"
Gina Leon Cas 0h28'54" 58d31'14"
Gina Lisanti Lyr 18h47'18" 35d44'26"
Gina Lola Peg 23h47'48" 14d56'6"
Gina Louise Coconato Crb 16h13'53" 30d41'39"
Gina Louise Gargano And 1h20'46" 33d57'4"
Gina Louise Heck And 1h43'38" 42d46'21"
Gina Lynn Armstrong Aql 19h25'19" 2d28'35"
Gina Lynn Martin Uma 11h55'5" 60d47'54"
Gina Lynnik Tau 3h51'51" 26d7'16"
Gina M Capitoni Lmi 10h28'43" 38d41'26"
Gina M. Carucci Aqr 22h47'2" -10d2'13"
Gina Macklin's Star of Excellence And 1h26'6" 48d53'16"
Gina Mae Gem 7h14'45" 33d22'5"
Gina Maria Del 20h41'24" 16d55'20"
Gina Maria Leo 11h33'1" 11d37'28"
Gina Maria Simeone Leo 11h7'37" -4d20'31"
Gina Marie Lib 15h41'36" -6d8'36"
Gina Marie Lib 15h16'54" -8d58'37"
Gina Marie Cap 20h13'24" -15d29'23"
Gina Marie Umi 14h11'16" 70d43'44"
Gina Marie Sgr 18h25'6" -33d1'24"
Gina Marie Ori 5h32'41" 2d26'20"
Gina Marie Cnc 9h3'47" 15d2'3"
Gina Marie Gem 7h13'0" 31d56'56"
Gina Marie Lyn 7h2'29" 50d50'5"
Gina Marie Her 17h31'13" 46d11'39"
Gina Marie #1 Aqr 23h3'35" -24d43'40"
Gina Marie Asaro, Esquire Sco 16h16'53" -16d31'8"
Gina Marie Barsema Cas 0h13'43" 54d33'18"
Gina Marie Candeliria Leo 11h47'4" 26d4'43"
Gina Marie Cesario Leo 10h47'46" 16d34'29"
Gina Marie Cocci Ari 1h52'30" 16d43'54"
Gina Marie DiPasquale Leo 9h51'5" 32d23'57"
Gina Marie Donato Lmi 10h45'30" 34d16'32"
gina marie faith Cnc 9h11'33" 10d48'8"
Gina Marie Ferraro Tau 4h19'59" 28d3'29"
Gina Marie Geiser Leo 11h13'46" -0d44'47"
Gina Marie Glaubke Lib 14h46'33" -9d12'40"
Gina Marie Gormley Leo 11h10'32" -0d17'56"
Gina Marie Hubinger And 2h30'41" 45d45'59"
Gina Marie Hunt And 1h46'35" 38d21'54"
Gina Marie Meehl, My Love Forever Tau 4h49'56" 22d26'55"
Gina Marie Perri And 1h9'50" 45d47'13"
Gina Marie Petrizzo Lib 15h43'25" -12d3'2"
Gina Marie Reccardi Crb 15h33'36" 30d12'28"
Gina Marie Rose Bianco Lib 14h54'16" -10d22'12"
Gina Marie Simoncelli-Arnold Cas 1h4'54" 61d54'13"
Gina Marie Staffierri Tau 4h17'32" 20d31'36"
Gina Marie Swan Tau 4h3'46" 24d18'8"
Gina Marie Wright Cam 4h35'43" 66d24'33"
Gina Marissa Golba And 23h8'51" 51d8'38"
Gina Martin And 2h12'18" 45d21'11"
Gina Meredith Umi 14h55'28" 73d10'47"
Gina Michele Aqr 20h59'43" -7d9'47"
Gina Michele Gerona Vir 14h6'55" -21d12'36"
Gina Michele Koester Sgr 19h8'56" -18d9'16"
Gina Michelle and James Peter Kosta Cyg 19h54'25" 36d17'15"
Gina & Minell 25 June 2005 Cnc 8h1'26" 18d23'10"
Gina Nichole Psc 23h56'31" -3d11'0"
Gina Nicole Gerwien Leo 11h21'6" -4d27'23"
Gina Nicole Mercieri Aqr 20h55'5" -12d6'16"
Gina Noel Nardecchia Cap 20h56'44" -15d2'57"
Gina Oberfrott Kirchweger And 0h32'20" 36d20'38"
Gina Our Bright Spot in the Sky Per 2h47'39" 50d13'21"
Gina Pamela Landinez Ari 2h15'51" 16d42'17"
Gina Paola Sametini Lib 15h47'58" -17d30'37"
Gina Passeri Crb 15h35'25" 33d28'17"
Gina Petrossian's star Cnc 8h21'56" 29d53'35"
Gina Preston Tau 5h31'49" 24d54'13"
Gina Quesito Smith Eri 3h57'14" -0d8'11"
Gina Rae Sco 16h9'43" -9d6'9"
Gina Rae Love Star Uma 11h59'16" 46d21'7"
Gina Riazzi Cam 4h14'37" 66d51'46"
Gina Romanelli Cnc 8h45'38" 28d11'17"
Gina Rose 11/23/04 11:16 pm Per 3h23'39" 47d47'23"
Gina Rose Patalano Sgr 18h59'18" -24d44'39"
Gina Sanchez Crb 15h26'47" 29d8'45"
Gina Santucci Uma 14h16'28" 58d44'41"
Gina Scarpo Turner Uma 10h6'3" 42d45'44"
Gina Showstopper And 0h59'45" 46d33'41"
Gina Sirard Cyg 20h26'27" 40d48'31"
Gina Suzanne Winstead Cyg 21h18'4" 55d19'49"
Gina "Svenia" Haukedahl Cam 5h48'39" 68d43'0"
Gina T. And 2h36'56" 43d27'53"
Gina Taddie Lyn 6h36'18" 56d38'49"
Gina Tallarita Sgr 19h15'55" -20d8'30"
Gina Terreri And 23h2'41" 48d10'12"
Gina the Precious Cyg 21h23'4" 50d32'50"
Gina Valencia Leo 11h52'57" 20d28'45"
Gina Vita Farinaccio And 23h15'11" 42d57'5"
Gina Vito Cas 1h21'9" 62d0'34"
Gina Woodruff Tau 4h6'14" 11d57'20"
Gina Zalowitz And 1h7'50" 42d1'40"
Gina, Lauren, Gabriel's Star Cap 20h17'4" -14d25'30"
Gina, Star of Truth & Beauty Cnc 8h51'10" 11d59'12"
Gina, Your Life Shines This Bright Lib 15h12'57" -4d29'44"
GINA1 Lib 15h21'19" -22d20'12"
Gina1020 Lib 15h41'41" -16d37'44"
GinaBlue Leo 11h4'27" 15d29'32"
GinaCole Leo 10h9'10" 22d14'20"
Ginalee Lep 5h31'8" -10d59'57"
Gina-Louise Grottke - 9.12.21 Aqr 22h49'11" -23d9'47"
Ginalp Ori 5h56'2" -0d3'54"
GinaMaria Grace Star Tau 4h13'29" 21d23'19"
Ginamarie Del 20h32'47" 10d49'5"
Ginamarie Letizia Aqr 22h8'36" -23d19'16"
Ginamaus Uma 9h53'46" 72d37'58"
GinaRyanTyler Ori 6h0'20" 10d43'2"
Gina's Light And 1h32'49" 36d57'29"
Gina's Medium Dipper Cnv 12h42'48" 44d15'2"
Gina's Muguppy Vir 11h38'59" 2d42'0"
Gina's Passion Ori 6h9'50" 11d52'54"
Gina's star Ari 3h18'8" 22d55'38"
Gina's Star Leo 11h15'8" 24d18'10"
Gina's star And 2h13'5" 39d16'27"
Gina's Star Peg 22h5'22" 36d13'2"
Gina's Star Lyn 6h18'29" 54d11'44"
GINA'S U of M 2006 Ori 4h48'51" 1d51'51"
Gineen Gallo Peg 22h22'48" 16d21'18"
Ginelle Leo 11h7'34" 16d19'29"
Gineska I. Rosado Peg 22h37'3" 11d7'34"
Ginette Gagné Del 20h47'44" 8d9'59"
Ginetteicus Psc 1h14'41" 24d50'54"
Ginevra Negrini Ori 6h17'49" 2d19'59"
Ginga Cnc 8h13'5" 14d17'40"
Gingell Rogers Vir 14h31'3" -0d19'30"
Ginger Lib 14h50'54" -4d48'45"
Ginger Lib 15h29'37" -20d33'57"
Ginger Aqr 23h52'56" -10d11'16"
Ginger Cas 0h52'55" 63d35'15"
Ginger Ari 1h57'56" 22d3'57"
Ginger Ori 5h29'12" 8d3'4"
Ginger Psc 1h9'57" 12d9'14"
Ginger And 1h18'32" 41d47'12"
Ginger And 1h2'54" 39d15'58"
Ginger Lmi 10h12'10" 35d46'46"
Ginger And 23h55'13" 42d46'5"
Ginger Lyn 7h3'16" 52d20'39"
Ginger Cas 0h19'3" 57d32'39"
Ginger Uma 10h12'15" 45d48'12"
Ginger 82 Aql 19h32'51" 12d30'38"
Ginger Al's Big 30! Uma 9h30'29" 44d57'39"
Ginger and Crystal Foehr Crb 15h31'45" 35d25'30"
Ginger and David Lyr 19h11'9" 27d52'23"
Ginger and Michael Uma 8h52'15" 59d21'14"
Ginger Barnes Cap 21h36'53" -22d44'23"
Ginger Belle Cas 0h47'0" 60d22'43"
Ginger Blankenship Com 13h4'45" 16d55'13"
Ginger Blossom Psc 1h20'17" 15d55'11"
Ginger Clara Karner Uma 13h3'27" 55d20'30"
Ginger Dawn Tau 3h51'31" 25d8'25"
Ginger Gabriel Lyn 7h33'19" 52d56'0"
Ginger Gueffroy Uma 10h28'11" 55d23'58"
Ginger Hall Vir 13h11'49" 1d39'9"
Ginger Hill Byers And 23h22'20" 51d47'45"
Ginger Kay Collins Uma 8h23'35" 72d20'4"
Ginger Keahi Allen Cas 1h19'23" 58d5'14"
Ginger Knapp & Family Lmi 10h27'9" 38d12'10"
Ginger Lee Kreigh Cnc 8h53'7" 10d9'31"
Ginger Leigh Sgr 18h10'23" -27d8'23"
Ginger Lorna Orner Cap 21h16'55" -19d52'3"
Ginger Lou Mace Collins Gem 7h40'24" 23d28'32"
Ginger Majewski Lmi 10h45'19" 34d12'54"
Ginger Marie Van Roekel Aql 20h20'58" 8d10'12"
Ginger Morel Vir 13h7'48" -15d43'31"
Ginger Moss Gem 7h0'23" 25d30'20"
Ginger Neal Mon 6h47'37" -0d13'33"
Ginger October 11, 2001 Cam 6h9'35" 67d31'2"
Ginger Rose Marie And 2h23'18" 50d35'24"
Ginger Sallee Mon 6h49'15" -0d36'19"
Ginger Smith Cnv 12h27'33" 33d29'7"
Ginger Suzanne Hoffman And 1h29'21" 35d5'8"
Ginger Thompson Gem 7h45'54" 24d55'13"
Ginger Thompson Ari 3h26'34" 20d18'2"
Ginger Traynor Steiner Cmi 7h33'56" 6d5'1"
Ginger Veronica Dunbar Cas 0h55'35" 59d12'12"
Ginger Vestal Cap 21h5'49" -19d18'47"
GingerDawn Aqr 23h30'31" -9d45'50"
Gingerlily Cnc 9h6'44" 6d53'41"
Ginger-Miles-Sammie And 2h22'2" 50d6'59"
GingeRon Uma 9h43'13" 47d19'38"
Ginger's Angel Star Cas 0h33'25" 55d2'33"
Ginger's Jewel Tau 3h42'40" 28d27'41"
Gin-Gin Fogelman Peg 23h5'18" 9d49'0"
Gin-Gin Sunshine Girl SWEET 16 And 0h15'52" 30d40'7"
Gingo Star Lyn 7h10'0" 49d44'45"
Gini Chuck Pitruzzello Tau 4h21'1" 24d38'2"
Gini Valentin Crawford And 23h20'15" 36d31'18"
Gini Woodrum Lyn 7h43'44" 35d53'3"
GiniSue'sMohawk Crb 15h23'12" 30d18'28"
Ginna Boo Psc 1h8'46" 4d11'19"
Ginna Chambers - Eye Street Star Cam 3h22'52" 60d45'12"
Ginnie Dooley Lib 14h55'46" -1d57'13"
Ginnie Mae Psc 1h10'36" 16d32'54"
Ginnie's Star Cas 0h17'30" 54d26'20"
Ginnstar And 0h48'15" 37d37'1"
Ginny And 1h39'22" 37d48'8"
GINNY Cnv 13h40'1" 35d35'57"
Ginny Gem 7h6'51" 33d37'14"
Ginny And 0h32'24" 34d54'52"
Ginny Lyn 7h21'47" 48d11'0"
Ginny Crb 15h55'36" 37d52'50"
Ginny Sgr 17h52'0" -17d0'9"
Ginny Altmeyer Cyg 20h1'26" 35d29'16"
Ginny and Dave Cyg 20h49'57" 44d20'56"
Ginny and Howard Cosner Crb 15h46'15" 27d6'51"
Ginny Cowart Crisler Tau 5h11'0" 17d46'16"
Ginny Davidson Cap 21h49'3" -11d50'30"
Ginny Elkins Vir 12h39'11" 0d39'58"
Ginny Ellen Brown, "ritz" Ori 5h25'59" -4d39'14"
Ginny Gail And 2h21'57" 50d31'36"
Ginny Gallaugher Com 13h8'58" 27d46'4"
Ginny Hayes' Shining Star Tau 4h31'49" 29d25'20"
Ginny Landle Psc 0h54'54" 33d24'30"
Ginny Lee Stegemiller And 1h20'26" 39d19'55"
Ginny Moses Lyr 18h53'53" 36d53'49"
Ginny Quirin Sgr 18h30'35" -26d59'23"
Ginny Ray Boo 14h27'44" 51d4'12"
Ginny Sampson Mon 6h45'53" 9d38'22"
Ginny Schmidt Peg 22h7'42" 10d4'2"
Ginny Stonecipher Tau 3h35'41" 4d11'22"
Ginny: The Shining One Cma 6h49'34" -16d39'21"

Ginny White
Per 2h49'59" 36d49'13"
Ginny wilkins
Lib 15h6'26" -6d57'5"
Ginny, Mark & Emma Grace Kruntorad
Lyn 8h8'45" 40d33'35"
Ginny071469
Cnc 9h11'41" 24d56'40"
Ginny218
Cas 0h38'23" 52d15'7"
Ginnyognive
Umi 16h56'39" 76d24'34"
Ginny's butterfly star of 'light'
Cas 0h50'9" 66d5'16"
Ginny's Friendly Star
Leo 9h52'58" 23d54'45"
Ginny's Saphire
Psc 0h55'47" 5d51'55"
Gino
Peg 0h13'46" 18d37'1"
Gino
And 2h3'34" 38d1'2"
Gino Agostinelli Jr.
Uma 10h9'49" 51d54'51"
Gino and Christine Cascio
Uma 13h39'26" 54d44'7"
Gino and Kerry
Sco 16h3'6" -9d43'4"
Gino Benedetti
Sgr 18h57'22" -34d26'8"
Gino Carraro
Cyg 20h29'17" 52d44'51"
Gino DiGiovacchino
Ori 5h56'43" -3d31'37"
Gino Dimitri Abel Bristow
Dra 18h58'12" 63d12'56"
Gino Giacomangeli
Uma 12h56'42" 62d28'48"
Gino Leon
Cas 1h15'49" 72d6'20"
Gino Luconi
Lib 14h54'2" -9d32'6"
"Gino" Martin
Ari 2h50'29" 15d52'23"
Gino Navaroli
Ari 2h3'40" 21d34'8"
Gino Pulice
Uma 8h15'24" 66d33'30"
Gino Roger Nathan Narboni
Cma 6h52'30" -12d27'9"
Gino Scialdone
Ori 4h52'28" 14d37'7"
Gino Thiago Schwarz
Per 3h18'11" 41d30'50"
Gintaras
Aur 6h13'44" 54d28'18"
Ginvera Marketing Ent.Sdn.Bhd.
Ori 5h34'51" 3d38'51"
Ginzburg, Mark
Lac 22h27'43" 49d18'49"
Gio
Per 4h8'53" 48d48'24"
Gioacchino Senese
Cam 3h31'54" 59d6'35"
Giobana Karenina Garcia Zarate
Ori 6h4'45" 0d15'24"
Giocondina
Cnc 9h4'24" 30d52'41"
GioCurryLaneSica2005
Uma 9h57'23" 57d49'11"
Gioele
And 2h15'43" 50d32'48"
Gioia
Dra 18h56'34" 50d23'54"
Gioia
Vir 12h37'58" 3d56'41"
Gioia
Cas 0h45'30" 60d16'2"
Gioia A.C.
Cam 7h3'14" 69d0'41"
Gioia di Giulia
Peg 22h43'38" 26d58'53"
Gioia Isabella Susan Iozzi
And 0h47'50" 29d36'27"
Gioia Marie Benenati
Vir 13h19'28" -7d46'15"
Giomarcus D'andre Laxamana Batac
Uma 13h48'50" 49d52'1"
GiòNuvola
Her 16h37'38" 24d13'52"
Giordan Ellie Gitchell
Umi 13h8'18" 73d43'28"
Giordana Reda
And 23h50'38" 41d28'49"
Giordano Mazzolini
Uma 9h21'42" 61d22'38"
Giorgia
Cas 23h14'36" 56d7'58"
Giorgia
Aqr 21h7'50" -13d16'17"
Giorgia
Umi 14h3'0" 74d11'43"
Giorgia
Sco 17h50'56" -36d1'54"
Giorgia
Lac 22h50'35" 37d39'43"
Giorgia
Aur 5h33'10" 32d53'56"
Giorgia
Peg 22h40'12" 27d33'41"
Giorgia
Ori 5h32'23" 14d2'38"
Giorgia
Vir 13h19'9" 12d46'45"
Giorgia Antonia Vourvahakis
Ari 2h20'44" 13d38'5"
Giorgia Bellini
Peg 22h21'19" 6d48'12"
Giorgia Kate Medici
Nor 16h9'39" -45d3'9"
Giorgia Tudico
Cnv 13h17'43" 33d42'29"
Giorgia-Foxy
Umi 15h53'2" 76d51'43"
Giorgiana Iliescu
Lyr 19h13'3" 43d27'32"
Giorgina Noo Noo 90
Ara 16h39'52" -53d3'18"
Giorgio
Cam 5h4'1" 79d27'25"
Giorgio
Aur 5h55'25" 32d57'37"
Giorgio
Vir 13h0'25" 5d47'25"
Giorgio Armani
And 23h16'8" 43d57'55"
Giorgio Banfi Jr
Aql 20h9'9" -0d18'43"
Giorgio Beretta
Aur 4h56'10" 37d49'8"
Giorgio e Generina 29.12.1962
Her 17h26'46" 32d54'52"
Giorgio Gallenzi
Crb 15h55'54" 32d24'14"
Giorgio Lo Stimolo
Sco 17h25'4" -30d33'37"
Giorgio Lorenzo Morganti
Cep 20h53'24" 62d18'33"
Giorgio Rufa
Vir 12h24'18" -8d30'40"
Giorgos & Hara
Del 20h34'18" 19d31'40"
Giovambattista (Johnny) Sussanna
Cru 12h29'17" -59d50'33"
Giovanna
Cam 4h36'40" 58d14'15"
Giovanna
Ari 2h18'58" 11d14'56"
Giovanna Carnevali
Aur 6h34'36" 32d49'37"
Giovanna Casola
Ari 3h26'28" 21d6'5"
Giovanna & Dustin 2007
Psc 23h27'31" 5d4'23"
Giovanna G. Bruno
And 2h20'38" 48d40'26"
Giovanna Germinario
Aqr 23h43'24" -12d6'40"
Giovanna Montesanti Gemignani
Ara 17h31'22" -47d14'2"
Giovanna Nobile
Cam 6h26'31" 69d37'14"
Giovanna Piccolo
Ori 5h59'48" 22d17'46"
Giovanni
Psc 1h7'34" 12d22'24"
Giovanni
Uma 11h54'51" 51d51'36"
Giovanni
Aur 6h28'18" 39d26'44"
Giovanni
Uma 8h12'45" 61d40'58"
Giovanni
Uma 12h32'16" 60d52'7"
Giovanni
Sco 17h45'11" -40d53'41"
Giovanni A. McNeil
Her 16h27'31" 24d8'40"
Giovanni Alexander Fernandez
Ari 2h18'26" 26d9'4"
Giovanni and Jovanni Waters
Cyg 20h23'58" 40d26'36"
Giovanni Antonio Rosas
Tau 3h27'54" 9d24'38"
Giovanni Battista Russo
Uma 9h4'40" 57d49'55"
Giovanni Berlingieri
Gem 7h31'26" 35d7'48"
Giovanni & Brigitta
Umi 16h4'17" 73d15'14"
Giovanni Charles Di Capua
Psc 0h20'19" 20d4'34"
Giovanni Christopher Cantini
Dra 17h50'46" 68d43'15"
Giovanni Corsa
Per 4h33'57" 48d47'8"
Giovanni Cortese
Cas 1h5'20" 63d0'45"
Giovanni De Marco
Lyr 18h50'22" 32d0'25"
Giovanni Di Crescienzo
Cyg 19h41'2" 34d38'48"
Giovanni e Leonora
Cyg 19h56'32" 31d22'7"
Giovanni Esposito Jr.
Uma 8h38'36" 55d21'59"
Giovanni Gagliano
Lyr 18h33'16" 33d50'20"
Giovanni Galacio
Cyg 20h58'14" 38d44'31"
Giovanni Gambino
Psc 23h36'59" 6d4'26"
Giovanni "Gio" Cavagnari
Ori 5h58'29" -0d42'0"
Giovanni Lancia
Cnc 8h41'12" 29d20'52"
Giovanni Loves Suzanne
Cnc 8h30'48" 31d59'35"
Giovanni & Madlen in love
Uma 12h24'28" 61d7'39"
Giovanni Mare
Umi 14h58'46" 74d36'35"
Giovanni Mario Branca
Aqr 22h40'7" 1d38'15"
Giovanni & Marisa Pasquino
Cyg 19h56'16" 32d24'31"
Giovanni Martinez
Lib 14h54'59" -2d7'38"
Giovanni Oliveri's Star
Leo 9h42'31" 32d12'24"
Giovanni Puricelli
And 1h55'58" 41d22'12"
Giovanni Rosser
Lib 15h0'3" -14d15'46"
Giovanni S. Castillo
Uma 13h58'0" 56d7'54"
Giovanni Sifuentes
Dra 18h52'52" 52d10'7"
Giovanni Vaccarello
Uma 8h39'17" 55d34'8"
Giovanni Viscogliosi
Uma 9h22'30" 65d11'17"
Giovanni, il mio Morino
Crb 16h21'19" 29d35'39"
Giovannia
And 2h38'36" 46d24'36"
GioVannie
Cnc 8h34'32" 10d39'56"
Giovanny
Vir 14h9'23" 2d55'17"
Giovi
And 23h39'41" 40d43'55"
Giovi
Dra 19h25'25" 61d4'0"
Giovia Chiara Deputato
Cnc 8h40'17" 32d26'0"
Giovi-Serendipity
Cas 0h57'23" 58d52'58"
Gipsy
Lac 22h25'44" 48d23'49"
Gipsy
Her 18h47'30" 19d37'42"
Girard Riehl
Her 18h13'20" 15d33'57"
Girasole
And 23h25'43" 52d28'59"
Girija Akash Samarth
And 23h49'0" 41d59'28"
Girish & Bindiya
Cyg 19h28'8" 51d27'32"
Girish Nagasamudra
Psc 1h17'19" 22d21'7"
Giriti Tamás csillaga
Uma 9h9'41" 46d43'25"
Girl
And 1h16'53" 35d29'40"
Girl
Psc 0h58'51" 3d21'16"
Girl Loves Boy Mas!
Psc 1h20'7" 28d17'17"
Girl Scouts of Troop 5462
Uma 12h34'54" 55d25'16"
Girlilla
And 2h20'11" 46d39'15"
Girllbert
Ori 5h20'36" 1d34'13"
Girls "12-25-05"
And 0h37'10" 34d51'51"
girls star
Tau 3h47'2" 25d53'47"
Girma
Peg 22h53'51" 32d22'43"
Giro "Joe" Capasso
Cep 2h31'41" 80d39'0"
Girr Cecil Mousekowitz Martinez
Leo 11h16'54" 1d21'19"
Girrel Lee Johnson
Aur 7h24'56" 40d4'10"
Gisa (Honey Bear) Arellano
Vir 12h24'11" 6d37'19"
Gisela
And 1h42'8" 49d26'0"
Gisela
Aqr 22h37'44" -0d37'57"
Gisela Behrendt
Cyg 20h56'45" 33d39'4"
Gisela Bertram
Ori 6h17'43" 2d10'15"
Gisela Lord
Aql 18h49'34" 9d57'40"
Gisela Möbius
Uma 11h2'51" 47d2'40"
Gisela N Witt
Cam 7h7'16" 71d26'25"
gisela schedy
Uma 11h53'2" 38d18'46"
Gisela Stachowski
Uma 9h0'21" 61d54'49"
Gisela Tavarez
Crb 16h15'55" 38d33'8"
Gisela Widmer
Umi 13h58'31" 76d3'38"
Gisela's Heart
Uma 11h32'43" 65d48'12"
Gisele
Sgr 18h58'4" -34d14'5"
Gisele Braceras
Vul 19h48'49" 21d57'12"
Gisele Chimara
Col 6h23'50" -37d1'6"
Giselle
Leo 10h26'32" 17d42'28"
Giselle
Cnc 8h49'26" 26d34'53"
Giselle
Leo 10h18'10" 9d26'12"
Giselle
Psc 1h9'12" 16d45'0"
Giselle
Cnv 12h48'5" 48d4'2"
Giselle Always and Forever
Vir 13h25'8" 12d19'51"
Giselle Ariane
Lib 15h23'46" -21d6'43"
Giselle Jonte Brown
Cap 20h14'4" -18d59'53"
Giselle Melchor
Cnc 8h35'17" 17d11'33"
Giselle Samantha Murillo
Ari 1h59'23" 14d29'53"
Giselle Star
Leo 10h54'43" 0d12'22"
Giselle's Sparkle
Sco 17h48'24" -39d33'30"
Gish's Star
Lmi 11h5'24" 33d3'33"
Gissel Sonnenbrot
Cap 21h30'36" -9d8'1"
Gissell Beatriz Gonzalez
Uma 11h20'52" 43d12'16"
Git Richter
Psc 1h22'29" 33d7'5"
Gita
Uma 11h27'6" 45d59'41"
Gitanilla-Ariadna Star
And 1h15'16" 46d49'20"
GITTA
Tri 2h19'10" 36d41'0"
Gitta Vass
Gem 6h41'16" 30d31'56"
Gitte, Angel of København
And 0h36'57" 28d22'1"
GITTI'S LUCKY STAR
Her 16h29'7" 30d50'15"
Giudice
Del 20h34'46" 19d4'24"
Giule
Peg 23h39'30" 27d48'49"
Giulia
Tau 6h0'48" 27d5'59"
Giulia
Ori 5h28'35" 11d18'29"
Giulia
Boo 14h4'11" 9d53'10"
Giulia
Ori 6h24'43" 14d21'54"
Giulia
Per 3h33'11" 49d20'22"
Giulia
Umi 15h4'41" 76d43'55"
Giulia 1.6.91
Her 18h29'37" 25d25'19"
Giulia Aievoli
Cas 0h27'31" 62d26'26"
Giulia Antelmi
Aur 6h14'46" 31d6'9"
Giulia Antonella Massei
Ori 6h25'29" 16d59'16"
Giulia Bellofatto
Cam 4h35'17" 66d20'10"
Giulia Bergamo
Cam 7h13'34" 84d48'38"
Giulia Borghi
Her 18h17'32" 28d36'16"
Giulia Brescancin
Umi 17h7'47" 78d37'2"
Giulia Cali
Umi 15h40'20" 67d56'17"
Giulia Camilla
Per 4h45'28" 43d43'8"
Giulia Castellani
Cam 4h37'46" 59d23'44"
Giulia Dall' Olmo
Peg 23h42'30" 11d14'5"
Giulia D'Amato
Per 4h33'40" 32d38'4"
Giulia Dell'Antico
Umi 17h41'17" 80d2'16"
Giulia (Giu-Giu) Tiramani
Uma 11h22'58" 59d17'0"
Giulia Mariani
Peg 21h30'12" 23d28'44"
Giulia Mezzetti
Aur 5h19'11" 48d48'39"
Giulia Palermo
Peg 22h9'7" 6d44'39"
Giulia Ratman
Cas 23h26'23" 52d52'37"
Giulia Sofia Milagros
Tau 3h42'32" 24d38'17"
Giulia'94
Cam 3h24'52" 60d13'39"
GIULIA-AALIYAH
Com 12h21'44" 27d36'53"
Giulian Keck Beaulieu
Lib 15h25'47" -20d34'53"
Giuliana
Lib 15h29'15" -4d31'38"
Giuliana
Aqr 21h47'27" 0d0'6"
Giuliana Martinelli Rosaia
Uma 9h16'11" 59d59'9"
Giuliana Marussi mamy blue
Her 18h16'20" 16d39'34"
Giuliana Olmo
And 23h19'11" 44d5'56"
Giulianna-Marie
Uma 12h4'45" 35d12'58"
Giuliano
Ori 6h17'17" 11d14'42"
Giuliano
Aqr 21h41'0" -8d21'56"
Giuliano S.
Aur 4h58'23" 36d21'7"
Giulio
Per 3h26'22" 49d25'38"
Giulio 27 luglio 1942
Leo 11h55'35" 25d53'22"
Giulio Camber
Peg 21h47'53" 25d57'29"
Giulio e Valentina
Ori 5h24'25" -4d25'9"
Giulio Spaccatrosi
Cyg 19h42'7" 28d46'59"
Giuly
Ori 6h0'28" 6d37'44"
Giuran
Cnc 9h20'57" 25d27'48"
Giuseppe
And 23h11'5" 37d0'43"
Giuseppe
Uma 13h31'20" 60d7'59"
Giuseppe Aldino
Gem 7h2'51" 22d39'13"
Giuseppe & Antonietta
Cyg 19h29'42" 53d44'11"
Giuseppe Antonino Colca
Mon 7h2'48" 5d44'52"
Giuseppe Campisi
Ari 2h0'10" 10d48'2"
Giuseppe Circognini
Aur 6h52'3" 43d44'46"
Giuseppe Crapanzano
Lyr 18h36'16" 30d5'33"
Giuseppe Ducco
Uma 10h59'4" 33d50'45"
Giuseppe Giovenzana-amare e volare
Aql 19h45'43" 7d38'26"
Giuseppe Grimaldi
Lep 5h54'34" -16d12'6"
Giuseppe Iellamo
Ori 5h50'18" 10d14'28"
Giuseppe Ippolito
And 2h20'52" 41d16'58"
Giuseppe Lorenzino MORETTI
Peg 21h53'7" 11d12'14"
Giuseppe Max Deleers-Certo
Tau 3h41'24" 22d15'33"
Giuseppe Petrone
Uma 13h3'32" 61d32'48"
Giuseppe Pierini
Cas 0h29'3" 62d22'14"
Giuseppe "Pino" Pirovano
Her 18h50'43" 23d58'51"
Giuseppe Romeo "IL MIO CAMPIONE"
Sct 18h30'29" -15d33'17"
Giuseppe & Rosalie Cangelosi
Cyg 21h22'35" 34d40'34"
Giuseppe Serpico
Cyg 21h23'31" 34d47'54"
Giuseppe The Pimp
Tri 1h44'24" 29d55'0"
Giuseppina Sole
Peg 21h54'2" 18d14'4"
GiusFil 65
Cam 3h20'2" 58d40'19"
Giusi's BIG star
Uma 9h15'55" 53d11'55"
Giustino Capodilupo
Sgr 18h59'54" -22d30'12"
Giusy
Per 3h7'10" 51d38'12"
Giusy
Ori 5h45'59" 1d56'15"
Giusy Romano
Her 18h34'27" 18d39'43"
Giusy Tinelli
Lyr 19h11'50" 38d56'57"
Give a Little Whistle
Umi 14h48'1" 78d21'49"
Giving you just a touch of Heaven..
Sco 17h35'59" -34d18'19"
Giz Casse
Aur 5h11'4" 42d0'52"
Giza Amor
Cnc 8h43'26" 31d7'25"
Gizela Stenzel
Ori 5h55'32" 17d37'41"
GIZELKA
Aqr 23h55'40" -10d53'57"
Gizella
Sco 17h51'59" -44d28'25"
Gizella
Ari 2h35'53" 14d35'59"
Gizelle Audrey Acio
Leo 11h53'59" 25d15'12"
Gizem "Gizzolisse" Kural
Cap 20h11'9" -15d28'9"
Gizem's Angel in the Sky
Sco 16h17'22" -29d41'51"
Gizmo
Vir 14h19'15" -19d24'35"
Gizmo
Tau 3h41'58" 8d50'13"
Gizmo
Cmi 8h10'10" 3d49'6"
Gizmo
Gem 6h53'33" 32d47'24"
Gizmo
Cas 0h8'41" 59d5'42"
Gizmo
Uma 11h34'28" 48d41'18"
Gizmo*05.09.92
Uma 9h33'39" 41d48'31"
Gizmo "Cowboy" Lamarre
Cmi 7h31'40" 7d22'28"
Gizzle
Leo 9h39'46" 28d37'3"
Gizzy
Ari 1h47'42" 24d26'26"
GJJG
Cyg 19h47'26" 43d30'47"
GL 5862
Aqr 23h5'53" -9d7'54"
G.L. Cole
Uma 10h36'40" 68d25'45"
GLACE
Per 2h24'34" 55d31'27"
Gladdiatrix ~ AKA Lisa Ann Gladd
Per 3h55'7" 37d20'34"
Gladiador
Her 16h34'44" 35d52'35"
Gladioli
Cas 0h35'41" 50d5'34"
Gladis Aimee Garcia
Cyg 19h49'23" 44d19'52"
Gladis Azar Saba June 7th, 1948
Gem 7h8'15" 34d15'29"
Gladis Cancino Moncada
Per 4h15'38" 31d58'52"
GLADSON
Uma 8h19'33" 62d54'5"
Gladwell
Uma 10h51'11" 49d5'55"
Gladwin Huber
Tau 5h58'24" 26d42'19"
Gladyce Haflund
Umi 17h22'53" 77d44'10"
Gladys
Uma 11h30'14" 60d41'59"
Gladys
And 0h43'58" 27d1'41"
Gladys
Psc 1h15'59" 18d14'34"
Gladys
Gem 6h40'38" 13d51'52"
Gladys
Dra 17h40'4" 51d57'58"
Gladys Adeline Tubinis
Ori 6h17'46" 3d16'42"
Gladys Bennet
Uma 10h3'11" 49d23'38"
Gladys Blair
Cas 1h47'46" 64d45'27"
Gladys Constance Arnaud
Sco 17h15'42" -39d16'23"
Gladys Cruz
Crb 15h28'24" 29d15'38"
Glady's D'Amato
Vir 15h3'11" 4d48'1"
Gladys Danielson
Cyg 21h44'29" 38d41'32"
Gladys Darling Doku
Gem 6h48'44" 31d52'14"
Gladys Demster
Aqr 22h36'1" 1d44'30"
Gladys Eskenazi
And 1h3'30" 41d52'46"
Gladys Eves
Uma 9h49'48" 49d58'15"
Gladys F. Carter
Tau 5h10'36" 17d38'9"
Gladys F. Olson
Aqr 21h6'48" -14d9'33"
Gladys Fierimonte
Uma 11h4'34" 54d35'0"
Gladys Garay
Cas 23h29'43" 53d30'12"
Gladys Garcia
Peg 22h1'50" 15d0'33"
Gladys Goh Ching Swi
Cra 18h50'10" -41d48'49"
Gladys Googins
And 0h49'24" 36d16'6"
Gladys "Gram" Williams
Sco 16h42'6" -37d49'15"
Gladys Irene Neale (Grandma)
Uma 10h22'12" 44d56'51"
gladys_jack_45_major
Uma 10h9'43" 46d45'19"
Gladys Joy
Aqr 23h24'51" -11d19'28"
Gladys June Vannausdle
Cas 3h32'12" 70d32'5"
Gladys L Rose
Her 16h45'47" 8d54'47"
Gladys Lawrence
Psc 1h24'1" 10d33'45"
Gladys Lee Morris
Cap 20h21'59" -11d57'3"
Gladys Marie
Lib 14h26'30" -20d11'40"
Gladys Mary Ogden
Peg 22h45'35" 26d59'21"
Gladys May Blain
Gem 6h33'5" 24d54'0"
Gladys "Nan" Gallagher
Cas 23h32'27" 53d43'23"
Gladys NASSIF TORBAYE
Cet 0h42'47" -0d10'24"
Gladys Noraine
And 0h40'29" 41d10'34"
Gladys Persaud
Leo 9h48'8" 28d44'50"
Gladys & Phil's 60th Celebration
Cyg 19h33'5" 53d21'17"
Gladys Porter
Lyn 7h39'30" 49d32'0"
Gladys Reinbold
Umi 4h29'23" 88d55'44"
Gladys Rose Hassler
Tau 4h17'12" 27d27'47"
Gladys S Jenkins
Per 4h9'10" 47d24'23"
Gladys Solarte
Tau 5h47'24" 17d19'23"
Gladys Teuke
Tau 3h45'7" 30d42'37"
Gladys Treat
Lep 5h14'45" -12d25'17"
Gladys Welch
Cyg 20h45'11" 32d39'59"
Gladys Wilson
Cap 20h28'6" -18d8'53"
Gladys Zuniga
Psc 1h2'53" 28d51'43"
Glaiza
Ari 2h40'20" 29d38'34"
Glam-Ma Anne Works
Uma 10h2'39" 49d58'46"
Glamorous Glynis
Sco 17h56'44" -30d10'25"
Glamorous Jennifer
Psc 23h57'27" -0d5'49"
Glan Lucas
Ori 6h11'27" 1d59'48"
Glanzlicht Carmen 56
Uma 12h4'36" 37d52'47"
Glaß, Steffen
Ori 5h7'55" 16d2'12"
Glatzel, Karl-Heinz
Uma 11h36'7" 33d18'33"
Glatzi
Lac 22h51'24" 54d5'32"
Glauriah
Sco 17h31'59" -32d48'30"
Glavij
Tau 5h56'34" 25d2'45"
Gleeson Ryan
Vir 13h21'40" 13d2'12"
Glen
Cnc 8h14'27" 21d59'33"
Glen A Huffman
Mon 6h51'22" -5d40'55"
Glen Alan Hipwell
Cap 21h41'46" -9d35'9"
Glen Allen Austin
Boo 14h32'50" 26d53'1"
Glen and Andrea Forever
Her 17h19'41" 20d29'50"
Glen and Burnell Russell
Uma 9h24'21" 42d49'9"
Glen Autobahn Bucher
Tau 4h14'5" 5d54'40"
Glen B.
Tau 4h42'2" 19d31'5"
Glen Baum
Lib 15h57'24" -18d16'38"
Glen D. Hammond
Per 2h52'15" 54d19'42"
Glen E. Murray
Crb 16h22'28" 36d49'53"
Glen Edison Banks
Cap 21h5'0" -18d47'4"
Glen Floyd Graves
Lib 15h13'38" -24d6'44"
Glen Franklin Johnson
And 0h34'20" 45d8'8"
Glen G. Bouchard
Per 3h20'5" 54d4'48"
Glen Hill
Uma 10h20'25" 57d8'17"
Glen Jeffrey Silverman
Cep 20h54'4" 63d15'25"
Glen & Jen Wilkins
Eri 3h57'52" -0d22'15"
Glen & Jo G. McNeill
Cyg 19h20'36" 28d59'3"
Glen M DaSilva's G-Star
Gem 6h53'25" 32d9'5"
Glen Max Burnham
Uma 13h27'1" 56d54'16"
Glen Maxwell Reid
Umi 13h57'11" 71d44'6"
Glen Michael Wllman
Lib 15h4'33" -15d54'55"

Glen Mitchell
Dra 18h27'32" 53d58'10"
Glen & Nicole Perry - Winkles
Pho 0h27'32" -52d36'12"
Glen P. King Jr.
Per 2h19'18" 56d45'47"
Glen R. Davis
Ori 5h22'3" 5d40'12"
Glen R Willis
Cep 21h6'24" 61d58'35"
Glen Richard Tweedie
Uma 11h0'40" 51d46'50"
Glen "Sparky" Hitchens
Tau 6h0'15" 24d40'0"
Glen & Stephie
Aqr 21h26'24" 0d11'39"
Glen Stewart Scotland
Sct 18h48'25" -9d28'32"
Glen Sug
Umi 16h32'8" 76d16'28"
Glen & Tina Peterkin
Cyg 21h47'51" 44d50'27"
Glen W Morris
Per 4h18'13" 50d11'31"
Glen Wade Box
Her 18h49'32" 25d18'10"
Glencora
Cyg 20h24'5" 47d48'19"
Glenda
Aql 19h53'43" 15d22'32"
Glenda
Mon 7h10'49" 1d16'55"
Glenda
Peg 21h26'58" 11d50'12"
Glenda
Vir 14h11'23" -4d44'49"
Glenda
Lib 14h58'24" -8d28'22"
Glenda
Uma 13h6'23" 52d49'27"
Glenda
Sgr 18h45'54" -35d15'59"
Glenda Adams
Cas 23h40'41" 56d2'4"
Glenda Carolyn Nordman
Lib 15h5'16" -5d12'19"
Glenda E. Mathews
And 0h44'1" 39d38'37"
Glenda Fay Hickman (Tomasiello)
Gem 6h42'43" 13d17'24"
Glenda Gayle
Leo 10h24'14" 8d55'33"
Glenda Granados
Lib 15h11'32" -4d12'45"
Glenda Hood
Leo 9h46'46" 31d59'26"
Glenda J. Diem
Lyn 6h59'15" 58d45'8"
Glenda J. Sims
Aqr 20h41'23" -11d38'59"
Glenda Jane Matheson
Psc 1h3'36" 7d4'11"
Glenda Jo Hall
Cap 21h42'19" -14d57'59"
Glenda Jo Houck
Umi 16h52'9" 79d52'47"
Glenda Johnston
Cru 12h31'6" -61d56'36"
Glenda Kay
And 23h18'29" 48d41'42"
Glenda L Johnson
Ori 5h35'16" 6d28'3"
Glenda Larsen Harrod
Uma 11h16'13" 42d22'32"
Glenda Lask
Leo 11h8'57" 25d20'30"
Glenda Lee
Tau 4h20'15" 4d33'12"
Glenda Lisette Colón
Gem 7h44'29" 33d17'11"
Glenda Lynn Cook Mea Astra
Lmi 10h48'44" 27d47'57"
Glenda Major
Leo 11h26'33" 14d53'26"
Glenda Malyon Coote 06/09/1956
Vir 13h23'43" -10d7'23"
Glenda Mary Majors Ramsey
Cnc 8h15'1" 15d13'7"
Glenda McClain Go Dale Jr
Sgr 18h42'39" -23d32'49"
Glenda Michelle
Cyg 19h55'59" 35d3'19"
Glenda Mocha
Gem 7h43'27" 20d59'18"
Glenda Reynolds Wilson
Leo 10h18'19" 23d4'46"
Glenda Salaam
Leo 11h17'18" -1d27'46"
Glenda Sue Dowdle
Cas 0h51'28" 47d39'47"
Glenda Taylor
Uma 10h11'45" 45d9'38"
Glenda Tessnear
Uma 11h4'44" 59d36'55"
Glenda The Yellow Rose Of Texas
Cas 0h36'6" 54d13'52"
Glenda Zametto
Uma 11h48'6" 57d34'36"
Glendaly Pabon
Lib 15h35'27" -18d10'8"
Glenda's Guardian
Per 3h1'5" 55d49'25"
Glenda's Light
Tau 4h34'43" 16d45'33"
Glendon L Fowler
Lib 15h6'39" -17d47'46"
Glenell
Cyg 20h10'42" 35d26'39"
Gleneta Jefferson
Cam 4h25'20" 66d32'28"
Glenister John Orton
Uma 9h32'11" 43d50'5"
Glenita E. Thurston Pensoneau
Uma 10h16'58" 54d26'36"
Glenjamin
Per 4h23'19" 41d1'6"
Glenn
Per 4h22'18" 35d7'47"
Glenn 4
Aql 19h19'38" 2d36'10"
Glenn A. Gregory
Sgr 18h36'32" -23d51'31"
Glenn A Kimball
Ser 18h16'14" -13d40'37"
Glenn A. Koch
Psc 23h34'24" 3d36'26"
Glenn A. Milner/MAGIC MAN 1/24/57
Aqr 23h5'50" -8d4'41"
Glenn Alan Halbur
Her 16h38'4" 43d38'0"
Glenn Allen Knutson
Uma 9h16'38" 56d16'0"
Glenn Allen Lagerman 1969
Sgr 20h21'28" -41d38'5"
Glenn and Florence
Cyg 19h59'5" 58d50'31"
Glenn and Jean Adams
Cnc 8h13'54" 7d41'29"
Glenn and Lisa Forever
Aqr 22h27'8" -6d28'54"
Glenn and Tina Snyder 09202003
Cyg 19h43'41" 34d4'32"
Glenn and Wendy McQuiston
Cyg 21h16'10" 52d55'25"
Glenn Arlen Olsen
And 2h11'38" 46d15'53"
Glenn Bell
Lyn 7h32'20" 36d12'50"
Glenn Bernard Kelly
Ori 5h20'27" 2d56'47"
Glenn & Brenee
Psc 1h3'40" 25d39'44"
Glenn Brookover & Darlene Corica
Cam 4h9'29" 74d23'50"
Glenn Brown
Ari 2h55'51" 18d17'10"
Glenn Buhr
Cep 21h30'11" 62d43'13"
Glenn Christopher Holcomb
Gem 7h35'15" 25d1'1"
Glenn Cooper Song Birds
Her 17h32'56" 14d43'27"
Glenn Curtis
Gem 6h33'33" 18d24'9"
Glenn D. Kimmel, Sr.
Uma 11h51'12" 40d41'46"
Glenn "Daddy" Campbell
Her 16h17'48" 47d1'36"
Glenn David Marshall
Cap 20h17'33" -12d57'34"
Glenn Davis Jordan
Lib 14h23'8" -12d18'6"
Glenn E. Stutts-The Law
Per 4h30'4" 47d4'32"
Glenn & Elaine Bolton
Cyg 21h21'9" 39d27'32"
Glenn Elmer Taylor, Sr.
Gem 7h15'48" 19d23'45"
Glenn & Emma Copella
Gem 6h42'57" 26d48'49"
Glenn Emory Knox
Ari 2h32'30" 26d25'41"
Glenn Erik Falkum
Per 4h27'17" 46d19'5"
Glenn Francis Eikenberg
Vir 13h25'11" 12d8'40"
Glenn G. Woods
Ari 3h20'24" 19d5'56"
Glenn Ganson
Cap 21h41'12" -15d8'56"
Glenn Genzel
Uma 11h31'1" 40d20'39"
Glenn Gerald Strawder
Boo 14h51'13" 19d6'46"
Glenn Gordon Batchelor
Gem 6h46'11" 28d22'13"
Glenn Greene
Uma 11h45'18" 29d33'59"
Glenn Grossman
Cnc 8h13'31" 32d42'10"
Glenn Henderson
Gem 7h45'26" 17d9'55"
Glenn I. Harvey
Ori 5h6'35" 4d2'53"
Glenn J Denning Tech Sargent WWII
Peg 21h43'33" 21d34'30"
Glenn J Stack
Her 16h47'24" 31d43'36"
Glenn Jerry Moore
Uma 10h40'21" 53d30'20"
Glenn & Jessica's Star
Uma 9h42'40" 60d25'41"
Glenn & Jodi Lotenberg Family
And 1h32'17" 48d42'50"
Glenn Keys
Cnv 13h57'30" 34d4'14"
Glenn LaDue
Her 18h16'33" 18d5'9"
Glenn & Linda Worthington
Cyg 19h57'44" 47d26'6"
Glenn Lobo
Cep 22h11'41" 68d42'33"
Glenn M. Clifton
Del 20h29'20" 12d36'36"
Glenn Michael Swiston
Tau 5h59'20" 28d3'23"
Glenn Myrrah Sellars
Uma 8h18'50" 71d39'11"
Glenn Newman
Uma 11h20'29" 53d34'24"
Glenn Nichalas Davis
Lib 15h6'31" -9d59'48"
Glenn & Nicole
Cnc 8h52'20" 11d27'36"
Glenn O. Helseth
Gem 7h10'1" 24d2'18"
Glenn R. Johnston
Sco 17h40'58" -36d40'14"
Glenn Rawnsley
Ori 6h3'3" 21d18'36"
Glenn Redemann
Per 2h32'4" 53d22'34"
Glenn Richard Knight
Uma 14h3'25" 55d6'31"
Glenn Rosazza
Psc 23h30'0" 4d56'37"
Glenn Rose
Per 4h15'32" 36d59'44"
Glenn Russell Beevor
Per 4h16'1" 35d0'28"
Glenn Ryan "RPM's Aurora"
Gem 6h50'33" 34d19'37"
Glenn & Sandra - 17 June 2006
Cru 12h29'43" -58d6'30"
Glenn Sangalli
Per 3h20'45" 46d19'19"
Glenn Sinacore
Umi 14h39'45" 73d50'10"
Glenn T. Shaw
Gem 6h1'52" 25d11'11"
Glenn the Great
Uma 11h10'1" 30d57'58"
Glenn Thomas Orsak
Tau 3h30'18" 17d6'59"
Glenn Trevor Mitchell
Cep 21h13'3" 65d53'36"
Glenn W White-'The Pink'
Ori 5h9'40" -0d15'9"
Glenn Walter Smith
Ori 5h38'5" 0d13'50"
Glenn Wayne Bisby
Her 17h49'3" 46d43'39"
Glenn William Horrigan D.D.S.
Per 2h18'43" 56d36'28"
Glenna Ann Bayer
Ori 6h19'45" 10d49'55"
Glenna Arminta Elkholy
Mon 7h17'26" -0d7'44"
Glenna Dorrough
Uma 11h21'7" 31d33'24"
Glenna Falkowski
Uma 10h3'12" 60d3'0"
Glenna J. Emel
Cas 1h25'13" 57d31'52"
Glenna Jean
Del 20h29'4" 16d36'39"
Glenna Mae "Ever Faithful One"
Cnc 8h36'5" 19d53'8"
Glenna Marie
Lac 22h20'17" 48d23'11"
Glenna (Penny) Payne
Cas 0h37'25" 52d2'12"
Glenna Tom Jacob Kornegay Johnson
Uma 14h27'24" 62d25'9"
Glenna Tuckett
And 0h35'17" 37d42'39"
Glennalee
Uma 14h28'37" 55d39'17"
Glenna's Star
Cnc 8h42'56" 30d42'36"
Glenne
Lib 14h49'17" -2d58'53"
Glennia Ruth Towne
Cyg 20h11'19" 51d1'24"
Glennie and Donald
Lyr 18h37'9" 39d15'54"
Glennie Henry
Crb 15h53'57" 29d3'48"
Glennis
Lyr 18h20'26" 36d49'43"
Glennis Lillian Hill
Cas 23h15'47" 59d49'42"
Glenn's Angel
Cen 13h14'18" -32d45'3"
Glenn's Glory
Per 2h56'2" 43d4'49"
Glenn's Light Shining Bright
Cru 12h27'14" -60d43'45"
Glenny T
Cep 21h13'58" 66d24'56"
Glen's Friendship Star
Aqr 21h59'35" -10d40'44"
Glenstar
Her 16h11'40" 11d56'25"
Glenwood C. Witsaman, Jr.
Lib 15h44'45" -19d39'0"
Glenwood Charles Butler
Uma 9h44'29" 58d13'38"
Glenys Coombes
Cas 2h5'1" 62d58'28"
GLIGA MEA
Sgr 19h14'19" -22d2'11"
Gliggle
And 23h3'50" 51d7'31"
Glimel
Cru 12h19'3" -56d44'35"
Glimmer of Hope
Psc 1h25'57" 6d27'22"
Glischinski
Her 18h13'46" 18d44'9"
Glitter Girl Krissy
Sgr 19h2'35" -28d37'33"
Glittering Lucrezia
Cru 12h5'39" -62d42'6"
GlitterStar
Cnc 9h4'15" 6d42'13"
GLJ247
Psc 1h5'34" 7d11'35"
GLMason
And 0h6'55" 46d49'19"
GL-MKTK Kruk
Lyn 7h36'20" 39d11'30"
Glo
Cyg 20h1'58" 35d5'43"
glo
Vir 14h36'10" -0d52'23"
Global Ordering
Uma 9h15'43" 49d50'49"
Globe
Umi 16h12'41" 82d11'15"
Globo
Lib 15h22'56" -7d19'54"
Glock II
Uma 9h8'53" 66d45'5"
Gloie
And 2h22'47" 45d26'28"
GloRaoul
Lyr 19h0'35" 38d55'9"
Glorfindel
Dra 17h36'49" 53d1'35"
Glori Dawn
Apu 16h55'49" -79d57'18"
Glori Marshall
Cam 4h36'34" 53d1'51"
Gloria
Cas 0h4'34" 53d24'12"
Gloria
Cyg 21h43'37" 44d51'14"
Gloria
And 1h56'50" 45d16'3"
Gloria
Lyn 7h58'27" 39d1'12"
Gloria
Uma 11h44'54" 43d36'4"
Gloria
Lyr 18h18'49" 32d12'2"
Gloria
Gem 8h7'44" 31d33'10"
Gloria
Psc 0h53'1" 32d54'20"
Gloria
Ari 2h9'40" 23d59'55"
GLORIA
Gem 6h54'55" 20d5'1"
Gloria
Lyr 19h17'12" 28d47'17"
Gloria
Lyr 19h15'6" 28d27'43"
Gloria
Com 12h39'9" 26d49'4"
Gloria
Cas 0h46'40" 61d54'4"
Gloria
Lyn 6h45'2" 59d34'55"
Gloria
Uma 9h59'35" 62d11'31"
Gloria
Sgr 18h12'31" -18d53'33"
Gloria
Lib 15h10'33" -21d47'22"
Gloria
Vir 14h14'48" -17d5'38"
Gloria 75
Umi 13h54'20" 72d45'28"
Gloria A. Berendse
Gem 6h53'9" 21d57'22"
Gloria A. Bivins, My Love
Leo 11h8'14" 26d10'1"
Gloria A Burns
Sco 16h41'33" -31d14'32"
Gloria A. Campos angel of sin
Sgr 19h11'55" -18d46'59"
Gloria Aiwasian
Cas 1h20'44" 57d58'20"
Gloria Amen
Crb 16h11'8" 30d40'48"
Gloria and Hank Brech
Cyg 20h14'40" 51d50'45"
Gloria and Nathan Reese
Vir 13h9'25" 2d56'52"
Gloria Andrea Murphy
Aqr 23h0'0" -9d53'51"
Gloria (Angel) King
Lyr 19h6'24" 47d13'30"
Gloria Angela Serrano
Ari 3h17'1" 27d59'59"
Gloria Ann Kuhns
Lyr 18h35'43" 37d15'9"
Gloria Ann Morra
Crb 15h37'22" 27d46'14"
Gloria Ann Simpson
Mon 7h32'16" -0d31'44"
Gloria Ann Wargo
And 23h24'1" 38d40'30"
Gloria Anne Cordova
Cnc 9h18'35" 8d32'39"
Gloria Anne Lavick
Uma 12h9'26" 49d30'34"
Gloria B. Casteel
Lib 15h15'54" -5d39'9"
Gloria B Gainer
Cap 20h46'50" -17d11'31"
Gloria B. Mouzon
Per 3h25'46" 36d53'32"
Gloria Berini
Cap 21h43'51" -8d58'58"
Gloria Blount
Mon 6h31'18" 6d34'46"
Gloria Brooke
Cas 23h29'31" 58d0'26"
Gloria C. Davis
Sco 16h8'46" -9d21'6"
Gloria Calcagno
Srp 18h12'6" -0d48'47"
Gloria Canlas
Lmi 10h26'20" 28d21'5"
Gloria Capdepon Bolin
And 0h43'9" 24d20'52"
Gloria Carroll Scollard
Sco 16h19'23" -11d36'7"
Gloria Cathrine
Ari 3h16'4" 28d29'31"
Gloria Choi
Cma 7h27'38" -15d16'59"
Gloria Coger
Leo 10h6'10" 22d15'55"
Gloria Cruz & Jaime Bautista's Baby
Lyr 19h15'54" 39d51'42"
Gloria Danna
Cas 0h52'37" 61d21'7"
Gloria Darlene Collins Riggle
Tau 4h2'34" 30d2'40"
Gloria DeNoi
Sgr 19h49'18" -22d2'57"
Gloria Diamond
Sgr 18h1'56" -32d2'19"
Gloria Dispenza Jones
Sco 17h56'21" -40d39'49"
Gloria Elizabeth Alston
Crb 16h11'15" 35d52'15"
Gloria Ellen
Uma 11h44'43" 44d35'32"
Gloria Faith Marie
Ari 1h52'30" 18d55'35"
Gloria Faye Hudnall
Per 3h53'27" 48d53'11"
Gloria Fontana Ruggeri
Sco 17h22'47" -41d48'22"
Gloria & Frank
Cyg 21h38'25" 50d39'52"
Gloria G. Burdette
Sgr 19h0'21" -24d13'28"
Gloria Garcia Nava
Cnv 12h40'56" 40d19'49"
Gloria Garcia von Kleinow
Lyn 8h20'5" 39d25'56"
Gloria Garing Massar
Cas 1h49'26" 64d38'20"
Gloria Gayden Corona
Cap 21h7'4" -24d53'0"
Gloria Gebhardt
Uma 9h54'42" 60d27'10"
Gloria Gill Hulsey
Cas 1h29'53" 63d10'1"
Gloria Godwin
Mon 7h19'27" -0d33'23"
Gloria Gonzalez
And 0h25'58" 43d1'46"
Gloria "Grandmom" DeLaski
Psc 1h1'25" 14d18'26"
Gloria - H & S Forever - LUT - CO
Cnc 9h8'8" 31d7'29"
Gloria Hailey 1st Love
Mon 6h57'41" -0d52'8"
Gloria Hamilton Toplitt
Uma 9h29'48" 66d59'6"
Gloria Hodge
Uma 10h12'29" 51d9'38"
Gloria Hulsey
Nor 16h19'27" -47d10'51"
Gloria I. Jones / Jan.21,1954
And 1h35'46" 48d18'6"
Gloria J. Camillo
Ori 5h20'33" 0d38'39"
Gloria J. Gail
Aqr 22h42'39" -0d58'23"
Gloria Jan 12/25
Uma 11h15'25" 30d34'57"
Gloria Jean
Tau 3h39'30" 16d8'29"
Gloria Jean
Cap 21h0'56" -18d45'39"
Gloria Jean Andersen
Psc 0h43'39" 14d53'55"
Gloria Jean Baird
Cap 20h55'1" -15d32'7"
Gloria Jean Boles Uphold
Gem 6h45'43" 28d24'15"
Gloria Jean Moloney
Lib 15h31'18" -27d11'42"
Gloria Jean Smith (Paul, Mack)
Aqr 20h43'56" 0d55'33"
Gloria & John Ricci
Uma 12h45'59" 58d7'11"
Gloria Johnson
Uma 9h58'16" 70d48'28"
Gloria Jon
Sco 16h11'0" -12d46'58"
Gloria Jones
Oph 16h24'52" -0d31'12"
Gloria Jones
And 0h12'15" 33d10'20"
Gloria K. Pitfido
Lib 14h50'43" -1d43'35"
Gloria Kirchoff
Lib 15h46'5" -18d3'24"
Gloria Larson
Gem 7h46'2" 27d51'27"
Gloria Lee
And 1h22'21" 48d0'6"
Gloria & Leo
Uma 10h36'6" 68d24'19"
Gloria Linda
Lib 15h46'53" -25d54'49"
Gloria Logan
Mon 6h31'19" 4d51'2"
Gloria Lou Donovan
Lib 14h50'29" -5d26'39"
Gloria Lynette
Ori 6h1'16" -0d42'14"
Gloria Lynn Shannon
Lmi 10h36'55" 35d40'4"
Gloria M. Butler
Cyg 21h15'25" 46d55'58"
Gloria Maira Cruz - BROWNIE
And 1h37'11" 43d11'34"
Gloria Margarita
Hya 9h1'14" -0d47'52"
Gloria Maria Gonzalez
Com 13h4'46" 18d22'27"
Gloria Martin
Aql 19h36'45" 7d43'20"
Gloria May Bass
Umi 14h30'1" 68d0'19"
Gloria McDermott Born: Apr 3/1926
Uma 11h32'58" 56d34'56"
Gloria McDonough
Cyg 19h49'5" 38d26'40"
Gloria Meza
Sgr 19h22'0" -18d10'38"
Gloria Mi Madre Querida
Aqr 22h37'41" -1d18'32"
Gloria Michelle Pena
Tau 3h40'21" 29d28'57"
Gloria Miller
Ari 3h6'18" 29d7'14"
Gloria Mina
Equ 21h13'42" 12d5'51"
Gloria Minerva
And 1h30'18" 48d45'23"
Gloria Mingone
Gem 6h34'0" 15d53'39"
Gloria Montoya
Cnc 8h37'34" 15d53'49"
Gloria My Love My Life My Destiny
Gem 7h22'22" 25d5'29"
GLORIA NAPPO
Cas 1h46'5" 64d58'35"
Gloria Norris Feldhusen
Crb 15h37'50" 38d25'57"
Gloria Northway
Tau 4h49'38" 23d1'18"
Gloria Occhiogrosso - My Lovely Mom
And 0h12'51" 27d45'25"
Gloria O'Donohoe
Tau 3h37'41" 16d53'6"
Gloria Patrica Hoff
Sco 17h54'49" -37d40'6"
Gloria Patrice
Cas 23h21'52" 57d10'8"
Gloria Patterson Presley
And 23h4'8" 52d18'41"
Gloria Perez
Lyr 18h35'48" 33d15'31"
Gloria Purcell
Sco 16h34'13" -43d10'18"
Gloria Purvis
Cas 23h27'55" 55d42'19"
Gloria R LaRose
Cas 0h46'46" 64d51'0"
Gloria Raphiel
Leo 10h32'59" 14d0'13"
Gloria Reid & Ronald L Sheard, Jr.
Sgr 19h27'31" -14d18'44"
Gloria Rowe
Crb 15h41'28" 26d50'53"
Gloria S. Evans
Vir 13h28'44" -14d9'22"
Gloria Silio
Cas 22h57'34" 57d21'51"
Gloria Sue Nelund
Tau 5h41'44" 17d7'13"
Gloria Sweet Angelo ~ Sept 28, 1928
Uma 11h41'28" 51d14'40"
Gloria Tetzlaff Kelly Brave & Kind
Uma 11h45'31" 60d46'16"
GLORIA - TITA
Cap 20h23'37" -15d32'1"
Gloria Tita Cardoza
Sco 16h51'49" -27d4'46"
Gloria Torres Rivera
Uma 12h57'54" 57d48'8"
Gloria Torres-Forastieri
And 23h23'16" 39d2'47"
Gloria Vandenberg
Cas 23h28'43" 57d48'23"
Gloria Victorine
Ori 5h58'0" 21d27'46"
Gloria Virtutis Umbra
Lyn 6h50'42" 52d56'35"
Gloria Yesenia Sanchez
And 0h19'39" 44d22'37"
Gloria Zataray
Gem 7h39'36" 24d0'10"
GloriaFrank
Vir 14h7'10" -16d20'14"
GloriaJustin
Uma 11h39'2" 51d4'5"
Glorialex
Cyg 21h34'46" 44d31'55"
Gloria-Lyn Bardugone
Gem 7h46'38" 33d20'3"
GloriaMouser
Cnc 8h34'49" 27d51'1"
Gloria's Butterfly
Sco 17h41'50" -41d11'46"
Gloria's Grace
Ori 5h34'24" 11d14'24"
Gloria's Loving Glow
Uma 9h7'38" 62d47'20"
Gloria's Wolf Star
Uma 11h21'43" 49d37'37"
Glorious Reagan
Psc 0h43'21" 14d28'10"
GLORITACUS
Leo 9h50'20" 7d12'33"
Glory
Uma 8h27'53" 68d25'1"
Glory
Cas 1h50'32" 63d29'28"
Glory B
Lyr 18h41'3" 34d56'4"
Glory Days 2005
Cam 4h17'4" 67d55'27"
Glory Jean
Cnc 8h48'3" 9d10'18"
Glory Nass
And 0h34'39" 36d2'22"
Glory of Love
And 1h27'8" 47d9'8"
Glorya Dawn Miles Reese
Peg 21h44'31" 13d43'50"
Glow
Lyn 6h42'4" 51d30'47"
Glowbabe72
Aqr 22h58'36" -15d21'10"
Glowing Ember
Lib 14h49'27" -18d34'19"
Glowing Gail & Donald Meredith
Cra 18h8'14" -37d5'30"
Glowing Guardians
Aqr 23h3'9" -15d11'9"
Glowing Kari
Gem 7h22'13" 31d53'44"
Glowing Leanna
Uma 8h17'45" 60d35'49"
Glowy
Uma 10h58'12" 61d36'6"
Glowy
Lyn 8h27'16" 45d11'19"
GLP3120
Psc 23h40'37" 6d16'27"
Glücksschtärn
Per 4h3'34" 47d42'48"
Glücksstern Barbara & Thomas
Uma 10h52'38" 41d29'2"
Glücksstern für Eva & Wolfgang
Ori 6h18'2" 14d24'53"
Glücksstern für Inge & Alfred
Uma 8h47'21" 61d36'4"
Glücksstern PETRA in Liebe Patrick
Ori 6h15'20" 15d40'25"
Glücksstern Stefan Katrin
Uma 8h47'36" 61d9'57"
Glücksstern von Anke und Matthias
Uma 12h2'37" 41d3'15"
Glücksstern von Sternchen & Bärchen
Uma 8h41'20" 54d28'17"

Gluecksstern fuer Lis und Christian
And 23h35'45" 43d6'48"
Glufke, Anna-Lena
Ori 6h21'16" 14d14'47"
Glug
Ari 2h57'23" 25d50'25"
Glusi, 05.10.1973
Umi 16h23'35" 77d37'39"
Glycerine
Sgr 19h6'55" -35d56'11"
Glyn
Peg 21h18'33" 17d11'41"
Glyn C. Roberts
Cnv 13h31'30" 36d49'20"
Glyn George Moore
Cnc 8h19'4" 13d44'26"
Glyn Stephen Reilly
Cep 21h40'43" 60d9'46"
Glyn Thomas
Per 4h6'30" 44d28'20"
Glynis Renee Brown
Lib 14h51'12" -3d21'32"
Glynis Shining Brightly Forever
Cyg 21h58'25" 55d7'4"
Glynis's Star
Cyg 21h8'35" 53d4'58"
Glynn & Julie & Family
Uma 8h41'12" 66d1'34"
Gma and Gpa Spencer
Lyn 7h26'14" 49d20'28"
G-Ma BevMo Beverly
Cyg 20h43'30" 39d35'38"
GMA LAURENE 1915 WLFTGC
And 0h25'2" 42d16'21"
G-man
Cnc 8h18'43" 9d27'29"
G-MART
Ari 2h15'6" 23d4'58"
GMCIMA 40
Nor 15h45'23" -59d55'18"
G.M.G. 02-01-1951 "Marty"
Aqr 21h1'45" -9d57'55"
GML Star
Cas 23h49'17" 57d13'9"
GML Strong Worthy Earthwarrior
Per 3h6'59" 51d55'2"
G.M.O. III
Cnc 9h9'47" 27d3'57"
GMO4Eternity
Aql 19h51'41" 9d6'41"
Gnanasothy
Umi 17h5'32" 76d33'24"
Gnarly Lisa
Sco 17h39'11" -40d8'29"
Gnartank
Lyn 8h53'48" 37d50'26"
Gnat's Guiding Light
Aql 18h54'54" -0d11'3"
Gneiss
Vir 12h16'15" -10d1'6"
gngpower2
Lyn 6h47'52" 52d52'44"
GNH
Per 4h14'26" 50d43'51"
Gnia
Uma 10h38'12" 55d42'20"
Gnoccola
And 2h38'12" 45d39'24"
Gnostic1
Lmi 11h3'50" 26d12'12"
Gnus Magnus
Sgr 18h29'1" -26d38'36"
Go Deo
Cap 21h39'52" -14d35'42"
Go Deo
Psc 0h53'47" 8d31'43"
GO DIAMOND
Lac 22h51'53" 52d31'2"
Go Mister In Perfect Time
Boo 15h22'46" 43d49'59"
Go Team
Aqr 22h15'18" -1d5'23"
Goad-hoshi
Tau 4h25'23" 21d21'10"
Gobbelonia
And 23h4'31" 47d13'38"
Göbel, Günther
Uma 11h39'39" 53d48'4"
Göbel, Jörg
Ori 5h52'5" 21d31'38"
Goblin
Cma 7h23'53" -30d14'29"
Gobon Dakit
Lyn 7h59'2" 58d58'11"
Gockel, Ralf
Uma 9h55'57" 65d6'25"
God Bless Mum
Lib 14h25'44" -15d21'34"
God Bless the Broken Road
Eri 4h33'30" -1d1'10"
God Bless The Broken Road
Tau 3h44'9" 21d44'2"
God bless the broken road
Lac 22h46'6" 50d47'7"
"god bless the broken road" tj & cb
Cyg 19h47'15" 35d10'30"
God Child of Mine
Tau 4h25'18" 12d57'13"
God is a Star
Lib 14h59'17" -18d15'17"
God Is Gracious & Worthy (GIGWOP)
Cet 2h3'59" -6d1'52"
God Loves Jack Arthur Bender
Cap 20h24'27" -11d29'37"
God*s child Goldie
Cyg 20h40'33" 41d42'54"
God Star Struck Me with Our Love
Sco 17h34'11" -40d18'34"
Godbaby Carolyn
Umi 14h59'22" 71d47'50"
GodbeTheGlory!Lovell+Connie=LK3&CEK
Cyg 21h35'44" 48d51'4"
Goddard Star ~ Kay
Cyg 19h30'3" 29d27'11"
Goddard Star ~ Les
Cyg 19h39'6" 32d33'52"
Goddess
Cyg 20h49'11" 31d44'7"
Goddess
Vir 12h8'16" -1d5'7"
Goddess Alexandria
Leo 10h49'50" 17d43'46"
Goddess April
Tau 4h14'39" 6d35'7"
Goddess April
Ari 2h5'1" 21d14'42"
Goddess Blessed
Lyr 18h53'29" 37d9'55"
Goddess Caitlin Gorden
Cap 21h44'42" -24d53'55"
Goddess Cathangela
Col 5h27'13" -31d48'2"
Goddess Christy
Cnc 8h42'30" 18d59'37"
Goddess Diane
Peg 22h15'47" 34d27'18"
Goddess Dina
Cas 0h53'52" 51d10'59"
Goddess Ellen
Leo 9h23'13" 7d20'21"
Goddess Izzie
And 0h25'11" 25d27'20"
Goddess Kalona
Cam 7h36'18" 66d43'46"
"Goddess" Lee Cimicata
Lib 14h52'43" -3d45'54"
Goddess Lene
Lyn 9h22'33" 38d16'18"
Goddess Linda
Aqr 21h42'43" -0d52'14"
Goddess Shona, Mother of Mason
Cnc 8h51'3" 13d19'38"
Goddess Terpsichore
Lib 15h10'49" -8d53'5"
Goddess Yael
Aqr 21h53'44" -8d21'35"
Goddesses Burnidge, Fetter, & Walsh
Cyg 19h48'55" 43d13'26"
GoddessTherese
And 0h16'46" 28d12'26"
Godelieve
Vir 13h11'45" 11d58'40"
Godess Booboo
Oph 17h51'20" -0d25'30"
Godfather Brian
Per 2h36'21" 54d29'42"
Godfather Louis Andreacchio
Her 17h56'54" 22d49'53"
Godfather Pants
Lyn 9h40'1" 40d17'9"
"Godfather Uncle Pete"
Uma 10h57'7" 41d15'49"
Godfrey
Lyn 7h47'11" 43d2'47"
Godfrey Goldsmith
Umi 13h19'42" 76d18'19"
Godfrey Joseph Sundmark
Sgr 18h11'6" -20d7'0"
Godfrey Pabo
Psc 0h57'52" 8d54'42"
Godfrey's Star
Uma 11h7'56" 31d16'13"
Godiva Guillory
Vir 14h39'35" 2d17'6"
godly paradise
Cap 20h15'57" -9d43'27"
Godmother & Aunt Blanche Deasy
Uma 12h9'12" 55d31'34"
Godmother J
Crb 15h36'27" 27d16'59"
God's Angel #13
Uma 10h40'41" 59d36'28"
God's Angel #14 Bales M. Brannon
Boo 15h20'27" 44d17'56"
God's Blessing
Lyn 8h0'50" 36d42'40"
God's Brightest Star, Trish
And 0h44'10" 36d3'8"
God's Child Amelia
Ori 5h23'22" 2d4'35"
God's Child; Ann R. Essenburg
And 1h13'47" 38d12'52"
God's Gift: Sisters-AmbyPattiLauren
And 0h46'38" 35d21'10"
Gods Gift to Jeanne
Ori 5h22'12" -0d14'15"
God's Gift, Amber
Sgr 19h20'4" -13d19'8"
God's Grace
Cyg 19h30'6" 29d57'3"
God's Harmony in Lana and Garet
Sge 19h37'53" 17d35'4"
God's Little Corner
Sco 16h39'8" -40d35'12"
God's Magic - Eric S. Jackson
Tau 4h14'10" 25d10'17"
God's New Angel Christopher Barrios
Umi 14h43'20" 73d36'1"
God's Special Angel Dan Bauer
Per 3h12'38" 54d8'20"
God's Special Angel ~ Heidi Cranmer
Lib 15h46'51" -11d10'31"
God's Special Angel Marie Mills
Sgr 19h29'34" -13d52'46"
God's Special Angel Marie Mills
Sgr 19h18'53" -13d56'3"
God's Special Angel ~ Richard Wolf
Leo 11h48'6" 11d36'17"
God's Sunshine
Ori 4h52'23" 11d46'1"
God's Sweetest Gift
Umi 15h28'11" 72d4'32"
God's Warrior - Adam Barr
Aur 5h16'9" 41d59'5"
God's Wenis
Lib 15h6'4" -8d15'25"
God's White Light-rs
Cnc 9h6'43" 12d57'46"
Godson - Oliver Robert O'Brien
Ori 5h12'6" 8d6'28"
Godzilla
Umi 15h44'56" 76d23'5"
Goettibueb Cedric Dal Balcon
Dra 16h48'33" 55d59'11"
Go-Gab for Gloria Pena
Uma 9h40'4" 66d30'27"
Gogegi
Cas 0h38'8" 58d9'13"
Gogi
Tri 2h26'8" 31d9'22"
GOGO
Uma 9h25'27" 69d6'2"
Goh Shiao En & Peng KangZhen
Tau 4h0'37" 8d44'53"
Gohar Pervez
Uma 9h27'51" 44d53'55"
Göhring, Ralf
Ori 6h6'45" 9d34'52"
GOIK
Cam 9h13'53" 78d54'46"
Gök, Zeliha
Uma 12h47'45" 53d44'55"
Gokben-Can
Sge 19h10'18" 18d58'41"
GÖKCE YANARDAG
Vir 13h14'21" -7d27'59"
Gokmen Gul
Cap 21h6'41" -16d15'22"
Golbahar
Leo 9h39'4" 32d16'14"
Gold Star Helen
Sco 17h46'51" -42d42'8"
Gold Star of Pat and Fred Robson
Sgr 20h28'29" -44d43'52"
Gold Star-Liz Tupper
Mon 7h52'49" -11d11'38"
Gold Star-Mark Ness
Per 2h45'28" 51d57'55"
Gold Star-Megan Jones
Cyg 21h45'43" 33d14'57"
Gold Star-Pat Pevan
Umi 18h38'26" 87d53'29"
Gold Star-Stefani Link
Lyn 7h8'13" 59d50'47"
Gold Star-Troy Lightfield
Uma 13h41'43" 51d3'54"
Gold William & Phyllis Pulver Star
Cyg 19h37'53" 30d45'30"
Golda Bear
Cyg 20h35'54" 52d40'6"
Golda Belle Allen Laas Morrison
And 23h2'33" 51d34'27"
Goldammer, Joshua
Ari 3h12'1" 18d51'33"
Golden
And 0h39'31" 23d25'21"
Golden Angel
Uma 13h17'42" 57d45'58"
Golden Anniversary - Jamal & Saba
Cru 12h23'50" -61d23'31"
Golden Ba & Dada
Lib 14h43'58" -22d58'6"
Golden Bhoy Chad Williams
Aql 19h19'0" 3d12'46"
Golden Boy
Uma 9h44'57" 49d12'19"
Golden Dream
Sge 19h41'55" 17d23'48"
Golden Glo
Gem 7h9'43" 16d24'15"
Golden Laineypie
Uma 11h29'17" 48d23'6"
Golden Lovely
Tau 4h12'3" 24d55'14"
Golden One's 30th
Vir 14h26'52" 3d13'47"
Golden Rose Loves Her ZRX Warrior
And 1h52'38" 45d43'51"
Golden Sands Hotel
Peg 22h3'26" 34d23'29"
Golden Son Jeremy
Gem 7h22'33" 30d12'12"
Golden Sunshine-XXV
Uma 9h24'48" 61d56'10"
Golden Valentine
Tau 3h27'8" -0d55'18"
Goldenrush Lindsay Luvbug
Cas 0h1'29" 53d52'54"
Goldenrush Softspoken Stephanie
Cas 1h14'7" 58d38'41"
Goldenyokko
Uma 11h13'26" 37d8'59"
Goldfish
Nor 16h6'1" -43d7'23"
Goldi
Gem 7h26'31" 28d22'56"
Goldie
Leo 11h35'38" 25d5'19"
Goldie
Equ 21h18'36" 8d24'16"
Goldie
Uma 9h42'14" 46d53'44"
Goldie
Cma 6h46'19" -28d31'24"
Goldie
Sco 16h9'0" -16d35'30"
Goldie
Uma 11h34'22" 56d9'28"
Goldie C. Fox
Vul 19h25'33" 26d25'18"
Goldie Forman
Her 18h32'20" 12d57'20"
Goldie K. Alvis
Dra 18h27'52" 77d15'0"
Goldie Levin
Uma 13h30'46" 55d30'8"
Goldie Mae Heaton
Crb 15h42'55" 27d4'6"
Goldie McCann
Pup 8h14'32" -21d57'50"
Goldie Olive Abrahams
Lib 15h17'51" -9d28'11"
Goldie The Kid Brown
Sgr 18h50'55" -20d35'3"
Goldie Virginia Benton King
Crb 15h51'59" 29d25'50"
Goldie & Yetta "A Twin Treasure"
Aqr 21h15'42" -8d59'44"
Goldielox3
Vir 13h11'25" 12d57'58"
Goldilocks
Dra 20h22'11" 67d14'13"
Goldin's Beacon
Tau 4h9'13" 15d40'13"
Göldner, Klaus-Reinhard
Lib 15h15'27" -10d34'44"
Goldorak
Cas 1h13'18" 68d41'36"
Goldsmith
Vir 12h42'6" 7d0'51"
Goldy
Ari 3h19'49" 29d16'25"
goldy33
Her 16h43'44" 17d55'27"
Golebiewski, Franz
Ori 5h2'8" 6d36'20"
Golfer's Paradise
Her 17h47'53" 42d39'9"
Goliath
Her 16h49'30" 35d13'39"
Golou Yang
Cap 21h40'42" -9d20'12"
Goma & Andreas
Tri 1h47'28" 30d3'18"
Gomathy Krishnamurthy
Lib 15h7'46" -4d10'9"
Gommy
Umi 15h42'44" 81d18'29"
Gommy/Pops
Crb 15h51'38" 34d39'52"
Gonçalo Alexiades Gonçalves Cerejeira
Ori 6h2'3" 20d55'33"
Goncalves
Cas 1h11'23" 65d46'54"
Gone Fishing
Tau 3h34'24" 24d59'32"
gongee doodle britches
Lyn 7h23'6" 44d39'28"
Gonne Leile Cotant
Cas 1h10'29" 65d50'47"
Gonsior, Andrea
Uma 10h53'22" 42d27'42"
Gonter
Sgr 18h15'33" -17d49'45"
Gontman
Leo 9h54'3" 9d46'8"
Gonzales Sierra Star
Crb 16h18'24" 38d40'6"
Gonzalo & Janice and Family
Uma 12h45'3" 54d40'58"
Gonzalo Mendez!!!
Cyg 21h29'23" 34d5'51"
Gonzalo Morales
Cma 7h12'50" -30d54'16"
Gonzalo Naval
Lib 15h27'20" -3d46'55"
Gonzalo TetúGptaEeedmcptpsDtgTG TaK
Vir 12h20'54" 11d37'23"
Gonzamel - My Light and Love
Gem 6h22'52" 18d27'25"
Gonzo
Vir 13h3'41" -11d14'6"
Goob
Vir 12h59'0" -12d33'49"
Goobanova
Per 4h6'53" 39d10'6"
Goober
Per 2h25'1" 55d49'44"
Goober
Tau 4h27'28" 23d44'31"
Goober
Leo 11h2'15" 3d27'25"
Goober
Leo 11h17'2" -2d42'32"
Goober
Dra 17h51'49" 59d33'30"
GooberFace
Lib 14h51'55" -16d53'25"
Goobie
Cap 21h29'16" -13d55'36"
"Gooby" - 1966
Cep 21h42'17" 56d47'10"
Goobybeana
Peg 23h20'55" 28d29'5"
Good Boy Brutus
Umi 13h37'40" 78d49'20"
Good Boy Davis
Umi 16h15'16" 74d50'47"
Good Boy, King!
Sgr 18h22'45" -30d45'31"
Good Charlotte
Lib 15h27'41" -4d37'27"
Good for Nothin
Dra 16h45'28" 66d38'31"
Good Morning Glory
Ari 2h21'36" 18d35'11"
Good Morning Sunshine
Psc 1h18'45" 12d36'15"
Good night moon
Aql 20h11'47" -0d21'44"
Good Son
Her 18h4'20" 22d7'14"
Good Together
Uma 10h51'31" 39d47'58"
Good Witch Barbara
Uma 12h3'54" 39d30'11"
Goodheart
Psc 1h36'26" 10d44'17"
GOODLOE'S GALAXY
Tau 4h32'46" 17d36'46"
Goodmorningbeautiful
Uma 13h46'33" 53d40'10"
Goodnight Cute Girl
Cap 21h19'14" -26d37'11"
Goodnight Elizabeth
And 1h37'47" 49d14'44"
Goodnight Irene
Gem 6h57'19" 35d2'35"
Goodnight Mark
Dra 17h39'4" 57d45'1"
Goodnight-Randall
And 2h1'12" 38d9'10"
GOODS
Leo 9h42'58" 26d37'20"
Goodwin
Psc 0h13'40" 0d42'38"
Goodwyn
Cyg 20h19'18" 47d56'57"
Goof
Gem 6h54'49" 24d12'1"
Goofety Goofball (Nathan)
Ori 5h32'29" 14d1'47"
Goofus and Gallant
Uma 8h35'35" 63d52'41"
Goofy
Leo 11h7'48" 23d53'30"
Goog
Ori 5h30'8" -0d0'39"
Google Infinity
Uma 8h25'42" 64d1'25"
Goomie's Heart
Vir 12h15'52" 12d6'13"
Goop
Sco 16h48'41" -32d51'21"
Goorgen
Cam 4h45'48" 63d57'1"
Goose
Sco 17h19'55" -43d32'23"
Goose
Ori 6h12'16" 5d42'46"
Goose
Lyn 8h15'2" 39d52'47"
Goose
Cyg 19h52'2" 33d8'54"
Goose and Monkey
Ori 4h50'42" 11d37'7"
Goose Love
Sgr 18h20'3" -24d32'57"
Gooseberry
Umi 14h36'54" 71d17'52"
Goosefish
Cyg 21h43'19" 43d57'59"
Goose's Goal
Sco 16h57'17" -39d3'18"
Goosey
Cas 0h25'33" 63d2'26"
Gootie
Leo 10h24'38" 26d32'9"
Gopi & Kate's Star
Gem 6h50'58" 34d24'28"
Göran Olsson
Uma 8h40'28" 51d29'24"
Goran Pekic
Peg 21h54'48" 17d55'23"
Göran Radalewski
Uma 8h52'50" 69d33'8"
Goranitestar
And 1h20'31" 47d10'40"
Gord Mar, 50th Anniversary
Uma 8h26'36" 62d17'11"
Gordan Ray Powers
Sco 17h3'48" -43d44'48"
Gorden Konieczek
Uma 14h28'22" 58d11'43"
Gordie
Lib 15h20'5" -27d23'5"
Gordie
Psc 1h13'32" 28d33'50"
Gordie Dad Williams
Her 16h57'55" 33d48'18"
Gordo
Tau 5h10'47" 23d37'4"
Gordo Red 25
Uma 11h3'16" 54d43'25"
Gordon
Cap 20h57'8" -21d3'2"
Gordon A. Morrisette
Per 2h42'35" 54d49'48"
Gordon Alfred Russell
Cru 12h45'58" -64d16'22"
Gordon & Alisa Gray 25 Anniversary
Umi 15h37'24" 74d52'41"
Gordon and June Meekin
Uma 8h50'6" 53d1'28"
Gordon and Mathilda Walker
Uma 9h42'0" 49d56'37"
Gordon and Ruth's courting star
Ori 5h25'8" -3d37'46"
Gordon Avery Pollock
Leo 9h30'36" 28d47'3"
Gordon Bassham
Aql 20h10'24" 6d24'18"
Gordon & Beverly Blevins
Uma 10h52'19" 55d17'26"
Gordon Bluestone
Aql 19h47'20" 11d23'46"
Gordon Brent Barrus
Per 3h51'24" 37d23'11"
Gordon Byron Hilty
Sco 16h31'1" -26d11'48"
Gordon C. Davis, Jr.
Cnc 8h8'26" 14d38'13"
Gordon C. Holtzer 12/15/1983
Uma 10h36'17" 59d56'10"
Gordon & Carole Burton united 1976
Umi 15h47'35" 74d32'55"
Gordon & Catherine Seaman
Ori 5h25'1" 1d55'38"
Gordon D
Crb 15h37'2" 31d23'15"
Gordon D. Smith
Her 16h41'3" 44d52'46"
Gordon D. Stock
Per 2h18'56" 54d39'15"
Gordon "Dad" Thorne, Sr.
Psc 1h9'52" 24d29'0"
Gordon Davenport
Oph 17h46'28" -0d20'22"
Gordon Dax Sturrock
Uma 9h10'53" 70d5'23"
Gordon Dickinson
Cep 21h52'24" 71d18'2"
Gordon Douglas Lowrie
Her 16h32'47" 45d34'5"
Gordon E. Henley
Boo 15h24'37" 38d14'14"
Gordon F. Ruddy
Vir 12h59'6" -7d53'13"
Gordon Franklin Campbell Jr.
Uma 11h23'13" 44d59'24"
Gordon Frederick Drage
Ori 5h12'6" 15d14'46"
Gordon George Mitchell, Jr.
Aql 19h6'50" -0d45'3"
Gordon GG Wong
Leo 11h17'40" 10d26'18"
Gordon Herget
Gem 7h26'8" 32d9'53"
Gordon John Bennett
Sco 16h50'29" -42d56'5"
Gordon John Burridge
Sgr 17h54'55" -29d45'13"
Gordon K Loucks, Jr.
Psc 0h22'21" 16d50'15"
Gordon & Kiiko Nehls 03/11/53
Psc 0h29'22" 5d13'10"
Gordon Lamberton
Umi 14h10'35" 69d40'55"
Gordon Lane
Aql 20h8'51" 7d51'36"
Gordon Lawitzke
Boo 14h52'0" 22d19'20"
Gordon Lee Coker Gilliam
Sco 16h5'58" -16d40'33"
Gordon Lloyd - One in a Million
Uma 8h12'8" 71d9'9"
Gordon Luong & Qiaoyue Hu
Uma 10h6'30" 56d6'22"
Gordon Marnoch
Cep 23h13'7" 73d49'18"
Gordon My Love
Her 16h35'20" 31d46'56"
Gordon "n" Chirstie
Peg 22h37'58" 7d52'40"
Gordon of Greenlake
Cas 1h43'21" 60d47'51"
Gordon Olson
Uma 12h9'19" 48d10'54"
Gordon Pasiciel
Ari 3h13'38" 28d20'32"
Gordon Philip Leeney
Her 16h52'28" 19d59'34"
Gordon R Barratt
Her 16h4'47" 16d57'16"
Gordon Ransome Padfield
Ori 5h39'12" 2d4'56"
Gordon Richard Carroll, Sr.
Uma 10h44'21" 71d2'13"
Gordon Richey
Her 17h20'40" 35d10'55"
Gordon Robert Conti Holton
Cap 20h32'12" -13d42'35"
Gordon & Rosalie MacDonald
Lyr 18h55'48" 33d24'52"
Gordon Russell Miller
Gem 7h39'55" 29d41'24"
Gordon Samuel Brown
Vir 14h49'43" 3d56'2"
Gordon Scott
Her 17h38'53" 25d26'16"
Gordon "Simply the Best" Hanson
Psc 23h11'0" -0d39'56"
Gordon- still shining at 60-Ruberg
Ori 5h31'26" 2d9'57"
Gordon Syverson
Per 3h13'36" 51d21'26"
Gordon Thomas James Taylor
Cru 12h36'7" -58d20'45"
Gordon Thomas Morgon b.1927
Uma 11h23'6" 40d11'50"
Gordon V. Hodde
Aql 19h51'23" -0d19'9"
Gordon W. Chase
Gem 6h32'29" 15d43'36"
Gordon Wendel
Cap 21h8'50" -16d9'25"
Gordon Woo
Cap 20h11'38" -17d37'45"
Gordon's Gazer
Cnc 8h16'59" 20d3'32"
Gordy
Tau 4h9'37" 4d28'54"
Gordy
Cet 2h14'50" -0d9'23"
Gordy
Cep 21h43'31" 62d15'19"
Gordy Sjelin
Per 4h15'48" 50d16'12"
Gorecki
Uma 10h32'23" 60d56'14"
Goretti Silva
Sco 17h58'29" -30d44'11"
Gorg
Cap 21h54'6" -9d13'48"
Görg, Thomas und Simone
Uma 10h18'44" 39d55'29"
Gorga Girl -Neenie's Star For Indie
Cru 12h19'17" -56d22'28"
Gorge
And 0h38'13" 23d53'11"
Gorgeous
Crb 15h36'23" 27d34'59"
Gorgeous
Cap 21h44'33" -14d41'0"
Gorgeous
Sgr 17h57'56" -18d47'19"
Gorgeous
Vir 13h25'45" -17d47'55"
Gorgeous
Aqr 21h49'32" -6d43'16"

Gorgeous
Vir 13h42'47" -1d21'56"
Gorgeous
Umi 13h56'3" 70d29'7"
Gorgeous Coincidence
Cyg 20h2'29" 33d56'28"
Gorgeous Debra Stones
Lmi 10h12'40" 28d58'15"
Gorgeous Edward L Hoener
Sco 17h55'9" -42d49'9"
Gorgeous Ellina
Cas 0h42'21" 65d50'54"
Gorgeous Genna
Lib 15h8'25" -4d18'18"
Gorgeous Grace
And 1h38'24" 42d25'21"
Gorgeous Grace
And 23h16'37" 51d31'13"
Gorgeous Graham
Uma 9h49'41" 51d26'25"
Gorgeous Jennifer Green
Uma 13h39'42" 49d4'57"
Gorgeous Jeri 65
Lyn 7h41'8" 49d56'49"
Gorgeous Krystin Johnson
And 0h44'7" 37d29'54"
Gorgeous Laura
And 0h21'47" 24d42'3"
Gorgeous Melinda Jayne
Vir 12h12'31" -0d54'16"
Gorgeous Michelle
Leo 11h27'59" 15d29'44"
Gorgeous Mr Thompson
Cyg 20h7'2" 32d23'9"
Gorgeous Reina N Y Cheng
Sgr 19h27'3" -35d4'50"
Gorgeous Sarah's Sparkle
Uma 9h32'43" 44d8'11"
(Gorgeous) Sherry Blackwell
And 23h9'58" 41d45'38"
Gorgeous Simone Coleman
Leo 11h3'28" 14d35'45"
Gorgeous Una Sang
Cra 18h26'47" -37d11'39"
Gorgeous X Tina
And 0h46'54" 23d34'10"
Gorgeous, forever & for always
Tau 5h47'19" 15d27'51"
Gorgey Love Muffy
Uma 11h15'13" 34d47'49"
Gorguito
Cap 20h48'12" -19d49'54"
Gorina Gor
Cas 1h53'39" 60d45'59"
Göring, Rita
Uma 13h47'47" 57d8'59"
Gorjez Walker
Aqr 21h49'33" -2d4'34"
Gorondolain "10-22-06"
Mon 6h33'0" 4d45'14"
Gorp
Cam 4h32'36" 66d53'24"
G.O.S. (George Oliver Skuse)
Umi 15h0'1" 77d5'11"
Gosia
Dra 17h55'22" 63d34'53"
Gosienka
Cap 21h11'55" -19d15'56"
Gosselin
Boo 15h35'14" 45d20'44"
Gossner Gertraud
Cam 5h55'46" 70d48'4"
Gotchu
Peg 21h34'21" 24d30'40"
GOTHA
Lyn 9h18'25" 37d39'19"
Gothic Gal
Psc 0h1'55" 8d53'19"
Gotkis
Mon 7h32'33" -1d0'59"
Gottenkind Laura Mumenthaler
Per 2h13'29" 54d25'28"
Gottfried Linss
Ari 3h20'54" 28d40'26"
Gotti Chrissi's Glücksstern
Uma 10h37'53" 64d12'24"
Gottschalk, Michael
Uma 9h44'37" 48d32'40"
Gottschalk, Steffen
Ori 5h5'48" 5d59'40"
Götz Lutterbey
Uma 8h20'50" 61d46'7"
Götz, Alfred
Uma 11h32'53" 39d40'7"
Götz, Hans-Norbert
Uma 11h51'28" 48d49'45"
Götze, Marko
Uma 8h50'8" 62d10'6"
Götzi
Uma 9h14'12" 58d16'39"
Goudschaal, Ude
Ori 5h36'29" -1d54'35"
Gouelle Ludovic
Tau 5h15'56" 20d39'46"
Govercia Family
Uma 11h37'53" 62d38'16"
Goy
Psc 0h16'22" 17d11'29"
Goz's Star
Cru 12h12'20" -57d25'15"
GP Ernie and GM Judy Hall
Cyg 20h21'13" 46d27'21"
GP&ROS
And 0h22'27" 46d10'9"
Gpa
Tau 4h37'14" 2d10'25"
Gpa Mikosz
Cam 3h56'31" 67d44'0"
G'pa Robert Fink
Cam 3h56'47" 66d58'33"
GpaDave
Ari 2h58'24" 18d54'7"
GPE Poggiaspalla
Leo 10h59'13" 7d11'23"
GPLBAWLPP
And 23h42'3" 41d40'39"
G.P.P.- My Little Rockstar
Cas 23h43'23" 54d49'4"
GQ
Vir 13h8'2" -16d54'46"
Gr Grandmas Marie Gladys Helen Baba
And 1h37'59" 45d28'42"
gra go deo
Ori 5h59'34" -0d45'21"
Gra Go Deo - W+K
Cyg 20h34'0" 35d16'8"
Gra Mo Chori
Tau 5h35'51" 27d23'53"
Grabowski
Psc 0h48'39" 16d2'51"
GRACE
Cnc 8h9'45" 7d18'20"
Grace
Vir 12h37'18" 2d51'51"
Grace
Gem 7h17'58" 24d59'12"
Grace
And 0h18'12" 31d41'42"
Grace
And 0h18'25" 31d48'4"
Grace
Leo 9h43'6" 25d58'33"
Grace
Ari 2h18'13" 24d59'20"
Grace
And 23h5'11" 35d42'16"
Grace
And 23h33'22" 36d7'34"
Grace
And 0h48'40" 41d14'53"
Grace
Uma 11h40'58" 41d45'53"
Grace
And 2h21'46" 42d0'55"
Grace
Uma 11h24'38" 37d24'48"
Grace
Lyr 18h44'7" 34d5'50"
Grace
Lyr 18h47'46" 31d25'42"
Grace
And 0h49'8" 36d39'49"
Grace
And 2h12'46" 51d2'57"
Grace
Cas 0h29'12" 51d52'2"
Grace
Umi 16h18'4" 80d48'58"
Grace
Psc 23h24'20" -3d11'14"
Grace
Sgr 17h59'3" -19d57'36"
GRACE
Sco 16h13'36" -10d49'53"
Grace
Uma 11h24'5" 58d1'24"
Grace
Cam 3h59'43" 64d18'19"
Grace
Uma 11h43'11" 60d28'13"
Grace
Cru 12h44'39" -57d53'31"
Grace
Cru 12h12'7" -62d5'39"
Grace
Sco 17h11'58" -32d37'24"
Grace
Sgr 19h29'41" -24d1'5"
GRACE 101
And 2h24'43" 46d10'39"
Grace - 2005 & Lilly - 2007
Cru 12h28'11" -60d48'30"
Grace 5
Lib 14h31'2" -18d58'50"
Grace "a heartbeat at my feet"
Cmi 7h47'41" 1d9'37"
Grace A. Hill
Peg 23h13'5" 17d50'8"
Grace A. Hudson
Crb 15h55'22" 30d37'37"
Grace Abigail
Crb 15h42'50" 32d27'52"
Grace Aimee
Psc 0h37'44" 12d27'29"
Grace Alcaraz
And 23h50'34" 42d33'39"
Grace Alden Pendleton
Uma 12h29'31" 59d57'49"
Grace Alex Khowaylo
Cas 23h58'48" 52d41'57"
Grace Alexandra Bowling
Lyn 8h19'46" 54d30'17"
Grace Alexandra Classon
Aqr 20h46'18" -14d19'29"
Grace Alice
And 1h16'14" 42d59'33"
Grace Alivia
And 0h1'43" 44d23'27"
Grace Amber Loretto
Ori 5h52'11" 18d42'44"
Grace Ana Emanoff
Cas 1h49'28" 64d56'7"
Grace Anca
Cas 0h42'36" 66d35'57"
Grace and Cyril
Uma 8h41'14" 56d15'35"
Grace and Erin's Special Star
Pav 18h34'23" -66d42'43"
Grace and Joel
And 1h32'32" 47d15'20"
Grace and Scott March 6, 2005
Uma 10h53'13" 39d53'41"
Grace and Travis
Leo 10h5'40" 25d24'25"
Grace Ann
Gem 6h45'50" 17d6'0"
Grace Ann
Aqr 20h58'6" 1d42'47"
Grace Ann
Sgr 19h43'4" -17d4'29"
Grace Ann Babcock-Taylor
Cas 2h23'54" 63d32'31"
Grace Ann & Clare Marie
Cyg 20h57'37" 37d39'32"
Grace Ann Dakis
Her 17h52'38" 31d13'25"
Grace Ann Diorio
Aqr 21h24'44" -0d6'35"
Grace Ann Goodman
And 1h2'6" 39d52'38"
Grace Ann Hlavaty
And 23h13'59" 43d58'5"
Grace Ann Ivancic
And 0h3'32" 44d30'25"
Grace Ann LaConte
Tau 5h1'15" 16d20'44"
Grace Ann Mancourt
Lyn 6h27'27" 60d48'53"
Grace Ann Matthews
And 2h14'32" 42d15'25"
Grace Ann Melillo
Lyn 8h28'48" 33d55'56"
Grace Ann O'Halloran
Aqr 22h13'51" 1d6'32"
Grace Ann Quartararo
Lib 14h57'19" -18d29'41"
Grace Ann Rolf
And 0h36'23" 43d40'27"
Grace Ann Taylor
And 1h27'55" 38d15'10"
Grace Ann Tennent
And 0h16'19" 41d0'54"
Grace Ann Wood
Peg 22h47'35" 14d2'20"
Grace Anna
Psc 0h56'1" 31d29'52"
Grace Anna Caccamo
Aqr 21h4'51" 1d30'34"
Grace Anna Schwarte
And 23h23'39" 42d53'17"
Grace Anna Warmington
And 0h2'58" 32d9'57"
Grace Anne Erwin
And 0h23'15" 31d11'57"
Grace Anne Wilson
Gem 6h36'20" 12d2'17"
Grace Annette Conway
Gem 7h24'38" 28d37'12"
Grace Armstrong
Vir 13h58'31" -8d21'12"
Grace Ash
Lib 15h16'1" -19d12'24"
Grace Aykroyd
And 2h14'38" 49d11'9"
Grace B. Levkovich
Cas 1h28'45" 57d44'43"
Grace Beatrice Reed
Cnc 9h8'47" 25d35'26"
Grace Bestland
Psc 1h15'55" 19d3'20"
Grace (Betty) Flanders
Lyr 19h10'59" 45d17'13"
Grace Borrows - Grace's Star
And 2h35'29" 39d21'32"
Grace Bryant
Cas 1h35'2" 60d59'12"
Grace Camille Tilyou
Ori 5h27'10" 3d21'40"
Grace Candida
Lyn 7h27'8" 56d54'7"
Grace Carbo
Cam 7h43'59" 67d46'39"
Grace Caroline
And 1h13'45" 36d8'58"
Grace Catherine Chiriatti
Gem 7h48'16" 25d40'12"
Grace Catherine Cunning
Ori 5h14'13" -7d22'26"
Grace Catherine Hawk
And 1h15'23" 43d24'52"
Grace Catherine McDaniel
Uma 10h42'0" 61d37'42"
Grace Catherine Yaklin
And 1h32'55" 50d6'14"
Grace Cerbone
Cnc 9h7'56" 14d52'58"
Grace Cheryl
Sco 17h48'51" -35d50'40"
Grace Clifford
Peg 22h31'34" 16d14'7"
Grace Coles
Uma 10h56'45" 49d11'53"
Grace Coogan
Leo 11h1'37" 10d0'56"
Grace Crawford
Aqr 21h41'1" 1d9'39"
Grace Crossley
And 1h18'1" 33d54'52"
Grace Crowley
And 1h41'27" 49d22'15"
Grace Curtis Debow
Lyn 9h17'42" 34d24'52"
Grace Daloisio
Crb 16h7'39" 38d40'4"
Grace De la cruz
Vir 13h27'19" 13d3'1"
Grace Dempster
Ori 5h26'51" -0d3'38"
Grace Diana Hobby
Umi 14h11'12" 73d43'24"
Grace Diva Matranga
Tau 4h19'55" 30d17'47"
Grace Drewes
Cnc 8h45'18" 28d5'26"
Grace Dudley
Sco 17h0'58" -34d30'31"
Grace E. Jacobs
Gem 7h39'46" 26d17'33"
Grace E. McFadden
And 0h44'59" 39d36'34"
Grace E. Smith
Gem 6h54'42" 21d51'39"
Grace Elaine Jeffers
Peg 21h42'46" 7d19'24"
Grace Eleanor
And 0h28'26" 25d10'1"
GRACE ELISABETH
Ari 3h12'41" 28d34'8"
Grace Elizabeth
And 0h27'2" 41d49'44"
Grace Elizabeth
And 1h32'43" 49d29'17"
Grace Elizabeth
Uma 9h27'42" 54d48'31"
Grace Elizabeth
Cas 1h9'46" 63d27'39"
Grace Elizabeth Amos
Aqr 22h7'1" -7d40'13"
Grace Elizabeth Attiyeh
Tau 4h2'11" 21d38'27"
Grace Elizabeth Austin
Vir 12h51'59" 8d44'1"
Grace Elizabeth Burgin
Lyn 7h4'30" 59d3'43"
Grace Elizabeth Cline Giblin
Cas 23h11'49" 56d2'22"
Grace Elizabeth Griggs
Uma 11h11'18" 35d23'9"
Grace Elizabeth Kelley
Vir 12h46'33" 5d57'57"
Grace Elizabeth Lachnit
Per 4h1'31" 43d47'34"
Grace Elizabeth Lahrmer
Cas 23h57'58" 56d44'17"
Grace Elizabeth Mahon
Crb 16h11'20" 28d16'25"
Grace Elizabeth Mandy
And 23h54'0" 33d50'33"
Grace Elizabeth Phelps
Tau 4h21'50" 18d36'11"
Grace Elizabeth Poppy Farnan
Tau 5h8'10" 24d47'2"
Grace Elizabeth Rood
Uma 11h43'52" 51d14'33"
Grace Elizabeth Salado
And 0h50'9" 39d36'38"
Grace Elizabeth Terry
Ori 5h13'10" 3d46'7"
Grace Elizabeth Villa
Lyn 8h0'31" 43d5'34"
Grace Elizabeth's Light
Uma 10h32'47" 48d56'5"
Grace Ellen Bordt
And 23h40'53" 46d37'16"
Grace Ellen Holmes
Psc 1h32'34" 15d4'7"
Grace Ellen Hutton
And 0h44'41" 42d30'51"
Grace Ellynn
Dra 20h17'24" 67d2'57"
Grace Emily Elliott
Cyg 19h57'3" 53d55'29"
Grace Emily Isabel See
Cas 23h35'18" 53d58'2"
Grace Evan Hernick
Cnc 8h51'58" 29d39'48"
Grace Evelyn
Cam 6h12'57" 78d46'58"
Grace Evelyn Bethray
Cru 12h12'23" -62d12'16"
Grace Evelyn Campbell
Vir 12h31'39" -10d27'5"
Grace Evelyn Homewood
Vir 13h21'18" -20d42'11"
Grace Evelyn O'Neill
Cas 1h0'49" 66d11'42"
Grace Evelyn Young
And 2h6'52" 39d7'14"
Grace Farrell
Uma 8h46'0" 68d17'44"
Grace Ferrara
Ori 5h15'49" 9d14'56"
Grace Fitzpatrick
And 23h48'12" 38d38'39"
Grace Frances
Crb 15h49'10" 37d55'18"
Grace Frances Nolfi Birth Star
Cnc 9h13'33" 11d44'15"
Grace G. Guevara
Lib 15h5'35" -1d7'53"
Grace Gardner
Cam 7h59'10" 69d41'34"
Grace Gaylord
And 0h44'12" 38d12'41"
Grace Geiger
Cas 1h38'26" 64d38'6"
Grace George Alzaibak
Gem 7h0'36" 29d35'20"
Grace Gibbs
Gem 7h25'7" 21d26'0"
Grace Giovanna Westol
Crb 15h37'20" 28d4'38"
Grace Goddard
Leo 11h16'22" 16d45'34"
Grace Grey
Cas 0h11'47" 63d39'21"
Grace Hamilton
Leo 10h24'50" 22d33'10"
Grace Hannah Kearney
Tau 4h10'32" 21d42'5"
Grace Harlow
Cas 0h15'48" 63d54'32"
Grace Hart
Cnc 7h59'7" 15d48'52"
Grace Heather
Psc 1h6'1" 31d37'8"
Grace Helen Hughes
Lib 15h25'50" -7d29'45"
Grace Heninger
Ari 2h38'48" 19d57'14"
Grace Holmes
And 2h10'7" 47d13'46"
Grace Imogen
And 0h58'9" 44d25'42"
Grace Imogen Purslove
And 23h32'52" 50d38'40"
Grace Isabella
Cas 1h40'30" 61d9'27"
Grace Isabella Wood
Uma 9h51'10" 55d11'35"
Grace J. Mennona
Cyg 20h9'47" 52d18'18"
Grace Javillo Mones
Cap 21h14'8" -26d34'6"
Grace Jessica Lombardo
And 23h27'6" 48d14'35"
Grace Jiexue Gerwin Gillach
Uma 9h38'56" 60d39'15"
Grace Joan Broderick
And 22h59'27" 41d22'28"
Grace & Jocelyn Stieb
Gem 7h25'14" 17d52'33"
Grace Johnson
And 1h3'30" 42d52'38"
Grace Kelly
Aqr 20h44'40" -8d53'23"
Grace Kelly Thornton
And 0h46'11" 25d28'17"
Grace Kennedy Minor
Uma 11h7'20" 61d18'12"
Grace Keough
Aql 19h46'46" -0d16'39"
Grace King Harris
Psc 23h56'53" -0d36'29"
Grace Kiyo Okazaki Inouye
Cyg 20h32'22" 52d54'14"
Grace Kleinman
Cas 2h1'5" 73d28'22"
Grace L. Masten
Aqr 23h4'33" -9d46'0"
Grace LaRee
Leo 9h31'22" 10d42'45"
Grace Laywood
And 0h33'46" 45d54'34"
Grace Lee Hoover, Born 01/17/1909
Cap 21h13'43" -27d6'20"
Grace Leigh Norman
And 1h38'5" 41d25'7"
Grace Lillian Greenman
Pho 1h20'34" -42d27'25"
Grace Lily Z
And 23h6'56" 35d25'51"
Grace Linder
And 2h22'41" 48d31'48"
Grace Lockhart
And 2h32'20" 45d38'29"
Grace Lombardo
Cnc 9h14'33" 8d43'46"
Grace Loralei Garber
Uma 11h44'31" 53d29'38"
"Grace Lorraine Reynolds"
Psc 1h20'46" 26d20'19"
Grace Louise
And 2h34'4" 44d14'4"
Grace Louise Paget
Ari 3h11'15" 18d50'58"
Grace Lowder
And 23h19'4" 52d13'46"
Grace Lynn
Lyn 7h14'29" 57d17'55"
Grace Lynn Murphy
Cas 23h54'11" 57d25'49"
Grace Lynn Pittman
Cyg 20h15'47" 52d29'51"
Grace Lynne Kesterson
And 23h41'7" 45d4'4"
Grace M. Ferreira
Ori 5h45'45" 6d52'58"
Grace MacKenzie Moore
And 23h59'34" 42d22'14"
Grace Madeline Parker
And 1h4'12" 42d44'11"
Grace Madison
And 23h11'41" 38d47'2"
Grace Marcie Rubano
Crb 16h1'25" 38d14'4"
Grace Margaret Harron
And 23h0'40" 51d21'1"
Grace Maria Eberhardt
Cas 1h13'44" 63d3'4"
Grace Marian Blackham Redmon
Cap 21h36'23" -15d27'2"
Grace Marie
And 23h11'8" 40d14'52"
Grace Marie Baumel
And 0h13'24" 35d18'45"
Grace Marie Claus
And 0h17'52" 21d45'16"
Grace Marie Cutean
Cap 21h13'33" -15d6'32"
Grace Marie Dukes
Dra 18h46'13" 56d7'38"
Grace Marie Mayo
And 23h0'38" 48d28'43"
Grace Marie Nelson
Tau 5h57'59" 27d22'25"
Grace Marie's "Angel" ~ Cecelia
Lyr 18h52'18" 33d13'29"
Grace Marie's Dream
Uma 9h44'31" 51d26'0"
Grace Marilyn Codd
Cas 0h43'56" 51d14'5"
Grace Markham Black
Crb 15h26'14" 31d43'25"
Grace Martha Whatmough
And 1h52'34" 46d40'1"
Grace Mary Fraser
Cnc 9h4'34" 17d44'50"
Grace Mary LoPiccolo
Lyn 7h36'34" 54d2'14"
Grace Mary Virginia Hasz
Cas 1h14'36" 68d30'52"
Grace May Caveney
Peg 23h59'23" 20d43'45"
Grace May Curtis
Tau 5h39'11" 27d25'59"
Grace McCleskey
Uma 8h13'21" 72d13'29"
Grace Melo
Vir 13h10'15" 11d42'53"
Grace Mendillo — The Best Zia
And 0h20'19" 22d40'24"
Grace Merrick Pugh
Aqr 23h5'2" -7d46'45"
Grace & Millie's Star
Gem 6h56'16" 28d55'12"
Grace Miriam Lane
And 0h26'0" 37d47'12"
Grace Monaco Orr *Princess*
And 2h8'48" 42d47'22"
Grace Moreno Cassells
Lyn 6h45'51" 51d45'49"
Grace Murillo
Uma 10h54'10" 56d59'50"
Grace Nardi Wood "Happy 50th"
Cap 20h34'16" -19d51'46"
Grace Niamh
Umi 13h20'44" 73d30'58"
Grace Nicole
And 23h43'45" 47d18'15"
Grace Nicole Anderson
Lyn 8h32'43" 44d10'4"
Grace Nicole Smith
Gem 7h32'36" 26d36'4"
Grace Nicole Vallone
Cyg 21h31'16" 34d4'45"
Grace Olivia
Cas 1h17'8" 54d58'8"
Grace Olivia B
Sgr 18h0'46" -29d34'6"
Grace Olivia Hackett
Cra 18h52'30" -41d35'57"
Grace Olivia Latter
And 0h6'20" 40d36'24"
Grace Opus Dei
Cru 12h17'49" -56d24'54"
Grace Orlando "My Shining Star"
Cas 0h3'51" 56d23'40"
Grace P. Blake
Peg 21h35'31" 21d59'41"
Grace P. Merrifield
Crb 15h48'4" 32d27'24"
Grace Pagliaro - SWEET-EST STAR
And 0h46'8" 29d6'31"
Grace Park Lane
Uma 8h25'16" 62d38'21"
Grace Patricia Young
Umi 16h47'12" 83d18'18"
Grace Payton Lennon
Sgr 17h59'35" -26d37'6"
Grace Pierias (Dove) Star
Per 3h48'41" 45d46'31"
Grace Piro Delia
Psc 1h14'27" 11d33'7"
Grace Pitetti
Crb 16h15'49" 38d59'33"
Grace Poppy Wakefield
Ori 5h54'30" 21d37'36"
Grace Robinson Komoroczy
Aqr 22h27'47" -0d6'6"
Grace Rolland
Uma 11h44'33" 45d35'45"
Grace "Rose" Karunanithi
Uma 10h22'31" 55d29'33"
Grace Ryan Abraham
Crb 16h10'33" 37d14'32"
Grace Sanders
Lyr 18h50'47" 32d13'48"
Grace Seitz
Sgr 19h12'30" -22d7'42"
Grace Shanti Connors
Sco 16h27'25" -30d24'9"
Grace Sheehan
Cas 0h19'48" 51d38'59"
Grace Shoemaker
Cas 23h43'0" 55d14'44"
Grace Shufelt's Super Star
Vir 14h3'52" -15d34'33"
Grace Sienna Honey Wilkinson
And 23h55'34" 39d44'6"
Grace Steranko
And 1h49'15" 41d26'38"
Grace Sylvia Green
And 2h2'46" 46d44'49"
Grace The Mother Star
Ori 5h38'55" -0d58'59"
Grace Therese Tanceusz
Psc 1h39'55" 18d24'12"
Grace Tin-Min Cheng
Sgr 18h42'18" -23d18'0"
Grace Tiva
Gem 7h21'31" 32d26'56"
Grace Tomaselli Sahyoun
Aqr 21h13'22" -9d17'40"
Grace Traynor
And 1h23'10" 39d58'7"
Grace Tung the Luna Star
Ari 1h59'35" 19d10'54"
Grace Valentine Dummar
Cnc 9h12'10" 32d1'33"
Grace Victoria
And 0h11'41" 46d24'16"
Grace Victoria Ellis
Lmi 9h52'50" 33d36'11"
Grace Victoria Iris Chick
And 0h33'38" 26d24'6"
Grace Violet Eilidh McCallum
Umi 14h39'28" 73d7'5"
Grace Waldron
Umi 17h10'26" 76d27'36"
Grace Welden
Uma 8h15'3" 60d11'42"
Grace Whaley Watkins
Gem 7h39'25" 27d12'23"
Grace Young Eun Liu
Com 13h13'27" 29d0'31"
Grace Yung Sau Leung
And 0h33'1" 24d43'35"
Grace Yung Sau Leung
And 1h46'37" 45d55'35"
Grace Yung Sau Leung
Cas 0h25'28" 61d7'21"
Grace Zhang
Sgr 18h37'42" -18d19'5"
Grace, stay golden indeed.
Lib 15h19'50" -6d2'52"
Gra-Cee
Uma 9h30'42" 42d20'1"
Gracee Lou Harrel
Cyg 21h44'41" 42d20'52"
Graceful Christine
Aqr 20h49'5" -12d7'9"
Graceful Paalua
Cas 23h21'14" 59d4'21"
Grace-Lynn Marie Janelli
Uma 14h6'43" 48d13'26"
Grace-Marie
Cam 4h11'36" 56d30'25"
Gracen Landry
Cyg 21h44'20" 39d31'16"
Grace's Dragon Star
Dra 19h24'24" 60d42'15"
Grace's Haven
Uma 10h26'55" 61d7'59"
Grace's Star
Sgr 18h14'19" -26d38'56"
Grace's Star
Leo 11h14'3" 24d3'55"
Grace's Wishing Star
And 1h3'22" 42d1'28"
Gracey Ireland
Tau 4h44'29" 19d57'16"
Gracey Lee Kelley
Aqr 22h33'39" 0d37'23"
Gracey Pearl
Uma 10h59'41" 62d47'29"

Gracey's Star
And 23h45'10" 48d25'43"
GraceyT_5
Cir 15h24'28" -56d47'22"
Gracia "Bella" Sham
Cam 5h31'38" 62d12'57"
Gracia Elizabeth Phinn
And 0h59'25" 45d52'1"
Gracia Rose
Vir 13h45'33" 3d2'1"
Gracie
Ari 3h7'34" 17d57'45"
Gracie
Tau 4h38'48" 20d55'59"
Gracie
Leo 11h18'1" 25d42'7"
Gracie
Leo 11h34'57" 25d58'52"
Gracie
Ari 2h45'42" 26d56'44"
Gracie
And 23h45'50" 41d44'57"
Gracie
Lyn 8h26'20" 46d23'58"
Gracie
And 1h21'41" 43d52'15"
Gracie
Umi 14h23'58" 67d12'35"
Gracie
Umi 14h56'50" 74d57'15"
Gracie
Vir 13h53'23" -11d42'56"
Gracie
Lib 15h14'28" -25d23'27"
" Gracie "
Cap 21h19'9" -23d56'24"
Gracie A. De Chirico
Cap 20h53'46" -15d14'28"
Gracie Ann Buckner
Lib 15h40'26" -29d6'24"
Gracie Ann Margaret Ridenbaugh-Karr
Tau 5h29'49" 24d35'34"
Gracie Anne Parker
And 2h22'31" 45d48'1"
Gracie Anne Randolph
Cnc 8h17'24" 32d6'7"
Gracie Chambers
And 1h43'17" 48d9'8"
Gracie Christine Straub-Berg
Sco 16h14'57" -16d19'51"
Gracie Dexter
Uma 13h54'49" 51d49'19"
Gracie Elaine
Sco 16h39'14" -29d32'32"
Gracie Elizabeth
And 2h0'22" 42d0'9"
Gracie Elizabeth Griffin
And 23h6'34" 51d8'29"
Gracie Fletcher
Per 3h10'35" 46d39'29"
Gracie Girl
Ori 5h33'24" -0d19'28"
Gracie Griffin
And 22h59'3" 52d14'33"
Gracie Harvey Rivera
Cyg 20h42'0" 34d46'49"
Gracie J. Livingston
Uma 11h33'31" 34d22'31"
Gracie Kay
Uma 8h41'48" 52d38'31"
Gracie Kay Landrigan
Lyr 18h47'36" 43d50'15"
Gracie Kennedy
Cap 21h29'56" -14d49'54"
Gracie Kevyn
Sco 17h53'37" -36d2'24"
Gracie Leigh
Leo 10h28'0" 24d42'16"
Gracie Leigh Rambin
Lib 14h24'39" -22d33'5"
Gracie Lou
Lib 14h35'46" -13d24'38"
Gracie Lou
Crb 15h24'4" 27d35'14"
Gracie LouAnn Grettum
Uma 10h23'27" 42d25'8"
Gracie Madeline Lange
Cru 12h41'53" -62d47'44"
"Gracie" Madison Grace Levy
Uma 11h15'1" 70d14'41"
Gracie Mae Binns
And 23h19'0" 47d33'47"
Gracie Maria
Leo 9h38'7" 29d50'18"
Gracie Marie
Sgr 18h8'13" -26d30'23"
Gracie Pearl Hough
And 0h17'58" 44d59'3"
Gracie the Beautiful
Sgr 18h17'27" -34d3'59"
Gracie Warsaw
Uma 9h17'48" 66d24'56"
Gracie050501
Tau 4h22'4" 26d39'54"
Graciela
And 23h49'55" 41d51'42"
Graciela
Lib 15h35'49" -8d56'21"
Graciela
Cra 19h6'16" -38d43'54"
Graciela
Sco 16h53'43" -41d59'8"
Graciela Eleta
Lyn 8h28'48" 51d28'49"
Graciela Godina
And 1h50'59" 46d47'4"
Graciela Quezada Overdorf
Vir 13h35'53" -6d54'5"
Graciela Villanueva Reyes
Ari 2h10'24" 11d22'54"
Graciemingmonggreen
Uma 8h37'30" 63d47'51"
Gracies Night Light
Aqr 21h39'19" 1d53'51"
Gracie's Smile
And 1h24'14" 40d52'53"
GRACIE'S STAR
And 23h0'22" 51d12'34"
Gracie's Twinkle Twinkle
And 0h15'20" 39d23'1"
Gracie's Yia-Ya
And 2h16'54" 48d50'34"
Graciosa Chaput
Per 1h51'7" 51d3'34"
Gracious
Lyn 7h38'49" 56d59'9"
Gracious Grace
Cru 12h45'30" -57d51'47"
Gracjan Adam zeTiger Lis
Cnc 8h54'4" 30d31'9"
Gracy Howenstine
Cap 21h51'52" -22d57'1"
Gracy Lillian Shivers
Mon 6h51'48" -0d29'16"
Gracyn Caroline Reed
Ari 2h6'31" 21d57'45"
Gracyn Elizabeth Woodley
Psc 0h9'6" 9d55'29"
Gracyn P. McCrary
Psc 1h12'48" 17d59'33"
Gradam
Umi 19h2'42" 87d4'43"
Graddy
Uma 11h36'38" 55d24'33"
Grade Beaute et Nobelesse
Sgr 18h6'6" -28d10'5"
Gradean Nooway
Ori 5h51'2" -3d27'43"
Graduate Amazing Aimee Jones
Aql 18h50'40" -0d54'52"
Graduate of the Year
Dra 15h13'45" 63d36'2"
Graduate Rodney Reary
Uma 11h53'45" 55d24'42"
Grady
Dra 17h19'4" 61d3'4"
Grady Alvin Prescott
Equ 21h16'48" 8d55'55"
Grady Bryce Murphy
Psc 0h55'17" 6d16'29"
Grady Casner's Star
Sco 17h53'29" -35d54'32"
Grady D Huron
Ori 6h18'30" 13d13'35"
Grady "G Man" Bass
Dra 17h21'54" 63d58'2"
Grady Goodlett Martin
Uma 10h55'48" 36d57'13"
Grady Helton
Psc 2h4'27" 9d25'52"
Grady Hubbard Robinson
Gem 7h23'26" 14d50'6"
grady logan scroggins
Lyn 8h48'56" 44d56'14"
Grady Mae
Lyn 8h15'19" 46d35'12"
Grady Scott Law
Per 3h50'22" 35d50'37"
Grady Tweed
Cam 4h26'28" 71d13'34"
GradyF6
Tau 3h42'14" 27d29'2"
Graeme Blench
Per 3h13'13" 42d37'4"
Graeme Hayes Jackson
Cyg 20h13'37" 40d5'15"
Graeme John Smithers - My Angel
Lib 15h39'2" -10d35'30"
Graeme Laslett - "G-Man"
Per 4h48'50" 43d35'15"
Graeme Macdonald
Dra 14h39'6" 55d30'16"
Graeme Paul Thompson
Tau 4h43'57" 20d31'8"
Graeme Robert Speck
Uma 11h25'29" 59d15'54"
Graeme Thornton
Per 4h9'33" 34d31'4"
Graeme W Briggs - Pater Supremus
Tau 5h10'53" 18d11'13"
Graeme Walsh
Tau 4h27'59" 15d57'48"
Graeme William Marley
Cep 2h10'33" 80d59'35"
Graeme "you're a star" Pohl
Cap 21h47'24" -15d26'9"
Graeme's Light
Cru 12h23'3" -58d5'22"
Gräfe, Michael
Uma 14h2'9" 50d46'21"
GraFelix
Uma 11h17'35" 34d27'52"
Grafton Bacardi
Lyn 6h33'1" 54d20'12"
Graham
Uma 10h4'47" 60d55'56"
Graham
Cyg 20h3'22" 45d48'0"
Graham
Per 3h22'45" 47d21'46"
Graham
Peg 21h28'7" 12d5'28"
Graham Albert Pearce
Her 17h43'38" 39d47'35"
Graham and Sue - Diligo Eternus
Cyg 19h40'8" 29d1'46"
Graham B. Thompson
Aql 19h56'26" 15d52'17"
Graham Betts
Psc 1h8'35" 29d54'34"
Graham Brennand Skinner
Cyg 20h41'17" 34d11'15"
Graham Brenton McKay
Ori 6h9'43" 5d52'6"
Graham & Brooke - Eternal Love
Cru 12h36'31" -62d46'32"
Graham Bruckner Garrison
Per 3h50'32" 40d3'43"
Graham Calvert
Cep 4h38'25" 84d7'11"
Graham Canny
Uma 9h35'10" 46d36'8"
Graham Clark Davies
Boo 14h28'13" 40d3'11"
Graham Clarke
Her 18h54'4" 21d56'36"
Graham D Newman
Uma 10h7'19" 49d28'5"
Graham David Allen
Dra 15h8'54" 55d38'33"
Graham David Walsh
Cep 20h41'4" 59d20'8"
Graham de Villiers
Cru 12h50'13" -60d52'31"
Graham Duffy
Aur 5h27'35" 36d35'14"
Graham Elliot Couret La Force
Cnc 9h17'39" 23d17'47"
Graham & Gemma Goodberry
Lmi 9h54'11" 39d43'53"
Graham H Cox
Per 4h43'23" 48d52'35"
Graham Harold Smith
Her 16h16'39" 6d44'11"
Graham Harrison
Uma 10h6'44" 50d8'6"
Graham Hunter Gher
Per 3h45'25" 43d59'27"
Graham & Jacquie
Cyg 20h7'36" 32d39'28"
Graham John Hurst
Uma 11h4'20" 42d20'45"
Graham Kovacs Herr
Uma 11h40'40" 39d19'53"
Graham Larkin
Per 4h27'59" 43d29'37"
Graham Louis Boyd
Sgr 18h3'16" -28d13'23"
Graham "Lum" Edwards
Tau 3h28'13" 9d6'3"
Graham - Luminous
Lib 14h59'25" -12d8'36"
Graham Marousek
Ori 5h58'11" 21d5'15"
Graham Michael Marcy
Boo 14h51'9" 54d35'45"
Graham Morgan
Her 17h2'33" 45d54'33"
Graham Mounsey Cutest Bum in Galaxy
Sco 17h28'47" -43d39'8"
Graham Nathan Cook
Gem 6h37'17" 26d6'57"
Graham Norman Walstrom
Dra 18h58'15" 70d56'31"
Graham Olsen
Umi 14h23'11" 75d23'15"
Graham Othmar Schulz 8/19/04
Lyn 6h18'3" 61d46'17"
Graham & Pauline
Lyr 18h45'32" 30d4'49"
Graham Pinto
Psc 23h16'10" -2d14'1"
Graham Pitman
Uma 9h20'12" 49d36'59"
Graham Reginald Loomes
Pho 1h21'46" -42d41'49"
Graham Robert Aspinall
Per 3h39'9" 39d49'25"
Graham Shai Radina
Aqr 23h12'58" -9d30'42"
Graham & Sheri Kortgaard
Ori 4h59'12" 11d15'0"
Graham Stephen
Lib 15h33'12" -16d20'45"
Graham Walker
Del 20h45'9" 11d8'27"
Graham Wason
Ori 6h0'0" 5d40'19"
Graham Wesley Naeseth
Aqr 22h19'45" -23d43'25"
Graham Wilson Smith
Her 17h46'27" 22d52'33"
Grahame D Callow Papa Star
Cep 22h15'31" 72d57'47"
Graham's Golden
Uma 12h5'20" 53d59'55"
Graham's HoverStar!
Umi 17h18'42" 84d55'55"
Graham's Star
Aqr 23h0'21" -7d35'19"
Graig Michael Spadoni
Ori 5h51'14" 2d21'26"
Grala, Marian
Ori 6h15'48" 13d20'2"
Gram
Cnc 9h9'1" 13d57'48"
Gram
Gem 7h10'17" 25d49'54"
Gram
Per 3h45'55" 34d5'31"
Gram
Cas 1h32'33" 60d28'38"
Gram
Sco 17h38'57" -32d41'6"
Gram & Bump's Star
Col 5h38'14" -33d7'34"
Gram & Grandaddy
Eri 3h56'40" -8d23'8"
Gram & Pop Crowder
Uma 11h44'46" 38d35'30"
Gram & Pop Patrone
Cyg 21h33'9" 55d10'21"
Gram Svajlenka
Cas 1h22'15" 52d21'52"
Grama Mom
Cas 23h23'59" 53d49'14"
Gramcrackers
Leo 10h27'0" 21d47'8"
Gram-Gram 77
Cnc 8h5'17" 15d22'34"
Gramie and Grampie Long
Uma 10h14'30" 48d11'56"
Gramie and Papa
Lyr 19h15'54" 29d48'20"
Gramma
Vir 13h39'13" -4d27'32"
Gramma Abshire
Tau 4h10'45" 23d35'50"
Gramma Amaparo Ornelas
Crb 15h34'21" 26d45'51"
Gramma Anne Scheeler
Vir 14h3'54" -3d22'48"
Gramma Doofy
Cas 1h25'56" 57d2'21"
Gramma Joan Endreson
Uma 8h40'49" 47d36'25"
Gramma Maki
Cnc 9h8'26" 31d34'18"
"Gramma" Mary J. Koss
Uma 11h22'45" 64d11'43"
Gramma Pat Demarest
Sgr 19h4'54" -30d9'51"
Gramma's Star
Ori 5h36'40" -0d21'34"
Gramma's Twinkle
Cnc 8h42'39" 8d43'0"
Gramme
Cyg 19h57'32" 31d9'0"
Grammie and Wyatt's Star
Sge 20h1'10" 19d35'46"
Grammie's Star
Cas 0h9'20" 58d18'33"
Grammione<\@>Heaven.com
Cas 0h21'10" 64d34'4"
Grammy
Cas 0h57'46" 61d32'55"
Grammy
Cas 2h50'17" 60d56'22"
Grammy
Cas 0h55'32" 57d22'14"
Grammy
Cas 1h27'27" 59d8'59"
Grammy
Her 16h57'45" 27d45'9"
Grammy
Tau 4h24'22" 27d10'42"
Grammy
Cnc 9h4'22" 24d35'11"
Grammy
Vir 13h18'49" 13d21'43"
Grammy and Grampy Tex
Sgr 18h2'5" -34d19'45"
Grammy and Grampy's Smile
Her 16h54'38" 33d23'26"
Grammy and Pappy Karcher
Del 20h36'16" 15d17'52"
Grammy (Carol Kindrick)
Lib 15h13'49" -4d5'9"
Grammy Carr's Star
Cas 1h19'19" 58d56'0"
Grammy & CrapDaddy
Cyg 21h13'18" 46d3'53"
Grammy DOT DOT
Per 3h0'59" 49d5'13"
Grammy Drowns Forever
Uma 9h7'58" 60d2'37"
Grammy & Grampy
Cyg 21h39'51" 46d30'34"
Grammy Joanne Elaine Baer
Uma 12h55'7" 58d42'2"
Grammy K
Per 3h27'1" 48d8'0"
Grammy Kathy
Ori 5h52'10" 10d12'2"
Grammy Linda Kay Hale Alter
Uma 11h25'50" 45d42'24"
Grammy of Loch Lloyd
Vir 13h20'43" 6d50'8"
Grammy P
Crb 15h46'4" 26d52'16"
Grammy & Papa Aubrey
Sco 17h46'5" -34d8'1"
Grammy Patty Taramaschi
Ori 6h17'31" 15d17'34"
Grammy Philly
Leo 10h15'56" 24d43'22"
Grammy Rebecca Breitman Jacoby
Cap 20h11'47" -12d44'20"
Grammy Ruth
Cas 1h12'28" 55d45'38"
Grammy Winner John R. Burk
Cap 21h7'5" -16d31'58"
Grammy YaYa Stirnaman
Uma 8h40'6" 47d50'11"
Grammy's Love
Psc 1h9'13" 27d51'51"
Grammy's Star
Cas 0h7'39" 55d55'47"
Grammy's Star Next to Claire & Drew
Oph 17h1'13" -7d51'52"
Gramochroi
Psc 1h17'55" 24d7'12"
Grampa Abshire
Tau 3h50'26" 14d0'52"
Grampa Meiwes
Aql 19h29'13" 14d21'48"
Grampa Preston
And 2h35'58" 44d45'44"
"Grampa" Steve Lee
Ori 5h37'50" -0d39'37"
Grampa Virg
Uma 12h1'39" 35d43'26"
Grampa's Star
Ori 6h11'14" 18d10'26"
Grampie Love
Her 16h46'58" 27d56'43"
Gramps
Per 2h54'57" 54d10'49"
"Gramps" John Salza
Cep 22h55'45" 77d42'41"
Gramps Laue
Vir 13h42'48" -10d43'51"
Gramps Power
Uma 8h50'30" 71d46'38"
GRAMPSY'S LITTLE RED
Cam 4h18'40" 69d5'21"
Grampy
Psc 1h26'10" 3d26'23"
Grampy - 7th December 1923
Lyn 7h12'10" 54d8'48"
Grampy Guitard
Ori 5h41'25" -4d42'4"
Grampy Richie Joyce
Per 3h36'46" 35d10'11"
Grampy Steve
Cep 22h11'30" 71d59'34"
Grampy Thiel
Uma 9h15'54" 67d26'26"
Grampy's Star
Her 17h57'52" 17d5'45"
Grams
Vir 12h56'0" 2d51'53"
Grams
Vir 12h1'18" 5d20'0"
Grams
Cas 1h55'47" 60d34'44"
Grams and Pops
Uma 11h13'42" 54d35'46"
Grams, Siegfried
Ori 6h14'4" 8d55'56"
Gramsey's Star
And 23h15'3" 48d29'37"
Gran and Gramps
Tri 2h10'41" 33d36'15"
Gran Dorris
Crb 15h33'6" 28d34'51"
Gran Forbes
Cas 0h20'27" 50d56'41"
Gran & Grandad McNeil
Uma 11h23'57" 71d30'14"
Gran (Janet Patricia Nosel)
Leo 9h42'19" 27d8'18"
Gran Nan "Our Guiding Light"
Cru 12h24'10" -57d26'20"
Gran Pamela Love Always Miss You
Col 5h55'15" -38d9'34"
Gran & Pop Woolnough
Cru 12h36'26" -56d6'28"
Gran & Popi's Place
Uma 11h30'13" 57d58'34"
Granata's World
Cap 20h20'49" -11d32'20"
Grand Annie
Sgr 18h48'10" -18d46'44"
Grand Barbados Beach Resort
Ori 5h53'18" 21d27'45"
"Grand" Bryson
Uma 9h2'54" 56d21'16"
Grand Dad
And 0h54'36" 43d48'5"
Grand - Daddy
Uma 11h38'51" 38d15'26"
Grand Four
Dra 17h0'49" 59d58'32"
Grand Mac
Aqr 23h4'22" -8d15'49"
Grand maman Irène
Cyg 19h40'5" 34d27'51"
Grand Martie
Cap 21h49'46" -18d0'5"
Grand Poobah
Uma 13h21'15" 55d11'29"
Grand Poussin
Cap 20h28'11" -27d2'53"
Grand Titan
Her 18h4'57" 33d14'14"
Grandad Alf
Uma 10h54'40" 70d36'20"
Grandad Bob
Uma 10h58'0" 67d6'57"
Grandad Bob
Cep 1h23'58" 83d14'13"
Grandad Bower
Per 3h52'17" 48d45'3"
Grandad Burke - Our Angel in the Sky
Uma 11h7'38" 58d8'26"
Grandad Dennis
Psc 0h46'46" 15d35'3"
Grandad Joe
Cep 21h30'30" 68d25'12"
Grandad John Burt - 80
Cep 22h37'18" 63d6'45"
Grandad Kelvin Johnston - Pappy's Star
Per 3h18'44" 40d31'26"
Grandad Laverie
Cep 22h4'58" 57d16'0"
Grandad Liam
Per 2h28'29" 51d48'50"
Grandad Norman Wrigglesworth
Ori 6h11'34" 15d25'38"
Grandad Richards
Cep 21h59'30" 66d28'26"
Grandad & Ryan's Little Stink Star
Vir 14h0'15" -16d0'59"
Grandad Stan
Vir 12h35'22" 1d40'14"
Grandad Tom & His Little Princess
Uma 9h9'27" 48d11'13"
Grandad Tony Cook "Cookus"
Uma 10h52'25" 43d48'17"
Grandad Vic
Aql 20h19'57" -3d17'44"
Grandaddy
Aqr 23h16'25" -14d10'46"
Grandaddy's Star
Her 15h56'33" 42d43'37"
Grandad's Light
Uma 9h18'54" 61d16'12"
Grandad's Star
Cep 21h26'52" 69d26'50"
Grandame Family Star
Ori 6h1'27" 20d50'55"
GRANDC
Aqr 22h40'38" -2d12'7"
Granddad
Cep 22h32'23" 69d7'48"
Granddad
Ori 5h28'42" 2d24'34"
"Granddaddy"
Cep 21h24'28" 62d24'11"
Granddaddy Bob
Vir 14h33'53" 1d23'8"
GrandDaddy Lipsey's Star
Cnc 8h28'42" 10d13'20"
Granddaddy Rowe
Her 16h41'39" 36d22'40"
Grande Arianis
Uma 9h26'50" 41d37'27"
Grande Oaks Nursing Staff
And 0h56'48" 38d58'4"
Grande Ruffy
And 0h44'26" 30d30'39"
Grandfather
Sgr 17h57'5" -20d30'14"
Grandfather Ken
Aqr 23h31'46" -13d58'46"
Grandfather's Gremlins
Aql 18h56'55" 8d18'31"
Grandma
Tau 5h42'36" 26d25'32"
Grandma
Cas 0h10'33" 62d55'41"
Grandma
Uma 9h16'52" 54d21'14"
Grandma Ada Delfa
Cas 1h15'13" 52d30'7"
Grandma and Grandpa
Cyg 20h4'31" 54d41'45"
Grandma and Grandpa Evans
Tau 5h55'13" 25d24'44"
Grandma and Grandpa Facius
Uma 11h27'27" 58d42'55"
Grandma and Grandpa Johns
Psc 1h46'13" 6d11'18"
Grandma and Grandpa Malstrom
Cyg 20h58'39" 43d9'1"
Grandma and Grandpa Sanderson
Aqr 22h5'4" 0d28'14"
Grandma and Grandpa Smith
Cyg 21h58'42" 54d24'6"
Grandma and Grandpa Wylie
Cnv 12h45'19" 39d11'55"
Grandma Ann
Uma 10h52'31" 63d56'33"
Grandma Ann
Cap 21h10'34" -20d7'22"
Grandma "Annie" Annette Schroeder
Cnc 9h20'46" 25d13'25"
Grandma Anoush
Lib 15h6'45" -1d9'49"
Grandma Askew
Cas 0h27'15" 61d45'51"
Grandma Audrey and Grandad Charlie
Cyg 20h20'0" 51d42'17"
Grandma Ausilia Chicki Casella
Lyr 18h37'54" 45d36'32"
Grandma Babs Prigozen Abrahams
Cas 1h27'31" 56d43'51"
Grandma(Barbara)&Grandpa(Dave)
Uma 11h14'26" 56d35'47"
Grandma Bea
Uma 11h34'32" 56d58'41"
Grandma Becker
And 0h49'8" 37d3'25"
Grandma Becker
Psc 1h3'48" 15d20'0"
Grandma Becky White
Aqr 23h36'5" -8d5'16"
Grandma Bell
Crb 15h19'57" 26d10'44"
Grandma Bertie
Vir 12h14'0" -1d12'57"
Grandma Betty Jane Stade
Vir 12h36'31" -11d5'24"
Grandma Betty Kretzler
Crb 15h38'23" 27d51'12"
Grandma Beverly
Per 4h12'50" 46d41'35"
Grandma Bif
Sgr 18h33'40" -21d51'59"
Grandma Birdie
Apu 14h26'35" -72d6'29"
Grandma BJ
Psc 1h30'32" 12d29'48"
Grandma Bobbie
Cnc 8h51'52" 26d14'22"
Grandma Carol Davis
Uma 10h0'17" 65d0'24"
Grandma Carolyn Martinez
Crb 16h18'4" 29d35'11"
Grandma Carr
Cas 0h22'31" 52d47'21"
Grandma Cerreta
Vir 12h39'14" 6d59'43"
Grandma Cheryl,Megan,Makenzie Star
Lyn 7h44'45" 39d29'31"
Grandma Cindy
Dra 18h55'45" 50d40'54"
Grandma Cindy's star
Sgr 18h4'37" -27d25'6"
Grandma Connie McKelvey
Cas 0h26'7" 54d35'53"
Grandma Cookie
Cas 0h25'36" 59d25'40"
Grandma Darlene L. Moore
Umi 15h13'41" 72d57'43"
Grandma Deane
Per 3h15'51" 50d17'4"
Grandma Diane McGinn
Uma 8h52'30" 61d3'30"
Grandma Dianna
Uma 14h1'41" 52d3'31"
Grandma Donna
Vir 13h16'0" 5d34'34"
Grandma Donna Marquez
Crb 15h52'14" 35d17'37"
Grandma Doris's star
Uma 11h36'0" 51d39'9"
Grandma Dorothy
And 0h16'13" 25d18'51"
Grandma - Dorothy Butler
Cas 1h59'50" 68d15'54"
Grandma Edy
Cas 1h9'31" 75d41'54"
Grandma Eileen Boland
Uma 9h28'28" 59d16'5"
Grandma Emilie Widi
Uma 12h28'22" 53d33'2"
Grandma Fanta
Psc 0h23'51" -0d24'5"
Grandma Fleszar
Aqr 22h53'17" -6d49'0"
Grandma Florence Meier
Per 4h6'11" 36d24'32"

Grandma Florie
Aqr 21h22'7" 1d54'2"
Grandma Frieda's Shining Star
Aqr 22h14'27" 2d36'26"
Grandma G.
Tau 5h39'39" 26d32'33"
Grandma Genny
Cyg 20h14'20" 49d18'59"
Grandma Gerry & Grandpa George
Cyg 21h19'37" 32d41'42"
Grandma "Gigi" Taylor
Cas 0h33'51" 53d44'36"
Grandma Ginny
Cas 23h34'21" 54d28'46"
Grandma Gladys
Uma 8h51'19" 54d1'3"
Grandma Goose
Vir 13h16'30" 5d38'8"
Grandma Gracie
Lib 15h49'41" -17d4'36"
Grandma & Grandpa
Cyg 21h44'8" 52d12'37"
Grandma & Grandpa
Cyg 20h53'48" 35d14'56"
Grandma & Grandpa Becker
Cyg 20h5'3" 40d7'46"
Grandma & Grandpa Brownlee
Uma 9h20'35" 53d53'18"
Grandma & Grandpa Dingle
Sge 19h37'51" 17d53'48"
Grandma & Grandpa Gehl of Kohler
Uma 11h14'23" 43d53'22"
Grandma & Grandpa Martinez's Star
Cyg 21h8'53" 48d13'56"
Grandma & Grandpa Moug
Per 3h20'17" 47d32'13"
Grandma & Grandpa O'Hare
Her 18h0'39" 17d27'57"
Grandma & Grandpa Paschke
Lib 14h53'33" -2d6'52"
Grandma & Grandpa Payne
Cyg 20h52'17" 50d25'39"
Grandma & Grandpa West
Crb 16h5'26" 36d13'40"
Grandma & Grandpa Wisniewski
Uma 9h54'3" 55d10'50"
Grandma Gray
Aqr 22h38'55" -5d14'57"
Grandma Gwen of the Galaxy
Cyg 21h1'42" 51d28'44"
Grandma Helen
Cnc 8h4'3" 20d54'35"
Grandma Helen
Col 5h42'42" -34d24'8"
Grandma Helen's Cookie Store
Ori 4h49'8" 11d32'54"
Grandma Henry
Cas 0h46'44" 58d5'1"
Grandma Hilma
Vir 12h37'18" -0d41'33"
Grandma Irene's star 44444
Leo 11h0'33" 3d7'53"
Grandma Jaimie
Cas 1h25'35" 62d4'9"
Grandma Jan
Cas 1h37'46" 63d37'19"
Grandma Jane
And 1h4'18" 34d41'59"
Grandma Janet Always Shining Bright
Cas 0h47'47" 53d27'54"
Grandma Jean
Lib 14h49'42" -17d28'47"
Grandma Jill
Cas 1h18'28" 54d47'6"
Grandma Joan Gross
Vir 15h2'31" 1d58'10"
Grandma Joan's Star
Ari 2h47'41" 10d43'47"
Grandma Johnson
Cas 0h38'55" 64d21'24"
Grandma JoJo
Uma 11h43'41" 41d3'24"
Grandma Jones
Aql 19h21'45" 6d26'29"
Grandma Joyce Montgomery Smith
Lyn 6h26'40" 54d35'41"
Grandma Judy/Grandpa Bill Nicholas
Cyg 20h19'19" 53d59'56"
Grandma K
Aql 19h37'12" -11d10'52"
Grandma Kae
Ari 3h12'35" 20d40'14"
Grandma Kathleen Ann Schneider
Lib 14h24'0" -13d10'9"
Grandma Kathy
Ari 3h11'55" 29d8'55"
Grandma Kathy
Uma 14h2'10" 52d24'26"
Grandma Kathy Jourdin
Lib 15h22'16" -15d52'1"
Grandma Keller
Per 3h48'34" 39d11'51"
Grandma King
Cap 21h44'24" -24d21'52"
Grandma Labonte
And 0h58'10" 40d42'27"
Grandma Lee and Pop/Pop
Cyg 19h44'50" 33d37'6"
Grandma Lila
Cas 1h20'45" 62d42'42"
Grandma Linda
Cnc 8h31'42" 25d33'43"
Grandma Linda Bear
Tau 5h39'56" 27d51'5"
Grandma Linda & Grandpa Stan
Crb 16h14'45" 37d9'40"
Grandma Linda Star
Psc 1h9'42" 27d57'3"
Grandma Lizbeth
Col 6h24'14" -37d10'24"
Grandma Lois
Uma 11h24'13" 54d45'42"
Grandma Lois Wilson
Sgr 18h17'24" -19d27'42"
Grandma Lola
Tau 4h25'9" 2d23'42"
Grandma Lorraine "Bubbles" Surufka
Leo 10h41'29" 13d54'4"
Grandma Lorraine McKenney
Lyn 6h20'20" 61d26'17"
Grandma "Lorraine" Mom
Uma 13h51'58" 58d17'45"
Grandma Lu Becotte
Cas 1h19'12" 63d45'59"
Grandma Macaroni
Uma 13h11'39" 58d26'59"
"Grandma Maggie"
Crb 16h20'46" 34d9'37"
Grandma Margaret
Ari 2h56'58" 26d37'5"
Grandma Marie
Aqr 23h3'42" -6d43'41"
Grandma Marjorie Lamphier
Uma 11h6'14" 56d51'44"
Grandma Marlow
Cas 1h27'30" 57d47'43"
Grandma Mary
Psc 1h9'3" 10d50'46"
Grandma Mary
Cap 21h7'51" -19d44'30"
Grandma Mary Ann Hafey Chandler
Uma 11h35'23" 48d9'31"
Grandma Mary Saidens
Ari 2h50'14" 13d23'3"
Grandma Mary's Light
Crb 15h42'35" 28d46'18"
Grandma Maxine
Peg 22h44'7" 26d59'24"
Grandma McAfee
Cep 21h32'46" 66d18'38"
Grandma Mickey always in our hearts
Cas 22h57'28" 58d53'23"
Grandma Millie Martin
Cas 1h46'10" 63d28'26"
Grandma Mugs
And 23h2'24" 42d48'34"
Grandma Netta
Per 3h22'42" 50d2'21"
Grandma - Our Shining Light in Life
Psc 1h13'40" 29d0'13"
Grandma & Papa McCurdy
Uma 11h12'11" 53d37'58"
Grandma Pat
Psc 1h17'8" 5d14'32"
Grandma Pat Pat
Leo 9h23'13" 7d48'32"
Grandma Peach
Uma 11h11'27" 49d34'56"
Grandma Peggy
Gem 6h53'34" 15d59'14"
Grandma Polly
Aqr 22h44'29" -4d50'15"
"Grandma Raindrop" Lorenzen
Cyg 19h39'26" 34d40'57"
Grandma Red
Cas 1h15'21" 57d13'0"
Grandma Rene
Lyn 7h57'0" 38d40'7"
Grandma Rene
Aqr 21h49'57" -1d8'38"
Grandma Rita
Lyn 8h2'10" 34d1'9"
Grandma Rita
Ari 2h24'38" 25d9'27"
Grandma Rosalie
Cnc 8h31'35" 24d53'0"
Grandma Rose
Pho 0h36'11" -41d41'56"
Grandma Ruth Star
Sgr 19h44'19" -15d28'49"
Grandma Sandy Watching Over Us
Sgr 18h30'57" -20d11'32"
Grandma Sheila
Cas 0h46'58" 53d55'20"
"Grandma" Sherry Burk
Crb 15h34'24" 27d4'11"
Grandma Shirley
Tau 4h37'29" 10d11'6"
Grandma Shultz
Tau 3h44'9" 29d0'21"
Grandma Smith
Cas 0h8'17" 57d23'33"
Grandma Spillie
Uma 12h3'28" 55d42'18"
Grandma Stacy
Lib 14h48'42" -3d43'8"
Grandma Star
Uma 14h5'56" 50d25'11"
Grandma Stookey
Gem 7h24'25" 26d28'39"
Grandma Susan Berg
Uma 10h46'22" 64d54'46"
Grandma Thorn
Cap 21h41'16" -12d58'42"
Grandma Titi "10-19-29 to 5-7-05"
Cas 0h26'17" 51d57'46"
Grandma Toni
Uma 14h18'36" 57d25'40"
Grandma Tucker Mace We miss & luv u
Uma 9h29'16" 54d42'59"
Grandma Veith
Ori 6h20'39" 7d28'11"
Grandma Weezie the Wonderful
Sco 16h39'28" -28d35'26"
Grandma Wendy
Sco 16h17'5" -17d23'55"
Grandma Z
Cap 20h28'14" -11d42'25"
Grandma-BDO&LO
Cas 1h56'10" 60d31'37"
Grandmaddie Steane
Ori 6h15'43" 6d24'28"
Grandma-Love
Cas 0h0'1" 57d10'24"
Grandmama
Cas 1h26'26" 66d57'45"
Grand-maman Fernande
And 23h51'34" 45d23'17"
Grand-Maman Pauline
Cas 22h59'13" 55d30'40"
Grand-maman Thérèse
Cas 1h19'18" 54d51'23"
GRANDMA'S ANGEL BOY
Lyn 7h44'56" 36d38'24"
Grandma's Babies Rainelynn & Chloe
And 23h2'20" 40d40'15"
Grandma's Guiding Light
Sgr 18h43'57" -30d6'46"
Grandma's Little Ballerina <3 KMS
Per 3h15'12" 52d10'7"
Grandma's Shining Light
Ori 5h35'8" 9d42'54"
Grandma's Star
Lyr 19h25'27" 38d25'46"
Grandma's Star
Cas 23h3'37" 59d9'49"
Grandma's Three Angels
Uma 11h40'9" 57d41'43"
Grandma's Wish
Aql 19h35'40" 12d2'3"
Grandmaster Wonik Yi
Aql 19h3'41" -0d45'9"
Grandmere Esther
And 0h20'18" 45d9'34"
Grandmom
Lyn 8h10'22" 35d7'22"
Grandmom Betty Schultz
Ori 5h49'10" -0d15'29"
Grandmom & Grandpop's Star
Cha 10h29'23" -76d6'30"
[Grandmom] KITTY Novicki
Ari 3h8'15" 29d52'16"
GrandMoM Liz
Sgr 19h30'3" -15d17'24"
Grandmom Nancy
Lyn 8h28'8" 56d25'53"
Grandmom Retzler
Ari 2h22'45" 24d16'19"
Grandmom Short
Psc 1h24'3" 29d12'13"
Grandmother Carol Lynn
Tau 4h4'33" 25d51'58"
GrandMother June
Gem 7h10'5" 27d27'41"
Grandmother Mary Willis
Leo 9h45'39" 24d40'27"
Grandmother's Star
Tau 4h28'35" 17d6'16"
Grandots
Mon 6h50'13" -0d9'7"
Grandpa
Ori 5h30'58" -3d45'25"
Grandpa
Cas 23h29'41" 55d14'31"
Grandpa
Cep 22h12'57" 62d36'19"
Grandpa
Cnc 8h6'43" 8d55'15"
Grandpa
Uma 12h0'17" 31d15'50"
Grandpa
Per 4h34'33" 44d33'15"
Grandpa Aaron Kane
Uma 9h27'41" 49d43'15"
Grandpa A.J. Moore
Umi 15h33'56" 76d40'32"
Grandpa Alan S. Reinig
Uma 11h58'23" 53d1'18"
Grandpa Albert
Sgr 19h43'25" -14d5'34"
Grandpa Ambro
Psc 0h12'34" 7d2'32"
Grandpa and Grace
Cyg 21h31'12" 44d42'51"
Grandpa and Grandma Bailey Star
Her 18h32'35" 20d13'3"
Grandpa and Grandma Haywood
Cyg 20h6'0" 35d47'10"
Grandpa and Grandma Whiteside
Uma 10h35'40" 42d23'25"
Grandpa and Grandma Wilson
Ori 4h56'15" 10d15'7"
Grandpa Bill
Aqr 22h0'20" 0d58'43"
"Grandpa" Bill Arabio
Cep 21h56'36" 62d41'34"
Grandpa Bill's Light
Cep 23h52'55" 84d48'32"
Grandpa Blackburn
Cyg 21h8'20" 53d15'21"
Grandpa Bob
Vir 14h8'36" -20d1'46"
Grandpa Bob
Per 3h35'23" 36d16'57"
Grandpa Bob
Her 16h27'59" 47d52'31"
Grandpa Bob Chase
Ori 5h52'41" 18d4'52"
Grandpa Briggs Love
Umi 13h4'55" 71d40'35"
Grandpa Bruce M. (Dagger) Mosier
Vir 12h30'9" 11d0'43"
Grandpa Bud
And 1h22'10" 45d9'9"
Grandpa Bud Pawlowski
Aql 19h34'28" 1d48'58"
Grandpa Charley
Psc 1h35'8" 14d0'53"
Grandpa Charlie Oehring
Umi 15h50'36" 79d2'51"
Grandpa Chuck
Tau 5h33'44" 26d20'26"
Grandpa Cliff
Cep 21h27'11" 64d37'29"
GrandPa Clyde
Her 16h17'57" 19d3'45"
Grandpa D
Per 3h43'32" 48d48'34"
Grandpa Dale
Cap 20h56'1" -20d14'17"
Grandpa Dale P
And 23h31'54" 47d17'50"
Grandpa Dan
Aql 19h29'42" 8d45'46"
Grandpa Dawson
Aql 19h50'22" -0d17'21"
Grandpa Demler
Umi 15h21'2" 71d59'45"
Grandpa Dosch
Uma 11h58'23" 40d23'53"
Grandpa Dunford
Uma 11h24'3" 54d22'50"
Grandpa Ed
Her 18h40'22" 22d18'36"
Grandpa Fisher
Cep 22h41'49" 72d22'53"
Grandpa Fleszar
Aqr 22h48'0" -8d58'41"
Grandpa Frank's Night Light
Umi 16h45'56" 78d39'41"
Grandpa Gerald Chambers, Sr.
Sco 16h9'26" -15d19'0"
Grandpa & Grandma
Cyg 19h43'53" 47d15'49"
Grandpa & Grandma Cone (Ron & Judy)
Cyg 20h31'30" 50d16'6"
Grandpa & Grandma Happy 50th
Her 16h22'6" 19d16'17"
Grandpa & Grandma Kim
Pho 0h28'14" -42d52'9"
Grandpa & Grandma Schau
Cyg 19h37'39" 35d35'15"
Grandpa & Grandma Steffenson
Eri 3h59'21" -9d49'28"
Grandpa Green
Ari 3h22'52" 27d26'10"
Grandpa Hal
Her 18h13'3" 41d30'34"
Grandpa Harry Radke
Lib 15h6'7" -6d57'12"
Grandpa Henry
Ori 4h58'10" 9d51'30"
Grandpa Hodgkinson
Cep 22h36'34" 73d30'40"
Grandpa Hutter
Pho 0h35'58" -41d50'39"
Grandpa Jack Aldridge
Sco 16h11'56" -15d5'31"
Grandpa Jacks Star VE3 FNJ
Sco 16h13'27" -13d36'27"
Grandpa Jay Eliezer
Ori 5h51'26" 20d34'45"
Grandpa Jim
Her 16h55'58" 34d30'3"
Grandpa Joe Hunnie
Cep 23h17'41" 76d35'55"
Grandpa Joe's Star "9-26-1941"
Lib 15h52'1" -6d56'21"
Grandpa John
Vir 14h6'24" -21d23'5"
Grandpa Jr
Cnc 9h16'45" 13d12'10"
Grandpa Kitchen
Lyn 6h29'14" 55d31'26"
Grandpa LaLa
Tau 5h45'40" 25d39'19"
Grandpa Lee
Peg 22h23'17" 20d59'38"
Grandpa Lou
Her 17h45'29" 47d21'30"
Grandpa Mac
Cap 20h33'57" -20d4'6"
Grandpa "Mac" Dad
Uma 14h7'51" 58d24'35"
Grandpa Mark
Ori 5h50'17" -0d48'46"
"Grandpa" Max Sills
Per 3h14'39" 53d13'56"
Grandpa Meier
Cep 21h43'49" 70d39'45"
Grandpa Mel Harkness
And 23h21'6" 43d47'10"
Grandpa Melvy and Granny Gloria
Uma 9h10'22" 60d57'45"
Grandpa Mike
Aql 18h50'17" -1d0'34"
Grandpa Mike's Wishing Star
Cnv 12h29'42" 32d19'50"
Grandpa Mo
Her 18h46'6" 22d13'7"
Grandpa Mouse
Umi 14h4'51" 69d58'20"
Grandpa & Nana Karseboom
Cyg 19h59'43" 30d18'42"
Grandpa Patrick
Uma 11h32'0" 56d9'53"
Grandpa Paul
Cep 0h1'57" 71d10'27"
Grandpa Paul Jimenez
Leo 11h21'33" 2d48'37"
Grandpa Phil
Ori 5h49'34" 7d30'46"
Grandpa Phil's Star
Aqr 22h54'16" -9d56'21"
Grandpa Pug Best Dad
Cnc 8h26'19" 10d59'47"
Grandpa Puzant
Aqr 22h42'29" 2d9'2"
Grandpa Ralph Engelbretson
Cep 21h26'41" 57d12'35"
Grandpa Ron Aasmundrud
Leo 11h9'15" 2d2'19"
Grandpa Rosa
Uma 10h3'14" 59d35'0"
Grandpa Roy
Cep 4h30'33" 87d24'46"
Grandpa Ruberry
Vir 11h58'17" -0d10'5"
Grandpa & Sabrina
Lyn 7h34'42" 45d49'37"
Grandpa Sal
Vir 13h46'21" 1d4'6"
Grandpa Soloperto
Lmi 10h48'9" 27d53'2"
Grandpa Spencer
Ori 5h32'41" 1d1'7"
Grandpa Stan
Per 4h21'9" 31d58'55"
Grandpa Steve
Her 16h20'41" 10d48'13"
Grandpa the Great - Stan Derby
Crb 15h38'34" 39d20'49"
Grandpa Tucholski 03/16/1932
Lib 14h34'0" -12d55'40"
GrandpaFoutch
Dra 17h44'8" 52d5'0"
Grandpa...Our Guiding Star
Lyn 7h29'3" 53d5'35"
Grandparent's Watching Star
Cyg 20h20'11" 43d59'58"
Grandpa's Star
Cep 22h2'20" 56d48'38"
Grandpa's Star - Tom Wilson
And 0h13'18" 27d29'34"
Grandpa's Starlight
Cep 22h23'28" 77d15'28"
Grandpa's World
Leo 11h28'15" 6d46'42"
Grandpop
Sco 17h23'26" -31d57'53"
"Grandpop" Andrew (Buddy) Rank
Leo 9h48'22" 28d3'39"
Grandpop Sonny Price
Cep 22h29'29" 66d0'26"
Grand-Roger
Lib 15h59'11" -9d45'43"
Grandrosie
Lib 15h55'44" -11d47'35"
Grandson's only "Pam"
Uma 8h37'35" 52d4'5"
Grandy
Sgr 19h18'41" -19d16'43"
Grange Brunner Gurr
Aqr 23h7'4" -15d39'55"
Grania Mary Maggio
Leo 11h31'29" 19d52'1"
Granite Boy - Alexander Ming Lee
Uma 9h48'6" 58d32'55"
Granitzer, Maria
Uma 8h54'44" 54d27'12"
GranJo
Vir 14h3'54" -14d20'35"
Granmart
Aqr 23h8'39" -9d19'42"
Granna & Grandpa's Golden Year!
Cyg 19h48'5" 48d15'45"
Gran-Nan
Sco 17h30'48" -41d16'29"
Granne
Sco 16h13'14" -11d42'4"
Granner
Gem 6h51'15" 31d50'0"
Grannie Jannie
Uma 8h56'51" 67d47'2"
Grannoids Star
Cnc 8h38'27" 8d2'43"
Granny
Tau 5h10'51" 18d2'42"
Granny
Peg 22h55'57" 22d7'54"
Granny
Sco 17h53'26" -38d15'13"
Granny
Sco 16h47'11" -32d58'23"
Granny and Papa Stan
Lyn 6h51'21" 51d42'11"
Granny Annie
Aqr 22h9'28" 0d19'3"
Granny Arlene Iris Maltz
Her 16h17'51" 49d9'7"
Granny Barb
Uma 8h53'35" 70d55'0"
Granny Borders
Uma 9h15'44" 53d40'57"
Granny/Bucka Star
Uma 9h4'27" 66d19'16"
Granny Burton
Cap 21h27'29" -23d18'28"
Granny Faye
Cap 20h30'10" -23d17'23"
Granny Goose
Lyn 7h32'52" 56d45'46"
Granny Harrell
Uma 9h26'43" 53d40'50"
Granny Ida Leta Terra
Lyr 18h48'59" 40d50'57"
Granny Jo's Star
Mon 6h49'29" 8d20'45"
Granny June
Gem 7h36'51" 33d7'37"
Granny Lavora J Ball
Sgr 17h48'48" -22d52'22"
Granny Long
Vir 14h8'50" -6d56'41"
"Granny" Marie Cook
Lib 14h44'48" -9d27'7"
Granny Marilyn
Cyg 21h12'15" 41d22'8"
Granny~Mom~Kristin
Gem 6h58'10" 17d9'34"
Granny Phyllis Stevens
Cas 2h25'16" 62d40'0"
Granny Wise Owl
Cas 23h44'28" 55d56'29"
Granny's Star
Peg 21h50'46" 15d56'46"
Granny's Wishing Star
Cas 23h10'29" 59d41'31"
GrannysLove
Aqr 22h42'23" -0d18'29"
Granpady
Gem 7h32'30" 26d10'2"
Gran's Glory
Cas 0h30'7" 62d44'15"
Gran's Star - Tyler's Grandma
Cas 0h33'27" 53d32'46"
Grant
Ari 3h25'42" 23d59'8"
Grant
Uma 8h16'48" 61d27'44"
Grant
Lib 14h51'19" -7d42'16"
Grant A Harkness 1 World Foundation
Her 18h30'42" 25d12'20"
Grant Allen Peretz
Leo 11h40'43" 14d25'4"
Grant & Amber
Cyg 20h27'51" 33d31'29"
Grant and Nancy's Star
Cyg 21h58'49" 48d38'19"
Grant Anthony Cusson
Cep 21h41'4" 59d9'35"
Grant Apkarian
Per 3h28'39" 48d31'41"
Grant Aram Killian
Sco 16h48'10" -33d43'54"
Grant (Beaky) Thompson
Ori 6h0'23" 14d24'55"
Grant C. Hilton
Boo 14h50'25" 34d20'48"
Grant Charles Putman
Per 3h14'42" 40d17'17"
Grant Connor Tebbe
Ori 5h56'35" 21d1'11"
Grant David Bailey
Uma 10h42'42" 47d13'26"
Grant Dominic Houle
Cap 20h42'23" -15d27'27"
Grant Dunbar
Her 16h35'16" 38d28'38"
Grant Edward
Aqr 20h54'11" -10d5'19"
Grant Edward Susman
Her 17h18'38" 19d23'11"
Grant Edward Zimmerman
Her 16h30'46" 47d13'47"
Grant Elliot Rheingold
Per 3h37'25" 49d40'22"
Grant Evan
Psc 0h1'35" -5d1'21"
Grant Gordon Stell - "Sanctuary"
Tau 5h3'7" 20d27'5"
Grant H.
Uma 10h3'11" 65d12'47"
Grant J Davis Birthday Star
Sco 17h57'23" -41d2'26"
Grant Kidd alias "Mr Skippy"
Uma 9h27'54" 45d15'23"
Grant Louis Sacco
Sgr 18h48'4" -33d11'55"
Grant McFerrin
Cap 20h36'52" -16d35'59"
Grant McKinnon Morgan
Lib 14h29'53" -19d19'0"
Grant Michael Kleman
Tau 3h46'29" 7d28'3"
Grant Michael Root
Cap 21h32'33" -18d45'15"
Grant My Precious Son
Ori 5h26'41" 4d25'47"
Grant Nicholas Giovannetti
Leo 10h11'52" 13d35'56"
Grant Nicholas Roser
Boo 14h21'38" 47d40'15"
Grant Noble Maxwell
Per 2h46'55" 44d40'18"
Grant Patrick Allen
Uma 10h7'34" 72d11'18"
Grant Patrick Boes
Cyg 19h53'18" 30d48'36"
Grant R. Mainland Bright Light
Tau 3h48'56" 20d5'29"
Grant R. Parsons
Lmi 10h44'55" 29d11'4"
Grant Richard Argall
Uma 12h39'19" 60d43'41"
Grant Rolfes Hackett
Her 17h49'45" 15d19'6"
Grant Ronald Noah
Ori 5h12'28" 6d11'50"
Grant Ross Hughes
Uma 11h34'37" 38d10'59"
Grant Scott Rood
Uma 11h32'16" 53d9'44"
Grant Simon
Lyn 6h55'57" 53d22'0"
Grant Steven Clabaugh
Uma 8h59'48" 70d19'59"
Grant Steven Solomon
Ori 5h19'18" 6d50'5"
Grant Taber
Aql 19h15'2" -7d45'11"
Grant & Tanya
Cru 12h53'38" -58d6'47"
Grant Thomas Stokke
Uma 9h12'55" 51d8'37"
Grant Warner Merrill
Psc 0h18'49" 3d54'24"
Grant William Turner
Psc 1h16'54" 21d55'50"
Grantham Wesley McWilliams
Per 3h16'48" 32d45'34"
Grants Boss Kelli Fitzgerald
Sco 17h7'3" -30d48'45"
Grant's Light
Sgr 20h17'1" -39d8'29"
Grants-Mimi,Gary,Bryan,Megan
And 0h37'10" 38d26'12"
Grant-Swee's-Zwick
Cap 20h28'0" -13d43'14"
Grantzau, Werner
Uma 11h38'28" 52d51'47"
Granville Bland Byrne III
Uma 13h52'29" 51d25'23"
Graphobia
And 23h21'16" 42d46'47"

Grasshopper
Ori 6h15'18" 9d10'21"
Grasshopper
Aqr 22h39'42" -4d52'1"
Grasshopper - Club Zürich 1886
Tri 2h9'50" 33d4'56"
Gräßler, Gisela
Uma 9h18'42" 44d17'25"
Grateful
Cas 0h27'12" 55d52'33"
Gratefully Jozzie
Cyg 20h25'33" 37d59'29"
Gratia's Pietra
Ari 2h6'33" 24d45'23"
Gratzer, August Alfred
Uma 10h39'54" 57d54'50"
Gravel Pants
Umi 21h10'29" 89d0'57"
Gravin
Sgr 19h46'38" -13d48'13"
Gravity Lover
Uma 12h18'21" 61d19'44"
Gravy
Uma 12h28'3" 56d58'34"
GRAVY - Biscuit's Beau
Ori 5h13'35" 8d12'43"
Grawnie
Cnc 9h6'32" 13d54'42"
Gray
Crb 16h3'27" 31d41'34"
Gray Chandler Layden
Ari 2h0'24" 14d25'26"
Gray Costner
Leo 9h22'16" 15d17'19"
Grayce Cunningham
Sgr 19h31'27" -15d3'2"
Graylen95
Uma 10h56'25" 56d25'46"
"Gray's Astronomy"
Ari 2h19'48" 14d5'59"
Gray's Nightlight
Umi 14h56'1" 78d12'58"
Grayson
Umi 16h25'30" 78d47'44"
Grayson
Sco 17h14'5" -44d2'52"
Grayson
Uma 9h23'32" 49d54'11"
Grayson Alesio
Her 18h43'18" 12d41'2"
Grayson and Reagan McClatchey
Uma 8h17'7" 69d32'44"
Grayson Benjamin Stark
Gem 6h22'8" 19d28'28"
Grayson Byczek
Aqr 22h8'51" 0d18'18"
Grayson Cassel
Uma 10h55'21" 65d38'5"
Grayson Clark
Boo 15h13'40" 42d14'50"
Grayson Dennis Zuromski
Cyg 20h5'46" 34d29'9"
Grayson Elizabeth Jenkins
Uma 9h44'32" 51d39'32"
Grayson Griffith
Psc 0h9'27" -0d9'53"
Grayson Guthrie Jones
Aur 6h56'35" 37d53'0"
Grayson Heather
Psc 0h50'56" 11d58'50"
Grayson Lawrence Schaeffer
Umi 12h6'31" 88d9'51"
Grayson Leigh
Cap 20h50'10" -24d53'6"
Grayson Lewis Coffin
Gem 7h18'24" 15d24'6"
Grayson MaCain
Uma 10h28'44" 54d35'52"
Grayson Nicole
Lib 15h5'38" -9d33'11"
Grayson Richard Hunter
Ori 5h33'25" 6d27'29"
Grayson Stone Quay
Cap 21h5'9" -25d11'54"
Grayson Thomas Stone
Ori 4h52'13" 2d27'17"
Grayson Ty Lamet
Cap 21h43'20" -23d29'38"
Grayson Wagner
Her 18h52'4" 22d14'33"
Grayson William Hutsey-Harwell
Dra 18h30'30" 63d46'43"
Grayson Xavier King
Com 13h3'36" 19d22'21"
Graysyn Lily
Uma 11h8'17" 66d35'28"
Grazel
Cyg 20h38'32" 42d35'58"
Grazette
Lyn 9h2'14" 41d42'19"
Grazia
Peg 21h25'2" 21d39'46"
Grazia 55
Ori 5h56'55" 11d40'57"
Grazia Bonavita
Cyg 20h34'8" 41d27'5"
Grazia Colatrella
Lmi 10h28'59" 30d29'46"
Grazia e Gilberto
Tau 3h40'53" 4d34'7"
Grazia Maria
Per 2h54'45" 44d10'46"
Grazia Montaleone
Lyr 19h19'1" 28d6'50"
graziaPgiorgio
And 23h56'12" 35d49'43"
Grazie Per Nove Anni
Cyg 20h48'0" 36d4'15"
Graziella
Ari 3h7'53" 29d24'9"
Graziella
Ori 5h44'36" -2d23'57"
Graziella
Ori 5h55'43" -8d22'22"
Graziella Carmelina
Lyn 6h41'41" 56d56'32"
Grazina *GG*
Cap 21h34'48" -18d33'3"
Grazyna and Tomasz Paduch
Cyg 21h43'59" 53d34'27"
Grealis
Gem 6h58'10" 22d32'57"
Great Auntie Nat
And 2h28'23" 46d29'22"
Great Big Bear
Uma 11h40'1" 57d1'54"
Great Dad of Barb Lainson
Cep 22h21'58" 72d47'27"
Great Daddy
Her 16h55'55" 34d46'2"
Great Day SA
Uma 11h42'21" 63d41'44"
Great Father Frost
Cep 21h34'21" 66d45'37"
Great Gram Hewitt
Crb 15h28'36" 29d46'41"
Great Gram Ruth Mustered
Cas 2h9'8" 59d11'32"
Great Grammy
Ori 5h39'48" 5d25'16"
Great Grandma Betty's Wishing Star
Uma 10h8'10" 48d58'44"
Great Grandma Eileen Mary Donohue
Umi 14h8'57" 72d36'27"
Great Grandma Johnson - Love Hunter
Vir 13h44'1" -3d29'56"
Great Grandma Sylvia Goodman
Cas 1h23'27" 59d57'20"
Great Guy Grampa
Her 17h13'28" 15d57'11"
Great Kate :)
Lib 15h27'6" -21d47'34"
Great Nanna Dodds
Cas 0h19'46" 51d32'5"
Great Nanny
Umi 16h8'2" 80d47'13"
Greater Auburn Community Chorus
Lyr 18h31'49" 37d28'3"
Greatest Dad In The Universe (Tom)
Ari 3h8'59" 29d50'39"
Greatest Tina
Cnc 8h55'25" 26d5'51"
Great-Grandpop Louis V. Barner
Ari 2h14'18" 24d41'58"
GREATmom
Aqr 22h34'4" -2d28'17"
Grecia Gardenia Chantong
Cas 23h33'41" 58d53'57"
Greco's Love, Hope and Faith
Her 17h20'19" 45d46'34"
Greek Goddess
Cas 1h6'55" 49d59'6"
Greeley and Chloe Togni Lucky Star
Uma 12h3'21" 51d50'43"
GREELY'S PLACE IN HEAVEN
Per 4h37'51" 32d9'17"
Green Acres
Psc 1h23'43" 5d1'28"
Green Dot Corporation
Cyg 20h40'3" 50d6'5"
Green Eyed Lady Lovely Lady
Cam 5h12'34" 65d11'15"
Green Eyes
Lyn 7h0'14" 54d48'9"
Green Eyes
Psc 0h49'46" 10d28'42"
Green Eyes
Cnc 8h19'4" 13d17'16"
Green Valley
Hya 9h24'5" -0d42'16"
Greene
Leo 11h11'10" -2d18'35"
Greene Wood Troop 28
Umi 15h56'45" 76d59'56"
Green-Eyed Sweetness: Alile
And 2h12'17" 50d3'12"
Greenie - Star Of Our Universe
Ari 3h7'10" 17d56'5"
Greenish Brown Female Sheep
Ari 2h51'49" 14d40'25"
Green's
Aql 20h2'21" -6d27'3"
Greer
Aqr 22h45'52" -0d42'6"
Greer Leslie Glenn Wells
Cas 23h43'11" 54d38'23"
Greerland ~ Where Fun Rules
Lyr 18h42'30" 35d16'31"
Greer's Star
Ori 5h36'20" 3d11'32"
Greg
Leo 11h2'36" 22d16'49"
Greg
Lmi 10h54'45" 25d22'57"
Greg
Her 17h35'40" 37d3'3"
Greg
Boo 14h46'53" 37d2'48"
Greg
Uma 8h19'23" 68d30'41"
Greg A. Stevenson
Aqr 21h44'21" -0d57'13"
Greg Adams
Ari 3h22'31" 28d0'52"
Greg and Adriane
Apu 17h25'57" -70d44'22"
Greg and Amanda
Lyr 19h8'53" 27d10'38"
Greg and April's Magic
Ari 2h56'34" 30d55'2"
Greg and Carol Mitchell Star
Aqr 21h21'35" -9d54'18"
Greg and Dani's Star
Cyg 19h38'50" 37d38'19"
Greg and Diane's Star
Uma 13h40'9" 57d32'12"
Greg and Dordis Hurlburt
Aqr 22h52'51" -21d21'9"
Greg and Kaitlyn Karcheski
Crb 16h4'17" 39d17'49"
Greg and Katie Alexander
Cyg 20h49'35" 42d34'57"
Greg and Linda's "Perfect Love"
Mon 6h51'34" -0d9'23"
Greg and Melissa Wade
Cyg 19h29'6" 52d30'2"
Greg and Polly Koontz
Sgr 18h29'19" -18d11'3"
Greg And Rachael Forever
Lyr 18h40'33" 34d30'55"
Greg and Sarah Haider
Cyg 20h15'7" 54d28'11"
Greg and Touriya
Gem 7h14'39" 17d10'22"
Greg Anderson
Aql 19h13'26" 6d49'32"
Greg Andrew Rikaart
Psc 1h28'51" 18d10'58"
Greg Augie Apr. 8
Cyg 19h33'55" 54d13'46"
Greg Beddow
Ori 5h8'5" 6d30'54"
Greg Bellinger
Her 18h31'17" 18d44'49"
Greg Boardman, Dr. DRC
Uma 10h54'3" 39d37'49"
Greg Bush "Ad Infinitum" 40th
Aql 20h5'44" 9d28'57"
Greg Campos
Aur 5h49'33" 46d51'18"
Greg Cappuzzo
Aql 19h4'33" 14d26'3"
Greg & Cara
Col 5h41'50" -35d9'15"
Greg Carere's Starfall
Umi 13h22'48" 73d44'58"
Greg Charles Tyransky
Aur 5h52'40" 51d29'16"
Greg & Christine's Star
Psc 1h24'56" 17d4'57"
Greg Corona
Cep 21h46'5" 61d14'48"
Greg Cottrell
Aql 19h9'41" -0d40'29"
Greg D. Krebs
Ind 20h43'39" -48d36'2"
Greg "Daddy" Kimball
Per 3h29'28" 52d12'9"
Greg "Danny" Cronin
Per 3h13'47" 52d41'7"
Greg & Darlene King
Lyr 18h47'16" 32d9'31"
Greg Desjardins Born: Sept 8/1957
Uma 11h31'53" 57d12'32"
Greg Dyer
Cir 14h5'9" -66d56'6"
Greg & Enisa
Cyg 20h4'9" 45d39'12"
Greg Fields
Tau 3h34'23" 2d35'17"
Greg Fowler's Freedom
Boo 15h9'55" 46d6'4"
Greg Fox
Cnc 8h25'56" 6d48'41"
Greg G. Malyska
Leo 10h44'6" 9d20'11"
Greg German
Dra 12h17'23" 72d39'55"
Greg Gomez
Tau 5h17'13" 21d21'21"
Greg Greenawalt
Uma 9h25'2" 72d44'5"
Greg Haggis Solomon 31.01.1985
Aqr 22h29'30" -7d13'34"
Greg Hazel
Tau 4h41'48" 16d39'3"
Greg Howard Comings
Psc 23h5'24" 7d33'56"
Greg *ILYFAE*
Vir 14h34'42" 5d3'35"
Greg Jacques
Tau 5h55'2" 26d3'33"
Greg & Jenny for Eternity
Vel 10h27'31" -42d25'33"
Greg & Johnique
Lyr 18h27'54" 32d27'51"
Greg Jones-my heart, soul and life
Lyn 6h51'1" 55d1'54"
Greg & Karen Harnden
Uma 9h16'9" 67d3'18"
Greg Kasey Turner
Uma 9h15'36" 47d59'37"
Greg & Kathie Tanner - forever
Cyg 21h50'53" 52d14'23"
Greg & Kathy Julius
Cyg 20h13'38" 50d36'51"
Greg & Katie Martin
Crb 15h33'38" 27d41'46"
Greg Kelly
Ori 5h25'35" 3d20'4"
Greg & Kelly together forever
Aqr 23h53'11" -13d15'3"
Greg & Kim Winters
Aql 20h8'40" 11d27'59"
Greg Koeck
Her 17h37'39" 41d13'39"
Greg & Kristi
Lmi 10h30'43" 38d10'13"
Greg & Kristy Collins
Cyg 20h4'53" 30d55'43"
Greg [l' étoile de nos vies]
Ari 2h47'12" 31d3'51"
Greg & Laina Saul
Cyg 19h43'41" 33d53'46"
Greg Lewis 77
Ori 4h46'51" 5d10'29"
Greg Liptac
Uma 11h16'37" 58d26'1"
Greg Marin
Crb 15h40'12" 35d29'15"
Greg McIvor
Aql 20h18'13" 8d53'10"
Greg Melbert
Per 4h8'52" 36d28'28"
Greg Mosley
Psc 1h25'45" 14d43'35"
Greg Mouat
Lyn 7h33'52" 36d44'37"
Greg N Weyant
Uma 9h6'21" 52d31'48"
Greg Nave
Per 4h28'5" 51d21'56"
Greg & Necia Always and Forever
Sco 17h27'17" -44d7'53"
Greg Nelson
Ari 2h47'45" 25d52'36"
Greg Nibbs
Cru 12h48'30" -60d29'6"
Greg O'Brien
Mon 7h49'21" -4d52'47"
Greg Paquin-Pacman
Umi 14h20'35" 84d16'58"
Greg & Pati Iseman
Sge 19h42'32" 17d14'8"
Greg Pentastar Points North
Uma 9h20'19" 68d45'13"
Greg Prather
Ori 5h24'5" 5d47'15"
Greg Putnam and A Sister's Love
Her 16h49'37" 34d6'4"
Greg "Red" Gilbert
Her 17h7'19" 34d14'39"
Greg Richard Moreschini
Eri 2h13'22" -53d17'53"
Greg Rittenhouse
Cma 6h16'17" -14d56'54"
Greg Ruhlander
Uma 9h46'4" 44d51'35"
Greg S. Phelps, My "Rock" Star
Vir 13h45'10" -10d29'0"
Greg Sellitti
Per 4h11'59" 45d5'27"
Greg senior
Uma 8h15'32" 61d16'6"
Greg Smith
Uma 11h16'39" 42d37'36"
Greg Sosebee
Aqr 22h18'20" -0d22'16"
Greg Soto
Equ 21h16'10" 9d4'9"
Greg Soto
Aql 19h7'39" 13d26'0"
Greg Steele
Cnc 8h15'10" 9d27'45"
Greg Steinbock
Uma 12h37'28" 59d19'55"
Greg Sterk
Per 3h37'7" 47d45'43"
Greg The Wiz Kid Benjegerdes
Leo 10h16'52" 21d19'19"
Greg Thomas
Aur 5h30'17" 40d27'18"
Greg Trnka
Leo 11h9'21" 8d39'15"
Greg Tucker
Per 2h51'36" 52d33'46"
Greg Vandeligt's Vision
Tau 3h54'12" 9d22'33"
Greg VanWinkle
Per 4h23'41" 42d26'42"
Greg W Melvin
Leo 11h31'26" 17d15'38"
Greg Waldron
Her 16h51'2" 37d0'18"
Greg Whitney Mitchell
Vir 12h42'47" 0d9'30"
Greg Williams
Sco 16h52'13" -42d47'19"
Greg Wohrle
Gem 7h12'14" 17d4'57"
Greg Wood
Her 17h6'27" 32d28'55"
Greg Zeller
Uma 13h23'19" 57d24'2"
Greg, Cindy, and Jersey Perkins
Uma 10h1'59" 68d26'19"
Greg, Madeline, Jason & Aliya Cone
Uma 11h39'40" 57d30'24"
Greg, The Super Spotter
Tau 5h27'18" 27d22'9"
Gregamus
Cep 20h42'42" 59d43'1"
GregAnn
And 2h22'14" 50d30'48"
Gregary Anderson - Rising Star
Lmi 10h30'46" 28d21'58"
GregDeborah_12/26/05
Cyg 21h30'12" 30d42'57"
Gregg
Aqr 23h8'36" -9d56'7"
Gregg Allan Bouchard
Aql 19h48'38" -0d29'5"
Gregg and Heidi Berg Forever
Uma 9h20'2" 67d55'38"
Gregg Balton
Crb 16h11'4" 39d19'49"
Gregg Bansavage
Aql 19h15'53" 3d18'2"
Gregg Cassano
Uma 11h34'2" 32d7'23"
Gregg DiNardo
Psc 1h22'31" 31d58'21"
Gregg Fox
Uma 9h24'45" 67d52'53"
Gregg Garon
Her 17h30'1" 33d50'16"
Gregg & Jan
Cyg 21h57'23" 49d16'43"
Gregg Joseph Bowen
Lyn 7h13'28" 59d2'46"
Gregg Klinzing *DADDY*
Per 3h20'2" 47d27'31"
Gregg Lee Workman
Boo 14h20'22" 43d12'53"
Gregg & Lynn
Cyg 19h50'45" 31d43'50"
Gregg Miller
Her 17h5'15" 32d1'37"
Gregg & Patti Bourland ~ 25 Years
Cyg 19h40'4" 30d51'29"
Gregg Reynolds
Uma 11h36'37" 29d13'27"
Gregg Scherban
Vir 12h42'30" 8d33'21"
gregg sean
Vir 13h10'44" 11d9'12"
Gregg T. Norman
Lib 15h20'44" -12d16'45"
Gregg Zimmerman
Vir 13h28'48" -0d1'24"
GreggAGenovese
Her 18h47'27" 20d58'4"
Greggan Elizabeth-Sarah Capps
And 1h38'27" 42d33'46"
Gregg's Star
Cep 2h29'58" 81d26'1"
GreggTara
Per 3h41'53" 49d29'55"
Greghan's Castle in the Sky
Per 2h44'29" 52d15'11"
Gregimus
Ori 4h50'1" 4d12'56"
Greg-n-Lina 2421
Umi 13h10'1" 69d32'13"
Gregochutney
Leo 11h29'18" 28d12'41"
Grégoire Leproust
Leo 10h54'6" 14d0'7"
Gregor Iain Bruce
Ori 5h33'37" -9d22'27"
Gregor Mario André
Uma 9h16'20" 68d49'55"
Gregor Nikolaus Gentschev
Lib 14h53'52" -6d26'56"
GREGOR & VESNA
Uma 12h43'4" 60d39'3"
Gregor Wolf
Ori 5h54'26" 17d40'5"
Gregorec
Ori 5h40'10" 0d16'53"
Gregorio
Per 3h22'53" 32d16'44"
Gregorio DiBetta
Cyg 20h17'33" 48d26'0"
Gregorio Luis Torales
Her 18h10'24" 48d3'10"
Gregory
Per 4h48'42" 45d43'12"
Gregory
Lmi 9h36'53" 34d11'0"
Gregory
Her 17h59'17" 14d52'56"
Gregory
Her 18h53'44" 23d9'20"
Gregory A. Hardy
Lyn 7h59'9" 57d46'50"
Gregory A Newman,
Peg 23h17'9" 17d8'4"
Gregory Aaron Pierson
Uma 12h12'43" 55d3'50"
Gregory Abplanalp
Ori 5h34'55" 9d52'26"
Gregory Adam Thorowgood
Uma 10h30'1" 44d32'2"
Gregory Aldridge Zander Jr.
Leo 10h27'30" 13d35'51"
Gregory Allan (Givit) Toomey
Cnc 8h49'8" 14d37'57"
Gregory Allen McCain
Psc 1h13'3" 17d2'40"
Gregory Allen Tonkovich
Cep 20h41'40" 65d31'1"
Gregory Allen Winterhalter
Cma 6h15'37" -14d5'1"
Gregory and Stacy Pucci
Lmi 11h0'5" 29d23'6"
Gregory and Susan's star
Lyn 7h55'52" 42d9'5"
Gregory Andrew Alexopoulos
Gem 7h35'56" 25d39'44"
Grégory Angelcat
Cas 23h44'46" 54d50'19"
Gregory Anthony Gines
Aqr 22h47'15" -6d21'16"
Gregory B. Bostick III
Uma 8h37'26" 57d39'43"
Gregory Baggett
Ari 2h44'6" 22d5'30"
Gregory Bernard Kirkland Jr.
Cyg 20h29'57" 57d22'28"
Gregory Botts
Per 4h9'38" 50d48'44"
Gregory Brian Hart
Boo 14h20'3" 45d27'54"
Gregory Brian & Sarah Jayne Forever
Cyg 20h38'59" 53d10'11"
Gregory Brian Steinhouse
Uma 11h10'17" 53d55'16"
Gregory Bryant Fales
Ori 5h32'4" -3d5'29"
Gregory "Bubba" Miller
Her 16h31'54" 37d56'45"
Gregory "Bubba" Rivera
Gem 6h39'51" 23d52'14"
Gregory & Carolyn Coburn's Star
Gem 7h34'34" 28d17'51"
Gregory Charles Peterson
Boo 15h3'39" 43d16'44"
Gregory Charles Prazenica
Ori 6h5'24" 17d40'46"
Gregory Charles Wilhelm
Lyn 6h57'49" 51d21'23"
Gregory Chee
Ori 5h18'53" -7d28'58"
Gregory Cordaro
Lyn 7h27'44" 55d4'37"
Gregory Croteau
Cyg 20h47'17" 37d21'42"
Gregory D. Young
Pho 23h46'45" -39d39'55"
Gregory Dacenko
Uma 10h26'21" 63d37'3"
Gregory Dakota Conklin
Her 16h47'54" 25d42'0"
Gregory Dale Bittle
Lib 15h7'13" -16d34'26"
Gregory Dale Haight
Her 18h18'41" 14d58'49"
Gregory Daughenbaugh
Sgr 19h52'57" -21d10'38"
Gregory David
Cap 20h24'27" -10d13'10"
Gregory David Potter
Psc 0h24'18" 7d40'47"
Gregory Davis Rhodes
Cap 20h24'38" -13d27'44"
Gregory Drew Wohrle
Gem 6h19'7" 24d4'51"
Gregory E. Creasy 9-14
Vir 12h20'4" 11d42'52"
Gregory E. LaBelle
Lib 15h43'25" -4d9'26"
Gregory E. Witt August 23, 1952
Per 4h41'18" 48d22'11"
Gregory Elijah Paeth
Uma 10h59'13" 44d44'30"
Gregory Evan Philactos
Uma 8h49'20" 70d58'30"
Gregory Francis Kehrig
Tau 5h36'59" 25d22'42"
Gregory Garland Haney 613
Aur 5h55'59" 42d52'3"
Gregory Garon
Sco 16h48'51" -31d34'31"
Gregory Gascon
Aqr 23h27'57" -17d42'0"
Gregory Gene Oblisk
Sco 16h40'10" -30d38'22"
Gregory Glazman
Cep 1h27'58" 80d25'5"
Gregory Glen Wharton
Peg 21h48'14" 9d23'1"
Gregory Glossy
Aur 5h58'18" 43d41'17"
Grégory Goulot 8/06/04
Aql 19h2'9" -8d19'18"
Gregory H. Gang: Love of My Life
Uma 10h48'53" 67d7'26"
Gregory Haralambopoulos
Cnc 9h16'43" 14d54'56"
Gregory Harrison
Aur 5h59'9" 32d48'54"
Gregory Harvey
Ori 5h42'44" 1d37'53"
Grégory Humbert
Cnc 8h6'49" 10d30'35"
Gregory J. Cecconi
Aql 19h6'19" -0d11'19"
Gregory J. Erickson
Boo 15h9'47" 48d18'31"
Gregory J Lucas
Ori 5h36'48" -1d40'45"
Gregory J. Mitchell
Lib 15h3'34" -1d42'39"
Gregory J. Ramage
Per 1h44'29" 54d33'13"
Gregory J. Sinclair
Uma 9h22'11" 43d25'30"
Gregory James Abrams
Cnc 8h20'18" 25d58'56"
Gregory James Blair
And 0h29'13" 42d36'28"
Gregory James Bonnen
Lib 15h35'5" -9d30'47"
Gregory James Gozzo
Cap 20h20'24" -19d44'47"
Gregory James Molamphy
Cap 21h58'20" -24d28'7"
Gregory James Patton, Jr.
Ind 21h40'39" -64d51'25"
Gregory James Vail
Cep 20h48'17" 64d21'15"
Gregory James Ward
Cas 2h28'46" 65d59'52"
Gregory John Benedict
Lib 15h39'34" -19d37'53"
Gregory John McKenna
Lyn 8h21'43" 58d22'56"
Gregory John Nail
Her 17h10'20" 36d5'58"
Gregory Joseph Wolf
Uma 13h2'51" 55d25'0"
Gregory Joshua Burton
Per 3h0'39" 45d58'58"
Gregory Jurczyk
Per 2h40'39" 54d0'20"
Gregory Karl Alexander, Jr.
Leo 11h5'52" 21d44'0"
Gregory Keith Maier
Tau 4h24'28" 30d10'39"
Gregory Kelley
Per 2h47'1" 46d5'13"
Gregory Kessler
Ori 5h38'17" 7d35'51"
Gregory Ksenych
Gem 6h29'3" 18d38'6"
Gregory L Cole
Per 3h16'7" 51d2'53"
Gregory L Roteik
Leo 9h36'14" 10d13'30"
Gregory L. Schweitzer
Ari 2h35'53" 26d35'46"
Gregory Lasater's Star
Per 2h32'50" 55d12'17"
Gregory Lawrence Maggio
Vir 12h15'38" 11d25'40"
Gregory Lee Bass
Ari 2h9'19" 23d39'59"
Gregory Lemarchal
Cas 2h50'54" 61d35'53"
Grégory Lemarchal
Lib 14h48'37" -4d44'9"
Grégory Lemarchal
Tau 3h38'28" 18d20'55"
Gregory Leon Walsh
Lmi 10h19'8" 31d44'31"
Gregory Liam Tripcony
Ori 5h12'29" 7d52'10"
Gregory & Lorraina Wieting
Uma 12h4'21" 52d59'22"
Gregory Louis Pipkin
Cap 20h36'55" -12d22'2"

Gregory Louis Wereski Sr.
Per 4h18'55" 44d27'26"
Gregory Lynn Harston
Boo 14h20'20" 23d2'13"
Gregory M. Sobotka
Leo 11h43'51" 26d19'59"
Gregory M. Vaughn
Sco 16h2'38" -17d18'45"
Gregory Mark Miller
Cma 6h35'21" -15d58'26"
Gregory & Marlene Gosser 12-12-1993
Cep 22h56'10" 72d1'34"
Gregory Martin Dugan
Vir 14h19'0" 3d25'4"
Gregory Martin Wiener
Gem 6h57'28" 14d10'12"
Gregory & Melinda Heffner
Lyr 19h3'18" 42d41'52"
Gregory Mercurio
Ari 2h35'43" 18d3'13"
Gregory Michael Finnerty
Sgr 18h22'43" -23d9'34"
Gregory Michael Fox
Lyn 8h23'41" 57d20'27"
Gregory Michael Hutson
Uma 8h16'57" 59d54'17"
Gregory Michael Moore
Cyg 19h56'55" 52d24'45"
Gregory Michael West
Boo 15h15'59" 45d44'3"
Gregory Monack
Aur 5h30'2" 47d45'41"
Gregory My Love Bug
Her 16h28'56" 45d36'58"
Gregory Myers
Per 3h5'51" 42d48'18"
Gregory N. Mitchell Jr.
Dra 17h40'50" 55d14'46"
Gregory Neil Sanders II
Cap 20h42'16" -16d37'47"
Gregory Neofitos
Sgr 18h19'6" -24d7'59"
Gregory Nicolas
Uma 9h35'5" 62d43'32"
Gregory Niggli
Com 11h59'26" 22d6'7"
Gregory Nolan Smith
Cap 21h42'27" -13d53'3"
Gregory O'Bannon
Per 3h42'12" 45d59'13"
Gregory Orville Brandenburg
Vir 13h45'32" -6d0'40"
Gregory Owen Ganser
Vir 13h2'51" 4d0'52"
Gregory P. Ochoa
Lyn 6h50'42" 56d18'5"
Gregory Pamela and Overton Boys
Tri 2h8'11" 33d2'1"
Gregory Papadopoulos
Uma 8h29'30" 61d55'49"
Gregory Paul
Per 4h49'25" 43d55'7"
Gregory Paul Watson
Leo 11h40'22" 15d33'24"
Gregory Peter Panayoti
Sgr 17h53'8" -29d31'31"
Gregory Phillip James Fuortes
Uma 10h31'46" 65d12'38"
Gregory Pittman
Aur 5h5'56" 47d36'38"
Gregory Prekupec
Ori 5h39'13" 8d56'48"
Gregory R. Auer
Sco 17h24'51" -42d47'46"
Gregory R. Nichol
Per 4h27'50" 45d40'35"
Gregory R. Specht
Dra 18h25'38" 57d37'52"
Gregory Rand
Uma 11h15'47" 58d19'31"
Gregory Randall Shifflett II
Lib 15h11'15" -18d6'58"
Gregory Ray
Uma 13h46'18" 61d58'26"
Gregory Richard Debert
Tau 5h29'52" 21d25'32"
Gregory Richard Lauersen
Uma 11h8'52" 62d30'53"
Gregory Richard Spencer, M.D.
Ori 6h16'55" 15d54'15"
Gregory Richter
Her 17h27'33" 18d52'8"
Gregory Robert Almas
Sgr 18h18'0" -25d40'15"
Gregory Robert Becker
Aql 19h49'53" -0d29'26"
Gregory Robert & Lisa Ann Tompkins
Ori 6h13'13" 3d44'22"
Gregory Robert Mueller
Sgr 18h10'17" -33d39'57"
Gregory Roberts Williams
Gem 6h49'5" 14d5'54"
Gregory Robin Chandler
Her 18h35'42" 17d44'7"
Gregory Roland Getty
Psc 1h13'26" 11d18'28"
Gregory Ross
Uma 11h11'16" 63d39'30"
Gregory Royce Robertson
Her 17h22'29" 27d50'57"
Gregory Rund
Per 4h24'59" 44d38'57"
Gregory Ryan Caplan
Sgr 19h11'54" -20d49'49"
Gregory S Fortner
Sgr 18h42'15" -16d58'20"
Gregory S. Snyder
Uma 8h35'39" 46d48'42"
Gregory Samp
Ori 6h2'10" -0d55'33"
Gregory Scott Bastow
Her 16h33'48" 32d6'54"
Gregory Scott Bresnen
Per 2h16'0" 57d11'12"
Gregory Scott Brown
Sco 17h24'20" -39d40'55"
Gregory Scott Hatfield
And 0h45'30" 40d15'46"
Gregory Scott Purdy II
Cap 21h24'40" -16d57'2"
Gregory Scott Schwan
Per 2h25'41" 55d14'57"
Gregory Shane Burke
Her 15h50'2" 39d48'56"
Gregory Shirin
Uma 12h1'0" 41d39'41"
Gregory Slavin
Cas 2h48'44" 65d28'40"
Gregory Smith
Boo 15h37'32" 47d18'7"
Gregory "Sonnie" Rasquinha
Tau 4h16'16" 29d21'44"
Gregory Stephen
Gem 6h50'50" 24d32'15"
Gregory Steve Hadjiyane
Gem 6h49'27" 34d35'19"
Gregory Steven Doan
Uma 9h13'14" 52d36'35"
Gregory Stewart Noelken
Ari 2h16'6" 26d2'6"
Gregory & Susan Moncrief 12-29-65
Lib 14h54'54" -7d27'3"
Gregory T. Dorsett
Her 17h59'39" 41d44'55"
Gregory Taylor (Our Love) Dumont
Ori 5h22'46" -8d4'20"
Gregory the Great (Greg Collins)
Leo 11h52'48" 23d54'12"
Gregory "The Russian Star"
Cep 21h28'11" 66d19'26"
Gregory Thomas McKeon
Tau 5h27'57" 24d26'24"
Gregory Thomas Swingle
Leo 9h56'6" 32d21'54"
GREGORY TWO-HAWKS BENJAMIN 2005
Dra 16h15'55" 56d7'53"
Gregory Vincent Shiah
Per 3h40'17" 47d58'21"
Gregory W. Gatto
Sco 16h39'20" -27d45'47"
Gregory Warden
Sco 16h42'13" -41d46'57"
Gregory Watson Schmidt
Ori 5h46'2" 5d50'37"
Gregory Weber Lang
Per 3h25'0" 38d28'5"
Gregory William Gray
Ori 6h7'42" 12d38'52"
Gregory William Paradis
Ari 3h13'9" 28d42'40"
Gregory William Summers, Jr.
Aqr 22h5'37" -0d42'56"
Gregory Worth Cochran
Her 17h3'8" 33d50'23"
Gregory-Jessica-Joshua-020607
Dra 16h39'58" 63d26'52"
Gregory...My Love...My Angel
Ori 6h19'56" 5d41'13"
Gregory's Girl
Tau 5h28'48" 21d59'56"
Gregory's glory
Uma 12h0'59" 46d50'34"
Gregory's Star
Vir 14h37'32" 2d48'56"
Gregory's Star
Gem 7h1'50" 25d12'20"
Greg's Candy
Sgr 18h58'36" -34d19'58"
Greg's Crummy Star (Greg Crum)
Uma 8h57'27" 56d14'56"
Greg's Song
Ori 6h1'28" -0d41'54"
Greg's Star
Aql 20h12'13" 15d12'14"
GregSteeleNothingLastFor everButUS
Per 3h39'26" 44d43'20"
GregWerner 60
Her 17h26'40" 31d16'51"
Greiner, Herbert und Christine
Ori 5h21'21" -9d9'55"
Greipl, Helga
Ori 5h20'14" 3d28'52"
Grejormel
Sco 17h19'16" -32d18'52"
Grelanna Will-child
Cnc 9h11'47" 32d39'59"
Greleon
Uma 11h39'32" 62d34'58"
Grell, Hartmut
Uma 10h41'24" 46d8'41"
Gremaf Reyes
Tau 5h18'43" 18d9'47"
GREMHOG
Aql 19h32'4" 7d50'20"
Grenda From Pocklooolooland
Cru 12h38'12" -59d21'0"
Grenville Curly Newton
Uma 11h31'51" 44d42'3"
Greta
Cyg 21h5'22" 47d41'47"
Greta
Peg 22h47'55" 3d36'7"
Greta
Cyg 21h5'22" 47d41'47"
Greta
Cam 12h59'13" 78d32'8"
Greta Elizabeth
Cam 3h52'55" 54d57'26"
Greta GC
Dra 18h33'8" 65d48'32"
Greta Gibson
Vir 14h4'27" -15d3'38"
Greta Gliessner
Lib 15h47'0" -25d32'58"
Greta Lavendol
Lyn 9h5'46" 38d6'40"
Greta Ling you shine so bright
Uma 9h6'30" 50d48'55"
Greta Morassi
And 0h17'34" 46d17'11"
Greta Naju Jargue
Ori 5h40'29" -7d49'1"
Greta Naselli
Cam 4h25'39" 54d31'37"
Greta Rose Munch
Vir 12h6'19" -9d39'14"
Greta Rose Tapley
Cas 1h38'19" 64d36'53"
Greta Simpson - gls - "Believe"
And 23h36'12" 47d31'44"
Greta White
Uma 13h9'50" 54d48'45"
Greta, My Shining Star
Vir 12h37'46" -8d9'20"
Greta's Life of Stars
Uma 10h16'51" 46d12'48"
Gretchell's Piece Of Heaven
Umi 13h23'52" 75d4'15"
Gretchen
Vir 12h44'28" -9d10'42"
Gretchen
Cas 0h42'36" 64d38'0"
Gretchen
Cas 1h23'20" 64d38'19"
Gretchen
Cas 1h39'50" 63d35'28"
Gretchen
Uma 9h1'5" 46d47'49"
Gretchen
And 0h35'53" 41d37'54"
Gretchen
And 0h22'42" 40d29'30"
Gretchen
Aqr 22h15'4" 1d24'3"
Gretchen
Cnc 8h37'28" 15d47'26"
Gretchen Ann
Ari 3h6'9" 29d12'36"
Gretchen Anne
And 0h31'52" 33d1'39"
Gretchen Anualao Smith
Sco 17h57'21" -30d25'54"
Gretchen Aufman
Sco 17h57'16" -37d44'16"
Gretchen Brunati
Ori 5h57'35" 6d50'49"
Gretchen & Dallas
Leo 9h26'28" 26d5'41"
Gretchen Elizabeth Reifeis
Umi 16h50'33" 79d55'14"
Gretchen Gilmore
Sco 16h41'22" -31d40'49"
Gretchen Gramlich
Lib 15h16'35" -26d16'38"
Gretchen "Grammy" Cumming
Cas 1h23'11" 56d50'23"
Gretchen Guendelsberger
Lyn 8h17'56" 45d51'47"
Gretchen Heiden
And 0h43'54" 45d24'12"
Gretchen Kay Gerke
Gem 6h23'38" 18d23'36"
Gretchen Krone
And 0h19'58" 36d5'22"
Gretchen Kurowski
Psc 1h37'56" 19d51'5"
Gretchen Louella Glaize
Ari 2h37'17" 21d18'34"
Gretchen Mac
Lyn 7h42'35" 50d27'56"
Gretchen Marie
Gem 6h34'24" 24d50'17"
Gretchen Olson ~Pure Sport~
Uma 12h57'38" 58d25'30"
Gretchen P
Tau 5h28'20" 19d33'39"
Gretchen Pumphrey
And 0h53'43" 38d47'0"
Gretchen Raquel
Peg 22h41'24" 28d23'41"
Gretchen Rucker 8/1/84-2/20/06
Leo 11h20'39" 14d21'43"
Gretchen Ruth Meisner
Tau 4h40'26" 22d14'52"
Gretchen Sael Sepulveda
Ari 3h28'42" 28d3'5"
Gretchen Talley
Cas 0h35'26" 64d49'15"
Gretchen the Great
Uma 10h6'51" 58d12'29"
Gretchen W. Hollis
Com 13h11'20" 17d50'39"
Gretchen Winkel
Lib 14h44'10" -17d42'28"
Gretchen1973
Oph 17h31'46" -0d16'45"
Gretchen's twinkle in the sky
Cyg 20h55'26" 44d53'58"
Grete Prets-Schwaiger
Gem 7h44'59" 20d17'33"
Gretel
Lyn 6h43'20" 51d22'35"
Grethe Lindholm
Uma 10h43'37" 57d4'35"
Grettell Aida 11122005
Cas 23h41'59" 57d28'8"
Grey
Aqr 22h46'46" -20d58'38"
Grey Arthur
Uma 14h3'56" 56d24'32"
Grey Beard
Aql 19h8'50" 12d5'4"
Grey, Reiner
Uma 10h9'46" 43d26'34"
Greysi
Crt 11h25'50" -8d33'26"
Greyson Arden MacLean
Her 16h23'15" 19d31'5"
Greyson Francis
Ori 5h57'56" 20d54'25"
Greyson Koichi Kendron Cervantes
Leo 9h37'30" 24d2'41"
Greyson Louis Fezatte
Crb 16h8'7" 37d54'11"
Greyson Reed Linnell
Crb 16h22'44" 36d42'30"
Grezdfez
Her 17h1'18" 13d7'23"
G.R.Hayden
Uma 11h42'21" 37d24'34"
Gribbins
Leo 11h18'23" 14d45'32"
Gricelda Agudo
Leo 10h24'59" 16d41'54"
Griff 2005
Aqr 22h3'56" 1d1'36"
Griffey
Uma 12h8'11" 51d6'3"
Griffin
Lyr 19h9'58" 38d25'41"
Griffin
Leo 11h13'16" 10d50'48"
Griffin
Lyn 6h48'17" 52d46'2"
Griffin Andrew Werner
Aqr 23h26'15" -9d25'22"
Griffin Baldwin Gher
Her 17h55'31" 21d58'54"
Griffin Benjamin Kessler
Aqr 22h35'24" -3d33'13"
Griffin Blane
Gem 7h42'27" 25d51'1"
Griffin Bolduc Cannon
Her 18h32'27" 19d46'38"
Griffin Brown
Umi 13h8'26" 74d43'39"
Griffin Cael Kearns
Del 20h44'23" 15d29'4"
Griffin Chappelle Carr
Ori 6h7'28" -0d13'33"
Griffin Craig Nelson
Her 16h26'19" 10d26'15"
Griffin David Miller
Sgr 19h41'55" -13d44'23"
Griffin Douglas Dykstra
Umi 14h25'34" 72d39'38"
Griffin Hawk
Gem 6h48'1" 28d41'20"
Griffin Hohepa Clark
Lmi 10h2'10" 40d0'39"
Griffin J. Herod
Leo 11h5'3" 21d18'26"
Griffin Joseph Loren Spencer
Per 2h43'54" 44d22'1"
Griffin Kane
Her 18h13'15" 29d9'10"
Griffin Langston Jones
Her 17h1'37" 30d13'37"
Griffin Lee Harwood
Vir 12h54'11" 7d33'24"
Griffin Lyons "Griffy's Star"
Sgr 19h50'19" -33d38'47"
Griffin Markwith
Cma 6h30'24" -11d55'26"
Griffin Matthias Webb Milner
Lyn 7h19'8" 49d33'14"
Griffin Merryweather Erno
Gem 6h31'54" 16d15'35"
Griffin Patrick Letko
Lib 15h16'46" -7d26'13"
Griffin Stephens Bunn
Sgr 18h49'12" -17d45'3"
Griffin Thomas
Umi 14h33'12" 79d53'11"
Griffin Thomas Mandirola
Leo 11h45'28" 24d37'25"
Griffin's Aerie
Uma 9h23'13" 62d13'3"
Griffin's Dad
Gem 7h26'17" 26d8'58"
Griffin's Light
Umi 14h24'54" 66d5'37"
Griffinstar04
Uma 11h43'54" 32d27'7"
Griffon
Vir 13h36'48" 1d9'14"
Griffon 14
Sgr 19h24'19" -12d38'43"
Grigor Balantzian *George*
Uma 11h18'5" 31d14'4"
Grillo
Her 18h44'39" 14d31'26"
grillz
Sco 16h14'18" -9d50'26"
Grim Nifty
Uma 12h2'48" 55d31'31"
Grimace
Crb 15h37'27" 37d27'43"
GRINGA
Ori 6h19'48" 19d34'20"
Gringo Largo
Boo 14h32'50" 28d8'2"
Gringoire - Matteo Setti
Per 3h3'43" 51d39'12"
Grins and Giggles
Cap 21h42'31" -13d0'56"
Grinsekatze Hebeisen
Her 18h14'14" 17d20'58"
Griselangel
Tau 4h35'42" 11d31'4"
Griselda Loza
Lmi 9h57'42" 34d25'30"
Griseldi Maria Corona Ortiz
And 1h54'50" 37d41'48"
Grisell Vazquez
Cnc 8h16'39" 22d39'10"
Griselle Rivera
Psc 1h21'54" 21d28'24"
Grismary
And 0h13'54" 26d31'37"
Grit Mause
Vir 12h20'28" -6d0'47"
Grit Schäfer
Uma 11h2'6" 57d26'52"
Gritter Borealis
Crb 15h25'31" 27d26'21"
Grizz (Carl Brownell)
Per 3h17'10" 51d53'15"
Grizzz
Psc 0h47'41" 17d12'40"
Grlee Siblex
Vir 12h5'26" -5d26'20"
Grma2Three
Vir 14h40'15" 4d59'51"
Grobys, Jürgen
Uma 8h24'40" 64d15'1"
GROG
Ori 6h13'19" 8d12'31"
Gröling, Willi
Cnc 8h55'1" 16d29'5"
Gromit
Cnv 12h31'58" 36d2'0"
Gromit
Umi 15h29'53" 71d10'38"
Gröning, Klaus-Peter
Uma 14h0'28" 55d18'58"
Groove Goddess (a/k/a Kerri-Ann)
Cas 23h11'10" 55d59'6"
Groovy Chick Janet (Jan Prince)
Cru 12h32'2" -59d53'36"
Gröschke, Reiner
Uma 9h56'23" 65d3'12"
Groß, Detlef
Uma 10h42'19" 44d27'14"
Groß, Günter
Ori 6h16'21" 11d4'31"
Gross, Siegmar
Uma 12h33'55" 58d52'21"
Große, Stefan
Uma 12h34'29" 58d41'26"
Grosseltern Nyffeler
Ori 5h59'59" 22d25'18"
Großer Frank-Steffi-Fine-Anna-Stern
Aqr 21h59'50" 0d14'17"
Grosser Tiger
Uma 9h18'45" 50d31'38"
Grosser, Oliver
Uma 11h59'17" 37d12'11"
Großi
Ori 6h12'52" 16d30'49"
Grosspapa
And 2h18'36" 50d10'33"
Grosss
Uma 10h52'37" 72d44'1"
Grouch 50
Uma 9h15'51" 68d42'55"
Ground Zero 8/5/05
Per 3h2'45" 52d17'25"
Groupie 1
Uma 12h3'29" 31d57'44"
Grove Havener
Leo 9h44'12" 6d45'21"
Grover
Lyn 8h11'14" 44d51'7"
Grover
Sco 16h2'54" -29d41'52"
Grover and Lucille Williams
Cyg 21h8'30" 35d34'28"
Grover Garrett
Sco 17h52'22" -33d19'54"
Grover Gorman Brittingham, Jr.
Dra 19h45'2" 70d3'30"
Grover Salancy
Ori 4h52'35" 11d8'23"
Grover Sloan
Ori 6h3'49" 5d45'27"
Grover's Girl
And 23h57'13" 36d44'25"
Grover's Star
Ori 6h10'18" 15d1'18"
Grow old with me . . . Happy 25th !
Psc 1h35'38" 8d38'52"
GRP-18
Cam 4h9'21" 68d55'46"
Grud
Lyn 6h44'41" 59d5'58"
Gru-Man
Cnc 8h47'32" 16d26'11"
Grumbleweed
Dra 17h35'2" 66d3'21"
Grumpa-Bapa6
Sco 17h15'32" -33d4'52"
Grumps Hawking 16.04.06
Cru 12h36'39" -58d27'34"
Grumps's Star
Uma 12h4'34" 60d13'20"
Grumpy
Dra 16h40'21" 56d37'53"
Grumpy
Uma 8h54'1" 49d41'9"
Grumpy Bennett
Cma 6h40'53" -17d39'32"
Grumpy Paisano Dad Ruscigno
Cnc 9h6'58" 30d38'12"
Grümschali
Vul 19h54'21" 23d2'24"
Gründel, Ida Dorothea Auguste
Ori 6h11'53" 13d23'41"
Grundmann, Robert-Frederik
Uma 14h3'1" 55d1'58"
Grunt
Lac 22h47'17" 36d39'4"
Grupo Elefante Mexico
And 2h11'7" 37d38'14"
Gryphy
Cap 20h24'50" -10d10'53"
Grzenda
Cap 21h53'8" -19d49'52"
G.S. Stevens
Uma 10h20'8" 47d29'55"
GSB (Girl Seddy Bear)
Uma 12h5'22" 34d2'11"
GSDeibert45
Uma 13h50'54" 52d20'38"
GSK Lifeguard of My Heart
Psc 1h16'17" 31d13'7"
gsnappyscap
Lmi 10h30'15" 33d59'27"
GSON
Tau 5h40'50" 20d4'50"
G-Spot, my favorite little guy.
Equ 21h18'18" 2d41'1"
GSRB
Ori 5h36'59" 3d17'35"
"G-Star"
Sco 17h30'37" -40d11'16"
GTF 1986
Uma 8h54'36" 66d52'6"
GTrineK
Psc 0h53'22" 4d32'46"
Gu Brath
Oph 17h34'24" 6d14'7"
Gu Hao-60th Birthday
Uma 11h32'5" 48d24'19"
Guadalupe
Leo 10h8'5" 21d21'58"
Guadalupe
Ser 15h20'39" -0d45'15"
Guadalupe Alonzo
Uma 11h44'21" 49d3'32"
Guadalupe Arellano
Aql 19h33'32" 9d40'44"
Guadalupe Ayala
Cnc 8h48'38" 13d0'14"
Guadalupe Brown
Cap 20h30'9" -22d42'28"
Guadalupe DeLeon
Uma 11h26'31" 44d37'44"
Guadalupe Moreno
Mon 7h30'46" -0d46'46"
Guadalupe Quiroga Ramirez
Uma 10h59'51" 53d20'50"
Guadalupe Rodriguez
Mon 6h45'29" -0d13'33"
Guadalupe V. Hernandez
Crb 15h39'42" 27d31'23"
Guadalupe Vera
Lyr 19h20'28" 34d14'11"
Guanghong Xu
Lyn 7h27'39" 52d4'53"
Guangxi Spirit Yi
And 23h13'8" 45d4'32"
Guapo Azul
And 0h39'12" 42d57'59"
"Guapo" Davis
Aur 7h17'59" 41d58'8"
Guard Dux
Lyn 9h4'1" 34d25'22"
Guardian
Gem 6h48'2" 21d50'46"
Guardian
Ari 2h53'19" 25d22'44"
Guardian
Uma 11h7'7" 64d2'14"
Guardian Angel
Leo 10h22'1" 25d33'23"
Guardian Angel
Cnc 8h29'25" 24d52'58"
Guardian Angel - Fred of Bedrock
Per 3h35'17" 36d31'30"
Guardian Angel Glen
And 0h1'59" 46d31'53"
Guardian Angel Nicole Ann Nogrady
Aqr 22h13'48" -13d25'22"
Guardian Angel Patsy
Mon 7h8'39" -0d26'1"
Guardian Angels
Cnv 13h51'53" 33d13'13"
Guardian Blue
Ori 6h10'17" -0d42'9"
Guardian Gray
Sct 18h58'16" -8d14'51"
Guardian of the Calendar
Cyg 20h33'8" 31d35'2"
GubStar
Eri 4h30'36" -32d27'10"
Gucci
Aur 5h21'10" 55d36'8"
Gucci Boy 1985
Cep 0h28'17" 82d51'7"
Gucia
Crb 16h0'28" 39d5'42"
Gudalovic
Per 3h22'24" 31d37'9"
Guddan's star
And 0h49'52" 23d9'55"
Gudhull
Cas 1h5'45" 69d29'13"
Gudrun
Ori 6h15'18" 15d33'15"
Gudrun
Gem 8h7'41" 31d32'35"
Gudrun & Marcel
Com 13h3'30" 15d7'1"
Gudrun Winkler Ennett
Leo 9h31'33" 28d50'14"
Guea-Yea Lian
Lib 14h42'41" -11d25'27"
Guendalina
Aur 6h28'39" 47d18'59"
Guendalina Tavagnacco
Aur 5h39'3" 32d34'18"
GUENHWYVAR
Ori 6h11'25" 8d53'41"
Guenola Larrere
Sco 17h50'20" -41d9'25"
Guenter Hahn 40
Her 18h21'22" 18d4'55"
Guenther & Marga
Gem 6h55'24" 17d55'14"
Guenther Trzenski
Uma 14h25'59" 58d48'34"
Guera
Sgr 20h4'38" -24d50'58"
Guerel Marion
Leo 10h1'6" 21d0'9"
Guerrier and Lauren
Cma 7h22'17" -11d53'59"
Guertin
Cyg 20h56'58" 48d39'26"
guess what
Aqr 23h45'9" -11d13'52"
Guess What...
Uma 10h31'57" 65d22'46"
Guess What Donasil
Aqr 22h1'3" -16d2'34"
"Guesswhat" - J & D
Eri 3h50'40" -21d9'44"
Guevel
Aqr 22h51'31" -9d53'40"
Guewen
Col 5h40'9" -29d27'47"
Guffin Gulch
Cep 20h41'51" 59d22'56"
Guggapook
Cas 23h57'26" 50d18'35"
Gughi
Crb 15h57'41" 38d32'2"
Guglielmino
Mon 8h3'31" -0d52'5"
Guglielmo
Umi 15h27'53" 72d45'35"

Guglielmo Costa
Ori 5h31'46" -8d2'36"
Gugliotti Family Star
Ori 5h26'58" -1d39'59"
Gugliuzza
Uma 10h25'28" 61d7'2"
GuguVic
Cnc 8h50'18" 30d54'14"
Guhert, Georg
Uma 9h4'51" 51d8'16"
Guichard St. Surin
Lyn 7h20'54" 52d34'3"
Guidance
Uma 11h30'27" 63d38'47"
Guidance
Tri 2h10'17" 34d52'4"
Guiding Angel
Mon 6h34'39" 1d46'37"
Guiding Light
Cnc 9h5'5" 14d43'42"
Guiding Light
And 23h33'44" 47d17'0"
Guiding Light for Ray and Allison
Cyg 20h56'55" 45d45'46"
Guiding Light Of Sarah Vaughan
Cru 12h17'45" -56d45'20"
Guiding Me Always
Cnc 9h11'53" 24d34'8"
Guiding Star
Uma 8h35'17" 56d29'58"
Guido Castillo
Crb 15h42'41" 38d2'29"
Guido Mentz
Uma 10h1'48" 44d55'15"
Guido Ottaviano
Peg 22h35'33" 18d49'8"
"Guido" Paul S. Bliss
Cyg 19h31'18" 53d4'43"
Guido R. D'Angelo
Cap 20h7'56" -16d34'7"
Guido Seffer
Uma 9h55'46" 50d25'10"
Guido - star of my soul
Ori 6h9'20" 15d40'22"
Guido Victor Regelbrugge
Her 17h13'31" 47d53'54"
Guidor
Sgr 18h31'12" -30d27'29"
Guido's Girl
Gem 7h12'28" 15d20'13"
Guido's Star
Leo 9h27'0" 27d18'59"
Guilainium
And 1h56'38" 35d54'6"
Guilhem Fontvieille
Sgr 20h25'35" -42d56'49"
Guilianna "Gigi" Mezzetta
And 0h59'40" 46d27'47"
Guilio Gambuto
Uma 12h27'11" 61d9'44"
Guilio & Tessie Franceschini
Cyg 20h25'51" 37d26'48"
Guillaume
And 1h44'46" 42d54'46"
Guillaume
Uma 9h10'28" 67d40'0"
Guillaume 21 août 2006
Leo 10h32'41" 21d45'24"
Guillaume Bouteiller
Uma 9h29'17" 43d2'21"
Guillaume Briand
Sco 17h28'8" -32d6'4"
Guillaume Couture
Lyr 18h29'19" 35d39'4"
Guillaume et Justine
Lyr 18h58'11" 26d3'45"
Guillaume Lanthier-Proulx
Uma 12h24'9" 52d49'50"
Guillaume Nebout
Cnc 8h52'50" 8d57'45"
Guillaume Pessy
Vir 11h44'38" 9d2'31"
Guillaume Richard
Dra 19h32'19" 68d1'44"
Guillaume + Sonya BouBou
Cyg 20h17'32" 51d10'48"
Guillerma *Guilly* Vallente
Gem 7h31'53" 14d56'18"
Guillermina Herman
Aql 19h56'52" 1d21'19"
Guillermina Valenzuela
Sco 16h26'41" -33d12'45"
Guillermo Alfonso Alondra 101042
Lib 15h10'22" -8d6'15"
Guillermo Augusto DeVeyga
Per 2h19'15" 57d8'13"
Guillermo Garza Martinez
Her 16h27'43" 7d48'12"
Guillermo Gonzalez
Uma 8h24'24" 64d26'10"
Guillermo IronMan Garsed
Dra 17h51'35" 67d50'20"
Guillermo Jesus Alvarez
Ari 2h47'49" 21d48'53"
Guillermo & Rosie Johanson
Lyr 18h49'10" 39d14'39"
Guillermo Sanclemente
Vir 12h11'33" -9d4'18"
Guillermo-Eva
Del 20h35'39" 13d56'56"
Guimi
Aql 19h28'9" 10d52'43"
Guin Matthews "Love Star"
Cep 21h24'31" 56d47'54"
Guinan Family Star
Lib 15h43'35" -20d31'59"
Guiness
Aql 19h51'49" -0d33'12"
Guinevere
Cas 1h50'6" 60d35'43"
Guinevere
Umi 15h21'48" 68d21'34"
Guinevere V
Tau 6h0'17" 25d31'37"
Guinness Aimone
Uma 10h28'14" 64d56'54"
Guiomar
Gem 7h31'21" 30d49'1"
Guirrermo Jose Corona
Her 18h7'22" 35d16'18"
Guiseppe Calderone
Uma 10h18'9" 40d59'15"
Guiseppi III
Per 4h48'10" 46d49'31"
Guissell Espinoza Fafalios
Her 17h26'30" 29d15'37"
Guitar Man 50
Per 3h9'47" 54d55'52"
Guitar Worship Dude 2021
And 1h1'43" 41d46'53"
Guitarman Mike
Lyr 18h49'22" 35d41'51"
Guitarrista Guille Martin
Cnc 9h20'28" 31d57'8"
Guitou
Lyn 6h34'56" 57d51'58"
Guiver
Cyg 21h37'41" 39d53'30"
GUJ
Aur 6h13'40" 33d39'45"
Guj
Aur 6h54'51" 44d54'42"
Gujarati
Crb 16h5'2" 39d36'58"
Gujjubear
Uma 11h32'4" 57d0'21"
GULAY
Uma 11h18'43" 46d45'59"
Gulay Ozdemir
Lyr 18h36'20" 35d54'41"
GULCAN
Lib 15h13'41" -23d32'28"
Gulchekhra L. Muradova
Psc 0h12'15" -2d2'6"
Gulcin
Per 3h45'37" 37d48'5"
Gülden
Lyr 19h8'32" 34d14'13"
Gulfstar
Lyn 9h16'32" 40d52'15"
Guli - the all most beautiful rose
Uma 11h0'29" 45d13'22"
Gulliver Pot of Gold Busier
Dra 18h44'0" 62d19'41"
Guls
Gem 7h17'43" 28d24'43"
Gülsah
Mon 6h35'50" 5d38'27"
Gülsah Spielmann, 04.02.1998
Lyr 18h29'36" 28d30'43"
Gulten Ozcelik
Cyg 21h15'42" 53d25'53"
Gulumjan, Roman
Uma 12h18'42" 58d1'37"
Gumba Joe
Cyg 20h12'8" 59d22'35"
Gumbee Land
Eri 3h24'7" -21d54'17"
Gumbo
Aql 20h13'31" 1d18'58"
Gumby
Sgr 19h24'47" -17d6'36"
Gumby Dean
Per 3h6'35" 53d24'26"
Gumma Fran
Ari 2h34'38" 27d16'50"
Gummi Bears
Sgr 18h19'1" -24d57'21"
Gummy
Ori 6h4'27" 18d25'27"
Gummy
Ori 6h10'31" 7d8'46"
Gummy Bear
Umi 16h50'30" 87d6'37"
Gummy Bear Star
Umi 15h56'35" 78d8'49"
Gump & Liz
Ari 2h17'21" 21d42'15"
Gump61433
Gem 7h30'51" 18d38'6"
Gumy
Sco 17h49'31" -40d59'44"
gunaikeio zvujio
Sgr 18h4'8" -28d4'56"
Gunda
Tau 3h54'34" 22d24'25"
Gundee
Uma 10h30'6" 43d3'32"
Gundi Hölscheid
Uma 10h2'18" 72d37'45"
Gundula
Gem 6h30'10" 26d2'57"
Gunel
Aql 19h35'38" 12d41'6"
Gunel
Crb 15h57'1" 34d46'14"
Gunhild au nom de notre Amour
Gem 7h24'43" 20d54'51"
Gunilla Blumchen Peter
Sgr 17h54'15" -30d2'51"
Gunilla's Star
Umi 13h56'42" 70d34'9"
Gunjan Doshi
Tau 3h55'31" 12d11'6"
Gunn Al-Sirius 01161926
Uma 9h31'31" 65d17'31"
Gunnar Douglas Rummel
Tau 5h45'37" 21d11'15"
Gunnar & Inger Carlsson
Cyg 21h1'59" 54d48'28"
Gunnar James Farrell Johnson
Psc 0h56'19" 8d34'13"
Gunnar Kelly
Cnv 12h36'13" 41d50'54"
Gunnar Pydde
Ori 6h21'24" 13d48'6"
Gunner
Cep 2h34'59" 84d28'26"
Gunner Bryan Campbell
Sgr 18h8'44" -30d8'31"
Gunner Campbell
Lyn 9h3'43" 36d55'12"
Gunner the Great
Psc 1h9'25" 20d51'12"
Gunnery Sergeant David Coronado
Cam 3h59'41" 70d25'47"
Gunnhild Cathrin
Lyn 7h53'38" 34d26'24"
Günni's Glücksstern
Ori 5h44'3" 2d15'44"
Gunnur Curmak
Aur 6h0'10" 47d13'56"
Gunny Hill
Crb 15h31'8" 30d59'23"
Gunter
Per 3h42'43" 49d22'11"
Günter
Ori 6h15'46" 9d32'13"
Günter 05.05.1972
Uma 12h31'18" 62d5'42"
Günter Alois Böhm
Uma 10h25'31" 40d4'9"
Günter Berger
Uma 10h37'26" 40d36'50"
Günter Bode
Uma 8h54'5" 62d38'32"
Günter Bonneik
Uma 9h36'50" 59d48'19"
Günter Conzelmann
Uma 11h31'53" 62d52'37"
Günter Dahms
Ori 5h55'20" 17d7'2"
Günter E. Pieper
Peg 22h26'50" 4d37'22"
Günter Emonds
Uma 10h6'28" 72d30'25"
Günter Geißler
Uma 9h33'37" 51d56'12"
Günter Holtmann
Uma 8h44'12" 52d26'42"
Günter Kaulfuß
Uma 10h27'25" 49d17'53"
Günter Michelfelder
Uma 8h18'27" 65d29'34"
Günter my Precious Star
Uma 10h4'10" 55d23'53"
Günter Pritzwein
Uma 11h19'39" 40d9'58"
Günter Reister
Uma 9h13'14" 58d27'16"
Günter Riedl
Uma 10h23'40" 40d35'33"
Günter Soltau
Uma 10h59'20" 69d37'16"
Günter Strupp
Uma 9h21'52" 47d56'22"
Günter Styra
Uma 11h15'26" 41d5'20"
Günter's Star
Sgr 19h11'20" -19d29'57"
Günther Langhammer
Uma 9h3'43" 69d56'35"
Günther Lehmann
Uma 9h11'27" 62d48'59"
Günther Poggensee
Ori 5h24'5" 7d21'26"
Günther & Sandra
Uma 9h42'50" 72d42'42"
Günther Wegert
Uma 11h28'36" 65d30'48"
Günther Wenth, 17.04.1956
Cyg 21h34'21" 35d28'47"
Günther, 10.09.1952
Cep 22h26'43" 66d59'36"
Günther, Engelbert
Ori 6h4'58" 10d31'6"
Günther, Ingrid
Ori 5h24'13" 13d55'38"
Günther, Sandra und Fabian
Sgr 19h18'25" -34d28'50"
Günther, Tobias
Uma 10h26'27" 67d32'47"
Gunther's Corner
Cap 21h43'38" -12d40'32"
Günti
Lac 22h50'11" 53d53'40"
Guo Qi
Psc 0h47'19" 18d7'27"
Guppy
Cam 7h50'49" 65d56'36"
Guppy
Cet 3h7'21" -0d11'23"
Gupy
Vir 12h49'50" 13d14'14"
Gurami Topchishvili National Hero
Uma 9h39'1" 71d42'27"
Guri L. Ravatn
Gem 6h41'42" 26d51'53"
Gurjyott Kaur Dhillon
Lmi 10h26'42" 37d26'17"
Gurksi Bär
Sgr 17h57'22" -17d10'45"
Gurleen Singh
Lyn 8h29'58" 50d14'38"
Gurminder Singh Rai
Dra 17h44'50" 65d52'3"
Gurpreet Singh (Guruji)
Cam 3h42'13" 64d13'28"
Gursharn
Dra 18h29'54" 71d38'10"
Guru Dad
Ori 5h27'12" 5d51'7"
Gurvan
Ori 4h50'39" -3d33'32"
Gus
Sgr 18h49'17" -30d23'12"
GUS
Leo 11h10'55" 1d5'36"
Gus
Vir 13h5'49" 9d20'8"
Gus
Per 2h40'0" 55d55'23"
Gus
Cnc 8h47'21" 30d18'53"
GUS - 31.12.1926
Cap 20h33'36" -9d2'36"
Gus Acuna
Gem 6h21'49" 26d24'4"
Gus and Lil Neilson
Uma 10h8'38" 69d25'10"
Gus Anders Barr
Boo 14h38'19" 51d35'42"
Gus Anthony Stavros
Her 18h54'17" 24d2'3"
Gus Boss
Ori 6h14'16" 16d17'17"
Gus Carameros
Cnc 9h16'58" 11d6'32"
Gus*Casas *IKT!*
Her 17h6'42" 31d59'35"
Gus' Getaway
Tau 4h34'6" 10d22'36"
Gus Green's Eastern Star
Cep 21h26'30" 61d51'56"
Gus Gustafson
Lmi 10h40'30" 27d39'35"
Gus Hull
Uma 10h7'20" 46d19'39"
Gus James Peters
Lib 15h38'50" -11d40'16"
Gus "Mr. Kosta"
Uma 9h59'23" 51d24'45"
Gus Palamides
Lyr 18h47'10" 30d36'0"
Gus Ridge
Ari 2h16'46" 26d37'36"
Gus Salcedo
Dra 19h2'58" 70d48'56"
Gus Spark
Tau 3h26'20" -0d10'13"
Gus the lopeared rabbit
Lep 5h13'18" -11d59'24"
Gus Turner
Aql 19h45'17" 3d33'29"
Gus Wallace Beaudry
Leo 10h31'17" 15d0'52"
Gus, Garrett & Brodie
Lyr 18h59'31" 33d38'53"
Gusanito
Sgr 18h20'30" -21d14'44"
GUSCHA-STERN
Crb 15h45'19" 28d18'59"
Gusette
Cnc 8h47'41" 32d47'0"
Gusi & Sarah
Ori 5h12'1" -0d3'48"
Guss
Uma 12h14'26" 52d40'23"
Gussie & Dutz
Ori 5h15'33" 2d51'19"
Gussie - Simple Treasures Are Rare
Pyx 8h46'46" -27d10'8"
Gussy Bunkster
Ori 6h15'0" 14d7'25"
Gustafson
Pho 0h59'52" -40d13'31"
Gustafson 1
Ori 5h56'1" -0d3'23"
Gustafson - Gervasi
Tau 5h35'1" 26d22'56"
Gustav 1
Lyn 6h59'49" 59d18'8"
Gustavo Annovelli
Vel 9h24'49" -52d10'52"
Gustavo Chandler Smith Sasis III
Her 18h24'43" 12d6'6"
Gustavo de Castilhos Pezzi
Dra 19h9'49" 60d4'59"
Gustavo E. Valdivia (Tavo 88)
Per 3h15'32" 31d29'7"
Gustavo Gigena
Cru 12h6'58" -55d50'49"
Gustavo Gorrochotegui
Cyg 20h6'49" 44d48'53"
Gustavo Grayeb Sousa
Cyg 20h0'31" 59d4'6"
Gustavo Manuel
Aql 19h9'3" -8d34'5"
Gustavo Rico
Ori 6h14'46" 15d35'40"
Gustavo Sanchez
Dra 18h55'55" 64d42'37"
Gustavo Serrano, M.D.
Boo 13h48'48" 23d23'59"
Gustavo's Star
Cap 20h26'4" -9d12'23"
Gustavson
Oph 17h29'10" 11d21'47"
Gusti Stoz
Umi 14h9'27" 74d16'30"
Gustie
Cet 2h46'8" -0d31'16"
Gusylvia
Cyg 21h38'27" 34d52'29"
Guta
Uma 12h28'47" 56d19'14"
GUTEVUSS LSVR 01.12.1942 VAIL IOWA
Cap 21h20'39" -18d9'19"
Güth, Heinz-Georg
Ori 6h13'9" 9d4'14"
Güthert, Toralf
Uma 10h39'24" 46d44'0"
Gutierrez
And 23h1'36" 46d55'46"
Gutknecht, Horst
Uma 11h35'23" 31d41'38"
Gutmueller 11 - Zachery William
Psc 22h51'25" -0d7'21"
Gutow, Bernd
Uma 13h53'55" 56d56'56"
Guttorm's Jewel
Aql 19h56'32" -0d57'37"
Gutzmann, Werner
Gem 7h31'0" 19d42'1"
Guy A Colella
Uma 14h14'58" 56d50'49"
Guy A. Gioia, Sr.
Ori 5h27'39" 5d28'0"
Guy A. Mastrion
Per 4h24'33" 48d56'14"
Guy and his Gal
Ori 6h17'5" 13d48'1"
Guy Anthony Patrick
Aql 19h45'41" -0d29'3"
Guy Bailey DeWeese
Aur 5h53'5" 37d19'3"
GUY BEN-GAL
Dra 19h57'49" 76d50'38"
GUY BOY
Her 17h40'55" 21d57'46"
Guy & Cindy Carlson 4/28/2000
Lyn 7h24'5" 53d45'55"
Guy Crossing & His Girls
Cru 12h10'41" -62d36'4"
Guy & Cynthia Mead
Cyg 21h14'17" 44d58'15"
Guy D. Briggs, TIA Board Chairman
Gem 7h42'48" 33d24'21"
Guy Daniel
Leo 11h15'31" -3d34'19"
Guy Delery
Uma 11h19'42" 35d15'18"
Guy Deyrolle
Cas 23h41'16" 51d12'2"
Guy Douglas Hall
Tau 4h16'20" 25d11'7"
Guy F. Chennault
Sco 16h9'56" -19d35'11"
Guy Francis II
Dra 18h16'35" 70d22'27"
Guy Gardener Stibal
Ori 5h40'42" 9d14'17"
Guy Girard
Gem 6h12'33" 23d49'41"
Guy Hargrove
Her 18h21'54" 21d6'31"
Guy Johnathan Boyer
Cap 20h30'45" -15d40'23"
Guy Laflamme
Umi 17h2'41" 77d8'53"
Guy Lon-Ho-Kee
Aql 19h33'55" 11d38'37"
Guy & Lura Brown
Umi 16h12'11" 73d38'23"
Guy Macdonald
Uma 11h50'35" 58d56'15"
Guy Man Morrison
Her 17h19'5" 27d18'56"
Guy Marcoux (a.k.a. Taz)
Aqr 21h20'19" -7d21'59"
Guy McKenzie Jr.
Uma 11h34'0" 64d23'24"
Guy Michael Couchman
Per 3h21'9" 41d48'43"
Guy Michael - "Lionheart"
Cru 12h35'30" -60d4'56"
Guy Paul Pollina
Peg 22h11'49" 7d15'23"
Guy Peloquin
Peg 22h32'54" 24d48'57"
Guy Robert Swalm
Tri 1h54'45" 30d6'32"
Guy Sanders #42
Per 3h44'21" 39d48'35"
Guy & Shantel
Cyg 21h42'55" 46d41'49"
Guy Shipe
Per 3h6'2" 52d19'41"
Guy Stanford Michael Coogan
Uma 9h24'41" 59d11'0"
Guy Stein Lesavoy
Vir 12h10'51" 11d57'11"
Guy T. Bucco
Cap 21h39'27" -8d54'45"
Guy William Hardin
Cap 21h45'37" -19d8'28"
Guyati
Ori 5h35'31" -3d22'47"
Guylaine Gustar
Cas 23h43'46" 55d12'13"
Guylaine Martel
Per 3h29'10" 33d57'46"
GuyTron-04
Per 3h21'41" 44d46'30"
Guyver Ullyart
Dra 12h34'34" 74d58'10"
Guyzee's Star
Cru 12h25'23" -58d50'7"
Guzal Kamilova
Uma 11h20'39" 31d57'9"
Guzzimoore
Her 17h17'33" 26d38'31"
Gvantsa Jishkariani
Sgr 18h34'18" -24d14'27"
GWA 143
Vir 14h1'31" -15d19'23"
GWD's ArticBlast
Sco 16h52'37" -34d55'30"
Gwelma Kennedy
Uma 9h46'43" 55d59'54"
Gwen
Uma 10h22'17" 58d48'35"
Gwen
Aql 19h27'31" -0d18'57"
GWEN
Her 18h19'19" 22d44'58"
Gwen
Gem 6h54'37" 25d12'42"
Gwen
Ori 5h58'34" 6d13'8"
Gwen A. DeHart - Light Of My Life
Per 4h35'8" 31d59'57"
Gwen Alto - 80 and still shining
Ori 5h34'12" -4d14'28"
Gwen Ananias
Lib 15h22'33" -5d28'10"
Gwen and Amanda
Aqr 23h32'19" -15d2'50"
Gwen Barr
Ori 5h27'8" -0d31'27"
Gwen Blanc
Com 12h42'21" 15d24'12"
Gwen Blazejewski Born: Oct 11/1965
Uma 11h30'25" 57d3'3"
Gwen Cole
Cas 1h17'50" 64d38'20"
Gwen Dillon
Ori 6h13'44" 2d52'44"
Gwen Flamberg
Peg 22h44'47" 21d49'43"
Gwen Flamberg
Cam 6h10'43" 60d42'45"
Gwen Gonzalez
Cnc 9h12'8" 13d6'36"
Gwen Good
Her 17h36'30" 32d27'30"
Gwen & Jim Fulwell's Home in Heaven
Uma 10h5'15" 56d20'55"
Gwen & John's 60th Anniversary Star
Cyg 19h43'39" 27d59'38"
Gwen Marie Gessinger
Cyg 21h51'9" 38d29'7"
Gwen Nichol Westfall
Uma 11h38'14" 53d20'35"
Gwen Scherbring
Com 12h40'48" 15d30'52"
Gwen Siewertsen
Cyg 21h17'50" 41d54'15"
Gwen Weaver
Peg 21h39'37" 7d20'48"
Gwenaëlle
Cyg 21h40'37" 49d27'37"
Gwenaëlle
Lib 15h31'28" -24d24'52"
Gwenaëlle 15/11
Cap 20h21'2" -15d51'58"
Gwenaëlle et Olivier
Aqr 23h23'52" -12d3'15"
Gwenda Souza Smith a.k.a Gwen
Cas 23h26'3" 53d10'50"
Gwenda-Katherine
Lib 14h57'3" -10d40'56"
Gwendalee
Leo 11h12'2" 4d55'55"
Gwendel Adell Webster
Crb 15h59'48" 28d55'41"
Gwendolen Williams
Cnc 9h8'40" 25d42'22"
Gwendoline May Blenkinsopp
Her 17h39'17" 38d18'26"
Gwendoline's Light of Love
Cyg 20h42'9" 46d30'1"
Gwendolyn
And 1h27'6" 49d0'23"
Gwendolyn
Aqr 23h11'22" -17d18'2"
Gwendolyn
Sgr 18h36'0" -21d46'1"
Gwendolyn
Psc 1h43'52" 5d11'26"
Gwendolyn Collins
Cnc 8h59'3" 29d23'40"
Gwendolyn Dianne Whetstone
Cam 4h6'55" 54d22'4"
Gwendolyn G. Traylor
Lyr 18h50'27" 33d56'6"
Gwendolyn Gibbons
Cas 1h24'42" 62d22'27"
Gwendolyn Goehringer
Cas 1h15'22" 62d20'25"
Gwendolyn Jane Urbas
Uma 8h37'35" 50d29'44"
Gwendolyn Jean
Ori 5h36'37" 11d12'19"
Gwendolyn Jean
Lib 15h15'27" -9d59'49"
Gwendolyn Rose
Psc 1h1'31" 31d39'43"
Gwendolyn Stalheim Martinson
Aqr 20h44'19" -2d31'44"
Gwendolyn Thelin Swanda
Psc 1h20'18" 31d38'33"
Gwendolyn Yael King
Uma 8h33'53" 65d16'37"
Gwendolynae Tauri
Cas 1h35'25" 58d2'44"
Gwendolyn's Dream
Dra 18h54'29" 71d22'44"
Gwenith
And 23h27'52" 40d35'58"
Gwenith Mary Griffiths
Cnc 8h31'51" 25d45'3"
Gwennie Scourfield
Peg 21h50'26" 36d15'11"
GwenTerryBishop65
Cas 23h35'25" 51d44'35"
GWFout
Lib 15h40'46" -28d38'58"
Gwiazda imienia Katarzyny Liebchen
Ori 5h49'40" 11d48'33"
GWIAZDA SYLWII I JACKA
Uma 10h1'7" 43d22'48"
Gwiazdeczka
Cyg 20h55'10" 47d19'54"
Gwiazdka Spelniania Zyczen Izy
Vir 13h39'52" 2d46'16"
gwiazdka z nieba dla kochanej Mamy.
Ari 2h21'42" 22d38'23"
Gwilym and Iris
Cyg 21h32'14" 38d8'53"
Gwinnette Missouri Meyers
Cru 12h2'47" -62d52'31"
Gwisc
Cas 3h16'18" 66d22'34"
G-Wiz
Uma 12h0'12" 39d56'16"
GWJW Allman
Tau 5h10'11" 18d5'24"
Gwladys, étoile de Bruguières
Uma 14h4'49" 61d13'17"
Gwondola-Kay
Sge 19h14'9" 18d52'37"
GWSAS
Crb 15h38'26" 37d26'14"
Gwydion-Maynard Sirois 7-25-04
And 0h30'33" 42d48'18"
Gwyn
Cyg 19h54'46" 31d22'42"
Gwyn
Cam 4h33'38" 59d23'22"
Gwynaeth
Mon 6h54'48" -0d38'12"
Gwyneth (Gwynnie) Elizabeth Vitek
Cas 2h48'58" 60d28'9"
Gwyneth Hope Hallmark
Cnc 9h20'2" 29d20'42"
Gwyneth Lee
Ari 2h39'34" 26d17'18"
Gwyneth Marie Herkert
Cam 5h58'59" 58d57'18"

Gwyneth Mitchell Price Thomas
Ori 5h53'3" 22d35'58"
Gwynn Johnson
Apu 15h25'9" -70d46'43"
Gwynn T. Swinson, SECC State Chair
Cyg 20h33'18" 41d21'36"
Gwynne Janette Harper
Uma 11h23'36" 69d39'57"
Gwynneth Ainslie Carter
Psc 1h21'52" 30d22'45"
Gwynneth R. Harrington
Peg 23h45'46" 26d52'30"
GXHughes
Ori 6h10'38" 7d38'6"
Gyamfi-Poku, Kvame
Ori 6h15'24" 14d22'0"
Gyaneeta
Lyn 7h4'20" 60d1'2"
Gylda
Umi 14h27'52" 75d25'8"
Gyler
Uma 12h26'34" 55d53'46"
Gym-a-nist AMB
Cas 1h17'50" 52d42'26"
Gymer
Psc 1h10'41" 18d13'2"
Gyna Denise
Ari 2h36'9" 24d30'41"
Gyoergy Pribliczki
Ori 5h53'27" 3d23'58"
Gyönyörû (Beautiful) Marie
Uma 10h25'34" 42d37'48"
György Ádám szerencsec-sillaga
Uma 9h55'9" 51d58'47"
Gyorgy és Kata Csillaga
Tau 5h49'21" 23d32'1"
Gyozike
Cap 21h18'58" -19d30'20"
Gypsy
Umi 14h41'28" 67d53'18"
Gypsy
Ori 5h38'3" 12d29'7"
Gypsy Kitty - Danielle Fortuna
Sco 17h17'22" -35d37'50"
Gypsy Porfirio
Umi 15h43'26" 75d3'57"
Gypsy Rawlings
Boo 14h40'34" 27d12'32"
Gypsy, Boo, Tess & Maxy - Much Loved
Col 5h10'46" -39d8'44"
Gypsy126
Aqr 20h52'50" -8d56'25"
Gypsy-Gitana
Peg 22h38'37" 11d56'9"
Gypsy's Dream
Lyn 8h29'57" 57d8'21"
Gypsy's Hero
Ori 4h52'39" 10d50'6"
Gypxy 1964
Lib 15h38'49" -24d32'39"
GYSGT David Laverne Eckerson
Sgr 17h52'47" -29d8'12"
Gyura László
Leo 9h50'50" 13d2'4"
Gyurcikám csillaga
Psc 1h15'44" 30d37'44"
Gyuri '67
Tau 5h30'55" 21d31'17"
Gyuri '85
Cas 0h45'33" 64d1'5"
Gyvan
Dra 17h10'41" 53d33'14"
H. and M. Wondergem
Lmi 10h21'18" 29d21'36"
H B TraditioNeal
Mon 6h20'53" -0d42'23"
H. Bruce Knox
Per 3h8'58" 43d26'36"
H Church Star Ladies and Survivors
Gem 7h49'24" 24d6'30"
H. D. Hollifield
Leo 11h55'37" 19d12'35"
H. D. Hudson Centennial Star
Uma 11h33'3" 53d20'47"
H. D. Shephard
Uma 10h18'25" 42d46'2"
H. David C. Gunderman
And 0h12'51" 37d50'40"
h*dew
Cam 7h1'48" 65d7'54"
H Dragon Heart
Sgr 18h51'38" -18d51'14"
H Edward Hanway
Umi 15h9'57" 68d3'32"
H. Franklin Baker
Sco 16h37'13" -32d23'44"
H. George Hamilton
Tau 5h33'50" 26d38'5"
H. Greg Weldy, Jr.
Per 3h13'34" 51d37'4"
H. H. Nichols
Aqr 22h35'21" 0d36'24"
H H Saum III
Vir 13h10'53" 8d2'23"
H. Herb Wittig
Lib 15h6'54" -27d39'29"
H. I. Sober, 5th Monkey Cycle
Uma 9h1'4" 52d28'16"
H & J
Cyg 21h21'32" 32d43'21"
H June Casteel & J T Stovall 1943
Umi 14h44'24" 68d9'24"
H L Babcock
Ori 5h18'41" -8d35'9"
H & L Sa
Cyg 20h30'52" 48d6'10"
H. L. Szilveszter édesany-ja: Lia
Uma 8h18'29" 72d18'2"
H. Lanean Hughes
Per 3h12'23" 45d5'10"
H. Lee Mitchell
Cep 20h40'23" 59d32'13"
H. Louis Chodosh, M.D.
Her 17h10'39" 35d26'39"
H. M. O. Celestial Mother
Cyg 20h12'21" 46d51'51"
H. Margarita
Uma 8h59'36" 55d22'6"
H. Michael O'Brien
Lyr 18h21'20" 30d25'40"
H & R Dooley's Irish Rose Star
Uma 8h40'56" 72d42'40"
H. Randy Cody
Aqr 21h46'22" -1d52'33"
H. Robert Guthrie
Ari 3h14'41" 29d8'22"
H. Russell Griffith (Russ)
Uma 9h45'8" 66d1'52"
H. S. James Neighbours
Cyg 21h26'59" 44d55'59"
H. Shockley Beyond the Call of Duty
Aql 19h42'51" 7d47'35"
H. Soufiane
Sco 17h25'48" -33d36'42"
H&T 2006
Ari 2h53'49" 29d42'50"
H. T. Erickson
Boo 14h36'0" 41d40'15"
H & TH
Leo 11h49'38" 19d0'36"
H W D -13
Vir 13h50'0" -2d45'56"
H. Woodie Lange
Her 16h19'36" 49d8'2"
H2R3
Her 17h8'52" 25d59'21"
Ha Hoang & Sangchi Tang
Cyg 20h56'6" 51d16'10"
Haaaalllle
Aql 19h34'26" 6d28'55"
Haag, Herbert
Uma 9h35'49" 58d39'16"
Haakon Charles Moninger
Vir 13h49'17" -12d8'27"
Haala Major
Boo 15h24'14" 34d58'53"
Haas Family
Aqr 20h51'38" -12d25'3"
Haas, Willi
Uma 11h18'21" 37d55'44"
Haase, David
Uma 10h43'21" 43d57'32"
Haase, Egon
Uma 10h53'54" 71d4'56"
Haase, Hartnut
Uma 10h40'43" 56d55'30"
Habakuk
Uma 12h1'11" 39d4'56"
Habbishaw Heaven
Leo 10h21'0" 26d49'17"
HABEBTY
Uma 9h5'21" 48d40'56"
Haberkamp, Hans-Josef
Uma 9h47'41" 72d49'35"
Haberkorn 1962
Crb 16h20'32" 37d30'56"
Häberlen, Ben
Uma 9h25'50" 63d51'16"
Haberman 122504
Uma 10h30'52" 51d51'58"
Habermann, Hans-Joachim
Uma 11h41'20" 29d38'40"
Habib 'Papik' Kazanchian
Aql 19h18'21" -11d25'29"
Habib, Fahima and Sana Nawid
Tri 2h13'53" 34d15'59"
Habiba and Ramy
Cyg 21h59'0" 54d34'21"
Habibi
Lib 15h6'51" -6d55'7"
HABIBI
Sco 17h1'38" -31d8'48"
Habibi Lizzi
Gem 6h46'46" 27d15'37"
Habibi M & M
Cyg 20h58'1" 32d42'59"
Habibteh Sandra
And 0h45'24" 38d45'0"
Habros
Lib 14h53'40" -7d18'16"
Hachinger, Maximilian
Sco 16h59'33" -39d1'42"
Häckel, Eberhard
Ori 6h19'3" 10d55'30"
Hackers Speedway
Aur 4h51'49" 32d50'2"
Hackstar 3
Cra 18h10'0" -39d29'56"
Hadassah
And 0h50'30" 41d50'11"
Haddad
And 0h16'0" 36d18'19"
Hadeel
Eri 4h22'52" -8d31'47"
Hadeel El Sourani
Uma 11h41'53" 38d5'41"
Hadleigh Maddox Dryden
Psc 1h16'2" 15d6'5"
Hadley Burcham Dolan
Lyr 18h21'48" 35d27'13"
Hadley Claire Allaria
And 1h42'41" 42d50'55"
Hadley Haven
Lyn 7h54'35" 52d57'16"
Hadley Jo Ellig
Cap 20h51'33" -18d2'41"
Hadley Jo Elsenbaumer
Ori 6h15'11" 7d35'28"
Hadley McLane Harris
And 2h18'36" 45d25'30"
Hadley "Millie" Moree
Lyn 8h43'20" 42d21'32"
Hadley Witherow
Aqr 23h27'48" -19d56'25"
Hadlock
Crb 15h35'24" 32d6'58"
Hadlock's Hurricane
Psc 0h45'38" 15d52'34"
Hadraniel Sunshine
Tau 5h17'6" 23d17'35"
Hadrian R. Katz
Leo 11h52'34" 25d4'4"
Hads
Cnc 8h45'47" 24d47'12"
Hadyn Jeffrey
Ori 6h6'8" 13d1'41"
Hae Ok
Cyg 20h39'46" 44d46'37"
HaeJung Kimberly Tonnesen
Leo 10h8'42" 15d6'26"
Haela Heather Melton Stapish
Dra 17h19'51" 65d30'37"
Haelee Cheyanne
Vir 14h35'17" 3d46'37"
Haeley and Luke Nardone
Uma 8h36'3" 64d42'52"
Hafida Nassall
And 1h41'35" 39d49'42"
Hafren Mia Vaughan
And 2h22'21" 39d4'42"
Hafsa Sufi
Lmi 10h59'9" 29d58'27"
Hagai Altman
Cyg 20h0'40" 37d31'19"
Hagan's Pointe
Cnc 8h13'49" 11d52'37"
Hagen Albus
Uma 11h59'7" 28d37'37"
Hagen Isaksen's Shining Star
Gem 7h47'44" 23d14'25"
Hagen Plötz
Uma 10h1'14" 64d51'39"
Hagen, Antja
Uma 9h16'6" 64d10'12"
Hagen, Werner
Ori 6h19'39" 16d47'19"
Haggman
Ori 5h38'43" -3d8'34"
Hagop Jack Baghdassarian
Lyn 7h36'49" 44d0'11"
Hagrid
Sco 17h44'48" -41d40'23"
Hahn, Edith & Harry
Uma 12h56'6" 54d27'9"
Hahnenkraft
Uma 8h56'13" 48d48'16"
Hahnia
Cep 21h27'46" 65d51'46"
H.A.H.S. Class of 2007
Lyn 7h47'19" 38d32'46"
Hai
Her 17h21'44" 15d45'31"
HAI
Vir 13h18'57" -3d4'16"
Hai & Alisa
Her 17h4'41" 33d24'33"
Hai Ming Hsia
Sgr 19h18'7" -13d7'53"
Haibing Liu
Oph 17h27'58" -0d23'46"
Haidee
Lib 15h30'43" -27d35'21"
Haider Murad
Cas 0h27'24" 62d24'57"
Haig John Kingston - 8.11.2005
Cru 12h20'28" -59d25'42"
Haig Kitishian
And 0h40'45" 45d34'29"
Hail Andrew
Lyn 8h8'51" 55d28'17"
haile
Lyn 6h27'24" 60d6'53"
Haile Rose Bohn
Crb 16h17'3" 28d4'17"
Hailee
Cam 6h44'18" 68d2'33"
Hailee Ava Pomeranz
And 2h38'27" 50d8'47"
Hailee Brooke Guyer
Ari 2h2'13" 15d39'48"
Hailee Elise Chappell
And 23h25'11" 48d44'45"
Hailee Hunter
Cas 1h34'38" 62d21'18"
Hailee Margret Sorum
And 1h43'16" 44d4'1"
Hailee Rose Freeman
And 0h2'46" 46d15'4"
Haileigh's Diamond
Cyg 21h15'49" 42d40'38"
Hailey
Cnc 8h9'20" 20d58'39"
Hailey
Lib 15h50'47" -9d21'51"
Hailey Alexa Anderson
Tau 5h53'6" 24d20'29"
Hailey and Tyler Wood
Cyg 19h44'13" 54d56'54"
Hailey Anne Giagnacovo
And 1h55'42" 41d53'29"
Hailey Autumn Randazzo
Lib 15h43'15" -17d26'50"
Hailey Baby
Uma 9h37'47" 70d36'54"
Hailey Berit Beck
Cyg 20h45'28" 33d11'24"
Hailey Bliss Velasquez
Sgr 18h57'41" -28d40'37"
Hailey Brazao
Lib 15h26'13" -15d8'25"
Hailey Brewer
And 1h31'31" 47d3'2"
Hailey Celeste Poteet
Psc 0h48'19" 19d50'6"
Hailey Coleen Gertz
And 0h17'22" 37d12'42"
Hailey D Fox
Uma 11h6'19" 35d20'16"
Hailey D. Rath (Halo)
Cas 0h24'23" 58d2'25"
Hailey Davidson
Sco 17h51'42" -35d38'14"
Hailey Elaine Carse
Ari 2h11'41" 23d5'30"
Hailey Elizabeth
Gem 7h11'5" 15d7'10"
Hailey Elizabeth
Psc 0h2'53" 6d33'55"
Hailey Elizabeth Bluml
Uma 11h31'27" 29d26'2"
Hailey Elizabeth Hall
Vir 14h0'5" -10d56'48"
Hailey Elizabeth Parks-Kunz
And 2h5'8" 45d18'33"
Hailey Elizabeth Stelzel 1992-2006
And 0h6'57" 35d0'52"
Hailey Emma (Kloy Kloy)
Uma 12h38'33" 59d1'46"
Hailey Filippone
Aqr 21h10'8" -10d19'28"
Hailey Gabrielle Bradley
Aqr 23h4'1" -7d27'29"
Hailey Grace Brown
Cap 21h40'21" -9d39'30"
Hailey Grace Dwyer
Leo 11h9'50" 22d3'6"
Hailey Hope
Umi 13h47'52" 87d20'31"
Hailey Jane Williams
And 0h20'49" 45d54'26"
Hailey Janea Raynor
Peg 23h49'25" 18d2'10"
Hailey Jo
Gem 6h57'43" 15d35'49"
Hailey Kathryn Osborne
Vir 12h34'32" -0d3'1"
Hailey LaRene Crook
And 2h22'21" 47d8'35"
Hailey Lashea
Crb 15h33'35" 26d59'44"
Hailey M. Denewiler
Dra 18h40'22" 83d43'12"
Hailey Madison
Lyn 7h16'26" 57d56'23"
Hailey Madison O'Rourke
And 0h42'3" 37d8'3"
Hailey Mae Scagliotti
Leo 9h30'6" 23d36'54"
Hailey Margaret Hrouda
And 23h46'19" 41d59'46"
Hailey Marie Albert
Ari 2h35'25" 22d0'15"
Hailey Marie Paramski
And 1h19'36" 37d56'8"
Hailey Marie Rumpca
Ari 3h16'27" 28d6'46"
Hailey Marie Smothers
And 2h27'35" 41d58'40"
Hailey & Matt's Star
Cyg 21h21'58" 44d36'12"
Hailey Morgan
Uma 9h37'54" 55d19'15"
Hailey Nicole Hahn
Gem 7h40'46" 22d4'12"
Hailey Nicole Stern
And 0h53'29" 37d9'29"
Hailey Nicole Zilch
And 1h0'31" 44d44'56"
Hailey- Papa's Girl
Ari 3h14'56" 29d23'53"
Hailey Pumpkin Pie Pelletier
Cap 20h37'1" -21d41'58"
Hailey Rae
Gem 7h17'1" 27d35'15"
Hailey Reese Borawski
Sco 17h45'14" -38d28'31"
Hailey Reese Hancock
Leo 10h45'15" 21d4'12"
Hailey Rose
Sgr 19h12'2" -23d41'6"
Hailey Rose Brennan
Cnc 8h19'16" 19d46'52"
Hailey Rose Gortakowski
Ori 5h31'7" 4d36'21"
Hailey Ruth Rinaldo
Aqr 21h16'50" -8d36'11"
Hailey Sianne Tuttle
Leo 11h33'47" 13d47'17"
Hailey Sophia
Uma 11h11'33" 52d55'37"
Hailey Sophia Kranis
And 2h18'2" 37d41'13"
Hailey Stiltner
And 0h39'6" 44d59'19"
Hailey Tushman
Uma 11h57'17" 35d54'30"
Hailey Vaughn
Vir 12h39'28" -5d4'15"
Hailey Woods
Sco 17h51'1" -38d33'56"
Hailey Yvette Doherty
Psc 2h0'25" 9d11'14"
Hailey-Joe Star of Possibilities
Lyn 7h37'13" 39d23'29"
Hailey's Inspiring Light
Ari 2h6'21" 24d12'57"
Hailey's Light
Leo 9h46'42" 13d27'39"
Hailey's Star
Lyn 9h11'30" 37d36'58"
Hailey's Star
Crb 15h43'3" 33d46'10"
Hailie
Psc 1h6'29" 22d24'25"
Hailie Jade
Aqr 23h52'23" -24d29'0"
Hailie Joanne
Tau 4h36'47" 27d44'0"
Hailie Mae
Boo 14h31'3" 52d29'7"
Hailie's Twinkling Light
And 0h13'39" 32d10'8"
Haim Saban
Her 17h10'56" 46d51'35"
Haim Sheli
Uma 8h50'10" 59d36'21"
Hainer Tibornak örök szerelemmel
Uma 11h9'31" 72d14'48"
Haines Joshua Tyler Wendell
Ari 2h32'36" 24d42'56"
Hainzl, Brigitte
Ori 5h27'41" 2d50'34"
HairGuRu
Aqr 23h19'27" -7d20'10"
HairryDad
Her 17h28'27" 45d48'23"
Hairy & Me
Cnc 9h11'0" 8d22'50"
Haixiang & Yichang
Uma 13h57'41" 62d18'10"
Hai-Ye Ni
Cyg 21h35'48" 35d21'1"
Haizon
Ari 2h33'58" 27d38'19"
Hajdar Divanovic
Ori 5h26'13" 9d5'39"
"HAJI"
Ori 5h56'26" -0d43'50"
Hajni Baba
Lib 14h43'17" -21d38'38"
Hajo
Uma 8h55'7" 48d34'42"
Hakan Yakin
Aur 5h3'37" 49d39'33"
Hakeber
Uma 9h43'55" 45d49'38"
Hakeemeh Heidari
Lib 15h24'13" -8d36'57"
Hakima
Umi 3h43'9" 88d39'21"
Hakuna Matata
Cen 11h19'22" -44d47'56"
Hakuna Matata Wrightson
Uma 11h44'19" 45d44'32"
Hal
And 0h43'40" 43d29'24"
Hal
Ori 5h37'21" 2d27'38"
Hal
Cnc 9h16'55" 25d37'56"
HAL
Ori 5h51'40" -8d22'8"
Hal and Katie LaFleur
Aur 5h24'8" 47d42'39"
Hal and Margaret North
Ori 5h22'49" 14d50'29"
Hal and Marlene Conrad
Aql 19h53'41" -0d24'30"
Hal & Betty Lou
Cyg 20h44'40" 37d31'33"
Hal Croak 9/13/1960
Her 18h53'45" 23d49'6"
Hal D. Morgan
Psc 0h47'47" 16d44'9"
Hal E Stickel
Aqr 22h30'33" 2d32'52"
Hal Hampton Jenrette
Tau 5h46'2" 21d24'55"
Hal & Judy DeMoss
Sge 19h50'26" 17d32'3"
Hal & Kathy Davis 25 yrs Together
Cyg 21h34'58" 36d58'19"
Hal loves Tracy
Psc 1h10'42" 12d37'26"
Hal Mills
Ari 2h21'16" 12d50'35"
Hal Ray Richardson
Vir 11h45'47" -0d48'50"
Hal Schwartz
Umi 17h34'28" 81d26'38"
Hal Wilson
Aqr 22h42'9" 2d11'20"
Hala
Vir 13h16'59" 4d53'19"
Hala baby
Aqr 22h7'8" -14d10'53"
Hala Poussin
Uma 8h34'11" 60d54'56"
Halasz
Aql 18h56'0" -0d44'17"
HalCin
And 0h31'36" 26d25'15"
Halcyn
Per 3h31'22" 50d21'35"
Halcyon Ember
Lyn 7h39'6" 35d40'10"
Hale and Rosie Schamel
Cyg 20h46'3" 44d31'56"
Halea 02042005
And 23h59'29" 48d9'46"
Haleah
Per 4h23'4" 34d45'22"
Halee Joy
Cnc 8h15'57" 10d7'25"
Haleh
Uma 9h15'17" 49d15'17"
Haleh Moustafa
Cnc 9h10'37" 26d37'5"
Haleigh and Marshall
Crb 15h34'9" 28d0'56"
Haleigh Crabb ~ November 30, 2006
Sgr 18h41'22" -34d6'5"
Haleigh Dawn Harrison
Vir 11h45'58" -6d5'34"
Haleigh Elizabeth Roberts
And 1h6'27" 35d33'10"
Haleigh Marie Friday Gottschalk
Crb 15h51'43" 37d23'56"
Haleigh Marie Gwin
Per 3h31'7" 45d5'50"
Haleigh Rose Shaw
Leo 10h50'27" 9d58'4"
Haleigh's LadyBug
Leo 11h24'36" 9d56'40"
Halema
And 1h2'2" 42d38'39"
Hale's Heart
Uma 10h25'25" 65d17'22"
Hale's Hollywood
Cas 0h22'32" 58d3'0"
Haley
And 23h4'54" 51d45'12"
HALEY
And 2h26'51" 48d50'20"
Haley
And 1h24'22" 40d19'54"
Haley
Uma 10h9'6" 44d40'0"
Haley
Uma 11h11'3" 30d8'4"
Haley
Tau 3h34'16" 21d30'46"
Haley
Leo 11h40'29" 26d33'1"
Haley
Com 12h24'31" 29d46'42"
Haley
Leo 9h35'40" 28d51'31"
Haley
Umi 9h48'59" 87d45'53"
Haley
Lib 15h45'56" -19d13'56"
Haley
Lib 15h49'7" -18d13'12"
Haley
Cap 20h29'53" -15d34'4"
Haley Adelle Hunter
And 0h18'0" 25d58'42"
Haley Aizlyn
Aqr 21h46'13" -2d1'4"
Haley + Angela
Aqr 20h51'44" -8d1'6"
Haley Anita Brousseau
Aqr 21h36'28" -7d42'51"
Haley Ann
And 1h38'48" 45d52'52"
Haley Ann Renneker
Leo 11h51'23" 10d57'16"
Haley Anne Knuepple
Vir 12h18'3" -0d54'48"
Haley Anne Snyder
And 23h46'16" 47d25'54"
Haley Au
Cru 12h16'1" -56d4'49"
Haley Autumn Agostinucci
Cap 20h25'44" -12d6'0"
Haley B
Ari 2h10'52" 23d51'0"
Haley Bettencourt
Umi 14h30'49" 75d37'22"
Haley Bopp
Ori 5h28'21" -3d46'52"
Haley Breann Gengler
Leo 9h37'17" 29d14'37"
Haley Breanne Stengle
Peg 22h29'56" 24d55'25"
Haley Brianne Scherr
Lmi 10h46'57" 28d23'40"
Haley Brooke
Cam 7h39'33" 65d38'45"
Haley Catherine O'Rourke
Aur 5h34'10" 40d18'28"
Haley Christine Cooper
Lmi 10h17'49" 28d32'47"
Haley Crystal Lenner
And 0h15'43" 27d33'18"
Haley D
Vir 13h25'13" -13d9'0"
Haley Dresdow
Cnc 8h36'23" 23d43'11"
Haley Elaine Kralik
Sco 17h41'36" -32d32'9"
HALEY ELISABETH MINIX
Psc 23h29'58" 7d14'39"
Haley Elizabeth
Uma 11h52'0" 39d19'8"
Haley Elizabeth Anderson
Ori 5h59'56" 17d44'59"
Haley Elizabeth Atchison
Uma 11h24'9" 33d30'37"
Haley Elizabeth Eastburn
Cam 4h10'55" 67d2'31"
Haley Elizabeth Nelson
And 0h8'12" 45d21'41"
Haley Elizabeth Wehr
Ori 4h58'26" 3d45'59"
Haley Ellen Smith
Uma 13h43'55" 56d29'20"
Haley Furtado
Lib 15h32'22" -9d34'7"
Haley Grace Wolfskill
Ori 5h47'2" 6d1'40"
Haley Hart
Vir 14h57'9" -0d8'28"
Haley Hesser
Cnc 8h10'45" 25d1'19"
Haley Holland
Ari 2h29'50" 19d42'21"
Haley Jaclyn Harding
Cap 21h38'0" -16d19'7"
Haley Jade
Cas 2h13'50" 64d9'46"
Haley Jade
Ari 2h59'30" 24d24'21"
Haley Jane
Psc 23h42'58" 6d33'20"
Haley Jean Mayo
And 0h16'32" 38d58'20"
Haley Jian Waw Timothy
Cam 6h0'44" 75d18'53"
Haley Jo Church
Umi 15h48'41" 86d35'56"
Haley Jolie
Peg 22h54'21" 8d19'8"
Haley Jordan Doniger
Umi 14h15'47" 72d44'27"
Haley Knutsen
Psc 1h43'7" 7d52'37"
Haley Lynn
And 0h19'42" 38d28'44"
Haley lynn prescott
Ari 2h41'5" 20d30'51"
Haley Lynn Shining Bright
And 0h19'50" 29d39'38"
Haley Lynn Wall
Lyr 18h51'35" 31d35'56"
Haley Mae Howard
Psc 1h10'9" 25d11'58"
Haley Maree Rich
Vir 14h18'32" 7d5'7"
Haley Marie
Peg 22h24'11" 32d28'53"
Haley Marie Jordan
Cap 21h10'48" -19d3'1"
Haley Marie Keyes
Ari 3h25'23" 29d52'4"
Haley Marie Mosch
Dra 18h13'17" 75d41'56"
Haley Marie Taylor
And 2h22'13" 43d24'51"
Haley Michelle
Gem 6h47'23" 23d2'37"
Haley Michelle Eazor
Aqr 20h39'44" -11d21'14"
Haley Miriam Doctor
Sgr 17h45'54" -24d52'28"
Haley Nichole Pereyda
Mon 6h52'4" 3d13'45"
Haley Nicole May
Cas 2h14'14" 64d7'20"
Haley Nicole Reh 2-2-1996
And 23h16'18" 45d20'25"

Haley Nicole Schultz
Cap 20h35'15" -22d46'26"
Haley Nicole Thierry
Cap 21h4'9" -19d4'29"
Haley Olivia Renee Gaskill
And 1h27'35" 37d26'47"
Haley Paige Hammer
Cyg 20h55'44" 32d28'44"
Haley Poarch
Peg 22h57'0" 31d57'27"
Haley Putnam
Crb 15h33'21" 36d4'57"
Haley Rainwater
Lmi 10h33'35" 32d15'51"
Haley Renee Brady
Umi 14h25'2" 74d33'36"
Haley Rose Marquardt
And 1h31'46" 48d42'8"
Haley Roy
Lib 15h49'41" -18d35'9"
Haley Ryan With Love
Lib 14h55'6" -15d24'53"
Haley Sauter
Lmi 10h34'49" 31d1'29"
Haley Seltenreich
Uma 11h16'33" 33d0'18"
Haley Smith
Ari 2h51'20" 27d50'53"
Haley Stearns
Cap 20h37'44" -22d14'53"
Haley Stuckmann
Uma 10h11'17" 57d43'34"
Haley Suzanne
Cir 15h17'12" -57d31'3"
Haley Sylvester
And 23h5'55" 45d54'53"
Haley Vatz
Cyg 20h19'8" 43d45'3"
Haley Wagner
Aqr 22h26'27" -2d47'16"
Haley Zaiden
Lib 15h11'53" -8d18'59"
Haley,Jacob,Keely
Lyn 8h54'44" 36d32'55"
Haley21097
Aqr 23h6'52" -7d41'9"
HaleyBear
Umi 15h10'1" 69d25'9"
Haleydactyl 4
Jumbopotamus
Tau 4h16'3" 20d33'3"
Haleydagit
Aur 5h45'22" 45d32'2"
Haleys Beauty
Sco 16h17'37" -16d32'8"
Haley's Big Store
Aqr 22h22'58" -23d19'32"
Haley's Comet
Cap 20h44'33" -21d21'10"
Haley's Dramatic Comet
Sgr 18h24'28" -26d2'17"
Haley's Heart
Per 2h15'44" 51d5'52"
Haleys' Heart
And 0h28'55" 27d12'35"
Haley's Heaven
Ari 3h15'29" 28d45'19"
Haley's Own
And 23h23'28" 41d50'14"
Haley's Shimmering Ray of Hope
Lib 15h35'57" -7d50'0"
Haley's Star
And 23h19'15" 47d29'14"
Haley's Star
And 0h48'14" 43d33'17"
Haley's Star For Life
Uma 11h35'57" 29d58'59"
Haley's Teddy Bear
Uma 9h13'17" 49d2'6"
Haley's Twinkling Star From T.C.
Uma 11h58'44" 63d35'35"
half-a-kasaba
Cyg 19h30'3" 29d56'23"
Half-n-Half Crispens
Sgr 19h28'1" -16d24'49"
Half's Heart
Col 6h0'4" -34d48'13"
Half-Way Dave
Dra 16h2'58" 68d2'50"
Hali
Cnc 8h35'32" 24d18'7"
Hali Ann
Uma 9h38'7" 48d44'57"
Hali Elizabeth Potters
Leo 11h37'5" -0d33'8"
Hali Raines
And 2h32'36" 44d6'35"
Hali Téa
And 23h15'25" 42d11'11"
Halia
Cnc 8h8'56" 27d17'51"
Halia Rain Hall-Harvey
Aqr 21h36'37" -2d42'44"
Halid-Roberto, 17.03.2002
Cas 0h49'36" 69d31'28"
Halie Annette
Uma 11h35'40" 49d7'4"
Halie Brooke "With All My Love"
Aqr 22h27'3" 0d47'38"
Halie Madelyne Bissell
Cap 20h22'25" -27d0'3"
Halie Marie McKenzie
Lib 15h31'35" -6d15'50"
Halie Norman
Cap 21h56'58" -14d54'36"
Halie's Star
Tau 4h0'44" 17d4'6"
HALIM KLAEB
Her 17h27'20" 26d18'40"
Halim Maya
Cyg 21h21'57" 38d36'18"
Halina
Umi 15h16'26" 79d27'8"
Halina Hildenbrand-Best
Mom Forever
Lmi 10h25'5" 38d5'20"
Halinki 18 urodzinki :-)
Love U
Uma 11h33'24" 53d32'20"
Hali's Heavenly 16TH
Sco 17h0'2" -31d40'51"
Halit & Simone para siempre
Uma 9h47'45" 45d49'53"
Hall - 1
Uma 10h0'17" 49d56'58"
Halle
Aqr 22h50'50" -9d2'28"
Halle Brook Billinghurst
Lyn 7h54'28" 58d16'14"
Halle Catherine Morana
Peg 21h26'17" 19d49'20"
Halle Elaine Kelsch
Cas 23h57'5" 56d53'54"
Halle Frances Lange
Gem 7h54'9" 15d16'29"
Halle Grace
And 23h46'39" 48d12'52"
Halle Hayes
Uma 9h56'13" 52d10'58"
Halle Hunt Stevens
Crb 15h27'47" 32d10'11"
Halle JoAnne Hardesty
Uma 12h3'35" 53d50'36"
Halle Marie Taylor
And 1h22'12" 38d36'17"
Halle Michelle Aten a.k.a. "Bootch"
Leo 11h25'4" 5d43'43"
Halle Murphy Gibson
Uma 9h44'50" 62d32'27"
Halle Noelle Burke
And 0h47'27" 22d21'51"
Halle Rueter's Shining Star
And 0h54'20" 37d51'8"
Halle Serena O'Conor
And 0h26'48" 44d19'22"
Halle, Eberhard
Uma 12h59'8" 53d6'8"
Halleh Afra
Lib 14h59'24" -3d18'5"
Hallei Alana Mleziva
Umi 16h16'9" 71d26'50"
Hallie's Little Mimi
Umi 14h41'2" 72d5'47"
Hallett's Happiness
Gem 7h34'36" 18d16'53"
Halley Caplan
Uma 9h23'59" 56d11'33"
Halli Dianne Keese
Cas 0h14'59" 58d12'47"
Halli Janae Caudell
Lib 15h3'33" -21d50'54"
Hallie
Cap 20h26'4" -9d26'49"
Hallie
Tau 3h31'59" 26d39'0"
Hallie Anne Wells
Sco 16h41'20" -32d14'40"
Hallie Danielle Hill
Psc 0h51'35" 28d7'3"
Hallie Elizabeth Hertzler
Sco 17h28'39" -42d23'0"
Hallie Elizabeth Watt
And 23h16'33" 42d29'19"
Hallie G. Sears
Cnc 8h26'18" 23d15'14"
Hallie Golden Taylor
Lib 15h44'41" -29d18'29"
Hallie "Jali" Rodarte
Lyn 8h48'10" 37d22'38"
Hallie Kaye Pratt
Lib 15h35'25" -26d43'10"
Hallie Osage
Aqr 22h32'56" -7d28'1"
Hallie Rae
Peg 22h28'57" 17d20'10"
Hallie Ryann Rhodes
Uma 9h58'1" 44d7'5"
Hallie The Priceless
Vir 13h9'47" 4d51'55"
Hallie's Comet
Lyn 7h30'10" 46d56'33"
Hallie's Dream
Vir 13h36'43" -7d50'8"
Hallie's Shooting Star
Uma 10h19'12" 68d6'47"
Halligan IV - I
Aqr 22h36'35" -16d45'5"
Halloween Pippin
Lib 14h57'9" -24d12'16"
Hall's Happiness
Lyn 8h41'38" 33d27'46"
Hallsboro 1927
Crb 15h33'11" 25d56'56"
Halo
Vir 13h8'1" 11d50'13"
Halo
Cap 21h58'57" -24d48'36"
Halo June
Sco 16h16'50" -19d3'1"
Haloa
Dra 18h33'55" 72d32'30"
Halpern, Sophie
Uma 11h59'38" 30d26'18"
Hal's halo
Uma 9h43'19" 47d16'55"
Hal's Star
Sgr 19h11'11" -21d56'29"
Halseeker
Vel 10h36'43" -54d10'34"
Halston Catherine Seton
Aqr 22h57'34" -4d59'17"
Halyn Thelma Lessard
Vir 13h35'55" -3d15'4"
Ham & Eggs
Her 18h5'50" 34d27'0"
Hamadryad
Vir 13h22'16" 11d59'25"
Hamann, Sandra & Dirk
Uma 12h7'5" 60d47'48"
Hamanusit
Uma 9h21'54" 47d27'48"
Hamdi and Nurten
Umi 14h34'59" 73d35'3"
Hamed The Angelic Velvet
Cep 22h3'5" 84d59'30"
Hamed, Mon Ange
Sgr 19h19'26" -21d55'9"
Hameed Nighat Saleemi
Aqr 22h43'4" -21d26'9"
Hamel Marion
Her 16h44'33" 28d56'38"
Hamers, Joachim
Uma 13h53'49" 56d10'49"
HaMi
Gem 6h1'10" 23d32'10"
Hamie
Gem 7h16'50" 13d37'42"
Hamilton
Per 3h4'43" 41d53'58"
Hamilton Bright
Cap 21h46'16" -11d8'34"
Hamish Gwekere Lunguela
Sgr 20h22'55" -42d50'43"
Hamish the Great McAlpine
Sgr 19h12'2" -21d6'57"
Hamley Cabrera
Lib 15h34'27" -11d42'54"
Hammer
Cap 21h19'43" -17d13'53"
Hammer
Uma 10h23'47" 57d21'10"
Hammer, Klaus
Ori 5h54'48" 21d47'17"
Hammer, Melanie
Uma 12h46'38" 61d22'27"
Hammer, Monika
Uma 9h42'14" 49d18'6"
Hammerhead
Sgr 19h21'57" -14d55'56"
Hammerich Family Star
Her 16h47'1" 36d36'44"
Hämmerling, Klaus
Uma 8h10'32" 63d45'35"
Hammer's Key
Leo 11h14'53" 12d9'23"
hammett
Uma 11h29'48" 62d52'15"
Hammonds Star
And 23h17'30" 42d55'9"
Hamner's Annabel in the Sky
Her 18h44'38" 19d3'26"
HAMP BURNETT
Uma 8h40'57" 61d35'9"
HAMPI
Crb 16h2'15" 33d55'24"
Hampton Alexander Poplin
Her 17h36'54" 36d50'33"
Hamsi
Cnc 8h25'7" 10d50'44"
Hamzy
Cyg 21h44'34" 53d42'45"
Han Ock Chang
Lac 22h24'10" 54d47'36"
HANA
Sgr 19h49'5" -22d57'21"
Hana
Leo 11h24'46" 13d28'21"
Hana
Gem 6h3'34" 25d11'17"
Hana Hamulic
Aqr 22h14'44" 1d6'26"
Hana Palazzotto
Cyg 21h4'42" 31d15'8"
Hana Starlion
Psc 1h8'52" 21d14'28"
Hana Vlasta Volavka
Mrozek
Del 20h38'6" 14d33'4"
Hanaa's Star
Uma 9h17'57" 47d44'20"
Hanah Mishel Bell
Gem 6h36'58" 15d23'12"
Hanan "Heno" Eid
Gem 7h18'51" 32d25'13"
Hanan Safagui
Ari 3h14'36" 21d14'52"
Hanan Salim
Cnc 9h21'53" 17d29'28"
Hanane Boutkhil
Aqr 20h47'30" -4d30'24"
Hanane (Noun my Baby)
Cyg 19h58'5" 38d50'47"
Hanani Christian Joy Taylor
And 23h3'31" 52d6'10"
Hanan's Star—The Jewel of the Nile
Aqr 23h54'13" -11d57'46"
Hana's Shining Star
Vir 14h13'43" -16d57'59"
Hana's Star
Tau 4h31'31" 16d27'55"
Hancl's Happiness
Gem 7h18'45" 20d6'24"
Hancock-Avila Lumina
Her 18h2'58" 24d32'58"
Hancocks
Tau 4h4'5" 16d9'2"
Hand in Hand
Cas 1h5'51" 48d53'9"
Hand to Hand
Cir 15h19'28" -58d40'13"
HandHammons
Lyr 18h49'7" 45d6'41"
"Handley"
Uma 10h32'22" 47d26'8"
Hands on a miracle... forever
Aqr 23h18'21" -13d28'13"
Handsome
Boo 14h47'5" 33d38'39"
Handsome
Ori 5h36'17" 5d57'42"
Handsome And The Princess
Lyr 18h40'11" 41d7'15"
"Handsome Cowboy"
Aur 5h50'43" 38d37'13"
Handsome Janssen
Per 3h6'59" 39d1'30"
Handsome (Joe Dinubilo)
Per 3h28'26" 48d26'53"
Handsome Mark
Leo 10h56'0" 10d22'56"
Handsome Prince Husband
Her 17h47'55" 40d41'29"
Hanedi Mansour
Gem 6h51'16" 22d17'15"
Haneen
Sco 17h51'7" -36d51'41"
Hanem Mohamed Hassan
Cas 2h5'25" 70d29'51"
Haney Love-Charles and Isabelle
Sgr 18h48'48" -20d24'17"
Hang Nguyen
Sgr 19h20'56" -26d19'32"
Hanh C.
Lib 14h34'50" -19d46'59"
Hanh Vo
Cnc 8h4'35" 12d13'55"
Hani
Umi 14h45'9" 74d5'1"
Hani and A J
Cap 21h35'2" -13d33'10"
Hania i Franek
Sgr 19h5'55" -22d5'40"
Hanicka Krejcova Kosarek
Aqr 22h38'2" -1d5'44"
Hanieh Shah
Cyg 21h32'54" 39d20'25"
Haniel's Insight
Lyr 18h48'20" 34d19'46"
Hanif-Mirza
Dra 18h26'11" 51d57'48"
Hanika
Leo 11h16'4" -2d57'19"
Hanisah Gani 1-11-1978
Sco 16h48'20" -27d58'34"
Hanita Saighal Walia
And 1h26'43" 34d43'39"
Hanizah Abdullah
Lmi 10h37'39" 31d51'32"
Hank
Cnv 13h46'6" 30d7'54"
HANK
Uma 10h27'50" 50d29'56"
Hank
Uma 9h34'43" 49d13'35"
Hank
Vir 12h22'34" 4d46'52"
Hank and Joanne
Cyg 20h34'16" 57d30'20"
Hank & Barbara Hicks
Lyn 6h41'8" 59d11'27"
Hank Cruz
Leo 9h45'13" 22d47'17"
Hank Cutler
Per 1h48'34" 51d14'10"
Hank Edwards Holland
Aur 4h56'1" 35d43'34"
Hank Espensen
Psc 0h37'17" 20d35'59"
Hank Falck
Boo 14h54'53" 22d47'31"
Hank G. Rule
Lyn 6h34'27" 56d44'24"
Hank Grenda
Gem 7h40'38" 22d16'3"
Hank Hall
Sgr 19h51'33" -11d46'3"
Hank Hearne The Eternal Light
Sco 16h45'31" -33d20'57"
Hank Loves Robyn
Cyg 20h59'39" 48d6'41"
Hank O'Neill (Molly's Loving Dad)
Uma 11h20'22" 52d59'37"
Hank POOPSI Gorecki
Gem 6h52'11" 12d25'19"
Hank & Stella by Starlight
And 1h25'58" 46d26'7"
Hank the Angry Drunken Dwarf
Uma 9h42'58" 59d3'43"
Hank the Tank
Cap 21h46'2" -14d28'18"
Hank Thompson
Vir 14h15'45" -13d8'26"
Hank Watkins
Per 3h11'21" 54d16'45"
Hanke Five
Cas 1h5'36" 60d15'4"
Hank's Shining Star
Leo 10h47'17" 15d30'46"
Hanky and Robin
Cyg 21h21'55" 34d24'33"
'Hanky Panky'
Dra 19h47'41" 61d23'29"
Hanna
Uma 12h57'48" 57d59'57"
Hanna
Lib 15h48'11" -19d32'13"
Hanna
And 1h50'45" 43d35'20"
HANNA
Cyg 19h38'6" 31d13'21"
Hanna
Leo 11h3'44" 16d13'26"
Hanna
Ari 2h53'27" 26d1'38"
Hanna~Adams Star
Cyg 19h37'40" 37d4'18"
Hanna Banana
Lib 15h7'3" -16d47'19"
Hanna Brunnerova
Ori 6h11'33" 12d3'42"
Hanna Claire
Leo 10h23'16" 24d46'43"
Hanna Cox
Lmi 10h44'0" 28d13'10"
Hanna Elizabeth
Cnc 8h51'59" 28d10'37"
Hanna Ella Viktoria Wedin
And 23h15'29" 51d36'42"
Hanna Emma Mae
Vir 12h46'10" 3d37'55"
Hanna Fay
Armstrong/Blackburn
Vir 12h43'53" -8d35'50"
Hanna Françoise Cook
Aur 5h23'24" 30d41'26"
Hanna Gibson
Sco 16h9'55" -18d18'54"
Hanna Heape
Del 20h42'12" 13d45'3"
Hanna Jasmin Götze
Uma 9h54'1" 51d33'30"
Hanna Jo Beautiful
Leo 10h8'16" 24d30'11"
Hanna Jolie Russell
Lyn 7h37'39" 47d5'58"
Hanna Kim
Cas 1h45'41" 61d47'53"
Hanna Lee
Sco 17h56'28" -40d57'13"
Hanna Magdalena 120354
Uma 11h5'10" 63d21'48"
Hanna Margaret Ralston
Peg 22h35'45" 15d56'23"
Hanna Maria
Uma 8h22'23" 68d17'1"
Hanna Martina
Lyr 18h30'2" 32d54'37"
Hanna Morgan Hughes
Sgr 18h10'45" -25d51'27"
Hanna Olsson
Cyg 21h21'7" 37d59'41"
Hanna Pride
Gem 6h32'5" 21d22'1"
Hanna Rebekah Gartshore Silk
Cyg 19h57'3" 30d20'33"
Hanna Rose
Ari 2h52'12" 26d1'24"
Hanna S. Elverum
Tau 3h40'33" 15d42'6"
Hanna Steiner
Peg 22h37'17" 8d46'32"
Hanna Whitaker
And 23h22'11" 39d33'9"
Hannah
Per 3h25'40" 48d26'6"
Hannah
Uma 9h42'7" 48d7'30"
Hannah
And 23h45'11" 48d10'18"
Hannah
And 23h33'14" 47d51'21"
Hannah
And 23h8'39" 51d5'26"
Hannah
Crb 16h11'11" 35d12'7"
Hannah
And 0h11'3" 34d57'55"
Hannah
Ori 4h50'14" 14d22'20"
Hannah
Psc 1h36'21" 27d35'31"
Hannah
Ari 1h53'22" 23d18'26"
Hannah
Cnc 8h12'1" 18d1'30"
Hannah
Leo 10h21'3" 24d21'48"
Hannah
And 0h10'54" 33d33'5"
Hannah
Tau 4h15'32" 26d56'32"
HANNAH
Cep 22h14'6" 68d53'17"
Hannah
Uma 11h11'22" 67d22'53"
Hannah
Lib 15h39'11" -18d53'56"
Hannah
Aqr 22h19'23" -14d15'10"
Hannah
Sgr 19h20'1" -12d9'4"
Hannah
Cap 20h14'57" -10d46'10"
Hannah
Lib 15h53'25" -4d20'50"
Hannah
Lib 15h10'44" -5d16'1"
Hannah 06/13/94
And 0h40'15" 44d46'49"
Hannah 18
And 0h27'55" 46d16'38"
Hannah Abigail Gooch
Uma 11h24'5" 41d43'27"
Hannah Adrienne Rogers
Ori 6h18'54" 15d35'33"
Hannah Aftergood
Sco 16h13'58" -22d2'19"
Hannah Alexandra
Ehrenberg
And 1h50'18" 45d9'37"
Hannah Alise
Lmi 10h13'9" 31d24'0"
Hannah Alison Latham
And 23h57'41" 42d40'58"
Hannah and David's Star
Sco 17h25'31" -35d27'49"
Hannah and Ian
Lmi 10h31'2" 36d59'54"
Hannah and Shayne Smith
And 0h58'59" 39d53'20"
Hannah Ashley Whiting
Cap 20h31'26" -20d52'3"
Hannah Autumn's Star
Lib 14h51'5" -1d38'5"
Hannah Ayleah Hatfield
Psc 1h25'49" 15d36'43"
Hannah Badger
Vir 12h27'37" -5d42'36"
Hannah Banana
Ant 10h17'10" -32d52'43"
Hannah Banana
Gem 6h34'49" 20d45'54"
Hannah Bea Finkelstein
And 0h28'31" 25d8'0"
Hannah Beatrice Luckman
And 23h44'27" 48d35'7"
Hannah Becky Alborough
And 23h19'42" 35d19'55"
Hannah Benjamin Tetro
Sgr 19h32'44" -16d52'34"
Hannah Beth Watson
Sco 16h6'0" -17d33'4"
Hannah Bethany Mauck
Lyn 7h36'50" 47d39'2"
Hannah Billings
Uma 11h22'41" 59d33'23"
Hannah Blake Whalen
Uma 10h58'1" 48d48'21"
Hannah Blanch
Cru 12h1'41" -62d34'55"
Hannah Bob Banana
Psc 1h22'33" 27d16'2"
Hannah Brisley
And 2h17'34" 37d38'26"
Hannah Brooke
Psc 1h28'22" 29d40'28"
Hannah Brown diZerega
Poulson
Vir 12h44'10" 5d35'57"
Hannah Butler
Vir 14h25'5" 2d30'45"
Hannah C. Havenar
'Beautiful'
Sgr 19h0'53" -32d39'1"
Hannah Cameron She-Ning Dykehouse
Vir 13h44'39" -8d32'36"
Hannah Carter-Schelp
Leo 9h41'34" 27d40'8"
Hannah Catherine Jackson
And 1h37'14" 41d3'51"
Hannah Catherine
Zarshenas
Leo 10h4'17" 26d24'19"
Hannah Chailyn Cochran
And 23h57'0" 45d7'29"
Hannah Chapman
Uma 11h26'36" 38d52'58"
Hannah Christine
And 1h42'19" 43d30'13"
Hannah Christine Harris
Psc 1h22'32" 32d43'41"
Hannah Christine Lee
Cnc 7h58'45" 15d43'2"
Hannah Claire Gatliff
Aqr 23h7'2" -5d40'38"
Hannah Coogan
And 0h36'57" 44d54'12"
Hannah Cooley
Uma 9h55'21" 60d59'30"
Hannah Curry
And 0h23'5" 41d56'44"
Hannah Danielle
And 2h10'7" 45d15'56"
Hannah Davis
Crb 15h51'32" 34d22'5"
Hannah Deanne Harvey
Lmi 10h43'37" 34d21'34"
Hannah Delaney Burke
Peg 23h31'24" 9d17'14"
Hannah & Derrick
Cyg 20h44'40" 52d20'19"
Hannah Dorothy Gidlewski
Cas 2h34'54" 65d2'30"
Hannah Duffy
Vir 11h56'30" 2d39'44"
Hannah Dyan Burtner's Star
Ori 5h15'20" 5d58'48"
Hannah Elaine Klingbeil
Mon 7h36'38" -0d36'4"
Hannah Elisabeth
Carbonneau
And 23h6'47" 49d26'45"
Hannah Elisabeth Shappell
And 23h7'11" 46d46'24"
Hannah Elisabeth Walent
Sgr 18h23'8" -27d45'14"
Hannah Eliza
Gem 6h32'4" 24d58'47"
Hannah Elizabeth
Crb 15h25'14" 25d59'23"
Hannah Elizabeth
Crb 15h37'28" 27d2'25"
Hannah Elizabeth
Aql 19h24'40" 7d59'9"
Hannah Elizabeth
And 2h19'12" 42d50'53"
Hannah Elizabeth Boehme
And 23h44'1" 46d0'39"
Hannah Elizabeth
Butterfield
Lyn 8h27'35" 54d56'47"
Hannah Elizabeth Cerullo
Tau 5h7'53" 26d59'58"
Hannah Elizabeth Cole
Psc 1h15'57" 23d57'14"
Hannah Elizabeth Flynn
And 23h31'47" 42d12'2"
Hannah Elizabeth Gately
Vul 19h51'4" 21d24'56"
Hannah Elizabeth Marshall
And 0h35'37" 27d21'18"
Hannah Elizabeth
McMonigal
And 23h44'27" 46d49'6"
Hannah Elizabeth
Montgomery
Per 3h29'27" 46d44'16"
Hannah Elizabeth Mora
Vir 13h44'38" 1d4'54"
Hannah Elizabeth Morgan
And 2h31'38" 38d33'43"
Hannah Elizabeth Reser
And 0h34'44" 46d25'38"
Hannah Elizabeth Sessa
Mon 6h50'47" -0d5'29"
Hannah Elize
Vir 12h13'15" -0d17'34"
Hannah Emily Blake
Uma 9h49'2" 63d31'10"
Hannah Engel
Cnc 8h19'49" 13d51'56"
Hannah Eriksson
Leo 9h25'1" 6d33'15"
Hannah Esther Rich-Byberg
Sgr 19h46'46" -15d49'0"
Hannah Faith
Vir 14h20'17" -0d30'28"
Hannah Faith
Lyn 7h28'44" 59d23'22"
Hannah Faith
Vir 12h17'43" 12d15'2"
Hannah Faith Little
Cnc 8h47'3" 28d32'59"
Hannah Fallon
Peg 23h54'25" 19d38'5"
Hannah Fauteux
Uma 11h16'43" 34d58'50"
Hannah Felicia Rodriguez-Jones
Cnc 8h19'24" 21d17'34"
Hannah Fonti
Lyn 6h48'38" 52d55'52"
Hannah Frances Hathaway
And 1h30'46" 49d32'13"
Hannah Frances Vaughan
And 1h15'33" 35d32'5"
Hannah Fryar * Daddy's Star
Lib 15h16'33" -5d37'17"
Hannah Georgianne Jones
Peg 22h31'25" 17d11'1"
Hannah Glosser
And 0h23'44" 24d16'15"

Hannah Gochenour
Her 17h2'37" 18d25'9"
Hannah Grace
And 0h19'17" 29d26'32"
Hannah Grace
Lyn 6h55'48" 59d37'26"
Hannah Grace Brown
Uma 10h22'6" 69d5'13"
Hannah Grace Dvorachek
And 0h38'30" 37d5'48"
Hannah Grace Hancock
And 0h36'37" 45d27'22"
Hannah Grace Horner
And 1h53'23" 36d57'18"
Hannah Grace Larroux
And 0h36'16" 31d44'33"
Hannah Grace McSwain
Tau 4h16'49" 5d37'8"
Hannah Grace Murphy
Hya 9h8'21" 3d45'27"
Hannah Grace Nixon
And 23h30'36" 37d51'36"
Hannah Grace Paye
And 0h4'48" 36d25'37"
Hannah Grace Robbins
Lib 14h51'32" -4d46'18"
Hannah Grace Rogers
Lib 15h10'4" -14d50'2"
Hannah Grace Thompson
Gem 7h37'34" 21d49'32"
Hannah Grace Tobin
And 0h31'26" 31d50'22"
Hannah Grace Trio
Cnc 8h16'45" 17d51'44"
Hannah Grace Ward
And 23h28'36" 48d25'42"
Hannah Gregor
Gem 7h2'5" 26d44'45"
Hannah Grubb
Ari 2h53'11" 27d34'19"
Hannah Guerino, in loving memory
Leo 9h55'27" 24d15'56"
Hannah Gutierrez
Lib 14h30'2" -8d50'15"
Hannah Hallman
Leo 9h24'30" 24d53'48"
Hannah Hansen
Tau 4h27'24" 14d14'47"
Hannah Hess
And 23h17'50" 48d8'0"
Hannah Hood
Psc 1h19'47" 14d43'20"
Hannah Hosmer
Cap 20h28'51" -22d28'19"
Hannah I. Tunnell
Gem 6h38'17" 13d45'10"
Hannah in the Heavens
And 1h22'28" 49d52'44"
Hannah J. Steinkamp
Leo 9h33'10" 14d17'22"
Hannah J. Williams
Cyg 20h18'8" 37d10'21"
Hannah James
Sco 17h43'44" -41d17'31"
Hannah Jane
And 23h1'32" 52d16'3"
Hannah Jane Edwards
And 2h38'6" 39d26'51"
Hannah Jane Ellis
Uma 11h43'40" 54d17'55"
Hannah Jane Keyser
Sco 16h11'10" -13d25'15"
Hannah Jane Woodall
And 1h17'5" 42d4'9"
Hannah & Jason 6.3.2005
Cyg 20h31'43" 36d39'6"
Hannah Jean Higgins
Uma 8h16'9" 71d20'6"
Hannah Jean Lenox
Ori 6h8'56" 15d35'38"
Hannah Jeanette
Psc 23h40'36" 1d13'46"
Hannah Jo Becker
Lib 15h6'45" -13d6'17"
Hannah Joanne Kramer
Crb 15h44'37" 26d43'50"
Hannah John
Cyg 20h10'18" 45d52'43"
Hannah Joy
Com 12h51'42" 17d26'3"
Hannah Joy Manaois
Uma 11h24'39" 36d8'31"
Hannah Joy Meckley
Sgr 18h44'41" -27d44'45"
Hannah Joyce & Olivia Kathleen
Lyn 7h40'39" 53d46'32"
Hannah K. Sowd
Cnc 8h0'8" 13d1'28"
Hannah Kaitlyn and Matthew Julian
Psc 1h14'14" 26d30'12"
Hannah Kate Parmar
Aqr 21h7'28" -9d50'9"
Hannah Katherine Boan Rutherford
Peg 22h29'54" 14d43'47"
Hannah Katherine Gibson
Lyr 18h49'8" 36d18'6"
Hannah Katherine Wolfe
Psc 23h58'7" -4d50'56"
Hannah Kathryn Glaze
Ari 2h26'12" 25d6'56"

Hannah Kay
Lyr 18h44'34" 39d55'29"
Hannah Keesy Barnes
Lib 14h24'35" -15d37'35"
Hannah Kelsey Price
Leo 11h8'37" 17d44'5"
Hannah Kirin
Cnc 9h4'21" 31d3'58"
Hannah Lauren Woods
Leo 10h24'39" 18d29'35"
Hannah LaVonne
And 2h29'0" 46d23'43"
Hannah Layne
Cnc 9h0'54" 18d29'11"
Hannah Lea Henderson
Peg 22h16'29" 7d3'45"
Hannah & Lee
Cyg 20h52'14" 49d38'24"
Hannah Lee
Ori 5h34'9" -5d10'19"
Hannah Leigh Jones
Gem 7h20'56" 31d49'30"
Hannah Lena
Uma 13h1'39" 53d46'8"
Hannah Li Aldridge
Cas 2h6'43" 63d57'55"
Hannah Lichtenberger
Crb 15h23'37" 31d34'34"
Hannah Lillioja
Cyg 20h36'34" 45d37'50"
Hannah Lindsay
Psc 0h46'59" 8d24'8"
Hannah Lois Young
Ari 2h14'38" 11d49'53"
Hannah Louisa Shapiro
Cam 6h15'51" 60d2'35"
Hannah Louise Barkby
Umi 15h17'15" 79d41'47"
Hannah Louise Beckett
And 1h41'40" 42d58'51"
Hannah Louise Carr
Umi 13h59'42" 70d20'53"
Hannah Louise Custons
And 23h49'48" 39d3'15"
Hannah Louise Edge
And 2h28'8" 49d37'59"
Hannah Louise Oldfield
And 23h55'18" 43d45'55"
Hannah Louise Scott "Weezy"
Lyr 18h36'4" 41d15'26"
Hannah Lupton
Ori 5h53'39" 7d6'16"
Hannah Lynn
Ori 6h5'2" 18d17'13"
Hannah Lynn Bogina
Uma 11h21'31" 45d33'37"
Hannah Lynn Gallagher
Ari 2h53'25" 27d5'25"
Hannah Madelaine Middleton
And 23h27'38" 42d52'3"
Hannah Madeleine Rappold
Cap 21h29'19" -16d48'22"
Hannah Mae
Cap 20h9'27" -10d35'3"
Hannah Mae - "Little Buddha"
And 23h24'5" 38d1'47"
Hannah Magnia
Lmi 10h26'38" 35d55'6"
Hannah Marie
And 23h29'52" 48d34'18"
Hannah Marie
Ari 2h49'14" 13d53'37"
Hannah Marie
Lib 15h26'56" -4d45'53"
Hannah Marie
Cap 21h32'56" -22d33'48"
Hannah Marie Anderson
And 23h46'16" 33d29'25"
Hannah Marie Bradshaw
Umi 14h20'24" 67d43'19"
Hannah Marie Broughton
And 23h44'26" 39d3'57"
Hannah Marie Burton
Uma 12h3'52" 42d33'47"
Hannah Marie Gould
Gem 7h52'12" 27d3'29"
Hannah Marie Grace
Cas 23h38'46" 53d59'26"
Hannah Marie Hawkins
And 0h56'52" 38d56'13"
Hannah Marie Hoffner
Cap 21h37'59" -17d30'51"
Hannah Marie Leake
Uma 8h31'41" 64d55'0"
Hannah Marie Rouse
And 2h22'46" 46d45'1"
Hannah Marie Scott
And 0h42'31" 25d21'13"
Hannah Marie Stassens
Psc 0h55'40" 4d28'33"
Hannah Marie Watson
And 1h54'31" 43d1'55"
Hannah Marie Young
Psc 0h0'16" -2d58'11"
Hannah Mary Orchison
Cas 0h21'57" 53d41'18"
Hannah Marye Hockaday
Crb 15h43'6" 26d13'53"
Hannah McCoy
And 0h35'8" 46d23'40"

Hannah McGraw
Lyr 18h44'23" 39d38'20"
Hannah McGuire
And 0h9'7" 44d27'54"
Hannah Means Angel
Gem 6h19'4" 23d36'54"
Hannah Michal Sanchez
Cnc 8h25'18" 17d52'39"
Hannah Michelle Medwid
Vir 12h38'14" 10d46'38"
Hannah Michelle Moy
And 0h19'47" 27d56'19"
Hannah Michelle Radner
Psc 0h35'11" 6d34'57"
Hannah Michelle Riles
Lyn 8h30'29" 56d2'49"
Hannah Michelle Rock
Pho 1h10'50" -39d53'6"
Hannah Michelle Snyder
And 2h17'2" 42d40'20"
Hannah Miller Lee
Tau 3h38'28" 24d27'3"
Hannah Mulkey
And 2h17'18" 47d22'22"
Hannah Munson
Ari 2h23'3" 25d26'6"
Hannah Murphy
Cas 23h45'19" 54d18'0"
Hannah Nadine
Cnc 8h12'14" 8d10'22"
Hannah Nichole Abell
Gem 7h30'18" 34d35'8"
Hannah Nichole Smith
Tau 5h40'30" 26d21'15"
Hannah Nicole Guardian Angel
Cyg 20h33'7" 40d22'49"
Hannah Nicole Hines
Tau 4h45'4" 23d9'52"
Hannah Noel
And 23h50'42" 41d25'34"
Hannah Noelle
And 0h5'50" 37d15'47"
Hannah Noelle Weiss
Sgr 19h6'20" -16d25'1"
Hannah Our English Sapphire Star
Eri 2h6'55" -53d19'45"
Hannah Paige Duncan
And 2h23'17" 47d56'14"
Hannah Patricia Peyton
Umi 15h22'57" 67d54'3"
Hannah Phillips
Cap 21h29'4" -15d43'38"
Hannah Pickar
Peg 22h46'27" 25d46'47"
Hannah Preister
And 1h17'4" 45d15'46"
Hannah Presley Davis
Lib 15h20'52" -23d28'26"
Hannah Pullen
Cru 12h15'36" -61d15'39"
Hannah Rae
And 1h49'14" 38d27'5"
Hannah Rae
Ari 2h27'20" 21d1'37"
Hannah Rae Freeman
And 0h5'17" 46d28'39"
Hannah Rae Kasper
Sco 16h16'41" -19d36'7"
Hannah Rae Miguel
Uma 11h49'9" 45d15'5"
Hannah Rae Woodruff
Hya 8h46'33" 4d55'11"
Hannah Rae's Own Star
Gem 6h15'39" 26d6'40"
Hannah Rain
Sco 17h52'46" -35d53'17"
Hannah Rain Krauss
Ari 2h44'21" 16d38'19"
Hannah Rainey
Mon 6h46'13" -0d59'18"
Hannah Randol
Vel 9h28'15" -42d57'5"
Hannah Ray Dubord (Little Peep)
And 0h12'24" 43d51'12"
Hannah Raychel Cooper
Umi 14h26'2" 69d59'13"
Hannah Rebekah Sparks
Lyn 6h51'55" 58d54'8"
Hannah Reisinger
Uma 12h26'20" 59d9'1"
Hannah Reneé Friend
And 23h17'41" 44d49'27"
Hannah Resha
Sco 16h45'33" -45d37'43"
Hannah Richards
Sco 17h52'4" -30d17'11"
Hannah Roberts
And 2h12'28" 50d2'31"
Hannah Roberts 13!
Her 16h36'17" 35d53'38"
Hannah Robertson Arndt
Ari 2h5'47" 18d7'35"
Hannah Rory
And 0h35'38" 36d15'23"
Hannah Rose
Cyg 21h10'49" 47d15'45"
Hannah Rose
Psc 1h20'47" 17d12'24"
Hannah Rose
Vir 12h45'8" 3d38'3"
Hannah Rose Beldue
Sco 16h18'3" -11d40'10"

Hannah Rose Cohen
Lib 15h25'12" -21d49'13"
Hannah Rose Leighton
Mon 7h34'2" -0d55'25"
Hannah Rose Mackay
Lib 14h49'25" -3d21'5"
Hannah Rose Pruett
Cas 1h9'39" 57d5'54"
Hannah Rose Warner
Uma 9h37'4" 42d36'45"
Hannah Roselle Spine
And 2h23'28" 47d54'14"
Hannah Rose's Dream
And 23h14'11" 51d40'2"
Hannah Roy
And 2h13'55" 45d39'44"
Hannah Ruby
And 1h12'51" 43d21'21"
Hannah Ruth
Ari 2h16'16" 17d28'41"
Hannah Ruth Arel
Cas 0h51'58" 63d44'36"
Hannah Ruth Hutchinson
Umi 17h26'27" 83d19'13"
Hannah Ruth Meade
Gem 7h26'50" 26d43'45"
Hannah & Samuel
Cas 0h39'44" 54d46'36"
Hannah Scheurer God's Angel #5
Gem 6h38'20" 16d0'26"
Hannah Shayne Pite
Aqr 22h34'52" -1d47'47"
Hannah Smallwood
Uma 8h43'56" 59d33'29"
Hannah Sophia Furda
Uma 12h13'38" 61d27'7"
Hannah Star
Gem 6h2'7" 24d46'26"
Hannah & Steve's Wedding Star
Cyg 19h50'8" 33d13'20"
Hannah Stidman
And 0h45'20" 28d23'32"
Hannah Strugnell
And 2h36'17" 44d59'14"
Hannah Sue
And 23h31'47" 45d59'20"
Hannah Sue DuFriend
Cyg 20h3'39" 31d48'44"
Hannah Susan Ewens
Cru 12h39'52" -60d7'22"
Hannah Suzan Press
Umi 15h21'53" 72d52'56"
Hannah Swacus
Sco 17h42'57" -39d47'17"
Hannah Tahhan
Aqr 22h3'57" 2d20'13"
Hannah Taylor Janes
Lyn 8h11'12" 46d24'45"
Hannah Valerie Ritter
Lib 14h22'14" -11d21'28"
Hannah W.
Aqr 21h40'59" 2d50'28"
Hannah Walid Taman
Uma 10h5'53" 51d48'50"
Hannah Walker
And 23h48'31" 34d46'29"
Hannah Walsh
Cyg 19h42'24" 33d31'41"
Hannah Welsh
Gem 7h45'5" 26d18'10"
Hannah Wert
Leo 10h59'14" 16d5'40"
Hannah Williams
Cas 0h20'52" 55d30'2"
Hannah Williams
Cas 23h20'41" 55d22'21"
Hannah Willow
Vir 13h10'15" 11d14'54"
Hannah Wimpee/Giltner
Lyr 18h47'51" 31d51'15"
Hannah Marie
Dra 19h15'21" 62d40'42"
Hannah, My Best Friend
Leo 11h15'13" 24d2'1"
Hannah1025
Sgr 18h23'24" -22d55'44"
Hannah...143...My Heart
And 0h3'13" 44d32'8"
HannahCH
Uma 13h44'31" 61d26'45"
Hannah-Joy Bird
Vir 13h5'56" -10d50'54"
Hannah-Leigh Elizabeth
Ari 2h49'57" 25d44'13"
Hannah-Nova
And 0h19'28" 26d42'14"
Hannah-Palma-Thérese Gajdosir
Vir 11h50'15" 6d51'36"
HannahRose
Tau 3h24'22" 17d17'35"
Hannah's Celestial Steed
Peg 21h58'6" 26d21'26"
Hannah's Domino
Cma 6h53'24" -23d26'12"
Hannah's Halo
Lib 15h34'45" -15d27'27"
Hannah's Halo
Uma 10h29'11" 54d36'14"
Hannah's Heaven
Gem 6h40'38" 17d50'16"
Hannah's Hero
Per 3h39'24" 43d50'54"

Hannah's Light
Uma 11h12'38" 33d3'34"
Hannah's Sparkle
Cyg 20h41'7" 39d8'49"
Hannah's Star
And 1h7'41" 34d23'36"
Hannah's Star
Gem 7h31'13" 24d47'0"
Hannahs Star
For 3h5'2" -27d19'9"
Hannah's Star
Cap 21h42'3" -24d35'31"
Hannah's Star
Cru 11h58'7" -61d58'36"
Hannah's Twinkling Star
And 1h42'36" 41d34'24"
Hannahs' Wish
Psc 0h9'57" 9d19'10"
Hannah's Wishing Star
Leo 10h19'40" 8d28'4"
Hanna-Leigh
And 1h25'33" 35d22'25"
Hanna-Rae Oddie
Cas 23h16'32" 56d7'57"
Hanna's Light
Cnc 9h18'18" 16d18'55"
Hanna.xxx
And 23h52'20" 41d49'29"
Hanne Ala-Rami
Uma 11h53'59" 42d51'45"
Hanne Damgaard Hansen
Cap 20h58'53" -27d22'50"
Hanne & Morten's stjerne
Vir 14h34'40" 7d19'40"
Hanne Reinhardt
Uma 9h21'59" 68d45'23"
Hanneke Winfree
Leo 9h55'33" 15d17'35"
Hannelore E. Benjamin
Uma 11h31'38" 52d48'56"
Hannelore Hecker
Ori 5h56'9" 12d52'56"
Hannelore Henderson
Peg 23h14'46" 17d26'19"
Hannelore Rosar
Ori 6h15'51" 10d57'48"
Hannelore Schmidt
Com 13h8'37" 17d26'24"
Hannelore Waitz
Cnc 8h6'33" 21d40'24"
Hannemann, Helmut
Ori 6h8'43" 16d23'18"
Hannes
Uma 12h50'48" 53d12'3"
Hannes der Oberschurke
Uma 9h2'4" 49d23'50"
Hannes Estl
Aqr 21h50'24" -2d59'11"
Hannes Floto
Uma 14h6'15" 54d39'39"
Hanni Gubler
Dra 18h43'36" 50d2'39"
Hanni Haddad
Uma 8h52'7" 71d23'5"
Hanni Kolb
Uma 8h58'31" 68d43'6"
Hanni Zobrist
And 23h5'46" 41d57'9"
Hanniball
Uma 10h6'9" 45d41'6"
Hannie
Gem 7h10'4" 33d42'29"
Hanno Ernst
Ori 4h48'57" -2d22'30"
Hanns Vollmer
Aur 5h44'5" 41d29'20"
Hanobe
Gem 6h44'33" 28d7'5"
~:+: hanonle :+:~
And 2h22'0" 41d57'14"
Hanoodi
Ori 5h13'32" 11d46'8"
Hanoosh
Aqr 23h26'59" -18d5'52"
Hanora
Mon 6h52'7" 8d30'24"
Hanover
Gem 7h49'20" 31d50'10"
Hans
Cas 0h46'45" 64d19'49"
Hans A. Knoben
Ori 6h18'6" 16d28'15"
Hans Aeschlimann
Lyn 7h29'19" 57d18'22"
Hans and Eva
And 1h38'12" 50d29'3"
Hans Barmettler
Crb 16h19'31" 38d27'13"
Hans Bergmann
Uma 8h14'57" 68d24'2"
Hans Bernd Weber
Uma 9h7'3" 49d56'50"
Hans Bohlein
Uma 11h9'28" 65d19'52"
Hans Bosch (Chulisnaquis)
Ori 6h15'14" 15d1'54"
Hans Brupbacher
Aql 19h57'30" 1d46'32"
Hans Clarin
Uma 8h15'35" 65d37'11"
Hans de Boer
And 0h18'22" 45d8'9"
Hans Dellert
Ori 6h23'42" 10d4'26"

Hans Diem
Cap 20h47'37" -16d13'43"
Hans Dutler
Tri 2h8'7" 33d19'18"
Hans E.
Cas 1h3'20" 62d24'0"
Hans Eduard Smodis
Uma 10h3'7" 52d12'49"
Hans Elias Fröhlich
Uma 8h55'7" 65d20'3"
Hans Forrer
Aur 4h56'0" 35d37'56"
Hans Franz Spiertz
Uma 12h33'36" 53d11'40"
Hans Garlichs
Aql 19h50'10" 2d34'10"
Hans & Gerti Tschallener-Senn
Cyg 21h53'5" 52d46'7"
Hans Giessel
Uma 11h14'53" 41d9'26"
Hans Günter Koch
Uma 11h12'4" 63d53'6"
Hans Handschuh
Com 13h12'1" 14d54'12"
Hans Henrik Nicolaisen
Sco 17h15'35" -42d37'5"
Hans Hopfenmüller
Uma 10h57'48" 59d14'49"
Hans Hörler
Lyr 18h42'53" 35d50'17"
Hans Hucker
Uma 9h58'45" 56d6'50"
Hans J. Hibbert
Uma 12h3'20" 31d27'56"
Hans Jespersen
Aql 19h47'55" -0d7'9"
Hans Joachim Groß
Uma 11h56'31" 30d58'5"
Hans Joachim Neumann
Vir 13h43'24" -17d57'36"
Hans Jochen Meuer
Ori 6h13'52" 7d50'53"
Hans Jürgen Graue
Uma 8h56'26" 48d43'14"
Hans Jurgen - Papa's Star Forever
Her 17h0'0" 16d40'59"
Hans Limon
Gem 6h54'12" 15d2'16"
Hans Müller
Uma 8h42'3" 66d55'7"
Hans Muth
Ori 6h20'0" 9d11'50"
Hans - my shining star - Alana
Vir 12h14'11" -0d56'49"
Hans N.Hansen
Her 17h27'28" 38d46'34"
Hans Peter Knöpke
Uma 9h8'47" 64d10'29"
Hans Peter Steierl
Uma 9h31'59" 63d56'59"
Hans & Rachel Klose
Pho 1h22'39" -42d58'23"
HANS SCHICK
Aur 5h50'43" 39d41'0"
Hans Schüller
Uma 11h0'56" 33d35'23"
Hans Schymura
Uma 9h13'17" 57d59'48"
Hans Stüsser
Ori 6h17'54" 3d42'42"
Hans Suter
Cas 0h6'58" 59d47'17"
Hans und Louise Hess-Streit
Aur 6h39'27" 39d3'33"
Hans Wendel
Ori 5h53'43" 6d47'23"
Hans Willi Neuls
Uma 9h26'20" 48d18'51"
Hans1976
Cap 21h47'49" -9d1'33"
Hansard
Lyn 8h7'4" 57d0'6"
Hans-Dieter Liss
Uma 8h47'12" 68d46'39"
Hans-Dieter Ritterhaus
Uma 9h55'5" 72d43'54"
Hansel, Heather, Brook, Zac
Ori 5h42'9" 1d8'43"
Hänsel, Peter
Cap 20h27'59" -16d55'11"
Hansenstar
Cnv 13h55'45" 37d47'53"
Hans-Hermann Aldenhoff
Ori 6h19'19" 7d39'0"
Hansi & Nici
Uma 8h57'22" 68d43'12"
Hansi & Rosey Star
Aur 5h54'58" 42d12'8"
Hansi Schwarzl
Ori 5h54'49" 12d33'58"
Hansibern
Cyg 20h53'5" 43d54'4"
Hans-Joachim Fritsche
Uma 8h42'44" 53d14'36"
Hans-Joachim Rasmus
Uma 10h33'20" 48d59'9"
Hans-Joachim Ullrich
Uma 10h8'52" 49d31'53"
hans-joli-stern
Uma 9h12'59" 64d22'18"

Hansjörg Kleinhanß
Uma 9h34'26" 63d0'22"
Hansjörg & Teresa Maissen
Cyg 21h41'40" 51d45'58"
Hans-Jorg Walter Rosler
Cnc 9h17'17" 14d8'26"
Hans-Jürgen Dittombée
Uma 8h27'45" 63d33'21"
Hans-Jürgen Drost
Ori 6h10'52" 6d29'51"
Hans-Jürgen Köning
Vir 14h3'42" -2d50'0"
Hans-Jürgen Kossatz
Ori 6h18'19" 5d46'4"
Hans-Jürgen Schulz
Uma 10h18'58" 48d11'14"
Hans-Jürgen Seewig
Uma 9h3'55" 60d19'28"
Hans-Jürgen Watzlaw
Ori 6h12'49" 7d29'44"
Hans-Jürgen Wild
Uma 11h43'53" 33d18'31"
Hans-Karl Seibert
Ori 4h49'46" 1d4'13"
Hansom Star
Her 17h32'0" 37d52'45"
Hans-Otto Appel
Uma 8h43'46" 68d48'21"
Hans-Peter Birkenbach
Uma 10h41'27" 71d34'59"
Hans-Peter Haas
Uma 10h47'21" 45d31'14"
Hanspeter Thiel
Umi 13h36'18" 70d12'52"
Hans-Peter Wöffen
Ori 6h19'51" 8d32'55"
Hans-Rudolf Daverio, 28.08.2003
Cep 22h34'1" 62d53'15"
Hans-Rudolf Matter
Cep 22h2'24" 68d46'55"
Hansruedi
Vul 20h20'41" 24d19'24"
Hansruedi
Per 4h37'34" 40d16'10"
Hansruedi Völlmin
Tri 2h14'33" 33d8'22"
Hansueli Meier Davos
Cas 0h23'50" 48d31'14"
Hansueli Winz
Tri 2h21'8" 36d59'10"
Hans-Ulrich Langer
Uma 11h1'5" 40d28'58"
Hans-Ulrich Schmidt
Ori 5h53'45" 6d41'3"
Hans-Werner Jung
Uma 8h16'1" 67d56'26"
Hans-Werner Koppelkamm
Uma 9h40'50" 41d40'15"
Hans-Werner Mydlar
Ori 6h17'34" 8d42'29"
Hansy
Peg 22h12'40" 6d42'52"
Hanz Lei
Tau 5h43'35" 23d27'8"
Hanz Macedonio - Rising Star
Uma 11h27'39" 61d59'58"
Hanzade Sultan
And 23h18'23" 41d55'55"
Hao-Jie 78
Dra 16h33'55" 62d25'37"
Haonan Nancy Yan
Lib 15h6'44" -10d44'18"
Hap-Nik 2005
Cyg 19h45'14" 42d19'51"
Happily Ever After
Her 17h24'16" 47d55'16"
Happily Ever After
Cyg 21h54'39" 52d24'47"
Happily Ever After
Crb 15h41'7" 37d26'7"
Happily Ever After
Vir 11h40'39" 8d8'19"
Happily Ever After
Col 5h7'36" -37d26'32"
Happily Ever After Begins Today
Lib 14h53'28" -6d52'39"
Happiness
Eri 3h36'36" -13d58'13"
Happiness
Cam 4h28'4" 66d18'20"
Happiness
Psc 0h9'58" 1d44'40"
happiness and smile forever
Leo 10h39'36" 14d13'4"
Happiness to Julie & Brian Miller
Cyg 20h57'23" 44d28'17"
Happiness727
Lmi 10h7'36" 38d53'7"
Happy
Umi 14h29'59" 71d10'56"
Happy 18th Abigail Elizabeth Sayer
And 23h43'3" 42d24'5"
Happy 18th Birthday Alicia Crawford
Umi 15h14'29" 74d14'23"
Happy 18th Birthday, Tony
Lyr 18h41'34" 31d41'7"

Happy 18th Fer-Fer August 25 2004
Uma 9h22'7" 56d46'10"
Happy 1st Anniversary Chad
Aqr 22h59'34" -10d54'51"
Happy 1st Birthday Ella Jane White
Lyn 6h56'56" 54d15'54"
Happy 1st Birthday Ryan N.Knipp
Lib 14h50'32" -1d44'15"
Happy 20th Birthday Katie!
Psc 1h45'14" 21d27'56"
Happy 21st Amber & Jeremiah Forever
Ari 2h47'1" 13d48'56"
Happy 22nd Birthday Erik
Cep 23h34'4" 78d18'40"
Happy 25 Anniversary Joe Libertore
Cyg 19h39'23" 34d1'22"
Happy 30th Anniversary
Cyg 19h39'49" 39d7'31"
Happy 30th Birthday Hols!
Cet 2h47'34" 4d19'33"
Happy 30th Birthday! Love Mike
Ari 3h8'24" 29d6'45"
Happy 30th Marlan & Judy Andrews
Crb 15h46'40" 29d48'56"
Happy 31st B-Day Dan
Cnc 8h2'51" 25d42'25"
Happy 35th Birthday, Reg
Aqr 22h13'57" -1d54'50"
Happy 40th Anthony Love Fi 17082007
Leo 10h26'49" 20d32'27"
Happy 40th Birthday Voortly
Cru 12h28'24" -60d12'24"
Happy 40th Randy Krauss!
Leo 9h54'54" 15d14'2"
Happy 50th B-Day Calvin
Uma 9h24'59" 65d23'39"
Happy 50th Christine Davies!
Cas 0h31'25" 65d48'21"
Happy 50th To A Star - Alan S. Gordon
Aql 19h47'21" -0d29'16"
Happy 60th Anniversary
Cap 20h54'7" -19d55'7"
Happy 94th b-day Narbada P. Tillak
Her 16h49'23" 37d29'48"
Happy Anniversary Nicholas & Sonya
Uma 10h7'31" 63d5'40"
Happy B Day-Dr. James F. Haw-Karli
Sgr 18h39'52" -33d7'12"
Happy Bar Mitzvah Irving
Uma 8h30'29" 64d40'53"
Happy Birthday Brittany
Tau 4h40'18" 18d3'14"
Happy Birthday Jess, Love Garrison
Uma 9h38'25" 65d13'12"
Happy Birthday Joe Baseball 88 2004
Uma 8h17'34" 60d45'19"
Happy Birthday Lauren
Tau 5h34'45" 26d12'30"
Happy Birthday Lindsey - Doug
Sco 16h11'8" -41d25'0"
Happy Birthday Lizzy
Cap 20h13'10" -14d38'10"
Happy Birthday Nancy Nichols Miller
Cas 0h46'48" 64d21'12"
Happy Birthday Nathan Huber
Her 18h5'49" 32d7'20"
Happy Birthday to My Buddy
Uma 10h23'37" 61d52'24"
Happy Birthday, April
And 1h43'52" 44d21'41"
Happy Birthday, Liz...SOM
And 1h24'41" 48d59'17"
Happy Boy
Ori 5h29'52" 5d51'25"
Happy Cartier
Gem 7h45'8" 32d0'30"
Happy Day Star Mom #1 <\@> Earth.Den
Ari 2h48'29" 25d16'58"
Happy Daze by Pop Pop
Peg 22h12'51" 33d26'36"
happy eighteenth, nurny!
Vir 12h19'25" 6d22'51"
Happy Ever After
Cyg 20h15'30" 30d33'28"
Happy Fathers Day~Lost Without You
Aur 5h51'3" 52d36'41"
Happy Harry
Her 17h22'8" 35d3'4"
Happy Harry
Umi 16h20'40" 70d50'36"
HAPPY HEIDI
Ori 5h14'33" -1d6'43"
Happy Hernandez
Uma 11h27'42" 47d54'45"
Happy Honey
Vir 13h15'44" -4d9'50"
Happy In Oz
Her 16h36'9" 31d31'3"
Happy Jack Ramsey Hecimovich
Uma 10h34'34" 59d29'20"
Happy John
Dra 17h39'8" 52d18'16"
Happy Little King
Crb 15h53'52" 34d4'31"
Happy. love. ours. always.Lisa&John
Aqr 22h20'55" -17d7'29"
Happy Planet of Booland
Uma 11h38'48" 38d40'56"
Happy Retirement #1 Daddy Harvey
Cep 21h29'21" 64d25'52"
happy smiling jenny
Lib 14h43'40" -16d9'53"
Happy Star 28
Sco 16h33'0" -29d44'55"
Happy together, forever!
Uma 8h48'31" 72d25'36"
Happy Trails
Lyn 7h43'53" 36d53'39"
Happy & True
And 23h0'57" 41d22'25"
Happy50thJoe
Per 2h39'2" 55d30'44"
Happybirthday! Rion!
Cap 20h37'8" -26d55'28"
Hapster
Uma 10h22'28" 56d30'21"
Haralambos A. Stamatatos
Uma 13h51'5" 50d34'40"
Harald
Uma 10h18'43" 47d40'42"
Harald & Christine Hemer
Cyg 21h27'55" 45d13'44"
Harald Dahms
Ori 5h11'31" 8d3'41"
Harald Eichhorn
Ori 4h47'0" 0d28'31"
Harald Fiedler
Uma 11h29'19" 61d16'3"
Harald Heise
Uma 8h42'21" 57d1'5"
Harald Kliche
Uma 13h31'31" 63d5'33"
Harald Michael
Uma 11h53'47" 33d26'6"
Harald Rauchleitner
Uma 10h35'23" 61d41'27"
Harald Richard Bregulla
Psc 1h25'23" 18d49'21"
Harald Seshun
Ori 5h57'48" 20d47'45"
Harald Simmel
Ori 5h17'20" 15d52'57"
Harald Tischer
Ori 5h15'58" 1d5'34"
Harald Wibbelmann
Uma 9h12'7" 56d13'41"
Haran's Love Nest
Ari 2h5'41" 23d30'16"
Harazin's Lighthouse
Lyn 8h8'9" 37d49'41"
Harbet
Cyg 19h51'48" 31d25'20"
Harborside Crestwoodstar volunteers
Ari 3h17'37" 23d32'10"
Harbour House
Cyg 19h40'5" 37d47'27"
Harby
Per 3h53'17" 42d30'10"
Hardcore Renee
Tau 4h28'51" 11d27'51"
Harder, Jörg
Uma 13h58'50" 57d10'25"
Hardev Sidhu
Lib 15h42'16" -18d40'21"
Harding's Bounty
Crb 15h47'16" 28d32'53"
Hardip Sighn "RAJ-BigDaddy-Gulati"
Her 16h34'26" 38d21'37"
Hardt, Johannes
Sco 16h50'53" -38d45'26"
Hardwired
Cnc 8h8'29" 17d13'1"
Hardy Kister
Uma 10h6'20" 63d38'42"
Hargis Hideaway
Eri 3h46'33" -0d23'4"
Harhay Star
Lib 14h55'20" -10d35'55"
Hari Nam Singh
Aur 5h57'57" 42d44'54"
Haribo Harry
Cep 22h46'11" 66d24'28"
HARIBO STAR
Uma 9h58'45" 42d41'20"
Harika
Tau 4h29'24" 28d27'24"
Harika Chandrasekhar
Sco 17h30'56" -44d38'18"
Hariklia
Sgr 17h49'30" -26d47'22"
Harilaos Leonardatos
Per 4h12'10" 50d42'19"
Harjit Singh
Uma 9h57'56" 59d56'31"
Harjot Samra
Cnc 8h26'14" 15d21'39"
Hark the Harrold
Her 17h22'51" 34d27'26"
Harkirat22
Lib 15h5'36" -5d39'1"
Harlan and Stephanie McCoy
Leo 11h7'56" 17d9'30"
Harlan Dale Miller
Umi 13h28'57" 74d0'29"
Harlan Doeg
Psc 1h17'29" 22d2'22"
Harlan Elkins Sr. Our Dad, Our Star
Gem 7h20'51" 30d52'14"
Harlan J. Jackson
Cap 21h41'40" -24d8'30"
Harlan Joseph Falejczyk
Uma 11h57'15" 62d9'42"
Harlan Stai
Uma 11h39'16" 31d58'22"
Harlene Krane
Cas 1h18'19" 57d27'37"
Harles I. Creech's Eastern Star
Her 16h50'28" 51d15'0"
Harley
Peg 23h38'43" 14d9'24"
Harley
Aql 20h5'52" 7d1'19"
Harley
Cam 4h30'43" 63d41'46"
Harley
Dra 18h19'49" 78d59'18"
Harley 37
Cap 20h41'23" -25d12'15"
Harley Alexander Ellis
Uma 10h17'19" 59d54'49"
Harley Connolly-Cole
Peg 21h57'0" 18d55'26"
Harley Cummings
Cnv 13h15'56" 40d54'50"
Harley D
Lyn 7h28'28" 50d24'58"
Harley Davidson Jack
Her 16h44'48" 46d51'20"
Harley Dawn
Psc 1h8'48" 4d40'56"
Harley & Dean
Lac 22h11'55" 37d37'32"
Harley Dockery
Umi 15h47'24" 71d24'41"
Harley Edgar Robbins
Her 17h25'28" 16d19'58"
Harley Elizabeth Bickerstaff
Vir 13h44'10" -18d1'10"
Harley E.Wilson
Her 17h23'38" 14d48'46"
Harley Hoffman Hunner
Her 17h50'19" 45d36'25"
Harley Hunter McLean
Umi 15h31'16" 75d5'7"
Harley Jae's Star
Per 4h28'39" 32d54'44"
Harley Joe Cool Killebrew
Cnv 12h32'50" 43d2'40"
Harley Keaton Grey
Per 2h13'47" 53d6'7"
Harley & Kim
Cyg 19h57'59" 30d36'47"
Harley Krause
Leo 11h23'15" 1d33'5"
Harley L. Cole
Uma 10h2'29" 56d33'36"
Harley L. Engel
Uma 12h43'9" 54d7'50"
Harley Loves Krystal Forever
Crb 15h17'18" 25d49'47"
Harley Michelle Martin
Sco 16h4'4" -25d13'5"
Harley Pattyann Mascolo
Lmi 10h35'44" 29d7'45"
Harley Pieter
Cru 12h37'11" -60d49'35"
Harley Reen Heath - Harley's Star
Umi 17h27'56" 80d19'22"
Harley & Susan
Uma 11h29'13" 61d59'44"
Harley Voss
Cma 6h49'17" -17d24'25"
Harley Waters
Aqr 22h4'48" -22d40'43"
Harley-Bear
Umi 14h1'5" 70d48'34"
" Harley-Dave " Komro
Lib 14h48'49" -11d53'28"
Harleykid
Sco 16h13'42" -24d48'40"
Harley's Comet!
Ori 5h32'38" 0d38'50"
Harley's Kitten
Cap 21h14'53" -22d21'47"
Harley's Star
Ori 5h9'21" 3d44'3"
Harley's Star
Cyg 20h24'12" 42d31'59"
Harli
Tau 5h32'31" 20d58'42"
Harlie
Umi 14h33'51" 76d3'48"
Harlie Nicole Umfleet
Gem 6h28'39" 26d48'23"
Harlie Rose
And 0h58'45" 46d8'29"
Harlock D Hero
Pho 23h34'59" -45d24'52"
Harlon
Cnc 8h21'53" 25d36'18"
Harlton G. Dunbar
Her 16h46'12" 43d54'24"
Harman Clan
Aur 5h47'17" 41d38'11"
HARMER
And 1h51'12" 46d10'25"
Harmina Tungal
Tau 5h53'1" 24d45'27"
Harmoni
Uma 11h34'46" 61d18'24"
Harmónia
Ari 2h31'47" 24d41'18"
Harmonia
Lyr 18h28'3" 40d47'23"
Harmonia
Uma 9h42'35" 44d12'58"
Harmonie Jenelle Hudgins
Lib 15h13'32" -7d9'4"
Harmonious Union
Cha 10h18'53" -78d34'46"
Harmony
Cap 20h49'36" -21d13'50"
Harmony
Lyn 8h34'58" 43d24'10"
Harmony
And 1h46'24" 38d9'12"
Harmony
Lyr 18h52'49" 31d50'28"
Harmony
Gem 7h2'31" 30d9'23"
Harmony
Uma 9h30'5" 51d49'1"
Harmony
Vir 14h21'27" 1d48'17"
Harmony Bench
Cas 1h28'37" 52d42'38"
Harmony Beth Dávila
And 0h58'43" 46d14'52"
Harmony Bird
Lib 15h57'36" -10d57'0"
Harmony Dee Montoya-Tafoya
Sco 16h18'23" -19d26'22"
Harmony Olson
Aqr 20h40'34" 1d17'15"
Harmony Rose
Cnc 8h2'57" 21d17'35"
Harmony's Guiding Light
Ari 3h9'53" 22d30'7"
Härning, Anne-Ute
Uma 14h0'26" 51d55'42"
Harold
And 23h50'6" 48d3'46"
Harold
Per 4h30'32" 42d40'52"
Harold
Ari 2h47'24" 27d35'20"
Harold
Aql 19h59'2" 8d2'36"
Harold
Ori 4h53'36" -0d0'23"
Harold
Cma 7h24'7" -29d23'36"
Harold Ahlers
Per 2h51'44" 54d12'59"
Harold Alfred Fors
Ori 5h34'16" 9d52'59"
Harold and Alice Bowen
Cyg 20h13'7" 35d7'8"
Harold and Hermoine Cambas
Sco 17h52'17" -35d11'28"
Harold and Karen Gerry
Aqr 22h33'3" 0d1'31"
Harold and Mary Dansdill Miller
Vir 13h11'45" 13d0'34"
Harold and Shirley Brodie
Uma 8h47'22" 51d3'11"
Harold A.Towle
Aqr 20h51'28" -10d30'54"
Harold B. Kennedy
Lyn 8h27'46" 45d19'58"
Harold Baird
Tau 4h42'46" 23d32'47"
Harold Bischoff-Dad and Grandad
Cyg 19h46'19" 36d21'43"
Harold Breeding, Sr.
Ori 5h46'7" 3d11'27"
Harold Brian Deats "Hank"
Aqr 21h49'24" -1d52'29"
Harold Brown
Cep 23h21'49" 76d58'34"
Harold Brown
Uma 11h25'19" 50d51'23"
Harold Bud Clark
Aur 5h16'25" 42d8'42"
Harold "Bud" Loyd
Per 3h32'25" 47d41'56"
Harold Butch Murphy
Ori 6h10'29" 15d2'25"
Harold C. Hammer
Cyg 21h14'11" 47d37'5"
Harold C. Hollenbeck, J.S.C.
Lib 15h22'44" -14d54'23"
Harold C. Mears
Ori 5h12'53" 12d55'34"
Harold Carl Breyer
Per 4h10'56" 34d21'1"
Harold Charles Caponigro
Uma 8h54'15" 61d10'46"
Harold Clarence Zank
Cep 20h36'29" 61d51'2"
Harold Custer Whiteside
Aql 19h53'40" 7d4'9"
Harold D. Ankeny
Per 2h55'21" 54d5'59"
Harold D. Mathews~75~2007
Aqr 22h11'51" -1d51'48"
Harold D. Smith 04-11-21
Uma 11h43'30" 62d29'51"
Harold Denney, Sr.
Aqr 21h40'17" -8d11'59"
Harold & Dorothy
Aql 19h23'8" -0d13'32"
Harold Douglas Stanfield
Uma 11h1'20" 69d33'48"
Harold E. and Frances P. Fitzgerald
Cyg 20h21'19" 42d33'30"
Harold E. Pearson
Uma 10h40'21" 59d12'24"
Harold E.and Marjory L. Garrett
Aqr 22h38'56" 0d11'50"
Harold Eric Leonard
Uma 13h30'21" 55d33'29"
Harold Eugene McElhinny
Lyr 18h45'25" 36d4'35"
Harold Evans
Sco 16h50'43" -41d38'0"
Harold F. Falter
Ori 5h15'15" -5d35'30"
Harold F. "Hutch" Hutchinson, Jr.
Dra 17h38'37" 54d57'15"
Harold F. Jacobs
Cas 0h21'57" 57d55'57"
Harold F. Wolfing, Sr.
Ori 5h11'50" 6d45'0"
Harold Francis Johnson
Per 4h12'20" 51d8'50"
Harold Galarneault Wagner
Ori 6h9'2" 15d33'41"
Harold George Scott
Ori 5h23'59" 7d52'15"
Harold H. Stege
Her 18h46'18" 25d20'5"
Harold (Hal) Floren
Ari 2h47'15" 27d21'56"
Harold Henry Bach, Jr.
Cnc 9h17'19" 8d5'27"
Harold Henson's Star
Her 18h33'30" 24d23'53"
Harold Hoss Owsley
Uma 11h24'59" 32d50'1"
Harold I love u forever Tammy
Per 3h13'9" 52d37'0"
Harold J. Hurley, Sr.
Her 18h46'37" 12d36'48"
Harold J. La Chapelle
Tau 4h1'16" 21d21'55"
Harold J. Schrade
Gem 6h57'29" 23d40'23"
Harold Jack Sartain
Her 17h26'51" 21d9'4"
Harold James Bateman
Uma 11h25'0" 59d58'1"
Harold James Boyle "Harry"
Gem 6h52'22" 14d13'28"
Harold James Sparks 04071967 M
Lyn 6h26'47" 57d36'12"
Harold & Jane Taubes
Cyg 21h45'5" 44d12'32"
Harold Joe Rush
Uma 12h14'54" 56d37'56"
Harold Joseph Welhouse
Uma 11h26'12" 65d47'39"
Harold K. Frye
Her 17h21'53" 35d54'12"
Harold Karlin
Uma 11h48'38" 29d29'20"
Harold & Katherine Grausam
Uma 10h29'25" 50d27'26"
Harold & Kaye Richardson
Cyg 21h30'13" 49d50'39"
Harold King
Vir 12h34'20" -9d51'5"
Harold L. Hendler
Vir 13h48'59" -8d19'11"
Harold L. Romanowski
Cnc 8h45'6" 26d17'42"
Harold L. Russell
Crb 16h9'31" 28d5'3"
Harold Leroy Carter
Aql 19h41'30" -0d47'19"
Harold Lloyd Damron
Cnc 8h35'15" 28d41'54"
Harold Makinson
Uma 11h29'55" 60d6'16"
Harold Morrow, Jr.
Her 17h47'37" 14d19'4"
Harold & Nancy Mizpah
Cap 21h42'18" -11d5'1"
Harold "Papa" Trice
Her 17h20'39" 23d36'16"
Harold Paul Loveall III
Leo 10h13'0" 23d0'48"
Harold 'Red' Frey, Jr.
Per 3h43'54" 45d30'52"
Harold Reynolds
Sco 16h57'11" -42d25'12"
Harold Richard Ziss "1921-2004"
Tau 3h37'37" 16d8'22"
Harold Robert Schoeck "Bob"
Aql 19h4'50" 15d18'27"
Harold & Roz 58th Anniversary
Gem 7h13'5" 16d51'22"
Harold S. Covell
Ari 2h32'13" 26d54'58"
Harold Satterfield Point Celestial
Psc 0h24'59" 2d45'17"
Harold Shavar Brahien Jackson
Gem 7h26'33" 33d27'17"
Harold Sheets
Sgr 19h11'26" -21d29'20"
Harold Shulke
Her 17h6'8" 31d1'13"
Harold Small
Uma 12h23'38" 58d45'53"
Harold Smith rocks <\@> 77 HSHGWB
Ari 3h9'3" 28d37'36"
Harold Snyder
Cep 22h25'19" 64d38'1"
Harold & Takako Iwamasa
Umi 15h32'31" 71d6'27"
Harold the Protector
Per 3h6'21" 42d2'15"
Harold Thomason - The Telescope Man
Leo 11h1'43" 19d47'34"
Harold Tretsch
Aur 5h52'24" 51d51'56"
Harold W Helmke, Jr
And 0h28'24" 42d22'51"
Harold W. Nix
Cep 20h38'27" 64d47'21"
Harold Warren Gerard Hanson
Tau 4h35'57" 27d42'21"
Harold William Brown
Crb 15h37'40" 31d16'46"
Harold William Daly
Ari 3h20'27" 28d46'33"
Harold William Willis
Cep 22h59'28" 70d16'19"
Harold Wolford
Uma 11h47'45" 38d4'18"
HaroldPierce
Aqr 22h38'35" 1d15'29"
Harold's Star
Ori 6h14'3" 8d57'12"
Haroldstar
Aql 19h31'21" 16d10'14"
Harout & Luiza
Leo 10h15'23" 15d21'52"
Haroutiun"Our Shining Star Forever"
Lib 15h53'4" -11d59'40"
Haroutyoun Sergey Simonyan
Lib 15h4'9" -20d37'40"
Harper
Ari 3h13'55" 28d40'19"
Harper Ann Christiansen
Aqr 21h2'45" 0d7'21"
Harper Beall Williams
Her 18h43'25" 19d9'26"
Harper Ellen
Lib 15h8'34" -7d0'27"
Harper Grace Neminski
Cap 21h17'33" -24d29'29"
Harper Kerrington
Aqr 21h54'55" 1d15'43"
Harper Law
Lmi 10h23'16" 33d14'9"
Harper Makena Rogers
Leo 10h42'47" 16d56'53"
Harper Mary Riley
Cap 21h23'4" -25d28'54"
Harper Rice
Per 4h39'40" 39d30'21"
Harper's Dad's Star
Ori 6h4'52" 18d31'17"
Harpers Trading (M) Sdn Bhd
Ori 5h34'37" 3d19'28"
HarQuinn 21915
Psc 0h50'10" 21d10'59"
Harraway All Star
Leo 11h24'26" 13d15'45"
Harraway Super Star
Leo 11h44'8" 12d34'46"
Harree Ethel Louise Siler, M.D.
Her 17h15'30" 19d12'10"
Harrell Dean Tindall
Her 17h12'29" 14d8'42"
Harri Ethan Chambers
Dra 16h21'22" 66d15'49"
Harri Hopp
Uma 10h8'47" 44d2'39"
Harri John Floyd Williams
Peg 22h50'43" 25d44'55"
Harrianne Pepper
And 0h2'15" 42d11'38"
Harriet
Aql 20h27'23" -8d16'9"
Harriet and Dick Sheaff
Cyg 20h16'30" 32d44'11"
Harriet and Jessie
Lyn 7h21'34" 59d19'21"
Harriet and Nial's Star of Love
Cyg 21h0'6" 46d8'36"
Harriet Ann
Peg 21h33'41" 11d16'58"
Harriet & Bob Whitlock
Uma 8h55'3" 47d13'29"
Harriet Cherry
Crb 16h12'25" 39d5'40"
Harriet Cohen Halpryn
Psc 23h12'21" 6d49'2"
Harriet Davies
And 0h41'20" 22d7'57"
Harriet Delina Durspek
Pic 6h49'56" -59d18'11"
Harriet Doniger
Her 17h17'57" 16d7'6"
Harriet Elizabeth (Betty) Moore
Tau 3h50'58" 25d27'34"
Harriet Ellen Smith
Cnc 8h47'47" 11d25'1"
Harriet Evelyn Davis
Ori 5h24'18" 6d15'29"
Harriet Foster Horton
Leo 10h19'8" 22d48'44"
Harriet Garvin
Uma 11h50'15" 31d6'29"
Harriet Grace Richardson
And 1h25'48" 34d25'35"
Harriet H. Amundsen
Cyg 21h50'12" 53d52'58"
Harriet - Isabel
And 2h19'30" 43d29'33"
Harriet K. Macchia
And 0h25'31" 33d8'58"
Harriet Kirkendall - Russell
Cap 20h38'24" -15d23'26"
Harriet Laub Levine
Uma 11h13'32" 56d13'42"
Harriet Lundberg
Uma 12h0'8" 65d5'38"
Harriet M. Hagerman
Lyn 7h59'38" 47d42'28"
Harriet M. Tandy
Cas 23h0'51" 58d9'43"
Harriet Marie Hinton
Ori 5h19'58" 5d41'14"
Harriet Matilda Long
And 0h18'43" 37d2'49"
Harriet Mendelovitch
And 1h8'30" 42d32'30"
Harriet Reuter (Hargwen)
Uma 10h40'48" 53d46'20"
Harriet Rotter
Vir 13h18'56" 6d14'23"
Harriet S. Kline
Gem 7h20'1" 27d2'55"
Harriet S. Makepeace
And 2h31'42" 43d24'23"
Harriet S Steele
Gem 6h7'45" 26d38'38"
Harriet Simon
Cas 3h15'19" 59d30'23"
Harriet Strugnell
And 2h35'31" 44d31'19"
Harriet "The Biddle" Morgan
Cyg 20h24'27" 52d38'39"
Harriet Todd Dindinger
Sco 16h10'43" -12d20'7"
Harriet Tunick Nathan
Tau 4h19'27" 18d32'25"
Harriet Waldstein
Ari 2h47'52" 28d58'16"
Harriet & Wally Jordan
Sge 19h50'27" 17d34'24"
Harriet Young Biondo
Leo 11h14'1" 7d29'5"
Harriet's Guiding Light
Uma 11h26'4" 64d50'6"
Harriet's Highlight
Cas 23h21'42" 58d35'52"
Harriett Elizabeth Phillips
Ari 2h54'31" 24d23'40"
Harriett Leeb
Per 3h16'25" 31d21'58"
Harriett Marie Coustinarious
And 2h30'24" 43d35'29"
Harriett & Sheldon Fuller
Cyg 20h6'12" 48d3'28"
Harriett Walton Morelan
Gem 7h2'20" 26d59'48"
Harriette Amity Jacobsen
Pav 20h45'10" -60d58'2"
Harriette Warshaw
Sgr 20h16'17" -43d47'50"
Harrington
Cep 20h43'46" 67d25'23"
Harrington
Ori 5h40'47" 9d13'51"
Harrington Family
Dra 19h34'42" 60d27'54"

Harris
Lyn 6h33'50" 57d50'15"
Harris
Psc 0h46'28" 8d4'50"
Harris
Her 17h3'59" 30d52'46"
Harris 06
Uma 10h58'24" 69d34'28"
Harris A. Desorda
Tau 3h33'11" 27d12'28"
Harris "Butch" Monson
Uma 8h13'51" 64d46'48"
Harris Christopher Stack
Lmi 10h21'16" 28d56'21"
Harris Connor Webster
Cnc 8h44'11" 8d34'15"
Harris Eisenhardt
Cyg 19h45'37" 34d0'33"
Harris Keen Kessler
Ori 5h57'46" 21d18'33"
Harris Letter
Leo 9h23'10" 19d19'27"
Harris Rutsky
Cnc 9h8'51" 27d32'47"
Harris' Sweetheart
Uma 9h24'27" 52d3'17"
Harris Vincent Fleisher
Ori 6h3'45" 19d1'46"
Harrison
Her 17h6'58" 32d44'24"
Harrison 060606
Cen 14h8'21" -42d2'26"
Harrison #1
Cap 21h2'26" -16d7'28"
Harrison Adam Cymbler
Psc 23h0'59" 7d55'17"
Harrison Allen Brunz
Sgr 18h32'40" -26d9'50"
Harrison and Taylor
Cap 20h25'57" -14d53'35"
Harrison Anthony
Lib 14h45'20" -15d25'27"
Harrison Anthony Farkas
Cap 20h18'5" -25d51'37"
Harrison Bachrack
Umi 15h26'20" 72d17'33"
Harrison Brian Hemsley
Uma 10h5'57" 50d1'45"
Harrison Cates Withers IV: "Sparky"
Psc 1h25'34" 24d30'58"
Harrison Charles Hickey
Cru 12h39'10" -60d57'39"
Harrison Daniel Jones
Per 4h9'34" 32d47'51"
Harrison David Marsocci
Per 2h50'39" 57d20'10"
Harrison Friel
Per 3h42'35" 42d51'17"
Harrison "Harry" Gray Mundy-Smith
Cru 12h36'36" -60d49'1"
Harrison 'Harry' Jones
Aql 19h50'36" 14d20'45"
Harrison Helton
Aql 19h49'48" 7d30'31"
Harrison Jack Suhr
Cen 14h10'2" -33d50'53"
Harrison Jacob Farina
Psc 1h17'28" 24d34'57"
Harrison James Boundy Humphrey
Cru 12h28'13" -60d37'24"
Harrison James Wise
Per 4h31'57" 39d11'30"
Harrison Jeffrey Greenberg
Per 3h25'57" 50d5'6"
Harrison Joel Smith
Her 17h58'14" 21d19'3"
Harrison John Chapman
Cas 0h48'14" 61d58'0"
Harrison John Prunty "The Happy"
Vir 14h32'57" 0d21'55"
Harrison Joseph Jess
Cru 12h29'12" -60d55'32"
Harrison Kross Carnevale
Per 4h0'4" 38d2'40"
Harrison Lee Peeples
Lib 15h15'49" -9d28'59"
Harrison Leonard Tucker
Her 16h12'57" 48d20'23"
Harrison Marsocci
Gem 7h35'52" 33d2'16"
Harrison Meyn
Cru 12h2'17" -62d15'42"
Harrison Michael Hardman
Her 17h51'5" 27d42'16"
Harrison Michael Lloyd
Uma 9h59'48" 45d39'52"
Harrison Miles Keith
Ari 2h28'46" 10d41'53"
Harrison Mitchell Ludwig
Cnc 8h0'53" 13d23'3"
Harrison Mitchell McArtor
Uma 10h42'5" 69d15'51"
Harrison Nicholas Haeflein
Vir 13h21'48" 12d5'32"
Harrison Oliver Barker
Cra 18h0'43" -37d8'32"
Harrison Oswald Gladden
Cnc 8h13'20" 29d18'30"
Harrison P Schultz
Uma 11h27'4" 55d25'45"
Harrison Paul Farr
Aur 5h24'23" 31d39'11"
Harrison Paul Weaver
Lmi 10h14'23" 31d55'25"
Harrison Perrie
Cru 12h21'39" -57d30'35"
Harrison Rogers Weber
Uma 10h1'20" 44d11'59"
Harrison Russell Jones
Ori 5h25'24" 2d11'9"
Harrison Sydney's Star
Per 4h5'57" 33d45'7"
Harrison Taylor Schaefer
Cep 21h33'28" 55d26'46"
Harrison Temple
Vir 15h7'1" 5d15'52"
Harrison Thomas Adams
Cen 13h26'48" -39d30'15"
Harrison Thomas Allen
Ori 5h19'30" -0d27'12"
Harrison Tyler Weaver
Gem 7h47'42" 34d58'35"
Harrison William Harbourt
Psc 0h36'48" 15d41'36"
Harrison William Ruiz
Uma 8h13'2" 71d46'47"
Harrison Winskill Dyson 29-09-03
Umi 13h30'17" 75d20'17"
HARRISONMYERS2004
And 23h12'50" 47d53'39"
Harrison's Guiding Hope
Per 3h49'9" 40d33'52"
Harrold and June
Cru 12h55'24" -59d32'15"
Harry
Lib 15h16'45" -7d16'52"
Harry
Uma 8h34'6" 68d17'40"
Harry
Umi 14h15'24" 65d58'10"
Harry
Lyn 7h53'14" 34d33'57"
Harry
Vir 13h20'46" 11d43'34"
Harry A. Blair
Cnc 8h25'0" 17d50'42"
Harry A. (Butchie) Davis Jr.
Uma 9h47'32" 45d49'16"
Harry A. Raines
Cep 21h30'42" 67d30'16"
Harry Adam Brooke
Ori 5h16'29" -4d38'9"
Harry Alfred Johnson Jr.
Tau 5h59'22" 25d29'10"
Harry Amphlett
Her 18h31'49" 21d43'17"
Harry and Christina
Cap 21h2'35" -15d12'53"
Harry and Doris Thomason
Uma 10h42'25" 49d29'24"
Harry and Dot Hix 60 years Strong
Cyg 20h2'23" 33d5'42"
Harry and Emma Fritts
Cyg 19h56'53" 30d7'32"
Harry and Irene Curran
Cyg 21h20'34" 49d33'12"
Harry and Jeremy
Uma 9h7'48" 47d34'5"
Harry and JoDean
Sge 19h12'5" 19d33'39"
Harry and Karen Hayes
Gem 6h50'39" 34d25'21"
Harry and Maryann Mayfield
Aqr 20h48'43" 2d14'32"
Harry and Monkey
Cap 21h27'22" -12d38'12"
Harry and Ramona
Cyg 21h31'38" 44d44'34"
Harry and Winifred Lennard
Cyg 19h49'43" 56d5'22"
Harry Andrew David Griffiths
Umi 17h6'32" 77d22'44"
Harry Anthony Thomas Lewis
Cep 21h45'31" 58d54'39"
Harry Benton Green
Ari 2h46'9" 26d4'48"
Harry Best
Uma 13h11'19" 61d36'34"
Harry Bill (Dad) Fifarek I Luv You
Her 16h50'57" 23d10'23"
Harry & Blue
Uma 10h31'2" 63d36'47"
Harry Bopp
Gem 7h33'13" 33d1'54"
Harry Bowden
Lib 15h45'11" -20d14'12"
Harry Brewer Davis
Ori 6h18'34" 9d32'44"
Harry Bridger - Harry's Star
Uma 9h56'11" 42d35'10"
Harry Brown
Per 3h31'26" 47d30'39"
Harry Bush
Uma 13h30'0" 58d11'59"
Harry Carr's Eternal Star
Col 5h9'39" -28d6'32"
Harry Cedric Samora
Uma 13h36'22" 51d23'17"
Harry Charles Pearce
Her 17h39'6" 18d49'22"
Harry & Charlie Davies-Carr
Her 18h19'17" 20d12'49"
Harry Christopher Hall
Psc 1h4'20" 25d55'27"
Harry & Colin Okorn
Umi 15h34'54" 69d50'4"
Harry Constantine Orbelian
Dra 19h30'44" 73d46'30"
Harry Dalling Trucking BPT.,CT.1933
Ori 6h17'13" 17d32'27"
Harry David Heddle Sayers - Harry's Star
Ori 5h57'28" 11d17'26"
Harry Davidow
Lib 15h31'57" -27d17'25"
Harry de Woeps - His Star
Gem 6h20'50" 20d37'3"
Harry Dennis Lynch
Cas 23h49'36" 57d53'9"
Harry Donald Betts Birthday Star
Cru 12h37'59" -59d45'6"
Harry Dowl Boteler
Aqr 22h29'20" -24d5'38"
Harry E. Jamison, D.D.S.
Ori 6h3'28" -3d49'13"
Harry E. Norton
Aur 5h53'49" 40d42'8"
Harry E. Zinn Sr.
Cap 20h50'56" -26d46'25"
Harry Edison Novinger
Ori 5h14'41" -0d10'10"
Harry & Edna Farrar
Cyg 19h51'6" 38d36'53"
Harry Edward Bellardini
Her 16h43'52" 47d27'32"
Harry Edward Collings
Ori 5h26'7" 14d13'22"
Harry Evans and Tina Brown
Cyg 20h32'43" 54d8'32"
Harry F. Doyle II
Uma 9h52'28" 58d1'38"
Harry Feder
Uma 11h34'28" 32d29'30"
Harry & Freda Per Ardua Ad Astra
Uma 8h41'28" 50d57'15"
Harry Frumerman
Cnc 9h9'0" 30d46'6"
Harry Fry
Cas 1h32'4" 68d15'39"
Harry G. Ladanye
Ari 3h18'26" 28d39'49"
Harry G Yeaggy
Lyn 9h2'45" 34d0'26"
Harry George Courts
Sgr 18h49'37" -29d6'31"
Harry George French
Dra 17h43'9" 62d9'59"
Harry George Poyser
Her 18h13'17" 28d10'53"
Harry Goldberg
Uma 10h21'8" 67d41'14"
Harry Greenhough
Cru 12h11'45" -62d11'49"
Harry Haaversen
Ori 5h32'1" -0d11'25"
Harry Hamilton
Ori 6h11'51" 15d51'51"
Harry Hannabuss
Cep 22h52'56" 65d39'21"
Harry Harold McCune
Uma 11h24'55" 54d34'59"
Harry Hegeman Johnson
Cep 21h40'13" 61d48'44"
Harry & Helen Lingerfelt
Cyg 20h22'36" 58d9'12"
Harry & Helen Umbriaco's Place
Cyg 20h8'34" 53d25'18"
Harry Herman Cottam V
Uma 11h50'27" 42d44'46"
Harry Herman Heuman
Ori 5h11'18" 5d16'31"
Harry & Hermione
Ori 6h8'46" 17d57'1"
Harry & Irene Brunson's Lucky Star
Cas 0h42'36" 66d10'39"
Harry J. Adams
And 0h22'37" 39d35'4"
Harry Jack Hughes
Leo 11h58'20" 20d24'56"
Harry James
Her 17h50'37" 19d41'25"
Harry James Briggs - Harry's Star
Uma 8h48'10" 59d4'46"
Harry James Cox
Cep 3h26'42" 81d14'19"
Harry James Currass - Harry's Star
Dra 18h25'56" 52d56'20"
Harry James Day
Dra 10h46'42" 79d29'3"
Harry James Lawrence
Sco 16h28'31" -36d38'5"
Harry James McAllister Addison
Peg 21h35'35" 26d44'8"
Harry James Mitchell Smith
Her 17h18'0" 13d58'41"
Harry Jefferson Shadbolt
Per 3h58'36" 33d18'4"
Harry John Hayden Grimmett
Per 3h17'21" 42d12'36"
Harry John Scleater - Harry's Star
Her 16h39'1" 17d4'37"
Harry John Stokes
Uma 10h52'30" 57d30'58"
Harry Joseph Graver, Sr.
Lib 15h5'53" -20d36'27"
Harry Joseph Womack
Her 18h18'38" 16d17'41"
Harry Jung
Gem 7h14'0" 33d6'12"
Harry Kaminski
Uma 11h22'51" 65d56'45"
Harry Kanavy
Ori 6h13'50" -1d33'16"
Harry Karl Alexander
Her 18h47'28" 20d13'49"
Harry + Kathy Abraham
Lib 14h52'38" -3d19'51"
Harry Kelly Thornton
Her 18h1'27" 21d3'53"
Harry Krakovsky
Per 3h31'29" 52d7'8"
Harry Krystalla
Gem 6h59'17" 15d11'7"
Harry & Kuquiz Scarborough
Uma 11h51'45" 30d20'47"
Harry L. Wilson
Ari 3h27'15" 27d18'35"
Harry Lamb
Mon 6h39'31" 7d2'21"
Harry LaMonda "80th & Retired"
Gem 7h56'16" 31d12'6"
Harry Larrison
Aur 5h36'44" 53d9'25"
Harry LaTour
Uma 10h28'4" 41d54'49"
Harry & Leah Traugott
Cyg 19h45'36" 45d52'20"
Harry Lee Gordon
Cnc 8h32'40" 31d5'53"
Harry Lewis Jenner
Vir 12h30'30" -5d53'2"
Harry Lewis Mello
Psc 1h6'33" 21d27'6"
Harry Lucas Rushman
Per 2h52'41" 45d31'13"
Harry Lumia
Uma 9h18'34" 42d50'43"
Harry M. Henderson
And 0h47'21" 38d51'25"
Harry M. Yockey
Pyx 8h37'50" -32d28'47"
Harry Martin Campbell
Umi 15h58'45" 75d16'46"
Harry Marvin Hakenson
Aql 19h44'34" -0d7'18"
Harry Marvin Womack
Ori 5h14'50" 5d40'35"
Harry Matthew Thornton - Harry's Star
Ori 5h24'9" 13d53'41"
Harry Mcdorman
Leo 9h33'45" 27d52'4"
Harry Meyers
Cap 21h44'27" -16d26'25"
Harry Michael John Rotheram
Per 3h30'26" 37d12'45"
Harry Milburn Dort, Jr.
Per 2h16'36" 55d51'3"
Harry Miller's Star
Uma 10h34'56" 50d48'55"
Harry Mitchell Hanna III
Tau 5h10'8" 17d48'24"
Harry Moeller
Cep 22h42'36" 77d42'19"
Harry Montana
Dra 18h47'12" 59d42'14"
Harry N. Eirich
Cam 3h58'41" 53d3'20"
Harry New
Cru 12h43'30" -57d30'12"
Harry Perkins
Uma 9h26'31" 58d1'33"
Harry Perry III
Sgr 17h58'45" -24d49'45"
Harry Peter
Cep 20h21'58" 60d53'18"
Harry Poskitt
Cap 20h30'55" -23d40'38"
Harry Potcake
Uma 13h5'47" 54d40'22"
Harry R. Pollard
Cep 23h26'54" 77d4'44"
Harry & Ruth Kline Star
Lmi 10h13'40" 28d9'43"
Harry S
Mon 6h47'28" 9d34'14"
Harry S. Bazemore
Psc 1h8'24" 28d8'27"
Harry S. Heart
Ari 1h48'30" 20d18'38"
Harry S. Ort
Tau 4h9'32" 28d23'2"
Harry Saint Germain
Gem 6h59'19" 18d9'7"
Harry Salvatore
Vir 14h10'42" -12d58'19"
Harry Sam Nolan
Cep 23h50'17" 74d13'17"
Harry Sebastian Hunter Hill July 92
Leo 9h24'15" 29d52'21"
Harry Segall
Boo 14h20'31" 19d6'45"
Harry + Seva
And 2h19'38" 39d8'2"
Harry Shrake
Cnc 8h47'24" 30d47'56"
Harry Star
Lib 14h52'14" -3d40'11"
Harry Steinman
Sco 16h38'59" -34d13'19"
Harry Stock
Uma 10h32'55" 58d11'37"
Harry & Sue
Tau 4h26'58" 12d16'28"
Harry T. Shamberger
Uma 11h45'41" 42d5'14"
Harry Taylor
Umi 14h15'42" 77d15'35"
HARRY the SCHNEIDER
Cep 1h35'12" 85d57'2"
Harry Theodore Holmberg
Uma 10h31'57" 51d42'41"
Harry Thompson's Star 2004
Uma 12h2'0" 37d23'6"
Harry Tyrrell's Star
Her 17h39'42" 38d7'41"
Harry V. Jorgensen
Ori 6h12'57" 13d38'26"
Harry W. Cline 10/15/24-8/17/03
Ari 1h49'1" 18d50'51"
HARRY W. KILLA
Psc 1h7'41" 26d40'49"
Harry W. Tretter
And 23h22'26" 51d21'44"
Harry W Walker - My Special Sparkler!!!!
Uma 11h2'52" 36d47'0"
Harry Walker
Dra 14h32'33" 55d19'3"
Harry Warner Hoff, Jr.
Sco 16h13'19" -10d16'35"
Harry Wilcock
Per 4h13'16" 49d9'56"
Harry William Petersen
Per 2h25'20" 54d24'17"
Harry Yates
Sco 17h31'49" -33d44'23"
Harry21
Dor 5h5'31" -66d3'5"
HarryandAngieArcherLove AlwaysNDW
Mon 6h30'8" 11d29'52"
Harrylda Anniversus 65 Major
Pho 0h4'36" -48d25'18"
Harry-Rujo
Sco 17h22'51" -41d25'23"
Harry's 2005 Starship Mercedes 190S
Uma 10h57'27" 33d53'51"
Harry's Halo
Lyr 19h5'49" 31d37'51"
Harry's Heart
Aql 19h35'34" 5d33'23"
Harry's Light
Her 18h22'4" 20d59'10"
Harry's Star
Peg 22h16'38" 19d44'29"
Harry's Star
Her 16h20'53" 14d33'20"
Harry's Star
Per 2h55'27" 40d47'26"
Harry's Star
Per 2h28'50" 51d8'33"
Harry's Star
Sco 16h17'49" -10d28'4"
Harsha Priya Jayanti
Boo 14h14'17" 42d5'16"
Harshal K Carter
Ari 3h21'53" 19d16'0"
Harsh's Star
Sgr 20h4'30" -20d14'24"
Hart- Love Dream Hope Faith
Cyg 20h26'37" 47d42'34"
Hart of Dixie Star
Cnc 8h23'40" 32d48'52"
Hart Schiess
Umi 13h42'27" 75d49'31"
Hart Star
Cap 20h15'55" -9d23'22"
Harte of my Heart
Cyg 21h32'33" 48d49'42"
Hartinger Mazzini
Mon 6h49'54" 6d36'22"
Hartman
Cap 21h38'11" -14d45'43"
Hartman021407
Tau 4h9'51" 6d2'17"
Hartmut
Ori 5h51'48" -0d41'23"
Hartmut Heske
Cru 11h56'46" -60d1'45"
Hartmut Menzel
And 23h46'42" 47d39'43"
Hartmut Skoeries
Ori 6h20'34" 16d30'57"
Hartmut Straßburger
Ori 5h58'30" 17d28'32"
Hartmuth Müller
Ori 6h15'49" 8d21'58"
Hartoon Der-Sarkissian
Uma 11h56'39" 34d45'17"
Hartrampf, Melina
Uma 9h26'39" 68d7'54"
Hart's Light
Cas 1h13'33" 53d57'59"
Hartwell
Uma 9h37'16" 48d29'17"
Hartwig Knetter
Uma 11h4'52" 63d26'15"
Hartzall on Guard
Ori 5h37'45" -2d29'58"
Haru Furuya
Lyn 7h50'15" 52d35'55"
Haruka, thank you for our new life
Sco 17h37'6" -35d34'39"
Haruki
Cap 20h37'31" -8d35'8"
Harumi
Sco 16h13'3" -25d15'20"
Haruna
Aqr 21h57'29" -4d10'19"
Harveer Singh Gill
Dra 20h25'36" 67d54'23"
Harvest Years Senior Center
Sco 17h50'21" -44d43'33"
Harvestere
Ari 2h38'55" 20d51'38"
Harvey
Boo 13h37'8" 18d23'53"
Harvey
And 2h37'48" 46d5'0"
Harvey Alexander Crighton
Cep 3h13'31" 83d39'10"
Harvey and Fina
Umi 15h13'55" 71d17'56"
Harvey Antony
Cep 21h54'31" 65d30'40"
Harvey Bennett Nassau
Aur 6h55'20" 44d1'53"
Harvey Cannova "For Our Journey"
Sco 16h16'22" -37d15'18"
Harvey Carr
Ori 5h0'20" -1d28'42"
Harvey Crihfield
Her 16h48'49" 8d50'54"
Harvey Dale Freeman
Lib 15h23'35" -6d19'18"
Harvey & Deanne Stone
Cyg 20h23'0" 57d9'56"
Harvey Declan Stretton
Her 18h10'9" 17d22'22"
Harvey Ernest George
Sgr 18h39'22" -16d35'37"
Harvey Fisher Branham
Cet 1h46'43" -8d29'27"
Harvey Gerstman
Per 3h53'1" 32d53'45"
Harvey Ginsberg
Uma 11h16'30" 30d44'38"
Harvey Harold Bird
Vir 13h10'21" 13d34'47"
Harvey Hauer
Boo 13h47'48" 11d15'10"
Harvey Hop
Aql 19h5'16" -0d45'14"
Harvey in the Sky with Diamonds
Cep 22h23'5" 61d58'6"
Harvey Jack Harris
Per 2h19'6" 55d55'3"
Harvey Jack Michael Smith
Per 2h56'30" 46d53'5"
Harvey Joseph Tetrick
Umi 16h13'7" 76d22'26"
Harvey L. Turnipseed
Uma 9h1'30" 50d19'2"
Harvey & Lever globetrotter star
And 1h26'40" 39d39'39"
Harvey Liebhaber, M.D.
Ori 4h51'33" -0d41'35"
Harvey & Louise Brown
Cyg 19h55'53" 33d26'26"
Harvey Mamon, M.D., Ph.D.
Cnc 8h46'38" 30d39'51"
Harvey Manger-Weil
Psc 0h42'27" 12d48'14"
Harvey Monroe Klein
Uma 10h27'26" 63d53'26"
Harvey P Vowell
Mic 20h57'13" -37d11'28"
Harvey Ross Jr
Cep 23h2'34" 70d29'55"
Harvey Sill - In Loving Memory
Oph 16h38'30" -0d50'48"
Harvey Star
Per 4h29'57" 42d59'7"
Harvey Tucker
Psc 0h44'40" 12d27'4"
Harvey W. Phelps
And 2h36'26" 46d46'2"
Harveybar12
Her 18h38'27" 15d42'56"
Harvimaz
Psa 22h50'17" -35d18'27"
harvizzy
Gem 6h39'4" 15d53'49"
"Harvo" Harvey Wendell Hertz
Crb 16h5'53" 27d56'55"
Härzli
Cep 22h56'34" 79d52'23"
Hasan Arat
Lib 15h4'6" -20d28'55"
Hasan 'Baby Face' Alaydrus
Sco 17h52'39" -39d48'46"
Hasana Ayasha & Karinda Akilah
Cyg 21h27'5" 47d51'49"
Hase
Uma 8h31'0" 68d43'31"
Hase liebt Bärchen für Immer & Ewig
Uma 9h56'9" 54d16'43"
Hasen-Schatz
Uma 10h4'54" 55d49'25"
Hashmukh Parmar
Per 2h29'43" 54d28'20"
Häsl Schnuggi
Ori 6h16'25" 15d44'47"
Häsl Schnuggi
Umi 13h9'29" 75d51'2"
Hasmik Bagramian
Ori 5h44'23" 11d28'35"
Hason's Heaven
Vir 13h20'7" -0d42'43"
Hassan
Cen 13h35'50" -38d27'22"
Hassan Afzal
Ser 15h55'36" 3d45'8"
Hassan Ali Safdari
Cma 6h41'2" -16d37'12"
Hassan Saleh
Her 18h24'30" 12d5'31"
Hassan & Suzana
Cyg 20h55'44" 34d24'13"
Hassan & Tiffany Grab Forever
Cnv 13h43'38" 37d38'55"
Hassan & Veronique
Uma 9h33'3" 59d55'39"
Hassiba Freiha
And 23h29'53" 47d28'44"
Hassie Clare Thurman
And 0h9'36" 34d33'12"
Hastings
Tau 4h31'37" 22d0'42"
Hastings
Gem 6h32'57" 16d57'51"
Hatak Intinkba Holitopa
Ori 5h51'11" 11d27'23"
HATCH
Uma 9h30'21" 42d58'31"
Hatem
Lyn 7h43'0" 36d59'19"
Hatem
Per 4h24'24" 46d38'49"
haTh nAri naTii
Pho 1h15'46" -44d39'39"
Hatley's Halo
Per 3h24'37" 36d15'31"
Hatshepsut Vega
Lyr 18h39'17" 39d26'25"
Hatteras
Cru 12h44'7" -56d6'10"
Hatthaya Paprakhon
Cam 3h42'24" 56d22'11"
Hattie and Leon Scott
Umi 16h19'59" 75d25'43"
Hattie J. Scarcliff
And 1h17'45" 37d58'7"
Hattie Jackson
Lyr 19h9'40" 37d16'51"
Hattie L. Kibler Bates
Gem 7h27'38" 32d47'54"
Hattie Manuel Fox
And 0h1'36" 44d42'39"
Hattie Miller
Leo 11h54'7" 11d57'7"
Hattie Rose Seagraves Swicord
Cap 20h49'12" -15d53'21"
Hattie Sorochen
Cas 1h23'49" 60d31'43"
Hattie's Bit of Heaven
Cas 1h16'34" 57d14'42"
Hattie's Heart
Lib 15h54'57" -12d54'9"
Hattie's Light
Crb 16h1'5" 32d33'25"
Hattie's smile
Cas 1h4'6" 53d50'26"
Hattie's Star
And 23h39'3" 47d58'11"
Hattie's Star
Cas 0h19'8" 52d16'43"
Haubold, Jens
Lib 15h18'22" -8d27'43"
Haubold, Ute
Uma 9h3'38" 51d37'54"
Haug, Peter
Uma 9h43'16" 49d27'59"
Haugus 2004
Cyg 21h8'16" 30d41'3"

Haui Tarshis
Psc 0h42'43" 18d1'10"
Hauke Christoph Kunst
Uma 11h40'16" 43d52'56"
Haunani
Leo 10h42'33" 7d22'55"
Haunani
Ser 15h22'31" -0d25'43"
Hausdoerffer Family
Legacy Star
Umi 7h4'48" 89d18'20"
Häusermann, Werner
Uma 12h57'29" 53d16'23"
Havaani Allegra
Peg 23h36'30" 9d40'28"
Have I Told You Lately I
Love You
Tau 4h15'15" 14d56'41"
Haven
Uma 10h27'33" 61d10'58"
Haven and Sarah
Lyn 8h52'49" 38d6'5"
Haven & Ashley Loved by
Gigi
Uma 11h16'34" 52d41'26"
Haven Grace Davis
Cnc 8h58'22" 18d57'27"
Haven Hughes-Miller
Uma 12h37'22" 58d4'8"
Haven Kathleen Lutz
Peg 22h1'28" 18d41'20"
Haven Kelsey Mills
Cap 20h37'53" -23d27'22"
Haven Lynn
Tau 4h35'21" 30d49'3"
Haven Marie Norris
Lyr 18h49'58" 38d49'54"
Haven Nicholas Mitchell
Psc 1h25'28" 26d40'20"
Haven Nicole Murauskas-
Clarke
And 1h31'43" 40d21'12"
Haven Skye
Lib 15h21'48" -17d40'54"
Haverford College Class of
2006
Ori 5h47'30" 8d14'30"
Haverkamp, Jörg
Gem 7h32'36" 20d2'37"
Havilah Henderson
Lyn 7h9'8" 59d3'9"
Havior
Gem 7h12'48" 24d19'28"
Havoc's Splendour
Ara 17h45'30" -47d1'3"
Hawaiian King
Per 3h12'9" 48d23'6"
HAWK
Vir 14h5'28" -19d29'58"
Hawk and Cindy
Uma 10h24'1" 40d57'3"
Hawk: Guardian of Dana L.
Paz
Peg 23h9'43" 28d54'17"
Hawk Star
Tau 4h21'26" 23d16'4"
Hawke Edson
Cap 20h43'26" -21d11'36"
Hawkes
Uma 14h18'25" 60d15'27"
Hawks Family Star
Uma 10h47'55" 45d47'33"
Hawley Swiecicki
Cnv 12h25'41" 47d5'18"
Häxli
Aur 5h30'29" 40d42'6"
Häxli Anika
Aql 19h35'19" 6d43'26"
Hay Now
Uma 10h8'52" 56d36'28"
Haya Hamad Ghanem
Shaheen AlGhanem
Vir 13h24'12" -5d24'6"
Hayam Ali Shaker
Tau 5h2'17" 19d4'3"
Haydar Sancak
Tau 4h40'45" 22d55'59"
Haydee
Ori 6h17'46" 16d15'6"
Haydee
Lyn 7h50'27" 53d6'24"
Haydee Alfonso
Cap 21h24'52" -16d23'2"
Haydee Perez
Vir 13h56'15" 5d41'23"
Haydee & Sammy
Vir 12h50'23" 0d59'17"
Hayden
Vir 12h58'16" 10d41'15"
Hayden
Ser 18h20'58" -13d40'32"
Hayden
Uma 13h46'41" 58d39'51"
Hayden Alexander
Ari 3h16'49" 27d42'46"
Hayden "Beasley" South
And 1h9'3" 38d50'24"
Hayden Bock
Sco 17h39'0" -37d27'58"
Hayden Boling
Her 16h44'24" 47d38'2"
Hayden Christensen
Ori 5h28'52" 0d13'23"
Hayden Christian 626
Cnc 8h13'15" 16d9'30"
Hayden Christian French
Uma 10h3'10" 59d42'10"
Hayden Clark Stevens
Darling Donnelly
Leo 11h27'50" 19d30'51"
Hayden Clark Wood
Cnc 8h0'19" 15d10'11"
Hayden Clive Jason Payne
Col 5h35'22" -33d41'38"
Hayden Cole Duppong
Cap 20h51'11" -15d10'55"
Hayden Costigan Moore
Uma 9h44'40" 57d35'54"
Hayden CS8705
Ori 5h44'10" 8d45'16"
Hayden Dale Aly
Uma 11h4'7" 48d9'42"
Hayden Daniel Duffey
Cap 21h9'29" -25d55'43"
Hayden Douglas
McMonigle
Her 17h29'21" 26d52'19"
Hayden Eagles
Cru 12h1'16" -62d4'18"
Hayden Fitts
Umi 15h15'45" 88d32'14"
Hayden Forrest Howell
Uma 10h21'8" 52d54'30"
Hayden Graham Schreiber
Gem 7h45'26" 31d21'21"
Hayden Grant Cooper
Cnc 8h25'34" 18d8'27"
Hayden Heaton Simmons
Tacoma
Cas 1h12'18" 49d26'2"
Hayden Jack Niehus
Sco 16h50'18" -33d22'8"
Hayden "Jake" Busot
Dra 18h22'23" 53d4'20"
Hayden James
Sgr 18h25'0" -24d26'27"
Hayden James Ficek
Uma 11h37'24" 59d28'9"
Hayden James Jones
Boo 15h34'54" 40d1'40"
Hayden Jeffrey Ragusa
And 0h40'22" 25d42'58"
Hayden Johnston
Tau 3h54'16" 9d31'36"
Hayden K. Conatser
And 23h33'12" 47d53'39"
Hayden Kevin Spence
Uma 10h40'38" 62d41'34"
Hayden Kirby
Lammermann
Sco 16h7'10" -15d34'32"
Hayden Kylie Wicks
Ori 5h49'21" 6d20'26"
Hayden Lance
Cra 19h3'39" -40d35'53"
Hayden Lauren Huber
Cap 21h20'53" -15d50'45"
Hayden Lee Gartner
Per 2h47'59" 55d37'51"
Hayden LeRue Pollock
Cap 20h34'19" -21d27'39"
Hayden & Logan Russell
Uma 11h30'33" 58d17'31"
Hayden Luke Christopher
Her 16h43'14" 18d42'18"
Hayden M Corl
Cnc 8h42'44" 19d37'23"
Hayden Maris Miller
Cep 20h40'16" 66d55'10"
Hayden Matthew
Uma 8h38'0" 53d14'16"
Hayden Michael Melycher
Umi 16h23'21" 76d9'59"
Hayden Michael Wagner
Vir 13h16'51" 8d26'7"
Hayden Moses
Aur 6h3'0" 47d37'47"
Hayden Patrick Howell
Lib 15h24'16" -3d53'11"
Hayden Quinn Smith
Psc 1h13'15" 27d23'20"
Hayden Reilly Glover
Sgr 19h15'29" -14d1'6"
Hayden Robert Evetts
Ari 3h27'47" 26d59'22"
Hayden Robert Kelly
Thomson
Cru 12h25'24" -60d29'29"
Hayden Robert Strenfel
Her 18h12'17" 31d38'19"
Hayden Scott Hauser
Psc 23h26'3" 3d9'54"
Hayden Scott Link
Ori 5h43'45" 7d31'0"
Hayden Tate Seal
Aql 19h50'31" -0d33'21"
Hayden Tyler
Umi 15h26'19" 82d56'20"
Hayden Young
Tau 5h4'57" 27d5'54"
Hayden's Lucky Star
Lyn 6h39'23" 54d19'10"
Hayden's Star
Umi 15h9'28" 77d27'47"
Haydn Cole Winnings
Leo 9h45'55" 24d44'35"
Haydn James Dawson
Cyg 20h30'16" 41d25'43"
Haydon Brown
Cnc 9h8'46" 31d10'24"
Haydon Mark Vaughan
Ori 5h51'29" 4d12'56"
Hayes Alan Heidinger
And 2h16'55" 47d16'25"
Hayes David Therrien
Ori 5h20'4" -4d49'12"
Hayes Lookout
Cam 3h50'45" 73d42'23"
Hayes Louis Nelson
And 0h57'56" 42d5'23"
Hayes Preston Rogers
Uma 10h4'45" 51d37'32"
HAYGRAYK
Lyn 7h34'30" 48d47'8"
Hayko & Narine
Hakopchikyan
Cyg 20h56'32" 32d19'33"
Haykush Nazaryan
Aqr 22h6'39" -0d25'28"
Haylee
And 23h14'47" 42d2'0"
Haylee
Tau 3h46'8" 27d7'15"
Haylee
Peg 22h52'48" 19d13'28"
Haylee Danyelle Mazzella
Cnc 8h16'10" 24d26'21"
Haylee Edna Colter
Crb 15h41'34" 26d17'44"
Haylee Faith
Col 5h42'17" -33d19'44"
Haylee February 24, 1996
Psc 0h53'2" 15d23'52"
Haylee Hunter Mannon
Aqr 22h41'14" -8d48'48"
Haylee Maree Dwyer -
29.09.1987
Cru 12h29'31" -60d10'8"
Haylee Olivia Balme
Sgr 18h22'1" -31d7'39"
Haylee Rae
Aql 19h22'35" 7d42'49"
Haylee Sue Rivers
Vir 12h41'3" -9d13'0"
HayleeAnn5
And 0h37'43" 33d23'40"
HayleeRosePearson
Psa 22h47'11" -32d34'14"
Haylee's Angel
Lib 15h32'24" -11d35'32"
Haylee's Heart
And 0h2'31" 40d2'7"
Haylee's Hope
Aqr 21h43'19" -8d11'24"
Haylei M Cudd
Com 13h10'55" 25d26'52"
Hayleigh Bop's Twinkle
Cap 20h51'9" -25d51'11"
Hayleigh Jo Smith
Vir 13h4'40" 11d45'32"
Hayleigh Pyn Lidbury
Cnc 8h33'22" 29d54'1"
Hayley
Peg 23h52'25" 11d10'12"
Hayley
Cyg 19h46'34" 34d37'13"
Hayley
Boo 15h38'6" 45d40'18"
Hayley
Cam 4h32'37" 59d5'44"
Hayley
Cap 21h46'15" -10d36'28"
Hayley
Lib 15h32'40" -16d39'57"
Hayley
Umi 16h52'4" 76d14'18"
Hayley Alana - Our
Precious Baby Girl
Cru 12h25'3" -60d35'20"
Hayley Ann
Gem 7h25'13" 30d49'40"
Hayley Anne Adamo
And 0h42'34" 26d32'41"
Hayley Barnes
Uma 8h43'47" 53d14'31"
Hayley Beth Morris
Gem 6h52'2" 33d29'36"
Hayley Brianne Ward
And 1h25'44" 42d39'30"
Hayley Bright
Gem 7h37'1" 23d32'0"
Hayley Burch CMP-GaMPI
Shining Star
Lyn 7h45'34" 47d51'9"
Hayley C. Biggens
Sgr 19h35'27" -38d3'45"
Hayley & David's Star of
Love!
Cyg 20h26'24" 52d55'50"
Hayley Dean - 29.09.1976
Lib 15h40'39" -18d29'37"
Hayley Devon
Gem 6h3'16" 22d50'46"
Hayley Elizabeth Feagley
Cyg 21h45'33" 47d46'6"
Hayley Fiskum
Umi 20h5'46" 88d34'36"
Hayley Fox Star
Vul 19h32'30" 26d15'44"
Hayley Girl
Leo 11h9'42" 6d43'3"
Hayley Grace
And 0h9'54" 43d45'15"
Hayley J. Page
Mon 6h46'49" -0d8'14"
Hayley Jane Brunz
Sgr 18h45'55" -26d57'28"
Hayley Jane Sherman
And 0h21'46" 46d22'16"
Hayley Jocelyn Elmore
Cas 0h5'54" 57d5'7"
Hayley Kruse
Leo 9h44'7" 21d12'37"
Hayley Louise Broad
Lib 15h4'40" -2d5'0"
Hayley Louise Ravenhill
And 2h7'6" 41d23'40"
Hayley Lynn
Tau 3h37'56" 27d43'49"
Hayley Marie Keithahn
And 2h14'10" 48d23'23"
Hayley Mazie Guittard
Tau 3h54'24" 28d21'0"
Hayley McNerney
And 0h20'46" 43d21'20"
Hayley & Michael O'Hora -
23.02.05
Cru 12h19'45" -63d37'37"
Hayley Michele
Leo 11h51'21" 21d53'21"
Hayley Michelle Love
Cyg 21h40'30" 51d12'23"
Hayley Minn's Star
Sge 20h9'12" 19d24'15"
Hayley Muse Douglass
Rupersberg
Tau 5h38'4" 22d41'45"
Hayley Nicole
And 2h9'47" 37d36'1"
Hayley Noel Zanetta
Cyg 19h58'26" 45d23'28"
Hayley Rae
Gem 7h26'35" 22d21'6"
Hayley Renee Curry
Cas 0h12'11" 57d4'40"
Hayley Scarlet Dodd
Eri 4h34'51" -35d47'21"
Hayley Theresa Jaqua
Lib 15h49'8" -5d36'51"
Hayley Watson
Ari 2h0'59" 19d53'18"
HayleyandJay-3742
Cyg 20h21'24" 46d23'41"
Hayleyann (Hoggi) Morton-
Joseph
Cep 22h0'41" 67d30'36"
Hayley's Comet
Gem 6h26'35" 24d28'33"
Hayley's Light
Cnc 9h4'35" 7d46'57"
Hayley's Sparkle
Cru 12h41'52" -57d30'59"
Hayley's Star
Peg 23h58'31" 20d48'38"
Hayley's Star
And 1h7'27" 38d24'59"
Hayley's Star
And 2h21'25" 42d18'19"
Hayley's Star for Many
Loved Years
Leo 9h44'22" 26d51'53"
Hayley's Wishing Star
Del 20h37'54" 15d19'55"
Hayli & Jerms
And 23h31'51" 47d30'19"
Haylie Anna
And 22h58'30" 48d38'45"
Haylie Cathryn Messina
Lyn 7h31'3" 36d12'23"
Haylie Elizabeth Mitchell
And 1h9'6" 38d16'31"
Haylie Marie James
Vir 13h1'22" -13d49'13"
Haylo
Vul 20h14'57" 22d40'49"
Haynes Family Star
Leo 11h7'15" -5d47'56"
HAYS 7
Ori 6h4'27" -0d49'37"
Hays- United Together in
Love
Crb 15h36'17" 31d47'40"
Hayseed
Aqr 22h25'35" -5d57'27"
Hayungs
Sco 16h6'40" -14d7'47"
H.A.Zapata
Uma 13h45'13" 55d8'0"
Hazard Meriden
Community
Lyn 8h11'14" 56d4'13"
Haze
Lib 15h45'0" -11d26'56"
Hazel
Umi 14h27'14" 74d22'57"
Hazel
Sgr 20h9'18" -38d50'0"
Hazel
Sgr 18h39'3" -23d32'11"
Hazel
Cam 4h15'37" 56d34'30"
Hazel
Lyn 7h5'56" 48d39'2"
Hazel
Cyg 21h42'22" 38d38'49"
Hazel
Tau 4h28'31" 15d4'30"
Hazel
Tau 4h3'2" 13d20'33"
Hazel 13
Lyr 18h36'53" 29d16'47"
Hazel A. Carlisle
Uma 13h36'49" 57d4'37"
Hazel and Julien Lefort Jr
Per 3h21'26" 41d35'25"
Hazel Ann Barnes
Cas 0h6'58" 53d17'16"
Hazel Ann Methe
D'Ingianni
Psc 0h48'20" 9d38'10"
Hazel Beatrice 1957
Cap 20h30'51" -24d11'31"
Hazel Borden
Cas 0h0'24" 56d59'56"
Hazel Castle
Aur 5h7'29" 33d29'1"
Hazel Charolette Blue
Dolecheck
Uma 13h39'50" 57d56'26"
Hazel Cordelia Hensley
Cap 20h15'28" -17d7'10"
Hazel Cottage
Cyg 21h18'17" 53d58'4"
Hazel D. V. Langille
Aqr 20h43'46" -1d28'53"
Hazel Dakers Menzies
Cyg 21h48'55" 43d8'33"
Hazel Dalziel
Uma 11h32'53" 56d37'14"
Hazel E. Adams 4/17/1921-
8/22/2006
Cas 0h44'57" 57d49'43"
Hazel Elardo
Cas 0h40'27" 64d33'40"
Hazel Elizabeth Kelley
Cyg 19h32'41" 28d26'52"
Hazel Gianatiempo
Umi 14h10'51" 67d30'0"
Hazel Grace Chag
And 23h25'45" 41d54'48"
Hazel Ida Way
Cnc 8h43'12" 7d42'28"
Hazel J. Flaherty
Sgr 18h32'17" -19d4'6"
Hazel Jane Tetrick
Ari 2h10'8" 23d51'1"
Hazel La Hood
Lyn 7h5'40" 45d57'20"
Hazel & Laz
Sgr 19h3'22" -12d24'9"
Hazel Lee Westervelt Orton
1925
Ori 5h42'18" 8d41'38"
Hazel M.
Aql 19h3'17" 1d33'52"
Hazel Marie (Hood) Hoffner
Lyn 9h1'56" 33d8'38"
Hazel Maxine Draeger
Uma 12h25'7" 60d25'6"
Hazel my Love
And 0h58'36" 45d55'57"
Hazel Patricia Moder
And 0h48'12" 42d14'2"
Hazel "Penny" Cecilia
Gocke
Cyg 21h25'13" 55d13'0"
Hazel Robison
Aqr 22h3'24" -15d8'44"
Hazel Rosalie Simpson
Uma 9h41'33" 42d8'27"
Hazel Rose Sell
Cas 1h25'41" 63d10'28"
Hazel Schuth
Lyr 19h15'40" 32d16'17"
Hazel Sievert
Crb 15h49'3" 34d10'37"
Hazel Smith
And 0h21'16" 27d57'32"
Hazel Supernova
Uma 12h39'52" 59d47'13"
Hazel T. Wilson 3-30-1911
Ari 3h8'56" 28d45'9"
Hazel Verna Parker Taylor
Cyg 19h53'3" 32d23'10"
Hazel Williams
Aqr 20h43'17" -12d25'12"
Hazel Worley Angel
Sgr 19h43'53" -22d50'16"
Hazel, Frazer & Rosie
Uma 8h57'39" 61d55'40"
Hazell Fallows
Cyg 19h47'45" 35d25'9"
Hazel-May
Gem 7h46'44" 29d18'6"
Hazelnut Kamila
Cru 12h23'6" -56d6'47"
Hazel's Birthday Star
Cas 3h13'44" 63d20'36"
hazels galaxy
Gem 7h3'24" 24d50'7"
Hazel's star
Cap 20h32'53" -17d59'39"
Hazel's Star
Umi 12h52'19" 86d26'26"
Hazem Khadidja Amara
Her 18h2'31" 28d19'19"
Hazen & Neeka
Leo 9h49'33" 7d32'14"
Hazy
Cma 7h20'39" -19d33'5"
HB
Tau 5h4'1" 24d57'53"
HB29
And 2h33'21" 40d52'31"
HBD2 Tooter
Uma 8h31'35" 71d19'48"
HBD3Tooter
Uma 13h36'26" 56d54'10"
HBD4 Tooter
Umi 14h33'48" 75d12'44"
HBD6 Munchkin
Umi 14h8'22" 72d22'44"
HBD7Munchkin
Umi 16h17'28" 77d17'23"
HBD8 Munchkin
Umi 14h20'29" 76d9'50"
HBNF
Cyg 20h22'36" 58d26'33"
H-Bomb Hot Rod 2005
Uma 9h4'54" 71d10'53"
HC ifetayo kabisa
Sco 17h29'5" -42d7'47"
HC-BC
Vir 13h20'37" -16d24'48"
H.D. Allen
Aqr 22h23'2" -5d46'21"
H.-D. Wiese
Uma 11h18'59" 47d57'3"
H.D.G.D.L.
Uma 8h26'22" 66d40'49"
H'dia Nie
Uma 11h52'29" 34d12'1"
HDK-Koster Princess Star
Cap 20h48'51" -15d39'31"
He Shenqun
Aur 6h5'58" 31d19'42"
Heaboo
Ari 2h36'34" 18d0'8"
head over feet
Uma 9h30'37" 42d52'29"
Headley Thomas "5-11-35"
Umi 17h1'5" 76d14'3"
Heahter Ann Bowman
Vir 12h14'8" -0d50'54"
healer
Vel 9h55'22" -51d55'33"
Healster
Tau 4h17'59" 27d4'43"
Health And Harmony
Cyg 20h22'32" 52d55'56"
Heart
Sgr 18h49'4" -16d16'37"
Heart Filled Box
Cyg 19h58'10" 35d19'17"
Heart in Hand
And 1h3'35" 42d2'7"
Heart Means Everything
Uma 10h7'34" 61d14'8"
Heart of Christy
Psc 1h35'15" 21d33'18"
Heart of Geoff
Cru 12h17'41" -56d33'38"
Heart of Gold
Cas 0h8'2" 50d17'57"
Heart of Jerrod
Cep 22h40'45" 76d47'57"
Heart of Marie
Gem 6h27'52" 24d45'30"
Heart of Marz
Uma 10h16'36" 49d26'45"
Heart of Melissa
Ori 6h16'46" 11d2'53"
Heart Of My Angel
Vir 13h18'8" -10d18'35"
Heart of the Night
Aql 19h51'23" 16d18'45"
Heart of the Sunrise
Per 3h42'51" 49d41'40"
Heart of Tonia
Psc 0h51'55" 17d0'47"
Heart of Tuesday
Aqr 22h26'14" -0d28'1"
Heart & Soul For Always
Ara 16h37'28" -47d37'59"
Heart You
Aqr 21h22'2" 1d16'6"
heartlock
Aqr 22h12'30" -8d25'48"
Heartmender the Blue
Cherry
And 1h35'16" 47d56'20"
Heart's Desire
Cyg 20h19'15" 54d28'47"
Hearts Desiree
Sgr 20h15'4" -39d5'14"
Hearts Ronnie
Her 18h53'44" 22d26'29"
Hearts-a-glow
Aur 6h30'3" 39d10'25"
Heath
Her 17h7'23" 28d25'58"
Heath
Cet 2h34'28" 8d38'37"
Heath
Dra 18h48'41" 63d48'45"
Heath
Cap 21h46'53" -18d5'24"
Heath and Kristin
Sco 16h23'8" -23d46'46"
Heath Andrew Snyder
Uma 8h43'6" 69d54'29"
Heath David Reichelt
Psc 0h24'54" 15d55'46"
Heath Erik Massey
Ari 3h5'11" 19d53'47"
Heath Hassler
Ori 6h18'53" 14d47'10"
Heath Hayes
Ori 5h26'48" 13d21'21"
Heath Matthews
Cru 11h56'29" -55d53'39"
Heath McLachlan
Cra 18h4'36" -37d12'6"
Heath Russell Behmer
Pho 0h13'7" -43d48'31"
Heath Ryan Naye
Ori 6h12'45" 15d29'43"
Heath William Kincade
Psc 0h54'21" 10d20'2"
Heathbar 02.20
Psc 1h21'37" 19d11'34"
Heather
Ari 2h32'37" 21d47'20"
Heather
Vir 12h38'50" 12d18'5"
Heather
Psc 0h37'38" 14d55'20"
Heather
Vir 12h13'35" 7d8'35"
Heather
Ori 5h21'57" 5d7'25"
Heather
Tau 4h34'26" 1d19'5"
Heather
Gem 7h19'12" 26d3'7"
Heather
Her 18h22'7" 15d50'4"
Heather
Gem 7h7'34" 15d22'48"
Heather
Gem 6h54'10" 20d36'45"
Heather
Leo 10h10'15" 15d21'3"
Heather
Lyn 8h2'22" 38d54'16"
Heather
And 1h18'55" 38d42'41"
Heather
And 0h27'8" 41d43'2"
Heather
And 0h41'0" 30d56'15"
Heather
And 0h50'46" 36d18'25"
Heather
Gem 6h54'10" 31d12'44"
Heather
And 2h25'52" 46d7'52"
Heather
And 23h27'50" 42d59'53"
Heather
Crb 16h15'7" 38d52'26"
Heather
Uma 11h46'39" 48d26'23"
Heather
Boo 15h12'52" 47d54'45"
Heather
And 23h23'34" 52d5'42"
Heather
And 23h15'49" 51d33'25"
Heather
Sco 17h27'8" -42d37'11"
Heather
Psc 0h1'21" 4d20'45"
Heather
Sco 17h30'19" -33d53'46"
Heather
Col 5h37'17" -31d31'38"
Heather
Col 5h38'6" -33d54'59"
Heather
Cma 6h28'38" -29d13'43"
Heather
Lib 15h13'1" -23d25'11"
Heather
Sgr 20h5'10" -25d16'55"
Heather
Dra 15h16'46" 62d21'27"
Heather
Cas 1h44'26" 63d56'17"
Heather
Cam 6h52'2" 67d9'0"
Heather
Cas 2h12'56" 62d39'20"
Heather
Cas 1h46'29" 61d13'22"
Heather
Cas 0h10'28" 61d14'17"
Heather
Cap 20h42'56" -17d20'58"
Heather
Vir 13h19'9" -19d34'36"
Heather
Lib 15h58'25" -17d25'5"
Heather
Cap 21h46'36" -11d34'17"
Heather
Aqr 23h47'53" -13d40'22"
Heather
Sco 16h8'48" -12d23'22"
Heather
Vir 14h18'7" -0d5'37"
Heather A Ditomaso
Aqr 22h30'18" -2d29'5"
Heather "A Giver at Heart"
Uma 14h22'49" 60d47'18"
Heather A. Horn
Lib 14h40'0" -9d20'39"
Heather A. Schermerhorn
Your Loved
Cyg 20h18'10" 46d43'39"
Heather & Abram
Vir 12h38'8" 8d17'27"
Heather Alexis Compton
Psc 23h11'57" 5d28'12"

Heather Alicia Harper
And 1h4'46" 37d56'23"

Heather Allison Knox
Ari 2h41'31" 27d15'50"

Heather and Brandon
Leo 9h52'53" 31d17'33"

Heather and Chris - Eternal Love
Uma 9h56'47" 48d30'54"

Heather and Garrett Bissell
Lyr 18h42'16" 30d51'15"

Heather and Jason's Eternal Flame
Vir 12h24'37" 0d4'33"

Heather and Jayson FOREVER
Cyg 19h59'56" 33d36'42"

Heather and John's Journey Star
Leo 9h54'1" 28d8'32"

Heather and Matthew Miller
Cyg 19h36'2" 53d10'29"

Heather and Matt's Star of Love
Cyg 19h56'58" 39d5'5"

Heather and Michael Fara
Cyg 20h28'3" 48d12'8"

Heather and Nick
Leo 9h46'41" 24d59'20"

Heather and Robbie
Gem 7h8'59" 34d41'25"

Heather and Roy
Cnv 12h37'10" 43d38'32"

Heather and Ryleigh McCright
Uma 11h33'9" 63d1'26"

Heather and Scott Donaldson
Cyg 20h16'1" 56d0'38"

Heather And Sebastian
Aqr 22h12'0" -0d39'51"

Heather and Shane Forvever
Ari 2h54'3" 21d26'44"

Heather and Timothy
Cnc 8h42'20" 14d48'58"

Heather and Todd for Evermore
Uma 14h19'22" 56d19'37"

Heather and Tony
Gem 6h23'49" 20d59'47"

Heather Andress Lewis
Per 3h48'36" 43d54'10"

Heather & Andrew Davis Eternal Love
Cyg 21h13'49" 46d8'1"

Heather Andryuk
Gem 6h52'58" 22d1'2"

Heather Angel
Lib 14h47'40" -10d28'3"

Heather Angel Lobello
Leo 11h14'16" 23d32'9"

Heather Ann
Leo 9h27'41" 14d16'30"

Heather Ann
Cyg 21h12'59" 46d16'23"

Heather Ann
And 0h42'35" 43d28'7"

Heather Ann
Gem 7h32'0" 31d32'57"

Heather Ann
Lyr 18h53'46" 36d3'32"

Heather Ann
Uma 11h58'30" 34d12'6"

Heather Ann
Cap 21h35'42" -18d45'52"

Heather Ann Burke Wolintos
Umi 16h14'53" 74d37'40"

Heather Ann Gluecklich
Uma 10h37'14" 40d34'43"

Heather Ann Halliday
Uma 10h26'8" 71d52'14"

Heather Ann Hill
Sgr 19h40'33" -12d25'3"

Heather Ann Howells
Lyn 8h24'43" 44d39'44"

Heather Ann Jacobs
And 0h30'46" 31d10'11"

Heather Ann Kingston
Sgr 19h5'41" -36d27'32"

Heather Ann Knipf
Uma 11h49'53" 34d44'15"

Heather Ann Koon
Peg 23h31'13" 17d50'14"

Heather Ann Lucarelli
Ari 3h17'19" 24d26'39"

Heather Ann Miller
Per 3h23'30" 47d27'30"

Heather Ann Moilanen
Lyn 8h13'58" 36d37'22"

Heather Ann Moran
Psc 0h52'34" 11d10'55"

Heather Ann Pickersgill
And 23h59'9" 45d34'2"

Heather Ann Solvig
Ari 2h15'4" 11d41'24"

Heather Ann Stewart
Cap 20h26'32" -21d16'29"

Heather Ann Summer Duvall
Uma 9h54'26" 58d17'39"

Heather Anna Coleman
Aqr 22h40'1" -19d30'24"

Heather Anne
Tau 4h19'59" 29d33'35"

Heather Anne
Cnc 8h46'34" 32d23'19"

Heather Anne Hoffman
Ari 2h45'59" 11d46'50"

Heather Anne Lavin
Sgr 18h38'8" -28d26'41"

Heather Anne Phillips
Cam 4h17'19" 56d31'47"

Heather Anne Simpson
Uma 11h5'36" 48d17'17"

Heather Annebelle
Ori 5h59'27" -0d4'17"

Heather Antoun Jones
And 1h2'36" 44d45'21"

Heather Arbuckle
Sgr 19h29'5" -17d56'37"

Heather Armstrong
Lyn 6h53'7" 52d5'44"

Heather Ashley
Ari 2h10'20" 25d51'54"

Heather Ashley 1126
And 1h3'8" 42d18'25"

Heather Ashley Davis
And 1h0'34" 36d35'3"

Heather Ashley Rounsaville
Cnc 7h59'2" 11d57'17"

Heather Athena Ipsico
Ori 5h59'29" 21d15'27"

Heather Atkinson
Sco 17h55'12" -41d44'26"

Heather "B"
Lyn 6h44'37" 54d0'15"

Heather B. & Ian
Sco 16h26'55" -32d33'11"

Heather B. Wilson
Cyg 21h57'24" 47d34'9"

Heather Baby
Del 20h21'25" 20d36'2"

Heather "Baby Girl" Harrington
Lyn 9h7'46" 33d53'31"

Heather Baehr
And 0h22'37" 46d20'22"

Heather Baker
Lyn 9h14'33" 33d17'33"

Heather Baker Born: Wed May 1/1974
Uma 11h31'12" 56d55'36"

Heather Barnett
Mon 6h42'52" 4d46'52"

Heather Barrett
Gem 6h39'43" 16d18'50"

Heather Bears Star
Aqr 20h38'22" 1d18'14"

Heather Beck
Uma 9h0'41" 62d49'47"

Heather Beckett
Cnc 8h46'57" 19d56'38"

HEATHER BIANCHI
And 0h19'19" 42d14'0"

Heather Bishop 28
Cru 12h56'5" -57d32'26"

Heather Bishop Barbour
Cap 21h50'23" -13d30'8"

Heather Blodgett
Cas 0h15'6" 52d39'21"

Heather Bolling
Tau 4h35'8" 23d22'15"

Heather Bowling
Sco 17h18'16" -32d35'17"

Heather Bradley
Cas 1h55'52" 62d5'3"

Heather Bram
Sgr 18h16'25" -27d13'28"

Heather & Brandon Irvin
Leo 10h27'14" 8d1'38"

Heather Brandow
Cyg 19h52'6" 31d30'51"

Heather BrieAnn Rabenold
Uma 11h7'6" 45d24'5"

Heather Brink Backus
And 1h29'46" 46d17'46"

Heather & Brody's Star
Cyg 20h39'37" 46d44'10"

Heather Brook Freeman
Leo 11h18'45" -2d14'17"

Heather Brooke
Uma 9h33'32" 71d16'2"

Heather Brooke
And 2h20'36" 49d2'38"

Heather Brooke Helm
Lib 15h7'26" -4d38'19"

Heather Brooke Smitherman
Lyn 6h42'30" 60d51'17"

Heather Buckley aka Lil Princess
Cap 20h57'51" -20d41'54"

Heather Bunch-Rindels my sweet love
Leo 11h19'56" 22d29'13"

Heather Burch
Sco 17h53'42" -40d37'42"

Heather Butler
And 1h12'30" 42d56'35"

Heather C. McAlea
Vir 13h28'12" 10d59'10"

Heather (Carmella) Graham
Lib 14h48'52" -15d51'25"

Heather Casey
And 2h1'31" 41d6'3"

Heather & Casey
Gem 7h26'40" 33d19'38"

Heather Charissa Brooks
Lyn 7h49'30" 38d13'55"

Heather Charmaine Imperatore
Cma 6h20'26" -22d17'38"

Heather Charnock
Dor 5h19'9" -59d18'32"

Heather Chase
Dra 15h46'58" 62d45'0"

Heather Childs
Cas 0h57'22" 58d51'16"

Heather Chin Chu Mc Laughlin
Lyn 8h4'15" 44d32'34"

Heather Christine
Uma 9h37'8" 50d49'41"

Heather Christine Viall
And 2h12'9" 38d23'31"

Heather Ciampi
And 23h35'42" 48d28'29"

Heather & Claudio de Oliveira
Crb 16h8'23" 38d55'31"

Heather Colston
Tau 5h38'0" 25d54'9"

Heather Cormier
And 0h44'53" 43d5'2"

Heather Cosgrove
Sgr 18h37'20" -22d59'14"

Heather Crowley
Ori 6h8'56" 20d46'45"

Heather Crowley
Her 17h21'45" 14d42'19"

Heather Cuddeback
Ari 2h44'48" 27d46'7"

Heather Cunningham
Tau 4h36'47" 23d14'14"

Heather D. Neese: Superstar
Peg 22h34'30" 29d29'40"

Heather Dalora Patron
Sco 16h48'23" -32d22'40"

Heather Dangerfield
Lib 14h48'26" -9d28'46"

Heather Daniel Our Little Princess
And 2h19'1" 46d26'58"

Heather Danielle May
Cap 21h6'45" -17d38'19"

Heather & David Wilner
Cam 3h54'50" 68d45'40"

Heather Dawn
Sgr 19h26'46" -16d16'16"

Heather Dawn
And 23h24'13" 37d54'17"

Heather Dawn
And 23h44'27" 37d8'3"

Heather Dawn
Tau 4h11'8" 4d58'39"

Heather Dawn Blankenship
Tau 4h35'27" 17d21'58"

Heather Dawn Dellinger
And 1h42'19" 43d58'21"

Heather Dawn Evans
Uma 9h22'28" 60d37'36"

Heather Dawn Hogan
And 23h4'16" 43d2'29"

Heather Dawn Holt
And 1h36'10" 42d52'41"

Heather Dawn Johnson
And 23h56'24" 43d0'43"

Heather Dawn Sullivan
Lib 15h15'54" -13d53'40"

Heather Dawn Watson
And 0h37'6" 36d3'3"

Heather De Haas
Ori 5h2'30" 10d8'23"

Heather Denise Bottum
Umi 16h0'18" 77d33'26"

Heather Dentoom
And 23h24'9" 47d22'1"

Heather Desautels-Germain
And 0h30'41" 39d22'5"

Heather Diane Dayhuff
And 0h38'14" 42d27'37"

Heather Diane Lyons
Lib 15h17'45" -27d31'58"

Heather DiDoneto
Cas 23h21'0" 54d50'24"

Heather Disberry
Gem 6h52'45" 22d4'40"

Heather D.J. Velvick
Cas 1h19'11" 63d37'38"

Heather Dodge
Ori 5h24'11" -0d35'32"

Heather Domantay
And 0h39'3" 36d40'25"

Heather Donahue
Uma 8h48'5" 60d48'49"

Heather Doucet
Gem 6h49'31" 30d13'20"

Heather Downing
Aqr 22h51'52" -7d47'59"

Heather Doyle
Mon 6h34'40" 8d24'42"

Heather Dreamed of Fairies
Vir 13h32'58" 1d42'17"

Heather Drewery
Aqr 22h33'1" -8d14'54"

Heather Dunham
Sgr 18h40'0" -28d30'53"

Heather Du-Val
Gem 7h26'14" 34d18'55"

Heather E. Dinneen
Aqr 21h54'59" 1d6'40"

Heather E. "Hodge" Noffsinger
Cap 20h13'41" -13d29'26"

Heather E. Regan
Gem 7h52'20" 19d10'42"

Heather Edseth
Cam 6h38'44" 68d10'5"

Heather Eggett
Aqr 22h18'9" -14d15'46"

Heather Egner
Lib 14h42'29" -17d51'54"

Heather Eileen Smith
Sco 16h48'11" -27d20'33"

Heather Elain
Uma 14h3'41" 51d32'27"

Heather Elaine
And 1h23'25" 48d18'59"

Heather Elaine
Uma 8h57'28" 56d44'35"

Heather Elisabeth Jane Quinn
And 23h19'0" 41d24'53"

Heather Elizabeth
And 2h36'21" 45d26'15"

Heather Elizabeth
Gem 7h43'24" 33d21'42"

Heather Elizabeth
Leo 9h42'32" 29d12'10"

Heather Elizabeth Castle
Cam 9h13'48" 78d51'29"

Heather Elizabeth Darby
Uma 8h40'24" 63d46'42"

Heather Elizabeth Guzzi
Ari 3h7'52" 18d7'59"

Heather Elizabeth Hall
Sgr 18h26'48" -31d40'31"

Heather Elizabeth Hubbs
Sco 16h44'22" -30d11'12"

Heather Elizabeth Jordan
Cap 20h26'33" -10d25'28"

Heather Emily Johns
Cap 21h33'21" -24d50'25"

Heather English
Vul 19h51'45" 22d6'20"

Heather & Eric Hullllooo Star
Tau 4h25'57" 2d37'32"

Heather Evet Woods
Uma 12h5'11" 38d24'31"

Heather F. Hancock
Hya 9h5'12" -0d37'44"

Heather F. Klepel
Vir 14h3'54" -13d28'1"

Heather Farnell
Sgr 19h8'20" -12d43'18"

Heather Feather
Cyg 20h40'39" 43d23'29"

Heather Fitzgerald
Cnc 9h14'54" 7d28'35"

Heather Fitzpatrick
Lyn 7h7'43" 49d13'44"

Heather Fraser
And 1h11'13" 35d38'23"

Heather Frey
Sco 17h52'12" -43d1'13"

Heather Gale
Ori 5h52'38" 22d46'24"

Heather Galloway
Tau 4h46'48" 27d4'52"

Heather Garcia
Per 4h4'8" 43d38'25"

Heather Geanine Brown
Cas 2h21'16" 63d19'17"

Heather Geary
Leo 9h41'31" 27d25'44"

Heather Gena
Uma 10h36'42" 70d22'0"

Heather George
Lib 15h50'18" -10d37'52"

Heather & Gio Cancelli
Sge 20h5'27" 17d14'0"

Heather Gondek
Aqr 22h25'9" -5d8'2"

Heather Gore
Peg 21h39'24" 23d35'40"

Heather Grace McAllan
Cra 17h59'55" -37d3'2"

Heather Gramza
Ari 2h4'28" 23d10'31"

Heather Grzech
And 2h21'37" 46d4'23"

Heather Hall
Peg 23h15'58" 31d53'20"

Heather Harris
Cyg 19h34'49" 29d7'23"

Heather "Heart Of The Night"
Cas 0h36'1" 57d40'4"

Heather Heebner
Uma 10h43'55" 42d15'1"

Heather Heida
And 0h43'52" 42d29'47"

Heather Heiser
Ari 3h20'30" 29d29'45"

Heather Hensiek, Love of My Life
Ori 5h28'32" 4d8'43"

Heather Herde
Ori 6h13'36" 17d47'12"

Heather Hines
Cas 0h20'39" 58d2'34"

Heather Hirschberg
Ari 2h0'41" 24d19'3"

Heather Hiwiller
Psc 0h40'59" 9d58'4"

Heather Hoffman
Lyr 18h47'0" 38d13'6"

Heather Holler
Ori 5h25'40" 2d37'15"

Heather Holly Hojnacki
Sco 17h49'19" -31d19'35"

Heather "Hoochie" Clark
Mon 6h43'46" -1d35'33"

Heather Hooke
Cnc 9h4'50" 31d48'36"

Heather Hughes
And 0h7'59" 40d24'4"

Heather Iandoli
Crb 16h15'38" 34d41'38"

Heather Illene
Cyg 20h55'34" 49d53'50"

Heather J. Gafka
Cyg 21h50'31" 42d44'29"

Heather J. Kinkaid
Cyg 19h42'0" 40d59'48"

Heather & Jamie 08132005
Leo 11h44'39" 26d51'27"

Heather Jantz
Psc 1h16'9" 17d21'1"

Heather Jean
Ori 5h55'18" 11d44'2"

Heather Jean
Psc 0h57'32" 3d47'46"

Heather Jean Tucker 1187
Uma 12h46'26" 54d22'46"

Heather Jean Winne Das
And 0h26'31" 25d47'2"

Heather Jeannine
Sgr 19h2'3" -14d57'22"

Heather Jo
Lib 15h27'46" -8d23'47"

Heather Jo
Psc 1h21'18" 32d39'8"

Heather Jo
Cyg 20h31'24" 31d9'49"

Heather Jo Roberts
Vir 13h45'59" -1d53'8"

Heather & Joe
Cyg 20h18'43" 51d58'34"

Heather Johnston
Psc 1h10'32" 14d23'8"

Heather Jones
And 2h19'40" 38d54'3"

Heather Jones Melvin
Sco 16h5'41" -12d24'25"

Heather Kahre
Aqr 22h52'29" -5d36'45"

Heather & Kam Will Shine Forever
Umi 15h28'3" 89d3'33"

Heather Kathleen Morningstar
Psc 1h5'21" 32d25'49"

Heather Kathleen Rybak
Crb 15h53'31" 35d8'34"

Heather Kelly Hedrick
Uma 10h31'38" 50d27'8"

Heather "Kiddo" Parry
Lib 15h45'8" -25d23'49"

Heather Kishel
Per 3h36'40" 43d19'27"

Heather Klodzinski
Tau 5h39'38" 21d20'22"

Heather Koppelman
Lac 22h22'33" 46d33'55"

Heather Korine Calvano
Cap 20h13'13" -16d56'34"

Heather Kramer Bolivar
Leo 9h37'47" 28d32'57"

Heather Kranz
Dra 17h0'34" 57d24'23"

Heather Kresh
Cen 12h50'8" -41d23'52"

Heather Kristen Sparrow
Gem 6h52'46" 26d33'32"

Heather Kristin Shiels
Cas 0h24'25" 47d58'11"

Heather L. Bartko
And 2h20'34" 50d5'16"

Heather L Johnson
Lyn 8h23'41" 54d46'16"

Heather L. Walker
Lyn 7h29'51" 48d52'13"

Heather La Rue
Gem 6h32'53" 26d9'20"

Heather Laaksonen
Cap 21h41'27" -14d36'57"

Heather Ladd Major
Sco 17h56'21" -36d33'12"

Heather Langenbach
Uma 10h8'15" 45d23'23"

Heather Laspino
Cap 20h14'9" -13d51'47"

Heather Lauderdale
Tau 5h26'50" 25d7'31"

Heather Laura Derring
And 1h50'4" 37d37'33"

Heather Lauren
And 1h18'30" 37d47'10"

Heather LaVern
Ari 2h4'17" 14d8'36"

Heather Leach
Cas 23h0'16" 53d25'52"

Heather Leah Keehn
Cnc 8h36'6" 14d59'5"

Heather Leann York
Ari 2h52'19" 28d20'5"

Heather Lee Crider
Lyn 7h27'12" 52d41'37"

Heather Lee DeSimone
Sco 17h36'39" -33d20'40"

Heather Lee Dowler
Mon 6h26'16" -0d30'36"

Heather Lee Eller
Cnc 8h32'57" 24d55'21"

Heather Lee Keel
Cyg 21h1'20" 46d6'56"

Heather Lee Price
Cap 20h7'3" -8d48'55"

Heather Lee Rose
And 2h34'12" 50d55'51"

Heather Lee Schofield
Uma 12h9'24" 56d56'0"

Heather Lee Wisniewski
Gem 7h11'21" 22d5'57"

Heather LeeAnn Craft
Psc 0h21'56" 5d44'12"

Heather Leigh
Cam 6h22'53" 62d57'8"

Heather Leigh Bush
Tau 5h24'34" 27d53'2"

Heather Leigh Haley D.O.
And 0h32'35" 32d3'54"

Heather Leigh Horn
Vir 11h42'29" -4d57'15"

Heather Licciardi
Cam 4h0'45" 57d53'5"

Heather Lin Aikey
And 23h2'40" 41d49'22"

Heather Lohrman
Uma 9h57'22" 50d14'25"

Heather Loram's Star
Cas 0h38'10" 52d20'52"

Heather Lorraine Owens
And 1h45'5" 41d42'21"

Heather Louise Hunt
And 2h32'2" 49d31'15"

Heather Louise LuLu Wall
Vir 13h47'52" -8d14'36"

Heather Love
Sco 16h9'59" -12d11'0"

Heather "Love"
Uma 11h8'51" 30d17'59"

Heather Loves Drew
Leo 10h15'15" 26d41'29"

Heather Lyn
And 0h39'9" 33d13'58"

Heather Lyn Roberts
Ori 5h50'36" 6d22'58"

Heather Lyn Sarofian
Gem 7h2'23" 24d38'22"

Heather Lyn Taylor
Ari 2h10'43" 23d32'20"

Heather Lyn Wenning
Psc 23h58'5" 7d5'30"

Heather Lynn
Psc 0h33'42" 8d49'39"

Heather Lynn
Vir 13h7'33" 13d1'10"

Heather Lynn
Tau 5h27'10" 28d32'12"

Heather Lynn
Tau 4h28'30" 28d16'30"

Heather Lynn
Lmi 9h28'58" 35d46'23"

Heather Lynn
Dra 17h57'27" 64d12'46"

Heather Lynn
Uma 11h0'23" 71d41'25"

Heather Lynn
Cas 23h16'34" 59d33'30"

Heather Lynn Davis
Sco 17h36'43" -32d3'11"

Heather Lynn Desilets
Cnc 8h17'54" 15d42'1"

Heather Lynn Gillis
And 0h40'41" 22d5'11"

Heather Lynn Griffin
And 0h57'17" 22d35'21"

Heather Lynn Hall
Srp 18h5'13" -0d21'41"

Heather Lynn Hanchett
And 0h46'28" 25d55'34"

Heather Lynn Hartmann
And 1h36'41" 44d30'35"

Heather Lynn Hayden
Ari 2h50'23" 22d29'31"

Heather Lynn Hornung
Cam 4h46'4" 72d56'15"

Heather Lynn Hughes
Aqr 21h13'56" 0d16'49"

Heather Lynn Jankowski
Lib 15h3'23" -26d26'51"

Heather Lynn Kennedy
Aqr 22h11'12" -2d55'10"

Heather Lynn Kono
And 0h42'12" 42d23'36"

Heather Lynn Koudelka
And 0h31'27" 42d24'32"

Heather Lynn Krugler
Cnc 8h58'58" 24d5'0"

Heather Lynn Markelz
Tau 5h36'10" 27d37'38"

Heather Lynn McCarthy-Cohen
Uma 13h37'51" 55d56'55"

Heather Lynn Noyes
And 1h32'27" 39d38'33"

Heather Lynn Pakstis
Cnc 9h6'25" 27d40'13"

Heather Lynn Paul
Uma 11h15'54" 63d32'26"

Heather Lynn Pham
Cas 0h37'7" 54d36'42"

Heather Lynn Ritts
Cnc 8h32'4" 27d48'12"

Heather Lynn Tisdel
Ari 3h21'3" 27d38'9"

Heather Lynn Walter, L.M.T.
Leo 10h20'17" 22d34'30"

Heather Lynn Welch
Cap 21h24'43" -27d6'46"

Heather Lynn Wilson
Cyg 19h46'43" 30d52'29"

Heather Lynnae
And 0h47'2" 38d57'19"

Heather Lynne Speck
Ori 6h14'42" 18d11'58"

Heather M Blankenship
Psc 1h27'33" 16d50'0"

Heather M. Brown
And 0h22'58" 30d59'9"

Heather M Fields
Aqr 23h23'34" -19d28'52"

Heather M. Hathaway
Uma 8h34'59" 49d54'50"

Heather M. Rousse "Violet Fire"
Cas 22h59'42" 58d4'51"

Heather Mack
And 2h25'52" 49d2'23"

Heather MacLeod
Lib 14h35'5" -17d16'23"

Heather MacNaughton
And 0h7'12" 39d43'34"

Heather Madere
Gem 6h55'49" 25d56'28"

Heather Makala
Aql 19h52'8" 12d55'40"

Heather Margaret Jamieson
And 1h1'35" 42d24'52"

Heather Maria
Leo 9h27'30" 7d12'41"

Heather Maria Robles
And 0h45'4" 23d23'57"

Heather Mariah Tatiana Rosenberger
Cyg 21h43'53" 51d13'7"

Heather Marie
And 23h51'47" 45d37'33"

Heather Marie
Cas 0h34'6" 54d11'54"

Heather Marie
And 0h39'57" 41d5'44"

Heather Marie
And 0h19'44" 36d42'49"

Heather Marie
Ari 3h7'40" 25d52'42"

Heather Marie
Her 18h48'0" 20d19'27"

Heather Marie
Psc 23h46'36" 5d18'56"

Heather Marie
Vir 12h36'1" 11d26'54"

Heather Marie
Lib 15h17'56" -16d2'51"

Heather Marie
Sgr 18h59'34" -14d9'48"

Heather Marie
Cap 21h40'57" -9d52'32"

Heather Marie
Lyn 7h19'4" 54d42'14"

Heather Marie
Lib 15h41'22" -29d54'3"

Heather Marie
Cap 20h55'33" -24d27'24"

Heather Marie Cathcart
Mon 6h52'2" -0d27'22"

Heather Marie Diers
Cap 20h19'37" -23d35'5"

Heather Marie Fitch Aguirre
Peg 22h38'35" 20d40'2"

Heather Marie Frazier
Cnc 8h50'54" 28d1'2"

Heather Marie Hess
And 0h32'19" 22d32'2"

Heather Marie Hibbs
Ari 3h12'19" 25d6'31"

Heather Marie Holt
Uma 11h36'40" 39d7'2"

Heather Marie Hunt
Aqr 21h49'36" -6d37'47"

Heather Marie Hurd
Crb 15h29'28" 28d18'4"

Heather Marie Joly
Ori 6h7'57" 17d55'0"

Heather Marie Knox
Ari 3h18'39" 21d22'55"

Heather Marie Lewin
Cas 2h36'47" 68d58'1"

Heather Marie Madrigal
Psc 0h54'34" 11d46'40"

Heather Marie Madruga
Leo 10h32'35" 26d11'31"

Heather Marie Malcolm (Mamas)
Aqr 22h47'49" -5d2'31"

Heather Marie McBride
Cyg 21h42'17" 54d27'54"

Heather Marie Mitchell
Leo 9h48'17" 9d46'24"

Heather Marie Pelikan
And 1h2'13" 41d2'24"

Heather Marie Pinchbeck
Crb 15h54'24" 37d31'15"
Heather Marie Roberts
Gem 7h10'59" 23d32'58"
Heather Marie Rosiak
Cap 21h5'49" -16d53'35"
Heather Marie Savadge
Cas 1h37'17" 65d17'17"
Heather Marie Schneider
And 0h27'18" 25d51'17"
Heather Marie Sievert
Uma 10h18'30" 44d6'15"
Heather Mathieson
Cnc 8h15'27" 23d21'28"
Heather May 50
Cra 18h33'40" -42d10'29"
Heather May Campbell
Ori 5h40'6" -0d19'0"
Heather Mayer
Tau 4h33'38" 20d3'13"
HEATHER MCDONALD
And 23h24'34" 50d43'57"
Heather McGowan
Cap 21h29'24" -16d11'5"
Heather Megan Fitzpatrick
Vir 12h2'2" 6d58'6"
Heather Meglino
Cyg 19h32'29" 53d17'23"
Heather Melissa
Lyn 7h45'32" 47d30'49"
Heather Melissa Sutphin
Cnc 8h46'47" 17d45'34"
Heather Metcalf
Sco 16h53'21" -40d49'25"
Heather & Michael Fenton
Umi 15h34'34" 73d36'45"
Heather & Michael's Silver Star
Cyg 21h21'32" 31d37'49"
Heather Michel Smith
Lyn 8h0'56" 46d31'18"
Heather Michelle Albright
Psc 0h21'59" 7d55'10"
Heather Michelle Armas Cruz
Leo 10h30'42" 25d44'24"
Heather Michelle Bohn "Bubbles"
And 1h47'50" 38d39'53"
Heather Michelle May
Cas 23h27'42" 57d34'39"
Heather Michelle's Star
Uma 9h15'38" 51d20'43"
Heather Millard
Lyn 7h34'2" 50d31'44"
Heather Morgan Damon
And 23h42'11" 47d55'31"
Heather Morgan Johnson
And 23h1'7" 51d38'41"
Heather Morgana Ward
Cnc 8h36'8" 7d37'16"
Heather Morris
Lyn 7h45'12" 49d8'10"
Heather Morrow
And 0h37'42" 31d41'1"
Heather Muir
Cas 0h18'3" 55d2'53"
Heather Muir-Peak
Ori 5h26'20" 10d48'17"
Heather "Mum" Smith
Sgr 18h4'0" -27d21'1"
Heather My Love 10-16
And 1h32'53" 46d30'55"
Heather My Princess
Lib 14h53'17" -3d2'48"
Heather N Caetano
Dra 14h50'1" 58d12'59"
Heather N. Crager
Vir 13h17'28" -5d4'10"
Heather N Edwards
Per 3h3'31" 55d8'33"
Heather N Hourigan
Cas 23h20'22" 55d38'45"
Heather N. Russell
Psc 1h32'27" 21d28'49"
Heather N. Zbikowski
Lib 14h51'37" -1d38'48"
Heather Nahaku Kalei
Ori 5h15'51" -6d48'16"
Heather Newald Weinhaus
Hya 9h9'57" -0d29'8"
Heather Nichelle Montanez
Tau 5h8'9" 18d26'12"
Heather Nicholle Lajoie
Uma 8h47'51" 55d16'28"
Heather Nicole
Del 20h35'0" 13d14'30"
Heather Nicole
Leo 10h24'57" 12d33'10"
Heather Nicole
Del 20h39'28" 15d49'30"
Heather Nicole
Ari 2h26'26" 27d39'14"
Heather Nicole
Cyg 21h44'36" 43d50'40"
Heather Nicole Bootskies
Leo 9h22'32" 8d50'21"
Heather Nicole Bramlett
Tau 5h51'23" 13d21'25"
Heather Nicole Cox
Lib 14h51'50" -2d41'33"
Heather Nicole Davis
Vir 12h10'50" 13d13'9"
Heather Nicole Foster
Lib 15h39'46" -17d39'36"
Heather Nicole Hlavaty
Cnc 8h22'52" 10d15'32"
Heather Nicole McGee
Cyg 21h9'29" 30d1'32"
Heather Nicole Page
Uma 10h33'32" 40d29'31"
Heather Nicole Riggs
Cyg 21h17'37" 43d38'5"
Heather Nicole Ringler
Aqr 22h9'58" -19d59'38"
Heather Nicole Sharpe Warner
Aqr 22h37'50" 0d47'27"
Heather Nicole Stallings
Sgr 18h34'36" -31d26'54"
Heather Nicole Stenglein
Leo 10h24'53" 26d17'25"
Heather Nicole Whetstone
Cap 21h15'56" -19d15'17"
Heather Nicole Williams
Cap 21h29'8" -16d3'36"
Heather Nihart
Gem 7h19'58" 25d43'54"
Heather Niki
Psc 1h32'56" 12d0'1"
Heather Nikki
And 23h16'41" 52d38'27"
Heather Noel
Psc 1h13'19" 24d10'54"
Heather Noël Howard
Cap 20h15'13" -17d1'47"
Heather Noelle
Lib 14h53'1" -15d52'57"
Heather Noelle Taylor
And 2h25'27" 50d0'35"
Heather Novak
Cha 10h41'20" -79d7'2"
Heather Nowak
Vir 13h21'17" 11d38'6"
Heather O'Connell
Umi 13h31'42" 70d58'44"
Heather Oraios
Per 3h45'36" 33d54'45"
Heather Osborne
Uma 10h53'24" 46d40'4"
Heather Osella
Tau 5h10'48" 17d53'6"
Heather Our Beautiful Dancer
Leo 11h37'5" 11d8'28"
Heather Parker
Leo 9h35'32" 10d0'19"
Heather Parrish
Ari 2h31'55" 13d24'12"
Heather Patterson
And 1h4'1" 46d36'15"
Heather Payton
Cyg 20h59'55" 47d38'48"
Heather Perry Weafer
And 2h2'34" 42d22'0"
Heather Phelps
Cnc 9h2'45" 8d37'28"
Heather Phillips
Gem 6h32'4" 26d0'39"
Heather Piersiak
Cas 0h51'20" 57d16'11"
Heather Pollard
Lib 15h15'46" -19d16'17"
Heather Powell
Her 18h32'52" 15d27'49"
Heather Pressler
Cyg 21h35'19" 32d59'52"
Heather Price
And 23h6'50" 52d18'10"
Heather Price
Hya 9h2'19" 0d13'7"
Heather "Pumpkin" Young
And 0h51'27" 40d2'53"
Heather R. Garrison
Dra 14h36'47" 56d29'34"
Heather Rachelle Martin
And 1h54'49" 42d8'59"
Heather Rackley
Lib 15h0'41" -12d24'11"
Heather Rae
Sgr 19h51'8" -27d53'29"
Heather Rae
Lyn 7h13'6" 47d34'37"
Heather Rae
Del 20h37'10" 19d56'6"
Heather Rae Buchanan
Dra 19h23'12" 67d50'28"
Heather Rae Lambert
Lib 15h51'50" -19d38'51"
Heather Raley
Peg 21h29'55" 15d41'48"
Heather Raye Banks
Cnc 8h44'50" 26d10'22"
Heather Rebecca
Vir 13h19'7" 11d33'46"
Heather Renee
Leo 11h49'31" 25d3'36"
Heather Renee
Crb 16h11'5" 29d21'54"
Heather Reneé
Sco 17h42'40" -33d38'39"
Heather Renee Boone
Ari 2h8'47" 24d52'3"
Heather Renee Forever Shining
Cap 21h58'15" -20d38'20"
Heather Renee Johnson
Tau 4h11'2" 4d39'27"
Heather Renee Mackenroth
Lyr 18h42'42" 30d7'35"
Heather Renee Mills
Del 20h47'24" 14d55'31"
Heather Renee Misutka
Dra 17h44'59" 57d9'32"
Heather Renee Rowe
Psc 1h6'25" 10d35'51"
Heather Renee Walker
Lyn 9h9'34" 42d59'49"
Heather Renee Welschmeyer
And 23h33'57" 41d46'25"
Heather Renee Wilde
Cnc 8h51'13" 8d43'26"
Heather Renee Yang
Dra 18h43'8" 50d1'0"
Heather Richards
Gem 7h24'45" 25d24'25"
Heather Richelle Jenkins
And 0h19'16" 36d18'26"
Heather Richey
Ari 2h22'1" 25d52'4"
Heather Richter
Ori 5h56'13" 21d3'53"
Heather & Rick Forever
Cyg 20h7'27" 54d25'43"
Heather & Rick Reardon Family Star
Peg 23h16'48" 31d33'30"
Heather Ritenour
Leo 9h29'35" 6d28'44"
Heather Roberts
Cas 23h36'20" 55d39'36"
Heather Robertson Simmons
Tau 4h22'16" 28d55'28"
Heather Robinson
Lyr 18h40'34" 31d47'46"
Heather Rose Arruda
Cnc 8h51'57" 30d27'27"
Heather Rose Forney
Cnc 8h23'22" 19d17'23"
Heather Rose Greene
Sco 16h16'33" -14d1'50"
Heather Rose Kearney
Uma 10h11'16" 59d28'16"
Heather Rose Peach
Psc 0h14'55" 0d15'8"
Heather Ruble
And 1h15'44" 49d24'3"
Heather S. Drummond, My Love
Lyn 7h52'20" 53d25'32"
Heather S. O'Neill
Crb 15h34'31" 33d6'3"
Heather S. Robertson
Aqr 22h16'16" -9d28'3"
Heather Saitas
Psc 1h46'35" 24d10'25"
Heather Schmidt
Uma 13h8'45" 53d55'40"
Heather Schwartz
Ari 2h33'9" 25d15'46"
Heather Scott
And 0h14'19" 28d57'39"
Heather Seagraves
Aqr 22h6'16" -3d20'6"
Heather Sguigna
Uma 11h25'52" 60d10'43"
Heather Shaffer
Peg 22h44'36" 26d48'21"
Heather Sheldon
Vir 13h17'48" 7d17'46"
Heather Shepherd
And 23h20'52" 45d27'13"
Heather Sherman
Ari 2h11'24" 27d16'28"
Heather Shewmake
Psc 1h0'52" 4d22'57"
Heather Shimokawa
Lep 5h40'46" -11d4'59"
Heather Shirley Barker
Cen 14h13'13" -33d6'51"
Heather Sills
And 0h28'9" 27d10'31"
Heather Sinchak
Sco 17h34'8" -42d5'43"
Heather Siordia
And 23h23'52" 47d34'21"
Heather Smart *Freedom*
Ari 3h10'19" 29d12'14"
Heather Sophia
Cas 23h38'14" 58d29'2"
Heather Squirrell & Matthew Rogers
Cyg 20h52'14" 30d23'2"
Heather Stacie
Ari 2h40'23" 30d58'58"
Heather Star Duff
Tau 4h6'17" 14d11'10"
Heather Star-X
Per 2h56'5" 54d59'51"
Heather Statum
Gem 7h5'56" 14d3'35"
Heather - Steve
Lyr 18h46'57" 36d29'15"
Heather Stewart
And 1h33'15" 49d11'59"
Heather Sue Radcliffe
Uma 11h48'34" 44d50'2"
Heather Sutton
Ori 5h1'4" 4d3'23"
Heather Suzanne Carter
And 23h35'41" 43d54'13"
Heather Suzanne Wojtecki
Psc 0h35'35" 8d45'12"
Heather Sweed
And 23h25'46" 49d49'34"
Heather Tatham
And 0h26'37" 26d58'49"
Heather Terri Denise
Vir 12h53'14" 1d32'0"
Heather Terry
Vir 13h27'7" 5d9'48"
Heather the beautiful
And 0h38'31" 45d0'31"
Heather The Feather
Gem 6h49'9" 13d4'55"
Heather ~ The Light in My Life
Cyg 19h29'30" 57d45'58"
Heather Torrez
Mon 6h35'48" 0d27'13"
Heather Trahan
Aqr 20h59'59" 0d26'35"
Heather Trautman
Umi 14h15'23" 69d33'48"
Heather Turner
Uma 11h43'8" 52d9'58"
Heather Tyler
Com 12h17'58" 27d51'27"
Heather Ulrey
Peg 23h25'9" 28d28'37"
Heather Vestfal
Cyg 20h9'44" 44d5'4"
Heather Viera
Leo 10h31'3" 26d9'15"
Heather Vohsing
Cas 0h25'11" 56d14'29"
Heather Wagner
Cyg 20h15'20" 35d46'19"
Heather Waugh's Star
Tau 5h3'9" 19d57'29"
Heather Weeks' Anniversary Star
Leo 10h59'37" 11d18'21"
Heather White, The Love Of My Life
Ari 1h52'42" 22d9'23"
Heather & Will 4Ever
Sge 20h4'35" 18d57'35"
Heather Williams ~ The Shining Star
Gem 7h17'43" 20d46'33"
Heather Wingfield
Aur 5h48'8" 45d19'6"
Heather Wise
And 23h2'2" 48d28'41"
Heather Wonderly 07
Lyn 6h55'7" 51d7'56"
Heather Woodward Gavin
Cam 5h30'20" 67d29'0"
Heather Yantes
Cyg 20h39'38" 50d8'38"
Heather Yeranoohie Bond
And 2h31'7" 48d2'54"
Heather Yuhasz
And 1h0'51" 35d43'33"
Heather Zehr
Sco 16h5'4" -13d50'35"
Heather Zielinski
Psc 1h13'16" 3d51'58"
Heather Zoma
And 1h12'34" 46d12'0"
Heather "Har" Potter
Uma 11h9'26" 42d17'3"
Heather, My love and life.
Sco 16h10'25" -12d52'4"
Heather, Taylor, and Ellie
Uma 9h38'57" 55d43'14"
Heather690000
Cas 0h17'17" 65d34'56"
Heather725
Leo 10h16'27" 8d57'31"
Heather88
Tau 5h33'36" 26d22'52"
Heatheran
Cir 15h6'48" -60d8'54"
HeatherAnne L'abbee
Sgr 18h2'47" -27d57'14"
Heather-Austin
Leo 10h24'2" 7d36'46"
HeatherB
Cap 21h50'16" -14d29'31"
HeatherHunni
Aqr 20h45'22" 2d5'38"
HeatherLovesRobinLovesBethsLove....
Cyg 20h0'41" 38d16'47"
Heatherly
Cnc 7h57'41" 11d24'15"
heatherly m stankeys star
Per 3h17'44" 42d30'47"
Heather-Lyn
And 1h20'29" 47d15'46"
Heather-Mae Trace
Cyg 22h1'12" 53d25'2"
Heather-Marie Jetté
Ari 2h3'50" 24d2'23"
HeatherMichelleWellington
Cap 20h37'52" -13d23'6"
Heather-O
Ori 5h35'29" -1d36'50"
Heather's Angel
Peg 23h26'3" 27d58'41"
Heather's Angel
And 2h27'33" 49d27'20"
Heather's Beauty
Cet 1h14'47" -0d3'3"
Heather's Birthday Star
Cru 12h18'8" -56d28'13"
Heather's Christmas Heaven
Cyg 19h39'38" 29d12'33"
Heather's Eye
Cap 21h40'31" -18d53'27"
Heather's Fairy Star
And 2h29'3" 45d37'36"
Heather's Gift
Psc 1h27'45" 10d56'18"
Heather's Glimmer
Leo 11h42'7" 14d30'17"
Heather's Heart
Uma 11h20'29" 68d11'0"
Heather's Hope
Cnc 8h48'29" 30d52'5"
Heather's Insatiable Beauty
Her 18h6'22" 25d5'9"
Heather's Love
Ori 5h5'1" 10d37'27"
Heather's Love
Aqr 22h36'50" -0d24'10"
Heather's My Heart, REGARDLESS
Vir 14h43'33" -0d9'15"
Heather's Pop Star
Leo 9h39'55" 12d8'12"
Heather's Shining Star
Uma 10h5'54" 45d37'30"
Heather's Smile
Tau 4h13'7" 26d52'32"
Heather's Smile
Lib 14h48'35" -9d27'30"
Heather's Star
Sgr 19h42'43" -14d8'20"
Heather's Star
Lib 14h50'0" -3d36'28"
Heather's star
Leo 9h58'7" 24d48'35"
Heather's Star
Psc 1h20'21" 23d46'55"
Heather's Star
Leo 11h40'37" 12d27'19"
Heather's Star
Ari 2h52'11" 17d56'43"
Heather's Star
Ori 5h31'43" 10d57'43"
Heather's Star
Ori 4h45'26" 0d29'18"
Heathers Star
Gem 7h43'20" 32d49'59"
Heather's star light in the heavens
Dra 19h28'13" 61d53'3"
Heather's "Stardust"
Uma 12h6'27" 49d44'37"
Heather's Twinkling Dream
And 0h37'53" 39d18'27"
Heather's Wonderland
Ori 5h46'15" 7d55'53"
Heather's Zephyr
Psc 23h48'26" 0d11'44"
HeathJigga
Cnc 8h28'7" 32d31'51"
Heath's Sweet Kaleena Marie
Gem 7h9'51" 25d54'48"
Heavan
Leo 11h15'36" 15d54'59"
Heaven 500
Cnc 8h51'38" 9d32'22"
Heaven and Earth
Cnc 8h46'21" 31d12'42"
Heaven Elisabeth
Leo 11h38'26" 25d28'27"
Heaven & George
Umi 15h1'34" 68d24'31"
Heaven Hill Distilleries, Inc.
Cam 7h0'51" 64d35'39"
Heaven Lee Ratliff
Lyr 18h45'53" 38d59'8"
Heaven Leigh Ram
Cyg 19h44'51" 37d36'16"
Heaven Mykel Richards
Lyr 19h12'35" 38d47'33"
Heaven Nadine Reinert
Sgr 19h22'43" -27d32'44"
Heaven on Earth- Christian and Sara
Dra 18h42'43" 72d27'49"
Heaven Rain
Gem 6h43'55" 20d32'43"
Heaven Rosenberg
Umi 17h22'41" 76d53'6"
Heaven Sent
Cep 21h5'6" 56d54'1"
Heaven Sent
Leo 11h49'29" 24d50'35"
Heaven Sent
Lyn 8h2'32" 35d12'12"
Heaven sent Fraser
Ari 1h56'25" 14d18'40"
HeavenLeah
Cnc 9h8'12" 31d57'8"
heavenLEY
Tau 4h8'37" 23d39'42"
Heavenly Amanda Matrecano
Cam 13h0'39" 78d34'26"
Heavenly Angel
And 0h58'30" 39d18'53"
Heavenly Bill
Psc 0h40'10" 18d13'7"
Heavenly Body : KJH 111
Ori 5h14'32" -0d58'2"
Heavenly Cal Bear
Uma 9h28'9" 49d14'57"
Heavenly Cathie, My Shining Star
Col 5h42'26" -42d2'54"
Heavenly Dancer
And 1h5'4" 37d31'58"
Heavenly Diane
Lyn 8h55'12" 41d9'2"
Heavenly Glory
Cyg 21h51'3" 41d33'46"
Heavenly Heather
Sco 16h59'31" -38d17'16"
Heavenly Heather Onties
Sgr 19h39'9" -14d53'33"
Heavenly Hindmarch
Ori 5h54'56" 7d55'47"
Heavenly Holcombs
Lib 14h49'16" -15d58'51"
Heavenly Hope
Crb 16h23'55" 28d43'18"
Heavenly Ivy
Psc 0h40'43" 20d39'44"
Heavenly Judy
Aqr 22h25'40" -1d52'25"
Heavenly Katie O'Grady
Sco 17h53'12" -35d55'49"
Heavenly Laura Lee
Mon 6h54'34" -0d20'6"
Heavenly Light of Mary Ann Smith
Cen 13h56'27" -60d54'13"
Heavenly Light Of Richard Rice
Sco 17h47'30" -41d18'58"
Heavenly Liza
Cas 1h17'58" 69d7'22"
Heavenly Manicotti #50
Tau 4h17'6" 27d48'8"
Heavenly Pamela Davis
Vir 12h12'22" -2d24'51"
Heavenly Pfer
Boo 14h26'1" 52d41'26"
Heavenly Randa
Lib 15h54'51" -17d24'21"
Heavenly Reminder
Lyr 18h42'13" 37d41'27"
Heavenly Spruijties
Cep 2h26'20" 81d59'21"
Heavenly Star of Grace
Ari 2h46'34" 12d5'31"
Heavenly Viau
Umi 16h22'3" 76d49'42"
HeavenlyMom121460
And 0h49'24" 43d42'2"
Heaven's Angel
Lyr 19h2'43" 31d55'46"
Heavens Angel Douglas R. Kielty
Sco 17h53'29" -40d26'35"
Heaven's Big
Cap 20h38'59" -17d59'42"
Heaven's Eyes
Lyn 8h55'44" 41d48'13"
Heaven's Gate
Cru 12h6'47" -58d25'50"
Heaven's Light (Elsie I. Iozzino)
Vir 13h59'20" -17d2'49"
Heaven's Open Door
Crb 16h0'43" 33d5'36"
Heaven's Reach
Cen 12h58'39" -30d37'55"
Heavens Slice
Uma 9h48'2" 66d14'46"
Heavens to Betty Mortimer
Lyn 8h4'9" 44d59'28"
Heavens-Heather
Sgr 18h26'1" -32d1'56"
HEAVON 4 TWO
Sco 16h10'55" -10d20'55"
Heba
And 23h6'36" 40d32'0"
HEBA YEHIA
Lyn 8h20'6" 42d57'22"
Hebron Lutheran Church, Blairsville
And 2h6'0" 42d37'40"
Hebron Lutheran Church, Madison, VA
Uma 9h30'53" 54d11'39"
Hebron T Hunt Jr
Ori 6h9'33" 7d46'2"
Hecke, Joachim
Uma 10h27'15" 65d40'50"
Heckerman Clan Constellation
Uma 13h53'16" 52d23'35"
Hector
Per 4h16'36" 41d55'23"
Hector
Cru 12h9'1" -61d15'57"
Hector & Amber
Uma 11h24'39" 52d59'43"
Hector Bryce Alexander Family Star
Her 17h26'54" 37d45'11"
Hector David Nieves
Per 3h20'8" 44d45'32"
Hector & Denna Fuentes Eternal Love
Cyg 20h18'43" 55d59'5"
Hector Enrique Sanz Morales
Aqr 22h6'50" -19d7'20"
Hector Gabriel
Cnc 8h2'4" 21d12'19"
Hector Isaac Martinez
Per 3h28'50" 47d19'19"
Hector & Katrina the skies are ours
Cnc 9h6'49" 28d2'14"
Hector L Rios Jr
Lyn 7h38'45" 37d33'22"
Hector Myinthan Jeyakaran
Cep 21h13'36" 61d49'31"
Hector Ponce
Cnc 8h56'38" 13d40'7"
Hector Riveroll
Aqr 23h4'38" -16d45'41"
Hector Vazquez
Psc 0h30'35" 9d3'46"
HectorApril
Leo 11h30'15" 17d11'37"
Hedda
And 1h46'56" 38d45'43"
Hedgehog's Laughter
Cyg 20h9'30" 36d22'0"
Hedi Scherer-Kamber
Aql 19h9'35" 6d1'10"
HEDRICK
Cyg 21h50'11" 38d28'20"
Hedrik
Sco 17h33'54" -41d47'56"
HEDWARD
Her 18h43'37" 20d24'46"
Hedy Schwartz
Umi 15h27'50" 74d31'58"
Hedy Wolpa
Uma 11h11'22" 36d7'32"
Hee Kyung Kim 2004
And 0h25'39" 27d54'57"
Heebee
Ori 5h41'43" -3d30'45"
HeeJunLUVSeungJoo
Dra 16h7'11" 64d27'33"
Heel-Miller
Cra 18h49'27" -39d51'11"
Heer aur Ranjha
Cnc 8h29'3" 15d28'54"
Hef
Cnc 8h44'46" 6d46'30"
Heffalump
Cas 1h29'25" 65d21'43"
Heffy (Heather Elizabeth)
Psc 1h36'42" 26d34'3"
Hegler, Andreas
Uma 9h8'7" 60d43'33"
Hehewuti
Sco 16h20'24" -12d16'47"
Hehrlein, Griedrich Wilhelm
Uma 10h20'35" 53d19'31"
Heiarii Amaru
Gem 7h8'56" 33d38'5"
Heide
And 0h40'48" 40d21'20"
Heide Kreutzer Boutell
Ari 3h18'8" 19d15'43"
Heidee Love
Vir 11h43'6" 5d39'41"
Heidemarie Krumpschmid
Tau 4h41'14" 5d0'49"
Heide-Regine
Uma 9h29'18" 44d43'3"
Heide's Hope
Lib 14h31'54" -10d6'2"
Heidi
Lib 15h25'10" -19d6'20"
Heidi
Vir 11h50'25" -4d35'55"
Heidi
Umi 16h53'26" 78d13'36"
Heidi
Vir 12h38'32" -5d20'4"
Heidi
Cas 1h27'58" 65d5'16"
Heidi
Cas 23h42'47" 55d58'8"
Heidi
Dra 14h36'52" 58d21'6"
Heidi
Sco 17h55'6" -37d46'6"
Heidi
Psc 0h26'44" 6d30'56"
Heidi
Sco 17h56'23" -37d24'1"
Heidi
Sgr 19h14'1" -24d10'34"
Heidi
Psc 0h53'53" 25d26'4"
Heidi
And 0h41'57" 40d54'30"
Heidi
And 0h44'31" 41d28'57"
Heidi
Cyg 20h15'26" 36d56'5"
Heidi
Uma 11h51'30" 31d30'11"
Heidi
Cyg 21h43'23" 43d58'12"
Heidi
Boo 14h10'14" 50d17'13"
Heidi
Uma 10h25'46" 51d46'53"
Heidi
And 23h19'53" 48d31'24"
Heidi
Gem 7h23'35" 14d8'19"
Heidi
Lmi 10h47'48" 26d46'22"

Heidi 020579
Peg 22h15'52" 11d56'23"
Heidi 071005
Aqr 22h13'53" 0d5'44"
Heidi 5/28/74
And 23h28'23" 42d19'49"
Heidi A Engel
Cyg 21h39'13" 38d46'11"
Heidi A. Schwarz
Aur 5h38'38" 33d10'4"
Heidi Aaron Rouse
Ferguson
Leo 9h24'52" 26d33'20"
Heidi Ann Potter
Vir 13h18'2" 2d14'2"
Heidi Ann Shuler Gordon
MQF12
Lib 14h22'33" -11d56'25"
Heidi Ann Wilhelm
Per 4h36'42" 39d40'26"
Heidi Annette Moore
Vir 13h27'49" -1d7'23"
Heidi bell
Aqr 23h5'47" -9d14'41"
Heidi Berkovitz "Wishing
Star"
Ori 5h35'47" 1d57'32"
Heidi Beth Drymer
Boo 14h32'6" 43d22'21"
Heidi Blake ~ June 28,
1964
Cnc 8h20'14" 6d48'9"
Heidi Boenisch
Ori 6h3'42" 21d20'46"
Heidi Brandenburg
And 1h33'33" 48d35'14"
Heidi Brianna Lee
Leo 11h41'1" 24d31'10"
Heidi Bruce
Ari 2h50'27" 15d5'37"
Heidi ; Busle 72
Mon 6h26'43" -4d15'56"
Heidi Casey
Sgr 18h26'1" -32d37'41"
Heidi Cauthon
Mon 6h47'23" 7d23'11"
Heidi Chrowl
Uma 9h4'1" 65d5'55"
Heidi Clare
Cnc 8h20'48" 32d8'24"
Heidi Deutsch
Gem 6h42'54" 26d59'33"
Heidi Dobson
Cam 3h34'10" 66d27'21"
Heidi Eileen
Psc 0h18'10" 2d46'7"
Heidi Franklin
Umi 15h0'54" 77d15'23"
Heidi Gail Collins
Leo 9h31'17" 29d29'1"
Heidi Grunzig
Uma 10h31'32" 51d26'29"
Heidi Gwerder
Cyg 20h10'24" 48d48'33"
Heidi Hager Holden
And 1h10'29" 37d2'2"
Heidi Haliniak
Sco 17h27'14" -37d31'6"
Heidi Hamilton
Lib 15h31'5" -7d1'46"
Heidi & Hans-Peter
Her 15h54'27" 42d51'51"
Heidi 'Hege' Valkenburg
Cru 12h1'22" -62d12'50"
Heidi Helen Hunter
Psc 0h57'4" 31d38'20"
Heidi Herrera
Tau 4h43'17" 28d32'48"
Heidi Jadwiga Aurora
Ruminski Moore
Uma 13h36'4" 62d20'39"
Heidi Janelle
And 23h21'6" 42d32'7"
Heidi Jean Graf
Cap 21h30'5" -9d24'38"
Heidi Jewell Bruster
Peg 22h35'6" 23d8'45"
Heidi Jo
And 0h56'26" 35d38'24"
Heidi Jo Wittig
Aqr 23h3'8" -15d32'10"
Heidi Johnson
Com 12h49'52" 13d23'52"
Heidi Jones Bucher
Tau 5h49'44" 16d11'11"
Heidi Kindorf
And 0h38'16" 37d26'20"
Heidi Klementovich
Cas 1h21'21" 64d6'3"
Heidi Kotansky
And 0h13'21" 43d16'17"
Heidi Kreitz
Uma 8h20'39" 66d54'56"
Heidi Lanette
Her 17h14'13" 15d44'3"
Heidi Lawler
Ori 6h13'25" 5d48'58"
Heidi Leanna
Com 12h13'57" 31d53'34"
Heidi Lee
Lyn 6h19'48" 56d36'29"
Heidi Lee Estrada
Peg 21h42'21" 24d11'47"
Heidi Leigh
Aqr 21h35'34" -0d28'37"

Heidi Leigh Gallagher
Ari 3h29'10" 20d38'20"
Heidi Lippuner
Cas 0h20'12" 53d44'35"
Heidi Lorraine Mangum
Cyg 20h27'3" 39d9'5"
Heidi Louise
Peg 22h2'45" 8d5'25"
Heidi Louise Grainger
Cyg 20h57'28" 46d45'22"
Heidi Louise Mitchell
Cnc 8h17'43" 22d21'32"
Heidi Lucy Burke (Boo)
And 1h20'41" 40d56'9"
Heidi Lyn Fire
Uma 11h5'10" 67d27'12"
Heidi Lynn Small
Sco 16h3'37" -21d26'11"
Heidi Lynne
Gem 6h35'24" 21d37'58"
Heidi M. Blough
Lyn 9h0'14" 39d3'26"
Heidi M. Schmidt
Mon 6h43'43" 5d57'34"
Heidi Mae Reisenauer
Ori 6h9'55" 6d55'26"
Heidi Mae Vredenburg
Sco 16h17'34" -28d35'33"
Heidi Manoleros
Cas 23h4'40" 55d57'9"
Heidi Marie
Ari 3h5'1" 16d0'51"
Heidi Marie
Leo 9h27'1" 25d59'19"
Heidi Marie
And 23h46'2" 40d57'14"
Heidi Marie Dawn
Ori 6h15'14" 6d39'32"
Heidi Marie Ertl
Cnv 13h40'15" 32d47'53"
Heidi Marie Parks
Vir 12h9'49" 3d59'28"
Heidi Marie Shaw
And 0h47'42" 35d26'56"
Heidi & Markus Hendel
Cyg 19h44'26" 35d19'11"
Heidi Mattmann
Gem 7h30'3" 20d24'51"
Heidi mein hellster Stern
des Universums
Her 18h13'11" 26d55'4"
Heidi & Michael
Leo 11h13'58" 7d54'22"
Heidi Middleton
And 0h27'59" 45d27'32"
Heidi M.L.
Cam 6h0'47" 62d9'34"
Heidi Morgan Prayon
Crb 15h45'10" 26d30'22"
Heidi Münzberg
Uma 9h7'14" 58d52'50"
Heidi "My Princess" Waye
Lyr 18h33'53" 33d0'25"
Heidi Nan Cohen
Ori 6h17'22" 15d22'7"
Heidi New
Gem 6h13'1" 22d43'25"
Heidi Norine Anderson
Tau 4h3'31" 22d26'49"
Heidi Orcholski: Green
Crayon 17
Uma 9h15'33" 54d54'48"
Heidi Orlean Hooker
Lyn 8h30'33" 57d21'17"
Heidi Pecoraro
Leo 9h52'29" 28d57'31"
Heidi & Peter
Cyg 20h46'19" 31d3'34"
Heidi Pierce
Mon 6h53'7" -1d35'9"
Heidi Rae
Psc 1h9'59" 28d35'12"
Heidi Rae Mashaw
Ori 5h55'12" 7d5'43"
Heidi Rauch
Ori 6h25'19" 10d33'18"
Heidi Rebecca Green
Cap 21h17'37" -27d7'4"
Heidi Renae Jones
And 0h23'10" 31d15'11"
Heidi Rodriguez
Cas 23h34'51" 54d53'24"
Heidi Schmeelk
Cas 1h39'34" 65d50'57"
Heidi Schneider-Schweizer
Uma 13h21'8" 62d49'38"
Heidi Smeder
Lmi 10h3'12" 39d56'37"
Heidi "Spidey" Delbridge
Leo 9h25'32" 28d56'59"
Heidi Starling
And 0h41'32" 38d6'24"
Heidi Steffanie Olmos
Ori 6h22'47" 14d17'54"
Heidi Steward - Loving
Mother & Wife
Cyg 21h37'15" 52d13'6"
Heidi & Tara. I Love You
Mom.
Peg 23h43'36" 24d24'14"
Heidi The Best
Tau 5h46'1" 15d59'30"
Heidi. The Love of my Life!
And 0h17'12" 44d2'16"

heidi und chris
Cnc 8h12'2" 7d47'3"
Heidi - Unsere Liebste, fuer
immer.
Uma 12h41'3" 59d40'28"
Heidi & Wayne Forever
Cyg 19h33'49" 29d24'37"
Heidi Weddle
Leo 10h53'1" 6d44'21"
Heidi Windish
Uma 13h36'20" 59d25'45"
Heidi Winterkorn Craemer
Cas 1h34'35" 61d57'24"
Heidi Wolfensberger/Forke
Uma 11h37'59" 35d58'59"
Heidi Wunderlin
Vul 19h24'16" 26d8'54"
Heidi´s Venus
Ori 6h15'49" 10d34'30"
HeidiDoug7805
Lib 14h52'25" -3d22'32"
HeidiKen
Sge 19h52'10" 18d50'56"
Heidiliz
Cnv 12h27'37" 37d30'38"
Heidina
Uma 9h28'43" 52d11'20"
heidiogandreas
Uma 9h30'41" 62d8'29"
Heidi's Beauty
And 0h33'13" 29d5'12"
Heidi's Countryside
Luncheonette
Vir 13h43'37" -20d3'40"
Heidi`s Kuschelbär
Ori 6h2'52" 9d27'49"
Heidi's Light
And 23h0'18" 45d12'32"
Heidi's Penguin
Sco 15h54'14" -21d54'55"
Heidi's Shining Star
Ori 6h16'35" 10d4'47"
Heidi's Star
Sgr 19h58'14" -30d15'57"
Heidi's Valentine Star
Lib 15h16'9" -11d2'1"
Heidi's Wispy Dream
Crb 16h14'20" 38d40'24"
Heiditta
Aur 5h29'53" 36d43'12"
Heido Karge
Uma 8h45'42" 63d24'10"
Heidy
Dra 17h48'13" 69d11'38"
Heidy
Ori 5h3'51" 5d23'6"
Heidy and Steven's Star
Tau 5h48'37" 22d24'9"
Heidy Brenneis
Boo 14h26'58" 41d34'20"
Heidy Carrasquilla Peña
Ari 2h42'6" 27d20'3"
Heidy's Diamond Jubilee
Star
Ari 2h38'38" 22d12'6"
Heike
Gem 6h56'16" 13d16'14"
Heike
Gem 7h51'39" 17d45'29"
Heike
Gem 8h6'5" 31d44'10"
Heike
Uma 14h6'24" 56d7'13"
Heike
Vir 12h26'41" -6d24'31"
Heike Buchmann-Meineth
Uma 9h29'14" 42d9'33"
Heike Marita
Aqr 22h41'3" -2d36'34"
Heike Schmid
Uma 10h25'13" 48d2'34"
Heike Tillmann
Uma 9h6'18" 60d54'41"
Heike & Uli
Ori 5h59'59" 22d50'16"
Heike und Peter
Uma 8h45'58" 58d0'44"
Heike Würl 04.05.1979
Uma 10h21'31" 44d8'37"
Heike, my Stäärn
Cyg 21h23'9" 49d0'40"
Heikes Stern
Uma 10h15'40" 43d0'34"
Heikki Blowsmen Gruner
Uma 11h47'48" 50d39'42"
Heiko
Uma 8h20'27" 67d43'25"
Heiko Bruns
Ori 6h12'4" 7d38'54"
Heiko Fialski
Uma 11h36'12" 35d37'13"
Heiko & Martina forever in
love
Uma 13h35'15" 60d34'21"
Heiko Ramspeck
Uma 8h23'30" 63d0'27"
Heilbryel
Cas 0h20'16" 57d10'52"
Heineken Défi Millénium
2005
Tau 5h45'23" 13d13'33"
Heinemann, Gitte
Uma 10h7'20" 54d27'57"
Heinemeier, Ulrich
Uma 12h14'3" 61d12'33"

Heinen, Friedrich
Uma 10h53'9" 43d58'11"
Heiner Drooff
Uma 8h59'17" 49d41'55"
Heinke, Helmut
Vir 12h53'45" 2d0'18"
HEINO
Sgr 17h52'57" -29d18'5"
Heino Leibersperger
Uma 8h38'2" 54d0'9"
Heino L'Estocq
Uma 9h48'29" 44d41'14"
Heino Winneburg
Uma 10h20'10" 40d4'55"
Heinrich Bjoern
Ori 5h22'44" 6d28'13"
Heinrich Grünewald
Uma 8h44'10" 67d51'3"
Heinrich Jakob Lamby
Uma 8h56'40" 60d28'29"
Heinrich Krüer
Uma 10h29'36" 48d1'59"
Heinrich Lenze
Her 18h41'26" 25d5'26"
Heinrich Schulte
Uma 9h32'34" 58d10'7"
Heinrich the Bull
Tau 4h23'52" 1d30'43"
Heinsch, Kristin & Pascal
Uma 10h8'27" 48d40'11"
Heintz
Peg 22h44'23" 10d2'6"
Heinz
Uma 11h38'9" 28d55'18"
Heinz
Her 16h17'44" 19d2'38"
Heinz
Uma 9h16'7" 47d17'2"
heinz
Per 3h40'28" 45d13'58"
Heinz Becker
Ori 4h45'47" 5d12'17"
Heinz Bertels
Uma 8h12'12" 65d47'21"
Heinz Bode
Uma 8h16'15" 68d47'50"
Heinz Bräuer
Uma 8h11'34" 67d25'51"
Heinz Butz
Uma 9h23'9" 47d41'34"
Heinz Colonius
Ori 5h11'51" -6d17'33"
Heinz Fischer
Uma 9h34'38" 59d23'5"
Heinz Fourberg
Uma 11h48'20" 38d28'48"
Heinz Gygax
Her 17h8'25" 19d21'53"
Heinz Harfst
Uma 11h12'36" 64d52'25"
Heinz Kellenberg, RB
Flawil
Cep 21h59'43" 72d39'45"
Heinz Kummer
Uma 9h37'37" 50d51'46"
Heinz Lorsbach
Uma 13h34'4" 59d56'38"
Heinz+Marie-Theres
Cas 1h59'44" 77d28'0"
Heinz Martschin
Tau 5h12'25" 25d0'28"
HEINZ Neuenschwander
Lyn 7h53'59" 34d15'31"
Heinz Pritzke
Ori 5h54'37" 7d18'5"
Heinz Robert Zurbrügg
Uma 9h24'13" 42d3'4"
Heinz Ross
Ori 6h24'13" 10d23'19"
Heinz Spickermann
Ori 5h47'21" 3d12'10"
Heinz Zilles
Ori 5h14'11" 8d43'56"
Heinz-Günter Seebach
Uma 14h11'57" 57d39'57"
Heinzinger, Katja
Ori 6h9'44" 9d9'38"
Heinz-Jürgen Otto
Uma 8h59'20" 47d35'9"
Heinz-Uwe Dettling
Ori 5h16'53" 0d48'58"
Heiratsstern Mandy &
Heiko
Uma 8h20'9" 61d33'41"
Heise, Daniel Lennart
Uma 10h56'50" 71d3'41"
Heiss, Gretel&Manfred
Uma 10h7'18" 58d33'26"
Heitland, Steffen
Uma 9h39'0" 67d20'37"
Heitmann, Ruth
Ori 5h0'49" 13d58'59"
Heitner-Weissmann
Uma 10h31'1" 64d37'54"
Heizmann, Andreas
Cap 20h38'24" -22d14'31"
Heizmann, Andreas
Uma 6h24'49" 20d16'48"
HeJa
Ori 6h13'33" 8d31'54"
Hekela 030406
Gem 6h45'36" 20d28'3"
Hela a Jara
Peg 22h40'2" 18d8'49"

Helaina Marie Battle
Uma 9h23'45" 61d13'47"
Helbing, Nicolas
Uma 11h11'56" 56d29'46"
Helbing, Wolfgang
Uma 10h7'40" 43d52'17"
Heldga & Rudy 50 Years of
Love
Cyg 19h46'1" 54d47'52"
Helen
Cyg 19h49'3" 52d44'52"
Helen
Cas 1h30'1" 63d32'4"
Helen
Cas 1h15'1" 62d19'4"
Helen
Cas 23h27'26" 54d59'7"
Helen
Uma 8h19'48" 71d38'6"
Helen
Cas 2h16'9" 74d23'24"
Helen
Sco 16h7'52" -8d54'25"
Helen
Ori 5h57'2" -2d14'3"
Helen
Cam 7h3'29" 76d42'49"
Helen
Vir 13h10'6" -2d15'25"
Helen
Pho 1h11'43" -47d30'47"
Helen
Uma 9h22'14" 42d27'35"
Helen
Lyn 8h27'51" 39d30'42"
Helen
And 2h8'32" 42d43'23"
Helen
Cnc 8h47'18" 32d18'44"
Helen
Tri 2h3'27" 34d24'38"
Helen
Cyg 19h50'14" 32d24'37"
Helen
Cyg 20h20'51" 30d45'37"
Helen
Lyr 18h49'23" 31d15'46"
Helen
Cas 0h19'50" 51d41'3"
Helen
And 23h34'49" 41d44'38"
Helen
Cas 0h23'1" 59d26'49"
Helen
And 0h36'3" 29d41'23"
Helen
Cnc 9h20'19" 19d37'35"
Helen
Tau 5h9'28" 18d19'57"
Helen
Ori 5h32'24" 5d17'55"
Helen A. Haberkorn
Cas 23h56'23" 59d11'55"
Helen A. O'Connor
Cas 0h40'59" 62d25'35"
Helen A. Sciorra
Cap 21h37'51" -15d37'43"
Helen Adams
Psc 0h29'28" 9d47'32"
Helen Alderman
Lib 15h6'29" -9d13'51"
Helen and Charlie Roy
Uma 11h49'54" 61d7'34"
Helen and Harold Max
Crawford
Vir 13h54'1" -17d2'44"
Helen and Jamie Small
And 0h49'41" 40d35'0"
Helen and Jason Frost
Cyg 19h55'40" 38d50'1"
HELEN AND JOHN
Psc 23h43'56" 6d58'12"
Helen and John Zalasky
Leo 9h44'31" 28d36'37"
Helen and Maurice Burke
Uma 12h14'39" 54d50'16"
Helen and Peter Motko
Psc 1h17'52" 31d25'44"
Helen and Taz
Cyg 20h14'2" 35d38'33"
Helen Ann Cox
Tau 5h53'2" 25d7'47"
Helen Ann Ishiyah
Psc 0h52'14" 16d14'17"
Helen Anne Hunter
Eri 4h35'44" -0d3'38"
Helen Apgar
Peg 21h25'53" 17d43'6"
Helen Ashton
And 1h54'40" 42d13'57"
Helen Assetto
Gem 8h6'19" 30d43'40"
Helen Avila
Psc 1h20'37" 10d1'49"
Helen Axelrood
Uma 10h26'31" 61d23'47"
Helen Babino
Uma 8h26'51" 69d9'44"
Helen & Baby Morgan
Umi 14h6'23" 67d59'10"
Helen Barnes
And 0h48'5" 30d53'13"
Helen Barnes & Michael
Lets Dance
Ori 4h49'58" 12d48'48"

Helen Bell Gordon 'Nana'
Uma 14h23'45" 57d15'14"
Helen Beyond Boundries
Lib 15h13'9" -3d53'16"
Helen Binnix
And 0h43'47" 41d26'11"
Helen Boehm
Lyr 18h29'10" 33d39'37"
Helen Bransby
And 0h4'1" 39d44'22"
Helen Breitenbach
Leo 9h40'58" 26d47'30"
Helen Briggs
Cyg 20h31'57" 45d28'50"
Helen Briyanda Jocelyn
Lib 15h36'3" -5d24'0"
Helen Brown
And 0h56'52" 35d6'46"
Helen Brown "1916-2006"
Cas 23h40'7" 53d46'54"
Helen Brown Hollis
And 1h35'2" 47d29'55"
Helen Buchman
Cas 23h54'50" 56d16'0"
Helen Butler
Lib 15h25'18" -11d47'10"
Helen Byus
Cap 21h32'49" -19d26'32"
Helen C. Duckham
Cas 23h37'52" 56d14'14"
Helen C. Hege
And 0h36'43" 43d13'11"
Helen C. Lazenby
Cnc 8h35'21" 29d15'26"
Helen C. Liu
Lyn 6h43'8" 57d44'12"
Helen Caecilia Chatigny
Crb 15h42'40" 33d23'16"
Helen Camisa Ardizzone
Lyn 8h1'56" 38d48'52"
Helen Cantor
Uma 11h52'1" 49d15'5"
Helen Carol Paul
Sco 17h56'31" -37d28'21"
Helen Case Korb "9-6-
2005"
Lyn 8h1'50" 55d52'13"
Helen Casey Collins
And 2h21'52" 45d49'50"
Helen Checki
Cam 6h45'43" 64d39'17"
Helen Cieslewicz
Crb 15h40'6" 36d23'40"
Helen Coble
Lyr 18h22'59" 46d9'23"
Helen Cole
Crb 16h24'1" 27d23'45"
Helen Collins
Col 5h13'24" -28d16'22"
Helen Corbett - 26 April
1920
Cru 12h2'4" -61d54'18"
Helen Craig
Mon 7h36'56" -0d20'54"
Helen Cutting Muntean "H"
Tau 4h31'36" 20d59'37"
Helen D. Hegerty
Umi 13h22'8" 73d23'5"
Helen D. Hernandez
Crb 16h23'12" 37d39'11"
Helen D. Wedel
Uma 8h53'10" 49d30'45"
Helen Dalby
Umi 14h46'44" 72d33'12"
Helen Dean
Peg 23h46'20" 11d5'52"
Helen Debra Eubanks
Leo 11h14'7" 15d25'41"
Helen DeLorenzo
Lyn 8h33'51" 44d1'0"
Helen Desmond
Ari 3h9'50" 25d59'4"
Helen Diana Chirigotis
Her 17h16'29" 48d49'33"
Helen Dimmock
Cas 1h2'0" 63d35'11"
Helen Dimulkas & Jeremy
Brenman
Uma 11h6'26" 72d7'4"
Helen Duffy Jones
Cap 21h40'12" -10d35'46"
Helen E. Brayton
Cyg 21h48'3" 43d53'28"
Helen E. Campbell, VMD
Sco 16h6'24" -10d28'54"
Helen E. Gutierrez
Cap 20h36'44" -19d21'31"
Helen E. Hansen 4/10/34-
2/25/07
Ari 1h56'16" 20d17'24"
Helen E Kristoff
Cas 2h25'51" 67d22'52"
Helen E Strand
Uma 11h55'4" 29d46'16"
Helen Elisabeth Blair Stuart
And 1h18'32" 42d50'46"
Helen Elise
Ari 3h13'6" 19d30'19"
Helen Elise Galloway
Aqr 20h50'45" -13d4'8"
Helen Elizabeth Albaneze
Fenton
Tau 5h44'36" 16d55'16"

Helen Elizabeth Evans
Anniversary
Vir 12h34'55" 2d52'19"
Helen Elizabeth Holmes
Cas 23h11'2" 54d51'14"
Helen Elizabeth Lamont
Aqr 22h14'24" -8d6'44"
Helen Elizabeth Russell
Cap 20h19'31" -16d8'38"
Helen Elizabeth Sharp
McDonald
Aqr 22h28'55" 0d50'29"
Helen & Ellsworth Piper
Aqr 22h37'55" -1d41'35"
Helen & Evan's Star
Uma 11h7'2" 30d43'27"
Helen Faye Rosenblum
Cnc 9h7'33" 7d34'53"
Helen Fenker Proteau
And 2h36'33" 49d53'22"
Helen Fenwick
Cyg 21h33'39" 48d46'49"
Helen Filipou - 8 February
1953
Cru 12h27'19" -60d47'24"
Helen Frances Weindorf
Comer
Tau 5h46'38" 22d19'23"
Helen G Austin
Lib 15h11'28" -21d29'22"
Helen G. Greenstein
And 2h15'12" 46d21'21"
Helen G. Nohr
Psc 23h40'10" 5d50'23"
Helen Gabrielle Smith
Lyn 6h59'21" 49d24'41"
Helen Galantich Richman
Cyg 20h9'17" 32d25'19"
Helen Gaughan
Cnc 8h6'56" 15d15'35"
Helen Gene Martin
Cas 0h15'46" 51d46'54"
Helen Germaine Tucker
Psc 1h28'39" 7d26'47"
Helen Gianferrara
Uma 11h26'30" 49d50'49"
Helen Gilbride
Ori 5h14'13" -9d12'57"
Helen Goldstein
Lyr 18h37'17" 39d46'13"
Helen Goodwin
Cra 18h1'22" -37d1'30"
Helen Gouveia
Crb 15h58'10" 35d7'30"
Helen Grace Kuhar
Aqr 23h2'42" -11d26'30"
Helen Grace Megson
And 2h23'13" 48d27'27"
Helen Gray Driver
Lib 14h53'55" -4d14'42"
Helen Guastaferri
Leo 9h26'3" 24d46'10"
Helen H. O'Brien
Uma 12h31'36" 53d2'15"
Helen H. Okamoto
Sco 16h5'46" -21d5'20"
Helen Haberkorn
Psc 1h4'27" 31d33'30"
Helen Hastings Cockrell
Psc 23h56'17" 7d33'23"
Helen Heuser Harwood
Uma 12h35'15" 53d18'57"
Helen Hicks
Uma 11h12'56" 44d27'19"
Helen "Honey" McVey
Uma 8h14'16" 70d5'43"
Helen Hope Hall Green
Vir 12h46'31" -9d48'6"
Helen Hunter
Ori 5h10'24" 15d16'31"
Helen I Baker
Aqr 22h8'55" -15d25'57"
Helen I. Jones-Bagg-Bortz
Lyn 6h41'49" 50d47'44"
Helen Isabelle Stander
Cas 0h32'58" 57d13'22"
Helen Isobel
And 1h23'22" 40d52'24"
Helen J. Fonner
Cas 23h24'0" 53d23'58"
Helen J Hunyady
Lib 15h7'5" -0d29'0"
Helen J. Trosper
Uma 10h43'0" 46d29'29"
Helen Jane Batchelor
Umi 13h56'51" 70d6'41"
Helen Jane Freeland
Psc 23h30'44" 4d19'3"
Helen Jane Wood
And 23h49'34" 45d9'10"
Helen Jean Cooke
And 1h8'49" 39d13'13"
Helen Jean McDonald
Cam 4h31'33" 53d40'46"
Helen Jean Reed-Poland
Crb 15h43'29" 32d4'37"
Helen Jeanette Long
Cas 1h32'43" 60d7'50"
Helen Jeffcoat
Aqr 22h42'33" 0d39'33"
Helen Jodzio
Uma 10h46'27" 49d11'10"
Helen & Joe Ippolito 85710
Cyg 21h25'44" 34d6'41"

Helen & Joe"Windy Acres" 11/28/1957
Cyg 21h24'52" 48d6'32"
Helen Johns BunStar
And 0h15'12" 34d44'41"
Helen Joyce Hunsaker
Lmi 9h23'45" 35d43'36"
Helen Joyce Kuck
Cas 23h17'0" 55d53'29"
Helen K. Hendricks
Sco 16h11'32" -12d10'8"
Helen Kate Ruane
Lyn 7h50'32" 38d41'22"
Helen Kay Daffern
And 2h5'57" 42d11'48"
Helen Keene
Cap 20h24'1" -20d23'20"
Helen Keller
Cas 0h59'16" 66d28'14"
Helen Kelly
Cyg 20h39'47" 38d26'4"
Helen & Ken Galvin
Uma 10h29'12" 51d39'31"
Helen Kiedaisch
Cap 20h56'42" -15d14'45"
Helen Krawczyk Matzkevich
Uma 8h34'56" 51d20'59"
Helen L. Filer
Aqr 23h14'28" -13d41'14"
Helen L. Hopkins "Aunt Honey"
Per 3h55'2" 44d4'1"
Helen L. Kulsha
Gem 7h49'58" 33d8'45"
Helen L. McBride
Cas 1h5'21" 63d42'8"
Helen L Pacyna Star
Vir 13h46'36" -2d6'38"
Helen L. Schwarzer
Cam 3h56'52" 53d23'55"
Helen LaRue
Crb 16h17'54" 35d0'11"
Helen Latta
Lyr 19h14'16" 27d11'14"
Helen Lee
Vir 14h11'46" -15d10'4"
Helen Lee
Uma 8h40'19" 68d26'16"
Helen Lehmenkuler
Leo 11h27'51" 14d33'3"
Helen Lessing
Eri 4h9'17" -12d26'29"
Helen Livelys' Granny Star
Umi 14h31'23" 71d2'43"
Helen Lockwood Van Zanen
And 0h42'57" 25d10'37"
Helen Lorraine Ingham
Psc 0h29'41" 4d25'33"
Helen Louise
Lyn 7h42'2" 38d54'56"
Helen Louise Gold (Seppala)
Uma 10h58'37" 37d2'22"
Helen Louise Morgan
And 2h35'17" 48d46'42"
Helen Loves Ivanh <30.06.05
Ori 5h36'57" -2d25'10"
Helen Loves Mark Forever & Always
Uma 12h53'9" 58d58'14"
Helen Lyons
Psc 1h21'48" 16d7'31"
Helen M. Corcoran
Cyg 19h45'48" 52d51'20"
Helen M. Galt
Cas 1h30'41" 61d18'50"
Helen Mae Anderson
Ari 2h16'20" 24d58'9"
Helen Mae Eckhart
Cnc 9h7'6" 30d22'27"
Helen Margaret Kenny 12/23/03
Uma 11h33'35" 50d18'27"
Helen Marian Williams
Lyn 8h2'24" 42d3'8"
Helen Marie
Gem 6h35'35" 15d8'36"
Helen Marie
Cnc 7h58'46" 12d46'49"
Helen Marie Boesenecker
Psc 1h6'51" 32d7'7"
Helen Marie Iglehart Kelley
Aqr 21h47'28" -7d44'59"
Helen Marie Long
Psc 0h48'58" 11d1'6"
Helen Marie Mason
And 1h36'2" 49d50'11"
Helen Marie Neiman Memorial Star
Cas 0h49'25" 60d4'54"
Helen + Markus
Ori 6h12'51" 21d2'40"
Helen Martha Larson
Lyr 18h57'23" 26d9'41"
Helen Martin
Crb 16h4'42" 36d6'35"
Helen Mary
Tau 4h17'29" 14d11'50"
Helen Mary Warner
Lib 15h5'2" -16d51'13"
Helen Mattison Wyatt
Cas 22h57'38" 57d20'38"
Helen Maxine Spittle
Cyg 19h57'19" 34d56'49"
Helen May Marens
Aqr 21h12'16" -8d0'38"
Helen May Reidelbach
Cyg 21h53'50" 51d33'3"
Helen McFarland
Crb 15h38'35" 32d57'33"
Helen Meyers
Psc 1h4'44" 10d31'19"
Helen & Michael Rettmuller
Cyg 21h40'24" 55d21'25"
Helen Miles
Cas 23h14'58" 58d34'51"
Helen Mitchell Dodderidge
Leo 10h37'0" 21d20'21"
Helen Moore
Uma 9h50'45" 55d18'28"
Helen Moore
Cas 2h13'52" 69d34'54"
Helen Morley
Vir 13h30'40" -13d42'54"
Helen Mother Grandmother & Friend
Cas 0h39'21" 57d39'51"
Helen Munday
Tau 3h52'9" 20d26'19"
Helen Murray
Tau 5h44'21" 26d20'53"
Helen (My Mom)
Sgr 18h48'16" -16d41'19"
Helen "Nana" Urban 7/3/35-8/8/04
Ori 5h30'45" 3d4'18"
Helen Natonski
Cyg 19h52'21" 32d47'45"
Helen Nevins
Uma 12h35'49" 55d19'32"
Helen Ng
Uma 9h44'13" 52d41'16"
Helen Nguyen
Lyn 7h5'50" 54d59'21"
Helen Nicole
Aqr 21h5'36" -8d43'13"
Helen Norton
Lmi 9h56'9" 35d14'50"
Helen Nowacki & Caregivers
Cyg 20h56'37" 46d28'7"
Helen O
Gem 6h10'52" 24d28'31"
Helen of our Hearts
Uma 10h45'2" 47d37'0"
Helen Olga Turton
Cas 1h13'11" 52d59'26"
Helen Olivia
Cyg 21h44'15" 36d55'2"
Helen Olson
And 0h22'34" 32d7'39"
Helen - Our Mom & Grandmom
Vir 12h6'55" 8d12'5"
Helen Oxner
Leo 9h23'6" 19d29'55"
Helen P
Aqr 20h55'11" -12d57'38"
Helen Pelt Mosure Elder
Cam 5h47'16" 56d35'21"
Helen Peralta
Cyg 20h6'10" 45d50'3"
Helen Phaneuf
Uma 9h53'57" 51d49'55"
Helen Popney Hamilton Memorial Star
Sco 16h7'25" -9d54'29"
Helen Posanke Anderson
Cas 0h37'11" 56d52'7"
Helen "Princess" Horvath
And 0h48'4" 46d27'25"
Helen Puralewski
Sco 17h44'54" -36d27'25"
Helen Quanstrum
Lib 15h14'34" -17d19'1"
Helen R. Yoesle
Aur 5h54'7" 52d1'9"
Helen Rachel Yeadon
Dra 16h59'13" 62d56'50"
Helen Randmäe
Vir 12h4'1" 1d0'57"
Helen Raybell
Cas 23h27'46" 53d5'32"
Helen Rayfield
Cas 0h12'16" 61d18'59"
Helen Rendlesham Burg
Cas 23h54'10" 57d47'14"
Helen Rhoda Alford
Cas 0h22'5" 53d53'23"
Helen & Robert
Vir 12h21'26" -6d51'44"
Helen Rook
Tau 4h14'36" 13d0'35"
Helen Rose Brotz
And 23h38'51" 49d3'3"
Helen Roslund
Lib 16h1'8" -16d23'15"
Helen Ross
Lyr 18h34'53" 27d49'59"
Helen Rossi
Crb 15h47'12" 34d4'48"
Helen Rouhana
Aqr 22h33'22" -1d45'22"
Helen & Rudy Rutenber
Sco 17h57'11" -30d13'31"
Helen Russell
Cas 0h28'46" 63d52'36"
Helen Ruth Rawson "Gan Gan" 1927
Gem 6h56'10" 19d58'52"
Helen Ruth Stone
Uma 9h14'4" 72d20'12"
Helen S. Caruso
Psc 22h52'36" 6d18'50"
Helen Sackler-Rosenberg
Aqr 21h13'48" 1d3'31"
Helen Sadeh
Cap 21h14'3" -25d22'46"
Helen Sadoff, Jeff's baby-doll star
Crb 15h31'59" 32d48'25"
Helen Sanker Landseadel
Sgr 18h15'10" -24d15'7"
Helen Scharschan
Crb 15h23'22" 28d36'31"
Helen & Scott Cunningham
Cyg 20h5'1" 35d3'1"
Helen Sealock
Cas 0h33'39" 55d24'45"
Helen Sealock
Cas 0h32'28" 54d4'45"
Helen Shelby Mitchell
Tau 5h44'17" 21d18'6"
Helen Sheppard
And 2h32'44" 37d45'16"
Helen - SHMILY
Gem 7h16'47" 32d38'26"
Helen Shoemaker
Ari 2h34'54" 18d47'19"
Helen Shoemaker
Uma 9h53'30" 59d39'23"
Helen Skirptunas
Lyn 7h58'39" 37d42'25"
Helen Sobal
Lyr 19h14'22" 31d29'12"
Helen Solensky
Gem 7h1'0" 33d36'32"
Helen Spano D'Amato *100th*
Vir 14h6'16" -11d44'11"
Helen Star
Peg 22h51'46" 21d45'0"
Helen Starbuck Betts
Crb 15h54'37" 33d27'17"
Helen Stayskal
Cap 20h21'53" -13d18'9"
Helen Stiles
Lib 15h42'22" -6d26'41"
Helen Stough
Uma 11h8'55" 55d26'52"
Helen Stuckey Barrett
Vir 12h47'25" 4d30'10"
Helen Stumpf-Marinelli
Cnc 9h14'22" 7d45'33"
Helen "Sue" Tweedie
Lyn 7h47'55" 35d59'3"
Helen T. Benjamin
Cas 23h48'21" 59d1'27"
Helen T. Fitzgerald
Uma 14h14'32" 57d51'26"
Helen T. McCaskill
Lyn 7h20'57" 56d43'41"
Helen Taccini
Crb 16h4'47" 35d23'43"
Helen Tamandong Germenis
Cas 0h24'38" 59d44'31"
Helen the Beautiful
Psc 1h16'32" 4d52'8"
Helen - The Greatest Mum In The World
Cas 1h18'3" 56d31'48"
Helen Todd
Cap 20h27'6" -17d25'34"
Helen Tohill's Star
And 23h28'26" 41d41'34"
Helen / Tony - Inseparable
Gem 6h47'25" 12d4'17"
Helen Tosney BA(Hons)
And 23h32'2" 36d21'2"
Helen Tyminski
Cyg 21h34'28" 39d37'27"
Helen V. Danahy
Cas 0h38'17" 52d47'13"
Helen Varble-Rautenbach
Ari 2h8'25" 23d26'35"
Helen & Victor Long
Cyg 20h25'12" 50d8'40"
Helen Victoria Hurst
Vir 12h45'17" 0d43'42"
Helen Viola Kiel
Uma 14h20'37" 54d58'4"
Helen Virginia Lee
And 0h54'20" 43d31'24"
Helen Vladessa Kolsevich
Leo 9h39'34" 26d56'26"
Helen Walker
Cyg 21h15'2" 47d40'30"
Helen Wallder
Cru 12h13'9" -60d55'40"
Helen Warrington Roberts Perry
Cam 4h13'4" 56d24'19"
Helen Wilkinson Allston
Uma 9h59'39" 68d38'6"
Helen Wilson
Cnc 8h10'24" 25d34'2"
Helen Winters Birthday Star
Cyg 20h38'8" 40d10'50"
Helen "Woodling" Stone
Leo 9h32'53" 7d2'20"
Helen Yaccarino
Cas 1h22'31" 61d0'20"
Helen Yvonne Humphris - Mum in a Million
Cas 23h20'29" 56d38'43"
Helen Z "Toots" Bunker
Vir 13h49'4" -12d20'19"
Helen Zhang
Ari 2h44'1" 30d20'41"
Helen Zuroff
Cas 1h42'59" 63d38'33"
Helen, the best mom in the world!
Cap 21h10'7" -15d30'30"
Helen, Victor & John Surman
Uma 9h7'44" 49d52'30"
HELENA
Uma 9h36'25" 48d49'13"
Helena
Cyg 20h2'43" 49d58'7"
Helena
Cas 2h46'4" 57d38'31"
Helena
And 23h39'55" 47d22'18"
Helena
Ari 2h12'14" 13d33'12"
Helena
Com 12h42'41" 13d27'11"
Helena
And 0h7'16" 30d2'44"
Helena
Lib 15h6'11" -16d41'46"
Helena
Cap 20h26'15" -13d28'59"
Helena
Cas 23h47'41" 53d52'35"
Helena
Uma 12h41'41" 60d51'19"
Helena
Uma 10h28'20" 61d23'16"
HELENA
Cra 18h29'54" -38d48'23"
Helena Adele - Light Shining Bright
Cru 12h9'37" -63d36'9"
Helena Alexandra Thomas
Lyn 7h28'44" 58d4'50"
Helena and Carl Yeaman
Cyg 20h0'42" 33d5'54"
Helena Anne Noel
Ori 6h20'53" 10d46'31"
Helena Audrey Anne Regensburg
And 23h37'25" 42d15'10"
Helena Ava
And 0h21'4" 42d26'2"
Helena Beale Wright Morris
Ari 3h11'13" 27d31'33"
Helena Bella
Uma 11h18'58" 29d56'21"
Helena Brooke Nelsen
Tau 3h52'49" 24d5'37"
Helena "Cuci" Ungaretti
Dor 4h19'52" -49d44'33"
Hèlèna Dupuy
Leo 11h6'51" -6d22'57"
Helena F. Maloney
Uma 10h27'51" 39d43'16"
Helena Felecita Hay
Ari 3h22'51" 27d51'53"
Helena Hanson
Cam 4h43'23" 55d45'46"
Helena Heurtel
Lyr 18h40'49" 34d5'23"
Helena Hickey
Gem 7h2'5" 20d56'52"
Helena Jean Walsh
Lib 14h35'16" -19d11'49"
Helena Lisiecka
Cru 12h37'31" -57d23'4"
Helena Maria Montambo
Cas 1h26'4" 61d3'58"
Helena Moon Clyne
Cas 1h31'19" 64d0'57"
Helena Nicole Dallas
Lyn 7h0'36" 46d9'59"
Helena Peltier
Cas 0h10'16" 53d22'4"
Helena Ru
And 23h7'21" 51d55'39"
Helena Ruby Frances Gonzalez
Leo 11h2'46" 9d20'23"
Helena Sias Witte
Per 2h29'16" 54d17'11"
Helena Slavsky
Cap 21h55'15" -9d46'14"
Helena Sundlun House
Psc 1h7'20" 28d58'33"
Helena & Ulf Hägg
Uma 9h5'35" 57d53'15"
Helena's Guiding Light
Cru 12h1'39" -62d2'32"
Helena's Light
Lib 14h58'7" -15d7'33"
Helena's Star
Lib 15h16'7" -9d50'2"
HelenClare Fahey
Cas 1h41'3" 60d19'12"
Hélène
Cap 20h36'29" -13d41'12"
Hélène
Her 17h47'24" 21d2'48"
Helene
Sgr 18h37'50" -35d19'24"
Hélène
Sgr 17h57'40" -22d40'39"
Hélène
Del 20h45'12" 19d11'43"
Helene
Cas 1h28'30" 58d6'51"
Helene
And 0h42'33" 41d53'25"
Helene Alexis toe Laer's 1st B-Day
Ori 5h30'8" 6d59'30"
Helene and Schmed Suspended
Per 3h41'22" 52d40'55"
Helene Bari O'Brien
Aqr 22h32'45" 1d40'37"
Helene Belmont
Cas 0h4'58" 56d47'44"
Helene Bergs
Sgr 18h53'7" -28d3'4"
Helene Bloom
Tau 5h13'19" 27d10'52"
Helene D. Thornton
And 0h28'45" 40d4'59"
Helene Daniel, Our Hero and Star
Vir 12h32'13" 12d10'29"
Hélène Defever
Cen 14h1'20" -44d10'3"
Hélène Diego Retailleau
Cas 0h27'2" 61d26'15"
Helene Eisenbud
Sgr 18h20'25" -32d52'58"
Helene En Or
Dra 17h42'6" 52d35'39"
Hélène et Lucie
Psc 0h55'57" 25d49'39"
Helene et Marc Eisenberg
Cas 1h11'5" 49d1'10"
Helene Forever
Vul 20h29'57" 26d24'13"
Helene Goldfine
Lyn 7h50'42" 36d28'55"
Helene Juliette Mickey Shambrook
Ari 3h25'28" 29d5'11"
Hélène Ladriere
Psc 0h39'14" 9d9'57"
Helene Marcus
Vir 14h37'40" 6d5'39"
Helene Margaret McLeod Cable
Uma 9h21'6" 44d23'14"
Helene & Mark
Ori 5h35'18" -1d27'13"
Hélène Maux - Granny
Sgr 18h15'45" -20d1'29"
Helene Mayer
Dra 16h16'42" 60d55'24"
Helene Paglaiccetti
Tri 2h28'21" 31d26'32"
Hélène Paquette
And 1h8'19" 42d2'49"
Hélène Quin
And 1h26'42" 49d45'32"
Helene Ronchetti
Tau 4h13'0" 27d16'7"
Helene Rosenberg Sacharow
Del 20h39'6" 15d18'56"
Helene S. Abramowitz, M.S.
Psc 1h8'24" 9d56'13"
Helene Saphire
Lib 14h54'5" -0d32'58"
Helene Shulkin
Ori 5h7'15" 8d59'3"
Helene T. Conlon
Leo 11h3'18" -0d17'22"
Hélène Urbanovsky
Psc 1h9'8" 23d12'43"
Helene & Viktoria Heiter
Uma 8h48'17" 69d23'3"
Helene Windish Fink
Uma 11h45'44" 33d12'3"
Helene Zukanovich
Lib 15h32'39" -4d41'22"
Helene-Kirk 143
Uma 8h13'11" 70d9'57"
Helene's Star
Sco 16h5'53" -10d0'3"
Heleni Sema
Psc 1h25'2" 17d51'55"
HelenJoseph
Sgr 18h23'33" -22d52'51"
Helenket B
Leo 11h24'17" 27d37'37"
Helen's Anniversary Star
Cyg 19h40'25" 30d49'33"
Helen's Bobby-Robert G. Rayve
Per 3h43'2" 44d51'27"
Helen's Halo
Tau 5h58'59" 25d7'48"
Helen's Shining Octogendrius
Cas 0h45'31" 63d48'45"
Helen's Star
Cyg 21h34'21" 54d47'7"
Helen's Star
Psc 1h22'40" 31d24'59"
Helen's Star
And 23h19'12" 41d38'4"
Helen's Star
Cas 0h41'15" 56d53'22"
Helenski
Aqr 22h19'21" -3d0'5"
HELENSKYE
And 22h58'8" 47d37'29"
Helga
Cas 0h49'42" 56d19'8"
Helga
Cyg 19h34'5" 54d3'36"
Helga
Uma 8h48'14" 61d4'55"
Helga Elisabeth Josefine Newald
Uma 8h34'43" 48d15'52"
Helga Güldner
Uma 8h42'0" 53d28'2"
Helga Helena Beekman
Leo 11h9'38" 4d56'36"
Helga Klara Luise Rodgers
Crb 16h18'27" 32d5'54"
Helga Kromp-Kolb
Cas 23h16'21" 55d6'16"
Helga Lehmann
Ori 5h59'10" -1d57'20"
Helga Leitner
Uma 10h22'20" 66d46'22"
Helga Lopresti
Cam 7h0'39" 69d3'29"
Helga * Muggi * Puppe * 14*05*1943
Uma 9h59'52" 72d49'8"
Helga Ott
Uma 12h2'26" 35d17'45"
Helga + Peter Simnacher "ZUHAUSE"
Ori 5h18'42" -4d52'58"
Helga Reinhard Berger
Uma 11h34'50" 29d0'14"
Helga & Richard Hoffmann
Uma 9h46'9" 53d33'3"
Helga Splinter
Uma 8h41'12" 63d50'29"
Helge Gabriel
Uma 10h34'46" 47d49'33"
Heli
Uma 14h2'29" 51d59'30"
Heli
Ori 4h59'29" 10d11'26"
Heli
Umi 16h13'32" 79d18'54"
Heli Alouf
Uma 11h36'37" 31d12'57"
Heli Constance
Del 20h24'47" 16d20'46"
H'elia
Vir 12h14'43" 8d47'50"
Hèlia
Lyr 18h20'26" 38d5'10"
Hèlia sempre meu amor
Tau 4h6'3" 25d46'46"
Helios
Lyr 19h10'6" 46d12'8"
Helioseleneos
Cam 4h15'45" 72d2'8"
Helix
And 0h32'35" 44d5'45"
Hella Jera
Tau 4h24'35" 23d18'32"
Helle Christensen of Denmark
Tau 3h34'36" 27d6'12"
Hellen Cobie
Cru 12h25'15" -58d56'53"
Hellen DevilsAngel
And 0h47'15" 42d7'0"
Hellen Elizabeth Rogers
Peg 21h31'19" 11d29'14"
Hellen Murray Jackson
Cyg 20h3'17" 39d52'49"
Hellen Pereira Menezes
Dra 9h44'57" 74d22'23"
HELLENNA
Cap 20h14'31" -22d15'18"
Heller Farms
Lyn 6h58'49" 52d15'19"
Heller, Katrin
Uma 13h12'37" 60d26'59"
Hellevi Marjatta Kaartinen Miller
Umi 16h51'22" 83d48'8"
Hellmann, Roger
Ori 5h36'21" 10d2'59"
hello. hello. hello.
Her 16h41'58" 44d41'30"
Hello!Pikachu102205
Cnc 8h26'31" 21d0'45"
Hello Sunshine
Cma 7h24'30" -29d14'56"
Helmig, Klaus
Cap 20h14'30" -24d5'20"
Helmut
Ori 6h17'5" 13d15'58"
Helmut A. Elsdörfer
Uma 10h4'21" 50d40'39"
Helmut Brieden
Uma 9h44'47" 72d47'52"
Helmut Elsner
Sco 16h9'33" -10d24'14"
Helmut & Inge Rieder
Lyn 7h17'15" 56d44'15"
Helmut Köhler
Uma 8h42'41" 48d16'4"
Helmut Mak
Uma 9h54'11" 41d51'57"
Helmut Muchow
Uma 10h5'58" 50d8'23"
Helmut Müller
Uma 9h52'42" 42d12'28"
Helmut Müller
Uma 10h57'26" 66d41'51"
Helmut Roth
Uma 9h20'28" 47d52'12"
Helmut Theysohn
Cyg 20h58'42" 46d58'10"
Helmut Wallenhorst
Uma 10h16'58" 42d16'23"
Helmut Willmann
Ori 5h35'43" 13d26'21"
Helmut Winkelströter
Uma 9h17'15" 53d53'1"
Helmut-Günter Massier
Uma 9h46'4" 61d18'17"
Helmuth Kugi
Her 16h27'30" 45d53'52"
Helmuth T. Billy, MD
Oph 17h21'35" 7d31'14"
Helmut's happy family
Uma 14h26'33" 57d44'10"
HELMUTT
Lib 15h19'3" -7d25'32"
Héloïse Fleischel
Gem 6h42'24" 24d21'52"
Heloueno2
Uma 11h29'52" 63d1'21"
Helping Find A Cure
Umi 15h45'6" 76d44'19"
Hely
Cnc 9h3'47" 26d5'32"
Hely Perttu
Uma 10h43'22" 56d10'44"
Helyn Bautista Tuquero
Vir 13h42'56" -5d3'51"
Hema
Ori 6h9'52" 20d44'1"
Hema Pyarilal
Tau 5h49'15" 26d17'45"
Hemali
Uma 8h55'24" 63d45'39"
Hemali & Premal
Uma 11h2'28" 52d18'55"
Hembrees Familius
Psc 1h39'47" 22d8'24"
Hemmalaya Tann
Uma 10h28'30" 62d11'32"
Hemotoggy
Uma 11h43'24" 37d8'27"
Hena
And 23h55'49" 34d56'44"
Hena Dodoy Drew
Sge 19h36'20" 16d53'0"
Hena Garcia
Cma 7h8'48" -25d41'58"
Henderson
Her 17h17'43" 34d4'6"
Hendricks Kendall Kim Tyson Tanner
Uma 13h47'31" 55d59'47"
Hendrik
Uma 9h31'19" 48d49'7"
Hendrik Gödje
Aqr 21h59'33" 1d18'40"
Hendrik Lassmann
Uma 9h40'43" 57d44'53"
Hendrix
Sgr 19h31'52" -29d19'29"
Hendy
Boo 14h16'52" 38d10'21"
Heneka
Cyg 20h50'10" 33d24'9"
Hengst, Ingrid
Uma 10h30'23" 41d32'52"
Heni és Péter Csillaga 07.07.07.
Cas 0h9'7" 54d37'36"
Henic302
Gem 6h59'30" 13d36'45"
Henley
Uma 11h7'31" 55d11'2"
HENMARUGILBERTE
Cyg 21h24'7" 35d47'16"
Henna
Ori 5h35'54" 3d6'33"
Henna (My Little Green Horn)
And 23h34'12" 35d42'18"
Henna Rashid
Vir 12h6'4" 7d10'11"
Hennepin
Cyg 19h30'3" 30d5'17"
Henner's Star
Per 3h8'1" 47d48'11"
Hennessey
And 1h19'2" 49d43'43"
Hennessy
Psc 0h36'16" 14d4'40"
Henning Christiansen
Uma 10h55'43" 57d49'42"
Henning Reith
Ori 6h23'5" 10d55'41"
HenningSuehrig
Her 16h15'35" 46d7'22"
Henry Arnold
Ori 6h17'26" 7d15'34"
HENNY
Uma 12h4'6" 29d25'12"
Henny AKA Andrew Henderson
Cep 20h44'4" 61d42'10"

Henny & Lars Brix
Uma 9h46'24" 65d28'27"
Henri & Annemi
Eri 4h25'18" -1d44'36"
Henri Becker
Uma 10h41'7" 58d10'57"
Henri Jude Phillips
Lib 15h16'53" -15d58'43"
Henri & Lydia
Psc 1h42'4" 21d49'29"
Henri Meiller
Her 17h19'34" 35d41'5"
Henri Ox Jean
Sgr 19h11'20" -21d56'39"
Henri Pierre de Villiers
Cep 21h58'51" 60d22'37"
Henri Salvador
Cnc 8h1'51" 27d36'48"
Henri Van Hove
Aqr 23h17'24" -8d46'7"
Henrie's Twenty-five
Cyg 19h38'27" 54d36'23"
Henriett Ercsei - "Puppa"
Pup 7h33'28" -48d45'18"
Henriett Sidney Sandor
Uma 10h8'28" 57d46'16"
Henrietta Bruno
Cam 4h8'58" 73d4'6"
Henrietta Greenlaw
Uma 10h30'18" 58d52'37"
Henrietta Jensen
Cam 12h8'48" 76d27'26"
Henrietta Margeret May-Keating
Psc 1h55'57" 6d3'57"
Henrietta Stacey Sands
Cra 18h4'17" -37d11'20"
Henriette
Col 5h6'0" -34d16'13"
Henriette Delphine Amelie
Cas 0h55'29" 51d50'4"
Henriette Vymazal
And 2h31'57" 46d41'22"
Henrik Prossegger Vraanes
Ari 2h48'52" 30d6'52"
Henrik Senius
Lyn 9h15'43" 39d27'26"
Henrike & George Geuther NY-FL-AZ
Tau 4h38'37" 28d25'35"
Henrique Fonseca
Pup 7h10'4" -39d45'43"
Henrique Igreja Dinis
Uma 10h33'37" 44d39'19"
Henry
Per 3h22'23" 44d56'53"
Henry
Aqr 22h4'34" -1d38'12"
Henry
Uma 9h54'5" 67d51'24"
Henry
Uma 12h56'2" 62d36'37"
Henry
Uma 8h39'26" 66d55'40"
Henry A. Redding III
Gem 7h4'57" 21d46'16"
Henry A. Waxman
And 0h33'21" 37d24'50"
Henry Alan Phillips
Cap 21h28'11" -16d18'17"
Henry Amos Daniels Jr.
Cap 21h30'23" -16d11'39"
Henry and Bessie's Star
Cyg 19h37'35" 28d17'14"
Henry and Irene D'Amario
Cyg 21h24'27" 52d8'57"
Henry and Jeanette Shaffer
Ori 5h1'33" -0d26'29"
Henry and Nancy Kissinger
Cyg 20h9'49" 51d40'42"
Henry and Nathalie Gemayel
Gem 6h38'42" 17d27'4"
Henry and Ruth Cyganiewicz
Umi 15h26'57" 72d57'37"
Henry Andrews
Per 3h34'3" 41d20'34"
Henry Ardini
Per 2h45'57" 53d35'9"
Henry Arnberg
Vir 13h58'32" -0d49'18"
Henry B. Brown
Per 4h7'28" 49d18'32"
Henry B. Hermanns
Umi 14h59'9" 76d48'34"
Henry Barnabei & Jean-Carlo Sanchez
Aqr 20h51'9" 0d6'8"
Henry Bear
Umi 17h36'0" 80d55'17"
Henry Becker
Uma 10h4'50" 62d42'33"
Henry Bergen Strahm
Ari 3h1'59" 25d14'56"
Henry Broadwell Watkins, Jr.
Aql 18h51'6" -0d46'20"
Henry C. Messatzzia
Psc 1h0'56" 9d20'30"
Henry Carl Meyer
Sco 16h9'26" -11d41'48"
Henry Carl Riechmann II
Uma 10h5'38" 64d28'24"
Henry Chance Aronofsky
Gem 6h38'21" 16d20'53"
Henry Charles Holle
Uma 12h10'59" 53d50'39"
Henry Charles Smit
Psc 1h32'57" 20d2'43"
Henry Chow
Dra 18h54'16" 72d56'28"
Henry Clark
Gem 7h14'3" 34d31'52"
HENRY CONVERSANO SPECTACULARIS
Her 18h36'24" 25d10'20"
Henry D Lowery
Uma 8h59'36" 50d20'46"
Henry Daniel Palmier
Ori 5h24'11" 13d43'45"
Henry David Evans
Aql 19h45'26" 8d10'35"
Henry Davies
Umi 14h24'38" 67d38'20"
Henry de Monpezat
Gem 6h24'35" 26d5'47"
Henry & Dolores Roehr
Cyg 21h25'58" 52d17'44"
Henry Douglas Horne Jr.
Per 3h42'39" 45d46'22"
Henry Duffett
Aql 19h51'51" -0d41'43"
Henry Dunbar - "Harry's Star"
Dra 19h5'25" 74d41'6"
Henry E. and Meldine M. Mitchell
Umi 14h8'31" 67d14'19"
Henry E. Sarnacki
Her 16h58'50" 30d9'19"
Henry Edgar McCone
Cyg 20h53'45" 35d43'40"
Henry Edward Campbell Kopp
Her 18h53'2" 24d6'7"
Henry Edward Urwin
Cnc 8h6'32" 11d49'50"
Henry Edwin Ozment
Cap 21h15'0" -15d0'6"
Henry Elfstrom
Gem 7h46'39" 26d2'58"
Henry et Léa Plée
Ori 6h18'27" 12d33'56"
Henry Evald Skytt
Lyn 8h10'35" 43d57'54"
Henry G McKenzie-My Hero-My Dad
Tau 4h28'39" 17d10'20"
Henry Geer Rogers II
Ori 6h10'48" 20d59'15"
Henry George Rogers
Peg 23h32'57" 17d3'8"
Henry George Ruaidhri Collins
Cam 7h3'5" 71d34'51"
Henry Giovannetti Rltr 2 the Stars
Aur 5h35'4" 45d15'0"
Henry Giroud
Gem 6h56'39" 14d46'29"
Henry Glasner
Psc 1h8'42" 19d4'0"
Henry Gluck
Per 4h11'8" 51d25'46"
Henry - Grandad the Great! - xxx
Ari 2h57'9" 21d35'43"
Henry Grzyb
Uma 10h16'27" 62d14'31"
Henry H. Schreiber
Cyg 21h22'24" 54d55'2"
Henry H. Thomas Sr.
Ori 5h55'27" 22d41'8"
Henry "Hank" Maniace
Vir 14h19'7" 0d10'30"
Henry "Hank" Peter Corda
Uma 11h39'44" 35d41'35"
Henry "Hank" Petruska
Boo 15h34'23" 43d6'51"
Henry "Hank" Shaffer
Uma 10h38'58" 65d14'32"
Henry Hawk Sandler
Tau 5h24'53" 25d54'56"
Henry Hilderbrand
Uma 11h17'7" 42d13'11"
Henry Hines MacGuire Munford
Uma 10h34'53" 50d31'4"
Henry Hollihead
Umi 16h0'3" 72d10'33"
Henry Hoven Maillet
Sco 17h52'54" -37d49'47"
Henry I. Reinert
Per 3h41'59" 38d8'9"
Henry J. Becker, Sr.
Psc 1h17'54" 16d12'22"
Henry J Boucher "1914-1996"
Tau 4h29'14" 15d25'33"
Henry J. Craig
Uma 10h19'9" 61d54'19"
Henry J. Malone "Hank"
Leo 9h39'0" 14d8'15"
Henry J. Phillips
Uma 9h32'25" 59d44'5"
Henry J. Urban D.D.S.
Uma 9h17'33" 60d51'57"
Henry James Colquhoun
Her 17h56'34" 24d54'47"
Henry James Greive
Sco 16h48'41" -35d43'51"
Henry Jay Baldwin
Cep 22h51'25" 71d7'22"
Henry Jay D'Auria
Uma 9h47'52" 52d5'47"
Henry John Adkins
Her 16h35'3" 30d51'43"
Henry John Capro
Per 2h48'37" 53d58'47"
Henry John & Erma Deutschendorf
And 1h16'19" 40d28'45"
Henry John Gartland
Cep 21h6'25" 58d38'20"
Henry John Steinberger
Dra 18h23'55" 59d4'46"
Henry Johnson
Per 4h23'36" 44d23'52"
Henry Johnson
Crb 16h19'28" 30d38'23"
Henry Jorgensen
Lib 15h30'35" -28d54'5"
Henry Joseph
Cru 12h18'40" -59d36'4"
Henry Josiah Lacher
Sco 17h39'48" -42d32'15"
Henry K.
Leo 11h41'28" 24d48'19"
Henry & Karen Bremmer
Uma 11h18'42" 63d13'16"
Henry Kotulski
Cyg 20h55'30" 51d29'31"
Henry Kramich
Cyg 19h46'25" 39d48'33"
Henry Krueger
Aur 6h17'35" 52d54'39"
Henry L. "Buddy" Pricer
Leo 9h27'6" 15d7'41"
Henry L. O'Cain
Sgr 18h9'28" -30d40'15"
Henry Lee Work
And 0h59'1" 45d50'37"
Henry Leonard James Scully
Per 4h7'49" 49d19'57"
Henry Lewis Gartland
Ori 6h1'50" 17d54'52"
Henry Llewelyn Dransfield
Per 2h25'9" 54d18'2"
Henry Loretto Coia III
Del 20h36'19" 15d15'51"
Henry Louis Pessina
Cep 22h14'49" 63d49'56"
Henry Luis Barea
Her 16h39'25" 7d36'0"
Henry M. Lerner
Aur 5h36'12" 46d30'47"
Henry M. Soule
Boo 14h48'5" 29d2'58"
Henry Major
And 23h29'41" 48d33'45"
Henry Mark Heaton
Lib 14h57'0" -1d5'27"
Henry Mark Higham
Cnv 12h36'23" 39d19'48"
Henry Markarian
Her 18h47'37" 22d6'3"
Henry Martin Ostrowski
Uma 9h56'21" 49d52'22"
Henry McCune Miller
Aur 5h21'20" 41d22'33"
Henry Meatloaf Charles Greco
Sgr 19h10'18" -20d48'17"
Henry Michael Hirsch
Psc 1h22'25" 14d53'57"
Henry Michael Viscardi
Umi 15h30'43" 72d5'5"
Henry Minis, Jr.
Her 17h23'15" 24d13'59"
Henry N. McDill
Tri 2h10'13" 33d56'0"
Henry N Plumstead
Tau 4h42'7" 24d10'47"
Henry Nathan Smith
Aqr 20h40'36" -1d56'56"
Henry Nathaniel Powell Heyburn
Sco 16h16'19" -9d44'52"
Henry Nguyen Loves Jenny Tran
Uma 9h44'47" 56d20'20"
Henry Norton Redding Jr.
Cnc 9h2'13" 21d8'48"
Henry Ostberg
Cnc 8h2'50" 25d58'20"
Henry Otto Gauger
Gem 6h47'6" 34d59'8"
Henry Pankauski
Aur 5h34'4" 45d40'18"
Henry Penner
Psc 1h24'44" 19d16'42"
Henry Peter Strauss
Tau 4h48'25" 21d31'40"
Henry Phoenix Gager
Lib 15h18'48" -19d27'35"
Henry Pollak Coy
Aqr 21h37'24" -2d56'40"
Henry "Poppie" Nied
Aur 5h41'6" 49d56'44"
Henry Rafalski
Sco 16h55'59" -33d2'38"
Henry Reynolds
Cep 1h22'49" 77d54'19"
Henry Richard Howeson
Psc 23h15'15" -0d42'0"
Henry Ritter Mortimer II
Leo 10h26'19" 12d11'16"
Henry Robert Kennedy
Cep 22h28'56" 62d34'11"
Henry Robert Simmonds
Per 3h22'23" 41d32'26"
Henry Robertson Aldrich
Lib 14h36'0" -24d29'26"
Henry Rudd Thornton
Lyn 7h57'29" 39d7'39"
Henry & Shannon Forever
Sge 19h51'26" 17d21'37"
Henry Shaw
Uma 14h3'43" 59d17'25"
Henry Stanley Bent III
Her 16h48'32" 32d32'1"
Henry & Stephanie Ruhl
Cyg 20h51'45" 43d0'17"
Henry Streed
Uma 8h44'18" 68d50'21"
Henry "Terry" Terranova
Ori 5h38'30" 1d17'38"
Henry Thomas Stringham Superstar
Cma 7h25'0" -23d15'12"
Henry Todd Kurtzman
Aqr 21h12'57" -13d23'38"
Henry Tyler Morton
Umi 10h5'25" 87d20'1"
Henry & Virginia Summers
Gem 6h41'9" 32d54'35"
Henry Waia Watson
Uma 11h15'56" 47d14'3"
Henry Waldo Golden, III
Ori 5h51'2" 3d21'41"
Henry Wallace Staub III
Tau 4h12'25" 25d16'22"
Henry Warren Zhong
Gem 6h38'19" 12d55'40"
Henry Westfall
Umi 16h22'1" 76d11'36"
Henry Wilco Kruger
Uma 11h45'11" 34d19'26"
Henry Williams
Umi 14h37'31" 82d59'23"
Henry Willis Lundy
Vir 12h48'12" -0d9'24"
Henry Zuk
Cep 21h48'42" 65d49'24"
Henryk
Uma 12h46'55" 53d52'15"
Henry's "Guiding" Star
Umi 14h1'55" 74d37'41"
Henry's special lite
Aqr 22h14'25" 2d36'33"
Henry's Star
Her 17h22'40" 35d16'47"
Henry's Star
Uma 12h26'8" 62d38'25"
Henry's Star
Sgr 18h3'32" -28d56'13"
Hentschel, Leonie
Sgr 18h2'51" -27d33'3"
Hentschel, Michael
Lib 15h45'54" -9d21'21"
Hentzel & Fischer forever
Tau 5h33'25" 24d10'52"
Henze, Jürgen
Uma 9h2'3" 50d20'13"
Henze, Miriam
Cnc 9h1'38" 16d7'28"
Her
Sco 17h20'15" -33d8'1"
Her Ladyship, Patricia A Mahoney
Ari 2h15'31" 10d39'52"
HERA
Boo 14h27'46" 52d11'43"
Hera
Lyn 7h12'11" 51d42'58"
Hera
Lib 15h16'10" -5d30'49"
Hera
Uma 11h25'53" 64d39'0"
Hera Hehr
Sco 17h17'20" -45d19'28"
Heraucourt, Markus
Uma 9h38'0" 54d8'29"
Herb
Hor 4h10'47" -41d43'54"
Herb
Per 2h30'5" 54d39'57"
Herb
Tau 5h56'32" 27d39'37"
Herb and Barb Clark
Cyg 21h27'25" 44d55'43"
Herb and Josephine Smiley
Ori 5h30'32" 1d18'27"
Herb & Betty
Gem 6h48'4" 32d53'48"
Herb & Carol Paustian
Sgr 19h8'16" -15d51'32"
Herb Christiansen
Ori 6h0'11" 10d26'1"
Herb & Diana Warner
Uma 10h14'54" 43d4'33"
Herb Gage (Syncom)
Boo 14h27'6" 14d37'16"
Herb & Kristi Seymour
Cyg 20h40'4" 53d51'10"
Herb Miller
Her 17h37'18" 32d28'26"
Herb Mosteller
Ori 6h8'49" 20d47'39"
Herb & Pilar
Cyg 19h44'16" 34d29'23"
Herb Rubin
Dra 16h53'16" 69d16'39"
Herb Weisman
Cep 22h15'45" 61d11'28"
Herber, André
Uma 12h59'17" 62d26'22"
Herbert
Uma 9h39'28" 54d58'7"
Herbert A. Schmidt Memorial Star
Ori 6h5'52" 16d3'0"
Herbert Arenz
Uma 11h11'37" 41d45'34"
Herbert Bagel Shines Forever
Uma 11h30'10" 45d17'34"
Herbert Clayton Hale Magney
Vir 13h11'45" -16d30'35"
Herbert Cohen "Herby"
Ori 5h56'9" 13d6'23"
Herbert Covil
Sgr 18h40'17" -16d34'44"
Herbert Curry Granger
Cyg 20h4'35" 48d19'25"
Herbert D. Brock
Lyr 19h26'45" 40d5'10"
Herbert Dixon
Per 2h58'46" 46d2'23"
Herbert "Dusty" Rhode
Tau 4h51'48" 22d41'28"
Herbert E. Flemming
Uma 8h10'11" 62d21'51"
Herbert E. Wood
Ori 5h7'34" -0d9'2"
Herbert Edwin Steinke Jr.
Uma 11h16'12" 38d56'9"
Herbert Friede
Uma 11h50'15" 28d44'25"
Herbert H. & Judith K. Frank
Tri 2h6'6" 32d12'37"
Herbert H Webb "Herbie's Star"
Aqr 22h56'37" -6d38'8"
Herbert Hoover Hill
Tri 1h49'19" 30d5'18"
Herbert Johannemann
Ori 5h59'24" 2d50'52"
Herbert Joseph Crane
Cyg 20h52'43" 31d59'2"
Herbert Lüthe
Ori 5h17'32" 1d35'12"
HERBERT MOSER - GLUCKSSTERN
Uma 9h55'28" 44d31'51"
Herbert & Muriel Nikula
Uma 10h54'35" 44d12'58"
Herbert Nöhsner
Ori 5h38'0" -4d4'6"
Herbert Press
Uma 9h54'47" 48d1'28"
Herbert Rohacsek
Cas 0h19'59" 53d36'39"
Herbert Rubinsky
Ari 2h35'23" 19d6'31"
Herbert Sands, First Magnitude Dad
Aql 19h48'11" -0d33'20"
Herbert Schorer
Uma 11h40'10" 29d14'6"
Herbert Simon
Aql 19h18'32" 4d10'12"
Herbert Spies
Uma 9h7'47" 58d24'48"
Herbert Sydney Gabler
Cep 23h29'47" 76d55'41"
Herbert T. Westley
Cap 20h55'2" -17d45'20"
Herbert Thomas Howland
Leo 11h23'4" 16d36'19"
Herbert V. Warne, Jr.
Her 16h39'32" 47d18'46"
Herbert Vesterling
Uma 10h11'19" 41d59'52"
Herbertz, Mia-Sofie
Uma 10h53'7" 66d24'3"
Herbie
Vir 13h24'20" 7d44'29"
Herbie Brown
Uma 13h20'53" 55d8'38"
Herbie & Rose Forever together xoxo
Apu 15h21'5" -71d5'27"
Herbie Watts
Lib 15h2'39" -1d30'9"
Herbold, Justin
Uma 10h7'30" 44d21'20"
Herby the Love Bug
Gem 6h48'6" 25d29'12"
Herci Marsden, Prima Ballerina
Cyg 20h39'56" 38d22'41"
HercsAngel Forever
Her 16h17'45" 6d47'1"
Herdis Kristine Jacobsen-Miller
Tau 3h56'12" 20d40'12"
Herdzina, Ulli, Lucas und Katrin
Ori 5h9'14" -0d59'22"
Here B Dragons Dad Mom Maria & Rob
Ari 3h11'9" 23d15'43"
Here Fishy Fishy
Psc 0h54'3" 29d0'5"
Here I Am, Jane Lee
Cas 23h10'59" 56d12'19"
Here We Are- Douglas & Megan
Uma 8h34'59" 63d36'55"
Here's to the Night....
Lyn 8h12'27" 45d51'3"
Here's Wishing: Eternal Imagination
Cap 21h30'44" -9d16'24"
Hereward Leofricsson
Lmi 9h54'13" 37d17'0"
HERGONOLI
Sgr 18h15'19" -16d10'53"
Hering, Gunter
Ori 5h55'9" 21d41'35"
Heriold (Herminia Silva)
Mon 7h55'48" -4d52'46"
Herkner, Andreas
Uma 10h50'29" 42d15'51"
Herky
Oph 17h14'2" -0d30'6"
Herm B.'s Star
Per 2h53'43" 54d50'43"
Herman
Cep 8h26'47" 86d0'0"
Herman
Sgr 18h22'53" -23d38'31"
Herman A. Cauley
Lyn 7h25'35" 49d42'1"
Herman and Nancy Taube
Vir 13h22'36" 6d3'24"
Herman Berkelaar
Vir 13h18'8" 12d48'47"
Herman E. Hart
Cnc 8h38'31" 29d13'8"
Herman G. Loaiza Jr.
Per 3h53'13" 42d51'29"
Herman Glenn Wiggins
Cma 6h35'56" -17d13'58"
Herman H Gutjahr
Sco 16h15'31" -9d25'15"
Herman Issac Smith
Aqr 22h18'39" 0d29'10"
Herman Lee Raines Jr.
Aqr 22h53'34" -15d7'33"
Herman Leonard Grimes
Ori 5h55'44" 3d22'30"
Herman Lyons
Her 18h29'56" 23d9'8"
Herman P. Berkelaar
Cep 22h35'31" 72d23'44"
Herman "Sonny" Rogers Jr.
Leo 11h50'40" 25d4'29"
Herman Tang
Gem 6h4'53" 24d25'8"
Herman Vanderbeek
Aqr 23h5'30" -11d23'59"
Herman William Edwards
Psc 1h27'58" 22d9'45"
Herman-Alice
Aql 19h36'13" 10d55'26"
Herman-Dieter Tollenaer
Uma 10h3'18" 44d44'36"
Hermann Hauer h²
Uma 9h48'0" 48d13'1"
Hermann Kaelin 18.03.1928
Uma 9h26'52" 46d26'24"
Hermann Lausträter
Ori 6h15'14" 5d43'31"
Hermann Leinen
Ori 6h7'25" 0d54'43"
Hermann "Mani" Dubach
Cas 23h36'27" 56d10'26"
Hermann Schimpel
Ori 5h56'14" 18d43'47"
Hermann Schneider
Her 17h33'8" 41d34'45"
Hermann Wiese
Ori 6h11'38" 8d4'41"
Hermann, Sebastian
Uma 11h11'16" 33d59'43"
Hermano daSilva Pereira, Jr.
Boo 14h40'36" 48d15'3"
Hermelinda Duran
Cap 20h26'25" -9d26'8"
Hermerath, Sabine
Uma 14h7'8" 55d45'6"
Hermes M. Furlin, Sr
Lyn 8h44'54" 35d34'52"
Hermie A. Ramos
Tau 5h46'1" 22d18'40"
Hermie Hirsh
Cap 21h2'34" -20d25'30"
Hermie McNerney
Aur 5h5'20" 37d23'19"
Hermilo Balderas
Ori 6h17'7" 14d27'29"
Hermina
Uma 8h26'6" 63d7'32"
Hermina Barbara Rausch Boeding
Lmi 10h46'36" 26d0'6"
Hermina Stracke
Cas 0h54'14" 54d20'58"
Hermine & Arnie Roussine
Lyr 18h28'20" 28d9'20"
Hermine Kit Lam
Uma 9h11'41" 66d21'3"
Hermine Viktoria
Uma 9h31'13" 69d1'49"
Herminia
And 0h56'26" 40d57'55"
Hermintrude Mary Elizabeth Simons
Cyg 20h39'43" 29d36'53"
Hermione Jade Bissonnette
Uma 9h15'20" 59d49'49"
Hermione Katrina
Lyn 7h53'27" 52d51'3"
Hermionie Eyre
Cyg 19h32'39" 28d26'53"
HERMMARY 06/18/1949
Lyn 7h41'3" 38d23'55"
Hermosa Princesa
And 1h56'13" 38d41'28"
Hermosa Rosa Santos
Cas 1h24'1" 59d8'15"
HERMOSA YVONNE
Leo 11h19'39" 20d1'39"
Hernandez Christelle
Uma 11h32'29" 60d37'17"
Hernandez - La Vida del Cielo
Uma 11h34'10" 48d12'4"
Hernandez - Stapleton
Dra 10h37'7" 75d23'3"
Herneet "Neetu" Kaur Rai
Lib 14h29'34" -23d48'32"
Hero
Ori 6h16'7" 16d12'11"
Hero and Prophetess
And 0h30'51" 35d46'28"
Hero- Brandon A. Gilcher
And 0h42'34" 24d49'54"
Hero of 2 dogs
Cnv 13h47'19" 30d3'3"
Hero Pasqualucci
Her 18h35'39" 15d32'11"
Hero special only shining star
Cnc 8h7'10" 22d33'25"
Herodotus Damianos
Uma 9h51'3" 62d49'53"
Héroe
Ori 5h57'5" 3d13'34"
Heron Pond
Umi 13h14'12" 73d34'19"
heron soaring
Cyg 20h25'24" 47d21'40"
Heron's Star
Uma 13h45'59" 54d13'9"
Heros' Heart, Nancys' Arthur
Ori 5h55'36" 21d11'21"
Herr Krantz [The Baby]
Umi 15h32'30" 71d30'2"
Herr M
Ari 2h57'54" 22d20'43"
Herrey, Anna & Philip
Ori 5h14'54" 5d39'54"
Herrmann & Katharina Schmid
Lac 22h50'9" 54d2'12"
Herrmann, Heinz
Uma 13h3'28" 60d33'44"
Herry Madison
Psc 0h50'21" 12d32'19"
Herschel Ryan Carithers
Uma 11h8'39" 60d44'36"
Hershé
Cas 0h40'56" 64d20'41"
Hershel Roell
Cnv 12h46'37" 38d28'46"
Hershey
Cas 1h36'40" 55d0'8"
Hershey
Cmi 7h40'59" 1d23'44"
Hershey Kiss II Stockwell
Cma 7h3'50" -26d30'57"
Hershey Morrison
Cma 7h26'24" -16d16'1"
Herta
Tau 3h25'22" 8d44'53"
Herta Hrdlicka
Uma 10h30'13" 48d37'8"
Herta + Willy Burri; 25.05. + 31.07.1932
Lyr 18h47'18" 39d56'55"
Hertha
And 0h16'7" 32d20'19"
Hertha Borovka
And 1h51'28" 44d26'49"
Hertta Katriina
Uma 8h38'56" 58d35'14"
Hertzler
Cap 20h38'34" -13d12'38"
Hervé
Cas 0h46'4" 54d42'57"
Hervé Bouillon-Tallendier
Ari 2h42'19" 15d54'35"
Hervé Laurent Cavatorta
Psc 1h8'54" 20d26'39"
Hervé Stoffel
Her 18h16'37" 21d3'43"
Hervé Vancompernolle
Ori 5h22'12" 9d15'9"
Herveline Case
Sgr 17h59'32" -18d6'50"

Herzele
Uma 8h45'56" 48d18'26"
Herzing, Melanie
Ori 5h2'42" 6d9'43"
Hesham & Maha.. always and forever
Cyg 21h14'30" 47d38'1"
Heshie Burak
Ari 1h58'42" 13d29'14"
Heshlaan
Col 5h45'48" -34d19'10"
Hesperus Jeremy Guyot
Cyg 21h19'9" 51d22'8"
Hesperus Luthien
Ori 4h47'51" -1d56'26"
Hess
Psc 1h6'39" 4d7'27"
Hessa Hamad Al-Attiyah
Sgr 19h4'12" -26d15'7"
Hessenbruch, Rolf
Ori 6h20'20" 10d16'35"
Hessler
Ori 5h56'27" 17d35'33"
Hesta Bailey, my Little Star
And 2h30'8" 45d48'23"
Hester Lynn
Del 20h52'40" 6d39'46"
Hester Marie
Ari 2h37'27" 27d57'2"
Hestia de Pascal et Florence
Cas 0h8'33" 57d9'28"
Hestia, Christie's star
Crb 15h45'54" 38d23'23"
Heston Adam Titzer
Her 17h42'20" 37d35'7"
Hetal S C
And 0h39'31" 26d34'19"
hetmanheath
Tau 5h13'25" 19d25'28"
Hetsel Brandon Minix
Ori 5h34'7" 8d43'16"
Hettie Darby
Umi 14h39'48" 88d22'18"
Hetty
Lmi 10h18'49" 37d24'27"
Hetty Ho Tran
Hya 8h48'55" 4d42'35"
Hetty Lee Stubbs
And 0h29'19" 38d39'25"
Heuser
Uma 8h59'50" 69d36'5"
Hev and Ferg
Tau 5h6'53" 16d21'25"
Hev Luna
Tau 5h47'3" 19d31'34"
Hevan
Ari 2h48'4" 27d6'10"
Heveling-Marceniuk, Katarzyna
Ori 5h27'47" 13d34'19"
Hevenly Haleakala
Eri 4h17'1" -33d10'45"
Hewbie
Cnc 8h45'21" 13d43'15"
Hewlett's 50th
Sgr 19h17'36" -15d36'23"
Hey Baby, Baby
Uma 10h35'21" 72d20'47"
Hey Dad always a Star David Grant
Cep 23h6'0" 72d8'41"
Hey Hi
Uma 11h25'13" 38d21'3"
Hey Jeff...You Drink!
Per 4h1'53" 44d39'57"
Hey Jude
Cep 22h9'43" 85d14'1"
Hey Jude
Cru 12h41'27" -58d21'10"
Hey Kevin...
Lib 15h45'37" -17d27'10"
Hey Renee
Psc 1h14'0" 24d56'0"
HEY SUZ
Tau 5h37'12" 26d0'1"
Hey Sweetness
Cnc 9h6'52" 25d18'39"
heydan
Leo 11h29'28" -2d42'33"
Heydecke, Hans-Joachim
Ori 5h5'56" 6d32'15"
Heydt, Franz-Herbert
Gem 6h27'18" 20d32'53"
Heyne, Anneliese
Sgr 18h24'46" -23d9'4"
Heyne, Cornelia
Ori 6h4'32" 10d40'54"
HeyokaShakti
Ori 6h14'13" 15d12'37"
Heywood G. Friedman
Per 3h36'18" 47d51'1"
Hezi Tali
Crb 15h30'26" 27d21'49"
H.F. Bunnell
Cnc 8h52'21" 18d32'38"
HG Shining Star++/
Peg 21h45'11" 24d14'26"
HGA90
Cnc 8h33'40" 11d16'58"
HGS85
Uma 10h38'19" 41d55'39"
HH Ganapathi Sachchidananda Swamiji
Lac 22h36'21" 48d53'14"
HHA III
Aqr 22h38'54" -0d45'31"
HHMoss
Cas 1h1'22" 61d27'3"
HHS Coach D
Dra 18h48'37" 51d6'33"
Hi Doc :)
Sgr 19h16'11" -24d49'0"
HI HO
Uma 13h48'41" 54d9'41"
Hi Jack
Ori 5h23'55" -3d0'3"
Hi-4Ever-J&J
Ori 6h17'12" 6d29'59"
Hiapo
Ari 2h20'51" 23d31'16"
Hiba 84
Cyg 20h41'15" 49d0'23"
Hiba Ladiki
And 1h31'57" 37d56'39"
Hiba & Mazen
Crb 16h8'26" 30d24'45"
Hiba Sarieddine
Her 17h41'39" 40d19'22"
Hickers R.M.
Ori 6h5'4" 5d51'50"
Hidden Lakes
Lyn 8h44'6" 44d46'55"
Hidden Treasures
Psc 2h1'26" 3d44'36"
Hideki H Hishikui-uchi Nasu-machi.
Sco 16h48'4" -41d29'24"
Hideo Soneoka
Lib 14h49'47" -18d11'50"
Hiding the Stars - Jennie Babin
Cma 6h54'15" -19d20'38"
Hi-Do
Lyn 7h31'21" 53d42'22"
Hiedi Honor Hall
Vir 14h57'13" 5d17'14"
Hiedi Marie Johnson 2/26/73
Psc 1h29'6" 6d15'40"
Hiedi "The key to my heart"
And 23h3'4" 45d37'30"
Hiedi's Song
Uma 11h25'31" 72d10'37"
Hiehie ipo hoku
Gem 7h7'22" 23d24'24"
Hielka
Psc 0h21'8" -2d5'55"
Hien Kim Luu
Cap 21h57'22" -19d52'52"
Hien Tran
Psc 0h32'19" 8d5'47"
Hiesha Nicole WilliamsDavis
And 23h40'4" 48d22'45"
HIGGALD
Lyn 6h43'52" 56d41'35"
Higginson Family
Sgr 17h57'51" -17d19'56"
High Above Big Wind
Uma 12h2'27" 32d4'21"
High Road Academy of Howard County
Uma 12h41'20" 59d8'33"
Higham;'s Heavenly Light
Aql 20h5'17" 2d53'11"
Highd
Dor 4h21'8" -55d58'40"
Highlander
Ori 5h46'6" 7d3'32"
Highlander
Cnc 8h2'18" 10d52'33"
Highlight Of My Life
Cru 12h38'22" -58d13'42"
Highnotes
Cap 21h47'46" -9d32'41"
highpocket
Lib 15h20'5" -5d36'38"
Higinio Hernandez
Aqr 22h33'32" 2d26'52"
Hihike
Sco 16h24'46" -32d30'45"
Hijie Hiahara Alperin
And 2h28'4" 48d45'35"
Hikari
Aur 6h1'43" 47d57'46"
Hikari
Uma 13h53'51" 59d31'53"
Hikari's Star
Cyg 19h58'1" 34d7'30"
Hilaire Daly Natt
Cra 18h18'0" -45d16'40"
Hilarah
Cas 1h6'56" 63d36'37"
Hilari Nicole Sloan
Uma 9h44'59" 49d41'19"
Hilaria
Leo 11h33'7" 17d3'1"
Hilarie Beam Zani
Vir 12h32'16" 2d51'20"
Hilary
Gem 7h10'41" 29d4'45"
Hilary
Lib 14h50'34" -6d32'0"
Hilary Alexander
Cyg 21h28'59" 51d10'40"
Hilary and David McCallum
Uma 12h10'23" 51d14'9"
Hilary and Justin
Leo 11h15'34" 25d23'25"
Hilary Andrea Johnman
And 1h32'4" 41d36'26"
Hilary Ann Duff
And 0h49'29" 39d49'55"
Hilary Ann Duff
And 0h52'5" 38d37'45"
Hilary Becker
Ori 5h9'56" -0d45'11"
Hilary & Billy
Vir 13h27'44" 11d51'54"
Hilary Cullen
And 1h21'52" 48d59'11"
Hilary Daniel
Sco 17h24'22" -31d52'16"
Hilary Dilary Doc
Leo 10h15'17" 17d1'4"
Hilary Dunn
Cam 5h40'17" 69d55'50"
Hilary E. Gaby
Aql 20h2'16" 9d29'15"
Hilary Elaine
Uma 11h48'3" 34d48'43"
Hilary Erin Williams
Her 18h27'36" 12d8'43"
Hilary Gibino
Vir 12h52'6" 12d46'24"
Hilary Griffiths
And 0h17'29" 34d1'21"
Hilary Haron
Vir 13h18'24" -13d52'13"
Hilary Jay Chambers
Leo 11h56'17" 26d34'11"
Hilary Jordan Fisher
Lib 15h3'15" -24d0'22"
Hilary Keppery
Lib 15h4'1" -9d21'22"
Hilary L Douglas (Beautiful Star)
Cas 1h38'34" 63d28'2"
Hilary Layne Schroeder
Gem 6h47'30" 26d19'16"
Hilary Lynn
Her 16h15'15" 42d36'40"
Hilary M. Crais
And 1h6'22" 48d38'10"
Hilary Nicole Martin
Peg 22h11'50" 6d18'27"
Hilary & Petrus Nepgen
Cnc 8h20'4" 28d48'17"
Hilary Pries
And 0h27'16" 40d25'10"
Hilary Regan, Superstar
Sco 17h30'47" -36d50'40"
Hilary Regan's Dreams
Sco 16h40'59" -34d29'15"
Hilary Schultheis
Lib 14h57'36" -0d44'9"
Hilary's Beautiful Star
Cas 1h13'16" 56d29'44"
Hilary's Star
Uma 14h2'20" 59d21'5"
Hilbert's Hope
Cam 4h3'9" 63d54'57"
Hilda
Umi 14h12'14" 70d33'25"
Hilda Aldunate
Vir 13h15'59" 5d27'2"
Hilda and Jerry Brosky
Cyg 21h31'27" 53d23'21"
Hilda and Joe Daily
Cyg 19h38'0" 51d33'50"
Hilda Anne Guenther Downing Paim
Leo 11h9'39" 22d49'4"
Hilda Birchall
Cas 1h50'54" 65d12'55"
Hilda Churchill
Cas 23h18'2" 55d57'33"
Hilda Clara Zacherl
Leo 11h27'3" 16d32'59"
Hilda & Drew Toshack
Cet 2h16'56" 6d26'22"
Hilda Hobma's Star
Uma 9h56'50" 71d55'27"
Hilda "Joyce" Holmes
Cas 23h26'59" 53d4'23"
Hilda Judith
Ori 5h13'26" 7d51'48"
Hilda & Leyton
Cyg 20h18'40" 49d37'6"
Hilda Louise
Lmi 10h6'49" 32d33'27"
Hilda M. "Bobbie" Fitzpatrick
Uma 11h39'6" 55d33'15"
Hilda Mae Wolfe
Lyn 8h18'5" 35d7'53"
Hilda Martha Ramirez de Marshel
Ari 3h18'11" 26d37'21"
Hilda Osborn Memorial Star
Cap 20h57'8" -19d35'19"
Hilda & Robert
Gem 6h47'56" 27d16'55"
Hilda Salloum
Tau 4h19'34" 17d17'41"
Hilda Star*k
Cas 22h58'7" 55d13'2"
Hilda Wilson
Crb 15h42'30" 26d47'8"
Hilda Youkhana
Aqr 22h36'26" -11d46'8"
hildalgo
Lyn 8h9'11" 39d3'55"
Hilde Hauff
Uma 12h4'58" 47d47'26"
Hilde L Thompson
Gem 6h55'52" 13d8'58"
Hilde Walter
Cas 23h4'4" 58d39'41"
Hilde Ward
Cas 0h54'39" 53d34'30"
Hilde Weist
Crb 15h42'37" 33d33'3"
Hilde Zawadzki
Cap 21h50'4" -16d4'17"
Hildegard Luise Sophie
Ori 4h56'3" 2d49'10"
Hildegard Paradiso
Dra 14h57'38" 55d54'26"
Hildegard Smith
Cnc 8h22'34" 9d23'52"
Hildegard Späth
Uma 11h5'11" 40d46'25"
Hildegard Therese Eberle
Sgr 19h55'19" -31d26'29"
Hild's Star
Lyn 9h7'53" 39d16'13"
Hileman Family Star
Gem 7h5'5" 31d59'52"
'Hileric'
Lyr 18h49'46" 35d7'28"
Hilfiger
Uma 9h56'35" 69d7'52"
Hilgartner
Ori 5h4'31" -0d25'37"
Hill 1994
Uma 11h51'52" 39d56'44"
HILL BOY Michael1950-2006 Adam-2006
Uma 10h1'47" 70d55'31"
Hilla
Her 17h48'58" 38d44'58"
Hillarie & Tom
Cyg 19h38'31" 31d52'32"
Hillary
Vir 12h14'49" 12d22'18"
Hillary
Vul 21h27'24" 26d38'27"
Hillary
Cap 21h24'8" -26d1'23"
Hillary 32
And 0h8'40" 44d59'55"
Hillary and Hugh Simmers
Aql 19h4'28" -9d22'39"
Hillary & Andrews Love
Cyg 19h41'34" 29d6'26"
Hillary Ann Sellers
Uma 10h26'4" 62d24'49"
Hillary Anna Jones
And 23h21'28" 42d12'28"
Hillary Blair
Crb 15h48'10" 38d6'13"
Hillary Charlene Rentrop
Uma 11h43'46" 49d37'16"
Hillary Corinne Johnson
Vir 12h59'23" 4d22'32"
Hillary Dawn Johnson
Cnc 8h2'20" 27d23'47"
Hillary Dawn Tennent
Cap 21h18'46" -14d59'22"
Hillary Giangregorio
Lyn 6h57'14" 52d24'28"
Hillary Goodman From Your Light
Leo 11h14'5" 4d40'27"
Hillary I Love You
And 0h49'24" 40d48'51"
Hillary Jane Thurman
Ari 2h20'7" 21d29'39"
Hillary Jo Jedlicka
Lmi 9h44'12" 37d54'5"
Hillary Joy Pietrucci
Dra 14h49'24" 59d50'25"
Hillary Kae Freitag
Cas 1h27'28" 53d7'23"
Hillary Louise Case
Uma 10h51'41" 42d24'22"
Hillary "Love Star"
Uma 11h3'57" 37d29'36"
Hillary Lynn
Cnc 8h5'34" 8d4'15"
Hillary Moskow
Lyn 9h3'38" 34d33'25"
Hillary Nikole Thomas
And 23h46'29" 45d34'27"
Hillary Paula Dixon; leo star*35
Leo 10h13'34" 17d26'12"
Hillary Peach
Sco 17h15'52" -32d51'31"
Hillary Rebecca Slapa
Cnc 8h58'39" 13d5'26"
Hillary Robin Rodgers
And 0h5'4" 45d9'31"
Hillary Rose Connor
Uma 8h47'3" 61d2'53"
Hillary Stich
And 1h16'1" 50d20'20"
Hillary Walker
M.S.M.F.I.L.Y.A & F
Uma 9h0'53" 57d56'55"
Hillary Zylick
Tau 4h37'29" 3d7'43"
Hillary's Graduation Star
Ori 5h51'4" 5d34'7"
Hillary's Smile
Ari 2h26'39" 23d45'54"
Hillbilly's Wishing Star
Lib 15h19'6" -5d53'7"
Hillebrandt, Stefan
Uma 12h56'19" 54d55'55"
Hiller, Joachim
Ori 6h2'23" 13d47'33"
Hillermann, Kathrin
Uma 10h12'4" 49d28'56"
Hillery Louise Brown
Dra 17h9'12" 60d54'47"
HillStar
Cyg 19h36'6" 29d50'29"
Hillster
Crt 11h34'32" -8d28'26"
Hillybug
Cam 4h25'30" 72d43'7"
Hilpert, Axel
Uma 8h28'22" 64d12'30"
Hiltmann, Hans
Ori 5h31'44" 10d0'50"
Hilton and Abigail
Cru 12h34'50" -60d5'32"
Hilton Howard Leach II
Tau 5h44'2" 14d7'17"
Hilton James Palmer
Sgr 18h25'48" -16d40'52"
Hilton-Stirling42
Crb 16h13'52" 38d18'59"
Hilyria
Psc 0h3'35" -0d22'4"
Himburg, Heinz-Michael
Uma 10h34'26" 66d20'10"
Hime Chan
Aqr 23h2'32" -14d11'44"
Himes Family Star
Aur 5h37'8" 48d31'29"
Himesama Jenn
Leo 10h29'14" 6d49'25"
Himml, Johann
Uma 9h11'7" 63d58'52"
Hina Vejay Patel
Peg 22h19'25" 23d16'24"
Hinatea Melody
Cyg 20h8'36" 45d23'54"
Hind Sidawi
Uma 11h32'2" 48d41'34"
Hinda Cecelia Khuen
Sgr 18h18'29" -16d36'56"
Hinda Cecelia Khuen
Sgr 17h53'46" -29d12'2"
Hindy Gluck
Gem 7h15'50" 21d37'28"
Hines Family
Umi 14h41'21" 75d16'17"
Hiniorite
And 0h18'7" 27d6'20"
Hinoki 2005.1.5.1506
Cap 20h8'10" -16d52'38"
Hinz, Wolfgang
Uma 10h56'1" 56d25'55"
Hipatia Escanio
And 0h17'2" 22d20'45"
Hippler, Dennis
Aqr 21h22'32" -0d14'32"
Hippocampus S & K
Uma 9h53'24" 64d53'51"
Hippopotamus - Jason Hung Shen Lu
Cep 23h19'30" 68d19'1"
Hiral
Uma 11h38'3" 53d44'58"
Hiram
Cam 3h52'35" 66d36'37"
Hiram Michael Gerickont
Aur 5h52'32" 46d55'0"
Hiren Ladwa
Cnc 8h53'50" 8d25'25"
Hiro Kuwana
Uma 11h24'40" 64d8'56"
Hiroki Ichikawa
Cnc 8h25'30" 22d46'7"
Hiroko
And 0h13'12" 36d8'57"
Hiroko Tarutani
Gem 6h48'41" 13d2'38"
Hiroko's Guiding Light
Cas 23h45'20" 54d19'43"
Hiromi
Cnc 8h12'3" 13d14'21"
Hiroshi
Aql 19h55'32" -0d27'55"
Hiroya & Yinni Love Forever
Cyg 20h45'16" 53d54'46"
Hirsch
Cap 21h39'35" -12d49'42"
HIRSCHI' S HALO
Cnc 8h42'35" 9d40'3"
Hirt, Michael & Christa - 17.12.1971
Ori 6h15'30" 11d11'48"
His desire
Cas 23h38'20" 53d16'42"
His Excellency Richard M. Weissman
Lyn 8h16'57" 41d30'56"
H.I.S. - H,S,C,M,R,E,BG,G,BB
Per 4h6'59" 45d0'4"
His Highness Fergus Miller
Cma 6h43'17" -20d53'42"
His name is Mr. Bill
Cap 21h55'16" -15d22'13"
Hisae
Hya 8h41'7" 5d14'57"
Hisashi Kitagawa
Sco 16h10'6" -28d37'22"
Hisham and Diana Forever
Cyg 20h32'58" 49d42'17"
Hitesh Patel
Per 3h34'0" 40d24'43"
hitomi
Leo 9h44'6" 11d31'6"
Hiwahiwa
Her 17h48'36" 42d52'5"
HJ 20/12/03
Cyg 19h54'57" 33d43'7"
H.J. St. Pierre
Srp 18h8'22" -0d6'32"
HJ6380
Pho 0h48'32" -52d33'6"
HjAaYsLoEnY
Uma 13h1'30" 62d1'34"
~Hjerte - Kinkajou~
And 0h22'39" 30d42'47"
hjomav
Cam 4h3'56" 65d58'45"
Hjomdsovid Jubilee
Lyn 8h7'27" 35d18'25"
Hjördis Eskola
Cyg 19h24'42" 53d26'21"
HJ's Star
Ori 5h50'11" -0d0'3"
Hlavac Persistent
Ori 6h17'59" 14d15'5"
Hlavac, Heinz
Ori 6h11'13" 8d50'28"
HLB42305ASB
Her 17h18'14" 46d32'41"
HLF-MEF2005
Aqr 21h7'19" 2d26'13"
HLHitsman
Gem 7h1'5" 29d59'27"
Hlirou
Per 3h1'20" 54d44'36"
Hlub
Umi 14h21'49" 68d25'52"
Hlynur & Berglind
Per 4h12'4" 45d55'30"
HM^{2}D
Dra 15h9'18" 56d20'26"
HMA 19740923
Ori 6h19'9" 9d36'43"
HMaaltety
Cyg 21h48'29" 36d16'39"
H-MAN
Ari 2h17'25" 17d16'8"
HMCS(SW/AW) Gabriel Cruz, USN-RET
Ori 5h27'53" -4d13'0"
HMegaStar
Cyg 20h22'42" 46d40'17"
HMG 04-28-1969
Eri 3h42'44" -0d9'9"
HMiller Brightstar
Cnc 8h41'9" 15d30'53"
HMREID
Cas 1h37'43" 62d0'23"
HMS186
Sco 16h51'54" -45d12'16"
H.N.D. MAX
Dra 18h26'49" 73d3'33"
h-n-i
Sge 19h49'38" 18d42'4"
H-n-J Duke
Cnc 8h29'7" 25d25'25"
Ho Bo
Vir 13h8'13" 12d27'1"
Ho Tsan Fai & Ellen O' Yang Oi Ling
Dra 16h37'24" 60d20'11"
Ho Wai Keung
Leo 11h5'33" 16d42'35"
Hoa T Kha
Ari 3h0'1" 29d14'22"
Hoang Nguyen
Cha 13h51'6" -76d5'26"
HOANGAL
Sco 16h43'33" -31d23'34"
Hobbes
Cep 21h39'46" 61d40'29"
Hobbes "Mini-man" Hawes
Cmi 7h25'11" 2d58'31"
hobbes star
Leo 11h29'46" 9d26'6"
Hobbit's Heart
Ori 5h33'51" -0d4'53"
Hobbs Karl Rumpca
Cnc 9h6'51" 31d35'21"
Hobday, Riley John - Born 15.2.2004
Cru 12h34'17" -60d54'4"
Hobe Cat
Psc 0h20'27" 17d43'32"
Hobie
Leo 9h25'23" 13d31'0"
Hobie Wallace
Sgr 19h40'58" -12d22'58"
Hobiger, Bernd
Uma 9h50'18" 67d17'16"
HOBI-STAR
Uma 11h27'27" 31d37'57"
Hobnob
Cmi 7h51'54" 6d2'51"
HoBo Love
Lyn 6h25'32" 60d8'8"
Hobo SKH
Lib 15h53'6" -17d51'16"
H.O.C.C - My personal 4 Stars
Ara 17h30'47" -47d41'29"
Hochleitner, Raimund
Uma 10h32'21" 42d17'47"
Hochrein, Fritz
Uma 11h20'59" 63d56'40"
Hochstein, Olivia
Uma 9h52'16" 46d12'32"
Hochzeit Nathalie&Thommy 25.09.2004
Uma 10h41'14" 40d4'12"
Hochzeitsstern Daniela & Alexander
Uma 9h8'48" 54d18'33"
hock
Ari 2h26'14" 17d45'47"
Hock Ju Edar Sdn Bhd
Ori 5h36'13" 3d35'12"
Hocke
Leo 11h45'1" 21d57'21"
Hocquart de Turtot
Leo 11h52'44" 24d32'49"
Hod
And 0h40'31" 30d6'44"
Hoda Mattar
Cas 0h24'26" 57d55'3"
Hodana K.
Aqr 20h43'29" -9d17'41"
Hodges Star
Cyg 19h58'34" 37d28'38"
Hoeck, Ulrich
Uma 9h2'16" 53d40'30"
Hoeflich
Uma 10h33'48" 60d49'12"
Hoel Ménard
Vir 14h57'48" 6d56'24"
Hoesing
Dra 19h44'23" 65d53'59"
Hofeditz, Uwe
Ori 6h17'13" 19d41'5"
Hofer Family
Sgr 19h47'29" -13d38'22"
Hofer, Karl-Heinz
Uma 11h36'28" 32d43'38"
Hoffman Construction Company
Per 3h12'9" 55d21'20"
Hoffmann
Ari 2h47'29" 26d48'17"
Hoffmann, Hermann
Gem 6h25'44" 19d12'34"
Hoffmann, Peter
Ori 6h3'48" 10d2'29"
Hoffmann, Thomas
Ori 6h20'47" 16d44'38"
Hoffmas - StefanHoffmann 30.04.1965
Uma 8h59'27" 60d27'7"
HOFFNUNG
Lac 22h3'29" 51d48'57"
Hoffzimmer, Helmut
Uma 9h23'38" 64d58'26"
Hofmann, Peter
Uma 9h15'5" 62d51'12"
Hog
Cma 6h40'6" -12d9'44"
Hogan Brown O'Brien
Her 18h2'49" 36d44'23"
Hogan Leland Carmichael
Cnc 8h51'37" 26d53'7"
Hogan McLaughlin
Psc 0h43'5" 14d57'10"
Hogan's Hope
Crb 15h34'6" 26d54'56"
Hogie
Uma 11h24'1" 54d23'4"
Höhle, Hans-Werner
Uma 12h46'39" 60d41'11"
Hohlfeld, Laura
Uma 9h20'2" 62d32'39"
Hohlfeld, Ulla
Uma 9h20'13" 63d53'7"
Hoi-Kai
Lmi 10h26'58" 35d18'14"
Hojda Star
Uma 12h5'7" 53d50'26"
Hokanson
Dra 19h43'57" 60d21'7"
Hokeo
Cap 21h47'42" -21d46'38"
Hokey
Cam 4h21'49" 73d16'45"
Hokshila
Psc 0h48'44" 4d43'30"
Hoksila
Tri 1h49'12" 34d6'18"
Hoku Chelsey & Barry
Cru 12h41'14" -55d55'22"
Hoku Christien
Gem 6h29'7" 21d5'46"
Hoku - Unser Stern
Uma 10h26'57" 40d36'15"
Hokuaonani
Cas 23h22'7" 59d40'15"
Hol*Leigh
Cra 18h50'40" -40d10'37"
HOLDEN
Cap 21h3'54" -16d4'26"
HOLDEN AKOPIANTZ ROBERTS
Leo 9h38'13" 28d5'44"
Holden Andre Taylor
Her 16h36'59" 23d17'48"

Holden Benjamin Miller
Her 16h41'44" 11d41'31"
Holden Bennett Nilsen
Per 3h7'16" 54d29'2"
Holden Brant Lund
Cnc 9h17'48" 15d0'51"
Holden James
Cep 20h42'13" 65d9'59"
Holden John Michael Seton
Vir 13h13'44" -21d40'23"
Holden Parker Lewis
Uma 11h27'49" 54d27'56"
Holden Reed's Star 2-11-93
Aqr 21h33'42" -0d51'40"
Holden Wayne Pyle
Aqr 21h5'4" 1d33'39"
Holding Hands
Ori 5h27'16" -3d59'52"
Holdyn Gene McDonal
Ori 6h17'57" 14d56'43"
Holger
Ori 5h8'11" 15d30'59"
Holger
Uma 12h7'41" 61d35'42"
Holger Boysen
Uma 11h8'35" 67d13'33"
Holger Härtel
Uma 11h35'49" 40d59'41"
Holger Iris
Ori 4h58'12" 10d9'12"
Holger Scheithauer
Boo 15h32'6" 45d45'38"
Holger Sieben
Uma 10h13'33" 64d2'28"
Holger Wülfken
Ori 6h20'39" 8d48'47"
Holgj
Cap 21h30'36" -13d34'33"
Holgy
Peg 21h40'39" 22d57'32"
Holidae Marie Swan
And 1h39'5" 49d17'43"
Holiday Michelotti
Uma 10h14'56" 42d22'42"
Holidaysburg Mike
Gem 6h57'44" 15d51'20"
Holkunium
Aql 19h55'58" 14d55'7"
Holl, Michael
Uma 10h12'40" 68d47'57"
Holland Amanda Spilker
Cnc 8h41'18" 7d18'21"
Holland Byrd
Uma 13h41'56" 58d23'50"
Holland & Jack Rossetto
Umi 16h58'47" 77d31'30"
Holland Olivia Holdsworth
And 2h11'26" 42d54'37"
Holland Olivia Stalcup
Tri 1h45'14" 33d32'41"
H.O.L.L.A.N.D WIMBY N AILED
Sgr 18h36'53" -17d53'29"
Holland's Rising Star
Cha 13h40'58" -77d39'29"
Hollee Hayden
Hya 9h5'1" -0d37'26"
Holley Marie Madderra
Sgr 18h14'35" -19d4'50"
Holli and Kiairra Lynn Bray
Her 16h14'49" 48d16'9"
Holli Ann
Lmi 10h50'5" 26d35'50"
Holli Elizabeth Pisarczyk
Lib 15h42'37" -3d53'39"
Holli Michelle Adams
And 0h24'10" 25d53'0"
Holli Nichole Knepper
Cas 1h12'2" 49d59'34"
Holli Rae Maloni
Vir 14h19'25" 0d9'16"
Holli Renae Hubright
Cnc 8h52'30" 31d47'31"
Holli Shirley
And 2h32'45" 43d55'45"
Holli Swick
Tau 4h32'11" 16d57'4"
Holliday
Leo 11h12'31" 11d51'28"
Hollie
Tau 4h22'52" 22d29'5"
Hollie
Aqr 23h56'9" -11d31'32"
Hollie Allen
Leo 10h28'32" 12d50'13"
Hollie Ann
Aqr 22h25'43" -5d50'44"
Hollie "Baby Doll"
Vir 11h40'27" 9d49'5"
Hollie Beth Lisle
And 0h17'25" 22d25'0"
Hollie Brooklyn Stewart
And 0h18'49" 31d5'29"
Hollie Calea Patton "My Girl"
Cap 21h35'2" -19d38'42"
Hollie Cook
Cnc 8h37'6" 8d1'59"
Hollie Elaine
And 23h18'50" 43d21'18"
Hollie Elizabeth Hixon
Cap 21h36'25" -22d49'56"
Hollie Holsbolsmelols
And 1h5'53" 47d24'36"
Hollie Janice Brown
And 2h6'43" 42d6'12"
Hollie Karen
And 1h55'47" 36d4'54"
Hollie Lenae Mann
Uma 13h54'9" 58d47'54"
Hollie Louise Hepworth
Boo 14h24'9" 46d2'31"
Hollie Love Sammy
Aql 19h3'46" 8d59'20"
Hollie Lynn
Cam 5h51'52" 69d30'57"
Hollie Marie Smith
Cas 1h18'4" 54d11'18"
Hollie May
And 23h23'47" 42d50'19"
Hollie Moon
Gem 6h45'4" 26d17'14"
Hollie Nicole Abrahams
And 23h29'34" 42d6'56"
Hollie Pincott
And 1h28'3" 43d36'22"
Hollie Rebecca Lunt
And 2h31'27" 46d29'47"
Hollie & Ryan's Star.
Cyg 20h58'39" 45d26'56"
Hollie Stevenson
Cnc 8h8'5" 29d15'4"
Hollie Susan Davies
And 23h12'11" 42d9'34"
Hollie Yount
And 0h25'12" 31d20'38"
Hollie's Dipper
Uma 14h3'28" 50d58'1"
Hollie's Nightlight
And 23h14'3" 48d16'28"
Höllinger, Hilkka
Uma 10h18'13" 65d25'2"
Hollins
Leo 11h24'16" 13d35'26"
HOLLIS
Psc 1h12'17" 32d31'6"
Hollis Alexa Polk
Cnc 8h45'49" 25d7'4"
Hollis Amalie Hyneman
Aqr 22h36'26" -0d2'22"
Hollis Bradwell
Cet 1h0'57" -0d43'48"
Hollis F. McLennan & Brian K. Lee
Cnc 8h2'28" 11d36'20"
Hollis Jay Newcomer
Aqr 22h12'10" -3d24'0"
Hollis Nicole Heckler
Leo 10h31'45" 26d9'3"
Hollister
Uma 10h40'13" 58d36'30"
Holloway
Psc 1h5'26" 32d35'8"
Holloway's Vesper Stella
Cnc 8h23'43" 26d36'44"
Holly
Gem 7h24'54" 24d37'29"
Holly
Uma 11h28'45" 28d58'53"
Holly
Gem 6h44'20" 21d47'0"
Holly
Tau 5h44'14" 22d28'28"
Holly
Tau 4h37'30" 7d34'33"
Holly
Psc 0h13'29" 8d23'35"
Holly
And 2h15'17" 37d54'36"
Holly
And 1h23'46" 40d19'23"
Holly
And 2h11'3" 41d30'5"
Holly
And 23h6'26" 51d26'23"
Holly
Cyg 21h46'39" 42d20'27"
Holly
And 0h18'51" 45d38'18"
Holly
And 2h22'37" 49d39'43"
Holly
And 1h17'53" 45d30'41"
Holly
Uma 14h24'35" 61d18'53"
Holly
Cas 2h20'31" 66d45'15"
Holly
Cas 0h39'59" 69d25'12"
Holly
Mon 7h23'21" -8d52'20"
Holly
Cap 21h11'29" -21d3'3"
Holly
Vir 13h27'45" -21d38'56"
Holly
Lib 15h42'21" -17d5'32"
Holly
Cap 20h35'42" -12d5'3"
Holly
Lib 14h46'23" -11d47'37"
Holly
Lib 15h7'41" -14d2'38"
Holly
Sgr 18h4'30" -27d19'5"
Holly
Psc 1h49'6" 6d24'4"
Holly
Sgr 19h43'18" -33d43'5"
Holly Alana Moylan
Vir 14h15'27" 4d29'18"
Holly and Cailan
Ori 5h11'49" 6d18'35"
Holly and Doug
And 23h23'19" 51d27'24"
Holly and Eric Sitting in a Tree...
Cyg 19h33'57" 51d56'29"
Holly and Grace's Guardian Angel
Aql 19h53'4" 5d7'2"
Holly and Ken
Lib 14h49'56" -0d47'27"
Holly and Kevin
Cap 20h28'42" -22d56'58"
Holly and Owen Dry
Aqr 21h6'42" -12d45'30"
Holly and Rob Shine Forever
Cyg 20h0'51" 57d7'13"
Holly and Thomas
Cyg 19h35'26" 30d8'8"
Holly Ann
Psc 0h58'17" 13d58'25"
Holly Ann Bryan
Lib 15h31'19" -25d22'21"
Holly Ann Bryant
Ari 1h53'39" 19d50'26"
Holly Ann Clancy
And 0h24'43" 32d33'52"
Holly Ann Clark
Ori 5h20'37" 6d44'46"
Holly Ann Copeland
And 23h43'57" 38d57'19"
Holly Ann Heggli
Aqr 22h40'36" -8d30'52"
Holly Ann Hodge
Psc 0h42'44" 9d58'45"
Holly Ann Koerbitz
Sco 16h25'31" -33d28'49"
Holly Ann Marie
And 0h1'45" 46d38'32"
Holly Ann Moore
Com 13h6'48" 18d27'53"
Holly Ann Papovich
And 2h7'34" 37d33'49"
Holly Ann Teasdale
And 23h54'59" 40d56'8"
Holly Ann Wenig
And 0h2'31" 38d33'46"
Holly Anne
Sgr 18h2'18" -28d13'5"
Holly Baker
And 23h20'3" 42d1'43"
Holly Ballantine
Col 6h17'27" -34d55'38"
Holly Batten
And 2h35'2" 49d18'14"
Holly Belinda
Cas 3h12'32" 58d45'15"
Holly Berry Harris
Tau 4h13'39" 9d6'30"
Holly Beth
Cap 21h39'28" -8d37'15"
Holly Beth Wiley
Cnc 8h43'19" 8d10'55"
Holly Bixby
And 2h15'8" 38d25'12"
Holly Brewster
Uma 10h6'29" 43d6'25"
Holly Bug
Lyn 8h4'19" 55d19'47"
Holly Calderado
Her 17h28'55" 27d44'34"
Holly Campano
Lyn 7h7'12" 58d49'0"
Holly Carlson
Cas 1h3'48" 49d41'33"
Holly "Carnie" Daly
And 0h40'35" 40d31'16"
Holly Caron
Cas 1h38'27" 65d20'9"
Holly Carpenter
Leo 9h59'36" 13d55'38"
Holly Celeste Cordero
Ori 6h18'18" 13d59'37"
Holly Christina Wright
Uma 10h28'35" 47d51'37"
Holly Claire Temperley
And 23h2'33" 45d9'56"
Holly Colette
Psc 1h10'4" 18d11'3"
Holly Crawford
Com 12h46'18" 16d15'42"
Holly Crowley Walsh
Sgr 18h36'35" -23d55'45"
Holly Cunningham
Cas 0h1'40" 53d26'31"
Holly D. Williams
And 1h27'57" 38d24'18"
Holly D'Addio
Lib 14h53'2" -1d8'17"
Holly Dahlia
Peg 22h54'10" 20d2'32"
Holly Danielle Doucette
Uma 12h24'41" 55d11'5"
Holly Davis Kurtzeborn
And 2h37'35" 44d58'54"
Holly Dawn Gessley
Ori 5h29'46" -1d10'7"
Holly Dawn Kingston
Cyg 21h26'6" 39d17'38"
Holly Dawson
Vir 12h46'14" -11d16'52"
Holly DeBel
Cap 21h51'18" -18d20'38"
Holly Dee Butler
Ori 6h17'13" 9d35'27"
Holly Denise Sigona
And 0h21'14" 28d31'2"
Holly Dory
And 23h16'12" 52d15'16"
Holly Driscoll
And 1h9'9" 42d26'29"
Holly E. Nobes
And 0h39'27" 37d59'2"
Holly Edith Lugnan
Cru 12h27'14" -60d17'14"
Holly Elise
And 2h19'0" 38d35'52"
Holly Elizabeth
Gem 6h54'43" 22d24'3"
Holly Elizabeth Accorsi
Leo 11h6'32" 0d48'27"
Holly Faith Shaffer
Uma 9h26'38" 41d45'56"
Holly Famularo A shooting star
Lib 15h16'8" -27d24'35"
Holly Farrell
And 0h11'26" 45d31'38"
Holly Femano
Cap 21h20'44" -21d20'38"
Holly Fiona Cunningham
Uma 9h10'4" 64d41'4"
Holly Frances Krenz
Sco 17h0'16" -33d58'20"
Holly Franks & AJ
Peg 22h30'4" 10d33'21"
Holly Fuller
Lyr 18h44'54" 35d15'20"
Holly Gal Reinders
Sgr 19h20'58" -27d58'10"
Holly Gilbertson
Del 20h53'0" 3d41'38"
Holly Goldsmith
Cap 20h19'16" -10d59'18"
Holly Gordon
Peg 22h23'50" 20d46'9"
Holly Gould
Ori 6h15'39" 18d4'3"
Holly Grace Jones
Cas 1h24'9" 65d52'25"
Holly Grace Osborne
And 1h27'22" 44d13'50"
Holly Grace Usher
And 23h56'23" 44d14'22"
Holly Groat
Uma 9h37'12" 57d48'12"
Holly Gwen Thompson
Sgr 19h48'28" -23d24'56"
Holly Haddad
Lyn 8h5'16" 40d52'13"
Holly Harris
Aqr 22h40'34" -1d44'19"
Holly Hemminger
Lyn 6h50'39" 54d56'21"
Holly Hyde
Cas 0h21'17" 55d59'25"
Holly * ILYFAE *
Vir 14h35'28" 3d53'51"
Holly Inger Monk
And 22h58'24" 51d36'10"
Holly & Jamie Two As One
Cyg 20h1'56" 47d14'22"
Holly Jan Scoby
Cyg 21h10'38" 46d49'32"
Holly Janelle Perez
Sgr 19h53'13" -42d47'35"
Holly Jaquel Sheppard
Cnc 8h39'12" 10d3'38"
Holly Jarvis
Cap 20h29'25" -11d24'30"
Holly Jean Nicholls
And 2h20'15" 37d23'50"
Holly Jean Watson
Sgr 18h35'54" -23d50'55"
Holly Jessica Hearn - Holly's Star
And 23h52'21" 46d18'6"
Holly Joanna
Leo 9h41'51" 29d42'38"
Holly Jones
Cas 23h31'4" 57d49'5"
Holly Joy
Cap 20h24'59" -24d3'7"
Holly K. Gartin
Vul 20h43'35" 27d54'22"
Holly Kacala Nebula
Ori 5h12'24" 11d40'40"
Holly Katherine
And 1h15'58" 41d29'21"
Holly Kay McClew
And 23h23'25" 47d21'54"
Holly Killian - Love - The Uncle Kev
Sgr 19h12'44" -24d0'34"
Holly Kishfy
Crb 15h44'40" 38d25'33"
Holly Kristine
Oph 16h54'49" -0d16'33"
Holly L. Tucker
Vir 12h43'17" 1d0'54"
Holly Laura Jackson
Cas 0h49'10" 69d41'5"
Holly Lee Owen-Pauers
Sgr 17h58'18" -28d36'51"
Holly Liles
And 23h56'9" 42d52'30"
Holly Louise
And 23h55'45" 38d36'58"
Holly Louise LaPrease
And 23h24'25" 42d33'53"
Holly (Love Bunny)
Lep 5h4'39" -17d17'12"
Holly loves you....always
Sgr 18h15'11" -19d45'26"
Holly Lynn
Vir 13h32'13" -2d20'35"
Holly Lynn
And 2h17'41" 47d36'3"
Holly Lynn
Gem 6h20'29" 19d18'10"
Holly Lynn Curtis
Cap 20h47'10" -21d2'46"
Holly Lynn Pancoast
Cas 0h53'12" 56d23'32"
Holly Lynn Schaefer
And 1h31'5" 47d52'15"
Holly Lynne Hamilton
Sgr 18h19'8" -33d1'20"
Holly M. Fought
Vir 13h40'52" 4d7'21"
Holly M. McGuffin
Gem 6h53'4" 22d15'1"
Holly M. Stirtz
Mon 7h22'24" -5d8'0"
Holly M. Walters
Mon 6h48'15" 7d56'30"
Holly Madrinich "I Love You"
And 0h37'44" 25d38'24"
Holly Marie
Per 2h23'58" 55d51'27"
Holly Marie
Lib 14h53'6" -11d16'2"
Holly Marie
Sco 16h57'22" -40d58'31"
Holly Marie Bashor 3/02/1992
Psc 23h26'28" 6d19'1"
Holly Marie Harrison
Cas 1h58'15" 64d43'33"
Holly Marie Perucci 8/2/1992
Leo 10h14'41" 15d0'8"
Holly Marie Redding
Gem 6h44'9" 13d22'59"
Holly Marie Taylor
Umi 14h44'17" 72d39'57"
Holly Mears
Lmi 10h9'12" 30d56'8"
Holly Mercurio
Uma 9h6'8" 59d31'27"
Holly Michele
Cap 20h8'27" -25d38'49"
Holly Michelle Becker
Uma 8h39'41" 70d0'18"
Holly Michelle Martin
Sgr 19h6'53" -21d5'16"
Holly Michelle Simpson
And 23h21'57" 41d37'38"
Holly Mizer
Aqr 20h39'29" 1d21'0"
Holly mon Coeur Enchante
Per 3h19'51" 32d40'37"
Holly Morgan
And 1h20'8" 49d2'5"
Holly Nebgen
Lib 15h5'46" -7d24'4"
Holly Nicole Diehl
Aqr 22h35'1" -16d38'15"
Holly Nicole Preston
Aql 19h45'55" 6d40'39"
Holly Noel
Vir 12h49'50" 6d41'38"
Holly Noel
Cap 20h57'8" -20d26'33"
Holly Noel
Lep 5h53'51" -11d47'44"
Holly Noel
Sco 17h17'52" -32d35'3"
Holly Noel Wilding
Cap 21h44'44" -12d6'24"
Holly Padgett
Gem 6h50'2" 19d59'15"
Holly Patricia Gorrell
Uma 12h44'12" 54d39'52"
Holly 'Pea' Tate
Ori 6h12'34" 20d49'23"
Holly "Piggly Poo"
Sgr 18h27'7" -30d35'18"
Holly Pyle
Peg 21h39'29" 21d39'21"
Holly R Bond
And 0h0'38" 37d3'12"
Holly R. Neises
Psc 1h18'6" 23d22'6"
Holly Rebecca Kuman
Sgr 17h52'33" -28d39'43"
Holly Rene' Hodgin
Lmi 10h14'12" 34d59'6"
Holly Renee Hutzell
Sco 16h40'14" -37d0'55"
Holly Renee Northup
Per 3h22'39" 55d6'32"
Holly Robbins
Per 3h8'31" 42d44'6"
Holly Robson
Cam 5h20'47" 65d59'46"
Holly Rose
Cam 3h42'31" 58d31'28"
Holly Rose
And 23h30'42" 42d1'41"
Holly Sandra Harrison
Cas 23h17'10" 56d5'18"
Holly Schrum
Ori 4h49'21" 1d28'48"
Holly Shepard
Cnc 8h14'10" 16d9'28"
Holly Shonk
Sco 16h23'12" -18d44'42"
Holly Smith
Lyr 18h49'19" 35d31'1"
Holly Stanislawska
And 1h2'13" 42d22'35"
Holly Stephens
Aqr 21h12'23" 1d43'53"
Holly Stone's Radiant Light
And 0h29'22" 32d50'2"
Holly Sugar
And 23h33'11" 36d26'14"
Holly Sunshine Perez
Cap 20h36'27" -16d11'5"
Holly Suzanne Hauser
Aqr 22h36'48" -23d4'41"
Holly Taylor Detrick
Aql 19h50'46" 10d54'14"
Holly Tift
Sco 17h56'30" -38d24'10"
Holly Truscott
Cru 12h5'31" -60d25'0"
Holly & Umair
Cyg 20h10'9" 35d12'21"
HOLLY VAIL
Cap 20h31'20" -23d30'0"
Holly Ventura
And 23h14'3" 43d36'5"
Holly Wilson
Cyg 19h58'18" 32d47'45"
Holly Wright The Star of Angels
Lyr 18h51'57" 44d13'20"
Holly Wyner
And 23h26'35" 48d37'50"
Holly Ziegenhorn
Sgr 18h14'49" -28d35'10"
Holly, Harper and Butch Fromm
Ari 3h5'15" 10d45'39"
Holly, Love Matt
And 1h33'6" 48d35'3"
Holly1932
Ori 5h34'21" 13d15'11"
Holly-Berries Loan Doan
Gem 7h44'53" 34d19'58"
Holly-Beth
Cma 6h13'48" -21d48'17"
Holly-Beth Kraft
Peg 21h27'48" 18d5'28"
Hollye
Tau 3h46'4" 5d18'9"
Hollye L. Smith
Lib 15h2'31" -11d57'19"
HollyJoel
Gem 6h48'15" 20d50'42"
HollyJoGriffith Our Brilliant Star
Lyr 18h43'34" 38d26'5"
Hollymaximus
Dra 19h5'20" 70d28'49"
HollymyLove
Ari 2h46'21" 13d53'10"
Holly-O
Cap 20h7'39" -8d43'16"
Holly's Comet over the Marginal Way
Gem 6h53'21" 32d37'24"
Holly's Destiny
Cas 1h14'38" 62d36'58"
Holly's Diamond
And 1h12'18" 44d35'30"
Holly's Heart
Sco 16h57'40" -41d37'24"
Holly's Light
Her 17h5'15" 33d44'38"
Holly's Love
Lyn 8h1'47" 41d25'11"
Holly's Perpetual Diamond
Leo 11h14'59" 16d48'30"
Holly's Smile
Uma 9h5'58" 60d8'1"
Holly's Stage Light
Uma 12h0'15" 46d27'44"
Holly's Star
And 23h41'44" 48d4'27"
Holly's Star
Umi 15h6'29" 68d32'46"
Holly's Star
Uma 10h41'12" 55d43'4"
Holly's Star
Lib 14h38'7" -18d2'56"
Holly's Star
Vir 13h2'13" -2d4'46"
Holly's Star
Sco 17h52'57" -35d53'23"
Holly's World (Holly Taylor)
Lyn 6h37'22" 56d56'33"
Hollywog
Cap 20h7'34" -18d33'13"
Hollywogs
Uma 13h20'5" 59d36'28"
Hollywood
Aql 19h42'11" -0d4'38"
Hollywood
And 0h25'41" 40d31'58"
Hollywood
Uma 11h53'58" 34d48'19"
Hollywood
Cnc 8h47'40" 16d50'38"
Hollywood
Vir 12h14'8" 12d48'16"
Hollywood HillBilly Hogette
Sco 16h10'0" -30d34'20"
Hollywood Jessica
And 23h18'51" 48d0'14"
Holmer Fischer
Uma 9h15'7" 48d5'30"
Holmes (aka Bubbles) Beltran Jr.
Ori 5h20'14" 4d36'15"
Holstafa Robo-Cop
Uma 8h37'45" 64d56'36"
Holt Michael Cole
Cap 21h37'2" -17d39'20"
Holtorf, Hans-Joachim
Ori 4h47'15" -0d52'20"
Holt's Star
Per 3h41'46" 44d47'11"
Höltschi, Peter
Uma 11h4'31" 37d55'51"
Holy Stars Rottie Lover
Sco 16h37'58" -28d40'52"
Holy Trinity 3rd Grade 2005
Uma 11h40'57" 57d43'38"
Holz, Conny
Ori 6h21'10" 14d41'4"
Holzapfel, Otto-Werner
Uma 12h1'5" 44d43'31"
Hölzemer, Hans Jürgen
Ori 5h7'50" 8d42'15"
Holzhauser Gudrun
Uma 12h24'34" 58d6'42"
Holzknecht, Claudia
Sgr 18h0'8" -27d41'3"
Homa Jessica Shadpour Michaelson
Ari 2h58'10" 12d42'46"
Home
And 23h24'18" 47d46'5"
Home
Umi 14h39'58" 77d8'50"
Home
Cap 20h37'47" -9d44'28"
Home 5683
Umi 15h59'36" 76d26'26"
Home Is Where The Heart Is *JGNKB*
Her 16h10'42" 24d50'35"
Homer
Boo 14h25'56" 27d29'30"
Homer
Uma 10h58'13" 55d18'25"
Homer 84-05
Gem 7h14'18" 32d56'9"
Homer E. Guthrie
Cmi 7h48'13" 1d7'40"
Homer E. Kelley, Jr.
Gem 6h54'14" 17d29'8"
Homer & Ingeborg-50th Anniversary
Pho 0h20'24" -42d45'54"
Homer & Mosolene Shaw
Cnc 8h54'38" 31d35'9"
Homer P. Angel
Aqr 20h44'24" -10d16'44"
Homer Ruhe of Ottawa, OH
Her 17h32'24" 48d43'50"
Homer Thomas
Vir 14h1'55" -15d47'0"
Homerion
Uma 9h20'47" 58d8'43"
Homeward Bound
Cha 10h19'38" -77d47'43"
Homie
Uma 8h29'42" 64d29'6"
Homie
Ari 2h10'56" 26d5'2"
Homie D
Sgr 19h38'32" -20d52'10"
Homölle, Rolf
Ori 5h0'44" 5d51'58"
Homy
Ori 5h49'33" -4d40'14"
Hon
Aqr 22h39'11" -5d39'21"
HON
Psc 23h26'28" 3d49'34"
HON
Her 16h9'50" 43d58'59"
Hon. E. Leo Kolber Superstar
Umi 15h47'55" 86d24'25"
Hon. James M. Scanlon
Cyg 20h57'28" 46d7'14"
Hon. Lance Gough
Cyg 21h39'5" 50d25'59"
Hon. Langdon D. Neal, Chariman
Cyg 20h55'29" 47d22'53"
HON / Mariana
And 0h21'4" 31d0'55"
Hon * Mernvia-Jean * JAC
Umi 13h56'58" 77d6'51"
Hon. Thomas R. Leach
Cyg 21h19'16" 46d7'33"
Hone Hone Chris & Myla
And 23h32'35" 37d48'46"
Honestly Perfect
Hys 1h40'4" -58d48'6"
Honestus
Psc 1h35'18" 23d55'38"

Honesty
Cep 22h30'51" 65d18'36"
Honesty Jason and Morgan
Cyg 19h41'13" 46d31'52"
Honey
Cyg 19h36'29" 47d1'31"
Honey
Cyg 20h30'55" 37d47'7"
HONEY
Cyg 20h48'31" 43d4'30"
Honey
Lyn 8h32'36" 45d10'19"
Honey
Peg 22h13'58" 7d5'16"
Honey
Aqr 20h50'7" 0d59'6"
Honey
Ori 4h55'28" 9d43'11"
Honey
Ori 5h8'32" 0d37'8"
Honey
Cas 2h50'33" 61d3'15"
Honey
Uma 14h5'53" 61d42'52"
Honey
Uma 13h14'49" 62d36'57"
Honey
Uma 12h37'31" 54d8'19"
Honey
Cap 21h22'41" -15d53'23"
Honey
Lib 15h29'50" -9d56'35"
Honey
Lib 14h23'42" -24d12'36"
Honey 1
Cas 23h52'38" 57d49'8"
Honey 4 Bunny
Cru 12h23'13" -57d0'27"
Honey Angel: Gladis Isabel Robles
Cnc 8h54'56" 19d2'1"
Honey Bear
Cnc 8h51'32" 17d19'11"
Honey Bear
Psc 0h29'40" 10d42'20"
Honey Bear
Boo 14h57'24" 50d30'9"
Honey Bear
And 0h34'39" 42d6'24"
Honey Bear
And 0h38'40" 43d22'21"
Honey Bear
Sco 17h2'27" -41d56'15"
Honey Bear
Uma 11h14'48" 58d53'44"
Honey Bear
Vir 13h32'20" -4d7'51"
Honey Bear Eyes
Umi 14h6'21" 78d27'57"
Honey Bear Kennedy
Uma 12h8'46" 60d41'19"
Honey Bear Palmer
Uma 10h59'48" 49d27'34"
Honey Bee
Crb 15h38'44" 27d1'40"
Honey Blossom
Sgr 18h13'15" -20d6'39"
Honey Bun
Psc 1h19'10" 6d22'9"
Honey Bunch
Lib 14h53'59" -1d50'6"
Honey Bunch
Ari 2h20'52" 25d40'52"
Honey Bunches
Ori 4h51'55" 0d33'17"
Honey Bunches (Mrs. Greg Forrest)
Cas 0h20'52" 61d52'55"
Honey Bunches of Oats
Aqr 23h23'56" -13d58'6"
Honey Bunches of Oats
And 0h31'6" 31d19'11"
Honey Bunny
Leo 11h23'36" 21d39'59"
Honey Bunny
Leo 11h36'54" 13d42'38"
Honey Bunny
Lep 5h14'16" -11d45'28"
Honey Bunny
Cas 1h5'47" 63d34'31"
Honey Bunny
Sco 16h27'2" -27d14'21"
Honey Bunny Dan
Tau 3h47'12" 9d24'26"
Honey Buns
Per 2h21'35" 52d29'45"
Honey Buns and Carletta Tinsman
Sco 17h28'56" -31d51'9"
Honey DEW
Sgr 18h56'41" -26d0'40"
Honey Farshad
Uma 10h52'57" 50d46'5"
Honey Feickert
Cyg 19h53'40" 39d5'11"
Honey Holly Cooper Fox
And 23h43'24" 36d5'55"
Honey Honey Honey My Honey
And 0h34'9" 36d54'9"
HONEY II
Leo 11h17'1" 11d42'15"
Honey Jimenez Mitchell
Uma 8h43'40" 56d35'4"
Honey Leigh Edwards
And 0h30'10" 25d56'12"
Honey Michael Horbanz
Uma 9h15'59" 51d43'15"
Honey Olivia Wade
And 1h22'4" 34d52'25"
Honey (OT ILY FEY S)
Cam 7h53'20" 72d29'16"
Honey - our little Angel
And 23h24'48" 51d14'16"
Honey Pot
Leo 9h22'6" 16d13'58"
Honey Sugar
Cam 6h9'51" 64d24'12"
Honey Sweet Dionne
Crb 15h36'5" 28d6'55"
Honey & The Moon Jenn 26-11-04 Adam
Cra 18h11'18" -40d14'36"
Honey & Walter's Piece of Heaven
Cyg 20h53'37" 46d42'35"
Honey Wife
Cas 1h22'6" 62d32'40"
Honey, Forever The Light Of My Life
Ari 3h27'56" 28d59'49"
HoneyB
Per 3h42'39" 36d37'33"
Honeybear
Cyg 21h33'10" 45d43'4"
Honeybear
Lib 15h1'46" -16d38'32"
HoneyBear-F
Tau 5h40'45" 24d31'55"
honeybear"kristin"star
Gem 6h51'30" 14d2'4"
Honeybee
Tau 3h38'28" 16d2'36"
HoneyBee
Gem 7h19'2" 34d46'59"
Honeybee
Umi 17h0'25" 86d33'59"
Honeybee
Cas 1h40'26" 61d0'12"
Honeybee Hideaway
Uma 12h33'53" 54d27'10"
HoneyBee621
Cnc 8h43'26" 8d2'54"
Honeybun
Cep 22h14'47" 59d8'45"
Honeybunny
Psc 1h7'23" 15d27'12"
Honeybunny
Cnc 8h10'0" 28d54'11"
HoneyBunny Diamond
Lib 15h59'19" -16d54'17"
HoneyBuns
Cyg 20h27'31" 46d37'36"
Honeycutts' Light Spectacular
Cyg 19h33'17" 31d10'54"
Honeydoo
Eri 3h42'31" -0d23'16"
Honeygirl
Cas 0h59'55" 62d25'22"
HoneyHoneyHoney
Peg 22h16'52" 24d16'34"
Honey-Joe Burnett
Gem 7h44'16" 15d55'57"
HoneyMomma
Vir 13h14'5" 6d33'47"
Honeymoon
Uma 9h46'41" 56d58'52"
Honeymoon - Willi & Brigitte Bieri
And 2h12'5" 41d0'18"
Honeypie's Star
Aqr 22h11'6" -3d0'57"
HoneyPoppie
Uma 8h53'11" 52d4'43"
Honey's Love
And 1h3'57" 37d15'45"
Honey's Star
Cnc 8h6'5" 18d31'25"
Honey's Star
Uma 14h15'19" 55d47'46"
Honey's True Love - Jerry
Uma 13h3'55" 59d46'19"
Hong
Vir 13h26'22" 13d22'15"
Hong Shan
Aqr 21h16'44" -8d3'17"
Hong Xi
Lib 15h0'25" -17d14'20"
Hong Yue
Cyg 19h33'58" 32d29'46"
Hong Zheng
Psc 0h18'42" 20d39'41"
HongVanTP
Lib 15h46'25" -5d3'26"
Honi Faye
Gem 6h32'30" 24d52'2"
Honi & Paul Forever
Ara 17h17'29" -52d13'13"
Honigbär
Cas 0h4'16" 52d48'13"
Honigkuchenseepferdchen
Cep 21h19'32" 66d24'13"
Honigmüsli
And 2h28'2" 44d11'3"
Honi-Mai Maisey
Peg 22h16'26" 19d51'39"
Honiola (Olga Libman)
Gem 6h25'36" 25d38'22"
Honkie Frederixon
Aql 19h4'10" 4d14'4"
Honkie1215
Ori 6h18'54" 13d38'18"
Honk's Forever Star from EY
Lib 14h42'57" -8d49'0"
Honoka
Tau 5h9'29" 16d17'11"
Honolulu Bor
Ori 5h55'55" 20d47'46"
Honor and Freya Kesteven
And 2h13'17" 48d48'59"
Honor Elizabeth Sarah
And 23h32'59" 46d23'32"
Honor Hailey Sewell
Uma 10h38'21" 45d13'15"
Honora Carmel Gillen
Cas 1h43'51" 61d47'11"
Honora June Larock
Cyg 20h56'0" 45d47'23"
Honorable Anthony J. DePanfilis
Her 18h42'9" 13d5'52"
Honorah Whittaker
Col 6h4'54" -42d41'27"
Honore
Psc 0h32'49" 12d11'46"
Honoré Le Conte de Poly
Cnc 8h17'1" 29d59'0"
Honorio
Aqr 20h55'22" -13d53'20"
Honors3B
Uma 9h30'57" 54d45'2"
Hontas
Psc 1h16'12" 16d8'50"
Honugirl
Cnc 8h18'11" 25d31'49"
Hoo Doo
Uma 9h20'8" 46d46'16"
Hooch
Uma 9h4'21" 49d14'28"
"Hook em Horns" Petty # 62
Boo 14h33'26" 12d52'4"
Hooker
Umi 14h23'4" 68d33'40"
HOOLEY
Umi 16h39'35" 79d57'23"
Hoolie RBR's 30
Lyn 7h38'10" 37d58'15"
Hoonie - Hunie
Lib 15h1'9" -16d36'51"
Hoonis
Cyg 21h37'20" 46d51'36"
"HOOP"
Uma 10h49'41" 55d50'54"
Hooper
Uma 12h33'48" 58d7'48"
Hooper & Nellie Collier
And 0h16'35" 29d51'39"
Hooperman
Her 17h19'57" 31d38'8"
Hoops-Harrison (souls as one)
Uma 9h36'1" 53d21'8"
HOOSIER6
Crb 15h52'18" 34d50'8"
Hoot/Unconditional Friendship
Ori 5h52'34" 7d9'19"
Hootchie Mama Linda
Crb 16h1'17" 38d49'5"
Hootersstern
Uma 8h53'19" 61d46'30"
Hootie-Scootie
Cru 12h39'47" -59d44'57"
Hooty McBoob
Cyg 20h31'24" 31d10'32"
Hoovens Realm
Uma 9h11'48" 59d13'1"
Hoover
Leo 10h21'37" 16d56'52"
Hoover091746
Vir 14h22'20" 3d6'41"
Hope
Leo 10h16'29" 6d56'1"
HOPE
Aql 20h18'35" 2d58'14"
Hope
Aqr 21h41'15" 0d43'4"
Hope
Peg 22h26'38" 3d35'13"
Hope
Ori 5h17'25" 8d7'8"
Hope
Aql 19h10'9" 11d13'55"
Hope
Gem 6h31'27" 12d32'56"
Hope
Ori 6h8'51" 13d29'36"
Hope
Gem 6h55'39" 16d54'35"
Hope
Tau 5h24'21" 26d56'31"
Hope
And 0h20'34" 32d26'59"
Hope
Cyg 20h42'43" 35d10'18"
Hope
Lyn 8h8'0" 41d30'50"
Hope
Lyr 18h52'6" 35d18'7"
hope
Lyr 18h45'27" 33d46'20"
HOPE
Boo 14h26'44" 41d45'25"
Hope
And 2h13'38" 45d33'33"
Hope
Per 4h45'14" 48d8'20"
Hope
Uma 9h10'14" 50d54'16"
Hope
Per 2h21'16" 55d59'46"
Hope
Cas 1h14'10" 58d11'29"
Hope
Lyn 7h42'24" 53d56'43"
Hope
Lyn 7h54'50" 52d53'2"
Hope
Cep 22h20'6" 58d41'48"
Hope
Dra 17h55'53" 56d50'12"
Hope
Cyg 21h14'47" 54d20'42"
Hope
Uma 12h3'12" 61d19'58"
Hope
Uma 9h31'34" 67d37'29"
Hope
Sco 16h4'40" -18d47'49"
Hope
Cap 20h20'31" -16d17'7"
Hope
Cap 21h54'28" -8d29'53"
Hope
Lib 15h29'42" -9d32'54"
Hope
Sco 16h10'25" -13d41'0"
Hope
Lib 15h32'38" -5d43'26"
Hope
Cma 6h57'0" -23d4'12"
HOPE
Sco 17h52'32" -36d27'46"
Hope 3996
Psc 1h43'59" 20d11'23"
Hope #5
And 0h26'53" 40d40'18"
Hope A. Gricar
Ori 5h28'31" 2d16'55"
Hope A. W.
Sgr 18h7'53" -17d8'51"
Hope Alexis
Lmi 10h39'27" 36d31'9"
Hope and Love!!
Umi 16h58'58" 79d24'18"
Hope Angelina
And 2h18'55" 44d52'35"
Hope Angelina
Psc 1h15'12" 25d53'3"
Hope Ann Fox
And 1h37'9" 38d35'11"
Hope Ashley Rideout
Ori 6h19'4" 18d45'31"
Hope BartgisKandhari
Psc 1h26'1" 13d25'36"
Hope Bernier
Peg 21h38'7" 15d33'45"
Hope Birmingham
Sco 17h33'42" -36d58'13"
Hope Carrell
Lyr 18h49'27" 38d20'25"
Hope Catherine Kersey
Peg 22h54'24" 31d20'18"
Hope Cernea's Star
Aqr 21h52'41" 0d17'53"
Hope Christin Ann Casale
Tau 3h29'43" 13d13'21"
Hope Cornish
Lyn 7h13'40" 56d23'59"
Hope Elise Donley
And 1h41'1" 47d52'24"
Hope Elizabeth
Tau 4h39'55" 0d48'24"
Hope Elizabeth Andersen
Tau 5h37'9" 25d47'57"
Hope Elizabeth Duncan
And 1h48'37" 42d53'39"
Hope Elizabeth Hemby
Gem 7h34'12" 29d46'28"
Hope Eliza-Marie Villarreal
And 0h26'41" 31d41'41"
Hope Elyse Hayes
And 1h59'23" 41d19'47"
Hope Eternal
Cyg 20h22'24" 30d59'43"
Hope Faith Love
Uma 14h6'36" 61d28'6"
Hope Fischer Racca
And 1h3'19" 40d33'5"
Hope for Alex & Sean
Sge 19h37'57" 18d4'33"
Hope for Andrew
Lyn 6h34'33" 54d16'26"
Hope Fuselier
Ori 5h34'58" 2d19'33"
Hope G.F.
Gem 6h53'27" 32d55'20"
Hope- I'll Love You Forever & A Day
And 0h58'57" 38d44'23"
Hope Isabella McClane
Cyg 21h46'41" 44d28'50"
Hope Jemima
Cas 23h22'8" 57d53'54"
Hope Leia Mary "Our Princess"
And 1h24'12" 47d43'14"
Hope Leilani Wehrle
Gem 7h43'23" 17d0'37"
Hope Lorelei
Cap 21h57'4" -21d47'32"
Hope Luckhardt-Flake
Leo 10h53'4" 14d33'24"
Hope Marie Blackwell
Lib 14h29'13" -10d14'20"
Hope Marie Boozer
Vir 13h4'25" -3d36'55"
Hope Marie Peralo
And 1h57'15" 43d56'19"
Hope Marie Robinson
Col 5h40'21" -28d43'3"
Hope Mary
And 23h15'52" 43d2'26"
Hope Noel
Psc 0h43'4" 15d19'56"
Hope of Anisa
Vir 11h58'42" 12d58'29"
Hope of Silverdragon
Uma 8h51'12" 60d39'9"
Hope of the Heart
Cyg 20h36'54" 52d15'26"
Hope O'Neill
Sco 16h7'15" -14d57'8"
HOPE our star
Vir 13h13'30" 7d0'3"
Hope Powell
Umi 15h41'7" 77d26'29"
Hope Reborn
Ari 3h15'38" 24d44'16"
Hope Rene' Williams
Uma 9h34'48" 68d29'43"
Hope Rogers, Hope Shines Eternal
Crb 15h39'17" 39d11'1"
Hope Star
Umi 19h22'14" 87d20'39"
HOPE & STRENGTH
Psc 0h41'19" 9d7'44"
Hope Swisher
Uma 13h3'53" 53d42'46"
Hope Ton
Lyn 7h11'53" 47d40'28"
Hope Wilbur
Lyn 9h1'10" 41d34'50"
Hope, Faith, Love and Strength
Uma 11h22'46" 48d44'20"
Hope, May It Always Sparkle for You
Aqr 23h9'11" -22d17'20"
Hope, Mother, Wife, and Friend
Lib 15h3'3" -0d33'23"
Hopeiness Skies in Bloom
And 23h37'7" 47d30'49"
hopelandic.4
Per 4h44'32" 49d13'4"
hopeless romantic
Her 16h46'27" 7d6'5"
Hope-Liberty
And 1h18'9" 41d14'43"
Höper, Elmar
Ori 6h19'51" 10d32'2"
Hopes and Dreams
Cap 21h31'30" -18d1'35"
Hope's Ray of Light
Gem 7h41'38" 33d47'36"
Hope's Star
Peg 21h50'43" 36d30'40"
Hope's Star
Leo 9h40'47" 7d10'20"
Hopestarius
Vir 12h54'53" 4d5'18"
Hopey Forever
Aqr 23h12'49" -10d28'11"
Hopey Lu
Uma 9h57'37" 51d30'57"
Hop-Hop-Shun(h)
Umi 15h11'36" 75d20'43"
Hopkins Family Star
Sex 9h41'40" -8d30'37"
Hopo
Ori 6h14'23" 20d53'17"
HopoCruzer31355
Umi 15h8'17" 69d40'7"
Hoppe, Eberhard
Ori 5h7'13" -1d43'46"
Hoppe, Lutz
Uma 10h40'56" 41d57'48"
Hoppelhase 310371
Sabine Brückner
Uma 12h26'12" 60d6'4"
Hopper
Aql 19h58'40" 13d43'38"
Hoppers Highlighter
Lac 22h21'41" 53d40'36"
Hopper's Inferno
Cyg 21h16'29" 46d35'18"
Hoppi
Umi 10h48'20" 88d41'50"
Hoppy
Cas 1h23'16" 52d14'44"
Horace Edward White Robertson
Leo 11h15'0" 6d48'47"
Horace & Glenda Reid - 6 June 1958
Col 5h49'20" -34d18'24"
Horace H Heidt
Lib 15h54'12" -12d31'39"
Horace Sherwin
Her 17h25'32" 28d6'6"
Horace Southward
Cen 13h31'37" -62d32'11"
Horace Stanley White 11
Aql 19h15'57" -0d0'39"
Horace Tiret
Sco 17h35'9" -39d19'39"
HoraceAllen 1920
Cas 1h8'20" 60d48'45"
Horacio Gonzalez
Lib 14h47'53" -10d54'47"
Horacio Reyes
Psc 1h11'42" 16d18'36"
HoranNickBarb
And 1h42'24" 44d29'15"
Horatio Purcell
Aur 5h52'43" 46d1'49"
Hörmann, Ingeborg Georgine
Uma 8h19'35" 68d3'20"
Hörmann, Marco
Uma 10h41'34" 54d28'14"
Hormoz Mottaghian-Milani
Per 4h12'57" 32d15'35"
Hörnchen
Uma 9h58'33" 42d39'1"
Hörnschemeyer, Albert
Uma 8h37'15" 58d59'6"
Horretta and Rodney Wilkins
Uma 11h26'35" 36d16'26"
Horse
Uma 11h14'17" 42d27'2"
Horsebaby
And 2h11'4" 49d27'30"
Horseshoe Lew - Lewis S. Frindt
Nor 16h17'20" -60d13'25"
Horst
Leo 11h21'54" -5d36'28"
Horst
Uma 10h42'59" 42d34'51"
Horst and Anneliese Kuhn
Dra 18h39'58" 50d20'5"
Horst Baganz
Uma 9h8'14" 63d34'1"
Horst Dieter Vogt
Ori 4h49'15" -2d13'7"
Horst Georg Bernhard
Ori 5h52'30" 7d42'23"
Horst Goetz
Uma 9h23'32" 49d35'16"
Horst Grossgebauer
Uma 9h56'41" 63d6'22"
Horst Grubert
Ori 6h16'10" 14d21'5"
Horst Habenicht
Ori 6h7'47" 7d54'28"
Horst Keilbach
Psc 0h32'27" 19d59'57"
Horst Kuhn
Ari 3h12'42" 21d10'53"
Horst Langner
Ori 5h10'13" 8d13'15"
Horst Merz
Uma 8h58'41" 61d11'41"
Horst Müller
Uma 8h36'17" 57d20'5"
Horst Oberbloibaum
Uma 10h34'41" 48d15'46"
Horst Papenfuß
Uma 8h18'40" 69d20'1"
Horst Peter Heyduck
Uma 9h32'14" 51d15'16"
Horst Pieritz
Uma 13h48'2" 58d41'30"
Horst Poellinger
Uma 13h18'9" 56d57'9"
Horst R F Nitzsche
Cap 21h22'31" -18d48'14"
Horst Radon
Ori 4h46'35" 1d8'24"
Horst Rekel
Psc 2h3'56" 9d2'39"
Horst Rosenau
Uma 10h16'52" 46d56'42"
Horst Sauter
Uma 10h33'25" 42d7'7"
Horst Schneider
Per 3h46'40" 51d10'2"
Horst Siegfried
Uma 8h17'43" 60d19'6"
Horst Tietze
Ori 6h17'8" 9d14'28"
Horst Trost
Uma 10h9'22" 51d51'31"
Horst Ulbrich
Uma 8h43'58" 50d55'10"
Horst Waclawek
Ori 5h54'59" -2d44'38"
Horst Walter Schmidt
Uma 9h9'37" 72d40'7"
Horst Zehrt
Uma 10h15'4" 46d7'20"
Horst-Dieter Jänicke
Ori 6h16'35" -1d6'28"
Horstl Jun. 4.2.1980
Uma 11h17'17" 38d31'47"
Hortense
Del 20h35'15" 14d58'25"
Hortense
Tau 4h38'38" 10d43'21"
Hortensia Bollettini
And 1h33'45" 41d33'56"
Hörtner, Patrizia
Uma 14h9'56" 56d45'43"
Horty's Halo
Lib 14h46'50" -0d54'3"
Horus
Uma 11h43'21" 50d47'47"
Horváth E. - Lily a sivatag rózsája
Cas 23h52'57" 58d44'0"
Horváth Levente 1987.01.27.
Cas 1h22'3" 68d39'6"
Hosanna and Alison Deforest
Cas 0h56'6" 63d24'48"
Hosanna O'Brien
Lmi 10h40'3" 24d56'58"
Hosanna Regina Mary Dina M Deforest
Gem 6h36'43" 21d52'18"
Hose-A
Leo 10h15'51" 22d4'49"
Hosea & Jenique Jones
Cyg 19h47'32" 56d51'23"
Hoseamandias
Aqr 22h14'29" 1d54'24"
Hosein Monzavi Father of New Ghazal
Lib 15h11'57" -27d50'0"
Hösel, Falk-Georg
Ori 6h20'54" 13d29'43"
Hoseymoni
Sgr 17h53'23" -16d26'11"
Hoshi Shoujo
Lyn 8h5'35" 34d49'30"
Hoshie
And 0h48'28" 37d55'10"
Hoshiko
Psc 1h26'21" 3d55'33"
Hoshiko-sama XXIII XIII Yori
Sco 17h28'53" -35d56'37"
Hoskins: Until the star fades xxxx
Cyg 20h5'12" 37d51'59"
Hospice of The Valley (Dobson Home)
Uma 11h30'46" 32d31'58"
Hospice of York Clients
Uma 8h28'53" 62d28'6"
Hossat Al Saud
Mon 7h6'9" -5d47'53"
Hossein Ghorbani
Umi 17h10'3" 80d53'11"
hoSTAR
Ari 3h21'40" 29d46'17"
Hostein Sylvie
Psc 1h3'28" 10d58'24"
Hostelet Emmanuelle
Lib 15h40'8" -15d49'18"
HosterBach
Lyn 7h50'12" 47d1'27"
Hostos Community College
Uma 11h21'47" 33d2'17"
Hot Chocolate
Cam 4h17'42" 62d40'4"
Hot Mama
And 0h43'32" 39d28'37"
Hot Momma
Leo 9h23'4" 11d6'20"
hot n steamy go0dness
Psc 23h11'6" 6d43'51"
Hot Rod Granny
And 2h21'52" 41d27'21"
HOT ROD LINCOLN
Pho 0h52'26" -57d39'55"
Hot Rod Mama
Cas 1h23'26" 62d42'17"
Hot Sauce, our beautiful friend
Uma 10h37'33" 44d48'58"
Hot Shot Guynes
Leo 10h11'22" 12d42'32"
Hot Stuff
And 2h35'22" 39d43'46"
Hot Tub
Ori 4h50'18" 5d51'5"
Hota-Pe
Dra 17h6'43" 68d2'37"
HotBoxer
Lmi 10h36'48" 37d54'41"
Hotboy Hartmann
Per 3h16'24" 44d20'12"
Hotel Waldhaus Sils-Maria
Uma 12h44'47" 56d28'50"
HotFire643
Lyn 8h57'33" 38d3'38"
Hotness
Cnc 8h28'50" 12d52'6"
Hots
Cmi 7h26'58" 8d57'48"
Hottie
Tau 4h11'8" 11d46'13"
Hottie
Crb 15h37'30" 27d18'31"
Hottie chappy69
Cap 21h3'36" -16d22'28"
Hottie Tricia
And 2h8'11" 43d25'50"
HottieScotty's BB Babycakes
Lyn 7h50'59" 53d38'51"

Hou Hsiao-Hua
Ari 2h30'23" 21d56'19"
Hou Ohana
Gem 7h5'20" 19d29'40"
Houbi Doubi
Cap 20h49'29" -14d58'4"
Houdini
Sgr 20h2'52" -27d18'45"
Houke Hoeksma
Per 2h39'21" 55d56'43"
Houligan
Boo 14h36'52" 32d4'43"
Hourez's Heart
Sco 16h42'43" -30d39'11"
Houry Jean Telvizian
Tau 4h23'48" 14d29'23"
House of Phoenix
Vir 12h8'49" -4d23'46"
Housh you are my "Shining Star"
Per 2h53'25" 56d5'52"
Houssam - Nadine 10 Sept 2004
Cyg 20h45'32" 40d30'36"
Houston
Peg 22h14'17" 32d56'13"
Houston
Her 17h59'4" 22d28'36"
Houston Butler
Aqr 22h57'10" -10d28'11"
Houston Chase Hickerson
Tau 5h2'39" 23d9'19"
Houston - Donovan
Aql 19h6'45" 3d6'24"
Houston Hard Rock Cafe'
Boo 14h52'43" 35d47'1"
Houston High School's Class of 1990
Crb 15h18'46" 29d39'22"
Houston Person
Sco 16h6'51" -17d10'3"
Houston Stone Yelland
Cep 22h30'7" 65d30'31"
Houston Whaley
Lyn 7h27'2" 54d52'41"
Houtan Dayhimi
Lyn 7h16'20" 58d47'14"
Hovatter -1
Uma 10h57'2" 60d10'10"
Hover and Pooh's Happy 39
Uma 8h32'43" 66d55'8"
Hovik & Lilit Abdunaryan
Mon 6h52'47" -0d14'58"
How do I love thee Bob, Let me count the ways
Cep 21h49'41" 62d54'53"
Howard
Vir 13h1'38" -3d40'21"
HOWARD
Leo 11h32'47" 26d39'31"
Howard
Ori 5h51'1" 12d16'46"
Howard
Per 4h29'56" 43d48'13"
Howard A. Grossman
Ari 2h6'41" 10d37'3"
Howard A. & Sandra M. Gaston
Cyg 20h43'42" 36d46'56"
Howard A. White
Her 17h8'4" 33d51'3"
Howard Aaron Pincus
Cyg 20h37'45" 53d49'21"
Howard and Joann Dail
Crb 16h7'37" 37d59'30"
Howard and Kim's Star
Cyg 19h46'40" 38d51'21"
Howard and Marion Imme
Cyg 21h57'45" 49d23'39"
Howard and Marjorie Brown
Cyg 20h42'23" 43d40'37"
Howard and Sandy
Vir 14h15'4" -15d59'55"
Howard Ann James
Del 20h40'39" 13d52'32"
Howard B. Adams
Sgr 19h44'22" -12d14'53"
Howard B. Lawson
Cap 20h22'14" -14d44'50"
Howard Benedict
Ori 5h19'56" 1d45'55"
Howard Bond Flower
Ori 5h21'41" 1d14'40"
Howard Booth Memorial
Uma 11h42'7" 61d39'18"
Howard Brown
Her 16h55'16" 17d55'10"
Howard Cohen, Star of our Family
Lyn 7h56'14" 42d11'18"
Howard Crain
Ari 2h23'54" 26d44'43"
Howard Crist
Ori 6h21'8" 14d19'29"
Howard Damon Vitkus
Boo 14h30'43" 19d51'42"
Howard Deshong IV
Her 16h38'51" 22d59'17"
Howard Dubman
Uma 8h53'5" 66d59'57"
Howard "Dusty" Rhodes
Uma 12h27'49" 52d29'40"
Howard E. Kane Esq.
Ari 2h23'15" 21d44'14"
Howard Egnor Ford
Crb 15h49'54" 37d54'31"
Howard & Erma Rust
Ori 5h41'57" 1d37'50"
Howard Eugene Kunder
Dra 18h53'10" 62d56'27"
Howard Faber
Cep 21h34'13" 65d51'31"
Howard Fisher
Uma 11h21'7" 60d28'11"
Howard F.Stiles Jr. 06/10/1941
Gem 7h31'36" 30d8'54"
Howard G. Bickel
Lyn 6h17'21" 55d3'10"
Howard G. Jones
Sco 16h15'36" -11d44'26"
Howard & Gale Samarin Anniversary
Gem 7h44'55" 26d21'19"
Howard Gerald's Damn Star,Y'all !!
Tau 3h43'12" 7d27'20"
Howard Glover
Cru 11h58'28" -59d38'27"
Howard Gurland
Ori 5h21'12" 3d15'21"
Howard H. Hall - Poppy
Vir 13h59'36" -20d49'29"
Howard H Lane
Ari 2h5'17" 18d43'45"
Howard & Heather Shipman
Per 3h31'16" 34d12'36"
Howard & Helen Wallace
Cyg 20h38'2" 39d30'34"
Howard Horcher
Her 16h27'2" 47d30'35"
Howard J. Pomroy
Ori 5h20'57" 0d35'42"
Howard Jay Fulfrost
Lyn 6h40'54" 54d24'22"
Howard Johnson
Per 3h9'44" 53d6'6"
Howard Jordan
Cma 7h4'11" -30d17'38"
Howard K Schroeder
Dra 16h57'29" 62d30'37"
Howard Kimmel
Lyn 8h30'43" 57d9'55"
Howard Kreitzman
Dra 19h54'56" 70d49'48"
Howard L. Goss
Cep 22h17'17" 72d35'1"
Howard Loves Susan
Per 4h28'23" 35d31'12"
Howard M Sartain
Per 3h27'28" 44d1'53"
Howard & Marie Hale
Cyg 20h54'34" 42d22'5"
Howard Michael Gordon
Uma 11h23'29" 42d36'9"
Howard Miskin
Aur 6h28'6" 34d36'22"
Howard My Friend Parrish
Sgr 19h15'49" -27d31'3"
Howard N. Goetsch
Boo 14h53'14" 25d35'7"
Howard Norman Stark
Umi 16h20'4" 72d23'43"
Howard O. Stafford
Ari 2h6'24" 23d2'45"
Howard Pearcy
Uma 11h30'50" 62d55'1"
Howard Price, A Fine Man.
Lyn 7h34'56" 37d27'38"
Howard R. Grifhorst III
Cap 21h29'0" -8d47'25"
Howard R. Hanna
Gem 6h41'12" 14d8'41"
Howard Ralph Geesman
Per 3h8'13" 51d2'6"
Howard Ray Shoemaker
Per 3h6'59" 38d18'58"
Howard Reinert
Cep 22h14'5" 72d28'54"
Howard Richard Safir
Aqr 23h48'36" -22d58'47"
Howard Rideout
Cep 22h18'11" 62d46'22"
Howard Riggs
Sgr 19h18'21" -14d42'6"
Howard Roy Liebowitz
Cap 21h7'27" -19d17'28"
Howard S. Oldham III 6-8-1965
Uma 14h15'25" 58d27'19"
Howard Sandberg
Per 3h52'21" 48d19'19"
Howard Scott Krin
Tau 5h40'52" 18d54'27"
Howard Scott Steng
Vir 13h6'13" 12d47'33"
Howard Sego Rorke
Ari 3h11'53" 27d25'10"
Howard Seth Lupowitz
Aur 5h57'14" 42d48'44"
Howard Settle
Cap 21h16'48" -25d37'37"
Howard Shapiro
Aur 6h2'30" 35d38'28"
Howard T Young
Aur 5h43'56" 52d31'58"
Howard The King
Aql 19h44'47" 6d40'27"
Howard W. Lahnemann
Lib 15h5'19" -15d27'50"
Howard, Vicki, Alex and Jeff Loff
Peg 23h27'40" 25d7'52"
Howard's Star
Aqr 21h12'58" 0d48'2"
Howard-The Peddler
Tau 4h44'9" 24d33'31"
HowardTricia
Cen 13h41'57" -43d10'51"
Howat-O'Reilly July 14th Paris 2007
Lmi 10h18'31" 37d16'57"
Howatto
Tau 3h33'1" 10d50'27"
Howdy Weaner
Aql 19h46'7" -0d13'31"
Howell Ray Fell Sr.
Her 17h35'29" 17d43'52"
HowellMarkel
Vir 11h49'52" 3d26'43"
HOWIE
Lyn 7h25'44" 50d24'19"
Howie & Allison
Ori 5h38'13" 3d6'37"
Howie and Angie Anne
Cyg 20h22'47" 52d38'57"
Howie Bayer
Cap 21h9'37" -14d49'8"
Howie / Dad / Grandpa
Cep 21h31'40" 63d53'59"
Howie Herndon
Her 17h7'55" 19d15'51"
Howie & Janet's Bon Jovi Star
Cyg 21h45'11" 40d24'24"
Howie Yatcilla
Per 3h11'59" 54d35'28"
Howling Acre's Khandu
Uma 11h35'21" 59d51'1"
How's Shining HYBJM Father's Day 07
Cep 20h36'58" 63d27'45"
Howwie
Tau 5h18'24" 20d33'41"
Hoyer, Jörg
Uma 9h47'38" 44d25'45"
Hoyi
Uma 11h38'49" 46d37'20"
Hoyt
Her 16h27'49" 48d25'14"
Hoyt Matthew Landis
Lac 22h50'20" 39d10'9"
Hoyt Stanford
Tau 5h9'0" 17d21'36"
Hoyt's Star
Gem 7h18'12" 19d42'22"
H.P. Iten
Cyg 21h27'49" 48d48'8"
HPDA:29:10:04:20:45:GMT :53.3:-1.5
And 2h17'21" 42d34'59"
H.-P.'s Zauberstern
Cas 0h2'34" 52d32'42"
Hrastnik, Horst
Uma 9h51'33" 65d43'7"
HRB 10-10-20
Umi 15h51'58" 76d31'37"
H.R.H.
Ori 5h31'11" 1d19'18"
HRH Alwaleed Bin Talal
Ori 5h17'36" 1d23'43"
HRH Marta
Cra 18h18'27" -41d2'1"
HRH Princess Rokhshan B. Wali
And 0h24'49" 32d43'30"
HRH Vada
Cap 20h38'12" -26d58'37"
HRH's Grandpa Hockey-Jack Rose
Crt 11h22'23" -17d48'18"
Hrimon Rakshit
Per 3h33'3" 44d42'11"
Hripsik Holo Shatikian
Leo 9h45'50" 21d49'45"
Hristu Bogdan Chepa
Leo 11h15'38" 3d20'41"
Hroberaht
Sco 16h34'10" -41d28'21"
H.R.S.-Maximus-V
Per 2h59'50" 40d32'32"
Hruska
Umi 15h24'57" 73d5'44"
HryuHryuMrik
Vir 13h24'24" -9d58'47"
HSIN-YI
Leo 10h41'32" 14d6'37"
Hsiuling Lu
Uma 11h46'23" 34d10'20"
HSS Bethany
And 22h58'17" 47d15'47"
H.S.S. - Heather Sameera Salame
Tau 4h46'51" 17d5'40"
Hsueh Chen
Dra 12h16'54" 72d17'58"
HSWF Volunteers
Uma 10h43'8" 51d35'8"
HTH FIND
Cru 12h9'25" -62d50'46"
Huang Yao Wei
Gem 6h42'16" 18d26'51"
HUB
Uma 10h9'31" 43d1'44"
Hub Bub
Dra 19h17'57" 67d26'33"
Hubba & Wifey
Tau 5h25'58" 27d12'15"
Hubbell
Uma 12h13'45" 62d25'32"
Hubbie
Aqr 22h24'54" -19d14'33"
Hubby Bubby Hennings
Her 17h17'26" 15d27'19"
Hubby & Tide
Uma 9h2'48" 62d48'32"
Hube-1-Kanube
Boo 14h53'3" 52d59'22"
Huber Ferenc
Leo 9h23'54" 11d44'46"
Huber, Hans
Uma 10h11'59" 55d39'54"
Huber, Siegi
Uma 13h36'50" 58d20'18"
Hubert
Uma 10h27'6" 66d47'8"
Hubert
Tau 5h58'6" 26d51'56"
Hubert A. Wudke, Husband, Dad & Opa
Vir 13h20'6" 4d22'39"
Hubert B. Haywood III
And 2h24'49" 46d35'34"
Hubert Bailey
Psc 0h39'23" 8d34'12"
Hubert Daniel McCormack
Ori 4h55'4" 2d45'29"
Hubert Eugene Bowen
Cnc 8h5'38" 8d14'58"
Hubert Finn Zappas
Per 3h49'8" 40d54'0"
Hubert Frederick Garner
Cep 22h0'17" 72d35'21"
Hubert Gray
Ori 6h17'31" 19d12'6"
Hubert Hahn
Uma 8h56'26" 56d12'27"
Hubert J. Lang
Uma 12h34'32" 59d42'21"
Hubert Lyle Hodges
Cnc 9h20'8" 32d35'38"
Hubert Pralitz
Per 3h40'21" 44d46'48"
Hubert Richard Pennings
Uma 11h51'51" 45d54'47"
Hubert Richter
Uma 10h47'27" 42d18'59"
Hubert Schiml
Ori 5h46'17" 12d21'50"
Hübner, Sabine
Uma 8h18'5" 63d55'52"
Hubschman's 50th Anniversary Star
Cyg 20h23'42" 32d56'48"
Huck
Lyr 18h43'57" 30d36'44"
Hucklberry
Aql 19h45'37" 2d23'18"
Huckleberry
Uma 10h38'46" 45d19'10"
Huckleberry
Uma 11h32'2" 62d55'24"
Huckleberry
Lyn 7h11'6" 57d5'27"
Huckleberry
Cap 21h40'27" -13d44'56"
Huckleberry
Lib 14h53'21" -2d41'47"
Huckleberry Hollow
Cnc 8h51'20" 28d48'3"
Huda
Vir 13h55'59" -20d39'50"
Huda Nebula
Tau 5h3'13" 24d57'25"
HUDELSON 2
Cyg 20h41'53" 33d53'38"
Hudson
Uma 11h22'32" 46d18'11"
Hudson
Sgr 19h54'54" -24d13'56"
Hudson and Maddie
Cma 7h27'12" -16d21'2"
Hudson Bay
Cmi 7h53'47" 12d56'2"
Hudson Blake Emerson
Uma 11h12'33" 55d20'40"
Hudson Bruno
Dra 17h21'18" 61d39'43"
Hudson Cline 1119
Lyn 8h5'47" 34d54'46"
Hudson Hunter Sheats
Crb 16h0'42" 34d34'23"
Hudson Jackson Sherwood
Leo 10h12'10" 26d12'5"
Hudson James Hallums
Umi 15h42'27" 73d2'26"
Hudson Kemper
Cnc 8h43'23" 16d2'41"
Hudson & Oak
Umi 15h26'52" 69d29'9"
Hudson Orion
Ori 6h24'42" 17d23'1"
Hudson Patrick Przybyski
Umi 15h37'12" 78d7'56"
HudsonDavidJohnson 10212005 1192005
Lmi 9h47'43" 39d56'54"
Hudson's Star
Aqr 21h12'57" 0d54'1"
Huepenbecker-Linzie
Uma 10h44'9" 61d52'8"
Huetta
And 1h47'21" 46d37'12"
Huey W. Spearman
Ori 5h50'16" 5d43'6"
Hüffeli
Crb 15h33'13" 38d18'14"
Huffman Constellation
Uma 9h55'11" 57d1'2"
Huffnickle
Gem 6h52'24" 32d54'6"
Huffooing Bear Star
Aql 19h29'37" -0d3'46"
Hufo
Cas 1h31'57" 69d14'24"
Hufriya Belle 71204 LRV
Crb 15h53'36" 27d6'24"
*hug
Psc 1h21'32" 32d54'10"
Hug Bug
Ari 2h8'30" 23d11'25"
Hug Toy
Uma 11h58'18" 46d6'20"
Hugale
Sge 19h25'12" 17d22'44"
Hugger Bugger
Umi 14h16'19" 75d25'29"
Huggle Tracy, Huggle
Ari 3h21'14" 20d51'55"
Huggy & Kitten
Cyg 19h59'27" 45d55'6"
HuggyBear!
Peg 22h3'48" 6d33'21"
Huggyville
Vir 12h16'34" 11d26'11"
Hugh Alan Amundson
Tau 4h21'51" 1d38'32"
Hugh B. Severs II M.D.
Aql 19h52'28" 11d49'14"
Hugh Bolton
Cep 22h11'43" 68d25'20"
Hugh Brien
Uma 9h21'15" 56d10'15"
Hugh Brockie
Umi 13h46'0" 71d18'31"
Hugh "Bud" Ritter
Uma 9h58'22" 57d55'43"
Hugh Cassidy
Lyr 18h50'0" 45d26'42"
Hugh Charles Macaulay, Jr.
Sco 17h17'37" -44d1'12"
Hugh D. and Gladys M. Anderson
Cyg 21h27'23" 30d55'27"
Hugh Damien Paul Melmore Scantlebury
Ori 6h18'57" 10d29'19"
Hugh Dinsmore
Per 2h42'30" 54d10'8"
Hugh Douglas Jamieson
Peg 22h30'31" 29d58'38"
Hugh Elwin Maingot
Aur 5h32'59" 47d27'3"
Hugh Francis O'Neill
Per 1h39'30" 54d32'17"
Hugh Frederick Lavery - Inspiration
Cru 12h26'59" -56d29'7"
Hugh Hefner
Uma 9h57'10" 59d54'48"
Hugh I. "Skip" Garrettson
Cnc 8h30'49" 12d36'4"
Hugh J Karr
Cep 23h25'56" 76d59'12"
Hugh Jackman - Star Of Our Hearts
Lib 14h54'3" -15d50'45"
Hugh Jarvis Dinteman
Lib 15h53'21" -10d54'28"
Hugh Joseph & Maureen Frances Quinn
Vir 13h16'1" 4d56'21"
Hugh Lambert
Ori 5h31'25" -4d19'22"
Hugh & Leslie Newton
Cyg 21h14'45" 47d18'0"
Hugh & Lois Adams
Uma 8h44'22" 51d34'51"
Hugh McDonald
Sco 17h24'13" -44d28'31"
Hugh Montgomery's Retirement Star
Uma 8h40'38" 49d40'48"
Hugh Owen
Cep 0h30'9" 79d1'25"
Hugh Owens
Gem 7h16'58" 20d22'22"
Hugh Patrick Cleary, Jr.
Her 17h21'16" 41d35'33"
Hugh Patrick Molloy Sr.
Cep 22h47'42" 74d21'30"
Hugh Patrick Ryan
Lyn 8h8'13" 46d48'35"
Hugh & Patty's Star
Cyg 20h25'1" 46d55'22"
Hugh Paul Fitzgerald 1969 - 2004
Cap 20h25'24" -15d43'4"
Hugh & Phyllis Proffitt
Uma 12h53'41" 56d37'0"
Hugh Pryce
Tau 4h26'53" 24d24'2"
Hugh & Rosemary Forever Love 8/51
Ori 6h8'52" -0d53'39"
Hugh Stewart
Lyr 18h27'39" 33d37'36"
HUGH STONE, JR.
Tau 3h40'52" 14d30'57"
Hugh Thomas McAulay
Cap 20h27'3" -9d58'3"
Hugh & Vera Kearns 45 Years of Love
Cyg 20h16'35" 45d43'57"
Hugh W. Johnston
Ari 2h41'18" 14d27'5"
Hugh Weyman Webb
Her 17h14'25" 47d4'0"
Hugh William & Edward Osborn L . Latta
Dra 16h0'9" 57d18'6"
Hughes Gang
Uma 10h41'58" 55d9'10"
Hughes - Squatrito
Ori 4h45'48" 3d58'36"
Hughie D 56
Cep 0h15'43" 73d49'56"
Hughie William Gilshan
Peg 21h16'8" 16d36'42"
Hugi
Vir 14h19'58" -0d15'26"
Hugo
Tau 5h51'33" 14d34'33"
Hugo
Per 2h44'52" 40d11'19"
Hugo A. Pinto III
Uma 8h15'43" 71d48'28"
Hugo A. Zelada Jr.
Her 17h26'3" 39d8'33"
Hugo Alberto Lara
Vir 13h18'59" 3d20'34"
Hugo & Alicia Hearns
Vir 12h50'53" 1d8'29"
Hugo and Jack Lauroesch
Uma 9h37'23" 61d49'58"
Hugo Castro
Cyg 19h41'21" 34d29'12"
Hugo Cincotta
Psa 21h32'58" -35d55'46"
Hugo & Concetta Sandolo
Cyg 19h46'45" 32d55'6"
HUGO CROSS
Her 16h49'41" 23d9'31"
Hugo David Woolley
Per 3h2'24" 56d48'2"
Hugo Flores - Shining Star
Aql 19h51'47" -0d49'48"
Hugo Frédéric Rovelon
Uma 8h34'44" 69d20'43"
Hugo Llopis
Uma 9h28'12" 49d49'6"
Hugo Luciano
Uma 11h33'46" 37d25'45"
Hugo Main Bochel
Umi 14h43'47" 67d31'3"
Hugo Martinez
Ari 2h20'27" 18d16'13"
Hugo Olvera
Lib 15h17'9" -4d27'13"
Hugo Omar Castillo
Uma 11h57'5" 32d16'9"
Hugo R.
Mon 6h31'32" 5d46'44"
Hugo Raffard
Uma 10h5'53" 44d49'15"
Hugo Samuel Amandus Gibson
Cyg 20h8'40" 56d52'15"
Hugo Thibodeau - 04-03-2002
Uma 11h31'13" 56d27'5"
Hugo Thorel
Cen 14h41'9" -64d26'55"
Hugo Wentzel
Ser 15h40'53" 10d30'58"
Hugolin Philippot
Cyg 20h9'7" 35d15'23"
Hugo's Hope
Mon 6h42'40" -10d21'21"
Hugs
Cyg 19h46'46" 33d58'34"
Hugs And Kisses
And 0h56'50" 38d49'14"
Hugs and Kisses
Lmi 10h33'8" 29d53'14"
Hugs & Kisses
Gem 7h24'28" 27d5'35"
Hugues Charles
Ori 4h53'18" 4d30'56"
Hugues de Planta
Her 16h49'30" 19d11'17"
Hugues Delahaie Riché
Cap 21h44'17" -16d57'8"
Huguette Nicolazo
Uma 9h40'44" 71d23'54"
Huhmum
Psc 0h11'19" 10d41'16"
Huhn, Heinrich
Uma 11h37'21" 32d41'52"
Hui Chun
Sco 16h43'7" -32d34'5"
Huisgen, Tiziano Maria
Uma 9h31'59" 49d32'2"
Hui-Wen Fu
Cas 0h24'59" 54d16'32"
Hulbear
Uma 9h23'55" 67d16'11"
Hulburd Family Star
Ori 4h48'57" 12d21'2"
Humaira
Lyn 8h47'19" 34d50'42"
Humaira
Cap 21h37'36" -12d54'37"
Humaira Ashraf & Shahid Siddiqi
Lib 15h14'26" -5d24'15"
Humble "Moma's Star"
And 0h21'6" 35d30'48"
Humenuh
Lib 14h37'7" -10d16'16"
Humera Ansar Ali
Ori 5h23'9" 4d33'34"
Humichan
Psc 0h21'28" 21d29'57"
Hummel, Werner
Ori 6h18'47" 10d5'38"
Humming Bird
Lib 15h24'37" -7d4'2"
Humming Bird Teddi Alvey
Lib 15h44'48" -8d45'2"
Hummingbird
Uma 8h10'38" 66d59'40"
Hummingbird
Leo 10h33'33" 26d16'31"
Hummingbird
Cyg 21h31'4" 51d37'59"
Humpherys
Ori 5h19'31" 15d55'15"
Hunalicious
Aqr 22h28'30" -19d7'1"
Huney Bear Huney Blossom Rose
Aqr 22h8'55" -3d47'48"
HuneypieDK
And 1h43'5" 50d33'40"
Huni Baby
Uma 13h40'1" 51d9'31"
Huni Two
Cap 21h15'45" -23d36'45"
Hunk Chak-ha Chee
Boo 15h38'47" 39d56'50"
Hunk & Effie
Uma 11h41'42" 49d12'23"
Hunky Wert
Psc 1h13'58" 27d4'44"
Hunna Bunna
Tri 1h48'59" 30d26'2"
Hunnies
Peg 23h24'40" 25d37'17"
Hunny and Pooh Bear
Del 20h37'17" 15d30'53"
Hunny Bear
Cyg 21h38'51" 40d48'42"
Hunny Buggles
And 2h32'23" 48d58'8"
Hunny Bunchy
Sco 16h13'35" -13d48'5"
Hunny Bunny
Lep 5h42'39" -12d6'49"
Hunny Bunny
Vir 13h20'49" 5d39'0"
Hunny Bunny (Jessica Hidalgo)
Sco 16h9'38" -17d49'8"
Hunny Nunny
Sgr 18h37'34" -18d23'23"
Hunny Trehan
Lyn 7h33'3" 36d37'37"
Hunt
Lyn 7h33'37" 38d20'12"
Hunter
Ori 5h36'14" 10d43'10"
Hunter
Ari 2h6'48" 21d37'32"
Hunter
Ari 3h10'11" 27d49'42"
Hunter
Uma 9h23'0" 64d17'52"
Hunter
Uma 11h5'38" 64d17'1"
HUNTER
Uma 14h8'55" 60d54'37"
Hunter
Uma 14h28'48" 57d9'39"
Hunter Alexander Goodwin
Ori 4h50'36" 11d43'31"
Hunter Amelia and Mom
And 0h13'1" 45d13'49"
Hunter and Candice
Ori 5h57'5" 7d21'24"
Hunter and Dublin
Peg 21h49'45" 21d26'45"
Hunter and Helena Dersch
Cas 1h15'15" 53d14'31"
Hunter Anthony
Ori 5h32'28" 4d59'53"
Hunter Austin
Sge 19h48'4" 18d32'50"
Hunter Austin Deel
Ori 5h31'16" -0d15'34"
Hunter Austin Hampton
Dra 19h38'59" 70d24'24"
Hunter & Brenda Dodge
Sgr 17h44'37" -26d19'25"
Hunter Brent Lile
Her 17h14'24" 26d55'6"
Hunter Brown
Peg 23h45'27" 28d16'10"

Hunter Chase Bedell
Dra 15h39'10" 58d10'47"
Hunter Chase Henry
Lyn 9h15'39" 34d25'42"
Hunter & Christy
Her 16h32'13" 32d43'1"
Hunter Clifton
Cru 12h21'18" -58d40'0"
Hunter Curtis Little
Aqr 20h40'18" 1d15'18"
Hunter David Jurenka
Uma 10h30'37" 64d24'28"
Hunter David Minner - Baby Angel
Psc 23h57'32" 3d1'24"
Hunter David Reed
Ori 6h2'2" 19d52'32"
Hunter David Schwartz
Her 18h11'14" 29d13'48"
Hunter Delaney
Sco 16h42'13" -44d5'30"
Hunter Derek Bradley
Ori 6h5'0" 18d7'39"
Hunter Douglas
Cma 6h38'47" -17d19'27"
Hunter Duke Husch
Umi 14h30'16" 79d53'11"
Hunter Evan
Gem 6h49'55" 33d0'56"
Hunter Forbes
Cyg 19h40'13" 38d47'57"
Hunter Galleher
Sco 17h33'26" -37d31'31"
Hunter Gavin Knorr
Lyn 7h56'31" 36d21'21"
Hunter George Ackerson
Ori 6h18'58" 2d14'15"
Hunter Grace Simpson
Aqr 22h37'43" -1d27'36"
Hunter Gregory Murchland
Aqr 22h27'28" -24d30'59"
Hunter Gregory Muth, BonBon & Poppy
Ori 5h35'9" 5d9'20"
Hunter Gruvman
Ori 5h51'25" 9d17'8"
Hunter Holstine
Her 18h49'59" 12d56'40"
Hunter Hudson
Ori 5h40'12" 1d21'20"
Hunter IV
Aqr 22h27'59" -18d11'24"
Hunter Jade
Ori 5h29'59" 2d26'55"
Hunter James Bandola
Vir 13h26'6" 12d20'55"
Hunter James McLachlan
Ori 5h49'9" 11d42'22"
Hunter John St. Pierre
Lib 14h50'12" -12d16'20"
Hunter John Stewart
Ori 5h35'52" -2d16'1"
Hunter Johnson
Leo 10h25'45" 14d29'32"
Hunter K. Conatser
And 23h49'7" 47d15'57"
Hunter Kendall Davis
Psc 1h17'12" 22d54'26"
Hunter Kordzikowski
Ori 6h9'1" 18d45'37"
Hunter Lee Boss
Cnc 8h47'20" 32d10'29"
Hunter Lee Cox
Ori 6h9'1" 6d0'17"
Hunter Lee Rowe
Sgr 19h38'12" -32d56'3"
Hunter Lily Stewart Bell
Crb 15h35'16" 38d8'34"
Hunter Logan
Ori 5h36'56" 12d22'29"
Hunter M. Brophy
Lib 15h36'18" -19d35'10"
Hunter M. Burgin
Pho 0h34'24" -41d59'46"
Hunter Mathew Petrich
Ori 6h1'45" 17d21'0"
Hunter Matthew Artzt
Ari 3h22'56" 21d59'23"
Hunter Merkle
Cap 21h45'5" -12d43'32"
Hunter Michael Joseph Headley
Sgr 19h39'19" -15d19'52"
Hunter Milo Smith
Uma 10h38'7" 65d33'16"
Hunter Nash Glickman
Tau 3h33'22" 12d30'21"
Hunter Nicole Palmer
Ori 5h47'58" -5d22'7"
Hunter Niedziejko
Lib 15h17'20" -12d22'24"
Hunter P. LaBarge
Umi 14h31'32" 73d34'55"
Hunter Patterson
Her 16h43'8" 34d50'30"
Hunter Paul Holton
Ori 6h16'27" 15d9'6"
Hunter Paul Ruel
Ori 5h20'55" 8d4'13"
Hunter Phoenix Jones
Psc 1h30'9" 11d2'24"
Hunter Powell
Ori 6h4'32" 18d24'20"
Hunter Reese
Uma 11h21'27" 33d16'15"
Hunter Reichers
Ori 5h49'58" 0d15'8"
Hunter Richard Harrison
Ori 5h38'39" -0d29'54"
Hunter Ronald Hales
Leo 10h48'12" 12d17'47"
Hunter Roy
Sgr 18h21'9" -31d56'6"
Hunter Ryan Alexander
Ori 6h4'46" -0d12'51"
Hunter Ryan Rebhan
Psc 23h26'54" 5d29'21"
Hunter Saikin
And 1h11'38" 34d58'41"
Hunter Samuel Javes
Vir 14h12'34" -9d57'10"
Hunter Scott
Uma 9h11'58" 67d26'9"
Hunter Taylor Lankford
Sco 16h13'50" -9d19'3"
Hunter Thomas
Ori 5h33'36" 4d58'54"
Hunter Thomas Gould
Lyr 18h58'26" 27d40'39"
Hunter Thomas Hanna
Aqr 22h38'48" -15d14'32"
Hunter Thomas Rodgers
Sco 16h6'44" -14d28'27"
Hunter Tope
Psc 0h29'26" 7d57'48"
Hunter Tushman
Cnc 8h11'51" 11d14'0"
Hunter Vale
Ori 5h18'19" -7d41'7"
Hunter Walsh
Ori 6h1'43" 18d3'24"
Hunter William
Tau 4h40'47" 22d21'27"
Hunter William Lassige
Uma 10h34'3" 67d17'30"
Hunter-Green
Crb 15h48'29" 38d35'33"
HunterJBB
And 23h34'43" 38d19'11"
Hunter's Heart
Sgr 18h21'38" -32d54'44"
Hunter's Heaven
Ori 6h18'7" 15d10'17"
Hunter's Star
Ori 5h38'55" 12d54'23"
Hunting Noah
Uma 8h15'0" 72d12'21"
Hunts Golden Star!
Cyg 21h10'55" 49d23'57"
Hup Seng Hoon Yong Brothers Sdn Bhd
Ori 5h35'15" 2d58'33"
Hupert
Psc 23h4'19" -1d26'20"
HUPFER
Peg 22h53'46" 12d31'19"
Hurit Catori
Aqr 22h33'8" -1d48'21"
Hurit Wikimak
Ori 5h56'25" 18d26'10"
Hurler-Millburgh
Uma 13h41'21" 52d55'30"
Hürlimann Fritz
Psc 0h46'34" 17d29'50"
Hurlock
Ori 5h29'25" 1d3'9"
Hurmet Unsalan
And 0h15'17" 45d43'0"
Hurricane
Her 16h34'57" 46d53'26"
Hurricane Jake
Leo 11h6'51" 23d0'59"
Hurtado
Aql 20h28'22" -0d19'53"
Husam Jihad Hussein Aqel
Ori 6h7'16" 18d55'36"
Husbando
Uma 10h48'13" 49d11'38"
Husbando Michael Ponzi
And 23h10'58" 48d24'33"
Huség & Szerelem csillaga-Gézusnak
Sgr 18h30'56" -16d40'11"
Hüseyin + Linda
Tri 2h39'14" 35d26'37"
Husker Du Rabbit
Leo 10h31'44" 11d20'46"
Husky Braving Elements
Uma 10h42'38" 46d11'37"
Husmeier, Joachim
Ori 6h8'20" 13d23'43"
Husna
Cnc 8h30'59" 13d33'49"
Husniye Siyahi
Leo 11h7'16" 20d23'8"
Hussam Hussein Salim
Cyg 20h20'23" 31d5'55"
Huston J. Mattson, Sr.
Lyn 7h30'43" 48d37'22"
Huszka Zsolt
Tau 3h32'56" 10d48'43"
Hutchinson Soulmates
Cyg 21h22'56" 50d8'38"
Hutch's special 44
Uma 9h28'35" 65d43'33"
Hüttermann, Georg-Wilhelm
Uma 10h26'48" 65d50'48"
Huw XXVII
Ori 5h42'23" -3d6'46"
HUWAIDA - my lovely loving wife
Uma 11h19'56" 37d27'46"
Huxley Charles Conner
Umi 16h27'30" 84d18'48"
Huxoll, Christian
Uma 11h18'9" 37d34'34"
H.W. Hemingway
Per 2h53'50" 51d33'58"
Hwa Tai Distribution Sdn Bhd
Ori 5h34'54" 3d33'41"
Hwa Thongs Bags Industries Sdn Bhd
Ori 5h33'5" 3d19'31"
H.W.Michael
Uma 13h35'4" 57d7'2"
H.X.G. Gillogly VII
Uma 11h49'24" 51d3'7"
Hyacinth
Cas 1h55'10" 63d50'55"
Hyatt
Uma 10h58'21" 68d50'36"
Hybodus
Umi 15h42'2" 77d35'20"
Hyde-Burnard-7
Psc 0h47'48" 8d30'39"
Hydro
Ori 5h33'44" 11d11'37"
HygCen
Peg 22h24'13" 29d37'58"
Hyllary Kay Merricks
And 0h36'50" 42d18'32"
Hylton & Janet Wilton
Peg 23h6'9" 26d48'31"
Hyman Golden
Ori 5h56'20" -0d2'34"
Hyménée
Uma 10h5'6" 59d27'49"
Hyo-Suk & Glenn's Endless Love Star
Mon 7h16'7" -0d44'19"
Hypoxi Dr. Norbert Egger
Uma 9h25'42" 54d13'52"
Hysen Veliu
Aqr 21h24'10" 2d0'4"
Hythsheika
Ori 5h32'6" -8d5'55"
Hyun Jin Kim
Aqr 21h2'18" 2d10'28"
Hyun "Julie" Shin
Cnc 8h46'14" 23d5'1"
Hyun June
Leo 10h37'7" 9d9'40"
Hyung Suk Chang
Cnc 8h33'34" 18d10'28"
Hyunjin Kim
Lyn 7h32'5" 43d52'43"
HyunJu Sung
Sgr 18h13'1" -30d49'59"
Hywel Robert Evans
Her 18h37'21" 18d2'22"
i<3marie
Leo 9h53'23" 19d51'25"
I. Ádám Péter, a gitáros
Uma 14h5'34" 51d31'1"
I AM
Cru 12h55'18" -64d25'20"
I AM
Lib 15h4'50" -1d0'51"
I am Bill Ladd
Aql 19h14'45" -7d51'56"
I am Destiny L(ove).
Cyg 21h38'47" 37d1'9"
I am Keeping You
Cyg 21h6'4" 43d52'56"
I choose you Al Boland!
Uma 11h26'37" 30d42'9"
I Do
Uma 11h28'43" 34d54'47"
I do it for you
Ari 3h17'3" 18d50'1"
i don't wanna miss a thing
Cyg 21h21'20" 52d2'5"
I Forever Do
Leo 10h37'18" 21d59'7"
I found my star in Del Mar 10/24/04
Ori 6h9'53" 14d39'31"
I Heart Kumey
Psc 1h21'14" 24d37'34"
I Hope You Dance
Apu 16h26'8" -77d55'22"
I. Jay Ford
Dra 19h27'46" 74d45'30"
I Know All About That
Gem 7h55'20" 17d5'20"
I ko 'ike maka
Lyr 19h11'35" 37d23'53"
I L T S Y I T N
Lyr 18h40'56" 47d34'29"
I L U 2 M 4 E
Uma 9h33'21" 63d4'30"
~*I "L" You*~
Cyg 21h31'40" 46d9'34"
I Lamp Kristi Cox
And 1h6'28" 46d46'8"
I Love Brittany With All My Heart
Gem 7h16'34" 23d52'50"
I Love Courtney
Ori 5h31'59" 1d0'53"
I love Crystal
Cyg 21h44'37" 48d9'20"
I Love Demi
Lmi 10h26'5" 36d55'18"
I Love Emily Vash
Cnc 8h46'0" 16d9'46"
I Love Erin for Eternity Star
Cyg 20h6'48" 32d11'54"
I love Hendrikje Louise Schalkwijk
Cnc 8h55'18" 16d33'45"
I Love Jared Timothy Morello
Lmi 10h24'1" 31d43'46"
I Love Jeanderson Jeannot
Sgr 19h46'34" -14d16'12"
I love Jeanne
Psc 0h43'10" 16d44'11"
I Love Joanna Rieke
And 1h15'21" 35d31'21"
I Love Jodi Lebar
Ori 5h31'47" 0d8'23"
I Love John Dixon!
Ori 5h19'31" 12d29'19"
I love Koreena
Psc 0h46'40" 20d53'50"
I Love Liz
Sgr 18h27'39" -18d10'19"
I Love Lucy
Sco 16h14'24" -14d9'44"
I love MaryAnn
Lyn 7h8'16" 52d16'34"
I Love Mollybeth
Uma 11h1'17" 48d16'11"
I Love Mommy
Leo 11h20'52" -5d32'8"
I Love My Ducky
Lyn 8h42'4" 34d17'3"
I Love My Hunnums - You Are My Star
Aql 18h52'7" -0d46'1"
I Love My Mom, Darlene Gelinskey
Lib 16h1'29" -16d9'29"
I Love My Piehole
Ari 2h39'49" 18d24'24"
I Love My Sister
Cyg 19h42'14" 37d22'56"
I Love My Uncle Danny Mills
Cep 20h52'5" 60d3'56"
I Love Natalie Batchen
Cas 0h20'51" 52d39'3"
I Love Nikki
And 0h49'13" 38d30'52"
I love no matter
Per 4h40'8" 40d14'2"
I LOVE PATZ
Boo 15h6'27" 14d45'2"
I Love Rebbecca
Lib 15h46'51" -24d6'45"
I Love Roger
Ori 5h44'26" 1d22'32"
I Love Sandman Bill Elkins
Psc 0h50'31" 6d34'6"
I Love Sandra Shirk
Tau 4h26'43" 13d4'7"
I Love Sarah
Boo 14h16'43" 47d7'4"
I Love Sean Jackson
Her 16h33'18" 42d21'46"
I Love Sharon
Crb 15h41'11" 37d5'38"
I love Stephanie and Lola!
Uma 8h55'4" 56d11'23"
I love u to the moon & stars & back
Vir 13h29'20" -6d31'39"
I love Walter Nova
Uma 12h7'14" 56d10'42"
I Love You
Uma 10h47'5" 58d29'17"
I love you
Uma 9h49'55" 53d16'48"
I Love You
Cyg 19h35'4" 31d4'46"
I Love You!
Per 3h24'12" 44d1'27"
"I Love You"
Aqr 22h10'6" 1d6'30"
I Love You
Leo 10h24'35" 9d11'42"
I Love You!
Sge 19h51'38" 17d46'2"
I Love You Always Mom Your Son Tim
Sco 17h17'20" -32d47'12"
I Love You Alyson
Lyn 6h51'22" 58d24'22"
I love you Amanda Riddle
Cas 0h27'2" 50d1'29"
i love you amy
Gem 7h6'35" 18d33'15"
I love you Amy
Ari 2h29'5" 10d44'40"
I Love You Andrea
And 22h59'46" 38d51'44"
I Love You Angel
Ari 2h53'21" 19d49'46"
I Love You Anna
And 0h26'43" 43d24'57"
I love you Anna McCudden
Lib 15h5'4" -6d41'45"
I Love You Ashley
Cyg 20h23'52" 36d42'3"
I love you beyond the stars!
Uma 12h1'47" 49d10'17"
I Love You Big C I believe in you.
Uma 9h31'27" 44d26'3"
"I Love You Bink" Lauren Arnold '05
And 0h13'44" 32d45'5"
I Love You Bubba
Uma 11h2'49" 39d38'29"
I Love you Christopher D Angelo CLJ
Ori 5h59'53" -0d45'11"
I Love you Christopher Fatur
Aqr 22h41'8" -2d19'57"
I love you Dad
Cnc 8h19'39" 10d43'31"
I love you Dad your sweetheart El
Psc 1h13'38" 27d22'52"
I love you David Anthony
Per 3h24'45" 46d52'49"
I love you Debbie Bock
Leo 10h38'56" 20d24'16"
I love you Denise
Cap 21h49'43" -13d5'17"
I love you Dorothy
Crb 15h51'7" 26d7'14"
I Love You Elisa Magalhaes
Gem 7h22'50" 34d3'39"
I Love You Every Day
Vir 12h41'39" 4d59'52"
I Love You - Fancy Face
Cap 21h47'7" -16d12'29"
I Love You Forever Colb
Cnc 8h20'58" 12d6'6"
I Love You Forever, Sarah Michelle!
Cyg 21h44'1" 45d43'36"
I Love You Hayley
Cnc 8h34'46" 8d59'56"
I Love You Huge
Per 3h34'48" 44d34'56"
I Love You I Like You And I Care
Cyg 20h28'37" 51d43'1"
I love you Ian - Kayla
Cru 12h55'53" -59d42'16"
I love you Izabela Wyrwich
Lib 15h39'32" -18d56'30"
I Love You James
Ori 5h50'12" 2d40'0"
I Love You Jason
Hya 10h7'54" -14d41'44"
I Love You Jenn
Psc 0h22'59" 18d11'18"
I Love you Jennifer Whitehead
And 1h4'9" 45d36'48"
I love you Jenny xxx
Cyg 21h55'18" 54d56'56"
I love you Joanne My beautiful star
Sgr 18h5'56" -29d7'11"
I Love You Just Because
Sgr 18h12'23" -31d38'24"
I Love You Just The Way You Are
Crb 15h47'52" 27d52'24"
I Love You Karyn
Aql 19h51'36" -0d36'0"
I Love You Katherine Barnes
Sco 17h6'48" -36d58'26"
I Love You Kiel
Cnc 8h25'24" 29d9'31"
I Love You Kimber, Happy Birthday
Lib 15h19'6" -19d21'39"
I love you Kirsten Robyn McCullough
Lmi 9h23'58" 39d5'57"
I love you Kristin Uselton forever
Psc 0h1'48" -1d47'8"
I Love You Kristine Love Boobs
Gem 7h30'49" 19d34'1"
I Love You Lacey
Psc 1h8'43" 24d56'10"
I Love You Laugh
Umi 15h23'16" 72d38'40"
I love you Lily P.
Her 17h0'4" 37d1'19"
I Love You Mallory
Mon 6h46'36" -3d18'18"
I Love You Mama, Mama I Love You
Ari 2h17'14" 25d54'39"
I Love You Meg!
Ori 4h54'54" 3d48'11"
I love you Megan Moiren
Psc 0h0'14" 9d47'57"
I Love You Michael John
Psc 1h4'23" 30d46'14"
I Love You Miss Katie
And 23h35'25" 38d36'52"
I Love You Mom and Dad.
Uma 9h6'31" 51d39'9"
I love you Mom and Dad
Uma 9h49'1" 55d27'28"
I love you MOMMY
Per 3h11'51" 52d24'34"
I love you Moore than anything!
Ari 2h19'50" 26d28'40"
I Love You More
Ari 2h38'29" 22d29'19"
I Love You More
Uma 9h54'50" 46d54'31"
I Love You More
Sgr 18h15'29" -22d17'15"
I love you more.
Cma 6h59'5" -28d49'46"
I Love You More Loomis!!
Gem 7h8'37" 18d13'24"
I Love You More Mom Star
Sco 16h3'48" -10d4'8"
"I Love You More Than..."
Lyr 19h16'57" 29d43'50"
I Love You More Than Chocolate
Boo 15h35'23" 47d10'24"
I Love You More Than The Galaxy!!!!
Vir 13h17'59" 12d37'0"
I Love You More Than Words Can Say
Cyg 21h45'59" 49d6'42"
I love you - My Gen
Ari 2h56'21" 26d54'39"
I love you my HBB so very much
Cap 21h8'53" -24d36'44"
I Love You Natalie Pahlavan
Uma 14h3'37" 61d45'56"
I Love You Nikki
Uma 8h37'36" 67d14'29"
I love you Nikola
And 0h32'12" 39d22'1"
I Love You Princess Anna!!!
Ari 2h15'22" 25d14'15"
I Love You Rik
Cyg 21h34'45" 34d31'6"
I Love You Rina. Happy Anniversary!
Cyg 19h32'8" 28d37'43"
I love you Ron
Vir 12h50'36" -0d21'53"
I Love You Roxiie
Ori 5h22'2" 0d19'7"
I love you smash
Cap 21h37'41" -17d38'28"
I Love You So Much Ashley
Gem 6h43'40" 13d39'43"
I love you Stacy
Aqr 21h53'22" 1d0'51"
I Love You Stephanie Ames, Robert
Lib 14h55'22" -5d49'26"
"I Love You" Susan
Uma 8h55'54" 58d41'57"
I Love You Ted
Cep 22h49'27" 71d5'1"
I Love You Teresa, Your Fiancè J.T.
Leo 10h55'10" 1d10'50"
I Love You THIS MUCH Mark Selbee!
Uma 9h45'32" 54d12'2"
I love you this much...T
Cyg 19h46'22" 44d42'29"
I love you Tinia.
Gem 6h20'21" 21d35'26"
I Love You To The Stars
Lib 15h0'42" -5d51'14"
I Love You Up To The Sky!
And 1h45'28" 49d56'50"
I love you way up to the sky!
Leo 10h44'6" 10d2'59"
I love you, Adam Kendall McClure.
Her 16h23'48" 11d11'18"
I LOVE YOU, CHARNETTA
Aqr 21h32'41" 0d14'42"
I Love You, Corey
Gem 7h10'18" 32d13'8"
I love you, E.R.W.
Cyg 20h33'48" 46d38'43"
I Love You, Karen
Leo 10h25'33" 13d32'58"
I love you, Mandy!!! (10-31-04) 143
And 1h40'6" 49d31'22"
I love you, Sean Daniel Vigil!
Her 17h51'44" 47d6'16"
I lovea ouu
Uma 11h15'33" 44d9'34"
I LoveUWAMHNB
Her 18h5'36" 17d56'21"
I Luff You Muffin
Cap 21h3'15" -18d48'30"
I made a wish and you came true!
Cnc 8h33'50" 28d5'38"
I made a wish, you came true.
Lmi 10h31'45" 34d46'48"
I Olivia Lampie Yia Panta
Cyg 21h20'35" 38d2'50"
I Petra tou Yanni
Cas 2h4'26" 62d37'11"
I Primi Cinque Anni
Psc 23h21'13" 7d22'0"
I promise I'll never forget
Tau 4h29'45" 21d38'47"
I promised you a star...
Aql 20h2'48" 13d56'7"
I Remember Mama-Forever a Star
Cas 2h6'41" 67d18'54"
I See Angels
Ari 2h11'38" 25d2'37"
I still Love You Heidi
Sgr 18h6'23" -17d30'24"
I swear by the Moon & the Stars-94
Gem 7h0'53" 14d44'36"
I vanya Ireth Arcamenel Mara Monkey
Crb 15h22'32" 25d59'28"
I. Ventura
Leo 9h33'35" 31d53'53"
I will always love you SKM, MKB <3
Lib 15h36'1" -20d2'34"
I Will Love You Always and Forever
Cyg 19h39'37" 30d59'5"
I will LOVE you for ever
Uma 9h44'12" 72d28'10"
I will Love you Forever
Lup 5h30'34" -16d47'24"
I Will Love You Forever And Always
Lib 15h55'43" -18d50'1"
I will Love You forever Jason Golf
Tau 4h0'14" 6d50'41"
I will Love you Forever Jen Brown
Cyg 20h16'49" 32d34'19"
I Will Love You Forever Nippy
Ari 3h23'17" 29d1'56"
I will love you forever Vaska
Cyg 21h41'51" 52d21'45"
I will love you forever...My Andrew
Lib 15h50'51" -11d54'53"
I Wish You Were Here
Umi 13h27'56" 73d24'53"
I Y E
Lib 14h44'15" -9d16'32"
i1Jo
Uma 8h40'41" 47d9'44"
I665 Scott Crossman
Aql 19h37'59" 14d30'54"
Ia Vang
Gem 6h47'24" 26d37'12"
IA Wikky
Uma 10h14'15" 56d39'34"
Iago
Lyn 7h44'20" 52d12'56"
Iago Cabeza Rodriguez
Uma 8h32'1" 63d42'19"
Iain Andrew Ferguson
Per 4h26'26" 35d7'50"
Iain Cummins Kennedy
Lib 14h52'46" -3d40'22"
Iain David
Cas 1h57'1" 60d55'22"
Iain MacBean
Eri 2h31'32" -40d0'59"
Iain Robert Muir
Psc 1h22'27" 31d10'17"
Iakovos
Uma 12h33'11" 57d52'6"
Iamsiri
Umi 16h51'22" 88d46'7"
Ian
Uma 9h12'52" 54d58'18"
Ian
Psc 0h16'44" 6d45'21"
IAN
Gem 6h47'15" 23d19'8"
Ian
Her 16h18'18" 25d48'42"
Ian 10
Lyn 6h25'47" 54d5'11"
Ian 22597
Lyn 8h4'33" 34d15'33"
Ian & Ada Gordon
Cyg 21h30'54" 54d40'18"
Ian Ainslie
Sgr 18h37'55" -28d11'58"
Ian Albert Proulx
Her 16h31'53" 44d28'4"
Ian Alexander MacKenzie
Cam 4h41'9" 72d20'49"
Ian Alexander Stout
Lib 15h36'2" -6d20'18"
Ian Allen Franasiak
Aqr 22h13'8" -8d48'53"
Ian Alvarez
Aur 6h18'9" 32d43'58"
Ian and Alex
Umi 13h23'41" 72d37'51"
Ian and Carly forever and always
Uma 9h26'42" 61d51'49"
Ian and Claire Kreczmer
Tau 3h32'18" 23d10'40"
Ian and Danielle
Cas 0h34'42" 54d36'42"

Ian and Jamie: Lifers
Sco 17h44'30" -44d54'58"
Ian and Kala's Star
Ori 5h9'26" 6d6'56"
Ian and Laura Ferguson
Cyg 21h6'4" 36d22'50"
Ian and Leah Forever
Uma 9h42'3" 60d11'3"
Ian and Rachel
Uma 11h34'53" 61d46'5"
Ian and Tanya's star
And 0h53'25" 39d44'42"
Ian Anderson
Per 3h1'32" 42d9'33"
Ian Andrew Brandenburg
Aqr 22h34'56" -1d48'47"
Ian Anthony McDonough
Aur 5h14'1" 46d58'32"
Ian Asher
Uma 9h23'33" 50d31'27"
Ian Barraclough - 25.12.1945
Cru 12h2'53" -58d18'14"
Ian Barrera
Uma 11h13'8" 38d56'50"
Ian Basil Harbinger Kane
Uma 9h22'14" 44d54'8"
Ian Bean
Ori 5h44'55" 5d50'55"
Ian Blesse
Lib 15h40'39" -11d46'37"
Ian Boczko
Per 3h50'10" 52d21'30"
Ian Bradley Olson
Ori 6h0'0" 21d18'16"
Ian Brendan Wrobel
Uma 11h23'14" 60d44'14"
Ian "Brocky" Brock
Psa 22h23'11" -30d29'5"
Ian C Anderson
Umi 16h2'47" 78d22'48"
Ian Campbell Whitney
Uma 9h29'37" 66d9'21"
Ian Charles Rowe
Cas 2h17'55" 68d52'47"
Ian Christopher Hughes
Aql 18h55'35" -0d4'31"
Ian & Clifton Rule
Ori 5h45'26" 6d57'15"
Ian Cochrane Loggie
Sco 17h40'40" -42d29'52"
Ian Crockford
Sgr 18h6'44" -34d14'40"
Ian Curtis Weaver
Vir 13h22'44" 4d24'8"
Ian David Hilgendorff
Her 18h3'16" 14d55'7"
Ian David Klinglesmith
Sco 16h11'31" -18d38'25"
Ian David Swenson
Cnc 9h21'9" 32d42'46"
Ian DeWitt 8.9.01
Uma 11h46'48" 53d48'40"
Ian Douglas Jolley
Ori 6h19'32" 7d56'15"
Ian Douglas Slater
Ari 3h13'41" 11d18'55"
Ian Drummond Hosea
Uma 11h7'46" 69d18'1"
Ian Dunn & Karen O'Brien
Crb 16h23'41" 37d46'34"
Ian Edward Murdock
Aur 7h15'29" 42d20'37"
Ian Erin Terry
Cap 21h14'47" -17d11'9"
Ian et Whitney Pour toujours
Apu 15h21'50" -71d30'54"
Ian Eugene Miller
Ori 6h10'58" 18d10'18"
Ian Fischer Barnum
Per 3h39'3" 45d25'52"
Ian Forever
Peg 23h16'40" 28d31'57"
Ian Foster
Equ 21h6'47" 12d1'34"
Ian Franklin aka Lovebug
Del 20h44'25" 5d17'47"
Ian Gabriel Varn Collins
Ari 2h42'5" 27d7'48"
Ian Glen Davidson 09051980
Ori 6h18'2" -0d24'11"
Ian Gore
Cen 14h29'58" -31d43'36"
Ian Graham
Cap 20h14'16" -14d41'0"
Ian Greatorex
Cep 20h52'25" 61d56'3"
Ian Greenblatt
Psc 1h3'49" 33d25'2"
Ian Gregory
Cap 20h17'36" -9d29'13"
Ian Hall
Vel 10h34'28" -50d3'7"
Ian Harrison
Leo 11h24'25" -3d38'45"
Ian Haynes Capozziello
Leo 11h22'27" 4d33'51"
Ian J Pavis
Cyg 20h57'43" 45d3'11"
Ian Jackson
Peg 22h34'16" 28d32'19"
Ian Jacob Staub
Her 18h4'46" 28d47'42"
Ian Jacob Taylor
Umi 16h24'14" 75d42'44"
Ian James MacIntosh
Ori 5h39'59" 1d4'42"
Ian James McDevitt
Gem 7h37'26" 32d3'28"
Ian + Jemma the perfect couple
Cyg 19h40'54" 33d44'39"
Ian & Joan Potter - "The Diamond Star"
Cyg 20h29'29" 30d28'51"
Ian Johnson Smithers
Aur 6h1'37" 47d38'43"
Ian Jones
Dra 16h15'10" 68d32'8"
Ian Joseph
Lyn 7h2'2" 48d30'46"
Ian Joseph Receveur
Uma 9h41'20" 54d6'26"
Ian Joshua Miller
Dra 15h47'46" 53d51'16"
Ian & Kara
Cyg 20h55'48" 46d46'7"
Ian & Kayla Gray
Crb 15h44'20" 26d18'34"
Ian Kilgannon 'The Love of My Life'
Per 3h36'5" 42d51'32"
Ian Ko and Kelly Chen
Vir 13h44'12" -5d12'9"
Ian Lantta O'Russa
Aqr 21h59'7" -8d56'21"
Ian Letnainczyn
Her 16h42'28" 47d23'24"
Ian Liebman
Leo 10h19'53" 8d8'42"
Ian Light
Per 2h44'46" 53d18'43"
Ian Livingston Urquhart
Sgr 18h38'48" -26d53'37"
Ian M. Dudley
Boo 14h14'58" 18d37'50"
Ian & Marianne Cusey
Mon 6h54'28" -0d25'38"
Ian Michael
Her 17h41'44" 47d8'24"
Ian Michael Connor
Lac 22h40'36" 50d26'17"
Ian Michael Harnett
Umi 14h31'48" 69d36'46"
Ian Michael Rosenberg
Leo 10h16'48" 10d25'28"
Ian Mitchell
Per 3h9'59" 55d29'16"
Ian N. Irons
Her 18h33'20" 18d15'14"
Ian Nathaniel Nelson
Her 17h48'12" 46d59'20"
Ian Nelson Burton
Tri 2h20'15" 35d9'21"
Ian Nigel
Gem 7h9'9" 34d59'40"
Ian Nightingale
Uma 10h13'28" 59d37'56"
Ian & Nikki - memoria in aeterna
Vel 9h2'23" -44d0'45"
Ian Norton
Umi 15h15'23" 86d17'36"
Ian Norton Redfield Star
Tau 5h22'43" 22d9'40"
Ian Nutter - (Nutts). 06.07.46
Dra 16h6'37" 67d48'56"
Ian of Caerphilly '68
Uma 11h25'42" 42d32'25"
Ian Patrick Kennington
Uma 10h33'25" 43d23'19"
Ian Patrick Riches
Boo 14h15'27" 16d51'58"
Ian Patrick Tait
Ori 6h14'59" 2d47'47"
Ian Pennington
Ori 5h12'49" 0d36'25"
Ian Peter Balchin Green
Cep 20h52'36" 67d9'26"
Ian Richard Robbins
Aql 19h43'19" 0d8'25"
Ian Robert Cady
Aql 20h16'32" 2d21'1"
Ian Robert Charles Hanhauser
Sco 17h8'30" -32d39'1"
Ian Robert Fisher - Love of My Life
Cyg 19h11'2" 51d5'40"
Ian Robert Loudermilk
Peg 23h35'3" 19d23'26"
Ian Robinson
Cep 23h54'40" 73d47'55"
Ian Robinson
Psc 23h22'11" -3d1'41"
Ian Ross Hampton
Cnc 8h30'13" 11d44'43"
Ian Sage Donley
Her 17h25'35" 28d42'5"
Ian Scott
Aur 6h22'20" 40d7'32"
Ian Scott 09/12/1953 - 13/01/2005
Per 2h14'13" 54d30'15"
Ian Shanley McNamara
Vir 14h16'2" 0d31'37"
Ian & Shanti: 1827 Days & Many More!
Pho 1h19'17" -55d21'8"
Ian & Sharon's Star
Peg 21h56'20" 19d49'3"
Ian Shea Yow
Umi 13h12'20" 70d14'5"
Ian Simpson of Bourtree Cottage
Uma 11h55'28" 45d59'6"
Ian Smith
Psc 0h26'39" 11d57'48"
Ian Snyder 08/16/2000
Leo 10h22'30" 9d24'52"
Ian Sterling 9/30/1982
Uma 11h58'44" 57d51'58"
Ian Stoner
Uma 11h45'43" 48d7'5"
Ian Stuart Mackie
Leo 9h30'50" 31d14'12"
Ian Sunrise06/25/88 Sunset12/27/05
Per 2h28'31" 52d15'3"
Ian Thomas
Her 16h35'45" 6d10'55"
Ian Thomas Pedersen
Lmi 9h44'48" 34d22'33"
Ian Thompson
Sco 16h16'18" -41d55'51"
Ian Tijo Ro
Umi 16h13'10" 73d15'53"
Ian & Tracey
Umi 14h42'25" 70d22'44"
Ian Varney's Dragon Home
Dra 17h46'24" 55d27'24"
Ian Walker Davis
Her 16h57'8" 37d1'19"
Ian William Boden
Aqr 20h45'55" 0d25'29"
Ian William Clark
Aur 5h25'31" 32d25'40"
Ian William Ehrlich
Cep 0h11'14" 76d9'37"
Ian Winter
Uma 10h18'31" 60d59'24"
Ian Woolley
Per 3h7'49" 42d17'32"
Ian Zubowicz
Per 3h19'22" 41d49'17"
Iane Montane
Psc 0h12'40" -4d35'44"
"ianne"
Cas 0h36'9" 53d38'7"
IANNIS SMARAGDIS
Uma 9h30'32" 45d39'8"
Iannuzzo
Crb 16h9'20" 30d54'16"
Ian's Flashlight
Lib 15h14'40" -20d40'57"
Ian's Kristi
Gem 7h21'7" 13d47'35"
Ian's Star
Ori 6h15'26" 9d51'50"
Ian's Star
Ori 6h16'28" 18d52'43"
Ian's Star
Lib 14h53'37" -2d22'49"
Ian's Star
Uma 10h37'58" 68d2'45"
Ianzano
Uma 11h48'14" 29d15'1"
I.B. "Bob" & Mary Chloe Roberts
Uma 13h58'35" 56d41'22"
Ibby
Vir 13h17'41" 6d45'45"
'Ibidem sub dio' - Michelle & Rhys
Cru 12h31'4" -59d50'14"
IBM
Peg 22h5'56" 6d34'36"
IBM/Visteon Intel Server Team
And 1h57'9" 39d5'5"
Ibonne del Carmen
Tau 3h41'41" 24d56'19"
Ibrahim Abdallah "angelus"
Uma 9h43'15" 60d22'26"
Ibrahim Al Husseini
Ori 5h11'49" 6d47'44"
Ibrahim and Najla Ar-Romaizan
Ari 2h19'7" 12d42'22"
Ibs
Uma 11h21'36" 46d52'57"
Ibtissam Hanna
Sgr 19h23'6" -28d51'53"
IBTREES
Cnc 8h52'6" 27d42'37"
Ibutuan
Ori 5h5'17" 3d56'56"
I.Bütyök - Vágó Balázs csillaga
Cas 23h34'18" 56d9'37"
IBV+JML
Cyg 21h55'47" 42d12'14"
IC30
Uma 8h35'47" 64d16'43"
Icabod The Star of Seperated Lovers
Gem 7h45'45" 20d39'3"
ICARUS
Ori 6h15'11" 16d7'52"
Icculus
Cam 3h57'4" 70d15'28"
ICE
Gem 7h43'33" 19d26'6"
Ice Box
Uma 10h34'50" 44d22'29"
Ice Dragon
Dra 17h45'38" 64d59'5"
Ice Lella
Ori 5h50'54" 2d53'45"
Ice Queen
Uma 12h59'54" 57d35'54"
Ice Skateing Goddess
And 0h16'57" 27d10'5"
Icel - The Sparkle of My Life
Cru 12h34'26" -58d44'11"
Icey
Lib 14h54'31" -3d49'44"
Ich Liebe dich,Anja Luthard
Uma 14h27'28" 57d46'39"
Ichiban Boshi Tagos
Cam 4h13'30" 67d0'41"
Icka
Sgr 18h35'26" -35d52'33"
Icka
Cnc 8h34'28" 19d0'4"
Ickle Baby Katie's Star
And 1h20'4" 33d48'16"
Ickles
Ari 2h46'6" 21d20'44"
Icky
Cep 1h50'14" 85d4'30"
Icodia
Cam 3h50'59" 68d9'58"
icpkjvt
Umi 13h10'44" 71d45'26"
ICYY
Cap 21h21'56" -17d36'16"
Ida
Ori 6h17'13" -0d41'24"
Ida
Uma 13h56'8" 57d35'56"
Ida A. Drummond
And 23h10'35" 42d10'12"
Ida Anbarian
Leo 9h40'17" 28d0'12"
Ida Angelico
Ori 6h6'51" -2d6'6"
Ida "Baby" L. Meryman
Crb 15h25'23" 28d54'33"
Ida Bee 601
Gem 6h36'42" 12d24'14"
Ida Brown Floyd
Umi 15h19'19" 70d0'43"
Ida Bruno
Ori 6h3'0" 18d45'11"
Ida Cecile LaRoche
Cas 2h50'45" 61d31'23"
Ida Cruz
Ari 2h2'22" 23d51'51"
Ida E. Joseph
Uma 12h10'42" 49d23'2"
Ida Eloise Tice Emerizy
Cnc 8h34'23" 23d39'22"
Ida et Daniel
Cyg 20h39'40" 44d52'12"
Ida & Fred
Cyg 19h36'52" 54d26'4"
Ida Giovanna Shaker
Ari 3h21'1" 28d5'54"
Ida Jane Britton
Uma 9h46'47" 65d28'4"
Ida Jane Mitsumori
Gem 7h6'27" 34d23'22"
Ida Judith Jerome
And 1h44'56" 42d0'58"
Ida Luisa Folkesson
And 0h56'26" 39d31'43"
Ida M. Raffaele-Montone
Lyn 8h10'1" 46d29'18"
Ida M. Russell
Uma 10h43'5" 50d34'42"
Ida Mae Allen
Aqr 20h52'47" -10d15'42"
Ida Mae Dey
Cnc 8h44'21" 30d59'29"
Ida Mae Lauf
Sgr 19h23'16" -22d41'10"
Ida Mae "Steen" Noble 7~1~1907
Cas 1h56'26" 67d57'2"
Ida Marguerite Armstrong AuWerter
Sgr 19h1'45" -30d22'8"
Ida Marie
Sco 16h44'43" -28d27'43"
Ida May Mixon
Tau 5h19'0" 23d20'3"
Ida Paini
Uma 9h45'46" 42d47'12"
Ida Radovich
Psc 1h3'3" 33d22'9"
Ida Rikhter Bright Star
Aqr 22h55'10" -3d57'14"
Ida Silver
Aur 5h12'57" 29d12'46"
Ida Spann
Cas 0h51'21" 60d44'43"
Ida West
Sgr 18h22'30" -21d49'28"
Idalia
And 1h38'6" 42d8'17"
Idalia Soto
Leo 10h53'15" 12d22'40"
Idalina & Antonio Goncalves
Uma 11h33'41" 41d3'2"
Idalis S.M. Heredia
Cam 6h10'22" 68d6'45"
Idamia
Peg 22h36'54" 19d8'38"
Idan Ofer
Boo 14h42'29" 19d23'9"
Idania Vanessa Mejia
And 23h35'9" 42d47'45"
Idara Essien
Aql 19h57'24" 10d58'1"
Iddo Yosha
Uma 10h14'13" 62d51'58"
Idea
Aql 19h29'2" -0d12'18"
Idealistic Tuna
Del 20h38'10" 15d24'53"
IDec
Gem 6h40'20" 25d18'19"
Idella's Unique Star of Peace
Uma 11h28'27" 62d23'20"
idem
Sco 16h13'50" -13d7'51"
Idem
Crb 16h0'25" 36d9'35"
Idget
Umi 17h34'4" 86d35'31"
Idgit
Del 20h34'7" 13d0'39"
Idgy
Leo 11h37'2" 26d21'39"
Idiat Azeez
Uma 12h43'45" 62d45'42"
Idil Tabanca
Leo 11h5'30" 20d20'18"
Idris Barzani
Uma 8h47'22" 46d37'46"
Idriys Emanuel Eyssallenne
Gem 6h48'29" 25d22'52"
IDS Rocks!
Psc 0h32'10" 21d0'17"
IEN's Aaron Chamberlain
Umi 16h22'1" 71d45'51"
IEN's Brian Gallof
Her 16h26'10" 11d42'42"
IEN's Courtney Crawford
And 23h44'12" 34d8'33"
IEN's Hyan Im
Dra 17h36'19" 69d35'8"
IEN's Johnson Lau
Per 3h40'30" 38d37'50"
IEN's Lily Man
Cyg 21h16'10" 45d6'27"
IEN's Linda Chen
Cas 23h37'52" 55d25'51"
Ieros Agapi Vanessa
Cnv 13h40'17" 34d9'1"
Ie-Ru Wang
Leo 11h31'40" 10d19'40"
Iesha Marie Patience Milford
Leo 9h42'27" 7d24'31"
Iesha Nichole Price
And 23h29'10" 36d26'56"
Iesha Nixon
Lyn 8h16'33" 41d31'43"
Iestyn
Her 18h54'18" 25d28'19"
Iestyn Mark Harrison Rowlands
Peg 23h47'25" 11d10'25"
I.F.
Leo 9h30'16" 32d28'50"
If only once in our lifetime
Cyg 19h39'6" 38d48'47"
If Tomorrow Never Comes
Cen 11h50'25" -52d21'46"
If you jump...I jump - VL & Kirbs
Ori 5h26'13" -0d30'46"
Ifat Pfister
Ori 5h53'10" 12d0'6"
Ifeoma Aitkenhead
Uma 11h29'51" 52d19'21"
Iffat Riaz
And 0h48'52" 25d27'56"
I.F.G-S
Mon 6h24'11" 8d21'24"
Ifigenia Anna Omiridis
And 23h15'6" 35d32'10"
ifj. Riczler Gusztáv diver
Cru 12h34'39" -58d0'46"
If-Maybe.us
Ari 2h53'23" 24d39'42"
Iford
Cyg 21h53'25" 52d47'19"
Ifra Nasir Khoso
Sgr 17h56'20" -28d26'40"
Ifty
Leo 11h15'57" 24d4'3"
Igaz szerelemmel Gergomnek!
Lib 15h32'16" -12d18'1"
Iggers' Shining Light
Lib 15h26'26" -5d32'3"
Iggy "The Polish Star" Szymanski
Lyn 7h23'48" 49d15'42"
Iggy's Stardust
Dra 19h47'40" 68d32'44"
Igi
Cnv 13h25'13" 31d9'59"
Igly
Ari 1h53'22" 12d53'7"
Ignacije Nikolic
Crb 15h58'3" 25d52'3"
Ignacio Adame
Lyn 7h38'10" 38d59'17"
Ignacio Avila Licon
Cyg 21h5'39" 54d30'10"
Ignacio de Satrústegui Aznar
Leo 11h6'4" 20d39'18"
Ignacio "Pop" Padilla
Uma 10h55'22" 62d12'39"
Ignatius "Iggy" Maniscalco
Cep 22h56'31" 64d2'52"
Ignazio Ciaccio
And 1h10'46" 41d55'27"
Ignazio "Tony" Gallo
Tau 4h55'36" 28d8'8"
IgnazioPaz
Ori 5h41'47" 1d39'41"
Ignious Yong Wai Hong
Psc 1h29'48" 13d36'9"
Ignis Micans Caeli Timothei
Gem 7h6'29" 11d1'15"
ignoci ingra
Ser 15h54'26" 7d11'19"
Igor
Ori 5h9'4" 15d52'51"
Igor Bekker
Dra 15h9'23" 58d33'11"
Igor Iskhakov
Vir 12h38'25" -4d35'33"
Igor&Julie
Cyg 21h22'15" 52d37'19"
Igor Klinchin
And 0h21'27" 27d0'21"
Igor Kogan
Her 16h12'16" 49d7'10"
Igor Livshits
Per 4h48'45" 45d44'8"
Igor Mjaskiwker
Uma 10h59'47" 51d14'21"
Igor Mordkovich
Vir 12h19'57" 12d14'15"
Igor Petar Dzadzic
Ori 5h52'0" 18d27'17"
Igor Puskelnik
Leo 11h5'1" 2d16'31"
Igor Salvi, notre étoile
Uma 9h37'2" 46d17'50"
Igor Shirley Narozniak
Aqr 22h34'15" -2d56'50"
Igor Trofimov
Vir 12h50'29" 6d57'10"
Igor Vatelman
Leo 10h12'6" 27d10'38"
Igor Walsh
Cma 6h48'39" -29d29'23"
Igor Y Lib
Del 20h43'15" 17d28'29"
Igor Zwilichowski
Uma 8h42'36" 68d41'10"
Ihab & Barbara
Lac 22h19'29" 47d1'0"
Ihlenfeld, Eckhard
Uma 9h46'49" 50d33'59"
Ihlenfeld, Petra
Uma 8h50'41" 48d42'44"
Ihle's Histoire D'Amour
Lyr 18h49'45" 33d17'54"
Ihli's Töffi
Tau 3h41'18" 27d37'13"
Ihsan Al Monzir
Her 16h58'51" 30d27'24"
IHSAN SCOTT
Aqr 21h7'23" -13d59'36"
Il Leggero de Luca
Her 18h11'26" 22d24'57"
iimer oben schauen
Sgr 17h59'11" -27d24'16"
Ik hou van je
Umi 15h7'10" 73d7'15"
IKA Jirina Butora-Franek
Cas 2h14'29" 73d37'52"
Ikaia-Purdy, Keith
Lib 15h18'18" -8d20'21"
Ikaika Noah
Her 17h17'46" 36d5'7"
Ikarus-Lamont
Peg 21h10'31" 18d7'49"
Ike and Amanda Always
Leo 9h47'56" 25d56'18"
Ike Messmore, The Sky Is The Limit
Aqr 23h41'21" -7d26'8"
Ikela Tenise Gilham
And 1h43'16" 45d50'34"
Ikelle Iturbe
Dra 19h19'25" 75d51'16"
Ikevis
Lyn 8h10'35" 40d14'18"
Ikey-Dog
Cma 6h42'55" -15d21'43"
Ikibin
Sgr 19h19'24" -21d45'35"
Ikikuku
Tau 4h27'52" 18d16'23"
ikkin retrop
Gem 6h43'7" 12d10'35"
Ikkle Elf
Umi 14h7'21" 74d30'13"
IKnewILovedU-Linh Vo&KhoaNguyen
Ari 2h19'3" 18d55'18"
Ikuisen Rakkauden Tähti
Cas 1h20'4" 51d58'28"
Ikuya "Eek" Kurita
Cnc 9h2'28" 13d23'51"
il battesimo di Philip
Cap 20h15'8" -9d33'58"
Il bello George
Psc 0h31'8" 18d25'16"
Il Bello Katie
Cnc 8h33'25" 8d12'7"
il campione del fiume
Her 18h9'15" 18d8'44"
Il Cuore Di Lauciello
Peg 22h47'40" 17d44'3"
IL MASSIMO
Cas 23h38'23" 57d8'27"
il mio amico migliore
Sco 17h43'12" -37d25'46"
IL MIO AMORE
Tau 4h1'4" 22d10'29"
il mio amore la mia anima gemella
Cnc 9h10'10" 25d3'33"
il mio amore, Samantha
And 23h24'25" 48d56'52"
Il Mio Angelo Piccolo
Umi 15h3'45" 75d17'43"
il mio principe
Peg 22h59'19" 25d37'51"
Il Mio Principe Caro
Cep 20h44'0" 62d25'8"
Il Mio Principe G. Calatagirone
Tau 5h5'13" 23d57'9"
il mio teresa bello
Cyg 21h34'56" 36d24'32"
Il Mylove, mia moglie, il Mio Reen
And 0h14'36" 32d25'32"
Il Nostri Amore e Sogni
Uma 13h43'19" 53d34'29"
Il Nostro Amore Sarà Sempre
Umi 16h38'3" 82d18'10"
Il nostro luce nel buio
Pho 0h8'57" -42d8'24"
Il Nostro Splendore
Uma 10h36'6" 70d9'18"
IL PRINCIPE AZZURRO
Cnc 8h22'43" 25d41'41"
IL SIGNORE - P. SIMINO
Vir 12h42'3" -9d24'41"
Il Tempismo e` Tutto
Cyg 21h43'43" 54d46'39"
Il vostro posto nel cielo
Cyg 20h21'36" 51d44'9"
Ila Hope Krauss
Ari 2h52'1" 26d38'52"
Ila Jean Dobyns
Vir 11h42'23" 10d9'50"
Ila Mae Nichols
Cap 20h48'42" -15d4'57"
Ila Mansukhani
Uma 9h39'31" 66d27'0"
Ila Pollard
Tau 5h49'58" 16d55'54"
Ilan
Cnc 9h8'16" 21d28'50"
ILAN
Uma 11h46'32" 46d11'10"
Ilan Caplan
Per 4h4'24" 34d30'9"
Ilana and Sam
Pav 19h17'48" -57d25'54"
Ilana Blitzer
Mon 7h16'41" -4d45'41"
Ilana Calderon
And 23h14'55" 50d41'25"
Ilana Danielle Caplan
Lib 15h2'7" -10d0'0"
Ilana Gershteyn
Sco 16h16'52" -21d4'45"
ILANA JADEN YULIS
Umi 14h24'37" 77d18'32"
Ilana Journo - 21st Birthday Star
Cyg 19h55'40" 33d30'5"
Ilaria
Lyr 18h42'40" 31d10'20"
Ilaria
Cyg 21h33'4" 34d0'44"
Ilaria
Cyg 20h6'37" 35d26'0"
Ilaria
Cas 2h7'4" 59d58'53"
Ilaria
Cam 3h42'50" 55d50'46"
Ilaria
Ori 6h2'59" 21d0'34"
Ilaria
Her 16h38'12" 29d38'36"
Ilaria
Mon 7h14'8" 0d53'26"
Ilaria
Umi 14h49'43" 88d12'15"
Ilaria 13*04*1979
Uma 11h23'2" 34d29'51"
Ilaria Alessio
Ori 5h40'28" 1d56'16"
Ilaria Bettozzi
Peg 22h52'27" 22d2'11"
Ilaria Dall' Olmo
Lyn 7h52'39" 45d6'38"
Ilaria Linda Mascia
Lyn 7h41'32" 59d2'47"

Ilaria Lisa
Aur 5h8'55" 40d3'23"
Ilaria & Lorenzo
Uma 10h1'37" 47d57'35"
Ilaria Martinelli
Her 17h39'38" 28d58'38"
Ilaria Persiano
And 1h22'5" 43d56'4"
ILARIO
Lac 22h20'37" 48d26'8"
Ilda Quizhpi
Sco 17h51'12" -33d13'55"
Ildi & Geri
Gem 6h30'5" 25d37'6"
Ildikó
Tau 3h56'46" 3d20'46"
Ildiko
Cas 0h56'9" 53d52'54"
Ildikó
Uma 11h1'6" 33d46'55"
Ildikó
Cas 1h56'58" 58d35'58"
Ildikó csillaga
Cnc 8h45'40" 23d47'23"
Ildiko Varga
Lmi 10h50'22" 35d58'10"
Ildy Lee
Uma 9h46'3" 55d18'36"
Ileah Olson
Ori 6h6'27" 18d25'39"
Ileana
Tau 4h38'19" 24d13'10"
Ileana
And 23h2'0" 42d31'13"
Ileana
Eri 3h7'46" -39d45'42"
Ileana
Cma 7h0'12" -11d51'17"
Ileana
Sct 18h53'42" -12d48'43"
Ileana Bumblebee Mendoza
Leo 9h28'48" 24d57'2"
ILEANA C.
Per 2h39'47" 42d16'32"
Ileana Doris Smith Brown
Crb 15h33'28" 28d31'19"
Ileana O'Connor
Cnc 8h20'51" 11d4'25"
Ilene
Tau 3h27'52" 17d31'55"
ILENE
Gem 7h15'9" 26d51'59"
Ilene
Tau 5h11'14" 26d32'55"
Ilene and Dom
Lyr 18h53'45" 43d37'25"
Ilene Emily Mae
Crb 15h53'56" 38d51'55"
Ilene & Gerson Field
Cyg 21h30'14" 39d14'55"
Ilene Grisi
Crb 16h14'41" 39d15'58"
Ilene Keyes
Uma 10h47'19" 51d57'38"
Ilene Lee
Cas 1h18'55" 64d47'36"
Ilene McDonald
Leo 11h1'2" -0d30'5"
Ilene My Love
Sgr 19h39'51" -13d4'46"
Ilene Schaffer
Aqr 21h48'21" -7d5'21"
Ilene Weller's Rainbow
And 23h25'34" 42d16'40"
Ilene's Eyes
Psc 1h40'2" 8d11'7"
Ilenia
Uma 14h26'1" 62d24'20"
ILENIA POVEGLIANO
Cyg 20h13'32" 50d22'23"
Ilenia Quaglia
Cyg 20h45'14" 32d40'27"
Iles-Gurney Twins
Uma 11h8'49" 59d44'26"
Ilgin & Erman
And 1h17'21" 36d54'28"
I-Li
Leo 9h40'41" 16d1'37"
Ili Nunez "Princess Peekaboo"
Lyr 19h14'31" 27d13'24"
Ilia
And 23h39'3" 48d33'25"
ILIA
Uma 12h14'9" 54d54'44"
Iliachtida
Tau 5h8'51" 24d29'5"
Iliah Natasha Perez
Crb 15h40'33" 32d4'17"
Iliana Alpha
Dra 14h59'30" 59d3'53"
Iliana Estee Weisberg
Ari 3h9'3" 12d32'42"
Iliana Rose Katsidis 13/01/07
Cru 11h57'0" -62d29'37"
Ilianna Goddess of Love
Del 20h32'57" 6d35'54"
Ilianna Queen of the Universe
Cas 0h57'25" 63d43'6"
Ilias N.
Her 17h52'40" 32d43'31"
Ilias Yiorgios Cholakis
Lyn 7h7'53" 59d19'56"
Ilike
Ari 2h15'54" 13d21'7"
Ilima's Hon Derek
Uma 13h13'40" 58d26'41"
Iliyana Lazarova
Tau 4h34'49" 17d1'49"
Ilizabeth Sarah
Ari 2h35'17" 21d48'11"
Ilja Kreis aus Soest Deutschland
Ori 5h16'7" 1d3'46"
ILKA
Sco 17h52'18" -35d51'45"
Ilkin Cheyenne Senner
Ori 5h16'3" -0d36'40"
I'll always be with u wheva u r-ALM
Gem 6h55'0" 14d5'30"
I'll Always Love You Louie
Per 3h11'32" 53d2'46"
I'll always love youTone, Luther
Aqr 22h25'0" 1d40'57"
I'll Look After You
Ari 2h47'37" 28d57'1"
I'll love you forever and always
Lib 14h49'39" -7d51'54"
Illa Stella Desuper
And 23h44'10" 46d25'4"
Illama
Uma 9h8'12" 71d58'0"
Illini Skip Alejos aka Skippy
Uma 10h5'16" 42d13'33"
Illogic love
Ori 5h35'48" 8d58'45"
IlluminaTing
Gem 7h24'40" 26d12'0"
Illusions
Leo 11h29'59" 3d4'35"
Illusive Muse
Boo 14h51'26" 44d6'11"
Illustrious Sir M. Starr Potentate
Per 3h13'33" 45d47'17"
Illustrious VeRonica
Sgr 18h2'16" -27d50'29"
iLMA
Lyn 7h27'5" 52d25'27"
Ilma Davina Chairani
Leo 10h36'54" 23d22'50"
Ilo
Her 18h56'49" 23d18'11"
Ilodia Caris Vogt
Leo 9h51'8" 22d51'37"
ILONA
Sge 20h20'1" 19d29'37"
Ilona
Psc 0h55'35" 27d21'43"
Ilona
Aql 20h13'35" 8d12'41"
Ilona
Uma 9h11'19" 53d28'20"
ILONA BEDNAREK
Crb 16h3'32" 36d19'34"
Ilona Erica
Lib 15h8'45" -10d28'27"
Ilona Kopelman
Sgr 17h52'9" -29d52'34"
Ilona Magdalena Dezentje
Cas 2h20'52" 72d47'52"
Ilona Seltmann
Uma 10h13'28" 46d15'28"
Ilona Tabar
Vir 13h26'12" -9d51'12"
Ilona Zielinski
And 2h10'31" 43d44'38"
Ilonka
Uma 11h2'1" 42d16'20"
Ilonka Wallace
Psc 1h8'54" 19d19'27"
ILove Susanalways
Cas 0h55'23" 58d47'28"
ILoveYouClaireHappy21st
Uma 10h36'3" 42d56'18"
Iloveyougcar
Crb 16h1'6" 25d58'8"
ILoveYouJessicaRaeGingrich
Ari 2h14'8" 11d31'39"
Ils C'M
Cas 0h21'50" 53d49'33"
Ilsa Impey
Umi 16h15'0" 84d15'9"
Ilse
Lyn 8h22'48" 55d41'20"
Ilse S Black
Uma 10h5'27" 48d18'6"
Ilse von Studzinski
Uma 9h42'53" 42d6'5"
Ilse Zimmermann
Ori 6h18'40" 5d59'45"
ILUDFT - the Emma & Mikkel Star
Cru 12h27'34" -60d58'58"
ILUM
Sgr 19h36'27" -14d52'37"
ILUM
Ori 5h56'54" 17d12'43"
ilurjohnson
Ori 5h9'54" 15d20'12"
Iluviel
Cap 21h47'37" -18d18'47"
ILuvUStef
Ari 3h25'46" 27d15'42"
ILUVYOU
Uma 11h40'4" 61d51'55"
Ilva Massignani
Uma 9h17'41" 57d50'30"
ILVDEM04
Uma 9h16'26" 55d28'51"
ILY
Lyn 6h50'23" 58d2'58"
ILY
Cnc 8h36'12" 28d41'29"
ILY
Leo 10h25'31" 21d38'51"
Ily Phelps
And 0h17'54" 31d11'51"
Ily von Kempen
Ori 6h19'14" 18d46'49"
Ilya, Roza, Nonna & Leanna Vinokur
Uma 11h41'42" 58d3'31"
ILYFAE
Boo 14h41'36" 10d40'19"
I.L.Y.M.
Cap 20h24'27" -10d57'14"
ILYMED - Colin & Diana
Per 3h34'7" 34d30'26"
Ilyssa Ashley Goodman
Crb 15h56'10" 26d40'58"
Ilyssa Shiri Bails
Dra 19h53'44" 70d54'37"
Ilze Vipulis
Crb 15h43'28" 30d6'39"
I'm gonna be around
Mon 7h30'28" -1d26'27"
I'm Jon Friggin Dively
Uma 13h48'4" 55d18'4"
I'm Loving you Sideways.
Sco 17h31'34" -38d0'29"
Im Meleth Le
Cas 2h18'20" 64d14'12"
"I'm So Glad You Got To See Me!"
Uma 13h27'31" 57d30'1"
I'm Sorry Amanda My Angel 'Star'
And 0h8'23" 32d30'52"
I'm Sorry please forgive me Danica
Uma 11h45'30" 61d29'58"
Im Soulbound
Uma 10h41'55" 56d1'46"
I'm Yours
Aur 5h36'51" 33d35'34"
IM4U
Peg 21h30'28" 15d38'17"
Ima Fay Mezzell
Leo 10h33'10" 23d28'28"
Imad Joudieh
Ori 6h18'49" 15d44'29"
Imadara
Lib 15h14'56" -22d10'4"
Imagin Ruth Thompson
Cru 11h59'36" -61d25'36"
Imagine: Susan Klesath's Star
Cas 1h29'43" 59d50'43"
Imagine...François-Xavier
Del 20h41'15" 9d43'19"
iMAJYN
Cra 18h1'9" -37d0'18"
Iman
Aur 6h4'21" 48d43'27"
Iman E. Holiway
Leo 9h45'31" 26d39'1"
Iman Victor Imhans
Vir 14h22'7" 1d25'6"
Imane
Ari 2h27'31" 18d59'5"
Imani Lewis-Thornton
Sco 16h6'35" -24d48'23"
Imani Mornike
Ari 2h22'45" 17d41'38"
Imani Naeem Murphy
Gem 7h33'20" 23d46'16"
Imanni
Cyg 20h40'52" 33d59'32"
Imants Laimons Brolis
Uma 14h4'4" 54d13'33"
IMAR
Lyn 7h59'23" 48d37'3"
Imberg, Anja
Uma 11h36'47" 28d26'2"
Imbrimia
Lac 22h36'52" 49d14'21"
Imelda
And 23h6'17" 51d7'40"
Imelda
Vir 13h22'18" 12d23'57"
Imelda A. Simmons
Uma 12h2'45" 63d46'46"
Imelda Lopez
Com 12h51'11" 22d33'11"
Imelda Nuno
Psc 0h23'49" 18d23'25"
Imelda, 26.10.1961
Her 16h42'32" 19d52'29"
Imhyt
Ori 6h7'46" 17d57'5"
Imiloa Pilialoha
And 0h37'39" 24d28'43"
I-Ming
Ori 6h15'6" 15d51'9"
Imitos De Oro
Ori 6h11'18" 9d12'5"
Imke Zerrenthin
Ori 5h20'2" 7d29'53"
Imlac100 JPB
And 1h42'15" 46d42'0"
Imma
Aur 5h39'22" 33d5'33"
Imma
Psc 1h22'21" 23d46'0"
Imma
Leo 10h23'37" 25d14'2"
IMMACOLATA
Cnc 8h28'27" 16d11'48"
Immaculate Angel
Sco 16h35'49" -29d23'48"
Immaculate Heart of Mary
Cae 4h55'34" -28d4'49"
IMMcKenna
Psc 1h9'5" 27d53'47"
Immortal Beloved
Cyg 20h34'48" 41d19'31"
Immortal Love
Boo 14h33'16" 42d15'22"
Immortal Marshall "The Sylver Star"
Tau 4h34'26" 9d13'9"
Immortal Rock
Lib 15h1'18" -14d20'9"
IMO Rawhide
Boo 14h34'30" 33d25'27"
Imogen
Col 5h51'9" -41d22'29"
Imogen and Evan
And 2h24'48" 49d11'51"
Imogen Ann
And 0h53'47" 23d59'27"
Imogen Bianca Gray
Cap 20h26'30" -22d55'47"
Imogen Catherine Johnson
Cas 23h24'25" 55d2'46"
Imogen Daisy Smith
Col 6h37'53" -34d38'11"
Imogen Darcy Brown
And 23h56'30" 42d11'12"
Imogen Elizabeth Pool
And 22h58'55" 36d54'11"
Imogen Erin Glass
And 1h56'2" 45d19'59"
Imogen Francesca Rogers - Imogen's Star
And 23h57'19" 45d22'27"
Imogen Holly Croucher
And 1h45'56" 36d52'5"
Imogen Jane Chandler
Cas 1h42'22" 63d5'12"
Imogen Lily Corvini
Cru 12h0'59" -63d42'42"
Imogen Loughlin
Leo 11h15'24" 5d34'23"
Imogen Louise
And 0h10'43" 47d35'42"
Imogen Margaret Muir Aitken
And 0h57'40" 43d47'32"
Imogen May Hicks
And 1h32'18" 41d3'46"
Imogen Olivia
And 2h0'58" 43d37'15"
Imogen Rachel Moore
And 2h23'10" 38d26'44"
Imogen Rose Cooper
Cyg 21h30'29" 35d19'5"
Imogen Rowan May Evans
And 23h52'16" 40d8'55"
Imogene Isaacs
Uma 10h1'42" 45d14'0"
ImoJean Brown Griffith
Uma 13h43'40" 57d42'24"
Imperator
Ori 6h14'14" 15d49'33"
"Imperial Princess" Ronda Marie
Tau 5h48'10" 13d51'15"
Impetuous Angel
Mon 7h17'54" -1d13'43"
Impossible Dream
Per 3h39'19" 36d12'51"
Impi Marie
Uma 12h40'31" 53d37'45"
Impressive Lonnie
Per 3h29'7" 47d47'52"
Imra & Stephenie
Tau 3h51'48" 27d23'53"
Imran Hasan Mansur
Lib 14h50'57" -13d54'23"
IMS
Cyg 20h32'28" 36d48'39"
IMT14U
Aqr 22h14'47" -1d19'27"
Imy
Cru 12h1'47" -63d13'42"
Imzadi
Lyn 9h2'34" 33d31'46"
In Aeternum Amor Sub Dio
Uma 11h59'56" 49d23'52"
In Celebration of Deb and Condor
Leo 10h15'24" 18d41'46"
In Concert 4 Ever Rod and Rachel
Uma 11h25'45" 34d19'2"
In everlasting memory-Ethel Murray
Uma 14h24'7" 61d6'47"
In ewiger Liebe für Brigitta von Tom
Aqr 22h50'58" -16d22'16"
in ewiger Liebe Janine und Chris
Ori 5h44'50" 3d0'28"
In ewiger Liebe Stefanie und Kaspar
Peg 21h42'44" 7d4'28"
In Fide Et In Bello Fortes G.M.C.
Per 4h4'21" 47d8'9"
In Gedanken bei Lisa Saal/Rohr
Ori 5h9'0" 6d41'9"
In Honor of Amy Cardiello
Sgr 18h53'24" -32d1'47"
In Honor Of Bernadette Finch
Vir 12h22'51" -3d26'38"
In Honor of Bill and Marie Gray
And 1h7'40" 38d33'36"
In Honor of Bob Fisher 1926-2005
Uma 9h10'32" 58d0'55"
In Honor of Deputy Wayne Koester
Sgr 18h49'24" -19d18'37"
In Honor of Julie & Henry Sullivan
Cyg 20h19'28" 47d5'42"
In Honor Of PCD Teachers 2005-2006
Uma 11h33'41" 56d6'57"
In Honor of Robert Lindemann
Per 3h22'50" 51d53'20"
In honor of Ruthanne
Uma 10h58'26" 53d2'30"
In Honor Of Sailor Fisher 1991-2004
Cma 7h22'29" -20d47'45"
In Honor Of Steve Scruggs
Lyn 7h13'14" 48d9'7"
In Liebe und Dankbarkeit euer André
Uma 11h22'6" 42d44'24"
In Lieu of A Love Enduring
Cyg 19h58'4" 30d14'45"
In love for Danielle
Sco 16h54'30" -38d27'49"
In Love & Loved Joan & Keith
Cyg 20h23'46" 30d44'27"
"In Love" Urs & Prisca Steiger
Lac 22h19'24" 48d22'20"
IN LOVE WITH YOU
Cyg 20h23'52" 47d54'2"
In Loving Memories of Vilma Alers
Vir 13h18'49" -13d40'53"
In Loving Memory AmberLialicejewell
Umi 14h29'17" 68d10'2"
In Loving Memory Barry Thomas Brown
Sco 17h25'2" -38d4'21"
In Loving Memory Katherine A Wilson
Uma 10h17'14" 48d51'24"
In Loving Memory of Alicja Polujan
Cas 1h11'52" 70d21'31"
In Loving Memory of Anne Phillips
Uma 9h29'0" 58d6'24"
In Loving Memory of Anyan
Cep 23h24'7" 70d49'22"
In Loving Memory of Birdo & Jack
Uma 10h29'13" 49d39'41"
In Loving Memory of Bryan Richards
Cyg 20h51'56" 48d53'50"
In Loving Memory of Charlie Morton
Her 17h13'55" 37d58'58"
In Loving Memory of Colin Nye
Uma 9h46'52" 42d20'24"
In loving memory of Colleen Cronin
Uma 11h38'27" 57d40'51"
In Loving Memory of Dad Malinchak
Uma 14h18'3" 56d24'31"
In Loving Memory of Dana Dandrea
And 0h45'34" 30d28'40"
In Loving Memory of Glen Whelpdale
Col 5h49'50" -34d6'36"
In Loving Memory of Greg Caldwell
Gem 6h30'32" 15d30'57"
In Loving Memory of Jadd Gilmore
Aqr 22h58'16" -11d2'1"
In Loving Memory of James F. Prell
Vir 14h36'7" 2d41'23"
In Loving Memory Of Joseph Krueger
Lib 15h7'35" -6d27'8"
In Loving Memory of Katie, our Angel
And 1h50'16" 46d39'41"
In loving memory of Kenny D.
Gem 7h14'57" 23d13'37"
In loving memory of Linda Inzero
Tau 3h43'58" 22d52'58"
In Loving Memory of Lisa Nones
Tau 3h39'6" 26d26'38"
In Loving Memory of Luke J. Stover
Cap 20h35'19" -19d39'26"
In loving memory of Michelle Lipski
Cas 1h19'2" 57d28'11"
In Loving Memory of Mick
Cep 0h23'16" 69d5'3"
In Loving Memory of Nikki Werner
Cmi 7h28'49" 10d27'2"
In Loving Memory of Norman & Jessie
Cru 12h29'7" -59d32'24"
In Loving Memory Of Operzenia Bates
Cas 23h7'54" 57d13'11"
In Loving Memory of Peter Obermaier
Dra 15h55'12" 52d2'1"
In Loving Memory Of Samuel Contino
Ori 5h57'27" 6d12'16"
In loving memory of Scott Coussens
Cru 12h36'57" -57d40'54"
In loving memory of Steve Brumfield
Uma 10h54'52" 44d12'28"
In Loving Memory Of Tom Buss
Per 4h7'49" 50d55'11"
In Loving Memory of Tony Hiller
Ori 5h28'13" 3d19'57"
In Loving Memory of Vera Carver
Per 2h28'18" 52d3'24"
In LovingMemory of Jak Leigh Andrew
Cru 12h41'26" -64d39'18"
In Lust & In Love, Forever & Always
Cnc 8h25'28" 13d32'54"
In Memory Jean Nardiello Fernicola
Uma 10h11'26" 70d53'49"
In Memory Jim Hill 8.08.46-22.11.05
Cru 11h59'3" -57d53'17"
In Memory - Michael Efton Olenchalk
Lib 15h13'57" -12d19'59"
In Memory of Adam Strain
Per 3h58'48" 38d8'41"
In Memory of Alex Lee Hill 8/10/87
Uma 10h53'8" 66d20'13"
In Memory of Anna Reece Ford
And 0h24'42" 27d29'9"
In Memory of Barry & Wilma Adkins
Cyg 21h16'44" 52d51'54"
In Memory of Ben Davis
Her 18h12'31" 21d53'57"
In Memory of Bernie Indyke
Ori 5h20'44" 6d52'13"
In Memory of Birdie McGinley
And 1h16'54" 34d6'46"
In Memory of Boo Kitty
Lyn 8h20'29" 50d48'41"
In Memory of Brooke Ashleigh
Cru 12h18'55" -57d5'48"
In Memory of Carson Milz
Uma 10h38'49" 62d24'4"
In Memory of Cassie
Col 5h48'21" -34d26'59"
In Memory of Cassie
Col 5h47'17" -35d34'12"
In Memory Of Chelsey R. Dimmitt
Cyg 19h48'18" 38d27'42"
In memory of Cle Cross
Uma 8h33'4" 64d14'14"
In Memory of Cole Tucker Willard
Her 17h53'25" 21d40'46"
In Memory of Dad - Shane's Star
Per 4h6'56" 46d51'26"
In Memory of Dan Manzella
Per 2h19'56" 52d40'18"
In memory of Dancer Samantha Ahn
And 0h22'23" 28d0'37"
In Memory of Dave Brown-Dear Friend
Uma 10h27'14" 48d38'53"
In Memory Of Davis Arthur Rollins
Uma 14h4'39" 53d24'58"
In Memory of Devin W. Deaton
Cep 21h55'57" 68d58'51"
In Memory of "Diana Marie Kautz"
And 0h41'51" 28d48'48"
In Memory of Diane E. Corey
Tau 4h18'10" 26d51'35"
In Memory of Dorothy Rose
Cru 12h34'11" -62d38'19"
In memory of Eddie
Aqr 21h7'17" -8d51'42"
In Memory of Edward Meinck
Her 18h28'25" 20d37'52"
In Memory of Eftyhia Mavridis
Col 5h46'31" -34d13'43"
In Memory of Elaine Weinberg
Cas 22h59'43" 53d46'11"
In Memory of Ellen Bradley Steel
Uma 9h16'33" 46d34'13"
In Memory of Eric J. Fries
Leo 11h5'26" 16d38'9"
In Memory of Evelyn Freiden
Crb 16h9'7" 36d10'22"
In Memory of Frank R. Crayton
Aqr 22h20'22" -6d51'56"
In Memory of Fred and Leila Hughes
Uma 12h36'49" 56d12'39"
In Memory of Fred and Willie Cap
Uma 12h45'20" 58d9'29"
In Memory of Gary Walsh - 1955 - 2006
Pho 1h50'50" -48d23'37"
In Memory of Gene Capola
Cap 20h31'51" -18d6'21"
In memory of George D. Dolphin
Uma 9h46'24" 43d17'5"
In Memory of Georgianna Tonelli
And 1h14'27" 45d34'15"
In Memory of Gramma Millie
Lmi 10h52'39" 26d8'50"
In Memory of Hayden Parker Sokolow
Cyg 19h48'58" 37d53'31"
In memory of Helen Turner
And 0h38'32" 37d5'20"
In Memory of Helena Russo
Cru 12h28'23" -59d27'35"
In Memory of Holy Ann Sneath
Per 4h5'51" 32d28'2"
In Memory of Jack Cooper Willard
Per 3h21'28" 44d28'3"
In Memory of James E. Belcher
Cep 21h22'12" 60d9'20"
In memory of Jerome & Joyce Goeden
Cyg 21h43'53" 33d22'9"
In Memory of Jessica Anne Robertson
Lyr 18h28'25" 30d51'53"
In Memory of KAT KLF JCY
Her 16h45'51" 24d56'30"
In Memory of Kenneth Conceicao
Aur 5h25'53" 45d23'10"
In Memory of Keryn Maree Leonard
Vir 14h5'55" -16d13'37"
In Memory of Kim Allen Tuttle
Lyr 18h44'33" 37d45'23"
In Memory of Kyle Weech
Per 4h11'26" 38d27'12"
In Memory of Liam Reukauf
Uma 9h14'42" 68d41'41"
In Memory Of Lindsay Mace
Cyg 19h47'8" 37d46'19"
In Memory of Logan Ashbaugh
Per 2h41'35" 57d9'43"
In Memory of Luthfi Hill
Uma 10h28'18" 45d8'26"
In Memory of Marcella Jana
Cru 12h57'5" -57d29'42"
In Memory of Maria & Alfio Di Carlo
Col 5h47'16" -28d50'44"
In Memory of Mario Limpias Terrazas
Vir 13h28'12" -10d19'39"

In Memory of Mary McIntyre
Gem 7h46'47" 16d8'1"
In Memory of Matt Jarvis
Aql 19h50'10" -0d4'34"
In Memory of Matt Myers
Psc 1h12'34" 27d48'23"
In Memory of Momma
Uma 8h33'31" 60d59'6"
In Memory of My Granny (Ninni)
Cas 0h10'54" 60d31'18"
In memory of my Pop Pop-We Luv U
Leo 10h10'40" 26d16'9"
In Memory of Nathan & Jaden Snider
Uma 9h44'16" 61d33'7"
In Memory of Neil Francis 4.02.1978
Aqr 21h18'27" -2d48'19"
In Memory of Nelson
Uma 13h55'32" 52d20'11"
In Memory of Our Angel Christy Lynn
And 0h40'19" 36d20'7"
In memory of our Dad - Matt
Cep 21h27'51" 64d34'58"
In Memory of Our Hero, Milt Johnson
Uma 12h36'28" 59d37'42"
In memory of Paige Emily Matthews
Cru 12h28'21" -56d2'27"
In Memory of Pat Trautman
And 2h0'35" 37d58'45"
In Memory of Patricia Nickel
Cru 12h11'16" -62d1'14"
In Memory of Paula Ward
Del 20h37'52" 15d35'53"
In Memory of Phyllis Frankel
Cas 0h37'26" 57d2'9"
In Memory of Ralph W. Young
Uma 10h56'20" 54d19'29"
In Memory of Raymond Majewski
Uma 8h30'39" 63d11'25"
In Memory of Rebecca Ann Cizewski
Gem 6h43'42" 17d25'15"
In Memory of Robert Amato
Uma 10h24'26" 60d48'10"
In Memory of Robert & Angeline Neal
Cru 12h56'15" -59d41'43"
In Memory Of Robert Donald Biddle
Ori 5h38'5" -0d59'19"
In Memory of Robert Kulevski
Sco 17h7'14" -40d2'45"
In Memory of Ron
Leo 11h19'48" 16d59'35"
In Memory of Rosemarie B. Ballard
Cas 1h28'19" 59d51'23"
In Memory of Sadie Jarvis
Col 5h36'32" -41d30'35"
In Memory of Samuel Stephen Field
Aql 18h47'30" 8d38'29"
In Memory of Scott James Mills
Cru 12h31'1" -59d45'54"
In Memory of Serna Lerner
Cas 23h5'16" 53d47'32"
In Memory of Shane Bomber Williams
Col 5h58'24" -35d22'17"
In Memory of Stephanie N. Glavin
And 23h10'21" 51d1'33"
In Memory of Sylvia Cagan
Cas 1h23'22" 60d58'5"
In Memory Of William A. McKinstry
Uma 10h35'46" 52d8'1"
In Memory Of William Ellis Bailey
Sco 17h41'6" -34d53'10"
In Night's Dark Sky, "Mary's Light"
Gem 6h52'0" 33d21'1"
In our dreams, in your eyes
Cru 12h1'56" -62d51'4"
In Rememberance of Pauline Clark
Ori 5h15'11" 6d14'35"
In Saecula Saeculorum
Dra 18h52'36" 50d7'11"
In Sky And My Heart Forever At Home
Vir 12h36'11" 12d0'1"
In The Air Tonight
Uma 11h36'54" 55d37'32"
In The Dark
Sgr 18h33'46" -22d4'7"
In The Sky Together George & Stacey
Cyg 20h49'1" 48d20'59"
In the Spirit - Deanna Kennedy
Lib 15h16'20" -22d2'16"
In Valerie & Torey Adkins Honor
Dra 10h9'30" 73d19'34"
Ina
Cep 21h49'21" 72d3'23"
INA
Uma 9h51'43" 68d23'47"
INA
Lyn 7h5'33" 52d22'9"
Ina
Tau 4h20'7" 9d26'15"
Ina
Lyr 19h12'41" 27d15'16"
Ina B.
Leo 11h10'18" 20d22'34"
Ina C. Billingkoff
Cap 21h40'26" -11d40'5"
Ina Fay Cheyney
Uma 8h56'3" 68d18'55"
Ina Hwang
Lyn 6h47'4" 50d39'42"
Ina Kay Hehl
Lyn 7h46'47" 48d22'15"
Ina Marie Wallen
Psc 1h8'41" 31d6'11"
Ina Nau
Cap 20h38'46" -18d15'36"
Ina R. Sinovoi
Tau 3h36'51" 29d42'0"
Ina & Ralf
Uma 11h37'42" 30d41'16"
Inaam Heavens Angel To The Stars
Cas 0h31'33" 63d7'52"
Inagi
Uma 8h39'1" 64d19'27"
Iñaki
Cnv 13h47'2" 37d39'7"
Inaki-Edurne 7-10-2004
Peg 22h13'36" 23d54'34"
Inalou
Cam 5h11'36" 66d9'35"
InaMichaelRios
Lyr 19h27'2" 41d34'27"
Inamo
Leo 10h21'13" 14d15'32"
Inamorata
Lyn 7h47'25" 52d34'52"
inamorato
Her 17h1'30" 13d45'43"
Inamoratos
Cyg 20h4'20" 33d2'6"
Inana will Twinkle
Vir 12h45'55" 6d40'29"
Inanna May Gibsone
And 23h10'20" 41d33'23"
Inara
Aql 19h6'14" -11d34'46"
Inara Grace
And 0h31'52" 33d24'7"
Ina's Hopes, Dreams & Happiness
Crb 16h11'10" 35d51'32"
Inas Rashad
Psc 23h21'39" 4d20'20"
INASTERN
Umi 14h8'6" 69d44'44"
Incantalupi
Leo 11h45'11" 20d41'33"
Incessant Love
Uma 11h47'38" 59d18'38"
incido amor
Ori 6h11'33" 3d29'11"
Incito of Gina
Lyn 8h1'6" 41d35'20"
" Incomparable " Mary Jones
Cas 1h41'42" 62d41'51"
Incredible Love Everlasting K&K
Aqr 22h59'16" -8d6'29"
Incubus
And 23h1'27" 50d37'52"
Inday Gazelle
Gem 6h55'6" 15d57'21"
Inday's Court
Cam 3h36'3" 63d32'26"
Indee
Ari 3h16'30" 23d10'11"
Indefinable Path
Her 17h53'14" 29d45'59"
IndelibleDream
Leo 9h44'10" 25d11'51"
Inden, Elke
Ori 5h2'22" 5d51'4"
Inder-Sneh
Dra 17h40'3" 62d18'47"
Indescribable
Leo 10h58'0" 11d24'18"
Indescribable Joy
Uma 9h35'8" 41d41'47"
Index
Umi 13h53'57" 69d47'33"
Indi Wirayudha Jegeg Bulan
Ind 21h13'2" -54d15'26"
India
Cas 0h14'12" 56d29'39"
India
Cnc 8h15'21" 7d26'21"
India A
Ori 4h52'32" 13d15'34"
India Hammond Waters
And 2h31'33" 49d3'52"
India Luce
Pav 18h48'16" -61d13'11"
India~Maher
Mon 6h25'23" -0d51'42"
India Malila Skye
Aqr 23h25'13" -19d44'9"
India Rose Godin
Mon 8h0'38" -0d27'49"
India Smith
Gem 6h2'35" 25d14'40"
Indian Princess Heather Sue Harris
Umi 14h13'1" 78d7'51"
Indiana Hope Jentjens
Aqr 22h6'17" -0d24'53"
Indiana Rose Winter
Cru 12h14'42" -60d18'30"
Indiana Sloan
Cma 6h56'22" -22d27'28"
Indiana Wayan Pratt
Uma 10h48'54" 65d42'42"
Indianna
And 1h55'0" 41d39'36"
Indianna Myers
Aql 20h5'21" 9d13'7"
IndianVette
Ari 2h41'55" 25d7'47"
India's Lucky Star
Umi 15h57'50" 75d13'54"
Indigo
Uma 9h31'6" 47d10'15"
Indigo Niki
Uma 11h11'46" 36d44'42"
Indigo Snow Waters
Sco 16h7'49" -12d47'7"
Indina, Viktoria
Uma 10h35'11" 65d38'38"
Indira
Peg 21h31'57" 15d35'50"
Indira
Aqr 22h14'52" 1d32'9"
Indira Leana
Ori 5h53'49" 6d44'16"
Indira Lucy Halder
Aqr 20h49'2" -0d16'30"
Indira The star that touches Heaven
Cyg 21h44'39" 44d38'5"
Indira76
And 2h15'50" 46d5'11"
Indira's Star (Estrella de Necesia)
Uma 10h58'57" 72d38'35"
IndiraSheehan
Lyn 8h11'43" 50d12'12"
Indivision
Ari 2h16'43" 10d47'31"
Indochine
Peg 21h30'9" 23d27'18"
Indra
Vir 12h19'53" 5d13'37"
Indra Medina
Sgr 19h15'22" -17d15'9"
Indra Rani
Uma 14h11'18" 58d4'27"
Indrani Muthukumaran
Lyn 7h36'42" 40d0'46"
Indri
Peg 23h35'40" 27d37'59"
Indser
Leo 9h45'11" 32d26'5"
indu
Ind 20h46'42" -46d39'28"
indu m lal
Gem 6h25'18" 23d37'5"
Indu Mathur
Cas 23h56'11" 53d24'33"
Indy
And 1h9'12" 45d42'7"
Ineke van der Linden
Crb 15h45'43" 36d49'47"
Inert Genius
Ari 3h10'43" 19d30'34"
Ines
Lyr 18h44'58" 29d32'37"
Ines
And 23h23'47" 35d19'39"
Ines
Cyg 21h1'29" 39d7'25"
Inés
Uma 9h49'55" 45d0'57"
Inès
Cen 13h31'20" -35d5'18"
Inés Arboli
Ori 6h17'32" 13d35'26"
Inès Baron
Umi 16h9'21" 83d9'14"
Ines Castejon Marcos
Uma 13h0'0" 59d10'1"
Ines Cervantes
Lyr 19h15'8" 26d26'38"
Ines Emilie Thorup
Vir 13h25'11" 12d8'25"
Ines hdsg
Lyr 18h36'38" 36d8'3"
Ines Margaret Mello
And 2h32'3" 48d57'54"
Ines Reig LLorens
Ari 2h19'42" 12d17'24"
Inès Romain
Cas 0h38'12" 51d31'0"
Ines Schubert
Uma 10h31'34" 49d1'40"
Ines - Souad Ghandour
Tri 2h32'7" 31d30'1"
Ines und Roland
Uma 11h14'55" 44d38'22"
Ines und Thomas
Peg 22h21'22" 3d1'48"
Inesa Petkeviciute
Aqr 22h59'8" -7d4'18"
Inese
Leo 11h4'18" 21d29'6"
Inese Delvere
Aqr 20h54'9" -10d51'26"
Inesella Pesante
Vir 13h18'57" 8d57'18"
inessa
Vir 12h36'27" 9d16'24"
Inessa Panova
Lyn 7h4'5" 45d58'12"
Inessa Sapin
Sco 17h56'7" -30d54'22"
Inez Brown
Sgr 20h11'56" -32d41'53"
Inez Garcia
Leo 11h1'52" 5d8'24"
Inez Marie Whitehead " Boots "
Boo 14h32'32" 44d41'54"
Inez Regan
Peg 21h35'39" 27d52'13"
Inez Ruth Mayer
Lyn 6h48'19" 51d3'6"
Inez Rutherford
And 0h24'58" 32d59'58"
Inez Stubblefield
Crb 15h30'2" 27d17'25"
Inez Van Lingen-Wife,Mom,Grandma
Ari 2h5'35" 18d20'15"
Infanis Armadillum
Uma 9h32'37" 64d11'4"
Infectious
Uma 9h45'29" 48d5'51"
Infiniment Moi
Tau 5h27'18" 25d40'57"
Infininest & Beyond
Eri 4h34'14" -4d41'59"
Infinitas
Cyg 20h20'16" 52d5'30"
Infinité
Psc 0h23'57" 15d56'22"
Infinite
Sgr 19h12'59" -33d40'25"
Infinite Abyss
Uma 8h22'44" 61d57'50"
Infinite Butterflies
Peg 23h30'43" 12d6'49"
Infinite Isabel
And 23h13'22" 51d45'49"
Infinite Love
Cyg 21h52'32" 47d1'39"
Infinite Love
Cyg 21h12'47" 40d40'4"
Infinite Love
Ari 2h14'45" 22d8'11"
Infinite Love - Chris & Natalie Green
Cyg 20h4'17" 51d56'45"
Infinite Love CJDT4205
Cyg 20h26'8" 52d22'16"
Infinite Love Jeremiah and Kristen
Lib 15h44'23" -17d2'51"
Infinite Possibility
Cam 4h38'47" 65d31'34"
Infinite Promise
Cyg 20h14'42" 49d28'32"
Infiniteum Geraldine Smith06
And 23h52'57" 48d3'26"
Infinite.....You and Your Love Are
Psc 1h22'33" 13d21'7"
Infiniti Rose Lee
Cas 0h37'42" 58d30'40"
Infinito
Ori 5h34'21" 3d25'11"
Infinitus Amor
Uma 8h15'26" 66d52'9"
Infinitus Ardor
Ari 2h28'3" 23d10'12"
Infinity
Cnc 7h55'51" 16d49'19"
Infinity
Cyg 19h21'53" 28d10'49"
Infinity
Aqr 21h35'43" 1d32'27"
Infinity
Psc 0h48'31" 8d45'56"
Infinity
Cas 0h22'35" 54d24'20"
Infinity
Cas 0h9'26" 53d28'18"
Infinity
Cyg 20h23'34" 45d36'19"
Infinity
Dra 18h24'34" 51d23'25"
Infinity
Uma 10h52'53" 46d32'58"
Infinity
Uma 9h49'55" 51d44'23"
Infinity
Uma 11h13'3" 51d5'49"
Infinity
Uma 11h14'26" 52d9'33"
Infinity
Uma 8h59'40" 48d6'55"
Infinity
Lmi 9h43'48" 37d50'42"
Infinity
Lyn 7h22'5" 44d33'6"
Infinity
Cyg 21h2'32" 30d41'40"
Infinity
Crb 15h51'55" 34d22'58"
Infinity
Uma 9h23'34" 64d37'28"
Infinity
Uma 9h23'0" 61d36'27"
Infinity
Uma 13h16'47" 56d9'9"
Infinity
Cep 5h24'21" 86d1'50"
Infinity
Lep 5h44'36" -13d41'40"
Infinity
Vir 13h21'55" -11d15'54"
iNFiNiTY
Ser 15h18'47" -0d9'50"
Infinity
Psc 0h41'6" 6d13'49"
Infinity
Sgr 19h58'16" -30d29'1"
Infinity
Psa 22h7'29" -28d41'44"
Infinity + 1
Sco 17h50'19" -35d24'24"
Infinity + 1
Dra 15h5'7" 63d13'23"
Infinity + 1
Ori 6h17'39" 9d6'49"
INFINITY and Now Beyond
Her 16h21'13" 48d9'3"
Infinity Dos
Aql 19h45'24" 14d49'6"
INFINITY - I love you Jamie & Avery
Umi 15h13'7" 74d58'10"
Infinity LLF 81305
Uma 11h41'42" 61d50'39"
Infinity Love
Gem 7h42'23" 24d4'15"
Infinity - This Is My Love For You
Umi 15h32'12" 72d20'34"
Infinity X
Leo 10h17'22" 18d31'26"
Infinity X 3
Her 18h44'2" 20d8'48"
Infinitys Love
Umi 19h29'17" 88d29'42"
Infohrm
Cru 12h42'48" -58d17'41"
Ing. František JUN, PhD.
Aql 19h55'6" 11d11'13"
Ing. Franz-Peter Bauer
Ori 5h42'15" 11d31'42"
Ing. Gerhard Morocutti
Uma 10h0'28" 51d23'9"
Ing. Peter BLAŠKO, CSc.
Aql 19h34'55" 13d33'16"
ING-042405
Lyn 6h45'42" 53d4'21"
Inga
Cma 6h44'4" -12d57'11"
Inga
Uma 10h31'33" 45d40'13"
Inga
Uma 10h35'24" 49d12'54"
Inga
Cnv 13h49'49" 36d36'47"
Inga
And 0h40'41" 37d16'40"
Inga "21st Birthday"
Cas 0h13'36" 51d29'23"
Inga Mara Stots
Leo 9h32'1" 29d19'32"
Inga Nyeng Benan
And 1h19'43" 45d26'11"
Inga's Star
Per 3h42'23" 39d37'26"
IngaWil
Dra 16h25'46" 63d7'38"
Inge Andersen
And 0h53'29" 37d14'51"
Inge Baumgartner 10.08.1940
Uma 9h31'38" 48d31'37"
Inge Berg
Uma 9h14'44" 69d45'10"
Inge Bodo 30- igster Hochzeitstag
Uma 11h3'49" 57d8'2"
Inge Dora
Vir 13h16'19" -0d58'9"
Inge + Giorgio, die besten Eltern des Universums!
Per 4h19'4" 34d41'38"
Inge Jaklin
Tau 5h45'12" 23d40'16"
Inge Krämer
Uma 11h35'54" 60d0'39"
Inge Metzger
Uma 11h5'18" 37d54'26"
Inge & Reiner
Ori 6h19'0" 10d38'37"
Inge Ruiseco
Psc 0h48'36" 11d31'32"
Inge und Alexi für immer
And 2h0'5" 38d9'22"
Inge Wong
Vir 13h5'23" -17d12'6"
Inge5151
Tau 3h45'6" 16d52'59"
Ingeborg Hallstein
Uma 11h40'52" 60d30'29"
Ingeborg Hammer
Ori 5h7'46" 1d14'25"
Ingeborg Janzen
Cas 1h37'7" 60d4'41"
Ingeborg Müller-Thiem
Uma 10h16'59" 70d14'31"
Ingeborg Runte Moran
Aql 19h0'56" -0d8'50"
Ingeburg Hromadka
Uma 10h12'23" 41d34'14"
Ingenium
Cyg 20h15'45" 50d9'21"
Inger
Cap 21h30'59" -21d7'27"
Inger Bergagård
Cas 1h16'11" 58d59'42"
Inger Marie Øksenholt
Cnc 9h0'33" 14d20'16"
Inggew9131971
Lyn 8h42'26" 43d9'45"
Ingheans
Uma 10h30'20" 68d0'27"
Ingibjorg Oskarsdottir
Cas 0h27'53" 66d54'4"
Ingmar
Uma 10h23'22" 43d55'57"
Ingmar Beins
Uma 10h13'40" 48d33'20"
Ingmar Pien
Uma 9h26'14" 52d34'53"
Ingo
Uma 11h19'9" 33d50'17"
Ingo Hoffmann
Uma 10h46'19" 63d35'49"
Ingo Oschmann
Dra 16h52'30" 55d25'20"
Ingo Pauly
Ori 5h56'12" 17d14'14"
Ingo Radusch
Uma 10h29'37" 41d43'21"
Ingo Schulz
Uma 9h31'32" 62d45'35"
Ingo Schütt
Uma 11h39'56" 49d10'52"
Ingolf Grimmer
Uma 9h32'10" 50d43'34"
Ingram
Her 17h15'39" 20d35'6"
Ingred Elizabeth
And 0h36'9" 28d20'19"
Ingrid
Gem 7h46'20" 25d37'14"
Ingrid
Ori 6h20'58" 11d10'48"
InGriD
Aql 20h1'12" 14d52'11"
Ingrid
Ori 4h53'28" 2d23'31"
Ingrid
Ori 6h16'31" 5d43'31"
Ingrid
Vir 12h8'19" 4d14'36"
Ingrid
Psc 1h40'40" 13d30'18"
Ingrid
Uma 11h58'29" 51d40'54"
Ingrid
And 1h45'52" 42d23'46"
Ingrid
Uma 8h22'20" 63d57'53"
Ingrid
Uma 12h57'56" 60d30'28"
Ingrid A. Hill
Crb 16h3'22" 33d23'20"
Ingrid Alicia Stabinski
Tau 5h33'54" 24d51'43"
Ingrid Amaris
Uma 14h3'19" 48d28'6"
Ingrid Ameli Hüttenrauch
Uma 8h26'19" 69d45'2"
Ingrid and Earl Hamilton
Cyg 21h36'35" 45d0'41"
Ingrid Ann
Pic 4h43'59" -46d36'38"
Ingrid Arlette Tielman Smith 2/16/53
Aqr 21h3'35" -12d25'52"
Ingrid Ava Schwarz
Crb 15h48'37" 27d32'40"
Ingrid Britt
Gem 6h50'16" 17d4'34"
Ingrid Carlsson P.A.
Lyn 8h13'34" 34d37'40"
Ingrid Castro-Rodriguez
Lmi 10h30'46" 36d57'16"
Ingrid Concepcion
Lyr 18h51'35" 36d41'24"
Ingrid Denisse Ayala Grosswiler
And 23h24'56" 49d45'10"
Ingrid Echeverry
Cap 20h30'9" -9d4'51"
Ingrid F - CM
Psc 0h59'49" 27d57'35"
Ingrid Helen Alter
Ari 3h13'1" 28d28'43"
Ingrid Labre
Uma 10h57'3" 38d46'26"
Ingrid Leon
Cnc 8h14'12" 20d39'21"
Ingrid ma Puce
Ori 5h6'52" 6d35'41"
Ingrid "mein Sternchen" Winkler
Ori 5h15'19" 6d24'25"
Ingrid Mercedes Bermudez
Leo 11h14'15" 6d16'52"
Ingrid mon petit ange adorée
Sgr 20h11'13" -36d54'28"
Ingrid; Mother of Bob and Ken
Cas 22h59'28" 58d47'51"
Ingrid Niklas
Ori 5h55'5" 21d59'26"
Ingrid Optimus Matris
Crb 16h18'33" 28d59'57"
Ingrid Paape Baburam Texiera
Mon 7h34'33" -0d49'49"
Ingrid Pawlytta
Uma 11h30'23" 40d59'38"
Ingrid Stähli
Lac 22h51'54" 53d21'4"
Ingrid Suprock
Sco 16h7'51" -10d51'59"
Ingrid Vargas (mi lok)
Cap 21h42'15" -11d56'47"
Ingridore Schautzuna
Lib 15h49'22" -18d12'9"
Ingrid's Eye
And 1h27'11" 43d50'3"
Ingrid's Star
Sco 17h37'16" -42d59'51"
Ingridskön Prinsessanealis
Crb 15h50'44" 36d34'1"
Ingrid-W
And 0h3'29" 36d43'9"
Inholte, Ralf
Uma 11h37'20" 31d57'19"
Iñigo Melchisidor Urquijo
Her 17h49'16" 32d16'56"
Inika Visser
Sgr 19h27'3" -42d58'49"
Inka Porath
Uma 10h37'7" 40d7'2"
Inke Deane - Tupperware
Cet 2h20'16" 8d9'12"
INKI
Uma 14h18'8" 55d19'39"
Inky
Aqr 23h41'36" -24d18'23"
Inlustris Lumen
Uma 11h33'9" 42d46'0"
Inmaculada Millán Álvarez
Col 5h11'46" -28d23'44"
Inmanius Martinius 1727
Mon 6h33'57" 7d38'41"
Inna
Uma 9h32'0" 57d51'0"
Inna Alexandrovna Tishkova
Cyg 21h39'35" 44d50'39"
Inna Khusid
Lib 15h43'30" -10d45'28"
Inna Krasnik
Cyg 20h20'35" 43d14'39"
Inna Kulikova
And 0h58'44" 41d8'42"
Inna Lazareva
Ari 2h55'10" 24d32'2"
Inna Nechipurenko Aciernos Red Star
Uma 11h24'6" 54d45'33"
Inna Ukrainian Goddess
And 0h29'32" 27d15'51"
Innamorato
Cam 4h3'11" 65d2'40"
Innamorato, Kly
Tau 5h58'20" 26d48'19"
Innes
Lmi 11h1'50" 28d37'16"
Innes
Sgr 18h40'15" -22d26'35"
Innes Murray Hendry
Umi 14h29'54" 68d11'31"
Innisfree
Cyg 20h20'1" 58d13'12"
Innishfael Luna
Cyg 20h12'58" 47d53'29"
Innita2005
Uma 10h29'20" 62d23'54"
Innocence - JCYX66
Uma 9h38'55" 50d32'39"
Innochka Sigizmyndovna
Gem 6h48'23" 32d16'23"
Inny Imon
Cnc 8h34'20" 32d42'24"
Inocencia G. Diaz Our Beloved Angel
Cas 0h51'40" 60d26'52"
Inolvidable Alska
Peg 23h13'56" 16d7'52"
Inome F.
Ori 4h51'46" 1d31'23"
Inova
Cap 21h35'45" -15d29'25"
InRoLove
Uma 14h2'32" 48d59'24"
Insaf
Ari 2h7'19" 26d30'42"
Insane Harley 2004
Uma 11h10'59" 34d35'28"

Insatiable Susie
Cyg 20h10'38" 39d19'19"
Inseparable
Uma 10h30'29" 64d37'4"
Inshallah
Cyg 20h1'41" 39d16'18"
Insieme
Lib 15h18'2" -17d47'59"
Insieme Pere Sempre
Cru 12h15'25" -62d59'19"
InSinkErator Evolution Series™
Uma 9h8'43" 67d19'59"
Insite
Umi 15h12'38" 76d22'6"
INSOMNIA
Uma 11h24'6" 31d16'18"
Inspiration
Peg 23h21'4" 32d4'46"
Inspiration
Lyr 18h49'44" 44d46'57"
Inspiration
Cyg 20h55'42" 42d42'55"
Inspiration
Peg 23h22'40" 15d58'53"
Inspiration
Ari 2h18'45" 14d22'31"
Inspiration
Pho 0h31'36" -43d16'58"
Inspiration Martha
And 23h17'10" 48d21'26"
"Inspiration" - Terry W Ibbotson
Psc 1h44'46" 12d58'18"
~InSpiRation~ Tom & Jean DeThomas
Uma 10h26'59" 47d56'55"
"Inspiration" Wendy Millman's Star
And 2h14'14" 39d29'27"
Inspiration45
Uma 12h4'5" 35d2'55"
Inspirational Spirit
Per 3h11'17" 42d51'35"
InspireWorks
Ori 5h28'1" 6d56'15"
Instigator
Cyg 20h55'22" 37d27'54"
Instigator and Butterfly 2006
Cyg 21h15'25" 52d35'30"
Integrated Tourism Star
Cru 12h51'1" -60d29'49"
"Integritas" - Through Renae's Eyes
Col 5h27'43" -35d22'47"
Integrity
Cep 21h40'11" 69d3'3"
Integrity's Faith
Uma 10h39'2" 49d37'26"
Intelliworks Orion
Ori 5h35'19" 9d11'46"
Intemporaliter
Ari 1h53'56" 20d5'39"
Intense
Umi 14h14'20" 72d50'18"
Intense Steve
Aur 5h50'14" 42d38'27"
Intensity Gail
And 23h24'47" 42d44'41"
Intention Point (Craig's Star)
Umi 16h26'29" 85d39'48"
Inter Nos In Aeternum
Cas 23h49'28" 56d50'8"
Intergalactic Instructor Erin
Lib 15h17'5" -6d23'2"
Intergalactic Tang Soo Do
Dra 16h36'55" 62d11'9"
Internally Flawless Winky
And 23h8'8" 50d50'50"
Interpretus Antarktos
And 2h26'38" 46d21'54"
Inti Lutrek
Vir 14h41'7" 5d23'26"
InTigrity
Aqr 20h46'24" -8d23'47"
Into the Mystic
Cas 1h23'31" 64d17'57"
Intrinsicate
Sgr 18h37'8" -16d56'55"
INVERS-Gruppe
Uma 14h23'46" 62d17'28"
Invictus
Uma 9h22'50" 63d44'4"
Invincible
Her 17h3'48" 31d59'14"
Invisible Fairy Tale
Mon 6h58'23" 2d37'41"
InWEnt
Uma 8h18'59" 59d57'0"
inz. Mieczyslaw Rus
Dra 18h10'31" 51d51'24"
Ioan Jinga
Aqr 21h40'36" -1d35'35"
Ioana
Ari 2h23'28" 15d29'56"
Ioana Mara
Cnc 9h14'5" 20d57'35"
Ioana_Moisescu78
Cyg 20h24'17" 34d32'41"
Ioannes Caelicola
Cap 21h37'11" -15d30'21"
Ioannes Tullius
Uma 12h41'30" 58d14'53"
Ioannis Anastasopoulos
Uma 9h27'15" 42d53'10"
Ioannis Brachos
Uma 11h35'30" 46d25'6"
IOANNIS CORCACAS
Sgr 20h19'49" -43d25'48"
Ioannis Haniotis
Leo 11h16'25" 21d4'49"
Ioannis Mandalas
Uma 10h19'40" 71d33'9"
Ioannis Masoutis
Uma 11h50'57" 43d45'27"
Ioannis Sotiropoulos
Uma 13h37'47" 51d48'52"
Ioannis Varotsos
Uma 13h47'55" 50d4'25"
Ioannou, Ilias
Lib 15h19'4" -11d4'5"
Iodgywtfuiogjyrwuanwwtlykj hmymtuwly
Pho 23h41'57" -42d25'53"
Iofi
Cyg 19h48'37" 33d26'29"
Iola Sausville
Uma 12h42'23" 57d31'35"
Iolanda
Aur 6h37'23" 31d23'11"
Iole
Sct 18h54'16" -12d46'51"
Ion SoulReble
Aql 19h43'30" 14d37'49"
Ione
Cas 1h56'9" 66d52'5"
Ioni
Com 13h11'56" 20d2'38"
Ionica
Sco 16h47'0" -41d55'19"
Ionie L Findley
Aqr 22h42'10" -0d51'27"
Ionis
Lac 22h29'14" 50d36'3"
IORINUS
Uma 8h55'51" 61d53'39"
Iorio Gianni
Her 16h42'42" 24d34'2"
Iosif Chain & Raissa Issaev
Psc 1h0'41" 26d17'21"
Iosif Uvaydov
Uma 12h1'15" 42d29'57"
i.o.u. one galaxy
Per 4h41'38" 46d42'4"
Ipa
Cnc 8h42'58" 17d13'23"
IPBNJU
Uma 13h45'9" 60d19'36"
Ipek & Tufan
Uma 11h10'38" 40d9'16"
Ipek Yesil
Ari 3h16'18" 15d51'55"
Iphaelon Jan
Psc 0h53'52" 13d14'36"
Ipheellike
Ari 3h23'40" 25d55'7"
Ipo Hoku Alexandria M Varga
Uma 11h19'48" 42d20'15"
Ippa
Per 3h25'26" 41d33'5"
Ippotis apo arthro Oros
Sgr 18h19'42" -16d53'47"
Iqbal M. Wali
Gem 7h4'18" 22d34'31"
I.R. Weinraub
Aql 18h56'43" -3d49'52"
Ira
Lib 15h11'20" -4d59'6"
Ira
Uma 10h45'3" 55d29'30"
Ira Citron
Umi 14h26'40" 66d59'50"
Ira Curtis Gunning
Cma 6h40'10" -17d33'36"
Ira Dankberg
Uma 11h12'52" 41d49'34"
Ira Jeffrey Marcus
Uma 11h54'24" 34d40'38"
Ira Ludwick
Cap 20h34'28" -20d32'49"
Ira Major
Cam 5h35'5" 58d26'48"
Ira Morgunovskaya
Ori 5h42'26" -4d20'25"
Ira "OPE" Mann
Sgr 18h14'57" -26d10'14"
Ira Reisman
Tau 4h33'34" 27d8'34"
Ira Rubinstein
Uma 10h21'33" 66d43'36"
Ira Tate
Umi 16h12'12" 80d43'7"
Ira the Star
Uma 10h26'28" 64d3'26"
Ira Weiss
Cnc 8h42'35" 16d11'11"
Ira Zimmerman
Psc 0h35'37" 9d18'48"
IRAD
Aql 19h16'1" 14d17'33"
Iraina
Cas 23h45'5" 52d41'18"
Iraqi Sweathearts,Tiffany & Daniel
Dra 16h31'27" 56d28'13"
Ira's Star
Per 3h30'1" 41d19'49"
Irasema Rodriguez
Aur 6h27'17" 34d13'10"
Irati Pisón
Leo 9h23'41" 8d42'32"
Irawan, Paul
Uma 8h51'56" 62d53'40"
Irazu
Sgr 19h27'51" -36d13'3"
Irchik
Umi 15h5'17" 77d16'48"
IRD. HB,TB,TB,NB,JB,DB,JD,L M,PR,PG
Uma 13h35'48" 49d25'35"
Ire
And 0h8'51" 39d43'9"
Irebio
Lyn 7h50'51" 45d4'12"
Ireck Humberto Briones
Gem 7h43'29" 29d52'17"
Ireilis
Aqr 22h37'20" -1d43'33"
Irelan Hope Inoshita
Aqr 21h53'41" 0d48'52"
Irelyn Hope Wilson
And 1h46'42" 42d32'25"
Iren Kazemi
Uma 8h28'47" 72d16'58"
Irena
Cas 23h34'54" 58d9'37"
Irena Diana Di Mauro
Uma 13h22'2" 55d12'6"
Irena Dimitrova
Lyn 7h18'7" 59d40'40"
Irena Lipshits&Vlad Cutler Forever
And 0h10'6" 40d36'5"
Irena Tintor
Aqr 22h6'41" -11d24'43"
Irena und Jürgen
Uma 11h11'33" 43d23'56"
Irenca
Cnc 9h4'46" 30d54'3"
Irenco
Uma 10h20'6" 42d45'32"
Irene
Crb 16h8'51" 33d36'27"
Irene
And 1h4'6" 46d30'45"
Irene
Uma 9h54'11" 49d1'38"
Irène
Tau 3h45'18" 8d54'18"
Irene
Leo 11h40'13" 10d45'34"
Irene
Gem 7h37'23" 27d31'38"
Irene
Cyg 19h51'42" 29d57'42"
Irene
Peg 23h28'37" 18d16'10"
Irene
Aqr 22h39'56" -19d29'3"
Irene
Aqr 22h37'34" -15d21'21"
Irene
Vir 12h57'5" -18d22'31"
Irene
Aqr 21h37'34" -5d58'9"
Irene
Uma 10h44'57" 59d20'7"
Irene
Cas 1h29'31" 62d8'37"
IRENE
Cam 6h30'26" 66d36'32"
Irene
Uma 12h18'0" 61d57'48"
Irene
Sco 17h19'58" -40d23'0"
Irene
Aqr 22h28'48" -22d32'44"
Irene
Peg 3h24'25" 41d31'58"
Irene 1915
Cas 0h58'12" 58d55'14"
Irene A. Brandt 1934-2005
Uma 11h30'40" 59d47'16"
Irene Aguiar Gallardo
Uma 9h54'49" 45d10'12"
Irene Alice
Sco 17h56'28" -42d58'38"
Irene Anastasia Tzinis
Gem 7h29'0" 30d39'39"
Irene and Dean Goodlaxson
Uma 11h23'50" 36d9'36"
Irene and Romeo Martin
Sco 16h15'17" -9d34'54"
Irene and Timothy George
Aqr 21h41'20" 2d30'40"
Irene Ann Miriello Kemper
Uma 10h31'42" 53d53'52"
Irene Armstrong
Uma 9h55'49" 58d12'57"
Irene Atkins
Aqr 20h47'53" 0d51'38"
Irene & Ben
Lyr 18h39'56" 39d50'57"
Irene & Benjamin Chasen
Cas 0h59'22" 54d10'15"
Irene & Bill S. Myers
Aqr 22h23'55" -3d56'59"
Irene Blythe
Cas 0h47'51" 67d10'30"
Irene & Boele
Cyg 20h26'32" 47d11'55"
Irene Bulick A M L A M Love Milton
And 23h23'46" 48d43'35"
Irene C. Forbotnick
Aur 5h41'51" 51d7'6"
Irene C. Smee
Crb 15h56'19" 25d40'6"
Irene Cole
Ori 5h43'43" 12d5'17"
Irene "Cricket" Collett Grindstaff
Aqr 20h46'15" 1d35'3"
Irene Cucci
Cap 20h41'34" -20d43'13"
Irene D. Anderson
And 0h13'36" 42d36'21"
Irene Deborah
Cam 5h39'25" 56d40'44"
Irene Desrosiers
Cas 1h37'35" 60d44'17"
Irene Diana Jacobs
Cas 0h29'52" 50d14'17"
Irene E. Jablonski
Ori 5h35'26" 10d52'5"
Irene E. Masserelli
Gem 6h35'52" 27d29'24"
Irene E. Steele
Cas 1h25'10" 63d0'30"
Irene Ebalobor Castaneda
Lib 15h18'8" -5d58'28"
Irene Eilinger
Ori 6h0'27" 14d4'10"
Irene Elizabeth Devellin
Gem 7h6'49" 19d48'32"
Irene Elizabeth Reid - Born 21-6-1946
Cas 23h56'26" 52d34'14"
Irène et Olive Ventrillard
Cap 20h15'19" -20d14'39"
Irene Evangelous
Sco 17h26'26" -42d24'21"
Irene Evelyn Jenkins Hammerslough
Tau 4h25'51" 23d37'9"
Irene F. Lagoo
Uma 14h20'45" 56d2'11"
Irene Fan
Sgr 18h29'24" -16d14'43"
Irene Fender Sisk
Cyg 20h52'3" 54d39'26"
Irene Fleming
Vir 13h11'5" -13d3'20"
Irene Francis Amaro
Uma 8h37'8" 60d17'26"
Irene Gladys Rivett
Uma 10h51'39" 64d50'6"
Irene Greenberger Yachbes
And 2h17'32" 41d44'38"
Irene Greenwood's Shining Light
Cas 1h38'58" 60d56'46"
Irene H. Schermerhorn
Leo 9h31'2" 31d23'39"
Irene Hall
And 23h11'36" 42d40'52"
Irene Hamilton Winner
Cas 3h24'17" 70d29'43"
Irene Hollerback Egan
Cas 1h49'26" 64d1'50"
Irene "Honey" O'Dwyer-Woods
Psc 1h19'34" 32d59'42"
Irene Hook
Uma 10h26'50" 40d46'38"
Irene Jackson
Cas 0h8'5" 55d33'6"
Irene James
Gem 6h17'36" 24d53'33"
Irene & Jared
Lib 15h40'34" -12d27'26"
Irene June
Sco 16h38'8" -28d36'45"
Irene June
Ori 6h6'55" 1d25'30"
Irene & Kay Hafen
Lyr 18h49'54" 31d50'44"
Irene Koch
Cas 1h18'28" 64d44'14"
Irene Kyriakoudes
Aqr 22h35'47" -8d1'39"
Irene Lake Drew
And 1h20'2" 34d43'8"
Irene Lapetina
Uma 10h26'28" 61d7'2"
Irene Leone
Umi 14h49'43" 88d12'15"
Irene Levine Ph.D.
Lib 14h30'16" -23d0'3"
Irene Lewis
And 0h47'55" 30d27'59"
Irene Lim somekinda wonderful star
And 0h25'35" 44d2'31"
Irene Louise Joraskie
Cas 0h37'13" 54d49'2"
Irene Love
Crb 15h53'53" 26d20'46"
Irene Lusk
And 0h30'28" 26d33'29"
Irene M. Chubb
Cas 1h3'56" 60d10'31"
Irene M. Sherry
Uma 13h41'46" 52d32'37"
Irene Mae Kassis
Cap 21h42'15" -11d7'47"
Irene Maggi
Uma 11h47'10" 52d45'48"
Irene & Mahlon Beels
Aqr 22h51'18" -11d20'43"
Irene Marie
Vir 14h16'19" -7d24'6"
Irene Mason
Uma 11h0'25" 51d28'25"
Irene Matz
Cyg 19h58'45" 42d32'40"
Irene Melice
And 0h36'5" 45d4'51"
Irene Miriam Ashley (Cita)
Uma 10h46'14" 69d14'10"
Irene Mitropoulos
Cas 0h20'3" 47d28'1"
Irene my love!
Aqr 21h44'38" -6d33'54"
Irene Nanny Tillery
Cnc 8h43'27" 16d31'45"
Irene Nicola
Psc 1h40'21" 20d47'46"
Irene - Our Star
Cas 1h57'18" 61d40'35"
Irene Patricia Ferrell
Cas 0h40'10" 53d30'33"
Irene Pauline Hoffman
Cnc 9h5'27" 31d24'41"
Irene Piech
Aqr 22h18'8" -3d12'53"
Irene Ping Xie
Psc 1h30'59" 14d15'40"
Irene Russell 27.12.1930 - 27.04.2006
Umi 14h6'37" 69d1'17"
Irene Ruth Aguayo Rivas
Cyg 20h38'57" 34d51'30"
Irene Rydz
Lyr 18h43'9" 40d33'19"
Irene S. DiMatteo
Cas 0h18'12" 60d42'23"
Irene Sainty
Eri 2h27'34" -45d10'54"
Irene San
Vir 11h38'44" 3d40'19"
Irene Sanchez
Aql 19h28'43" 6d25'32"
Irene Sienko (Mom is a real star!)
Cas 1h59'31" 61d56'31"
Irene Siriusilla
Ori 6h0'36" 3d58'5"
Irene Sophia Romero
Leo 11h42'0" 26d45'57"
Irène Stocker
Aur 5h20'35" 43d27'33"
Irène Stübing
Cyg 20h53'26" 48d35'14"
Irène Stübing
Her 16h33'41" 13d42'41"
Irene Stumbo Eagler
Peg 22h55'27" 31d21'14"
irène super amore
Crb 15h38'32" 29d26'32"
Irene T. Lynch
Sgr 18h18'59" -16d55'25"
Irene Tau
Psc 1h15'26" 25d20'13"
Irene The Dancing Queen
Cap 20h51'11" -22d1'43"
Irene Thomas - The Star of Irene
Cyg 21h40'4" 35d31'49"
Irene Topliss
Gem 7h3'4" 14d59'48"
Irene Varol
Aqr 21h18'47" -11d1'48"
Irene Virginia Tomczak
Lib 14h47'2" -24d43'58"
Irene Volz
Ari 2h12'37" 21d16'24"
Irene Walker
Umi 14h54'27" 71d55'28"
Irene Walters
Crb 15h44'52" 28d53'16"
IRENE WENGER
Umi 15h35'40" 72d56'0"
Irene White "Best Grandma"
Uma 11h37'32" 40d47'19"
Irene Woldanski
Lyn 7h39'20" 54d19'16"
Irene Wright
Peg 23h58'38" 10d51'9"
Irene Wymbs
Tau 3h27'32" 5d8'41"
Irene Yuan
Sgr 18h58'44" -31d11'21"
Irene Zanette
Sco 17h22'50" -45d7'3"
Irene, 01.01.2003
Umi 14h5'6" 68d56'3"
Irene, Light Of The Heavens
Uma 14h23'54" 56d18'1"
IreneBob65
Vir 12h42'7" 4d49'18"
Irene-Jasmine-Rojas-5000
Leo 11h20'52" 20d43'16"
Irene's Odyssey
Pho 0h35'31" -41d46'5"
Irene's Silver Star
Cnc 8h21'2" 14d45'26"
Irene's Star
Lib 15h3'38" -2d0'1"
Ireta & Fred Elkins
Lib 15h20'2" -21d32'54"
Irfan, Rizwan and Sunny Ahmed
Umi 14h0'55" 73d33'25"
IRIA
Ori 5h28'46" 3d13'7"
Iriada
Uma 10h16'27" 55d37'10"
Irina
Sgr 17h47'51" -21d42'42"
Irina
Ori 6h8'27" 20d11'9"
Irina
Ari 3h27'43" 27d2'15"
Irina
Lmi 9h32'58" 37d4'52"
Irina
Gem 7h44'9" 32d38'8"
Irina
Lyn 7h31'32" 51d27'8"
Irina Anatolyevna Evstigneeva
Lyn 7h37'34" 58d27'27"
Irina and Minas Bagyan
Gem 6h57'39" 18d21'28"
Irina Ashkinazi
Lyn 8h8'31" 59d3'50"
Irina Astra 9
Cap 21h44'31" -22d31'2"
Irina Baburov
Tau 5h12'41" 16d35'3"
Irina Batkova
Tau 4h20'34" 26d55'29"
Irina Budileanu
Cnc 8h8'22" 14d41'51"
Irina Cornetta
Tau 5h55'12" 25d24'41"
Irina Genya
Lib 15h49'47" -19d58'4"
Irina Gouchtchina
Lyr 19h1'8" 32d40'52"
Irina Kuznetsova
Lib 15h44'36" -23d52'29"
Irina Lapina (Kisiun)
Cyg 20h52'4" 35d1'36"
Irina Lipovich
Vir 12h41'9" 4d28'48"
Irina Malladi
Aqr 22h30'15" 1d56'59"
Irina Mijatovic Vranes
Peg 22h41'9" 18d11'22"
Irina Mitnik
Aql 19h46'9" -0d10'32"
Irina Munarova (Munroy)
Lyn 6h19'12" 56d18'7"
Irina Petoeva
Cam 7h8'14" 78d46'2"
Irina Puzankova
Aqr 23h40'27" -4d23'14"
Irina Shlafman
Leo 9h37'58" 26d9'40"
Irina Stanescu
Lyn 9h10'16" 34d48'51"
Irina starbar Popova
Cyg 20h17'2" 40d2'56"
Irina Syrtsova
Leo 10h51'33" 15d47'1"
Irina V.
And 23h12'21" 47d57'17"
Irina Volchok
Ari 2h55'19" 30d24'54"
Irina Yanovskaya
Gem 7h11'24" 28d26'12"
Irina's 45th Birthday
Tau 3h50'59" 18d4'55"
Irine Price Memorial
Ari 2h11'24" 22d44'29"
Irini Christodoulides
Uma 10h33'40" 58d15'34"
IRINI VASSOS
Lib 15h12'48" -13d39'58"
IRINKA18
Uma 9h24'42" 61d7'26"
Irinushka
Cyg 19h57'51" 52d31'28"
Iris
Cas 1h59'37" 66d45'28"
Iris
Cam 7h33'40" 66d41'31"
Iris
Uma 12h41'22" 60d55'43"
Iris
Sco 16h5'56" -11d32'20"
Iris
Ari 2h9'48" 23d48'11"
Iris
Tau 4h49'50" 20d12'55"
IRIS
Vir 15h4'24" 4d8'12"
Iris
Mon 6h47'35" 3d6'40"
IRIS - 25.04.01
Tri 2h38'50" 33d45'7"
Iris 731
Leo 9h50'45" 14d16'18"
Iris and David Forever
And 0h25'4" 31d20'2"
Iris Beth Camay
Cap 20h26'22" -12d35'3"
Iris Cabrera
Vir 14h14'26" 3d42'10"
Iris D. Reeder
And 2h34'6" 43d31'36"
Iris DeLeon Denizac Nadal
Cas 2h15'43" 64d28'5"
Iris Dewese Dake
Cas 0h7'26" 61d28'46"
Iris & Eddie Jupp
Cyg 20h15'18" 51d18'58"
Iris Elizabeth Frome
Lib 15h25'2" -22d51'46"
Iris Ellen
Lmi 10h32'6" 37d54'10"
Iris Ellen Edwards Rees
Cas 23h43'1" 57d31'9"
Iris Emelyne Westerman
And 23h14'0" 37d16'17"
Iris "Empress" Casillas
Cas 23h26'16" 57d19'37"
Iris Gail
Ari 2h33'34" 25d23'47"
Iris & George
Cyg 19h39'59" 44d5'8"
Iris Geraldine Corey Peacock
Uma 9h9'37" 54d44'16"
Iris Giove
Cas 1h2'11" 63d43'56"
Iris Joanna Bryant
Ori 5h26'43" 2d46'32"
Iris Krasnow
Vir 12h35'59" -7d58'39"
Iris Lynette
Sco 17h41'11" -40d12'47"
Iris Margaret Kelly
Cas 23h0'5" 54d9'3"
Iris Mary Moody Ferguson
Uma 11h29'17" 45d38'25"
Iris May
Cyg 21h45'15" 54d34'42"
Iris May E. Infante
Lib 15h46'43" -6d19'39"
Iris Medina
Leo 11h1'16" -2d37'24"
Iris Mischelle
And 0h13'16" 25d23'12"
Iris Montoya
Cas 23h47'48" 56d16'12"
Iris Parks
Per 4h24'4" 43d26'21"
Iris Rabinowitz March 17, 1937
Psc 2h4'51" 6d37'13"
Iris S Vest My Love
Uma 11h10'22" 40d1'50"
Iris Saint-Albin
And 23h47'14" 35d41'0"
Iris Smith
Gem 6h37'51" 14d12'16"
Iris Varnadore Brooks
Uma 9h59'19" 62d58'8"
Iris Virginia Kissane
Psc 1h25'26" 23d37'38"
Iris Wasserstein
Lyr 18h48'54" 35d37'17"
Iris Westphall
Sco 17h24'55" -31d36'47"
Iris-Andrea
Lac 22h45'30" 53d56'38"
Irisday
Ori 5h33'33" -7d58'18"
Irish
Her 17h0'31" 46d30'29"
Irish
Her 17h19'54" 26d49'35"
Irish Angel : Kathryn Logan
Cap 21h17'34" -16d8'26"
Irish Daisy
Gem 6h29'15" 13d49'55"
Irish Dungca
Vir 13h0'58" 11d38'36"
Irish Eyes
Mon 6h47'32" 7d45'56"
Irish Eyes
Lmi 10h28'37" 36d31'6"
Irish Eyes
Lyn 8h1'15" 41d55'53"
Irish J. McCarty
Ori 5h13'44" -0d31'56"
Irish Night on Waikiki
Gem 6h34'21" 22d0'21"
Irish Prince - Bad Boy Byrnie
Sco 17h38'39" -42d18'45"
Irish Princess
Lib 15h29'8" -4d23'3"
Irish princess
And 0h51'12" 37d26'36"
Irish "Tiger"
Her 18h15'50" 14d23'53"
IrishLee
Crb 15h17'56" 28d58'3"
Irishred
Lyr 18h42'40" 35d16'10"
Iriss Rainbow Goddess
Cas 0h50'24" 57d28'42"
Irit
Lyn 7h53'25" 35d32'30"
Irizze
Ari 2h21'44" 26d42'44"
Irja
Tri 2h6'46" 32d18'49"
Irleen Lydia Bowen
Aqr 22h21'23" -1d53'50"

Irma Angelica Benavides
Tau 5h40'30" 25d42'10"
Irma Ann
Cap 21h46'21" -17d16'23"
IRMA BARAJAS
Sgr 17h55'2" -28d21'23"
Irma Cleopatra Serna
Lib 15h28'42" -7d45'42"
Irma Colon Morales
Tau 5h1'55" 18d34'27"
Irma E. Fleming
Umi 18h27'58" 88d48'59"
Irma Jean Slavick
Peg 21h35'40" 26d50'30"
Irma Iesa
Sco 17h48'24" -39d16'5"
IRMA MARTINEZ
Cnc 8h20'25" 12d4'57"
Irma Miestrella Shall Forever Shine
Cyg 21h41'13" 31d52'30"
Irma Miller
Uma 9h49'1" 46d36'56"
Irma Mischler
Ori 5h59'34" 3d34'45"
Irma & Oscar Forever
Aqr 22h50'39" -8d6'26"
Irma S. Finis
Uma 8h38'10" 55d36'30"
Irma Strickland Suggs
Cap 21h37'55" -18d26'52"
Irma Sunshine Rodriguez
Cas 1h17'59" 65d53'17"
Irma und Hans Hess
Aur 6h37'59" 39d12'53"
Irma Wuertz - Mom and Shining Star
Per 2h38'44" 55d50'53"
IrmaJean Carr
Sco 17h52'7" -30d55'37"
Irmak Okullari
Uma 11h0'54" 42d58'46"
Irma's Love
Gem 6h46'47" 33d6'13"
Irmgard
Leo 10h7'42" 24d4'50"
Irmgard Körnchen
Ori 5h58'14" 21d35'54"
Irmgard Leidig
Cas 23h41'49" 54d16'11"
Irmgard Müller
Uma 14h13'41" 59d6'15"
Irmgard und Christian
Mon 7h7'10" -7d21'12"
Irmin Kelman
Cas 0h24'48" 65d17'3"
Irmtraut Karin Antonucci
Aqr 20h50'56" 2d10'47"
IROCHKA
Sco 17h21'24" -37d55'5"
Irochka Briller
Leo 11h47'38" 19d36'6"
irocjr
Cnc 8h36'27" 17d5'13"
Iron Mike 7064X
Ori 6h16'33" 10d45'16"
Ironcire777
Cmi 7h41'28" 0d31'17"
Ironman
Cap 20h33'38" -18d27'2"
Ironman - Eric Adair Fontaine
Per 4h37'32" 41d47'3"
Ironman Gil Mancilla
Her 17h18'40" 23d19'34"
Ironstar
Sco 17h21'17" -41d25'35"
IRONWIENS
Umi 14h58'40" 76d29'44"
Ironwill
Uma 12h45'32" 56d53'32"
Irra
And 23h24'52" 47d4'40"
Irresistible
Aur 5h30'45" 47d59'50"
Irresistible Cawse
Leo 10h13'5" 20d52'32"
Irresistible Jody
Lyn 7h47'17" 58d51'48"
Irresitible Isabella Taylor
Lmi 10h31'39" 34d46'47"
irryst
Aur 5h17'43" 28d51'31"
IrSa.8607
Uma 9h15'5" 69d57'33"
IrUxI
Cnc 8h3'16" 22d12'27"
Irv
Cep 23h38'18" 77d54'55"
Irv "Rosie" Rosenthal
Ari 3h13'41" 22d34'48"
Irvette Ione
Cas 1h41'8" 58d0'44"
Irvin
Vir 13h21'50" 4d29'54"
Irvin Anderson 10/20/1920
Lib 15h0'9" -6d52'25"
Irvin Clark
Umi 14h45'31" 73d16'33"
Irvin Noel Soto Crespo
Sgr 19h21'46" -18d11'3"
Irvin V. Deggeller
Per 3h24'23" 32d15'7"
Irvin Zager
Uma 11h10'30" 52d27'23"
Irving Barnowsky
Lyn 8h31'4" 41d11'53"
Irving Bronstein
Boo 14h30'6" 18d34'59"
Irving C. Mensch
Cas 1h55'17" 60d35'59"
Irving Greenfield Alger, Jr.
Uma 10h59'53" 48d52'6"
Irving Gubin
Ori 5h35'34" 3d23'20"
Irving Herbst
Her 18h28'13" 21d2'31"
Irving Hottle
Uma 11h39'48" 53d24'37"
Irving Liebenthal
Per 3h9'24" 54d26'26"
Irving "Red" Dorfman
Aur 5h9'50" 48d8'33"
Irving & Yvonne Curtis
Tau 3h26'18" 7d27'18"
Irwin and Miriam Blose
Crb 16h15'58" 31d14'59"
Irwin Garfinkle
Gem 6h40'36" 24d36'10"
Irwin H. Sher
Dra 17h36'56" 55d50'18"
Irwin Jay & Rebecca Marklin Fine
Gem 6h14'30" 25d3'6"
Irwin Jay Solomon
Cnc 9h19'12" 26d18'3"
Irwin Konokoff
Cyg 19h49'52" 39d10'51"
Irwin Roe Saulsbury
Lib 15h5'39" -21d0'17"
Iryna
And 23h22'4" 42d52'53"
Iryna Hladun
And 2h20'13" 48d54'58"
Iryna Oliver
Uma 8h23'56" 61d56'7"
Iryna Yufa
Cas 0h17'26" 61d57'13"
Irys Dawn
Vir 13h34'42" -14d22'2"
ISA
Aqr 22h48'25" -14d0'32"
Isa
And 0h6'6" 43d21'26"
I.S.A
Crb 16h9'11" 28d37'19"
Isa
Peg 21h30'13" 23d30'54"
Isa
Tau 5h52'25" 23d56'27"
ISA
Leo 10h16'9" 15d57'14"
Isa
Ori 4h44'8" 1d27'32"
Isa & Alex 2002
Aqr 21h27'7" 1d2'48"
Isa Alexandra Rodrigues
Lyn 7h40'51" 40d49'41"
Isa and Peter - The Wedding Star
Cyg 20h18'16" 46d14'37"
Isa Hassan
Umi 16h26'29" 70d57'42"
Isa Luis Ada
Uma 10h59'59" 33d55'42"
Isa Olev
Uma 9h29'0" 65d1'17"
Isaac
Cnc 8h26'17" 7d10'17"
Isaac
Ari 2h39'35" 26d29'9"
Isaac Alan Vellutini
Vir 13h2'55" -20d43'59"
Isaac and Danielle
Crb 15h34'55" 33d0'28"
Isaac and Katherine Perkins
Lyn 8h23'59" 41d35'14"
Isaac Andrew Ault
Boo 14h37'6" 51d36'48"
Isaac Anthony Amador
Lyn 7h59'20" 38d3'24"
Isaac Aren "Picachu" Koffa
Tau 4h17'42" 29d14'16"
Isaac Bazbaz "Bag"
Cap 20h27'32" -12d14'50"
Isaac Bennett Newlin
Her 18h29'43" 23d26'19"
Isaac Britt Conner
Aql 20h5'55" 7d34'35"
Isaac Cameron North
Uma 9h11'43" 66d0'47"
Isaac Charles Fowler
Gem 7h16'22" 23d32'15"
Isaac Conrad Westly
Sgr 18h10'21" -25d26'56"
Isaac Daniel
Vir 13h3'56" 12d26'0"
Isaac Daniel Greenlaw
Aqr 22h19'6" 1d31'39"
Isaac Dean Chapman Bordfeld
Her 17h43'2" 37d13'47"
Isaac Elijah Jacobson
Gem 6h40'38" 30d21'14"
Isaac Frederick Dancy
Her 17h28'17" 38d54'59"
Isaac Frye
Per 2h33'56" 53d24'58"
Isaac Gregory Carlson Cordova
Vir 13h41'2" 3d0'14"
Isaac Haiden Ritchie
Ori 5h21'51" 6d50'5"
Isaac Halen Faynsud
Gem 7h10'50" 17d12'32"
Isaac Henderson Butler
Aqr 21h52'6" -7d14'58"
Isaac Henry Tarker
Gem 7h38'53" 28d11'32"
Isaac James McKay
Col 5h32'15" -41d23'7"
Isaac Jennifer
Umi 15h22'14" 72d24'54"
Isaac Jesse Leon
Per 4h3'15" 46d38'34"
Isaac Joseph
Vir 14h7'42" -2d52'58"
Isaac Laframboise
Uma 11h25'7" 60d23'53"
Isaac Lee Joyce
Uma 14h20'4" 61d36'8"
Isaac Lucas Rutter
Cep 21h57'27" 60d46'59"
Isaac Matthew Lethbridge "11-07-03"
Her 16h45'17" 28d33'51"
Isaac Michael Ross
Umi 16h45'39" 83d18'40"
Isaac Neil Griffith
Cep 2h47'32" 83d31'2"
Isaac Newton Farrell
Psc 1h36'37" 11d9'9"
Isaac Nicholas Tirado
Ori 5h10'58" 11d40'55"
Isaac Peraza
Her 16h38'53" 37d48'56"
Isaac Peter
Her 18h11'30" 22d49'40"
Isaac Raijman
Dra 17h16'48" 64d47'43"
Isaac Raymond Holland
Per 4h3'11" 44d39'22"
Isaac Samuel Tiscoe
Ori 6h14'21" 8d0'0"
Isaac Stockdale
Lib 15h25'3" -9d52'25"
Isaac Stroud, Sr.
Her 18h7'5" 36d10'53"
Isaac "The Monkey"
Ori 4h53'41" -2d47'24"
Isaac Theoden Archer
Per 4h31'21" 32d47'4"
Isaac Thomas Mattich
Lib 15h6'1" -1d39'17"
Isaac Van Horn-Morris
Cas 23h37'20" 61d50'14"
Isaac Walter Nathaniel Paredes
Col 5h48'56" -33d4'6"
Isaac Wesley Pinkerton
Her 17h37'57" 28d19'53"
Isaac William Circle
Aql 19h54'3" -0d7'49"
Isaac Zale
Uma 11h32'30" 29d24'12"
Isaac & Zoe Basker
Lyr 19h27'28" 37d42'25"
Isaacs' Star
Ori 6h18'14" 10d46'56"
Isaac's Star
Sgr 18h49'58" -28d39'53"
Isaah Carrizosa
Psc 1h2'41" 11d23'33"
Isaak Volynsky
Boo 14h34'44" 35d40'23"
Isaak Volysky
Umi 15h34'16" 75d59'5"
Isabeau
And 0h48'29" 40d22'57"
Isabel
Lyr 18h27'18" 37d8'24"
Isabel
Cas 1h28'17" 52d12'58"
Isabel
Mon 6h32'4" 7d41'43"
Isabel
Gem 6h51'53" 17d2'0"
Isabel
Aqr 22h34'36" -0d11'43"
Isabel
Cap 20h31'42" -12d14'46"
Isabel
Lib 15h14'30" -22d28'10"
Isabel
Cas 23h45'42" 54d15'51"
Isabel 05
Sgr 18h12'26" -29d16'26"
Isabel A. Alvial
And 23h10'57" 42d43'42"
Isabel Alexandra Reed
Tau 4h40'26" 15d39'42"
Isabel Ameri
Ari 2h2'39" 24d6'7"
Isabel Ann
And 0h52'43" 39d43'3"
Isabel Ann Sightler
Cam 4h37'12" 68d2'15"
Isabel Anna-Rose
And 2h20'56" 46d17'40"
Isabel Banegas
Uma 11h29'58" 46d14'4"
Isabel Beatrice
Sgr 19h42'41" -14d53'30"
Isabel Bixby Carey
And 23h47'7" 41d50'57"
Isabel C.
Ori 5h19'42" 6d12'28"
Isabel Cayla
Ari 3h13'36" 29d8'32"
Isabel Cecile
Lmi 10h40'14" 32d33'27"
Isabel Charlotte Brennan
And 23h36'27" 36d12'10"
Isabel Charlotte Challenor Ludwig
Cru 12h56'31" -58d49'40"
Isabel Charlotte Warren
And 0h58'33" 45d49'29"
Isabel Christensen
Cas 0h8'10" 56d50'51"
Isabel Coleman
Sgr 19h36'7" -13d29'25"
Isabel Consuelo Vazquez
Gem 7h19'14" 31d25'19"
Isabel Cristina
Gem 7h3'7" 24d15'4"
Isabel Cuevas - A star
And 0h23'47" 32d14'44"
Isabel D. Ayers
And 0h16'43" 29d23'12"
Isabel Dailey
Sco 16h41'4" -39d4'10"
Isabel Daisy Lipman - Isabel's Star
And 0h0'8" 41d5'35"
Isabel Daisy Thompson
And 1h40'56" 41d51'58"
Isabel Gallagher
And 2h14'19" 46d41'54"
Isabel Giovanna's Star
And 0h4'17" 32d33'22"
Isabel Gomes Gerber
Peg 21h55'34" 19d45'8"
Isabel Grace
Lyr 18h29'49" 38d30'28"
Isabel Grace Brooker
And 1h10'23" 46d37'36"
Isabel Grace Christie
Vir 12h5'41" -0d19'13"
Isabel Grace Hammond
And 1h24'10" 49d14'57"
Isabel Grace Longson
Cyg 20h3'59" 31d56'21"
Isabel Grace Maki
Cap 21h7'13" -21d44'4"
Isabel Gracie
And 1h7'59" 46d4'4"
Isabel Grady
Leo 9h43'41" 26d54'51"
Isabel Harley Keim
Cyg 21h33'6" 36d18'36"
Isabel Hutchins Cahill
Lib 14h54'13" -6d13'47"
Isabel - Imi
Cep 22h5'57" 61d21'26"
Isabel J. Fee
Lyn 7h24'58" 50d11'34"
Isabel Jade
Uma 10h46'24" 69d23'17"
Isabel & Jenny
And 23h7'54" 46d9'14"
Isabel Juliana
Tau 5h13'23" 23d3'26"
Isabel Julie McMenamon
Lyr 18h50'27" 44d40'23"
Isabel Kate Prejean
Peg 22h2'56" 19d7'21"
Isabel Kay Ross
Cap 21h1'52" -20d44'43"
Isabel Kirstie Munroe
Cyg 21h25'31" 31d21'59"
Isabel Lake Cope Parsons
Lib 15h11'32" -11d57'52"
Isabel Laurel Obhof
Cas 0h38'7" 52d37'5"
Isabel Leia Phemayotin
Psc 0h52'56" 27d55'1"
Isabel Lorien Zimmerman
Sco 17h45'15" -35d52'20"
Isabel Lozano Lopez
Cap 20h11'55" -13d42'32"
Isabel Lucia Barrett
Cas 23h40'22" 53d1'59"
Isabel & Marcel
Vir 12h37'33" 4d23'52"
Isabel Margarita Lara
Crb 15h43'11" 31d57'22"
Isabel Maria Lopez
Aqr 21h51'33" -6d54'13"
Isabel Marie Nelson
And 0h21'31" 38d44'36"
Isabel McInnis
Cap 20h16'54" -11d0'23"
Isabel McKie
Cas 2h27'20" 72d47'28"
Isabel Meram Urbancic
Sco 15h53'51" -20d38'44"
Isabel Mersky
Uma 11h38'6" 64d9'51"
Isabel Mitchell
And 23h28'2" 48d21'37"
Isabel Noemi Barberi
Cas 0h50'47" 53d16'40"
Isabel Norwood
Cas 2h51'32" 61d51'35"
Isabel "Papi's Star"
Ori 5h14'4" 1d2'49"
Isabel Parra
And 23h7'14" 39d55'51"
Isabel Pitaro
Lyn 8h20'16" 39d45'14"
Isabel Q&O
Sco 16h7'32" -21d57'12"
Isabel Rose Maya
Cnc 9h4'39" 13d40'30"
Isabel Rose Zurowski
And 2h34'19" 46d39'6"
Isabel Sara
And 2h32'2" 49d22'8"
Isabel Saunders
And 1h16'57" 45d14'56"
Isabel Stanley
Uma 9h9'49" 49d15'19"
Isabel Stusser
Dra 17h26'11" 68d15'29"
Isabel T-B
And 0h49'28" 36d59'53"
Isabel Victoria
Lyn 9h19'42" 36d35'34"
Isabel Victoria Manzano Carrero (Keka)
Vul 20h38'49" 26d59'19"
Isabelbrett
Vir 14h6'41" 1d51'53"
Isabelica
Del 20h37'23" 10d34'1"
Isabelinha
Uma 13h26'9" 52d56'8"
Isabelisse
Lyn 6h51'58" 55d54'53"
Isabelita
Lyn 7h20'31" 56d41'52"
Isabell Claire Wick
Cnc 8h5'1" 7d5'29"
Isabell Marie
Cas 1h6'50" 66d15'32"
Isabella
Cas 1h29'30" 62d33'18"
ISABELLA
Cas 23h16'34" 54d27'51"
Isabella
Umi 15h31'56" 71d35'36"
ISABELLA
Cap 21h35'6" -16d24'54"
Isabella
Cap 21h50'4" -17d57'27"
Isabella
Vir 13h39'26" -7d43'4"
Isabella
Lib 15h16'46" -7d22'6"
Isabella
Lib 15h6'53" -29d4'19"
Isabella
Sgr 19h0'26" -30d38'14"
Isabella
Sco 17h49'57" -44d2'0"
Isabella
Ori 4h49'56" 7d23'8"
Isabella
Lmi 10h52'27" 27d0'5"
Isabella
Leo 11h47'26" 27d21'32"
Isabella
Boo 14h46'4" 25d2'14"
Isabella
And 0h41'26" 29d58'19"
Isabella
Leo 11h48'25" 19d22'47"
Isabella
And 0h39'33" 31d20'7"
Isabella
Crb 16h18'55" 37d1'37"
Isabella
Cyg 19h44'53" 33d40'57"
ISABELLA
And 1h47'45" 43d0'17"
Isabella
And 2h16'5" 38d39'41"
Isabella
Cas 0h56'41" 47d4'52"
Isabella
Lyn 7h2'3" 45d39'28"
Isabella
And 23h17'16" 42d55'15"
Isabella Adele Martorelli
Aqr 22h11'22" 0d56'52"
Isabella Amanda Steinley
Sgr 17h57'29" -18d0'13"
Isabella Angelina Ruiz-Ramos
Com 12h46'29" 27d49'51"
Isabella Ann
And 1h18'1" 45d41'2"
Isabella Ann
Uma 9h57'59" 58d40'59"
Isabella Anne Tyrrell
Cnc 7h59'17" 11d3'58"
Isabella Annie Scott
And 0h17'28" 45d30'43"
Isabella Annie Thornton
And 0h59'35" 45d53'47"
Isabella Arianna Taylor-Bowes
Gem 6h58'46" 18d33'58"
Isabella Ashton Novelli
And 1h21'57" 43d34'41"
Isabella Audrey Eierweis
Col 5h24'45" -39d28'16"
Isabella Baliva
Cru 12h8'51" -62d57'28"
Isabella Bärtsch
Sge 20h1'38" 18d30'42"
Isabella Bondi
Sco 16h28'44" -29d37'49"
Isabella Bouquio
And 2h8'39" 39d8'23"
Isabella Bravo
Cnc 9h3'49" 8d42'32"
Isabella Breanne's Star
Tau 5h51'16" 14d20'44"
Isabella Brianna
Cyg 19h54'53" 36d56'19"
Isabella Brown
And 0h36'42" 31d23'51"
Isabella Calderai
Cam 5h55'24" 56d30'2"
Isabella Carlin Lee
Tau 5h53'54" 25d28'54"
Isabella Carolina Zumaya
Cap 20h27'31" -12d5'0"
Isabella Castaneda
Crb 16h17'27" 30d16'8"
Isabella Caulboy
And 0h56'44" 39d24'56"
Isabella & Christian Rossini
Umi 17h0'31" 78d8'27"
ISABELLA CHRISTINA GRACE HEROLD
Aqr 22h4'48" -18d47'51"
Isabella Christina Greco
Gem 7h9'23" 15d39'56"
Isabella Claire Ricciardi
Cas 0h9'53" 57d30'59"
Isabella Claire Stockwood
Tau 3h55'17" 6d2'9"
Isabella Claire-Cron Miceli
Dra 16h23'20" 57d41'34"
Isabella Corvino
And 1h19'56" 48d20'10"
Isabella Cristina 08042005
And 2h36'13" 49d58'12"
Isabella Curello
Dra 19h19'52" 66d15'21"
Isabella DelBasso
And 2h33'15" 49d46'12"
Isabella Dianne Gulczynski
And 1h46'33" 45d2'40"
Isabella Doris Mahaney
Cas 1h17'7" 57d22'46"
Isabella Elizabeth Trapp
Cnc 8h36'13" 27d47'37"
Isabella Emma
Lib 14h54'13" -1d29'52"
Isabella Erin Girdler
And 23h39'58" 35d11'26"
Isabella Eva Wood
Vir 11h50'50" -4d29'44"
Isabella Filomena Aslanian
Lac 22h23'3" 51d36'11"
Isabella Flora Grace Edwards - Izzy's Star
And 23h20'6" 37d37'10"
Isabella Fox Bruketta
And 0h59'6" 43d18'37"
Isabella Francesca Stacy Camilleri
And 23h8'6" 51d9'6"
Isabella Francesca Vaccari
And 0h13'5" 25d59'12"
Isabella Gabriella
And 23h14'7" 41d46'13"
Isabella Garraway
Cas 0h4'55" 53d17'43"
Isabella Gia Shin Lo
And 1h12'17" 36d28'34"
Isabella Glassman
And 23h37'25" 47d26'2"
Isabella Grace
And 23h16'12" 49d19'44"
Isabella Grace
Leo 10h35'31" 20d18'10"
Isabella Grace
Vir 13h55'1" 2d39'26"
Isabella Grace Alana
Uma 11h29'1" 64d57'13"
Isabella Grace Banks
And 23h30'4" 38d9'16"
Isabella Grace Bianco
Crb 16h19'3" 33d49'30"
Isabella Grace Chapman
Lyn 8h5'26" 34d33'7"
Isabella Grace DePasquale
And 2h31'23" 45d40'21"
Isabella Grace Gray
And 2h5'31" 41d22'47"
Isabella Grace Guarino
Vir 13h12'3" 7d8'48"
Isabella Grace Huber
And 23h4'1" 47d11'10"
Isabella Grace - Isabella's Star
And 2h12'14" 49d34'57"
Isabella Grace Louise Newton
And 2h21'40" 42d29'38"
Isabella Grace Nannini
Psc 1h25'25" 22d43'55"
Isabella Grace Orlando
And 1h43'21" 36d32'9"
Isabella Haley Silver
Leo 9h54'38" 31d6'16"
Isabella Harriet
Vir 13h51'33" -9d42'38"
Isabella Hope
Aqr 22h29'29" -7d19'6"
Isabella Hope Goring
And 23h5'38" 40d8'23"
Isabella India Simons
And 1h27'35" 49d13'25"
Isabella Jade Cupo
Sco 16h50'10" -40d33'14"
Isabella Jade Sterling
Sgr 17h58'17" -27d0'42"
Isabella Jae Boodt
Vir 12h9'36" -6d40'9"
Isabella Jayne
And 23h37'11" 50d16'14"
Isabella Jean Forte
Lib 15h48'33" -5d41'52"
Isabella Jean Gonzalez
Cnc 8h55'25" 13d31'50"
Isabella Jessica Marie Beham
Gem 7h19'46" 31d1'57"
Isabella Jordan Kennedy
Aur 5h17'42" 43d56'12"
Isabella Josephine Hoener
Cas 0h37'25" 59d2'35"
Isabella Kathleen Gioia
And 0h56'53" 38d46'10"
Isabella Kayla
Ari 3h9'13" 21d25'44"
Isabella Lee
Com 12h12'55" 31d24'6"
Isabella Lorée Sundberg
Peg 22h27'50" 32d15'9"
Isabella Loren Compres
Cnc 8h27'45" 26d22'36"
Isabella Louise Shay
Cnc 8h39'36" 30d46'27"
Isabella Luisa
Lmi 10h46'4" 34d34'12"
Isabella Lynne Ramos
Ari 2h47'18" 27d16'51"
Isabella M Hunter
Cas 0h13'1" 57d20'12"
Isabella Madison Cavanaugh
Cyg 19h34'1" 32d10'5"
Isabella Mae
Psc 1h22'25" 25d8'41"
Isabella Maisie Samuel
And 1h14'10" 42d24'13"
Isabella Maiuri-Winter
Cap 20h32'38" -22d46'44"
Isabella Margaret
Lyn 9h11'31" 39d55'54"
Isabella Maria
And 23h58'33" 34d54'14"
Isabella Maria
Uma 11h51'13" 47d10'18"
Isabella Maria
Sgr 18h49'30" -16d7'42"
Isabella Maria Angelica Jobo
Cnc 9h3'44" 30d19'21"
Isabella Maria Carcelli
Cas 0h51'5" 57d18'25"
Isabella Maria French
And 0h3'6" 33d26'41"
Isabella Maria Okuniewska
Tau 5h31'38" 21d57'41"
Isabella Maria Tagliareni
And 1h32'41" 46d17'29"
Isabella Marie
Cam 4h3'34" 56d24'25"
Isabella Marie
Peg 23h38'22" 23d46'55"
Isabella Marie
Cap 21h21'33" -27d10'48"
Isabella Marie Cruz
Dra 20h20'23" 68d59'33"
Isabella Marie Kluis
Tau 3h49'41" 24d24'54"
Isabella Marie Lupi
Leo 10h8'22" 20d24'56"
Isabella Marie Scozzola
And 0h13'48" 26d59'27"
Isabella Marie Soon
And 0h45'25" 30d5'11"
Isabella Marie Wren
Crb 15h42'33" 31d28'29"
Isabella Mary Barr
Sgr 19h19'7" -16d7'14"
Isabella May - Born 22nd May 2005
Pho 23h43'49" -55d51'20"
Isabella May Newman
Cnc 8h13'6" 15d17'48"
Isabella Mia Curcio
And 0h12'59" 43d11'25"
Isabella Mina Rivera
Lib 15h5'31" -11d3'3"
Isabella Mola
Umi 17h8'36" 83d27'9"
Isabella Mosca
Aqr 23h36'23" -15d28'17"
Isabella Mya
Ari 3h2'16" 25d21'52"
Isabella Nai'a O'Kalani Sloan
Ori 5h20'36" -0d17'1"
Isabella Nicole
And 23h58'25" 42d39'26"
Isabella Noelle
And 1h2'11" 43d38'28"
Isabella Nykita Cartier
Cam 3h35'14" 58d57'43"
Isabella Palmiero
Gem 7h6'14" 23d15'31"

Isabella Peck August 2, 1906
Cas 1h17'20" 55d58'35"
Isabella Pilar Zoll
Ari 2h37'23" 28d20'11"
Isabella Pizzingrilli
Psc 1h51'10" 5d57'40"
Isabella Prime
Ori 5h59'19" 6d3'26"
Isabella Prince Crane
Sgr 18h34'22" -23d44'16"
Isabella Quinn Vanslette
Psc 1h2'57" 27d19'14"
Isabella R. Bickhaus
Vir 13h25'50" 5d5'36"
Isabella Rachel LaFerrera
Cas 23h47'55" 62d1'56"
Isabella Rachel Popofsky
Umi 15h19'30" 67d30'16"
Isabella Rae Hall
Vir 13h28'45" 6d40'4"
Isabella Rago
Cas 0h15'18" 65d32'44"
Isabella Raquel
Umi 14h38'26" 74d59'35"
Isabella Rebecca Woods
Uma 10h45'48" 54d13'40"
Isabella Reese Wright
Sgr 19h23'56" -30d18'52"
Isabella Renato
Mariateresa Renato
And 2h35'3" 48d49'39"
Isabella Renee
Cam 7h13'27" 77d5'1"
Isabella Renee Palanca
Sco 16h49'54" -44d15'22"
Isabella Ricchiuti
Lib 15h13'6" -19d6'40"
Isabella Rogers
And 23h49'5" 46d5'42"
Isabella Rosa
Sgr 17h51'58" -29d46'51"
Isabella Rosa Mercado - Nini's Star
And 0h23'37" 33d37'43"
Isabella Rosano
Cap 20h25'37" -13d15'55"
Isabella Rose
Sco 16h52'28" -44d49'44"
Isabella Rose
Ari 3h28'1" 24d18'32"
Isabella Rose
Com 12h47'33" 16d48'30"
Isabella Rose
And 23h38'56" 47d7'38"
Isabella Rose
And 23h28'55" 41d16'5"
Isabella Rose
And 0h42'32" 44d32'38"
Isabella Rose
Gem 7h40'22" 31d39'45"
Isabella Rose Balestrieri
Cyg 21h58'12" 47d1'30"
Isabella Rose Bottiglieri
Tau 3h47'29" 28d8'56"
Isabella Rose Colon
And 0h29'59" 32d42'10"
Isabella Rose Crespo
Cas 2h20'12" 63d13'32"
Isabella Rose Lennon
And 23h44'47" 41d33'8"
Isabella Rose Marrocco
Gem 6h18'13" 23d2'35"
Isabella Rose Scalesse
Lyn 7h58'41" 51d9'34"
Isabella Rose Troup
And 0h24'51" 41d31'53"
Isabella Rose Vivilecchia
Cyg 20h43'31" 52d23'46"
Isabella Rose White
Gem 6h20'8" 27d33'34"
Isabella Ryan Bailey
Ari 2h24'8" 26d4'3"
Isabella Scarlet Winkler
Psc 0h40'2" 14d37'2"
Isabella Scarlett Troup
Lib 15h18'29" -6d6'5"
Isabella Seren Murphy
And 23h46'39" 37d46'25"
Isabella Serenity
Uma 8h18'2" 69d25'24"
Isabella Shay Besse
Cnc 9h8'32" 27d56'18"
Isabella Sikiric
Cyg 21h28'48" 39d44'10"
Isabella Sky Shobe
Dra 17h26'52" 65d13'37"
Isabella Sofia Carrerou
And 23h37'20" 41d23'36"
Isabella Sofia Lucas
And 23h2'16" 51d3'32"
Isabella Solange Davidson
Crb 15h46'2" 38d9'58"
Isabella Spiezio 231287
Ori 5h35'55" 14d59'48"
Isabella Stavreski's Star 04/10/2002
Cru 12h25'37" -57d49'50"
Isabella Storm
Lib 15h46'4" -26d9'16"
Isabella Susan Murray
And 23h50'59" 41d34'20"
Isabella the Rising Star
Ari 3h21'13" 27d17'39"
Isabella "The Warrior Princess"
Sgr 18h37'22" -17d51'53"
Isabella Treleaven - Our Beauty
Psa 22h19'26" -30d48'55"
Isabella Tucker
Crb 15h38'43" 37d43'23"
Isabella und Karin
Dra 16h26'56" 63d7'9"
Isabella Van-den-Rijse
Cyg 21h16'45" 52d53'19"
Isabella Victoria Leah Gilbert
And 0h1'19" 40d24'12"
Isabella Von Frohlich
Lib 15h45'39" -18d55'58"
Isabellae Prima Lux
Sco 17h54'4" -44d14'13"
Isabellalove
Cyg 19h48'14" 33d32'21"
IsabellaMaria
Lib 14h54'32" -3d45'39"
Isabella's Christmas Star
Uma 11h19'48" 39d7'46"
Isabella's Rainbow Angel Star
Ara 17h10'17" -55d30'43"
Isabella's Shining Star
And 0h57'58" 40d42'21"
Isabella's Wishing Star
And 0h1'9" 33d56'17"
Isabelle
Lyn 7h33'43" 41d27'31"
Isabelle
Her 17h30'40" 39d59'16"
Isabelle
And 1h7'9" 46d3'13"
Isabelle
Cas 1h23'24" 58d4'32"
Isabelle
Tau 5h54'25" 24d22'29"
Isabelle
Crb 16h22'5" 28d30'52"
Isabelle
Her 18h33'8" 21d34'25"
Isabelle
Leo 10h21'44" 10d39'10"
Isabelle
Aqr 22h32'26" -14d14'56"
Isabelle
Vir 13h3'29" -15d20'35"
Isabelle
Dra 14h51'35" 58d33'30"
isabelle
Cep 22h33'4" 66d56'15"
Isabelle Aimée
Ori 5h50'39" 9d48'56"
Isabelle Alexis Macker
Psc 0h27'21" 11d43'46"
Isabelle Ann Lewis
Cam 7h46'22" 64d15'7"
Isabelle Anna Casey-Holowka
Aqr 22h29'49" 0d57'7"
Isabelle Anne Cantle
Cru 12h20'53" -56d51'4"
Isabelle Anne Laure
Ori 5h11'10" 15d48'58"
Isabelle Anne Olson
Tau 4h33'49" 21d27'2"
Isabelle Artis
Ari 3h5'39" 29d42'56"
Isabelle Baeza
Vir 13h27'9" -16d27'59"
Isabelle Boulay.. Ste Félicité
Her 18h39'15" 24d54'38"
Isabelle Cano et Raphaël Maupetit
And 23h9'26" 38d48'45"
Isabelle Carmen Hollie Bryan
And 1h22'11" 43d31'50"
Isabelle - Cédric
Uma 14h11'38" 56d40'53"
Isabelle Christine Guna
Cas 0h28'35" 51d39'49"
Isabelle Claire Doetsch
Psc 1h11'2" 22d41'36"
Isabelle&Dario
Uma 8h55'58" 48d37'10"
Isabelle Desjardins
Ori 5h32'51" 3d36'59"
Isabelle Dumas 21-10-1980
Her 15h53'22" 39d55'7"
Isabelle Elizabeth Howerton
Tau 4h36'28" 24d8'50"
Isabelle Ellen Guest
And 1h55'20" 46d48'22"
Isabelle Erica Hills
And 0h40'35" 40d27'36"
Isabelle Frances Levine
And 1h31'57" 45d0'28"
Isabelle Gingras
Cas 0h34'29" 61d35'1"
Isabelle Gollas
Uma 13h12'9" 52d28'42"
Isabelle Grace
And 0h17'46" 44d32'12"
Isabelle Grace
Ari 3h16'6" 15d15'23"
Isabelle Grace
Sgr 19h53'18" -12d2'6"
Isabelle Grace Chevalier
And 2h35'36" 40d8'33"
Isabelle Grace Shearer
And 23h43'11" 42d21'24"
Isabelle Hannah Kuhn
Umi 16h12'26" 73d15'10"
Isabelle Holzach
And 23h29'20" 47d6'37"
Isabelle Hope
Cas 2h27'2" 67d47'55"
Isabelle Iverson
And 1h42'11" 43d30'44"
Isabelle Jacklyn Barragan
Aql 19h31'19" 7d36'8"
Isabelle Jane Haag
And 23h42'43" 44d45'2"
Isabelle Jean
And 0h21'12" 32d1'41"
Isabelle Joan Griffin
And 0h19'54" 44d25'31"
Isabelle Kahatra Singh
And 23h28'54" 38d1'21"
Isabelle Kathleen Raine
And 1h24'20" 50d13'9"
Isabelle Kiffel
Aqr 20h39'29" -10d11'52"
Isabelle Leueen May
Leo 9h31'52" 29d23'16"
Isabelle Lhopiteau
Psc 2h2'54" 6d27'33"
Isabelle Lily
And 23h36'5" 48d35'40"
Isabelle Louise Holah - Tilly's Star
Ori 5h54'18" 17d14'30"
Isabelle Lucia Giaquinto
Gem 7h6'36" 32d7'51"
Isabelle Margaret Tolputt
And 1h28'19" 34d18'31"
Isabelle Maria Flores
Sco 16h52'11" -32d43'49"
Isabelle Marie Orth
Cyg 19h44'50" 47d30'9"
Isabelle Mary
Psc 0h56'55" 19d5'58"
Isabelle Mary Louise Hewett
And 1h29'21" 47d13'56"
Isabelle Mills
And 23h11'50" 42d40'35"
Isabelle Monteferrante
Cap 20h28'22" -16d0'6"
Isabelle Niamh Franklin
And 23h1'44" 51d48'11"
Isabelle Patricia Rauh
Aqr 22h22'20" 1d54'21"
Isabelle Phan Speer
Lyn 7h51'49" 53d9'38"
Isabelle Poinsignon
Uma 11h57'0" 33d27'27"
Isabelle Pooja Eiden
Gem 7h32'16" 27d6'52"
Isabelle Poppy Spencer
And 1h57'18" 46d57'13"
Isabelle poupée d'ange
Peg 21h28'12" 7d39'27"
Isabelle Rebecca - Belle
Leo 9h50'28" 21d29'53"
Isabelle & René
Crb 15h45'57" 29d47'34"
Isabelle Rose
And 23h4'50" 50d41'56"
Isabelle Roseanne - our Little Princess
And 2h35'20" 50d33'24"
Isabelle Schaff
Umi 16h24'10" 71d21'15"
Isabelle Silverman
Cas 1h25'12" 62d44'35"
Isabelle St-Laurent
Ari 3h10'27" 30d56'1"
Isabelle & Sylvain Olivier
Sco 17h54'1" -32d33'11"
Isabelle Tartaglione
Cap 21h43'30" -18d50'3"
Isabelle Weulersse
Ari 2h50'47" 11d4'17"
Isabelle Whitehead
And 23h32'1" 38d31'7"
Isabelle Zamblera
Her 16h20'20" 19d37'57"
Isabelle-Matou
Tau 4h42'25" 22d56'40"
Isabelle's Light
Cas 0h59'17" 63d42'30"
Isa-belle-Ssandra
Gem 6h46'41" 23d37'41"
Isabel's Inspiration
Cyg 20h20'22" 56d7'46"
Isabel's Leroy
Psc 1h34'12" 7d34'5"
Isabel's Star
Ori 5h57'6" 11d37'54"
Isabel's Star
And 23h49'49" 33d52'8"
Isac & Mathilde Fournier 08.12.04
Boo 13h48'4" 12d5'46"
Isadora
Gem 6h36'41" 14d9'43"
Isadora
Cnc 7h56'24" 17d34'38"
Isadora Anne Mauricia
Vir 13h24'54" 11d56'57"
Isadora Lucile Powell
Uma 9h7'47" 46d51'56"
Isadora Sandlin
Cam 4h45'59" 59d59'23"
Isaiah
Per 3h32'38" 41d2'31"
Isaiah Carvalho
Per 2h6'40" 56d51'6"
Isaiah David Bishop
Per 4h7'49" 37d21'44"
Isaiah Green
Uma 11h8'1" 71d9'54"
Isaiah James-Michael Livesey
Umi 9h57'59" 87d22'43"
Isaiah Joe Martinez
Her 18h20'26" 22d10'51"
Isaiah Joseph Cuevas
Ori 5h16'25" 15d13'4"
Isaiah Joseph Musuta
Cap 21h54'38" -21d16'57"
Isaiah Kyle Calloway
Psc 0h24'18" 10d33'28"
Isaiah Lee Young, Jr.
Psc 1h37'49" 22d12'45"
Isaiah Manuel Penuela
Psc 0h14'40" 8d50'17"
Isaiah Micah Nino Landron
Cap 21h29'16" -15d30'50"
Isaiah Morava
Boo 15h0'46" 49d59'37"
Isaiah "Mr" Holt Jr.
Lib 15h36'22" -17d59'11"
Isaiah Patrick
Aql 19h47'48" -0d32'4"
Isaiah Paul
Per 3h8'5" 52d13'20"
Isaiah & Payton Cardilicchia
Uma 11h9'11" 37d31'49"
Isaiah Raven Urbina "03/17/1999"
Psc 2h1'19" 5d58'28"
Isaiah Richard Serrato
Sgr 19h5'22" -28d20'27"
Isaiah Tutor
Uma 13h23'54" 56d50'47"
Isaiah Yetter
Gem 7h43'34" 24d56'46"
Isaiah Zahid Husain
Leo 9h30'28" 16d0'54"
Isaiah, light of our lives
Uma 13h50'43" 55d45'1"
Isaias Argueta
Cyg 19h35'20" 28d53'15"
Isaias Ibarra
Cap 21h37'8" -22d7'35"
Isaias Joaquin Rios
Boo 15h30'36" 40d43'49"
Isak
Cas 1h24'39" 67d50'34"
Isalia Nunez
Cnc 9h19'9" 11d34'10"
Isalinda - Mi Rayito de Luz!
Lyn 7h16'4" 59d8'49"
Isaline Meylan
Umi 14h24'46" 76d40'47"
Isalyn Navarro Colón
Gem 7h3'35" 14d17'17"
Isalys
Per 4h18'35" 49d50'42"
IsaT aime FabN
Ori 6h16'40" 15d0'41"
Isay's Star
Sco 16h6'39" -10d34'34"
Isbitzki, Joachim Josef
Uma 8h21'9" 67d30'24"
Isebel Gurrola
Umi 15h32'32" 78d11'26"
Isebelle Grace Jones
And 0h56'20" 35d59'10"
Isel Leontine Pollock
Leo 9h37'10" 27d59'56"
Isela
Cap 20h9'41" -17d20'22"
Isela Castillo
Sco 17h12'37" -40d13'25"
ISH
Ari 2h48'52" 26d23'31"
Isha and Bryan
Cyg 20h15'29" 54d24'19"
Isharidah Ishak
Cyg 19h37'57" 35d17'21"
Isharna
Cra 18h43'0" -38d28'47"
I.S.H.B.Y.
Vir 13h48'11" 0d25'12"
I-Sheng Chu (baby)
Vir 13h12'21" -14d37'13"
ISHI
Lib 14h49'30" -3d24'15"
Ishi-Israel
Cyg 20h9'22" 52d43'44"
ishiperdayfulla
Cnc 9h6'54" 13d58'48"
Ishtar
Cru 11h59'48" -59d47'19"
Ishtar 3
Peg 22h34'45" 29d3'20"
Ishtar Mi Estrella Brillante
Dra 17h48'19" 63d2'42"
Isiah Hunt
Uma 11h5'58" 36d24'25"
Isiah Miguel Olivas
Her 17h14'31" 39d15'25"
Isidar Mithrim - Sapphire Star
Lyn 6h31'39" 56d59'43"
Isidoro Femia
Cap 21h44'50" -11d6'0"
Isidre i Eva
Ori 5h55'12" 17d15'23"
Isidro Rodriguez
Uma 9h19'43" 60d47'39"
Isilwen
Uma 11h15'48" 30d8'20"
ISIS
Gem 7h14'24" 31d59'46"
Isis
Lmi 10h58'26" 29d13'27"
Isis
Lyn 7h9'33" 55d32'29"
Isis Blue
Gem 6h49'20" 12d31'47"
Isis Bunny
Cyg 21h29'50" 54d49'3"
Isis Doris
Vir 11h37'26" -0d13'48"
Isis Estrella Haslem
Lib 15h42'43" -29d30'44"
Isis Francisca Wilhelmina Luijks
Uma 11h41'23" 52d30'10"
Iskandar N. Ghobril
Ori 6h5'35" 13d21'0"
Isla Bell Wood
Vul 19h45'57" 26d11'36"
Isla del Cielo
Cnc 9h16'30" 16d3'5"
Isla Frances Nangle
And 0h15'50" 48d5'48"
Isla Frances Staveley
And 0h39'10" 36d53'23"
Isla Louise Zimmerman
Cyg 19h46'36" 35d33'32"
Isla Madeline
Mon 6h49'39" -2d45'1"
Isla Maggie Eleanor Williams
And 1h25'18" 34d3'13"
Isla Presley
Uma 9h55'35" 48d43'53"
Isla Ramage
And 2h33'52" 38d15'55"
Isla Rose McIntosh
And 1h50'0" 45d16'17"
Islamorada Baby!!
Uma 12h1'0" 56d1'57"
Island In The Sky
Lyn 7h48'43" 53d49'58"
Island Inn Hotel
Cyg 19h26'58" 52d51'5"
Island Villas
Cam 4h30'39" 67d22'29"
Islay Anne Poskitt
Lmi 9h57'5" 37d15'56"
Isle of View
Cyg 20h25'0" 37d6'12"
Islewolf
Lup 15h13'5" -31d51'44"
Islic, Andreas
Aqr 21h50'18" -1d37'25"
Islwyn Haydn Thomas
Cep 20h54'14" 60d3'35"
Ismael Antonio Scott Martinez
Aql 19h22'35" -1d53'11"
Ismael Emmanuel
Her 17h12'19" 34d2'14"
Ismael Merari and Isai Jacob Mejia
Uma 10h19'49" 66d22'58"
Ismaelia
Her 16h28'52" 15d2'1"
Ismary
Cap 21h21'15" -15d14'1"
Ismary Esther Williams
Uma 13h47'45" 52d24'50"
Ismene Papamatthaiou
Vir 13h20'30" -17d52'11"
Ismenia Suhail Molina
Vir 13h27'31" -21d52'2"
Ismeri
Aql 20h1'49" 10d17'40"
Iso
Peg 22h0'23" 17d8'3"
Isobel
Umi 14h52'34" 81d7'3"
Isobel Ada Gosling - Isobel's Star
And 1h13'15" 44d33'48"
Isobel and Mark's Wedding Star
Cyg 21h36'21" 39d32'45"
Isobel Grace Busby - Isobel's Star
And 2h29'1" 48d55'6"
Isobel Jackson's Dream
And 1h2'3" 46d25'29"
Isobel Jade Noon
And 1h18'18" 45d39'12"
Isobel Josephine Peachey - Isobel's Star
And 23h39'36" 36d24'19"
Isobel Louise Griffiths
And 2h13'49" 48d8'12"
Isobel Nancy Joanne John
And 23h29'12" 43d5'14"
Isobel Olivia
And 0h55'13" 45d56'19"
Isobel Poppy
And 0h20'4" 38d47'0"
Isobel Ruth Khan
And 2h32'17" 46d54'4"
Isobella Scarlett Mayo
And 1h52'7" 40d50'52"
Isobelle
Cas 0h31'4" 55d23'42"
Isobelle Grace Weaver
And 2h5'52" 46d37'32"
Isobrite
Sgr 19h39'58" -14d53'17"
Isolde
Aqr 22h36'19" -14d14'4"
Isolde
Uma 11h0'22" 43d1'25"
Isolde Altmann Helgert
Ori 5h22'11" -7d36'25"
Isolde Regina Hennessey
Lyn 8h20'1" 43d33'4"
Israel
Cap 20h39'0" -16d51'37"
Israel Elias Parra
Umi 16h20'27" 70d21'32"
Israel Izzy Gallihugh
Leo 10h15'35" 21d57'42"
Israel Klabin
Leo 10h27'0" 9d22'11"
Israel Phillips
Her 17h58'50" 23d42'24"
Issa
Cas 3h4'40" 61d33'7"
Issabel Nicole Serrano
Cnc 8h23'8" 24d55'24"
Issabella
Leo 9h28'15" 11d49'23"
Issabella Cimadevilla
Sgr 20h28'26" -42d58'22"
Issabelle Nakibuuka Clark
Dra 20h39'49" 70d33'48"
Issac Hall Jr.
Uma 10h55'16" 45d51'23"
Issac Yelchin "4-14-97"
Umi 14h56'49" 73d54'42"
Issam
Cyg 21h54'8" 50d53'40"
IsseyMiyakexxx "Kym"
And 23h42'24" 43d2'52"
Issues & Fires
Psc 23h3'22" 5d23'59"
Issy
Vir 12h39'8" 0d36'19"
Ista Wanzi Francis
Cam 12h28'18" 77d32'32"
Istanbul Ekonomi
Uma 11h34'15" 46d19'19"
iStar
Lyn 7h5'14" 50d7'57"
Istinova
Sco 17h50'18" -31d55'3"
Istvan Nemeth
Sco 16h29'5" -36d25'46"
Istvan, Sara
Ori 5h15'32" -8d23'44"
Isuf Zukaj
Uma 11h57'15" 36d6'45"
Isyss Kayla~Michelle Riley
And 22h59'40" 52d18'7"
"It is what it is"-March 2005
Umi 17h10'44" 80d10'24"
"It just feels right."
Cyg 21h12'20" 35d42'39"
Ita (Andrea Salome Hernandez Peña)
Crb 15h41'15" 32d29'27"
Itala Marcella Azzarelli
Aql 19h37'8" 13d56'8"
Italian Boy & Greek Girl
Cyg 21h51'5" 53d0'44"
Italian Princess
Uma 9h20'14" 68d21'16"
Italian Princess
And 22h58'29" 40d25'46"
Italiano Sandrine
Del 20h36'48" 14d45'5"
Itchy
Dra 15h21'2" 59d1'44"
Iter Bryony
And 2h38'18" 48d48'1"
ITERA
Ari 3h13'46" 28d38'31"
Itiel
Vir 13h14'21" 11d9'11"
ITLO
Peg 22h39'15" 29d19'37"
Ito Yuuma
Leo 10h59'34" 4d26'9"
ITRACO
Uma 8h55'45" 67d49'16"
ITROICA V
Lyn 6h19'48" 57d44'25"
iTs A gReAt DaY tO bE aLiVe!
Tau 3h59'49" 28d21'5"
Its a Super Star!
Aqr 22h39'48" 1d5'9"
It's a Wonderful Life Samantha
Ori 5h24'49" 5d22'5"
It's All About M.eghan E.dwards
Cap 20h37'1" -23d6'54"
It's Ok to be Different.
Aqr 21h25'42" -4d13'4"
It's Ours
Cyg 20h48'26" 52d12'22"
It's Something That We Do
Uma 11h44'10" 37d20'4"
It's Ta-Rue
Tau 4h43'46" 9d32'29"
It's Your Love
Per 2h46'14" 54d11'27"
It'sDa'TopO'DaBoatMon
Lib 15h39'43" -5d43'5"
ITSME
Umi 14h53'44" 68d54'12"
Itta
Cap 20h28'22" -23d7'9"
itty bitty
Umi 14h41'51" 73d17'6"
Itty Bitty Baby Bunny
Psc 1h36'4" 17d40'22"
Itzel Gabriela Hicks
Vir 11h48'4" 3d34'38"
Itzi Te Amo Por Siempre
Cam 7h33'9" 79d4'2"
"Itzy"
Cru 12h23'24" -57d8'52"
IU American Catalan Newyorker
Cyg 19h57'24" 31d42'6"
iubirea noastra
Sgr 19h41'59" -14d35'33"
Iubirea Vietii Mele
Uma 10h51'57" 40d29'37"
Iukekini
Ori 5h51'3" 6d13'54"
Iulia
Umi 15h34'47" 67d33'19"
Iulia Alexandra Andrews
And 23h39'12" 33d57'13"
Iuna
Lyn 8h37'45" 39d22'48"
Iura Deborah
Aur 5h36'8" 33d2'53"
Iva
Ari 2h4'2" 11d32'19"
* Iva *
Cas 1h19'32" 69d4'50"
Iva Alexis
Lmi 10h45'7" 33d46'46"
IVA aviva REUVEN
Leo 11h44'53" 25d33'18"
Iva Hopayhmah Crandall
Oph 17h19'35" -0d33'32"
Iva May Warner
Lyr 18h19'15" 39d29'57"
Iva Melissa Henson
Cap 21h43'1" -12d34'51"
Iva Petrovic
And 23h24'41" 52d54'30"
Iva Potocka SVK
Lib 15h10'56" -5d6'12"
Iva Renée Olive
And 23h21'56" 51d39'56"
IVA (shekinah)
Lib 14h57'49" -4d8'54"
Iva Strljic
Cas 1h19'45" 51d25'37"
Iva Tobias
Leo 10h55'19" -3d6'21"
Iva Wolfe
Ari 2h15'24" 22d32'16"
Ivan
Per 4h50'4" 46d31'1"
Ivan
Her 18h6'23" 49d4'20"
Ivan
Cyg 19h55'10" 32d31'53"
Ivan
Ori 5h53'52" -8d17'15"
IVAN
Uma 8h20'44" 64d55'0"
Ivan 1
Her 18h47'20" 20d5'15"
Ivan - 22.01.
Peg 0h8'28" 18d39'14"
Ivan Avendano
Uma 10h29'17" 68d16'30"
Ivan Bruce
Cep 22h6'19" 72d1'21"
Ivan Dale Driver
Psc 1h27'33" 5d12'49"
Ivan Earl Bohlender July 1986
Leo 11h22'2" 19d38'56"
Ivan & Elsie
Cyg 20h28'56" 47d16'18"
Ivan F Ingraham
Her 17h21'16" 34d4'2"
Ivan Galliver
Uma 8h26'7" 70d6'50"
Iván Guisado Martín
Cyg 19h56'46" 50d32'49"
Ivan I. Nesch
Uma 10h29'18" 49d13'54"
Ivan Jay Yarrow
And 23h58'30" 33d40'15"
Ivan Jenkins
Per 4h13'18" 45d8'22"
Ivan Johnson 10/17/1976
Lib 15h9'17" -10d15'2"
Iván José Lucas Souto
Cnv 12h53'37" 41d35'41"
Ivan Kavazovic IVANKO 1.8.1983.
Cas 0h0'22" 53d2'33"
Ivan Keith Briscoe
Uma 12h3'27" 38d37'22"

Ivan Kirov
Tau 4h28'59" 27d36'25"
Ivan Klima
Uma 11h47'19" 28d34'53"
Ivan Koval
Gem 6h23'29" 21d47'53"
Ivan Lee Koh Lim
Ari 2h34'17" 12d52'55"
Ivan LEGENDS
Ori 6h17'7" 3d44'6"
Ivan Medina
Lib 14h50'22" -4d9'56"
Ivan Panisse
Aql 19h35'43" 12d13'52"
Ivan Previtali
Cyg 19h33'3" 44d43'28"
Ivan Romo Rodriguez
Cnc 8h25'33" 20d4'43"
Ivan Ruiz' ethernal shining star
And 0h45'23" 25d42'20"
Ivan S. Phelps
Lyn 7h10'16" 55d25'48"
Ivan Saltarelli
Cyg 21h32'4" 35d53'35"
Ivan Sean McMillian
Per 2h48'38" 53d7'33"
Ivan Stanley
Cap 20h27'34" -12d49'33"
Ivan Subbotin
Lyn 8h22'4" 44d22'8"
Ivan Szitron
Gem 7h27'48" 26d26'15"
Ivan Tchernenko
Per 4h19'8" 51d12'8"
Ivan The Cat
Lyn 9h10'37" 37d54'28"
Ivan the Terrible
Dra 19h38'17" 72d48'24"
Ivan Vaclaw
Ori 6h5'56" 13d52'57"
Ivan W. Sanchez & Christal L. Redd
Cas 0h47'9" 56d20'56"
Ivan Weatherbee, Jr.
Vir 13h35'9" -13d22'46"
Ivan Wilt
Ari 2h51'41" 28d25'16"
Ivana
Uma 13h31'47" 60d9'56"
Ivana Cacciatori
Gem 7h14'20" 21d5'6"
IVANA - DIANA
Boo 14h17'57" 13d29'35"
Ivana Dusickova
Uma 8h49'29" 65d28'1"
Ivana ElpEz
Sco 17h24'10" -30d28'31"
Ivana la mia mimimma
Uma 9h45'38" 49d53'24"
Ivana Marie
And 1h25'29" 47d14'31"
Ivana Tubic
Mon 8h4'7" -0d57'52"
Ivana Veljkovic
Sgr 18h9'35" -22d1'41"
Ivana Walter
Uma 9h23'28" 47d10'52"
IvaNardeen
Cnc 8h52'2" 30d41'37"
IvaNathan
Uma 10h37'8" 55d52'16"
Ivancica & Ivanic Jandric
Cyg 21h23'29" 37d14'0"
Ivancik
Uma 11h6'10" 61d33'22"
Ivanka & Hans Dörig
Umi 17h5'57" 78d24'47"
IVANKO - DR. IVAN SHVACHUK
Lac 22h18'13" 52d2'11"
Ivan-Myrtle Prendergast
Sgr 18h14'39" -17d50'13"
Ivann
Boo 14h53'29" 20d57'30"
Ivanna Hunt
And 23h56'53" 47d29'14"
Ivanna Ortiz
Aqr 23h28'39" -9d15'24"
Ivano Ruffo
Per 3h23'2" 39d25'31"
Ivanovic, Anton
Ori 6h14'26" 2d40'58"
Ivan-Valles
Dra 17h39'46" 50d48'40"
Ivasleighan
Pyx 8h47'29" -20d21'0"
Ivelisse
Vir 15h2'53" 6d15'57"
Ivelisse
Leo 11h25'45" 13d0'53"
Ivellise Vazquez
Lib 14h37'13" -11d21'9"
Ivens
Per 3h22'12" 42d45'52"
Ivernia Henderson
Gem 7h4'3" 26d37'17"
Ives Coco Chisman Duffy
And 23h37'38" 41d46'29"
Ives S. Chinea - My Mighty Eros
Her 17h14'22" 49d28'47"
Iveta Krizkova
Ari 3h16'32" 24d43'30"
Ivetta
Psc 1h0'4" 15d57'27"
Ivetta Manukyan
Lib 15h2'16" -6d55'37"
Ivette
Lyn 6h48'20" 53d21'53"
Ivette Luna
Sco 17h49'55" -37d2'36"
Ivetteliz Negron
Cam 4h23'47" 73d18'16"
Ivey and Tiffy
Lyn 7h36'1" 45d25'17"
Ivi Dale Mooney
Uma 11h36'48" 42d50'46"
Ivi Juarez Corazon de Alex
Crb 15h44'7" 34d7'35"
Ivica & Sandra forever in love
Cas 1h13'56" 69d12'1"
Ivis Abella
Ari 2h14'17" 22d17'55"
Ivis & Dorothy Morris
Lmi 10h4'9" 34d38'3"
IVK
Uma 9h15'18" 59d56'59"
Ivka
And 2h3'29" 42d21'43"
Ivllon Elaine
Oph 17h52'21" -0d14'6"
Ivna
Boo 15h2'32" 13d10'10"
Ivo
Leo 11h42'54" 26d26'30"
Ivo Amado
Lyn 8h3'15" 48d47'50"
Ivo HK - Mistral
Psc 1h7'38" 29d51'4"
Ivo Obrist
Vul 20h30'25" 26d33'47"
Ivo Pitra
Lib 14h46'26" -0d35'29"
Ivo, occhi di velluto
Crb 16h11'20" 29d1'34"
Ivoire
Cas 0h23'7" 59d23'11"
Ivomas
Psc 1h10'17" 32d58'0"
Ivonett
Ori 5h55'21" 22d15'24"
Ivonne
Tau 5h23'55" 19d45'17"
Ivonne
Psc 0h9'29" 11d20'22"
Ivonne & Dan
Cyg 19h34'46" 53d0'22"
Ivonne "Denny" Savone
Uma 11h35'2" 43d50'9"
Ivonne & Jeff Wadleigh
Mon 7h22'29" -0d18'27"
Ivonne M. Ortiz
Psc 0h42'18" 7d15'46"
Ivonne & Marc
Ori 6h4'53" 13d55'7"
Ivonne Marie Bates
Sco 16h11'13" -15d32'14"
Ivonne Pfaff Tuero
Col 6h24'26" -36d9'56"
Ivor Davies
Ori 6h17'35" -0d29'56"
Ivory & Patrick
Uma 9h31'7" 49d4'48"
Ivory Rose
Gem 6h50'2" 22d21'28"
Ivory Sophia
Umi 15h9'50" 78d28'27"
IVY
Ori 5h11'24" -1d0'32"
Ivy
Vir 14h11'18" 1d16'30"
Ivy
And 22h59'55" 41d5'20"
Ivy
Lyn 8h39'39" 34d6'42"
Ivy ADO
Uma 11h20'49" 54d38'59"
Ivy and Albert Kyle
Mon 6h25'53" 8d15'29"
Ivy Angel Gopaul
Tau 4h43'53" 19d5'27"
Ivy Beth Browne
Cas 23h33'49" 56d16'34"
Ivy Bilka
Ori 6h18'11" 14d44'39"
Ivy Camille Mullins
Singleton I
Psc 1h7'48" 21d42'54"
Ivy Dinan
Lyr 19h14'21" 39d34'35"
Ivy Green
Leo 11h39'14" 10d19'39"
Ivy Haddaway
Crb 15h23'37" 31d53'7"
Ivy Hannah Halpern
Gem 7h6'52" 34d8'5"
Ivy "Jack's Star"
Uma 14h0'41" 49d45'28"
Ivy Jane
And 23h50'34" 41d37'6"
Ivy Josina
Vir 12h45'15" -10d17'14"
Ivy[muscles]
Sgr 18h5'46" -17d39'14"
Ivy Perdue
Cyg 20h7'40" 47d8'36"
Ivy Rose Kwiatkowski
Com 12h41'53" 17d49'37"
Ivy Rose Long
Vir 12h38'43" 0d56'4"
Ivy Sarah Goldfarb
Cyg 20h40'33" 53d59'57"
Ivy Speck
Cam 4h39'55" 55d40'19"
IvyHays
Srp 18h29'19" -0d40'22"
IvyJustin1-10-2005
Cas 1h6'37" 60d19'41"
Ivy's Shining Star
Uma 12h15'35" 57d38'24"
Ivy's Star
Ari 2h21'10" 19d30'11"
Iwaki Hiroyuki & Waka
Leo 9h29'10" 21d14'54"
Iwanaga Star
Per 3h45'8" 32d10'27"
Iwona
Vir 12h33'23" 12d31'17"
Iwona Pruszynska
Psc 1h20'48" 30d51'47"
Iwona R.
Cap 21h42'18" -8d45'50"
Iwona's Star
Gem 8h7'26" 31d35'37"
Ixon's Pride
Uma 11h20'2" 40d38'30"
Iyana Kaiman
Uma 11h7'4" 63d52'14"
Iyanna Lynn Blankenship
Leo 10h56'48" 7d13'46"
IYSAISDMOIRNE
Uma 9h18'11" 56d48'16"
Iz
Ari 2h52'32" 16d49'25"
Iza Bean
Lyn 7h21'3" 56d18'21"
Iza et Ewen
Umi 14h22'32" 75d44'27"
Izaak James Edward Harding
Her 18h51'48" 21d51'57"
Izaak Michael Even
Her 17h43'48" 16d17'59"
Izabeau Potter
Uma 14h25'25" 56d45'42"
Izabel Anna Ratsey-Woodroffe
Uma 9h23'52" 50d2'15"
Izabela Nedzka
Aqr 21h4'57" -9d33'36"
Izabela R.
Ari 2h20'18" 13d46'25"
Izabella
Uma 9h47'46" 47d15'14"
Izabella
And 0h6'36" 47d2'17"
Izabella
Lyn 6h57'8" 44d35'58"
Izabella
And 1h1'10" 41d11'3"
Izabella Eva Grlica
Cas 0h2'16" 56d19'19"
Izabella Gonzalez
And 0h37'22" 36d9'30"
Izabella Irene Carsey
Cyg 21h35'34" 35d6'44"
Izabella Katarzyna Derda
Sco 16h12'0" -11d58'32"
Izabella Logsdail
And 0h22'23" 36d18'40"
Izabella Lorraine Ireland(IzzyGirl)
Tau 4h33'43" 14d21'21"
Izabella Maria Garcia
Vir 14h7'6" -16d49'10"
Izabella Rose
And 2h31'23" 50d30'14"
Izabella Victoria Villanos
And 1h9'59" 35d12'30"
Izabelle Lorraine Vazquez
And 23h11'57" 39d24'58"
Izaiah Kenney
Uma 11h31'8" 46d53'21"
Izaskun
Her 17h46'13" 39d55'10"
Izel Ibarra
Dra 18h41'7" 52d21'41"
Izeyah Anthony Yenter
Uma 11h50'57" 29d27'10"
Izia
Cas 0h27'45" 62d24'25"
Izik Youker
Uma 9h51'11" 69d49'22"
izlbee
Vir 12h37'59" 0d27'59"
Izsákné Wágner Edina & Izsák András
Cas 23h56'28" 58d8'14"
Izzabella McBeth Bloom
Lib 15h40'34" -29d41'5"
Izzat Hazem
Gem 7h32'58" 15d32'52"
Izzie's Pixie Dust Angel
Peg 22h20'28" 23d6'19"
Izzy
Cnc 8h22'58" 26d44'24"
IZZY
Cet 2h24'6" 9d19'49"
Izzy
Aqr 22h26'11" 0d19'32"
Izzy
And 2h26'53" 50d28'27"
Izzy
Lib 15h22'3" -6d37'50"
Izzy
Lib 15h12'37" -7d21'6"
Izzy G
Cnc 9h5'51" 22d31'10"
Izzy Gobioff's Star
Gem 7h31'9" 15d28'58"
Izzy Rodgers
Tau 5h42'57" 26d47'48"
IzzyB
Aqr 21h49'56" -4d2'20"
Izzy-Shine - Isaac John Amsterdam
Lib 15h21'33" -3d58'39"
J
Umi 14h10'16" 69d28'40"
J.
Ari 3h23'27" 23d45'20"
J
Lyr 19h8'24" 26d28'18"
J
Her 17h28'23" 24d24'47"
J
Leo 10h17'59" 15d30'2"
J & A
Leo 10h45'35" 13d51'32"
J A G's DB
Aql 19h49'12" 8d50'8"
J. A. Rowell Jr.
Lib 15h28'11" -11d38'47"
J A T I K
Pho 1h5'56" -40d51'54"
J. Alexander Jarvis
Uma 8h26'4" 61d30'8"
J. Allan Tinkham's Star
Aqr 21h38'22" -1d3'1"
J and E
Apu 13h59'17" -71d31'0"
J and G
Aqr 22h58'20" -19d4'7"
J and J
Cyg 20h49'18" 34d18'41"
J and J Butler 11/12/05
Sco 17h47'42" -37d38'24"
J and J Foehr
Lmi 10h40'39" 24d3'41"
J. Andrew Nie
Ori 6h13'19" 15d38'4"
J. Arthur Loignon
Per 2h54'21" 41d28'8"
J. B.
Leo 9h38'34" 27d14'2"
J. B. & Dodge Pixels Forever Young
Pyx 8h47'17" -30d52'16"
J B Frankosky
Per 4h14'50" 45d22'41"
J & B Rasmussen
Cyg 21h14'27" 35d47'53"
J. B. Shepelwich
Dra 18h30'41" 55d5'6"
J. Benjamin Witter
Gem 6h54'35" 31d47'43"
J Bird
Aql 19h5'53" 3d0'8"
J Boy's All
Aur 5h14'53" 29d19'32"
J. Brayton Fogerty
Aqr 22h44'33" -19d49'52"
J. Brooks Brown
Aql 19h30'44" 11d15'24"
J & C
Aqr 22h55'11" -17d57'26"
J. C. "Chuck" Conley
Cnc 9h19'12" 27d12'55"
J & C Gordon
Cnc 8h14'51" 26d9'18"
J C Jacob
Uma 12h8'59" 54d4'15"
J & C Star of Love
Cyg 20h29'35" 46d32'9"
J C Wahlberg
Psc 1h2'47" 31d44'23"
J. Chad Roth
Psc 1h23'58" 31d20'16"
J. Christopher Pardini
Lyn 7h9'40" 58d33'57"
J. Corcoran, DO JD-Peaceful Warrior
Her 16h26'15" 24d56'22"
J. Curtis Cree
Ori 6h5'13" 13d37'29"
J & D 3/18/06
Cyg 20h46'35" 42d55'32"
J D Weidner, III
Ori 5h31'56" 2d16'44"
J. Dale Wainwright
Ori 5h14'53" 8d19'15"
J. Danyale Little
Sgr 18h19'47" -20d26'59"
J. Darlene Enfield
Gem 7h23'47" 25d39'41"
J. Daryl McLaughlin
Uma 13h41'49" 51d17'54"
J. De Cardenas
Lib 15h1'26" -6d8'35"
J. DeLois Stewart
Uma 11h23'13" 45d31'29"
J Diggy
Cnc 8h41'8" 24d53'56"
J*Dizzle
Psc 1h13'59" 7d1'30"
J. Douglas Brown
Her 16h24'45" 9d0'0"
J. E. Halonen
Aql 19h51'24" -0d35'43"
J. E. Hood.s Star Catcher
Ari 2h19'13" 25d41'41"
J*E*M
Cam 4h13'6" 66d31'21"
j e m ison
Mon 7h19'29" -0d26'5"
J Foley 05/11 C Cunha
Tau 5h38'42" 26d42'28"
J. Ford
Uma 13h50'9" 61d53'29"
J Franklin Presley
Crb 15h38'14" 33d43'15"
J & G 25 Grace Under Pressure
Uma 8h31'45" 67d59'16"
J G B
Cas 1h18'14" 51d11'55"
J. Garwood Family
Aqr 22h34'17" -5d4'58"
J. Golino
Uma 12h14'2" 54d28'46"
J Grush
Uma 10h20'32" 45d9'50"
J Guadalupe Perez - Lupe
Sgr 18h21'24" -27d14'54"
J - Hawk
Ari 2h53'12" 28d6'59"
J. Ira Harris
Uma 8h24'15" 64d2'35"
J & J
Dra 17h10'38" 57d11'9"
J & J
And 0h46'56" 36d53'26"
J&J amerò per sempre
Gem 7h30'49" 34d39'58"
J&J Forever and Always
Cyg 19h46'26" 37d37'12"
J & J Forever In The Stars
Pho 0h31'6" -55d57'16"
J J Holt
Cra 18h35'54" -41d25'53"
J & J Kreeger -39
Sge 19h16'12" 18d38'9"
J & J Malgo "40"
Uma 9h31'0" 57d45'32"
J. J. Russell
Uma 9h26'21" 65d21'14"
J J SPELLER
Leo 9h36'17" 31d55'26"
J J Superstar
Her 17h41'51" 37d48'22"
J J Tunnadine
Umi 13h24'46" 73d28'30"
J. J. Wade
Her 17h20'54" 37d3'46"
J. Jackson
Cet 2h46'3" -0d32'45"
J. Jay
Her 16h40'58" 28d34'9"
J. Jay Lee
Per 2h18'0" 51d46'26"
J Jazz
Dra 16h7'7" 55d3'39"
J&K FOREVER
Vir 15h6'40" 4d29'42"
J. Keith Petty
Cam 3h59'54" 69d15'55"
J. Kenneth Menges, Jr.
Crb 16h9'57" 28d14'4"
J. Kim Prather
Umi 15h19'54" 85d5'19"
J. Kirk Douglas Conners
Her 16h36'44" 31d15'1"
J K's Destiny
Cyg 19h32'52" 51d48'53"
J & K's Palace
Cyg 19h39'27" 34d29'23"
J+L= A&F Camaleon
Ori 6h19'19" 2d35'9"
J. L. Acosta
Cnc 8h47'2" 8d3'55"
J. L. Castor
Ori 6h7'43" 9d18'34"
J. L. Fletcher
Uma 8h50'49" 57d25'17"
J&L Giacoppi
Cyg 21h29'35" 51d46'3"
J L S 0203
Aqr 21h12'47" -11d1'39"
J & L Snyder - Unwavering Love
Lib 15h17'32" -7d10'4"
J. L. Very
Gem 6h36'12" 22d57'25"
J. La Verne C.
Cnc 8h54'43" 27d35'46"
J. LANE Purcell
Aur 5h22'26" 32d40'37"
J. Lyle Roland
Cap 21h40'8" -14d19'55"
J & M
Lib 14h25'32" -9d26'8"
J&M
Cam 4h14'54" 53d4'20"
J & M 01-17-2004
Cnc 8h17'14" 19d22'55"
J&M&B Werner
Sge 20h1'20" 18d7'45"
J & M Moore's Family Star
Cru 12h30'31" -55d42'5"
J M P - A S P 04
Uma 13h14'5" 63d0'46"
J M Panatier
Uma 11h27'52" 51d38'27"
J M S Cody
Umi 14h45'59" 71d43'1"
J. Marie Denzer
Leo 11h13'4" 24d4'54"
J. Mark Harper & Jill Fortner
Cyg 20h18'52" 52d49'7"
J. Mark Staley
Lib 15h46'13" -29d37'57"
J Marren
Sgr 18h21'0" -16d32'53"
J. Mello's Fire
Dra 18h57'27" 50d25'19"
J. Michael
Ari 2h33'16" 17d48'42"
J. Michael Higginbotham
Leo 10h5'27" 13d45'21"
J. Michael Simon
Sco 17h55'50" -41d53'16"
J Montoya
Cru 12h29'19" -55d43'21"
J. Mumethius
Her 17h48'51" 36d1'56"
J&N
Leo 10h40'33" 9d49'21"
J. N. Cooper
Cep 1h45'13" 80d29'9"
J n S
Aqr 22h54'51" -8d30'59"
J. Nana Grote 14
Gem 6h30'52" 27d42'24"
J. Nichole 12-06-74
And 1h34'44" 41d49'59"
J. O. King, Jr.
Uma 10h43'25" 62d45'37"
J O S Centennial
Gem 6h29'30" 19d28'47"
J O S L
Ori 5h57'18" 12d13'1"
J Owen Dryden
Lyn 7h39'8" 46d53'52"
J. P. B. No. 1
Lib 14h27'23" -12d33'37"
J P COMOLA
Ori 5h41'59" 1d22'58"
J P Rosciglione
Uma 8h35'49" 48d32'6"
J. Pancake
Per 4h21'20" 35d28'32"
J. Patrick
Cam 3h32'13" 56d14'56"
J. Patrick Flippin
Her 17h53'49" 18d10'35"
J. Peter Williams
Sco 16h27'50" -30d58'25"
J R Bundfuss
Peg 21h39'54" 26d43'8"
J & R Crabb
Cyg 20h28'48" 44d1'49"
J. R. Cummard
Per 4h14'7" 47d32'52"
J&R Foreverrr
Ori 6h15'41" 10d28'14"
J R Hudson, Grandpa of the Czar
Per 3h12'45" 54d53'38"
J. R. Kukla
Cnc 8h4'32" 13d10'12"
J. R. Richter
Leo 11h44'54" 26d56'17"
J. Ray
Sgr 19h39'29" -31d40'25"
J Richard C Sainsbury
Ori 5h41'8" -0d23'28"
J. Richard Lancaster
Gem 6h22'1" 18d43'1"
J Ro
Psc 1h20'30" 17d46'16"
J. Robert Bullock
Sgr 19h51'37" -41d26'39"
J. Robert Schweikert, M.D.
Psc 0h56'50" 15d54'45"
J Rock
Aqr 23h8'58" -7d17'59"
J. Rod Cameron
Aql 19h9'52" 6d47'33"
J. Ronald Spence, Ph.D.
Her 17h38'44" 37d1'19"
J. Rosa Fleming
Her 17h18'52" 20d35'20"
J. R.'s Everlasting Rose
Cyg 21h23'48" 51d21'39"
J&S Eternal Light
Tau 3h51'12" 8d23'41"
J. Sadowsky Multiage Guiding Light
Uma 11h22'56" 32d42'9"
J/Schultz-65
Boo 14h47'21" 22d47'41"
J Scott
Ari 3h15'17" 20d58'23"
J Smoov & JJ Star
Ari 2h26'46" 27d38'58"
J Square
Umi 13h4'8" 74d42'34"
J. Stephen Parks
Lib 15h22'47" -8d0'33"
J~sto
Leo 10h6'39" 22d49'46"
J & T: Five Years and Counting...
Car 8h9'31" -58d50'41"
J T Lawson
Uma 11h40'1" 28d30'36"
J*T Reed
Equ 21h16'19" 10d45'24"
J T Z
Psc 0h18'25" 7d23'21"
J. Thomas Holland
Boo 14h53'46" 38d54'8"
J Tim Walsh
Sgr 18h21'58" -32d9'15"
J. Timothy Porea, IV
Umi 14h14'40" 75d48'31"
J. Victor A. Collomb
Umi 16h3'18" 83d23'12"
J. W. Diver
Sgr 19h29'2" -17d58'30"
J Zacarias
Uma 9h37'49" 52d54'14"
J. Zuber
Her 17h17'13" 16d8'8"
J, M & M always together
Ori 5h52'4" 18d23'35"
J2
Tau 5h46'48" 22d39'36"
J^2
Vir 12h43'10" 12d2'18"
j3ngerale
Uma 11h39'6" 35d25'56"
J4F111
Her 16h54'19" 30d39'47"
J9
And 23h4'23" 51d27'2"
J-9
Uma 10h5'36" 48d20'46"
J-9
Cam 3h55'2" 64d14'0"
J9383J
Vir 13h56'50" -21d48'18"
J.A. Drago
Cap 21h4'49" -21d51'16"
Ja Man Ja
Cas 1h0'28" 57d27'36"
J.A. "The Pumper"
Cnc 8h59'21" 23d34'4"
J.A. van Draanen van den Burg
Aqr 20h55'3" -11d59'3"
J.A. World
Uma 10h38'26" 50d18'26"
J.A.A.B Sanguinetti Family
Uma 13h46'13" 56d1'23"
Jaabir Syed
Cnc 8h13'2" 16d48'30"
Jaamir Alexander Love
Psc 1h36'49" 27d31'36"
JAAN
Umi 15h5'54" 73d49'9"
Jaana Luik
Cnc 8h12'23" 15d19'45"
jaano nick dha
Cyg 20h16'20" 50d46'40"
Jaan's Priti
Cas 1h25'33" 55d22'25"
Jaarne 19.11.2003
Ori 6h18'29" 11d7'55"
JaB4E
Cnc 8h32'13" 32d22'18"
Jabba Love
Uma 9h49'48" 66d22'15"
Jaber
Ori 5h5'58" 4d36'17"
Jablonski, Klaus
Uma 9h56'29" 65d21'23"
JaB-PHRoSTiLL M
Psc 0h47'37" 8d41'25"
Jabsen
Uma 11h8'41" 34d23'56"
JabZ 01021934
Uma 9h0'57" 62d49'44"
JAC
Dra 18h32'28" 59d53'52"
Jac Attack
And 0h16'47" 27d37'9"
Jac Jones
Umi 22h56'39" 89d27'25"
Jac & Roc
Cyg 20h43'6" 45d32'49"
Jac Venza
Cap 20h18'2" -23d35'16"
JAC5506
Dra 17h2'24" 58d52'12"
Jacalyn Christine Endres
Lib 15h40'3" -3d52'38"
Jacamelle
Cru 12h12'49" -62d24'33"
JacCon 06
Ori 5h34'27" -1d27'20"
Jace
Cap 20h46'20" -18d38'34"
Jace Edward Raley
Lyn 7h11'40" 44d35'44"
Jace Emiliano Panis
Her 17h38'27" 27d42'13"
Jace Homec
Uma 11h31'59" 42d46'40"
JACE - Jason Miles Cohen
Cnc 9h8'38" 32d35'35"
Jace Kaine Postoak
Boo 14h18'32" 16d43'4"
Jace Robert Serrano
Lyn 8h15'32" 56d20'8"

Jace William Hunter
Vir 12h54'50" -1d56'6"
Jaceb Liam
Crb 15h40'4" 26d37'56"
Jacek Jakobczak
Dra 17h25'0" 51d14'39"
Jacek Kobus (Happy Jack)
Lyn 6h32'44" 60d55'1"
Jacelyn
And 0h52'38" 38d26'43"
JACEMN
Uma 10h15'55" 54d6'57"
Jacen William Jeremay
Lyn 7h57'7" 34d40'3"
JACES 7
Lyn 9h9'1" 36d44'17"
Jace's Hole
Aqr 22h19'12" -2d9'27"
Jacey
Lib 15h35'3" -4d37'1"
Jacey
Ari 1h59'12" 13d9'59"
Jacey D Davis
Her 17h52'15" 22d27'0"
Jacey Domoto
Lyn 7h49'56" 38d53'4"
Jacey Marie Fugate
Psc 0h21'4" 6d31'31"
Jacey, The Unforgettable Star
Gem 6h34'40" 17d3'33"
Jacglen
Cra 18h6'16" -39d17'16"
Jaci
Dra 16h23'51" 58d37'5"
Jaci
Psc 0h59'18" 25d43'25"
Jaci
Her 16h32'2" 47d27'39"
Jaci Gaddis
Psc 1h22'38" 10d8'26"
Jaci Leigh Martin
Lmi 10h31'53" 33d7'50"
Jaci Sullivan
And 0h44'58" 42d27'47"
Jaci Timko
Cam 4h3'4" 54d44'56"
Jacie Couse
Cnc 8h54'36" 8d1'40"
Jacie Jo Junkin
Sco 16h50'25" -27d4'23"
Jacie Schulman
Sex 10h38'28" -0d44'47"
Jacin
Gem 6h43'43" 16d56'25"
Jacinda
Cnc 8h30'6" 7d11'3"
Jacinda Rose Feldman
Gem 6h44'37" 30d59'36"
Jacinthe Grandmaître
Her 16h8'30" 24d9'25"
Jack
Her 18h39'7" 22d44'40"
Jack
Tau 5h38'26" 24d6'56"
J.A.C.K.
Tau 4h27'37" 29d52'28"
Jack
Leo 10h14'18" 18d43'51"
Jack
Her 18h44'19" 16d1'22"
Jack
Vir 13h43'43" 1d53'22"
Jack
Ori 5h13'32" 7d46'27"
Jack
Gem 7h20'4" 33d3'53"
Jack
Aur 5h49'14" 33d5'46"
Jack
Aur 6h15'52" 33d16'36"
Jack
Per 3h17'10" 53d4'50"
Jack
Boo 15h13'0" 41d33'21"
Jack
Cyg 21h16'6" 43d7'51"
Jack
Cas 1h0'46" 49d54'5"
Jack
Leo 11h4'2" -0d36'14"
Jack
Cep 1h31'1" 78d40'45"
Jack
Cep 0h15'42" 78d0'29"
Jack
Umi 15h25'57" 69d9'49"
Jack
Cep 20h44'54" 63d58'8"
Jack
Aqr 23h25'43" -18d52'45"
Jack - 02.21.1957
Psc 1h31'42" 13d47'54"
Jack 11 and Jackie Banta
And 2h35'55" 45d47'31"
Jack 133
Sgr 19h9'23" -13d5'47"
Jack A. Demetree II
Cep 20h51'34" 65d4'33"
Jack A Kasperski
Uma 11h23'3" 30d29'19"
Jack Aaron Chambers - Jack's Star
Gem 6h2'3" 26d9'39"

Jack "Ace" Hugh Cottrell
Uma 10h42'35" 52d42'36"
Jack Adelman
Aur 6h53'48" 37d36'12"
Jack Adrian Chaston
Sgr 19h26'52" -41d29'35"
Jack Alan Bishop - Jack's Star
Lmi 10h5'0" 28d15'42"
Jack Alberto Oldham
Per 4h0'7" 32d4'57"
Jack Alexander Campbell
Uma 9h9'27" 61d20'9"
Jack Alexander Christie
Per 2h55'36" 45d42'31"
Jack Alexander Harvey - Jack's Special Star
Per 3h18'2" 55d12'39"
Jack Alexander Hayes
Her 16h33'51" 12d35'50"
Jack Alexander Littlewood
Her 16h20'24" 17d10'1"
Jack Alfie Plowright
Cep 20h29'22" 63d13'16"
Jack Allen Carbone
Cap 20h29'10" -12d19'38"
Jack Allen Hogan
Uma 10h51'32" 69d44'34"
Jack and Alice
Cyg 20h31'45" 50d9'12"
Jack and Anne Pleasance
Uma 10h28'24" 67d44'42"
Jack and Beth Birli
Her 18h3'20" 17d45'33"
Jack and Cathy Oates
Sge 19h41'29" 17d51'22"
Jack and Charlotte Royal
Sco 16h56'47" -37d47'24"
Jack and Emma Wilson
Umi 15h35'44" 77d29'39"
Jack and Estelle
Cyg 19h53'45" 30d36'33"
Jack and Gloria Normoyle
Gem 6h40'39" 16d28'7"
Jack and Ivana, God's gift to me
And 0h21'43" 42d9'10"
Jack and Jadyn's Star
Cnc 8h35'37" 31d4'36"
Jack and Janet Wilson
Tau 3h37'22" 28d45'12"
Jack and Jenn's Star of Love
Cnc 8h28'22" 23d17'53"
Jack and Keley
Lyr 18h47'52" 39d58'51"
Jack and Koniko Chan
Cyg 19h58'43" 30d53'11"
Jack and Lexi
Sco 15h53'54" -21d38'18"
Jack and Loretta
Cyg 20h2'13" 37d54'18"
Jack and Maggie's Star
Per 3h37'29" 33d55'14"
Jack and Mary Schliessmann
Cyg 21h35'38" 39d15'50"
Jack And Megan's Star
Umi 15h41'42" 71d23'22"
Jack and Melinda
Crb 16h2'40" 27d39'15"
Jack and Nadine
And 0h19'16" 32d58'25"
Jack and Peggy Miller
Eri 3h48'22" -0d19'17"
Jack and Renee
Cyg 19h58'16" 30d56'20"
Jack and Rose
Dra 19h31'26" 72d58'35"
Jack and Roz Kaplan
Cyg 19h41'53" 54d26'4"
Jack Andrew Hambleton
Per 3h47'23" 36d57'5"
Jack Andrew Raimo
Sco 16h11'22" -22d24'14"
Jack Andrew Welch
Dra 18h54'11" 59d45'56"
Jack Andrew Wilson - "Superstar"
Uma 9h44'7" 42d38'24"
Jack Anthony Draycott
Per 3h19'17" 40d40'34"
Jack Anthony Hobaica
Uma 12h20'29" 58d46'30"
Jack Anthony Jarrett
Ori 5h51'59" 3d47'59"
Jack Anthony Veiga
Per 4h42'49" 49d33'22"
Jack Archie Cantle
Cru 12h20'32" -57d25'11"
Jack Artenstein
Uma 10h35'59" 53d32'16"
Jack Arthur Gergets
Her 16h5'46" 46d22'5"
Jack Artrup
Ori 4h49'14" 11d49'25"
Jack Ashworth
Aql 19h53'29" 4d54'42"
Jack Augusta Reynolds
Umi 15h22'30" 68d32'26"
Jack Austin Robertson
Her 16h11'31" 4d10'54"
Jack Avent
Ori 5h10'9" 12d24'2"

Jack Axelrood
Uma 10h31'48" 62d48'21"
Jack B Jones
Per 2h17'15" 57d0'6"
Jack & Barb Milaskey, Earth Angels
Gem 7h48'32" 27d51'58"
Jack & Barb Rexworthy Together Again
Cru 12h5'7" -63d39'46"
Jack & Barbara Forever
Uma 8h59'12" 72d19'49"
Jack & Barbara Hayes
Psc 0h22'32" 9d53'33"
Jack Barber
Her 17h34'28" 35d23'32"
Jack Barker
Gem 6h40'28" 15d9'44"
Jack Baronett
Psc 1h11'10" 28d15'45"
Jack Basile
Aqr 21h59'43" -14d53'5"
Jack Bellis
Sgr 19h14'2" -17d16'43"
Jack Benjamin Nash
Her 18h56'17" 16d14'40"
Jack Bennett Clapper
Ori 5h15'32" 6d22'22"
Jack Benninger
Uma 11h8'5" 55d25'40"
Jack Bergen Vermedahl
Ari 3h6'37" 16d12'37"
Jack & Betty Ennen
Cyg 19h37'23" 27d55'30"
Jack & Betty Haggerty
Lyr 19h22'43" 38d3'51"
Jack & Beverly Degnan
Aql 19h58'18" -0d6'39"
Jack Binns
Gem 6h42'1" 12d17'6"
Jack Blackjack Santos
Uma 11h28'45" 48d8'10"
Jack Bolton
Uma 11h40'3" 57d6'32"
Jack Boydle
Ori 6h4'47" 14d16'28"
Jack Boytim
Vir 12h8'14" 11d7'56"
Jack Bradley Ethan Aaron
Her 17h19'2" 38d52'49"
Jack Brady
Her 17h55'5" 28d57'34"
Jack Brancel
Per 3h42'27" 52d22'22"
Jack Brennan
Uma 9h32'25" 69d38'10"
Jack Brent Schuermann 9/8/1997
Vir 12h57'50" 0d55'3"
Jack Brian Baker
Cen 12h30'30" -50d31'59"
Jack Brockten Levy
Her 17h40'18" 36d36'37"
Jack Buffett Kelly
Aur 5h44'0" 42d19'19"
Jack Buiskool
Uma 8h56'56" 60d14'28"
Jack Burgoine
Uma 12h7'31" 47d43'37"
Jack C. Huggins
Tau 5h36'16" 18d3'49"
Jack Caelen Dowse
Ori 6h16'4" 16d32'19"
Jack Cahill
Aql 19h49'46" 6d22'28"
Jack Carl West "Baby Jack"
Lib 15h39'15" -18d58'38"
Jack Carlson
Uma 10h14'2" 70d47'43"
Jack Carter
Sgr 18h43'33" -27d22'44"
Jack Carter
Her 16h46'53" 30d32'51"
Jack & Cathy Tracy
And 0h19'40" 45d12'17"
Jack Chambers - Jack's Star
Per 4h2'46" 43d17'47"
Jack Chandler
Aql 19h8'28" -10d21'39"
Jack Chandler Zalewski
Psc 0h49'5" 16d0'39"
Jack Charles
Col 5h13'23" -38d55'51"
Jack Charles Bond
Cap 21h54'18" -13d58'46"
Jack Charles Grable
Umi 14h22'38" 76d17'12"
Jack Charles Miller
Tau 5h29'51" 21d20'25"
Jack Chastain
Ori 6h16'24" 9d25'54"
Jack Cheetham
Her 16h16'27" 43d54'44"
Jack Christopher
Cep 2h3'20" 85d34'33"
Jack Christopher Berger
Psc 1h21'1" 26d19'53"
Jack Christopher Liwienski
Sco 16h46'55" -26d28'15"
Jack Christopher Street
Dra 16h58'56" 52d58'27"
Jack Cillian Conway
Her 17h49'26" 17d10'51"

Jack Clark
And 23h25'54" 42d0'29"
Jack Clements
Psc 1h35'5" 11d5'33"
Jack Clemons "CTR" Star
Aur 5h32'37" 41d28'10"
Jack Clift
Uma 8h54'19" 71d4'30"
Jack Collins
Her 17h47'5" 46d29'58"
Jack Colties
Aur 5h52'50" 40d54'29"
Jack Connor Speakman
Her 18h2'43" 29d2'52"
Jack Corbett
Sex 10h14'2" -0d52'56"
Jack Corbette Nicholls
Cru 12h34'4" -57d23'44"
Jack Corwin Whitener
Lib 15h2'23" -5d55'50"
Jack Cox
Sgr 18h5'2" -31d23'23"
Jack Craig
Equ 21h15'44" 9d0'1"
Jack Crow
Tri 1h40'28" 29d25'11"
Jack D Green
Psc 1h21'32" 16d31'38"
Jack D. Rickly
Uma 9h34'54" 67d19'17"
Jack Dale Thomas
Uma 10h11'12" 52d36'10"
Jack David Connolly
Sgr 17h54'1" -29d57'32"
Jack David Jones
Uma 12h22'35" 56d26'20"
Jack David McAuliffe - 7/06/58
Cnc 8h38'20" 8d22'42"
Jack David Messinger
Leo 9h26'6" 11d36'13"
Jack David Nahgahgwon
Ori 5h19'56" 3d49'16"
Jack David Stephens
Uma 14h5'29" 50d56'1"
Jack David Watson
Ari 2h51'15" 18d49'41"
Jack Davis Weisskopf
Her 18h46'36" 22d27'45"
Jack Dawson Bridges 5-3-99
Tau 5h26'4" 23d5'24"
Jack & Debra Ordner
Crb 15h50'51" 35d33'48"
Jack Denton Paris
Lib 15h21'39" -10d27'47"
Jack Devaney
Aur 5h22'51" 52d30'54"
Jack "DOC" Hoffpauir, D.V.M.
Vir 12h3'28" 3d7'57"
Jack Dolle
Cap 21h42'1" -15d53'50"
Jack Donald Lewis
Cru 12h23'57" -58d31'21"
Jack Donald Smith
Aqr 23h0'17" -7d46'58"
Jack Douglas Becker
Uma 9h45'25" 56d10'35"
Jack Dowding
Vir 12h31'41" 5d47'53"
Jack Dowling Lalor
Lib 15h7'10" -13d1'27"
Jack Dunlavy
Cnc 8h47'20" 25d44'54"
Jack Duplantis
Cep 22h10'53" 60d58'45"
Jack Dylan
Del 20h54'42" 9d3'40"
Jack Dylan Burgos Watt-Roy
Uma 9h24'30" 44d7'48"
Jack Dylan Hirsch
Crb 15h46'38" 31d40'48"
Jack Dylan Williams
Sco 17h13'1" -30d7'30"
Jack E Adams
Ari 3h0'58" 18d22'22"
Jack E. Brown
Uma 10h23'16" 66d35'19"
Jack E. Crow
Lmi 10h44'55" 28d7'32"
Jack E. Leedom
Per 3h13'52" 46d44'26"
Jack E. Olson
Tau 5h11'39" 18d24'6"
Jack E. Paige
Sgr 19h10'30" -19d37'47"
Jack E. Robinson
Lib 15h22'11" -28d2'30"
Jack E. Vickers
Per 3h31'58" 48d10'43"
Jack E. Walker
Per 2h46'50" 54d0'41"
JACK EBERT
Her 18h5'49" 17d29'6"
Jack Edsill
Sgr 20h5'35" -24d47'6"
Jack Edward
Psc 1h31'58" 24d0'13"
Jack Edward Byers
Gem 6h46'42" 18d28'24"
Jack Edward Colton
Per 4h22'22" 34d44'35"

Jack Edward Lindsell 17022006
Cra 18h47'59" -40d34'34"
Jack Edward McGaha
Per 4h38'4" 40d12'17"
Jack Edward Sullivan
And 1h49'22" 45d34'47"
Jack Edward Upton
Per 4h37'10" 52d36'24"
Jack & Eileen Clark
Uma 11h43'54" 47d56'2"
Jack Elliot's Wishing Star
Sco 17h37'38" -41d56'42"
Jack Ellis Norris
Vir 14h10'49" -17d15'37"
Jack & Elsie
Cyg 20h28'17" 33d6'36"
Jack Entwistle
Cep 21h26'31" 55d34'51"
Jack Epper
Ori 6h0'34" 10d44'45"
Jack Eugene Jagger
Cep 3h35'15" 86d34'13"
Jack Evan
Vir 12h35'25" 12d27'19"
Jack F. Lasday
Psc 1h25'2" 3d4'11"
Jack F. Mertz
Uma 13h52'34" 49d27'8"
Jack Falvo III
Cap 20h23'37" -9d49'36"
Jack Farmer
Aqr 20h50'56" 1d43'27"
Jack Farren
Gem 6h47'29" 16d15'47"
Jack Fendrick
Boo 14h37'26" 17d48'52"
Jack Finlay Burlikowski
Uma 11h50'16" 43d44'6"
Jack Fisher
Her 17h16'28" 34d25'44"
Jack Flaherty
Leo 9h33'3" 27d49'16"
Jack & Florence Cecil
Cyg 19h27'50" 28d53'28"
Jack Forrester
Sgr 18h55'24" -20d29'30"
Jack Frances Christian
Uma 8h20'30" 66d44'5"
Jack Franzman
Gem 7h19'50" 24d19'44"
Jack G. Bitterly - Inventor
Uma 11h6'59" 34d6'51"
Jack G. Mowers
Ari 3h25'49" 26d23'38"
Jack & Gail Stoakes
Cyg 21h16'8" 44d30'25"
Jack Gatto
Ori 6h5'23" 17d19'17"
Jack George Buswell - Jack's Star
Peg 21h36'16" 25d35'32"
Jack Gibbons
Per 3h13'34" 53d6'9"
Jack & Gladys Kemp
Cyg 20h11'10" 46d41'30"
Jack Gordon Lahar
Cep 22h7'18" 56d22'39"
Jack Grafman
Her 18h17'42" 21d51'29"
Jack Grayson Ellis
Cap 20h45'15" -21d25'33"
Jack Green
Ori 5h5'8" 3d51'53"
Jack Greenslade
Uma 10h22'33" 43d26'3"
Jack Gregory Mc Innes
Leo 11h10'9" 11d32'41"
Jack Gresser
Her 17h44'50" 14d47'54"
Jack Gunnels
Cep 2h26'35" 81d19'36"
Jack Gutowski
Her 16h44'1" 15d47'19"
Jack Gutowski
Uma 10h29'49" 50d58'26"
Jack H. Brooks, Sr.
Sco 17h47'11" -40d25'22"
Jack H. Corn "Molaris"
Umi 13h38'46" 88d41'10"
Jack H. Scaff, Jr., M.D.
Gem 7h33'9" 14d37'33"
Jack Halper
Psc 0h33'10" 3d50'34"
Jack Harper
Cap 20h28'8" -11d33'20"
Jack Harrison also Da
Cap 20h52'18" -25d47'5"
Jack Hartford
Cep 21h39'19" 64d7'31"
Jack Haughney
Cyg 19h37'22" 39d18'4"
Jack Hayes Lights the Links and Sky
Gem 7h13'46" 31d57'37"
Jack Henry Schweitzer
Sgr 19h7'52" -31d0'5"
Jack Henry William Andreson
Sco 17h58'23" -31d59'11"
Jack Herrle
Ari 3h28'30" 19d57'19"
Jack Higgins
Leo 11h37'50" 26d16'53"

Jack Hinkins
Ori 4h55'0" -3d32'53"
Jack Holden Canosa
Aqr 22h50'43" -7d14'20"
Jack Holden Caulfield
Per 3h32'18" 44d53'0"
Jack Holmes
Umi 14h45'37" 77d40'27"
Jack Hornor
Aqr 21h1'12" 0d39'56"
Jack Houston Gipson
Sco 16h45'40" -26d31'46"
Jack Howard Stewart
Cen 12h35'57" -51d19'11"
Jack Hyman Chaet
Uma 8h58'19" 58d22'36"
Jack Iverson
Lyn 6h48'25" 61d26'17"
Jack J Cicotta
Her 18h43'29" 17d23'5"
Jack Jace Gooden
Cra 18h0'37" -37d8'55"
Jack & Jacey
Cyg 20h44'49" 35d51'32"
Jack Jack
Sgr 17h54'46" -29d53'35"
Jack Jackson
Gem 6h41'17" 16d50'27"
Jack James
Her 17h2'50" 12d47'3"
Jack James Redhead
Umi 15h19'19" 77d39'49"
Jack Jasey Davidson
Cen 13h42'29" -41d14'57"
Jack Jeanes (Dad/Pop Pop) 90TH
Vir 12h32'7" -9d38'44"
Jack & Jeanne 1949
Lyr 18h31'30" 36d31'0"
Jack & Jeannie Jacques
Cyg 21h24'6" 53d6'28"
Jack Jesse James - 12.10.1999
Cru 12h29'40" -60d32'17"
Jack & Jessica, Love Shines Forever
Cyg 19h57'47" 31d53'22"
Jack John
Dra 18h37'49" 55d27'0"
Jack Joseph Applefeld
Per 4h7'23" 35d58'36"
Jack Joseph Bandola
Vir 13h11'47" 12d22'34"
Jack Joseph Powell
Vir 12h54'48" 6d48'4"
Jack Joseph Towers
Gem 7h25'14" 17d3'20"
Jack Joseph Valentine Williams
Per 4h35'11" 36d10'9"
Jack & Joyce - Shining Together
Cru 12h1'7" -61d31'3"
Jack & Juanita Jackson's Star
Uma 11h30'38" 47d23'54"
Jack & Julia Church
Cyg 20h37'10" 54d51'46"
Jack Jury
Aqr 23h0'36" -10d0'12"
Jack K. Kelley
Aur 5h18'48" 42d28'44"
Jack Keating
Her 18h40'51" 18d26'37"
Jack & Keith's Star
Cnc 8h46'38" 27d23'5"
Jack Kelly
Uma 12h5'12" 62d33'42"
Jack Kendrick Johnson
Uma 8h57'2" 50d2'30"
Jack Kenneth
Lib 15h35'47" -9d34'0"
Jack Kenneth Lind
Umi 14h47'51" 77d0'58"
Jack Kerrigan
Ari 2h33'13" 30d48'2"
Jack Kiviat
Ari 2h34'22" 13d31'40"
Jack Kjentvet
Gem 6h32'39" 26d34'5"
Jack Koenig
Uma 10h47'54" 47d20'31"
Jack Kramer
Gem 6h48'30" 17d37'2"
Jack L
Eri 3h42'52" -0d16'22"
Jack L. Brackin
Mon 6h53'30" -0d22'2"
Jack L. Greider (Bi-Pa)
Tau 4h22'4" 23d15'6"
Jack L. Moore
Vir 13h30'13" -8d35'38"
Jack L. Thorne
Umi 14h37'7" 70d25'44"
Jack Labarge
Uma 8h22'0" 61d52'10"
Jack LaBruere Hurst
Leo 10h20'9" 23d18'56"
Jack Lane Claver
Pho 1h18'38" -46d4'7"
Jack Lark IV
Per 3h8'9" 51d25'14"
Jack Lark, IV
Uma 13h40'51" 55d25'7"

Jack Larry Joseph Phelan
Lib 14h52'52" -0d55'57"
Jack Larry Nunez (BEST DAD EVER)
Lib 15h29'0" -8d39'28"
Jack Laurence Sedlock
Lyn 7h53'34" 49d6'36"
Jack Lawless
Cyg 20h42'45" 46d32'2"
Jack Le Quesne
Cyg 19h56'24" 46d48'36"
Jack Leib
Her 17h40'34" 27d24'6"
Jack Leo Sweda
Lmi 10h42'30" 26d8'59"
Jack Leo Wilson
Ari 2h12'52" 11d50'42"
Jack Lewis Ramey
Per 3h5'19" 43d48'16"
Jack Likely's Star
Per 3h24'57" 49d51'24"
Jack Logan
Lyn 7h11'27" 51d8'13"
Jack Logan
Dra 16h6'43" 57d30'22"
Jack Loughridge Wargo
Aqr 22h7'52" 2d1'31"
Jack Louis Lester
Umi 12h57'54" 86d22'55"
Jack M. Datin
Cep 20h46'14" 61d20'8"
Jack M. Troop
Aql 19h58'16" -0d28'15"
Jack Madison Abernathy
Ari 3h5'46" 15d1'22"
Jack Maestas
Uma 9h47'33" 69d56'44"
Jack Magee
Per 4h9'47" 33d47'2"
Jack Mann
Crb 15h45'55" 29d30'25"
Jack Mann
Lib 14h56'22" -0d40'19"
Jack Manning
Her 18h8'12" 16d29'35"
Jack March Purdy
Uma 13h58'14" 51d18'35"
Jack & Marcia
Uma 9h4'24" 60d32'18"
Jack Mark Darlington
Cyg 19h37'28" 39d42'17"
Jack Martin Engelhardt
Uma 13h9'25" 57d23'1"
Jack Mason Mead
Uma 11h5'15" 44d56'4"
Jack Mathias
Lyn 7h0'34" 45d59'42"
Jack Matthew Cuddeback
Aql 19h41'6" 3d56'9"
Jack Matthew Dudhill
Ari 3h22'53" 19d13'41"
Jack Matthew James - Jack's Star
Per 3h34'15" 35d54'31"
Jack Matthew Sconziano
Cnc 8h58'19" 9d38'15"
Jack & Mavis Waldron
Dra 17h59'46" 66d37'47"
Jack McCafferty Morris
Umi 16h38'16" 81d22'26"
Jack McCarthy
Ori 6h17'6" 9d27'36"
Jack McCleary
Sco 16h8'25" -14d42'18"
Jack McCormick
Uma 10h23'39" 69d6'52"
Jack McCormick
And 23h36'17" 48d21'36"
Jack McDonald
Uma 10h15'38" 46d33'38"
Jack McGillowey
Uma 12h13'14" 52d56'27"
Jack McKenney
Cnc 8h22'3" 22d24'25"
Jack McKenney Arthur
Lyn 7h43'8" 42d14'15"
Jack McLaughlin
Cnc 8h29'12" 31d10'40"
Jack McLaughlin
Leo 10h14'30" 7d19'43"
Jack Mead
Umi 16h8'12" 78d19'10"
Jack Michael Fridrikson
Ari 3h18'58" 29d31'44"
Jack Michael Gutting
Ori 5h44'45" 3d23'47"
Jack Michael Kirages
Gem 7h22'11" 17d44'1"
Jack Michael Kirakosian
Vir 13h19'48" -22d27'28"
Jack Michael Mitchell
Gem 6h43'59" 28d30'26"
Jack Michael Riva
Lmi 10h17'5" 30d39'54"
Jack & Michelle Evans
Cyg 19h50'25" 33d15'17"
Jack Mills
Aql 19h29'47" 4d34'0"
Jack Mills Jr.
Ori 5h29'28" 3d24'15"
Jack Mitchell Hered
Uma 14h8'47" 62d12'24"
Jack Mitchell O' Connor
Uma 9h45'59" 43d43'50"

"Jack" Monroe
Ori 5h26'22" 6d15'24"
Jack Montague
Tau 4h45'31" 19d5'53"
Jack Moreland's Star
Her 17h25'15" 28d5'5"
Jack M.T. Haroutunian "01/02/2005"
Lyn 7h34'6" 39d25'28"
Jack Munro's Twinkler
Umi 16h39'4" 77d21'17"
Jack M.W. Phelps
Oph 16h24'54" -0d2'57"
Jack (My Eeyore)
Lib 15h16'32" -10d37'50"
Jack Myers
Her 18h56'9" 13d13'42"
Jack Myles Hric
Aur 6h18'35" 32d28'20"
Jack N Parker - Our Wise Three Men
Cru 12h4'10" -60d46'35"
Jack & Nancy Douglas
Cyg 20h35'38" 43d46'52"
Jack Nash
Uma 10h26'10" 63d38'33"
Jack Neuer & Jean Neuer
Vir 13h20'43" -12d56'16"
Jack Newbold
Cep 20h42'46" 63d54'28"
Jack Nicholas Pawlich
Cnc 8h24'47" 32d1'8"
Jack Noel
Vir 11h42'30" 9d53'34"
Jack Nolan Brooks III
Per 2h49'13" 57d6'48"
Jack O. Bakker
Uma 9h24'5" 72d31'15"
Jack Oliver Boss
Gem 6h21'7" 20d27'37"
Jack Oliver Burns
Ori 6h8'58" 7d13'59"
Jack P. Brickman
Her 17h36'38" 33d51'16"
Jack P. Facundus "Dr. Jack"
Her 17h40'36" 30d37'27"
Jack Padraig Quinlan
Lib 14h24'5" -18d5'11"
Jack "Papa" Ventling
Uma 11h33'19" 45d2'38"
Jack "Pappaw" Smith
Uma 9h9'14" 59d25'33"
Jack Parker
Per 2h43'20" 51d9'34"
Jack Partis
Uma 12h1'41" 65d26'14"
Jack Patrick Simmon
Aqr 22h40'29" -12d6'2"
Jack Paul James (LDJ)
Vir 12h9'48" 12d58'23"
Jack + Peg True Love 4-50
Cyg 20h49'30" 43d54'2"
Jack Perry - Jack's Star
Uma 11h56'31" 37d43'18"
Jack Peter Arthur Orgill
Umi 17h22'54" 86d17'14"
Jack Peter Chedzey
Per 4h10'39" 44d54'58"
Jack Peter Ray Rushman
Her 18h53'11" 21d55'35"
Jack Peter Tony Berry
Gem 6h53'3" 22d14'27"
Jack PH Chin
Aql 19h26'33" 10d59'28"
Jack Phelan
Cep 21h46'15" 71d7'48"
Jack Phillipps
Cap 21h32'51" -19d30'42"
Jack "Pop Pop" Green
Cma 6h41'26" -18d6'55"
Jack Porter
Aql 19h53'0" 1d55'59"
Jack Porter Stone
Uma 12h2'28" 50d41'6"
Jack Prentice
Ori 5h22'6" 3d46'59"
Jack Presley Riggs
Tau 4h49'23" 22d29'28"
Jack Preston Perry's Special Star
Gem 7h24'28" 25d35'4"
Jack Purton
Leo 9h25'45" 10d53'45"
Jack Quincy Warden
Aql 19h51'5" 13d42'17"
Jack R. Bescherer
Her 16h34'7" 35d8'30"
Jack R. Dalton
Lib 15h7'26" -19d13'58"
Jack R Jellison
Umi 15h29'48" 71d56'25"
Jack R Stratton
Cam 4h9'44" 68d53'38"
Jack Randal "Big Bear" Phillips
Uma 11h19'47" 49d53'34"
Jack Rau
Aur 5h34'57" 53d49'50"
Jack Rene Perkins
Lyn 6h28'26" 56d2'34"
Jack Reuben Kaplin
Cnc 8h43'2" 22d45'45"
Jack Rhead & Elizabeth Sue Lecheler
Cyg 19h36'11" 53d7'8"
Jack Richard George Pearce
Per 2h57'48" 45d29'46"
Jack Robert Daniel
Ori 5h23'18" 6d0'35"
Jack Robert Lenihan
Cep 21h47'58" 64d7'31"
Jack Robert Morrow
Umi 14h42'28" 75d37'10"
Jack Robert Nelson
Leo 10h40'48" 13d54'21"
Jack Robinson
Uma 13h7'56" 56d31'50"
Jack Rodgers
Vir 12h43'50" 1d47'3"
Jack Rodgers
Boo 15h30'51" 46d59'29"
Jack Roland
Sgr 18h14'3" -29d11'20"
Jack Roman Smith
Ori 5h38'27" 6d12'13"
Jack & Ronan Doyle
Uma 12h9'28" 62d24'43"
Jack Ross Toledo
Ari 2h51'16" 17d30'58"
Jack Rudolph Rains
Her 17h10'9" 32d50'11"
Jack Russell Gilcrease
Aql 19h43'23" -0d2'22"
Jack Ryan Berube
Cep 20h52'24" 68d2'14"
Jack Ryan Harrington
Uma 9h55'14" 42d21'9"
Jack Ryan Pesso
Ori 6h8'29" 13d55'26"
Jack Ryan Pozner
Psc 0h15'54" 4d16'24"
Jack Savant
Ori 5h44'20" 2d21'59"
Jack Schmittel
Cep 22h16'26" 70d46'42"
Jack Schmitz
Cep 0h3'46" 72d44'46"
Jack Shafer
Uma 9h59'26" 45d41'45"
Jack Shaw 5.17
Uma 10h24'45" 48d58'10"
Jack Sheffield Stalcup
Her 16h31'48" 26d6'48"
Jack Sher
Cru 12h28'45" -59d19'15"
Jack Shertzer
Cnc 8h40'20" 31d5'5"
Jack Shurr
Cap 20h11'36" -17d34'59"
Jack Sitts
And 23h18'36" 48d41'16"
Jack Solitro
Cru 12h4'55" -60d0'43"
Jack Spoleti
Uma 11h31'27" 63d9'42"
Jack Stanley
Cep 0h21'47" 77d39'43"
Jack Steiner
Her 17h21'54" 35d49'14"
Jack & Stella Forever
Sco 16h58'39" -38d25'24"
Jack Stephen Koets
Aql 19h6'52" -9d47'20"
Jack Sterling Klusmann
Per 3h42'52" 43d47'48"
Jack Stone
Tau 3h49'17" 21d22'11"
Jack Stuart Pike - Jack's Star
Umi 13h26'13" 75d2'25"
Jack Sullivan
Her 17h55'53" 23d20'57"
Jack Swafford
Aql 19h16'40" 0d9'8"
Jack T. Flaherty
Per 4h19'29" 51d9'18"
Jack Taylor Shapiro
Cap 21h1'13" -19d19'37"
Jack Tayseer Bahouth
Uma 8h58'5" 56d37'25"
Jack Terrence O'Keeffe
Cra 18h0'2" -37d28'41"
Jack The Brat
Leo 9h29'48" 28d37'22"
Jack Thomas Borowiec
Ari 3h12'8" 12d8'20"
Jack Thomas Brockhagen
Psc 1h4'25" 29d23'28"
Jack Thomas Foley
Per 3h34'31" 45d48'11"
Jack Thomas Jayaraman
Ori 6h11'47" 6d4'44"
Jack Thomas Kelly
Per 2h19'1" 51d40'16"
Jack Thomas Liley Son Brother- Frog
Cru 12h18'39" -56d35'57"
Jack Thomas Oehlschlager
Cap 21h25'55" -15d21'23"
Jack Thomas O'Hara
Ari 2h43'23" 30d31'5"
Jack Thomas Rankin
Uma 12h9'12" 47d51'23"
Jack Thomas Richmond - Jack Thomas' Star
Cru 12h16'32" -63d55'21"
Jack Thomas Scambler - Jack's Star
Cep 21h50'38" 58d50'7"
Jack Thomas Schwanke
Umi 16h38'0" 81d53'43"
Jack Thurlow
Ari 1h57'24" 21d7'51"
Jack Tilghman Wojciechowski
Cnc 8h14'26" 12d55'24"
Jack Timothy Foster
Uma 9h7'51" 63d37'12"
Jack Todd
Her 17h27'35" 18d28'50"
Jack & Tricia Beeler
Crb 16h15'21" 31d3'12"
Jack Tucker Clark
Her 18h43'25" 24d43'35"
Jack Tyler
Uma 11h33'40" 39d59'36"
Jack Tyler Lambert
Leo 11h26'14" 9d19'42"
Jack Tyler Thoms
Her 16h36'52" 48d9'41"
Jack V. Adams
Aqr 23h10'45" -7d35'9"
Jack V. Ferguson
Vir 13h25'49" -4d8'32"
Jack Vail Fedor
Psc 1h8'44" 23d34'38"
Jack Varon
Leo 10h4'47" 24d16'37"
Jack Verland Mattila
Lyn 6h20'40" 54d50'30"
Jack Virgo Watching Over You
Per 3h48'14" 34d56'22"
Jack Vuillemot
Ari 2h7'15" 12d50'22"
Jack W. McNerney
Tau 4h24'52" 18d22'0"
Jack W Sanders
Cen 13h38'51" -38d3'42"
Jack W. Sidwell
Uma 11h29'45" 46d8'26"
Jack Waskowsky
Uma 12h0'18" 37d32'49"
Jack Wayne Russell, Sr.
Ori 5h38'16" 1d58'10"
Jack Wayne Shaffner
Uma 8h13'27" 59d46'0"
Jack Weinstein
Cma 6h49'59" -28d26'42"
Jack Whitaker
Per 4h21'26" 50d27'40"
Jack William Cawte
Per 2h57'58" 45d28'42"
Jack William Davies
Vir 14h48'54" 6d7'14"
Jack William Dunkerley
Cyg 19h52'9" 32d41'8"
Jack William Goodwin
Ari 2h54'23" 12d38'10"
Jack William Goodwin
Tau 5h35'41" 22d54'0"
Jack William Higham
Lmi 10h19'2" 31d32'44"
Jack William Joyce
Uma 10h49'35" 40d36'10"
Jack William Marshall Leitch
Peg 21h37'13" 18d52'1"
Jack William Morton
Ori 4h53'35" 2d16'15"
Jack William Provan - Jack's Sparkle
Ori 6h14'7" -4d1'52"
Jack William Schroeder
Aqr 21h43'38" -7d59'19"
Jack William Scott
Crb 15h52'54" 32d9'47"
Jack William Seibert
Tau 5h56'8" 24d51'10"
Jack William Spittal
Dra 17h11'33" 55d53'5"
Jack William's Star
Per 4h13'42" 35d49'58"
Jack Willoughby
Uma 11h51'26" 33d13'47"
Jack Willoughby
Ari 3h20'2" 28d34'57"
Jack Wilson
Per 3h24'53" 44d0'34"
Jack Wilson
Per 4h33'25" 49d40'18"
Jack Winter
Ori 6h12'30" 15d9'1"
Jack Woldert
Boo 14h37'18" 29d3'59"
Jack Womack
Uma 10h39'8" 48d20'30"
Jack Yates
Aql 19h12'52" 6d0'43"
Jack Zanette
Sgr 19h43'20" -24d15'46"
Jack, Daisy & Michael Rogan
Dra 17h45'42" 62d54'48"
Jack, Melinda, Madison, & J.T. Fox
Cap 20h43'6" -20d38'38"
Jack, Snow Owl, Crans
Uma 11h37'46" 63d39'27"
Jack, Tesoro mio
Leo 11h11'3" 13d11'39"
Jack08031924
Leo 10h41'20" 6d58'55"
Jack-Al
Crb 16h9'13" 35d59'55"
JaCKal 1982
Aur 7h14'6" 42d44'57"
Jackal 94
Oph 17h6'0" -0d31'43"
Jackalopus Majoris
Aqr 22h51'23" -19d53'56"
Jackamore
Lib 14h34'56" -18d57'3"
JackAsh
Vir 13h42'31" 2d2'53"
Jackay Hickman
Aqr 22h23'49" -3d8'35"
JackD
Uma 8h32'57" 68d5'28"
JACKDUFFY
Sgr 18h36'2" -23d56'49"
Jackee Miller
Ori 5h53'54" 18d22'35"
Jackee Nowicke
Uma 9h27'33" 66d24'9"
Jäckel, Jesko
Ori 6h8'46" 16d11'17"
Jackeline Bulnes
And 2h18'43" 50d41'35"
Jackeline Ma
Sco 16h14'50" -20d54'39"
Jackeline Rodriguez Reyes
Lib 15h23'30" -26d17'18"
Jackeline's Jewel
Ari 3h17'40" 22d21'40"
Jackeline's Star
Sgr 18h11'15" -18d40'53"
Jackelyn Marie
Psc 1h17'38" 16d28'6"
Jacki
Aqr 21h51'37" -1d10'39"
Jacki and Scott Hester
Uma 13h18'9" 61d3'43"
Jacki "Angala" Knutson
Umi 16h18'17" 76d24'3"
Jacki Jobe
Cyg 20h46'44" 30d46'33"
Jacki Wildman Wales
Ori 6h2'20" 13d52'31"
Jacki Zellers
Aqr 21h54'35" -1d29'55"
Jackie
Aqr 22h56'32" -6d30'41"
Jackie
Cap 21h34'32" -21d6'57"
Jackie
Cap 20h51'25" -18d55'45"
Jackie
Aqr 22h0'48" -11d27'3"
Jackie
Sgr 19h11'27" -13d23'57"
Jackie
Sgr 19h3'47" -14d20'12"
Jackie
Uma 11h23'3" 62d36'43"
Jackie
Lyn 6h36'45" 61d20'11"
Jackie
Cyg 19h54'33" 57d4'17"
Jackie
Cma 6h33'8" -28d21'0"
JACKIE
Sgr 19h22'8" -26d49'58"
Jackie
Sgr 17h55'32" -29d28'41"
Jackie
Sgr 18h9'49" -31d19'15"
Jackie
Cnc 9h7'4" 12d40'8"
Jackie
Vir 12h8'3" 7d40'13"
Jackie
Vir 13h11'19" 12d10'37"
Jackie
Ari 2h3'45" 14d42'34"
Jackie
Ari 2h22'2" 11d49'13"
Jackie
Leo 9h30'42" 16d48'44"
Jackie
Del 20h36'47" 16d2'39"
Jackie
Psc 1h34'8" 26d55'0"
Jackie
Psc 1h0'21" 26d59'26"
Jackie
Ari 3h8'22" 26d3'22"
Jackie
Ari 3h20'51" 29d15'39"
Jackie
Tau 5h17'29" 26d37'12"
Jackie
And 0h54'34" 38d14'11"
Jackie
Uma 9h33'24" 42d14'15"
Jackie
Lyr 18h39'52" 36d56'16"
Jackie
And 0h31'29" 36d59'58"
Jackie
Cnc 8h39'40" 32d34'25"
Jackie
Cnc 8h38'46" 31d14'2"
Jackie
Leo 9h36'55" 31d28'21"
Jackie
Leo 9h53'45" 30d42'55"
Jackie
Lyn 8h27'32" 37d0'1"
Jackie
And 0h45'56" 45d41'58"
Jackie
Lyr 18h43'9" 39d40'10"
Jackie
Uma 13h56'22" 52d4'5"
Jackie 3
Leo 11h41'59" 21d42'27"
Jackie A. Sherman
And 0h37'16" 31d44'8"
Jackie and Aaron
Cyg 20h45'48" 37d25'3"
Jackie and Andrew
Cap 21h9'24" -17d18'32"
Jackie and Brian's Star of Promise
Ori 6h5'32" 10d10'13"
Jackie and Clyde for Eternity
Vir 14h5'33" 6d40'37"
Jackie and Desirae
Cae 4h39'37" -37d5'6"
Jackie and Jared's Star (10/12/06)
Cyg 19h42'20" 47d55'22"
Jackie and Nick
Umi 16h18'18" 76d42'32"
Jackie and Tom Parker
And 23h13'4" 48d38'9"
Jackie Anderson
Cap 21h57'13" -24d36'12"
Jackie Angell-Clouser
Cap 21h35'14" -17d42'45"
Jackie Baby
Mon 7h13'59" -0d45'16"
Jackie Belcik Alger
Cnc 8h59'54" 24d48'25"
Jackie "Beluga Bear" Walley
Aqr 23h1'59" -13d30'13"
Jackie Borcherding
Cam 7h48'50" 64d1'47"
Jackie Boy
Sgr 18h30'11" -16d24'10"
Jackie Brighton
Cra 18h55'41" -40d32'58"
Jackie Brown
Uma 11h30'15" 58d12'35"
Jackie Brown
Ari 2h47'11" 25d52'9"
Jackie Catz
Cyg 20h2'12" 59d37'50"
Jackie (Chicklet)
And 1h15'17" 38d46'5"
Jackie & Cliff Lee
Vir 13h34'52" 2d15'46"
Jackie Composto
Cas 0h1'46" 57d15'6"
Jackie Cowper
Gem 6h58'43" 14d50'5"
Jackie Coxhead
Lyn 8h26'35" 42d16'20"
Jackie Daddona
Tau 4h48'22" 22d58'30"
Jackie Datino
Gem 6h54'5" 22d20'46"
Jackie Dawkins - "Jackie's Star"
Cas 0h55'34" 74d44'40"
Jackie Denise Stewart
Cmi 8h5'27" -0d7'21"
Jackie Denton
And 0h38'13" 41d53'42"
Jackie & Dougie Walker
Cyg 21h10'55" 46d29'20"
Jackie Doyle
Tau 5h47'23" 27d52'20"
Jackie Ducci
Sco 16h42'5" -30d58'2"
Jackie E. Robinson
Boo 15h38'0" 47d34'15"
Jackie East
Cas 1h57'0" 64d11'7"
Jackie Ernestine Cole Wright
Psc 1h47'47" 5d50'6"
Jackie Eschbaugh
And 0h15'0" 38d58'16"
Jackie Estep
And 1h42'58" 42d12'58"
Jackie Faye Perry
Psc 23h43'53" 4d10'18"
Jackie Fellows
And 0h54'40" 41d21'29"
Jackie Forbes-Olson
Sco 17h41'1" -44d51'34"
Jackie Fraser
Sgr 18h58'4" -27d54'50"
Jackie Fryer
Lyr 19h8'23" 26d32'13"
Jackie G Light Bearer
Cas 23h19'34" 58d36'41"
Jackie Garvin
Cyg 19h51'53" 31d5'12"
Jackie George
Aqr 21h51'47" 2d5'19"
Jackie Giuffrida Star
Lyn 6h57'33" 59d28'47"
Jackie Gomes Darosa
Sco 16h14'1" -12d20'42"
Jackie Gonzales
Leo 11h12'11" 2d43'49"
Jackie Graham, Educator
Uma 12h31'59" 58d52'31"
Jackie Grippo
Gem 6h1'21" 24d44'26"
Jackie H Stamos
Aql 19h2'46" 1d30'27"
Jackie Harrison
Umi 14h14'56" 69d41'38"
Jackie Hemsley
Per 4h14'55" 46d44'47"
Jackie Hutton
Cas 0h3'11" 59d46'55"
Jackie Iles
And 23h25'31" 35d51'35"
Jackie & James Country Star
Cyg 21h27'38" 46d30'57"
Jackie Jennings
Uma 10h59'36" 63d10'34"
Jackie & Jim Sechrist
Cnc 8h44'5" 19d18'50"
Jackie Jo Piggush
Sco 17h51'49" -35d50'3"
Jackie & Joey Best Friends Forever
Uma 8h35'41" 64d0'25"
Jackie & John xox Mason
Cru 12h17'7" -57d44'39"
Jackie Jones
Lyn 7h39'12" 57d52'59"
Jackie Keiper
Aqr 22h20'25" 1d39'28"
Jackie Lee Burkett
Lib 15h6'51" -16d51'53"
Jackie Leigh Briggs
Mon 7h17'28" 0d53'11"
Jackie Leo Fantabulous
Leo 10h52'47" -1d10'47"
Jackie Levine
Cnc 8h34'57" 32d33'46"
Jackie Louise
Uma 11h58'46" 64d29'29"
Jackie Lumenosa
Crb 15h46'17" 33d11'17"
Jackie Lupo
Cnc 8h46'45" 27d33'40"
Jackie Maas-Hull
Lib 14h51'35" -0d42'36"
Jackie MacAskill
Cas 1h35'35" 61d8'12"
Jackie MacPherson
Tau 4h27'51" 10d20'45"
Jackie Mae Murphy
And 23h18'13" 47d17'12"
Jackie Mah
Cas 23h42'53" 57d33'7"
Jackie Maki
Lyr 19h2'34" 33d39'6"
Jackie Marsh
Cas 23h37'30" 55d28'23"
Jackie Martin
Psc 1h13'50" 24d50'33"
Jackie Mitchell
Uma 10h35'20" 62d54'32"
Jackie (Mum) and Les Clarke
Cas 1h20'10" 57d19'7"
Jackie Murdock
Del 20h25'5" 9d40'58"
Jackie Murray
Umi 4h10'10" 88d49'13"
Jackie "My Little Guppy" Perez
Psc 0h57'12" 19d54'48"
Jackie (My Love)
Aqr 23h38'0" -22d53'6"
Jackie N Porter
Cas 2h11'43" 69d50'30"
Jackie Oliver's Love
Vir 12h56'17" 3d55'12"
Jackie Overmeyer
Cas 1h13'54" 62d48'37"
Jackie Perris Porcho
Crb 16h9'11" 34d40'12"
Jackie Perris-Porcho
Peg 23h29'50" 23d23'17"
Jackie Perrotta
Cas 0h33'31" 62d42'17"
Jackie Phillips Star
Ari 2h15'26" 27d24'57"
Jackie Powell God's Angel #9
Cap 20h40'21" -17d18'23"
Jackie Rae
Aqr 21h33'11" 1d45'18"
Jackie Rae
Uma 11h0'8" 51d42'17"
Jackie Rae Hull Thomas
Vir 13h7'15" -15d8'39"
Jackie & Rene
Cyg 21h14'35" 43d18'13"
Jackie Roberts
Aqr 21h0'1" 1d15'47"
Jackie Robinson
Cas 23h56'32" 56d32'59"
Jackie Rose
Her 16h49'39" 35d18'4"
Jackie Rose - My Brightest Star
Cru 12h29'35" -58d47'47"
JACKIE ROTH
Uma 10h31'9" 48d15'31"
Jackie S
And 23h54'43" 37d55'51"
Jackie Slater-Biwer
Lyn 8h5'10" 38d16'44"
Jackie Smitrovich
Uma 11h0'47" 48d20'59"
Jackie Sortino
Cas 0h27'18" 51d13'16"
Jackie Spairana
Cap 20h32'33" -13d36'32"
Jackie Sue Chandler
Gem 6h40'9" 23d29'16"
Jackie Susan Thompson
Ari 2h8'1" 23d17'29"
Jackie 'Teen Angel'
Mon 6h50'14" -0d33'9"
Jackie "Thaumastos"
Uma 8h37'43" 59d57'57"
Jackie Thompson
Cas 0h52'34" 52d37'36"
Jackie Tractenberg
Cyg 19h38'15" 28d23'29"
Jackie Traylor
Lmi 10h17'28" 36d43'20"
Jackie Vinkler
Gem 6h45'22" 23d48'14"
Jackie Vu, my One Love
Lib 15h4'9" -7d58'42"
Jackie Wahl Love Always Leighland
Mon 7h3'29" 4d27'33"
"Jackie" William Peryam
Uma 11h59'30" 59d27'48"
Jackie Withers Smith
Leo 10h34'9" 9d49'46"
Jackie Woodville Thompson
Uma 10h59'2" 45d8'6"
Jackie Zaldaña
Lyr 18h42'7" 37d18'29"
Jackie, my favorite cuddle buddy
Crb 16h3'31" 26d4'2"
Jackie080804
And 2h24'21" 45d7'45"
Jackie11
Uma 11h1'27" 41d25'25"
Jackiebo and Ruthie
Cyg 19h56'40" 41d51'45"
jackiedennis(boobear)
Psc 1h12'17" 10d2'46"
JackieK1998
Leo 11h13'10" 15d53'56"
Jackieland
Ari 2h51'11" 28d29'55"
Jackielyn L.
Psc 1h7'48" 27d56'10"
JackieO.524
Aqr 23h19'9" -18d23'3"
Jackie's Aura III
Vir 12h27'45" 2d15'8"
Jackie's Game
Lyr 18h54'40" 43d17'3"
Jackie's Hidden Village
Lib 15h38'25" -18d2'14"
Jackie's LilStarr
And 2h5'4" 39d38'22"
Jackie's Lucky Star
Cnc 8h52'33" 6d50'44"
Jackie's Mom
Sco 16h41'44" -28d49'3"
Jackie's Star
Srp 18h26'7" -0d33'49"
Jackie's Star
Ori 5h26'39" 13d27'58"
Jackie's Star
Peg 23h46'49" 27d49'30"
Jackie's Star
Lmi 10h39'26" 38d53'59"
Jackie's Star
Gem 7h26'29" 35d13'10"
Jackie's Star
And 23h17'34" 51d55'15"
Jackies Ultimate Vacation
Lyr 18h50'32" 33d35'0"
Jackii 1-24-1922
Tau 3h52'11" 23d27'19"
Jackineum Katelyneus
Crb 15h48'16" 26d12'23"
JacKit
Uma 9h28'29" 46d27'28"
Jack-Jouk
Cru 12h25'44" -57d52'55"
Jacklen Mary Smith
Cas 1h7'34" 61d48'26"
Jackline & Vincent Sanfiorenzo
Ori 6h15'13" 15d14'17"
Jacklyn A. Klubek
Aqr 21h31'40" 2d27'5"
Jacklyn Colvard
Cas 0h36'57" 63d58'0"
Jacklyn Danielle Hedrick
Tau 3h50'0" 20d36'8"
Jacklyn Elena Farinas
Del 20h38'51" 16d28'23"
Jacklyn & Griff Blatz
Cyg 19h42'36" 34d4'7"
Jacklyn Kay
Aqr 21h53'58" -7d10'19"
Jacklyn Kim Young
Psc 1h22'56" 16d54'27"
Jacklyn Palmer
Psc 1h21'5" 33d10'12"

Jacklyn R. Hammer
Lyr 18h43'48" 30d15'32"
Jacklyn Sue Stairs
And 1h5'46" 34d26'40"
Jacklyne Laverghetta
Ari 2h36'25" 14d52'16"
Jacklyn's Star
Lib 15h31'6" -19d27'2"
JackO Hudson
Per 3h23'1" 50d56'16"
Jackpot Philip
Leo 9h58'8" 16d34'17"
Jackqueline
Psc 1h19'30" 16d50'50"
Jackqueline
Cas 1h57'32" 61d57'23"
Jackqueline Ann
Cam 3h58'6" 67d50'21"
JackRTyler
Lyn 6h54'9" 54d3'18"
Jack's 12-Pointed Star
Uma 10h5'44" 70d20'14"
Jack's 25th Anniversary Star
Her 17h33'15" 14d17'27"
Jack's Celestial Kenai Dream
Sgr 19h46'38" -12d35'11"
Jack's Christening Star
Umi 15h51'22" 79d15'7"
Jack's Cupid
Sco 16h10'11" -17d3'8"
Jack's Daddy - Adam Robert Smith
Gem 7h28'40" 26d28'2"
Jack's Dot
Per 3h5'5" 51d52'58"
Jacks Funes, The Brightest Love
Tau 5h44'53" 19d0'20"
Jack's Lil Angel
Pho 1h51'34" -44d21'25"
Jack's Matthew 6:19
Uma 11h23'41" 54d25'39"
Jack's Quince
Per 3h21'58" 42d22'47"
Jack's Soaring Star
Cep 22h33'21" 62d43'50"
Jack's Star
Uma 8h26'36" 72d38'29"
Jack's Star
Dra 16h42'52" 66d17'41"
Jack's Star
Per 2h52'52" 38d34'8"
Jack's Star
Per 4h43'45" 41d15'59"
Jack's Star
Uma 10h45'30" 51d15'8"
Jack's Star
Psc 0h58'51" 13d53'13"
Jack's Tally-Ho
Boo 14h40'36" 38d1'20"
Jack's Wishing Star
Per 3h11'18" 53d9'20"
Jackson
Sge 20h13'50" 18d9'35"
Jackson
Uma 12h44'5" 55d44'1"
Jackson
Cap 21h10'15" -18d19'11"
Jackson
Sco 17h53'23" -36d40'8"
Jackson Allard Li Jones
Leo 11h53'1" 19d48'6"
Jackson and Van
Per 3h18'0" 48d5'12"
Jackson Andrew Lalos
Ari 2h19'33" 11d7'15"
Jackson Andrew Milinovich
Uma 10h39'17" 39d40'1"
Jackson Anthony Williams
Lib 15h23'22" -16d14'6"
Jackson Asteri
Per 3h10'27" 54d33'42"
Jackson Benjamin Barry
Per 4h34'29" 41d19'33"
Jackson Bradley Sams
Her 17h36'47" 33d13'52"
Jackson Brian Gallagher
Psc 22h52'38" 5d23'34"
Jackson Bunker Hobbs
Leo 11h48'28" 26d15'2"
Jackson C. Pearce
Cnc 8h43'39" 22d16'10"
Jackson Carter - Our Star Is Born
Cru 12h28'19" -59d56'3"
Jackson Chad Rosvall
Leo 11h48'34" 26d6'12"
Jackson Clancy Scruggs Grandson
Sco 17h53'39" -35d52'4"
Jackson Cole Adler
Cnc 9h9'26" 30d22'1"
Jackson Connor Peddicord
Cep 22h55'41" 70d49'21"
Jackson Conroy
Cru 12h5'53" -61d39'30"
Jackson Cook
Uma 9h7'5" 65d38'43"
Jackson Cooper Rowe
Per 3h17'45" 54d15'28"
Jackson Cooper Tarpley
Uma 8h51'49" 46d34'1"
Jackson Crawford Sirmon
Ori 5h27'41" 0d27'2"
Jackson D. Wiegand
Uma 12h3'50" 39d58'6"
Jackson Dance
Cam 6h37'25" 65d54'32"
Jackson Dance Center - 7/25/03
Cyg 20h9'43" 33d27'20"
Jackson Daniel
Ori 5h9'44" 1d47'34"
Jackson Dean Savich
Aur 5h46'21" 32d23'21"
Jackson Douglas Shedd
Aqr 22h5'52" -1d46'46"
Jackson Eric Court
Cru 12h27'22" -55d54'17"
Jackson Family Memorial Star
Aql 19h34'32" 5d46'26"
Jackson Fisher Gallimore
Gem 6h37'2" 16d47'1"
Jackson Foster
Sgr 18h35'45" -28d57'48"
Jackson Gray
Aur 5h48'7" 45d1'54"
Jackson Gray Kourouklis
Leo 10h1'40" 22d23'14"
Jackson Grey
Gem 7h49'43" 32d20'33"
Jackson Guy Kimble
Lyn 7h40'3" 57d45'17"
Jackson Hall Williams
Her 17h40'10" 16d7'24"
Jackson "Hambone" Hunt
Lyn 9h0'2" 38d25'16"
Jackson Henry
Lyn 7h47'42" 49d4'22"
Jackson Henry Bello
Ari 3h12'7" 27d30'56"
Jackson Henry Born
Sgr 19h11'42" -18d27'47"
Jackson II
Ori 5h38'48" 9d5'59"
Jackson James Wolintos
Umi 16h20'0" 77d23'40"
Jackson Kai Vieira
Cyg 21h9'44" 47d39'57"
Jackson Keith Hundelt
Ori 6h18'37" 19d7'4"
Jackson Kent Saller
Tau 3h30'22" 8d11'22"
Jackson Kole Benjamin
Mon 7h0'47" -0d51'39"
Jackson Lane
Aql 19h11'58" -7d49'46"
Jackson Lavery Gilbert
Umi 14h26'25" 75d35'8"
Jackson Lee Graves
Umi 15h41'45" 72d19'54"
Jackson Matthew Trumbull
Tau 5h53'13" 23d58'14"
Jackson Meyer
Aur 5h52'22" 52d7'52"
Jackson Mitchell Atterton
Lib 15h33'28" -12d13'35"
Jackson Moffatt
Her 18h47'35" 16d2'7"
Jackson Monroe Aven
Tau 4h34'56" 16d56'54"
Jackson Montgomery Bloomer
Vir 13h25'49" -0d11'43"
Jackson Murphy
Ori 6h20'19" 9d19'0"
Jackson Noah Bloom
Cnc 8h41'56" 29d27'25"
Jackson Noah Caron
Aql 19h19'0" 15d7'21"
Jackson Patrick Allegra
Leo 10h10'15" 21d8'22"
Jackson Paul
Her 17h21'58" 36d5'29"
Jackson Paul Garlitz
Dra 17h37'36" 58d51'2"
Jackson Peter
Ari 2h45'43" 14d15'0"
Jackson Philip Widdecombe
Sgr 19h25'55" -15d1'31"
Jackson Ray Scheradella
Tau 4h48'24" 19d22'38"
Jackson Rice
Aql 18h50'47" 8d50'58"
Jackson Riley Edward-Todd
Cap 20h19'6" -16d32'48"
Jackson Riley Mentzer
Ori 6h3'14" 9d59'44"
Jackson Robert
Aur 5h33'10" 53d27'51"
Jackson Robert Smithers
Aqr 22h37'9" -2d2'18"
Jackson Roy Chicklis
Aql 19h16'19" 3d16'54"
Jackson Rudolph Lechelt
Uma 11h31'26" 33d4'3"
Jackson Russell Hill
Uma 10h36'57" 59d43'11"
Jackson Ryan Bennett
Cnc 9h19'42" 28d39'27"
Jackson Ryan Shea
Per 3h33'2" 48d15'19"
Jackson Scott Charles Cline
Aqr 22h10'39" -4d30'47"
Jackson Seymour Hardiman
Uma 10h44'30" 72d1'8"
Jackson Stephen Dobbins
Cyg 19h39'27" 33d29'40"
Jackson Story
Cmi 7h8'56" 4d18'11"
Jackson T. Godwin (UB 5)
Ori 5h49'26" 8d14'44"
Jackson Theodore Cooke
Boo 14h45'56" 38d35'16"
Jackson Thomas
Ori 4h50'46" 1d33'57"
Jackson Thomas Boehm
Lib 15h35'43" -9d36'51"
Jackson Thomas Marchand
Aur 5h59'37" 42d14'51"
Jackson Thomas Marston
Leo 11h28'52" 9d3'9"
Jackson Tyler Richards
Gem 7h17'13" 15d46'11"
Jackson Tyler Tucker
Uma 8h56'25" 67d59'21"
Jackson Tyler Weir
Ari 2h40'59" 20d40'8"
Jackson Valentine Sellitto
Aqr 23h0'3" -5d19'56"
Jackson Vincent McCormick
Sgr 18h24'58" -22d55'13"
Jackson Wallace Lemon
Her 17h48'14" 22d26'0"
Jackson William Schroeder
Sco 17h56'43" -42d38'35"
Jackson Young Olney
Aql 19h29'42" 10d38'42"
Jackson-Guy
Tau 5h7'29" 22d39'22"
Jacksonian Hope
Lyn 8h42'53" 34d4'43"
Jackson's Layer
Lyr 19h1'5" 34d0'17"
Jackster Bear
Uma 11h21'7" 31d11'11"
Jackus Lanoius B.
Leo 11h4'45" -0d55'32"
Jacky
Aql 20h24'19" -8d0'48"
Jacky
And 0h44'22" 41d54'0"
Jacky
Gem 6h21'16" 18d38'20"
Jacky and Joe Travis Anniversary
Cyg 20h27'32" 48d2'5"
Jacky Cheung Hok Yau
Cnc 8h19'40" 10d6'2"
Jacky Giordano
Uma 10h51'48" 39d50'36"
Jacky Giordano
Uma 11h8'56" 61d34'10"
Jacky Jr., Sydney, Ashley Crincoli
And 0h35'4" 39d36'15"
Jacky Lai Jiun Loong
Aqr 22h44'1" -0d45'30"
Jacky P. Lam
Umi 15h21'18" 67d53'56"
Jacky R. Smith
Cam 7h47'3" 65d42'25"
Jacky Rousselet
Umi 14h9'18" 73d7'34"
Jackybean - Jack Noell Burket
Psc 1h36'20" 17d49'30"
Jacky's Place
Tau 4h18'53" 17d40'48"
Jaclene Hodges
Lyn 8h23'43" 54d28'5"
Jaclin Ruth Collins
Cyg 20h4'1" 33d57'53"
Jaclyn
Her 17h37'31" 48d4'59"
Jaclyn
Ari 2h5'26" 19d45'37"
Jaclyn
Ari 2h20'54" 19d36'29"
Jaclyn
Cas 23h45'27" 56d26'20"
Jaclyn
Vir 14h15'24" -15d54'13"
Jaclyn A. Andersen
Vir 11h40'43" 3d22'13"
Jaclyn Albright
Cnc 8h52'1" 13d39'54"
Jaclyn Allsup
Lyn 8h3'59" 51d20'24"
Jaclyn Amber Churilla
And 23h1'9" 48d4'19"
Jaclyn Ann Jacobs
Cyg 21h48'6" 43d50'43"
Jaclyn Anne Calarie
And 0h41'20" 44d11'20"
Jaclyn Arcy and Bradford Bailey
Gem 7h20'49" 17d10'14"
Jaclyn Christopher
Uma 9h26'1" 64d49'13"
Jaclyn & Christopher Winkler
Sgr 17h59'50" -29d59'14"
Jaclyn Connors
Vir 12h8'5" 11d26'44"
Jaclyn Elizabeth Berlin
Sco 17h8'2" -40d24'11"
Jaclyn Fay
Cyg 21h24'23" 33d0'0"
Jaclyn Gale
And 1h40'28" 42d47'28"
Jaclyn Griffin
Cnc 8h54'7" 27d9'12"
Jaclyn Hamelka
Aqr 22h17'18" -2d48'49"
Jaclyn Ittel
Lyn 7h32'12" 39d31'56"
Jaclyn Joy
And 0h21'40" 39d9'45"
Jaclyn K. Malleck
Cas 0h47'26" 61d30'18"
Jaclyn Lea Hunt
Uma 11h51'23" 45d57'52"
Jaclyn Lee Sorlie
Tau 4h51'23" 26d35'28"
Jaclyn Lorraine Albanese
Lyn 8h58'48" 39d34'56"
Jaclyn M. Zito
Per 3h17'2" 49d4'2"
Jaclyn Marie
Gem 7h48'50" 31d28'55"
Jaclyn Marie Fisher
Sgr 17h54'5" -17d30'53"
Jaclyn Marie Klaffka
Vir 13h16'52" 12d32'54"
Jaclyn Marie Locktosh-Kawaii-Luna
Cap 21h35'30" -9d10'6"
Jaclyn Marie Micaela Lucidonio
Tau 4h27'9" 8d58'34"
Jaclyn Michetti
Cas 1h36'50" 65d28'10"
Jaclyn Monique Turner
Sco 17h23'42" -35d34'19"
Jaclyn Morgan Bonilla
And 0h25'9" 43d2'43"
Jaclyn Mowrer - Beautiful Angel
Psc 1h15'43" 15d59'41"
Jaclyn Nicole Mesa
Ari 2h20'34" 12d40'58"
Jaclyn Ochs
Cap 21h35'27" -19d46'34"
Jaclyn Paige Valentas
And 2h21'8" 44d42'50"
Jaclyn Rae Sperduto
Umi 15h7'56" 70d16'24"
Jaclyn Ray Whitman
Ori 6h4'15" 21d0'9"
Jaclyn Reneé Kucera
Sco 16h17'17" -9d11'39"
Jaclyn Reyes Barrera
Ori 5h20'27" 7d1'9"
Jaclyn Samantha
Per 3h24'55" 50d6'6"
Jaclyn Sara
Cap 21h58'33" -21d35'36"
Jaclyn Shaheen
Cas 0h23'42" 54d23'21"
Jaclyn Snow
And 0h42'28" 34d58'14"
Jaclyn Stewart
And 0h14'43" 28d28'52"
Jaclyn Suzanne Browning
Ori 5h7'31" 7d54'2"
Jaclyn Templin
And 0h39'26" 31d22'22"
Jaclyn Thatcher
Vir 13h19'10" -0d15'53"
Jaclyn Veillette
Lib 16h1'23" -7d24'35"
Jaclyn Wilson
Cap 20h53'28" -24d31'20"
Jaclyn, Mirabile Visu
And 0h56'28" 38d28'2"
Jaclyne & Richard Frazier 8/13/05
Ori 5h36'34" -0d28'17"
Jaclynn
Psc 0h18'39" 8d41'2"
Jaclynn Marie Gerdts
Gem 6h7'38" 25d39'15"
Jaclyns Faith
Cyg 21h27'54" 30d55'50"
Jaclyn's Hope
And 23h1'0" 40d45'13"
Jacob
Per 3h18'34" 52d20'44"
Jacob
Uma 9h34'21" 50d26'45"
Jacob
Ari 2h17'28" 25d14'44"
Jacob
Her 18h23'10" 18d34'6"
Jacob
Ori 5h8'34" 5d53'12"
Jacob A. Hinkle
Her 16h48'36" 38d31'4"
Jacob Aaron Corr
Boo 14h49'5" 53d34'24"
Jacob Aaron Glorioso
Psc 1h33'25" 15d25'29"
Jacob Aaron Leach
And 0h15'10" 42d23'53"
Jacob Ace Gallman
Ari 2h32'3" 27d28'7"
Jacob Adam Israel
Dra 19h20'45" 75d27'35"
Jacob Alan Bashawaty
Uma 10h1'45" 68d0'52"
Jacob Alan Schlaud
Psc 23h24'33" 4d59'26"
Jacob Alan Stocks
Umi 15h28'27" 67d37'50"
Jacob Albert Gilchrist
Uma 12h29'17" 53d18'19"
Jacob Alexander Arroyo
Per 3h11'0" 44d38'18"
Jacob Alexander Berner
Boo 14h16'50" 45d39'18"
Jacob Alexander Carswell
Gem 7h41'13" 25d10'50"
Jacob Alexander Condiff
Her 15h51'35" 44d50'30"
Jacob Alexander Crenshaw
Cru 12h48'58" -56d19'19"
Jacob Alexander Crissman
Aql 19h46'50" 8d25'41"
Jacob Alexander Leibowitz
Cap 21h38'53" -11d5'42"
Jacob Allan
Boo 14h26'36" 18d34'6"
Jacob Allen
Her 16h20'59" 43d32'12"
Jacob Allen 040798
Per 3h45'54" 49d49'41"
Jacob Allen Heinitz
Aqr 23h29'7" -16d5'4"
Jacob Allen Inman
Gem 7h28'56" 17d13'24"
JACOB ALLEN JAIMES
Sgr 17h56'9" -27d34'32"
Jacob Allen Rinard
Vir 12h9'16" -4d44'19"
Jacob and Alyssa
Cyg 20h25'34" 47d53'54"
Jacob and Ivy Forever
Sge 19h33'44" 16d51'45"
Jacob and Jonathan Lord
Cas 1h51'36" 60d49'42"
Jacob and Kelly Brown
Cyg 21h57'27" 48d17'50"
Jacob and Mary Jones
Cyg 21h44'5" 36d11'37"
Jacob Andrew Carlson
Lyn 7h46'5" 41d59'21"
Jacob Andrew Nelson
Uma 9h1'42" 57d0'5"
Jacob Andrew Weaver
Her 16h43'34" 15d25'51"
Jacob Angel Boy Willman
Tau 5h45'39" 22d28'18"
Jacob Anthony Casilio
Her 17h0'37" 37d57'13"
Jacob Anthony Kempster
Uma 10h10'19" 44d11'8"
Jacob Archie Staats
Ari 2h53'56" 26d36'33"
Jacob Armstrong Our Love Is Eternal
Psc 1h10'56" 32d35'12"
Jacob Armstrong Reynolds
Uma 13h54'10" 53d7'52"
Jacob August Kuhn
Her 18h17'21" 29d11'18"
Jacob B. Nielsen
Lyn 8h40'13" 38d53'53"
Jacob Benjamin Pressman
Uma 10h5'34" 46d51'6"
Jacob Boehs
Ari 2h15'8" 12d5'38"
Jacob Bradley Bohrer
Lib 15h9'12" -15d59'49"
Jacob Bradley & Carrie Briana
Gem 6h58'40" 13d7'31"
Jacob Brennan Hatton's Star
Cet 1h41'30" -9d5'39"
Jacob Brettargh
Cen 13h49'26" -40d29'36"
Jacob Brian Brody
Per 3h14'57" 53d20'14"
Jacob Brian Gillig
Her 17h3'23" 14d19'29"
Jacob & Britni Forever My Life
Uma 9h5'56" 53d2'37"
Jacob Brower
Per 3h29'50" 32d10'59"
Jacob Carl Heavlin
Vir 13h18'11" -22d25'9"
Jacob Carl Smith
Cep 22h28'39" 73d2'49"
Jacob Case
Aur 5h46'20" 45d29'26"
Jacob Castanho
Gem 7h30'0" 24d46'38"
Jacob Chambers McCrum
Psc 1h51'7" 5d1'10"
Jacob Charles Friedrich
Dra 17h7'59" 52d29'52"
Jacob Christensen
Cru 12h1'53" -63d13'54"
Jacob Christian Duncan
Lyn 8h59'58" 44d14'50"
Jacob Christian Sauer
Per 3h40'11" 48d4'38"
Jacob Christopher Beaver
Lib 14h23'26" -17d18'1"
Jacob Christopher Moffett
Ari 2h10'53" 27d45'8"
Jacob Clare
Tau 4h0'47" 27d5'33"
Jacob Clay Kreis
Psc 1h7'52" 22d9'28"
Jacob Cohen
Psc 1h5'37" 30d23'9"
Jacob Cooper Hydovitz
Peg 22h41'56" 17d29'46"
Jacob Crawford
Cap 21h13'14" -16d31'58"
Jacob D. Rittenhouse
Psc 1h32'51" 9d25'54"
Jacob D. Spencer
Per 4h38'36" 40d39'17"
Jacob Dagnino
Aur 5h5'22" 34d44'21"
Jacob Dale Heinzman
Aqr 21h13'18" 2d22'7"
Jacob Daniel Castaneda
Aql 19h3'56" -0d55'8"
Jacob Daniel Franco
Vir 12h30'44" -0d6'22"
Jacob Daniel Gundrum
Uma 11h56'48" 39d10'27"
Jacob Daniel Sargeant
Dra 11h24'25" 77d42'16"
Jacob "Dash" Gurreri
Cap 21h46'37" -12d15'54"
Jacob David Buol
Cep 21h37'6" 66d20'13"
Jacob David Falica
Per 2h56'42" 41d21'33"
Jacob David Jackson
Vir 14h5'18" -10d55'15"
Jacob David Reeve
Umi 13h55'22" 73d36'55"
Jacob David Wrona
Sco 17h29'12" -41d53'8"
Jacob Dean Brown's Special Star
Ori 5h42'17" 11d22'17"
Jacob Dempsey Ubamos
Leo 10h25'4" 6d52'57"
Jacob Dodgen Fuller-NAVY
Cnv 13h40'26" 41d57'7"
Jacob Domas
Cyg 20h30'21" 30d3'59"
Jacob Donald Decker
Uma 10h39'53" 57d0'29"
Jacob Donavon Garland
Uma 13h13'18" 53d38'40"
Jacob Douglas Kope
Tau 4h2'29" 22d57'45"
Jacob Duckworth
Ori 5h38'30" 2d20'57"
Jacob Dylan [Cubby] Lawson
Aql 18h58'42" -0d5'7"
Jacob Dylan Mreen
Leo 10h9'57" 26d7'6"
Jacob Dylan Rhea
Cnv 12h53'40" 42d38'11"
Jacob E Carbajal
Her 16h35'45" 46d57'49"
Jacob Earl Dunaway
Aql 19h40'57" 3d41'15"
Jacob East
Eri 4h35'27" -15d39'12"
Jacob Edward Monhart
Tau 5h45'31" 16d54'28"
Jacob Edward Rueth
Uma 11h37'21" 41d20'17"
Jacob Eli Stolker
Uma 8h42'42" 53d0'32"
Jacob Ellis Mason
Uma 8h59'52" 59d15'17"
Jacob Ellis Wilde
Uma 11h32'58" 57d7'3"
Jacob Essex Wood
Uma 11h30'51" 56d42'58"
Jacob Ethan Antoney Bray
Ori 6h3'55" 17d6'23"
Jacob Eugene Savage
Her 16h31'45" 11d1'56"
Jacob Evan Lamar Gonzalez
Ori 6h6'3" 0d11'16"
Jacob Evertt Oriet
Cep 22h24'50" 61d51'40"
Jacob F. Oleshansky: Orbiter I
Gem 7h50'0" 16d44'33"
Jacob Franchi
Tau 5h32'7" 21d36'15"
Jacob Francis
Vir 12h44'46" 7d17'33"
Jacob Franco
Dra 17h43'4" 55d11'18"
Jacob Fry
Ari 2h39'43" 30d36'59"
Jacob Gage
Her 17h55'42" 22d57'49"
Jacob Garretson
Uma 9h35'19" 42d29'55"
Jacob Gary Marsden
Tau 3h31'33" 4d46'36"
Jacob Gheorghe Mihalca
Psc 0h24'59" -4d28'41"
Jacob Golladay
Sco 17h53'48" -36d33'30"
Jacob Gregory Castrellon
Aql 18h49'29" -0d5'36"
Jacob Guerra
Gem 7h29'58" 20d35'24"
Jacob Gwin
Per 3h9'59" 56d20'14"
Jacob Hardcore Shairs
Cap 20h20'39" -10d44'20"
Jacob Haven Schreiner
Her 17h34'40" 35d58'46"
Jacob Hayashi
Uma 11h41'53" 39d23'39"
Jacob & Heather
Umi 15h23'11" 75d3'31"
Jacob Hendrickson
Ori 6h1'5" 19d55'47"
Jacob Henri Zogaib
Her 16h52'27" 17d4'29"
Jacob Henry Herzog
Per 3h1'5" 46d27'19"
Jacob Hesse
And 1h18'49" 49d17'52"
Jacob Hugh Joseph Hower
Tau 5h22'9" 18d3'37"
Jacob Ian Muller
Lyr 18h38'25" 39d23'36"
Jacob ("Jack") Walter Baur
Her 18h7'54" 48d43'57"
Jacob "Jake" Donald Anderson
Umi 15h35'30" 68d12'38"
Jacob "Jake" Omdahl
Uma 11h28'20" 30d44'34"
Jacob James Dygan
Her 17h41'14" 35d25'59"
Jacob James Neltner
Her 16h37'4" 37d31'25"
Jacob John Bassett
Uma 11h5'28" 67d26'33"
Jacob John Bennett
Per 2h58'32" 41d53'1"
Jacob John Kay - Jakey's Star
Del 20h30'57" 20d9'29"
Jacob John Nagel II
Ori 5h54'22" 6d19'30"
Jacob John Rushford
Leo 9h45'49" 28d3'38"
Jacob & Jordan Lindberg
Cyg 20h31'53" 35d18'20"
Jacob Joseph
Crb 15h43'38" 26d23'5"
Jacob Joseph Narotzky
Vir 13h24'38" 12d21'49"
Jacob Joseph Smith
Ari 2h40'2" 26d27'46"
Jacob Joshua Shapell
Vir 13h57'43" -12d16'49"
Jacob Kelly
Sgr 19h37'55" -12d30'3"
Jacob Kent Rife
Cnc 8h16'10" 32d47'39"
Jacob LaValley
Lib 15h39'11" -14d50'40"
Jacob Lear
Aqr 22h57'44" -10d52'18"
Jacob Lee
Aur 5h54'53" 52d0'8"
Jacob Lee Cagal-Bley
Per 4h14'45" 51d13'42"
Jacob Lee Ewert
Gem 7h40'26" 25d11'47"
Jacob Lee Hodgson
Aqr 23h27'41" -14d27'47"
Jacob Lee Klassen
Ari 3h13'9" 28d38'24"
Jacob Lee Wempe
Peg 21h47'4" 13d12'49"
Jacob Lee William
Aqr 23h0'19" -7d36'18"
Jacob Lenin Portillo
Uma 9h25'56" 49d25'46"
Jacob LeRoy Metzger
Gem 7h13'54" 22d40'35"
Jacob Lester Jaremko
Tau 5h50'39" 27d8'28"
Jacob Liam
Cas 0h47'29" 61d46'4"
Jacob Lillioja
Ori 5h50'4" 6d43'28"
Jacob Lipsey
Sgr 18h11'53" -19d38'19"
Jacob Lloyd Gilson
Lib 14h53'16" -8d10'51"
Jacob Logan Davis
Uma 14h23'55" 61d10'58"
Jacob Loran Dessert
Aur 6h29'37" 40d58'38"
Jacob Loves Kristina
And 2h23'24" 46d57'19"
Jacob Lowell Alexander Burlap
Cyg 20h14'40" 47d17'16"
Jacob Luke Watts
Uma 13h5'24" 56d27'9"
Jacob Luke Zuba
Cyg 19h58'14" 36d17'52"
Jacob Lynn Grossman
Uma 9h47'32" 55d43'39"
Jacob M. Ford
Lyn 7h38'3" 36d52'21"
Jacob Mabee
Dra 12h34'48" 73d11'16"
Jacob Malachi Oberle-Criqui
Lyn 8h10'23" 50d13'46"

Jacob Manus
Aql 19h5'8" 15d39'51"
Jacob Marcus Frye
Ori 5h47'28" 7d27'30"
Jacob Marcus Gerhart
Umi 15h26'15" 71d13'58"
Jacob Marlow
Boo 14h37'57" 33d42'56"
Jacob Martin Gordon
Cyg 20h58'25" 42d39'5"
Jacob Martin Kahn
Aql 19h30'27" 2d54'20"
Jacob Martin Kahn
Lib 14h23'44" -18d35'6"
Jacob Matthew Anagnostis
Per 3h52'59" 51d53'58"
Jacob Matthew McClain
Dra 18h24'49" 58d8'16"
Jacob Matthew Roberts
Lib 15h2'43" -11d7'58"
Jacob Matthew Strommer
Per 3h4'30" 51d11'54"
Jacob Matthew Walters
Ori 5h28'33" -1d27'23"
Jacob Max
Aqr 22h32'50" -1d51'5"
Jacob Maybe
Dra 19h57'50" 75d42'54"
Jacob McCaugney Miller
Ori 5h55'46" 12d24'34"
Jacob Michael Campbell
Cam 4h23'44" 74d57'55"
Jacob Michael Derby
Her 18h35'35" 15d27'8"
Jacob Michael Espindola
Tau 5h57'2" 27d52'33"
Jacob Michael Henkemeyer
Gem 7h24'15" 33d8'26"
Jacob Michael Litwin
Aql 19h46'12" 7d47'17"
Jacob Michael Loggins
Uma 9h36'23" 68d30'21"
Jacob Michael Moore
Uma 11h0'25" 38d56'39"
Jacob Michael Rehor
Ori 6h8'36" 9d17'13"
Jacob Michael Rousey-Noel
Cnc 9h8'34" 30d24'6"
Jacob Michael Schuette
Lib 15h31'48" -19d29'30"
Jacob Michael Sforza
Ori 4h49'45" 5d15'23"
Jacob Michael Thompson
Her 16h37'52" 26d24'46"
Jacob Michael Zakas
Vir 12h26'52" 2d18'14"
Jacob Miguel Romero
Sgr 19h40'52" -13d49'31"
Jacob Miller
Her 18h56'37" 16d49'53"
Jacob Mitchell
Per 3h24'38" 46d52'50"
Jacob Mockabee
Vir 12h11'3" -9d34'18"
Jacob Moore
Gem 7h50'2" 31d9'24"
Jacob Morton Kirkland
Ori 5h25'7" -4d33'15"
Jacob Nathaniel King
Per 3h45'38" 37d32'47"
Jacob Neil Boger
Peg 22h6'26" 7d41'43"
Jacob Newhouse
Aur 5h10'2" 41d10'29"
Jacob Nicholas Pfeiffer
Cnc 9h9'7" 16d15'28"
Jacob Nicholson 1-30-94 / 8-29-04
Her 16h47'28" 33d36'59"
Jacob Nicolas Berninger
Uma 12h9'10" 61d9'17"
Jacob Noah Barela
Uma 11h27'57" 50d1'39"
Jacob Nyland Trudeau
Ori 5h42'0" -0d36'57"
Jacob Oliver Soloway
Cnc 8h11'41" 32d7'44"
Jacob O'Neill Eddy
Sco 16h9'45" -10d55'23"
Jacob Owen Schmidt
Cam 7h41'50" 81d52'49"
Jacob P. Gibeault
Per 3h24'20" 48d47'21"
Jacob Patrick
Uma 9h1'31" 52d12'40"
Jacob Patrick Aherne
Psc 1h41'49" 4d25'32"
Jacob Patrick Kirkner
Ari 3h2'34" 26d11'45"
Jacob Patrick McClain
Her 16h16'12" 22d57'2"
Jacob Patrick Murphy
Boo 14h41'46" 14d40'22"
Jacob Patrick Ritter
Her 17h18'42" 18d35'49"
Jacob Paul
Vir 14h14'32" -16d48'58"
Jacob Paul Cox
Dra 17h1'40" 62d56'0"
Jacob Paul Dosch
Lyn 8h31'45" 45d17'57"
Jacob Paul Green
Ari 2h41'51" 25d7'26"
Jacob Paul Hampson
Per 2h22'12" 56d20'47"
Jacob Paul Hoover-Koehler
Aur 6h30'32" 38d53'28"
Jacob Paul Nowell
Sco 17h54'8" -35d44'54"
Jacob Paul Orlando July 21, 1997
Uma 9h23'10" 47d23'10"
Jacob Paul Rheault
Vir 14h17'27" 2d15'42"
Jacob Perlson
Equ 21h18'5" 8d52'55"
Jacob Peter DeMaio
Cnc 8h50'50" 31d54'5"
Jacob Peter Duane Claus
Her 17h16'41" 19d6'43"
Jacob Peter Whibley
Ori 5h55'3" 22d0'34"
Jacob Philip
Sco 16h47'51" -41d33'29"
Jacob Pitre
Aur 5h12'57" 42d54'11"
Jacob Poillucci
Aql 19h46'37" -0d29'55"
Jacob Pope
Her 18h37'5" 15d24'20"
Jacob Quinlan
Per 3h40'50" 38d1'20"
Jacob R. Wayson
Her 18h44'45" 20d25'38"
Jacob Ray Cutright
Aql 20h4'42" 10d32'26"
Jacob Ray Torano
Dra 17h26'10" 69d42'12"
Jacob Reece Johnson
Gem 6h21'24" 26d50'55"
Jacob Reid Mantooth
Psc 1h33'39" 11d47'2"
Jacob Remmey Brown
Cep 23h3'27" 59d46'40"
Jacob Reuben Benfield
Her 18h52'29" 20d49'19"
Jacob Richardson Barrock
Umi 16h59'35" 77d24'19"
Jacob Riley
Tau 5h28'5" 26d24'57"
Jacob Riley Greene
Ari 1h58'4" 14d2'10"
Jacob Riley Stewart
Gem 7h11'42" 33d42'18"
Jacob Robert Grech 08/07-09/27/2004
Cep 23h6'7" 80d21'46"
Jacob Robert Hannah
Uma 8h9'27" 63d47'5"
Jacob Robert Owens
Lib 15h20'18" -5d2'32"
Jacob Robert Sheridan
Sgr 19h46'13" -17d14'43"
Jacob Robert Slaughter
Vir 13h21'23" 4d22'51"
Jacob Rooks
Psc 23h50'15" -3d10'21"
Jacob Ross Coulter
Uma 11h1'27" 53d11'9"
Jacob Rourke
Crb 16h21'11" 31d51'4"
Jacob Roy Chapman
Ori 5h30'10" 9d47'20"
Jacob Rubin
Her 17h28'43" 47d42'41"
Jacob Russell Quase Rhodes
Gem 7h27'56" 18d37'11"
Jacob Ryan Buckwald
Per 3h21'11" 41d50'39"
Jacob Ryan Castro
Crb 16h21'24" 28d59'44"
Jacob Ryan Hamilton
Ori 5h13'6" 15d58'12"
Jacob Ryan Manis
Her 17h59'50" 46d39'24"
Jacob Ryan Neubauer
Aur 6h56'29" 39d6'50"
Jacob Ryan Osborne
Gem 7h13'8" 33d43'0"
Jacob Ryan Young
Cap 20h54'49" -16d46'39"
Jacob Ryan's Belt
Gem 7h37'22" 31d9'57"
Jacob Sam
Uma 13h28'48" 52d54'11"
Jacob Sampson
Aqr 22h45'47" -23d9'13"
Jacob Samuel Adessky
Sgr 18h19'42" -32d13'19"
Jacob Samuel Neal
Sgr 19h14'31" -25d41'13"
Jacob Santoro
Aur 5h28'47" 45d42'4"
Jacob Schoap
Uma 13h21'36" 52d52'14"
Jacob Scott Jones
Her 17h44'52" 35d55'38"
Jacob Scott Kersh
Ori 5h29'36" 3d29'59"
Jacob Scott Micheals
Aur 5h55'24" 36d42'9"
Jacob Scott Slater's Star
Cra 18h19'24" -41d8'8"
Jacob Scott Thompson
Ori 4h46'2" 4d37'55"
Jacob Sean Davis
Uma 9h26'44" 45d1'18"
Jacob Sean Smith, Sarasota, Florida
Ori 5h41'17" 2d16'49"
Jacob Shaun Haddad
Ori 5h22'46" 8d53'49"
Jacob Sidney Lake
Umi 15h27'45" 76d51'28"
Jacob Singer
Tau 3h53'2" 17d44'55"
Jacob Smith
Cnc 8h29'55" 25d13'14"
Jacob Solomon Barron
Her 18h7'19" 31d48'57"
Jacob Soto
Umi 16h33'43" 76d36'19"
Jacob & Stephanie Levin
Cyg 21h27'39" 35d59'32"
Jacob Steven Beck
Lmi 10h31'22" 31d6'24"
Jacob Stevenson Miles
Lyr 18h39'40" 39d39'36"
Jacob Stone Collins
Aqr 22h34'35" -1d39'11"
Jacob Stuart Trindell
Per 2h27'29" 55d9'4"
Jacob Thomas
Aql 19h45'36" 12d28'36"
Jacob Thomas Baurle
Ori 5h30'21" 3d6'2"
Jacob Thomas Braymiller
Lyn 8h29'23" 39d20'32"
Jacob Thomas Compton
Leo 10h20'27" 14d35'42"
Jacob Thomas Fein-Ashley
Gem 7h40'47" 34d27'28"
Jacob Thomas Hollerbush
Uma 10h22'54" 53d34'20"
Jacob Thomas Kenneth Fawdry
Uma 11h32'16" 55d57'48"
Jacob Thomas Ochwat
Cnc 8h51'10" 31d43'18"
Jacob Thomas Pereira
Vir 12h40'55" 0d54'45"
Jacob Thomas Putland
Ori 5h19'56" 15d0'34"
Jacob Thomas Sternoff
Vir 12h43'32" 7d26'22"
Jacob Ton Nguyen
Cnc 8h23'47" 24d52'7"
Jacob Townley Reed
Cap 20h53'45" -15d26'26"
Jacob Travis Rauch
Her 17h50'20" 47d44'28"
Jacob Tyler Riha
Uma 11h41'47" 47d27'18"
Jacob Vann Gurwell
Crb 15h52'56" 35d48'50"
Jacob Vasquez
And 0h18'59" 27d3'53"
Jacob Vincent Roth
Cam 7h30'59" 63d42'3"
Jacob W. McGrath
Sgr 19h16'7" -21d5'40"
Jacob W. Wagner
Umi 15h43'9" 85d40'44"
Jacob Wade Kent
Aql 19h6'28" 2d53'23"
Jacob Wayne Faulkenberry
Sco 17h51'28" -36d44'11"
Jacob Weldon Miller
Ori 5h21'39" 4d14'30"
Jacob Wetterling
Lyn 7h39'52" 52d54'35"
Jacob William
Dra 17h29'19" 56d47'5"
Jacob William
Psc 0h30'36" 13d9'48"
Jacob William Flaxman
Gem 7h50'31" 27d6'26"
Jacob William Granger
Vir 14h9'26" 6d14'57"
Jacob William Korducki
Boo 14h30'40" 51d44'1"
Jacob William Liedtka
Umi 14h33'9" 75d17'56"
Jacob William Werdean
Vir 12h23'29" -5d20'41"
Jacob Wismans
Per 2h38'36" 56d5'55"
Jacob Zogaib
Leo 9h37'35" 11d10'50"
Jacoba Marinos
Aqr 22h37'35" 2d26'54"
Jacob-Bear
Sco 17h26'7" -38d40'48"
Jacobi David Caudwell
Uma 11h33'20" 34d47'50"
Jacobo Levy Alcahe
Her 16h29'15" 48d36'32"
Jacob's Christening Star
Umi 14h40'14" 76d26'32"
Jacob's DreamSeeker
Uma 10h5'10" 55d23'49"
Jacob's Guardian Star
Per 4h19'55" 35d1'41"
Jacob's Guiding Light
Cep 4h48'18" 84d21'30"
Jacob's Guiding Star
Per 2h52'23" 56d57'14"
Jacob's Heart
Umi 15h4'47" 78d47'51"
Jacob's Heavenly Light
Lyn 7h48'15" 57d39'52"
Jacob's Hoku
Ari 3h27'0" 23d22'42"
Jacob's Journey
Her 16h54'54" 36d3'59"
Jacob's Ladder
Per 4h35'13" 35d12'16"
Jacobs Ladder
Her 18h34'38" 17d58'18"
Jacob's Light
Umi 15h38'12" 69d9'58"
Jacob's Ryan
Uma 13h59'9" 59d11'15"
Jacob's Son Shine
Ori 5h34'51" 1d53'36"
Jacob's Special Star
Psc 1h18'26" 26d26'34"
Jacob's Star
Ori 5h7'13" 8d15'18"
Jacob's Star
Psc 0h27'44" 18d6'21"
Jacob's Star
Uma 8h57'34" 53d38'10"
— Jacob's Star — for Jake Nelson
Vir 13h48'4" -12d2'47"
Jacob's Star - Our Shining Light
Cru 12h41'20" -63d54'26"
Jacob's Wishing Star
Uma 12h33'20" 57d35'43"
Jacobson Family Star
And 1h46'27" 49d11'39"
Jacobus
Aqr 21h48'55" -2d51'41"
Jacodi
Sco 16h14'38" -29d6'25"
Jacole
Umi 14h34'9" 74d37'43"
Jacopo e Laura 27/07/2005
Cam 9h15'21" 75d35'1"
Jacopo Romano
Peg 23h57'14" 22d18'53"
Jacopo Trulli
Umi 16h59'32" 83d20'39"
JACOT9
Vir 12h22'42" -6d19'12"
jacovy
Uma 11h52'37" 54d1'16"
JACOYA
Aur 5h39'20" 40d23'35"
Jacqeline Schnuppe
Uma 8h10'7" 65d39'40"
Jacqi Joy
Del 20h34'59" 15d56'2"
Jacqie & Carol
Uma 11h17'28" 39d41'45"
Jacqua Larry
Uma 10h25'44" 63d49'0"
Jacqualine
Aqr 22h2'18" -0d51'56"
Jacqualyn Jade
Sgr 19h48'30" -12d47'5"
JacquCluzel YvetPiétrolongo
Ori 5h48'31" 5d44'0"
Jacque
Lyn 8h25'2" 33d46'9"
Jacque
And 23h17'48" 47d23'29"
Jacque
Sgr 18h46'8" -29d15'44"
Jacque and Mary Reitz
Umi 15h24'4" 73d13'33"
Jacque Ellison - YaYa
Uma 8h54'11" 63d24'47"
Jacque Frolich
Ori 5h59'33" 22d37'42"
Jacque "Golden Lady"
Cas 0h23'9" 56d43'1"
Jacque Lowenstar
Sgr 19h17'19" -14d7'33"
Jacque Ratz
Cap 21h48'5" -17d55'16"
Jacque & Ronald
Vir 12h34'17" 11d22'4"
Jacque Shelton
Tau 4h26'35" 14d31'12"
Jacque Tollen Coxe
Lib 14h54'0" -20d9'45"
Jacquelene
Ori 6h23'25" 15d45'22"
Jacquelin Fabris Georgens Clain
Gem 7h12'6" 26d44'31"
Jacquelin Marie Pitre
Vir 11h57'8" -0d9'38"
Jacquelin Michelle Presbaugh
Aqr 20h42'8" -10d17'25"
Jacquelina Furfaro Brown
Cnc 9h1'50" 23d0'55"
Jacquelina Rose Luna
Cas 23h5'1" 55d57'6"
Jacquelinas
Cas 0h32'9" 64d5'2"
Jacqueline
Cas 0h13'19" 63d23'11"
jacqueline
Uma 11h9'22" 55d58'31"
Jacqueline
Uma 8h50'16" 55d23'18"
Jacqueline
Cam 7h28'36" 60d44'55"
Jacqueline
Cas 2h18'14" 63d6'15"
Jacqueline
Umi 13h39'47" 72d17'21"
Jacqueline
Aqr 21h10'24" -11d42'23"
Jacqueline
Aqr 22h55'53" -10d37'56"
Jacqueline
Vir 12h22'9" -6d52'42"
Jacqueline
Sgr 18h7'50" -28d49'39"
Jacqueline
Aqr 23h16'3" -22d9'4"
Jacqueline
Cru 12h16'27" -64d0'30"
Jacqueline
Sco 17h56'17" -37d49'57"
Jacqueline
Tau 4h24'38" 28d25'10"
Jacqueline
Leo 10h30'56" 19d31'6"
Jacqueline
Ari 2h53'51" 25d34'18"
Jacqueline
Psc 23h1'14" 5d57'16"
Jacqueline
Cnc 8h48'25" 6d39'24"
Jacqueline
Vir 13h14'51" 11d14'4"
Jacqueline
Leo 9h30'42" 12d38'23"
Jacqueline
Ori 5h54'15" 16d36'27"
Jacqueline
Uma 9h54'17" 45d38'49"
Jacqueline
And 1h4'56" 45d34'47"
Jacqueline
Lyn 8h57'4" 37d41'25"
Jacqueline
And 0h23'6" 41d32'11"
Jacqueline
And 1h19'54" 41d20'47"
Jacqueline
And 0h43'2" 41d1'55"
Jacqueline
And 0h42'35" 43d47'33"
Jacqueline "#1 Mom" Frisina
And 0h5'7" 44d50'44"
Jacqueline 10/13/1982
Lib 15h38'11" -10d39'23"
Jacqueline A.
Gem 6h32'27" 21d15'41"
Jacqueline A. Knox
Ori 5h54'33" 21d9'27"
Jacqueline A. Rose
Gem 6h40'15" 22d37'30"
Jacqueline Adele's Shining Star
Aqr 22h44'49" -23d52'26"
Jacqueline Adorno - Jacque Star
Gem 6h46'35" 33d15'57"
Jacqueline Alexis Bolte
Uma 12h56'25" 62d40'1"
Jacqueline and Gerry
Dra 17h57'57" 79d7'56"
Jacqueline and Julianne O'Donnell
Cyg 20h45'29" 51d56'45"
Jacqueline and Vanessa
Peg 23h34'0" 18d18'25"
Jacqueline Angelina
Leo 10h37'43" 7d38'31"
Jacqueline Ann
Leo 10h41'2" 6d31'44"
Jacqueline Ann Elizabeth Catalano
Lib 15h7'23" -4d48'50"
Jacqueline Ann Heckathorne Beach
Umi 15h23'10" 71d27'9"
Jacqueline Ann Hyland
Cas 23h34'16" 55d15'35"
Jacqueline Ann Safrah
Tau 3h56'22" 21d24'35"
Jacqueline Anna Berner
And 0h20'10" 29d35'13"
Jacqueline Anne Barrett
Cas 1h31'45" 62d13'50"
Jacqueline Anne Reed
And 2h0'54" 40d26'55"
Jacqueline Anne White
And 1h7'23" 41d31'36"
Jacqueline Ann's Wild Irish Rose
Gem 6h31'11" 15d40'43"
Jacqueline Antionett Hopkins
And 23h10'47" 48d36'49"
Jacqueline Barnard
Cap 21h50'41" -15d21'18"
Jacqueline Barr
Ori 5h28'22" -0d23'10"
Jacqueline Baumann White
Tau 4h22'27" 13d54'18"
Jacqueline Benson
Umi 14h28'6" 85d11'19"
Jacqueline Berry
Tau 3h31'3" 12d19'52"
Jacqueline Boehni
Ori 5h41'13" -3d35'15"
Jacqueline Bonner
Aqr 22h17'34" -8d34'33"
Jacqueline Brook Puleo
Lib 14h26'16" -12d11'10"
Jacqueline Bruning
Tau 5h49'19" 22d18'40"
Jacqueline Byrd A Star For My Star
Aqr 21h4'3" -11d39'39"
Jacqueline Carrillo
Vir 12h35'46" 6d17'46"
Jacqueline Catherine Palumbo
Ari 3h4'44" 29d6'51"
Jacqueline Christine Stark
And 23h18'39" 43d13'38"
Jacqueline Chroptek
Cyg 21h35'35" 53d46'5"
Jacqueline Coivous
Tau 3h52'58" 17d41'28"
Jacqueline Colette
Vir 13h54'2" -7d48'58"
Jacqueline Combes Cahayag
Dra 15h27'4" 61d12'37"
Jacqueline Coon
Lyn 7h27'4" 56d48'30"
Jacqueline Costa-Lapouge
Uma 10h24'16" 70d36'50"
Jacqueline Cotton
Per 2h16'42" 51d25'32"
Jacqueline Cramer
Cap 20h23'0" -22d53'32"
Jacqueline D. & James R. Fleming
Cnc 8h49'43" 14d44'59"
Jacqueline Danielle
Cnc 8h36'27" 29d20'40"
Jacqueline & David - Together Forever
Cru 12h23'46" -58d20'42"
Jacqueline Delores Henrich
Crb 15h35'26" 38d25'56"
Jacqueline Detombe
Vir 14h52'21" 4d53'17"
Jacqueline DeVelle Polier
And 1h46'4" 44d3'45"
Jacqueline Diane Hodge
Leo 10h35'20" 13d27'2"
Jacqueline DiLeo
Uma 9h54'44" 55d45'8"
Jacqueline Dorosz
Aqr 20h55'43" 0d47'52"
Jacqueline Downey Festa
Ori 6h16'2" 15d45'32"
Jacqueline Eck
Cas 0h0'11" 57d0'43"
Jacqueline Elaine Flanigan
And 0h47'36" 38d34'5"
Jacqueline Elezaj
Cnc 9h11'41" 30d41'32"
Jacqueline Elizabeth Huck
And 1h55'0" 43d7'23"
Jacqueline Elizabeth Wilke
And 2h23'34" 39d21'36"
Jacqueline Ellen Pozar
Her 17h53'6" 27d30'48"
Jacqueline Eltgroth 4-23-1920
Tau 4h8'4" 27d10'27"
Jacqueline Faith Brennan
Lyn 7h24'2" 53d13'56"
Jacqueline Faye Ferguson
Hya 9h14'49" -0d31'12"
Jacqueline Francesca
Aqr 23h11'52" -12d21'46"
Jacqueline Goelz
Cas 1h33'23" 63d20'26"
Jacqueline Goorsky
Gem 7h40'11" 16d14'14"
Jacqueline Grace Friar
And 1h30'55" 48d50'50"
Jacqueline Graeme Tiddy
Cas 0h9'0" 53d26'14"
Jacqueline Grübli 09.06.1961
And 1h2'3" 38d48'33"
Jacqueline Guerrero
Cap 21h41'14" -21d30'54"
Jacqueline Gunther
Cap 20h49'12" -15d59'44"
Jacqueline Hall - Jack's Joy
Cas 23h46'13" 56d39'19"
Jacqueline Hecht White
And 0h49'3" 40d35'24"
Jacqueline Hernandez
Cam 6h22'59" 68d32'2"
Jacqueline Holt
Tau 3h42'35" 4d19'33"
Jacqueline Hornik
Dra 17h42'39" 62d39'43"
Jacqueline Ivory Burt
Umi 15h30'52" 74d51'38"
Jacqueline J. Baby Gonzalez
And 0h19'44" 21d44'39"
Jacqueline "Jackie"
Ari 3h22'32" 26d3'25"
Jacqueline Jane Conaway
Vir 14h23'38" -2d46'56"
Jacqueline Jane Miles
Aqr 23h39'42" -6d46'7"
Jacqueline "Jaz" Zepatos
And 2h21'22" 49d55'33"
Jacqueline Jean Sichenzia
And 23h37'2" 47d0'6"
Jacqueline Jessup Skinker
Cas 1h17'10" 54d42'6"
Jacqueline Joseph Ryan
Aur 5h15'21" 42d51'23"
Jacqueline & Josephine Mitsdarffer
And 2h24'52" 41d44'45"
Jacqueline Joy Valenciana
Cru 12h50'28" -60d12'29"
Jacqueline Joyce Odneal
Vir 12h1'57" 5d45'34"
Jacqueline & Justin
Cyg 20h21'2" 43d35'24"
Jacqueline K.
Cnc 8h51'13" 25d39'1"
Jacqueline K-84
Vir 12h17'34" 11d13'39"
Jacqueline Kasen
Sgr 17h57'33" -17d6'32"
Jacqueline & Kevin Frank
Cyg 20h21'25" 54d46'12"
Jacqueline Kevorkian Wilking
Sgr 18h57'5" -35d29'21"
Jacqueline L. Clock, née Mahoney
Lyn 7h18'9" 47d48'51"
Jacqueline L. Holmes
Vir 13h16'45" -2d35'25"
Jacqueline L Sorelle
Lyn 6h32'21" 59d5'56"
Jacqueline Laura
Lyn 7h58'21" 37d33'15"
Jacqueline Leah Tomten
And 1h32'36" 42d53'45"
Jacqueline Lee Johnson
Gem 7h7'52" 15d39'2"
Jacqueline Leigh Esperti
Uma 9h6'19" 61d14'39"
Jacqueline Leigh Jordan
Sgr 18h53'13" -16d49'37"
Jacqueline Lenora Howell
Sgr 20h4'49" -35d12'35"
Jacqueline Lila
Cru 12h30'34" -58d52'59"
Jacqueline Linda Blincow
And 2h21'42" 38d47'23"
Jacqueline Loftis
Boo 15h28'55" 44d48'5"
Jacqueline Lois Epstein
Lyn 7h51'55" 54d42'43"
Jacqueline Lombardo
Cnv 13h35'18" 30d42'43"
Jacqueline Margaret Anne Linnane I
And 2h8'34" 43d2'17"
Jacqueline Marie
Cas 0h40'22" 64d20'35"
Jacqueline Marie
Uma 8h29'31" 70d30'32"
Jacqueline Marie Burr
Psc 0h18'55" 15d30'33"
Jacqueline Marie Dominguez
Cnc 8h19'36" 21d31'44"
Jacqueline Marie Dussia
Leo 10h17'44" 21d58'59"
Jacqueline Marie Ellis Hall
Tri 2h23'20" 32d18'2"
Jacqueline Marie Forsling
Cnc 8h3'21" 22d12'38"
Jacqueline Marie Guerrero
Lib 15h18'54" -9d17'46"
Jacqueline Marie Hekker
Leo 11h5'31" 3d22'1"
Jacqueline Marie Morton
Uma 10h33'7" 39d34'8"
Jacqueline Marie Sessa
Leo 10h19'37" 14d37'51"
Jacqueline Marie Walrond
And 0h34'28" 34d40'31"
Jacqueline Marie Watts
Uma 11h55'53" 63d22'28"
Jacqueline Marie Willis
Cnc 9h5'25" 24d57'12"
Jacqueline Marisa Armas
Sgr 18h19'52" -18d31'6"
Jacqueline Marisa Del Pilar
Lib 15h57'28" -19d26'23"
Jacqueline Marmol
And 0h58'23" 46d12'43"
Jacqueline Marquez
And 0h44'49" 41d49'15"
Jacqueline Mary O'Halloran
Ser 18h31'38" -0d51'15"
Jacqueline May
Tau 5h33'52" 24d35'23"
Jacqueline Melo
Tau 3h50'49" 24d33'48"
Jacqueline Merando
And 23h32'51" 42d45'41"
Jacqueline Michele Loiselle
Dra 17h58'24" 52d34'17"
Jacqueline Michelle Cecil
Lyr 19h16'14" 32d24'31"
Jacqueline Michelle Krueger
Tau 4h14'5" 27d13'50"
Jacqueline Michelle Mawla
Ori 5h6'56" 7d40'6"
Jacqueline Mo Anam Cara
Crb 15h35'49" 31d31'35"
Jacqueline Muriel
Sco 16h14'36" -11d36'52"

Jacqueline Newman Mylett Psc 0h11'55" 3d10'6"
Jacqueline Nicole Brazeal Gem 6h7'3" 26d3'27"
Jacqueline Nicole Church Aqr 21h40'14" -1d42'54"
Jacqueline Noel Soule Henriksen Sgr 18h5'32" -23d26'8"
Jacqueline Noelle Betro Uma 10h9'11" 47d29'18"
Jacqueline Noelle Dinglasan Peg 22h39'31" 27d19'14"
Jacqueline P Tau 4h37'31" 12d13'21"
Jacqueline Patrice Hatfield Cyg 21h56'35" 48d9'44"
Jacqueline Patricia Smyth Uma 8h24'44" 63d15'1"
Jacqueline + Patrick Ori 6h1'45" 9d59'49"
Jacqueline Pfister Cas 0h48'7" 60d20'19"
Jacqueline Popinko And 23h14'3" 36d47'36"
Jacqueline Rae And 23h0'14" 41d41'7"
Jacqueline Rae Gonzalez Cnc 8h39'13" 8d12'31"
Jacqueline Raffaele Psc 1h24'46" 6d31'26"
Jacqueline Rene' Menger Aql 19h42'32" -0d0'37"
Jacqueline Rodgers Vir 13h7'13" 6d27'18"
Jacqueline Romero Nieves Crt 11h33'5" -7d45'47"
Jacqueline Rose Ouellette Mon 6h45'11" -0d4'43"
Jacqueline Rose Polden And 23h15'54" 49d4'11"
Jacqueline Rose Sardina Bohm And 1h18'42" 45d40'47"
Jacqueline Rose Wiegard Cap 20h49'53" -21d23'44"
Jacqueline Ruma Her 17h50'11" 32d4'44"
Jacqueline Ruth Alberts And 2h23'43" 50d12'20"
Jacqueline Salazar Ari 2h24'52" 18d0'15"
Jacqueline Sarah Jones Tau 4h5'43" 0d44'19"
Jacqueline Sendra And 23h21'28" 48d9'24"
Jacqueline Shaughnessy Cas 1h29'33" 63d47'28"
Jacqueline Simpson Lmi 10h30'41" 35d14'31"
Jacqueline Solvino Dalessio Crb 15h48'54" 26d24'59"
Jacqueline Spörri Cas 1h55'9" 69d14'32"
Jacqueline Starr Lewis Cmi 7h29'17" 7d4'29"
Jacqueline Sue Parks Cnc 8h1'53" 10d58'37"
Jacqueline T Nelson Tau 3h50'53" 16d7'8"
Jacqueline Talbert Luminary Lib 14h52'39" -1d55'27"
Jacqueline Tanner And 2h8'37" 37d56'12"
Jacqueline Taylor Crisafulli Sgr 18h35'42" -22d47'0"
Jacqueline Theresa Quinn And 0h14'57" 30d1'12"
Jacqueline Weir And 2h19'20" 43d44'46"
Jacqueline Weyand Psc 0h17'8" 8d19'49"
Jacqueline Wheeler And 1h51'8" 46d32'22"
Jacqueline & Yves Umi 14h49'21" 69d24'9"
Jacqueline Yvette Caouette Sgr 18h43'45" -17d15'46"
Jacqueline.C.Forever Aqr 22h55'50" -23d59'46"
Jacqueline's Grace Sgr 18h28'47" -17d3'8"
Jacqueline's Shining Star Vir 13h24'15" -2d56'2"
Jacqueline's Star 2+2=3 Cnc 9h5'36" 16d14'8"
Jacqueline's Tranquility Sgr 18h44'27" -18d52'42"
Jacquelyn Cyg 21h34'58" 44d48'15"
Jacquelyn Cyg 20h36'27" 51d11'19"
Jacquelyn 143 And 23h13'49" 38d58'22"
Jacquelyn Adams Leo 11h6'16" 20d29'53"
Jacquelyn Alexis Naomi Johnson Vir 13h28'0" 11d10'56"
Jacquelyn Ann Lindemann Uma 11h19'59" 47d28'53"
Jacquelyn Bickle Gem 7h22'46" 26d20'15"
Jacquelyn Butler Cas 1h35'46" 61d13'55"
Jacquelyn C. Bennett Uma 10h17'26" 56d14'50"
Jacquelyn Cecilia Cole Cnc 8h47'41" 31d57'39"
Jacquelyn Christine Williams Crb 16h18'28" 32d28'7"
Jacquelyn Cross And 1h30'13" 50d7'2"
Jacquelyn Driesslein Uma 10h43'20" 43d15'4"
Jacquelyn Dwelle Bates Cap 21h57'4" -9d9'12"
Jacquelyn Elizabeth Crb 15h51'26" 36d45'28"
Jacquelyn Esparza Crb 15h41'30" 31d30'23"
Jacquelyn Janette Hunt Vir 14h16'40" 1d27'58"
Jacquelyn Jean Tau 5h20'13" 19d12'54"
Jacquelyn Joy Knie Dra 17h53'12" 57d0'52"
Jacquelyn Kelsey Tri 1h57'15" 31d5'0"
Jacquelyn Kyndal Wilgus 6/5/02 Gem 7h44'9" 33d20'1"
Jacquelyn Lee Cornell Vir 14h20'11" 1d7'15"
Jacquelyn Lee Flowers Leo 10h55'56" 11d42'43"
Jacquelyn Lee Flowers Leo 10h59'12" 11d37'3"
Jacquelyn Lezette Umi 14h33'56" 75d11'7"
Jacquelyn Lou Edwards Cyg 20h43'26" 32d22'6"
Jacquelyn M. Veuleman Sco 16h40'15" -32d1'35"
Jacquelyn Marie Gem 7h25'59" 32d9'59"
Jacquelyn Marie Bekech Barry Lib 14h53'26" -0d41'26"
Jacquelyn Marie Mullen And 23h17'3" 41d1'58"
Jacquelyn Marie Negri Uma 11h17'15" 39d8'41"
Jacquelyn Marie Newberg And 2h36'56" 50d30'46"
Jacquelyn Marie Schneider And 2h29'27" 43d42'41"
Jacquelyn Marie Stahl Cyg 19h46'43" 33d20'8"
Jacquelyn Marie Wallenberg Uma 9h17'21" 71d13'25"
Jacquelyn Melissa Kotnarowski Leo 11h34'14" -1d55'28"
Jacquelyn Monique Ori 5h8'1" 6d22'25"
Jacquelyn Moore Cas 2h24'36" 65d28'39"
Jacquelyn Owen Sgr 19h29'10" -41d44'56"
Jacquelyn Ranae Shibilsky Lyr 18h18'58" 32d36'1"
Jacquelyn Renee Anderson Sco 16h7'59" -16d35'17"
Jacquelyn Rose Bonnenfant 8/1/1938 And 2h22'59" 50d5'12"
Jacquelyn Scott Cas 1h4'39" 54d11'51"
Jacquelyn Smith Uma 10h38'8" 42d48'35"
Jacquelyn Taylor Uma 13h41'7" 50d22'38"
JacquelynDean Lyr 18h58'15" 27d2'58"
Jacquelyne Uma 8h44'16" 66d32'6"
Jacquelyne Cecelia Berkheimer Tau 5h12'12" 18d54'7"
Jacquelyne Mayce Collard Sgr 19h48'47" -25d21'50"
Jacquelynn M. Stanley-Rivera And 2h20'17" 47d49'23"
Jacquelynn Mae Bezik Lib 15h8'20" -12d57'1"
Jacquelynn Michelle Querry Lib 15h31'25" -11d36'39"
Jacquelynn Nicole Cas 23h1'8" 59d14'35"
Jacquelynn Nicole Moore Tau 4h22'43" 24d17'41"
Jacquelynn Rose Stratakis Cap 21h55'35" -22d9'57"
Jacques Sct 18h52'40" -11d32'20"
Jacques Leo 11h34'3" 26d20'9"
Jacques Alexander Christian Quinn Hibbs Uma 9h16'9" 72d16'5"
Jacques Antoine Bahuaud Cas 0h1'52" 53d6'25"
Jacques Blulle Cam 3h43'29" 64d7'27"
Jacques Brouhet Gem 6h47'31" 35d21'21"
Jacques Desprez Cas 0h52'20" 50d32'42"
Jacques Elie Fortin Ori 6h13'8" 2d12'50"
Jacques Francois Kessler Per 3h43'31" 45d41'42"
Jacques Hainard Uma 10h37'20" 58d27'29"
Jacques Hélaine Ari 3h4'10" 19d17'45"
Jacques Jolivet Lib 15h41'22" -18d35'50"
Jacques Laroue Aqr 23h22'35" -19d8'22"
Jacques le gars dans les étoiles Ori 6h4'45" 21d29'9"
Jacques Lecler Umi 16h7'13" 70d13'8"
Jacques Legrée And 23h0'16" 37d20'12"
Jacques & Louise Paul-Hus Uma 8h37'18" 72d18'19"
Jacques Méthot Ari 3h5'25" 17d57'6"
Jacques Morly aka DADA Her 18h6'51" 30d37'39"
Jacques Olivier Ori 5h18'6" 0d1'18"
Jacques Robinson Sgr 18h19'33" -19d7'31"
Jacques Thévenaz Ori 5h7'23" -0d44'32"
Jacques Tremblay Aql 19h56'28" 12d2'31"
JacQuetta Clayton Cam 12h57'32" 79d51'47"
Jacqui Psc 1h7'11" 3d19'13"
Jacqui Leo 11h35'50" 18d31'39"
Jacqui and Ian's star And 1h26'4" 44d21'38"
Jacqui and Tom's Star Cas 23h7'14" 58d41'1"
Jacqui Barrett Cnc 9h19'0" 14d54'53"
Jacqui Jean Sgr 17h53'11" -29d50'33"
Jacqui King And 0h17'40" 45d30'8"
Jacqui Margaret Uma 9h46'50" 41d56'16"
Jacqui Ooi Col 5h59'0" -35d11'32"
Jacqui Pyper Cru 12h3'1" -61d57'14"
Jacqui Taylor Sco 16h15'18" -16d53'3"
Jacqui Vivanco Lmi 10h41'45" 38d38'59"
Jacquie Fiorini And 0h30'4" 36d37'36"
Jacquie Kerner Cas 0h46'24" 57d8'21"
Jacquie Martell Crb 15h51'0" 26d50'21"
Jacquie's Star And 0h51'9" 44d21'46"
Jacquiline Moore Gem 6h56'55" 18d13'15"
Jacqui's Star Lib 14h49'32" -15d22'26"
Jacquita Bonita And 0h18'5" 26d14'0"
Jacqulyn Dee Karp Leo 10h37'0" 18d8'9"
Jacs Tuesday Tess Durberville Sco 16h8'42" -13d50'10"
Jacson T. Long Oph 16h41'5" -0d53'1"
JACW Uma 10h42'18" 57d59'40"
Jacy Crb 16h9'5" 37d47'20"
Jacy Lundberg And 23h26'59" 47d18'36"
Jacy Rose Crb 15h50'37" 25d51'13"
Jacy Roselius Umi 15h34'41" 86d59'35"
Jäd Lib 15h8'37" -7d58'13"
JADA Uma 13h35'45" 56d5'0"
Jada Cnc 8h1'38" 18d35'53"
Jada Ann Rokisky - Raulerson Uma 10h21'37" 58d18'5"
Jada Bella Frances Hillman Uma 11h18'53" 30d33'52"
Jada Beth Aqr 20h43'8" 1d44'24"
Jada C Sco 17h22'19" -36d54'40"
Jada Elizabeth Harris Cnc 8h40'19" 7d12'19"
Jada Faith Besso Uma 11h19'58" 37d15'33"
Jada Grace Patton And 0h38'46" 34d55'27"
Jada Jones Ori 5h4'4" 15d16'27"
Jada Mae And 0h10'59" 34d28'20"
Jada Marie Wilson Ari 3h9'56" 27d32'12"
Jada Nicole Psc 0h42'46" 17d29'23"
Jada R. Stebenne Crb 16h18'29" 33d52'2"
Jada 's Star Ori 6h8'28" 2d9'25"
Jada 's Star Ori 6h0'52" 5d52'3"
Jada Shruthi Srikanthan And 0h8'37" 29d17'7"
Jadaboo And 23h11'10" 47d54'45"
Jadaboo And 23h14'29" 48d44'11"
Jadah Hettinger Ari 2h14'32" 20d36'8"
Jadah Weaver - Shining Star Cam 4h27'2" 62d28'3"
Jada's Star Vir 14h1'29" -7d13'21"
Jade Sco 17h5'56" -36d43'18"
Jade Ori 6h3'2" 13d14'54"
Jade Vir 12h45'45" 6d36'37"
Jade Gem 7h5'49" 26d42'12"
Jade Leo 11h27'19" 28d18'0"
Jade And 0h1'43" 48d40'31"
Jade Per 3h22'18" 42d33'17"
Jade A. Girton Tau 4h12'53" 24d55'10"
Jade Alexa White Aql 18h58'59" -4d23'43"
Jade Alexendra Aql 19h7'42" 5d25'46"
Jade Alexis And 23h54'11" 40d46'58"
Jade Amanda Pollock Cru 12h10'7" -60d38'1"
Jade & Andrew - Backflips & Fireworks Psc 1h0'46" 29d1'30"
Jade Angelina Moffett And 1h19'56" 42d56'54"
Jade Ann Maracic Lib 15h14'6" -4d58'49"
Jade Athena Emerald Dixon Ori 6h3'9" 18d38'45"
Jade Audrey Eunjung Kim Sco 16h13'38" -16d56'40"
Jade & Billy Cyg 19h36'28" 30d29'42"
Jade Brennan 16/2/76-01/01/2006 Cra 18h35'37" -40d33'18"
Jade & Bruce - Together Forever Cru 12h38'29" -57d46'34"
Jade Burgener Cam 3h21'43" 56d11'36"
Jade Chasm - Wedding - Stuart/Mary Ann Cru 12h26'22" -61d54'25"
Jade Chevalier Peg 21h32'12" 9d41'1"
Jade Christen Ari 2h16'40" 22d41'19"
Jade Ellaine Sellings Pho 0h8'42" -44d52'23"
Jade Eryn Rose Aqr 22h4'42" -7d53'7"
Jade Evan Aqr 22h20'34" -2d10'42"
Jade Faith Phillips Ari 3h6'6" 22d48'1"
Jade Fernandes Lib 15h9'50" -13d59'43"
Jade Frota Uma 10h43'21" 46d31'52"
Jade Gallant Cas 23h43'21" 58d35'19"
Jade Georgia Lib 15h40'28" -28d27'30"
Jade Gray And 2h31'38" 46d37'27"
Jade - Halima Uzuri And 23h39'6" 38d29'45"
Jade Hannah Cas 0h58'13" 48d31'10"
Jade Hayward And 23h27'20" 47d44'9"
Jade Ka-Lee VanDertuin Ari 3h10'33" 20d19'16"
Jade Louise Wilson And 0h12'25" 48d29'31"
Jade Loves Galenn Lyr 18h54'53" 33d27'51"
Jade Mackenzie Meads Uma 9h45'14" 61d2'57"
Jade Martinez Psc 1h41'31" 19d4'14"
Jade Martinus Psa 22h32'3" -30d3'2"
Jade McGough And 0h54'36" 44d13'24"
Jade Melanie Ori 6h1'21" 9d29'3"
Jade Melody Dickey Lmi 10h36'0" 30d51'32"
Jade Michelle D. Psc 1h25'17" 13d1'5"
Jade Morgan I Love you more Lyn 6h53'24" 51d41'26"
Jade Myburgh Lib 15h27'50" -11d44'21"
Jade Nevaeh Getman Gem 7h27'47" 15d51'57"
Jade Renae Cra 18h9'20" -40d0'43"
Jade Riley Hausmann Dra 16h1'22" 64d59'2"
Jade Ryen Galt Dra 17h18'2" 62d2'46"
Jade S.Y. Chang Aqr 22h5'25" -0d27'28"
Jade Terese Pugsley Sgr 19h31'50" -41d27'24"
Jade Van Lafitte Lib 15h13'9" -9d52'38"
Jade Victoria Washington Her 17h3'14" 31d1'51"
Jade Ya Xuan Tau 5h40'35" 26d7'8"
Jadeaspect Ori 4h48'26" 11d47'52"
Jaded Night And 0h37'37" 37d39'5"
Jadeline Ann Mullikin Mon 6h48'51" -0d8'35"
Jaden Umi 15h25'5" 78d20'5"
Jaden Umi 14h41'9" 74d25'17"
Jaden Vir 12h33'24" 11d3'1"
Jaden Abrams Papas Per 4h20'56" 48d14'18"
Jaden Anine Stock And 23h2'35" 45d1'57"
Jaden Daufeldt Frank And 1h14'50" 50d4'13"
Jaden David Ori 5h21'10" -8d11'13"
Jaden D'Shae Turner Gem 6h41'55" 21d27'4"
Jaden Elijah Umi 14h17'39" 72d53'32"
Jaden Elizabeth Kalb Cyg 20h46'1" 50d32'44"
Jaden Eric McCorkle Ari 2h7'7" 25d5'48"
Jaden Grace Heller Sco 17h51'41" -37d51'6"
Jaden Gray Crossley Lyn 7h31'7" 38d4'41"
Jaden Ky Ori 6h18'46" -0d35'5"
Jaden Leigh Garner Uma 8h51'49" 69d8'20"
Jaden Marie Sco 17h50'56" -44d50'44"
Jaden Mark Ari 3h4'5" 25d15'9"
Jaden McClellan Lyn 8h25'26" 33d41'0"
Jaden McKay Gibson Cap 21h46'15" -14d11'19"
Jaden Noelani Linares Psc 2h1'14" 5d41'12"
Jaden Philphot~Outstanding Student Peg 23h47'30" 25d54'8"
Jaden Roger Cam 3h41'39" 55d23'47"
Jaden Rosenhain Umi 15h42'2" 79d23'1"
Jaden Saadig Collier Uma 11h21'25" 39d41'5"
Jaden Scott Nanney Cnc 9h18'39" 13d50'14"
Jaden Skylar Bennett And 0h0'51" 41d8'53"
Jaden T. Mirza Mohammad-Ali Naraghi Cap 21h57'52" -11d30'58"
Jaden Taylor Gravelding Lib 14h34'1" -17d26'47"
Jaden Taylor Welsh Lyn 7h38'55" 47d15'18"
JaDene Scott Jones Cas 1h0'45" 56d39'27"
Jaden's And 0h10'16" 34d9'27"
Jaden's Dreammaker Aqr 22h42'5" 1d8'41"
JadenSky Cyg 19h59'11" 43d5'9"
Jade's Derbstar Ari 3h4'2" 19d47'19"
Jade's Star Umi 15h10'11" 75d33'32"
Jadeumuah Cnc 8h43'33" 7d3'10"
Jadhira Cnc 9h9'56" 31d42'34"
Jadia Uma 10h15'38" 51d24'5"
Jadiana Psc 1h45'26" 6d29'15"
JADIE Ari 2h48'25" 27d35'55"
Jadma Aqr 22h18'54" -3d6'21"
Jadon Aql 19h41'32" 13d1'27"
Jadon Christopher Vaughan Sco 16h47'33" -27d5'55"
Jadon Curtis Sanchez Uma 10h16'21" 45d32'59"
Jadon Elliot Jackson Gem 7h38'40" 28d7'17"
Jadon Francois Michel Lib 14h49'45" -9d22'31"
Jadon S. Mitchell Sgr 18h13'56" -19d16'8"
Jadon's piece of the universe Lmi 10h40'12" 36d8'42"
Jadra Crt 11h16'31" -24d25'21"
Jadranka Mihic Cru 12h20'45" -61d32'50"
Jadranka-Ivanka And 2h2'43" 46d7'41"
Jadyn Aleczandria McLain Psc 1h57'42" 7d44'42"
Jadyn Alexia Uma 11h30'47" 58d33'44"
Jadyn Paige Johnson Cnc 8h37'58" 24d4'21"
Jadyn Simmons Uma 11h46'38" 65d30'23"
Jadyn Sue Marie Sell Per 3h30'52" 42d27'32"
Jadyn Thomas Ashcraft Lib 14h51'43" -2d27'31"
Jadyn Trivett Per 3h36'19" 48d35'14"
Jadyn Zane Smith Umi 13h7'18" 74d50'11"
JadynLaila And 0h20'35" 24d59'38"
Jadzeea's Night Light (JDE) Lyn 7h14'15" 56d50'30"
Jadzia Dax Tieman Cnv 12h47'34" 44d18'29"
Jadzia Denkiewicz Cap 21h42'44" -16d46'55"
Jadzlaida Avestruz nee Paraja Folk Tau 5h46'20" 26d59'14"
Jae & Bette Edwards Uma 10h11'21" 52d7'21"
Jae H. Chung Gem 6h48'5" 14d3'30"
Jae J. Jang Vir 13h26'53" -12d57'11"
Jaeden Mason Michael 12/16/03 Per 3h21'47" 47d31'27"
Jaeden N Pev Leo 11h26'50" 0d57'3"
Jaeden Starnes Her 16h57'53" 31d55'3"
Jaedon Isaiah Alberts-Wheeler Equ 20h59'31" 5d29'21"
Jaedyn Latrell Pittman Lib 15h8'19" -6d1'42"
Jaedyn Nicole Antle Ari 3h25'39" 29d24'13"
Jaeger Lyn 8h8'46" 39d45'1"
Jaeggi's Celestial Yin-Yang Srp 18h12'53" -0d21'6"
Jael Chac And 0h31'1" 27d50'20"
Jael Fischer Per 4h19'29" 35d7'37"
Jael Guzman Cnc 9h2'32" 28d22'27"
Jaelen McKenley Woodard Boo 14h31'43" 40d14'10"
Jaelize Boogiebites Psc 23h53'33" 5d25'16"
Jaelyn Love Geib Cnc 9h6'1" 8d14'41"
Jaelyn Marie Psc 23h5'14" 6d24'1"
Jaelyn McCurdy Sgr 18h3'48" -25d45'7"
Jaelyn Thurner Peg 23h39'54" 10d16'40"
Jaenelle Contreras Cnc 8h57'47" 23d9'9"
Jaepa Cap 21h12'45" -20d6'32"
Jaerigell Vir 12h43'35" 12d20'29"
J.A.ESCAMILLA Ari 3h0'4" 21d30'31"
JaEse<\@>DocShi Cep 21h10'51" 58d3'20"
Jafar Irshaidat Cap 21h40'13" -8d39'49"
Jafar M. Pierre Cep 21h45'29" 60d55'11"
Jafcor Ori 5h31'33" -5d14'23"
Jaffar Sgr 19h14'20" -18d38'35"
Jaffe Worley Vir 14h6'29" -16d10'3"
Jaffrey A. Adams Ari 2h51'26" 22d25'57"
JaG Tau 4h32'46" 19d15'34"
JAG Peg 23h46'19" 29d45'29"
JAG Umi 17h22'23" 77d9'9"
JAG Uma 13h47'16" 61d57'35"
Jag Chellam Cas 23h20'45" 55d20'47"
JAG My Special Love, Kimosobe Umi 14h58'14" 72d50'58"
Jag Rall Per 2h37'10" 53d51'30"
Jagdfeld, Anne-Maria Ori 5h42'23" 7d21'11"
Jagdiep Dhonsi Peg 23h58'13" 27d40'15"
Jagdip Singh Ghuman Lib 15h37'8" -8d10'49"
Jagella, Peter Ori 6h13'13" 16d35'0"
Jägermeisterlein Remy Cnv 12h45'13" 43d8'25"
Jagger Ori 6h10'46" 15d2'37"
JaggeR Ori 4h54'9" 10d49'49"
Jagger 66 Ori 5h34'55" -0d57'31"
Jagger C. Brown Her 18h6'33" 19d40'36"
Jagger Doss Thompson "Star Is Born" Uma 9h52'8" 53d27'35"
Jagger Risk Sgr 19h27'47" -14d17'8"
Jaggie Per 2h43'35" 50d35'27"
Jagjeet Gill Cam 4h24'13" 63d43'28"
Jago Gem 6h56'50" 18d43'57"
JAGO 95-05 Sco 16h43'36" -31d39'29"
Jagoda Vir 13h17'17" 6d56'53"
JAGS Sgr 18h4'14" -28d8'20"
Jagunne Mist Cam 3h47'12" 61d53'17"
jagusha Her 16h38'52" 6d25'59"
J.A.H. Uma 11h33'16" 64d15'0"
Jahaira Galarza Sco 17h7'47" -40d40'36"
Jahan Khubani Gem 7h17'18" 19d57'54"
Jahangir and Jane's Faith in Love Apu 15h19'1" -71d55'31"
Jahanshir Babadi Tau 3h55'0" 17d39'51"
Jahlani Farai Alamu Uma 11h30'29" 60d43'0"
Jahn, Gunter Uma 8h19'5" 68d6'50"
Jahn, Michael Uma 9h43'19" 72d49'45"
Jahna Elaine Merle Lugnan Cru 12h23'5" -58d32'53"
Jahnavi Jane Stackhouse Cam 3h39'47" 59d55'16"
Jahneen Nadeu Crb 16h16'55" 33d24'56"
Jahs Love Tau 4h12'51" 26d24'29"
Jah's Star Uma 12h57'8" 57d45'53"
Jahvon Sco 17h38'15" -40d56'17"
Jai Gem 7h7'38" 26d37'40"
Jai Ori 5h52'48" 21d2'40"
Jai Dillard Sanford Aqr 21h23'3" -9d23'49"
Jai & Drita Cyg 21h34'25" 44d55'6"
Jai K. Reed-Sweetpea Psc 1h11'25" 21d34'9"
Jai Stephen Cru 12h28'58" -58d51'4"
Jai Tomas Causby Cru 12h36'49" -59d14'35"
Jaianca And 0h25'55" 26d53'42"
Jaida Lee Jacobie Cnc 9h4'8" 30d41'2"
Jaida Lotus Fiji Tau 3h53'4" 26d45'34"

Jaida Monet Marrero
Uma 9h35'59" 44d11'58"
Jaidalee
Cnc 8h29'35" 11d34'19"
Jaidan James
Uma 10h11'41" 59d29'2"
Jaidan L. Swanson
Psc 23h54'56" 1d46'41"
Jaidan Swanson
Psc 1h3'49" 30d44'2"
Jaide Ashli Dixon
Leo 10h18'5" 26d53'20"
Jaiden
Lib 15h3'12" -0d42'47"
Jaiden Amber Kelly
Leo 11h3'21" 8d43'0"
Jaiden David Lopez
Psc 1h33'47" 27d35'12"
Jaiden "JJ"
Ari 3h20'19" 28d59'9"
Jaiden John Connor
Umi 15h54'27" 74d47'25"
Jaiden Lee Shute (Peanut) (I.E.)
Leo 10h35'40" 18d5'30"
Jaiden Lynn Reid
Cnc 8h58'17" 23d59'7"
Jaiden Lynne Algren
Tau 4h27'43" 22d34'58"
Jaiden Melcher
Gem 6h39'52" 24d53'2"
Jaiden Michael
Sco 16h15'24" -14d34'52"
Jaiden Renea Fenton 1-5-2004
Cap 20h48'40" -15d57'34"
Jaiden Sachin Patel
Cnc 8h21'48" 12d49'41"
Jaiden's Wish
Sgr 20h10'33" -39d6'25"
Jaidison Turrill
Peg 22h9'21" 5d57'57"
Jaidon Bleu Pena
Cam 4h0'13" 68d30'42"
Jailene Joanne Adelle Smith
Lyr 18h30'16" 41d32'16"
jaim
Ari 2h20'27" 22d0'36"
Jaime
Ari 2h18'39" 21d19'9"
Jaime
Vir 12h59'42" 7d17'6"
J'aime
Ori 5h11'18" 1d18'47"
Jaime
Peg 21h44'59" 27d31'26"
Jaime
Lyn 8h32'41" 51d10'14"
Jaime
Uma 10h0'53" 46d50'1"
Jaime
Uma 9h34'7" 45d27'46"
Jaime
Uma 9h1'32" 50d23'32"
Jaime
And 0h3'28" 44d7'36"
Jaime
Sco 16h6'8" -15d0'56"
Jaime
Sco 16h7'41" -10d37'33"
Jaime
Lib 15h6'48" -2d39'18"
Jaime
Umi 16h48'23" 75d5'57"
Jaime Adam Fuentes
Aql 20h8'0" 9d53'22"
Jaime & AJ
Cas 1h6'29" 61d52'1"
Jaime Alexander Garcia
Tau 4h25'59" 19d5'16"
Jaime Allyn Crain
Sco 16h47'10" -31d14'42"
Jaime and Konner Stowe
Lyr 18h50'47" 33d52'21"
Jaime Ann Tupper
Vir 14h43'51" 2d53'22"
Jaime Anne Cohen
Cas 1h18'20" 58d44'48"
Jaime & Brian
Cyg 20h23'27" 39d4'10"
Jaime Brooke
Cas 0h23'0" 53d36'9"
Jaime Brooke Ghorieshi
Aur 6h55'11" 38d15'23"
Jaime Bruder
Ori 5h39'27" -0d34'41"
Jaime Buck
Uma 11h25'7" 46d50'18"
Jaime Caviola
Lib 14h28'18" -11d16'37"
Jaime Clemente ~ Faith - Hope
Vir 11h45'9" 8d4'36"
Jaime Contreras
Uma 8h23'4" 64d6'59"
Jaime Couture
Lyn 9h10'9" 33d40'19"
Jaime & Dave's Star
Lyr 18h49'23" 37d12'4"
Jaime Del Poignee
Aqr 21h8'39" -14d25'9"
Jaime Denise Zimerman
Lyr 18h30'54" 31d43'52"
Jaime Donovan, All My Love Forever
Aqr 20h47'37" -7d21'19"
Jaime E. Fromson
Aqr 22h45'32" 2d9'47"
Jaime Elizabeth Danielowski
Ari 3h19'16" 29d51'46"
Jaime Elizabeth Presley
Lib 15h16'6" -23d40'45"
Jaime Elizabeth Yeomans
And 2h37'38" 46d49'19"
Jaime Foster
Aql 19h41'16" -0d3'1"
Jaime Francisco Lagos Garcia
Per 4h22'57" 32d12'6"
Jaime Gonzalez
Lyn 6h55'39" 54d57'53"
Jaime I. Eichenbaum
Lyn 7h32'35" 42d15'23"
Jaime "Jaimes" Enion
Pho 0h6'22" -51d48'39"
Jaime Jane Mush
Uma 12h32'29" 62d58'22"
Jaime & Janine Jordan
Boo 14h58'40" 24d20'49"
Jaime & Joel's Wedding Star
Crb 16h9'48" 38d55'12"
Jaime Kathleen Ruark
Lyn 7h8'44" 58d56'26"
Jaime Klemz
And 0h38'57" 45d11'9"
Jaime L. Staniszeski
Cyg 20h58'56" 45d13'54"
Jaime L Trivette
Her 18h35'28" 15d38'53"
Jaime Lee
Ari 2h39'30" 14d3'11"
Jaime Lee
Sgr 19h30'37" -28d45'40"
Jaime Lee Bee Ching
Cnc 8h2'27" 10d9'39"
Jaime Lee Fusaro
Lib 14h59'3" -24d6'32"
Jaime Lee Jones
Leo 9h43'28" 30d45'29"
Jaime Leigh Drebit
Sgr 19h24'51" -21d6'13"
Jaime Leigh L'Heureux
Uma 12h17'49" 61d13'2"
Jaime leigh Murdock
Gem 7h19'10" 19d8'15"
Jaime Lim S.C.
Sco 17h20'15" -32d15'29"
Jaime Lowe
Aqr 21h32'39" 0d4'14"
Jaime Lyn
Lib 15h5'40" -22d54'51"
Jaime Lyn Coffman
Lib 14h32'19" -11d4'3"
Jaime Lyn Slatt
Uma 8h47'0" 61d44'51"
Jaime Lynette
And 23h37'32" 39d20'29"
Jaime Lynn
Uma 11h33'1" 44d33'11"
Jaime Lynn
Gem 7h39'25" 26d29'43"
Jaime Lynn
Aqr 22h11'16" -2d3'0"
Jaime Lynn
Umi 16h52'22" 76d15'33"
Jaime Lynn
Sgr 18h20'32" -29d53'32"
Jaime - Lynn
Sgr 19h11'7" -36d24'47"
Jaime Lynn Brehm
Gem 6h21'31" 21d28'35"
Jaime Lynn Jablonski
And 1h41'20" 41d51'39"
Jaime Lynn Kelsch
Ari 2h15'26" 11d7'22"
Jaime Lynn Martinez
Uma 9h10'20" 54d26'4"
Jaime Lynn McGrew
Cyg 19h49'31" 31d34'49"
Jaime Lynne Boyle
Cyg 20h5'46" 42d15'43"
Jaime Lynne Helt
Cas 0h16'3" 59d51'18"
Jaime Lynne Stecker
Ari 2h59'34" 18d21'37"
Jaime Marie JMR
Lib 15h34'53" -8d28'16"
Jaime Marie Thies
Vir 14h42'2" 2d45'35"
Jaime Mata
Ori 5h26'53" 3d1'1"
Jaime Maurice Wood
Ari 2h23'39" 18d15'49"
Jaime Nicole
And 0h48'21" 37d47'36"
Jaime Nicole Milano
Cas 0h55'21" 56d32'52"
Jaime Nicole Paddock
Per 4h41'50" 49d19'22"
Jaime Nicole Paddock
Per 4h46'56" 49d5'46"
Jaime Noelle Decker
Psc 0h48'0" 9d57'59"
Jaime Oni
Uma 9h3'30" 59d10'9"
Jaime Reedholm
Lmi 10h10'28" 35d33'0"
Jaime Reyes Fernandez
Her 17h27'2" 27d19'47"
Jaime Reynoso
Uma 11h39'10" 44d28'44"
Jaime Rische
Psc 1h20'41" 14d29'31"
Jaime Rodriguez "I Love You"
Leo 11h3'30" 16d41'36"
Jaime Rose Cardello
Lib 14h53'4" -2d12'14"
Jaime Sanders
Psc 1h3'45" 32d45'31"
Jaime Siobhan Marshall - 18 Oct 1979
Cru 12h39'22" -60d39'27"
JAIME SLANEY
Vir 12h42'25" 1d11'46"
Jaime Starr
Dra 19h42'3" 75d16'7"
Jaime "Sweetness"
Sgr 18h45'11" -18d9'48"
Jaime Tecza
Ori 6h15'54" 14d29'1"
Jaime & Terry
Lib 15h45'1" -5d25'58"
Jaime V's Star
Del 20h28'9" 18d19'35"
Jaime Whitten
Dra 18h20'40" 74d15'48"
Jaime, Lizzie & Kadence Osborn
Uma 13h20'30" 61d26'47"
Jaime, the Light of my Life Always
Cnc 8h17'3" 20d24'31"
Jaime-Brian
Uma 10h55'39" 50d24'51"
Jaimee
Cnc 9h3'27" 12d37'12"
J'Aimee A. Louis
Sgr 18h14'1" -18d10'16"
Jaimee Alexandra Navarrete
Gem 7h2'36" 30d59'57"
Jaimee Bottger
Vir 12h30'38" -0d11'45"
Jaimee Ellen Goad
And 2h35'46" 44d29'50"
Jaimee Lowe
Leo 10h43'32" 10d5'58"
Jaimee McCrimmon
Cru 11h57'8" -59d26'6"
Jaimee Nicole Goodrich
Cnc 8h29'41" 7d17'58"
Jaimee's Guiding Light
Cas 2h40'15" 57d52'59"
Jaime-Leigh Matty
And 2h36'40" 49d44'34"
Jaimelyn
Lib 15h38'40" -3d54'1"
Jaimelyn Whitni Jacome
Leo 9h34'33" 25d39'57"
Jaime's Christmas Star
And 2h37'27" 44d9'31"
Jaime's Lucky Ladybug
Cam 5h3'43" 76d16'42"
Jaime's Shining Star
Psc 0h50'2" 15d58'53"
Jaime's Star
Ori 5h34'29" 4d9'55"
Jaimesyn Delilah
Sco 17h29'20" -40d30'13"
JAIMEY CUTRELL
Cnc 8h59'28" 31d33'57"
Jaimi
Ori 6h7'40" -0d42'50"
Jaimi Brooke Maloff
And 0h15'49" 39d24'24"
Jaimie Catherine Griffin
Sco 17h48'49" -41d11'27"
Jaimie Denise Kos
Tri 2h21'20" 36d51'11"
Jaimie Elizabeth Algate
Vel 8h14'14" -44d52'9"
Jaimie Janell Rosaguera
Leo 11h54'22" 19d32'21"
Jaimie Johnson
Vir 12h25'8" 0d35'54"
Jaimie Johnson-Kline
And 0h5'44" 40d33'0"
Jaimie Lee
And 0h0'24" 45d52'11"
Jaimie Leigh
Leo 11h44'37" 26d16'57"
Jaimie Lynn
Leo 10h15'56" 25d42'57"
Jaimie Lynn Ferretti
Uma 11h40'55" 61d55'43"
Jaimie Lynn Frey
Gem 7h12'6" 34d34'48"
Jaimie Marie Walker
Lib 14h37'44" -9d54'40"
Jaimie Olivia Star
Lyn 7h33'26" 49d20'20"
Jaimie Owen
Umi 13h46'18" 74d7'4"
Jaimie Phillips
Aqr 23h23'38" -9d52'39"
Jaimie Rachel Elizabeth Rivera
Aur 5h10'10" 30d19'27"
Jaimie Reilein
And 2h16'4" 46d0'48"
Jaimies Star
Lyr 18h58'26" 28d6'18"
Jaina
Col 6h11'52" -40d57'47"
Jaina M. Johnson
Ari 2h40'16" 28d31'32"
Jaina Marie Rusnell
Sco 16h26'39" -33d43'36"
Jaine
And 0h35'37" 33d49'25"
Jaira my Love
Psc 23h56'46" -4d11'41"
JairdanSarah
Ori 6h17'0" 10d28'27"
Jaired Riley - Always in my heart
Uma 11h33'49" 64d19'0"
Jaisal G. Sanger
Ori 4h53'2" 11d10'34"
Jaisey
Sco 16h22'6" -26d43'47"
Jaisy-Venus
Cmi 7h23'28" 10d52'58"
Jaiya Gonsalves Ellis
Uma 11h24'6" 34d56'30"
Jajuba
Gem 7h42'18" 15d57'32"
Jak Cameron
Psc 1h8'50" 32d32'24"
JAK*KMB
Uma 10h44'31" 56d10'19"
Jak Owen Harper - Jak's Star
Aql 19h6'4" -7d0'10"
Jakaaf "4-29-06"
Ari 2h33'27" 26d19'29"
Jakaeden
Cas 1h28'8" 58d35'11"
Jakarleon
Lyn 8h44'16" 40d17'48"
Jake
Cnv 12h26'42" 33d26'40"
Jake
Her 17h7'35" 32d44'55"
Jake
Cnv 12h28'22" 46d53'44"
Jake
Her 17h21'56" 26d59'25"
Jake
Uma 8h39'33" 67d52'31"
Jake
Sgr 18h6'35" -27d10'24"
Jake
Sco 16h54'26" -41d56'3"
Jake 99
Cep 22h20'14" 61d59'49"
Jake Adam Shaw
Sgr 18h48'56" -21d38'59"
Jake Alexander Frydman
Aqr 22h51'25" -15d19'39"
Jake & Alisha
Aqr 22h37'17" 0d4'7"
Jake and Becky
Her 18h11'38" 29d22'7"
Jake and Jesse
Vir 13h28'8" -15d2'42"
Jake and Kelly
Uma 11h9'44" 45d58'54"
Jake and Lynn
Cam 7h38'57" 81d18'31"
Jake and Maggie
Cyg 20h48'13" 31d41'43"
Jake and Nicki's Star
Per 4h7'25" 42d30'47"
Jake and Pippa Longley
Uma 10h28'24" 45d31'47"
Jake Andres Cammarata
Boo 14h42'12" 18d37'59"
Jake Andrew Allen
Psc 1h7'53" 29d22'49"
Jake Andrew Wohld
Uma 12h5'4" 33d14'39"
Jake Angelo
Aur 6h34'29" 39d4'41"
Jake Anthony Keebaugh
Uma 9h59'39" 60d1'47"
Jake Arvonio
Uma 11h27'9" 59d28'34"
Jake Austin
Vir 14h17'21" 0d21'18"
JAKE BARDING
Her 16h21'54" 48d46'46"
Jake Bartes
Per 3h16'57" 51d41'19"
Jake Baumgarten
Her 16h45'4" 46d38'0"
Jake Bellis
Boo 14h47'20" 30d31'46"
Jake Brian Harrison
Uma 8h31'34" 71d13'57"
Jake Briscoe
Lyn 6h30'40" 58d7'0"
Jake Brown
Cru 11h56'26" -57d22'46"
Jake Cangiano
Dra 16h10'49" 51d56'9"
Jake Christian Sendaydiego
Per 3h34'36" 44d26'30"
Jake Cohen
Leo 9h41'17" 29d1'32"
Jake Coleman
And 0h47'18" 40d26'31"
Jake Craig Watson
Per 4h13'54" 37d12'52"
Jake D. Hellman
Uma 14h28'9" 59d21'40"
Jake "Daj'a" Medeiros Middleton
Aql 19h35'57" 11d44'50"
Jake David
Uma 8h15'59" 64d14'12"
Jake David Port
Lyn 8h23'45" 51d43'54"
Jake David Sieling
Boo 14h42'9" 51d17'41"
Jake & Dott Schwaner Family
Aql 19h50'9" -0d14'7"
Jake Douglas Humberstone - Jake's Star
Ori 4h51'20" -3d13'34"
Jake Edward Hayden
Dra 19h30'56" 69d56'9"
Jake Elliot Faulkes
Del 20h34'45" 6d49'5"
Jake Elliott
Aur 5h32'12" 45d46'29"
Jake Erik Pettersen
Uma 10h38'9" 65d41'2"
Jake Erik Pettersen August 19, 2002
Leo 10h15'16" 16d3'31"
Jake Evan Trembly
Per 4h48'39" 43d37'0"
JAKE ~ Forever Our Shining Star!
Her 17h11'50" 33d48'18"
Jake Garrison
Tau 4h10'35" 27d21'37"
Jake Gelvin
Her 16h29'28" 22d23'15"
Jake Graham Edwards Owen
Col 5h58'33" -36d47'14"
Jake & Grandad's Star
Aql 19h43'10" 4d30'15"
Jake H. Newcomer
Aqr 22h37'44" -0d36'21"
Jake Hallonquist
Dra 19h20'12" 72d4'12"
Jake Hamelburg
Sgr 19h30'56" -17d0'9"
Jake Harding
Peg 22h9'25" 19d32'32"
Jake Harley Davidson
Uma 10h27'2" 66d16'52"
Jake Harrison Mangini
Per 4h17'23" 40d26'50"
Jake Harvey Hewitt Sullivan
Her 18h44'3" 14d3'36"
Jake & Jill Forever
Cyg 20h1'56" 33d32'6"
Jake Joseph Kennedy
Aqr 22h7'14" 2d32'24"
Jake Joseph Kennedy
Umi 17h23'56" 82d19'14"
Jake Kelly
Her 18h53'18" 23d19'39"
Jake Kiyokane
Mon 7h32'11" -0d45'47"
Jake Lannin
Her 16h31'10" 45d57'24"
Jake Levine
Her 16h13'59" 46d52'48"
Jake Lewis Evans - Jake's Star
Her 17h41'58" 40d44'38"
Jake Liam Garrett "I Love You"
Mon 7h36'25" -0d29'52"
Jake Liu
Lib 14h47'7" -24d53'19"
Jake Lynch
Cma 7h17'51" -11d15'44"
Jake Maass
Cnv 12h44'52" 39d3'58"
Jake Madsen
Gem 7h41'38" 32d17'36"
Jake Marca
Aqr 22h12'54" 0d25'12"
Jake Martin Babigian
Ori 5h14'46" -0d41'30"
Jake Michael Heale
Ari 3h8'9" 29d26'0"
Jake Michael Johnson
Umi 15h16'37" 71d35'21"
Jake Michael van der Vlugt
Boo 14h38'35" 32d7'48"
Jake Michaeldon Edgar II
Ari 2h2'20" 19d57'58"
Jake Monroe Friedman
Ari 3h7'18" 25d46'20"
Jake Montgomery Conti
Cep 21h34'39" 66d18'24"
Jake Mosier
Cnc 8h48'21" 17d47'39"
Jake 'N' Jess
Tau 4h29'50" 18d34'46"
Jake & Nannie's Forever Love Star
Cap 20h22'15" -13d39'1"
Jake Nathan Mercer
Umi 13h44'2" 73d41'55"
Jake Neiman
Lmi 9h48'35" 35d12'20"
Jake Nhan Kuzyk
Umi 16h21'53" 73d30'56"
Jake of Diamonds
Cep 22h33'23" 66d12'48"
Jake Orans
Cnc 8h15'59" 25d11'5"
Jake Pearce Kenny
Lmi 10h41'23" 24d51'20"
Jake Pearson - Jakey Boy's Star
Her 18h31'13" 14d23'28"
Jake Pietersen
Col 6h32'38" -33d42'50"
Jake Quentin Walden
Uma 10h25'23" 54d1'48"
Jake & Rachel
Ori 5h59'13" 21d3'54"
Jake Rector
Lyr 18h45'49" 32d32'1"
Jake Rhys
Per 2h59'42" 42d52'20"
Jake Ridgway
Uma 11h21'34" 64d27'23"
Jake Robert Smith #4
Aur 5h41'2" 46d21'27"
Jake Robert Westaway
Peg 23h57'16" 25d1'20"
Jake Robert Widman
Per 4h33'23" 31d17'46"
Jake Rodgers
Ori 5h30'36" 5d54'15"
Jake Ryan
Uma 8h39'2" 52d59'22"
Jake Ryan Lauro
Cnc 8h36'3" 24d0'36"
Jake Ryan Parry - Jake's Star
Her 18h43'53" 13d35'19"
Jake Sanders
Lyr 18h54'51" 32d29'21"
Jake & Sarah Yunker
Cyg 20h42'24" 43d0'35"
Jake Scot McDonald
Ari 2h32'21" 25d34'58"
Jake Scott
Ari 2h43'41" 27d41'45"
Jake Skelton
Uma 9h48'31" 58d16'20"
Jake Smith
Uma 10h18'0" 41d54'41"
Jake Stan Parkes
Umi 16h19'25" 73d39'23"
Jake Star
Leo 11h23'59" 7d9'54"
Jake Szlakowski
Cma 7h11'15" -25d11'3"
Jake Taylor Pontz
Leo 10h57'4" 8d41'1"
Jake Textor
Her 16h34'9" 34d24'54"
Jake The Snake
And 0h51'31" 39d58'58"
Jake Thomas Alan Thomson
Uma 9h2'8" 59d0'37"
Jake Thomas Farnsworth
Cep 21h57'36" 63d26'53"
Jake Thomas Martin Woodhouse
Per 4h34'5" 39d5'42"
Jake Thomas Mayo
Tau 5h34'15" 19d50'51"
Jake Thomas Pike Warren
Aur 5h46'31" 42d6'30"
Jake Thomas Sullivan - Jake's Star
Umi 13h13'59" 75d44'51"
Jake Thorn
Per 4h49'43" 41d34'9"
Jake & Tiff Goodin Est. 11-20-2004
Lyr 18h46'47" 36d9'59"
Jake Tyler
Dra 15h7'54" 59d54'39"
Jake Tyler Reuter
Psc 1h26'33" 2d50'1"
Jake Vitrofsky
Ari 2h15'13" 25d10'24"
Jake W Carlson
And 0h47'30" 35d54'33"
Jake W. Sherrill
Uma 13h39'54" 48d58'15"
Jake Walklett
Cep 20h12'31" 61d38'34"
Jake Wallen Franzen
Vir 12h6'2" 6d29'5"
"Jake Walters Bright Night Light"
Psc 1h45'29" 5d25'48"
Jake Whiting
Gem 7h28'32" 20d29'21"
Jake William Bredin
Per 4h25'0" 49d53'52"
Jake William Galloway
Per 4h18'37" 45d13'11"
Jake Wilson Miller
Uma 8h31'45" 66d40'54"
JakeCHill
And 2h25'57" 45d14'15"
JakeJohns
Her 18h9'33" 22d33'20"
Jakelinne Leon
Lyr 18h51'43" 39d29'30"
Jake's Asteroid B-612
Ari 2h10'0" 25d16'58"
Jake's Aurora
Aql 19h29'19" 3d59'31"
Jake's Bar Mitzvah Star
Umi 14h22'16" 75d16'3"
Jake's Bright Star
Her 17h53'39" 17d17'2"
Jake's Light
Cnc 8h40'23" 31d11'43"
Jake's Light
Umi 13h25'42" 70d54'25"
Jake's Star
Lmi 9h41'14" 38d22'58"
Jakestar
Dor 5h6'45" -69d31'21"
JAKESTAR96
Aqr 22h30'47" -1d1'42"
Jakester's Sparkle
Aql 18h50'24" -0d36'5"
JakeSumner & KatieWiegand FOREVER
Ari 3h9'28" 11d17'4"
Jakey
Leo 9h22'16" 10d35'31"
Jakey's Star
Per 3h7'53" 47d36'57"
Jakhye Jonathan Wilkins
Uma 14h16'23" 57d15'19"
Jakiratso
Umi 15h15'17" 74d57'55"
Jaklin Dauterman
Cyg 21h55'44" 49d10'19"
JakLlyWu
Sco 16h28'52" -40d5'19"
jaklyi
Lyr 18h41'17" 34d59'31"
Jakob Alexander Stawowy
Dra 17h30'56" 67d23'34"
Jakob Brett Gates
Lib 15h13'52" -15d36'5"
Jakob Dächert
Ori 5h14'14" -6d24'19"
Jakob Dyer Archibald Niece
Crv 12h20'18" -18d7'50"
Jakob Haag
Ori 6h16'51" 7d29'15"
Jakob Hartnagel
Ori 6h18'40" -1d22'22"
Jakob + Jenny / Hammer
Del 20h46'53" 15d12'24"
Jakob Leon's Stern
Leo 10h21'33" 24d4'7"
Jakob Loves Sarah
Sgr 19h16'9" -15d24'28"
Jakob Lowell Langholz
Aur 5h5'11" 33d34'53"
Jakob Nankervis
Leo 10h25'34" 13d36'55"
Jakob Palmer Anderson
Cap 20h49'51" -20d36'51"
Jakob Peter
Cnc 9h7'3" 32d9'38"
Jakob Yankauskas
Uma 11h26'0" 38d12'35"
Jakober, Reinhold
Uma 9h2'37" 52d6'20"
Jakobs, Nicole
Uma 9h41'30" 48d46'25"
Jakov Mesenzhnik
Uma 11h54'35" 46d15'13"
J.A.K.S.
Mon 7h15'10" -0d43'53"
Jakub Jedlicka
Uma 9h53'41" 50d20'7"
Jakye Jackson
Uma 10h16'55" 63d33'30"
JaLa381
Per 3h18'34" 46d13'3"
Jalagi Udoda
Her 17h27'54" 36d39'21"
Jalal Shahda
Mon 6h50'42" -6d31'40"
Jalan
And 0h36'52" 45d26'59"
JALANA
Lib 15h15'58" -24d6'18"
Jalana Jean Palma
Cyg 19h47'5" 31d26'40"
Jalana Palma
Tri 2h26'23" 35d57'8"
Jaleesa Charelle
And 0h30'5" 27d18'21"
Jaleesa Collins
Vir 12h17'27" 13d6'10"
Jaleina Dee Aboud
And 1h5'37" 43d50'50"
Jalen
Gem 7h3'10" 10d54'15"
Jalen Rose
Cnc 8h42'58" 28d3'46"
Jalena Christina
Cap 21h36'0" -13d31'24"
Jalene
Aqr 23h3'33" -7d12'15"
Jalexis
Sco 16h53'3" -38d36'29"
Jalila
Psc 0h57'50" 8d9'30"
Jalina Jezabel Gutierrez
Sgr 18h57'6" -26d1'9"
Jalina Wolfe
Lyr 19h10'31" 26d28'29"
Jalissa - Our Precious Princess
Cra 18h19'41" -42d17'19"

Jalmer Ray Mattila
Ori 5h33'21" 1d38'39"
Jalona Lenae's Star
Uma 9h33'50" 67d20'18"
Jalosu
Cnc 9h8'17" 16d17'50"
Jaltarich
Cru 12h37'4" -58d21'37"
Jalynn Browning
Gem 7h33'12" 26d0'14"
Jalynn Marie
Lib 14h38'30" -12d8'18"
"JAM"
Uma 8h43'21" 72d41'33"
JAM
Ori 6h11'52" 7d13'58"
J.A.M. Forever
Dra 20h11'44" 69d39'58"
J.A.M. Loves R.M.M.
Cyg 20h15'25" 44d56'18"
Jama
Peg 22h30'8" 8d2'53"
Jamaal Kenyatta Soweto McCoy
Cnc 8h59'58" 27d40'49"
Jamaal Solomon Jeremiah Womble Jr.
Sco 17h25'48" -34d31'47"
Jamaica me Crazy
Ori 5h36'18" -3d14'53"
Jamal
Uma 10h36'21" 45d11'53"
Jamal Allen Polk
Cnc 8h36'34" 14d41'49"
Jamal Dimashk
Col 5h54'10" -33d50'13"
Jamal Jafar Mohamed Badr Al-Kathiri
Aql 19h18'28" 0d25'53"
Jamal Lewis
Umi 14h52'11" 72d17'1"
Jamal "Yohance" Richardson
Cep 23h5'13" 83d32'39"
Jamar Laster
Crb 16h22'21" 38d17'0"
Jamar "My Star Forever"
Uma 9h38'17" 72d7'50"
JAMAR TEODOSIO
Uma 12h0'2" 39d48'58"
Jamarkoe
Uma 12h8'28" 51d11'49"
Jamarlia Yassin Warner
Lyn 9h14'50" 37d33'19"
Jamas Solo
Ori 4h50'29" 11d20'55"
Jambery
Uma 10h18'54" 55d28'19"
JamCin : James and Cindy Bedard
Men 5h22'52" -73d39'51"
J.A.McLinden
Cas 1h2'18" 63d21'12"
Jamealya Holt
Uma 13h45'37" 54d55'55"
Jamee
Peg 21h46'17" 7d42'39"
Jamee Mitchell
And 0h26'20" 42d53'26"
Jameel - Helow Mouna
Per 3h17'41" 53d57'6"
jameg
Leo 11h24'31" -0d36'13"
Jameilah Torres
Lib 14h24'29" -18d57'19"
Jameixel ILFP
Uma 12h11'22" 53d13'22"
Jamela
Gem 6h54'13" 33d22'59"
Jamele F. Toolis
Lyn 6h48'52" 54d0'40"
Jamelen
Cra 18h53'56" -40d32'56"
Jameran
Uma 8h46'55" 47d1'1"
Jamery
Psc 1h19'55" 26d52'6"
James
Leo 11h43'38" 27d19'20"
James
Leo 11h41'26" 13d51'57"
James
Psc 0h39'46" 9d3'39"
James
Aqr 22h28'56" 1d36'43"
James
Leo 11h4'27" 3d13'7"
James
Per 4h23'3" 50d34'57"
James
And 1h5'33" 46d6'39"
James
Sco 16h58'41" -38d52'17"
James
Aqr 23h12'34" -20d23'22"
JAMES
Uma 10h5'14" 56d51'54"
James
Aqr 22h23'34" -14d34'13"
James
Psc 23h57'19" -4d24'57"
James A. Bannister
Vir 13h16'43" 6d14'25"
James A. Berta
Lyn 7h39'29" 48d46'25"
James A. Carnes III
And 1h35'10" 41d20'29"
James A. Dare
Uma 11h31'59" 50d56'26"
James A. Dasaro
Ori 4h46'17" 1d52'8"
James A. Davis
Aur 5h37'56" 49d30'47"
James A. DeAtley
Her 17h59'50" 23d52'47"
James A. Evers
Cep 0h23'32" 72d12'20"
James A. Fitzgerald
Uma 11h23'32" 64d18'25"
James A. Gjesvold
Tau 4h42'23" 21d55'49"
James A. Greco
Lib 14h49'22" -17d17'37"
James A. (Jim) Valentino
Vir 13h37'54" -3d29'6"
James A. Kennedy, Sr.
Uma 9h45'54" 46d34'24"
James A. Kiphart Jr.
Uma 8h16'6" 67d1'19"
James A. MacIntyre #1 Dad
Uma 12h5'24" 63d49'35"
James A. & Mary E. Moran
Psc 0h35'53" 19d43'3"
James A. McInerney
Lac 22h34'8" 50d26'24"
James A. Miller
Dra 16h38'43" 63d8'10"
James A. Moore
Her 18h36'56" 21d43'59"
James A. Needham
Leo 10h21'26" 24d50'38"
James A. Reynolds
Lib 14h53'45" -0d37'8"
James A. Robertson III
Cep 20h48'5" 68d51'1"
James A. Schmieder
Lib 15h11'21" -20d33'8"
James A Smitley
Aql 19h45'16" 15d29'36"
James A. Todd
Uma 13h36'24" 59d43'45"
James A. Wors, Jr.
Ori 6h11'52" 12d0'49"
James Aaron
Vir 13h20'47" 6d42'10"
James Abry
Lyn 7h57'30" 56d14'13"
James Adam
Sco 16h7'8" -8d29'25"
James Adam Gilliatt
Her 17h49'19" 22d25'15"
James Addison Trillo
Aur 5h8'20" 36d5'5"
James & Agnes Flynn
Cyg 20h39'27" 38d15'46"
James Alan Galley
Cep 2h17'17" 81d4'3"
James Alan Kase, Sr.
Cep 22h41'11" 74d49'15"
James Alan Keller
Boo 14h40'45" 37d0'2"
James Alan Massey
Per 2h53'8" 50d40'26"
James Alan Mazurkiewicz
Cyg 19h35'47" 30d3'35"
James Alan Perna
Lib 15h8'0" -7d59'5"
James Alastair Tain Teasdale
Per 3h1'25" 46d28'8"
James Albert Bochert
And 0h46'19" 44d25'56"
James Albert Miller
Ori 5h12'56" 6d55'25"
James Alburn Omps
Aqr 20h45'43" -8d49'37"
James Alcorta Garza
Boo 14h16'39" 24d43'49"
James Alden Harpell
Lyn 7h51'47" 56d55'45"
James Alderfer
Leo 11h46'44" 22d23'58"
James Alexander
Uma 10h51'56" 45d56'34"
James Alexander Bartlett
Ori 6h17'45" 1d26'25"
James Alexander Crowe
Crv 12h41'5" -20d6'38"
James Alexander Harris
Pho 1h51'8" -43d8'58"
James Alexander Herbert
Per 3h4'10" 45d28'5"
James Alexander Irving "Little Jim"
Ori 6h17'8" 18d53'3"
James Alexander 'Jim' Bacon-15.5.50
Cru 12h21'29" -61d24'8"
James Alexander North Star Kenning
Sgr 19h17'54" -16d38'47"
James Alexander Stewart 100626
Uma 8h49'58" 63d22'31"
James Alfred Dobner
Cep 21h16'3" 83d3'59"
James Alfred Hamilton (Jim)
Uma 13h29'20" 59d36'46"
James & Ali Tully
Cyg 21h16'37" 38d12'6"
James Allen Caza
Cnc 8h37'20" 16d4'41"
James Allen Cottam
Sco 16h56'8" -38d37'2"
James Allen Dame, Sr.
Cnc 8h31'53" 23d7'50"
James Allen Forester 9/22/1926
Cyg 19h37'20" 30d5'41"
James Allen Moody
Uma 8h45'16" 57d53'48"
James Alley IV
Leo 9h56'58" 19d36'25"
James Allington Reid II
Gem 6h40'5" 23d41'23"
James Altomare
Her 17h15'49" 20d40'54"
James Alvy Fowler IV
Vir 15h4'2" 6d20'26"
James Amihyia-Marsden
Her 18h25'45" 15d15'59"
James Amos Yaeger
Leo 10h46'37" 12d15'59"
James & Anastacia's Soul-Mate Star
Crb 16h3'31" 35d39'31"
James and Amy's Star
Aqr 20h51'36" 0d37'37"
James and Arika Hoscheit
Leo 10h12'9" 14d55'58"
James and Betsy Hannan
Umi 15h16'54" 76d38'3"
James and Bonnie Clemency
Per 3h18'38" 32d35'6"
James and Carolyn Williams 143
Leo 9h50'32" 22d12'19"
James and Cristen
Uma 10h15'16" 69d36'53"
James and Cynthia Lenox
And 1h12'11" 38d36'48"
James and Donna Rensel forever
Cyg 21h46'26" 49d30'21"
James and Emma Liley's Wedding Day
Cyg 20h5'2" 37d14'21"
James and Eve
Ori 6h17'36" 10d47'1"
James and Hayley Hamilton
Per 3h51'11" 34d0'5"
James and Janet Carr
Leo 10h41'22" 8d22'38"
James and Jean Rice
Lyr 18h29'27" 28d22'29"
James and Jenna Parker
Umi 14h37'49" 75d0'0"
James and Jennifer Long
Crb 15h34'14" 33d25'47"
James and Jennifer Meddaugh
Crb 15h50'26" 36d2'42"
James and Kelly Olson
Uma 8h13'22" 70d4'43"
James and Kristina
Uma 10h30'6" 47d10'32"
James and Kyle Laing's Star
Gem 6h29'39" 27d9'17"
James and Leslie Allen
Crb 15h31'20" 32d14'22"
JAMES AND LINDA PATTERSON
Sco 17h22'54" -32d11'58"
James and Lucille Schroeder
And 1h2'26" 45d27'21"
James and Marcella Tracy
Umi 15h18'49" 76d45'53"
James and Mary Alice Funkhouser
Cyg 20h52'0" 47d6'40"
James and Mary Lorenzen
Uma 13h20'44" 57d28'59"
James And Maxine Talbot
Crb 16h2'28" 30d17'12"
James and Melanie
Lyn 7h52'33" 38d56'4"
James and Michelle
Sco 16h1'45" -21d52'17"
James and Naomi Garcia
Umi 16h18'9" 75d26'38"
James and Nicole
Uma 8h47'23" 50d32'8"
James and Nikki Cottrell Family
Lyr 18h54'23" 31d12'20"
James and Nikki's Eternal Light
Cnc 8h33'32" 11d22'39"
James and Oma 51
Lib 14h53'24" -1d53'32"
James and Pamela Perrins
Uma 8h46'9" 52d41'40"
James and Rachel Hammond
And 1h45'3" 36d34'4"
James and Shannon Gill
Cnc 9h6'59" 32d8'48"
James and Shannon Smith Wedding Day
Uma 11h19'52" 36d16'2"
James and Shirley Schaad
Cas 0h30'49" 54d12'54"
James and Simone White
Her 18h54'41" 24d10'36"
James and Summer Eternal Love
Cir 14h59'44" -66d30'18"
James and Tanya Martin
Cyg 19h49'51" 31d8'28"
James and Yuridiana
Cyg 20h7'54" 47d54'56"
James & Andrea
Tau 5h31'15" 26d17'47"
James Andrew Bodin, Jr.
Umi 13h28'21" 71d55'36"
James Andrew Coco
Cap 20h13'31" -25d27'26"
James Andrew Crandall
Her 17h1'9" 46d25'57"
James Andrew McKinney
Gem 6h59'59" 13d21'31"
James Andrew Tomasiello
Psc 1h41'31" 24d36'28"
James&Angel05
Cap 21h31'43" -14d59'11"
James Angelo Guerin
Umi 15h58'5" 79d5'37"
James Angelos Aitken
Lyn 7h40'29" 53d4'3"
James & Angie's Anniversary Star
Cyg 21h18'9" 44d18'10"
James Anthony Broaddus
Gem 7h22'44" 22d52'23"
James Anthony Creighton
Col 5h30'5" -34d33'32"
James Anthony Ertan Moore
Gem 6h30'26" 19d58'56"
James Anthony Hutchings
Tau 4h12'33" 4d36'40"
James Anthony Katz
Cap 20h58'47" -26d35'14"
James Anthony Lee Wooten
Per 3h45'11" 43d58'30"
James Anthony Mulcahy
Uma 10h8'15" 67d38'8"
James Anthony Ollier - Jimmy's Star
Per 4h3'2" 42d8'22"
James Anthony Radosevich
Her 16h17'25" 19d13'50"
James Anthony Thomas, Jr.
Aql 19h37'14" 7d39'48"
James Anthony Walsh
Her 17h35'26" 21d39'17"
James Anthony Wilson
Uma 11h47'15" 46d43'26"
James & April Perkins
Cyg 21h37'16" 47d36'15"
James Arjun Schmulen
Umi 15h12'27" 69d14'25"
James Arron Fage
Ari 2h40'3" 27d15'11"
James Arron Nicholas
Aqr 22h8'44" -2d22'29"
James Arthur
Peg 22h21'48" 33d39'27"
James Arthur Alexander Jr.
Pho 0h31'34" -43d38'21"
James Arthur Duffy
Her 16h30'34" 14d15'27"
James Arthur "Little Man" Sullivan
Lyn 7h53'28" 57d16'45"
James Arthur Mepham II
Gem 7h38'38" 20d23'29"
James Arthur Netterstrom
Psc 0h13'46" 7d50'25"
James Arthur Rayburn
Lmi 9h51'22" 36d49'36"
James Atwood
Leo 9h41'34" 11d49'13"
James August
Uma 11h5'8" 66d28'59"
James Aurora, Jr.
Sco 17h18'6" -36d23'53"
James B. Clifford
Psc 23h6'3" 1d55'32"
James B Codlin
Sco 16h46'19" -30d34'25"
James B. Culliton
Sco 16h6'43" -18d59'24"
James B. Fuller
Uma 12h4'37" 42d50'49"
James B. McGee
Cep 22h18'1" 63d33'7"
James B. McGrath 4th
Leo 11h13'54" 15d33'39"
James B. Moses, Jr.
Her 17h22'56" 38d30'43"
James B. Rudberg
Ari 2h43'1" 26d49'26"
James B. Spicer
Boo 14h42'13" 28d12'23"
James B. Spicer
Aql 19h0'59" 14d45'10"
James Babcock
Aql 19h30'7" 11d12'14"
James Bachmeier
Lmi 10h16'21" 31d56'12"
James Barber
Lyn 7h54'47" 51d55'11"
James Barcelona "04/06/47"
Ari 2h16'34" 21d59'52"
James Bargone
Uma 10h16'23" 60d22'8"
James Barson
Psc 0h55'40" 8d38'37"
James Barson
Psc 23h32'52" 4d6'29"
James Baum
Aql 19h2'1" -0d3'1"
James "Bease" White
Cyg 20h7'9" 56d0'40"
James Benoit
Psc 0h43'8" 10d3'40"
James Benson Davis, Sr.
Per 2h44'2" 53d8'20"
James Benton Plummer
Sex 10h28'13" -0d51'51"
James Berard
Her 17h52'45" 14d25'50"
James Berger LaBarre
Uma 11h40'39" 32d44'12"
James Berlin Dean
Sgr 19h20'52" -22d20'12"
James Bernard Augustine
Uma 10h50'43" 56d33'17"
James Bernard Hall Jr.
Per 3h34'2" 43d45'45"
James Bernard Larson
Leo 10h49'20" 18d30'6"
James Bernard Palmier
Per 4h39'52" 49d15'50"
James Berry
Per 3h29'48" 47d38'32"
James Bertrand II
Lyn 6h36'36" 58d48'2"
James Best
Lyn 8h48'36" 46d30'59"
James Biggi
Aur 5h49'2" 46d2'44"
James Bissell Woodcock
Gem 7h6'36" 33d32'1"
James Blake Harnagel
Cyg 21h32'22" 49d30'7"
James Blake Yokum
Lmi 9h36'50" 36d34'12"
James Blankenship
And 23h12'14" 42d49'39"
James Bliven
Uma 10h5'41" 46d32'35"
James Bonacci
Lmi 10h1'35" 37d41'23"
James Bond
Aql 19h46'10" -0d36'24"
James Bone
Cap 21h29'10" -13d23'57"
James & Bonnie Heidenberger
Lyr 18h33'37" 36d21'27"
James 'boo boo bear' Coulen
Cep 20h9'11" 60d34'50"
James Bookhamer Sr.
Per 2h43'58" 57d10'14"
James Bouchard
Tau 5h49'42" 13d14'0"
James Boyd Holley
Ori 5h30'32" 1d23'21"
James Boyd Ihle
Lyn 7h38'45" 51d34'44"
James Boyd Lamb
Gem 7h1'19" 27d48'59"
James Bradford Webster
Uma 11h5'33" 36d25'27"
James Bradley Bates
Per 2h37'34" 56d8'13"
James Bradley Burkett
Ori 6h15'35" 14d54'30"
James Bradley Burkey
Dra 19h18'20" 60d21'35"
James Braniff
Lyr 19h12'11" 37d27'11"
James Braydan
And 0h36'8" 31d5'56"
James Brent Willis 10-11-1958
Lyn 8h4'29" 56d53'32"
James Brett Keller
Per 3h25'58" 47d26'7"
James Brian Bates
Aqr 22h40'21" -12d1'0"
James Brian Manghan
Her 16h17'41" 6d32'47"
James Brian Tibbit
Boo 14h45'59" 33d26'7"
James Brittain Ferguson
Ari 2h7'42" 17d42'40"
James Bruce Gilliland
Psc 23h58'58" 5d7'52"
James Buchanan
Her 16h31'16" 32d54'36"
James Buford Paul
Vir 12h55'16" -11d1'7"
James Burns
Cep 22h34'21" 74d57'10"
James Burns Lees
Umi 3h57'59" 88d39'54"
James Burns Lees
Tau 3h42'54" 16d22'55"
James Burr Hall
Lac 22h25'19" 42d46'9"
James Bynum Knoderer
Boo 14h41'34" 34d46'37"
James Byron Ward
Lib 15h13'23" -18d47'1"
James C Burge
Ari 2h47'38" 15d26'50"
James C Coulthard
Leo 10h54'43" 14d40'13"
James C. Darling
Lyr 18h35'30" 39d47'57"
James C & Evelyn O Ross
Cyg 21h31'8" 52d32'52"
James C. Jensen Sr.
Ori 5h53'46" 21d35'44"
James C. Maxwell
Cep 22h24'26" 63d8'47"
James C. Neal
Hya 9h13'15" -0d16'9"
James C. Newcom
Per 3h17'17" 52d44'9"
James C O'Connor
Uma 10h31'30" 49d13'25"
James C Ostendorf
Umi 15h22'36" 68d51'25"
James C. Robbins
Leo 11h43'12" 26d8'57"
James C. Ross
Per 4h37'33" 39d15'54"
James C. Wright
Her 17h48'1" 40d18'10"
James C Yelland
Umi 13h24'59" 72d41'38"
James Callum Jones
Her 18h47'58" 14d3'35"
James Calvin Barbery III
Aql 19h34'12" 6d38'26"
James Canfield Small
Tau 4h5'2" 30d57'39"
James Cannavino
Sgr 17h58'13" -17d16'35"
James Capper
Cru 12h17'49" -56d18'12"
James Carl Bruse
Ori 6h7'19" 7d25'13"
James Carl Porter
Uma 11h16'24" 58d28'4"
James & Carolyn Frenkil
Cyg 21h19'16" 54d19'51"
James Carroll Greenan
Sgr 19h39'43" -14d16'48"
James Cartwright - Jay's Star
Per 3h35'59" 49d16'20"
James Casey
Cep 21h13'49" 58d59'50"
James Caula Donnelly
Her 17h28'54" 43d0'10"
James Chapple
Cep 22h45'34" 60d49'56"
James Charles Mueller (aka Teddy)
Leo 9h37'35" 27d16'18"
James Charles Terbush
Cnc 8h52'45" 26d26'22"
James Charles Tipton
Ori 5h22'5" 2d19'7"
James "Chief" King
Cap 21h46'55" -11d28'58"
James Childers
Aql 20h9'21" 10d34'21"
James Christian Martin Sr.
Uma 11h4'17" 44d39'34"
James & Christina Munro (10/07/1965)
Cyg 19h38'6" 35d38'33"
James Christopher Grant - Jim's Star
Uma 13h26'30" 53d18'5"
James Christopher Santos
Aqr 22h50'55" -5d26'43"
James Ciarlante
Uma 10h3'42" 61d21'39"
James & Clare DePinto
Cyg 20h35'49" 34d21'12"
James Clark
And 0h28'16" 32d56'31"
James Clay Hankla
Lib 15h44'28" -8d29'7"
James Clayton Hunter
Ori 4h48'31" 11d5'30"
James Clifford Todd
Her 18h52'34" 23d42'23"
James Coburn
Uma 11h26'13" 39d5'22"
James Cody Hilliard
Boo 14h23'2" 16d7'18"
James Cole Durham
Vir 13h22'55" 11d35'46"
James Comfort
Ori 6h18'24" 13d54'35"
James Cordero
Uma 13h26'58" 56d49'57"
James Courtney
Cru 12h37'26" -57d54'48"
James Coyne aka Pop
Sgr 19h23'29" -20d15'51"
James Craig Christianson
Vir 12h39'20" 1d25'54"
James Cruz II
Cep 21h48'50" 59d0'2"
James Cuthill Ogg
Per 2h13'49" 51d27'1"
James D. Botte
Her 16h21'0" 49d36'34"
James D. Buckley
Cnc 8h52'45" 7d57'30"
James D. Connor
Cap 21h49'50" -24d4'50"
James D. Curran
Gem 6h25'13" 18d32'47"
James D. Fulton
Sgr 18h13'59" -26d23'25"
James D. McDermott
Uma 10h20'19" 51d51'58"
James D. Miller husband to Patricia
Aqr 22h19'26" -16d38'56"
James D. Reed
Lib 14h53'59" -5d41'9"
James D. Richardson
Vir 13h46'3" -10d15'31"
James D. Trumble
Lib 14h23'57" -11d1'0"
James: Dad of the Universe
Cas 0h10'47" 60d39'2"
James Damien
Per 3h9'39" 45d45'46"
James Daniel Alfandre
Ori 5h53'46" 20d45'53"
James Daniel Berru
Ori 5h4'30" -1d13'52"
James Daniel Bowe
Per 2h26'23" 55d19'49"
James Daniel Brear
Per 4h45'49" 49d0'2"
James Daniel Burke
Equ 21h16'13" 10d59'29"
James Daniel Major
Per 3h37'18" 46d22'20"
James Daniel Reeves "2-97-87"
Ori 5h18'43" 1d41'42"
James Darryl Hunt Shining Star
Ori 5h58'51" 17d17'0"
James Daryl Pitts
Her 16h50'38" 17d11'18"
James David
Psc 0h35'47" 13d27'10"
James David Auslander
Ari 2h9'23" 22d56'31"
James David Carr
Cep 20h53'12" 59d37'33"
James David Edie
Lib 15h12'59" -25d31'58"
James David Hamer
Per 4h27'57" 34d14'20"
James David Lee Russell
Sgr 18h24'56" -23d4'6"
James David Rathke
Psc 23h40'35" -1d8'11"
James David Shepherd
Lyn 8h9'5" 58d54'0"
James David William Brooks
Uma 10h42'7" 72d46'23"
James Davidson Heffernan Jr.
Vir 11h45'50" -5d14'11"
James Davies-Bonner
Umi 14h32'21" 82d13'54"
James Dean Stoy
Aql 18h46'5" 8d53'29"
James Dean Zickler
Uma 13h54'28" 49d38'15"
James DeAngelis
Cep 22h17'43" 61d1'44"
James Debney
Per 4h26'11" 34d45'51"
James Declan O'Hara
Her 17h58'30" 23d18'2"
James Deon Reed
Equ 21h7'45" 11d45'14"
JAMES DI GUISEPPI
Lyn 7h27'48" 57d14'9"
James Dignam Grady
Ori 5h42'53" 0d33'30"
James Dillon Dellobuono
Dra 16h6'29" 61d30'27"
James Dobson Branham "Granddaddy"
Umi 12h24'7" 87d6'19"
James Don Montgomery
Ori 5h34'23" -4d30'14"
James Donald Catchen, I
Psc 0h9'56" 7d22'4"
James Donald Deans
Aql 19h17'52" 9d17'23"
James Donald (JD) Latta
Uma 8h37'9" 65d37'40"
James Donald Rice
Ari 1h51'26" 17d43'11"
James Doran
Her 17h39'44" 36d55'47"
James & Doreen For Eternity xxx
Cyg 19h45'51" 50d17'10"
James Dorsey, Sr.
Cyg 21h13'34" 47d9'9"
James "Doug" Schneeberger
Ori 6h3'23" 16d55'58"
James Douglas
Cru 12h26'17" -56d58'48"
James Douglas Catania
Tau 5h40'11" 26d26'54"

James Douglas Connor Psc 23h48'44" 7d6'7"

James Douglas Foulkes Sco 16h14'19" -33d26'4"

James Douglas Haney Forever Shines Ori 5h33'45" 1d56'46"

James Douglas Stapleton Per 4h47'55" 47d42'54"

James Dylan Setzer Gem 7h16'38" 29d5'39"

James E. Aiken Ori 5h55'20" 21d11'38"

James E Beckum Ari 2h16'47" 23d29'48"

James E. Bjork, Jr. Ori 5h22'21" 1d9'3"

James E. Bowers Lib 15h27'11" -8d19'10"

James E. Davis "PaPaw." Uma 11h16'10" 40d13'36"

James E. Duncan Per 3h47'11" 50d47'19"

James E. Dymacek Psc 1h35'37" 21d22'18"

James E. Hamilton "2-17-1996" Uma 10h58'44" 45d47'11"

James E. Hanlon III 6-14-1952 Gem 6h47'17" 23d59'23"

James E. Hathaway Leo 10h43'50" 10d52'41"

James E. & Janet S. Parks Lib 14h58'29" -14d45'39"

James E. Lewis, Jr. Ori 5h19'25" 6d2'29"

James E. Lyons Boo 14h31'14" 32d32'38"

James E. Martin son of Paul+Flavia Per 4h45'48" 43d42'55"

James E. McCannell Lib 15h4'49" -19d5'58"

James E. McPheters Leo 11h16'51" 13d27'18"

James E. Novitzki Aql 20h16'55" 1d10'40"

James E. Osborne Cnc 8h29'46" 23d42'27"

JAMES E. PEMBERTON, JR. Uma 9h17'25" 65d44'31"

James E. Perrier Ari 2h53'6" 30d0'18"

James E. Reynolds Lib 15h39'31" -14d20'7"

James E. Schlink Cap 20h45'54" -24d31'2"

James E. Solberg Cap 21h1'25" -22d8'25"

James E. Summers Per 2h36'37" 53d22'49"

James E. Tennant Her 16h13'59" 24d44'44"

James E. Vasil Her 16h43'57" 37d54'46"

James E. Whitaker And 23h19'29" 47d59'2"

James Earl Finnegan Uma 9h57'53" 52d15'23"

James Earl Hicks Leo 11h7'2" 20d37'7"

James Earl Hines Ori 5h4'29" 2d30'17"

James Earl Peters Gem 7h38'29" 25d43'16"

James Eastwood Cnc 8h41'42" 13d34'37"

James Edison Mullins Ori 5h30'54" 2d0'7"

James Edward Vir 12h12'48" -0d54'34"

James Edward Bowers Psc 0h44'38" 15d41'58"

James Edward Carbine IV Ari 3h13'33" 21d3'50"

James Edward Christen 04-03-1941 Ari 2h16'58" 12d2'1"

James Edward Daly Sgr 18h48'46" -27d45'26"

James Edward Denbow Uma 9h9'0" 58d40'17"

James Edward Denzin Per 3h16'20" 44d32'58"

James Edward Gaston Per 2h26'51" 54d16'52"

James Edward Maltby - Jim's Star Ori 6h18'55" 16d25'37"

James Edward Mullen Lyn 7h47'54" 37d22'16"

James Edward Murray III Ori 6h7'48" 13d52'49"

James Edward Newman Cnc 8h32'42" 30d38'16"

James Edward Palovchik Uma 12h5'32" 54d56'48"

James Edward Patrick - "Pooh's Star" Umi 14h15'26" 69d42'26"

James Edward Peppers Tau 4h17'51" 7d4'41"

James Edward Roycroft Uma 10h31'14" 67d46'6"

James Edward Salisbury Cep 21h42'31" 65d55'20"

James Edward Smiley III Her 16h45'27" 31d32'41"

James Edward Snow Her 18h47'22" 20d9'22"

James Edward Todd Healer Uma 9h22'51" 67d5'54"

James Edwin Brooks Dra 15h56'27" 55d17'46"

James Edwin Dorsey Tau 5h37'2" 25d35'52"

James Edwin Todd Leake Per 3h14'4" 43d28'40"

James Eichenberger Aur 5h27'58" 53d25'55"

James & Elanor Uma 13h55'12" 54d49'15"

James Eldon Floring Ori 5h36'3" 7d5'4"

James Eldon Russell Cep 22h37'50" 57d54'15"

James Eldon Schwandner Her 17h40'17" 40d8'31"

James & Elizabeth Aur 6h33'33" 34d34'49"

James Elliot Knox, III Peg 22h50'54" 24d4'4"

James Ellis Uma 9h49'45" 61d59'3"

James Elmore Cep 22h13'0" 60d16'24"

James Emil Wortman III Her 18h46'33" 22d9'17"

James Emmanuel Morris Psc 1h9'25" 28d36'26"

James Ernest Carmine Uma 10h43'44" 67d57'5"

James Ernest Mangrum Uma 10h50'57" 44d28'36"

James Espinosa Gem 6h58'46" 18d31'13"

James Estep and Lauren Cunningham Ori 4h47'17" -0d33'3"

James et Simone Ori 5h54'46" 2d59'38"

James Everette Lucas III Aur 5h25'33" 39d50'3"

James F. Baumann Cap 20h17'9" -10d36'23"

James F Benigni Lyn 7h41'32" 52d43'2"

James F. Draude "Star Of My Life" Uma 8h59'54" 58d30'17"

James F. Egan, Jr. Her 17h17'15" 35d17'40"

James F. Frounfelker Cnc 8h56'10" 7d55'29"

James F. Glennon Aql 19h50'17" 9d48'38"

James F Lynch Cap 20h37'27" -21d8'21"

James F Nilon, Jr Psc 0h49'5" 11d30'43"

James F. O'Shea Aql 19h22'25" -0d30'22"

James F. Poyerd Per 2h41'36" 55d1'26"

James F. Roberts, Jr. Aql 18h50'3" 8d40'48"

James F. Vaughn, Sr. Ori 5h33'44" 0d21'21"

James F Whittington from his Family Her 17h43'22" 22d49'41"

James F. Wright, Jr. Tau 4h27'33" 17d56'36"

James Fenwick Sct 18h54'47" -4d18'58"

James Ferguson Hamilton Ori 5h25'42" 9d7'14"

James Finley Martin Her 17h29'1" 28d31'30"

James Floyd Akins Gem 6h29'59" 26d35'47"

James Foley Uma 9h40'52" 67d58'28"

James Forbes Winter IV And 1h41'12" 45d22'9"

James Foreman Lmi 9h29'43" 34d8'42"

James' Forever Bright Tau 4h44'45" 24d45'25"

James Foster Dra 10h21'20" 77d56'14"

James Foster Scott IV Ori 4h53'39" -0d4'57"

James Fouche Lyr 18h43'38" 37d41'50"

James Francis Bailey Ari 2h13'31" 25d27'57"

James Francis Cunniff Cep 0h24'40" 73d30'58"

James Francis Jorgensen Psc 0h15'42" 1d14'11"

James & Francis Laidlaw Gem 6h3'50" 24d25'50"

James Francis Mannakee Uma 12h3'34" 36d46'16"

James Francis O'Brien Ori 5h28'40" -4d20'16"

James Francis Pitney Lib 15h59'17" -9d49'41"

James Francis Spaulding Ori 5h32'44" -0d51'21"

James Francis Stroshine Ari 3h16'11" 29d8'41"

James Francis Turton Cru 12h26'52" -56d21'44"

James Franey Langkop Tau 3h44'6" 25d11'55"

James Frank DeCarolis Lib 15h4'28" -6d22'41"

James Frank Pitman Vir 14h56'39" 5d13'6"

James Frank Webber Cep 22h15'52" 61d43'26"

James Franklin Conley II Uma 8h43'2" 48d55'18"

James Franklin Giffen Uma 11h26'54" 59d58'57"

James Franklin Pasley Gem 6h5'57" 25d34'37"

James Franklin Truan, I Love You Her 17h22'47" 37d53'8"

James Franklin Wallace 3 Uma 8h55'58" 66d2'29"

James Fred Ezell Uma 10h42'29" 51d27'38"

James Frederick William Rowe Vir 13h16'50" -3d3'46"

James Frederick Wilson Jr. Her 17h51'52" 24d14'19"

James "Fud" Kilpatrick Vir 13h12'18" 11d13'18"

James G. Fagan II Aur 7h30'33" 43d50'21"

James G. Gutman, Jr. Aql 19h48'57" -0d36'40"

James G. Hopper Cet 2h37'58" 8d2'21"

James G McHale - The Omega Star Uma 9h0'55" 59d16'28"

JAMES G. PAYNE Cep 22h58'36" 75d57'37"

James G. Tennison, Sr. Sco 16h47'46" -25d4'45"

James Gaeton Lesanti Her 16h54'5" 36d13'44"

James Gaherity Aqr 23h33'10" -18d3'59"

James' Galaxy Quest Ori 5h8'52" -0d2'35"

James Gannett Aqr 22h37'46" -9d41'22"

James Genovese (Jim) Sgr 18h0'40" -23d26'35"

James George Foggin Cam 4h2'55" 53d14'49"

James Gerard Flynn Her 17h3'53" 19d50'21"

James Gerard Hurley Gem 7h51'50" 33d52'27"

James Glenn Messmore Psc 0h59'30" 25d35'51"

James & Gloria Uma 10h30'37" 47d46'40"

James Goddeeris Ori 5h34'34" -0d35'59"

James Gordon McKinley IV Leo 11h10'6" 22d21'42"

James Graham Jardine 60th Birthday Tau 4h13'29" 25d45'46"

James (Grandpa) Milligan Cap 20h32'45" -13d53'0"

JAMES GRAY TAFT BERRY Cam 3h45'55" 68d34'35"

James Grayson Bridgeman Uma 11h39'27" 45d50'42"

James Gregory Smith Ari 1h56'32" 22d0'47"

James Gribling And 23h3'17" 52d33'5"

James Grogan Tau 5h11'23" 17d59'17"

James Gronewald Vul 21h18'47" 24d49'41"

James Gunnard Freeburg Per 2h13'3" 54d19'6"

James "Gypsy" Haake Dra 17h15'22" 53d13'27"

James H. Beebe Ori 6h15'14" 15d59'38"

James H. Brewer, Jr. Leo 10h18'23" 24d30'34"

James H. Byrdwell Leo 10h40'56" 17d52'26"

James H. Cash Cep 0h0'27" 77d31'58"

James H. Costello Ori 6h22'16" 16d47'42"

James H. Dulin Cep 2h25'33" 82d57'28"

James H. "Mac" McFarland Cnc 8h21'28" 22d35'33"

James H. Mahoney Uma 13h36'3" 55d2'44"

James H. Mittelman Sgr 18h31'41" -27d6'22"

James H. Morehouse Tau 3h54'56" 3d56'59"

James H. Morrow Jr. Per 4h9'12" 37d52'54"

James H. Rosenthal Vir 12h50'46" -1d24'14"

James H. White Sgr 18h17'15" -17d19'47"

James H White Sgr 18h7'48" -26d33'58"

James Hal Austin Uma 11h21'27" 43d52'31"

James Haley Vir 14h13'38" -16d45'24"

James Haley 1915 Lib 15h9'33" -8d20'1"

James Hall - Pip Umi 14h3'45" 69d45'32"

James Halligan Aqr 22h10'3" -2d13'23"

James Hambright Vir 12h31'36" -0d40'55"

James Hamilton Brookhouser Uma 13h37'48" 50d5'47"

James Hannegrefs Uma 9h50'21" 60d23'20"

James Hansen Uma 13h51'22" 51d29'9"

James Hanson Hale Ser 16h10'57" -0d20'18"

James Haro Aqr 22h38'51" -1d39'20"

James Harold O'Bryan Jr. Cap 20h54'38" -25d8'5"

James Haron & Rachel Doherty Aql 20h5'5" -0d20'20"

James Harper Buckley Sgr 19h32'31" -37d54'1"

James Harry Ebel Gem 7h46'55" 15d1'13"

James Harvey Moody aka Papaw Dooter Her 17h33'20" 36d32'3"

James Haskell Herrell "Big Jim" Ori 5h46'7" 6d31'22"

James Hayter Ori 5h16'33" -0d14'0"

James & Helene Armstrong Cyg 20h22'36" 52d31'22"

James Hemingway Vir 12h12'18" 10d50'4"

James Henderson Sealy-Thompson Per 3h11'20" 46d30'46"

James Henry Gala Uma 11h52'55" 63d13'18"

James Henry Hamilton-Smith Cnc 8h14'21" 13d2'53"

James Henry - Jim's Star Cep 2h17'12" 81d16'18"

James Henry Morris Tau 4h30'19" 18d31'31"

James Henry Price Srp 17h59'10" -0d29'35"

James Henry Seller Uma 10h46'36" 68d24'49"

James Herbert Mead (Yimmels) Cma 6h49'31" -28d9'1"

James Herbig Ori 6h14'58" 9d2'18"

James Hetrick Leo 10h24'15" 17d39'17"

James Hetzel Her 17h4'16" 31d43'26"

James Higgins Her 16h19'15" 20d37'26"

James Hilton Aur 5h34'0" 45d45'2"

James Hiram Beasley IV Sco 17h23'1" -42d17'29"

James Holdford Sgr 18h3'27" -22d36'58"

James Holladay Gordon Dra 18h53'24" 72d16'17"

James Homiak Uma 12h4'45" 39d43'33"

James Hovanec Cnc 8h44'1" 32d48'6"

James Howard Blake Mon 6h52'14" 8d42'4"

James Howard Gibert III Sgr 18h40'7" -33d29'33"

James Howard Loop Leo 11h8'52" -0d5'4"

James Howard Tolley Cep 23h36'50" 72d5'27"

James Howard Warren Sanders And 1h30'6" 48d27'7"

James Hoyle Cmi 7h17'31" 8d54'13"

James Hubert Barlow Tau 5h43'27" 20d13'27"

James Hudson Crb 15h33'36" 28d25'10"

James Hudson Bearden Aql 18h52'24" -0d42'7"

James Hugh Brincefield, Jr. Aql 19h50'40" 7d19'36"

James Hultgrien, Jr. Sgr 18h24'59" -33d59'27"

James Hunter Clark Sgr 19h26'15" -17d7'58"

James Hunter Gaffney Uma 10h46'37" 62d47'36"

James *I had a grunch* Muse Lyr 18h34'34" 34d2'24"

James I. Keane Sco 16h19'48" -11d2'13"

James Ian Derbacher Per 4h49'53" 50d29'46"

James Ira McCormick III Boo 14h27'50" 19d50'5"

James J. Fitzpatrick IV Leo 9h40'1" 27d11'30"

James J. Flanagan Dra 9h43'59" 76d53'40"

James J. Flanigen And 23h4'0" 37d4'29"

James J. Keehan Cap 20h44'10" -19d48'32"

James J. Kuzma Uma 8h35'28" 54d42'58"

James J. Lansing Jr. Per 3h32'35" 43d25'47"

James J & Lynne P Foster Uma 11h15'12" 52d34'2"

James J. McDermott, Jr. Uma 8h38'9" 58d16'38"

James J. Middlebrook Ser 15h58'37" -0d34'29"

James J. Miller Ari 2h23'15" 25d55'18"

James J. Mulligan Cet 1h28'31" -0d18'46"

James J. Murray Ori 4h59'17" -0d14'57"

James J. Murt Aqr 21h31'29" 1d40'19"

James J. O'Mara Sr Cep 23h7'38" 70d11'14"

James J. Pennell Gem 7h55'31" 33d5'0"

James J. Porretta Cam 4h11'31" 70d37'53"

James J. Sadowski Ari 3h13'11" 11d37'59"

James J. Sailer Eri 4h27'6" -0d22'54"

James J. Vigilante Cnc 8h29'16" 31d29'56"

James Jaafil Cnc 8h40'34" 24d31'48"

James Jackson Jr. Ori 5h3'34" 15d7'24"

James Jackson King Aql 19h52'24" -0d27'34"

James (Jamie) Alan Echols Ari 2h37'28" 21d11'54"

James "Jamie" Tracy White Ori 6h23'17" 14d4'46"

James & Jane Borsare Cyg 20h0'17" 44d28'40"

James Jeffrey Stacey Sgr 20h0'15" -33d31'53"

James Jen Fei Na Cen 14h34'13" -60d4'18"

James Jerrall Wise Per 2h39'9" 52d7'47"

James & Jillianne Pup 7h39'9" -48d46'4"

James 'Jim' Gerard Carrigan Col 6h11'3" -40d18'59"

James "Jim" Lusk Ori 6h16'39" 14d10'49"

James "Jim" McDonald Vir 14h1'50" -7d3'12"

James "Jim" Richard Prescott Uma 11h31'16" 56d16'29"

James "Jim" T. Roberts Uma 9h2'49" 63d19'18"

James Jimbo Belli Cep 21h33'27" 63d32'58"

James "Jimbo" Keith Kearns Leo 10h24'15" 21d53'43"

James&Jimena Aqr 21h37'44" 1d54'53"

James "Jimmie" Jersie Ori 5h21'8" 4d7'57"

James "Jimmy" Dalton Psc 23h46'41" 4d55'35"

James "Jimmy K" Katsempris Cam 4h25'22" 64d45'42"

James -Jimmy- Neilon Leo 11h24'6" 12d19'45"

James "Jimmy" P. Lovely Uma 8h50'38" 69d10'44"

James "Jimmy" Smith Uma 11h45'46" 36d43'22"

James Joesph Antongiorgi My Tuffie Per 3h19'52" 48d39'36"

James John Berenato Del 20h43'27" 16d10'43"

James John Flavell Per 4h39'32" 39d58'36"

James John Lewandowski Vir 12h52'45" -7d25'44"

James John Nelson II Lmi 10h3'53" 29d22'9"

James John O'Hara Per 2h52'3" 45d15'15"

James Johnson Ori 6h7'9" 18d26'17"

James Johnson Jr. Ari 2h13'17" 25d52'3"

James Johnson Jr. Ari 2h44'17" 26d35'13"

James Jonathon Drummond Her 16h30'34" 45d50'13"

James Jones Uma 8h20'25" 66d21'23"

James Joseph Vir 13h49'29" -19d33'5"

James Joseph Ashby Sgr 19h37'13" -12d39'38"

James Joseph Brady IV Sco 17h36'44" -42d31'39"

James Joseph Bruce Cyg 19h40'4" 32d3'6"

James Joseph De Palma Cap 20h8'46" -14d13'54"

James Joseph Deibel Cnc 8h13'49" 32d2'44"

James Joseph Donahue Cyg 19h56'10" 30d27'31"

James Joseph Duggan " Dad " Uma 11h57'16" 57d46'53"

James Joseph Edwards Ari 2h57'2" 21d23'21"

James Joseph Haynes Umi 15h33'37" 75d8'37"

James Joseph Hughes Ori 6h16'57" 14d56'40"

James Joseph Janota Tau 3h37'18" 25d3'1"

James Joseph King Aql 19h47'24" -0d57'10"

James Joseph Knauer Lyr 18h30'41" 38d19'35"

James Joseph Lipski Per 2h24'42" 52d28'21"

James Joseph Maycroft Tau 4h29'41" 21d58'54"

James Joseph O'Gormley Per 4h43'55" 39d18'10"

James Joseph O'Neill Uma 11h49'53" 32d47'9"

James Joseph Pelletier Aur 5h51'25" 47d21'1"

James Joseph Quinn Ari 2h39'15" 22d41'11"

James Joseph Robb Dra 15h27'53" 58d19'53"

James Joseph Schrage Vel 9h21'58" -38d23'49"

James Joseph Shimmin Per 4h49'7" 45d29'28"

James Joseph Williams Uma 10h55'7" 49d17'16"

James Joseph Williams III Umi 14h23'38" 78d11'6"

James Joseph Wood Cnc 7h58'7" 13d40'8"

James Jr Her 17h53'1" 23d13'33"

James Julian Dessau Tau 5h48'27" 27d8'30"

James Junior Margeson Per 4h20'35" 36d53'23"

James Justin Byrne Uma 10h51'22" 60d40'45"

James K. Avera Uma 10h52'31" 57d15'53"

JAMES K. HUTHWAITE Sgr 18h42'25" -32d48'20"

James K Jake Johnson Ari 2h10'23" 22d49'44"

James K. Marias Del 20h16'42" 13d19'51"

James K. Twist Aqr 22h31'39" 0d39'14"

James Kahokuleleokalani Aken Cap 20h18'36" -19d43'32"

James & Kaleigh Uma 11h28'3" 58d54'43"

James & Kathryn Ford Cyg 20h29'44" 46d0'17"

James Kaufman And 1h2'14" 46d18'23"

James Keaton Her 16h34'10" 33d45'21"

James Kelly "Jimmy" Jarvis Leo 10h20'5" 13d40'21"

James Kelly Todd Uma 10h58'3" 48d42'29"

James Kemper Agnew Aur 5h53'39" 50d56'35"

James Kenneth Cox "75" Leo 11h2'0" -4d58'42"

James Kenneth Strobel Pho 0h57'41" -51d27'42"

James Kenneth Wallerstedt Ori 5h22'13" -7d16'13"

James "Kieran" Luke Rosica Cep 21h46'40" 61d52'8"

James Kilbane Sgr 18h43'42" -21d44'27"

James Kim Uma 13h20'39" 62d52'48"

James Kimbrough Stapleton Tau 5h20'6" 19d19'45"

James King Ori 5h8'16" 3d51'1"

James King Elliott Lyn 7h8'21" 58d51'51"

James Kirby Meyer-Mary Allis Meyer Cyg 20h47'54" 46d53'39"

James Konetski Aur 5h49'55" 54d20'49"

James L. Byrd "Dad" Uma 9h4'31" 60d26'23"

James L Courson IV - My RockStar Aqr 22h11'29" -14d28'25"

James L Delamorton Psc 22h58'50" 7d23'12"

James L. Fann Gem 7h12'7" 14d4'54"

James L. Faulkner 2/24/48 Psc 23h14'32" -1d57'47"

James L. Fults, Jr. Equ 21h13'59" 9d59'40"

James L. Gardner, M.D. Her 18h10'26" 42d7'41"

James L. Holt Per 4h21'16" 52d31'29"

James L. Nauser Cep 21h41'52" 61d44'27"

James L. Rice Umi 17h44'35" 80d32'10"

James L. Smith Aqr 21h41'45" 0d17'40"

James L. Taylor Aqr 20h42'26" -7d45'14"

James L. Vance Uma 9h5'41" 48d26'13"

James L. Ware III Gem 6h18'58" 24d5'14"

James L. Watson Boo 14h10'56" 32d22'4"

James Lackey Leo 11h32'29" 27d25'51"

James Lanza Gem 7h39'17" 20d55'43"

James Lapine Per 4h7'12" 51d25'5"

James Larry Allen Psc 0h12'57" -0d24'33"

James Laskey Ramirez Tau 4h16'1" 19d20'3"

James&Laura Cyg 21h11'29" 53d31'14"

James Lawless - Jamsey's Star Cep 20h50'56" 57d16'49"

James Lawrance Armstrong Psc 23h24'23" 1d21'2"

James Lawrence Case III Gem 7h19'42" 23d10'45"

James Lawrence Fingland Cap 21h7'32" -24d30'16"

James LCD Tingley Aqr 22h20'9" 2d24'17"

James Lee Umi 15h48'35" 76d46'17"

James Lee Bass Ari 2h13'49" 24d51'30"

James Lee Hollis Her 17h33'9" 37d20'34"

James Lee Reynolds Her 17h26'50" 25d41'1"

James Lee Rinehart Boo 14h42'7" 34d30'28"

James Lee Shields, Jr. Lyn 8h56'29" 34d12'46"

James Lee Williamson Uma 9h48'2" 63d17'41"

James Leo Scott II Aqr 21h45'34" -3d6'28"

James Leo Topper IV Psc 22h57'9" 4d51'35"

James Leon French, Sr. Tau 5h45'50" 17d10'46"

James LeRoy Beeson Cnc 8h41'39" 24d38'15"

James Leroy Eickhoff Her 16h49'30" 41d48'21"

James LeRoy Hinich Umi 13h10'20" 69d34'8"

James Lester Murray Psc 23h51'44" -3d15'22"

James Levandowski Uma 12h41'58" 61d40'30"

James Levitt And 0h19'39" 27d43'37"

James Lew McIlvaine Lyn 7h51'5" 44d51'0"

James Liam Brady Boo 14h39'9" 37d40'1"

James Libby "My Superstar" Sco 16h7'35" -13d13'54"

James & Linda Brown Sgr 18h47'31" -20d39'53"

James & Linda Christou Cru 12h10'5" -61d40'59"

James Linsenmeyer Ori 5h48'4" 7d23'25"

James "Little Big Boots" Bruno
Tau 4h34'8" 27d56'25"
James Livingston Fagan
Ari 3h24'55" 29d31'27"
James Logan Gerhard
Aqr 23h17'39" -18d44'13"
James Lopez
Ser 15h56'23" 7d29'3"
James Lorenc
Vir 12h35'41" -8d18'34"
James & Lori
Eri 3h51'41" -0d30'35"
James Loves Dena Slater
Lib 14h53'5" -3d7'7"
James Luca Richardson
Umi 17h41'26" 80d51'21"
James' Lucky Star
Dra 20h26'17" 72d49'17"
James & Luke
Lyn 8h7'28" 39d3'6"
James Lynch
Per 3h9'10" 51d41'54"
James Lynn Allen
Boo 13h50'5" 16d27'6"
James M. Bickhaus
Sco 17h43'24" -32d49'1"
James M Bobrowski
Uma 10h4'6" 52d18'25"
James M. Carter
Cep 20h42'24" 68d35'53"
James M. Gilliam Jr.
Aur 5h51'52" 53d10'28"
James M. Grace
Aql 19h23'23" -10d35'11"
James M. Hollifield
Ari 3h15'20" 29d15'59"
James M. Jacobsen
Aur 6h4'32" 53d46'48"
James M. Jordan, Jr.
Her 17h23'43" 36d41'12"
James M. Kapsa
Per 3h38'23" 46d50'52"
James M. Kennedy, Jr.
Uma 10h5'42" 67d8'39"
James M. Logan
Lib 15h8'7" -26d27'45"
James M. Lundy
Aqr 23h0'19" -9d12'18"
James M. Macpherson
Aur 5h56'29" 36d25'12"
James M. McCuiston Jr.
Vir 12h9'2" 13d0'18"
James M. Morris
Cnc 8h34'6" 17d2'4"
James M. Sellgren
Aur 5h47'57" 43d52'52"
James M. Senape
Uma 11h54'21" 42d44'35"
James M. Souby Oquirrh Star
Lib 15h9'30" -27d0'44"
James M. Squiccirini
Tau 4h1'25" 23d16'16"
James Macdonell
Vir 12h29'13" -4d20'43"
James MacLain Adams
And 23h19'29" 47d56'34"
James Madere
Ari 2h56'42" 25d1'54"
James Madison Calvert III
Uma 10h54'17" 61d27'27"
James Madison Martin
Cap 20h42'4" -25d53'21"
James Madison Ruble
Aqr 21h46'45" -1d42'4"
James Malone Taylor
Uma 13h59'32" 57d58'27"
James & Mandy
Cas 0h43'19" 62d5'28"
James Manning
Aql 19h43'41" 5d4'5"
James & Marcia Riley
Cyg 21h59'11" 55d7'57"
James Marcinko III
Vir 12h47'46" 6d12'18"
James Marcus Moore
Per 4h13'46" 41d36'14"
James Margiotta
Boo 14h29'43" 51d36'41"
James Mark Newbould
Her 17h11'6" 15d0'9"
James Marshall
Ori 6h15'52" 13d56'26"
James Marshall Smith
Lib 14h52'3" -10d25'53"
James Marsters
Leo 9h47'18" 7d48'26"
James Martin
Her 18h45'57" 21d5'36"
James Martin
Cap 21h39'21" -14d46'16"
James Marvin Hamilton
Lyr 19h13'20" 45d18'25"
James Mary Kane
Leo 11h3'33" -1d30'35"
James Matt Foster
Pho 0h33'11" -42d14'34"
James Matthew Ciminelli
Dra 18h53'20" 63d0'32"
James Matthew Eby
Ori 5h46'8" -10d29'13"
James Matthew Jagielski
Sco 17h29'31" -44d56'37"
James Matthew Letton
Cep 22h47'46" 74d19'42"
James Matthew Mackay
Lyn 8h1'55" 39d27'33"
James Mauricio
Srp 18h27'41" -0d29'44"
James & May
Ori 6h3'36" -0d45'2"
James McAllister Jr. (Jim)
Uma 8h19'2" 72d23'18"
James McCliggott
Per 2h35'9" 53d27'29"
James McClintic
Her 18h34'2" 25d59'5"
James McClure
Aql 19h12'33" 2d32'37"
James McClurg Mathers
Cep 20h44'50" 63d3'0"
James McGuff
Lyn 7h38'35" 36d28'16"
James McLaughlin
And 23h31'26" 48d12'44"
James McLaughlin
Per 3h26'36" 45d13'43"
James McNicol
Aqr 20h52'45" 2d14'9"
James McVey
Ori 4h57'57" -0d7'32"
James & Megan
Cyg 21h28'1" 51d10'9"
James Meketa
Her 18h10'6" 27d21'30"
James & Melodee's Special Star
Umi 15h59'44" 76d23'28"
James Menegoz
Cep 23h37'55" 78d6'1"
James & Meredith
Cma 7h20'20" -19d17'9"
James Merrel Allison
Cap 20h23'22" -24d32'2"
James Michael
Psc 0h11'27" 4d34'43"
James Michael
Umi 13h24'32" 70d30'8"
James Michael
Crb 15h30'10" 27d9'6"
James Michael Allen Glandon's Star
Aql 18h42'22" -0d12'33"
James Michael Anderson
Mon 6h31'53" 9d39'16"
James Michael Bosh
Cyg 19h53'8" 51d40'51"
James Michael Bush
Per 4h49'27" 43d48'23"
James Michael Castelli
Sco 16h6'26" -19d25'0"
James Michael Collins
Vir 12h34'34" 12d0'54"
James Michael Cox
Aur 5h35'20" 33d44'29"
James Michael Coyne Memorial Star
Aqr 22h13'32" 0d35'39"
James Michael Cunningham
Vir 12h30'6" 7d40'15"
James Michael D'Aquino
Aur 6h26'25" 42d55'41"
James Michael David Myre
Ori 5h26'5" 4d43'9"
James Michael East
Boo 15h38'25" 41d12'31"
James Michael Evans Star
Sco 16h9'30" -19d32'58"
James Michael Everett
Ori 5h37'36" 4d7'21"
James Michael Everett
Boo 14h15'49" 20d45'20"
James Michael Favaloro
Uma 11h16'1" 39d14'34"
James Michael Frank
Her 17h23'0" 38d22'32"
James Michael Grau
Aur 5h15'8" 42d12'12"
James Michael Henson
Cnc 8h45'59" 26d47'39"
James Michael Hilz II
Her 17h12'6" 35d58'48"
James Michael Holden Jr
Psc 1h11'37" 12d5'12"
James Michael Joseph O'Leary
Her 18h6'48" 17d30'6"
James Michael Keane
Uma 10h1'28" 68d4'51"
James Michael Love
Ari 2h47'34" 14d59'46"
James Michael McGlinn
Per 2h51'8" 53d49'30"
James Michael Miller
And 0h20'4" 41d59'4"
James Michael Mulhern
Ari 3h27'45" 22d29'41"
James Michael Mullen
Boo 14h35'48" 26d28'46"
James Michael Pabis
Cyg 20h29'15" 34d36'22"
James Michael Proctor III
Lmi 10h12'48" 28d34'5"
James Michael Taylor
Uma 9h22'1" 61d15'21"
James Michael Thibault
Cnc 8h52'59" 28d7'0"
James Michael VanGilder 459
Gem 7h10'39" 25d32'37"
James Michael Walker
Eri 4h39'0" -0d53'27"
James Michael Weller
Uma 11h43'20" 36d42'41"
James Michael Young
Dra 18h26'35" 73d25'55"
James Michael Zee
Dra 12h0'54" 70d8'27"
James Micheal
Lyr 18h38'41" 30d31'18"
James Middleton Marsh
Ori 5h59'12" 11d32'54"
James Miller
Gem 7h4'37" 26d47'47"
James Milman Raum
Aqr 22h23'3" -23d24'14"
James Milne Diplock
Sgr 18h7'15" -27d16'16"
James Minor Sr. my eternal light
Leo 9h43'17" 24d52'17"
James Mischler
Cnc 9h12'8" 14d16'59"
James Mitchell Holloway
Cru 12h25'7" -61d4'31"
James Molenari
Uma 9h57'53" 46d0'19"
James Monroe Gasho
Uma 9h2'51" 60d10'13"
James Monroe Parnell
Psc 1h7'36" 23d53'29"
James Mooney
Ari 3h3'43" 25d45'2"
James Moore and Danielle Ficker
Umi 1h50'48" 89d32'0"
James Morgan
Her 17h31'36" 36d27'10"
James Moriarty
Cyg 21h47'22" 47d3'0"
James Morrell
Uma 10h19'17" 69d46'7"
James Morrissey
Cep 0h2'57" 75d19'22"
James Murphy Shapiro
Boo 14h53'4" 21d13'1"
James Murphy, our Angel
Gem 7h16'3" 27d11'17"
James "My Angel" Soloway
Cyg 19h43'25" 28d24'12"
James my loving husband
Her 18h42'8" 20d12'13"
James (My Romeo)
Ori 5h24'1" -0d26'30"
James N Angela Orfe
Uma 11h16'27" 43d33'30"
James N Hart-Poppy's Star
Cru 12h22'23" -57d9'56"
James Nappi
Her 18h20'1" 20d58'26"
James Nathan Adams
Her 15h53'11" 45d3'10"
James Naylor
Uma 9h18'32" 54d0'9"
James Neil Loves Leah Rastiello
Cyg 19h39'53" 30d46'29"
James Neville Pasley
Vir 11h44'48" 3d51'12"
James Nicely Sr.
Aql 18h49'4" -0d55'8"
James Nicholas Frampton
Her 17h22'32" 16d31'48"
James Nicholas Jimas
Gem 7h1'27" 21d50'17"
James & Nicole Wood
Crb 16h15'35" 32d43'31"
James Nissen
Ori 5h22'11" 7d9'52"
James & Nora Church
Cyg 20h18'43" 33d16'24"
James Norman Serphos
Cep 21h30'20" 65d55'55"
James Nunzio Costa
Leo 11h25'30" 9d25'25"
James O. Bailey
Aqr 23h22'6" -19d26'9"
James O. Reitan
Cyg 21h19'34" 43d57'11"
James Oliver Gardiner
Peg 22h16'49" 17d2'35"
James Olson
Lyn 7h52'5" 51d6'47"
James Opgenorth
Dra 20h4'8" 74d24'18"
James Orand Ewing
Ori 5h26'10" 1d3'50"
James Orion Orlopp
Ori 4h53'31" 12d13'47"
James Orville Jordan
Her 17h36'11" 44d39'43"
James Oscar Benson
Cep 22h23'28" 61d57'58"
James Oscar Taylor
Aql 19h46'40" -0d1'33"
James Oswald Halverson
Vir 12h43'14" 5d0'28"
James Owen Fox
Leo 10h38'3" 18d7'25"
James P. Cima, D.C.
Ori 6h3'30" 14d4'57"
James P. Conley
Aql 19h36'26" 3d30'23"
James P. Devlin
Lyn 6h46'6" 52d44'11"
James P. Flynn
Aqr 21h12'37" -9d9'32"
James P Jone
Cnc 8h48'25" 28d44'20"
James P. Kreindler
Sco 17h34'57" -45d17'23"
James P. Monohan, Jr.
Per 2h40'46" 57d26'18"
James P. Naifeh
Cep 22h44'50" 60d0'58"
James P. Neely
Cep 22h18'22" 62d36'29"
James P. O'Donnell
Gem 7h35'10" 30d53'57"
James P. Remick
Uma 14h25'26" 59d50'32"
James P. Sanfilippo (JP)
Lib 14h47'31" -5d12'18"
James P. Sciackitano
Aur 5h55'21" 42d34'52"
JAMES P SINDLINGER
Dra 17h3'7" 57d4'15"
James Padraic
Ari 2h21'30" 12d0'44"
James Pagano, Jr.
Tau 5h38'24" 27d16'43"
James Palmer
Com 13h2'22" 24d37'17"
James Palmer
Ori 4h45'16" 4d26'56"
James Palmer & Elenora Brown 4ever
Leo 10h21'37" 23d28'3"
James Parker Knowles
Sgr 18h24'23" -22d9'41"
James Parker McCulley, M.D.
Cep 20h59'46" 62d19'0"
James Patrick Bolenbaugh
Tau 4h12'40" 29d8'44"
James Patrick Cippoletti
Her 16h36'34" 22d42'23"
James Patrick Cippoletti
Cep 20h44'39" 64d19'12"
James Patrick Flynn
Leo 11h36'20" 22d41'55"
James Patrick Huber
Gem 7h20'36" 34d12'11"
James Patrick Joyce
Ari 1h55'8" 18d13'30"
James Patrick Kimball
Aqr 22h9'59" -8d29'15"
James Patrick Lavin Jr.
Dra 20h30'41" 67d50'40"
James Patrick Marshall
Ori 5h50'36" -0d17'17"
James Patrick Mortell
Ari 1h54'30" 18d39'53"
James Patrick O'Donnell
Gem 6h22'17" 23d50'59"
James Patrick O'Toole
Ari 2h6'33" 23d27'40"
James Patrick Perks
Psc 1h19'19" 25d37'30"
James Patrick Reeve
Ori 6h15'44" 15d29'14"
James Patrick Ronsky
Sco 17h35'59" -36d32'24"
James Patrick Sewell
Boo 14h32'49" 24d43'38"
James Patrick Sheehan, Jr.
Ori 5h30'36" 3d44'25"
James Patrick Sidoti
Per 3h58'24" 36d24'39"
James Patrick Urciuoli
Ari 2h9'44" 24d8'26"
James Patrick's Green Lantern
Sgr 18h43'3" -25d1'5"
James Patterson
Aqr 22h50'56" -12d25'58"
James Paul
Her 16h44'23" 14d28'50"
James Paul Duncan
Ori 6h4'21" 11d2'48"
James Paul Keener Memorial Star
Uma 9h54'17" 43d7'1"
James Paul Monegan
Cyg 20h9'53" 38d49'52"
James Paul Mullins
Uma 10h34'31" 46d18'11"
James & Pauline Marler
Aql 19h36'22" -0d7'36"
James Pavelin 29
Cep 23h36'41" 83d18'1"
James Pealer
Uma 10h41'15" 51d56'48"
James Percival Ray 09/06/1937
Cep 22h40'46" 76d28'4"
James Peter Crosby
Aur 5h51'53" 52d29'26"
James Peter Emburey
Sco 17h18'22" -33d1'40"
James Peter Fowden
Cep 20h50'29" 62d38'25"
James Peter McCarthy
Uma 11h20'41" 51d13'42"
James Pharo
Per 3h46'14" 48d50'31"
James Philip Christian II
Cap 20h19'50" -8d49'58"
James Philip Prack
Ori 4h51'51" 2d12'59"
James Pippin
Cyg 21h21'30" 31d4'49"
James Pitman
Cyg 21h27'1" 36d18'56"
James Portmann
Boo 14h33'37" 17d50'55"
James Prescott Brandon
Uma 9h45'32" 48d46'0"
James Prior
Lyn 8h8'40" 55d50'41"
James Procario
Cep 21h33'48" 64d21'38"
James Prowse Mansion in the Sky
Her 16h39'11" 26d47'39"
James Pufahl
Per 4h15'11" 40d34'24"
James Purcell
Per 3h46'37" 41d42'43"
James Quesnell
Leo 10h20'50" 17d44'26"
James R. Bailey
Leo 9h56'25" 31d30'59"
James R. Bentley
Aql 19h49'58" 4d23'20"
James R. Bowen
Aql 20h3'13" -0d50'16"
James R Clark Jr
Cep 22h11'20" 61d45'38"
James R. Coates, Sr.
Aqr 21h51'59" -0d51'12"
James R. Cuca
Aur 5h27'42" 43d3'21"
James R. Fike, Pilot Extraordinaire
Vir 14h19'32" 0d1'35"
James R. Garland
Gem 6h31'12" 23d48'23"
James R. Hanlon
Uma 11h5'26" 37d15'6"
James R. Hildreth
Per 2h26'24" 56d57'54"
James R. Lewis
Cap 20h24'21" -10d21'48"
James R Lewis, Jr.
Tau 4h13'5" 12d39'59"
James R. McNelis
Uma 11h2'28" 69d3'10"
James R. Morton
Per 4h32'29" 33d31'15"
James R. Mrizek, Jr.
Lib 15h16'0" -5d38'30"
James R. Petersen
Vir 13h14'26" 4d54'7"
James R. Sands
Sco 17h49'3" -41d37'47"
James R. Schmalz
Psc 0h28'47" 4d43'2"
James R. Schroeder/Brenham, TX
Cyg 19h47'2" 38d40'40"
James R. Tibbetts, Sr.
Cyg 19h56'12" 31d31'23"
James R. Wells
Ari 3h12'32" 28d40'3"
James R. Widmayer
Vir 12h43'8" 12d29'7"
James Ragan
Lyn 9h1'52" 39d1'2"
James Ramey
Ori 5h25'36" 2d41'37"
James Randall Barrett
Gem 6h21'38" 21d35'18"
James Randall Flewellen
Ori 5h39'13" 0d2'25"
James Randall Persky
Per 3h14'23" 46d42'40"
James Randall Taylor "Randy"
Ori 5h42'19" 2d20'39"
James Randell Moody
Cap 20h11'2" -23d58'9"
James "Randy" Noble
Dra 18h44'47" 53d13'31"
James Ray
Cet 1h33'23" -8d59'20"
James Ray Siegmund
Uma 11h27'41" 29d55'58"
James Ray Thomas
Cnc 8h47'36" 21d55'36"
James Raymond Arena
Umi 16h40'32" 79d19'44"
James & Rechell Against the World
Aqr 22h40'14" -11d21'31"
James Reese Ford
Per 4h45'29" 46d36'50"
James Reposkey
Cep 22h14'17" 73d43'28"
James Richard Douglas
Her 16h29'15" 48d25'49"
James Richard Farmer
Uma 9h50'5" 53d39'21"
James Richard John Ravenhill
Cep 21h23'22" 56d59'58"
James Richard Pastorik
Cnc 8h41'10" 31d31'2"
James Richard Pfirrmann
Ori 5h26'5" 3d14'52"
James Richard Roberts
Tau 4h39'25" 11d48'23"
James Rieck
Vir 13h35'40" -2d39'58"
James Riley Williams
Psc 0h40'56" 8d0'0"
James Ringer
Her 17h25'2" 47d17'8"
James Robbins
Aur 5h53'33" 50d57'14"
James Robbins
Per 3h42'48" 45d18'59"
James Robert
Boo 14h28'19" 18d5'42"
James Robert
Psc 1h24'47" 27d33'30"
James Robert
Cep 1h52'59" 81d56'22"
James Robert "Bob" Griffin
Per 2h37'46" 53d45'15"
James Robert Brooks
Sco 16h45'4" -40d23'34"
James Robert Brown
Tau 3h48'8" 30d23'12"
James Robert Browning
Cep 21h10'1" 81d54'2"
James Robert Caprarella
Her 16h44'6" 47d6'36"
James Robert Clarke
Sgr 18h5'59" -27d57'53"
James Robert Collins
Cam 7h11'26" 75d16'48"
James Robert Coons
Her 18h51'20" 24d42'37"
James Robert Croy
Ori 6h8'51" 8d33'29"
James Robert David Sayer
Cas 23h53'56" 48d52'0"
James Robert Draughon
Cap 21h0'2" -20d49'55"
James Robert Feddick Shining Wisdom
Uma 9h26'32" 52d22'4"
James Robert Francis Hudson
Peg 22h13'6" 12d3'2"
James Robert Harter
Aqr 21h18'55" -5d27'1"
James Robert John Rollinson
Dra 15h13'56" 58d46'23"
James Robert Lyons
Psc 1h35'10" 26d6'14"
James Robert Martin
Sgr 18h50'16" -16d44'38"
James Robert Morgan
Tau 5h39'7" 22d16'33"
James Robert Navar
Ori 5h53'24" 12d14'48"
James Robert Orr Worlds Best Hubby
Vir 14h32'38" -2d4'10"
James Robert Patillo
Gem 7h6'57" 17d1'50"
James Robert Silverman
Dra 16h54'20" 60d33'39"
James Robert Stanton
Aur 6h53'54" 37d38'54"
James Robert Vakis
Sgr 19h9'40" -23d15'34"
James Robert Vaughn
Aqr 23h20'47" -18d58'14"
James Robert Wallick
Ari 2h39'36" 29d54'40"
James Robie Ray IV
Gem 7h28'58" 24d29'23"
James Roby Smith
Vel 9h3'9" -38d14'51"
James Robyzad Williams
Cas 23h1'35" 55d30'15"
"James" Rockin with Tight Pants
Uma 10h25'36" 39d33'8"
James Rocks Whiteford II
Lib 15h45'35" -17d17'9"
James Rod Davis
Uma 11h16'49" 30d40'18"
James Rogers Pritchard
Aur 5h56'35" 36d55'26"
James Roland Lindstrom
Uma 11h9'44" 59d38'14"
James & Rosalind Dixey
Cyg 19h56'1" 40d10'35"
James Rosenberger
Her 17h25'41" 32d54'20"
James Ross Woods
Aqr 22h0'28" 0d59'3"
James Roviezzo
Lac 22h52'0" 52d50'45"
James Roy
Cnc 8h24'2" 31d57'45"
James Roy DelGado
Per 4h29'8" 43d12'14"
James Roy Manley
Ori 6h15'56" 16d1'13"
James Russell Alexander Riley
Cam 4h31'17" 57d15'8"
James Russell Cook
Tri 2h37'44" 31d30'38"
James Russell Kirts IIII
Umi 15h32'6" 72d44'1"
James Rutland
Uma 8h19'23" 63d47'58"
James Ryan Church 517 78 3874
Uma 13h23'25" 56d23'56"
James Ryan Coker
Sgr 20h13'44" -36d41'55"
James Ryan Kress
Ori 6h11'8" 9d16'17"
James Rynell Brown
Cnc 8h37'34" 23d48'39"
James S. Bradley
Lib 15h34'29" -7d32'40"
James S. Gallagher
Ori 5h34'30" 9d12'49"
James S. Mack
And 23h27'12" 48d55'48"
James Sabatini
Tau 4h38'26" 8d24'19"
James Sachs
Aqr 22h24'31" -1d2'19"
James & Saira
Aur 5h27'17" 45d55'41"
James Sanchez & Alexandra Montiel
Cyg 20h49'24" 36d40'9"
James Saneda
Her 18h22'37" 18d17'55"
James Santos Castagnozzi
Tau 5h26'49" 22d26'52"
James Saponaro
Her 17h23'30" 36d46'19"
James Sarantopulos
Ari 2h13'31" 23d59'2"
James Schlick
Cnc 8h37'10" 30d43'52"
James Schuyler Miller
Ori 5h20'19" -1d4'42"
James Scott
Cap 20h41'16" -25d21'9"
James Scott Connelly
Aqr 21h52'45" 0d45'37"
James Scott Ferguson
Gem 7h13'32" 14d36'7"
James Scott Frampton
Per 3h47'19" 31d52'18"
James Scott Glenn
Per 4h23'28" 40d48'1"
James Scott Porter
Uma 11h15'9" 53d11'9"
James Scott Thompson
Per 4h14'23" 51d15'7"
James "Seamus" Salvatore
Dra 18h6'17" 70d10'29"
James Shaheen
Uma 10h52'22" 65d47'8"
James Shaw & Agnes DeBusk Farquhar
Crb 16h12'28" 27d37'53"
James Sheehy "Your Personal Star"
Cam 4h24'31" 63d39'11"
James Sherman 3/3/73
Psc 1h10'47" 16d52'18"
James Sidney Rosinbaum, Sr.
Uma 8h38'37" 52d5'0"
James Simmons
Her 17h48'47" 40d20'0"
James Simon from Rebecca De Falco
Her 18h36'37" 22d11'38"
James Smith
Pyx 8h54'35" -32d30'35"
James Sotirakos
Ori 6h10'6" 16d11'25"
James "Sparky" Renick
Uma 11h32'13" 62d41'13"
James Spence Bishop III
Gem 6h10'0" 25d6'10"
James Spencer
Sco 17h54'17" -35d52'7"
James Spotts
Cam 3h58'32" 66d49'53"
James Stanley Murray
Per 3h11'14" 52d6'8"
James' Star
Her 18h20'39" 22d49'39"
James' Star
Leo 11h0'13" 4d3'2"
James & Stephanie
Psc 1h21'18" 10d58'37"
James Stephen Cooney
Her 17h33'20" 45d16'44"
James Stephen Dewberry
Ari 2h20'18" 23d58'37"
James Stephen Kurz
Aql 19h16'7" 3d36'36"
James Steven Springer
Ori 6h17'41" 16d7'15"
James Stevenson
Per 3h14'18" 47d19'35"
James Stokes Hagan
Uma 9h9'10" 46d54'24"
James Stonewall Myatt, Jr.
Vir 13h6'55" 5d20'23"
James Strickler
Per 3h16'18" 51d48'48"
James Stuart Shoemaker
Aqr 22h37'9" 2d19'21"
James Stucchio
Aqr 23h35'34" -18d56'36"
James Sumter Hanna, II
Uma 11h17'8" 44d3'13"
James - Superstar-McKernan
Her 17h0'50" 14d29'50"

James & Svetlana
Cap 21h22'29" -22d49'17"
James Swartz
Cnv 13h49'53" 33d2'48"
James SY Wong - ah jim
Gem 6h32'15" 12d5'2"
James T.
And 2h36'46" 46d18'20"
James T. Adkins II JT
Cyg 21h14'15" 45d22'19"
James T. Arnold, Jr. a.k.a Superman
Lyn 6h30'14" 53d56'38"
James T. Dunn, Jr.
Uma 12h0'16" 44d34'36"
James T. Erwin Sr.
Uma 10h41'54" 70d11'6"
James T. Evelti
Cnc 9h4'1" 18d23'12"
James T. Gorman
Uma 13h6'31" 58d2'49"
James T Haines
Vir 15h8'22" 5d58'51"
James T Higgins
Her 17h20'31" 45d59'43"
James T. Hughes
Uma 11h33'55" 40d31'30"
James T. Lee
Aqr 22h1'25" -13d12'11"
James T. Malone
Cyg 21h33'8" 35d46'18"
James T. Plattner
Cnc 8h44'34" 31d12'5"
James T. Todd
Lyn 7h39'56" 52d32'45"
James T Wright & Albert L Patterson
Her 16h46'14" 32d35'56"
James T Young
Ori 5h44'26" 11d19'39"
James & Tamara Windsor 2006
Cyg 21h57'49" 46d28'51"
James Taylor DuBois Glase
Sco 17h54'2" -32d3'58"
James "Teddy" Holt
Uma 11h21'27" 63d22'16"
James Thomas Appleyard
Per 1h50'1" 51d19'33"
James Thomas Branciforte
Sgr 19h16'55" -20d43'38"
James Thomas Brazil
Psc 1h19'11" 14d23'46"
James Thomas Clark
Ari 2h9'14" 13d8'43"
James Thomas Conaty
Uma 9h47'57" 71d9'34"
James Thomas Copley
Crb 15h49'15" 31d24'51"
James Thomas Devaney
Sco 16h11'20" -9d37'2"
James Thomas Downing
Per 4h9'19" 50d33'14"
James Thomas Guida
Uma 13h25'3" 57d42'32"
James Thomas Johnson
Her 16h59'33" 22d16'37"
James Thomas M Loves Audrey Ann H
Sgr 19h40'32" -13d0'58"
James Thomas Massetti
Aql 19h0'23" -10d14'24"
James Thomas McDonald
Umi 13h14'40" 89d38'16"
James Thomas Miars
Her 17h13'52" 15d44'58"
JAMES THOMAS PION
Tau 5h50'39" 27d20'8"
James Thomas Shadwell
Leo 11h45'21" 26d8'7"
James Thomas Sumpter
Pho 1h33'11" -41d22'17"
James Thomas Walter
Umi 16h22'31" 75d42'30"
James Thompkins
Cap 21h52'36" -24d24'52"
James Thompson
Her 16h8'21" 46d52'43"
James Thomson 4/4/39 UK
Cam 4h33'26" 66d50'6"
James Tice Rollins
Ori 4h59'55" 10d22'29"
James Timothy Stevenson
Cet 1h29'44" -0d31'24"
James Tobias Wells Ritter
Sco 17h19'19" -32d58'31"
James Todd McRae
Sco 16h12'57" -16d19'28"
James Todd's Lakin Five
Per 4h20'20" 44d56'37"
James Tomlinson
Ori 6h7'11" 20d46'0"
James "Tommy" Greene
Uma 10h52'56" 69d44'37"
James Tong
Ori 5h3'4" 10d15'1"
James Tony and Rita Carolyn Kidd
Tri 1h56'10" 34d26'2"
James Travis Tucker
Her 17h11'18" 48d8'46"
James Trevathan
Sco 16h46'21" -32d49'12"
James Tuljus
Ari 2h25'53" 25d21'0"
James Tyler Stroupe
Ari 2h40'16" 26d18'30"
James "Ullie" Allen
Her 16h36'10" 49d0'28"
James V. Leszczynski XXXOOO
Her 16h19'28" 48d12'37"
James V. Patti
Per 3h19'31" 47d10'8"
James V. Sparrow
Oph 17h15'14" 0d36'42"
James V. Spignesi, Jr.
Cap 21h1'51" -18d3'31"
JAMES VACA, JR.
Her 16h58'42" 33d9'47"
James Van Gee
Cap 21h34'31" -17d18'59"
James Vander Velde *Jimmy's Star*
Lib 14h55'47" -10d12'27"
James Verlin Armstrong
Lyn 6h47'9" 56d29'28"
James Vernon Anthony Postle
Per 4h28'45" 49d47'48"
James & Veronica Nagel
Lyr 19h2'12" 34d55'3"
James & Vikki 2 Hearts 1 Love
Ori 6h11'2" 19d25'28"
James Villanneva
Cyg 20h21'40" 47d40'59"
James Vincent DiStasio
Vir 12h51'46" 12d35'1"
James Vincent Franco
Her 17h39'18" 33d31'52"
James Vincent Herrmann
Lib 14h51'20" -17d43'15"
James Vincent Le
Lmi 9h28'59" 33d37'12"
James Vincent Robibero (10-7-1985)
Uma 8h43'25" 46d43'20"
James Vincent Young Sr.
Ori 5h39'35" 7d57'56"
James Vines
Her 18h6'30" 24d22'42"
James Vito
Her 17h20'44" 36d31'8"
James Vollendorff
Leo 11h23'16" 22d33'30"
James W. Barnes
Her 16h42'51" 28d44'26"
James W. Brown
Psc 1h6'42" 25d58'3"
James W. Dunn is a star Dad.
Lib 14h51'25" -4d4'45"
James W Evans
Lyn 7h37'49" 39d36'56"
James W. Ferris
Umi 14h4'40" 72d43'51"
James W. Franklin, Sr.
Uma 10h4'26" 49d24'58"
James W. Hendrickson
Leo 11h57'57" 20d14'57"
James W. Hippler
Psc 0h31'49" 9d4'40"
James W. Kennedy III
Uma 10h34'8" 46d2'19"
James W. Milstead 2002
Lmi 10h14'5" 39d58'35"
James W. Osborn "Ozzy's Light"
Cas 23h26'24" 54d34'49"
James W Southgate & Lauren M Smith
Cas 2h21'6" 73d10'49"
James W. Taylor
Ari 2h53'53" 19d12'9"
James Waldroup
Sgr 19h5'33" -31d52'26"
James Walker
Cep 22h6'21" 57d10'2"
James Walker Young
Ari 2h11'31" 21d39'5"
JAMES WALLACE
And 23h6'4" 48d35'25"
James Wallace Pikey, III
Sgr 18h18'18" -23d21'12"
James Walter O'Neil
Sgr 19h44'0" -12d31'49"
James Ward Davis
Ari 2h23'36" 25d14'30"
James Ward Love
Uma 8h25'13" 70d57'22"
James Warren Bowker
Ori 6h0'29" 0d41'56"
James Warren Greer
Ori 5h24'6" 1d2'38"
James Wayne Holmes
Ari 2h21'33" 25d47'51"
James Wayne Voeller
Uma 8h36'25" 46d54'35"
James Wert
Ori 5h58'38" 20d56'32"
James Wesley Patterson
Crb 15h36'31" 33d8'13"
James Wesley Terrell
Cas 23h54'57" 54d14'45"
James Wesley Tunnell
Uma 13h51'40" 52d2'1"
James Wessman
Tau 3h51'20" 10d11'1"
James Wetheral Caulkett 6/22/1916
Cnc 8h14'10" 6d59'43"
James Wiking Nordhem Stevens
Cnc 8h35'1" 22d39'11"
James Wilkins
Cnc 8h29'15" 23d58'27"
James William
Cap 21h53'22" -18d4'48"
James William Adams III
Ari 2h27'30" 24d14'18"
James William Carruth
Ari 3h3'27" 15d49'14"
James William Carruth
Leo 9h29'33" 9d29'4"
James William Dudek II
Her 17h29'36" 31d31'2"
James William Fahrlander Sr.
Leo 9h36'15" 29d17'14"
James William Garner
Aqr 23h15'54" -8d55'0"
James William Heaslet
Boo 14h42'19" 36d56'8"
James William Kempfert
Cap 21h44'2" -9d48'11"
James William Marple
Cep 22h16'4" 55d55'19"
James William McCaig
Uma 8h41'30" 47d40'37"
James William Needham
Uma 11h1'49" 38d39'2"
James William Perling
Gem 7h8'50" 26d13'36"
James William Plahitko
Aqr 23h1'35" -9d24'17"
James William Talbot
Her 16h31'44" 25d41'15"
James William Tankersley
Per 3h44'31" 45d43'16"
James William Ullrich
Cep 21h47'12" 60d40'3"
James Wilson
Uma 11h21'13" 60d56'2"
James Wilson Kennedy - 14.11.2003
Cru 12h34'28" -61d59'32"
James Wolfe
Ori 5h41'10" -0d42'31"
James Wolfgang Donahue
Per 3h0'56" 54d30'31"
James Wolfgar Ramsey
Lyn 7h54'15" 39d8'12"
James Wood Jr.
Uma 8h49'13" 62d51'12"
James Woodman Morse
Her 16h34'24" 32d3'22"
James Wyant Rappaport, Jr.
Tau 5h44'50" 25d56'3"
James Yanskey
Boo 14h41'56" 54d30'8"
James Yoder
Uma 10h7'59" 64d57'30"
James York
Per 2h21'54" 52d20'10"
James Yoshioka
Her 18h5'52" 38d27'16"
James Young Major
Aqr 22h39'46" -1d46'43"
James Zac Rainey
Uma 10h3'38" 52d52'43"
James Zangrilli
Per 2h47'31" 53d25'20"
James, Amber & Maria Alejandra
Cnv 12h42'45" 34d58'10"
James, Bethany & Ruby
Cyg 20h44'50" 38d12'52"
James, Kyle, Brock & Sarah
Uma 9h4'44" 61d6'0"
James, My angel shining star.
Per 4h45'14" 46d39'42"
James, my soul mate forever
Gem 7h23'9" 24d13'50"
JAMES1224
And 22h59'46" 48d2'54"
JamesCarlOsborne.1944.12.02USA
Per 1h48'25" 51d15'4"
Jamesgaard-Lauseng
Ori 6h5'54" 20d2'45"
Jamesia
And 0h24'8" 27d40'55"
jamesk34
Lib 15h21'52" -21d47'0"
Jameson
Cen 13h26'23" -37d51'32"
Jameson
Her 17h14'24" 27d31'7"
Jameson Creed Absher
Lib 15h0'55" -6d16'51"
Jameson Edward O'Hara
Cyg 20h59'56" 32d25'35"
Jameson Henry Keeves
Aur 5h47'40" 50d1'11"
Jameson Loves Mommy.
Vir 15h3'23" 5d0'36"
Jameson O'Neill Maher
Her 18h16'32" 23d26'2"
JamesR.Andrews,Jr(Jimmy)six2274
Cnc 8h28'37" 7d1'32"
James's Little Jaelen
Peg 22h28'0" 24d37'3"
James's Yellow Star
Per 3h14'13" 42d50'50"
Jamesstar
Per 4h38'25" 42d6'18"
Jamesy
Tau 4h0'29" 30d13'36"
Jamey Halke
Uma 12h41'12" 59d10'15"
Jamey Masi
Tau 3h46'58" 8d49'7"
Jamey Stiles
Lyn 7h51'55" 38d54'18"
Jamey Woo
Uma 9h28'51" 49d1'28"
Jami 731
Ori 6h10'56" 15d32'16"
Jami Bezerra
Cas 0h18'51" 61d15'24"
Jami Dawn McDaniel
Her 17h0'20" 30d34'24"
Jami & Derek Kuglin
Ori 5h59'49" 6d23'31"
Jami Hergert
Vir 12h55'49" -8d52'35"
Jami Hickerson
Uma 9h30'26" 48d56'50"
Jami Javaid
Aqr 21h59'16" -7d38'40"
Jami K. Vitale
And 1h43'45" 45d6'31"
Jami L Smith
Cam 4h0'31" 68d36'18"
Jami Labree
Cas 0h21'29" 50d36'41"
Jami Lea
Gem 6h50'47" 23d15'28"
Jami Leah
Lyn 8h30'15" 35d21'34"
Jami Leighann Frazier
Gem 7h49'2" 21d27'6"
Jami Lynn
Cyg 21h17'54" 46d10'27"
Jami Lynn Jones
Lyn 6h58'45" 51d0'5"
Jami M. Falchetta, Christmas, 2005
Cas 0h20'30" 54d56'34"
Jami & Marc
Ori 5h36'56" -0d19'27"
Jami Marie
Per 4h6'3" 34d21'36"
Jami Marie Ellis
Lyn 7h56'46" 56d44'41"
Jami - My Heaven On Earth
Ari 2h26'15" 25d37'52"
Jami Ober Gan
Cyg 19h53'5" 37d17'21"
Jami Poole, My Angel.
Sco 17h26'58" -37d0'33"
Jamie
Sco 17h22'12" -32d47'18"
Jamie
Sco 17h56'55" -30d51'57"
Jamie
Psa 22h21'8" -31d26'27"
Jamie
Uma 11h59'57" 59d14'58"
Jamie
Uma 11h32'18" 52d38'57"
Jamie
Cas 1h22'3" 64d48'1"
Jamie
Cam 4h16'9" 68d31'23"
Jamie
Umi 15h24'14" 74d0'48"
Jamie
Dra 20h40'10" 67d40'0"
Jamie
Uma 11h21'59" 60d24'26"
Jamie
Umi 14h23'19" 77d54'36"
Jamie
Cam 5h24'24" 78d5'22"
Jamie
Cap 20h34'24" -9d29'26"
Jamie
Sco 16h4'53" -17d38'23"
Jamie
Cap 21h7'54" -21d54'26"
Jamie
And 1h20'18" 44d3'2"
Jamie
Lmi 10h45'37" 39d14'38"
Jamie
Psc 0h55'2" 33d38'31"
Jamie
Gem 7h25'1" 29d8'18"
Jamie
And 0h24'42" 32d31'41"
Jamie
Cyg 19h32'42" 28d38'17"
Jamie
Com 12h14'45" 28d57'31"
jamie
Vir 11h37'53" 2d44'22"
Jamie
Ori 5h14'11" 8d16'17"
Jamie
Ori 6h14'59" 15d16'0"
Jamie
Ori 6h14'59" 15d27'54"
Jamie
Ori 6h20'22" 14d42'55"
Jamie
Mon 6h48'7" 7d56'51"
Jamie 3149
Ori 5h37'19" 8d6'54"
Jamie 3MAIC
Peg 22h48'1" 22d15'28"
Jamie A. Collins
And 0h41'54" 43d38'57"
Jamie A Zubak
Cnc 8h1'58" 22d11'33"
Jamie & Aaron Cooper
Cyg 20h21'58" 54d32'14"
Jamie Adams - I Love You Mom
Vir 13h17'39" -20d7'49"
Jamie Alexander Love
Ori 6h21'41" 16d37'52"
Jamie Alexandra Grimsley "Allie"
Sco 17h43'31" -40d55'47"
Jamie Alicia Golden
Lib 15h43'27" -29d12'2"
Jamie Allene Ross
Sgr 18h48'2" -30d23'53"
Jamie Allene Ross
Sgr 19h41'17" -17d20'9"
JAMIE ALLISON
And 23h23'33" 47d52'28"
Jamie and Angie
Umi 14h15'39" 68d48'2"
Jamie and Austin Rae
Aql 20h8'53" 11d31'0"
Jamie and Carl Weber
Lmi 10h30'48" 38d54'48"
Jamie and Cecy Robson
Cyg 19h52'32" 37d28'49"
Jamie and Chris
And 1h32'40" 44d52'42"
Jamie and Danny
Aur 6h1'3" 48d1'18"
Jamie and Darren
Col 5h44'40" -31d10'48"
Jamie and David
Ori 5h46'36" 6d31'54"
Jamie and Frankie Forever
Cyg 20h2'32" 56d52'47"
Jamie and Kathy
Cap 21h9'2" -15d52'19"
Jamie and Michelle Nooyen
Cyg 21h32'41" 52d49'32"
Jamie and Mindy
Umi 15h18'55" 73d49'55"
Jamie Ann Hradesky
And 1h7'11" 42d45'8"
Jamie Ann Thomas
Gem 6h46'47" 26d41'14"
Jamie Anne
And 0h25'50" 43d3'33"
Jamie Anne Andrews
Sco 16h5'10" -11d8'28"
Jamie Anne Hirdler
And 2h9'50" 42d59'0"
Jamie Anthony McIntyre
Per 4h15'55" 51d9'33"
Jamie & Ashley Forester
Crb 15h45'0" 33d48'43"
Jamie Ashlynne
Cnc 8h8'18" 7d30'33"
Jamie Austin Wade Gaffin
Cyg 19h58'45" 42d15'36"
Jamie Azelle Schlatter
Peg 22h39'2" 15d39'6"
Jamie Baird
Ari 3h22'11" 27d21'38"
Jamie Barrie
And 1h43'17" 45d32'13"
Jamie Bear
And 0h42'32" 38d2'59"
Jamie 'Bear' Rafferty
Umi 15h22'56" 79d59'50"
Jamie Beckler-Farrar
Cnc 8h50'42" 20d53'45"
Jamie Bencivenga
Cnc 8h19'58" 32d45'41"
Jamie Benefield
And 2h21'30" 50d1'10"
Jamie & Bianca Dallas 101106 Eternity
Col 6h2'15" -30d13'38"
Jamie Blair
And 1h2'33" 44d17'14"
Jamie "Bonita" Moncur
Uma 12h4'15" 50d45'35"
Jamie & Brad
Cyg 20h56'25" 39d47'2"
Jamie & Brian Swanson
And 0h38'11" 46d12'24"
Jamie Brior
Umi 15h13'54" 86d39'12"
Jamie Brooke Irwin
Sgr 18h49'55" -16d16'30"
Jamie Brooke Schulman
Vir 12h11'37" 12d31'20"
Jamie Carol Harris
Ori 5h24'11" -0d3'7"
Jamie Cash
Crb 15h20'42" 32d22'55"
Jamie Christina
Dra 17h42'21" 50d48'45"
Jamie Christina
Tau 3h48'23" 3d13'9"
Jamie Christopher Vincent Gallagher
Per 4h13'26" 42d56'8"
Jamie Ciaran Payne
Cep 21h27'44" 66d6'9"
Jamie Clough
Gem 7h4'56" 24d12'59"
Jamie Colleen Stanton
Ori 5h57'40" -0d45'27"
Jamie Coughlin
Vir 13h31'3" -18d35'59"
Jamie Curti
And 22h57'52" 37d19'23"
Jamie D Clarke Baby Girl
Ari 2h44'24" 28d5'10"
Jamie Dalton 2005 Luv The Bullock's
Umi 15h23'57" 76d7'30"
Jamie Daniel Maines
Gem 7h10'22" 34d8'57"
Jamie Deal
Cap 20h21'55" -15d2'42"
Jamie DeCicco
Lib 14h53'27" -1d17'39"
Jamie Denise
Vir 12h18'3" -4d15'11"
Jamie Denys Sargeant
And 23h48'33" 38d3'33"
Jamie Diane
Tau 4h13'20" 1d5'36"
Jamie Diane Tisdale
Sgr 17h57'58" -26d50'21"
Jamie Dickel
Uma 8h56'18" 64d44'37"
Jamie Dixon
Sco 16h12'37" -9d34'50"
Jamie Dodd
Her 16h44'37" 34d30'49"
Jamie Downing
Aqr 22h30'23" -1d35'44"
Jamie Dunn
Lib 15h3'20" -1d42'9"
Jamie E Frank
And 0h51'47" 44d24'20"
Jamie Elaine
Uma 11h46'26" 55d28'4"
Jamie Elizabeth Carter
Aqr 22h4'20" 1d54'54"
Jamie Elizabeth Elmhirst
Vir 14h1'49" -16d47'28"
Jamie Elizabeth Valentine
Cas 1h46'35" 61d10'39"
Jamie Ella
Lib 15h5'27" -1d5'21"
Jamie F. Drennen
Sgr 20h6'8" -41d14'57"
Jamie Faulk
Aqr 22h31'57" -4d26'0"
Jamie Finnerty
Peg 21h50'15" 7d23'43"
Jamie Flynn
Umi 16h43'30" 77d21'21"
Jamie Fontes
Uma 10h46'8" 47d0'41"
Jamie Frala Cooper
Aqr 22h11'3" -2d56'18"
Jamie Freshour
Lib 15h9'59" -2d29'55"
Jamie Frost
Uma 11h38'28" 46d29'55"
Jamie Fuller
Aqr 22h29'21" 0d18'29"
Jamie G
Uma 11h27'33" 59d42'41"
Jamie Gath
Uma 11h23'37" 53d7'41"
Jamie George William Walton
Per 4h33'52" 40d10'52"
Jamie Gilpatric Hopper
Sgr 19h39'31" -15d18'19"
Jamie Gordon
Cas 0h47'14" 60d44'11"
Jamie Gorton
Lib 14h53'24" -1d7'54"
Jamie & Greg Wedding Star
Cyg 21h40'36" 49d31'59"
Jamie Guardia
Col 5h56'5" -34d45'42"
Jamie Guest
Umi 17h9'59" 80d58'35"
Jamie Gunn
Cep 20h39'23" 55d28'20"
Jamie Guy Williams - Jamie's Star
Dra 17h25'41" 57d28'28"
Jamie Hanlon
Her 17h8'3" 27d34'28"
Jamie Harris loves Josh Arnold
Psc 1h22'46" 27d49'56"
Jamie Heard
Aqr 20h46'36" -9d10'30"
Jamie Hinman's Smile
Psc 0h54'0" 10d5'13"
Jamie Hipp
Cam 4h10'43" 68d3'51"
Jamie Holland
Leo 10h34'53" 15d40'8"
Jamie Hope
Sgr 17h58'38" -18d4'37"
Jamie Hopkins
Peg 22h2'10" 7d57'47"
Jamie Hot Mom
Crb 15h46'30" 32d7'3"
Jamie I. Harris I love you Dad
Umi 13h53'47" 69d47'37"
Jamie J
Lyn 8h33'14" 47d9'5"
Jamie J. King
Ari 2h15'7" 26d12'57"
Jamie J. Tomasz
Gem 6h45'51" 19d38'14"
Jamie Jack Fryer
Gem 7h2'19" 28d0'26"
Jamie Jackson
Uma 9h1'29" 56d39'45"
Jamie James Wisner
Psc 0h27'30" 4d58'5"
Jamie Jameson
Uma 10h58'59" 49d45'3"
Jamie Jammers
Uma 11h40'11" 53d47'42"
Jamie Jaynes
Umi 15h30'46" 70d5'15"
Jamie Jensen
Lib 15h47'2" -25d44'34"
Jamie (Jnr) Miller
Cap 20h48'40" -25d26'38"
Jamie Jo
Peg 23h51'50" 18d9'41"
Jamie Jo Lebrun
Ari 2h7'27" 23d14'9"
Jamie Jo Pacino
Crb 15h36'17" 34d29'38"
Jamie Joseph Rice
Cap 20h22'45" -13d17'25"
Jamie & Joshua
Ori 5h36'41" -4d25'47"
Jamie June
And 2h6'53" 43d19'5"
Jamie Kathleen Reed
Gem 7h21'15" 30d46'49"
Jamie & Kathy Stares Forever
Crb 16h14'44" 38d41'46"
Jamie Katy Hu
Ari 2h46'30" 27d41'51"
Jamie Kay Tanner my universe
Ari 2h13'47" 23d58'37"
Jamie Knabel
Uma 13h30'31" 59d6'28"
Jamie Kyle
And 0h48'56" 35d47'13"
Jamie L. Burkholder
Lmi 9h50'50" 39d32'46"
Jamie L. Cochran
Her 17h57'36" 27d9'1"
Jamie L. Galati
Vir 11h57'22" -8d57'20"
Jamie L. Gallardo
And 23h0'25" 48d34'29"
Jamie L. Preston
Umi 11h46'32" 88d38'59"
Jamie L. Rowe
Mon 7h36'31" -0d48'0"
Jamie Lafave
Uma 10h17'14" 43d33'26"
Jamie Lange
Umi 16h53'23" 79d29'29"
Jamie Laraine Stewart
And 2h33'43" 43d51'48"
Jamie Lauren Nash 143
Leo 10h14'17" 24d58'6"
Jamie Lauren Stephens
Uma 10h7'13" 60d59'26"
Jamie Lea Caracci
Lyn 8h12'15" 42d40'58"
Jamie Lea Robinson
Uma 10h27'13" 52d59'59"
Jamie Lee
Uma 14h26'36" 59d15'50"
Jamie Lee
Cam 4h25'53" 67d1'54"
Jamie Lee
Umi 15h28'46" 73d44'51"
Jamie Lee
Psc 0h26'14" 11d26'2"
Jamie Lee Groth
And 2h17'5" 45d19'32"
Jamie Lee Mangione
Tau 4h1'31" 23d44'46"
Jamie Lee Philippi
Vir 12h33'33" 3d9'59"
Jamie Lei Bruhn
Cyg 21h7'2" 47d47'4"
Jamie Leigh Harris
Gem 7h15'12" 33d46'55"
Jamie Leigh Hilliker
Cyg 19h42'6" 29d28'33"
Jamie Leigh Jordan
Gem 7h30'59" 32d39'42"
Jamie Leona Annelle Myrick-Duckett
Psc 1h30'5" 22d42'36"
Jamie Letourneau
Cam 5h1'9" 65d36'48"
Jamie Liam Bryant
Lib 16h0'46" -13d34'15"
Jamie Liam David Colin Burgess
Dra 18h25'28" 53d50'45"
Jamie Lindberg Gould
And 23h14'9" 41d27'25"

Jamie Loughlin
Tau 5h39'12" 26d43'31"
Jamie Louise Jenson
Dra 18h23'38" 58d35'12"
Jamie Loves Jenn
Ari 3h27'35" 27d58'30"
Jamie Loves Tim Bible
Tau 5h51'58" 24d1'6"
Jamie loves Tina
Cnc 8h4'30" 18d6'44"
Jamie Lu
Her 17h16'51" 17d26'47"
Jamie Lyn Arnold
Uma 8h57'53" 61d23'15"
Jamie Lyn O'Malley's Awesome Star
Lib 15h26'27" -14d18'37"
Jamie Lynch
Tau 4h0'10" 27d37'33"
Jamie Lynn
Cnc 8h40'29" 20d53'11"
Jamie Lynn
Del 20h31'18" 8d1'29"
Jamie Lynn
Ari 2h14'32" 16d26'47"
Jamie Lynn
Leo 11h18'12" 10d5'54"
Jamie Lynn
And 2h26'50" 50d27'10"
Jamie Lynn
And 1h32'5" 48d10'26"
Jamie Lynn
Uma 11h52'4" 32d13'9"
Jamie Lynn
Uma 12h2'8" 38d59'38"
Jamie Lynn
Cas 1h23'32" 68d57'43"
Jamie Lynn Borkgren
Sco 16h33'38" -28d54'25"
Jamie Lynn Bresley
And 1h41'21" 38d32'22"
Jamie Lynn Campbell
Cap 20h25'52" -11d38'52"
Jamie Lynn Church
Cap 20h8'16" -20d4'31"
Jamie Lynn Colaneri
And 23h44'30" 34d1'26"
Jamie Lynn Crozat
Cmi 8h6'16" -0d18'34"
Jamie Lynn Dowell
Ori 4h59'36" -0d50'55"
Jamie Lynn Edwards
Sco 16h13'20" -12d45'27"
Jamie Lynn Hathcock
Cap 21h39'56" -18d25'21"
Jamie Lynn Hawley
And 0h58'42" 40d2'19"
Jamie Lynn Holewski
Uma 13h7'45" 60d12'20"
Jamie Lynn Hopkins
Lmi 10h22'39" 33d0'22"
Jamie Lynn II
And 2h35'41" 44d3'56"
Jamie Lynn Jacks
Lyn 6h27'52" 59d10'19"
Jamie Lynn Klimowicz
And 1h5'58" 45d16'54"
Jamie Lynn Kramer
And 2h33'23" 44d46'17"
Jamie Lynn Larkin
And 23h47'21" 48d21'30"
Jamie Lynn McDonald
Sco 17h34'32" -33d33'52"
Jamie Lynn Murray
Leo 11h53'11" 22d14'59"
Jamie Lynn Pace-DeGuara
Tau 3h35'22" 22d18'7"
Jamie Lynn Rawson
Vir 12h30'57" 6d16'55"
Jamie Lynn Roberts
Uma 11h2'58" 41d58'52"
Jamie Lynn Sell
And 1h28'37" 47d12'10"
Jamie Lynn Snyder
Lyn 6h25'46" 56d48'51"
Jamie Lynn Tatum
And 0h16'35" 29d5'19"
Jamie Lynn Theriault
Gem 6h57'23" 35d13'8"
Jamie Lynn Wade
Cap 21h38'58" -8d27'49"
Jamie Lynn Yeager
Cnc 8h36'47" 19d24'10"
Jamie Lynne
Lib 15h45'41" -18d18'48"
Jamie Lynne Morgan
Lmi 10h42'48" 23d40'22"
Jamie Lynne Shea
And 1h0'25" 44d46'0"
Jamie Lynne Sisler
Aql 19h38'59" -10d20'59"
Jamie Lynne Teklits
And 23h22'24" 43d4'38"
Jamie Lynn's Smile
Uma 8h35'35" 59d49'33"
Jamie M. Herskowitz, my son
Per 2h40'52" 52d45'7"
Jamie M. Wayne
Ori 5h2'42" 13d53'18"
Jamie Maria
Ori 6h9'24" 9d19'13"
JAMIE MARIE
Ari 2h5'15" 23d13'30"
Jamie Marie
Lup 15h31'7" -38d45'25"
Jamie Marie Chavez
Cnc 8h37'2" 17d44'43"
Jamie Marie D'Amico
Ori 5h38'35" 3d2'37"
Jamie Marie Pitts
Crb 15h39'47" 32d30'3"
Jamie Marie Poe
Aqr 22h16'37" -2d18'36"
Jamie Marie Sacino
Leo 11h4'27" 22d5'35"
Jamie Marie Sprouse
Psc 1h32'34" 12d29'29"
Jamie Marie Wilbur
Uma 8h37'21" 64d50'27"
Jamie Marinaro
Psc 1h6'50" 32d16'48"
Jamie Marr
Tau 3h28'42" -0d15'30"
Jamie Massaro
Uma 9h36'9" 44d39'10"
Jamie McCulloch
Gem 7h15'21" 22d32'38"
Jamie McPeck
Lyn 8h34'7" 34d11'51"
Jamie Meadows
And 23h19'50" 42d48'55"
Jamie+Melissa Shining Together 4Eva
Cru 12h32'55" -61d14'46"
Jamie & Michael Morrison
Umi 15h48'19" 72d15'24"
Jamie Michelle
Uma 11h3'23" 69d39'46"
Jamie Michelle
And 23h14'44" 48d19'36"
Jamie Michelle
Uma 9h55'41" 49d25'9"
Jamie Michelle Sivinski
And 2h24'25" 45d36'48"
Jamie Miller (The Love of My Life)
Cnc 8h17'46" 24d47'36"
Jamie Millis
Lib 15h4'37" -1d22'55"
Jamie Murray
Tau 5h18'6" 18d27'59"
Jamie Murray Mitchell
Per 2h13'59" 54d37'40"
Jamie N. Myrick
Aqr 22h23'35" -6d21'24"
Jamie n Sarah
Leo 11h16'21" -4d9'12"
Jamie Neil Morrison
Uma 9h15'32" 70d5'8"
Jamie Nicole Sunhine
Sco 17h21'20" -31d47'29"
Jamie Nicole Upton
Cnc 8h35'43" 22d27'33"
Jamie Ortega
Cyg 21h19'33" 46d8'33"
Jamie Oscar Evers - my star
Lyn 7h32'49" 36d45'31"
Jamie Paradise
And 23h34'46" 36d57'18"
Jamie Paul Jolly
Umi 19h51'36" 87d37'9"
Jamie Paul Pipkins
Aql 20h18'1" -0d8'6"
Jamie Paul Rendall
Uma 9h9'24" 52d55'8"
Jamie Peer
Sgr 18h52'28" -17d11'38"
Jamie Perry
Ori 5h13'44" 2d35'51"
Jamie Philip Kelly
Umi 16h27'39" 85d27'23"
Jamie Phillips
Psc 1h18'6" 3d18'2"
Jamie Pierce
Cap 20h21'39" -9d41'52"
Jamie Pion-Ritter
Tau 5h26'9" 20d3'12"
Jamie Powers
Sgr 19h53'57" -12d29'0"
Jamie Quackenbush
Tau 5h39'26" 26d1'1"
Jamie Rae
Crb 15h46'22" 33d16'59"
Jamie Ragsdale
Uma 12h22'29" 54d16'20"
Jamie Ray Diane Scott
Lyn 6h40'34" 55d51'48"
Jamie Raymond Clayton
Col 5h43'32" -28d12'31"
Jamie Reed Saxena
Gem 6h31'42" 26d44'37"
Jamie Regina Tapia
Ari 2h52'1" 24d33'51"
Jamie Renee
Cnc 9h7'9" 31d1'43"
Jamie Richardson-Hardie
Aqr 22h9'47" -3d57'33"
Jamie Riggs
Peg 23h53'56" 17d55'29"
Jamie Rivera
Uma 8h35'12" 66d26'15"
Jamie Robert Lyons
Ari 2h40'14" 30d48'4"
Jamie Rose
Gem 7h28'5" 31d3'17"
Jamie Rose Brids
Uma 12h57'7" 62d46'39"
Jamie Rose Isaak
Tau 4h37'38" 22d36'1"
Jamie Rowan Duffin
Cnc 9h19'38" 11d13'27"
Jamie Ruth Hill
Cas 1h25'56" 60d43'57"
Jamie Ryan
Ari 2h7'48" 15d43'16"
Jamie Ryan
Uma 11h43'26" 43d5'18"
Jamie S. Prewitt
And 1h14'28" 45d13'25"
Jamie S. Weisberg
Tau 5h53'4" 27d47'55"
Jamie Satterwhite
Gem 7h52'25" 29d26'38"
Jamie Scherer Berndt
And 0h33'3" 35d20'45"
Jamie Schnittjer
Ori 5h24'20" 9d1'3"
Jamie Shannon
Umi 14h26'5" 68d31'43"
Jamie & Sharon Butterfield
Cyg 19h45'18" 33d1'23"
Jamie Sheppard
Lyn 7h40'51" 47d53'35"
Jamie Shough
Ari 3h7'31" 20d9'31"
Jamie Squared
Uma 13h42'26" 57d21'43"
Jamie Star
Cyg 20h29'17" 52d21'50"
Jamie Stephen O'Reilly
Ari 2h54'31" 28d40'25"
Jamie Steven Bell
Umi 16h23'57" 72d22'26"
Jamie Stevenson
And 23h2'2" 48d6'53"
Jamie Stricchiola
And 0h19'0" 37d28'5"
Jamie Sue
And 23h23'28" 50d35'40"
Jamie Sue
Psc 1h8'43" 27d26'35"
Jamie Sue Rebella
Lmi 10h30'59" 30d6'18"
Jamie T. Choi
Lib 15h35'10" -24d36'21"
Jamie Tarabay
Pho 0h17'15" -41d19'39"
Jamie Taylor- aka: Mom & Meme
Cyg 19h35'59" 28d25'58"
Jamie Trisztan Horton
Tau 3h31'19" 17d23'16"
Jamie Turitz
Sco 16h7'53" -8d58'39"
Jamie Turnbull
Uma 9h44'57" 62d43'57"
Jamie Twisssay Birkenberger
Pup 8h4'50" -34d41'53"
Jamie Tyler South 4/7/06 - 5/1/07
Cru 12h26'8" -61d14'30"
Jamie Victoria Croker Cline
Leo 10h31'0" 14d22'58"
Jamie Vincent
Tau 5h48'21" 27d4'18"
Jamie Wiesner-Ortega
Tau 4h29'5" 29d40'38"
Jamie William Townes
Uma 12h26'41" 55d40'45"
Jamie Witsen
Cap 21h46'37" -24d2'49"
Jamie Wittenberg
Lib 15h33'25" -6d16'13"
Jamie, Mary-Veronica
Leo 11h15'48" 11d44'55"
Jamie0704
Cnc 8h49'21" 8d3'15"
Jamie7511
Lib 14h57'11" -7d34'19"
JamieBear
Ori 5h34'6" 9d42'47"
Jamie-Leigh
Peg 22h39'41" 21d9'38"
Jamielily
Cru 12h25'47" -58d43'51"
Jamiell
Psc 1h20'23" 27d12'23"
JamieL.Osio2005
Lmi 10h7'50" 30d17'55"
Jamie-Louise Hawkins
And 23h38'25" 41d43'27"
JamieLove071803
Cas 23h36'21" 53d34'32"
Jamie-Lyn
Leo 9h26'50" 17d11'44"
JamieMarie
Tau 4h16'31" 23d22'43"
Jamienne
Cyg 20h11'55" 56d14'56"
Jamie's Angel Star
And 0h55'31" 40d40'23"
Jamie's Butterfly
Gem 6h32'14" 14d6'40"
Jamie's First Star to the Left
Crb 16h4'26" 30d54'55"
Jamie's Forever...Townsend's Love
Leo 10h34'24" 23d38'52"
Jamie's Jewel
Cas 23h45'52" 51d22'56"
JAMIE'S LITTLE SAM
Ari 2h44'35" 30d5'15"
Jamie's Love
Uma 8h41'22" 64d0'25"
Jamie's Obeda
Cet 1h14'18" -0d29'8"
Jamie's Star
Uma 8h57'21" 61d57'30"
Jamie's Star
Cru 12h54'44" -59d20'2"
Jamie's Star
Sco 17h57'53" -30d18'8"
Jamie's Star
Per 2h53'53" 56d1'12"
Jamie's Star
Ori 6h7'0" 18d21'47"
Jamie's Star
Cnc 8h55'57" 7d59'45"
Jamie's Star
Aqr 21h26'16" 2d2'40"
Jamie's Sunshine
Gem 6h38'29" 27d50'32"
Jamil Ahmed Malik
Dra 17h5'25" 55d26'0"
Jamil & Ghida
Her 16h46'25" 31d12'45"
Jamil Lucy
Ari 2h11'59" 13d1'20"
Jamila
Ori 5h29'44" 1d18'59"
Jamila
Cnc 8h12'19" 13d4'0"
Jamila Zakiya Thema
Vir 12h59'4" -21d41'5"
Jamilah Abdelhag
Sco 15h54'0" -25d44'48"
Jamille Cucci
Vir 13h20'38" -0d9'48"
Jamillet Cortez
Tau 4h35'50" 9d31'42"
JamilovesAnthony..Forever & always
Sco 16h12'41" -9d55'24"
Jamilynn Ember King
Gem 7h25'44" 14d17'53"
Jamin Kristan
Gem 6h58'24" 19d18'32"
Jamina Jana Kauer
Peg 23h44'20" 13d19'48"
Jamina Vasbinder
Peg 22h8'28" 8d0'18"
Jami's Beauty
Tau 4h22'41" 18d43'2"
Jami's StarLight
Lyn 8h48'19" 45d51'51"
Jamison Adara Lientz
Uma 9h12'11" 50d45'13"
Jamison Covington
Leo 10h28'40" 17d38'17"
Jamison Daniel Pangilinan
Her 18h55'2" 19d4'26"
Jamison Irving Braniff
Leo 10h55'34" -1d12'57"
Jamison "Jamie" Keith Rish
Per 3h42'54" 43d53'9"
Jamison Kathleen Traub
Ari 3h29'14" 23d49'40"
Jamison Moore
Crb 16h10'33" 39d7'37"
Jamison Renae Erwin
Crb 15h50'5" 28d4'10"
Jamison Yttri
Umi 14h55'33" 75d54'27"
JamisonDrewCupp11-18-1987
Her 17h47'23" 35d11'14"
Jamison's Brilliance
Dra 16h7'25" 62d11'9"
Jamity
Cyg 20h25'45" 46d43'18"
Jamly
Cap 20h47'45" -15d38'5"
JamMel
Dra 16h33'13" 52d41'41"
Jammel
Uma 11h4'34" 48d45'35"
Jammer
Cmi 7h38'22" 0d57'45"
Jammi Lynn Arlington
Ari 2h3'37" 23d24'4"
Jammie Apodaca
Gem 6h57'13" 26d15'29"
Jammie Castillo
Uma 9h18'57" 45d50'16"
Jammie Fuller
Sgr 17h58'3" -18d6'41"
Jammie McCale Phi Sigma Chi Alumni
Umi 14h43'43" 72d51'30"
Jammie's Baby Pie
Psc 1h25'12" 17d28'50"
Jammin' Jamie
Uma 9h22'1" 65d23'35"
jammine
Cas 0h9'3" 56d17'8"
Jamo
Umi 14h7'20" 68d43'49"
Jamo
Uma 12h29'36" 55d48'34"
Jamoaimay
Umi 15h53'22" 79d58'36"
Jamon Elliot Johnson
Cnc 8h51'12" 25d51'48"
Jamona
And 2h7'36" 41d16'26"
Jamone M. Lewis Jr. My Son My Angel
Uma 10h31'49" 42d13'50"
Jamonix
Lib 14h25'39" -17d48'53"
Jamoo
Cyg 20h52'26" 31d52'48"
Jamp-4
Her 17h25'54" 20d8'33"
Jamquel
And 0h7'8" 43d57'31"
Jamshan Amour
Ori 5h58'26" 18d18'9"
Jamus Lizdosis
And 0h10'12" 31d19'7"
Jamy
Aqr 23h23'44" -20d17'42"
Jamye Cappetta Shah
Aqr 23h0'14" -22d29'27"
Jamye Lee Henderson
Cnc 8h8'43" 22d18'14"
Jamye's Hope
Uma 10h37'18" 48d38'17"
Jan
Uma 9h19'34" 50d10'14"
Jan
Uma 13h55'36" 49d17'51"
Jan
And 23h22'25" 47d15'56"
Jan
Per 3h22'50" 45d16'34"
Jan
Lyn 8h49'54" 41d49'7"
Jan
Aur 6h34'23" 35d2'43"
J.A.N :)
Lyr 19h9'30" 27d17'3"
Jan
Ari 2h31'53" 20d11'18"
Jan
Sco 15h55'37" -22d10'52"
Jan
Vir 11h49'34" -2d57'55"
Jan
Aqr 22h51'31" -7d13'2"
"Jan"
Uma 12h12'34" 54d26'9"
Jan
Uma 8h43'13" 58d0'1"
Jan
Uma 13h43'59" 56d25'32"
Jan Alan Kline
Leo 11h33'42" 17d45'4"
Jan Alexander
Lyn 6h53'55" 59d1'31"
Jan Alexander Saward
Per 4h9'56" 42d33'35"
Jan and Harry Anniversary Star
Crb 15h40'19" 30d4'14"
Jan and Jules Pellerin Special
Cyg 21h54'36" 45d57'29"
Jan and Steve Frisch
Cyg 20h31'46" 43d29'54"
Jan - Ann - Guiding Lights Forever
Leo 10h10'23" 11d40'56"
Jan Barry Carroll
Lib 15h11'3" -9d37'11"
Jan Booth
Cap 21h37'13" -17d32'13"
Jan & Brian Arnett - Best Bosses
Cru 12h18'42" -57d25'27"
Jan Bright
Lyn 7h2'42" 53d48'8"
Jan Broadwater
Mon 6h42'23" 2d43'23"
Jan Brown - 2004
Peg 22h19'26" 17d33'3"
Jan Brozek
Lyn 8h44'15" 34d25'46"
Jan Bühler
Cyg 21h29'12" 50d12'51"
Jan Burger
Tau 3h31'51" 27d54'21"
Jan Busch
Lyr 19h9'34" 43d5'55"
Jan C. Dunner
Lib 15h0'0" -15d37'36"
Jan Carlton Childress
Cnc 8h16'46" 23d46'29"
Jan Castells i Sanllehi
Cap 20h34'10" -18d32'22"
Jan Cornfeld/ Daddy
Psc 1h1'56" 33d3'17"
Jan Correll Coleman's Star
Lmi 10h39'35" 37d27'37"
Jan & Danuta Czubiak
Lmi 10h28'32" 29d42'58"
Jan Dawson
Lyn 6h50'43" 56d51'31"
Jan Dornseif
Lyn 7h28'50" 52d32'25"
Jan E. Adams
Aqr 22h33'30" 0d18'30"
Jan "El Cuki" Velez
Uma 9h25'17" 50d7'57"
Jan Elowe
Aqr 22h12'23" -4d8'23"
Jan "Ember" O'Rourke
Psc 22h55'48" 7d30'23"
Jan Etchells
Cas 23h46'45" 56d33'58"
Jan F. Dunlap & Sandi Hinton
Sco 16h40'16" -31d0'20"
Jan Grauer
Gem 6h33'28" 19d40'38"
Jan Gregory-Geoffrey
Vir 13h52'37" -20d30'8"
Jan Haines
Ori 5h23'25" 13d41'39"
Jan Heaton
Lmi 10h15'22" 31d3'29"
Jan Hedstrom Wyatt
Ari 2h7'17" 14d41'35"
Jan Hildebrand
Cep 3h53'50" 85d45'38"
Jan Hinkle Brainard
Peg 22h33'47" 3d6'49"
Jan i Joasia
Aqr 22h36'52" -0d52'41"
Jan Jan
And 23h44'22" 47d49'9"
Jan Jepson
Aqr 23h20'26" -19d25'5"
Jan K. Davidson
Lib 14h52'37" -2d20'34"
Jan Keltner
Cas 0h35'58" 53d52'3"
Jan Kenneth Widaman
Lib 14h24'9" -16d56'24"
Jan Kersten
Uma 9h23'18" 57d11'27"
Jan Kingma
Lib 15h37'33" -17d1'13"
Jan Kline
Cap 21h56'38" -17d43'46"
Jan Konecny
Ari 3h4'3" 25d35'48"
Jan Krieger
Uma 8h31'32" 63d19'8"
Jan L Golden 08.07.2005
Leo 9h25'55" 30d58'27"
Jan L. Hume Holter
Lyr 18h47'22" 36d42'15"
Jan Larsen
Uma 9h34'21" 61d2'20"
JAN - LENA - MAREN
Crb 16h15'13" 36d53'54"
Jan Libby French
Cas 1h5'49" 62d54'8"
Jan Lindsey
Tau 4h21'7" 26d25'39"
Jan Louise Brown-Dunaway
Uma 11h21'19" 51d18'9"
Jan Maas 10.000 days young
Uma 11h26'1" 58d53'11"
Jan Marie
Ori 5h6'18" 1d9'57"
Jan Marie Peterson
Ari 3h7'16" 11d53'53"
Jan Marie Tillman
Vir 12h42'23" 2d0'41"
Jan May
And 0h14'13" 43d8'55"
Jan McAlister
Cas 1h31'16" 61d11'21"
Jan McCaskill
Lmi 10h44'38" 27d47'43"
Jan Michael Fordham
Sco 17h47'11" -42d58'21"
Jan Miles
Cas 23h22'17" 59d27'21"
Jan Morgan
Cap 21h5'0" -27d1'6"
Jan "My Heavenly Star" Turlick
And 2h35'27" 46d17'12"
Jan Oberröder
Uma 13h57'13" 59d29'17"
Jan O'Brien
Gem 6h43'40" 17d59'53"
Jan of Santorini
Sco 17h20'53" -37d57'5"
Jan Olaf Jörg Hensel
Uma 10h32'28" 47d37'47"
Jan Oppliger
Vul 21h7'49" 27d59'8"
Jan & Paul Constable
Cyg 19h53'47" 52d8'39"
Jan Perry
Lyn 8h0'47" 51d34'27"
Jan Persson, Jr.
Leo 10h12'15" 26d44'54"
Jan Porter
Uma 10h48'43" 46d36'32"
Jan Potter
Uma 11h33'4" 61d49'36"
Jan Reine
Cas 1h18'16" 54d18'15"
Jan Saltsman
Uma 12h56'23" 61d11'47"
Jan Scarpella
Her 17h48'30" 38d54'52"
Jan Schär, 05.02.1984
Lyr 18h46'57" 40d45'20"
Jan Segnitz, M. D.
Cas 1h18'19" 57d20'27"
Jan Selman
Ori 5h38'55" -0d2'38"
Jan Synnamon
Lyn 9h11'24" 36d25'10"
Jan Tubiolo
Gem 6h57'7" 24d8'12"
Jan Victor Wawszczak
Cyg 21h1'22" 46d13'53"
Jan W. Wise
Uma 11h47'6" 63d58'11"
Jan White
Uma 14h11'59" 55d18'56"
Jan Wilfried Meyer
Uma 11h44'12" 61d21'22"
Jan Wißler
Ori 5h59'54" 7d12'48"
Jan Wolfe-McAlister
Cas 23h49'57" 62d6'27"
Jan Yarnell
Gem 6h58'41" 15d9'44"
Jan Yuille
And 23h53'9" 35d33'42"
Jan Zerek Catherwood
Ari 1h53'4" 19d6'49"
Jana
Ori 5h31'17" 10d49'40"
Jana
Peg 22h34'16" 22d5'40"
Jana
Her 18h12'13" 18d21'43"
Jana
Crb 15h58'29" 38d32'24"
Jana
Uma 8h46'40" 48d38'12"
Jana
And 2h22'48" 45d46'4"
Jana
Cas 1h8'7" 63d23'16"
Jana Baker
Umi 14h50'44" 74d47'18"
Jana Belot
Ori 5h35'34" 2d58'1"
Jana Bieger
Sco 16h42'5" -26d34'51"
Jana Buder
Cyg 19h19'21" 53d10'54"
Jana Caranante Davis
Uma 12h2'33" 41d23'37"
Jana&Christin
Uma 8h11'45" 67d28'0"
Jana Connor 243
Lib 15h18'20" -4d19'33"
Jana Corien Dimmig
Cnc 8h36'58" 24d24'19"
Jana Escuin Trigo
Cap 20h42'46" -21d52'11"
Jana & Frantisek Jiricek
Cyg 19h42'42" 43d16'40"
Jana Gail Marshall
Psc 1h9'27" 11d13'16"
Jana Gibson
Sco 16h58'23" -38d30'36"
Jana Joseph
Mon 6h49'4" 7d37'39"
Jana Katrina Schenk
Umi 13h23'5" 87d0'51"
Jana Kimmel
Sgr 19h13'4" -25d18'20"
Jana Larsen Peters
Cap 21h3'11" -25d1'54"
Jana Lee
Ori 6h14'3" 15d0'50"
Jana Lee
Uma 11h31'53" 45d20'51"
Jana Leigh
Ari 2h11'46" 22d55'2"
Jana Lynn Smith
Uma 10h3'29" 68d41'30"
J'ana M. Rogers
Uma 8h52'32" 56d28'45"
Jana Marie
And 1h46'32" 50d18'49"
Jana Marie Head
Tau 4h31'47" 6d20'38"
Jana Marie Sneddon
Cyg 19h39'22" 39d20'28"
Jana Rebecca Wark
Cnc 8h53'19" 22d39'51"
Jana Reder-Münnich
Uma 8h38'56" 54d55'44"
Jana Schlieter
Uma 8h44'22" 53d36'29"
Jana Simmons
Uma 11h15'28" 58d42'25"
Jana Sinclair Bagwell
And 23h19'56" 41d57'33"
JANA SINDLINGER
Dra 16h53'58" 57d23'25"
Jana Tuscher
Sgr 18h23'26" -32d18'46"
Jana Yonemura
Lyn 7h58'30" 44d37'17"
Jana, Marc-Jordan, Uwe
Ori 6h9'18" 16d50'55"
Jana, My Sweet Girl
And 23h4'12" 47d50'7"
Janabellah
Sgr 19h25'15" -28d11'40"
Janabeth Fleming Taylor
Cnc 9h12'19" 25d3'6"
Janadan
Mon 6h27'7" 9d6'40"
JaNae
Tau 4h34'33" 6d47'22"
Janae Bry'Ann Grissell
Lyn 8h40'48" 34d7'18"
Janae Carmella
And 1h56'15" 38d18'53"
Janae Jean Almen Smith
Tau 4h40'10" 24d4'25"

Janae Michelle Seignious-
Shields
Ari 3h18'3" 24d38'24"
Janae Nafziger Un Cadeau
du Ciel
Lib 14h37'19" -9d12'58"
Janae Souza
Mon 6h44'44" 7d30'34"
Janaebigwatson
Lib 14h45'16" -11d18'15"
Janaeh'
Uma 10h20'7" 49d26'35"
Janai
Cas 0h45'39" 73d58'10"
Janai Watkins and Binh Ly
Lib 15h41'14" -5d38'54"
Janaki Patel
Lyn 7h4'40" 58d10'29"
Janaki Rajan-Hancock
Uma 9h5'3" 48d2'41"
Janaki Rana
Gem 6h55'12" 24d50'5"
Janalea
Psc 1h35'19" 10d16'44"
Janalee
Uma 9h37'46" 44d16'19"
Janalee Julana Hubbard
Lyr 19h23'10" 37d55'45"
Janani
Gem 7h11'55" 22d18'19"
Janara
Tau 3h44'3" 23d27'18"
Janard Andrew
Her 16h9'11" 48d41'4"
Jana's Glücksstern
Uma 8h26'46" 61d28'37"
Jana's Pegasus Sadalbari
Earth Star
Peg 22h44'43" 24d40'5"
Jana's Smile
Sco 16h7'19" -14d33'15"
Jana's Starr
Cap 21h13'20" -19d45'41"
Janay Beverly
And 23h10'44" 50d47'42"
Janay Rasey
Aql 19h17'14" -0d9'5"
Janaya Reeves
Vir 13h18'57" 6d8'20"
Jancie Rose Robbins
Ori 5h33'6" -1d15'21"
Jancy
Lib 14h34'40" -10d29'32"
Janda
Peg 23h43'1" 18d49'3"
JANDA
Uma 9h25'42" 50d0'10"
Janda Drew
Cap 20h54'10" -24d52'36"
Janda Karen Peanut Love
Umi 16h32'1" 76d28'28"
Jan-David Soutar
Vir 13h36'38" -3d22'54"
Jandi Lynn Searle
Sgr 18h56'4" -22d5'38"
Jandindili und Bauro -
01.02.2004
And 2h1'3" 46d15'15"
JandJ Love Forever 1994
Per 4h24'31" 35d16'20"
JANDK
Cam 4h8'14" 66d13'48"
JandNs Bear
Gem 6h23'57" 21d48'23"
Jandolyn Alyse
Ori 5h35'49" -0d14'17"
Jandow, Christina
Uma 9h27'7" 48d45'56"
JANDY
Tau 5h53'16" 24d59'22"
Jandy
Umi 16h16'22" 74d37'10"
Jane
Uma 12h51'12" 62d30'41"
Jane
Cas 1h48'9" 60d20'32"
Jane
Mon 6h41'14" -6d24'13"
JanE
Cap 20h53'29" -19d30'13"
Jane
Sco 16h13'45" -11d20'30"
Jane
Leo 9h24'8" 29d24'46"
Jane
And 0h42'52" 22d35'50"
Jane
Psc 0h8'26" 8d24'57"
Jane
And 23h4'46" 51d37'7"
Jane
Cyg 21h59'14" 51d59'57"
Jane
And 23h21'21" 47d11'23"
JANE
And 0h32'39" 37d26'44"
Jane
Cyg 20h8'55" 30d34'17"
Jane
Boo 14h45'41" 36d50'51"
Jane
And 2h29'40" 44d41'35"
Jane
And 0h5'10" 44d55'45"
Jane A. "Gabby" Valencia
Del 20h44'55" 15d22'17"
Jane A Giambattista
Vir 12h12'37" -0d54'42"
Jane A. Maurer
Cas 0h59'55" 63d42'53"
Jane A. VanZant
Cyg 20h58'15" 36d55'35"
Jane Allen
Vir 13h0'14" -12d4'21"
Jane Allen Colby
Cas 0h19'12" 47d48'4"
Jane Allison Frick
Cas 1h12'24" 57d5'6"
Jane Amy
And 23h6'50" 46d1'0"
Jane and Andrew -
Wedding Day
Cyg 20h59'31" 37d9'56"
Jane and Jen - Always
Aqr 23h19'46" -10d42'51"
Jane and Jim
Peg 23h51'46" 19d50'22"
Jane and John Pinel
Cnc 8h34'17" 13d58'56"
Jane and Neal Grunstra
Cyg 20h45'56" 40d11'22"
Jane and Richard
McDonald
Ori 5h50'41" 9d12'18"
Jane Ann Barnett Kelly
Leo 9h46'42" 27d55'54"
Jane Ann Bohmer
Psc 0h56'54" 26d54'28"
Jane Ann Shih
Gem 7h14'27" 20d17'8"
Jane Ann Wilhite
Mon 6h38'11" 10d46'10"
Jane Ash
Ori 5h8'54" 15d16'35"
Jane B. Converse
Cnc 8h43'9" 18d4'27"
Jane Baber
Psc 23h25'12" 3d50'43"
Jane Bagley
And 1h18'58" 40d45'17"
Jane Baltadonis
Leo 10h57'25" 6d3'46"
Jane "Beanie" Flint
Uma 11h18'58" 64d27'10"
Jane Becker Kantor
Cnv 12h35'7" 43d34'2"
Jane Bennett
Cnc 8h35'57" 31d44'36"
Jane Bernard
Tau 5h33'52" 21d58'3"
Jane & Bob Smith 50th
Anniversary
Cyg 20h19'22" 45d56'59"
Jane Bodeau
Ori 6h8'40" 3d33'49"
Jane Bonanno
And 1h54'27" 42d15'44"
Jane Braboy
Gem 6h27'43" 17d2'1"
Jane Bryce Wilson
Psc 0h50'48" 9d55'3"
Jane C.18
Lyr 18h54'37" 36d28'18"
Jane Calista
Psc 1h4'52" 23d44'5"
Jane Cardona Valdez 07-
13-2003
Leo 10h17'22" 22d45'12"
Jane Caroline
Aqr 23h1'24" -14d24'27"
Jane Catherine Hannon
Vir 11h49'14" 2d41'41"
Jane Catherine Rachel
"Love Star One"
Vir 12h16'37" -0d57'23"
Jane/Charles Branson
Tau 4h28'51" 29d38'56"
Jane Christian
Lib 15h43'15" -6d45'48"
Jane Clair
Sco 16h7'28" -8d51'4"
Jane Corlette
Cas 1h29'29" 65d44'52"
Jane Coulson Aldrich
Uma 13h27'21" 61d24'58"
Jane D. Harker
Sgr 18h42'0" -31d10'39"
Jane Delores Oakley-
Clifford
Aqr 21h48'57" -6d8'9"
Jane Dowling
Cam 7h11'4" 73d27'50"
Jane E
Sco 17h38'25" -40d50'11"
Jane E.
Tau 5h39'45" 25d59'58"
Jane E.
Psc 1h6'39" 11d55'51"
Jane E. Bartsch
And 0h14'37" 25d8'28"
Jane E Moss
Gem 7h13'35" 32d44'51"
Jane Eileen Baldry
Lib 15h19'56" -9d37'48"
Jane Elizabeth
Sco 16h10'44" -11d8'43"
Jane Elizabeth
Cap 20h46'12" -19d38'31"
Jane Elizabeth
Cas 2h2'30" 62d26'50"
Jane Elizabeth
And 2h27'48" 47d43'48"
Jane Elizabeth Brennan
Murphy
Tau 5h54'33" 26d51'15"
Jane Elizabeth DePasqual
Ori 6h7'58" 6d41'7"
Jane Elizabeth Dick
Cap 21h44'20" -12d58'57"
Jane Elizabeth Eastaff
Cas 23h33'37" 57d4'30"
Jane Elizabeth Jackson
Uma 11h57'40" 31d0'46"
Jane Elizabeth Lubs
Cas 0h49'25" 59d24'35"
Jane Elizabeth Marsden
Gem 7h16'31" 16d1'4"
Jane Elizabeth Mitchell
Cyg 21h35'5" 40d43'11"
Jane Elizabeth Palmer -
"Lady Jane"
Peg 21h36'5" 26d8'50"
Jane Elizabeth Phee
Del 20h34'37" 14d35'58"
Jane Elizabeth Reitz
Vir 12h53'51" 12d9'44"
Jane Elizabeth Schroeder
Lib 15h2'9" -9d36'52"
Jane Elizabeth Seys
Tau 4h13'30" 26d36'25"
Jane Elizabeth Townsend
And 0h24'57" 28d12'19"
Jane Ellen Conway
Per 2h55'16" 44d15'15"
Jane F. Coolican
And 23h3'16" 48d24'27"
Jane F. Mullen
And 23h42'36" 45d37'7"
Jane Fairhurst Blakey
Cyg 20h45'25" 35d22'58"
Jane Fikslin
Cyg 20h7'54" 36d38'9"
Jane Fitzharris 12/30/1920-
3/26/04
Cas 0h25'24" 64d53'42"
Jane Frangos
Cnc 8h43'6" 12d58'13"
Jane Frishcosy
Mon 6h44'44" -0d8'12"
Jane G. Akin
Lyr 18h48'21" 44d47'28"
Jane Graham Brown
Macdonald
Cyg 20h57'45" 31d12'6"
Jane H. Conti
Sgr 18h53'10" -32d28'37"
Jane H. Hopkins
Lib 15h54'11" -17d37'11"
Jane Haggerty Our Star Of
Strength
Cyg 20h38'3" 53d56'9"
Jane Hankins
Cas 0h12'18" 50d18'24"
Jane Hankle
Cas 1h25'46" 60d21'48"
Jane Hatton
Cep 23h4'28" 72d19'0"
Jane Hauk Garza
Com 12h47'8" 30d38'39"
Jane Hawking
Cas 0h10'49" 55d6'43"
Jane Hays
And 0h43'26" 35d1'4"
Jane Heather Wilson
Vir 13h20'53" -0d4'12"
Jane Hill Pritchard Stanley
Vir 14h18'56" 3d38'55"
Jane Hilldigard Bernal
Uma 9h46'52" 63d2'31"
Jane Hinz-Bowser
Crb 15h35'26" 32d21'16"
Jane Hughes 19th Hole
Uma 8h29'48" 65d28'16"
Jane (Janie-Bird)
Ronhovde
Psc 1h28'31" 27d35'3"
Jane JinKyung Ha
Tau 4h13'39" 14d24'43"
Jane Jorvina
And 2h10'17" 38d13'46"
Jane Juel Swan child of
God
Uma 13h47'3" 49d23'16"
Jane K. Wilson Family Star
Tri 2h38'9" 33d55'6"
Jane Kathleen Spomer
Spielman
Cas 1h12'49" 69d8'32"
Jane Kimberly
Sco 16h13'30" -15d13'28"
Jane Kolomiyets
Cyg 21h3'32" 47d32'22"
Jane Larkworthy
Lyr 18h28'55" 32d7'48"
Jane Larkworthy
Mon 6h41'9" 4d47'19"
Jane Lavery
Cnc 9h10'36" 15d45'22"
Jane Lemiuex
Psc 1h42'35" 14d24'1"
Jane Lindskold
Vir 14h26'46" 1d32'47"
Jane Lockhart
Uma 9h59'33" 46d4'33"
Jane Louise Davies
Tau 5h46'23" 16d59'37"
Jane Louise Pawlowicz
Sgr 18h1'6" -27d49'16"
Jane M 2005
Cas 23h22'36" 54d53'10"
Jane M. O'Brien
Psc 1h28'8" 21d16'43"
Jane Maisie
Gem 6h52'59" 30d13'50"
Jane Manausa
Cnc 8h59'8" 15d5'45"
Jane Marchie-Gregan
Cap 21h40'16" -18d21'53"
Jane Margaret Griffiths
Psc 1h47'4" 9d51'16"
Jane Mariam Easo
Lib 15h6'3" -10d14'54"
Jane Marie
And 2h23'50" 45d7'57"
Jane Marie Cook
Lyn 8h1'16" 51d14'57"
Jane Marie Craigg
Cyg 20h57'47" 31d18'59"
Jane Marie Lant
Aqr 22h23'2" -5d23'7"
Jane Marie McCarthy
Peg 22h37'41" 17d56'46"
Jane Marie Ostash
Cas 1h8'7" 49d20'46"
Jane Marie Zelinsky
Cas 0h46'25" 53d43'50"
Jane Martin
Uma 8h44'1" 54d43'21"
Jane McAnulty... Who
Loves Ya Baby
Leo 11h7'10" 3d21'17"
Jane Menees
Uma 11h3'54" 67d6'52"
Jane & Michael Rudes
Cnc 8h8'58" 29d21'16"
Jane Milgram: Wife, Mom &
Grandma
Cas 1h7'4" 59d29'8"
Jane "Mimi" Ball
Uma 9h41'22" 47d9'41"
Jane (Mommy)
Tau 4h24'42" 28d6'50"
Jane Montgomerie
Tau 5h46'26" 21d18'25"
Jane Muir Sawyer
Per 4h48'47" 46d48'41"
Jane Mulcare
And 0h26'56" 32d52'44"
Jane - My Angel
And 23h27'14" 36d42'9"
Jane Newman
And 2h32'2" 41d27'40"
Jane Olivia Zimbalist
Uma 11h38'23" 37d42'1"
Jane Park
Lyr 19h19'17" 33d43'43"
Jane Peacock-Panos
Cas 1h28'47" 57d23'45"
Jane Pendleton Blauvet
Uma 9h57'20" 43d4'1"
Jane & Phil Forever
Lib 15h20'9" -9d41'13"
Jane Pope Blake
Com 13h6'32" 27d14'55"
Jane Pufky Nesbitt
Crb 16h2'27" 36d44'26"
Jane Purdy
Uma 12h23'56" 54d7'58"
Jane Q
Lyr 18h42'17" 34d45'29"
Jane R. Goldring
Crb 15h36'46" 31d19'7"
Jane Rankin
And 2h21'4" 45d8'40"
Jane Reid
Lyn 7h35'39" 38d5'27"
Jane Rex
Aql 19h12'53" 7d57'6"
Jane Risebury
And 23h53'33" 35d26'0"
Jane Roberta Melampy
Cas 0h9'56" 57d2'11"
Jane Romanek
Cep 22h2'57" 69d35'3"
Jane Ross
Cam 3h21'48" 66d4'27"
Jane R.Summers
Tau 4h38'28" 22d33'44"
Jane S. Lipscomb
Mon 6h54'33" -0d29'7"
Jane Salverio
Del 20h49'9" 16d4'25"
Jane Samuel - Star for Life
Tau 3h39'52" 28d45'0"
Jane Serues
And 1h48'17" 45d39'48"
Jane Siehl Moore
Aqr 22h18'39" -2d25'41"
Jane Sunshine Duong
Cnc 9h4'3" 24d37'55"
Jane Sykes and Lisa Sykes
Leland
Uma 11h36'43" 46d55'40"
Jane T. Hansen
Cas 23h43'34" 59d52'28"
Jane "The Mayor" Frisina
Cas 0h22'8" 50d2'55"
Jane Tiernan
Leo 9h55'56" 16d38'1"
Jane Tupper
Aql 20h18'50" 3d57'47"
Jane Turner Bedell
Cas 1h42'23" 64d19'54"
Jane Valentine
Uma 9h28'27" 52d37'26"
Jane Vella
Cnc 9h7'3" 23d55'32"
Jane W. Lee
Leo 11h53'17" 15d35'39"
Jane Walker Gray
Uma 11h16'21" 53d19'22"
Jane Walton
Uma 10h31'17" 47d26'2"
Jane Ward
And 1h4'8" 41d9'34"
Jane Watkins
Peg 21h44'5" 27d42'16"
Jane Webster's 40th
Birthday Star
Cyg 20h31'4" 38d54'32"
Jane Weiss
Tau 5h55'16" 26d1'44"
Jane Westerhoff
Cas 1h1'18" 49d30'34"
Jane Whitehead
Gem 6h50'34" 15d35'12"
Jane Wolfe
Crb 16h9'49" 34d38'7"
Jane Woll Johnson
Eri 4h40'0" -27d24'38"
Jane²
Cyg 20h5'55" 32d39'50"
JaneandJackEiler
Gem 7h8'34" 32d50'11"
Janeane Haines
Cas 0h52'7" 60d31'13"
Janeane Marie Tuttle
Aqr 22h30'5" -2d3'35"
Jane-Anne's Cosmic Twin
Mon 7h31'29" -0d20'35"
Janeczek Family
And 0h44'46" 37d33'49"
JaNed
Lib 15h35'14" -23d47'32"
JaneDeany42
And 1h24'20" 47d39'21"
Jane-Dee Reyes
Cam 4h39'8" 58d8'7"
Janeen Blythe Watkins
Cap 20h10'35" -9d18'8"
Janeen Michele McKenzie
Uma 13h23'10" 57d18'44"
Janeenly
Uma 11h44'55" 63d16'33"
janejacq
Lib 14h52'54" -13d43'50"
Janek 27.02.1972
Psc 0h44'0" 18d22'44"
Janel
Psc 0h49'19" 11d40'28"
Janel
And 1h57'52" 46d15'58"
Janel
Lib 15h33'26" -7d18'2"
Janel Anise Mcintosh
Lib 15h5'21" -23d55'26"
Janel Evelyn
Cap 20h15'4" -21d53'28"
Janel Foote
Uma 9h19'34" 65d39'36"
Janel Lee
Ari 2h32'19" 24d8'50"
Janel Louise Marks
Raymond
Psc 1h15'38" 7d26'31"
Janel Marie
Tau 4h29'23" 19d38'6"
Janel Marie Fontaine
Lib 15h8'41" -8d31'32"
Janel Marie Kochuk
Uma 11h45'51" 57d26'42"
Janel Melina Kaui Malate
Lmi 9h30'47" 34d20'52"
Janell
And 0h21'58" 41d13'46"
Janell
Ari 2h13'8" 16d57'28"
Janell Aguda
And 0h12'32" 44d40'5"
Janell Blaylock
Mon 7h53'40" -2d12'14"
Janell Deann Smith
Ori 5h30'2" 5d9'30"
Janell James
And 0h24'20" 43d45'42"
Janell Kozette Schroeder
Cap 21h39'5" -12d24'21"
Janell M. Fielding
Cyg 20h48'22" 33d30'49"
Janell Malnory
And 2h3'54" 42d6'35"
Janell Renee Gardner
Leo 11h18'7" 16d58'1"
Janell Stoller
Cnc 8h43'51" 17d23'47"
Janell: The Center of my
Universe
Gem 7h39'49" 32d41'49"
Janella Foster
Uma 10h32'21" 51d4'52"
Janelle
And 1h12'38" 42d16'5"
Janelle
Sgr 19h50'59" -12d13'35"
Janelle
Uma 11h31'37" 58d41'47"
Janelle
Pho 0h43'59" -41d38'51"
Janelle
Cap 21h9'6" -25d21'32"
Janelle Bullock
Crb 15h19'57" 29d29'31"
Janelle Calado
Cma 7h13'58" -14d34'6"
Janelle Elizabeth Little
And 2h7'22" 38d52'46"
Janelle Eve Amsley
Her 17h46'55" 47d16'11"
Janelle Fischer
Leo 10h42'17" 9d4'0"
Janelle & Josh's Love
Cnv 13h19'13" 42d40'18"
Janelle Katherine Tompkins
And 23h14'48" 51d47'3"
Janelle Kay Engles
Leo 11h57'28" 21d14'57"
Janelle Khan - My One And
Only
Aqr 23h6'33" -21d17'9"
Janelle Kim Hottes
Aqr 23h46'55" -3d53'10"
Janelle & Landrum
Cap 20h17'27" -19d5'3"
Janelle LaPlante
Psc 0h43'50" 18d50'10"
Janelle Leila Manuel
Sco 17h57'26" -40d49'20"
Janelle Louise Woodward
Cra 18h6'49" -37d10'9"
Janelle Lynn
Cyg 19h36'4" 32d39'32"
Janelle Lynn Kennedy
Umi 15h14'54" 73d34'31"
Janelle Marie
Umi 14h57'47" 73d3'49"
Janelle Marie Jahnke
Cnc 8h52'59" 7d8'40"
Janelle Mary
Gem 6h46'12" 34d35'31"
Janelle Maurene
Ari 2h30'59" 23d8'41"
Janelle & Michael
Uma 11h11'21" 47d44'3"
Janelle Nicole Lutz
Leo 10h24'33" 13d43'56"
Janelle Rader
Peg 23h7'53" 9d42'4"
Janelle Reiko Ka'ipolani
Hirata
Gem 6h50'5" 14d5'9"
Janelle Ryder
Sgr 19h32'6" -18d39'27"
Janelle Shea
Lyn 8h46'25" 43d40'46"
Janelle Smallfry Andree
Aqr 22h38'36" -1d9'50"
Janelle Wilson - A Star
Leo 9h29'2" 16d5'47"
Janelle, Leslie's mom
Cap 20h23'58" -12d33'58"
Janelle's Beauty
Sco 17h52'58" -37d16'23"
Janello
Ari 3h15'27" 25d4'58"
Janely Perez
Aqr 22h59'33" -8d7'48"
Janelz
And 0h50'13" 46d39'17"
Janene Austin
Cam 3h46'2" 53d45'31"
Janene Lee
Lmi 10h22'44" 38d50'57"
Janene Renee Gwin
Per 2h42'16" 54d14'51"
Janenn Sanders
Umi 14h50'53" 69d45'9"
Janers
Sco 16h8'32" -13d18'50"
Jane's Blessings
Cas 1h55'16" 65d0'23"
Jane's Forever Mardi Gras
Star
Crb 15h53'46" 27d26'49"
Janes Kitchen
Aqr 22h11'53" 0d6'32"
Jane's Kitten Man
Gem 7h4'20" 14d31'37"
Jane's Star
Mon 6h45'58" 8d12'48"
Jane's Star
Cnc 8h9'19" 21d40'44"
Jane's Star
Cas 0h25'47" 59d24'54"
Jane's Star
Lmi 10h37'51" 30d35'58"
Jane's Star of Lovelight
Cap 20h25'52" -20d29'44"
JANE'S WORLD
Uma 10h55'36" 37d27'17"
Janess Nicole Livecy
Aqr 23h7'36" -20d48'20"
Janessa
Aqr 22h30'57" -1d55'44"
Janessa Danielle
Cnc 9h6'54" 19d40'40"
Janessa Nikole Miller
Vir 13h34'29" -11d12'3"
Janet
Lyn 7h1'13" 59d34'28"
Janet
Cas 23h37'9" 59d24'57"
Janet
And 0h17'27" 31d3'21"
Janet
Ori 5h3'33" 13d46'28"
Janet
Lmi 10h34'7" 31d30'54"
Janet
Crb 15h45'48" 31d24'56"
Janet
Leo 9h43'35" 31d40'40"
Janet
And 1h1'5" 36d44'4"
Janet
And 1h18'40" 41d34'48"
Janet
And 1h1'0" 40d2'33"
Janet
And 23h9'49" 40d25'44"
Janet
And 23h17'28" 43d24'1"
Janet
And 2h24'49" 45d52'6"
Janet A. Crosby
Lib 14h53'27" -16d54'19"
Janet Adams
Crb 15h47'23" 38d26'42"
Janet Ali
Uma 11h39'7" 36d44'13"
Janet An Inspiration To All
And 23h6'43" 42d28'33"
Janet And Brook Forever
Cyg 21h10'40" 46d30'9"
Janet and Jazzy
Sgr 19h54'53" -34d51'18"
Janet and Jonathan's Love
Aqr 22h0'15" -15d13'18"
Janet Ann
Cas 1h0'51" 62d18'5"
Janet Ann McKiernan
Cap 21h4'18" -18d41'38"
Janet Ann Nickerson
Vir 14h5'27" -19d55'6"
Janet Ann Quirico
Cnc 8h33'9" 22d52'57"
Janet Ann Tracey
Cas 2h30'18" 65d59'13"
Janet Asher
Uma 10h7'56" 47d59'47"
Janet Auman
Cas 1h30'1" 57d8'55"
Janet B. Carl
Vir 13h3'10" 11d23'53"
Janet B Gibbs
And 0h18'46" 44d41'18"
Janet Barden
Sco 16h34'40" -34d16'23"
Janet Beauteous Bryant
Sgr 18h6'3" -27d19'7"
Janet Bemis Harris
Sgr 19h6'19" -31d33'33"
Janet Bernice Gauchas
Tau 5h10'56" 22d19'26"
Janet Bistrek Waddington
Uma 13h44'1" 58d44'10"
Janet B.J.J.P. Mincks
Cas 0h58'35" 64d19'14"
Janet & Bob Barbee June
24, 1955
Cyg 21h18'3" 42d33'5"
Janet Boyd Parry
Cas 0h59'59" 55d41'12"
Janet Bryan Adams
Peg 22h20'54" 21d30'59"
Janet Burns
Leo 9h38'31" 27d29'24"
Janet Byron Rogacki
And 1h16'27" 45d24'57"
Janet C. Imhoff
Lib 14h55'11" -17d14'21"
Janet C Tutalo
And 0h51'11" 39d31'38"
Janet Callif
Cyg 20h49'52" 35d59'35"
Janet Carlson Freed
Peg 21h31'28" 12d40'20"
Janet Case
Ori 4h53'42" 12d35'6"
Janet Catalano
Cas 1h7'2" 65d35'26"
Janet & C.H. Knowles
Shining Bright
Crb 15h36'7" 35d53'32"
Janet Chapin Polack 7-5-
1923
Lyr 18h25'55" 32d59'44"
Janet Christine Lucas Fain
(Oma)
And 0h24'34" 45d24'43"
Janet Ciabattari
And 0h44'35" 43d46'52"
Janet Claire Graziano
Psc 1h22'51" 24d8'1"
Janet Cooper (My Coop)
Cas 1h24'13" 64d2'34"
Janet Cornelius
Equ 21h12'19" 5d7'55"
Janet D. Britland
Sgr 18h46'48" -36d40'58"
Janet Darlene
Aqr 22h54'53" -4d30'4"

Janet Davidson Uma 12h8'10" 50d46'6"
Janet Davis Sex 10h39'34" -0d19'36"
Janet Dearo Lyn 7h42'41" 57d38'45"
Janet Dee Anderson Psc 1h38'4" 17d45'44"
Janet Delnero Cyg 21h8'52" 53d41'38"
Janet Denise Dew Cyg 19h58'36" 31d27'25"
Janet DiVizio Boehm Aqr 22h15'49" 1d34'11"
Janet Doetsch Cnc 9h6'35" 16d59'45"
Janet Domelle Sgr 18h31'50" -17d10'51"
Janet Dowling Lyn 8h31'14" 55d5'25"
Janet E. McGee Cas 0h26'51" 61d19'59"
Janet Easley Cap 21h31'7" -19d50'0"
Janet Edwards Uma 13h11'27" 58d21'32"
Janet Elaine Boehler Sgr 18h32'5" -17d33'11"
Janet Elaine Herrmann And 1h20'29" 43d18'56"
Janet Elaine Shea And 1h43'52" 45d40'25"
Janet Eleanor Boswell Gem 6h39'12" 20d47'56"
Janet Elizabeth Bell And 23h13'20" 49d8'58"
Janet Elizabeth Garton Lib 15h23'16" -10d58'35"
Janet Elizabeth Haberl Cas 0h58'47" 55d48'45"
Janet Elizabeth Lee Cas 2h18'51" 73d21'4"
Janet Elizabeth Newman Eri 3h52'24" -7d55'19"
Janet Elizabeth O'Connell Mon 6h49'55" -0d29'4"
Janet Ellen Pearman Vir 12h36'41" 3d53'46"
Janet Evangelina Sainsbury And 23h28'53" 35d49'38"
Janet F. Bryan HJM Psc 0h28'29" 14d37'44"
Janet Fara "always with us" Uma 12h11'0" 60d26'25"
Janet Ferguson Uma 12h5'11" 63d11'27"
Janet Ferrell Cyg 21h14'13" 45d52'39"
Janet Fields Vir 13h26'46" 4d24'22"
Janet Fisher Star Crb 15h51'43" 39d12'22"
Janet Fleming And 0h47'19" 37d24'12"
Janet Fontenot And 2h2'51" 37d25'34"
Janet Frankl-Lockwood And 23h20'47" 47d58'50"
Janet Frigerio Cas 23h27'36" 53d19'13"
Janet from "Out There" And 23h36'1" 48d25'5"
Janet Gale Johnston Gem 6h15'23" 25d32'19"
Janet Gard Cornebise Cyg 19h51'39" 36d16'11"
Janet Gayle Sgr 18h17'16" -27d54'35"
Janet Gildea Lyn 7h28'4" 56d34'21"
Janet Gloria Henze Leo 11h38'37" 17d49'15"
Janet Gomez Crb 15h33'0" 33d52'54"
Janet Gray Aqr 22h36'6" 1d22'14"
Janet Grist Uma 10h44'19" 59d33'53"
Janet Handyside Crb 15h44'2" 26d7'49"
Janet Harding Vir 14h11'32" -7d50'14"
Janet Hatcher My Guiding Light Uma 12h21'5" 54d42'21"
Janet Hill Lyn 8h45'37" 43d30'19"
Janet Hoehn And 23h27'26" 41d21'48"
Janet Holland Cas 23h20'55" 56d39'7"
Janet I. Holcomb Umi 15h23'4" 71d17'43"
Janet Iacampo Uma 8h53'12" 66d24'17"
Janet & John Burns forever Cyg 20h24'0" 37d14'38"
Janet & Justin's Star Uma 8h41'15" 52d38'47"
Janet K. Greene Cas 0h35'13" 52d44'23"
Janet K. Mayer Mon 6h46'7" -5d55'59"
Janet Katherine Miller Lyr 19h27'53" 41d8'1"
Janet Kent Cyg 19h58'26" 30d1'26"
Janet Kim And 23h38'49" 41d39'27"
Janet Kunka Sgr 17h59'22" -24d30'32"
Janet L. Malatesta Cap 20h29'5" -13d25'3"
Janet L. Sojka Leo 9h34'31" 27d58'59"
Janet L. Stamato Lyn 7h7'8" 50d52'42"
Janet L. VanderSYS Leo 10h13'28" 27d3'15"
Janet L. Yameen Ari 2h13'49" 11d52'14"
Janet LaFlouf Ari 3h11'24" 27d52'2"
Janet Lanchbery Cas 23h17'38" 54d23'48"
Janet Lang Lib 15h3'20" -0d31'51"
Janet Laura Meyers Cyg 21h39'16" 31d51'34"
Janet Lee Sgr 19h29'19" -17d58'22"
Janet Lee Albertson Aqr 22h4'43" 1d17'4"
Janet Lee Alverman Cas 1h28'21" 63d19'16"
Janet Lee Pitman And 0h45'22" 43d0'42"
Janet Lee Schoenhoff Her 16h52'55" 32d56'19"
Janet Lee Shibao Cas 0h32'38" 55d30'11"
Janet Leigh And 23h7'59" 42d13'27"
Janet Leigh Brown Parrish Vir 13h32'34" -9d23'23"
Janet & Lenny Mancini Cyg 20h35'41" 53d47'31"
Janet Li Lyr 18h28'58" 36d27'21"
Janet light of my life Sgr 19h18'46" -26d20'7"
Janet "Little B" And 1h11'22" 41d57'8"
Janet Lou Stanton Cnc 8h16'37" 8d36'18"
Janet Lough Aqr 23h2'32" -8d31'7"
Janet Lynn And 0h50'5" 40d2'33"
Janet Lynn And 2h19'15" 47d55'46"
Janet Lynn Brintnall Uma 10h56'47" 63d55'56"
Janet Lynn Bush Vir 14h32'4" -2d27'37"
Janet Lynn Dorman Smart Aqr 22h42'58" -2d4'33"
Janet Lynn Morton Uma 13h44'11" 59d16'52"
Janet Lynn Nease Tau 4h31'21" 27d56'48"
Janet Lynn Norwood Cas 1h6'57" 61d40'14"
Janet Lynn Steinert Crb 16h11'0" 36d1'47"
Janet Lynn Yucker Uma 10h20'30" 67d25'53"
Janet Lynne Moline Cas 0h38'10" 54d28'13"
Janet Lyric Narish Lyr 18h53'23" 39d44'13"
Janet M Casbeer Peg 23h30'16" 15d26'17"
Janet M. Escudero Cnc 8h19'36" 9d21'13"
Janet M. Iaci Srp 18h10'13" -0d0'29"
Janet M. Koinzan Vir 14h16'35" 0d34'46"
Janet M Lane And 1h3'2" 36d0'4"
Janet M. Staples Uma 10h15'6" 43d2'40"
Janet M. Zalewski Cyg 21h45'18" 42d13'43"
Janet MacKay Lib 15h31'36" -26d29'9"
Janet Madeleine Morgan Halford And 1h23'28" 39d46'36"
Janet Maggi Cárdenas Aqr 20h42'41" -13d39'30"
Janet Mann And 23h18'19" 51d37'51"
Janet Manson Vir 12h41'41" 7d0'28"
Janet Marie Cyg 20h8'55" 35d44'4"
Janet Marie Cyg 21h34'40" 30d17'7"
Janet Marie And 0h16'20" 34d25'58"
Janet Marie Mon 7h11'29" -1d44'52"
Janet Marie Beach Per 3h10'0" 40d30'51"
Janet Marie Hoey Cap 20h17'34" -14d22'7"
Janet Marie Lemon Finn Sgr 19h8'1" -32d47'4"
Janet Marie Naples Uma 8h42'32" 57d59'35"
Janet Marie Price Lyn 8h36'39" 36d54'53"
Janet Marie Slivka-Sungail Lyn 8h9'30" 37d50'26"
Janet Marie's Beauty Cyg 20h31'32" 43d9'5"
Janet Marion Cyg 21h15'16" 40d27'26"
Janet Marlene Walker Del 20h33'56" 18d13'9"
Janet Marotta ~ April 8, 1948 Ari 2h47'58" 27d32'43"
Janet Mary Jones Aqr 22h25'37" -24d49'22"
Janet Menh Voong Cnc 8h24'5" 14d45'55"
Janet Miller Leo 9h30'21" 27d46'12"
Janet Miller's Love and Guidance Uma 8h48'6" 50d28'58"
Janet Morgan Johnson Leo 11h7'42" 20d14'8"
Janet Morse Grimsley Uma 11h3'39" 56d25'28"
Janet my love And 0h29'15" 26d52'19"
Janet My Love Cyg 20h23'0" 50d3'11"
Janet Myers And 2h6'17" 42d45'35"
Janet N. Magan Vir 13h40'6" 1d40'2"
Janet Ney Uma 14h15'27" 61d58'3"
Janet Niemi (Dearest Mom) And 1h22'49" 33d57'47"
Janet Norris Cyg 21h21'27" 35d57'32"
Janet "Our Shooting Star" Cas 1h25'31" 55d21'15"
Janet P. Spaulding Vir 13h53'53" -10d22'1"
Janet Palazzotto Tau 5h33'39" 27d52'30"
Janet Paola Ferrarese Cnc 8h50'58" 12d18'32"
Janet Patricia Frye Olson Hood-BF Cas 1h30'24" 68d10'12"
Janet Pauli Crb 16h4'28" 39d0'41"
Janet Peterson-Williams Leo 11h20'19" 7d18'54"
Janet R. A. Lara And 22h59'26" 40d31'31"
Janet Raeside Lockie Haldane Smith Per 3h11'53" 41d37'42"
Janet Read Uma 10h50'5" 48d45'34"
Janet Reiko Lincicome Cnc 8h36'53" 14d51'16"
JANET ROSE "NENA" Cyg 19h38'59" 30d13'55"
Janet Ross Tau 3h30'24" 18d1'35"
Janet Rye And 0h47'25" 26d43'24"
Janet S Herman WOMD Psc 0h43'40" 14d49'11"
Janet S. Mayo Tau 4h29'26" 16d37'1"
Janet S. Miller Cap 20h53'5" -18d14'27"
Janet S Nankin Cnc 8h41'55" 14d28'18"
Janet S. Plum And 23h22'57" 48d44'23"
Janet Sarnoski And 0h22'56" 35d52'52"
Janet Skidmore And 23h31'41" 42d47'59"
Janet Stanton Vul 19h25'13" 26d29'28"
Janet Stuckey Aqr 22h51'31" -7d13'5"
Janet Sue Oph 17h7'7" -0d10'29"
Janet Sue Leo 10h51'33" -0d19'5"
Janet Sue Psc 1h21'54" 7d35'38"
Janet Sue Dunmoyer And 2h8'28" 43d40'39"
Janet Sue Inglis Brown Ari 2h2'9" 23d52'26"
Janet Sue Press Aqr 22h35'50" -3d4'41"
Janet Susan Gage Vir 13h41'36" 3d27'17"
Janet Susan Viker Psc 0h17'55" 15d19'40"
Janet Tindell And 0h39'2" 38d14'49"
Janet Tremaine Cnc 8h52'55" 11d48'54"
Janet Trevino Cam 4h3'2" 69d1'34"
Janet Twamley "Gramma Star" And 23h37'11" 46d27'14"
Janet V. Wilson Lyn 7h45'6" 43d6'42"
Janet Valerie Ohlhausen Uma 12h13'3" 62d7'49"
Janet Vincent And 23h9'29" 46d58'3"
Janet W. Becker Uma 8h42'6" 58d15'41"
Janet W. Podboy Lyn 6h46'5" 50d41'26"
Janet Wagner Crb 16h13'40" 38d49'52"
Janet Waldron Cas 0h24'22" 56d19'21"
Janet Warbington Gem 7h4'15" 24d52'4"
Janet Westfall Lib 14h49'31" -4d54'36"
Janet Zajac Grote And 23h20'26" 48d10'30"
Janet Zenna Cyg 21h47'14" 50d3'7"
Janet52 Sco 17h29'0" -42d40'21"
Janeta Blas Cnc 9h5'17" 7d23'0"
Janeth Danixia Campos Uma 13h22'28" 56d3'11"
JANETH91101JHON Sco 16h22'7" -33d6'47"
Janet's Heart Leo 11h30'40" 11d1'51"
Janet's Star Cam 4h33'7" 56d35'43"
Janet's Star Sgr 18h48'12" -29d24'42"
Janet's World Aql 19h54'56" 12d28'37"
Janett Gem 7h22'58" 32d48'4"
Janetta Powelson Her 17h5'47" 30d40'14"
Janette And 1h20'49" 34d34'9"
Janette Lmi 10h41'52" 25d50'29"
Janette Aurora Cano Lib 15h21'59" -18d20'25"
Janette Dominique Duran Cnc 9h6'11" 31d21'57"
Janette E Acosta And 1h23'50" 43d16'39"
Janette Horton And 1h39'15" 50d19'47"
Janette K. Ratley And 1h7'46" 43d38'19"
Janette Kathleen Ori 5h38'53" -2d33'23"
Janette Kelly Kincaid Cas 0h1'42" 59d25'23"
Janette Martinez Cnc 8h1'34" 10d8'6"
Janette Our Shining Star Cyg 21h52'36" 42d43'21"
Janette "Princess" McCartney And 1h30'45" 41d17'24"
Janette Steigerwald Ori 5h28'35" 8d52'43"
Janette & Stephen - to the Moon ... & Back Cyg 19h36'20" 30d9'19"
Janette Teodora Concannon Cnc 8h0'20" 12d58'19"
Janette Whited Hya 8h44'46" -11d6'3"
Janette,"LA UNGIDA" Psc 1h6'56" 20d51'31"
Janetzy Sco 16h10'34" -14d57'56"
JANEVA Cnc 8h45'26" 16d19'24"
JaneW Ser 15h51'44" 20d28'33"
Janey 33 Leo 9h39'24" 30d38'37"
Janey BMama Walker Lib 15h24'4" -4d1'21"
Janey Charters Cyg 21h28'5" 38d4'37"
Janey Luong And Michael Kwan Cnc 9h7'55" 32d28'56"
Janey & Maree Blair 16-08-1970 Pho 0h32'11" -49d28'50"
Janey Woo And 2h7'46" 41d27'21"
Janey, John, Amelia, Isabel & Oliver Cnc 9h7'51" 9d32'50"
Janey's Blessing Gem 7h5'30" 31d53'38"
Janforehand and J.EdwardswitzerJr. Lyn 6h28'50" 57d9'19"
Jang Hsin Hui Hys 1h45'49" -61d18'21"
Jango Vir 13h6'31" -18d57'15"
Jani Elizabeth Makowiec Chaborek Sco 16h5'14" -15d6'48"
Jani Lannenmaki Vul 20h34'44" 23d9'7"
Jani Lindholm Dra 19h17'59" 66d39'35"
Jani Novotny Lib 14h25'52" -21d44'35"
Jani Teli Cnc 8h9'22" 14d14'15"
Jania Ka'iulani Talakai-Nichols Sco 17h55'7" -40d25'1"
Jania Saipale Gem 6h51'23" 25d3'3"
Janialma Lyr 18h48'49" 39d17'39"
Janic Com 12h46'22" 29d34'19"
Janic Baumgartner Tau 4h11'16" 28d4'4"
Janice Ari 2h32'26" 26d36'49"
Janice Tau 4h27'43" 15d45'9"
Janice Vir 12h9'8" 3d32'9"
Janice Vir 12h46'34" 4d46'7"
Janice Crb 15h49'24" 38d32'19"
Janice And 23h11'34" 48d0'41"
Janice Cas 1h57'48" 68d0'43"
Janice Cas 1h39'18" 61d6'2"
Janice <3 Twiggy Sco 17h31'56" -36d20'34"
Janice A. O'Brien, D.V.M. Vir 13h59'27" -15d41'21"
Janice A. Thibodeau Leo 9h27'11" 19d8'28"
Janice Adams Podhorsky Sgr 19h19'4" -17d39'27"
Janice Aguila Lyn 7h59'56" 33d49'15"
Janice Aileen Chapman Ori 5h26'29" 9d0'52"
Janice Alvarenga Cas 1h35'50" 72d40'20"
Janice Amedia April 14, 1925 Ari 2h33'42" 30d0'45"
Janice Amelia Fabry Cas 2h36'48" 67d33'1"
Janice Amstadt - Shining Star And 0h31'4" 25d26'2"
Janice and Geneva Cyg 20h29'14" 47d37'44"
Janice Ann Cap 20h11'30" -11d10'34"
Janice Ann Shiver Crews Uma 9h22'18" 72d13'15"
Janice Anne Cnc 9h20'13" 30d19'34"
Janice Anne Poirier Cyg 21h55'3" 51d3'3"
Janice Ann's 50th Birthday Cnc 9h10'36" 19d45'55"
Janice Antoinette Leon Uma 11h45'21" 35d25'17"
Janice Avery Pizzotti Lyr 19h16'57" 33d38'57"
Janice Ayers And 0h39'0" 37d2'55"
Janice B. Graham Leo 10h48'2" 8d55'6"
Janice Bader's Star Sgr 17h54'47" -28d57'37"
Janice Ball Cyg 21h6'41" 52d49'52"
Janice Barnwell Sgr 18h46'55" -17d1'38"
Janice Barr Ori 5h29'3" -0d50'13"
Janice Beavan Cyg 20h12'53" 40d33'9"
Janice Beck Cap 20h27'27" -15d20'4"
Janice Bell Sgr 19h19'15" -27d23'7"
Janice Bercovich-Reid Vir 13h32'17" -17d17'40"
Janice Beth Lyr 18h32'20" 31d57'7"
Janice Blank Cas 0h25'14" 62d6'8"
Janice Blu Witheridge - Monticello Cyg 21h30'54" 45d52'40"
Janice Brennan Uma 11h48'39" 62d21'56"
Janice Brown Findlay And 23h27'41" 42d1'50"
Janice "Bucky" Parker Cas 1h3'9" 65d8'33"
Janice Burgess "Gammy's Star" Uma 11h22'21" 65d45'25"
Janice C. Affonso Sgr 18h50'3" -33d18'22"
Janice C. Kreamer Crb 15h56'22" 30d27'44"
Janice Capps Gem 7h49'30" 18d34'23"
Janice Catherine Clements And 0h40'43" 30d55'38"
Janice Cimbalo Aqr 21h32'25" 2d18'15"
Janice Conklin Cam 7h40'26" 79d5'41"
Janice Crewe Monks Psc 1h33'5" 8d56'56"
Janice Dawn Lyn 8h10'4" 52d10'30"
Janice Deshaun Leo 10h30'21" 18d46'39"
Janice Diane Goldstein Aqr 22h30'0" -22d53'26"
Janice Dianne Hunter Tau 4h6'3" 24d48'22"
Janice DiPietrantonio Leo 9h29'28" 19d30'27"
Janice E. Astor Vir 13h24'52" 4d4'35"
Janice E. Humphrey Sco 17h51'4" -41d49'45"
Janice Eileen McDearman Vir 12h22'51" -4d0'28"
Janice Elaine Ephron Oph 17h42'27" 6d37'55"
Janice Elaine Turek Cas 3h4'1" 63d57'48"
Janice Elisabeth Curry Tau 4h6'47" 28d15'59"
Janice Elizabeth Ari 3h21'55" 29d5'26"
Janice Everett Ari 2h33'8" 25d27'11"
Janice Farr Lac 22h27'4" 43d11'32"
Janice Fay Niklas Aur 5h43'22" 38d14'30"
Janice Ferguson Tau 4h6'11" 28d48'52"
Janice Fifer Cma 7h20'53" -19d40'52"
Janice Fornaciari Aqr 20h41'0" -8d53'4"
Janice & Frazer Golden Anniversary Cyg 19h46'38" 33d30'27"
Janice Gail Vir 13h27'51" 12d19'57"
Janice Garrod Sterrett Cnc 9h4'46" 32d45'42"
Janice Gerber Gem 7h27'50" 32d13'13"
Janice Giles Cas 1h0'57" 54d21'38"
Janice Girouard And 1h14'41" 34d58'20"
Janice Goline And 0h52'21" 38d23'20"
Janice H. Blevins Com 13h5'16" 18d47'45"
Janice Hale Sgr 20h4'51" -25d37'37"
Janice Hayflich Crb 15h30'23" 28d30'25"
Janice Henley And 0h19'7" 29d37'59"
Janice Herma Ari 1h47'58" 18d57'5"
Janice Higgins Ori 6h1'50" 13d16'47"
Janice Hope Ferguson Lyn 7h32'48" 42d2'41"
Janice Irene Mills Williams Ari 3h12'21" 23d57'53"
Janice J (Frey) Huddleston Gem 6h32'21" 23d14'43"
Janice Jeanne Uma 10h10'17" 70d55'39"
Janice "JJ" Furst Aqr 22h2'11" -0d14'36"
Janice June (Princess) Tau 5h9'51" 18d22'22"
Janice Kardos Cnc 8h9'19" 17d41'31"
Janice Kay Calhoun Gem 6h46'48" 32d8'30"
Janice Kay Durrill Lyn 9h18'36" 33d59'5"
Janice Kay Epstein Ari 2h21'0" 19d28'50"
Janice Kay Hillenburg Tau 5h26'3" 21d6'4"
Janice Kaye Cnc 8h8'22" 12d49'58"
Janice & Kirk Packard And 23h15'26" 44d8'45"
Janice Koop And 0h29'41" 41d49'31"
Janice L. Edwards-Snell Cnc 8h25'27" 12d40'5"
Janice L. Holmes Heimberger And 1h38'15" 43d0'24"
Janice L. Porter - Heavenly Star Cap 20h33'58" -17d24'42"
Janice Lambert Zuza Lib 15h40'12" -19d57'35"
Janice LaVanWay Sgr 18h43'14" -24d45'5"
Janice Leady Tau 4h35'33" 12d38'37"
Janice Lee Abate Sgr 18h34'16" -16d59'50"
Janice Lee Weiss Gem 7h2'18" 16d58'6"
Janice Lewis Psc 1h24'53" 16d42'48"
Janice Lewis "Nangirl" Ori 5h26'44" -0d15'57"
Janice Lillian Francisco Cnc 8h9'16" 8d48'35"
Janice Little Black Bear 59 Gem 6h26'17" 27d31'37"
Janice Lucia Psc 1h14'3" 26d45'42"
Janice Lusk Crb 15h50'34" 34d6'21"
Janice Lynn Doughty Sgr 19h33'30" -30d43'38"
Janice M. And 23h21'53" 44d52'43"
Janice M. Adams Sco 17h22'17" -42d12'29"
Janice M. Marinaccio Cas 0h50'34" 64d8'37"
Janice M Noe Lmi 10h28'5" 38d21'43"
Janice Malin Peg 22h23'44" 31d9'15"
Janice Margaret Noe Cyg 21h42'19" 43d2'39"
Janice Marie Cam 4h35'58" 53d31'25"
Janice Marie Craig Cas 0h13'29" 50d18'37"
Janice Marie Fulkerlicous 35 Cnc 9h20'57" 18d51'2"
Janice Marie Sawyer Aqr 23h52'9" -4d48'40"
Janice Marie Vendola Vir 12h41'35" 4d47'51"
Janice McGill Psc 0h43'20" 3d50'30"
Janice McNamara Cas 1h42'23" 61d41'22"
Janice Mercadante Aqr 23h31'6" -8d48'50"
Janice Millar Aqr 22h5'29" 1d38'52"
Janice Miller Lyn 7h22'29" 53d19'26"
Janice Muriel Brewer Lyr 18h47'7" 31d10'14"
Janice My Love Lyr 18h45'55" 35d46'56"
Janice O'Dwyer And 23h39'55" 47d27'7"
Janice Pantano And 0h24'47" 40d3'2"
Janice Renata Wassink Oskam Cas 0h56'38" 58d17'30"
Janice Ricki Singer Aqr 22h37'19" -24d27'32"
Janice Rider Lyr 18h46'55" 37d56'44"
Janice Sousa Tau 5h59'16" 22d54'1"
Janice Stadler Cap 21h19'24" -26d45'50"
Janice Stephanie Psc 1h8'17" 32d38'36"
Janice & Steve's Love Cyg 20h52'49" 36d26'34"
Janice Stipp Uma 8h36'37" 59d57'18"
Janice Tolman Olson Gem 6h43'28" 30d26'35"
Janice Tracy Garneau Lake Uma 11h49'20" 56d48'54"
Janice W. Wilson Umi 14h54'21" 71d31'41"
Janice Williams And 1h41'56" 39d18'23"
Janice Willis Ari 2h27'0" 19d13'12"
Janice-Elsie Lyn 7h16'12" 52d14'45"
Janice's Dream Upon A Star Gem 6h50'13" 31d41'36"
Janice's Stars (C.N.B.) Umi 14h13'11" 68d17'20"
Janice's Wish Lyn 8h24'26" 40d42'27"
Janick Sgr 19h19'56" -21d54'30"
Janick Stephan Järmann Her 16h41'20" 6d26'36"
Janicka And 0h40'56" 44d37'14"
Janicse Huff Cnc 8h19'31" 14d5'45"
Janie Ori 5h18'51" 3d45'3"
Janie Cnc 8h17'55" 20d58'27"
Janie Leo 10h56'27" 19d55'6"
Janie And 23h39'4" 34d27'12"
Janie Lib 15h43'11" -17d9'4"

Janie
Lib 14h53'52" -12d25'42"
Janie
Uma 12h14'9" 53d41'0"
Janie
Sco 17h38'37" -44d53'55"
Janie Baglien
Cas 0h24'10" 59d8'37"
Janie Berry
Ori 5h16'14" 0d27'41"
Janie & Bud Deal
Sge 19h50'53" 18d50'57"
Janie Bynum
Uma 9h24'3" 69d5'14"
Janie Carol
Sco 17h23'52" -35d52'48"
Janie D
Cam 4h30'55" 57d21'13"
Janie Elizabeth Fannin
And 0h29'7" 42d51'1"
Janie Elizabeth Nunn
Crb 15h53'20" 39d1'41"
Janie Hernandez "My Love"
And 1h4'31" 44d18'25"
Janie I. Howard
Srp 18h8'35" -0d11'41"
Janie Louise
Sgr 18h43'0" -28d49'34"
Janie Lynn
Uma 9h45'13" 61d31'7"
Janie Lynn Ball
Mon 6h42'56" -0d0'57"
Janie M. Behm
Vir 14h8'16" -16d46'24"
Janie Marie Catlin
Cnc 8h29'51" 24d21'4"
Janie Marie Fischer
Tau 5h56'3" 23d42'29"
Janie Pimentel
Cyg 20h2'42" 37d31'17"
Janie Roselli
Uma 10h3'56" 49d39'8"
Janie Ruth
Sgr 18h7'6" -27d45'19"
Janie Speltz
And 23h21'33" 42d56'47"
Janie Teresa Chauhan
Cnc 8h25'15" 6d58'53"
Janie Vaile
And 2h38'14" 44d26'46"
Janiece
Cyg 21h57'36" 39d23'13"
Janie-In-The-Sky
Leo 10h22'10" 12d55'26"
Janie's Star
Lib 15h3'14" -6d58'48"
Janie's Starrbrite
Gem 7h6'24" 28d51'25"
Janifrance Douaire
Uma 11h29'35" 59d32'8"
JaNiJu
Lyn 7h2'45" 45d23'56"
Janik Spattz
Uma 8h37'38" 50d54'44"
Janin Blanco
Vir 13h30'35" 3d7'56"
Janina
Tau 5h47'43" 23d57'33"
Janina
Uma 13h0'33" 58d46'42"
Janina
Vir 14h43'15" -0d1'41"
Janina Bettinelli
Umi 15h4'34" 67d35'45"
Janina Joiner
Eri 4h42'1" -0d57'48"
Janina Monica Barlowski D'Abate
And 1h13'23" 38d48'42"
Janina N
Cap 21h49'10" -13d40'58"
Janina Sophie
Tau 5h8'23" 25d13'12"
Janinaleo
Leo 11h19'35" -0d19'32"
Janine
Aqr 20h43'31" -13d50'10"
Janine
Sco 16h10'12" -26d2'14"
Janine
Ari 2h20'58" 24d47'20"
Janine
Ari 2h8'35" 23d35'37"
Janine
Psc 1h35'51" 8d51'49"
Janine
And 0h5'25" 35d46'18"
Janine
And 1h15'48" 33d59'2"
Janine
Lyr 19h26'55" 38d13'59"
Janine
Uma 9h22'34" 49d53'43"
Janine Allison Finlay
Aqr 22h30'16" -1d36'43"
Janine and Eric
Gem 6h22'51" 21d3'45"
Janine & Angelo's Eternal Star
Lib 14h51'34" -2d2'40"
Janine Anne Cook
And 0h33'42" 25d9'25"
Janine Bivona
Sgr 18h45'31" -16d56'28"
Janine Boros
Cra 18h53'34" -41d55'2"
Janine & Bülent
Uma 9h53'51" 48d20'28"
Janine Chantal Albert
Uma 10h50'14" 71d30'39"
Janine Elisabeth Walda
Psc 0h24'31" 15d57'36"
Janine H. Varnum Dunson 1/24/1963
And 2h15'27" 50d8'42"
Janine Hausammann
Ori 5h7'30" 5d27'46"
Janine Le Dourner Delattre
And 0h37'51" 27d35'41"
Janine Leigh Kelker
Gem 7h45'50" 30d21'15"
Janine Lynn
Uma 9h23'1" 50d8'39"
Janine Martin
Lib 15h2'8" -7d21'40"
JANINE MAZZA
Lmi 10h19'10" 37d55'37"
Janine McShane
And 2h38'20" 50d8'56"
Janine Michelle
Sgr 18h0'0" -27d18'9"
Janine Nash
Psc 0h54'17" 10d54'34"
Janine Nedd
Sco 16h18'28" -11d57'4"
Janine O'Toole
And 23h14'46" 49d31'29"
Janine Pollak
Her 18h54'9" 24d28'22"
Janine Rascher
Cep 21h33'6" 63d5'12"
Janine Ratcliffe
Gem 6h38'16" 14d40'43"
Janine Rose
Cas 0h25'10" 61d26'17"
Janine Scott
Gem 7h51'1" 31d36'3"
Janine Starr
Sco 16h54'54" -38d27'38"
Janine van Twest
Psc 0h58'24" 9d55'31"
Janine Weber
Ori 5h55'19" 21d32'52"
Janine Whitlock
Uma 9h57'9" 58d53'30"
Janine Widmer
Ori 5h17'3" 0d16'16"
Janine, 12.02.1974
Cam 6h21'46" 77d40'57"
Janine's Beautiful Twinkle
And 0h56'15" 23d35'41"
Janine's Star
And 1h19'38" 35d21'6"
Janire
Per 4h44'34" 36d49'24"
Janis
Sco 16h13'34" -13d30'41"
Janis
Cam 7h51'24" 72d20'12"
Janis
Psc 0h46'20" 7d5'54"
Janis A Piantedosi
Leo 10h35'29" 14d43'25"
Janis Bauser
Umi 14h8'27" 67d11'56"
Janis Blair "04/20/1957"
Cas 0h16'53" 54d38'4"
Janis Burnside
Ori 5h56'13" 6d42'23"
Janis Castillo
And 22h59'8" 47d0'28"
Janis D Terhune
Sgr 18h27'24" -16d11'25"
Janis Fritschi
Uma 11h13'49" 30d1'49"
Janis is Fifty
Crb 15h28'0" 30d8'21"
Janis Kristine Carter
Mon 6h38'46" 0d57'25"
Janis Lee
Cyg 20h31'7" 51d45'3"
Janis Levine-Loving Mother
Ori 6h5'14" 21d26'52"
Janis Littel
Cap 20h52'15" -14d35'8"
Janis Lynne Tatomer
Ori 6h10'10" 18d15'43"
Janis M. McGhee
Lyn 7h1'24" 57d37'59"
Janis Machala
Cas 0h44'47" 61d50'54"
Janis Marie's Baby
Lyn 8h15'49" 34d41'43"
Janis Marlene Rillo
Vir 12h59'51" 11d14'11"
Janis May
Uma 9h27'30" 41d38'36"
Janis Payne Ballard
Cas 0h46'21" 56d53'53"
Janis Perren
And 1h40'43" 48d32'12"
Janis R. Pence
Boo 14h12'10" 28d5'0"
Janis Rumba
Crb 15h51'38" 36d12'31"
Janis Shoss Block
Lib 15h28'8" -9d46'53"
Janis Trebick
Tau 4h23'56" 9d32'17"
Janive
Sge 19h45'59" 17d16'42"
Janiz Lee Antonio
Lib 15h49'16" -12d31'49"
JANJAMIN
Uma 8h51'27" 47d2'31"
Janjan's Star
Psc 1h24'39" 16d46'22"
Janka
Leo 10h43'18" 16d3'28"
Jankel
Ari 2h31'41" 27d3'16"
Janki
Cyg 21h31'21" 34d46'39"
Janley's Journey
Lyn 6h57'40" 52d57'32"
Jan-Luca
Uma 10h8'34" 47d1'50"
Janmarie
Mon 6h49'26" 11d31'16"
Jan-Michael Niemann
Uma 11h8'12" 71d25'46"
JANN
Tau 4h51'15" 19d49'48"
Jann Margaret Wilson
Psc 1h32'33" 4d55'32"
Jann R. Montemagno
Aql 19h30'47" 3d43'41"
Janna
And 0h42'20" 41d7'47"
Janna
Vir 13h50'25" -16d30'52"
Janna and Kyle
Uma 11h6'8" 41d52'35"
Janna and Stan
Sco 16h6'34" -12d53'42"
Janna Bear
Ori 5h31'7" -7d57'42"
Janna Bennett
Aqr 22h40'18" -2d40'37"
Janna Bowden-Paine
Vir 12h29'27" 3d31'26"
Janna Brooke Crutchfield
Tau 5h10'12" 21d38'50"
Janna Ekstrom
Cas 0h19'49" 51d11'19"
Janna Kaye
Vir 12h6'51" -11d5'41"
Janna Lee
Gem 6h20'43" 23d4'29"
Janna Marie
Sco 16h17'4" -16d22'29"
Janna Marie Alves
And 1h7'3" 43d45'15"
Janna Noelle Hardy
Aqr 23h38'0" -14d32'33"
Janna Rose Chilson
And 0h23'28" 41d26'13"
Jannabeth Mackenzie Garrison
Umi 14h50'3" 72d4'37"
Janne Bjerkeset
Cas 1h21'47" 54d49'12"
Jannea Sandker
Umi 14h28'9" 74d25'17"
Janneke van Haperen-de Man
Aqr 23h5'49" -9d4'35"
J.Anne.Rackham
Lib 15h59'45" -12d34'3"
Jannet Caroll
Cap 20h25'15" -18d44'13"
Jannette A. Mendiola
Cap 21h13'1" -26d54'26"
Jannette Lopez
Lib 15h47'12" -14d21'42"
Jannette und Norbert
Uma 8h47'47" 62d45'18"
Janni
Uma 10h17'44" 50d32'21"
Janni
Crb 15h41'35" 29d6'9"
Jannie
Leo 10h9'9" 17d18'38"
Jannie Mae JMB102565
Sco 17h51'8" -36d16'28"
Jannik Rosenthal
Tau 3h38'35" 26d14'1"
Jannine DeMarie
Lyn 6h45'34" 57d14'25"
Jannine & Robert
Peg 22h43'58" 25d26'44"
Jannis Nelle *20.12.2001
Ori 6h14'9" 16d11'12"
Jano
Uma 14h13'8" 56d4'41"
Jano Pilavdjian
Cnc 8h41'45" 27d36'36"
JaNonda Jean Parks Collins
Gem 6h47'25" 32d38'30"
JanOrs
Sco 17h4'10" -30d41'49"
Jànos & Brigitta
Leo 10h11'57" 10d49'13"
Janos Pataki
Cep 22h2'14" 64d2'55"
JanRae
Cyg 21h25'6" 52d15'51"
Jan's Ocean Lights
Del 20h41'54" 15d17'33"
Jan's Star
Tau 3h42'13" 24d29'57"
Jan's Star
Cyg 21h35'29" 47d31'26"
Jan's Star
And 23h21'6" 46d35'57"
Jan's Star *9-23-1954*
And 1h12'13" 33d47'57"
Jans Wife
Vir 14h8'4" -11d15'50"
Jan's Wishing Star
Sgr 19h3'32" -34d22'16"
Janse, Hajo
Uma 11h0'44" 39d55'8"
Jansen Michael Adkins
Aql 20h20'25" 1d6'18"
Jansen Stijn
Lib 14h46'13" -13d12'57"
Jansen, Dieter
Uma 11h17'27" 54d23'59"
Jansen, Ralph
Uma 12h47'9" 62d29'17"
Jansen-Spurney Spectacular
Ori 5h33'34" 9d35'51"
Janssen, Thomas
Ori 6h17'29" 16d6'13"
Janstar
Leo 10h13'33" 16d45'44"
Janstar
Crb 15h58'32" 33d3'44"
January
Gem 6h50'30" 17d25'32"
January Anne
Cnc 7h58'58" 11d39'14"
January Courtney Penne
Uma 10h14'7" 45d0'50"
January J. Gill "1/27/1981"
Vir 12h29'7" 2d3'8"
January Valdes Heath
Mon 7h18'11" -0d19'32"
JanuaryJo
Umi 13h54'54" 70d30'57"
January's Song
Lib 15h12'59" -6d56'26"
Janum
And 23h2'53" 35d37'36"
Janus
Ori 5h29'55" 1d9'53"
Janus Henlam
Car 9h55'54" -66d7'40"
Janusz
Cyg 21h15'2" 43d14'28"
Janusz Gerth
And 23h36'47" 47d49'40"
Janusz & Susanne
Ori 5h4'49" -0d0'8"
Janvi & Arun
Tau 5h26'29" 25d19'52"
Janyce Cuccio
Ari 2h9'20" 24d18'56"
Janyfers star
Per 4h23'40" 33d49'18"
Janz Pacific
And 1h36'2" 49d24'52"
Janzen Agtaguem
Cnc 8h22'29" 9d16'17"
Janzen Reese Arellano
Aqr 21h44'47" -3d29'59"
JaOIReDxANder
Lyr 18h52'8" 34d0'44"
Japhy
Ori 5h43'30" 8d37'36"
"Japiplatz"
Tri 2h17'36" 36d53'5"
J-Aponte-1
Umi 15h56'6" 72d36'37"
jaqeb
Uma 9h26'12" 46d12'10"
Jaqua Sweetie
Cnc 8h36'38" 8d7'52"
Jaquelincita Lara
Lyr 18h51'53" 34d27'39"
Jaqueline 41106
Aqr 22h45'14" -6d46'1"
Jaqueline Edwards
Leo 10h30'12" 23d23'39"
Jaqueline & Henri's Wedding Star
Cyg 21h26'15" 39d22'27"
Jaqueline "Jackie" Sokol
Uma 13h40'57" 52d17'27"
Jaqueline Jackster
And 2h37'21" 50d22'25"
Jaqueline Martha Brooks
Sgr 18h31'33" -26d15'11"
Jaqueline Probert "Nanna"
Umi 16h22'37" 87d58'25"
Jaqueline Rose Freed
Lyn 6h57'9" 44d47'43"
Jaquelyn M Simonson
Ari 2h53'56" 29d31'22"
Jaquelyn Peaches Mcquillan
Sgr 19h11'7" -17d15'35"
Jaqui Parsons a Star at 40
Del 20h32'54" 4d48'5"
Jaqui Sadler
Gem 6h39'44" 13d55'20"
Jaquline Kay Kluge
Cap 21h21'41" -15d35'54"
JAR
Leo 10h39'28" 7d39'45"
J.A.R.
Cyg 19h48'9" 39d20'30"
Jara
And 0h29'50" 42d8'2"
Jarad Anthony Damiens
Lib 14h52'21" -4d36'22"
Jarbreeah - our two star are one
Ori 5h10'57" 3d59'30"
Jardin Leleux
Leo 11h5'34" 16d10'56"
Jarecki's Silver
Ori 5h1'8" 12d57'2"
Jared
Psc 23h28'26" 4d41'56"
Jared
Psc 1h10'35" 24d7'17"
Jared
Her 16h48'51" 17d57'9"
Jared
Her 17h20'39" 34d8'55"
Jared Alan Barton
Her 16h40'26" 37d34'22"
Jared Alan Helms
Cep 21h25'56" 64d30'25"
Jared Alexander Colvin
Uma 10h31'57" 66d28'57"
Jared & Allison
Cyg 20h38'34" 38d20'52"
Jared and Amy
Lyr 18h49'7" 31d42'34"
Jared and Jaime's Star
Cyg 21h32'20" 45d46'8"
Jared and Jessica Forever Always
Crb 15h33'11" 31d12'10"
Jared and Mandy's Star
Ori 5h5'10" 10d55'40"
Jared and Paige Forever
Her 17h17'36" 46d42'12"
Jared and Tori's Dream Star
Her 17h45'37" 48d0'14"
Jared Andrew Cenziper
Lib 14h54'40" -2d41'21"
Jared Andrew Phillips
Uma 9h2'2" 63d46'43"
Jared Arthur Anderson
Cnc 8h50'19" 31d16'18"
Jared / Aunt Fran
Uma 12h37'6" 58d20'35"
Jared Austin Stonecipher
Gem 7h9'18" 25d46'16"
Jared Benjamin Lenner
Tau 3h55'14" 27d11'30"
Jared Bitner
Ori 5h13'34" 4d51'52"
Jared Blake Call
Her 17h21'25" 17d34'53"
Jared "Boo" Weaver
Her 17h48'14" 28d3'41"
Jared Brian Masi
Boo 15h41'13" 39d36'50"
Jared Brosky
Cnc 8h16'50" 18d58'35"
Jared C. Elliott
Her 17h45'59" 36d58'36"
Jared Christensen
Hya 8h34'30" -11d13'52"
Jared Christopher Miller
Tau 5h24'38" 28d7'3"
Jared "Cowboy" Swaim
Lyn 8h38'22" 33d55'13"
Jared "Dada" Leon
Cap 20h38'14" -15d6'22"
Jared David
Uma 9h23'11" 58d39'12"
Jared & Emily
Cyg 19h48'18" 29d59'39"
Jared Foro
Sgr 17h58'44" -29d59'20"
Jared Francis Eagle Trepepi
Cyg 21h55'59" 48d15'34"
Jared G Breaux 1982
Vir 13h22'53" -12d47'25"
Jared Glazer
Per 3h19'40" 53d20'17"
Jared Grant Klein
Ari 2h38'3" 30d30'32"
Jared Guthrie
Ori 5h18'11" 5d21'59"
Jared Hartsock
Aur 5h49'39" 53d44'1"
Jared Heath Nadel
Cnc 8h16'14" 18d6'49"
Jared Hodges McCollough
Peg 22h31'14" 28d13'8"
Jared Houle
Aql 20h8'58" 3d21'30"
Jared Isaac/JareBear
Uma 9h9'1" 52d20'47"
Jared James Hostetler
Cyg 21h19'52" 54d51'27"
Jared Jensen
Uma 10h15'37" 46d1'48"
Jared Joseph Grassi
Sgr 18h46'19" -28d6'55"
Jared Joseph Rogers
Gem 7h45'33" 33d34'29"
Jared K. Neal
Her 16h31'39" 33d50'7"
Jared K. Stern
Cap 20h23'49" -10d31'42"
Jared Kaiman
Uma 10h45'55" 61d44'4"
Jared Kevin Steele
Per 3h28'22" 47d20'30"
Jared Lael Widman
Lyn 8h40'33" 39d42'18"
Jared Lee
Sgr 18h50'21" -25d8'52"
Jared Leigh Hicks
Psc 1h1'1" 27d28'40"
Jared Leto 30 Seconds to Mars
Cap 21h2'49" -26d18'32"
Jared Matthew Davis
Sgr 19h16'12" -34d10'43"
Jared Matthew Forman
Sgr 18h24'45" -32d38'41"
Jared Matthew Shoupe
Uma 10h26'22" 56d20'17"
Jared McBride"My Heart and Soul"
Her 17h40'10" 39d11'34"
Jared & Meredith Miller
Cyg 20h17'37" 46d5'18"
Jared Michael Abelow
Uma 11h50'39" 31d42'25"
Jared Michael Bruno
Crb 16h4'7" 39d33'43"
Jared Michael Hargrove
Aqr 22h17'17" 0d53'33"
Jared Michael Hoppenfeld
Sco 16h21'44" -25d23'7"
Jared Mikel Maronde
Uma 9h3'0" 68d41'0"
Jared & Monica Khan
Cyg 21h57'23" 54d31'43"
Jared Orian
Sco 17h52'57" -35d39'21"
Jared Paules
Her 17h13'35" 38d20'36"
Jared Pelletier
Uma 8h55'0" 50d14'51"
Jared Peter Reynolds
Uma 13h36'13" 52d40'8"
Jared Phelan - Our Star In The Sky
Cra 18h1'5" -37d11'53"
Jared Robert Toevs
Gem 6h50'36" 23d35'12"
Jared Sandridge
Ori 5h23'46" 7d52'20"
Jared Scott Beach
Aql 19h48'27" -0d18'1"
Jared Scott Rodriguez
Crb 15h52'18" 35d17'43"
Jared Scott Sussman
Aqr 20h55'33" -8d13'0"
Jared "Skippy" Goff
Cep 23h1'38" 70d32'23"
Jared Smith
Aqr 23h18'43" -7d9'25"
Jared Smith
Per 3h35'35" 41d18'39"
Jared Spencer Ritter
Lib 15h34'51" -8d17'3"
Jared Thomas Anderson
Cnc 8h37'6" 30d46'45"
Jared Thompson Kurzawa
Sco 17h52'21" -35d45'9"
Jared Tyler
Lib 15h58'30" -14d47'38"
Jared Tyler Labial Ho
Leo 11h21'49" 17d52'11"
Jared V. Petruso
Ari 2h11'45" 27d1'58"
Jared Victor
Per 4h22'5" 48d20'3"
Jared Walker Vaughn
Uma 10h2'31" 46d45'27"
Jared Walsh
Cyg 19h47'27" 33d22'36"
Jared Wayne
Psc 0h32'52" 13d29'2"
Jared Wayne Meeker
Cap 20h35'2" -25d19'0"
Jared Week's Good Luck Charm
Her 16h52'0" 16d53'46"
Jared White
Sco 16h55'17" -38d28'58"
Jared Who Swims in the Stars
Ori 5h33'12" -0d24'46"
Jared's Scale
Dra 15h18'5" 64d4'59"
Jared's Star
Uma 11h18'36" 60d25'28"
Jared's Star
Umi 16h57'3" 76d47'58"
Jared's Star
Boo 14h37'15" 19d2'5"
Ja-Rem
Uma 11h9'32" 70d19'47"
Jaremkewicz, Orest
Uma 12h52'17" 54d47'14"
Jaren Boyd Ririe
Ori 6h6'34" 15d33'2"
Jaren-30
Uma 10h18'30" 44d58'42"
Jaret Crawley
Her 17h0'2" 24d19'31"
Jareth Zephir Pierpont
Sgr 18h31'47" -16d3'27"
Jarglo
Ari 1h56'38" 21d28'32"
Jarilyn
Lyr 19h18'38" 32d53'9"
jar-k-9
Uma 10h28'41" 45d41'8"
Jarl Alfred Oscar Gibson
Peg 21h36'27" 21d58'16"
Jarladon
Lyn 7h51'19" 46d21'15"
Jarlehag
Uma 11h54'12" 64d55'50"
Jarmila Solc
Tau 5h53'18" 25d22'13"
Jarno Ziegler Hüppi
Uma 9h46'34" 56d33'11"
Jarod
Uma 14h3'24" 57d22'45"
Jarod Ackerman
Ari 2h24'21" 24d4'19"
Jarod Dale Zikmund
Her 16h18'18" 19d13'47"
Jarod Joseph Murphy
Ori 5h22'55" 2d48'12"
Jarod Kenneth Mitchell
Her 16h31'40" 45d47'49"
Jarod "Lemon" Smith
Uma 10h58'36" 36d29'0"
Jarod Mark Mariska
Aql 19h20'38" -0d4'47"
Jarod Robert Wills 08/22/1998
Dra 15h32'33" 64d20'4"
Jarod Tyler Edwards
Aqr 22h19'12" -3d37'17"
Jarod W. Trafton
Uma 8h53'47" 54d36'33"
Jarom Jay Heusser
Her 18h44'54" 18d56'10"
Jaromill
Tau 3h36'29" 10d41'35"
Jaros in the Sky
Dra 19h18'52" 61d35'34"
Jaroslav Josef Jasan Havlin
Cam 3h49'15" 74d8'30"
Jaroslaw D.
Ori 6h12'24" 6d32'39"
Jaroslaw Monkiewicz
Uma 10h6'32" 62d50'23"
Jarrad Paul Grahame
Pho 0h35'27" -50d37'20"
Jarrad Thornton "JMT" 04/06/1989
Cru 12h30'11" -62d34'11"
Jarrad's Star
Aqr 22h38'20" -17d24'10"
Jarred
Leo 11h42'12" 27d0'43"
jarred 21898
Aqr 22h27'11" -2d7'58"
Jarred Belford
Umi 14h12'55" 67d15'10"
Jarred Charles Collins
Aqr 22h31'33" -2d0'50"
Jarred Douglas Qualls
Ori 5h56'7" 19d26'22"
Jarred Meade
Aur 5h41'2" 50d15'38"
Jarred my "star" my "stud" Houston
Lyn 8h30'8" 38d27'5"
Jarred Nicholas Wells
Her 16h31'36" 49d27'49"
Jarred R. DeWees
Cap 20h14'13" -8d54'28"
Jarred27
Aql 19h49'14" -0d19'55"
Jarrel N. Willey
Uma 9h8'33" 67d13'24"
Jarrelyn Karen Martin
Leo 9h56'43" 13d59'59"
Jarren "Petie" Moser
Uma 8h44'51" 50d52'40"
Jarret Titcombe
Cru 12h17'49" -56d53'45"
Jarret Tobe Detwiler
Uma 10h19'24" 54d12'0"
Jarret's Bubbles
Her 18h8'0" 16d52'52"
Jarrett
Uma 9h18'41" 46d51'30"
Jarrett and Shanno's star
Lmi 10h28'4" 33d20'33"
Jarrett Brandon Richburg
Per 3h34'0" 46d23'21"
Jarrett Kenny Estep
Per 4h33'20" 41d21'50"
Jarrett Lee Stanley
Uma 8h35'2" 48d48'53"
Jarrett Rush
Cnc 8h15'0" 30d42'47"
Jarrett's Star
Cnc 9h14'37" 8d12'30"
Jarrick Starnes
And 0h42'40" 41d23'11"
Jarrid Christopher Crespo
Ari 2h34'34" 26d22'51"
Jarrid L. Amico
Per 3h27'34" 51d35'18"
Jarrod
Leo 11h23'55" 10d52'22"
Jarrod & Aiden's Shining Angel
Cru 12h33'32" -59d53'48"
Jarrod Alan Kuehl
Cep 22h10'44" 66d40'51"
Jarrod Alexander Tidaback
Ori 5h11'2" 15d11'3"
Jarrod - Always & Forever - Linda
Pyx 8h52'56" -31d58'28"

Jarrod and General
Her 16h53'13" 29d42'25"
Jarrod D. Payne
Aur 5h16'28" 31d17'51"
Jarrod Duane
And 0h18'58" 45d42'59"
Jarrod Edward Jonas
Aql 20h7'18" 15d12'1"
Jarrod Joseph Drew, Sr.
Leo 9h36'59" 27d20'23"
Jarrod Lanclow
Ori 5h25'52" 14d39'19"
Jarrod Michael Bresky
Lib 15h33'4" -9d18'59"
Jarrod Pinson
Ori 5h57'47" 17d35'51"
Jarrod Robert Kay
Uma 8h34'7" 65d55'57"
Jarrod Rodriguez
Ori 6h18'47" 7d16'4"
Jarrod & Sonia
Cyg 19h48'15" 29d47'30"
Jarryd E T Barrah- love you always
Dor 4h59'6" -65d39'23"
Jaruwan Senfeld
Peg 21h50'59" 13d40'31"
Jarvis Family
Uma 9h27'28" 72d36'56"
jaryndevin
Ori 5h47'50" 1d10'8"
Jas
Ori 6h14'59" 15d8'26"
JAS
Sgr 19h3'24" -20d36'47"
Jas i Malgosia
Gem 6h45'30" 14d19'33"
Jas Palmer
Uma 11h21'27" 28d27'22"
JAS33
Gem 7h4'27" 23d15'31"
JAS619
Cam 4h22'20" 68d13'38"
Jasa Lee-Noland Snyder
Aql 19h42'20" -0d6'24"
Jasani Major
Tau 3h31'22" 21d19'25"
Jasco
Peg 23h43'2" 15d24'38"
jascon360
Uma 12h40'37" 62d12'10"
Jase
Vir 13h10'9" 12d18'53"
Jase Alan Katzenberger
Her 17h54'30" 14d35'22"
Jase & Alli's - Love's Shining Light
Cru 12h29'21" -58d48'0"
Jase Dawson Wells
Lyn 8h12'40" 35d33'48"
Jasen Weitekamp
Her 17h23'27" 14d23'8"
Jashica Princess Ta
Cnc 8h48'16" 26d4'9"
Jasi and Karen
Cnc 9h11'46" 22d54'47"
Jasi Schatz
Per 4h23'11" 46d53'28"
Jasi the third
Per 4h14'19" 44d12'12"
Jasin Parker's Star of Wishes
Aqr 22h51'21" -21d9'50"
Jasio Jakubczuk
Lyn 7h4'19" 54d22'22"
JasJenGavinEmaleeMiller-JJGEM
Ori 5h8'52" 6d15'39"
Jasjit Kaur Ahluwalia
Uma 11h25'38" 37d22'26"
Jasley Yusely Salguero
Cap 20h48'57" -16d9'21"
Jaslin
Dra 19h41'9" 61d33'27"
Jasmadia
Her 17h2'3" 44d6'8"
JasMat
Uma 10h7'25" 48d33'14"
Jasmeet Barn
And 1h24'35" 39d30'16"
Jasmia
Lib 15h21'33" -21d27'9"
Jasmijn
And 2h35'22" 49d32'27"
Jasmin
Cas 1h8'32" 49d6'12"
Jasmin
Uma 10h8'11" 47d24'4"
Jasmin
Cyg 21h32'44" 36d42'48"
Jasmin
Lmi 10h47'42" 34d10'10"
JASMIN
Vir 13h19'13" 6d27'6"
Jasmin
Umi 16h38'1" 76d3'52"
Jasmin
Umi 14h51'19" 86d19'17"
Jasmin
Uma 13h15'59" 62d16'50"
Jasmin
Sco 16h48'30" -26d4'3"
Jasmin Buntzler
Per 4h6'5" 34d37'12"
Jasmin Deboo
Peg 21h54'40" 8d54'28"
Jasmin Feichtinger
Uma 10h25'15" 69d50'5"
Jasmin Gardner
Cra 18h44'11" -41d0'25"
Jasmin & Hansjörg
Her 18h12'29" 42d3'21"
Jasmin (Jas71406)
And 23h10'9" 49d43'1"
Jasmin "Jassie" Maria Lipat
Lyn 8h4'45" 39d4'41"
Jasmin Lees
Peg 22h28'17" 7d40'2"
Jasmin mis Engeli
Her 17h33'41" 14d19'17"
Jasmin Mokbel
Cnc 9h21'57" 17d36'18"
Jasmin Moxie St. Claire
Crb 15h35'59" 26d20'26"
Jasmin Oberschelp
Uma 9h52'27" 49d13'9"
Jasmin & Patrick
Dra 17h26'40" 63d14'16"
Jasmin Schmid
Lyr 18h47'4" 33d9'16"
Jasmin Simmons
Cru 11h57'25" -62d53'50"
Jasmin Soraya Meier
Sge 20h4'5" 17d22'32"
Jasmin/Venera
Uma 14h19'23" 59d13'40"
Jasmin Violet Pederson
And 0h15'55" 37d4'4"
Jasmin Vollenweider
Dra 9h37'10" 73d29'14"
Jasmin, mein Sternchen
Uma 12h58'5" 52d47'37"
Jasmina Budisic Zahir
Sgr 19h9'45" -21d26'22"
Jasmina Radjevic
And 0h29'11" 44d19'8"
Jasmina und Marcel
Uma 14h6'15" 48d26'38"
Jasmine
Uma 13h42'34" 49d14'20"
Jasmine
And 23h19'48" 47d11'38"
Jasmine
Cyg 20h21'26" 44d28'9"
Jasmine
And 23h2'36" 40d25'10"
Jasmine
And 2h26'29" 46d41'19"
Jasmine
Per 4h44'50" 49d35'20"
Jasmine
And 0h14'1" 44d20'11"
Jasmine
And 0h25'24" 41d22'25"
Jasmine
And 0h36'54" 42d5'32"
Jasmine
Cnv 12h37'57" 44d7'42"
Jasmine
Lyn 9h12'33" 37d11'35"
Jasmine
Com 13h8'21" 29d32'15"
Jasmine
Cnc 8h15'13" 25d22'28"
Jasmine
Gem 6h42'57" 23d13'12"
Jasmine
Vir 15h8'50" 6d45'31"
Jasmine
Peg 22h15'25" 7d27'11"
Jasmine
Lib 15h43'57" -9d5'21"
Jasmine
Aqr 23h3'30" -10d2'34"
Jasmine
Lyn 7h8'26" 59d54'1"
Jasmine
Sgr 18h19'31" -29d5'17"
Jasmine Alina Pena
Peg 22h26'34" 29d24'26"
Jasmine Altagracia Santana My Angel
Sgr 17h46'46" -21d33'52"
Jasmine Amber Herald
Mon 7h15'39" -0d54'51"
Jasmine Astra-Elle Marston
Sco 16h10'45" -41d18'14"
Jasmine Bradley
Lyn 6h28'55" 57d10'52"
Jasmine Breeze Mathis
Ori 6h16'2" 14d14'19"
Jasmine Briana
And 0h13'17" 44d59'5"
Jasmine "Carissima"
Aql 19h32'40" 6d46'42"
Jasmine Castaneda
Peg 22h42'39" 27d34'36"
Jasmine Chean
Aqr 23h9'24" -17d0'53"
Jasmine Danielowski
Uma 10h17'49" 58d14'23"
Jasmine De Sisto
And 23h42'3" 38d24'26"
Jasmine Double Brew
Cma 6h27'34" -15d30'7"
Jasmine Edelle Ross
Lyn 7h54'57" 34d7'44"
Jasmine Eleanor
And 1h28'11" 48d24'21"
Jasmine Elizabeth Binney
Aql 19h51'50" -0d17'7"
Jasmine Elizabeth Oliver
Sco 16h11'34" -16d3'57"
Jasmine Ellie Smith
Aqr 21h58'6" -7d50'43"
Jasmine Food Corporation Sdn Bhd
Ori 5h33'30" 3d42'14"
Jasmine Grace Hill
Cru 12h25'13" -56d51'17"
Jasmine Helena Blyth
Lyn 7h14'9" 48d24'51"
Jasmine Heung
Gem 7h46'55" 24d34'49"
Jasmine Hinkley
Uma 13h42'32" 56d34'16"
Jasmine J. Blair
Uma 12h22'11" 60d51'35"
Jasmine "Jessie" Biggins
Gem 6h50'28" 34d42'21"
Jasmine John Sioui
Boo 15h38'6" 45d35'42"
Jasmine Judith Sturgess Tasker
And 2h10'29" 38d46'33"
Jasmine K. Dixon
And 2h37'29" 45d19'35"
Jasmine Katerina
And 23h42'24" 34d1'31"
Jasmine Kiah Major
And 23h19'21" 52d5'53"
Jasmine Kymberly Disgdiertt
Lib 14h38'59" -16d54'1"
Jasmine Lau
Cnc 8h34'24" 27d34'45"
Jasmine Leah
And 2h12'28" 43d28'36"
Jasmine Leilani Yelle
Cyg 19h31'31" 53d5'2"
Jasmine Lewellen
Lib 15h47'18" -12d14'12"
Jasmine Linette Cruz
And 23h38'25" 42d46'1"
Jasmine Lothien
Ori 6h14'12" 13d16'33"
Jasmine Lynn Hawkins
Crb 16h4'52" 28d28'4"
Jasmine Maldonado
Lib 14h56'3" -4d11'30"
Jasmine Maria Martinez
Mon 6h15'53" -9d12'41"
Jasmine Maria Rega
And 1h17'33" 43d8'57"
Jasmine May
Leo 11h36'39" 23d56'23"
Jasmine May McCormick
And 23h11'54" 44d28'33"
jasmine medina
Tau 4h39'2" 22d20'57"
Jasmine Michael (I Love you)
Cas 0h33'24" 61d13'53"
Jasmine Monet Smith
And 0h55'56" 34d19'46"
Jasmine "My Little Princess"
Com 12h42'28" 17d25'44"
Jasmine N. Conway
Cas 1h28'22" 64d32'34"
Jasmine Nicole Streeter
Aqr 23h39'45" -8d25'38"
Jasmine Noel Pacheco
And 2h21'34" 49d25'59"
Jasmine Olivas
Cam 5h29'40" 56d19'43"
Jasmine Olivia Bold
And 23h15'51" 52d10'27"
Jasmine Olivia McGURRELL
Ori 6h8'27" 17d51'25"
Jasmine Onano
Gem 6h44'56" 35d19'26"
Jasmine Osman-Lambrinos
Gem 7h0'38" 20d46'54"
Jasmine Peanut & Tank Rios
Per 2h52'24" 33d35'17"
Jasmine Postorino
Umi 14h22'16" 77d51'7"
Jasmine Ramirez
Uma 11h30'1" 56d15'24"
Jasmine Robles
Crb 15h44'58" 32d44'35"
Jasmine Rochelle Sissenstein
Leo 10h4'54" 13d19'37"
Jasmine Rose Garcia
And 0h44'35" 37d35'32"
Jasmine Sabrina Ferrence
Cyg 21h49'33" 47d55'57"
Jasmine Salome
Vir 13h34'58" -9d46'4"
Jasmine Schatz's Star "Happy 13th"
Cap 20h41'27" -17d31'30"
Jasmine Scheuer
Gem 6h9'3" 26d11'56"
Jasmine Silva-Reyes
And 0h45'10" 30d31'27"
Jasmine Star
Uma 8h56'5" 53d24'50"
Jasmine Star
Umi 15h0'48" 70d11'8"
Jasmine Starr Burris
Vir 14h0'58" -15d34'9"
Jasmine Sweet 16
And 23h20'35" 43d38'47"
Jasmine Sydney Maskell
Dra 20h18'49" 62d39'3"
Jasmine Taylor Young
Lib 15h7'11" -4d23'11"
Jasmine Trachsler
Uma 10h59'1" 36d12'37"
Jasmine Trey DeGuevara
Leo 10h10'45" 20d55'13"
Jasmine Trudo
Cyg 21h8'8" 46d58'5"
Jasmine Wallace 05/18/1989
Uma 11h52'12" 34d48'31"
Jasmine West/16
Lib 14h48'52" -3d2'45"
Jasmine, Isabella & Nicholas
Tri 2h8'42" 34d27'8"
Jasminélyla
Cyg 22h2'22" 50d55'35"
Jasmine's Fire
Gem 6h52'16" 21d17'25"
Jasmith
Sgr 19h31'26" -13d27'20"
Jasmyn Anissa
Psc 1h14'54" 26d21'53"
Jasmyn Cruz
Crb 15h50'38" 36d48'21"
Jasmyn Joy
Lyn 6h33'46" 56d25'22"
Jasmyn Michelle
Aqr 22h41'51" 0d36'42"
Jasmyne R. Harris aka "da Princess"
And 0h43'57" 42d34'13"
Jasna
Lyr 19h3'28" 27d55'33"
Jasna Janicijevic
Cap 21h44'29" -10d49'57"
Jasna Stanivuk
And 1h8'22" 46d17'1"
Jasnad
Lib 15h41'8" -19d17'33"
JasNate
Leo 9h25'57" 17d29'26"
Jaso, The Great White Hunter
Uma 8h48'24" 47d0'40"
Jason
Per 2h23'54" 51d5'19"
Jason
Her 16h18'14" 44d5'6"
Jason
Per 2h12'17" 52d55'35"
Jason
Her 17h37'22" 33d31'55"
Jason
Leo 11h6'19" 21d20'54"
Jason
Ori 5h53'50" 17d30'46"
Jason
Her 17h4'17" 25d9'7"
Jason
Leo 10h1'32" 22d56'47"
Jason
Ori 5h29'39" 14d6'50"
Jason
Ori 5h49'13" 9d23'39"
Jason
Ori 5h31'58" 0d22'49"
Jason
Ori 6h14'31" 15d53'24"
Jason
Sco 16h6'7" -17d25'2"
Jason
Aqr 22h9'8" -2d49'16"
Jason
Dra 18h31'1" 56d52'22"
Jason
Cep 0h19'31" 81d22'13"
Jason
Sco 16h30'20" -31d17'15"
Jason A. Alcorta
Ori 6h11'59" 3d13'24"
Jason A. Reinhart
Cnc 8h48'29" 15d17'24"
Jason A. Scott
Her 18h45'55" 21d11'13"
Jason A. Weinberg
Uma 9h35'30" 44d16'9"
Jason Adam Cooper
Her 17h45'14" 44d40'51"
Jason Adam Schechter
Her 17h44'29" 33d24'32"
Jason Adams
Her 18h38'46" 17d44'48"
Jason Adrian Zechiel
Sgr 18h37'7" -35d7'22"
Jason Alan Baird
Gem 6h58'22" 13d35'18"
Jason Alcorn
Cep 22h50'30" 60d7'40"
Jason Alexander Clark
Ari 3h10'43" 29d2'43"
Jason Alexander Mushorn
Tau 4h25'3" 25d34'48"
Jason Alfred
Boo 14h33'25" 52d32'14"
Jason Allan Perkins
Cep 21h47'35" 63d14'16"
Jason Allen Bohn
Her 17h7'38" 32d56'55"
Jason Allen Glover
Boo 14h47'25" 33d29'17"
Jason Allen Scott
Uma 10h50'45" 68d59'32"
Jason Allen Wales
Gem 6h56'6" 14d56'27"
Jason & Amanda
Lyr 18h53'33" 32d2'50"
Jason & Amanda
Cyg 20h18'48" 55d24'38"
Jason & Amber's Light Forever
Ori 5h10'35" -0d48'27"
Jason & Amy Nolff
Cyg 19h38'39" 47d23'24"
Jason and Alyssa
Vol 7h22'46" -64d20'57"
Jason and Amy Always and Forever
Cnc 8h35'44" 16d11'42"
Jason and Carissa Jacobs
Cnc 8h46'55" 25d50'36"
Jason and Elise for all time
Cru 12h8'42" -63d44'54"
Jason and Jessica
Cap 21h12'49" -19d42'11"
Jason and Jordan Forever
Cyg 20h41'39" 37d19'2"
Jason and Julie's Wedding Day Star
Cyg 21h38'46" 48d52'54"
Jason and Katie Chilcoate
And 1h45'36" 43d24'2"
Jason and Katie Suter
Cyg 21h44'36" 43d56'21"
Jason and Kelly
Cyg 20h30'46" 34d11'25"
Jason and Leasha
Lyr 18h16'39" 33d39'59"
Jason and Margaritta Schiavello
Leo 11h0'4" 24d49'29"
Jason and Mary
Ori 5h20'37" 6d15'44"
Jason and Melanie Forever
Ari 2h14'27" 21d2'38"
Jason and Melissa Neef
Leo 9h50'38" 21d48'49"
Jason and Melissa's Star
Leo 11h48'55" 19d53'38"
Jason and Paige
Cap 21h0'33" -19d53'55"
Jason and Patty
Sge 19h47'52" 18d49'34"
Jason and Paula's Eternal Love Star
Lyr 18h52'14" 35d7'38"
Jason and Rachel Davis
Lib 15h12'12" -4d20'35"
Jason and Regan Watson
Sge 19h51'41" 18d53'6"
Jason and Sarah DeChiaro
Crb 16h22'12" 30d24'22"
Jason and Terri Randall
Gem 7h37'53" 28d1'56"
Jason and Tonya Tostrup
And 1h4'5" 45d55'16"
Jason Andrew Culos
Lyr 19h19'15" 29d22'22"
Jason Andrew Dantonio
Psc 0h25'20" 6d55'22"
Jason Andrew Franklin
Uma 9h25'29" 60d0'41"
Jason Andrew Gibbs I
Ari 2h26'52" 24d6'1"
Jason Andrew Messier Jr.
Psc 1h6'45" 33d29'20"
Jason Andrew Suggs
Ori 4h51'58" 8d12'35"
Jason Andrew Vela
Uma 12h8'28" 50d0'10"
Jason & Angela
Umi 15h12'44" 77d55'36"
Jason Anthony Putnam
Leo 11h42'50" 24d43'28"
Jason Arthur Musick
Cnc 8h48'47" 31d40'14"
Jason Arthur Towers
Uma 10h12'39" 61d7'9"
Jason Arthur Wigand
Ori 4h57'40" 15d34'29"
Jason B. Dua
Her 17h7'40" 32d28'10"
Jason Baby love of my life
Umi 9h52'8" 85d56'50"
Jason Bagley
Ori 5h54'4" 13d55'27"
Jason Baltimore
Cma 6h50'32" -12d22'15"
Jason Barnoff I love you love Emily
Vir 14h33'36" 6d2'56"
Jason Beau Morgan
Ori 5h22'53" 1d18'12"
Jason Beaver
Leo 11h4'47" 3d10'31"
Jason & Becky
Ari 2h20'0" 17d30'14"
Jason & Bek Levy - When 2 Become 1
Col 5h28'40" -35d10'0"
Jason Belen
Uma 8h46'53" 59d8'40"
Jason Bennett
Uma 10h45'31" 62d0'59"
Jason Bernard Runell
Ori 5h58'47" 6d10'36"
Jason Bickley
Sgr 20h8'1" -27d59'26"
Jason Bigger
Psc 0h50'37" 16d20'19"
Jason Bland
Vir 13h18'38" -17d51'35"
Jason Blue Boultinghouse
Lmi 10h25'29" 33d17'47"
Jason Boyd & Ehren Nowell Mullins
Psc 1h28'5" 27d36'55"
Jason Bradley (B-boy) Bender
Ori 5h50'20" 10d25'1"
Jason Brant Fausey
Leo 11h47'5" 22d21'32"
Jason Brian & Jordan James Peters
Umi 14h52'46" 72d32'43"
Jason Brian Mundinger
Uma 10h46'54" 40d20'0"
JASON BROWN
Aur 5h51'1" 47d56'28"
Jason Bruce Lyon
Ari 1h53'12" 11d33'10"
Jason Brunette
Tau 4h38'3" 19d56'22"
Jason Bryan Bonetti
Crb 15h45'36" 26d55'21"
Jason C. Bunting
Ser 16h19'57" -0d29'25"
Jason C. Getka
Umi 14h29'47" 75d21'33"
Jason C. Hall
Psc 0h21'41" 11d54'15"
Jason C. Kratz
Leo 11h39'50" 26d56'30"
Jason Calhoun Muse, Jr.
Her 16h38'18" 34d2'43"
Jason Carrol Buchanan
Lyn 7h42'49" 53d20'8"
Jason Cavanaugh
Tau 4h42'8" 22d34'2"
Jason Charles DeCaro
Dra 17h12'57" 58d58'3"
Jason Charles Kenney
Leo 11h3'50" 22d32'57"
Jason Charles Lockwood
Per 3h28'2" 51d11'0"
Jason & Chelsea
Cyg 20h45'48" 46d46'2"
Jason Cheney
Boo 14h43'33" 37d2'54"
Jason (Chester) Ballenger
Her 18h40'40" 20d13'50"
Jason Christides
Cnc 8h42'26" 31d51'48"
Jason Christopher King - I love you
Ari 2h39'21" 14d41'55"
Jason Christopher Land
Gem 6h54'33" 22d49'59"
Jason Christopher Lewis
Aql 19h42'37" 11d51'55"
Jason Christopher Rupp
Her 18h34'43" 19d1'28"
Jason Christopher Thomas
Her 17h45'58" 37d4'6"
Jason Christopher Townsend
Tau 4h21'44" 26d47'23"
Jason Claire Burbas
Aqr 21h3'11" -10d4'2"
Jason Clifford Williams
Per 4h24'23" 45d13'3"
Jason Cody Walker
Umi 17h0'52" 81d7'55"
Jason Cooper Ward
Aqr 22h8'7" -9d34'47"
Jason Corbett
And 1h1'58" 34d28'1"
Jason Corkins
Psc 1h20'29" 32d42'33"
Jason Coulas
Gem 6h10'56" 22d31'20"
Jason Cronin
Vir 12h22'8" 11d25'18"
JASON D
Tau 4h47'53" 24d18'34"
Jason D. Lewis
Uma 8h41'22" 50d20'12"
Jason Daniel Brink
Tau 5h26'10" 18d52'1"
Jason Daniel Cool
Ser 16h14'10" -0d19'26"
Jason Daniel Epling
Her 17h16'45" 32d18'43"
Jason Daniel Haigh
Umi 16h20'50" 77d42'17"
Jason Daniel Hatcher
Ari 2h35'51" 12d58'43"
Jason Daniel Holland
Tau 4h35'33" 18d45'17"
Jason Daniel Laino
Vir 12h27'30" -6d43'19"
Jason David Burnett
Del 20h47'37" 10d36'52"
Jason David Jennings
Uma 10h36'31" 63d42'57"
Jason David Klotz
Cnc 9h11'44" 32d32'31"
Jason Dean Claice
Boo 14h46'6" 36d37'51"
Jason & Deanna
Tau 5h43'51" 28d1'27"
Jason & Debbie's Eternal Love Star
Cru 12h27'40" -58d0'40"
Jason DeMatteo
Per 3h19'49" 51d14'30"
Jason & Devonn
Sge 19h23'43" 18d5'40"
Jason Digby
Ori 6h19'45" 13d43'28"
Jason Doering
Tau 4h35'26" 21d18'4"
Jason Douglas Forsyth
Sco 16h10'36" -13d15'36"
Jason Dunn
Uma 10h7'5" 45d26'30"
Jason Dutile
Her 17h43'11" 45d28'42"
Jason E. Fish
Ori 5h50'2" 8d54'36"
Jason .E. Folker
Dra 14h51'25" 56d29'31"
Jason E. Kuerth
Cyg 21h45'28" 46d28'40"
Jason E. Phillips
And 2h34'20" 40d1'10"
Jason E. Wolff
Leo 9h48'45" 31d0'30"
Jason Easthope
Aur 5h53'44" 53d21'32"
Jason Edward Barrigar
Ori 5h50'50" 6d57'24"
Jason Edward Dayhuff
And 0h53'39" 39d36'42"
Jason Edward Helmreich
Boo 14h53'24" 18d47'17"
Jason Edward Kee
Ori 6h18'13" 15d26'6"
Jason Edward Lee
Lyn 6h59'56" 51d11'3"
Jason Edward Moseley
Ori 5h24'15" 3d45'18"
Jason Edward Setser
Her 17h36'23" 36d29'27"
Jason Elder Faughn
Uma 10h8'19" 56d42'5"
Jason Elhers
Uma 12h5'45" 54d48'46"
Jason Eliseoduaine Vela
Uma 8h38'41" 61d14'38"
Jason Eric Bresmon Bubba 1979-2004
Tau 4h9'49" 28d3'47"
Jason Eric Savoy
Lac 22h23'50" 52d1'23"
Jason & Erin - Forever - 8.10.05
Cap 20h28'53" -23d33'55"
Jason Evans
Cep 1h56'34" 84d50'53"
Jason F. Blackshear
Cyg 19h56'31" 41d32'33"
Jason F. Southerland
Lib 15h39'50" -18d50'52"
Jason Forline
Ori 5h51'53" 7d7'55"
Jason Frank Harvey
Uma 8h59'13" 68d27'44"
Jason Frank Lake
Cnc 8h49'31" 14d9'50"
Jason G. Grams
Ari 2h8'20" 22d28'46"
Jason G. Lewis
Dra 18h56'16" 56d6'21"
Jason Gan
Cru 11h57'24" -55d46'55"
Jason Garret Prim
Cnc 8h42'58" 22d4'47"
Jason Garrett
Per 3h32'56" 49d24'47"
Jason Gembra
Her 17h25'8" 38d26'16"
Jason George Kennedy
Uma 10h51'33" 68d57'48"
Jason Glen Meader
Lib 14h42'6" -11d22'17"
Jason Glenn Dela Vega
Cap 20h22'51" -11d47'10"
Jason Graves
Uma 10h44'43" 50d3'13"
Jason Green
Her 16h54'5" 17d33'14"
Jason Greg Merritt
Uma 10h7'53" 55d37'8"
Jason Haines-Weddle
Sgr 17h56'16" -25d47'13"
Jason Hamblin
Cyg 21h20'55" 32d46'58"
Jason Harris Fox
Cnc 8h30'58" 16d13'43"
Jason Hart
Ori 5h9'21" 13d0'53"
Jason Hayden Blades
Psc 1h23'10" 27d38'11"
Jason & Hema Owens
Cyg 19h54'46" 33d18'47"
Jason Hinojosa and Family
Per 3h30'59" 45d3'9"
Jason Homesley and Kristin Thompson
Vir 12h43'34" 5d19'7"

Jason Howard Wheeler II
Sco 17h55'42" -42d18'43"
Jason Hunley
Her 16h35'3" 33d32'39"
Jason Hunter
Ori 5h38'28" 5d45'22"
Jason & Idalmis
Uma 11h29'58" 34d31'55"
Jason Ide
Cas 1h19'0" 60d22'21"
Jason Introwitz
Her 17h43'56" 44d12'30"
Jason J. Perkins
Gem 6h59'15" 31d1'36"
Jason J. Propsom
Cap 20h47'31" -21d26'38"
Jason J. West
Ori 5h41'42" 11d50'11"
Jason James Lenartowicz
Uma 8h50'54" 53d3'8"
Jason James Pickford
Aqr 23h38'54" -17d59'30"
Jason James Semperger
Her 16h17'36" 16d53'28"
Jason James Stofregan
Her 18h5'10" 31d53'38"
Jason James Wagner
Aqr 22h13'42" -2d56'35"
Jason Joseph Baker
Ori 5h15'22" 2d52'21"
Jason Joseph Menteer
Aql 19h11'26" 5d41'14"
Jason Joseph Suess
Her 18h26'28" 14d36'55"
Jason & Julie
Leo 11h20'6" -3d56'45"
Jason & Julie '04
Vir 14h37'21" -0d1'21"
Jason K. Davidson
Leo 11h24'2" 16d57'24"
Jason Kai Yuen Chan & Iris S W Leung
Umi 14h50'10" 69d19'44"
Jason Karel Bandish
Per 2h47'3" 53d37'45"
Jason & Karen Kline
Cyg 20h9'3" 46d32'11"
Jason & Karla Eriksson
Cyg 21h19'7" 40d7'21"
Jason Karns
Cap 20h16'51" -22d59'47"
Jason Karras
Leo 9h33'55" 7d14'25"
Jason & Karyn Schwartz
Gem 6h44'20" 30d46'56"
Jason Keith Green Sr. (J.J.)
Uma 8h41'0" 70d23'20"
Jason & Kelly
Ori 6h4'16" 17d16'1"
Jason Kempston
Leo 9h42'25" 28d57'12"
Jason Kenneth Martino
Her 17h48'6" 33d40'43"
Jason Kenneth Martino Jr.
Gem 7h43'19" 19d33'3"
Jason & Kristen-Soulmates Forever
Cyg 20h18'24" 43d0'57"
Jason & Kyle Quillen
Uma 8h33'39" 65d35'48"
Jason L Beal
Sco 17h53'51" -35d46'27"
Jason L. Crews
Crb 15h33'14" 29d10'24"
Jason L. Dubble
Cnc 8h32'51" 31d35'23"
Jason L. Fletcher
Ori 6h14'48" 15d12'6"
Jason Lapuz: Friend, Rock & Angel
Cyg 20h42'53" 34d33'57"
Jason Lawrence Schwartz
Cnc 8h49'35" 13d44'49"
Jason Lee
Leo 11h35'43" 9d27'5"
Jason Lee
Ari 3h28'36" 22d12'22"
Jason Lee Anthony Murach
Her 17h48'8" 33d48'35"
Jason Lee Fisher
Uma 10h27'11" 66d25'0"
Jason Lee Hill
Uma 11h1'49" 35d2'39"
Jason Lee Lambert
Vir 12h59'16" 12d26'33"
Jason Lee Norwood
Uma 10h42'34" 47d5'41"
Jason Lee Patterson
Her 18h22'39" 23d23'20"
Jason Lenice
Per 3h4'43" 48d38'22"
Jason Leo
Aqr 22h0'38" -20d18'34"
Jason Liam Jackson
Boo 14h31'16" 20d43'47"
Jason & Lindsay's 1st Anniversary
Cyg 21h35'3" 52d30'41"
Jason & Lisa Daoust
Cyg 20h54'4" 34d50'17"
Jason Llewelyn Cleavley Redman
Dra 18h20'12" 78d47'23"
Jason Logan Baird
Her 17h2'53" 14d25'0"
Jason & Lori
Cyg 21h27'15" 46d44'1"
Jason Loves Ashley
Ori 5h56'49" 11d39'54"
Jason Loves Laura
Ari 2h16'55" 25d37'22"
Jason M. Baines
Cap 21h26'40" -15d13'18"
Jason M. Cameron
Per 3h4'23" 37d42'10"
Jason M. Comardo
Sco 17h35'59" -34d38'26"
Jason M Egenski & Seren J Egenski
Cyg 19h38'18" 35d54'37"
Jason M. Rodrigues
Sgr 18h29'49" -27d17'17"
Jason M. Vincent
Ori 5h54'52" 21d0'49"
Jason Madsen
Uma 10h57'36" 68d38'46"
Jason Maloy
Per 3h9'1" 56d53'10"
Jason Manning
Cma 7h25'7" -15d24'46"
Jason Marc Ayers My Love Forever
Cyg 21h44'40" 53d47'34"
Jason Marcoux
Dra 20h21'49" 67d25'9"
Jason Mark Wooster
Ori 5h14'49" 5d45'48"
Jason Martin
And 0h42'41" 26d52'37"
Jason Martino Jr.
Her 16h29'54" 44d45'31"
Jason Matteo Bollino
Psc 23h1'33" 5d41'54"
Jason Matthew Lutz
Sgr 18h1'14" -27d33'1"
Jason Matthew Mccaughey
Psc 0h18'44" 9d6'30"
Jason Matthew Sands
Dra 17h5'4" 57d32'38"
Jason Matthew Soltys
Umi 16h25'37" 78d17'11"
Jason Matthew Walters
Cep 22h8'37" 65d28'16"
Jason Mazzochi
Ori 5h42'35" 2d47'37"
Jason McGrath Nilsen
Ari 3h16'32" 27d49'25"
Jason Means
Uma 11h47'59" 44d41'28"
Jason & Megan
Vir 14h4'6" -9d3'23"
Jason & Megan Land
Cyg 20h57'7" 30d1'29"
Jason & Mel Ferguson 6/5/06
Cru 12h30'4" -61d45'13"
Jason & Melissa Anne
Cyg 21h41'41" 34d3'25"
Jason Mestre
Aur 6h21'51" 40d42'32"
Jason Michael
Gem 7h13'31" 28d3'9"
Jason Michael
Ori 6h4'28" 5d21'6"
Jason Michael
Sgr 19h47'35" -27d51'23"
Jason Michael 2004
Cap 20h37'52" -14d19'53"
Jason Michael Adams
Cap 21h53'30" -22d10'54"
Jason Michael Boccia
Crb 16h23'54" 30d36'28"
Jason Michael Chehova
Cnc 8h47'10" 14d22'28"
Jason Michael Clotfelter
Aql 19h10'16" -2d48'3"
Jason Michael Duncan
Her 16h34'55" 12d53'53"
Jason Michael Ertzberger
Tau 4h17'37" 26d58'25"
Jason Michael Fisher
Lib 15h50'36" -20d18'30"
Jason Michael Jedlicka
Uma 8h14'42" 61d55'47"
Jason Michael McKee
Her 17h30'24" 44d51'52"
Jason Michael Mrosla
Uma 8h50'29" 63d53'3"
Jason Michael Nietfeld
Uma 8h48'54" 65d18'2"
Jason Michael Reynolds
Lib 15h0'27" -22d22'9"
Jason Michael Schulze
Tau 4h29'47" 28d54'25"
Jason Michael Sell
Ari 2h30'26" 24d36'33"
Jason Michael Weaver "Weavs"
Lmi 10h39'23" 26d54'42"
Jason Michael Wilson
Tau 4h27'35" 20d8'37"
Jason Michael Witucki
Tau 4h1'19" 18d29'38"
Jason & Michelle
Cyg 21h45'56" 52d11'1"
Jason Milewski
Dra 18h53'32" 52d17'27"
Jason Mitchell
Per 3h9'24" 43d20'36"
Jason Mitchell
Sgr 19h32'44" -12d54'32"
Jason Molina
Sco 16h7'48" -9d2'8"
Jason Moreno N Isela Rosado 4 ever
Lac 22h47'18" 37d46'7"
Jason Naddox & Kelley Mann
Cyg 19h44'27" 34d54'25"
Jason Nicholas Parrish
Per 1h47'7" 51d24'49"
Jason & Nicole Nunes
And 0h23'43" 43d19'4"
Jason Nile Poeth
Dra 18h22'59" 72d10'25"
Jason Nye
Aql 19h19'15" 9d3'32"
Jason Owen Doherty
Sgr 18h1'50" -16d3'12"
Jason Page
Her 17h20'55" 32d51'4"
Jason Parker Bregman
Uma 12h19'52" 56d0'35"
Jason Paterson's Star
Sgr 17h54'41" -23d46'6"
Jason Patrick Arnold
Cyg 19h54'36" 33d11'23"
Jason Patrick Newman
Leo 9h41'13" 27d30'9"
Jason Patrick Patterson
Gem 7h29'25" 19d17'8"
Jason Patrick Rickart
Cap 20h54'8" -15d9'29"
Jason Paul
Tau 5h45'35" 22d27'49"
Jason Paul Alvarado
Dra 15h10'43" 61d54'15"
Jason Paul Aufdenberg
Aqr 21h12'59" -8d42'17"
Jason Paul Cross
Ari 2h20'46" 14d14'19"
Jason Paul Freeman
Vir 11h39'31" 4d20'30"
Jason Paul Wade
Ari 2h53'13" 29d28'7"
Jason "Peepers" Begonia
Her 17h18'59" 32d9'44"
Jason Pettus
Ori 5h15'13" -0d44'1"
Jason Phillip Auxier
Her 16h42'18" 18d14'57"
Jason Polack
Uma 10h29'8" 41d5'54"
Jason R. and Kristen E. Hoyt
Peg 23h0'39" 30d44'48"
Jason R. & Asheley C. Bonney
Cyg 21h44'9" 46d38'34"
Jason R. Bamford
Cep 23h16'46" 71d28'42"
Jason R. D'Engenis
Ari 2h57'46" 14d2'48"
Jason R Earl
Tau 5h6'18" 24d56'31"
Jason R. Herman
Cap 20h49'23" -19d5'4"
Jason R. Hughes
Gem 6h22'27" 17d37'54"
Jason R. Richmond
Uma 9h35'50" 42d12'1"
Jason R. VanHaren
Gem 6h53'25" 17d51'54"
Jason R VanTassel
Cyg 20h9'20" 35d19'32"
Jason & Rachel Swejkowski
Lyn 6h46'9" 53d34'25"
Jason Ralph LeMay Jr
Her 16h32'54" 46d48'53"
Jason Randy Hunt
Per 3h7'8" 38d35'46"
Jason Ray Edwards (Artist)
Aqr 21h3'4" -10d16'32"
Jason Raymond Michel
Uma 9h8'51" 62d26'1"
Jason & Rebecca
Tau 4h39'22" 12d9'23"
Jason Reed
Vir 15h1'52" 4d38'18"
Jason René Barrios
Uma 13h17'27" 59d46'58"
Jason & Renee Forever
Cyg 20h14'6" 50d36'30"
Jason Rich
Sgr 19h5'6" -17d4'41"
Jason Richard Collins
Ori 5h27'6" 3d0'44"
Jason Riel
Sco 17h53'48" -36d32'2"
Jason Robert Gootee
Her 17h44'33" 32d0'30"
Jason Robert Lee Morrissey
Cap 20h39'48" -19d3'31"
Jason Robert Tharp
Sgr 17h55'43" -20d58'0"
Jason Roberts
Ari 1h58'14" 19d6'0"
Jason Roxas Cruz
Vel 9h14'50" -44d16'35"
Jason "Rudy" #62
Tau 5h44'17" 21d2'12"
Jason Russell Lyman
Uma 11h47'46" 34d35'18"
Jason & Ruth Trussell
Cyg 20h40'27" 51d52'56"
Jason Ryan Gaffney
Cnc 8h47'41" 31d11'16"
Jason Ryan Oberweis
Tau 4h6'2" 23d45'8"
Jason Ryan Peters
Gem 7h34'33" 27d28'2"
Jason S. Ashby
Her 17h37'30" 36d15'48"
Jason Sager
Tau 3h52'44" 22d19'27"
Jason Samuel
Uma 12h36'19" 61d47'50"
Jason & Sarah Back
Crb 15h46'12" 27d10'54"
Jason Saunders
Umi 17h23'6" 76d0'27"
Jason Schattauer
Per 3h12'7" 55d33'51"
Jason Schmidt Our Shining Star
Psc 0h37'18" 10d20'32"
Jason Schoch, MA.
Her 17h10'52" 31d54'27"
Jason Schultz
Uma 11h28'36" 52d41'44"
Jason Scott
Uma 9h12'8" 66d59'47"
Jason Scott Byerly
Uma 10h44'34" 39d29'57"
Jason Scott Crane
Her 16h40'14" 5d1'8"
Jason Scott Crane
Cep 22h18'59" 60d25'3"
Jason Scott Ferron
Cra 18h48'34" -37d9'31"
Jason Scott Thomson
Boo 14h53'34" 54d48'2"
Jason Selby Young
And 0h54'11" 39d24'35"
Jason & Shannon
Cyg 19h55'53" 41d9'29"
Jason & Shaun Hatcher
Ori 5h38'12" -1d18'8"
Jason Sherman
Lib 15h45'49" -18d17'28"
Jason & Sheryl Shaw
Cyg 20h15'15" 52d15'34"
Jason Shoop - Rising Star
Cma 6h38'7" -15d26'45"
Jason Singer
Uma 11h37'4" 55d27'21"
Jason Sizeland's Star
Cnc 8h34'12" 32d23'35"
Jason Skywalker
Aqr 23h6'34" -6d46'24"
Jason Sowinski
Her 17h34'25" 36d40'38"
Jason Spaulding
Uma 10h22'31" 57d53'48"
Jason Spencer Smith
Cnc 8h3'36" 9d4'8"
Jason Spice
Cen 13h33'38" -37d36'52"
Jason Starlight 12-19
Sgr 18h9'53" -22d19'45"
Jason Sundar
Lib 15h18'15" -20d24'46"
Jason T. Craig ~ Timeless Dreams
Psc 1h8'21" 27d50'59"
Jason T Griffeth
Cnc 8h14'42" 21d17'0"
Jason T. Smolen
Lyn 7h3'23" 47d15'38"
Jason T. White
Aql 20h4'18" 14d55'12"
Jason "Tag" Finkes
Vir 13h42'20" 5d8'30"
Jason "Tate" Prater
Uma 10h37'52" 41d13'51"
Jason Tate Smith
Cep 4h29'48" 81d20'17"
Jason the Boy
Her 17h15'59" 45d40'3"
Jason The Great
Umi 17h36'39" 82d56'13"
Jason The Just
Her 17h51'19" 22d1'27"
Jason Theobald
Aqr 22h48'7" -10d7'46"
Jason Thomas Basilicata
Her 16h13'40" 48d29'36"
Jason Thomas Brzozowski
Aqr 22h33'53" -0d33'1"
Jason Thomas Pinto
Leo 10h16'10" 25d51'25"
Jason Thomas Silva
Aqr 22h58'20" -6d48'32"
Jason Thomas Wicker
Leo 10h10'45" 11d28'27"
Jason Thomas, Our Shining Star
Her 16h13'11" 48d8'8"
Jason Todd + Jodene Beck
Uma 10h31'57" 59d4'2"
Jason Travis
Cyg 21h54'33" 41d16'47"
Jason Triantafyllidis
Per 2h45'23" 51d53'28"
Jason & Trish Pfeiffer
Uma 10h44'40" 39d22'44"
Jason & Trish Soulmates Forever!
Lib 15h54'45" -13d57'8"
Jason Tropf
Her 17h26'9" 17d26'30"
Jason Tucker Hebb
Psc 1h15'54" 26d39'16"
Jason Tyler Bagliani
Sgr 18h14'58" -29d11'46"
Jason Tyler Brock
Vir 13h28'59" 11d14'31"
Jason Tyler Foreman
Sco 17h26'19" -38d8'45"
Jason Tyler Kennedy
Ari 2h48'36" 25d23'38"
Jason Valcarcel
Aur 6h0'49" 38d18'43"
Jason & Valerie Lilly
Cas 1h20'19" 57d24'25"
Jason Vaughan Sutton
Crb 15h41'36" 32d43'34"
Jason VonKundra
Vir 11h44'43" -0d52'26"
Jason W Budd
Psc 0h9'46" 10d1'31"
Jason W. Clinch
Uma 12h1'6" 38d59'18"
Jason W. Sheckells
Aqr 21h57'27" 0d56'14"
Jason W Trusty ~ 12/11/2005
Aqr 22h24'59" -20d43'20"
Jason Wade Cato
Vir 12h11'29" 11d38'18"
Jason Wallace
Ori 6h7'53" 16d24'36"
Jason Wayne Sharp
Psc 0h51'36" 5d22'29"
Jason Wesley Ivey
Uma 14h18'1" 58d25'27"
Jason White Eagle
Her 18h44'32" 21d2'39"
Jason Widoff
Boo 14h17'44" 29d23'7"
Jason Wieand
Aur 6h30'3" 35d11'52"
Jason Wiemer
Cas 0h38'14" 52d2'41"
Jason Willard Danford
Boo 13h52'25" 11d10'57"
Jason William
Leo 10h14'47" 17d36'45"
Jason William Cooper
Psc 1h55'5" 6d23'52"
Jason William Heaton
Psc 0h22'17" 10d15'46"
Jason William Kolts
Aqr 22h29'49" -2d39'58"
Jason William Reedy
Uma 10h20'35" 51d46'33"
Jason William Smith
Uma 11h43'18" 49d28'26"
JASON WILSON (SUPER-MAN)
Gem 7h30'2" 33d22'55"
Jason Ying 16/07/1978
Cnc 8h49'54" 15d11'0"
Jason Yuan
Aur 6h26'30" 48d42'53"
Jason Zachary Elk
Umi 15h34'6" 80d9'16"
Jason Zaman
Dra 18h21'12" 54d36'33"
Jason Zocchi loves Robin Martindale
And 23h23'19" 48d39'57"
Jason Zukas
Lib 14h33'18" -10d58'55"
Jason, Jack, Georgina & Joe Coe
Peg 0h0'26" 18d36'20"
Jason, Kelly & Tom Gasteiger
Cru 12h25'37" -58d32'12"
Jason, My Special One and Only
Aqr 22h9'26" -1d38'47"
Jason, Vicky, Keisha & Connor
Cra 18h49'28" -39d53'35"
Jason-2
Dra 19h9'35" 61d24'19"
Jason41477
Ari 3h14'27" 11d54'41"
Jasonator
Tau 4h32'43" 25d52'32"
JasonD.Wood-RockStar-LoveSarah4ever
Ori 5h22'3" 13d58'25"
Jasonelias
Uma 11h31'29" 43d57'59"
JasonHillaryKing
Lib 14h30'59" -10d13'28"
JasonJulieLucasElise..Eternally One
Lyr 18h30'52" 35d38'49"
Jason-Paul Trevor Cooke
Psc 0h41'19" 19d12'7"
Jason's and Rachel's Star
Aqr 21h50'53" -5d46'33"
Jason's Angel Bailey
Ori 5h13'24" 11d35'3"
Jason's Immortal Pearl
Aqr 22h32'26" -2d23'50"
Jason's Ladybug
Umi 16h10'15" 70d11'4"
Jason's Love
Sgr 17h52'28" -28d37'17"
Jason's Star
Lib 15h0'6" -19d58'1"
Jason's Star
Cnc 8h50'21" 26d9'13"
Jason's Star
Cyg 20h57'32" 34d17'13"
Jason's Star
Cyg 21h48'2" 46d9'24"
Jason's World
Psc 1h41'52" 6d12'45"
Jason's Yin Star
Her 16h38'57" 44d19'24"
Jaspa
Boo 13h49'27" 10d26'39"
Jasper
Vir 13h36'52" 1d9'43"
Jasper
Leo 11h7'42" 1d48'51"
Jasper
Cma 7h24'47" -33d3'10"
Jasper
Cma 7h14'12" -14d57'43"
Jasper "Archie" Collins
Hya 10h20'49" -16d13'10"
Jasper Bennett Conigrave 11 05 04
Pyx 9h2'39" -27d36'20"
Jasper Earl Barber
Her 17h19'39" 34d50'46"
Jasper Eddison
Per 4h9'6" 43d56'55"
Jasper Henshaw Wise
Her 17h9'20" 18d48'51"
Jasper Joe & Deltona Debbie, 4/3/04
Sge 19h49'10" 18d42'32"
Jasper Johns
Tau 4h45'18" 26d16'37"
Jasper Jones Kemp
Tau 3h30'56" 13d17'12"
Jasper Joseph Ricchio
Aqr 22h36'0" -3d28'0"
Jasper Matrisciano- Horwitz
Dra 17h25'50" 56d16'16"
Jasper McLellan (Our Star)
Lyn 7h46'29" 41d15'32"
Jasper Spencer Hare
Her 17h25'16" 14d19'5"
Jasper Sugarpie Honeybunch
Cma 7h24'1" -33d5'11"
Jasper Tang Ung
Uma 14h15'41" 61d43'58"
Jasper William Gary Franklin
Peg 22h8'56" 31d1'36"
Jasper Wren, Anisa and Philip Wren.
Dra 18h55'37" 69d24'47"
Jasper's Star
Vir 14h17'50" -9d42'47"
JASPIN
Sgr 18h20'59" -23d9'36"
Jaspreet
Sco 17h22'12" -40d48'7"
Jaspreet Balh
Uma 8h29'8" 66d44'12"
Jasri
Leo 10h26'35" 23d40'30"
Jassica Kay Dillon
Cap 21h57'53" -9d3'28"
Jassim Yousif Al Nassar
Per 4h12'31" 31d23'20"
JASSS
Uma 9h38'20" 53d10'47"
Jastara
Uma 13h52'50" 48d57'8"
Jastin
And 0h26'17" 41d15'18"
Jastina
Aql 20h14'1" 5d21'29"
Jaston Thomas Gray
Boo 13h50'32" 16d34'13"
Jasydney Estella
Lyn 6h34'21" 56d53'23"
Jathinon
Per 4h42'50" 46d42'4"
Jatin (Honey)
Cyg 19h21'30" 28d56'35"
Jatin Kumar
Cnc 8h6'25" 21d4'14"
Jatinder Chohan
Uma 10h42'8" 54d17'58"
Jatinder Kaur Sandhu
Tau 5h9'54" 21d11'57"
Jaton
Peg 22h21'28" 7d20'49"
Jaume Torra Canelles
Psc 0h59'16" 16d2'59"
JAV
Gem 6h51'4" 33d44'10"
Javal S. Davis
Umi 15h38'51" 73d9'17"
Javan and Brandy
Umi 15h37'20" 73d16'11"
Javan Daniel
Vir 14h42'25" 3d46'22"
Javanna Elliott
Aqr 20h51'17" 0d17'24"
Javaun Tillacy Mitchell
Uma 14h13'4" 55d46'58"
Javen Charles
Umi 15h19'17" 69d51'30"
J-Avery
Tau 5h46'37" 17d5'50"
Javi
Leo 11h2'2" 0d30'50"
Javi y Sonia
Ori 5h38'31" -1d54'3"
Javid Fynn Lapp Yoder
Uma 10h15'2" 64d46'45"
Javier
Leo 10h54'33" 7d22'19"
Javier Allegue Cardenas
Dra 18h19'7" 75d51'3"
Javier Beltran
Cam 5h32'45" 71d25'28"
Javier Carrazana (Daddy)
Sco 16h15'37" -32d34'30"
Javier Castaneda - Rising Star
Lac 22h20'0" 53d38'28"
Javier DeJesus Lopez
Lib 14h56'44" -16d50'20"
Javier Diaz
Lib 15h51'48" -18d8'5"
Javier Garcia Merinero
And 1h12'53" 46d10'18"
Javier Huca & Linda Mirelez
Mon 7h48'42" -1d48'45"
Javier & Karina Garcia
Cyg 20h55'7" 34d57'7"
Javier Las Heras
Tau 3h32'57" 23d31'33"
Javier M. Alvarez Jr.
Ori 4h51'27" 1d34'54"
Javier Marin
Ori 5h28'2" 1d19'43"
Javier Morales
Boo 14h55'34" 52d27'5"
Javier Moreno Valle
Gem 7h9'28" 16d53'37"
Javier & Paolo Adevoso
Gem 6h41'7" 15d24'42"
Javier Perea Serna
Cap 21h11'3" -22d18'32"
Javier Perez-Giraud
Cap 21h14'57" -25d3'46"
Javier & Rahel
Ori 5h56'23" 21d30'1"
Javier Rosado
Lib 15h45'30" -9d36'59"
Javier Soto Alonso
Cnc 8h7'36" 22d42'54"
Javier Tiberius
Cas 23h5'47" 57d5'36"
Javier Torres
Dra 17h57'4" 50d52'44"
Javier W Cruz
Sco 16h5'23" -15d32'3"
Javier y Lolita
Dra 17h22'10" 51d39'10"
Javier y Nati
Cap 20h46'15" -21d17'28"
Javi's Star
Per 4h34'6" 34d30'8"
Javito
Uma 12h57'4" 54d35'39"
Javon Gaymon
Uma 9h48'30" 48d35'42"
Javon Xavier-Lee Smith
Leo 10h26'13" 20d10'29"
Jawad
Ari 3h18'44" 28d42'19"
Jawad Haidari
Sgr 18h22'24" -32d59'18"
Jawad & Shamina
Cyg 20h35'11" 54d31'4"
jawikai
Pho 0h49'6" -40d38'49"
JAWS
Ari 2h50'8" 27d27'28"
JAWS
Leo 9h26'13" 14d43'44"
Jax
Cam 3h22'25" 58d1'12"
Jax
And 1h9'27" 36d26'2"
Jax
Dra 17h15'7" 56d37'24"
Jax
Cas 23h29'58" 52d56'54"
Jax
Uma 11h22'18" 60d41'47"
Jax
Cap 21h30'38" -18d34'0"
Jax Charles Ross
Per 2h41'0" 56d23'59"
Jax Hoppy Hopkinson - My Tree Fairy
Peg 23h55'39" 8d45'56"
Jax Star Esq.
Sgr 18h16'9" -19d14'45"
JAXON
Cap 21h50'20" -11d55'24"
Jaxon
Aqr 22h23'57" -3d13'9"
Jaxon Aiden Day
Cap 20h22'32" -14d25'8"
Jaxon Blayne Jenkins
Ori 5h7'13" 5d27'32"
Jaxon Boyd McMonigle
Her 18h5'21" 19d13'57"

Jaxon C. Baker
Per 3h14'38" 51d17'11"
Jaxon Charles
Aqr 21h39'1" 1d47'29"
Jaxon Graves
Ori 6h8'49" -0d58'31"
Jaxon Richard Saren
Uma 11h30'17" 31d3'8"
Jaxon Vai Sutton-Sinapati
Cru 12h42'37" -64d28'54"
Jaxsen York Dodd
Tau 4h23'35" 24d8'52"
Jaxson
Her 17h44'9" 23d8'2"
Jaxson
Uma 10h41'2" 63d31'5"
Jaxson Avery Gray
Tau 4h1'56" 15d51'47"
Jaxson Jeffrey Lehnen 1st Birthday
Lib 15h28'56" -16d41'19"
Jaxson Richard Griffith
Cyg 19h53'24" 31d58'47"
Jaxson Tyler Harper
Per 3h27'50" 49d25'3"
Jay
Tau 4h19'47" 9d18'32"
Jay
Her 16h33'11" 15d19'36"
Jay
Ori 5h55'11" 21d54'4"
Jay
Lib 15h29'37" -8d32'31"
Jay
Uma 13h33'15" 53d52'5"
Jay
Uma 11h48'55" 58d1'48"
Jay
Men 4h58'35" -75d4'41"
Jay Alford
Cen 13h23'23" -39d10'52"
Jay & Alisa Wilcox
Cyg 20h38'50" 45d40'35"
Jay Allsman
Cma 7h21'4" -15d43'19"
Jay and Denise Keenan
Cyg 20h17'37" 55d0'32"
Jay and Diane McMullen Star
Cyg 20h18'42" 38d18'19"
Jay and Esther
Aur 6h23'17" 41d36'30"
Jay and Granda's Star
Per 2h12'6" 52d42'57"
Jay and Jack's Star
Cyg 21h57'14" 46d27'11"
Jay and Kristin Forever!
Cyg 19h41'13" 42d40'34"
Jay and Lesli's Star
Cnc 8h23'2" 19d50'54"
Jay and Maria Perez
Lyn 7h37'12" 39d32'7"
Jay and Meg
Uma 11h19'23" 43d55'24"
Jay and Patty Cook
Uma 9h27'54" 42d27'31"
Jay and Patty Cook
Crb 15h17'51" 30d37'31"
Jay and Rea's eternal love
Cyg 20h9'46" 45d53'1"
Jay and Tracy 2004
Umi 15h10'58" 75d37'31"
Jay B. Goldberg
Ari 2h58'33" 21d1'3"
Jay B. Morrall
Cyg 20h52'48" 31d58'51"
Jay Bastian
Per 3h21'15" 45d23'6"
Jay Bateman
Cep 22h14'35" 59d20'36"
Jay Battles
And 0h57'5" 42d36'55"
Jay - Beau Coates
Dra 20h38'0" 71d19'56"
Jay Bernstein - Star Maker
Per 4h8'38" 47d0'2"
Jay Bradford Staley
Cap 21h29'34" -10d2'46"
Jay Brady
Aqr 23h1'27" -7d14'58"
Jay Brame
Dra 19h2'3" 74d10'44"
Jay Bretz
Lib 15h24'31" -5d5'4"
Jay Brian Rappaport
Psc 1h39'30" 24d57'1"
Jay C. Foggy
Cyg 20h31'34" 41d28'11"
Jay Callahan
And 23h7'6" 48d28'41"
Jay Carlson
Cyg 19h53'52" 31d45'17"
Jay Carol Clark
Leo 11h14'44" 21d20'58"
Jay Clark Miles "Jay Bird"
Boo 14h7'42" 34d27'56"
Jay Cruz
Lib 15h7'19" -5d15'22"
Jay&Dana71198
Vir 11h56'48" 9d36'7"
Jay David Cherski
Per 2h24'39" 54d41'22"
Jay David Weinstock
Uma 10h46'37" 43d44'59"
Jay Davis
Uma 10h29'36" 58d24'16"
Jay Davis
Uma 10h20'54" 59d3'0"
Jay Dimock
Tau 3h24'32" 16d36'18"
Jay & Donna; One Within This Star!
Cyg 20h51'43" 48d30'18"
Jay Douglas Roberts
Leo 10h25'15" 18d9'24"
Jay E. Baldwin
Ori 5h42'17" -0d49'27"
Jay Effross
Ori 5h18'48" 6d48'31"
Jay Egipciaco
Ori 5h35'22" -1d53'22"
Jay Ellenburg
Aur 4h57'48" 37d25'55"
Jay Eygji
Cnc 9h8'25" 13d51'55"
Jay Foreman
Ori 5h52'2" 11d15'25"
Jay Gallup, one in a million
Uma 10h6'23" 59d4'28"
Jay Glen Kendall Cook
Tau 4h24'0" 26d50'6"
Jay Greso
Peg 22h49'53" 26d12'16"
Jay Hallam
Aur 6h36'35" 38d50'48"
Jay Hammock 07/31/1973
Leo 10h1'8" 22d37'36"
Jay Hansen
Uma 13h40'2" 49d48'59"
Jay & Heather's Fire
Lyn 8h24'47" 43d42'34"
Jay Henry Silverstein
Sgr 18h54'28" -35d14'4"
Jay Ingram
Cnc 8h17'0" 15d27'11"
Jay & Irene
Aqr 22h21'48" 2d16'53"
Jay Jay
Per 4h0'23" 32d13'31"
Jay jay
Her 17h28'14" 31d7'53"
Jay & Jay
Cyg 21h28'34" 52d20'57"
Jay Joseph Athey
Ori 6h9'25" -0d5'40"
Jay Kerry Wallerstedt
Ori 5h30'32" -4d44'36"
Jay L. Padratzik
Her 16h33'28" 14d9'8"
Jay Lamark
Cet 1h11'47" -0d39'52"
Jay Leff
Ori 5h27'19" -0d38'34"
Jay Lightfoot
Vir 12h37'26" -0d38'2"
Jay Lininger
Per 4h16'15" 45d12'41"
Jay Loves Adehl
Cru 12h35'26" -59d6'49"
Jay Luchun
Vir 13h37'33" -15d21'49"
Jay M. Goldthwaite
Cnc 9h16'41" 25d47'53"
Jay Macey Rosenblum
Tau 5h37'3" 22d45'5"
Jay & Madonna's Star
Vir 12h35'54" 10d47'38"
Jay & Margaret Robinson
Dra 17h35'48" 55d40'50"
Jay & Marissa Stein
Gem 6h30'0" 25d3'47"
Jay Mascovich
Cep 22h10'47" 67d2'38"
Jay McCain
Boo 13h50'54" 12d15'12"
Jay McCarthy
Uma 11h39'5" 50d5'9"
Jay & Melinda Glass
Uma 9h42'41" 56d21'30"
Jay Morganstern
Gem 6h51'50" 26d28'28"
Jay "Mr. Hollywood" Bernstein
Cep 4h7'40" 87d50'26"
Jay (My King)
Cep 22h9'54" 53d31'46"
Jay&Newman
Gem 6h3'36" 25d45'35"
Jay & Nikki
Ori 6h6'27" 18d19'31"
Jay Norris
Aql 19h7'44" 1d56'52"
Jay O. Darling
Dra 18h32'3" 75d35'36"
Jay Over the Bay
Her 18h46'45" 19d52'9"
Jay Pee's Star From Tami
Her 17h38'57" 28d41'21"
Jay Quincy Johnson
Ori 5h27'2" 3d28'44"
Jay R Smith-The Brightest Star
Gem 7h42'29" 30d34'23"
Jay & Ray "Feafa"
Cyg 19h22'2" 53d44'13"
Jay Reed Overholser
Her 18h7'0" 48d45'27"
Jay Robert Sanders
Sco 16h4'28" -28d3'42"
Jay Robinson
Psc 22h52'45" 6d34'13"
Jay Rosson I love U
Tri 1h54'39" 28d45'29"
Jay Ruban
Uma 11h56'36" 29d55'26"
Jay Ryan Powell
Ori 5h33'28" -1d9'26"
JAY SALDIVAR
Her 15h56'4" 44d57'23"
Jay Sanjay Shah
Her 16h33'0" 36d50'5"
Jay Schwartz
Uma 11h49'3" 52d58'25"
Jay Sheahan
Leo 9h38'8" 27d12'49"
Jay Silverman Aka Jaybird
Tau 4h10'7" 22d22'34"
Jay Sonshine's 50th
Leo 10h1'23" 23d32'51"
Jay Stailey #50
Lmi 9h40'59" 33d48'40"
Jay Steele
Cnc 8h19'19" 25d37'30"
Jay Steven Hoyle
Her 17h8'59" 30d40'45"
Jay Superman Valentin
Ari 2h50'52" 29d28'36"
Jay "The Captain Of My Heart"
Tau 4h0'9" 17d58'22"
Jay — The Love of My Life
Cep 21h23'49" 56d23'2"
Jay Unruh
Cnc 8h58'5" 22d38'6"
Jay Vance Smith
Tau 5h43'5" 28d5'46"
Jay Vasquez
Ari 3h5'59" 22d34'33"
Jay Vladimir Judkins
Sco 16h8'48" -14d0'13"
Jay W. Greenstone
Vir 13h35'5" -17d12'40"
Jay William Lilledahl
Cep 21h20'12" 58d9'37"
Jay William Pearce
Per 3h39'23" 39d20'15"
Jay William Warren
Her 17h51'43" 25d1'2"
Jaya
And 1h47'16" 49d14'26"
Jaya Kutti
Sgr 19h15'53" -16d50'8"
Jaya Vaswani
Vir 11h48'20" 4d56'23"
Jayaa
Ari 2h39'59" 19d43'39"
JayaDevi Harilal
Leo 11h26'0" -4d9'32"
JayandDanielle
Cyg 20h39'2" 47d28'59"
Jayant Ullal Nayak
Ari 3h13'5" 27d37'29"
JayBee
Crb 15h43'9" 32d55'56"
JayBee
Cnc 8h23'5" 30d29'40"
JayBeeJay
Lib 14h53'53" -6d46'51"
JayBird
Vir 13h33'36" -18d55'24"
jaybird
And 23h38'26" 47d41'57"
Jaybird
Psc 1h9'45" 29d48'23"
Jaybird
Aql 19h10'34" 15d16'43"
Jaybird
Ori 5h54'42" 21d25'2"
JayBird
Ori 6h3'56" 15d49'57"
Jayca M. Fleming
And 1h11'50" 42d20'21"
Jayce
Mon 6h45'51" 3d28'31"
Jayce Cassidy Vrbka
Crb 15h38'41" 25d58'49"
Jayce Charles Rains
Uma 10h5'13" 56d39'42"
Jayce David Perry
Sgr 19h17'53" -13d49'4"
Jayce Michael Scholler
Umi 15h24'58" 71d43'30"
Jayce Parker McElwee
Uma 10h58'32" 61d50'50"
Jayce Rasaan Brown
Cap 20h18'54" -14d39'42"
Jaycee
Cep 23h53'54" 68d41'21"
jaycee
Tau 5h31'48" 20d0'54"
Jaycee
Peg 23h11'14" 13d14'25"
Jaycee Rose
Tau 5h47'33" 26d12'0"
Jaycie O'Brien
Uma 9h3'38" 56d55'3"
Jayda
Crb 16h4'58" 36d22'44"
Jayda Grace Castillo
Gem 7h20'31" 33d22'53"
Jayda Jourdin
Dra 15h29'2" 60d22'53"
Jayda Madison
And 0h12'16" 28d15'49"
Jayda Mundon
Eri 4h41'34" -16d12'17"
Jayda Rose's Lightning Bug Star
Cam 6h55'1" 79d55'58"
Jaydan Michael Thomas Israel
Uma 11h51'4" 62d27'1"
Jayde
And 0h29'23" 43d47'15"
Jayde Christine
And 0h58'24" 39d24'1"
Jayde Gearing
Sco 16h9'55" -19d9'37"
Jayde Laura McAuslan
Psc 0h38'22" 8d36'54"
Jayden
Lyn 7h36'50" 37d49'2"
Jayden
Lib 15h55'24" -17d54'26"
Jayden Alexander Pace
Vir 14h41'9" 3d51'10"
Jayden Bree Prestholdt
Cyg 20h2'23" 45d56'16"
Jayden Broms Daube
Sgr 19h27'59" -15d11'0"
Jayden Chancey Rogers
Vir 14h34'11" 4d1'16"
Jayden Dean Folckemer
Her 17h0'47" 39d47'14"
Jayden Dermer
Uma 11h36'45" 51d51'59"
Jayden Desjardins Born: Oct 15/1995
Uma 11h31'4" 57d13'37"
Jayden Dominic
Ori 5h32'24" -4d21'10"
Jayden First Born
Leo 10h11'30" 16d57'51"
Jayden - Forever My Star
Vir 12h0'40" -0d4'42"
Jayden Graham Sherman
Lib 14h28'45" -16d8'17"
Jayden Hawke Escover
Lib 15h17'59" -9d35'45"
Jayden Hurst
Uma 13h16'53" 62d23'57"
Jayden Isaac Wali
Sco 16h52'56" -32d48'11"
Jayden Joel Harris "J.J."
Ari 3h14'41" 16d13'54"
Jayden Jose Rios
Her 18h1'32" 20d58'49"
Jayden Joseph Lopreiato
Vir 14h1'8" 6d24'17"
Jayden Julia Pappas
Aqr 21h42'59" -1d42'3"
Jayden Kenohi Sperry Exzabe
Cap 20h23'57" -14d6'9"
Jayden Lee Finkelstein
And 2h18'57" 45d50'34"
Jayden Lynn Schatzer
Aqr 21h31'25" 0d31'58"
Jayden Matthew Dubanewicz
Her 18h35'57" 22d28'16"
Jayden Matthew Scott Barella
Boo 14h35'35" 52d2'20"
Jayden Michael Samonte
Tau 3h26'22" 9d43'37"
Jayden Paul
Her 16h19'1" 44d28'21"
Jayden Ricky Sweeting
Umi 14h5'51" 74d25'7"
Jayden Riley DeLuca
Uma 10h49'46" 44d22'53"
Jayden Rose Durfee
Sgr 18h43'13" -27d47'31"
Jayden Shane Hall
Cru 12h24'28" -56d56'58"
Jayden Terhune Scott
Aqr 21h47'55" -6d52'16"
Jayden Tyler Opper
Her 17h26'25" 16d8'43"
Jayden Victor Fulgencio
Ari 2h47'5" 25d51'50"
Jayden Wayne Mathis
Her 16h19'11" 9d12'38"
Jayden William Colwell
Cap 21h55'0" -14d37'36"
Jayden William Cory
Psc 1h11'36" 27d0'53"
Jayden Young, You Brightin My Soul
Ori 6h19'36" 5d59'59"
Jaydene Hennessy
Col 6h9'35" -35d58'21"
"Jaydens Beauty"
Gem 6h45'43" 20d49'34"
Jayden's Love
Col 5h56'33" -30d34'44"
Jayden's Lucky Star
Umi 16h10'7" 74d5'19"
Jaydie
Lib 14h57'26" -4d55'49"
Jaydie Lynn King
Lib 15h43'43" -7d57'17"
Jaydin Talbert Rossiter
Leo 9h31'56" 27d50'39"
Jaydnmadisonbissig
Cap 20h17'16" -9d11'12"
Jaydon Kyle Porter
Per 3h30'14" 37d16'38"
Jaydy
Cam 4h14'23" 66d4'38"
Jaye
Dra 17h59'13" 51d55'5"
Jaye Alexandra
Cas 0h15'4" 62d29'18"
Jaye and Amy
Aqr 22h24'35" -0d1'33"
Jaye Emily Flegg - My Guiding Light
Cap 20h35'56" -18d13'30"
Jaye Pozzobon Prescott
Uma 11h32'22" 57d15'18"
Jaye Thompson
Uma 8h47'20" 53d36'41"
Jaye, My Wife My Love
Sgr 18h22'2" -23d43'25"
Jaye436
Tau 4h21'29" 12d0'36"
Jayelle Danielle
Uma 11h24'15" 68d52'21"
Jayelynn
Sgr 18h24'32" -32d8'26"
Jay-Glen Hockemeier
Ori 6h9'18" 15d50'57"
Jayla
Dra 15h13'41" 56d53'59"
Jayla
Vir 13h9'36" -21d57'23"
Jayla Kyann
Leo 11h7'52" 13d57'42"
Jayla Weigart
Cam 5h43'54" 61d44'14"
Jaylee
Lib 15h18'44" -23d36'54"
Jaylee
Peg 22h21'25" 24d14'24"
JAYLEE ALASKA
Psc 1h1'13" 12d16'55"
Jayleen
And 23h58'30" 34d40'15"
Jaylen Michael
Lyn 7h0'18" 51d46'54"
Jaylene
Lib 15h34'7" -27d22'28"
Jaylene
Sco 16h13'20" -14d44'57"
Jaylene Raedell Stiner
And 1h42'54" 38d59'58"
Jaylen's Mom
Cas 1h43'56" 60d17'11"
Jaylin
Cap 21h0'58" -16d27'37"
Jaylin Aliza Owen
And 1h16'41" 37d51'12"
Jaylin Kiara Sonnleitner
And 0h21'3" 26d19'1"
Jaylin Star
Dra 16h6'8" 54d3'47"
Jaylinn Nicole
Aqr 21h43'13" 1d17'48"
Jayln Star
Cap 21h45'57" -22d24'16"
Jayme
Vir 13h36'54" 1d20'57"
Jayme Ann Alaimo
Cnc 8h17'21" 31d28'31"
Jayme Bisi
Aqr 21h19'27" -9d30'22"
Jayme Day
Aql 19h36'53" 5d3'41"
Jayme Felice Pierce
Mon 6h47'12" 8d33'49"
Jay-me Huguette Maxine Bradford
Ori 5h39'34" 11d30'11"
Jayme Kaye
Sgr 19h44'13" -13d33'9"
Jayme Lee Quick
Leo 11h49'22" 22d49'38"
Jayme Loves Greg Forever
Tau 5h55'46" 27d21'13"
Jayme Lyn
Del 20h42'30" 10d14'2"
Jayme Lynn
And 2h10'44" 43d57'45"
Jayme Lynn Totora
Uma 11h22'44" 50d25'57"
Jayme M. "Missy" Rodriguez
Cyg 20h15'11" 42d34'5"
Jayme Michelle Cameron
Ori 5h19'14" 7d24'2"
Jayme Montemayor
Gem 7h18'42" 14d20'49"
Jayme Nicole Karp
Sco 16h13'3" -12d0'4"
Jayme Patricia Campagnola
Sgr 18h15'13" -23d46'45"
Jayme Richards
Tau 4h41'44" 22d44'6"
Jayme Rose
Ari 2h7'22" 15d58'38"
Jayme Rose Wagner
Uma 10h32'7" 40d57'5"
Jayme Sue Fink
And 23h15'29" 47d33'38"
Jayme Uhl
Tau 5h7'21" 27d11'2"
Jayme Weber
Cas 2h26'33" 63d29'37"
Jayme Wessen
Lib 15h33'9" -8d22'28"
Jaymee Marie Heuton
Sco 17h11'37" -30d38'27"
Jaymee, I Will Love You FOREVER
Oct 21h23'22" -75d48'2"
Jayme's Little Chattahoochee
Aqr 20h52'28" -9d32'15"
Jay-Mi Amor, Mi Estrella-Natalie
Sco 17h46'28" -40d35'3"
Jaymi E. Peña
And 23h13'15" 43d11'31"
Jaymie Lyn Tetreault
Leo 10h37'12" 13d14'48"
Jaymie Lynn Jones
Uma 8h45'22" 60d35'9"
Jaymie Reel
Leo 11h23'13" 17d32'36"
Jaymin Anthony Illas
Aql 19h25'44" -11d35'52"
Jaymi's Destiny
And 0h30'53" 36d51'3"
JayMo
Pup 6h46'14" -42d50'5"
Jayn
Vir 12h32'9" 11d19'45"
Jayna
Gem 6h20'3" 26d19'49"
Jayna Beth Dixon
And 2h21'8" 48d18'37"
Jayne
Per 3h25'25" 45d11'34"
Jayne
Ori 5h12'41" -0d52'18"
Jayne & Andy's Star
And 23h54'49" 34d37'32"
Jayne Beving
Vir 13h2'14" -4d0'4"
Jayne C. Veslov
Del 20h45'6" 3d5'16"
Jayne & David 060603
Cyg 21h15'55" 47d24'51"
Jayne Davies
Umi 16h6'46" 79d20'11"
Jayne Dennis
Cnv 13h40'55" 35d21'2"
Jayne "Ducky" Dietz
Cyg 20h11'42" 36d42'0"
Jayne Elizabeth Lucas
Ari 3h19'31" 21d3'46"
Jayne Frances Moran
Aqr 21h56'11" 1d26'58"
Jayne Harper
And 0h30'41" 31d34'0"
Jayne Heather Williams
And 23h27'9" 38d46'57"
Jayne Holbrough
Cas 1h14'36" 55d56'31"
Jayne Louise Langdale
Cas 23h16'7" 54d40'5"
Jayne Lynne Chaffin
Aur 6h15'39" 45d48'11"
Jayne Mae
Cnc 8h12'28" 16d30'1"
Jayne Marie
Uma 9h48'32" 51d56'47"
Jayne Marie
Ant 10h48'33" -35d51'3"
Jayne Mawson
Cas 0h2'38" 59d41'59"
Jayne - my superstar
Tau 3h48'3" 10d37'31"
Jayne Senecal Suffolk Law 2007
And 23h12'59" 45d57'45"
Jayne Tarpy
And 1h4'30" 46d24'25"
Jayne & William Herrera Xmas star
Cyg 21h30'3" 45d38'2"
Jayne-Marie Jackson
Sgr 17h53'40" -17d34'20"
Jayne's Light
Pho 1h6'52" -46d37'59"
Jayne's Promise
Cas 23h30'29" 51d48'24"
Jayne's Star
Cyg 21h16'40" 46d28'36"
Jayne's Star
And 0h41'38" 40d16'50"
Jayne's Star
Cru 12h27'11" -63d4'45"
Jayne's Star
Cas 23h3'43" 59d23'45"
Jaynie Marie Lopez
Aqr 23h12'12" -8d46'26"
jayNsway
Ori 6h1'31" 17d51'48"
Jaynyn Raye Goddard
Ori 5h55'37" 3d11'31"
Jay's Allument Josie
Leo 10h20'30" 7d58'6"
Jay's Astellas Star
Sgr 19h48'40" -12d25'10"
Jay's Blue Heaven
Umi 13h37'22" 76d6'49"
Jay's Blue Star
Aqr 22h25'43" 1d52'28"
Jay's Diamond in the Sky '05
And 2h25'33" 48d49'57"
Jay's Love
Eri 4h28'27" -3d11'4"
Jay's Piece of the Cosmos
Ori 5h31'0" -1d42'1"
Jay's Star
Dra 17h36'29" 53d56'50"
Jay's Star
Leo 11h16'19" 12d5'41"
Jay's Wishing Star
Per 3h35'39" 48d59'31"
Jayse's Star
Ori 5h35'15" -2d13'34"
Jayshree
Aqr 23h3'26" -9d8'24"
Jayshree Kansara
Ari 3h11'53" 26d4'28"
' Jayson '
Cyg 19h53'0" 36d47'25"
Jayson Erik
Aql 19h5'21" -0d26'4"
Jayson Ferrell Carraway
Aql 19h52'19" -0d36'34"
Jayson J Haydon
Aql 19h49'24" -7d21'16"
Jayson John Thomas Foye
Psc 22h59'56" -0d38'52"
Jayson Jon Ott
Cas 0h35'35" 64d15'49"
Jayson M. DeZuzio
And 1h14'23" 38d0'24"
Jayson Michael Stockdale
Psc 1h28'5" 28d31'35"
Jayson ODonahue
Lib 15h22'16" -5d58'21"
Jayson P.
Cnc 8h30'30" 29d28'28"
Jayson Phillips Smith, Jr.
Umi 15h26'2" 71d45'43"
Jayson Raine
Aqr 20h55'6" -7d5'34"
Jayson's Hope
Gem 6h44'30" 33d6'35"
Jayster's Star
Dra 18h49'19" 68d45'6"
JayTon
Lyn 6h30'0" 54d10'9"
Jaz
Leo 9h36'50" 30d50'14"
Jaz
Lmi 10h28'56" 36d38'28"
Jaz
Uma 9h24'46" 42d48'51"
"Jaza Star" - Jasmine Elisha Bluck
And 2h31'26" 38d30'4"
Jazalyne DeVito
Vir 12h21'59" 11d26'35"
Jazan
Ori 5h37'30" 1d29'7"
Jazline
Leo 11h27'32" 3d33'28"
Jazlyn
Cnc 8h46'51" 12d14'57"
Jazlyn Leana Jew
Leo 10h32'58" 21d28'7"
Jazman Banks
Uma 11h12'14" 45d39'20"
Jazmin
Vir 11h38'37" 6d47'9"
Jazmin Adriana
Aqr 22h59'14" -7d59'57"
Jazmin Alexis Roper
Vir 12h50'9" 5d49'46"
Jazmin Almeida
Psc 1h33'39" 18d2'56"
Jazmin Cheyenne
And 2h35'13" 44d13'12"
Jazmin Franco
Uma 10h24'10" 48d0'52"
Jazmin Marie Martinez
Psc 1h10'3" 25d0'5"
Jazmin Nicole Little
Sco 16h5'58" -12d25'14"
Jazmin Sky Wilson
Dra 17h18'58" 62d29'54"
Jazmina
Del 20h34'41" 15d8'4"
Jazmine
Cas 1h39'32" 63d28'0"
Jazmine C. Hayes
Gem 7h48'46" 26d42'23"
Jazmine De jesus Contreras
Vir 11h51'57" 6d56'4"
Jazmine Diaz
Leo 10h41'10" 8d56'30"
Jazmine Hope
Cam 4h26'41" 66d46'57"
Jazmine Katinka
Cas 3h15'13" 57d29'56"
Jazmine Marie
And 0h48'10" 42d11'31"
Jazmine Rivers
Eri 3h43'24" -2d26'44"
Jazmine - Tribute by Robert Heward
Psc 23h15'35" 5d47'27"
Jazmine Webster
Lib 14h59'44" -11d45'40"
Jazminita
Dor 5h8'1" -57d51'17"
Jazmon
Sgr 19h4'47" -15d50'9"
Jazmyn Dawn Alberts Wheeler
Oph 18h1'57" 4d0'36"

Jazmyn Lynn Wiley
Cep 1h36'15" 80d8'2"
Jazmyn Sweet
Leo 10h59'8" 15d8'1"
Jazmyne
Aqr 20h40'16" -1d8'9"
Jazmyne 54
Gem 7h44'38" 30d58'35"
Jazmyne Deann
Tau 3h55'5" 1d42'40"
Jazmyne Rose
And 23h58'36" 41d32'11"
JAZZ
Cyg 19h57'40" 37d41'31"
Jazz
Ori 5h52'16" -3d40'20"
Jazz Bouzon de St Jean Dillac
Cap 20h59'36" -24d58'51"
Jazz & Chazz
Cyg 20h28'28" 47d11'3"
Jazz Man
Ori 5h59'52" 14d2'9"
Jazz McArdle
Umi 16h24'30" 85d6'33"
Jazzelle
Per 2h49'29" 37d1'20"
Jazzmin Grace Morrison
Cru 12h51'10" -64d32'30"
Jazzy
Cma 6h48'16" -14d29'48"
Jazzy Angel
Lyr 19h14'8" 33d28'5"
Jazzy J
Vir 13h14'35" 11d43'59"
JAZZY JAZZ
Lib 15h40'44" -6d15'51"
Jazzy-Golden Pooh Bear
Lyn 8h37'3" 45d35'36"
JB
Gem 6h46'5" 28d9'29"
JB~1017
Sgr 19h5'47" -23d22'52"
JB Brown
Uma 10h5'23" 71d2'19"
J.B. Gerlach III
Her 16h46'18" 25d55'10"
JB & KC Burdsall 19th Anniversary
Aql 20h10'25" 8d0'9"
JB Langley
Her 17h18'11" 25d52'8"
JB LoveLife HK
Cas 2h21'19" 59d58'23"
J.B. Raftus
Ori 5h12'16" 9d18'43"
JB SB 04-03-04
Sco 17h54'34" -35d50'46"
JB Squared
Cam 4h9'49" 68d26'35"
JB2
Umi 16h3'9" 77d17'50"
JB-B Kalck+Inès-nès Tran Dinh
Uma 12h2'0" 37d14'15"
JBCAM
Crb 16h19'52" 29d29'36"
JB-Day
Ari 2h44'53" 24d39'37"
J.B.E. & S.J.K. Forever
And 1h20'9" 47d15'53"
Jbear
Sco 17h17'36" -43d33'59"
J.Beck star
Lib 15h54'7" -11d56'57"
J-Belle Amour
Uma 11h39'27" 31d50'53"
JBF Jr.
Vir 13h14'29" 4d52'19"
JBFox
Ori 5h13'22" -0d53'0"
JBHunt
Vir 13h26'23" 13d24'37"
J-bird
Vir 13h21'48" 9d30'28"
J-Bird
Aqr 22h17'31" 0d37'24"
J-Bird's Dream
Uma 12h36'58" 58d51'45"
JBM My Brightest Star
And 1h11'0" 34d25'42"
Jbma3470
Lyn 8h7'25" 57d7'37"
jbnjenny
Cnc 8h19'29" 8d26'24"
J.B.O.
Cnc 9h11'16" 10d29'45"
J-BO
Her 18h52'42" 23d20'26"
JBooneGR
Uma 11h22'6" 38d22'59"
J-BOP 28
Lyn 7h1'26" 51d25'15"
JB's Superstar
Vir 14h42'18" -1d10'19"
jbug 8467
And 1h2'22" 41d32'41"
J.Butterfly
Cyg 20h19'28" 47d23'6"
jbzollo
Umi 15h57'32" 77d38'37"
JC
Cen 13h12'26" -49d37'3"
JC 11
Cma 6h59'27" -18d59'25"
JC^2
Uma 10h38'15" 53d27'31"
JC and MC
Cyg 21h13'42" 44d57'57"
JC Binder
Cnc 8h22'53" 27d44'29"
JC & Bruce FY06 Competitive Battle
Uma 12h18'1" 55d2'58"
J.C. Byrd
Psc 0h14'52" -1d49'13"
J.C. French
Lyn 8h43'42" 35d40'31"
JC Harmon
Ori 5h41'13" 3d5'45"
JC Hendry
Aqr 21h49'23" -3d17'43"
JC Leslie
Sgr 19h32'32" -37d58'36"
JC Love You Forever LC
Ari 2h42'16" 28d5'41"
JC Shinebox Sisk
Dra 19h20'25" 64d26'33"
JC - That's Amoure
Cra 18h4'9" -43d37'59"
JC weatherman
Cnc 9h0'53" 31d27'5"
JC White
Cyg 19h47'29" 29d23'16"
JCAH 2001
Leo 11h55'25" 27d38'11"
J.C.B. 1
Cyg 20h49'7" 30d37'44"
JCB 29
Gem 7h9'49" 16d7'44"
JCFellure
And 2h29'56" 44d1'46"
JCGoode121357
Cep 21h44'32" 66d23'24"
JCGS The madman
Ari 3h6'47" 18d43'12"
JCKennedyColwell
Uma 9h16'15" 48d30'3"
JCL 21
Cnc 8h0'56" 11d42'24"
JCM1
Uma 9h1'46" 70d8'29"
JCmb4
Per 3h17'34" 43d48'38"
J.C.M.R. Petrie Star
Uma 9h22'16" 47d52'11"
JCPrinslooMariannePrinsLoveStar
Psc 0h15'3" -1d54'52"
JCR&DJM
Lyn 8h3'32" 55d44'29"
J-Crab 062557
Uma 10h37'16" 55d36'57"
Jcrstar7
Sco 16h16'21" -12d1'17"
JCruz14
And 22h59'4" 39d9'19"
J.C.S (AKA: JACK) MY LIFE, MY LOVE!
Per 4h17'2" 51d37'56"
jc's star
Tau 4h14'22" 21d51'33"
JC's Talented Woman
Cas 1h39'43" 65d52'22"
JCsquared
Crb 15h32'52" 29d31'7"
JCVICTOR-IA
Her 18h53'20" 25d24'2"
JCW-143
Gem 7h19'56" 25d14'21"
J.D.
Boo 14h27'29" 18d8'8"
J.D
Lyn 7h21'3" 49d42'31"
JD and RP Jamieson
Uma 10h33'1" 66d11'56"
J.D. Boone
Lup 15h27'32" -32d17'52"
JD Butler
Uma 11h5'36" 64d12'46"
J.D. Duffy's Shining Star
Uma 8h25'36" 60d56'4"
J.D. Hart
Aqr 22h20'10" 1d50'34"
JD Hoffman 5
Gem 7h42'27" 34d50'26"
J.D. Hohman
Ori 5h24'39" -4d32'28"
J.D. Kittinger
Tau 4h20'37" 2d19'14"
JD & Laurie - 22 Ditto
Cyg 20h5'19" 37d46'7"
JD & MC: Love, Fidelity, Friendship
Tau 4h30'51" 19d4'1"
JD Northrup
Ori 6h15'39" 15d44'2"
J.D. Pegg
Psc 0h12'12" 6d20'16"
J.D. Stanley
Tau 4h13'11" 16d25'1"
J.D. Throgmorton
Vir 13h17'20" 12d57'35"
JD Wiggin
Tau 3h49'51" 15d6'44"
JD57
Sco 17h4'13" -35d35'43"
JD622
Ori 5h36'20" -1d5'20"
JDM
Lib 15h7'49" -5d12'58"
J-Doodles
Psc 23h18'49" 1d55'13"
JDP416
Psc 1h10'41" 23d20'26"
J.E. Hutchison
Her 17h1'37" 13d17'26"
Je t'adore
Aqr 21h1'30" -11d47'47"
je t'adore tre's
And 0h19'38" 23d15'9"
Je t'adore, Barbara Zawisla
Cas 0h15'36" 63d26'41"
Je T'aime
Uma 9h29'10" 57d28'7"
je t'aime
Lib 14h52'57" -4d23'37"
Je t'aime
Umi 15h20'18" 81d9'3"
Je t'aime
Sco 15h54'50" -22d35'26"
Je T'aime
Leo 10h4'20" 20d40'48"
Je t'aime
Vir 13h26'23" 6d51'58"
Je t'aime
Per 3h49'17" 34d21'28"
Je t'aime
Lmi 10h15'25" 32d22'38"
Je t'aime à la lune et au dos.
Leo 10h14'53" 15d21'4"
Je t'aime Anette
Psc 1h27'34" 16d41'34"
Je t'aime Babacar Louis Camara
Cyg 21h51'21" 50d4'40"
Je T'aime Beaucoup
Cyg 19h27'33" 53d13'18"
Je t'aime Maman
Tau 4h6'34" 20d55'1"
Je t'aime SNC
Eri 3h50'45" -0d23'3"
Je t'aime~Te amo~Ich liebe Dich
Cas 23h34'59" 52d20'32"
Je t'aime toujours Hunter
Del 20h49'8" 11d13'44"
Je t'aime Whitney
Cyg 19h35'24" 37d51'13"
Je t'aime, je t'aime, je t'aime
Uma 8h38'15" 56d54'31"
Je Vous Aime
Leo 11h44'38" 24d0'34"
Jeafra
Uma 12h2'17" 45d39'38"
Jeah
Lyn 7h54'10" 34d2'15"
Jean
And 0h51'2" 39d21'19"
Jean
And 1h47'28" 41d58'11"
Jean
Cas 0h20'7" 58d6'14"
JEAN
Cnc 8h44'24" 22d55'32"
Jean
Gem 7h17'3" 29d17'10"
Jean
Cnc 9h10'39" 15d8'36"
JEAN
Tau 4h35'48" 19d41'36"
Jean
Ori 5h21'24" 0d48'1"
Jean
Aqr 20h58'52" -12d42'47"
Jean
Cap 21h4'27" -26d11'39"
Jean #1 Allstar
And 1h49'16" 41d30'41"
Jean A. Bielitzki
Uma 9h50'14" 45d27'41"
Jean Abrahamian Martin
Lib 14h54'24" -2d34'9"
Jean Acaster
Cas 23h23'49" 55d25'50"
Jean Addison
And 1h41'8" 39d55'19"
Jean & Agnes' Star
Lyr 18h43'53" 34d6'32"
Jean Alain Boumsong - Boomy -
Col 6h2'24" -28d23'35"
Jean And Gerry Hagstrom's Star
Uma 8h51'25" 69d19'26"
Jean and Mark Bell
Del 20h57'52" 12d56'26"
Jean and Reuben Edwards
Umi 15h2'21" 80d2'59"
Jean and Vivian's Love Shines
Ari 2h48'49" 22d5'10"
Jean Andersen
Eri 4h9'28" -0d7'41"
Jean Ann
Leo 9h46'22" 31d56'35"
Jean Ann "Chip" Miller
Lyr 18h41'36" 37d8'30"
Jean Ann Decker
Gem 6h14'38" 26d34'3"
Jean Ann Dorsey
Lyr 18h52'25" 37d11'8"
Jean Ann Fitzsimonds
Cas 1h29'7" 60d57'22"
Jean Ann Kimerling
Her 18h2'7" 32d18'57"
Jean Ann Remus
Aql 19h18'57" 4d54'6"
Jean Ann Sturdevant
Tau 5h47'37" 22d40'50"
Jean Anne Attwell
And 1h56'9" 36d20'13"
Jean Anne Hedelius
Tau 4h17'3" 13d7'24"
Jean Ann-Marie
Leo 11h42'58" 25d35'54"
Jean Arbucke Johnson
Uma 10h26'43" 45d54'16"
Jean Armitage
And 1h33'59" 41d2'19"
Jean Arnett
And 1h42'2" 50d18'27"
Jean Arthur
Vir 13h13'17" 4d18'3"
Jean B. Cardoza
Cam 6h28'43" 64d59'43"
Jean B. Weber
And 0h45'45" 45d26'43"
Jean Bair
Uma 11h13'42" 67d49'44"
Jean Baker Sigford
Uma 11h57'45" 44d19'31"
Jean Baptiste St.Pierre, III
Uma 12h1'27" 54d50'46"
Jean Barrick Fortunatus
Leo 10h18'23" 20d54'0"
Jean Beaujean
And 1h3'35" 41d48'12"
Jean Belanger
Uma 11h22'45" 58d31'10"
Jean Belizar
Gem 7h15'23" 23d51'37"
Jean Bell
Psc 0h43'35" 16d23'26"
Jean Bergstrom
Cas 23h30'27" 51d9'17"
Jean & Bernie Boyce
Aql 18h58'22" -0d27'44"
Jean & Bert's Golden Wedding Star
Uma 8h33'30" 62d36'23"
Jean & Bill
Uma 10h21'24" 43d31'58"
Jean Blosser
Cas 1h46'29" 65d33'39"
Jean Bowden
Cas 3h13'20" 59d57'35"
Jean Brink
Cap 20h15'14" -15d33'10"
Jean Brisson (POUPI)
Pho 0h37'19" -39d37'58"
Jean & Bruce Graham - 02.08.1941
Cru 12h19'17" -62d34'28"
Jean Bruneau
Cep 21h28'12" 55d52'36"
Jean Brunell Guyon
Tau 4h19'17" 18d37'0"
Jean Budd Edge
Ori 5h36'41" 5d4'4"
Jean Butler - Rising Star
Vul 19h56'50" 22d51'2"
Jean Carey
Lyr 18h52'47" 44d45'37"
Jean Cargill Sleet
Eri 4h35'57" -20d14'49"
Jean Carol McKenna
Ori 6h13'43" 8d57'49"
Jean Catherine
Ori 5h36'48" -2d25'26"
Jean Chang
Leo 10h50'19" 6d23'28"
Jean Charles André ESPERANDIEU
Uma 10h40'48" 71d44'24"
Jean(Chink) and Jim(Wish) Ippolito
And 1h26'43" 41d13'19"
Jean Christophe Deville
Cap 20h10'35" -14d19'18"
Jean Christopher
Ori 4h58'41" 15d13'15"
Jean & Chuck Faulkner
Uma 9h58'19" 69d16'2"
Jean Cocozza
Tau 5h9'13" 23d11'37"
Jean Collins
Sco 17h52'18" -35d2'58"
Jean Compton
Crb 15h37'55" 27d48'15"
Jean Corbett
Crb 15h31'59" 30d41'4"
Jean Coumbaras
Uma 13h51'8" 48d50'12"
Jean Cygan Cruickshank Oswald
Uma 10h2'10" 48d53'38"
Jean Da Luz
Vir 14h16'30" -5d25'11"
Jean Davis
Lyn 8h26'20" 36d25'35"
Jean de Faramond
Uma 8h24'50" 62d9'35"
Jean de Servien - Kenwood
Cep 20h39'38" 58d42'4"
Jean Delcel 05.10.03
Gem 6h29'33" 25d27'53"
Jean Delores Smith
Cas 23h18'23" 59d38'32"
Jean Desilets
Lmi 9h57'49" 38d38'31"
Jean DeVoy
Cap 20h18'46" -19d52'31"
Jean E. Iavelli
Gem 7h13'18" 16d4'55"
Jean E. McNelis
Uma 9h47'59" 42d48'49"
Jean E. Nitschke
Cnc 8h59'56" 6d42'4"
Jean Elizabeth Parcher
Leo 9h23'32" 29d15'3"
Jean Elizabeth Whittaker
Sgr 18h3'26" -32d1'15"
Jean Ellen Booth
Sco 16h56'5" -38d12'20"
Jean et Rose
Uma 13h41'49" 52d32'34"
Jean Ewen
Lyn 6h31'41" 57d8'5"
Jean F. Larroux, IV
Crb 15h38'55" 26d25'35"
Jean Figorito
Uma 10h25'54" 69d11'39"
Jean Fischer
Ori 5h15'54" 7d4'55"
Jean Fleming
Cas 1h18'53" 51d11'24"
Jean Foggon
Cas 23h17'18" 59d27'10"
Jean Francis Roberts Murgatroyd
Tau 5h46'20" 28d4'38"
Jean François Guérard
Col 5h20'28" -39d16'9"
Jean François User
Uma 8h22'18" 62d2'3"
Jean & Françoise
Uma 11h19'37" 61d11'19"
Jean G. Imdorf "Mom-Mom"
Uma 10h49'13" 47d57'31"
Jean G. Kosche
Uma 11h38'52" 49d15'38"
Jean Galanis-Sexton
Tau 5h40'30" 25d56'13"
Jean & George Via star
Lmi 10h9'9" 29d57'10"
Jean Gertner
Uma 10h15'5" 42d10'34"
Jean Ghislain mon amour
Uma 9h52'13" 65d34'15"
Jean Godfrey-June
Umi 15h8'56" 71d18'15"
Jean Godfrey-June
Com 12h41'2" 13d59'4"
Jean Grau's 90th B-Day Star
Vir 13h20'58" 9d13'30"
Jean Gregoire
Lyr 18h31'6" 30d51'5"
Jean Haddad Suez
Her 18h24'59" 17d10'38"
Jean Hammer
Uma 11h57'23" 58d43'4"
Jean Hansen of Felton
Cnc 8h31'57" 27d30'42"
Jean Helena
Com 12h18'30" 28d31'45"
Jean Hennah
Cru 12h17'41" -59d0'2"
Jean Henriette
And 23h48'59" 41d44'44"
Jean Henry
And 23h49'19" 35d37'13"
Jean Hensley
Uma 9h8'51" 50d50'12"
Jean Horton
Tau 5h25'9" 27d56'17"
Jean Hughes The Wildling
Lib 15h5'3" -6d10'13"
Jean I. Boyd
Cas 1h12'21" 63d2'54"
Jean Ivy (Nan)
Cas 23h51'14" 57d31'0"
Jean & Jack Karr
Lyr 18h38'10" 34d11'15"
Jean & Jack Wales
Uma 11h2'48" 41d6'26"
Jean & John
Leo 11h10'51" 16d6'32"
Jean (John) Serbu
Tau 4h27'9" 23d31'28"
Jean Jones
Leo 11h44'27" 24d19'27"
Jean & Julie Hughes
Tau 4h27'40" 29d15'38"
Jean K. Tessmann
Ori 5h31'34" 8d4'37"
Jean K VIII
Vir 13h23'59" 13d37'3"
Jean Kimbrough
Tau 4h51'33" 29d20'39"
Jean Knight
Aql 20h5'54" 11d51'47"
Jean Kouremetis, Light of my Life
Cyg 19h36'34" 30d13'31"
Jean Krum (Peanut)
Umi 14h23'36" 75d40'41"
Jean Krygrowski
Umi 15h31'19" 74d55'17"
Jean L. Hashem
Uma 9h36'22" 52d10'13"
Jean L. MacLeod Batson
Crb 15h48'2" 27d30'16"
Jean L Schumann
Uma 11h32'37" 59d50'3"
Jean Ladutke
Gem 7h2'30" 17d32'22"
Jean Larizza
Uma 9h48'0" 67d31'10"
Jean Le Bohec
Del 20h40'55" 19d53'53"
Jean Leimomi Maile Luning
Aqr 22h26'51" -5d40'39"
Jean Leo
Leo 9h40'52" 27d36'40"
Jean Leslie Freed
And 23h12'20" 51d43'30"
Jean Lorenti
Tau 4h36'36" 27d59'15"
Jean Lorraine Wyns
Aqr 22h31'13" -4d30'16"
Jean Louis Lespagnol
Cma 6h21'10" -21d53'43"
Jean Louise Harpeng-Rice
Sco 16h12'48" -14d23'53"
Jean Louise Hopkins
Vir 13h45'54" -10d29'45"
Jean Louise Lonsinger Hildebrandt
Cas 0h10'28" 54d2'12"
Jean Lucille Sellars
Peg 21h42'39" 27d37'0"
Jean M. Amico
Leo 9h40'56" 29d6'54"
Jean M. Crowley 40th Birthday
Lib 15h19'27" -6d2'28"
Jean M. & Eric A. Hoffrage
Cyg 20h21'57" 56d4'12"
Jean M Fornell
Aql 19h55'42" 14d31'26"
Jean M. Jastrzebski
Uma 10h13'8" 63d1'34"
Jean M. Morris
And 2h21'17" 49d1'29"
Jean M. Moyer
And 1h18'21" 34d51'47"
Jean M. Renner ~ Friend & Grandma
Cam 5h10'28" 62d48'56"
Jean Mannhaupt
Dra 16h31'44" 66d2'55"
Jean Marcia Lamson Davis
Aqr 22h5'45" 2d20'24"
JEAN MARGARET
Tau 5h46'37" 16d11'6"
Jean Margaret - Eighty Sparkling Years
Cru 12h43'17" -56d36'51"
Jean Marie
Psc 1h19'58" 21d4'58"
Jean Marie
And 0h56'12" 35d27'45"
Jean Marie
Lmi 10h11'35" 30d53'30"
Jean Marie Antoinette Iadarola-JNRR
Leo 9h54'55" 19d10'19"
Jean Marie Cash
Lib 15h36'33" -17d40'51"
Jean Marie Comstock
Lib 15h14'37" -15d19'36"
Jean Marie Concetta Raio
Sgr 18h6'20" -30d40'57"
Jean Marie Davidson
Psc 23h6'3" 6d9'27"
Jean Marie Drost
Tau 4h37'27" 1d11'46"
Jean Marie Duffy
Uma 10h39'27" 67d3'17"
Jean Marie Epps Richardson
Sgr 18h28'9" -26d33'15"
Jean Marie Jacoby 6-12-1955
Gem 6h23'12" 27d45'2"
Jean Marie Louis
Lib 15h4'45" -3d29'31"
Jean Marie McBain
Boo 14h55'34" 48d38'54"
Jean Marie Mills Lewis
And 0h56'3" 40d5'54"
Jean Marie (Neana) Dubuc
And 0h55'16" 43d16'0"
Jean Marie Newman
Aqr 22h11'58" -0d22'23"
Jean Marie Payne Beautiful
Crb 15h43'53" 25d45'15"
Jean Marie Rioux
Uma 10h5'59" 52d23'30"
Jean Marie Sillick
Cap 20h11'52" -17d30'59"
Jean Marie Sinclair
Cnc 8h21'16" 17d29'38"
Jean Marie Slaugh
Gem 6h4'0" 24d44'53"
Jean Marie (Zellen) Dewan
Sgr 18h47'33" -22d23'8"
Jean Marsicovete
Ori 5h30'19" -0d11'37"
Jean Mary Joy
Peg 21h30'45" 15d54'53"
Jean Mary Kenny
And 0h11'18" 45d44'49"
Jean Mason Cohen
Tau 5h57'44" 25d48'33"
Jean Mawart
Uma 9h25'37" 48d48'24"
Jean McBride Pate
Cnc 9h4'10" 24d5'15"
Jean McDonald
Cyg 19h57'45" 32d8'4"
Jean McGillan Sweeney
Cas 1h37'11" 61d9'54"
Jean Michele Parry Amo
Aqr 22h40'54" 1d29'19"
Jean Mikota
Cas 1h27'2" 57d5'20"
Jean Miller (Granna)
Ari 2h49'17" 21d13'55"
Jean Mitchell
Umi 13h4'50" 75d12'47"
Jean "Moe" Gagnon
Leo 9h28'44" 13d26'42"
Jean Montano
Sco 17h46'44" -33d3'36"
Jean Murray
And 23h24'22" 46d58'22"
Jean Musker - Our Mum
Cyg 20h10'58" 39d16'16"
Jean Nickels
Psc 1h13'39" 28d59'51"
Jean Nicole
Cap 21h34'5" -12d49'20"
Jean O'Hanrahan
And 23h54'36" 34d49'30"
Jean Oldham Kaspar
Cet 0h52'0" -0d29'12"
Jean One
Ari 2h5'43" 21d39'18"
Jean P. Lash
Per 2h58'41" 51d34'15"
Jean Paddington
Umi 15h12'42" 85d39'32"
Jean Palmer
Lyn 8h9'25" 42d31'13"
Jean Pants
Cnc 8h49'48" 32d18'48"
Jean Paul Lupori
Per 4h24'56" 52d40'21"
Jean Paul Perreault
And 0h19'21" 25d18'56"
Jean Phyllis Bartell Lalagos
Cyg 20h37'29" 35d45'32"
Jean Pierre
Uma 11h58'20" 41d0'57"
Jean Pierre Chanial
Cnc 8h28'56" 15d38'54"
Jean R. Solomon
Lyr 18h34'38" 27d37'5"
Jean Raudenbush
Aql 19h36'43" 10d46'48"
Jean René Bernaudeau
And 23h41'1" 42d8'46"
Jean Renee Martin Schwermer
Lyr 19h6'35" 43d2'51"
Jean Riach & David Walker's Memorial Star
Cyg 21h32'4" 32d46'22"
Jean & Richard H. Battista
Cyg 19h24'40" 51d41'1"
Jean S. Schroeder
Cap 21h11'59" -21d53'47"
Jean Santomero
Cas 0h17'8" 51d55'23"
Jean Shirley Hughes
Crb 15h49'51" 27d34'10"
Jean Singagliese
Uma 10h2'51" 63d30'22"
Jean Sipora Spurgeon
Peg 22h24'43" 22d23'46"
Jean Smith
And 0h42'5" 42d11'47"
Jean Smith
Uma 12h42'35" 54d51'29"
Jean Solis
Leo 10h10'11" 17d55'54"
Jean Stewart Quigley
Uma 11h14'38" 32d57'15"
Jean Strayhorn Baynham
Cas 0h21'36" 57d52'22"
Jean Stucki
Uma 13h49'27" 48d2'33"
Jean Swan
Psc 1h26'18" 11d44'34"
Jean T. Green
Vir 12h9'45" 3d22'17"
Jean & Tony Grisz
Crb 16h15'33" 30d56'29"
Jean & Tullis
Lyn 7h14'31" 53d26'56"
Jean Turner Rushing Harper 8/11/27
Cyg 20h43'12" 36d8'43"
Jean Two
Ari 2h0'24" 21d6'29"
Jean Tytla
Crb 15h36'47" 37d53'33"
Jean V. Watts
Lep 6h0'32" -13d50'6"
Jean Vanden Driessche
Leo 9h29'54" 17d35'44"
Jean Veronese
Cyg 19h49'3" 33d12'22"
Jean W. Woodman
Leo 11h19'47" -2d3'39"
Jean Walrand
Vir 12h8'8" 3d16'52"

Jean - We Love you Mom
Cas 0h25'55" 65d15'33"
Jean White
Uma 9h32'58" 68d5'58"
Jean Williams forever a STAR!
Aql 19h32'12" 11d32'53"
Jean Williams Pierce
Ari 2h31'45" 25d46'39"
Jean Wilson
Cyg 21h50'10" 45d19'51"
Jean Wood
Cep 20h54'55" 68d1'49"
Jean Wright - born July 18th, 1951
And 23h14'33" 50d43'14"
Jean Wyer Batug
Cas 1h39'45" 67d26'59"
Jean Y. Prickett
Cam 4h16'39" 55d47'4"
Jean Yvonne Morrison
Psc 0h41'33" 17d50'38"
Jean Zeckendorf
And 23h13'17" 51d16'43"
Jeana Bell
Srp 18h8'24" -0d3'48"
Jeana Elaine
Lyn 6h48'0" 56d42'54"
Jeana Groose
Sgr 19h40'56" -16d15'29"
Jeana Morgan
Lyn 7h13'5" 51d56'54"
Jeana Renee
Lib 15h41'57" -7d53'58"
Jeanamika
Gem 6h51'47" 17d22'6"
JeanAnn
Cam 4h12'50" 54d47'24"
JeanAnn
And 1h36'24" 42d27'39"
JeanBiatekCox
Gem 7h14'57" 17d23'41"
Jean-brice Ca.Ve
Ori 5h12'3" 0d54'14"
Jean-Claude Canat
Uma 10h37'5" 40d53'18"
Jean-Claude Coutant
Mon 7h29'44" -0d19'33"
Jean-Claude Duhamel
Cas 0h19'52" 47d27'16"
Jeane
Sco 16h6'25" -8d46'56"
Jeane Wessels
Uma 10h19'34" 40d36'52"
Jeanell Santee 122381
And 2h37'40" 44d10'47"
Jeanelle Marie Montez
Psc 1h9'35" 13d9'12"
Jeanepan
And 2h30'48" 45d18'59"
Jeaners16
And 0h27'19" 30d20'34"
Jeanetta
Uma 11h48'41" 64d2'53"
Jeanette
Sgr 18h17'15" -23d8'46"
Jeanette
Tau 4h31'22" 3d35'2"
Jeanette
Vir 13h19'39" 6d1'33"
Jeanette
Ari 2h25'28" 18d58'44"
Jeanette
Cas 0h39'50" 57d1'22"
Jeanette
Cas 1h34'20" 57d56'43"
Jeanette
Cyg 21h33'29" 32d15'43"
Jeanette "100"
Cyg 20h45'21" 37d37'2"
Jeanette A. Duffy
Lyn 7h35'42" 40d0'43"
Jeanette and Bernie
Lyn 6h54'41" 53d37'47"
Jeanette and Rod
Leo 9h29'15" 12d17'51"
Jeanette Andrea Kopaczewski
Lyn 8h36'21" 33d18'46"
Jeanette Behrens
Cam 4h39'25" 65d12'38"
Jeanette & Bonnie Brillhart
Per 3h27'10" 43d15'40"
Jeanette C Brox
Uma 10h48'33" 46d29'17"
Jeanette Cheatham #1 Mom
Aqr 22h51'11" -22d22'28"
JEANETTE CHRISTINA GONZALEZ
Psc 1h26'9" 5d36'57"
Jeanette Clark
Ori 5h49'14" -2d27'53"
Jeanette D. Dedvukaj
Uma 13h46'11" 53d17'34"
Jeanette D West
Cnc 9h5'1" 31d4'21"
Jeanette E. Fernstrom
Cas 0h41'18" 57d31'36"
Jeanette Elizabeth
Uma 13h41'3" 61d31'50"
Jeanette Feldpausch
Ori 6h11'14" 15d42'42"
Jeanette Ferguson Nestor
Ari 2h47'36" 20d28'32"

Jeanette Flournoy
Cam 3h54'4" 56d14'18"
Jeanette Fouste
Uma 9h26'23" 45d47'13"
Jeanette Kahala
Aql 19h47'31" 3d57'57"
Jeanette L. Allen
And 23h11'11" 50d49'45"
Jeanette Lawrence
Cap 21h15'13" -16d34'48"
Jeanette Levine
Lyn 8h28'3" 39d7'15"
Jeanette Lockridge
Cas 0h33'28" 54d4'39"
Jeanette Lorinz
Cap 21h52'51" -20d1'56"
Jeanette Lynn Snell
Vul 19h49'0" 22d52'40"
Jeanette Lyons
And 0h45'0" 26d34'11"
Jeanette Malecki
Ari 3h17'23" 29d7'5"
Jeanette Marie Nicholson
Psc 1h38'26" 20d40'20"
Jeanette Marie Recco
Aqr 21h3'35" 2d24'56"
Jeanette McGowan
Aqr 23h6'2" -7d52'28"
Jeanette Moorer
Lib 14h47'34" -8d3'7"
Jeanette Murray
Uma 13h42'53" 57d56'57"
Jeanette ("Nettie") Leah Tollefson
Cnc 8h8'34" 16d30'28"
Jeanette Nicole Wear
And 0h40'24" 31d23'41"
Jeanette Peck
And 1h56'15" 45d10'44"
Jeanette Regina Brillhart
Cas 1h24'36" 52d8'18"
Jeanette Renee Niles
Sgr 19h9'44" -15d5'3"
Jeanette Shaffer
Uma 10h40'28" 67d59'3"
Jeanette Shifflett
Gem 6h16'19" 22d57'7"
Jeanette Siegel My Loving Wife
Cap 21h43'17" -13d34'2"
Jeanette (S.M.)
Cru 12h26'0" -57d39'57"
Jeanette Sonnenberg
Cas 23h20'15" 53d51'35"
Jeanette Tayamen
Cas 23h27'11" 56d31'56"
Jeanette Taylor
Uma 10h24'30" 59d51'25"
Jeanette Teitelbaum
And 0h49'6" 41d53'17"
Jeanette Tetlow
Uma 11h13'13" 58d46'41"
Jeanette & Verdun Boots
Cyg 20h1'34" 30d8'24"
Jeanette Wadsworth
Uma 8h17'52" 71d59'30"
Jeanette Wieczorek
Gem 7h18'46" 31d20'36"
Jeanette Woodward
Lyr 18h53'14" 43d36'10"
Jeanette0404
Mon 6h35'22" 10d52'46"
Jeanette's Eternal Promise
Leo 9h42'20" 31d54'26"
Jeanette's Joy
Cnc 8h14'36" 23d24'52"
Jean-Francois Deschenes
Per 3h46'53" 51d12'54"
Jean-Frédérik Martin
Cep 20h47'53" 60d35'33"
Jean-Guy Bigras
Dra 17h50'52" 67d8'31"
Jeani Thomson, Purple Hearts
Ori 6h9'14" 6d58'9"
Jeanic Thibeaudeau-Guérin
Cas 0h36'51" 58d27'36"
Jeanie
Cnc 9h4'15" 9d10'5"
Jeanie
Tau 4h14'17" 16d6'14"
Jeanie
And 2h21'26" 49d8'49"
Jeanie
And 23h24'51" 51d46'7"
Jeanie Bear
And 0h46'4" 36d13'48"
Jeanie Clinton
Aqr 23h29'56" -7d16'44"
Jeanie Dotlich
Lyr 18h25'48" 31d32'51"
Jeanie Goodnature
And 0h29'46" 45d24'34"
Jeanie Hershey
Lyr 19h0'5" 42d39'34"
Jeanie Lacey Weglarz
Lyn 8h8'13" 45d37'40"
Jeanie Rachael Brooks
Gem 6h36'39" 14d7'58"
Jeanie27
Peg 22h45'52" 21d29'36"
Jeanina Boone
Aqr 23h48'17" -15d2'4"
Jeanine
Uma 10h32'6" 68d12'50"

Jeanine
Psc 23h55'43" 7d32'4"
Jeanine
And 0h45'37" 42d32'37"
Jeanine
And 0h25'4" 38d25'35"
Jeanine Carolyn
Aql 19h33'16" 10d42'2"
Jeanine Caruso
And 2h24'55" 45d23'34"
Jeanine Cerveny
And 1h17'13" 41d56'12"
Jeanine Cirillo
Cas 0h13'42" 61d26'49"
Jeanine Cumiskey
Uma 10h6'39" 60d43'38"
Jeanine Gilmore
Umi 15h35'33" 71d14'10"
Jeanine Hackman
Lyr 18h30'40" 36d21'20"
Jeanine Heckman
Uma 13h37'44" 56d30'38"
Jeanine Hydrick
Vir 12h29'23" 6d5'30"
Jeanine Lynn
And 0h37'37" 37d3'54"
Jeanine M. Green
Dra 14h45'54" 59d13'4"
Jeanine Marie Furia
Sgr 18h33'26" -36d27'34"
Jeanine Mary
And 0h22'42" 38d40'2"
Jeanine Michele Mitchell
Aqr 22h31'47" -0d13'55"
Jeanine Needles
And 0h39'23" 45d52'53"
Jeanine Petitpas
Uma 10h1'24" 67d7'19"
Jeanine R. Kroeger
Cap 21h38'2" -19d43'48"
Jeanine Rubino
And 23h10'55" 43d26'19"
Jeanine Tweedie
Uma 13h16'58" 63d4'54"
Jeanine Walling
Cyg 19h55'5" 36d22'59"
Jeanine Willis
Leo 10h59'49" 14d17'55"
Jeanine's Shining Star
Tau 4h9'34" 15d30'15"
Jeanine's Star of Love
Col 5h42'31" -31d54'47"
Jeanious Imbecilious
Vir 14h49'42" -20d42'23"
Jean-Jacques Blaise
Ori 4h46'11" -3d39'50"
Jean-Jacques Haye
Tau 4h28'8" 28d48'12"
Jean-Jacques Shingling
Sco 17h51'57" -36d24'32"
jeanjenks04131956-09172006
Ori 6h21'6" 16d25'4"
Jean-Louis Avril
Cas 0h55'32" 54d20'56"
Jean-Louis Le Quintrec
Cnc 7h58'27" 10d51'9"
JEAN-LUC FEUZ
Boo 14h32'48" 42d18'44"
Jean-Luc Fourniou
Lib 14h51'40" -11d57'41"
Jean-Luc Matussiere
Peg 21h29'8" 9d45'29"
Jean-Marc Humberset
Leo 11h44'2" 23d56'2"
Jean-Marc Laglasse
Uma 10h11'23" 44d8'37"
Jean-Marc Rioux
Cep 0h11'46" 69d38'2"
Jeanmarie
And 1h31'19" 46d52'18"
JeanMarie
Tau 4h39'21" 21d42'54"
JeanMarie Danese
Lyn 6h32'20" 57d27'45"
Jean-Marie Primaux !!!
Cam 5h21'27" 67d19'57"
Jean-Marie Zimoch
Aqr 21h7'7" -8d56'37"
Jean-Melissa
Umi 15h58'28" 78d41'20"
Jeanna
Vir 12h56'37" 6d50'18"
Jeanna
And 1h47'18" 38d1'2"
Jeanna Adams
Cap 20h27'54" -22d52'19"
Jeanna D. Viola
Sco 16h18'22" -11d2'30"
Jeanna lee
Aqr 23h5'2" -8d47'32"
Jeanna Lee Carden
Ari 2h44'22" 16d3'24"
Jeanna Marie Harvey
Ori 5h8'43" 3d1'17"
Jeanna Middleton
Cas 1h25'54" 61d11'5"
Jeanna Nicole Pagnotta
Cap 21h49'38" -11d41'14"
Jeanna Ravelo
Gem 7h44'35" 33d43'22"
Jeanne
And 1h6'11" 37d40'39"
Jeanne
Dra 18h43'13" 51d25'55"

Jeanne
Com 12h19'15" 27d55'51"
Jeanne
Crb 15h51'54" 25d48'20"
Jeanne
And 0h14'9" 32d26'42"
Jeanne
Ari 3h12'23" 29d40'25"
Jeanne
Cnc 8h10'36" 24d39'26"
Jeanne
Ari 2h6'55" 24d17'22"
Jeanne
Lib 15h7'27" -21d3'31"
Jeanne
Vir 14h6'2" -7d49'44"
Jeanne
Vir 13h35'24" -4d10'49"
Jeanne
Lib 14h53'26" -4d21'29"
Jeanne
Uma 10h14'12" 54d54'35"
Jeanne
Uma 10h3'1" 53d30'45"
Jeanne
Uma 10h8'56" 54d47'16"
Jeanne
Uma 10h37'18" 60d47'18"
Jeanne 2
Cyg 21h31'10" 49d32'13"
Jeanne 92815
Uma 14h3'50" 60d26'20"
Jeanne Adele Bear
Uma 11h30'40" 40d39'3"
Jeanne Amber
Leo 9h26'8" 26d9'37"
Jeanne and Bruce Marks
Uma 11h59'6" 36d25'41"
Jeanne and Carl Schulze
Cyg 21h27'47" 39d45'21"
Jeanne and Carlos
Cap 21h15'6" -15d45'1"
Jeanne and Liborio
Tau 6h0'2" 27d16'27"
Jeanne and Paul
Sge 19h11'55" 16d52'6"
Jeanne Ann O'Donnell
Vir 13h21'30" -13d21'12"
Jeanne Anne Sheets Bender
Leo 9h34'41" 27d35'56"
Jeanne B. Goodwin
Uma 11h29'57" 54d28'51"
Jeanne Bean
Lib 15h46'8" -19d17'9"
Jeanne Bishop
Lib 14h50'21" -14d41'47"
Jeanne Bourdon
Lyn 6h18'33" 60d49'6"
Jeanne Bug
Dra 12h41'12" 73d2'34"
Jeanne Burrill Raymond
And 1h5'26" 45d44'35"
Jeanne Byrd
Lmi 10h11'37" 37d11'2"
Jeanne Byrd
Leo 10h48'39" 13d57'9"
Jeanne C. Beard
Uma 11h39'46" 47d51'9"
Jeanne C. Varnum Estes 6/19/1964
And 2h19'30" 48d45'35"
Jeanne Carney
Leo 9h29'48" 28d37'40"
Jeanne Carolyn Abrams
Sgr 18h8'21" -27d52'1"
Jeanne Claire Eagleson
Cap 20h23'6" -24d55'59"
JEANNE COGNATO
And 2h23'11" 48d37'38"
Jeanne D Eyre
Sco 16h42'54" -32d30'54"
Jeanne d'Arc L.M.
Vir 13h18'7" 4d16'21"
Jeanne Dawes
Cyg 20h42'6" 35d29'36"
Jeanne Delia Lucé Sciubba
Lib 14h33'55" -9d59'45"
Jeanne Delia Sciubba
Lib 14h49'43" -2d13'5"
Jeanne Demore
Psc 1h26'19" 11d33'47"
Jeanne Dodson
Aqr 22h4'4" 0d16'54"
Jeanne E. Day
Cas 0h16'19" 52d25'6"
Jeanne Eloise
Aqr 23h33'46" -13d18'6"
Jeanne Emma Gressier Segall
Aqr 20h48'12" -8d18'24"
Jeanne Fanelly Coyne
Lib 14h47'36" -6d44'9"
Jeanne Filbert - Shining Star
And 2h18'40" 45d11'50"
Jeanne Flowers
Lyn 7h33'40" 53d15'58"
Jeanne Frances Free - Happy 80th Mum.
Tau 3h39'17" 28d28'7"
Jeanne G Chiaccio
Cas 2h8'14" 62d26'46"
Jeanne Gable
Cyg 20h25'15" 44d26'48"

Jeanne Gable and Geoff Anders
Uma 10h13'55" 66d8'12"
Jeanne Garon
Dra 16h39'3" 58d53'15"
Jeanne Hargrove
Uma 10h9'42" 48d42'56"
Jeanne & Harold
Cyg 20h33'10" 41d13'0"
Jeanne Hazel Nunes
Uma 11h12'38" 46d7'5"
Jeanne HMD Our Star Mother of 3
And 0h15'50" 27d51'43"
Jeanne Hopp
Ori 6h10'2" 15d8'12"
Jeanne L. Chandli
Sex 10h49'3" -0d11'58"
Jeanne La Barre Jones
Lyn 9h13'59" 42d30'47"
Jeanne LeBold
Lib 15h4'22" -12d47'49"
Jeanne M Howard
Cam 3h51'45" 59d51'13"
Jeanne M. Purtill
Aql 19h57'13" 10d40'27"
Jeanne Madonna
Cyg 20h1'18" 37d11'31"
Jeanne Mantell
Sco 17h10'57" -43d11'28"
Jeanne Marie
Sgr 19h56'12" -25d7'19"
Jeanne Marie
Sgr 19h14'6" -22d15'4"
Jeanne Marie
Cam 4h13'21" 70d47'58"
Jeanne Marie
Per 2h50'50" 53d18'3"
Jeanne Marie
Uma 13h54'13" 50d6'50"
Jeanne Marie Beaver Stefanik
Uma 8h25'25" 65d24'12"
Jeanne Marie Connor
Lyn 7h33'33" 54d7'59"
Jeanne Marie Farley
Psc 23h33'30" 4d42'57"
Jeanne Marie Giacinto
Gem 7h41'39" 32d0'14"
Jeanne Marie Hoff Neuman
Cas 0h40'20" 57d19'8"
Jeanne Marie Jacqueline Lenfest
Sgr 19h38'48" -13d31'27"
Jeanne Marie Silva
And 0h1'50" 39d11'37"
Jeanne Marie Stephens Izant
Vul 20h19'50" 23d57'44"
Jeanne McCarthy
Aqr 22h14'18" -1d47'33"
Jeanne McNamara
Aur 6h34'2" 34d21'14"
Jeanne My Love
Ori 5h56'1" 6d16'56"
Jeanne Nicole Giles
And 2h20'19" 47d55'37"
Jeanne Niewoehner
Cas 0h30'0" 65d17'45"
Jeanne Ouellet - My North
Umi 12h9'38" 88d36'48"
Jeanne P. Steele "Mom"
Peg 22h21'50" 4d34'30"
Jeanne Palmer Goodwin
Vir 12h53'5" -8d51'12"
Jeanne & Papa Gearhart
Cyg 21h38'9" 32d5'33"
Jeanne Patricia Bundrick
Tau 4h23'52" 17d17'23"
Jeanne Razzano
Cnc 8h22'14" 20d25'55"
Jeanne Ro
Tau 5h15'45" 27d49'17"
Jeanne Rolland
Umi 13h51'22" 70d57'39"
Jeanne S. Einselen
Crb 15h27'13" 30d43'23"
Jeanne Schumann
Ori 5h53'56" 7d23'56"
Jeanne Schwarzentraub
Aqr 21h47'32" -7d7'55"
Jeanne Seese
Lyn 7h55'18" 38d42'15"
Jeanne Sewell
Cyg 21h57'19" 48d43'51"
Jeanne Sobine, Sister and Aunt
Crb 16h11'48" 39d3'16"
Jeanne the Bean
And 1h36'25" 47d9'31"
Jeanne Weaver
Cap 21h51'46" -14d25'14"
Jeanne Williams
Tau 3h37'32" 29d9'24"
Jeanne Williamson Shelley
Cnc 8h11'8" 25d19'20"
Jeanne Z.
And 2h25'32" 49d44'46"
Jeanne-Marie Zusy Cote
And 1h7'15" 36d26'50"
Jeannene Allen
Leo 9h44'51" 6d46'42"
Jeannene Annette Tanielian
Uma 9h21'19" 61d47'4"

Jeanne's star
Lib 15h16'30" -6d13'31"
Jeanne's Tender Twinkle
Sco 16h10'19" -10d19'3"
Jeannet et Fleur
Uma 8h26'41" 61d8'5"
Jeannette
Cap 21h32'39" -23d39'1"
Jeannette
Gem 6h48'34" 23d3'19"
Jeannette
Gem 6h55'23" 22d22'15"
Jeannette
Cnc 8h17'43" 32d20'33"
Jeannette
Lyr 18h40'57" 38d8'10"
Jeannette
Uma 11h37'55" 45d30'49"
Jeannette 2004
Uma 9h7'35" 55d36'13"
Jeannette 9-22
Uma 8h57'16" 50d40'43"
Jeannette Brenner
Her 17h23'56" 36d18'2"
Jeannette Carlisle
Sgr 18h28'45" -16d23'24"
Jeannette & Dominik
Umi 14h1'30" 69d29'32"
Jeannette "Jan" Goff
Ori 4h46'44" 9d3'22"
Jeannette Johnson
Lyn 6h46'39" 61d41'57"
Jeannette Lteif
Lib 15h53'22" -20d15'31"
Jeannette M. Chandler "My Red"
Leo 11h17'47" 22d48'45"
Jeannette M. Gould-Schmidt
Cas 23h31'28" 52d55'40"
Jeannette Marie
Uma 11h50'44" 50d15'20"
Jeannette Marie
Cas 1h19'0" 56d29'8"
Jeannette Marshall
Aql 19h15'20" 15d13'8"
Jeannette Maureen Morris
Ori 6h17'8" 14d58'21"
Jeannette "Memere" Lecuyer
Uma 9h46'4" 51d50'46"
Jeannette Miriam
Cap 20h34'55" -27d7'29"
Jeannette Nanavati
Sco 16h9'46" -22d59'2"
Jeannette R. Maass
And 1h23'4" 45d24'17"
Jeannette Riopelle
Cas 1h16'54" 67d8'28"
Jeannette R.O.C. 716
Dra 17h39'2" 53d18'53"
Jeannette Rydzewski
Cnc 9h7'41" 25d23'4"
Jeannette Sanchez
And 23h4'51" 38d15'55"
Jeannette & Sebastian Droese
Cap 21h57'6" -14d11'26"
Jeannette Seemann
Aqr 21h45'4" -0d27'8"
Jeannette's Light
Cas 2h14'20" 72d26'20"
JeanneYoho
And 23h21'33" 44d35'53"
Jeanni
Vul 19h31'30" 25d59'33"
Jean-Nicolas et Vanina
Uma 9h40'45" 45d38'18"
Jean-Nicolas Maltais
Uma 13h29'28" 62d0'37"
Jeannie
Cas 23h51'22" 53d27'21"
Jeannie
Vir 13h33'19" -17d34'26"
Jeannie
And 23h25'30" 37d56'3"
Jeannie
Crb 15h53'51" 34d41'6"
Jeannie
Lmi 10h41'39" 30d54'29"
Jeannie
And 0h21'42" 26d13'47"
Jeannie
Leo 10h1'47" 16d36'52"
Jeannie Allison Blake
And 2h17'53" 44d57'28"
Jeannie Anderson
Leo 10h26'45" 17d52'53"
Jeannie Anderson
Ari 2h0'42" 24d38'57"
Jeannie Charlee Jorg !!!
Cyg 21h15'58" 42d43'59"
Jeannie & Charles Maroosis
Crb 15h50'45" 35d40'57"
Jeannie Cross
Uma 13h21'52" 60d0'11"
Jeannie Glisson-Davis
Umi 14h57'8" 76d59'14"
Jeannie Grzelak
Leo 10h51'23" -2d29'14"
Jeannie Jones
And 22h59'16" 39d17'6"
Jeannie Kelso
Psc 23h19'31" -1d50'42"

Jeannie Kim
And 1h4'18" 36d15'22"
Jeannie Lan-Thi Boris
Psc 1h23'20" 17d34'44"
Jeannie "Lil' Piece of Heaven"
And 0h1'28" 45d12'19"
Jeannie Louise Thomas
Cas 1h19'57" 63d28'12"
Jeannie Lynn Parrott
And 22h58'17" 50d15'30"
Jeannie Lynn Ulmer-Boehme
Ori 5h52'40" 12d18'46"
Jeannie M. Hauner
Gem 6h3'27" 21d57'38"
Jeannie "Ma" Foss
Uma 8h25'38" 63d26'28"
Jeannie Marie
Uma 12h16'40" 62d56'15"
Jeannie Mia-Auerbach
Ari 2h33'56" 25d5'5"
Jeannie & Mike Smith
Cyg 20h46'34" 37d0'14"
Jeannie Morrison
Lib 15h39'17" -5d1'30"
Jeannie Nicole Guzman
Leo 10h53'30" 13d59'36"
Jeannie Richardson
Vir 13h35'18" -2d34'29"
Jeannie Rose
Mon 6h45'14" -0d22'36"
Jeannie Sills' Star
Srp 18h16'49" -0d16'20"
Jeannie & Ted Scalise
Peg 22h10'36" 10d31'53"
Jeannie & Tim Somos Siempre Juntos
Uma 9h19'7" 60d40'7"
Jeannie Van Dyk
Mon 7h1'49" -9d11'49"
Jeannie Wade
Dra 16h19'41" 52d21'13"
Jeannie Wadley
Cyg 21h10'42" 41d28'26"
Jeannie-Girl
Cas 1h16'21" 68d20'26"
Jeannie's Jewels: Adam, Mona &John
Ori 5h47'55" 6d22'49"
Jeannie's Lilypalooza
And 2h19'12" 48d50'39"
Jeannie's Twinkle
Uma 13h52'29" 56d57'30"
Jeannine
Cas 23h48'46" 56d24'27"
Jeannine
Sco 16h17'8" -11d28'7"
Jeannine
And 0h29'32" 45d21'9"
Jeannine
Boo 15h21'5" 49d3'51"
Jeannine
Psc 23h20'37" 3d33'8"
Jeannine All My Love Always
Leo 9h29'53" 31d49'53"
Jeannine Bassetti
Cas 0h29'7" 63d58'38"
Jeannine de Port Lyautey
Peg 21h39'48" 27d55'31"
Jeannine et Philippe
Leo 10h11'41" 13d24'6"
Jeannine Horvath
Uma 13h38'5" 54d0'34"
Jeannine Kay Brockert
Cas 0h25'51" 52d56'9"
Jeannine Lambeth
Peg 22h38'13" 26d37'53"
Jeannine Levy
Aqr 22h10'55" -14d9'25"
Jeannine Louise Murrell
Cap 21h53'45" -20d19'29"
Jeannine Marie Daleas
And 2h5'29" 37d46'20"
Jeannine Marie Keasley
Cas 0h26'58" 62d50'32"
Jeannine Miceli Martin
And 23h19'43" 49d53'42"
Jeannine*Midnight Shopper*
Gem 6h46'4" 19d19'11"
Jeannine Monique Pedroza
Leo 9h56'21" 21d30'17"
Jeannine Nelson
Cap 20h10'38" -15d25'41"
Jeannine Renee - Misty - Sophie
Her 16h36'6" 14d48'57"
Jeannine Vogler
Cas 1h31'19" 71d48'43"
Jeannine Wade
Psc 0h28'9" 3d40'10"
Jeannine Willis
Vir 11h38'32" -4d54'5"
Jean-Noël Bosserelle
Vir 13h12'55" 11d52'43"
Jeannthe
Dra 17h58'13" 59d43'43"
Jeanny
Cep 21h41'8" 71d44'14"
Jean-Paul Ear & Lucie Mai-Truc Ho
Del 20h29'22" 16d40'58"

Jean-Paul Genevrier
Uma 11h27'5" 38d32'5"
Jean-Paul & Rosina Couture
And 0h28'54" 38d18'47"
Jean-Paul Saija
Cas 23h30'9" 58d41'23"
Jean-Philip Delbano
Aqr 22h19'51" -12d53'5"
Jean-Pierre
Her 18h6'26" 29d14'12"
Jean-Pierre Barbault
Lyn 7h41'2" 53d25'9"
Jean-Pierre Boladian II
Vir 13h45'6" -8d3'7"
Jean-Pierre Chanet
Uma 10h10'7" 44d55'11"
Jean-Pierre Marie Capredon
Leo 10h17'15" 20d13'30"
Jean-Pierre Smigielski
Ori 5h53'22" 2d47'36"
Jean-René Tocqueville
Uma 10h33'9" 40d26'13"
Jean's Angel
Aqr 22h58'50" -7d55'7"
Jeans Girls
Lyn 7h42'28" 38d24'51"
Jean's Golden Star
Gem 6h3'5" 25d42'48"
Jean's Little Bit of Heaven.
Leo 10h6'33" 16d59'16"
Jean's Shining Light
Crb 16h8'42" 33d17'44"
Jean's Shining Star
And 0h27'25" 39d32'31"
Jean's Shining Star
Sco 17h46'47" -42d35'57"
Jean's Star
Uma 10h17'15" 66d51'23"
Jean's Star
Crb 16h18'58" 30d11'59"
Jean-Sébastien Barbe
Cep 20h47'10" 65d18'56"
JeanV
Lyn 7h56'39" 58d1'12"
Jeanvo
Vir 14h15'54" 4d36'11"
Jean-Yves
And 23h29'29" 44d55'38"
Jear Bear's Twilight
Uma 11h18'21" 45d44'53"
Jeara
And 0h26'44" 32d31'16"
Jearsionn
Agr 22h2'54" -3d9'8"
J-East Trading Sdn Bhd
Ori 5h31'46" 2d54'5"
Jeb
Lib 14h57'0" -4d21'31"
(JEB) Janice & Eugene Barlow
Crb 16h8'12" 34d10'48"
JEBB
Umi 15h11'8" 81d10'49"
Jebbie
Aqr 22h5'8" -15d56'58"
Jebeney Turrill
Mon 6h52'34" 8d12'59"
Jebrille Walls
And 2h4'25" 42d25'2"
JEC1961
Ari 2h55'36" 10d30'28"
Jeca bebica - moja zvezdica
Cas 22h59'9" 58d41'39"
Jeca Pereca
Cam 4h19'23" 56d57'59"
Jecca's Piece of the Heavens
Aqr 21h41'25" -1d56'1"
Jecedes
Dra 16h14'9" 54d33'35"
"Jed"
Lib 15h20'59" -7d13'58"
Jed
Uma 10h56'5" 38d46'46"
Jed Coleman Cornelison
Sco 16h10'37" -16d3'30"
Jed Hudson
Psc 0h27'24" 4d26'13"
Jed Independence
Dra 20h27'23" 70d22'39"
Jed Rosenzweig & Meghan O'Neil
Cyg 19h25'53" 54d51'2"
Jed W. Bernstein
Her 17h46'58" 47d42'16"
Jed Wayne Mayes
Lib 15h21'30" -12d55'39"
Jed Wells 76
Her 17h58'33" 29d21'18"
Jedadiah
Ori 5h4'14" 3d45'41"
JedDawg
Cma 6h50'2" -16d31'35"
Jedde
Umi 14h4'27" 69d0'45"
Jeddediah Del Mar Hazard
Aur 5h17'58" 35d58'34"
Jedediah Seth Fancher
Cep 21h6'30" 57d16'12"
Jedeen
And 0h10'55" 41d3'17"
Jedge
Ori 6h19'59" 18d48'57"
Jedi 76
Per 2h59'55" 40d52'34"
Jedi de Coeur
Cas 0h56'7" 53d56'35"
Jedi James
Ari 2h34'30" 28d39'45"
Jedi Knight Chris
Aqr 21h8'25" -13d31'22"
Jedi Master Samuel Brett Smith
Lib 15h14'58" -22d46'33"
"Jedi Master Teffet" Tevfik
Uma 10h44'12" 54d21'27"
Jediah Darrell Nathaniel Webb
Lmi 10h28'40" 37d58'17"
Jedidiah
Vir 13h54'44" -6d36'16"
Jedidiah and Lauren
Cyg 20h23'58" 48d54'41"
Jedina moja ljubav Ivana
Cas 23h14'46" 54d16'23"
Jedrus
Sco 16h9'33" -13d14'55"
Jed's Star
Cap 20h15'50" -10d1'49"
Jeena's Crown
Aur 7h30'3" 43d24'40"
Jeenie Beenie
Cas 23h28'57" 51d21'15"
Jef50
Uma 10h39'2" 55d0'30"
Jef7rey Scott Hildner
Lib 15h57'46" -20d6'5"
JeFa
Sco 17h50'19" -44d37'0"
Jeff
Uma 12h18'4" 53d55'46"
Jeff
Uma 12h27'36" 62d26'44"
Jeff
Uma 11h13'26" 32d39'20"
Jeff
Her 18h37'9" 20d9'40"
Jeff
Tau 5h53'11" 24d44'52"
Jeff~60
Dra 17h27'34" 50d44'52"
Jeff Akins
Cep 21h51'32" 69d29'47"
Jeff Allan Rupprecht
Cap 21h38'28" -20d18'27"
Jeff Ambrose
Ori 6h15'33" -0d21'52"
Jeff and Ashley Martin
Cyg 21h17'0" 51d45'58"
Jeff and Cary's Love
Uma 9h36'9" 52d23'31"
Jeff and Crystal Doerr
Cyg 19h47'54" 38d5'16"
Jeff and Danie forever
Her 18h35'52" 22d38'50"
Jeff and Donna's 25th Anniversary
And 0h20'35" 30d41'52"
Jeff and Erin Abbott
Cyg 22h1'48" 53d31'24"
Jeff and Gai Forever 26/01/2007
Cru 12h28'31" -60d9'2"
Jeff and Jamie
Cyg 21h28'16" 51d58'1"
Jeff and Jean Grossman
Uma 11h34'47" 57d37'53"
Jeff and Jenn~ Forever Lovers
Crb 15h58'51" 35d32'39"
Jeff and Jodi Harris
Cyg 19h48'39" 29d42'38"
Jeff and Julia Miller
Uma 10h57'59" 42d27'54"
Jeff and Karen's Anniversary Star
Cyg 20h30'20" 37d35'34"
Jeff and Kate
Crb 15h44'50" 26d46'10"
Jeff and Kathy
Cma 7h11'30" -27d51'19"
Jeff and Kenna
Cap 21h41'44" -17d53'58"
Jeff and Laurie
Cap 21h7'0" -19d39'37"
Jeff and Lisa
Cap 21h28'25" -22d45'58"
Jeff and Lubna Roberts
Pho 1h13'0" -43d40'36"
Jeff And Marci, Amelia Island
Gem 6h56'22" 18d48'42"
Jeff and Maria Munro
Ari 3h5'47" 28d58'52"
Jeff and Mona
Cap 20h24'18" -26d7'26"
Jeff And Nancy
Cyg 20h21'44" 55d59'44"
Jeff and Newland Campbell
And 0h15'35" 26d58'39"
Jeff and Sandra Worstell
Lyn 8h16'29" 58d52'0"
Jeff and Tammy Always and Forever
Her 18h40'23" 13d9'40"
Jeff and Terry in the Stars Forever
Ori 5h17'1" 4d31'18"
Jeff and Tracey
Uma 11h41'45" 39d54'22"
Jeff Anderson and Pole' Jensen
Cyg 20h24'15" 57d41'8"
Jeff Arthur
Sco 17h30'58" -38d27'15"
Jeff B
Dra 18h47'40" 75d21'31"
Jeff Balcer
Ori 4h51'2" 14d6'3"
Jeff Baxter
Psc 1h15'23" 18d8'43"
Jeff Belair
And 0h16'56" 46d21'54"
Jeff & Beth Ellen Gallino
Cyg 20h20'54" 52d45'15"
Jeff & Betsy Cole
Lyn 8h33'3" 45d1'52"
Jeff 'Bubs' Wade
Psc 1h39'23" 20d26'0"
Jeff C. Langston
Dra 17h9'15" 60d12'55"
Jeff Calvin
Ari 2h44'36" 27d42'29"
Jeff Cantamessa
Sco 16h7'32" -16d5'9"
Jeff Canter
Uma 9h0'43" 72d23'54"
Jeff + Celina Forever
Cyg 21h24'16" 43d58'43"
Jeff Chandler Woods
Ari 1h58'3" 21d53'51"
Jeff & Christine White
Cyg 19h55'29" 33d21'48"
Jeff Cohen
Vir 12h48'58" 2d20'34"
Jeff Comment
Uma 9h21'34" 61d34'4"
Jeff & Cristina's Sexy Star
Cyg 21h21'13" 46d58'1"
Jeff D Martin
Dra 16h40'57" 74d2'17"
Jeff *DADDY*
Leo 11h54'49" 20d2'16"
Jeff Davis, number 44
Aql 19h50'49" -0d30'33"
Jeff Divito
Crb 15h51'29" 30d57'23"
Jeff & Donna's Journey To True Love
Sgr 18h30'30" -16d12'32"
Jeff Edward McCormick
Ari 3h19'34" 27d50'18"
Jeff Ellner
Per 3h27'3" 33d48'14"
Jeff Elms
Her 17h4'3" 16d10'14"
Jeff Farr
And 0h25'53" 37d32'0"
Jeff Ferguson
Vir 14h52'8" 0d43'30"
Jeff G Eisenmann
Ori 6h5'25" 17d24'10"
Jeff Grater
Aql 19h30'18" 7d55'40"
Jeff Graydon
Cap 20h33'22" -14d7'3"
Jeff Greenberger (Doc)
Dra 19h27'47" 59d14'27"
Jeff Greenberger (Doc)
Ori 5h33'52" 3d37'20"
Jeff Greenberger (Doc)
Per 3h11'5" 54d5'45"
Jeff Hamil
Per 3h28'1" 43d30'4"
Jeff Haslam
Tau 4h40'57" 20d59'28"
Jeff & Heather
Tri 2h30'36" 31d34'17"
Jeff Hilsher
Uma 10h37'45" 61d6'23"
Jeff Hiscock
Uma 8h40'46" 68d0'10"
Jeff Horn
Uma 11h11'32" 61d46'50"
Jeff Howard
Tau 4h46'34" 19d29'36"
Jeff J. Petersen
Ori 5h32'17" 6d47'51"
Jeff + Jamie Prei
Lyn 7h53'7" 39d30'48"
Jeff & Jamie's Unchained Melody
Cyg 20h23'35" 56d4'36"
Jeff & Janet Filippi
Eri 4h48'48" -14d22'44"
Jeff Jarka
Aql 19h38'9" 4d23'46"
Jeff Jenkins
Dra 17h5'5" 63d8'18"
Jeff & Jill forever
Boo 14h36'58" 34d35'18"
Jeff & Joanne 25 Years And Counting
Cyg 21h16'39" 46d3'9"
Jeff & Joel
Psc 0h12'50" 5d43'24"
Jeff Johnson
Ari 1h57'29" 22d49'14"
Jeff Johnston
Cnc 8h13'3" 24d58'48"
Jeff Jordan
Cap 21h42'46" -20d42'40"
Jeff & Judy Stone's Star
Psc 0h9'49" 4d8'50"
Jeff & Kara Kreer ~ Eternal Love
Leo 10h12'22" 15d24'53"
Jeff & Kaylee
Cyg 20h43'30" 43d43'48"
Jeff & Keli
And 23h19'56" 44d38'15"
Jeff Kesler
Tau 4h17'41" 9d21'7"
Jeff Kinney-You completed my heart!
Her 17h21'44" 27d59'39"
Jeff Knutson
Uma 11h47'12" 47d6'41"
Jeff Kreidenweis
Leo 10h53'52" -1d4'0"
Jeff & Kristen Forever 12-25-2002
Ori 5h27'25" 3d17'53"
Jeff Kubarych
Gem 7h42'45" 33d52'57"
Jeff & Kym
Gem 7h37'3" 31d59'25"
Jeff & LaDonna Amberg
Uma 9h47'49" 51d27'18"
Jeff Lee Cox
Uma 11h42'9" 29d27'53"
Jeff Letellier
Her 17h9'31" 36d55'5"
Jeff Lewsadder
Ori 4h53'49" 1d30'41"
Jeff & Linda
Cen 13h28'8" -37d34'17"
Jeff & Linda Barnes
Cyg 20h45'54" 52d11'27"
Jeff Lindenman
Ari 2h45'36" 20d33'23"
Jeff Linhares
Aur 5h25'4" 33d25'46"
Jeff Lochow
Her 17h46'14" 38d26'7"
Jeff Long
Her 17h18'46" 35d47'45"
Jeff Loves Mary
Cyg 21h25'39" 54d59'27"
Jeff Lumia
Uma 13h57'22" 56d26'7"
Jeff Lyzenga
Cyg 19h42'25" 36d46'15"
Jeff M. Sharp
Uma 8h57'5" 54d7'0"
Jeff Macedo
Cnc 8h22'23" 10d31'14"
Jeff Matala - Rising Star
Ori 6h17'58" 19d18'21"
Jeff & Melissa Krutzfeldt
Dra 10h3'14" 77d51'22"
Jeff Melson
Agr 20h55'11" -11d14'42"
Jeff Mesplay
Cap 21h9'44" -25d38'34"
Jeff Meyers
Cep 22h1'27" 60d12'44"
Jeff Mikutis
Boo 14h31'50" 40d47'48"
Jeff Milano-Johnson
Uma 9h3'57" 68d36'16"
Jeff Millard
Peg 21h50'50" 26d53'4"
Jeff Morgan
Cep 20h46'40" 69d27'10"
Jeff Morris
Psc 1h23'28" 16d52'53"
Jeff Morris
Lac 22h17'36" 49d11'19"
Jeff Myers
Ori 5h37'19" -1d11'29"
JEFF NEWCOMBE
Lyn 7h48'14" 46d13'7"
Jeff Niemi
Aur 5h19'48" 48d49'28"
Jeff Nitschke
Dra 18h40'36" 62d31'26"
Jeff Orswell
Uma 11h50'59" 46d35'57"
Jeff Paige
Cas 23h53'24" 61d25'53"
Jeff Pardey creator Ken Ashi Do
Dra 20h21'59" 67d5'41"
Jeff Parks
Boo 14h39'24" 33d26'36"
Jeff Paul Davis
Per 2h6'43" 54d32'28"
Jeff & Peggy
Sge 19h15'34" 18d53'22"
Jeff Peterson
Cep 21h50'14" 63d20'41"
Jeff Pinkner #1 Daddy
Her 18h5'42" 38d54'3"
Jeff Piper-In Loving Memory
Dra 12h31'13" 71d17'55"
Jeff Post
Cas 23h40'37" 51d54'25"
Jeff & Purusha
Ori 5h35'42" -0d35'39"
Jeff Pynes
Cmi 7h32'0" 10d34'0"
Jeff Ramsdell
Lyn 7h11'4" 58d49'51"
Jeff Rauen
Boo 15h0'9" 34d59'8"
Jeff Reynolds
Cru 12h31'14" -59d14'35"
Jeff Roisum
Her 17h35'49" 35d51'44"
Jeff Romain - Our Shining Star Here
Per 3h5'2" 53d30'21"
Jeff "Romy" Scales, Our Matchmaker
And 0h34'9" 37d28'37"
Jeff Runyon
Gem 7h24'41" 25d25'9"
Jeff Ryan Hollinger
Cnc 8h40'29" 18d34'44"
Jeff Ryan Julio
Lib 14h55'16" -10d16'52"
Jeff Sage
Per 3h20'10" 51d57'7"
Jeff Salvadore
Her 16h42'54" 33d1'29"
Jeff & Sarah 2-2-01Ruth1:16
Uma 12h39'26" 62d49'3"
Jeff Satterlee
Her 17h27'48" 33d46'25"
Jeff Satterlee
Boo 15h23'16" 40d19'6"
Jeff Sewecke
Sco 17h46'30" -40d32'50"
Jeff Sherin aka Crystal St. Clair
Uma 11h4'39" 68d31'17"
Jeff Shetler
Uma 8h12'36" 65d20'29"
Jeff Smith
Sco 16h12'12" -14d6'22"
Jeff Smith
Her 17h33'37" 31d57'21"
Jeff Sparrow
Ori 5h52'6" -0d41'17"
Jeff Storm Harkavy
Cru 12h5'11" -63d0'45"
Jeff Strieker
Per 3h18'31" 56d49'34"
Jeff "Sweetie" Olson
Umi 20h48'15" 89d18'38"
Jeff "The Rocketman" Shoemake
Cap 21h32'21" -17d32'36"
Jeff Turk
Uma 9h38'56" 71d3'40"
Jeff Wells
Tau 4h10'55" 10d40'47"
Jeff White
Dra 19h23'32" 59d28'58"
Jeff Wiltzius
Her 17h46'54" 44d51'31"
Jeff Wylie Schultz
Ori 5h45'35" 1d21'6"
Jeff, My World is Brighter
Nicci
Her 16h46'55" 33d15'43"
JeffandMeg0925
Lib 15h22'21" -4d8'1"
Jeffanie
Cyg 19h36'49" 53d44'5"
Jeffblair
Uma 11h27'14" 58d24'37"
JeffC #1
Agr 22h38'24" -0d38'9"
Jefferson B. Leinberger
Leo 9h51'23" 8d27'16"
Jefferson K Hobbs
Her 18h54'29" 20d57'0"
Jefferson Luz
Cnc 8h47'22" 12d23'42"
Jeffersons Everything named Angela
Col 6h9'9" -32d41'49"
Jeffery
Her 18h11'35" 38d8'54"
Jeffery Alan Wright
Sgr 18h5'21" -27d41'37"
Jeffery Allan McCormack The 1st.
Her 17h18'47" 15d23'0"
Jeffery And Erin Forever
Lyr 18h44'36" 41d13'58"
Jeffery B. Eargle
Vir 12h46'40" 6d33'51"
Jeffery Carr
Lib 14h25'57" -21d16'59"
Jeffery Coia
Aql 20h11'28" 14d47'27"
Jeffery Cononver
Uma 9h4'30" 60d34'11"
Jeffery D. Jarosiewicz
Leo 10h9'53" 26d36'32"
Jeffery D. Rosine
Her 17h42'48" 20d4'44"
Jeffery David Miller
Psc 23h25'18" 3d8'34"
Jeffery David's Star
Dra 17h26'4" 51d3'0"
Jeffery Derek Sakundiak
Crb 16h9'36" 37d56'58"
Jeffery Dwight
Sgr 19h37'25" -15d21'35"
Jeffery Foster Llewellyn
Gem 7h46'40" 31d36'54"
Jeffery G. Harrison
Per 3h2'27" 47d42'0"
Jeffery James
Ori 5h7'33" 4d2'41"
Jeffery James Turner Jr.
Boo 14h38'47" 22d14'7"
Jeffery James Wiltberger
Her 17h46'41" 37d5'24"
Jeffery John Racine
Tau 5h21'59" 18d0'26"
Jeffery Joseph Kozak
Lib 14h36'31" -10d29'17"
Jeffery Keith Beck
Ser 15h16'19" -0d33'35"
Jeffery L. Kitchen, Esq.
Uma 10h27'29" 45d34'40"
Jeffery L. Peterson
Pic 5h50'43" -56d15'50"
Jeffery Lynn
Psc 0h38'8" 15d38'10"
Jeffery Mathews
Cap 20h34'10" -20d0'5"
Jeffery Neil Vortherms
Lib 15h11'0" -15d24'12"
Jeffery Noah Green
Uma 8h39'23" 49d10'30"
Jeffery O. Childers
Lib 15h0'24" -17d46'53"
Jeffery Potter
Tau 5h25'0" 22d41'53"
Jeffery Ray Fox "Number One Dad"
Her 16h32'37" 42d12'28"
Jeffery Scott Ball
Her 17h54'46" 14d43'36"
Jeffery Scott Bracewell Jr.
Her 17h29'53" 38d53'40"
Jeffery Scott Buckley
Per 3h11'48" 45d21'17"
Jeffery Thomas Nesmith
Sgr 19h23'42" -16d11'13"
Jeffery "Tiger" Walker
Lyn 8h42'46" 44d46'16"
Jeffery Wade Laird
Lyn 8h26'16" 54d18'15"
Jeffery, Emily, Rachel, Oreo Talpas
Peg 22h9'27" 31d35'49"
Jeffery's Light
Sgr 19h0'13" -18d49'7"
Jeff-JMan-Meyer love IS C/O 2005
Her 17h39'55" 23d7'54"
Jeffra Elizabeth Stafford
Lyr 18h30'13" 36d31'3"
Jeffrei
Sgr 18h33'1" -32d40'51"
Jeffrey
Sgr 18h25'40" -18d1'5"
Jeffrey
Vir 13h40'35" -4d4'11"
Jeffrey
Uma 12h11'54" 62d7'32"
Jeffrey
Her 16h4'46" 45d18'37"
Jeffrey A. Berns Christmas Star
Leo 11h29'59" -6d8'10"
Jeffrey A. Bursey
Ari 2h43'58" 28d54'47"
Jeffrey A. Fowler - Go Blue!
Uma 10h37'48" 54d35'49"
Jeffrey A. Gavitt
Cap 20h55'27" -14d47'30"
Jeffrey A Lowe
Ori 5h26'3" 2d35'45"
Jeffrey A. Medinas
Lib 15h23'49" -5d35'46"
Jeffrey A. Modell
Her 16h31'46" 46d28'45"
Jeffrey A. Wegner
Sco 17h51'25" -40d5'58"
Jeffrey Adam King
Boo 15h18'16" 48d0'17"
Jeffrey Adam Martin
Tau 5h31'26" 23d50'40"
Jeffrey Adams Underhill
Sgr 17h57'28" -29d47'9"
Jeffrey Alan
Ori 5h20'27" -1d30'20"
Jeffrey Alan Davis
Per 2h21'47" 55d21'12"
Jeffrey Alan McLean
Sgr 18h25'37" -27d55'16"
Jeffrey Alan Willis
Uma 8h13'50" 69d3'1"
Jeffrey Alexander
Her 17h41'49" 17d53'5"
Jeffrey Allan Sumpter
And 0h17'55" 27d2'8"
Jeffrey Allen
Ori 4h54'7" 10d18'46"
Jeffrey Allen Hallowell
Aql 19h39'27" 10d5'48"
Jeffrey Allen Howard
Leo 9h37'48" 27d5'48"
Jeffrey Allen Kassel
Leo 10h57'58" -0d8'6"
Jeffrey Allen Korman
Leo 10h57'24" 11d48'27"
Jeffrey Allen Lee
Uma 12h19'32" 62d46'21"
Jeffrey Allen Millard
Gem 6h59'46" 12d41'7"
Jeffrey Allen Nesbit
Sco 17h51'9" -39d38'21"
Jeffrey Allen Price
Sgr 18h6'13" -27d14'56"
Jeffrey Allen Woerner
Dra 16h32'58" 64d54'9"
Jeffrey Allister Block
Uma 10h46'33" 52d44'1"
Jeffrey and Amy O'Brien
Tau 5h44'25" 25d44'32"
Jeffrey and Annamarie Hastings
Ara 17h27'0" -47d11'4"
Jeffrey and Candace Hutton
Lyr 19h18'15" 28d4'59"
Jeffrey and Carla Penwarden
Ori 5h38'39" 12d55'33"
Jeffrey and Lynette
Col 6h13'9" -42d35'13"
Jeffrey and Mary Sadler
Lac 22h26'15" 47d34'18"
Jeffrey and Robin Wechsler
Col 5h31'42" -34d28'9"
Jeffrey Andrew Whitehead
Sgr 19h37'18" -13d13'22"
Jeffrey Andy Thompsen
Aur 5h47'41" 54d59'37"
Jeffrey B. Prince
Lib 15h50'18" -19d14'40"
Jeffrey Barrington
Cap 20h53'34" -15d3'52"
Jeffrey Bashan Thomas
Uma 10h54'48" 44d1'25"
Jeffrey Bernard Schmidt
Sco 17h40'26" -41d55'43"
Jeffrey Blake Bourgeois
Cnc 8h42'51" 23d46'57"
Jeffrey Bolte
Dra 19h33'33" 61d4'17"
Jeffrey "Boo" Scott Suddjian Rangel
Uma 8h59'19" 46d33'51"
Jeffrey Brendan Spengler
Psc 1h40'39" 18d35'40"
Jeffrey Brewster Norwood
Cyg 20h37'16" 33d57'44"
Jeffrey Burge
Vir 12h11'58" 12d14'4"
Jeffrey Burkhart
Lyn 7h6'26" 48d30'24"
Jeffrey C. Davis
Per 2h54'30" 53d58'44"
Jeffrey C. Stauffer
Vir 12h52'41" 2d56'39"
Jeffrey C. Wilkins
Sgr 18h28'35" -27d6'35"
Jeffrey Canlon Clark
Her 16h52'56" 16d54'49"
Jeffrey Cap Morgan
Aql 19h10'6" -7d43'30"
Jeffrey Carl Wilhelm
Cap 20h44'7" -25d54'56"
Jeffrey & Carolina's Star Bright
Crb 15h39'43" 27d39'27"
Jeffrey & Caron Simpson
Cam 4h30'0" 66d47'17"
Jeffrey Charles
Uma 10h21'11" 55d25'13"
Jeffrey Charles Bennett
Ori 6h2'19" 18d37'5"
Jeffrey Charles Castro
Her 18h47'2" 16d25'4"
Jeffrey Chattin
Ori 5h8'35" 12d38'22"
Jeffrey Christensen
Cnc 9h6'39" 27d45'29"
Jeffrey & Christina-Always&Forever
Uma 9h13'48" 69d19'40"
Jeffrey Cobb
Psc 1h21'14" 29d56'57"
Jeffrey Conrad
Agr 21h0'17" 1d38'4"
Jeffrey Cornelius
Ori 5h37'4" 0d24'28"
Jeffrey Craig Benson
Per 3h49'40" 32d20'10"
Jeffrey D. Barnes
Gem 6h48'1" 15d39'1"
Jeffrey D. Lamb
Cnc 9h11'18" 17d15'17"
Jeffrey Dalton Bowers
Sco 17h58'23" -37d57'46"
Jeffrey Daniel Matthews
Per 3h3'39" 53d55'31"
Jeffrey David Burd
Tau 3h50'46" 14d22'12"
Jeffrey David O'Neil
Uma 11h11'37" 62d10'29"
Jeffrey David Sabatino
Cep 22h1'33" 73d3'47"
Jeffrey Dean Clark
Uma 11h22'5" 29d10'44"
Jeffrey Dean Gamble
Sgr 19h24'7" -21d43'0"
Jeffrey Dean Hibbs
Psc 1h9'16" 28d7'25"
Jeffrey Dodd Krull
Cap 20h26'53" -12d21'9"
Jeffrey Donner Glossip
Tau 4h21'15" 24d9'44"
Jeffrey Douglas Bullock
Per 3h23'7" 41d14'41"

Jeffrey Douglas Eloi
Cep 23h10'53" 72d3'11"

Jeffrey Douglas Lupo
Sco 17h0'16" -38d53'27"

Jeffrey D.Woods
Her 17h49'44" 47d56'50"

Jeffrey Edward Arnold
Her 17h10'45" 35d5'39"

Jeffrey Edward Daniels
Tau 5h0'51" 16d37'48"

Jeffrey Edward Smith
Uma 13h44'3" 56d38'15"

Jeffrey Ellis-my soldier & soulmate
Ori 5h35'34" -1d41'24"

Jeffrey Evans Jackson
Her 17h17'7" 28d10'1"

Jeffrey Featherston
Ori 5h11'26" 5d42'57"

Jeffrey Fichner
And 0h39'44" 36d27'22"

Jeffrey Fossen
Cnc 9h17'16" 6d29'43"

Jeffrey Frederick
Ori 5h37'48" 0d2'20"

Jeffrey Gaeto
Sgr 19h52'7" -17d39'46"

Jeffrey Gerard Brooks
Leo 9h40'33" 27d43'51"

Jeffrey Glenn
Ari 2h15'27" 26d4'50"

Jeffrey Goldman
Her 16h40'7" 35d39'21"

Jeffrey H Worth
Tau 5h20'39" 26d21'38"

Jeffrey Hague
Cap 21h39'22" -21d11'1"

Jeffrey Hamrick
Cnc 9h14'39" 31d52'31"

Jeffrey Harris Drake
Psc 1h40'5" 2d38'7"

Jeffrey Heeley
Cma 7h22'57" -15d5'1"

Jeffrey Hottensen
Vir 13h21'50" -11d4'57"

Jeffrey Huffman
Cep 2h29'36" 80d59'3"

Jeffrey Iken
Gem 6h23'33" 21d11'45"

Jeffrey Irvine
Uma 10h3'52" 59d26'11"

Jeffrey J. Macedo
Leo 9h51'22" 28d20'13"

Jeffrey J. Metz
Her 16h43'5" 29d54'50"

Jeffrey J. Pettit
Her 18h48'21" 21d58'35"

Jeffrey Jackson Parette
Uma 12h1'43" 57d15'28"

Jeffrey Jaguar Spira
Vir 13h29'44" -16d25'36"

Jeffrey James Liebowitz
Uma 10h25'33" 39d27'45"

Jeffrey James Morgan
Ori 5h47'21" 6d13'33"

Jeffrey James Ortagus
Cap 20h29'38" -13d17'18"

Jeffrey Jason Jacobs
Her 18h19'8" 26d58'19"

Jeffrey Jay Miller 52
Sco 17h8'42" -42d54'29"

Jeffrey "Jeffers" Michael Aubry
Per 2h56'49" 40d59'10"

Jeffrey & Jennifer
Tau 4h19'36" 18d7'37"

Jeffrey Jerome Massie
Cap 20h28'48" -13d42'29"

Jeffrey "J.J." Harford
Uma 13h54'19" 56d37'45"

Jeffrey John Chateau Jr.
Cap 21h10'18" -16d52'23"

Jeffrey John Loundy
Per 3h28'19" 48d57'59"

Jeffrey John Mossor
Aqr 22h25'36" -5d3'23"

Jeffrey John Standefer
Cnc 8h54'40" 11d48'14"

Jeffrey Jon Wielock
Per 3h12'47" 52d14'2"

Jeffrey Jordan Marsh
Her 17h52'43" 28d18'45"

Jeffrey Joseph Fuka Planet Explorer
Uma 9h29'20" 58d54'12"

Jeffrey Keith Bell
Per 3h24'53" 48d13'0"

Jeffrey Keith Harvey
Uma 10h20'57" 51d12'59"

Jeffrey Kofi Brown
Uma 10h3'48" 69d48'44"

Jeffrey Krouldis
Lac 22h55'45" 51d12'2"

Jeffrey Kurtz
Her 17h9'49" 32d30'51"

Jeffrey Kyle Pettey
Psc 0h0'24" -2d49'32"

Jeffrey L. Crandall
Uma 10h35'35" 62d14'25"

Jeffrey L. Golaszewski
Sco 17h53'2" -44d22'0"

Jeffrey L. Hicks
Uma 10h35'36" 66d31'54"

Jeffrey L. Minor
Lyn 6h40'31" 53d6'30"

Jeffrey L. Monroe
Leo 11h56'27" 23d31'15"

Jeffrey L Piehl
Psc 0h47'4" 12d58'54"

Jeffrey L. Weeks Jr.
Aqr 21h18'9" -0d56'25"

Jeffrey L. Wiener
Her 16h22'27" 44d27'40"

Jeffrey LaCroix
Dra 16h40'38" 52d54'43"

Jeffrey Laitinen
Cep 22h11'57" 73d39'45"

Jeffrey Laurence Ellis
Cen 13h27'13" -41d11'32"

Jeffrey Lee Cyran
Sco 17h56'44" -30d4'28"

Jeffrey Lee Dutt
Her 16h44'12" 21d14'33"

Jeffrey Lee Miracle
Sgr 18h59'36" -14d51'50"

Jeffrey Lee Rogers
Lib 15h5'43" -0d51'17"

Jeffrey Lee Wilson
Cyg 19h33'33" 28d6'20"

Jeffrey Leonard Bayse
Vir 13h21'48" -0d56'18"

Jeffrey Levell Parker
Per 3h8'2" 43d42'28"

Jeffrey Lim Garcia
Her 17h47'18" 36d7'59"

Jeffrey Lynn Foss
Sgr 18h3'21" -28d53'34"

Jeffrey Lysle Fluharty
Boo 14h38'35" 50d36'23"

Jeffrey M. Farwell
Tau 4h8'28" 6d40'19"

Jeffrey M. Queen
Sgr 19h22'9" -35d2'44"

Jeffrey M Rashid
Her 17h36'43" 37d25'54"

Jeffrey MacDonald
Lib 15h54'13" -10d35'26"

Jeffrey Macedo
Aur 6h11'44" 53d33'36"

Jeffrey Mark
Lib 15h23'28" -15d10'36"

Jeffrey Mark Jr.
Lib 15h21'20" -16d12'35"

Jeffrey Mark Wittig
Sco 16h12'1" -9d20'26"

Jeffrey Marshall
Per 4h17'23" 48d59'5"

Jeffrey Marshall Payn
Uma 13h56'45" 60d59'57"

Jeffrey Martin
Psc 1h36'20" 18d8'0"

Jeffrey Masters
Crb 15h42'42" 26d22'29"

Jeffrey Mathew Cahill
Per 4h19'19" 50d49'41"

Jeffrey Meyer Turntables
Uma 8h25'46" 66d41'35"

Jeffrey Michael Cooperstein
Sgr 19h1'32" -12d36'24"

Jeffrey Michael Gasper
Lmi 10h24'47" 33d55'37"

Jeffrey Michael Gibson
Uma 13h50'6" 56d59'37"

Jeffrey Michael Gold
Vir 13h32'13" -16d3'35"

Jeffrey Michael Peterson
Uma 8h52'24" 52d24'33"

Jeffrey Michael Spain
Gem 7h49'10" 19d43'20"

Jeffrey Michael Squires
Psc 0h28'16" 15d49'25"

Jeffrey Michael White
Leo 10h25'49" 15d5'16"

Jeffrey Michaels
Lyr 18h36'22" 39d50'51"

Jeffrey Mmmm SvenDahl, M.D.
Cep 22h25'47" 65d28'13"

Jeffrey Molloy
Sco 16h21'33" -25d38'44"

Jeffrey Moran
Cep 0h18'53" 79d5'54"

Jeffrey Musante
Sge 19h26'21" 18d37'52"

Jeffrey Narkiewicz's guardian star
Cnc 8h20'2" 14d33'41"

Jeffrey Neal Rhame
Uma 9h23'46" 46d10'52"

Jeffrey Neil Van Horn
Uma 8h52'49" 50d35'14"

Jeffrey Noel Shields
Aql 19h45'6" -0d6'19"

Jeffrey Nolan Chicoski
Psc 23h17'30" 6d48'54"

Jeffrey Oakes
Leo 11h15'4" -0d37'12"

Jeffrey P. Holley
Aur 5h27'54" 36d25'45"

Jeffrey P. Katon
Psc 0h42'33" 10d48'21"

Jeffrey P. Kilner
Her 16h55'29" 35d23'19"

Jeffrey P. Manley
Aql 19h11'31" -0d8'16"

Jeffrey P. Mills
Ari 2h40'8" 26d47'39"

Jeffrey P. Wetzel
Cep 22h44'23" 74d58'23"

Jeffrey Palmer Turner, "QJ57"
Cnc 9h8'11" 32d16'6"

Jeffrey Patrick Campbell
Her 16h55'49" 15d44'2"

Jeffrey Paul And Jessica Lavonne
Uma 11h25'56" 30d9'22"

Jeffrey Paul Janness
Uma 12h45'50" 52d35'40"

Jeffrey Paul Mihali Mush
Uma 12h39'35" 61d2'2"

Jeffrey Paul Nadeau
Psc 1h9'4" 28d47'42"

Jeffrey Perez
Cep 23h33'21" 75d49'41"

Jeffrey Peter
Cap 21h33'25" -23d21'40"

Jeffrey Philip's Star
Dra 19h22'27" 78d42'3"

Jeffrey Plotkin
Aqr 22h27'19" -2d37'27"

Jeffrey Pomeroy's Hunting Star
Ori 5h27'37" 5d4'46"

Jeffrey Powell Smalley
Cma 6h42'10" -12d39'40"

Jeffrey R. Corveau
Aur 5h53'14" 52d39'13"

Jeffrey R. Pennington
Psc 1h53'0" 7d46'37"

Jeffrey Ralph Miller
Dra 17h37'33" 67d11'51"

Jeffrey Ray Weaver
Tau 5h38'51" 18d21'37"

Jeffrey Rita
Ori 5h44'19" 3d37'6"

Jeffrey Robert Colianni
Ari 2h19'45" 11d1'14"

Jeffrey Robert Dady
Psc 1h16'54" 18d59'34"

Jeffrey Robert Goss
Per 4h13'19" 47d36'47"

Jeffrey Robert Thompson
Cap 20h34'26" -9d38'2"

Jeffrey Robert Willett
Per 4h11'51" 36d30'7"

Jeffrey Ross Hayet
And 2h34'32" 40d1'28"

Jeffrey Rothrock
Per 3h11'10" 41d23'17"

Jeffrey Ryan Kalnas
Leo 11h6'21" 8d58'11"

Jeffrey S. Alexander
Lib 14h51'43" -1d33'39"

Jeffrey S Gudenkauf
Cap 20h28'1" -24d14'5"

Jeffrey S. Hipschman
Uma 8h52'0" 52d25'55"

Jeffrey S. Mann je t'aime
Psc 22h56'48" 4d35'2"

Jeffrey S. Wallace
Ori 5h44'4" 1d47'4"

Jeffrey Salvadore
Per 3h30'8" 47d35'17"

Jeffrey Salvadore
Dra 16h33'20" 52d6'27"

Jeffrey Scott Conway
Her 17h47'0" 34d26'3"

Jeffrey Scott Johnson
Per 3h9'17" 39d13'35"

Jeffrey Scott Johnson
Her 17h47'29" 49d4'52"

Jeffrey Scott Pounds
Aql 19h46'34" 13d54'37"

Jeffrey Scott Rapkin
Lib 15h25'44" -3d43'24"

Jeffrey Scott Senter
Her 18h1'14" 21d30'32"

Jeffrey Scott Tardiff
Sco 17h42'39" -32d3'19"

Jeffrey Scott Wilson
Her 17h48'40" 41d33'38"

Jeffrey Sheldon Hawk
Uma 13h26'4" 59d3'38"

Jeffrey Snicker Snackers
Vir 12h27'54" -6d25'19"

Jeffrey Spielvogel
Per 4h29'4" 32d39'49"

Jeffrey (Sput) Duay
Sex 10h24'0" -5d2'11"

Jeffrey Stephen
Sgr 19h33'2" -12d53'42"

Jeffrey Stephen Isler
Lib 15h27'33" -24d28'13"

Jeffrey Stephen Martinez
Tau 4h13'52" 17d50'58"

Jeffrey Steven Dunlava
And 0h25'20" 41d49'33"

Jeffrey Stockham Gibbs
Tau 4h30'51" 29d6'39"

Jeffrey Stone Huntoon
Cep 22h42'42" 65d43'7"

Jeffrey Stovall March 24th, 1982
Peg 22h26'40" 11d25'1"

Jeffrey Stuart Wightman
Her 16h50'0" 27d14'59"

Jeffrey Sturgis
Ori 5h52'59" 21d46'54"

Jeffrey T. Brock
Ori 6h11'24" -2d5'25"

Jeffrey T. Noblet, Jr.
Vir 13h25'21" 9d1'22"

Jeffrey Taylor
Ari 2h36'34" 13d58'53"

Jeffrey Thomas Judd
Ari 3h11'6" 23d2'57"

Jeffrey Thomas Kroll
Tau 4h38'16" 28d10'22"

Jeffrey Thomas Miller
Ari 3h16'25" 28d5'27"

Jeffrey Thomas Sodeman
Her 16h37'37" 29d10'25"

Jeffrey Thomas Wyand
Leo 11h30'31" 10d23'21"

Jeffrey Todd Pieters
Sco 16h2'54" -10d35'20"

Jeffrey Tolliver
Per 3h40'31" 36d4'10"

Jeffrey W. Pool
Ori 5h47'49" 9d18'4"

Jeffrey W. Scarpa
Boo 14h36'13" 33d35'18"

Jeffrey Wade and Sheena Robyn A/F
Cap 20h52'52" -26d12'6"

Jeffrey Welter
Aqr 21h1'51" 2d24'48"

Jeffrey Willem Hendrik De Wit
Ari 2h6'25" 24d52'9"

Jeffrey William
Uma 9h44'54" 61d59'25"

Jeffrey William Dean
Per 3h10'23" 56d25'12"

Jeffrey William Herlache Jr.
Ori 5h14'35" 3d11'22"

Jeffrey William Milosch
Uma 11h43'5" 30d23'43"

Jeffrey Zola
Uma 9h34'10" 64d56'38"

JeffreyRobertLovesNicoleDanielle
Tau 5h12'9" 26d41'13"

Jeffrey's Song
Leo 11h43'4" 11d35'10"

Jeffreys Star
Leo 9h40'34" 28d39'18"

Jeffreystar
Ori 6h5'24" 21d12'16"

Jeffri W. Bantz
Ari 2h15'4" 27d16'30"

JeffRo Mai pu'uwai mai 'oe
Lac 22h42'28" 54d3'36"

Jeffros Star
Umi 15h36'12" 84d58'27"

Jeffry Leo Bates
Lib 16h1'14" -7d21'22"

Jeffry Michael Allan O'Neill
Leo 9h23'30" 30d48'50"

Jeffry Paul Zander
Her 17h9'27" 34d2'2"

Jeffs'
Ori 5h21'36" -1d27'10"

Jeff's Brecken Bear
Umi 13h57'6" 75d58'51"

Jeff's Castle
Ori 5h29'26" 3d48'0"

Jeff's Girls
Tau 4h24'33" 10d6'31"

Jeff's Jetson
Aql 19h53'51" -0d57'53"

Jeff's shining star
Tau 4h6'21" 8d52'1"

Jeff's Star - Leave Only Bubbles
Cnc 8h55'55" 30d37'2"

Jeff's Virgo Seeking
Vir 14h2'28" -0d50'15"

Jeff-The-Star
Psc 1h14'50" 28d23'53"

Jeg Elsker De
Uma 9h9'19" 66d6'39"

Jegan
Lyn 7h54'25" 41d50'41"

Jegen, Tina
Uma 8h20'51" 62d39'48"

Jegglin
Cyg 20h11'41" 50d39'58"

Jehad Al-Sarraj
Ori 5h20'21" -0d25'45"

JEHAN 16
Vir 12h12'22" 11d15'41"

Jehanara Tejpar
And 22h57'54" 39d36'4"

Jehanne Mabilais
Peg 21h32'9" 8d27'41"

Jehanzeb Khan
Ori 6h15'8" 2d44'24"

Jehne, Günter
Uma 9h1'48" 63d8'14"

Jehvon Javie
Uma 9h26'37" 48d49'15"

Jeidy
Uma 10h15'12" 52d31'42"

Jeising, Mechthild
Uma 13h16'25" 61d27'58"

jeka1913
Crb 15h56'47" 26d0'10"

Je-Ka-Ju
Tau 4h27'3" 25d42'23"

Jelani Naeem White
Uma 9h44'11" 56d49'35"

Jelena
Uma 10h36'58" 57d12'47"

Jelena
Umi 13h4'30" 69d21'17"

Jelena Bjelica
And 1h50'45" 46d9'35"

Jelena Brkic - Moja Zvezda
Cyg 21h47'54" 47d56'45"

Jelena Gapeyeva
Cap 21h47'3" -13d19'54"

Jelena & Goran
Lmi 10h13'44" 37d45'29"

Jelena "Mäusle" Schmid
Uma 10h12'49" 49d51'43"

Jelena Pejic - 26.12.2005.
Cyg 19h52'23" 52d29'21"

Jelena - Sara 31/08/05
And 1h16'42" 50d25'32"

Jelena Tanasic
Per 4h21'16" 46d29'49"

Jelena Tarbuk
And 23h34'1" 48d23'24"

Jelena und Dany forever
Per 4h47'6" 40d20'26"

JELENA, 12.09.1973
And 1h18'31" 43d27'43"

(JELEP) Joseph & Linda Powell
And 23h25'38" 42d37'11"

Jelica i Cedomir Jovanovic
Ori 5h53'7" 20d47'17"

Jelina
Lyr 18h51'53" 38d21'21"

Jellibi
Uma 13h55'10" 49d4'41"

Jello with a -Y
Lyn 7h42'35" 36d37'58"

Jelly
Uma 9h52'58" 61d16'11"

Jelly
Cas 23h18'5" 54d32'25"

Jelly Baby
Pho 0h50'7" -52d7'31"

Jelly Bean
Psc 0h37'12" 11d44'25"

Jelly Bean
Vir 13h25'54" 13d10'24"

Jelly Belly 23 - Cunha 2007
Uma 10h21'17" 48d57'51"

Jelly Loves Jimmy
Per 3h17'13" 41d28'58"

Jellybean
Tau 3h31'45" 8d23'39"

Jellybean
Tau 4h31'42" 13d6'56"

Jellybean
Ori 5h34'21" 10d53'13"

Jellybean
Crb 15h48'46" 27d9'38"

Jellybean
Cap 21h49'23" -24d13'46"

Jellybean
Lib 15h17'7" -27d9'46"

Jellybean
Cam 4h29'22" 73d13'59"

Jellybean
Umi 14h18'41" 76d30'58"

Jellybeans Star Forever InOurHearts
Cru 12h17'30" -56d57'40"

Jelodaco
Per 3h20'9" 43d31'0"

Jelz
Lib 15h44'42" -14d28'10"

J.E.M.
Cas 1h6'40" 61d22'53"

J.E.M.
Ori 5h59'35" 22d44'37"

Jem
Ori 6h1'32" 6d52'15"

Jem Dragon
Cra 17h59'56" -37d6'59"

Jem Jem
Cas 0h52'49" 51d38'31"

Jema Khan
Cyg 19h38'26" 34d20'50"

Jemdayanes
Del 20h34'32" 10d36'13"

Jemell's Angel
Cru 12h28'37" -59d48'47"

JEMENI
Gem 7h44'41" 20d29'30"

Jemerika
Sco 16h34'11" -30d20'3"

JEMFAM
Umi 16h28'53" 77d16'44"

Jemilbra
Lib 15h0'44" -11d57'3"

Jemily1
Lyn 7h1'42" 51d17'28"

Jemima Kate Hargreaves
And 23h19'55" 40d5'56"

Jemima Mackay
Cyg 19h32'45" 31d45'53"

Jemine Bryon
Per 2h47'53" 40d14'48"

Jemison Tayla Margaret Kennedy
Aqr 22h57'52" -10d11'31"

Jemma
And 23h59'20" 35d8'21"

Jemma
And 23h36'4" 45d25'55"

Jemma and Cristian
Cyg 19h40'15" 32d5'26"

Jemma & Andrew Campbell
Ara 17h22'40" -56d14'39"

Jemma Fiorini
And 0h33'24" 39d19'1"

Jemma Louise Allan
Aqr 23h0'24" -14d22'55"

Jemma Louise Asprey
Del 20h21'26" 9d44'4"

Jemma Ruby Hainge
And 23h13'51" 52d11'59"

Jemma Stretch
And 1h57'37" 42d18'46"

Jemmar
Cyg 21h38'20" 42d6'55"

Jemmarie Xiomara
And 0h38'56" 27d51'24"

Jemma's Smile
Cru 12h1'31" -62d59'35"

Jemmi
And 0h44'49" 35d57'24"

Jemmi Aquino
Cnc 9h3'39" 20d37'3"

Jemonie - soul sisters
Cru 12h0'22" -64d19'16"

Jem's Star
Tau 5h37'37" 21d1'56"

Jemsi
Leo 11h2'27" 5d55'12"

jemstar
Vir 11h51'21" 6d6'9"

Jen
Vir 13h19'41" 5d7'14"

Jen
Ori 5h12'50" 0d57'28"

Jen
Psc 23h44'22" 7d31'32"

Jen
Cnc 8h21'48" 12d26'37"

Jen
Ori 6h4'44" 11d38'24"

JEN
Leo 9h34'18" 28d21'10"

Jen
Crb 16h12'25" 35d6'48"

Jen
Lyn 6h52'19" 52d3'25"

Jen
Uma 8h47'32" 47d46'49"

Jen
And 23h20'29" 48d18'15"

Jen
Sgr 19h2'57" -29d0'0"

Jen
Ori 5h54'33" -0d42'13"

Jen
Aqr 23h9'27" -8d6'31"

JEN
Aqr 22h3'37" -0d31'26"

Jen 10/04/03 Roger
Leo 11h35'53" 26d1'27"

Jen A. McCormick
Ori 5h28'18" 2d27'17"

Jen Albright
Uma 10h56'43" 48d50'31"

Jen and Baby Carlos Forever
And 0h43'47" 39d6'50"

Jen and Becki 12/12/06
Cyg 21h25'2" 34d56'32"

Jen and Gee
Cyg 20h50'36" 46d11'13"

Jen and Matt Forever
Cyg 21h27'59" 51d35'55"

Jen and Nick's star of LOVE
Cyg 21h30'7" 47d55'4"

Jen and Pete 1st Anniversary
Ara 17h23'15" -47d56'29"

Jen and Rich Sawyer
Sco 16h8'15" -18d1'0"

Jen and Ryan
Apu 16h47'40" -70d28'6"

Jen and Steve Darling
Lmi 11h6'30" 25d11'13"

Jen and Wes
Sge 20h4'20" 17d44'27"

Jen 'Aura' Krezel
Leo 11h48'16" 21d3'10"

Jen B Ian
Vir 14h42'9" 2d59'28"

Jen (Baby girl)
Ori 6h10'23" 8d49'52"

Jen & Bob's Stargaze
Sgr 19h22'1" -23d34'44"

Jen & Brian Forever
Psc 1h4'1" 4d25'21"

Jen & Brian Valentine's Day 2007
Cyg 20h39'17" 33d59'26"

Jen Byrne
Vul 20h38'14" 27d29'21"

Jen Carter is special. I love You.
Vir 12h36'45" -9d12'36"

Jen Chau
Lib 15h35'36" -7d39'59"

Jen & Dave - Love of a Lifetime
Cyg 21h29'39" 34d24'7"

Jen & David
Cyg 21h27'11" 35d31'39"

Jen Dimichele
Sco 16h43'46" -31d51'41"

Jen Doss
Crb 15h48'39" 27d59'59"

Jen Doyle - "Light of My Life"
And 23h27'13" 37d14'29"

Jen Durval
Lep 5h9'31" -14d11'46"

Jen & Dwayne
Psc 23h12'52" -1d58'34"

Jen Eckhaus
Sco 16h4'22" -11d24'8"

Jen Elyse
Uma 11h16'55" 35d55'27"

Jen Fenstra
And 23h9'39" 35d53'53"

Jen Filli
Lac 22h26'15" 52d3'30"

Jen Fish aka "The Grene Ham"
Ari 2h58'35" 26d23'16"

Jen Gaskell's Rainbow Skittles
Cas 2h5'18" 70d58'44"

Jen Grinspan
And 23h39'45" 48d29'5"

Jen Ingram
Cnc 8h56'40" 10d11'32"

Jen & Jay Simon
Tau 5h8'45" 28d35'52"

Jen Jen
Del 20h33'41" 15d0'34"

Jen Jen
Ori 5h25'37" 8d57'59"

Jen Jen
Cap 20h24'14" -20d58'37"

Jen Jen
Lib 15h24'25" -5d12'11"

Jen "Jenisis" Samoranos
Aqr 22h24'49" -9d19'41"

Jen - Jennifer Ann
Leo 9h38'8" 28d16'12"

Jen&Jeromy5/2/1998
Cyg 19h33'47" 54d20'0"

Jen & Josh
Cyg 21h30'23" 53d28'34"

Jen & Ken Becker
Cyg 19h56'37" 31d30'20"

Jen Keyes
Ori 6h8'22" 9d15'0"

Jen Larkin
And 1h1'57" 41d42'49"

Jen Loveless
And 1h29'23" 49d50'56"

Jen Loves Owen
Ari 2h15'13" 23d10'38"

Jen Loves Wael Forever
Per 4h49'42" 40d1'36"

Jen M. Sweet
Cam 4h20'45" 70d55'29"

Jen M Sweet
Cam 3h58'54" 68d29'28"

Jen MacLennan
Umi 5h17'48" 89d29'20"

Jen Marie
Cnc 9h8'38" 9d52'54"

Jen Mark Karafilis
Lib 14h56'8" -20d41'59"

Jen Marsh
Lib 14h50'32" -5d9'54"

Jen Meredith Lane
And 2h32'1" 45d22'10"

Jen Merin
Crb 15h59'20" 38d36'42"

Jen Michelle-30
Her 16h37'1" 35d48'46"

Jen "My Answer"
And 1h16'51" 43d0'27"

Jen & Nate
And 1h45'29" 50d0'23"

Jen Nguyen
Cru 12h25'49" -58d17'2"

Jen 'Peaches' Lundy
Lyn 7h59'25" 57d14'32"

Jen Pfeiffer
Uma 9h33'6" 47d17'6"

Jen Pollock
And 23h2'27" 48d20'25"

Jen Pryor
Cap 20h12'30" -24d44'54"

Jen Romney
And 1h12'37" 41d27'15"

Jen Rox
Leo 9h30'0" 9d5'32"

Jen & Ryder Allen (Goshorn)
Sco 16h8'3" -17d32'51"

Jen & Sandy
Cyg 19h44'48" 32d14'55"

Jen Shade
Lyn 6h28'10" 60d51'50"

Jen Smith
Lib 14h51'48" -2d34'54"

Jen Stevens - Our Shining Star
Dor 5h0'8" -68d27'0"

Jen Tatarcyk
Psc 0h49'47" 16d14'42"

Jen the Angel
And 23h39'26" 38d13'51"

jen thy
And 1h42'34" 45d14'10"

Jen & Tim's Twinkle
Psc 1h7'47" 19d28'12"

JEN TMQ
Ori 6h17'18" 14d11'43"

Jen Vadella
Cyg 19h45'26" 32d11'53"

Jen, 9/22/82 - My Beautiful Love
Cas 0h39'22" 65d47'41"
Jena
And 1h19'40" 37d17'38"
Jena
And 0h52'33" 41d4'31"
Jena
Psc 1h18'55" 11d10'17"
Jena
Leo 11h11'55" 20d21'57"
Jena and Matt
And 1h45'11" 49d0'16"
Jena C. Minor
Cam 7h19'49" 63d18'38"
Jena (Jean) Ferratier
Crb 15h39'55" 35d46'0"
Jena Koch
And 23h46'35" 40d7'23"
Jena Lanai Marie Nicolini
Lyn 7h47'26" 46d39'45"
Jena Le Reber
Cyg 21h41'27" 32d28'41"
Jena Marie Arendt
Leo 11h13'45" 12d27'46"
Jena Marie the only one for me
And 23h29'52" 48d13'12"
Jena Michele
Ari 1h59'50" 12d52'5"
Jena Noel Brown
Dra 18h36'50" 56d45'52"
Jena Noel Stuteville
Cyg 21h28'13" 36d41'17"
Jena Robinson
Cyg 20h31'59" 50d12'19"
Jena Waggett
Cnc 8h29'7" 19d26'18"
Jenae "Angel Girl" Phillips
Gem 6h25'5" 26d16'13"
Jenae M. Pitts
Ori 5h33'17" 9d27'21"
Jenae Michelle Graham
Tau 4h36'3" 12d6'15"
Jenae Reane Mayfield
Sgr 19h9'18" -30d7'12"
Jenai Nicole Miller
And 23h26'14" 44d1'56"
Jenan Daoud
Lyn 7h47'14" 42d3'21"
Jena's Light
Ari 2h57'19" 26d19'33"
Jena's Shinning Hope
Cap 21h57'42" -22d50'35"
Jena's Star
And 1h25'11" 44d51'55"
Jena's Star
Uma 8h39'18" 48d58'15"
JenaStar
Cas 1h38'55" 66d33'23"
JenBear
Cet 3h11'54" -0d9'49"
Jency Nicole Hernandez
Psc 1h17'50" 16d41'31"
JENDOG
Crb 15h24'3" 31d30'58"
JENDY
Per 2h46'31" 51d20'8"
Jeneath Omega Tordil Quick
Cam 4h22'33" 63d56'33"
Jenee Rosh
Lyn 7h12'47" 54d44'0"
Jeneice
Umi 15h16'41" 69d16'54"
Jenel Bartee
Uma 11h22'45" 48d2'6"
Jenell
Lyn 8h7'24" 39d29'12"
Jenelle Alexander
And 0h29'57" 43d8'42"
Jenelle Ann Raba
And 23h24'34" 48d8'54"
Jenelle M. Meloun
Gem 7h1'32" 15d55'39"
Jenelle Marie
Cyg 21h40'47" 32d55'50"
Jenelle Marie Sergi
Aqr 22h25'4" 0d23'16"
Jenelle Marie Zielie
Cas 23h1'10" 53d46'59"
Jenelle Nicole Griego
Tau 5h51'13" 26d10'52"
Jenelle Rose
Uma 10h7'58" 65d36'55"
Jenelle "Supercutie" Woodard
And 0h51'41" 44d12'3"
Jenelle Wilson
Cnc 8h57'33" 9d30'46"
Je'Nen's Amber Glow
Lib 14h56'40" -5d51'16"
JenEric Petrick
Lyn 7h5'51" 51d21'23"
Jenessa
Sgr 18h27'23" -25d5'57"
Jenevaeh
Sco 16h40'55" -29d9'6"
Jenevieve
Psc 0h22'20" 19d26'30"
JENFI
Lib 15h18'50" -15d20'11"
Jeng-Tyng Hong
Umi 14h18'29" 77d36'53"
Jeni
Cap 20h21'31" -9d53'58"
Jeni
Gem 6h46'55" 23d33'55"
Jeni
Ori 5h59'39" 20d59'18"
Jeni
And 1h13'3" 46d2'17"
Jeni
Lyn 7h54'39" 34d17'10"
Jeni Clifton
Cnc 8h42'57" 31d14'3"
Jeni De'Ath
Cyg 20h33'8" 46d14'1"
Jeni Justine Fry
Ari 3h22'11" 29d12'38"
Jeni M. Sylcox
Ari 3h8'29" 29d27'24"
Jeni Rose
Tau 4h15'50" 19d26'30"
Jeni Teresa
Sco 16h17'22" -42d11'49"
Jeni the love of my life
And 22h59'47" 50d52'42"
Jeni West
Uma 10h8'19" 61d34'42"
Jenia Arshakian
Crb 15h42'8" 31d25'16"
Jenica Ellen Guarisco
And 0h3'51" 45d12'41"
Jenica Marie My Love
Umi 15h37'59" 73d6'5"
Jenice's Sparkle
And 0h0'39" 47d51'27"
Jeniefer L. Benefield
Cyg 20h58'5" 46d40'33"
Jenifaaa
Vir 12h11'3" 12d6'14"
Jenifer
Del 20h32'30" 7d23'15"
Jenifer
Psc 1h11'35" 29d17'26"
Jenifer
And 1h9'24" 38d51'6"
Jenifer Ammons *Mom*
Aqr 22h42'21" 0d35'8"
Jenifer Ann
Tau 5h29'59" 25d7'29"
Jenifer Aurora Aguirre
Her 16h21'7" 45d36'4"
Jenifer Eccles
Cnc 8h42'50" 23d4'51"
Jenifer Elizabeth
Sco 16h18'0" -10d45'57"
Jenifer Etta Bourne
Tau 3h45'17" 28d47'9"
Jenifer Helton
Vir 12h5'55" 4d14'44"
Jenifer Hunter
Tau 5h57'5" 25d13'47"
Jenifer Johnson (J.J)
Vir 11h51'54" 9d51'47"
Jenifer Kat Krizmanic
Vir 12h47'41" 12d18'40"
Jenifer Katherine Grace
Tau 3h46'57" 11d1'22"
Jenifer Lynne Stoltz
Uma 10h8'10" 46d48'55"
Jenifer Rebecca
Cnc 9h6'51" 19d57'6"
Jenifer Robin Foss
Sco 16h7'30" -12d57'24"
Jenifer Sue Valentine (831)
Cas 0h49'28" 61d21'26"
Jenifer Susan Garner
Cas 1h26'3" 61d38'44"
Jenifer Truan
Cas 0h4'58" 57d18'13"
Jeniffer Mei-Lan Mas
Lmi 9h49'54" 38d24'47"
Jeniffers22
And 2h13'44" 50d16'41"
Jenika Loriane Fodge
Aqr 21h35'39" 0d58'46"
Jenilee Bates
Ari 2h19'51" 23d36'4"
Jenilee Jaber
Cnc 8h23'44" 6d47'1"
Jenilee Petit
Her 17h36'21" 16d53'50"
Jenilynn Ann
Ori 4h43'54" 1d24'39"
Jenimoy Eternal Light
Cru 12h36'11" -61d3'10"
Jenine Marie
And 23h12'46" 51d38'33"
Jenise
Vir 13h16'28" -4d52'18"
Jenissa
And 0h21'27" 40d43'19"
Jen-izzle
Cas 0h42'34" 52d32'51"
Jenjira
Vir 13h52'3" -6d20'0"
JENJOHN92256
Dra 14h48'21" 54d59'16"
Jenjulie
Cam 4h23'51" 63d43'2"
Jenkz
Lib 15h6'26" -7d19'26"
Jen-Liz-Nic
Cas 0h47'16" 64d10'54"
JenLonn
Ori 5h22'40" 3d20'21"
JenMitchR03102001
Umi 14h31'33" 72d46'10"
Jenn
Lac 22h17'23" 53d13'31"
Jenn
Aqr 22h43'49" 1d43'3"
Jenn
Ori 6h5'40" 9d29'28"
Jenn
Ori 6h4'5" 10d49'20"
Jenn
And 0h21'53" 25d32'59"
Jenn
Psc 1h38'9" 27d49'19"
Jenn
And 0h28'38" 29d22'4"
Jenn
Cas 1h23'10" 53d44'24"
Jenn
And 23h8'15" 47d8'38"
Jenn
And 23h25'48" 47d50'6"
Jenn
And 1h22'44" 47d30'15"
Jenn
Lyr 18h35'56" 41d20'5"
Jenn 17
Lib 15h26'34" -24d53'10"
Jenn and Ariel
Sco 17h51'50" -35d55'2"
Jenn and Fluke Forever!
Cyg 20h45'50" 44d27'24"
Jenn Beau
Sco 16h14'13" -12d8'9"
Jenn Causby
Lyn 7h23'38" 46d35'38"
Jenn Cosio
Ari 2h43'0" 28d21'40"
Jenn Dias
Uma 9h32'3" 44d11'37"
Jenn DiClaudio
Tau 5h39'23" 26d46'3"
Jenn E. Dixon Halifax Nova Scotia 1980
Uma 9h29'12" 60d33'12"
Jenn East
Cas 1h34'56" 61d14'12"
Jenn Elcock
Ari 3h4'59" 27d38'7"
Jenn Fortune
Cap 21h26'58" -23d1'31"
Jenn & Gavin Newman
Cyg 21h58'41" 49d45'31"
Jenn & Gord McCubbin
Lyn 7h35'58" 35d44'4"
Jenn Harvey, Beautiful Artist
Vir 13h37'32" -13d26'2"
Jenn & Jake
Psc 1h25'52" 5d55'8"
Jenn Kline
Lib 15h9'29" -23d11'9"
Jenn Kolada
And 0h46'26" 40d46'13"
Jenn L. Marr
Leo 11h29'8" 20d53'9"
Jenn Lewis
Sgr 19h10'9" -21d50'52"
Jenn Lin 22
Ori 6h20'30" -1d20'23"
Jenn My Sunshinee
Leo 9h29'52" 31d39'20"
Jenn Noll
Uma 8h31'25" 67d57'14"
Jenn Rankin
Uma 9h43'30" 61d21'53"
Jenn Renee
Sco 17h15'19" -44d10'41"
Jenn Rosen, LJCDS, Class of 2005
Uma 11h33'48" 56d56'55"
Jenn Smith
Cas 0h19'21" 61d39'28"
Jenn Star
Sgr 18h15'47" -24d32'44"
Jenn Tim James 4 ever
Leo 9h33'52" 32d6'50"
Jenn & Tom
And 2h10'43" 42d6'0"
Jenn Vincenti
Uma 12h0'41" 47d0'14"
Jenn Voelker
Uma 13h1'40" 58d1'24"
Jenn & Will - October 13th, 2001
Cyg 20h19'41" 43d3'21"
Jenn Windsor
Gem 7h14'14" 33d27'11"
Jenn Windsor
Tri 1h55'15" 33d53'31"
Jenn, Kendall & Lisa's Jewel
Umi 13h9'12" 73d26'8"
Jenn, My Cosmic Girl
Ari 2h59'45" 19d18'52"
Jenn,u are the hi-LIGHT of my life.
Psc 1h28'56" 16d14'14"
Jenna
Psc 0h49'30" 17d10'30"
Jenna
Ari 3h26'49" 20d48'13"
Jenna
Tau 4h23'48" 18d10'13"
Jenna
Leo 11h15'5" 13d42'15"
Jenna
Leo 10h33'35" 13d47'57"
Jenna
Cnc 9h1'13" 7d59'2"
Jenna
Ari 2h18'43" 12d10'58"
Jenna
Aqr 22h28'17" 2d14'51"
Jenna
Leo 11h14'6" 20d33'20"
Jenna
Her 18h9'8" 17d16'14"
Jenna
Tau 4h33'10" 24d33'43"
Jenna
Com 12h21'42" 26d2'24"
jenna
Lyr 19h7'54" 26d55'21"
Jenna
Psc 1h6'46" 32d24'38"
Jenna
Lyn 7h35'34" 36d29'12"
Jenna
Crb 16h1'25" 33d24'41"
Jenna
Crb 16h20'57" 36d7'43"
Jenna
And 0h16'59" 43d33'9"
Jenna
Cyg 20h42'41" 35d40'29"
Jenna
Crb 16h15'19" 38d36'53"
Jenna
And 1h20'0" 45d30'42"
Jenna
And 2h37'50" 50d18'27"
Jenna
Her 16h6'27" 49d10'33"
Jenna
Uma 10h58'21" 55d25'5"
Jenna
Lyn 7h14'20" 53d39'56"
Jenna
Ori 5h35'56" -3d10'53"
Jenna
Umi 16h47'36" 82d18'13"
Jenna
Vir 13h30'24" -14d19'15"
Jenna
Lib 15h43'55" -18d12'18"
Jenna
Vir 13h40'20" -20d10'25"
Jenna
Sco 16h55'32" -42d16'10"
Jenna 21
And 23h59'44" 35d7'21"
Jenna A. Leary
Cyg 20h10'51" 51d20'23"
Jenna aka jbdancer
Aqr 20h40'2" 0d42'3"
Jenna Alizabeth Butler
And 23h6'39" 48d59'44"
Jenna Allyson's Star
Cnc 8h16'44" 30d55'4"
Jenna and Anna
Cas 0h15'27" 56d28'17"
jenna and mike
And 2h19'39" 46d36'41"
Jenna Angela Buckley
Sco 16h54'47" -38d16'9"
Jenna Angelynn
And 2h9'42" 44d29'10"
Jenna Ashley Abdalla
Crb 16h4'0" 35d34'39"
Jenna B. L. Summers
And 2h36'22" 50d0'36"
Jenna "Banina" Toler
Psc 0h24'30" 20d20'8"
Jenna Barber Shines
Uma 12h45'50" 60d3'24"
Jenna Bear
Psc 0h58'0" 19d12'40"
Jenna Benna
Gem 6h21'2" 18d31'42"
Jenna Beth
Ari 1h59'13" 22d3'29"
Jenna Blumie 2006
Cap 20h16'39" -13d35'46"
Jenna Brugger
Cyg 19h54'14" 31d37'57"
Jenna Cambria
Ori 6h5'17" 20d55'28"
Jenna Christine
Lib 15h50'2" -19d52'27"
Jenna Conforti
And 0h38'20" 27d35'5"
Jenna Conley
Sgr 18h2'34" -27d34'1"
Jenna Constance Walker
Vir 12h28'19" 10d38'9"
Jenna Corren Ramsey
Sgr 19h39'43" -13d51'3"
Jenna Dean
Cru 12h48'22" -59d38'34"
Jenna Dillon
Her 17h3'17" 34d33'51"
Jenna Dorothea Peak
Sco 17h53'32" -42d58'57"
Jenna Dove
And 1h10'58" 37d59'43"
Jenna Elaine Bachert
Aqr 22h44'27" 0d40'55"
Jenna Elisabeth Ritter
And 0h37'37" 28d33'37"
Jenna Elizabeth
And 2h30'41" 39d25'51"
Jenna Elizabeth Catanzaro
Sco 16h58'6" -35d47'26"
Jenna Elizabeth Martin Davis
Sgr 18h29'57" -15d56'17"
Jenna Elizabeth Nicole Childress
And 0h32'44" 41d56'24"
Jenna Elizabeth Simmon
Aqr 22h39'55" -15d18'19"
Jenna Elizabeth's Star
And 0h53'21" 43d4'48"
Jenna Elmer
Crb 15h22'2" 28d24'57"
Jenna Florence Mathieson
And 1h17'3" 49d50'59"
"Jenna" Frost
And 2h31'28" 45d19'17"
Jenna Grace Humphrey
And 0h17'20" 29d14'47"
Jenna Grace McDaniel
Gem 6h37'51" 15d18'10"
Jenna Hamilton
Lyn 6h54'22" 55d49'17"
Jenna Is Amazing
Leo 11h33'2" 22d16'18"
Jenna "JB" Sheehan
And 23h28'38" 44d27'22"
Jenna Jenkins
And 2h12'54" 49d6'31"
Jenna Johnson
Cas 1h25'47" 68d30'0"
Jenna Joy
Crb 15h45'40" 32d43'29"
Jenna K. Elliott
Psc 0h4'8" -5d15'12"
Jenna K Rizzo
Cap 21h41'55" -18d45'22"
Jenna Karnatski
And 0h17'29" 32d0'0"
Jenna Kate Conniff
Gem 6h47'26" 25d32'29"
Jenna Katelyn Krueger
Dra 16h3'53" 58d11'42"
Jenna Kay Morin
And 1h4'13" 46d56'15"
Jenna Kelly Saunders
Lib 14h51'42" -4d31'6"
Jenna Kim Marie
Lib 14h56'33" -0d56'48"
Jenna Kinghorn & Morgan Conrad Jr.
Lyn 8h0'37" 41d22'25"
Jenna- Kristine
Sgr 18h12'23" -27d45'56"
Jenna L. Meyer
And 0h40'16" 40d42'9"
Jenna L. Nichols
Aqr 20h39'37" -2d2'25"
Jenna Lane
Gem 6h7'20" 27d31'51"
Jenna Layne Hooter
Sco 16h41'15" -30d57'17"
Jenna Lea
Cas 23h27'8" 56d49'2"
Jenna LeAnn Schaefer
And 1h16'47" 37d31'39"
Jenna Lee
Crb 16h12'7" 37d57'37"
Jenna Lee
Peg 21h32'26" 15d21'0"
Jenna Lee Paparozzi
Cap 20h21'10" -25d23'49"
Jenna Lee Zamelsky
Lyn 7h19'47" 59d25'47"
Jenna LeeAnn
And 0h46'9" 36d50'41"
Jenna Leigh 1
Lib 14h48'23" -12d21'31"
Jenna Leigh Hall
Ari 1h53'7" 17d40'59"
Jenna Leigh Mullins
Uma 11h43'32" 47d55'17"
Jenna Leigh VanDyne
Col 5h55'27" -27d40'31"
Jenna Leigh Yeager
Lib 15h5'41" -6d59'32"
Jenna Lewis Chappell
Sco 16h55'16" -34d6'7"
Jenna Lily Jane Mason
Her 17h4'7" 32d19'37"
Jenna Lopez
Ari 2h37'27" 13d7'7"
Jenna Lyn
Leo 11h51'23" 26d9'37"
Jenna Lyn
Cap 21h41'43" -11d40'34"
Jenna Lyn Milford
And 2h19'1" 50d29'7"
Jenna Lynn
And 1h27'3" 44d38'4"
Jenna Lynn
Cas 23h46'27" 60d57'12"
Jenna Lynn
Sgr 17h54'6" -28d48'8"
Jenna Lynn Aurora
Gem 6h36'1" 22d59'47"
Jenna Lynn "Baby J"
Ari 2h25'3" 25d58'44"
Jenna Lynn McKay
Lib 14h51'55" -1d38'0"
Jenna Lynn McKinney
And 23h26'33" 47d38'28"
Jenna Lynnae
And 23h40'26" 45d27'53"
Jenna Lynne Ferguson
Sco 17h53'44" -31d31'43"
Jenna M. Ianuzzi
Psc 23h0'48" -0d47'6"
Jenna M Kopp
Sgr 17h59'38" -29d56'20"
Jenna Mae
Leo 11h15'36" 14d19'53"
Jenna Mae Bees
Cas 0h37'27" 51d28'50"
Jenna Mae Smith "naIL'em dUdet"
Leo 10h59'48" 17d39'30"
Jenna Mae's Star
Vir 15h7'29" 4d48'45"
Jenna Mann
Ari 2h6'50" 23d8'43"
Jenna Mari Yoshimura
Vir 12h17'12" 5d23'2"
Jenna Marie
Leo 9h46'23" 29d17'33"
Jenna Marie
Sco 16h58'2" -32d21'16"
Jenna Marie
Cma 6h50'9" -14d30'42"
Jenna Marie Carpenter
Sco 17h49'13" -42d44'0"
Jenna Marie Dunsmore
Lib 14h53'10" -2d30'7"
Jenna Marie Faith Ross
And 23h14'35" 51d51'39"
Jenna Marie Goins
Sgr 19h4'23" -25d57'4"
Jenna Marie Holmes
Sgr 18h54'48" -16d58'33"
Jenna Marie Krosch
Cam 3h16'37" 59d19'53"
Jenna Marie Miller
Leo 10h34'37" 9d36'39"
Jenna Marie Moser
Cas 2h17'8" 64d19'34"
Jenna Marie Ruffler
Leo 9h43'53" 29d31'12"
Jenna Marie Street
Cnc 8h24'34" 24d9'9"
Jenna Marie Taono
Cnc 9h11'13" 21d2'39"
Jenna Marie's Star
Cap 21h39'39" -9d22'34"
Jenna Marissa
Her 16h24'6" 45d9'47"
Jenna McLaughlin
And 0h55'44" 40d59'1"
Jenna Mclaughlin
Ori 5h29'58" -0d6'50"
Jenna McNicholas Beacon of Light
Lib 15h36'47" -7d25'16"
Jenna Meade
Cam 5h59'40" 58d14'9"
Jenna Michelle
Leo 9h53'15" 30d23'22"
Jenna Michelle and Chris O'Neal
Her 17h23'26" 38d36'57"
Jenna Michelle Hutchinson
Cap 21h4'22" -22d21'10"
Jenna Murphy
And 23h54'50" 35d54'19"
Jenna Nicole
Cam 3h22'44" 64d7'21"
Jenna Nicole
Sgr 18h28'31" -24d7'3"
Jenna Nicole
Sgr 18h13'15" -34d16'9"
Jenna Nicole Courtney
And 0h29'53" 42d8'1"
Jenna Nicole Creasy
And 0h39'1" 29d46'55"
Jenna NiCole Long
Leo 9h39'4" 10d19'46"
Jenna Nicole Samuelson
And 2h22'1" 42d54'54"
Jenna Noelle Ushkowitz
Lyn 7h48'46" 56d0'20"
Jenna N.P.
And 23h48'43" 47d32'29"
Jenna P. Keadle
And 0h50'30" 43d2'26"
Jenna "PJ" Kerner
Cap 20h7'19" -9d20'26"
Jenna Quitugua Kayyali
Lyn 6h17'47" 57d50'55"
Jenna Rae
Ori 5h52'21" -0d20'1"
Jenna Ray Knight
Cnc 9h13'36" 29d37'41"
Jenna Raykova
Uma 14h6'7" 48d25'44"
Jenna Robb
Peg 21h43'1" 23d27'37"
Jenna Rose
And 2h11'47" 39d13'46"
Jenna Rose
Gem 6h42'36" 30d38'31"
Jenna Rose
Cru 12h40'8" -60d35'30"
Jenna Rose Kayleigh Foran
Aqr 21h52'12" -0d3'50"
Jenna & Scott
Cyg 21h57'9" 52d51'53"
Jenna Sims
Ari 2h35'39" 18d47'17"
Jenna "Skippy" Thompson
Aql 19h13'33" 7d30'37"
Jenna Sperry
Lib 14h58'48" -24d28'27"
Jenna Staup
And 2h31'18" 44d48'46"
Jenna Sue
Vir 13h25'22" 11d28'45"
Jenna Susan
Her 17h24'11" 48d8'1"
Jenna Templeton
And 23h36'6" 43d57'1"
-Jenna- The Devine Goddess Of Love
Tau 5h43'20" 18d54'12"
Jenna (The Teacher)
Ari 2h18'9" 23d32'29"
Jenna Theresa Scalise
Lyn 6h39'20" 56d54'28"
Jenna Townsend
Aqr 22h6'56" -0d53'9"
Jenna Trujillo
Tau 5h32'39" 18d31'13"
Jenna Uhe
And 1h40'12" 43d26'34"
Jenna Watson
And 23h43'43" 46d42'41"
Jenna West
Vir 12h16'51" 12d2'4"
Jenna Wheeler Clark
Vir 13h12'32" -0d54'0"
Jenna, My Special Gem
Sco 16h15'26" -14d56'7"
jennaferlyn & allen
Leo 10h10'47" 20d20'18"
Jennah DiLauro
Cyg 19h35'54" 29d22'36"
Jennalee
And 2h33'52" 40d20'1"
Jennalise Hall
Uma 11h36'57" 42d54'46"
Jennalyn 12
Cyg 21h36'56" 52d45'21"
Jennamarie9
Psc 1h26'35" 16d16'1"
JennAndron IV
Mon 6h37'13" 11d9'2"
Jenna-Piper
Cmi 7h32'20" 10d31'24"
JennaRob
Sco 16h10'18" -13d42'5"
Jenna-Rose
And 1h17'53" 38d45'26"
Jenna's Diamond
And 2h22'57" 50d10'28"
Jenna's Little Star
Uma 11h45'36" 41d33'16"
Jenna's Star
And 1h27'15" 43d43'30"
Jenna's Star
And 23h9'18" 51d46'26"
Jenna's Star
Tau 4h37'40" 21d15'9"
Jenna's Star
Tau 4h18'22" 4d26'7"
Jenna's sweetness star
Psc 0h34'41" 14d28'18"
Jenna's Wishes
Cap 20h28'51" -14d17'34"
Jennasea
Mon 6h47'54" 7d57'9"
Jennatron
Aqr 23h31'12" -13d37'46"
Jennavieve Elizabeth Schrage
Vir 13h52'36" -2d7'57"
Jennay Bugg
Uma 9h8'36" 54d35'24"
Jennbrez
Aqr 22h26'9" -5d15'53"
JennDamere05
Vir 12h36'37" -0d24'3"
JennDipper
Leo 10h45'27" 16d56'35"
Jenne Diane Naper
Lyn 7h34'42" 35d40'9"
Jenne Lynn Anderson
Mon 6h30'12" 1d36'12"
Jenne S Frisby
Cnc 8h53'23" 13d35'24"
JENNE'E
Her 15h52'7" 43d25'24"
JenneeLee 2-24-83
Psc 0h52'48" 27d14'18"
Jennefer L. Filauri
Lyn 7h38'11" 45d57'43"
Jennell Carcich
Umi 15h59'56" 78d4'39"
Jennell Lenea
Cnc 8h53'30" 32d11'32"
Jennelle
And 1h15'1" 37d9'57"
Jennelle Koerner
Lib 14h50'11" -3d17'13"
Jennelle Kristin
Cas 23h45'59" 59d52'50"
Jenners14
Tau 3h29'56" 7d1'23"
Jennessee Lorenzo
Ari 3h6'49" 29d24'22"

Jennet and Matthew Singleton
Cyg 20h56'48" 47d58'10"
Jennettacless
Ari 2h15'7" 23d22'47"
Jennette Renea Henderson
Cas 1h15'31" 63d52'47"
Jennfier Ainsworth
Sgr 18h3'42" -21d4'23"
JennGer
Cyg 20h13'50" 52d0'47"
Jenni
And 0h30'56" 29d50'33"
Jenni
And 0h36'28" 24d18'35"
Jenni
Psc 1h12'37" 28d14'41"
Jenni
Gem 7h6'24" 23d45'59"
Jenni
Srp 18h28'31" -0d18'3"
Jenni & Allen
Cyg 21h52'11" 53d37'29"
Jenni Angel 24
And 0h43'45" 44d2'22"
Jenni Ann Harmon
Leo 11h38'29" 21d2'14"
Jenni Baker
And 0h55'28" 36d10'46"
Jenni Beth Browning
Cnc 8h36'16" 19d32'31"
Jenni Brice
Leo 10h20'31" 16d30'37"
Jenni Brooke Ranes
Sco 17h29'58" -38d9'30"
Jenni Decker-Shelton - Tupperware
Mon 6h33'9" 8d54'41"
Jenni Dillman
And 0h12'4" 41d7'56"
Jenni E Greer
Gem 6h59'45" 13d34'23"
Jenni Eckstrom
Gem 6h46'26" 17d22'6"
Jenni Elisabeth
Lyr 18h31'30" 39d0'6"
Jenni Espinal My Little Butterfly
Cyg 20h9'52" 48d22'38"
JENNI GARDNER
Gem 7h40'23" 15d55'7"
Jenni Girl's Light
Uma 12h34'38" 62d4'18"
Jenni Houser: Grandma's Bright Star
Cas 0h47'43" 54d40'47"
Jenni & Jim Gevry
Uma 9h15'4" 49d31'4"
Jenni Jo Garber
Leo 11h1'58" 15d31'2"
Jenni Joyce Fitzgerald
Ari 2h8'5" 24d14'53"
Jenni Kuhn
Cas 1h20'35" 61d52'15"
Jenni Leigh
Tau 4h29'58" 2d54'41"
Jenni Leigh Brunson
Vir 13h34'19" -1d49'23"
Jenni Marie
Uma 11h54'1" 49d40'5"
Jenni Maureen
Cyg 19h42'39" 40d50'27"
Jenni Meloy
Gem 6h44'42" 16d45'51"
Jenni - Minni
Her 18h15'48" 19d2'25"
Jenni mon princesse
Cyg 20h5'33" 37d26'35"
Jenni Nickels
Ari 3h15'35" 29d57'24"
Jenni Ottobre "A 21st Star"
Sco 17h17'2" -45d30'19"
Jenni Rose
Ori 6h15'58" 15d45'19"
Jenni & Sam - Love Eternal
Col 5h22'22" -42d14'48"
Jenni Schmid
Uma 11h26'13" 54d32'2"
Jenni Schroeder & Bradley Gold
Vir 13h55'20" -8d50'53"
Jenni Spenny
And 0h25'50" 29d44'34"
Jenni Stack
Cas 0h43'52" 63d10'35"
Jenni Wren
And 0h37'42" 37d25'0"
Jenni's Hochzeitsstern
Uma 9h19'42" 72d15'34"
Jennia Loree
Sgr 19h9'4" -16d38'5"
Jennibean
Psc 0h56'5" 29d26'14"
Jennibear
Uma 9h29'42" 64d58'47"
Jennica Ann Rudzik
Gem 6h41'47" 20d41'26"
Jennica Denice (Buttons)
Cap 21h36'38" -13d41'23"
Jennica June Church
And 1h25'58" 48d23'33"
Jennie
And 23h25'19" 47d8'23"
Jennie
And 0h22'13" 38d28'9"
Jennie
Tau 4h40'57" 17d15'56"
Jennie
Peg 23h47'48" 10d0'33"
Jennie
Ori 6h19'5" 9d58'37"
Jennie
Ori 5h57'29" 9d37'29"
Jennie
Lib 14h50'22" -3d41'21"
Jennie
Lyn 6h45'21" 57d32'6"
Jennie
Sco 17h56'11" -41d12'39"
Jennie
Sco 16h15'22" -28d26'43"
Jennie
Cap 20h28'54" -23d57'38"
Jennie and Pete's Star
Cyg 20h13'34" 38d10'25"
Jennie Aquino
Aqr 22h16'59" -18d54'54"
Jennie Barnes
Uma 13h53'39" 50d27'47"
Jennie Berkson-Edelstein
And 0h41'24" 43d14'15"
Jennie Capasso
Uma 10h20'54" 44d50'20"
Jennie Catherine Dickovitch
Vir 13h3'22" 12d31'32"
Jennie Cheng
Sgr 18h55'21" -31d32'4"
Jennie Crane
Cas 1h29'37" 64d35'47"
Jennie Elizabeth Curry
Cyg 20h18'12" 34d16'55"
Jennie Elizabeth Jacobs
Uma 13h49'37" 48d39'49"
Jennie Frances Mell
Ari 2h40'26" 29d27'56"
Jennie Francis
Cas 0h27'30" 50d1'16"
Jennie G.
Cap 21h43'14" -13d51'39"
Jennie Gambino
Cas 0h49'57" 63d11'47"
Jennie Gamble
Uma 10h43'30" 60d4'2"
Jennie Gazetas
Lib 15h49'47" -19d14'33"
Jennie Giger
Crb 16h6'25" 37d15'44"
Jennie Hall Jerome
Ari 2h16'22" 23d31'50"
Jennie Haynes
Vir 13h43'48" -1d15'7"
Jennie Heston
Cap 21h10'52" -26d21'25"
Jennie & Justin always & forever
And 23h37'52" 47d56'28"
Jennie Katharine Boyd
Lmi 10h19'9" 34d40'30"
Jennie Kawahara
Ari 2h26'34" 10d31'28"
Jennie Kaye
Eri 4h5'47" -0d35'15"
Jennie Kobayashi
Uma 11h56'3" 36d44'26"
Jennie Laird
Cam 4h20'19" 63d30'53"
Jennie Lynn Reeves
Tau 5h27'29" 26d58'47"
Jennie Marie
Uma 9h27'15" 53d47'16"
Jennie Marie Behrens likes girls.
Ari 2h40'22" 27d45'49"
Jennie Marie Lucci
Gem 6h29'50" 12d40'53"
Jennie Mikay
Cas 1h42'15" 62d5'20"
Jennie & Mike 6-1-2001
Cyg 21h51'46" 47d13'19"
Jennie Monaco
Psc 1h26'12" 23d4'11"
Jennie Morris
Ori 5h52'4" 5d57'28"
Jennie MySpace Novak
Vir 13h11'2" -17d0'44"
Jennie Napoli
Lyn 8h11'21" 58d48'4"
Jennie Penning
And 23h2'56" 41d35'58"
Jennie Pie
Tau 3h58'52" 4d27'6"
Jennie Portalatin
Cas 1h37'41" 68d26'27"
Jennie Quinby 100
And 22h59'0" 47d5'50"
Jennie Rebecca Mayer
Lib 15h6'0" -9d45'9"
Jennie Ross Wild
Uma 9h13'44" 65d11'12"
Jennie Weaver
Peg 21h55'49" 11d46'18"
Jennie Wren (J.A.Forster)
Cas 23h56'29" 59d55'38"
Jennie, Craig's Shining Star
Aur 5h11'8" 42d14'11"
JennieDrew Kidwell
Cyg 20h41'40" 31d25'12"
Jenniekathryn
Tau 4h44'21" 23d36'30"
Jennie-Mae
Mon 7h54'51" -10d21'4"
Jenniepants
And 0h39'7" 37d13'57"
jennie-wren
Cas 0h36'53" 55d46'17"
Jennifer
Per 2h42'13" 52d32'38"
Jennifer
Lac 22h27'57" 49d9'29"
Jennifer
And 23h31'36" 47d21'38"
Jennifer
And 23h21'55" 38d9'31"
Jennifer
Cyg 21h44'33" 41d47'45"
Jennifer
And 23h28'5" 44d49'1"
Jennifer
And 23h29'48" 42d38'3"
JENNIFER
Cyg 20h13'29" 41d38'13"
Jennifer
And 2h17'35" 50d36'31"
Jennifer
And 2h25'39" 49d3'58"
Jennifer
And 2h36'36" 45d11'12"
Jennifer
And 2h33'17" 46d1'26"
Jennifer
And 1h5'28" 37d27'13"
Jennifer
Tri 2h16'35" 31d57'54"
Jennifer
Gem 7h5'16" 32d41'42"
Jennifer
And 1h16'7" 41d45'12"
Jennifer
And 1h40'31" 38d10'14"
Jennifer
And 2h3'4" 38d39'19"
Jennifer
And 2h11'5" 38d13'46"
Jennifer
And 0h59'0" 37d56'27"
Jennifer
And 0h47'58" 39d21'58"
Jennifer
And 0h39'48" 44d10'41"
Jennifer
And 0h46'40" 41d9'29"
Jennifer
Tau 4h36'11" 24d22'51"
Jennifer
Tau 5h54'19" 25d9'47"
Jennifer
Tau 3h52'18" 28d15'44"
Jennifer
Gem 6h44'43" 23d1'10"
Jennifer
Gem 6h22'42" 26d23'52"
jennifer
Leo 9h47'0" 27d8'29"
Jennifer
Psc 1h8'53" 28d32'59"
Jennifer
Ari 2h11'21" 23d37'51"
Jennifer
Peg 22h19'31" 16d20'46"
Jennifer
Gem 7h22'59" 16d40'19"
Jennifer
Gem 7h17'7" 19d31'49"
Jennifer
Gem 6h16'25" 22d26'25"
Jennifer
Psc 0h52'44" 17d14'54"
Jennifer
Aql 19h31'21" 8d3'32"
Jennifer
Tau 4h46'39" 18d18'27"
Jennifer
Tau 4h26'28" 17d44'40"
Jennifer
Tau 4h34'43" 18d12'46"
Jennifer
Tau 4h0'33" 18d46'22"
Jennifer
Ori 6h10'41" 15d37'14"
Jennifer
Her 17h53'35" 14d29'2"
Jennifer
Leo 10h34'19" 7d40'56"
Jennifer
Tau 4h28'4" 2d26'59"
Jennifer
Ori 5h15'33" 6d48'36"
Jennifer
Cmi 8h1'38" 1d28'34"
Jennifer
Ori 5h12'39" 12d37'17"
Jennifer
Psc 23h17'38" 0d41'32"
Jennifer
Del 20h56'6" 7d11'37"
Jennifer
Aqr 21h42'24" -0d11'45"
Jennifer
Vir 14h18'45" -4d43'56"
Jennifer
Vir 12h32'51" -1d10'55"
Jennifer
Lib 15h2'57" -0d33'58"
Jennifer
Mon 6h52'51" -0d4'37"
Jennifer
Mon 7h21'59" -0d27'7"
Jennifer
Psc 0h6'9" -0d46'47"
Jennifer
Lib 15h38'13" -12d14'39"
Jennifer
Cap 21h49'40" -13d24'15"
Jennifer
Lib 15h55'19" -18d21'48"
Jennifer
Cas 23h13'40" 54d36'44"
Jennifer
Cas 1h58'2" 62d15'13"
Jennifer
Uma 9h10'24" 63d26'54"
Jennifer
Uma 9h31'9" 60d36'24"
Jennifer
Cas 1h2'59" 69d35'43"
Jennifer
Uma 10h16'38" 67d30'36"
Jennifer
Cam 7h22'46" 70d35'53"
Jennifer
Dra 18h7'6" 70d29'2"
Jennifer
Umi 16h12'30" 72d6'32"
Jennifer
Cap 21h27'25" -24d44'3"
Jennifer
Cap 20h27'30" -26d22'52"
Jennifer
Sco 17h58'8" -37d28'50"
Jennifer
Sco 17h21'57" -31d43'58"
Jennifer
Sgr 19h17'13" -28d47'10"
Jennifer
Sgr 18h43'19" -24d7'53"
Jennifer
Sco 16h52'15" -43d44'29"
Jennifer
Sco 17h53'5" -38d16'19"
Jennifer
Sco 17h35'41" -39d2'2"
Jennifer #1
Cnc 9h14'44" 15d43'44"
Jennifer 1
And 23h19'57" 51d27'31"
Jennifer "12-16-91"
Cam 6h29'59" 62d30'28"
Jennifer 1994 I Love You
And 2h25'30" 45d41'3"
Jennifer 2005
Lmi 9h51'50" 41d22'45"
Jennifer 25
Cas 0h25'56" 56d47'38"
Jennifer 912
Cam 5h32'35" 66d53'3"
Jennifer A. Beauchamp
Sgr 18h20'54" -27d46'14"
Jennifer A. Boyce
Sco 16h43'18" -32d34'21"
Jennifer A. Brisse
Gem 6h26'49" 23d6'9"
Jennifer A. Hanyecz
Uma 11h35'3" 48d2'38"
Jennifer A Johnson
And 0h42'15" 42d21'13"
Jennifer A. Mendoza
And 0h16'21" 42d46'9"
Jennifer A. Salemme
Psc 1h33'56" 7d45'23"
Jennifer A Sanchez
Leo 11h13'7" 22d20'30"
Jennifer A. Zubyk
Cnc 8h34'19" 13d13'14"
Jennifer Abbot Bautz
Lib 14h49'18" -0d51'41"
Jennifer Ackerman
Sco 16h10'20" -10d55'32"
Jennifer Adams, my beautiful wife
Cas 0h32'14" 51d43'22"
Jennifer Adele Rothenberg
Gem 7h7'2" 33d46'58"
Jennifer Agnes Yoon
Uma 8h55'0" 47d23'37"
Jennifer (aka: Mommy)
And 0h40'44" 33d39'57"
Jennifer Alaine Brown
Sgr 19h41'26" -14d16'25"
Jennifer Alexandria
Sco 16h14'44" -10d42'53"
Jennifer Alexis May
Psc 1h20'48" 4d44'16"
Jennifer Alice Conners
Tau 5h44'31" 23d34'55"
Jennifer Alice Welchman
And 1h45'11" 48d55'33"
Jennifer Alison Wozniak
Tau 5h10'23" 25d14'15"
Jennifer Allen
Sco 16h49'45" -26d27'41"
Jennifer Allison Floyd
Uma 13h49'37" 51d59'20"
Jennifer Amanda Beltz
Aql 19h49'33" 16d3'2"
Jennifer Amera
And 0h14'39" 42d12'4"
Jennifer Ammons
Lyn 8h36'59" 38d51'51"
Jennifer and Alan
Cyg 20h44'18" 33d28'51"
Jennifer and Ava Lee
Leo 11h0'56" 17d51'1"
Jennifer and Brandon
Cyg 20h13'13" 32d34'52"
Jennifer and Carie May 19, 2006
Cas 23h2'28" 55d11'4"
Jennifer and David
Sco 17h58'3" -34d57'34"
Jennifer and Dean Forever
Cyg 20h9'19" 58d28'8"
Jennifer and James McGinn
Col 5h31'54" -34d44'14"
Jennifer and Jason
Sco 16h11'10" -10d31'3"
Jennifer and Jimmy
Umi 16h17'29" 76d55'57"
Jennifer and Joseph
Cyg 20h30'10" 56d14'56"
Jennifer and Joshy
Cyg 21h27'35" 37d13'9"
jennifer and kevin
Cyg 21h39'56" 36d9'16"
Jennifer and Michael Mitra
Cap 21h52'41" -14d19'22"
Jennifer and Michael Reich
Sge 19h12'32" 18d55'18"
Jennifer and Oleg Bortman
Sge 19h35'19" 19d0'19"
Jennifer and Richard's Star
Cyg 21h49'5" 41d38'57"
Jennifer And Troy
Uma 9h21'51" 72d53'24"
Jennifer Andrews
Lib 15h45'9" -6d34'11"
Jennifer - Angel - Star
Uma 11h25'39" 37d0'7"
Jennifer Angelique
Gem 6h42'18" 18d12'40"
Jennifer & Angi
Peg 21h34'0" 24d57'41"
Jennifer Ann
Leo 9h45'38" 27d16'57"
Jennifer Ann
Ori 5h31'59" 4d46'18"
Jennifer Ann
Lyn 7h0'45" 47d30'30"
Jennifer Ann
Mon 6h51'47" -0d26'28"
Jennifer Ann Agnew
Tau 4h2'46" 26d36'39"
Jennifer Ann Allor
And 2h37'10" 44d9'57"
Jennifer Ann Barrett (Peanut)
And 23h26'34" 44d48'10"
Jennifer Ann Bentley Birthday Star
Cas 1h18'20" 68d13'38"
Jennifer Ann Bowling
Psc 1h20'34" 2d45'47"
Jennifer Ann Bullock
Cnc 8h58'2" 23d35'34"
Jennifer Ann Cioffi
Leo 11h46'47" 20d40'31"
Jennifer Ann Cioffy
Leo 11h10'55" -2d10'32"
Jennifer Ann Crecente
Aql 20h5'14" 1d28'6"
Jennifer Ann Cullinan
Aqr 22h57'16" -10d57'0"
Jennifer Ann Elizabeth Steinher
And 1h8'59" 38d14'47"
Jennifer Ann Engriser
Leo 11h6'16" 21d50'20"
Jennifer Ann Fletcher
And 0h21'41" 46d2'2"
Jennifer Ann Glass
And 0h42'28" 27d24'51"
Jennifer Ann Horton
Sgr 18h48'42" -17d3'29"
Jennifer Ann Huskey
Cas 0h46'50" 51d51'4"
Jennifer Ann Johnson
Uma 11h23'51" 30d0'41"
Jennifer Ann Justice
Uma 10h22'37" 41d50'28"
Jennifer Ann Kelley
Aqr 21h47'36" -2d13'22"
Jennifer Ann Lavine
Gem 6h51'13" 20d47'47"
Jennifer Ann Majors
And 0h34'52" 23d34'37"
Jennifer Ann McFarland
Uma 10h37'53" 46d26'6"
Jennifer Ann Middleton
Leo 10h27'27" 15d9'8"
Jennifer Ann Milot
Lib 15h27'35" -13d40'0"
Jennifer Ann Nordstrom
Cap 20h41'9" -19d4'30"
Jennifer Ann Pommerenk
Sco 17h55'54" -37d55'21"
Jennifer Ann Rivera
Leo 11h7'2" 22d26'19"
Jennifer Ann Sitton
And 0h41'24" 36d31'50"
Jennifer Ann Stevens
Tau 4h21'4" 24d43'48"
Jennifer Ann Stone
And 23h24'37" 49d1'41"
Jennifer Ann Timbers
Gem 7h13'8" 28d20'49"
Jennifer Ann Toby Hayse
Col 5h32'52" -34d3'55"
Jennifer Ann Townsend
And 1h40'56" 43d48'23"
Jennifer Ann White
Gem 7h15'19" 32d8'50"
Jennifer Ann Wreski
And 23h48'12" 33d30'23"
Jennifer Ann Wright
Ori 5h7'57" 5d21'33"
Jennifer Anne Carr
And 1h43'41" 37d48'16"
Jennifer Anne Contapay Asercion
Vir 13h1'27" 5d59'51"
Jennifer Anne Garner
Ari 2h46'56" 27d15'19"
Jennifer Anne Harrison
Cap 20h15'11" -23d16'31"
Jennifer Anne & Jonathan Calvin
Leo 9h53'50" 13d48'46"
Jennifer Anne Linsenmeyer
Vir 12h44'9" 5d26'59"
Jennifer Anne Mickunas
Vir 14h9'56" -1d14'25"
Jennifer Anne Smulow
Ari 2h3'36" 20d40'43"
Jennifer Anne Wolf
And 1h37'24" 48d17'43"
Jennifer Anne Wong
Cas 23h36'36" 53d2'17"
Jennifer Arellano Culbertson
Aqr 22h49'22" -10d44'37"
Jennifer Ashlee Hatfield
Gem 7h43'17" 17d59'17"
Jennifer Ashley Heyden
Leo 11h10'5" 9d43'44"
Jennifer Ashley Roth
And 2h18'29" 47d52'8"
Jennifer Ashley Smith
Aqr 21h40'28" -1d16'55"
Jennifer Atkins
Aqr 21h56'27" -0d23'55"
Jennifer Augspurger
Sco 16h47'21" -34d18'30"
Jennifer B.
Cap 21h42'26" -18d43'19"
Jennifer B. Callinan
Psc 1h27'48" 32d3'59"
Jennifer Baby Girl Dooley
Cnc 8h53'53" 13d54'46"
Jennifer Bailey
Gem 6h43'5" 24d25'0"
Jennifer Bailey
And 23h26'23" 47d23'6"
Jennifer Baker
Gem 7h42'20" 17d58'54"
Jennifer Barley
Cas 23h36'12" 54d19'52"
Jennifer Bates
Leo 9h27'10" 14d33'1"
Jennifer Baum
Ari 2h9'33" 23d59'4"
Jennifer Beals
Gem 7h43'30" 16d13'24"
Jennifer Behnke
And 0h49'9" 37d11'45"
Jennifer Bell
Com 12h39'20" 27d51'37"
Jennifer Bell
Lib 15h33'13" -21d20'33"
Jennifer (Benapfl) Schmitt
Mon 6h53'27" -0d19'58"
Jennifer Bengtzen
Aqr 21h12'32" -11d30'52"
Jennifer Bertolani
Lmi 10h37'37" 31d35'5"
Jennifer Beth Levine
Sco 16h7'42" -10d31'52"
Jennifer Bethany Powell
Uma 11h14'8" 36d45'41"
Jennifer & Betty
Cyg 19h41'3" 51d27'14"
Jennifer Bever's Eternal Guide
Crb 15h35'18" 32d36'45"
Jennifer Biondo
Sco 17h56'30" -30d26'13"
Jennifer Bloomfield
Cnc 8h18'39" 9d22'3"
Jennifer Boniface
And 2h37'57" 43d26'23"
Jennifer Boulos- Kryptonite
Gem 7h19'38" 25d50'47"
Jennifer Boyd
Cra 19h3'15" -39d34'23"
Jennifer Bradshaw
Apu 17h9'50" -79d19'6"
Jennifer Brand Loves Darren Geiger
Cas 22h57'23" 54d31'56"
Jennifer Brandon
Cnc 8h5'49" 15d53'54"
Jennifer Brandon
Cnc 8h36'17" 20d28'24"
Jennifer Branton
Lib 14h57'26" -15d55'12"
Jennifer Bray
And 1h23'30" 40d42'8"
Jennifer Breezy Marie White
Cap 20h29'12" -9d53'45"
Jennifer Brickner
Sco 17h50'8" -39d4'10"
Jennifer Bridden
And 0h1'52" 41d5'29"
Jennifer Bridget's Star
And 23h47'29" 36d14'3"
Jennifer "Bright Eyes" McCrary
Lyn 6h55'56" 57d58'48"
Jennifer Brooke Hedrick
Sgr 18h35'9" -30d29'17"
Jennifer Brooke Kasper
Aqr 22h34'40" -0d32'52"
Jennifer Brooke Lambert
Uma 11h19'38" 36d30'15"
Jennifer Brown
Com 12h21'35" 28d27'58"
Jennifer Brown McDonald
Uma 10h26'10" 69d26'34"
Jennifer Brown Star
And 1h15'10" 38d4'56"
Jennifer Bryce Clay
Leo 9h38'38" 13d4'12"
Jennifer Buchanan
Aqr 21h13'29" -3d38'15"
Jennifer Burley
Ari 2h45'18" 18d48'29"
Jennifer Bylciw Crisson
Lib 15h4'31" -22d1'19"
Jennifer C. Damboragian
Leo 9h36'28" 29d34'55"
Jennifer C. Huggins
And 0h2'33" 41d15'55"
Jennifer C. Pelham
Crb 15h25'4" 31d33'7"
Jennifer Calderon
Crb 16h10'6" 33d38'32"
Jennifer Calvanese "1970-2001"
Umi 14h27'44" 73d58'57"
Jennifer Camp
Cnc 9h8'26" 14d26'42"
Jennifer Carbonell
Psc 0h41'52" 10d23'9"
Jennifer Carlie "Baby"
Leo 11h19'29" 15d48'9"
Jennifer Carlo
Uma 10h14'29" 50d4'4"
Jennifer Carmen Fenech
Cas 0h37'47" 62d34'48"
Jennifer Carnes
Psc 1h5'23" 32d35'15"
Jennifer Carol Lee
And 1h21'25" 37d48'46"
Jennifer Carter Carnes
Sgr 18h4'1" -27d4'42"
Jennifer Carter Pritchard
Uma 9h21'10" 43d4'17"
Jennifer Cassady
Ari 1h59'59" 23d34'21"
Jennifer Castellanos
And 23h55'4" 46d29'6"
Jennifer Castro Leon
Ari 3h28'44" 22d11'33"
Jennifer Catherine
Gem 7h12'38" 32d31'22"
Jennifer Catherine Boehle
Psc 0h55'0" 29d41'36"
Jennifer Catherine Cardella
Cyg 20h20'45" 58d9'2"
Jennifer Cathleen Houseman
Aqr 21h35'32" 2d21'13"
Jennifer Cathrine Brown
Cnc 8h29'55" 22d30'45"
Jennifer Cauthen
Cma 6h47'43" -16d32'2"
Jennifer Caye's Star
Ori 6h10'28" 15d1'29"
Jennifer Cebrian
Vir 12h43'5" 4d1'25"
Jennifer Cerrone-Shining Star
And 1h9'33" 41d42'24"
Jennifer Chambers
Ari 3h4'27" 18d43'24"
Jennifer Chan
And 22h59'32" 52d2'7"
Jennifer Chan
Cas 23h39'27" 52d34'0"
Jennifer Chapa
Lib 15h26'18" -7d52'58"
Jennifer Chelsea Heshon - Jennifer's Star
And 0h6'43" 39d45'34"
Jennifer Chiao
Lyn 6h44'57" 52d33'26"
Jennifer Chin
Psc 1h48'51" 3d50'53"
Jennifer & Chris
Uma 11h51'9" 31d35'11"
Jennifer Christine Dammann
Cnc 9h13'10" 32d24'20"
Jennifer Christine Fauver
Cnc 8h2'31" 26d25'38"
Jennifer Christine Johnson
Leo 10h4'47" 26d34'33"
Jennifer Christine Rhodes
Uma 11h55'34" 41d7'46"
Jennifer Christine Rogers
Psc 1h25'44" 17d3'13"

Jennifer Christine Sireci
Gem 7h6'54" 31d16'43"
Jennifer Chung - Lil' Piggie
Aqr 22h56'19" -8d18'23"
Jennifer Clair
And 1h53'34" 36d22'32"
Jennifer & Clarence Carr
Cyg 19h49'4" 37d58'0"
Jennifer Clement's SuperStarFlare
Vir 13h36'42" 4d22'25"
Jennifer Coen (Princess)
And 2h19'18" 47d13'35"
Jennifer Cohen
Ari 3h2'49" 13d40'55"
Jennifer Coleman
Apu 14h41'8" -74d9'43"
Jennifer Colleen Hashbarger
Tau 4h38'40" 24d43'4"
Jennifer Colony
Aqr 20h47'6" -12d41'6"
Jennifer Connolly
And 23h48'4" 36d59'51"
Jennifer Cooper
Cas 1h24'38" 51d48'35"
Jennifer Cooper
Sco 16h59'43" -44d35'15"
Jennifer Cordaro
Lyn 7h36'39" 57d28'16"
Jennifer Correia
And 23h11'42" 42d59'13"
Jennifer Correll
Cnc 8h46'55" 30d32'56"
Jennifer Crabtree
Oph 16h25'34" -0d7'42"
Jennifer - Craig
Cap 20h43'13" -25d42'6"
Jennifer Craig
Ori 5h52'18" 21d0'53"
Jennifer Crandall
Cas 0h30'33" 50d52'22"
Jennifer Cruz
And 2h38'41" 46d23'21"
Jennifer Cuvin
And 1h15'17" 46d37'19"
Jennifer D. Asher
And 1h9'17" 38d48'42"
Jennifer D. Elbert
And 0h28'52" 43d19'3"
Jennifer D Murray
And 0h29'20" 29d38'58"
Jennifer D. Nolt
Dra 16h7'34" 57d10'46"
Jennifer Dadson
Cnc 8h44'14" 28d48'31"
Jennifer Dale
Lyn 7h33'17" 42d34'39"
Jennifer Damron
Ari 2h18'12" 23d31'29"
Jennifer D'Angelo
Aqr 23h53'2" -7d6'39"
Jennifer Danielle Goulet
Sgr 19h12'58" -17d45'51"
Jennifer Danielle Honeycutt
And 0h43'52" 43d51'25"
Jennifer Daniels
Tau 4h8'7" 12d58'4"
Jennifer Davis
Leo 10h36'41" 15d33'51"
Jennifer Davis
Leo 9h27'33" 25d5'9"
Jennifer Davison
Vul 19h4'38" 24d12'37"
Jennifer Dawn
And 23h25'11" 46d46'47"
Jennifer Dawn Andrade
Psc 0h41'24" 6d51'27"
Jennifer Dawn Cotton
Cyg 19h35'57" 41d36'50"
Jennifer Dawn Harzke "16"
Sco 16h26'38" -25d45'13"
Jennifer Dawn Lee Miller
And 0h59'27" 38d12'17"
Jennifer Dawn Lynch
Sgr 18h54'8" -16d32'45"
Jennifer Dawn Neugent
And 1h55'37" 38d26'11"
Jennifer Dawn Rakow
Cas 1h28'3" 62d22'47"
Jennifer Dawn Rheeling
Cap 20h37'22" -21d2'59"
Jennifer Dawn Wilke
Ari 3h6'59" 16d13'16"
Jennifer Deane Stuppy
Ori 6h16'23" 14d5'32"
Jennifer Debra Frangipani
Sco 17h40'26" -40d5'12"
Jennifer Decker
Lyn 7h31'34" 49d55'57"
Jennifer Decorus Angelus
Dra 18h51'39" 50d9'29"
Jennifer Decrey
Cap 21h41'11" -13d50'39"
Jennifer DeJohn - "Sweet-Pea"
And 23h6'24" 48d6'36"
Jennifer Delaney
Com 12h33'43" 17d28'5"
Jennifer Dempsey
And 23h17'56" 47d32'50"
Jennifer DeMuria
Oph 17h19'36" -23d10'48"
Jennifer Denaro
Cam 6h38'26" 64d13'44"
Jennifer Denise
Leo 11h12'15" 24d16'34"
Jennifer Denise Beesley
Lib 15h5'5" -4d32'40"
Jennifer Denise Moore
And 1h7'30" 43d26'44"
Jennifer Denise Short
Cap 20h31'1" -13d46'51"
Jennifer Denise White
Uma 13h47'35" 52d19'4"
Jennifer Dennis
Uma 9h13'54" 53d48'24"
Jennifer Derrick
Ori 6h13'20" 15d6'21"
Jennifer Diane
Cap 21h49'25" -13d11'22"
Jennifer Diane Adair
Uma 11h54'37" 28d22'22"
Jennifer Diane Kalinowski
Cap 20h35'15" -23d36'32"
Jennifer Diane O'Brien
And 23h39'40" 46d56'25"
Jennifer Dieterle
And 1h2'26" 42d28'14"
Jennifer Dill Reinhard
Crb 15h38'15" 27d18'12"
Jennifer DiMario
And 23h6'40" 49d22'20"
Jennifer Dobbins
Cyg 21h59'34" 47d13'54"
Jennifer Donnelly
Cas 1h12'34" 53d58'13"
Jennifer Dooms
Cas 2h40'34" 63d17'2"
Jennifer Dorothy 4503
Ori 5h27'9" 2d20'31"
Jennifer Drago
Lyn 8h27'8" 39d0'0"
Jennifer Dru Arney
Cnc 8h2'40" 21d39'23"
Jennifer Dryden
And 23h0'41" 51d18'11"
Jennifer Dudley, Mon Amour Mon Tout
Lib 15h0'21" -17d55'29"
Jennifer Dugan-Saghir
Psc 0h38'1" 6d58'16"
Jennifer Dullaghan
Cnc 9h21'18" 11d20'24"
Jennifer Dunlap
Lib 15h8'36" -6d33'33"
Jennifer Duttry
And 0h43'54" 36d43'24"
Jennifer E. Hart
Gem 6h4'0" 25d18'25"
Jennifer E. Mendez
Lib 14h57'26" -9d50'28"
Jennifer E. R. Gibson
Ari 2h54'44" 24d45'25"
Jennifer Eaton
Gem 6h28'57" 25d4'2"
Jennifer Edith Kemmerly
Sgr 19h33'59" -18d41'5"
Jennifer Egbert
Vir 12h38'59" 7d53'29"
Jennifer Eilis Duffy
And 22h58'51" 39d32'41"
Jennifer Elaine
Sco 16h20'21" -27d56'21"
Jennifer Elaine Liepis
Dra 15h54'29" 54d47'32"
Jennifer Elise Davis
Ori 6h5'54" -0d39'11"
Jennifer Elise Poston
Aqr 22h8'47" -9d22'40"
Jennifer Elise Stothart
Mon 6h28'0" 9d4'54"
Jennifer Elizabeth
Tau 3h59'49" 22d22'58"
Jennifer Elizabeth
Tau 4h17'17" 16d56'8"
Jennifer Elizabeth
And 2h23'35" 48d23'39"
Jennifer Elizabeth
Sgr 18h7'0" -16d40'0"
Jennifer Elizabeth
Cas 23h5'28" 58d19'0"
Jennifer Elizabeth Ang
Sco 16h57'34" -37d0'1"
Jennifer Elizabeth Budge
Uma 12h0'23" 48d34'57"
Jennifer Elizabeth Byrne
And 0h17'37" 26d17'29"
Jennifer Elizabeth Iacono
Cas 1h28'43" 62d38'19"
Jennifer Elizabeth Madigan's Star
Cas 0h26'55" 64d41'53"
Jennifer Ellen Broadhurst
Ori 6h18'22" 15d23'57"
Jennifer Ellen Buchan
And 22h59'16" 41d23'24"
Jennifer Ellingboe
Ari 3h24'49" 29d13'5"
Jennifer Elyse Abeli
And 23h8'13" 49d9'48"
Jennifer Emily
Sgr 18h5'35" -27d3'5"
Jennifer Endsley
Cas 1h31'57" 56d52'18"
Jennifer & Eric Smith
Cyg 21h12'42" 40d36'47"
Jennifer & Erik for Eternity
Cyg 20h34'29" 58d26'10"
Jennifer Erin Ballenger
Cma 6h50'15" -15d41'51"
Jennifer Erin Bruno
Tau 4h16'16" 29d20'43"
Jennifer Erin Valentine
And 0h46'52" 33d44'51"
Jennifer Ert Schmeiser
Cnc 7h59'9" 11d47'38"
Jennifer Espinueva Valdez
Ori 6h19'34" 7d15'22"
Jennifer Evans
Lyn 8h49'14" 34d19'16"
Jennifer Eve Smithers
Cyg 21h9'54" 45d31'46"
Jennifer F. Tinker
Aqr 22h21'42" -8d53'14"
Jennifer Faith Schwarz
Cas 0h55'51" 47d44'32"
Jennifer Faith Soneboulam
Gem 7h35'20" 29d45'4"
Jennifer Federle
Crb 15h31'7" 32d3'20"
Jennifer Fidotta
Ari 2h34'15" 25d59'26"
Jennifer Fields Palbicke
Leo 11h12'4" 16d38'58"
Jennifer Filauro
Dra 17h0'50" 63d4'34"
Jennifer Fithen
Cyg 20h2'48" 47d0'26"
Jennifer Flynn
Aql 19h10'26" 14d14'26"
Jennifer Fong
Leo 11h12'4" 20d31'37"
Jennifer Fong
Ari 3h16'49" 28d15'15"
Jennifer Forever
Mon 7h44'59" -1d57'50"
Jennifer Frances Esposito
Lib 14h59'11" -17d29'51"
Jennifer Frances Simon R.N. BSN
Cas 1h34'27" 62d13'36"
Jennifer Fry
Lib 14h32'35" -24d41'43"
Jennifer G. Clarke
Vir 12h19'47" -9d17'23"
Jennifer Gail
Psc 1h40'49" 6d28'7"
Jennifer Gail Kettner "Lola"
Gem 7h44'23" 33d30'16"
Jennifer Garcia
Vir 13h19'55" 5d43'48"
Jennifer Garcia
Cap 21h33'16" -17d18'20"
Jennifer Garrett
And 22h58'14" 52d26'21"
Jennifer Gattis's "Mystic Star"
Tau 3h52'41" 24d48'12"
Jennifer Gayle Swanson
Cas 1h12'41" 72d9'57"
Jennifer Gentile
Gem 6h36'52" 24d18'33"
Jennifer Giovinco
Leo 9h43'51" 21d22'5"
Jennifer Giusti
Vir 13h20'42" -18d2'7"
Jennifer Glorioso
Uma 14h27'26" 56d13'20"
Jennifer & Glory
Uma 10h37'15" 42d38'55"
Jennifer Goco
Leo 10h11'51" 15d49'41"
Jennifer Goldsmith
Lyn 9h12'35" 33d42'26"
Jennifer Goldstein
And 1h48'27" 39d46'28"
Jennifer Goldstein
Tau 3h43'3" 15d49'15"
Jennifer Gomez
Sgr 18h1'2" -19d52'5"
Jennifer Gonzalez
Ori 5h3'35" 5d18'42"
Jennifer Gonzalez
Gem 7h16'33" 33d8'39"
Jennifer Goode-Chris' Favorite Star
Psc 0h15'57" 6d30'37"
Jennifer Gooderham
Umi 13h32'13" 74d53'9"
Jennifer Gora
Lyn 8h5'7" 41d5'48"
Jennifer Gower
Ari 2h52'24" 19d11'9"
Jennifer Grace
Crb 15h32'26" 31d37'24"
Jennifer Grace
Lyn 7h20'38" 46d7'21"
Jennifer Grace Farinella
Uma 10h14'8" 45d30'1"
Jennifer Graham
Leo 10h53'9" 12d48'55"
Jennifer Grawien
Tau 4h25'24" 7d46'43"
Jennifer Gray
Lyn 7h53'40" 53d22'50"
Jennifer Gray
Uma 8h42'20" 53d13'34"
Jennifer Grayson
Uma 13h53'33" 48d21'3"
Jennifer Green
And 0h33'40" 32d45'20"
Jennifer Greenberg
Cam 4h44'53" 73d20'44"
Jennifer & Gregory Gulick
Sge 19h40'48" 18d8'40"
Jennifer Griffith
Ari 2h48'13" 14d35'53"
Jennifer Grynewicz - Shining Star
Uma 8h51'13" 46d50'13"
Jennifer Guastella
Tau 4h14'30" 29d16'57"
Jennifer H
Cas 0h26'51" 48d40'57"
Jennifer H. Brown
Tau 5h27'20" 25d43'56"
Jennifer H. Langlois
And 0h5'9" 35d18'53"
Jennifer Haan
And 23h31'2" 48d25'46"
Jennifer Haddock
Cnc 8h31'35" 10d47'53"
Jennifer Hahn
Cet 3h14'45" -0d47'27"
Jennifer Hall
Aqr 20h41'50" -2d9'29"
Jennifer Hallett
Cnc 8h46'14" 23d49'49"
Jennifer Han
Cmi 8h1'26" -0d16'24"
Jennifer Harris
Cas 0h42'7" 66d26'12"
Jennifer Harris
Vir 14h21'13" 2d1'24"
Jennifer Harroff
And 23h17'43" 43d7'21"
Jennifer Hathcock
Cnc 8h43'46" 22d26'0"
Jennifer Hatton's 21st Birthday
And 23h14'32" 37d18'50"
Jennifer Hayes
Lib 15h21'41" -4d47'45"
Jennifer Hedges
And 2h19'20" 38d43'31"
Jennifer Helen Wood
Cnc 8h35'45" 28d0'53"
Jennifer Helene White
Sgr 18h48'25" -33d54'23"
Jennifer Henderson
Crb 15h28'47" 29d47'31"
Jennifer Hernandez
Gem 6h56'6" 22d1'45"
Jennifer Herrera
Ori 6h19'20" 2d37'56"
Jennifer Hess
Ari 2h37'9" 25d30'14"
Jennifer Hixenbaugh
Cyg 19h53'5" 59d35'38"
Jennifer Holden
And 2h25'5" 49d54'31"
Jennifer Holmes
Vul 19h26'30" 25d23'24"
Jennifer Holmes
Dra 17h2'21" 57d31'5"
Jennifer Hope Wells
And 0h49'40" 40d47'13"
Jennifer Houghtaling Neal
And 1h45'45" 42d28'47"
Jennifer House
Cas 0h19'36" 54d50'56"
Jennifer Hudec
Aqr 22h58'55" -21d49'50"
Jennifer Hurley
Cas 23h2'58" 54d43'40"
Jennifer Innes
Lyn 7h22'11" 45d32'27"
Jennifer Irene Baldwin
Vir 14h6'35" 2d41'50"
Jennifer Irene Bellmore
And 23h46'43" 33d51'4"
Jennifer Irene Ostromecki
Psc 0h10'21" 7d37'46"
Jennifer is Ryan's Love
Vir 12h13'37" -1d45'0"
Jennifer J.
Mon 6h46'18" -0d55'15"
Jennifer J
Ari 2h13'6" 22d51'21"
Jennifer J. Hale
Leo 10h10'51" 26d22'5"
Jennifer J. Hill
Cas 2h22'54" 63d44'53"
Jennifer J. Pfister
Psc 0h52'34" 16d9'43"
Jennifer J. Reath
Psc 0h14'24" 10d31'9"
Jennifer J. Varner
Cam 5h16'52" 62d23'17"
Jennifer Jackle
Ari 2h37'0" 24d47'16"
Jennifer & James
Aqr 22h52'50" -6d56'20"
Jennifer James Marcelle Gilliam
Lmi 10h25'10" 30d45'13"
Jennifer Jamielyn
And 0h38'30" 31d35'42"
Jennifer Jaqua, The Best Mom EVER!
Uma 11h19'50" 67d48'4"
Jennifer Jarrett
Del 20h33'36" 7d1'9"
Jennifer Jean Boyd
Leo 11h23'43" 25d53'26"
Jennifer Jean Lasker
Cas 0h48'39" 53d22'44"
Jennifer Jean Marretta
Gem 7h46'24" 33d41'7"
Jennifer Jean Mays
Psc 1h26'55" 18d19'52"
Jennifer Jeanne Campbell
Lib 15h46'53" -7d56'40"
Jennifer Jeanne Huckleberry-Wincher
Cnc 8h48'45" 28d54'57"
Jennifer Jene'
Sco 16h11'24" -10d37'43"
Jennifer Jill Massey
And 0h32'48" 30d28'38"
Jennifer Jill Strickland
Uma 11h19'19" 51d42'2"
Jennifer Jillson Harrington
Dra 16h5'23" 57d43'7"
Jennifer Jimenez Bravo Grulkey
Aqr 23h18'42" -12d38'37"
Jennifer Jo
Sco 17h28'20" -45d22'8"
Jennifer Jo Bowling
And 0h59'30" 39d57'54"
Jennifer Joan Brio
Psc 1h22'15" 20d18'6"
Jennifer & Joe 4 Ever
Cyg 20h25'41" 52d22'38"
Jennifer Johnson
Tau 4h37'10" 18d15'2"
Jennifer Johnson
Ari 3h18'26" 27d53'53"
Jennifer & Jon Estes
Uma 11h10'45" 67d6'10"
Jennifer Jones
Lyn 6h51'47" 52d13'35"
Jennifer Jordan Copeland
Sco 16h50'9" -26d44'23"
Jennifer Josephine
Ari 2h47'46" 20d50'12"
Jennifer Joy
Cam 5h19'48" 58d3'38"
Jennifer Joy Martin
And 1h4'43" 39d5'10"
Jennifer Joy Steider
Vir 12h49'48" -8d26'55"
Jennifer Judi Macke
Cyg 19h34'49" 54d51'55"
Jennifer Jungshin Rim
Leo 11h41'32" 14d25'55"
Jennifer Juniper
Cnc 9h13'5" 25d38'14"
Jennifer Jurczynski
Vir 13h29'0" -6d30'10"
Jennifer & Justin
Uma 13h46'22" 58d55'59"
Jennifer K
Sgr 19h45'37" -13d25'33"
Jennifer K. Spencer
And 0h25'36" 44d17'0"
Jennifer & Kaitlyn's Star
And 0h55'30" 38d31'11"
Jennifer Kaleialoha
Sco 16h52'28" -26d55'15"
Jennifer Kallas
And 23h14'10" 41d42'2"
Jennifer Kanon Rick Forever & Ever
Vir 13h37'27" 4d47'51"
Jennifer Karl
Vir 13h22'9" -0d7'56"
Jennifer Karr's Gazing Star
Vir 11h57'41" -0d4'59"
Jennifer Kate
And 1h19'19" 48d56'38"
Jennifer Kate Woods
And 1h5'0" 41d38'17"
Jennifer Katherine Jamieson Fulton
Lmi 9h30'28" 35d4'41"
Jennifer Katherine Lee
Cas 23h53'27" 54d46'29"
Jennifer Kathleen Belter
Cyg 20h5'44" 48d32'9"
Jennifer Kathleen Husk
Uma 8h51'39" 64d28'37"
Jennifer Kathryn Sylva
And 1h18'17" 45d21'4"
Jennifer Kawahara
Crb 15h47'25" 35d46'36"
Jennifer Kay
Uma 11h53'10" 34d25'38"
Jennifer Kay
Uma 11h19'2" 54d17'13"
Jennifer Kay
Aqr 22h36'3" -23d3'43"
Jennifer Kay Hicks
Gem 7h27'28" 17d51'48"
Jennifer Kay Jones
Uma 12h37'58" 57d24'42"
Jennifer Kay Lockridge
Aqr 21h53'52" 1d3'23"
Jennifer Kay Schmidt
Com 13h5'52" 29d8'50"
Jennifer Kaye
Cyg 20h10'34" 36d40'53"
Jennifer Kaye Dickinson
Cas 23h41'56" 53d21'30"
Jennifer & Keith Haines 08-25
Umi 2h35'10" 88d47'59"
Jennifer Keli Mathis
Vir 13h39'36" 5d2'51"
Jennifer Keller
Dra 17h29'45" 55d3'9"
Jennifer Kelly
Sco 16h8'3" -13d33'42"
Jennifer Kemper
Aqr 22h46'42" -18d10'55"
Jennifer - Ken Jr.
Tau 4h44'7" 23d48'28"
Jennifer Kepler
Vul 19h22'23" 25d18'7"
Jennifer Kern, GaMPI Shining Star
Lyn 8h2'38" 40d23'44"
Jennifer Kidonakis
Cyg 19h53'3" 50d29'56"
Jennifer Kirstine
Gem 7h0'9" 24d0'54"
Jennifer Klein
And 23h32'26" 41d6'9"
Jennifer Koplin
Gem 6h45'16" 33d25'43"
Jennifer Kotch
Lyn 9h4'37" 38d56'59"
Jennifer Kovacevich
Psc 23h53'43" 1d0'29"
Jennifer Kozlowski
Tau 4h30'10" 16d57'34"
Jennifer Kristen Attaway
And 0h48'19" 40d48'44"
Jennifer Kristina Kuhn 19.11.1981
Uma 11h44'26" 37d36'54"
Jennifer Kyle Barringer
Leo 9h23'44" 26d41'34"
Jennifer Kyzer
Ari 2h2'11" 22d51'7"
Jennifer L.
Uma 9h33'3" 49d50'42"
Jennifer L. Alekson / "Babbers"
Ari 3h6'46" 19d31'50"
Jennifer L. Brase (Happe)
Lyr 18h48'57" 43d25'44"
Jennifer L. Chavez
Sgr 19h6'37" -31d5'43"
Jennifer L. Chu, M.D.
Sco 16h6'15" -17d18'10"
Jennifer L. Covell
Mon 7h52'30" -0d21'58"
Jennifer L. Hawley
Cas 23h31'24" 53d31'46"
Jennifer L. Hayek
Psc 0h2'30" 0d23'29"
Jennifer L. Hunt
Vir 12h45'1" 0d53'24"
Jennifer L. Kane
Cyg 20h5'35" 59d7'6"
Jennifer L. King
Vir 13h10'0" 1d12'4"
Jennifer L. McKay
Crb 16h23'45" 36d46'29"
Jennifer L. Neilson
Cnc 9h11'30" 16d38'28"
Jennifer L. Posen
Lyr 19h17'28" 29d30'49"
Jennifer L Rockwood
Uma 13h26'12" 54d29'21"
Jennifer L. Rocque
Lib 14h33'33" -17d21'15"
Jennifer L. Salyers
And 0h45'52" 38d28'29"
Jennifer L. Smith
Vir 12h31'5" -0d44'55"
Jennifer L. Soukup
Cnc 9h17'39" 8d30'56"
Jennifer L Sparks
Gem 7h24'41" 33d26'49"
Jennifer L Swenson
And 23h37'1" 47d2'38"
Jennifer L. Villacci
And 1h5'26" 45d44'35"
Jennifer L Wilson
Cnc 9h14'7" 16d17'29"
Jennifer L. Wilson (Bunny)
Lep 5h16'6" -11d44'47"
Jennifer "LadyBug" Davis
Tau 3h50'9" 22d38'0"
Jennifer Lafontaine
And 23h59'7" 46d42'2"
Jennifer LaMoureaux
Pho 1h8'33" -48d2'51"
Jennifer Lane
Psc 0h53'33" 25d25'54"
Jennifer Lane
Ari 2h33'6" 10d59'49"
Jennifer Lant - "Babygirl"
Vir 13h38'6" 0d36'46"
Jennifer LaSarge - Light of My Life
Cas 23h5'2" 53d21'29"
Jennifer Lasker
Uma 11h50'45" 64d12'44"
Jennifer Lauben
Gem 6h42'51" 25d22'13"
Jennifer Laura Lumb
Tau 6h0'51" 27d52'9"
Jennifer Lauren
Leo 9h26'30" 32d57'47"
Jennifer Lauren Cialone
And 0h15'12" 45d53'17"
Jennifer Lauren Cialone
Aqr 23h52'38" -3d26'37"
Jennifer Lauren Godby
Cap 20h45'28" -20d44'53"
Jennifer Lauren Halloran
Tau 5h50'29" 16d25'42"
Jennifer Lauryn Le
Lib 14h49'55" -2d14'6"
Jennifer LaVonne Nowak
Lyn 7h23'43" 54d45'49"
Jennifer Layne Davis
Lib 15h36'56" -19d54'29"
Jennifer Le
And 0h27'4" 33d30'10"
Jennifer Lea Druhan
Tau 4h23'36" 24d49'31"
Jennifer Leah
And 0h2'50" 45d17'8"
Jennifer Leanne Bjerk
And 0h51'49" 41d28'54"
Jennifer Lebrethon
Ari 2h59'33" 21d57'59"
Jennifer Lee
Gem 7h1'49" 27d7'30"
Jennifer Lee
Lyn 8h32'30" 42d14'0"
Jennifer Lee
Cap 20h24'37" -18d50'8"
Jennifer Lee
Cam 4h1'27" 67d45'59"
Jennifer Lee
Sco 17h0'28" -31d38'5"
Jennifer Lee Adams
Gem 7h44'31" 20d39'54"
Jennifer Lee Boling
Vir 13h43'49" -0d11'43"
Jennifer Lee Chrucky
Cnc 8h13'49" 16d44'43"
Jennifer Lee Derr's Starr
Cas 1h47'55" 62d45'24"
Jennifer Lee East Johnson
Sco 17h58'13" -42d12'22"
Jennifer Lee Farnkopf
Ari 2h30'54" 11d33'55"
Jennifer Lee Feist
Tau 5h49'13" 24d14'52"
Jennifer Lee Fritz
Tau 3h51'34" 17d34'18"
Jennifer Lee Ganoe
Gem 7h49'31" 32d44'55"
Jennifer Lee Gray
And 23h20'32" 42d4'20"
Jennifer Lee Hill
Uma 10h27'33" 56d21'53"
Jennifer Lee Joyal
Sco 16h16'4" -12d10'49"
Jennifer Lee Macri
And 23h20'1" 49d2'49"
Jennifer Lee Mayhew
Tau 4h39'37" 24d45'28"
Jennifer Lee Nechodomu
Gem 7h42'50" 31d58'22"
Jennifer Lee Newby-Haskin
And 0h35'32" 45d56'42"
Jennifer Lee Olivares 22
Tau 4h9'48" 25d26'43"
Jennifer Lee Phillips
Her 17h2'21" 33d4'52"
Jennifer Lee Predieri
And 2h6'42" 42d7'25"
Jennifer Lee Prestianni
Sgr 18h46'37" -28d54'36"
Jennifer Lee Roth
Ari 2h29'33" 19d25'37"
Jennifer Lee Schnurr
Sgr 18h1'12" -27d49'45"
Jennifer Lee Scolnick
Cnc 9h5'19" 32d14'54"
Jennifer Lee Sibley
Psc 1h20'43" 25d18'51"
Jennifer Lee St. Croix
Ari 3h12'26" 27d45'50"
Jennifer LeeTyler
Ori 6h9'12" 15d26'13"
Jennifer Leigh
Sgr 18h7'21" -26d36'31"
Jennifer Leigh
Mon 6h54'31" -0d20'32"
Jennifer Leigh Angell
Uma 11h11'11" 28d36'53"
Jennifer Leigh Bond
Ori 5h34'34" 1d50'18"
Jennifer Leigh Case
Psc 0h51'2" 13d6'5"
Jennifer Leigh Deslatte
And 0h17'40" 30d56'50"
Jennifer Leigh Dillon
And 2h16'28" 45d45'4"
Jennifer Leigh Giuliano-Slaughter
Ari 2h41'10" 27d42'17"
Jennifer Leigh Guy
Ori 5h15'46" 8d13'4"
Jennifer Leigh Haan
Vir 12h45'5" 1d34'10"
Jennifer Leigh Plaugher
Aql 19h2'10" 16d8'11"
Jennifer Leigh Rizk
Cap 20h33'39" -13d34'36"
Jennifer Leigh Ruffing
Lyn 7h43'25" 41d43'16"
Jennifer Leigh Youngs
Lyn 6h44'39" 53d30'16"
Jennifer Leona Gerischer
Vir 13h27'41" -4d48'0"
Jennifer Lesley Allen
Cas 1h25'55" 72d14'37"
Jennifer LiLy Andrews
Gem 7h4'33" 15d36'51"
Jennifer Lin Simonetti
Gem 7h20'53" 34d8'30"

Jennifer Linh Reiter
Cet 2h4'24" -14d34'58"
Jennifer Linn
Cnc 8h40'10" 24d53'40"
Jennifer Linville will you marry me
Leo 10h42'1" 14d34'51"
Jennifer Lisa Patch
Ari 2h20'56" 18d17'4"
Jennifer L'Italien
Crb 16h15'40" 38d22'43"
Jennifer Loiseau
Cap 21h40'42" -9d31'54"
Jennifer Long
Mon 6h46'49" -5d24'33"
Jennifer Lopez
Ori 6h17'11" 6d42'22"
Jennifer Loreen
And 23h26'24" 48d1'0"
Jennifer Lorene Stillwell
Uma 9h5'39" 49d44'3"
Jennifer Lorien
Lib 14h49'4" -2d11'43"
Jennifer Lorraine
Lyr 18h46'31" 41d10'45"
Jennifer LouAnne Clark
And 0h59'22" 40d27'27"
Jennifer Louise
Lib 15h37'22" -11d22'38"
Jennifer Louise Alvord
And 22h58'20" 40d47'33"
Jennifer Louise Gibbs
And 23h8'13" 50d46'16"
Jennifer Louise Lang
Cas 0h26'2" 61d52'31"
Jennifer Louise McGarty
And 0h44'26" 22d18'38"
Jennifer Louise Perkins
And 0h50'43" 41d41'41"
Jennifer Loves Efrain Forever
Vir 12h51'57" 12d7'18"
Jennifer Loves Robert
Sgr 18h8'39" -27d52'45"
Jennifer LSchubbe's Piece of Heaven
Lup 15h37'25" -40d56'34"
Jennifer Lucas
Lib 14h47'39" -11d31'9"
Jennifer Lupo 10/18/75
Lyn 6h35'1" 60d0'20"
Jennifer Lyn
Vir 13h1'5" 11d7'36"
Jennifer Lyn
Crb 15h43'16" 37d45'58"
Jennifer Lyn Abbott
And 0h53'2" 39d4'4"
Jennifer Lyn Gallacher
And 1h17'13" 37d15'4"
Jennifer Lyn Lipari
Gem 6h49'39" 26d35'11"
Jennifer Lyn Oden
Uma 10h59'21" 45d33'20"
Jennifer Lynn
Cyg 20h21'55" 50d43'17"
Jennifer Lynn
And 0h3'41" 47d48'55"
Jennifer Lynn
Lyn 8h14'32" 34d4'6"
Jennifer Lynn
Cyg 19h52'58" 32d49'5"
Jennifer Lynn
Cyg 19h47'23" 32d17'31"
Jennifer Lynn
Lyn 7h40'2" 38d43'21"
Jennifer Lynn
Gem 6h30'38" 27d7'37"
Jennifer Lynn
Ari 3h6'0" 29d39'44"
Jennifer Lynn
Tau 5h55'39" 25d21'54"
Jennifer Lynn
Crb 15h32'7" 26d32'1"
Jennifer Lynn
Gem 6h26'37" 21d0'57"
Jennifer Lynn
Cnc 8h22'20" 22d26'43"
Jennifer Lynn
Ari 2h39'34" 24d50'26"
Jennifer Lynn
Ari 2h36'34" 27d26'18"
Jennifer Lynn
Psc 1h8'6" 11d34'32"
Jennifer Lynn
Psc 0h10'34" 10d33'3"
Jennifer Lynn
Dra 19h23'4" 61d4'56"
Jennifer Lynn
Lyn 8h20'43" 52d46'17"
Jennifer Lynn
Uma 11h19'29" 54d35'46"
Jennifer Lynn
Cap 21h56'41" -21d13'20"
Jennifer Lynn
Sco 16h4'44" -16d21'21"
Jennifer Lynn
Lib 14h48'16" -0d50'46"
Jennifer Lynn
Lib 15h4'1" -1d7'58"
Jennifer Lynn
Lib 15h19'59" -6d45'31"
Jennifer Lynn
Pho 1h19'30" -54d24'47"
Jennifer Lynn 12
Cap 21h11'9" -23d23'54"

Jennifer Lynn Alaimo
Leo 10h26'50" 6d45'33"
Jennifer Lynn Anderson
Cas 23h0'48" 55d52'24"
Jennifer Lynn Ascenzi
Sco 16h17'45" -11d44'45"
Jennifer Lynn Barbee
Cap 20h27'44" -14d51'19"
Jennifer Lynn Beebe
And 1h11'30" 37d2'38"
Jennifer Lynn Billingsley
Leo 11h7'24" 5d40'37"
Jennifer Lynn Bono Pujol
Psc 1h1'28" 4d0'34"
Jennifer Lynn Bradley
And 23h9'9" 47d22'45"
Jennifer Lynn Brendel
Vir 12h47'27" -10d23'59"
Jennifer Lynn Brown
Cas 2h51'21" 60d11'21"
Jennifer Lynn Brown
Leo 9h29'36" 28d33'31"
Jennifer Lynn Chrusniak
Cnc 8h43'54" 18d50'47"
Jennifer Lynn Clayton "Angel Eyes"
Uma 13h37'27" 53d57'28"
Jennifer Lynn Coleman
Ari 2h18'15" 23d22'8"
Jennifer Lynn Davis
And 23h25'11" 46d11'45"
Jennifer Lynn DeHarts Star
Sco 16h17'59" -29d46'54"
Jennifer Lynn Dillow
Vir 13h43'28" 2d34'34"
Jennifer Lynn Dolph
Uma 12h4'4" 32d5'1"
Jennifer Lynn Edinger
Cas 0h8'12" 56d49'6"
Jennifer Lynn Emanuel
Aqr 22h37'59" -2d29'39"
Jennifer Lynn Frisbie
Sgr 20h5'11" -35d7'26"
Jennifer Lynn Garner
Uma 11h57'12" 46d56'14"
Jennifer Lynn Gregory
Uma 10h6'45" 62d51'26"
Jennifer Lynn Hammond
Cas 22h59'59" 55d47'14"
Jennifer Lynn Hanna
Sco 17h53'22" -36d0'35"
Jennifer Lynn Harrington
And 0h13'5" 29d30'18"
Jennifer Lynn Hawkins
Psc 1h35'7" 18d12'20"
Jennifer Lynn Henderson
And 1h25'4" 37d45'56"
Jennifer Lynn "Jennae" Hayes
Uma 11h48'50" 47d18'52"
Jennifer Lynn Jones
Lib 15h20'3" -8d27'28"
Jennifer Lynn Kelly
Lyr 18h42'4" 40d3'27"
Jennifer Lynn Kephart
Lyn 8h10'37" 41d29'37"
Jennifer Lynn King
Peg 22h5'49" 9d39'21"
Jennifer Lynn Konopa
Uma 11h19'41" 46d23'13"
Jennifer Lynn Lippincott
Sco 17h26'43" -38d59'33"
Jennifer Lynn Loftus
Leo 11h30'27" 17d19'10"
Jennifer Lynn Maddox
Uma 10h8'29" 61d49'33"
Jennifer Lynn Mahan
Lib 15h44'32" -18d36'0"
Jennifer Lynn Marie Toro
Sco 16h24'3" -32d42'11"
Jennifer Lynn Masson
Leo 10h14'59" 12d10'47"
Jennifer Lynn McCormick
Cap 20h57'47" -21d44'24"
Jennifer Lynn McDonnell
Leo 10h59'51" 15d37'49"
Jennifer Lynn Mitchell
Cap 21h54'3" -8d25'37"
Jennifer Lynn Moriarty
Ari 2h53'3" 21d58'49"
Jennifer Lynn Nall
Aqr 21h20'16" -5d25'26"
Jennifer Lynn Newman
Cas 0h33'59" 48d2'20"
Jennifer Lynn Oates
And 0h27'32" 28d29'47"
Jennifer Lynn Ohsman
Lib 15h55'15" -19d37'49"
Jennifer Lynn Ondo
Ori 5h5'53" 14d16'59"
Jennifer Lynn Page
Ori 5h50'45" 5d57'48"
Jennifer Lynn Patterson
Cnc 9h6'35" 29d48'31"
Jennifer Lynn Poole
Cas 1h29'2" 69d48'25"
Jennifer Lynn Powers
Sco 17h28'11" -39d18'10"
Jennifer Lynn Rachael Porile
Cas 23h40'26" 54d52'53"
Jennifer Lynn Ryan
Sgr 20h0'0" -32d18'47"
Jennifer Lynn Sample
Aqr 21h33'52" -1d54'34"

Jennifer Lynn Savoroski
Sgr 18h56'54" -29d16'38"
Jennifer Lynn Schnee
Vir 11h49'23" 9d5'50"
Jennifer Lynn Shelton
Psc 0h48'56" 16d57'38"
Jennifer Lynn Shewmaker Monson
Aur 5h56'1" 32d46'13"
Jennifer Lynn Skidmore
And 23h37'4" 35d45'3"
Jennifer Lynn Sliwa
Aql 19h7'56" 2d31'47"
Jennifer Lynn Sullivan
And 23h15'16" 36d24'18"
Jennifer Lynn Swart DeMarco
Aqr 22h8'19" -22d40'58"
Jennifer Lynn Sweet
Ari 2h23'36" 27d46'25"
Jennifer Lynn Sylvester
Lib 14h54'11" -0d35'10"
Jennifer Lynn Taylor
Umi 15h32'33" 73d56'53"
Jennifer Lynn Tedseco
Tau 4h17'0" 1d41'13"
Jennifer Lynn Thompson
Aqr 22h47'48" -17d5'59"
Jennifer Lynn Thornton
Per 3h42'22" 48d45'17"
Jennifer Lynn Villarreal
Mon 6h46'41" 6d44'23"
Jennifer Lynn Watz
Cas 1h12'59" 58d12'15"
Jennifer Lynn Wernitznig Sebena
Cas 0h22'30" 59d13'40"
Jennifer Lynn Williams
Ari 3h11'40" 29d38'15"
Jennifer Lynn Williams
Lyn 7h54'43" 52d33'58"
Jennifer Lynn Wilt-Ritter
Lib 14h38'56" -17d45'11"
Jennifer Lynn Wimberly Ackerman
Gem 7h31'8" 33d44'32"
Jennifer Lynn Zientek
Gem 6h6'28" 27d18'33"
Jennifer Lynne
Cnc 8h37'52" 15d14'2"
Jennifer Lynne
Leo 10h33'59" 13d28'19"
Jennifer Lynne
Gem 7h2'16" 34d4'8"
Jennifer Lynne
Uma 10h35'29" 54d29'28"
Jennifer Lynne Catton
Vir 12h16'53" 3d23'52"
Jennifer Lynne Orr
Ari 2h29'26" 27d36'49"
Jennifer Lynne Short
And 23h5'18" 40d50'12"
Jennifer Lynne Tarantino
And 23h36'40" 48d52'24"
Jennifer Lynne Whitlock
Cyg 20h59'1" 45d35'42"
Jennifer M. Coulter/Jigamon
Cam 5h46'22" 58d21'47"
Jennifer M. Davidson
Sco 16h25'7" -37d12'51"
Jennifer M. Leon
Cnc 9h6'59" 11d58'42"
Jennifer M. Norton
Uma 13h39'8" 59d5'41"
Jennifer M Ortiz
Cmi 7h19'3" 0d13'39"
Jennifer M. Penick
Psc 1h43'21" 13d28'25"
Jennifer M Peterson
Cnc 9h6'51" 28d2'33"
Jennifer M Rowe
Psc 0h51'10" 8d34'39"
Jennifer M. Trego
And 0h28'47" 40d6'57"
Jennifer M Valdez
Aqr 23h46'19" -17d9'7"
Jennifer M. van den Bosch
Lib 14h41'21" -24d45'8"
Jennifer M. Waldo
Sco 17h17'20" -44d5'44"
Jennifer MacDill 17
Sco 16h58'0" -39d48'53"
Jennifer MacNaughton
Tau 4h29'25" 20d6'57"
Jennifer MacNaughton
And 0h11'19" 29d37'54"
Jennifer Mae
Gem 7h44'49" 30d51'56"
Jennifer Mae Johnson
Mon 7h22'19" -0d29'12"
Jennifer Mae Lischak
Tau 5h58'8" 24d31'22"
Jennifer Maggi
Ori 5h34'14" -5d28'16"
Jennifer Major
Lyn 7h41'4" 37d53'14"
Jennifer Maley
Psc 23h49'33" -0d3'28"
Jennifer Mallory
Aqr 21h11'13" 0d5'26"
Jennifer Manalac Garcia
Lib 15h39'38" -17d54'57"
Jennifer Mankey
Vir 13h1'0" 10d54'2"

Jennifer Margaret McGowan
Cas 1h43'1" 64d8'25"
Jennifer Margaret Passmore Sadler
And 23h33'5" 48d8'18"
Jennifer Maria Borski
Gem 6h9'23" 26d55'58"
Jennifer Maria Patel
Sco 16h11'52" -15d51'58"
Jennifer Marie
Sgr 18h50'17" -19d2'19"
Jennifer Marie
Lib 14h50'41" -2d14'25"
Jennifer Marie
Uma 12h0'6" 63d46'41"
Jennifer Marie
Lyn 7h33'59" 58d42'6"
Jennifer Marie
Leo 9h43'39" 21d38'54"
Jennifer Marie
Vir 12h15'27" 9d50'24"
Jennifer Marie
And 1h45'20" 49d5'6"
Jennifer Marie Aguliaro
Gem 7h4'25" 33d43'20"
Jennifer Marie Bailey
And 0h54'35" 38d34'17"
Jennifer Marie Beany
Aql 19h59'39" 8d55'22"
Jennifer Marie Buol
And 1h6'44" 44d37'23"
Jennifer Marie Calvo
And 0h18'36" 32d16'46"
Jennifer Marie Cassell
Mon 6h28'14" 9d12'47"
Jennifer Marie Dahmer 12:24 A.M.
Aqr 22h6'18" 0d45'33"
Jennifer Marie DeTalvo
Cas 1h23'36" 52d12'14"
Jennifer Marie Dobek
And 0h18'0" 44d0'29"
Jennifer Marie Embrey
Aqr 22h19'37" 0d34'28"
Jennifer Marie Emery
Lyn 7h34'38" 56d47'53"
Jennifer Marie Flanders
Cnc 9h7'36" 31d5'7"
Jennifer Marie Frickman
Tri 1h40'50" 29d56'3"
Jennifer Marie George
Sco 16h13'39" -13d37'24"
Jennifer Marie Golinski
And 23h25'46" 47d36'14"
Jennifer Marie Gonzalez
Peg 22h21'45" 8d59'26"
Jennifer Marie H 69
Leo 10h41'10" 14d54'39"
Jennifer Marie Lopez
Leo 11h39'47" 17d42'44"
Jennifer Marie Lytle
And 0h18'30" 43d59'39"
Jennifer Marie Morris
Mon 6h44'41" -0d22'14"
Jennifer Marie Myles (j.a.d.a)
Cyg 20h21'41" 58d15'23"
Jennifer Marie Piatt
Umi 14h33'3" 73d27'7"
Jennifer Marie Pierson
And 0h27'52" 41d25'53"
Jennifer Marie Rodriguez
Cap 20h58'41" -25d58'17"
Jennifer Marie Smith
Cep 21h46'48" 58d48'48"
Jennifer Marie Storm
Umi 14h53'13" 75d35'57"
Jennifer Marie Thomas
Cen 13h43'21" -31d15'17"
Jennifer Marie Tynon mpd
Cas 0h42'21" 65d55'26"
Jennifer Marie Vitcak
Uma 9h38'27" 65d56'51"
Jennifer Marie Whitenight
Cyg 20h16'15" 37d33'27"
Jennifer Marie Wilson
Vir 12h7'37" 2d41'33"
Jennifer Marie's Tinkerbell
And 1h25'9" 42d40'58"
Jennifer & Mark
Cyg 20h27'44" 49d44'37"
Jennifer Marley
Sgr 18h27'51" -24d35'56"
Jennifer Marquis
Tau 3h40'15" 29d22'28"
Jennifer Mary
Aqr 22h59'48" -7d58'8"
Jennifer Mary Schindler
Lyn 9h36'12" 40d9'17"
Jennifer Mary Theresa Rosinski
Ari 3h24'13" 28d34'47"
Jennifer Mary Zak
Aqr 22h26'2" -0d48'30"
Jennifer Mason Gray
Aqr 22h0'54" -8d44'42"
Jennifer Masters
Umi 16h17'19" 71d48'33"
Jennifer & Max ~ Amor Vincit Omnia
Cyg 20h5'19" 45d32'30"
Jennifer & Max living on the Stars
Lyr 18h55'20" 33d15'17"

Jennifer Mayes
And 2h24'48" 45d52'34"
Jennifer Mayle
And 1h11'44" 42d45'19"
Jennifer McCaughey-Bradley
And 2h33'13" 45d35'27"
Jennifer McDiffett
Sgr 18h35'54" -23d58'16"
Jennifer McDougall
And 0h39'57" 41d23'35"
Jennifer Mcquillan
And 0h15'8" 22d11'6"
Jennifer Meghan Wilson
Psc 0h37'8" 19d26'55"
Jennifer Melissa
Cas 0h13'19" 57d14'10"
Jennifer Melissa Ann Lunn
And 0h40'5" 28d16'3"
Jennifer Meng
Uma 9h48'10" 48d40'59"
Jennifer Meyer
Leo 11h22'20" 15d53'22"
Jennifer Micah
Aqr 21h32'33" 1d17'28"
Jennifer & Michael
Crb 16h0'32" 33d51'48"
Jennifer & Michael Villiers
Cyg 21h36'20" 47d37'25"
Jennifer Michaels
And 0h32'13" 25d3'33"
Jennifer Michele Hunter
Dra 16h24'19" 58d33'47"
Jennifer Michele Testa Clayton
Ari 2h32'48" 17d42'50"
Jennifer Michelle
And 1h43'39" 47d52'38"
Jennifer Michelle Callis
Mon 6h44'42" -0d6'55"
Jennifer Michelle Daniels
Sco 16h13'38" -16d44'29"
Jennifer Michelle Danna
Uma 9h40'46" 50d26'43"
Jennifer Michelle Drake
Cyg 20h26'12" 31d30'3"
Jennifer Michelle Kaspar
Uma 10h33'5" 69d58'28"
Jennifer Michelle Kohlmyer
Sex 10h22'14" 4d58'48"
Jennifer Michelle Smith
Lib 14h52'29" -12d44'20"
Jennifer Michelle Stanek
Cma 7h9'30" -22d44'0"
Jennifer Michener
Vir 14h3'25" -15d16'5"
Jennifer Michie
Cnc 8h2'45" 11d17'4"
Jennifer Mihye Oh
Cap 21h51'6" -23d37'51"
Jennifer Mike
Dra 19h20'47" 61d13'58"
Jennifer Minervini
Sco 17h39'8" -33d7'26"
Jennifer Miranda Martin
And 1h4'28" 41d12'38"
Jennifer Mitchell
Gem 7h16'0" 33d54'58"
Jennifer Moase
And 2h22'20" 48d55'1"
Jennifer Mold
Mon 6h51'56" 6d16'58"
Jennifer Molenda
Uma 10h55'50" 58d20'26"
Jennifer Mollison
And 0h32'16" 31d50'39"
Jennifer Moore
Aqr 22h15'23" 1d57'52"
Jennifer Moretz
Lmi 10h30'6" 36d5'41"
Jennifer Morris
Gem 7h14'1" 20d2'2"
Jennifer Müller
Uma 9h58'32" 48d11'21"
Jennifer Murhon
Lib 14h51'1" -2d16'9"
Jennifer Murray
Psc 1h52'43" 7d28'34"
Jennifer Musser (Jenny)
Uma 9h32'4" 50d8'54"
Jennifer Musso
And 0h8'44" 47d34'47"
Jennifer My Mahal
Vir 14h28'22" 0d18'28"
Jennifer My Perfect Love
Uma 10h23'9" 45d18'5"
Jennifer Myers
Lib 15h5'54" -17d9'23"
Jennifer N. Gong
Vir 14h8'29" -11d24'43"
Jennifer N. Zito
And 23h38'10" 48d4'41"
Jennifer Nations-Pelton
Mon 6h46'43" -0d9'43"
Jennifer Neeley
And 1h5'4" 33d52'54"
Jennifer Neese
Gem 6h26'58" 17d48'42"
Jennifer & Neil
Per 3h14'6" 36d46'35"
Jennifer Nelson
And 23h1'51" 36d5'33"
Jennifer Nenos
Crb 15h55'10" 32d51'17"

Jennifer Ngo
Sco 16h10'45" -11d39'31"
Jennifer Nguyen
Uma 13h20'24" 55d25'54"
Jennifer Nichole Dick
Uma 9h6'5" 54d38'13"
Jennifer Nichole Faber
Ari 2h33'55" 25d16'3"
Jennifer Nichole Kniceley
Uma 11h4'51" 37d26'34"
Jennifer Nichole O' Hare
Lib 14h53'58" -17d34'9"
Jennifer Nichols
Gem 6h48'34" 26d30'21"
Jennifer Nicole
And 2h12'42" 50d20'12"
Jennifer Nicole Baldwin
Leo 11h44'18" 19d58'51"
Jennifer Nicole Banks
And 23h28'38" 39d24'40"
Jennifer Nicole Carey
And 1h5'6" 41d13'21"
Jennifer Nicole Cramer
Lyn 8h32'13" 34d1'41"
Jennifer Nicole Miller
And 23h55'51" 36d4'23"
Jennifer Nicole Pickford
Car 7h32'35" -52d19'53"
Jennifer Nicole Potts
Leo 11h26'42" 12d47'15"
Jennifer Nicole Waters
Ari 2h35'1" 27d1'10"
Jennifer Nifkin/Howes
Ori 6h21'31" 15d4'46"
Jennifer Noel Colgan
Uma 12h0'57" 31d51'59"
Jennifer Noll
Uma 9h29'53" 44d35'19"
Jennifer O
Uma 8h41'35" 57d49'53"
Jennifer O Connor
Gem 6h37'33" 22d20'56"
Jennifer Odette Robbins
Cap 21h35'51" -18d43'53"
Jennifer Olivia Travis Eldredge
Ari 3h15'49" 29d2'37"
Jennifer One
Cas 0h22'27" 56d43'15"
Jennifer Osmanson
Vir 12h48'19" 2d14'1"
Jennifer Our Mom
Cyg 20h52'40" 31d26'22"
Jennifer Ousley
Uma 11h46'42" 53d38'7"
Jennifer Owen
And 0h41'30" 31d17'59"
Jennifer Padula
Lib 14h59'23" -15d49'1"
Jennifer Pagano
Her 17h21'15" 34d27'46"
Jennifer Paige
And 0h55'5" 37d26'34"
Jennifer Paige Barclay
Com 12h32'43" 16d45'20"
Jennifer Paige McPherson
Dra 16h57'28" 63d48'47"
Jennifer Palafox
And 2h28'29" 43d23'34"
Jennifer Palm
Sgr 19h26'53" -15d25'49"
Jennifer Park
Sgr 18h52'56" -32d37'33"
Jennifer Patino Blod
Tau 3h54'56" 5d45'50"
Jennifer Patricia Alexander
Ari 3h5'42" 28d30'28"
Jennifer & Patrick Schumacher
Ori 5h58'59" 6d51'53"
Jennifer Patterson
Uma 9h52'27" 43d41'12"
Jennifer Paul
Cap 20h29'28" -9d30'3"
Jennifer Pearce
And 22h59'41" 40d51'40"
Jennifer Pearce
Tau 5h40'14" 26d39'42"
Jennifer Pearson
Cas 2h31'56" 60d34'43"
Jennifer Pearston
Lib 14h53'47" -0d46'49"
Jennifer Pelucca
Lib 15h32'21" -4d41'35"
Jennifer Pena
Crb 15h42'34" 32d7'45"
Jennifer Pequignot
Sco 17h32'11" -39d6'3"
Jennifer Peterson
Del 20h30'11" 7d17'12"
Jennifer Peterson
Vir 12h53'32" 12d42'38"
Jennifer Phillips
Gem 7h29'48" 24d56'27"
Jennifer Pikes Piece of Heaven
And 0h31'8" 23d39'29"
Jennifer 'Pip' West
Cnc 8h53'29" 29d22'51"
Jennifer Piscopiello
Cap 21h41'20" -17d57'36"
Jennifer Platt Kienitz
Ori 5h35'38" 13d23'37"
Jennifer Plett
Cnc 8h44'33" 23d10'33"

Jennifer Ponder
Uma 12h19'0" 53d51'49"
Jennifer Pratt
Psc 2h0'25" 4d20'18"
Jennifer Presciti
Aqr 23h19'57" -10d27'25"
Jennifer Presnell Is Very Special
Ori 4h45'11" 7d13'13"
Jennifer Price
Vir 12h43'48" 1d48'47"
Jennifer "Princess" Essaian
Cap 20h35'38" -12d37'7"
Jennifer Prior
Her 17h13'59" 18d15'34"
Jennifer Pullen **JLP**
Leo 11h9'14" 21d32'49"
Jennifer Purvis
Crb 16h6'29" 35d38'49"
Jennifer Quinn
Tau 3h53'15" 14d54'39"
Jennifer R. Ballew
And 1h56'45" 45d6'15"
Jennifer R Gora
Cnc 8h24'41" 15d39'35"
Jennifer R. LaRosa
Lmi 10h51'21" 27d48'30"
Jennifer R. Morris-Shining Star
Dra 15h30'5" 55d27'47"
Jennifer R. Mullins
And 2h18'37" 47d48'34"
Jennifer R Szoltysik
Sco 17h20'46" -38d27'2"
Jennifer R. Terranova
Uma 11h19'39" 31d6'50"
Jennifer R18
Cap 20h28'56" -12d7'3"
Jennifer Rachel Kernoff
Lib 15h3'33" -1d6'9"
Jennifer Rae
Cap 21h43'25" -19d34'39"
Jennifer Raelynn Henry
Mon 6h29'48" 9d5'54"
Jennifer Rahall
Psc 0h52'43" 26d29'43"
Jennifer Rahn
Psc 1h54'56" 5d31'1"
Jennifer Ramirez
Tau 5h48'49" 20d6'35"
Jennifer Ramsey
Ori 6h22'33" 16d40'45"
Jennifer Rebecca
Sco 16h10'30" -12d47'46"
Jennifer Rebecca Blanzy
And 23h11'12" 41d3'51"
Jennifer Rebekah Nutt
Sco 15h56'46" -21d45'56"
Jennifer Redepenning
Lyn 7h57'32" 37d51'3"
Jennifer Reiley
Cnc 9h4'13" 31d5'59"
Jennifer Renda
Per 2h27'58" 55d13'4"
Jennifer Rene McLaughlin
Vir 12h50'39" 1d27'55"
Jennifer Renee
And 23h41'57" 37d26'52"
Jennifer Renee
Uma 13h43'4" 60d15'26"
Jennifer Renee Anderson
Vir 13h21'58" 3d38'5"
Jennifer Renee Buchanan
Cas 23h25'25" 58d39'36"
Jennifer Renee Glimpse
And 2h13'32" 43d13'36"
Jennifer Renee Hope
And 0h16'51" 36d0'19"
Jennifer Renee Lee
Gem 6h42'42" 24d27'46"
Jennifer Reneé Thomas
Lyn 7h8'42" 51d55'10"
Jennifer Renia Love
Uma 9h53'41" 54d0'7"
Jennifer Resnick
Lyn 8h0'26" 43d15'30"
Jennifer Reye Nielson
Ari 2h12'52" 23d58'59"
Jennifer Rhian Valentine
And 2h23'2" 42d12'50"
Jennifer Rice
Aqr 22h21'54" -13d2'50"
Jennifer & Richard Brown "Forever"
Cyg 21h40'49" 46d2'54"
Jennifer Riebeling
Cnc 8h40'53" 25d36'3"
Jennifer Riede
Gem 7h25'32" 35d1'48"
Jennifer Riggins
Cnc 9h17'15" 10d17'54"
Jennifer Riggs-Sauthier
Cyg 20h26'21" 39d11'42"
Jennifer Robinson
Lyr 18h47'41" 38d5'25"
Jennifer Robyn McKittrick Syme
Col 5h23'27" -28d56'35"
Jennifer Rock
And 23h40'40" 47d50'7"
Jennifer Rodrigues
Ari 2h36'19" 14d58'25"
Jennifer Rome
Lib 14h32'19" -10d52'14"

Jennifer Romeo & Gina Mazzuca Cyg 20h36'12" 48d1'41"
Jennifer Ronneburger Lmi 10h49'53" 28d31'30"
Jennifer Rose Tau 5h42'14" 24d44'31"
Jennifer Rose Ari 3h1'17" 18d18'22"
Jennifer Rose Lyn 6h47'15" 50d34'43"
Jennifer Rose Cnc 9h10'13" 31d49'19"
Jennifer Rose And 1h42'17" 42d35'16"
Jennifer Rose Sco 16h16'21" -11d49'32"
Jennifer Rose Lib 15h58'53" -5d10'25"
Jennifer Rose Dra 16h49'45" 64d7'17"
Jennifer Rose Uma 10h4'20" 60d36'25"
Jennifer Rose Adkins Cnc 8h1'25" 17d28'26"
Jennifer Rose Allan Crb 15h48'42" 37d17'54"
Jennifer Rose Callighan And 1h55'1" 37d51'32"
Jennifer Rose Cortez Cyg 19h51'33" 36d17'19"
Jennifer Rose Daily Crb 15h50'13" 34d8'59"
Jennifer Rose Ecker Lib 15h45'48" -26d38'35"
Jennifer Rose Nations-Ovalle And 0h16'33" 29d59'30"
Jennifer Rose Siciliano And 2h29'16" 50d0'52"
Jennifer Rose's Heavenly Body Lyn 7h57'6" 43d20'15"
Jennifer Rowan Repa And 0h30'21" 27d18'46"
Jennifer Rubeck And 0h22'31" 44d0'42"
Jennifer Ruff Lmi 10h41'29" 23d38'43"
Jennifer Rushing Tau 5h47'36" 17d17'46"
Jennifer Ruth Conley Lib 15h55'56" -7d12'49"
Jennifer Ruth Nelson And 2h27'36" 42d47'47"
Jennifer Ruth Thoma Uma 9h17'18" 52d4'13"
Jennifer Ryan And 2h17'47" 47d39'34"
Jennifer Ryan Tau 4h50'4" 23d34'31"
Jennifer Ryan Hanzlik Tau 5h5'7" 16d32'57"
Jennifer S. Fresnedi Cyg 19h52'50" 35d42'25"
Jennifer S. Griffin Cap 20h32'22" -19d59'7"
Jennifer S. Kovacevich Psc 23h24'52" 1d38'2"
Jennifer Sabotage And 2h12'39" 44d28'47"
Jennifer Saenz Cnc 8h17'8" 23d38'56"
Jennifer Salarzon Vir 13h31'34" -6d21'2"
Jennifer Saleh Mon 7h0'44" 7d12'41"
Jennifer Salinas Cnc 8h47'53" 21d35'54"
Jennifer & Sam Cap 21h36'28" -16d6'27"
Jennifer Sanati And 23h48'26" 34d48'58"
Jennifer Sanford Aql 19h33'52" 11d50'34"
Jennifer Santilli Gem 7h15'2" 32d37'0"
Jennifer Saporito Leo 9h48'37" 22d23'38"
Jennifer Sattler Lyn 8h56'7" 39d47'49"
Jennifer Saucedo And 1h4'4" 45d43'39"
Jennifer Schneider "Jenn's Star" Ari 2h15'19" 26d28'28"
Jennifer Scholwin Vir 14h39'43" 3d53'40"
Jennifer Schramm Ori 5h27'9" 5d16'49"
Jennifer Schreck Sco 16h38'14" -30d36'48"
Jennifer Scoggins Ori 6h0'19" 10d27'25"
Jennifer Scorchess Cas 1h20'23" 65d10'1"
Jennifer Scott And 23h28'46" 45d10'29"
Jennifer Seals And 0h23'18" 39d3'54"
Jennifer Sears (Mason) Tau 3h34'24" 24d30'51"
Jennifer Serrano-Galvan Cyg 20h1'13" 35d8'54"
Jennifer Shaffer Ori 6h10'42" 9d6'45"
Jennifer Shannon Cohen Mon 6h45'55" -0d7'32"
Jennifer Sicuro Uma 12h50'17" 52d31'36"
Jennifer Sirkin Dra 15h32'40" 64d28'48"
Jennifer Skelton Sco 17h57'36" -33d53'20"
Jennifer Skinner Crb 16h4'35" 35d54'47"
Jennifer Skolochenko Uma 8h37'35" 66d50'45"
Jennifer Skrabutenas Tau 3h59'45" 21d32'39"
Jennifer Skuse And 1h15'32" 42d12'12"
Jennifer Smith Uma 11h14'25" 48d39'50"
Jennifer Smith Gem 6h50'46" 13d46'22"
Jennifer Sonja Smith Sgr 19h19'27" -14d44'37"
Jennifer Specht Uma 11h30'4" 51d15'50"
Jennifer Stadt Cap 20h19'46" -11d39'19"
Jennifer *Star Dancer* Roeske And 1h47'37" 39d1'58"
Jennifer Star Jennings Sco 17h55'8" -30d26'47"
Jennifer StarBaby Lloyd Cap 20h53'11" -24d2'6"
Jennifer Steele Tau 4h25'34" 17d51'53"
Jennifer Sterchi Lyr 18h30'28" 36d38'16"
Jennifer Stewart Dra 18h38'5" 61d29'56"
Jennifer Sturgis Leo 10h17'57" 13d28'21"
Jennifer Sue Lib 15h45'58" -18d41'20"
Jennifer Sue Butler 143 Leo 11h18'30" 24d2'33"
Jennifer Sue Gullett Leo 11h37'25" 19d48'6"
Jennifer Sue - My Eternal Love Pho 0h2'51" -47d51'9"
Jennifer Sue O'Neil Cas 1h53'58" 61d44'10"
Jennifer sue Zundel's star Aqr 22h51'18" -23d37'39"
Jennifer Sullivan Mon 6h30'9" -1d1'33"
Jennifer Sung Mon 6h52'46" -0d30'30"
Jennifer Sung Ori 4h56'42" 1d41'9"
Jennifer Sur Tau 5h27'13" 25d37'5"
Jennifer Susan Gem 6h43'55" 19d0'54"
Jennifer Suzanne Hughes Leo 11h12'10" 24d5'46"
Jennifer Sweatman Per 2h16'49" 51d43'6"
Jennifer Sweet Vir 15h6'10" 3d45'21"
Jennifer "Sweet Pea" Constant Uma 11h34'6" 40d22'11"
Jennifer Swinburg Sco 16h6'40" -17d15'15"
Jennifer Taddeo 061567 Uma 12h3'39" 37d38'44"
Jennifer Talamas And 0h19'2" 29d46'50"
Jennifer Tamara Nhlane Leo 9h30'52" 26d35'6"
Jennifer Tan Ong Cnc 8h43'23" 11d6'2"
Jennifer Tatangelo And 2h23'12" 38d5'30"
Jennifer Tattershall Vir 13h16'1" -2d58'37"
Jennifer Taylor Uma 14h25'50" 56d50'53"
Jennifer Teresa Webb Gem 7h1'27" 28d9'34"
Jennifer Terrell Smith And 2h16'7" 38d28'26"
Jennifer Thamm Sgr 19h6'3" -17d53'54"
Jennifer The Beautiful Lyn 8h50'58" 40d19'28"
Jennifer "The Dozer" Reitz Aqr 23h36'7" -8d39'26"
Jennifer Therese Drobish Ori 5h45'56" 7d24'18"
Jennifer Thomas Tau 5h6'25" 28d0'0"
Jennifer Thompson Mon 7h31'22" -0d21'17"
Jennifer & Tine Cyg 20h48'43" 31d16'33"
Jennifer & Todd Hall Umi 16h4'29" 84d44'34"
Jennifer Torain-Jones And 0h3'9" 40d30'33"
Jennifer Toscano Lmi 10h31'57" 33d29'15"
Jennifer ToTo Cas 0h21'9" 57d31'24"
Jennifer Tung Vul 21h2'7" 22d58'41"
Jennifer Turner Cnc 8h4'4" 17d45'49"
Jennifer Uhl Eri 4h37'54" -0d58'17"
Jennifer Unell Ray Ori 5h32'28" -6d12'40"
Jennifer Vail Umi 16h19'37" 75d22'37"
Jennifer Van Caemerbeke Psc 1h43'9" 12d47'44"
Jennifer Vergados Uma 9h52'55" 51d40'29"
Jennifer Villalva Lyn 7h18'25" 57d47'43"
Jennifer Vlachos And 1h24'4" 40d38'2"
Jennifer Vona Sgr 19h38'40" -35d28'45"
Jennifer Vos Vir 14h5'36" 2d56'50"
Jennifer Vrana Leo 11h51'43" 21d0'9"
Jennifer Wackerman Gem 6h53'59" 32d38'50"
Jennifer Wadland Cyg 21h30'22" 37d52'2"
Jennifer Wall Psc 1h23'51" 17d34'15"
Jennifer Wallace And 0h19'51" 40d52'58"
Jennifer Ward Uma 10h25'53" 61d41'57"
Jennifer Warner Tri 2h26'50" 30d52'52"
Jennifer Watkins "Love Of My Life" Lyr 19h13'25" 45d52'3"
Jennifer Webb Aqr 21h18'28" -10d50'9"
Jennifer Weibert Sgr 18h13'5" -20d1'17"
Jennifer Weissman Lyr 18h35'24" 32d4'31"
Jennifer Wells Leipham Gem 7h3'43" 32d30'32"
Jennifer Whitehead, Jones Painumkal Ori 6h13'7" 15d45'0"
Jennifer Whitney Spitler Aqr 21h47'25" 0d31'54"
Jennifer Williams And 0h42'58" 27d15'12"
Jennifer Williams Lib 15h17'35" -20d11'32"
Jennifer Willms Psc 0h15'39" 17d46'40"
Jennifer Wilson Cam 4h5'48" 55d33'51"
Jennifer Wnorowski Lyn 6h25'15" 61d23'44"
Jennifer Wolfenden Psc 0h19'24" 5d51'23"
Jennifer Wolowich Cap 21h49'26" -10d24'12"
Jennifer Won Sgr 19h17'29" -14d31'33"
Jennifer Wong Mei Yoke Ori 5h38'40" -3d50'14"
Jennifer Wyatt Cyg 21h56'31" 52d3'28"
Jennifer Wyn Hall Lib 15h42'55" -14d14'44"
Jennifer Yeaman Cas 23h54'6" 53d56'45"
Jennifer Yee Psc 23h14'59" 3d31'53"
Jennifer Young Uma 12h45'43" 62d10'53"
Jennifer Zapor Psc 0h52'44" 31d14'6"
Jennifer Zaun Cas 0h16'55" 52d47'34"
Jennifer Zerolis - The Wookie Psc 23h6'11" 7d54'47"
Jennifer Zissou And 23h35'54" 44d3'33"
Jennifer, A. J. and Sloopy Tri 1h54'27" 30d15'20"
Jennifer, A. J. and Sloopy Uma 12h22'0" 58d42'56"
Jennifer, The Great! Gem 6h42'31" 23d2'16"
Jennifer,Alvaro,Eric,Ryan Viquez Lyn 9h14'14" 33d32'13"
Jennifer42482 Tau 4h6'52" 27d33'28"
Jennifer-Ann Gibbons And 0h44'39" 32d0'26"
Jenniferis Genesis Gem 7h48'14" 32d52'49"
Jennifer-K And 23h33'6" 47d55'42"
Jennifer-Lee Vir 13h17'36" 9d8'46"
Jennifer-Lyn Gwaley Sco 17h28'10" -45d31'59"
JenniferLynne And 23h26'7" 49d34'4"
JenniferMarieHardy-Harris Cnc 7h57'33" 12d4'0"
Jennifer's 21st Birthday Star And 23h46'15" 41d20'43"
Jennifer's 50th Birthday Star Cnc 9h7'18" 24d26'12"
Jennifer's Angel Sgr 18h55'35" -29d46'46"
Jennifer's Barry Aqr 21h39'9" -0d0'8"
Jennifer's beauty And 0h18'54" 37d0'50"
Jennifer's Celestial Crecent Crb 16h16'7" 32d41'30"
Jennifer's Court Cas 23h30'4" 55d17'41"
Jennifer's Dawn Leo 9h45'32" 20d10'7"
Jennifer's Destiny Lib 15h23'46" -12d6'3"
Jennifer's Dream Star Vir 12h33'51" 8d17'53"
Jennifer's Dreaming Star Sgr 19h42'16" -14d22'8"
Jennifer's Eternal Light And 23h3'0" 47d55'49"
Jennifer's Heart Psc 1h50'56" 8d6'22"
Jennifer's Joy Cas 1h41'3" 62d58'2"
Jennifers Love Uma 10h41'44" 49d19'37"
Jennifer's Love And 0h20'9" 44d35'41"
Jennifer's Lovely Diamond Leo 9h45'35" 7d55'44"
Jennifer's Mark Uma 13h12'42" 54d58'38"
Jennifer's muah star Per 3h48'1" 43d54'5"
Jennifer's Pixie Star And 0h25'7" 32d14'0"
Jennifer's Place In Heaven Cnc 9h14'31" 28d12'24"
JENNIFER'S PYEWACKET Ari 3h6'15" 28d9'52"
Jennifer's Shining Star And 0h33'41" 28d14'50"
Jennifer's Shining Star Vir 14h6'58" 4d49'35"
Jennifer's Shining Star Uma 10h35'24" 64d52'47"
Jennifer's Sir Lmi 9h54'44" 33d56'3"
Jennifer's Smile And 23h49'40" 43d31'30"
Jennifer's Star Lyn 7h56'27" 48d58'35"
Jennifer's Star Cyg 21h14'53" 45d58'26"
Jennifer's Star Crb 16h10'56" 36d15'16"
Jennifer's Star Tau 4h39'14" 3d23'35"
Jennifer's Star Cmi 7h28'22" 9d46'11"
Jennifer's Star Ari 2h9'19" 24d30'44"
Jennifer's Star Cap 20h33'20" -19d36'54"
Jennifer's Star Ori 5h30'20" -3d8'37"
Jennifer's Star; Always & Forever Aqr 22h33'36" -2d1'43"
Jennifer's Star of Everlasting Love And 2h19'37" 47d13'24"
Jennifer's Wish And 23h59'54" 48d3'18"
Jennifer's World Uma 10h3'9" 60d0'22"
Jenniffer Gem 7h10'14" 32d35'31"
Jenniffer Kiefer's Star of Strength Psc 23h52'20" -0d4'32"
Jennifier St. Germain Psc 1h14'48" 26d40'52"
Jennilee Psc 0h54'50" 28d10'48"
Jennina Sco 17h17'0" -32d53'54"
Jenni's Flute And 0h12'5" 35d43'22"
Jennis George Richard III Ori 6h4'9" 5d13'41"
Jennise Ashley Lib 14h52'25" -14d28'11"
Jennisima Loving XMaStar Nectaria Tau 4h41'2" 5d45'49"
Jennivee Psc 0h48'22" 12d39'17"
Jenn-Matteo Cnc 9h17'1" 9d59'29"
Jenn-O Psc 1h22'34" 17d6'14"
Jennody Ori 5h32'44" 1d29'47"
Jenn's Confirmation Star Cas 0h35'0" 57d21'54"
Jenns Shooting Star Per 3h17'57" 42d30'15"
Jenn's Star Per 4h33'11" 40d35'40"
Jenn's Star And 1h5'41" 37d5'45"
Jenn's Star Cap 20h16'1" -15d11'45"
Jenn's World Cas 2h23'59" 71d37'1"
Jenny Umi 13h55'56" 69d32'32"
Jenny Umi 14h2'27" 70d41'9"
Jenny Uma 11h9'21" 65d41'47"
Jenny Uma 10h14'39" 56d29'1"
Jenny Lyn 7h9'7" 58d33'30"
Jenny Lib 15h50'57" -16d59'15"
Jenny Aqr 21h4'3" -8d43'13"
Jenny Cap 20h32'33" -13d4'59"
Jenny Aqr 22h57'50" -6d39'30"
Jenny Lib 15h12'55" -5d49'41"
Jenny Lib 15h5'27" -3d59'0"
Jenny Col 5h50'22" -32d14'21"
Jenny Aur 7h13'29" 41d57'20"
JENNY Uma 10h37'1" 43d24'59"
Jenny And 0h42'15" 40d59'16"
Jenny And 1h30'37" 39d41'13"
Jenny And 1h46'56" 44d0'26"
Jenny And 1h37'33" 43d46'8"
Jenny And 23h16'56" 47d45'19"
Jenny And 23h20'26" 48d41'23"
Jenny And 23h3'41" 47d51'40"
Jenny Uma 9h54'32" 46d31'16"
Jenny And 1h17'13" 46d28'42"
Jenny Lyn 8h7'18" 45d30'21"
Jenny Aur 5h47'25" 48d43'59"
Jenny Lyr 19h22'53" 37d31'6"
Jenny Vir 13h21'36" 4d9'14"
Jenny Vir 12h42'38" 0d29'9"
Jenny Aqr 22h15'54" 2d24'59"
Jenny Psc 1h6'46" 14d1'0"
Jenny Psc 1h31'10" 14d50'58"
Jenny Ari 2h18'33" 17d3'51"
Jenny Tau 4h35'33" 20d52'50"
Jenny Ori 6h18'29" 14d43'30"
Jenny Leo 10h14'0" 14d51'17"
Jenny And 0h32'58" 28d44'7"
Jenny Ari 2h19'51" 25d35'9"
Jenny Tau 5h54'30" 25d29'23"
Jenny Gem 7h16'24" 23d43'49"
Jenny Leo 10h14'55" 26d37'28"
Jenny 1922 Cas 0h28'3" 57d0'22"
Jenny 20011958 Crb 15h48'41" 34d52'52"
Jenny 711 Per 2h52'39" 48d36'24"
Jenny A O Brien Ari 1h56'9" 13d23'47"
Jenny & Alexander Vir 13h53'43" -12d10'38"
Jenny Alpert And 23h15'1" 51d56'24"
Jenny and Brian's Wedding Star Cyg 20h55'19" 34d43'6"
Jenny and Glen Star Sco 17h52'56" -35d40'40"
Jenny and Yuno Eri 3h54'5" -0d5'10"
Jenny & Andrew's Wedding Star Cyg 20h18'48" 41d20'37"
Jenny Andry Lib 15h5'1" -15d10'47"
Jenny Angel And 2h30'26" 43d47'20"
Jenny Ann And 0h54'42" 35d16'1"
Jenny Ann And 1h56'1" 45d12'32"
Jenny Ann Armstrong Dra 17h59'9" 52d42'45"
Jenny Ann Craig Cyg 19h51'33" 35d54'15"
Jenny Ann Spatz Sco 16h16'28" -13d25'27"
Jenny Anne Tran Cnc 8h48'58" 32d32'55"
Jenny B Cas 23h56'56" 52d44'20"
Jenny B. Bermema Lib 15h20'49" -12d56'27"
Jenny "Baby Girl" Sugar Lib 15h26'47" -21d11'30"
Jenny Bailly Lyn 7h45'39" 39d0'10"
Jenny Balin Psc 1h19'6" 16d13'57"
Jenny Barry Gem 6h32'41" 21d5'44"
Jenny Bear Cap 20h27'16" -13d40'45"
Jenny Bear Aqr 22h37'9" -0d16'41"
Jenny Bear Umi 15h1'16" 74d9'18"
Jenny Bear Psc 1h9'38" 5d11'58"
Jenny Bee Uma 10h18'48" 50d51'26"
Jenny Bell Aqr 21h9'13" -13d10'26"
Jenny Belle Mere Cnc 8h14'57" 15d56'5"
Jenny & Bill Esson Cyg 20h19'44" 52d37'55"
Jenny Boo Star Lmi 10h40'24" 27d49'10"
Jenny & Brad Leo 11h28'33" 16d3'56"
Jenny Brooke Burgess And 0h32'7" 28d43'55"
Jenny Cento Cas 23h1'17" 55d29'54"
Jenny Cepero Cas 1h41'27" 61d11'3"
Jenny Chapman And 2h18'52" 38d55'57"
Jenny Cheevers Uma 10h23'10" 51d2'14"
Jenny Chick Cas 0h0'8" 59d56'18"
Jenny+Christoph die Liebe für Immer Uma 13h26'5" 57d7'45"
Jenny Clary *Always Remember* Gem 6h50'18" 16d53'9"
Jenny Coad And 1h31'5" 49d15'51"
Jenny Coad Cyg 20h11'16" 44d42'40"
Jenny Conn Cyg 20h16'54" 32d36'4"
Jenny Crow Uma 8h50'19" 65d18'43"
Jenny Cynthia Davis Cnc 9h8'44" 25d3'50"
Jenny D Leo 9h41'5" 28d24'41"
Jenny Danley Cnc 8h54'50" 12d32'13"
Jenny & David Pho 0h38'13" -41d9'3"
Jenny Douglas Peg 22h20'52" 4d45'51"
Jenny Driver Aql 19h48'24" 7d26'27"
Jenny Duffy Aqr 22h25'44" -7d15'35"
Jenny Earles And 0h45'9" 36d7'45"
Jenny Elizabeth Cyg 21h13'42" 55d2'14"
Jenny Elizabeth Grimes Lmi 10h37'19" 30d47'50"
Jenny Ellisen Leo 11h29'52" 25d54'22"
Jenny Encarnacion<3 Lmi 10h31'21" 37d32'2"
Jenny Eun Chang Cru 12h39'19" -57d59'15"
Jenny Faye Lyr 18h42'34" 37d19'14"
Jenny Fine Lyn 7h35'24" 49d41'5"
Jenny Flora Marine Continente Uma 10h10'14" 56d28'1"
Jenny Forever Lib 15h35'52" -14d25'11"
Jenny Foster Crb 15h30'53" 28d24'37"
Jenny Furrer Vir 13h7'57" -11d18'7"
Jenny Glock Uma 8h46'31" 51d19'23"
Jenny Greer Aqr 22h24'24" -0d23'17"
Jenny Gresham Doyen Cap 20h33'11" -17d51'35"
Jenny Griffith Lyr 19h19'42" 29d32'9"
Jenny Hatfield Uma 8h36'9" 60d29'24"
Jenny Hicks Psc 0h48'22" 7d26'46"
Jenny Hsin Ying Wu Ari 2h58'27" 18d25'11"
Jenny Hyo Lee: My Star And 23h42'17" 45d38'54"
Jenny Irizarry Dra 16h11'45" 54d42'5"
Jenny Jadelyn Tran Leo 10h27'23" 21d54'59"
Jenny Jones Cyg 19h52'37" 31d32'7"
Jenny Jones Aqr 22h56'55" -18d59'54"
Jenny Jungbluth Uma 9h15'49" 64d59'22"
Jenny K Lyn 7h31'35" 48d37'50"
Jenny Kam 1987 Ari 2h17'10" 24d19'55"
Jenny "Karate Kid" Sobecke Umi 14h22'37" 88d25'21"
Jenny KB Cnc 8h13'42" 6d44'7"
Jenny Kisslan Aqr 21h4'24" -10d6'3"
Jenny KK Cheung Vir 13h18'5" 4d32'43"
Jenny KL Aqr 22h55'44" -9d25'6"
Jenny Kodofakas Tau 4h7'29" 7d12'27"
Jenny (Ko-u Ipo) Sco 17h47'51" -41d33'17"
Jenny & Kurt 4Eva Uma 9h33'30" 54d26'3"
Jenny Kwan Cas 23h35'52" 53d20'57"
Jenny L. Garcea Sgr 18h0'48" -18d4'50"
Jenny Lacombe Psc 23h46'1" 5d36'6"
Jenny Lam Lyn 7h10'9" 54d10'27"
Jenny Lavoie Cyg 19h19'2" 53d50'34"
Jenny Lee Ori 5h26'39" 5d59'12"
Jenny Leigh Del 20h52'28" 2d59'41"
Jenny Leigh Ori 5h36'43" -5d6'48"
Jenny Leigh Anderson Vir 13h24'31" 13d21'31"
Jenny Leigh Wallace Gem 6h54'51" 14d24'23"
Jenny Len Cap 21h45'57" -18d10'35"
Jenny Lennon Cyg 20h13'22" 38d12'4"
Jenny. Light of my life. Phil x x x Eri 4h15'23" -24d51'17"
Jenny Liu Cyg 21h16'11" 54d42'5"
Jenny Lorien Zimmerman Psc 0h25'26" 10d55'40"
Jenny Lynn Aqr 22h42'2" 2d2'57"
Jenny Lynn Her 18h4'26" 21d19'1"
Jenny Lynn And 0h7'12" 44d0'6"
Jenny Lynn Dra 18h26'48" 73d53'6"
Jenny Lynn Astor Peg 22h20'6" 14d43'30"
Jenny Lynn Berry Uma 11h26'14" 32d37'48"
Jenny Lynn Conley And 1h14'40" 38d14'5"
Jenny Lynn Kanouff And 23h38'23" 42d30'24"
Jenny Lynn Rehm Sgr 19h0'5" -35d43'57"
Jenny Lynne Karalus Leo 11h41'34" 25d32'48"
Jenny M Noe Uma 10h40'33" 48d21'26"
Jenny Mac Aur 5h39'45" 53d2'57"
Jenny Mahvi Crb 15h17'58" 28d12'49"
Jenny Margaritha del Rocio Uma 9h8'16" 48d7'38"
Jenny Marie Leo 9h47'27" 22d3'35"
Jenny Marie Umi 13h56'41" 77d47'5"
Jenny Marie A Guerrero And 1h40'52" 42d42'25"
Jenny Marie Alexander Sgr 18h39'49" -36d10'55"
Jenny McCarthy Aqr 21h4'46" 0d45'10"
Jenny McCawley Sco 17h54'10" -36d28'19"

Jenny McCue
Gem 6h20'25" 21d53'38"
Jenny McCullough
And 1h45'26" 43d16'21"
Jenny Melnyk
And 23h22'55" 41d44'30"
Jenny Mitchell MacDonald
Aqr 22h39'12" -22d13'2"
Jenny Moats
And 23h23'22" 48d7'42"
Jenny Murkins
Cas 0h29'33" 50d23'29"
Jenny Murkins
Cas 3h14'41" 66d55'20"
Jenny Murphy
Cas 0h21'28" 51d10'2"
Jenny - My Ladybug
Cyg 21h25'34" 30d33'29"
Jenny My Love
Cyg 19h56'21" 45d46'36"
jenny nacimiento
And 0h11'53" 45d41'3"
Jenny Nicole Hammons
Sgr 18h15'57" -18d39'2"
Jenny P
Aqr 21h1'57" -12d53'15"
Jenny P. Tran
Cas 0h31'49" 60d14'7"
Jenny Palazzo
Tau 3h40'29" 21d8'5"
Jenny Park
Leo 11h25'49" 23d51'48"
Jenny Paul
And 0h21'53" 27d52'57"
Jenny Peanutbutter
Psc 0h27'11" 19d5'53"
Jenny Phebe
Com 12h37'53" 28d8'25"
Jenny R. Kain
And 23h13'47" 42d30'38"
Jenny R & Sarah L
Tau 4h10'31" 5d23'28"
Jenny Rae Lattin
Cnc 8h21'41" 13d29'51"
Jenny Ramirez
And 0h34'9" 30d47'59"
Jenny Ramone
And 23h44'5" 37d58'4"
Jenny Ray Perry "best mom ever!"
Lyn 7h11'26" 52d26'59"
Jenny Rebecca Hubert
Cyg 20h4'16" 37d16'53"
Jenny Rios Maldonado
Peg 22h17'19" 6d7'2"
Jenny Ritchie
And 23h19'29" 47d33'32"
Jenny Ritter
Her 17h24'47" 37d6'26"
Jenny Rose
Cas 0h21'6" 62d7'1"
Jenny Rose
Lyn 8h22'23" 56d47'54"
Jenny Rowe
And 0h5'47" 37d22'27"
Jenny Rules The World
Sgr 18h6'47" -34d5'26"
Jenny & Ryan
Cyg 21h20'45" 35d43'40"
Jenny S. Williams
And 1h52'11" 35d38'10"
Jenny Salvador
Sgr 18h12'54" -22d9'53"
Jenny Sanchez
Ori 5h55'46" 7d0'7"
Jenny Sangpasanesoul
Lib 15h10'34" -9d57'33"
Jenny Schwenke
Cyg 21h52'46" 53d7'3"
Jenny Seymour
Sco 16h54'12" -30d43'16"
Jenny Shadini Zabaneh
Vir 12h41'55" 0d44'28"
Jenny Shipton
And 0h33'49" 25d31'11"
Jenny Showalter
Tau 4h45'59" 19d5'4"
Jenny Silva
Lmi 10h59'55" 31d24'45"
Jenny Smith
And 1h45'9" 42d32'17"
Jenny Snookums Pribula
Vir 13h8'55" 12d7'18"
Jenny Spear
Lib 15h54'16" -11d5'25"
Jenny Spencer
Aql 19h57'15" 8d55'47"
Jenny Star
Vir 13h31'3" 1d20'1"
Jenny Stevens
And 2h14'40" 45d51'59"
Jenny Sue
Psc 0h31'46" 15d45'0"
Jenny Sue Smith/Trouble
And 1h22'44" 40d43'20"
Jenny Sue's Eyes
Lyn 9h6'18" 45d10'48"
Jenny Superstar
Mon 6h44'8" -0d7'52"
Jenny Susan
Uma 11h19'8" 65d21'59"
Jenny Swarbrick
Lyr 18h29'46" 31d49'52"
Jenny Sweeney
And 2h20'55" 50d12'53"
Jenny - Sy - Katie
Peg 21h55'38" 33d53'21"
Jenny T
Crb 16h10'14" 37d56'57"
Jenny the messenger
Tau 4h30'25" 20d19'47"
Jenny "The Sparkle of Tulsa's Eye"
And 0h17'25" 26d44'29"
Jenny "The Sparkle of Tulsa's Eye"
And 0h11'51" 26d25'15"
Jenny Toth
Cas 0h49'35" 50d29'53"
Jenny Tuck
Leo 10h0'4" 15d16'36"
Jenny Viccars
Dra 19h41'13" 70d24'58"
Jenny Wang
Leo 11h6'23" 15d9'59"
Jenny West
Boo 14h26'12" 48d45'8"
Jenny Wilson
Del 20h38'39" 15d20'21"
Jenny, Martin & Paul Redl
Cas 0h23'54" 55d28'17"
Jenny-311
Psc 0h39'56" 19d15'41"
Jennyann Marie Francis
Cas 23h23'25" 53d36'19"
Jennybabe
Ori 5h30'36" 5d5'4"
Jennybabe
Ori 5h27'45" 5d0'50"
Jennye Marie Grobmyer Dulaney
And 2h16'42" 39d10'54"
Jennyface bablazin'
Uma 9h46'42" 67d15'31"
Jennyfer Fairy Princess
Sco 17h39'8" -37d40'48"
Jennyfer Jennings
Cas 1h34'43" 60d43'29"
JennyFur
Vir 14h6'41" -15d26'53"
Jennykris
Col 6h0'16" -31d52'24"
Jenny-Lynn Borgatti
Vir 13h21'12" 12d43'16"
Jenny-Lynn Somma
Psc 1h17'17" 20d33'37"
Jenny's Bright & Beautiful Star
Leo 9h55'8" 24d56'14"
Jenny's Guiding Light
Lib 15h23'8" -7d36'5"
Jenny's Piece of Heaven
Psc 1h14'13" 26d41'51"
Jenny's Sparkle
Vir 13h35'18" -10d26'54"
Jenny's Star
Cas 0h59'10" 63d32'21"
Jenny's Star
Tau 4h44'10" 21d58'10"
Jenny's Star
Cyg 21h32'47" 40d52'19"
Jenny's Sugar Plum Fairy
Leo 10h13'19" 15d54'41"
Jenny...Will You Mary Me?
Cyg 21h21'53" 50d13'18"
Jenoli
Cru 12h19'53" -57d33'0"
Jenon Tintern
Cep 23h27'53" 87d18'36"
Jenovanna
Uma 11h26'53" 53d38'5"
JenPaws 07
Aqr 22h3'9" 1d57'33"
JenRay 25
Lib 15h37'35" -17d18'13"
Jens
Ori 5h54'42" 12d24'22"
Jens
Ori 5h2'42" 6d21'8"
Jens
Uma 11h25'59" 29d27'42"
Jen's Akela
Cma 6h43'41" -16d40'18"
Jens and Jennifer
Her 16h50'38" 35d27'5"
Jens Becker
Uma 11h1'19" 65d13'15"
Jens Busch Junior
Sco 17h56'40" -30d7'28"
Jens Curtis Egeland
Per 2h43'15" 52d47'42"
Jens Daisy Mae
Mon 6h50'37" 7d26'50"
Jen's DragonStar
Dra 19h40'39" 66d10'1"
Jens E.
Lyn 7h56'30" 42d48'36"
Jen's Eternal Light
Ari 2h53'54" 14d14'49"
Jens Frederick Egeland
Per 2h49'46" 53d58'19"
Jens Helbig
Ori 6h19'21" 6d13'28"
Jen's Jam
Her 17h10'55" 36d23'35"
Jens "Jimbo" Keyes
Uma 11h2'6" 72d4'31"
Jens Klein
Ori 5h13'22" 15d12'4"
Jen's Little Star
Aqr 22h21'17" -7d10'32"
Jens Mårten Ingemar Larsson
Ori 6h23'3" 15d1'19"
Jen's Pebble
Cyg 21h12'47" 51d46'29"
Jens Peter
Uma 12h57'0" 52d42'6"
Jens Pietrusky
Uma 9h30'53" 47d1'44"
Jen's Place
Lib 15h13'42" -22d22'2"
Jen's Star
Ori 5h32'47" -2d11'44"
Jen's Star
And 0h55'8" 46d26'28"
Jen's Star
Ori 5h44'31" 3d36'35"
Jens Teuber
Uma 10h13'32" 49d57'31"
Jens und Franzi
Uma 8h39'39" 59d21'20"
Jens, mein Kuschelbär
Uma 10h19'6" 68d38'9"
Jens, meine Sternschnuppe
Uma 10h3'56" 46d35'12"
Jensen Boy's
Boo 14h34'12" 52d0'9"
Jensen Callum Houghton
Umi 17h22'22" 85d57'28"
Jensen Jack Glenn
Vir 13h23'18" -15d6'10"
Jensen Stewart
Cru 12h27'20" -59d11'45"
Jensen William Seculer
Lib 15h24'36" -12d0'51"
Jensenator
Her 17h25'5" 35d9'5"
Jenski
Ori 5h50'32" 2d14'22"
Jenson James Burgess
Uma 11h21'45" 29d14'6"
Jenson John Astbury
Per 3h0'40" 44d1'4"
Jenson815
And 1h37'4" 50d16'6"
Jenstar
Vir 13h20'26" 11d38'4"
Jenstar
Vir 12h32'37" 12d6'5"
JenStar25
Leo 9h42'1" 27d53'32"
Jensyn Bella
Leo 11h28'23" 18d34'1"
Jentel Marie Adams
Uma 8h35'41" 57d17'27"
JentheGrea
Psc 1h25'42" 14d1'9"
Jentri Furniss
And 0h40'4" 43d48'11"
Jentri Skyler Benbrook
Ori 5h40'25" -0d4'13"
JenWen - 1
Umi 23h40'23" 89d31'10"
Jenya Sagalchik
Sco 17h31'29" -44d7'39"
Jenyfer Willis
Lib 15h49'44" -18d16'21"
Jenzie
Per 4h31'44" 39d28'56"
Jenzy
Psc 1h8'40" 23d28'4"
Jeona Sanders
Cyg 20h17'21" 43d20'30"
Jeorge Barbour
Her 17h23'46" 22d32'49"
Jeoung Suk Kern
Uma 11h11'20" 55d51'19"
j-e-p
And 0h24'46" 38d57'37"
J.E.P. and J.A.K. Forever
Psc 0h13'27" 4d8'50"
Jer and Anna - 2yrs I love you
Cyg 21h18'58" 41d17'9"
Jer Bear
Umi 16h43'41" 83d44'25"
Jer Bear's Blu Rayne
Uma 8h15'56" 65d6'3"
Jera Kay Lammy
Vir 14h43'37" -4d51'24"
Jerad's Star
Uma 9h55'55" 48d0'16"
Jerald and Audrey White
Cyg 20h41'45" 55d0'9"
Jerald and Colleen Watt
Per 2h38'4" 55d10'14"
Jerald D. Melrose
Ori 5h21'52" 7d27'49"
Jerald E. Smith
Uma 9h35'39" 50d32'49"
Jeraldine A. Council
Uma 13h5'14" 59d43'50"
Jeraldine "Jeri" Smith
Vir 12h23'46" 8d59'47"
Jeralynn Strong
Crb 15h23'30" 30d47'3"
Jeramie Pahlman - July 29, 1977
Cyg 19h36'49" 28d7'43"
Jeramy Michael Reimer
Her 18h35'40" 20d16'11"
Jeramy Ybarra
Her 17h23'48" 47d14'31"
jerang
Lyn 8h22'3" 56d13'2"
Jerard Thomas Carney
Dra 17h9'55" 57d15'5"
Jere and Scott Forever
Lyn 8h26'6" 43d6'20"
Jere D. Trejo
Leo 9h22'55" 10d4'39"
Jere J. Schoettmer, Ph.D.
Gem 7h43'55" 33d41'41"
Jere Jeannoel
Vir 12h45'2" -1d3'53"
Jere&PatriciaKyle
Lyn 8h48'58" 33d13'46"
Jered Wray
Gem 6h52'11" 29d56'44"
JEREJESS
Sge 19h24'39" 18d18'54"
Jerelind Mysay Forever
Lib 14h35'29" -17d48'56"
Jeremaine "Mimin" Santiago
Uma 8h31'27" 67d29'9"
Jereme Roy Katzel
Boo 14h48'40" 36d6'49"
Jeremiah
Per 3h19'9" 45d47'58"
Jeremiah
Gem 7h14'11" 28d7'52"
Jeremiah
Vir 12h17'29" 3d7'58"
Jeremiah Ashukian
Umi 15h8'25" 77d35'32"
Jeremiah Byron Sanders
Aqr 23h5'55" -13d54'23"
Jeremiah & Clarisse
Lyr 18h40'2" 38d26'15"
Jeremiah Claude Barnes Sep 3, 1998
Umi 15h4'19" 75d26'59"
Jeremiah David Reed
Aqr 21h52'25" -3d10'56"
Jeremiah D.Penland
Ori 5h47'22" 3d30'30"
Jeremiah George Dellas
Her 17h6'20" 27d53'5"
Jeremiah J. Price
Sco 17h7'48" -35d56'56"
Jeremiah John
Lib 15h8'28" -10d40'8"
Jeremiah Johnson
Sco 16h48'14" -32d39'22"
Jeremiah Kellogg
Cet 1h9'8" -19d10'0"
Jeremiah Kenneth Jones
Ori 6h8'23" -0d30'47"
Jeremiah Leary
Aqr 22h31'26" 1d13'25"
Jeremiah Lucas Rayne
Psc 1h44'0" 6d3'23"
Jeremiah Mahlone Shepard
Sgr 19h37'49" -15d50'7"
Jeremiah N. Hansen
Per 3h5'30" 56d22'50"
Jeremiah Reynoso
Sgr 18h39'40" -31d5'39"
Jeremiah Ryan Shelton
Ori 6h17'18" 10d40'10"
Jeremiah Scott
Dra 18h19'28" 49d42'40"
Jeremiah Scott Eisenhardt
Ori 5h3'57" 9d31'33"
Jeremiah Scout Visnouske
Aql 19h46'40" -0d36'2"
Jeremiah Shane Conrey
Aqr 23h1'43" -11d34'31"
Jeremiah Shane Culbreth
Cap 20h26'52" -12d27'2"
Jeremiah Stewart Olney
Her 17h25'5" 25d32'28"
Jeremiah Thomas Oestreich
Aqr 22h48'51" -21d11'11"
Jeremiah William Lawrence
Aql 19h40'11" 6d31'41"
Jeremiah's Love
And 2h19'47" 48d25'19"
Jeremias-Begleitstern
Lyr 18h54'43" 26d34'24"
Jérémie Gariépy-Théoret
Uma 9h0'3" 72d10'20"
Jeremie Yvon Fournier
Uma 12h16'15" 62d27'18"
Jeremily
Lyn 8h3'41" 57d7'34"
Jeremi's Star
Vir 12h41'59" 3d34'25"
Jeremy
Aqr 21h34'51" 2d3'3"
Jeremy
Ori 6h11'8" 8d49'1"
Jeremy
Leo 11h48'49" 21d23'28"
Jeremy
Leo 10h28'38" 15d22'28"
Jeremy
Leo 9h50'43" 15d26'50"
Jeremy
Lyn 6h59'18" 47d23'57"
Jeremy
Uma 10h35'47" 46d55'9"
Jeremy
Uma 11h42'14" 51d59'17"
Jeremy
Boo 14h43'4" 32d58'16"
JEREMY
Lac 22h44'46" 52d38'57"
Jeremy
Cap 20h50'34" -16d7'31"
Jeremy
Sco 16h16'43" -10d22'11"
Jeremy
Aqr 23h1'40" -5d53'30"
Jeremy
Ori 5h24'45" -0d39'53"
Jeremy
Lib 15h8'6" -29d54'35"
Jeremy
Sco 17h42'57" -44d44'8"
Jeremy
Scl 23h38'55" -37d30'20"
Jeremy A. McDougall
Sco 17h38'51" -41d19'22"
Jeremy Adam Chizewer
Her 17h9'53" 30d24'28"
Jeremy Alexander Davis
Lib 15h3'25" -6d54'55"
Jeremy Allan Anderson
Uma 8h40'43" 51d21'14"
Jeremy Allen
Aql 19h23'53" 4d11'44"
Jeremy Allen Brown
Ori 5h7'30" 4d18'8"
Jeremy & Amanda Roberts
Ori 5h1'28" 3d53'4"
Jérémy Amelin
Cnc 9h6'26" 14d29'57"
Jeremy and Amy Robinson
Ori 4h51'36" 4d28'31"
Jeremy and Briana Forever
Cyg 19h40'39" 38d21'18"
Jeremy and Crystal Ramos
And 1h17'41" 37d33'56"
Jeremy and Dawn Tirado
Cyg 19h23'57" 51d8'48"
Jeremy and Eva Love in the Sky
Aqr 21h45'2" -0d7'24"
Jeremy and Melissa
Uma 8h37'46" 62d30'57"
Jeremy Andrew Lewis
Aqr 22h35'31" -4d8'29"
Jeremy Blake White
Gem 6h31'36" 16d28'50"
Jeremy Bloom
Ori 6h11'19" 3d23'4"
Jérémy Brunat
Cas 0h26'36" 59d44'25"
Jeremy Bryan Hicks Jr.
Uma 9h42'56" 59d10'58"
Jeremy Bryson Retino
Aqr 23h10'20" -7d52'18"
Jeremy Budz
Per 3h33'17" 35d22'32"
Jeremy Calavrias
Cru 12h25'57" -61d13'24"
Jeremy Caldwell Simmons
Per 3h10'25" 49d25'7"
Jeremy Calvin Perry
Ori 5h16'11" 6d12'37"
Jeremy Chapman
Dra 17h0'18" 69d30'16"
Jeremy Charles (Jem's Star)
Aqr 22h40'35" 2d2'4"
Jérémy Chavaudra
Ori 4h46'49" 3d58'39"
Jeremy Cole
Psc 23h55'25" 7d4'3"
Jeremy Craig
Aqr 22h45'26" 1d15'5"
Jeremy Crosslin
Lyn 9h10'0" 33d15'40"
Jeremy D. Ross
Her 15h50'41" 39d40'42"
Jeremy D. Smith
Cnc 8h39'35" 32d21'57"
Jeremy Daniel Hawley
Uma 9h45'23" 60d40'19"
Jeremy Daniel Hinz
Aur 5h17'29" 41d54'26"
Jeremy Daveau
Boo 15h1'2" 47d38'31"
Jeremy David Coleman
Lib 14h47'17" -6d41'17"
Jeremy David Doak
Per 2h52'36" 52d14'30"
Jeremy David Price
Ari 1h51'25" 17d49'25"
Jeremy David Tuell
Her 18h10'15" 15d34'4"
Jeremy Deal
Vir 14h17'59" 1d26'24"
Jeremy Dean Sheppard "BEAR"
Vir 11h53'54" 3d47'20"
Jeremy Delon
Her 16h41'24" 46d20'28"
Jeremy Desiree Correia
Uma 10h17'6" 53d12'46"
Jeremy Dom Sisko Dawson
Ari 3h2'44" 20d53'8"
Jeremy Douglas Bowman
Aqr 21h49'9" -1d16'1"
Jeremy Drew Fontanez Sr.
Uma 11h46'42" 46d41'20"
Jeremy Elmer Kerl
Ari 2h18'8" 24d38'54"
Jeremy Elvis Pearce
Crb 15h55'46" 27d8'47"
Jeremy Espy
Per 3h15'35" 46d31'43"
Jeremy Ethan Fox
Per 2h37'42" 55d51'51"
Jeremy Grebe
Lib 15h51'52" -19d24'19"
Jeremy Guapito Gibbs
Ori 5h58'21" 18d38'21"
Jeremy Holmes
Dra 18h10'38" 75d18'49"
Jeremy J. Englert
Umi 13h36'12" 72d10'40"
Jeremy J Kemp
Uma 12h45'28" 63d9'25"
Jeremy J. Maumus
Gem 7h50'0" 24d39'19"
Jeremy Jackson
Uma 11h58'19" 54d59'38"
Jeremy & Jaclyn 4always
Cyg 20h33'49" 52d22'4"
Jeremy James Hortman
Uma 8h59'19" 55d54'10"
Jeremy James Maley
Lib 15h4'17" -1d3'42"
Jeremy James Siipola
Uma 10h14'49" 45d50'22"
Jeremy & Jayna Will Shine Forever
Ori 5h26'12" 3d51'41"
Jeremy "Jer-Bear" John Morales
Cnc 8h36'50" 24d46'57"
Jeremy John
Cap 21h39'1" -24d24'52"
Jeremy John Glen
Gem 7h24'14" 18d0'7"
Jeremy John Simon
Her 16h44'2" 45d5'0"
Jeremy Joseph Knost
Aql 19h9'18" 13d12'32"
Jeremy Joseph Murray
Ori 6h11'59" 16d41'20"
Jeremy Joseph Neves
Lib 15h25'55" -7d58'23"
Jeremy Karlovits
Psc 0h22'37" 12d6'44"
Jeremy Keith Gray
Sgr 19h15'10" -17d13'42"
Jeremy Kenneth Weiss
Sco 17h38'48" -30d16'1"
Jeremy & Lacey Just Remember Faith
Umi 15h58'36" 75d8'0"
Jeremy Laddin
Cep 21h31'39" 63d46'28"
Jeremy Lakso
Cep 22h28'17" 62d45'25"
Jeremy Lebowitz
Her 17h20'26" 45d9'41"
Jeremy Lechtzin
Ori 6h7'36" -0d20'14"
Jeremy Long
Uma 10h17'49" 71d59'30"
Jeremy Loves Kimberly
Her 18h42'56" 19d13'1"
Jeremy Lynch Blvd.
Cyg 20h49'13" 46d5'46"
Jeremy M. Hooper
Uma 11h22'38" 32d5'29"
Jeremy & Maria Moore
Ari 2h41'45" 26d26'36"
Jeremy Mark Zeller
Cep 1h27'6" 81d19'34"
Jeremy Martin Bucher
Lyn 7h56'28" 47d54'20"
Jeremy Mathers
Cnc 9h14'8" 27d52'30"
Jeremy Matthew Duran
Lyn 8h10'53" 52d8'43"
Jeremy Matthew Miles
Her 18h7'52" 27d4'30"
Jeremy McCoy
Her 18h44'13" 21d16'28"
Jeremy & Meghann
Cyg 21h58'43" 53d0'24"
Jeremy Michael Foster
Cap 20h14'19" -15d15'57"
Jeremy Michael McCannon
Uma 11h40'22" 41d2'53"
Jeremy Michael Miller
Peg 22h26'25" 24d59'47"
Jeremy Michael Ruiz
Umi 15h17'50" 67d51'30"
Jeremy Michael Weber
Cep 20h7'49" 60d46'51"
Jeremy Mink
Cnc 9h17'58" 11d12'31"
Jeremy & Mom
Cyg 20h51'19" 48d39'53"
Jeremy Money
Her 17h35'26" 32d12'5"
Jeremy "Monkey" Archibald
Ori 5h48'57" 9d8'0"
Jeremy Morgan
Sgr 18h29'54" -22d8'34"
Jeremy Patrick Gullette
Psc 0h57'3" 28d56'45"
Jeremy Paul Longyear
Uma 8h16'39" 68d10'29"
Jeremy Pession
Peg 23h48'21" 28d8'1"
Jeremy Petty
Aur 7h2'41" 35d28'58"
Jeremy Polychronopoulos
Cnv 13h45'13" 31d37'3"
Jeremy PS
Dra 19h17'12" 74d52'47"
Jeremy R. Dobbe
Uma 9h23'20" 71d16'19"
Jeremy R. Henley
Leo 9h24'9" 30d59'37"
Jeremy Reed Thompson
Aur 5h54'9" 53d14'57"
Jeremy Richard Best
Uma 8h41'58" 65d52'23"
Jeremy Robert Cicolino
Vir 13h34'41" -7d30'55"
Jeremy Robert Smith 26-9-2004
Lib 15h40'28" -10d37'37"
Jeremy Rochon
Lyr 18h29'29" 35d58'33"
Jeremy Rourke
Ari 2h43'58" 26d10'49"
Jeremy Ryan Pitcher Born: Jan 22/1989
Uma 11h31'10" 57d34'14"
Jeremy Salois
Per 4h26'52" 50d55'6"
Jeremy Santo Garofalo
Cep 0h5'0" 80d16'39"
Jeremy Scott Arney
Her 17h13'4" 35d58'43"
Jeremy Scott Stretz
Her 17h36'22" 36d38'18"
Jeremy Shawn Rohrs JX
Per 3h42'46" 52d1'50"
Jeremy Slocombe
Cyg 20h53'17" 47d48'29"
Jeremy Sosman Ham
Aql 19h7'31" 0d55'32"
Jeremy Spencer Watson
Per 4h22'1" 39d38'23"
Jeremy (stallion)
Dra 19h3'15" 68d43'59"
Jeremy Sterling 6/6/1986
Uma 11h14'12" 61d3'59"
Jeremy Steven
Crb 15h53'34" 27d11'49"
Jeremy Steven Devault
Ori 5h33'7" -0d27'25"
Jeremy Syth
Aqr 23h3'11" -12d34'39"
Jeremy The Bamf
Ori 4h53'48" 4d20'42"
Jeremy The Expert Discenza
Uma 12h38'16" 58d6'5"
Jeremy Thomas
Gem 6h41'53" 16d9'31"
Jeremy Thomas
Gem 6h48'16" 34d11'5"
Jeremy Todd Nickel
Per 3h21'15" 42d20'19"
Jeremy Todd Williams
Cnc 9h10'27" 15d21'10"
Jeremy Tyler Gibson
Leo 11h29'45" 18d47'19"
Jeremy V. Olson
Sgr 18h26'3" -18d17'45"
Jeremy W. Bailey
Gem 7h4'56" 34d26'48"
Jeremy Walter Reynolds
Ari 2h37'36" 25d53'34"
Jeremy Wayne Melton
Ori 5h28'1" 2d15'3"
Jeremy Wayne Newman
Ari 2h9'4" 14d31'18"
Jeremy Webb
Cep 21h37'10" 60d14'54"
Jeremy Whitehead
Cnc 8h20'25" 32d57'8"
Jeremy Williams
Her 17h43'8" 36d43'3"
Jeremy Williams
Her 17h19'56" 29d10'6"
Jeremy Wolsten-Croft
Cep 21h55'21" 58d57'41"
Jeremy Wood
Gem 6h55'43" 29d30'4"
Jeremy Wright
Cma 6h37'33" -25d32'58"
Jeremy Yearsley
Per 4h11'36" 51d54'42"
Jeremy Zappettini Pearson
Mon 7h36'30" -0d41'23"
Jeremy, Salena & Paytyn Miller
Eri 4h43'14" 0d12'8"
JeremyAndKristen143
Cyg 19h48'27" 37d3'3"
Jeremy-Jacob Alexander Gray
Lib 15h29'25" -27d39'14"
Jeremy's Beacon
Vir 13h48'4" -5d34'22"
Jeremy's Journey
Cep 0h9'10" 79d43'53"
Jeremy's Light
Uma 8h54'3" 58d29'57"
Jeremy's Light
Her 16h48'20" 41d35'44"
Jeremy's peace
Cnc 9h18'58" 8d44'21"
Jeremy's Star
Ari 3h28'18" 21d50'55"
Jeremy's Star
Ari 1h54'54" 20d38'33"

Jeremy's Sweet Jennifer Cnc 8h54'55" 26d21'19"
Jerene And 23h43'2" 46d57'42"
Jerene Lee Vaagen Ari 3h8'38" 19d55'15"
Jeret Lee Tadtman Sgr 18h33'39" -28d17'39"
Jeri Aql 19h31'44" 10d1'17"
Jeri Ann Hansen Gem 7h19'16" 31d0'27"
Jeri Frances Berry Cnc 9h4'52" 24d35'47"
Jeri Garner Psc 1h8'8" 8d27'18"
Jeri Girl Lyn 8h35'44" 33d18'49"
Jeri Hatcher Ari 2h35'24" 25d11'46"
Jeri J. Schweikert Lib 15h12'50" -14d27'18"
Jeri Lou Coffey Uma 10h34'31" 50d28'24"
Jeri Lou Wilber And 2h32'18" 39d28'9"
Jeri Lynn Lib 15h26'53" -6d11'30"
Jeri Lynne Spiero Sco 16h4'40" -16d22'5"
Jeri Nicolas Leo 11h6'54" 10d30'36"
Jeri of Dana Point Uma 11h17'44" 35d20'3"
Jeri Ryan Psc 0h22'14" 18d41'29"
Jeri & Scott's Unity Star Vir 11h57'53" -9d48'45"
Jeri Woods And 23h21'47" 42d25'50"
Jerian,John,Samantha and Daniel Ori 5h7'34" 15d4'20"
Jeriann 50th Tau 5h30'53" 22d1'52"
Jeriann Firestone Uma 12h46'6" 56d28'46"
Jerica Uma 12h14'49" 53d33'20"
Jerica Curtin Tau 5h43'59" 27d26'52"
Jerica M. Jenke Vir 13h43'30" -7d17'21"
Jerica Renee Cross And 1h20'25" 35d22'18"
Jerica Rhodes And 23h14'3" 48d4'24"
Jerica's Star And 2h9'28" 39d4'54"
Jericho Aari Armstrong-Van Vleck Uma 10h17'34" 63d48'32"
Jericka Joyce Correia And 0h52'0" 44d49'59"
Jeridith Aql 19h41'25" 13d25'40"
Jerie L. Gant Peg 23h37'39" 18d38'37"
Jerilyn Aql 20h5'31" -9d52'27"
JeriLynn Sue Koehler Lmi 10h42'40" 33d12'42"
Jerin Basin Lib 14h51'8" -6d20'42"
Jeris Cep 20h40'15" 63d55'40"
Jeri's World Dra 18h40'17" 67d6'8"
Jerk Face Jasmine Cnc 8h44'10" 7d34'55"
JerKat Srp 18h38'2" -0d27'54"
Jermain Wedderburn Sco 17h17'1" -44d9'6"
Jernigan Psc 23h51'24" 5d17'16"
Jerod MB-HMD Gem 7h33'16" 21d52'13"
Jerod Michael Fulmer Her 18h11'18" 18d37'14"
Jeroen en Jennifer van der Meer Uma 13h58'16" 52d19'59"
Jerold and Dorothy Beeve Vir 12h40'23" 1d41'16"
Jeroldine 50 Tau 3h58'39" 17d21'45"
Jerome Leo 10h13'43" 14d42'56"
Jérome Lyn 7h44'9" 39d17'52"
Jerome A Hayford Aur 5h50'14" 53d15'14"
Jerome Allen Shreiner Uma 9h13'0" 49d30'19"
Jérôme Amar Uma 12h9'33" 53d40'33"
Jerome Anuszkiewicz Ori 5h38'26" 11d54'40"
Jerome Best Aur 5h50'43" 48d39'26"
JEROME BROWN Sco 16h16'26" -41d27'39"
Jérôme Bulteau Com 12h44'12" 15d58'8"
Jerome Charles Edmond Psc 0h46'6" 16d33'34"
Jerome Elisah Purvis lll 7/26/1973 Leo 11h21'36" 16d26'31"
Jérôme et Cynthia Ori 5h44'58" 11d49'3"
Jerome Grace Cyg 19h50'53" 38d22'45"
Jérôme Hélène Salomé Vouvet Ori 5h12'47" 6d53'10"
Jérome Hoffmann Cyg 19h45'17" 55d42'12"
Jerome Horton Cep 20h38'39" 63d15'41"
Jerome J. Viola Per 3h35'2" 51d38'5"
Jerome John Callahan Boo 14h49'1" 26d23'51"
Jerome & Kathleen Lech Mon 7h20'26" -0d53'53"
Jerome Kopmar Vir 13h33'42" 6d33'31"
Jerome Lanier Dra 16h41'48" 56d33'25"
Jérôme Leclercq Lyn 7h54'31" 56d27'24"
Jerome Leo Gostkowski Gem 6h35'51" 17d37'21"
Jerome Li Guaiguai Tau 4h26'49" 23d18'53"
Jerome Loubere Uma 9h12'14" 61d13'8"
Jérôme Lugrin Mon 7h4'47" 5d2'58"
Jerome Lukas Psc 0h34'19" 7d49'22"
Jerome M. Kern Ori 5h3'18" 5d27'31"
Jerome Majeski Uma 12h41'36" 55d44'31"
Jérôme Müller Uma 12h18'6" 59d41'32"
Jerome P. Oberman Lib 15h37'18" -17d16'51"
Jerome P. Viola Per 3h51'49" 51d0'41"
Jerome Radwin Lib 14h52'28" -0d55'45"
Jerome Roman Pellowski Uma 11h8'15" 72d29'30"
Jerome S. Moore Aur 5h11'25" 33d12'26"
Jerome Saxton Her 17h34'58" 36d22'13"
Jérôme Scherer Umi 13h19'47" 73d20'23"
Jerome Schulman And 0h49'21" 37d46'43"
Jerome Simon Cnc 8h7'32" 15d9'0"
Jerome Smolar Uma 9h51'11" 45d42'51"
Jerome Steven Alvarado Cap 21h57'33" -17d53'47"
Jerome Vernon Wilkinson Lib 14h53'37" -11d54'46"
Jerome Victor Jarger Her 17h52'33" 29d4'17"
Jerome Weitzner Cep 22h6'8" 53d24'44"
Jerome Wells Uma 10h57'59" 65d32'21"
Jérôme, envie de folie à la vie! Aql 20h15'56" 0d13'32"
Jerome's Boo Cap 20h8'53" -18d5'30"
Jeromy Morris Shining Hope Her 17h17'46" 35d51'33"
Jeromy W. Pierroz Tau 5h14'5" 23d19'12"
jeronimo Cas 2h0'11" 69d27'33"
Jerrad Keith Fowler Ari 2h31'49" 21d58'14"
Jerral Brooks Miller Cap 20h26'7" -10d49'25"
Jerrell Streeter Aur 5h10'50" 36d19'50"
Jerrene Johannesen Uma 11h57'58" 39d58'10"
Jerret Dale Knight Her 17h41'30" 21d28'41"
Jerrett Stark 743 Tau 4h41'10" 23d10'18"
Jerri Ann Deavers Her 16h24'11" 44d19'51"
Jerri Berry Sgr 18h33'30" -24d49'48"
Jerri Casat Uma 11h58'43" 59d1'0"
Jerri D. Sgr 17h56'44" -18d1'37"
Jerri Grell (Lady High Pockets) Uma 11h32'51" 64d8'35"
Jerri Kim Thomas Sco 17h51'59" -36d4'13"
Jerri Kitchen Uma 10h59'59" 35d3'22"
Jerri Lea Shaw Umi 14h34'59" 74d0'24"
Jerri Lou Cap 20h23'47" -12d44'3"
Jerri Lynn Macke Lep 5h50'31" -12d34'41"
Jerri & Marcus Bush And 0h38'42" 39d5'53"
Jerrica Tau 4h14'35" 16d13'47"
Jerricus Duanus Leo 11h55'23" 21d36'13"
Jerrie Uma 11h12'52" 56d2'52"
Jerrie 1976 Cnc 8h41'20" 9d8'37"
Jerrika Mahala Cap 20h26'42" -10d49'58"
Jerrimae Uma 11h14'48" 33d27'53"
Jerrimy John Wilkins Her 17h30'17" 43d22'26"
Jerrod J Riggleman Cnc 8h22'44" 29d52'33"
Jerrold Gould Psc 1h10'45" 9d59'23"
Jerron Our Shining Light Leo 9h27'1" 11d45'9"
Jerry Psc 0h20'32" 17d16'36"
Jerry Aqr 22h18'5" 1d57'52"
Jerry Boo 14h30'35" 40d46'21"
Jerry Uma 13h59'15" 49d27'35"
Jerry Her 17h15'11" 48d22'2"
Jerry Cyg 20h10'33" 47d33'16"
Jerry Uma 10h6'7" 43d2'14"
Jerry 2000 Uma 12h16'43" 57d47'28"
Jerry 50 Tau 3h55'19" 19d28'14"
Jerry A. D'Ambra Sr. Her 16h11'11" 46d55'18"
Jerry A. Graham Cyg 20h28'51" 43d49'36"
Jerry A Poff,Light of My Life 25yrs Uma 12h6'21" 50d11'31"
Jerry A. Yankaskas Uma 9h3'58" 67d8'34"
Jerry A. Zahn Per 3h14'33" 52d41'5"
Jerry Abramson Her 17h14'28" 19d59'7"
Jerry Acheson Gem 7h16'0" 21d59'10"
Jerry "AKA Butch" & Michelle Curry Aqr 22h18'21" -0d21'3"
Jerry Allen Boles Ori 5h32'59" 4d11'8"
Jerry Allen Schroyer Umi 14h46'46" 82d58'58"
Jerry and Eunmi Deutsch Cyg 19h55'53" 31d24'47"
Jerry and Georgia Kearney Her 16h45'2" 47d50'36"
Jerry and Joni Miskovic Leo 11h19'50" -3d55'54"
Jerry and Julie Green Lyr 19h24'23" 31d28'19"
Jerry and Linda Bruckheimer Uma 10h18'10" 44d47'35"
Jerry and Myra Shiplett "12-20-70" Sgr 18h18'26" -21d0'8"
Jerry and Phyllis Meyer Cyg 20h44'6" 46d17'25"
Jerry and Sherry Walker Uma 9h3'52" 59d26'20"
Jerry and Tiffany Appley Ever After Umi 15h32'45" 73d57'42"
Jerry Bartholome, My Beloved Psc 23h58'30" -0d30'50"
Jerry Bartholow's Unity Star Ari 2h33'35" 18d32'16"
Jerry Bassin Sgr 18h26'32" -34d32'7"
Jerry Beck "Awesome Daddy" Boo 14h53'49" 22d46'33"
Jerry Belcher And 2h37'35" 46d18'1"
Jerry Bernius Uma 9h46'11" 58d23'35"
Jerry Bonino Per 2h42'39" 54d4'43"
Jerry Brinster Her 16h22'28" 5d45'36"
Jerry Brown Cnc 9h11'32" 10d31'45"
Jerry Camina Ori 5h21'40" 4d11'31"
Jerry Carletti Leo 9h49'34" 22d58'3"
Jerry Carter & ET Home Uma 11h30'58" 58d25'57"
Jerry Cellon Aur 5h25'41" 33d36'41"
Jerry Chafee Uma 13h57'8" 56d25'20"
Jerry Cheung Aql 19h11'20" 13d10'6"
Jerry & Cristy Cyg 21h46'46" 54d6'37"
Jerry D. Headlee Cep 23h0'26" 70d30'57"
Jerry D Strand Ori 4h49'44" 2d55'33"
Jerry D. Williams Leo 9h48'33" 20d45'55"
Jerry 'Dad Rocks' Zaborowski Crb 16h13'6" 38d38'8"
Jerry Daniel Bryant Jr Cnv 13h27'17" 32d1'51"
Jerry David Baldwin Cyg 20h4'52" 38d5'30"
Jerry Davis Fogleman Ori 5h37'36" 1d21'23"
Jerry Davis Porter Lyn 6h58'58" 45d49'24"
Jerry Dean Amaro Ori 5h37'45" 8d35'24"
Jerry Dean Collis Dra 20h25'0" 63d11'6"
Jerry DelGaudio Uma 8h42'47" 49d27'32"
Jerry & Diane Self Cyg 21h26'8" 34d32'34"
Jerry Dianne Craig Crystel Rebecca Cru 12h46'8" -64d35'32"
Jerry Diekmann Uma 11h59'15" 62d38'30"
Jerry Don Calloway Per 4h16'37" 36d48'33"
Jerry Douglass Peg 23h15'17" 33d17'12"
Jerry Doyle Lewis, Jr. Ori 5h20'55" 5d4'25"
Jerry Dozier Aql 19h45'5" -0d35'47"
Jerry Drew Cozby Aur 5h8'51" 46d53'41"
Jerry DuShane Gem 6h47'34" 13d20'51"
Jerry Edward Secor Psc 1h26'37" 22d42'16"
Jerry Edwin Fralick Umi 15h13'48" 71d23'35"
Jerry & Evie Levitz Cyg 20h54'13" 40d8'41"
Jerry Foster Uma 10h4'42" 65d18'28"
Jerry Friend Leo 10h19'46" 6d44'29"
Jerry Geronimo Radja Lib 15h24'23" -27d35'52"
Jerry Giles Ori 5h7'46" 9d21'33"
Jerry Goodwin III Tau 5h52'11" 25d21'22"
Jerry Greenlees Her 17h12'10" 31d46'30"
Jerry Hagan Aqr 21h17'18" 2d9'20"
Jerry Halsband Uma 11h28'55" 33d48'41"
Jerry (H.M.P) Christie Cep 0h17'21" 69d36'4"
Jerry Holbrooks Ori 4h48'2" 3d50'32"
Jerry Holland Per 3h33'9" 50d14'20"
Jerry Howard Uhlig Dra 18h29'2" 59d26'8"
Jerry Irving Ori 5h43'3" 8d43'2"
Jerry J. & Daniel P. Fleming Uma 10h33'39" 57d0'31"
Jerry James House Psc 0h51'24" 12d46'4"
Jerry (Jod) Martin Boo 14h23'36" 47d16'21"
Jerry Joseph Ari 3h11'30" 18d21'21"
Jerry Joyce Uma 10h6'24" 47d42'3"
Jerry & Joyce Dunning Sge 19h51'50" 17d22'44"
Jerry Jurgens And 0h46'38" 33d23'25"
Jerry & Kim Schwamn Cyg 19h59'35" 56d58'39"
Jerry Kouzmanoff Cep 20h54'6" 57d33'48"
Jerry Kozlowski Uma 11h13'55" 49d29'31"
Jerry L. Ballew Uma 8h49'44" 56d18'1"
Jerry L. Carbone & Mary E. Carbone Cyg 19h33'48" 47d47'52"
Jerry L. Guetschow, Jr. Aqr 21h5'3" -11d25'1"
Jerry L. Todd Aqr 23h5'18" -9d19'59"
Jerry Laine Jr. 4-30-75 Ori 6h18'23" 9d29'28"
Jerry LaVon Pace And 23h2'59" 48d4'7"
Jerry Lee Sco 16h10'8" -13d39'33"
Jerry Lee Vir 13h15'50" -2d27'19"
Jerry Lee Hall Her 17h21'42" 16d54'12"
Jerry Lee Houck Lmi 9h44'7" 34d34'54"
Jerry Lee Kronquists's Shining Star Aql 19h16'2" -0d4'24"
Jerry Lee Walker, Jr. Gem 7h23'42" 24d18'29"
Jerry Lee Werner Uma 10h28'14" 44d49'54"
Jerry Lett Per 4h45'59" 36d31'48"
Jerry Lynn Biddinger Umi 17h8'26" 79d29'22"
Jerry Lynn Gibson Aur 5h53'48" 55d25'20"
Jerry Lynn Shenep Lib 14h51'0" -3d59'48"
Jerry & Lyra - Lasting Love Eri 3h48'48" -10d42'26"
Jerry M. Durkis Sgr 20h24'31" -42d11'35"
Jerry M. Shiota Per 3h24'41" 52d26'29"
Jerry Mac Curry Psc 1h34'49" 21d51'49"
Jerry Maggi Ori 5h20'14" 6d38'58"
Jerry Malone Cep 22h34'13" 57d54'17"
JERRY & MARIE'S EVER-LASTING STAR Cyg 21h0'25" 44d50'5"
Jerry & Marilyn Newman Cyg 20h24'59" 46d14'49"
Jerry & Mary Martin Canada Dra 18h18'18" 63d40'38"
Jerry May Uma 11h9'25" 71d38'0"
Jerry McCarthy Cyg 20h27'43" 55d2'55"
Jerry Mead Uma 9h3'16" 56d7'26"
Jerry Michael Wise Uma 11h53'7" 43d48'13"
Jerry Monkarsh Uma 10h18'53" 46d52'26"
Jerry Montgomery Cyg 20h57'2" 36d5'56"
Jerry Moylan Ari 2h33'27" 22d41'19"
Jerry My Dream Come True Her 17h9'57" 31d58'35"
Jerry N. Corley Uma 11h33'18" 62d55'5"
Jerry Neal Blair Uma 8h50'44" 51d49'10"
Jerry Neal Harms Pho 0h33'40" -42d13'38"
Jerry & Nunzia Uma 10h18'52" 45d56'57"
Jerry O Uma 9h52'7" 55d13'54"
Jerry Ossie Digman Dra 16h10'3" 55d23'41"
Jerry Ozell Hancock Sgr 19h36'10" -15d56'44"
Jerry & Pam Register Her 17h53'27" 14d27'48"
Jerry Parsons Uma 10h27'3" 65d35'24"
Jerry Patrick Miller Tau 4h50'35" 22d20'30"
Jerry Patton "Our Star Gazer" Sgr 18h14'59" -19d44'31"
Jerry & Penny Now and Forever Star Uma 9h21'41" 47d27'15"
Jerry Peters Ori 5h33'22" 4d17'4"
Jerry "Prince Charming" Shrovnal Ori 6h17'7" 20d50'38"
Jerry Punches Cep 22h42'4" 62d1'35"
Jerry Ray "BEAR" Fedynik Ari 2h55'25" 28d15'12"
Jerry Ray Pridgen Srp 18h8'21" -0d19'19"
Jerry Reggie Rupp Uma 13h46'55" 57d23'2"
Jerry & Rita Potter Umi 16h9'14" 75d46'49"
Jerry Robert Kyle Uma 8h25'25" 61d17'55"
Jerry S. Oakes Per 3h6'8" 44d13'4"
Jerry Sarnowski Per 4h36'21" 48d50'36"
Jerry Schoenherr Cep 20h48'58" 59d26'29"
Jerry Shaw Her 16h47'5" 36d24'14"
Jerry Shellabarger Ori 5h20'52" 3d54'0"
Jerry Silberman Uma 13h20'22" 56d24'5"
Jerry Skinner Aql 19h33'7" 10d21'15"
Jerry Skotland Ori 6h6'8" 20d47'48"
Jerry Smialek Boo 14h34'10" 54d27'28"
Jerry Sol Cupples Boo 13h38'32" 21d32'32"
Jerry Sr. Lib 15h50'25" -10d6'49"
Jerry Star Aql 19h54'21" -0d29'48"
Jerry Stover Ori 5h40'3" 1d20'29"
Jerry Theodore Zoll Ari 2h11'52" 24d53'16"
Jerry Thomas Kelley Tau 4h14'45" 24d52'8"
Jerry Thorn Uma 13h33'6" 56d57'48"
Jerry Tone Aql 19h0'7" -0d39'1"
Jerry V. Reynolds Leo 10h57'8" 16d5'49"
Jerry W. Carter Sco 16h11'50" -14d54'19"
Jerry W. Cashon Uma 10h41'46" 40d10'33"
Jerry W. Hall Leo 10h22'30" 27d0'44"
Jerry W. Kier Her 18h30'16" 21d18'37"
Jerry W. Levin Lac 22h21'25" 47d33'38"
Jerry W. Taylor Aur 5h37'1" 48d41'41"
Jerry Walker Lib 14h58'28" -22d19'43"
Jerry Walkup Cap 21h39'54" -15d33'23"
Jerry Wayne DeLano—Our Star Tau 5h45'14" 16d54'22"
Jerry Wayne Mitts Uma 11h1'5" 37d25'18"
Jerry Wayne Watson Leo 9h34'36" 12d8'30"
Jerry Weber Vir 12h21'1" -6d31'27"
Jerry Westlund Leo 10h24'9" 26d4'2"
Jerry Woods Uma 11h17'44" 54d9'51"
Jerry, Claire and Oscar Cyg 21h10'14" 47d8'16"
Jerry,Pat,Christopher&Zachary Bowen Per 2h59'45" 55d20'23"
jerryandlisapalmer Cyg 20h27'31" 49d49'10"
JerryBeth Gem 7h18'16" 18d39'15"
Jerry's 75th Bash Ori 5h22'9" -4d34'55"
Jerry's Eternal Ride Uma 11h29'6" 50d0'35"
Jerry's Light Crb 15h58'20" 31d47'52"
Jerry's Light Uma 13h21'29" 55d49'39"
Jerry's Shining Stars P,Z,H & J Her 18h9'27" 18d17'59"
Jerry's Star Per 4h4'44" 42d59'44"
Jerrys Star Lyr 19h13'42" 37d37'43"
Jerry's Star B.B.B. 010203 Per 3h52'52" 33d51'42"
Jerry's Star B.B.B 010203 Ori 5h32'23" 14d18'19"
Jerry's Stars-Niki,Sami,Jami,Alexa Per 3h25'23" 50d20'32"
Jerry's Wishing Star Leo 10h23'39" 10d0'28"
Jerrystar Her 18h27'32" 18d28'0"
Jer's Babygirl Shelley Lib 15h38'40" -17d18'54"
Jer's Star Lib 15h4'15" -9d47'49"
Jersey And 0h0'33" 40d49'28"
Jersey Ben Her 16h23'19" 11d2'57"
Jersey Girls Cas 0h35'27" 59d2'57"
Jersey Grace Cabana Per 4h30'33" 31d48'56"
Jersey Jeff Wewers Aur 5h38'14" 49d57'21"
Jersey's Shining Star Psc 1h23'24" 17d20'45"
JerseystarMom Ari 2h35'14" 21d26'55"
JERSHIRTERLAR Lyn 6h44'4" 60d6'14"
Jerus Reign Cam 4h19'17" 67d4'31"
Jerusalem Sky Hartley Vir 13h26'9" -6d14'20"
Jery Carl Tairua Ori 5h41'15" -1d27'22"
Jeryl D. Kershner Ori 5h54'4" 22d41'53"
Jeryl Gould Ori 6h16'15" 18d17'3"
Jerzee Cap 21h34'20" -24d31'57"
Jerzey Cora Beyak Gem 7h12'23" 32d7'25"
Jerzy Wielowiejski Uma 10h24'25" 47d56'31"
Jerzyk Jamroziak Cyg 21h7'43" 30d20'10"
Jes Lac 22h48'49" 49d11'57"
JES Psc 1h8'0" 7d30'30"
jes Lib 15h5'28" -28d35'53"
JES Psc 1h7'7" 7d12'9"
Jes Cru 12h26'52" -60d47'56"
Jes 9 Cas 0h26'48" 62d21'12"
Jes Is The Best Tau 3h26'13" 16d6'50"
Jesanu Gem 6h45'33" 30d17'1"
Jesc Plutt Lib 14h48'11" -15d25'53"
Jeseika Lea Gem 7h36'44" 17d35'48"
Jeselle Cap 20h47'8" -22d10'30"
Jesenia Marina Rodriguez Aqr 21h43'21" -3d52'4"
Jesh and Jehn Sehjal Peg 23h12'8" 27d6'24"
Jesi Carolyn Cas 1h0'0" 62d32'30"
Jesi Gaines Lib 14h30'16" -13d22'43"
Jesi Jane Major Vir 13h33'53" -22d19'37"
Jesi Yost Aql 19h43'55" -0d0'34"
Jesica Gem 7h30'47" 26d37'0"
Jesica Ann And 0h48'21" 38d45'58"
Jesica Christian Nor 16h11'25" -57d16'17"
Jesica L. Milton Gem 7h9'36" 31d34'38"
Jesica Lynn Ori 5h53'24" 18d3'12"
Jesica Marie Uma 12h5'54" 45d54'26"
Jesika Kuzuoka Aqr 22h22'41" 0d48'47"
Jesika Meghan Hubbard Ori 4h53'16" 3d32'31"
Jesika Salt Tau 3h48'39" 24d40'58"
Jesi's Jordan Cap 21h54'41" -9d13'25"
Jeska And 0h15'43" 43d39'50"
Jeska Day Crb 16h18'16" 32d36'49"
Jes-n-Chuck's Piece of Heaven Ori 5h39'18" 5d31'51"
Jess Leo 9h48'26" 12d40'17"
Jess Tau 4h31'14" 27d10'18"
Jess Crb 15h37'24" 36d31'48"
Jess Per 3h27'22" 43d16'37"
Jess And 2h35'22" 49d16'22"
Jess And 23h13'13" 41d51'38"
Jess And 23h23'48" 43d22'33"
Jess Aqr 23h43'35" -10d18'56"
Jess Lyn 7h56'30" 58d15'24"
JESS Uma 9h19'26" 66d33'28"
Jess and Adam - Forever and Always Uma 9h27'55" 44d30'23"
Jess and Bren's Star Cnv 13h32'30" 37d51'3"
Jess and Cel Murillo Lyn 8h4'43" 34d52'42"
Jess and Dave Forever Uma 13h48'15" 48d35'27"
Jess and Ken's Lucky Star Sge 20h3'50" 18d8'25"
Jess and Oliver's Star And 2h37'23" 44d55'56"
Jess and Tim's Star Of Eternity Lib 15h18'30" -9d14'55"
Jess Ashton Cnc 9h1'12" 22d42'39"
Jess Baker Lmi 10h24'30" 28d39'38"

Jess Be
Gem 7h18'39" 18d2'57"
~*Jess~BFF<3~11:11~Make A Wish*~
Psc 0h23'13" 3d38'0"
Jess Dileo
And 2h36'52" 49d29'33"
Jess Elizabeth
Ori 5h21'6" 1d3'47"
Jess Emily Aldridge
And 23h41'14" 35d24'38"
Jess Franklin McCormick
Cnc 8h10'40" 24d58'1"
Jess Fuhriman
Uma 9h11'59" 46d58'56"
Jess Jenkins
Vir 11h47'48" 9d24'1"
Jess Jo
Aqr 23h31'32" -13d11'16"
Jess Maryea
Uma 9h46'7" 58d16'23"
Jess Michael Miller
Ari 2h31'49" 11d46'36"
Jess Mo
Ari 2h58'5" 25d55'13"
Jess Murray
Tau 4h43'11" 27d37'50"
Jess & Poppy
Aql 19h22'55" -0d53'45"
Jess' Radiant Ava
Cas 22h57'47" 54d35'46"
Jess Robert Feathers, Jr.
Cap 21h40'0" -15d18'36"
Jess Smoky Rickard
Tau 5h58'5" 24d29'58"
Jess Stahr
Lib 15h21'59" -6d22'25"
Jess - The Brightest Star
Cru 12h0'39" -61d45'38"
Jess Weixler
Cyg 19h34'54" 52d29'37"
jess "wifey"
Cas 1h19'35" 54d39'12"
Jess, Babefish
Lyr 18h39'34" 30d49'54"
Jessa
Cnc 8h14'42" 32d44'34"
Jessa and Joe Hussey Star
Cyg 19h36'10" 31d38'47"
Jessa Bratman Pynes
Cap 20h59'7" -16d5'32"
Jessa Greiwe
Cas 0h32'29" 66d14'15"
Jessabella & Thomas Bradley
Cyg 20h0'9" 30d35'23"
Jessafanopiessica
Aqr 21h53'39" -0d49'37"
Jessalyn Belle Stytzer
Sgr 19h10'53" -13d47'47"
jessamerson
Uma 10h37'24" 47d11'2"
Jessamine France Dirain
Lib 15h47'54" -3d52'17"
Jessamy
Sco 16h49'29" -31d4'21"
Jessamyn Suzanne Felker
Cap 20h38'56" -20d36'44"
Jessc
Cnc 8h50'36" 31d6'11"
JessCAngel
Sco 16h54'3" -43d50'23"
JessDa
Lyn 7h51'54" 37d51'51"
Jesse
Uma 11h29'24" 43d37'29"
Jesse
Her 16h34'15" 34d15'56"
Jesse
Per 3h29'24" 45d26'38"
Jesse
Ori 4h50'7" 1d7'53"
Jesse
Cnc 8h27'6" 11d49'27"
Jesse
Cap 20h16'11" -11d38'40"
Jesse
Uma 13h29'47" 56d43'32"
Jesse A. Taylor's Star of Oden
Ori 5h57'14" 20d53'32"
Jesse Addison Davis
Psc 1h10'20" 7d29'44"
Jesse Allen Manning
Lib 15h9'47" -14d52'13"
Jesse Allen Sprague
Gem 6h51'27" 26d53'26"
Jesse & Amy's Star
Per 4h23'55" 43d17'45"
Jesse and Jennifer
Uma 11h30'12" 62d23'10"
Jesse and Lois Turner's True Love
Gem 7h40'49" 31d4'26"
Jesse and Louise Ford
Uma 10h12'47" 72d20'24"
Jesse and Mandy Queck
Uma 12h4'59" 56d28'54"
Jesse and Ruth Hayes
Psc 0h52'53" 25d1'33"
Jesse and Stephen Nathan's 30th
Cyg 21h24'1" 36d28'15"
Jesse And Tobi
And 2h18'0" 38d36'53"
Jesse Ayala
Ari 2h31'47" 25d34'28"
Jesse Boy Blue
Tau 4h29'38" 20d48'23"
Jesse Brandon Hobbs
Psc 1h39'51" 24d59'20"
Jesse Brasco Knox, Jr
Sgr 18h6'32" -27d41'31"
Jesse Bruner
Per 3h17'28" 45d21'47"
Jesse Buttons - Our Angel In Heaven
Cru 12h34'50" -59d28'57"
Jesse Bytautas
Cnc 8h51'45" 29d40'56"
Jesse C. Gutierrez
Ari 2h41'37" 20d40'43"
Jesse Carlos
Her 17h20'25" 28d25'13"
Jesse Caruso
Per 3h27'32" 48d37'45"
Jesse Citrowske
Umi 14h45'38" 75d11'44"
Jesse Clark P
Sco 17h58'9" -31d10'12"
Jesse Clifford Johannsen
Lyn 7h15'46" 58d37'27"
Jesse Cole Davis
Umi 15h3'5" 71d51'58"
Jesse Curtis Phillips
Vir 13h49'58" 5d53'42"
Jesse Cyman
Lib 14h53'44" -2d48'52"
Jesse Daniel Ricks
Ari 2h34'21" 13d42'41"
Jesse Daniel Tillery
Tau 3h45'24" 14d52'57"
Jesse David Galligan
Uma 11h58'47" 44d2'3"
Jesse David Garman 8/29/06 - 9/8/06
Umi 15h42'40" 83d56'12"
Jesse David Pino
Leo 10h58'5" 17d7'27"
Jesse & Dee: Until Oranges After
Cyg 20h11'20" 57d34'38"
Jesse Dominick Sidor
Lib 15h23'38" -15d3'50"
Jesse Dominique Brian Bushong
Gem 6h43'36" 26d21'39"
Jesse Donaldson Jr.
Lib 15h24'12" -26d28'44"
Jesse Douglas Rowe
Cnc 8h51'12" 13d5'37"
Jesse E. Coe
Cap 20h9'14" -24d30'39"
Jesse Elijah Oppenheim Lee
Her 16h34'50" 35d16'30"
Jesse Eugene Butler
Tau 4h52'42" 29d57'58"
Jesse Furnas
Oph 16h59'11" -0d44'49"
Jesse Gene Levetta
Per 4h10'2" 34d12'28"
Jesse Gonzalez
Cam 4h23'35" 70d27'59"
Jesse Gregson
Cru 12h29'59" -61d47'22"
Jesse Guerra
Ari 2h49'44" 16d38'49"
Jesse H. Cloud III
Lyn 7h51'41" 47d6'47"
Jesse Henry Aldridge
Uma 9h13'6" 57d23'37"
Jesse Howe
Uma 8h57'42" 55d21'59"
Jesse J. C. Cracknell
Leo 10h39'42" 8d10'49"
Jesse J. McGee Sr.
Her 17h30'19" 26d49'11"
Jesse J. Merone
Aqr 22h40'11" -18d3'7"
Jesse J. Rego
Lyn 7h58'39" 54d12'39"
Jesse Jacobs
Uma 13h54'46" 54d37'38"
Jesse Jade James Conroy
Lyn 7h25'53" 59d15'9"
Jesse Jake Warren
Umi 16h47'24" 82d29'23"
Jesse James Terreson
Umi 14h38'23" 68d39'18"
Jesse James Turk
Lmi 10h29'32" 37d21'25"
Jesse James & Wyatt James Melton
Umi 15h37'43" 79d12'53"
Jesse & Jane
Cyg 20h46'41" 35d37'40"
Jesse & Jessica, Always & Forever
Ori 5h10'24" 8d40'1"
Jesse Jo Nyersh
Cra 18h52'10" -37d19'33"
Jesse Jones
Aql 19h46'37" 3d2'59"
Jesse Joseph Coburn
Cas 23h16'27" 55d54'1"
Jesse K.
Vir 12h44'55" 4d58'53"
Jesse Kais Kierans
Sge 19h57'23" 16d44'39"
Jesse Knudsen Larson
Uma 11h27'7" 40d43'42"
Jesse Kohlmann Colin Bauman
Umi 16h52'31" 75d32'48"
Jesse Kostecki
Cma 6h42'31" -12d54'50"
Jesse Kuroski
Uma 13h41'58" 61d31'38"
Jesse Lee
Uma 11h0'57" 61d24'0"
Jesse Lee Breeden
Ori 5h23'29" 5d7'29"
Jesse Lee Case
Uma 10h57'45" 57d9'47"
Jesse Lee Howell
Aqr 20h46'5" -3d27'56"
Jesse Lee Rosener
Tau 4h20'54" 27d42'6"
Jesse Leslie Love
Gem 6h59'14" 17d8'16"
Jesse Lhotka
And 23h41'52" 47d16'55"
Jesse Lockamy
Uma 10h34'45" 40d13'54"
Jesse & Madeline
Uma 11h22'13" 39d15'3"
Jesse Marshall Benham
Lib 15h9'3" -19d54'14"
Jesse Martin Hernandez
Lib 14h49'52" -0d56'1"
Jesse Martin Tuten
Gem 6h56'58" 14d44'6"
Jesse McKinney
Lmi 10h24'56" 36d29'55"
Jesse Michael Chacon
Aqr 23h27'15" -8d32'20"
Jesse Michael Cross
Uma 9h9'19" 46d36'39"
Jesse Michael Renaud
Ori 6h3'43" 20d18'7"
Jesse Mitchell
Aqr 22h14'55" -19d30'57"
Jesse Mullins
Lyn 7h32'44" 53d19'20"
Jesse Muse
Lib 14h59'53" -14d57'18"
Jesse My Sun
Tau 5h49'0" 16d28'38"
Jesse Myrick
Peg 22h22'2" 6d56'20"
Jesse & Natalie
Cyg 21h33'47" 53d15'2"
Jesse Nicholas Deratto
Lib 14h39'37" -24d41'12"
Jesse Nicholas Kefallinos
Cru 12h33'13" -56d9'59"
Jesse & Nicole 10/03/03
Per 2h53'21" 55d36'25"
Jesse "Our Prince" James
Ori 6h12'14" 3d19'8"
Jesse P.
And 2h16'44" 47d47'15"
Jesse Plisch
Ori 5h26'55" 1d44'34"
Jesse Quintana The Pizza King
Sgr 18h25'42" -32d34'38"
Jesse R. Graw, Jr.
Lyn 7h49'56" 41d50'56"
Jesse Radice-Hyde - The Shining Light
Cru 12h36'3" -60d47'7"
Jesse Ray Eyman III
Leo 9h23'11" 16d9'18"
Jesse Reuben Jacobs
Cnc 9h21'56" 32d22'42"
Jesse Robert Sommer
Lib 15h36'34" -19d44'8"
Jesse Romeo Witmer
Aqr 21h0'29" 1d59'29"
Jesse Rose
Psc 1h41'18" 17d55'29"
Jesse RYAN James Hernandez
Her 16h39'43" 39d59'56"
Jesse Ryan Lee
Her 16h41'40" 18d46'17"
Jesse Saldivar-Honey Bunny
Vir 13h12'28" 5d24'43"
Jesse Saul Bookspan-14
Dra 17h24'3" 68d0'34"
Jesse Scott
Ori 6h4'54" 13d18'19"
Jesse Sockman
Leo 10h44'10" 11d30'56"
Jesse Spade
Cyg 20h30'47" 36d46'7"
Jesse Spears
Gem 7h39'27" 22d52'45"
Jesse Stephen Powers
Vir 12h16'16" 1d56'29"
Jesse T. Lowe
Lyr 18h42'51" 30d55'54"
Jesse T. Turon
Aqr 22h57'6" -20d36'14"
Jesse Tapia
Cyg 20h59'49" 43d3'58"
Jesse (The Boy) Jakel
Ori 6h0'46" 19d38'20"
Jesse Thomas Neff
Aqr 21h0'12" 1d1'56"
Jesse Thompson
Sgr 19h50'25" -12d20'22"
Jesse Turner
Lyn 7h34'21" 53d4'31"
Jesse Vega
Uma 8h36'43" 53d12'29"
Jesse Viggiano
Psc 0h52'12" 27d48'57"
Jesse Virginia
Cyg 19h36'46" 52d2'40"
Jesse Wayne Swisher
Vir 14h2'18" -16d8'1"
Jesse William Davie
Crb 16h9'15" 27d7'4"
Jesse William Sheldon
Per 4h35'47" 40d46'6"
Jessea Macaluso
Crb 15h46'27" 31d55'58"
Jesseca
Aql 19h27'57" 4d56'11"
Jesseca Caryn Spittle
Tau 5h29'24" 18d44'16"
JesseFrancis
Cet 0h55'2" -0d35'21"
Jesselee Angeles
And 23h22'20" 47d6'50"
Jessemae
And 2h35'59" 49d7'22"
Jessemi
Lyr 19h3'47" 36d19'28"
Jessen, Christian
Ori 5h13'44" 0d21'13"
Jessenia "Our Star" Ruiz
Uma 11h25'35" 33d40'50"
Jessenia Rodriguez
And 0h30'24" 36d40'52"
Jesseroooo M. 90
Psc 0h51'17" 20d20'52"
Jesse's 40
Leo 9h39'29" 28d24'24"
Jesse's Beautiful Mommy Melissa
Per 2h55'48" 51d1'33"
Jesse's Bright, Smiling Star
Cap 20h38'51" -13d23'22"
Jesse's Girl
Uma 9h20'48" 57d7'49"
Jesse's Hitch
Aql 19h36'36" 8d44'9"
Jesse's Jo
Psc 1h27'21" 16d11'51"
Jesse's Perfect Mom
Aqr 22h24'53" 0d7'52"
Jesse's Star
Vir 13h20'23" 4d34'9"
Jesse's Star
Gem 7h2'35" 16d15'39"
Jesse's Star
Cru 12h21'50" -58d26'44"
Jesse's Wonderful Mom
Aqr 22h21'58" -2d15'50"
Jesseth7
Umi 13h39'29" 77d25'17"
Jess-Girl
And 1h11'11" 43d6'40"
JesSher
Aql 19h55'21" 15d33'42"
Jessi
Tau 4h5'37" 8d21'48"
Jessi
Lyn 9h14'4" 43d10'52"
"Jessi"
Lmi 10h44'54" 37d19'9"
Jessi
Cas 1h28'50" 51d4'26"
Jessi
Cap 21h36'3" -16d10'39"
Jessi
Cas 23h50'18" 56d59'19"
Jessi
Sgr 18h0'7" -23d0'44"
Jessi
Sgr 18h0'16" -26d30'25"
Jessi Abshire
Tau 4h47'53" 20d53'8"
Jessi and Colin's Star
Per 3h33'19" 34d18'5"
Jessi and Dennis in love forever
Uma 12h33'27" 60d56'34"
Jessi & André
And 23h40'46" 38d45'42"
Jessi Ann West
Aqr 22h10'6" -2d43'40"
Jessi Eckstine
Umi 14h11'59" 67d31'12"
Jessi & Jesse Carter
Leo 10h19'5" 21d23'32"
Jessi May
Ori 5h34'0" -1d5'8"
Jessi Michelle Boney
Gem 6h29'18" 20d35'28"
Jessi Rae (S.Q.)
Vir 13h27'13" 12d9'16"
Jessi Renae Kovars
Gem 6h56'30" 19d57'49"
Jessi Renee Dygan
And 0h57'7" 39d52'38"
Jessi Vega
Lyn 9h36'42" 40d22'22"
Jessia
Crb 15h43'59" 29d4'33"
Jessianna
And 2h25'47" 50d18'51"
Jessibella
Cnc 9h6'49" 32d37'52"
Jessica
Cnc 9h4'0" 30d52'54"
Jessica
Lyn 8h29'8" 34d51'31"
Jessica
Gem 7h46'21" 31d0'59"
Jessica
Ari 2h38'54" 31d9'57"
Jessica
And 1h21'51" 36d15'43"
Jessica
And 0h51'52" 35d14'46"
Jessica
Lmi 10h24'38" 33d13'10"
Jessica
Crb 16h23'45" 30d41'43"
Jessica
Cyg 19h46'9" 35d54'20"
Jessica
Crb 16h13'10" 35d20'54"
Jessica
Lyn 8h40'58" 40d5'14"
Jessica
Uma 10h43'43" 40d16'18"
Jessica
Lmi 10h49'46" 37d49'29"
Jessica
And 2h8'36" 41d45'58"
Jessica
And 0h46'27" 42d54'0"
Jessica
And 0h54'34" 38d26'24"
Jessica
And 0h42'26" 37d31'11"
Jessica
And 1h40'42" 42d29'19"
Jessica
And 1h39'38" 42d3'58"
Jessica
And 1h13'24" 41d52'46"
Jessica
And 0h58'1" 43d8'23"
Jessica
And 1h47'38" 37d55'38"
Jessica
And 2h36'56" 50d28'15"
Jessica
And 1h45'17" 49d25'55"
Jessica
And 1h55'12" 46d48'13"
Jessica
And 1h23'55" 47d0'7"
Jessica
And 23h5'47" 40d16'51"
Jessica
And 23h5'11" 42d34'22"
Jessica
And 23h30'35" 44d31'40"
Jessica
And 23h22'39" 42d41'3"
Jessica
Cas 0h25'24" 50d13'11"
Jessica
Cyg 20h14'24" 44d28'57"
Jessica
Lyr 18h50'55" 44d49'32"
Jessica
And 23h17'49" 47d14'13"
Jessica
And 23h42'27" 47d29'28"
Jessica
And 23h22'37" 51d35'39"
Jessica
Cas 23h29'28" 51d27'13"
Jessica
And 22h59'6" 47d47'29"
Jessica
Cyg 21h57'15" 48d27'3"
Jessica
Cyg 20h6'19" 48d16'6"
Jessica
Uma 10h50'59" 50d25'7"
Jessica
Peg 23h34'34" 25d15'3"
JESSICA
Peg 22h49'44" 23d25'41"
Jessica
Peg 22h53'0" 25d37'39"
Jessica
And 0h21'35" 30d25'4"
Jessica
Gem 7h20'25" 29d34'8"
Jessica
Leo 11h2'10" 22d4'3"
Jessica
Cnc 8h41'30" 15d32'0"
Jessica
And 0h15'3" 25d21'1"
Jessica
Her 18h44'31" 20d35'5"
Jessica
And 0h32'38" 28d10'19"
Jessica
Vir 13h26'57" 8d4'5"
Jessica
Gem 7h40'47" 14d18'24"
Jessica
Ari 2h53'53" 19d8'39"
Jessica
Tau 3h38'11" 17d42'9"
Jessica
Tau 4h4'19" 18d21'20"
Jessica
Tau 4h36'33" 20d35'14"
Jessica
Aql 19h27'1" 9d17'54"
Jessica
Aql 19h3'36" 14d31'12"
Jessica
Psc 1h31'5" 12d1'36"
Jessica
Aqr 22h27'44" 2d10'0"
Jessica
Peg 22h25'36" 7d13'30"
Jessica
Psc 0h39'10" 14d56'21"
Jessica
Psc 0h45'48" 10d22'8"
Jessica
Psc 0h36'16" 8d52'16"
Jessica
Psc 0h11'38" 8d24'27"
Jessica
Vir 12h43'48" 0d29'43"
Jessica
Ori 5h9'27" 6d13'47"
Jessica
Dra 19h51'46" 79d41'35"
Jessica
Cam 7h40'48" 83d7'28"
Jessica
Crt 11h21'33" -14d34'13"
Jessica
Cap 20h57'10" -17d36'33"
Jessica
Aqr 22h28'49" -22d12'10"
Jessica
Lib 14h33'57" -17d1'5"
Jessica
Sco 16h10'1" -18d28'17"
Jessica
Cap 20h25'18" -12d53'1"
JESSICA
Aqr 23h5'50" -8d35'36"
Jessica
Lib 15h54'53" -9d44'45"
Jessica
Umi 13h50'35" 73d31'59"
Jessica
Cas 2h50'19" 61d20'10"
Jessica
Cas 2h34'19" 66d8'48"
Jessica
Cas 1h9'53" 64d3'15"
Jessica
Cyg 19h45'26" 57d51'13"
Jessica
Uma 9h41'37" 54d7'46"
Jessica
Lyn 7h28'35" 57d13'40"
Jessica
Lyn 7h16'20" 53d43'38"
Jessica
Uma 13h15'14" 53d38'9"
Jessica
Aqr 23h44'26" -18d40'25"
Jessica
Lib 15h44'22" -28d23'0"
Jessica
Sco 16h25'35" -31d5'41"
Jessica
Aqr 23h47'0" -22d54'55"
Jessica
Psc 0h56'53" 3d6'5"
Jessica
Sco 16h59'14" -41d55'17"
Jessica 16-06-2006
Cas 3h12'13" 61d15'47"
JeSsiCa <<3
Gem 7h30'12" 29d31'24"
Jessica 827
Lyn 7h6'10" 52d23'27"
Jessica A. Coffin
Lib 15h29'11" -20d52'42"
Jessica A Desanctis
Tau 5h36'24" 27d44'54"
Jessica A. Kempf
Ori 6h18'7" 14d54'12"
Jessica A. Leinen's Birthday Star
And 1h11'30" 43d35'48"
Jessica A Mawdsley
And 2h21'5" 50d4'33"
Jessica Abrams Gibbs - JAG
Lib 15h55'23" -9d33'5"
Jessica "Akachan" Mai Sample
Sgr 18h27'33" -19d41'7"
Jessica Alcantar
Sco 16h5'35" -15d30'52"
Jessica Alejandra Sanchez
Sgr 19h14'18" -13d50'29"
Jessica Alexis Anne Singh
Psc 1h41'1" 23d44'36"
Jessica Alice
Sco 16h30'36" -28d2'18"
Jessica Alison Kern
Ari 3h17'51" 19d28'28"
Jessica Allyson Richman
And 1h20'47" 50d5'49"
Jessica Amber Diament
And 23h48'49" 34d37'25"
Jessica Amelia Tomasiello
Cnc 8h31'27" 10d9'9"
Jessica Amelie Hopkins
Col 6h2'13" -41d25'19"
Jessica and Armando
Uma 14h13'17" 58d18'37"
Jessica And Brandon Forever
Cyg 20h34'5" 34d57'44"
Jessica and Brenda
Vir 12h38'0" 0d27'10"
Jessica and Chad's OURS
Lib 14h29'47" -9d47'0"
Jessica and David Van Gilder
Cam 4h18'59" 71d51'53"
Jessica and Jared
Cyg 20h17'58" 51d33'35"
Jessica and Jesse 4ever
Lac 22h5'56" 51d32'42"
Jessica and Jonathan Landis
And 0h32'53" 46d18'46"
Jessica and Josh's Star of Love
Cnc 8h35'45" 16d34'53"
Jessica and Justin Harmon
Per 4h39'53" 42d1'25"
Jessica and Lydia Puryear
Cnc 9h9'15" 8d12'38"
Jessica and Patrick
Cyg 20h18'3" 47d11'47"
Jessica and Paulo's Valentine Star
And 0h40'53" 42d41'23"
Jessica and Thomas Carey
Gem 6h55'46" 14d38'21"
Jessica and Wesley's Star
Gem 6h43'1" 15d43'24"
Jessica Anderson
Gem 7h26'24" 15d47'10"
Jessica Anderson Bianco
Gem 7h26'9" 23d11'50"
Jessica Andres
Sco 17h6'35" -30d30'54"
Jessica Ann
Sgr 18h6'15" -27d1'41"
Jessica Ann
Gem 6h52'29" 27d42'49"
Jessica Ann
And 0h8'40" 29d4'7"
Jessica Ann
Cnc 8h47'29" 11d40'55"
Jessica Ann
Tau 5h51'6" 16d59'14"
Jessica Ann
Psc 0h50'21" 14d0'33"
Jessica Ann
Psc 1h1'17" 13d54'56"
Jessica Ann
And 1h6'14" 44d41'8"
Jessica Ann
Lyr 19h25'35" 38d13'54"
Jessica Ann
Crb 16h14'44" 38d24'45"
Jessica Ann Bouchard
Gem 7h28'50" 19d47'45"
Jessica Ann Cole
Vir 15h6'3" 4d8'17"
Jessica Ann Dickinson Skeen
Uma 10h22'37" 63d57'0"
Jessica Ann Gaissert
Psc 1h39'31" 11d3'20"
Jessica Ann Gardner
Cnc 9h4'43" 32d28'59"
Jessica Ann Groft
Uma 9h34'54" 46d6'8"
Jessica Ann Ham
Lyn 6h25'19" 55d31'36"
Jessica Ann Hejl
Tau 4h11'8" 29d4'46"
Jessica Ann Heller
Leo 10h23'51" 22d10'21"
Jessica Ann Jacobi
Uma 13h29'26" 54d58'33"
Jessica Ann Jones
Pho 0h16'8" -49d21'8"
Jessica Ann Kohlmyer
Pyx 8h59'24" -32d16'5"
Jessica Ann Lynch
Cnc 8h37'12" 21d55'55"
Jessica Ann Mancourt
Her 17h15'18" 27d26'17"
Jessica Ann Markovich
Ari 3h12'4" 11d36'1"
Jessica Ann Moffett
Psc 0h34'1" 7d59'40"
Jessica Ann Owens
And 0h12'50" 26d27'55"
Jessica Ann Reddoch
Cnc 8h51'10" 27d39'15"
Jessica Ann Riley
Cap 20h24'40" -27d1'7"
Jessica Ann Smith
Aqr 20h49'38" -7d51'9"
Jessica Ann the most beautiful star
Uma 11h2'38" 34d7'18"
Jessica Ann Trice
Leo 11h32'42" 20d50'20"
Jessica Ann Woolfolk
Ari 3h3'47" 18d0'59"
Jessica Anne
Peg 23h36'4" 21d48'57"
Jessica Anne
Lyn 7h45'36" 36d32'37"
Jessica Anne
Sgr 18h48'40" -34d33'15"
Jessica Anne Benes
Uma 9h26'29" 56d27'27"

Jessica Anne Cioffi
"Pumpkin"
And 1h28'34" 49d14'8"
Jessica Anne Hayner
Tau 4h39'3" 23d15'18"
Jessica Anne Larson
Cnc 8h43'35" 15d58'40"
Jessica Anne Lashbrook
Cyg 20h31'48" 52d47'40"
Jessica Anne Lawson
And 0h23'23" 31d28'23"
Jessica Anne Miller
Aqr 23h55'16" -11d25'1"
Jessica Anne Pelasky
Sco 15h53'12" -22d56'1"
Jessica Anne Prater
Dra 10h11'37" 78d30'38"
Jessica Anne Quinn
Cru 12h18'28" -57d15'52"
Jessica Anne Rooke
Uma 12h33'3" 55d49'14"
Jessica Anne Walby
And 23h25'20" 35d37'46"
Jessica Anne Wellard
And 0h41'4" 37d51'45"
Jessica Ansell
Peg 22h51'28" 9d1'6"
Jessica Aracely Sanchez
Tau 5h54'52" 28d14'48"
Jessica Arletta Miller
Peg 22h28'14" 26d3'7"
Jessica Ashley Browder
And 1h48'12" 35d53'29"
Jessica Ashley Scutt
And 0h12'46" 26d37'11"
Jessica Ashley Seabeck
Aqr 22h40'34" 2d0'17"
Jessica Auleta
Sgr 18h20'27" -17d0'2"
Jessica Aust
And 23h8'10" 42d16'15"
Jessica Austin
Cyg 21h43'38" 52d19'27"
Jessica Avery
Uma 11h11'42" 29d59'47"
Jessica Azzaro
Tau 3h59'18" 11d3'48"
Jessica Babcock
Tau 5h39'14" 25d49'4"
Jessica "Baby Girl" Powers
Vir 13h39'38" 0d39'31"
Jessica "Baby J" Lease
Dra 17h8'23" 64d21'53"
Jessica Ballas
Cas 0h11'42" 51d34'15"
Jessica Bambridge
And 23h0'50" 41d51'33"
Jessica Barbato
And 1h25'30" 48d30'45"
Jessica Barrett - 13th
Birthday Star
Cru 12h29'21" -61d1'10"
jessica barrett + jeff sass
And 23h18'12" 49d27'47"
Jessica Bathauer
Lac 22h29'23" 42d22'7"
Jessica Baugher
And 0h11'54" 46d7'26"
Jessica Bayer
And 0h40'59" 38d41'48"
Jessica Bayless
And 0h35'34" 28d28'2"
Jessica Bean
Cnc 8h50'20" 25d37'25"
Jessica Beatrice DeMaio
Sgr 18h39'4" -21d51'53"
Jessica Beaudrot
Leo 11h28'46" 11d13'13"
Jessica (Beautiful)
Sco 17h52'35" -36d14'51"
Jessica Belle
Umi 16h37'47" 77d34'37"
Jessica Beloin
Umi 17h11'8" 75d6'49"
Jessica Beran
Cam 4h0'35" 58d3'6"
Jessica Berke
Psc 1h34'22" 3d38'42"
Jessica Bertin (Peep)
Psc 0h19'21" -3d5'22"
Jessica Bice
Gem 6h58'3" 15d3'18"
Jessica Blackler
Cas 23h53'4" 56d15'29"
Jessica Blackwell
Mon 6h25'15" 9d19'14"
Jessica Blair
Cnc 8h18'35" 17d51'31"
Jessica Blair
Cyg 20h35'15" 37d22'6"
Jessica Blair Szymanski
Uma 14h15'42" 60d23'28"
Jessica Blancher
Cas 23h52'36" 50d24'25"
Jessica Blankenship
Sco 16h50'8" -26d46'48"
Jessica Block
Gem 7h51'33" 16d14'28"
Jessica Blythe
Sco 17h2'51" -32d59'15"
Jessica Bowen
Cam 5h58'26" 56d4'13"
Jessica Bradwell
Cnc 9h13'8" 13d47'35"
Jessica & Brandon's Star
Tau 5h52'58" 14d39'41"
Jessica Brassell
Ori 4h56'11" 3d3'36"
Jessica Brett
And 0h11'5" 38d41'5"
Jessica Brianne Tucker
Uma 9h1'31" 64d2'3"
Jessica Brianne Wills
And 1h17'43" 35d59'0"
Jessica Briggs
Ari 3h26'32" 20d38'21"
Jessica Bright
And 0h44'5" 43d20'47"
Jessica Bright Eyes
Cap 21h32'8" -17d38'40"
Jessica Brisley
Cas 0h25'40" 56d33'25"
Jessica Brito
Cnc 9h8'23" 10d49'31"
Jessica & Brody
Cyg 19h39'30" 50d39'12"
Jessica Brovont
Psc 1h21'7" 5d23'46"
Jessica Brown
Gem 7h6'50" 27d16'1"
Jessica Brown - Bugg
Cas 0h9'10" 56d13'56"
Jessica Brummert
Uma 12h21'38" 59d48'29"
Jessica Bryant
Vir 12h13'34" 11d46'53"
Jessica Burns Lepore
Cas 1h47'33" 60d51'47"
Jessica Burrow
Vir 14h31'10" 3d1'9"
Jessica Bybel
And 0h14'56" 46d46'13"
Jessica C. Albers
Lyn 7h32'35" 39d6'42"
Jessica C. Cook
Mon 6h46'1" 7d43'47"
Jessica C. Thomas
Sex 10h36'58" -0d11'13"
Jessica Cable
Lyn 7h38'1" 55d17'50"
Jessica & Callum
Uma 13h46'24" 47d58'52"
Jessica Candace
Cyg 20h2'1" 38d10'55"
Jessica Carlo
Cap 21h50'41" -20d32'35"
Jessica Carolyne Hundl
Cas 2h18'45" 69d38'21"
Jessica Carrington
Vir 13h25'50" -15d18'15"
Jessica Carter
Lyn 7h34'19" 53d22'25"
Jessica Casey Gray
Leo 11h45'12" 25d50'35"
Jessica Cassie
Lib 14h52'59" -0d47'19"
Jessica & Cassie's Star
Cyg 19h19'2" 52d56'58"
Jessica Cat Barton
Cnc 9h17'37" 16d49'19"
Jessica Catherine
Uma 8h22'14" 60d32'23"
Jessica Catherine
Laughton
And 0h28'30" 46d17'42"
Jessica Cavallaro
Gem 6h1'44" 28d0'28"
Jessica Chahir
Sgr 19h9'14" -15d3'38"
Jessica Chaney's "Sweet
16"
Sgr 18h15'56" -20d26'58"
Jessica Chang
Aql 19h33'5" 8d4'23"
Jessica Chapman
Lyn 6h24'53" 57d11'20"
Jessica Chilicki
Cam 6h1'31" 69d3'57"
Jessica Christensen
Hya 8h19'1" -11d14'23"
Jessica Christiana Golding
Cam 5h39'13" 63d28'45"
Jessica Christine
And 1h6'36" 45d45'37"
Jessica Christine Alexander
Ari 3h7'23" 14d30'3"
Jessica Christine Citrino
Lyn 7h23'59" 57d39'15"
Jessica Christine Henzel
Lib 15h8'53" -4d58'11"
Jessica Christine Nesbit
Sco 17h29'52" -45d20'28"
Jessica Christine Ostley
And 0h32'3" 45d17'59"
Jessica Christine Panther
Aur 5h55'6" 32d14'53"
Jessica Churcher
Cyg 21h16'38" 47d29'39"
Jessica Clark
Crb 15h43'4" 26d29'59"
Jessica Clark - a Shooting
Star!
Umi 14h21'11" 73d35'45"
Jessica Clause
Cap 21h32'36" -20d28'14"
Jessica Colman
Per 3h34'55" 44d31'57"
Jessica Colon
Aqr 22h18'30" -6d15'49"
Jessica Colotti
Aql 19h16'59" 7d15'33"
Jessica Combs
And 0h59'4" 36d9'47"
Jessica Conn
And 1h8'1" 33d56'10"
Jessica Connie-Marjorie
Woodin
Cap 20h59'12" -21d55'40"
Jessica Contrares
And 1h35'41" 44d46'48"
Jessica Cooley
And 0h50'46" 40d34'25"
Jessica Cornelia Mulle
Uma 10h21'38" 61d2'7"
Jessica Costanza
And 23h39'13" 34d4'39"
Jessica Courtney
Uma 10h35'50" 59d39'45"
Jessica Cowing
Leo 9h24'27" 20d13'2"
Jessica Coy
Cru 12h1'8" -61d54'38"
Jessica Cristina
Sandmann, 11.08.2002
Uma 11h12'23" 61d7'48"
Jessica Crocker
Sgr 19h5'56" -29d3'24"
Jessica Crowe
Cnc 8h9'6" 22d25'50"
Jessica Cruz
Leo 11h4'19" 20d38'10"
Jessica Crystal Shaelin
Gunther
Sgr 18h59'48" -31d11'11"
Jessica Cuddles Pippi Pop
Pei - 381
Cyg 20h30'39" 53d56'52"
Jessica Curran
Ori 6h2'44" 19d9'31"
Jessica D. Allan
Cyg 21h32'7" 39d8'40"
Jessica D Davis
Aqr 22h51'55" -9d18'57"
Jessica D Hancock
Cyg 19h38'35" 28d5'8"
Jessica Dains Davis
Lyn 7h47'39" 41d35'24"
Jessica & Damian
Cyg 20h9'56" 33d59'5"
Jessica DaMore
Uma 9h54'44" 58d10'22"
Jessica Dangler12
Cyg 20h0'54" 52d40'57"
Jessica Danielle Farrell
And 23h22'32" 47d15'42"
Jessica Dauphinais
And 1h47'3" 38d17'53"
Jessica David Taylor
Gem 6h39'27" 18d41'33"
Jessica Dawn
Peg 22h22'6" 4d51'49"
Jessica Dawn
Ori 6h13'17" 15d22'20"
Jessica Dawn Barnes
And 2h22'40" 47d54'40"
Jessica Dawn Guillot
And 1h14'36" 48d58'36"
Jessica Dawn Louk
Uma 8h51'0" 61d39'49"
Jessica Dawson
And 0h19'39" 46d30'14"
Jessica De La Cruz
Psc 0h9'26" 5d13'28"
Jessica Death
Del 20h37'35" 13d49'6"
Jessica Deckinger
And 1h1'22" 45d37'54"
Jessica Dee Scott 11-01-
1995
And 1h44'32" 40d53'48"
Jessica Delane Davis:
Patience
And 0h42'37" 44d15'40"
Jessica Delgado
Gem 7h16'37" 20d7'35"
Jessica Dellorso
Leo 11h38'49" 27d23'49"
Jessica Denise Mallett
Leo 10h17'47" 12d53'32"
Jessica DePaolo and David
Barr Star
And 23h12'23" 43d46'51"
Jessica Dewey
Sgr 18h45'9" -31d25'10"
Jessica Diamond Ramsay
Aqr 22h35'31" -1d45'45"
Jessica Diane Savich
Cap 20h40'5" -26d6'9"
Jessica Dillon
Peg 22h14'20" 7d3'5"
Jessica Dittenbir
Tau 5h31'26" 27d13'2"
Jessica D'nee
Leo 10h12'43" 24d16'18"
Jessica "Dolface" Organ
Dra 16h25'48" 63d35'45"
Jessica Donahue
Aqr 22h12'43" 1d12'29"
Jessica Douglas
Rhodericks Charm
Sco 17h45'36" -34d17'1"
Jessica Downey
Sgr 17h56'15" -27d48'50"
Jessica Duarte
Cas 1h50'23" 61d12'6"
Jessica Duarte
And 0h35'15" 31d8'55"
Jessica Duby
And 2h17'17" 50d8'20"
Jessica Duvall "Monkey"
Lib 14h37'53" -9d26'44"
Jessica Dyer
Gem 7h34'2" 26d10'51"
Jessica E. Biederman "My
Bebez"
Gem 6h56'12" 19d46'30"
Jessica E. DeMaria
Cas 0h16'47" 58d12'38"
Jessica E. Smith
Cnc 8h55'58" 18d12'54"
Jessica E. Stickle
Lyr 18h35'24" 35d50'39"
Jessica E. Williams
Sco 16h15'32" -27d15'31"
Jessica Egan
Tau 3h49'16" 22d59'8"
Jessica Elaine
Vir 14h0'46" -8d18'36"
Jessica Elizabeth
Dra 16h15'54" 58d23'58"
Jessica Elizabeth
And 0h41'15" 26d21'23"
Jessica Elizabeth
And 23h32'22" 41d41'48"
Jessica Elizabeth Artigliere
Ari 3h19'58" 28d40'38"
Jessica Elizabeth Data
Cyg 19h58'3" 50d57'21"
Jessica Elizabeth Langston
Sgr 19h36'14" -27d7'24"
Jessica Elizabeth Mackin
Leo 9h33'54" 28d6'52"
Jessica Elizabeth Omelian
Uma 11h10'12" 47d26'28"
Jessica Elizabeth Reddick
Sco 17h28'16" -44d57'55"
Jessica Elizabeth Wilson
Glover
And 0h48'41" 22d22'6"
Jessica Ellen Gould
Cas 1h34'17" 63d52'8"
Jessica Ellie
And 2h28'55" 50d3'59"
Jessica Ellie Reid
And 2h37'45" 45d33'39"
Jessica Elsasser
Leo 9h24'18" 12d14'35"
Jessica Emily Jobson
And 23h33'23" 41d17'40"
Jessica Emily Medeiros
And 1h22'23" 43d18'24"
Jessica & Emma Mahoney
And 23h38'14" 41d28'57"
Jessica & Emma Tyler, TT
Loves You
Umi 14h0'0" 71d50'29"
Jessica Eppley
Uma 9h57'6" 62d52'16"
Jessica Erin
Sco 17h52'28" -43d35'16"
Jessica Erin Dean
And 1h46'14" 45d34'42"
Jessica Erin Porter
Lyn 8h9'53" 34d40'30"
Jessica Erstad "1-17-04"
And 1h38'15" 42d35'25"
Jessica Estrada
Tau 5h21'24" 23d43'55"
Jessica et Jérémy
Umi 14h42'3" 81d33'0"
Jessica Evelyn Howze
Aqr 22h4'45" -1d9'43"
Jessica Everlast
Hys 1h59'28" -74d12'32"
Jessica Evon Love
Tau 5h7'27" 20d58'0"
Jessica Faith
Leo 10h12'59" 12d53'25"
Jessica Faith Hanna
Ori 6h17'48" 10d11'34"
Jessica Faraone Mennella
Crb 16h11'22" 34d48'32"
Jessica Faraone Mennella
Umi 14h19'1" 67d45'42"
Jessica Feldman
Cam 4h40'34" 52d45'46"
Jessica Fevrier
And 2h8'26" 46d48'24"
Jessica Fiene
Lib 14h50'8" -3d18'9"
Jessica Finley
Gem 7h10'18" 18d23'7"
Jessica & Fiori
Leo 10h13'19" 26d36'5"
Jessica Flege - Shining
Star
And 23h0'57" 42d44'23"
Jessica Flynn
Leo 10h11'54" 22d7'14"
Jessica Foley
Uma 9h19'3" 58d46'3"
Jessica Folgueras
And 0h19'51" 32d39'43"
Jessica - Forever Beautiful
Aqr 23h31'48" -8d16'27"
Jessica Fortes
And 23h18'25" 46d58'16"
Jessica Frances Lifman
Aqr 22h18'52" 1d18'49"
Jessica Fraser
Leo 9h44'5" 17d33'48"
Jessica Frederick
Aqr 22h6'57" 0d54'30"
Jessica G. Steenmeyer
Ari 3h11'14" 26d41'2"
Jessica Gabrielle Jeanne
FEY
And 23h39'1" 33d33'15"
Jessica Gallagher
Sco 16h9'37" -15d44'7"
Jessica Gallina
Ori 6h6'58" 21d18'10"
Jessica Garcia
Cas 1h18'43" 62d10'56"
Jessica Gardepe
Ari 3h17'28" 12d24'29"
Jessica Gayle Johnson
And 0h57'1" 41d25'38"
Jessica Geist
Cyg 19h48'8" 33d26'33"
Jessica Gennell McCoy
Lyn 8h23'20" 55d2'9"
Jessica Gentian McAdams
Cyg 20h17'57" 43d45'58"
Jessica Gerber
Sgr 17h58'40" -28d11'27"
Jessica Germata
And 0h32'18" 41d32'28"
Jessica Geronimo
Gem 6h21'41" 26d8'38"
Jessica Gibson
Her 16h48'53" 36d6'50"
Jessica Giddens
And 23h40'13" 41d25'41"
Jessica Gilbertson
Cnc 8h41'40" 11d6'13"
Jessica Giles
Cnc 8h8'18" 8d48'33"
Jessica Gilkison
Tau 5h6'3" 24d35'46"
Jessica Giordano
Psc 1h20'42" 10d12'4"
Jessica Gladys Morgan -
"Jessica's Star"
And 1h14'13" 34d44'33"
Jessica Gleeson
Ori 6h6'40" 14d23'51"
Jessica Glynn
Vir 11h56'39" -8d57'1"
Jessica "Goddess"
Steindorff
Gem 7h12'36" 23d49'39"
Jessica Golden
Ari 2h40'10" 14d47'44"
Jessica Gonia
Cap 21h49'38" -24d44'41"
Jessica "Gorgeous"
Sandrelli
And 0h55'12" 46d39'0"
Jessica "Gorgious" Wall
Sco 17h3'56" -33d11'59"
Jessica Grant
Tau 4h1'38" 18d24'10"
Jessica Gray
Cas 1h30'37" 60d59'52"
Jessica Gregorek
Sco 17h45'37" -45d10'43"
Jessica Gribbin
Ori 5h24'57" 6d7'15"
Jessica Grier
Dra 16h43'24" 69d21'37"
Jessica Guardino
And 0h57'35" 22d35'54"
Jessica Gurdemir
Lyn 8h28'45" 40d13'33"
Jessica H.
Crb 16h11'31" 34d32'59"
Jessica H B
Cas 0h20'33" 51d22'33"
Jessica Hager
Vir 12h16'59" 8d0'34"
Jessica Hahn
Cnc 9h11'2" 30d59'6"
Jessica Haley Stavinoha
Gem 6h46'57" 12d10'34"
Jessica Hall
Sco 16h11'42" -12d58'55"
Jessica Halleman
Cas 22h58'10" 56d27'31"
Jessica Hallowell
Cyg 19h55'16" 35d39'35"
Jessica Hamilton
Uma 11h57'58" 41d53'19"
Jessica Hamlyn
And 23h8'49" 48d49'23"
Jessica Hannah Boff
Ari 3h27'5" 21d47'24"
Jessica Harris
Lyn 7h46'57" 53d9'46"
Jessica Harris
Lyn 6h21'51" 59d10'39"
Jessica Hartnett
And 0h53'20" 45d43'50"
Jessica Hastings
Cyg 20h36'14" 37d2'25"
Jessica Haughton
Tau 5h25'35" 19d5'56"
Jessica & Hayley Borino
And 2h0'29" 37d53'58"
Jessica Heaps *Our Star*
Lib 14h26'54" -18d33'2"
Jessica Hegge
Psc 0h52'4" 5d4'19"
Jessica Helena
Uma 10h20'43" 55d19'0"
Jessica Helene Smith
Ori 5h7'38" 6d1'4"
Jessica Henderson
Leo 11h19'37" -0d6'23"
Jessica Hendon
Lib 15h16'17" -23d21'22"
Jessica Henrickson
And 0h43'24" 40d35'13"
Jessica Hensley
Sgr 18h11'0" -31d51'44"
Jessica Hernandez
And 23h17'52" 52d54'10"
Jessica Herrick
Cam 5h47'28" 59d1'55"
Jessica Hines
Del 20h39'12" 15d18'49"
Jessica Hobson
Dra 20h40'47" 72d31'16"
Jessica Honer
Umi 15h14'36" 75d45'14"
Jessica Honey Jane Miller
Gem 6h36'42" 17d6'40"
Jessica Hoover
Cas 23h20'5" 58d6'44"
Jessica Hope McCarthy
And 23h46'27" 48d28'9"
Jessica Hounsell
Cas 23h59'8" 65d41'27"
Jessica Humphrey
Cas 23h41'33" 54d35'58"
Jessica I love you!! From
Dominic
Ori 5h52'44" 20d19'45"
Jessica I. Quintana
Leo 9h37'59" 26d50'43"
Jessica Ingrid Taylor
Lyr 18h33'42" 26d14'6"
Jessica Irene Campbell
Ari 2h11'25" 21d52'49"
Jessica Isabel
And 23h25'39" 42d16'45"
Jessica Ivory Dunn
Uma 11h5'21" 35d44'8"
Jessica J.
And 1h9'41" 39d53'31"
Jessica J
And 0h47'58" 45d35'3"
Jessica Jacob
Cam 4h35'5" 53d52'29"
Jessica Jade Pruitt
Ori 5h39'24" 3d0'21"
Jessica "Jaibird" Goulet
Tau 3h46'11" 27d2'43"
Jessica Jamaica Moon
Gem 6h55'1" 22d46'25"
Jessica Janay Forbing
And 1h7'43" 45d8'11"
Jessica Jancewicz
Uma 11h23'25" 40d57'12"
Jessica Jane
Uma 9h47'2" 54d20'56"
Jessica Jane
Cra 18h6'10" -38d24'11"
Jessica Jane Mu
Aqr 22h53'35" -11d51'3"
Jessica Jane Wimpee
Gem 6h48'3" 12d52'44"
Jessica Jane Wood
And 2h22'4" 43d0'8"
Jessica Janine Smiley
And 0h47'43" 40d54'15"
Jessica Jarosky
Uma 9h31'47" 42d37'6"
Jessica Jarvis
Cyg 20h37'7" 39d50'17"
Jessica Jean Coello
And 0h18'17" 28d43'46"
Jessica Jean (JJ) Werner
Cap 21h38'49" -17d35'48"
Jessica Jean King
Lyr 18h32'18" 37d6'58"
Jessica Jean Olsen
And 0h46'24" 37d44'20"
Jessica Jean Rock
Gem 6h48'13" 34d6'17"
Jessica Jennifer Hendrix
Sgr 18h42'56" -35d34'44"
Jessica Jennings
Gem 7h28'23" 16d26'11"
Jessica Jill Sellers
Ori 5h50'28" 18d17'2"
Jessica Joan
And 0h7'58" 42d40'1"
Jessica Joanne
Cas 1h13'34" 54d32'30"
Jessica Joanne Patton
Ari 2h41'0" 26d24'48"
Jessica Joanne Rau
Ari 2h15'38" 23d6'37"
Jessica Johanna Morris
Umi 15h51'19" 74d9'14"
Jessica Johnston
Ori 5h17'10" 0d8'55"
Jessica Johnston (J Rock
Major)
Cas 1h24'10" 64d7'45"
Jessica Jones
Mon 6h26'19" 8d48'1"
Jessica Jones
Tau 4h31'57" 28d37'7"
Jessica Jordan Green
Tau 4h16'14" 27d20'40"
Jessica & Josh
Cyg 21h25'59" 45d50'27"
Jessica Joy
Cap 20h9'44" -24d59'24"
Jessica Joy DiFelice
Cyg 19h32'4" 29d15'6"
Jessica Joy Marshall
Ori 5h7'24" 4d5'36"
Jessica Joy Renee Avatar
Mauldin
Cyg 19h45'27" 33d39'55"
Jessica Jurkosky
Leo 11h17'5" 18d10'14"
Jessica & Justin
Sco 16h40'57" -29d24'25"
Jessica Justine
Klotunowitch
Vir 12h36'51" 11d36'39"
Jessica JZ
Lyn 8h54'42" 38d2'14"
Jessica K.
Cam 5h56'38" 60d7'8"
Jessica K. Leone
Ori 5h37'56" -0d23'55"
Jessica K Scott is Brents
true love
Psc 1h32'37" 12d50'9"
Jessica Karalash
Sco 16h37'47" -37d20'35"
Jessica & Karl
Cyg 19h57'21" 39d57'58"
Jessica Kaspardlov
Sco 16h8'23" -18d5'29"
Jessica Katherine
McBrearty
Crb 15h37'53" 27d38'28"
Jessica Kathryn
Lmi 10h36'5" 31d8'17"
Jessica Kay Bobo 11-27-
1987
And 2h33'20" 43d13'26"
Jessica Kay Marshall
And 1h8'12" 37d23'57"
Jessica Kay Perez
Eri 3h44'44" -0d7'29"
Jessica Kay Singletary
Gem 6h21'2" 18d7'43"
Jessica Kay's Star
And 2h35'48" 50d12'11"
Jessica Kendzor
Sgr 20h21'52" -35d7'15"
Jessica Keri Bluth
Uma 9h27'20" 45d0'41"
Jessica & Kevin Weigand
Cyg 21h56'35" 37d15'10"
Jessica Kindred
Cra 18h46'14" -38d18'34"
Jessica Kirkwood
Cap 20h28'42" -11d37'52"
jessica kligerman
Cas 1h44'45" 61d49'52"
Jessica Koran Correll
And 23h14'30" 40d58'34"
Jessica Korpela
Mon 6h28'52" 8d27'45"
Jessica Kotzmoyer
Uma 12h59'52" 55d13'47"
Jessica Kreps
Cap 21h48'0" -13d12'12"
Jessica Kristen
And 0h59'13" 40d26'13"
Jessica Krystal Haros
Gem 6h58'26" 15d33'13"
Jessica Kunkel
Lyn 8h28'37" 33d31'30"
Jessica Kwicien
Ari 3h2'39" 27d51'3"
Jessica & Kyle
Leo 11h52'28" 21d56'40"
Jessica Kylie Mauk
And 0h40'18" 42d52'4"
Jessica L.
Sco 16h52'47" -41d33'4"
Jessica L. Barcafer
And 0h52'34" 39d55'29"
Jessica L. Bullen
Uma 10h23'25" 45d4'42"
Jessica L. Coleman
Gem 6h42'2" 22d56'26"
Jessica L. Dickerson
And 0h35'26" 38d14'34"
Jessica L Gordon
Per 3h33'58" 39d48'23"
Jessica l Heaton
Lib 15h30'14" -21d9'18"
Jessica L. Osborn
Vir 12h18'51" 4d16'5"
Jessica L. Pavich
Vir 14h2'28" -7d49'10"
Jessica L. Vitale
Lib 15h9'47" -20d48'21"
Jessica L. Wagoner
Tau 4h30'43" 17d28'59"
Jessica La Preciosa Prima
Psc 0h17'39" 21d30'39"
Jessica Lacoste Hebert
Cnc 8h46'21" 23d40'50"
Jessica LaCroix
And 1h47'16" 41d24'53"
Jessica LaCroix
And 1h17'13" 50d26'21"
Jessica Ladawn Rogers
Gem 6h46'55" 19d34'57"

Jessica Laedermann Del 20h32'44" 4d57'4"
Jessica Laine Harris Cyg 19h47'36" 48d14'7"
Jessica Lane Vir 13h34'25" 12d5'34"
Jessica Langebrake's Star Lyn 6h34'38" 56d17'21"
Jessica LaPlaca Cnc 8h9'27" 26d36'31"
Jessica Lauren McGhie And 1h29'28" 35d14'6"
Jessica Lauren Salter And 23h16'28" 42d1'57"
Jessica Lavendol Lyn 8h5'52" 35d29'50"
jessica lavine Per 4h33'11" 39d51'32"
Jessica Lawson Lyn 7h44'57" 35d43'21"
Jessica Le' Psc 1h19'42" 24d12'27"
Jessica Lea Lindgren Cas 1h41'59" 61d21'24"
Jessica Lea Thomas Sco 16h47'57" -26d55'8"
Jessica Leah Mason Psc 0h2'41" 8d12'47"
Jessica Leah Pierce Sgr 18h30'53" -17d31'37"
Jessica Leah Southers Aqr 22h31'2" -5d40'12"
Jessica Leah Wilson And 0h22'37" 40d4'51"
Jessica LeAnn Bargainear Del 20h33'44" 14d12'26"
Jessica LeAnn Burch Lyn 7h57'8" 38d57'41"
Jessica Leask Cas 0h7'27" 54d6'30"
Jessica Lee And 23h32'39" 41d6'37"
Jessica Lee Uma 11h10'52" 38d23'12"
Jessica Lee Com 12h16'33" 32d7'54"
Jessica Lee Cnc 8h5'33" 13d34'34"
Jessica Lee Cas 1h58'3" 61d30'10"
Jessica Lee Cas 1h19'35" 61d51'2"
Jessica Lee Cra 18h34'22" -42d54'34"
Jessica Lee Cook Sco 17h54'47" -44d58'18"
Jessica Lee Dover Aqr 23h54'3" -10d51'50"
Jessica Lee Lamprecht - Hawley Cap 21h23'57" -18d3'5"
Jessica Lee Ludwig Ari 2h52'8" 29d38'18"
Jessica Lee Madonna And 0h14'26" 40d12'34"
Jessica Lee Novak And 23h23'46" 43d23'32"
Jessica Lee Smoose Gem 7h28'56" 33d20'31"
Jessica Lee Snider Lyr 18h44'38" 39d58'23"
Jessica Lee Vigil Vir 13h43'25" 3d13'18"
Jessica Lee Wingham And 0h47'46" 35d25'0"
Jessica Lee Ziegler Cyg 19h51'14" 38d21'27"
Jessica Lehman Collard Dra 20h40'56" 69d54'21"
Jessica Leigh Lib 15h8'21" -17d55'31"
Jessica Leigh Mon 6h44'5" -0d5'10"
Jessica Leigh Cnc 9h8'14" 32d45'1"
Jessica Leigh Ori 5h42'9" 8d24'11"
Jessica Leigh Crb 16h7'25" 26d34'11"
Jessica Leigh Baryla And 0h42'57" 42d27'38"
Jessica Leigh Harris Cap 20h26'6" -13d32'29"
Jessica Leigh Hessman And 0h22'46" 42d21'52"
Jessica Leigh Johnson And 0h25'5" 41d58'49"
Jessica Leigh Lachiver And 0h38'56" 34d40'19"
Jessica Leigh Lonsdale Ari 2h39'6" 19d47'15"
Jessica Leigh McQuillin Cas 1h10'20" 52d32'7"
Jessica Leigh Mullis And 0h26'21" 43d43'52"
Jessica Leigh Schindler Lib 15h30'43" -24d6'22"
Jessica Leigh Stewart Ari 3h10'50" 27d30'44"
Jessica Leighton Moore Lyn 6h24'40" 55d52'38"
Jessica Lembo And Kristin Kenyon Lib 14h30'0" -17d39'54"

Jessica Leong Cet 1h5'15" -20d31'25"
Jessica Li Cas 23h9'35" 58d52'6"
Jessica Lilly And 23h24'2" 51d50'1"
Jessica Liniece Storer And 23h20'35" 49d7'32"
Jessica Logan Vir 12h30'32" 3d13'34"
Jessica Lombardi Aqr 22h12'13" -3d29'30"
Jessica Long Cas 2h22'21" 73d55'50"
Jessica Lopez And 2h22'42" 46d7'16"
Jessica Loraine Cunningham Gem 6h25'57" 21d46'52"
Jessica Lorton Muse And 1h9'51" 45d31'59"
Jessica Louise Brouff Cru 12h9'45" -59d37'31"
Jessica Louise Collins And 23h26'3" 39d52'41"
Jessica Louise Vernon And 1h17'11" 45d22'28"
Jessica Louise Vicars Cnc 8h17'6" 19d34'28"
Jessica Louise Wagstaff And 0h13'45" 31d58'55"
Jessica Love Aql 19h31'31" 8d39'55"
Jessica Loves Michael Forever Psc 1h20'28" 33d22'30"
Jessica Lucente Per 4h23'55" 52d39'37"
Jessica Lucio Cnc 8h47'29" 10d2'54"
Jessica Lucy Fisanick Psc 1h3'20" 17d17'37"
Jessica Luise Sanabria Vir 13h35'30" 2d1'46"
Jessica Luna Vir 14h33'7" -3d55'36"
Jessica Lyn Vir 13h42'29" -10d54'10"
Jessica Lyn Gore Lmi 9h31'30" 34d7'25"
Jessica Lyn Jandreski And 23h5'26" 49d57'39"
Jessica Lyn Mcelroy Leo 10h41'12" 16d33'16"
Jessica Lyndsay Rosenblum Leo 11h43'3" 26d20'3"
Jessica Lynn Tau 3h54'56" 24d29'42"
Jessica Lynn Gem 6h17'35" 22d23'6"
Jessica Lynn And 0h26'17" 26d56'40"
Jessica Lynn Ori 5h11'52" 6d39'7"
Jessica Lynn Ori 5h12'10" 15d27'25"
Jessica Lynn And 23h6'4" 48d28'3"
Jessica Lynn Uma 14h1'36" 48d26'46"
Jessica Lynn Uma 10h23'45" 47d30'18"
Jessica Lynn Uma 10h53'51" 51d44'8"
Jessica Lynn And 2h26'46" 48d8'35"
Jessica Lynn Boo 14h44'6" 34d19'4"
Jessica Lynn And 0h39'49" 37d52'51"
Jessica Lynn Lyn 8h5'6" 43d2'4"
Jessica Lynn Lyn 9h29'54" 41d12'31"
Jessica Lynn Aqr 22h20'20" -6d9'43"
Jessica Lynn Vir 13h53'28" -21d48'44"
Jessica Lynn Cas 2h16'50" 67d49'23"
Jessica Lynn Uma 9h23'24" 66d33'40"
Jessica Lynn Uma 10h34'13" 66d31'21"
Jessica Lynn Cyg 19h27'27" 58d4'28"
Jessica Lynn Cap 20h47'54" -25d15'2"
Jessica Lynn Alameda And 0h43'32" 39d11'30"
Jessica Lynn Ballard Uma 9h16'47" 58d12'19"
Jessica Lynn Barber And 0h16'31" 32d49'28"
Jessica Lynn Beaulne Umi 16h2'53" 76d22'25"
Jessica Lynn Bergener Uma 13h39'17" 57d37'49"
Jessica Lynn Boyce Lyn 8h10'48" 39d59'12"
Jessica Lynn Bush Ori 6h6'35" 10d16'15"
Jessica Lynn Byers Cnc 8h25'31" 28d36'57"

Jessica Lynn Cass Cnc 7h58'47" 15d9'59"
Jessica Lynn Chioco And 23h18'9" 44d10'49"
Jessica Lynn Dingman Cam 7h2'22" 70d29'49"
Jessica Lynn Estrada Gem 7h53'7" 29d58'58"
Jessica Lynn Fekany Uma 10h27'17" 44d49'1"
Jessica Lynn Flaxington Vir 12h49'23" 12d50'42"
Jessica Lynn Flyte Uma 12h48'56" 60d32'49"
Jessica Lynn Foley Sco 16h13'51" -15d40'54"
Jessica Lynn Gagnon Aqr 22h15'20" -3d31'12"
Jessica Lynn Griner Uma 9h19'48" 61d6'47"
Jessica Lynn Hayward Cam 4h20'18" 67d57'24"
Jessica Lynn Higgins Per 3h48'24" 48d54'46"
Jessica Lynn Hunter And 1h29'11" 39d14'19"
Jessica Lynn Janz Lib 15h38'21" -9d35'26"
Jessica Lynn Klein Uma 10h9'44" 51d4'4"
Jessica Lynn Litten Lib 15h55'0" -17d18'7"
Jessica Lynn Matuszak Aqr 22h37'38" -2d3'53"
Jessica Lynn McCullough Tau 5h37'46" 25d51'19"
Jessica Lynn Murphy And 1h18'34" 38d55'22"
Jessica Lynn Parron Peg 22h40'49" 31d51'47"
Jessica Lynn Patton Cas 1h10'43" 55d40'42"
Jessica Lynn & Paul Morris, Jr. Cas 0h44'48" 61d56'49"
Jessica Lynn Pellegrino And 1h20'1" 36d20'8"
Jessica Lynn Peterson And 0h58'26" 45d42'42"
Jessica Lynn Quinones Uma 9h55'9" 43d52'41"
Jessica Lynn Rachfalski And 23h12'35" 47d48'40"
Jessica Lynn Racila Vir 11h56'34" -0d7'54"
Jessica Lynn Ray And 23h35'23" 49d47'47"
Jessica Lynn Read Com 12h48'36" 25d51'7"
JESSICA LYNN REED Ari 2h49'21" 25d15'51"
Jessica Lynn Robertson Lep 5h49'32" -21d45'46"
Jessica Lynn Rodriguez Leo 9h26'51" 11d48'29"
Jessica Lynn Scalabrini Cyg 21h45'18" 54d29'47"
Jessica Lynn Shewfelt Cnc 8h29'44" 14d59'46"
Jessica Lynn Trahan Ari 2h19'34" 24d44'1"
Jessica Lynn Vos And 1h24'19" 35d11'41"
Jessica Lynn Wayte Cnc 9h9'33" 21d48'49"
Jessica Lynn Westcott And 0h39'43" 40d39'43"
Jessica Lynn Woodworth Uma 13h9'1" 57d13'32"
Jessica Lynn Yocom Ori 5h38'55" -5d10'6"
Jessica Lynn Yori Vir 13h56'19" -15d59'43"
Jessica Lynn Zar Sco 17h16'36" -43d43'31"
Jessica Lynne Gaskill Uma 9h31'50" 58d44'37"
Jessica Lynne Roof Lib 15h30'14" -27d45'50"
JESSICA LYNNE SCHUETTE Cap 20h57'52" -21d8'11"
Jessica M. Adviento Ori 5h56'7" 7d43'48"
Jessica M Bartol Sgr 18h32'0" -33d19'12"
Jessica M. Benning Ori 5h46'40" 8d14'3"
Jessica M. Cardenas And 1h17'33" 38d40'21"
Jessica M. Escalera And 0h33'21" 45d17'22"
Jessica M. Garcia Vir 12h42'45" 3d2'55"
Jessica M Garcia Taylor Cnc 9h10'53" 6d39'4"
Jessica M Letona Lyr 18h44'18" 39d30'49"
Jessica M. Masis Cam 3h53'48" 70d14'0"
Jessica M Modlin And 23h23'51" 48d43'11"
Jessica M. Palys And 1h58'59" 46d6'24"

Jessica M. Pearson Tau 4h0'35" 10d51'36"
Jessica M Riechert Tau 3h47'45" 6d13'27"
Jessica M. Story Aqr 22h24'18" -24d39'23"
Jessica M Wirick Lib 15h6'37" -8d23'20"
Jessica Mabel Campodonico Crb 16h9'42" 35d30'54"
Jessica Mae Leo 9h33'22" 29d18'24"
Jessica Mae Martinez Tau 3h32'44" 24d57'30"
Jessica Magna Dra 17h40'28" 52d22'6"
Jessica Mai Williams And 1h16'43" 33d58'3"
Jessica Maire Castillo And 1h47'19" 41d52'18"
Jessica Major Per 4h16'22" 45d1'38"
Jessica Maliza Hopkins Cas 0h19'42" 56d48'53"
Jessica Mamon Leo 9h59'52" 20d23'11"
Jessica Maree Cabral - Purple Angel Vir 13h48'5" -11d15'19"
Jessica Margaret Adissi Lyn 9h13'58" 33d1'54"
Jessica Margareta Falis Cnc 8h34'57" 32d20'38"
Jessica Margit Puckett Peg 22h42'56" 24d32'55"
Jessica Margret Joan Bell And 23h9'38" 51d33'46"
Jessica Marie Crb 16h5'16" 39d6'30"
Jessica Marie Ari 3h2'16" 30d35'12"
Jessica Marie Cnc 9h8'54" 27d45'59"
JESSICA MARIE Ari 2h13'0" 17d26'12"
Jessica Marie Lib 15h15'29" -16d20'15"
Jessica Marie Uma 10h34'22" 66d11'18"
Jessica Marie Uma 11h45'26" 61d7'18"
Jessica Marie Lyn 7h45'45" 55d52'53"
Jessica Marie Lib 15h24'26" -25d45'43"
Jessica Marie 10-18-86 Lib 15h20'41" -27d34'23"
Jessica Marie Almaraz Cap 20h28'26" -13d2'42"
Jessica Marie Badalian Tau 5h40'3" 21d37'31"
Jessica Marie " Beautiful Angel " Her 18h3'47" 15d1'32"
Jessica Marie Berry And 0h37'3" 32d2'42"
Jessica Marie Blaney And 23h16'33" 49d22'58"
Jessica Marie Boyce Pho 1h47'10" -47d47'36"
Jessica Marie Brislee Uma 11h47'5" 30d3'45"
Jessica Marie Burkett Aqr 20h48'7" -1d9'49"
Jessica Marie Carlisle Sco 16h44'35" -42d55'43"
Jessica Marie Carrol Cam 5h46'36" 58d43'47"
Jessica Marie Clark Ari 2h16'40" 13d20'37"
Jessica Marie Darnell Lib 15h6'4" -22d55'59"
Jessica Marie Deleon "Always Luv U" Ari 3h19'45" 19d53'28"
Jessica Marie Drunasky Leo 11h2'49" 14d21'30"
Jessica Marie Dyer Tau 3h44'53" 26d8'32"
Jessica Marie Ewing Vir 13h6'53" 12d43'49"
Jessica Marie Fletcher And 2h21'32" 38d40'45"
Jessica Marie Hess And 0h11'10" 40d14'22"
Jessica Marie Hightower Cam 5h25'49" 62d1'41"
Jessica Marie Hoak Leo 10h16'45" 16d40'31"
Jessica Marie Holmberg Cap 20h36'40" -20d3'46"
Jessica Marie Hughes Crb 15h38'29" 29d2'39"
JESSICA MARIE ILU And 2h13'24" 44d25'17"
Jessica Marie Kleya And 2h18'49" 44d59'4"
Jessica Marie Landry Per 2h52'50" 38d33'7"
JESSICA MARIE LITTLE-JOHN SWEET 16 Cap 21h24'27" -19d20'58"
Jessica Marie Ney And 0h44'55" 23d39'22"

Jessica Marie Norman Uma 11h35'31" 46d42'32"
Jessica Marie Paris And 1h32'42" 48d34'14"
Jessica Marie Pierschbacher Cyg 19h44'6" 34d0'56"
Jessica Marie Pollnow Cyg 20h0'20" 44d9'4"
Jessica Marie Roberson Uma 9h45'19" 46d18'24"
Jessica Marie Smith Cnc 9h12'23" 26d46'47"
Jessica Marie This Vir 13h44'1" -19d45'24"
Jessica Marie Tucker And 23h0'26" 43d4'56"
Jessica Marie Wassif Leo 11h20'35" 3d53'26"
Jessica Marie Watcke Lib 15h18'48" -4d7'54"
Jessica Marie Wetch Cap 21h56'7" -19d39'41"
Jessica Marie Woodward Cam 5h3'48" 72d0'34"
Jessica Marissa Ari 2h33'39" 26d46'10"
Jessica Marquez Sco 16h11'7" -23d46'18"
Jessica Martin Cap 21h5'19" -24d56'24"
Jessica Mary Drabyk Lyn 8h33'15" 34d49'17"
Jessica Mary Talty Kerr And 1h0'59" 45d2'5"
Jessica Matherne Lyn 7h59'30" 42d52'19"
Jessica Matilda Lib 15h9'18" -26d43'58"
Jessica Matlin Del 20h14'27" 15d55'3"
Jessica Maureen Lajiness Gem 7h21'35" 22d46'19"
Jessica May Sgr 18h57'44" -23d10'28"
Jessica May Frost And 1h28'36" 35d29'47"
Jessica May Harding And 1h30'39" 40d48'44"
Jessica Mazzola Sge 20h19'16" 18d49'30"
Jessica Mazzurco Uma 11h8'41" 49d40'30"
Jessica McCarson Tau 5h39'58" 25d12'53"
Jessica McClure Lyn 7h57'11" 41d19'35"
Jessica McComb Leo 11h40'22" 26d33'48"
Jessica Mcdonald "Top" Lyn 7h53'29" 47d33'30"
Jessica McHugh Uma 8h57'9" 57d57'45"
Jessica Meade - Shining Star And 1h20'23" 49d49'53"
Jessica Medina Sco 16h21'6" -18d13'13"
Jessica Melanie Rainey Vir 13h15'33" 13d5'23"
Jessica Mendes Cnc 8h7'59" 25d8'46"
Jessica Meredith Messenger And 0h47'18" 38d25'38"
Jessica Meringer And 23h15'30" 44d51'47"
Jessica Merriott Crb 15h31'8" 32d32'24"
Jessica Meyers Lib 14h30'29" -18d41'20"
Jessica Mi Amor Cnc 9h10'10" 28d57'16"
Jessica Michaut Mon 6h40'31" -1d50'25"
Jessica Michele Psc 1h28'49" 10d29'7"
Jessica Michele Lauria And 1h30'31" 49d8'16"
Jessica Michelle Lyn 8h48'16" 43d46'26"
Jessica Michelle Cnc 8h2'38" 26d13'4"
Jessica Michelle Ponder Gragg And 0h25'53" 41d7'17"
Jessica Milne And 1h7'20" 46d25'33"
Jessica Mitchell Cyg 21h24'27" 31d6'17"
Jessica Mitchell Umi 14h54'48" 89d11'54"
Jessica Mitchell Umi 16h10'24" 74d49'26"
Jessica Mohr Sco 17h41'36" -36d28'43"
Jessica Mollie & Jacob James Bennett Gem 6h25'37" 19d16'52"
Jessica Monique Desso And 0h14'42" 46d51'40"
Jessica Montana Brock Uma 10h46'0" 58d27'17"
Jessica Montano Her 16h36'47" 26d19'29"

Jessica Moore Cas 1h16'1" 62d1'44"
Jessica Morgan And 23h44'6" 37d10'14"
Jessica Morgan 4 Psc 1h8'30" 3d20'58"
Jessica Morgan Trentacosta Lib 15h19'18" -11d39'35"
Jessica Morton Ori 6h2'26" 18d54'33"
Jessica Muenker And 23h15'17" 40d50'57"
Jessica Murphy Lyr 19h17'20" 29d44'31"
Jessica Murphy Tau 3h33'21" 13d48'51"
Jessica Musuta And 0h41'7" 40d27'57"
Jessica Muszynski Del 20h33'19" 8d16'24"
Jessica My Angel Cas 23h51'40" 52d42'20"
Jessica My Angel I Love You Uma 8h18'53" 67d7'26"
Jessica my hearts desire Cap 20h31'2" -16d29'57"
Jessica My Love Leo 10h14'20" 12d10'37"
Jessica My Love Lyr 18h55'3" 35d5'42"
Jessica (My Love) Cherchio Ori 5h23'18" 2d7'50"
Jessica My Love Przybyla Gem 6h18'39" 22d6'24"
Jessica My Mon Cheri And 0h33'17" 31d42'18"
Jessica my Wootzen Cap 21h18'20" -16d20'49"
Jessica Myers Vir 12h49'16" 5d13'0"
Jessica Myers And 23h12'17" 51d46'0"
Jessica N. Burke Psc 23h25'12" 7d35'22"
Jessica N Lout Gem 6h15'41" 26d47'53"
Jessica N. Reynolds Lyn 8h43'44" 38d17'35"
Jessica N Skyeler Eri 3h54'36" -10d33'59"
Jessica N. Swifka Cap 21h49'41" -24d7'23"
Jessica Nail Tau 4h38'19" 22d34'42"
Jessica Natofsky Lyn 7h4'9" 52d10'29"
Jessica Nemeth & E.J. Gagola Cyg 20h12'21" 36d43'13"
Jessica Newton Uma 9h25'20" 58d8'10"
Jessica Nguyen Cap 21h21'54" -14d44'57"
Jessica Nguyen And 1h41'13" 37d29'36"
Jessica Niamh Smith And 2h12'26" 49d13'12"
Jessica Nichol Davis Cap 20h51'25" -15d15'0"
Jessica Nichole Cap 20h29'57" -16d32'7"
Jessica Nichole Howard Cas 0h18'12" 55d36'2"
Jessica Nichole Preble And 0h1'57" 39d25'58"
Jessica Nicole Cnc 8h20'26" 25d1'23"
Jessica Nicole Cnc 8h32'33" 28d28'16"
Jessica Nicole And 0h43'10" 28d34'50"
Jessica Nicole Vir 14h4'37" -17d8'3"
Jessica Nicole Cap 20h38'49" -13d10'7"
Jessica Nicole Lib 15h8'16" -27d50'36"
Jessica Nicole Bass Ari 2h41'24" 26d55'28"
Jessica Nicole Borm Lyn 6h52'27" 57d41'33"
Jessica Nicole Branham Leo 10h21'25" 21d22'20"
Jessica Nicole Callan Sco 16h4'2" -13d37'45"
Jessica Nicole Felder Ari 2h8'49" 13d45'10"
Jessica Nicole Hayes Cyg 21h42'49" 53d4'30"
Jessica Nicole Johnson And 0h35'44" 41d17'41"
Jessica Nicole Painter Tau 3h41'37" 5d50'30"
Jessica Nicole Pappas Leo 11h33'20" 27d29'4"
Jessica Nicole Perkins Cyg 21h8'40" 48d28'12"
Jessica Nicole Raynor Peg 23h22'47" 17d9'7"
Jessica Nicole Ruiz And 1h22'36" 39d16'34"

Jessica Nicole Sasaoka And 2h29'1" 42d48'23"
Jessica Nicole Schmid Ari 2h22'59" 24d52'45"
Jessica Nicole Turner Crb 16h7'51" 38d23'46"
Jessica Niermann Sgr 19h47'2" -31d22'6"
Jessica Nino Lyn 8h7'7" 36d56'6"
Jessica Noel Cnc 8h10'56" 15d3'39"
Jessica Noell Pinna And 0h7'39" 33d35'46"
Jessica Noelle Vir 13h4'12" 12d50'41"
Jessica Nora Smith Gem 7h25'56" 26d40'38"
Jessica Norene Prouty ~ "Piglet" Vir 12h19'32" -8d55'14"
Jessica Nova Leo 11h23'28" 23d40'25"
Jessica Nowak Lyn 7h20'31" 50d21'35"
Jessica Ochoa Cyg 20h59'35" 32d28'20"
Jessica Oerly Cap 20h33'58" -16d55'12"
Jessica O'Kane Cas 1h21'20" 63d41'54"
"Jessica Olivia Alexandra Star" And 1h44'41" 41d15'8"
Jessica P. Delph Uma 11h7'31" 47d39'10"
Jessica P Garcia And 0h23'43" 33d44'43"
Jessica P. Jovillar Ori 5h49'36" 8d31'13"
Jessica Paige Leo 11h15'36" 14d36'29"
Jessica Paige Salzman Ari 2h38'18" 14d5'53"
Jessica Paige Stitt Cnc 8h4'14" 6d47'52"
Jessica Palmer Leo 10h30'38" 18d33'7"
Jessica Parsons And 0h47'48" 42d5'10"
Jessica Passi - Jessica's Star And 0h4'11" 46d31'40"
Jessica Pate's Star Sco 17h21'34" -42d3'12"
Jessica Patricia Arnold Psc 0h42'53" 15d51'52"
Jessica Pena Cnc 7h57'33" 12d41'35"
Jessica Penguin Vir 12h27'4" 11d8'26"
Jessica Peoples Leo 10h22'3" 23d44'21"
Jessica Perez Gem 7h15'57" 15d35'28"
Jessica Perkins Sgr 18h13'27" -28d9'45"
Jessica Perri Kirsch And 23h17'42" 35d19'42"
Jessica Perry Aqr 21h35'42" 0d36'47"
Jessica Perry Newell Aqr 22h15'35" 0d22'49"
Jessica Peterson Cas 0h10'30" 54d14'27"
Jessica Pfieffer Sgr 18h29'14" -27d56'22"
Jessica Phylieche Auldridge Lib 15h10'21" -5d45'43"
Jessica Pickford Cas 0h19'18" 51d22'31"
Jessica Piper And 23h41'42" 48d8'1"
Jessica Piper And 23h35'0" 45d0'13"
Jessica Player's Star And 2h13'20" 38d54'34"
Jessica Pokorni Cnc 8h32'9" 25d23'21"
Jessica Poulin Aqr 20h57'55" 1d6'46"
Jessica Presbie Vir 14h12'49" 5d30'0"
Jessica Prestifilippo Aqr 21h35'6" 1d49'49"
Jessica Prime Ori 4h47'6" -3d5'36"
Jessica Pujol And 0h50'42" 37d58'15"
Jessica Quintero's Special Star Sco 17h54'5" -41d4'45"
Jessica Quiroga Psc 0h42'36" 14d16'51"
Jessica R. Sanders Ori 5h36'44" 10d4'46"
Jessica R. Williams Favier And 1h13'0" 40d17'16"
Jessica Rabbit Ari 2h46'18" 21d50'42"
Jessica Rachel Ari 2h20'10" 11d0'54"
Jessica Rachel Reed Vir 13h14'42" -4d38'51"

Jessica Rae
Mon 6h26'26" -3d24'40"
Jessica Rae
Cra 19h5'2" -38d28'42"
Jessica Rae
And 2h8'27" 45d41'27"
Jessica Rae 12/6/03
Sco 16h41'55" -30d49'1"
Jessica Rae Cornejo
And 2h27'9" 42d34'38"
Jessica Rae Stoner
And 23h15'9" 52d9'11"
Jessica Rain Bucy
Leo 10h51'42" 19d54'4"
Jessica Rainwater
Crb 15h50'16" 26d38'59"
Jessica Randone
Ori 5h14'9" -1d21'33"
Jessica Ray
Ari 3h24'46" 26d43'38"
Jessica Ray Deedon
Per 1h44'11" 54d26'30"
Jessica Read
Sco 16h19'35" -42d18'21"
Jessica Reba
Vir 14h1'28" -8d18'57"
Jessica Reeping - Light Of My Life
Uma 10h51'33" 63d24'50"
Jessica Reis Labombarde
Vir 13h29'52" 5d20'46"
Jessica Rene
Psc 0h21'19" -4d27'55"
Jessica Rene Cole
Psc 0h45'47" 14d27'20"
Jessica Renee
Gem 6h45'10" 15d38'21"
Jessica Renee
Ari 2h38'5" 28d1'56"
Jessica Renee
Lyn 6h49'25" 55d24'3"
Jessica Renee Anderson
And 0h45'58" 43d34'1"
Jessica Renee Benner
Sgr 18h18'43" -29d20'9"
Jessica Renee Mitchell
Mon 6h32'31" 7d36'36"
Jessica Renee Musselman
Ori 6h11'46" 9d15'1"
Jessica Renee Oostindie
Ari 2h16'55" 21d4'10"
Jessica Renee Wenzel
Vir 14h36'16" 3d41'41"
Jessica Renee Zambon
Sco 16h42'6" -28d54'51"
Jessica Repasky
Leo 11h9'15" 4d0'41"
Jessica Rhea Cook
Cnc 8h22'14" 22d47'54"
Jessica Richard
Uma 9h53'21" 66d33'50"
Jessica Richards
Lyr 19h0'47" 41d47'35"
Jessica Ritchey
And 23h10'52" 43d2'50"
Jessica Ritzman Hanson
Psc 0h30'35" 8d18'54"
Jessica Rizzolo
Psc 1h46'24" 13d43'31"
JESSICA RIZZUTO DILLON
Crb 15h20'6" 27d1'47"
Jessica Rochelle Luczak
Tau 5h35'10" 17d27'15"
Jessica Rolon
And 23h36'27" 38d47'43"
Jessica Rose
And 23h2'17" 51d35'51"
Jessica Rose
Uma 9h55'4" 43d35'41"
Jessica Rose
And 0h8'56" 28d43'56"
Jessica Rose
Gem 7h30'7" 20d33'34"
Jessica Rose
Cas 2h22'1" 68d4'11"
Jessica Rose
Lib 14h50'33" -1d3'28"
Jessica Rose
Lib 15h36'59" -17d34'38"
Jessica Rose
Lib 15h16'34" -23d36'40"
Jessica Rose Dorr
Ari 3h29'18" 25d41'11"
Jessica & Rose Farmer
Vir 12h56'46" -16d17'42"
Jessica Rose Johnson
Cru 12h39'54" -60d14'16"
Jessica Rose Webb
Vir 13h12'26" -20d24'36"
Jessica Rose Wheeler
And 2h32'54" 38d29'57"
Jessica Rosengrant 831
Leo 10h28'35" 12d49'38"
Jessica Rose's Love
Peg 21h16'27" 18d49'45"
Jessica Ross
Cru 12h14'11" -62d21'9"
Jessica Rowan
Gem 7h43'32" 25d17'46"
Jessica Rowe
Gem 6h45'12" 19d13'51"
Jessica Rowe
Cam 3h22'40" 60d5'5"
Jessica Ruth Joy Snock
Leo 11h49'8" 26d58'0"
Jessica Ruth Myers
Tau 4h13'5" 23d52'32"
Jessica Ruth Rood
And 23h9'12" 40d20'46"
Jessica Ryan
Cru 12h4'21" -62d33'45"
Jessica S. Lindsey
Uma 10h34'40" 54d32'37"
Jessica S. Seagraves
Psc 1h9'18" 28d48'18"
Jessica Salas
Sco 16h5'40" -11d28'58"
Jessica Saldutte
Gem 7h11'33" 21d44'5"
Jessica Samantha Madris
Lib 15h15'46" -20d29'32"
Jessica Sanders
Cas 23h16'38" 55d49'32"
Jessica Sandrini Rogue
Lyn 7h7'45" 45d5'2"
Jessica Sara
Ori 6h14'37" 15d18'48"
Jessica Sarah Hauck
Uma 12h4'17" 40d51'58"
Jessica Schofield
Umi 15h25'37" 72d57'55"
Jessica Schollenberger
Cas 23h15'52" 55d18'33"
Jessica Schutz
Tau 4h12'51" 24d41'44"
Jessica Schwab
Cas 23h46'30" 57d33'54"
Jessica Scott Neyman
Lyn 7h14'53" 48d28'16"
Jessica Shae Doggen
Cru 12h3'20" -61d40'42"
Jessica Shannon Brown
Gem 6h22'12" 24d10'12"
Jessica Shannon Scott
Sgr 18h31'13" -16d46'11"
Jessica Shannon Ware
Cnc 8h20'8" 14d15'54"
Jessica Shea Warren
Ori 6h17'29" 10d21'6"
Jessica Shoemaker
Ori 6h15'5" 16d23'32"
Jessica Shook
Vir 14h16'27" 5d49'29"
Jessica Shortis
Sco 17h22'15" -38d54'4"
Jessica Shutiva
And 0h31'50" 29d36'17"
Jessica Sierra
Sco 16h17'1" -11d4'14"
Jessica Sierra Yow
Cas 1h38'41" 63d7'6"
Jessica Smiech
Uma 13h15'3" 53d41'47"
Jessica Smith
Crb 15h31'28" 27d23'8"
Jessica Smyth
Aqr 22h7'13" -1d36'19"
Jessica "Sneaks" Kalivoda
Vir 14h21'35" 2d34'48"
Jessica "Sniffles" Camacho
Sgr 18h39'25" -36d1'26"
Jessica Solene Belkin Heys
Lyn 7h0'59" 51d12'48"
Jessica Sophie Anne Taylor
And 2h34'41" 46d55'44"
Jessica Spring Thomas
And 0h0'21" 35d2'49"
Jessica "squishy" Herb
And 2h12'23" 45d56'1"
Jessica Staffeld
Gem 6h39'28" 14d37'19"
Jessica Stansbie
Cnc 9h13'51" 6d56'32"
Jessica Starcevic
Uma 11h45'27" 33d9'30"
Jessica Starr Dean
Uma 11h28'21" 39d27'32"
Jessica Sten
Cnc 9h6'0" 8d7'7"
Jessica Stephanie Caisse
And 23h19'35" 47d55'2"
Jessica Stephens
And 2h27'29" 49d32'7"
Jessica & Steven Forever and Always
Dra 18h58'22" 55d12'32"
Jessica Stiles
Uma 10h51'9" 56d0'52"
Jessica Sue (Williams) Kneisley
Mon 6h57'47" -0d50'46"
Jessica Summer Dotsey
And 0h39'54" 32d42'6"
Jessica Sunshine
Psc 0h43'52" 7d16'26"
Jessica sunshine Walker
Sco 16h28'40" -26d20'32"
Jessica Susan Coons
Leo 11h39'5" 24d54'44"
Jessica Swagel Rodriguez
Vir 12h48'59" -11d24'32"
Jessica Sweeney
Ari 2h8'16" 12d18'25"
Jessica Szopinski Star
Cas 0h16'2" 54d44'58"
Jessica T. Thorngren
Aql 19h6'43" 14d54'43"
Jessica Tabor Matthews
Crb 15h27'20" 30d36'46"
Jessica Tart
Ori 5h11'39" 6d41'12"
Jessica Tate's Star
Leo 11h1'50" 15d46'57"
Jessica Taylor
Cap 21h34'52" -19d0'56"
Jessica Taylor
Cas 1h27'38" 67d40'35"
Jessica 'The Captain' Ducrou
Psc 1h54'29" 2d49'2"
Jessica Thompson
Ari 2h14'4" 12d47'11"
Jessica Thorne-Booth
Uma 10h9'8" 44d1'54"
Jessica Tilton
Cas 0h1'1" 54d18'46"
Jessica Titus
And 23h15'6" 43d16'38"
Jessica Trevino
And 23h25'26" 42d19'54"
Jessica Trimboli
Ari 3h14'27" 28d33'9"
Jessica Troth
And 0h29'15" 41d29'23"
Jessica Tucker
Leo 11h34'44" -5d21'15"
Jessica Tyler
Lyn 7h15'52" 57d18'12"
Jessica Tyler Cohen
And 23h3'57" 41d49'6"
Jessica Usbeck
Lac 22h44'9" 36d58'7"
Jessica Usher
Ari 3h26'25" 20d26'17"
Jessica V. Gomez
Vir 13h20'8" 7d20'35"
Jessica Vahey Rappaport
Ari 3h10'59" 29d20'14"
Jessica Valdivia
Uma 9h0'24" 59d4'41"
Jessica Valerie Smith
Cas 23h46'32" 59d17'25"
Jessica Van Note
And 23h13'25" 40d55'51"
Jessica Vega
And 23h21'56" 43d19'59"
Jessica Vega
Psc 1h29'42" 13d28'20"
Jessica Velez
Lib 15h45'25" -24d6'16"
Jessica Victoria Janner Major
Crb 15h37'36" 30d51'21"
Jessica Virginia Rivera (J.V.R)
Cnc 8h39'6" 19d18'18"
Jessica Vitela
And 1h19'30" 43d1'20"
Jessica Wainner
Sgr 19h13'33" -34d18'12"
Jessica Walrath
Cyg 19h32'18" 31d57'59"
Jessica Walters
Lmi 10h30'58" 36d26'43"
Jessica Ward
Aqr 23h53'22" -12d12'49"
Jessica Waterlilly-0 Diaz
And 0h39'42" 33d0'45"
Jessica Weller
Leo 11h5'13" -5d6'25"
Jessica Wellner
Lib 15h23'43" -4d23'52"
Jessica White
Crb 16h8'14" 30d10'4"
Jessica Wiegel
Sco 17h38'10" -37d34'36"
Jessica Wigginton
Leo 11h45'21" 21d8'24"
Jessica Willacy
Leo 11h4'59" 15d18'55"
Jessica Wisniewski
And 1h5'49" 45d1'10"
Jessica Wolf
Uma 10h30'23" 47d16'54"
Jessica Wood - The Light of My Life
Gem 7h18'14" 24d12'32"
Jessica Woyce
Sco 17h53'41" -35d51'29"
Jessica Wurtzebach
Ari 2h6'40" 22d49'17"
Jessica Ximena
Leo 11h20'23" 7d14'30"
Jessica Y. Richardson
Cyg 19h45'44" 32d11'50"
Jessica Yang
Sco 17h55'15" -37d29'24"
Jessica Yoon
Lyr 19h14'7" 26d55'30"
Jessica Yvonne Johnson
Cas 0h43'38" 51d52'10"
Jessica Zakhem
Vir 13h15'37" 3d58'57"
Jessica Zeller
Sgr 18h0'5" -27d4'14"
Jessica, ma belle princesse
And 0h4'35" 41d15'33"
Jessica, My Love
Lyn 6h17'46" 60d11'26"
Jessica, My One True Love
Sco 17h13'41" -41d6'6"
Jessica, My Shining Star
Cap 21h40'18" -9d16'49"
Jessica, My Sweet Heart
And 0h55'29" 34d50'0"
Jessica041503
Ari 2h47'13" 18d33'19"
Jessica820
Leo 11h17'22" 16d55'42"
Jessica-Bella
Aqr 21h39'47" 2d27'50"
Jessicah
Lyn 8h39'42" 38d8'15"
JessicaJason
Cyg 19h56'16" 30d28'18"
jessicalauren
Sco 16h55'59" -38d49'13"
JessicaLeighAlbright
Ori 6h20'37" 7d15'4"
Jessica-Lyn
Gem 7h39'42" 33d0'53"
JessicaM
Aqr 22h53'13" -5d38'44"
JESSICA-MARIE
Gem 6h36'17" 17d32'15"
Jessicara
Ori 6h20'9" 18d55'2"
Jessica's angel
And 0h43'56" 44d5'53"
Jessica's Angel
And 1h43'11" 43d34'23"
Jessica's Bright Light
Crb 15h27'19" 25d37'32"
Jessica's Dawn
Cas 0h28'40" 57d4'35"
Jessica's Dawn
Sco 17h46'44" -42d57'57"
Jessica's Guiding Light
Gem 7h29'53" 34d51'55"
Jessica's Halo
Sgr 18h53'11" -33d7'15"
Jessica's Heart
Cyg 21h34'24" 46d20'9"
Jessica's Heaven
Psc 1h24'41" 5d47'42"
Jessica's Heavenly Window
Per 4h48'43" 46d45'40"
Jessica's Kip
Tau 5h47'27" 21d25'37"
Jessica's Light
Lyr 19h17'32" 28d27'21"
Jessica's Mica Mica
Tau 3h33'20" 25d38'49"
Jessica's Piece of Heaven
Dra 19h25'41" 60d5'57"
Jessica's Sanctuary
Uma 11h20'16" 48d49'42"
Jessica's Shining Star
Vir 13h41'53" -5d52'12"
Jessica's shining star
Aqr 20h49'59" -10d57'2"
Jessica's Shining Star
Sgr 18h26'15" -27d21'47"
Jessica's shining starlight
Uma 11h39'50" 55d10'39"
Jessica's Smile
Uma 11h4'58" 47d44'49"
Jessica's Star
And 22h58'25" 45d55'9"
Jessica's Star
And 1h23'20" 40d53'17"
Jessica's Star
And 0h32'10" 38d7'5"
Jessica's Star
And 2h8'29" 41d51'43"
Jessica's Star
Psc 0h44'32" 15d17'50"
Jessica's Star
Vir 14h45'15" 2d10'27"
Jessica's Star
Vir 14h51'55" 4d25'47"
Jessica's star
Uma 11h0'1" 57d26'6"
Jessica's Star
Cas 23h45'27" 58d17'55"
Jessica's Star
Cas 23h7'56" 56d12'9"
Jessica's Star
Uma 12h33'25" 61d55'38"
Jessica's Star
Uma 8h26'8" 66d5'46"
Jessica's star
Cap 20h21'9" -11d18'39"
Jessica's Star
Aqr 22h0'48" -15d7'57"
Jessica's Star
Cap 20h38'10" -20d51'34"
Jessica's Star
Srp 18h20'51" -0d16'49"
Jessica's Star
Aqr 21h47'33" -1d11'3"
Jessica's Star
Sco 17h57'57" -36d24'18"
Jessica's Star
Cru 12h38'1" -57d23'46"
Jessica's Star So Sweet
Lib 15h40'10" -6d40'32"
Jessica's Stride
Peg 22h25'59" 28d22'28"
Jessicas Tulip
Leo 11h26'35" 16d27'35"
Jessica's Willum in the Sky
Cnc 8h49'52" 31d16'7"
Jessica's Wishing Star
Ari 2h25'49" 25d20'38"
Jessica-S2
Uma 11h36'4" 55d14'39"
Jessicca Miller
Lib 14h38'14" -14d44'48"
Jessie
Umi 15h36'56" 85d13'33"
Jessie
Uma 14h2'4" 58d21'20"
Jessie
Uma 10h28'44" 67d6'44"
Jessie
Sco 16h51'15" -41d7'25"
Jessie
Crb 16h8'32" 26d44'34"
Jessie
Leo 11h42'28" 25d59'55"
Jessie
Tau 4h23'42" 27d48'23"
Jessie
Gem 7h20'19" 29d23'8"
Jessie
Cnc 8h35'55" 8d3'57"
Jessie
Leo 11h1'18" 9d15'20"
Jessie
And 1h39'22" 41d29'46"
Jessie
And 0h4'40" 44d25'7"
Jessie
Cas 0h8'52" 56d23'40"
Jessie
Cam 4h10'26" 59d1'22"
Jessie
Per 3h4'34" 54d55'38"
Jessie
And 0h19'38" 46d24'11"
Jessie
And 23h7'31" 41d35'36"
Jessie
And 23h32'46" 38d44'38"
Jessie & Alana Wilson
Cru 12h14'27" -63d40'25"
Jessie and Joe 10-26-03
Aqr 22h53'21" -15d19'46"
Jessie and Nic
Uma 11h9'59" 38d27'5"
Jessie Ann Koegel
Uma 9h43'28" 51d25'0"
Jessie B
Tau 5h8'33" 17d56'17"
Jessie B. Bishop
Vir 13h17'53" 5d24'57"
Jessie Barlow
Sco 17h56'40" -31d23'50"
Jessie Bell
Ari 3h20'38" 29d21'26"
Jessie Bellard
Lyn 6h23'57" 55d16'46"
Jessie Blue Elfgifu
And 23h32'54" 41d54'23"
Jessie Brunner
Psc 23h16'31" 6d59'47"
Jessie C. Smit
And 23h31'31" 42d39'49"
Jessie Cable
Her 17h59'36" 21d45'48"
Jessie Campbell Moen
Uma 10h40'14" 53d56'43"
Jessie Caroline Weidman
Cas 1h57'3" 63d2'54"
Jessie Catherine you are MY forever
Vir 12h43'48" -9d3'58"
Jessie Catriona Rose Meek
Lyn 8h7'14" 36d31'52"
Jessie Clifton
Cru 12h5'53" -56d29'23"
Jessie Cole
Psc 1h22'49" 16d17'54"
Jessie Cronan
Uma 10h30'14" 62d47'18"
Jessie D. Wilson
Oph 16h6'8" -3d44'23"
Jessie Dreyer (Porage)
Dra 19h5'13" 61d37'50"
Jessie Elaine Edgar
Vir 11h46'12" -5d0'41"
Jessie Ewing
Cap 21h0'50" -21d51'6"
Jessie Frost
Uma 11h35'26" 47d59'30"
Jessie Fuji Hamelin
Cas 23h44'34" 55d44'0"
Jessie Fyfe Macmath Bungarz
Ari 1h57'40" 22d10'10"
Jessie Heywood
Sco 16h46'31" -41d23'5"
Jessie Hiltz
Psc 1h7'6" 32d47'42"
Jessie in the sky with diamonds
Cnc 9h3'59" 30d47'39"
Jessie Ivey
Aqr 22h13'9" -1d39'40"
Jessie James George
Leo 11h8'26" 20d38'8"
Jessie James Goldsmith
Cnv 12h43'18" 39d44'13"
Jessie & Jerame
Crb 15h21'25" 31d10'54"
Jessie Jo Boswell
Uma 10h41'15" 47d21'21"
Jessie & Joe
Cyg 20h25'51" 48d34'53"
Jessie & Keith Herron
Lyr 18h37'28" 26d21'25"
Jessie Kientzy
Uma 11h28'20" 41d11'50"
Jessie Labree
Cas 23h17'55" 54d26'42"
Jessie Lee
Cnc 8h12'16" 10d27'35"
Jessie Lee Greenfield
Tau 4h26'53" 22d33'49"
Jessie Lee Meyer
Lib 14h50'54" -0d42'2"
Jessie Lee Reagan
Tau 5h22'29" 27d11'35"
Jessie Leigh Mitchell
Cas 22h59'9" 58d11'24"
Jessie Lin Bush
Crb 15h26'16" 28d18'26"
Jessie Lou
Tau 3h30'0" 26d12'24"
Jessie Lou S. Gatchalian
Vir 15h8'39" 0d51'38"
Jessie Lovano
Ori 5h30'44" 3d12'1"
Jessie Lovie Watt
Uma 11h20'21" 33d19'12"
Jessie Lynn
Lyr 19h11'1" 27d0'8"
Jessie M. Miller
Lib 15h23'55" -5d32'10"
Jessie M. Stimpson (Giggles)
Cyg 21h22'12" 33d45'14"
Jessie Mae
Uma 13h51'23" 50d54'45"
Jessie Major
Uma 11h16'9" 31d21'5"
Jessie Marie
Leo 9h48'17" 21d58'11"
Jessie Marie Baker
Leo 11h49'47" 10d33'54"
Jessie Marie Ellison
Cas 0h46'18" 54d16'2"
Jessie Marie Evans
Sgr 17h58'14" -17d16'13"
Jessie May - 09.01.1985
Cru 12h28'39" -58d52'25"
Jessie Michelle Kaas, 20 March 1984
Uma 9h59'44" 64d31'50"
Jessie Mikami
Dra 18h33'23" 61d49'55"
Jessie Miller
Ari 2h5'5" 18d25'30"
Jessie Moonwitch
Lup 16h4'37" -35d2'5"
Jessie Murray
Vir 13h27'44" 5d42'30"
Jessie My Shining Star
Peg 22h59'0" 9d52'29"
Jessie Nicole Lusero - How Special?
Uma 10h49'5" 45d35'35"
Jessie Pépin
Sco 17h22'47" -42d25'55"
Jessie Preston
Ari 2h44'54" 16d37'6"
Jessie R. Schutte
Lmi 10h11'48" 29d51'17"
Jessie Raye
Leo 11h14'5" 14d11'43"
Jessie Rene Russell "Pooky"
Tau 5h3'14" 24d2'0"
Jessie Revel
And 23h7'7" 35d24'53"
Jessie Rose Pierick
Leo 9h41'30" 26d8'22"
Jessie RubenStar
Sgr 18h31'10" -20d3'42"
Jessie Sampson "AKA" Atailia
Cas 0h17'27" 52d20'16"
Jessie Savin
Uma 10h20'57" 47d45'15"
Jessie Sclater
Cnc 8h47'31" 27d15'57"
Jessie Smith
Uma 14h27'54" 61d54'4"
Jessie Snyder McCormack
Leo 9h41'48" 32d35'32"
Jessie Starr
Cnc 8h18'20" 14d53'46"
Jessie "stinky" Wendell
Aql 20h11'40" 13d48'27"
Jessie Taylor-Wood
Cas 0h54'21" 56d8'38"
Jessie The Princess Kebert
And 23h16'57" 47d47'33"
Jessie Tinman
Peg 23h44'18" 14d51'34"
Jessie Wood
Uma 14h22'30" 56d13'5"
Jessie Y. Pumphrey
Cyg 19h53'52" 31d43'44"
Jessie Yeh
Cnc 9h9'54" 32d32'7"
JESSIE Z
Cap 21h45'1" -9d30'54"
Jessie & Zane Walls
Lib 15h9'47" -20d42'1"
Jessie, I love you bro.
Per 4h20'36" 44d2'8"
JessieCamp
Leo 9h28'16" 11d9'6"
Jessie-James
Cyg 20h8'43" 34d36'3"
Jessie-Lynn Menna
Vir 13h23'44" -17d30'28"
Jessie's Angel
And 23h5'8" 41d57'20"
Jessie's Boy - Jim McColl truly loved
Uma 11h8'53" 40d25'47"
Jessie's Guardian Night Angel
Uma 8h43'6" 64d52'59"
Jessie's Ninja Star
Aqr 22h45'27" 0d36'40"
Jessie's Piece of heaven
Leo 10h14'18" 13d23'54"
Jessie's Smile
Uma 11h47'51" 35d13'38"
Jessie's Star
And 0h27'42" 43d29'6"
Jessies Star
Cnc 8h24'7" 11d40'3"
Jessie's Star
Psc 1h45'53" 25d42'33"
Jessie's Star
Vir 13h24'4" -7d8'6"
Jessie's star
Lib 14h53'38" -1d51'38"
Jessie's Star
Sco 17h58'14" -39d57'34"
Jessie's Star LMF
Ari 2h18'13" 26d24'55"
Jessie's Wish
Peg 21h38'24" 10d40'52"
Jessika
Psc 23h58'51" 7d11'50"
Jessika Ann Hayes
Cyg 19h35'43" 34d6'56"
Jessika J Curti
And 0h32'54" 35d18'21"
Jessika Manning
Cas 23h34'28" 51d40'36"
Jessika Sicignano
Aur 6h35'55" 38d50'51"
Jessika Twila Chapman
And 1h52'11" 41d47'37"
Jessikah and Ross Forever
Cyg 20h20'46" 34d10'37"
Jessikia
And 0h17'21" 27d17'23"
Jessilyn Lacy Sellars
Lac 22h33'36" 50d41'37"
Jessinko
Uma 9h56'53" 66d20'29"
Jessipoo
Uma 8h9'4" 60d23'56"
Jessirome
Cyg 19h15'8" 53d52'58"
Jessi's Angel
Lmi 9h35'49" 36d2'36"
Jessi's Light
Lyr 18h42'4" 31d6'4"
Jessi's Place in My Heart
Vir 13h57'56" -19d44'12"
Jessi's Star
Vir 13h3'28" -4d54'21"
Jessi's Star
Psc 1h11'45" 6d17'30"
Jesska Blue
Cap 20h17'2" -19d19'45"
Jess's Oasis
Col 5h30'47" -42d23'28"
Jess's Shadow
Cru 12h11'9" -64d24'39"
Jess's Star
Cnc 8h9'55" 7d50'3"
Jesstopher 10-22-2006
Dra 16h45'27" 63d58'19"
Jessy
Lib 15h15'26" -8d26'0"
Jessy
Ori 5h24'39" 2d5'20"
Jessy
Ari 2h25'4" 26d56'23"
Jessy
Ari 3h14'33" 28d36'57"
Jessy A. Wilson
Per 2h16'11" 51d13'49"
Jessy Bouvet
Sco 16h43'29" -42d34'39"
Jessy Critten
Uma 10h40'34" 47d17'12"
Jessy de Jussy
Umi 10h22'31" 88d15'49"
Jessy Girl
Leo 11h17'47" 11d10'0"
Jessy Kay
Lmi 10h33'55" 37d55'7"
Jessy Keating
Vir 14h41'24" 2d14'0"
Jessy Soto
Ori 5h48'39" 5d41'55"
Jessy-Bee
Psc 23h39'56" 5d5'18"
Jessye
Lyr 18h45'28" 28d42'1"
Jessye Serene Callahan
Aqr 22h2'38" -7d47'16"
JessyJames
Cyg 21h19'58" 44d59'23"
JessyLove
Ori 5h53'19" 18d27'21"
Jessy's
Tau 4h23'7" 9d4'16"

Jessy's Star Cyg 21h24'32" 39d48'39"
Jesteen Gutierrez Lyn 8h33'51" 53d44'1"
Jester Psc 0h43'5" 16d11'12"
Jestina Joan Ferrante Lyn 7h12'51" 51d0'9"
Jestine Lib 15h21'0" -17d31'2"
Jesus Cep 22h0'6" 68d26'20"
Jesus Cyg 20h25'8" 34d26'27"
Jesus Ari 3h20'42" 19d36'13"
Jesus Ori 5h43'29" 6d39'51"
Jesus and Shirley Carreon Uma 9h33'52" 64d6'3"
Jesus Ayala-Pena Her 17h9'5" 30d57'59"
Jesus & Carmen Cyg 20h26'18" 43d47'47"
Jesus Christ, The Son of God Cep 22h46'50" 64d5'30"
Jesus Esteban Valenzuela Sgr 19h12'15" -22d14'57"
Jesús Gamez Echevarria Aur 6h8'44" 53d31'57"
Jesus Gustavo Roman Gem 7h17'0" 22d33'57"
Jesus L. David Aql 18h51'27" -0d5'16"
Jesus Lucio Jr. Uma 9h1'46" 57d0'35"
Jesus Marcial Arellano Boo 14h42'23" 26d2'46"
Jesus(mijo)SuarezTeAmod eaquiaPluto Uma 9h39'18" 43d40'0"
Jesus Nieves Ori 5h43'48" -5d27'56"
Jesus Rodrigo Fonseca Leo 9h43'21" 14d2'16"
Jesus S. & Tokuko Linda Dizon Cyg 20h2'17" 31d49'17"
Jesus Sablan Her 17h41'57" 26d52'18"
Jesus, Jocelyn, & Jesse Saucedo Sgr 19h18'22" -17d13'26"
Jesusy Mon 6h39'48" 8d37'24"
Jesyca Nicole Datri Lyn 8h20'57" 34d2'46"
Jet Per 3h6'44" 38d24'26"
Jet Erykah Cru 12h36'28" -55d53'19"
JET - Jake Ethan Thomas Her 18h30'5" 16d48'24"
Jet & Joyce Sco 17h53'47" -40d10'36"
Jet Lindahl Lyn 7h42'20" 53d55'41"
Jet Van Der Wall Cas 0h14'43" 52d48'59"
JeTaime Cyg 20h35'35" 44d43'19"
Je'taime FEAEL Uma 13h7'18" 62d30'37"
Jet's Star Peg 23h27'44" 24d34'39"
Jetsca Uma 11h40'52" 54d23'24"
Jett Cas 23h46'58" 52d43'11"
Jett and Grace Jamison Uma 11h19'7" 29d31'24"
Jett Boyd Ori 5h41'51" 2d13'33"
Jett Kane Dilger Sgr 18h6'5" -27d3'5"
Jett Mulrooney Cru 12h49'3" -60d47'39"
Jett Odachowski Cap 20h16'0" -11d6'41"
Jett Stephen Humphrey Aug 1, 2004 Leo 11h34'36" 1d12'54"
Jetta Bug Sco 16h10'59" -29d2'25"
Jetta Darrow Cyg 21h38'17" 53d6'57"
Jette Jorgensen Leo 11h17'4" 18d0'40"
Jetti Uma 8h14'6" 66d56'34"
Jetti Umi 14h11'16" 71d59'23"
Jett's Star Aql 19h39'21" 15d4'56"
Jettstar Blue Lyr 18h42'48" 40d36'28"
Jevon Cordell Blakemore Uma 12h2'38" 57d56'50"
JEW Forever and Ever SLW Cyg 21h30'29" 52d12'1"
JEWA Boo 14h42'28" 14d19'10"
Jewel Crb 15h17'27" 27d6'1"
Jewel Crb 15h32'9" 28d12'9"
Jewel Uma 8h37'14" 52d34'30"
Jewel 21 Ori 5h50'18" 6d3'17"
Jewel Black Uma 10h41'3" 40d2'38"
Jewel Eliza Hancock Gem 6h49'40" 31d5'40"
Jewel for Dara Umi 17h8'5" 81d23'30"
Jewel of Mellisa And 0h31'23" 36d13'27"
Jewel Okwechime And 23h43'37" 35d41'15"
Jewel Over Vegas: The Jenal Nicole And 2h10'20" 42d20'24"
Jewel65 Pho 0h41'30" -40d58'55"
Jewelium Ori 5h20'50" 8d23'40"
Jewell Leo 10h2'59" 12d28'31"
Jewell And 23h54'55" 41d32'48"
Jewell A Province Dra 17h50'42" 55d48'12"
Jewell Ann Lagger Lyn 8h7'22" 34d33'44"
Jewell Elder Ori 5h44'15" -2d39'7"
Jewell Mae Lewis Tau 4h36'49" 12d11'49"
Jewell Price Covey Cap 20h35'22" -10d53'54"
Jewell Williams Vir 13h17'40" 4d44'54"
Jewels Ari 2h48'1" 14d55'36"
Jewels Tau 4h41'44" 24d0'14"
Jewels Ori 6h8'58" 17d56'7"
Jewels Uma 11h30'43" 36d3'2"
Jewels Cyg 21h53'34" 49d41'19"
Jewels Sgr 18h57'0" -34d38'49"
Jewels Alexa Davis Cap 21h43'28" -12d53'43"
Jewels Bai Image Peg 23h1'20" 30d37'0"
Jewel's Light of Love Aql 19h57'45" -6d19'58"
Jewels & Sean hbk Sgr 19h26'17" -22d41'11"
Jewelya Marie Brown-Perras Umi 14h41'42" 67d59'42"
Jewelz Sco 16h58'20" -38d41'14"
Jewl of the Night Sky Cap 20h23'21" -9d28'21"
Jewl of the Universe Tau 4h39'46" 27d31'30"
Jewlz Feris And 1h10'36" 43d4'6"
Jezabelle Lib 14h54'3" -3d31'6"
Jezebel Uma 11h43'48" 42d6'43"
Jezreel Karen Ninalga Psc 1h14'50" 5d0'59"
Jezza's Star - Jeremy T Price Lib 15h51'59" -18d49'44"
Jezzika Vir 13h19'44" -13d36'54"
J-F Deotto Per 2h50'23" 49d36'17"
JF & KS Cyg 20h12'48" 51d23'45"
JFM,DOC,1939 Tau 5h51'5" 27d33'0"
JFS 4.23.50 Tau 3h34'17" 17d50'16"
JG & JW - Together Even When Apart Sgr 19h50'4" -13d13'29"
jgatmk Cyg 21h57'42" 39d42'0"
JGiardina_Uncle Jim_<3CLASS of 2005 Cyg 19h42'34" 29d5'39"
JGL, Inc. Vir 13h9'29" 5d8'21"
JGL-CHOC-CHIP Per 3h45'3" 47d43'6"
JH33 Cas 22h58'27" 57d39'49"
Jha'Nae Brie Stoffer Lmi 10h13'9" 31d52'53"
Jhane Mary Pappenfus Uma 8h54'1" 55d56'18"
<-= JHBH: a diamond in the sky => Vir 13h43'14" -9d36'50"
Jhenn And 23h45'3" 33d55'57"
JhiemBecito Tau 4h35'13" 30d10'15"
Jhina Crb 15h45'33" 36d11'23"
JHK Jackie Harden Kimbrough Tau 4h33'5" 14d32'39"
Jhonn Ramirez Dra 18h32'7" 55d51'54"
Jhony Evaristo Cap 21h27'29" -10d41'11"
JHS Uma 9h24'33" 62d24'36"
JHS,shine bright for all to see Umi 14h25'30" 78d35'15"
JHT Oneiros 1 Crb 16h2'43" 29d16'17"
jhugs Crb 15h44'3" 26d57'54"
Ji Hyun Lee Psc 1h8'20" 25d39'10"
Ji yi Tau 4h20'15" 29d24'40"
Ji Ying Lyn 7h6'44" 59d24'33"
Jia And 1h40'19" 36d28'46"
JIA Jinbo Uma 8h21'23" 61d48'21"
Jia Kennie Blackwell And 1h55'38" 38d47'14"
Jian Xing Liu Gem 7h6'52" 20d56'24"
Jiang Lijuan Sco 17h18'4" -35d35'31"
Jianna Barone Sartorio Psc 1h14'57" 30d50'51"
Jiap F. Sco 16h20'42" -25d34'6"
Jiaur Rahman Ori 6h16'1" 13d20'55"
Jiaxue Hu Her 18h42'45" 19d47'19"
Jibran Cep 22h56'52" 59d35'44"
JiCE Uma 13h41'4" 49d15'7"
Jiderra Angelica Barry And 23h4'7" 48d11'28"
Jido And 0h46'43" 35d41'7"
Jiefen & Dan Forever in the Cosmos Sco 17h20'1" -42d32'34"
Jigga Gem 7h36'23" 26d25'25"
Jiggerlorius Leo 11h47'14" 24d55'28"
Jiggers Uma 11h45'24" 45d10'14"
Jiggo Col 5h27'20" -39d48'10"
Jihae Tau 4h29'20" 5d37'48"
Jihane Cas 0h30'0" 50d57'34"
Jihane Bachir, "Jano" Lyr 18h51'28" 26d20'12"
Jihwan Uma 9h33'43" 45d9'21"
JIHYE the true love of Matt Psc 0h47'12" 16d39'49"
JiJi Cornish Volz Aqr 21h59'39" -9d19'9"
Jijounette Del 20h46'39" 5d26'15"
JIL Sco 16h10'23" -20d36'33"
Jil Renee Maddox Cap 20h55'43" -21d35'51"
Jila, Haheh, & Omid Fard Uma 11h16'9" 69d15'4"
Jilana & Rudy Crb 15h20'47" 32d22'36"
Jilbearte Marie Lep 5h13'9" -11d16'49"
Jilda Henderson Lyn 9h6'53" 40d45'57"
Jilda Marnell Ari 1h59'46" 13d29'5"
Jill Leo 11h3'18" 5d44'42"
Jill Cnc 8h55'19" 10d38'2"
Jill Vir 12h45'14" 10d22'6"
Jill Gem 7h23'22" 27d30'16"
Jill Gem 7h18'39" 25d45'3"
Jill Ari 1h50'59" 23d5'50"
JILL Lyn 8h9'14" 44d42'4"
Jill And 23h41'53" 45d49'24"
Jill Lyr 18h47'6" 44d9'27"
Jill Lyn 8h33'37" 48d16'29"
Jill And 2h22'12" 45d45'40"
Jill And 2h11'55" 45d4'0"
Jill Aqr 22h52'54" -20d28'23"
Jill Cas 2h24'52" 74d53'31"
Jill Cam 5h0'58" 62d6'16"
Jill Uma 8h47'33" 53d19'12"
Jill Sco 16h32'35" -35d55'6"
Jill A. Gallagher Cap 21h10'56" -19d8'44"
Jill A. Jones And 1h44'47" 41d7'30"
Jill A. McConnell—Exceptional Mom! Cas 0h37'40" 55d29'42"
Jill A Nelson Peg 22h5'47" 19d44'48"
Jill Allison Del 20h44'37" 15d41'8"
Jill and Jeffrey, Panitch/Snyder Cyg 21h34'34" 49d16'39"
Jill And Lenny - Forever And Always Cyg 20h11'16" 43d46'1"
Jill and Matt's 1 Year Anniversary Lyr 18h31'9" 35d55'16"
Jill and Shay's Star Cap 20h25'46" -18d50'51"
Jill Anderson Lyr 18h41'3" 30d3'49"
Jill Angelia Campbell Gem 6h51'0" 19d27'55"
Jill Angelina And 0h46'2" 45d42'46"
Jill Ann And 23h35'52" 38d20'15"
Jill Ann Camarce Natividad Oph 16h44'37" -0d48'11"
Jill Ann Johnson Leo 10h44'37" 8d6'47"
Jill Ann Karwales Fink Gem 6h49'30" 27d30'32"
Jill Ann Kemble Cyg 21h25'16" 53d9'9"
Jill Ann Lewis-Rowder Com 12h33'45" 27d42'52"
Jill Ann Stall Vir 13h23'32" 12d26'7"
Jill Anne St. Denis Cap 20h55'33" -25d7'39"
Jill & Anthony September 18, 2004 Cyg 20h54'11" 46d46'51"
Jill Arshan Sco 17h51'34" -36d4'14"
Jill Ashley Cnc 8h50'48" 7d5'52"
Jill B. Wiscomb...Our Shining Star Cas 0h54'44" 54d0'48"
Jill Ball Cra 18h15'46" -41d3'59"
Jill Beechly Lib 14h42'44" -24d23'45"
Jill Bergeron Uma 11h10'1" 52d0'5"
Jill Betkowski Uma 11h10'11" 67d31'11"
Jill Biechler King Ori 6h8'49" 15d28'36"
Jill Blume Cas 23h4'18" 56d43'28"
Jill C. Paschal Vir 13h16'12" -21d40'4"
Jill Caitlin Cnc 8h28'13" 26d26'45"
Jill Catherine Adams And 0h9'46" 32d0'18"
Jill Christine Hurley Aqr 22h30'2" -2d6'47"
Jill Colleen Galbreath Cnc 8h54'0" 25d51'30"
Jill Constantine And 23h2'17" 47d22'10"
Jill Cooper Cas 0h16'37" 57d25'9"
Jill Crepaldi Her 18h23'42" 21d50'23"
JILL D. MARSHALL Vir 13h1'27" 10d0'54"
Jill DeGuilio Sco 17h15'9" -44d26'58"
Jill Dewey Cnc 8h21'50" 18d37'55"
Jill Diann Platt Gem 6h36'56" 25d9'7"
Jill Dianne Hardy And 1h18'24" 42d49'10"
Jill Dionisio Aqr 21h17'26" -8d50'31"
Jill Dooley Umi 15h50'21" 73d56'19"
Jill Easton Vir 11h43'14" 7d13'25"
Jill Ellen Polgar Uma 9h42'24" 48d44'37"
Jill Erica Combe Gem 6h25'46" 21d57'48"
Jill Erika McCary "Jem" Aqr 22h31'18" 1d5'35"
Jill Ford's Star Srp 18h7'19" -0d12'26"
Jill Frances Jackson Lyn 9h9'28" 38d45'38"
Jill Franks Cas 1h54'11" 69d21'34"
Jill Garnsey Mon 6h42'17" -0d25'27"
Jill Garrecht Cas 23h1'41" 53d37'28"
Jill Giesing Ari 1h53'37" 18d40'53"
Jill Gonzalez And 1h52'44" 38d43'14"
Jill Harpt Mon 7h1'1" 6d58'16"
Jill Headrick Cam 4h19'3" 54d43'32"
Jill Henrikson's Family Star Ori 5h14'37" -0d45'33"
Jill Ilene Montemuro And 22h59'32" 51d54'22"
Jill Jamgochian Lib 15h0'40" -22d6'51"
Jill & Jason Climer Umi 15h23'2" 75d51'31"
Jill & Jason - January 18, 2004 Uma 10h29'58" 44d7'12"
Jill & Joey 2006 Vir 12h39'11" -9d48'49"
Jill Jones * Silvers Umi 15h5'43" 68d53'53"
Jill Jose Crb 15h46'7" 35d2'13"
Jill Kathleen Porter Hutchison And 0h41'34" 26d41'50"
Jill Kathryn Harwell Cyg 20h11'57" 48d7'4"
Jill Kelsey Pup 7h45'40" -22d26'27"
Jill Kiley And 0h20'27" 42d39'10"
Jill Kleinfeldt And 0h50'57" 40d39'42"
Jill Kristine Murphy Leo 9h23'57" 21d40'52"
Jill Krol Aqr 21h39'14" 2d15'48"
Jill L. Allen-Margolis Cap 21h49'35" -24d12'38"
Jill Laura Creedon Ari 2h59'27" 22d22'31"
Jill Leah Ortega Psc 0h44'19" 10d45'8"
Jill Leanne Treen Her 16h30'44" 13d57'7"
Jill Lee Gem 6h42'5" 14d40'28"
Jill Leslie Narowetz Cyg 20h16'43" 38d5'59"
Jill Loree Mitchell Tau 5h12'53" 16d31'58"
Jill Lowen Cyg 19h37'28" 31d51'38"
Jill Lynne Kelly Sgr 18h37'18" -34d42'43"
Jill M. Castle Cap 20h26'20" -12d43'4"
Jill M. Espelin Sgr 18h28'13" -26d52'28"
Jill Maddison Feigelman Tau 4h21'3" 26d29'54"
Jill 'Maiden' Ashurst And 0h1'35" 37d0'34"
Jill Margrethe Fritzo Ari 3h4'32" 22d32'11"
Jill Marie Cnc 8h10'14" 10d5'5"
Jill Marie Cet 1h7'27" -0d28'57"
Jill Marie AKA Tracy LaVonne And 2h10'48" 45d55'30"
Jill Marie Armas Sgr 18h38'32" -18d15'42"
Jill Marie & Michael Robert Gibbon Per 2h35'59" 56d10'31"
Jill Marie Milosch Uma 8h47'16" 55d23'31"
Jill Marie Nyenhuis And 0h35'25" 45d28'21"
Jill Marie Schisler Cyg 20h3'42" 51d33'16"
Jill Marie Varney Ari 3h27'57" 27d24'26"
Jill Marlene Uma 9h55'46" 42d12'26"
Jill McShane Cas 1h29'7" 66d24'25"
Jill Meredith Panitch Lyn 7h7'26" 59d57'37"
Jill Michelle Conger Uma 11h48'8" 45d37'30"
Jill Milberg Cap 21h45'36" -19d36'9"
Jill Miura Uma 11h10'16" 62d14'49"
Jill Mongey Psc 0h54'29" 29d15'35"
Jill Nagy Anderson Cas 23h8'50" 55d35'20"
Jill Negrete Sco 16h7'5" -18d25'35"
Jill Oester ~ My Girl Cyg 20h52'12" 48d34'38"
Jill Owen Lib 15h16'17" -6d19'52"
Jill P McCarthy's Star Cyg 19h29'11" 54d5'16"
Jill Patricia Davis Uma 9h26'35" 43d4'9"
Jill Pianelli Cyg 21h48'45" 36d8'52"
Jill Pollhammer Ari 2h14'12" 23d49'7"
Jill Rae Enright Leo 10h35'59" 17d37'8"
Jill Rene Gem 7h0'4" 23d8'50"
Jill Robin Sgr 19h31'0" -35d33'30"
Jill Roelie Uma 13h16'43" 58d49'24"
Jill Rutherford Cyg 21h45'56" 48d26'0"
Jill Ryan Crb 15h32'59" 32d4'13"
Jill Sandie- RD's Shining Star Lyn 8h50'26" 41d56'44"
Jill Sarah Evans - Just Jill Lmi 10h11'40" 32d31'26"
Jill Sarapata Cas 0h15'41" 55d57'56"
Jill Sheena Lockyer Cyg 19h56'54" 56d58'9"
Jill Shmoo Goslee And 1h16'54" 42d21'20"
Jill Sierra And 0h15'37" 37d10'52"
Jill Simmons And 23h20'59" 47d21'19"
Jill Stephens Mon 7h35'54" -0d25'34"
Jill Stiner Cap 20h12'46" -10d2'0"
Jill Susan Page And 23h51'56" 48d7'50"
Jill Suzanne Ari 3h12'12" 30d7'50"
Jill Suzanne Atchison West Lib 15h6'8" -0d29'18"
Jill Suzanne Toney Uma 14h22'49" 57d5'15"
Jill the sexy Lyn 8h27'6" 39d8'45"
Jill Thomas Lyn 7h43'25" 53d45'36"
Jill Thompson And 0h34'36" 37d28'29"
Jill Trnavsky Psc 1h1'10" 11d51'27"
Jill Tulloch Lib 15h57'8" -18d16'30"
Jill Tuttle Lyn 8h14'8" 42d36'21"
Jill Van Kempen Psc 1h15'8" 27d3'22"
Jill Vest Gem 7h33'53" 26d26'13"
Jill Violet Hogan Uma 12h53'15" 59d45'51"
Jill Walsh Tau 5h54'19" 26d11'56"
Jill Wignall And 0h4'4" 32d48'38"
Jill Wikstrom Gem 7h6'40" 20d51'3"
Jill, Alan & Delana Tri 2h24'40" 35d36'15"
Jill, For Now, For Always, Forever Umi 14h46'49" 79d42'19"
Jill, The Fire In My Heart Dra 19h36'20" 70d28'3"
Jill, the shining star of FCC And 2h23'3" 38d1'40"
Jillana Grant Raddall Tichenor Ori 5h55'57" 18d23'23"
Jillana Rae Treadway And 2h30'56" 44d12'28"
Jillanne DeBari And 1h27'51" 40d4'31"
Jillasta Gem 6h33'56" 20d50'32"
JillB Sco 16h12'46" -11d36'27"
Jillene K. Psc 1h5'9" 32d42'54"
Jillene Rose And 0h15'6" 46d44'19"
Jilli Bean Ari 2h36'5" 24d34'30"
Jillian Gem 7h39'47" 17d31'23"
Jillian Ari 3h16'17" 26d35'26"
Jillian Leo 10h42'26" 7d8'41"
Jillian Cnc 8h47'41" 12d56'29"
Jillian Ari 2h32'34" 18d25'41"
Jillian Psc 1h18'2" 15d28'42"
Jillian And 23h52'29" 40d53'21"
Jillian Lyr 18h49'38" 37d48'0"
Jillian Gem 7h14'59" 33d26'56"
Jillian And 1h23'20" 44d41'13"
Jillian Cap 20h15'5" -17d49'51"
Jillian Umi 15h28'8" 82d55'52"
Jillian Uma 10h25'21" 67d32'47"
Jillian Cas 1h37'8" 60d50'49"
Jillian Pho 0h18'31" -48d57'21"
Jillian Sco 17h30'51" -39d45'37"
Jillian Cap 21h13'43" -27d7'45"
Jillian 6375 Sgr 19h11'26" -12d35'27"
Jillian Alexander Brower Peg 22h54'39" 12d31'6"
Jillian Alexis Katke Cap 20h13'53" -12d12'59"
Jillian Alise Hansen Uma 10h58'0" 50d30'9"
Jillian Amanda Lazzara And 23h46'35" 48d40'18"
Jillian Amistoso Cyg 20h55'43" 47d10'18"
Jillian Ann Psc 0h30'21" 17d23'24"
Jillian Anna Mahen Uma 12h4'4" 46d21'36"
Jillian Anne O'Shaughnessy Cannon Ori 6h8'0" 19d53'28"
Jillian Aubrey Sgr 19h59'59" -25d0'10"
Jillian Beth Airington Aql 18h53'53" -0d31'25"
Jillian Braun Cam 8h26'17" 83d9'4"
Jillian C. And 1h7'43" 43d6'58"
Jillian*Caellyn*Jason*Loving Family Uma 10h6'19" 43d54'17"
Jillian Clute Crb 15h24'45" 30d57'47"
Jillian Dade Rice Cnc 9h2'55" 19d34'13"
Jillian Dawn Converset Ori 5h48'53" -2d13'39"
Jillian DeAnna Aaron And 2h32'29" 45d6'51"
Jillian DeCamp Vir 11h59'7" -8d21'54"
Jillian Denise Tarpley Psc 23h52'5" 4d51'45"
Jillian Elizabeth Charroud Vir 12h43'6" 4d36'41"
Jillian Elizabeth Savko Lyn 7h2'47" 57d6'25"
Jillian Ella Bauco Lib 15h55'56" -17d32'45"
Jillian Ellen Leo 9h30'54" 11d43'50"
jillian emmalexa Cyg 19h39'36" 34d52'37"
Jillian Grace Uma 9h34'32" 46d30'48"
Jillian Grace Mulholland Dra 16h13'44" 53d46'11"
Jillian Grace O'Connor And 23h16'29" 48d44'21"
Jillian Herzberg Uma 10h3'16" 59d25'2"
Jillian Honey Ayran Sco 17h27'51" -31d49'40"
Jillian Hoss Sco 17h42'46" -37d34'49"
Jillian Isabella Cra 18h44'54" -41d28'30"
Jillian Jamruszka Vir 13h10'24" -6d5'53"
Jillian Jewell Vir 11h48'24" -6d19'42"
Jillian "Jill" Erica Keating Cas 23h24'14" 56d16'21"
Jillian Joy Sagi Psc 0h36'30" 14d41'1"
Jillian Korgeski Cas 1h22'38" 55d14'33"
Jillian Kristine Boda Uma 11h33'39" 47d34'34"
Jillian Kristine Lacey Sco 17h49'24" -39d24'11"
Jillian La Rocque Star Gem 7h1'19" 16d53'49"
Jillian Lane And 1h16'33" 49d4'47"
Jillian Lang And 1h10'32" 35d56'16"
Jillian Leigh Harris Aqr 21h6'52" -12d40'43"
Jillian Leigh Sandrey And 23h14'48" 49d24'3"
Jillian Lynn Snyder Gem 6h52'22" 32d37'4"

Jillian M. Ross
Vul 20h40'23" 22d13'10"
Jillian Mae Hamrick
Agr 20h55'13" -8d29'43"
Jillian Mare
Agr 22h30'19" -2d55'1"
Jillian Margaret Parry
Agr 21h20'47" -1d38'52"
Jillian Mariah Klassen
Vir 12h14'25" 10d39'3"
Jillian Marie Fleury
Lyn 8h16'36" 39d3'55"
Jillian Marie Kyle
Vir 13h20'57" 7d39'7"
Jillian Marie Wisniewski
Mon 6h45'41" -4d54'19"
Jillian & Michael, Forever in Love
Cyg 20h16'50" 32d21'24"
Jillian Michelle Ann Bucola
And 1h59'1" 39d12'33"
Jillian Michelle English
Agr 22h12'1" -21d19'25"
Jillian Murray
Cas 23h59'55" 58d4'48"
Jillian my Angel
Leo 10h14'43" 25d45'19"
Jillian N. Conroy's Shining Beauty
And 1h20'1" 36d21'13"
Jillian Nichole Kirts
And 2h16'33" 43d33'49"
Jillian Olive Cook
Leo 10h51'20" 18d56'43"
Jillian Olivia
Oph 17h23'15" 8d16'24"
Jillian Pancini
And 0h43'54" 22d5'29"
Jillian Pasco
Lmi 10h56'35" 27d19'2"
Jillian Pauley
Dra 18h54'41" 76d49'40"
Jillian Peterkovich
Tau 5h59'13" 27d10'47"
Jillian Peterson
Dra 18h21'32" 52d44'18"
Jillian R. Joyce
And 0h39'29" 34d18'28"
Jillian Rachel Comeau
Her 17h34'24" 47d37'54"
Jillian Rae
Sco 17h37'45" -44d36'50"
Jillian Rae Courlas
Lyn 7h45'14" 51d20'19"
Jillian Rae Graham
Vul 19h52'14" 24d27'5"
Jillian Raye
Cap 21h45'19" -14d21'39"
Jillian Renee Lapolla
Cam 4h16'19" 66d28'41"
Jillian Robinson
Tau 4h24'45" 29d31'38"
Jillian Rose
Cyg 19h51'33" 36d4'40"
Jillian Rose Fadeley
Lyn 8h8'7" 34d40'0"
Jillian Ryan Collins
Leo 9h40'7" 26d28'39"
Jillian S. Wilay
Gem 6h50'25" 34d12'0"
Jillian Spano
Cas 0h43'16" 65d32'5"
Jillian Star
And 0h41'19" 41d1'15"
Jillian (star) Anesta
And 23h17'32" 52d15'8"
Jillian Stevenson
Lib 15h46'39" -9d25'4"
Jillian Stover
Dra 18h49'25" 73d26'15"
Jillian Sutton
Cap 20h42'17" -25d57'37"
Jillian Terese Heinbockel
Lmi 9h48'4" 39d1'49"
Jillian (the David and Aryn Star)
And 0h25'56" 43d32'21"
Jillian Tufano
And 23h24'11" 47d20'37"
Jillian Vito
Psc 1h16'38" 31d56'5"
Jillian Wickett
Tau 5h40'15" 27d51'52"
Jillianne
Lyn 8h8'37" 37d44'55"
JillianRobin
Cyg 20h4'3" 48d8'29"
Jillian's Buttercup Shines Forever
And 0h33'51" 26d34'43"
Jillian's Heaven
Psc 0h11'32" 0d6'41"
Jillian's Jewel
Cap 20h58'4" -19d49'3"
Jillian's Jewel
Uma 9h16'17" 52d24'8"
Jillian's Jewel
Lyn 7h20'47" 46d31'7"
Jillian's Light
Cap 21h48'21" -14d51'31"
Jillian's Skylight
Uma 10h43'34" 55d45'28"
JilliansMysticalMystery
And 2h6'31" 43d42'12"

JilliBean
Sco 16h9'26" -25d58'26"
Jillie Ree
Lac 22h54'2" 38d35'7"
Jillien Rachelle Barnhill
Umi 17h11'31" 76d56'27"
Jillima F. 27.12.2003
Cas 0h5'23" 52d30'34"
JillJill
Ari 2h23'22" 12d10'12"
Jillou
Ara 17h46'18" -48d30'56"
Jill's 65th Birthday Star
Cas 22h59'38" 54d41'10"
Jill's Grace
And 0h21'59" 24d45'53"
Jill's Guiding Light
Cyg 21h59'50" 51d22'39"
Jill's Radiance
Cnc 9h21'21" 26d52'47"
Jill's Star
Tau 5h34'55" 21d48'58"
Jills86
Sgr 18h9'43" -28d38'56"
Jillstar
Cas 23h58'6" 60d6'12"
Jilly
Tau 5h22'44" 19d43'1"
Jilly
Tau 4h21'34" 26d48'5"
Jilly 33
Gem 7h26'31" 31d30'11"
Jilly and Steve's Wedding Day
Cyg 20h12'30" 52d57'32"
Jilly Bean
Vir 11h43'21" -3d3'15"
Jilly Bean
Crt 11h35'22" -16d25'45"
Jilly Bean
And 1h56'55" 36d27'3"
Jilly Bean
And 23h52'22" 45d24'12"
Jilly Bean
Cas 0h52'32" 52d7'3"
Jilly Bean Cleveland
Crb 16h7'42" 34d50'51"
Jilly Bean Smile
Lyr 18h52'32" 36d16'22"
Jilly Hartman
Uma 13h40'24" 52d15'40"
Jilly Love
And 2h17'51" 48d30'21"
Jilly Willy
Sgr 18h11'1" -17d24'41"
JILLYBEAN
Vir 13h14'53" -4d10'52"
JillyBean
Leo 10h16'40" 23d31'33"
Jillybean
Del 20h24'8" 9d40'16"
Jillybean Denise, Love of my Life
Leo 9h22'37" 10d56'49"
Jillyn Rachelle
Peg 21h48'54" 9d28'44"
Jillynn
Psc 0h54'33" 4d30'3"
Jillynne
Ari 2h9'39" 22d56'35"
Jilly's Twinkle
Crb 15h36'52" 38d26'31"
Jim
Cyg 20h53'34" 47d14'27"
Jim
Psc 23h29'48" 7d57'35"
Jim
Leo 9h22'15" 12d58'42"
Jim
Sco 17h56'54" -36d35'38"
Jim
Cap 21h26'56" -21d56'6"
Jim
Agr 20h42'15" -9d3'35"
Jim
Uma 14h3'53" 56d29'22"
Jim
Lyn 6h23'56" 56d37'25"
Jim
Uma 9h53'51" 72d31'42"
Jim & Albina
Umi 16h8'6" 76d23'23"
Jim Alexander
Uma 11h58'42" 64d54'13"
Jim and Alice Linton
Agr 21h58'17" 1d5'47"
Jim and Ann Hustead
Sge 20h4'34" 18d30'56"
Jim and Betty Murphy
Cyg 20h46'54" 33d19'3"
Jim and Bev Forever
Tel 18h14'20" -49d4'46"
Jim and Carol's Star
Ori 5h34'43" -1d15'0"
Jim and Carolyn Whelan
Crb 15h44'9" 27d55'56"
Jim and Cheryl
Lib 14h50'29" -9d21'51"
Jim and Debbie Blalock's Lucky Star
Cyg 20h31'45" 41d18'27"
Jim and Debby's Anniversary Star
Umi 16h53'47" 76d52'45"

Jim and Donna Imer
Umi 15h30'56" 74d22'24"
Jim and Elizabeth McAlister
Cyg 21h24'19" 38d45'17"
Jim and Evelyn's Star
Cas 1h19'45" 67d17'22"
Jim and Ginny Cheney
Sgr 18h16'19" -17d36'19"
Jim and Helen McCaul
Sge 20h5'13" 17d40'28"
Jim and Jean
Gem 7h9'42" 22d44'30"
Jim and Jen
And 23h15'9" 44d51'47"
Jim and Jen
Boo 14h20'51" 51d37'52"
Jim and Jo Hicks
Per 3h31'4" 44d26'12"
Jim and Joanne Barbati
Cyg 20h20'17" 54d46'41"
Jim and Judy
Her 16h57'54" 24d22'15"
Jim and Judy Star
Uma 14h13'38" 58d14'0"
Jim and Julie
And 23h28'47" 42d30'40"
Jim and Kari Kohlhaas
Uma 11h18'36" 38d30'55"
Jim and Lauren Forever
Uma 10h34'1" 67d26'48"
Jim and Lauren Merryfield
Crb 15h47'50" 26d31'53"
Jim and Leeann Iacino
Cyg 19h50'44" 32d4'53"
Jim and Lisa Wood
Cyg 20h39'47" 44d57'55"
Jim and Marla West
Uma 11h32'10" 56d47'32"
Jim and Mary Jo Short
Lyn 7h56'29" 54d56'50"
Jim and Nancy
Uma 10h33'21" 64d1'16"
Jim and Nicole
Cyg 20h38'4" 37d2'41"
Jim and Noreen Perrone
Cyg 20h51'30" 40d24'23"
Jim and Norma Groves
Umi 16h1'22" 75d47'0"
JIM and NORMA THOMAS
Eri 4h22'44" -0d52'56"
Jim and Rosemary Calvert
Gem 7h11'47" 28d4'8"
Jim and Rosemary Whitright
Cyg 21h36'1" 55d17'33"
Jim and Sandie Stewart
Cyg 19h47'35" 37d34'23"
Jim and Sandy Parrish
Lyr 18h32'45" 36d6'18"
Jim and Sharon Angus Forever Love
Agr 22h48'17" -6d51'15"
Jim and Sharon Miller
Uma 9h44'2" 66d39'26"
Jim and Shelly Forever
Uma 10h36'53" 48d51'44"
Jim and Tera Gall
Dra 17h33'44" 58d15'37"
Jim and Terri Raleigh
Cyg 20h24'19" 47d46'51"
Jim and Tina DeHarts Superstar
Vir 13h18'16" -20d55'58"
Jim and Wanda Gerlits
Cyg 20h5'59" 42d23'2"
Jim & Arlene
Cyg 21h38'23" 48d15'14"
Jim Azbell
Cnc 8h25'26" 12d48'23"
Jim Baker
Per 3h17'44" 51d4'23"
Jim & Barbara Raden
Cyg 21h44'35" 52d38'24"
Jim Becker
Cen 13h28'5" -37d19'53"
Jim & Beth Duff - Wedding Star
Vir 14h38'22" 3d52'27"
Jim "Big Daddy" Ball
Uma 9h41'59" 47d19'19"
Jim Biles
Aql 19h43'51" -0d8'2"
Jim "BO" Jackson "Dad"
Uma 9h40'53" 72d51'17"
Jim Bob
Ori 6h13'15" 3d43'42"
Jim Bond
Her 17h39'36" 36d21'50"
Jim Bonde
Tau 5h59'21" 27d52'7"
Jim & Bonnie Leitzel
Lyr 19h6'49" 42d27'36"
Jim & Bonnie Mush
Uma 8h34'13" 71d30'42"
Jim Bowe
Boo 14h44'33" 27d38'24"
Jim Brannies
Tau 5h13'14" 23d9'25"
Jim Bremer
Her 16h35'25" 34d44'45"
Jim Bridgewater
Boo 14h57'47" 51d41'6"
Jim Bubba Morris Wells
Boo 14h20'0" 16d13'3"

Jim "Buckets" Manning
Per 3h14'5" 52d9'11"
Jim Buckless
Aur 5h39'31" 49d29'13"
Jim "Budweiser" Furletti
Umi 15h49'41" 73d42'41"
Jim Cairns
Tau 5h30'34" 26d32'48"
Jim Cannon
Ori 6h8'15" 13d40'57"
Jim & Carol Rink
Dra 13h20'4" 66d4'59"
Jim & Catherine DeFranceschi
Cam 13h18'12" 78d54'28"
Jim Chappine ~ AllMyLove, AllMyLife
Cep 23h38'40" 74d58'23"
Jim Chiu
Cep 22h38'16" 77d51'43"
Jim & Christine's Lucky Paris Star
Cyg 20h14'9" 50d13'23"
Jim & Claire Cunningham
Cyg 19h48'44" 33d21'1"
Jim Clark
Boo 14h40'55" 40d30'27"
Jim Clarke
Uma 11h39'20" 31d56'55"
Jim & Claudia Sesso ~ Fifty Years
Cyg 19h37'30" 31d31'9"
Jim Cloud
Gem 6h52'3" 21d27'23"
Jim Collas
Sgr 19h58'14" -43d6'49"
Jim Crow
Cep 0h8'48" 66d59'51"
Jim D. Norwood Jr.
Her 16h29'28" 13d35'6"
Jim D. Ross
Cyg 19h25'59" 47d6'12"
Jim (Dad) Costa
Aur 6h31'21" 46d8'11"
Jim & Dalton,Bev,Jeff & Cindy Bond
Uma 10h18'19" 67d39'21"
Jim Danbert's Star
Crb 15h49'59" 28d35'4"
Jim Darr
Her 16h44'5" 15d22'49"
Jim Davis
Boo 15h37'29" 42d35'21"
Jim Deanne Hill
Cyg 19h50'9" 39d8'18"
Jim & Debbie Always
Uma 11h51'8" 57d17'44"
Jim & Debbie Hodder
Cyg 20h35'2" 58d57'48"
Jim Delebak
Cep 20h26'42" 60d25'39"
Jim DeMaio
Her 16h31'44" 46d4'5"
Jim & Diana Stum
Cyg 21h36'37" 45d47'43"
Jim & Diane Sheets
Her 17h53'40" 24d0'33"
Jim Dickens
Ari 2h31'57" 25d20'50"
Jim Dodge
Umi 14h56'19" 69d57'28"
Jim & Doris Sehm March 9, 1957
Uma 9h47'44" 68d45'7"
Jim & Dorthy Hardin
Uma 8h34'41" 62d56'59"
Jim Dudley
Aql 20h0'26" -0d29'11"
Jim E. Tysinger
And 1h45'11" 50d7'1"
Jim Engelhardt & Linda Proskow
Cyg 19h59'55" 52d21'37"
Jim & Enid
Cyg 20h40'46" 47d11'45"
Jim FiFi Forester
Ori 5h58'44" 21d57'2"
Jim Filgate
Tau 3h58'45" 30d9'55"
Jim Franklin
Aql 20h1'40" 12d8'15"
Jim Freeman
Gem 7h1'58" 21d45'22"
Jim G. Wolf
Dra 19h19'25" 59d16'25"
Jim Gallagher
Uma 10h6'41" 50d49'41"
Jim Galli
And 0h34'2" 43d31'47"
Jim & Gayle Bouskos
Cen 13h37'54" -37d51'36"
Jim Glass
Sgr 18h16'19" -19d35'42"
Jim Godbey
Gem 7h17'4" 27d29'43"
Jim Gometz's Shining Star
Leo 9h45'2" 28d50'50"
Jim Grady
Cep 0h15'59" 71d26'36"
Jim Guida 50
Ori 5h55'15" 8d47'48"
Jim Guilfoil
Psc 1h25'18" 27d26'32"
Jim & Gwen
Tau 4h34'44" 20d9'1"

Jim Gwilt's Star
Tau 4h39'5" 5d42'59"
Jim Hair, Sr./Granddaddy
Gem 6h25'6" 20d50'57"
Jim Hamrick
Sco 16h26'31" -28d13'31"
Jim Hardway
Dra 17h32'39" 60d34'22"
Jim Hardy
Ori 5h42'14" -0d39'58"
Jim Headley
Ori 5h50'31" 2d9'4"
Jim Heber
Per 3h51'22" 37d6'14"
Jim Hockensmith
Tau 5h34'42" 27d53'13"
Jim Hosmer
Uma 12h28'44" 60d6'19"
Jim I'm amazed by You. Love4ever
Boo 14h42'3" 32d31'44"
Jim Jenkins' Universal ONENESS Star
Her 16h33'47" 38d41'29"
Jim & Jess Bennett
Cyg 20h47'22" 42d20'52"
Jim Jill Kelly Emily & Saturn Davis
Uma 9h5'26" 56d25'8"
Jim Johnson
Uma 10h39'16" 53d45'32"
Jim Joyce "Big Heart"
Lyn 8h9'12" 53d14'38"
Jim & Joyce Vinita
Lyr 18h40'47" 31d0'0"
Jim "JP" Potter
Leo 9h56'22" 20d28'57"
Jim Kalis
Uma 11h11'50" 30d24'33"
Jim & Katherine Bedenko
Uma 10h36'5" 64d25'22"
Jim Kavanagh
Her 18h47'2" 15d28'12"
Jim Keeney
Psc 23h34'44" 0d7'48"
Jim Kenny
Ori 5h57'10" 18d22'40"
Jim & Kim Walt
Sge 20h5'56" 18d38'48"
Jim Kirwan
Lyn 8h0'13" 57d13'12"
Jim Klaren
Boo 14h28'53" 15d13'55"
Jim Kubiak
Lib 14h29'56" -24d22'56"
Jim L. Umberger
Uma 10h31'42" 51d6'43"
Jim LaBau
Lyr 18h55'19" 39d13'57"
Jim & Laurie, Once In A Lifetime
Leo 10h29'40" 23d2'4"
Jim & Leah - The sparkle of my life
Sco 17h28'23" -35d33'59"
Jim & Leanne Diefenbach's 19th Year
Sgr 18h39'11" -26d48'37"
Jim & Leslie Everitt Steve & Alison
And 0h48'44" 35d48'21"
Jim & Lexi
Umi 13h22'21" 86d52'9"
Jim & Linda Thorndike
Per 3h8'8" 47d3'23"
Jim & Liz
Cru 12h23'50" -61d44'56"
Jim & Lois Constant
Vir 11h44'30" -2d11'50"
Jim Long
Per 3h13'23" 55d42'7"
Jim Love
Her 16h46'48" 43d17'45"
Jim Lundgren
Sco 17h0'17" -40d12'43"
Jim Lupo
Uma 10h56'4" 45d53'9"
Jim MacMahon
Ori 5h34'46" 4d25'48"
Jim & Margie Barefield
Uma 9h33'49" 64d11'33"
Jim & Martha Calvinperez
Lib 15h48'0" -18d29'52"
Jim Mauro & John Seidel "21+"
Cap 21h25'43" -23d27'17"
Jim McBurney - John 3-16
Ori 5h44'5" 11d19'55"
Jim McCann's Eastern Star
Per 4h14'27" 45d0'58"
Jim McCune
Her 17h26'10" 24d43'11"
Jim McGuire 1924 - 2000
Per 4h41'53" 37d36'48"
Jim Milan
Leo 10h26'1" 12d49'5"
Jim Miller
Cma 6h58'49" -15d53'36"
Jim Moise
Sgr 19h16'42" -22d15'44"
Jim Muller
Vir 12h26'48" 0d29'15"
Jim Murphy
Psc 0h52'40" 17d15'40"

Jim Murphy
Uma 13h57'0" 61d35'5"
Jim Murphy Legacy 97
Ori 5h21'45" 0d35'3"
Jim Musgrove
Uma 10h0'36" 48d47'33"
Jim My Angel From God
Umi 15h26'28" 70d30'58"
Jim Nance
Crb 15h30'10" 27d53'43"
Jim Napier
Lyn 8h28'3" 50d26'58"
Jim & Nell Turners Shining Spirits
Gem 6h50'38" 22d59'59"
Jim Nerney
Vir 14h43'15" 2d53'1"
Jim Niederpruem
Lib 14h41'19" -24d33'25"
Jim Nothhouse
Per 3h21'34" 36d7'13"
Jim O'Leary
Cep 1h52'8" 81d58'38"
Jim O'Shea
Gem 6h52'11" 32d37'29"
Jim Overstreet
Lyn 7h11'26" 56d57'56"
Jim Packer
Uma 10h36'48" 40d11'22"
Jim "Papa" Geib
Psc 1h37'19" 10d47'49"
Jim Pari, my eternal love
Her 16h14'14" 13d58'24"
Jim & Pat's Lucky Christmas Star
Umi 15h56'49" 78d34'42"
Jim & Patty Standring
Uma 8h54'32" 53d17'3"
Jim Paxton
Agr 22h28'45" -3d47'4"
Jim Pendry
Gem 7h37'7" 22d5'58"
Jim Perisho
Cep 22h17'28" 57d34'49"
Jim Pfitzinger
Per 2h51'40" 51d18'59"
Jim Piechowiak
Leo 10h24'39" 21d3'15"
Jim "Poppy" Arnold
Cep 22h2'14" 67d30'17"
Jim Profitt
Per 2h45'43" 52d54'34"
Jim Pugash Family Star
Uma 9h40'1" 55d48'44"
Jim & Rachael Waterman
Vir 14h19'4" 1d38'25"
Jim Ray Rainwater Jr.
Sco 16h18'50" -8d38'34"
Jim Reichers
Ori 5h53'50" 0d24'50"
Jim Richard Nelson
Cep 23h25'13" 77d25'58"
Jim Riley
Uma 9h53'36" 54d58'53"
Jim & Rita Purkis
Cyg 20h3'6" 48d4'8"
Jim Ritterhoffs Quixotic Star
Per 2h20'53" 56d9'34"
Jim Roberts - My father & friend
Her 17h53'38" 47d51'6"
Jim & Rod McMullen-Oksuita
Psc 0h52'59" 3d5'49"
Jim Rosebrough
Cap 21h42'26" -13d46'7"
Jim & Ruby Klimek
Uma 9h36'27" 63d48'27"
Jim Schnoering
Boo 14h24'43" 45d57'58"
Jim Shapy
Uma 10h53'4" 54d38'11"
Jim & Sharon
Lyr 18h47'55" 40d3'59"
Jim Shook
Aur 5h39'33" 49d30'8"
Jim Southern
Aql 18h46'24" 9d18'7"
Jim Sr.
Boo 14h19'57" 19d4'24"
Jim Stratton
Tau 4h32'20" 18d57'37"
Jim Stuberg
And 2h18'59" 47d26'20"
Jim & Susan Tyte
Lib 14h23'20" -14d41'57"
Jim & Susie Maus
Cyg 20h48'50" 30d28'37"
Jim "Sweetheart" Manlutac
Uma 8h36'46" 53d15'47"
Jim Tam
Aql 19h46'59" 12d36'6"
Jim & Tarah Lu's Girl
Cas 23h33'29" 58d8'14"
Jim Teller
Her 17h38'57" 20d15'43"
Jim & Terri Leftwich
Cyg 19h55'35" 30d15'52"
Jim & Tina
Lyr 18h52'40" 44d8'20"
Jim&Tina Whitworth 30th Anniversary
Uma 8h15'8" 72d20'33"

Jim & Tommy, Father & Son I Luv U
Dra 19h48'16" 65d58'11"
Jim Tozzer
Leo 10h8'55" 14d58'37"
Jim Trears
Peg 0h2'34" 24d35'18"
Jim Uchytil
Ori 5h26'34" 13d45'10"
Jim "Villanova" Hodges
Uma 8h31'9" 61d16'25"
Jim W Clause Happy Birthday
Per 3h18'23" 48d3'13"
Jim Walls
Uma 9h5'5" 49d54'0"
Jim Warren
Cep 20h29'7" 60d13'47"
Jim Warych
Per 3h6'50" 42d13'14"
Jim Watt
Per 4h17'26" 45d13'44"
Jim "Weasel" Serrine
Her 16h51'26" 23d10'8"
Jim Weaver's Smokin' Hot Super Star
Ser 18h17'27" -13d43'43"
Jim & Wendy Marshall
Umi 14h59'11" 67d58'2"
Jim Wilson
Ari 1h53'29" 17d48'20"
Jim Wilson
Aur 5h38'1" 50d3'2"
Jim Woolway
Ori 4h52'6" 11d45'43"
Jim, A Little Is Enough
Umi 18h24'37" 87d4'7"
Jim, Joy, Jay Raynor
Uma 11h58'34" 47d30'31"
Jim, Kay, Jay and Taylor Chitty
Cap 21h0'6" -17d2'38"
Jim, Kelly And Austin Partin
Tri 2h22'14" 35d54'35"
Jim, Kerri, Mary Crescent, Anna
Per 3h14'7" 31d35'12"
Jim, TLOML
Per 3h26'26" 47d47'38"
Jima
And 23h5'11" 40d22'7"
JimaLea
Umi 17h2'52" 79d31'33"
Jiman Khosravan
Ari 3h23'43" 28d55'27"
JimAnni Nuestra Estrella Eternal
Cyg 20h49'4" 35d36'18"
Jimaria
Uma 10h54'57" 60d43'43"
Jimber
Leo 10h22'39" 22d27'51"
Jimbo
Gem 6h31'15" 25d55'11"
Jimbo
Vir 13h7'19" 2d0'11"
Jimbo
Psc 1h13'10" 20d47'10"
jimbo
Per 3h26'52" 52d49'9"
Jimbo
Cap 21h42'9" -17d36'3"
Jimbo Stewart
Tau 5h27'42" 21d11'14"
Jimbo The Clown
Per 2h43'2" 56d17'47"
Jim-Bob Acker
Leo 9h41'26" 13d22'56"
Jimbo's Big Honkin' Star
Her 16h23'8" 23d35'35"
JimCheri 40
Uma 11h49'38" 60d59'55"
Jimena Villalonga Abascal
And 2h9'24" 42d32'25"
Jimenez Familia
Uma 9h40'45" 53d41'53"
Jimer McCusker
Boo 15h7'48" 40d56'30"
Jimi Scozzaro III
Umi 15h21'22" 74d5'15"
Jiminet
Sco 17h42'41" -42d22'14"
Jimi's DeDe
Leo 11h11'55" 22d3'37"
JimJim
Cma 6h54'37" -20d10'40"
JimJimJimmityJim
Cap 21h13'43" -15d50'8"
Jimm
Cap 20h56'1" -26d13'42"
JimMariannaGreen
Sge 19h26'17" 17d28'9"
Jimmer Shimmers
Agr 22h53'10" -12d11'32"
Jimmette O'Briant
Her 17h14'44" 30d39'19"
Jimmie and Delberta Steele
Her 16h26'10" 5d15'15"
Jimmie Anne Haisley
Dra 18h32'36" 60d10'29"
Jimmie B. Horton
Aql 19h50'13" -0d31'32"
Jimmie Bear
Ori 5h52'29" 20d21'13"

Jimmie C. Lynch Jr.
Her 16h33'5" 29d25'38"
Jimmie Diane Bolinger Pyburn
Vir 13h43'5" 7d7'25"
"Jimmie" James F. Rogers
Vir 11h51'1" 9d7'27"
Jimmie Jones
Sco 16h13'33" -18d26'19"
Jimmie Joseph Glynn
Gem 6h39'8" 19d46'40"
Jimmie K. Carey
Her 16h44'7" 47d55'37"
Jimmie L. Saxton
Cnc 9h17'32" 32d35'13"
Jimmie Lee Campbell
Sco 16h40'50" -32d12'53"
Jimmie Nelda Bishop
Tau 4h18'3" 7d45'8"
Jimmie Newsome
Dra 19h19'21" 64d36'19"
Jimmie R.
Ori 5h58'24" -0d40'50"
Jimmie R. Bailey, Jr.
Leo 10h38'13" 18d49'45"
Jimmie Tom McDonnell
Her 17h19'27" 29d25'25"
Jimmie&WayneTabor50thAnniversary
Leo 11h25'11" 14d12'39"
JimmieFrank
Cyg 19h55'21" 45d31'25"
Jimmy
Her 17h8'14" 33d36'13"
Jimmy
Her 17h10'7" 33d54'8"
Jimmy
Tau 4h8'58" 16d8'7"
Jimmy
Ori 5h23'36" 2d33'23"
Jimmy
Tau 4h27'2" 7d19'6"
Jimmy
Tau 3h28'6" 2d27'48"
Jimmy
Gem 6h32'36" 24d26'17"
Jimmy
Boo 14h50'22" 22d27'39"
Jimmy
Her 18h18'18" 15d5'23"
Jimmy
Aqr 22h12'41" -4d42'35"
Jimmy
Psc 23h14'22" -2d2'3"
Jimmy
Sgr 18h13'59" -20d8'12"
Jimmy
Aqr 20h51'5" -9d3'35"
Jimmy
Cap 20h29'27" -12d34'13"
Jimmy
Lyn 7h19'51" 58d1'0"
Jimmy
Uma 11h38'3" 59d59'22"
Jimmy
Sgr 17h54'39" -29d29'27"
Jimmy
Cru 12h44'48" -55d50'58"
Jimmy and Amanda Gleason
Cyg 20h14'13" 55d29'9"
Jimmy and Angel Forever
Sco 16h56'24" -42d8'11"
Jimmy and Carolyn Pike
And 1h41'33" 49d2'41"
Jimmy and Cathy Hunter
Cyg 20h8'54" 59d17'34"
Jimmy and Jenny and Cole Potter
Ori 5h27'45" 10d18'52"
Jimmy And Laura Two Souls One Heart
Aqr 23h2'40" -5d29'28"
Jimmy and Mandi's
Psc 1h7'59" 14d20'49"
Jimmy and Mary Adair
Cyg 19h37'20" 40d36'13"
Jimmy and Natalie's Star
Cyg 20h13'13" 34d52'46"
Jimmy and Rachel Murphy
Cyg 20h32'42" 37d23'33"
Jimmy and Shawna Morris
Ori 6h13'24" 16d15'10"
Jimmy and Shelley
Ori 4h54'51" 7d35'4"
Jimmy Antony Wade
Uma 10h37'25" 49d1'16"
Jimmy Barnette
Her 16h35'8" 38d26'14"
Jimmy Bateman
Aur 5h50'8" 53d14'11"
Jimmy Ben
Her 18h5'3" 18d0'15"
Jimmy Bond & Mariya Grygorash
Uma 10h37'39" 60d12'53"
Jimmy Bond & Mariya Grygorash
Umi 14h17'52" 70d42'39"
Jimmy Boney
Her 18h15'1" 23d48'54"
Jimmy Byrd
Lmi 9h36'8" 38d58'42"
Jimmy Cancilleri
Gem 7h19'0" 24d54'1"
Jimmy & Caroline
Eri 4h28'48" -1d43'12"
Jimmy Chambers
Ari 2h14'18" 25d31'31"
Jimmy Charles Parker
Her 16h41'56" 6d45'48"
Jimmy Cline Franklin
Ori 5h33'23" 3d14'26"
Jimmy Dale Marrs Jr.
Ori 5h38'58" 6d26'8"
Jimmy Dale Mayes
Per 3h9'16" 53d18'3"
Jimmy Dawn Carruth, Jr.
And 0h24'25" 25d6'13"
Jimmy Dean Coombs - Daddy
Her 17h18'40" 28d4'58"
Jimmy & Debbie Essex
Cyg 21h33'54" 37d28'55"
Jimmy DeCecco "Checkers"
Uma 8h24'13" 67d24'42"
Jimmy Doyen Cox
Sgr 17h45'18" -19d58'47"
Jimmy Doyle Pugh II
Tau 5h50'19" 17d15'7"
Jimmy Forrester - Gentleman Pirate
Vel 9h19'40" -47d12'49"
Jimmy Fred
Her 18h6'50" 47d45'46"
Jimmy Fullan
Per 3h11'42" 53d53'51"
Jimmy Gee - Eternal Optimist
Gem 6h39'11" 14d41'50"
Jimmy Glidewell
Cap 20h57'51" -18d48'21"
Jimmy Griffiths
Uma 11h6'44" 36d43'31"
Jimmy H King
Her 16h42'41" 35d49'3"
Jimmy Hahn
Ari 2h8'4" 23d14'40"
Jimmy Hayes
Her 17h12'10" 27d19'32"
"Jimmy James" James M Rowlands
Ori 6h21'6" 10d44'47"
Jimmy Jenkins - Superstar
Cap 21h37'4" -9d5'14"
Jimmy & Jessie McLaren
Uma 9h16'14" 57d20'25"
Jimmy Jim
Uma 10h16'25" 68d17'15"
"Jimmy Joe" Johnson
Cap 20h31'54" -14d50'47"
Jimmy Joe Montcalm
Del 20h41'23" 15d49'24"
Jimmy Jr. 7/10/1990
Umi 16h21'39" 75d49'53"
Jimmy & Katie St. John for Eternity
Her 16h49'1" 35d33'24"
Jimmy Lee
Gem 6h20'52" 18d40'37"
Jimmy Lee
Lib 15h33'10" -18d47'13"
Jimmy Lee Taylor
Gem 6h23'55" 27d40'3"
Jimmy LeGrande
Her 17h41'22" 16d40'49"
Jimmy/Lissa
Her 17h0'13" 29d23'26"
Jimmy Loves Cyndi
Cas 1h21'35" 68d55'46"
Jimmy Loves Katie Forever & Always
Cyg 21h30'3" 38d42'9"
Jimmy Loves Monica For All Eternity
Cnc 7h59'57" 17d10'7"
Jimmy Loves Rachel
Vir 13h14'5" -19d49'8"
Jimmy M. Casanova
Her 17h22'58" 15d15'49"
Jimmy & Marisa
And 23h0'23" 48d42'40"
Jimmy Miller - Star Maker
Her 17h29'1" 21d31'27"
Jimmy/Monique
Sgr 18h37'1" -27d36'55"
Jimmy Newman & Mary Muller
Cyg 19h38'47" 39d3'43"
Jimmy O'Keefe
Tau 5h56'27" 25d1'37"
Jim-my one & only true love
Psc 23h47'57" 6d24'50"
Jimmy Osiek
Psc 1h48'29" 9d30'40"
Jimmy & Patty Forever Shining Star
Lyr 19h15'18" 26d33'30"
Jimmy Pelliccia, Jr. 2/15/04
Psc 0h20'24" 15d8'44"
Jimmy & Phyl Kaplan
Crb 15h34'56" 28d31'32"
Jimmy Pines
Ori 5h49'18" 1d24'37"
Jimmy Pitt
Cnc 9h16'54" 30d22'34"
Jimmy Pollard
Ori 5h46'52" 2d19'51"
Jimmy Powell
Uma 12h1'43" 37d51'14"
Jimmy R. Jamison's Bright Spot
Dra 17h39'40" 64d41'3"
Jimmy R. Reade
Her 17h47'43" 47d53'17"
Jimmy Ray
Cap 21h37'5" -23d34'6"
Jimmy Reeder
Per 2h21'1" 54d47'27"
Jimmy Robertson
Leo 10h18'3" 15d39'39"
Jimmy Ryan
Ori 5h56'2" 14d16'5"
Jimmy & Sarah
Cnc 8h31'13" 25d43'43"
Jimmy Setrakian
Her 16h36'42" 34d53'0"
Jimmy Sharkey
Vir 13h56'44" -0d53'9"
Jimmy *SKY* Ludacka
Her 16h24'2" 23d53'20"
Jimmy Spann
Aqr 23h47'29" -8d47'28"
Jimmy Spurs
Per 3h15'48" 44d3'57"
Jimmy & Stephanie
Lib 15h19'59" -7d23'52"
Jimmy & Sue Bedard
Cyg 21h36'29" 41d0'58"
Jimmy "The Butcher" Smith
Psc 1h19'49" 31d34'58"
Jimmy The Cokeman Nelson
Cyg 21h46'25" 52d34'40"
Jimmy THE DEAL McNeil
Per 2h33'6" 56d12'59"
Jimmy The Drummer
Tau 3h33'44" 9d52'28"
Jimmy+Therese
Uma 13h48'27" 47d55'0"
Jimmy & Thuy
Her 16h52'36" 42d45'58"
Jimmy Trejos
Sgr 19h0'51" -21d42'27"
Jimmy Turner, Jr.
Tau 4h24'50" 12d39'53"
Jimmy~Vicki...ALWAYS & FOREVER
Del 20h52'19" 13d21'53"
"Jimmy Ward - Shining On"
Uma 11h30'30" 56d25'42"
Jimmy Wasko
Lyr 18h49'36" 34d13'4"
Jimmy Watkins
Uma 11h31'39" 46d52'31"
Jimmy Whitten
Cyg 20h9'59" 50d38'38"
Jimmy Wyatt
Cap 20h50'36" -17d45'55"
Jimmy Z
Cap 20h32'43" -15d54'11"
Jimmy, Georgie, and Gianna
Uma 11h23'40" 62d40'52"
Jimmy, Jenna and Jillian Staiti
Uma 10h56'50" 67d55'29"
Jimmy, Jimmy, Jimmy
Cnc 8h31'21" 32d50'17"
jimmy070443
Aql 19h45'10" -0d3'8"
JimmyChonga
Boo 15h6'41" 34d9'21"
Jimmy-Django
Ori 5h32'8" 10d1'31"
Jimmy's Candle
Sco 16h16'43" -8d37'54"
Jimmy's ilyvm
Lyn 8h29'6" 41d0'35"
Jimmy's Star
Per 3h41'54" 44d3'7"
Jimmy's Star 2/7/64 - 1/3/01
Aqr 23h55'6" -24d1'33"
Jimmy's Star 62847
Ori 5h54'23" 12d5'5"
Jimmy's Winking Star
Ori 6h14'50" -3d48'5"
Jimmy'sGirl
Sgr 19h36'25" -33d9'58"
Jimothy
Cap 20h26'0" -11d5'5"
Jims April Day
Leo 9h42'45" 27d24'55"
Jim's Baby Girl Crystal
Ari 2h50'43" 27d54'46"
Jim's Birthday Star
Crb 16h1'29" 37d15'59"
Jim's Bright Maggie
Lib 15h28'12" -24d50'19"
Jim's Dragonfire
Dra 18h56'21" 61d4'7"
Jim's Dream
Mus 12h32'34" -68d24'49"
Jim's Jewel
Ari 2h57'36" 25d11'17"
Jim's Kiss
Tau 5h51'41" 28d23'23"
Jim's Shining Paula
Tau 3h24'10" 19d2'0"
Jim's Skinned Knee Award
Ori 6h7'40" 5d54'58"
Jim's Special Star
Tri 1h57'54" 30d27'14"
Jim's Star
Cnc 8h52'2" 31d54'6"
Jim's Star
Per 4h39'53" 44d13'3"
Jim's Star
Aql 19h17'35" 0d24'0"
Jim's Star: Heaven Sent
Aur 5h46'43" 52d53'47"
Jim's treasure
And 23h10'15" 51d24'51"
Jim's Very Own Star
Lyn 7h54'52" 42d16'36"
Jim's Wishing Star
Cnc 8h44'49" 7d1'43"
Jimsik
Aqr 22h15'3" -4d38'23"
JimStar
Her 16h38'18" 11d58'35"
Jin Hwa Kim Ariel Hofman
And 1h34'59" 47d7'50"
Jin Star
Boo 15h24'18" 38d4'25"
Jina Lee
Leo 10h12'45" 10d31'56"
Jinan Ellahib
Cas 1h24'34" 51d54'18"
Jinarhett Duncan
Ari 2h16'33" 12d36'45"
Jinbo Jia
Cyg 21h33'9" 32d46'27"
Jinbo Sun
Gem 7h8'57" 23d11'13"
Jinda
Cam 4h9'28" 68d26'52"
Jindalay Garnett
Mon 6h43'18" -0d21'23"
Jinelle
Vir 12h34'54" 3d16'6"
Jinelle
Tau 5h34'52" 18d40'44"
Jinetian P143
Uma 11h42'2" 43d19'51"
Jinfeng Liu
Sco 16h40'59" -34d53'18"
Jing Tian
Uma 10h18'36" 47d40'0"
JingJing
Aqr 22h34'28" 1d51'12"
jingjing
Sgr 17h56'7" -29d5'57"
Jingy
Leo 9h33'45" 28d8'19"
Jinky
Uma 11h37'43" 37d53'5"
Jinna
Uma 10h35'5" 44d44'46"
Jinna
Aqr 21h40'17" 0d30'22"
Jinnae Nicolette
Cas 1h41'24" 65d3'18"
Jinnell Sharlee Killingsworth
Umi 13h43'2" 72d35'57"
Jinnie and Greg
Mon 7h17'43" -0d4'37"
Jin'Sen
Dra 18h32'32" 63d21'13"
JinShan
Vir 12h59'31" -14d7'56"
Jinshil
Uma 9h3'58" 57d16'11"
Jinwon Park
Uma 11h12'48" 40d42'43"
Jinx
And 0h15'34" 31d59'56"
Jinx
Uma 11h4'19" 61d2'0"
jinya
Gem 7h21'40" 34d26'56"
JIOIAMIA
Lyn 9h15'7" 37d57'58"
Jiovanna Meraz
Aqr 22h11'51" -2d41'0"
jirafa orgullosa
Boo 15h0'35" 13d4'14"
Jireh Martillo
Per 3h10'33" 53d37'50"
Jirene
Umi 13h40'42" 73d23'14"
Jirik a Bozenka Slukovi
Cyg 20h47'46" 45d35'18"
Jiritah
Uma 8h21'14" 70d37'36"
Jirsten
Uma 10h32'31" 63d22'54"
Jirtania
Del 20h36'43" 18d26'28"
JISSEL
Ari 3h26'0" 26d47'42"
Jistarrr
Uma 10h28'10" 61d2'33"
Jitka
Cam 3h47'47" 63d59'11"
Jitka Heinzova
Sco 17h13'4" -34d8'36"
Jitka, mein Stern
Uma 9h28'59" 41d31'45"
Jitterbug
Ari 2h16'5" 12d36'52"
Jitters
Gem 6h17'50" 21d40'29"
Jiu Jitsu
Del 20h41'7" 13d59'17"
Jivani Devi
And 1h58'20" 42d58'59"
Jiwon
Vir 13h21'23" 13d7'40"
Jiwon and Ustino
Cyg 21h34'52" 52d27'27"
JJ
Cyg 21h31'3" 48d56'5"
JJ
Gem 7h24'48" 31d7'8"
J.J.
Tau 4h4'22" 17d58'40"
JJ
Tau 3h43'31" 22d58'52"
J.J.
Lib 15h8'35" -8d3'33"
J.J. Bassetti
Vir 14h26'38" 0d44'38"
JJ "Boo-kie" Routh
Sgr 18h9'40" -31d2'1"
JJ Buster Brown
Leo 10h4'4" 22d48'49"
J.J. Carroll
Per 3h37'8" 44d55'55"
JJ Cole
Umi 14h24'44" 75d26'28"
J.J. Gaughan
Lyn 8h8'49" 42d36'8"
JJ Gus
Ori 6h17'46" 10d16'34"
JJ Mckibben
Lmi 10h19'21" 38d10'53"
JJ Minor
Crb 16h2'15" 29d34'40"
JJ & SS Forever
Umi 14h26'13" 73d47'29"
J.J. Wagener
Leo 9h31'31" 29d9'0"
JJ Woodruff
Sco 16h10'24" -15d57'47"
Jj,Dm,Ja,Bn MOCNY's Night Diamond
Cyg 19h48'50" 52d35'55"
JJ60
Dra 18h49'50" 55d14'31"
J.Jay Heiden loves Erin Troxell
Crb 15h36'8" 26d19'49"
JJB
Leo 10h22'30" 27d10'57"
JJC and KMH sail the sky forever
Leo 9h45'44" 26d0'1"
JJCWYAIL
Leo 11h28'22" 13d13'38"
JJDobosh
Cnc 8h20'47" 22d8'16"
JJF24
Sgr 19h57'38" -17d2'4"
JJHNAT
Uma 9h33'20" 41d50'52"
JJJ Schlacks
Sco 17h56'9" -40d16'42"
JJMELLO-27
Cap 20h43'24" -15d7'21"
JJMS40
Vir 12h16'44" 12d59'41"
JJNRS 16-2-96
Boo 14h15'34" 41d43'22"
jjoanbarone
Uma 8h52'14" 49d40'51"
JJRICATEU
Uma 9h40'38" 64d50'36"
JJ's Angel
Cyg 19h39'6" 38d4'37"
JJ's Angel
Uma 10h48'23" 43d19'44"
JJ's Burning Star of Love
Sgr 19h31'59" -31d40'39"
JJ's Star
Aql 18h45'9" -0d33'1"
JJTyr7701
Peg 21h37'50" 21d7'50"
JK 57-60
Aur 5h15'8" 29d37'35"
JK Forever
Cyg 21h45'52" 49d52'43"
JK Robideau
Lib 15h32'18" -25d0'12"
JK2162003637
Aqr 22h10'45" -16d11'29"
JKAK2005
Tau 3h55'32" 24d28'30"
JKButterfield8953
Leo 9h34'18" 26d30'21"
J'Kee
Uma 13h40'0" 57d11'6"
JKJ O'Connor Family Star
Lyn 7h46'10" 53d42'51"
JKM 4984
And 1h14'58" 39d11'24"
JkrSbp511
Sgr 18h23'30" -32d21'29"
JKSANDMDS
Uma 13h14'28" 57d47'58"
JL
Gem 7h41'53" 23d21'45"
J.L. Huff Jr
Aql 19h32'17" 4d33'32"
JL04251972
Tau 5h34'10" 27d10'27"
Jl4ine
Uma 9h57'32" 57d56'44"
jlarceri
Umi 14h50'10" 74d34'40"
JlarsenSB/CA
Lmi 9h59'1" 35d45'10"
JLB
Sgr 18h34'4" -33d40'50"
J.L.B.
Lib 15h46'13" -28d27'59"
JLB friend and healer
Ori 6h4'33" 21d15'59"
J.L.Caster75
Sgr 19h11'59" -15d57'46"
JLD 082105 MK
Cyg 20h29'47" 53d21'28"
JLE & MPV
Aql 20h10'57" 6d15'55"
JLECHD
Cap 21h37'19" -14d8'21"
J'lexhew Viencek
Sco 16h38'12" -25d8'1"
JLH1964
Cam 4h0'35" 70d26'2"
JLH918
Vir 12h3'35" 5d17'51"
JLH-JMC
Ori 5h27'19" 8d1'28"
JLHouk 1938
Ori 5h17'35" 11d41'59"
JLK 44
Per 4h45'36" 49d38'15"
JLK/PJE823
Umi 16h3'24" 75d57'57"
J.L.M.
Vir 13h15'25" -19d3'31"
JLM...24...06/05
Gem 7h45'24" 25d25'46"
JLMayne123056
Cyg 20h16'21" 49d45'29"
JLN & JMB
Cyg 20h31'42" 36d30'31"
J-Lo
Lyn 8h25'55" 45d46'43"
JLO2001
Uma 11h24'26" 51d4'34"
Jlona
Cas 0h11'20" 52d25'7"
JLP - 10/14/1981
Mon 6h48'44" 7d6'52"
JLPJLP2
Uma 13h45'58" 60d53'7"
JLS 143 6*5*03
And 0h32'45" 39d7'4"
JLT011064
Her 17h22'29" 46d49'15"
JLUV and the LOVE FAMILY
Ori 5h56'29" 17d40'43"
JLV
Oph 17h43'29" -0d5'50"
J-lynn
And 0h58'41" 38d29'46"
JM
Lyn 9h17'31" 44d34'26"
J.M. Castori
Sgr 19h50'2" -34d56'57"
JM & Christina Denaut
And 1h17'19" 45d54'2"
JM Corazon Para Siempre...xoxo
Lyn 7h1'42" 47d33'6"
JM Julier
Crb 16h0'45" 27d29'33"
JM & SM
Cyg 20h19'40" 53d53'30"
JM1080
Per 2h26'8" 55d24'41"
J-Mae
Aur 5h32'1" 39d15'12"
JMAR
Uma 9h22'0" 66d32'36"
J-ME
Ari 3h6'44" 28d53'18"
jmeasrutsa
Cas 0h38'53" 55d45'10"
JMG-1109
Uma 12h24'47" 56d14'39"
JMGreco02042004
Uma 11h54'27" 30d20'30"
JMH & MRW - PROM 2007
Cnc 8h35'24" 28d13'7"
JMH26
Sgr 17h57'4" -17d53'26"
J-michael
Her 17h6'28" 27d3'10"
JMKaminski
And 2h17'33" 47d36'18"
JMMoats_1984
And 23h39'55" 47d34'17"
JMO
Her 18h36'2" 17d44'48"
JMole
Sgr 20h4'10" -12d31'48"
JMorganK3
Tau 4h3'13" 22d25'51"
J.M.Ramsay
Her 17h57'20" 14d57'24"
JMS - Greatest Mother In Universe
Gem 6h44'39" 18d29'46"
JMS : Jolie Merveille Suprême
Aqr 22h8'0" -24d26'46"
JMSH-SCR
Lyn 7h28'55" 50d21'45"
JMss 2B 4 EVER
Peg 22h21'24" 19d15'14"
JMT
Aqr 21h17'30" 2d1'24"
JMT Doyle 091331
Vir 12h47'29" 6d17'17"
jmteas1
Psc 0h33'49" 19d5'37"
JMVH1024
Sco 16h11'44" -13d9'12"
jmwest 98
Crb 16h10'0" 28d24'5"
Jmyeriah Jamal
Lyr 18h54'18" 32d11'29"
Jn Phillipe & Renefred Maingrette
Aqr 23h40'42" -13d21'15"
JnA
And 23h23'30" 41d24'9"
Jnet
Lyn 6h52'2" 51d27'35"
JNG07
Sco 17h35'22" -42d30'55"
J-n-J Eternal Love 10-23-04
Ori 6h0'24" 21d42'37"
JnJ's
Cyg 20h14'43" 48d46'53"
JNM12800
Sgr 18h55'0" -29d57'53"
JNR08302001
Lyr 19h13'38" 26d35'26"
JnUoLalhE
Umi 16h51'10" 79d18'54"
Jo
Ori 5h34'24" -10d27'53"
"Jo"
Uma 9h24'28" 69d46'41"
Jo
Umi 13h12'35" 69d24'52"
Jo
Ori 6h2'41" 18d0'24"
JO
Cnc 8h44'30" 19d1'59"
Jo
Mon 6h29'40" 8d21'19"
Jo and Andy - 1st Anniversary Star
Cyg 21h43'11" 44d15'49"
Jo and Bill "5-22-55"
Cyg 21h13'49" 36d38'28"
Jo and Danny, my friend from heaven
Ari 2h20'2" 18d33'36"
Jo and Harry Good
Cyg 19h48'18" 35d47'26"
Jo Ann
Vir 11h39'24" -0d49'0"
Jo Ann Catherine Weickert Fluegeman
Umi 15h22'59" 71d10'44"
Jo Ann Elizabeth Peña
Gem 6h21'50" 23d47'40"
Jo Ann Fingerhut
Gem 7h29'41" 28d47'59"
Jo Ann G. Arnold
Cap 20h56'10" -24d32'56"
Jo Ann H. Austin
Cas 1h39'37" 63d5'13"
Jo Ann Justus
Crb 15h43'54" 29d9'17"
Jo Ann Kerr
Tau 3h31'23" 7d39'4"
Jo Ann Marie Von Elsner
Ari 1h49'32" 19d36'36"
Jo Ann Novotny
Sgr 17h57'43" -16d59'15"
Jo Ann Proctor
Lyn 8h20'7" 34d42'23"
Jo Ann (Smith) DeFeudis
Vir 12h48'39" 7d51'38"
Jo Ann Taylor
Leo 11h35'25" 25d22'40"
Jo Ann Tennant
And 23h41'37" 45d1'55"
Jo Ann Woerndle
Uma 11h37'1" 46d36'19"
Jo Anne Lantagne Beaulac
Crb 15h59'6" 35d39'14"
Jo Arends
Crb 16h10'7" 33d45'59"
Jo Baboo
Lac 22h21'26" 47d8'17"
Jo Beth Eads
Crb 15h29'7" 29d53'38"
Jo Bird
Uma 10h6'43" 45d5'47"
Jo Bondonese
Vir 13h28'28" 12d21'45"
Jo Brantley
Lyn 8h32'29" 35d5'20"
Jo Carol and Bobby Elliott
Cyg 19h51'21" 30d25'42"
Jo Clancy
Gem 7h32'13" 35d3'47"
Jo & Danny Hales
Cyg 21h29'8" 33d44'1"
Jo & Darren Galea
Col 5h30'8" -38d24'22"
Jo Deacon
Uma 13h31'43" 58d36'31"
Jo Dolan's Light
Cru 12h42'39" -56d28'35"
Jo Edna Boldin
Sgr 20h22'51" -41d53'33"
Jo Ellen
Ari 2h7'30" 25d50'54"

Jo Ellen and Wayne
Uma 9h35'18" 54d51'38"
Jo Ellen Baker
Cnc 8h18'31" 19d7'13"
Jo Ellen Savage
Vir 14h22'46" -3d43'19"
Jo Federonich
Cnc 9h0'31" 29d42'24"
Jo Feiler
Cet 1h13'45" -0d15'21"
Jo & Gareth
Cyg 21h45'27" 55d12'18"
Jo Greaux Dervan
Gem 6h44'46" 32d24'43"
Jo Hans Lang
Ori 6h18'8" 9d56'36"
Jo Hassan's Piece of Heaven
Cra 18h4'33" -37d8'5"
Jo Highland's Special Place
Cnc 9h9'12" 16d27'53"
Jo Jo
Ori 5h54'10" 12d20'53"
Jo Jo
Ori 5h27'0" 3d40'58"
Jo Jo
Lmi 10h37'12" 36d11'26"
Jo Jo
Cru 12h56'41" -59d12'55"
Jo Jo Krueger
Uma 11h40'34" 65d17'54"
Jo Jo Magoo
Tau 5h40'31" 26d28'38"
Jo Jo OrangeBlossom
Tau 4h56'29" 19d19'35"
Jo Joslin Gossett
Mon 7h22'19" -0d28'33"
Jo Katz - ROW 2005
Leo 9h27'21" 31d10'0"
Jo Lernout (Not From Hauspie)
Psc 1h24'22" 29d43'13"
Jo Macpherson
Cyg 19h37'35" 54d27'15"
Jo Majors
Tau 5h51'45" 24d48'26"
Jo Mama
Uma 13h52'33" 56d55'58"
Jo Mannes
Uma 12h14'9" 53d25'48"
Jo Marie Byrd
Lib 14h54'13" -8d32'48"
Jo Morast's White Star in the Sky
Ari 1h55'23" 18d32'6"
Jo Morgan Jackson
Uma 8h48'59" 49d1'32"
Jo & Murray 60th Anniversary Star
Cyg 20h28'47" 32d29'29"
Jo & Nath - 05/02/05
Cru 12h13'20" -61d39'0"
Jo Ruth Patterson
Lib 14h47'34" -4d9'42"
Jo Schwartz
Umi 14h45'50" 69d54'40"
Jo Shea
Vir 12h50'0" 7d56'31"
Jo So Bright
Uma 10h28'50" 64d43'19"
Jo Stimpson - Always In My Heart
Lyn 6h52'7" 53d55'17"
Jo the Dragon
Dra 11h22'42" 77d38'18"
Jo Van Meter III
Uma 10h49'20" 52d45'7"
Jo Wallace Schneider Thru Eternity
Uma 11h26'46" 60d15'52"
Jo, Everything happens for a reason
Uma 11h22'1" 53d9'49"
Joachim
Lyr 19h3'8" 27d29'10"
Joachim Bauch
Uma 8h50'8" 69d22'34"
Joachim Eduard Gustav
Uma 10h32'32" 72d8'50"
Joachim Gabka
Psc 0h36'34" 6d44'6"
Joachim Hacker
Uma 11h32'24" 35d55'51"
Joachim Handschug
Uma 11h58'21" 50d10'22"
Joachim Heinrich Hans Werner Barella
Cma 7h1'58" -17d58'14"
Joachim Janos
Uma 10h58'39" 33d30'13"
Joachim Kölschbach
Ori 6h23'18" 15d58'54"
Joachim Majowski
Uma 8h26'48" 69d11'52"
Joachim Poppek
Uma 9h5'17" 59d32'42"
Joachim Seifert
Uma 9h47'52" 56d10'46"
Joachim Stratmann
Uma 9h20'0" 55d27'55"
Joan
Uma 11h47'21" 61d32'12"
Joan
Crb 15h30'25" 27d13'39"
Joan
And 2h31'1" 43d30'6"
JOAN
Uma 11h19'3" 50d1'40"
Joan
Cas 1h18'58" 57d25'24"
Joan
Cas 1h26'17" 52d2'51"
Joan
And 1h51'13" 46d43'39"
Joan 1931 Didier for 75th
Aqr 22h9'57" -1d9'48"
Joan A. Boettcher
Lyn 9h23'59" 40d8'46"
Joan A. Gallichant
Sgr 19h36'40" -13d58'40"
Joan Allen
Crb 15h56'20" 34d51'31"
Joan and Carl
Vir 11h41'19" -5d42'26"
Joan and George Thompson
Per 4h48'2" 39d28'44"
Joan and Howard Jones
Uma 10h23'44" 71d35'46"
Joan and Jan Unger
Sco 16h9'52" -9d9'22"
Joan and John * Together Forever
Cyg 20h50'2" 41d26'46"
Joan and Justin, Together Forever
Cas 1h7'5" 61d33'3"
Joan and Linda
Lyn 7h59'52" 34d50'33"
Joan and Richard Forever
Cyg 21h55'25" 48d53'17"
Joan and Ross Hansen
Psc 23h59'23" 10d9'54"
Joan Angela
Tau 5h47'48" 21d8'46"
Joan Ann Olsen
And 2h24'53" 46d55'18"
Joan Ann Walsey
Sgr 17h58'34" -29d17'13"
Joan Aurelie Bengtson
And 1h45'46" 44d17'35"
Joan Avis Jepsen
Cas 0h11'56" 53d19'36"
Joan Barbara Turner
Uma 11h5'9" 33d33'42"
Joan Bebe Raffe
Leo 10h37'9" 14d45'56"
Joan Beh DeVries
Cnc 8h21'22" 22d15'16"
Joan Benninger
Cas 0h10'50" 54d12'20"
Joan & Bill
Gem 7h18'34" 19d13'58"
Joan & Bill Skedd
Cyg 19h52'31" 31d49'49"
Joan Bills
Cas 23h32'58" 53d14'5"
Joan Blair Morones
Cyg 20h41'26" 51d21'54"
Joan Bouthillier
Uma 10h38'2" 51d3'28"
Joan Bradley
Cyg 19h48'37" 37d36'3"
Joan Burrows
Uma 9h47'59" 62d21'12"
Joan C. Bosshammer, AKA Toofy
Crb 16h9'54" 32d15'42"
Joan C. Flaherty
Ant 9h35'47" -25d54'9"
Joan C. Leodori
Uma 9h22'38" 50d15'24"
Joan C Niemer
Tri 2h2'13" 32d11'27"
Joan Caffrey
Psc 0h13'43" 5d33'31"
Joan Callaway
Her 17h30'24" 32d3'27"
Joan Campopiano-Powe
Cnc 9h9'19" 27d31'33"
Joan Carol
And 1h19'53" 38d11'31"
Joan Carol Thomas Merkel
Cas 1h37'58" 65d31'34"
Joan Carta
Cas 1h8'26" 63d33'11"
Joan Cetin
Cas 1h18'22" 64d28'15"
Joan & Charles Boyle
Uma 8h29'40" 61d1'17"
Joan Christine Niemer
Crb 15h39'43" 29d19'30"
Joan Christnacht
Uma 11h42'53" 59d54'34"
Joan Christopherson
Cyg 20h24'14" 47d50'5"
Joan Ciura
Mon 6h51'22" -0d53'28"
Joan Claire
And 1h12'50" 42d16'26"
Joan Claire Parisi
Sco 16h3'30" -14d1'47"
Joan Collette Zelten
Cas 23h47'56" 50d1'17"
Joan Comito
Cas 1h46'12" 60d30'45"
Joan Constance & Hiram Ernest
Cyg 20h26'49" 30d45'28"
Joan Cronin McGrade
Vir 13h53'30" -5d17'47"
Joan D. Maud
Sco 16h9'49" -11d18'52"
Joan D. Specker
Cam 6h31'50" 68d30'1"
Joan Dale
Cas 23h2'38" 57d55'14"
Joan & Dave Curto
Cyg 19h54'18" 30d40'46"
Joan & Dave's Eternity
Lyr 18h27'54" 31d35'8"
Joan & Dean
Lib 14h50'45" -23d46'31"
Joan Del Clark
Cnc 8h18'54" 25d27'29"
Joan Delores Morgan
Tau 4h5'32" 7d28'57"
Joan Deshefy
Leo 9h33'10" 28d38'41"
Joan Doherty *MOM*
Leo 10h17'31" 25d20'15"
Joan Dominicis
Sco 16h17'43" -16d22'33"
Joan & Doug
Cyg 20h2'25" 31d38'55"
Joan E.
Vir 13h17'48" 3d15'45"
Joan E
Aqr 22h5'27" -0d32'2"
Joan E. Gibson
Crb 15h29'58" 32d27'33"
Joan E. Hemus
Cas 0h39'7" 71d7'25"
Joan E. Mallen
And 0h40'24" 45d56'6"
Joan E Siravo
Cyg 21h42'26" 46d0'15"
Joan E Tharpe
Cas 1h44'17" 60d41'43"
Joan E Wolf
Uma 10h56'25" 46d5'35"
Joan Ebbecke *MOM*
Lib 15h21'7" -13d11'51"
Joan Edith Trow
And 0h12'16" 25d20'22"
Joan Eileen
Cyg 19h34'18" 48d25'16"
Joan Eleanor Gislason
Uma 11h26'43" 60d25'43"
Joan Elizabeth
Aqr 23h25'3" -18d46'54"
Joan Elizabeth
Her 17h14'30" 48d31'21"
Joan Elizabeth Thompson
Ari 1h59'14" 22d36'21"
Joan Elizabeth Winters
Crb 15h48'30" 27d22'57"
Joan Ellen Gabel/Creampuff
Psc 1h10'47" 10d7'48"
Joan Ellen Leskowsky Washburn
And 1h5'18" 38d1'14"
Joan Emberger Strong
Cnc 8h48'14" 13d37'55"
Joan & Ernie Dewald
Cyg 19h59'0" 34d24'15"
Joan F. Brigante
Uma 13h43'17" 51d2'38"
Joan F. Karasick
Psc 1h42'21" 27d56'1"
Joan F. Lawrence
Cas 0h45'24" 62d20'11"
Joan F. Martino
Tau 5h37'40" 27d17'56"
Joan Falk
Aqr 23h54'12" -11d22'1"
Joan Farrell Koss
Cas 0h25'43" 53d8'32"
Joan Fatton
Umi 14h37'50" 69d58'17"
Joan Fernan Lilly
Cas 1h36'43" 64d28'52"
Joan Fonda
Aqr 22h19'23" -15d27'7"
Joan Foster
Ari 2h55'7" 24d59'2"
Joan Frances Irving
Uma 11h3'33" 45d28'47"
Joan & Franco's Star
Cas 1h30'40" 60d8'25"
Joan Fuske
Aqr 22h31'19" 1d4'27"
Joan Geilow
Cas 23h22'40" 53d48'6"
Joan & Gerald Paley
Lyn 7h7'42" 59d8'25"
Joan Geraldine Plater
Crb 15h42'4" 36d19'46"
Joan Gill
Uma 10h55'29" 61d36'39"
Joan Goodman
Lyr 19h12'48" 45d31'41"
Joan Gough
And 23h2'29" 35d42'33"
Joan Grace Nolan
Ari 2h59'53" 12d2'14"
Joan Grace Shuttleworth Whitley
Lib 14h51'20" -1d0'38"
Joan Griebel Reynolds
Leo 9h38'12" 20d48'54"
Joan Guy
Gem 6h51'33" 14d31'23"
Joan H. Baez
Cas 1h5'1" 56d28'46"
Joan H. Kozel, In Our Hearts & Sky
Umi 14h58'23" 88d55'6"
Joan H. Seaman
Cap 21h53'5" -18d3'36"
Joan Haas
Lyr 18h27'3" 31d3'40"
Joan Hamilton Lohnes
Aqr 22h0'16" 2d5'19"
Joan Harvey's Star 70
Psc 0h37'39" 7d35'12"
Joan Hilda Ethel Flint
Col 5h22'9" -32d7'38"
Joan Hindley-Meals
Sco 17h18'24" -42d7'35"
Joan Hussey
Vir 13h46'26" -12d46'13"
Joan I Montse
Lyn 7h21'35" 58d45'59"
Joan I. Overman
Del 20h36'21" 14d11'8"
Joan Inkson
Cas 1h56'18" 62d24'16"
Joan Irene Sirota
Uma 12h22'30" 61d48'43"
Joan & Jean Wagner
Sco 17h56'57" -31d36'24"
Joan Jett
Vir 13h24'42" -8d10'29"
Joan K. Molloy
Uma 10h1'49" 59d19'35"
Joan Kathleen McNulty
Cap 20h59'57" -15d2'13"
Joan Kathleen Race
Cas 0h22'44" 50d47'58"
Joan Korton
Leo 11h27'38" 0d56'8"
Joan Kostreva
Cap 20h17'52" -23d33'22"
Joan Kriegler
Cyg 20h39'43" 53d29'57"
Joan L. Butler
Uma 13h59'43" 53d1'46"
Joan L. Gerrish
Tau 5h9'9" 18d38'8"
Joan L. Kowalski
Cyg 19h29'44" 57d38'55"
Joan La Bayne
Crb 15h34'20" 32d22'31"
Joan Lawless
Cas 23h54'42" 59d43'44"
Joan Lawrence
Cyg 20h53'41" 49d4'36"
Joan Lechtur
Crb 16h5'11" 38d30'10"
Joan Leslie Beldegreen
Uma 11h11'47" 57d2'45"
Joan Linsey Pitts
Uma 9h16'11" 50d5'50"
Joan Louise Marshall
Sco 17h44'24" -42d34'44"
Joan Lynch White
Leo 11h1'49" 10d48'32"
JOAN LYNETTE ZIDEK
Gem 6h52'2" 30d52'49"
Joan M. Batchelder
Ori 6h19'30" 15d15'2"
Joan M. Bradley
And 2h14'38" 37d54'46"
Joan M. Haas
Vir 12h10'11" 11d59'55"
Joan M Lulf
Sgr 18h56'27" -31d18'50"
Joan M. Nelson
And 23h8'9" 51d17'31"
Joan Marie
And 2h23'2" 47d34'48"
Joan Marie
And 2h36'54" 43d56'23"
Joan Marie Albright
Cap 21h45'24" -9d22'44"
Joan Marie Alverson
And 0h22'16" 38d48'4"
Joan Marie Cook
And 0h43'45" 39d55'18"
Joan Marie Dehmlow
And 1h28'51" 46d11'35"
Joan Marie Goffinet
Cyg 20h27'53" 45d11'35"
Joan Marie Jacoby 6-12-1955
Gem 6h41'12" 30d58'4"
Joan Marie Kane
Umi 15h6'48" 84d58'42"
Joan Marie Lourdes - Your Chosen Name here
Peg 21h10'36" 19d51'25"
Joan Marie Montovani
Aql 19h14'10" 2d48'46"
Joan Marie Moore
Psc 0h8'0" 4d55'44"
Joan Marie Purvis
Psc 0h19'56" -5d5'17"
Joan Marie Reiniger
Cam 5h4'13" 72d15'20"
Joan Marie Schuh
Cam 4h7'26" 54d57'43"
Joan Marissa Newland
And 1h5'26" 42d18'15"
Joan Mary Jacobs
Ari 2h36'15" 24d36'57"
Joan Mary Rosati
Sco 16h8'40" -19d2'58"
Joan Mary Rose (nee Hambling)
Aqr 20h55'41" 0d41'16"
Joan Mason
Leo 10h37'33" 6d41'58"
Joan May Stockwood 0208
Aqr 22h12'1" 0d48'41"
Joan May Ward
Cas 0h12'50" 52d18'56"
Joan Mazzi
Psc 2h3'5" 7d0'52"
Joan McGuriman
Vir 14h43'44" -5d59'40"
Joan McGurk
Aql 19h16'8" 14d46'6"
Joan Mink Star
Sco 16h11'11" -13d21'22"
Joan Minty Thompson Forever Star
Cru 12h31'12" -59d20'48"
Joan Mollenhauer
Cas 23h31'42" 58d7'34"
Joan Moran
Cyg 19h38'18" 36d7'22"
Joan Muns I Casals
Leo 9h46'33" 30d12'24"
Joan Myra
Ari 2h39'35" 29d50'26"
Joan N. "Joanie" Dascoli
Sgr 19h24'8" -42d27'18"
Joan P. Arminio
Ari 2h1'8" 18d49'39"
Joan Parker Dembek
Sgr 19h20'16" -16d32'29"
Joan Paton
Cas 0h8'12" 55d24'5"
Joan Patricia Quinlan
And 23h9'28" 42d24'42"
Joan & Paul Taylor
Vir 12h20'24" 3d13'31"
Joan Peaches Hubbert
Sgr 18h54'7" -28d12'49"
Joan Pearson Jopson Bayfield
Lyn 7h33'30" 49d24'44"
Joan Perry Vincunas
Ori 5h23'38" 2d0'51"
Joan Peterson
Tau 5h35'16" 18d26'47"
Joan Porter
Cas 1h16'13" 54d5'23"
Joan Potter Levinson
Ari 3h20'45" 29d33'1"
Joan Puddy
Cyg 20h27'41" 31d18'10"
Joan R. Edwards
Cap 21h33'30" -16d28'29"
Joan R. McLeish
Cnc 8h2'9" 26d10'16"
Joan Rachel
Aqr 21h1'57" -10d35'57"
Joan Radtke
Dra 18h41'52" 65d25'10"
Joan & Randy
Cyg 20h21'18" 54d24'40"
Joan Ray Nicolay
Aqr 22h26'15" -1d2'38"
Joan Rice
Uma 11h8'33" 69d22'39"
Joan Rich
Uma 13h2'59" 61d18'11"
Joan & Richard Platt
Cyg 21h33'33" 51d13'6"
Joan Rispoli
Cap 20h39'0" -15d44'25"
Joan Rivelis Becker
Lyr 18h46'8" 32d30'41"
Joan Roberts
And 0h47'25" 34d56'0"
Joan Roberts - Donlin - Peter
Cnc 8h39'54" 31d9'2"
Joan Romanelli
Lib 15h54'36" -8d18'34"
Joan Romano-Pilla
And 0h29'26" 37d24'57"
joan/rose
Aqr 22h54'26" -7d51'4"
Joan S Sneddon
Aql 19h19'53" 11d17'13"
Joan & Sal Belissimo
Lib 14h51'40" -3d32'18"
Joan Sayek
And 2h23'48" 45d47'41"
Joan Schwarzbach
And 1h31'3" 42d51'31"
Joan Sellers
Uma 10h37'1" 61d58'44"
Joan Shelby VanAtta
Cas 0h5'26" 50d25'3"
Joan Skiba
Cas 1h35'3" 62d16'24"
Jo-An Skowronek
Vir 12h40'11" 12d28'9"
Joan Squared
Cas 1h20'6" 58d5'39"
Joan Steadman's Stellar Body
Cap 20h28'59" -11d59'58"
Joan *Stella* Nolan
Uma 9h44'55" 67d59'10"
Joan Stiber Our STAR Teacher
And 0h27'5" 41d33'23"
Joan Stubbings
And 0h27'28" 37d1'53"
Joan Suzanne Serchuck
Ari 2h36'2" 29d20'41"
Joan T. Queen: Watch Over Mayberry
Cas 0h25'2" 54d9'51"
Joan T. Quinlan-Wielock
Aqr 21h3'11" 0d20'28"
Joan Tantum Kelly
Ari 2h9'14" 23d12'14"
Joan THE MATCHMAKER Elder
And 2h16'49" 50d31'8"
Joan Theis ... My "Joanie Fair"
Vir 11h52'13" 8d17'27"
Joan Theresa Ragan
Lyn 7h14'58" 59d35'32"
Joan Theresa Sronce-ROmero
Cas 1h44'22" 63d32'33"
Joan Thompson
Sgr 17h58'3" -30d1'0"
Joan Tiska Kennedy
Cnc 8h54'13" 23d58'2"
Joan Trozzi
Cyg 20h4'56" 33d48'48"
Joan V. Dobbs, M.D.
Cma 6h38'40" -15d12'22"
Joan Valentine
Aqr 23h2'52" -7d32'12"
Joan Vindett
Cyg 20h4'5" 56d35'15"
Joan Waldman
Cas 1h3'42" 60d51'17"
Joan Weisman
Crb 16h7'47" 28d1'53"
Joan Wickings
Lyn 6h24'46" 59d0'8"
Joan Willson
And 23h52'7" 44d15'37"
Joan Zane
Ari 2h34'53" 19d30'9"
Joan, My Stability
Cam 4h43'48" 54d27'53"
Joan, The Tracy's North Star
Lmi 10h31'43" 32d49'36"
JOANA
Cas 3h14'29" 71d10'49"
Joana
Cap 20h40'10" -22d23'4"
Joana Pavlova
Lib 15h30'31" -16d9'2"
Joana Rudiakov
Cyg 20h12'5" 37d8'58"
Joana Zayas
And 1h3'30" 35d28'30"
Joananel Gonzalez Wishing Star
Uma 9h34'40" 55d17'52"
Joandi's Brilliance
Uma 11h46'2" 36d29'59"
Joandon
Sgr 18h29'33" -17d3'14"
Joane Cornell
Psc 1h14'17" 27d41'52"
Joani
Ori 5h3'52" -1d44'27"
Joanie
Cas 23h45'38" 51d49'7"
Joanie
Crb 16h8'5" 38d11'14"
JOANIE
And 2h35'39" 50d3'33"
Joanie
Per 3h23'13" 45d13'56"
Joanie Aileplay
Aql 19h2'17" 9d0'35"
Joanie Anne Bright
And 1h48'3" 40d57'12"
Joanie Benoit
Mon 7h41'41" -8d27'29"
Joanie Critchley
Tau 4h21'33" 24d31'44"
Joanie Detwiler's Star
And 2h25'18" 46d33'6"
Joanie G
And 0h48'19" 36d50'26"
Joanie Graham
Sex 9h43'1" -0d24'22"
Joanie K. Bevington
Umi 18h9'45" 87d34'57"
Joanie Lee
Psc 1h44'39" 13d30'38"
Joanie LeMieux
And 0h18'15" 26d21'1"
Joanie Loves Chachie
Cyg 21h22'32" 38d49'34"
Joanie Lynn
And 23h24'33" 42d31'2"
Joanie Marie Laurer
And 0h46'58" 44d26'56"
Joanie "Mommy" Cleary
Cas 1h22'23" 63d48'37"
Joanie Riley
Per 3h25'46" 45d27'29"
Joanie Sompayrac & Tony Grossi
Lyr 18h37'29" 36d8'41"
Joanie Wardell - Dad's Little Girl
And 0h54'7" 38d58'11"
Joaniekins
And 2h18'59" 46d41'46"
Joanie's Star
Peg 21h39'11" 21d4'6"
Joanie's Star
Leo 11h56'2" 18d45'53"
Joanie's Star
Crb 15h55'4" 29d57'53"
Joanie's Star
Cap 20h38'24" -13d49'12"
Joanie's Star of Hope
Gem 6h29'20" 24d3'13"
JoaniMac
Uma 10h28'54" 56d8'36"
Joanmylove Shesmystarius
Cnc 9h10'30" 8d21'24"
JOANN
Psc 0h54'35" 17d0'18"
Joann
Tau 3h34'58" 26d40'51"
JoAnn
Leo 10h46'51" 17d15'13"
Joann
And 2h24'29" 50d11'27"
Jo'Ann
And 1h15'0" 45d36'18"
JoAnn
And 1h54'27" 46d48'33"
JoAnn
And 0h50'47" 45d26'2"
JoAnn
And 23h14'21" 43d42'46"
Joann A. Imus
Lmi 10h14'33" 31d21'58"
JoAnn - Angel in the Sky
Sco 16h10'52" -11d3'21"
JoAnn Ballard
Cmi 7h58'59" -0d13'45"
JoANN Bates
Cam 6h34'0" 62d43'9"
Joann Bennett Sanford (Joann Baby)
Sgr 19h9'20" -18d14'33"
JoAnn Bodall
Crb 15h47'41" 26d35'7"
Joann Brzozowski
Cru 12h34'59" -58d10'53"
JoAnn Cates
And 0h48'4" 37d33'18"
Joann Cochrane
Lyr 18h30'59" 28d13'30"
Joann Costarella
Cep 21h39'55" 71d52'27"
Joann Delellis
Lyn 6h56'45" 50d41'20"
Joann DiGiovanni
Ari 3h26'28" 19d45'43"
JoAnn Donato Corry
Cyg 21h22'1" 39d31'19"
Joann Espinoza
Uma 11h5'52" 37d9'50"
JoAnn Fink
Uma 11h11'52" 67d37'43"
Joann Foley
Leo 10h18'16" 26d54'54"
Joann Gambino
Sco 16h6'10" -12d28'29"
Joann Grace O'Rorke
Uma 8h52'51" 65d58'31"
Joann Gray
Peg 21h42'24" 23d16'37"
JoAnn Handley
Mon 7h15'51" -0d40'42"
JoAnn Hanshew
Uma 11h57'7" 63d42'29"
Joann Hare
Uma 10h54'55" 46d26'40"
Joann Haring
Lyn 7h42'59" 40d58'32"
JoAnn Hattie Ward
Uma 13h51'29" 49d21'44"
JoAnn Heath's Eastern Star
Cas 0h38'28" 52d20'14"
Joann Hentkowski Wojtaszek
Sco 16h3'1" -11d57'51"
JoAnn Hoeth
And 0h23'13" 35d46'34"
Joann "Jo"
And 23h19'38" 37d43'59"
Joann Jordan
Sgr 17h54'11" -28d43'10"
JoAnn "Josie" Castle
Cap 21h32'23" -24d47'38"
Joann Karasinski
Crb 15h27'44" 31d9'50"
Joann Kathleen Dallow
Sco 17h4'58" -36d0'22"
Joann Kim
Aqr 22h48'25" -21d40'19"
Joann Koontz
Lyn 9h38'29" 40d5'1"
Joann Kuehn
Cas 1h32'8" 64d53'16"
JoAnn L. Woltz
Aqr 23h38'4" -7d27'29"
JoAnn Lang
Ori 6h11'23" 20d47'9"
Joann Lombardo
Leo 9h27'4" 15d44'37"
JoAnn M. Clark
Cas 1h12'43" 57d54'14"
Joann M. McHugh
Cyg 19h55'53" 32d51'38"

JoAnn M. Rooney
And 23h12'27" 47d43'26"
Joann Mae Antonio
Peg 0h3'5" 13d30'36"
Joann Marie Aznavorian
Ari 2h36'9" 27d49'37"
Joann McGuin
Cyg 19h31'1" 48d49'10"
Joann Michalkowski
Cnc 8h47'5" 32d15'29"
Joann Mitchell
Cas 1h17'6" 62d47'51"
Joann & Mordechai
Cnc 9h17'7" 25d52'1"
JoAnn Nakagawa
And 0h45'2" 37d0'9"
Joann P Bakay
Lib 14h49'36" -21d4'31"
JoAnn Pecenka
Peg 22h14'21" 6d20'1"
Joann Queeney
Uma 12h48'42" 60d39'52"
JoAnn Rees
Umi 16h8'17" 74d30'32"
JoAnn Ricks
Her 17h4'5" 28d14'38"
JoAnn Roberta Fegan
Sgr 19h43'54" -20d8'45"
JoAnn Rose
Ori 5h16'2" -8d28'16"
JoAnn S Minchey
Crb 15h44'6" 25d47'52"
JoAnn Schorah
Leo 10h23'25" 25d5'16"
JoAnn Smith
And 1h37'8" 48d4'3"
JoAnn Tartaglia
Ari 2h6'39" 15d42'18"
Joann "Tootie" Scelsa
Cas 1h34'15" 62d45'29"
JoAnn & Vic Blanco
Boo 14h33'1" 19d13'30"
Jo-Ann Vultaggio
Ari 2h17'34" 21d26'2"
Joann Weber
And 0h9'16" 40d31'37"
JoAnn Wood
Lyn 8h18'20" 41d58'23"
JoAnn092350
Lib 14h59'30" -9d58'16"
Joanna
Cap 20h29'40" -11d59'43"
Joanna
Sco 16h17'27" -18d4'32"
Joanna
Cas 0h39'48" 61d58'2"
Joanna
Crb 15h19'50" 30d38'25"
JoAnna
Uma 9h57'26" 45d5'17"
Joanna
Leo 10h31'26" 9d50'50"
Joanna
Uma 11h9'2" 29d10'55"
Joanna and Edward Lamolinara
Cyg 21h21'18" 53d20'25"
Joanna and Larry's Wedding Star
Cnc 8h54'38" 26d56'47"
Joanna Austin Mcleod
And 1h25'23" 41d45'43"
Joanna Banana
And 0h33'44" 42d59'54"
Joanna Beukema
Cnc 8h9'41" 7d34'59"
Joanna Biggers
Aqr 22h25'33" -4d1'27"
Joanna Bolter in our Hearts Forever
Ori 5h47'39" 5d44'22"
Joanna Buchanan
Cas 0h2'13" 56d30'58"
Joanna Buettner
Cas 0h32'59" 59d5'36"
Joanna Bunches
Lyn 6h43'49" 52d30'49"
Joanna Burlison
Lmi 10h41'2" 26d43'40"
Joanna Claire Young
And 0h41'10" 35d40'17"
JoAnna & David Chmielecki ~Always~
Cyg 21h40'53" 39d42'32"
Joanna Dee Nash
Cnc 8h19'46" 17d30'47"
Joanna DeMaria
Cap 21h20'25" -16d1'14"
Joanna Demartino
Cyg 19h35'27" 35d17'17"
Joanna Dittrich
And 1h38'44" 48d34'6"
Joanna Doguiles
Uma 11h38'22" 51d8'3"
Joanna E. Ruvolo
Cas 1h43'9" 64d45'8"
Joanna Elizabeth Kouchich
And 0h7'22" 39d15'10"
Joanna Elyzabethe Finnegan
Mon 6h26'51" -0d38'14"
Joanna & Jason Simpson
Cyg 20h25'35" 44d50'11"
Joanna Jenkins
And 0h22'8" 42d2'22"
JoAnna & Jerome
Lyr 19h12'41" 45d53'9"
Joanna Jimenez
Uma 11h13'40" 51d0'28"
Joanna "Jojo" McCandlish
And 0h7'43" 44d36'9"
Joanna L. Sullivan
Tau 3h47'25" 29d24'31"
Joanna Lara
Com 12h22'43" 26d11'47"
Joanna Leigh
Cap 21h14'43" -19d45'49"
Joanna Leigh Hudale
Lyn 7h43'10" 57d51'10"
Joanna Leigh Valentas
And 2h25'22" 48d41'40"
Joanna Lewis Lane Forever
Gem 6h30'27" 16d34'4"
Joanna Lynn
Psc 0h17'55" 15d22'6"
Joanna Lynn Presti
Cas 23h2'34" 57d18'22"
JoAnna Lynn Robinson
Psc 0h52'26" 31d43'52"
Joanna Lynn Tammaro
Uma 9h48'8" 70d57'16"
Joanna Lynn Thiede
Tau 5h18'44" 22d51'36"
Joanna Marcinkowska
Lyn 7h3'25" 48d17'17"
Joanna Margaret Galman
Psc 23h28'47" 6d54'8"
Joanna Marie
Tau 5h9'44" 23d27'2"
JoAnna Marie Howsden
Uma 10h23'15" 42d32'9"
Joanna Marie Vinoski
Lib 15h18'10" -18d37'53"
Joanna Marina
Sco 16h18'16" -13d10'3"
Joanna Marino
Cnc 8h0'40" 20d11'18"
Joanna Mary Constantino Carlson
Lib 14h54'8" -3d5'19"
Joanna Mary Deakin
Lmi 9h55'6" 38d20'48"
Joanna Mateo
Uma 9h10'34" 48d17'49"
Joanna Minaya
Leo 11h14'46" 15d7'9"
Joanna Morales
Lib 15h24'2" -12d46'50"
Joanna Morgan
Aur 6h57'33" 38d28'5"
Joanna- My love, best friend, life
Gem 6h49'33" 24d32'34"
Joanna Nadia Quigley
Uma 11h35'10" 35d28'14"
Joanna Nicole Siino
Cnc 8h11'48" 13d6'45"
Joanna Nicole Tufexis
And 23h59'24" 42d15'50"
Joanna Peruskie
Sco 17h27'23" -38d12'50"
Joanna Peterman-Hagan
And 0h30'32" 38d44'57"
Joanna R Jones
Uma 8h48'31" 64d36'41"
Joanna Reaves Mincke Henry
Psc 23h44'53" 6d21'13"
Joanna Repsold
Leo 10h21'20" 14d48'11"
Joanna & Robert Lucas
Cyg 21h37'57" 35d48'39"
Joanna Seroczynska
Cam 4h28'14" 53d41'51"
Joanna Shaughnessy
Cas 0h46'58" 64d20'34"
Joanna Sotomura
Ori 5h32'27" 10d51'59"
Joanna Stephanie Beltran
Crb 15h50'54" 34d6'18"
Joanna & Tom
Vir 12h47'35" -0d17'17"
Joanna Truong
Uma 13h10'14" 60d5'26"
Joanna Turner Bisceglio
Ori 5h52'10" 18d10'0"
Joanna Velasco
Lyn 7h15'15" 49d28'23"
Joanna Walsh
And 2h28'10" 39d11'50"
Joanna Wells
Sco 16h9'58" -18d9'45"
Joanna Xerri
Cas 1h9'23" 63d34'28"
JoAnna Young 1982
Cas 1h47'8" 66d24'3"
Joanna111042
Uma 14h23'18" 55d2'11"
Joanna-Mom 2 Zachary and Jacquelyn
Cnc 8h10'17" 7d35'18"
Joanna-My Promising Star of Light
Cap 21h52'53" -19d41'28"
Joanna's Angel
Crb 15h51'16" 36d39'56"
Joanna's Love
Cap 20h54'41" -22d1'51"
Joanna's love... Gary's light
Gem 7h33'21" 30d8'2"
Joanna's Star
Sco 16h5'5" -17d1'43"
Joanna's Star
Umi 17h10'43" 78d0'20"
Joanna's Star
Uma 9h11'54" 61d23'11"
Joanna's Star
Sco 17h50'36" -36d20'30"
JoAnna's Wishing Star lug3305
And 2h14'1" 50d1'55"
Joanna'sLoveStarJames
And 1h35'25" 45d11'24"
JoAnne
And 23h29'37" 42d11'28"
Joanne
Crb 15h34'58" 39d18'0"
Joanne
Cas 0h18'14" 56d33'32"
Joanne
And 23h1'32" 48d5'6"
Joanne
Crb 15h37'17" 34d4'22"
Joanne
And 1h46'56" 43d24'22"
Joanne
And 1h46'57" 37d35'23"
Joanne
And 0h42'58" 39d29'31"
Joanne
And 23h55'26" 34d47'0"
Joanne
Lyn 7h27'59" 44d16'14"
Joanne
And 2h17'48" 42d16'37"
Joanne
Vir 13h19'32" 11d30'41"
Joanne
Ari 3h8'3" 29d21'24"
Joanne
Mic 21h9'57" -28d17'11"
Joanne
Uma 8h28'0" 60d49'51"
Joanne
Lyn 7h43'37" 58d1'47"
Joanne
Ori 5h54'4" -1d54'27"
Joanne
Sgr 19h6'27" -21d16'40"
Joanne A. Ahrens
Apu 16h16'52" -74d41'58"
Joanne A Deslandes
Crb 15h30'48" 26d58'5"
Joanne Akar
And 1h19'42" 40d39'47"
Joanne Allayne Hawkins
And 23h20'51" 42d45'46"
Joanne and Linwood Fredericksen
Umi 15h20'5" 82d10'54"
Joanne "Angel" Royston
Peg 22h41'3" 14d21'25"
Joanne Biocca
Uma 8h40'36" 59d56'1"
Joanne Blain
Aqr 22h25'31" -24d41'41"
Joanne Boylan
Cas 1h22'31" 62d29'52"
Joanne Brawner
Vir 12h12'34" 5d8'35"
Joanne Byrne
Sgr 19h34'50" -28d2'55"
Joanne C. Aird
Leo 9h35'40" 24d51'52"
Joanne D. Quinones
Vir 13h35'6" -12d59'5"
Joanne Dean
Peg 21h50'36" 10d34'27"
JoAnne Doucet
And 0h30'39" 38d37'12"
Joanne Duran
Uma 8h12'13" 71d18'51"
Joanne E. Pomada
And 0h20'39" 31d53'2"
Joanne Elizabeth
Psc 0h17'3" 17d18'21"
Joanne Ellery - 21st
And 1h41'50" 45d18'1"
Joanne Esposito Covais
Cnc 8h6'4" 20d13'52"
Joanne (Fancy Face)
And 0h44'20" 40d7'24"
Joanne Ferguson - Superstar!
Equ 21h8'15" 9d43'4"
Joanne Ferris
Cyg 19h52'14" 31d36'8"
Joanne Fiona Troth - Mummy's Star
And 1h21'10" 35d11'38"
Joanne Fisher
And 0h33'39" 24d0'21"
Joanne Foster
Cam 5h16'54" 74d43'22"
Joanne Frances Allen
And 1h16'6" 42d18'27"
Joanne G. Conners
Cas 23h4'15" 53d51'21"
Joanne Garabedian
And 0h34'33" 37d1'58"
JoAnne Goulet
Cas 23h31'29" 57d10'3"
Joanne Goulet's Star
Cas 1h52'58" 61d11'3"
Joanne Grzyb
And 0h49'42" 40d10'44"
Joanne Guariglia Bartolomeo
Uma 13h38'44" 56d38'55"
JoAnne Gubrud
Tau 4h28'16" 19d9'27"
Joanne H. Kidd
Ori 5h36'39" 3d45'53"
Joanne Hackney
Aqr 23h5'48" -6d45'38"
Joanne Harper
And 0h32'51" 25d53'4"
Joanne Harrison
Psc 0h39'26" 6d46'26"
Joanne Holland
And 0h14'2" 29d58'47"
JoAnne Hunter
Uma 11h33'42" 38d44'9"
Joanne James
Mon 6h50'12" 7d37'19"
Joanne Jensen
Sgr 18h17'49" -28d0'9"
Joanne "Jody" Amarino
Gem 7h13'6" 19d42'44"
Joanne Johnston Boyer
Vir 13h12'26" 10d48'26"
JoAnne JW1 41205
Tau 5h33'40" 19d30'54"
Joanne Kientz
Cnc 8h56'6" 16d6'1"
Joanne Krol
And 23h19'7" 40d37'43"
Joanne Kruk
Sgr 18h1'9" -30d50'16"
Joanne L Breault-Whitmore, R.N.
Cap 20h24'7" -13d28'59"
Joanne Lajam
Sgr 19h44'37" -14d58'52"
Joanne Lavin
Cyg 19h37'5" 38d31'33"
JoAnne Lee
And 1h37'54" 42d53'30"
Joanne Lee McCubbin
And 1h3'22" 42d13'3"
Joanne Leischuck
Ori 6h9'32" 19d54'46"
Joanne & Les
And 1h55'58" 43d1'42"
Joanne Lilian Macklin
Cas 23h54'55" 50d2'47"
Joanne & Linoy
And 1h8'7" 41d52'5"
Joanne Loposky
And 2h8'34" 42d55'31"
JoAnne Louise Parks
Uma 8h41'0" 69d44'3"
Joanne Louise York
Tau 5h44'43" 17d39'3"
Joanne Lucille Conroy/Barry
Per 3h22'6" 44d17'28"
Joanne Lunardini
Sgr 18h52'41" -19d29'27"
Joanne Lynne
Uma 10h7'52" 46d41'32"
Joanne M. Berbower
Crb 15h52'4" 28d26'41"
Joanne M Levine
Tau 4h41'3" 19d32'24"
Joanne M. Messier
And 1h30'26" 39d47'28"
Joanne M. Pawinski
Psc 0h18'21" 0d8'27"
Joanne M. Rapisarda nee Seward
Leo 9h53'14" 21d45'1"
Joanne Marie
Ori 6h7'8" 17d22'18"
Joanne Marie
Col 5h13'39" -28d10'7"
Joanne Marie Brastow
Uma 11h30'34" 64d29'9"
Joanne Marie Lewis
Ari 2h2'54" 22d39'25"
Joanne Marie O'Brien
Lyn 6h54'33" 59d31'19"
Joanne Marie Radich Durkin
Vir 13h30'54" -0d13'53"
Joanne Marie Sturgeon
Uma 13h5'40" 57d51'1"
Joanne Marot
Gem 6h46'17" 23d10'12"
Joanne Mary Costello
And 23h42'2" 42d14'45"
Joanne Meehan, always a Star
Per 3h45'36" 48d56'17"
Joanne Meshboum
Dra 18h10'6" 78d10'9"
Joanne Mickett
Lyn 8h34'58" 57d41'55"
Joanne (Mom) Lajam
Uma 8h54'37" 71d6'20"
Joanne Murdoch
Ari 3h16'25" 19d40'33"
Joanne - "My Joie de La vie"
Leo 11h38'19" 26d2'53"
Joanne Myers
Cas 23h26'26" 52d8'13"
Joanne N. Chapman
Tau 4h42'30" 22d44'48"
Joanne Nirchi
Lyr 18h40'21" 31d32'41"
Joanne O'Laskey
Tau 4h31'18" 18d35'26"
Joanne Patricia Smith
Cap 21h35'18" -19d2'1"
Joanne Patricia Thomas
Lib 15h34'4" -8d14'42"
Joanne Pitcher Born: Feb 23/1962
Uma 11h32'30" 57d43'50"
Joanne (Posh) White
And 23h27'30" 46d52'45"
Joanne Pratley
Aur 6h30'35" 34d42'21"
Joanne Richmond
Ari 2h1'43" 14d36'14"
Joanne & Robert Buonfiglio
Lib 15h3'9" -4d1'49"
Joanne Roessner
And 0h32'31" 32d2'54"
Joanne & Ronald Spears Sr.
Crb 16h24'43" 27d22'35"
Joanne S. Wiviott
Uma 8h27'18" 67d3'20"
Joanne Saad
Cas 0h56'48" 53d34'46"
Joanne Sawatzke
Cas 0h2'27" 56d31'14"
Joanne Schmidt
Dra 18h40'20" 58d30'0"
Joanne Sierra
Cnc 8h59'43" 11d26'53"
Joanne Spix
Dra 19h26'40" 63d45'13"
Joanne & Steven
Cyg 21h25'34" 30d58'40"
Joanne Susan Turack
And 23h18'44" 47d33'15"
Joanne Svoboda
And 0h20'6" 41d55'32"
Joanne Szczygiel
Aqr 22h13'16" 0d53'47"
Joanne T. Drummond
Ori 4h50'28" 13d18'23"
Joanne Tadevich
Vir 15h7'4" 4d29'17"
Jo-Anne Tayag
Cyg 19h54'57" 36d36'23"
Joanne Theresa Monahan-Crehan
Uma 10h22'10" 70d7'23"
Joanne Tiamo Mia Amo're
Cas 1h8'47" 56d27'13"
Joanne Varga
Leo 11h3'26" 4d25'2"
Joanne Ventura
Psc 1h24'6" 27d34'54"
Joanne Victoria Todd
And 0h49'7" 32d0'45"
Joanne W
Uma 13h33'40" 59d6'27"
Joanne Walling
Aqr 23h48'21" -15d51'43"
Joanne Wolfe
Gem 6h49'38" 26d38'22"
Joanne, Beacon of Beauty.
Tau 3h41'56" 25d58'29"
Joanne, my dreamer
Ori 6h21'42" 14d0'45"
Joanne051072
Lib 15h32'30" -23d12'39"
Joanne2121
Sco 16h16'0" -14d30'45"
JoanneHoffman
Uma 9h45'12" 43d44'25"
JoannelLoveYou
Cnc 7h56'38" 13d17'31"
Joanne's Geordie Wishing Star
And 23h54'47" 36d38'47"
Joanne's Light
Umi 5h9'28" 89d26'17"
Joanne's Rising
Sco 17h51'3" -35d48'13"
Joanne's Star
Oph 16h37'48" -0d33'6"
Joanne's Star of Hope
And 23h23'45" 51d52'28"
JoAnne's Valentine
Sgr 19h1'19" -33d3'0"
Joann-Heaven's Angel
Cnc 8h20'19" 17d27'16"
Joannie
Cas 0h27'4" 54d56'35"
Joannie Almodovar
And 23h20'28" 39d21'5"
Joannie Halls
Ori 5h45'39" 3d40'0"
Joannis
Leo 11h5'11" 17d21'55"
Joann's Eternal Beauty
Crb 15h49'30" 26d5'0"
JoAnn's Eternal Light
Psc 0h53'32" 27d54'40"
Joann's Eternal Love
Psc 0h2'14" 0d43'14"
Joann's Star
Gem 7h51'16" 32d20'1"
Joan's Brilliant Star
Cas 23h28'39" 59d13'2"
Joan's Crown
Cyg 20h47'39" 39d25'39"
Joan's Everlasting Light in Heaven
Cnc 9h0'29" 18d49'17"
Joan's Light Shines Down
And 22h58'47" 37d2'17"
Joan's Place in the Sky
And 2h9'14" 42d56'26"
Joan's Place in the Sky
And 2h9'50" 41d56'29"
Joan's Star
Cas 23h45'58" 56d1'31"
Joan's Star
Cas 1h27'50" 65d41'3"
JoAnthony
Lyn 7h41'38" 35d47'44"
Joany Sze
Tau 4h37'37" 22d39'14"
Joany Thibodeau - 04-04-2000
Uma 11h32'2" 57d1'22"
Joao Antonio Dias da Fonseca
Gem 6h38'19" 19d50'19"
Joao Antonio Galego Da Cunha
Uma 8h24'22" 66d11'33"
João & Carla
Peg 21h39'11" 22d41'6"
Joao Daniel Silva Pereira
Uma 9h48'5" 68d40'37"
joao luiz silva
Psc 0h6'28" -1d31'0"
João Pedro Leal Avelar Dias
Her 17h58'22" 24d25'26"
Joaquim Aguilar Barbany
Cyg 21h48'28" 36d27'20"
Joaquin
Cyg 20h26'14" 44d55'59"
Joaquin Alejandro Nunez
Uma 9h29'22" 55d6'3"
Joaquin and Agnes Sueiro
Sge 19h13'21" 19d43'16"
Joaquin Augustus Auger
Ari 2h27'47" 21d48'16"
Joaquin Boutsen
And 2h30'18" 50d8'16"
Joaquin Eros
Gem 6h17'40" 23d9'54"
Joaquin Gomez
Vir 14h49'26" 3d51'48"
Joaquin Goquiolay
Tau 5h46'30" 27d35'18"
Joaquin Joseph Machado
Leo 11h49'51" 20d24'32"
Joaquin Rafael Hurtado
Cep 21h33'7" 61d25'31"
Joaquin SanMiguel
Cep 22h26'47" 62d0'13"
Joaquin Valdez
Aqr 21h45'10" -0d30'27"
Joasia's Star
Cas 23h40'23" 62d31'5"
Job
Leo 9h35'53" 17d49'44"
Job Daniel Branning
Lmi 10h38'55" 23d34'38"
Jobe
Cap 20h51'19" -26d47'9"
Jobe - Our Little Angel
Sgr 18h0'10" -28d11'3"
JoBeth
Lyn 7h28'14" 59d2'9"
JoBeth Banas
Tau 3h36'28" 13d26'48"
Jobeth & Miguel
Umi 15h57'55" 72d21'7"
Jobeth Miller
Uma 13h32'24" 57d55'9"
Jobeth's Angel
Lyr 18h23'47" 44d17'34"
Jobey's Star
Cyg 19h54'52" 35d33'15"
JoBo's Love
Lyr 19h11'32" 45d17'20"
Jobyanne
Aqr 21h5'15" 2d22'18"
Joc Palm
Uma 10h19'40" 53d39'14"
Jocelin Padilla
Cap 20h32'18" -17d20'37"
Jocelin Padilla
Cap 20h32'29" -17d15'54"
Jocelyn
Cap 21h2'59" -20d34'33"
Jocelyn
Dra 18h46'40" 80d3'16"
Jocelyn
Cas 0h31'14" 62d56'5"
Jocelyn
Leo 11h0'47" 0d32'52"
Jocelyn
Com 12h18'22" 23d47'32"
Jocelyn
Lyr 19h13'19" 26d54'35"
Jocelyn
And 23h51'49" 42d42'46"
Jocelyn
And 2h30'52" 43d21'50"
Jocelyn 04
And 2h16'49" 48d34'51"
Jocelyn Alexis Provo
Cnc 8h38'57" 30d25'19"
Jocelyn Allura Altman
Psc 1h6'33" 28d34'18"
Jocelyn Ashlee Rivera
Ori 6h4'1" 15d44'26"
Jocelyn Ashley Nastasi
Psc 0h55'36" 18d23'41"
Jocelyn Elizabeth
Lib 15h17'45" -16d49'18"
Jocelyn Elyse Booth "Sweet Pea"
And 0h42'40" 38d40'7"
Jocelyn Eve Mortimer
Ari 2h21'26" 12d10'26"
Jocelyn Grace Davis
Her 16h51'21" 32d10'0"
Jocelyn H. Sicat JHS-5
Cap 20h50'59" -21d37'30"
Jocelyn Hemond
Crb 15h18'30" 27d56'39"
Jocelyn Joanna Zink
Gem 7h18'45" 19d23'49"
Jocelyn L. Maragno
Cam 5h14'1" 60d3'17"
Jocelyn Larae
And 23h31'9" 41d19'56"
Jocelyn Lauren Topf
And 23h48'29" 41d48'43"
Jocelyn Louise Ancona
Uma 13h6'19" 53d46'34"
Jocelyn M. Cruzen-Beauregard
Crb 15h37'34" 31d34'38"
Jocelyn M. Ellis
Aur 6h14'59" 31d57'27"
Jocelyn M. Viola
Sco 16h9'40" -10d46'42"
Jocelyn Maribel
And 23h10'29" 44d15'43"
Jocelyn Marie
Cap 21h53'5" -18d31'13"
Jocelyn Marie
Sgr 18h20'41" -19d57'57"
Jocelyn Marie Jordan Pugurwag
Umi 15h14'26" 73d16'21"
Jocelyn Mary Broughton
Aqr 22h22'31" -7d24'36"
Jocelyn My Light
Cnc 8h12'49" 29d45'40"
Jocelyn Player
Uma 12h1'49" 40d26'3"
Jocelyn Racine
Aqr 22h23'1" -6d1'28"
Jocelyn Renee
And 0h1'16" 40d50'52"
Jocelyn Rodriguez
Aqr 23h4'2" -7d6'29"
Jocelyn Rose
Aqr 22h58'30" -11d1'39"
Jocelyn Rose Oceguera
And 1h50'22" 41d10'43"
Jocelyn Skye
Uma 13h20'49" 57d7'20"
Jocelyn Sloane Fital
Sgr 19h57'29" -30d27'17"
Jocelyn Valentina
Lmi 10h40'49" 31d5'8"
Jocelyne
And 23h14'26" 35d17'28"
JOCELYNE GLINZ
Umi 13h15'59" 88d14'11"
Jocelyne Yvette Renée Duret
Ari 3h23'40" 19d26'35"
Jocelynn Raine Lindgren
And 23h22'39" 47d14'2"
Jocelynne Angelique
Cam 7h19'14" 63d0'8"
Jocelyn's Star
Lib 14h53'40" -6d16'59"
Jochelle
Psc 23h30'5" 2d24'30"
Jochen Eckhardt
Uma 9h49'23" 56d12'49"
Jochen Kristopher
Lmi 10h17'15" 34d36'20"
Jochen Lange
Ori 6h15'2" 20d42'16"
Jochen Sauer
Psc 0h40'37" 11d9'15"
Jochen Schühle
Per 3h10'25" 43d3'44"
Joci
Lyn 7h44'2" 44d39'9"
JOCI
Col 5h42'16" -33d18'46"
Jocinda
Cap 21h49'51" -11d47'38"
Jock
Cma 6h45'38" -17d23'33"
Jock McDivot
Lyn 6h43'9" 54d2'21"
Jock & Sheena Hill
Cyg 19h54'50" 31d11'3"
Jocko
Uma 9h2'47" 57d35'27"
Jocovo
Umi 16h17'17" 74d19'6"
Jodans
Lib 14h53'11" -5d31'39"
Jodayne
And 22h58'21" 51d30'41"
Jodee
And 1h3'55" 44d17'39"

Jodee "Kissy" Renick
Uma 11h25'53" 44d29'26"
Jodee Lee
Lmi 10h31'19" 34d37'42"
Jodee "Leet Penguin" Steenberg
Cep 22h4'21" 69d42'28"
JoDee Lynn
Uma 9h39'5" 55d33'42"
Jodee Nicole Rose
Leo 10h25'58" 24d48'20"
Jodell Susan Marsh
Psc 1h21'27" 16d4'36"
Jodella
Gem 7h7'11" 30d27'59"
Jodene Le Clus
Cap 20h25'31" -18d15'46"
Jodi
Lib 15h7'58" -16d15'48"
Jodi
Vir 13h47'10" -4d33'50"
Jodi
Sgr 18h51'22" -30d47'35"
Jodi
Cam 3h54'9" 55d5'23"
Jodi
Crb 16h19'50" 38d29'12"
Jodi
Aqr 22h40'53" 2d11'46"
Jodi
And 0h19'11" 30d39'57"
Jodi
Ari 1h47'17" 23d25'28"
Jodi and Dave
Uma 11h45'18" 43d36'19"
Jodi and Elisha Forever
Cnc 8h50'3" 21d52'4"
Jodi and Mike Puterbaugh
Cyg 21h44'49" 53d50'46"
Jodi Ann Engbrecht
Eri 3h45'11" -0d27'33"
Jodi Arlene McGee
And 23h50'22" 45d10'10"
Jodi Bartman
Ori 6h10'59" 16d4'27"
Jodi Colleen Rayburn
Cap 20h23'9" -22d47'24"
Jodi Collis
Crb 15h36'44" 38d50'12"
Jodi Crawford
Cru 12h46'25" -64d39'32"
Jodi Creek Larson
Aqr 22h24'58" -4d5'47"
Jodi Daily
Sgr 18h42'29" -24d9'11"
Jodi & Dave McLaren
Cyg 21h30'39" 33d43'51"
Jodi Donnelly
Uma 13h34'19" 62d18'17"
Jodi Elizabeth Doughty
And 23h11'5" 42d50'27"
Jodi Estep
Crb 15h38'43" 33d18'35"
Jodi Fellner
And 0h23'2" 27d24'11"
Jodi Fleissig
Sgr 19h44'44" -16d45'49"
Jodi Forever
Crb 16h6'51" 38d1'25"
Jodi & Fred Poust
Del 20h39'55" 15d23'38"
Jodi Gibson
Lyn 6h49'59" 51d29'50"
Jodi Hale
Lib 15h44'39" -5d33'17"
jodi judith
Ari 1h47'39" 19d34'44"
Jodi K. Rusignola
Lyn 19h18'7" 34d17'20"
Jodi Katharine Allison James
Her 17h55'0" 28d51'33"
Jodi Keating
Leo 10h51'53" 12d47'38"
Jodi Lane Sullivan
Uma 9h32'38" 67d47'4"
Jodi Lea Bagwell
Ori 5h33'4" 4d34'38"
Jodi Lea Reed
Cas 23h43'56" 55d0'35"
Jodi LeAnn Mattson
Leo 9h43'56" 21d35'13"
Jodi Lee
And 23h22'47" 43d42'44"
Jodi Lee Conner
And 23h43'47" 43d9'56"
Jodi Lee McNamara
And 0h27'25" 37d43'17"
Jodi Lee Roush
Her 17h13'25" 15d42'17"
Jodi Lee Smith
Leo 10h53'46" 0d42'33"
Jodi Leigh Ayers
Ari 3h26'1" 23d49'3"
Jodi Louise French
And 23h31'58" 41d47'54"
Jodi Lynn
Cyg 20h12'32" 35d15'8"
Jodi Lynn
Lyn 8h35'15" 36d51'42"
Jodi Lynn
Gem 6h22'18" 18d38'16"
Jodi Lynn Gordon
Dra 18h51'50" 56d8'26"
Jodi Lynn Newbold
Aql 18h43'22" -0d1'16"
Jodi Lynn VandenHull
Leo 11h7'28" -0d15'12"
Jodi Lynn Warshaw
Lib 15h12'57" -18d2'31"
Jodi Lynn Wiemers
Vir 13h25'7" 3d30'10"
Jodi Lynne 2004
And 0h43'16" 39d34'33"
Jodi Mae
And 1h15'58" 50d18'2"
Jodi Marie Meads
Uma 9h19'30" 63d28'38"
Jodi Min Lilla Svala 1962
Leo 11h42'58" 24d31'36"
Jodi Mylie
Uma 10h4'57" 69d47'39"
Jodi Odie
Ari 2h53'0" 27d15'42"
Jodi Polser
Uma 8h38'54" 47d28'14"
Jodi Primm
Tau 4h13'0" 26d21'38"
Jodi Redman
Lyn 7h41'4" 45d40'32"
Jodi Roberts
Umi 15h20'55" 69d32'16"
Jodi Rodar
Umi 16h45'24" 75d55'15"
Jodi Schumacher
Uma 10h14'41" 71d44'50"
Jodi Seever
And 1h15'55" 48d45'53"
Jodi Shipplett
Cnc 8h4'47" 11d3'27"
Jodi Steinhart
Ori 5h35'34" 7d14'0"
Jodi Swiatkowski
Gem 7h19'41" 19d13'59"
Jodi T. Pucillo
Uma 13h10'54" 56d43'47"
Jodi Walls
Umi 14h12'18" 65d29'11"
Jodi Wilshire
Uma 9h59'25" 42d29'31"
Jodi Wooten
Crb 16h16'47" 35d5'53"
Jodi, Carlos , Anthony, Danyka
Tau 3h27'54" 17d37'51"
Jodi, The love of my life
Lib 14h51'1" -2d16'15"
Jodi-Annabelle Lenz
And 2h2'37" 46d21'44"
Jodie
Lmi 10h26'13" 35d18'57"
Jodie
And 0h8'59" 44d14'16"
Jodie
Ari 3h15'16" 28d35'13"
Jodie
Ori 5h18'31" -4d4'4"
Jodie
Sco 16h57'10" -38d39'0"
Jodie Allison Nestman Buhse
Cnc 8h50'32" 26d29'21"
Jodie and Johnny's Star
Ari 3h8'31" 20d54'54"
Jodie and Paige Hukill
And 23h20'41" 48d26'25"
Jodie Ann
Cnc 8h17'5" 7d52'59"
Jodie Anne Hyde
And 0h34'1" 42d27'14"
Jodie Bella
And 0h3'33" 44d39'55"
Jodie Beth
Ari 2h12'4" 15d35'21"
Jodie Bruhn
Ari 2h25'19" 24d45'24"
Jodie Danielle Kathleen Petty
Ari 2h50'48" 11d0'18"
Jodie Diane Mogensen
Cas 0h33'10" 54d12'42"
Jodie Groneck *MOM*
Leo 11h26'39" 17d7'37"
Jodie Hill
And 23h37'29" 47d57'21"
Jodie Hohenstein
Cap 21h11'49" -17d32'20"
Jodie Hood
Ori 5h7'26" 8d43'50"
Jodie & Jalie
Uma 8h58'0" 50d37'35"
Jodie & Justin - Together For Eternity
Ara 17h46'47" -48d45'24"
Jodie Kathryn Robbins
Cas 23h32'7" 56d57'22"
Jodie Kingston
Peg 22h27'32" 10d24'26"
Jodie L. Manuel
Sgr 18h36'15" -23d52'19"
Jodie Lynn
Cam 4h20'53" 78d59'7"
Jodie Lynne
Cnc 8h19'1" 10d14'12"
Jodie Marie McIntosh (Jodes)
And 1h54'7" 41d34'46"
Jodie Marie Pickel
Sgr 18h58'11" -34d45'20"
Jodie Marie Wild
Uma 11h28'4" 60d51'56"
Jodie Melinda
Cru 12h17'48" -57d3'2"
Jodie My Love Always and Forever
Uma 9h46'49" 66d21'39"
Jodie Owen Bullard
Umi 14h20'21" 69d52'21"
Jodie Reid
And 23h20'39" 52d11'3"
Jodie Renae Burns
And 0h44'35" 38d40'36"
Jodie Rowlett
Cru 12h47'13" -57d3'56"
Jodie Steiner
And 23h9'16" 40d38'10"
Jodie Weatherill's Star
Sge 20h9'11" 17d28'34"
Jodie Wigham
And 23h44'22" 33d48'14"
Jodie's Light
Pho 2h23'36" -45d57'45"
Jodies Star
Uma 12h7'32" 49d48'21"
Jodie's Star
Gem 6h52'55" 15d48'26"
Jodie's' 'This One Here'
Umi 13h45'52" 71d54'2"
Jodi-n-Steve
Cyg 20h37'8" 33d50'10"
Jodi's
Sco 17h47'41" -42d32'48"
Jodi's Eye
Cen 13h49'40" -59d20'36"
Jodi's Star
Tau 4h38'9" 19d5'37"
Jodi's Star
Vir 12h11'58" 5d13'16"
Jodi's Stellar Dream
Gem 6h35'56" 12d37'19"
Jodrell Brewster 55
Uma 13h48'51" 61d55'53"
Jody
Mon 7h40'16" -2d2'21"
Jody
Ori 5h35'40" 5d26'31"
Jody
Gem 7h6'28" 27d24'24"
Jody A. Crum
Uma 9h20'19" 61d15'41"
Jody Adametz
Aqr 22h26'35" -23d0'41"
Jody and Keith Mullins
Boo 14h22'5" 49d4'58"
Jody Ann Shea's Guiding Star
Tau 5h7'48" 25d45'46"
Jody Berry
Uma 9h39'49" 44d15'33"
Jody Bingham
Crb 16h0'30" 26d16'51"
Jody Buffalo
Cas 23h43'34" 56d3'18"
Jody Butterfly Lawson
Psc 23h23'38" 6d42'27"
Jody Byron James Noble
Lyn 8h23'47" 47d25'37"
Jody Chad Price
Boo 15h36'16" 45d43'55"
Jody Cheyenne Marie Taylor Branizor
Ori 5h8'32" 12d27'58"
Jody Christine Browns
Cap 21h28'29" -18d47'47"
Jody & Clark Kennedy
Cru 12h11'19" -60d39'53"
Jody Cramer
Lyn 6h35'55" 55d43'40"
Jody DelaPaz
Tau 4h37'58" 27d28'55"
Jody Edward Adkins
Lmi 10h21'5" 35d35'57"
Jody Fisher
And 23h44'11" 39d8'43"
Jody Hartman
And 2h38'8" 46d7'19"
Jody Lee Bowie
Lmi 10h40'15" 32d14'17"
Jody Lee Cable LMTE
Ari 1h58'58" 20d28'1"
Jody Lee Mayes
Lib 15h58'52" -9d17'31"
Jody Lee Miles, my life, love hope.
Per 2h15'22" 54d37'58"
Jody Lee Weisser
Ari 2h52'55" 28d37'26"
Jody Lee, The Rebel
Hya 9h34'11" -0d42'51"
Jody Louise Serewicz
Umi 7h55'30" 88d46'17"
Jody Lynn
Tau 4h18'24" 2d36'46"
Jody Lynn Black
And 2h25'53" 48d41'4"
Jody Lynn Ryan of Ruth & Jack Kydd
Gem 6h46'4" 32d36'42"
Jody Mae Davis
Uma 10h54'14" 64d56'59"
Jody Martin
Cas 0h12'47" 62d43'29"
Jody Moro
Cas 1h18'33" 57d28'19"
Jody Olea
And 1h43'41" 46d8'22"
Jody Peel
Her 16h21'38" 10d27'2"
Jody "Puffer" Leigh Roberts
Gem 7h18'54" 25d9'2"
Jody Puterbaugh 80th Birthday Star
Psc 1h18'18" 25d51'59"
Jody Reid
Uma 11h51'3" 47d57'38"
Jody Strickland and Bob Marcynyszyn
Lyr 19h26'8" 38d48'58"
Jody Walker, Daughter of God
And 2h26'9" 49d19'49"
Jody Wayne Sikes
Ori 5h38'48" 5d34'42"
Jody Winkenbach, My Love Forever
Crb 16h9'42" 36d34'12"
Jody Winslow McGovern
Cnc 8h27'25" 23d5'4"
Jody's Grandchildren
Cnv 13h58'49" 31d59'15"
Jody's Guiding Light
Tau 5h39'3" 22d14'21"
Jody's Star
Ori 6h4'17" 12d34'42"
Jody's Star
Her 18h54'55" 23d21'23"
Jody's Star
Lyr 18h20'54" 38d44'49"
Jodz & Han - Best friends forever
Cyg 19h39'23" 29d41'47"
Joe
Aur 5h18'57" 28d53'47"
Joe
Cnc 8h9'30" 31d25'22"
Joe
Uma 8h43'19" 55d12'7"
Joe
Umi 13h53'0" 72d30'18"
Joe
Cet 0h43'43" -0d26'24"
Joe
Umi 19h51'27" 86d22'35"
Joe
Umi 16h53'48" 83d41'17"
Joe
Lib 15h31'21" -4d13'53"
Joe
Sgr 18h12'46" -20d1'13"
Joe A. Coty
Sco 16h17'30" -12d17'45"
Joe A. DeTommaso
Uma 11h55'7" 41d50'2"
Joe A. Franco
Aur 5h39'49" 42d41'2"
Joe Ahdoot
Her 16h34'59" 42d2'5"
Joe Allen Bailey
Per 4h48'26" 40d1'36"
Joe and Anita
Cru 12h2'27" -60d41'8"
Joe and Barbara Kunovic
Per 2h15'14" 51d37'32"
Joe and Brenn
Leo 11h43'10" 26d49'51"
Joe and Bridge
Cyg 20h15'13" 55d11'47"
Joe and Charlotte Yeglic
Umi 15h29'9" 74d37'2"
Joe and Christina
Cyg 21h21'34" 37d59'55"
Joe and Hazel Temple
Ari 3h25'45" 19d52'30"
Joe and Heather make a Forever Love
Sgr 19h11'52" -15d37'34"
Joe and Jeanette Bear
Uma 11h20'28" 53d2'5"
Joe and Jodie
Sge 19h28'28" 17d6'49"
Joe and Joey
Vir 13h29'57" -16d13'29"
Joe and Josephine Pereira
Boo 14h38'39" 29d30'44"
Joe and Karen Anatra
Sge 20h6'39" 19d37'47"
Joe and Luca's mummy Nicola
Cas 3h35'59" 70d11'3"
Joe and Marceline - 8/14/05- Forever
Cyg 21h12'0" 47d1'43"
Joe and Mary McQuesten's Star
Lyr 19h16'34" 29d8'23"
Joe and Patty Schuster
Uma 9h14'9" 59d52'11"
Joe and Reni
Cyg 21h45'19" 50d55'10"
Joe Anderson
Ori 6h15'33" 14d30'7"
Joe & Ann June 18, 1955
Cyg 19h28'40" 46d58'56"
Joe & Anna
Cyg 19h45'1" 30d19'33"
Joe Aragona: A star for 50 years
Leo 9h44'50" 29d33'3"
Joe Aronica
Uma 9h45'9" 56d31'33"
Joe & Ashley Miller
Uma 11h21'4" 37d40'18"
Joe B
Dra 17h34'25" 61d39'31"
Joe Bass
Cap 20h13'45" -25d38'12"
Joe & Bea Sanches
Uma 13h47'16" 51d0'17"
Joe "Beb" Dorsey
Aqr 23h9'38" -7d52'2"
Joe Bell
Lib 14h48'4" -13d47'28"
Joe Bethune's Angel Star
Lyr 18h54'33" 28d14'44"
Joe/Betty Martin 50yrs of Fam. Joy
Cyg 21h59'4" 50d46'55"
Joe & Betty Santoleri
Cyg 21h18'25" 44d3'29"
Joe Bob Droke
Aql 20h2'44" 13d53'41"
Joe Bonene & Joyce Shaw
Dra 16h0'56" 62d15'0"
Joe Borg
Sco 16h46'3" -25d16'56"
Joe Boss memorial Star
Uma 10h7'38" 46d38'14"
Joe BouRamia
Uma 8h44'58" 68d24'16"
Joe Bowen
Sge 19h51'34" 17d32'49"
Joe Bowman
Aur 5h38'20" 44d10'45"
Joe Brimacombe
Dra 15h7'27" 62d47'18"
Joe Burriola
Lib 15h37'48" -8d37'45"
Joe Bustos
Psc 1h32'43" 9d52'11"
Joe C. 2, husband of wisdom - 50
Per 3h1'56" 47d27'5"
Joe Calabrese
Her 16h47'30" 4d48'51"
Joe Calvin Yerby
Aql 19h14'4" 13d24'2"
Joe Caporicci
Cep 22h36'23" 66d38'59"
Joe & Cari Faso
Gem 7h41'4" 26d47'35"
Joe Carlin
Uma 10h20'55" 61d30'55"
Joe & Carmenza Weizenblut
Uma 11h31'22" 37d33'44"
Joe & Cassie Loran
Leo 11h21'12" -6d16'47"
Joe Castanho
Gem 7h25'2" 23d47'1"
joe+chelsea's
Cyg 19h46'13" 56d16'3"
Joe "Chovey" Antonacceo
Aur 5h55'57" 53d28'50"
Joe Christopher Balls
Umi 14h55'53" 75d37'48"
Joe Cindric Jr.
Uma 9h10'10" 68d47'29"
Joe Clark
Uma 12h26'21" 53d44'53"
Joe Clark
Uma 12h6'10" 45d56'2"
Joe Coleman
And 0h24'59" 42d9'33"
Joe Costa
Sgr 18h24'35" -22d50'59"
Joe Crosby
Eri 3h10'28" -7d47'50"
Joe D. Church
Cma 7h6'46" -30d40'19"
Joe Daniel Barton
Dra 17h39'5" 59d9'12"
Joe Dario Santoyo
Her 17h53'48" 25d41'44"
Joe Darren Beard (Joe's Star)
Her 17h41'1" 33d2'54"
Joe David Simmons
Gem 6h58'43" 27d42'1"
Joe & Dayna Forever
Cyg 20h22'59" 55d57'7"
Joe DeGise
Tau 5h52'3" 25d29'29"
Joe Desimone
Cnc 8h52'42" 32d21'26"
Joe Diaz
Per 2h48'8" 51d49'23"
Joe Donohoe
Cep 20h48'28" 62d53'4"
Joe Dowling
Aql 19h50'38" -0d34'3"
Joe "Dragon" Crocker
Dra 10h58'9" 78d33'22"
Joe E. Pompa, Sr.
Her 18h57'10" 15d31'25"
Joe Elekes, Jr.
Lib 15h12'36" -4d0'3"
Joe & Ellen Harris - wed 8/11/1956
Psc 0h35'41" 20d48'51"
Joe Elliott Oliver
Her 16h33'31" 19d7'50"
Joe & Erica
Sgr 18h1'20" -27d56'19"
Joe et Michelle-Amor Semper
Umi 15h23'59" 75d38'50"
Joe Evan Thomas
Per 1h35'52" 52d3'30"
Joe Everett
Uma 9h48'3" 44d57'53"
Joe F. Brosius
Psc 1h7'53" 10d3'12"
Joe Fairchild
Aur 5h45'43" 50d2'39"
Joe Falcone The Atlantic City Man
Cep 22h44'19" 74d2'24"
Joe Fekete
Tau 4h46'4" 23d56'26"
Joe Felix Rowland
Sco 17h10'46" -38d22'38"
Joe Fernandez
Dra 18h50'0" 54d51'10"
Joe Fernandez
Ari 2h53'0" 13d23'15"
Joe Fitzgerald: 16th March 1974
Psc 1h35'8" 10d46'35"
Joe Florio
Leo 10h18'2" 24d18'16"
Joe Flynn Board
Cep 22h37'43" 56d56'59"
Joe Garagiola
Dra 19h33'17" 68d3'43"
Joe Gleason
Uma 9h47'24" 57d31'26"
Joe Greathead 1
Cyg 19h43'18" 32d19'43"
Joe Guarino
Sco 17h24'37" -41d34'7"
Joe Gullo
Per 3h49'10" 36d10'38"
Joe Hanley
Cap 20h37'23" -18d0'0"
Joe Hanratty
Tau 3h45'15" 28d36'2"
Joe Hanson
Leo 9h31'51" 10d5'20"
Joe Hardy
Boo 15h37'47" 45d34'38"
Joe Harvard
Cap 20h41'7" -25d41'58"
Joe Harvey, Jr.
Boo 14h46'36" 29d45'54"
Joe Haskins and Ginny Haskins
Leo 10h4'53" 13d7'7"
Joe Hawkins
Tau 5h2'57" 16d13'18"
Joe Hazan
Ari 2h53'37" 26d0'46"
Joe Helowicz
Per 4h41'22" 43d51'30"
Joe Hillhouse
Cyg 19h53'20" 33d31'12"
Joe Hot Set-up
Leo 11h6'50" 27d5'59"
Joe Jackson's Place in the Heavens
Ori 6h17'50" 14d52'4"
Joe & Jan
Uma 14h24'20" 61d5'3"
Joe & Jessica Harris
Cyg 19h41'40" 39d7'56"
Joe Joe
Leo 11h25'11" -0d38'2"
Joe Joe Ornelas
Cnc 8h43'39" 18d40'22"
Joe John Foy
Per 3h35'51" 48d7'5"
Joe John Perry
Cyg 20h33'6" 47d9'53"
Joe John Truzzolino
Ari 3h15'36" 25d58'55"
Joe & Joy ~ Truly, Madly, Deeply
Uma 9h32'29" 68d51'49"
Joe & Joyce ~ 30 Years Together
Vir 13h16'55" 12d20'24"
Joe "JRan" Randall Jr.
Uma 8h45'24" 69d59'46"
Joe & Karina Zavala
Ari 3h23'9" 22d0'5"
Joe & Kathy 32
Cyg 21h36'12" 37d44'31"
Joe Kazmierczak
Her 17h14'8" 26d53'56"
Joe & Kelsey Clements
Ori 5h30'17" -5d21'15"
Joe Keshishian
Cep 22h58'55" 82d48'43"
Joe Killacky Family
Sgr 19h16'9" -19d59'38"
Joe & Kim - 6th January 1996
Ori 5h35'27" -1d54'24"
Joe Klempay
Ari 2h30'9" 19d35'18"
Joe Kodba
Cap 20h8'23" -14d35'43"
Joe Krantz
Vir 13h52'45" -12d13'4"
Joe & Kristee
Lmi 10h36'53" 38d29'48"
Joe & Kristen
Cyg 21h27'51" 51d28'7"
Joe & Kristin Meehan
Lib 14h50'12" -5d54'55"
Joe L. Velasquez
Uma 9h18'11" 43d4'53"
Joe Lewis - An Ornish's Best #15
Uma 13h52'10" 56d31'9"
Joe Lindbloom
Her 17h42'1" 47d52'44"
Joe Lockwood, Loved Forever
Ari 3h17'15" 19d16'26"
Joe Longthorne
Per 4h35'15" 33d37'31"
Joe Lorenzen
Cep 23h18'47" 77d7'15"
Joe&Lori Piscopio"Our Wedding Star"
Uma 11h33'26" 58d11'25"
Joe & Lorraine
Sgr 18h19'49" -31d15'42"
Joe & Lorraine Amico
Sge 19h44'58" 18d15'10"
Joe Loves Jennifer
Sco 16h20'30" -30d21'14"
joe loves kathy
Cap 20h10'14" -9d44'1"
Joe Loves Sarah
Sco 16h12'30" -12d33'45"
Joe Lucas
Umi 16h17'49" 80d17'30"
Joe Lumia
Uma 8h48'36" 47d55'36"
Joe M. Riester
Vir 14h17'32" 1d47'57"
Joe Mac
Sco 17h51'59" -36d20'43"
Joe Mantell
Sgr 18h39'11" -33d10'10"
Joe Marciano
Per 4h21'8" 50d41'32"
Joe & Margie Angelica
Sco 16h53'49" -34d55'57"
Joe & Marie Cipolla
Cap 21h34'7" -13d58'35"
Joe Marola
Gem 7h31'44" 26d28'50"
Joe Masters
Ari 2h48'36" 13d43'30"
Joe Matthew Hannigan
Per 3h35'5" 37d6'31"
Joe McElhiney
Cep 21h5'32" 57d50'53"
Joe Michael Palermo
Ori 5h33'32" 2d50'42"
Joe Miller
Leo 10h21'53" 8d52'25"
Joe Miller
Cep 21h12'37" 74d49'12"
Joe Montalvan 9/23/36
Lib 15h21'56" -6d31'15"
Joe Mora
Vir 13h39'3" -0d9'40"
Joe Moretti
Leo 9h43'10" 31d13'7"
Joe "Mr. Hoops" Weisbrodt
Gem 7h27'7" 20d18'3"
Joe N. Franco
Aql 19h45'26" 4d41'15"
Joe N Jones
Psc 0h38'34" 6d44'3"
Joe n Mya "My Shinning Star"
Lyn 7h46'27" 53d12'41"
Joe & Net
Gem 7h3'1" 30d12'24"
Joe Nichols
Sgr 18h1'56" -27d56'54"
Joe & Nicole
Cyg 19h47'46" 38d51'33"
Joe Nosari
Sgr 18h4'59" -23d59'53"
Joe Novosedlik
Cep 2h16'10" 85d24'13"
Joe Nunley
Dra 19h49'20" 70d7'6"
Joe O
Sgr 18h0'3" -35d11'45"
Joe O'Bryan
Gem 6h42'57" 19d52'49"
Joe Oppedisano
Leo 9h44'46" 7d58'34"
Joe Oscar Hill III
Aur 6h32'21" 33d54'22"
Joe Packard's Quiet Place
Aur 5h41'43" 33d43'58"
Joe Paglia
Uma 10h36'15" 69d14'18"
Joe Paige...Love
Cas 23h10'39" 56d10'34"
Joe & Pat Scheidt Eternal Love
Per 3h46'8" 38d57'22"
Joe & Pat's "50th"
Cyg 19h35'25" 31d16'55"
Joe & Peggy O'Connor
Aql 19h57'56" 8d38'41"
Joe Pereira
Her 16h20'5" 45d6'18"
Joe Perrone
Her 17h49'57" 45d46'44"
Joe Perry
Her 17h28'50" 26d25'1"
Joe Pezzino "The Bookie"
Her 18h44'52" 21d33'14"

Joe Pietro
Lib 15h10'59" -6d21'0"
Joe Podhurchak
Vir 13h4'48" 11d23'3"
Joe Pons - "The Patriarch"
Peg 23h41'10" 10d16'47"
Joe Prasil
Lib 14h55'57" -5d7'38"
Joe Presloid
Uma 8h51'13" 56d49'18"
Joe Rainbow
Peg 22h35'58" 20d47'20"
Joe Ray
Cam 3h56'9" 67d31'18"
Joe Re
Her 15h52'51" 44d38'24"
Joe Reilly
Uma 11h52'30" 37d17'20"
Joe Rice
Ari 2h49'50" 28d30'42"
Joe Roberts
Her 18h44'27" 25d44'17"
Joe & Rosa Chibnik
Cyg 20h18'40" 44d4'27"
Joe Rufino
Her 16h12'5" 23d55'24"
Joe Ruggiero
Crb 15h48'35" 27d9'39"
Joe Salibi
Cas 2h19'41" 59d13'46"
Joe Sandahl
Aql 20h14'45" 7d56'58"
Joe Santiago
Her 17h43'1" 18d52'5"
Joe Sapere &Amputees Across America
Her 16h45'41" 8d55'41"
Joe Saynavong
Cnc 9h7'38" 11d32'33"
Joe & Sharon
Cyg 19h46'26" 38d3'34"
Joe & Sharon Cade Love Always
Cas 23h15'9" 55d47'30"
Joe Shoemaker
Boo 14h45'42" 36d0'45"
Joe Simon
Lyr 18h33'10" 32d53'18"
Joe Skaife
Com 12h46'22" 27d58'21"
Joe "Smoopy" Mason
Uma 9h42'20" 50d4'40"
Joe Soap
Per 4h36'59" 32d50'45"
Joe Solomito, Sr.
Cep 22h26'23" 77d48'6"
Joe Sproul's Star Gazer
Aur 5h19'57" 41d44'16"
Joe & Steffani Grande
Lmi 10h41'51" 38d43'1"
Joe Sucharda
Her 16h31'38" 46d55'23"
Joe (Super Constellation) Politi
Boo 15h22'35" 43d11'47"
Joe Sweeney
Aur 5h22'47" 39d12'44"
Joe & Tammy Foryan
Aur 5h27'42" 32d20'11"
Joe Taverner, Love Always Cindy
Gem 7h38'38" 32d45'5"
Joe "The Juggler" Welling
Lib 14h42'44" -8d43'40"
Joe "The Sicilian" Restivo
Cmi 7h26'12" 8d27'26"
Joe Thirty
Cnc 8h2'19" 14d41'5"
Joe Todd Campbell
Cnc 8h18'2" 15d34'41"
Joe Toohey
Psc 1h10'23" 26d6'2"
Joe Valente
Cnc 7h59'0" 14d59'48"
Joe Vallette
Uma 12h32'8" 58d3'16"
Joe van Akker's first birthday star
Aqr 23h55'20" -18d40'15"
Joe Vidal
Cap 20h21'0" -10d31'53"
Joe Vilane
Cyg 21h12'45" 45d12'54"
Joe & Virginia Golden Anniversay
Uma 9h36'55" 56d57'11"
Joe W. Kelly Jr.
Uma 10h55'24" 50d9'28"
Joe W. Kirkland
Cep 20h49'14" 68d56'38"
Joe W. Maze
Aqr 22h18'56" 1d47'22"
Joe Waguespack "My Shining Star"
Uma 8h35'5" 63d27'48"
Joe Walker Sidgwick
Umi 16h17'46" 70d21'55"
Joe Wesley Punches (Joeronimo)
Psc 0h50'10" 28d52'57"
Joe & Whitney
Tri 2h26'2" 30d50'12"
Joe Will Love Yvonne 4 Ever.
Cyg 20h15'31" 43d54'38"
JOE (William Joseph) CALI
Psc 0h25'48" -2d17'4"
Joe Williams
Per 4h29'6" 36d26'27"
Joe Wing
Leo 9h40'8" 26d51'41"
Joe, Gian and Giovanni
Cnv 13h11'4" 45d6'54"
Joe, in loving memory of Catherine
Cnc 8h24'13" 14d19'39"
Joe, Katie, Michael & Thomas
Cas 23h10'49" 58d49'29"
Joe, my one and only, I love you
Her 16h20'43" 47d55'38"
Joe2e
Her 17h27'28" 35d0'22"
Joe40
Ori 6h13'54" 6d4'12"
joeaim
Her 16h8'32" 46d26'4"
JoeB&CarolJ~InLove4Ever
Mon 6h45'15" 11d18'6"
JOECAM
Uma 9h32'17" 65d38'23"
JoeDelaoLovesYuriBazan: EndlessTruLv
Cap 20h38'9" -20d3'38"
JoeDeray
Sgr 18h17'14" -28d24'53"
JOEDI
Psc 1h14'56" 30d29'17"
Joedrea
Uma 10h5'29" 49d12'15"
JOEGE
Uma 11h38'9" 33d34'0"
Joehoney
Her 16h59'46" 38d3'12"
Joeinne Emmette Armga
Lyn 7h42'17" 39d33'31"
Joel
Ari 3h2'10" 12d51'22"
Joel
Psc 1h32'36" 12d12'29"
Joel
Lmi 10h39'5" 28d48'58"
Joel
Her 18h52'28" 21d30'8"
Joel
Sgr 19h45'41" -15d23'25"
Joel
Ori 5h5'43" -0d12'21"
JOEL
Lac 22h42'59" 52d53'28"
Joel 30 May
Gem 7h29'0" 18d44'47"
Joel A. Belaire
Uma 9h43'10" 45d23'55"
Joel A. McCullough
Her 17h39'38" 16d41'38"
Joel Alexander Growney
Ori 5h8'37" 15d19'41"
Joel Alexander Meadows
Lyn 7h10'53" 51d11'43"
Joel Alonzo Lowell
Aur 6h15'35" 52d17'52"
Joel and Amber's "Forever Star"
Psc 1h24'28" 14d28'13"
Joel and Clare Gorski
Tri 2h11'4" 34d12'21"
Joel And Heathers Star
Cyg 20h34'21" 36d37'26"
Joel and LaShawn soul mates TLF
Lmi 10h5'20" 36d54'42"
Joel and Lisa
Ari 2h37'49" 11d55'48"
Joel and Sam Always and Forever
Cyg 20h13'57" 52d25'13"
Joel Andrew Lardner - shining forever
Cru 12h28'2" -60d48'26"
Joel Aram ~ Señor Moustache ~
Per 3h33'3" 47d12'43"
Joel B. Mayer
Lyn 8h17'29" 56d12'16"
Joel Batule
Tau 4h22'42" 24d9'59"
Joel Beak
Ser 18h30'5" -0d14'7"
Joel Camarena
Boo 14h41'43" 34d46'56"
Joel Charles Wilson
Pho 0h58'8" -52d54'2"
Joel Daniel Marcus
Per 3h34'48" 33d9'59"
Joel Dauterman
Aql 19h24'6" -10d37'35"
Joel David
Uma 11h5'45" 68d37'34"
Joel David Walker
Cyg 21h42'26" 47d56'29"
Joel Demulder
Cas 0h19'50" 51d37'17"
Joel E. Andrews
Psc 1h40'58" 5d13'35"
Joel E. Staubs
Aql 19h18'55" -0d3'58"
Joel Edgar
Cep 22h16'54" 68d21'40"
Joel Edward Simchick
Sco 16h42'9" -40d44'45"
Joel Ethan Tate
Umi 15h9'40" 67d44'11"
Joel Gabriel Santos
Umi 14h35'39" 75d35'46"
Joel Garrison
Lib 14h52'31" -15d25'40"
Joel George Blackledge
Umi 14h44'33" 71d33'9"
Joel Glassborow
Cen 11h45'27" -59d12'40"
Joel Goski
Aqr 22h20'26" -9d50'35"
Joël Graille
Her 18h16'6" 15d25'41"
Joel Henry Baron
Sgr 18h2'55" -18d14'21"
Joel Hetrick
Psc 1h28'32" 17d4'35"
Joel Hill's Christening Star
Umi 16h25'17" 74d46'29"
Joel I Steele
Aqr 21h48'53" 2d13'14"
Jo-El In Memory For "Cosmo"
Per 3h0'41" 52d50'24"
Joel Jacob Stone
Ori 5h58'43" 6d28'16"
Joel Jordan Maiman 3-6-1943
Tau 3h48'5" 26d9'46"
Joel Keith Snyder
Ari 3h18'46" 30d43'7"
Joel Klinger
Psc 0h51'51" 27d37'32"
Joel L. Marnette
Aql 20h18'56" 2d26'22"
Jöel "Le pètit prince"
Aur 6h15'41" 45d23'21"
Joel Marcus Williams
Ori 5h56'46" 12d23'53"
Joel Martin Cohen 4-14-33
Umi 14h26'27" 75d18'49"
Joel Matthew Ficek
Uma 11h17'13" 60d7'8"
Joel Michael Dunham I
Sco 17h1'52" -41d39'50"
Joel Montgomery
Uma 9h59'26" 57d45'26"
Joel Osamu Dominie
Cap 20h45'40" -18d4'58"
Joel Patrick
Her 16h26'34" 49d37'46"
Joel Patrick Buckingham
Cep 22h20'51" 67d39'29"
Joel Pederson
Ori 6h9'24" 15d16'36"
Joel Pesqueira
Cep 23h20'55" 81d6'17"
Joel Price
Uma 8h37'26" 53d29'35"
Joel Robert Behr
Uma 9h58'40" 59d58'54"
Joel Robert Weissman
Per 2h20'37" 56d6'49"
Joel Rosado
Lib 15h12'22" -14d42'35"
Joel Rose
Her 16h33'14" 34d25'28"
Joel Rosenstock
Her 17h37'24" 31d47'8"
Joel Santino
Per 3h49'54" 47d43'46"
Joël Sebazungu
Ari 3h13'50" 18d45'35"
Joel & Sharon's Shining Star
Tau 5h20'7" 25d16'40"
Joel Spiro
Uma 10h18'5" 59d48'6"
Joël Stefan Mägli
And 0h17'43" 45d44'55"
Joel Stephen
Dra 15h7'8" 60d28'0"
Joel Stephen Parker
Tau 4h4'53" 5d3'44"
Joel T. Goski
Aqr 20h41'23" -11d9'40"
Joel Thomas Dummett
Psc 1h5'13" 11d13'36"
Joel Totherow "My Baby"
Cep 22h57'11" 70d48'23"
Joel Trester Bouslog
Cap 21h31'19" -16d6'56"
Joel Wayne Merriman
Aur 5h37'40" 44d7'47"
Joel & William's
Uma 11h40'41" 30d23'13"
Joela Dos Santos
Tau 3h25'9" 18d56'57"
JoelAnn
Cyg 19h46'41" 37d8'7"
Joelconn
Ori 5h53'26" 21d57'6"
Joeleen
Psc 0h43'7" 9d55'49"
Joeleen Hughes
And 23h25'14" 46d55'20"
Joeleen L. Mellars
Leo 11h14'2" 8d54'29"
Joelene
Cru 12h6'18" -61d7'52"
Joelene Ann Bergonzi
Aqr 20h59'40" -12d44'31"
Joeli Lynn Mothershead
Leo 10h36'56" 20d59'26"
Joelie
Aqr 23h26'54" -19d53'57"
Joelila 13 (Joel Benjamin Rushton)
Aql 19h29'39" 3d40'46"
Joella
Cyg 20h10'7" 40d31'12"
Joelle 05
Leo 11h34'36" -6d15'41"
Joelle 14.10.67
Ori 5h55'29" 11d32'53"
Joelle and Vincent
Lib 15h6'53" -3d48'16"
Joelle Anne Kairys
Psc 1h12'42" 31d32'50"
Joelle Anne Snyder
And 0h43'12" 40d2'37"
Joëlle Berends
Umi 14h3'33" 67d14'10"
Joëlle Karam
Cyg 19h35'37" 29d50'53"
Joelle Khoury Aouad
And 23h29'58" 47d24'19"
Joelle M. Wood
And 23h14'39" 43d5'42"
Joëlle Mansuy Delcroix
Vir 11h40'28" -0d58'23"
Joelle Martel
Tau 4h29'18" 1d38'48"
Joelle Novak
Ori 6h6'0" 18d52'24"
Joelle Peri Jonas
Uma 10h31'8" 51d6'55"
Joelle R Zielin
Aqr 23h53'38" -10d7'6"
Joelle Stephanie White
Crb 15h39'32" 33d56'6"
Joelle Swistak, My Love
Lyn 7h55'3" 54d45'33"
Joelle Vincenzo Macrino
Vir 11h42'30" -3d13'35"
Jo-Ellen Gibson (Auntie Jo)
Leo 11h41'6" 21d5'5"
Joellen J. Zitch
Cap 20h59'57" -20d27'38"
JoEllen Martin
Cnc 7h57'33" 11d9'34"
Joelle's jewell
Ari 2h4'6" 14d10'43"
Joelley
Ari 2h7'19" 24d57'2"
JoEllyn Ciani
Lyn 7h14'5" 52d10'8"
JoelMarieMcDevitt
Lyn 7h44'7" 54d46'13"
Joel's Inspiration
Sgr 18h21'0" -32d53'20"
Joel's Star
Sgr 19h16'21" -14d28'59"
Joelstar
Equ 21h8'44" 11d52'35"
Joelu
Cyg 20h59'53" 45d12'24"
Joelyn Denice
Sgr 19h22'22" -41d49'58"
Joely's Angel
And 23h34'33" 41d55'17"
JoeMarieAnnMarie Jauregui-Theodosis
Aqr 22h5'12" -0d44'54"
JoeMooner
Sco 16h10'31" -16d10'54"
Joene C. Kemp
Aqr 21h37'8" 0d6'54"
Joerlings
Her 16h39'46" 34d26'43"
Joern + Sabine
Cyg 20h43'22" 36d23'26"
Joero
Cas 0h28'38" 63d9'21"
Joe's Adored Sebastian
Cmi 7h30'23" 7d42'43"
Joe's Bar & Grille
Leo 9h27'55" 12d49'8"
Joe's Destiny 2017
Aur 5h39'33" 49d46'17"
Joe's Dream
Cap 20h13'7" -13d13'37"
Joe's Edge
Peg 21h59'53" 17d11'17"
Joe's Star
Per 4h49'56" 43d13'54"
Joe's Star
Cnc 9h4'47" 32d19'50"
Joe's Star
Uma 11h59'39" 58d16'57"
Joe's Star
Cru 12h25'10" -55d45'48"
Joe's Star thats to fat to do stuff
Psc 1h18'15" 24d3'7"
Joe's Treasure in the Sky
Cap 20h31'1" -26d29'50"
Joesph Edwin Joyner
Ori 5h26'42" 1d40'37"
Joesph Francis Salmon Sr.
Cep 22h12'41" 60d18'34"
Joesph M. Kovach
Lib 14h57'4" -18d25'49"
Joesph M. Kraus
Cyg 20h30'15" 31d41'27"
Joesph Paul
Uma 11h41'47" 47d50'50"
joespomicmac
Tau 3h39'0" 27d53'48"
JoeStar
Uma 9h10'35" 56d45'54"
Joet
Sgr 18h3'35" -21d46'26"
JoEtta G Abo
Uma 11h5'42" 48d27'13"
Joetta & John Furrer
Lyr 18h52'22" 31d50'42"
Joetta Waterman
Com 12h31'48" 23d31'42"
Joette
Sgr 18h27'23" -19d5'21"
Joette the October Diamond
Cnc 9h8'7" 12d9'36"
Joette Ward
Lyr 18h42'52" 40d4'47"
Joey
And 1h33'15" 46d25'26"
Joey
Per 1h41'50" 54d29'17"
"Joey"
Gem 7h42'41" 33d26'2"
Joey
Aur 5h36'3" 40d34'16"
Joey
Uma 11h18'12" 39d13'5"
Joey
Ori 5h11'59" 12d2'52"
Joey
Boo 14h53'33" 23d32'6"
joey
Lmi 10h48'27" 28d44'14"
Joey
Leo 9h59'57" 17d39'15"
Joey
Aqr 23h9'7" -7d58'46"
Joey
Lib 15h31'3" -12d1'40"
Joey
Aqr 22h22'15" -2d59'11"
Joey
Uma 9h36'15" 54d31'59"
Joey
Lyn 6h20'17" 57d57'12"
Joey
Uma 12h43'47" 57d24'51"
Joey
Uma 10h39'14" 56d46'7"
Joey
Cep 23h1'23" 71d9'58"
Joey
Umi 15h6'15" 72d56'51"
Joey
Umi 14h55'20" 73d32'15"
Joey
Uma 10h12'33" 66d59'21"
Joey
Uma 8h54'35" 61d51'54"
Joey *2006*
Tau 4h20'49" 28d4'43"
Joey Alan Statlander
Cap 21h5'59" -17d10'49"
Joey Allen Smeen
Uma 8h54'48" 50d27'31"
Joey and Brandi Roberts
Sge 19h44'59" 19d0'8"
JOEY AND CHRISTINIE
Aqr 21h12'51" -8d31'50"
Joey And Kara Star
Umi 15h31'30" 74d27'4"
Joey and Sabrina Forever
Her 17h51'54" 29d1'21"
Joey Andrew Gomez
Per 2h48'16" 53d48'13"
Joey & Angie Always
Sge 19h44'42" 17d37'47"
Joey Arditi
Ori 6h18'57" 14d41'57"
Joey Baby
Gem 6h47'22" 17d47'40"
Joey Balls and Snaps Clams
Del 20h15'56" 9d43'56"
Joey Bear
Sco 16h7'29" -12d13'29"
Joey Bear Luce
Leo 10h11'11" 22d12'4"
Joey Bell
Uma 11h35'36" 28d43'56"
Joey Berard
Ori 6h5'0" 9d51'58"
Joey BoBo David
Umi 16h18'34" 76d51'12"
Joey Brown
Uma 9h50'5" 55d20'42"
Joey Burke
Leo 10h13'1" 15d41'28"
Joey Carl Gorringe
Per 3h10'8" 56d33'12"
Joey Caronna
Cma 6h33'28" -29d53'40"
Joey Colt Crow
Uma 12h1'2" 50d43'32"
Joey D and Mikey W
Uma 9h44'34" 66d34'56"
Joey D Sclafani
Aqr 23h51'57" -17d18'29"
Joey Di Santo
Per 4h6'38" 44d10'24"
Joey DiMonte
Ori 5h43'27" 1d45'24"
Joey Dorn
Cnc 8h41'16" 18d48'41"
Joey DuMouchel
Sgr 18h59'7" -30d5'35"
Joey Errichiello
Uma 12h40'20" 61d2'52"
Joey Fitch
Cnc 9h13'11" 7d45'25"
Joey Flesner
Dra 12h55'9" 72d2'39"
Joey Francis
Umi 15h38'16" 77d45'27"
Joey Fuschetto
Her 18h39'30" 21d13'28"
Joey G.
Sco 16h6'8" -15d6'1"
Joey Girl
Lyn 7h28'59" 45d1'46"
Joey Green
Aql 19h43'22" -0d7'6"
Joey Harrington
Uma 12h33'22" 58d35'23"
Joey Hitchcock
Peg 22h10'36" 7d13'47"
Joey Hyland
Ori 5h4'26" 5d16'9"
Joey & Jenn Forever and a Day
Cyg 20h6'55" 35d19'8"
Joey Jowett
Uma 8h26'32" 62d36'10"
Joey Kaiser "God's Angel Among Us"
Cma 7h10'15" -18d41'42"
Joey Kosnick Jr.
Uma 11h2'45" 49d25'38"
Joey<\@>Leo.Star
Uma 9h34'46" 56d23'10"
Joey Letourneau
Uma 9h56'9" 43d21'28"
Joey Loves Colleen
Uma 13h1'12" 60d3'34"
Joey Loves Hayley Miller
Ari 3h20'8" 29d34'53"
Joey M. Reece
Lac 22h24'41" 49d5'4"
Joey Mandel
Sco 16h19'32" -10d58'49"
Joey Martin
Ori 4h51'28" 13d14'2"
Joey McManus
Lyn 7h32'17" 36d46'10"
Joey Meisinger The Little Man
Uma 11h30'50" 58d33'53"
Joey Mendes
Uma 11h15'55" 57d44'44"
Joey Mendonca Ramos
Sco 17h37'7" -33d15'0"
Joey Mishel
Aur 5h45'27" 49d25'41"
Joey Muffin Cakes
Aqr 21h44'49" -0d31'48"
Joey "My Bird" Williams
Aql 19h31'20" 0d0'8"
Joey My Love
Sgr 18h33'22" -26d11'28"
Joey "My One and Only True Love"
Tau 5h44'54" 19d56'46"
Joey Patricia Weatherill
Aql 19h35'57" 3d2'6"
Joey Patrick Malone
Cep 22h59'13" 66d54'18"
Joey R. Summers
Her 17h45'49" 14d57'9"
Joey Randall Cunningham
Lyn 9h18'8" 38d44'10"
Joey Ryan Lagner
Cnc 8h45'44" 18d17'53"
Joey Sammon
Per 3h52'18" 39d37'49"
Joey Saporito
Her 17h14'25" 33d51'46"
Joey & Shauna Maisto
Uma 10h32'31" 68d15'14"
Joey Sikes
Ori 5h1'16" 11d13'28"
Joey Supak
Leo 11h9'48" 14d29'51"
Joey Tabak
Sgr 19h40'39" -13d21'23"
Joey Taylor
Ari 3h12'2" 29d16'25"
Joey Valvano
Aqr 22h53'32" -17d15'4"
Joey Vincent DeMarco-LaPlante
Her 17h34'16" 47d35'36"
Joey Vitale
Her 17h10'41" 32d34'6"
Joey Williams
Gem 7h6'49" 29d0'24"
Joey Zuccarello
Vir 11h42'37" 5d53'40"
Joey, my Lovie
Sgr 18h57'20" -33d43'29"
JoeyBear
Umi 16h9'27" 85d43'52"
Joey-Fish
Uma 10h35'11" 58d4'2"
Joey-oey-oey
Dra 19h20'16" 64d51'32"
Joey's 50th
Cap 20h30'8" -10d47'24"
Joey's Faith
Ari 3h17'38" 28d55'48"
Joey's Nightlite over Marginal Way
Cnv 13h50'41" 38d5'16"
Joey's Star :)
Cyg 20h1'43" 39d45'59"
Joey's Star
Dra 13h25'35" 67d56'29"
Joey's Star
Cru 12h7'35" -58d53'53"
JoeyTamara
Aur 5h25'0" 47d38'25"
Johamy
Cyg 20h37'37" 58d33'7"
Johan Gremaud
Cyg 20h9'3" 48d56'40"
Johan Iskandar
Lib 14h51'49" -1d34'53"
Johan Lysiak
Cen 13h39'40" -43d27'4"
Johan Peter & Sarah Jane Hansen
Cyg 21h18'15" 32d32'29"
Johana & Valentina
And 1h0'25" 45d58'15"
Johanka
Cyg 21h57'51" 50d49'32"
Johann
Uma 8h42'36" 47d57'14"
Johann Brunnhuber
Uma 11h13'38" 64d46'44"
Johann Daxner
Ori 6h22'21" 15d23'10"
Johann Erich Teutsch
Uma 11h56'24" 55d4'8"
Johann Friedrich Zubrod
Uma 9h2'21" 59d18'59"
Johann J. Heuser
Psc 0h44'16" 5d53'35"
Johann Michael Rieger Jr. "Mike"
Uma 12h16'29" 62d21'40"
Johann Pellkofer
Uma 10h29'1" 48d53'12"
Johann Rettenwender
Ari 2h18'21" 12d49'35"
Johann Rettenwender
Umi 16h7'41" 79d57'46"
Johann Sagerschnig
Ari 3h17'10" 28d39'48"
Johann Schiessel
Cnc 8h28'53" 29d23'28"
Johann Sonntag
Uma 9h58'42" 56d58'37"
Johann und Johanna Niederreiter
Ori 6h20'28" 6d19'36"
Johanna
Equ 21h12'15" 8d20'20"
Johanna
Gem 6h45'21" 23d16'45"
Johanna
Gem 6h44'12" 24d12'27"
Johanna
And 1h34'21" 41d9'26"
Johanna
Aur 6h32'14" 38d33'49"
Johanna
Uma 10h22'44" 41d55'35"
Johanna
Dra 19h23'56" 64d25'59"
Johanna
Cam 7h57'9" 65d20'31"
Johanna
Cep 22h1'28" 72d39'57"
Johanna
Sgr 18h10'51" -19d33'16"
JoHanna
Lib 15h22'52" -22d5'39"
Johanna
And 1h15'51" 48d20'41"
Johanna Allen
Cas 0h59'47" 56d32'44"
Johanna and George
And 23h18'50" 47d37'10"
Johanna Angel Gardner
Cap 20h28'7" -23d33'9"
Johanna Augusta
Lyn 8h7'46" 37d37'8"
Johanna Chica
Sgr 18h8'5" -18d2'0"
Johanna De Simone
Sgr 20h4'28" -30d11'29"
JOHANNA DEGRAAF ROELEVELD
Leo 10h29'52" 20d48'38"
Johanna Dora Strouse
Uma 11h38'14" 50d20'0"
Johanna Eloa Canfield
Uma 13h44'56" 57d48'58"
Johanna G. Gonzalez
Cyg 22h0'31" 50d22'23"
Johanna Hermine
Umi 15h14'14" 77d32'4"
Johanna J. Zakrzewski
Sgr 19h38'26" -13d30'23"
Johanna Jane Smith
Ari 3h2'24" 25d45'46"
Johanna Kate
Peg 22h24'52" 20d44'35"
Johanna Katharina Helene
Uma 9h33'53" 63d49'23"
Johanna Kotterba
Uma 12h5'51" 46d19'54"

Johanna Lacy
Del 20h33'19" 4d1'23"
Johanna Lucas
And 1h3'6" 45d19'40"
Johanna M. Paltrineri
Cep 23h22'38" 79d31'16"
Johanna Marie Aspiras
Cyg 20h7'58" 35d29'52"
Johanna Marie Ledell
Lyr 19h23'38" 37d33'47"
Johanna Marie Swope
Uma 10h5'33" 45d35'29"
Johanna My Reason
Heitritter
Ari 2h16'11" 25d14'13"
Johanna Ochoa
Cas 1h37'44" 64d57'40"
Johanna Perez
Crb 16h13'33" 33d2'40"
Johanna Reinbolz
Gem 7h28'40" 19d12'36"
Johanna S B Nani
Ari 2h51'24" 27d40'57"
Johanna Schleser
Pho 0h16'56" -53d49'27"
Johanna Sedeborg
Cas 0h32'32" 61d12'37"
Johanna Star
Uma 10h39'7" 48d23'35"
Johanna "Tina" Almaraz
Aql 19h15'40" 9d9'31"
Johanna Y. Escobar
Aqr 20h52'27" -3d12'33"
Johannah Red Sorensen
Peg 22h13'33" 8d22'3"
Johanna-Mehdy
Uma 11h7'16" 33d11'5"
Johanna-Rain Freya
Heimberger
And 1h43'54" 42d16'45"
Johanna's Birthday Star
Mon 6h50'54" -5d57'28"
Johanna's Sparkle
And 23h37'53" 43d34'50"
Johanne
And 2h38'34" 41d8'32"
Johanne Comtois
Vir 13h2'35" -20d3'47"
Johanne Girard
Sgr 19h53'36" -29d1'37"
Johanne & Richard Conry
Cyg 20h18'26" 38d46'11"
Johanne St-Germain
Uma 8h38'37" 58d40'56"
Johannes
Lyr 18h45'38" 34d2'59"
Johannes Altmann
Aur 5h22'53" 30d2'16"
Johannes de Waard
Uma 10h1'4" 63d41'29"
Johannes Deeg-Karamujic
Uma 10h22'36" 54d26'29"
Johannes Grommes
Uma 11h16'50" 39d55'36"
Johannes Henke
Ori 6h19'14" 9d50'48"
Johannes Heppe
Uma 11h39'26" 36d0'55"
Johannes Nowak
Uma 8h45'33" 57d27'1"
Johannes Riedl
Cnv 13h32'10" 50d4'6"
Johannes Steffens
Dra 11h12'55" 77d51'52"
Johannie
Uma 8h14'28" 66d27'44"
Johlean
Uma 11h27'3" 63d13'29"
Johlise Jerenee
Uma 11h11'36" 51d48'8"
John
Lyr 18h35'9" 39d47'21"
John
Crb 15h17'20" 30d30'17"
John
Lyn 7h34'10" 36d22'5"
John
Aqr 21h41'46" 0d47'8"
John
Psc 1h12'2" 11d26'45"
John
Leo 11h38'26" 17d5'1"
John
Gem 6h5'20" 27d4'12"
John
Uma 8h10'57" 67d20'54"
John
Cep 20h55'53" 68d35'2"
John
Cap 21h4'3" -19d40'40"
John
Sgr 18h54'17" -18d48'59"
John – My Knight in
Shining Amour
Cap 21h3'14" -16d25'44"
John A. Bauman
Lyn 9h8'2" 37d24'39"
John A. Boyd Jr.
Cep 22h34'7" 60d32'54"
John A. Canta
Aqr 21h38'30" 2d17'21"
John A. Carbo
Ari 2h44'21" 14d24'40"
John A. & Connie Smith
Bird
Ori 5h13'0" 1d12'57"
John A. DeVito
Uma 11h8'45" 68d51'11"
John A. Ellis, Jr.
Psc 1h43'19" 13d16'13"
John A. Foltz
Aql 19h35'39" 12d10'2"
John A. Frank
Her 17h8'1" 34d25'32"
John A. Garvin
Leo 11h41'21" 21d21'22"
John A Horti
Sco 16h42'58" -25d37'16"
John A. Izzo, Jr.
Sco 16h54'42" -42d6'32"
John A. "Johnny" Kuriger
Sco 16h8'18" -16d39'6"
John A. Kell
Boo 14h30'47" 29d47'54"
John A. LaBarbera
Cnc 9h8'47" 30d22'53"
John A. Lahoski
Uma 9h23'31" 62d36'45"
John A. Lane
Psc 0h42'13" 17d53'2"
John A. Large, Sr (in memory of)
Uma 12h0'25" 30d30'14"
John A. Larkin
Lmi 10h24'16" 35d28'9"
John A. Lytle
Aql 19h8'6" 5d59'23"
John A. Malone, MD
Ori 5h55'26" 17d40'15"
John A. McArthur III
Dra 18h47'51" 63d58'19"
John A. Meek
Tau 4h4'59" 5d22'50"
John A. Middleton
Aur 7h30'11" 41d5'31"
John A. Naney
Lmi 10h50'2" 31d50'11"
John A. Richardson
Tau 5h56'43" 26d2'2"
John A Schafer's Shining
Light
Vir 15h9'20" 0d28'27"
John A Scott, Sr-Much
Love Forever
Per 3h40'4" 45d53'27"
John A Smith
Ori 5h58'59" 2d31'20"
John A. & Tobe K. Martinez
Cyg 21h30'42" 49d14'7"
John A. Traficante
Ori 5h36'49" 1d37'14"
John A. Vento II
Cep 22h19'48" 64d0'15"
John A. Weisz
Pho 23h30'4" -47d49'57"
John A. Willard
Ari 2h23'50" 12d25'52"
John A Wright (speck)
Cep 21h17'33" 66d48'29"
John A. Yager
Gem 6h31'47" 15d39'44"
John Aaron Hancock
Dra 17h7'48" 54d13'49"
John Adam Adams
Gem 7h48'9" 32d6'47"
John Adam Schepp
Tri 2h7'16" 33d18'55"
John Adam Shaner
Lyn 7h37'22" 52d10'8"
John Adams
Cep 3h29'16" 81d4'34"
John Adams Norton III
Umi 14h42'48" 75d26'55"
John & Afra Wilder The
Mabrouk Star
Cru 11h59'48" -58d53'19"
John aka Jacques Barton
Puls
Leo 11h30'41" 23d16'55"
John Alan
Aur 7h27'38" 39d37'3"
John Alan Cootes
Cru 12h44'14" -64d6'51"
John Alan Harbinson
Per 2h10'8" 54d31'53"
John Alan Isley
Aqr 22h47'49" -15d57'49"
John Alan Loretta
Sgr 19h10'20" -20d55'54"
John Alan Williams
Sco 16h36'36" -29d21'54"
John Alatopoulos
Uma 8h34'52" 47d13'16"
John Albanese
Sco 16h51'21" -38d39'38"
John Albert O'Brien
Cep 0h17'53" 67d27'33"
John Alec Ferrell
Cyg 19h38'48" 34d21'43"
John Alexander
Lib 15h20'38" -8d36'8"
John Alexander Fairbanks
Vir 13h18'3" 3d54'48"
John Alexander McClain
Cmi 7h28'45" 5d45'46"
John Alexander Murdoch -
Miss You
Uma 10h55'14" 72d13'37"
John Alexander Patterson
Lyr 18h45'11" 32d1'29"
John Alexander Richard
Pfirrmann
Sco 16h5'21" -16d26'23"
John Alexander Rossetto,
Jr.
Leo 10h24'17" 18d58'55"
John Alexander Walker II
Her 17h25'41" 36d21'47"
John & Alexandra Tierney
Uma 11h17'50" 60d48'45"
John & Alexia Williams
Cru 12h4'45" -59d30'25"
John Alfred
Crb 16h19'14" 30d4'9"
John Alfred Fountain
Cru 12h34'20" -62d16'27"
John Allan Mekkelsen
Ari 2h24'5" 21d55'18"
John Allan Reynolds
Uma 9h39'54" 55d17'7"
John Allan Staab, Jr.
Cru 12h33'52" -56d11'7"
John Allen "AL-P" Pickett,
Jr.
Uma 9h29'14" 46d30'26"
John Allen Conway, IV
Her 16h42'9" 29d1'19"
John Allen Crenshaw
Per 1h43'33" 54d30'55"
John Allen Davis
Gem 7h8'55" 16d6'34"
John Allen Gipson
Her 17h40'55" 33d51'22"
John Allen Johnson
Her 17h22'34" 35d30'56"
John Allen Slabaugh
Uma 9h15'55" 51d18'24"
John Alston Logue
Aql 19h42'16" -0d6'5"
John Anatoliy
Vir 13h2'28" 4d1'12"
John and Alma - Written in
the Star
Aqr 23h4'10" -3d56'13"
John and Angell Fontenot
And 1h24'51" 40d58'54"
John and Anna Mae Gallice
Leo 9h42'8" 26d47'8"
John and Aspen Allen
Ori 4h46'57" 3d16'4"
John and Bea: True Stars
Forever
Leo 11h43'57" 25d59'44"
John and Betty Matulovich
Love Star
Peg 22h50'12" 19d7'34"
John and Brandee Love
Lives Forever
Cyg 20h43'51" 52d25'26"
John and Carol Carney
And 2h12'31" 38d26'44"
John and Carol Sapienza
Cas 1h41'30" 63d41'30"
John and Carrie Simmons
Cyg 19h58'12" 47d7'0"
John and Catherine Leslie
Leo 11h33'47" 26d21'35"
John and Catherine
McInnis
Cyg 21h21'56" 38d20'51"
John and Cathy Wojewoda
Cyg 20h51'9" 33d14'52"
John and Charlee Super
Star
Cyg 19h47'8" 30d33'59"
John and Cheryl
Cyg 21h33'55" 47d6'41"
John and Cheryl Unity
Potentate
Cyg 19h38'28" 34d7'22"
John and Christine
Tau 5h45'53" 15d30'36"
John and Christine Maier
Vir 11h49'9" 9d11'52"
John and Christine Wilkens
Aqr 23h42'17" -15d29'7"
John and Danielle De
Jesus
Tau 4h27'24" 18d28'22"
John and Deirdre Joyce
And 2h17'14" 42d27'15"
John and Delores Sauers
Uma 9h33'11" 58d26'59"
John and Diane Winner
Aqr 23h50'38" -10d56'36"
John and Donna's Cosmo
Kostesich
Cyg 19h42'57" 41d4'11"
John and Edel Stone
Psc 1h28'47" 17d9'7"
John and Elaine April 2006
Cas 23h6'28" 58d59'17"
John and Elaine Kofoed
Uma 13h43'26" 53d39'51"
John and Elizabeth Dines
Boo 15h5'7" 49d53'42"
John and Elysse Fleece
Cyg 20h48'20" 53d43'25"
John and Faye 47 The
Anniversary
Cyg 21h40'20" 46d14'15"
John and Freddy Howell
Aql 19h52'0" 1d16'26"
John and Garnet
Uma 8h58'35" 58d46'26"
John and Goldie Chor
Cyg 21h45'18" 44d13'50"
John and Hels Febbrari in
love 4eva
And 23h44'27" 46d33'45"
John and Jackie
Umi 17h9'57" 75d46'7"
John and Jackie Nelson
Uma 8h41'33" 51d24'46"
John and Jane Cadman
Col 6h16'6" -35d6'31"
John and Janet Smith
Cyg 21h13'22" 47d18'7"
John and Jillian - I love you
Leo 9h22'31" 10d21'40"
John and Jillian's star
Cyg 19h50'21" 33d41'52"
John and Jo
Tau 4h38'14" 19d24'56"
John and Joan
Cyg 20h16'14" 46d20'48"
John and Joanne
Carothers
Dra 17h17'25" 53d0'23"
John and Jody 8172002
Cyg 19h31'47" 30d37'2"
John and Judy McCurry
Cyg 21h54'53" 46d32'56"
John and June Hoye
Vir 12h42'25" 8d51'0"
John and Kara DeFelice
Deeley
Cnc 8h40'42" 22d54'11"
John and Kelly
Uma 9h44'35" 55d24'6"
John and Kimberly Duclos
Cyg 20h10'1" 31d37'29"
John and Leslie 2-19-05
Crb 16h22'47" 32d56'31"
John and Lisa
Cam 3h19'33" 67d11'35"
John and Lois, 50 years
Ori 5h30'48" -0d17'7"
John and Mandy
Umi 18h50'17" 86d15'59"
John and Mandy
Umi 14h23'40" 70d49'58"
John and Mandy's 4-ever
Star
Her 17h35'18" 32d24'11"
John and Margaret Gordon
Lmi 10h3'30" 33d24'48"
John and Margaret Healy
Cyg 20h19'17" 52d16'0"
John and Maria
Cyg 20h2'45" 39d8'16"
John and Marie Spezia
Cyg 19h11'3" 51d11'38"
John and Marilyn Kennedy
Per 3h49'57" 41d54'20"
John and Marjorie White
Uma 10h43'19" 57d32'56"
John and Martha Satterfield
Uma 10h38'55" 51d12'49"
John and Mary McGovern
Per 4h7'21" 40d42'23"
John and Maureen Dean
Lyn 9h6'4" 34d18'46"
John and Meliezza Walker
Per 3h35'43" 32d49'3"
John and Miriam
Cyg 21h6'34" 47d41'0"
John and Niamh Bowler
Cyg 20h13'16" 37d6'29"
John and Nicole Forever
And 2h22'37" 47d15'46"
John and Nikki
Cyg 20h5'10" 37d27'58"
John and Nina
And 1h36'44" 40d6'54"
John and Noreen Pavlos
Uma 9h39'54" 50d42'34"
John and Patricia Salisbury
Cyg 20h11'23" 55d25'25"
John and Patti Cournoyer
Cap 21h30'24" -19d26'42"
John and Patti Forever -
05/09/05
Cap 21h12'10" -19d53'23"
John and Pattie's Heavenly
Place
Peg 21h39'28" 24d42'28"
John and Pearl McMullen
Per 4h27'14" 34d37'47"
John and Phil Smith Lamb
Sco 16h43'51" -32d31'56"
John and Raylyn Eubanks
Uma 12h9'7" 47d11'56"
John and Rhonda
McKenzie
Lib 15h25'38" -21d36'13"
John and Rose Forever
Star
Cru 12h52'36" -57d37'43"
John and Sandra Wareham
And 0h50'15" 41d26'9"
John and Sarah
Vir 12h47'15" 11d8'33"
John and Shalina
Sparkman
Vir 13h53'39" -21d50'17"
John and Stephanie
Sgr 19h19'25" -17d10'10"
John and Summer's First
Anniversary
Ori 5h29'44" 10d26'10"
John and Susan Alfano
Sco 16h10'58" -11d25'6"
John and Teresa
Ind 20h46'43" -46d53'6"
John and Viola St. John
Family
Uma 8h51'46" 47d17'8"
John and Virginia
Middleton
Cyg 20h36'7" 50d12'5"
John and Zina Bash
And 0h43'31" 23d50'3"
John Anderson
Cep 21h51'10" 56d10'40"
John Andresini
Tau 3h49'21" 28d22'32"
John Andrew
Per 3h43'53" 37d26'31"
John Andrew Cross
Lyn 7h40'59" 35d48'44"
John Andrew Fannin, II
Cnc 9h6'47" 28d21'6"
John Andrew Garlans, IV
Aql 19h6'23" -9d40'0"
John Andrew Hamilton II
Sgr 17h57'31" -29d58'36"
John Andrew Sianez
Aqr 22h30'0" -1d55'14"
John Andrew Stoich
Per 3h43'27" 50d51'29"
John Angel Janis
Per 3h59'9" 33d6'7"
John & Anita Buting
Psc 0h48'44" 13d49'0"
John & Anita Thrussell
Cyg 21h46'24" 44d20'12"
John & Ann Harris
Uma 10h22'10" 66d30'47"
John & Annie Chowns
Cyg 20h7'59" 35d24'18"
John Annoni
Aql 19h58'4" -10d0'29"
John Anthony
Her 17h15'46" 39d16'9"
John Anthony Andriella
Psc 0h9'48" 11d17'37"
John Anthony Branthwaite
Per 4h9'29" 39d44'31"
John Anthony DeMarco III
Vir 12h32'6" -0d0'53"
John Anthony Fiato
Aql 19h8'2" -0d47'30"
John Anthony Gulla
Leo 10h12'52" 14d48'1"
John Anthony Henry, Jr.
Vir 12h44'2" 4d54'30"
John Anthony Hope
Her 16h39'25" 37d45'50"
John Anthony III
Sgr 17h56'22" -29d19'38"
John Anthony "Jay"
Sampietro
Per 4h0'35" 35d8'46"
John Anthony Lucas
Ori 5h26'8" 2d32'17"
John Anthony Moore Sr
Dra 17h41'32" 57d39'33"
John Anthony Plumley, Jr.
Her 17h55'33" 49d33'23"
John Anthony Scarkino
Cap 20h23'6" -12d36'51"
John Anthony Travasso
Lib 14h41'26" -17d11'24"
John Anthony Watson - My
soulmate.
Dra 15h38'20" 63d0'26"
John Archer Hatch
Cep 23h13'52" 72d48'28"
John Arthur Charles Hilliker
Cep 22h45'41" 66d57'43"
John Arthur Cruz
Tau 4h24'7" 20d34'4"
John Arthur Fregeau
Per 3h22'36" 52d3'21"
John Arthur Hughes
Uma 9h22'11" 65d19'48"
John Arthur Marchena
8/5/1956
Leo 9h49'26" 32d19'35"
John Arthur Miller
Boo 14h42'55" 19d10'55"
John Arthur Rensing
Uma 11h44'16" 30d42'49"
John Attarian
Sgr 19h28'23" -12d42'20"
John Author McEvoy
Equ 21h21'6" 9d22'27"
John Awong
Cet 1h32'38" -3d42'54"
John B
Cap 20h17'49" -13d1'35"
John B
Uma 11h21'11" 48d0'58"
John B. Dawson
Psc 1h18'52" 31d14'33"
John B. Flynn "Jack"
Uma 10h34'30" 51d19'27"
John B. La Due IV
Sco 16h8'43" -13d36'25"
John B. Norton
Per 3h9'22" 51d46'58"
John B. Renton
Cap 21h11'16" -18d2'15"
John B. Sigel
Her 17h52'44" 24d4'23"
John & Baby Sister Kimby
Uma 13h57'55" 61d34'15"
John Bachino
Her 16h37'57" 39d1'17"
John Bachor, Günther
Cnc 9h14'27" 15d33'36"
John Bagdon
Crb 15h33'20" 28d22'14"
John Bailey
Aql 19h50'52" 8d15'46"
John Baillie
Peg 22h39'1" 8d10'36"
John Baker
Aur 5h0'43" 49d21'38"
John Banker
Ori 5h52'30" 18d22'39"
John & Barb, 30 Years
Until Forever
Cnc 8h40'12" 17d40'55"
John Barnwell Fishburne
Per 3h36'9" 46d40'35"
John Barone
Her 18h38'25" 19d9'9"
John Barrand's Dream
Psc 23h4'32" 8d2'31"
John Barrie Mellonby
Per 3h13'0" 32d23'41"
John Barringer Steever
Per 3h4'29" 41d44'50"
John Barry Yates 26
November 1935
Sgr 17h53'32" -29d29'40"
John Battlestarny
DiPalermo
Aql 19h51'42" 10d39'25"
John Bebbling
Vir 13h0'9" -1d29'29"
John Bechtel
Per 2h56'29" 54d50'31"
John Beering
Psa 22h43'43" -26d23'21"
John Bellew
Per 2h29'13" 56d59'45"
John Bello
Aql 19h41'4" -0d7'44"
John Belongia
Cep 22h15'47" 65d20'49"
John Bencivenga
Uma 11h8'31" 30d57'6"
John Benjamin Fein-Ashley
Gem 7h29'29" 32d38'17"
John Benton Travers
Boo 14h45'26" 51d5'57"
John Benton Weaver III
Ori 4h50'11" 4d5'50"
John Berkley Helm
Ari 3h16'15" 19d22'8"
John Bernard Clennell
Per 4h18'52" 42d45'28"
John Bernhard McLain
Her 17h19'2" 17d0'0"
John Berry Gosnell
Ori 6h20'23" 2d7'57"
John & Betty Cox
Ori 5h33'58" -0d19'50"
John Bevan Chuck
Her 17h7'20" 31d48'40"
John Biafore Jr.
Cep 21h20'10" 60d6'37"
John Bickford Henderson
Per 4h48'5" 49d10'25"
John bigjohn616 Palmiero
Gem 6h44'39" 33d6'26"
John Blackwelder
Aur 5h50'29" 53d39'40"
John Blackwell & Jeannie
Armstrong
Her 17h54'2" 21d23'59"
John Blaschke
Her 17h46'34" 36d44'11"
John Blee
Sco 17h9'57" -32d13'12"
John Blodgett Hearty
Vir 12h9'16" -8d37'35"
John & Bobbie Baker
Sex 10h46'35" -6d10'12"
John Bolf
Psc 0h49'24" 9d54'3"
John Bonsignore
Tau 4h30'58" 27d10'39"
John Boone Arnold
Uma 8h38'46" 66d48'43"
John Borum
Per 2h39'22" 54d38'25"
John... boy in the sky...
Umi 14h29'48" 67d46'32"
John Bradford Fritz
Gem 6h5'23" 22d18'4"
John Bradley Mangrum
Her 17h44'11" 37d32'1"
John Bradshaw Layfield
Uma 11h47'56" 60d41'20"
John Brady May
Psc 0h7'45" 11d25'38"
John Brandon Jarriel
Her 17h12'1" 28d31'23"
John Brandon Kirkman
Ari 3h16'52" 21d23'9"
John Brandt
Cru 12h38'45" -59d48'35"
John & Brandy
Umi 14h24'26" 65d42'47"
John Brasel "10-11-47"
Lyn 7h1'17" 51d18'29"
John & Brenda
Uma 12h4'51" 53d41'41"
John Brennan
Leo 9h35'39" 31d18'6"
John Bret-Harte
Uma 8h36'17" 54d6'28"
John Bright
Per 4h11'53" 33d14'37"
John Brisslinger
Uma 12h53'37" 58d59'30"
John Brodylo
Vir 13h15'35" 13d36'18"
John Bruce Reid
Aqr 23h0'16" -20d46'10"
John Bruzzano
Uma 11h48'47" 53d49'37"
John Bryan Philpott
Uma 10h12'8" 43d28'9"
John Bubnis
Leo 9h31'8" 12d10'36"
John Buchanan
Crb 16h23'9" 34d52'1"
John Buchanan
Uma 10h8'9" 54d18'32"
John Buckland
Cep 0h21'50" 78d51'33"
John Buggins - Ruby Star
Cyg 21h15'45" 47d0'18"
John "Bullet" Standingdeer
Her 16h37'4" 36d52'27"
John Burnell Southall
Per 3h0'10" 54d40'52"
John Bussell
Uma 11h56'20" 44d33'46"
John *Buzzy* Bollman
Uma 9h29'40" 60d47'49"
John C. Albanese 8-25-
1933
Aqr 23h12'42" -21d57'49"
John C. Boyd
Ori 5h46'31" 2d53'13"
John C. Bussani
Aqr 22h1'3" 2d7'10"
John C. Callow
Sco 17h25'38" -42d18'46"
John C. Campbell
Lib 14h42'11" -18d8'47"
John C. Christie
Boo 14h25'44" 45d9'59"
John C. Dikeman
Sco 17h46'46" -31d28'29"
John C. Galloway
Lyn 7h8'58" 50d53'25"
John C. Grinnell Jr.
Ari 1h57'40" 24d8'41"
John C. Healy
Uma 8h34'2" 59d49'1"
John C. Heymann
Boo 14h17'53" 17d5'21"
John C. Lange
Lib 15h51'39" -12d32'20"
John C. Ledbetter
Ori 4h54'11" 14d30'57"
John C. Moyer
Uma 10h48'32" 62d31'22"
John C. Nax
Uma 11h55'4" 35d26'10"
John C Shy
Lyn 8h13'44" 45d51'4"
John C. Suares
Uma 8h34'3" 47d5'40"
John C. Wildt
Tau 5h45'53" 13d42'11"
John C. Zagarella,
O.Praem.
Vir 14h20'56" -0d27'28"
John Cabral
Leo 11h42'40" 24d49'21"
John Caldwell
Vir 12h25'20" 1d24'23"
John Caldwell
Uma 10h42'26" 70d48'45"
John Calvin
Aql 20h1'19" -8d45'23"
John Calvin Elmes II
Ori 6h12'28" 17d11'39"
John Campbell
Lib 15h45'49" -19d12'59"
John Caravello
Per 2h18'9" 55d10'38"
John "Cariño" Kelley
Uma 10h58'46" 49d6'44"
John Carl Fisher
Her 16h28'28" 49d9'41"
John Carl Liguori
Aqr 23h1'5" -5d31'46"
John Carl Mahoney
Lib 15h16'52" -18d48'11"
John Carleton Taylor, Jr.
Leo 11h18'42" 13d14'8"
John Carlile, Jr.
Her 18h26'42" 23d35'51"
John Carlin
Her 16h36'50" 38d25'55"
John Carneiro
Ari 2h53'38" 28d1'38"
John & Caroline King
Lib 15h12'46" -6d14'5"
John Carr
Psc 1h26'34" 10d2'1"

John Carrington Eggleston
Lib 14h51'50" -8d12'2"
John Carroll
Gem 7h0'0" 18d28'28"
John Carter Fidkin
Per 2h49'55" 51d27'28"
John Cavan
Aqr 23h8'41" -8d34'0"
John Cervini,PHD
Sgr 18h54'38" -34d21'46"
John Chandler Hawley
Cep 22h43'13" 67d11'1"
John Chapman Marshall
Tau 4h43'47" 27d37'45"
John Charles Casserino
Sco 16h8'20" -17d24'21"
John Charles Collins
Peg 21h12'3" 12d23'28"
John Charles Culp
Lmi 10h16'24" 32d53'50"
John Charles Fragassi
Cep 5h1'48" 82d10'3"
John Charles George Spatig
Per 2h55'42" 45d53'18"
John Charles Gough
Her 18h47'39" 16d23'46"
John Charles Gray Jr.
Ori 6h9'4" 2d39'52"
John Charles Gray Sr.
Per 4h46'5" 50d31'17"
John Charles Malone
Cep 22h11'45" 61d34'1"
John Charles Polkey
Umi 17h8'41" 76d45'30"
John Charles Rosenberg
Lib 14h25'42" -21d40'37"
John Charles Russell's Bright Light
Lib 15h34'17" -3d53'47"
John Charles Stryker
Uma 9h54'46" 48d42'41"
John & Charlotte Lee
Cap 20h25'5" -15d42'6"
John Cheetham
Uma 10h8'38" 58d36'50"
John Chiappetta
Per 4h15'11" 47d4'21"
John Chojnacki
Per 3h12'37" 42d41'45"
John Christian Harrington
Per 4h6'6" 43d52'57"
John Christian Schilhab
Her 18h21'21" 22d48'50"
John & Christie
Psc 1h6'39" 33d5'56"
John & Christine DeCorte
Ori 4h47'55" 10d32'37"
John Christodoulides
Uma 10h37'21" 59d36'31"
John Christopher Angel
Tau 3h57'37" 22d29'18"
John Christopher Donahue
Ori 4h55'0" 3d1'19"
John Christopher Grigsby
Leo 10h8'35" 24d30'29"
John Christopher Jordan Rock
Pho 0h35'41" -56d9'16"
John Christopher Krause
Tau 4h11'1" 7d48'44"
John Christopher Lewis
Sgr 19h1'0" -28d36'28"
John Christopher Reyes
Boo 14h39'54" 54d26'56"
John Christopher Stielper
Her 17h20'28" 32d46'19"
John Christopher Valvardi
Sgr 18h55'33" -18d48'22"
John*Clare Star
Per 3h5'13" 56d51'22"
John Clarence LaPaille
Cnc 8h17'28" 16d43'20"
John Clark III
Cnv 13h47'27" 30d9'51"
John Clark O'Dell
Lib 15h35'58" -9d36'28"
John Clark Sampson
Her 16h48'55" 18d57'7"
John Clarke
Cru 12h19'53" -57d33'54"
John Claude Horton
Ori 5h54'55" 21d50'43"
John Claxton Ockenden 70th Birthday
Her 16h53'43" 34d32'58"
John Clifford Burrow
Vir 14h51'37" 5d29'15"
John Clinton Self
Aur 5h44'52" 49d33'14"
John Cloud
Ori 6h15'57" 9d29'11"
John Cochran
Ori 5h15'24" -7d43'34"
John Cogley
Lib 14h51'40" -5d11'3"
John Coleman
Cep 22h55'57" 71d0'16"
John & Colleen Hopkins
Cyg 21h38'30" 44d48'24"
John & Colleen Palmer
Uma 8h16'30" 67d1'14"
John "Colonel" Strong
Dra 16h48'44" 58d4'39"
John Conner
Vir 12h6'59" 11d21'16"
John Connor Minich
Cap 21h57'47" -12d46'35"
John Conrad Fuchs, Jr.
Cnv 12h47'26" 43d25'23"
John Conrad Hanks
Ari 2h36'59" 25d26'34"
John Constantine Rigas
Lmi 9h59'2" 38d47'58"
John Convery, Jr.
Sgr 17h55'23" -30d0'30"
John Coogan
Uma 9h5'12" 50d43'22"
John Cooley
Ari 1h51'11" 20d59'13"
John Costigan
And 1h20'9" 37d28'52"
John Cote
And 1h49'42" 41d42'32"
John Cowper Barford
Tau 5h53'3" 25d31'1"
John Coyle
Lac 22h40'25" 48d46'26"
John Crawford Spinks
Cep 20h25'17" 61d41'34"
John Crummy's Lucky Star
Vir 13h21'19" -13d2'31"
John Culetsu
Per 3h16'23" 31d13'22"
John Cullen Lancaster, Jr.
Aql 19h12'23" -0d51'16"
John Curran, M.D.
Her 17h10'6" 19d24'40"
John Curtis Giacometto
Cep 22h41'15" 73d7'36"
John Curtis Steffen
Ori 5h18'0" 11d47'52"
John Curtis Walker
Sct 18h50'41" -12d52'9"
John Curtis, Jr.
Aqr 22h37'20" 1d57'27"
John D. Barbour Sr. "Johnny"
Lyn 8h46'36" 44d33'5"
John D. Barnes = My Angel Dad
Uma 13h10'14" 60d43'21"
John D. Beeson
Her 18h47'30" 14d35'24"
John D. Buckley
Dra 14h46'28" 56d21'18"
John D Charboneau
Her 18h47'30" 21d42'48"
John D. Curran
Dra 10h21'41" 75d30'40"
John D. Davis
Mon 6h37'1" -6d38'40"
John D Davis
Leo 9h22'17" 7d8'23"
John D. Focht
Cnc 9h9'56" 11d7'37"
John D. Goolsbee
Ori 6h10'37" 15d3'7"
John D. Helgesen
Per 3h21'5" 44d36'31"
John D. Heyer
Cnv 12h46'5" 44d12'56"
John D. Kaprat
Ari 2h13'30" 22d33'2"
John D. Kelly
Vir 13h49'39" -16d38'14"
John D. Maher, Jr., Daddy
Her 16h40'47" 48d49'22"
John D. Moore
Cnc 8h6'43" 6d53'29"
John D. Norsworthy
Per 3h37'44" 49d16'22"
John D. Smith
Gem 6h33'16" 21d43'12"
John D. St. John
Uma 11h12'57" 59d38'40"
John D. W. Mullis
Sco 16h8'46" -12d37'30"
John D Willbanks
Her 17h53'9" 14d45'25"
John Dachiw
Ori 6h1'12" 5d10'13"
John "DAD" Upton
Cam 4h16'7" 58d1'28"
John "Daddles" Love
Per 4h28'29" 43d23'8"
John Daddono
Pyx 8h37'34" -20d28'44"
John 'Daddy' Ballard
Gem 7h13'31" 19d50'33"
John & Dale Fuqua
Ori 5h35'56" 3d37'45"
John & Dalena
Cyg 19h49'11" 38d47'31"
John Daniel
Crb 16h0'44" 31d38'24"
John Daniel Boyd Grant
Cyg 21h9'18" 51d57'55"
John Daniel King
Cep 22h31'1" 73d26'19"
John Dankovchik
Lib 15h38'4" -26d8'48"
John - Dapper - Gibbs
Cap 21h40'17" -19d7'57"
John David
Lib 15h43'10" -24d27'45"
John David Adachi-'67
Lib 15h8'51" -16d3'17"
John David Adams
Per 4h20'32" 41d5'1"
John David Alkire
Her 17h16'37" 43d6'15"
John David Curry "Peterpan"
Lib 14h39'35" -10d37'54"
John David Elvin
Sco 17h41'4" -40d13'16"
John David Hartney, My Bunnylove
Lep 5h18'51" -21d57'35"
John David Khawam
Vir 14h55'8" 2d45'6"
John David Lackey
Leo 11h45'39" 19d31'57"
John David Leroy Hanston
Cep 21h10'12" 66d28'28"
John David Moe
Sco 16h17'19" -14d57'0"
John David Parker
And 0h12'26" 33d8'0"
John David Ramirez
Cyg 20h24'33" 49d24'43"
John David Riggins
Ori 5h31'43" 2d25'24"
John David Ruge
Uma 14h12'23" 60d58'29"
John David Samuelson
Per 3h45'54" 41d22'46"
John David Santos
Cnc 8h20'30" 14d3'45"
John David Shamus Wrotniewski
Cru 12h41'49" -58d18'2"
John David Vogelzang
Sgr 18h22'19" -34d22'7"
John David Wilson
Cap 21h14'46" -19d50'18"
John Davidson
Lac 22h43'1" 53d14'51"
John Davila
Del 20h39'32" 15d26'8"
John Davis Turman
Lyn 7h47'49" 53d39'22"
John D.Barbour Sr.
Boo 15h21'18" 49d16'48"
John De Canto
Cep 22h18'31" 72d3'30"
John De Dickson
Aql 20h28'44" -0d17'12"
John Deanna's Knight
Cet 1h4'35" -19d28'30"
John Deatherage
Peg 22h50'0" 22d41'1"
John & Debbie Scully - 35 Years!
Sge 19h25'17" 18d10'0"
John DeFabio
Ari 3h15'29" 22d21'10"
John DeGrazia Til I Call Your Name
Aql 19h22'35" 7d57'24"
John DeLucia-Areola Constellation
Cyg 19h37'52" 28d20'32"
John Denison
Vir 13h43'41" -5d21'13"
John Dennis Ryan III
Gem 7h16'49" 28d10'29"
John Derek Rogers
Cap 20h36'1" -10d54'15"
John Devlin and Frank Rayment
Uma 10h16'23" 60d14'6"
John Dimitriadis
Lyn 7h37'56" 57d19'55"
John Dixon
Aur 5h19'29" 31d20'14"
John Dolloff
Uma 12h15'4" 58d59'14"
John Donovan - Sleepy - 07061942
Cru 12h27'26" -59d47'4"
John Doranda
Gem 7h1'53" 17d38'20"
John & Dorthea Gray
Peg 22h29'37" 6d29'18"
John Douglas 4261931
Tau 4h36'29" 17d11'59"
John Douglas Bell
Ori 5h24'11" 14d7'25"
John Douglas Campbell
Sct 18h46'44" -5d17'14"
John Douglas Clark
Cyg 20h28'47" 35d47'25"
John Douglas Deans
Uma 9h39'19" 69d37'9"
John Douglas Emerson
Vir 13h5'25" -2d20'22"
John Douglas Haycock
Uma 8h14'3" 69d22'8"
John Douglas Regulus
Lmi 10h12'17" 35d16'46"
John Douglas Tobin, Sr.
Tau 3h32'20" 19d23'24"
John Douglas Vance
Her 16h37'33" 34d50'56"
John Drain
Per 2h27'20" 51d58'14"
John Drakopoulos
Dra 17h41'5" 53d41'53"
John Dreux
Ari 2h18'19" 25d22'57"
John Drew Frederickson
Sco 17h45'4" -44d33'30"
John Drew Laurence Nolan
Vir 13h14'4" 11d12'1"
John Duane Brooks - JBBEERMAN
Gem 7h58'50" 32d6'6"
John Duddley Holman
Cru 12h8'0" -57d25'2"
John (Duffer) Flynn
Lib 15h54'41" -19d27'43"
John Dugan 1913-1989
Uma 11h46'1" 34d41'4"
John Duhe
Gem 7h20'55" 29d24'5"
John Dunkin Hartley
Ori 6h3'39" 19d40'3"
John E Bowers
Leo 10h28'24" 22d29'41"
John E. Brandon
Gem 7h10'34" 28d22'44"
John E. Brocato
Ari 3h17'51" 25d0'21"
John E Bulaga Jr"Never Forget" 9/11
Per 3h17'6" 56d17'8"
John E. Capps
Her 17h37'13" 16d18'50"
John E. Carbaugh Jr.
Uma 11h54'47" 62d57'5"
John E. Cross
Uma 11h0'37" 54d37'42"
John E. "Eric" Czernec
Per 2h51'45" 54d9'30"
John E. Harris
Ori 6h9'11" 20d0'29"
John E. Heaner IV
Lib 15h16'51" -5d22'39"
John E Hewitt Jr
Per 4h18'36" 39d28'41"
John E. Kavelak, Jr.
Vir 13h21'47" 12d53'10"
John E Lonnborg July 28th,1999
Tau 3h58'25" 10d25'39"
John E. McDonald, Jr.
Cnc 9h20'21" 27d38'9"
John E. Miller
Ari 2h34'55" 14d31'19"
John E. Morse
Uma 10h58'50" 48d41'41"
John E Oliver
Leo 11h40'54" 20d7'59"
John E. Phillippi
Sco 16h12'1" -30d51'52"
John E. Ramirez
Her 17h19'22" 16d55'11"
John E. Saeger
Cnc 8h50'0" 19d17'44"
John E. Shank
Cap 21h34'10" -14d16'6"
John E. Tipton
Uma 11h32'3" 63d42'16"
John E. White
Uma 12h4'41" 58d39'0"
John Earl Crothers
Aqr 22h28'9" 1d23'48"
John Easey
Ori 5h35'39" -1d36'15"
John Easterling
Cen 13h25'7" -33d21'5"
John Edward
Del 20h46'17" 3d42'0"
John Edward Aversa
Leo 10h19'54" 13d21'7"
John Edward Fink
Cnc 8h18'35" 22d58'45"
John Edward Gorrill
Cep 23h9'27" 71d11'30"
John Edward Hackett
Aql 20h1'51" 10d42'5"
John Edward Kostic
Dra 20h12'11" 62d14'10"
John Edward Lyman
Uma 11h37'50" 57d17'55"
John Edward Morgan
And 23h11'1" 51d29'22"
John Edward My Love
Tau 5h47'47" 17d38'44"
John Edward Price
Her 17h14'3" 36d26'46"
John Edward Shields III
Aur 5h12'28" 28d47'14"
John Edward Thomas
Her 17h41'10" 15d10'4"
John Edwin Blades Stuart
Her 16h43'16" 8d38'58"
John Edwin Presson
Per 4h45'21" 48d26'22"
John Edwin White
Sco 16h28'10" -42d9'1"
John & Eileen
Umi 17h1'56" 79d27'23"
John Eldh
Ori 5h40'59" 1d9'25"
John Elis Crosslin
Uma 10h59'41" 56d17'13"
John Ellsworth Barnett
Uma 11h20'11" 33d6'27"
John Ellsworth Vernon
Ari 2h52'25" 26d22'34"
John Elmer Lawton
Aur 5h25'38" 39d44'12"
John Elmer Osbo
Uma 12h44'28" 56d51'52"
John Emilio
Vir 12h45'48" 12d46'26"
John Emrich
Per 3h17'40" 54d28'20"
John & Erica
Her 16h59'4" 29d23'29"
John Erik Robert Thurman
Lib 15h40'43" -19d24'53"
John Ethan
Lib 15h1'58" -0d40'54"
John Eugene Powell
Psc 1h9'55" 10d14'26"
John Evan Lyman
Her 16h58'44" 37d8'36"
John Everett Benton's Star
Aqr 23h14'45" -13d5'17"
John Everhart
Ari 2h39'34" 20d0'8"
John Eyre
Leo 11h30'23" 10d32'27"
John F. Adams
Lib 15h6'18" -5d45'4"
John F. Aldrich, Jr.
Cep 22h25'24" 73d24'43"
John F. Antaya
Cap 20h17'14" -19d18'7"
John F. Ashley
Gem 7h31'0" 34d21'59"
John F. Beucler
Lyn 8h3'40" 34d43'4"
John F. Brueck
Tau 5h50'11" 27d22'58"
John F Butler
Cep 22h41'38" 65d40'36"
John F. Capalbo
Boo 14h6'26" 20d59'3"
John F. Christensen
Per 3h15'43" 45d29'15"
John F. Ciurczak
Lib 15h46'43" -20d58'55"
John F Cook
Aql 18h46'40" 8d54'47"
John F. Dobzenski III
Lib 15h12'52" -5d47'58"
John F. Gartner
Boo 14h37'56" 53d46'0"
John F. Gillen
Tau 4h27'25" 20d22'44"
John F. Grebb
Per 3h3'20" 55d0'26"
John F. Isolano
Leo 10h17'46" 24d4'28"
John F. (Jack) Graffam
Lib 15h41'47" -17d19'39"
John F. "Johnnie Red" Lettiere
Uma 11h48'26" 60d24'34"
John F. Keane
Uma 9h56'43" 48d30'21"
John F. Kennedy
Aql 19h52'44" -0d58'37"
John F. Klein
Leo 11h38'52" 26d7'8"
John F. Lockhart "1945-2006"
Psc 0h16'53" -4d29'57"
John F. Lynch, Jr.
And 0h33'18" 39d17'56"
John F Neifert III "Beloved Son"
Ori 5h34'38" 1d40'8"
John F. Opel
Aur 5h6'48" 47d28'14"
John F. P. Clark
Cep 21h40'5" 55d50'55"
John F. Peil "A Great Man"
Per 2h21'16" 51d44'33"
John F. Schilling
Ori 6h4'56" 18d30'59"
John F. Smetana
Lib 15h17'11" -11d45'24"
John F. Sprada
Uma 10h24'50" 42d25'53"
John F. Theuer
Her 18h14'53" 29d19'1"
John F. Van Aken Jr.
Peg 21h12'6" 15d52'19"
John F. Vito
Cap 20h33'10" -21d7'47"
John F Vito
Cap 20h59'52" -17d37'27"
John F. Walsh
Lib 15h26'16" -9d56'34"
John F. Widdifield
Her 17h36'5" 16d33'26"
John Falchetta
Cma 7h12'8" -32d34'43"
John Fallon "Johnny"
Aur 5h52'37" 46d42'48"
John Farr
Uma 9h42'32" 70d21'25"
John Farver
Lyr 18h46'4" 31d47'9"
John Fazio Winters
Cnc 8h45'27" 26d35'4"
John Ferrara
Umi 17h1'28" 89d23'40"
John Field
Uma 11h51'14" 49d9'26"
John Findlay
Cyg 21h56'55" 48d13'24"
John Fisher Boyd
Uma 9h54'40" 44d24'42"
John Fitton
Lyn 8h25'36" 42d26'54"
John Flenady I Love You
Sgr 19h39'59" -14d27'10"
John Foster
Aql 19h48'10" -0d2'48"
John Foster
Cap 20h54'8" -24d43'15"
John Foster
Gem 6h35'33" 22d36'18"
John Foster Houseal
Ori 5h29'38" 10d41'57"
John Fox
Lyn 6h49'42" 51d10'42"
John Francis Conlon
Leo 10h41'58" 7d28'14"
John Francis Duffy, Sr.
Uma 10h48'47" 63d58'45"
John Francis Larkin
Cyg 20h7'27" 52d19'25"
John Francis LeMieux
Per 3h18'2" 48d51'42"
John Francis Nemanich
Psc 1h24'24" 13d9'4"
John Francis Reed
Cyg 19h38'47" 32d3'11"
John Francis Stewart III
Lib 15h3'20" -1d28'17"
John Francis Wright Jr.
Uma 10h54'13" 50d26'15"
John Frank
Uma 10h42'42" 50d57'51"
John Frank Bell
Aql 19h51'30" 10d47'48"
John Frank Grova
Her 17h16'39" 20d24'43"
John Frank Kasper
Cap 21h44'54" -19d25'0"
John Franklin
Sgr 18h9'21" -27d39'34"
John Franklin Lee III
Leo 10h14'31" 9d40'58"
John Franklin Noble IV
Uma 12h5'31" 52d14'32"
John Fred Gourrier
Umi 15h14'51" 69d36'28"
John Frederick Allman
Ori 6h7'15" 16d33'58"
John Frederick Convery
Del 20h33'16" 15d24'18"
John Frederick Dormady
Gem 7h10'46" 17d3'19"
John Frederick Nare
Per 4h40'33" 40d18'4"
John Frederick Rotchford Jr.
Leo 9h38'7" 28d3'27"
John Frederick Schultz III
Vir 13h14'48" -20d24'15"
John Frederick Zinn Anderson
Her 16h28'26" 11d26'11"
John Fredrick Menear
Leo 11h57'36" 23d21'37"
John Frerichs
Ori 5h37'46" -2d1'50"
John Fuller
Uma 8h11'50" 65d31'43"
John Fuller's Falcon Fire
Ori 6h17'5" 10d50'38"
John Furlow
Sgr 18h21'47" -29d37'47"
John G. Brodak
Ari 2h8'13" 23d0'1"
John G. Campbell
Aur 5h23'52" 39d44'44"
John G. Grzywacz
Cap 21h6'43" -15d15'4"
John G. Harkins
Aur 5h34'50" 33d22'11"
John G. Lipscomb
Gem 6h54'20" 24d10'1"
John G. Makrakis
Psc 1h8'15" 27d59'3"
John G Miller
Gem 6h21'49" 21d45'22"
John G. O'Neill
Cep 21h36'40" 69d2'57"
John G. Reinhart
Umi 16h32'30" 76d54'56"
John G Turck
Cep 22h30'5" 74d19'27"
John Gagliardi
Per 3h30'9" 45d24'36"
John Garrett
Ori 5h53'34" 6d11'8"
john gast
Cap 20h30'15" -26d41'45"
John Gates
Ari 2h34'22" 20d58'26"
John Gaudio
Cep 22h36'44" 69d4'46"
John Geddings
Cas 0h20'17" 56d27'47"
John George Karavas
Cnc 8h30'8" 11d33'0"
John George Sundbye II b. 1/19/1931
Her 17h41'24" 27d23'17"
John Gerald Meagher,MD
Sco 16h15'25" -8d29'3"
John Gerald Spierto-Diodato
Uma 8h45'14" 64d4'59"
John Gerard
Cru 12h28'39" -60d8'52"
John Gersper
Aur 6h25'28" 51d44'18"
John Gessey
Cep 20h46'58" 63d1'58"
John Giannouli Skordos
Leo 11h5'52" 21d7'2"
John Gibbins
Cep 0h9'52" 70d41'56"
John Gilmore Meyer, Jr.
Her 17h21'57" 27d39'29"
John & Ginny Star
Mon 8h3'28" -0d51'55"
John Glover Garwood
Aql 19h11'32" -9d25'26"
John Godfrey
Cru 12h49'3" -60d49'58"
John Golden
Per 4h15'8" 51d6'22"
John Goodwin
Per 4h24'26" 49d58'30"
John Googins
Her 18h13'8" 28d43'44"
John Gordon Swanson
Cyg 20h12'41" 39d51'2"
John Gorman
Cep 22h28'33" 60d25'29"
John & Grace Macaskill - 25.6.1946
Uma 8h44'26" 70d0'16"
John Graesser's Arrow of Light
Ori 5h47'59" 8d44'33"
John Graham Evans
Dra 18h5'1" 76d28'39"
John Graham Williams
Cep 20h56'10" 83d3'26"
John Grandison Underhill
Leo 11h33'15" 7d24'1"
John Gregory Cottingham
Cyg 19h42'18" 32d36'23"
John Grenfell
Aql 19h2'3" 15d41'18"
John Grubiss
Tau 4h42'33" 25d47'53"
John Guerrasio
Uma 10h36'6" 66d23'15"
John Guggino
Per 3h29'59" 51d30'33"
John Gurino, Jr.
Leo 9h51'19" 31d37'48"
John Gustavson
Uma 11h27'22" 59d35'0"
John H
And 23h58'32" 37d50'15"
John H. "Bill" Powell
Umi 15h2'5" 70d27'12"
John H. Chatham
Uma 9h36'51" 42d39'33"
John H. Dorr
Per 4h20'9" 35d18'30"
John H Eberwein
Cmi 7h24'20" 1d57'11"
John H. Feth
Her 16h37'1" 18d17'27"
John H Hanson
Cap 21h25'48" -25d35'9"
John H Muller & Cheryl M Buchanan
Dra 19h46'1" 62d48'55"
John H. Nogales
Aql 19h41'36" -0d7'31"
John H. Parks, MD, Founder AAP TDP
Vir 11h54'30" 3d14'55"
John H. Robinson Jr.
Psc 1h46'42" 5d45'42"
John H. Ryan
Vir 12h19'29" -3d32'39"
John H. Schriever
Uma 11h18'19" 30d48'20"
John H. Stella
Ari 2h30'30" 11d5'23"
John H. Stella
Aql 20h22'35" 0d24'38"
John H. Wohlfarth
Boo 14h45'19" 38d33'58"
John Hailey
Her 17h56'40" 29d55'19"
John Haley
Aur 5h27'17" 46d54'14"
John Hall
Uma 8h49'19" 61d58'8"
John Halsey
Ser 18h21'41" -14d32'15"
John Hamilton Bottoms
Cap 20h52'26" -23d23'21"
John Hamilton Campbell
Aql 19h51'1" 16d20'41"
John Hammel
Cep 22h4'19" 56d16'7"
John Hands a true "star"
Per 4h44'4" 40d41'24"
John Hanna
Her 17h10'17" 32d4'54"
John Hanson
Aqr 22h28'7" -1d8'54"
John Hardin
Uma 11h21'34" 67d44'55"
John Harson
Her 18h54'4" 24d6'39"
John Hartmann
Ori 5h39'26" 1d29'18"
John Hart's Star
Tau 5h37'5" 25d46'8"

John Hassan
Aur 5h26'7" 44d52'23"
John Hatch
Uma 14h25'36" 62d12'52"
John Hauschildt
Gem 7h46'13" 27d14'40"
John Healey
Cep 21h53'53" 64d41'19"
John Healy
Cep 21h19'3" 59d53'1"
John Heintz
Cru 12h1'34" -58d16'31"
John Heise's Wish
Sgr 19h15'52" -29d11'20"
John & Helen Karaszewski
Uma 8h59'7" 67d5'31"
John Henrik Andersen
Per 3h42'35" 43d45'43"
John Henry
Her 17h42'40" 20d21'51"
John Henry
Lib 15h3'4" -1d10'34"
John Henry
Cap 20h16'57" -21d24'1"
John Henry
Sco 16h41'51" -36d5'26"
John Henry Castillo
Her 18h51'31" 23d5'33"
John Henry Cunningham
Lyn 8h4'15" 41d3'3"
John Henry Debes IV
Sgr 19h0'56" -13d23'44"
John Henry Forrest - 2 April 1965
Boo 15h30'35" 45d2'40"
John Henry Hall
Ori 5h52'55" 3d54'35"
John Henry Kipnis
Gem 6h41'37" 31d44'14"
John Henry Laird, Jr
Cap 21h49'22" -10d8'35"
John Henry Mattei
Gem 7h27'27" 25d36'25"
John Henry Pollnow
Her 16h12'4" 14d15'56"
John Henry Ross
Tau 5h52'36" 23d30'18"
John Hesling and Carly Steinhauer
Cyg 19h37'26" 30d4'10"
John Hill
Lmi 9h41'54" 34d5'46"
John Hlinka
Cep 23h6'13" 71d0'21"
John Hoerdt
And 23h10'21" 42d49'3"
John Hollins
Uma 11h29'48" 58d49'53"
John Homer Sanders
And 1h8'58" 36d33'58"
John Honrycha
Her 17h41'36" 29d12'19"
John Howard Brackett Sr.
Umi 15h36'21" 74d30'29"
John Howard Caflisch
Her 16h37'12" 23d35'31"
John Howard Hess
Uma 8h43'5" 61d21'54"
John Howard Wilson
Sgr 18h33'14" -36d20'13"
John Howie
Ori 5h51'46" 7d29'36"
John Hozzian
Uma 12h4'20" 62d5'34"
John H.Thomas v1vrv2
Aqr 22h0'41" -16d53'57"
John Hudgens Pescador De El Rio
Eri 4h24'54" -2d34'11"
John Hudson
Uma 10h9'33" 41d59'47"
John Hugh O'Banion, Sr.
Her 16h21'5" 22d27'37"
John Hummel
Uma 13h2'36" 59d4'53"
John Hunsucker
Leo 10h54'54" 16d1'44"
John I Jackson
Crb 16h3'59" 37d2'35"
John Igielski
Cep 3h55'24" 81d34'12"
John Ignatius Bacile
Cap 21h33'9" -14d55'54"
John III
Cnv 13h21'47" 46d44'16"
John Illingworth Bradbury
Uma 10h28'3" 53d51'59"
John & Inger Nicoletti
Boo 14h38'38" 24d36'34"
John & Irene Klimis
Cyg 19h30'59" 30d46'35"
John Irwin James
Tau 3h34'44" 4d47'16"
John Italia, Ph. D.
Vir 12h58'3" 6d57'37"
John Ivor Hubbard
Per 4h41'32" 46d51'35"
John J. Austin, Jr.
Sgr 18h56'2" -34d40'50"
John J Ayre
Uma 11h10'39" 71d52'37"
John J. Brogan, III
Ari 2h41'23" 18d46'30"
John J. Buchanan, Jr.
Per 3h34'36" 46d58'21"
John J. Burton
Aql 19h29'24" -0d19'42"
John J. Carras
Her 17h7'21" 24d8'31"
John J. Chandler
Tau 5h41'40" 25d32'53"
John J Coopersmith Sr.
Aur 5h47'46" 54d18'52"
John J. Daly Jr. and Mary N. Daly
Cyg 19h24'25" 50d5'29"
John J. Diamond, Jr. "Digger"
Cep 23h55'40" 69d35'26"
John J Downey
Psc 1h8'12" 27d31'8"
John J. Duggan
Sgr 18h26'7" -32d57'50"
John J. Gordon
Cep 0h5'45" 67d32'58"
John J. Hallinan
Ari 2h47'16" 10d33'49"
John J. Hansen
Cnc 9h10'28" 12d8'25"
John J Hogan
Psc 1h31'25" 33d22'57"
John J. "Jack" Haughney III
Umi 15h4'42" 81d27'31"
John J. Jacksen Jr.
Ari 2h51'46" 28d11'9"
John J. Karl
Aur 5h52'24" 51d56'27"
John J. Kaye
Lib 15h6'31" -1d32'25"
John J Kraus
Lyn 8h38'21" 45d20'1"
John J. Laundry Jr.
Aur 6h36'51" 35d3'35"
John J. Macaluso
Her 16h52'8" 33d52'45"
John J. Masington
Aur 5h51'1" 55d41'10"
John J. McGrath
Uma 11h16'23" 34d39'43"
John J. Morris, Jr.
Sco 16h50'1" -33d40'34"
John J. Murphy
Cas 1h23'20" 65d40'17"
John J. Murphy Jr.
Cyg 20h32'32" 45d20'5"
John J. O'Neill
Per 3h6'35" 54d43'52"
John J. Patterson
Uma 11h21'47" 30d48'28"
John J. Pauly
Per 3h39'10" 52d14'0"
John J. Picone March 31, 1918
Ari 1h54'20" 23d4'6"
John J Piotrowski Wings of Freedom
Aql 19h45'10" 8d9'30"
John J. Pisani
Vir 14h31'31" -0d42'15"
John J. Pytel
Lyn 7h40'50" 37d26'21"
John J. Ryan Jr.
Ori 6h6'31" 21d13'6"
John J Schmid
Aqr 21h5'29" -14d1'50"
John J. Shade III
Per 3h44'1" 41d22'39"
John J Snyder III—John J Snyder IV
Per 3h45'18" 48d38'39"
John J. "Sonny " Kovatch
Lyn 7h30'33" 35d38'17"
John J Sr. & John J Jr. Albanese
Per 3h0'14" 54d37'37"
John J. Sullivan Jr.
Per 3h36'12" 45d31'47"
John J. Tickner
Lyn 8h13'55" 33d23'41"
John J. Vittorio
Uma 13h41'41" 49d51'19"
John J Woods 111 "Woodsy"
Ori 5h34'38" -0d24'11"
John "Jack" Daly
Uma 9h24'54" 58d27'32"
John *Jack* Davis
Per 4h29'12" 34d55'25"
John Jack Dutnall
Tau 5h57'12" 25d16'7"
John 'Jack' F. Hecker
Uma 11h9'30" 34d21'13"
John "Jack" Faherty
Ori 5h29'25" 2d38'37"
John "Jack" Heigelmann
Uma 12h34'31" 57d34'47"
John "Jack" McCarthy
Per 3h21'25" 47d45'11"
John "Jack" Serra
Leo 9h58'8" 10d48'49"
John (Jack) W. Roberts
Aql 19h21'25" -7d14'5"
John Jackson Professor Emeritus
Cnc 9h1'43" 19d15'45"
John Jacob Cramer
Ori 5h12'50" 1d1'55"
John James Bordoy
Uma 12h0'35" 61d25'22"
John James Boyle 60
Lib 15h5'56" -1d37'34"
John James Brunelli
Tau 3h50'58" 30d42'34"
John James Errington
Cep 22h3'49" 56d58'47"
John James Lynch
Ori 5h32'6" -1d31'29"
John James McAvoy
Cep 20h39'10" 64d22'52"
John James Starr
Leo 11h42'21" 24d18'55"
John James' twinkle-twinkle little
Lmi 10h10'28" 39d27'0"
John James Washam, Jr.
Ari 2h7'9" 26d19'53"
John James Weitort
Lmi 10h56'2" 27d33'27"
John & Jan 37th Wedding Anniversary
Cyg 20h28'10" 37d6'52"
John & Jane
Uma 9h19'26" 41d45'18"
John & Janice King - 50 Golden Years
Cyg 19h43'31" 32d32'56"
John & Janice Magee: 30yrs Together
Per 3h14'21" 45d14'9"
John & Jaquline Brown
Cyg 19h37'21" 28d30'43"
John Jaramillo
Lac 22h38'32" 48d56'49"
John Jay Menesini
Gem 7h49'45" 18d11'10"
John Jay Thomas
Gem 6h56'52" 14d56'18"
John Jeffery Meade
Cap 20h56'4" -18d6'55"
John & Jennifer Moore
Lyr 18h46'56" 36d53'32"
John Jesus De Sousa
Pyx 8h33'34" -24d41'26"
John Jesus Velasquez
Sco 16h55'50" -38d20'5"
John JK Kantzas
Cep 22h18'4" 68d25'25"
John & Joel Crocitti
Cru 12h46'13" -64d31'42"
John Joel Lorr
Sgr 17h57'5" -19d0'22"
John Joesph Paul Alfano
Vir 13h9'38" 8d44'36"
John Joha
Sex 10h26'6" -0d51'48"
John "Johnny Boy" Collins
Boo 14h53'53" 17d39'49"
John Joseph Baggio
Psc 23h3'59" 6d7'36"
John Joseph Baker 111
Aql 20h36'8" -0d5'51"
John Joseph Betz
Leo 9h34'12" 13d55'52"
John Joseph Browne
Aql 19h14'32" 6d50'47"
John Joseph Burke
Lib 15h41'22" -26d2'6"
John Joseph Crawford
Uma 10h17'21" 56d42'50"
John Joseph Dombek JR
Cyg 20h26'6" 30d49'4"
John Joseph Gawlinski Jr.
Per 4h46'24" 42d11'35"
John Joseph Gruba
Cep 6h26'2" 86d28'6"
John Joseph Haneklau
Cnc 8h29'1" 26d43'37"
John Joseph Heaton
Cep 20h43'2" 61d47'18"
John Joseph James, a true hero
Her 16h11'29" 22d32'58"
John Joseph Kelly, Sr.
Equ 21h10'4" 11d17'7"
John Joseph Kissane
Cep 22h54'19" 76d0'38"
John Joseph LaRusso
Aur 5h52'45" 53d33'45"
John Joseph Ligotti
Ori 5h43'7" 2d47'28"
John Joseph Madeline
Tau 4h46'2" 25d47'51"
John Joseph O'Hagan
Uma 10h25'14" 41d13'17"
John Joseph Osborne
Crb 15h26'4" 28d55'27"
John Joseph Owens
Psc 23h59'4" -4d34'27"
John Joseph Polefka
Her 17h32'4" 44d8'3"
John Joseph Raymonds
Leo 9h23'44" 10d12'47"
John Joseph Renfrow
Mon 7h49'56" -2d22'8"
John Joseph Riel
Uma 13h59'57" 57d11'11"
John Joseph Roddy III
Aur 5h41'36" 28d53'26"
John Joseph Shingler
Sco 16h40'42" -37d8'8"
John Joseph Spirito
Sco 16h33'3" -25d58'5"
John Joseph Sweeney
Boo 14h13'43" 50d0'5"
John Joseph Trela
Cyg 21h57'38" 52d55'48"
John Joseph Turk, Jr.
Tau 4h22'42" 28d24'13"
John Joseph Turner
Ori 5h26'20" 10d19'46"
John Joseph Wertzberger
Cyg 19h59'51" 43d20'34"
John Joseph Whelan
Ori 6h2'0" 11d1'15"
John Joseph Wilcox
Aur 5h59'0" 33d3'18"
John & Josephine Magnasco
Cyg 19h33'52" 50d53'57"
John & Joyce
Cyg 21h44'13" 45d23'27"
John & Joyce Cuffarri
Cyg 20h47'45" 31d59'34"
John Jr.
Gem 7h42'47" 32d44'2"
John Jr Mitchell
Per 4h25'24" 49d20'41"
John Julian Freiermuth
Ori 5h55'33" 20d52'42"
John Junior - God's Angel
Peg 22h27'46" 30d6'48"
John K 1
Her 17h18'33" 36d8'14"
John K Eglinton
Uma 11h5'36" 72d44'12"
John K. Hotze
Per 3h43'4" 44d47'3"
John K. Prihoda
Uma 13h43'28" 61d35'41"
John K. Taylor Sr.
Psc 0h55'51" 27d11'1"
John Kadin Gazzawaway
Her 17h21'1" 15d48'4"
John Karl
Cep 22h35'7" 61d35'56"
John Kasmer
Uma 9h36'8" 61d17'44"
John Kath
Ori 5h47'24" 6d38'31"
John & Kathleen
Her 17h20'46" 34d33'7"
John Kathleen McGondel Bampa Gamma
Uma 9h31'10" 67d5'11"
John & Kathy Hooper
Cyg 19h54'37" 32d30'30"
John Kawalkin
Sco 17h24'15" -42d9'12"
John Keed deGroot
Aur 5h13'58" 42d26'23"
John Keith Wise
Uma 10h54'16" 71d53'4"
John Kell
Uma 10h52'39" 66d44'5"
John Kempf
Ori 5h13'51" 6d14'13"
John Kennedy Pittman Jr.
Ori 5h56'51" 6d40'44"
John Kenneth ~K.L.
Uma 11h26'10" 62d42'36"
John Kerby Lathrop
Uma 9h31'41" 57d51'24"
John Ketchmark
Dra 15h53'8" 58d8'34"
John Kevin O'Brien
Lib 15h37'0" -16d59'8"
John Kim
Aur 5h3'46" 34d30'18"
John & Kim Goldinger
Cyg 21h15'45" 52d35'12"
John & Kimberly
Cyg 20h51'1" 45d45'31"
John King
Ori 5h18'31" 5d49'16"
John Kiyoko Hammond
Cap 21h14'13" -15d42'58"
John Klismet
Uma 10h25'12" 55d3'24"
John Knight
Per 3h1'1" 50d24'46"
John Knueppel, My son, My Buddie
Crb 16h7'1" 26d23'10"
John Kondas seventrees
Lib 14h55'6" -2d2'59"
John Konrad
And 23h26'42" 48d36'16"
John Koopman
Cep 23h23'54" 77d1'40"
John Kopera
Psc 1h21'9" 29d4'16"
John Korinko
Lyn 7h38'54" 57d0'54"
John Korner
Per 3h7'23" 53d26'13"
John Kozicki, Jr.
Gem 7h40'57" 33d52'47"
John "Krako" Guthrie
Her 17h20'7" 34d12'24"
John Kress
Leo 11h44'8" 24d40'49"
John Kuepper
Leo 11h22'20" 26d1'44"
John L. and Mary McFarland Liccini
Ori 6h0'29" 13d52'36"
John L. Baldwin
Cep 22h20'41" 62d17'22"
John L. Brown Celtic Warrior
Ori 5h24'21" 4d1'45"
John L. Cordeiro
Uma 9h12'51" 64d19'22"
John L. Daw
Aql 19h52'50" -0d12'29"
John L. Emhoff
Her 17h52'41" 22d20'44"
John L. Folino, Jr.
Her 18h10'17" 18d7'43"
John L. Hartman, Jr.
Lib 15h3'36" -26d30'2"
John L. Hoormann Jr.
Cap 20h12'38" -25d56'28"
John L. Johns
Boo 14h30'42" 17d44'40"
John L. Kalal
Ari 2h12'10" 24d1'20"
John L. Kaster
Uma 10h45'21" 53d36'38"
John L. Kelley, Jr. "JJ"
Umi 15h40'4" 79d51'12"
John L. Marshall
Sco 16h57'12" -11d18'24"
John L. Muraglia
Her 17h23'59" 16d41'1"
John L. 'Skip' Moroz
Her 16h40'5" 6d49'58"
John L. Tierney
Leo 10h31'13" 21d58'41"
John L. Wolken
Her 17h20'46" 34d7'21"
John Lacey's Destiny
Cyg 20h49'40" 31d38'56"
"John Lad"
Umi 17h43'19" 82d15'23"
John Lage Dewey
Gem 6h37'59" 18d28'25"
John & Lalla Porter
Lyr 19h18'32" 29d1'17"
John Landers
Uma 11h33'31" 57d39'1"
John Landis
Cnc 8h44'35" 20d17'23"
John Lane Bennett
Aql 19h26'17" 3d13'10"
John Laney
Vir 14h19'33" -6d16'39"
John Lange
Per 3h7'25" 47d17'20"
John LaRossa
Cap 20h59'45" -19d30'28"
John LaSorsa
Gem 7h43'13" 20d5'9"
John Lauer
Cyg 21h8'59" 36d30'1"
John & Laura
Cap 21h57'29" -21d46'4"
John & Laura Dalton - Forever
Cap 21h34'22" -14d31'40"
John & Laura, Forever
Cyg 20h10'56" 31d56'48"
John Laurila, My Love
Her 16h22'50" 23d26'46"
John Lawless
Sco 16h58'28" -40d59'38"
John Lawrence Bates
Uma 9h21'44" 61d17'6"
John Lawrence Schekeloff 23.10.84
Cru 12h37'59" -58d18'36"
John & Leah
Uma 9h11'45" 59d54'43"
John Leddy
Lyn 7h41'36" 38d59'30"
John Lee Davenport
Her 17h38'49" 18d12'16"
John Lee Gabbert
Lib 14h52'48" -1d26'53"
John Lee Whyte
Per 3h16'59" 43d43'4"
John Leiby II
Her 17h16'7" 46d20'51"
John Lenthal Eells
Sgr 17h55'56" -17d27'53"
John Lentz
Per 3h12'20" 53d29'26"
John Leonard Anthony Griffin
Per 2h38'23" 56d5'6"
John Leonard Appling
Ari 2h31'28" 25d11'46"
John Leonard Appling
Ari 2h42'50" 26d50'47"
John & Leslie Kammerman
Cyg 20h18'23" 53d57'8"
John Lester Rhodes
Lib 14h52'30" -10d46'6"
John Leverett
Psc 1h31'54" 11d48'13"
John Lewis Koch
Cep 22h13'6" 66d29'19"
John & Lia DeFelice
Ari 2h23'33" 18d5'57"
John Liam Moran
Cnc 8h4'4" 17d7'45"
John Lilla
Aur 6h52'4" 41d40'5"
John & Linda Austin
Cyg 21h13'5" 47d13'37"
John & Linda Doyle
Cnc 8h40'33" 15d46'9"
John Lindenmeyer
Ori 5h37'46" 13d1'0"
John Lipp and Shi Ying
Umi 16h39'55" 84d9'18"
John Lloyd Brunson
Cep 21h44'19" 55d53'35"
John Lloyd Morris
Cru 12h35'3" -58d32'39"
John Loper
Lib 14h51'46" -10d4'40"
John Lori & Julian Camden
Uma 9h54'40" 62d19'8"
John Louis Orfe
Uma 13h43'47" 58d45'22"
John Louis Teitloff, Jr.
Ori 6h9'24" 3d41'32"
John & Louise O'Brien
Cyg 19h43'8" 36d36'28"
John Lovatt
Cep 23h51'26" 84d3'49"
John Loves Jessica
Dra 17h26'2" 67d15'11"
John Lucian Sills
Sco 16h7'51" -11d43'17"
John & Lucille Kenning
Lyr 18h53'50" 34d49'11"
John & Luisa Puzder
Lyr 19h22'49" 37d30'24"
John Luke Maher
Cap 20h9'53" -25d24'58"
John Lunde
Per 3h2'59" 46d43'39"
John Lurcock
Her 16h48'57" 40d43'43"
John Luu - Shining Star
Cep 22h43'14" 72d19'59"
John Lyle Waldroop
Sgr 18h49'57" -19d34'59"
John & Lynn's Star for Always
Per 3h43'44" 45d54'33"
John Lyons Buckley IV
Del 20h34'18" 13d5'55"
John M. Ammerman
Vir 12h23'11" -5d40'3"
John M Barbery
Gem 6h45'37" 16d28'44"
John M. Baur
Aur 5h6'42" 34d31'6"
John M. Cannistra Jr
Uma 10h55'36" 67d38'44"
John M. Erdelac
Aqr 23h54'39" -4d35'27"
John M. Gilliam
Psc 0h23'48" 6d25'32"
John M. Gilmore
Her 18h52'22" 24d20'55"
John M Grinstead
Cnv 12h24'1" 46d26'23"
John M. Ireland
Gem 7h15'3" 23d44'43"
John M. Kenney
Aqr 23h2'27" -8d1'14"
John M. McCurdy III
Lib 14h47'36" -11d15'5"
John M. Moore
Sgr 18h57'3" -31d47'38"
John M. Moratelli
Ori 6h1'20" 15d31'55"
John M. Norris
Aql 19h49'9" 3d39'7"
John M. Skeffington, Jr.
Leo 10h22'43" 26d13'23"
John M. Zerolis - Crown Prince
Leo 10h14'21" 13d49'40"
John M. Zomp
Umi 14h12'10" 76d24'46"
John Mac
Lib 15h51'57" -18d37'9"
John Mac
Lib 14h33'22" -17d57'32"
John "Mac" McLaughlin
Per 3h10'21" 53d30'42"
John MacArthur Burness
Cep 22h35'12" 65d33'16"
John & Maddy White
Cyg 20h37'33" 42d28'30"
John Maemori
Lib 14h45'32" -9d5'33"
John Maines
Sco 17h12'22" -39d54'31"
John Malcolm Dedinsky
Lup 15h39'46" -32d26'37"
John Mallery
Per 3h44'4" 43d53'14"
John Mancuso, My Baby Boy
Her 15h52'9" 42d41'28"
John Mandracchia
Ori 6h12'51" 14d20'24"
John & Margaret Clough for Eternity
Cra 18h2'11" -37d15'29"
John & Margaret Shevlin
Cyg 19h38'40" 28d6'17"
John Mari
Boo 13h51'42" 12d24'52"
John & Marilyn
Sco 16h13'31" -16d16'4"
John Mark Gatling
Leo 11h14'20" 14d43'28"
John Mark Ponder Husband & Father
Per 3h41'16" 34d16'56"
John & Marlys Achartz
Umi 15h15'42" 72d1'12"
John Marshall Fernberger
Ori 5h27'59" 1d42'37"
John Marshall Gill
Aqr 21h45'32" -7d22'42"
John Marshall Sonntag
Uma 8h55'53" 60d8'54"
John Martin
Uma 9h28'31" 63d50'17"
John Martin
Uma 11h6'22" 42d18'32"
John Martin Riiska
And 1h14'52" 45d43'38"
John Martin Smyth
Cep 21h24'57" 66d37'50"
John Marvin Herrick
Per 2h37'30" 54d50'36"
John & Mary Hires
Cyg 20h19'9" 35d3'4"
John & Mary Kay Entsminger
Lyr 18h39'13" 37d37'23"
John & Mary Tuegel Forever
Uma 11h54'53" 34d30'1"
John & MaryBeth Reiss
Cyg 19h43'7" 53d45'0"
John & MaryEllen Blanchford-Whalen
Cyg 21h20'47" 32d58'31"
John Masephol's P.I.T.S. Stop
Her 16h47'48" 46d42'50"
John Mason Ammerman
Umi 16h33'43" 78d7'46"
John Mather Harris
Sco 16h5'11" -22d8'42"
John Matthew Benda
Lib 15h28'42" -23d10'52"
John Matthew Pimley
Her 17h47'36" 29d9'44"
John Matthew Raymonds
Aqr 21h40'9" -5d51'4"
John & Maureen
Umi 16h23'8" 81d16'24"
John Mazza
Ari 3h18'41" 19d37'29"
John McBride
Leo 9h35'53" 28d40'5"
John McCarthy
Uma 8h56'37" 52d21'15"
John McColm Thomson
Ori 4h47'18" 5d28'53"
John McCurrie
Aqr 21h59'40" -13d28'17"
John McElmurry Campbell
Tau 5h22'14" 18d17'48"
John McElroy "50"
Cep 1h41'49" 77d38'15"
John McGraw
Cas 0h23'59" 51d38'28"
John McIntyre Sandles
Uma 10h38'16" 58d50'37"
John McKenzie
Aql 20h9'0" 10d47'3"
John McLaughlin
Per 3h8'33" 57d5'50"
John McMath
Uma 13h35'46" 48d18'14"
John Megown Mayhall IV
Crb 16h3'59" 33d38'4"
John & Melinda Ann Hacker
Cyg 19h31'42" 52d0'51"
John & Melissa - Soul to Soul
Cru 12h37'15" -58d24'47"
John Mellish
Ori 5h24'23" -0d37'40"
John Merkel
Ori 4h48'17" 14d15'46"
John Mersek
Lyn 7h50'36" 43d4'14"
John Mezhir
Her 17h15'42" 42d9'56"
John Miceli
Her 16h14'0" 46d5'52"
John Michael
Ori 5h49'45" 9d17'14"
John Michael
Gem 7h53'28" 15d13'40"
John Michael
Ser 18h21'34" -13d39'4"
John Michael Baker
Cap 21h4'41" -16d20'28"
John Michael Barnette
Equ 21h20'43" 8d9'3"
John Michael Barry
Psc 1h27'30" 10d55'3"
John Michael Bartol, Jr.
Lmi 10h46'21" 37d49'11"
John Michael Bauer
Uma 9h26'7" 43d30'46"
John Michael Doody
Uma 11h13'44" 64d29'49"
John Michael Fehil
Her 17h38'4" 37d28'15"
John Michael Fitzgerald
Per 4h38'53" 40d46'11"
John Michael Francis
Per 3h31'50" 31d15'33"
John Michael Gilliam
Leo 11h31'57" 0d39'15"

John Michael Harmon Per 4h16'41" 51d12'53"
John Michael Hyatt Uma 12h9'3" 60d55'21"
John Michael Lee Thomas Her 17h39'45" 16d32'3"
John Michael Mahon Flannery Aqr 22h31'2" 2d0'46"
John Michael Manning Her 18h56'10" 20d9'25"
John Michael McGee Per 4h19'34" 44d12'9"
John Michael Melchiori Sco 16h17'29" -10d35'44"
John Michael Murray Sgr 19h53'40" -29d11'18"
John Michael Pereira Her 17h12'45" 26d59'10"
John Michael Pirruccio Tau 4h4'14" 27d24'6"
John Michael Pottage Cnv 13h45'33" 30d18'5"
John Michael Ruiz Her 17h19'33" 17d23'37"
John Michael Sprafka Cep 22h32'25" 71d52'20"
John Michael Stauffer Gomez Ori 4h50'51" -0d0'33"
John Michael Suozzo Lib 14h50'21" -14d49'28"
John Michael Sykes Uma 8h57'37" 65d45'37"
John Michael Turner Per 4h12'46" 48d54'50"
John Michael Vincent Berrellez Her 17h10'13" 33d46'40"
John Michels Aqr 21h30'0" 2d24'50"
John Mike Azar Sco 16h25'54" -33d49'51"
John Mikula Uma 11h14'21" 54d53'46"
John Miles Dawson Sco 17h42'23" -37d21'43"
JOHN MILLER'S STAR Ori 5h53'16" 21d56'19"
John & Millie DiBiase 2-24-74 Lyr 18h52'16" 36d41'32"
John Milton Kennedy Cir 14h44'10" -63d53'45"
John Milton Schulz Uma 8h19'20" 62d50'52"
John Mino Sr. Per 4h26'15" 34d12'55"
John Minter Bowyer Uma 11h0'39" 49d9'4"
John Mitchell Lindsay Cas 1h24'19" 59d22'12"
John Molly Sco 16h16'53" -27d50'45"
John & Monica's Sterlng Silver Star Cyg 21h28'52" 47d5'25"
John "MonkeyBum" Hallett Uma 11h30'43" 56d51'35"
John Montelione Boo 14h37'24" 23d54'18"
John "Moon" Mullen Per 3h24'35" 48d4'2"
John Moreira Jr. Gem 7h18'54" 21d46'44"
John Morgenstern III Gem 7h23'49" 31d42'31"
John Moriarty Ari 2h15'58" 23d15'43"
John Muller Tau 5h48'42" 26d19'33"
John Mullett's Lucky Star Gem 7h48'25" 33d6'55"
John Munson Moore Cap 20h38'44" -17d11'40"
John Murphy Dra 18h56'13" 70d12'57"
John Murren Sco 17h28'6" -35d6'15"
John my Diamond in the Dark Ori 6h20'1" 19d1'0"
John My Forever Love Be Happy Ari 3h27'8" 20d2'26"
John N. Diana, M.D. Uma 11h32'27" 42d32'56"
John 'n' Marg 31.12.04 Cyg 21h23'55" 50d38'58"
John N. Stefano Per 2h54'25" 45d27'38"
John Nathan Mon 6h35'48" 7d59'34"
John Nazarian Lac 22h54'57" 38d35'37"
John Nee Cep 22h38'47" 85d35'43"
John Neely Dowden Cyg 20h24'10" 35d23'17"
John Nelson Taylor Jr. Gem 7h24'48" 24d58'34"
John Nicholas Elston Lyn 8h49'47" 34d33'29"
John Nicholas Heffernan Her 18h38'28" 12d17'19"
John Nichols 2005 Pro Bono Star Ori 5h38'28" 2d30'3"
John Nicolas Her 17h39'43" 29d1'45"
John Niegocki Ari 1h58'35" 13d15'31"
John & Nina Uma 8h33'56" 65d59'41"
John Nobles Ori 4h54'48" 1d12'14"
John "NoNo" Nichols Aqr 21h51'3" -0d15'52"
John Norman Holland And 0h20'27" 28d7'7"
John Norman Nilsen Boo 14h14'52" 28d3'11"
John Novaks Star Sgr 18h54'43" -30d15'47"
John O. Lee Leo 9h31'58" 24d59'54"
John O. Yeider, Jr. Her 18h35'52" 15d31'31"
John O'Brien Cap 21h20'41" -15d37'46"
John Ochs Cep 20h40'27" 63d49'48"
John O'Connor Vir 13h7'4" -5d17'15"
John O'Connor's Sparkling Shamrock Uma 9h47'18" 48d13'34"
John O'Donoghue Per 2h34'2" 52d51'2"
John Ohriner Cnc 8h36'20" 23d25'36"
John Oliver Mumford Uma 8h48'20" 62d40'36"
John Omar Aur 5h36'20" 49d21'51"
John Orin McGuire II Uma 13h33'53" 54d36'12"
John Orr Gem 6h51'43" 22d52'49"
John Outzen Ari 3h29'9" 26d56'47"
John P. Bardong Leo 11h3'54" 5d23'9"
John P. Cloutier Her 16h30'25" 46d38'6"
John P. Connolly Jr. Her 16h29'5" 42d32'43"
John P. Dick - Our Dadda Sgr 19h13'48" -23d25'10"
John P. Fitzpatrick Uma 11h35'32" 53d52'42"
John P. Fleckenstein Uma 10h28'2" 63d7'15"
John P Hansmeyer Uma 8h41'17" 65d10'15"
John P. Hernon Uma 10h53'6" 39d41'32"
John P. Hogan Lyn 7h13'41" 55d6'49"
John P Lowney Jr & Daune E Nobriga Uma 10h23'15" 57d4'59"
John P McFadden Leo 10h20'9" 11d48'2"
John P. Mosser, Sr "Double Eagle" Uma 12h40'42" 56d27'25"
John P. Venturella Vir 12h19'47" 12d7'53"
John P. Willey 3/30/45-10/6/01 Ari 1h53'48" 17d19'30"
John Padraic Lavelle Her 16h3'42" 47d23'25"
John "Pap" Muldoon Psc 0h14'34" 11d34'56"
John "Papa" Eyler Per 3h29'50" 35d46'56"
John (Papa) Hemmings Vir 14h0'55" -16d1'2"
John Papp Aql 19h11'8" 14d49'55"
John Pappas Her 17h32'7" 34d13'51"
John Parker and Valerie Ann Cap 20h47'18" -20d9'14"
John Passamonte Cnc 8h2'54" 14d19'12"
John Pat McDonald Leo 10h12'22" 21d43'52"
John Patone Lyn 7h13'42" 50d45'27"
John & Patricia's Ruby Star Cyg 19h55'26" 57d54'56"
John Patrick Ori 6h6'30" 10d42'2"
John Patrick Bartel Tau 3h51'36" 15d16'30"
John Patrick Collier Sgr 18h36'52" -16d29'31"
John Patrick Duval Sgr 18h53'10" -20d26'37"
John Patrick Dylan Robinson Uma 8h39'55" 71d17'14"
John Patrick Flanagan Her 17h11'23" 34d54'10"
John Patrick Green Sgr 18h21'52" -32d37'34"
John Patrick Higgins Her 17h48'49" 30d47'46"
John Patrick Kennelly Vir 13h32'22" 4d41'22"
John Patrick Konzelman Sgr 19h6'48" -19d15'58"
John Patrick Maguire Lmi 10h7'7" 28d13'30"
John Patrick Moriarty Psc 23h6'12" 0d57'47"
John Patrick Neal Umi 16h16'27" 77d36'51"
John Patrick Pollis II Aur 5h5'47" 40d37'7"
John Patrick Ward Gem 6h44'5" 23d40'28"
John Patrick Winchester Uma 8h38'10" 50d44'19"
John Patten Aur 5h40'30" 43d5'9"
John Patterson Crb 15h56'48" 39d30'55"
John Patterson Aqr 22h37'21" -0d8'23"
John Patula Cep 22h32'48" 74d2'42"
John Paul and Kathleen Maczko Col 6h24'8" -33d41'58"
John Paul Anderson Uma 9h55'14" 59d0'24"
John Paul Arrascue Uma 11h35'28" 36d1'39"
John Paul Bascelli III Uma 11h40'15" 59d46'4"
John Paul Beech Ori 5h44'34" 1d18'49"
John Paul Bertolina Lmi 9h28'33" 36d21'56"
John Paul Bocchi Psc 0h52'32" 7d53'19"
John Paul Brody Lyn 7h54'6" 34d37'24"
John Paul Corchado Per 3h28'7" 38d56'14"
John Paul Cunningham Per 3h21'10" 41d30'14"
John Paul Dallas Gem 7h41'35" 24d43'37"
John Paul DeJoria Cep 3h22'28" 81d57'4"
John Paul Delobel Lib 15h55'7" -16d54'45"
John Paul Dickmeyer Psc 1h13'37" 11d45'15"
John Paul Duffy Umi 15h9'45" 73d40'54"
John Paul Edward Breen - Jack's Star Cas 2h20'18" 62d7'56"
John Paul Elliott Her 16h17'51" 44d46'49"
John Paul Fowler Gem 7h31'47" 26d11'12"
John Paul Gidos Cep 23h22'41" 77d18'33"
John Paul Hinklin Lmi 10h20'40" 35d43'44"
John Paul I 7165 Ori 4h48'30" 11d30'44"
John Paul "Jack" Jones III Ari 2h33'24" 26d33'49"
John Paul Jones Aqr 23h36'5" -19d42'37"
John Paul Kaminski, Jr. Cap 21h13'9" -15d57'14"
John Paul Morrison II Her 16h40'31" 35d27'25"
John 'Paul Muad'Dib' Dawson Cyg 19h55'0" 44d34'21"
John Paul Painchaud Sr. Tau 4h19'20" 26d50'53"
John Paul Philippe Lamb Sco 16h42'10" -41d32'18"
John Paul Romero III Cnc 9h12'3" 31d17'48"
John Paul Tiffin Lmi 10h44'3" 32d27'25"
John Paul Totels Psc 1h0'24" 24d52'58"
John Paul Varagnolo Sco 16h14'53" -11d42'40"
John Paul Webb Uma 11h56'58" 42d30'23"
John Paulk Sgr 19h51'3" -12d23'51"
John Paul's Rising Star Her 18h23'4" 17d12'29"
John Paulson Her 16h41'45" 36d31'49"
John "Peanut" Glasgow Psc 1h32'36" 3d42'46"
John Perry Psc 1h15'29" 23d23'44"
John Peshoff Per 3h25'13" 49d0'51"
John Peter Bodie Per 3h6'10" 55d57'20"
John Peter Drewal Cnc 8h50'29" 12d19'4"
John Peter Falcon Gem 7h34'9" 28d7'44"
John Peter Ferraro Uma 12h36'0" 54d13'15"
John Peter Gibson Per 4h32'50" 33d25'52"
John Peter Gidders' Star Ori 5h18'58" -0d40'17"
John Peter Kass And 2h20'43" 46d23'35"
John Peter Kieffer Cnv 12h42'47" 40d57'10"
John Peter Lafferty IV, M.D. Oph 17h5'24" -13d19'33"
John Peter Lazar born Sept. 7, 2006 Vir 13h18'9" -19d57'7"
John Peter Lucia Sgr 18h27'19" -22d14'31"
John Peter Rose III Uma 10h36'22" 49d48'17"
John Peter Tardibuono Per 2h43'46" 52d58'50"
John Philip Bass Cep 22h47'16" 72d27'21"
John Philip Cosentino Uma 14h18'47" 62d9'6"
John Philip Scattergood Her 18h12'4" 23d51'53"
John Phillip Davies Peg 23h52'0" 10d50'14"
John Phillips Ari 1h59'20" 21d54'16"
John Picard Her 17h14'2" 26d56'42"
John Picini Sr. Her 17h59'46" 22d5'21"
John Pickell Tau 4h39'50" 22d56'8"
John Piero DeRose Her 16h8'36" 24d35'56"
John Piper Psc 1h18'34" 30d42'3"
John Piraneo Jr. Cyg 20h51'28" 46d18'13"
John Plassio Umi 14h34'22" 71d35'16"
John Pleban Uma 10h46'38" 68d26'42"
John Pollard Jr. Aur 5h43'55" 49d16'26"
John "Pop" O'Rourke Cep 23h41'13" 74d31'2"
John Poplett Ori 5h0'15" 10d11'26"
John Porette Hya 9h23'49" 0d23'26"
John Presley Uma 10h8'5" 46d37'9"
John Price Dra 16h44'16" 64d57'3"
John "Puffy" Ludford Peg 21h33'11" 15d47'27"
John Pulsonetti Loved Husband/Dad Cep 22h34'10" 58d2'36"
John Puskarich Cap 21h9'26" -23d46'52"
John Quick Per 4h16'51" 43d52'58"
John Quillen*The Love Of My Life* Cyg 20h59'22" 47d43'9"
John Quinn Cep 20h48'21" 60d1'15"
John Quinn Ward Umi 15h16'30" 70d36'1"
John Quintin Doroquez Cnc 9h6'2" 32d11'23"
John R. Bell II Gem 7h1'12" 20d31'20"
John R. Butler, Sr. Cap 21h58'38" -9d18'53"
John R. Cisco Sco 16h45'44" -40d3'8"
John R Deho Tau 5h51'9" 17d54'27"
John R. & Esther Lewis Umi 15h28'47" 73d0'24"
John R. Fee Sgr 18h29'8" -18d55'38"
John R. Haecherl Boo 15h38'52" 39d45'46"
John R. Honeycutt Per 3h15'25" 45d49'16"
John R Kretsch "Jonrok" Boo 14h33'20" 22d17'55"
John R. Lucas Crb 15h44'3" 37d19'26"
John R & Marcelina Morgan & Family Cyg 21h43'41" 45d46'20"
John R. Moxon, Jr. Per 3h46'8" 44d1'29"
John R. Portelance Uma 9h14'59" 69d45'36"
John R. Quirk Her 17h45'53" 46d19'36"
John R. Shaw Uma 10h39'25" 66d39'46"
John R. Tidwell Ori 5h39'11" 2d51'58"
John R. Vercher III Ari 2h43'45" 14d12'38"
John Ralph Pointer Boo 14h32'53" 37d7'43"
John Rancier Owens Dra 17h49'49" 59d15'20"
John Randall Zink Ari 2h52'56" 14d17'1"
John Randol Sanfilippo Per 3h25'35" 48d7'2"
John Randolph Bergmann 3/13/2004 Psc 0h49'26" 17d53'53"
John Raph Newton; always our Star Leo 9h32'28" 13d11'48"
John Rawley Hessick Per 2h54'19" 54d51'49"
John Ray Cep 20h38'27" 61d39'18"
John Ray Young Ori 5h52'44" 22d23'34"
John Raymond Brennessel Ori 6h11'38" 19d35'40"
John Raymond Ferguson Gem 6h31'16" 16d37'29"
John Raymond Hautzinger Sco 17h26'55" -40d24'6"
John Raymond Lamb Lib 14h52'38" -5d21'30"
John Raymond Sorenson Per 3h12'7" 44d49'19"
John Red Cru 12h27'7" -58d22'43"
John Reeves Cyg 19h38'6" 35d15'30"
John Reid Quattlebaum Per 3h14'50" 51d38'4"
John & Reinelda Cwalinski Uma 11h20'2" 53d6'5"
John & Renee's Shining Star Cyg 20h27'49" 37d16'13"
John Renzullo, amare e la forza Cep 22h10'47" 61d13'39"
John Rich Ori 6h9'13" 14d30'49"
John Richard Begg, Jr. Umi 16h14'33" 76d24'36"
John Richard Edwards Jr. Aqr 21h20'33" 0d29'20"
John Richard Frieders Her 17h46'21" 40d59'25"
John Richard Kempf Lyn 7h17'17" 49d0'59"
John Richard Montgomery Lib 15h32'41" -9d27'3"
John Richard Rivers Dra 20h28'49" 67d45'51"
John Richard Samuels Jr. Her 17h45'54" 35d9'38"
John Richard Schneider Ari 2h16'9" 16d51'9"
John Richard Spann IV Her 17h39'21" 16d48'14"
John Rick Umi 16h25'46" 77d39'1"
John Ricky Spires Boo 13h42'29" 14d40'42"
John Riley Ari 2h43'23" 29d52'9"
John Riley Her 16h37'18" 26d0'11"
John Riley - Wicky Cru 12h36'31" -59d41'22"
John Rimmele Umi 15h20'49" 74d53'26"
John & Rina Rastetter Cas 0h18'49" 57d36'5"
John Robert Per 2h18'38" 51d45'56"
John Robert Bowman Aql 20h4'16" 10d34'55"
John Robert Bowser Cyg 20h11'13" 30d17'2"
John Robert Corcoran Per 3h12'37" 54d1'5"
John Robert Dorr III Umi 17h9'47" 78d30'2"
John Robert Goheen Boo 14h35'39" 19d9'35"
John Robert Green Sge 19h43'37" 17d54'56"
John Robert Harvey Cep 21h28'36" 83d4'9"
John Robert Johnson Her 17h46'32" 44d12'35"
John Robert Logan Sco 17h2'11" -40d17'6"
John Robert Logue Her 17h1'37" 19d47'5"
John Robert Luttjohann Her 16h31'30" 32d35'32"
John Robert Maries Gem 7h28'51" 18d25'50"
John Robert Miller Vir 14h16'28" 3d36'11"
John Robert Moody Her 16h28'38" 46d38'4"
John Robert Nelson Leo 11h45'35" 15d19'19"
John Robert Orr Leo 10h27'7" 20d38'34"
John Robert Schuyler Her 16h55'2" 31d52'35"
John Robert Simonin III Lib 15h6'32" -6d39'33"
John Robert Small Vir 12h40'6" 1d5'34"
John Robert Waslin Ari 3h20'20" 24d31'49"
John Robert Watson Boo 15h9'4" 48d59'33"
John Robokos Gem 7h3'51" 28d49'32"
John Rodacker, Jr. Aqr 22h42'1" 0d11'34"
John Rodarte Her 17h43'37" 16d1'20"
John Rodriguez Crt 10h59'37" -13d48'33"
John Rogalski Cru 12h9'16" -56d3'59"
John Rogers & Alycia Malloy Cyg 20h24'17" 54d49'19"
John Roland Wolsey Cru 12h34'49" -58d40'23"
John Roller Cep 22h6'57" 53d47'13"
John Roman and Vanessa Revard Gem 6h1'11" 24d24'29"
John Romanucci Her 17h55'50" 14d54'47"
John Ronan Jr. Ari 3h7'41" 19d27'50"
John & Rose Ari 2h40'23" 28d34'50"
John Rose Uma 10h18'35" 64d38'29"
John Rossbach Per 3h45'4" 45d8'25"
John Rothwell Lyn 7h19'38" 54d6'5"
John Roulston Halford Dra 20h12'0" 69d4'31"
John Roy Lee Sr Lac 22h50'57" 50d8'47"
John Royall Ori 5h19'59" 12d12'20"
John Rudy Kondor Her 17h11'29" 23d4'58"
John Rundquist Cyg 19h44'28" 36d38'18"
John Rushby Cep 21h25'53" 63d51'21"
John Russell Meers Psc 1h34'2" 5d14'30"
John Russell Youngblood Uma 11h35'47" 50d36'19"
John Russo Lyr 18h29'35" 36d39'38"
John & Ruth Bricker Cyg 21h27'1" 34d58'21"
John & Ruth McLay Aqr 23h11'55" -22d27'40"
John Ryan and Tabitha Lynn Hill Peg 23h50'37" 23d43'26"
John Ryan Daniel Floreani Ori 5h53'4" 10d34'40"
John Ryan Grassby Aql 19h16'53" 0d24'18"
John Ryan Kennelly Aqr 21h10'56" -3d12'45"
John Ryan Pike Sco 16h12'14" -10d46'17"
John S. and Penny Wilson Cyg 19h36'33" 30d19'44"
John S. Black Per 3h5'47" 53d35'22"
John S. Bragan Uma 10h48'6" 63d38'1"
John S. Ciappenelli (Fahto) Uma 9h12'23" 62d53'37"
John S. Corbett Leo 11h15'10" 2d47'41"
John S. Drugas Her 16h51'56" 37d30'56"
John S. Griswold Boo 14h41'25" 26d18'32"
John S. Herlihy - "Our Rockstar" Aqr 22h32'0" -10d0'53"
John S. Holden Jr. Uma 11h27'9" 62d30'36"
John S. Knight Per 3h20'29" 43d18'14"
John S. Lewis Gem 6h56'46" 13d14'53"
John S. Marshall, Jr. Per 2h52'16" 53d25'9"
John S. & Phyllis Diloreto Guthrie Lyn 7h17'20" 57d43'34"
John S. Wargo Gem 7h22'23" 30d40'59"
John - Sabine Cyg 21h54'31" 41d40'39"
John Salvador Ellis Her 17h27'49" 36d7'48"
John Salvatore Bartolotta III Sgr 19h42'50" -11d51'46"
John Samuel Gynan Cnc 8h8'40" 20d27'50"
John & Sandra Mitchell 12/28/1963 Cap 20h37'18" -10d53'2"
John & Sandra ~ You brighten my day Sge 19h36'55" 17d33'11"
John & Sandy Sloan Cyg 21h38'19" 36d56'50"
John Sanfilippo & Family Cyg 20h11'5" 40d48'20"
John Sanford Skidmore And 23h31'17" 49d5'8"
John Sang-Heun Kim Gem 7h9'50" 26d34'43"
John Sann Umi 16h20'8" 77d14'19"
John & Sarah's Shining Star Cyg 20h8'8" 45d7'20"
John Schleiff Cap 21h41'24" -23d54'8"
John Schmidt Pho 0h39'35" -46d19'26"
John Schriner Transue Aur 5h43'2" 46d10'44"
John Schubert McPhail Uma 11h2'30" 36d43'57"
John Scoggins Cyg 20h39'28" 54d38'29"
John Scott Fancher Sco 16h58'18" -38d37'26"
John Scott Hampshire Uma 8h42'9" 48d46'23"
John Scott Webb, Jr. Sgr 18h19'25" -30d19'9"
John Scrivines Cep 20h49'4" 64d55'53"
John Seabury Farrell First Birthday Lac 22h26'54" 49d43'3"
John Sebastian Brigiotta Dra 19h49'7" 70d13'34"
John Sebastian Danvoye Sco 17h45'33" -37d44'44"
John Segginger Aql 19h30'54" 4d2'57"
John Seldon Nelson Her 18h11'11" 28d51'20"
John - Semper Lucidos Cen 11h33'10" -48d39'3"
John Setiabudhi - Jansz And 1h34'32" 39d56'12"
John Seymour Cowen Dra 18h23'30" 52d52'25"
John Sherman Estey Lyn 7h17'27" 47d24'18"
John & Sherry Dryer Uma 9h7'46" 67d48'21"
John Sheyon Uma 9h18'47" 53d56'16"
John Shillington Lib 15h27'42" -6d24'42"
John Shuler Aur 7h16'4" 42d22'15"
John & Siew Huang Ryan 12/5/07 Leo 10h27'30" 15d17'27"
John Sikes Gibbs, III Psc 1h20'58" 18d4'11"
John Silver Trugler Uma 10h50'50" 63d1'5"
John Simkin Cyg 21h31'55" 48d47'28"
John Simone Cyg 20h37'29" 37d20'35"
John Simpson Tau 3h40'39" 27d53'37"
John Smith (Smithy) True Blue Friend Cru 12h37'19" -59d55'8"
John Speight Cep 22h1'56" 65d11'15"
John Spencer Harrison Aiken III Cap 21h16'18" -16d4'16"
John Sperni Cep 3h30'43" 87d14'53"
John Spooner Ser 15h28'37" -0d8'22"
John "Squirrel" Cunningham Sco 16h10'43" -11d33'56"
John Sr. "Dad - Pops" Uma 11h30'18" 34d17'44"
John & Stacy Billings Cyg 19h37'7" 29d58'48"
John Stanley Madden, IV Umi 15h25'9" 71d9'23"
John Stanley Trella Psc 1h40'8" 5d28'55"
John Starczynski Cnc 8h49'28" 31d39'13"
John Starman Shelton Cyg 19h39'15" 28d24'35"
John Stefek Leo 11h3'0" 21d11'51"
John Steinkamp Ari 2h47'41" 15d38'46"
John Stepasiuk Uma 11h44'3" 64d53'53"
John Stephan Karas Cep 4h42'7" 82d23'8"
John & Stephanie Cyg 20h58'34" 44d56'58"
John Stephen Lowery Psc 0h49'31" 16d53'30"
John Stephen O'Neil Psc 1h10'58" 10d26'32"
John Steplar "Your Guiding Light" Per 2h50'4" 42d57'28"

John Steven McEntire
Ori 5h12'16" 3d36'27"
John Stevens Powell
Cep 22h57'58" 73d3'43"
John & Stevie-Lee
Pepperell 07/04/2007
Cru 12h15'15" -57d24'54"
John Stewart
Uma 9h57'40" 52d24'30"
John Stewart Nielsen
Boo 15h1'21" 41d49'43"
John Stewart Raeside
Per 3h24'9" 39d20'34"
John Stillings
Per 2h24'21" 56d52'15"
John Stockton
Ori 5h18'55" 1d50'37"
John Stoner
Her 18h56'54" 16d47'20"
John Storey
Uma 9h3'10" 47d16'27"
John Stroud
Her 16h56'16" 16d49'44"
John Stuart Frazer
Cep 20h41'39" 56d48'12"
John Stuart Phillips
Sco 17h51'34" -37d27'6"
John Stypulkoski
Leo 11h12'54" 17d24'35"
John Sullivan
Her 16h46'46" 38d25'28"
John Sullivan
Per 2h48'40" 42d18'32"
John Sullivan Sr.
Uma 12h11'53" 62d30'27"
John Swenson
Lib 15h4'9" -6d23'9"
John Szabo's Star
Uma 13h57'6" 58d59'12"
John T. Baum
Ori 4h52'11" -0d11'29"
John T. Cook
Sco 16h47'3" -31d25'40"
John T. Elder
Uma 10h28'59" 58d7'49"
John T. Ferrara
Leo 9h32'20" 14d25'56"
John T. Holzum
Cet 3h8'22" -0d9'49"
John T. Hummingbird
Sgr 18h34'38" -22d36'19"
John T. Masklee
Lib 15h37'17" -19d9'49"
John T. Messere
Sco 16h6'20" -12d51'22"
John T Moore
Cep 21h28'14" 63d57'13"
John T Piazza
Aur 7h23'47" 40d3'57"
John T. Schmidt III
Vir 13h1'25" -7d49'18"
John T Swanick
Cep 23h19'42" 75d30'10"
John T. Vucurevich
Ori 6h0'2" -0d46'39"
John Taggart
Umi 15h31'12" 76d0'49"
John Tatore
Cnc 8h31'0" 11d59'49"
John Ted Oliver
Aql 20h21'2" 3d8'23"
John & Tegan Barber
Leo 9h31'6" 16d40'40"
John (Terry) Grage
And 23h20'8" 44d38'47"
John Terry's Proverbs
11:25 Star
Gem 6h36'11" 17d38'30"
John the Baptist's Eternal
MoJo
Cnc 9h14'29" 10d40'33"
John the Beloved
Per 4h26'17" 33d12'59"
John "the Don" Simmons
Leo 11h44'17" 22d40'2"
"john the welder"
Per 3h29'42" 36d40'14"
John Theodore Schmick
Crb 15h54'12" 39d25'14"
John Theotokas
Uma 11h14'31" 39d19'46"
John Thomas Boy
Cnc 9h21'21" 27d6'33"
John Thomas Bramer
Psc 1h13'43" 17d49'54"
John Thomas Canada
Ori 6h13'45" 9d12'56"
John Thomas Danco
Her 18h56'18" 23d40'6"
John Thomas Gerold, Jr.
Tau 5h46'33" 22d40'32"
John Thomas Iodice
Uma 9h25'52" 61d34'30"
John Thomas Jones
Per 2h44'11" 53d59'21"
John Thomas Joyce
Ari 1h56'40" 18d47'45"
John Thomas Manning
Tau 5h27'28" 24d20'37"
John Thomas Matlock
Umi 16h8'55" 81d35'47"
John Thomas Oakley
Ori 6h4'33" 9d49'42"
John Thomas Stubler
Vir 13h15'24" -15d31'37"
John Thomas Thomas
Aql 19h36'30" -1d55'7"
John Thomas Thomas
Aql 19h30'18" -1d42'0"
John Thomas Ungefug
Uma 12h1'37" 44d46'57"
John Thomas Vinson
Cap 20h17'54" -23d43'35"
John Thomas Ware
Per 4h49'25" 49d59'22"
John Thomas Willis
Ori 5h27'9" 0d45'22"
John Thornton
Uma 13h24'53" 54d36'40"
John Tiffin Patterson
Umi 15h14'3" 71d21'1"
John Timothy Alexander
Houlihan
Uma 11h24'25" 51d14'6"
John & Tina Berte 50th
Anniversary
Uma 11h7'40" 41d20'50"
John & Tina Douma
Uma 11h29'33" 60d21'1"
John & Tish Burns
Cyg 20h7'3" 55d56'33"
JOHN TODESCHINI
Lac 22h40'43" 53d11'1"
John Toomey's Star
Umi 16h19'17" 81d27'47"
John Torpey
Tau 4h47'18" 21d52'22"
John Torres - Little Lover
Man
Her 16h44'53" 24d38'4"
John Travis Nuckolls
Sco 17h4'20" -44d5'12"
John Tremblay
Ori 5h39'33" -3d31'30"
John Trevor Aurther
Hiscock
Cas 23h40'8" 54d19'19"
John Trevor Field
Cep 0h18'28" 67d35'11"
John & Tricia Morrin Lloyd
Vir 14h30'50" -2d53'55"
John Troy Hubbard
Her 17h20'57" 38d0'36"
John Turman Bosstick
Per 3h45'1" 45d2'54"
John Tyler
Lyr 18h58'57" 46d59'43"
John Tyler Wanner
Ori 6h0'42" -0d1'33"
John Udle
Sgr 19h36'19" -23d2'7"
John Urso
Leo 11h41'8" 25d50'53"
John V. Boeker
Her 17h26'17" 39d7'43"
John V. Jr. & Regina F.
Carrigan
Cyg 20h10'40" 49d6'19"
John V. Lombardo, Jr.
Umi 14h20'42" 75d34'16"
John V Petrovic
Ari 3h22'33" 28d16'26"
John V. Stewart
Umi 17h12'16" 85d36'43"
John V. Walsh, Jr.
Boo 14h24'1" 34d27'4"
John Vahid AmirAbbassi
Boo 14h37'26" 37d48'4"
John Valente
Per 3h17'17" 53d17'1"
John Valentine "Johnny V."
Carroll
Cnc 9h8'33" 21d52'34"
John Van Kempen
Cnc 8h24'29" 10d28'54"
John Vandenberg
Lyr 18h25'40" 37d55'32"
John Vandermale &
Jocelyn Malette
Cyg 21h12'30" 51d44'32"
John Velasco
Leo 9h39'52" 11d53'12"
John Victor Beyer
Psc 1h49'27" 8d19'46"
John Victor Shea, Jr.
Ori 6h2'6" 17d34'41"
John Victor Thoner Rush
Her 17h40'0" 20d9'2"
John Villesvik
Per 4h39'40" 40d45'37"
John Vincent Catalano
Uma 11h33'24" 37d2'17"
John Vincent Compono
Her 16h11'16" 41d58'23"
John Vincent Diorio
Ari 3h22'7" 29d48'22"
John Vincent Krunis
Uma 8h42'44" 58d23'27"
John Vincent White
Her 18h10'10" 22d47'57"
John Vipulis
Crb 15h37'42" 29d12'24"
John W. Allan
Sgr 19h25'52" -15d57'6"
John W. Armstrong Jr.
Lyn 8h0'39" 40d21'24"
John W. Bolish
Uma 11h34'30" 54d48'29"
John W. & Diane
VanDervoort
Cyg 21h35'15" 33d34'21"
John W. Dockstader
Her 17h24'39" 27d36'26"
John W Harthan
Psc 1h24'15" 18d19'55"
John W. Hopkins
Leo 10h13'43" 18d48'20"
John W. Horton, Jr.
Her 18h15'51" 15d49'59"
John W Hubbard
Cyg 21h14'0" 44d56'25"
John W. Jernstedt
Sgr 17h53'24" -29d5'2"
John W. Keith
Cep 22h53'48" 71d47'37"
John W. Lefebvre
Per 4h4'29" 36d18'38"
John W. Leggett III
Gem 7h33'11" 30d23'9"
John W. Lyle
Per 2h28'5" 55d23'30"
John W. Murphy, Jr.
Cep 2h32'38" 81d24'5"
John W. Potter
Her 17h45'57" 23d3'16"
John W. Roland, Jr.
Her 18h1'38" 27d11'36"
John W. Rowley
Ari 2h51'7" 13d8'32"
John W. Waller, III
Vir 13h18'21" 6d10'42"
John W. Zerolis - The
Diplomat
Leo 10h10'13" 20d1'47"
John Wade Shiver
Per 4h33'54" 32d43'2"
John Wallace Dadswell
Uma 9h5'1" 62d37'41"
John Walter
Per 4h43'16" 40d0'30"
John Walter Keith Thorne
Cep 4h35'31" 83d13'53"
John Walter Silva
Tau 4h28'12" 1d34'4"
John Wanto
Her 17h11'48" 32d3'51"
John Warling
Aur 5h16'1" 36d5'59"
John Warren Acampa,
M.D.
Uma 11h37'46" 36d37'16"
John Warren Downer;
Fearless Hunter
Ori 4h50'11" -0d41'49"
John Warren Malanik
Cnc 8h21'34" 10d31'26"
John Wasilewski
Aur 5h54'23" 53d40'29"
John Wayne Andrew
Schaller 08051932
Lyn 8h12'4" 56d39'28"
John Wayne "Jack"
Pangborn
Psc 23h44'11" 7d21'26"
John Wayne Spivey
Sco 17h40'9" -32d12'14"
John Weaver
Ori 5h20'11" -5d27'0"
John Weeks
Aur 5h26'44" 33d49'18"
John Weidemuller, Juan in
a Million
Sco 16h51'16" -42d44'28"
John Wells
Sgr 18h5'24" -22d35'40"
John Welz
Uma 11h35'19" 52d20'38"
John Wemyss - The Star of
Weemo
Uma 10h16'29" 55d8'53"
John Wendell Rutledge
Uma 9h38'1" 67d40'55"
John Werner Kluge
Oph 16h43'48" -0d51'59"
John Wesley Kendall I II III
IV V
Cep 0h23'13" 72d12'17"
John Wesley McKinney
Cnc 8h24'27" 10d7'58"
John Wesley Morgan
Faulkner
Gem 6h22'47" 21d5'44"
John Wesley Worley
Boo 13h50'52" 10d0'57"
John Weston Loving
Husband/Father
Ori 5h40'6" 1d18'19"
John Whatmough
Per 3h25'8" 36d7'55"
John Whited
Hya 8h39'54" -12d53'59"
John Wight
Cas 22h57'41" 54d35'35"
John Wilcox
Uma 11h15'30" 55d57'51"
John Wiles
Cep 22h49'36" 66d48'40"
John Wilford
Uma 8h15'11" 70d22'57"
John William Aiken Gallo
Psc 0h17'6" 20d38'16"
John William Alderton
Per 4h41'15" 47d11'37"
John William Alexander
Matthews
Psa 21h51'27" -34d23'48"
John William Banser,Jr.
Uma 8h56'27" 57d13'26"
John William Barrington
Gem 7h23'37" 25d41'23"
John William Bell
Per 3h43'11" 44d51'45"
John William Bonnett
Per 2h57'47" 40d31'39"
John William Bredbenner
Cep 22h42'36" 60d41'5"
John William Byrnes
Ara 17h43'25" -48d39'13"
John William Chappelow
Uma 8h53'32" 48d44'51"
John William Daniels
Her 17h38'26" 35d28'19"
John William DeCoudres
Ori 5h29'47" 10d50'22"
John William Eckman III
Aqr 22h32'15" -23d49'21"
John William Edward
Gardiner
Cap 20h24'50" -27d16'35"
John William Enzman, Jr.
Her 16h24'37" 25d19'15"
John William Eppich 11-15-
1916
Ori 5h41'53" 2d8'12"
John William Franklin Jr.
Psc 1h1'15" 14d56'59"
John William Hendrix II
Aqr 22h25'32" 2d28'9"
John William Hendrix Sr.
Cyg 20h5'0" 52d50'11"
John William Leach
Aqr 23h1'27" -7d13'18"
John William McDonald
Gem 6h51'26" 23d35'13"
John William Mulvihill
Umi 14h28'10" 75d14'14"
John William Murray
Cep 20h42'28" 64d13'26"
John William O'Day Jr.
Leo 9h44'41" 27d46'34"
John William Parkinson
Her 16h57'9" 39d27'14"
John William Powell
Cru 12h34'39" -62d45'52"
John William Richmond
Franklyn Hole
Lep 5h34'31" -14d53'4"
John William Simpson
Ara 17h10'48" -54d3'19"
John William Titmas
Cyg 20h5'38" 51d32'13"
John William Toffelmire
Cnc 9h4'47" 7d31'47"
John William Trey Wilson
III
Aql 19h51'32" -0d36'27"
John William Vaccaro
Tau 4h16'51" 24d50'56"
John William Walker
Her 17h7'56" 33d31'22"
John William Warren
Leo 11h17'34" 24d43'29"
John William Warren III
Aql 19h13'29" 6d12'57"
John William Zorbas
Her 17h17'2" 35d31'8"
John Wisecup
Cet 2h45'10" 7d18'59"
John Wiseman
Uma 10h25'22" 51d23'2"
John Wojtanowski
Psc 1h41'22" 18d24'48"
John Wright
Psc 0h39'58" 14d56'16"
John Wright
Psc 0h47'53" 7d49'28"
John Wyatt Ball
Boo 14h44'57" 14d40'27"
John Wylam
Per 3h7'2" 51d6'50"
John X. Brooksbank
Ari 2h47'12" 17d17'1"
John x Pilot
Gem 6h7'41" 23d59'29"
John Yessir Ford
Per 3h42'55" 37d30'59"
John~ You Are My Shining
Star
Her 18h31'36" 23d6'4"
John Zaniello
Cnv 12h41'39" 48d18'50"
John Zarco
Cnc 9h0'30" 19d6'17"
John Zehnder
Lyr 18h36'13" 31d42'49"
John Zohn "Indescribable"
Cyg 21h52'53" 53d44'57"
John, Becky and Caitlin
Sullivan
Ori 6h11'49" 16d44'8"
John, Betty, Trish
Weidemuller
Uma 11h32'14" 32d42'56"
John, Bonnie + Mandi
Mateka
Lyr 18h51'53" 31d56'23"
John, Clara and Joanne
Kassmier
Lyr 18h47'7" 37d12'32"
John, Dina, Hope, & JJ
Goodenbury
Mon 6h54'17" 9d13'31"
John, Frank and Paul
Figucia
Cyg 20h48'58" 37d18'4"
John, God's Gracious One
Ori 4h52'40" 3d53'48"
John, Jane, Patrick, Ellie &
Bradley Daw
Aql 20h8'29" -0d37'47"
John, Jenny and Jillian
Durand
Tri 2h8'34" 34d52'12"
John, Jr.
Per 3h42'55" 49d2'48"
John, Tara, and Deuce
Uma 9h16'35" 64d7'40"
John143Mary
Uma 10h58'41" 59d12'59"
Johna Marie Wettstein
Gem 7h16'28" 32d34'13"
Johna Pierson, #1 Love
Cap 20h9'15" -20d8'37"
JohnAKinn
Aqr 22h59'6" -9d58'59"
JohnandJaime
Sco 17h56'12" -30d22'1"
JohnAshEliz
Tri 2h22'53" 36d0'0"
Johnathan
Uma 9h30'2" 68d43'4"
Johnathan A. Hernandez
Gem 7h15'21" 16d55'59"
Johnathan and Raqual
Smith
Cyg 20h50'56" 37d31'14"
Johnathan Andrew
Barkdale
Ori 5h28'30" 0d31'13"
Johnathan Andrew Guthrie
Per 2h12'31" 56d14'6"
Johnathan Andrew Mars
Cnc 8h19'13" 20d30'37"
Johnathan Caraballo
Cep 22h28'35" 62d32'35"
Johnathan David
Cep 20h24'7" 61d19'46"
Johnathan Dhuey Shaline
Psc 1h44'34" 11d46'2"
Johnathan Edward Born
Uma 11h49'42" 56d31'29"
Johnathan & Jenelle forev-
er........
Cnc 8h53'55" 30d10'15"
Johnathan & Jennifer
Cyg 21h31'40" 53d25'6"
Johnathan & Jessica
Ori 5h56'55" 17d31'35"
Johnathan Joseph John
Uma 10h48'59" 45d22'15"
Johnathan Matthew
Wallace
Cap 20h52'40" -26d26'2"
Johnathan Patrick McHenry
Uma 11h25'13" 38d44'29"
Johnathan Paul Mucinskas
Cep 21h33'31" 64d10'42"
Johnathan R. Prado
Her 17h38'56" 36d46'5"
Johnathan Thomas
Uma 11h13'50" 46d24'51"
Johnathan Wellington
White
Per 3h35'22" 44d3'54"
Johnathan William Riutzel
Leo 9h27'11" 17d47'56"
Johnathan Wingfield Bunch
Per 2h41'4" 52d48'41"
Johnathan's Wishing Star
Lyn 7h40'44" 37d43'40"
Johnathon Barber
Aqr 23h54'13" -4d17'44"
Johnathon Cale Lewis
Shines Forever
Per 2h20'35" 51d36'30"
Johnathon David Hoppe
Sgr 19h20'15" -13d45'55"
Johnathon Eric Brousseau
Her 16h44'35" 13d9'7"
Johnathon & Fiona Brock
Cyg 21h53'56" 50d46'15"
Johnathon Garret Pole
Tau 4h41'59" 18d30'20"
Johnathon Grauer
Leo 11h55'29" 19d2'23"
Johnathon Wayne Asbury
Her 16h31'18" 35d52'8"
Johnbo278
Per 4h27'16" 44d5'32"
Johnboy
Sgr 19h22'35" -12d35'17"
John-Boy
Cru 12h37'13" -60d21'23"
JohnBRoman
Cen 13h45'52" -42d20'28"
johncarle
Cap 21h28'38" -18d36'25"
JohnCarlo Thrush
Sco 15h56'11" -22d12'41"
JohnCath Gazing Star
Crb 16h2'32" 30d56'29"
JohnCindyGillianCaitlin
Lyr 18h27'25" 36d20'46"
JohnCollier
Ori 5h36'58" -1d55'55"
John-David and Amy Chan
Bamford
Uma 10h50'50" 67d37'2"
Johnetta L. Shablack
Lyn 6h55'58" 56d12'13"
Johnetta's Joy
Uma 9h1'54" 72d25'31"
Johnetta's Joy
Uma 11h51'27" 42d52'43"
Johnette Mannato
Cnc 9h13'56" 32d1'58"
JohnFrances
Uma 9h38'18" 54d34'24"
JohnHughes53
Cep 3h45'24" 81d12'32"
Johni Cerae Boyd Bellamy
Lyn 6h38'6" 54d3'48"
Johnifer
Lmi 10h25'49" 32d35'43"
Johnita
Mon 6h36'40" 3d54'19"
Johnithan Alexander
Per 4h42'42" 49d18'51"
JohnJCoxSr
Leo 11h33'34" 13d46'4"
JohnJosephNeedham
8161994 10:24 AM
Ori 5h14'54" 9d6'48"
John-Marc Knight
Per 3h7'14" 52d7'24"
John-Marie
Uma 10h55'6" 69d2'13"
JohnMarleau
Umi 14h35'38" 72d44'19"
Johnna
Uma 11h40'29" 45d43'10"
Johnna 92785
Dra 20h15'52" 62d3'8"
Johnna and Brian Armijo
Ori 5h22'39" -5d35'47"
Johnna B Jones
Cap 21h0'35" -17d11'23"
Johnna "Beeface" Foley
Leo 10h56'56" 12d24'29"
Johnna Grady Rosenthal-
Vincenti
Aqr 21h20'6" 1d59'16"
Johnna Jean Pick
Cnc 8h53'15" 28d17'2"
Johnna Lynne
Sgr 18h31'47" -34d31'59"
Johnna Marie
Uma 10h8'50" 49d10'51"
Johnna Noel
Cyg 20h40'12" 45d19'24"
Johnnie
Dra 12h43'44" 67d9'58"
Johnnie (Angel Mom)
Lyons
Ori 5h35'47" -2d32'30"
Johnnie Blair
Uma 8h58'31" 61d34'32"
Johnnie Bray
Uma 13h42'40" 58d34'57"
Johnnie Cunningham
Dra 16h27'0" 58d30'28"
Johnnie & Deni Woleuwich
Cyg 21h36'55" 49d35'18"
Johnnie Douglas Collins
Leo 10h10'30" 16d33'16"
Johnnie Duren
Uma 9h7'16" 64d41'50"
Johnnie Faye Kimberling
Ari 3h9'4" 19d8'6"
Johnnie H. Pope, Jr.
Ori 5h21'25" 2d14'2"
Johnnie & Jackie
Polechronis
Leo 11h19'46" -3d53'9"
Johnnie Jones
Leo 9h28'40" 12d21'52"
Johnnie Kay
Uma 11h42'10" 32d52'44"
Johnnie Lee Mixon
Sco 16h40'10" -34d13'42"
Johnnie Lynn McKown
Lib 15h27'51" -15d2'14"
Johnnie Newsome
Cap 20h37'0" -19d37'26"
Johnnie Riley Cannon
Per 3h22'14" 32d5'6"
Johnnie Sue
And 0h28'25" 32d44'58"
JohnnieMae Shipman
Cas 1h24'33" 62d6'33"
Johnno
Uma 10h56'44" 34d1'7"
Johnny
Her 17h8'0" 30d37'3"
Johnny
Aur 5h41'55" 48d48'0"
Johnny
Her 17h15'27" 27d29'22"
Johnny
Cnc 8h11'32" 20d17'28"
Johnny
Psc 0h54'50" 7d28'19"
Johnny 3
Lib 15h4'35" -15d49'59"
Johnny 5
Uma 10h18'33" 55d56'32"
Johnny 5
Cnc 8h3'29" 22d28'3"
Johnny A. Jennings
Aur 6h34'10" 36d50'34"
Johnny Alberts
Aql 19h45'46" 16d15'40"
Johnny Alex Yono
Boo 14h32'1" 27d16'36"
Johnny and Ann Hart
Wedding Star
Cyg 20h19'48" 46d30'38"
Johnny and Deirdre
McLaughlin
Cyg 19h49'16" 33d28'19"
Johnny and Diane's Star
Psc 0h7'19" 10d0'15"
Johnny and Karen
Shalhoub
Ori 5h23'33" 3d11'36"
Johnny and Lindsey 2006
Uma 9h13'21" 55d35'2"
Johnny and Shirley Boyd
Lyr 18h58'36" 27d5'45"
Johnny and Tess
Tri 2h9'10" 34d24'6"
Johnny Angel
Cyg 20h59'53" 33d39'45"
Johnny "ANGEL" 71785
Aql 19h48'16" 2d10'48"
Johnny Aura
Ori 5h32'8" -0d8'4"
Johnny B
Cap 21h35'27" -13d47'2"
Johnny B Goode
Vir 13h17'56" 12d50'49"
Johnny "Badd" Campoli
Ari 2h48'41" 27d19'37"
Johnny Barna
Her 16h41'28" 36d50'54"
Johnny Beaulieu
Uma 10h26'5" 65d59'48"
Johnny Blue Eyes
Per 4h35'34" 40d33'3"
Johnny "Boo" Bocchino
Her 17h47'56" 45d11'35"
Johnny Boy Anderson
Cyg 20h13'37" 56d18'7"
Johnny Boy's Star
Eri 2h40'35" -41d49'29"
Johnny Brackett
Ari 2h37'14" 26d8'39"
Johnny Bum
Her 18h34'56" 17d31'6"
Johnny Burleson Parks
Sgr 20h8'53" -39d38'19"
Johnny C Hood
Lmi 10h12'45" 28d41'12"
Johnny Carlos
Sco 17h54'43" -36d26'36"
Johnny Champion
Cep 22h10'44" 53d56'24"
Johnny Chan
Her 18h35'18" 15d10'37"
Johnny Clarence Williams
Jr
Aql 19h39'17" 4d3'45"
Johnny Cochran McBee,
Jr.
Lmi 10h38'53" 32d2'35"
Johnny Cooney
Aur 5h58'1" 42d51'35"
Johnny D.
Her 17h34'57" 18d25'58"
Johnny Davis
Cyg 19h40'55" 38d47'47"
Johnny & Dean Barnes
One Year
Psc 1h55'3" 8d42'9"
Johnny Dewayne Lewis
Ari 2h18'45" 21d17'41"
Johnny Dixon
Tau 4h10'10" 13d46'14"
Johnny Drew Pilla
Psc 1h33'33" 27d39'57"
Johnny Ed
Leo 10h11'58" 25d48'50"
Johnny Edward Mancu
Her 16h21'57" 43d38'28"
Johnny Edward Smith
Cyg 19h57'39" 46d48'13"
Johnny Farr
Ori 5h18'26" 15d49'13"
Johnny Fazz
Cep 22h30'46" 61d58'2"
Johnny Flamingo
Dra 17h3'39" 53d11'57"
Johnny Football Head
Uma 11h40'58" 63d36'51"
Johnny - Forever Our Star
Cru 12h35'3" -58d46'41"
Johnny G Jr.
Her 18h45'44" 20d33'24"
Johnny G Our All Star
Uma 11h7'34" 44d56'4"
Johnny Gaines Bennett
Uma 11h28'18" 38d25'36"
Johnny Gerard Odum
Tau 3h26'41" -0d9'33"
Johnny Glowatch
Cep 21h16'11" 61d19'39"
Johnny Hollywood & Rogue
Psylocke
Ari 2h47'51" 27d48'28"
Johnny J
Lib 14h51'13" -8d22'34"

Johnny Jew
Cap 21h40'39" -17d0'23"

Johnny (JT) 180
Umi 14h49'41" 80d43'18"

Johnny Kliewer
Her 17h9'26" 35d45'59"

Johnny La-La
Her 16h19'40" 48d44'42"

Johnny Lee Bailey
Sgr 19h52'25" -28d22'24"

Johnny Lee McKissick III
Cap 21h49'45" -13d28'55"

Johnny Lee Welch, Jr.
Sgr 17h54'22" -29d55'12"

Johnny Lynn
Mon 7h21'32" -0d24'46"

Johnny M. Phan
Cnc 8h50'29" 20d28'23"

Johnny Mack Lockard Jr.
Her 17h37'5" 18d25'30"

Johnny Martha
Tau 5h29'18" 20d19'21"

Johnny Martindale - Shining Star
Her 17h2'22" 16d37'48"

Johnny Mika!
Leo 11h16'9" -2d12'16"

Johnny Mike 49
And 0h43'24" 39d27'46"

Johnny Monk
Her 18h56'28" 24d14'8"

Johnny - Mr. E
Aur 5h46'26" 33d34'38"

Johnny Peluso
Her 16h48'38" 39d38'45"

Johnny Purga
Umi 14h16'39" 66d25'28"

Johnny Ray
Leo 10h30'48" 8d37'56"

Johnny Red
Ori 6h11'2" 12d40'31"

Johnny Robert Harris
Uma 9h15'53" 56d35'45"

Johnny Rocco Moeschl
Uma 10h47'22" 40d40'10"

Johnny Rock Star
Lib 15h25'44" -22d41'38"

Johnny Russo
Aqr 20h42'26" -7d45'4"

Johnny Scops
Sco 16h46'29" -32d52'44"

Johnny & Sibere
Cru 12h45'57" -56d33'53"

Johnny Steele 2/16/59
Umi 14h15'26" 71d33'4"

Johnny Taylor
Gem 6h44'42" 16d3'29"

Johnny Thunder
Vir 14h42'36" -2d17'30"

Johnny U Sears aka Bam Bam
Sgr 18h32'6" -16d3'9"

Johnny Upchurch
Her 16h24'19" 19d35'25"

Johnny Vegas
Pho 1h38'49" -47d40'43"

Johnny W. Hall IV
Lib 14h49'0" -1d59'45"

Johnny Walsh
Uma 9h32'39" 69d10'56"

Johnny Warren
Cru 12h56'49" -58d32'47"

Johnny Warrington
Uma 10h43'6" 69d38'17"

Johnny Whoops DeJonge
Her 16h27'47" 35d28'35"

Johnny Widdifield
Ori 5h55'30" 18d46'59"

Johnny Windsor
Psc 22h56'11" 3d18'48"

Johnny Wooldridge
Cas 1h8'47" 55d30'26"

Johnny Z
Pho 1h34'50" -50d13'18"

Johnny, Our Hero! We Love You!
Per 3h22'52" 50d16'17"

JohnnyAbaco
Tau 4h39'37" 19d57'9"

johnnyangel01
Tau 5h52'57" 15d10'30"

JohnnyBoy
Her 17h7'22" 32d40'22"

johnnycook
Uma 8h32'2" 65d6'47"

Johnnydiaz
Her 16h41'54" 15d11'19"

JohnnyPooh
Lyr 18h41'46" 34d17'21"

Johnny's Daddy
Pyx 8h48'29" -22d34'32"

Johnny's Sea of Tranquility
Lib 14h34'10" -18d13'7"

JOHNPATRICK
Gem 7h13'39" 19d39'30"

John-Paul Slovick
Umi 14h34'59" 77d40'57"

John-Paula-Kimberly
Uma 8h28'42" 63d24'34"

JohnRay
Psc 1h56'3" 6d35'58"

John-Raymond Bartolomeo
Psc 0h52'25" 15d33'15"

John-Ryder
Psc 0h52'46" 28d0'0"

Johns Baby Gibson
Umi 16h30'50" 77d43'5"

John's Black Pearl
Pyx 8h31'21" -21d55'19"

John's Bright Light
Aql 19h21'0" 6d50'28"

Johns Family Fleur de Quatre
Cra 18h0'27" -37d12'32"

John's Guiding Light
Tau 5h32'3" 20d19'33"

John's Inugami
Vir 12h48'44" 7d31'26"

John's Light Forever Burns
Her 17h25'58" 34d30'38"

John's Love for Paula
Cnc 8h17'33" 18d56'47"

John's Star
Ori 6h8'35" -0d35'56"

John's Star
Uma 10h16'49" 61d27'26"

JOHN'S STAR "DOUGLAS"
Lyn 9h12'29" 42d26'6"

John's Star of Faith, Hope and Love
Per 4h34'19" 37d55'30"

John's Thor
Cnv 13h43'5" 32d0'44"

John's Wish
Cnc 8h38'8" 8d58'51"

Johnson
Uma 8h43'22" 52d3'37"

Johnson Family
Tau 4h29'38" 23d57'12"

Johnson & Johnson 4/2/05
Cyg 21h14'2" 31d48'20"

Johnson75
Sco 16h12'5" -13d19'59"

Johnson's Star
Vir 12h46'58" 2d20'17"

Johnston Eternal Star
Lyn 6h34'18" 58d12'51"

JohnStrainLovesAshleySieving
Sco 16h49'16" -32d39'24"

johnsx
Sco 17h15'37" -38d40'49"

Johnsy Bautista-Anderson (Angel)
Gem 7h25'43" 33d7'12"

JohnThomas Few
Per 3h8'30" 51d43'53"

Johny & Barbara
Equ 21h0'49" 12d9'33"

Johny Breco
Ori 5h6'23" 4d0'14"

Joi Rae Wilkins
Leo 9h33'16" 28d22'1"

Joi Taheera-Samaiyah Dunbar
Lib 14h25'11" -17d21'5"

Joi, star in my eyes
Psc 1h1'40" 9d56'6"

Joice (Turkey) Terry
Dra 16h44'31" 69d26'27"

Joie
Cyg 19h38'57" 29d19'25"

Joie Rose DelGiorno
Umi 14h28'48" 73d58'43"

Joined in Life - Jim & Mel Whipps
Cru 12h38'10" -60d40'20"

Joji
Umi 14h27'5" 73d14'27"

JoJo
Lib 15h57'6" -9d30'56"

JoJo
Sgr 18h0'17" -28d1'44"

Jo-Jo
Tau 5h33'6" 27d59'4"

JoJo
Gem 7h33'9" 26d32'2"

JoJo
Gem 6h50'59" 20d40'8"

JoJo
And 0h49'0" 22d5'36"

Jo-Jo
Leo 9h31'48" 11d14'39"

JoJo
Lyn 7h19'1" 50d28'36"

JoJo
Uma 9h36'22" 48d28'11"

JoJo
Uma 9h50'33" 41d33'41"

JoJo Bailey-Dallas
Cas 23h53'38" 50d31'45"

Jojo Fegan
And 2h21'18" 41d9'5"

jojo houd
Cam 4h4'22" 53d7'35"

JoJo Hwang
Umi 15h34'49" 73d59'53"

Jojo love Fabio 16.11.2003
Per 4h32'55" 39d25'49"

JoJo Loves Karbomb
And 0h39'45" 31d29'56"

Jojo - Major
Uma 11h15'38" 50d45'8"

JoJo - Sugarbear - Durden
Cap 20h25'56" -13d43'0"

JoJo the Beautiful
Del 20h33'58" 12d32'59"

JOJOGAL
Sco 17h38'15" -38d53'45"

Jojo's Booger
Tau 3h47'55" 13d0'33"

Jojo's Star
Umi 16h26'18" 75d37'32"

JOJO'S STAR
Dra 15h25'29" 55d34'19"

JoJo-Sooty
Lyn 7h34'10" 56d48'49"

Joker
Ari 3h18'25" 19d18'19"

Joker
Aur 5h32'31" 43d15'58"

JoLa
Lmi 10h22'35" 36d18'37"

Jola (Mamuja)
Cnc 8h52'57" 32d14'59"

Jolaina's Celestial Presence
Leo 9h41'10" 12d39'38"

Jolan K Valo
Uma 9h31'11" 44d57'1"

Jolanda + Gerold Welte
Boo 14h38'37" 11d31'57"

Jolanda "Joli" Gerber
Her 16h44'14" 42d49'8"

Jolanda Lazecki
Umi 17h7'11" 75d15'52"

jolanda mühle
Cep 20h47'9" 80d23'33"

Jolane Morgan
Aqr 22h34'29" -16d58'49"

Jolanta
Uma 10h21'6" 45d11'23"

Jolanta (Fire of our burning Love)
Uma 11h26'11" 35d26'25"

Jolanta Kalinauskiate Little
Lmi 10h48'33" 26d5'25"

Jolanta Soltys
Vir 13h12'14" -11d31'30"

Jole
Peg 21h39'45" 20d39'55"

Jole' Amber Dominguez Midyette
Uma 13h52'53" 61d41'24"

Jolea
Lib 15h6'44" -0d57'35"

Jolean
Aqr 21h4'57" 0d57'57"

JoLee Buckner
Uma 8h34'5" 50d28'19"

Joleen
And 0h1'46" 42d52'7"

Joleen
Sco 17h31'2" -39d47'45"

Joleen Marie
Sco 16h5'48" -11d34'4"

Joleen & Steve
Tau 3h52'40" 24d42'22"

Joleen Tala Ricci
Uma 9h30'31" 61d26'28"

Joleene Kristen Uticone
Vir 14h59'34" 6d0'57"

Joleen's Twinkling Spirit
Cas 1h19'4" 65d31'44"

Jolena Rose McCarty
Cyg 20h28'3" 35d0'25"

Jolene
Cnc 8h20'24" 6d38'21"

JoLene
Gem 6h42'25" 12d32'50"

Jolene <3 Mike
Leo 11h30'8" 19d45'39"

Jolene Chalk
Cnv 13h44'59" 30d54'21"

Jolene Consuelo
And 0h19'18" 43d10'36"

Jolene Fay
Lep 5h20'11" -11d8'40"

Jolene Goucher Animal Whisperer
Cyg 20h10'30" 40d44'35"

Jolene Green
Leo 9h32'0" 14d15'56"

Jolene Hoffman
Mon 8h4'50" -0d27'44"

Jolene Hopkins
And 0h0'15" 42d20'23"

Jolene Jessica Gayle Sloan
And 0h16'4" 26d49'25"

Jolene Jordaan
Cnc 8h20'26" 9d54'44"

Jolene & Joshua
Cyg 20h20'37" 36d33'10"

Jolene Kay Hatton
Crb 16h9'26" 38d39'20"

Jolene Kostohryz
Tau 4h27'44" 18d29'46"

Jolene Lamphere
Sco 17h31'14" -44d54'16"

Jolene M Fisher
Dra 16h21'57" 53d30'18"

Jolene Mary "Jellybean"
Cas 0h7'12" 53d46'10"

Jolene Mullenix
Gem 7h32'56" 29d36'19"

Jolene Munch
Psc 1h16'41" 27d14'54"

Jolene P. Petrella (Jo's Star)
Gem 6h5'14" 22d13'45"

Jolene Westerling
Gem 6h50'59" 34d25'54"

Jolene28
Ari 2h45'53" 24d47'31"

Joletta Sberna
Cas 1h39'59" 64d54'38"

Jolhi Helena Clements
Cra 19h1'53" -39d43'36"

Joli
Oph 17h11'15" -0d6'36"

Jo-Li
Per 2h27'49" 52d5'55"

Jolia Amber Rodgers
And 23h19'38" 46d57'23"

Jolie A Hoffman
Aql 19h32'53" 14d34'4"

Jolie Alana Hiatt
And 23h18'57" 43d54'34"

Jolie Amandine Delmotte
Lib 15h28'41" -29d50'5"

Jolie Bell
Gem 6h54'57" 33d39'39"

Jolie Dava Kaner
Uma 11h45'38" 31d24'9"

Jolie Delphine
Uma 13h38'27" 53d56'45"

Jolie Dubinski
Cyg 20h55'8" 31d54'14"

Jolie June
And 0h57'17" 40d35'50"

Jolie Lauren Mary Sullivan
And 23h38'44" 38d10'30"

Jolie Lourdes Dubois
Uma 10h30'6" 62d27'30"

Jolie Lynn Johnson
Uma 13h18'18" 60d20'48"

Jolie Maeva
Umi 12h26'23" 89d9'48"

Jolie Valtierra
Psc 0h44'34" 19d19'48"

Jolie Vaughn
Lib 15h8'21" -10d43'50"

Jolika '85
Cas 0h48'16" 65d10'23"

JoLinda Ewing
Cyg 19h58'16" 39d57'24"

Joline
Ari 2h41'29" 14d32'56"

Joline Alpha of Pegasus
Peg 22h35'40" 22d29'17"

Joline LeBlanc
Cnc 8h50'58" 32d52'56"

Joline Marie
Cap 21h29'16" -17d42'56"

Jolita & Rolandas Jankauskas
Sgr 19h18'23" -19d1'9"

Joly Fanny
Cap 20h27'50" -18d27'27"

Jolyn G. Post
Aql 20h17'1" 1d12'21"

Jolyne Madhany
Uma 9h47'45" 41d54'21"

JoLynn Berg
Leo 9h41'16" 25d36'36"

Jolynn Cheston
Leo 11h24'33" 5d28'40"

Jolynn "Hummingbird"
Mon 6h35'33" 8d43'14"

Jolynn Marie Bartlett
And 1h52'27" 42d37'54"

Jolynn's Wishing Star
Lyr 18h48'12" 38d33'44"

Jolyon James Rush
Her 17h41'36" 19d10'4"

Jolyon Purton
Aqr 22h34'50" -15d33'36"

Jomaine Chan Mei See
Mon 6h39'50" -5d52'15"

Jomana
And 2h16'27" 50d56'28"

Jomar - Werzowa Wedding Star
Dra 19h42'46" 64d55'56"

JoMarie
Crb 15h46'48" 32d32'10"

Jomark NHMD
Lmi 9h42'40" 35d8'25"

Jomayra *Jomi*
Uma 10h24'47" 61d24'57"

Jombi Kid & Kati
Sco 17h23'12" -43d6'1"

JoMiAbMaSo
Sgr 18h28'0" -16d35'33"

Jomist Wedding Star
Col 5h42'18" -36d58'28"

Jomoma
Uma 14h10'19" 60d55'35"

Jon
Uma 14h1'31" 52d46'22"

Jon
Cnc 7h57'16" 10d5'52"

Jon A. Manafort
Cnc 9h20'46" 13d43'58"

Jon Alan Lamendola
Cep 0h22'24" 69d48'34"

Jon Alex Merkle
Uma 11h2'26" 38d10'33"

Jon Alexander Cooke
Lyn 6h29'18" 57d47'1"

Jon Allison
Her 17h15'39" 15d30'55"

Jon & Alyssa
Apu 15h31'21" -71d39'47"

Jon and Amanda
Her 17h0'16" 32d15'44"

Jon and Amy Forever
Cyg 20h53'21" 35d4'59"

Jon and Amy Seaton
Gem 6h46'48" 33d47'44"

Jon and Andrea
Psc 0h34'50" 8d26'40"

Jon and Bonnie's Fifteen Months
Ari 2h47'20" 25d44'38"

Jon and Brittany Forever Valentines
Vir 12h6'12" -5d22'58"

Jon and Heather
Per 3h19'34" 46d7'4"

Jon and Holly Cunningham
Leo 11h5'15" 23d30'26"

Jon and Kara Lindstrom
Cyg 21h23'21" 52d36'10"

Jon and Kristina
Crb 15h32'49" 29d49'35"

Jon and Shari Allen
Sco 16h13'13" -12d31'30"

Jon & Andrea * True Love
Lyr 18h44'41" 34d53'42"

Jon & Anna 10-3
Tau 5h21'20" 26d52'58"

Jon Bickford and Brian Baum
Aqr 22h31'38" 1d28'52"

Jon Bockwich
Leo 11h36'50" 9d10'9"

Jon Brit Mills
Uma 9h38'41" 53d50'4"

Jon Broder - Shining Star
Aql 19h40'54" -9d29'44"

Jon Brown's Great Ball Of Fire
Sgr 19h55'58" -44d12'25"

Jon Bruce Kenyon
Umi 15h28'31" 69d7'3"

Jon Brusco
Dra 18h53'2" 52d48'4"

Jon Budar
Per 4h16'50" 51d16'35"

Jon C Bengtson
Aqr 23h2'49" -10d36'43"

Jon Chan
Tau 4h38'43" 3d51'12"

Jon Christian Evensen
Aql 19h41'25" 14d56'56"

Jon Corey
Her 18h35'44" 19d28'58"

Jon Curtis Webber
Ari 2h19'24" 11d8'55"

Jon & Danielle
Cyg 20h21'47" 34d16'29"

Jon Davis Horton
Ori 5h43'38" 5d24'26"

Jon Dean
Crb 15h56'24" 36d5'56"

Jon Erik Wolfskill
Ori 5h27'50" 6d19'36"

Jon Ferber, the Brightest of All
Lyn 7h27'14" 52d56'55"

Jon Franz
Sgr 17h47'41" -16d40'53"

Jon Frederic Snow
Cnc 8h49'17" 9d24'43"

Jon Frederick Warner
Ari 2h13'25" 26d4'25"

Jon Frye
Her 17h33'46" 33d43'0"

Jon Genevieve Dean
Pho 1h38'56" -44d16'32"

Jon Giambalvo
Her 17h15'13" 26d29'34"

Jon Gudmundsson
Cap 20h29'26" -21d26'32"

Jon Gustafson
Uma 8h56'58" 52d28'4"

Jon H. Cochrane
Boo 14h26'42" 19d55'21"

Jon Hay 60
Uma 11h57'22" 52d54'4"

Jon Hebert
Vir 13h23'4" -1d23'55"

Jon & Heidi Bleich
Cyg 20h47'58" 47d4'57"

Jon Henry IV
Sco 16h51'17" -38d34'40"

Jon Howell
Lyn 7h54'38" 38d58'39"

Jon Hunsinger
Dra 20h21'37" 62d59'30"

Jon Hunter Kimberlain
Cnc 7h56'50" 16d20'15"

Jon & Jamie
Umi 14h34'5" 77d13'42"

Jon + Jenna - Forever Means Forever
Lyr 18h50'3" 36d5'48"

Jon Joel Lindskog
Her 17h35'9" 37d15'15"

Jon Jon
Lib 15h36'15" -10d43'40"

Jon Kaminski
Cap 21h40'56" -18d14'31"

Jon Keith Pruett
Her 17h29'23" 16d6'25"

Jon Kevin Abdoney
Vir 14h41'34" 3d54'7"

Jon Klicker
Per 3h24'8" 44d6'37"

Jon L. Sartor
Cnc 8h47'53" 9d13'36"

Jon LaBarron Walker II
Gem 7h3'46" 22d21'7"

Jon Ladra
Uma 11h0'13" 40d57'37"

Jon Languasco
Cnc 9h5'57" 23d20'2"

Jon Latorella's Guiding Light
Uma 10h49'16" 64d57'3"

Jon Lee Marcussen
Per 2h39'19" 56d51'57"

Jon Loves Brenda Forever
Cyg 21h31'4" 46d42'8"

Jon Lucas
Aql 19h30'6" 13d27'44"

Jon Manafort-Independance Day 1946
Dra 19h56'22" 72d45'44"

Jon Marcus Ross
Ara 17h32'24" -46d7'44"

Jon Mayer
Lyn 8h7'7" 56d52'49"

Jon & Melissa Menke
Cyg 20h12'54" 35d24'0"

Jon Michael Bartlett - "JonBoy"
Per 3h25'16" 35d37'56"

Jon Michael Schutte
Uma 12h6'46" 54d30'44"

Jon Michael West
Her 18h34'41" 22d2'52"

Jon Mikel Aguirre
Cnc 9h14'21" 11d35'2"

Jon Mosteller's 50th!
Aqr 23h12'54" -9d39'24"

Jon Muehlenkamp
Sgr 18h40'52" -26d47'22"

Jon Nabors
Sgr 19h38'11" -14d36'13"

Jon & Natalie
Cyg 20h32'21" 47d12'47"

Jon & Nicole Karch
Lyr 18h48'34" 38d46'11"

Jon Orr
Cnc 8h28'18" 13d16'13"

Jon P. Henderson
Ori 5h56'48" 6d14'12"

Jon P La Prad
Cap 21h21'12" -21d38'17"

Jon Papamihail's Star
Ari 1h48'5" 17d31'14"

Jon Paul
Ori 6h16'13" 14d57'35"

Jon Paul Mitchell
Cap 20h23'42" -14d42'22"

Jon Postiglione's Wishing Stella
Psc 0h10'56" 8d6'23"

Jon R Conway
Sgr 18h5'27" -23d26'1"

Jon R. Fink
Sgr 18h3'9" -17d17'15"

Jon & Rachel
Col 5h30'20" -30d54'23"

Jon & Rachel
Cyg 21h17'29" 42d44'0"

Jon Rampling's Mum Sue
Col 6h17'26" -41d6'33"

Jon Richard Garrison
Ori 5h8'44" 9d10'31"

Jon Robert Tonne - Phone Guy 13
Aqr 23h38'25" -9d27'59"

Jon Rocco Marchi Mauro
Boo 14h43'56" 32d9'43"

Jon Rodney Allums
Per 3h45'4" 43d52'48"

Jon S.
Her 17h19'31" 31d4'46"

Jon S.
Lmi 10h13'48" 28d41'44"

Jon & Sam
Cyg 20h16'6" 52d29'44"

Jon Sammy
Cam 7h53'59" 64d8'50"

Jon Scott
Uma 11h47'13" 43d58'46"

Jon Scott Fixler
Her 16h34'38" 37d58'40"

Jon Simon
Cen 13h26'53" -35d12'26"

Jon Speiser Loves Andrea Harding!!!
Ori 6h4'47" 15d8'33"

Jon & Stacey Forever
Cyg 20h56'2" 46d40'23"

Jon Szabo
Aql 19h46'15" -0d5'0"

Jon T. Schuetz
Cep 23h14'35" 71d58'39"

Jon Taylor Valentino
Per 3h8'32" 43d57'42"

Jon:The Blood
Gem 7h3'48" 25d18'40"

Jon Todd
Ari 3h29'0" 27d4'14"

Jon Tristan Jerman
Leo 10h5'29" 16d39'12"

Jon W.
Cyg 20h35'15" 35d54'23"

Jon West
Per 3h44'4" 43d44'52"

Jon & Yesi's Hope
Cyg 21h23'32" 47d44'41"

Jona
Cnv 13h54'30" 36d37'54"

Jonae Nicole Lloyd, A Gift from GOD
And 1h0'2" 45d7'52"

Jonah
Cyg 21h33'2" 47d37'44"

Jonah
Aqr 22h10'55" 2d16'44"

Jonah
Cru 12h38'16" -62d36'29"

Jonah Andrew Morss
Per 3h17'22" 49d49'0"

Jonah Ashston Gipson
Gem 7h42'1" 16d9'17"

Jonah Caillier
Cnc 8h30'5" 27d42'34"

Jonah Cox
Psc 0h53'33" 19d0'42"

Jonah David Bouchard
Lmi 10h25'16" 37d55'16"

Jonah Drew
Cnc 8h49'3" 14d16'29"

Jonah Duane Brey
Uma 12h26'46" 59d45'10"

Jonah Edward
Crb 15h44'7" 27d37'35"

Jonah & Emily
Ori 5h4'5" -1d7'58"

Jonah Everett
Her 17h39'55" 36d56'33"

Jonah Folbe
Cet 1h7'46" 0d26'40"

Jonah Greyson Sena
Lib 14h57'24" -21d43'54"

Jonah Jericho Abreu
Umi 17h36'45" 81d32'26"

Jonah Joseph Hughes
Ori 5h41'51" 1d25'35"

Jonah K. Ross
Aur 7h3'52" 37d42'45"

Jonah Lee Couch
Lmi 10h27'44" 37d57'19"

Jonah Lillioja
Her 16h42'34" 36d56'33"

Jonah Michael Cervenec
Ori 5h27'26" 3d4'5"

Jonah Michai Stockwell
Ori 5h35'30" -5d32'53"

Jonah Parkes Casella
Ari 2h51'58" 19d15'13"

Jonah Renee Kitchen
Aqr 22h27'12" -12d38'50"

Jonah Tracy
And 23h23'24" 47d14'19"

Jonah William Åke Murden
Her 18h45'20" 15d9'38"

Jonah William Ammons
Cas 0h31'35" 63d10'13"

Jonah Wry Pomfrey
Her 17h55'2" 23d54'23"

Jonahail
Cyg 21h44'36" 42d45'22"

Jonalee
Crb 16h14'35" 33d18'19"

Jonalisonstarward
Cyg 19h41'8" 39d2'20"

Jonana Leah
And 0h48'50" 46d30'27"

Jonas
Her 18h49'56" 14d28'4"

Jonas
Uma 13h42'39" 58d1'1"

Jonas Abbou
Aql 19h32'52" 11d37'45"

Jonas Abram Houston
Leo 11h29'39" 2d28'9"

Jonas Beymer
Cap 20h34'2" -11d5'54"

Jonas Brothers
Lyr 18h49'14" 38d52'21"

Jonas Bucher
Per 3h8'2" 51d21'49"

Jonas Holte
Her 17h40'36" 31d42'41"

Jonas Istad
Her 16h59'35" 23d38'45"

Jonas Marc
Cyg 19h35'3" 44d52'34"

Jonas Shining Star
Gem 7h39'47" 32d49'12"

Jona's Star
Peg 22h19'35" 5d56'32"

Jonas William Raiha
Vir 13h23'1" 12d59'25"

Jonatan Cerrada Moreno
Lib 14h30'44" -11d48'18"

Jonatan Romero
Ori 4h49'49" 4d52'45"

Jonathan
Ori 5h18'44" 1d30'7"

Jonathan
Aqr 22h42'4" 0d36'46"

Jonathan
Ori 4h48'41" 13d37'0"

Jonathan
Ori 6h25'8" 14d23'59"

Jonathan
Ari 2h25'57" 19d40'55"

Jonathan
Peg 22h34'40" 19d39'19"

Jonathan
Her 16h59'39" 32d45'5"

Jonathan
Uma 10h31'25" 40d31'21"

Jonathan
Aur 5h42'54" 45d58'18"
Jonathan
Her 18h6'12" 49d3'50"
Jonathan
Uma 9h32'36" 46d55'55"
Jonathan
Uma 9h57'29" 47d13'36"
Jonathan
Aqr 23h51'38" -10d19'42"
Jonathan
Sgr 17h59'58" -17d25'44"
Jonathan
Lyn 7h55'4" 57d21'46"
Jonathan 12/6
Sgr 20h25'45" -42d58'41"
JoNathan 42
Her 16h41'16" 13d43'1"
Jonathan A. Giardini Memorial Star
Her 16h28'10" 19d25'46"
Jonathan A. Zizak - JAZ
Her 17h30'41" 16d55'10"
Jonathan Aaron
Her 16h26'13" 48d57'13"
Jonathan Adam Burkhart
Uma 11h19'58" 71d24'5"
Jonathan Adams
Cap 20h21'28" -14d10'10"
Jonathan Agazaryan
Sco 16h43'37" -33d37'49"
Jonathan Alan Mullins
Lib 15h38'47" -12d10'14"
Jonathan Alexander Klunder
Cyg 20h22'35" 59d15'40"
Jonathan Alexander Wezowicz
Cep 21h25'31" 63d52'50"
Jonathan Allard Mattlage
Cap 21h28'54" -23d15'9"
Jonathan and Chase FF
Cnc 8h4'56" 25d3'21"
Jonathan and Crystal
Leo 10h17'16" 21d1'11"
Jonathan and Edwin
Leo 11h39'48" 23d31'57"
Jonathan and Krista
Vol 7h37'3" -64d50'2"
Jonathan and Laura
Uma 11h4'56" 63d33'58"
Jonathan and Sharie
Lyr 18h35'27" 26d13'45"
Jonathan Andrew Batey
Cru 12h29'31" -61d12'53"
Jonathan Andrew Chang
Vir 13h11'51" -0d54'8"
Jonathan Andrew Rodeghero
Cet 1h16'19" -0d41'46"
Jonathan Andrew Weldon
Dra 19h58'52" 71d26'12"
Jonathan Anthony Bauer
Lib 15h15'9" -12d16'32"
Jonathan Aurthor Swack
Her 16h11'4" 49d23'24"
Jonathan Ayala
Gem 7h34'25" 30d1'4"
Jonathan Ball
Peg 21h44'52" 27d57'43"
Jonathan Bennett Sims
Uma 11h45'41" 42d45'26"
Jonathan Blake Hayward Shemmell
Psc 1h19'7" 11d37'56"
Jonathan Blake Yow
Boo 14h26'35" 52d57'56"
Jonathan Boyd, Abba Extraordinaire
Gem 6h44'5" 14d49'40"
Jonathan Brandon Dominic Solis
Aql 18h50'41" -0d45'26"
Jonathan Brenn CEO
Ori 6h12'13" 2d46'24"
Jonathan Bruce Germain
Her 18h47'16" 21d8'17"
Jonathan Brunelle
Per 3h7'14" 49d2'51"
Jonathan Brzostowski
Leo 10h59'37" 5d30'53"
Jonathan Burras
Boo 14h31'17" 33d52'41"
Jonathan Byers
Cyg 21h34'35" 34d0'58"
Jonathan C. Arroyo F.
Lmi 10h44'8" 30d23'9"
Jonathan Cadena
Per 3h5'2" 47d8'7"
Jonathan Cantú
Ori 6h4'38" 10d21'27"
Jonathan Careri
Psc 23h57'9" 5d59'26"
Jonathan Carl Fineo
Her 18h54'37" 26d5'52"
Jonathan Charles Halpin
Vir 14h56'59" 5d16'49"
Jonathan Charles Riggin
Pho 0h54'9" -53d8'25"
Jonathan Chavez
Per 3h47'8" 41d34'35"
Jonathan & Christina Rhude Scott
Cyg 19h48'7" 33d25'23"
Jonathan & Christy
Cyg 19h27'59" 53d29'20"
Jonathan Clifford Brewer
Sgr 18h2'51" -28d2'22"
Jonathan Clifford Dorich
Sco 16h44'24" -33d2'5"
Jonathan Craig Kennedy
Per 3h37'46" 40d13'28"
Jonathan Currie
Umi 16h0'2" 79d1'43"
Jonathan D. Louth
Uma 8h26'55" 60d36'16"
Jonathan D. Steele
Her 17h39'5" 34d50'10"
Jonathan D. Swain
Ori 5h53'47" -0d47'47"
Jonathan Dagnicourt
Aqr 22h2'21" -14d17'27"
Jonathan Dan Gray
Boo 14h45'53" 30d26'49"
Jonathan Daniel Betts
Ari 2h16'18" 23d21'26"
Jonathan Daniel Greene
Tau 5h36'14" 25d32'35"
Jonathan Daniel Lyons
Cru 12h38'59" -58d49'8"
Jonathan Daniel Petrinitz
Vir 12h39'32" -7d21'18"
Jonathan Daniel Poirier
Aqr 20h55'24" -13d26'35"
Jonathan Daniel Willenberg
Uma 10h3'58" 53d36'1"
Jonathan David
Cas 23h43'44" 61d36'55"
Jonathan David Bate
Lmi 10h6'47" 39d37'56"
Jonathan David Bowman
Aur 5h28'29" 33d29'52"
Jonathan David Carney
Gem 6h55'48" 22d11'1"
Jonathan David Didden
Uma 11h9'25" 65d52'37"
Jonathan David Eidinger Weinbaum
Ori 6h15'11" 13d31'36"
Jonathan David Hicks
Per 3h48'41" 42d23'24"
Jonathan David Marchese
Per 2h36'34" 56d34'14"
Jonathan David Philip Steven
Crb 16h22'48" 38d57'31"
Jonathan David Pittard
Dra 16h50'13" 60d38'17"
Jonathan David Rogol
Ori 5h34'10" 2d58'20"
Jonathan David Roth
Dra 16h56'41" 74d30'51"
Jonathan David Vollmuth
Lib 15h4'10" -2d38'7"
Jonathan Dean Adams
Uma 11h2'9" 52d9'10"
Jonathan DeBoer
Uma 11h22'23" 56d37'12"
Jonathan Dickerson
Cnc 8h11'23" 9d1'52"
Jonathan 'Double J' Jarvis
Vir 13h9'10" 11d36'3"
Jonathan Douglas White
Uma 9h19'33" 42d27'16"
Jonathan Duren Pack
Uma 9h20'51" 50d47'2"
Jonathan E. Rose
Uma 12h27'12" 56d55'14"
Jonathan Earl Bodine
Aql 20h9'54" -0d36'26"
Jonathan Edward De Larme
Cap 21h58'24" -9d7'4"
Jonathan Edward Fleming
Lyn 8h30'21" 33d43'2"
Jonathan Edward Huttner
Per 2h56'18" 46d58'4"
Jonathan Edward Pelz
Uma 11h22'59" 48d20'54"
Jonathan & Erica O'Brien
Crb 16h18'28" 38d59'34"
Jonathan Francis Ring
Gem 7h15'11" 16d17'16"
Jonathan Frederick May
Aqr 22h26'27" -2d10'26"
Jonathan Frederick Williams
Uma 11h38'9" 52d1'40"
Jonathan Friedrich Behrend Eaton
Ari 2h12'59" 23d11'5"
Jonathan G. Fernandez
Vir 12h23'55" -3d2'49"
Jonathan G. Flick - My Shining Star
Del 20h35'52" 7d21'16"
Jonathan G. Wallace
Cap 20h50'59" -15d59'42"
Jonathan Gabriel Vance
Uma 9h34'18" 45d30'28"
Jonathan Geoffrey Conway
Sgr 18h55'22" -17d32'45"
Jonathan George
Vir 13h19'44" -17d11'20"
Jonathan George
Cep 22h26'39" 67d0'16"
Jonathan Gonzalez
Sgr 18h2'54" -27d32'6"
Jonathan Gosselin
Per 3h30'8" 39d34'27"
Jonathan Graham
Aql 19h16'34" -0d2'48"
Jonathan Gremos 15.05.04
Tau 5h9'56" 17d53'33"
Jonathan Guy Tunnell
Uma 13h36'55" 56d43'39"
Jonathan H. Ortiz
Her 17h39'4" 39d16'19"
Jonathan Hauger
Uma 9h54'26" 72d1'37"
Jonathan Hogsett
Per 4h22'21" 39d27'59"
Jonathan Hollis Redding
Psc 0h17'25" 7d39'9"
Jonathan Horvath
Leo 10h15'1" 24d47'55"
Jonathan Howard Hayes
Cyg 19h29'17" 54d14'56"
Jonathan Hume Crowther
Cep 21h33'0" 64d39'36"
Jonathan Hunter Campbell
Ori 5h43'50" 7d53'7"
Jonathan Irwin Martin
Aur 5h51'41" 47d13'12"
Jonathan James
Her 18h43'3" 20d20'16"
Jonathan James Ellenburg
Per 3h1'17" 42d45'53"
Jonathan James Whitlow
Ori 5h27'29" 2d49'13"
Jonathan Jason Dooley
And 1h54'59" 38d55'27"
Jonathan Javier Coria
Sco 16h12'7" -9d11'24"
Jonathan Jay Michalczak
Vir 13h18'22" 12d52'44"
Jonathan Jenkins
Gem 7h32'8" 17d49'36"
Jonathan Jose Thurman
Cep 21h48'49" 67d12'54"
Jonathan Kaiyan Hubbard Scoggins
Tau 4h45'3" 21d41'25"
Jonathan Kale Tabar
Vir 14h46'23" 5d19'10"
Jonathan Kardos
Aql 20h7'43" 3d57'22"
Jonathan Keith Staso Celestial Grad
Her 18h32'56" 17d23'54"
Jonathan Keith Sutherland
Psc 0h48'20" 5d44'32"
Jonathan & Kelly Anniversary Star
Sco 17h1'29" -35d53'28"
Jonathan Kingston Keen
Lib 15h36'54" -25d20'53"
Jonathan Kody Pietsch
Tau 4h38'49" 15d6'12"
Jonathan Kountis
Her 16h29'32" 4d40'12"
Jonathan Krasinski
Vir 14h12'30" -8d54'40"
Jonathan Kyle Cureton
Cnc 8h18'14" 9d23'43"
Jonathan Kyle Ramirez
Her 16h34'7" 28d16'8"
Jonathan L. Pool
Per 3h47'37" 51d22'14"
Jonathan LaPointe
Umi 13h22'14" 74d34'0"
Jonathan Lee Frommelt
Her 17h31'48" 36d27'56"
Jonathan Lee's Star
Cyg 20h36'34" 34d57'31"
Jonathan Len Vowell 21
And 2h33'47" 50d30'15"
Jonathan Leslie Reeves
Uma 9h11'23" 56d48'38"
Jonathan Levi Gowern Martin
Uma 9h27'7" 69d55'53"
Jonathan Levine
Lac 22h53'58" 51d41'37"
Jonathan Lewis
Her 17h48'59" 42d6'50"
Jonathan Logan
Boo 14h27'22" 18d21'55"
Jonathan Lynch
Uma 11h50'43" 32d4'1"
Jonathan M. Hauser
Aur 5h7'55" 34d23'30"
Jonathan M. Howard
Aur 6h28'8" 48d35'42"
Jonathan M. Maddox
Lyn 8h2'48" 54d15'40"
Jonathan M. McLaren JR.
Sgr 18h7'53" -26d32'8"
Jonathan M. Sneedse
Ori 5h45'41" 0d43'58"
Jonathan Mahal
Cnc 9h14'30" 24d1'3"
Jonathan Malara's Super Smiley Star
Her 18h5'29" 22d40'29"
Jonathan Marc Dragul
Cep 22h15'53" 61d29'7"
Jonathan Martin Hooper
Uma 9h0'36" 47d32'41"
Jonathan Matthew Long
Cyg 20h2'54" 33d49'51"
Jonathan Matthew Wiacek
Boo 13h48'6" 10d2'50"
Jonathan Max Imsland
Lib 14h48'23" -3d20'52"
Jonathan Meyers
And 0h32'41" 43d1'15"
Jonathan Michael Allen
Uma 9h46'9" 49d25'54"
Jonathan Michael Carpenter
Vir 14h41'22" 2d33'7"
Jonathan Michael Davis
Vir 13h46'26" 2d30'41"
Jonathan Michael Engelhardt
Uma 9h55'52" 66d31'33"
Jonathan Michael Groboski
Uma 12h37'43" 56d48'49"
Jonathan Michael Resch II
Cap 21h43'44" -10d43'58"
Jonathan Michael Stepp
Cyg 21h50'41" 36d36'41"
Jonathan Mickiewicz
Lmi 10h19'55" 39d10'58"
Jonathan Milton
Lyr 19h10'5" 26d23'38"
Jonathan Minogue
Sgr 18h7'9" -31d59'0"
Jonathan Montgomery Wisler
Dra 17h35'13" 65d32'15"
Jonathan Morales
Her 17h27'27" 46d34'26"
Jonathan Murcia
Cep 22h17'11" 74d10'17"
Jonathan Murphy
Ori 6h24'30" 11d4'47"
Jonathan - My Shining Star
Ori 5h36'27" -4d12'13"
Jonathan Myers Gregory
Her 17h32'38" 46d11'31"
Jonathan Napert
Her 17h58'7" 29d16'34"
Jonathan & Nicole
Gem 7h40'47" 19d9'58"
Jonathan Nigel Carter
Uma 10h48'23" 47d0'38"
Jonathan Norman
Uma 11h15'6" 72d1'7"
Jonathan O. Teeter Sr.
Leo 10h10'38" 18d5'15"
Jonathan Ortiz & Sayra Sandoval
And 0h32'1" 38d10'39"
Jonathan P. Wilkinson
Her 16h17'22" 45d30'54"
Jonathan Patrick Assur
Per 2h56'23" 41d39'35"
Jonathan Patrick Phillips
Aql 19h43'17" 2d36'1"
Jonathan Patrick Reed Grabow
Cep 23h4'7" 78d3'12"
Jonathan Paul Brou
Tau 5h5'10" 23d25'44"
Jonathan Paul Cooper
Per 2h53'5" 46d13'28"
Jonathan Paul Dean Allen
Sco 16h38'11" -41d19'36"
Jonathan Paul Galiano
Leo 11h32'48" 26d10'30"
Jonathan Paul Levine
Uma 11h53'8" 39d25'33"
Jonathan Paul Lewinski
Cap 20h25'19" -13d53'36"
Jonathan Peter Carmina
Cyg 21h13'4" 43d42'59"
Jonathan Philip Smith
Cas 1h8'13" 60d10'43"
Jonathan Piercy
Ori 5h52'9" 7d7'24"
Jonathan Pierre Verchiens
Uma 11h22'31" 54d28'23"
Jonathan P-Nutty Pavlis
Sgr 18h23'45" -25d30'17"
Jonathan Preas Street
Tau 4h30'50" 21d59'49"
Jonathan Prescott Brown
Cnc 8h1'44" 11d32'24"
Jonathan Quetania's Quasar
Ari 3h4'23" 28d21'31"
Jonathan Quezada
Tau 3h46'34" 30d24'42"
Jonathan R. Biedell
Her 17h41'36" 34d3'53"
Jonathan R. Brust
Sco 17h50'2" -42d18'35"
Jonathan Resler
Her 17h18'11" 15d39'21"
Jonathan Richard Zimmerman
Her 17h19'36" 27d21'14"
Jonathan Richards
Her 17h29'55" 25d45'40"
Jonathan Ricky Hiers
Cas 23h39'53" 53d27'51"
Jonathan Rico Nazzaro
Sco 16h8'54" -10d6'53"
Jonathan Robert Gyngell
Per 2h54'27" 42d3'51"
Jonathan Robert Holland
Per 4h16'31" 52d16'13"
Jonathan Robert Noble
Psc 1h11'46" 28d48'12"
Jonathan Robert Peters
Cep 0h1'38" 72d56'12"
Jonathan Roberto Quinn
Her 17h37'26" 22d8'57"
Jonathan Rodriguez
Gem 6h13'42" 23d4'26"
Jonathan Rohman
Gem 6h48'37" 34d18'49"
Jonathan Ross Ingalls
Uma 13h13'8" 53d50'54"
Jonathan Ruiz
Ari 2h9'51" 16d11'2"
Jonathan S.
Tri 2h24'51" 30d38'31"
Jonathan Samuel
Cyg 20h15'7" 40d9'3"
Jonathan Samuel Bemis
Vir 14h50'24" 5d29'24"
Jonathan Scott Teraoka
Tau 5h35'9" 22d5'40"
Jonathan (Snuffy) Taurins
Leo 9h44'48" 31d41'0"
Jonathan Sonnet - Wilson Boschiero
Eri 4h25'55" -0d26'38"
Jonathan Sorenson
Per 3h9'2" 43d11'28"
Jonathan Spencer Burton
Sgr 19h57'40" -30d13'24"
Jonathan Starr
Psc 1h13'33" 22d40'59"
Jonathan Stewart Wilkins
Her 18h22'40" 29d19'41"
Jonathan Stringer
Ori 5h29'21" 2d25'56"
Jonathan "Sweet Ace" Armstrong
Uma 11h54'31" 59d9'50"
Jonathan T Kuhn
Del 20h48'43" 17d37'46"
Jonathan T. Work
Her 18h18'40" 18d8'18"
Jonathan Thomas Donaghey
Per 3h6'44" 52d5'19"
Jonathan Thomas Rozek
Psc 1h18'0" 16d8'29"
Jonathan Thomas Shannon
Leo 10h15'2" 24d18'51"
Jonathan Todd Melloy
Tau 5h33'59" 24d36'18"
Jonathan Tomlinson
Cyg 21h30'26" 42d41'47"
Jonathan Trant
Dra 18h51'4" 61d54'58"
Jonathan Tucker
Cas 1h9'27" 52d40'48"
Jonathan Unseld
Uma 11h55'1" 48d53'38"
Jonathan Vick
Her 17h38'20" 36d17'3"
Jonathan Vincent Allen
Cnc 8h42'37" 31d40'18"
Jonathan Vincent Weed
Aql 19h40'22" 12d13'50"
Jonathan Vordermark
Boo 13h46'39" 24d58'47"
Jonathan Watson
Her 16h31'51" 31d48'1"
Jonathan William Dekkers
And 0h58'58" 36d55'55"
Jonathan Williams McCrane
Her 17h23'5" 37d44'6"
Jonathan Wood
Leo 11h9'48" 17d3'44"
Jonathan Yarborough
Ori 5h36'55" 0d3'30"
Jonathan Yenzer
Per 3h25'57" 35d42'50"
Jonathan Zappola
Cap 20h11'31" -16d47'23"
Jonathan's Bahamut
Vir 11h54'52" -4d46'10"
"Jonathan's Eternal Nightlight"
Tau 4h37'6" 27d50'45"
Jonathan's First Christmas
Uma 11h26'14" 60d44'34"
Jonathans Glow
Lyr 18h50'59" 44d51'49"
Jonathan's Star
Ori 6h7'47" 17d21'48"
Jonathon
Ori 5h48'31" 7d29'52"
Jonathon
Dra 16h7'0" 51d38'54"
Jonathon
Lib 15h15'40" -10d17'39"
Jonathon Austin Hubbard
Dra 16h39'18" 63d41'21"
Jonathon Brockman the Penguin Boy
Boo 14h32'40" 21d25'11"
Jonathon Bryson Ynwa
Tau 5h11'35" 17d54'32"
Jonathon Chase Hall
Vir 12h11'15" -4d54'0"
Jonathon Chason
Ori 6h20'53" 14d21'27"
Jonathon E. Yule
Cep 22h22'19" 62d40'16"
Jonathon Edwin
Per 1h45'56" 54d25'18"
Jonathon Eugene Treasure
Lyn 8h2'46" 42d28'30"
Jonathon Gordon Phillips
Tau 4h20'27" 12d42'49"
Jonathon (Jack) Perone
Per 3h46'44" 43d51'22"
Jonathon James Burks
Tau 3h31'50" 24d45'59"
Jonathon Kevin Aupperle Jr.
Per 2h29'57" 56d36'49"
Jonathon Ly
Ari 2h1'3" 14d10'39"
Jonathon Michael Keys
Uma 8h26'4" 61d29'39"
Jonathon Moffitt
Crb 15h24'19" 25d46'37"
Jonathon Neumann
Gem 6h37'14" 16d49'58"
Jonathon Patrick Ransom
Ori 5h28'58" 1d33'24"
Jonathon Philip Wright
Uma 12h23'20" 57d42'6"
Jonathon Poli
Leo 9h57'32" 16d20'35"
Jonathon Robert Alfiero
Lyn 8h25'36" 55d12'48"
Jonathon Ryan Brown
Sgr 18h58'27" -28d52'55"
Jonathon Spinks
Uma 10h28'38" 50d20'53"
Jonathon Thomas Wallace
Aur 5h39'25" 48d55'20"
Jonathon Weedman
Vir 14h5'12" -11d51'57"
Jonathon William Rex Dyer
Ari 3h12'56" 18d57'6"
Jonathon Yocum
Per 2h53'6" 45d10'42"
Jonathon's Light
Sco 17h11'28" -33d23'29"
Jonathons Racer (Jonathon R. Yates)
Crb 15h45'51" 32d50'51"
Jon-Charles Emory Teel
And 1h33'27" 43d6'54"
Jonda Lynn McGill Engle
And 0h36'12" 41d44'15"
JonDawn Feldman
Aql 19h2'55" -0d47'21"
Jone Gamble's Shining Star
Cam 4h31'26" 64d46'44"
Joneen Marie Cook
Sgr 18h12'48" -34d20'7"
Jonelee
Umi 14h17'45" 66d2'40"
Jonelica
Cam 3h41'36" 58d34'17"
Jonell D. Humphrey
Uma 8h11'38" 68d18'1"
JoNell R. Westhues
Per 3h45'12" 49d3'4"
Jo'nelle
Uma 9h27'13" 67d40'0"
Jonelle Lynn Stever
Her 17h9'27" 30d44'42"
JoNell's Desire
Peg 21h40'27" 7d16'11"
Jones Arthur Dickerson
Uma 9h47'5" 62d43'22"
Jones Girl
Uma 11h27'0" 59d55'10"
Jones' Kazoo Beacon
Umi 15h36'45" 77d11'6"
Jones R. Weitzman
Leo 11h13'0" 8d3'48"
Jones Wedding Star
Cyg 20h53'16" 48d38'12"
Jonessa
Lyn 6h51'48" 61d12'39"
Jonesy
Uma 11h43'54" 45d32'17"
Jonette Noel Pettyjohn
Sgr 18h24'49" -35d47'12"
Jong Il Park
Mon 6h52'18" -0d53'11"
Joni
Leo 11h18'31" -0d41'54"
Joni
Sco 16h6'5" -17d46'32"
JONI
Sgr 18h57'1" -31d14'20"
Joni
Cnc 8h36'11" 8d4'45"
Joni and Ricky Gardner
Uma 9h16'56" 66d44'43"
Joni B.
Vir 13h23'34" -17d38'22"
Joni B.
Ori 5h36'0" 13d38'52"
Joni Bartlett
Uma 10h47'9" 64d4'11"
Joni Caggiano
Cap 20h22'13" -9d37'3"
Joni Cailyn
Aqr 21h15'5" -11d56'0"
Joni Chase
Aqr 22h9'39" -3d20'24"
Joni Gage
And 1h45'43" 50d24'46"
Joni Henderson
Del 20h40'34" 14d33'33"
Joni Jones
Crb 15h35'15" 31d30'36"
Joni Kay
And 23h19'13" 44d31'1"
Joni Kerley
Leo 11h4'49" 5d16'3"
Joni Kristen Howard
Her 18h39'13" 19d10'16"
Joni LaMont
Ari 2h22'16" 18d8'46"
Joni Lampl
Cam 3h35'1" 63d10'44"
Joni Layden Waldron
Leo 9h34'52" 28d59'25"
Joni Lucas
Uma 11h32'33" 54d44'29"
Joni Lynn
Lyn 7h54'12" 56d3'51"
Joni Lynn McCalman Stephens
Cnc 8h46'7" 27d14'34"
Joni Mae Staple
Ari 2h39'40" 21d54'58"
Joni Mahoney
Aur 5h39'24" 33d41'41"
Joni Marie Fincher
Tau 4h33'3" 28d26'44"
Joni Marie Perry
Com 12h29'0" 25d18'10"
Joni Montee
Her 18h30'30" 18d23'43"
Joni MS Francis
Ari 3h24'44" 28d31'51"
Joni Rae Katz
Apu 15h26'17" -75d28'7"
Joni Rene'
And 0h31'28" 37d48'49"
Joni Russo
Aqr 22h44'51" -0d57'48"
Joni Sharrah
And 1h17'52" 47d30'13"
Joni Summitt
Gem 7h17'18" 22d9'9"
Joni Tindall
And 23h52'37" 39d24'25"
Joni Tracey Taylor
And 0h50'57" 37d21'26"
Joni Wilders 2005
Gem 7h23'33" 32d52'46"
Joni Yoswein
Cyg 19h55'7" 30d59'6"
Joni39
Aql 19h52'2" 1d33'30"
Jonie
And 0h20'32" 33d13'27"
Joni-roseann
Gem 7h38'22" 17d36'33"
Joni's Eternal Light
Leo 10h25'10" 18d53'6"
Joni's Utopia
Dra 18h52'27" 50d6'35"
Jon-Jon Lott's Star
Aqr 21h41'19" -3d29'32"
Jon-Kely Cassara
Uma 9h40'43" 61d51'48"
Jonlawman
Sco 17h49'58" -37d29'6"
Jon-leslie
Aqr 21h38'29" 1d35'43"
Jon-Michael
Gem 6h21'18" 18d22'7"
Jonmichael
Aqr 21h59'39" -9d52'9"
Jon-Michael C. Peterson
Sco 15h53'20" -24d43'21"
Jonmichael Mantelli
Per 2h36'56" 52d29'19"
Jon-Michael Ray Drawdy
Gem 6h44'51" 15d4'34"
JONNA
Ari 2h56'51" 26d29'44"
Jonna Akin
Vir 12h18'11" 12d23'44"
Jonna Annette Upson
Mon 6h36'5" -0d18'17"
Jonna Asbury McClain
Lmi 10h0'46" 37d8'39"
Jonna Belle
Gem 7h16'9" 22d39'11"
Jonna Goodan
Per 3h42'45" 45d25'56"
Jonna Lynn Sondall
Crb 15h43'48" 32d46'39"
Jonna Marie Drake
Cap 20h13'35" -10d36'44"
Jønna og Stinemor
Per 2h56'2" 50d44'33"
Jonna Seal-Lark
Tau 5h48'26" 25d53'50"
Jonna-Flash
Gem 7h25'36" 32d6'49"
Jonnamy
Dra 19h0'44" 61d26'12"
Jonnell Michael
Tri 2h21'4" 33d10'13"
Jonni
Leo 11h43'16" 20d56'17"
Jonnie Lol
Cas 1h26'27" 55d39'59"
Jonnie Wilson Tancock
Umi 13h44'19" 70d54'56"
Jonni's Light
And 0h52'41" 40d50'32"
Jonny
Leo 10h7'30" 26d12'17"
Jonny
Leo 10h54'44" 12d8'11"

Jonny and Christina O'Leary
Cnc 8h52'38" 31d12'5"
Jonny Axtman and Family
Uma 10h27'28" 47d29'8"
Jonny Bier
Tau 4h15'54" 22d18'19"
Jonny & Brittney Forever Love Star
Cyg 19h47'12" 52d13'56"
Jonny Kay
Cap 21h35'16" -16d32'4"
Jonny-My-Jonny
Gem 6h57'43" 16d6'50"
Jonny's Star
Cyg 21h23'13" 50d23'46"
Jono Hardy
Dra 19h43'29" 61d42'26"
JonOncale4-11-2001
Ari 2h21'1" 22d49'55"
Jon-Paul Ian Dayton
Cep 23h19'42" 82d15'19"
Jonpaul Miller
Cap 20h33'36" -24d2'10"
Jonpaul Wesley Ursick
Her 16h41'34" 44d57'55"
Jon's Hope
Psc 0h48'10" 15d29'52"
Jon's Light
Leo 10h22'24" 6d45'24"
Jon's Love
Cnc 8h43'0" 22d10'24"
Jon's Mark
Uma 9h30'14" 65d0'9"
Jon's Star In Honor Of His Dad, Jim
Psc 23h30'20" 7d20'50"
Jonty Yamisha
Vir 15h8'21" 4d52'50"
Jonty's Star
Umi 14h4'56" 77d55'5"
Jo-nus
Per 3h47'58" 41d37'57"
Jonya Danea Cooke
Uma 10h21'21" 53d18'17"
Joo Yon Kim
Psc 1h35'59" 23d35'11"
JooHee Kang
Umi 13h19'37" 75d43'30"
Jooi Brothers Sdn Bhd
Ori 5h37'43" 2d58'37"
Jools
And 23h34'3" 42d11'11"
Joon Rose Minus
And 2h6'27" 41d43'10"
Joon Suh
Crb 16h23'51" 33d13'5"
Joona
Uma 9h32'14" 64d57'44"
Jooooo/sh and leeee/yo
Uma 9h32'51" 60d20'32"
Joop van den Ende Superstar
Umi 16h58'25" 78d0'48"
joost
Ari 3h5'27" 29d42'39"
Joost, Lars
Vir 11h49'14" 3d34'39"
Jooyeon Lim
Uma 11h43'17" 29d31'45"
Joplin Star Scarlet
And 1h35'9" 47d29'11"
Jor and Laur's Star
Col 5h31'48" -34d29'42"
JORA Star aka " You & Me" Star
Cap 21h48'45" -19d41'7"
Joram Nicholas William 7.30.00
Leo 10h28'59" 12d46'32"
Jordan
Cnc 9h10'40" 13d27'30"
Jordan
Cnc 8h20'15" 13d41'30"
Jordan
Tau 5h47'45" 21d59'29"
Jordan
Vir 12h38'52" 6d12'16"
Jordan
Ori 4h52'24" 12d52'34"
Jordan
Ori 4h48'44" 13d9'45"
Jordan
Psc 1h23'2" 26d12'11"
Jordan
Leo 11h28'28" 17d3'27"
Jordan
Cyg 19h38'12" 37d41'30"
Jordan
Dra 17h18'25" 51d5'44"
Jordan
Uma 11h41'16" 41d47'51"
Jordan
Sgr 18h29'41" -16d46'47"
Jordan
Aqr 20h39'24" -2d0'55"
Jordan
Ser 15h22'59" -0d42'40"
Jordan
Umi 13h43'18" 73d25'34"
Jordan
Dra 18h7'56" 67d32'39"
Jordan
Cep 0h0'43" 67d49'47"
Jordan
Lyn 7h23'30" 58d29'54"
Jordan A. Hyde
Lyr 18h51'22" 31d48'22"
Jordan Acosta
Uma 9h45'24" 44d26'37"
Jordan Adelynn Wells
Peg 21h30'22" 15d30'30"
Jordan Alana Cates
Ori 5h56'0" -0d41'33"
Jordan Albert Guggenheim
Cap 20h56'9" -19d44'39"
Jordan Alecia Osborn
Lac 22h45'12" 50d25'10"
Jordan Alexander Bruck
Umi 16h32'51" 88d29'51"
Jordan Alexis
And 0h17'14" 27d44'56"
Jordan Alexis
Tau 4h13'22" 18d28'4"
Jordan Alexis Kidd
Lib 15h0'21" -10d29'59"
Jordan Alice McNeal
Uma 12h9'20" 49d23'22"
Jordan and Ashton Elliott's Star
Uma 9h52'51" 68d12'35"
Jordan and Carmen
Cap 21h34'12" -15d21'25"
Jordan and Laura
Cyg 20h25'56" 58d57'0"
Jordan and Lindsey together forever
Cap 21h35'50" -8d33'55"
Jordan and Tyler McFall
Cap 20h17'46" -12d4'52"
Jordan Andrew Jackson
Psc 1h26'31" 2d50'24"
Jordan Ann Sherman
Cnc 8h5'47" 27d30'21"
Jordan & Annastasia 22 August 2004
Cyg 20h12'37" 49d6'34"
Jordan Anthony Rowe
Tau 4h23'24" 30d43'16"
Jordan Ari Kramer
Lib 14h53'24" -7d40'46"
Jordan Arthur Stephen Jacko
Per 3h14'30" 42d51'4"
Jordan Ashley Essenburg
Aql 19h59'32" -0d26'43"
Jordan Axelle Butler
Eri 4h27'1" -0d31'21"
Jordan B. Steed
Leo 10h12'46" 24d52'18"
Jordan Banks
Gem 6h27'26" 23d14'54"
Jordan Beard
Aqr 21h9'52" -13d23'43"
Jordan Bond
Per 4h7'20" 43d54'5"
Jordan Branzburg Sick
Uma 10h16'55" 42d32'9"
Jordan Breen
Lib 15h24'7" -27d38'30"
Jordan Brianne
Ari 2h59'22" 26d8'22"
Jordan Brice
Ari 2h29'14" 18d58'8"
Jordan Brice Robertson
Dra 17h44'24" 50d53'27"
Jordan Brittany Dowd
Aur 5h38'35" 38d29'28"
Jordan Brooke Hutsell
And 0h47'33" 45d12'49"
Jordan & Chase Baker
Her 18h38'4" 19d41'57"
Jordan Chiappetta
Uma 12h41'41" 60d20'26"
Jordan Christine Hern
Gem 6h37'47" 27d53'53"
Jordan Christopher Graham
Uma 11h55'4" 63d45'43"
Jordan Claire Carroll
Vir 11h59'12" -10d19'0"
Jordan Clark
Dra 15h29'25" 62d28'59"
Jordan Clark Hawley
Her 16h52'9" 13d30'31"
Jordan & Clyde
Umi 15h2'5" 74d9'40"
Jordan Copeland
Cyg 21h54'15" 54d37'41"
Jordan Daniel Teixeira
Leo 9h40'34" 30d44'35"
Jordan Dante Baum
Uma 12h5'2" 29d24'43"
Jordan David
Tau 3h48'44" 27d50'21"
Jordan David Block
Umi 15h14'25" 77d5'4"
Jordan David Boath
Cru 12h33'26" -63d13'29"
Jordan David Mandelbaum
Psc 1h33'8" 21d52'13"
Jordan Dean
Cnc 8h45'58" 14d32'30"
Jordan Dean Burden
Ari 2h36'39" 13d0'22"
Jordan Denzil Meyers - 11.09.1997
Cen 13h21'47" -57d50'40"
Jordan DiLauro
Cyg 21h28'1" 39d24'49"
Jordan Dwane
Umi 15h12'1" 71d34'53"
Jordan Elaine Barnes
Lib 15h24'16" -19d16'59"
Jordan Elaine Davidson
Uma 9h57'0" 47d43'39"
Jordan Elisa Sitea
Ori 6h1'23" 6d26'33"
Jordan Elizabeth Aiello
Uma 9h17'40" 55d13'41"
Jordan Elizabeth Conley
Cyg 20h3'13" 33d15'4"
Jordan Elizabeth Fade
Cyg 20h57'8" 33d44'45"
Jordan Elizabeth Greenwald
Lib 14h50'15" -0d33'48"
Jordan Elizabeth Martinez
Lib 14h55'0" -13d28'32"
Jordan Elizabeth Williams
Ari 3h12'51" 21d14'55"
Jordan Emelia Reeves
Mon 6h44'45" -0d21'5"
Jordan Emilea Richey
Cap 21h40'14" -12d29'13"
Jordan Emily Cox
Lyn 7h15'13" 56d57'44"
Jordan Emily Jeter
Ari 3h23'11" 18d45'38"
Jordan Emily Ryz
Uma 10h37'46" 43d1'47"
Jordan Eric Schoem
Ori 6h18'48" 9d33'27"
Jordan Eusebio Verrier
Lyn 7h17'55" 57d19'22"
Jordan Flaherty
Aur 5h40'28" 46d8'21"
Jordan Flowers
Sgr 18h15'13" -25d46'48"
Jordan - Forever in our Hearts
Cru 12h7'8" -64d36'41"
Jordan Francis Esser (5-14-04)
Tau 4h34'26" 19d45'23"
Jordan Frank
Cas 0h15'0" 57d18'48"
Jordan Garcia
Her 17h29'5" 39d38'56"
Jordan Goslee
Sgr 20h4'22" -43d51'54"
Jordan Hamilton Bullard
Lyr 18h53'12" 32d27'24"
Jordan Hartnett
Dra 16h56'34" 60d46'55"
Jordan & Holly
Ori 5h59'34" 22d28'0"
Jordan Husted
Uma 10h29'30" 44d47'23"
Jordan Isaac Nelson
Tau 3h45'12" 22d57'11"
Jordan J. Janov
Aqr 22h42'34" -21d47'23"
Jordan Jane
Cnc 8h10'15" 16d9'8"
Jordan Jane Tomczyk
Lyn 9h20'29" 40d2'43"
Jordan Jeffrey Wiltey
Tau 4h50'54" 28d53'21"
Jordan Jud
Sco 17h21'46" -42d16'3"
Jordan Kailani Aoki
Gem 6h49'26" 32d56'52"
Jordan Kaiser
Per 3h31'22" 40d12'40"
Jordan Karamujic
Uma 11h26'47" 52d39'52"
Jordan Kate 1
Uma 11h58'7" 56d24'40"
Jordan Kaufman
Uma 10h41'44" 41d20'53"
Jordan Kay
Aur 5h47'15" 40d11'17"
Jordan Kettering
Lib 15h50'47" -20d17'43"
Jordan King Worlds Best Daddy
Vir 13h22'57" 1d49'33"
Jordan Klipp
Leo 11h6'54" 10d0'9"
Jordan Kyle Begraft
Tri 1h40'36" 29d52'37"
Jordan L. Bleile
Umi 14h26'6" 75d21'1"
Jordan Lai
Ari 2h33'33" 27d52'21"
Jordan Lamm
And 0h36'37" 27d1'8"
Jordan Leah
Aqr 22h26'20" -5d12'44"
Jordan Lee
Cyg 21h44'57" 53d4'17"
Jordan Lee Boettcher
Leo 11h26'50" 21d47'28"
Jordan Lee Boettcher
Leo 11h52'40" 19d3'1"
Jordan Lee McMullen
Vir 12h27'52" -5d50'17"
Jordan Leigh
And 0h30'58" 36d17'36"
Jordan Leigh Killmon-Welch
Aql 19h21'34" -7d47'58"
Jordan Lindsay Waterhouse
Crb 15h55'31" 31d30'47"
Jordan Lindsey
Mon 6h32'49" 7d3'30"
Jordan Louise Roberson
And 0h17'32" 45d34'6"
Jordan Love
Sgr 18h44'36" -18d26'11"
Jordan M. Williams
Boo 14h45'57" 26d7'59"
Jordan Mac's shining star
Gem 7h11'47" 25d29'3"
Jordan Madison Lenahan
Cnc 9h4'11" 31d35'12"
Jordan Madison Minnick
Aqr 23h3'17" -12d58'42"
Jordan Mari
Uma 9h55'43" 44d2'26"
Jordan Marie
Peg 22h10'36" 11d52'22"
Jordan Marie
Sco 17h29'20" -45d33'23"
Jordan Marie Boyer
Cap 21h40'47" -14d51'42"
Jordan Marie Quickstad
Cap 20h37'45" -10d47'34"
Jordan McWatters
Mon 6h47'55" 7d21'46"
Jordan Michael
Lib 14h23'47" -10d59'0"
Jordan Michael
Aqr 21h47'28" -5d31'40"
Jordan Michael Church
Her 18h23'0" 19d13'15"
Jordan Michael Kolnick
Leo 11h44'28" 13d11'12"
Jordan Michael Ruttenber
Lmi 10h23'47" 38d56'44"
Jordan Michelle
And 2h14'30" 45d8'50"
Jordan Michelle
Ari 2h50'35" 22d21'47"
Jordan Michelle Nelson
Ori 6h16'31" 16d45'37"
Jordan & Miranda
Cru 11h59'11" -64d34'57"
Jordan Mucci
Uma 10h10'39" 54d2'28"
Jordan - My shining star
Cap 21h15'2" -17d3'28"
Jordan & Natasha
Cyg 21h12'47" 32d3'53"
Jordan Neely Martin
Peg 23h13'59" 25d59'37"
Jordan Newington
Tau 5h49'3" 18d59'12"
Jordan Nichole 16
Aqr 21h49'15" -7d0'24"
Jordan Nicole Cooper
Leo 10h44'43" 16d46'41"
Jordan Nicole Morelock
Sgr 17h54'0" -26d17'49"
Jordan Noelle
Cap 21h9'31" -15d54'45"
Jordan Orion Krause
Ori 5h47'54" 8d53'1"
Jordan Paige Aucompaugh
Gem 6h55'18" 15d46'37"
Jordan Paige Gardner
Ari 1h57'17" 13d55'36"
Jordan Partridge
Dra 17h2'51" 58d22'7"
Jordan Peter
Aqr 22h39'48" -8d14'49"
Jordan Peter Williams
Sco 17h0'22" -36d20'0"
Jordan Phyllis Jacqueline Chandler
Lib 14h53'25" -3d32'53"
Jordan R. Holt
Cap 21h2'24" -16d4'29"
Jordan R Magnusson
Sco 16h17'51" -12d32'43"
Jordan Rae Weaver
Crb 16h13'58" 25d47'8"
Jordan Rebecca Diaz
Cnc 8h18'45" 17d38'20"
Jordan Reese Antongiorgi My Angel
Per 3h41'59" 46d20'2"
Jordan Reichers
Ori 5h46'35" -3d21'42"
Jordan Renee' Gadoury
Psc 0h7'56" 2d43'40"
Jordan Robert
Uma 9h29'16" 56d42'54"
Jordan Robert Landry
Per 2h51'13" 52d22'48"
Jordan Robertson Rude
Tau 4h23'19" 22d54'47"
Jordan Rose Hewitt
And 2h25'46" 45d49'9"
Jordan Ryan Woodruff
Leo 10h54'45" 0d31'5"
Jordan S. Brell
Cru 12h48'20" -60d6'2"
Jordan Salyer
Aqr 20h39'7" 1d48'44"
Jordan Samantha
Leo 10h51'59" 16d59'20"
Jordan Schapiro
Per 3h33'38" 32d12'37"
Jordan Seltzer
Cyg 20h45'29" 33d40'44"
Jordan Sha Arnett
Lib 15h4'3" -6d44'16"
Jordan Shante'lle
Ori 4h53'45" 3d37'9"
Jordan Sharon Pierce
Mon 8h10'56" -0d35'28"
Jordan Shea McCafferty
Lyn 8h41'28" 33d17'3"
Jordan Sienna Ernst
Vir 14h27'27" 3d55'58"
Jordan Souder
And 0h40'47" 28d29'31"
Jordan Stoodley
Uma 10h14'6" 46d27'1"
Jordan Strother's Wish Upon a Star
Uma 11h12'22" 68d20'49"
Jordan Sund
Sgr 19h8'22" -25d12'15"
Jordan T. Coughlan
Cnc 8h24'20" 27d55'58"
Jordan Taylor Wilkes
Peg 21h55'50" 18d16'55"
Jordan Terrance Blake Jenkins
Eri 3h48'22" -0d52'3"
Jordan "The Warrior" Crawford
Aqr 22h12'43" -1d7'52"
Jordan Thomas Marston
Psc 0h35'21" 12d58'11"
Jordan Tirillo
Gem 7h34'49" 30d47'55"
Jordan Trinity Call
Sgr 18h23'36" -23d0'31"
Jordan Vassallo
Aql 19h59'50" 11d24'12"
Jordan Victoria Carroll
Del 21h0'6" 16d26'50"
Jordan Victoria Jimenez
And 0h34'25" 46d31'20"
Jordan Watkins
Ori 5h43'3" 8d0'9"
Jordan Wattenburger
Cet 2h43'54" -0d27'15"
Jordan Wayne Beaupre
Aqr 23h53'29" -10d53'29"
Jordan William
Umi 15h19'6" 70d24'35"
Jordan Woda
Uma 10h55'41" 63d11'0"
Jordan Yuter
Tau 4h8'45" 23d41'2"
Jordana
Ari 2h15'20" 24d2'47"
Jordana Alexis Monasebian
Umi 15h42'32" 72d59'6"
Jordana Rose
Cnc 9h15'29" 31d15'33"
Jordana Rose Ross
Cnc 9h10'37" 30d30'24"
Jordana Sofia Gasparinho
And 0h2'50" 43d37'6"
JORDANI
Aql 18h41'37" -0d38'7"
Jordanina
Dra 18h51'30" 59d26'11"
Jordanium Prime
Lyn 8h53'7" 33d3'22"
Jordan...my best friend and love
Boo 15h8'39" 24d22'51"
Jordanna
Ari 3h15'21" 15d39'42"
Jordanna
Gem 7h12'51" 32d4'37"
Jordanna Ball
Gem 7h42'26" 31d43'52"
JordanPatrick
Gem 7h41'56" 23d11'13"
Jordan's Antipop
Cap 20h56'55" -20d42'38"
Jordan's Everlasting
Psc 0h34'56" 17d3'23"
Jordan's Grandma-Susan Gayle Graves
Cnc 8h28'52" 26d10'48"
Jordan's Jobsite
Umi 13h38'39" 74d26'55"
Jordan's Light
Lib 15h3'46" -12d50'28"
Jordan's Ma
Lib 15h20'53" -7d22'14"
Jordan's Star
Uma 8h52'1" 53d0'32"
Jordan's Star
Pho 0h38'24" -46d19'20"
Jordan's Star
Cnc 8h48'19" 27d17'18"
Jorde
Mon 6h46'11" 11d26'22"
Jordee
Lyn 7h32'2" 43d18'18"
Jorden Neuleich
Sco 17h17'54" -32d54'34"
Jorden's Bell
Psc 0h23'53" 7d59'7"
Jordi
Ori 6h18'31" -1d40'56"
Jordi Roig Castells
Ari 1h47'55" 21d45'19"
Jordia Claudan
Cyg 20h35'11" 46d43'34"
Jordie and Stace
Uma 11h24'29" 59d58'5"
Jordie C
Sco 17h8'39" -38d32'40"
Jordie Yaniel Guzman
Uma 10h21'42" 61d48'46"
JordiMeritxell
Sgr 19h7'39" -24d43'52"
Jordin Geovany Scott
Dra 19h27'13" 63d29'16"
Jordin's Star
Cam 4h9'36" 53d15'3"
Jordis Rylee
Cra 18h38'12" -39d50'34"
Jordon
Sco 16h52'58" -42d56'18"
Jordon
Ori 4h50'47" 11d8'58"
Jordon Isaiah
Vir 14h29'9" -1d1'2"
Jordon Joshua Sandefer
Lib 15h41'48" -15d19'48"
Jordon Kennedy Burroughs
Sco 16h53'14" -38d48'54"
Jordy
Tau 4h17'14" 26d48'54"
Jordyn
And 0h40'21" 45d9'32"
Jordyn
Lib 15h15'41" -27d12'26"
JORDYN
Cap 21h34'12" -11d21'26"
Jordyn
Umi 16h56'34" 77d48'38"
Jordyn
Cam 4h42'24" 62d0'18"
Jordyn
Dra 16h28'15" 57d32'58"
Jordyn Alexis Hurtubise "Pumpkin"
Cap 21h0'53" -26d51'26"
Jordyn Brynn Black
Uma 13h42'27" 61d14'35"
Jordyn Emma Elkan
And 2h31'37" 47d49'44"
Jordyn Faith Mandigo
Lib 14h49'1" -2d21'4"
Jordyn Fisher
Cam 7h30'13" 70d22'11"
Jordyn Haylee Ross
Gem 7h40'42" 23d50'13"
Jordyn Johnson's Star
Gem 7h48'3" 30d50'28"
Jordyn K. Lipply
Mon 6h52'25" -0d8'6"
Jordyn Katherine Lapointe
Lib 15h6'21" -2d21'55"
Jordyn Lei Moey
Gem 7h9'27" 23d45'48"
Jordyn Lynn
And 2h20'15" 48d39'20"
Jordyn M. Albright
Lmi 10h16'30" 29d30'41"
Jordyn Mekenna Chrisco
Aqr 22h11'36" -2d27'49"
Jordyn Michelle Grice
Aqr 22h34'53" -0d48'47"
Jordyn Michelle's Star
Umi 14h58'29" 71d58'46"
Jordyn Mosley
Lmi 10h8'47" 34d17'8"
Jordyn Nicole Miller & Grampy Ron
Uma 11h30'31" 55d25'18"
Jordyn Nicole Myers
Lyr 18h43'47" 39d17'58"
Jordyn Noelle Stanoch
Sgr 19h32'40" -18d3'23"
Jordyn R. Sansavera
Lyr 18h22'54" 38d8'31"
Jordyn Shea Paulson
And 0h34'29" 39d2'34"
Jordynn Ella Campbell 10/25/2002
Sco 17h51'51" -41d16'15"
JORDYNROSE
Sco 16h6'27" -11d2'31"
Jordyn's Star
Psc 23h49'49" 7d4'26"
Jordy's Star
Cru 12h6'15" -61d49'8"
Joren Clowers
Ari 3h7'46" 29d4'10"
Jorene (Nana) Metcalfe
Tau 4h51'19" 15d44'55"
Jorepana
Gem 7h17'36" 29d32'4"
Joretta Titus-Barnett
Uma 13h10'15" 58d40'28"
Jörg
Uma 8h19'23" 60d41'24"
Jörg Ahrens
Uma 14h25'29" 61d24'21"
Jörg Borowietz
Uma 10h12'19" 50d30'16"
Jörg & Colette's LUCKY STAR
Uma 8h51'58" 67d33'52"
Jörg Hassel
Uma 9h11'47" 62d38'39"
Jörg Hochuli, 09.11.1967
And 0h44'25" 41d31'12"
Jörg Kälin
Tri 2h28'34" 34d3'13"
Jörg Lier
Uma 12h3'16" 55d33'52"
Jörg Mattern
Uma 10h17'34" 40d15'53"
Jörg mein Schatz
Uma 8h31'17" 68d39'6"
Jörg Mühlfeld
Ori 6h17'52" 8d55'0"
Jörg Niemann
Uma 8h42'2" 67d3'48"
Jörg Schubert
Uma 9h0'33" 64d34'19"
Jörg Tautrim
Ori 4h46'40" 5d9'15"
Jörg Tessmer
Ori 6h12'38" 6d45'24"
Jörg und Valerie
Uma 11h21'53" 43d37'39"
Jörg Vetter
Ori 6h20'14" 8d40'52"
Jörg Vohwinkel
Uma 8h50'44" 64d28'39"
Jörg Wulf
Ori 6h19'3" 9d7'22"
Jörg-Andreas Sausel
Ori 6h19'56" 15d16'34"
Jorge
Vul 21h8'12" 23d33'3"
Jorge Alejandro Balbela
Ari 2h8'34" 17d21'53"
Jorge and Olga Forever
Leo 10h17'11" 12d30'1"
Jorge Aurelio Tetzpa Ramirez
Sgr 18h1'28" -27d37'45"
JORGE Baby Jorge
Uma 9h40'20" 69d37'12"
Jorge Ceballos
Cyg 20h5'18" 38d29'24"
Jorge Corona's Harley In The Sky
Cep 22h9'49" 59d55'16"
Jorge Enrique Moran-Loqui
Her 17h26'7" 30d31'47"
Jorge Enrique Nuñez Wolff
Boo 14h35'15" 52d53'19"
Jorgè Enrique Zabala
Her 17h24'4" 33d50'36"
Jorge Escamilla
Ari 3h13'32" 28d38'46"
Jorge Eudocio Aracena
Cru 12h36'26" -58d8'15"
Jorge Figueroa
Lep 6h2'3" -14d29'58"
Jorge Hernandez
Cep 22h17'47" 62d50'55"
Jorge L. Lugo Fiol
Lyn 6h27'13" 56d58'38"
Jorge & Loredana
Cyg 20h6'52" 54d5'13"
Jorge Luis Andrade Fernandes
Psc 0h28'47" 8d20'9"
Jorge Luis Figallo
Cas 23h0'39" 56d54'25"
Jorge Luis Perez
Aur 5h53'33" 35d51'54"
Jorge & Martha
Lyr 18h34'33" 37d56'29"
Jorge Morrison
Sco 17h51'36" -35d39'38"
Jorge Nuñez, Jr.
Sco 17h16'45" -34d10'53"
Jorge O. Uriarte
Gem 7h31'57" 33d29'7"
Jorge Patricio Michel
Uma 13h24'20" 55d40'53"
Jorge Rabaza
Per 3h33'20" 33d55'51"
Jorge Reyes
Umi 16h28'4" 86d34'40"
Jorge Rimblas - Orion Consulting, Inc.
Ori 5h10'58" -0d2'15"
Jorge Riofrio
Aql 20h14'53" 4d19'18"
Jorge Rivera
Per 2h38'12" 55d45'3"
Jorge Rodriguez
Cnc 8h51'18" 13d45'41"
Jorge Romero
Ari 2h37'13" 25d50'26"
Jorge Vidal Aranda
Boo 14h54'57" 25d21'41"
Jorge Webster My Heart
Ori 6h12'18" 15d13'31"
Jorge y Claudia
Cyg 20h24'12" 55d15'22"
Jorgelina Flores
Cra 18h15'28" -37d13'59"
Jorgen Aaron Singer
Lib 14h33'44" -14d59'51"
Jorgenson
Uma 11h6'24" 46d26'41"
Jorgie April McKinlay
Cru 12h24'15" -57d29'58"
Jorgurany
Vir 12h54'13" -0d39'42"
Jori Bertomen
And 23h45'26" 48d8'26"
Jori Bonselaar
Ori 5h44'29" -4d54'19"
Jori Leigh Andrus
Ari 3h21'30" 19d48'39"
Jorie Kendall Clawson
Uma 10h47'8" 56d48'46"

Jorinne Antoinette Jackson
Leo 11h30'42" 12d46'6"
Joris Daedalus ter Meulen Swijtink
Lyr 18h27'57" 39d37'39"
Joris Dominic Bieg
Uma 8h51'1" 55d52'30"
Joris Rousset
Leo 11h0'52" 11d37'34"
Joris Zahn
Uma 14h2'25" 50d33'37"
jorisofie
Boo 14h44'38" 32d41'38"
Jorja and Bill
Sgr 19h41'54" -13d20'58"
Jorja Bleu Michaels
Tau 5h52'34" 26d14'19"
Jorja Emily Walker
And 23h28'37" 37d52'26"
Jorja est Natus
And 2h35'50" 47d21'2"
Jorja Hayley Buttriss
Psc 2h1'5" 8d15'31"
Jorja Hope - 28/03/2004
Cru 12h10'43" -62d3'48"
Jorja Mae Luther
And 1h32'7" 46d50'43"
Jorja Mei-Li Townson
Peg 21h35'27" 15d33'47"
Jorja Nicole Osborne
Dra 10h9'37" 73d0'0"
Jorja Rae Barrett * Sweetpea *
Leo 10h29'57" 19d40'17"
Jorjah & Jazmins Star
Cyg 19h48'18" 34d6'16"
jor-k-11
Uma 11h18'43" 31d35'37"
Joronda Baxter
And 23h28'34" 47d8'15"
Jory Alexander Owen
Her 17h28'31" 34d55'30"
Jory Fine
Tau 5h46'38" 16d37'56"
Jory Hammons
And 23h38'35" 47d27'51"
Jory Murphy
Lmi 10h47'5" 27d46'17"
Jo's 18th Birthday Star
And 23h0'17" 52d28'30"
Jo's Angel
Cas 23h31'59" 55d50'15"
Jo's Elmo
Cma 6h46'35" -12d36'13"
Jo's Serendipitous Star
Sgr 18h13'30" -17d53'39"
J.O.'s Sky Space
Ori 5h18'20" -7d43'53"
Jo's Star
Uma 11h42'23" 45d40'9"
Josafel Dominic Ananto
Gem 6h49'54" 14d9'1"
Josah R.C.
Uma 10h26'58" 68d5'22"
Josalan M. Giavanna
Aur 5h29'30" 51d54'46"
Josalyn
Dra 17h28'26" 63d26'13"
Josana Mendes
Cru 12h15'19" -64d30'15"
Josander Paki Paki
Uma 13h7'4" 56d13'47"
Josann
Sco 16h11'23" -13d12'47"
Josanna Morningstar-Reece
Lyn 7h56'29" 45d32'34"
Josanne
Uma 11h4'45" 48d24'46"
Joscelyn Reed
Cas 0h20'18" 61d25'22"
Jose
Dra 9h38'43" 79d29'14"
José
And 23h25'38" 48d20'1"
Jose
Ari 3h16'14" 15d23'11"
Jose 22-7-1978
Vir 14h7'29" -2d57'40"
Jose A. Figueroa
Tau 5h59'28" 26d25'10"
Jose A. Filpo
Sco 16h11'57" -19d10'31"
Jose Abel Espinoza
Gem 6h9'17" 26d45'20"
Jose Alexander Lopez
Her 17h24'2" 16d0'25"
José Alexander Sanchez
Psc 0h52'48" 13d41'49"
Jose & Alicia Garcia Anniversary
Aqr 23h17'36" -7d42'34"
Jose and Kelli
Leo 9h23'25" 9d23'54"
Jose and Tanya
Cyg 19h38'16" 35d6'41"
José Angel Ocasio
Uma 12h1'23" 60d34'43"
Jose Angel Rodriguez Forever
Aql 19h36'27" 4d9'53"
José Antonio Manuel Martinez
Crb 16h21'22" 38d12'18"
José Antonio Rodriguez
Cep 21h18'11" 66d22'45"
José Antonio Rodriguez Rodriguez
Cas 1h17'2" 72d3'27"
Jose Antonio Torres
Tau 5h57'56" 24d36'42"
Jose Arana
Ori 5h32'49" -0d18'53"
Jose Aurelio Hernandez
Cnc 8h26'26" 17d56'26"
Jose Balmaceda Bayquen
Uma 11h39'24" 30d49'6"
Jose Castiella chico del screto
Her 16h40'34" 26d13'45"
Jose Castilleja
Cep 22h19'12" 72d50'27"
Jose Castro
Her 17h48'11" 41d56'19"
Jose Cisneros
Per 3h9'16" 43d5'25"
Jose Correia "12-25-60"
Uma 13h50'58" 54d58'53"
Jose Corrigan
Sco 16h33'54" -28d40'39"
Jose De Jesus Covarrubias, FLAQUITO
Sgr 18h31'13" -26d54'15"
José Enrique Morones
Psc 0h27'54" 11d19'59"
Jose F. Graves II
Aqr 22h28'1" -14d48'37"
Jose Foradada III
Per 3h24'53" 45d32'16"
José Gamez "para mi Amor galactico"
Boo 14h43'27" 30d1'47"
José Garrett Kessler
Tau 3h45'29" 27d5'48"
Jose George Witt
Per 3h42'19" 48d0'26"
Jose H. Sotolongo-Pla
Leo 10h11'9" 12d55'53"
José i Maria
Cyg 19h32'51" 32d25'15"
José Ignacio Barrera
Aql 19h15'37" 14d57'56"
José Ivo De Morais
Sco 17h32'13" -39d25'21"
Jose Jesus Suarez
Per 3h30'30" 47d15'9"
José "Johito" Rivera Valladares
Ori 5h47'2" 5d2'13"
José Juan (Joshua) Perales Castro
Psc 23h29'2" 5d13'7"
Jose L. Rangel
Uma 11h22'28" 41d51'48"
Jose L Riera
Per 2h43'4" 54d31'13"
Jose Lorenzo Diaz Suarez
Cma 6h56'41" -23d23'41"
José Luis Alcantara de la Rosa
Cep 21h44'19" 62d23'9"
Jose Luis Arenas Campana
Gem 6h51'53" 13d31'57"
Jose Luis Cubas
Leo 10h46'9" 18d51'39"
Jose Luis Fraile Rodriguez
And 0h55'6" 23d32'18"
Jose Luis Fuller
Aqr 22h16'36" -1d18'36"
Jose Luis Herrera
Ori 5h35'6" 4d55'59"
Jose Luis Miranda
Vir 13h42'48" -18d54'31"
Jose Luis Murillo Bagundo
Aqr 23h6'11" -11d53'12"
Jose Luis Rambo Rivera
Ori 6h17'3" 11d4'58"
José Luis Soto III
Sgr 18h8'32" -16d41'45"
Jose Luis Valdez Jr.
Cep 21h58'30" 65d29'35"
Jose Luis Verbera
Vir 13h40'26" 7d6'3"
Jose Luz Soriano Sosa
Sgr 19h6'49" -32d10'38"
Jose M. Peralta
Sge 19h59'34" 16d39'18"
José Manuel Blasco Lasierra
Gem 6h53'25" 15d50'19"
Jose Manuel Euvin
Her 17h26'9" 26d40'44"
Jose Manuel Fernandez Gil
And 1h21'30" 40d26'6"
Jose Manuel Franco Lopez
Tau 4h32'28" 8d23'18"
Jose Manuel Machin
Vir 13h37'30" 4d30'44"
jose manuel toledo
Aqr 20h46'39" -8d44'29"
José María del Arco
Dra 17h40'19" 54d49'37"
José Maria del Arco
Dra 18h21'2" 48d52'3"
Jose Maria Hiceta Porguez
Lib 14h49'0" -8d19'11"
Jose Maria Suarez
Lyr 19h2'6" 42d37'18"
Jose Mario Valdes Gallo
Boo 14h57'38" 50d20'9"
Jose Martin Pacheco
Psc 23h35'34" 6d50'30"
Jose Miguel Franco 29.07.1983
Cas 1h2'13" 62d6'30"
José Miguel Garau
Boo 13h48'6" 11d7'31"
Jose Miguel Marin Medina
Uma 10h46'39" 50d46'42"
José Moya
Dra 16h12'38" 55d34'35"
Jose Noel Ortiz
Aur 5h38'17" 49d9'20"
Jose & Olga Aguirrebeitia
Sco 17h18'48" -43d56'12"
José Orlando Schnider
Cas 0h55'19" 54d18'29"
Jose Ortega
Uma 8h38'6" 52d26'41"
Jose "Pumpkin" Rodriguez
Her 17h51'17" 48d32'59"
Jose Ramón
Umi 14h41'26" 78d28'53"
Jose Ramon Peinado, JR
Sgr 19h24'34" -16d34'50"
Jose Ramon Perez
Tau 4h16'6" 13d30'12"
Jose Ramon Ramos
Ori 5h22'12" 13d13'33"
Jose Ramon y Hortensia Abreu
Lyn 7h44'43" 56d40'59"
Jose Raphael Ferraz Esposito
Psc 0h23'25" 19d22'44"
Jose Rios
Leo 11h35'7" 9d34'19"
Jose Roberto Arruda
Cen 14h33'2" -61d2'46"
Jose Roberto Arruda
Pyx 9h3'5" -26d44'49"
Jose Rodriguez
Aql 19h15'18" -9d54'54"
Jose Rodriguez
Sge 19h50'44" 18d45'1"
José Rodriguez Bermudez
Ari 2h31'12" 27d45'7"
Jose Rojas
Per 2h48'52" 39d38'15"
Jose Romero Rodriguez
Uma 11h35'52" 29d35'38"
José Soares Ribeiro
Uma 8h20'41" 62d7'6"
Jose T. Hurtado
Gem 7h31'24" 17d47'10"
Jose "the sparkle in my eye"
Vir 14h56'28" 7d18'54"
José Toniann Scarpa Timothy King
Uma 12h14'10" 61d2'38"
Jose Trinidad Ibal Rodriguez
Gem 7h15'32" 17d38'41"
Jose Vicente Dasí Ferrandis
Peg 21h48'36" 27d38'24"
Jose y Patricia
Aur 4h49'18" 35d55'5"
JOSEAA 1974
Cyg 21h51'0" 50d37'27"
Josean & Stephanie Marrero
Cnc 8h18'18" 13d9'15"
Joseann
Lib 15h7'38" -11d16'50"
José-Antonio
Ori 6h8'43" 14d45'15"
JoseChristaTeAma
Aqr 22h6'12" -13d5'22"
Josée
Cas 1h32'55" 67d34'32"
Josee Boivin
Sco 17h6'55" -38d20'10"
Josée Boudreault
Umi 11h34'35" 86d39'46"
Josée De Santis-Chow
Leo 11h1'20" -0d0'17"
Josée Dompierre
Umi 16h48'44" 76d19'4"
Joseé Gélinas
Cas 23h18'27" 58d36'10"
Josée Landry
Aqr 23h26'23" -21d32'25"
Josef A Barmoha Cuban Wonder
Cam 3h59'31" 70d12'12"
Josef A. Müller
Ori 6h4'59" 19d12'31"
Josef Anetzberger
Uma 10h48'5" 42d20'26"
Josef & Anita Choate
Ori 6h13'36" 19d23'9"
Josef & Barbara
Ori 5h22'34" 3d7'29"
Josef Bochatz
Ori 6h22'48" 16d42'42"
Josef Elias Koller
Per 4h20'27" 51d10'7"
Josef Földes
Aqr 20h49'12" -5d57'36"
Josef Freser (Dide) 14.10.1932
Cru 12h25'20" -57d28'55"
Josef Geisenhofer
Uma 8h58'36" 57d23'33"
Josef Goergen
Dra 17h27'0" 50d52'14"
Josef Haas
Cas 22h59'30" 54d48'56"
Josef Heinlein
Uma 8h37'23" 54d11'4"
Josef Koch
Uma 8h55'18" 60d59'29"
Josef Kriyan De Leon
Uma 9h3'59" 67d47'48"
Josef Krystek
Uma 11h32'57" 37d34'16"
Josef Meyer
Uma 11h28'24" 31d45'38"
Josef Oerding
Uma 11h52'0" 30d29'29"
Josef Prinz
Uma 13h40'5" 59d27'6"
Josef Reum
Per 3h46'32" 40d28'13"
Josef S. Opela
Cyg 19h56'28" 53d37'51"
Josef Schaider
Uma 11h22'13" 38d46'0"
Josef Schauer "Superstar"
Umi 13h26'2" 74d45'32"
Josef Schwitter
Dra 20h10'47" 72d30'22"
Josef Sedlmair
Uma 8h22'13" 62d28'4"
Josef Sikiric
Cyg 20h41'10" 34d51'16"
Josef Siwczyk
Uma 12h2'39" 45d44'2"
Josef the Honey Bear
Ari 2h33'59" 20d46'32"
Josefa Dias Zachariadhes
Vir 12h8'30" 5d18'37"
Josefa Salinas 20 Years In Radio
Lib 15h13'21" -6d29'27"
Josefina
Aqr 22h33'20" -2d35'4"
Josefina
Psc 1h24'54" 32d10'0"
Josefina Galvan
Lyn 6h37'51" 57d25'12"
Josefina Karpecki
Gem 7h35'29" 23d30'22"
Josefina Lara Zavala
Crb 15h40'12" 33d43'14"
Josefina Pidilla
Uma 11h15'9" 61d59'26"
Josefina Silver
Peg 21h39'20" 27d55'12"
Joseline
Eri 4h26'2" -3d34'21"
Joseline 2342
Ori 4h50'12" -2d59'55"
Joselito Cocoy Catalan
Dra 17h35'4" 55d37'32"
Joselito Padua Jr.
Cyg 19h50'42" 51d36'11"
Josener H. Aguinaldo, Sr.
Uma 11h44'10" 35d10'22"
Josep Barrera
Gem 7h23'56" 31d1'59"
Josep Blanch
Del 20h42'13" 4d45'41"
Joseph
Psc 1h37'39" 13d47'28"
Joseph
Ori 5h28'14" 2d54'45"
Joseph
Leo 10h13'42" 13d26'16"
Joseph
Cnc 8h58'10" 28d33'14"
Joseph
Gem 7h19'19" 24d26'22"
Joseph
Cnc 8h52'50" 23d4'15"
Joseph
Aur 5h51'30" 36d2'45"
Joseph
Lac 22h30'41" 49d9'0"
Joseph
Per 3h24'8" 53d2'7"
Joseph
Uma 13h29'25" 55d7'33"
Joseph
Umi 13h20'53" 76d13'23"
Joseph
Cma 6h43'51" -14d3'5"
Joseph
Lib 14h48'54" -1d2'58"
Joseph
Sgr 19h42'55" -14d25'43"
Joseph A. Brewer
Cep 22h58'11" 63d24'42"
Joseph A. Caprera
Leo 11h23'20" 25d18'19"
Joseph A. Coviello
Per 3h27'17" 35d31'24"
Joseph A. Cunard
Uma 9h26'25" 68d31'13"
Joseph A. D'Amico
Tau 3h31'14" 20d58'46"
Joseph A. Fitzkee
Ori 5h47'40" -0d58'55"
Joseph A. Galluccio
Uma 9h23'0" 56d49'17"
Joseph A. Haas, M.D.
Cnc 8h37'12" 24d4'37"
Joseph A. Hardy III
Aur 5h48'7" 49d32'21"
Joseph A Hinni 50th Birthday Star
Aqr 21h9'7" -14d15'32"
Joseph A. Howard II
Lyn 8h0'35" 40d3'57"
Joseph A Leone III
Per 3h54'37" 37d45'23"
Joseph A. Marchese
Cnc 9h12'0" 21d26'22"
Joseph A. Mendoza 77286
Cnc 8h45'47" 7d23'59"
Joseph A. Otterstedt
Sco 16h9'19" -11d37'37"
Joseph A. Parziale
Sgr 18h6'36" -27d40'41"
Joseph A Patalano
Sgr 18h52'54" -20d25'34"
Joseph A. Patton
Cnc 8h23'42" 24d46'9"
Joseph A. Rappa
Sgr 18h58'11" -27d54'56"
Joseph A. Remsik
Her 17h44'31" 37d34'40"
Joseph A. Rinaldi
Aqr 22h38'56" 0d49'38"
Joseph A. Santana
Ari 2h38'37" 19d49'56"
Joseph A. Santangelo III
Umi 16h11'50" 78d27'52"
Joseph A. Satrape
Vir 12h36'57" 8d28'46"
Joseph A. Thrain
Psc 0h40'45" 20d29'57"
Joseph A. Tolino Sr.
Per 1h50'9" 47d54'37"
Joseph A. Tomenello
Uma 10h10'19" 44d7'55"
Joseph A. Verica
Gem 7h25'54" 29d31'21"
Joseph A. Vincuillo, Jr.
Vir 12h9'6" 11d15'4"
Joseph A. Ziegler
Boo 13h45'7" 18d1'31"
Joseph Aaron
And 0h50'26" 45d13'53"
Joseph Abdo Maatouk
Leo 11h13'34" -1d15'35"
Joseph Aguilar
Psc 0h23'45" 12d46'7"
Joseph Alan Chuter
Per 3h2'43" 54d2'20"
Joseph Albenese
Aql 19h7'8" 2d52'26"
Joseph Albert Centrone
Per 1h38'18" 54d23'56"
Joseph Alex Larson
Uma 12h0'36" 46d21'4"
Joseph Alexander
Crb 15h30'6" 29d17'18"
Joseph Alexander Awsiukiewicz
Lyr 18h50'36" 37d26'5"
Joseph Alexander Gorry
Umi 15h50'26" 82d18'4"
Joseph Alexander Zdrojewski 1996
Aqr 22h23'40" -15d54'25"
Joseph Alfonso Cetrulo
Uma 10h4'38" 67d21'15"
Joseph Alfred Prive'
Del 20h36'51" 14d9'58"
Joseph Allen Dale Guidry
Cnc 9h6'5" 11d10'54"
Joseph Allen Perry Birthday Star
Sco 16h11'57" -20d10'36"
Joseph Aloysius Coyle
Ori 5h55'17" 17d16'57"
Joseph Amoruso
Her 18h8'32" 28d30'45"
Joseph and Anna Rotante
Aql 19h25'29" 0d11'9"
Joseph and Ashley
Cyg 21h35'39" 47d50'5"
Joseph and Dulce Bruner
Cyg 19h42'37" 36d56'46"
Joseph and Helen
Gem 7h46'51" 32d18'8"
Joseph and Ida Auerbach
Cnc 8h39'41" 24d31'25"
Joseph and Margie Anheier, 1/19/53
Cyg 20h34'25" 30d49'54"
Joseph and Mercy Metherate
Crb 15h57'18" 34d20'4"
Joseph and Michelle Cavallaro
Uma 11h54'27" 36d22'48"
Joseph and Mildred Roselli
Ori 6h8'42" 13d58'25"
Joseph and Sandra Augustine
Cyg 19h40'41" 34d41'51"
Joseph and Stefanie Wackler
And 0h36'54" 38d27'22"
Joseph Anderson Glover "Joey"
Per 3h26'52" 45d8'26"
Joseph Andrew
Ori 5h52'3" 12d25'59"
Joseph Andrew Bathgate
Aur 6h27'58" 35d4'25"
Joseph Andrew Colianni
Tau 4h14'7" 28d39'43"
Joseph Andrew Gallo
Cep 22h44'6" 67d50'20"
Joseph Andrew Green
Aqr 22h55'9" -16d30'52"
Joseph Andrew Koval IV
Boo 14h55'58" 35d30'14"
Joseph Andrew Lombardo
Sgr 18h19'26" -16d36'48"
Joseph Andrew Teller
Psc 2h1'33" 6d13'45"
Joseph Angelo Zangri, Jr.
Her 17h39'4" 40d53'32"
JOSEPH ANN
Crb 15h54'43" 31d55'49"
Joseph & Anna Batchelder
Cyg 21h51'29" 42d29'51"
Joseph Anthony
Umi 15h33'41" 79d45'11"
Joseph Anthony
Cas 1h21'40" 64d51'56"
Joseph Anthony Abruzzo
Aqr 22h17'47" -22d47'57"
Joseph Anthony Brooks, Jr.
Per 2h41'12" 53d42'28"
Joseph Anthony Bukartek Jr.
Psc 1h35'36" 11d2'8"
Joseph Anthony Cristelli
Tau 5h49'42" 22d53'51"
Joseph Anthony D'Adamo
Uma 8h34'13" 65d54'38"
Joseph Anthony DeCarlo, Jr.
Uma 12h29'8" 61d10'52"
JOSEPH ANTHONY DEGLOMINI
Lib 14h42'14" -19d51'6"
Joseph Anthony Emanuele
Gem 7h6'32" 14d1'30"
Joseph Anthony Facciola
Per 3h29'28" 51d44'27"
Joseph Anthony Gallart
Sco 17h6'9" -38d19'53"
Joseph Anthony Gilchrist
Ori 5h12'11" 5d53'40"
Joseph Anthony Gutierrez
Cyg 20h2'31" 56d31'51"
Joseph Anthony Hanson
Per 4h10'32" 50d48'56"
Joseph Anthony Hill Jr.
Per 4h33'30" 40d39'1"
Joseph Anthony III
Aqr 20h54'27" -8d53'52"
Joseph Anthony Mager
Leo 9h36'50" 7d3'14"
Joseph Anthony Maiatico
Uma 9h12'46" 55d10'29"
Joseph Anthony Marchena 8/5/1956
Leo 11h33'9" 2d35'20"
Joseph Anthony McHugh
Uma 12h15'3" 55d1'9"
Joseph Anthony Montanaro
Lyn 8h30'58" 34d49'52"
Joseph Anthony Morrison
Cep 21h48'38" 56d8'44"
Joseph Anthony Pascarelli
Col 5h8'31" -35d40'55"
Joseph Anthony Pedry
Tau 4h27'2" 16d56'26"
Joseph Anthony Peter Menna
Cnc 8h48'30" 27d54'43"
Joseph Anthony Petrak
Her 17h4'1" 37d34'47"
Joseph Anthony Puglisi
Cyg 21h5'32" 39d0'49"
Joseph Anthony Schwalbach
Uma 9h0'4" 57d0'23"
Joseph Anthony Sernio Sr.
Per 2h28'22" 55d58'56"
Joseph Anthony Tapper
Psc 0h28'38" 16d19'50"
Joseph Anthony Vazquez
Her 17h41'51" 44d14'33"
Joseph Anton Wetzel
Uma 9h35'51" 59d2'39"
Joseph Archer's Little Star
Her 17h29'35" 21d34'52"
Joseph Arleo
Leo 10h36'40" 18d57'9"
Joseph Armetta
Cnc 8h49'2" 31d34'22"
Joseph Arthur Roper
Aqr 21h37'56" -5d52'8"
Joseph Arthur Rosati
Aqr 21h59'47" -9d46'21"
Joseph Attardi
Cyg 19h37'3" 29d17'32"
Joseph & Audrey Smolinski
Uma 11h12'36" 29d45'1"
Joseph B. Kreckel
Sgr 18h18'54" -28d45'56"
Joseph B. Lammers
Cyg 19h40'53" 48d1'29"
Joseph B. Psalmonds, Jr.
Ari 2h25'6" 11d31'17"
Joseph & Barbara Crawshaw
Cyg 19h59'29" 38d49'1"
Joseph Bart Bates
Per 3h24'48" 48d53'21"
Joseph Bart Bates
Aql 19h50'21" 15d58'15"
Joseph Bedkowski, Jr.
Vir 13h55'52" -8d19'48"
Joseph Belfiore
Her 18h28'55" 21d29'50"
Joseph Bendoraitis
Ori 5h56'37" -0d10'17"
Joseph Benjamin Baratta, MD
Lyn 7h35'16" 37d28'50"
Joseph Benjamin Delmonico
Aur 5h26'54" 53d8'51"
Joseph Bernard Lange
Crb 15h57'10" 31d58'36"
Joseph Besch
Gem 6h34'26" 12d22'37"
Joseph Black
Her 17h8'56" 38d20'13"
Joseph Bobb
Sco 16h5'6" -12d20'34"
Joseph Bono
Psc 1h11'13" 31d47'51"
Joseph Borriello
Aur 5h34'44" 43d59'41"
Joseph Bou Rached (yoso)
Uma 11h58'15" 36d46'42"
Joseph Bradbury
Cep 22h40'14" 61d4'24"
Joseph Bradley Clermont
Aqr 23h23'52" -14d20'27"
Joseph Brett Woodruff
Sge 19h49'26" 18d42'35"
JOSEPH BREVETTI
Her 18h45'10" 21d23'39"
Joseph Broughton
Cep 22h51'41" 70d17'34"
Joseph Browne
Cep 22h3'10" 68d15'1"
Joseph Bryan Harvey
Per 3h11'0" 50d59'38"
Joseph & Brynne Violi
Cas 0h55'52" 62d34'2"
Joseph "Buddy" Mauriello
Uma 10h10'58" 53d40'59"
Joseph Bulone
Sgr 20h1'43" -22d18'6"
Joseph Burke Flanagan, JR.
Her 18h24'15" 18d36'52"
Joseph Bushek
Uma 9h5'13" 57d34'38"
Joseph Buta
Her 16h37'0" 34d17'22"
Joseph C. Arters
Sco 16h15'15" -8d28'36"
Joseph C. Baht
Uma 9h53'26" 65d40'27"
Joseph C. Morabito
Dra 16h36'55" 60d23'49"
Joseph C. Sink
Sgr 19h1'2" -34d21'37"
Joseph C. Townsell
Ari 3h1'9" 25d45'30"
Joseph Cabral, Jr.
Aqr 22h5'4" -4d25'21"
Joseph Caccavale
Peg 22h47'38" 15d12'23"
Joseph Camarra
Cap 20h15'50" -13d53'47"
Joseph Caperna 2
Leo 9h39'10" 27d3'30"
Joseph Capizzi
Per 4h47'13" 45d28'44"
Joseph Capobianco
Uma 9h40'40" 64d59'31"
Joseph Capps Sherrill
Aur 5h9'24" 40d49'11"
Joseph Cardamone, Sr.
Leo 11h34'10" 16d35'16"
Joseph Carl Riechmann
Uma 11h11'26" 50d14'7"
Joseph Carnie Patton II
Boo 14h42'38" 52d41'38"
Joseph Carolus
And 2h21'42" 39d41'10"
Joseph Charles Belardinelli Jr.
Cyg 21h55'16" 53d24'36"
Joseph Charles Brummel
Uma 10h4'18" 46d3'27"
Joseph Charles Capucini
Leo 9h43'41" 30d12'52"
Joseph Charles Daly
Cep 21h12'44" 61d3'28"
Joseph Charles Edwards
Ori 6h9'40" 8d21'58"
Joseph Charles Gillick
Vir 12h56'55" -7d38'41"
Joseph Charles Honeysuckle
Uma 11h20'57" 35d58'9"
Joseph Charles Nophut
Tau 5h9'25" 27d46'20"
Joseph Charles Papariello, II
Cep 2h29'43" 83d46'23"

Joseph Charles Papariello, III
Cep 21h33'29" 59d50'58"
Joseph Charles Shong
Psc 0h10'55" -0d43'18"
Joseph Chase Gillis
Her 17h52'56" 18d29'19"
Joseph Chinosi
Uma 8h10'49" 62d44'15"
Joseph Christopher
Aur 6h28'55" 51d35'9"
Joseph Christopher John Davoli
Leo 9h34'8" 23d27'0"
Joseph Christopher Lombardi
Per 4h21'5" 39d36'45"
Joseph Christopher Plumley
Ari 2h40'53" 21d25'7"
Joseph Christopher Tamagna
Cnc 9h3'59" 30d47'58"
Joseph Christopher Wachendorfer
Vir 13h47'29" -5d0'17"
Joseph Ciavolella
Cap 21h49'23" -22d16'51"
Joseph Clarence Wohlschlaeger
Her 18h46'29" 12d50'19"
Joseph Clay Stovall, Sr.
Ori 5h28'35" 1d30'13"
Joseph Clifford Smith
Uma 10h52'22" 64d37'36"
Joseph Clyde Skinner III
Umi 13h48'15" 78d26'41"
Joseph Coffin and Jin Hong Family
Uma 11h31'27" 30d0'44"
Joseph Cohen
Tau 4h41'38" 28d8'13"
Joseph Columtino
Ori 5h59'16" -0d12'46"
Joseph Conrad Torpey Toole
Gem 6h38'6" 15d44'28"
Joseph Constant
Ari 2h18'46" 12d48'0"
Joseph Covelli
Cep 22h32'30" 72d36'49"
Joseph Cunningham
Gem 7h46'1" 23d7'11"
Joseph D. Laszlo, MD.
Per 3h16'44" 47d36'54"
Joseph D. Massucci
Uma 10h10'51" 43d48'32"
Joseph D. Pullen
Her 18h0'19" 22d58'51"
Joseph D Weiss
Cam 4h1'31" 67d25'0"
Joseph "Dad" McGrady
Cap 21h36'47" -13d10'28"
Joseph D'Adamo
Leo 10h59'39" 18d42'33"
Joseph Dadoush
Vir 13h14'9" -6d23'53"
Joseph Daks
Sgr 18h17'18" -32d31'20"
Joseph D'Alessandro
Uma 9h3'49" 67d43'49"
Joseph Daniel
Sco 17h57'6" -37d7'50"
Joseph Daniel Benavidez
Aur 5h33'31" 45d32'30"
Joseph Daniel Dowsing Joyce
Dra 15h37'47" 58d42'18"
Joseph Daniel Hubball
Peg 21h49'38" 8d4'51"
Joseph Daniel Kuehl
Boo 15h9'16" 41d30'0"
Joseph Daniel Watts - Joe's Star
Per 3h58'9" 32d53'5"
Joseph & Danielle
Vir 13h25'13" 11d43'23"
Joseph & Danielle Biafore
Dra 16h6'55" 63d11'44"
Joseph Darrell Goff, Sr.
Ari 2h11'38" 26d7'46"
Joseph David
Per 3h13'8" 51d12'15"
Joseph David Courtney - Joseph's Star
Dra 17h16'49" 52d39'11"
Joseph David Frederick
Cep 22h20'20" 73d6'3"
Joseph David Licameli (JD)
Vir 13h29'3" -6d30'59"
Joseph David Rubino "Joe" BestTwin
Cnc 8h37'52" 31d59'9"
Joseph David Ter Louw
Ori 6h21'23" 14d18'31"
Joseph David White
Leo 10h35'58" 13d24'11"
Joseph De Feo 5 Firey Years.....
Leo 10h5'54" 21d1'10"
Joseph Dean Scarpa
Aur 5h42'12" 42d15'49"
Joseph Debelak
Dra 14h51'32" 58d46'11"
Joseph Dee, Jr.
And 23h0'50" 50d53'20"
Joseph Delango
Uma 11h46'26" 34d51'9"
Joseph Denney
Sgr 18h51'35" -21d39'16"
Joseph DePalma
Uma 8h12'46" 61d2'16"
Joseph DePiano
Aur 5h45'21" 46d20'41"
Joseph Devine
Pho 1h32'48" -41d29'21"
Joseph "Dewey" Ywuc
Ori 6h6'24" 5d43'28"
Joseph 'Dick' Higgins
Aqr 21h38'1" 1d42'38"
Joseph DiMatteo
Ari 3h22'47" 28d20'5"
Joseph Dinolfo
Cep 21h27'11" 64d33'13"
Joseph DiNuzzo
Ori 5h43'46" -0d5'40"
Joseph Domenic
Boo 14h51'17" 50d31'18"
Joseph Dominic
Cnc 9h6'26" 12d59'25"
Joseph Dominque Michel Magna
Uma 8h12'34" 70d48'26"
Joseph Dorchak
Aur 5h53'9" 48d29'37"
Joseph Douglas Pearson
Uma 12h4'49" 32d16'26"
Joseph Douglas Powers
Psc 0h37'58" 13d1'21"
Joseph Doyle Engle
Cap 20h49'48" -21d4'15"
Joseph Dylan Richards
Umi 13h21'43" 72d27'51"
Joseph Dyson William Gore III
Psc 1h6'51" 28d13'6"
Joseph E. Brandt
Cas 0h32'59" 62d35'35"
Joseph E. Burchick
Uma 8h52'4" 50d51'1"
Joseph E. Byrne
Her 17h19'15" 45d46'26"
Joseph E. Ceratt
Uma 11h30'9" 29d24'19"
Joseph E. Dilsaver
Per 4h24'4" 32d41'51"
Joseph E Faith II
Ari 2h14'1" 25d52'35"
Joseph E. Fay
Cap 20h29'23" -10d55'1"
Joseph E. Garoppo "Joppo"
And 23h39'29" 48d23'26"
Joseph E Lemmons
Dra 17h43'15" 54d3'19"
Joseph E. Longo
Cap 20h10'27" -18d8'9"
Joseph E. Murray 1939 - 2005
Uma 9h53'0" 53d11'5"
Joseph E. Willey 6/7/14-5/1/45
Ari 1h54'6" 20d3'5"
Joseph Edmund Maddigan
Sco 16h17'2" -9d7'25"
Joseph Edward Bruney
Aqr 22h38'54" -0d25'45"
Joseph Edward Burt
Sco 17h58'41" -39d6'57"
Joseph Edward DiPietri
Gem 7h39'31" 32d36'46"
Joseph Edward Dowler
Umi 16h0'27" 74d41'51"
Joseph Edward Glaser
Ari 1h59'31" 17d37'15"
Joseph Edward Hebbeln
Her 17h59'34" 27d13'46"
Joseph Edward Krause
Umi 16h0'6" 74d17'45"
Joseph Edward Murray
Ori 5h27'35" 1d55'2"
Joseph Edward Smith
Her 17h48'55" 22d34'41"
Joseph Edward Talbot III
Uma 11h2'30" 68d2'56"
Joseph Edwards
Her 17h46'33" 16d16'29"
Joseph Egli
And 0h11'41" 25d52'53"
Joseph Elise
Sco 17h40'25" -36d19'55"
Joseph Ellis
Umi 13h35'33" 72d55'44"
Joseph Ellsworth Baker
Vir 13h28'48" 3d26'37"
Joseph Elvert
Ser 18h21'16" -13d38'52"
Joseph & Emily Bedell
Cyg 21h14'22" 44d39'9"
Joseph Ernest Haviland III
Per 3h33'1" 47d53'20"
Joseph Ernest Visco
Aqr 23h48'48" -12d34'27"
Joseph Esposito
Aqr 22h40'47" 0d30'46"
Joseph Eugene McCoy
Vir 14h5'17" -8d30'59"
Joseph Eugene Phillips
Ori 6h3'7" -0d16'51"
Joseph Evan Peat
Psc 1h1'48" 27d34'5"
Joseph F and Evelyn M Pugliese
Uma 13h31'24" 57d45'8"
Joseph F. Cussio
Ori 5h33'48" 10d51'54"
Joseph F. DeMaria
Per 3h22'18" 52d6'33"
Joseph F. Lynch
Cep 20h43'2" 61d51'2"
Joseph F. Rogozinski
Per 4h30'54" 40d44'55"
Joseph F. Sarra Sr.
Cep 21h33'10" 57d56'57"
Joseph F. Shaw
Lup 15h36'40" -39d0'0"
Joseph F. Weitlauf
Her 17h36'45" 46d10'3"
Joseph Fadziewicz
Cep 21h34'36" 64d18'53"
Joseph Falchetta Jr.
Uma 11h4'47" 63d30'44"
Joseph Family Angel
Col 5h23'41" -29d46'6"
Joseph Fegan - Forever Above.
Cep 20h34'47" 61d30'56"
Joseph Floyd Giacometto
Aur 5h38'55" 40d34'8"
Joseph Forcellina & Hope Forcellina
Uma 12h54'46" 57d32'52"
Joseph Frances Ragni
Per 3h44'17" 51d32'56"
Joseph Francis Charlow
Uma 8h27'16" 68d57'23"
Joseph Francis Dudek
Ori 6h9'29" -0d38'8"
Joseph Francis Grosso
Ori 5h30'15" 1d52'0"
Joseph Francis Lyons, Sr.
Cap 21h57'43" -22d40'36"
Joseph Francis Raymaker
Per 2h50'7" 53d9'37"
Joseph Francis Zuccalmaglio
Dra 16h39'22" 64d1'2"
Joseph Frank
Ori 6h19'2" 15d9'25"
Joseph Frank Risi
Vir 13h28'49" -6d36'40"
Joseph Franklin Ballew III
Ori 5h34'39" 13d14'25"
Joseph Franklin Moore, Jr.
Lib 14h43'23" -24d28'47"
Joseph Franklin Romes
Cyg 20h53'5" 37d8'1"
Joseph Franklyn McMullen
Umi 15h49'7" 76d49'43"
Joseph Franks
Cnc 8h52'44" 17d48'33"
Joseph Frederick Poertner 1934-2006
Ori 6h6'6" 11d57'37"
Joseph Furci
Uma 9h0'26" 48d21'54"
Joseph G. Ciacci
Ori 6h12'53" 15d34'13"
Joseph G. Ferreira
Aqr 20h39'42" 0d42'17"
Joseph G. Visciano
Psc 23h24'35" 5d23'48"
Joseph Gabriel Orlando
Per 4h15'10" 36d0'27"
Joseph Gabriel Toro
Sgr 19h32'55" -13d50'30"
Joseph Gangloff
Tau 5h51'15" 17d41'55"
Joseph Gardner Hardin
Ori 5h30'58" -0d35'16"
Joseph George Breder
Cap 20h48'55" -24d31'3"
Joseph George Iozzi
Aqr 22h4'42" 1d2'47"
Joseph George Kehoe .- Anchors Away
Uma 11h11'10" 70d30'24"
Joseph George Tolley - Joseph's Star
Her 18h14'32" 21d14'16"
Joseph Gerald Teresa
Leo 10h45'41" 18d21'36"
Joseph Gervais
Psc 1h23'19" 27d41'30"
Joseph Gibson Blackmer
Leo 11h58'14" 27d30'44"
Joseph Gibson Sayers Snr. 21.06.1937
Pup 7h44'52" -28d19'44"
Joseph Giglio
Cnc 9h11'37" 31d39'56"
Joseph Glade Repasky
Her 17h16'29" 28d58'18"
Joseph Gogolinski
Cap 20h27'53" -23d17'52"
Joseph Golemme
Per 2h22'59" 57d10'8"
Joseph Gordon Alexander Cullen
Psc 0h53'42" 18d53'42"
Joseph Gordon Atherton
Aqr 21h54'6" 1d31'24"
Joseph Gorman
Her 17h21'39" 47d57'49"
Joseph Graham Ewing
Ari 2h10'17" 27d23'30"
Joseph "Grandma & Grandpa's Star"
Her 17h26'24" 32d23'46"
Joseph "Grandpa" Schmidt
Uma 10h10'44" 57d28'7"
Joseph Graziano
Uma 8h36'34" 58d36'2"
Joseph Guerra
Cap 21h55'31" -16d21'19"
Joseph Guzzardo
Aur 5h29'57" 47d46'25"
Joseph H. Herold
Cnv 12h42'39" 48d12'12"
Joseph H King 11
Cap 21h26'29" -25d35'23"
Joseph H. Levine
Her 17h52'57" 19d38'14"
Joseph Hadjijoseph
Cep 22h17'30" 72d56'59"
Joseph Hagen - Tupperware
Per 3h39'52" 32d42'13"
Joseph Haig Boyd
Ori 5h53'48" 11d26'19"
Joseph Haley
Tau 4h39'25" 15d5'54"
Joseph Haze Arciniega
Aqr 22h12'12" -1d53'54"
Joseph Hazen
Her 18h49'20" 21d20'31"
Joseph Heitkemper
Ori 5h15'21" 7d52'16"
Joseph Henry Daly
Ari 1h57'21" 13d12'27"
Joseph Henry Draheim Jr. 01201981
Aqr 22h53'20" -22d48'55"
Joseph Herskovic
Per 3h22'15" 38d3'15"
Joseph Hester
Sgr 18h45'20" -21d29'23"
Joseph Horvat - Shine over us always
Aqr 21h46'36" 1d0'4"
Joseph Hunter
Ori 4h49'16" 1d12'50"
Joseph Hutnik Jr.
Cep 20h50'22" 59d27'24"
Joseph I. Ciacco
Cep 22h31'31" 74d43'51"
Joseph Iacopella
Psc 23h50'40" 5d26'37"
Joseph Ikaika Wai'ole Kali
Ari 2h20'58" 11d44'6"
Joseph Introna
Aur 5h23'48" 30d4'14"
Joseph & Iris
Cnc 8h35'38" 9d48'27"
Joseph is so sexy, a star was born.
Her 18h7'35" 22d39'50"
Joseph Isadore Ruks
Gem 7h35'0" 27d26'1"
Joseph & Ivan - Sent with Love, Bev
Uma 11h29'34" 64d10'36"
Joseph J. Alioto
Her 16h20'47" 17d59'46"
Joseph J. Cozzone Sr.
Lyn 8h3'24" 49d16'1"
Joseph J. Farina
Lyn 7h40'27" 41d29'8"
Joseph J Fenty, Jr
Her 18h43'48" 15d35'7"
Joseph J. Hudome
Per 2h33'52" 56d11'0"
Joseph J. Jados, Jr.
Uma 11h4'35" 61d42'5"
Joseph J. Martin
And 0h59'29" 44d12'44"
Joseph J. Merlino Sr.
Ori 5h49'49" -0d19'47"
Joseph J. Sapronetti - has my heart
Ori 5h43'41" 6d32'36"
Joseph J. Sisca, Jr. 20 Yrs HVB BDB
Cep 21h47'54" 64d30'45"
Joseph J. Sochoka
Ori 5h1'16" 9d58'56"
Joseph J. Ventura Jr.
Her 18h35'37" 15d25'39"
Joseph Jake Parsons - 'Baba Jo'
Umi 15h43'22" 74d48'14"
Joseph James
Leo 10h19'21" 23d0'17"
Joseph James Aceta
Sco 16h14'1" -11d48'29"
Joseph James Jenkins
Uma 11h21'26" 44d15'47"
Joseph James Jr.
Uma 9h32'53" 72d45'49"
Joseph James Lee
Per 3h12'5" 47d47'46"
Joseph James Limbrick
Her 18h47'24" 20d57'9"
Joseph James Parker, III
Psc 23h34'50" 0d57'9"
Joseph James Rasmussen
Cep 21h22'39" 64d49'53"
Joseph James Thompson
Her 16h35'44" 29d20'24"
Joseph James Walden
Tau 5h36'45" 22d49'34"
Joseph & Jamie
Cnc 9h17'58" 17d0'54"
Joseph & Janice Correia
Her 17h33'45" 43d21'14"
Joseph Jayden Dafoe
Tau 5h37'43" 25d43'31"
Joseph Jefferson Hennesy, IV
Per 3h4'10" 55d43'18"
Joseph Jeffrey Saltzgaber
Leo 9h38'40" 10d38'9"
Joseph Jerome Traylor
Lyn 7h51'21" 54d12'54"
Joseph "JoeP" Pajot
Cep 22h33'50" 58d5'29"
Joseph "Joey" Marquez
Uma 10h6'4" 45d31'41"
Joseph "Joey" W. Pascale
Uma 8h53'27" 62d51'37"
Joseph John casella
Cyg 20h14'53" 59d42'15"
Joseph John Chopski
Pyx 8h56'24" -32d33'45"
Joseph John Curley
Umi 14h35'18" 69d55'18"
Joseph John DeAngelo
Ori 6h20'43" 6d16'41"
Joseph John Hinton
Sgr 19h31'29" -37d51'16"
Joseph John (Joe) CHOP-SKI
Lib 15h22'30" -28d21'21"
Joseph John Kucic
Lyn 8h9'23" 56d53'51"
Joseph John Parkes
Mon 6h42'49" 1d21'0"
Joseph John Pereira
Leo 11h29'23" 22d58'46"
Joseph John Ragusa
Per 3h23'15" 45d28'32"
Joseph John Rodriguez
Leo 10h10'15" 24d52'13"
Joseph John Tromp
Uma 12h3'32" 51d18'54"
Joseph John Willman (Dad)
Ari 3h17'44" 19d36'31"
Joseph John Zewiski
Aqr 23h6'39" -10d15'48"
Joseph Jonathan Sherman
Aql 20h2'5" 10d33'56"
Joseph Jordan
Aur 7h11'48" 42d2'2"
Joseph "Josh" Super
Sco 16h13'34" -16d38'55"
Joseph JP Spitz
Cnc 8h25'30" 9d6'6"
Joseph Jr. Albert Dippong
Leo 10h30'38" 9d1'20"
Joseph Jude Love
Uma 11h35'52" 35d55'48"
Joseph Jude Peter Parr
Pho 23h52'51" -46d0'52"
Joseph Julio Rodio, Jr.
Vir 12h14'33" 6d52'23"
Joseph K. Miller
Her 18h37'13" 18d52'55"
Joseph Kaden Halloran
Vir 13h4'42" 12d49'7"
Joseph Kalvin
Per 2h40'58" 54d3'52"
Joseph Kamionka
Lyn 7h40'16" 35d46'58"
Joseph & Karen Cathey
Cyg 20h1'34" 52d7'55"
Joseph Karl Baranowski
And 2h28'46" 45d58'14"
Joseph & Kathleen LaLonde
Per 4h26'45" 46d25'30"
Joseph Kehr
Cyg 21h12'40" 44d51'33"
Joseph Kelley
Lup 14h45'15" -51d17'41"
Joseph Kellogg
Vir 12h51'43" 0d30'51"
Joseph & Kendall
Psc 23h1'6" 7d48'1"
Joseph Kenneth Medrano
Her 17h42'0" 27d8'17"
Joseph " Keo" Feldman
Aqr 22h1'47" -4d37'54"
Joseph Kevin Kline
Psc 1h24'34" 17d57'1"
Joseph Kopera
Per 3h6'27" 42d13'4"
Joseph Korte Rosenberg
Cnc 8h26'30" 10d33'19"
Joseph Kratlian
Aql 19h50'16" 13d32'9"
Joseph Krauthamer & Robert Krauthamer
Aql 18h59'45" 5d49'19"
Joseph Kronenwetter
Umi 13h46'42" 75d23'22"
Joseph L. Cofer
Psc 1h18'55" 16d53'30"
Joseph L Donati
Dra 17h26'54" 51d28'40"
Joseph L. Feniak
Ari 2h17'28" 24d34'32"
Joseph L. Mastin
Ori 5h35'3" 3d0'32"
Joseph L. Molis II
Aqr 21h35'54" 1d30'9"
Joseph L. Stefanoni
Ori 5h53'8" -8d19'15"
Joseph L. Viola
Her 17h44'32" 47d58'47"
Joseph la Mia Stella Brillante
Sgr 18h47'59" -28d42'29"
Joseph Labonte
Leo 9h28'37" 30d31'41"
Joseph Lantin
Her 16h31'25" 39d2'51"
Joseph Larragy
Tau 4h34'25" 28d0'34"
Joseph Lawrence
Her 17h34'3" 48d44'16"
Joseph Lawrence Bolton's Star
And 0h11'51" 45d48'8"
Joseph Lawrence Yott
And 0h27'34" 38d46'35"
Joseph Lee Cobb's 5th Anniversary
Uma 11h4'32" 70d9'35"
Joseph Lee DeSimini
Cnc 9h13'26" 29d33'18"
Joseph Lee Devon
And 23h47'14" 47d21'54"
Joseph Lee Thompson
Aur 5h24'16" 44d20'58"
Joseph Leighton Nase
Psc 0h0'14" 0d13'12"
Joseph Lemm
Leo 11h2'9" 9d29'38"
Joseph Leo
Sco 16h5'19" -12d10'19"
Joseph Lewis Clarke
Umi 13h22'28" 71d3'59"
Joseph Lightner Hutchison
Sco 17h51'14" -35d52'4"
Joseph & Linda Waggoner
Lyr 18h35'8" 36d36'56"
Joseph & Lisa Anne Naame
Umi 15h35'52" 73d0'9"
Joseph Lloyd Gavin
Per 2h11'18" 55d54'38"
Joseph Lloyd Lyons III
Sco 16h26'34" -26d57'59"
Joseph & Louisa Guzzo
Del 20h38'22" 15d45'46"
Joseph Loza
Lib 15h22'9" -5d32'27"
Joseph Lubas
Ari 2h54'23" 27d47'39"
Joseph Luc
Aqr 21h8'29" -12d22'31"
Joseph "LuLu"
Her 18h39'28" 15d25'59"
Joseph Lutz
Cnc 7h55'31" 13d9'35"
Joseph Lynn Jackson
Psc 1h4'18" 18d4'11"
Joseph Lynn Jackson
Psc 0h27'25" 6d39'38"
Joseph M. Demblowski
Aur 5h45'0" 54d59'51"
Joseph M. Dolbee
Vir 12h21'58" -8d41'52"
Joseph M. Filosa
Psc 0h49'1" 17d7'50"
Joseph M. Fung
Ori 5h50'31" 22d39'15"
Joseph M. Novosedliak
Aur 5h53'53" 52d36'20"
Joseph M Sheeler
Cnc 9h12'18" 17d10'45"
Joseph M. Wyban Jr.
Umi 15h33'24" 71d34'47"
Joseph Magarian
Sgr 19h7'13" -13d44'49"
Joseph Malec
Cyg 19h59'34" 33d21'42"
Joseph Mannello
Boo 14h43'50" 23d15'20"
Joseph Mantegna
Per 4h21'27" 47d52'44"
Joseph & Maraid McEnery
Crb 16h6'27" 30d44'28"
Joseph & Margaret Pisuzko
Uma 10h48'25" 39d58'29"
Joseph Mariano DeVito III
Sco 17h42'5" -37d58'43"
Joseph Mario Basciano
Ori 5h35'43" -5d36'27"
Joseph Marion Kahn
Sco 16h10'49" -17d9'34"
Joseph Mark Trohman
Vir 13h7'2" 11d39'18"
Joseph Martarano
Leo 11h21'57" 1d51'50"
Joseph Martin Citrin
Cep 21h25'26" 64d44'31"
Joseph Martin Murphy
Her 17h48'58" 48d37'55"
Joseph Martin Waller
Per 2h59'35" 41d24'16"
Joseph & Mary Amodei
Cyg 21h12'16" 30d19'12"
Joseph & Mary Spiteri
Uma 11h46'57" 50d42'9"
Joseph & Mary-Margaret DelliCarpini
Ori 6h11'28" 18d54'42"
Joseph Mastanduno
Cyg 21h11'13" 45d36'41"
Joseph Matthew Bentley
Lib 15h27'20" -9d54'30"
Joseph Matthew Gilbert
Her 17h30'32" 30d21'24"
Joseph Matthew Krimm
Sco 16h7'37" -14d14'27"
Joseph Matthew Thomas Rodetsky
Aqr 22h57'51" -9d52'18"
Joseph McCarrol, Jr
Sgr 18h33'26" -22d18'38"
Joseph McElroy
Per 4h42'35" 46d26'9"
Joseph Megna
Ori 5h39'29" -0d31'42"
Joseph & Mellisha Moore
And 23h16'18" 43d52'53"
Joseph Melvin Parrette
Uma 9h52'21" 69d25'36"
Joseph Mercado, Sr.
Tau 5h11'30" 20d9'34"
Joseph Merlo
Boo 14h35'38" 34d22'51"
Joseph Mettee
Her 16h56'5" 15d38'14"
Joseph Michael
Lac 22h22'48" 53d1'39"
Joseph Michael
Sco 17h56'14" -31d41'43"
Joseph Michael Ales III
And 0h45'55" 38d59'53"
Joseph Michael Archer
Uma 10h31'40" 44d59'46"
Joseph Michael Arfre
Ari 3h7'13" 27d59'23"
Joseph Michael Bauer
Uma 11h2'15" 58d49'0"
Joseph Michael Beatty
Uma 12h4'32" 35d16'43"
Joseph Michael Bilodeau $.99
Ori 5h14'16" 7d11'26"
Joseph Michael Blaine
Sgr 18h51'17" -18d16'24"
Joseph Michael Chicoski
Gem 7h36'36" 26d55'21"
Joseph Michael DiAdamo
Psc 0h54'4" 31d36'40"
Joseph Michael Dwyer
Hor 4h7'4" -40d45'16"
Joseph Michael Fasnacht
Aur 6h14'48" 47d29'22"
Joseph Michael Finnegan
Aur 5h44'51" 49d22'24"
Joseph Michael Hightower
Cep 23h41'39" 72d38'27"
Joseph Michael Hogan
Per 2h44'3" 54d16'10"
Joseph Michael Janas
Umi 15h26'43" 71d43'29"
Joseph Michael Kane
Cnc 8h48'37" 14d54'24"
Joseph Michael Malthe
Aqr 22h44'6" -9d22'7"
Joseph Michael Matatyaou
Her 16h49'0" 25d20'56"
Joseph Michael Moore
Her 16h50'12" 28d0'23"
Joseph Michael Negri
Cnc 8h7'36" 20d41'21"
Joseph Michael Padula
Cep 3h56'2" 80d34'33"
Joseph Michael Palaggi
Ori 5h36'18" 4d33'31"
Joseph Michael Payne
Aqr 23h18'15" -10d36'7"
Joseph Michael Pelo
Her 17h38'55" 21d12'35"
Joseph Michael Pisa, Jr.
Aur 6h4'44" 36d7'26"
Joseph Michael Reyes...Joey
Aqr 22h45'17" 1d1'49"
Joseph Michael Rhoda
Leo 9h36'43" 27d24'50"
Joseph Michael Riley
Aur 5h46'27" 41d55'21"
Joseph Michael Rominski
Per 2h55'12" 53d52'58"
Joseph Michael Schwartz, Jr.
Leo 11h18'1" 24d41'21"
Joseph Michael Suhayda
Cyg 20h26'2" 51d5'14"
Joseph Michael Tremarco
Uma 12h34'55" 52d44'11"
Joseph Michael White III
Cnc 8h35'36" 15d47'42"
Joseph Michael Wilson
Uma 11h12'29" 33d43'56"
Joseph Michael Wood's Star
Lmi 11h5'4" 29d19'30"
Joseph Michalski
Cap 20h9'44" -26d8'52"
Joseph Micheal Santos
Per 3h26'17" 44d49'19"
Joseph "Mickey" Mantz
Psc 1h27'23" 29d31'18"
Joseph Milton Newell
Umi 14h51'30" 77d24'48"
Joseph Minervino
Cep 22h56'16" 57d30'42"

Joseph Mitchell Gault Ori 6h12'23" 1d19'9"
joseph monroe burton sr Uma 9h12'55" 59d2'28"
Joseph Montana: My heart and soul Uma 12h27'22" 53d10'35"
Joseph Moretti Cecchi Her 17h10'59" 38d5'49"
Joseph Morgan Hendrickson Sco 17h57'50" -30d55'18"
Joseph Motto Lib 15h18'26" -5d36'7"
Joseph Mulroy Gem 7h10'50" 34d21'14"
Joseph Murray Aql 19h28'8" -0d41'46"
Joseph N. Barbati ~ October 2, 1963 Lib 14h49'13" -3d6'13"
Joseph N. Gleba Per 2h52'10" 40d12'6"
Joseph N. Impastato III Leo 9h48'14" 7d9'9"
Joseph N. Schmitt Psc 1h10'14" 31d17'35"
Joseph Naime of Maximum Wizdom Cae 4h51'5" -28d44'54"
Joseph Nathan Bello, Jr. Cnv 12h45'2" 40d1'58"
Joseph Nathanial Conway Per 4h49'45" 41d44'25"
Joseph Negler Aur 5h46'5" 48d57'31"
Joseph Neil Anderson Lib 16h0'24" -15d16'30"
Joseph Neil Smith Tau 4h34'58" 19d46'59"
Joseph Nguyen Cap 20h17'20" -11d10'42"
Joseph Nicholas Ori 5h36'5" 5d40'16"
Joseph Nnaemeka Onwuta Uma 11h31'2" 58d25'11"
Joseph Nucci Her 17h50'39" 28d12'55"
Joseph O. Stumpf Ori 6h25'33" 17d9'22"
Joseph Odierna " April 3, 1940" Ori 6h3'52" 19d47'15"
Joseph "Ollie" de Matos - 04.04.1960 Ori 5h21'23" -2d1'42"
Joseph "Ollie" de Matos - 04.04.1960 Cru 12h22'30" -63d23'21"
Joseph Onorati Her 17h12'7" 34d4'50"
Joseph Otto Behnke Per 3h18'7" 53d33'23"
Joseph P. Carmody Aqr 21h36'44" 0d12'59"
Joseph P. Gavron Vir 13h13'18" 3d33'46"
Joseph P. Greelish, III Cyg 19h38'3" 29d48'48"
Joseph P. Imbriani, Jr. Lib 14h44'38" -17d4'17"
Joseph P. Kelton Lib 14h48'24" -9d18'57"
Joseph P. Kirkner, IV Her 17h55'37" 22d13'51"
Joseph P. O'Hara Ori 5h53'27" -2d44'40"
Joseph P. O'neill Per 3h59'39" 49d19'12"
Joseph P. Puppy David Cnc 8h11'45" 27d8'33"
Joseph P. Taulbee Aur 5h5'25" 47d0'1"
Joseph P. Urso Sco 16h18'29" -10d32'4"
Joseph Panattieri Lyn 8h30'27" 34d15'39"
Joseph Papasergio Gem 7h44'57" 33d14'11"
Joseph "Pappy" Homsy Her 18h1'42" 14d23'4"
Joseph Parisi Aur 5h48'32" 50d59'29"
Joseph Pascal Bedard Per 3h49'29" 33d34'51"
Joseph Pasquale Deon Umi 14h39'39" 67d59'48"
Joseph Pastor Ari 2h18'2" 23d58'32"
Joseph & Patricia Psc 0h55'51" 17d7'7"
Joseph & Patricia Cornetta's 50th Cyg 21h12'10" 42d14'32"
Joseph Patrick Gem 7h42'27" 28d31'53"
Joseph Patrick Edward McGrath Leo 11h26'43" 11d11'3"
Joseph Patrick Mathews Her 18h10'32" 16d25'38"
Joseph Patrick Serdar Uma 9h54'28" 63d46'45"
Joseph Paul Tau 4h31'1" 30d0'30"
Joseph Paul Bailey Sco 16h5'28" -17d27'0"
Joseph Paul Carroll II Tau 4h34'11" 28d43'56"
Joseph Paul Conti Leo 11h22'42" 12d48'11"
Joseph Paul Ehnen Her 16h6'16" 47d46'49"
Joseph Paul Faderl 2:05a.m. 7lb.7oz Ari 1h59'23" 20d12'53"
Joseph Paul Leddy Aqr 21h10'17" 0d37'23"
Joseph Paul Nixon Cen 13h40'54" -40d40'43"
Joseph Paul Pellicane Tau 5h43'37" 25d45'35"
Joseph Paul Sanderson Jr. Aql 19h13'1" 6d18'10"
Joseph Paul Smith Uma 14h19'23" 58d56'52"
Joseph Paul Tomlinson Umi 13h31'33" 70d2'50"
Joseph Paul Veraldi, III Cnc 8h19'36" 28d22'59"
Joseph Paul William Varley Uma 8h51'34" 47d5'19"
Joseph Pehar Her 16h34'18" 36d8'0"
Joseph Pemmaraju Dakin Her 17h12'16" 35d25'48"
Joseph Pennacchio Uma 10h25'23" 49d3'49"
Joseph Peplinski Umi 14h40'38" 73d38'1"
Joseph Perez Her 17h46'24" 39d16'42"
Joseph Perrone Uma 10h1'12" 53d48'51"
Joseph Peter Affinito Uma 9h27'30" 71d13'45"
Joseph Peter McMahon Per 2h21'47" 55d43'25"
Joseph Philip John Ori 6h10'55" 9d25'1"
Joseph Phillips Leo 9h25'37" 29d9'13"
Joseph Phillips Zahornacky Cap 20h18'13" -10d48'58"
Joseph Pinder Cep 23h2'33" 87d2'29"
Joseph Pischel Lib 15h32'39" -26d21'44"
Joseph Pontoriero Sco 17h29'9" -35d29'37"
Joseph "Pop-Pop" Orfetel Ori 5h24'7" -8d25'35"
Joseph Porter-Mackrell Her 16h11'54" 3d46'22"
Joseph Poulin (honey bun) Umi 4h36'57" 88d44'40"
Joseph Prescott Lowry Her 17h49'59" 39d15'57"
Joseph Puzzo Per 3h20'22" 51d40'57"
Joseph R. Albi Jr. Lib 14h54'28" -2d43'23"
Joseph R. Cress Uma 11h35'21" 55d52'14"
Joseph R. Dupre, M.D. Her 18h21'14" 23d3'20"
Joseph R. Mascio Her 18h31'58" 17d30'22"
Joseph R. Mounts Uma 11h45'29" 49d34'35"
Joseph R. Otto Tau 5h51'4" 17d16'58"
Joseph R. Ourada Her 17h40'33" 36d28'30"
Joseph R. Sadowski Cep 22h36'51" 73d58'26"
Joseph R. Spisak "Joey" Uma 11h7'32" 50d40'37"
Joseph R. Vaughan, Jr. Aqr 21h46'18" -2d33'55"
Joseph R. Wheeler Psc 1h13'37" 28d42'53"
Joseph R. Zuppardo Peg 22h20'59" 12d6'18"
Joseph Raccuia, M.D. Aqr 23h19'48" -18d41'53"
Joseph & Rachel Decosimo Uma 9h0'28" 47d23'14"
Joseph Rafaniello Aql 19h40'6" 11d0'33"
Joseph Rand Uma 11h22'18" 61d13'24"
Joseph Randall Olsen Ari 2h33'5" 12d1'50"
Joseph Raso Ori 6h4'23" 11d5'20"
Joseph Ray Ganzermiller Vir 12h51'32" 12d49'56"
Joseph Ray Kiesel Cnv 14h2'50" 38d10'56"
Joseph Ray King Vir 12h51'6" 5d33'11"
Joseph Raymond Jones Sgr 19h40'53" -16d18'4"
Joseph Raymond Michalski Uma 11h17'36" 47d30'7"
Joseph Raymond Pedersen Lyn 8h46'7" 36d24'43"
Joseph "Rayo" Scollo Uma 11h29'17" 45d56'16"
Joseph Read Joyce, Jr. Cyg 20h10'26" 56d59'2"
Joseph Read Pritchard Per 4h27'4" 43d31'36"
Joseph Reburn Sgr 18h11'22" -20d31'51"
Joseph Reginald Williams Cas 1h25'28" 67d16'52"
Joseph Reihel Ori 6h2'45" 14d5'0"
Joseph Renner Uma 10h6'10" 68d51'54"
Joseph Rhoda Sco 17h9'1" -43d48'31"
Joseph Ribeiro Uma 13h49'3" 54d10'58"
Joseph Richard Ferrell Pho 0h44'40" -41d12'10"
Joseph Richard Latch Uma 11h40'59" 60d3'35"
Joseph Richard Lumadue Cnc 8h36'35" 31d47'35"
Joseph Richard Pozzuoli Her 16h18'33" 20d5'27"
Joseph Richard Sacks Cyg 20h8'6" 46d5'44"
Joseph Richard Stickland Vir 13h14'33" 5d34'50"
Joseph Richards Boo 14h44'26" 36d36'0"
Joseph Richardson Pegler Uma 11h8'29" 50d0'21"
Joseph Roark Aql 19h17'15" -8d12'33"
Joseph Robert Bradley Sco 17h29'6" -33d28'40"
Joseph Robert Caruso Cyg 19h59'0" 39d27'45"
Joseph Robert Cutinella And 23h54'39" 39d28'59"
Joseph Robert Grillo aka Just Joe Psc 1h24'36" 15d7'7"
Joseph Robert Johnson Cap 21h35'19" -15d25'59"
Joseph Robert Pallardy Cep 20h10'16" 60d9'53"
Joseph Robert Pappenfus Uma 12h50'50" 60d8'6"
Joseph Robert Reeder DeGrange Cyg 21h29'45" 39d42'43"
Joseph Rocco Biondo Vir 12h18'56" 12d31'16"
Joseph Rocco Danubio Cep 22h59'17" 71d2'27"
Joseph Roderick Chidester And 2h20'55" 46d0'48"
Joseph Rodriguez Valdez Uma 13h27'34" 56d15'28"
Joseph Romeo Vachon Sgr 18h5'35" -18d49'32"
Joseph Ronald Butler Her 17h49'17" 27d35'12"
Joseph Rosario & Clara Walker Cyg 20h0'18" 53d41'29"
Joseph & Rose Keller Cnc 7h56'34" 19d10'46"
Joseph & Rose Lumia Uma 8h45'7" 59d4'3"
JOSEPH & ROSEMARIE PINTO Uma 11h52'35" 60d26'34"
Joseph Ross Vickery, Jr. Uma 14h2'50" 54d22'10"
Joseph Royal Earle Gem 7h51'12" 17d41'35"
Joseph Rucci Sr. Lyn 7h31'56" 38d40'49"
Joseph Russo Peg 23h54'3" 27d29'30"
Joseph Ryan Aqr 21h52'47" -8d10'34"
Joseph Ryan Millward Dra 17h50'33" 64d22'10"
Joseph Ryon Gross Uma 9h39'46" 50d21'22"
Joseph S. Colonna Superstar Daddy Cnc 8h45'25" 13d37'35"
Joseph S. Middleton Psc 1h21'27" 5d59'55"
Joseph S. Nolasco Ari 1h58'14" 20d54'42"
Joseph S. Rosenberg Col 5h43'4" -32d7'49"
Joseph S. Sled "Our Shining Star" Tau 4h24'34" 0d35'47"
Joseph S. Zalegowski Tau 4h34'41" 19d39'23"
Joseph "Salami" Noguera Umi 15h6'2" 76d20'21"
Joseph Salee Aur 5h44'44" 41d41'50"
Joseph Salvatore Chiaramonte Gem 6h32'21" 25d50'33"
Joseph Samuel Gordon Uma 9h14'58" 57d16'54"
Joseph Samuel Potts And 0h44'16" 36d42'12"
Joseph Saverio Serrano, Jr. Tau 5h14'19" 28d0'53"
Joseph Scales Uma 13h56'21" 55d17'37"
Joseph Schieberl Umi 15h20'8" 74d36'31"
Joseph Schmidt Per 3h47'56" 43d47'59"
Joseph Schulman Her 17h35'5" 44d19'52"
Joseph Sciarra IV Tau 4h39'7" 22d58'32"
Joseph Scott Conklin Cyg 20h46'58" 44d17'46"
Joseph Scott McPike Vir 12h37'11" -7d19'57"
Joseph Scott "Scotty" Smith Umi 15h25'56" 71d15'55"
Joseph Scrivani Vir 14h5'52" -9d19'27"
Joseph Seikali Lib 15h41'26" -17d50'1"
Joseph Shannons 3 Cnc 8h11'19" 25d5'32"
Joseph Sharp Umi 15h14'4" 76d44'44"
Joseph Shawn Chizanskos Ori 4h52'5" 5d11'19"
Joseph & Sheila Always & Forever Cyg 20h21'15" 43d35'16"
Joseph Sime Vandervalk Per 3h35'1" 35d4'59"
Joseph Simpson Cru 11h58'47" -63d4'16"
Joseph Snyder Sgr 18h11'52" -18d49'54"
Joseph Souto Amado, Jr. Sco 17h16'41" -31d57'15"
Joseph Spennato Leo 9h22'38" 15d50'32"
Joseph Spike Mackenzie Umi 14h38'52" 71d0'3"
Joseph Sr. & Irene Catallo Uma 13h18'5" 54d48'41"
Joseph Stanley Parker Cam 4h44'31" 59d28'58"
Joseph Stanzione Cas 0h5'48" 53d0'20"
Joseph + Stephanie Uma 10h18'23" 44d33'15"
Joseph Steven Fanelli Her 18h54'43" 25d25'2"
Joseph Steven Jaggers Aqr 22h42'33" -23d21'24"
Joseph Stogner Her 18h6'21" 34d43'5"
Joseph Story Per 3h13'30" 52d27'19"
Joseph Stuart Cullip Umi 15h41'20" 67d37'56"
Joseph Sullivan Uma 11h59'26" 59d25'7"
Joseph Sung Canlas Cen 13h45'2" -46d4'22"
Joseph Sydney Johnston Her 18h37'11" 15d42'42"
Joseph Sydney Johnston, Jr. Ori 5h32'27" 3d17'16"
Joseph T. Albrizio, Jr. Vir 12h17'20" -0d55'55"
Joseph T. Baroz Cep 23h16'48" 75d45'52"
Joseph T Brown Ari 2h48'10" 25d35'6"
Joseph T. Dixon Tau 5h52'33" 14d44'12"
Joseph T Hill Lup 15h41'53" -32d38'13"
Joseph T. Larsen Cas 0h16'17" 63d12'40"
Joseph T. Loweree Boo 14h39'34" 33d59'21"
Joseph T. McNamara Aur 6h32'7" 41d20'20"
Joseph T. Trentacosta Jr. Uma 9h13'47" 56d9'22"
JOSEPH TAAFE Boo 14h43'4" 43d56'55"
Joseph Tabasco Leo 11h47'16" 24d26'26"
Joseph & Tamala Correia Lyr 19h9'46" 26d36'26"
Joseph Tennant Per 3h21'18" 39d43'36"
Joseph Terrence Loney Per 2h50'51" 44d56'13"
Joseph Thaddeus Nagy Uma 11h17'18" 53d25'56"
Joseph Thalimer Umi 14h23'39" 75d24'15"
Joseph & Theresa Ciecierski Family Cyg 20h5'9" 45d49'30"
Joseph Thomas Ciminelli Cyg 21h30'18" 31d41'44"
Joseph Thomas Giannotti Uma 8h34'16" 49d15'45"
Joseph Thomas Haroutunian Leo 11h56'35" 26d2'54"
Joseph Thomas Hawthorne Her 17h50'26" 21d35'40"
Joseph Thomas Maritato Uma 10h42'4" 49d34'20"
Joseph Thomas Monello, III Leo 9h58'15" 22d45'6"
Joseph Thomas Perdue Lyn 6h36'52" 56d21'7"
Joseph Thomas Phlipot Her 18h2'46" 26d14'12"
Joseph Thomas Schena Vir 12h32'20" 6d8'34"
Joseph Thomas Stone Per 4h12'16" 34d15'34"
Joseph Thomas Suty Her 18h35'58" 21d47'5"
Joseph Thomas Sylcox Per 4h5'34" 44d21'34"
Joseph Thomas Vatterott Uma 8h44'15" 56d37'0"
Joseph Thomas Visslailli II Uma 9h57'55" 67d28'23"
Joseph Thomas Weddeke Aur 6h30'47" 35d36'5"
Joseph Thomas Wilson Per 2h38'46" 51d48'33"
Joseph Thomas Worthy Her 16h36'7" 8d6'22"
Joseph Thomson Per 4h24'10" 41d5'24"
Joseph Tim Byrne Her 16h22'27" 6d21'32"
Joseph Timothy Monahan Uma 9h53'56" 69d19'22"
Joseph Toner Per 2h45'48" 53d29'11"
Joseph Troman Per 4h12'7" 43d2'56"
Joseph Tyler Macy Per 3h18'34" 51d58'37"
Joseph Tyler Podlas Sco 17h32'52" -38d50'38"
Joseph V. Barba Lib 15h26'35" -14d10'18"
Joseph V. Dugoni Cyg 21h40'22" 40d27'31"
Joseph Valentine Hainey Uma 13h41'9" 55d25'53"
Joseph Valento Uma 8h20'31" 61d16'9"
Joseph Valerie Leger Uma 14h16'24" 56d3'22"
Joseph van der Laan Aql 19h40'53" -0d25'3"
Joseph Vervaet Gem 6h47'40" 33d52'41"
Joseph Vierra Amaro Ori 5h28'47" 7d23'0"
Joseph Vincent James Incorvaia Sgr 18h7'36" -31d36'27"
Joseph Vincent Pici Uma 12h15'54" 55d13'38"
Joseph Vitale Lyn 7h26'11" 53d13'49"
Joseph W. Anderson Lmi 9h41'20" 38d22'38"
Joseph W Andruskiewicz Uma 11h14'0" 52d6'0"
Joseph W. Bails Aur 5h28'46" 40d57'56"
Joseph W. Bixby Umi 15h32'32" 71d9'44"
Joseph W. Hutcheson Sco 16h14'44" -11d3'1"
Joseph W. Majewski Aur 5h50'20" 54d30'17"
Joseph W Marty II Lyn 7h2'56" 56d30'12"
Joseph W. Orsini Tau 5h54'28" 28d17'0"
Joseph W. (Poppie) Sabol Leo 9h42'5" 26d41'14"
Joseph W. Posker Psc 23h22'29" 6d11'41"
Joseph W. Price, II Cep 22h12'47" 60d54'18"
Joseph W Ryan Her 17h57'27" 21d53'52"
Joseph W. Ryan Ori 6h1'33" 17d33'31"
Joseph W Testa Cep 22h11'20" 61d0'3"
Joseph W. Zullo Her 16h34'28" 16d27'3"
Joseph Walter Kempka Mon 6h54'51" -0d11'31"
Joseph Warren Uma 12h2'4" 53d23'54"
Joseph Wedow Aqr 23h21'49" -19d41'14"
Joseph Weisen Uma 11h45'38" 61d22'54"
Joseph West Talbert Mon 6h34'18" 10d25'55"
Joseph Wilkes Eagar Uma 11h25'6" 53d10'40"
Joseph William Aaron Kearney Umi 16h4'8" 73d51'47"
Joseph William Arenth Lib 15h12'11" -17d12'14"
Joseph William Bailey Uma 11h38'16" 34d52'59"
Joseph William Bastek Vir 13h7'2" 7d39'48"
Joseph William Espinola Sr. Per 3h49'53" 41d58'13"
Joseph William Espinosa Her 16h10'3" 47d14'2"
Joseph William Klunder Sco 16h9'16" -12d59'56"
Joseph William Madsen Ori 6h8'17" 20d45'15"
Joseph William Martuscello, Jr. Uma 11h10'14" 34d10'37"
Joseph William McLaughlin Ari 1h52'41" 19d46'50"
Joseph William Moranto Lyn 7h45'23" 36d19'25"
Joseph William Namath Her 17h20'32" 24d58'58"
Joseph William Ramage Per 4h17'22" 45d18'34"
Joseph William Robinson 10 25 1932 Boo 14h49'38" 36d31'57"
Joseph William Schwartz Umi 16h14'27" 75d46'14"
Joseph Winshman Cap 20h13'8" -22d57'46"
Joseph Wissman Uma 11h45'38" 43d54'6"
Joseph Woodrow Anderson Vir 13h49'59" -7d41'36"
joseph x spellman Lyn 7h43'29" 45d20'17"
Joseph Zaidler Per 4h10'14" 34d5'49"
Josepha Gawron Aur 5h41'20" 39d56'1"
Josepha Schiffman Leo 11h37'58" 11d26'50"
Josephina Magdalena Aleida de Jong And 23h40'27" 38d22'20"
Josephina Maria Olagues Sco 16h35'47" -30d34'0"
Josephine Col 5h42'4" -33d3'19"
Josephine Sco 17h18'55" -32d5'2"
Josephine Sco 17h54'6" -38d26'33"
Josephine Vir 13h52'38" -7d41'16"
Josephine Aqr 21h13'49" -8d42'19"
Josephine Cam 4h51'15" 63d3'46"
Josephine Cas 2h16'29" 64d10'50"
Josephine And 1h20'10" 48d9'21"
Josephine Cam 3h39'24" 53d5'28"
Josephine And 23h51'32" 48d0'46"
Josephine Uma 9h20'22" 44d26'25"
Josephine Peg 22h1'16" 35d16'30"
Josephine Psc 1h24'31" 32d9'4"
Josephine Vir 13h14'29" 11d19'14"
Josephine Leo 11h20'32" 15d50'35"
Josephine 1962 Per 4h21'8" 43d38'1"
Josephine A. Truzzolino Ari 3h3'50" 19d28'45"
Josephine Acosta And 0h27'15" 36d2'48"
Josephine Adelle Stringer Leo 11h25'35" 0d18'20"
Josephine Andrea Ori 5h54'48" 18d7'48"
Josephine Anita Cas 1h42'27" 61d21'47"
Josephine Ann Pilate Lyn 6h41'53" 57d32'4"
Josephine Anna DiLeonardo Uma 8h39'44" 49d25'3"
Josephine "Aunt Jo" Shelman Uma 11h55'42" 61d57'49"
Josephine Bubala Lyn 7h32'48" 58d43'21"
Josephine Campese Uma 9h41'36" 45d36'38"
Josephine 'Chi-Chi' Mosier Ari 2h35'12" 14d41'58"
Josephine Chlöe Louise Rogerson Cas 1h2'33" 49d24'59"
Josephine Corbin Everett Cas 1h47'36" 60d31'10"
Josephine DeBuino Lynch Rupert Ari 2h6'47" 24d0'19"
Josephine DeFriece Wilson Cnc 8h16'45" 11d37'56"
Josephine Dickson Centennial Star Sco 16h16'36" -13d14'59"
Josephine Eileen Madrid Sco 16h9'32" -10d21'30"
Josephine Elefante Vir 13h29'42" 13d18'14"
Josephine Ellen Kendall Vir 13h20'35" 7d8'31"
Josephine Eve Duckworth Ari 2h11'51" 26d14'24"
Josephine Farrow Psc 1h29'55" 26d52'46"
Josephine Fazio Cyg 19h41'0" 31d34'7"
Josephine Frediani Cas 0h34'11" 63d27'53"
Josephine Gascoigne Per 4h6'56" 47d8'5"
Josephine Gerena Per 4h6'50" 33d54'24"
Josephine Giancana Aqr 21h35'46" 1d52'18"
Josephine Godfrey Beers Cas 0h53'18" 63d37'12"
Josephine Grabowski And 23h27'38" 43d1'56"
Josephine Grace Parker And 2h35'0" 47d37'46"
Josephine Harrison Cas 23h6'36" 54d51'57"
Josephine Hau Gem 7h15'21" 24d17'15"
Josephine Hernandez And 0h15'9" 35d2'16"
Josephine Hope - Josie's Star And 0h2'25" 41d2'21"
Josephine Illingworth-Law And 23h23'14" 52d26'44"
Josephine Jamieson Cap 21h46'34" -13d17'43"
Josephine Joan (Cookie) DeFruscio Sgr 19h23'27" -27d53'40"
Josephine Johnson Gahn Sco 17h12'17" -44d54'55"
Josephine "Jojie" Oliver And 2h16'13" 41d39'2"
Josephine Jordan Mahoney Star Tau 3h35'26" 29d35'33"
Josephine Kayastha Gem 7h28'31" 34d37'5"
Josephine Kenyon Lib 14h27'34" -19d38'2"
Josephine L. Graffeo Cep 22h18'12" 64d55'40"
Josephine Latino And 0h29'59" 43d24'18"
Josephine Letourneau Lyn 8h6'55" 52d48'53"
Josephine Lugara Dor 4h22'43" -49d11'33"
Josephine Lynn- God's Birthday Gift Umi 15h17'43" 68d44'32"
Josephine Lyttle McProud Vir 11h48'17" 5d48'9"
Josephine M. Pustelniak Feeny Crb 15h35'46" 26d29'59"
Josephine M. Raffa Uma 8h23'12" 63d56'0"
Josephine Mae McSpadden Aqr 22h21'13" -24d19'57"
Josephine Marie Aql 19h23'37" -0d27'59"
Josephine Marie Hannan Gem 6h18'55" 27d38'58"
Josephine Mayerick Lyn 8h47'3" 39d6'14"
Josephine Mellor Uma 9h31'38" 69d52'58"
Josephine & Michael Cyg 19h46'29" 56d42'57"
Josephine Monteleone Ori 6h12'51" 6d20'5"
Josephine Mosketti Cam 4h18'59" 55d23'5"
Josephine Nash Dougherty Uma 11h5'29" 37d12'10"
Josephine Nicola-Sosie Pops-17-9-04 Vir 11h38'51" 4d46'4"
Josephine Pacifico Uma 10h42'4" 47d46'26"
Josephine Palmer Lyr 19h10'47" 27d22'5"
Josephine Panzer Sco 17h59'1" -39d18'1"
Josephine Price-Smith Ari 2h44'25" 27d18'57"
Josephine Ramos And 23h8'42" 39d58'35"
Josephine Randazzo "Class of 2006" And 0h33'48" 30d43'3"
Josephine Reynolds Peg 22h30'41" 21d38'22"
Josephine Rose-May And 1h28'44" 33d50'22"
Josephine Ruiz Gonzalez Com 12h29'17" 27d5'0"
Josephine Sellitti Per 3h23'46" 46d55'55"

Josephine Sinclair
And 23h1'3" 37d25'22"
Josephine Sorbera
Her 16h44'47" 46d39'23"
Josephine Suarez
Cam 7h30'53" 63d16'26"
Josephine Tate Duncan
Crb 16h13'45" 34d7'4"
Josephine Tayana Reyes
Psc 0h45'3" 20d36'22"
Josephine Teresa Murdoch
Cas 0h48'38" 56d33'51"
Josephine Tubbs
Uma 8h50'26" 57d34'19"
Josephine V. Myrick
Crb 16h8'17" 35d3'38"
Josephine Vida del Rosario
Leo 10h46'47" 8d5'38"
Josephine Yakobiszyn
Mon 6h46'3" 7d12'2"
Josephine Zann
Cam 5h7'6" 69d59'9"
Josephine-Dianora
Dra 18h56'53" 69d21'23"
Josephine's Shining Star
And 0h15'37" 46d38'50"
JosephRyan
Cnc 9h8'54" 7d13'39"
Joseph's Class Star
Ori 5h48'47" 5d50'53"
Joseph's Love
Sge 20h4'0" 16d56'22"
Joseph's Pride
Dra 16h13'22" 54d46'12"
Joseph's Star
Peg 22h40'45" 16d42'31"
Joseph's Zenith
Umi 4h26'49" 88d26'18"
JOSEPHUS
Umi 15h17'16" 71d19'51"
Josephyne E. Corsi
Sgr 18h10'11" -35d59'55"
Jose's Eternal Pluto
Sco 16h3'53" -8d23'41"
Jose's Guiding Light
Sco 17h18'3" -41d14'15"
Jose's Xterra of Ecstasy
Tau 4h34'31" 30d49'0"
Josette
Crb 16h13'18" 37d41'54"
Josette
And 23h19'51" 50d50'52"
Josette
Uma 9h5'58" 60d45'58"
Josette
Uma 13h35'41" 55d40'25"
Josette Halegoi
Lib 15h2'30" -26d58'53"
Josette Miles
Sgr 19h22'18" -22d43'0"
Josh
Ori 5h54'51" -0d43'51"
Josh
Boo 15h24'47" 44d7'7"
Josh
Uma 11h9'25" 43d0'36"
Josh and Alison Hille
And 2h9'28" 43d0'41"
Josh and Andria
Leo 10h19'44" 12d13'54"
Josh and Anna
Cyg 20h45'12" 37d56'3"
Josh and Anne Forever
Lyr 19h2'51" 32d8'51"
Josh and Bridget's Star
Her 17h35'38" 40d51'41"
Josh and Cindy
Cyg 21h14'48" 44d57'51"
Josh and Clare Zealous
Cyg 21h26'26" 32d44'38"
Josh and Holden McNeel
Per 3h20'29" 39d11'39"
Josh and Kyla Gust
Gem 6h50'18" 32d16'2"
Josh and Leanne forever
Lyn 7h51'50" 38d25'6"
Josh and Linsey Forever
And 1h51'13" 46d5'16"
Josh and Lisa's Star
Cyg 19h37'53" 32d17'41"
Josh and Megan Vaught
Ari 3h2'35" 20d3'15"
Josh and Meghan's First
Anniversary
And 0h33'21" 33d40'3"
Josh and Michele's Star
Her 17h2'48" 17d34'41"
Josh and Nichole's Eternal
Love
Cyg 19h19'9" 46d55'7"
Josh and Nicole's
Ori 6h7'1" 9d43'43"
Josh and Rachel Hyatt
Cas 0h41'56" 58d20'54"
Josh and Samantha
Cyg 21h45'38" 52d51'19"
Josh and Sara Lusk
Forever and Ever
Cyg 20h13'56" 41d2'40"
Josh & Ashley
Leo 10h59'1" 1d59'3"
Josh & Ashley Woods
Uma 10h58'24" 53d34'9"
Josh Baird
Uma 11h33'46" 65d27'29"
Josh Baker GaMPI Shining
Star
Her 16h40'41" 34d54'39"
Josh Benoit
Lmi 10h40'24" 23d51'14"
Josh & Beth Hopkins
Her 16h23'28" 13d34'3"
Josh Billiot
Ori 5h58'30" -0d39'59"
Josh & Brita
Ari 2h46'11" 27d37'17"
Josh Burkhardt
Uma 8h51'3" 54d38'52"
Josh*Buttface
Cep 23h5'27" 71d37'1"
Josh & Carla
Sge 20h1'33" 18d28'55"
Josh Carmean
Aqr 23h55'53" -13d33'44"
Josh Chavers
Umi 14h49'17" 73d28'43"
Josh Christopher
Sco 16h45'34" -41d17'19"
Josh Davis
Ori 5h32'34" 5d44'58"
Josh Dolman
Per 2h26'29" 57d10'17"
Josh Ebrahemi
Uma 11h30'28" 47d20'13"
Josh Edward Sullivan
Umi 14h50'17" 71d8'42"
Josh Engelkemier
Her 17h28'23" 39d0'2"
Josh Ervin
Lib 14h57'57" -1d52'39"
Josh Ferro & Autumn
Ferro-Walters
Cyg 20h55'16" 46d40'27"
Josh Gavin Innes
Umi 13h21'8" 69d34'23"
Josh Glick
Ori 5h37'32" 1d50'41"
Josh Goffena
Leo 10h34'38" 14d10'43"
Josh Hagan
Aur 5h24'0" 39d23'40"
Josh Harrison
Aqr 22h37'19" -19d25'47"
Josh & Jeree's 2R1
Cyg 19h30'28" 53d54'49"
Josh & Jessica Evans
Cyg 20h14'20" 49d7'22"
Josh & Joanna
Cyg 19h23'25" 51d15'12"
Josh & Joyce's Timeless
Love
Lyr 19h12'47" 26d34'1"
Josh & Katia
Cnc 8h39'11" 16d47'42"
Josh & Katie Fenska
Sge 19h42'26" 18d56'39"
Josh Kenneth McGhee
Uma 8h45'16" 68d20'43"
Josh Lewton
Crb 16h19'40" 30d22'4"
Josh & Lisa Houle
Sge 20h3'2" 18d14'37"
Josh Loves Jen
Uma 10h30'13" 66d1'59"
josh maxwell
Aur 5h45'47" 38d57'5"
Josh & Meredith John
Uma 11h24'21" 56d48'50"
Josh Monier
Ori 6h3'41" 13d32'17"
Josh my star in Heaven
Aur 5h16'7" 30d52'28"
Josh n Chelle
Cyg 20h12'12" 32d33'9"
Josh Neckelsky
Boo 14h46'6" 17d59'33"
Josh & Neysa
Ara 17h29'7" -47d37'15"
Josh Nickels
Ari 2h42'3" 25d26'30"
Josh & Nicole 11-27-99
Uma 9h23'4" 63d9'16"
Josh & Nicole Karasow
Cyg 19h35'15" 51d45'36"
Josh Orenstein
Aqr 20h48'54" -8d27'8"
Josh Parker Hart
And 23h11'34" 41d54'10"
Josh Patterson Loves Nikki
Worth
Cnc 9h10'28" 19d6'52"
Josh Pringle & Aleshia
Grubbs 4ever
Vir 12h44'44" 2d29'58"
Josh R.
Aqr 21h48'43" -0d23'31"
Josh Ray
Cep 21h49'23" 74d8'31"
Josh Rubin
Her 17h43'9" 15d16'22"
Josh Rubin
Per 4h14'45" 51d32'37"
Josh Russell
Dra 17h14'8" 59d10'35"
Josh Salcines
Gem 7h22'19" 29d58'13"
Josh Scott Baker
Col 5h31'32" -35d27'3"
Josh & Shelly's
Uma 8h32'44" 65d29'39"
Josh Sloan (Infinity x
Infinity)
Ori 5h36'26" -2d4'30"
Josh Stone
Cam 4h6'12" 71d30'53"
Josh Stowers, I love you
Her 17h17'17" 30d22'3"
Josh & Susan 010202
Cyg 20h37'3" 41d40'12"
Josh Swimmer Star Hengel
Ari 2h8'17" 20d53'23"
Josh Tabak
Sco 16h12'54" -9d40'6"
Josh the Twinkle of
Nannies Eyes
Vir 13h25'5" 9d23'45"
Josh & Toni's
Tumblemomma
Srp 18h6'48" -0d50'3"
Josh & Trisha
Cep 21h30'12" 63d56'8"
Josh Turner
Sgr 18h32'11" -27d4'28"
Josh Unger
Sco 16h8'58" -15d10'37"
Josh Wartchow
Cnc 8h39'36" 16d22'16"
Josh Wheeler / Stinky
Her 18h14'45" 16d19'5"
Josh Wilde
Her 18h14'12" 27d31'3"
Josh, Jamie, Love
Her 16h49'22" 30d4'58"
Joshan Nevzat Ertan
Sgr 18h6'7" -28d0'4"
Joshanna
Sco 17h47'26" -39d54'36"
Joshaua Lee
Cnv 12h24'31" 37d46'25"
Joshea
Peg 22h26'17" 6d24'19"
Joshia Burke
Uma 8h46'22" 72d40'43"
Joshica Major
Uma 9h46'19" 46d19'55"
Joshie
Leo 11h36'21" 21d56'5"
Joshie Jack
Cas 1h22'29" 58d8'48"
joshily
Cnc 9h13'53" 21d48'54"
Joshka Immortalis
Pho 0h35'59" -50d13'4"
Josh's Angel, Emily Brooke
Tau 4h31'0" 12d5'19"
Josh's Flame
Dra 15h17'16" 57d32'34"
Josh's HomeRun
Uma 13h23'18" 61d55'24"
Josh's Princess
And 2h35'58" 45d43'47"
Josh's Smile
Ori 4h53'49" 12d49'24"
Josh's Star
Per 4h20'20" 52d46'25"
Josh's Star
Uma 12h3'59" 51d21'14"
Joshu Lawrence Goebeler
Sco 17h24'30" -42d40'27"
Joshua
Sco 17h14'11" -43d44'48"
Joshua
Cha 11h4'5" -82d58'24"
Joshua
Sgr 17h55'50" -29d37'4"
Joshua
Uma 9h17'49" 69d53'18"
Joshua
Cep 22h57'9" 64d17'32"
Joshua
Cep 0h18'19" 77d58'17"
Joshua
Umi 14h48'54" 67d47'49"
Joshua
Uma 13h55'5" 52d33'54"
Joshua
Sgr 18h26'12" -15d59'30"
Joshua
Vir 13h11'25" -22d25'18"
Joshua
Cap 20h30'17" -10d49'54"
Joshua
Umi 13h42'45" 77d16'25"
Joshua
Boo 15h28'43" 48d52'9"
Joshua
Per 3h16'19" 52d40'29"
Joshua
Per 3h33'10" 49d34'36"
Joshua
Uma 8h34'16" 49d57'50"
Joshua
Her 17h13'31" 38d29'43"
Joshua
Cyg 20h9'41" 32d28'36"
Joshua
And 0h55'51" 39d28'41"
Joshua
Ori 5h54'50" 11d42'16"
Joshua
Ari 3h1'49" 10d40'4"
Joshua
Vir 14h53'1" 4d39'33"
Joshua
Ori 5h42'18" 3d18'1"
Joshua
Her 18h5'7" 14d28'29"
Joshua
Ari 2h47'30" 27d53'54"
Joshua
Lmi 10h44'26" 26d31'18"
Joshua
Leo 9h36'11" 27d38'4"
Joshua
Com 13h6'48" 26d45'27"
Joshua A. Folman (Dork)
Gem 7h2'8" 20d36'58"
Joshua A. Helms
Her 16h41'6" 35d5'40"
Joshua A. Hurt
Ori 5h51'50" -8d19'53"
Joshua A. McKay
Cep 22h9'59" 56d54'23"
Joshua A Vargas
Cnc 9h6'45" 21d3'24"
Joshua A. Woods
Uma 10h23'13" 61d34'6"
Joshua Aaron George
Stribblehill
Dra 15h27'52" 56d44'20"
Joshua Adam Cohen
Psc 1h1'12" 19d9'46"
Joshua Adam Dewey Price
Her 17h59'13" 23d53'46"
Joshua Adam Ewert
Gem 6h31'34" 22d14'24"
Joshua Adam Gordon
And 23h15'48" 42d17'43"
Joshua Adam Levine
Cap 21h10'1" -19d39'30"
Joshua Adams, A Shining
Star
Lmi 9h53'42" 33d9'33"
Joshua Ainsworth-Elkington
Per 2h10'31" 54d9'8"
Joshua Alan Murphy
Ori 5h50'52" 9d37'19"
Joshua Albert
Uma 10h30'23" 46d42'6"
Joshua Alec Freedline
Vir 12h13'33" 9d23'42"
Joshua Alejandro Holgin
Equ 20h59'52" 10d43'57"
Joshua Alexander Bias
Uma 8h27'25" 64d17'2"
Joshua Alexander De Losa
Cru 12h18'33" -59d41'15"
Joshua Alexander
Herschell
Uma 8h22'3" 72d46'5"
Joshua & Alexander Munir-
Fawcitt
Cnv 12h44'17" 42d34'9"
Joshua Alexander Muresan
Cyg 20h1'8" 37d36'31"
Joshua Allan
Vir 14h26'34" 0d56'22"
Joshua Allan Jorgensen
Her 16h14'7" 44d37'56"
Joshua Allen Ellis
Psc 23h46'31" 3d23'51"
Joshua Allen Hanson
Uma 11h17'33" 63d57'26"
Joshua Alton Romines
Lib 15h21'45" -20d38'28"
Joshua Alton Duncan
Lib 15h5'44" -20d6'12"
JOSHUA ANANT JOLLY
Sgr 18h14'38" -20d3'7"
Joshua and Angela
Cyg 21h27'43" 43d48'35"
Joshua and Ashley West
Lyr 18h46'46" 39d54'34"
Joshua and Ayren Keller
Lib 14h54'7" -3d9'18"
Joshua and Cory
Reichenbach
Uma 11h19'7" 54d56'4"
Joshua and Hollie Schuff
Ori 5h24'34" 8d52'41"
Joshua and Jennifer
Sgr 18h37'49" -28d6'55"
Joshua and Marcy
McCaskill
Gem 7h22'56" 27d1'26"
Joshua and Rebecca
Ori 5h30'32" 7d14'26"
Joshua and Shalaine
Forever
Cyg 20h27'51" 54d20'7"
Joshua and Tracy
Cyg 21h22'45" 43d17'34"
Joshua Andrew Blythe
Sco 17h56'36" -37d3'54"
Joshua Andrew Forness
Uma 8h58'39" 56d5'34"
Joshua Andrew Hartwick
Per 4h35'49" 44d59'46"
Joshua Andrew Koza
Cyg 19h54'14" 38d36'1"
Joshua Andrew Long
Cas 1h2'5" 49d52'32"
Joshua Anthony Barlow
Her 17h34'45" 16d59'16"
Joshua Anthony Bryson -
21.11.2004
Cru 12h14'5" -62d23'5"
Joshua Anthony & Jacob
Daniel
Psc 1h17'8" 21d9'9"
Joshua Antony Bosence
Cep 22h55'47" 71d8'45"
Joshua Apollo Lambert
Ari 3h6'15" 15d16'44"
Joshua & Ashli Foarde
Scl 0h43'0" -28d13'43"
Joshua Ashton Culbert
Boo 14h30'59" 50d9'14"
Joshua Augustine
Col 5h31'45" -27d33'18"
Joshua Avery
Aur 6h6'24" 46d22'23"
Joshua Avery Spector
Per 2h59'20" 46d23'13"
Joshua Baron Watson
Aql 20h2'17" -2d23'52"
Joshua Barrows
Cep 22h56'45" 60d26'12"
Joshua Batchelder
Uma 12h43'41" 59d27'19"
Joshua Beaver
Dra 19h13'8" 66d24'14"
Joshua Beck
Uma 11h25'48" 31d6'10"
Joshua Ben Taylor
Per 2h51'30" 46d24'18"
Joshua Benjamin
Per 4h44'54" 39d13'29"
Joshua Benjamin John
Sillick
Her 18h30'39" 24d59'6"
Joshua Bernard Galvan
Tau 4h23'11" 14d53'22"
Joshua Bernstein
Uma 11h47'21" 45d36'1"
Joshua Bibby
Uma 11h8'59" 38d46'48"
Joshua Blewitt
Umi 17h3'14" 79d29'55"
Joshua Brandon Paige
Cas 23h58'38" 60d53'43"
Joshua Brandon Stuckey
Aqr 22h0'51" -10d12'5"
Joshua Brett Sackstein
Tau 3h57'28" 25d10'22"
Joshua Bruenning
Aur 5h42'5" 40d53'52"
Joshua Burk
Mon 6h39'19" 4d38'12"
Joshua Burke Keith
Ori 5h54'28" 18d39'5"
Joshua Cafasso Conroy
Cyg 19h49'53" 36d50'18"
Joshua Cahill
Ori 5h38'11" -3d15'18"
Joshua Caine
Ari 2h18'36" 12d18'59"
Joshua Caleb Longdin
Her 17h24'14" 25d58'43"
Joshua Camron Rabow
Ari 1h54'44" 18d44'54"
Joshua Carl Rumpca
Cnc 9h10'30" 14d52'7"
Joshua Carrick
Gem 7h56'2" 28d58'2"
Joshua Chapman
Uma 11h25'10" 36d45'17"
Joshua Chase Patterson
Gem 6h38'32" 15d20'3"
Joshua & Chelsea
Goodness
Vir 13h58'9" -8d9'16"
Joshua Chidlow
Gem 7h17'19" 14d38'15"
Joshua Christian Sjoberg
Her 18h21'49" 21d21'46"
Joshua & Christie Lum
Cyg 21h59'8" 50d41'53"
Joshua Christopher David
Harris
Per 4h15'35" 31d50'13"
Joshua Clint Finley
Uma 9h54'50" 62d45'16"
Joshua Cocklin 21
Uma 8h43'36" 50d46'59"
Joshua Colby Wright
Her 17h32'34" 33d52'19"
Joshua Connor Allan
Towse
Uma 11h6'32" 71d4'18"
Joshua Cook
Aqr 21h50'50" -0d49'10"
Joshua Cook
Cyg 21h17'23" 42d43'38"
Joshua Creed
Psc 0h6'18" 5d48'24"
Joshua D. DaSilva
Uma 9h27'42" 43d19'10"
Joshua D. McCrary
Sco 16h25'15" -32d49'12"
Joshua D. Scott
Her 17h33'30" 37d35'12"
Joshua Dale
Umi 14h16'19" 70d10'22"
Joshua Daniel
Uma 11h9'58" 61d38'54"
Joshua Daniel Adams
Aqr 20h43'46" -3d0'2"
Joshua Daniel Brandt
Tau 4h31'8" 17d9'57"
Joshua Daniel Findley
Sgr 20h2'5" -44d20'20"
Joshua Daniel Maddison
Dra 19h41'29" 61d23'20"
Joshua Daniel Manders
Dra 18h29'19" 50d57'45"
Joshua Daniel Michalec
Ori 5h38'55" 6d37'6"
Joshua Daniel Price
Aqr 23h53'51" -4d49'26"
Joshua Daniel Smith
Leo 9h31'18" 12d0'9"
Joshua Daniel Stine
Her 17h48'14" 44d19'9"
Joshua Daniel Tongs - 21
June 1999
Cru 12h38'37" -59d25'7"
Joshua David
Dra 18h30'2" 72d35'20"
Joshua David
Ori 6h9'33" 14d33'50"
Joshua David
Her 17h49'18" 21d57'33"
Joshua David Adair
Ari 2h0'46" 20d22'51"
Joshua David Barboza
And 0h35'24" 32d59'3"
Joshua David Berdugo
Lmi 10h22'20" 28d49'18"
Joshua David Cordone
Uma 11h9'54" 42d10'21"
Joshua David Cotton
Leo 11h45'36" 11d28'34"
Joshua David Erlenbusch
"3-22-01"
Lyn 6h45'23" 61d14'36"
Joshua David Gardner-
Hyden
Aqr 22h46'2" -4d55'31"
Joshua David Gurbal
Sgr 17h56'55" -25d57'35"
Joshua David Harton
Cnc 8h38'16" 31d14'49"
Joshua David Hurttgam
Cyg 19h52'14" 36d12'12"
Joshua David Jessie
Keyser
Leo 9h29'35" 28d5'44"
Joshua David Lauzer
Uma 10h35'7" 58d59'23"
Joshua David McMullen
Uma 12h38'12" 62d36'35"
Joshua David Nygaard
Her 16h33'10" 34d40'8"
Joshua David Peña
Cep 22h59'35" 75d42'51"
Joshua David Penland
Ori 5h50'30" 6d37'48"
Joshua David Sanders
Ari 2h8'19" 17d8'25"
Joshua David Talley
Cyg 21h29'21" 51d36'23"
Joshua Davis
Her 18h43'4" 20d11'9"
Joshua & Dayton Sylvester
Sge 20h5'37" 16d31'53"
Joshua Dean Averitt
Vir 14h40'49" 5d8'10"
Joshua Dean Mangrum
Uma 10h39'58" 43d17'32"
Joshua Dennis Harding
Her 17h6'28" 32d51'28"
Joshua Desmond Bartle
Peg 23h21'22" 32d53'15"
Joshua Dickenson
Ori 6h20'3" 7d11'36"
Joshua Dickey
Aql 20h18'21" -2d59'45"
Joshua Dominic Culpepper
Per 3h19'52" 53d34'22"
Joshua Don Mason
Her 16h39'56" 31d57'46"
Joshua Dorsey Thompson
Psc 0h38'16" 7d44'0"
Joshua Dupy
Her 17h9'41" 32d46'58"
Joshua Earl Philpott
Tau 3h46'39" 19d39'54"
Joshua East
Cru 12h1'45" -62d12'32"
Joshua Eddington's Star
Uma 14h13'30" 57d44'54"
Joshua Edward Johnson
Psc 23h22'27" 2d22'14"
Joshua Edward "Starz"
Rosario
Lac 22h48'12" 42d49'47"
Joshua Edward Taylor
Gem 6h56'14" 25d12'57"
Joshua Eli
Umi 16h25'53" 72d50'43"
Joshua Eli Colson
Lyn 7h27'36" 50d25'29"
Joshua Elvert
Psc 1h18'43" 18d3'29"
Joshua Eric Tomlinson-JET
Star
Uma 9h44'39" 51d58'38"
Joshua Eugene Schoeller
Vir 13h15'4" 11d17'11"
Joshua Eun Miller
Vir 13h20'43" -13d22'58"
Joshua Evan Mitchell
Uma 10h6'56" 47d40'54"
Joshua Evanick
Her 17h26'24" 45d24'21"
Joshua Everett Mallers
Aur 5h54'3" 36d57'38"
Joshua Ewers
Per 2h23'4" 57d10'9"
Joshua F. Barber
Uma 11h6'43" 44d57'14"
Joshua Fairchilds
Her 18h16'48" 14d47'4"
Joshua Ferguson
Cnc 8h51'37" 27d26'36"
Joshua Filsoof
Cep 22h9'51" 61d33'45"
Joshua Fionn Keyes
Umi 15h52'55" 79d5'31"
Joshua Forinash
Cyg 20h2'38" 31d24'33"
Joshua Foster
Uma 10h4'58" 47d59'8"
Joshua G. Boyer
Aqr 22h41'8" 0d53'0"
Joshua Gabriel Jacobs
Her 17h27'13" 34d35'41"
Joshua George Head
Ori 5h33'16" -8d54'20"
Joshua Gilbert Acuna
Dra 15h12'25" 57d10'41"
Joshua Gillis loves Katrina
Elliott
Sco 17h53'26" -35d48'14"
Joshua Grace
Col 5h52'0" -35d14'0"
Joshua & Gramie Star
Umi 15h6'50" 76d19'31"
Joshua Greene
Cep 20h47'17" 65d33'12"
Joshua Gregory
Aqr 22h39'48" -0d11'49"
Joshua Grey Eason
Eri 4h24'39" -2d5'47"
Joshua Gross & Gerald
Brinkman
Uma 8h10'54" 66d55'2"
Joshua Gutman
Vir 12h19'50" 12d6'36"
Joshua Hawkey
Uma 10h23'19" 47d41'51"
Joshua Hayes
Tri 2h5'46" 34d16'28"
Joshua Hayes
Uma 12h6'43" 58d31'34"
Joshua & Heather Lane
Mon 7h16'24" -0d48'25"
Joshua Isaac
Per 2h48'16" 51d58'51"
Joshua Isaiah Barbarossa
Gem 7h13'19" 29d18'4"
Joshua Ivan Williams
Cen 13h31'56" -38d10'28"
Joshua J. Dickinson 1975
Psc 23h21'51" 4d36'42"
Joshua J. Jordon
Dra 20h3'29" 70d16'20"
Joshua J. Saneda
Her 16h36'40" 36d54'6"
Joshua J. Twyman
And 0h56'50" 39d17'23"
Joshua Jacob
Her 16h35'39" 24d12'22"
Joshua Jacob Riehlman
Leo 11h18'21" 23d21'14"
Joshua James
Cnc 8h25'59" 10d9'37"
Joshua James
Aqr 22h32'52" -2d4'54"
Joshua James Anderson
Per 4h20'55" 36d56'30"
Joshua James Beldue
Sco 16h8'1" -12d5'59"
Joshua James Bond
Cru 12h41'31" -57d56'59"
Joshua James Chadwick
Sgr 19h46'7" -12d18'54"
Joshua James Garsea
Cyg 20h36'27" 54d22'26"
Joshua James Green
Ari 2h47'59" 27d43'4"
Joshua James Johnson
Sgr 18h3'53" -28d10'15"
Joshua James Lambert
Dra 19h29'40" 69d12'53"
Joshua James Lambert
Umi 15h37'26" 72d11'26"
Joshua James Lynne
Ari 3h2'31" 21d39'36"
Joshua James Perrett
Dra 17h33'37" 67d41'9"
Joshua James Randell
Per 4h49'36" 42d51'45"
Joshua James Taskovics
Sco 16h55'31" -43d11'58"
Joshua James
VanNieulande
Cnc 8h32'15" 23d30'26"
Joshua James Warren
Lyn 7h58'58" 57d19'15"
Joshua Jarred Chisholm
Lib 15h15'8" -21d49'59"
Joshua Jason Francis
Uma 9h24'19" 62d24'33"
Joshua Jay
Ori 6h15'28" 14d46'16"
Joshua Jay Farrar
Psc 23h59'0" -4d26'48"
Joshua Jay Sawford
Umi 13h39'55" 74d19'26"
Joshua Jeffrey Fowler
Uma 12h6'29" 52d22'25"

Joshua & Jennifer
Cyg 20h28'58" 49d50'58"
Joshua Jeraac Fout
Dra 15h41'17" 58d31'44"
Joshua Joel Sawyer
Psc 1h36'30" 21d31'7"
Joshua & Joelle's Love is 4 Ever
Cnc 8h49'28" 30d44'43"
Joshua John
Ori 5h31'36" 2d12'8"
Joshua John Arroyo
Per 3h24'33" 49d43'14"
Joshua John Fleming
Vir 13h35'50" 12d21'43"
Joshua John Monczewski
Her 17h7'59" 34d51'11"
Joshua John O'Brien
Cru 12h18'13" -62d59'4"
Joshua John Roger Tucker
Psc 1h35'40" 9d43'43"
Joshua John Rulla
Uma 10h43'17" 50d16'44"
Joshua John Wynn Lightbown
Per 3h36'52" 39d16'39"
Joshua Johnson Hanks
Aur 6h22'49" 32d25'48"
Joshua Jon
Cnc 8h51'20" 13d48'18"
Joshua Jon-Michael Wheeler
Sgr 19h18'50" -21d54'24"
Joshua Joseph
Cnc 8h46'10" 30d42'39"
Joshua Joseph Banville
Sco 16h52'29" -37d15'1"
Joshua " J.T." Thomas Cullers
Uma 9h17'42" 48d3'17"
Joshua K. Zimmerman
Sgr 18h6'9" -27d59'50"
Joshua Kash
Umi 19h12'59" 86d17'18"
Joshua Kashat
Uma 9h37'5" 46d53'45"
Joshua Keith Bowlby
Uma 10h12'29" 55d28'29"
Joshua Kenneth Crow
Umi 16h51'41" 80d32'43"
Joshua Kevin Steele
Per 3h29'21" 43d21'54"
Joshua Khurin
Aqr 22h1'23" -17d8'55"
Joshua Koskie
Cap 21h42'23" -14d13'23"
Joshua Kostelecky
Vir 11h44'50" -3d50'47"
Joshua & Kristy
Lib 15h17'11" -4d54'17"
Joshua Kyle Junier
Ori 5h17'14" 5d48'37"
Joshua L. Jackson
Her 17h52'8" 49d56'18"
Joshua L. Levin
Cma 6h41'32" -18d4'24"
Joshua L. Talley
Ari 2h51'59" 19d6'43"
Joshua Lane
Vir 14h37'3" 4d58'11"
Joshua Launer
Per 3h10'52" 53d0'38"
Joshua Laverne Ryan
Uma 9h47'19" 46d33'46"
Joshua Lee
Her 18h3'37" 14d25'8"
Joshua Lee Fink
Umi 14h21'21" 72d46'21"
Joshua Lee Hess
Sgr 19h21'41" -13d7'14"
Joshua Lee Nathe
Sco 17h8'46" -31d24'13"
Joshua Leland
Per 3h16'32" 54d10'12"
Joshua Lennon Pierce
Ori 5h52'55" -0d39'50"
Joshua Leonard Pearlstein
Aqr 21h47'38" -1d36'24"
Joshua Levitt
Gem 6h54'1" 33d40'44"
Joshua Lewis-Stone
Gem 6h53'1" 17d34'21"
Joshua LittleBear Swank
Her 17h30'10" 46d18'15"
Joshua Logan Plank
Vir 14h12'59" 2d16'40"
Joshua Louis Goodman
Per 3h6'54" 53d57'20"
Joshua Louis Witham Trayford
Cnc 8h47'28" 21d31'32"
Joshua Lunt
Ori 5h56'51" 11d40'26"
Joshua (Mark John) Michaylow
Her 17h39'54" 40d46'58"
Joshua Mark Robbins
Uma 10h43'38" 50d24'4"
Joshua Martinez
Tri 2h15'29" 35d21'26"
Joshua Mathiuet
Cam 6h17'46" 67d9'39"
Joshua Matthew
Cap 21h34'4" -14d46'50"
Joshua Matthew
Her 17h22'21" 36d40'19"
Joshua Matthew Purton
Ori 6h18'21" -1d26'37"
Joshua Matthew Spittal
Cep 22h38'29" 65d4'10"
Joshua Matthew Wimer
Ori 6h8'24" 6d49'24"
Joshua Maxwell Eaves
Leo 9h44'0" 27d19'14"
Joshua Mayberry
Tau 4h25'3" 21d43'38"
Joshua Mayes
Tau 4h36'19" 25d43'0"
Joshua McCulloch
Cru 12h21'54" -58d21'6"
Joshua McGrew
Ori 5h48'9" 2d36'47"
Joshua Michael Ault
Cnv 13h44'8" 30d38'12"
Joshua Michael Burke
Uma 10h40'20" 63d38'15"
Joshua Michael Derbyshire
Cep 20h54'34" 62d12'20"
Joshua Michael Fraser
Ari 2h19'47" 14d59'55"
Joshua Michael Frederick Hill
Her 16h47'12" 4d57'38"
Joshua Michael Hagler
Srp 18h24'23" -0d5'29"
Joshua Michael Herring 2/17/1988
Aqr 22h29'46" -1d41'23"
Joshua Michael Jones
Uma 11h47'48" 54d2'11"
Joshua Michael Myers
Her 17h50'54" 29d51'12"
Joshua Michael Stanton
Her 18h43'27" 17d34'6"
Joshua Michael Thomas
Vir 14h19'19" 5d16'13"
Joshua Michael Weber
Cnc 8h43'57" 7d29'15"
Joshua Michael Wilcox aka Bud
Her 17h49'31" 46d26'9"
Joshua Micheal Zwart
Lyr 19h4'25" 47d12'27"
Joshua Mikel Baxter
Cnv 12h50'45" 38d16'44"
Joshua Miller
Tau 4h56'28" 25d8'30"
Joshua Miller and Samantha Miller
Umi 15h43'10" 79d49'48"
Joshua Minato
Sgr 18h49'14" -20d17'48"
Joshua & Misha
Cyg 19h42'36" 37d9'47"
Joshua Mitchel Lynen
Ori 5h28'20" 2d34'43"
Joshua Mitchell
Cnc 8h31'50" 32d39'49"
Joshua Mitchell
And 1h16'44" 46d40'12"
Joshua Monk
Uma 8h52'49" 56d44'57"
Joshua Mooney
Cyg 21h21'2" 32d42'28"
Joshua Moun
Cru 12h3'49" -62d34'23"
Joshua Myles Bauer
Leo 9h40'32" 26d23'7"
Joshua Nathaniel Cheshire
Her 17h44'15" 36d9'21"
Joshua Nicolas DeAses, Jr.
Uma 11h19'51" 29d8'38"
Joshua Noah Sky Williams
Sco 16h8'39" -20d2'38"
Joshua P Ort
Her 17h42'21" 34d18'50"
Joshua Paul Cashier
Lib 14h41'40" -11d0'38"
Joshua Paul Harris
Ari 2h41'32" 27d17'14"
Joshua Paul Jackson, Man of Valor
Her 16h40'26" 22d18'7"
Joshua Paul Porter
Lib 14h50'29" -1d1'4"
Joshua Paul Rogers
Vir 13h19'1" -21d55'32"
Joshua Paul Sheridan
Per 3h26'55" 48d19'4"
Joshua Perry Johnson
Her 17h25'15" 36d22'22"
Joshua Perry Ortiz
Sco 16h50'57" -34d44'12"
Joshua Peter Christopher Benes
Gem 6h48'44" 34d16'40"
Joshua Phillip Earlenbaugh
And 0h13'51" 33d43'37"
Joshua Pickering
Cnc 8h57'8" 29d28'31"
Joshua Quinlan
Per 3h55'4" 38d56'41"
Joshua R Bordelon
Per 4h21'22" 33d42'41"
Joshua R. Jett
Cap 21h6'12" -15d2'42"
Joshua R. Kindler
Ori 5h55'52" 21d8'47"
Joshua R. Stubblefield
Tau 4h17'29" 29d44'34"
Joshua Ray
Ori 5h56'25" 22d31'19"
Joshua Ray Rogers
Her 17h59'47" 29d24'39"
Joshua Raymond Rain
Aqr 22h39'4" -0d36'26"
Joshua Reagan Walderbach
Cyg 19h38'35" 38d1'43"
Joshua Reese Pruitt
Uma 14h4'15" 57d52'47"
Joshua Riestra
Psc 0h8'23" 6d20'36"
Joshua Ringleb
Her 18h36'52" 17d20'49"
Joshua River Netanel
Eri 4h29'3" -2d23'28"
Joshua Robert Geary
Tau 5h43'13" 23d25'4"
Joshua Robert Hermonat
Tau 5h54'22" 28d19'18"
Joshua Robert Hewitt
Aur 5h28'23" 29d49'3"
Joshua Robert Kenyon
Ari 3h6'49" 18d42'55"
Joshua Robert Scheels
Dra 19h8'59" 78d28'13"
Joshua Robert Smith
Ori 5h58'6" 17d9'16"
Joshua Robert Winans
Uma 14h11'16" 55d22'50"
Joshua Ross Houston
Cnc 8h29'6" 18d59'42"
Joshua Roy
Aur 5h48'28" 41d57'14"
Joshua Rubin
Her 18h31'50" 23d25'4"
Joshua Russell Lyman
Her 16h42'46" 32d12'46"
Joshua Ryan Carr
Aqr 23h18'32" -13d8'42"
Joshua Ryan Rappaport
Cnc 9h3'57" 29d33'46"
JOSHUA SAI LOVES AMI MIO
Per 3h6'28" 51d53'13"
Joshua Sam Schpok
Gem 6h4'46" 23d27'58"
Joshua Samuel 19
Ori 5h57'1" 21d9'6"
Joshua Schmidt
Her 17h31'42" 44d36'41"
Joshua Scott Anderson
Boo 14h3'33" 18d43'30"
Joshua Sebastian
Her 17h19'52" 28d3'55"
Joshua & Serena Buckley
Vir 12h36'20" 11d38'41"
Joshua & Shannon
Cyg 21h36'29" 50d48'1"
Joshua Shepherd
Uma 9h13'26" 57d53'2"
JOSHUA SPARAGE
Cep 22h39'20" 65d39'15"
Joshua Stanley Howell
Leo 11h25'8" 6d19'5"
Joshua Stanley White
Leo 10h51'53" 14d6'56"
Joshua Stephen Cookson
Leo 10h44'4" 17d12'21"
Joshua Stephenson
Sgr 18h10'52" -17d20'20"
Joshua Steven C.
Aqr 22h9'43" -1d1'24"
Joshua Steven Derosier
Per 3h49'5" 37d40'58"
Joshua Steven James Bell
Ori 4h50'29" -3d11'47"
Joshua Stewart Linden
Cyg 20h2'43" 43d45'21"
Joshua Stokes
Her 18h25'4" 17d7'56"
Joshua Stone
Cet 2h27'43" 9d10'18"
Joshua Stone (star)
Per 3h46'20" 49d54'52"
Joshua "Sweetness" Rogers
Per 4h35'1" 42d10'10"
Joshua Taggart Miller - "Coach"
Cas 0h38'21" 55d30'37"
Joshua Tague
Her 17h21'46" 27d50'5"
Joshua Taj Mandara
Cnc 8h20'29" 30d52'58"
Joshua Taylor Chastain
Uma 11h39'15" 51d10'12"
Joshua Taylor Hall
Cep 23h8'29" 70d14'58"
Joshua Taylor Robertson
Ori 5h46'22" 1d3'35"
Joshua the Promise Star
Her 16h36'44" 38d20'33"
Joshua Theodore Roth
Umi 14h34'23" 72d41'5"
Joshua Thomas Baumgarden
Uma 13h58'8" 61d55'28"
Joshua Thomas Edwards
Tau 4h11'31" 13d4'3"
Joshua Thomas Gordon - Joshua's Star
Per 3h33'19" 36d2'52"
Joshua Thomas Jacques
Cep 20h47'38" 61d22'30"
Joshua Thomas Keene
Per 2h57'55" 46d23'57"
Joshua Thomas Lowery
Uma 11h19'51" 56d53'15"
Joshua Thomas Pantalone
Umi 15h47'4" 78d11'29"
Joshua Thomas Perry
Lyn 7h50'9" 42d6'28"
Joshua Thomas Skelly
Per 3h47'15" 43d10'1"
Joshua Thomas Spencer
Umi 15h25'46" 71d50'33"
Joshua Thomas William Seed
Cyg 21h30'27" 37d57'52"
Joshua Thomas Zoch
Pho 1h5'49" -41d9'49"
Joshua Timothy Melcher
Ari 3h4'17" 19d53'36"
Joshua Timothy Walker
Aql 19h43'0" -0d10'17"
Joshua Trigg - Joshua's Star
Per 3h9'37" 45d10'11"
Joshua Troutman
Umi 14h44'54" 73d11'1"
Joshua & Tyler Yoder
Aql 19h43'21" 3d19'53"
Joshua Urso
Lib 14h59'43" -10d4'48"
Joshua Valliant Fears
Uma 11h29'24" 62d16'7"
Joshua Victor Caron
Ori 5h11'17" 6d19'6"
Joshua Von Stetten
Aqr 22h50'59" -4d13'19"
Joshua W. Mooney
And 23h51'48" 48d36'42"
Joshua Walsh
Gem 6h26'29" 21d30'37"
Joshua Weinberg and Kristy Burch
Vel 9h45'56" -45d2'31"
Joshua William Cook
Ari 2h35'41" 19d5'46"
Joshua William Loewe
Uma 9h32'18" 52d21'11"
Joshua William Pyke
Her 17h20'37" 34d7'27"
Joshua William Ross
Cru 12h44'57" -57d31'41"
Joshua William's Name Day Star
Aql 19h32'53" 11d48'20"
Joshua Wolf Smith
Ori 6h17'16" 14d35'40"
Joshua Woods
Uma 13h13'13" 60d43'45"
Joshua Yankauskas
Uma 11h28'56" 40d41'37"
Joshua Yates Clanton
Ori 5h27'55" 1d50'27"
Joshua you brighten our lives
Cmi 7h50'21" 0d6'44"
Joshua, light of our lives
Uma 13h43'53" 54d11'14"
Joshua, Vanessa & Sean Smith
Lmi 11h3'51" 24d52'25"
Joshua,Benedikt Wehinger
Peg 21h30'46" 15d32'17"
JoshuaGeorgePowers1012 79Celestial
Lib 15h27'37" -27d54'6"
Joshuah Christopher Samuel Parnell
Umi 4h21'24" 89d12'58"
JoshuaJA
Ori 5h36'7" 7d24'9"
Joshua-Jacqueline
Cyg 19h47'12" 34d39'58"
JOSHUA-KRIS
Cma 6h43'50" -17d31'31"
Joshua-Marc Tanenbaum
Vir 13h9'27" 12d0'56"
Joshua's Christening Star
Uma 10h32'33" 47d30'8"
Joshua's Jenniffer
Ori 5h40'31" -0d10'5"
Joshua's Jolt
Ori 4h50'55" 5d3'26"
Joshua's Own
Aql 20h7'34" -7d14'4"
Joshua's Piece of Paradise
Pho 1h35'47" -41d27'45"
Joshua's Place in Heaven
Cep 2h4'26" 80d58'51"
Joshua's Play Land Star
Psc 0h43'43" 16d47'48"
Joshua's Smile
Her 16h59'59" 29d42'8"
Joshua's Star
Per 2h9'39" 55d9'59"
Joshua's Star
Sgr 18h42'30" -25d19'22"
JoshuaZCoven
Cap 20h30'47" -18d24'39"
Joshy
Lib 14h49'19" -20d45'7"
Joshy Poo
Ori 4h56'38" 3d0'54"
Joshy Washy
Tau 5h34'23" 24d6'0"
Josi
Vir 13h34'14" -1d4'51"
Josi Sue
And 1h15'31" 49d29'54"
Josiah
Per 3h50'30" 41d37'13"
Josiah Brant Harsell
Ori 5h23'49" -4d32'24"
Josiah Gatlin Cochran
Umi 15h27'48" 70d55'17"
Josiah Howard Kiefel
Psc 0h17'7" 8d33'4"
Josiah Parker Gutowski
Her 16h36'55" 39d1'17"
Josiah Ramon Alvillar-Torres
Umi 15h26'31" 72d9'12"
Josiah Schick
Per 3h29'23" 38d41'4"
Josiah Zachary Cook
Psc 0h46'15" 15d42'17"
Josiah's Reach
Sgr 19h35'40" -14d4'39"
Josiane Aurore Bryan
Gem 7h30'27" 25d56'33"
Josiane Paschoarelli
Crb 16h15'46" 36d26'42"
Josie
And 23h40'22" 39d2'1"
Josie
Cam 4h29'40" 56d41'21"
Josie
Cyg 20h22'49" 58d50'26"
Josie A. Castellanos
And 1h30'9" 49d33'16"
Josie Anne Oliver
And 23h26'7" 47d53'10"
Josie Caroline
And 1h51'12" 37d29'10"
Josie Carroll
Cru 11h58'45" -62d5'21"
Josie Gardiner
Uma 11h56'54" 53d42'8"
Josie Holly Hammond
And 0h49'11" 25d46'34"
Josie "Jewel of Light"
Cnc 8h20'27" 26d46'9"
Josie & Julius Gladysz
Per 3h13'16" 46d53'29"
Josie Kate
Sco 16h4'4" -17d10'49"
Josie Lee Weickert
Psc 0h34'8" 8d14'25"
Josie Lynette Creasy
Leo 10h53'51" 0d3'8"
Josie O Reilly
Uma 12h24'3" 52d39'23"
Josie Paige Solomon
And 1h39'7" 42d44'30"
Josie Rae Dittmer
Vir 12h7'32" -9d13'42"
Josie Rosas
Uma 11h52'35" 47d33'3"
Josie Scott
Umi 16h24'19" 77d54'7"
Josie Sepulveda
And 1h12'15" 42d11'16"
Josie Tait
Lyn 8h8'46" 37d54'52"
Josie West Garey
Gem 7h48'7" 34d46'5"
Josie Wieberg
Leo 10h46'36" 18d3'40"
Josiephine Joao Elliott
Sco 16h9'22" -11d34'27"
Josif Vons
Leo 11h0'4" -0d37'35"
Josip Fuduric
Uma 9h25'42" 47d38'10"
Josito JL
Cap 20h13'4" -8d57'42"
Joslyn
Aqr 22h8'7" -0d55'42"
Joslyn Eileen Christie
Leo 10h8'3" 20d10'17"
Joslyn My Mommy Kreh
Cnc 8h48'11" 30d21'25"
Joslyn Nicole Morrison
Crb 15h37'48" 32d30'25"
Joslynn Rosalia 82205
Crb 16h1'58" 38d3'0"
Joss Pegram
Umi 16h18'35" 71d52'48"
Josse Vargas
Tau 3h31'58" 18d19'50"
Josselyn-Tina
Cnc 8h19'33" 8d33'44"
Jossie Holland
Sco 16h38'58" -35d0'54"
Jossie's World
Mon 6h52'51" -0d18'2"
Jossy John- Agnus Dei
Per 2h15'32" 56d7'42"
Jost Langenfeld
Lyn 7h46'30" 36d24'46"
Jost, Jean
Uma 11h2'2" 47d44'2"
Josue
Her 18h38'10" 15d37'52"
JOSUE'
Cen 11h48'55" -52d19'51"
Josue A. Ortiz
Cap 20h25'6" -12d2'43"
Josue Ernesto Valencia
Lyn 8h12'20" 56d6'41"
Josue' M. Lugo
Aql 19h45'0" 6d14'28"
Josue Torres
Vir 14h18'47" 1d0'17"
JOSWICK925
Cnv 14h0'4" 36d51'50"
Josy
Sge 20h3'14" 18d44'23"
Josy
Tau 4h41'8" 22d49'32"
Josy Süssmeier
Uma 8h32'40" 61d20'44"
Josy Wigzell
Lib 15h5'52" -5d37'35"
Jouaux, forever my love
Sco 17h56'47" -30d44'33"
Jouline Schindler's Star
Umi 13h36'38" 76d39'48"
Joumana Aram
Lib 14h59'9" -15d36'41"
Joumana Sarkissian
And 1h47'11" 39d21'14"
JOURDAN
Aql 20h7'13" 7d55'32"
Jourdan Belle-Marie
Srp 17h59'25" -0d8'9"
Jourdan Leigh Rettig
Ari 3h13'29" 27d20'31"
Jourdes Reagan Jewell Wullbrandt
Aqr 22h7'41" -0d26'15"
Journee Lluvia Grace Levi
Ari 2h46'44" 25d5'52"
Journey
Cyg 19h36'4" 29d24'39"
Journey
Cnv 12h50'13" 37d58'43"
Journey Across the Blue Bridge
Eri 3h51'47" -0d35'2"
Journey Elizabeth Nicole Ann
Cnc 8h50'49" 28d36'14"
Journey Michelle Thomas
Sgr 18h58'21" -26d48'39"
Journeyman
Per 3h21'25" 52d7'49"
Journey's End
Cru 12h25'56" -58d0'12"
Journi
Lib 14h51'50" -3d20'1"
Jousha Our Christmas Star
Per 3h3'43" 40d28'44"
Jovan Owimrin
Cap 21h28'34" -22d42'24"
Jovana 44
Sco 16h6'28" -15d54'28"
Jovana Bijelic - 09.11.1981.
Peg 23h59'1" 28d37'26"
Jovana-Rae Pumehana Shigetani
Aqr 21h39'34" -4d37'51"
Jovanny E. Guillen
Uma 11h58'51" 55d56'11"
Jovelli
Cru 12h32'26" -59d26'18"
Jovena
Psc 23h4'39" -0d59'51"
Jovi McDeavitt
Gem 6h51'6" 32d22'6"
Jovial Dualism
Sgr 18h48'4" -27d49'26"
Jovica & Sne▯ana
Uma 9h53'54" 48d47'51"
Jovie Isobel Maree Ellis
Cru 12h37'26" -58d28'50"
Jovie Kathleen Cordahl
Vir 13h17'49" 12d39'22"
Jovin
Sco 16h12'45" -9d9'0"
Jovita Soto
Tau 3h49'42" 8d18'18"
Jowena
Psc 1h20'51" 16d12'3"
JO.W.M.B-35-42-1989
Sgr 19h41'0" -21d42'23"
Jowynne Khor
Vir 12h35'4" -10d40'13"
Joy
Aqr 21h43'9" -4d6'28"
Joy
Cam 13h59'56" 79d42'36"
Joy
Lyn 7h29'20" 52d39'46"
Joy
Cas 1h39'30" 61d6'34"
Joy
Umi 15h4'11" 72d40'22"
Joy
Umi 13h54'37" 71d36'38"
Joy
Psc 1h17'22" 20d47'31"
Joy
Ari 2h56'1" 22d18'28"
Joy
Tau 3h38'15" 16d36'31"
Joy
Psc 1h46'29" 8d22'56"
Joy
Psc 0h42'46" 8d7'16"
Joy
Mon 6h29'22" 4d41'38"
Joy
Her 17h16'43" 27d4'10"
JOY
Gem 6h19'31" 23d13'43"
JOY
And 0h18'39" 25d28'26"
Joy
Cas 0h59'25" 54d24'25"
Joy
Cyg 21h37'7" 42d34'45"
Joy
And 23h29'58" 44d29'26"
Joy
Psc 1h17'13" 30d59'45"
Joy
Cyg 19h51'0" 33d31'44"
Joy
And 1h44'18" 37d39'6"
Joy
And 1h46'26" 41d43'40"
JOY 1218
Sgr 19h16'36" -26d19'10"
Joy Amrhein
Cnc 8h45'12" 32d15'4"
Joy and Bobbe Andersen-Runions
Cyg 19h59'21" 38d18'25"
Joy and Denise forever !
Tau 3h52'19" 24d51'33"
Joy and Glory
Lib 15h23'18" -16d20'43"
Joy and Jim Reinknecht
Cyg 21h6'52" 46d58'45"
Joy and Phil
Cyg 19h50'48" 54d26'58"
Joy Ann Fisher
Aqr 23h0'18" -7d47'41"
Joy Ann Sullivant Mathews
Gem 7h35'50" 16d20'10"
Joy Anna VanCleave
Gem 7h14'12" 30d38'14"
Joy & Anthony - Inifinity & A Day
Pho 23h39'52" -41d52'27"
Joy Antoinette Eckmann
And 1h23'26" 50d24'48"
Joy Belle "Jacie" O'Neal
Lyn 8h17'3" 46d14'0"
Joy Belles
Sco 16h3'42" -23d22'22"
Joy & Bob
Cyg 20h40'17" 45d42'49"
Joy Brusca
Crb 15h47'7" 34d15'7"
Joy Buller
Vir 15h6'55" 0d11'9"
Joy Button LaFleur
Joy'sDancingStar
Ari 3h0'7" 19d44'51"
Joy Corban Shelton
Cyg 19h59'19" 46d35'30"
Joy Coughtrey
Lyr 18h35'0" 39d23'44"
Joy Cristina Hubbard
Sgr 19h45'4" -14d27'15"
Joy & Dave's Star
Per 4h50'43" 50d26'15"
Joy Eadie
And 0h3'19" 33d52'26"
Joy Elaine
Cra 18h4'7" -37d29'23"
Joy Elisha
And 0h43'21" 44d54'31"
Joy Elizabeth Cole Losee
Cyg 20h15'34" 38d34'11"
Joy Ellen Curry
Lmi 10h0'32" 38d5'15"
Joy Engebretson
Dra 16h44'55" 55d57'12"
Joy Etta Levack Molloy
Ari 2h14'15" 18d7'27"
Joy Evelyn Herzog
Peg 22h29'10" 28d58'14"
Joy Factor
Cnc 8h15'1" 29d1'43"
Joy Gera
Per 4h14'9" 48d35'17"
Joy Hearno Hyman
Cas 23h40'50" 51d50'12"
Joy IV
Cnc 8h45'58" 11d18'14"
Joy & John Sorrells Starbright
Cyg 21h44'57" 50d22'36"
Joy K Perry
Leo 9h23'39" 19d46'37"
Joy Kearney
Leo 9h56'32" 27d57'17"
Joy Kenney
Leo 9h31'34" 11d40'9"
Joy Kenney
Crb 15h33'47" 31d22'12"
Joy Kittrell
Gem 7h27'35" 15d22'3"
Joy Kudia "Joy's Shining Star"
Aqr 22h5'58" -1d9'56"
Joy L Bamberg
Per 2h50'39" 51d2'3"
Joy Lane
Cam 7h32'38" 76d42'12"
Joy (Light of My Life)
Aqr 22h50'31" -8d24'6"

Joy Lill
Sgr 18h48'23" -26d52'0"
Joy Linda Williams
Psc 0h48'26" 21d22'21"
Joy Loshigian 10 lbs. 2 oz. 21"
And 2h19'59" 45d49'44"
Joy Lynn Salvo Eck
Cnc 8h43'4" 12d34'42"
Joy Lynne Turner
Aqr 23h13'56" -6d52'18"
Joy Margaret Barnes
Cas 0h21'43" 59d13'10"
Joy Marie
Lyn 6h28'16" 54d18'32"
Joy Marie Boudreaux
Sgr 19h0'23" -31d28'11"
Joy Marie Graham
Uma 9h40'37" 56d40'17"
Joy Maxine Miller
Cas 23h9'4" 56d19'32"
Joy McCauley
Cas 1h7'55" 49d28'8"
Joy McCusker
Cam 5h51'19" 56d29'53"
Joy & Mircea
Lmi 10h39'16" 23d32'24"
Joy Moglovkin
Cas 1h34'50" 63d22'1"
Joy Newberry Boschen
Ori 5h36'47" -1d32'54"
Joy Nicholas
Cam 3h49'10" 67d43'34"
Joy of life and Eternal Love
Vir 13h56'30" 0d4'2"
Joy of Love...50 years
Cnv 12h40'56" 46d11'49"
Joy of My Life - First Anniversary
Cas 0h21'31" 61d38'12"
Joy O'Leary (The Best Mum) - 24/08/1960
Vir 13h11'27" 4d21'14"
Joy Page
Uma 13h19'38" 57d21'14"
Joy Pervis
Uma 14h5'50" 56d20'29"
Joy Peters
Cyg 20h50'8" 47d21'59"
Joy Rabbit
Cap 20h30'52" -13d18'29"
Joy Song Owen
Sgr 18h10'10" -35d5'21"
Joy Steinmeyer
Lyn 8h46'58" 33d32'19"
Joy - The Image of Beauty
Ari 2h33'22" 13d44'49"
Joy Thomas
Com 12h41'52" 17d31'21"
Joy to the World
And 1h41'57" 42d44'58"
Joy To The World
Cap 20h7'20" -18d47'2"
Joy & Todd
Uma 10h29'15" 67d5'16"
Joy Turner
And 0h5'59" 39d12'19"
Joy Warner, Director of CCS
Lib 15h15'58" -19d23'13"
Joy Wong
And 23h55'3" 48d7'36"
Joy Wynn
Psc 1h42'46" 16d36'2"
Joya Iris Odessa
Cas 1h37'40" 68d17'47"
JoyBearPatKatGav
Uma 8h47'5" 48d41'45"
Joyce
Cyg 19h55'10" 37d20'54"
Joyce
Lyr 18h36'33" 31d23'19"
Joyce
Leo 10h36'26" 22d11'29"
Joyce
Cnc 9h13'2" 19d30'34"
Joyce
Her 17h40'9" 15d34'31"
Joyce
Cyg 20h0'4" 53d19'16"
Joyce
Cyg 21h28'35" 53d47'14"
Joyce
Uma 12h9'59" 54d0'37"
Joyce
Mon 7h55'32" -0d50'47"
Joyce
Psc 0h39'26" 6d2'50"
Joyce A. Currier
Crb 15h42'23" 39d23'10"
Joyce A. Gonzalez
Tau 4h9'40" 17d32'58"
Joyce A. Hayashi
And 1h10'38" 38d17'8"
Joyce A. Robinson
Tau 4h46'26" 17d4'13"
Joyce Adelaide Koch
Com 13h11'14" 28d11'38"
Joyce Allen
Umi 13h53'17" 69d24'31"
Joyce Amy Hughes
Lep 5h7'18" -11d12'55"
Joyce and Bernie Gammonley
Cnc 9h11'31" 17d17'52"
Joyce and Bob Farrell
Per 3h2'15" 54d17'52"
Joyce and Derrick Sholl Golden Star
Cyg 20h59'47" 32d56'28"
Joyce and Ken Rice
Cyg 21h44'59" 55d16'41"
Joyce and Robert Baldwin, Jr.
Aqr 22h38'29" 0d25'17"
Joyce Andy
Uma 9h27'47" 67d5'5"
Joyce Ann
Sgr 19h0'20" -34d42'39"
Joyce Ann
And 23h12'42" 50d44'37"
Joyce Ann 58
And 23h39'49" 50d39'50"
Joyce Ann Crowe
Peg 23h0'49" 26d7'1"
Joyce Ann Johnson
Ari 3h11'4" 28d4'5"
Joyce Ann Miller
Cas 0h29'31" 63d5'17"
Joyce Ann Rochelle
Tau 3h43'20" 3d21'27"
Joyce Ann Walker Boozer
And 0h55'3" 41d55'53"
Joyce Ann Welch
Cam 3h29'5" 57d9'29"
Joyce Arlene
Gem 6h16'19" 24d46'44"
Joyce B Lockwood
Uma 11h49'46" 31d54'4"
Joyce Benedetto
Sco 17h15'20" -32d55'47"
Joyce Bissonnette
Vir 14h39'35" -3d58'19"
Joyce & Bob Dolan 50th Anniversary
Cyg 21h15'32" 47d44'11"
Joyce & Brian's "Golden Star"
Cyg 19h47'32" 35d1'51"
Joyce Brown
And 0h4'26" 45d16'29"
Joyce Brusgulis
Aqr 20h40'43" -9d4'50"
Joyce Burkoff's Star & Deli
Ari 2h26'54" 25d1'58"
Joyce C Flynn
Lyn 7h40'13" 39d8'15"
Joyce C. Herman
Cam 6h19'24" 68d5'25"
Joyce C. Schubkegel
Uma 8h43'40" 66d7'50"
Joyce C. Stiltner
Lyr 18h43'19" 35d1'32"
Joyce Capuano
Uma 9h1'20" 50d38'57"
Joyce Carol
Uma 10h47'45" 49d5'47"
Joyce Caughron Rhodes
Crb 15h40'26" 28d4'27"
Joyce Cecilia Maya
Ori 5h35'2" -2d53'6"
Joyce Chan
Lyn 6h47'39" 52d42'3"
Joyce Chen
Tau 5h45'10" 23d18'37"
Joyce Criner's Eternal Light
Per 2h50'45" 38d25'2"
Joyce Dascoulias
Umi 15h29'58" 69d54'13"
Joyce E. Dufault
Sgr 19h5'34" -34d12'22"
Joyce E. Reed
And 23h25'32" 41d27'27"
Joyce Eberhardt Alexander
Cnc 8h9'16" 15d21'0"
Joyce Elaine
Cnc 8h18'11" 14d27'56"
Joyce Elaine Jansen
Lib 15h44'50" -23d19'58"
Joyce Elaine King
Ori 6h11'51" 14d36'54"
Joyce Elaine Lee
Tau 4h16'11" 22d38'17"
Joyce Elaine Lowe
And 0h44'7" 25d39'12"
Joyce Elaine Smith 02/06/1942
Aqr 20h56'51" 0d50'42"
Joyce Elaine Sweeney
Ori 5h56'23" -0d25'26"
Joyce Elayne Cool
And 0h27'11" 29d14'20"
Joyce Elizabeth Jones
Lyr 18h48'8" 37d48'2"
Joyce Elizabeth Kirby
Cas 23h49'27" 53d59'15"
Joyce Elizabeth Watson
Cas 2h14'37" 62d48'39"
Joyce Ellen
Uma 10h34'5" 39d25'48"
Joyce Ellen Peck
Mon 6h51'6" -0d9'23"
Joyce Emily Seichepine
And 1h15'8" 43d4'12"
Joyce Emma Dube
And 1h17'37" 34d58'7"
Joyce Empson
Uma 11h16'26" 72d39'3"
Joyce Enders WBGAM
Lib 15h25'39" -23d0'39"
Joyce & Ernie Cassidy 65 Years
Per 3h8'14" 33d20'2"
Joyce Evelyn Hart Bronow
Gem 7h19'55" 24d46'45"
JOYCE EVON
Cnc 8h25'32" 29d56'36"
Joyce Ewing
Lyr 18h36'28" 34d53'29"
Joyce/Freedom
Vir 13h32'45" -2d45'15"
Joyce G. Anderson
Gem 6h47'24" 25d14'57"
Joyce Gale
Crb 15h50'57" 29d7'34"
Joyce Gasper
Aqr 21h36'54" 1d56'6"
Joyce Gharghour
Uma 12h4'39" 51d14'13"
Joyce Grace Susan
Ori 6h9'15" 7d4'28"
Joyce Green
Cas 3h14'36" 63d25'35"
Joyce (Gunner) Hutchens
Aql 19h32'29" -11d10'6"
Joyce H. Tustin
Uma 11h41'21" 29d51'57"
Joyce Harriet Luhm
Lib 15h7'14" -10d32'40"
Joyce Harrison - Joyce's Star
Col 6h23'35" -36d27'49"
Joyce Healy
Ari 1h47'22" 19d18'58"
Joyce Heneberry
Aqr 22h11'17" -10d39'28"
Joyce H.H.H.
And 1h2'50" 38d46'46"
Joyce Hills
Cas 0h9'8" 54d30'34"
Joyce Holmes
Aql 20h28'36" -8d21'36"
Joyce Huang
Leo 9h33'0" 15d26'21"
Joyce Ione
Cas 0h51'37" 63d44'33"
Joyce & Irvine Finlay's Ruby Star
Cyg 19h43'2" 39d15'57"
Joyce Irwin
Cas 23h18'2" 62d34'39"
Joyce Jerman Bell
Ori 5h18'39" 0d34'25"
Joyce JP Fossum
Cas 1h42'40" 67d20'59"
Joyce Kay Crumrine
Uma 12h56'14" 59d37'12"
Joyce Ken
Sco 17h55'27" -30d54'27"
Joyce Kezerle
Cap 20h25'38" -13d35'33"
Joyce L. Mills
Ari 2h12'37" 18d16'2"
Joyce L. Smith
Cyg 20h20'24" 52d44'37"
Joyce Larsen
Uma 9h50'14" 55d15'38"
Joyce Lawton
Cas 23h32'40" 58d30'36"
Joyce Lee
Lyn 7h51'21" 46d56'18"
Joyce Lee Coelho
And 23h13'35" 47d15'38"
Joyce Lee Hunt
Uma 10h1'5" 44d5'2"
Joyce Lee Ramsdell
Aqr 22h55'58" -17d21'55"
Joyce Leutink
And 1h6'48" 44d53'30"
Joyce Luk
Aqr 22h49'32" -10d40'32"
Joyce Lynne Lacey
Umi 10h16'8" 87d36'53"
Joyce M. Rivera
Vir 14h22'0" 0d12'38"
Joyce M. Schofield
Cam 4h6'16" 53d23'17"
Joyce MACDonald Stewart;Sis Jeanie
And 23h46'3" 46d38'57"
Joyce Mae Strand
Cyg 21h27'26" 37d8'10"
Joyce Maiti Mikiugu
Cas 0h19'15" 51d15'57"
Joyce Mamos
Tau 4h9'57" 27d25'56"
Joyce Marie
Gem 6h39'32" 14d0'42"
Joyce Marie
Vir 12h58'3" 11d10'15"
Joyce Marie Rowley
Psc 23h20'5" 4d3'40"
Joyce Marie Watson
Sco 16h35'47" -29d44'17"
Joyce Marquardt
Aqr 22h13'3" -2d44'51"
Joyce Martin Loughridge
Crb 15h38'9" 26d23'18"
Joyce & Marvin's Golden Star
Cyg 19h44'56" 28d18'41"
Joyce Mary Kerr
Cas 0h51'39" 57d3'57"
Joyce & Matthew - Beargros Chewgros
Cnc 8h39'2" 29d42'47"
Joyce Monk
Sco 16h46'23" -32d55'48"
Joyce 'n Jimmy
Lyr 18h44'6" 31d18'24"
Joyce Nelson
Sgr 18h35'18" -22d30'43"
Joyce Norton Burrows
Psc 0h49'32" 12d5'54"
(Joyce) Paula Berrier
Cas 1h11'52" 62d31'36"
Joyce Pettit
Uma 11h14'55" 61d42'8"
Joyce Phillips~Nana Shines Bright
Vir 13h38'23" 1d38'58"
Joyce Pritchard
Aql 19h6'37" 14d53'57"
Joyce Queary
And 23h59'28" 42d42'58"
Joyce & Raymond Pickering 1942/2002
Ari 2h42'35" 30d5'39"
Joyce Rebecca Young
Uma 11h39'53" 47d3'9"
Joyce Rosen
Gem 6h59'51" 27d56'2"
Joyce Rowe's Star
Crb 15h52'21" 35d19'54"
Joyce Scheffer-Glaze
Cas 23h20'50" 57d58'57"
Joyce Schleuniger
Her 18h23'51" 12d34'17"
Joyce Severson
Crb 15h48'56" 37d30'48"
Joyce Smith
Umi 15h9'7" 84d25'37"
Joyce Snyder, Mom & Grammie
Leo 9h22'25" 15d40'40"
Joyce Spector's Light
And 1h29'23" 37d18'39"
Joyce Steel
Cas 23h59'16" 49d40'50"
Joyce & Steve 070707
Cnc 8h14'30" 22d13'51"
Joyce & Steven Nelson
Umi 16h16'16" 80d53'8"
Joyce Sykes
Cas 0h19'48" 55d49'32"
Joyce & Tony's Special Star
Cyg 20h4'23" 54d7'6"
Joyce Toth Bennett
And 23h41'38" 44d34'6"
Joyce V. Richards
Vir 11h54'54" 0d2'15"
Joyce V. Richards
Vir 11h49'15" 1d15'9"
Joyce V. Richards
Vir 11h47'10" 1d54'41"
Joyce Vordahl
Lib 14h54'55" -0d46'36"
Joyce Weiss
Cas 1h39'18" 63d53'52"
Joyce & William Ballard
Cyg 21h20'0" 51d30'13"
Joyce Wilson Jurgensen
Ori 5h43'5" 5d29'42"
Joyce Womack
Crb 15h34'28" 32d0'29"
Joyce Workman
Cas 1h2'3" 64d22'48"
Joyce Yi-Ha Leung Ho
Aqr 21h50'8" -2d42'36"
Joyce Yvonne Grubbs
Cyg 20h18'52" 48d28'8"
Joyce Zirbes
Crb 15h38'17" 37d59'18"
Joyce, la mejor madre
Cas 0h50'4" 64d5'30"
Joyce-Best Wife, Mother, and Nana
Cas 1h22'16" 57d20'6"
Joyce-Jessica
And 23h59'40" 39d29'43"
JoyceKReynolds
Cap 20h14'28" -16d1'46"
Joyce-Lee 2004
And 2h22'28" 48d52'42"
Joycelyn
Ori 5h29'28" 1d31'28"
Joycelyn Anita McKay Crumpton
Sco 17h3'39" -39d58'39"
Joyce's Nitelite
Crb 15h36'1" 31d5'39"
Joycie
Leo 11h14'8" -2d15'38"
Joydancer
Cnc 8h52'44" 30d23'2"
JoyElizabeth My Best Friend My Wife
Uma 10h30'15" 47d44'20"
Joye's Dreams 381
Lib 15h47'3" -12d55'43"
Joyeux Anniversaire Femme Perle
Cen 12h31'44" -34d36'13"
Joyjoy
Vir 12h46'16" 0d44'21"
Joy-Joy
Leo 10h15'50" 19d25'49"
Joy-Joy Emerson
Uma 11h36'46" 39d55'0"
Joy-Leigh Ray
Psc 23h50'38" -0d18'45"
Joylyn Boynuince
And 0h33'12" 27d25'11"
Joylynn
And 23h23'5" 41d18'26"
Joylynn M. Trinidad
Cap 21h37'42" -12d9'0"
Joynoel Bermingham
Tau 3h37'56" 27d25'57"
Joyous Sun Ray & Demeter Isis
Cma 6h45'24" -18d15'8"
JoyRay
Uma 14h24'47" 62d19'58"
Joys Beartiful Brown Eyes
And 23h44'15" 34d54'38"
Joy's Brilliant Star
Tri 2h22'25" 34d43'39"
Joy's smile
Ori 6h5'16" 10d54'59"
Joy's Snow White
Her 17h49'50" 47d7'33"
Joy's Star
And 0h32'50" 42d52'58"
JOY'S STAR
Ori 5h32'15" 6d34'14"
Joy's Valentine
And 0h32'59" 32d53'40"
JoySmiley
Uma 9h21'4" 42d55'24"
JoyWillYouMarryMeCharlie
Lib 15h20'1" -8d49'47"
Jozef Jana
Cnc 8h27'35" 17d17'3"
Jozes Hajdukiewicz
Uma 13h37'56" 50d19'4"
Jozes Kieltyka
Uma 11h55'4" 39d19'4"
Jozlynn AnnMarie Stewart
Leo 9h41'19" 27d52'24"
József 40.
Aqr 21h59'4" -12d13'51"
Jozso 1987
Psc 0h6'10" -0d58'26"
JP
Aqr 21h45'5" -1d7'32"
J.P.
Sco 15h52'39" -20d32'24"
J.P.
Cas 0h41'29" 65d23'13"
jp
Uma 13h19'22" 58d47'4"
JP
Gem 6h46'3" 27d47'47"
JP
Cyg 19h58'2" 52d2'23"
J.P. 1996
Gem 7h8'55" 34d29'43"
J.P. and Francine Wright
Lyr 18h31'36" 36d1'33"
J.P. G.W. W.F. S.M.F.
Uma 10h31'53" 59d30'28"
JP Haynes Nothing Short of Wondrous
Vir 13h10'15" -17d8'24"
JP & Helen
Lyr 19h0'20" 43d2'48"
JP Hollingsworth
Her 17h44'10" 48d19'37"
JP "Jacqueline Paulette" Eckstrom
Cnc 8h58'33" 15d51'36"
JP Kruer....a star in loving memory
Vir 12h38'12" -0d18'9"
"JP" MY LOVE
Gem 6h59'19" 15d21'40"
JP N&J
Gem 7h15'32" 17d28'32"
J.P. Perkerewicz
Aqr 21h36'46" 2d5'13"
Jp Schmidlein & Alana Rome
Uma 11h19'16" 44d57'23"
JP Sekinger
Uma 8h41'41" 55d28'0"
JPa
Sgr 19h21'57" -27d41'55"
JPA911 Happiness Star
Umi 14h34'46" 73d43'40"
JPC 15761.6
Ari 2h16'41" 14d42'48"
JPDAD0551
Her 18h7'19" 48d14'39"
JPh Raimbault
Vir 11h42'14" -1d35'15"
J-Philippe et Nathalie Lagache
Uma 11h58'26" 63d25'56"
JPitt KPitt
And 1h55'21" 46d33'28"
JPJ60
Sgr 18h59'37" -35d19'53"
JPJefferisELSudbury - TrueLoveForever
Cra 18h49'32" -40d49'8"
JPK
Her 17h2'31" 33d5'12"
JPM0747
Aql 19h0'21" 5d14'34"
jp.ninie.for.eternity
Cep 22h46'48" 71d37'20"
JPony
Tau 4h11'45" 17d10'14"
JPPITTSII
Uma 11h51'51" 48d30'5"
JPri2
Ari 2h23'58" 11d23'16"
JP-Rock
Boo 15h17'50" 41d37'59"
JP's Dearest Ser Huey
Cnc 9h10'21" 16d28'9"
JP's Star OYea
Cnc 8h0'24" 12d48'43"
jpsbp
Her 17h33'27" 26d42'3"
JQ
Psc 0h56'58" 33d20'48"
JQ and NN
Cyg 20h27'25" 47d21'56"
JR
Psc 0h1'0" 9d59'11"
JR
Uma 10h35'25" 63d20'1"
J-R Barnes
Her 18h27'54" 12d7'12"
JR Christina 2007.7.7
Ori 5h51'45" 21d7'4"
J.R. Courage
Cen 13h10'7" -43d38'22"
JR DEEMER
Lyn 8h53'28" 41d45'27"
J.R. Dennehy
Dra 18h51'35" 53d54'47"
Jr Elias Cardenas
Cep 22h23'5" 61d27'38"
Jr. G.
Psc 0h29'22" 5d6'39"
JR Hughes
Uma 10h24'40" 45d39'0"
Jr. N J.
Lib 15h16'39" -9d14'26"
J.R. Rumpler
Ori 5h34'19" 5d26'0"
J.R. & Sally Forever
Uma 11h48'1" 35d5'18"
J.R. "Superboy" Lynn
Her 17h7'58" 17d22'10"
JR The King
Cep 22h12'33" 54d10'14"
Jr Wildfire
Aql 19h35'32" 3d16'19"
JRA & JAN...All my wishes come true
Cyg 19h42'37" 53d14'39"
jrb
Ori 6h5'12" 17d15'48"
JRC + SRW = Perfect Fit
Uma 12h55'19" 54d35'17"
Jre
Uma 8h31'6" 63d49'54"
jrenteria712
Umi 14h13'42" 68d52'2"
JRGramigna
Sco 17h50'3" -40d10'55"
J-Rod
Ori 6h7'40" 16d15'59"
JRP1981
Uma 11h41'14" 50d48'6"
Jr's 4.0
Uma 11h19'47" 45d10'7"
JRSentz 03-09-95
Psc 23h4'22" -0d38'28"
J.R.Swanson 3-26
Ari 2h12'25" 25d8'7"
JRW
Cam 4h25'0" 69d6'29"
JRW & SBB ~One Love~
Ori 5h4'16" 4d23'7"
JRW-III
Psc 1h3'29" 25d52'54"
J.R.Wright
Per 2h38'13" 55d42'5"
JS
Sgr 19h40'12" -16d0'20"
J.S. Kessel
Eri 4h22'11" -0d17'53"
JS mori eta-naru aiku
Ari 2h41'33" 26d18'44"
J's Star
Psc 23h28'43" 5d1'28"
J's Star
Cru 12h39'44" -60d33'22"
J's Starman
Ori 5h22'57" 2d9'37"
Js2Hh4E
Lib 14h51'38" -1d37'21"
Jsabella Graf
And 23h16'13" 48d28'50"
Jsabella Huber
Com 11h58'26" 20d18'11"
JSCAMMB of the Frisco Brubakers
Gem 7h1'2" 31d49'26"
J-S-Deland-E-Thibaudeau-02-21-2007
And 23h58'15" 34d25'10"
JSE
Tau 4h1'22" 16d7'53"
J-SEvans
Peg 22h2'14" 19d14'41"
JSF 143
Sco 16h52'6" -41d54'36"
JSlettvet
Cas 1h20'44" 61d21'41"
J.S.L.L.
Tau 4h6'16" 27d46'19"
J.S.M
Sgr 18h34'12" -22d40'48"
JSM Robot
Uma 12h43'1" 54d23'58"
JSpark's Brilliance
Gem 7h21'52" 25d47'22"
J-STAR 22
Cyg 20h52'49" 32d42'40"
J.Suliszka
Cyg 19h38'55" 34d49'59"
J.T. Carricarte (The Last Romance)
Cep 22h28'22" 69d21'55"
J.T. Chambers
Leo 10h17'27" 17d1'39"
JT Dolan
Ari 3h20'30" 27d55'15"
J.T. Hunter
Peg 21h48'58" 15d2'53"
Jt & Katie EvErLaStInG LoVe!
Cyg 21h35'50" 47d46'36"
JT & Kelli
Gem 7h45'28" 21d32'34"
J.T. - Majestic Miracle
Uma 9h28'56" 47d56'50"
J.T. Mayo
Per 3h23'11" 50d17'27"
JT & Mel - Always
Eri 3h44'58" -0d52'16"
JT Parr
Her 18h16'30" 15d32'11"
JT Wilburn
Sgr 19h49'47" -28d18'30"
JT, Teri, Court + Chandler Mateka
Del 20h42'18" 15d26'59"
JT11706
Lyn 7h31'46" 43d34'51"
jtand6
Uma 11h58'20" 59d6'28"
JTA's Wandrin Star
Cep 0h59'21" 79d23'9"
JTG + KEG
Cap 20h58'55" -19d46'10"
JTGTDS7707
Ori 4h49'7" 2d45'32"
JT-IV
Vir 13h5'54" -21d49'21"
JTK2003
Cnc 8h56'0" 24d11'9"
JTOSH
Aqr 22h30'26" -1d48'40"
JTP - 6/10/1985
Peg 22h13'8" 6d8'35"
JTR
Her 17h14'22" 28d36'58"
JTRussell
Uma 8h47'25" 69d3'55"
JT's
Uma 12h51'36" 62d39'43"
J.T.'s Fire
Lib 15h35'2" -5d15'29"
JT's Yhetti Star
Dra 17h14'50" 65d33'28"
J.T.TannenbaumT.L.N.D.
Boo 14h46'7" 28d33'8"
Ju ju
Cnc 8h13'51" 8d56'11"
Ju Ju
And 1h48'46" 37d36'16"
Ju Ju
Umi 13h54'47" 78d5'57"
Ju Ju Bees
Sco 17h56'11" -31d3'38"
Ju86
Cas 1h59'23" 69d4'22"
Juan
Sgr 19h47'6" -14d39'15"
Juan A. Martinez's "Amazing Heart"
Her 16h44'2" 39d46'23"
Juan Amador Rodriguez
Uma 9h34'31" 61d58'26"
Juan and Aimee's Star
Ori 6h23'50" 13d17'33"
Juan and Dawn Garcia Twentyfive
And 2h34'9" 39d56'6"
Juan and Tess Arroyave
Cyg 19h47'50" 43d25'20"
Juan Andres Cardenas Gonzalez
Umi 14h37'26" 86d43'23"
Juan Andres Ruiz de Alda
Gem 6h44'15" 22d14'5"
Juan Antonio Gallego Cordero
Gem 6h49'10" 16d52'37"
Juan Antonio Vecino
Gem 7h43'49" 32d49'2"
Juan Armando Araujo
Lib 15h17'49" -20d34'47"
Juan (Austin) Aguirre
Ori 5h3'56" 2d42'19"
Juan Bonilla
Ori 5h42'59" 3d41'58"
Juan Burciaga
Her 16h41'39" 28d29'18"
Juan Camilo Lopez
Uma 9h47'36" 61d15'17"
Juan Carlos
Ari 1h50'12" 18d37'20"
Juan Carlos Cales
Her 17h8'48" 14d47'24"

Juan Carlos Cruz Arrayz
Pho 2h15'34" -45d39'24"
Juan Carlos Dumas
Per 2h17'43" 57d6'10"
Juan Carlos Gonzalez
Boo 14h44'23" 18d50'43"
Juan Carlos Gonzalez's Family Star
Sge 19h44'47" 18d51'11"
Juan Carlos Hernandez
Umi 15h20'22" 68d51'37"
Juan Carlos Ibarra, Jr.
Tau 5h34'5" 19d25'44"
Juan Carlos Molina-Kennedy
Lac 22h23'43" 47d58'52"
Juan Carlos Motamayor
Her 17h58'51" 20d12'4"
Juan Carlos Orozco
Sgr 18h49'20" -20d16'33"
Juan Carlos Rivera Rodriguez
Sco 17h51'29" -35d48'28"
Juan Carlos Solis Rosas
And 23h16'18" 43d44'12"
Juan Carlos Suarez
Cen 13h38'17" -41d7'18"
Juan Carlos Zamora
Boo 14h25'22" 14d24'52"
Juan Ceja
Psc 1h14'20" 3d34'32"
Juan Colo'n
Del 20h33'11" 17d1'32"
Juan Daniel Hernandez-Mombiedro
Cap 21h54'37" -11d14'18"
Juan Daniel Sanchez
Ori 5h22'47" 2d23'22"
Juan Diego Perez
Dra 18h26'5" 66d15'4"
Juan D'vaughn Pierre
Gem 6h46'3" 13d44'11"
Juan Enrique Mejia
Uma 10h39'31" 68d7'25"
Juan Fernandez Mendez
Peg 21h47'36" 25d42'16"
Juan Francisco Fernandez Muñoz
Lyn 8h24'15" 34d49'25"
Juan G. Santiago
Cyg 19h48'56" 31d40'11"
Juan Gerardo Gonzalez Lagoa
Ori 5h16'58" -6d44'1"
Juan - God has been gracious
Sco 17h4'20" -37d16'48"
Juan "Gonzo" Gonzales
Vir 14h49'5" 5d20'45"
Juan Hilario Perez Cardona
Sgr 18h27'4" -17d31'21"
Juan & Jennifer's Christmas Star
Lib 15h12'47" -16d39'56"
Juan Jesús
Del 20h23'40" 9d40'13"
Juan Limasa III
Dra 16h33'51" 67d48'3"
Juan Loves Susan
Cam 5h31'14" 66d14'23"
Juan Lu and Cedar Lee
Cap 20h9'50" -20d25'46"
Juan M. Ortiz
Psc 23h57'56" 6d7'23"
Juan & Maria
Cyg 21h22'36" 39d19'27"
Juan Miranda
Tau 3h41'5" 11d21'23"
Juan Morales
Sgr 18h16'10" -19d38'14"
Juan Navarro
Vir 13h56'9" -7d50'45"
Juan Pablo
Boo 14h15'54" 28d24'26"
Juan Pablo Barreto
Ari 2h9'41" 22d57'1"
Juan Patino
Cap 20h19'5" -9d46'53"
Juan Quintanilla
Dra 17h59'23" 56d45'3"
Juan Ramon Sanchez
Ari 2h15'7" 26d24'15"
Juan Ratita Blanca Pasos
Per 3h13'9" 52d33'17"
Juan Rivera
Lyn 8h19'39" 44d47'15"
Juan Roberto Vega
Ori 5h21'41" 3d21'35"
Juan Serano
Uma 10h0'32" 71d11'45"
Juan Servin Fonseca
Leo 10h56'22" 15d14'41"
Juan Tello
Tau 5h13'5" 25d10'16"
Juan, Luna y Maite
Psc 1h9'37" 11d33'36"
Juana
Lib 15h3'51" -29d53'11"
Juana Alers Duprey
Uma 11h32'54" 60d28'2"
Juana Casillas
Sco 16h19'59" -17d30'52"
Juana Maria Ruiz
Aqr 23h31'59" -17d15'18"
Juana Martinez - Shining Star
Lyn 7h26'45" 45d34'55"
Juana McClain
And 2h31'32" 43d32'57"
Juana Ramos Santiago Reynoso
Uma 11h36'51" 62d43'5"
Juanchi de la Rua
Tau 5h47'26" 17d23'22"
Juanchi y Mari
Cyg 20h23'35" 54d23'41"
Juanita
Mon 6h46'52" -0d47'35"
Juanita
Her 16h48'3" 33d47'4"
Juanita A. Urioste
Uma 11h56'33" 42d48'20"
Juanita Aviles
Uma 11h9'17" 33d23'41"
Juanita Bokano
And 1h31'7" 34d3'21"
Juanita Carpenter
Sgr 18h17'31" -23d10'21"
Juanita Flores
And 23h16'17" 41d47'45"
Juanita G. Wilson
Crb 15h37'15" 38d5'8"
Juanita Guadarrama
Cap 20h49'51" -14d39'50"
Juanita & Harvey Meir
Cyg 21h13'24" 45d46'7"
Juanita "Jenny" Bachiller
Ori 5h27'24" 6d27'21"
Juanita Joan
Vir 13h41'37" 3d45'12"
Juanita Lucille Turin
And 22h59'21" 50d48'16"
Juanita Luisa
Cyg 21h10'37" 47d23'31"
Juanita Mae Vaught
Vir 14h35'34" 5d57'10"
Juanita "Mammaw" Stegner
Cas 1h35'49" 60d25'21"
Juanita Marie Johnson
Vir 11h44'35" -4d28'43"
Juanita Martin
Cas 22h59'59" 53d47'39"
Juanita Minter
Cap 20h18'28" -21d27'1"
Juanita New
Ari 3h6'35" 27d47'47"
Juanita "Nita" Louise Choate
Psc 1h37'59" 24d8'47"
Juanita Odell Reeves Laminack
Cnc 8h32'41" 23d46'18"
Juanita Pettigrew
Cas 0h42'1" 60d40'31"
Juanita Pierce
Cas 0h57'7" 50d37'24"
Juanita Rachael Bashaw
Cap 20h48'32" -26d1'53"
Juanita Ray Monegan
Her 17h55'2" 29d39'19"
Juanita Robinson
Aqr 22h15'37" 1d2'23"
Juanita Sanchez
Per 3h25'16" 44d2'7"
Juanita Schoenmann
Ari 2h7'59" 18d32'35"
Juanita Tremor Tabayoyong
Uma 11h17'11" 57d5'53"
Juanita Trostle
And 23h22'42" 42d32'45"
Juanita Who Is Much Loved
Tau 3h43'21" 15d50'8"
Juanita's Butterfly Star I L Y 4 E
Cyg 20h22'29" 36d29'53"
Juanmy
Vir 13h45'43" -18d16'31"
JuAnnaJor
And 0h26'52" 31d29'32"
Juaquin Enrique Cordero
Her 16h16'0" 45d45'51"
Juasito
Psc 23h10'7" 4d28'37"
Jubbley Mummy's Star
Cas 0h52'45" 56d17'0"
Jubel D' Cumpiano
Tau 5h7'16" 25d39'16"
Jubilee
Leo 11h15'41" 11d22'29"
Jubilene
Lmi 10h39'6" 32d34'26"
Juborovia
Cas 23h26'40" 57d48'6"
Jubri
Cyg 21h11'3" 44d15'10"
Juchan
Sgr 18h51'33" -19d27'26"
Juchert, Hans-Jürgen
Ori 5h12'57" 7d13'40"
Juchiban Support
Gem 7h4'35" 10d30'33"
Jucobie
Sgr 19h37'19" -16d24'1"
Jucrystaldy
Aqr 21h53'12" -3d13'16"
Jud
Cnv 12h38'13" 39d19'45"
Judah
Psc 2h1'47" 8d20'47"
Judah Saul Greenwald
Cep 21h52'11" 58d28'22"
Judah White
Del 20h46'49" 11d42'56"
Judamel
Ori 5h51'15" 9d5'45"
Judd Bazzel
Cnc 8h48'51" 14d29'46"
Judd Moul 50
Aqr 21h59'10" -3d35'21"
JuddNova 2004 Jeff Melinda Garrettw
Cyg 21h36'54" 39d17'47"
Jude
Uma 9h58'1" 47d35'3"
Jude
Cnc 9h9'23" 31d40'15"
Jude
Cam 12h39'27" 76d27'20"
Jude & Abigail
Cyg 19h35'43" 31d33'14"
Jude and Laurie Dressler
Cyg 21h14'51" 31d33'43"
Jude Anthony Vanover
Gem 7h23'12" 32d1'47"
Jude Cameron McKechnie
Vir 13h33'19" -9d10'6"
Jude Caymin Cordova
Uma 8h32'29" 72d17'59"
Jude "Cookie" Collett
Psc 23h28'0" 5d33'53"
Jude Daniel Oliver
Psc 1h27'26" 18d40'34"
Jude David Norman
Uma 9h38'47" 70d3'49"
Jude Elias Rowell
Ori 4h48'11" 14d11'12"
Jude Elizabeth Thomas
Uma 9h20'24" 58d19'11"
Jude Francis Lang
Ori 5h36'58" -10d46'57"
Jude Gorecki
Sco 16h25'19" -40d57'38"
Jude Henry Lucas
Aur 6h5'9" 37d26'9"
Jude Kurdy
Ari 2h36'28" 22d3'13"
Jude Lee
Cyg 20h29'1" 52d31'2"
Jude M. Cochran
Uma 11h48'57" 32d46'1"
Jude Samuel Stratford
Per 2h56'53" 55d19'26"
Jude Symonds Keener
Lyn 6h35'55" 55d44'11"
Jude T. Wanniski
Uma 9h32'16" 57d22'31"
Judee Marie
Cyg 20h19'17" 51d24'43"
Judeestar
Eri 4h15'31" -9d44'55"
Jude-McCartney-Gray
Umi 13h46'7" 69d48'45"
Judes - 18.3.1996
Gem 6h27'52" 25d0'49"
Jude's Special Wishing Star
Sco 17h45'9" -41d27'27"
Judey Star Calcaterra
Cyg 20h1'19" 40d39'44"
Judge Carl Bryson Jones
Cnv 13h45'53" 36d42'49"
Judge Frank P. Walsh
Cru 12h17'54" -56d47'0"
Judge Kathleen M. Dailey
Leo 9h55'37" 15d7'36"
Judge Richard W. Millard
Cep 21h41'44" 64d7'35"
Judge Roger C. Plichta
Cnc 8h9'20" 24d31'48"
Judge Walker Thomas
Ari 1h49'27" 21d17'45"
Judi Baby Davey
And 23h0'21" 41d27'21"
Judi Behan Sweet
Psc 1h20'17" 12d28'50"
Judi Dekart *You Are My Star*
Cas 2h10'14" 62d18'4"
Judi Dempsey
Crb 15h36'23" 27d6'7"
Judi French
Cas 1h21'59" 63d16'41"
Judi Galie
Leo 9h42'30" 12d44'49"
Judi Jacobsen
Lac 22h27'22" 50d26'56"
Judi Kay Fryer
Cnc 8h46'55" 31d34'58"
Judi Maguire
Leo 9h31'20" 11d35'19"
Judi McQueary
Vir 11h37'57" -5d0'23"
Judi My Love
Aqr 22h37'30" -2d3'19"
Judi & Oscar
Cyg 20h27'32" 56d10'48"
Judi Rase
Tau 5h53'19" 25d10'12"
Judi Yingst
Leo 10h12'18" 11d35'28"
Judianne Flores
Mon 6h50'15" -0d5'31"
Judie Ganek
Leo 9h52'5" 28d50'2"
Judie Gooch
Ori 6h9'15" -3d59'49"
Judie Lynn Thompson
Crb 16h6'33" 28d26'9"
Judiebear
Tau 3h46'5" 20d38'32"
Judie's Dream - ER
Uma 9h39'50" 69d1'44"
Judiet Bailey
Sco 17h13'38" -43d28'4"
Judimom1
Cap 21h21'22" -27d21'27"
Judi's Star
Leo 11h36'33" 27d28'36"
Judi's Wish
Leo 10h34'24" 8d53'34"
Judi'sNine
Gem 7h46'32" 31d20'31"
Judit
Ori 6h19'37" 15d30'22"
Judit & András = 100
Uma 12h4'13" 57d31'36"
Judit Blahut
Psc 0h55'6" 27d0'52"
"Judit szerencsecsillaga"
Tau 3h54'18" 0d31'16"
Judith
Leo 9h41'26" 19d23'1"
Judith
And 1h24'54" 41d39'24"
Judith
And 0h55'38" 38d16'15"
Judith
Per 4h38'13" 40d13'18"
Judith
Cas 0h44'42" 53d33'49"
Judith
Sct 18h47'44" -12d33'15"
Judith
Pho 1h18'48" -48d14'30"
Judith
Cnc 8h5'20" 22d42'58"
Judith A. Knickle
And 0h56'37" 41d13'10"
Judith Adeline`
Cnc 8h1'15" 12d19'51"
Judith Adrienne McClaughlin
Per 4h41'47" 45d20'5"
Judith Aide Soto
Psc 1h15'25" 15d47'20"
Judith Allen-Ferguson
Ari 2h2'42" 12d34'13"
Judith Allyn Shapiro
Peg 22h12'47" 23d45'44"
Judith and Pat McDonagh
Cyg 20h38'31" 34d27'37"
Judith Angel Flühmann
Lyr 18h50'2" 33d18'31"
Judith Anlauf
Cyg 21h19'26" 32d53'9"
Judith Ann
And 0h56'31" 39d9'24"
Judith Ann
Lyn 7h59'31" 43d3'16"
Judith Ann
Leo 9h28'49" 25d57'48"
Judith Ann
Leo 10h9'33" 15d40'15"
Judith Ann
Aqr 20h39'59" -9d14'51"
Judith Ann
Aqr 22h19'36" -7d40'10"
Judith Ann Ashbee
Lmi 10h38'7" 32d50'6"
Judith Ann Bullard
Sco 16h45'46" -27d40'55"
Judith Ann Butcher
And 1h11'43" 45d2'37"
Judith Ann Chebli
Cas 1h47'50" 61d35'33"
Judith Ann Davis
Aqr 22h32'26" -0d26'2"
Judith Ann Fischer
Peg 21h36'21" 26d50'52"
Judith Ann Fisher
And 1h31'43" 49d12'26"
Judith Ann Jaggers
Cas 1h28'1" 57d11'57"
Judith Ann Kiely
Sgr 17h53'13" -17d2'35"
Judith Ann Kirk
Cyg 20h41'18" 54d50'24"
Judith Ann Lathrop
Lyn 7h7'13" 52d40'6"
Judith Ann Poppleton
Uma 11h23'22" 61d5'31"
Judith Ann Porter Born June 4, 1951
Leo 9h52'57" 23d30'18"
Judith Ann Recklie Robertson
Cnc 8h46'24" 17d16'43"
Judith Ann Richardson Bunney
Uma 10h22'17" 44d18'52"
Judith Ann Stahlman
Lyn 7h34'21" 49d22'57"
Judith Anne Connell
And 23h39'8" 48d23'17"
Judith Anne Geatches
Lyn 6h32'26" 56d16'42"
Judith Anne Hildre
Sco 16h12'0" -11d19'18"
Judith Anne Jackson
Gem 7h12'27" 14d32'34"
Judith Anne Whitmire
Mic 21h4'11" -39d43'11"
Judith Ann's Star
Vir 11h39'36" -1d46'31"
Judith B.A. Walker
Dra 18h26'54" 74d45'55"
Judith Billings
Pho 0h39'12" -39d56'31"
Judith Bolton Sells
Cas 0h11'18" 55d49'8"
Judith C Gibson
Leo 11h21'9" 16d12'41"
Judith Carol
Uma 10h18'4" 65d42'6"
Judith Carol Edlinger
Ari 2h26'6" 24d35'54"
Judith Cataldo
Aqr 21h16'39" 1d37'21"
Judith Cloutier
Cyg 20h23'4" 48d27'9"
Judith Collins
Uma 11h3'4" 36d9'2"
Judith Coralie et Thalys
Cma 6h36'45" -24d37'40"
Judith Dawn
Sgr 17h54'49" -17d3'51"
Judith Denise O'Neill
Aqr 22h13'18" -0d7'20"
Judith Diane Sasse
Cap 21h55'48" -14d59'56"
Judith Diane Warren
And 23h6'32" 49d23'44"
Judith Douthit
Crb 15h49'30" 36d30'58"
Judith Driscoll
Cas 0h40'39" 53d33'0"
Judith E Chason
Cyg 21h29'10" 31d59'1"
Judith E. Stewart
Cas 1h43'16" 64d38'20"
Judith Elaine
Lib 15h7'27" -6d49'12"
Judith Elaine Osborne
Tau 5h32'56" 24d39'18"
Judith Elizabeth
Gem 6h21'42" 19d46'8"
Judith Ellen Antisdel
And 1h32'5" 48d13'9"
Judith Ellen Caldwell
And 1h54'18" 46d1'47"
Judith Ellen Rutowski
Uma 10h2'6" 66d48'8"
Judith Eloise
And 1h29'14" 48d59'16"
Judith és István Szerelemcsillaga
Uma 10h58'27" 69d14'28"
Judith Eva Nagy
Pho 0h7'22" -45d0'22"
Judith Fores
Tau 3h36'54" 27d50'44"
Judith Green
Cas 23h58'51" 66d2'43"
Judith Hahn
Crb 15h34'36" 37d46'31"
Judith Henry
Cap 21h41'15" -15d37'55"
Judith Hill McNally
Sgr 17h53'48" -29d56'46"
Judith Hoffmann
Ari 3h13'47" 12d23'30"
Judith Holland
And 23h45'35" 34d21'57"
Judith Hui San Tsang
And 23h38'47" 42d31'44"
Judith Irene
Ari 2h40'43" 17d7'55"
Judith Izzo
Lyn 7h55'24" 50d27'23"
Judith J Arevalo
Gem 7h20'6" 16d37'32"
Judith J. Ledgerwood
Cep 21h43'19" 64d6'15"
Judith Jacqueline Goodrich
Uma 11h16'38" 48d27'53"
Judith Jacqueline Weissman
Sco 16h5'24" -11d33'7"
Judith Jane Anderson
Gem 6h29'45" 26d27'49"
Judith K. Brelsford
Del 20h38'52" 15d28'43"
Judith K. Brendal
Cap 21h35'17" -13d47'47"
Judith K. Vogt
Psc 22h55'1" 2d38'51"
Judith Kenrick
Cas 1h55'1" 61d50'25"
Judith Kinser Schoonover
And 0h57'47" 37d0'27"
Judith L Tierney
Dra 15h9'52" 55d25'28"
Judith Laird Newman "Judy"
Uma 10h41'57" 70d31'44"
Judith Lay Brewer Davis 02/27/1941
Ori 6h14'58" 8d26'17"
Judith Lee Hendrix Brumfield
Ari 3h23'29" 16d11'53"
Judith Letticia Tyree Stinson
Dra 17h44'33" 56d23'45"
Judith Lisa
Lyn 8h32'28" 35d29'11"
Judith L.Kurlander
Lyn 7h50'53" 50d8'30"
Judith Lynette Latta
Cru 12h30'25" -63d0'26"
Judith Lynn Dziedzic
Tau 4h35'36" 18d47'52"
Judith Lynn O'Connell
Psc 0h18'5" 15d21'3"
Judith Lynn Paul
Cas 0h0'38" 56d37'4"
Judith M. Folmar
Cas 1h12'49" 64d47'12"
Judith M. French
Aqr 22h24'10" -5d44'4"
Judith M. Kirchner
Lib 15h50'22" -19d6'52"
Judith Macke
Lep 5h48'52" -11d2'57"
Judith Manczuk Krucki
And 0h39'28" 24d14'59"
Judith Margaret
Sgr 20h26'45" -41d51'34"
Judith Margo Elfenbein
Crb 15h32'59" 32d37'31"
Judith Margolin
Sgr 18h1'43" -26d47'29"
Judith Martin
Cyg 19h40'33" 29d17'50"
Judith Mary Batty
Cas 0h26'5" 60d43'4"
Judith Mary Bruer
Lac 22h21'37" 53d49'24"
Judith May Jackson
Gem 7h13'32" 23d11'35"
Judith McCarty
Cas 23h21'44" 55d5'48"
Judith McNaulty Star
Per 3h1'53" 47d54'38"
Judith Menger
Aqr 22h23'52" -6d55'21"
Judith Muneca Casado
Gem 7h4'14" 34d46'38"
Judith Norman
Cap 21h25'17" -22d9'23"
Judith Oakley
And 2h1'58" 38d39'31"
Judith P Sabree
Sgr 18h3'34" -28d2'50"
Judith Patricia Markowski
Psc 0h44'25" 9d24'36"
Judith Pilkington
Com 13h22'49" 24d11'56"
Judith R. Linke
Uma 11h14'28" 62d40'51"
Judith Rabold Nicholas
Aqr 20h49'14" 1d1'10"
Judith RL
Ari 2h33'51" 25d12'35"
Judith Rose
Lmi 10h10'32" 41d5'45"
Judith Rose Aquilina
Cnc 8h39'17" 30d45'30"
Judith Ruth Jackson
Leo 10h33'20" 16d6'26"
Judith S. & Charles L. Suplee, Jr.
Uma 11h24'37" 64d56'13"
Judith S. Shaffer
Uma 8h39'28" 72d10'20"
Judith Santos
And 0h40'47" 41d44'53"
Judith - Semsa 2
Sgr 18h29'34" -27d34'23"
Judith Simmons
Cet 1h25'37" -0d33'26"
Judith Sims
Cnc 8h15'14" 11d56'21"
Judith Stelter Meline
Cnc 9h17'40" 32d10'30"
Judith Swanson
And 1h1'54" 37d36'58"
Judith Swanson
And 0h43'7" 43d14'22"
Judith T. Detig
Lib 15h36'8" -17d55'30"
Judith Tegan
Psc 1h55'13" 5d35'53"
Judith "The Heroine" Valeski
Mon 7h15'52" -3d49'50"
Judith Viola
Cas 0h55'50" 57d0'50"
Judith Watts Roberts
Sco 15h55'21" -20d33'2"
Judith Weinstock
Uma 10h26'56" 43d18'26"
Judith Young
And 0h29'18" 26d52'15"
Judith Yvonne Mullin
Vir 12h31'20" 12d2'18"
Judith Zimmerman, My Sunshine
Cas 0h46'30" 64d25'37"
Juditha
Leo 11h0'25" 23d44'48"
JudithD
And 1h44'33" 49d11'53"
Judith's Baby Faith
Uma 11h37'41" 63d29'52"
Judith's Diamante
Sgr 19h15'49" -34d51'18"
Judith's 'North' Star
Pyx 8h38'36" -20d45'18"
Judith's star
Psc 1h25'43" 16d23'3"
Jud's Nitelite ~ Glows for Fahs
Lyn 7h27'27" 56d40'56"
Judson Alexander Guice
Per 3h46'33" 33d58'56"
Judson Freeman, Jr.
Ori 5h33'15" 8d0'39"
Judson G. Shelton
Cap 21h18'25" -17d19'0"
Judson "Jud" Miller
Ori 5h29'6" 6d34'37"
Judtih Rose
Cas 0h16'15" 51d57'9"
Judy
And 1h45'33" 48d18'59"
Judy
Uma 11h15'57" 45d3'8"
Judy
Uma 12h1'41" 40d30'49"
Judy
Lyn 8h31'33" 40d47'10"
Judy
Leo 11h26'23" 28d3'24"
Judy
Her 17h27'38" 20d29'52"
Judy
Aqr 23h18'24" -12d40'25"
Judy
Uma 9h57'19" 64d25'2"
Judy 2/18 & Tim 9/18 Puckett
Uma 8h27'34" 63d7'20"
Judy 92956
Crb 16h18'49" 36d40'33"
Judy Adler - A Shining Star
Cru 12h27'40" -59d45'34"
Judy Aiello
Cas 2h17'30" 62d30'47"
Judy and Ed Garland
Lyn 7h57'40" 42d50'35"
Judy and Henry Walley
Cyg 21h28'10" 31d38'4"
Judy and Mark
Cas 2h6'2" 63d48'26"
Judy Anderson
Cyg 20h41'34" 38d18'18"
Judy Ann
Gem 6h44'13" 22d57'30"
Judy Ann
Uma 11h3'25" 58d52'32"
Judy Ann Ayer
Cas 0h28'34" 54d59'13"
Judy Ann Battalini 4/20/1939
Uma 11h31'6" 56d40'18"
Judy Ann Chisum Korry
Uma 11h40'24" 36d48'3"
Judy Ann Fowler
Mon 6h48'57" -0d22'1"
Judy Ann Garrison
Cyg 19h59'30" 48d14'7"
Judy Ann Haney
Lib 14h56'31" -24d21'25"
Judy Ann Leewright
Sco 17h16'3" -32d31'24"
Judy Ann Magnuson-Silva
Uma 9h33'7" 66d8'10"
Judy Ann Mueller - 06/03/1953
Gem 6h58'8" 21d34'8"
Judy Ann San Fratello
Cnc 8h11'30" 28d22'57"
Judy Ann Towhill
Cas 23h7'32" 56d34'50"
Judy Anne Hamilton
Lyn 9h32'41" 39d26'23"
Judy Anne Scholl
Sgr 19h40'51" -15d5'59"
Judy Ashmore Pearson
Ori 6h4'52" 10d37'54"
Judy Asmussen
Cyg 19h46'39" 37d18'31"
Judy Beidle
And 1h41'25" 44d42'14"
Judy Bellerose
Psc 0h43'31" 9d47'4"
Judy - Beloved Mother
Lib 15h6'14" -7d32'9"
Judy Blue Eyes
Leo 10h42'20" 7d22'23"
Judy Blue Eyes Rogers
Leo 11h56'28" 21d49'13"
Judy Boop
Gem 6h33'19" 17d54'35"
Judy Booty
Psc 1h37'19" 3d48'12"
Judy Boris
Cas 0h56'13" 56d53'56"
Judy Breen Philbin
Psc 1h13'32" 23d42'46"
Judy Brickel
Sgr 19h56'27" -28d11'52"
Judy Bryant
Sco 15h55'24" -21d4'13"
Judy "Butterfly Angel" Wilson
Lyr 18h50'31" 35d5'47"
Judy C. Agno
Crb 15h46'28" 29d28'59"

Judy C Norton Uma 10h23'31" 49d31'41"
Judy C Welsh Leo 10h34'7" 17d28'41"
Judy Carpenter Lyr 18h46'16" 44d29'34"
Judy Corcoran Cas 1h38'10" 63d49'20"
Judy Corona Sco 16h30'54" -29d47'14"
Judy <Da Cutie> Howell Uma 9h25'22" 44d46'40"
Judy Dacey Vir 13h19'29" 11d11'12"
Judy Davis Uma 9h14'43" 49d42'33"
Judy Dayton Mon 7h59'35" -0d33'18"
Judy Del Rosso Aqr 22h47'36" -19d41'13"
Judy Delois Hicks Ari 2h17'12" 22d58'53"
Judy & Dennis King Cnc 9h7'44" 17d20'37"
Judy DeVito Cas 23h45'25" 61d12'12"
Judy Doris Roulette Cas 1h20'40" 57d12'15"
Judy Dworin Star of Peace Sgr 18h57'34" -30d4'4"
Judy Elaine Bliss Dra 15h51'58" 54d12'56"
Judy Elaine McDonald Ari 2h18'20" 23d37'31"
Judy Eleanor And 0h54'12" 38d6'52"
Judy Ellen Per 4h36'41" 33d15'19"
Judy és Zsolt Örök Szerelme Tau 5h38'15" 18d53'15"
Judy(Ethyl)Jones Lyn 6h53'21" 50d48'35"
Judy Flesch Felperin Ori 5h32'8" 14d37'37"
Judy G. Minter Mon 7h9'23" -0d54'25"
Judy Gale Genshock Cap 20h49'53" -19d2'58"
Judy Gale Jones And 0h44'43" 45d24'38"
Judy Gale Sparky Gritz Leo 9h28'14" 31d45'18"
Judy Garner Clason Sco 17h52'58" -36d16'10"
Judy Gee Lyn 9h3'53" 42d5'10"
Judy Giacobbe And 0h23'20" 42d6'54"
Judy Girl Vir 12h17'50" -11d29'24"
Judy Giuliano Aqr 22h55'13" -15d54'20"
Judy Glenn Tau 3h41'54" 29d21'39"
Judy Goldberg Lib 14h52'56" -5d18'32"
Judy Grace Uma 9h30'21" 62d44'26"
Judy Graebner Cyg 19h36'57" 48d40'15"
Judy Griffith Tau 5h46'42" 25d35'20"
Judy Hanley Ari 2h42'55" 26d57'11"
Judy Hatch Col 5h55'57" -36d4'16"
Judy Holland Cas 0h17'8" 56d28'22"
Judy & Hon Uma 11h26'22" 44d26'57"
Judy Humphrey Cas 0h23'42" 50d49'44"
Judy in the Sky Crb 15h49'6" 33d45'18"
Judy Irene Freeman Uma 8h9'53" 60d8'23"
Judy Isaac Lyr 18h47'9" 41d11'15"
Judy & Jerry Cohen's Star Uma 11h55'58" 59d8'26"
Judy Jo Leo 11h15'40" -1d46'6"
Judy Jo Stokem Cap 21h37'7" -14d50'47"
Judy John And 0h24'3" 45d39'15"
Judy Johns Sgr 17h51'1" -25d39'47"
Judy Johnson Horton Sco 16h6'43" -25d38'28"
Judy K. Ferguson And 1h16'30" 49d11'10"
Judy K. Harmon Lib 15h33'16" -18d24'46"
Judy K. Whatley Vir 13h14'46" 4d29'45"
Judy Kalista, Wonderful Mother Sgr 18h47'19" -18d29'29"
Judy Kardinal And 23h12'46" 47d38'33"
Judy Karole Sco 16h3'9" -22d10'55"
Judy Katz Cas 0h41'8" 64d31'53"
Judy Kay And 0h20'24" 31d13'14"
Judy Kay Listenberger Uma 14h4'23" 58d48'18"
Judy Kay Stolting Moore Galpin Psc 1h9'56" 4d31'54"
Judy Kingston Tau 4h42'53" 24d41'19"
Judy Kiss Stevens And 0h26'1" 43d22'43"
Judy Kuhn Cyg 21h27'49" 33d11'40"
Judy L Hammons Psc 0h39'9" 17d48'48"
Judy L King Lib 15h17'11" -11d39'22"
Judy Le Mon 7h1'14" 8d3'51"
Judy LeeAnn Hill-Universes Best Mom Her 17h16'48" 22d30'32"
Judy Lee's Shining Star Tau 5h53'45" 24d55'45"
Judy Lorraine Henry Aql 19h43'22" -0d2'11"
Judy Louise Davis Psc 0h17'35" 0d2'46"
Judy - Love of my life Crb 16h14'56" 31d3'29"
judy loves jeff Ori 6h23'20" 16d26'51"
Judy Lucille 8/22/54 Leo 11h35'25" 1d9'28"
Judy Lynn Cyg 19h50'40" 33d19'55"
Judy Lynn Psc 1h18'23" 5d39'51"
Judy Lynn Sco 16h53'25" -35d28'7"
Judy Lynn Epstein Cap 21h25'34" -24d38'36"
Judy Lynn Foster Uma 10h5'25" 48d44'54"
Judy Lynn McCaleb Psc 0h19'26" 17d2'59"
Judy Lynn McRobb Ori 4h45'49" 12d48'0"
Judy Lynn Shearer Cap 21h35'6" -16d29'45"
Judy Lynn Trabulsi Ori 5h13'39" -0d50'44"
Judy M Vir 13h21'49" -8d8'46"
Judy M. Bashara Cyg 20h16'32" 54d26'46"
Judy M. Helm And 2h11'56" 44d24'14"
Judy M. Whitley Uma 10h27'52" 40d24'28"
Judy Mae Mal Lyn 8h20'23" 42d7'39"
Judy Malisa Snyder Cnc 9h19'57" 10d17'48"
Judy Marie And 2h26'13" 45d1'56"
Judy Mazzoil Michael Pho 0h34'19" -43d0'4"
Judy Michel Uma 10h14'45" 48d38'8"
Judy Mikula Uma 11h45'21" 51d54'5"
Judy Miller Cas 0h11'50" 54d9'31"
Judy Miller Her 16h17'19" 7d1'43"
Judy (Mississippi) WILLIAMS, Dr.J. Lac 22h49'54" 49d15'59"
Judy "Mom" Wegerer And 0h22'15" 28d29'10"
Judy Montoya Crb 16h8'45" 35d28'4"
Judy Moore Cam 5h48'9" 66d49'46"
Judy Mullowney Barlow Cas 1h31'22" 62d27'1"
Judy Murphy Aqr 22h38'21" -17d39'34"
Judy Newnes Murphy's Shining Star Uma 11h6'28" 64d26'10"
Judy & Otto Kucera Jr. Cyg 19h16'30" 51d24'26"
Judy - Our Superstar Mum - Thompson Cru 12h32'31" -62d51'6"
Judy Papineau Aql 19h17'31" -0d0'13"
Judy Patricia Miller And 2h35'56" 45d6'57"
Judy Price Cmi 7h35'48" 9d37'16"
Judy Provost And 2h11'5" 37d40'57"
Judy Quigley Turnell Psc 1h8'58" 14d52'44"
Judy Rabotte - Shining Star Lmi 10h7'10" 38d41'34"
Judy & Ramon Cyg 19h56'37" 47d11'4"
Judy & Ray Papesh Per 3h46'22" 49d36'9"
Judy Renee Haley Cyg 21h30'13" 51d40'38"
Judy & Rick Shapiro Cyg 19h37'37" 51d26'14"
Judy & Ron's 50th Anniversary Star Cra 18h7'21" -37d2'23"
Judy Rosales Sco 16h5'28" -14d47'19"
Judy Rose Redmond Cas 1h0'30" 61d21'2"
Judy S. Williams Uma 12h0'44" 41d11'38"
Judy Sartell Cas 0h13'47" 55d41'19"
Judy Schaar Balliet Cyg 19h45'4" 30d18'21"
Judy Schelkun & Family Cas 1h40'58" 64d4'41"
Judy Sue ("Shopgirl") Cas 1h46'9" 71d28'32"
Judy Tackett Gem 7h34'51" 28d5'42"
Judy Tieman Lyn 8h0'32" 44d11'9"
Judy & Tim Cyg 21h43'56" 50d18'42"
Judy Tribble Cas 0h59'20" 60d32'52"
Judy Van Rest Vir 13h48'14" -18d29'6"
Judy Viny Peg 21h43'7" 20d51'49"
Judy Violet Uma 10h29'0" 58d0'3"
Judy Vrablic And 1h31'37" 39d41'12"
Judy & Wavel Hunt & Valen Aqr 23h17'1" -10d56'52"
Judy Werner Uma 10h32'28" 42d52'31"
Judy Whipple Lyn 8h17'57" 52d19'58"
Judy Wilder Uma 8h49'50" 71d28'53"
Judy Willard Tau 4h10'7" 22d56'40"
Judy Williams Cnc 8h48'47" 27d35'14"
Judy Williams Stogniew Vir 11h38'34" 3d57'54"
Judy Wilson Leo 11h47'51" 16d32'39"
Judy Yates Her 17h29'3" 28d23'40"
Judy Yaworski Cnc 8h56'22" 11d39'22"
Judy Z Ari 2h35'37" 26d14'40"
Judy Zika Dra 18h4'59" 71d23'21"
JUDY ZUBISKY Lib 14h29'40" -24d7'49"
Judy, the chunkiest mom ever! Leo 9h28'57" 17d0'48"
Judyann Taveras Ari 3h13'4" 27d45'2"
JudyBen 140 Cyg 21h16'50" 47d39'3"
JUDYBLU Crb 16h13'16" 30d20'29"
judykennedy3858 Psc 0h16'8" 18d40'47"
Judy's Bear And 0h38'14" 37d52'28"
Judy's Diamond of Love Cas 0h35'55" 54d42'36"
Judy's Eyes Cas 23h35'5" 52d7'15"
Judy's Fantasy Del 20h42'10" 17d6'54"
Judy's Heart And 0h27'13" 25d56'10"
JUDY'S JASMINE Cma 6h48'22" -14d51'32"
Judy's Joy And 2h36'37" 45d50'26"
Judy's Shining Star Gem 7h9'33" 27d33'15"
Judy's Shining Star Ori 5h36'31" 2d39'51"
Judy's Shining Star Cap 21h43'57" -22d30'47"
Judy's Star Cap 20h47'27" -23d41'17"
Judy's Star Sco 16h47'54" -38d8'11"
Judy's Star Lmi 10h5'18" 32d46'7"
Judy's Star of Love and Memories And 0h31'30" 40d55'15"
Judy's Star - " WOW !" - 21.10.52 Lib 15h36'6" -5d48'31"
Judy's Tiny Angel Lyr 18h51'30" 31d41'56"
Judy's Unions Uma 9h13'31" 59d27'5"
Judy's White Light Lyn 6h39'1" 56d53'2"
Judystewartstarteacher Lyn 7h41'28" 38d8'13"
Judyth Curlybird Cyg 19h47'9" 34d3'19"
Juel Lmi 10h41'34" 33d34'46"
Juell Chase Solaegui Her 18h39'42" 24d48'54"
Jueneta Boyle Ori 5h39'12" 3d39'9"
Jue-piter Cnc 8h46'13" 17d57'19"
Juerg Walder Cas 1h6'32" 56d40'48"
Juergen Leo 10h52'52" -4d44'32"
Juese Comer Vaughn Uma 12h23'49" 60d35'51"
jüge-sinaida Her 18h21'17" 12d16'1"
Jughead Ori 6h13'31" 15d27'52"
Jughead The MIGHTY Ori 6h21'54" 10d13'35"
Juhász Attila Uma 9h13'22" 70d45'18"
Juhi Bagaitkar Uma 10h30'52" 47d7'47"
Juhi Jaisinghani Gem 6h22'40" 25d16'34"
Juhi's Daddio! Leo 11h52'51" 22d46'37"
Juhos Mónika Csillaga Uma 10h38'3" 61d51'36"
Juice *My Shining Star* Uma 10h38'32" 60d21'18"
Juicey77 Lyn 8h24'35" 34d29'5"
Juicy esto perpetua Lmi 10h7'33" 40d19'39"
Juicy J Cnc 8h29'17" 30d58'8"
Juicy Wrigley Gum 43-93 Mon 7h36'51" -0d39'13"
Juila Kohout Mon 7h0'22" -5d58'48"
Juile Madrid Sgr 18h10'0" -18d37'22"
Juius Durham Per 3h13'2" 41d13'58"
Jujeeb Sgr 18h56'22" -20d10'29"
JuJu And 23h57'50" 38d37'49"
JuJu & JaX Inseparable, even in space! Cru 12h2'55" -63d47'8"
JuJube Adamo Aql 19h41'25" 12d7'20"
JUJUbeast Sgr 18h59'17" -25d46'4"
Juju's Lily Sun Lyr 19h19'33" 38d10'7"
JULA Uma 9h33'42" 68d13'37"
Jula Jane Pho 2h4'23" -41d31'20"
JulaBeth Uma 14h3'7" 58d2'7"
Julaine Cabot Peg 22h23'21" 4d38'48"
Julaine's Universe Sgr 19h14'42" -22d17'17"
Julaugust And 0h49'27" 39d29'33"
JulBud Peg 23h44'3" 24d24'5"
Jule Aqr 23h2'56" -9d17'16"
Jule Uma 8h46'54" 60d48'22"
Jule And Joshua Steele Cyg 19h40'30" 34d42'29"
Jule of the Sky Cnc 8h32'43" 18d53'19"
Jule Thomas Faulkner Her 18h38'23" 18d22'11"
Juleah and Jon Crb 15h27'45" 32d27'53"
Julee Sgr 18h29'28" -27d17'31"
Julee Heckermann Crb 15h41'32" 29d31'3"
Julee Marie & Jonathan Robert Leo 9h49'10" 7d58'30"
Julee Paige Tolk Sgr 18h57'15" -27d13'50"
Julee, Anthony & Britany Silvernail Uma 10h8'39" 46d10'31"
Juleen M. Edwards "Bug" 06/22/85 Uma 11h51'59" 61d56'21"
Julen Iturbe Lyn 7h47'37" 58d36'28"
Julene Nicole Cas 2h13'29" 64d33'16"
Jules Uma 9h8'37" 62d6'12"
Jules Umi 15h33'0" 69d53'1"
Jules Uma 8h53'40" 53d51'11"
Jules Leo 11h10'4" -2d8'56"
Jules Leo 11h20'39" -3d32'47"
Jules Lib 14h54'23" -2d29'1"
JuLeS Sco 16h34'0" -26d32'34"
Jules Cap 20h8'54" -24d23'26"
Jules Aqr 22h38'12" -23d32'29"
Jules Leo 11h15'24" 14d9'56"
Jules And 23h50'57" 42d22'55"
Jules Crb 16h10'44" 36d44'56"
Jules Aqr 21h44'43" 1d57'11"
Jules Ori 5h25'58" 13d25'30"
Jules Ori 4h55'2" 5d35'32"
Jules Vir 11h49'0" 2d36'47"
Jules Ari 2h10'51" 24d43'20"
Jules' Angel Tau 4h7'54" 22d34'54"
Jules ASCH Cnc 8h38'47" 28d54'2"
Jules Brenneis Cas 0h56'53" 57d15'33"
Jules' Constellation Gem 7h5'48" 15d23'4"
Jules & Drew Ridley - Lovers 4 Ever Pho 2h23'54" -40d25'52"
Jules Grosswald "Happy 80th" Sgr 18h42'14" -29d36'36"
Jules - in memory of Julie Keighran Cru 12h50'13" -57d47'4"
Jules Kreisberg Lyn 7h46'24" 58d19'6"
Jules Le Grand Dra 17h41'34" 65d44'41"
Jules loves Ed And 0h21'11" 26d0'56"
Jules Mabie Lyr 18h54'3" 39d42'50"
Jules Morgan Brantley Lyr 19h20'29" 39d3'35"
Jules of the sky Cnc 8h41'31" 16d20'47"
Jules Star of Life Tau 4h39'28" 23d22'59"
Jules & Steve Fraser-Burton Cyg 20h5'17" 55d23'58"
Jules - The Gem Of My Heart Cru 12h19'33" -56d24'45"
Jules Welch Uma 11h23'56" 52d33'32"
Jules - Who will always shine bright. Vir 13h26'42" -9d13'13"
Jules, Vicki and Immy Umi 15h32'25" 75d9'1"
Jules713 Cnc 9h19'10" 8d18'15"
Jules.E Mon 6h47'18" 6d59'56"
JulesK Cas 0h58'8" 69d54'4"
Julespops Cnc 8h2'11" 15d19'51"
Julesy Gem 7h45'29" 34d41'37"
Julesy Lib 15h32'24" -21d44'4"
julhub11 Uma 9h25'35" 57d8'38"
Juli Lyn 8h5'36" 52d46'27"
Juli Umi 16h16'42" 77d24'19"
Juli Uma 9h42'24" 47d14'57"
Juli 1978 Lib 15h55'36" -14d20'39"
Juli and Jason Lain Dra 18h45'44" 59d25'11"
Juli Ann Aviles And 0h7'10" 35d36'40"
Juli Ashton Crb 15h36'51" 36d34'31"
Juli Cheramie Lyr 18h26'38" 32d41'27"
Juli E. Freimanis Vir 14h6'12" -16d4'51"
Juli Moore Lyn 7h31'2" 59d21'29"
Juli Reeves Smith, Mother of I & A Cnc 8h54'12" 26d21'11"
Juli Urban And 0h33'56" 39d48'41"
Julia And 0h31'9" 42d23'2"
Julia And 23h4'35" 35d29'54"
Julia Uma 11h44'46" 38d12'6"
Julia Lyr 18h42'38" 31d31'13"
Julia And 0h44'36" 31d34'13"
Julia Psc 1h18'42" 31d18'18"
Julia Cas 1h22'40" 56d35'17"
Julia Cas 0h30'46" 55d48'37"
Julia Cam 3h43'57" 57d23'43"
Julia And 23h52'26" 48d11'56"
Julia And 23h38'19" 47d18'44"
Julia Cyg 21h52'44" 49d41'38"
Julia Cyg 20h6'31" 39d50'38"
Julia And 23h19'21" 42d24'31"
Julia Ari 2h57'14" 27d27'6"
Julia Ori 5h57'58" 22d0'41"
Julia Leo 9h54'50" 18d28'0"
Julia Ori 5h55'52" 2d26'50"
Julia Psc 0h45'53" 8d24'17"
Julia Ori 6h10'24" 8d48'0"
Julia Psc 1h12'11" 19d29'11"
Julia Lyn 7h15'26" 58d26'40"
Julia Uma 11h20'8" 59d32'59"
Julia Uma 11h12'53" 54d10'51"
Julia Uma 9h51'37" 55d3'14"
Julia Cas 23h46'49" 55d36'59"
Julia Cas 23h7'55" 56d13'25"
Julia Cas 23h4'3" 55d42'58"
Julia Umi 16h3'5" 72d23'27"
Julia Cam 7h40'3" 63d57'25"
Julia Lib 15h56'2" -10d17'28"
Julia Mon 7h52'2" -3d9'5"
Julia Psc 23h6'40" -0d8'52"
Julia "Celebrate My Life" And 23h46'52" 45d46'57"
Julia 13 Cam 3h55'58" 67d22'59"
Julia A Reeves Vir 12h25'6" 1d23'52"
Julia Abigail Cnc 8h19'3" 7d36'33"
Julia Addison Holdsworth Tau 3h41'31" 27d22'7"
Julia Agahan Cnc 8h28'36" 17d27'17"
Julia Alex Umi 14h10'58" 77d47'28"
Julia Alice DeVita Cnc 8h1'48" 12d26'22"
Julia Allison Auer And 23h23'48" 38d30'1"
Julia Alyse Uma 10h6'50" 59d32'46"
Julia and Claire And 2h22'7" 43d3'19"
Julia Angelina Oskwarek Vir 11h47'46" 7d26'28"
Julia Ann Per 3h2'14" 46d59'6"
Julia Ann Box Crb 15h53'36" 26d35'2"
Julia Ann Breaux And 0h56'31" 40d55'52"
Julia Ann Curcio Gem 6h48'59" 31d42'21"
Julia Ann Golod And 0h42'3" 36d4'10"
Julia Ann Groin Lib 14h51'32" -3d26'21"
Julia Ann Hoffman Vir 11h50'1" -6d5'16"
Julia Ann Keck 21st Birthday Star Sgr 19h8'41" -30d21'59"
Julia Ann Lazzara Cas 1h18'34" 53d51'24"
Julia Ann Magness Cas 0h12'48" 51d22'54"
Julia Ann Rowley Cap 21h8'51" -25d44'36"
Julia Ann Teeple Sco 16h17'21" -13d54'19"
Julia Ann Vega Cas 1h18'10" 54d13'34"
Julia Anne Tizzio And 23h1'11" 41d41'0"
Julia Anne Warnick And 23h3'48" 48d14'23"
Julia Anne Zernell Lyn 7h3'33" 54d19'30"
Julia Antoinette Huck And 1h57'21" 45d19'3"
Julia Antoinette Silecchia Leo 9h27'57" 31d33'28"
Julia Arreola Oph 17h19'33" -23d1'45"
Julia B. Willis Uma 12h5'48" 41d55'6"
Julia Bailey + Gregory Garland Cyg 20h24'52" 48d5'54"
Julia Bargiel And 0h51'40" 36d54'13"
Julia Bell Cas 1h38'49" 62d16'3"
Julia Benson Cas 0h58'20" 49d25'14"
Julia Berlina Sharp And 2h31'22" 49d27'35"
Julia Besner And 2h29'31" 40d48'43"
Julia Beth Cohn Peg 22h58'31" 25d45'46"
Julia Blackwell And 23h6'0" 49d48'26"
Julia Blanchard Dra 14h56'43" 57d18'28"
Julia Bogoyavlenskaya-Ad Gloriam! Sco 17h27'47" -42d52'20"
Julia Brackett And 0h52'35" 24d6'0"
Julia Bree Gold Tocco Cap 21h46'4" -11d49'24"
Julia Brennan Aqr 22h37'36" 1d42'37"
Julia Brianne Murillo Cnc 9h10'4" 7d47'28"
Julia Brooke Rock Gem 7h0'57" 18d0'5"
Julia Caitlyn Roppolo Cyg 20h29'32" 34d11'46"
Julia Calcerano Cas 1h20'24" 68d30'32"
Julia Capanna And 0h36'48" 27d12'58"
Julia Caradec Sco 17h46'7" -39d30'28"
Julia Carriveau Dra 18h56'56" 61d3'0"
Julia Catherine Tri 1h40'29" 29d36'50"
Julia Catherine Leo 11h21'43" 15d32'12"
Julia Catherine And 1h23'57" 48d33'28"
Julia Catherine-Marie Mills Vir 14h28'15" 1d13'1"
Julia Cavanaugh Lyn 9h13'6" 34d31'9"
Julia Chapman Selman Tau 4h39'36" 19d7'41"
Julia Christine And 23h26'37" 38d40'58"
Julia Christine Johnson And 0h53'44" 39d48'12"
Julia Christine Noonan Uma 13h20'28" 58d45'59"
Julia Claire Hubbell Huntington Ari 2h47'25" 28d56'5"
Julia Claire Price Aqr 22h43'30" 0d12'2"
Julia Connor Lyn 8h28'15" 43d19'54"
Julia Dano Krisfalusi Psc 1h27'11" 16d26'20"
Julia Danvers Gem 6h2'25" 25d25'14"
Julia & Dave Mockler Wedding Star Crb 16h1'26" 37d20'59"
Julia Davis Linthicum Aql 20h10'55" 14d49'36"
Julia Dawn Bolton Umi 15h22'41" 68d1'12"
Julia De-Vall, Shining on Peg 21h59'33" 36d22'12"
Julia Devi Cas 1h43'50" 63d30'9"
Julia Diane Heidtman And 1h25'34" 50d5'15"
Julia Doran And 2h16'55" 46d46'20"
Julia Dreisbach Lyn 7h42'48" 57d6'39"
Julia E. Mowery Lyr 18h52'42" 43d14'45"
Julia E. Phillips Aqr 22h37'38" -18d39'51"
Julia E. Robinson Tau 5h36'25" 20d37'14"
Julia E. Tower Crb 15h45'54" 35d47'55"
Julia Earl McGlothlin Uma 11h29'0" 49d46'41"
Julia Earle And 0h10'8" 33d0'8"
Julia Eda And 0h45'19" 39d36'47"
Julia Eichenberger Oph 16h50'48" -6d35'46"

Julia Elisabeth Reid Uma 8h54'47" 70d27'9"
Julia Elizabeth Cas 1h39'44" 62d53'6"
Julia Elizabeth Aql 18h42'6" -0d3'36"
JULIA ELIZABETH Cap 20h27'29" -9d52'37"
Julia Elizabeth And 0h14'19" 45d19'47"
Julia Elizabeth Bauer Peg 23h39'13" 15d27'25"
Julia Elizabeth Cerverizzo Ari 2h52'43" 28d49'19"
Julia Elizabeth Fanelli Cas 0h39'19" 60d22'47"
Julia Elizabeth Foster Lyr 18h39'6" 28d24'32"
Julia Elizabeth Graeber Tau 4h35'48" 11d33'43"
Julia Elizabeth Kane And 0h48'28" 41d35'41"
Julia Elizabeth Levine Aur 6h56'32" 38d8'13"
Julia Elizabeth Mitchell Vir 15h1'49" 1d53'40"
Julia Elizabeth Stran And 2h18'1" 46d19'42"
Julia Elizabeth's Star Cap 20h30'40" -20d31'47"
Julia Eva Gähler And 23h13'56" 43d5'49"
Julia Evelyn Uma 11h20'2" 60d22'24"
Julia F. Miller Sco 17h30'19" -45d18'13"
Julia Faith And 1h45'43" 38d8'44"
Julia Flaherty Leo 9h41'33" 6d41'36"
Julia Flora McLellan And 2h28'39" 44d17'27"
Julia "Fordtner" Gem 6h48'11" 24d38'50"
Julia Formosa Peg 22h42'47" 21d27'18"
Julia Frances Sgr 19h17'37" -20d56'8"
Julia Francis Donald Brennan Leo 11h35'46" 8d32'20"
Julia G. Curry Ori 5h29'26" 5d24'30"
Julia Gabrielle Valois And 1h57'41" 46d26'35"
Julia Giba Fernandez And 1h18'30" 46d6'52"
Julia Goldman Tau 5h52'34" 25d6'57"
Julia Grace Com 12h48'13" 30d40'12"
Julia Grace Howard And 23h30'45" 48d23'36"
Julia Graham Leo 11h41'17" 25d5'49"
Julia Graham Humeke And 0h16'36" 29d51'42"
Julia Green Cyg 19h30'56" 32d13'35"
Julia Guerin Pic 5h51'52" -48d26'35"
Julia "Gules" Vir 12h2'24" -4d11'30"
Julia Gulia Lyn 7h39'35" 53d46'43"
Julia Gulia Sgr 18h33'33" -36d22'39"
Julia Hanley And 1h39'38" 42d39'27"
Julia Hannah Fein-Ashley Gem 7h37'38" 32d49'45"
Julia Hauser Uma 9h40'42" 63d49'12"
Julia Heitin Tau 5h9'8" 22d56'45"
Julia Helene Eagan Cyg 21h45'54" 53d7'32"
Julia Hiller Aqr 22h23'34" -23d34'0"
Julia Hilyard Cyg 19h44'23" 42d6'59"
Julia Hrecinic Gem 7h46'8" 34d56'27"
Julia Huntley Powell Vir 12h23'11" 12d0'41"
Julia Imboden Her 16h43'4" 6d25'0"
Julia Irene McGuire Gem 7h0'37" 20d24'56"
Julia Isabel Peg 21h41'55" 18d51'58"
Julia Isabelle Ittig Mon 7h30'53" -1d32'39"
Julia - Jadyn's Mom Sco 16h16'36" -12d4'4"
Julia K. Fox Cnc 8h27'30" 14d36'10"
Julia Kannel Cyg 19h48'18" 32d7'53"
Julia Katelyn And 0h10'57" 43d52'37"
Julia Kathleen Celano Cyg 21h32'19" 33d28'26"
Julia Kathrin Bethe Gem 7h10'12" 28d33'33"
Julia Kaye Tau 5h49'38" 14d12'34"
Julia Kest And 23h32'45" 48d3'8"
Julia Kologe Aqr 22h7'24" -7d34'19"
Julia Kreisz Cnc 8h19'9" 20d27'32"
JULIA KUDRYA Cas 1h7'34" 61d21'24"
Julia L. Brayton And 2h36'1" 49d14'34"
Julia L. Millard Lib 15h43'3" -17d34'17"
Julia l'angelo Cyg 19h57'22" 51d42'18"
Julia Lau Sco 17h43'49" -41d35'37"
Julia Laurin Hall Ori 5h38'39" -2d26'50"
Julia - Lea ~ Nana's Love Cap 21h50'58" -8d52'20"
Julia Ledee Cap 21h47'57" -9d28'58"
Julia Lee Chapman Sgr 19h36'8" -13d6'24"
Julia Leith Orr Psc 0h59'53" 27d3'44"
Julia Lopez And 2h34'30" 50d17'11"
Julia Loren Blum Dra 19h24'10" 76d43'42"
Julia Louise Hargrove Aqr 22h34'29" 1d21'39"
Julia Louise Orlando Ori 5h30'5" -0d46'18"
Julia Love Aql 19h42'12" -0d24'11"
Julia Lynn Ori 6h13'22" 15d19'3"
Julia Lynn Ari 2h15'26" 26d57'1"
Julia Lynn Arwood Vir 14h34'17" -0d30'37"
Julia Lynn Kirsch Peg 23h16'39" 20d59'52"
Julia M Mailhiot Cap 20h18'21" -23d53'14"
Julia M. Zindel And 23h23'20" 42d24'22"
Julia MacKinnon Cas 0h32'59" 54d5'49"
Julia Madison Fargo Aqr 23h10'55" -9d39'22"
Julia Mae Breed Cap 21h44'13" -21d25'37"
Julia Mae "Dena" Winburn Uma 9h35'17" 41d27'4"
Julia Margaret Nussbaum And 23h46'29" 45d12'12"
Julia Marian Laughlin Vitale Peg 22h17'8" 11d56'47"
Julia Marie Cep 3h29'54" 82d54'19"
Julia Marie Cas 2h13'9" 64d17'44"
Julia Marie 2004 Leo 10h7'35" 14d0'35"
Julia Marie Courtney Ori 6h15'58" 9d30'53"
Julia Marie Gagliardi Cnc 8h30'12" 25d38'20"
Julia Marie Howard Cnc 9h17'17" 19d8'28"
Julia Marie Jenkins Uma 11h0'5" 50d6'57"
Julia Marie Steggerda Ari 2h12'2" 25d51'31"
Julia Marotta Ori 6h19'43" 18d54'55"
Julia Martini Gem 6h52'32" 34d13'25"
Julia Mary Hughes Aqr 22h7'28" 0d42'28"
Julia Massey-Rogers Crb 15h33'56" 26d24'19"
Julia Mateja Sgr 17h57'53" -26d6'48"
Julia May 708 Vir 12h37'29" -9d34'30"
Julia McDonie Tau 3h49'59" 15d40'23"
Julia Megia And 0h52'40" 40d38'7"
Julia Melnikov Cas 1h17'32" 50d40'30"
Julia Mihalko Tippie 01/29/99 Aqr 20h53'29" -12d25'3"
Julia Millstein Lyn 8h17'40" 36d49'38"
Julia Miranda Johnson And 0h15'37" 29d39'50"
Julia Moana Pho 0h58'28" -49d15'13"
Julia Mogilev Cas 23h6'30" 55d39'21"
Julia Mrak Vir 13h48'10" -19d28'37"
Julia my love Lib 14h48'28" -9d44'41"
Julia Nevaeh Bardaro Lib 14h49'43" -2d18'3"
Julia Nichole Locke Lib 15h26'21" -3d49'29"
Julia Nicole Mon 6h55'10" -7d53'44"
Julia Nicole Sco 17h30'54" -37d59'53"
Julia Nicole Needhammer Vir 14h30'33" 1d37'33"
Julia Nikolaevna Gravois Psc 1h40'22" 8d29'43"
Julia Nile Cas 0h54'19" 62d8'32"
Julia Ossey Sgr 18h9'39" -34d46'0"
Julia Ostrom Per 4h33'4" 32d52'36"
Julia Paige And 2h19'45" 48d58'20"
Julia Panaro And 22h59'15" 41d10'53"
Julia & Peff Lyn 7h30'1" 59d24'43"
Julia Perry Leo 11h42'12" 23d58'34"
Julia Persova And 1h6'6" 35d32'3"
Julia Petrucci Sco 17h33'38" -44d18'38"
Julia Pinheiro Raposo Lib 15h21'35" -12d49'39"
Julia Pulcher Astrum Ab Caelum Lib 14h48'57" -7d30'24"
Julia R Cardona Gem 6h51'43" 33d8'26"
Julia Rachael Sgr 18h50'25" -27d11'19"
Julia Rain Aqr 22h31'34" -1d0'24"
Julia Reagan Salm Del 20h27'19" 20d50'31"
Julia Remus Umi 14h29'36" 69d20'24"
Julia Renée Near And 2h12'45" 43d51'23"
Julia Renee Wright Dra 18h35'55" 77d33'45"
Julia Revillion Vir 13h20'18" 3d32'21"
Julia Rivera Cnc 8h41'28" 17d49'23"
Julia Robyn James Lyn 7h0'46" 51d37'5"
Julia Rogers And 23h41'29" 42d15'40"
Julia Romero Uma 11h24'22" 64d33'16"
Julia Rose Aqr 22h48'37" -10d12'45"
Julia Rose Crb 15h45'43" 31d33'0"
Julia Rose Leo 9h32'54" 27d17'55"
Julia Rose Diamond Crb 15h45'32" 32d35'16"
Julia Rose Evans Lmi 10h37'7" 37d59'47"
Julia Rose Forte Lib 15h32'1" -6d38'28"
Julia Rose Joseph And 0h39'38" 45d17'16"
Julia Rose Servin Aql 18h56'25" -0d53'1"
Julia Santos Mon 6h32'39" 8d34'32"
Julia Sara Krawiecki-Gazes Aqr 21h21'41" 2d17'37"
Julia Sasano Miller Sgr 18h30'57" -16d52'28"
Julia Schläpfer Lac 22h45'50" 54d3'27"
Julia Schmidt-Goldberg Cas 1h40'33" 62d48'50"
Julia Schwartz Lib 15h51'12" -16d15'44"
Julia Scott Crb 16h17'5" 31d50'19"
Julia Shack Sackler Lyn 9h15'50" 37d23'53"
Julia & Shannon's Twinkle Sgr 18h17'40" -18d1'49"
Julia Shing Hui Chan Cnv 12h48'48" 41d9'20"
Julia Shouldice Uma 14h17'10" 60d41'54"
Julia Sigri Pappas Sco 15h51'44" -26d2'42"
Julia Simon And 1h52'37" 45d8'49"
Julia Sky Leo 9h42'38" 28d28'30"
Julia Skye Landy Tau 4h11'10" 14d24'48"
Julia Sloan Donnell Lib 15h18'39" -4d10'23"
Julia Smithson Lyn 7h43'11" 56d17'51"
Julia Sophia Reichel Uma 11h59'51" 50d40'30"
Julia Stahl Leo 11h31'51" 10d17'16"
Julia Star Lmi 10h31'53" 31d3'57"
Julia Sutherland Cnc 8h45'26" 19d57'48"
Julia T. Cobarruviaz Cas 23h23'26" 58d7'4"
Julia Tempongko Cnc 8h48'0" 17d49'54"
Julia Teresa Mundt Sammaritano Lib 15h40'55" -17d48'59"
Julia Termine "Julia's Jewel" Cas 1h45'18" 61d34'7"
Julia The Beautiful And Bright Aqr 22h19'52" -12d35'0"
Julia Thompson Cep 7h39'59" 85d58'16"
Julia Ty Goldberg Aqr 22h43'56" -5d32'46"
Julia V. Dyachuk Lib 14h50'57" -1d58'51"
Julia Valderrama Ari 1h58'24" 18d1'3"
Julia Valentina Maxwell Crb 16h13'53" 36d28'1"
Julia Vashti Muse Cole Leo 10h22'14" 14d5'58"
Julia Veronica Werbel Gem 7h39'39" 24d9'11"
Julia Whitesell Leo 11h11'46" 6d37'43"
Julia Whyel Cas 23h41'46" 58d46'49"
Julia Yvette Ferraro Ari 2h8'0" 21d9'32"
Julia, Jane, Jay, Jim Ori 6h5'38" 15d43'2"
JuliaB10212003 Tau 3h59'2" 24d24'9"
Julian Ori 6h10'50" 15d30'25"
Julian Ori 5h8'23" 6d39'41"
Julian And 1h49'25" 42d1'37"
Julian Her 16h46'30" 48d35'29"
Julian Dra 17h41'22" 59d9'12"
Julian Umi 16h44'31" 76d35'15"
Julian For 2h49'47" -35d12'31"
Julian A. Landin Aql 19h24'41" 8d39'55"
~Julian and Anna Forever~ Cyg 20h34'50" 58d8'26"
Julian Buitrago Uma 10h8'16" 70d31'51"
Julian Casanova Her 18h10'7" 23d59'51"
Julian Chasin Cyg 20h52'45" 41d16'49"
Julian Christian Attardi Umi 14h50'20" 70d2'52"
Julian "Dada" Rivera Tau 4h42'24" 8d50'24"
Julian Daniel Linegar Leo 9h55'55" 29d15'33"
Julian Dean The Love Bug Psc 0h25'18" 8d18'43"
Julian Demetrius Per 3h20'1" 44d23'46"
Julian Dominik Uma 9h40'58" 47d37'14"
Julian Douglas Conrad Sco 16h44'42" -31d57'10"
Julian Dylan Fairchild Sgr 19h47'23" -14d39'27"
Julian E. McCann Uma 9h31'31" 45d11'27"
Julian Edgardo Uma 11h35'29" 54d44'8"
Julian Edward Alderfer Jr. Her 17h46'35" 48d15'58"
Julian Garrison Marius Hlozek Lyn 7h11'55" 52d58'42"
Julian Gary Clarke - Jules' Star Ori 6h19'32" 15d32'57"
Julian Gray Moffatt - My Maverick Cru 12h7'47" -60d57'42"
Julian Grollo Tau 5h49'25" 21d7'15"
Julian Harold Sparrow Ori 5h56'51" -0d47'9"
Julian Hideki King Umi 15h8'24" 79d35'55"
Julian Hiram Crb 16h6'18" 37d40'20"
Julian Innes Fader Per 4h34'44" 31d24'50"
Julian " Jake " Michael Arnau Boo 14h33'1" 34d2'48"
Julian James Fiano Lib 14h49'59" -8d0'7"
Julian James Lizzio Cnc 8h49'41" 24d38'11"
Julian James McMahon Uma 9h6'21" 58d31'54"
Julian Jason Offman Umi 16h58'29" 77d50'33"
Julian John De Mattia 25.05.67 Gem 7h18'10" 23d55'58"
Julian John Michael King Uma 9h34'39" 42d36'15"
Julian Joseph Gerard Caruso Cru 12h20'12" -56d52'11"
Julian Joseph Thorvund Lmi 10h45'19" 34d19'37"
Julian Kilkenny JKLM And 23h1'17" 35d32'15"
Julian Konstanty Bleess Umi 15h29'43" 73d3'0"
Julian Lopez Uma 12h7'37" 60d2'29"
Julian Luther Boland Psc 23h18'18" 3d28'15"
Julian Marie Aqr 22h24'29" -3d21'24"
Julian Maurice Ori 5h54'50" 17d26'17"
Julian Michael Augugliaro Hash Uma 11h39'33" 43d3'43"
Julian Michael Orth Uma 13h55'16" 53d34'42"
Julian & Nicky Gething Cyg 21h55'34" 43d24'57"
Julian Philip Santorelli Sco 16h17'48" -10d11'32"
Julian Reyes Vir 14h18'29" 2d19'35"
Julian Richard Aur 5h30'52" 44d22'21"
Julian Richard Nathaniel Briggs Cyg 21h53'40" 39d15'17"
Julian San Angel Menchaca Umi 14h30'10" 76d47'33"
Julian Schaffhouser Uma 10h55'16" 56d55'11"
Julian Sebastian King Gem 6h36'53" 14d19'44"
Julian Senderey Ori 5h32'1" 2d18'22"
Julian Shaheen Cnc 9h14'45" 17d3'13"
Julian Timms 21st Star - love Mum Cru 12h33'22" -59d55'34"
Julian von Nagel Per 3h9'13" 43d22'58"
Julian Wooten Johnson, III Cnc 7h56'3" 11d9'39"
Julian Zydan Kahalley Aqr 21h53'38" -2d29'9"
Juliana Lyn 7h18'1" 59d24'5"
Juliana Vir 14h17'12" 1d33'6"
Juliana Gem 7h13'29" 26d17'18"
Juliana Cyg 19h46'26" 35d1'15"
Juliana Lyr 18h46'40" 30d24'8"
Juliana Alvarez Vir 13h44'6" 4d23'43"
Juliana and Brett Her 16h30'48" 13d26'3"
Juliana Barfield Lib 14h54'24" -9d27'39"
Juliana Crysler Umi 15h46'20" 84d51'51"
Juliana Dandach Uma 11h21'11" 45d7'24"
Juliana & Jens Uma 10h0'14" 42d4'22"
Juliana Katherine Walton Uma 12h59'32" 61d17'56"
Juliana Khosasi Boehmer And 1h21'29" 38d58'52"
Juliana Lopez Cas 0h21'22" 55d47'2"
Juliana Lorraine Robles Cap 21h33'2" -15d37'42"
Juliana Marie Bakker Vir 14h36'55" 2d23'10"
Juliana Marie Lopez Sgr 18h4'27" -23d35'54"
Juliana Marie Ricigliano Cnc 8h15'41" 22d9'3"
Juliana Mejia Cas 2h33'23" 67d51'45"
Juliana Michelle Our Shining Star Leo 10h27'53" 18d53'16"
Juliana Nicole Rotondo Cyg 20h35'37" 47d26'20"
Juliana Omrae Uma 8h50'27" 65d54'34"
Juliana Patrice Staab Cas 0h26'11" 62d3'55"
Juliana Pearson Aqr 22h28'59" -20d48'59"
Juliana Rochelle Alvarez Tau 4h14'55" 25d5'34"
Juliana Rojel-Esquibel Cap 21h45'22" -13d22'50"
Juliana Romano And 0h4'27" 40d41'28"
Juliana Ulrych And 23h16'6" 48d30'19"
Juliana Vargas e Daniel Svec Leo 9h23'19" 10d34'27"
Juliana Vasquez Griffeth Cyg 20h17'27" 53d43'16"
Juliana Veronica Maruncic Ari 3h10'11" 14d21'47"
Juliana Vivian Hadjiyane Cap 20h19'10" -9d5'30"
Julian-Alexander Breuner, 04.08.2003 Ori 5h0'50" 6d12'15"
Juliane Crb 16h12'28" 35d48'27"
Juliane and Chad: Forever One Uma 11h12'55" 51d7'53"
Juliane and Kris And 0h19'22" 45d3'36"
Juliane Araneo Cas 1h13'21" 51d0'22"
Juliane & Laurent Cas 0h7'14" 56d42'7"
Juliane Segner Ori 6h8'2" -3d49'11"
Juliani Aalias Rosales Lib 14h48'23" -7d15'9"
JuliaNicoleSmith God's special gift Uma 13h31'14" 53d33'15"
Julian-Kalob Her 18h6'43" 47d29'42"
Juliann (CuteGirl) Ori 6h7'53" -0d17'29"
Juliann Elizabeth Holzer Gem 7h20'42" 14d55'1"
Juliann Ellan Dickerhoff Ori 6h6'12" 18d49'50"
Juliann Marie Gem 7h32'47" 21d27'17"
Julianna And 1h9'43" 42d44'59"
Julianna Asia Cowan Ari 3h15'49" 27d16'11"
Julianna Brooke Lapham And 0h18'0" 35d52'42"
Julianna DeMarco Uma 11h30'14" 59d0'20"
Julianna Grace Cas 1h34'44" 66d11'34"
Julianna Lei Selig Lyr 19h17'1" 28d38'31"
Julianna Marcella Her 17h59'4" 49d47'40"
Julianna Marie Giorgio Cam 6h20'7" 67d40'14"
Julianna Marie Holt Aqr 23h19'38" -10d50'45"
Julianna Marie Lichtenstern Gem 6h47'39" 27d39'55"
Julianna Martin And 23h43'9" 37d30'54"
Julianna Matura Read And 0h36'39" 44d25'47"
Julianna Moon Mon 6h48'47" -0d7'20"
Julianna Osuch Aqr 22h24'57" 1d50'0"
Julianna Porpora And 1h15'53" 41d22'16"
Julianna Rizzitelli Cyg 20h13'35" 55d29'27"
Julianna Roxana Sarmiento Psc 0h34'24" 7d15'4"
Julianna Salvatore And 0h33'33" 36d17'27"
Julianna Schlosser Lmi 10h4'3" 29d13'20"
Julianne Leo 11h46'14" 26d14'20"
Julianne Vir 15h0'5" 6d44'29"
Julianne Tau 3h50'20" 0d25'5"
Julianne Cnc 8h45'29" 32d6'22"
Julianne Cyg 21h21'52" 45d58'58"
Julianne Cam 3h39'42" 57d11'45"
Julianne Lib 14h46'55" -11d13'26"
Julianne Lib 15h28'50" -19d24'28"
Julianne Douglas Sco 16h57'29" -41d52'24"
Julianne Franks Lib 15h33'27" -19d57'48"
Julianne Griswold Cyg 20h27'43" 52d24'45"
Julianne Lovejoy Vir 12h6'19" 10d49'12"
Julianne Lucero Woodall Cas 0h50'5" 50d44'48"
Julianne Marie Sgr 19h55'39" -33d6'18"
Julianne Marron Wise Sgr 19h48'48" -14d15'15"
Julianne Regina Cuellar Cas 23h45'33" 56d22'26"
Julianne Sherwin - My True Love Vir 13h35'51" -3d48'26"
Julianne "Susie" Malstan Sad Ari 2h17'22" 13d6'59"
Julianne The Love Star Lib 15h1'54" -1d5'17"
Julianne (Tweetsie) Manchester Cap 20h38'20" -19d58'56"
Julianne Victoria Anastasia Lake Uma 10h16'43" 47d27'30"
Julianne Villeneuve Umi 4h11'44" 88d59'8"
Julianne's Star Aqr 22h12'54" -2d35'40"
Julianne's Star Cnc 8h6'19" 25d44'26"
Julianne's Star Psc 1h8'21" 26d54'38"
Julian's Star Aqr 22h40'33" -1d22'22"
juliapiera And 2h25'5" 47d49'57"
Julia's Eye Cnc 8h35'29" 7d56'27"
Julia's Grandma Doris Cas 23h59'33" 56d53'16"
Julia's guiding light memories JFG Cnc 8h53'58" 13d27'31"
Julia's Light Cas 1h40'44" 65d10'25"
Julia's Light Ori 5h28'22" -0d33'26"
Julia's Lucky Star Sgr 19h27'32" -41d38'54"
Julia's Rider in the Sky Psc 23h36'44" 4d52'16"
Julia's Shining Light Cnc 9h0'54" 17d43'20"
Julia's Shining Star Cnc 8h8'15" 18d14'40"
Julia's Shining Star And 2h12'48" 45d31'4"
Julia's Star Umi 14h5'29" 75d46'45"
Julia's Star Vir 14h3'31" -19d56'11"
Julia's Star Uma 13h34'22" 54d39'38"
Julia's Star of Hope Peg 21h34'40" 22d16'6"
Julia's Wedding Star Lyr 18h49'20" 33d54'52"
Julibean Her 17h36'22" 32d20'31"
Julibear Vir 15h5'16" 3d53'8"
Julie Psc 23h6'28" 7d28'59"
Julie Ari 2h34'0" 13d48'14"
Julie Cnc 8h21'2" 14d12'36"
Julie And 0h32'3" 22d0'39"
Julie Her 17h55'56" 16d27'19"
Julie Leo 9h33'44" 17d17'28"
Julie Cnc 8h13'7" 29d59'17"
Julie Cnc 8h45'50" 27d8'19"
Julie Tau 5h58'12" 28d21'45"
Julie Ari 3h28'48" 27d42'19"
Julie Leo 9h43'51" 23d53'16"
Julie Cyg 19h44'0" 28d9'36"
Julie Lyr 18h55'53" 26d45'35"
Julie Lmi 10h0'49" 30d27'2"
Julie And 0h20'55" 35d34'59"
Julie And 0h43'25" 41d24'31"
Julie And 1h44'33" 42d12'33"
Julie Lyn 9h41'22" 40d2'37"
Julie And 1h1'22" 45d18'25"
Julie Cyg 21h8'43" 43d42'3"
Julie Cyg 19h51'0" 39d8'42"
Julie And 23h43'5" 45d14'22"
Julie And 23h15'56" 47d6'42"
Julie Lyn 7h46'24" 59d27'26"
Julie Cas 23h47'28" 54d48'11"
Julie Cas 23h26'25" 54d58'42"
Julie Dra 15h25'7" 57d32'56"
Julie Cas 1h27'40" 67d12'0"
Julie Cas 2h39'41" 67d1'1"
Julie Umi 14h12'14" 78d43'12"
Julie Mon 6h51'26" -7d15'7"

Julie
Leo 11h12'52" -1d32'8"
Julie
Lib 15h42'27" -28d25'35"
Julie
Sco 17h44'40" -37d18'1"
Julie 0417
Cas 1h35'30" 61d29'36"
Julie 5-30-92
Gem 7h38'0" 23d15'6"
Julie A. Dutcher
And 0h35'46" 36d26'26"
Julie A. Ginley
And 0h11'43" 43d59'29"
Julie A Harris
Agr 23h12'11" -8d3'34"
Julie A. Sechrist
Cas 1h23'51" 69d27'54"
Julie A Trossbach
Gem 7h52'1" 32d42'3"
Julie Abigale
And 2h23'53" 50d0'1"
Julie a.k.a. My Sugarbear Star
Lmi 9h55'8" 36d20'2"
Julie Alice
Ori 5h35'28" -0d38'7"
Julie and Ben - True Love Forever
Cyg 21h13'34" 44d6'18"
Julie and Bill Forever
Leo 11h26'48" 18d7'0"
Julie and Brandon Fordham
Lep 5h16'43" -21d46'50"
Julie and Daniel Clark
Cas 1h29'26" 52d10'59"
Julie and Don Jenkins
Her 17h13'6" 20d22'4"
Julie and Edward Kershaw
Dra 15h11'33" 55d12'48"
Julie and Gary McKernan
Cyg 21h51'5" 45d48'23"
Julie and James
Ari 2h36'40" 14d12'12"
Julie and Kenny
Cnc 9h0'50" 19d33'0"
Julie and Rich " First True Love"
Agr 23h35'53" -8d58'25"
Julie and Richard
Umi 15h22'50" 71d49'23"
Julie and Rick Sinkfield
Lyn 8h37'33" 34d12'30"
Julie and Steve Markovich
Cyg 19h27'6" 54d29'45"
Julie and Theo
Cnc 8h18'56" 13d22'21"
Julie Ann
Mon 6h32'54" 7d43'1"
Julie Ann
Vir 13h0'42" 7d11'17"
Julie Ann
Cnc 8h59'45" 16d31'23"
Julie Ann
Tau 4h30'45" 26d33'9"
Julie Ann
Lyn 8h26'46" 33d8'18"
Julie Ann
And 23h58'33" 33d3'47"
Julie Ann
Lyn 8h44'33" 40d20'13"
Julie Ann
Uma 10h49'45" 42d25'43"
Julie Ann
Cam 3h57'25" 54d10'4"
Julie Ann
And 2h19'11" 45d54'17"
Julie Ann
Vir 13h39'36" -17d23'0"
Julie Ann
Sgr 18h48'13" -20d36'54"
Julie Ann
Sco 16h44'25" -40d8'51"
Julie Ann Bishop
Lib 15h8'20" -10d44'56"
Julie Ann Cosco-Deputy
Cap 20h12'29" -24d14'6"
Julie Ann Couch Star
And 0h40'46" 32d32'59"
Julie Ann Dorney
And 0h43'29" 45d10'28"
Julie Ann Druxman
Ari 3h16'7" 29d2'49"
Julie Ann George
Uma 9h5'23" 49d58'37"
Julie Ann Gibson
Cyg 19h38'40" 32d0'13"
Julie Ann Giorgio
Agr 20h45'46" 1d57'53"
Julie Ann Goff Brown
Mon 6h48'30" 6d47'3"
Julie Ann Hansen
And 2h23'58" 41d59'59"
Julie Ann Killingsworth
Psc 1h2'28" 8d28'49"
Julie Ann Knicklebine
Uma 13h40'53" 48d19'55"
Julie Ann Lemieux
Umi 14h41'51" 73d8'35"
Julie Ann McMullen
Psc 1h37'46" 20d5'49"
Julie Ann Miers
Sco 17h53'47" -35d55'57"
Julie Ann Morris Barry
Psc 1h28'38" 28d40'6"
Julie Ann Page
Vir 13h30'16" 1d54'22"
Julie Ann Ramos
Sgr 19h44'18" -12d3'56"
Julie Ann Rothe
Ari 2h58'29" 20d13'35"
Julie Ann Skaggs
Gem 7h2'36" 32d38'35"
Julie Ann Smith
Lib 15h31'0" -24d18'57"
Julie Ann Sosa
Cnc 8h2'21" 8d0'31"
Julie Ann Troyer
Cap 21h31'26" -16d25'30"
Julie Ann Trumbo
Leo 10h20'11" 25d37'31"
Julie Ann Williamson
Vir 12h46'13" 3d52'48"
Julie Ann Wise
Crb 16h21'59" 34d43'6"
Julie Ann, I Love You...Michael
And 23h36'16" 46d55'42"
Julie Anna Shaw
And 1h40'27" 48d25'42"
Julie Anne
And 23h13'26" 49d12'18"
Julie Anne
Sgr 19h34'4" -36d22'15"
Julie Anne Catherine Donovan
Gem 6h28'40" 22d40'13"
Julie Anne Hillaby
Cas 0h47'56" 62d38'55"
Julie Anne Lackey
And 23h2'46" 51d4'21"
Julie Anne Lister - 21.11.1955
Sco 17h44'18" -41d35'23"
Julie Anne May
Mon 6h49'13" -0d33'45"
Julie Anne Murray
Lmi 10h1'52" 30d4'33"
Julie Anne Popovich
Uma 9h11'37" 64d43'22"
Julie Anne Waters
Gem 7h28'4" 32d51'10"
Julie Annette Goans
Lib 15h7'23" -14d49'23"
Julie Anya
And 0h7'45" 31d16'50"
Julie & Arthur Kondos
Ara 17h26'15" -55d12'16"
Julie B Herbert
Cam 4h22'17" 69d20'37"
Julie baby
Sgr 19h11'7" -12d47'39"
Julie Baldwin
Cas 23h36'31" 58d5'7"
Julie Ballantyne
Mon 8h7'9" -1d6'47"
Julie Barnes-Lattimer
And 0h29'30" 39d39'33"
Julie Barr
Ori 5h28'26" -0d19'3"
Julie Beach Star
And 1h20'34" 50d1'6"
Julie "Bear"
Cyg 21h50'4" 51d46'7"
Julie Beaulieu
Leo 10h22'2" 16d25'26"
Julie *Beauty Queen*
Leo 10h24'45" 27d2'24"
Julie Bellant
And 23h55'56" 34d42'4"
Julie Bellinder
And 0h28'55" 42d12'54"
Julie Berns
Uma 8h35'27" 71d8'16"
Julie Beth Vadner
Uma 9h1'15" 48d51'34"
Julie Bindelglass
Cas 0h26'54" 61d3'39"
Julie Bird
Vir 12h48'31" 1d26'52"
Julie Blackman
Lib 15h52'50" -5d35'43"
Julie Blair Padgett
Ori 5h41'56" -0d16'40"
Julie Boledovich
Uma 11h55'43" 53d9'11"
Julie Bonselaar
Cas 0h57'56" 64d0'17"
Julie Boo Boo Bear Russell
Umi 15h21'38" 70d56'58"
Julie Bowman
Cas 1h25'3" 51d32'37"
Julie Boyer's Star Abby
Sco 17h51'8" -36d41'20"
Julie & Brian Shumard
Per 3h8'56" 46d18'33"
Julie Brown
Uma 11h13'56" 56d15'21"
Julie Bulloch Bunge
Tau 5h44'32" 25d52'23"
Julie (Butter) Gauthier
Sgr 19h0'2" -30d28'53"
Julie Buzzurro
And 0h36'10" 35d3'6"
Julie C. Brooks
Psc 0h57'32" 15d9'42"
Julie Callaghan
Umi 14h12'50" 68d27'41"
Julie Canter
Cyg 21h30'56" 42d38'43"
Julie Caouette
Cyg 20h41'38" 51d20'13"
Julie Carolyn
Cru 12h13'31" -60d30'5"
Julie Catherine Rubel
Uma 9h36'8" 63d24'0"
Julie chiquitita Blaszczyk
Cnc 8h57'40" 24d11'14"
Julie Chircop
Leo 11h37'38" 22d22'16"
Julie Christine
Agr 22h20'44" -0d32'39"
Julie Clayton "1942-2005"
Ari 2h28'34" 14d25'24"
Julie Colomby
Agr 22h52'49" -15d36'59"
Julie Comer
Lyn 6h19'0" 61d53'33"
Julie Connor
Her 16h32'13" 30d42'23"
Julie Cooke
Sgr 18h4'48" -17d20'45"
Julie Cordero
Uma 13h51'37" 54d44'55"
Julie Cotton
Ori 6h18'24" 8d37'5"
Julie Cress
Cyg 19h40'34" 33d46'21"
Julie D. Cintron
Agr 22h12'4" -1d28'40"
Julie Daniels (Juliette)
And 23h13'27" 44d8'51"
JULIE "DARLIN" STANTON
Lyn 7h50'36" 45d0'20"
Julie Dawn
Her 16h29'13" 44d1'10"
Julie Dawn Barlow
Cas 1h28'58" 61d23'19"
Julie Dawn Layell
Vir 11h55'32" -6d10'30"
Julie De Las Vegas
Tau 5h48'6" 18d59'8"
Julie De Leon
Dra 18h54'13" 55d16'42"
Julie Della Ratta
Tau 4h29'56" 16d4'20"
Julie Dermer
Lyn 7h54'5" 34d23'29"
Julie Dernoncour 24/05/2007
Cas 23h48'16" 50d43'1"
Julie Descamps
Uma 11h55'48" 34d46'7"
Julie Desloover
Ori 5h50'21" 22d49'9"
Julie Do
Ori 5h20'47" 6d7'19"
Julie Dorsett
Cnc 8h52'10" 28d29'46"
Julie Downie
Peg 22h42'50" 27d46'52"
Julie E. Johnson
Leo 11h56'0" 21d21'31"
Julie E. Sherwood "The Kitten Star"
Ari 3h27'38" 21d38'19"
Julie Edmundson
Tau 3h57'38" 21d5'26"
Julie Eileen Haugen
Agr 21h47'22" 1d52'42"
Julie "El Pollo Loco" Kim
Tau 5h48'30" 23d7'29"
Julie Elizabeth
Leo 11h43'15" 27d0'11"
Julie Elizabeth Davies
Lyn 7h32'14" 56d16'14"
Julie Elizabeth Goodwin
Cas 1h0'50" 67d27'21"
Julie Elizabeth Matthews
Leo 9h51'9" 32d6'10"
Julie Elizabeth Topoozian
Uma 9h43'48" 44d33'0"
Julie Elizabeth Ward
Agr 22h41'49" -18d21'7"
Julie Ellen
Cas 1h27'17" 72d3'50"
Julie Elliott
Vir 14h6'33" -14d58'44"
Julie Emson
Cyg 21h56'0" 39d12'30"
Julie & Esther's Star
Cas 0h15'21" 51d59'50"
Julie Faddoul
Lib 15h58'30" -12d31'43"
Julie Farley
Lmi 10h30'46" 37d14'25"
Julie Fecske
Lyn 8h20'47" 40d11'48"
Julie Finer
Gem 7h17'2" 32d56'1"
Julie Follete
Lib 15h23'40" -9d46'47"
Julie for ever
Umi 15h56'47" 80d31'47"
Julie Forbes (ILU)
Cas 0h37'0" 62d42'1"
Julie Fraley's Shining Alta
Cnc 9h9'36" 25d42'0"
Julie Fransisco
Lib 14h49'30" -3d48'54"
Julie Fritchtnitch
Uma 11h53'18" 33d13'18"
Julie from the BX
Sco 16h12'0" -10d2'20"
Julie Gae Jones "12/20/1956"
Sgr 19h12'59" -21d8'2"
Julie Gaffney
And 0h22'38" 42d44'18"
Julie Galvan
Cnc 8h19'13" 9d20'34"
Julie Garczynski
Cnc 8h7'32" 7d7'54"
Julie Gillispie - Shining Star
And 23h38'11" 44d5'51"
Julie Goldstein
Lyn 8h4'2" 58d23'29"
Julie Grace 26364
Del 20h30'56" 17d18'18"
Julie Griffin
Lib 14h36'15" -12d27'29"
Julie & Guillaume
Uma 14h17'16" 60d34'14"
Julie Hamelin
Psc 1h37'7" 13d52'54"
Julie Hamilton
And 1h24'27" 34d31'20"
julie harris <3
Uma 10h9'36" 54d59'17"
Julie Harrison
Ari 2h9'15" 25d8'9"
Julie Hawley
Tau 4h8'43" 14d0'48"
Julie Hemming's 50th Birthday
Mon 6h57'46" 9d46'58"
Julie Hemsley
Lmi 10h58'0" 27d17'57"
Julie Higdon Segal
Crb 16h3'47" 34d40'32"
Julie Hillary
Agr 22h7'56" -15d35'8"
Julie Holder's Piece of Heaven
Cru 12h15'35" -59d5'59"
Julie Holmes
Dra 16h34'35" 65d3'39"
Julie Huynh
And 23h28'30" 42d22'50"
Julie I McBride
And 1h36'8" 49d13'53"
Julie I. Stone
Sco 16h44'42" -29d52'25"
Julie Ibach
Psc 1h23'14" 11d0'22"
Julie J.
And 1h42'18" 45d3'1"
Julie J. Allen
Umi 17h39'53" 86d33'22"
Julie J. Smith
Ori 6h20'49" 14d43'39"
Julie Jack's Birthday Star
And 0h4'19" 39d58'5"
Julie Jensen
Cap 21h49'20" -21d2'7"
Julie&Jérôme
Vir 14h12'20" 6d10'39"
Julie "Jewels" Kafskey
Ari 2h12'27" 22d0'36"
Julie "Jujee" Willis
Sco 17h3'19" -33d21'54"
Julie *Jules* Sigal
Cap 20h10'53" -9d17'46"
Julie Juls Dewees
Leo 11h23'37" 13d22'5"
Julie (JuVixx) Creager Munford
Tau 4h16'13" 30d49'34"
Julie Karen My Love and My Life
And 0h31'49" 28d38'36"
Julie Kates
Cas 0h36'9" 69d40'42"
Julie Kellner
Crb 16h12'18" 36d28'56"
Julie Kikla
Psc 1h10'52" 11d26'27"
Julie Knapp
And 23h47'16" 35d34'52"
Julie Knapp
Lyn 7h45'37" 38d33'11"
Julie Knox
Leo 11h7'23" 13d17'57"
Julie Kristine Edeus
And 23h22'42" 42d23'37"
Julie L. Collier Thompson
Lmi 10h53'17" 26d44'19"
Julie L. Frishman
Agr 22h35'3" -12d29'25"
Julie L. M. The Star Everlasting
Ari 2h39'21" 28d21'12"
Julie Lagana
Lyn 6h57'43" 47d4'10"
Julie Laura Jones
And 1h19'29" 47d44'29"
Julie Lavoie
Crb 16h15'54" 37d37'27"
Julie Lee Clifton
Leo 10h18'39" 16d59'39"
Julie Lee Kelin
And 0h49'23" 37d4'40"
Julie Léger
Umi 15h34'13" 85d56'23"
Julie Lehua McKeague
Lyr 19h14'4" 27d7'36"
Julie (lil boss) Lopreiato
Umi 16h43'12" 76d35'52"
Julie Lindell
Sco 16h28'49" -26d34'8"
Julie Lockman
Cyg 20h11'58" 33d36'19"
Julie Louise
Pho 0h37'31" -48d56'49"
Julie "Love"
And 0h45'51" 39d16'14"
Julie loves Paul & Paul loves Julie
Uma 13h26'46" 53d40'12"
Julie Lynette Hogan
Tau 5h9'6" 18d23'14"
Julie Lynn
Cnc 7h58'2" 14d38'7"
Julie Lynn Belvoir
Cnv 12h20'8" 48d47'55"
Julie Lynn Johnson
Cam 4h7'9" 57d21'28"
Julie Lynn Schweikert
Psc 1h7'3" 22d19'26"
Julie Lynn Sullivan
Sco 17h33'44" -40d9'58"
Julie Lynn Yore
And 0h43'43" 39d3'13"
Julie Lynne
Agr 22h11'14" 0d14'32"
Julie Lynne Delatorre
Psc 0h38'21" 6d13'38"
Julie Lynne Strauss
Vir 13h36'52" -1d38'13"
Julie M Johnson
Ari 2h47'47" 10d34'51"
Julie M. Lopez
Psc 1h20'27" 15d28'23"
Julie Madeline Zeller 2-26-1994
Lyn 7h46'49" 47d14'31"
Julie Maheu
Lep 5h55'49" -12d7'20"
Julie + Marco = Amour Eternel
Umi 13h7'35" 71d17'50"
Julie Marie
Vir 13h27'41" -4d1'30"
Julie Marie
Gem 7h39'7" 15d52'17"
Julie Marie Andress
Lmi 10h26'39" 37d55'54"
Julie Marie Cabral
Tau 4h18'52" 4d55'47"
Julie Marie Eisan
Crb 15h54'6" 28d45'22"
Julie Marie Frank
Lyn 7h31'42" 35d30'12"
Julie Marie Heilgeist
And 0h4'52" 34d41'54"
Julie Marie Koerperich
And 23h45'1" 36d50'40"
Julie Marie Letton
And 1h25'23" 49d8'56"
Julie Marie Milosevich
Tau 3h57'55" 22d22'1"
Julie Marie Morrison (the baby)
And 0h45'1" 36d12'48"
Julie Marie Redman
And 1h0'57" 40d49'5"
Julie Marie Rooswinkel
Uma 13h22'7" 61d29'13"
Julie Marie Rucinski
Cap 20h27'30" -18d35'3"
Julie Marie Thompson
And 0h29'57" 27d3'12"
Julie Marie Wiejak
And 23h16'4" 47d54'35"
Julie Marschner
Peg 22h43'21" 2d52'59"
Julie Mary Abe
Cas 23h35'6" 56d10'56"
Julie Massari
Cyg 21h47'6" 50d11'54"
Julie Mayer
Tau 4h42'36" 8d9'4"
Julie McAvinn Shines Forever
Cas 2h43'39" 57d46'20"
Julie McDanel
Tau 4h48'41" 26d19'27"
Julie McDermott
Lib 15h8'40" -5d58'49"
Julie McKenna
Cyg 20h17'32" 49d28'58"
Julie McKinstry-Harvey
Cyg 20h35'2" 58d12'28"
Julie McPowell
Cnc 8h46'33" 19d58'42"
Julie Melissa Floyd
Lyn 7h6'53" 60d34'42"
Julie Metcalf
Cas 3h30'46" 70d10'28"
Julie Michael
Lyr 18h49'28" 35d5'35"
Julie Michelle DiVita
And 23h11'0" 52d20'16"
Julie Michelle Merz
Vir 14h27'22" 7d6'24"
Julie Michelle Saunders
And 2h11'49" 50d8'54"
Julie Michelle Wood
Cyg 19h38'17" 34d39'2"
Julie & Mike's Wedding July 16,2005
Umi 15h19'47" 74d37'40"
Julie Mildred Cooke
And 24h0'0" 45d23'27"
Julie Minnich
Tau 5h55'17" 25d11'19"
Julie mon étoile je t'aime
Agr 21h23'1" -10d8'9"
Julie Monkey
And 0h16'3" 25d11'27"
Julie Monster
Sgr 19h7'59" -27d41'5"
Julie Monty - Tupperware
Crb 15h31'0" 27d19'44"
Julie Moran
Uma 10h7'12" 52d8'6"
Julie Morant
Uma 11h9'34" 28d36'26"
Julie Mørch Jensen
Tau 4h28'7" 19d58'39"
Julie Morgan
Lyn 8h4'7" 42d5'41"
Julie Morissette
Umi 17h10'19" 83d22'47"
Julie Moss Johnson
Gem 6h54'21" 12d56'15"
Julie Murphy
And 23h12'51" 52d8'28"
Julie "My Everything" Thomas
Ari 1h52'46" 22d3'31"
Julie My Love
Cyg 19h45'44" 39d14'50"
Julie My Precious
Lyn 8h30'31" 33d24'15"
Julie Myers
Vir 12h18'48" 11d50'9"
Julie Napoletano
Uma 12h23'46" 59d10'18"
Julie & Nathan Forever
Cyg 20h47'23" 47d44'13"
Julie Naughton
Cyg 21h38'6" 50d11'20"
Julie Neoterikos Anatello Isaggelos
Cyg 20h1'36" 31d47'56"
Julie Newall
Cas 23h16'38" 55d9'48"
Julie O'Brien
Uma 11h53'20" 62d9'12"
Julie Olga
Psc 1h47'46" 5d5'43"
Julie O'Malley
Cas 1h34'35" 57d57'46"
Julie Oram - "My Truly"
Cyg 20h45'7" 35d14'23"
julie p
And 0h34'22" 31d3'53"
Julie Pannos
Cyg 21h39'30" 36d56'59"
Julie Peasgood
Cas 0h36'1" 60d8'31"
Julie Pederson
Lac 22h24'11" 43d33'31"
Julie Penman
And 2h34'3" 50d58'23"
Julie Poe
Lyr 18h38'35" 44d59'19"
Julie Powell
Dra 16h42'40" 56d11'59"
Julie Quisnam Ego Diligo
Cas 23h21'58" 55d50'19"
Julie R Colehower
And 1h10'2" 44d45'43"
Julie R. Richey
Lyn 6h45'2" 55d31'1"
Julie Rayburn Fowler
Gem 7h7'35" 24d34'1"
Julie Raye
Vir 13h24'38" 12d54'23"
Julie Reed
Cas 0h23'18" 56d54'58"
Julie Renae Runcie
Dra 18h36'30" 59d12'46"
Julie Rene Farley
Leo 11h18'4" -6d40'46"
Julie Reverberi
Uma 8h21'42" 62d6'31"
Julie & Richard Bickerton
Her 18h52'53" 21d14'41"
Julie Riggs
Her 16h26'14" 24d41'8"
Julie Roberson
Dra 17h57'42" 51d42'24"
Julie Robin (My Jewel)
Cnc 9h6'54" 23d25'57"
Julie Rockstar
Vir 13h5'11" -15d35'48"
Julie Romeo's Special Star
Uma 8h19'58" 71d33'16"
Julie Rosalyn Lavin
Agr 20h54'29" -8d26'3"
Julie Russel
Cas 23h41'51" 53d5'28"
Julie & Ryan Hershey
Leo 9h49'47" 8d8'5"
Julie S
Cru 12h26'12" -58d25'20"
Julie Sharon Whitt
Cas 0h2'13" 59d12'42"
Julie Shea
Sgr 18h2'59" -28d1'45"
Julie Shorack
Uma 13h57'5" 51d13'0"
Julie Shriner Kontsis
And 23h54'1" 35d28'1"
Julie Siegel
Sgr 18h39'42" -30d12'19"
Julie Sino-Shekhman
Sgr 17h53'25" -22d31'2"
Julie Sitts
Lyn 8h20'27" 40d59'25"
Julie Skibski
Dra 19h18'44" 74d12'58"
Julie - Sladkii
Cyg 19h32'55" 30d40'12"
Julie Smith
Umi 15h42'1" 82d55'1"
Julie Solange Odette Labergère
Uma 11h30'34" 60d38'5"
Julie Spears Luke's Only Star
Leo 10h24'21" 13d58'44"
Julie Speziale
Sge 19h55'55" 19d17'21"
Julie "Splendiferocious" Murray
Uma 11h42'52" 64d53'27"
Julie Spychalski and Cricket Tryon
Cyg 20h58'48" 31d18'1"
Julie Star
Cas 0h15'18" 62d34'10"
Julie Stronberg
Lyn 7h55'15" 56d26'44"
Julie Taidy
Leo 9h32'12" 32d26'20"
Julie Tempest
Tau 5h50'18" 17d16'25"
Julie Teresa
Ori 5h33'31" 8d24'53"
Julie the Stoolie Averna
Vir 12h3'47" -7d42'20"
Julie Therese Crawley
Crb 15h18'7" 26d28'37"
Julie Tiao
Leo 11h15'59" 19d26'8"
Julie Truong Mentrum
Psc 1h23'24" 26d10'17"
Julie V. Payne
Uma 12h6'6" 65d39'5"
Julie Valdez
Mon 7h36'25" -0d35'13"
Julie Van Waeyenberghe
Uma 9h32'27" 48d34'0"
Julie Vanderslice
And 23h20'40" 47d53'1"
Julie Weinman
Uma 8h37'1" 60d36'41"
Julie Weir - 10-08-58
And 23h49'25" 41d33'29"
Julie Wells
And 23h13'21" 44d16'20"
Julie Weston - Eternally Yours
Pho 0h48'20" -56d36'34"
Julie Winston
And 2h17'48" 41d36'11"
Julie Wittet
Gem 7h51'18" 15d0'1"
Julie Wulie
Cyg 20h26'30" 45d13'52"
Julie Yarosch
Cas 23h26'41" 51d45'8"
Julie Yoon
Leo 11h30'31" 14d37'19"
Julie You Light Up My Life
Uma 12h0'13" 45d22'3"
Julie, Best Friend Forever
And 23h47'28" 39d51'11"
Julie, Dylan, BreAnn
Tri 2h10'11" 34d15'28"
Julie, In Our Hearts Forever...
Sgr 17h52'27" -28d55'5"
Julie, Tyler, Rahni Dog & Rex Dog
Cru 12h36'24" -61d7'12"
Julie,Tom & Joe
Per 3h9'59" 38d56'18"
JulieAileen1
Leo 9h36'59" 7d0'34"
Julieann
Vir 13h45'55" -20d18'29"
Julieann
Sgr 18h50'42" -18d46'5"
Julieann Amanda Ward
And 2h36'48" 43d47'18"
Julieann Hoffman
And 1h30'16" 49d47'16"
Julieann Killingbeck
Lyr 19h20'9" 29d13'15"
Julieann Marie Fegan
Tau 4h21'6" 26d13'11"
Julieann Vanni
Leo 11h13'26" 17d14'35"
Julieanna Ann Mullins
And 0h48'39" 42d32'56"
Julieanne
Ori 6h9'4" 8d27'15"
Julieanne Richard
Tau 4h46'57" 21d45'51"
JulieAnne19031977
Ori 5h40'24" -2d27'39"
Julieanne's Star of Grace
Tau 5h47'56" 16d37'41"
JulieBharuchaTrivedi
Sgr 18h22'14" -22d35'34"

Juliebug Aql 19h42'33" -0d0'33"
JulieJustinTaylorMasonDrezdinHirsch Uma 11h43'58" 41d44'24"
Juliemar González Cap 21h35'55" -14d31'48"
Julien And 23h0'27" 52d55'43"
Julien Lyn 7h52'58" 38d20'34"
Julien Aubrey Boyance Lyn 7h46'55" 39d47'47"
Julien Bezerra Leo 10h5'22" 26d15'21"
Julien Bouvet Oph 18h17'44" 9d18'34"
Julien Briat Cas 0h56'35" 54d10'5"
Julien Delaitre Her 16h29'0" 5d41'25"
Julien et Laetitia Uma 8h57'8" 53d21'31"
Julien F Umi 13h39'3" 78d8'9"
Julien Hamel Peg 21h37'21" 26d46'34"
Julien J. Napoli Tau 3h48'24" 2d52'3"
Julien je t'aime Lep 6h9'49" -18d49'59"
Julien&Laetitia,Gillot-Fallone Cas 23h39'21" 54d41'45"
Julien & Loraine Forever Uma 8h47'44" 46d39'49"
Julien Madison Fuselier Umi 15h13'12" 81d17'50"
Julien Naginski's Lucky Star Cnc 8h9'11" 32d50'52"
Julien & Pepere Goupil Uma 9h46'37" 42d46'36"
julien polla And 1h21'39" 43d11'42"
Julien Sorel Vir 13h26'25" 9d39'40"
Juliena Chiara Albert Uma 8h9'27" 59d48'33"
JulieNJoe65 Tau 5h8'5" 28d0'46"
Julienne Aqr 22h9'44" 1d2'38"
Julienne Sharpe Equ 21h18'51" 7d31'32"
Julienne Skye Sco 17h19'21" -32d44'26"
Julien-Sandro Braunwalder Ori 6h7'40" 21d11'49"
Julieofmyheart Lib 15h27'55" -11d11'45"
Juliepie And 0h35'5" 29d24'9"
juliepie Lyn 7h40'28" 51d4'8"
Julierevs And 23h12'53" 47d26'28"
Julie's Bright Star Lyn 6h33'37" 55d11'45"
Julie's DreamCatcher Sco 16h15'0" -11d13'37"
Julie's Half Mile Psc 1h16'0" 32d49'57"
Julie's Heart in the Sky Peg 22h37'16" 10d56'30"
Julie's Husband Sco 17h14'42" -43d45'27"
Julie's Jewel In The Sky And 0h20'37" 31d6'45"
Julies Momskra Cnc 8h33'41" 23d52'9"
Julie's Shining Heartlight Ori 5h4'41" 9d28'2"
Julie's Star Psc 0h49'49" 16d3'58"
Julie's Star Tau 4h25'41" 17d16'28"
Julie's Star Leo 10h40'53" 8d38'31"
Julie's Star And 0h26'50" 33d19'3"
Julie's Star Leo 9h23'25" 32d45'14"
Julie's Star Dra 16h34'57" 51d39'37"
Julie's Star Sgr 19h4'5" -35d7'57"
Julie's Star Pho 1h54'16" -44d4'24"
Julie's Star Eri 3h47'56" -39d0'51"
Julie's Star Aqr 23h30'54" -15d44'8"
Julie's Star Vir 13h7'41" -4d25'49"
Julie's Starlight Lib 15h11'46" -6d20'57"
Juliessa Marie Talavera And 1h26'15" 43d23'7"
Juliet And 2h25'53" 46d25'23"
Juliet Cyg 19h35'44" 39d50'10"
Juliet Mon 6h50'8" 8d33'58"
Juliet Uma 12h2'20" 53d28'47"
Juliet Psc 0h57'23" 3d16'8"
Juliet 828486 Crb 16h7'16" 35d7'28"
Juliet and Randy-IWLYF-7-08-06 Lib 15h55'18" -9d30'39"
Juliet Anne Uma 11h1'39" 34d14'13"
Juliet Anne And 22h59'47" 39d39'27"
Juliet Berrettini And 23h28'21" 42d38'42"
Juliet Birch Cyg 19h40'17" 35d30'33"
Juliet Bullock Cas 2h21'3" 70d54'38"
Juliet Hay Cyg 19h33'33" 32d7'35"
Juliet Iannoli-Ballard Uma 10h52'47" 72d43'29"
Juliet loves Charles Lib 14h52'10" -1d0'0"
Juliet Lynn Ari 3h26'26" 26d56'42"
Juliet Lyons Gray And 0h20'37" 28d0'17"
Juliet Mansfield-Clark Cas 0h47'4" 58d6'13"
Juliet Marie Graves Cas 2h33'48" 62d5'58"
Juliet Marie Petruzelli Peg 22h9'15" 31d49'52"
Juliet Pearl Baker Lmi 10h22'2" 36d54'22"
Juliet Perez Vir 12h6'27" 0d50'2"
Juliet Rene Aqr 21h41'16" 1d45'57"
Juliet Revis Vir 14h17'42" -12d48'24"
JULIET ROSSETTI Aqr 22h14'48" 1d37'7"
Juliet Theresa Giroux Lyr 19h13'1" 27d16'38"
Juliet & Tiffany Leo 9h48'40" 26d0'47"
Julieta And 0h11'49" 44d32'27"
Julieta Perez Aqr 23h39'36" -7d19'18"
Julieta-Loved And Missed Cnc 8h47'58" 17d40'32"
Julieth Crb 16h17'19" 26d45'46"
Julietta Lib 15h30'23" -9d42'58"
Juliette Lib 15h6'1" -3d29'31"
Juliette Umi 15h12'16" 82d34'22"
Juliette Sgr 19h9'9" -24d33'27"
Juliette Col 6h19'22" -33d50'2"
Juliette Peg 21h51'15" 27d16'14"
Juliette Gem 6h6'37" 25d25'14"
Juliette Her 16h41'23" 10d44'30"
Juliette And 1h58'36" 47d0'58"
Juliette Per 3h23'49" 49d16'25"
Juliette Adele Pleso Cyg 19h13'58" 52d37'49"
Juliette Antonia Paterno Ari 3h22'20" 29d40'29"
Juliette Brittain Cas 23h42'36" 53d9'30"
Juliette Chombart Uma 9h34'42" 42d24'26"
Juliette Clagett McLennan Cnv 13h53'28" 37d54'54"
Juliette Edjour Amour Eternel Sco 17h50'51" -38d53'4"
Juliette Huckel Crb 16h21'57" 30d11'12"
Juliette Jo Berg Gem 7h6'57" 34d25'25"
Juliette & John Lyr 18h47'59" 36d39'11"
Juliette L. Curley Uma 10h53'27" 55d51'59"
Juliette Louise Peace Uma 10h54'16" 35d8'27"
Juliette M. Goulet Tau 5h46'10" 22d8'41"
Juliette Marie Sgr 19h14'4" -35d33'34"
Juliette Martin 07.10.1923 Cas 23h25'19" 54d49'59"
Juliette Nahide Uma 8h37'12" 62d40'19"
Juliette Reid Gem 6h16'37" 24d35'48"
Juliette ULN Vir 12h45'48" 6d50'27"
Juliette Van De Ven Psc 1h18'47" 24d33'25"
Juliette Vigne Cas 0h38'59" 61d47'55"
Juliette, 18 mars 2006 Psc 0h12'19" 9d45'50"
Julija Koscejeva Cas 23h27'54" 55d12'48"
Julija Laventa And 2h5'36" 43d43'19"
JULIJANA - 211063V And 1h51'11" 39d51'5"
Juli'mari Vir 14h14'32" 6d52'15"
Juli-Mary Fraser Beatty Uma 11h52'25" 43d2'32"
Julio Aql 19h13'47" 3d12'29"
Julio Tau 5h59'6" 23d58'33"
Julio Lib 15h28'55" -5d53'16"
Julio A. Irizarry And 23h57'28" 42d26'22"
Julio A. Larrea Cam 5h44'32" 78d2'37"
Julio Aspe Viñolas Ori 5h14'54" -0d37'6"
Julio Cañada Bellido Leo 10h46'55" 22d42'58"
Julio Cesar Velez Aur 5h17'17" 41d55'38"
Julio Figueroa Leo 11h23'55" 27d10'3"
Julio Lemus Boo 14h57'20" 25d32'12"
Julio loves Yami And 2h25'15" 46d45'57"
Julio Maggi Cep 22h49'41" 72d54'52"
Julio Miguel Leo 11h35'10" 14d30'20"
Julio Pena, Jr. Vir 13h3'18" 13d59'40"
Julio "Pookie" del Rio Ori 5h7'45" 3d51'26"
Julio R. Marrero Crb 15h39'9" 38d58'9"
Julio Rivera Burgos Uma 9h53'47" 55d26'42"
Julirina Uma 12h45'41" 52d41'9"
Julironyphialynne And 1h58'8" 35d43'0"
Julisa's Warm Heart Tau 5h43'58" 18d18'28"
Julissa Cnc 8h28'32" 13d58'10"
Julissa Ori 6h0'42" 17d46'10"
Julissa Sgr 18h14'23" -24d55'53"
Julissa Carmona Cap 21h46'38" -10d38'1"
Julissa Iris Vir 14h6'18" -20d56'23"
Julissa James Gem 7h23'23" 33d36'29"
julius Leo 11h10'25" 16d30'50"
Julius Cep 22h8'30" 57d3'1"
Julius Col 5h21'4" -38d37'3"
Julius Christian Taylor Tau 4h43'0" 24d59'24"
Julius Christian Taylor Tau 4h21'42" 11d26'15"
Julius & Dwan Santiago Cyg 21h24'27" 30d45'43"
Julius Gantman Sgr 19h21'30" -22d31'1"
Julius Granata Aql 20h18'52" 1d59'22"
Julius Joel Gotlieb Cnc 8h53'54" 8d7'49"
Julius Landell-Mills Uma 10h11'19" 64d58'7"
Julius Leavitt Aql 19h51'10" -0d52'6"
Julius May Her 18h33'57" 13d18'47"
Julius Pinter Lib 15h30'55" -6d36'17"
Julius Zarnofsky Aqr 20h41'15" -7d29'24"
Juliya Pavlova Uma 8h13'42" 63d25'46"
Jullian Theodore Kambouropoulos Gem 8h1'4" 28d4'45"
Jullisa Star Peg 21h10'44" 19d58'2"
Juls Gem 7h2'6" 31d43'47"
Juls in the sky Lyn 8h43'41" 41d21'29"
July 1, 1984 Ara 17h29'51" -47d25'1"
July 1st Uma 10h15'6" 68d57'52"
July Angel Umi 13h52'5" 69d57'23"
July M Lozano Lyn 7h37'2" 58d32'37"
July Sky Cat Dra 18h34'58" 64d26'53"
July Uittenbogaard Dra 19h49'26" 61d15'34"
Julyan Cnc 8h10'32" 10d21'12"
Julz Psc 1h11'45" 11d22'7"
JULZ And 2h28'55" 41d58'59"
Julz And 1h42'8" 46d43'0"
Julz Vir 13h23'50" -8d29'6"
Julz Hoku Uma 12h43'33" 59d14'26"
Julz Sully And 0h52'9" 40d15'35"
Julz61 Leo 11h37'34" 23d55'44"
JulznJess Cyg 21h10'59" 34d6'40"
Juma Uma 14h1'12" 56d40'46"
Jumamas precious stars JarBriHayMac Cyg 19h42'1" 33d14'24"
Jumana Psc 0h44'36" 19d22'54"
jumanji Lyn 7h56'43" 40d22'16"
Jumbo & Aki Tsuruta Cyg 20h12'21" 49d31'24"
Jumi Cnc 8h37'26" 29d16'22"
JuMic Tau 3h27'58" 17d47'56"
Jumpin For Joy Ari 2h41'50" 20d51'12"
Jumpin' Joe Petroni Her 16h32'54" 23d56'36"
Jumpywolf Sco 16h29'6" -25d21'30"
Jun and Stella Iglupas Crb 15h33'53" 37d30'5"
Jun Li Gem 6h38'26" 19d40'46"
Jun Love Star Psc 0h39'31" 3d30'0"
Juna Cnc 8h24'19" 30d52'45"
Juna, Adria, Kaleen Ori 4h54'22" 10d9'12"
Junann Toll-Lopez Cyg 21h55'56" 49d49'38"
Junaquist Joy Peg 22h6'9" 12d20'29"
Junction Joe Uma 8h56'22" 47d27'9"
June Gem 7h8'43" 20d56'44"
June And 0h29'53" 27d26'56"
June Col 5h8'22" -34d54'16"
June Cas 23h36'1" 54d20'18"
June Cap 21h25'52" -17d53'54"
June 18 Leo 10h23'4" 23d20'43"
June A. Anderson McElhinny Lyr 18h51'37" 36d55'8"
June A. Ramos Ari 2h52'52" 27d20'26"
June Albiez Uma 9h23'58" 68d26'46"
June and Barney's Gold Anniversary Lyn 7h59'55" 49d41'3"
June and Jerry Broyles - Forever Lyn 8h0'9" 34d24'54"
June and Joe Roshe Per 3h26'43" 44d22'33"
June and Orville Mock Lyr 19h2'7" 33d14'2"
June Angeline Norton Babcock Ari 2h44'9" 31d0'21"
June Ann Malina Sgr 17h54'38" -29d18'36"
June Ann Ritch Crb 15h26'30" 27d3'2"
June Ann's Shining Star Psc 0h25'59" 0d20'36"
June Azzinaro Col 5h46'56" -29d24'21"
June B Studley Gem 7h28'40" 33d32'17"
June Ballard Cnc 8h20'53" 22d38'14"
June & Bernard Barwick Cyg 20h10'44" 50d46'1"
June Blaine Gem 7h14'37" 21d24'15"
June Brennan Craig Gem 7h49'25" 25d12'28"
June Bug Ori 5h29'11" 8d5'39"
June Bug Her 16h53'59" 36d27'30"
June Bug Lyn 7h13'48" 58d39'37"
June C. Mahoney Sco 15h53'46" -20d25'2"
June Carmen Luebcke (Daisy June) And 23h19'48" 42d25'28"
June Chabot Uma 9h18'34" 43d39'54"
June Collier Beyer Pic 5h44'34" -47d46'15"
June Delores Lynne Lacey Umi 10h19'29" 88d24'13"
June Dolman - "J60 DOL" Aur 7h9'8" 38d6'59"
June Doremus And 23h32'7" 43d3'37"
June Edith Darby Cas 3h28'25" 75d7'46"
June Edmondson Cnc 8h35'15" 17d9'44"
June Elery Hennebry Cyg 19h46'41" 36d40'58"
June Elizabeth Cas 0h28'51" 55d49'10"
June Elle Sco 17h31'25" -42d39'23"
June Ellis Cnc 9h20'8" 19d57'56"
June Frye Cyg 19h51'9" 33d36'49"
June Gallagher Sco 17h16'24" -38d45'21"
June Guerra Ori 5h34'51" 9d24'40"
June Hammer Cnc 9h3'45" 31d35'47"
June Hertha Gorton Tillson Vir 14h28'47" 4d7'37"
June Irene Jennings Uma 10h51'4" 64d0'27"
June & John Sgr 19h2'8" -30d56'43"
June & Johnny Dra 18h35'36" 58d12'25"
June Johnson Aqr 22h27'26" -23d22'53"
June Kent-Bailey Cas 0h55'34" 58d57'39"
June Kimbel Uma 13h0'11" 53d38'13"
June Kitson Uma 13h42'34" 52d1'9"
June Landry/Domenick Ori 5h59'26" 20d58'55"
June Larsen Davis Lyn 7h44'33" 47d7'37"
June Lillian Beasley Cru 12h30'28" -60d41'14"
June "Lucky Star" Gladfelter And 0h29'14" 44d34'58"
June M. and Alvis S. Hardy Cyg 19h57'54" 35d17'56"
June M. Fish Cyg 19h56'47" 44d50'7"
June M. Vanourek Cam 5h6'13" 65d15'8"
June Marie Gilmore (Pooh) Leo 10h15'9" 21d28'13"
June Marie Green Vir 13h9'46" 11d24'34"
June Mary Giannelli Ari 2h0'13" 18d59'53"
June McCartney Cas 23h1'35" 53d30'18"
June McCraw 6 Uma 9h36'28" 67d34'35"
June Merlo Crb 16h18'1" 30d39'44"
June "Mom" Wheatley Cas 1h12'25" 51d29'9"
June 'Moon Beam' Baker Gem 7h18'12" 14d33'44"
June (My Mum) Per 3h19'46" 41d16'35"
June O'Connor And 0h32'45" 38d12'3"
June Oney Lyr 18h40'25" 35d50'23"
June Pappas Cma 6h22'49" -29d43'59"
June Pauline Chabot And 0h45'15" 30d21'27"
June Pawlotsky *A Hero* Per 2h45'30" 56d23'10"
June Pearl White Cas 23h13'38" 55d47'18"
June Pierson Gem 7h42'49" 16d45'43"
June Potter Lucier And 0h43'22" 36d32'22"
June Q. Burgess Cnc 8h4'43" 13d37'36"
June Rose Cnc 8h46'27" 16d14'49"
June Rosemary Johnson Crb 15h46'44" 32d15'9"
June Sember Vir 14h6'38" -20d41'7"
June Star Tau 5h20'7" 17d9'39"
June Sylvia And 0h32'50" 46d12'5"
June T. Brennan Aqr 21h59'33" -15d12'11"
June The Teaching Star Peg 21h36'6" 20d46'27"
June & Timothy Raymond Peg 21h44'32" 23d29'20"
June Twilight Ori 5h16'55" 15d18'56"
June Vivian McCarthy Gem 7h24'14" 33d42'28"
June W. Hoeflich, Board Chairperson Crb 15h32'30" 37d40'6"
June Warren Cas 23h20'59" 56d8'49"
June Wright Cnc 8h19'51" 13d12'19"
June, Best Mom&Wife in the universe Cnc 8h51'4" 9d57'15"
June23*88*Leiff Koenigs*August10*06 Uma 11h23'23" 59d59'41"
Juneau-Zmuda Cyg 20h33'15" 40d33'43"
Junebug Lyn 8h25'35" 33d42'21"
Junebug Gem 6h43'2" 29d27'7"
JuneBug Sge 19h51'24" 18d41'10"
June-Bug (1st wedding anniversary!) Cyg 20h10'13" 35d25'38"
Junebug In The Sky With Diamonds Crb 15h37'35" 30d53'20"
JuneBug2006 Gem 6h12'3" 22d48'20"
Juneifer Gem 6h41'25" 17d52'57"
June's Angel Lyn 8h5'37" 52d33'33"
June's Angel Ori 5h35'18" -2d7'56"
June's Angels Uma 12h4'51" 41d33'47"
June's Light Cnc 8h28'35" 32d59'25"
Junes Norwell Aur 5h33'28" 45d39'9"
June's Pearl Cas 23h51'37" 56d20'2"
June's Wish Upon A Star And 0h1'53" 39d28'4"
Jung Kim Tau 3h43'46" 9d37'35"
Jung Soon Lib 15h12'50" -21d51'42"
Jung Wan Lyn 7h21'12" 45d6'38"
Jung, Daniel Ori 6h19'19" 13d45'33"
Jung, Herbert Uma 9h6'4" 51d45'30"
Jung, Michael Ori 6h17'0" 19d55'54"
Jung, Vladimir Uma 10h19'11" 50d0'3"
Junge, Timo-Manuel Aqr 21h26'45" 0d23'52"
Jungle Eyes Cam 4h28'27" 66d43'35"
Jungle Fighter Per 3h25'15" 47d21'39"
Jungledaisy Umi 14h4'4" 66d44'53"
Junia Clarke Cru 12h19'12" -56d48'15"
Junia & Ted Doan Love Forever Cyg 19h46'26" 36d1'13"
Junichiro & Yoko Aqr 22h30'45" -7d50'30"
Junie Lyme' Fifarek And 1h48'34" 42d40'25"
Junior Gem 6h21'26" 18d40'43"
Junior Her 16h57'46" 29d52'28"
Junior Aqr 22h47'47" -7d41'4"
Junior Sco 15h59'4" -20d47'13"
Junior Umi 14h19'59" 66d55'47"
Junior Ara 17h27'9" -47d56'18"
Junior - Angel of Destiny Per 4h49'17" 39d31'25"
Junior Roadman Umi 14h26'58" 68d11'40"
Junior's Milky Way Vir 11h54'51" 7d8'52"
Juniper Del 20h35'43" 11d53'18"
Juniper "Juni" Quaglia Uma 13h32'9" 53d24'52"
Juniper Spring Tree And 23h58'14" 39d57'8"
Junius & Berneice Millard Uma 10h53'28" 44d41'34"
Junius Disarson Rosanes Lyn 6h39'34" 56d43'14"
Junker, Hermann Uma 13h38'52" 62d14'23"
Junko Hanjo Hoshi Umi 16h16'9" 72d51'52"
Junkochan Gem 7h5'9" 33d11'13"
JunKris Uma 13h49'42" 49d42'16"
junne Psc 0h32'8" 8d57'51"
Junnita Nelson And 0h58'8" 39d16'44"
Juno Hsu Ori 5h31'11" -4d46'26"
Junod's Dreamland Psc 0h48'49" 17d6'38"
Jun's star Psc 0h19'17" 12d2'30"
Junto Por Las Estrellas Her 18h7'25" 48d25'8"
Juntos Para Siempre Lyn 8h41'45" 34d53'49"
Juntos Para Siempre Mark and Amy Cyg 20h0'14" 44d36'35"
Juntos somos uno Sco 17h29'37" -30d30'29"
Junue Gem 6h19'52" 25d42'44"
Junwan & Jianhua Yuan Forever Vir 14h14'6" -9d59'51"
Jupa Aqr 22h8'29" -14d36'9"
Jupiter Uma 10h30'21" 57d4'53"
Jupiter Joe Cap 21h56'42" -22d16'30"
Jupiter - Nicole rose Tau 6h0'7" 28d23'30"
JUPPI FOREVER Uma 11h7'20" 37d46'27"
Juraj Sipko Psc 1h10'47" 16d16'14"
Jurate M. Bucha Aql 19h44'19" -0d6'26"
Jurek the Great Uma 11h40'56" 31d2'57"
JUREMA PAULUS Uma 12h0'23" 58d0'24"
Jürg Cyg 19h36'49" 31d54'15"
Jürg Aur 6h28'26" 35d7'4"
Jürg Bernegger Ori 5h8'16" 4d58'59"
Jürg Egli Crb 16h23'40" 36d16'55"
Jürg G. Fischer (Firg) Cnv 13h58'19" 35d33'55"
Jürg Hirzel Dra 19h10'57" 66d35'29"
Jürg Lohri-Spengler Aql 20h4'41" 11d58'12"
Jürg Rothen Crb 15h40'0" 25d55'38"
Jurga Mardosaite Gem 6h51'51" 22d52'29"
Jürgen Uma 10h1'35" 61d21'20"
Jürgen Albrecht Uma 8h28'25" 63d59'0"
Jürgen Barthel Uma 10h29'52" 47d58'17"
Jürgen Bettels Uma 11h33'33" 33d3'17"
Jürgen Böye Ori 6h19'12" 10d6'20"
Jürgen Fritz Uma 10h6'25" 63d40'8"
Jürgen Gattner Uma 8h54'8" 63d36'10"
Jürgen Gebhardt Uma 10h59'0" 58d3'35"
Jürgen Heyn Uma 8h57'21" 68d32'39"
Jürgen Hoffmann Uma 11h18'34" 41d46'22"
Jürgen Isermann Uma 11h41'49" 65d29'30"
Jürgen Kaiser Uma 8h52'55" 59d17'20"
Jürgen Klose Ori 4h51'16" 0d31'9"
Jürgen Kolter Ori 5h28'17" 8d47'5"
Jürgen Kömen Uma 11h43'30" 31d44'40"
Jürgen Krügel Vir 13h39'51" -18d35'16"
Jürgen Lindemann Uma 9h39'42" 41d59'25"
Jürgen Lloyd Uma 11h42'36" 38d30'25"
Jürgen Massong Ori 5h44'54" 12d5'53"
Jürgen Meisezahl Uma 10h8'32" 62d43'26"
Jürgen & Monika Uma 10h46'44" 72d15'34"
Jürgen Mühle Uma 11h10'25" 42d2'14"
Jürgen Pallokowski Uma 9h20'40" 42d0'40"
Jürgen Pruess Ori 5h12'29" -6d47'6"
Jürgen Radke Uma 11h0'15" 58d20'56"

Jürgen Rögner
Uma 9h12'58" 70d52'45"

Jürgen Rückheim
Uma 8h42'44" 62d49'9"

Jürgen Rühlmann
Ori 5h48'32" -3d59'56"

Jürgen Schnell
Uma 9h47'50" 68d29'40"

Jürgen Schwarzenau
Ori 5h57'5" -3d10'57"

Jürgen Vogler
Uma 9h50'8" 68d27'14"

Jürgen Weiss
Ori 5h58'38" 12d19'34"

Jürgen Weißkirchen
Ori 5h40'23" 9d14'11"

Jürgen Wiedenmann
Uma 10h12'28" 43d32'20"

Jürgen Woldemar Kröhnert
Uma 13h4'26" 53d58'6"

Jürgen-Peter Beier
Uma 8h13'42" 70d3'15"

Jürgens Glücksstern
Uma 8h52'54" 49d22'37"

Juri
Cyg 21h5'10" 45d0'59"

Juri
Cam 7h47'13" 84d28'23"

Juri
Cyg 21h5'10" 45d0'59"

Juri De Marzi
Ser 18h16'5" -15d29'2"

Jurifran*Eve
Uma 11h32'18" 38d45'37"

Jurika
Aqr 21h23'25" 1d59'22"

Juris Bialecki
Cmi 7h32'34" 4d16'24"

Jurni Abigail Hurd
Peg 21h40'31" 15d32'7"

Jurni Anderson
Uma 8h47'44" 63d10'51"

Jurnie
Dra 17h31'0" 58d23'51"

Jusep i Mònica
Ari 3h6'52" 17d48'31"

Jussara "The Star of Deep Feelings"
Cyg 20h52'1" 54d10'48"

Jussie James
Psc 1h11'32" 29d38'34"

Just a Little One!
Cas 0h25'24" 60d49'38"

"Just Be a Light"
Uma 13h51'24" 55d58'45"

Just Because
Uma 10h40'21" 58d49'19"

Just Dick Jefferson
Her 16h49'51" 16d36'29"

" Just Doug " Douglas Mark Doran
Her 16h32'7" 47d16'31"

Just Flying High
Psc 0h38'50" 8d17'29"

Just For Dawn
Leo 10h26'35" 17d22'54"

Just For SB
Psc 23h9'51" 1d13'42"

Just for the two of us
Cyg 20h57'20" 38d44'23"

Just for you
Uma 10h32'44" 48d39'58"

Just Frank
Cep 22h31'6" 60d17'57"

Just Glide My Sculli
Cru 12h22'27" -57d10'34"

"Just Hannah"
Peg 23h10'27" 17d5'4"

Just Imagine My Love
Lib 15h42'53" -10d46'28"

Just Imagine This
Uma 14h0'6" 48d58'32"

Just Jake "03/01/2005"
Psc 1h19'55" 26d57'49"

Just Jay
Cnc 8h35'41" 23d37'43"

Just Joe "9/17/1931 - 1/22/2005"
Uma 8h22'32" 61d58'28"

Just Know
Her 18h19'16" 23d18'56"

Just Married!
Per 3h0'8" 56d8'20"

just married Angelika & Daniel
Uma 9h42'12" 55d35'8"

Just Norman
Leo 9h49'19" 28d15'9"

Just One Look
Uma 11h26'33" 63d53'46"

Just One Wish
Uma 10h21'53" 69d37'52"

Just Peachie
Per 3h31'41" 44d39'48"

Just Smile
Ori 5h27'33" 9d55'18"

Just Us
Uma 12h9'39" 55d40'37"

Just us...
Lib 14h51'15" -2d34'10"

Just You and I
Cyg 20h56'10" 36d33'8"

Justaphine
Ori 5h57'11" 17d51'26"

JustChel Eternity
Cyg 20h18'56" 51d36'6"

Justeanie
And 23h25'26" 48d39'16"

Justeen
Cyg 20h1'30" 38d38'15"

Justeen Kaytlyn Gamboa
Gem 7h30'2" 15d0'28"

Justeen The Epitome Of Friendship
Cnc 9h8'15" 26d13'6"

Justella Marie Wiernicki
Uma 9h48'42" 42d7'3"

Justen Alexander Horvath
Uma 9h20'27" 59d21'42"

Justena
Leo 10h17'16" 26d54'27"

Justess Pletcher
Uma 10h31'3" 62d19'54"

Justice
Aqr 21h54'41" 0d42'0"

Justice Armani
Gem 7h38'51" 34d11'26"

Justice Kachadurian
Cma 6h55'23" -21d30'1"

Justice Mills
Per 3h8'6" 53d43'56"

Justice Salvatore R. Martoche
Boo 15h18'57" 42d5'14"

Justice Susi
Uma 12h1'1" 38d41'54"

Justin
Uma 10h24'18" 40d37'25"

Justin
Vir 12h10'9" 7d28'21"

Justin
Ori 5h58'13" 6d0'31"

Justin
Vir 12h50'21" 12d38'38"

Justin
Her 17h2'50" 12d59'46"

Justin
Crb 15h30'52" 29d4'4"

Justin
Tau 4h45'2" 26d21'38"

Justin
Leo 10h42'17" 15d20'47"

Justin
Psc 1h22'26" 26d39'56"

Justin
Sgr 19h9'59" -13d9'18"

Justin
Sgr 18h6'54" -16d40'44"

Justin
Sgr 18h35'42" -16d39'6"

Justin
Umi 5h3'47" 88d38'5"

Justin
Uma 8h15'35" 69d6'27"

Justin
Cep 22h25'23" 62d54'14"

Justin
Uma 12h38'24" 54d32'25"

Justin
Uma 12h48'46" 57d28'47"

Justin<3Klair Forever
Ari 3h19'40" 27d40'1"

Justin 919
Cma 6h58'18" -11d17'46"

Justin A. and Crystal R. Stephenson
Crb 15h33'42" 25d50'4"

Justin A. Bodman — "My Sweetie"
Leo 10h23'14" 25d44'36"

Justin A. Cochran
Cam 3h50'10" 68d0'17"

Justin A. Heller
Uma 9h13'36" 51d23'32"

Justin Aaron Campbell
Peg 21h32'26" 17d52'10"

Justin & Alaina
Cnc 8h4'9" 15d42'59"

Justin Alan Dangerfield
Umi 15h15'17" 77d18'45"

Justin Alexander
Per 3h48'29" 43d55'2"

Justin Alexander Kentish Star
Cap 20h34'30" -9d4'19"

Justin Allan Markwell
Uma 10h54'26" 52d56'21"

Justin Allen Pedersen
Sgr 19h45'30" -13d12'36"

Justin & Alyssa
Ori 5h52'2" 3d1'10"

Justin & Amanda
Ori 5h53'3" 21d28'32"

Justin Amandus Stegemann
Her 16h32'26" 33d40'16"

Justin and Alexis "Our True Love"
Tau 5h52'1" 16d6'19"

Justin and Allayne
Uma 11h44'11" 50d49'27"

Justin And Beth
And 23h53'5" 37d26'38"

Justin and Christina
Eri 3h51'11" -0d57'43"

Justin and Courtney Alvarez
Cnc 8h38'27" 18d16'53"

Justin and Jessica
Her 17h56'10" 14d56'29"

Justin and Kalee, Forever in Love
Lyr 18h50'21" 43d31'53"

Justin and Kristy: Friends Forever
Uma 13h54'14" 48d44'28"

Justin and Lucia
Cyg 20h23'7" 47d5'41"

Justin and Lynsie, Forever & Always
Ori 5h54'4" 3d53'59"

Justin and Michael Moore
Her 17h46'14" 43d21'6"

Justin and October Gonzales
Cyg 21h16'47" 37d6'55"

Justin and Steph
Aqr 23h3'33" -8d12'46"

Justin Andrew Guay
Ori 6h9'6" -3d51'25"

Justin & Annabelle Johnstone
Dra 18h55'26" 58d48'6"

Justin Anthony
Uma 11h34'8" 56d12'42"

Justin Anthony Barulli ~ 10/26/83
Lac 22h26'55" 48d11'55"

Justin Anthony Salas
Hya 9h31'22" -0d12'7"

Justin Arnold
Cnc 8h45'55" 13d3'57"

Justin August Wren
Vir 13h33'14" 4d37'3"

Justin B. Riddle - Sir
Sco 17h51'24" -31d55'58"

Justin "BB" Michaels
James Gonzalez
Psc 0h41'14" 8d25'50"

Justin Bellows
Ori 4h59'34" 12d21'52"

Justin Bennett - Codeman
Her 16h58'17" 27d42'14"

Justin Blackman
Sct 18h51'29" -9d41'5"

Justin Boles
Lac 22h17'35" 50d35'57"

Justin Bradford Bailey
Uma 9h47'18" 59d35'14"

Justin Brian Jacobsen
Cap 20h52'49" -16d1'51"

Justin Brian Jacobsen
Sgr 18h46'7" -28d37'17"

Justin Broeker
Aql 19h26'26" 4d35'26"

Justin Bruce
Leo 11h26'13" 3d58'30"

Justin "Bruce" Kirby II
Sco 17h55'1" -38d15'52"

Justin Burger
Tau 3h38'10" 28d22'19"

Justin Bush
Dra 20h16'20" 62d34'27"

Justin Carlisle Tuten
Sgr 19h55'40" -32d11'28"

Justin Carter
Her 16h38'22" 14d43'21"

Justin Castellano
Her 16h49'56" 21d40'30"

Justin Chase
Per 4h38'13" 39d44'38"

Justin Christopher Carroll
Ari 3h28'18" 22d30'16"

Justin Clay Henley
Crb 15h33'7" 27d6'45"

Justin Cole Worley
Ari 2h3'22" 21d8'58"

Justin Cole Worley
Cep 21h59'49" 60d56'8"

Justin & Corinn
Umi 15h25'11" 74d32'24"

Justin & Courti
Cru 12h2'7" -62d59'37"

Justin Crabtree
Boo 14h33'35" 34d35'0"

Justin & Crystal Knowles
Uma 12h4'31" 49d3'55"

Justin Curtis January 29, 1991
Ori 6h2'29" 13d31'8"

Justin D
Ori 5h56'8" 22d41'9"

Justin D. Gilley
Uma 11h49'35" 42d35'4"

Justin Daniel Fowler
Aur 6h28'40" 34d46'55"

Justin Daniel Pistininzi
Sco 17h15'3" -43d41'21"

Justin Daniel Tate
Ori 5h34'7" 0d42'19"

Justin Daniel Vitters 11/12/03
Lib 15h9'52" -16d45'34"

Justin & Danielle
Lmi 10h29'46" 37d30'30"

Justin & Danielle Forever
Lyn 8h10'21" 41d9'16"

Justin David
Per 2h53'46" 49d7'38"

Justin David Braun
Sgr 17h55'47" -18d6'44"

Justin David Brooks
Leo 11h44'24" 23d21'23"

Justin David Kemp
Sco 16h52'14" -40d14'0"

Justin David Palkovich
Uma 11h57'59" 51d1'22"

Justin Davie 8
Tau 5h13'56" 22d30'51"

Justin Davis Herman
Per 3h27'39" 45d26'3"

Justin Dreese
Uma 11h47'7" 56d15'40"

Justin Dreyer - Shining Star
Aur 7h28'4" 42d6'28"

Justin Duane Aines
Per 3h16'43" 43d8'14"

Justin E. Weickert
Per 3h22'47" 50d56'49"

Justin Edward DeSpain
Cnc 8h42'34" 14d37'2"

Justin & Elizabeth Ramedia
Cyg 21h15'50" 32d24'49"

Justin Eric Allen
Ari 2h31'11" 27d31'1"

Justin Everett Prening
Cap 21h41'25" -10d35'24"

Justin F. "Jay" Carlson
Her 17h19'30" 38d4'57"

Justin Fairchild
Umi 14h40'30" 70d31'50"

Justin Fleitz
Gem 7h39'54" 26d34'43"

Justin Fletcher Segrest
Sgr 18h38'50" -30d31'37"

Justin Frederick Fricke
Psc 0h38'48" 8d32'21"

Justin French
Cma 6h49'49" -27d52'0"

Justin Friedrichsen
Her 18h32'14" 24d9'17"

Justin Fulkerson
Ori 6h11'18" 19d35'37"

Justin Gage McCullom
Uma 8h49'0" 70d20'43"

Justin Galhier Guthrie
Cyg 21h28'10" 50d30'42"

Justin Gallant
Tau 3h47'11" 9d16'39"

Justin George Webber
Her 18h9'48" 37d0'13"

Justin Gerard Brandmeyer
Umi 15h20'58" 72d59'53"

Justin Gesford
Ori 6h10'25" 18d48'21"

Justin & Gina
And 0h15'16" 26d30'28"

Justin Glenn Castillo
Vir 12h50'38" 3d34'46"

Justin Goldfarb
Per 4h6'26" 33d52'55"

Justin Grant LeLaCheur
Leo 10h36'39" 12d36'29"

Justin Gruber
Her 16h12'7" 48d37'9"

Justin Hall
Per 2h41'8" 40d36'33"

Justin Halsey Moore
Ari 2h33'16" 21d50'18"

Justin (Handsome Jack) Lunsford
Cap 21h24'20" -14d39'22"

Justin Harres
Her 18h3'34" 18d7'40"

Justin Hauck
Ari 2h44'45" 28d26'4"

Justin Heath Young
Lmi 10h30'23" 37d20'3"

Justin Henderson
Lyn 8h15'51" 54d4'20"

Justin Hester
Per 4h24'37" 43d18'7"

Justin Hurlston Arnold
Dra 18h19'19" 67d30'17"

Justin Idol
Dra 18h56'17" 59d53'41"

Justin J. Roller
Psc 1h23'48" 27d48'36"

Justin Jade Sharphead
Lib 15h36'32" -7d21'24"

Justin James
Sco 17h52'33" -41d42'21"

Justin James Heck
Per 3h17'30" 45d34'41"

Justin James Hobday
Her 18h31'27" 24d4'50"

Justin James Keeney
Aqr 21h52'24" -5d46'11"

Justin James Reid
Leo 10h13'52" 25d58'17"

Justin James Rennilson
Ori 4h58'11" -1d53'46"

Justin ~JD~Repuhn
Aur 5h14'2" 42d24'30"

Justin & Jennifer~Always & Forever
Cyg 21h20'59" 51d3'26"

Justin Joseph
Umi 15h26'10" 77d2'8"

Justin Joseph Wendt
Per 3h39'19" 46d46'34"

Justin JUBBY Breen
Cap 21h19'46" -19d54'7"

Justin Kaplan
Gem 6h51'0" 32d2'38"

Justin Karl Phillippi
Per 3h14'55" 53d56'32"

Justin Keele
Her 18h24'7" 12d44'20"

Justin Keith Harmon
Gem 6h25'43" 23d22'34"

Justin Keith Scott
Tau 4h47'56" 18d38'50"

Justin Kendall
Ori 5h37'52" -0d49'51"

Justin & Kerry Bialik
Cas 0h32'44" 63d24'33"

Justin Klinksiek
Aql 19h52'10" -0d12'6"

Justin Knox "Jay" Burgin
Cma 7h10'4" -21d37'32"

Justin Kupinski
Dra 16h24'49" 61d36'11"

Justin Kyle Dailey
Sco 15h51'8" -20d41'19"

Justin L. Mayer
Leo 10h50'0" 10d42'33"

Justin LaNier Payton
Aql 18h56'49" -0d40'58"

Justin Lee
Uma 11h59'1" 47d41'7"

Justin Lee McElfresh
Lyr 18h41'6" 38d50'30"

Justin Lee Shrake
Aqr 22h7'48" -22d42'10"

Justin Lee Sollner
Gem 7h39'33" 31d48'41"

Justin Lee Stauffer
Per 3h23'44" 34d2'50"

Justin Ley 23
Lmi 10h56'30" 27d16'6"

Justin&LilJ-My Moon Ropers-Love AJ
Pho 1h0'35" -50d6'44"

Justin & Lisa Atwell Always Forever
Uma 14h6'34" 61d50'57"

Justin Lloyd - My Adventurer
Pyx 8h40'22" -21d20'27"

Justin Loomis - July 6, 1983
Uma 10h26'50" 65d33'34"

Justin Loves Britt
Cap 21h40'22" -8d55'8"

Justin M. Cartwright
Umi 15h27'17" 76d19'50"

Justin M Ferreras
Cap 21h31'41" -15d10'57"

Justin M. (husband and father)
Her 17h8'14" 35d43'28"

Justin M Madson
Her 17h6'11" 26d36'25"

Justin M May
Sco 17h3'34" -43d49'31"

Justin M. Mayer
Boo 14h37'21" 23d58'43"

Justin M McAbee
Umi 17h24'29" 78d50'19"

Justin M. McCarty
Her 17h40'15" 39d32'40"

Justin M. Mirabile
Vir 13h24'23" -15d9'45"

Justin M. My Shooting Star Forever!
Her 17h43'47" 36d53'16"

Justin M Prather AKA Bucket
Cnc 8h23'29" 25d33'13"

Justin & Malinda Mowrey "6-5-2004"
Cyg 20h48'53" 35d38'14"

Justin Marc Abrams
Boo 15h10'46" 44d15'47"

Justin Mark Earnest
Cyg 20h5'20" 35d8'29"

Justin Mark Reichs
Sco 17h51'24" -36d16'42"

Justin Marshall Armitage
Her 16h47'5" 37d14'7"

Justin Marvel
Her 17h2'0" 29d39'12"

Justin Matthew & Allison Grace
Uma 8h53'25" 61d11'30"

Justin Matthew Daniel
Ori 5h41'27" 2d16'15"

Justin Mazoway
Her 18h1'40" 17d29'45"

Justin McNutt
Uma 8h39'52" 51d25'33"

Justin Mezera
Gem 7h36'18" 22d43'15"

Justin Michael Barton
Lib 15h19'48" -5d53'44"

Justin Michael Braddock
Aqr 22h30'44" 2d24'15"

Justin Michael Caron
Aur 5h15'42" 29d23'16"

Justin Michael Cynor
Boo 15h12'26" 49d17'15"

Justin Michael Heman
Ori 5h24'5" 5d37'59"

Justin Michael Herb
Leo 10h14'45" 14d8'35"

Justin Michael Leninsky
Uma 11h7'23" 53d58'5"

Justin Michael Ray
Lib 15h35'25" -19d40'41"

Justin Michael Solberg
Lib 15h5'52" -0d34'34"

Justin Michael Tanner
Ori 5h46'36" 2d27'49"

Justin Michael Torres
Uma 12h2'44" 39d27'0"

Justin Michael Walters
Sco 16h12'52" -13d38'4"

Justin Michael Williamson
Boo 14h30'42" 19d51'57"

Justin Michael Wood
Gem 6h52'41" 27d37'42"

Justin Mickelson
Umi 20h54'39" 89d5'47"

Justin Mikal Coffey
Sgr 19h49'40" -12d18'46"

Justin Mish
Sco 17h12'46" -44d10'16"

Justin Monks
Umi 14h10'33" 74d42'52"

Justin Nelson
Her 18h39'51" 25d19'13"

Justin Nielsen
Cyg 21h38'45" 39d15'35"

Justin Nolan Perlman
Ori 5h15'39" -4d22'12"

Justin Nurrish
Cru 11h59'11" -63d4'34"

Justin Odom
Uma 11h14'4" 68d18'37"

Justin O'Neill
Cap 21h53'39" -19d51'13"

Justin Owen McComiskey
Uma 11h9'58" 49d33'3"

Justin Patrick Clement
Vir 12h40'8" 3d21'42"

Justin Patrick Cruz
Her 16h35'53" 47d1'42"

Justin Patrick Keoni Souza
Ori 5h21'6" -6d55'29"

Justin Patrick Merkle
Umi 14h41'0" 76d35'4"

Justin Patrick Woodham
Psc 0h46'20" 17d33'17"

Justin Paul
Aqr 20h55'33" 0d16'25"

Justin Paul Jeffre
Psc 1h6'9" 14d2'34"

Justin Paul Ricciarelli
Cap 21h10'19" -19d55'32"

Justin Paul Shooter Rogers
Her 16h46'13" 49d8'12"

Justin Pelkey's Hope
Ori 6h21'22" 14d51'48"

Justin & Penelope McClenny
Cyg 21h54'51" 49d36'59"

Justin Pete
Sco 17h16'36" -43d20'8"

Justin Peter
Her 17h41'57" 48d38'4"

Justin Peter Szwandrak
Lib 14h49'9" -18d26'15"

Justin Pierre "Bärchen" geb. 6.10.1996
Sco 17h15'8" -32d25'1"

Justin Plain Jones
Ari 2h2'42" 24d4'23"

Justin Plein-Schmitt
Aqr 21h1'51" 1d33'31"

Justin Quintana
Vir 13h47'54" 4d48'19"

Justin R. Rought
Aqr 23h51'11" -22d33'35"

Justin R Souza
Vir 13h17'33" 4d25'31"

Justin & Rachel Glantz
Cyg 21h30'5" 51d33'14"

Justin Ray Baker
Umi 15h14'22" 84d1'24"

Justin Reed Martenka
Dra 19h37'9" 65d31'7"

Justin Regli
Uma 9h43'28" 62d29'6"

Justin Riffle
Uma 9h18'42" 47d41'48"

Justin Roase
Vir 13h23'54" 13d16'14"

Justin Robert Balfour
Cas 1h35'37" 60d11'8"

Justin Robert Brown
Cnc 8h25'33" 17d14'48"

Justin Robert Etheridge
Uma 12h5'30" 59d7'36"

Justin Robert Fehl
Per 4h30'0" 31d47'25"

Justin Robert Michael Komine
Tel 18h33'24" -45d51'50"

Justin Robin Waters
Ori 5h41'34" 0d13'33"

Justin & Robyn
Del 20h35'41" 16d45'49"

Justin Rodriguez
Uma 8h47'41" 71d36'40"

Justin Ross Walker
Lyn 6h43'50" 56d30'45"

Justin Russell
Her 17h12'55" 29d10'13"

Justin Russell Goetz
Her 16h8'31" 48d46'9"

Justin Ryan Henkel
Per 3h20'50" 41d47'37"

Justin Ryan Misura
Cyg 19h50'17" 30d35'3"

Justin Ryan Mumley
Tau 5h4'50" 24d57'13"

Justin S. Johnson
Cnc 8h37'57" 19d32'59"

Justin S. Morace Star
Per 3h47'10" 50d48'49"

Justin S. Wenzl
Vir 13h27'45" 6d55'17"

Justin Samuel
Gem 7h41'14" 20d25'8"

Justin Samuel Scott
Ori 5h3'6" -3d30'9"

Justin & Sarah
Cyg 20h22'45" 37d15'26"

Justin Schillinger
And 0h25'6" 35d1'39"

Justin Scott Murray
Leo 10h12'20" 17d38'57"

Justin Scott Propst
Ori 5h57'21" 20d58'50"

Justin Smetak
Uma 9h18'25" 67d56'26"

Justin Sonntag
Her 17h41'47" 37d0'30"

Justin Speegle
Aql 19h0'51" -0d11'19"

Justin Stapleton God's Angel #3
Leo 10h7'1" 11d45'21"

Justin & Stephanie
Sge 19h40'48" 17d53'0"

Justin Steven Voorhees
Ori 5h56'54" 16d55'26"

Justin Sundance Gosvener
Vir 12h43'16" 6d26'31"

Justin T. LonCavish Beloved
Psc 0h1'47" 8d47'49"

Justin T. LonCavish Our Son
Psc 0h4'21" 8d56'46"

Justin T. Scarberry
Vir 12h48'24" 6d45'2"

Justin T. Schaefer
Her 17h36'58" 40d40'39"

Justin T. Schmidt
Cnc 8h52'12" 31d20'19"

Justin+Tanya's Star
Sgr 17h53'22" -24d44'36"

Justin Taylor
Her 18h49'27" 20d14'30"

Justin:The Beginning of Eternity
Sco 16h6'41" -15d3'51"

Justin - The One I Love
Aqr 22h52'56" -11d45'6"

Justin Thomas
Uma 11h22'31" 59d8'16"

Justin Thomas
Psc 23h21'44" 2d46'17"

Justin Thomas Black
Umi 14h48'0" 73d51'52"

Justin Thomas Roppolo
Cyg 21h30'7" 36d24'39"

Justin Thomas Tralli
Her 16h40'25" 31d11'47"

Justin Todd Burns
Uma 10h14'31" 67d24'32"

JuStiN TyLeR
Ari 3h23'17" 24d32'34"

Justin Tyler Berfield
Psc 0h42'34" 7d5'5"

Justin Tyler Schwandt
Uma 8h19'33" 72d38'8"

Justin W. McMurray
Eri 3h46'57" -3d57'22"

Justin Wade Lunning
Uma 8h57'31" 69d24'17"

Justin Wade Wallweber
Her 18h11'23" 16d4'31"

Justin Walker Conklin
Aqr 21h28'10" -0d6'55"

Justin Wayne Mueller
Sco 16h6'1" -17d51'10"

Justin Weber
Crb 16h15'15" 28d9'57"

Justin Weber
Tau 5h51'49" 16d56'8"

Justin Wells
Cma 7h3'33" -21d48'12"

Justin & Wendy - A New Life Together
Cen 11h28'50" -53d8'11"

Justin William Bissett
Aqr 22h45'28" -17d25'30"

Justin William Bissett
Per 2h12'34" 57d27'36"

Justin William Cassingham
Ori 5h46'35" 7d8'56"

Justin Winston Gosnell
Aur 6h22'18" 34d24'34"

Justin Wood
Ori 5h48'5" 2d16'27"

Justin Wright HateThrill
Ori 5h31'38" 2d26'40"

Justin Yirka
Uma 10h2'24" 59d8'11"

Justin - You Light Up My Life
Cru 12h56'20" -59d27'52"

Justin Zumsteg
Aur 5h36'55" 39d36'57"

Justina
Psc 0h0'11" 4d39'6"

Justina Andrea Moore
Ori 5h37'16" 3d19'50"

Justina Lynn
Uma 8h48'6" 51d17'45"
Justina Marie
And 0h37'35" 43d22'59"
Justina Marie Selby
Uma 10h41'29" 52d18'2"
Justina's Star
Lyr 18h54'6" 36d41'59"
Justine
Lyn 8h4'47" 33d44'23"
Justine
And 1h19'45" 41d30'39"
Justine
Lyn 8h50'57" 41d16'6"
Justine
And 23h18'44" 47d14'38"
Justine
Crb 15h46'8" 39d29'26"
Justine
Crb 16h13'50" 37d58'43"
Justine
Peg 22h37'57" 11d31'51"
Justine
Psc 1h27'33" 22d45'28"
Justine
Ari 2h16'40" 25d13'30"
Justine
Sco 16h56'24" -32d47'24"
Justine
Cap 20h20'8" -13d16'20"
Justine
Aqr 22h27'18" -6d58'52"
Justine
Vir 14h13'27" -9d45'51"
Justine 1436
Cas 23h33'35" 51d14'29"
Justine Alethea
Ori 4h55'56" -0d10'56"
Justine Alexandra Luth
And 1h2'4" 36d12'54"
Justine Alsop
Cas 2h29'40" 67d56'2"
Justine Angelina Locke
Cas 0h36'51" 59d22'42"
Justine Anne
Dor 5h36'17" -63d53'37"
Justine Boler 14-2-1986
Aqr 22h37'54" 2d35'23"
Justine Braun
Vir 14h17'50" -13d9'21"
Justine Clarke
Lyn 8h15'19" 44d47'21"
Justine E. Keller (D.O.B. 9/5/17)
Vir 14h4'24" -15d41'6"
Justine Elisabeth Boerst
Cyg 20h10'12" 44d1'58"
Justine Ferdjani
Crb 15h43'12" 35d17'13"
Justine - Forever True
Cru 12h19'16" -57d25'27"
Justine Gonzales
Peg 21h49'10" 7d13'10"
Justine Hendron
Lyn 7h50'52" 57d26'2"
Justine Higgins
Aqr 22h15'21" -19d35'30"
Justine Hoffman
Ari 3h27'0" 22d59'0"
Justine "Jess" Baldasare
Lib 14h57'27" -5d55'56"
Justine Lane Czarecki
Lib 15h14'46" -13d4'2"
Justine Lo
Lyn 9h6'42" 39d57'49"
Justine Louise Sinclair
Cru 12h39'21" -61d5'14"
Justine & Marc
Uma 11h36'11" 62d29'47"
Justine Marie
Cas 0h22'39" 58d46'48"
Justine Marie Mantua
Lyr 18h34'40" 32d28'52"
Justine Martins
Vir 14h8'53" -4d21'56"
Justine Melissa
Her 16h25'27" 45d9'56"
Justine & Michael's Wedding Star
Cru 12h20'52" -57d32'33"
Justine Nicole Hochberger aka Teeny
Gem 6h10'12" 25d39'17"
Justine Nicole Shivers
Leo 11h19'20" 5d0'23"
Justine "Rainbow" Schiro
Gem 7h17'9" 25d41'15"
Justine Rose Quattlander
Psc 1h25'35" 24d17'43"
Justine S Cooper
Cas 0h3'56" 50d12'9"
Justine Schmid
And 23h34'24" 36d3'24"
Justine Susan Clarke
Cap 21h49'5" -20d31'8"
Justine "Tenie" Smith
Cyg 19h38'34" 29d51'9"
Justine Yvonne Pappas
Sgr 18h25'14" -26d26'52"
Justine-Manon
Cas 0h11'26" 53d30'52"
Justine's Christmas Star
Leo 10h36'43" 21d21'5"
Justine's Escape
Uma 11h28'45" 61d49'47"
Justine's Jewel
And 23h46'47" 34d15'20"
Justine's Sea Green Star
And 2h15'59" 46d26'20"
Justinia Cobbuscus Oysterlatte
Tau 5h33'33" 17d13'13"
Justin-Kumi
Ori 6h11'0" 8d50'4"
Justin's
Cas 1h37'42" 62d20'3"
Justin's Baby Girl
Cap 21h44'26" -9d18'14"
Justin's Everlasting Jewel
Leo 11h12'8" 24d50'38"
Justin's Gallactic Empire 4-10-1995
Per 3h34'6" 50d48'27"
Justin's Guiding Light
And 0h53'19" 38d44'28"
Justin's Jewel
Aur 5h43'38" 41d46'13"
Justin's Ma
Lib 15h10'3" -7d26'12"
Justin's Star
Vir 12h45'34" -1d16'56"
Justin's Star
Aqr 22h51'28" -8d42'54"
Justin's Star
Her 17h54'41" 49d0'50"
Justin's Star
Psc 1h35'22" 17d5'46"
JUSTIN'S "STAR 21"
Her 18h24'59" 15d37'39"
Justinus Rex
Per 3h26'31" 32d20'49"
Justis
Cnc 9h10'46" 8d29'36"
Justito "cookielips" Ordonez
Her 17h23'52" 17d0'26"
JUSTIVAN
Cnc 8h28'17" 22d8'22"
Just'n Mey
Cnc 9h6'55" 28d33'50"
Juston
Tau 4h9'44" 16d17'26"
JustShell
Her 17h4'47" 13d28'50"
Just-us
Cyg 19h30'6" 28d56'52"
Justus
Uma 11h19'54" 71d1'21"
Justus Taaffe
Ori 6h9'32" 3d24'56"
JUSTUS TWOFEATHERS
Aqr 23h5'1" -7d58'34"
Justus Whalen
Tau 4h10'27" 2d55'1"
Justy Love
Tau 5h45'24" 22d25'17"
Justyn Michael
Umi 15h15'35" 68d0'42"
Justyn Philip Joyce
Cap 20h25'53" -10d43'0"
Justyna
Umi 15h50'45" 73d9'23"
Justyna
Uma 10h54'22" 65d27'9"
Justyna
Ari 3h13'3" 29d7'20"
Justyna Bielska
Lib 15h24'25" -9d23'58"
Justyna G.
Lyn 7h25'22" 44d38'8"
Justyna Jochym
Lib 15h56'36" -7d29'58"
Justyna "V. J"
Cyg 20h2'4" 35d34'34"
Justyne Butler
Sgr 18h23'44" -23d9'14"
Jutta
Uma 9h35'47" 48d38'34"
Jutta Dauba
Cnc 9h16'18" 18d21'40"
Jutta G. Schwarz
Uma 11h7'44" 29d17'43"
Jutta & Peter
Uma 11h27'2" 31d38'46"
Jutta Rudolph
Lyn 9h10'20" 35d28'40"
Juuso's Star of Luck
Umi 12h37'19" 87d54'29"
JUV
And 1h9'53" 45d39'49"
Juvern
Cep 20h42'31" 61d47'14"
Juvie Lyn Bernales
Sco 17h56'12" -40d49'16"
Juwelz
Cnc 9h12'43" 14d30'41"
Juying Gao
Cnc 8h44'50" 20d57'24"
Juzt My Star
Col 6h29'54" -41d29'23"
JUZZROMS
Aqr 22h43'34" -21d3'50"
Jvan Hugentobler
Sge 19h54'52" 17d47'36"
JVera
Uma 11h48'6" 43d50'47"
J.W. & Deborah Hall
Dra 19h13'26" 75d29'28"
J.W. Harrington
Ari 3h9'29" 28d57'20"
J.W. & H.P. 4th Month of Bliss
Uma 10h28'12" 64d25'22"
JWG
Vir 13h30'25" -12d45'54"
JWH-1913
Psc 1h8'0" 24d21'16"
JWINZ
Tau 4h3'52" 8d45'22"
JWN (Juden,Wolcott,Nicolo)
Uma 9h53'1" 44d13'14"
JW's Dream Chaser
Uma 12h43'15" 59d49'8"
Jybri Iman
Cap 20h23'53" -12d22'27"
Jyl Dougherty
Crb 15h37'26" 33d37'30"
Jyles, We Love You Forever
Vir 11h39'52" 2d52'21"
Jyll Cooperstein
Gem 6h23'46" 21d26'16"
Jyllian & Peter White
Lmi 10h8'44" 32d57'24"
Jyme Beth Shaw
Uma 11h29'23" 59d13'54"
Jymmi Aleseychuck
Leo 10h59'53" 5d57'35"
Jynell 1986
Cam 5h48'45" 68d26'26"
Jyoti - The Light in my Life...
Psa 22h4'23" -32d20'59"
Jyoty Parkash
Cap 21h55'37" -10d23'4"
Jytte
Cas 2h49'57" 60d11'23"
JZing1
Tau 5h36'13" 22d50'19"
K<\@>
And 0h6'21" 45d14'25"
K
Cyg 20h59'7" 43d17'22"
K
Lib 14h35'28" -18d5'21"
K 0829 Recollections of Summer
Vir 13h26'54" -0d15'17"
k <3 c
Per 2h44'38" 56d18'57"
K & A Promise Star
Cyg 20h33'14" 46d47'48"
K A T, Jr.
Dra 19h9'4" 70d16'6"
K. Amber Runion
And 1h45'3" 38d52'34"
K and T 1995
Lyn 8h18'33" 35d26'29"
K & B - November 23rd 1974
Cru 12h11'38" -62d54'23"
K. Blazak
Cma 6h41'14" -16d55'35"
K Bob Burks
Leo 11h40'17" 18d11'43"
K. Brett Wilburn 5/17/2002
Per 3h8'38" 41d50'20"
K. C.
Ari 2h59'4" 25d11'4"
K & C Bishop's Reigning Star of RCP
Her 18h55'31" 13d36'1"
K C Wong
Dra 16h30'48" 54d54'5"
K+D 1120
Leo 11h5'32" 21d6'35"
K+D Forever
Gem 6h13'5" 23d39'35"
K & D Walsh
Cyg 20h2'19" 53d53'27"
K Dawg
Sgr 19h30'10" -12d55'54"
K - Diggy
Uma 11h10'43" 49d37'6"
K Dub & Bern Richardson
Cyg 21h58'53" 49d27'47"
K. Faith
Aqr 20h44'38" 1d32'41"
K. Hieger
Uma 8h30'35" 64d13'33"
K Hop
Sco 16h30'56" -27d21'54"
K. I. B.
Cnc 8h39'54" 13d31'4"
K & J
Tau 4h37'47" 25d24'59"
K. J.
Cru 12h25'36" -60d6'35"
K. J. McLaughlin
Umi 16h13'33" 74d5'40"
K. Jasienowski
Cap 20h36'41" -11d21'0"
K. Jay
Ori 6h2'8" 11d10'57"
K. Jeffrey Eriksen Ph.D.
Cas 0h8'20" 59d46'46"
K Joy
Cas 0h59'44" 61d15'53"
K&K
Ari 2h25'14" 19d8'42"
K & K
Cnc 9h2'2" 18d7'35"
K&K Forever
Tau 4h12'15" 16d25'9"
K & K Werbelow 34th Anniversary
Cyg 20h20'4" 54d31'52"
K+L Bordlemay
Cyg 20h31'18" 39d34'5"
K L C
Umi 10h2'52" 86d34'25"
K L Peterson
Uma 11h48'50" 36d53'49"
K&L Westermann
And 23h24'50" 40d40'42"
K Laugh, the best BIG ever!
Ari 2h28'6" 26d8'27"
K. Lynn Taylor / M. Jay Kornblatt
Sgr 18h39'11" -34d23'19"
K & M
Lib 15h47'44" -7d35'42"
K & M
Cyg 19h49'36" 56d45'25"
k & m
Leo 10h56'59" 0d46'0"
K & M
Cyg 19h36'57" 31d1'24"
K. N. Roseberry
Lib 14h25'8" -19d19'54"
K Oda Smile 57
Vir 11h45'33" 1d19'4"
K & P 7-3-1992
Sex 9h50'59" -0d47'6"
K. P. W.
Aqr 22h47'54" -23d34'41"
K. Pannenbacker & A. Jordan
Lyr 18h27'20" 29d15'51"
K Poo
Uma 9h40'29" 46d3'50"
K & S Geigle
Per 3h35'28" 34d27'31"
K & S Meuer 25th Anniversary Star
Ori 5h34'58" -0d27'58"
K Squared
Cas 1h1'52" 60d18'46"
K. Star
Cam 6h39'50" 66d3'3"
K T Baier
Umi 14h23'43" 78d7'43"
K. T. Hanford
Lyn 7h6'52" 51d10'3"
K T K Trinity
Uma 8h28'31" 67d43'57"
K Thuraisingam
Lib 15h23'1" -11d13'27"
K - TIE
Aqr 21h57'44" 0d23'11"
K. T.'s 16th "Better Believe It!"
Lmi 10h43'57" 38d15'41"
K & V 1st Anniversary Star
Mon 7h46'53" -3d32'32"
K Ylan Joy Marshall
Lyn 9h41'34" 40d45'7"
K Z Star
Ari 2h4'47" 19d31'48"
K1A9T9I5E
Lyn 8h9'35" 56d7'57"
K2
Uma 13h40'2" 57d21'35"
K2 Kim Ferguson - Angel of Peace
Vir 13h54'41" 2d0'2"
K2 Rules
Cas 0h7'11" 58d39'59"
K23
Ori 5h35'59" 10d44'22"
K2K
Cap 20h52'30" -25d22'12"
K6LAE
Psc 1h1'41" 4d3'37"
K8 the Great
Cas 1h33'6" 57d5'7"
K8E
Leo 9h27'56" 27d7'35"
K8trix
And 2h16'42" 42d52'49"
KA
Leo 11h31'17" 12d21'6"
Ka Borio & Ka Epay Macalintal
Cyg 19h58'59" 57d41'32"
Ka Ja
Peg 23h10'47" 25d58'23"
Ka Tet
Gem 7h26'19" 34d23'38"
K.A. Tony Davies
Cep 21h54'35" 60d15'3"
Ka Ying Xiong
Uma 9h32'5" 47d42'32"
Ka2
Cap 20h28'9" -26d8'37"
ka7ozo
Gem 7h25'39" 30d50'28"
Kaaday Kalat
Uma 12h45'11" 61d20'36"
Kaafta's Spellcaster
Sco 16h46'36" -31d43'35"
Kaáli Zsombor
Sco 17h27'54" -32d51'53"
Kaan Ata Sen
Leo 10h40'15" 7d6'9"
Kaan Erenler
Her 18h20'45" 20d30'39"
Kaare Albert Hansen
Lyn 9h22'10" 33d50'55"
Kaarina Kivilahti
Uma 10h32'22" 61d12'37"
Kaavya
Vir 14h33'11" 6d50'26"
Kaayla062305Greg
Gem 6h42'8" 19d31'52"
K.A.B. Luminary
Aur 5h56'9" 36d57'52"
KABADA
Uma 13h41'57" 56d43'52"
Kaban Mathias
Leo 11h23'28" 5d59'12"
KABBOY "STAR" EIPHNH
Psc 1h41'58" 5d36'23"
Ka-Be Java
Per 3h23'10" 45d35'24"
KABEHAAN X
Uma 10h31'54" 68d32'13"
Kabelitz 07 Bluestar
Uma 13h12'30" 53d27'2"
Kabir Seth
Aql 19h22'55" -0d50'17"
KABLAFS
Tau 5h46'57" 22d16'44"
Kaca
And 1h12'26" 46d31'22"
Kacee Everlasting
Uma 13h31'42" 56d44'34"
Kacee Jeanne Herrick
And 2h36'36" 45d23'59"
Kacee Lynn Smith
Uma 8h56'47" 65d13'38"
Kacey
Uma 8h29'3" 65d9'19"
Kacey
Lyn 7h31'25" 38d31'35"
Kacey A. Bean
Sco 16h14'13" -14d35'15"
Kacey and Rita's Star
Tau 5h23'58" 19d11'19"
Kacey Brooke VanDuzer
Lyn 7h39'9" 41d43'20"
Kacey Dean
Leo 9h41'57" 17d37'0"
Kacey Elizabeth
Lyr 18h51'22" 33d12'35"
Kacey Lynn
Gem 6h39'54" 16d29'36"
Kacey Marie Taylor
Boo 14h19'12" 45d44'16"
Kacey Roessler
Uma 9h36'55" 48d27'55"
Kacey Wolfe's Star
Mon 6h51'40" 7d10'8"
Kachna
Cas 0h15'43" 54d51'16"
Kaci Caroline
Sco 16h5'0" -29d33'14"
Kaci Danae
Peg 22h56'43" 30d47'16"
Kaci Leigh Marter
Aqr 21h36'6" -0d38'50"
Kaci Maddison
And 0h39'27" 29d27'40"
Kaci Marie Toomey
And 23h17'34" 52d12'23"
Kaci McKenna
Leo 11h11'28" 22d3'21"
Kaci Rae Svornik
And 0h6'9" 45d36'26"
Kacia
Uma 11h43'11" 50d15'42"
Kacie
Lib 15h31'16" -24d53'54"
Kacie Elizabeth Harrison
Peg 0h0'48" 26d48'1"
Kacie Flynn
Psc 0h17'51" 20d8'17"
Kacie Grace
Mon 6h44'36" -0d5'29"
Kacie Janae
And 0h49'58" 45d10'34"
Kacie Madden Brown
And 1h18'17" 42d4'15"
Kacie Mae Butler
Sco 16h8'27" -10d14'18"
Kacie Marie Kesselak
Lib 15h4'23" -0d59'35"
Kacie Michelle
And 23h32'20" 48d55'11"
Kacie Michelle Craft
And 1h19'44" 42d58'44"
Kacie Parker
Sco 17h40'33" -37d5'23"
Kacie's fun-time party star
Cap 21h41'31" -16d29'54"
Kackee's Star
Umi 14h20'37" 78d12'16"
Kacy Dee
Ori 5h30'53" 4d9'58"
Kacy Elizabeth Slattery
Cas 0h25'32" 60d36'59"
Kacy Lynn Potemra
Sgr 18h1'43" -27d7'37"
Kacy's Dreams
Dra 16h54'42" 52d17'19"
Kadaisha LaDonna Simon
Sgr 18h15'17" -35d55'31"
kadalin
Ari 2h52'22" 19d4'56"
Kadalla, Jens
Uma 13h22'52" 62d28'36"
Kaddi
Leo 11h25'23" -3d31'21"
Kade John Scott Malcolm
Ori 5h31'2" 13d26'0"
Kade Michael
Sco 16h13'37" -10d7'32"
Kadee Newberry
Leo 10h16'35" 15d22'27"
Kaden
Sgr 18h26'33" -16d46'2"
Kaden Alan Pearson
And 23h12'46" 48d47'26"
Kaden Cesped
Umi 15h33'23" 71d34'33"
Kaden Elton Raley
Umi 14h32'15" 74d48'41"
Kaden Faragher
Ori 5h36'21" -8d55'1"
Kaden Grover
And 0h56'19" 39d23'32"
Kaden I Munoz
Vir 14h41'3" 4d45'24"
Kaden J. Breger
Uma 12h5'3" 31d30'39"
Kaden Jefferson Davis
Eri 4h13'25" -30d49'4"
Kaden Kalstrom
Oph 17h5'51" -0d53'53"
Kaden Michael Carson
Uma 9h28'37" 45d56'50"
Kaden Nickolus Taylor
Ori 4h51'33" 13d43'12"
Kaden Ray Berry
Vir 15h10'46" 5d13'20"
Kaden S. A. Rodriguez
Uma 9h50'41" 55d50'30"
Kaden William Huber
Ori 5h4'41" 13d17'43"
Kaden Xavier
Dra 17h45'23" 55d44'10"
Kadence Michelle Watts
Vir 13h32'24" -13d10'53"
Kadence Nora Kim
And 0h11'53" 46d31'12"
Kadence Olivia Marut
Lib 14h51'18" -3d45'31"
Kaderstemli
Umi 14h10'15" 65d46'26"
Kadesh
Tau 4h46'14" 22d49'0"
Kadi Dianne Ballard
Aql 19h18'46" 5d36'40"
Kadia Kariann
Lib 14h58'50" -5d3'9"
KADIAN
Dra 17h51'36" 53d17'18"
Kadie Janette Waller
Lib 15h4'24" -4d16'29"
Kadie Rose Cain
Dra 9h29'52" 76d46'4"
Kadijah Nur
Umi 14h15'29" 75d56'30"
Kadima 11/05/01
Umi 16h23'14" 73d40'11"
Kadin E. Hoven
Per 3h17'44" 53d15'35"
Kadire & Louie
Cyg 19h46'18" 56d29'15"
Kado
Cyg 21h40'32" 46d25'58"
Kadri-Liis
Cap 21h46'47" -17d0'7"
Kady
Per 3h26'9" 45d3'1"
Kady and Tamia
Uma 10h30'23" 64d1'0"
Kady Louise Smith
Leo 11h12'21" -0d2'31"
Kady Morgan
Uma 11h55'59" 60d47'59"
Kady (Planet Kathryn) 31-8-2005
Cru 12h39'41" -57d54'56"
Kaeden & Jalen
Lyr 18h46'13" 40d42'29"
Kaedence Channon McCarthy
And 2h29'48" 41d24'17"
Kaedi Marie Stoddard
Peg 21h41'29" 26d5'55"
Kaedys
Uma 11h16'55" 52d29'51"
Kael
Her 17h30'22" 46d21'29"
Kael Alexander Pearson
Crb 15h29'16" 30d18'8"
Kael, For Eternity
Uma 11h28'52" 59d3'33"
Kaela
Vir 12h52'5" 11d29'8"
Kaela Anne
And 22h57'56" 50d59'23"
Kaela E. Maroney
Ari 3h12'56" 28d44'31"
Kaela Godfrey
Cas 0h24'31" 61d31'43"
Kaela Marie Bergen
Vir 13h5'25" 9d11'36"
Kaela N Devin
Sco 17h50'41" -38d15'59"
Kaela Rose and Meghen Rae
Umi 15h38'9" 77d55'1"
Kaela-Marie Alicia Bardillon
And 1h40'8" 38d18'27"
Kaelan William Traub
Uma 10h55'32" 44d35'32"
Kaeleb Aedan McCarthy
Umi 15h45'35" 78d53'36"
Kaeleer
Umi 10h40'10" 89d9'25"
Kaelem Michael Wagner
Per 3h9'29" 53d44'54"
Kaeli Ann Abfall
Cnc 8h27'28" 6d37'49"
Kaeli Mae Slingerland
Cas 1h52'23" 63d28'12"
Kaelie Makenzi Hooper
Vir 13h24'26" 13d39'52"
Kaelin Nicole Berry
Uma 9h26'31" 42d17'0"
Kaelin Rhea
Dra 12h46'48" 76d0'45"
Kaelin's Promise
Gem 6h49'19" 18d7'33"
Kaeli's Star
Gem 7h21'21" 17d45'56"
Kaelyn
And 1h9'33" 36d23'21"
Kaelyn Annalisse Jones
Cas 1h17'54" 62d21'40"
Kaelyn Deanna Edwards
Vir 11h47'38" 5d1'10"
Kaelyn Fergeson
Mon 6h49'8" 3d8'43"
Kaelyn Lauren
Cas 1h20'57" 53d18'16"
Kaelyn Marie
Lyn 7h6'20" 61d17'5"
Kaelyn Marie Knox
Leo 11h14'55" 24d3'28"
Kaelyn Nicole Tyler
Cnc 9h7'26" 23d32'5"
Kaelyn Paige Oeth Riddle
Gem 6h45'40" 21d1'2"
Kaelyn Renae' Garrett
Tau 4h21'18" 12d6'34"
Kaelyn Rose
Aqr 21h38'23" 2d12'55"
Kaelyn Victoria
Peg 22h16'9" 16d2'23"
Kaelynn and Matthew Vollmer
Cyg 20h21'7" 54d29'12"
Kaelynn Faith Cheney
Ori 5h49'9" -2d49'27"
Kaenid Vestna
Lyn 7h27'30" 53d10'52"
Kaenishea & Andrew's Star
Ori 5h53'13" 12d22'19"
Kaera's-fireball
Uma 13h23'31" 57d7'24"
Kaetlynn Liesel
Sgr 18h11'41" -20d1'0"
Kafi
Uma 8h38'18" 58d58'58"
KAFLA 17th july
Lyr 18h49'6" 30d20'17"
Kaflea Snoelle
Aqr 22h13'17" -0d29'16"
Kafrica
Uma 11h18'0" 50d2'59"
Kafsleg Cazap
Uma 12h25'58" 58d40'5"
kage
Lyn 7h57'29" 45d36'37"
Kahealani
Leo 11h17'37" -3d10'16"
Kahi Pili Ola
And 1h11'19" 47d9'17"
Kahla Mae Getschman 05/04/87
And 0h50'3" 46d39'7"
Kahla Nicole
Umi 15h18'59" 67d30'58"
Kahlana Barfield
Vul 21h0'40" 22d47'53"
Kahle Erlanne Kuronen
Lib 15h52'3" -17d2'22"
Kahle Toothill
Cru 12h16'18" -57d4'38"
Kählert, Wolfgang
Uma 12h58'8" 53d33'19"
Kahlia Cherise
Tau 4h34'48" 8d54'53"
Kahlil Piscopo
Sgr 18h28'51" -24d44'41"
Kahlila Isel Torres
Cyg 19h44'38" 55d45'55"
Kahlo Romaria Aureliana Garrison
Lib 15h10'59" -27d42'47"
Kahlua
Aqr 22h16'36" -3d1'41"
Kahlua Jean
Cma 6h51'12" -15d57'57"
Kahoku
Sco 17h30'8" -40d22'7"
Kahu
Del 20h19'5" 9d40'45"
Kahuna
Uma 8h33'9" 71d53'48"
Kahunatutu
Lyn 7h42'22" 46d0'59"
Kai
Ari 2h23'13" 17d59'26"
Kai
Uma 11h26'19" 54d24'47"
Kai Arden
Pho 0h23'52" -47d23'1"

Kai B. Parker
Aur 5h35'58" 44d11'45"
Kai Barry Rae Suz Shel
Greg Lachie
Pho 2h23'58" -45d38'41"
Kai Burrell
Her 18h20'8" 18d58'34"
Kai Christopher Sims
Per 4h32'7" 31d38'0"
Kai den Hertog - 11.06.2002
Cru 12h54'38" -60d53'13"
Kai Diego
Cap 21h45'58" -11d42'15"
Kai Forrest Skoloff
Psc 0h52'25" 15d33'6"
Kai Ieuan Rogers
Umi 13h51'16" 77d54'20"
Kai Joseph Burke
Leo 9h40'54" 31d17'33"
Kai Lani Pence
Umi 13h17'5" 70d33'11"
Kai Louis Bristow
Uma 9h22'41" 59d20'12"
Kai Maddox Weiss
Vir 13h10'56" 13d23'42"
Kai Malaika Cannon Hill
Lib 15h29'0" -4d40'30"
Kai Masson - Perrault
Sco 17h44'39" -38d55'6"
Kai McCabe
Sgr 18h46'57" -17d50'50"
Kai Morningstar
Uma 12h11'15" 54d50'39"
Kai Newton
Umi 9h42'19" 87d15'5"
Kai Pascal
Cas 0h39'39" 69d37'29"
Kai & Paul McCormack
Lib 15h26'20" -6d34'51"
Kai Reichardt Hoffman
Tau 4h29'55" 21d0'32"
Kai Sidney
Per 4h25'45" 34d38'1"
Kai Stephenson
Ara 17h21'19" -52d11'14"
Kaia
And 0h13'43" 46d25'0"
Kaia Ann Hollingsworth
Gem 7h33'32" 27d31'27"
Kaia Fox
Boo 14h42'57" 23d15'41"
Kaia Hedvig Engelsrud
Lib 14h53'31" -2d18'50"
Kaia Marie Johnson
Sco 17h52'19" -41d48'24"
Kaia Roisin Hval Flynn
And 1h47'27" 36d34'0"
KaiAni Day
And 1h34'59" 39d10'2"
Kaiden
Psc 0h55'27" 28d46'58"
Kaiden Christopher
Per 2h22'58" 57d23'38"
Kaiden Craig Swyney 12-05-2002
Her 16h43'38" 42d34'37"
Kaiden Daniel Hess
Umi 15h51'38" 78d1'33"
Kaiden David
Uma 9h8'21" 55d49'1"
Kaiden Herbert Norton
Cru 12h32'34" -62d37'1"
Kaiden Riley
Cnc 9h19'58" 9d18'29"
Kaidon Ray Pugh
Uma 9h39'49" 55d11'28"
Kai`enna Rain de la Pena
Cyg 21h40'0" 51d22'28"
Kaiet Cian Moreno
Uma 13h42'16" 57d24'35"
Kaigan LeMaster
Dra 17h20'20" 59d16'15"
Kai-Kai
Lib 14h33'40" -24d41'29"
Kaikua'ana Rachel
Vir 14h15'2" -6d18'37"
Kaikua'ana Tara
Cnc 8h57'10" 18d56'44"
Kaila
Uma 10h41'11" 65d36'7"
Kaila and Carlos
Lib 15h49'21" -5d34'53"
Kaila Fewster
Uma 11h29'31" 58d28'37"
Kaila Kristina Hoppe
Tau 5h11'43" 23d4'25"
Kaila Marie Wheeldon
Lyr 19h11'58" 26d30'12"
Kaila Rae Ioane
Uma 10h45'53" 50d11'44"
Kaila Rhianne
And 0h13'58" 28d2'30"
Kaila Uribe
Uma 9h30'28" 49d6'15"
Kaila Walker
Pup 7h39'11" -21d39'33"
Kailani Beth - 11 March 2006
Cru 12h25'35" -58d51'54"
Kai-Lani Kelly
Umi 15h14'49" 69d32'1"
Kailani Ramey
Sco 17h27'17" -43d17'32"
Kailani Villavicencio
Peg 22h41'57" 12d5'55"
Kaila's Eye
And 0h9'6" 39d58'8"
Kaile Marie Burnett "Love Always"
And 1h44'5" 48d59'53"
Kailee
Leo 11h52'48" 23d29'38"
Kailee And Brown Mommy
Cnc 8h14'5" 25d18'17"
Kailee Andrews
Cam 4h6'8" 66d2'41"
Kailee AnnMarie Hill
And 0h34'0" 34d26'39"
Kailee Ann-Marie Templemire
Lib 15h12'34" -6d23'57"
Kailee Breanne Phillips
Crb 15h26'44" 28d10'32"
Kailee J
Ari 2h56'55" 25d56'3"
Kailee Love
Psc 0h47'9" 20d34'43"
Kailee Mae Verdeyen
Tau 5h7'12" 18d1'37"
Kailee N. Minor
Lmi 10h52'35" 36d37'16"
Kaileigh
Gem 6h49'28" 22d53'36"
Kailey
Peg 22h52'12" 32d39'51"
Kailey
Lyn 8h5'20" 50d56'15"
Kailey
Lyr 18h54'39" 43d55'3"
Kailey Ava
Leo 9h37'8" 22d26'11"
Kailey B. Bowles
Lib 15h13'45" -15d3'35"
Kailey "Babe" Moran
Psc 23h30'13" 4d32'48"
Kailey Bear
Ori 5h39'19" 1d38'32"
Kailey Beth
Cnc 8h42'56" 6d46'26"
Kailey Edder
And 1h1'21" 38d23'13"
Kailey Elizabeth Bates
And 1h26'18" 49d2'51"
Kailey Elizabeth Coleman
Uma 9h22'43" 59d29'40"
Kailey Elizabeth Walker
Cae 4h44'25" -45d14'50"
Kailey Elle Epstein
Peg 22h16'12" 34d45'2"
Kailey Gayle
Cam 7h55'19" 71d6'57"
Kailey Hebb
Leo 11h55'58" 26d18'55"
Kailey Jade Lozano
Vir 14h4'45" -15d52'50"
Kailey Madison Granger
Cnc 9h1'30" 9d43'50"
Kailey Nicole Suplinskas
Per 3h38'33" 35d46'12"
Kailey Rose Betancourt
Cap 20h18'6" -14d22'25"
Kailey Rowlands
Leo 9h38'34" 31d59'27"
Kailey1
Her 17h38'24" 15d54'15"
Kaili Moser
And 0h18'32" 43d3'14"
Kailie
And 2h37'29" 43d28'57"
Kailie Amanda Rhines
And 0h39'3" 37d43'51"
Kailie Jade Nottingham
Cas 1h12'29" 55d17'27"
Kailum's Beauty
And 2h28'7" 43d27'8"
Kailyn
Tau 3h31'42" 9d51'6"
Kailyn Christine
Ari 3h17'5" 27d50'27"
Kailyn Rhea
Cap 20h25'18" -26d47'52"
Kailyn Rionne Bufford
And 1h17'48" 50d4'16"
Kaimana
Vir 12h16'11" -4d2'30"
Kaimana
Lyn 7h40'27" 52d59'28"
Kaimen Stremel
Peg 22h18'29" 8d54'41"
Kain
Gem 7h34'29" 24d8'39"
Kain Magick (Wicca Star)
Peg 22h11'44" 7d26'15"
Kain Michael McGill
Umi 14h40'3" 73d56'19"
Kaindl, Walter
Cnc 9h2'35" 16d35'17"
Kaine John Arnott
Ari 3h12'49" 31d3'37"
Kaine Knight VII
Cma 6h40'16" -17d0'12"
Kaine & Nic - Forever & a Lifetime
Cru 12h42'27" -56d20'13"
Kainen
Sgr 18h36'38" -24d3'12"
Kaing Lao Loves
Constance Matsumoto
Vir 13h11'42" -3d50'58"
Kainoa Yukio Endo
Per 4h21'25" 36d15'32"
Kain's Star
Sgr 18h7'35" -20d6'14"
Kaioli
Lyn 7h36'12" 53d8'18"
Kaipo Michelle McKeague
Crb 16h12'3" 35d3'20"
Kaira
Tau 4h36'1" 28d38'56"
Kaira Nicole Knox
Lyn 7h45'14" 36d0'39"
Kairi Rain
Gem 6h35'14" 12d50'31"
Kaisa Shirren Heywood
Tau 5h37'37" 24d5'48"
Kaiser Bear
Uma 9h37'10" 66d9'21"
Kaiser J. Sneen
Umi 15h40'16" 68d10'42"
Kaiser Karch
Cyg 19h35'52" 33d43'57"
Kaiser LDAK
Crb 16h9'24" 39d32'24"
Kaiser Star
Umi 15h20'41" 72d59'19"
Kaiser Steven Potter
Cnc 9h3'3" 27d38'3"
Kaiser, Axel
Uma 10h19'10" 65d40'59"
Kaishek Vang
Cnc 8h18'36" 12d8'55"
Kait & Jem 4ever 8-20-03
Sgr 19h18'46" -13d53'6"
Kaiti & Andrew
And 2h28'22" 50d19'12"
Kaiti Bug
Lib 15h32'1" -19d27'19"
Kaitland Royce Volpe
Mon 6h12'55" -7d10'26"
Kaitlin
Lyn 7h58'0" 39d42'56"
Kaitlin
Tau 4h1'17" 7d35'37"
Kaitlin A Green
Sgr 19h10'40" -33d59'18"
Kaitlin Alexandra
And 0h57'19" 22d43'2"
Kaitlin and Sean Forever
Uma 10h51'12" 50d26'8"
Kaitlin and Shawn - Eternity Star
Cyg 20h22'51" 56d34'38"
Kaitlin & Andy
Per 3h24'58" 45d2'3"
Kaitlin Ann
Psc 0h52'43" 15d54'0"
Kaitlin Ann Baumann (Pumpkin)
Ori 5h49'28" 9d2'27"
Kaitlin Aubrey
And 0h18'14" 32d40'12"
Kaitlin D. Matz
Uma 11h31'6" 38d28'11"
Kaitlin Elizabeth Brynes
And 2h13'50" 41d44'45"
Kaitlin Elizabeth Hess
Vir 14h14'32" -14d14'14"
Kaitlin F. Vicari
Lib 15h17'37" -5d25'50"
Kaitlin Hope Christie
Gem 7h43'44" 25d31'31"
Kaitlin Ione
Uma 9h6'43" 64d43'45"
Kaitlin Leigh Jarvis
Cam 3h50'23" 67d51'17"
Kaitlin Longworth
Cas 1h30'9" 61d57'30"
Kaitlin Loves Derrick
Cyg 20h26'43" 52d45'13"
Kaitlin M. Leidy
Gem 7h3'48" 19d33'52"
Kaitlin Marie
And 1h12'34" 46d43'24"
Kaitlin Marie Arkebauer -1 Year Old
Cnc 8h52'34" 27d49'1"
Kaitlin Marie Reuter
Cyg 21h1'1" 42d55'41"
Kaitlin Marie Sorto
Vir 12h52'4" 10d41'29"
Kaitlin Mary Mulvihill
Tau 5h46'1" 25d42'1"
Kaitlin McIntyre
And 2h26'20" 46d29'5"
Kaitlin McKenzie Nichols
Tau 4h34'57" 15d18'11"
Kaitlin Melissa Clark
Uma 12h2'39" 43d17'7"
Kaitlin Michelle Kruser
Cap 20h25'1" -13d12'0"
Kaitlin Nevaeh Peck
Cap 20h27'41" -14d34'22"
Kaitlin Nicole Tucker
Ori 5h41'33" 11d33'16"
Kaitlin & Pamela Adams-Tanney Star
Cas 3h13'23" 61d34'38"
Kaitlin Patricia Rees
Gem 6h28'27" 20d40'49"
Kaitlin Rebecca
Uma 8h49'45" 58d35'43"
Kaitlin Rochelle Ertle
Lib 14h47'42" -24d10'46"
Kaitlin Rose Sidlauskas
Cru 12h40'54" -62d16'29"
Kaitlin Venables
And 1h3'0" 47d10'21"
Kaitlin White
Crb 15h32'40" 28d21'32"
KAITLIN WILSON (SUPERGIRL)
Leo 9h43'37" 27d46'36"
Kaitlin's Heart Changed The World
Dra 17h22'10" 51d21'45"
Kaitlin's Star
Ori 5h55'18" 18d21'43"
Kaitlin's Wish
And 1h33'50" 48d33'17"
Kaitlin's Wish
And 0h49'10" 38d40'30"
Kaitlyn
Lyr 18h55'28" 35d40'58"
Kaitlyn
And 0h44'15" 27d2'42"
Kaitlyn
Ari 3h19'46" 24d3'9"
Kaitlyn
Peg 22h3'32" 16d4'49"
Kaitlyn
Ori 6h16'37" 10d39'33"
Kaitlyn
Sco 17h29'49" -42d22'18"
Kaitlyn
Sco 16h4'48" -8d40'0"
Kaitlyn
Lib 14h24'43" -18d58'49"
Kaitlyn Alyse
Ori 5h54'2" 12d40'6"
Kaitlyn Andrea Gilmore
Tau 3h25'36" 11d21'27"
Kaitlyn Ann Akins
And 0h8'13" 36d48'16"
Kaitlyn Ann Lutz
Uma 11h39'56" 51d58'35"
Kaitlyn Ann Wegner
And 0h19'55" 44d23'44"
Kaitlyn Annemarie "Cecilia" Barney
Her 18h56'6" 23d50'13"
Kaitlyn Ashley
And 23h15'10" 50d53'51"
Kaitlyn Ashley Bright
And 22h59'31" 41d13'9"
Kaitlyn Barningham
Aqr 22h21'21" -4d48'35"
Kaitlyn Berry
Tuc 22h46'49" -56d44'42"
Kaitlyn Blakley
Cas 1h25'0" 53d19'45"
Kaitlyn & Brennan
Sco 16h16'59" -10d16'15"
Kaitlyn Brooke
Lyn 7h50'56" 39d0'55"
Kaitlyn Clare Reilly
Gem 6h52'56" 16d55'47"
Kaitlyn Colleen Westland
Sgr 18h23'4" -17d0'19"
Kaitlyn D. Smith
Lib 15h5'53" -4d39'21"
Kaitlyn Dakota King
And 0h41'35" 29d24'44"
Kaitlyn Dorothy Hassard
Cap 21h42'12" -11d5'13"
Kaitlyn E. Budzowski
Cyg 21h56'13" 49d53'55"
Kaitlyn Elias
Sco 17h37'2" -43d11'21"
Kaitlyn Elise Sheppard
Uma 10h26'49" 66d40'44"
Kaitlyn Elizabeth
And 0h22'3" 25d42'49"
Kaitlyn Elizabeth Anness
Cas 1h59'30" 60d19'42"
Kaitlyn Elizabeth Bavirsha
Umi 14h25'57" 68d57'40"
Kaitlyn Elizabeth Burnett
And 23h9'3" 39d59'20"
Kaitlyn Elizabeth Taylor
Ari 2h8'10" 17d26'56"
Kaitlyn Ella Doyle
And 1h22'47" 43d9'39"
Kaitlyn Fraser
Leo 11h26'45" 5d7'33"
Kaitlyn Gail Curtin
Uma 9h52'14" 60d16'59"
Kaitlyn Gail Curtin
Sgr 18h59'52" -32d40'25"
Kaitlyn Golden
Aqr 22h25'56" -4d34'54"
Kaitlyn Grace Whittick
And 1h36'24" 41d10'31"
Kaitlyn Hart
Ari 2h16'37" 23d15'34"
Kaitlyn Hoskovec
Sco 17h23'2" -42d50'44"
Kaitlyn Isabelle Ransome
Gem 7h15'39" 26d50'36"
Kaitlyn Isabelle Troester
Tri 2h2'25" 33d20'23"
Kaitlyn Jane "Yetta" Helf
And 23h13'7" 51d23'53"
Kaitlyn Karen Seymour
Leo 10h27'42" 17d5'18"
Kaitlyn "Katie" Palmer
And 0h21'0" 38d48'42"
Kaitlyn Laura Hoppe
Psc 0h0'1" -0d16'42"
Kaitlyn Lauren Kofler
Uma 11h10'1" 29d47'52"
Kaitlyn Layne
And 0h29'33" 38d10'6"
Kaitlyn Leigh Wunsch
Lib 15h5'0" -0d55'57"
Kaitlyn "Lil Baba" Rezny
Ori 6h2'34" 16d53'40"
Kaitlyn Louise Meyrose
And 2h24'51" 50d11'24"
Kaitlyn M. Cotton
Gem 6h30'7" 20d58'3"
Kaitlyn Manion
Cnc 8h51'55" 29d42'45"
Kaitlyn Marie Benes
Vir 14h0'27" -7d40'51"
Kaitlyn Marie Clark
Leo 10h19'26" 13d31'53"
Kaitlyn Marie Connolly
Ori 6h10'39" 7d27'43"
Kaitlyn Marie D'Alessio
Lyr 19h8'20" 27d8'34"
Kaitlyn Marie Dante
And 2h27'45" 43d49'34"
Kaitlyn Marie Druckreier
Uma 12h5'56" 60d22'31"
Kaitlyn Marie Ferencz
Uma 9h24'5" 63d51'16"
Kaitlyn Marie Freyer
Sgr 18h58'46" -25d23'44"
Kaitlyn Marie Hartman
Aqr 21h39'36" 0d42'13"
Kaitlyn Marie Price
Cyg 20h30'50" 37d20'58"
Kaitlyn Marie Timblin
And 2h13'54" 45d50'11"
Kaitlyn Marie Wayne
And 23h25'34" 51d19'50"
Kaitlyn Marie Witak
Ari 2h26'9" 19d43'20"
Kaitlyn Marie Zile
Lyn 8h20'33" 39d6'19"
Kaitlyn Marisa Smith
Ari 2h34'5" 24d56'13"
Kaitlyn Mary Eddy
And 0h33'9" 45d3'56"
Kaitlyn Mercedes Marin
Mon 6h49'56" -0d17'28"
Kaitlyn Michelle Bridgeforth
Cap 21h38'18" -15d34'7"
Kaitlyn Moore
Lyn 7h37'0" 41d32'3"
Kaitlyn Needham
And 2h15'36" 45d41'15"
Kaitlyn Nicole Alarie
And 1h33'42" 47d35'57"
Kaitlyn Nicole Cunningham
And 2h38'24" 40d57'23"
Kaitlyn Nicole Hernandez
Uma 10h29'51" 51d27'56"
Kaitlyn Nicole Kaufman
And 1h16'21" 43d5'45"
Kaitlyn Nicole Mann
And 0h34'1" 27d47'50"
Kaitlyn Nicole Sabino
Cap 21h43'56" -10d57'5"
Kaitlyn Nicole Vandenburg
Peg 22h20'33" 28d20'35"
Kaitlyn Nikole Simpson
Uma 8h36'22" 59d56'3"
Kaitlyn Noelle
Umi 13h35'31" 72d37'26"
Kaitlyn November Dyck
Umi 4h41'49" 89d5'3"
Kaitlyn O'Connor
And 1h6'9" 44d30'52"
Kaitlyn Patricia Murphy
Uma 10h33'40" 67d54'44"
Kaitlyn Potashnyk
Umi 15h45'10" 70d8'35"
Kaitlyn Quiroz
And 1h56'25" 41d51'40"
Kaitlyn Rae Bauer
Gem 6h46'30" 23d25'30"
Kaitlyn Reagan Miller
Mon 7h33'6" -1d24'19"
Kaitlyn Renee Richards
Umi 14h17'59" 87d38'51"
Kaitlyn Rial
Cap 21h45'42" -13d3'50"
Kaitlyn Rose
Lmi 9h28'41" 33d36'6"
Kaitlyn Rose Ellis
Lyr 18h37'29" 39d9'8"
Kaitlyn Rose Green
Sgr 19h11'55" -20d17'36"
Kaitlyn Rose Lock
Del 20h41'41" 17d12'2"
Kaitlyn Rose Staley
And 1h50'20" 38d34'42"
Kaitlyn Ryan Wyatt
Cam 3h48'54" 59d4'18"
Kaitlyn Ryleigh Eiring
Sgr 19h39'3" -12d47'32"
Kaitlyn Sara
And 23h6'28" 47d50'36"
Kaitlyn Sechrist
And 0h8'25" 45d32'19"
Kaitlyn Solis
Cas 1h23'33" 64d49'52"
Kaitlyn Tamily Lera
And 2h12'19" 38d37'58"
Kaitlyn Veronica Bradley
And 23h46'10" 37d1'21"
Kaitlyn Wendling
Vir 15h3'26" 3d46'46"
Kaitlyn Zimmerman
Uma 8h15'20" 65d59'8"
Kaitlynn
Cas 2h38'39" 66d44'29"
Kaitlynn
And 23h18'21" 51d12'59"
KaitLynn Cierra
And 0h18'53" 29d6'50"
Kaitlynn Elizabeth Hurt
Leo 9h48'56" 6d57'51"
Kaitlynn-S1
Uma 11h31'47" 56d4'51"
Kaitlyn's Home
Sge 19h56'19" 19d37'20"
Kaitlyn's Light
Vir 12h48'42" 5d51'8"
Kaitlyn's Paradise
Uma 9h36'3" 69d39'0"
Kaitlyn's Star
Uma 10h7'55" 61d13'15"
Kaitlyn's Sweetest Star
And 23h12'49" 51d34'38"
Kaity Hagmier
Cas 1h4'55" 65d15'8"
Kaityn
Mon 7h40'38" -7d29'59"
Kaiulani
Uma 9h44'0" 63d55'44"
Kaiy Kwong
Eri 3h56'53" -10d55'53"
Kaiya May
And 23h30'57" 42d32'9"
Kaiya Rae
Uma 11h33'3" 61d29'56"
Kaiya Rayvin - Akoda Rain
Sco 17h38'8" -36d36'17"
kaizhou zhu
Psc 1h35'58" 5d54'21"
Kaj & Alan's Eternal Star of Love
Ori 5h14'49" 11d18'32"
Kajinaga Daishi
Aqr 22h0'7" -18d45'26"
Kajoli
Uma 8h35'39" 59d30'50"
Kajsia LaVae
Cam 5h22'38" 65d0'11"
Kajus Januskis Walsh
Gem 7h20'47" 34d40'34"
KAK
Aql 19h5'42" -11d1'55"
Ka-Ka
Lyn 8h17'53" 41d37'15"
Kakarina
Cas 23h55'1" 53d47'34"
Kakl
Psc 0h53'2" 18d45'48"
Kako Mochi
Tau 5h18'13" 23d12'58"
Ka-Kun of star
Cap 20h32'53" -21d40'8"
Kal Brar
Uma 10h4'43" 61d31'21"
Kal Kirschenman
Mon 6h46'30" -0d57'1"
Kal Patrick Lawrence
Lib 15h12'50" -7d5'4"
Kala Cieciorka
Uma 8h17'27" 61d34'3"
Kala King
Cas 0h27'46" 50d45'52"
KALA (Kymberley Anna Lisa Allcock)
Umi 4h2'42" 88d41'29"
Kalachaland G. Daryanani
Cep 20h43'20" 60d28'53"
Kalair Myrick
Uma 11h32'31" 48d50'33"
Kalan Shael Kalynka
Uma 11h30'14" 57d25'37"
Kalani
Uma 10h36'20" 67d53'29"
Kalanstin_6_24_07
Lyn 9h6'33" 35d27'10"
Kala's star
And 0h16'3" 29d54'45"
Kalawika
Uma 9h31'30" 68d48'1"
Kalcjum A P
Cas 0h15'59" 54d40'56"
Kalder, Helga
Uma 11h26'28" 36d19'27"
Kaldizar
Cnc 8h37'16" 24d46'34"
Káldor Mihály
Uma 9h59'8" 46d53'8"
Kale
Ori 5h52'35" 21d4'21"
Kale Nocona
Ari 2h49'50" 15d48'55"
Kalea May Velasco-Cosare
Tau 4h52'20" 25d37'56"
Kaleb
Her 18h22'26" 18d12'0"
Kaleb
Her 17h29'1" 14d36'16"
Kaleb Curry
Aur 6h7'6" 48d19'11"
Kaleb Daniel Lovings
Per 3h28'45" 32d19'11"
Kaleb Garza
Aur 5h46'24" 49d31'6"
Kaleb J Lyonnais
Umi 14h18'42" 74d16'33"
Kaleb Martin Lee
Cep 22h22'34" 70d58'36"
Kaleb Michael
Aqr 23h15'12" -10d31'27"
Kaleb Michael Hatch
Her 16h13'28" 49d29'14"
Kaleb Michael Schemke
Cep 1h37'10" 80d48'55"
Kaleb Nathaniel Barton
Sco 17h50'46" -38d38'1"
Kaleb R. James
Tau 5h35'48" 26d42'55"
Kaleb Stephens Raley
Aqr 21h53'19" -8d11'22"
Kaleb's Star
Cas 0h1'46" 53d54'51"
Kalee
Ari 2h49'49" 27d54'51"
KaLee Ann Thorp
Ari 3h0'56" 28d53'19"
Kalee Chrissanthos
Aqr 22h36'44" -5d46'17"
Kalee Green
And 0h21'14" 43d25'27"
Kalee Hodges
Aqr 21h1'59" -12d12'52"
Kaleen
Sco 16h58'5" -38d16'9"
Kaleen Elizabeth Pezzuti
Crb 16h11'44" 36d31'42"
Kaleena
Sgr 19h13'30" -33d3'22"
Kaleena Anne Hughes
Lmi 9h52'52" 39d0'58"
Kaleh the Brilliant
Cyg 19h25'4" 47d1'37"
Kalei
Cmi 7h38'17" 7d5'43"
Kalei Diane
Aqr 22h18'44" 1d50'15"
Kalei Marie Eddinger
Mon 6h34'7" 0d53'32"
Kaleigh
Lyn 8h51'28" 36d5'40"
Kaleigh Ann
Del 20h30'53" 18d18'55"
Kaleigh Michelle
Tau 4h20'12" 3d30'42"
Kaleigh Morgan Bailey
Cnc 8h27'59" 16d40'32"
Kaleigh Nicole
Leo 11h23'10" 15d41'55"
Kaleigh Varney
Lib 15h8'6" -5d25'32"
Kaleigh's Graduation Star
And 1h16'58" 45d39'53"
Kaleigh's ILY Star
Ari 2h48'19" 15d47'45"
Kaleionehu
Sco 16h43'14" -26d59'35"
Kal-el McGraw
Gem 7h5'25" 29d35'0"
Kalen
Umi 15h51'27" 79d49'0"
Kalen Chi Johnson
Per 4h24'42" 40d44'45"
Kalen Josef Coe
Aur 5h32'40" 37d58'2"
Kalen Kristine Dorman
Cyg 20h16'25" 32d37'59"
Kalena
Gem 7h45'40" 31d47'12"
Kalena
Umi 9h23'39" 86d24'37"
Kalena Lynn Willard
Cam 6h22'33" 63d51'56"
Kaleonahe O Hau'ula Burke
Cru 12h46'40" -62d43'37"
Kalev Nathan Pallares Feingold
Ori 6h21'4" 10d59'27"
Kalévellic
Dra 15h0'6" 63d14'6"
Kaley
Lib 16h0'26" -9d30'18"
Kaley
Gem 7h9'39" 32d9'0"
Kaley
Cas 2h10'14" 59d59'24"
Kaley Ashton
And 0h41'6" 31d26'54"
Kaley Daigle
Uma 11h20'29" 38d4'20"
Kaley Faith Basford
And 1h17'6" 49d4'16"
Kaley Kiermayr
Aql 20h36'41" -0d23'19"
Kaley Nicole
Ari 2h36'10" 13d12'24"
Kaley Oestreicher
Psc 0h48'30" 9d3'43"
Kaley Rae Fellows
And 1h14'58" 50d22'29"
Kaley Rose 3-19-1995
Del 20h32'1" 6d50'46"
Kaley, Jacklyn, Ericka, & Kira
And 23h22'24" 38d14'47"
Kaley-Major
Dra 10h17'2" 79d45'29"

Kaley's Star
And 0h38'31" 35d46'8"
Kali
Lyr 19h7'50" 32d54'32"
Kali
Cyg 20h13'7" 48d52'30"
Kali
Leo 10h39'29" 9d45'18"
Kali
Leo 11h31'59" 24d32'51"
Kali
Leo 11h18'6" 22d25'2"
Kali
Aqr 22h30'18" -1d55'17"
Kali
Umi 14h8'16" 74d23'28"
Kali
Sgr 17h58'30" -25d23'36"
Kali A. Ferrell
Lyn 8h47'5" 44d33'3"
Kali and Aaron
Leo 11h2'31" 4d30'28"
Kali and Richie's Eternal Love Star
Cnc 8h48'48" 14d36'27"
Kali Ann
Vir 13h40'34" -6d10'30"
Kali Brandt
Crb 16h1'47" 37d19'22"
Kali Cooper
Cap 21h55'8" -12d37'57"
Kali Ilea Spelts
Lib 15h28'54" -7d47'46"
Kali Kay Knudson
Lib 15h42'59" -25d30'0"
Kali Kluba
Lyn 7h23'23" 57d36'12"
Kali L. Peterson
Vul 20h25'22" 27d50'49"
Kali MacLeod - 13th Birthday
Lib 15h32'28" -19d40'48"
Kali Marie
Psc 1h3'25" 26d28'16"
Kali Marie Carr
Sgr 18h39'38" -17d36'55"
Kali ... "My Reason"
And 1h47'18" 49d55'19"
Kali Rae
And 0h10'58" 32d45'12"
Kali Rose Trowt
Sgr 19h36'26" -24d43'52"
Kalia Patricia Lynch
And 2h18'26" 43d12'31"
Kaliamurthy & Prabavathy
Psc 1h21'21" 24d57'45"
Kaliann Rose Fox
Gem 6h4'0" 23d50'22"
Kaliappan
Vir 13h34'54" -12d30'22"
Kalicia Team Dr. Höhne
Uma 11h20'59" 70d4'37"
Kalico
Cas 23h31'40" 52d29'1"
Kalie Ann Mashburn
And 1h39'47" 36d53'51"
Kalie Rae Pathuis
Lib 15h23'4" -9d26'22"
Kalika Alicia Weathers
And 1h8'39" 43d9'28"
Kalin L Sergott
Lib 15h19'22" -20d44'21"
Kalina
And 0h46'55" 39d35'10"
Kalina
Gem 6h27'27" 22d54'31"
Kalina Danae Jazuk
Lyr 18h48'14" 39d11'48"
Kalina L. Barnes
And 0h16'17" 27d37'5"
Kalinna's Star
And 0h22'2" 45d2'42"
Kalinowski Star - Florence
Cnc 8h44'49" 31d35'32"
Kalinowski, Viktor
Uma 8h57'8" 61d49'28"
Kali's Star
Aqr 22h54'39" -8d22'32"
Kali's Star
Gem 6h51'25" 28d11'45"
Kali's Twinkle
Peg 22h9'8" 6d36'17"
Kalista Marie
And 0h17'4" 28d25'22"
Kaliyah
Ori 5h39'1" -2d53'16"
Kaliyah Jordyn Desiree Brite
Psc 1h16'59" 14d59'39"
Kaljar
Ori 5h29'17" 2d37'23"
Kalk, Klaus-Rainer
Ori 4h57'47" 2d58'26"
Kalkbrenner, Andreas
Ori 5h33'38" 10d9'36"
Kallan MacFarlane Palmer
Per 3h8'33" 55d47'4"
Kalley Marie Vincent
And 0h35'46" 31d38'57"
Kalli
Tau 5h37'0" 25d24'2"
Kalli
And 0h39'49" 41d8'42"
Kalli Jean Swenson
Lyn 7h34'52" 38d25'44"
Kallias
Lyn 8h42'4" 45d1'33"
Kallie
Crb 15h54'41" 26d42'42"
Kallie and Dillon's Star
Her 18h32'14" 15d28'23"
Kallie Anna Meyer
And 1h47'38" 43d25'41"
Kallie Fitzpatrick
And 23h30'0" 47d15'22"
Kallie Parisa Kalliche Weber
Lyn 6h58'13" 46d59'18"
Kallie Sue Fraley
And 1h21'49" 47d31'10"
Kallie's Valentine
Vir 13h38'3" -18d13'22"
kalliope
Per 3h42'58" 37d11'29"
Kalliope
Vir 13h19'33" 6d14'28"
Kalliope - Ourania
Ori 6h16'25" 14d46'13"
Kalliopi
Tau 5h41'19" 24d18'59"
Kallista
Aql 20h13'6" 5d49'18"
Kallisti
Cyg 19h56'51" 40d36'40"
Kallisti - Claudia
Gem 6h54'10" 22d8'17"
Kallistos Holly
Leo 10h57'12" 0d58'49"
Kalliswood Calypso (Kallie)
Cma 7h2'38" -31d0'47"
Kallstrom Diaz
Sco 16h16'54" -18d16'3"
Kallum Aron Higgins
Uma 11h57'18" 36d57'35"
Kallum Young
Per 3h53'5" 48d59'2"
Kalman Alexander
Cep 21h36'50" 63d57'29"
Kalmár Péter
Psc 0h55'4" 26d53'2"
Kalmd Kuethe
Per 3h25'41" 47d9'54"
Kalob Lee Douglas Manke loveGrandma
Lib 15h7'13" -2d38'18"
Kalon
Cnc 8h11'40" 25d9'27"
Kalos Spiros
Vir 13h21'9" 4d47'57"
Kalronz
Ari 3h7'52" 18d20'54"
Kalsy
Uma 9h37'45" 69d25'56"
Kalumbe
Hya 10h33'58" -26d46'26"
Kalusch, Stephanie
Uma 9h11'30" 58d10'5"
Kalvin Oldenburg
Cap 21h38'48" -16d50'22"
Kalvin1994
Gem 6h30'0" 16d51'0"
Kalyan Chakravarthi Kolli
Gem 7h15'27" 19d10'33"
Kalyl F Borowski
Sco 17h25'57" -42d13'33"
Kalyn
Gem 7h52'46" 33d13'44"
Kalyn Eternally
Sgr 20h11'19" -31d23'54"
Kalynn Dupre
Cas 0h54'20" 51d41'44"
Kalynn Elizabeth Bronson
Aqr 21h9'51" -10d3'36"
Kalynn Kerekes
And 23h28'38" 48d32'7"
Kalynn Onara Lee
Sco 17h53'41" -35d18'11"
Kalynneia
Cas 1h19'42" 55d10'35"
Kalyushonok Super Star
Leo 10h44'36" 16d33'4"
KAM
Uma 12h53'43" 63d12'59"
Kam Saquins
Ari 2h33'22" 25d20'5"
Kama
Ori 5h35'56" -6d15'56"
Kamal
Vir 13h13'49" 5d12'20"
Kamal and Natasha Pallan
Cyg 20h47'31" 51d44'4"
Kamal Benkaci
Lyn 7h4'59" 51d11'47"
Kamal El Tayyeb
Vir 12h46'50" -6d24'36"
Kamal Nour, Kouki
Ori 6h7'11" 13d38'21"
Kamal Valambhia
Ari 3h14'19" 23d31'13"
Kamala
Vir 14h41'53" 3d21'32"
Kamala Bernstein
Uma 13h49'12" 57d28'56"
Kamalani
Lyr 19h7'45" 26d20'28"
Kamali T.
Lyr 19h1'59" 41d40'10"
Kamalini
Leo 10h58'10" -0d46'29"
Kamalrukh Sethna
Umi 15h35'17" 71d31'1"
Kamara
Cnc 8h22'49" 8d10'22"
KamBooYa
Tri 2h40'32" 33d53'15"
Kambrie Rose Bartley
And 2h9'48" 38d51'4"
Kambriel
Tri 1h57'56" 34d48'57"
KAMC
Cyg 21h14'28" 44d2'35"
Kamden Rich
Lib 15h27'10" -26d16'26"
Kamdyn
Uma 8h59'25" 49d57'38"
Kamee Joe Kinyon
Aqr 22h33'17" -11d25'22"
Kameha Erin Umphrey
Ori 6h8'36" 8d45'6"
Kameha Umphrey
Lyr 19h7'21" 30d37'48"
Kamel Ouali
Sgr 18h12'1" -27d38'6"
Kamelia Mihova
Cas 1h22'56" 72d54'23"
Kameo Smith
Uma 13h39'43" 52d54'22"
Kameron
Dra 19h9'24" 70d11'14"
Kameron Scooter Warren
Lib 15h7'36" -24d18'11"
Kameron Taylor Elliott
Umi 14h19'49" 84d42'13"
Kameron Townsend
Ari 2h9'9" 24d44'38"
Kameroni
Ari 2h9'19" 25d8'1"
Kameryn Dunahoo
Uma 11h11'24" 29d7'26"
Kamesh Kumar Khaitan
Ori 6h5'56" 21d21'9"
Kami
Tau 4h34'8" 20d53'16"
Kami
Cam 6h26'4" 77d52'30"
Kami
Uma 9h22'7" 69d26'6"
KAMI
Lac 22h47'51" 54d17'31"
Kami 1992
Leo 11h44'10" 25d0'24"
Kami Elizabeth
Sgr 18h59'39" -33d50'35"
Kami Krieger
Uma 8h36'46" 58d18'54"
Kami Lou Lindberg
Leo 11h29'10" 16d18'59"
Kami Midkiff
And 0h36'59" 36d3'53"
Kami Nagami. 143.
Psc 1h3'29" 2d56'52"
Kami Powell
Sco 17h7'6" -38d18'48"
Kamiesha
Lib 14h26'51" -9d25'58"
Kamil
Aur 5h52'46" 52d55'41"
Kamil Tariq Dixon- Love
Ori 6h4'17" 17d54'59"
Kamila
Lmi 10h43'51" 34d28'17"
Kamila
Umi 15h3'0" 71d21'44"
Kamila Kurbanova
Cas 23h32'30" 58d6'58"
kamiland
Sco 16h9'26" -17d48'12"
Kamile Cilli, Queen of the Universe
And 2h26'17" 43d8'16"
Kamilla
And 0h11'21" 32d45'53"
Kamille K.
Her 16h46'50" 31d36'12"
Kamille LaDawn Stoor
Uma 11h37'52" 62d42'16"
Kamini
Gem 6h41'11" 24d25'40"
Kamini Lucknauth
Cnc 8h48'0" 11d23'36"
Kaminie
Lyn 7h31'54" 53d21'3"
Kaminski
Cam 4h37'24" 69d20'30"
KamiSue
Aqr 22h43'52" -19d46'31"
Kamla Devi Ramlakhan
Lyr 19h13'8" 26d31'33"
Kamlesh Amin
Sgr 18h51'5" -18d45'43"
Kamm, Harald Klaus Jürgen
Psc 1h17'39" 28d47'22"
Kamm, Jörgen
Ori 5h53'1" 21d31'16"
Kammy
Uma 10h19'16" 61d28'15"
Kamonchanoke Saisamorn
Ari 2h36'31" 26d23'20"
Kamo's Angel
Umi 13h21'9" 69d37'59"
KAMPED
Ari 3h17'25" 28d14'50"
Kampfstern Doris
Uma 12h18'26" 57d49'15"
"Kampourakis"
Cas 0h31'10" 54d53'15"
Kampuchea
Cam 3h33'36" 54d36'4"
Kamradt
Her 16h57'16" 50d25'14"
Kamran
Per 4h50'0" 41d3'31"
Kamran U Qureshi
Leo 10h17'33" 26d28'12"
Kamren Rose Perkins
Cru 12h40'39" -56d5'19"
Kamrie Jade
Peg 23h17'32" 17d23'37"
Kamron Barron
Sgr 18h10'25" -34d15'58"
Kamryn Jade Price
Her 16h27'36" 44d58'21"
Kamryn Kamalani Marks
Vir 13h19'59" -16d46'54"
Kamryn Kristine Edwards
Vir 14h14'58" -11d19'20"
Kamryn Lane McDonald
Uma 12h46'44" 57d52'55"
Kamryn Reayah Anderson
Cnc 8h29'34" 32d29'39"
Kamryn Rose Burns
Aqr 22h8'22" -0d42'33"
Kamryn Skye
Uma 14h9'31" 57d20'26"
Kamy Kemp
Psc 2h1'22" 6d20'6"
kana143
Cnc 8h14'36" 11d3'25"
Kanafchian
Uma 11h49'46" 42d23'31"
Kanako. M
Leo 9h41'40" 8d21'54"
Kanan Jasse Victorio Vieira
Cyg 21h30'48" 45d2'23"
Kanani
Cap 20h35'52" -17d47'20"
Kandace Ann
Lyn 6h44'4" 53d40'29"
Kandace Elyce
Gem 7h28'9" 19d27'47"
Kandace Lauryn Richmond
And 0h31'33" 39d17'52"
Kandace Nora Katchuk
Sgr 18h8'57" -27d4'4"
Kandace "Sunshine" Baker
Lep 5h7'59" -12d37'8"
Kandee
Sco 16h33'25" -31d22'23"
Kandess Floyd
Ari 3h20'47" 27d29'28"
Kandi
Cnc 8h12'52" 15d32'41"
Kandi Ann Deerfield
Lyn 6h28'39" 56d44'24"
Kandi Kirkpatrick
And 0h37'40" 38d57'30"
Kandi Lynn Pyle
Mon 6h54'38" -0d31'2"
Kandice and Michael
Cyg 20h1'27" 35d46'52"
Kandice C. Haynes
Ori 6h14'15" 15d29'40"
Kandice Flaherty
Psc 0h55'9" 2d49'40"
Kandice Rene Newkirk
And 2h34'7" 45d22'13"
Kandice Walsh I Love You
Cas 23h53'45" 53d56'8"
Kandida Chale Wells
Cap 20h41'4" -19d1'21"
Kandie L. Sims
Gem 7h11'54" 27d40'27"
Kandie, Allan, Tim, Malyna
Uma 8h58'49" 51d43'16"
Kandon Allen Riley
Her 16h49'18" 36d20'57"
Kandy Cordero
And 0h30'58" 39d44'22"
Kandy Kane
Cnc 8h10'59" 7d58'19"
Kandy Lee Jones
Leo 10h13'19" 21d11'4"
Kandyce
And 0h57'3" 43d12'6"
Kandyce V. Kahn
And 2h17'56" 49d40'54"
Kane Blanchard Cippoletti
And 23h12'41" 47d14'50"
Kane George Tuckwell 11.04.1989
Boo 15h19'12" 47d49'57"
Kane Hines Owings
Her 16h10'14" 48d9'45"
Kane Maloney - My Shining Star
Cap 21h58'22" -20d6'14"
Kane Michael Ashe
Tau 5h7'28" 27d26'56"
Kane Michael Funke
Boo 14h43'49" 52d23'12"
Kane Rable
Her 17h17'33" 47d14'32"
Kane Robert Kuester
Her 17h32'53" 44d53'6"
KANE THOMAS MECOMBER
Uma 10h30'19" 49d43'16"
Kane's Light of Love
Leo 11h56'57" 13d31'29"
Kanevsky-Janus 2y 3m 18d
Uma 12h39'43" 58d35'43"
Kang Family
Mon 6h52'23" 7d52'44"
Kanga Pooh
Uma 13h5'56" 60d6'2"
Kangaroo
Psc 1h8'39" 22d59'22"
Kangaroo Brooks
Sco 17h33'46" -41d44'22"
Kangourou dentellier
Leo 9h23'39" 17d2'17"
Kani Sangare
Uma 11h44'17" 40d16'37"
KaNi Star
Leo 11h24'10" 17d34'31"
Kanika Chawla
And 0h8'20" 41d53'14"
Kanishtha
And 23h27'24" 47d57'5"
Kanitz, Wolfgang
Uma 12h3'10" 52d24'28"
Kanji
Uma 9h43'19" 41d49'15"
Kanji Bhimani
Cap 21h19'49" -22d32'18"
Kankabear
Uma 11h21'24" 34d0'39"
KANN VERWENDET WERDEN
Ori 6h22'4" 10d49'38"
Kanna O Burch
Lib 15h34'34" -6d41'19"
Kannitha, My Hunie Bunches of Oats
Sgr 18h33'11" -28d1'2"
Kannon Robert Testerman
Dra 17h21'34" 60d13'57"
Kanoelani
Uma 11h13'9" 50d7'9"
Kanoknapa Sue-Jia
Umi 15h45'11" 82d23'57"
Kanoko
Aqr 21h57'50" -4d55'19"
Kanon Ando
Dra 18h26'9" 51d33'5"
Kanoudles
Cyg 20h34'19" 31d5'37"
Kanouff
Uma 8h53'9" 51d34'12"
Kansas Dan
Ari 1h59'35" 14d0'29"
Kansas M Vinson
Psc 0h8'36" -5d13'2"
Kant
Ari 1h57'40" 19d26'4"
Kantorova, Rita
Uma 9h12'4" 62d42'16"
kanuckahubbles
Cnc 9h10'18" 28d14'58"
Kanunik's Red Baron
Her 18h29'22" 12d33'23"
kanwal imtiaz
Cyg 19h37'9" 30d20'44"
Kanye, Touch the Sky
Scl 0h9'33" -28d4'55"
Kanzis
Sco 17h15'19" -44d59'10"
Kao Malaysia Sdn Bhd
Ori 5h37'31" 3d6'17"
Kaochan
Lib 15h30'36" -8d0'42"
Kaola Bear and Bunny Rabbit
Sgr 18h36'24" -23d52'57"
Kaori Yao
And 0h46'13" 33d14'59"
Kaori's Star
Lmi 10h25'30" 34d55'34"
kaos dreaming
Lib 15h5'50" -13d47'3"
Ka-Pants
Ori 4h47'57" 8d54'12"
Kapeish
Mon 7h2'26" -8d57'5"
Kapetanakis
Ori 5h31'55" 7d15'23"
KAP-I
Uma 13h35'33" 51d24'37"
Kapinos
Uma 8h10'39" 60d2'39"
Kapitän Kerk
Uma 11h37'40" 53d59'26"
Kapitza, Roland Helmut
Ori 6h3'18" 13d30'33"
Kaplanidou Eva
Lib 15h2'57" -24d50'8"
Kaplan's Star
Uma 11h49'30" 39d21'51"
Kapoustkina Irina
Ari 3h14'24" 15d39'6"
Kapp Perry
Vir 12h53'38" 7d1'20"
Kappa
Mon 6h20'23" -4d11'47"
Kappa Delta Chi
Aql 19h33'48" 14d2'48"
Kappa Delta Chi
Lmi 10h12'6" 33d16'24"
Kappa Kappa Kappa, Inc.
Uma 9h23'35" 61d50'0"
Kapp's Light
Her 18h54'7" 24d40'41"
Kaprice il mia bella
Cnc 8h27'42" 21d7'0"
Kara
Gem 7h22'16" 33d27'57"
Kara
And 23h49'4" 34d5'8"
Kara
And 23h16'15" 41d6'46"
Kara
Cas 23h46'8" 58d10'37"
Kara
Vir 12h55'2" -4d5'35"
Kara
Lib 15h1'26" -10d29'6"
Kara
Lib 15h14'39" -27d31'53"
Kara Alice
Tau 3h56'41" 10d14'46"
Kara and Brandon Forever
Sco 16h35'52" -29d58'0"
Kara and Bryan
Cyg 19h42'38" 52d20'29"
Kara and Dougie
Cyg 21h28'32" 53d27'31"
Kara and Jacob's Star
Ori 5h52'29" 21d10'58"
Kara and Lilianna's Star
And 0h14'5" 26d37'7"
Kara Angelbear Signorelli
Lib 15h22'55" -6d11'24"
Kara Ann
Psc 0h50'50" 10d0'4"
Kara Ann
And 0h42'16" 46d22'58"
Kara Ann Goidosik
Gem 6h47'34" 20d53'48"
Kara Ann Mabee Ex Adyto Cordis
Aqr 22h12'24" 0d19'57"
Kara Ann Orange
And 2h17'16" 47d36'58"
Kara Ann Wilhite
Gem 6h49'39" 32d15'18"
Kara Anne Herson
And 0h28'13" 42d3'41"
Kara Aubriann Kidd
And 0h40'32" 39d55'33"
Kara Bailey Berkemeier
Cap 20h27'49" -12d3'6"
Kara Baker
Cam 5h49'50" 57d41'2"
Kara Beara
Uma 13h17'14" 62d15'28"
Kara Beth
Cnc 8h59'24" 31d38'31"
Kara Beth Buggey
Lib 14h52'26" -3d37'28"
Kara Beth K.
Vir 14h10'14" -13d15'52"
Kara Beth MacKinnon
Ori 5h41'24" -2d22'37"
Kara Bounds Gilbreath
And 0h21'13" 32d12'46"
Kara Brock
Her 17h55'30" 22d21'27"
Kara Burris
Mon 7h25'20" -5d32'41"
Kara Buysse
Vir 11h51'1" 4d56'15"
Kara Charlie
Cap 20h32'26" -17d0'27"
Kara Clancy McGrath
Lyn 7h47'10" 47d47'42"
Kara Connell
Oph 17h38'7" -0d34'15"
~Kara~ "Cor Leonis"
Cnc 8h49'47" 14d53'15"
Kara E. Lee
Vir 14h15'15" 4d17'33"
Kara Eliza Cunningham
Psc 23h29'28" 5d6'23"
Kara Elizabeth Lynch
Cnc 8h46'44" 21d57'50"
Kara Gastelum
Crb 15h49'3" 26d50'10"
Kara Hokulani
Leo 11h16'0" 18d42'56"
Kara Jean
And 0h0'21" 42d47'10"
Kara Jellybean Patterson
Ari 2h17'22" 24d48'55"
Kara Jo
Aql 20h22'35" 4d6'58"
Kara Jo Price- A star has been born
Psc 1h13'25" 27d34'37"
Kara Johnson - Peter's Star Teacher
Umi 14h26'28" 70d57'55"
Kara Juhlin my loving daughter
Cas 1h29'44" 61d41'42"
Kara Kaylene Holland
Sgr 18h14'57" -20d6'31"
Kara Kijanka "Boo Boo"
Cnc 8h42'31" 12d56'15"
Kara Kitterman
Lib 14h37'39" -17d54'34"
Kara Kristen Schroeder
Mon 5h59'58" -7d35'53"
Kara Kristine
And 2h23'49" 47d53'46"
Kara L. Day
And 0h42'5" 33d42'38"
Kara Ladelle Scruggs
Crb 15h45'45" 27d45'23"
Kara Lane "Karaboo"
Peg 23h20'2" 28d51'39"
Kara Laughlin
Sco 16h10'29" -10d28'55"
Kara Lea
And 1h8'39" 37d30'1"
Kara LeBlanc's Robbie Loves U Star
Cnc 8h41'34" 31d34'20"
Kara Lee
Vir 12h31'57" 11d3'48"
Kara Lee Stone
Uma 11h21'52" 29d36'6"
Kara LeighAnn Knight
And 0h23'28" 31d59'7"
Kara Liedel
Aql 19h22'36" -10d55'19"
Kara Lily McLaren
And 0h4'58" 44d36'37"
Kara Lyndall Pholi
Cru 12h34'53" -64d12'1"
Kara Lynette Stegall
Lyn 8h27'17" 53d5'44"
Kara Lynn
And 1h33'0" 43d58'2"
Kara Lynn Davis
Hya 9h34'45" -0d29'50"
Kara Lynn Hess
And 0h53'43" 37d45'24"
Kara Lynn Russell
Cnc 9h13'8" 17d15'45"
Kara Lynne
Gem 7h10'13" 34d5'2"
Kara Mackenzie Tolbert
Aqr 22h37'9" -24d46'1"
Kara Marie
Aqr 22h4'40" -1d13'59"
Kara Marie
And 1h49'51" 43d12'51"
Kara Marie
And 0h45'58" 45d33'44"
Kara Marie
And 23h24'10" 45d52'32"
Kara Marie Akgulian
Tau 4h44'50" 22d3'54"
Kara Marie Anderson
Cap 21h32'22" -8d59'10"
Kara Marie Kirby
Cam 5h57'38" 60d14'21"
Kara Marie Zdimal
Tri 2h37'50" 33d54'32"
Kara McCready
Lyn 7h33'23" 53d13'7"
Kara McNulty
Lyn 8h3'5" 40d24'41"
Kara Mercolino
Cyg 21h30'36" 46d2'13"
Kara Michelle
Lib 14h38'37" -23d12'12"
Kara my principessa-My star 4eva JT
Col 5h20'15" -40d24'36"
Kara My Sweet Granddaughter
Uma 11h13'11" 52d52'52"
Kara N. Parrish
Psc 0h20'8" 15d52'43"
Kara Nebula 5
Leo 9h59'51" 20d56'13"
Kara Nicole Care-bear Effertz
And 23h20'6" 49d19'50"
Kara Nicole Pasquale
Cnc 8h44'38" 32d10'24"
Kara Nicole Patterson
And 1h25'23" 44d48'47"
Kara Nowak
Ori 5h8'0" 4d2'50"
Kara Olivia Latham
Sco 17h34'18" -35d53'20"
Kara Penney
Aqr 21h9'44" -3d5'19"
Kara Pizzo
And 2h26'54" 49d0'4"
Kara R. Morgan
Cnc 8h48'11" 14d29'19"
Kara Renae Carter
Aqr 23h45'56" -14d38'15"
Kara Seaburg
Ori 5h42'55" 7d1'17"
Kara Starr
And 23h54'0" 34d48'56"
Kara Stewart
Lib 14h51'1" -1d47'52"
Kara Sunflare Eriksen
Ori 6h4'35" 18d56'34"
Kara Sweeny's Star
Cam 3h32'19" 58d43'24"
Kara Taylor Lamm
Lib 15h31'41" -23d32'49"
Kara: The "Star" in my Life
Lib 15h23'55" -6d58'6"
Kara (The star in my sky)
Lib 15h55'17" -19d0'49"
Kara Theis
Sco 16h12'52" -10d51'34"
Kara Wethy
Ari 3h20'22" 29d16'26"
Kara12
Cas 1h7'32" 61d6'39"
Karaan Anne Gibbs
Cas 1h24'36" 51d17'45"

Kara-Beth Croft
And 2h3'40" 35d40'9"
Karagan
Tau 4h12'28" 11d32'24"
Karah
Lyn 9h21'14" 34d26'30"
Karah Ann Cummins
Vir 14h18'33" 6d51'39"
Karah Ellen Novak
And 1h9'46" 36d55'50"
Karah Grayson
Psc 1h8'27" 29d28'25"
Karah's Star
Crb 16h13'52" 35d55'26"
Karakwah
And 23h4'54" 42d41'27"
Karalaxy
Leo 11h6'27" 14d1'2"
Karalea Ann Godsey
Lmi 10h35'39" 32d17'21"
Kara-Lee Chiemi Maeda
Ari 2h50'35" 18d32'1"
Karaleen (Little Red) Fletcher
Lib 15h40'49" -5d15'14"
Karalyn Louise Aprill
And 0h13'38" 35d0'15"
Karam
Ori 6h5'47" 13d48'16"
KARAMA
Aqr 22h39'21" -3d45'13"
Karamel
Lyr 18h45'33" 34d29'41"
Karamellitiikeri
Cas 23h33'34" 58d9'31"
Karamia
Lyn 7h14'28" 59d34'20"
Karamia
Uma 9h41'42" 55d24'7"
Karan
Com 12h23'22" 17d23'39"
Karan and Rick
Aql 19h48'14" 11d12'35"
KARAN KAY
Leo 11h45'18" 21d45'23"
Karan Romaine
Crb 15h51'5" 26d30'43"
Karaoglan, Joshua A.
Uma 9h49'15" 47d14'1"
Kara's Bubie & Danny's Babydoll
Cyg 20h12'18" 54d26'39"
Kara's Love
And 0h40'10" 33d49'15"
Kara's Love
Leo 10h23'34" 14d11'46"
Kara's Prairie Dog
Lib 15h11'44" -4d46'22"
Kara's star
Aqr 21h2'55" -12d7'20"
Kara's Star
Cma 6h44'56" -28d47'57"
Kara's Star
Vir 14h48'2" 2d33'37"
Kara's Wishing Star
And 1h4'39" 38d49'7"
Karat DeLeon
Ori 5h44'36" 11d25'59"
Karate Kid DiLorenzo
Tau 5h52'46" 14d40'58"
KaraTom777
Cam 5h21'32" 63d31'34"
Karbachinskiy Grigoriy
Crb 15h47'29" 34d17'35"
Karbarfuzzywag
Gem 6h50'9" 14d25'23"
Karbela Reza
Tau 5h14'27" 17d48'26"
Karch Alexander Molnar
Cep 4h8'32" 81d55'38"
Kárcis Brúnó
Tau 4h23'56" 9d51'45"
Kar-D
Lyn 6h27'39" 60d19'47"
Kare Bear
Uma 11h34'15" 63d55'4"
Kare Bear
Psc 0h55'18" 26d43'56"
Kare Bear's Lucky Star
Leo 10h41'38" 13d27'51"
Karebear
Cnc 8h12'36" 9d46'14"
Kare-Bear
Crb 15h39'55" 26d52'19"
KareBear
Umi 15h3'52" 84d32'34"
Karebearmar
Psc 1h43'31" 5d48'53"
Karee Devon Love
Mon 6h50'34" -0d27'39"
Kareem Radwan
Uma 9h51'55" 52d37'51"
Kareem Saleh "One in a million"
Uma 12h39'11" 58d38'43"
Kareem Saleh "One In Infinity"
Sco 16h13'3" -13d34'5"
Kareen Nadine Bolger-Synan
Aqr 21h37'52" -7d39'24"
Kareen Welt
Sex 9h51'31" -0d35'7"
Kareena Dua
Psc 1h10'55" 26d43'29"
Kareena M Mccauley
Cam 3h50'12" 55d7'15"
Karel
Boo 14h33'57" 52d38'18"
Karel Joseph Antonin Vlach
Cep 22h46'23" 64d37'59"
Karel Madeleine DesMarais
And 1h15'14" 45d54'6"
Karel Schliksbier Badger Fan
Aqr 22h13'12" -2d9'54"
Karele
Mon 6h45'35" 8d58'15"
* Karen *
Ari 2h22'38" 27d30'33"
Karen
Gem 7h9'17" 18d55'58"
Karen
Ari 2h50'49" 28d43'3"
Karen
Tau 5h52'18" 26d28'6"
Karen
And 1h40'31" 47d19'20"
Karen
Aur 6h29'12" 47d46'59"
Karen
And 0h7'19" 47d29'29"
Karen
Cyg 20h37'57" 38d48'21"
Karen
Cas 0h52'28" 57d5'9"
Karen
Cyg 21h33'18" 46d51'53"
Karen
And 0h54'15" 37d9'44"
Karen
Cyg 20h45'57" 31d57'9"
Karen
Uma 11h17'3" 43d31'54"
Karen
Lyn 7h34'12" 39d1'18"
Karen
Lyn 7h49'1" 41d25'22"
Karen
Aqr 22h44'4" -0d56'53"
Karen
Lib 15h19'50" -5d26'11"
Karen
Lib 15h13'2" -13d45'27"
Karen
Cas 1h39'21" 64d28'32"
Karen
Cas 23h3'39" 58d47'1"
Karen
Uma 12h46'12" 54d58'45"
Karen
Lyn 6h48'10" 53d38'5"
Karen
Sco 17h46'20" -41d38'28"
Karen 080672
Uma 9h57'55" 43d11'48"
Karen 440
Psc 0h33'22" 5d53'17"
Karen A.
Cap 20h38'45" -13d51'54"
Karen A. Beardsley
Crb 15h55'24" 38d52'1"
Karen A Braun
Cam 4h46'53" 69d56'57"
Karen A. Bryant
And 0h26'50" 32d46'48"
Karen A. Kelly
Uma 12h7'24" 50d38'22"
Karen A. LOVES Peter J. Callahan
Cyg 20h53'13" 36d51'12"
Karen A. Mayhew
Tau 3h40'6" 22d58'53"
Karen A. Mercer
Gem 7h32'46" 19d43'33"
Karen - A Star For Your Crown!
Cyg 20h37'49" 41d36'55"
Karen A. Tomapat
Cas 0h0'34" 56d51'9"
Karen Acaster's Star
Ser 15h52'54" 22d54'10"
Karen Adams
Eri 3h59'42" -38d44'15"
Karen Adams - The Light of My Life
Tau 4h28'10" 21d12'21"
Karen Adele Foglesong
Umi 15h4'31" 67d39'39"
Karen & Adrian Taylor's Anniversary
Cyg 20h24'19" 31d12'48"
Karen Alchin
Cra 18h10'12" -39d39'21"
Karen Alicia Cavanagh
Lyn 8h15'28" 39d21'13"
Karen Amanda Lefas
And 1h11'0" 38d11'57"
Karen Amanda Reitman
And 1h9'41" 37d29'47"
Karen Amidon
Cnc 8h25'42" 11d5'4"
Karen and Carl
Cyg 19h46'7" 29d30'6"
Karen and Craig Pease
Uma 10h3'21" 47d17'43"
Karen and Garry Mraz
Cyg 19h38'35" 51d42'25"
Karen and Jay
Crb 15h38'19" 33d16'33"
Karen and Mac Siegel Star
Uma 10h43'27" 41d49'14"
Karen and Michaels Star 3-21-1998
And 23h5'52" 40d18'12"
Karen Angela Vella Kelley
Lib 15h23'7" -27d6'46"
Karen Anita
Cas 0h13'30" 57d13'53"
Karen Ann
Vir 13h5'51" 10d46'16"
Karen Ann
Sgr 19h16'50" -30d26'32"
Karen Ann
Lyn 6h27'41" 57d59'42"
Karen Ann
Aql 20h37'41" -7d49'4"
Karen Ann
Cap 21h14'12" -19d35'22"
KAREN ANN 39
Sco 16h38'29" -33d25'23"
Karen Ann Brown
Ari 2h34'5" 25d26'18"
Karen Ann Church's Star 7777777
Dra 14h34'2" 59d2'37"
Karen Ann Dupont
Tau 5h46'36" 24d11'45"
Karen Ann Emde
Leo 9h44'33" 26d52'8"
Karen Ann Flaherty
Vir 13h22'50" 8d2'1"
Karen Ann Franklin "Apple Blossom"
Cnc 8h10'22" 18d27'22"
Karen Ann Kluge
And 2h30'19" 38d8'54"
Karen Ann Leschke
Aqr 21h35'53" 0d39'11"
Karen Ann Maraccini
Peg 22h36'48" 27d1'58"
Karen Ann Menard
Cas 0h32'11" 59d16'40"
Karen Ann Richards
Lyn 8h24'6" 55d55'18"
Karen Ann Rogers-Fulton
Crb 15h47'18" 34d29'7"
Karen Ann Smith-Ryabchenko
Lyr 18h32'7" 41d29'39"
Karen Ann Snyder
Tau 3h42'10" 6d6'39"
Karen Ann Stafford
Sco 17h53'2" -42d49'13"
Karen Ann Tomlinson
Uma 11h25'56" 45d38'17"
Karen Ann Waring
Cas 23h27'7" 52d8'17"
Karen Ann Webb the Pooh Ma
Uma 14h9'1" 56d52'49"
Karen Ann Woods Moritz
Lib 15h47'45" -18d21'33"
Karen Ann Yaksic Foissotte
Leo 9h51'35" 13d31'4"
Karen Ann Zalewski
Sco 16h48'44" -26d41'18"
Karen Anne Haberlan
Sgr 19h31'41" -38d27'43"
Karen Anne Heron
Sco 17h51'4" -37d8'24"
Karen Anne Pilecki
Cam 3h37'20" 67d40'41"
Karen Anne Wallace
Sgr 18h59'58" -19d2'21"
Karen Annette
Psc 23h3'15" 7d32'1"
Karen Antoine
Uma 9h30'41" 47d25'52"
Karen Antonow
Her 16h58'6" 30d56'30"
Karen Anvelt
Cas 23h5'37" 58d54'3"
Karen Askew
And 1h6'33" 46d2'35"
Karen Avis Coles
Gem 6h44'2" 23d32'11"
Karen A.Weyant
Leo 9h34'24" 27d32'22"
Karen B. Dunlap
Ori 6h15'37" 15d5'51"
Karen B. Luther
Cyg 20h23'53" 48d46'47"
Karen Baley
Psc 1h21'23" 32d33'38"
Karen Ball
And 23h27'35" 41d24'26"
Karen Barnett
Sgr 19h10'32" -22d20'11"
Karen Barrett
Vir 13h6'24" 10d57'10"
Karen Barth Menzies
Sgr 20h25'56" -30d50'14"
Karen Basurto Salomon
Gem 6h57'25" 19d6'31"
Karen Baumert Marks
Aqr 22h6'35" -2d38'22"
Karen Baur
Cap 20h34'18" -19d58'30"
Karen Beam
And 23h1'25" 48d47'47"
Karen Beckwith Culver
Uma 12h48'51" 58d20'56"
Karen Beheyt
Dra 19h45'25" 65d48'42"
Karen Bellamy
Cas 2h46'55" 71d27'26"
Karen Bennett
Ori 6h2'23" 20d11'54"
Karen Berard The Ma Star
Her 18h39'0" 15d24'15"
Karen Bermudez
Crb 15h46'18" 35d55'37"
Karen "Best Mom in the World "Koch
Aqr 23h8'48" -14d53'55"
Karen Beth
And 0h26'19" 42d54'5"
Karen Beth Sheres
And 1h16'23" 46d40'39"
Karen Bishop
And 23h5'49" 51d2'20"
Karen Bloomberg Potter
Uma 9h25'55" 67d26'1"
Karen Bluemle
Psc 1h22'12" 31d24'48"
Karen Bootz
Vir 14h41'24" 3d48'20"
```Karen Borealis
Tau 5h30'34" 18d22'12"
Karen Boyer-Crain
Cas 0h25'1" 59d38'20"
Karen Braly
Crb 15h27'15" 26d31'7"
Karen & Brent - Make A Wish
Cru 12h8'35" -62d54'13"
Karen Brown Spivack 1950-2002
Uma 10h10'36" 46d49'38"
Karen Buckingham U R My Superstar
Sex 10h34'10" -0d15'34"
Karen Bullman Finley
Lib 15h11'3" -11d17'47"
Karen Bunny Passtn Renino
Sgr 18h21'39" -32d46'35"
Karen Burton
Cap 20h49'9" -15d34'32"
Karen Buschardt-Pisarczyk
Gem 7h24'6" 20d42'11"
Karen C. Loehr
Cas 23h11'13" 55d17'52"
Karen Campbell
Ari 2h16'46" 14d3'52"
Karen Casey Joslyn
Lyn 7h45'41" 56d36'31"
Karen Catherine
Mon 8h3'26" -0d43'19"
karen center of my universe
Lib 14h56'31" -13d18'52"
Karen Chan Capece
Ori 4h48'54" 4d35'23"
Karen Chapman
Lyr 19h11'56" 27d36'2"
Karen Cheever
Dra 19h22'26" 68d53'26"
Karen Chenowith - A Friend at "50"
Uma 10h26'29" 68d25'4"
Karen & Chris Forever
Leo 9h59'17" 7d13'26"
Karen Christiansen
Uma 11h43'54" 36d58'22"
Karen Christine Jouwstra
Cap 20h36'23" -17d39'25"
Karen Christine Michael
Lib 15h5'32" -23d49'48"
Karen Christine Westbrook
Sgr 18h21'26" -23d9'15"
Karen Cianci Brennan
Cas 23h34'0" 55d3'52"
Karen Clarkson's Christmas Star
And 23h41'28" 37d51'56"
Karen Coates
Cas 0h18'37" 61d10'59"
Karen Collura
Cas 23h25'19" 56d9'52"
Karen Commins
Tau 4h30'25" 22d9'37"
Karen Cortes
Psc 1h8'52" 18d7'57"
Karen Cotton
Lib 14h54'7" -13d34'50"
Karen Crumley
Mon 7h22'6" -0d59'45"
Karen Cushwa
Psc 0h25'49" 7d25'14"
Karen D. Berkowitz
Cas 3h12'31" 61d0'2"
Karen D. Blakey
Ori 5h53'31" 6d45'49"
Karen D. Drauss
Cas 0h12'45" 63d11'32"
Karen D. Majors
Umi 15h32'42" 72d45'21"
Karen Danielle Engle
Psc 1h19'16" 15d48'18"
Karen & Darren McCabe
Cyg 19h55'36" 54d47'17"
Karen & Darren Snowball Forever
Ara 17h2'32" -58d43'19"
Karen & David Hardman Ruby Star
Cnv 13h38'24" 31d4'24"
Karen Davis
Com 12h46'50" 16d0'44"
Karen Dawn
Tau 4h14'22" 2d59'9"
Karen Day
Lmi 9h25'35" 37d19'44"
Karen Day Frescas
Cam 4h57'23" 52d58'33"
Karen Dear's Sparkle
Sco 17h57'19" -40d2'5"
Karen Decorus
Cru 12h17'41" -56d57'41"
Karen DeFilippi
Cap 20h44'0" -21d14'2"
Karen Denise Corda
Uma 11h34'43" 36d50'5"
Karen Devenney
And 1h3'41" 42d45'5"
Karen Diane
Ari 2h9'42" 12d58'41"
Karen Dianne
Aqr 23h4'45" -11d40'42"
Karen Dickens
Leo 11h0'56" 22d39'47"
Karen Diehl
And 0h35'31" 46d2'9"
Karen DiGregorio Walker
Uma 9h8'18" 47d55'32"
Karen Driskill King
Cap 21h32'56" -19d19'12"
Karen Dudleston
And 23h30'4" 48d14'20"
Karen E. Bennett
Lib 14h28'52" -20d2'42"
Karen E. Bolick
Leo 11h17'48" 16d43'48"
Karen E. Page
Sco 16h27'21" -28d36'58"
Karen E. Sinatra
And 1h44'39" 49d25'36"
Karen Earle Smith
Uma 10h5'24" 60d39'13"
Karen Eileen Anderson
Ari 2h39'58" 14d11'44"
Karen Ekkelund Petersen
And 1h32'31" 40d12'56"
Karen Elaine
And 23h17'32" 43d47'48"
Karen Elaine Johnston Price
Her 17h6'34" 32d51'49"
Karen Eliska Drozdiak
And 1h0'11" 45d23'47"
Karen Elizabeth
Lyn 6h52'4" 56d4'0"
Karen Elizabeth
Lib 14h52'35" -4d52'4"
Karen Elizabeth Dyer, My Bae
Cap 21h12'34" -15d42'40"
Karen Elizabeth Hagerty
Uma 9h54'7" 58d42'7"
Karen Elizabeth Urry
Col 6h27'59" -33d15'46"
Karen Elizabeth Walker
Gem 6h47'42" 20d40'24"
Karen Elizabeth White
Cas 1h21'6" 61d9'29"
Karen Ellen
Lmi 10h21'47" 37d36'5"
Karen Ellen Sandifer
Mon 7h22'14" -0d21'30"
Karen Fahy
Cyg 19h47'12" 29d33'34"
Karen Fair
Cas 1h52'5" 61d51'36"
Karen Faye Angel Eyes
Leo 10h18'10" 25d7'27"
Karen Feldman
Lyn 7h37'59" 48d20'52"
Karen Fern Stoloff
Cas 23h26'46" 52d57'50"
Karen Fialkowski
And 23h8'57" 39d6'20"
Karen Fisher ~ World Class Angler
Sco 17h52'45" -35d56'32"
Karen Flanders-Angel of Inspiration
Cyg 19h14'57" 51d4'27"
Karen Fliss
Lib 15h36'22" -9d56'39"
Karen Ford
Sgr 19h55'35" -11d49'21"
Karen Foreman
Cas 0h48'45" 56d25'41"
Karen Francis Streb
Ari 2h15'9" 16d40'38"
Karen Frantz
Gem 6h33'7" 19d41'28"
Karen Gail Skoler
Cas 1h27'17" 57d59'28"
Karen Gardner
And 23h40'10" 47d58'23"
Karen Genesse
And 23h56'48" 43d24'12"
Karen Gillespie
Ori 6h2'5" 20d47'24"
Karen Gillespie
Cas 23h40'46" 56d20'54"
Karen Giuntoli
Ori 6h25'28" 17d13'50"
Karen Givner Heinly
Uma 9h32'32" 62d23'15"
Karen Glenn Stella Splendoris
Ori 5h7'2" -0d37'7"
Karen Goeltz
Cyg 21h51'48" 49d38'52"
Karen Goodkind
Uma 8h19'41" 70d7'5"
Karen Goucher - 10.06.1960
Gem 6h49'25" 24d9'26"
Karen Graff
Cas 0h41'22" 51d17'13"
Karen Graham
Cru 12h27'56" -59d44'5"
Karen Grantham
And 0h45'34" 37d15'13"
Karen & Greg 86
Sgr 18h36'36" -32d35'57"
Karen Guarasi
Aqr 20h42'22" -0d5'31"
Karen Guyer Permenter
Psc 1h15'36" 15d3'36"
Karen H. Bentfield
Lyn 7h29'44" 45d2'26"
Karen Habanek & Family
And 2h10'26" 42d35'17"
Karen Han
Lib 15h11'3" -16d31'22"
Karen Harris - 16 March 1984
Psa 21h33'47" -36d13'19"
Karen Helen Conlon "Star"
Lmi 10h32'37" 38d6'33"
Karen Hiddemen
Aur 5h18'39" 43d53'21"
Karen Higgins
Psc 23h23'26" 2d27'24"
Karen Holbrook
Cyg 19h57'5" 37d38'38"
Karen Honey
And 0h29'52" 40d40'8"
Karen Hope
Gem 6h33'11" 22d12'7"
Karen Horn
And 23h22'36" 44d25'42"
Karen Horsfall
Cyg 20h34'20" 37d30'22"
Karen Horsley
And 0h12'23" 37d10'31"
Karen Horton A Courageous Lady
Crb 16h12'39" 37d45'18"
Karen+lain
Cyg 20h57'55" 39d3'51"
Karen Illona Mocek Jones
And 1h20'40" 40d36'56"
Karen Imelda
Psc 1h8'10" 5d45'48"
Karen in the Sky
Lyn 7h33'37" 54d4'30"
Karen In The Sky With Diamonds
Vir 12h2'31" -0d5'33"
Karen Isemann Fetty
And 0h28'17" 44d22'16"
Karen J. Baker
And 2h15'45" 49d59'5"
Karen J. Haman-Miller
Aqr 22h14'52" -14d13'46"
Karen James
Cas 23h25'26" 56d8'29"
Karen Jane Reed
Lmi 10h6'36" 38d54'29"
Karen Jane Thomas
And 0h12'17" 41d43'30"
Karen Janzen
And 2h5'17" 43d11'10"
Karen Jean
And 23h40'4" 47d34'55"
Karen Jean
Lib 15h11'32" -6d37'18"
Karen Jean Hathaway
Lib 15h9'50" -7d36'41"
Karen Jean O'Connor
And 0h22'2" 37d36'37"
Karen Jean Stoddart
And 1h3'2" 35d5'46"
Karen Jean Teresa Skalecki
Cap 21h54'48" -9d3'56"
Karen Jeanne & Matthew Robinson
Cru 12h31'23" -59d23'13"
Karen Jerrett
Cnc 8h14'35" 7d37'37"
Karen Jessica Marissa Millie Piazza
Ari 2h38'43" 13d5'46"
Karen Jo
Uma 11h38'49" 45d54'57"
Karen Jo Allen
And 2h18'45" 50d35'19"
Karen Jo Cooper-Green
Vir 12h17'54" 10d56'48"
Karen & John Together, Forever
And 2h31'21" 38d27'14"
Karen Johnson
Cas 1h36'50" 67d55'15"
Karen Joy Canning
Aqr 22h29'16" -1d0'20"
Karen Joy Helmstetter
Psc 1h9'9" 32d42'58"
Karen Joy Quinn
And 0h14'13" 32d27'59"
Karen Joy Sheaffer
And 2h37'36" 39d39'32"
Karen Joy Smith
Cas 1h36'32" 67d30'45"
Karen Julia Smith
And 2h38'2" 37d42'20"
Karen Kaminski
Ari 2h35'51" 17d59'7"
Karen Kari Kenny Kane Jesse Brill
And 0h47'39" 34d54'31"
Karen Kathleen
Sco 17h46'22" -33d40'12"
Karen Kaufman
Psc 0h48'34" 8d59'33"
Karen Kay Panarra
Tau 3h32'17" 21d26'8"
Karen Kayser
And 0h54'6" 37d56'46"
Karen & Keith Fraher
Ori 5h20'20" 0d5'10"
Karen Kimberly
Sgr 18h14'35" -29d16'15"
Karen Kirk
And 23h7'31" 36d18'46"
Karen Klick / Karebear
Cam 4h10'47" 69d18'39"
Karen KMP Pawlowski
Lyn 7h35'22" 36d19'22"
Karen Korecek
And 0h41'21" 45d28'17"
Karen Krawchuk
Sgr 19h14'31" -21d1'5"
Karen Kremser
And 0h36'14" 37d28'53"
Karen Kristin
And 0h5'19" 45d12'52"
Karen Kwasha Jacober
And 23h28'42" 48d15'7"
Karen L Davis
Aqr 22h10'53" 2d4'26"
Karen L. Diekhans
Uma 11h26'12" 60d38'23"
Karen L. Gibson
Cas 23h47'27" 53d11'4"
Karen L. Kaufman
Lyn 8h3'2" 38d25'37"
Karen L. Mack
Psc 22h54'49" 5d39'17"
Karen L Peck
Cas 22h59'21" 55d44'45"
Karen L. Rudolph
Uma 11h29'38" 40d28'21"
Karen L. Samenow
Gem 7h16'55" 21d56'59"
Karen "La La" Paluzzi
Lib 15h55'25" -11d3'13"
Karen LaCapra
Cap 21h1'16" -19d11'24"
Karen LaMarine
Lib 15h24'22" -20d39'49"
Karen Lancaster
Ari 2h51'11" 28d57'14"
Karen Langton Hill McLaughlin
Uma 12h17'1" 59d16'52"
Karen Lau
Leo 9h44'25" 28d32'15"
Karen Layne
And 0h16'45" 28d47'36"
Karen Lebb
Cam 5h27'37" 67d56'4"
Karen Lee Ashcraft
And 0h4'44" 45d0'22"
Karen Lee Galley
Uma 8h49'31" 69d52'58"
Karen Lee Holler
Lyn 7h35'11" 58d46'32"
Karen Lee Letton
Lyr 18h52'20" 43d59'38"
Karen Lee Pacheco
Uma 11h16'51" 62d2'5"
Karen Lee Sherie Ellen
Cap 20h40'51" -20d55'42"
Karen Lee Stewart
Cas 23h36'9" 53d9'54"
Karen Lee Stiles
Tau 4h0'17" 21d18'33"
Karen Lee-This Star Plays 70s Music
Lyn 8h49'44" 46d0'57"
Karen Leigh Jaramillo
Cap 20h21'14" -16d14'6"
Karen Leigh (Pike) Matarangio
Sge 19h29'52" 18d25'32"
Karen Leighann
Per 3h24'19" 44d38'28"
Karen Lesley
Cas 1h55'50" 65d3'29"
Karen Lesley Marshall
Tau 4h10'28" 4d51'33"
Karen Leslie Compagno
Psc 1h31'56" 8d5'8"
Karen Leslie Wilkinson
Per 3h3'46" 54d14'30"
Karen Leung Ting Ting K<\@>R
Cnc 9h1'45" 15d16'22"
Karen Liat Chandally
Sgr 19h39'11" -14d36'6"
Karen Lind Clark
Cyg 21h18'17" 43d18'2"
Karen Lindsey Palmer
Sco 16h5'48" -16d59'37"
```

Karen Liva
Vir 12h24'11" -5d7'31"
Karen Lo
Aqr 23h14'30" -19d9'3"
Karen Lois Pritz
And 23h42'24" 35d26'50"
Karen Longwell
Cas 0h27'23" 63d46'45"
Karen Lopez
And 0h31'24" 42d6'36"
Karen Louise
Leo 11h46'59" 25d2'9"
Karen Louise Goetzinger
Vir 11h50'14" 3d45'21"
Karen Louise Hafer
Lib 15h43'58" -17d0'41"
Karen Louise Marsh
Leo 11h53'9" 21d47'51"
Karen Louise Roling
Gem 6h42'14" 13d23'51"
Karen Louise Wilson
Cyg 20h59'50" 32d38'1"
Karen Louise Windle
Vir 13h29'21" -21d10'45"
Karen Lucille Kitzis
Lib 14h57'44" -2d21'36"
Karen Lunny
Per 3h18'57" 48d35'22"
Karen Lyn Ellithorp
Uma 12h37'16" 53d2'1"
Karen Lyn Gallant
And 0h25'2" 31d33'53"
Karen Lynette Simpson
And 1h7'33" 36d13'21"
Karen Lynn And Baby
Gem 6h37'31" 21d7'31"
Karen Lynn Annis
Ari 2h35'16" 27d54'7"
Karen Lynn Breazeale Baker
Vir 12h49'24" 11d41'57"
Karen Lynn Childers Davis
Uma 11h34'5" 61d45'15"
Karen Lynn Cottrill
Lyr 18h30'29" 36d46'50"
Karen Lynn Gray
Uma 9h29'13" 48d36'14"
Karen Lynn Marie Sampson
Vir 13h12'54" -20d14'14"
Karen Lynn McLaren
Aqr 23h18'54" -18d11'5"
Karen Lynn Miller
Ari 3h19'44" 16d0'22"
Karen Lynn Quattone
Lib 15h36'38" -12d9'59"
Karen Lynne Baynard
Psc 1h7'28" 28d32'54"
Karen M
And 1h52'42" 36d8'39"
Karen M.
Cas 23h24'54" 59d1'25"
Karen M Bala
Her 17h40'51" 15d7'21"
Karen M Burke
Lib 14h51'56" -2d10'5"
Karen M. Jordan
Ari 2h49'45" 15d45'11"
Karen M. Kehoe Clancy Star
Ori 5h58'56" 22d48'9"
Karen M. Wigger
Cas 0h33'36" 57d0'3"
Karen MacClain
Cas 23h23'27" 58d3'20"
Karen MacWilliams
And 0h39'16" 45d42'5"
Karen Mae Magnuson ~ 08-29-1944
Vir 13h23'7" 13d31'31"
Karen Mahoney Merritt
Sco 16h48'56" -32d59'58"
Karen & Marc
Uma 10h41'19" 56d25'47"
Karen Maria Tyler
Crb 15h50'16" 36d50'4"
Karen Marie
Cnc 9h6'20" 31d21'59"
Karen Marie
Tau 4h0'29" 19d20'41"
Karen Marie
Ari 2h28'43" 18d9'34"
Karen Marie
Gem 6h18'4" 26d10'26"
Karen Marie
Psc 1h46'38" 5d20'8"
Karen Marie Agnes Brown Nicewander
Uma 10h9'28" 59d18'29"
Karen Marie Caruso
Vir 13h22'18" 5d3'4"
Karen Marie Choate
Aqr 21h38'42" -1d19'19"
Karen Marie Hering
Psc 0h21'56" 3d32'53"
Karen Marie Magro
And 1h43'52" 42d18'31"
Karen Marie Marlin
Ori 5h23'29" 7d26'52"
Karen Marie Moore
Cas 0h53'55" 58d35'14"
Karen Marie Shaffer
And 1h24'57" 46d13'15"
Karen Marie Tompkins
Cnc 9h4'52" 13d38'1"

Karen Marie's Light
Cam 3h30'57" 55d1'52"
Karen Marks
Mon 7h21'42" -0d37'22"
Karen Marshall
And 1h56'45" 39d7'39"
Karen Martingello
Sgr 18h48'30" -16d17'49"
Karen & Marvin Block
Cyg 21h42'53" 31d3'24"
Karen Mary Livingston
Tau 4h20'21" 16d55'52"
Karen Maudsley - Mum's Star
Cas 0h20'9" 50d6'56"
Karen Maureen Berkowitz
Lib 15h7'20" -0d31'53"
Karen Maxville
Sgr 19h33'25" -14d52'41"
Karen Mc Ilwain
Cnc 8h18'16" 10d13'24"
Karen McAulay
Her 16h31'25" 5d21'6"
Karen McCabe Krasner
Cap 21h35'31" -10d47'21"
Karen McKee
Lib 14h33'59" -13d21'15"
Karen Medina "The Star Of My Life"
Uma 9h30'13" 69d11'40"
Karen Melissa
Cnc 8h46'52" 24d8'44"
Karen Michaelis
Aqr 22h48'19" -21d1'2"
Karen Michele Hardy Beal
Her 18h29'42" 12d14'32"
Karen Michelle
Cnc 8h37'16" 31d43'3"
Karen Michelle Bryant
Cap 20h57'42" -22d7'44"
Karen Michelle Bryant
Sco 16h40'13" -25d13'9"
Karen Michelle Long
Sgr 19h33'10" -17d9'2"
Karen Michelle Loveless
And 23h30'6" 42d29'25"
Karen Michelle McKune
And 2h8'15" 38d31'28"
Karen Michelle Pushea
Cas 1h21'10" 57d5'4"
Karen Michelle Welch
Vir 12h57'41" 12d26'28"
Karen Michie's Cosmic Magic
And 1h47'49" 36d20'19"
Karen Mitchell
Sgr 19h23'25" -26d7'17"
Karen Mobley
And 23h22'58" 47d33'55"
Karen Mohan
And 23h15'24" 42d17'14"
Karen Molero Star
Dra 9h39'2" 74d45'27"
Karen "Mom" Gough
Lyn 7h26'59" 53d2'39"
Karen Morales
Tau 5h45'22" 19d54'58"
Karen Mortensen
Psc 1h9'53" 7d6'59"
Karen Mullen - Out of this World!
Per 4h4'35" 47d24'57"
Karen Muram
Cnc 8h56'58" 13d3'57"
Karen My Love
Psc 0h34'54" 8d20'46"
Karen My Love
And 23h47'40" 45d26'6"
Karen My Love & Treasure
Cas 1h30'41" 62d45'4"
Karen Myers
Peg 22h7'53" 18d43'16"
~Karen N Mooky's Star~
Lyn 8h50'30" 46d28'11"
Karen Napoletano
Ori 5h47'8" 7d41'40"
Karen Newell Byrd Braznell
Sco 16h40'19" -28d55'48"
Karen Nichlos
Gem 7h47'15" 24d39'18"
Karen Nina Dizon
Cap 20h57'22" -21d22'28"
Karen Norman "65"
Cnc 8h46'36" 6d47'59"
Karen Nyby Holt
Psc 1h21'10" 11d6'6"
Karen Olson
Aql 19h53'3" 14d22'8"
Karen O'Shea
Sco 16h12'23" -18d3'51"
Karen P. Hensley
Uma 8h59'47" 72d23'38"
Karen Patrice Krutish
Sco 17h18'37" -40d37'16"
Karen & Peter Kaminski
Cam 3h43'36" 58d37'39"
Karen Peters
Leo 11h36'26" 22d16'10"
Karen Phoenix Diep
Cnc 9h20'48" 10d27'7"
Karen Princess Of Scotland
Tau 4h5'4" 16d2'24"
Karen Prunczik
Cnc 7h59'40" 14d30'19"

Karen Puccini Guevara
Gem 7h33'19" 28d0'23"
Karen "Pudge" Bollnow
And 1h36'32" 40d51'19"
Karen Quast
Gem 6h31'8" 20d23'37"
Karen R. Cowland
Leo 9h27'23" 15d30'11"
Karen R. Ialacci
Psc 0h19'3" 0d6'59"
Karen Ragno Griffith
Cas 1h14'56" 63d29'16"
Karen RaNae Stephens
Com 12h14'42" 26d29'49"
Karen Re
Her 16h44'35" 42d51'27"
Karen Reap
Uma 9h41'44" 68d23'31"
Karen Reynolds
And 0h25'2" 25d50'1"
Karen Riera
Lmi 10h46'50" 28d5'8"
KAREN RKC
Tau 5h32'26" 28d17'44"
Karen Rodriguez
Tau 4h27'54" 23d1'12"
Karen Rogers Green
Vir 13h43'5" 3d48'45"
Karen Romanowitch
Mon 6h46'21" 10d59'19"
Karen Rose
Uma 10h33'1" 70d10'33"
Karen Rosebrough
Sco 16h19'49" -11d58'38"
Karen Rossetto Frick
Psc 1h9'43" 27d12'27"
Karen Rush Elko
Uma 11h15'29" 51d51'12"
Karen Ruth Rummler
Lyn 7h48'4" 38d23'46"
KAREN RYCHEL
Cap 21h1'0" -20d49'20"
Karen S.
Sco 17h54'39" -37d56'40"
Karen S. Alexander
Her 18h18'47" 28d13'30"
Karen S Chang
Lyn 8h13'28" 47d39'8"
Karen S Clifford
Vir 13h3'35" 13d38'43"
Karen S. Hedrick
And 23h42'22" 45d31'46"
Karen S Northwood 1781969
Aql 19h36'13" 12d20'56"
Karen S. Peltz-Gelazius
Leo 11h14'22" 14d58'36"
Karen S. Tytgat
Del 20h18'15" 9d40'51"
Karen & Sandy Berenberg
Cyg 20h53'24" 46d45'29"
Karen Sastri Dindial
Tau 4h18'18" 9d33'50"
Karen Schnell-Goldberg
Sgr 18h36'32" -21d4'41"
Karen Scott Jackson
Uma 8h40'10" 72d52'56"
Karen & Scott's 12th Anniversary
And 0h36'8" 39d12'30"
Karen Seiter
Uma 10h38'29" 66d59'9"
Karen Shipp
Vel 9h26'44" -40d34'10"
Karen Siegler
Aqr 22h18'57" -2d46'49"
Karen sincerus quod decorus
Sco 16h6'50" -15d54'51"
Karen - Star of Kiermor
Cas 1h27'5" 53d37'43"
Karen Stauffer
Cnc 8h47'34" 19d44'28"
Karen Steele
Psc 1h13'50" 27d14'55"
Karen Stevens
And 2h14'44" 50d24'14"
Karen Stewart
Vir 13h26'38" 11d52'10"
Karen Stewart Mega Star
Peg 21h55'24" 10d24'30"
Karen Strong
Aqr 22h16'9" 1d31'0"
Karen Sue
Mon 5h57'42" -7d19'17"
Karen Sue Artman-Von Dane
Uma 11h14'31" 63d58'35"
Karen Sue Breese
Uma 8h47'58" 66d51'28"
Karen Sue Degen (Wife/Mom/Nana)
Cas 1h41'58" 63d34'46"
Karen Sue Johnson
And 0h48'33" 37d1'20"
Karen Sue Manderick
Lib 15h16'15" -12d9'48"
Karen Sue Roy
Crb 15h37'32" 30d6'56"
Karen Sue Tautenhahn
Aqr 22h53'46" -16d28'20"
Karen Sundmaker
Cas 1h30'28" 61d22'48"
Karen Susan White
Lyr 18h52'13" 44d35'24"

Karen SW
Aqr 22h33'33" -0d14'1"
Karen Swinburn - My Sweetness
And 23h52'56" 35d17'21"
Karen Szweda
Cas 1h21'8" 60d32'20"
Karen TarBush
Tri 1h57'16" 30d33'18"
Karen Taubmann Johnson
Leo 9h27'7" 25d36'6"
Karen the Goddess Vincent
Lib 15h36'53" -10d43'35"
Karen the Special Bean!
Peg 22h50'24" 26d8'34"
Karen The Young Duck
Cas 23h22'45" 59d41'58"
Karen Thompto
Ori 5h45'34" 11d53'9"
Karen Thornbro
Vir 13h46'44" -6d23'40"
Karen Tiger Pants Nulty
Lmi 10h42'58" 27d21'19"
Karen Timmons
Tau 5h56'54" 27d23'10"
Karen Tommy Rea
Lib 15h3'20" -1d39'30"
Karen Trana Teachers Star
Uma 10h26'9" 55d9'1"
Karen Tucker
Ori 5h57'38" 21d13'46"
Karen Unruh-Wahrer Stellar Friend
Uma 8h47'53" 46d35'29"
Karen Valeo
Cnv 13h15'53" 41d19'53"
Karen Veltre
Gem 7h16'49" 33d52'37"
Karen Verrinder
Lib 15h33'58" -6d27'9"
Karen Villagonzalo
Gem 7h40'0" 14d14'52"
Karen Wallace Sorenson
Cap 21h37'21" -8d56'39"
Karen Walter
Psc 22h51'31" 3d25'23"
Karen Walton
And 1h34'59" 50d0'35"
Karen Wasser
Per 2h46'23" 53d20'45"
Karen Weathers
Cap 21h8'49" -20d35'53"
Karen Webber DaMore
And 23h28'19" 48d34'10"
Karen Werner Rung
Per 2h5'40" 56d19'10"
Karen Wertenberg
Crb 15h48'2" 25d47'47"
Karen White's Valentine Star
Ori 5h36'4" -2d27'4"
Karen Wierer
And 1h41'24" 37d8'8"
Karen Williams
Mon 6h11'19" -7d17'10"
Karen Williams
Ori 6h0'38" -3d47'46"
Karen Wilson
Aqr 22h5'9" -4d12'59"
Karen Wolf
Lyr 18h33'46" 32d12'40"
Karen Wolfe
And 23h45'46" 34d45'18"
Karen Woolf
Aqr 22h32'13" 0d35'50"
Karen Wright
Cas 23h16'53" 54d17'38"
Karen Yamsuan
Ari 3h25'56" 23d36'36"
Karen Yazmin Bustamante Benson
Umi 13h54'46" 75d36'18"
Karen Yvonne Rhoades
Uma 9h26'34" 65d41'28"
Karen "Zenith" Gumber
Aqr 23h49'46" -15d25'8"
Karen, My Love
Cas 23h14'33" 55d40'47"
Karena
Oph 17h22'24" -22d30'56"
Karena Bianca Groom
And 23h19'51" 50d39'41"
Karena Dowling-Dunny
Cas 23h55'5" 56d52'43"
Karena Ellen
Vir 13h3'2" 12d40'1"
Karena KK
Gem 6h29'40" 25d1'29"
Karena Nicole
Vir 12h45'30" 12d30'58"
KarenFranzese's Glow
And 2h36'26" 43d50'2"
Karenjo Golob Heronemus
Sgr 19h8'31" -33d4'34"
KarenLee
Cyg 19h36'46" 28d35'36"
Karenna Borealis
Crb 15h50'25" 36d32'5"
Karen's Angel Star
And 23h28'56" 42d30'28"
Karen's Cartref
Mon 6h50'37" -0d48'52"
Karen's Cellar Door
Tau 5h35'31" 26d33'9"

Karen's Eyes
And 23h41'53" 36d28'15"
Karen's gift from Adam
Cap 20h58'49" -20d52'9"
Karen's Joy
Cap 20h58'15" -20d6'32"
Karen's Light
Cap 21h54'11" -18d6'57"
Karen's Special Star
Cyg 21h24'51" 30d35'20"
Karen's Star
Cas 0h3'31" 53d19'20"
Karen's Star
And 23h10'41" 47d8'49"
Karen's Star
Cam 3h55'51" 55d46'27"
Karen's Star
Uma 11h40'50" 46d36'49"
Karen's Star
Del 20h25'10" 20d14'2"
Karen's Star
Aqr 22h2'22" -21d2'28"
Karen's Star
Pho 0h18'0" -47d19'59"
Karen's Star
Sgr 17h59'28" -24d46'45"
Karen's Star
Sgr 19h1'16" -23d27'0"
Karen's Star - Beautiful Blue Eyes.
Del 20h24'56" 10d18'10"
Karen's Treasure
Mon 8h4'34" -0d47'15"
Karen's Woossa Star
Dra 18h42'5" 70d19'24"
Karensa
Cap 21h8'17" -17d35'27"
Karen-S.B.-50
Tau 3h49'25" 24d43'29"
Kare'N'Shel KarenKai
Eri 3h42'11" -0d10'58"
Karenstar
Crb 15h55'41" 34d58'57"
Karenza Nadine Gray
Cen 13h0'28" -32d18'57"
Karenza Nadine Gray
Cen 12h46'9" -40d0'22"
KARERIC
Uma 12h38'19" 60d56'11"
Karey
Cam 4h19'43" 63d0'42"
Karey Brook
Ari 2h52'4" 18d43'41"
Karey Everett Bruner
Uma 10h37'34" 52d52'39"
KARF
Uma 9h24'31" 67d13'46"
Karhino
Cas 0h19'2" 56d43'38"
Kari
Cyg 20h51'21" 40d48'20"
Kari
And 1h39'11" 49d34'6"
Kari
And 1h20'3" 45d2'4"
Kari
And 0h44'3" 35d46'5"
Kari
Peg 22h0'52" 36d1'15"
Kari
Tau 5h52'41" 13d54'37"
Kari
Vir 12h16'34" 4d38'41"
Kari
Crb 16h9'4" 27d10'40"
Kari
Dra 16h20'5" 68d24'56"
KARI
Cra 18h49'18" -38d20'20"
KARI
Sco 17h30'55" -32d37'37"
Kari A. Hedden
Tau 4h35'48" 17d11'21"
Kari Angela Gopaul
Crb 15h34'10" 30d25'26"
Kari Ann
Cas 1h20'7" 64d59'43"
Kari Ann
Sco 16h14'45" -11d56'33"
Kari Ann Ellis
Uma 11h35'54" 34d19'17"
Kari Ann Skeval
Leo 10h48'35" 9d38'32"
Kari Anne Marshall
And 1h14'41" 41d22'0"
Kari Anne Merrifield
Vir 12h39'0" 0d18'11"
Kari April Spann
Psc 0h37'56" 11d38'7"
Kari Beth Sammon
Lyr 18h50'18" 36d19'38"
Kari Blankenship
Psc 1h38'2" 27d55'11"
Kari Dav
Aqr 21h26'33" 2d0'12"
Kari Day Macdonald
Tau 4h34'30" 11d30'34"
Kari Denise Nelson
Leo 11h41'36" 22d10'48"
Kari Dorflinger
Per 3h11'5" 49d20'45"
Kari Edwards
Ari 2h42'21" 26d24'49"
Kari Elizabeth
Cnc 8h12'20" 15d59'0"

Kari Elizabeth Allen
Ori 6h5'0" 19d32'55"
Kari Eternal
Vir 12h22'59" 7d48'12"
Kari Hardesty
Sco 17h21'40" -39d40'40"
Kari - Highest Distinction Graduate
Cnc 9h6'7" 30d53'51"
Kari J. McCall
Dra 17h14'44" 66d47'56"
Kari J. Podboy
Aqr 22h6'12" -10d22'11"
Kari & Jay Allen
Cmi 7h24'53" 8d4'18"
Kari Jeanne McGrath
Sgr 19h25'3" -41d46'29"
Kari Jo
Cnc 8h21'5" 12d42'19"
Kari K
Sco 17h12'27" -40d57'7"
Kari Kathryn Taylor
Uma 9h56'19" 58d6'0"
Kari L. Agosta
Cnc 8h24'12" 23d2'17"
Kari L. Gibson
Uma 8h58'12" 48d48'46"
Kari "Lil Booty" Duchyns
Boo 14h37'21" 51d43'14"
Kari Lynn
Crb 15h53'56" 39d31'36"
Kari Lynn Mango
Gem 7h18'27" 20d20'8"
Kari M. Kutcher
Leo 9h25'21" 30d52'54"
Kari Martinez "My Shining Star"
Umi 14h33'57" 72d41'16"
Kari Michelle
And 0h22'8" 46d33'13"
KARI MICHELLE
Leo 9h56'22" 13d32'17"
Kari Mills Corbin
Tau 5h58'29" 26d0'25"
Kari Mozer
Lyn 6h38'35" 59d1'53"
Kari Neilson
Ori 5h35'26" -5d15'11"
Kari Nelson
Aqr 22h7'44" -0d53'44"
Kari Nicole Racey
Lyr 18h37'51" 30d6'26"
Kari Rose Jablonski
And 23h38'13" 37d22'20"
Kari Shaw
Cyg 19h39'17" 30d20'36"
Kari Solanskey
Lib 15h54'42" -3d52'41"
Kari Sue and Keira Lyn Parker
Ori 6h12'12" 7d36'28"
Kari - The twinkle in my eye
Cas 23h28'18" 55d43'53"
Kari Youngblood
Ser 15h19'24" -0d49'58"
Kariann Marie
Del 20h19'22" 8d54'32"
KariAnne
Ari 3h16'33" 29d23'14"
Karie
Uma 8h51'16" 54d1'40"
KARIE ANN
Cap 20h16'31" -13d44'48"
Karie Ann Lent
And 2h13'20" 37d58'12"
Karie Ann Mason
Gem 6h41'52" 26d15'0"
Karie "Bear"
Uma 11h22'15" 35d2'47"
Karie Daniel
Psc 1h46'21" 19d20'20"
Karie Hovey
And 0h17'18" 40d11'16"
Karie Mumtaz
Sco 17h25'20" -32d27'13"
Karill
Uma 12h8'0" 58d3'52"
Karilynn Oi
Sge 19h41'8" 17d26'9"
Karim
Her 18h39'30" 25d41'58"
Karim Assef
Uma 9h49'53" 69d50'35"
Karim Bensid
Her 16h30'35" 23d26'30"
Karim & Dana Harb
Vir 14h36'28" 7d13'4"
Karim Masarweh
Uma 11h54'22" 45d38'46"
Karim the most Generous
Tau 4h6'37" 30d0'55"
Karima
Gem 6h37'42" 15d12'8"
Karima Tahri
Cap 21h18'33" -26d9'16"
Karim's Heart
Mon 7h31'24" -7d7'28"
Karin
Umi 13h12'15" 87d51'52"
Karin
Uma 9h28'38" 66d31'5"
Karin
Uma 9h30'27" 64d31'51"

Karin
Uma 12h11'0" 53d30'29"
Karin
Uma 8h41'9" 54d27'8"
Karin
Sgr 19h2'44" -34d2'5"
Karin
Ori 6h1'59" 21d12'53"
KARIN
Psc 1h8'13" 28d25'39"
Karin
Leo 9h26'23" 24d47'41"
Karin
And 2h4'4" 42d54'50"
Karin
Uma 12h6'6" 45d17'32"
Karin
Cas 0h58'57" 53d31'50"
Karin 2241990
Psc 0h23'47" 6d15'49"
Karin Amstutz
Cas 1h9'32" 63d27'44"
Karin and Jim
And 0h11'45" 37d16'54"
Karin Andreen
Cap 20h34'43" -9d59'36"
Karin Anita
Psc 1h44'48" 22d2'23"
Karin Bätscher
Crb 16h22'30" 27d19'18"
Karin Brede
Ari 2h20'26" 14d12'46"
Karin Brigitte
Uma 9h8'48" 50d22'11"
Karin + Bruno Schleiss - Leu
Uma 9h54'13" 67d17'21"
Karin Bühler
And 23h16'54" 48d3'20"
Karin & Christian
Cas 23h36'25" 51d39'14"
Karin Coheley
Dra 17h2'26" 54d3'40"
Karin "Colibri" Sprecher
Sco 17h52'13" -30d28'58"
Karin Frankenstein
Uma 11h39'15" 29d2'52"
Karin Gentile
And 23h39'12" 39d3'59"
Karin Gesse-Ettinger
Gem 7h54'10" 15d11'41"
Karin Götz
Crb 15h33'57" 37d17'1"
Karin Haid - Scambor
Uma 8h52'15" 49d14'26"
Karin Heiss
Uma 11h56'54" 32d4'29"
Karin Heß
Ori 6h17'17" 15d42'51"
Karin Hilbert
Uma 11h14'7" 40d11'46"
Karin & John Wanamaker
Crb 16h9'43" 38d54'36"
Karin Joline
Cas 1h26'26" 65d0'13"
Karin Klokow
Uma 13h5'11" 59d46'37"
Karin Maria Töpfer
Uma 9h2'55" 67d50'20"
Karin Marlis Sand Lynch
Cap 21h29'5" -16d3'36"
Karin Matz
Tau 3h51'43" 28d52'32"
Karin Meador-Frans
Uma 10h57'5" 67d14'1"
Karin Michelle Lynch 06-24-1988
And 0h44'39" 29d11'14"
Karin Mock
Cyg 21h28'55" 39d59'3"
Karin Moosche
Ori 5h58'37" 17d23'56"
Karin Mugavero
And 0h12'2" 25d38'5"
Karin & Oliver Scheuner
Cas 23h50'45" 50d17'57"
Karin Rochelle
Sgr 20h18'36" -35d11'46"
Karin Rupperts Birthday Gift
Aqr 21h47'4" -7d15'48"
Karin Ruth Hagstrom
Lib 15h0'28" -20d45'36"
Karin Schicke-Gerisch
Uma 11h38'25" 32d59'21"
Karin Scotti
Uma 14h13'51" 55d32'57"
Karin Sondergard
Uma 10h25'34" 52d8'9"
Karin Spears
Sco 16h4'45" -17d3'31"
Karin Vauthier
Cnc 8h46'43" 11d55'35"
Karin Woods
Leo 11h36'44" 22d59'31"
Karin Ziskind
Lyr 19h8'35" 27d21'30"
Karina
Cnc 8h24'29" 10d36'23"
Karina
Lyn 9h16'51" 45d42'2"
Karina
And 23h26'40" 46d56'46"
Karina
Aqr 23h0'44" -9d42'2"

Karina
Umi 16h22'44" 72d32'49"
Karina
Cru 12h24'26" -62d32'18"
Karina Allysa Emeric
Ari 3h7'37" 12d27'16"
Karina and Marty
Cyg 21h47'26" 52d44'43"
Karina & Andrew
Tau 4h29'29" 17d31'7"
KARINA ARDITE
Uma 12h46'53" 55d22'3"
Karina Carrera
Ari 2h49'16" 17d50'49"
Karina Celiz 4/20
Tau 5h44'2" 22d48'32"
Karina Chumacero
And 2h27'36" 43d28'35"
Karina Dawn
Mon 6h39'56" -0d9'36"
Karina De Melo Bacellar
And 2h23'47" 39d27'28"
Karina E. Contreras
Sgr 17h58'46" -16d58'8"
Karina G. Hermosa
Ari 1h56'9" 21d28'40"
Karina Gabriela Malke
Cas 0h33'50" 48d1'47"
Karina Gabrielle
Ari 2h13'11" 25d0'50"
Karina Garcia
Cas 1h12'53" 50d8'34"
Karina Janine
Crb 16h8'35" 37d54'56"
Karina Jolie Gonzalez
Crb 15h48'36" 26d35'44"
Karina L. Baker
Ari 2h45'21" 28d5'8"
Karina Leiva
Tau 4h38'27" 29d6'57"
Karina Lija Moritis
Cyg 20h12'39" 38d56'4"
Karina Majana Quinn
Cas 23h32'4" 51d57'9"
Karina & Marco
Lmi 10h27'26" 33d39'41"
Karina Margaret Cannelli
Cam 12h55'26" 76d49'51"
Karina Maria Ceglinski
Ari 3h18'58" 28d14'27"
Karina Marie
And 2h22'19" 42d36'29"
Karina Marie Valen
And 0h32'31" 28d48'23"
Karina Marion Roessig
And 2h12'40" 50d4'15"
Karina Nayr Ortiz Ramos
Cnc 8h13'46" 12d39'23"
Karina Robles Figueroa
Cnc 8h34'11" 10d28'18"
Karina S. Lau
And 0h29'53" 38d27'39"
Karina Stambouliah
Car 9h53'59" -66d35'42"
Karina & Steve's Forever Star
Cyg 20h7'18" 40d21'35"
Karina Tanuwijaya
Sgr 19h2'52" -35d23'3"
Karina Vity Dahl
Aqr 21h37'3" -3d52'27"
Karina W.
Psc 1h12'33" 5d48'59"
Karina Xochitl
Peg 22h22'3" 8d19'2"
Karina, Christian & Andrea
Per 3h25'22" 44d21'5"
Karina28
Uma 11h59'41" 36d35'4"
KarinaEganValarieJeanette CindyAlSin
Tri 2h6'33" 33d50'17"
Karinas Luminarias (Karen's Star)
Cap 20h13'13" -14d59'54"
Karina's Star
Lmi 10h40'2" 35d59'20"
Karinas Star
Cyg 20h14'39" 51d23'40"
Karinda
Tau 5h14'37" 27d48'1"
Karine
Cnc 8h19'17" 23d42'35"
Karine
Crb 16h2'14" 29d21'28"
Karine
Sgr 19h48'31" -13d9'35"
Karine
Hya 8h14'31" -2d6'36"
Karine
Sgr 19h16'56" -27d45'25"
Karine Avoinet
Ori 5h2'53" -0d3'14"
Karine Ayvazyan
Cap 20h38'21" -23d30'12"
Karine Babaian
Crb 15h52'26" 33d35'27"
Karine Dufour 27.09.72
Umi 17h2'24" 76d11'53"
Karine et Esteban
Cas 0h1'56" 54d13'26"
Karine Neros
Uma 8h38'56" 56d30'48"
Karine Robichaud (Q C) 15-04-86
Uma 11h22'49" 60d34'14"
KARINE VAREILLE
Umi 15h12'25" 84d47'53"
Karing Kara-Leigh
Cnc 8h13'57" 25d24'4"
Karinka
Lib 15h19'7" -7d8'5"
Karinne Renee' Hunt
Lmi 10h36'27" 30d45'51"
Karinne Versot
Aqr 22h0'43" -18d37'58"
Karin-Renee04
Lib 15h11'53" -15d39'51"
Karin's Star
Psc 0h49'45" 14d42'6"
Karins Topper
Leo 11h27'55" 22d43'38"
Karin-Schnausi
Aql 19h34'57" 7d25'52"
Karinsita
Psc 1h29'36" 32d43'37"
Karinzinha
Ari 2h22'24" 21d18'51"
Karis
Lmi 10h39'4" 34d24'26"
Karis LeTard
Sco 17h36'51" -37d56'35"
Kari's Sparkle
And 0h10'25" 33d41'45"
Karisa
Vir 13h5'28" 3d0'37"
Karisa Jean Burkhalter
Leo 11h36'18" 23d25'16"
Karisa Lynn Darrow
Lib 14h55'54" -18d49'51"
Karishe
And 1h24'32" 43d9'27"
Karishma
Aqr 21h15'12" 1d27'23"
Karishma Kajal Misri
Tau 5h42'23" 25d13'25"
Karissa
Cnc 8h11'5" 28d24'52"
Karissa
Gem 6h54'57" 13d13'54"
Karissa
Cra 18h27'53" -42d14'50"
Karissa Ann Spring's Lucky Star
Cas 23h57'33" 56d41'37"
Karissa Courtney Patterson
Mon 6h30'35" 4d17'24"
Karissa Knox
Cnc 9h15'47" 29d37'21"
Karissa Kristen Lania
Cap 20h11'10" -9d18'27"
Karissa Krueger
And 1h9'54" 44d53'2"
Karissa Lee
Leo 11h38'48" 24d6'40"
Karissa Louise
And 1h20'23" 47d30'14"
Karissa Marie Samuels
Uma 9h50'46" 65d32'5"
Karissa Marie Schwartz
Gem 7h29'1" 15d17'0"
Karissa Oi
Cet 1h5'49" -18d50'2"
Karissa Vickery
Lyn 8h14'30" 34d54'30"
Karisstar
Aur 5h30'24" 47d52'11"
Karki
Lib 15h6'54" -0d51'27"
Karl
Dra 19h1'42" 61d7'11"
Karl
Umi 14h16'3" 67d48'45"
Karl A. Zeggert
Gem 6h4'23" 25d35'47"
Karl Adventia Rodrigues - Happy 40th
Cru 12h57'11" -59d17'48"
Karl Alan Williams
Leo 9h23'59" 27d1'36"
Karl Albert Ziemba
Lmi 9h44'24" 35d50'56"
Karl and Amy
Gem 7h32'56" 31d47'31"
Karl and Beth Grgurich
Cma 6h19'13" -30d32'32"
Karl and Sandy
Vir 13h46'46" -6d47'39"
Karl Beilstein
Aur 6h23'21" 48d27'34"
Karl Callsen
Ari 2h44'51" 20d53'59"
Karl Cushey
Aql 20h10'31" 9d3'49"
Karl D. Augenstein
Her 18h26'0" 12d29'49"
Karl D. Hayberg
Sgr 19h47'8" -23d5'24"
Karl Dale Pitchford
Ari 1h56'47" 12d58'13"
Karl David Doyle
Cep 22h24'24" 62d7'40"
Karl David Hainsworth
Uma 8h56'54" 64d20'17"
Karl Davies
Dra 15h5'48" 64d4'23"
Karl Derenthal
Ori 5h54'59" 6d55'35"
Karl Doherty Schroeder
Ori 5h22'51" 6d40'51"
Karl & Dori Kukawa
Uma 11h43'50" 57d19'6"
Karl E. Tenney
Sgr 19h30'56" -13d8'4"
Karl Echele
Uma 8h53'12" 60d47'34"
Karl F. Bayer Knights Brilliance
Aqr 23h37'12" -18d48'19"
Karl Fischer
Cep 2h10'55" 81d18'49"
Karl Frederick Fallenius
Dra 17h59'42" 62d41'15"
Karl Gaßmann
Ori 5h56'21" 12d32'31"
Karl & Georgina Sell
Cyg 20h13'28" 30d1'7"
Karl Haas
Her 17h49'21" 41d25'42"
Karl Hal Jensen
Her 16h38'18" 20d18'51"
Karl Heinz Breuer
Sco 17h31'12" -44d13'56"
Karl Heinz Ehrenmann
Cas 1h54'38" 64d34'12"
Karl Heinz Steindl
Uma 12h11'27" 53d13'21"
Karl Hermann Hutzler
Uma 9h57'36" 55d50'12"
Karl Hyslop 5.6.78 My Shining Light
Cru 12h8'26" -61d52'51"
Karl & Ivy - I Love You, Forever...
Ori 6h19'52" 19d20'37"
Karl Jason Bates
Cep 1h4'13" 84d29'41"
Karl John Delaney
Cyg 21h54'41" 46d58'48"
Karl Jordans
Ori 6h23'51" 15d48'50"
Karl Justin Packer
Leo 9h45'41" 32d47'7"
Karl K. Burdette (Paw Paw)
Cnc 8h35'34" 24d15'53"
Karl Karlson
Cas 1h3'31" 61d24'52"
Karl Katzenberger
Leo 11h15'40" 4d10'33"
Karl & Lisa Taylor-Love Everlasting
And 0h6'1" 38d11'8"
Karl loves Gem
Cyg 20h11'46" 38d0'49"
Karl M Koshlap and Anne Kressmann
Uma 11h20'15" 43d36'18"
Karl Martin Petermann, Sr.
Cyg 21h44'10" 41d20'32"
Karl Maurice Gallant Jr.
Her 17h11'7" 34d48'5"
Karl Mayrschofer
Uma 10h50'55" 44d50'46"
Karl & Michelle
Boo 14h21'25" 18d48'53"
Karl Miller
Per 3h30'49" 34d20'35"
Karl & Mom
Lyn 9h7'57" 39d28'4"
Karl Norbert Merz
Uma 12h35'31" 61d48'47"
Karl O. Ott
Uma 11h5'22" 37d6'8"
Karl P. Gutwalt
Lib 15h12'8" -5d51'24"
Karl Pfüller
Uma 9h28'1" 51d58'47"
Karl Pinard - My True North
Cnc 8h53'43" 13d36'37"
Karl Robert Kirschner
Uma 10h55'44" 47d46'58"
Karl S. Guers
Per 3h42'17" 48d22'44"
Karl Schenck and Jen Necelis
Uma 9h41'38" 63d41'32"
Karl "SeaWolf" Moyed
Leo 11h53'54" 25d48'54"
Karl Steiger *The Malfated*
Per 3h7'45" 52d50'29"
Karl Strugalla
Uma 8h43'33" 48d0'28"
Karl & Susan Swope
Cyg 20h0'53" 33d26'20"
Karl & Suzie Smith
Ori 5h36'0" -1d3'40"
Karl "The Star Of My Life"
Aqr 22h54'10" -5d51'36"
Karl Thomas Halko Jr.
Cap 21h32'58" -9d24'8"
Karl Ulrich Menges
Uma 11h49'47" 29d16'50"
Karl Underkoffler
Uma 11h55'19" 42d3'16"
Karl (Väti) Gnehm
Vul 21h8'5" 22d48'44"
Karl/Vera Ruppert
Uma 12h52'7" 53d53'56"
Karl & Vikki Zajac
Leo 10h40'9" 7d23'35"
Karl W. Hardy
Boo 14h54'32" 20d59'24"
Karl Waymon Beetz
Her 16h39'24" 33d56'4"
Karl Zentgraf
Uma 8h56'44" 63d14'13"
Karla
Sgr 18h29'12" -18d38'24"
Karla
Aqr 23h5'29" -5d3'41"
Karla
Aqr 22h2'57" -0d46'5"
Karla
Lyr 18h28'50" 37d46'6"
Karla
And 23h25'7" 47d6'20"
Karla
Com 12h45'4" 17d17'54"
Karla
Ari 2h34'29" 24d49'1"
Karla
Tau 5h47'57" 17d1'34"
Karla & Alex
Vir 14h12'29" -19d31'36"
Karla and Gary's True Love Star
Ori 5h26'56" 3d23'42"
Karla and Jamie
Lyr 19h12'18" 46d15'11"
Karla and Peter
Umi 15h32'34" 75d24'43"
Karla and Sean
Lyn 7h41'49" 38d54'58"
Karla Ann Overholt
Cas 23h52'57" 56d26'16"
Karla Anne Trifelitti
Sgr 19h8'4" -12d51'16"
Karla Benitez
Vir 12h22'36" -6d27'17"
Karla Carter
Sgr 17h50'26" -28d48'57"
Karla & Chris
Leo 10h48'21" 17d42'40"
Karla Dolby
Cru 12h11'17" -60d56'18"
Karla Dudley
Cyg 19h40'32" 29d43'0"
Karla Dunn
And 0h29'2" 45d26'19"
Karla Ellis
Vir 11h44'40" 9d27'0"
Karla Gleva
Gem 6h51'6" 22d35'44"
Karla Jane
And 0h46'37" 42d21'50"
Karla Jean
Lyr 18h38'47" 46d52'55"
Karla Juliane Renton
Lyn 9h7'38" 33d50'46"
Karla Kay
Uma 11h39'58" 52d13'51"
Karla Kaye
Lyn 6h35'15" 58d40'48"
Karla Kirby
Lmi 10h9'46" 34d41'22"
Karla Kristine Reuter
Cnc 8h47'49" 31d23'36"
Karla Lee
Com 13h13'39" 17d56'50"
Karla Louise
Sgr 18h18'23" -22d54'4"
Karla Manuel Scarlet Diego Rodrigo
Crb 15h48'29" 27d31'52"
Karla Marie Pollock
Tau 5h24'59" 22d18'30"
Karla May Taccad
Cam 6h31'58" 69d44'11"
Karla Noel
And 0h14'26" 43d13'7"
Karla Ricard
Cas 3h4'42" 64d43'34"
Karla Rose Lund
Lyr 18h39'10" 31d27'42"
Karla Ross
Mon 7h6'3" -7d25'10"
Karla Sherene
Oph 16h43'6" -0d50'11"
Karla Stewart "Star Of My Life"
Psc 0h11'21" -0d37'15"
Karla Veronica Bahamonde
Psc 1h36'0" 17d48'11"
Karla Wilson
Tau 5h21'34" 18d16'8"
KARLA Y TONY FOREV-ER
Lib 15h7'57" -6d59'59"
Karla Yazmín Cardenas Lopez
Leo 9h28'44" 26d1'11"
Karla z Mornsteinu-Zierotina
Leo 9h24'4" 10d58'18"
Karla's Precious Star
Gem 6h20'27" 18d1'17"
Karla's Star
Cas 0h22'47" 57d36'21"
Karla's Tranquility
Ori 6h8'29" 20d59'16"
Karla-Z
Lyn 7h53'56" 57d14'0"
Karl-Dietmar Cohnen
Uma 11h15'52" 63d56'49"
Karlea Rae Craft
And 1h15'51" 41d37'52"
Karlee
Aqr 22h45'49" 1d16'52"
Karlee & Frank
Ari 2h2'17" 23d55'7"
Karlee Kaye
Uma 11h37'25" 52d14'49"
Karlee Mika Kimura
Lib 14h51'32" -5d30'12"
Karleen Stephan Ballard
Ori 6h6'25" 13d52'45"
Karleen Thompson
Gem 6h48'57" 15d33'46"
Karleigh's Star
Uma 8h27'36" 63d32'56"
Karlen Dorsey Flagg
Tau 5h33'8" 20d37'36"
Karlen Evins
Ari 2h21'4" 24d0'4"
Karlene M. Reyes
Sco 16h42'8" -32d46'57"
Karlene Sneden
And 1h33'48" 45d39'22"
Karley Faith Hoagland
And 23h39'11" 49d52'6"
Karley Landry King
Ari 3h22'43" 22d4'26"
Karley Mummert
Tau 4h34'2" 24d31'51"
Karley Rasmussen
Vir 12h42'9" -7d50'13"
Karlforever
Uma 11h14'17" 46d58'48"
Karlfried Tusselmann
Uma 9h22'24" 49d47'47"
Karl-Günther Glocke
Uma 10h3'46" 68d31'58"
Karl-Heinz
Sco 16h36'7" -29d44'56"
Karl-Heinz Baar
Ori 6h24'13" 10d48'44"
Karl-Heinz Bauer
Ori 5h9'6" 0d16'54"
Karl-Heinz Hasselbach
Uma 9h39'27" 58d49'56"
Karl-Heinz Haubennestel
Uma 10h7'54" 53d6'11"
Karlheinz Löffelmann
Uma 10h51'40" 45d45'9"
Karl-Heinz Sick
Uma 10h6'34" 44d21'41"
Karli
Uma 9h36'57" 57d55'33"
Karli
Uma 10h28'0" 69d12'14"
Karli
Aqr 21h40'20" -0d44'49"
Karli Camp
Lyn 7h35'49" 36d59'44"
Karli Dana Fortunato
Cam 5h8'55" 64d51'30"
Karli Elene Baker
Sgr 18h24'42" -34d7'24"
Karli Elizabeth Lyn Wandless
Uma 10h16'20" 57d7'11"
"Karli" Kühl
Uma 9h35'34" 50d30'15"
Karli Marie Nichols, Our Princess
And 0h46'42" 43d46'11"
Karli Rae Fannin
Sgr 19h11'37" -22d8'53"
Karlia Ann
Cas 22h59'53" 55d0'46"
Karlian
Mon 7h8'40" -0d25'44"
Karli-Anna's Diamond in the Sky
Cru 12h29'16" -60d14'55"
Karliayn Hittle
Gem 7h40'38" 16d16'3"
Karlie
Aqr 22h19'14" 0d41'33"
Karlie Alexis Walsh
And 23h3'2" 39d26'51"
Karlie Breeden
And 1h42'40" 38d26'3"
Karlie E Tamura
And 0h5'55" 45d6'13"
Karlie Jo Kaska
Lib 15h23'18" -5d53'5"
Karlie King
Aur 5h8'28" 46d58'5"
Karlie Lois Garner
Leo 9h46'5" 28d50'17"
Karlie Marie
Cnc 7h59'0" 13d17'57"
Karlie Scott Kersey
Cnc 8h40'11" 10d47'58"
Karlie Shea Miller
Gem 7h29'38" 15d8'38"
Karlie Stenerson
Cap 21h31'42" -12d34'54"
Karlie Vale
Lyn 9h23'24" 40d25'17"
Karlina A. Escobar
Gem 7h9'25" 15d14'54"
Karlita Ortiz
Psc 1h23'34" 21d8'36"
Karlitto
Uma 10h44'49" 57d6'53"
KarlMathiasBoyd
Gem 7h38'42" 33d54'27"
KarloJoha & Familie
Uma 8h44'18" 48d14'23"
Karlos
Cep 21h30'2" 59d2'59"
Karlotta Allen
Lyn 7h3'15" 51d0'1"
Karlow Yervant Krikor
Sgr 18h11'22" -26d23'47"
Karl's Two Bells
Aqr 22h54'57" -16d7'59"
Karls, Anton Theodor
Lib 15h53'16" -4d7'19"
Karlsson, Ylva Elisabeth
Uma 9h13'14" 51d51'30"
KarLu365
Lyn 7h45'14" 42d12'57"
Karl-Udo
Lib 15h38'47" -24d15'16"
Karly
And 23h12'39" 43d59'56"
Karly and Michael Forever
Cyg 19h40'23" 33d55'51"
Karly "BG" Arong
And 0h25'50" 31d43'1"
Karly Frances Kolden
Cnc 8h41'16" 17d25'57"
Karly Gay
Lyn 7h44'13" 35d56'18"
Karly Jade Torgerson
Cnc 8h46'19" 31d4'46"
Karly Jo
And 1h8'31" 42d52'13"
Karly Kaj
Leo 11h45'24" 26d2'5"
Karly Marie Hunter
Uma 8h58'24" 67d33'55"
Karly Michelle Marie Frame
Her 17h15'59" 34d42'14"
Karly Rich
Cyg 21h26'12" 39d59'41"
Karly Sheehan
And 1h21'36" 48d52'15"
Karlyann Kayden
Sco 17h52'36" -30d18'49"
Karlye Fader
Aql 19h1'39" 17d20'34"
Karlye Ghrist
Lib 15h7'34" -3d4'52"
Karlye Green
Cyg 20h38'51" 45d55'8"
Karlyn & Alan
Cyg 19h56'59" 32d28'45"
Karlyn Christina Warley
And 23h11'45" 47d2'35"
Kar-Lyn Grace
Uma 10h3'2" 48d20'37"
Karlyn Ivy
And 23h31'41" 45d47'1"
Karlyn Koh
Del 20h41'18" 18d4'41"
Karm
Leo 11h0'26" 2d39'52"
Karma
Tau 3h58'41" 5d49'5"
Karma
And 23h7'2" 48d26'19"
Karma
Per 4h23'29" 52d39'43"
Karma
Umi 14h4'4" 77d22'21"
Karma
Cma 7h24'47" -24d30'3"
Karma Igou
Gem 7h53'25" 18d50'31"
Karma Kimball
Mon 8h6'46" -0d54'44"
Karma Leigh Allen
Gem 6h52'44" 16d54'28"
KaRmA St. LoU
Sge 19h53'52" 18d48'6"
Karma, Anne, & Blake Bowers
Lyn 6h50'37" 61d47'21"
Karma's Joy & Blue Passion
Cnc 8h51'35" 28d40'45"
Karmella
Ori 5h30'36" 7d25'12"
Karmen Ehman
Mon 6h34'23" 3d18'15"
Karmen Faith Stegall
Psc 0h55'5" 17d17'3"
Karmen Isayan
Uma 10h45'33" 53d11'15"
Karmen Krystal Hampton
Uma 10h27'27" 40d38'2"
Karmin Joy
Ori 5h22'1" 6d38'3"
Karn Duke
Ori 5h36'5" 11d4'51"
Karn Ethan Hollis
Cep 21h38'8" 64d17'32"
Karna Dene Valen
Uma 9h55'43" 56d47'20"
Karnen
Umi 15h16'57" 70d45'35"
Karnolt, Michael
Uma 8h15'2" 72d40'58"
Karol Ann
And 1h19'57" 46d57'19"
Karol Ann & Don
Sge 19h40'9" 18d55'2"
Karol Anne Ilagan
Psc 0h6'33" 6d16'15"
Karolien & Christophe
Cyg 19h43'32" 33d57'24"
Karolina
Ari 2h37'43" 22d9'22"
Karolina Aleksiejczuk
Cap 21h44'53" -10d39'7"
Karolina and Jeff's Star
Sco 16h59'35" -38d50'47"
Karolina Anna Warzocha
Ori 6h22'7" 11d4'22"
Karolina & Armin
Leo 9h45'20" 32d6'47"
Karolina Krakowiak
Leo 10h8'31" 23d3'27"
Karolina Stokowska
And 23h41'37" 36d46'20"
KAROLINARAY
And 2h8'3" 42d31'37"
Karoline
Leo 11h40'8" 22d18'56"
Karolinka Sladki
Psc 23h55'32" 1d52'53"
Karolinka Zema
Vir 12h31'15" 10d43'50"
Karolyn
Sgr 17h53'5" -17d46'58"
Karolyn Elise Niskanen
Vir 13h54'37" 6d12'28"
Karolyn Francis Buttle
Cyg 21h58'53" 55d1'3"
Karolyn Leslie Ed Whitson
Uma 8h43'40" 59d36'25"
Karolyn Marie Barsamian
Vir 13h16'58" 5d10'19"
Karoo
Cma 7h14'10" -17d36'48"
Karoon
Dra 15h17'49" 59d8'51"
Karra Nance
And 0h27'31" 44d51'10"
Karrah Monica
Lyn 7h0'14" 48d42'42"
Karri Ann
Cyg 20h53'17" 54d51'24"
Karri Keller
Cam 3h49'40" 56d13'4"
Karrie
Uma 14h26'17" 62d16'2"
Karrie E Waterman "princess"
Gem 7h39'48" 32d6'13"
Karrie Ellen Hargreaves
Gem 6h29'11" 16d15'21"
Karrie Hughes
And 23h53'52" 44d1'39"
Karrie & London A&F
Aqr 21h0'23" 0d30'29"
Karrie Louise Bender
Cnc 8h36'17" 13d31'53"
Karrie Lynn Johnson
And 0h30'13" 41d47'32"
Karrie Pierce
Leo 11h5'35" 17d46'31"
Karrie-Marie
Tau 3h31'14" 9d42'56"
Karrieta W. Schreiber
Cyg 21h10'17" 48d25'48"
KARRIN
Pav 20h7'20" -71d34'45"
Karrin Patricia Huynh
And 2h17'55" 48d47'52"
Karron Hart Miller
Uma 9h48'25" 54d1'41"
Karsen Amy Meerkreebs
Mon 6h52'57" -0d55'19"
Karsjens, Hildegard
Uma 8h37'56" 63d59'22"
Karsten Reker
Uma 9h30'34" 57d51'55"
Karsten van den Steenoven
And 23h53'12" 43d18'26"
Karsten Wendt
Ori 6h21'21" 10d14'21"
Karsyn Brooke Altman
Lib 14h59'35" -17d40'48"
Karsyn Marie Turner
Uma 11h50'30" 59d44'22"
Karsyn Taylor
Crb 16h12'13" 34d14'16"
Karter Julian
Cnc 8h41'15" 24d9'25"
Karthik Mani Patibandla
Umi 15h49'4" 80d25'0"
Karthik Sekhar
Cnc 8h55'5" 25d14'6"
Karun
Umi 14h16'4" 78d7'9"
Karuna Nandkumar
Vel 9h55'30" -43d23'17"
Kary
Gem 7h23'10" 32d57'10"
Kary And MaryEllen
Cyg 19h37'38" 31d53'46"
Kary Groom
Cas 2h7'4" 62d30'40"
Kary Kaltenbronn
Mon 6h46'28" -0d5'6"
Kary Knudslien
Psc 1h5'41" 7d2'46"
Kary Lynn
Tau 3h47'46" 28d14'26"
Kary Taylor
And 0h37'59" 45d8'40"
Karye Denise Ghaderpanah
Sco 17h33'19" -42d29'39"

Kate & Scott's Wedding Love Star
Cru 12h37'12" -57d12'54"
Kate Seibert
Leo 9h54'19" 21d17'49"
Kate Sharkey Christopher
Leo 11h45'23" 20d30'3"
Kate Shining Star of David
Vir 13h27'40" -9d13'33"
Kate Shiver
Tau 4h40'37" 28d3'56"
Kate Simons
Cas 2h19'2" 68d47'48"
Kate Sinead McKenna
Cnc 9h2'11" 29d49'52"
Kate Sloan
Mon 6h53'18" -5d45'31"
Kate Sloane Lieberman - K8 the K9
Sgr 19h48'47" -15d23'33"
Kate Steciw
Gem 7h15'11" 20d40'59"
Kate Stephany
Uma 9h31'19" 67d3'16"
Kate Stoddart
Cra 18h41'46" -39d9'12"
Kate Stromberg
Lib 15h21'14" -14d19'28"
Kate "Sunny Jim" Boyce
Leo 10h57'34" 14d3'3"
Kate Superstar
Peg 22h18'31" 17d37'14"
Kate T
And 23h54'37" 48d40'37"
Kate Taylor
Aqr 22h2'47" -14d22'11"
Kate Taylor-Ily
Lyr 18h32'33" 38d43'22"
Kate The Baby Doll
Psc 0h51'18" 10d45'40"
Kate "The Great" Donovan 10/18/97
Lib 14h38'46" -9d23'30"
Kate Tomkinson
And 23h34'44" 46d28'43"
Kate Turner
Cas 1h58'54" 60d59'48"
Kate Vanderbach
Gem 6h18'9" 25d29'27"
Kate Venz
Sco 16h16'51" -13d20'53"
Kate Walldorf
Tau 4h15'20" 7d10'8"
Kate Watts
Cru 11h56'51" -63d27'39"
Kate Winberry
Sco 16h13'44" -25d17'38"
Kate Winton
Psc 0h42'8" 17d33'38"
Kate With Melanie
Cnc 8h34'8" 32d40'50"
Kate Zajicek, D.C.
Crb 16h3'4" 34d59'42"
Kate, My Princess
And 1h56'2" 45d39'22"
Kate58
Sco 17h29'31" -41d54'45"
Kateczar
Cyg 19h39'16" 33d50'56"
KateDan
Ori 6h4'20" 12d37'58"
Katee Ann
Per 3h20'38" 32d34'3"
Kateen's Encore
Uma 8h34'35" 72d1'5"
Katelee Marie Hart
Lib 15h5'58" -17d56'13"
Katelin Amber Lenczuk
Cas 1h22'28" 56d46'23"
Katelin Ann Sweeny
Dra 16h51'28" 68d32'17"
Katelin Elizabeth
And 1h44'55" 43d5'41"
Katelin Facer
Del 20h45'34" 14d29'47"
Katelin Nicole
Cnc 8h13'59" 11d43'26"
Katelin Radcliff
Cas 1h34'15" 57d19'2"
Katelin's Kisses
Psc 1h6'30" 14d47'46"
Katelyn
Tau 4h16'42" 13d10'42"
Katelyn
Cnc 8h27'22" 26d27'18"
Katelyn
Peg 22h39'31" 29d38'24"
Katelyn
Cas 0h52'49" 50d40'38"
Katelyn
And 1h37'3" 47d26'40"
Katelyn
Gem 7h1'54" 35d11'0"
Katelyn
Her 16h41'35" 35d5'11"
Katelyn
Lmi 10h29'41" 36d18'15"
Katelyn
Sco 17h31'26" -42d15'16"
Katelyn A. Provencher
And 23h55'59" 33d20'59"
Katelyn Alyssa Jeffcoat
Cnc 9h3'46" 32d31'24"
Katelyn Anastasia
And 1h37'57" 47d34'0"
Katelyn Anastasia's Star
Sgr 19h0'40" -17d15'40"
Katelyn and John
Cyg 19h42'2" 38d50'10"
Katelyn Anh Vo
Vir 13h7'55" -1d48'54"
Katelyn Ann Bolt
Tau 4h24'39" 19d17'15"
Katelyn Ann Burdette
Ori 5h54'7" 6d49'15"
Katelyn Ann Nowacki
Psc 1h32'11" 4d31'30"
Katelyn Anne Campbell
Gem 7h28'2" 24d36'32"
Katelyn Arlene
Uma 8h52'56" 59d46'59"
Katelyn Blaise Steeley
And 0h58'28" 44d0'24"
Katelyn Breanna Tener
And 0h11'19" 33d10'7"
Katelyn Bussard
Cnc 8h39'37" 24d44'23"
Katelyn Casper
Cap 20h42'34" -22d21'14"
Katelyn Chambers
And 23h48'28" 45d52'25"
Katelyn Charlotte Richardson
Uma 11h24'44" 63d49'46"
Katelyn D. Lopez
Aqr 20h46'23" -11d57'20"
Katelyn Danielle
Uma 14h25'32" 56d38'56"
Katelyn Deanna Wood
Crb 15h52'17" 25d53'19"
Katelyn Dieterle
And 1h47'47" 39d7'52"
Katelyn Elizabeth
Ari 2h3'14" 18d22'50"
Katelyn Elizabeth
Sgr 18h25'28" -16d54'41"
Katelyn Elizabeth Conner
And 2h11'56" 39d32'0"
Katelyn Elizabeth DeDea
Gem 6h40'38" 31d9'32"
Katelyn Elizabeth Devries
And 2h23'10" 47d13'30"
Katelyn Elizabeth Milheim
Dra 18h23'17" 74d22'41"
Katelyn Elizabeth Thomas
Cnc 8h34'38" 30d56'47"
Katelyn Elizabeth Wolfram
Cap 21h46'57" -9d0'40"
Katelyn Erica Lilly
Lib 15h52'3" -11d43'33"
Katelyn Faith Dayana Stamberger
Leo 11h42'26" 15d50'7"
Katelyn Fiona Rose
Cru 11h56'27" -58d42'14"
Katelyn Frank
And 0h33'52" 34d1'10"
Katelyn Gall
Cru 11h57'29" -61d59'55"
Katelyn Grace Coombs
Sgr 18h54'40" -27d1'45"
Katelyn Grace Gillian Beacon
And 23h16'59" 35d26'19"
Katelyn Grace Ingle
Leo 11h43'11" 25d55'22"
Katelyn Hartigan
Cnc 8h18'53" 20d6'28"
Katelyn Herzog
Cas 0h39'44" 58d4'22"
Katelyn Ivy
Umi 16h5'45" 83d25'49"
Katelyn Jane Vander Heide
And 1h47'57" 40d44'7"
Katelyn Jean King
Vir 14h38'27" -2d14'32"
Katelyn/Jennifer - My Sister's Star
And 1h13'54" 36d31'21"
Katelyn Joy
Vir 13h52'38" -20d43'57"
Katelyn Justine Forgue
And 0h58'42" 46d8'46"
Katelyn Lee Burley
And 23h21'5" 47d43'24"
Katelyn Lick
Crb 15h34'45" 37d27'27"
Katelyn M. Pavlik
Srp 18h18'5" -0d6'56"
Katelyn Mackenzie
Lmi 10h46'26" 27d2'39"
Katelyn Mae Williams
Cma 7h25'9" -24d30'24"
Katelyn Marie
Uma 12h9'29" 55d25'27"
Katelyn Marie
Vir 11h42'2" 7d18'47"
Katelyn Marie
And 1h15'27" 45d37'29"
Katelyn Marie Arone
Tau 3h32'16" 21d34'43"
Katelyn Marie Blaszkowski
Leo 11h38'35" 20d24'48"
Katelyn Marie Kent
And 0h28'48" 32d27'28"
Katelyn Mary Lewis Kasseroler
And 0h45'15" 37d31'58"
Katelyn Mary Takes
Cas 1h23'4" 62d47'0"
Katelyn McElearney
And 2h13'26" 46d34'18"
Katelyn Miette LaMorte
Psc 0h16'19" 8d27'56"
Katelyn Nicole Eddis
Lib 15h50'49" -16d34'48"
Katelyn Nicole Stengle
Aqr 22h42'27" 1d17'26"
Katelyn Rebekah vonDiezelski
Lib 15h38'47" -28d54'12"
Katelyn Renee Bell
Leo 10h16'52" 11d28'52"
Katelyn Reynolds
And 0h43'50" 42d3'1"
Katelyn Rose
Crb 15h36'1" 35d3'15"
Katelyn Rose Humenik
And 23h29'10" 48d1'31"
Katelyn Rose Mauriello
Leo 10h55'27" 17d51'50"
Katelyn Rose Smith
Dra 16h47'30" 56d19'4"
Katelyn Schwartz
Cas 23h34'48" 52d9'6"
Katelyn Segarra
Her 18h1'52" 25d21'45"
Katelyn Snell (Cimaria)
Tau 5h14'41" 24d28'37"
Katelyn Williams
And 23h9'29" 50d40'11"
Katelyn With A "K"
Lib 15h26'22" -5d44'56"
Katelyn Yoho
And 1h16'21" 45d44'56"
Katelynn
Lyn 8h43'15" 35d37'46"
Katelynn Ann Castaldo
Lyn 7h7'9" 51d4'35"
Katelynn Christine Guccione
Lyr 18h57'19" 26d6'27"
Katelynn E. Kelliher
Lib 15h20'50" -26d48'28"
Katelynn Faye
Lyr 18h37'38" 34d24'34"
Katelynn Marie Stout
Aqr 21h12'19" -9d12'52"
Katelynn Moore
Vir 14h17'18" -16d44'59"
Katelynn Volpigno
Cnc 8h43'33" 32d8'5"
Katelynne Marie Kratz
Lyn 9h22'12" 41d7'13"
Katelynn's Star
Cyg 19h39'56" 28d18'41"
Katelyn's 1Luv
Aqr 21h24'51" -7d48'16"
Katelyn's Faith
Cas 0h49'18" 50d39'4"
Katelyn's Punkin Girl Star
Aqr 23h24'21" -13d9'27"
Katelyn's Star of Serenity
Cas 0h35'3" 53d54'8"
Katemilly Alvaravo
And 1h25'31" 47d39'22"
"Kater Bug" Kaylee E.M. Gaston
And 0h27'37" 29d12'5"
Katerchen
Ori 6h25'14" 10d25'48"
Katere
Ari 2h46'44" 25d52'38"
KATERI
Cnc 8h31'0" 9d27'12"
Kateri
Dra 15h1'50" 62d46'8"
Kateri Ann
And 23h54'57" 45d28'51"
Kateri Emilia Estrella
Mon 7h10'46" -1d34'16"
Kateri Wendt
Cap 20h36'36" -20d46'20"
Katerina
Uma 9h20'2" 65d44'30"
Katerina
Cas 22h58'29" 58d23'16"
Katerina
Cnc 8h31'27" 9d42'56"
Katerina
Aqr 22h16'42" 1d29'48"
Katerina
And 0h56'16" 23d16'12"
Katerina
Gem 6h51'58" 16d53'1"
Katerina
Tau 5h36'42" 28d16'17"
Katerina Bashurina
And 23h14'21" 52d32'19"
Katerina Demetriou
And 2h29'58" 38d45'22"
Katerina E. Perdue "TODD"
Cyg 19h47'33" 41d27'21"
Katerina Gavrielidou
And 1h57'39" 40d55'35"
Katerina Hajjar
Aqr 22h7'49" -12d50'56"
Katerina Kreyndlin
Cap 21h3'19" -17d15'35"
Katerina Mandreka
Uma 11h3'31" 36d33'58"
Katerina Margaret Sitaras
Tau 5h47'18" 25d39'48"
Katerina Petropoulou
Uma 9h3'17" 72d18'9"
Katerina Thallasomatousa
Dra 18h38'22" 52d31'45"
Katerina Tsemberlidou
Uma 9h36'9" 46d24'28"
Katerinka
Vir 12h1'30" 8d14'1"
KateRose Teitelbaum
And 23h41'46" 47d29'9"
KateRuby
Uma 11h21'23" 46d59'24"
KATES
Ori 6h4'6" -0d37'4"
Kate's Absolute Devotion
Aqr 22h31'57" 1d21'22"
Kate's Birthday Star
Vir 11h49'5" 10d10'6"
Kates Brightest Star In My Night
Uma 9h38'23" 46d38'35"
Kate's Daddy's Star
Sco 17h56'26" -41d3'23"
Kate's Dream
Ori 5h25'24" -9d17'27"
Kate's Heart's Content
Cyg 20h56'45" 30d28'16"
Kate's Luminary #1
And 2h32'15" 45d37'39"
Kate's Star
Cyg 19h59'14" 30d37'26"
Kate's Star
And 1h16'13" 34d9'18"
Kate's Star
Ari 3h20'3" 29d6'22"
Kate's Star
Leo 9h49'59" 24d53'45"
Kate's Star
Leo 9h37'45" 22d3'3"
Kate's Wish
Vir 12h51'42" 8d16'13"
Kate's Wishing Star
Cnc 9h2'6" 9d3'49"
Kates World
And 23h12'51" 52d8'43"
Kates1918
Cap 21h29'18" -19d34'49"
Katesand
Lyn 7h46'33" 43d5'6"
Katetopia
Mon 6h46'6" -0d26'4"
Katey
Aqr 22h33'6" -22d14'25"
Katey
And 0h23'35" 35d27'26"
Katey
And 23h10'11" 50d19'59"
Katey
And 23h33'9" 38d24'0"
Katey B - Baby Doll
Psc 1h53'11" 2d59'14"
Katey Blair
Leo 10h15'44" 13d36'47"
Katey Deaton
Ari 2h13'10" 24d3'58"
Katey MAK
Ori 5h50'38" 3d17'3"
Katey Marie
Aqr 21h23'5" -7d36'47"
Katey Mortellaro
Ori 6h7'22" 18d41'36"
Katey Scarlett Biggam
Cas 1h5'10" 62d23'34"
Kath 60
Cas 1h26'38" 63d52'37"
Kath and Jer's Love Star
Uma 11h17'56" 58d3'4"
Kath & Barry's 40th Anniversary Star
Cru 12h29'34" -59d21'46"
Kath Blenkinsop
Gem 7h4'27" 11d14'29"
Kath Conroy
Cas 23h58'13" 50d0'9"
Kath & Des' Celebration of Hearts
Col 6h24'50" -35d34'0"
Kath forever shining
Vir 13h15'44" 2d32'12"
Kath Groves
Cas 1h26'16" 56d8'48"
Kath Walshe
Vel 8h51'5" -46d19'33"
Kath, you're always in our thoughts
Lyr 19h17'33" 28d7'39"
Kathaleen 47
Ori 5h7'1" 15d10'59"
Kathalena
Uma 12h21'46" 58d50'53"
Kathali
Gem 6h52'36" 21d42'35"
Kathandi
Uma 13h35'25" 59d9'8"
Kathandra
Lyn 7h1'30" 57d57'55"
Katharina
Lyn 7h33'0" 56d55'56"
Katharina
Ori 6h18'19" 10d28'59"
Katharina
Cas 0h56'28" 56d22'55"
Katharina
Uma 10h11'50" 51d31'56"
Katharina 64
Cas 1h3'58" 54d0'37"
Katharina Abler
Ori 6h23'2" 16d6'0"
Katharina & Andreas Wiehrdt
Lup 15h37'44" -40d16'52"
Katharina Emelien
Umi 9h5'19" 88d5'21"
Katharina Krieser
Uma 10h32'36" 48d52'27"
Katharina Kristensen
Ari 3h3'51" 29d35'38"
Katharina Obertrifter
Uma 9h28'8" 56d34'54"
Katharina "Pokey" Laus
Cas 1h34'52" 57d29'44"
Katharina Tschiggerl
Aqr 21h19'28" -10d59'7"
Katharina Weber
Cep 0h26'31" 67d53'58"
Katharina-Ingrid 13.11.1998
Uma 10h5'11" 64d21'48"
Katharine Alexis McGorty
Uma 9h19'50" 69d53'44"
Katharine and David's Wedding Star
Lyr 18h27'3" 32d57'34"
Katharine and Kenneth
Cyg 19h50'43" 30d18'7"
Katharine Andrews Maxwell
Cas 1h33'42" 58d17'37"
Katharine Anne Pelton
Vir 14h6'59" -7d52'2"
Katharine Bailey Merritt
And 0h12'19" 46d32'24"
Katharine Berg
Lyr 18h26'49" 38d13'25"
Katharine Bridget Comiskey
Uma 11h15'51" 50d32'46"
Katharine Cooper 30
Aqr 23h22'20" -17d58'27"
Katharine Douglass Hesmer
Ari 2h11'47" 23d41'12"
Katharine Drexel Brown
Sco 17h16'1" -30d44'47"
Katharine Fox
Tau 4h32'29" 11d44'21"
Katharine Harrison
Gem 7h40'19" 20d21'30"
Katharine J. Hahn
Cyg 20h56'52" 36d10'26"
Katharine K
And 23h16'5" 48d27'37"
Katharine Kindervatter
Gem 7h24'36" 24d47'35"
Katharine Lily Haddon
Aqr 22h19'6" -24d10'57"
Katharine Lynn Redmond Angeltoall
And 0h35'34" 41d3'36"
Katharine Mae Rhoten
And 1h21'24" 44d46'36"
Katharine Marie
Per 4h50'2" 46d44'30"
KATHARINE MARIE McCANN
Mon 6h49'55" 7d56'52"
Katharine Moyer
Cap 20h45'10" -21d11'13"
Katharine Nelson
Ser 15h30'47" 17d46'53"
Katharine Rhodes
Lib 15h44'21" -9d36'36"
Katharine Stafford
Vir 12h30'45" -11d21'15"
Katharine "Star Light" Quinn
And 1h42'11" 45d46'26"
Katharine Steele
And 0h45'58" 25d22'47"
Katharine Thalberg Stirling
Lyn 7h29'17" 49d22'26"
Katharine Wolfson
And 0h13'7" 33d38'33"
Katharine Wolfson
Psc 23h29'4" 3d32'0"
Katharine's Star
Ori 5h47'44" 6d44'48"
Katharine's Star
Gem 7h17'20" 14d36'47"
Katharos
Cnc 8h3'22" 6d40'25"
Katharos
Cap 21h37'50" -11d13'18"
Katharos Lupa
Sgr 18h6'33" -27d43'48"
Kathe Diamond
And 2h15'2" 37d33'29"
Kathee Grace
Cap 20h7'20" -17d39'47"
Katheen Karen Mcshane Larson
Psc 0h26'9" 16d0'54"
Katheet
Cnc 8h43'14" 8d11'52"
Katherene Johnson Latham
Lib 15h4'0" -25d29'23"
Katherin
Psc 1h29'4" 19d59'1"
Katherina Jalbert
Cas 23h47'45" 54d9'23"
Katherina S. Avanessova
Gem 6h45'15" 22d1'42"
Katherine
Boo 14h43'50" 17d32'0"
Katherine
Psc 1h23'51" 27d24'5"
Katherine
Ari 3h14'33" 27d19'13"
Katherine
Peg 22h25'27" 11d14'32"
Katherine
Tau 4h40'10" 17d12'24"
Katherine
Leo 11h34'10" 14d6'5"
katherine
Cnc 8h12'39" 6d38'38"
Katherine
Leo 11h25'41" 3d30'57"
Katherine
And 1h27'25" 39d24'33"
Katherine
And 2h18'20" 43d1'10"
Katherine
Lyr 18h50'3" 32d22'42"
Katherine
And 0h35'4" 45d51'40"
Katherine
Uma 11h52'7" 49d12'16"
Katherine
Uma 14h4'59" 57d20'4"
Katherine
Lyn 7h17'19" 59d12'41"
Katherine
Cam 7h2'14" 72d36'47"
Katherine
Sgr 19h19'34" -21d47'24"
Katherine
Sco 16h26'56" -18d42'6"
Katherine
Aqr 23h6'13" -6d41'5"
Katherine
Aqr 22h20'50" -24d17'23"
Katherine
Sco 17h25'42" -41d56'0"
Katherine A. Button
And 23h21'26" 41d16'55"
Katherine A. Reynolds
Aqr 20h49'51" -13d57'4"
Katherine Adams Booth
Psc 0h34'24" 9d1'27"
Katherine Adams, Nick's Angel
And 1h26'26" 40d42'31"
Katherine Adele Cornett
Leo 10h43'49" 18d13'35"
Katherine Alexandria Whitenack
And 23h47'40" 45d18'58"
Katherine Alexis Tolompoiko
Ari 2h16'36" 22d49'59"
Katherine Alfredo
Pho 0h57'53" -44d19'13"
Katherine Alice Matthews
And 0h44'7" 46d19'19"
Katherine Alis Magarian
Aqr 21h7'45" 0d51'0"
Katherine Allison Sipple
Cas 1h15'18" 58d56'54"
Katherine Allison Thompson's Star
Umi 15h11'7" 70d18'56"
Katherine Anastacia Murphy
Aqr 20h58'10" -10d38'1"
Katherine Anastasia Jones
Uma 13h30'54" 53d13'40"
Katherine and Brad's Star
Uma 10h34'52" 70d41'12"
Katherine and Jed McLaughlin
Leo 11h28'39" 9d56'42"
Katherine and Joseph Williams
Uma 9h8'4" 56d43'51"
Katherine and Kenneth Coleman
Uma 9h13'22" 63d36'3"
Katherine and Michael Visconti
Cyg 21h34'44" 39d41'9"
Katherine Ann
Tau 5h9'48" 23d47'33"
Katherine Ann Barone
Ari 3h21'51" 30d26'18"
Katherine Ann DeRouen
Uma 10h54'29" 49d17'32"
Katherine Ann Harden
Sgr 18h26'58" -18d8'18"
Katherine Ann Joyce
And 23h21'27" 48d51'53"
Katherine Ann Leitch
Leo 9h48'54" 27d15'20"
Katherine Ann Obrecht (Kate)
Cap 20h37'48" -13d34'14"
Katherine Ann Pistillo
Cas 1h41'51" 60d47'40"
Katherine Ann Riley
Uma 9h20'59" 47d29'15"
Katherine Ann Shaub
Cnc 8h41'30" 15d49'49"
Katherine Ann Smith
Per 2h47'24" 41d14'42"
Katherine Ann Torres Miranda
And 0h51'30" 45d26'11"
Katherine Ann Tranberg
Sco 16h10'38" -9d44'10"
Katherine Annabel Ellis
And 1h23'11" 44d24'6"
Katherine Anne
Uma 11h50'38" 49d48'20"
Katherine Anne D'Aquila
Vir 13h58'54" -9d36'38"
Katherine Anne Dineley
Ori 6h1'40" 19d9'38"
Katherine Anne Faith Busiek
And 0h11'24" 39d55'26"
Katherine Anne Falula St.Pierre
Leo 11h32'28" 20d54'34"
Katherine Anne Fernandez
Lyn 7h13'20" 55d38'54"
Katherine Anne Janice Kane
Psc 23h51'8" -0d13'33"
Katherine Anne Lunde
And 0h35'52" 45d2'40"
Katherine Anne Place
Cas 1h41'25" 63d4'40"
Katherine Anne Reynolds
Her 17h51'16" 46d17'48"
Katherine Anne Savage
Cam 3h26'5" 61d54'21"
Katherine Anne Smith
Lyn 7h35'15" 49d3'32"
Katherine Anne Tannahill
Sco 16h55'14" -42d1'37"
Katherine Anne Young
Ori 5h54'54" 13d2'8"
Katherine Ansley
Sgr 18h10'6" -31d50'8"
Katherine Araceli Urrutia
Psc 23h32'5" 1d48'29"
Katherine Aubrey MacLeod
Tau 5h27'22" 17d0'58"
Katherine Baluha
Vir 13h44'6" -20d12'29"
Katherine Beattie
Cas 0h17'49" 47d28'18"
Katherine Beecher (Tassie)
Vir 12h34'21" -1d10'38"
Katherine Belle
Gem 7h11'6" 17d53'6"
Katherine Berdecia
Vir 13h42'50" -17d5'17"
Katherine Block Wolpert
Cas 0h11'10" 53d33'20"
Katherine Bonacorsa
Leo 9h31'5" 27d30'7"
Katherine Boone Cody (Kitty)
Vir 13h31'52" 5d45'51"
Katherine Breedlove
Aqr 21h7'58" -7d17'19"
Katherine Brooke
And 23h29'25" 47d52'33"
Katherine C. Ryan
Lib 14h30'12" -10d46'27"
Katherine Cadwallader Douglass
Sco 17h47'26" -40d14'2"
Katherine Cerasoli
And 1h28'11" 45d30'54"
Katherine Cheng
Com 12h28'1" 28d28'55"
Katherine Chesley
Uma 10h54'45" 58d21'46"
Katherine Church
Gem 7h13'15" 32d26'25"
Katherine Claire Gautreaux
Leo 11h52'47" 22d9'25"
Katherine Claire Gayeski
Lib 15h20'25" -4d47'0"
Katherine Clare Bail
Aqr 23h35'8" -12d56'11"
Katherine Coletta Rospert
Uma 12h0'43" 53d10'39"
Katherine Cotton Elizabeth Hammond
And 0h10'59" 27d48'33"
Katherine Danene
Cnc 8h19'19" 23d57'5"
Katherine D'Avanzo
And 2h34'33" 37d36'41"
Katherine David
Gem 7h22'21" 26d7'31"
Katherine Davidson
Lib 15h37'21" -10d42'9"
Katherine Davis
Oph 17h7'42" -0d15'44"
Katherine Dawne Dunn
Psc 0h15'45" 12d8'30"
Katherine Dean
And 23h1'0" 48d9'0"
Katherine Deleon
Vir 14h27'52" -0d39'49"
Katherine Desiree' Wingate
Ari 2h18'51" 13d50'12"
Katherine DeWinter
Lib 15h47'57" -5d23'21"
Katherine DeWitt Baker
Gem 7h33'41" 23d29'12"
Katherine Diane
Cas 1h38'14" 69d10'29"
Katherine Dickson Baxter
Cas 1h42'35" 67d59'47"
Katherine DiRenzo
Psc 0h40'7" 7d50'12"

Katherine Dodds Cas 0h43'10" 56d48'28"
Katherine E. "Kathy" Jobin Cas 1h38'17" 68d42'41"
Katherine E. Kloss Gem 7h8'48" 34d14'53"
Katherine E. Landau Crb 15h43'56" 27d50'13"
Katherine E. Rodack Gem 6h49'57" 34d5'26"
Katherine Elaine Meadows And 0h28'37" 28d47'56"
Katherine Elizabeth Lmi 10h26'51" 28d27'46"
Katherine Elizabeth Vir 12h35'15" 4d43'23"
Katherine Elizabeth Cnc 8h9'32" 13d21'13"
Katherine Elizabeth Uma 9h22'32" 43d48'22"
Katherine Elizabeth And 1h36'22" 41d31'28"
Katherine Elizabeth Abbey Sgr 18h7'10" -16d48'47"
Katherine Elizabeth Bridget Beebie Cnc 8h47'50" 32d31'34"
Katherine Elizabeth Ernstrom Leo 11h35'49" 24d10'48"
Katherine Elizabeth Foster Vir 12h9'33" -0d6'9"
Katherine Elizabeth Gessner Sco 16h47'32" -35d15'40"
Katherine Elizabeth Hebert And 2h36'29" 46d5'59"
Katherine Elizabeth Henry Aqr 20h41'10" -7d22'52"
Katherine Elizabeth Higgins Cnc 8h46'42" 32d11'39"
Katherine Elizabeth Hilliard Psc 1h20'20" 16d58'53"
Katherine Elizabeth Hovey Peg 21h49'34" 7d38'42"
Katherine Elizabeth Johnson Cyg 19h52'37" 33d14'32"
Katherine Elizabeth Justice Sgr 19h15'4" -13d43'53"
Katherine Elizabeth Khais Sco 16h8'54" -31d19'29"
Katherine Elizabeth Miele Leo 9h59'26" 12d55'25"
Katherine Elizabeth Mumpower Gem 7h24'37" 24d19'33"
Katherine Elizabeth Pacheco Vir 11h46'20" 4d54'37"
Katherine Elizabeth Podjan Cam 3h24'13" 64d43'21"
Katherine Elizabeth Reese Vir 14h27'46" 5d9'30"
Katherine Elizabeth Veatch Sgr 18h20'50" -17d21'7"
Katherine Elizabeth Werner Tau 4h9'9" 2d29'26"
Katherine Elizabeth Wettstain And 0h47'33" 22d14'22"
Katherine Elizabeth Willkens Sco 17h41'31" -41d19'12"
Katherine Ellane Dilks Ari 2h42'32" 28d41'4"
Katherine Emily Bear Aqr 22h40'34" -15d47'19"
Katherine "Emily" Mawson Cas 1h44'5" 63d31'33"
Katherine Esther Dagermangy Eri 3h16'45" -9d19'49"
Katherine F. Osabe Cyg 21h50'49" 52d1'37"
Katherine - Face of Woman Spirit Crb 15h46'19" 27d49'35"
Katherine Faith And 0h28'4" 31d40'44"
Katherine Falco Cas 0h26'9" 57d20'30"
Katherine Fay Geddes Cnc 9h12'49" 17d6'41"
Katherine Ferguson Aqr 22h54'56" -8d26'33"
Katherine Finfrock Lac 22h21'55" 46d38'14"
Katherine Flores Tau 5h29'53" 27d6'41"
Katherine Forster Cas 23h39'41" 54d31'30"
Katherine Frances Koebel Leo 11h28'19" 27d22'30"
Katherine Frost Cas 1h9'37" 56d35'39"
Katherine Funkhouser/Jerry Psc 1h1'40" 4d34'8"
Katherine Geissler Cap 20h9'15" -17d22'10"
Katherine Giambrone Ari 2h13'56" 10d38'2"
Katherine Gomez Lmi 10h44'37" 26d10'53"
Katherine Gonzales Ari 3h10'42" 28d36'1"
Katherine Gorman Cnc 8h15'32" 31d32'22"
Katherine Grace And 1h35'45" 38d20'5"
Katherine Grace Gem 6h33'14" 24d43'14"
Katherine Grace Butler Tau 5h12'29" 23d39'30"
Katherine Grace Kedeshian Eri 4h12'54" -7d15'58"
Katherine Grace Kefallinos Cru 12h35'32" -57d53'37"
Katherine Grace Lamont Aqr 22h56'53" -5d0'7"
Katherine Grace McClain Cas 1h34'13" 57d41'0"
Katherine Grace Pham Crb 16h14'48" 32d43'45"
Katherine H. Logan Uma 9h49'25" 51d48'33"
Katherine Halley: Katie's Star Cap 20h7'43" -21d39'42"
Katherine Hardman-McPhillips Leo 9h53'37" 22d2'32"
Katherine Haye Walker Cas 23h34'55" 51d26'30"
Katherine Henriott-Alsobrooks Sco 17h52'23" -35d42'42"
Katherine Herries And 0h15'12" 27d22'43"
Katherine Herron Cas 23h14'10" 59d12'2"
Katherine Holden And 23h29'50" 41d15'23"
Katherine Howland King Cap 20h25'0" -14d59'7"
Katherine Inez Thornton Her 18h47'51" 20d30'7"
Katherine Irene Leo 10h59'30" 6d24'23"
Katherine Isabella Dombrowski Uma 10h37'58" 57d8'33"
Katherine J. Parks Sco 16h4'40" -18d22'15"
Katherine Jameson Moores Sco 16h51'43" -44d10'49"
Katherine Jan Clark Lyn 7h24'42" 44d24'39"
Katherine Jane Cardenas Gem 6h52'36" 33d37'56"
Katherine Jean Crb 16h15'3" 33d17'38"
Katherine Jean Vir 12h42'15" 12d5'51"
Katherine Jeanette Cuhsnick And 2h18'56" 46d56'49"
Katherine Jo Lib 15h0'5" -8d23'44"
Katherine Jones Uma 10h2'21" 59d52'1"
Katherine & Joshua And 2h6'44" 42d36'49"
Katherine Joy Del 20h40'43" 3d31'14"
Katherine Joy Uma 8h39'13" 58d43'49"
Katherine Joyce Bowskill Uma 11h58'30" 56d42'18"
Katherine Julia And 23h6'38" 50d55'14"
Katherine K Glombicki Lib 14h45'0" -11d10'41"
Katherine Kasia Milewski Aqr 21h52'11" 2d13'33"
Katherine "Kat" Galotti Sgr 18h42'53" -25d50'27"
Katherine Kelly Sco 17h18'16" -32d41'36"
Katherine Khoshaba Benson Cas 1h5'41" 61d57'55"
Katherine & Kiara Bancroft Psc 1h38'59" 9d8'44"
Katherine Kokkos Cas 1h0'17" 61d24'39"
Katherine Kristine Leo 9h55'36" 8d15'27"
Katherine L. Cerullo Cap 20h8'19" -17d5'7"
Katherine L. Krug Ari 2h8'57" 23d15'11"
Katherine L. Saia Gem 7h13'34" 31d35'31"
Katherine Lapham Cnv 12h39'2" 44d15'9"
Katherine Lauren Taylor Jones Psc 1h26'25" 4d19'1"
Katherine Laurette Dato Gem 7h38'29" 21d34'31"
Katherine Laurie Manchester Ari 2h12'33" 25d0'31"
Katherine Lawlor And 1h7'42" 33d57'44"
Katherine Leah Vitale Crb 15h38'20" 25d47'31"
Katherine Lee Dane Vir 12h39'49" 7d39'59"
Katherine Lee Harrington Gem 6h37'45" 14d53'51"
Katherine Lee Madelaine Becker And 0h14'14" 44d2'27"
Katherine LeGrand Collins Lib 14h52'52" -2d5'24"
Katherine Leigh Parker Leo 10h22'6" 23d35'1"
Katherine Lena Allison Vir 13h14'12" 13d46'3"
Katherine Lindmark And 0h38'47" 40d47'47"
Katherine Linea Jones And 1h41'34" 49d51'46"
Katherine Linehan And 0h4'15" 40d15'16"
Katherine Livingston Avera Gem 7h59'13" 32d54'52"
Katherine Lloyd O'Bryan And 0h26'46" 37d1'3"
Katherine Lorraine Ari 3h24'21" 22d6'3"
Katherine Louise And 2h7'11" 45d31'52"
Katherine Louise Neville Sgr 19h51'37" -27d57'17"
Katherine Louise Stewart Uma 12h1'17" 35d50'54"
Katherine Loutas Psc 23h2'43" 6d37'12"
Katherine Lucy Terry And 0h31'29" 30d25'12"
Katherine Lyndsey Webb Lep 5h30'41" -23d3'21"
Katherine Lynn Lyn 7h53'43" 48d10'44"
Katherine Lynn Sweet Aqr 21h15'10" -14d11'40"
Katherine Lynne Bojarski Sgr 18h51'44" -27d42'54"
Katherine Lynnwood Lyn 6h45'57" 50d41'47"
Katherine M. Collier Psc 1h19'30" 31d50'58"
Katherine M Licciardello Dra 15h6'10" 62d0'26"
Katherine MacKenzie Cody Lib 14h53'6" -2d23'12"
Katherine Madison Orr Vir 15h10'24" 5d48'51"
Katherine Maex World Lyr 18h58'32" 33d33'51"
Katherine Margaret Simon Ori 5h37'21" 7d58'53"
Katherine Marguerite Cap 21h24'32" -15d43'7"
Katherine Maria Ferrello Aqr 21h9'20" 0d48'28"
Katherine Maria Schwinn Dra 19h49'16" 72d56'29"
Katherine Marie Umi 13h53'17" 77d56'51"
Katherine Marie Sgr 18h21'39" -24d47'4"
Katherine Marie Ari 2h36'57" 13d3'7"
Katherine Marie And 23h13'31" 49d22'58"
Katherine Marie Barta Sco 16h6'23" -10d31'29"
Katherine Marie Battaglia Leo 11h16'1" 25d31'2"
Katherine Marie Elder Cas 23h59'57" 52d53'34"
Katherine Marie Flynn Cas 1h50'1" 62d1'37"
Katherine Marie Hays Cas 1h21'48" 68d0'59"
Katherine Marie Kazarian Tau 5h45'33" 24d37'21"
Katherine Marie Leland Ori 6h8'10" 16d23'38"
Katherine Marie Mason Cas 0h18'43" 62d52'11"
Katherine Marie Rogers And 23h1'2" 50d41'41"
Katherine Marie Swarm Lib 14h30'54" -22d33'14"
Katherine Marie Woodruff Gem 7h30'31" 32d21'10"
Katherine Marie Yturralde Aqr 20h39'11" 0d43'33"
Katherine Mary And 1h49'52" 45d11'23"
Katherine Mary Johnson Sco 16h52'2" -42d5'41"
Katherine Mary Paras Cam 5h10'58" 60d22'27"
Katherine Mary Silver Psc 0h16'28" 11d38'49"
Katherine M.Burlett Ari 3h18'46" 29d5'18"
Katherine McCoy Hudson And 0h57'27" 40d8'59"
Katherine McKenzie Eubanks 10:17pm Leo 9h27'53" 23d14'59"
Katherine Meredith Spencer Uma 9h32'23" 46d11'47"
Katherine Mi Gorda Ramirez Tau 4h36'12" 17d11'51"
Katherine Michelle Kuhn de Batres And 23h4'41" 40d44'12"
Katherine Miller Bryant And 0h29'49" 40d52'24"
Katherine Mills Cap 21h38'48" -10d5'37"
Katherine Morgan Potts Leo 9h46'19" 31d49'37"
Katherine Moritz And 1h12'21" 42d8'16"
Katherine Morrow Uma 8h26'44" 72d17'50"
Katherine mule ear Flatt Lyn 7h41'5" 37d4'16"
Katherine Muryn Gem 7h42'44" 33d44'1"
Katherine Natale Cam 6h39'43" 63d48'3"
Katherine Nice Cas 1h41'59" 60d11'17"
Katherine Nicholson Com 12h50'17" 30d15'33"
Katherine Nicole Castro Leo 10h40'53" 17d29'29"
Katherine Nicole Leigh Kennedy Tau 3h59'24" 9d34'24"
Katherine Nicole Martinez Cas 1h37'20" 61d16'58"
Katherine Nicole Rieske Tau 5h52'31" 27d1'2"
Katherine O. Mills Leo 10h30'16" 8d5'58"
Katherine O'Brien And 1h22'17" 45d46'18"
KATHERINE OKINGA Sco 17h13'54" -44d2'7"
Katherine Overmyer-Cotton Cap 20h20'32" -24d54'50"
Katherine Owen and James Rosenthal And 1h28'58" 47d31'47"
Katherine Page And 0h57'43" 45d14'10"
Katherine Paige Stockberger Gem 6h29'18" 23d14'11"
Katherine Parker Peterson Cas 1h45'8" 65d51'12"
Katherine Perchik And 2h18'59" 48d14'0"
Katherine Pereira Cyg 21h53'37" 48d5'57"
Katherine Petronellus Lmi 10h59'33" 32d31'15"
Katherine Phan Gem 7h20'58" 31d55'7"
Katherine Pincus Aqr 22h44'16" -1d48'18"
Katherine Poland Vir 12h33'35" -9d52'40"
Katherine Quigley Uma 8h34'58" 64d17'14"
Katherine R Boswell Leo 10h39'43" 14d51'1"
Katherine R. Lowisz Cam 4h31'51" 53d52'45"
Katherine R. Murphy Cap 20h28'55" -14d8'50"
Katherine Ramsey's Wishing Star Vir 13h26'29" -0d56'47"
Katherine Rea O'Dell Cas 23h41'8" 59d43'43"
Katherine Renae Riha And 1h42'6" 46d54'1"
Katherine Roaseau Dra 16h43'27" 54d28'27"
Katherine Roberta Dodge (Dear Star) Lyr 18h53'49" 36d39'29"
Katherine Roberts Cyg 20h2'56" 48d0'55"
Katherine Rodack Uma 12h5'51" 45d10'37"
Katherine Roscoe Lep 5h20'30" -12d0'37"
Katherine Rose Uma 9h30'22" 47d50'47"
Katherine Rose And 23h26'45" 37d33'17"
Katherine Rose Ori 5h27'23" 7d45'37"
Katherine Rose Craft Cnc 8h40'42" 24d57'9"
Katherine Rose Goris Sgr 19h50'39" -28d19'3"
Katherine Rose Halpern Cnc 8h55'16" 31d44'57"
Katherine Rose Jones Cas 0h12'33" 62d33'44"
Katherine Rose Kelly And 23h18'15" 52d14'46"
Katherine Rose Leathers Psc 1h6'41" 28d57'12"
Katherine Rose Lumia Uma 12h59'36" 61d3'46"
Katherine Rose Moore Gem 7h14'13" 24d8'41"
Katherine Rose Park Leo 9h27'57" 26d12'11"
Katherine Rudloph Lyn 8h18'55" 43d59'39"
Katherine Rush Hook Vir 14h9'46" -18d32'45"
Katherine Ruth Ribaul Peg 22h6'19" 11d14'49"
Katherine Ruth Stevens (Stretch) Del 20h33'28" 13d6'45"
Katherine Sheehan Lyn 9h4'50" 36d57'2"
Katherine & Sheridan LaPorte Uma 8h35'25" 54d9'33"
Katherine Shipley And 0h43'26" 27d5'0"
Katherine Slapko Gem 7h15'24" 25d51'6"
Katherine Speencer And 2h10'45" 44d42'11"
Katherine Stafford-Stack Leo 11h29'52" -4d51'27"
Katherine Steinmann And 1h37'25" 38d0'57"
Katherine Straface Vir 13h19'12" -7d57'42"
Katherine Suzanne Roehm Uma 11h12'2" 33d37'13"
Katherine T. Hairfield Sgr 18h15'6" -19d19'32"
Katherine Taylor Doniger Sgr 19h30'8" -13d25'49"
Katherine Thames Kidd Sge 19h50'28" 18d11'39"
Katherine "The Bright" And 0h9'5" 44d44'44"
Katherine "The Great" Thorsnes Cnc 8h33'31" 23d38'5"
Katherine Tod Arshalous Norris Aqr 22h43'50" 0d52'29"
Katherine Tsang Cas 1h31'39" 67d53'53"
Katherine V. Lawlor Haynes/Mom Cas 1h37'43" 60d2'17"
Katherine (Vargo) Byrnes Cyg 20h40'44" 46d17'39"
Katherine Virginia Rumburg Crb 15h46'5" 28d29'44"
Katherine Walker Fangio Cyg 20h20'19" 31d11'4"
Katherine Watford Vir 13h34'20" -18d24'46"
Katherine Webster Leo 10h43'19" 14d43'15"
Katherine Wells My Mother, My Angel Leo 11h3'0" 4d59'47"
Katherine Wilson Clark Lmi 10h49'32" 33d40'10"
Katherine Yount Vir 12h37'1" -8d5'9"
Katherine Zerebiec Aqr 22h39'52" -17d16'31"
Katherine, Ryan Zimmerman June 06 Uma 9h18'5" 59d11'21"
KatherineAnneKahler12070 3LoveSEERWN Cyg 20h29'29" 32d13'32"
Katherine's Dream Cru 12h19'30" -57d22'1"
Katherine's Maxinator Peg 22h12'58" 7d34'42"
Katherine's Serenity Uma 12h44'0" 62d8'32"
Katherine's Shining Star Ari 2h21'7" 12d42'3"
Katherine's Wishing Star Cnc 8h35'49" 7d46'29"
Kathern B., Mama, Grandma Vir 14h39'16" 3d5'38"
Katheros de Sollers Firmus Sco 17h17'49" -39d27'12"
Katheryn And 1h5'15" 36d59'4"
Katheryn Alison Young Hya 9h25'54" 5d15'50"
Katheryn Ann Johnson Ari 2h43'54" 28d1'9"
Katheryn Ann Russo Cnc 8h38'50" 8d58'25"
Katheryn L. Fariss Peg 23h25'17" 11d48'26"
Katheryn L. Weaver Peg 22h13'7" 24d5'15"
Katheryn Lynn Mace Ori 5h55'1" 7d11'21"
Katheryn R. is loved by Michael R. Vir 14h9'57" -12d33'6"
Katheryn Rachael St.Amant Psc 1h9'56" 18d20'35"
Katheryn Schrack Cnv 12h48'57" 39d2'23"
Katheryn T. Alesi Leo 11h21'4" -0d30'38"
Kathi Sco 17h50'30" -37d27'48"
Kathi And 0h6'52" 43d47'53"
Kathi Vir 14h42'26" 3d13'56"
Kathi A Pull's Star Ori 5h38'20" -1d22'5"
Kathi F. Hammer Uma 13h56'20" 48d38'14"
Kathi Girl Ori 5h58'50" 11d31'52"
Kathi Hamel Psc 0h49'51" 15d16'26"
Kathi J Lauria Crb 16h4'24" 35d48'30"
Kathi Kaufman Ari 2h10'24" 25d29'29"
Kathi Kelly - GaMPI Shining Star Cep 22h6'45" 60d54'8"
Kathi M. Nash Tau 4h5'39" 8d55'1"
Kathi McEwan Psc 22h51'26" -0d18'31"
Kathi Spradley Lewis Sgr 17h52'45" -17d14'2"
Kathi Steiger Ori 6h15'49" 9d29'57"
Kathi Vail Com 13h13'22" 19d22'22"
Kathi Wharton Lyn 9h7'25" 38d5'14"
Kathi Whitley Mon 6h53'32" -0d54'33"
Kathi, ich liebe Dich - Dein Dani Uma 14h17'20" 57d58'46"
Kathia Uma 8h27'35" 72d55'25"
Kathie Ann Tarpey And 0h44'36" 31d34'41"
Kathie Breen Davison Cas 0h10'42" 61d5'55"
Kathie Fouré de Prat-ar-Rouz Tau 3h41'16" 18d5'2"
Kathie Gerritzen Sgr 18h5'2" -16d55'13"
Kathie Gilberti And 23h10'42" 48d38'3"
Kathie J Cap 20h35'55" -18d19'43"
Kathie June Meredith Sgr 20h23'16" -42d51'27"
Kathie Louise Lib 15h36'23" -8d13'46"
Kathie Makings Cap 21h5'52" -27d10'26"
Kathie Marie Corby Lewis Ori 5h28'41" 2d29'26"
Käthie-Marie & Susann Reiche Uma 10h18'20" 68d56'17"
Kathie's Love Psc 0h35'59" 10d40'27"
Kathie's Star Uma 13h16'36" 53d56'51"
Kathie's Star - Nikki's Light Gem 7h6'49" 18d40'36"
Kathi's Dream Cap 20h26'30" -13d27'31"
Kathleen Cap 21h54'15" -14d49'27"
Kathleen Vir 12h8'37" -1d46'29"
Kathleen Vir 14h15'8" -13d37'12"
Kathleen Vir 12h40'9" -6d25'37"
Kathleen Uma 10h3'37" 56d34'33"
Kathleen Lyn 7h29'47" 55d8'41"
Kathleen Cas 0h31'22" 61d43'25"
Kathleen Cas 1h2'36" 61d21'57"
Kathleen Uma 10h42'1" 69d30'32"
Kathleen Uma 9h35'0" 70d0'59"
Kathleen Lib 15h35'39" -24d49'27"
Kathleen Aqr 21h58'45" 1d42'37"
Kathleen Aqr 22h17'54" 1d55'23"
Kathleen Her 18h28'21" 13d38'27"
Kathleen And 23h18'37" 52d8'23"
Kathleen And 23h46'23" 37d43'26"
Kathleen Lyn 8h37'57" 34d3'51"
Kathleen Cnc 8h22'24" 32d23'33"
Kathleen Cyg 19h40'28" 32d4'1"
Kathleen Crb 16h15'20" 33d16'37"
Kathleen Lyn 8h37'48" 40d32'20"
Kathleen 1951 Cas 1h14'18" 57d30'46"
Kathleen A. Adamson Cnc 9h7'39" 31d46'32"
Kathleen A. Cronk Cas 23h36'12" 54d27'25"
Kathleen A. Machala Sgr 19h22'1" -16d32'0"
Kathleen A. Russo Lyn 7h58'8" 47d56'17"
Kathleen Ally Lib 14h52'55" -6d49'28"
Kathleen Ames Vir 13h19'49" 7d45'41"
Kathleen and Cary Alpha one Uma 9h24'31" 42d18'18"
Kathleen and Dennis Cyg 19h59'28" 31d3'15"
Kathleen and Jim Buckley Cyg 21h12'14" 40d27'9"
Kathleen and Mauricio Kossler Cyg 19h43'33" 39d59'25"
Kathleen and Michael McManus Tau 4h21'32" 14d59'4"
Kathleen Ann Vir 14h7'8" -12d31'2"
Kathleen Ann Avila Cap 20h20'59" -11d22'45"
Kathleen Ann Bosett - 1933-2003 Ori 5h8'35" 6d19'45"
Kathleen Ann England Per 3h45'13" 41d27'4"
Kathleen Ann Grove Tau 3h32'5" 23d51'3"
Kathleen Ann "Kat" Waker Cas 0h24'58" 54d59'8"
Kathleen Ann La Borde And 1h49'9" 38d59'37"
Kathleen Ann Leo Uma 11h6'49" 53d51'27"
Kathleen Ann Ross Ari 3h9'23" 10d56'27"
Kathleen Ann Saiano Vir 13h12'0" -16d16'27"
Kathleen Ann Slocum (Granny) Uma 13h40'56" 53d19'47"
Kathleen Ann Snyder Ori 5h54'50" -2d23'49"
Kathleen Ann Titley Cas 1h15'56" 67d44'57"
Kathleen Anne Conway Muhr Cap 21h32'20" -19d8'37"
Kathleen Anne Mee Aqr 22h53'21" -20d10'31"
Kathleen Anne September 10th, 1965 Vir 11h54'57" 9d35'37"
Kathleen Anne Wilson Leo 9h50'6" 31d41'7"
Kathleen Balk Mon 6h42'3" 5d31'11"
Kathleen Barbara Latif Umi 14h17'14" 74d21'14"
Kathleen Barbee-Herzog Aqr 22h15'1" -8d49'15"
Kathleen Barenes Ori 5h38'15" -5d12'3"
Kathleen Barilone Crb 15h21'6" 29d22'18"
Kathleen Bars Uma 10h20'34" 42d46'43"
Kathleen Bayless And 0h47'17" 40d38'28"
Kathleen Beckwith Worth Uma 9h43'46" 67d14'38"
Kathleen Berg's Ball of Gas Uma 8h43'22" 53d17'57"
Kathleen Bogen Lyn 8h18'3" 58d35'36"
Kathleen Bowes Psc 23h21'55" 3d56'9"
Kathleen Boyland "Kitty's Star" Tau 5h35'46" 25d32'27"
Kathleen Brey Gem 7h48'35" 31d54'31"
Kathleen Browne Lyn 9h14'28" 36d54'2"
Kathleen Brunkhorst Vir 12h23'46" 0d3'22"
Kathleen C. Kriner Tau 5h38'12" 25d46'11"
Kathleen C. Maggi Aqr 22h9'39" -16d12'25"
Kathleen C. Matthews Vail Lac 22h56'27" 51d34'4"
Kathleen C. Pulley Aqr 22h36'11" 0d15'34"
Kathleen C. Sager And 23h56'38" 42d2'11"
Kathleen C. Sandbakken Peg 22h28'37" 30d55'1"
Kathleen Cabrera Cas 0h56'38" 56d13'48"
Kathleen Carole Carlino And 23h22'57" 41d48'19"
Kathleen Casper Dra 19h22'27" 59d24'54"
Kathleen Charlier Crb 15h42'32" 36d21'9"
Kathleen Christine De Leon Tau 4h7'6" 27d23'16"

Kathleen Clayton
Cet 1h50'29" -0d38'44"
Kathleen Collins
Cas 0h23'39" 48d16'30"
Kathleen Connell
Leo 11h46'35" 21d28'38"
Kathleen Cooney Lisa
Cas 1h28'4" 62d48'49"
Kathleen Coulombe
Ari 2h12'13" 22d53'44"
Kathleen Czajkowski
Cha 10h32'4" -76d50'48"
Kathleen D. Battaglia
Gem 7h46'17" 33d7'39"
Kathleen D. Gagne
And 1h45'23" 45d54'5"
Kathleen D. Sexton
Cap 21h36'13" -18d33'27"
Kathleen Davis
Vir 13h26'48" -13d3'33"
Kathleen DeAnne Herrin Ridolfi
Cas 0h31'12" 54d47'31"
Kathleen Deborah Moran
Cas 1h4'41" 60d17'9"
Kathleen DeSanti Haefele
Vir 12h47'6" -6d48'54"
Kathleen "Dixie" Mee
Crb 15h47'28" 26d5'9"
Kathleen Donohue
Cas 0h25'42" 58d10'54"
Kathleen Dunn
Lyn 7h32'23" 35d59'24"
Kathleen E. Barth
Lyr 18h46'23" 34d59'39"
Kathleen E. Fisher
Cnc 8h3'8" 15d9'3"
Kathleen E. Grebb
Cas 0h56'16" 53d48'49"
Kathleen E. Terceiro
And 0h42'25" 40d47'25"
Kathleen Eldridge 2000-5
Cep 21h23'52" 62d45'0"
Kathleen Elizabeth Anastacia Horan
Cap 21h11'10" -16d35'19"
Kathleen Elizabeth Mulnix
Cap 21h18'53" -25d51'16"
Kathleen Elizabeth Panteleakis
Uma 11h0'27" 57d51'10"
Kathleen Elizabeth Rayner
Aqr 22h42'24" 1d56'23"
Kathleen Elizabeth Ruckdeschel
And 0h29'4" 37d17'50"
Kathleen Elizabeth Zimmermann
Uma 12h29'51" 52d49'36"
Kathleen F. Smith
Sco 17h6'36" -34d31'19"
Kathleen Frances
Uma 10h17'22" 70d59'20"
Kathleen Frances Watson
Her 16h15'19" 46d55'43"
Kathleen Francis
Uma 8h20'59" 62d40'56"
Kathleen Francisco
Gem 7h15'38" 32d23'11"
Kathleen Garrett Page
Per 3h48'29" 46d25'20"
Kathleen Gehrke
Leo 11h11'32" 16d45'35"
Kathleen Gentz Dowling
Car 10h31'42" -60d25'53"
Kathleen Gilling
Per 3h17'48" 47d33'57"
Kathleen Ginty
And 22h58'58" 48d42'10"
Kathleen Gladkowski
And 23h47'13" 35d11'51"
Kathleen Glaubinger
Uma 12h25'14" 54d38'26"
Kathleen Gorman - Mum - "Mrs Woman"
Cas 0h31'29" 50d47'1"
Kathleen Grace
Cap 20h55'49" -27d10'10"
Kathleen Green
And 2h34'59" 43d27'37"
Kathleen & Gregory
Lyr 18h51'55" 44d22'29"
Kathleen Guthrie
Gem 7h43'55" 16d17'17"
Kathleen H
Vir 13h59'36" -0d51'6"
Kathleen H. Messey
Cas 0h47'23" 58d5'46"
Kathleen Hanson-Gough
Pho 0h47'18" -49d3'48"
Kathleen Harrington
Lyn 6h57'20" 56d7'41"
Kathleen Hatch 20 Years HVB
Uma 13h19'45" 56d54'8"
Kathleen Hooper Luten
Mon 6h48'4" -0d12'15"
Kathleen Hope Judah-Myers
Lmi 10h26'38" 28d59'49"
Kathleen Hopely
Psc 0h9'3" 12d18'19"
Kathleen Howard Guzaj
Ori 5h27'29" 6d22'20"
Kathleen Hughes
Cas 0h0'27" 53d13'27"
Kathleen Hughes
And 1h29'4" 39d46'41"
Kathleen Husko
Sco 16h7'15" -14d23'33"
Kathleen Hynson
Ori 5h0'38" 10d7'10"
Kathleen I. Roy
Gem 7h13'54" 26d6'56"
Kathleen J. Alvanos
Cnc 8h57'45" 16d6'47"
Kathleen J. De Salvo
Gem 7h28'41" 14d23'42"
Kathleen J. Dumas
Sco 16h49'31" -33d38'52"
Kathleen J. Kunkle
And 0h24'29" 33d8'4"
Kathleen Jackson
And 2h26'40" 46d46'59"
Kathleen Jean Ryczek
Lib 15h18'1" -5d51'7"
Kathleen Jentry
Per 3h36'19" 45d25'32"
Kathleen Joyce Smith Northcutt
Ari 2h58'35" 25d17'23"
Kathleen Judisky
Uma 11h44'27" 65d30'6"
Kathleen K Miller
Cap 21h29'49" -23d41'16"
Kathleen K. Touset
Gem 7h30'13" 26d45'48"
Kathleen Kaialani Askew-Wilson
Cap 20h56'23" -25d20'34"
Kathleen "Kathi" Patricia McGuire
Ori 5h53'56" 17d3'48"
Kathleen "Katie" Lorraine Eckerman
Ari 2h19'34" 14d30'5"
Kathleen "Katie" Riley
Sco 17h57'36" -41d31'48"
Kathleen Kaufman
Tau 5h27'39" 25d2'56"
Kathleen Kaye Cantu
Umi 16h13'15" 78d2'29"
Kathleen Kaylor-Vance
Aqr 22h39'19" -12d5'32"
Kathleen Kellaigh
Cnc 8h32'5" 9d49'51"
Kathleen Kent Thornton
Cep 21h35'12" 59d15'28"
Kathleen Kerins
Sco 17h17'25" -37d7'43"
Kathleen & Kevin Owens
Tau 4h18'32" 3d6'8"
Kathleen Keyes Gallagher
Ari 2h0'41" 20d36'59"
Kathleen Kirkwood Samuel
Lyn 7h33'13" 57d10'48"
Kathleen Klein
Ari 2h53'0" 18d57'26"
Kathleen Kooker Meinck
Cyg 21h57'27" 45d36'16"
Kathleen Krawetzky
Cas 2h28'57" 66d23'0"
Kathleen L Kane ki ki
Sco 17h56'32" -30d27'3"
Kathleen L. Perry
Uma 8h9'1" 62d2'23"
Kathleen L. Pinkett
And 2h33'22" 49d8'48"
Kathleen L. Simon
And 0h19'15" 46d19'37"
Kathleen Lannon-Brewer
Lib 14h50'22" -6d23'13"
Kathleen Larson McShane
Psc 1h30'44" 4d34'19"
Kathleen Laura Smith
Vir 11h51'42" 9d14'12"
Kathleen Leah
Cap 20h33'46" -24d33'2"
Kathleen Lehew
Ari 2h37'32" 29d16'33"
Kathleen Leota Churchill
Mon 6h53'4" -0d42'3"
Kathleen "Lil Mom" Brush
Lib 15h11'8" -4d6'11"
Kathleen Lila Petrak
And 1h20'56" 49d28'7"
Kathleen Logan
And 2h9'31" 43d26'14"
Kathleen Lombardi
Uma 9h8'42" 53d23'38"
Kathleen LoPresto
Lyn 7h37'12" 49d32'54"
Kathleen Louise
Psc 0h3'42" 7d31'46"
Kathleen Louise
Uma 8h35'48" 72d57'56"
Kathleen Louise Stearns
Aqr 21h34'45" 1d9'27"
Kathleen Louise Walk
Peg 21h26'27" 17d45'48"
Kathleen Loveland
Gem 7h1'6" 13d14'31"
Kathleen M. Beauman
Tau 5h59'35" 23d51'31"
Kathleen M. Collins
Cnc 8h44'4" 16d22'42"
Kathleen M Hughson
And 0h55'6" 34d51'33"
Kathleen M. Junikiewicz
Lyn 9h13'25" 41d55'28"
Kathleen M. L. Wright
Crb 16h14'44" 36d0'50"
Kathleen M. Liberatore
Ari 2h18'34" 26d27'17"
Kathleen M Rapkin
Psc 0h37'4" 6d54'19"
Kathleen M. St.John-Brough
Leo 9h26'16" 12d56'15"
Kathleen Macechak
Uma 9h5'33" 71d10'45"
Kathleen Mackenzie Fletcher
Vir 13h22'27" 11d31'7"
Kathleen Mae Kazolas 8/8/62-1/27/81
Leo 9h35'6" 32d10'31"
Kathleen Malloy
Oph 17h32'31" 6d47'55"
Kathleen Marea
Sco 16h14'35" -12d6'37"
Kathleen Margaret Sullivan
Cyg 20h32'30" 37d8'57"
Kathleen Marguerite Rickey
Peg 21h58'24" 11d47'32"
Kathleen Marie
Ari 3h8'26" 12d38'18"
Kathleen Marie
Crb 16h18'50" 27d46'39"
Kathleen Marie
Cnc 8h50'12" 31d10'26"
Kathleen Marie
Lyn 7h0'56" 45d19'39"
Kathleen Marie
Cas 0h50'52" 69d28'59"
Kathleen Marie
Cam 5h58'34" 61d42'59"
Kathleen Marie
Psa 21h34'52" -27d27'38"
Kathleen Marie Benefiel
Ori 5h8'5" 9d42'50"
Kathleen Marie Campbell
Cnc 9h13'1" 7d53'14"
Kathleen Marie Fellenz
And 23h16'15" 42d9'41"
Kathleen Marie Fults
Mon 6h26'34" 8d18'3"
Kathleen Marie Jiminaro
Aqr 21h54'55" 1d8'30"
Kathleen Marie MacLeod
Sco 16h6'31" -11d13'11"
Kathleen Marie Madding
And 0h39'48" 41d33'19"
Kathleen Marie Marks
Leo 11h42'40" 11d10'27"
Kathleen Marie May
Lyn 9h10'57" 37d25'10"
Kathleen Marie McDonnell
Cyg 21h34'43" 30d39'34"
Kathleen Marie Neary
Crb 15h47'19" 34d47'20"
Kathleen Marie Nohrenberg
Sgr 20h9'46" -39d43'50"
Kathleen Marie Ritchie
Cyg 19h50'2" 33d30'7"
Kathleen Marie Sande
And 2h27'10" 47d16'53"
Kathleen Marie Stehr Kennedy
Ari 3h28'59" 23d21'1"
Kathleen Marie Stone
Psc 1h0'38" 29d31'53"
Kathleen Marie Townsend
Leo 11h35'47" -6d38'11"
Kathleen Marrion
Lib 14h22'51" -17d55'9"
Kathleen Martin
Cyg 20h42'48" 34d57'16"
Kathleen Mary
Gem 6h53'12" 20d31'15"
Kathleen Mary
Sgr 18h10'23" -27d54'27"
Kathleen Mary ** Big K
And 0h55'39" 42d50'8"
Kathleen Mary Cain
Sco 17h28'53" -45d20'48"
Kathleen Mary Coolong
Pho 1h18'4" -41d10'48"
Kathleen Mary Elizabeth Kilgore
And 23h16'46" 47d56'30"
KATHLEEN MARY KLEIN
Cnc 8h36'34" 15d44'17"
Kathleen Mary Meehan
Psc 1h23'26" 15d15'5"
Kathleen Mary Neill
Gem 7h45'25" 15d57'9"
Kathleen Mary Windle
And 1h40'16" 42d29'46"
Kathleen May Gibb
Cas 1h8'47" 51d5'9"
Kathleen McCarthy
Vir 12h46'17" 5d16'29"
Kathleen McCormick Gilbert
Crb 16h8'11" 37d29'3"
Kathleen McFadden
Del 20h33'43" 12d16'14"
Kathleen McGaughan
Tau 5h52'58" 24d21'0"
Kathleen McGuire
Ori 5h4'14" 7d28'49"
Kathleen McHale
Sco 17h54'21" -42d46'36"
Kathleen McMaster
Tau 5h46'46" 28d3'49"
Kathleen Meuer-Wolf
Cma 7h24'40" -24d43'41"
Kathleen Michele Lewis
Crb 15h44'10" 31d53'35"
Kathleen Michelle Preston
Cas 0h34'33" 59d43'58"
Kathleen Molino
And 2h30'38" 50d6'1"
Kathleen *MY MINI*
Sgr 18h22'17" -23d44'50"
Kathleen N. Buckley
And 23h19'18" 51d16'31"
Kathleen N. Smith
Ori 5h39'4" -0d12'34"
Kathleen Nannini
Ari 3h23'46" 27d45'9"
Kathleen Nealon
Vir 14h3'1" -8d1'24"
Kathleen Nicole Burke
Sco 16h3'17" -19d23'12"
Kathleen Nicole Kendall
Gem 6h47'3" 32d10'50"
Kathleen Noell Geary
Sgr 19h16'40" -21d51'50"
Kathleen O'grady-Bell
Her 16h53'33" 32d23'1"
Kathleen O'Reilly
And 1h41'32" 42d55'27"
Kathleen P. "Ár nGrá" Quinn
And 23h14'16" 49d25'46"
Kathleen P. McIntire Stellar Home
Ari 2h35'6" 25d14'48"
Kathleen Patricia
Gem 7h28'21" 22d20'15"
Kathleen Patricia Devamey
Cnc 8h16'5" 11d17'15"
Kathleen Patricia Fitzsimons Liedel
Psc 1h11'55" 4d41'18"
Kathleen Patricia Holmes
Cas 0h37'55" 64d15'4"
Kathleen Patricia Michael Noonan
And 0h42'9" 30d38'30"
Kathleen Perrins' 90th Birthday Star
Cas 0h0'16" 54d11'20"
Kathleen Piccinich
Mic 21h15'35" -31d38'7"
Kathleen Ping-Sien Reichenberger
Uma 9h57'17" 64d55'29"
Kathleen Portatadino
Leo 10h23'24" 6d47'39"
Kathleen Prime
Aqr 22h44'2" -6d30'36"
Kathleen Quinnell & John Greninger
Del 20h48'27" 13d12'58"
Kathleen R. Arthur
Cas 23h38'21" 53d51'13"
Kathleen R. Shea
Cyg 20h13'57" 42d19'58"
Kathleen Rae Kinamon
Srp 18h34'6" -0d32'57"
Kathleen Rasmussen's Star
Ari 2h31'21" 21d50'12"
Kathleen Reagan
Cnc 8h44'28" 31d40'34"
Kathleen Reddish
And 1h26'48" 43d11'14"
Kathleen Redkatt Greck
Gem 7h48'32" 34d23'13"
Kathleen Renee Lacy
Cas 0h44'23" 47d53'53"
Kathleen Robertson
Lyn 9h7'10" 37d59'6"
Kathleen Rood
Ari 3h7'44" 30d24'5"
Kathleen Rose Miller Nailling
Ari 2h6'55" 26d15'18"
Kathleen Ruble
Uma 11h49'37" 63d27'42"
Kathleen Ruggio Cunniff
Cas 1h34'41" 63d50'27"
Kathleen Russo
Leo 10h31'9" 17d15'4"
Kathleen Rutkowski
Lib 15h11'9" -27d58'35"
Kathleen Ryan
Cas 0h45'9" 64d21'59"
Kathleen S. "Kathy" Raccio
Peg 23h24'14" 17d56'27"
Kathleen Sabbatino
Uma 9h23'33" 55d13'52"
Kathleen Sanders
Cap 21h4'15" -20d42'6"
Kathleen Sarah Mattice
Cas 1h49'11" 62d43'34"
Kathleen Sgroi Ackerman
Uma 9h27'50" 68d24'44"
Kathleen Shanahan
Ori 5h21'47" -4d16'9"
Kathleen Stallman
Tau 3h52'42" 20d10'33"
Kathleen Stefano McGoldrick
Umi 14h13'30" 73d34'1"
Kathleen Stipcevich
Gem 6h33'29" 26d0'49"
Kathleen Sue Parks
Psc 23h27'26" 4d53'35"
Kathleen Susanne McQueen
And 2h28'29" 50d3'24"
Kathleen T. Vuong
Tau 3h43'11" 16d23'8"
Kathleen Teeples
Cas 1h19'0" 52d7'48"
Kathleen - The Best Mom Ever
Cas 1h48'37" 64d30'28"
Kathleen Theresa (Linehan) Hughes
Sco 17h24'51" -37d10'44"
Kathleen Therese Connelly
Cyg 20h15'43" 37d8'2"
Kathleen Therese Lazzini
Vir 12h52'55" 9d9'37"
Kathleen Tilton
Lib 14h51'12" -3d14'4"
Kathleen "TINKER" Quirke COX
Vir 13h20'25" 13d3'28"
Kathleen Tornblom
Mon 6h56'22" 2d43'31"
Kathleen Tuatha De Danann
Psc 0h21'54" 16d19'54"
Kathleen V. Capps
Tau 4h13'4" 15d11'15"
Kathleen Vallejo
Sgr 18h49'34" -28d59'38"
Kathleen W Orr
Psc 1h3'43" 14d0'35"
Kathleen W. Stephens
Uma 8h42'30" 52d8'59"
Kathleen Walden
Lyr 18h49'59" 44d23'54"
Kathleen Walder
Gem 6h8'12" 22d45'9"
Kathleen Walsh
Uma 9h42'24" 72d41'28"
Kathleen Washa
Cap 20h31'19" -23d3'43"
Kathleen Webb
Ari 2h34'4" 25d58'58"
Kathleen Wisecup
Lib 15h43'29" -6d29'55"
Kathleen Worthington
Lyr 18h49'20" 32d31'55"
Kathleen Wynne
Lyn 7h49'13" 52d57'40"
Kathleen Zimmerer
Sco 16h25'49" -32d5'8"
Kathleen, Regina de maneo
Crb 15h31'19" 25d44'25"
KathleenCarbonaro
Uma 10h11'25" 42d25'33"
KathleenCeleste a.k.a. Babydoll
Uma 9h3'12" 57d48'11"
Kathleenia Elisabetha
Aqr 22h39'19" 1d54'24"
KathleenPrice1021
Leo 10h34'49" 23d36'44"
Kathleen's Heart
Gem 6h3'9" 26d37'49"
Kathleen's Heaven
Cas 0h58'34" 63d20'33"
Kathleen's Morgan
Cyg 19h27'1" 54d4'12"
Kathleen's Place in the Heavens
Cnc 9h6'37" 27d22'54"
Kathleen's Star
Cyg 21h47'39" 40d53'18"
Kathleen's Star
Per 3h31'7" 45d11'44"
Kathleen's Star
Aqr 23h50'41" -11d0'23"
Kathleen's Star
Col 5h53'59" -38d5'34"
Kathleen's Triumph
Cnc 9h3'27" 32d9'37"
Kathleen's Wish
Lyn 9h0'53" 39d1'24"
Kathleen's-T-J-63
Psc 0h17'58" 17d6'40"
Kathlena R. Ruiz
Leo 10h43'43" 17d44'44"
Kathline " Kat " Davis
Umi 15h23'45" 75d18'38"
Kathlyn Ann Janson Rider
Lib 15h53'17" -6d36'48"
Kathlyn *Broadway* Pogue
Sco 16h8'15" -19d4'17"
Kathlyn Crews
And 1h10'9" 42d25'7"
Kathlyn Jane Mitchell
Leo 11h10'13" 10d2'29"
Kathrin
Ori 5h53'57" 16d32'46"
Kathrin
And 1h3'36" 42d19'15"
Kathrin
Uma 8h52'57" 58d49'27"
Kathrin
Cep 22h21'55" 67d3'40"
Kathrin Bachrack
And 0h25'58" 44d14'2"
Kathrin Hess-Gribble
And 0h52'54" 46d29'4"
Kathrin Lashek
Uma 11h20'22" 38d42'22"
Kathrin Marie Campbell
Mon 6h43'59" 5d50'40"
Kathrin Richter
Uma 14h18'50" 62d19'7"
Kathrin Stacher
Cep 22h6'40" 62d7'39"
Kathrin Stacher
And 23h25'22" 44d4'16"
Kathrin & Steven Antonelli
Cyg 21h38'45" 51d59'42"
Kathrin und Alex
Uma 10h25'30" 39d31'39"
Kathrina K. West
Cyg 21h29'40" 49d8'47"
Kathrine Alison Fleming
Lib 15h21'55" -17d59'47"
Kathrine C. Caccavale
Mon 6h46'56" 7d25'48"
Kathrine Charlize Kohl
Cas 0h34'38" 57d58'55"
Kathrine Mary Ksen
Psc 0h43'29" 3d46'21"
Kathrine Meer
Psc 1h19'29" 17d25'18"
Kathrine Nichole Henson
Lib 15h47'38" -5d39'43"
Kathrine Rose
Cas 1h4'24" 61d32'24"
Kathryn
Lyn 7h12'57" 55d38'38"
Kathryn
Uma 10h46'52" 72d1'13"
Kathryn
Aqr 23h50'59" -10d51'15"
kathryn
Sgr 19h49'42" -35d20'39"
Kathryn
Vir 15h6'59" 5d7'10"
Kathryn
Psc 0h59'29" 9d58'31"
Kathryn
Ori 5h37'6" 11d16'15"
Kathryn
Cnc 8h33'41" 16d6'48"
Kathryn
Psc 1h9'48" 29d26'5"
Kathryn
Tau 4h43'44" 23d9'46"
Kathryn
Tau 5h5'10" 25d24'50"
Kathryn
Cyg 21h52'9" 45d21'49"
Kathryn
Lyn 7h47'57" 49d26'35"
Kathryn
And 0h33'32" 44d30'4"
Kathryn
And 1h3'20" 37d49'24"
Kathryn
And 0h22'25" 42d44'8"
Kathryn 05
Peg 21h42'30" 24d19'11"
Kathryn 44
Lib 15h4'57" -10d59'11"
Kathryn A. Boyle
Crb 16h20'0" 38d52'54"
Kathryn A. Byars
Mon 6h42'3" 6d16'13"
Kathryn Addie Lerche
Vir 14h34'39" 6d2'14"
Kathryn Alaine
Gem 6h49'55" 14d37'14"
Kathryn Alden Glavin
Ari 2h7'5" 25d5'49"
Kathryn Alexandra
Uma 10h54'2" 33d41'30"
Kathryn Alyssa McDaniel
Cap 20h29'4" -12d51'33"
Kathryn Amanda Hollenbeck
And 0h41'30" 28d5'52"
Kathryn Amanda Sullivan
Del 20h46'42" 13d16'0"
Kathryn Anastasia Beals
Aqr 21h45'42" -1d57'56"
Kathryn and Andy Hooper
Del 20h36'19" 14d38'55"
Kathryn and Bruce Green
Cyg 20h42'20" 39d39'58"
Kathryn and Elizabeth's Star
And 1h9'37" 45d9'31"
Kathryn and Keith's Cosmo
Sgr 19h39'14" -35d2'34"
Kathryn and Leslie Stangel
Uma 11h11'10" 52d48'56"
Kathryn Angela Reeve
Cap 21h28'40" -23d21'27"
Kathryn Anita Lindell
And 1h49'7" 38d1'0"
Kathryn Ann
Uma 10h44'52" 46d23'35"
Kathryn Ann
Psc 22h53'17" 2d35'40"
Kathryn Ann
Uma 9h51'0" 55d1'16"
KATHRYN ANN BEISER #23 10/24/1988
Uma 13h55'53" 56d44'3"
Kathryn Ann Bradford
Umi 16h25'14" 73d31'6"
Kathryn Ann Comune
Sco 16h7'39" -9d18'32"
Kathryn Ann Davis
Cas 0h15'59" 51d57'41"
Kathryn Ann Dusek
Cap 20h33'1" -14d3'11"
Kathryn Ann Eisenhower
Tau 4h3'47" 27d5'35"
Kathryn Ann Haust
Uma 14h23'7" 55d56'57"
Kathryn Ann McGinnis
Crb 15h49'43" 33d29'13"
Kathryn Ann Porter-Romelotti
Mon 7h16'43" -0d6'21"
Kathryn Ann Sharp
Lib 14h29'58" -10d40'26"
Kathryn Ann Silvera
Lyn 8h10'33" 45d22'59"
Kathryn Ann Wakelin
Umi 9h26'19" 86d16'57"
Kathryn Ann Webb
Com 12h25'22" 18d37'16"
Kathryn Anna Holbrook Maurais
Lib 15h38'55" -20d0'35"
Kathryn Anne
Cyg 20h2'8" 32d26'37"
Kathryn Anne Courier Scammell
Uma 14h24'58" 61d55'36"
Kathryn Anne Falbo
And 1h34'17" 48d43'56"
Kathryn Anne Reynolds
Uma 10h48'47" 61d10'17"
Kathryn Anne Schultz
Cap 21h57'42" -18d27'33"
Kathryn Anne Smiley
Lib 14h54'36" -2d20'47"
Kathryn April Tuscano
Lyn 8h46'49" 36d52'7"
Kathryn Arnold
Cas 23h17'16" 54d53'30"
Kathryn & Ashley Chauvin U5U
Cnc 8h29'4" 11d25'6"
Kathryn Ashley Woods
Uma 11h17'29" 53d58'24"
Kathryn Audrey Zannino
Psc 1h28'40" 21d41'30"
Kathryn Baggott "Aunt Babe"
Crb 15h38'14" 28d15'11"
Kathryn Bagust
Sco 16h46'2" -43d28'12"
Kathryn Barwick
Uma 9h1'7" 50d22'47"
Kathryn Belateche 4-10-1998
Ari 2h14'53" 20d43'31"
Kathryn Bender
Uma 11h11'11" 36d16'7"
Kathryn Beth
Vir 13h18'4" 5d26'30"
Kathryn Blais
Tau 4h14'53" 17d6'43"
Kathryn Blake Bjurberg
Cyg 19h43'19" 29d25'48"
Kathryn Bock
Cru 12h25'45" -61d2'41"
Kathryn Bomar
And 2h0'0" 38d25'24"
Kathryn Bubba
Uma 10h0'56" 61d42'20"
Kathryn "Button" Briden
Ari 3h12'40" 29d5'0"
Kathryn Carmela Della Fera
Sgr 18h11'57" -27d26'57"
Kathryn Cawiezell
Sex 9h50'19" -0d31'37"
Kathryn & Charles
Sge 20h6'55" 21d3'36"
Kathryn Charlotte Huntington
Tau 4h35'38" 26d55'40"
Kathryn Christine Kuipers
Tau 3h45'14" 28d27'32"
Kathryn Chrysostom
Aqr 21h35'8" 1d25'26"
Kathryn Claire
Leo 11h29'40" 8d45'12"
Kathryn Claire Smith
Leo 11h25'18" 0d21'1"
Kathryn Constance White
And 0h1'5" 35d27'27"
Kathryn Copello
Cyg 20h10'9" 52d9'16"
Kathryn Curtis Chisholm
Cas 0h35'57" 55d33'29"
Kathryn D Wriston
Lyn 8h8'2" 56d54'8"
Kathryn Damon - My True Love
Cnc 8h1'58" 10d34'12"
Kathryn Danielle Jones
Cnc 8h21'59" 21d20'26"
Kathryn Diane Toombs
Vir 13h42'54" -6d54'21"
Kathryn Dorothy Huscher Johnston
Psc 1h10'0" 26d23'22"
Kathryn Douglas Infinitus LubEre
Lib 14h51'39" -1d52'19"

Kathryn E. Dothage Lib 15h37'11" -18d54'25"
Kathryn Eileen Maloney (Sugarbear) Cas 2h12'58" 64d37'53"
Kathryn Elaine Needham Cnc 9h0'17" 28d36'18"
Kathryn Eleanor Reynolds Sgr 18h11'45" -20d49'23"
Kathryn Elisabeth Tau 3h32'0" 9d51'29"
Kathryn Elizabeth Vir 14h10'20" 4d7'39"
Kathryn Elizabeth Ari 3h13'17" 28d47'26"
Kathryn Elizabeth And 1h3'35" 45d22'9"
Kathryn Elizabeth Lmi 10h47'35" 36d0'18"
Kathryn Elizabeth Badal Mon 6h49'27" -0d10'26"
Kathryn Elizabeth Crowley And 23h22'30" 45d23'42"
Kathryn Elizabeth Douglass And 0h15'44" 27d40'14"
Kathryn Elizabeth Garland Lac 22h24'14" 49d33'55"
Kathryn Elizabeth Glessner Lib 14h50'1" -1d32'12"
Kathryn Elizabeth Griswold And 0h18'20" 38d57'50"
Kathryn Elizabeth Inman Ari 3h15'14" 23d39'8"
Kathryn Elizabeth Lantrip Sgr 18h48'7" -30d23'11"
Kathryn Elizabeth Milam Psc 0h40'22" 8d31'45"
Kathryn Elizabeth Morgan Ori 4h51'40" 4d54'19"
Kathryn Elizabeth Munro Uma 10h37'36" 69d5'26"
Kathryn Elizabeth Nilles And 1h55'32" 37d9'42"
Kathryn Elizabeth O'Leary Uma 13h29'45" 52d39'22"
Kathryn Elizabeth Slaughter Psc 1h21'39" 19d18'39"
Kathryn Elizabeth Soloway Ori 6h0'56" 21d7'6"
Kathryn Elizabeth Thomas Aqr 23h53'22" -14d1'59"
Kathryn Elizabeth Tinsley Porter Leo 11h8'5" 22d52'0"
Kathryn Elizabeth Williams Cyg 21h59'13" 52d45'17"
Kathryn Erin Levasseur Aqr 22h26'39" -1d41'8"
Kathryn Esther Viveiros Ari 3h20'7" 28d32'58"
Kathryn Ethyl Tau 3h30'50" 17d48'29"
Kathryn Evans Uma 13h58'8" 59d37'32"
Kathryn Eve & Rufino Eugene Cyg 20h10'49" 30d56'15"
Kathryn Faherty Ori 5h35'21" -2d17'40"
Kathryn Feehley Crb 16h10'45" 37d52'30"
Kathryn Feeney Gaughan Miller Lyn 9h6'59" 34d3'33"
Kathryn Fettkether Ori 5h18'46" 9d15'22"
Kathryn FILIPOVICH Tau 4h32'49" 21d4'44"
Kathryn Frances Dodds Lmi 10h39'42" 26d40'42"
Kathryn Gilfoil Drouant Cyg 19h36'9" 33d51'28"
Kathryn Grace Holmes Crb 15h39'29" 28d49'44"
Kathryn Grace Schwartz Sco 16h36'33" -30d41'12"
Kathryn Granger Gem 7h29'10" 20d13'6"
Kathryn Helen Gatiss 28.5.73 Gem 6h55'32" 15d41'1"
Kathryn Helen Swint Uma 11h11'51" 53d47'6"
Kathryn Hope Bond Uma 11h56'0" 48d24'19"
Kathryn Iconomopulos Peg 21h20'3" 17d33'53"
Kathryn Irene Cap 21h25'42" -16d15'34"
Kathryn J Shaw Cyg 21h56'49" 45d20'38"
Kathryn Jane Bulis Ari 2h47'52" 24d27'18"
Kathryn Jane Kuziak Lmi 10h43'55" 34d25'31"
Kathryn Jane Long Leo 11h22'44" 20d22'18"
Kathryn Jane Norrgard-Grossman Lyn 8h14'21" 39d56'49"
Kathryn Jasperse Cnc 8h18'2" 15d50'19"
Kathryn Jo Bailey Cap 21h47'29" -18d23'28"
Kathryn Jo Paskorz Gem 7h49'55" 19d49'12"
Kathryn Joy Heim Cnv 12h36'6" 45d22'38"
Kathryn "Kathy" L. Twitty Aqr 21h44'45" -1d17'55"
Kathryn Katie Flynn Lyr 18h50'49" 31d48'18"
Kathryn Kay Corrodi Lib 15h1'17" -11d17'26"
Kathryn Kay Green Tau 4h34'52" 20d7'24"
Kathryn Kelleher Sturm Uma 10h17'58" 44d35'37"
Kathryn Keller And 0h37'42" 42d30'43"
Kathryn Kelsey Tau 5h25'54" 22d23'30"
Kathryn Kleps And 23h0'41" 47d3'14"
Kathryn Kunkel Psc 1h20'8" 26d6'36"
Kathryn Larimore Cam 5h5'2" 66d1'54"
Kathryn Latrelle Dayhuff And 0h48'10" 35d19'16"
Kathryn Laughlin Cas 1h22'11" 64d36'45"
Kathryn Lauren Pearson And 1h18'20" 45d51'25"
Kathryn Lavina Noyer Uma 12h19'28" 55d35'40"
"Kathryn" Ledson Cra 18h0'27" -37d5'6"
Kathryn Lee Per 2h27'4" 54d15'40"
Kathryn Lee Arrington Sgr 19h29'17" -30d50'13"
Kathryn Lee Wisor And 0h43'26" 40d44'12"
Kathryn Linwood Rawls Per 2h25'35" 54d55'49"
Kathryn Locke Brown Psc 0h44'50" 10d36'19"
Kathryn Lorraine Scheps "Katie" Lyr 18h27'12" 36d27'7"
Kathryn Louise Sco 16h49'53" -27d19'11"
Kathryn Louise Lyn 7h11'27" 56d31'34"
Kathryn Louise Sherman Cas 0h15'53" 62d43'37"
Kathryn Louise Tallman Gem 7h25'7" 31d45'57"
Kathryn Louise Valiquette Uma 8h27'8" 63d9'3"
Kathryn Luanne"ICE Kitty" Archibald And 2h30'7" 39d5'28"
Kathryn Lynn Hoffman Sgr 17h53'15" -17d26'13"
Kathryn Lynn Olczak Tau 4h37'40" 4d18'23"
Kathryn Lynn Zwick Cas 23h51'4" 58d17'48"
Kathryn Lynne And 0h54'44" 41d43'10"
Kathryn Lynne Lyons And 0h45'24" 45d29'32"
Kathryn M. Di Camillo Cap 21h11'27" -16d28'8"
Kathryn M. LaBarge Umi 14h44'14" 73d55'16"
Kathryn M. (Lindy) Bazner 1-23-75 Aqr 22h54'54" -10d14'12"
Kathryn Mae Mills Ari 2h0'50" 14d33'48"
Kathryn Margaret Frey Dra 17h11'54" 59d49'0"
Kathryn Marie Aqr 22h24'19" -7d51'10"
Kathryn Marie Sgr 20h23'35" -35d9'24"
Kathryn Marie Bueker Sco 16h24'26" -18d41'21"
Kathryn Marie Clark And 1h5'2" 36d49'33"
Kathryn Marie Gnech Cnc 8h32'33" 14d41'51"
Kathryn Marie Long And 0h42'10" 22d8'50"
Kathryn Marie McGuinness Uma 10h16'59" 49d37'38"
Kathryn Marie Orsino Psc 1h9'17" 29d49'8"
Kathryn Marie Waterbury Uma 9h55'8" 72d7'13"
Kathryn Mary Cardone And 0h44'31" 21d58'30"
Kathryn Mary Dulaney Lyn 8h1'9" 38d44'19"
Kathryn Mary Egger Vir 13h11'16" 5d22'40"
Kathryn Mary Engelhardt Uma 9h25'55" 60d13'56"
Kathryn Mary Martel Slvrmncrb And 23h31'6" 42d43'59"
Kathryn Mary Presley Troxell Cnc 8h13'7" 8d25'17"
Kathryn Mary Ronningen Ari 3h9'25" 27d48'27"
Kathryn May Nightingale And 23h12'51" 48d18'16"
Kathryn May Nightingale Tri 2h25'13" 30d9'10"
Kathryn McKenzie Hohmann Cap 20h27'30" -19d51'15"
Kathryn Megan David Leo 9h45'19" 32d25'11"
Kathryn Merkel Uma 11h21'18" 47d57'25"
Kathryn Michelle Ori 6h18'40" 10d52'56"
Kathryn Michelle White Ori 6h10'14" -0d5'26"
Kathryn Milnes Lyn 8h43'7" 33d3'57"
Kathryn Miltenberger Uma 10h27'19" 62d2'54"
Kathryn Modaff Uma 11h9'2" 30d38'14"
Kathryn Monsour Sco 16h23'17" -36d23'44"
Kathryn Moseley Lyn 7h46'18" 37d55'48"
Kathryn Nicole And 0h53'11" 34d52'23"
Kathryn Nicolosi And 2h8'29" 42d16'47"
Kathryn Nora O'Malley Ori 5h22'41" -5d31'52"
Kathryn Pamela Golden And 0h28'45" 37d7'25"
Kathryn Patricia Ori 5h58'1" 18d24'59"
Kathryn Patricia Cap 21h41'59" -12d52'29"
Kathryn Patricia Cap 21h42'39" -10d38'23"
Kathryn Patton Sgr 18h55'38" -18d56'43"
Kathryn Penn Psc 1h19'53" 27d18'8"
Kathryn Pryor Uma 11h31'18" 60d3'5"
Kathryn Real Poindexter And 1h29'37" 39d59'45"
Kathryn Rebecca Ari 2h30'44" 26d34'35"
Kathryn Rebecca Jones Cas 0h5'21" 52d53'58"
Kathryn & Richard Denise Cyg 21h36'24" 41d15'2"
Kathryn Rose 1965 And 1h43'11" 46d3'32"
Kathryn Rose Wehrmeister Sco 16h55'23" -40d51'11"
Kathryn Rose's Little Star And 23h36'19" 47d40'57"
Kathryn Roys Cen 13h49'57" -46d0'10"
Kathryn Rudderow-Stites Crb 15h39'17" 27d56'52"
Kathryn S. Fuller Cas 0h9'2" 53d26'49"
Kathryn S. Smith Cam 5h49'31" 64d25'38"
Kathryn Schledwitz Lewis Vir 13h33'43" 12d16'52"
Kathryn Shankle Per 3h47'47" 45d52'42"
Kathryn Sheppard-Irwin Vel 9h26'18" -52d27'59"
Kathryn Smith Per 4h31'52" 43d55'19"
Kathryn Socha And 0h19'32" 37d33'27"
Kathryn Spivey Boo 15h9'25" 13d7'43"
Kathryn & Steven Mendez Uma 11h40'33" 49d20'55"
Kathryn Strickland Mon 6h33'7" 7d45'38"
Kathryn Suzanne Cavanah Umi 9h21'10" 86d12'14"
Kathryn Suzanne Martin Vir 12h54'56" 6d7'39"
Kathryn Swenson And 23h30'0" 38d53'28"
Kathryn Taylor Leo 10h35'46" 12d58'18"
Kathryn the Great Cyg 20h10'32" 51d47'38"
Kathryn The Princess And 2h13'58" 47d30'12"
Kathryn Theresa Femino Ari 2h22'34" 23d48'22"
Kathryn Theresa Hue Leo 11h57'24" 14d59'4"
Kathryn Thomson And 23h26'23" 47d15'33"
Kathryn Tissiman - Mum & Grandma Cas 23h16'59" 54d38'25"
Kathryn Valentine Cas 1h42'24" 63d20'45"
Kathryn Valido Lib 15h31'10" -12d54'23"
Kathryn von Steiger Cas 0h38'51" 48d15'34"
Kathryn Waggoner Gem 6h28'50" 22d21'54"
Kathryn Ward And 23h39'10" 47d25'21"
Kathryn Wasserman Davis Psc 0h54'40" 25d11'53"
Kathryn White Crb 15h37'44" 37d51'4"
Kathryn Whitman Cas 0h20'7" 50d30'44"
Kathryn Wilfreda Diggs Cap 20h29'10" -14d21'13"
Kathryn Woodcock Cas 0h43'5" 51d3'54"
Kathryn Yancey #1 Cmi 7h29'31" 7d22'2"
Kathryn Youso Mon 6h49'12" -4d54'19"
Kathryn Zilke Vul 21h23'47" 26d35'20"
Kathryn-1935 Aqr 21h48'34" -0d6'35"
Kathryna Hlyniansky Sco 17h46'43" -37d26'1"
Kathryn-Anne Alexandria Lewis And 2h33'32" 44d49'57"
Kathryne Elizabeth-Lee Wood And 1h4'27" 42d17'14"
Kathryne L. Hayde Dra 16h43'19" 54d3'58"
Kathryne Sophia McCreary Aqr 23h2'45" -9d16'55"
Kathryne Trunzo Spera Tau 3h59'30" 8d52'47"
Kathryne Y Z Ari 2h26'0" 18d39'37"
kathrynfantasia04061989 Ari 2h6'1" 13d42'51"
Kathryn's Beauty And 0h25'17" 42d45'11"
Kathryn's Love Vir 13h41'49" -6d10'43"
Kathryn's Star Cyg 21h22'11" 32d46'11"
Kathryn's Star And 2h23'10" 46d50'59"
Kathryn's Star in the sky. Aql 19h23'31" -9d56'20"
Kathryn's Star of Hope Per 1h37'50" 54d27'30"
Kathy Cyg 20h28'49" 46d41'8"
Kathy Uma 9h36'3" 49d13'30"
Kathy And 0h26'51" 42d50'2"
Kathy Uma 10h36'10" 40d42'31"
KATHY Tau 5h42'6" 24d39'29"
Kathy Lib 14h52'8" -7d57'29"
Kathy Aqr 22h57'56" -9d34'57"
Kathy Vir 13h13'3" -22d9'8"
Kathy Leo 11h9'40" -4d8'58"
Kathy Ori 5h2'13" -1d24'14"
Kathy Cep 2h50'27" 84d6'54"
Kathy Uma 11h27'29" 64d41'41"
Kathy Psc 0h9'48" 3d24'25"
Kathy 12 vert émeraude Cap 20h22'29" -26d38'10"
Kathy A. Longo Sex 10h40'7" 2d59'51"
Kathy A Positive Bright Beginning Leo 10h14'18" 23d15'28"
Kathy A. Stude Uma 8h35'48" 71d40'59"
Kathy Alaniz, "Our other mom" And 1h9'7" 42d40'38"
Kathy (all my love) ma wee darlin And 23h18'4" 51d3'3"
Kathy Altman Lyn 7h13'33" 52d39'48"
Kathy and Don Forever Lib 15h13'47" -25d20'59"
Kathy and Ed Rossi Cyg 21h17'40" 52d4'21"
Kathy and Joseph Zappulla Lyr 18h29'30" 36d14'44"
Kathy and Steve Tau 4h15'17" 25d25'50"
Kathy Anderson Dra 17h11'33" 68d30'27"
Kathy Ann Sabo Lib 15h8'39" -3d54'8"
Kathy Ann Trimble Wilson Cas 0h32'7" 69d59'59"
Kathy Ann Wagner Peg 22h42'26" 29d48'43"
Kathy Archer Lmi 10h22'45" 32d36'34"
Kathy Arroyo Cyg 19h47'41" 33d27'50"
Kathy B. Eastham Ori 5h32'39" -0d17'56"
Kathy Beauregard And 23h22'28" 48d36'25"
Kathy Bierl Lmi 10h41'51" 33d52'8"
Kathy Blackburn I love you Cnc 9h4'12" 22d46'18"
Kathy Boo Leo 11h8'25" 21d3'1"
Kathy Brady & Ron Adamietz Uma 8h14'44" 65d13'44"
Kathy Brooks Johnson Crb 15h49'0" 37d38'39"
Kathy Cairns Shepherd Sco 16h14'54" -12d41'51"
Kathy Callanan And 1h7'3" 42d29'47"
Kathy Carpenter Lyn 8h3'26" 55d12'53"
Kathy Carter Mon 6h52'10" -0d29'20"
Kathy Casey Mon 7h17'54" -6d3'22"
Kathy Cherry Uma 8h29'13" 61d54'20"
Kathy Coffman Craib Ari 2h16'5" 24d33'7"
Kathy Conners Uma 14h28'47" 57d7'41"
Kathy Conrod Cas 0h32'36" 61d38'21"
Kathy Costello Psc 0h7'46" 6d1'47"
Kathy Couet Lyn 7h43'10" 50d10'30"
Kathy Crawford And 2h25'18" 50d16'31"
Kathy Cunningham And 0h8'24" 44d4'5"
Kathy & Dan Cyg 19h55'54" 32d1'2"
Kathy Davis Uma 10h32'44" 50d44'44"
Kathy Demoors Vir 13h13'13" 5d35'41"
Kathy Dillinger Cas 0h20'36" 61d59'56"
Kathy DiPietropolo Sco 16h43'49" -28d30'2"
Kathy Doan Leo 10h27'57" 6d33'39"
Kathy Donovan Is 60 Years Old! Cam 5h8'57" 64d48'52"
Kathy Dove Allen 08/09/1956 Umi 15h46'26" 76d14'5"
Kathy Drummond Cas 0h28'23" 57d33'46"
Kathy Drysdale Uma 13h28'49" 58d3'7"
Kathy Dunn's Star Cam 5h59'5" 68d36'8"
Kathy E. Koglin Cyg 19h22'47" 53d46'1"
Kathy E. Silvey And 2h35'46" 43d44'55"
Kathy Ebersolds Star Lyn 6h57'43" 48d35'31"
Kathy Ehrhard Cam 4h28'50" 54d32'46"
Kathy Engelhardt Lib 15h0'23" -10d1'53"
Kathy Ferdinandsen Cam 7h50'29" 61d45'34"
Kathy Fincham & Nathan Van Stell Aql 19h8'16" -0d2'19"
Kathy Flood Virnig Aqr 22h35'18" -4d57'36"
Kathy Flores Lib 14h50'41" -2d2'18"
Kathy Freeman Ari 2h50'8" 12d43'47"
Kathy Frisk Crispin Lib 15h17'9" -27d53'10"
Kathy Fronk Ori 6h4'4" 4d33'20"
Kathy G Cas 1h37'3" 68d55'6"
Kathy Hacker Cnc 8h22'31" 11d29'34"
Kathy Harper's B-day Valentine Star Aqr 21h7'45" -3d17'28"
Kathy Hedgepath And 0h37'48" 39d4'36"
Kathy Helen Loera Vir 13h16'8" 2d6'52"
Kathy Henke Psc 0h27'31" 7d46'16"
Kathy Henry Iacopi Jane Lib 15h23'51" -4d46'11"
Kathy Hogan Uma 8h28'41" 66d2'22"
Kathy Hollman Umi 14h40'7" 68d21'40"
Kathy Howard Cnc 8h7'48" 18d30'58"
Kathy Hull Cas 1h17'44" 60d51'51"
Kathy I Tausevich Leo 9h38'37" 31d9'24"
Kathy Irene Palmer Uma 11h30'44" 56d41'22"
Kathy J. Ziemba And 2h19'1" 42d48'33"
Kathy Janese Smith Psc 23h21'3" 5d5'15"
Kathy Jean Lmi 10h46'41" 31d15'33"
Kathy Jean Farris And 0h18'33" 35d12'48"
Kathy Jo Leo 11h57'9" 26d56'53"
Kathy Jo Marlow Cnc 8h37'14" 15d52'55"
Kathy Joe Ari 3h25'43" 21d53'34"
Kathy Joe Hodges Lyn 7h8'42" 48d35'59"
Kathy & John 1st Anniversary Star Umi 14h17'40" 73d1'7"
Kathy Jones Cam 6h6'32" 60d26'13"
Kathy Kalitta Ari 2h25'34" 24d12'27"
Kathy Kapps King Tau 3h51'56" 24d9'55"
Kathy Kennedy Lyn 9h3'8" 41d52'42"
Kathy (Keter) Jo Scheiern Sgr 18h9'3" -21d23'12"
Kathy Koerner Lyn 7h9'50" 56d31'53"
Kathy Konjura Leo 9h22'27" 14d24'22"
Kathy Kotronakis Aql 19h57'3" 9d5'2"
Kathy Kunc Uma 11h37'23" 35d30'41"
Kathy L. Riggs Lyn 6h37'3" 57d57'6"
Kathy Landwermeyer Tau 5h2'54" 24d35'17"
Kathy Launer Tau 4h34'20" 28d54'27"
Kathy Lawday Lyn 8h23'36" 55d44'20"
Kathy Lawson's Star, She is Loved! Tau 4h35'8" 16d43'35"
Kathy Lindstrom Pho 0h20'23" -39d50'25"
Kathy Louise Olson Uma 13h33'44" 62d38'47"
Kathy Lynn Lib 14h22'54" -17d8'23"
Kathy Lynn Crb 15h46'37" 26d51'2"
Kathy Lynn Cornett And 0h42'43" 31d38'54"
Kathy Lynn Honaker Lib 15h51'15" -16d55'20"
Kathy Lynn Moyer Cas 1h12'37" 62d22'15"
Kathy M. Sco 16h12'31" -10d24'15"
Kathy M. Armeno Uma 9h35'34" 42d1'24"
Kathy M. Diaz Lib 15h5'53" -4d44'13"
Kathy M. Tully Sco 16h10'24" -10d32'27"
Kathy Mace And 0h33'22" 38d19'54"
Kathy Macon Uma 9h43'14" 49d29'38"
Kathy Mae Uma 11h43'19" 60d14'26"
Kathy Mallon Per 3h31'39" 49d24'36"
Kathy Mallon & Family Tau 5h15'27" 18d18'28"
Kathy Marie 1 Leo 11h5'47" 7d46'26"
Kathy Marie Combs Per 4h17'42" 39d31'40"
Kathy May "Our Hero" And 1h34'3" 45d15'39"
Kathy McCann Col 5h51'20" -36d5'8"
Kathy Mcgrane Lyn 6h28'53" 54d12'9"
Kathy McKenna Ari 2h57'40" 25d2'12"
Kathy McKnight And 23h24'0" 47d33'29"
Kathy McPherson Cam 5h22'3" 60d21'57"
Kathy Menard My Angel Lyn 8h55'39" 36d43'33"
Kathy Michini And 0h56'42" 38d38'8"
Kathy Mickus Lyr 18h45'43" 35d50'54"
Kathy & Mike Quattrone Cyg 21h30'1" 53d51'58"
Kathy Millburg Cyg 20h32'54" 36d58'50"
Kathy Milliser Tau 4h10'10" 24d26'22"
Kathy My Beacon In Life Lib 15h47'34" -28d56'58"
Kathy "Nana" Dial Cas 0h30'22" 55d49'31"
Kathy Nelson Tau 4h24'49" 15d41'15"
Kathy Neman "Never Forget" Uma 12h3'22" 62d50'0"
Kathy Nicaj Gem 7h43'15" 34d13'28"
Kathy Nickle Lyn 8h11'53" 36d21'51"
Kathy Nickoloff And 2h34'17" 37d45'46"
Kathy Nori - Believe Cas 0h34'25" 50d39'44"
Kathy O'Neil Ingram Cas 0h38'56" 61d50'52"
Kathy P Cas 0h23'45" 63d1'46"
Kathy Pattman Cas 0h10'20" 53d27'39"
Kathy Pena And 23h3'52" 50d51'30"
Kathy Polizzi And 0h32'31" 45d50'30"
Kathy R. - My Eternal Sojourn Per 3h15'44" 52d37'10"
Kathy Ratey, The Star of My Life Cyg 21h38'14" 52d9'53"
Kathy Reiter Tau 4h24'41" 27d10'16"
Kathy Renee Umi 21h9'30" 88d52'22"
Kathy & Richard Psc 0h56'12" 29d36'56"
Kathy & Rick Sge 19h37'24" 16d51'4"
Kathy Riley, Angel Extraordinaire Lyn 7h45'29" 38d38'50"
Kathy Ritter Cas 1h23'45" 55d56'35"
Kathy Rose Lyn 7h27'24" 53d29'48"
Kathy S. Vir 13h59'23" -8d28'28"
Kathy Sammons Ori 6h3'8" 17d59'17"
Kathy Sande, Angel of the stars Uma 8h32'6" 64d15'19"
Kathy Sandoval Ari 2h51'33" 29d24'47"
Kathy Santschi And 0h22'54" 41d36'32"
Kathy Scherer Cam 6h38'48" 62d59'0"
Kathy Schlau Gem 7h16'13" 24d37'52"
Kathy Schmaltz And 23h48'13" 47d11'34"
Kathy Severtson Cyg 20h1'39" 56d32'12"
Kathy Shenice Coleman Lyn 8h13'41" 34d20'29"
Kathy Skiadas Cru 12h27'27" -60d46'29"
Kathy Slicis And 23h28'41" 48d40'3"
Kathy Smeltzer Cas 0h12'2" 52d13'33"
Kathy Smith And 2h27'46" 49d1'33"
Kathy St. Clair Gem 6h47'9" 23d1'49"
Kathy St. Pierre Psc 1h10'30" 28d22'21"
Kathy Stanwick Mon 6h45'33" 6d53'56"
Kathy & Steve's Star Uma 11h34'13" 58d30'45"
Kathy Stewart Tau 3h23'47" -0d1'58"
Kathy Sue Bailey Avicola Lyn 7h34'48" 39d8'1"
Kathy Sue Johnson Lib 14h27'52" -19d16'34"
Kathy Sue McClain Hya 9h22'28" 3d29'38"
Kathy T. Cnc 8h16'29" 15d32'55"
Kathy The Viqueen Grabowski Lmi 10h27'34" 32d57'59"
Kathy Tinker Rockwell Cnc 8h11'49" 20d59'9"
Kathy & Tony's Star Cyg 20h27'38" 33d37'34"
KATHY VADALA And 23h46'10" 34d19'11"
Kathy Van Buskirk Lib 14h52'3" -18d58'8"
Kathy van Neck Dra 20h32'35" 75d16'37"
Kathy Verett Cnc 8h37'46" 10d36'31"
Kathy Waitt And 23h20'22" 47d42'12"
Kathy&Ward 2005-25 HappyAnniversary Lyn 8h28'43" 39d28'58"
Kathy Wilmarth Psc 0h45'36" 18d47'14"
Kathy Wilson Uma 9h19'26" 56d57'21"
Kathy with a K Crb 15h30'18" 28d55'14"
Kathy Wolsh Cas 1h18'58" 62d30'58"

Kathy- World's Best Grandma
Crb 15h38'11" 26d35'3"
Kathy Zyrlis
Sco 17h58'28" -30d22'5"
Kathy, The Vision
Vir 13h7'9" 3d43'15"
Kathy, will you marry me?
Uma 8h38'37" 67d2'32"
Kathy316
Psc 0h52'20" 27d12'32"
Kathyana Gomez-Martinez
Ori 5h31'25" 4d11'58"
Kathyann
And 1h15'45" 41d48'34"
Kathyanne
Gem 6h45'26" 26d18'31"
Kathyleen Natividad Taylor
Leo 11h3'20" 16d58'49"
Kathy-Lee's Cubs
Uma 11h26'30" 46d55'45"
KathyLouEtheridge20thAnniversary
Aql 19h17'17" 5d7'18"
Kathyrn Hope So Cole: My Katta
Ori 5h55'52" 11d31'46"
Kathyrn's Kosmo
Cap 21h23'41" -18d9'43"
Kathy's Bear Star
Uma 8h53'36" 70d56'22"
Kathy's Butterfly
And 2h21'7" 50d29'29"
Kathy's Cup of Tea
Cyg 20h51'25" 35d20'12"
Kathy's Eternal Eye
Sco 17h24'50" -42d19'28"
Kathy's Grace
Ari 2h4'53" 20d53'20"
Kathy's Heavenly Star
Vir 14h28'54" 6d38'33"
Kathy's Krystal
Lib 15h5'52" -28d0'5"
Kathy's Light
Psc 0h53'54" 9d16'55"
Kathy's Little Star
And 2h19'57" 43d40'58"
Kathy's Obsession Starry Eyed Larry
Vir 11h56'54" -0d6'7"
Kathy's Pink Star
Peg 22h34'43" 32d33'20"
Kathy's Place
Cas 0h55'14" 50d47'15"
Kathy's Shadow
Cas 0h47'37" 74d11'16"
Kathy's Song
Per 3h55'5" 37d2'34"
Kathy's special blessing
And 2h22'39" 46d11'15"
Kathy's Star
Cas 0h45'46" 51d41'35"
Kathy's Star
Uma 8h50'38" 48d1'56"
Kathy's Star
Uma 10h16'53" 50d53'8"
Kathy's Star
Leo 9h41'5" 32d25'39"
Kathy's Star
Leo 11h7'27" 20d3'10"
Kathy's Star
Tau 5h38'28" 23d40'26"
Kathy's Star
Uma 9h13'8" 64d6'41"
Kathy's Strawberry
Ari 2h42'41" 25d47'16"
KATI
Ori 6h20'57" 7d8'46"
Kati
Cnc 8h36'54" 13d41'26"
Kati
Uma 13h55'30" 49d34'25"
Kati
Uma 8h31'21" 60d5'14"
Kati
Cas 2h50'38" 60d15'22"
Kati
Sco 16h13'37" -13d10'51"
Kati
Aqr 23h11'40" -21d55'28"
Kati
Sgr 19h36'41" -23d40'14"
Kati Annis
Crb 16h12'51" 37d14'27"
Kati B.
Cnc 8h44'24" 31d3'58"
Kati Buie
Umi 14h49'14" 71d18'3"
Kati George
Mon 7h20'20" -0d35'43"
Kati Gray
Aqr 23h19'52" -14d46'31"
Kati Gray
Psc 1h21'3" 16d21'39"
Kati J
Cap 21h39'38" -12d20'4"
Kati Jo Provost
Sco 16h12'15" -12d13'35"
Kati Lyn Niemeth
Ori 6h4'14" 0d23'33"
Kati Lynn Wheeler
Cnc 8h52'17" 25d22'5"
Kati (néni) Szerencsecsillaga
Cnc 9h20'50" 22d44'1"
Kati Nicole
Per 3h42'30" 46d28'35"
Kati Rachae
Uma 9h3'51" 49d3'47"
Kati Sebök
Cru 12h46'23" -58d41'49"
Katia
Peg 23h12'11" 13d25'52"
Katia
Peg 22h26'27" 21d13'12"
Katia 14Nov73
Her 17h8'24" 19d5'0"
Katia Cagossi
Her 18h22'38" 21d11'29"
Katia et Julien
Umi 14h59'40" 76d43'26"
Katia G. Broomes
Ari 2h46'42" 27d49'31"
Katia Gissele Chavez
Ori 6h10'47" 15d36'50"
Katia Joy
Crb 15h53'0" 37d26'20"
Katia Lestarr Allen
Gem 6h27'2" 25d50'51"
Katia Maria Szwejbka
Lyr 18h31'38" 32d50'7"
Katia Melita Carrilho
And 1h22'40" 49d55'27"
Katia Meniconi
Cam 5h32'3" 71d23'48"
Katia Penna
Aur 5h10'11" 41d2'18"
Katia Sandra Elena
Cas 1h2'38" 56d58'29"
Kati-Alex
Cyg 21h33'18" 32d15'31"
Katianna R. Nardone
Ari 2h32'39" 18d34'38"
Katia's Star
Ori 5h40'56" 7d40'59"
Katie
Ari 3h4'48" 13d57'5"
Katie
Psc 0h39'7" 7d55'8"
Katie
Aql 19h54'12" 6d3'11"
Katie
Ori 6h19'33" 2d34'36"
Katie
Vir 13h46'56" 0d15'27"
Katie
Leo 11h35'50" 24d13'23"
Katie
Lmi 10h59'18" 26d21'5"
Katie
Cnc 9h6'41" 28d54'41"
Katie
Peg 23h46'40" 25d50'10"
Katie
Ari 2h39'15" 29d42'17"
Katie
Ari 2h19'36" 25d17'58"
Katie
Psc 1h43'0" 27d58'1"
Katie
And 0h22'46" 26d13'15"
Katie
Cnc 8h10'20" 21d56'43"
Katie
Gem 7h23'4" 15d20'54"
Katie
And 1h36'50" 42d8'14"
Katie
And 0h47'11" 38d49'11"
Katie
And 0h19'53" 43d17'49"
Katie
Per 4h4'18" 41d31'30"
Katie
And 2h9'53" 42d35'53"
Katie
Lyn 9h7'15" 38d37'48"
Katie
Lmi 9h45'43" 40d16'40"
Katie
Uma 10h0'3" 44d55'53"
Katie
And 0h26'44" 36d1'10"
Katie
And 0h42'58" 33d28'9"
Katie
And 1h3'23" 35d49'51"
Katie
And 23h56'8" 45d27'48"
Katie
And 23h17'30" 52d10'17"
Katie
And 23h14'29" 52d17'36"
Katie
And 23h11'10" 48d42'31"
Katie
And 2h20'43" 49d53'45"
Katie
Cyg 19h34'29" 44d58'3"
Katie
Hor 3h12'24" -49d39'28"
Katie
Cra 19h4'5" -39d21'5"
Katie
Sgr 18h36'30" -22d31'43"
Katie
Cma 7h5'42" -25d35'46"
Katie
Aqr 22h8'42" -0d24'13"
Katie
Sgr 19h32'45" -12d11'24"
Katie
Lib 14h53'26" -11d51'32"
katie
Lib 14h52'6" -9d21'40"
Katie
Lib 15h37'57" -18d58'42"
Katie
Sco 15h54'53" -20d42'4"
Katie
Cas 23h48'46" 61d35'59"
Katie
Uma 10h38'6" 68d54'49"
Katie
Uma 10h53'23" 69d41'40"
Katie
Cas 2h23'38" 66d10'17"
Katie
Cas 23h51'22" 57d3'7"
Katie
Uma 13h27'3" 56d2'53"
Katie
Lyn 8h28'40" 57d58'4"
Katie 09
Dra 15h39'56" 61d4'31"
Katie 143
Uma 8h40'8" 61d39'12"
Katie A. Bardong
Aqr 23h43'48" -11d14'46"
Katie A. Warren - 1437
And 2h19'54" 46d29'56"
Katie Ai's Birthday Star From J & J
And 0h44'34" 21d58'40"
Katie Alexandra
Gem 7h30'37" 28d51'2"
Katie Amanda Kelly
And 0h52'10" 43d49'43"
Katie and Anthonelli
Uma 13h24'4" 62d50'43"
Katie and Brandon Barks
Cnc 9h15'10" 8d27'15"
Katie and Brian Thatcher
Leo 10h37'13" 22d39'30"
Katie and Craig
Uma 10h37'2" 43d42'30"
Katie and Dan
Lyr 18h43'50" 39d24'32"
Katie and Eric Star
Uma 10h41'11" 61d13'1"
Katie and Garrett forever
Gem 7h7'10" 32d41'35"
Katie and Graham Thompson
Cyg 20h22'48" 31d54'40"
Katie and Jim Zehr
Cyg 20h17'2" 53d6'20"
Katie and Kevy's Love Star
Del 20h36'16" 14d23'58"
Katie and Lisa
Cyg 19h44'8" 33d26'37"
Katie And Nick's 1 Year Anniversary
Cnc 8h57'14" 29d1'43"
Katie and Tim's Wedding Day
Cyg 20h3'31" 55d7'56"
Katie & Andrew Forever
Sgr 19h4'54" -17d34'47"
Katie Angel
Tau 4h8'10" 6d15'29"
Katie Ann
Uma 10h29'55" 40d7'3"
Katie Ann
Lyn 8h19'55" 43d17'34"
Katie Ann
Cas 1h17'3" 56d55'14"
Katie Ann Bushey
Psc 1h47'48" 9d7'0"
Katie Ann Fuller - My Light
Gem 7h1'3" 19d19'57"
Katie Ann Gabbert
Sco 17h11'37" -33d46'21"
Katie Ann Hunter
Vir 12h9'53" 10d37'53"
Katie Ann Jenkins
Uma 9h21'23" 49d58'52"
Katie Ann Keller
And 0h35'27" 42d20'29"
Katie Ann Lyons
And 0h3'27" 44d8'48"
Katie Ann Mullen, lickle diamond
And 1h8'39" 45d45'16"
Katie Anna
Vir 12h47'40" 8d51'34"
Katie Anne Dove
And 2h1'20" 38d49'22"
Katie Anne Hughes
Sco 16h10'5" -35d2'6"
Katie Anne Jones
Vir 12h4'38" 2d4'22"
Katie Anne Kelley
Psc 1h9'7" 17d47'15"
Katie Anne Pope
Psc 1h19'34" 17d7'24"
Katie B 92
Psc 1h13'21" 14d42'45"
Katie B. Loftin
Tau 3h40'12" 19d14'21"
Katie Baby
Psc 1h3'5" 11d36'6"
Katie "Baby Girl" Hoyle
And 0h44'47" 38d28'42"
Katie Bagshaw
And 2h27'33" 50d20'6"
Katie Balk
Gem 7h6'0" 32d55'36"
Katie Bamber (nêe Perkins)
Cas 1h56'46" 61d49'13"
Katie Barker
Sco 16h8'31" -11d15'34"
Katie Bauer
Cap 21h7'51" -17d6'49"
Katie Bear
Cap 20h24'16" -9d45'14"
Katie Bear
Ori 6h12'44" 15d28'16"
Katie beauty as bright as this star
And 23h39'32" 46d28'6"
Katie Beckstrom
Lib 14h24'17" -19d21'24"
Katie Bedard
Lyr 18h48'15" 31d42'19"
Katie Bell
Cra 18h13'42" -40d46'25"
Katie Benson
Psc 1h21'1" 4d29'30"
Katie Bergdale
Cas 23h5'5" 57d59'37"
Katie Beth
Lib 14h31'29" -10d32'9"
Katie Beth
Uma 11h55'29" 36d23'54"
Katie Beth Pierce
And 0h22'17" 29d13'31"
Katie Beth Spier
And 1h6'27" 43d47'5"
Katie Birger
Ori 6h12'56" 18d43'4"
Katie Bjorgaard
Leo 9h47'16" 22d5'23"
Katie Blevins
And 0h13'10" 29d10'54"
Katie Bond
Vir 12h33'57" 0d41'56"
Katie Boozer's Beacon
Cnc 9h10'41" 28d15'28"
Katie Bopp
Vir 13h19'28" 13d25'59"
Katie Braganza
Lyn 6h38'44" 54d11'55"
Katie Bratley 4eva xD
And 1h16'4" 50d5'52"
Katie Breanna Laumann
Uma 12h54'30" 55d55'3"
Katie Bretzke
Cap 20h28'17" -10d54'37"
Katie Brideau
Pyx 8h48'58" -24d42'35"
Katie Bright
Lyn 7h37'9" 44d21'20"
Katie Brooke Buckwalter
Lyn 6h53'23" 55d46'23"
Katie Brown
And 1h54'26" 36d19'14"
Katie Brumley
Leo 10h11'35" 10d27'49"
Katie Bug
Ori 6h17'26" 14d38'7"
Katie Bug
Lyn 7h6'33" 61d37'1"
Katie Bug
Sco 16h13'14" -18d7'9"
Katie "Bug" Lynn Barry
Tau 3h48'39" 23d46'42"
Katie Bug's Star
Cap 21h41'19" -11d18'17"
Katie Burke
Ori 6h2'19" 9d52'48"
Katie Burke
Cyg 20h14'30" 47d12'30"
Katie Burr
And 0h54'59" 22d8'47"
Katie "buster" Ann
Tau 3h44'6" 30d26'26"
Katie Byrne
Cen 13h49'29" -41d54'51"
Katie Cafaro Livingston
And 1h8'33" 44d30'26"
Katie Calpin
Aqr 23h18'19" -18d25'17"
Katie Capello 22
Cas 1h13'38" 57d50'49"
Katie Carroll
Psc 1h28'36" 27d50'23"
Katie Carter Conlon
And 23h17'2" 51d39'52"
Katie Cassell
Cnc 8h37'24" 20d55'33"
Katie Cassidy
And 23h5'39" 44d59'49"
Katie Cassidy
And 1h58'48" 41d34'22"
Katie Cassidy Higgins
Aqr 22h8'44" 1d11'30"
Katie Cato Star
Pho 2h15'13" -45d27'44"
Katie Cecci
And 0h59'42" 43d20'24"
Katie Cecile Danna
Cam 3h59'5" 65d22'40"
Katie & Chad-Forever in the Heavens
Leo 9h57'10" 7d40'2"
Katie "Cheeks" Kirch
Uma 9h32'1" 46d2'14"
Katie Cheyenne
And 23h13'19" 52d49'45"
Katie & Chris Snee
Cyg 21h36'51" 46d44'14"
Katie Christine Nordvik
And 0h44'2" 25d24'18"
Katie Chrysler
Cam 6h36'10" 75d56'7"
Katie Cinnamon Fraser
Cam 3h56'40" 54d44'16"
Katie Ciurczak
Uma 11h17'54" 29d38'9"
Katie Clarke
Vir 11h57'31" -6d30'44"
Katie Clarke the reason is you
Sgr 17h56'45" -17d4'48"
Katie Click
Uma 8h54'24" 52d49'35"
Katie Colbert
Uma 10h33'58" 70d55'0"
Katie Corona Australis
Cra 18h7'24" -41d9'30"
Katie Cosenza
Cap 20h33'10" -9d53'13"
Katie Coughlin
Gem 7h42'22" 31d30'15"
Katie Crossett
Cas 1h30'36" 61d55'24"
Katie Cutshall
Uma 12h4'19" 39d3'14"
Katie D
And 1h44'18" 44d39'52"
Katie Danner
And 1h12'34" 36d7'59"
Katie Davis
And 1h42'2" 42d1'25"
Katie Dennett
Lyr 18h44'37" 40d51'36"
Katie Diffenderffer
Uma 10h19'50" 68d33'39"
Katie Ditzel
Psc 0h15'43" -2d52'23"
Katie Dubia
Vir 13h46'30" -14d0'29"
Katie Dupre
Sco 16h28'0" -31d35'58"
Katie Elisabeth McKenna
And 1h29'43" 44d2'59"
Katie Elise Bolus
Leo 11h18'55" 24d15'42"
Katie Elise Tominaga
Sgr 18h26'19" -16d44'55"
Katie Elizabeth
Lyn 7h33'38" 37d57'54"
Katie Elizabeth
And 1h3'59" 33d52'26"
Katie Elizabeth Ayer
And 23h12'51" 51d54'57"
KATIE ELIZABETH DAUD
And 1h29'52" 35d31'3"
Katie Elizabeth Foy
Sgr 17h54'14" -21d42'48"
Katie Elizabeth Fuhrman
Vir 13h59'53" 4d53'32"
Katie Elizabeth Leslein
Lib 15h35'17" -24d20'11"
Katie Elizabeth Munshower
Peg 21h52'57" 14d25'31"
Katie Elizabeth Pilgrim
Vir 14h42'32" -1d3'9"
Katie Elizabeth Rock
And 1h31'11" 39d26'59"
Katie Elizabeth Webb
Sco 15h57'20" -25d5'57"
Katie Ella Mary Sillick
Umi 15h48'18" 74d13'3"
Katie Ellen Genchi
Tau 3h59'19" 25d42'11"
Katie Ellen Young
Lyn 7h43'35" 47d40'39"
Katie Elliott
And 0h30'17" 26d3'56"
Katie Eloise
Cam 4h19'9" 66d9'50"
Katie & Emily
Cnv 13h56'0" 32d15'25"
Katie Erin
Tuc 23h15'3" -57d17'1"
Katie Erin Sarnowski
Lib 15h11'24" -27d49'17"
Katie Faduski
Crb 16h11'39" 39d6'35"
Katie Fais
Ari 2h18'27" 24d46'3"
Katie Faith
Sco 16h6'8" -10d20'24"
Katie Faugno
Lyn 8h39'4" 34d50'14"
Katie Ficca
And 0h14'11" 45d18'48"
Katie Finn
And 2h15'33" 46d14'49"
Katie Flood
Crb 15h39'55" 36d42'14"
Katie Forster's Star
Lyn 7h41'29" 42d49'26"
Katie Fox
And 1h18'46" 50d23'40"
Katie Frances Challinor
And 23h42'57" 41d28'36"
Katie Funge
And 23h22'35" 43d4'26"
Katie Funk
Ari 3h3'15" 14d23'13"
Katie Fussell
Leo 10h14'41" 12d2'4"
Katie G.
Cas 0h36'42" 57d10'22"
Katie Gaffney
Cas 2h40'45" 61d57'9"
Katie/Gavin Radike's Shining Star
Aqr 22h33'46" 1d39'35"
Katie Gay
Gem 6h56'21" 22d23'24"
Katie Gene Coble
Psc 1h32'30" 17d49'11"
Katie Gerber
Sgr 18h4'7" -28d23'26"
Katie Gilbey's Birthday Star
Psc 0h53'53" 18d1'8"
Katie Girl
Gem 6h46'7" 23d14'26"
Katie Girl
Cnc 8h50'25" 30d33'57"
Katie Girl Zuidema
And 0h18'55" 42d55'39"
Katie Glenn
Cap 21h1'28" -17d2'3"
Katie Grace Blackwell
Leo 11h11'24" 8d17'1"
Katie Grace Marshall
Ari 2h15'47" 24d14'40"
Katie Grdina
Psc 1h22'11" 25d17'34"
Katie Greer
Aqr 22h36'25" -2d27'46"
Katie Gysler
And 23h20'39" 49d12'43"
Katie Hale Ratzan
And 23h10'22" 44d43'21"
Katie Hales
Cas 0h55'21" 73d25'22"
Katie Hammitt
Lyr 18h25'32" 39d35'34"
Katie Hart
And 0h56'33" 41d47'59"
Katie Hathaway Gordon
And 0h36'3" 43d10'23"
Katie Healy
Lib 15h12'54" -14d5'16"
Katie Hefelfinger
Cyg 20h45'43" 37d4'54"
Katie Heilenday
Psc 0h4'32" 10d36'41"
Katie Hewitt
Ori 5h56'10" 6d54'47"
Katie Hewko (Squishystar)
And 2h31'49" 38d27'38"
Katie Hill
Uma 9h25'24" 42d6'23"
Katie Hill
Aqr 22h0'46" -14d25'37"
Katie Hobbs
Uma 11h18'49" 31d6'55"
Katie Hoffmann
Cap 20h22'34" -23d26'53"
Katie "Hollywood" Woodruff
Psc 1h39'11" 7d32'32"
Katie Hope Baker
And 0h9'16" 44d4'30"
Katie Hoy 2007
Psc 0h37'21" 16d44'32"
Katie I.L.Y.
And 0h17'52" 44d24'51"
Katie Is The Most Beautiful Star
Ari 2h28'56" 26d34'30"
katie j
Gem 7h13'54" 16d52'35"
Katie J
Tau 5h23'26" 23d8'1"
Katie Jacobson
And 2h30'43" 44d44'9"
Katie Jan Gale
Cnc 8h14'19" 16d27'0"
Katie Jane
Ori 5h39'7" 12d5'59"
Katie Jane Piggott
And 23h30'23" 38d13'58"
Katie Jane Schaffel
And 0h21'17" 25d31'20"
Katie Jane Winchester
Cas 0h56'1" 47d40'40"
Katie Javanaud - Katie's Star
Oph 16h26'49" -21d27'8"
Katie Jayne Reith
And 23h52'42" 39d12'41"
Katie Jean
Cap 21h33'49" -9d23'29"
Katie Jean
Cas 1h36'54" 68d21'37"
Katie Jean
Lyn 7h46'20" 53d56'14"
Katie Jean Little TMBF FMRS
Uma 10h56'13" 62d56'11"
Katie Jean Rincavage
And 1h53'7" 38d31'12"
Katie Jeffryes
Lmi 10h1'39" 40d19'34"
Katie & Jeremy's Start of Eternity
Sge 20h3'19" 16d52'18"
Katie Jo
Ori 5h53'25" 7d26'15"
Katie Jo
Cas 1h4'35" 48d56'5"
Katie Jo
Lib 14h47'5" -24d57'51"
Katie Jo Baby
Vir 12h54'30" 12d34'52"
Katie Jo Cooper
Tau 5h56'18" 24d22'15"
Katie Jo Guild
And 23h24'29" 40d50'56"
Katie Jo Sprague
Ari 2h46'15" 27d50'26"
Katie Johnson
And 0h28'56" 42d56'24"
Katie Johnson
Umi 16h7'6" 74d48'29"
Katie Johnson
Cap 20h45'19" -21d32'15"
Katie Jordan
Sco 17h56'34" -30d39'22"
Katie Judy Anna Smith
And 2h13'44" 37d28'46"
Katie June Ouderkirk
Sco 17h35'42" -38d7'6"
Katie Justine Black
Gem 7h38'58" 34d49'33"
Katie Kalna
Cap 20h38'16" -10d22'7"
Katie Kelly
Cnc 9h7'56" 31d47'59"
Katie Kelly DeSieno
Tau 5h56'30" 23d40'32"
Katie Kennedy
Lyn 7h48'53" 53d47'10"
Katie Kent
Mon 6h42'10" -5d58'41"
Katie Klara
Ori 5h58'24" 17d55'58"
Katie Kye
Cnc 8h27'45" 9d34'14"
Katie L. Alford
Leo 9h59'4" 20d50'5"
Katie L. Wilson
Psc 1h5'14" 31d38'13"
Katie Laine Wrublesky
Uma 10h30'40" 47d51'4"
Katie LaMarre
Sgr 18h24'29" -32d32'3"
Katie Larkin
Leo 10h16'26" 16d52'5"
Katie Lasater's Star
Per 2h40'30" 56d27'48"
Katie Leander
Leo 9h56'32" 21d24'7"
Katie Lee
Cas 1h28'57" 66d20'17"
Katie Lee & Nathaniel Gilbert Shaw
Cyg 19h39'49" 56d2'54"
Katie Leigh Anders
Lyr 18h49'55" 42d53'47"
Katie Leigh Barrows
Tau 5h32'32" 26d43'54"
Katie Leisegang
Cra 18h47'43" -39d26'53"
Katie & Lexi Shurts
Peg 23h36'44" 22d40'28"
Katie Lindsay
Aur 5h41'0" 54d50'19"
Katie & Lionel For Ever
Ari 2h10'52" 26d16'10"
Katie Lockett
Cas 0h28'9" 50d59'39"
Katie Lockett
Eri 3h48'46" -0d45'3"
Katie & Logan Forever
Cyg 21h35'27" 49d28'1"
Katie Lopatowski
Leo 11h50'39" 25d24'1"
Katie Louise
Lyn 7h44'42" 42d3'35"
Katie Louise 2/28/84-7/6/05
Lyn 7h26'30" 48d47'46"
Katie Louise Bateman
Cas 0h59'35" 66d34'23"
Katie Louise Haffie
And 23h21'49" 38d16'52"
Katie Louise Le Grove
And 1h55'42" 40d14'50"
Katie Louise - My Little Star
And 1h54'55" 37d45'53"
Katie Louise Piper
Tau 5h46'19" 17d11'42"
Katie & Luciano
Gem 6h54'10" 21d57'19"
Katie Luise Geer
And 23h13'52" 52d17'9"
Katie Lux Lucis
Psc 0h58'54" 26d44'24"
Katie Lynn
Lmi 10h40'29" 23d46'47"
Katie Lynn
Tau 5h53'30" 27d17'18"
Katie Lynn
Leo 10h57'58" 6d57'50"
Katie Lynn
Per 3h10'29" 53d52'47"
Katie Lynn
Lyr 18h41'51" 43d14'9"
Katie Lynn
Lyn 7h51'10" 41d59'24"
Katie Lynn
Cas 0h47'30" 60d50'28"
Katie Lynn Armstrong
Vir 13h13'24" -22d10'11"
Katie Lynn Byars
Cnc 8h14'23" 18d21'10"

Katie Lynn Connin
Lib 15h28'26" -8d12'1"
Katie Lynn Cross
Uma 10h4'34" 67d1'52"
Katie Lynn Dickel
Vir 13h11'48" 12d22'19"
Katie Lynn Gustafson
Vir 12h2'23" 4d19'47"
Katie Lynn Monks
Gem 7h9'54" 14d20'37"
Katie Lynn "Pepper" Schnaiter
And 23h17'54" 47d4'52"
Katie Lynn Romano
Gem 7h42'6" 34d25'4"
Katie Lynn Walker
Uma 13h11'34" 57d27'39"
Katie Lynnet scito te ipsum
Lyn 7h16'5" 49d9'26"
Katie Lynn's Star
Uma 11h16'15" 54d22'20"
Katie M Bouchard
Psc 1h25'42" 31d43'28"
Katie M. Dethlefsen
Aqr 23h2'57" -12d58'8"
Katie Madeline Santamaria
Ori 6h4'5" 19d5'1"
Katie Mae
Lyn 7h42'45" 41d49'13"
Katie Mae
Lib 14h48'20" -15d31'0"
Katie Maguire
Psc 1h40'11" 7d8'12"
Katie Mai Claire Hughes
Umi 14h40'30" 69d30'3"
Katie Malay
Lyn 8h57'59" 35d55'43"
Katie Marie
And 0h44'23" 44d45'23"
Katie Marie
And 2h11'55" 39d19'33"
Katie Marie
And 23h16'52" 48d35'47"
Katie Marie
Leo 11h15'20" 10d48'54"
Katie Marie
Lyn 7h23'8" 55d9'53"
Katie Marie Alcott
Cas 1h37'18" 66d56'37"
Katie Marie Calhoun
Lib 15h4'22" -5d47'4"
Katie Marie Dirkes
Cnc 8h34'15" 8d40'46"
Katie Marie Felty
Cap 21h49'5" -13d27'44"
Katie Marie Fitch
Tau 4h45'50" 19d50'50"
Katie Marie Gates
Cas 1h26'30" 62d46'52"
Katie Marie Hanna
Sgr 18h44'17" -29d37'13"
Katie Marie Johnston
Cap 21h55'48" -14d16'48"
Katie Marie Rowland
Vir 12h9'25" -7d56'3"
Katie Marsala
Aqr 20h50'17" -12d23'8"
Katie Mary Leonardi is Amazing
And 0h16'50" 44d12'16"
Katie Mary Walker
And 1h14'45" 45d46'21"
Katie Mashburn Meredith College '06
Vir 13h25'18" -9d45'37"
Katie Maurer
Dra 16h7'49" 53d20'59"
Katie Maurizi's Fire
And 2h2'19" 39d24'10"
Katie May
And 2h29'27" 40d18'9"
Katie May
And 2h28'8" 49d38'24"
Katie May
Cas 0h25'23" 60d39'41"
Katie May
Cap 21h23'37" -17d39'6"
Katie May Wilkens
And 0h12'54" 30d1'59"
Katie Maycroft
Sgr 18h12'52" -19d18'7"
Katie May's Star
And 2h7'24" 39d19'43"
Katie McDonald
Ari 3h13'0" 28d59'41"
Katie McDowall
And 23h14'14" 41d52'21"
Katie McMeekin
Lmi 9h27'9" 34d49'50"
Katie Mcteer
Uma 14h15'31" 57d57'57"
Katie Meagan Holihan
Peg 23h59'31" 29d58'0"
Katie Melinda
Lyr 18h49'15" 37d33'48"
Katie Menish
Peg 22h52'12" 21d44'26"
Katie Mi Lady
And 0h51'0" 37d0'15"
Katie Michel
Uma 8h33'44" 60d23'9"
Katie Miller
Psc 0h9'22" -1d47'37"
Katie Milligan
And 23h14'3" 46d46'29"
Katie/Misty Maloney
Per 3h25'18" 44d10'31"
Katie Mitchell Smith
Lyr 19h18'16" 35d17'13"
Katie Mohen
Peg 23h41'35" 30d4'41"
Katie Moran
Cnc 9h8'40" 30d57'47"
Katie Morse
Leo 11h30'56" 13d5'54"
Katie Muedeking, Best Mom Ever
Per 2h51'15" 50d43'28"
Katie my Guiding Light
Psc 0h35'37" 3d19'40"
Katie "My Souls Desire"
Lib 15h59'51" -19d42'3"
Katie Myers
Cap 21h0'8" -17d11'39"
Katie N' Jeff 12-20-2003
Cyg 20h22'41" 48d0'48"
Katie Nadine Brimer
And 0h29'29" 31d40'14"
Katie Nadine Tucker
Vul 19h45'14" 28d48'42"
Katie & Neil's Wedding Star
Cyg 20h52'23" 49d27'58"
Katie Nichole Richey
Aqr 22h1'44" -1d57'39"
Katie Nicole
Uma 9h26'20" 61d39'16"
Katie Nicole Sample
Ori 5h55'57" 21d36'15"
Katie Nicole Vasey
And 0h14'46" 48d29'36"
Katie Niezgoda, La Donna di Dio
Gem 6h55'47" 21d50'12"
Katie O'Banner of Nevers
Cas 1h2'11" 51d38'5"
Katie O'C
Uma 10h8'55" 61d22'1"
Katie O'Dell
Lyr 19h20'38" 34d6'46"
Katie O'Donnell
And 23h41'4" 44d48'18"
KATIE OGAREK - A STAR IS BORN
Psc 1h33'47" 27d31'21"
Katie O'Neill
Umi 10h27'33" 88d38'56"
Katie O'Rourke
Sgr 18h13'44" -19d18'39"
Katie Osteen
Cap 21h32'28" -15d5'57"
Katie Owens
And 2h17'34" 47d3'30"
Katie P. Green
Leo 11h8'44" -1d5'58"
Katie Painter
And 0h2'5" 44d32'45"
Katie Parenti
Vir 13h45'53" -5d59'26"
Katie Parkin
And 1h28'38" 49d47'29"
Katie Parrott
Ori 6h7'56" 5d56'33"
Katie Patricia Moroney
Lyn 8h25'28" 33d46'0"
Katie Penman
And 2h13'21" 48d13'43"
Katie & Peter's Magical Star
Cyg 19h39'13" 44d58'1"
Katie Peterson
Leo 11h15'0" 16d29'2"
Katie Petkus
Gem 7h12'47" 23d7'26"
Katie Philpot
Cyg 21h52'4" 37d14'30"
Katie Pika
Lyn 7h53'13" 53d31'40"
Katie Pisino
Aqr 21h52'41" 0d54'1"
Katie Pleski
Vir 13h12'41" 3d2'29"
Katie Pokres
Ori 4h59'52" -0d22'10"
Katie Pullin
Uma 9h15'2" 51d17'12"
Katie Punkstar Reynolds
Vir 13h18'35" 5d31'57"
Katie "Pure Beauty" Hoffman
Cyg 20h47'38" 30d38'18"
Katie Putnam
Ori 6h18'1" 10d16'38"
Katie Queen of Magic
Tau 4h19'43" 24d50'44"
Katie R. Crow
Lyn 7h39'21" 36d58'7"
Katie Rae
Psc 1h20'42" 18d26'27"
Katie Raeasis
Uma 11h19'3" 50d38'56"
Katie Rebecca
Gem 7h16'46" 32d9'51"
Katie Reid
And 0h37'18" 30d53'32"
Katie Renee
Ori 5h8'17" 6d40'34"
Katie Renee "Pookie" Schmitz
Ori 5h57'14" -2d22'17"
Katie Rockefellow
Cas 0h9'57" 59d42'37"
Katie Rose
And 1h37'21" 42d49'21"
Katie Rose
Ori 5h23'33" 14d2'49"
Katie Rose
Ari 2h8'22" 23d3'18"
Katie Rose Crawford
Cyg 20h56'51" 39d31'9"
Katie Rose Fox
Lyn 7h17'34" 55d18'23"
Katie Rose Simon
And 0h16'59" 32d8'21"
Katie Rosenberger
Sco 17h25'59" -38d26'4"
Katie Rosenkranz
Tau 4h43'9" 26d41'6"
Katie Rotundo
And 0h48'29" 40d56'50"
Katie & Ryan
Cyg 21h47'30" 48d41'19"
Katie Sadler
Gem 6h54'47" 28d57'24"
Katie Scarlett Wells
Del 20h55'12" 6d21'22"
Katie Schneider
And 23h43'53" 45d44'14"
Katie & Sean 11-11-05
Psc 22h51'25" 3d36'42"
Katie Sheridon
And 23h3'22" 50d44'11"
Katie Simpson
Cnc 9h8'25" 7d4'36"
Katie Smelas
Uma 9h21'15" 49d24'10"
Katie Sparkles
Cap 20h21'24" -12d18'38"
Katie Sredniawa
And 0h11'34" 29d3'42"
Katie Star
Sgr 19h37'10" -24d5'36"
Katie Stardust - You light up my Life
And 23h45'37" 35d4'57"
Katie Stephens
Sco 17h39'33" -45d26'41"
Katie & Steve
Cyg 19h44'20" 36d9'58"
Katie & Steve Johnson
Cyg 20h46'30" 44d59'59"
Katie Stevens
Lmi 10h29'46" 34d44'20"
Katie Stokes
Gem 7h32'34" 29d35'45"
Katie Stone
Sgr 19h13'45" -13d39'41"
Katie Stripling 04
Vir 13h53'54" -7d45'45"
Katie Sue
Col 5h57'16" -34d28'57"
Katie Sue
Vir 11h52'7" 8d36'43"
Katie Sukow
Leo 11h5'26" 11d3'31"
Katie Sunshine Princess Whitsett
Gem 6h30'36" 17d55'13"
Katie Tardio
Aqr 21h40'28" 2d30'35"
Katie "TaTa" Biritz
Lib 15h58'53" -5d44'19"
Katie Teresa Green
Her 16h29'5" 45d35'53"
Katie Terry
Her 17h3'58" 28d8'9"
Katie The Clown
Lib 15h40'36" -10d26'10"
katie the incredable
And 2h24'35" 46d40'49"
Katie Thomas
Cap 21h9'54" -17d38'2"
Katie Tillthestars Hadrunaway
Apu 16h7'59" -76d42'25"
Katie & Timmy
Cyg 20h35'40" 43d5'7"
Katie Trew
Crb 15h38'40" 39d3'35"
Katie & Tristran Shephard
Del 20h36'52" 15d27'30"
Katie Trusty
Tau 4h5'34" 23d49'5"
Katie Twinkletoes Fuss
Ori 6h13'0" 9d40'13"
Katie Utke
Boo 15h24'47" 46d28'49"
Katie Vacura
Vir 14h57'16" 4d28'11"
Katie Vail
Umi 16h59'0" 75d49'24"
Katie Vaughan
And 23h20'43" 48d15'12"
Katie Vi Bohatch
And 0h43'13" 31d47'41"
Katie Vogelgesang
And 1h9'49" 38d59'30"
Katie Von Yoakum
Ari 2h11'21" 24d8'10"
Katie Von Yoakum
Ari 2h12'57" 23d30'6"
Katie Walters
Uma 9h15'26" 46d43'40"
Katie Walworth -Dama de Noche-
Psc 1h0'38" 14d54'19"
Katie Weber
And 23h50'51" 47d56'32"
Katie Weeden and Keith Gerarden
Gem 7h32'7" 23d25'55"
Katie Welling
Uma 10h36'2" 59d32'44"
Katie Westmoreland
And 0h23'1" 32d23'41"
Katie White
Uma 12h47'7" 58d25'22"
Katie Williams
Ari 3h12'58" 19d56'13"
Katie Wilson
Sgr 18h14'51" -19d50'41"
Katie Wipfler
Psc 1h21'2" 12d12'40"
Katie Wojo
Gem 8h6'57" 30d4'19"
Katie Woo
And 0h54'41" 41d13'56"
Katie Young
Crb 16h22'3" 39d25'40"
Katie Zalenski
Cap 20h24'9" -18d1'23"
Katie Zey
Aur 5h56'34" 44d12'59"
Katie Ziehme
Cyg 19h56'33" 31d23'13"
Katie Zinda Ingram
Ori 6h0'51" 10d36'30"
Katie Zobec
Lib 15h9'6" -6d36'7"
Katie Zwolak
Aqr 22h5'10" -9d21'36"
Katie, I love you my angel
Cap 20h22'25" -23d9'37"
Katie, Kelly, Kristin, Mandy, Sarah
Crb 16h18'8" 34d17'10"
Katie, Love
Dra 17h20'1" 62d0'0"
Katie, meus decorus astrum
Cas 23h56'58" 50d12'54"
Katie, My beautiful softball star
And 23h35'26" 40d5'33"
Katie, Ross, and Ellie's Star
Tri 2h8'9" 33d4'27"
Katie, The JHU Star
Lib 15h27'43" -17d14'12"
Katie,Meagan & Sean Parsons
Umi 15h8'35" 87d37'46"
Katie5Wishart
Tau 4h29'20" 21d18'12"
Katieanna Groody
Vir 11h47'41" -6d13'10"
Katiebear
Psc 1h9'8" 10d17'19"
Katie-Bug
And 0h58'59" 41d14'45"
Katie-Bug
And 23h8'52" 51d13'33"
Katiebug
Sgr 18h37'14" -31d14'9"
KatieBug81
Lyn 7h23'30" 56d37'32"
Katiebug's Star
And 0h38'20" 21d42'29"
Katiedid
Psc 23h22'15" 5d39'22"
Katiedoll
Cas 0h39'29" 61d3'57"
KatieJames
Aqr 22h14'37" -1d57'12"
Katie-Kate
Ari 2h13'31" 25d41'29"
KatieLinn
Lmi 10h28'43" 37d56'6"
Katie-N-Scott
Cap 20h33'47" -19d5'40"
Katie's Charisma
Aqr 20h58'39" 1d16'25"
Katie's Diamond
Vir 12h48'38" 3d31'16"
Katie's Diamond
Her 17h31'20" 41d30'39"
Katie's Diamond in the Sky
Lup 15h48'4" -34d7'21"
Katie's Dream
Aql 19h20'6" 9d1'14"
Katies Eyes
Psc 1h4'24" 32d48'12"
Katie's Guiding Light
Vir 13h48'28" 2d34'59"
Katie's Heart
Cnc 8h8'55" 15d17'15"
Katie's Heaven
Cas 22h58'9" 53d53'10"
Katie's Hope
Uma 8h25'24" 62d26'1"
Katie's Hope
And 0h55'48" 35d37'1"
Katie's Illumination
Gem 7h43'16" 17d5'27"
Katie's Kaleidoscope
Her 17h8'6" 36d38'39"
Katie's Love Star
And 0h46'49" 38d38'52"
Katie's Lyric
And 23h0'29" 48d40'51"
Katie's Place to Call Home
Cap 20h49'50" -18d44'19"
Katie's Purity
Uma 11h21'35" 63d57'26"
Katie's Shady Shining Star
Lib 14h59'4" -18d40'33"
Katie's Shining Star
Uma 10h14'58" 52d27'51"
Katie's Shining Star
Vir 12h45'24" 3d34'10"
Katie's Shining Superstar
Leo 11h5'22" 21d47'27"
Katie's Smile
Tau 3h36'12" 22d41'53"
Katie's sparkle
And 1h28'3" 47d32'43"
Katie's Sparkle
And 0h43'4" 30d48'32"
Katie's Star
Cnc 8h48'5" 31d53'14"
Katie's Star
And 0h18'31" 43d10'26"
Katie's Star
And 2h37'36" 49d0'0"
Katie's Star
And 23h30'22" 44d46'40"
Katie's Star
Tau 5h17'37" 28d0'50"
Katie's Star
Leo 9h25'43" 26d0'35"
Katie's Star
Leo 10h57'24" 16d46'46"
Katie's Star
Leo 11h37'53" 22d15'4"
Katie's Star
Vir 13h18'23" 13d22'16"
Katie's Star
Ori 6h21'6" 10d38'22"
Katie's Star
Cnc 7h57'15" 12d8'26"
Katie's Star
Aqr 21h36'53" -0d21'56"
Katie's Star
Umi 15h31'21" 75d5'49"
Katie's Star
Umi 14h56'35" 74d53'42"
Katie's Twinkle
Uma 11h23'36" 30d37'49"
Katie's Vermont
Lib 15h23'18" -19d24'57"
Katie's Vibe
Psc 1h11'3" 18d20'38"
KatieSammy BFF
Umi 15h31'56" 73d10'25"
KatieSue
Lmi 10h19'4" 35d25'19"
KatieVanessaNg
Lyn 6h48'36" 51d50'28"
Katigan M Leoness
Aqr 23h34'10" -23d35'37"
Katika Zuri Ndoto
Cet 2h42'12" -0d32'9"
Katim
Lyn 7h45'15" 41d53'2"
Katina
Lyn 8h36'51" 33d37'6"
Katina
Leo 11h3'10" -2d34'41"
Katina Alexander
Crb 15h48'50" 37d6'9"
Katina & Daryl Hebert
Tra 16h15'49" -62d15'34"
Katina Emelia Anthony
Tau 4h16'40" 19d20'36"
Katina Jenell Vasquez
Cas 0h18'22" 57d57'7"
Katina Moore
Leo 11h32'44" 25d16'24"
Katingo Amoure Von Gulde Eterna
Uma 10h43'13" 62d52'45"
Katinka
Pho 0h37'16" -41d18'32"
Kati's Star
Vir 12h45'29" 12d7'17"
Katiuscia
Aur 5h56'6" 32d10'14"
Katja
Tri 2h27'27" 31d43'25"
Katja
Uma 10h18'27" 49d12'37"
Katja
Ori 4h57'33" 10d25'18"
Katja
Uma 9h6'35" 68d54'1"
Katja
Cas 23h3'28" 55d54'0"
Katja
Lyn 7h18'20" 52d52'21"
Katja
Uma 13h0'24" 53d53'11"
Katja & Joe
Vul 20h27'42" 27d12'4"
Katja&Marcel
Umi 17h6'15" 79d53'28"
Katja Martin Rettberg
Lib 14h51'15" -4d48'37"
Katja Pia, 22.06.1960
Ori 6h14'16" 20d41'46"
Katja Rasmussen
Uma 12h40'25" 57d4'22"
Katja Schäfer
Uma 8h26'4" 61d14'37"
Katja Susanna
Cap 20h27'50" -20d20'58"
Katja & Timo
Uma 10h24'50" 39d57'8"
Katja Williams
Cnc 8h3'9" 24d13'39"
Katka-30
Psc 1h33'25" 18d43'27"
Katla 4
Lib 14h46'34" -17d29'37"
Katleen Wolter
Uma 11h47'42" 32d1'23"
Katlena
Tau 4h31'45" 19d12'5"
Katlin Bess
Aqr 23h4'17" -5d15'7"
Katlin Brooke
Cyg 20h40'25" 53d21'46"
Katlin Charles Cravatta
Gem 7h38'45" 21d30'33"
Katlin Christine Tippie
Cap 20h52'15" -19d25'16"
Katlin Danielle Hickson
Ori 5h36'25" -6d47'12"
Katlin Michelle Whitesel
Cnc 8h33'33" 24d36'43"
Katlin Michelle Whitesel
Uma 12h10'27" 51d44'28"
Katlin Morgan Woodburn
Gem 6h30'10" 25d56'12"
Katlin Swadosh
Ori 6h8'56" 7d8'55"
Katlin Sweet Baby Girl King
Gem 6h34'17" 12d54'7"
Katline R.C
Cnc 8h25'17" 30d18'34"
Katlyn
Gem 7h12'48" 24d5'28"
Katlyn
Cas 0h39'6" 61d18'20"
Katlyn Ann Diana Johnson
Lib 15h1'10" -3d26'6"
Katlyn Ann Leathers
Uma 12h0'13" 31d49'47"
Katlyn Anna Rego
Gem 7h45'11" 32d48'8"
Katlyn Archer
Lyr 19h9'34" 45d11'1"
Katlyn Elise
Aqr 22h53'57" -11d19'7"
Katlyn Elizabeth
Vir 13h17'46" 6d54'26"
Katlyn Owens
Lyn 7h31'0" 48d33'50"
Katlyneu
Cas 1h30'43" 62d30'12"
Katlynn & Aaron Star
Umi 15h24'3" 73d13'31"
Katlynn Caroline Begraft
Vul 20h18'11" 22d49'43"
Katmandu
Uma 13h20'41" 56d8'51"
Kato Bachus
Uma 11h6'7" 47d47'19"
Kato Wilson
Cam 4h9'28" 68d33'11"
Kat-Ory
Pho 23h42'23" -52d7'23"
Katrin
Gem 7h39'34" 32d24'43"
Katrin and Alen
Cnc 9h12'8" 7d52'59"
Katrin Dirk
Uma 10h0'2" 46d14'15"
Katrin Grün
Uma 10h24'10" 41d2'49"
Katrin Olschnegger I Love You
And 0h40'24" 35d46'30"
Katrin Scherz
Ori 5h55'38" 12d44'33"
Katrin "Spocki" Loock
Aql 18h57'49" 15d46'28"
Katrin Tust
Lib 15h38'18" -14d17'44"
Katrin und Sascha
Uma 12h20'13" 58d4'17"
Katrina
Dra 16h1'21" 58d35'47"
Katrina
Uma 8h44'22" 70d43'45"
Katrina
Sco 16h15'41" -10d4'17"
Katrina
Psc 0h1'11" 1d55'29"
Katrina
Sco 16h52'4" -27d6'38"
Katrina
Vul 19h31'58" 23d33'20"
Katrina
Gem 6h6'24" 26d45'10"
Katrina
Psc 1h27'10" 11d0'49"
Katrina
Tau 5h13'21" 19d8'2"
Katrina
Psc 1h21'7" 16d44'14"
Katrina
Lyr 18h50'10" 36d54'41"
Katrina
Cyg 19h44'27" 30d7'56"
Katrina
And 0h48'12" 39d16'32"
Katrina Amber Longobardi
And 1h18'21" 34d33'49"
Katrina and Joey's I LOVE YOU STAR
Sgr 17h49'34" -27d19'40"
Katrina Ann (Fisher) Newcomb
Sco 17h9'49" -38d20'15"
Katrina Ann Rome
Umi 13h44'40" 78d0'5"
Katrina Anne
Cyg 21h27'49" 51d16'42"
Katrina Bella
Tau 4h43'16" 23d32'22"
Katrina Berndt
And 1h25'39" 39d21'50"
Katrina Campbell
Lib 15h38'48" -19d59'58"
Katrina Carol (Kaycee)
Peg 22h22'9" 15d32'42"
Katrina Chersicla
And 1h34'50" 49d3'58"
Katrina Clarissa Pascual
Aqr 22h2'18" -13d0'55"
Katrina Cutter
Leo 11h0'15" 18d0'52"
Katrina Danelle Monk
Cyg 19h44'48" 40d17'37"
Katrina Davis
Com 12h29'39" 26d58'36"
Katrina & Dion's Christmas Star!
Ori 5h35'9" -3d39'26"
Katrina E. Kenward
Lmi 10h42'56" 25d51'47"
Katrina Eldridge
Uma 12h57'49" 61d29'47"
Katrina Elise Elizabeth Webb
Ari 2h48'29" 28d14'11"
Katrina Eugene
Cnc 8h46'51" 23d9'11"
Katrina Frey
Leo 10h20'12" 11d55'39"
Katrina Holloway
Lib 15h29'31" -26d11'44"
Katrina Hrubiec
Cyg 20h54'50" 31d0'50"
Katrina ingen Columb
And 0h2'57" 34d21'13"
Katrina Jane Hauck
Cen 13h2'29" -47d22'44"
Katrina JoAnn Haugen
Peg 23h34'49" 13d37'23"
Katrina Juarez
And 0h48'39" 39d33'56"
Katrina L. Gilbert
Psc 0h37'12" 8d9'7"
Katrina L Murrell
Lib 15h9'17" -5d25'10"
Katrina Louise Magdol
And 0h46'7" 40d52'12"
Katrina Louise Vernon - Mummy's Star
Cas 23h2'12" 55d38'33"
Katrina Lynn Cloutier
Ori 6h8'49" 20d46'35"
Katrina Lynn Cooley
Psc 0h30'25" 7d12'41"
Katrina Lynn Knapic
Sgr 19h30'52" -37d52'33"
Katrina Lynn Walter
Sgr 19h12'29" -35d22'19"
Katrina M. Pedrini
Ari 1h57'17" 19d9'9"
Katrina Maney
Aqr 21h33'42" 2d16'57"
Katrina Maree
Leo 11h9'56" -2d19'4"
Katrina Marie
Lib 15h41'59" -16d54'42"
Katrina Marie
Uma 11h32'17" 63d26'31"
Katrina Marie
Lyn 7h22'6" 50d11'28"
Katrina Marie Grabbe
Aql 19h14'13" 1d55'33"
Katrina Marie Sheely
Tau 4h25'46" 19d24'25"
Katrina Marie Smart
And 1h0'50" 46d11'7"
Katrina McGough
And 23h25'54" 46d55'39"
Katrina Michelle Payton
Gem 6h32'30" 22d40'7"
Katrina Neefus
Aur 6h16'39" 51d32'47"
Katrina Nelson
Sco 16h41'43" -32d33'12"
Katrina Neran
And 0h40'6" 43d44'47"
Katrina Nick
Uma 9h21'39" 66d59'53"
Katrina O
Psc 1h20'5" 14d21'42"
Katrina "Pearl of Wisdom"
Cru 12h25'45" -62d19'43"
Katrina Pickard
Lyn 7h27'31" 45d27'6"
KATRINA R#1
Mon 6h48'29" 7d28'24"
Katrina Rachelle Weaver
Col 5h38'13" -31d36'27"
Katrina Rae
Ori 5h59'37" 21d11'51"
Katrina Raquel Lazenby
Cnc 8h53'19" 14d46'57"

Katrina René
Ori 6h13'29" 15d7'33"
Katrina Renee
And 23h49'16" 42d11'52"
Katrina Roberts
And 2h5'42" 43d0'3"
Katrina Ruth Daoud
Uma 10h58'25" 48d36'58"
Katrina * Ryan * Fred
Uma 11h47'6" 59d59'40"
Katrina Savage
And 23h35'40" 36d47'4"
Katrina Sheree Tom
Cnc 8h42'26" 22d15'52"
Katrina Smith
Psc 1h18'10" 26d37'21"
Katrina Starling
Boo 15h5'15" 40d46'36"
Katrina Starr (12/29/80-4/9/05)
Uma 8h38'1" 57d8'7"
Katrina Stepikura
Gem 7h37'13" 14d3'16"
Katrina Susan Lehman
Aqr 22h35'39" -1d10'39"
Katrina Torelli
And 1h33'43" 48d30'13"
Katrina - Untainted Star
Col 6h3'45" -30d19'20"
Katrina Valencia
And 2h38'14" 38d56'22"
Katrina W. Gnat
Psc 0h38'55" 9d22'23"
Katrina Wallace
Vir 13h17'43" -20d48'52"
Katrina Zanfirov
And 0h19'39" 34d36'50"
Katrina, Mei-Lan Chen (AMK)
Uma 10h37'30" 70d26'33"
Katrinaballerina
Lyr 19h11'59" 26d15'21"
Katrina-Jason
Sco 17h6'3" -35d47'44"
Katrina-Lyn Nicolle Laws
Cnc 8h11'34" 25d12'57"
Katrina's Brent
Lyr 18h53'42" 36d59'23"
Katrinas Karma
Uma 10h2'41" 70d44'31"
Katrina's Little Piece of Heaven
Crb 16h3'13" 30d50'2"
Katrina's Papilio
And 23h39'16" 33d9'13"
Katrina's shining light
Tau 4h17'50" 20d42'33"
Katrina's Smile
And 0h52'32" 40d20'22"
Katrina's Star
Ori 6h11'40" 9d11'36"
Katrina's Star
Vir 12h49'27" -10d31'42"
Katrina's Star
Vir 13h42'6" -1d45'58"
Katrine Kolberg
And 22h59'38" 50d39'12"
Katrine Rachel Levy
Gem 7h28'23" 23d40'45"
Katrine-Snut
Ori 5h16'19" -1d52'13"
Katrizzle
Lib 14h54'19" -1d56'37"
Katryn Pichette
Vir 14h56'10" 3d20'34"
Katryna Lynn Baker
Ari 3h21'52" 29d34'9"
Katryna Seania
Gem 7h4'36" 30d46'39"
Kats
Sco 16h10'35" -11d14'10"
Kats Meow
Tau 5h39'52" 24d14'58"
Kat's Star
Gem 7h50'38" 28d36'40"
Kat's Star
Uma 11h11'39" 36d46'33"
Kat's Star
Mon 6h48'30" -0d17'39"
Kat's Uncle Chris
Uma 9h26'59" 44d35'32"
KatsMark
Sco 17h55'23" -37d9'20"
Katsouni
Uma 9h22'30" 69d21'30"
KATstar
Lyn 9h6'1" 41d28'59"
Katstar
Ori 6h18'21" 10d9'5"
Katsu Hakoda
And 0h41'0" 25d31'17"
Katsu & Tomo's Shin' Star
Gem 6h24'32" 25d2'3"
Katsue
Cap 20h9'4" -14d27'39"
Katsumoto
Her 17h18'45" 46d15'49"
KATT
Uma 8h57'19" 56d42'10"
Katt Kopecky
And 0h33'21" 37d57'2"
Katteryne & Juan Manuel
Cyg 20h53'53" 46d26'6"
Kattie Pirkle
And 0h13'46" 42d7'43"

Katts Preshus
Lmi 10h33'37" 35d32'37"
Katty Munchkin
Sgr 19h10'10" -24d52'11"
(Katty&Ronnie) Eternal Love
Cyg 20h48'41" 47d18'44"
Katura Crum
Vir 12h33'28" -9d40'34"
Katurah Desiree Carter
Aqr 21h4'22" -10d2'28"
Katusha
Uma 13h48'50" 53d49'5"
Katy
Aqr 23h11'51" -19d47'17"
Katy
Cas 0h5'44" 59d3'46"
Katy
Boo 14h34'26" 39d5'7"
Katy
And 0h2'3" 37d36'50"
Katy
And 0h27'13" 27d5'2"
KATY
Psc 23h25'30" 3d27'30"
Katy
Aql 20h11'18" 4d31'50"
katy_03
Oph 17h10'48" 2d56'48"
Katy Alexandra
Cas 2h11'15" 65d22'59"
Katy Barber
Cnc 9h12'31" 21d57'40"
Katy Becker
Leo 11h0'38" 3d41'11"
Katy Bobby Wolf & Lux's Family Rock Star
Mon 6h48'14" -6d49'55"
Katy Boone
Mon 7h32'45" -0d39'33"
Katy Chin "Ping Ying" Peng
Ari 2h56'13" 27d46'0"
Katy Christy
Cnc 8h34'2" 14d2'57"
Katy Elgie
Cas 1h44'50" 65d18'30"
Katy Elizabeth
Lib 15h7'10" -23d9'50"
Katy Elizabeth Smith
And 2h28'44" 45d43'28"
Katy Emma Bowling
Peg 22h54'35" 15d16'31"
Katy & Eric Sandoval
Uma 11h23'49" 56d1'13"
Katy Gilbert
Aqr 22h53'23" -5d48'18"
Katy Goldfarb
Lmi 10h33'35" 38d4'37"
Katy Grace Payne
Aql 20h8'41" -1d1'49"
Katy Grau
Uma 13h27'4" 53d6'17"
Katy Hankins
And 23h16'46" 41d16'2"
Katy Kal
Leo 9h24'13" 14d38'9"
Katy "KatyByrd" Masuga
Uma 9h26'42" 49d48'19"
Katy Kryger
Dor 4h43'54" -54d58'54"
Katy Lane White
Leo 10h43'8" 19d37'5"
Katy Langdon
Peg 21h42'54" 25d53'41"
Katy Lee Doherty
Cap 20h48'4" -15d59'57"
Katy Leigh
Umi 13h4'9" 74d24'32"
Katy Leigh Piechura
Psc 1h32'4" 17d39'31"
Katy Li
Gem 6h50'35" 33d58'54"
Katy Lynn Lund
Cnc 8h34'30" 31d31'49"
Katy Marie Burns
Vir 13h10'50" 12d27'10"
Katy McCoy
Leo 10h27'59" 22d43'2"
Katy Michelle Williams
Cap 21h57'38" -17d35'32"
Katy Peterson
Ari 2h27'0" 26d25'23"
Katy Q Star
And 23h16'16" 49d33'55"
Katy Rose
Uma 9h8'59" 53d0'39"
Katy Saban
Uma 8h48'39" 53d29'52"
Katy Shepherd
Gem 7h8'51" 21d19'8"
Katy- Shining In Our Hearts Forever
Cyg 21h44'16" 50d23'32"
Katy Skinner
And 23h17'16" 43d12'6"
Katy Sue
Gem 7h13'13" 24d7'14"
Katy Susan Sherlock
And 23h1'22" 50d41'9"
Katy Tortorici
Sgr 18h54'36" -29d52'54"
Katy Traurig
Cma 7h12'26" -18d13'29"
Katy Victoria Rutledge
Uma 9h8'13" 51d18'34"

Katy Williams
Gem 6h28'20" 25d56'57"
Katy, veux-tu m'épouser ? Fréd
Umi 14h20'41" 70d29'16"
Katya
Leo 9h49'23" 28d5'33"
Katya
Ori 5h27'41" 7d4'28"
Katya and Clement
Cyg 20h35'38" 33d16'57"
Katya Antoinella Davis
Aql 19h5'38" -6d2'49"
Katya Grace
Sgr 18h15'27" -24d21'58"
Katya Li
Crb 15h50'47" 35d23'48"
Katya Marie Monarski
Dra 17h59'30" 68d57'50"
Katya Selene
Psc 0h17'9" 17d40'28"
Katya Shcherbakova
Aqr 21h9'2" -14d16'7"
Katya Watelski
And 1h4'19" 38d24'2"
Katya Yuryev
Leo 11h21'17" 3d57'36"
Katyapoo
Uma 12h51'47" 57d16'22"
KatyBoBrunoSampson
Tau 5h57'52" 24d56'25"
Katybug
Peg 23h54'16" 28d44'51"
Katy-Bug
Mon 6h48'13" 7d28'16"
Katybug
Cas 0h45'10" 60d9'18"
Katy-did
Ori 4h56'32" 9d51'59"
Katydid
Gem 7h29'8" 34d10'25"
katykat
Cas 0h57'37" 48d22'39"
KatyLeonardo
Lib 15h11'36" -4d21'30"
Katy's Guardian 12/04/2003
Lyr 18h51'39" 35d25'35"
Katy's Little Star
And 23h24'25" 50d42'34"
Katy's Star
And 1h26'5" 49d43'28"
Katy's Star
Lib 14h49'42" -4d6'2"
katz
Leo 10h47'15" 15d27'3"
Katzer Kati Örök Szerelmem
Uma 10h20'8" 60d24'49"
Kaufman
Lib 15h9'22" -3d34'44"
Kaulahea
Vir 13h42'14" -17d1'29"
Kaulmann, Anja
Uma 8h48'42" 57d52'29"
Kaulukukui-Duerr
Cyg 20h5'10" 42d12'24"
Kaus Medius Pathak
Sgr 18h19'31" -28d11'6"
Kaushik "Kanu" Mankad
Cam 5h0'56" 57d26'16"
Kauto Star
Peg 22h52'33" 9d57'18"
Kaveeta Peetra Sukhdeo
Oph 17h14'26" -0d54'9"
Kaveh
Cnc 9h16'54" 22d53'35"
Kaveman
Tau 3h55'24" 17d54'46"
Kavin (King) Pike
Cep 22h8'16" 58d9'47"
Kavita Bahl
Ari 2h44'41" 28d53'6"
Kavita Chadee
And 0h43'12" 41d20'33"
Kavita Jayant Gaglani
Uma 8h31'2" 63d15'8"
Kavita's Star
And 1h57'12" 42d32'28"
Kavitha Vijayan
Psc 23h28'12" 3d41'15"
KAVL
Uma 11h30'16" 62d52'29"
KAW + LDSJ
Uma 11h59'25" 37d30'34"
Kawabataya
Uma 10h31'36" 66d43'19"
Kawandeep Virdee
Crb 16h23'33" 29d17'16"
Kawehelani Auld
Vir 14h15'8" 5d9'48"
Kawkaw Serge Liajli Al Hayatte
Ori 5h13'4" -0d27'56"
Kay
Uma 12h25'6" 61d10'45"
Kay
Cas 1h25'25" 64d28'29"
Kay
Psc 0h45'1" 7d58'40"
Kay
Peg 23h52'48" 24d52'23"
Kay
And 23h44'3" 33d51'18"

Kay
And 1h16'39" 34d15'33"
Kay
Per 3h28'18" 49d20'58"
Kay Achtelik
Uma 9h23'32" 49d27'11"
Kay and Gates Scoville
Cyg 19h59'21" 31d32'49"
Kay and Joe's Star
Boo 14h51'36" 21d2'43"
Kay and Mike's Rocco Star
Cap 20h16'1" -11d17'38"
Kay and Moochie
Sco 16h7'25" -16d54'4"
Kay Angelica
Dra 20h11'41" 64d26'45"
Kay Angus's Birthday Star
Sco 16h56'32" -38d2'2"
Kay Anne
Peg 23h37'8" 13d40'7"
Kay Aycock
Leo 9h52'35" 17d40'7"
Kay B. Koglin
Cyg 19h27'26" 51d57'39"
Kay Beck
Vir 13h18'32" 5d24'2"
Kay Boo Boo
And 2h29'36" 38d17'50"
Kay C. Aycock
Sco 17h52'41" -30d26'7"
Kay Carpenter ~ "Princess of Stars"
Sgr 18h14'55" -18d56'43"
Kay Cooley
Cas 1h8'12" 61d46'45"
Kay DeFrancesca
Equ 21h10'55" 11d34'33"
Kay Elizabeth Estes Gravette
And 0h41'46" 33d4'6"
Kay Foth
Sco 16h52'43" -36d32'16"
Kay Frances
Lyr 18h27'20" 32d47'53"
Kay Francis Proberts 30-12-1998
Cru 12h50'47" -57d44'24"
Kay & George Anderson
Umi 15h14'2" 68d53'48"
Kay Gould
Psc 1h16'38" 8d32'40"
Kay H. LaBau
Lyr 18h37'33" 34d3'43"
Kay Hommel
Uma 9h27'44" 45d22'45"
Kay Joyce
Lmi 9h59'34" 38d39'0"
Kay Kay
And 0h17'40" 45d38'12"
Kay Keen
Gem 6h56'22" 15d31'43"
Kay Kime
Per 4h7'55" 34d59'34"
Kay Klinkenborg
Peg 23h47'48" 11d4'30"
Kay Kohler
Crb 15h42'59" 28d47'14"
Kay Koop
Ori 6h3'37" 18d42'58"
Kay Laurel
Psc 1h10'55" 15d17'17"
Kay Lile
Crb 16h13'43" 34d41'46"
Kay Lin
Srp 18h12'5" -0d16'41"
Kay Lorraine
And 2h19'26" 43d7'4"
Kay Lorraine Hansen
Aqr 21h55'50" 1d14'53"
Kay Lyn Robinson
Cyg 20h30'47" 59d8'0"
Kay Lynn
Lib 15h5'58" -8d29'33"
Kay Lynn
Vir 13h26'1" 3d40'15"
Kay Lynn Dykes
Aqr 22h52'38" -6d3'29"
Kay Lynne Bowers
Cas 0h10'43" 55d31'51"
Kay May
Cas 1h30'50" 63d35'8"
Kay McDonald
Peg 23h10'43" 17d23'58"
Kay McGarvey
Lyn 9h7'46" 33d53'57"
Kay Miller
Sco 17h52'24" -36d0'48"
Kay Moffet
Peg 22h43'4" 5d27'20"
Kay Mortellaro
Cyg 19h51'7" 36d43'13"
Kay Mutton
Cas 23h45'9" 55d38'51"
Kay Pagano
Vir 14h37'23" 3d3'13"
Kay & Paul March 5, 1944
Uma 9h22'19" 58d25'27"
Kay Ross
And 0h42'52" 45d19'31"
Kay Roycroft
And 1h22'5" 47d3'7"
Kay Scharch
Sgr 18h27'22" -16d56'30"
Kay Small's Star
Psc 0h53'19" 27d52'14"

Kay Sobczak
Per 3h50'48" 48d48'10"
Kay Star
Mon 7h2'23" -6d15'19"
Kay Stauder
Cnc 8h37'23" 7d47'10"
Kay & Steve Knee
Crb 15h46'12" 37d32'31"
Kay Stirling
Ind 21h40'27" -70d59'12"
Kay Strong
Tau 5h18'32" 28d6'49"
Kay T. LaBau
Lyr 18h47'58" 35d3'38"
Kay Witt
Cnc 8h42'44" 12d4'1"
Kaya
And 1h35'44" 46d35'51"
Kaya
Cnv 12h28'19" 47d11'37"
Kaya
Ori 5h38'42" -1d1'11"
Kaya Amber Rowan
And 23h6'31" 52d11'0"
Kaya and Connie 1946 Love Endures
Uma 11h12'54" 67d23'6"
Kaya Glick
Umi 15h10'25" 87d22'16"
Kaya Grace
Vir 14h3'49" -13d5'29"
Kaya Jade
Del 20h58'3" 12d8'57"
Kaya Lilley
Ori 5h26'45" -4d8'11"
Kaya Rose
Umi 14h4'43" 76d28'16"
Kaya Samantha Brewer
Cru 12h0'2" -64d17'41"
Kayan
Mon 6h45'35" -0d25'37"
Kayana Daniella Benson
Cas 23h10'1" 59d16'15"
Kayanasadat Dakhilitabatabaei
Uma 12h49'28" 55d24'43"
Kayanna King
And 2h28'3" 41d28'54"
Kaya-Rose Anela o' kalani Cabilao
And 2h8'48" 45d36'14"
Kaybaby
And 0h36'4" 32d57'27"
Kayce
Uma 10h7'30" 46d11'36"
Kayce
Cap 20h33'11" -23d40'11"
Kayce Lynn Clark
Cap 20h35'24" -12d1'51"
Kayce Marriner
Lmi 10h29'50" 36d2'53"
KayCee
Cas 0h38'58" 50d48'56"
Kaycee E Owens My Angel On Broadway
Vir 14h37'45" 1d35'46"
Kaycee Grace Richman
Ori 6h1'21" 17d47'1"
Kaycee Lyne A. Doria
Leo 10h11'3" 10d26'1"
Kaycee Lynn Sexton
Sco 16h8'16" -16d20'53"
KayDan Star
Per 3h38'1" 46d3'13"
Kaydee
Umi 16h35'38" 77d12'35"
Kayden Cole
Vir 13h26'10" 13d39'24"
Kayden Dylan Boen
Dra 19h46'0" 63d40'49"
Kayden Michael Long
Cnc 8h2'29" 14d50'1"
Kayden Ray Charlie Stewart
Per 4h21'17" 49d8'8"
Kayden Ray Levin
Peg 22h14'39" 32d19'33"
Kaydence Andrea
Cyg 21h11'26" 45d12'52"
Kaydence Darlene Johnson
Cas 2h12'10" 66d19'39"
Kaydence Lynn
Lmi 9h25'21" 37d3'59"
Kaydence Marie
Lib 14h58'27" -9d57'17"
Kaydence Marie Krause
And 1h38'56" 41d47'56"
Kaydree Mya
Lib 15h27'18" -6d35'53"
Kaye
And 0h40'52" 30d18'26"
Kaye and Mort
Vir 13h5'20" -20d51'55"
Kaye Francis Brown-England
Sgr 19h7'15" -25d52'16"
Kaye Robert Shazsam
Cru 12h26'9" -58d42'13"
Kayelon
Cnc 8h52'41" 24d26'25"
Kayhan Tabrizi
Cap 20h56'59" -20d49'45"
Kayin Duran Pereirra
Her 18h28'45" 16d18'0"

Kayin Jameson Adeyemi
Uma 10h8'49" 54d50'7"
Kayla
Uma 11h1'54" 66d5'10"
Kayla
Uma 12h10'10" 62d53'52"
Kayla
Aqr 21h22'16" -13d31'18"
Kayla
Umi 23h53'13" 88d44'16"
Kayla
Pho 0h30'29" -40d29'39"
Kayla
Sco 17h31'57" -40d48'39"
Kayla
And 0h36'8" 27d57'18"
Kayla
Ori 6h1'49" 21d16'41"
Kayla
Leo 11h4'8" 16d49'33"
kAyLa
Gem 6h44'47" 24d5'16"
Kayla
Gem 6h53'4" 23d42'18"
Kayla
Tau 4h17'48" 23d52'44"
Kayla
Her 17h25'40" 27d27'49"
Kayla
Vir 12h48'35" 5d31'27"
Kayla
And 0h51'6" 34d37'28"
Kayla
Tri 2h30'22" 36d22'1"
Kayla
Cnc 8h45'50" 31d31'19"
Kayla
Cyg 19h38'27" 32d20'39"
Kayla
And 1h44'9" 42d45'37"
Kayla
And 2h10'7" 43d50'11"
Kayla
Uma 11h34'17" 38d52'43"
Kayla
And 23h59'40" 47d47'11"
Kayla
And 2h37'2" 49d0'27"
Kayla
And 0h14'20" 45d46'15"
Kayla
Crb 15h37'9" 37d40'19"
Kayla
Cyg 20h37'18" 40d59'49"
Kayla A Acosta
Uma 10h57'41" 60d54'50"
Kayla Adeline Fox
Uma 8h31'46" 69d29'27"
Kayla and Keith
Cyg 20h28'9" 55d36'8"
Kayla and Kelina's Pooky
Cnc 9h11'8" 31d16'50"
Kayla and Nate Henise
Cyg 20h30'9" 33d4'12"
Kayla and Steve Nadal
Cyg 20h18'52" 55d6'2"
Kayla Ann
Her 17h50'7" 34d55'9"
Kayla Ann
Cnc 9h9'3" 7d50'45"
Kayla Ann Cameron
Aqr 23h49'59" -19d30'33"
Kayla Anne Corso
Uma 10h21'7" 43d59'23"
Kayla Anne Elstien
Psc 1h19'29" 24d25'33"
Kayla Ann-Marie Myers
Leo 10h48'25" 19d44'51"
Kayla Arroyo
Lyn 7h12'15" 49d21'20"
Kayla Ashley You Light Up My Life
Sgr 19h19'26" -22d46'8"
Kayla B
Crb 15h46'19" 37d2'36"
Kayla Baumann
Uma 11h48'34" 42d55'36"
Kayla Beth
Cyg 20h50'43" 34d16'48"
Kayla Beth Stewart's Star
And 23h41'20" 42d42'54"
Kayla Bliss Czajkowski
Tau 4h57'5" 16d49'53"
Kayla Blurr Talls
Cnc 8h50'29" 8d19'28"
Kayla Bree Bauer
Aqr 21h30'27" 2d23'50"
Kayla Brown
Umi 16h10'51" 75d0'11"
Kayla Chiara Mills
Cet 1h1'8" -0d24'55"
Kayla Christine Friedrichsen
Ori 5h32'21" 14d41'53"
Kayla Christine My little beautiful
And 0h8'36" 43d42'53"
Kayla Cody
And 1h15'17" 36d59'4"
Kayla Coleman
And 1h41'52" 43d18'33"
Kayla & Corey Forever
Cyg 21h13'42" 42d41'37"
Kayla Cullins
Vir 12h49'34" -1d26'34"

Kayla Danielle
Cam 4h4'36" 55d27'48"
Kayla Danielle McGrew
Sgr 18h51'16" -29d57'8"
Kayla Danielle Sanders
Cyg 20h3'38" 40d29'10"
Kayla Darlene
Mon 6h36'20" 8d8'13"
Kayla Darlene Dondero
And 23h28'59" 47d8'18"
Kayla Dawn
Cap 20h20'6" -11d48'46"
Kayla Destany Marie Vandever
Sgr 18h15'41" -17d25'30"
Kayla Diane Floyd
Del 20h34'49" 5d53'38"
Kayla E. Marrufo
And 23h46'56" 42d55'33"
Kayla Elizabeth Canale
Sco 16h50'56" -28d7'37"
Kayla Elizabeth Clemente
Vir 14h12'36" -8d17'26"
Kayla Elizabeth Machleit
Ari 3h14'54" 27d53'11"
Kayla Elizabeth Manning
Psc 1h5'53" 13d9'59"
Kayla Elizabeth Montgomery
And 23h29'31" 45d6'37"
Kayla Elizabeth Morrison
And 0h18'10" 43d45'47"
Kayla Ellena Thomsen
Aqr 22h51'30" -4d1'21"
Kayla Emily Grace Clayton "Muffin"
Uma 13h34'21" 55d2'44"
Kayla Evon Cline
Uma 9h30'22" 47d7'39"
Kayla Faith
Gem 7h37'14" 33d54'27"
Kayla Fekel
Cnc 8h32'40" 28d49'43"
Kayla Frost
Ori 4h58'55" -0d21'9"
Kayla Gatto
And 1h22'45" 34d34'10"
Kayla Glover
And 1h7'24" 34d57'29"
Kayla Grace Gutkoski
Sco 17h5'35" -37d38'17"
Kayla Hall
And 0h49'23" 40d43'17"
Kayla Heather Peter
Cyg 20h7'46" 31d41'20"
Kayla Hensley
Lyn 7h53'28" 58d4'49"
Kayla Hildebrandt - Forever Shining
Col 5h50'41" -41d37'23"
Kayla Hunt
And 2h22'22" 49d51'57"
Kayla "Idiot" Rodriguez
Tau 4h29'57" 12d4'23"
Kayla Irene
Vir 12h12'2" 5d28'49"
Kayla Jane Wenrich
Umi 13h31'18" 74d7'10"
Kayla Jean
Tau 5h58'11" 27d20'1"
Kayla Jean
Leo 11h2'17" 20d40'16"
Kayla Jean Bay
Leo 10h32'2" 17d48'34"
Kayla Jean Brigham
And 1h12'17" 38d49'12"
Kayla Jean Higginbotham-Mathews
Cas 23h44'51" 55d20'31"
Kayla Jennifer
And 2h33'16" 39d44'9"
Kayla Jo Coffman
Uma 10h46'50" 57d8'52"
Kayla Joanne
Aqr 22h54'44" -11d23'34"
Kayla Johanna Hunt
Aqr 22h28'40" -23d39'7"
Kayla Jones
Cam 4h17'13" 56d15'22"
Kayla Jones
And 0h24'11" 29d1'17"
Kayla Joy
Aqr 22h6'46" 0d51'28"
Kayla Joy Miklya
Ari 2h28'56" 12d56'5"
Kayla Joyce Capener
Sgr 19h25'38" -13d5'41"
Kayla June Rice
Cas 1h41'57" 60d21'15"
Kayla K. Diaz Grandchild of IGD
Cas 0h39'31" 56d57'5"
Kayla Kramer
Crb 16h23'30" 32d26'7"
Kayla Kristin Pryor
Sco 16h24'56" -28d18'1"
Kayla Kristina Cox
And 0h26'13" 27d30'58"
Kayla Krodinger
Tau 4h25'6" 20d43'11"
Kayla Krystle
Aqr 22h36'22" -0d26'1"
Kayla L. Jusko
Cap 20h10'27" -10d3'40"

Kayla L. Williams
Aqr 22h13'21" 0d23'12"
Kayla La' Sha Osborne
And 0h35'9" 28d8'10"
Kayla Lea
Lyn 8h13'9" 43d23'4"
Kayla LeeAnn Houskeeper
Psa 22h26'0" -31d32'12"
Kayla Leigh Smolka's
Magical Star
Ari 3h10'3" 18d45'54"
Kayla Lilley Shirlaw - 25th
Dec 06
Cap 20h55'50" -20d31'25"
Kayla Lillie Chisley
Lib 15h49'7" -19d10'48"
Kayla Long
Gem 6h54'23" 12d48'59"
Kayla Loren Wise
Lep 5h34'53" -12d36'53"
Kayla Louise Star
And 0h51'25" 39d58'24"
Kayla Lynette Miller
Uma 11h51'32" 49d25'6"
Kayla Lynn McCall
Cap 21h21'25" -23d56'46"
Kayla Lynn Seymour
Cnc 8h44'42" 25d23'52"
Kayla M. Bartell 03-16-89
Psc 1h12'2" 6d42'10"
Kayla M. DeLeo
Ori 5h29'35" 6d3'37"
Kayla Madison Newbill
Uma 11h22'4" 40d19'53"
Kayla Mae Acker-Berkich
Gem 6h42'25" 30d55'44"
Kayla Marae De Vincent
Cnc 8h24'17" 7d13'48"
Kayla Margaret Mason
Boo 13h54'10" 10d23'14"
Kayla Maria
Tau 4h38'28" 8d41'6"
Kayla Marie
Lmi 10h51'51" 28d54'13"
Kayla Marie
Psc 1h26'57" 28d17'58"
Kayla Marie
And 1h41'14" 43d16'59"
Kayla Marie
Uma 9h27'30" 51d32'16"
Kayla Marie
Vir 13h41'49" -15d11'54"
Kayla Marie
Cas 23h42'24" 57d17'32"
Kayla Marie Adsit
Cnc 8h27'23" 28d48'11"
Kayla Marie Bridges
Uma 10h25'59" 62d11'52"
Kayla Marie Dominick
Lyr 19h8'4" 26d58'7"
Kayla Marie Floyd
And 0h20'41" 27d43'34"
Kayla Marie Gonzalez
Gem 7h56'58" 31d48'53"
Kayla Marie Hallisey
Lib 15h21'41" -16d23'50"
Kayla Marie Holt
Uma 11h40'25" 53d33'40"
Kayla Marie Husby
Vir 13h6'48" 11d59'11"
Kayla Marie Martin
Sco 17h8'38" -31d8'58"
Kayla Marie Mathers
Vir 12h24'33" 8d59'56"
Kayla Marie Matheus
And 2h23'24" 47d55'52"
Kayla Marie McGowan
Aql 19h17'46" 7d15'2"
Kayla Marie Miller
Uma 10h52'32" 59d18'41"
Kayla Marie Raiser
Sco 16h27'16" -25d30'24"
Kayla Marie Riedel
Psc 1h10'42" 28d7'20"
Kayla Marie Shifflett
Lib 15h9'21" -18d10'33"
Kayla Marie's Star
Cru 12h15'19" -58d41'18"
Kayla Martin
And 2h15'17" 46d52'36"
Kayla Mary Donnelly
And 2h30'13" 46d12'17"
Kayla & Maya Guimaraes
Aqr 22h7'42" -0d29'43"
Kayla McKenzie
And 23h26'53" 43d28'47"
Kayla Michelle
And 0h34'55" 26d19'46"
Kayla Michelle Hughes
Ori 6h14'1" 6d23'45"
Kayla Morgan
Mon 7h45'41" -8d55'47"
Kayla Nadine Kraus
Her 16h31'24" 14d23'45"
Kayla Nicole
Cyg 20h55'9" 42d19'11"
Kayla Nicole Alexander
And 0h16'54" 29d38'41"
Kayla Nicole Griffin
Lyn 7h2'59" 45d18'33"
Kayla Nicole King
Del 20h46'56" 16d12'19"
Kayla Noel III
And 0h27'5" 40d34'37"

Kayla Ohlsson
Cap 20h21'8" -10d45'15"
Kayla "Our Family's Star"
And 1h49'49" 47d24'31"
Kayla Pankow
Her 17h54'43" 26d35'23"
Kayla Parton
Gem 6h32'57" 24d29'36"
Kayla Phillips
Sco 16h4'14" -11d55'8"
Kayla Rae Bless Mommy's
Sunshine
Gem 6h59'26" 32d23'19"
Kayla Rae Corley
Lib 15h26'11" -6d49'34"
Kayla Rae Higgins &
Robert Lee Ginn
Lib 15h45'31" -19d33'52"
Kayla Raquel
Sco 16h5'47" -14d36'15"
Kayla "Red" Pickett
Uma 8h9'40" 60d6'33"
Kayla Rená Long
Lib 15h50'34" -19d44'12"
Kayla Renae
And 0h22'56" 36d3'11"
Kayla Rene Shick
Tau 4h17'47" 24d20'0"
Kayla Reneé
Cnc 8h26'30" 19d16'0"
Kayla Renee Vaughn
Mon 7h57'22" -5d1'24"
Kayla Ricketts
Lyn 8h20'33" 57d41'9"
Kayla Roberta Altobelli
Uma 11h25'45" 50d12'12"
Kayla Rose Casale
And 23h10'55" 47d22'6"
Kayla Rose Osmon
Aqr 23h46'8" -19d54'15"
Kayla Rose Scarpaci
Peg 23h3'55" 17d5'14"
Kayla Rose Uitenbroek
And 0h47'45" 44d31'13"
Kayla Ryann Kroge
And 1h17'33" 49d26'11"
Kayla S
Cas 0h20'24" 57d15'36"
Kayla Saadeh
Aql 20h8'18" -0d54'27"
Kayla Simone Stephens
Tau 3h45'32" 19d37'39"
Kayla Smith
Gem 6h7'1" 24d8'5"
Kayla Smith Bozek
Uma 10h26'18" 53d34'36"
Kayla St. Denis
And 1h7'56" 37d32'29"
Kayla Sue Taylor
Uma 8h36'33" 71d15'23"
Kayla Tinoco
Boo 14h58'58" 14d29'34"
Kayla Uscilowski
Uma 11h26'59" 38d46'7"
KaylaChelle
Uma 12h55'58" 57d45'5"
Kayla-Cherie Mary Bourg
Lyn 8h25'10" 37d6'22"
KaylaFrances
Uma 10h38'41" 56d39'5"
Kaylah Gwyn
Aqr 21h44'51" -2d35'21"
Kaylan Ackles
Tau 4h37'13" 6d36'47"
Kaylan Barrego
And 23h9'59" 47d23'14"
Kaylan Brynna Ann Iverson
And 2h9'9" 44d24'56"
Kaylan Nicole
And 1h33'36" 42d40'14"
Kaylan Noel Goerge
Cyg 21h16'41" 42d57'16"
Kaylana Cherelle Padre
Ari 2h52'54" 28d8'48"
Kaylaneum
Sgr 18h52'19" -18d53'37"
KayLanni
Psc 0h36'55" 5d20'53"
Kaylan's Star
Cha 10h48'32" -79d7'52"
Kayla-Pie
Vir 14h31'14" -4d47'16"
Kayla's and Colt's Star
Lib 14h53'46" -3d40'40"
Kayla's Cupid
Lyr 18h39'46" 37d27'8"
Kayla's Light
And 1h47'34" 38d11'30"
Kayla's Shining Star
Gem 7h2'38" 21d7'33"
Kayla's Shining Star
Ori 5h36'31" 13d16'41"
Kayla's Star
Cas 0h21'5" 63d59'49"
Kayla's Star
Uma 8h18'28" 64d26'30"
Kayle Kidan
Crb 15h34'52" 33d46'6"
Kayle Wood
Psc 1h6'31" 33d1'57"
Kaylea Grace
Her 18h15'4" 18d41'46"
Kaylee
Vir 13h26'19" 12d42'36"

Kaylee
Lyr 18h53'46" 31d45'58"
Kaylee
And 1h12'25" 43d51'57"
Kaylee
Uma 11h22'6" 43d49'16"
Kaylee
Lib 15h15'50" -9d8'28"
Kaylee Adele Guldin
And 1h45'9" 46d3'20"
Kaylee Alexandra
And 22h58'15" 51d21'37"
Kaylee Angone
And 23h54'28" 41d16'6"
Kaylee Brooke Criner
Lmi 10h37'57" 35d3'28"
Kaylee Cartwright
Crb 15h16'28" 25d59'51"
Kaylee Christine Moran
Cnc 9h19'43" 18d7'34"
Kaylee D
Mon 6h31'22" 10d28'5"
Kaylee D White
Sgr 18h28'34" -17d33'7"
Kaylee D'ann Marie Everett
Cnc 9h1'23" 15d18'29"
Kaylee Elise Rae
Psc 23h49'50" 5d21'9"
Kaylee Elizabeth Quinn
And 23h20'12" 52d27'2"
Kaylee Grace
And 2h30'44" 40d28'53"
Kaylee Grace Friesen
And 0h51'12" 40d22'56"
Kaylee Grace Merkle
1:33pm
And 23h43'31" 46d34'13"
Kaylee Grace Spencer
Sgr 18h14'8" -19d11'48"
Kaylee Green
Umi 18h16'39" 86d43'49"
Kaylee Hawkins
And 1h39'48" 44d44'4"
KAYLEE HOPE CONLIN
Aqr 22h44'58" 1d0'53"
Kaylee Hope Franz
Cam 6h59'32" 64d38'43"
Kaylee Isabel Torres
Leo 9h35'56" 27d41'14"
Kaylee Jo Tigner
And 0h46'19" 43d24'52"
Kaylee Kimbril Sears
Ari 2h49'43" 17d16'53"
Kaylee Krueger
And 0h38'53" 32d39'41"
Kaylee Logan
Vul 19h29'6" 26d3'8"
Kaylee Marie
Peg 22h40'13" 28d39'46"
Kaylee Marie
Uma 8h36'48" 62d7'44"
Kaylee Marie Agster
And 23h13'12" 52d19'29"
Kaylee Marie Bonner-
Overstreet
Lyn 8h11'59" 55d37'0"
Kaylee Marie Daniels
Tau 5h10'38" 24d17'22"
Kaylee Marie Epperson
Vir 14h16'2" 1d53'29"
Kaylee Marie Goodwin
And 1h20'39" 38d28'31"
Kaylee Marie Hoffer
Lib 15h10'36" -4d32'39"
Kaylee Marie Neff
Dra 17h42'4" 55d30'28"
Kaylee Marie Waters
Uma 14h19'5" 56d42'24"
Kaylee Mary Perchal
Vir 12h53'43" 4d41'49"
Kaylee Mary White
And 0h40'2" 40d58'12"
Kaylee May Bradley
Tau 3h58'28" 6d52'42"
Kaylee May Keating
Gem 7h3'21" 32d27'8"
Kaylee McDermott
Cas 0h54'2" 59d29'12"
Kaylee Megan
Wrigglesworth
And 1h14'34" 45d43'30"
Kaylee Morgan Fearis
Lib 14h56'16" -18d17'43"
Kaylee Nicole Steele
Sgr 18h6'10" -32d25'3"
Kaylee Rattie
And 2h24'4" 47d4'46"
Kaylee Reese Buescher
Vir 12h58'37" 4d51'17"
Kaylee Robyn Christiansen
Dra 19h34'4" 60d45'29"
Kaylee Rose
Cnc 9h11'40" 30d49'4"
Kaylee Rose
And 1h57'27" 41d6'2"
Kaylee Rose Swann
Gem 7h5'0" 29d28'42"
Kaylee Schultz
Crb 15h44'16" 39d27'41"
Kaylee Seamann
Gem 7h28'57" 21d7'27"
Kaylee Teresa
And 0h19'8" 28d4'12"

Kaylee Warner and Liz
Effinger
Cyg 19h41'30" 54d52'25"
Kaylee Wells' Christmas
Star
Lyn 7h19'23" 49d53'31"
Kayleeanna Nkauj Hlis
Vang
And 2h8'15" 37d35'3"
Kayleebug
Sco 17h18'12" -32d48'18"
KayleeGiraud
Cas 1h39'46" 64d16'20"
Kaylee-Jo Maureen Pepe
Cnc 8h43'27" 18d35'16"
Kayleen R. Mitchell
Uma 11h43'14" 57d11'31"
Kayleen Schaefer
Mon 7h54'9" -2d30'45"
Kaylee's Bella Stella
And 0h50'56" 42d16'18"
Kaylee's Night Light
Cap 20h23'37" -12d7'29"
Kayleieh Dian Wollam/Erin
J Joseph
Lyr 19h14'44" 26d24'34"
Kayleigh
Sgr 18h50'6" -18d34'30"
Kayleigh
Vir 12h23'11" -7d17'6"
Kayleigh
Umi 15h26'56" 72d32'59"
Kayleigh Ann
Cam 5h19'27" 64d10'22"
Kayleigh Ann Elizabeth
Tau 4h7'41" 23d8'41"
Kayleigh Ann Leventhal
Tau 4h21'49" 17d26'54"
Kayleigh Athena
Lyn 7h40'11" 36d42'42"
Kayleigh Barnette
Ari 2h41'26" 25d16'43"
Kayleigh Cheyenne
Koyama
Vir 13h48'29" 4d55'34"
Kayleigh Denae Woodall
Cap 21h13'24" -16d37'22"
Kayleigh Faye Jenkins
And 1h8'6" 44d32'35"
Kayleigh Hamersly
Dra 19h16'1" 75d0'59"
Kayleigh Lambert-Gorwyn
Cas 23h49'50" 57d28'58"
Kayleigh Louise Bell
And 2h33'8" 50d31'45"
Kayleigh Marie
Psc 23h45'21" 1d20'43"
Kayleigh Michele Logan
Ori 6h12'7" 18d30'31"
Kayleigh Rose
Ori 6h9'1" 18d21'4"
Kayleigh Rose Kilgore
Her 18h6'39" 24d7'29"
Kayleigh Rose Tinsley
Cnc 9h17'4" 13d41'40"
Kayleigh S. Primavera 8-
30-2001
Vir 15h2'48" 4d11'16"
Kayleigh Snook
Leo 11h4'48" 0d57'4"
Kayleigh Tanya
Chamberlain
And 0h33'49" 45d23'55"
Kayleigh Walsh
And 0h17'14" 34d17'17"
Kayleigh's first Christmas
And 1h16'2" 48d54'37"
Kayleighs Foreverland
Aqr 22h47'28" -5d33'5"
Kayleigh's Star
Gem 6h59'18" 25d4'25"
Kaylen
Lyn 7h59'7" 36d29'39"
Kaylene
Crb 15h47'28" 33d16'46"
Kaylene's Sparkle
Col 5h20'12" -34d46'27"
Kayley
And 23h19'48" 47d45'59"
Kayley Archuleta
Crb 15h37'36" 39d7'30"
Kayley Jayne Eastwood -
27.03.1987
Uma 11h50'53" 47d19'36"
Kayley Kolleen
And 2h19'20" 47d57'57"
Kayley Ryan Sonognini
Smith
Lyr 19h11'39" 37d16'32"
Kayley Wynn Doyle
Cam 7h8'27" 73d37'59"
Kayli Lynn Sexton
And 2h23'36" 47d1'14"
Kaylie Alyssa Wrather
And 1h7'43" 36d58'17"
Kaylie Ann Maloney
And 1h3'7" 46d37'17"
Kaylie Grace
And 1h29'27" 44d40'21"
Kaylie Margaret Chudyi
And 2h27'11" 47d1'56"
Kaylie North
Aql 19h33'8" 9d52'4"
Kaylie's Star
Ari 3h21'23" 27d36'22"

Kaylin Beth
And 0h41'5" 27d48'4"
Kaylin Corinn
Cnc 9h5'18" 27d6'27"
Kaylin Kristine Flaxman
Gem 7h44'0" 27d14'26"
Kaylin Kristine Tucker
Leo 11h34'43" 22d51'47"
Kaylin Linnemann
Umi 14h53'3" 79d28'36"
Kaylin Liu
Ari 3h8'54" 11d4'48"
Kaylin Marie Falkner
Peg 22h11'37" 14d37'56"
Kaylin Morrisey Barrow
Dra 19h40'29" 60d3'42"
Kay-Louise
Uma 10h29'11" 55d15'46"
Kaylyn
Com 13h11'25" 29d3'14"
Kaylyn Bielatowicz
Sco 17h40'51" -44d51'7"
Kaylyn Elaine Warren
Uma 8h54'23" 67d46'31"
Kaylyn Elizabeth Jones
Sco 17h45'45" -32d2'47"
Kaylyn Louzetta Berrett
Cap 21h38'35" -11d11'31"
Kaylyn Oertly
Leo 10h31'39" 17d34'9"
Kaylyn Shumate
And 0h21'50" 28d46'40"
Kaylyn Zvonek
Crb 15h25'1" 28d41'36"
KayLyne Marie McCormack
Cnc 9h9'8" 25d34'48"
Kaylynn
Ari 1h51'34" 17d33'57"
Kaylynn
Tau 3h48'33" 5d40'19"
Kaylynn
Psa 22h10'53" -25d4'25"
KayLynn Brooke
Peg 23h28'15" 22d11'17"
Kaylynn Chelsea Sheets
Psc 1h38'18" 17d31'59"
Kaylynn Marie
And 23h23'31" 48d3'38"
Kaylynn Rose Ellis
Sgr 19h38'11" -13d54'40"
Kaylynn Washnock
Her 18h45'24" 20d10'19"
Kaylyn's Complexus
Psc 1h9'57" 6d37'16"
Kayna Jo Kloke
Per 3h8'57" 40d55'0"
Kayo 1941
Aql 20h12'0" -0d40'40"
Kayo Carroll
And 0h31'40" 39d37'48"
Kayoko and Masahiko
Umi 13h43'17" 73d49'34"
Kay's Christmas Star
Cnc 8h29'36" 15d51'21"
Kay's Star
Psc 0h51'20" 15d12'13"
Kay's Star
Uma 10h45'27" 56d7'43"
Kay's Star
Uma 12h10'30" 57d13'1"
Kaysha
Tau 3h49'25" 30d41'3"
Kaysie Anne Henderson
And 23h43'14" 35d50'3"
Kayson Amare Welch
Tau 5h32'8" 22d31'54"
Kayson Wade Jorgensen
Aqr 23h13'35" -14d56'38"
Kayte Junior
Lyn 8h26'47" 33d26'38"
Kaytie I. Fournier
Uma 9h45'10" 49d48'0"
Kaytlin
Cyg 20h34'12" 38d16'51"
Kaytlin LeMier
Cyg 21h24'35" 52d22'24"
Kaytlin Marie
Umi 17h39'19" 80d58'16"
Kaytlin Smith
Aqr 22h37'48" -2d35'55"
Kaytlyn Michelle Worner
Vir 13h17'11" 4d14'3"
Kaytlynn Renee Ellis
Vir 12h41'13" 4d58'54"
Kaytren
Pav 19h18'27" -67d43'36"
Kaytren
Pav 19h27'27" -56d51'6"
KayVar
Uma 13h24'12" 58d12'51"
KAZ
Uma 11h25'34" 54d40'29"
Kaz
Lmi 10h18'52" 35d48'20"
Kaz Buchanan
Cnc 9h21'25" 6d36'46"
Kaz Panda 1957
Psc 1h51'41" 2d46'52"
Kaz Star
Aql 19h37'14" 13d12'33"
Kazer Alex
Del 20h38'52" 15d24'29"
Kazimiera Kozlowski
Uma 8h21'57" 61d32'54"

Kazimierz Witold
Pomykalski
Psc 0h54'6" 13d19'42"
Kaz's Reach
Cas 23h28'0" 55d24'9"
Kazue Okauchi Mundy
Gru 21h49'10" -36d49'33"
Kazuki Meguro
Lib 15h38'18" -19d20'59"
Kazuko Eddy
Lyr 18h49'6" 31d22'31"
Kazumi
Cap 20h8'28" -16d53'0"
Kazuo Mizuno
Lib 15h21'45" -28d17'0"
Kazuya Umeki
Vir 13h14'26" 7d59'42"
Kazy Morning Star
Ori 6h15'16" 21d8'19"
KAZZ
Uma 11h29'42" 38d1'49"
Kazza and Duggy!
Cyg 21h25'57" 50d4'59"
Kazza Lang
And 1h14'10" 46d25'23"
KB
Uma 10h39'37" 46d43'29"
K.B
Dra 16h44'15" 52d37'36"
KB_11
Cir 15h6'20" -57d15'54"
KB Verus Vindexicis
Col 5h50'13" -34d6'22"
K-Bear
Cam 6h34'27" 68d10'36"
KB.Eternity.04.02.05
Cyg 19h51'42" 32d10'3"
K-Bomb
Cap 21h1'30" -27d29'45"
KBR1986
And 1h8'16" 42d36'33"
KBrnBrite4ElMo
Aql 19h36'45" -0d15'19"
kBrown
Ari 3h21'52" 27d45'43"
KB-SNAZ
Leo 10h23'12" 23d32'2"
KC
Leo 9h29'19" 27d55'35"
K.C.
Leo 11h56'33" 26d7'12"
KC
Vir 13h27'14" 11d32'15"
KC
Cnc 8h14'46" 8d47'32"
KC
Cnv 12h26'18" 33d43'45"
K.C.
Cyg 21h55'9" 52d24'21"
K.C.
Eri 4h40'17" -0d2'4"
KC
Lyn 6h20'34" 60d16'34"
KC 11-15-2005
Vir 13h12'2" -17d29'1"
K.C. and Me
Ori 6h17'44" 14d40'51"
"KC" Cassandra A. Jordan
And 2h4'13" 38d7'26"
"KC" Cassandra A. Jordan
And 2h3'33" 37d35'49"
K.C. Falls Fantastic 50th
Birthday
Per 3h41'34" 45d45'1"
KC Friedman
Tau 5h19'34" 25d10'20"
KC Hunter Johnson
Cep 23h9'42" 79d22'50"
KC - Loridanimichelle - JB
Cyg 20h56'22" 34d16'48"
KC Magrabi
Uma 9h11'31" 53d21'12"
K.C. Ng
Uma 10h31'55" 66d45'0"
K.C. "Poozy" & E.M. "Papa"
Forever
Sgr 19h9'33" -16d7'23"
KC X 2.
Ari 2h37'6" 27d5'59"
KC6768
Oph 17h4'2" -0d10'0"
K.C.A. McKiterick
Lyn 8h23'1" 35d9'18"
K.Carnevale & M. Stines
Cyg 20h13'7" 54d17'16"
Kchito 5
And 1h41'40" 39d0'16"
kcm+smeb
Sco 16h59'19" -36d1'25"
KCM1940
Cnc 8h2'13" 23d55'24"
KCMS
Cnc 8h19'42" 10d45'36"
KcmyKC Rogers
Sco 17h56'35" -42d37'34"
KC's starr
Dra 15h14'1" 55d7'16"
K.D.
Peg 22h19'59" 6d9'48"
K.D./D.M.
Gem 6h43'19" 16d52'26"
K.D. Keating
Tau 4h9'42" 21d27'58"
KD5EVW
Uma 11h42'48" 38d23'56"

KDAR
Lib 14h39'45" -10d54'11"
KD-AS
Uma 14h26'21" 60d26'35"
Kdea
Tau 5h34'45" 20d49'20"
Kdk9601
Cap 20h26'39" -10d27'43"
KDL 2007
Uma 11h34'0" 61d38'42"
KDM050757
Tau 5h1'35" 16d9'39"
Ke aloha o Pakelika a me
Kelilina
Cas 0h10'42" 59d20'19"
Ke JoTL 1000
Uma 10h18'25" 40d0'16"
Keagan Daniel
Psc 0h43'9" 8d22'57"
Keagan John Leigh
Ori 6h18'57" 8d28'25"
Keagan Matthew
Altenbaumer
Uma 11h53'12" 53d48'48"
Kealey
Umi 15h5'30" 74d46'33"
Kealey Skye Hall
Cru 12h51'7" -63d53'12"
Kealey Zane Pertiller
Lib 15h22'45" -23d36'35"
Kealley Heller
Cnc 8h19'36" 19d40'42"
Kealley May
Cnc 8h50'3" 30d19'39"
Kealy Michele
Aqr 22h53'34" -20d12'34"
Kean Michael Meoli
Gem 7h12'18" 27d42'22"
Keana
Gem 7h52'7" 27d42'56"
Keana
Cam 4h20'35" 57d8'55"
Keana Kristine Quintero
Cnc 9h10'34" 27d44'11"
Keana O'Kalani
KaiKaiNaAlii DeCoite
Com 12h41'23" 28d39'25"
Keanna Briles
Leo 11h11'16" 24d35'19"
Keanna Lee 02-08-02
Aqr 22h12'28" 2d11'20"
Keanu Charles
And 23h42'3" 41d5'23"
Keanu Morgan
Ori 6h14'49" -1d37'23"
Keanu Reeves
Vir 14h33'8" 1d0'37"
Keara McAndrew
Cyg 20h46'14" 42d51'3"
Kearney
And 2h22'42" 46d35'47"
Kearney
Lib 15h4'40" -21d42'6"
Kearny/Peg Bennett
Cas 0h54'23" 59d16'54"
Kearonita1
Ori 6h5'51" 9d48'34"
Kearstin Sue Fulmer
Cyg 19h54'40" 33d5'41"
Keasha
Lyn 8h38'6" 43d16'57"
Keath "Snickerdoodle"
Patterson
Vir 13h47'50" 1d3'56"
Keating 3
Uma 11h40'7" 58d36'47"
Keaton Allmen
Ori 5h7'55" 15d34'16"
Keaton Eric Piper
Her 16h35'21" 28d52'8"
Keaton Lancaster
Boo 15h38'8" 48d35'51"
Keaton Spencer Bosse
Lib 15h1'11" -9d50'22"
Keaton Steven Milerowski
Leo 10h18'41" 25d54'18"
Keaton's
Uma 9h25'20" 53d10'12"
Keazy & Kyo-ka
Tau 5h43'47" 21d31'44"
Keb - Victoria's Favourite
Cep 22h34'58" 66d57'4"
Keca L Carpenter
Sco 16h8'25" -11d35'26"
Keceli Meszaros Matyas
Cep 22h30'6" 72d0'48"
Kecia "Butterfly" Reyes
Lib 15h9'5" -6d9'5"
Kecia Simone Browne
Tau 4h59'10" 16d33'6"
Keddy L Ball-Davidson
Cnc 9h14'1" 23d22'25"
K.Edgette
Gem 6h30'22" 21d41'49"
KEDRIK & MARLENE
Uma 12h59'37" 52d34'25"
Kee
Sgr 19h12'55" -12d53'23"
Kee
Lyr 18h50'4" 36d7'20"
Kee Ganstar
Boo 14h30'20" 28d59'7"
Kee Kee's own special
place
Lmi 10h22'10" 34d28'47"

Keefe Thomas
Col 6h34'33" -33d30'38"
Keefer's Star
Lib 14h50'56" -2d11'24"
Keegan
Cru 11h59'12" -60d37'8"
Keegan
Ori 6h17'40" 13d10'4"
Keegan Anthony McLachlan
Lib 15h30'20" -16d21'26"
Keegan C. Raprager
Per 3h30'35" 46d50'16"
Keegan Christopher Cook
Her 17h40'4" 44d57'22"
Keegan David Chevrier
Tau 4h2'19" 22d5'37"
Keegan Edward Grazioli
Umi 15h14'33" 75d6'51"
Keegan Grant
Uma 13h41'51" 54d11'10"
Keegan Jay Grovert
Sgr 18h46'22" -26d55'41"
Keegan Joseph Jackson
Ari 2h46'2" 21d1'26"
Keegan Joseph Thurber
Ori 6h13'48" 8d49'10"
Keegan Leith
Tau 5h37'1" 18d58'17"
Keegan Mateasen Sanders
Ori 6h22'14" 13d50'34"
Keegan Matthew Duncan
Uma 12h51'8" 56d7'48"
Keegan Peter Kienbaum
Uma 9h6'26" 65d56'49"
Keegan Randall Lambert
Aql 19h40'25" 13d41'46"
Keegan Thomas Verse
Cru 12h29'49" -57d53'53"
Keegan's Star
And 0h21'40" 38d13'19"
Keeghan Harley Smith
Uma 13h46'22" 61d57'4"
Keegstrd
Umi 14h29'36" 68d30'30"
KeeKee
Tau 4h7'54" 25d53'40"
Keekeeree
Sco 17h12'3" -34d17'48"
Keeker
Cam 4h4'46" 65d41'21"
Keeker the Beautiful
Aqr 22h42'43" -3d2'32"
Keela
Gem 7h18'13" 33d59'28"
Keela Burgess
Cam 5h14'0" 68d3'12"
Keela McKim
Crb 16h3'4" 34d46'39"
Keelan Eve Feighery
Cnc 8h29'57" 24d25'23"
Keelee Ilynn Shultz
And 23h44'46" 47d31'56"
Keeler 36
Uma 9h48'59" 57d24'57"
Keeley Louise Ayre
And 23h4'9" 42d25'24"
Keeley Tang
Cas 0h55'30" 50d30'6"
Keeley's Star
And 2h19'35" 47d54'45"
Keelie Addison Schneider
Aqr 21h54'56" 1d32'30"
Keelie Ann Stark
Vir 13h44'42" -6d20'43"
Keelin McGrath
Cas 1h17'20" 54d18'33"
Keelin Sinead Reilly
Lyn 7h2'32" 50d41'54"
Keelin07202004
Lyr 19h8'1" 27d1'37"
Keeling's Light
Uma 11h49'9" 42d24'10"
KEELTRICK "7-20-1962"
Uma 9h22'50" 57d24'6"
Keely
Vir 12h23'48" -3d41'20"
Keely Abbott's 21st Star
Cas 1h22'41" 52d36'38"
Keely Amber Hall
Uma 10h57'43" 66d7'25"
Keely Ann
Uma 11h27'16" 52d57'1"
Keely Ann
Sgr 19h13'43" -20d14'38"
Keely Jean Kirkpatrick
Cap 20h17'42" -23d51'31"
Keely Paige
Uma 12h34'37" 58d14'4"
Keely T. Stone
Peg 21h57'53" 25d34'50"
Keely Tyler Fleming
Sgr 19h25'48" -32d24'30"
Keely Watt
Uma 12h34'0" 62d44'46"
Keely Wynne Parslow
Vir 12h43'48" 2d44'22"
Keely, Mother of Brittany
Uma 12h39'19" 58d27'38"
Keelyn
And 0h43'9" 24d51'4"
KeelynJustin
Cyg 21h59'34" 48d47'3"

Keemo & Chloe Lisotta-Brookstein
Uma 13h39'20" 58d3'51"
Keena
Uma 13h44'42" 55d53'12"
Keena
Ari 2h47'47" 13d56'27"
Keena Astemborski
Cap 20h41'56" -21d18'39"
Keenan Benjamin DuBois
Lib 15h16'56" -25d36'51"
KEENAN BRIAN BOYD IGHB
Sgr 17h52'4" -29d55'14"
Keenan Joseph Healey
Dra 18h27'25" 54d41'51"
Keenan Paul
Leo 10h24'6" 23d2'51"
Keene Reyman
Per 3h12'12" 41d56'26"
Keenebaby
Ori 5h33'18" 8d1'14"
Keenen Cole Merrick
Ori 5h36'35" -1d55'42"
Keenon's Star
Uma 9h9'25" 71d49'32"
Keenyn Kenrick Henderson-Yendrys
Uma 9h46'54" 58d44'11"
Keep Joan Happy
And 0h45'32" 30d26'52"
Keep Me in Your Pocket, Kelly
Gem 7h39'11" 27d26'55"
Keep Shining Lil Star!
Umi 16h19'57" 82d58'13"
Keep Smiling Maree
Ari 2h30'56" 21d2'20"
"Keep the Light" Stephen & Lynne
Psa 21h51'12" -32d41'28"
Keep Watching Over Me
Ari 2h43'14" 15d23'41"
Keep Your Dreams Alive
Ori 5h26'44" 6d30'8"
Keeper of Karen's Light
Uma 10h13'50" 50d15'45"
Keeper of my Stars, Joseph Robert
Her 17h27'3" 39d40'11"
Keeper Of Our Hearts***BCD + RBP
Uma 9h37'18" 45d28'28"
Keeper of the Stars
Lyr 18h54'23" 32d23'43"
Keeper of the Stars
Cap 21h31'39" -17d2'7"
Keep-it-Natural
Uma 9h21'12" 64d15'6"
Keera Barker
Peg 0h7'11" 15d38'42"
Keeran's Desire
Sgr 19h4'41" -27d36'54"
Kees and Corrie DeKievit
Tri 2h24'3" 35d9'29"
Kees F. Stahl
Cyg 20h9'46" 38d23'24"
Keesha and Lea Mills
Cyg 19h57'40" 48d3'49"
Keesha Tucker
Cra 18h48'4" -38d26'3"
Keeth Furan
Ori 5h35'29" 12d16'16"
Keetie
And 2h35'7" 45d25'2"
Keeton, Emilie, & Hinson Chesnutt
Tri 1h54'18" 29d21'9"
Keeton-Spence
Crb 16h12'8" 32d10'18"
Keeya's Star
And 0h56'48" 46d11'44"
Kefani2005
Ori 6h18'5" 8d58'10"
Kegan Michael Pettit
Cas 0h33'57" 63d30'21"
Kegan Thompson
Hya 8h43'46" 5d17'20"
Kegsy
Uma 10h6'44" 60d1'18"
Kei
Uma 11h55'59" 48d34'36"
Kei & Nozomi Endless happiness
Cnc 8h50'45" 15d40'38"
Keia Simmons, Star to Wish Upon
Aqr 22h38'16" -22d9'40"
Keicha Myers
Ori 5h17'14" -1d47'7"
Keifer John Schlekeway
Cep 22h50'37" 66d15'29"
Keighley
And 23h43'53" 48d25'18"
Keighley Ashkenazy
And 23h52'17" 38d47'24"
Keiki
Cas 0h25'51" 51d56'2"
Keiko
Aqr 22h36'2" -21d5'9"
Keiko Hayasaka
Lib 15h0'6" -17d24'52"
Keiko Ishino
Sco 17h24'37" -37d33'19"

Keiko Noguchi
Aqr 20h58'38" 1d5'40"
Keiko Son
Uma 10h44'50" 66d24'22"
Keiko'N'CJ 4eva
Gem 6h3'8" 25d19'12"
Keil Johnson & Lisa Iverson
Vir 11h53'5" 3d59'19"
Keil, Wilfried
Ori 6h3'46" 9d58'46"
Keila Guzman
Gem 6h46'19" 14d45'34"
Keilana's Kiss
And 23h5'25" 36d14'9"
Keilani Hupe
Vir 13h46'36" -12d43'33"
Keilee Marie Salvatore
And 1h57'8" 43d8'40"
Keilian Emily Rudisill
And 23h30'59" 48d22'10"
Keily Ann
Ari 2h58'38" 26d21'8"
Keina Bo Beena
Uma 11h36'3" 44d35'55"
Keina Wagner
And 1h51'13" 44d28'48"
Keir Vivienne Marticke
Lib 14h59'56" -9d25'12"
Keira
Ari 3h1'40" 28d0'58"
Keira
Ari 2h56'25" 26d11'3"
Keira Adcock
Cas 1h31'34" 71d50'35"
Keira Aili Flynn
Uma 12h7'17" 54d4'58"
Keira Ann Thomas
Lib 15h8'21" -4d51'53"
Keira Ashley Gallagher
And 1h30'46" 50d13'12"
Keira "bean" McCulloch
Gem 6h4'26" 27d2'16"
Keira Beth Mendez - Keira's Star
And 0h28'17" 25d50'44"
Keira Bradbury
And 23h12'56" 39d28'58"
Keira Christine
Boo 14h44'26" 32d25'36"
Keira Christine Courtney
Uma 13h40'54" 53d56'38"
Keira Dawn
Cam 3h34'16" 58d13'12"
Keira Dian
Gem 7h45'45" 32d16'24"
Keira Eleanor Franklin
And 23h55'7" 42d39'30"
Keira Elizabeth
And 1h29'34" 47d31'30"
Keira Elizabeth Moss
And 2h35'29" 48d44'36"
Keira Elizabeth Washkau
Vir 12h42'17" 12d2'2"
Keira Ellie Louise Benfield
Umi 14h31'39" 77d25'12"
Keira Hanae DuBois Swisher
Uma 12h36'37" 59d27'1"
Keira Jane Bosland
Lib 14h51'9" -3d8'28"
Keira Kesia Luevano
Gem 7h33'33" 16d26'36"
Keira Leigh
And 23h2'53" 51d31'4"
Keira Louisa
And 1h57'45" 36d27'40"
Keira Marie McBrine
And 23h5'54" 48d13'7"
Keira Matty
And 23h3'36" 48d40'49"
Keira May Howley
Cyg 20h30'46" 35d0'53"
Keira Morgan O'Sullivan
And 1h32'11" 47d6'38"
Keira Nickolena Lynch
Ari 2h58'7" 28d25'57"
Keira Rae
Lib 15h42'2" -3d56'32"
Keira Raelyn Ruiz
And 0h21'17" 38d38'3"
Keira Rose King
Sgr 19h17'8" -25d1'46"
Keira Sage Holley
Gem 7h31'37" 15d32'5"
Keira Smith
Crb 15h46'32" 33d28'15"
KEIRA SOLEIL WRIGHT
Umi 15h34'37" 72d9'17"
Keira Tiggardine
And 0h28'39" 26d12'1"
Keiran and Katies Promise Star 21
Cyg 20h56'31" 37d1'13"
Keiran James Blaszczyk
Leo 9h59'56" 16d56'31"
Keiran Ross Whitefield
Ari 3h7'45" 12d3'10"
Keira's Christening Star
And 1h22'38" 49d49'58"
Keira's Star
Cas 23h18'31" 58d53'5"
Keirsten Frazier
Sco 16h4'5" -17d55'3"

Keirston Picard
Sco 16h6'42" -10d23'46"
"Keisa's Star" Guarding You Always
And 23h15'4" 47d38'13"
Keisha
Lyr 18h45'48" 39d51'9"
Keisha
Lib 15h14'37" -6d4'41"
Keisha Alexander
Uma 11h37'16" 52d21'46"
Keisha Elizabeth Kamery
Per 2h53'50" 40d14'11"
Keisha King
Lyn 6h31'40" 56d44'50"
Keisha Kinney
Crb 15h36'40" 36d15'53"
Keisha Montelongo
Psc 1h24'0" 17d37'43"
Keisuke & Mai
Psc 1h1'37" 21d45'37"
Keith
Vir 12h7'21" 11d20'17"
Keith
Gem 6h26'46" 19d51'48"
KEITH
Uma 11h32'54" 37d12'49"
Keith
Her 17h31'42" 39d29'20"
Keith
Dra 17h17'34" 63d59'50"
Keith
Cap 21h58'45" -19d58'27"
Keith
Sgr 19h30'35" -37d59'11"
Keith A. Crosby
Sco 17h50'15" -37d35'51"
Keith A. Western
Leo 11h5'31" 0d23'57"
Keith Aaron Scull
Psc 23h7'9" 6d27'45"
Keith Alan Christen
Umi 15h30'53" 71d40'30"
Keith Alan Novorr
Aql 19h57'59" 11d30'48"
Keith Alan Roach
Cas 1h27'11" 57d32'57"
Keith Alan Westbrook
Tau 3h34'56" 11d6'42"
Keith & Alice
Crb 15h52'26" 36d27'16"
Keith Allen Anderson
Aql 19h28'39" 7d24'57"
Keith Allen Brokenshire
Cyg 20h18'43" 34d31'49"
Keith Allen Hunt
Ser 18h19'40" -13d39'57"
Keith and Cara Forever -N-Always
Tau 4h31'39" 21d8'41"
Keith and Claire McQuhae
Cyg 20h39'58" 44d50'18"
Keith and Jessica
Boo 14h30'33" 43d25'32"
Keith and Judy Hanf
Lmi 10h4'54" 37d21'31"
Keith and Keri
Cyg 20h30'44" 52d17'29"
Keith and Lori Forever
Umi 15h24'31" 74d49'10"
Keith and Lynn Sottile
Cyg 19h57'14" 58d56'56"
Keith and Lynne Gilfillan
Cas 23h6'58" 58d57'31"
Keith and Mary Ellen Page
Sgr 18h40'23" -16d26'2"
Keith and Maryann's Star
Lyr 18h47'13" 43d13'59"
Keith and Meghan Feusner
Cyg 21h29'26" 35d2'6"
Keith and Michel Schulz
Leo 11h26'55" 13d50'45"
Keith and Michelle Leary
Cyg 19h46'9" 36d1'54"
Keith Andrew Miller
Cyg 20h31'13" 32d9'41"
Keith Andrew Priola
Dra 16h52'38" 59d4'47"
Keith & April's Eternal Love
Sgr 18h36'45" -23d54'56"
Keith & Ashley Pitcher
Cyg 21h15'46" 44d59'59"
Keith Baldwin Christmas 2006
Cep 20h49'43" 57d54'57"
Keith Bett
Cep 0h15'31" 73d54'50"
Keith Bogus
Uma 13h44'21" 58d19'38"
Keith Bradshaw B.J.C. L.J.B. D.J.B
Uma 12h13'24" 59d43'17"
Keith Branson
Aur 5h33'41" 41d36'44"
Keith Brent Albelli, Jr.
Ari 3h11'19" 28d38'13"
Keith Broeckel
Vir 13h40'42" -3d23'15"
Keith C. Gorton
Psc 0h25'50" 7d1'39"
Keith C. Jones
Leo 11h12'56" 5d9'8"
Keith Caldwell
Sco 17h54'16" -36d37'23"

Keith "Chippy" Fitch
Cyg 21h13'47" 36d29'10"
Keith & Christy Brevard
Cyg 20h18'10" 34d59'8"
Keith Cook
Uma 11h39'17" 31d54'30"
Keith Corey Walker
Leo 10h6'50" 20d15'26"
Keith D. Martin
Sgr 18h29'10" -16d32'50"
Keith Darwin
Sgr 18h16'35" -24d37'12"
Keith David Krieg
Uma 9h49'21" 59d34'8"
Keith David Shearer
Psc 23h17'5" 7d47'5"
Keith Davies
Uma 9h57'32" 69d1'29"
Keith Diaferia
Aql 19h32'49" 2d35'59"
Keith E Gonyou
Uma 11h1'14" 64d37'32"
Keith E. Minor
Cep 3h58'2" 86d35'1"
Keith E. Ramsey
Crb 15h52'37" 37d29'15"
Keith E. Revoir
Uma 10h49'47" 51d53'39"
Keith E. Shanks-Reece
Ori 5h21'36" 6d31'52"
Keith Edgerton Ph.D
Cnc 8h31'3" 9d42'40"
Keith Edward Jones
And 1h1'41" 38d39'9"
Keith Edward Parks
Cmi 7h27'33" 11d49'18"
Keith Eldridge
Her 16h35'9" 21d5'36"
Keith Emmett Connors
Her 17h2'26" 36d29'8"
Keith Eugene Heinzman Jr.
Her 16h44'39" 32d58'6"
Keith Falasco "My Shining Star"
Uma 10h54'56" 53d58'25"
Keith Faulkner Voyles, Jr.
Ori 6h5'25" -3d57'53"
Keith - Forever and a Day
Per 2h22'21" 55d20'3"
Keith Francis Lutzen, Jr. 7:35 p.m.
Tau 4h35'18" 20d48'10"
Keith Frank Nutting
Cep 22h28'41" 84d53'48"
Keith Fred "Flames" Keller
Sco 17h18'10" -40d55'57"
Keith Frederick
Psc 0h32'6" 3d5'42"
Keith Gilbert Brianna Anisah Isis
Aqr 22h18'52" 0d22'53"
Keith/Gina
Cyg 21h55'10" 53d36'26"
Keith Griffin
Cnc 8h58'43" 29d18'47"
Keith Hallissey
Her 17h48'57" 48d17'17"
Keith Harris Family Star - Auckland
Cru 12h47'59" -60d38'52"
Keith Heika
Per 4h7'39" 35d7'14"
Keith Henry Novak
Dra 19h21'8" 59d7'51"
Keith Heustice
Cep 23h5'25" 70d53'45"
Keith Hilles-Pilant
Aur 5h15'53" 41d9'57"
Keith Holliday
Aur 5h28'18" 46d26'40"
Keith Honey
Her 16h16'10" 25d46'18"
Keith Hugh Hill
Ori 6h13'47" 15d6'57"
Keith Humble
Uma 10h10'19" 53d37'5"
Keith Hurlic
Tau 4h37'6" 13d14'2"
Keith "Hypno Man" O'Neill
Tau 4h36'27" 14d33'39"
Keith Ireland
And 2h34'43" 45d27'32"
Keith J. Holmes
Uma 10h35'57" 68d21'18"
Keith J. J. Koeb
Dra 20h22'3" 67d8'29"
Keith J. Krach
Ari 2h36'10" 22d22'55"
Keith J Shanabrough
Crb 16h15'17" 37d8'43"
Keith James Gates
Vir 12h40'19" 6d20'13"
Keith James Martin
Per 3h8'54" 52d52'20"
Keith Jeffrey Holmes
Uma 9h44'58" 49d3'10"
Keith Jobson
Per 4h14'22" 40d49'54"
Keith John
Ori 5h38'24" -0d33'8"
Keith Joseph Gerard
Psc 1h28'39" 23d34'16"
Keith Jr.
Uma 9h55'12" 47d56'33"

Keith Judge
Uma 11h13'13" 45d0'5"
Keith "Kid" Crenshaw
Sco 16h7'32" -9d50'6"
Keith & Kim Moser
Uma 11h52'34" 35d26'49"
Keith Kocan
Uma 12h6'51" 46d45'37"
Keith L. Beasley
Dra 17h49'2" 67d26'11"
Keith L. Butler
Uma 11h41'54" 42d18'16"
Keith L. Olsen
Psc 0h34'54" 20d42'38"
Keith L. Urban
Sco 16h24'14" -25d12'45"
Keith Lacey
And 2h20'21" 45d0'23"
Keith Lambourne
Lib 14h41'3" -18d17'6"
Keith & Lesley Harrop
Cas 2h50'56" 57d59'41"
Keith - Light of my Life
Cep 22h12'14" 63d57'3"
Keith Lightfoot
Aqr 20h42'23" -11d45'26"
Keith Limburg
Her 17h41'16" 45d22'15"
Keith M. Eckel
Lib 14h57'51" -5d7'28"
Keith MacKillop #1 Dad In The Universe
Uma 8h27'8" 62d6'12"
Keith Maclean Currie
Aql 19h15'54" -0d7'2"
Keith MacPherson's Heavenly Body
Cru 12h15'50" -63d38'23"
Keith & Mandi
Cyg 19h45'47" 35d31'27"
Keith Marchant
Ori 6h9'16" 21d8'21"
Keith&Marcia Buchanan Gleneira 2004
Cru 12h44'29" -57d58'26"
Keith Maroney - m'athair mór
Cep 21h34'58" 64d34'47"
Keith Marsh
Per 4h24'13" 47d44'22"
Keith Martin
Ori 6h22'42" 14d27'1"
Keith Martin Davis
Boo 15h31'46" 49d48'34"
Keith Matthew Malinao
Per 3h36'22" 49d9'21"
Keith McDaniel, Sr.
Psc 1h44'31" 12d48'17"
Keith McGlinchey
Sco 17h57'39" -42d57'11"
Keith McKenzie
Tau 5h2'29" 24d41'3"
Keith Mearns
Cyg 20h27'13" 33d16'17"
Keith Mellon-Butler
Her 17h15'20" 27d44'22"
Keith Melvin Aspin
Uma 10h33'30" 64d34'1"
Keith Miller
Tau 4h44'42" 19d1'34"
Keith Moore
Leo 11h35'28" 14d45'31"
Keith Myron Wright
Cap 20h21'24" -12d34'54"
Keith N. Cook
Her 16h28'0" 46d22'57"
KEITH - NYPH8001 - FPFD317
Lib 15h27'52" -19d6'51"
Keith O'Donnell
Cep 21h26'47" 63d46'19"
Keith Oldfield
Aur 5h3'6" 40d5'11"
Keith Pagan
Ori 6h1'31" -0d11'36"
Keith Patrick -Loving You Eternally
Col 5h24'34" -30d1'30"
Keith Perryman Joines
Vir 13h18'36" 11d40'40"
Keith R. Mires
Psc 1h17'11" 19d22'46"
Keith Raymond Tofts
Aur 5h13'28" 36d38'47"
Keith Renshaw Siddall
Per 2h52'13" 45d11'38"
Keith Robert Harrington
Aqr 21h47'12" -0d41'54"
Keith Robert Wegner
Crb 16h5'57" 38d28'37"
Keith Roper
Aqr 21h43'45" -5d52'46"
Keith Rossman
Per 3h9'43" 56d30'49"
Keith Russell Little
Per 3h2'28" 41d30'24"
Keith Ryan Nintzel
Ori 6h6'20" 10d0'3"
Keith S. Ducotey
Vir 13h5'46" 7d57'47"
Keith & Sandra Weser
Cyg 19h45'14" 35d25'53"
Keith Scribner
Cnc 8h44'20" 31d39'20"

Keith Smallwood Fraser
Uma 9h19'36" 50d6'46"
Keith Sullivan & Elizabeth
Cyg 21h43'38" 43d53'37"
Keith Swinehart
Cap 20h52'24" -17d42'59"
Keith T. Miller
Sgr 19h49'14" -12d26'17"
Keith & TaShena's Star of Love
Cyg 20h30'55" 32d51'23"
Keith Tatler
Uma 12h6'25" 53d5'36"
Keith Teague Chapman
Ori 5h21'25" 8d8'27"
Keith Terriault's Star
Aur 6h35'55" 34d17'36"
Keith Thomas
Her 18h29'35" 12d6'47"
Keith Thomas Dressel
Per 3h9'34" 51d26'27"
Keith Thomas Galvin
Ori 5h54'28" 1d1'3"
Keith Thomas Hayes
Uma 10h49'49" 53d42'19"
Keith Thumper Hinson
Ori 6h4'42" 6d21'5"
Keith Towells
Lmi 10h33'7" 37d40'27"
Keith Urban
Sco 16h4'51" -21d9'8"
Keith & Val-Started Out With A Kiss
Ori 5h38'24" -0d20'45"
Keith VonRapacki
Uma 9h25'29" 51d38'56"
Keith Walker
Ori 5h24'35" -0d45'50"
Keith Wesley Cameron
Lib 15h10'24" -21d57'42"
Keith William Marcola
Cep 23h4'43" 79d29'17"
Keith Wolos
Vir 12h56'57" -4d56'59"
Keith Wyatt Coley
Cap 20h59'14" -21d22'10"
Keith, Kara, Kim & Kyle Armstrong
Uma 11h30'43" 44d20'55"
Keith25
Her 17h39'41" 17d47'48"
KeithClareMichelleHeather
And 1h1'37" 37d10'43"
keithloriekaleighcaseyde-von"family"
Uma 10h38'47" 64d13'54"
KeithlovesShannon
Vir 13h14'35" 12d22'7"
Keith's Flying Circus
Ori 4h52'50" 10d57'22"
Keith's Heavenly Light
Uma 10h51'48" 71d8'24"
Keith's Light
Her 18h52'43" 23d23'39"
Keith's Star
Cep 0h28'54" 83d41'22"
Keith's Wishing Star
Tau 3h48'53" 17d54'14"
Keka
Gem 6h21'34" 18d9'46"
Kekai Malu
Cap 20h50'54" -25d27'5"
Kekaulaheaokeaholoa
Dra 16h45'19" 53d2'19"
Kekealaniokana'auao
Aqr 22h50'39" -13d2'34"
Kekoa Haines
Uma 11h31'48" 57d36'47"
Kel
Aqr 21h58'33" -8d18'2"
Kel Bel
Aqr 22h38'45" 1d2'10"
Kel Bugz
Psc 0h54'4" 9d0'37"
Kel Kel
Lib 15h19'0" -20d2'26"
Kela
Cnc 8h50'26" 31d12'51"
Kelah Leshae
Tau 3h31'6" 7d45'19"
kelalex07
Lyn 6h26'21" 60d43'41"
Kelandel
Uma 8h21'54" 61d33'44"
Kelaris
Lyn 6h47'6" 51d52'43"
KELBIE05
Gem 7h15'21" 19d25'16"
Kelby & Aidan Stach
Sco 17h51'52" -35d53'49"
Kelce Bre Sanford
Uma 9h48'21" 61d23'5"
Kelcea
Uma 10h19'20" 55d6'31"
kelci
Gem 6h22'43" 19d57'8"
Kelci Leigh Grzywnowicz
And 23h33'22" 47d12'36"
Kelcie
Uma 12h26'2" 55d35'42"
Kelcie Linn Harvey
And 1h14'20" 41d29'57"
Kelcie Sonora Clark The Bug
Sco 16h8'31" -15d31'17"

Kelcy Sullivan
Lib 14h59'31" -15d55'53"
Keldi Arnold
And 22h59'7" 38d47'35"
Keli
Ori 5h3'28" 2d4'17"
Keli and Andy
Per 2h11'47" 56d26'40"
Keli Ann Heiple
Uma 9h39'50" 62d40'21"
Keli Ann Rosado
Uma 14h2'1" 54d23'16"
Keli Blackwell
Cam 6h0'7" 60d54'50"
Keli Marie Jordan
Gem 6h33'53" 27d9'34"
Kelian
Uma 10h2'26" 45d39'23"
Kelie Anna 3/11
Psc 0h31'54" 19d53'59"
Keliiokalani RKM 6463
Gem 6h44'43" 21d33'18"
Kelin Salvador
Ori 6h17'12" 5d40'38"
Kelin Stanley "My Warrior Princess"
Ori 5h29'45" 0d29'25"
kelindaRolon709
Per 2h41'8" 56d36'10"
Kelise Amatorious Remedium Astram
Cas 1h0'31" 69d52'57"
Kelisha Bartholomew
Lyn 7h27'4" 46d27'56"
Kell
Cap 21h33'39" -17d56'15"
Kell Bell
Uma 10h38'39" 41d55'14"
Kell Bella
Cap 20h35'59" -20d11'42"
Kella
Lyn 6h31'39" 61d36'42"
Kella Kuka dos Santos
Cnc 9h8'27" 31d42'53"
Kella Renee
Lyn 8h21'23" 46d32'53"
Kellan
Tau 4h26'29" 23d19'51"
Kellbell
Cnc 8h24'12" 22d54'6"
Kellbell
Crb 15h43'4" 25d36'51"
Kellbell
Lyn 7h22'54" 56d37'26"
Kellcie Snyder
Uma 8h32'40" 63d57'35"
Kelldog
Cma 7h1'2" -23d15'55"
Kellee
Ori 5h54'0" 21d7'35"
Kellee
Aqr 22h8'11" 0d56'38"
Kellee
Her 17h43'11" 33d49'49"
Kellee Bo Belly
Lib 15h37'59" -18d33'48"
Kellee Craft
Cap 20h32'33" -18d4'52"
Kellee Diane
Ori 6h7'57" -0d21'10"
Kellee Jane Barnett
Cas 0h6'20" 57d17'23"
Kellee "KK" Hogan
Uma 11h36'37" 42d29'59"
Kellee Murchison Bennett
Uma 9h44'21" 53d26'15"
Kellee Nicole 1
And 0h49'31" 42d16'6"
Kellee's Light
Vir 13h18'53" -14d26'42"
Kellee's Star
And 1h9'22" 39d39'40"
Kellen
Leo 11h44'17" 27d12'4"
Kellen
Tau 4h20'2" 26d13'40"
Kellen Allayne Duffey
And 0h42'18" 39d4'1"
Kellen Campbell Dickinson
Uma 9h24'48" 62d50'16"
Kellen Carter Scott
Lyn 7h47'8" 47d53'28"
Kellen Kjer Flanigan
Ori 4h47'26" -0d2'5"
Kellen Kristhina Ferreira de Souza
Ari 1h47'40" 19d6'3"
Kellen McNeal's Shining Star
Uma 9h46'57" 57d36'24"
Kellen Michael Ortiz
Aqr 22h40'10" 2d20'38"
Kellen Michael Symmes
Psc 0h26'8" 19d2'9"
Kellen Tate Koller
Leo 10h53'28" 14d58'53"
Kellen William Fitzgerald
Cnc 9h5'45" 32d14'27"
Keller
Gem 6h55'22" 23d22'41"
Keller Collins
Her 16h13'22" 47d15'53"
Keller Raye Hylleberg
Aqr 21h52'32" 0d42'46"
Keller, Dirk
Aqr 20h44'7" -6d54'54"
Keller, Herold
Ori 6h18'32" 9d32'20"
Keller's Barbeque
Psc 1h20'46" 7d15'59"
Kelley
Lyn 7h34'56" 53d0'24"
Kelley
Uma 9h20'19" 67d18'40"
Kelley
Ari 2h48'48" 20d21'35"
Kelley
And 23h24'46" 43d3'3"
Kelley
Cnc 8h45'48" 32d7'48"
Kelley A. Lane
Uma 10h11'45" 69d5'43"
Kelley A. Trull
Lyn 7h47'32" 47d48'44"
Kelley Amanda Carter
Ari 2h53'23" 22d17'26"
Kelley and Jim Forever
Cyg 21h17'13" 46d9'28"
Kelley Ann Strasser
Cap 20h52'26" -27d0'44"
Kelley Anne Bowman
Uma 9h35'15" 47d14'50"
Kelley Anne Oakley
Cas 23h36'27" 52d48'48"
Kelley Barlow
Tau 5h47'32" 22d26'31"
Kelley Bean
Lib 14h48'39" -1d49'13"
Kelley Carter Sides
Cnc 8h16'15" 28d8'29"
Kelley Charise Grimes
Cnc 8h28'31" 17d31'13"
Kelley Christine McCabe
And 0h37'38" 25d16'19"
Kelley Colleen Kays
Lyn 8h35'9" 38d42'55"
Kelley Danielle Hickle
Cnc 8h23'34" 13d6'52"
Kelley DeAnn
Psc 1h12'13" 11d36'9"
Kelley Dearest
Umi 15h26'50" 72d17'56"
Kelley Duncan
Aqr 22h52'31" -17d19'24"
Kelley Earp
Cnc 8h19'17" 15d7'43"
Kelley Elizabeth Humphreys
Uma 9h12'42" 58d8'24"
Kelley Elizabeth Werden
Lyn 7h33'44" 47d52'41"
Kelley Exe
Lmi 10h10'29" 28d14'15"
Kelley Jennifer Boggs
Leo 11h45'16" 24d55'30"
Kelley Jo
And 22h59'16" 51d4'26"
Kelley Lee
Gem 6h44'8" 13d47'24"
Kelley Lyn Siefert
And 23h12'47" 43d8'58"
Kelley Lynn Dwyer
Aqr 22h14'42" -22d31'38"
Kelley McArthur
Vir 13h3'41" 12d59'1"
Kelley McCaffrey
Umi 16h34'29" 80d8'41"
Kelley McCarter
Tau 3h42'27" 19d7'52"
Kelley Michelle Enoch
And 0h45'55" 31d31'42"
Kelley Minogue
Tau 4h10'38" 27d39'47"
Kelley Morrell
Ari 2h13'50" 22d47'52"
Kelley Overton
And 0h47'38" 40d20'11"
Kelley Reeves
Sco 16h8'35" -12d58'54"
Kelley Rhodes
Lib 14h40'5" -24d41'35"
Kelley Shaddock
And 23h49'3" 45d9'18"
Kelley Sliauter
And 0h39'59" 28d42'20"
Kelley Sue Conniff
Cam 5h34'17" 62d53'29"
Kelley T. Cheney
Lyn 7h44'43" 36d16'32"
Kelley Thorpe
Leo 10h52'51" 22d49'46"
Kelley Wirths
Lyn 7h13'20" 57d53'37"
Kelley-Ann
Psc 0h50'27" 3d54'10"
Kelley's Love
Vir 14h12'21" -12d1'25"
Kelley's Lucky Star 1
Lib 15h5'13" -12d11'5"
Kelley's Rainbow
Boo 14h44'44" 53d17'35"
Kelley's Star
Uma 9h26'0" 57d22'44"
Kelley's Star
Uma 13h17'3" 58d4'53"
Kelleys Wishing Star
Tau 5h57'54" 24d40'31"
Kellhy Faraggi
Cas 23h32'53" 51d39'18"
Kelli
Crb 16h10'41" 36d42'58"
KELLI
And 0h47'51" 40d59'27"
Kelli
And 1h9'22" 43d47'19"
Kelli
Gem 6h10'45" 27d33'2"
Kelli
Psc 23h2'17" 8d2'26"
Kelli
Dra 12h25'57" 72d27'19"
Kelli
Sco 17h11'29" -44d3'57"
Kelli A. Karns
Gem 7h3'27" 25d8'46"
KELLI ANN
Cap 21h39'31" -11d14'0"
Kelli Ann
Vir 12h14'13" -9d8'42"
Kelli Ann Clover
Leo 11h11'2" -1d14'17"
Kelli Anna Butkovich
Sgr 17h56'54" -27d48'6"
Kelli Anne Beerer
Cap 21h18'52" -16d15'59"
Kelli Anne Lugo
Gem 7h31'12" 25d56'39"
Kelli Barber
Ori 5h24'7" -4d37'45"
Kelli Bridget Conneely
Sco 17h16'30" -32d6'11"
Kelli Britz
Cam 5h50'6" 61d25'20"
Kelli Cummings
And 1h48'29" 41d1'49"
Kelli D. Chapman
Uma 9h37'15" 60d27'59"
Kelli Elizabeth
Cnc 8h43'44" 15d49'41"
Kelli Giddish
Ari 3h20'33" 25d56'50"
Kelli Gilliland Forever Has My Heart
And 2h27'40" 42d16'32"
Kelli Grace
Cap 21h8'31" -16d23'9"
Kelli Hebert
Leo 11h32'29" -1d40'23"
Kelli Holleman
Gem 6h36'23" 13d13'35"
Kelli Ivy
Umi 14h57'53" 72d22'46"
Kelli Jean Bryant
Uma 10h42'52" 43d23'5"
Kelli Jeanne's star
Vir 11h39'50" 4d48'29"
Kelli Jo Berry 05
Ori 5h18'4" 0d38'6"
Kelli Jo McNemar
Cas 0h14'52" 57d3'22"
Kelli Jo Thompson
Sgr 19h46'34" -13d39'43"
Kelli K Bird
Cnc 8h47'9" 32d43'9"
Kelli Ketzler
And 0h53'49" 44d16'8"
Kelli Kristine West
Mon 7h12'20" -1d26'24"
Kelli Kusak
Vir 11h54'55" -3d19'59"
Kelli Lee
And 23h34'28" 43d7'30"
Kelli Leslie
Aqr 21h10'5" -3d42'6"
Kelli Lindberg
Gem 6h47'14" 19d2'37"
Kelli Lyn Cole Celestial Forever
Uma 9h35'13" 44d32'52"
Kelli Lynn Biondich
Lyn 6h52'0" 56d17'45"
Kelli M. Doyle
And 0h40'8" 40d0'31"
Kelli Macauley
Leo 10h0'10" 20d19'22"
Kelli Marie
Per 2h58'22" 47d11'1"
Kelli Marie
Uma 10h20'13" 52d14'51"
Kelli Marie Lakeman
Cyg 20h22'56" 39d12'35"
Kelli Marie Sullivan
Psc 1h12'51" 26d46'28"
Kelli Marie Vannatta
Lyn 7h17'38" 48d34'8"
Kelli Mariy Harlan
Leo 9h25'9" 26d57'57"
Kelli Martin
Com 11h58'49" 26d24'9"
Kelli Martin
And 2h15'24" 43d41'45"
Kelli Martisch
Psc 1h11'34" 32d20'28"
Kelli McKenzie
And 1h18'52" 49d28'20"
Kelli Middleton
Lyn 8h23'14" 34d19'55"
Kelli Morgan
Cam 4h17'32" 62d51'7"
KELLI MY ANGEL
Lyn 6h21'17" 59d27'25"
Kelli Nicole Crane
Ori 6h11'55" 15d28'51"
Kelli Parris
Cam 3h46'47" 55d22'26"
Kelli Priest
Peg 21h22'15" 19d22'13"
Kelli Renee
Sgr 18h50'29" -16d28'18"
Kelli Rynders
Lyn 6h54'23" 51d47'21"
Kelli Samantha Thomas
Aql 19h41'37" -0d11'31"
Kelli Samantha Thomas
Aql 19h40'31" -0d55'46"
Kelli Superstar Evans
Psc 1h43'13" 5d31'52"
Kelli Taylor
Umi 15h54'53" 71d5'14"
Kelli Turpin
Cas 0h21'42" 59d57'5"
Kelli Victoria
Cas 23h33'31" 53d43'41"
Kelli Wakefield
Crb 16h9'52" 32d51'44"
Kelli Walter "Class of 2006"
Uma 11h38'52" 53d24'38"
Kelli Wolff
Leo 9h36'33" 32d41'0"
Kelli Wright & James Dickerson
Cyg 19h38'6" 42d50'40"
Kelli530
Gem 6h26'32" 27d5'54"
Kellian
And 1h12'53" 43d3'27"
KelliDreamStar
Per 4h25'48" 34d44'44"
Kellie
Crb 15h39'35" 33d55'10"
Kellie
Lyr 18h51'33" 33d48'31"
Kellie
And 23h25'38" 38d56'45"
Kellie
Cnc 8h15'22" 15d35'10"
Kellie
Mon 6h37'38" 10d58'14"
Kellie
Cas 2h20'23" 67d45'31"
Kellie
Cam 7h2'40" 64d20'57"
Kellie A Seymour
Sgr 18h12'14" -20d29'0"
Kellie and Anthony
Sgr 19h24'38" -19d36'13"
Kellie Ann
Vir 13h27'43" -21d32'19"
Kellie Ann Jureka
Tau 3h49'57" 18d58'45"
Kellie Ann Shimansky
Uma 10h2'52" 53d33'59"
Kellie Anne Holland: The Silly Girl
And 0h34'21" 31d35'45"
Kellie Bartram
Cas 1h20'21" 62d11'50"
Kellie Blackmon
Cnc 8h50'30" 27d7'37"
Kellie & Brandon
Cyg 19h34'10" 31d52'59"
Kellie Connolly
Cru 12h43'55" -60d52'26"
Kellie & Corrie MacLean
Gem 7h29'29" 33d44'43"
Kellie Elizabeth McCartney
Gem 6h45'41" 15d22'41"
Kellie Faye Massey
Lyr 18h49'45" 37d11'4"
Kellie Gene
Uma 11h36'8" 64d4'34"
Kellie Guglielmelli
Psc 1h30'8" 18d1'6"
Kellie Hoa Hoang
Tau 5h41'51" 24d7'45"
Kellie I Love You 4/19/90 Love Tim
Cas 23h6'19" 54d9'36"
Kellie J. Gilbert
Uma 9h0'26" 47d3'30"
Kellie J Rollman
Aqr 23h49'30" -10d14'30"
Kellie Jean Archer
Psc 1h7'12" 10d48'31"
Kellie Jo Penn
Cyg 19h32'34" 31d57'32"
Kellie Jo Wyeth-Shankel
Ori 5h56'35" -0d12'55"
Kellie Josell
And 1h17'17" 45d33'13"
Kellie Leigh Barger
Sco 16h39'58" -28d17'57"
Kellie Lynn
Psc 1h23'41" 23d33'43"
Kellie Lynn Wiackley
Uma 13h30'15" 55d22'35"
Kellie Marie Bennett
Mon 7h35'43" -0d56'24"
Kellie Marie Datin
Psc 0h15'54" 15d22'7"
Kellie (My Maid of Honour) xx
Vul 18h59'11" 23d8'46"
Kellie Pamela
Lyr 18h50'47" 44d39'26"
Kellie Rae Walker
Cnc 9h7'32" 31d37'15"
Kellie Star
Ori 5h18'28" 8d58'24"
Kellie Stence
And 0h55'56" 42d46'6"
Kelliebelle
Ori 6h15'22" 11d2'1"
Kellie's Caelesitis Flammae
Peg 23h20'12" 32d50'28"
Kellie's Glisten
Sgr 19h21'27" -15d56'6"
Kellie's Star
Umi 15h45'16" 74d36'0"
Kellie's Star-Light
Ori 5h22'41" 13d50'23"
Kelli-Marie
Sco 16h12'16" -14d25'2"
Kelli's Hope
Gem 7h22'55" 24d44'26"
Kelli's Monkey
Aur 7h23'7" 40d29'12"
Kelli's Star
Gem 6h49'5" 28d23'21"
Kelli's Star
Leo 10h44'43" 14d41'7"
Kelli's Star
Lib 15h34'39" -8d0'16"
Kelli-sue Catherine Conine
Cas 1h18'37" 63d2'34"
Kellogg EMP63
Ori 6h4'52" 13d51'2"
Kellster
Lyn 8h23'45" 33d56'8"
Kelly
Cnc 9h3'22" 32d0'20"
Kelly
And 23h38'26" 36d57'1"
Kelly
Uma 9h40'23" 47d39'10"
Kelly
Uma 11h5'0" 51d44'32"
Kelly
Cyg 21h16'45" 46d59'49"
Kelly
Uma 13h24'4" 52d29'52"
Kelly
Cas 0h47'7" 53d25'38"
Kelly
Per 3h14'30" 54d18'50"
Kelly
And 23h1'24" 51d20'48"
Kelly
And 22h58'20" 47d28'8"
Kelly
Ori 6h16'24" 14d16'19"
Kelly
Leo 9h26'25" 8d42'40"
Kelly
Cnc 9h7'40" 13d58'15"
Kelly
Peg 22h22'53" 9d30'9"
Kelly
Ari 2h36'31" 19d41'43"
Kelly
Ari 2h34'21" 18d24'25"
Kelly
Aql 19h8'59" 2d28'17"
Kelly
Lyr 18h38'24" 29d8'33"
Kelly
Crb 16h7'42" 25d56'28"
Kelly
Ari 2h56'42" 24d40'52"
Kelly
Cas 23h29'48" 58d10'15"
Kelly
Uma 9h32'29" 57d46'3"
Kelly
Uma 8h19'25" 59d49'42"
Kelly
Uma 8h16'59" 68d37'6"
Kelly
Uma 11h48'39" 61d51'11"
Kelly
Eri 3h37'1" -21d59'41"
Kelly
Aqr 22h8'18" -12d41'3"
Kelly
Cap 20h23'28" -13d57'22"
Kelly
Cap 20h31'34" -12d30'34"
Kelly
Sgr 18h40'54" -21d44'6"
Kelly
Cap 21h4'56" -15d11'49"
Kelly
Lib 15h46'32" -17d26'0"
Kelly
Lib 15h31'42" -6d15'20"
Kelly
Lib 15h35'52" -6d40'54"
Kelly
Lib 15h12'8" -27d52'45"
Kelly
Sgr 18h12'47" -27d23'46"
Kelly
Sco 17h37'38" -34d35'15"
Kelly
Ara 17h48'25" -47d14'59"
Kelly 9-23-2000 Kevin Tetrick
Lib 15h3'4" -6d30'53"
Kelly A. Campos
Peg 22h26'28" 33d51'8"
Kelly A. Carney
Leo 9h52'39" 22d22'3"
Kelly A. Harper
Cnc 8h35'29" 30d57'14"
Kelly A. Hudgens
Sgr 19h34'24" -27d26'34"
Kelly A. Martin KellzBellz
Gem 6h47'17" 24d54'48"
Kelly A. McGovern
Per 4h36'16" 40d16'8"
Kelly A. Ryan
And 23h30'50" 48d11'32"
Kelly Abba
And 2h37'40" 50d13'35"
Kelly Agnes Slater
Cas 0h49'32" 61d48'11"
Kelly Aine Leahy
Cyg 20h32'19" 54d6'47"
Kelly Altobelli
Uma 9h26'22" 43d14'34"
Kelly Amanda Smith
And 23h10'8" 45d42'50"
Kelly Amanda White
Cyg 20h50'9" 35d54'28"
Kelly and Adrian's Wedding Star
Cyg 20h46'15" 34d52'23"
Kelly and Austin's Love
Her 16h47'24" 25d26'42"
Kelly and Ben together
Cyg 20h35'34" 34d53'28"
Kelly and Brent
Boo 14h24'3" 13d28'50"
Kelly and Chris
Uma 11h10'41" 51d31'16"
Kelly and Christian Richards
Cnc 7h59'2" 15d50'33"
Kelly and Dave
Uma 10h45'1" 60d22'38"
Kelly and Kaye Mayo
Pup 7h46'33" -13d37'17"
Kelly and Lauren Witnesses Always
Uma 8h53'43" 54d35'5"
Kelly and Rob
Uma 9h7'19" 62d47'0"
Kelly and Rob forever
Cyg 19h51'34" 33d36'51"
Kelly And Scott Forever
Cnc 8h30'18" 15d46'19"
Kelly and Sean
Per 3h20'47" 46d2'29"
Kelly Anderson December 4, 1961
Peg 22h7'54" 31d50'18"
Kelly & Andy Smuskiewicz
Cyg 20h36'34" 51d7'0"
Kelly Ann
And 23h11'33" 51d4'10"
Kelly Ann
And 23h10'0" 51d36'13"
Kelly Ann
And 23h17'7" 47d42'29"
Kelly Ann
And 1h47'6" 38d59'57"
Kelly Ann
Crb 15h42'26" 36d25'21"
Kelly Ann
Leo 11h41'57" 16d16'33"
Kelly Ann
Cnc 9h4'31" 24d30'23"
Kelly Ann
Vir 12h36'18" 6d5'18"
Kelly Ann
Ari 2h10'34" 14d56'15"
Kelly Ann
Dra 15h21'55" 59d38'54"
Kelly Ann
Psc 1h20'2" 7d23'7"
Kelly Ann Beahan
And 23h39'8" 47d49'24"
Kelly Ann Bonnar
Cnc 8h14'51" 11d49'23"
Kelly Ann Burden
Gem 7h18'7" 13d26'5"
Kelly Ann Caferty
Vir 12h27'27" 12d17'25"
Kelly Ann Fannon
Cnc 8h52'39" 27d2'44"
Kelly Ann Fowler
And 23h21'56" 52d26'39"
Kelly Ann Johnston
Psc 1h23'17" 14d48'6"
Kelly Ann Kloss
Psc 1h10'6" 31d59'30"
Kelly Ann Martin
Vir 14h41'16" 4d52'47"
Kelly Ann May
Crb 15h46'56" 27d30'21"
Kelly Ann McMillian
Crb 16h3'30" 26d6'34"
Kelly Ann Muro/Adore Her
Uma 11h44'21" 64d34'43"
Kelly Ann Noack
Psc 1h18'41" 24d44'34"
Kelly Ann Norvell
Leo 11h31'52" 14d32'31"
Kelly Ann Pfeiffer
Lyn 6h58'18" 58d22'26"
Kelly Ann Potter "Light Of Life"
And 0h47'15" 37d54'9"
Kelly Ann Renstrom
Crb 15h49'50" 30d21'36"
Kelly Ann Rodger, 1985
And 23h57'37" 42d41'10"
Kelly Ann Rosa
Leo 10h30'17" 24d45'53"
Kelly Ann Saine
Sco 16h49'52" -43d35'47"
Kelly Ann Scow
Cas 0h54'42" 62d4'7"
Kelly Ann Sweinhart
Leo 10h30'26" 17d38'23"
Kelly Ann Westhusing
Ori 5h12'15" -0d2'33"
kelly ann williams
Aqr 23h20'49" -19d20'33"
Kelly Ann Zani
Leo 9h49'42" 27d49'21"
Kelly & Anna
Crb 15h50'22" 36d37'33"
Kelly Anne
And 1h43'13" 50d32'37"
Kelly Anne
Uma 13h9'59" 58d57'56"
Kelly Anne Braun
Tau 5h41'23" 19d26'45"
Kelly Anne Burch
Crb 16h8'40" 36d44'25"
Kelly Anne Carsey
Lyn 9h42'42" 40d18'56"
Kelly Anne CastaNeda
Lyn 8h17'28" 34d22'35"
Kelly Anne Leimer
Psc 2h1'35" 4d25'1"
Kelly Anne Parker
Cyg 20h48'9" 49d36'1"
Kelly Anne Rohde
Cap 20h27'14" -11d50'11"
Kelly Anne Ross
And 0h52'22" 41d3'10"
Kelly Anne Wilcox
Cam 4h26'15" 57d35'56"
Kelly Anthony Yeager
Her 18h51'40" 24d11'4"
Kelly Antoinette Hoyt-Clairmont
Gem 7h30'11" 31d11'40"
Kelly April Denis-Pierce
And 1h17'22" 46d7'42"
Kelly Archambeault
Gem 7h16'38" 13d27'46"
Kelly Arnold
Peg 22h47'50" 26d15'10"
Kelly B
Cyg 21h11'0" 45d40'35"
Kelly Baker
And 23h45'9" 36d42'24"
Kelly Barons
Cap 20h27'40" -13d24'15"
Kelly Barras
Cas 1h41'19" 63d36'44"
Kelly Bear
Leo 11h31'51" 17d45'32"
Kelly (Beba) Scordos
Leo 9h37'14" 28d18'16"
Kelly Beckman-Crabtree
Lib 14h44'28" -11d7'31"
Kelly Bella's Star
Sgr 19h55'48" -28d25'28"
Kelly Belle
And 2h9'50" 46d44'22"
Kelly Belly
Cas 0h7'34" 56d18'28"
Kelly Belly
Sco 16h16'48" -14d48'53"
Kelly Belly Olson
Cma 6h55'39" -27d22'37"
Kelly Berger
Cyg 20h48'7" 31d28'51"
Kelly Boardwolf Panfil
Crb 15h36'8" 26d7'26"
Kelly Boman
Lib 15h2'2" -23d39'11"
Kelly Bradford Greer
Tau 5h4'55" 19d25'16"
Kelly Brennan
Uma 11h43'57" 44d31'15"
Kelly Brianne Clarkson
Per 2h46'18" 54d3'21"
Kelly Brittney Sung
And 0h5'47" 47d42'0"
Kelly Brock
Lyr 18h49'47" 40d53'17"
Kelly Burress
Uma 13h11'11" 55d9'15"
Kelly "C B" Kohler
Aqr 22h40'43" -0d7'3"
Kelly C. Keller
Cyg 21h13'43" 45d38'52"
Kelly Campbell
Lmi 10h25'16" 28d36'35"
Kelly Canales
Ari 3h8'49" 21d26'46"
Kelly Caniglia
Vir 14h28'33" 2d4'50"
Kelly Carney-Rogers
And 2h36'21" 43d44'32"
Kelly_Casperson
Gem 7h49'15" 30d52'39"
Kelly Cassandra
And 0h52'25" 43d38'56"
Kelly Castro
Crb 15h51'29" 35d11'27"
Kelly Chirico
Lmi 11h4'19" 29d13'0"
Kelly Choy & Pop Pop
Mon 7h9'33" -0d57'30"
Kelly & Chris Jackson
Cyg 20h13'35" 39d13'54"

Kelly Christine Lib 14h53'14" -3d41'5"
Kelly Christine Linn Cap 21h12'12" -15d23'9"
Kelly Clark Winebrenner Ari 3h20'44" 22d1'27"
Kelly Coenenberg Lyn 7h50'34" 57d49'34"
Kelly Collins Ori 5h41'26" 1d28'1"
Kelly "Coochi" Frank Vir 14h21'5" 0d31'7"
Kelly Cooley Aqr 22h32'29" -3d5'15"
Kelly Cooper Vir 13h39'58" -1d49'43"
Kelly Correy Mon 6h45'53" -0d9'6"
Kelly Cristine Gem 7h43'58" 25d46'40"
Kelly Cudney Sgr 18h21'48" -31d18'7"
Kelly Cunningham Vir 14h39'22" 3d53'54"
Kelly Cusick Crb 15h54'1" 27d13'12"
Kelly D. Butcher Ari 3h11'33" 28d39'32"
Kelly D Fernandez Lyn 8h36'16" 36d16'51"
Kelly & Daniel Magre: A Love Light Cyg 21h31'22" 45d59'36"
Kelly Darrin & Felice Antoinette Cyg 21h15'58" 46d47'58"
Kelly Davies | 28-11-1985 Sgr 17h56'52" -23d51'53"
Kelly Dawn Leo 11h24'48" 13d37'13"
Kelly Dawn Brown Vir 12h7'48" -2d45'11"
Kelly Dawn Herrod Lyn 9h4'9" 46d23'43"
Kelly DeAngelis Tau 4h3'59" 9d59'39"
Kelly Denise Cnc 8h9'24" 15d11'11"
Kelly Denise Peterson Vir 14h29'43" 1d25'22"
Kelly Diane Uma 9h54'51" 55d46'38"
Kelly Diane Enos Leo 11h20'1" 14d45'39"
Kelly Diane Long Sgr 18h12'0" -32d49'52"
Kelly Diane Russo Umi 14h30'23" 73d59'16"
Kelly Dianne Scott And 0h14'47" 29d47'41"
Kelly Dianne White Fisher Lib 15h45'1" -16d59'36"
Kelly Ditrich And 1h45'6" 37d41'59"
Kelly Dunham Ari 3h8'47" 18d3'49"
Kelly Dyan Maze ~ Amour Vrai And 23h14'25" 47d2'56"
Kelly E. Brogan Tau 5h23'47" 27d2'18"
Kelly E. Wolf Lyr 19h12'17" 26d26'53"
Kelly Eirinn And 0h29'41" 28d31'40"
Kelly Elaine Stunda Ari 3h28'27" 20d51'41"
Kelly Eldridge (Owen + Mark) Cyg 19h21'17" 29d47'29"
Kelly Elizabeth Uma 12h6'14" 53d50'45"
Kelly Elizabeth Burch Ari 2h32'13" 26d51'20"
Kelly Elizabeth Correll Cam 3h46'50" 55d7'46"
Kelly Elizabeth Faison Aqr 22h36'43" -4d13'41"
Kelly Elizabeth Hobbs Uma 10h43'42" 49d3'23"
Kelly Elizabeth Key's Star Leo 11h27'13" -4d30'22"
Kelly Elizabeth Ott Tau 4h12'20" 17d5'50"
Kelly Elizabeth Wilson And 2h29'51" 48d40'48"
Kelly Elizabeth Ziebell 72188 And 0h9'57" 42d4'9"
Kelly England And 22h58'44" 47d49'16"
Kelly Ennis' Star Gem 7h19'10" 13d52'48"
Kelly Erickson Dra 19h45'38" 65d17'56"
Kelly Erin McDaniels Star Tau 4h26'38" 20d53'19"
Kelly Etta Jerosch (Owen) msw Lmi 10h9'17" 31d23'13"
Kelly Everett Gem 7h50'31" 33d57'13"
Kelly Exley Ari 2h13'57" 25d24'2"

Kelly Family Eri 4h12'27" -11d56'41"
Kelly Family BIX 7 Star Her 17h20'38" 15d58'40"
Kelly Fay Peg 22h21'12" 23d41'9"
Kelly Fischbach And 2h19'59" 40d40'32"
Kelly Flores Sco 16h58'30" -36d2'22"
Kelly Forever Mine And 0h32'11" 38d3'47"
Kelly Fort I Vir 13h49'54" -4d50'25"
Kelly France Uma 11h33'45" 63d37'1"
Kelly Frances Malaguti Uma 13h38'59" 50d59'12"
Kelly Fresh And 0h30'35" 30d16'27"
Kelly Freund Uma 10h34'39" 66d47'28"
Kelly Friend Mon 6h54'59" -0d35'40"
Kelly G Manning Lyn 7h38'18" 54d4'35"
Kelly Gail Postlethwait Lyn 6h30'25" 57d43'20"
Kelly Gassie Lyn 7h51'9" 38d21'36"
Kelly Gencarelli Gem 7h17'44" 32d49'57"
Kelly Geraghty Leo 11h57'56" 28d7'50"
Kelly Gorgeous Ori 5h16'39" -7d48'20"
Kelly Gorgeous Donovan Vir 15h1'8" 4d15'29"
Kelly Gormley Uma 10h14'52" 67d55'47"
Kelly Got Dra 17h9'51" 59d4'55"
Kelly Grace Koers Sco 16h13'35" -15d7'56"
Kelly Grossbauer Uma 14h5'51" 49d20'20"
Kelly Guzzi Uma 10h27'41" 59d52'19"
Kelly Gwynne Schroeder Cas 0h24'8" 57d30'54"
Kelly H. Su Sco 17h1'57" -39d8'37"
Kelly Hairrell Lyn 6h22'14" 61d3'34"
Kelly Hale Peg 22h37'17" 8d17'31"
Kelly Hale Tau 5h27'20" 27d43'1"
Kelly Harmon Miller Sco 16h55'47" -36d7'32"
Kelly Havig And 23h31'21" 35d39'2"
Kelly Henzl Uma 13h19'12" 57d31'28"
Kelly Heyniger Cas 1h25'41" 63d51'3"
Kelly Hidano Aql 19h28'1" 8d13'32"
Kelly Hildreth Vir 14h13'42" -16d18'31"
Kelly Hilton's Star Dra 16h55'11" 61d57'18"
Kelly Hoban Psc 1h16'49" 16d13'37"
Kelly Holcomb Leo 11h3'9" 21d1'38"
Kelly Hopper And 0h10'49" 36d50'8"
Kelly & Howard Umi 15h19'47" 74d1'19"
Kelly Ireland Frick Cet 2h43'50" -0d32'6"
Kelly Irene Gatsakos And 2h12'47" 43d3'8"
Kelly Is Like Whoa Cyg 21h56'54" 43d2'24"
Kelly J Bradley Peg 22h32'53" 25d21'18"
Kelly J Causey Wonderful Star Ori 5h7'34" 8d58'32"
Kelly J. Dickens Aur 5h33'15" 43d49'30"
Kelly J. Ross Cas 23h55'57" 57d43'49"
Kelly James Lib 15h24'27" -4d45'50"
Kelly Jane And 0h15'56" 31d7'17"
Kelly Jankowski Cas 23h9'53" 54d50'42"
Kelly Jayne Pho 2h19'57" -40d53'43"
Kelly Jean Uma 13h45'22" 54d11'52"
Kelly Jean Tau 5h47'27" 26d10'7"
Kelly Jean Psc 1h15'19" 22d4'7"
Kelly Jean Houston Tau 4h18'57" 17d11'35"
Kelly Jean Laughery Uma 10h25'17" 72d24'18"
Kelly Jean Osborn Sgr 18h17'4" -35d22'34"

Kelly Jean R Psc 1h9'35" 2d57'2"
Kelly Jennell Ari 2h30'56" 22d25'33"
Kelly & Jim Normandin Crb 15h20'13" 31d23'19"
Kelly Jo Lyr 18h52'18" 33d17'3"
Kelly Jo Cap 20h21'3" -12d40'52"
Kelly Jo and Jeff Chase Cyg 20h7'19" 30d40'31"
Kelly & John Lyr 18h32'19" 37d2'9"
Kelly & John Mabusth Uma 12h41'18" 55d21'29"
Kelly Johnson And 1h2'59" 42d43'23"
Kelly Jones Gem 7h30'11" 20d34'39"
Kelly Jones Ari 2h11'27" 24d46'59"
Kelly Jordan Her 18h40'48" 25d17'50"
Kelly Jordan Lyn 7h29'29" 48d52'43"
Kelly Joyce And 0h45'6" 39d45'36"
Kelly Joyce Gendreau Ori 5h29'59" 2d36'38"
Kelly Kae O'Brien Lib 14h52'36" -6d41'16"
Kelly Kathleen Ari 1h53'5" 23d18'49"
Kelly Kathleen Mack Cyg 21h53'49" 47d23'33"
Kelly Kay And 23h44'53" 44d47'42"
Kelly Kay Ari 3h15'28" 19d21'32"
Kelly Kay Mathias Crb 16h19'59" 38d52'17"
Kelly Kicer Ensminger Mon 6h56'29" 9d47'5"
Kelly Kizer Cnc 8h43'41" 22d40'28"
Kelly Klima Cam 4h34'59" 61d34'20"
Kelly Koerber Uma 9h42'49" 58d41'53"
Kelly Kuhn's Star of Love Sco 17h22'22" -41d51'6"
Kelly Kyaw Aung Psc 1h5'40" 33d5'1"
Kelly L. Crabtree Sco 16h7'56" -17d52'51"
Kelly L Hall And 0h20'47" 41d6'48"
Kelly L. Lyons Star Cas 1h5'44" 49d11'27"
Kelly L Sorenson Aqr 20h41'27" -9d21'31"
Kelly Lahr Lyn 7h17'47" 51d9'23"
Kelly Laine Manahl Ari 2h40'21" 21d34'59"
Kelly Lamb Ensor Crb 16h14'16" 38d7'4"
Kelly Leann Dadisman Ori 5h13'33" 9d6'4"
Kelly Lee Uma 11h31'48" 39d29'39"
Kelly Lee Cap 21h10'26" -19d50'45"
Kelly Lee Berger Leo 10h24'6" 19d22'7"
Kelly Lee Bonnville Umi 15h25'32" 72d2'39"
Kelly Lee Cannon Uma 11h21'41" 33d37'24"
Kelly Lee Elliott And 1h7'32" 42d40'21"
Kelly Leigh Bails And 23h32'41" 46d53'48"
Kelly Leigh Marion Tau 5h3'6" 20d0'57"
Kelly Leigh Miller Cnc 8h44'46" 21d22'57"
Kelly Lockwald Uma 9h56'28" 70d26'7"
Kelly Lorraine Ryan Aql 19h8'40" 13d0'28"
Kelly Lou Quillin Umi 15h13'5" 69d36'58"
Kelly Louise Adams Cra 19h2'37" -37d21'8"
Kelly Louise Allen And 1h34'2" 36d59'9"
Kelly Louise Broome And 1h29'11" 38d47'10"
Kelly Louise Cunningham Uma 9h39'17" 42d31'26"
Kelly Louise Green And 0h3'51" 44d16'2"
Kelly Louise Hodges And 2h3'28" 42d23'41"
Kelly Louise Oldroyd Cas 23h43'24" 52d34'7"
Kelly Louise Santry And 2h22'3" 41d20'2"
Kelly Loves Robbie Her 16h43'56" 14d49'33"
Kelly Lyn Gem 7h22'57" 19d49'6"

Kelly Lyn And 1h0'42" 41d13'53"
Kelly Lyn Houze Ari 2h36'42" 25d24'58"
Kelly Lynn Ori 6h18'21" 18d48'23"
Kelly Lynn Psc 0h34'53" 9d21'20"
Kelly Lynn Cas 1h17'51" 57d27'50"
Kelly Lynn Lyn 6h26'59" 59d24'53"
Kelly Lynn Sgr 19h37'59" -13d53'57"
Kelly Lynn Anders And 1h23'1" 44d37'5"
Kelly Lynn Cowell Cas 0h41'33" 54d32'12"
Kelly Lynn Griffin Heckman Sco 17h33'47" -36d57'14"
Kelly Lynn Hass Cap 20h15'48" -14d51'35"
Kelly Lynn Hogan Cas 22h57'13" 57d43'40"
Kelly Lynn Hutcherson Cyg 20h18'55" 37d23'0"
Kelly Lynn Phillips Crb 15h36'14" 39d2'34"
Kelly Lynn Richardson And 0h47'59" 40d30'2"
Kelly Lynn Roach And 23h14'11" 51d5'13"
Kelly Lynn Sexton Lyn 8h28'20" 51d6'27"
KELLY LYNNE NEIDORF Uma 9h45'1" 60d15'34"
Kelly Lynne Noddle Cyg 21h48'2" 43d55'27"
Kelly M. And 23h59'26" 36d46'11"
Kelly M. D. D'Agostino Uma 11h6'17" 63d15'7"
Kelly M. Grey Star Cas 1h40'23" 60d46'33"
Kelly M. Rangel Ori 5h47'27" 8d13'13"
Kelly Mackenzie Knight Tau 5h5'47" 20d58'1"
Kelly Mackey Sgr 18h32'0" -19d54'27"
Kelly Mahone Sgr 19h42'35" -14d27'1"
Kelly Mairin O'Donnell Sgr 19h55'52" -41d19'32"
Kelly Maria Durbin Ori 5h46'5" 12d29'45"
Kelly Marie Leo 11h26'51" 0d5'41"
Kelly Marie Per 3h28'16" 44d52'7"
Kelly Marie Gem 6h47'19" 34d32'40"
Kelly Marie And 23h51'56" 38d0'2"
Kelly Marie Sgr 18h55'10" -17d38'41"
Kelly Marie Cas 1h52'54" 65d1'4"
Kelly Marie 42684 Tau 4h22'51" 19d49'18"
Kelly Marie Allen And 23h8'59" 51d36'37"
Kelly Marie Colson Mon 7h5'51" -6d11'31"
Kelly Marie Goring And 1h39'29" 40d46'16"
Kelly Marie Harin Vir 13h17'43" 5d48'17"
Kelly Marie Henson And 1h47'55" 39d56'9"
Kelly Marie Lenehan And 1h46'46" 43d8'43"
Kelly Marie Makinson And 0h3'42" 42d54'16"
Kelly Marie Nubbie Ari 2h12'58" 24d8'8"
Kelly Marie Phillips Vir 13h14'49" -2d54'56"
Kelly Marie Vita Cnc 8h9'46" 32d1'45"
Kelly Mark Psc 1h8'54" 32d45'48"
Kelly Marshall Phi Sigma Chi Alumni Umi 15h56'39" 82d8'49"
Kelly Martinez Aqr 23h1'24" -8d33'44"
Kelly Maureen Fitzgibbon Aqr 21h10'33" -8d44'57"
Kelly Mays Uma 11h59'36" 35d5'29"
Kelly McAvoy Psc 1h0'15" 6d57'58"
Kelly McBey Aqr 22h36'43" 2d20'33"
Kelly McCarthy Cyg 20h11'47" 33d59'40"
Kelly McClure Lmi 10h29'59" 36d43'41"
Kelly McCollum Vir 12h27'21" -4d14'18"
Kelly McDonnell Vir 13h16'36" 4d38'58"
Kelly McGarvey And 23h6'37" 51d39'3"

Kelly McGinn Lyn 7h36'47" 39d13'27"
Kelly McKinney Vir 12h22'28" -1d46'59"
Kelly McNabb Cam 3h48'24" 65d19'12"
Kelly McNamara And 0h1'19" 39d24'35"
Kelly Meek Cra 18h12'21" -37d12'37"
Kelly Megan Lemmons Her 17h39'41" 16d48'59"
Kelly Mesko Lib 14h57'32" -9d18'39"
Kelly Meyer And 0h44'0" 26d49'29"
Kelly Meyer And 22h58'24" 48d6'21"
Kelly Michelle And 23h15'6" 42d5'55"
Kelly Michelle And 0h37'30" 27d44'31"
Kelly Michelle Aqr 20h40'11" -11d38'35"
Kelly Michelle Anderson Ari 1h57'20" 21d12'56"
Kelly Michelle Cohen Lyn 7h35'29" 48d33'11"
Kelly Michelle Dukes Cap 20h40'35" -14d42'52"
Kelly Michelle Flagg Aqr 21h14'55" 1d21'34"
Kelly Michelle Griffin Sco 17h50'13" -37d18'59"
Kelly Michelle Smith Uma 13h54'13" 61d46'52"
Kelly Miller Tau 3h24'32" 10d2'42"
Kelly Misotti Lyn 7h10'51" 61d39'18"
Kelly Mitchell Gem 7h10'48" 27d27'17"
Kelly Moore-2005 Tau 5h48'53" 14d22'45"
Kelly Morgan And 0h24'26" 33d7'12"
Kelly Mou And 2h35'39" 45d14'4"
Kelly Murphy And 23h19'53" 42d52'24"
Kelly Murphy Psc 0h24'33" 4d36'41"
Kelly my love Ori 6h10'32" -3d59'52"
Kelly - My Princess Psc 0h40'25" 15d36'57"
Kelly Myong Dahle Her 18h12'47" 41d17'3"
Kelly Neil And 23h17'41" 48d43'4"
Kelly Nicole Lyn 6h58'1" 58d48'56"
Kelly Nicole Chin And 1h28'55" 50d35'16"
Kelly Nicole Haynes Leo 11h50'27" 25d15'12"
Kelly Nicole Perry Ori 6h5'10" 9d46'10"
Kelly Nicole Sullivan Ori 5h37'0" -2d12'18"
Kelly Noel Sgr 18h19'45" -32d57'9"
Kelly Noel Spurr Sgr 18h34'26" -18d6'13"
Kelly Noelle Sgr 19h33'0" -23d16'51"
Kelly Norris Cas 0h27'46" 51d53'5"
Kelly Notch Gem 6h43'20" 18d30'11"
Kelly O'Leary Lyr 18h49'13" 30d54'15"
Kelly O'Malley And 0h9'49" 46d28'13"
Kelly O'Reilly Ori 6h10'51" 19d32'46"
Kelly Osburn Psc 1h25'16" 12d46'43"
Kelly Outlaw Tate Vir 12h15'0" 9d25'24"
Kelly P. Meddaugh Her 18h37'5" 15d23'43"
Kelly Page Cuthbertson Uma 9h21'12" 44d15'16"
Kelly Palcziewski Preston Uma 10h26'29" 64d57'38"
Kelly Pearson And 0h54'1" 46d26'19"
Kelly Pellerin Uma 10h0'36" 59d16'28"
Kelly Pence Vir 12h16'37" 1d45'5"
Kelly Pentsch Leo 10h22'14" 10d19'31"
Kelly Pieters Uma 11h43'14" 41d15'6"
Kelly Polonus Luminous Lyn 9h2'40" 35d50'48"
Kelly Pressley's Eternal Diamond And 2h29'19" 41d34'46"
Kelly Priutt Her 17h5'10" 41d1'15"
Kelly Pruett Barnhardt Vir 13h0'21" -1d36'43"

Kelly Quirke And 0h28'45" 21d42'19"
Kelly R. Underwood Cas 1h3'28" 62d37'52"
Kelly Rae And 0h6'6" 43d49'59"
Kelly Rae Lyr 19h14'10" 33d22'31"
Kelly Rae Condiff Crb 15h36'22" 39d29'16"
Kelly Rae Galusha Cas 0h30'43" 51d8'6"
Kelly Rae Sommerfeldt Cyg 19h26'17" 36d4'29"
Kelly Rae Soots Sgr 17h54'37" -28d49'42"
Kelly Rafael Stefany Uma 14h13'20" 61d40'31"
Kelly Ralston Uma 9h3'44" 52d57'51"
Kelly Randal Brady Cap 21h37'8" -19d44'23"
Kelly Rankine Gem 7h4'22" 10d44'9"
Kelly Rebecca Gem 6h57'41" 31d6'22"
Kelly Reiko Vir 12h15'17" 10d42'56"
Kelly Renee Varesio And 23h35'51" 45d3'3"
Kelly Ripp A Crb 16h8'11" 31d3'31"
Kelly Roach Hackett And 2h28'35" 43d25'36"
Kelly Robert Keeton Dra 18h37'4" 53d29'0"
Kelly Roberts Cap 21h7'48" -19d37'51"
Kelly Roberts Tau 3h50'0" 0d55'19"
Kelly Robertson of Sandyglass Court Cru 12h48'52" -62d12'40"
Kelly Roderick Cas 1h32'11" 68d38'46"
Kelly Rodgers Tau 4h37'20" 17d53'27"
Kelly Rose Cnc 8h16'1" 28d29'55"
Kelly Rose Lib 14h46'26" -11d31'16"
Kelly Rose Garcia And 0h34'29" 30d21'20"
Kelly Rose Klumpp And 0h26'41" 26d57'19"
Kelly Rose Stonehouse Ari 3h12'13" 20d25'22"
Kelly Ruiz Cyg 21h41'37" 43d50'23"
Kelly Ruth Anderson Aqr 23h29'52" -23d13'51"
Kelly S. Fowler Cap 21h15'22" -26d23'41"
Kelly S. Givens Sco 17h52'2" -30d24'10"
Kelly S. Sauder Cas 1h11'0" 59d41'33"
Kelly Savin Lyn 8h13'1" 54d23'9"
Kelly Scott Bradley Tau 4h36'30" 26d38'40"
Kelly & Sean Cyg 21h35'35" 51d8'55"
Kelly & Sean November 4th 2005 Cra 18h5'29" -38d17'57"
Kelly Servedio Ari 3h13'27" 15d52'21"
Kelly Shanahan And 23h20'1" 44d48'30"
Kelly Shanley Uma 12h19'37" 57d9'19"
Kelly Shannon Martin (Hort) Sco 15h56'24" -22d23'39"
Kelly Shereck And 0h48'45" 36d38'19"
Kelly Shows' Open Arms Star And 0h31'28" 25d22'8"
Kelly Sigro And 23h52'15" 41d15'39"
Kelly & Simon Greenwell Cyg 21h49'2" 47d17'37"
Kelly Smrz Umi 16h27'45" 77d9'50"
Kelly Squirrel Ratfield Leo 11h8'11" -2d3'21"
Kelly (Sqwurrel) King Leo 10h11'35" 13d56'27"
Kelly Staar Ori 5h28'8" 1d49'16"
Kelly Starr And 23h17'56" 41d55'38"
Kelly St.Clair Readyhough Leo 10h54'38" 15d9'4"
Kelly Sterling Cyg 21h37'35" 54d9'27"
Kelly Sterling McLaen Shining Star Lib 15h31'35" -9d8'28"
Kelly & Steve's Promise of Love Cyg 20h14'19" 55d19'15"

Kelly Stewart Cas 1h5'52" 63d39'8"
Kelly Stewart Sco 16h42'52" -29d34'51"
Kelly Sue Cole Cnc 8h36'52" 9d46'26"
Kelly Sue Kesar And 2h11'48" 43d49'14"
Kelly Sue McNabb And 0h35'50" 44d25'10"
Kelly Sullivan Per 3h39'34" 44d12'23"
Kelly Sullivan Dra 19h26'56" 59d16'50"
Kelly Sundgren Lyr 18h41'36" 30d42'50"
Kelly Sura Vir 14h2'18" -16d9'16"
Kelly Surgalski Vir 12h49'30" -10d53'44"
Kelly T Jones Aqr 23h1'7" -8d18'13"
Kelly Taylor And 1h18'39" 41d57'8"
Kelly Taylor Crb 15h28'21" 29d4'29"
Kelly The Beautiful Vir 12h29'19" 4d28'26"
Kelly; The Brightest Star in my Sky Cyg 20h35'13" 48d14'45"
Kelly "The Girl of My Dreams" Cyg 19h57'55" 51d29'29"
Kelly Theresa Coughlin Leo 11h34'29" 27d55'13"
Kelly Thorne Muzzi Aql 19h28'48" 6d7'46"
Kelly Tobias Cyg 21h55'20" 52d42'18"
Kelly & Tony Cyg 19h36'1" 30d31'34"
Kelly Tracy Cnc 8h33'38" 16d49'32"
Kelly Vanlaeken Lyn 6h59'34" 48d13'46"
Kelly Vogt Sco 16h7'33" -18d31'42"
Kelly Wallace Uma 11h59'44" 51d36'8"
Kelly Walsh Uma 11h50'38" 43d42'31"
kelly waterbury Umi 17h39'19" 87d17'12"
Kelly Watkins Aqr 22h14'35" 1d21'47"
Kelly Werhane Althoff Ari 2h50'7" 26d32'5"
Kelly Wessel Sgr 18h5'26" -28d1'16"
Kelly Wilson Lyn 6h37'55" 57d3'5"
Kelly Wilson Leo 9h31'5" 31d16'18"
Kelly Wing Ari 2h34'39" 13d50'33"
Kelly Wood-Stille Cyg 20h33'53" 42d39'22"
Kelly Woolwine Tau 4h39'15" 19d47'16"
Kelly Wooten Tau 5h33'48" 26d49'54"
Kelly Wucherpfennig Gem 6h27'27" 23d1'7"
Kelly Wyatt And 0h46'47" 35d55'11"
Kelly, Chelsea, Blaine Ori 6h7'45" 17d50'10"
Kelly, Kori, D.J. & Doug Dawkins Uma 12h3'14" 46d6'29"
Kelly, my guiding light Cnc 8h52'33" 31d54'55"
Kelly, My Love Uma 11h50'12" 50d35'1"
Kelly, my love and my life Aqr 23h34'15" -7d48'20"
Kelly512 Uma 10h12'35" 69d39'26"
Kelly-Anne Bickhart And 23h57'58" 45d11'57"
KellyAnne's (a) Star Psc 1h42'51" 7d33'56"
Kellye Lawry Dra 16h29'34" 58d3'53"
Kelly-Elizabeth Sgr 18h0'5" -27d35'2"
Kellyellie Lyn 6h50'11" 51d3'50"
KellyGator And 1h10'21" 42d32'27"
KellyJo Sgr 19h18'51" -26d26'36"
KellyJo Sco 17h57'6" -41d5'54"
KellyJo Lyn 7h3'43" 55d34'38"
KellyKat Psc 0h48'47" 16d29'11"
Kellykins Psc 0h21'26" 10d2'29"
Kellykins And 1h52'8" 46d21'17"
KellyKurz Vir 13h25'8" -14d1'38"

KellyLynn
Mon 6h46'1" -0d29'14"
KellyMarleneicaBrighticus
Lib 15h1'11" -17d59'12"
Kellyn Palladino
Sgr 20h4'34" -26d9'27"
KellyO102384
And 0h42'14" 40d27'10"
Kellypher
And 23h14'39" 48d44'37"
Kelly's 1st Wedding Anniversary!
Del 20h37'17" 15d19'32"
Kellys Angel
Sco 17h17'12" -32d53'55"
Kelly's Astellas Star
Cnc 8h49'16" 28d12'58"
Kelly's Beauty
Aqr 22h14'45" -1d19'58"
Kelly's Birthday Star
And 1h41'42" 37d4'18"
Kelly's Chief
Sco 17h55'27" -39d58'40"
Kelly's Desperado Star
Sco 17h42'0" -42d20'3"
Kelly's Dream
Cas 0h6'1" 57d21'3"
Kelly's Eyes
Tau 5h10'47" 18d6'12"
Kelly's Heart
Her 18h33'28" 19d51'6"
Kelly's Ketch
Vir 13h23'26" 11d25'4"
Kelly's life will forever shine.
Cyg 20h30'55" 48d25'27"
Kelly's Light
Leo 9h54'29" 8d53'36"
Kelly's Light
Umi 14h11'24" 69d19'9"
Kelly's loves Molly Peta and Lauren
Uma 13h47'24" 57d25'44"
Kelly's Piece of Heaven
Uma 10h8'36" 44d8'24"
Kelly's Reut
Uma 11h33'13" 44d17'52"
Kelly's Shining Light
Cap 20h31'41" -27d14'18"
Kelly's Sole Mate
Mon 6h53'16" -0d3'29"
Kelly's Song
Cnc 8h48'20" 25d14'16"
Kelly's Star
Gem 6h24'11" 27d48'19"
Kelly's Star
And 0h8'47" 28d48'6"
Kelly's Star
Psc 1h10'6" 11d56'37"
Kelly's Star
Vir 13h8'32" 5d23'48"
Kelly's Star
Uma 11h31'5" 44d12'55"
Kelly's Star
Sgr 18h36'21" -23d50'35"
Kelly's Wish
Sco 16h49'33" -27d9'20"
Kelly's Wish
Lyn 8h46'40" 36d32'55"
Kelly's Wishing Star
Gem 6h50'34" 31d29'51"
Kelly's Wishing Star
Lmi 10h46'52" 34d22'34"
Kellystar
Lyr 18h36'57" 30d6'11"
Kellystar
Lib 15h31'52" -29d35'6"
KellyStar12162005
Aqr 22h3'29" -18d1'31"
KellyTaylor
Vir 13h44'18" -7d21'59"
Kelly-Ute
Uma 9h46'55" 55d47'16"
Kel-Mark 1
Dra 16h56'20" 52d51'6"
KelMikAvi Light
Cyg 21h12'2" 35d37'19"
Kels
Sgr 19h45'48" -42d28'9"
Kels ~N~ D.B.'s "Wedding Star"
Vir 14h24'21" 2d38'12"
Kelsa Gratereaux
Cnc 8h13'54" 30d25'51"
Kelsa Kuchera
Vir 12h58'47" 7d28'41"
Kelsea
Uma 9h43'7" 51d0'50"
Kelsea Addison
Lyn 9h8'21" 42d56'26"
Kelsey
Lmi 10h15'48" 39d2'52"
Kelsey
Gem 7h6'12" 31d2'27"
Kelsey
And 23h0'31" 52d4'58"
Kelsey
Tau 4h35'9" 11d43'51"
Kelsey
Tau 4h24'14" 17d45'40"
Kelsey
Leo 11h50'16" 17d56'48"
Kelsey
Dor 4h24'34" -49d31'3"
KELSEY
Psc 0h28'14" 6d1'41"
Kelsey
Cas 23h28'54" 53d48'16"
Kelsey
Uma 10h32'51" 59d0'34"
Kelsey
Dra 16h46'0" 64d37'2"
Kelsey
Lib 14h51'15" -2d28'7"
Kelsey
Lib 15h27'59" -21d32'47"
Kelsey Akers
Cap 20h58'30" -17d20'36"
Kelsey Amber Heavlin
Ari 3h9'0" 29d25'12"
Kelsey And Dave
Cyg 21h35'18" 46d40'32"
Kelsey And Eric
Lib 15h26'16" -4d32'46"
Kelsey and Eric Always & Forever
Uma 10h21'16" 48d50'35"
Kelsey and Wesley
Vir 11h49'41" 9d42'2"
Kelsey Ann
Tau 4h17'12" 14d39'5"
Kelsey Ann Little Star
Aqr 22h46'44" -0d46'28"
Kelsey Ann Marin
Sgr 19h16'12" -30d25'2"
Kelsey Anne McHugh
Tau 4h24'30" 16d12'28"
Kelsey Anne O'Connor
Aqr 21h58'21" 0d53'44"
Kelsey Anne's First Birthday Star
And 1h56'10" 37d53'14"
Kelsey Blythe Peronto
Aql 19h18'46" -0d35'1"
Kelsey Boettcher
Uma 11h23'57" 53d22'57"
Kelsey Boland
And 2h20'48" 39d56'36"
Kelsey Brandes
Vir 13h25'0" -0d27'6"
Kelsey Brooks
And 0h53'8" 39d51'34"
Kelsey C.
Crb 16h12'13" 37d31'0"
Kelsey Candelmo
Tau 4h30'49" 29d1'0"
Kelsey Castanho
Gem 7h35'35" 25d38'49"
Kelsey Clare Welch
Uma 10h22'20" 45d0'45"
Kelsey Danielle Johnson
Tau 4h34'43" 19d55'36"
Kelsey Dawn Castaneda
Ari 2h36'25" 20d9'7"
Kelsey Dawn Six
And 0h14'0" 41d24'40"
Kelsey Dean
Ori 5h32'36" 1d2'47"
Kelsey Dee Wohlman
And 1h11'21" 41d52'40"
Kelsey Devich
Uma 11h56'57" 58d49'49"
Kelsey Devich
Uma 10h24'21" 72d10'25"
Kelsey Diane
Ari 2h14'28" 26d55'50"
Kelsey Dodd
Crb 16h11'5" 37d15'2"
Kelsey Eaves
Lib 15h43'2" -15d59'59"
Kelsey Elise Zorich
Sgr 19h52'50" -28d37'58"
Kelsey Elizabeth Blades
Ari 2h10'48" 26d38'54"
Kelsey Elizabeth Cumming
Sgr 18h53'54" -20d52'33"
Kelsey Elizabeth Luckett
Cnc 8h46'23" 32d4'46"
Kelsey Erin Granger
And 23h46'1" 45d47'22"
Kelsey Fowler
Lib 14h53'30" -6d43'40"
Kelsey Grace
Cnc 8h3'39" 24d23'46"
Kelsey Griffin Pogonyi
Umi 10h58'51" 89d40'47"
Kelsey Hanna Hendrix
Gem 7h3'23" 15d42'33"
Kelsey Horsington
Uma 9h28'28" 55d0'12"
Kelsey Irish
And 22h58'16" 50d49'26"
Kelsey J Dosen
Uma 12h39'8" 60d34'55"
Kelsey Josephine Syrett
And 23h30'24" 48d42'55"
Kelsey "Karter" Sullivan
And 1h27'21" 48d33'56"
Kelsey & Kevin Together Forever
Uma 9h9'6" 55d0'26"
Kelsey "k-g" Shaul
Lib 15h8'13" -5d33'12"
Kelsey Kjos
Cnc 8h41'11" 32d45'31"
Kelsey Klaver (Kstar)
And 1h35'36" 40d1'52"
Kelsey Krentz
Ori 5h36'29" 7d56'23"
Kelsey L. Cox
And 1h18'16" 45d22'32"
Kelsey Laine Ross
Lib 14h29'14" -19d13'16"
Kelsey Lancaster
Uma 11h46'27" 45d22'4"
Kelsey Lauren
And 2h36'28" 45d44'53"
Kelsey Lauren Edwards
Leo 10h14'12" 16d7'10"
Kelsey Leak
Peg 22h34'50" 11d48'22"
Kelsey Lee
Psc 1h10'30" 19d53'37"
Kelsey Lee Nordvik
And 0h43'56" 22d0'8"
Kelsey Lee Primm
Lmi 10h33'40" 36d46'24"
Kelsey Leigh Box
Lyn 7h43'5" 48d48'15"
Kelsey Leigh Lindberg
Ari 2h32'29" 25d41'44"
Kelsey Liann
Leo 11h39'20" 18d23'54"
Kelsey Liles
Uma 11h29'1" 65d42'15"
Kelsey Lynn
Uma 8h25'59" 64d10'7"
Kelsey Lynn
Uma 9h43'11" 57d0'56"
Kelsey Lynn Bigden
And 0h29'30" 44d32'31"
Kelsey Lynn Hunt
And 2h24'17" 49d31'3"
Kelsey Lynn Kramer
Uma 12h52'17" 59d19'21"
Kelsey Lynn Martin
Uma 10h0'15" 70d50'5"
Kelsey Lynn Sayler
Uma 10h35'2" 64d32'24"
Kelsey Lynn Srednicki
And 0h36'5" 36d8'2"
Kelsey Mae Walker
Gem 7h53'39" 14d12'50"
Kelsey Marie
Vir 15h9'41" 3d19'19"
Kelsey Marie
Ari 3h23'47" 28d54'27"
Kelsey Marie
Sco 16h10'17" -30d42'27"
Kelsey Marie Allen McQuade
Sco 16h6'31" -8d53'47"
Kelsey Marie Blindt
Cyg 21h37'53" 32d5'11"
Kelsey Marie Clawson
Uma 10h20'10" 55d32'32"
Kelsey Marie Gallagher
And 0h45'27" 45d38'5"
Kelsey Marie Langworthy
Tau 4h26'20" 22d46'7"
Kelsey Marie Mccalla
Ori 5h43'30" 0d3'53"
Kelsey Mathouser
Cyg 21h38'59" 47d28'5"
Kelsey Meidell
Cas 0h27'47" 57d13'16"
Kelsey Merrill
Uma 9h25'39" 56d9'20"
KELSEY MI AMOR
Boo 14h48'59" 50d41'46"
Kelsey Murphy "My Little Princess"
And 23h27'39" 48d9'39"
Kelsey Nicole
Ori 6h14'59" 15d12'49"
Kelsey Nicole Akers
Crb 15h39'28" 26d45'43"
Kelsey Nicole Hines
Cap 21h0'53" -27d26'38"
Kelsey Nicole Swyney 4-30-2005
And 0h33'53" 41d16'56"
Kelsey Nicole Thaysen "sissy"
And 0h30'44" 29d12'24"
Kelsey Noel Reitenauer
Leo 11h4'39" 6d28'9"
Kelsey Peterson
Cam 5h20'6" 66d50'4"
Kelsey Pettengill
And 2h12'13" 42d53'26"
Kelsey RaNae
Vir 12h2'41" 3d55'36"
Kelsey Rebecca Hoj Hattam
Tau 5h56'33" 23d5'46"
Kelsey Renee Hammond
And 22h59'41" 47d32'20"
Kelsey Richmond
Gem 7h23'38" 33d26'49"
Kelsey Roberts Grandfather's Angel
Mon 6h54'2" -0d25'48"
Kelsey Rose
And 0h54'16" 36d55'21"
Kelsey Rose Rincavage
And 23h25'48" 43d39'18"
Kelsey Rose Shenton
Ari 3h10'23" 26d53'58"
Kelsey Samuel
Cyg 21h18'33" 45d14'19"
Kelsey Schmutz Kleeman
Uma 10h30'17" 53d4'40"
Kelsey Schnabel
Cmi 7h24'35" 8d4'11"
Kelsey Seerden
Cnc 8h51'50" 14d3'22"
Kelsey Shannon Harrington
Lyr 18h37'43" 30d16'19"
Kelsey & Shetlinn, Love Forever
Uma 13h51'47" 48d38'44"
Kelsey Siefert
Mon 6h25'21" -4d1'48"
Kelsey Sixteen
Sgr 18h21'35" -23d6'55"
Kelsey Slone
Mon 6h53'56" -0d43'57"
Kelsey Taylor Roth
And 23h17'10" 48d4'43"
Kelsey the Loveable Princess
Aqr 22h11'9" -9d16'46"
Kelsey Tinoco
Sgr 19h30'54" -37d36'35"
Kelsey Whelan
Uma 10h51'58" 70d47'51"
Kelsey Wickett
Ari 3h20'16" 29d24'59"
Kelsey Williams
Cap 21h10'15" -25d50'34"
Kelsey's Guiding Light
And 0h8'48" 31d42'2"
Kelsey's Nova
Uma 10h25'37" 55d58'45"
Kelsey's Star
Cas 23h8'47" 54d28'44"
Kelsey's Star
Uma 10h32'25" 67d51'38"
Kelsey's star
Uma 8h56'6" 63d21'4"
Kelsey's Star
Cap 21h52'54" -9d59'24"
Kelsey's Star KDD
Lib 15h13'16" -21d43'13"
Kelsey's Star of Love
Tau 4h13'20" 29d49'59"
KELSHER
Lyn 6h43'52" 57d23'28"
Kelsi Amanda Robinson
And 0h15'22" 30d26'0"
Kelsi Brandt
Crb 15h55'35" 37d52'7"
Kelsi CupCake 21
And 2h15'48" 47d13'55"
Kelsi Falvey
And 23h50'26" 46d56'57"
Kelsi Hirai
Vir 14h9'17" -17d16'5"
Kelsi Johanna Brennan
And 22h58'21" 52d49'21"
Kelsi Kay Lile
Gem 7h52'51" 29d44'27"
Kelsi Lynn Weaver
Cnc 8h15'24" 17d36'7"
Kelsi Noel Nibbana
Cap 20h25'12" -19d29'50"
Kelsi Ryan Mitchell Happy Birthday
Tau 4h13'23" 27d53'6"
Kelsi White
And 0h4'28" 40d52'39"
Kelsie Ann
Aqr 22h37'31" 0d31'22"
Kelsie Atkinson
Psc 1h13'4" 20d27'8"
Kelsie Baker
Uma 11h58'58" 31d6'27"
Kelsie Blue
Umi 16h23'53" 78d40'57"
Kelsie Bo
Uma 9h34'46" 69d51'34"
Kelsie Burt
Cnc 8h16'51" 25d6'39"
Kelsie Dawn Burleson
Dra 18h58'3" 60d14'28"
Kelsie Emerick
Lib 14h54'9" -5d34'56"
Kelsie Greene
Lyr 18h46'53" 38d37'9"
Kelsie Jean Reder
Cnv 12h44'46" 40d59'27"
Kelsie Lauren Burlingame
Ari 2h20'55" 27d7'13"
Kelsie Roche - soon to be Briley
And 1h36'48" 48d30'29"
Kelsi's Love
Sco 17h20'27" -39d32'2"
Kelsi-You're the Best Star
Eri 4h26'38" -1d2'42"
Kelson Lynch
Cma 6h34'57" -15d46'20"
Kelson Mark Rastigue "Little Lion"
Lmi 10h22'35" 36d10'19"
Kelson Robert Meredith
Dra 10h31'46" 74d42'3"
Kelson Weber
Her 16h31'44" 47d4'36"
Kelstar
Lyn 8h0'42" 41d18'40"
Kelstar
Aqr 20h42'1" -12d47'59"
Kelstar
Cru 12h26'21" -60d54'2"
Kelstar 40
Cru 12h39'35" -58d49'17"
Kelsy Ann Gonsowski
And 0h26'10" 31d5'25"
Kelsy Nicole
Lyn 8h0'44" 38d32'52"
Kelsye Taylor
Cap 20h7'51" -24d29'26"
Keltec
Ori 6h12'26" 6d3'34"
Keltie
And 0h39'45" 43d6'8"
KelTom
Leo 11h14'45" 3d32'58"
Kelton Louie Braxton
Uma 9h28'2" 60d42'35"
Kelvin
Ori 5h30'20" -0d49'7"
Kelvin
Her 18h23'50" 18d32'52"
Kelvin and Ina Bramadat
Cas 23h32'47" 58d0'57"
Kelvin Dylan "K.D." Montgomery
Lib 15h7'43" -22d51'13"
Kelvin Easter
Aur 6h8'29" 36d53'3"
Kelvin J. Monzon
Her 18h12'1" 16d7'18"
Kelvin John Richards
Per 4h27'2" 35d37'40"
Kelvin Richardson
Umi 15h11'14" 82d48'14"
Kelvin Von Roenn
Per 3h30'26" 48d16'37"
Kelvin Zane King
Ori 5h30'14" 8d58'11"
Kelvingrove Art Gallery & Museum
Peg 23h56'15" 27d32'13"
Kelvins Way
Pyx 8h53'20" -26d42'33"
Kelvis
Cap 20h28'32" -24d28'14"
Kelvyn Parsons 1965
Dra 16h2'31" 56d32'24"
Kelyce2233
Sgr 19h14'29" -18d12'15"
K.E.M. 30
Lyn 7h53'8" 37d33'52"
Kemal Has CINGILLIOGLU
Uma 9h28'38" 63d43'27"
KEMATOSA
Uma 12h57'47" 59d4'17"
Kemba Chambers
Uma 9h59'23" 52d3'42"
Keményi Rita csillaga
Gem 6h29'16" 25d23'22"
KEMO
Cnc 9h11'50" 16d37'40"
KemPaula
Cyg 21h7'7" 51d34'35"
Kempenich, Sonja
Lib 14h38'55" -9d32'40"
Kemper, Rolf
Uma 10h18'36" 39d41'23"
Kempken, Dirk
Cnc 8h57'46" 16d12'19"
Kempton
Tau 5h47'9" 24d4'22"
Kempy
Ara 17h46'8" -46d59'33"
Ken
Psc 0h9'19" 7d4'51"
Ken
Leo 10h30'50" 13d28'26"
Ken
Uma 12h3'36" 39d26'16"
Ken
Cnc 8h45'3" 31d50'46"
ken
Her 16h58'24" 31d35'51"
Ken
Aur 5h12'32" 48d7'44"
Ken
Cyg 19h52'31" 42d34'19"
Ken , Mona Dunning
Col 5h33'16" -28d33'40"
Ken & Allice Dean "Dee" Carroll
Cap 20h29'22" -12d50'58"
Ken & Amy
Vir 12h38'54" 11d16'24"
Ken and Angie Forever
Sge 19h28'38" 18d16'0"
Ken and Anita
Uma 10h2'3" 41d43'10"
Ken and Carol Zimmerman
Cyg 20h42'3" 43d35'35"
Ken and Christine Rowland
Cyg 19h52'59" 38d40'38"
Ken and Claire
Cyg 21h52'3" 47d53'52"
Ken and Emily Mask
Psc 23h52'15" 7d4'52"
Ken and Emma, Love Always
Cyg 19h54'17" 52d37'20"
Ken and Joan 11KJLV22
Cyg 21h45'6" 40d35'28"
Ken and Kelly's Star
Cyg 19h40'56" 44d6'49"
Ken and Lisa
Per 3h12'34" 50d16'55"
Ken and Melody
Umi 15h55'26" 87d59'45"
Ken and Sigrid Bulloch
Cyg 20h20'45" 58d29'24"
Ken and Steph's MollieK
Cyg 21h22'5" 34d45'8"
Ken and Tammy
Cyg 19h48'17" 33d37'12"
Ken and Vivian Duva
Uma 8h38'32" 68d52'27"
Ken Andrew Grant
Cep 21h34'58" 59d28'28"
Ken & Audrey Collins
Cas 0h1'3" 54d18'21"
Ken Austin
Psc 1h45'14" 5d8'24"
Ken+Barb=Love4Ever
Cyg 20h13'17" 47d48'2"
Ken (best friend,soul mate, lover)
Gem 6h53'5" 16d53'58"
Ken Bloch
Cep 21h49'6" 62d46'43"
Ken Breidenstein
Per 4h16'33" 51d16'34"
Ken Cajet
Aql 20h10'9" 3d47'40"
Ken Carroll, the Honorary Rose
Aql 19h32'59" 12d13'4"
Ken Ciaglaski
Uma 10h36'33" 40d22'5"
Ken & Cindy
Cep 0h9'36" 69d35'15"
Ken Cook
Vir 12h58'24" -14d5'20"
Ken~Del~Ya
Lyr 19h21'36" 38d52'15"
Ken Desmarais
Lib 15h27'12" -23d6'2"
Ken & Dottie Ann
Lyn 8h55'47" 38d4'13"
Ken Duncan
Her 16h20'40" 44d35'53"
Ken Ferestad
Eri 4h39'2" -0d51'40"
Ken Fluchr
Aqr 22h42'16" 0d42'26"
Ken Geary
Ori 5h22'6" 1d33'24"
Ken Goldwyn's Star
Cma 6h42'2" -17d56'53"
Ken Hazelwood - Forever Shining
Nor 15h54'41" -53d58'39"
Ken Heissler (Pumpkin)
Ori 5h55'56" 18d26'49"
Ken Hosterman
Her 18h26'12" 12d12'35"
Ken Isaac
Uma 12h42'0" 52d35'7"
Ken & Jackie <3
Lyn 6h59'19" 48d34'7"
Ken & Jerri Morton
Cyg 19h51'8" 37d7'17"
Ken & Joan Focken
Cyg 21h24'58" 45d21'37"
Ken Jones
Aqr 22h9'24" -3d23'37"
Ken & June - Together Forever
Uma 10h53'26" 60d51'50"
Ken & Katie Conrick
Aql 19h23'26" 15d19'20"
Ken & Kelly Kennerly
Cyg 20h57'23" 46d27'26"
Ken Kelly, My Forever Love
Psc 1h45'41" 24d43'16"
Ken Kessler
Tau 5h28'28" 27d37'52"
Ken Kilgour
Her 16h33'59" 46d44'33"
Ken Knecht's Star
Cep 22h22'51" 62d57'56"
Ken Kochenour
Lyn 7h50'1" 56d50'18"
Ken Kugler
Dra 16h3'33" 67d43'6"
Ken LaBau
Lyr 18h40'25" 35d22'34"
Ken Laub
Dra 17h38'31" 51d43'59"
Ken & Laura Silver Anniversary
Cyg 20h41'4" 50d50'52"
Ken Leonard Carolina's Dad
Uma 8h32'30" 70d18'40"
Ken & Leontine "For Ever Always"
Apu 15h26'16" -70d28'53"
Ken & Lindsay
Cyg 19h49'2" 43d22'51"
Ken & Lisa Fukayama
Ori 5h2'38" 4d13'16"
Ken & Lori
Cyg 21h24'40" 44d3'33"
Ken loves Bonnie
Uma 9h31'15" 62d46'9"
Ken Loves Carrie
Umi 14h14'10" 76d6'5"
Ken loves Maryanne forever
Cru 12h29'29" -59d23'58"
Ken LSW
Psc 1h35'7" 14d28'11"
Ken Ly
Cep 0h9'0" 66d44'10"
Ken M. Jastrzebski
Uma 10h46'57" 62d8'11"
Ken "Mac" McDougall
Leo 9h42'13" 28d5'44"
Ken Maggi
Ori 5h16'4" -8d26'17"
Ken Marisseau
Sgr 18h43'36" -18d44'11"
Ken Matheis Sr.
Uma 9h25'23" 67d20'29"
Ken Mathews
Ori 6h4'3" 10d44'55"
Ken McCalvey
Sco 16h53'47" -42d22'1"
Ken Millard
Cep 22h6'34" 57d18'27"
Ken Miller
Peg 22h39'50" 18d10'51"
Ken Moon
Ori 5h24'52" 2d57'57"
Ken Morgan
Cep 22h51'34" 71d37'55"
Ken New
Per 2h46'46" 52d24'54"
Ken Ohara
Uma 10h43'24" 58d18'30"
Ken Pachulski
Per 4h37'40" 31d7'30"
Ken & Peggy - 05/13/98
Cyg 21h13'58" 42d53'34"
Ken Presland's Star - Grandad+Maddy+Eva
Cep 23h47'18" 81d19'41"
Ken & Priscilla
Sgr 18h58'7" -29d36'57"
Ken Ray
Vir 12h28'45" 11d58'44"
Ken Rod
Aql 19h32'12" 12d33'15"
Ken Rowland
Per 4h30'2" 39d47'32"
Ken Ruth Hugo
Cnc 9h16'9" 30d14'51"
Ken & Sandy Wright
Lyr 19h2'47" 42d52'17"
Ken & Sara Sonntag Star
Uma 9h23'21" 65d34'18"
Ken Schaaf
Uma 8h38'55" 56d11'37"
Ken Scharringhausen
Her 17h7'27" 46d23'19"
Ken Scott
Vir 12h41'46" 2d50'53"
Ken Semboshi
Aqr 22h6'15" -7d17'23"
Ken Sexton
Cru 12h33'45" -61d36'30"
Ken Speiser's Watching Over All
Uma 11h48'15" 48d3'42"
Ken Stanley Sikorski RAJ
Aql 20h2'59" 10d19'2"
Ken Startz
Cap 21h18'1" -17d21'9"
Ken Sturdevant
Her 16h17'30" 25d24'0"
Ken Sutin
Her 16h50'15" 48d30'14"
Ken Telgen
Uma 11h43'37" 31d19'17"
Ken & Teresa
Vir 12h11'57" 12d8'55"
Ken Toomey - World's Best Dad
Her 18h26'30" 26d56'13"
Ken & Tricia Robertson
And 1h7'11" 43d45'45"
Ken Tuttle
Cnc 8h29'15" 16d56'5"
Ken Waller
Aqr 22h25'2" -19d11'42"
Ken Weathers
Lyn 7h53'40" 51d4'21"
Ken Wheatley Star
Umi 14h1'41" 71d16'51"
Ken Wood
Cap 20h36'39" -20d29'39"
Ken Wood
Cma 6h39'40" -26d13'23"
Ken Yiu
Uma 11h13'11" 51d22'36"
Ken & Yvette
Sgr 17h53'36" -28d6'12"
Ken Zatlin
Her 17h12'4" 34d13'58"
Ken, Jennifer, Buddy, Phoebe
Vir 12h16'37" 3d20'15"
Kena Lynn Hamman
Leo 11h22'48" 16d50'10"
Kena Williams
Cap 21h39'21" -15d22'25"
Kenai Rose Weinhold
Crb 15h41'6" 27d53'44"
Kenan Pratt
Psc 0h46'21" 9d50'10"
Kenan Yildiz
Leo 10h59'39" -0d18'7"
Kenange Anniversary Star
Cra 18h38'11" -43d16'23"
Kenay Shane
Umi 16h8'53" 73d48'46"

KENBECKY 3-7-11
Psc 0h12'59" 6d31'20"
Kenbo's Rocket
Ori 6h10'54" -3d55'40"
KenChristDougNickChristina Bogarts
Uma 8h12'4" 71d56'57"
Kenda
And 23h49'57" 45d2'30"
Kendahl Brooke Youngs
Lyn 9h2'34" 39d55'54"
Kendal Elise
Gem 7h8'16" 14d35'43"
Kendal Mae Abramson
And 1h50'19" 45d25'23"
Kendal Sharwell Smith
Ori 5h58'54" 20d46'11"
Kendall
Ori 5h31'11" 4d56'46"
Kendall
Per 3h41'47" 45d30'54"
Kendall
Cyg 20h29'12" 54d23'20"
Kendall
Cap 20h36'46" -17d3'28"
Kendall
Sco 17h34'10" -33d39'1"
Kendall Amelia Bauer
Uma 9h40'49" 60d15'20"
Kendall Angela
Peg 21h35'48" 26d25'26"
Kendall Anne Williams
And 0h39'23" 27d41'52"
Kendall Ashley Haff
Mon 6h49'55" 8d32'36"
Kendall Aubertot
Uma 11h49'19" 62d1'14"
Kendall B. Combs
Uma 13h43'22" 52d13'32"
Kendall "Banana" Leech
Lyr 18h49'53" 31d20'57"
Kendall Brinn Zimmerman
Uma 11h43'51" 34d27'18"
Kendall Capito
Equ 21h16'2" 12d27'13"
Kendall Christine
Psc 1h29'30" 16d1'55"
Kendall Christine Quick
Her 17h11'54" 24d51'46"
Kendall Elizabeth
Cyg 20h13'32" 33d10'8"
Kendall Grace Kanakanui
Vir 13h52'13" 3d50'39"
Kendall Grace Kraft
Mon 7h16'57" -0d8'3"
Kendall Hayes
Psc 1h8'37" 29d50'36"
Kendall Irene Gaura
And 0h31'10" 25d3'42"
Kendall Jaclyn Rivera
Ari 2h38'37" 25d40'16"
Kendall Joseph Oliver
Cas 0h7'2" 55d5'21"
Kendall K.
Uma 10h1'8" 43d11'25"
Kendall Karasawa
Cyg 19h50'43" 38d55'12"
Kendall Lauren
Aqr 22h28'28" -0d46'37"
Kendall Lauren Shea
Uma 10h12'47" 46d24'58"
Kendall Linn Ross
Uma 11h8'12" 30d40'56"
Kendall Malone Wagner
Aql 19h44'56" 2d32'1"
Kendall Marie Berg
And 23h54'39" 34d34'52"
Kendall Marie Sarantinos
Sco 17h47'36" -36d27'47"
Kendall Morgan
Leo 10h13'15" 18d23'25"
Kendall Nicole Haworth
And 2h8'32" 39d17'28"
Kendall Nicole Wise
Leo 10h13'24" 24d1'19"
Kendall Ray
Sco 16h57'40" -44d11'56"
Kendall Reese Morgan
And 1h17'35" 38d37'57"
Kendall Reese Wendland
Sco 16h11'44" -16d34'48"
Kendall Rice
Uma 9h23'0" 43d24'32"
Kendall Rose Falkner
And 1h16'23" 41d12'2"
Kendall Rose: Love for All
And 0h14'48" 42d7'20"
Kendall Roy Jones
Lyn 8h26'36" 42d56'9"
Kendall Schmiedel
Umi 14h25'32" 72d3'36"
Kendall Shira Weber
Lyn 6h57'41" 53d12'2"
Kendall Suppes
Lyn 7h17'10" 58d14'29"
Kendall Vera Lockett
Sgr 18h21'42" -29d28'45"
Kendall Zellmer
Uma 11h9'10" 67d7'29"
Kendall's Kiss
And 23h16'50" 51d24'12"
Kendall's Star
And 1h43'31" 36d8'57"
KendallTillery
Cas 1h16'17" 62d42'25"
Kendal's "Super Star" Olympaid
And 0h17'40" 44d39'8"
Kendell Davey
Tau 5h47'10" 20d35'57"
Kendell Lee Wright
Ari 3h5'36" 14d7'53"
Kendell Ray Lambert
Tau 5h16'0" 19d10'53"
Kendelle Stanton Menzo
Uma 8h31'24" 64d22'30"
Kendell's Everlasting Light
Sgr 18h10'37" -22d2'56"
Kendi Fanchin
Leo 10h11'38" 11d0'14"
Ken-Diand Kaye
Sgr 18h11'56" -31d37'52"
Kendle and Robin Bowler
Her 17h48'21" 22d48'11"
Kendle Elizabeth Kidd
Uma 8h23'57" 71d10'44"
Kendol - Rainbow
Mon 6h56'37" -0d53'53"
Kendon Richards
Dra 16h1'28" 62d48'4"
Kendor
Uma 8h46'35" 58d39'48"
Kendra
Aqr 22h24'22" -7d25'29"
Kendra
Sco 16h47'24" -27d6'2"
Kendra
Sco 17h48'24" -42d44'8"
Kendra
Ari 3h22'39" 25d12'7"
Kendra
Psc 1h8'4" 26d53'9"
Kendra
Tau 5h40'13" 21d16'48"
Kendra
Psc 0h56'45" 31d8'45"
Kendra
And 1h6'44" 45d27'26"
Kendra Alicia Gutierrez
Cyg 19h44'36" 30d54'6"
Kendra Bencun
Umi 16h47'36" 78d40'56"
Kendra Butler
Cyg 21h7'54" 47d14'13"
Kendra Charmel Dodd
Cap 20h58'40" -19d24'21"
Kendra Cherise
And 1h23'0" 36d28'17"
Kendra Creighton
Ari 3h21'2" 27d48'1"
Kendra Danielle
And 1h52'28" 37d42'37"
Kendra Danielle Johnson
Dra 18h52'30" 52d1'12"
Kendra Dawn Bryant
Vir 12h39'37" 0d45'6"
Kendra Feemster
And 1h40'6" 45d35'45"
Kendra Fisher
Aqr 22h22'13" -8d37'17"
Kendra Humphries
Uma 9h6'19" 47d11'25"
Kendra J Clements
Vir 13h38'37" 3d24'52"
Kendra J. Curry
Mon 7h58'20" -6d37'59"
Kendra J. Hinkley
And 23h16'39" 44d20'50"
Kendra Joan "Buckwheat" Richards
Cyg 19h40'15" 37d12'18"
Kendra Kading
Lib 14h28'37" -11d56'8"
Kendra Kay Sowers
Uma 8h26'7" 67d39'24"
Kendra Larkin
Leo 10h54'21" 5d25'33"
Kendra Lea Overturf
Vir 13h25'54" 5d1'0"
Kendra Lee
Lyn 7h27'23" 49d8'11"
Kendra Leigh Weaver
Cnc 8h51'40" 18d53'22"
Kendra Linn
Ori 6h17'47" 9d3'35"
Kendra Lynn
Lyn 8h4'4" 34d44'16"
Kendra Lynn Hinkley
Ari 3h25'33" 29d31'55"
Kendra Marie Campbell
Vir 13h5'27" 0d14'24"
Kendra Marie Olmscheid
Mon 7h45'4" -3d40'56"
Kendra & Matthew Streit
Cas 0h39'0" 57d26'21"
Kendra McGarvey / Chubbs
Lyn 7h26'28" 50d9'25"
Kendra McMahon - The Lady
And 0h3'37" 34d13'43"
Kendra Michele Allen
Uma 11h1'22" 39d6'50"
Kendra Michelle Guldan
Leo 10h28'20" 17d36'18"
Kendra "My Shining Star"
Peg 21h28'22" 15d53'36"
Kendra N Osness
Cnc 9h10'35" 15d25'35"
Kendra Nichole Reese
Leo 9h35'32" 29d17'0"
Kendra Pearson
Lib 14h50'40" -2d18'54"
Kendra Renee
Leo 11h1'35" 16d53'50"
Kendra Ryanne Winegar - I Love You!
Ori 5h56'8" 7d6'41"
Kendra Vivian Bocinsky
Ari 2h20'57" 13d20'14"
Kendra Yancey #12
And 0h52'56" 38d29'3"
Kendra'd Spirit
And 0h38'0" 33d5'43"
Kendra's Star
Sco 16h11'36" -31d40'34"
Kendra's Starr
Ori 5h41'5" 6d27'2"
Kendrick Alfonzo Gray
Uma 8h28'45" 63d3'35"
Kendrie Dawn Escoe
Aql 20h10'15" 11d28'1"
Kendy
Uma 8h49'16" 48d16'45"
Kendy Goodwin
Ori 5h29'15" 14d19'38"
Kendyl Lynn Nester
Psc 0h14'33" 8d8'51"
Kendyl Taylor Mygatt 22 Feb 1990
Psc 23h51'36" -0d18'41"
Kendyll MacKenzie
And 22h58'48" 37d27'36"
Kendyll Ryan Bergh
Her 16h58'40" 33d13'44"
kenearn 178
Uma 9h46'52" 44d26'56"
Keness
Cru 12h53'56" -56d10'37"
Keness Ebony Prince
Cma 6h18'34" -30d32'28"
Keneth Bryant Morrison
Psc 1h1'9" 8d9'2"
KENGO 0131
Aqr 23h36'54" -13d25'51"
kenhua
Uma 11h45'5" 45d3'30"
Keni Lynn
Cnc 9h11'18" 15d56'36"
Kenia Huete -02/05/1983-
Cas 0h30'58" 60d11'23"
Kenia Maria Peña
Dra 19h44'23" 61d44'53"
Kenikila Nani, u', maika', Hula
Cap 20h27'39" -11d40'8"
Kenin Michelle Zaharias
Sco 16h48'3" -28d20'57"
Kenita
Leo 10h10'47" 23d0'37"
Kenita Bonita
Cyg 21h42'54" 43d55'36"
Kenj
Umi 16h45'13" 79d32'52"
KenJac1
Uma 8h52'15" 56d18'39"
Kenji Smile
Leo 9h51'29" 12d24'28"
Kenji Tseem Muaj Vang
And 0h35'8" 46d20'14"
KenKendra "Forever In Love Star"
Cyg 20h33'53" 35d13'48"
Kenley Paige
Uma 10h48'40" 52d13'18"
Kenn Beatson
Uma 9h11'21" 56d38'51"
Kenn Manzerolle
Cep 22h58'0" 72d25'37"
Kenn Youngar
Cap 21h17'37" -19d16'6"
Kenna
And 23h17'40" 48d4'45"
Kenna and Lauren Tornow
Cyg 19h44'50" 55d35'42"
Kenna and Lola
Cas 23h30'8" 54d55'15"
Kenna Byrne
Her 18h47'28" 20d17'52"
Kenna Ingrid Duncan
Cas 23h39'11" 59d42'1"
Kenna Jean
Lib 14h48'52" -17d33'6"
Kenna Jean
Lyr 18h49'19" 33d59'41"
Kenna Jo
Lib 14h30'33" -14d17'58"
Kenna Johanna
Lib 14h48'47" -4d22'53"
Kenna Lynn Monahan
Cap 20h59'41" -19d33'46"
Kenna Paige Lane
Psc 23h49'21" 6d24'3"
Kenna Pressley Kron
Umi 15h36'7" 80d33'24"
Kenna Speights
Ori 5h35'55" 1d11'35"
Kenna Thomas,CMP-GaMPI Shining Star
Cyg 21h52'35" 39d9'31"
Kennady
Tau 3h39'42" 30d27'2"
Kennady Ann
And 23h59'47" 36d29'26"
Kennard Bishop Torrence Sr.
Aql 19h25'46" 10d25'27"
Kennatalia
Uma 10h49'8" 61d48'45"
Kennaty
Per 3h37'12" 34d21'8"
Kennedi Brooklyn
Boo 15h34'23" 45d17'51"
Kennedi Danielle Biondi
And 0h14'27" 30d46'38"
Kennedi Qiana Miles
Cap 21h47'1" -23d29'45"
KENNEDI TENNILLE
Sco 17h38'37" -41d50'0"
Kennedy
Uma 8h53'40" 71d43'5"
Kennedy A. Jackson Gallegos
Dra 18h33'19" 75d23'45"
Kennedy Allen
Gem 7h44'10" 33d36'43"
Kennedy Ann Elizabeth Huber
Uma 8h12'51" 66d21'54"
Kennedy Campbell
Peg 21h34'35" 22d0'12"
Kennedy Catherine
Cnc 8h54'59" 32d21'33"
Kennedy Elizabeth Snyder
Cas 0h19'46" 61d18'53"
Kennedy Gammage
Uma 9h40'50" 54d59'43"
Kennedy Grace Snyder
Lyn 7h1'15" 52d29'35"
Kennedy Grace Ziglar
Cnc 9h6'35" 25d58'32"
Kennedy Hilton Phillips
Uma 10h59'58" 62d13'28"
Kennedy Jane Miller - 1st Communion
Cmi 7h7'18" 9d32'23"
Kennedy Lane Pittsley
Psc 1h43'36" 9d50'35"
Kennedy Laurel
Uma 12h7'7" 50d36'22"
Kennedy Leigh White
Tau 5h8'0" 21d22'2"
Kennedy Marie
Sgr 18h25'32" -20d57'33"
Kennedy Marie Gerber
Crb 16h13'6" 34d39'4"
Kennedy Morris
And 22h57'57" 51d56'38"
Kennedy Nicole Hairston
Mon 8h10'19" -0d37'48"
Kennedy Rae
Psc 23h30'7" -1d11'11"
Kennedy Rose
Sgr 18h32'19" -26d51'58"
Kennedy Selina Howard
And 0h43'10" 31d5'37"
Kennedy Star
Ari 3h21'33" 27d29'43"
Kennedy Teagan Daddy TCT Star
Per 3h2'27" 41d36'9"
Kennedy's Shining Star
Cnc 8h30'15" 26d10'58"
Kenneith T. Collins
Aur 5h39'9" 39d32'29"
Kenneth
Boo 15h1'25" 41d32'57"
Kenneth
Her 18h39'42" 15d48'14"
Kenneth
Ori 5h23'58" 1d54'34"
Kenneth A. Huff
Uma 13h55'21" 50d50'19"
Kenneth A P Gear
Ori 5h56'45" 7d47'23"
Kenneth A Richards
Cep 22h59'47" 61d1'25"
Kenneth & Adyson
Vir 12h26'58" 12d15'32"
Kenneth Alan Martin
Cas 0h24'38" 58d1'4"
Kenneth Alan Roberts
Per 3h4'31" 38d2'17"
Kenneth Alan Roberts
Uma 13h40'31" 55d11'21"
Kenneth Allen Wenzel II
Vir 13h21'5" 3d19'50"
Kenneth Alvin Bishop
Aql 19h4'53" -0d0'50"
Kenneth always love Crystal
Sgr 18h32'49" -26d14'31"
Kenneth Amadeo Giacopazzi
Ari 2h5'22" 24d0'34"
Kenneth Amaro Heart Star
Tau 5h8'25" 18d28'20"
Kenneth and Felicia Kuffell
Cyg 21h27'59" 45d29'13"
Kenneth and Holly's Star
Uma 8h44'39" 55d3'32"
Kenneth and Kathy Wagenbrenner
Peg 22h34'29" 33d38'54"
Kenneth and Lela
Uma 9h1'52" 49d13'44"
Kenneth and Margarita's Star
Cap 21h11'17" -17d45'49"
Kenneth and Melicia Brewer
Lib 15h46'48" -21d6'22"
Kenneth and Tiffany Nolla
Peg 22h51'48" 26d54'45"
Kenneth Anthony Caldera~ 03/10/1975
Uma 12h12'1" 58d45'51"
Kenneth Arnold Brent
Uma 13h39'37" 52d7'9"
Kenneth Ashley Taylor
And 0h27'3" 32d41'43"
Kenneth Avery Woods
Sgr 18h49'3" -17d38'17"
Kenneth B. Franklin
Uma 10h37'25" 68d6'16"
Kenneth Baker
Boo 15h27'37" 40d48'50"
Kenneth Bauer Jr.
Ori 6h23'57" 16d27'4"
Kenneth Bert Langille
Cas 23h44'54" 51d46'59"
Kenneth Bertschinger
Lib 15h34'42" -6d46'25"
Kenneth Bodnar
Her 17h25'54" 35d11'43"
Kenneth Britt
Gem 7h52'42" 25d17'21"
Kenneth Brooks
Sco 16h12'57" -10d18'39"
Kenneth Burnett
Vir 12h2'17" 3d25'11"
Kenneth C Bungarz, Jr.
Cap 20h34'15" -20d13'3"
Kenneth C. Crow 1-24-1961
Aqr 22h38'14" -24d25'9"
Kenneth C. Murray III
Aur 5h42'55" 48d14'51"
Kenneth C. Rebello
Crb 15h43'13" 26d42'39"
Kenneth C Small
Aqr 23h6'53" -16d34'43"
Kenneth C. Whalen
Cyg 19h52'56" 31d18'11"
Kenneth Cameron
Sgr 18h10'56" -22d36'33"
Kenneth Carter Superstar
Psc 23h50'5" -0d18'22"
Kenneth Casey
Ari 2h51'2" 12d32'26"
Kenneth Cavin 1 of 2 Angels at 50
Cru 12h23'50" -57d33'58"
Kenneth Charles Barden
And 1h28'40" 34d11'9"
Kenneth Charles Phillips
Uma 9h15'48" 50d0'2"
Kenneth Clarence Miller
Vir 12h46'19" 11d37'22"
Kenneth Clayton Groves
Cnc 8h54'54" 19d42'59"
Kenneth Clock
Cep 22h3'8" 67d49'52"
Kenneth Coryell
Lyr 18h43'52" 37d58'44"
Kenneth Craig McKinley
Her 17h52'50" 25d7'36"
Kenneth D & Betty A Baxter
Cyg 20h4'22" 40d44'17"
Kenneth D. Campbell
Leo 11h56'36" 19d35'4"
Kenneth D. Marx
Uma 8h53'54" 67d15'13"
Kenneth D. Paganetti
Lib 15h41'3" -6d32'26"
Kenneth D. Stern
Per 3h6'17" 52d29'36"
Kenneth Dale Anderson-noony
Gem 7h29'49" 25d45'27"
Kenneth & Darlene Dickens 2005
Uma 11h8'32" 30d49'45"
Kenneth David Castro
Uma 10h8'6" 46d58'15"
Kenneth David Stuber - My Star!
Uma 10h42'12" 55d34'58"
Kenneth Davis Rodgers
Her 18h16'29" 28d45'11"
Kenneth Dean Smith
Leo 9h48'7" 27d48'6"
Kenneth DeHart
Sgr 18h14'45" -18d13'10"
Kenneth Delano Parrish, Jr.
Her 18h53'23" 23d40'14"
Kenneth Dills
Psc 0h25'12" 17d26'26"
Kenneth Dollhopf
Uma 9h49'33" 53d43'13"
Kenneth Douglas Mills
Uma 11h28'43" 59d12'6"
Kenneth Doyle
Cep 21h48'7" 61d19'25"
Kenneth Dresser
Uma 14h12'39" 58d18'20"
Kenneth E. and Laura Yvonne Holmes
Cnc 8h26'13" 12d13'29"
Kenneth E. Brunot
And 23h35'48" 45d20'25"
Kenneth E. Duncan, III 7/17/83
Uma 10h30'24" 65d40'32"
Kenneth E. Honig
Lib 14h34'7" -20d2'52"
Kenneth E. Laswell
Dra 19h11'43" 65d31'41"
Kenneth E Sands
Her 17h6'29" 32d11'31"
Kenneth E. Vargas
Her 18h25'1" 13d16'17"
Kenneth Earl Daugherty, Sr.
Cep 20h25'52" 61d1'19"
Kenneth Earl Hall
Sco 17h49'55" -35d44'35"
Kenneth Earl Webb
Del 20h34'42" 12d51'4"
Kenneth Edward Elberson
Uma 10h56'39" 71d26'3"
Kenneth Edward Keough
Ori 5h22'23" 14d8'49"
Kenneth Edwin Severine
Psc 0h4'13" 2d37'33"
Kenneth & Elisa Sands
Uma 12h7'10" 51d45'26"
Kenneth Eugene Hunt Sr.
Per 3h34'3" 48d8'20"
Kenneth Eugene Weber
Lib 14h39'26" -15d56'53"
Kenneth Everett Couey
Her 17h10'25" 25d43'1"
Kenneth Fernstrom
Her 18h32'18" 18d2'25"
Kenneth Fink
Uma 9h26'5" 42d26'31"
Kenneth France
Her 17h30'25" 40d27'34"
Kenneth Francis Salvi, 11-15-1970
Uma 10h59'37" 35d17'19"
Kenneth Franklin Armijo
Ori 5h30'11" -2d43'14"
Kenneth Friedberg
Per 3h27'52" 43d37'35"
Kenneth G Athmann
Sco 16h11'20" -10d16'1"
Kenneth G. Caldwell
Sgr 18h56'1" -32d57'29"
Kenneth G. "Ken" Thompson
Ari 2h14'49" 26d37'59"
Kenneth George Peter Eddington
Gem 7h20'12" 23d42'32"
Kenneth Gervase Williams - Ken's Star
Uma 10h2'49" 41d44'11"
Kenneth Gherardini
Cep 22h17'12" 61d27'57"
Kenneth Gnat
Psc 1h4'48" 30d48'25"
Kenneth Graham Patrick Warwick
Umi 16h29'41" 81d13'9"
Kenneth H. Godstrey 9/21/1933
Per 3h17'47" 42d41'3"
Kenneth H. Kane
Boo 14h6'59" 43d10'46"
Kenneth H. Seymour
Gem 6h45'26" 18d20'50"
Kenneth Hagen Jr.
Her 17h40'47" 36d20'30"
Kenneth Harb
Lib 15h7'21" -1d12'47"
Kenneth Harlan Constant
Peg 0h4'36" 17d40'1"
Kenneth Herbert Boze Kelsoe
Boo 15h4'42" 26d50'21"
Kenneth Herbert Matthews
Cnc 8h22'59" 21d25'9"
Kenneth & Hope Broomer
Gem 8h0'3" 32d19'35"
Kenneth Isaac Parker III
And 0h46'28" 37d51'51"
Kenneth J. Cahill
Her 15h51'27" 47d41'4"
Kenneth J. Freiberg
Cep 20h41'52" 55d26'12"
Kenneth J Gibbs
Uma 12h14'9" 58d59'21"
Kenneth J Kemp
Boo 14h34'15" 30d51'15"
Kenneth J. Kleinert, Jr.
Uma 11h31'21" 57d45'29"
Kenneth J. Kleinert, Sr.
Uma 11h50'13" 58d28'56"
Kenneth J. Leppold
Ari 2h20'59" 16d23'5"
Kenneth J. Lustgarten
Her 17h17'45" 19d8'37"
Kenneth J. Schwinn
Uma 12h9'51" 60d0'34"
Kenneth J. "Sharky" Donoghue
Leo 10h8'39" 12d41'48"
Kenneth J. Shinn
Cyg 21h25'39" 35d44'11"
Kenneth J Terry
Sgr 19h17'47" -18d15'55"
Kenneth James D'Angelo
Leo 10h29'27" 19d31'40"
Kenneth James (Kenny) Greer
Ori 5h33'25" 10d52'54"
Kenneth James Outland
Psc 0h47'42" 16d35'13"
Kenneth James Waruszewski
Uma 9h35'33" 71d28'23"
Kenneth Jameson
Ori 5h58'52" 20d57'26"
Kenneth John Kampa, No.1 Dad
And 2h37'28" 46d24'35"
Kenneth John Kelly
Gem 7h32'41" 15d42'58"
Kenneth John Primising
Her 17h53'47" 49d48'42"
Kenneth John Rock (Rockman) 11-27-53
Her 17h19'16" 27d51'20"
Kenneth Joseph McGowan
Sco 16h3'24" -9d37'30"
Kenneth Juan Harper
Sco 15h52'10" -21d31'58"
Kenneth K. Klein
Her 17h2'28" 14d11'50"
Kenneth Karl Albert
Per 3h36'31" 46d45'28"
Kenneth "Kenny" David Kirkland
Cyg 21h44'59" 33d21'7"
Kenneth "Kenny" King
Uma 11h48'27" 63d39'11"
Kenneth King
Leo 11h24'59" -0d5'50"
Kenneth Koopersmith
Cep 21h35'5" 67d22'3"
Kenneth Kramer
Aql 19h17'53" 14d19'59"
Kenneth Kress Rogers
Leo 10h22'25" 16d18'46"
Kenneth L. Anderson
Boo 14h44'19" 50d41'45"
Kenneth L. Forsythe, Sr.
Aur 6h33'25" 35d10'12"
Kenneth L Hyde
Lyn 8h39'55" 45d54'31"
Kenneth L. Slezak
Tau 5h49'38" 16d59'8"
Kenneth L. VanOordt
Leo 11h11'1" 9d30'41"
Kenneth Lasky
Per 3h15'55" 53d10'37"
Kenneth Lavern Saul
Her 16h21'35" 44d32'28"
Kenneth Lawrence Behrens
Aql 19h5'22" -0d13'40"
Kenneth Lawrence Pardoe
Cyg 19h54'53" 37d4'2"
Kenneth Lee Fox
Cyg 20h31'50" 49d20'30"
Kenneth Lee Newport
Dra 18h4'56" 71d56'9"
Kenneth Lee Richards
Gem 7h8'15" 21d28'55"
Kenneth Leo Bostwick
Cap 21h34'29" -22d55'54"
Kenneth M. Brown Jr.
Her 17h14'38" 19d51'41"
Kenneth M. Koller
Aur 5h50'4" 51d15'57"
Kenneth Manley Waggoner II
Tau 4h22'41" 26d24'36"
Kenneth Marcell
Tau 5h18'54" 24d15'36"
Kenneth Mark Shindle
Aqr 21h24'25" 1d52'43"
Kenneth Marquis
Her 17h24'29" 14d52'57"
Kenneth Mattus
Per 3h9'43" 42d54'2"
Kenneth & Maudine Lantz Family Star
Cyg 20h27'0" 46d40'17"
Kenneth Mecham
Per 2h37'11" 55d17'55"
Kenneth Michael D'Elia
Lyn 7h24'1" 48d33'6"
Kenneth Michael Routt
Lyn 6h45'9" 57d19'50"
Kenneth Moon
Mon 7h1'4" -8d34'0"
Kenneth Moreira
Lyr 18h15'13" 39d42'35"
Kenneth Morris Smith
Aql 19h27'36" 13d34'18"
Kenneth Munson
Gem 7h8'48" 23d16'48"
Kenneth Nicholas Franke
Tau 4h0'13" 7d47'20"
Kenneth Norval Luman
Equ 21h14'32" 10d18'5"
Kenneth Olden,PH.D.,SC.D.,L.H.D.
Ori 6h19'45" 14d56'42"
Kenneth O'Neill
Crb 16h2'29" 33d27'20"
Kenneth Ostrega
Cnv 13h56'2" 35d3'0"
Kenneth P Kehle
Uma 8h59'25" 69d52'36"
Kenneth P. & Madrene G. DeHart
Umi 14h28'47" 72d4'26"

Kenneth Patrick Hickey
Lac 22h10'26" 52d51'31"
Kenneth Patterson
Uma 13h51'11" 48d12'16"
Kenneth Paul Weel
Cap 21h34'29" -14d5'24"
Kenneth & Peggy Howell
Uma 10h52'28" 43d19'2"
Kenneth Peter Mendoza
Lyn 6h35'40" 55d16'45"
Kenneth Phillip Hirschfeld
III
Psc 1h2'37" 10d21'54"
Kenneth Phillips
Vir 11h58'12" 3d39'59"
Kenneth Powell
Uma 8h59'23" 70d17'59"
Kenneth Preston Morris
Cap 21h36'14" -15d0'15"
Kenneth R. Bell
Boo 14h35'57" 27d26'29"
Kenneth R. Halleran
Aqr 23h29'34" -13d57'44"
Kenneth R. Hammond
Her 17h15'57" 15d52'46"
Kenneth R. Keller
Leo 9h41'41" 31d17'7"
Kenneth R. Klein
Cap 21h40'7" -10d11'40"
Kenneth R. Morris
Aur 6h9'19" 50d19'58"
Kenneth R. Sellai
Uma 11h15'12" 63d15'18"
Kenneth R. Trietch
Ari 2h53'36" 10d58'52"
Kenneth R. Wash
Vir 12h47'2" 3d55'14"
Kenneth Rasco
Aqr 21h59'23" 0d53'28"
Kenneth Ray Dresh
Cnc 8h7'43" 8d28'59"
Kenneth Ray Ludwig
Ori 6h13'7" 6d43'5"
Kenneth Ray Martin
And 0h8'59" 45d45'52"
Kenneth Ray Thomas
Aql 19h23'10" -0d1'13"
Kenneth Ray Vieira
Cyg 19h39'23" 52d39'29"
Kenneth Raymond
Lmi 10h32'34" 32d57'6"
Kenneth & Rebecca
Forever!
Cru 12h24'51" -63d34'52"
Kenneth Redcliff
Cep 21h49'33" 63d44'5"
Kenneth Refugio
Sco 17h57'12" -30d30'29"
Kenneth Rich 12/26/82
Cap 21h25'38" -19d1'39"
Kenneth Richard Pugsley
Aqr 22h8'48" 1d49'0"
Kenneth Rieter
Uma 8h29'57" 63d42'21"
Kenneth Robert Rausa
Aqr 21h18'4" -7d46'16"
Kenneth Robert Thomas
Bowen
Tau 3h59'45" 23d54'22"
Kenneth Roland Shanks
Per 4h38'54" 50d53'39"
Kenneth Ronald Jacobsen
Cas 0h24'32" 52d3'31"
Kenneth Russ
Uma 11h5'57" 40d39'0"
Kenneth S. Breckenridge
Sr.
Vir 13h19'28" 11d49'18"
Kenneth S. MacDonald
"KensStar"
Ori 5h9'36" 12d25'58"
Kenneth S. Marvin, M.D.
Aqr 22h43'33" -0d55'46"
Kenneth Scott Cosgrove
Uma 11h18'19" 53d37'32"
Kenneth Sean Cleveland
Lib 15h27'8" -10d0'47"
Kenneth Seeberg Bronkie
Aql 19h56'55" -0d30'21"
Kenneth & Shirley Harris
Aql 19h27'10" 0d25'59"
Kenneth Smith Huffer
Vir 12h50'55" 12d1'20"
Kenneth Speicher
Sgr 18h58'0" -20d2'56"
Kenneth Spivey
Leo 11h29'42" 21d59'23"
Kenneth Stanley Sikorski
Aql 20h11'4" 6d57'42"
Kenneth Stephen Onsa
Cnc 8h41'6" 31d53'43"
Kenneth Stevenson
Aqr 22h15'26" -2d13'32"
Kenneth T. Salzer
Lyn 8h8'36" 56d37'54"
Kenneth Taylor Sax
Her 18h34'54" 20d15'12"
Kenneth Teise
Lup 15h37'9" -46d57'13"
Kenneth Thomas Slattery
Ori 5h39'36" -0d12'22"
Kenneth Tyler
Leo 11h44'38" 17d47'25"
Kenneth Vincent Tacelli
Lib 15h22'33" -9d59'6"
Kenneth W Carpenter Jr
Psc 23h9'55" 1d41'58"
Kenneth W. Fardie
Aql 19h50'52" -0d34'26"
Kenneth W. Gunderson
Lyr 18h45'14" 33d48'42"
Kenneth W. Harlan
Cap 20h26'12" -10d36'11"
Kenneth W. Hilse
Cap 20h35'46" -22d45'34"
Kenneth W. Nikels
Gem 6h54'29" 27d42'6"
Kenneth W. Ross
Her 17h6'52" 30d54'50"
Kenneth W. Young, Sr.
Lac 22h26'47" 46d35'35"
Kenneth Wade Klier
Her 17h10'31" 32d47'14"
Kenneth Watson
Uma 9h34'20" 60d7'46"
Kenneth Wayne Cook
Leo 11h21'54" 22d17'42"
Kenneth Wayne Martino
Her 17h51'40" 49d12'36"
Kenneth Wayne Moretz Jr.
08/08/1997
Leo 9h48'15" 25d1'43"
Kenneth Wayne Powell,Sr.
Vir 13h32'37" -0d54'1"
Kenneth Wayne Rumbaugh
Cet 1h0'24" 2d1'5"
Kenneth Wayne Stewart
Psc 1h13'31" 25d52'26"
Kenneth Whatman
Tau 5h41'11" 26d24'36"
Kenneth William Gall
Ari 3h8'48" 29d15'53"
Kenneth Wright Cameron
Uma 10h22'37" 65d55'0"
Kenneth Young - Shining
Star
Her 17h40'14" 31d6'46"
Kenneth Zipovsky
Leo 11h7'43" 20d40'0"
Kennetha "Kittie" Marie
DeGrange
Cyg 19h52'44" 51d40'37"
Kenneth's Star
Dra 17h23'7" 62d25'57"
Kenney E. Howe
Ara 17h16'20" -46d55'48"
Kenney's Star
Aqr 21h31'23" 2d23'14"
Kennisa Richele
Cas 23h22'18" 57d46'5"
Kennon David Champlin
Vir 12h19'41" 8d28'15"
Kennon's Star
Dra 15h3'24" 60d35'42"
Kennth Lee Eppelheimer
Her 18h45'59" 20d14'42"
Kenny
Her 18h21'0" 21d6'53"
Kenny
Gem 6h55'24" 17d12'20"
Kenny
Gem 6h40'3" 29d12'37"
Kenny
Ori 6h20'58" 7d37'57"
Kenny
And 23h32'40" 40d30'58"
Kenny
Per 4h26'47" 51d18'48"
Kenny
Per 4h36'48" 37d37'42"
Kenny
Dra 17h28'27" 55d49'38"
Kenny
Sgr 19h6'30" -30d35'23"
Kenny & Ali
Uma 13h52'5" 54d53'24"
Kenny Alligood
Cmi 7h35'20" 5d15'2"
Kenny and Amanda's
Forever
Cyg 20h19'0" 39d48'35"
Kenny and Andrea's
Starlight Dreams
Tau 5h48'2" 26d2'15"
Kenny Anderson
Sco 16h55'36" -39d1'52"
Kenny Bear
Leo 9h23'20" 27d2'8"
Kenny Bernstiel, Jr.
Ori 6h5'6" 21d15'35"
Kenny Bertin
Per 3h9'10" 50d14'54"
Kenny "Big Daddy"
McDougall
Uma 11h47'23" 58d51'2"
Kenny Bourg
Ori 5h25'56" 2d26'10"
Kenny Brown
Uma 13h2'38" 58d1'14"
Kenny Broxton
Uma 11h49'29" 60d31'23"
Kenny C. & Mary G.
Her 18h41'19" 20d21'46"
Kenny Campbell
Lac 22h14'43" 48d13'36"
Kenny Campbell
Dra 19h28'14" 64d7'21"
Kenny & Carla
Sge 19h46'45" 18d22'4"
Kenny Carrillo
Her 16h34'27" 13d34'50"
Kenny Cook
Uma 11h32'12" 32d39'5"
Kenny Crabtree
Uma 10h56'32" 53d40'29"
Kenny DeMaio Jr.
Lyn 7h37'11" 36d12'41"
Kenny Egor
Per 4h46'31" 50d14'45"
Kenny Fahy
Aqr 22h34'7" -0d59'51"
Kenny Ferrie
Lib 15h4'52" -27d17'29"
Kenny Fischer
Tau 5h8'3" 18d37'57"
Kenny Glim
Her 17h33'5" 37d1'26"
Kenny Gordon, Jr.
Gem 6h20'54" 26d56'1"
Kenny Houng
Dra 16h55'22" 60d23'23"
Kenny Hunter
Ori 6h2'51" 11d9'12"
Kenny Jamrog
Cap 20h47'12" -17d40'5"
Kenny Joe Linton
Cnc 8h43'5" 7d0'22"
Kenny Kelly
Aql 19h37'21" 8d38'20"
Kenny Kerr
Cap 20h14'55" -17d31'3"
Kenny Lejnieks
Ori 6h16'34" 14d14'25"
Kenny Lex
Ari 3h16'0" 29d22'2"
Kenny Loves Julianne
Cyg 20h51'22" 34d49'23"
Kenny Loves Kerri Mostest
Dra 19h6'9" 68d16'17"
Kenny Lowrie
Cep 2h28'44" 82d35'57"
Kenny M. Schauer
Uma 11h14'45" 70d19'18"
Kenny Mac
Psc 0h10'36" 4d35'30"
Kenny Martel
Her 16h48'57" 29d27'27"
Kenny "Max" Conroy
Cyg 19h57'25" 57d52'19"
Kenny & Melissa
Sge 19h26'32" 18d21'50"
Kenny M.Loves Sheryl
McCormick
Cyg 20h34'54" 38d17'3"
Kenny Mon Coeur
Her 16h14'0" 44d8'23"
Kenny & Monica Benoit
Gem 7h17'46" 15d49'27"
Kenny Moore
Psc 1h34'57" 12d12'0"
Kenny Nash
Uma 11h10'58" 34d19'53"
Kenny of the Cay
Her 17h23'48" 20d16'45"
Kenny Palmer
Her 17h40'30" 35d46'17"
Kenny Penny
Sco 17h37'11" -41d35'39"
Kenny ~ Phil & Gary's All
Star Dad
Ori 6h14'6" 18d23'37"
Kenny Ponitz
Ari 2h48'21" 15d25'34"
Kenny & Sallie Knox
Uma 11h19'47" 55d15'45"
Kenny & Sherrie
Woodard's 12th Year
Cas 0h35'42" 59d16'28"
Kenny Smith
Per 2h14'15" 54d23'2"
Kenny Sparks
Dra 14h48'23" 57d51'2"
Kenny Stellate
Lib 15h2'22" -14d27'55"
Kenny Sue
Lib 15h53'33" -18d12'33"
Kenny & Susan - Forever
Col 6h24'59" -41d57'21"
Kenny Sweeney
Cyg 20h4'49" 37d53'23"
Kenny Swinford
Boo 14h36'31" 23d8'26"
Kenny T Just For You
Uma 11h55'47" 34d36'42"
Kenny the Bear
Uma 8h51'58" 69d29'52"
Kenny-&-The-Kids' Star
Dra 19h10'2" 68d35'40"
Kenny Y. Kawamoto
Vir 13h2'53" 6d21'13"
Kenny, My Shining Star
Per 2h49'13" 44d32'23"
Kennyboy - Pups
Uma 9h52'11" 68d28'18"
Kenny-QT
Aqr 20h40'13" -11d36'50"
Kenny's Curiosity
Cnc 8h14'40" 9d27'16"
Kenny's Freedom Star
Tau 5h39'4" 26d38'37"
Kenny's love Cheryl
Cap 20h30'23" -12d26'39"
Keno
Cap 21h30'5" -14d9'40"
Kenozah
Cap 20h7'51" -11d8'34"
KenRea
Psc 1h11'53" 16d2'42"
KenRhonda4/10/07
Lyr 18h54'23" 32d6'49"
Kenrick Erin Nobles
Uma 11h12'3" 39d48'12"
Ken's Cosmic Hideaway
Uma 13h33'19" 61d22'46"
Ken's Peg 'O My Heart
Leo 9h39'7" 26d54'14"
Ken's Perfection
Her 17h22'46" 39d36'0"
Ken's Shining All-Star
Peg 21h35'5" 6d33'14"
Ken's Star
Cep 20h53'1" 59d59'4"
Ken's Star 17.7.29 Love
Forever Bev
Sco 16h49'15" -27d44'44"
Ken's Vision
Vir 12h4'54" 6d24'9"
Kensey Lowe (Our Baby
Girl)
And 23h30'50" 41d50'38"
Kent A. Roseman " My
Shining Star"
Her 17h47'15" 38d47'47"
Kent and Liz
Sgr 18h38'22" -28d2'58"
Kent and Susie's Split-
Apart Star
Cyg 20h17'52" 47d5'7"
Kent Bastin
Per 3h17'52" 48d33'15"
Kent Douglas Chastain
Tau 4h13'41" 17d38'57"
Kent & Emily Owens
Eri 2h46'33" -41d0'19"
Kent Errol Kline
Uma 12h3'3" 50d50'50"
Kent Forever
Aqr 21h12'10" 0d14'15"
Kent Gambill
Ori 5h38'47" 2d12'55"
Kent Grey Bader
Equ 21h19'27" 6d42'35"
Kent Hall
Per 2h52'17" 56d52'31"
Kent Hammond Stockdale
Cru 12h29'54" -57d43'53"
Kent Hoi loves Kathy
Pacyga
Lib 16h1'18" -6d30'34"
Kent ***I Love You More***
Ori 6h10'57" 8d41'37"
Kent Johnson
Her 17h9'24" 33d54'18"
Kent Jon Udy
Leo 9h28'18" 29d15'35"
Kent Kreider
Her 17h56'55" 19d47'39"
Kent Lockhart
Ari 2h32'49" 12d39'55"
Kent Michael Cousins
Her 16h7'56" 17d23'35"
Kent Michael Smith
Her 18h22'17" 21d50'43"
Kent Mikel Kerr
Her 16h39'9" 6d44'44"
Kent & Nancy Noyes
Cnc 8h44'49" 7d7'51"
Kent Noyes
Lib 14h52'20" -3d48'24"
Kent P and Jenny Lou
Ewart
And 0h18'32" 27d53'12"
Kent Rubright "Happy 25th"
Cep 23h5'48" 75d26'18"
Kent S. Aiken
Per 3h9'36" 42d54'15"
Kent Spencer Goulden
Gem 7h5'47" 31d45'19"
Kent Stevinson
Cas 0h21'7" 54d46'53"
Kent W. Thompson
Peg 22h50'13" 15d45'11"
Kent Wade Ponder Shines
Forever
Per 3h39'54" 42d3'15"
Kent Warren Koch
Aqr 21h46'10" -1d27'31"
Kentaro & the child of kiku-
mi takumi
Leo 9h28'1" 21d16'21"
Kentiffany
Uma 8h56'28" 50d47'58"
Kenton and Mary Yoder
Per 3h12'11" 55d39'7"
Kenton James Onofrychuk
-S'ayapo
Ori 6h16'33" 1d46'42"
Kenton & Lorraine's Star of
Hope
Lyr 18h47'26" 37d7'11"
Kenton Pattie
Her 18h21'18" 28d28'54"
Kenton Scott Shaw
Sco 17h55'36" -44d25'18"
Kenton V. Te Lindert
Uma 10h36'1" 62d54'47"
Kenton Wentz
Aur 6h29'56" 51d41'43"
Kentster
Ori 4h53'31" 1d36'51"
Kentucky Tri-Delta
Aql 19h40'17" 2d53'53"
Keny Blanco-Mazur
Crb 15h23'45" 29d22'46"
Kenya D'Ann
Sco 16h51'43" -38d17'27"
Kenyaka
Tau 3h24'53" 16d34'21"
Kenyatta Bernard Gowdy
Ori 5h20'21" 3d18'23"
Kenyetta Steel
Lyn 9h15'25" 40d14'18"
Kenyia Tilson
And 2h37'53" 43d48'23"
Kenza Khachani
Tau 5h17'35" 18d13'4"
Kenza Margaoui
Lib 15h39'5" -6d42'24"
Kenzie
Aqr 20h48'30" -7d9'37"
KENZIE
Ori 6h15'11" 15d54'13"
Kenzie Alexis Smith
Peg 21h52'50" 18d9'37"
Kenzie Burnett
And 0h38'29" 21d51'7"
Kenzie Cox
And 0h17'27" 28d25'6"
Kenzie Jo Wheeldon
Crb 16h6'26" 34d57'20"
Kenzie Lauren Odell
Sco 16h36'45" -34d45'26"
Kenzie McCormick
Umi 16h23'49" 75d22'18"
KenzieK
Leo 11h44'36" 22d32'50"
keo_1011
Uma 11h44'49" 52d39'20"
Keon Heather Kyan
Lyn 7h9'24" 52d37'10"
Keon Latrel Conerly
Tau 5h24'25" 27d31'35"
Keona Watkins
Vir 11h52'5" -4d43'0"
Keonaona Burke
Cru 12h45'0" -60d9'35"
Keoni
Lmi 11h3'59" 27d2'1"
Keoni Leon
Vir 12h56'49" -5d6'27"
Keoni Morgan
Lyn 7h21'5" 44d48'23"
Keonte S S Glover
Cap 20h23'52" -10d41'21"
Keplers Forever
Cnc 8h34'8" 27d59'7"
Keppelazarus
Cnc 9h7'0" 30d35'51"
Keppler, Manfred
Ori 6h20'8" 10d24'24"
KER
Vir 12h12'55" 10d49'30"
Kera
Ari 2h47'4" 25d51'40"
Ke-Ra 93
Cyg 20h48'41" 35d22'44"
Kera Rochelle 880418
Uma 11h10'56" 43d58'58"
Kérastase
And 0h7'9" 42d13'18"
KerBear
Ari 3h24'15" 26d16'52"
KerBear
Cas 23h54'11" 57d39'0"
Kerbel, Sergej
Ori 6h8'18" 9d4'5"
Kered & Freya
Dra 18h56'29" 52d28'51"
Kered & Freya's Wish Star
Gem 6h58'6" 16d59'7"
Keren
Sgr 18h10'54" -27d28'58"
Keren June Ott
Cap 20h28'3" -9d19'35"
Keren's Star
And 2h13'38" 42d35'43"
Kerényi
Cap 20h37'20" -21d28'53"
Kerezsi Pálma
Uma 13h55'32" 49d56'18"
Keri
And 2h7'5" 38d37'52"
Keri
Gem 7h42'1" 32d12'16"
Keri
Tau 4h24'24" 17d53'4"
Keri
Cap 21h37'2" -9d44'10"
Keri
Cap 21h42'25" -12d37'30"
Keri A Flynn
Gem 6h34'47" 24d17'0"
Keri and Ed's Hope
Lyn 7h5'58" 50d47'30"
Keri and Mike Fowler
Gem 7h4'57" 25d8'31"
Keri and Robbie Robbins
Lyr 19h26'59" 37d59'53"
Keri Angela Fico
And 23h13'57" 51d37'40"
Keri Ann
Sco 17h54'44" -38d11'49"
Keri Ann DeRoss
Leo 10h43'27" 17d52'52"
Keri Ann Hebert
And 2h23'25" 49d30'52"
Keri Ann Johnson
Uma 13h32'29" 56d2'20"
Keri B - 21
Ori 6h16'34" -0d25'30"
Keri Brandon Justin Chris
Uma 8h34'22" 66d6'11"
Keri Brothers
Cap 20h17'49" -11d43'0"
Keri Craig Lee (Australia)
Cen 13h58'4" -60d15'21"
Keri ete strella
Lib 14h48'56" -15d51'15"
Keri Joy McGowan
Uma 10h52'59" 57d47'39"
Keri Lee
Aqr 20h40'15" 0d19'19"
Keri Loves Chris
Sco 17h30'15" -32d55'55"
Keri Lyn Daniele
And 0h36'6" 38d21'54"
Keri Lyn Lenosky
Gem 6h35'43" 20d20'16"
Keri Lyn Martin
Vir 13h12'44" -20d44'35"
Keri Lynn
Leo 11h41'2" 22d13'19"
Keri Lynn
Gem 7h18'4" 13d25'28"
Keri Lynn Shedden
Cyg 21h10'13" 46d1'57"
Keri Lynn Wells
Sgr 18h6'45" -26d44'3"
Keri & Marc Messer
And 0h39'11" 27d56'27"
Keri Nowak
Cas 1h40'36" 63d6'55"
Keri Rene Painter
Mon 6h41'24" -0d24'45"
Keri Romano
Cas 2h11'23" 64d56'20"
Keri Romine
Uma 13h57'48" 53d59'0"
Keri Taylair Zimmerman
And 23h55'8" 33d45'9"
Keri The Keally
Crb 15h38'31" 35d3'43"
Keri Thorand + Andy
Gärtner
Uma 9h28'26" 62d31'41"
Keri Wilson "Kerbear"
Umi 15h20'31" 77d9'16"
Keri Yapaolo
Cnc 9h11'9" 11d12'4"
Keri, Cherished Daughter
Gem 6h22'10" 18d19'37"
Kerial
Ori 5h24'45" 2d58'34"
Keriann
Ori 5h53'31" 12d9'16"
Keriann Pilsbury
And 2h26'29" 47d19'21"
Kerianne Gavyn Vince
Leo 11h16'55" 17d48'46"
Keridwen
Uma 10h1'8" 47d15'3"
Kerilynn
Psc 1h24'13" 24d27'40"
Kerilynparhunepouranios
(K4A)
And 2h22'6" 41d18'34"
Kerina's Pearl
Uma 8h24'30" 70d3'31"
Kerinea
And 1h1'1" 45d18'23"
Kerin's Love
Cas 23h42'3" 57d6'1"
Keris' Christening Star
And 2h35'52" 49d37'20"
Kerish's Radiant Aurora
Crb 15h30'27" 29d49'43"
Keristar
Ori 5h16'3" 6d17'40"
Kerkow Family Star
Ori 6h16'24" 14d42'35"
Kermeth Engel
Uma 9h24'30" 46d13'59"
Kermit
Del 20h47'44" 13d54'58"
Kermit Allen Taylor
Her 17h20'58" 37d43'3"
Kermit Daniel Sarber II
Cnc 8h20'45" 15d22'14"
Kermit J. Schulz
Cnc 8h23'46" 22d1'22"
Kernads
Uma 13h32'12" 56d27'9"
Kernick
Uma 8h54'47" 63d39'29"
Kerpes, Michael
Uma 11h29'59" 31d27'10"
Kerra Elizabeth
And 0h18'58" 42d46'36"
Kerrbear
Lyn 6h47'27" 56d41'56"
Kerrey "The Love of My
Life"
Aqr 20h44'42" -7d42'8"
Kerri
Lib 15h47'28" -16d53'28"
Kerri
Lyn 8h46'8" 37d50'8"
Kerri
Cnc 8h14'49" 31d12'57"
Kerri
Psc 0h54'55" 27d11'53"
Kerri
Peg 22h32'0" 21d51'11"
Kerri
Cnc 8h26'20" 22d34'52"
Kerri
Tau 5h39'42" 20d23'8"
Kerri Allen
Leo 11h6'37" 21d58'30"
Kerri Allen
And 2h1'34" 37d36'11"
Kerri Ann McCall
Vir 13h39'25" -18d21'18"
Kerri Ann Thibault
And 1h48'27" 42d18'48"
Kerri Anne
Ori 5h43'35" 0d20'53"
Kerri Anne Layton.
Tau 4h30'42" 6d32'26"
Kerri Anne Schekeloff -
31.12.1949
Cru 12h36'12" -59d53'49"
Kerri Ann's Dreams
Psc 23h13'14" 6d49'46"
Kerri Bella Morrin
And 1h39'44" 47d35'43"
Kerri C. Houser
Cas 0h54'54" 57d7'35"
Kerri "Care Bear" Nyers
Peg 23h19'47" 32d42'39"
Kerri Diane Glover Rose
Sgr 18h57'52" -34d57'50"
Kerri Driscoll
And 0h54'9" 36d38'5"
Kerri Floyd
Uma 10h58'32" 61d59'49"
Kerri Frizzi
Sco 16h50'59" -45d28'8"
Kerri Janelle Wilson
Crb 16h9'46" 32d19'9"
KERRI JAYNE SEARS
Cyg 19h51'52" 55d10'46"
Kerri Jean
Aqr 23h50'9" -23d6'54"
Kerri Jean De Graff
And 0h31'20" 29d27'25"
Kerri Jo Read Lane
Uma 11h17'40" 31d33'26"
Kerri Kaplan
Uma 8h29'43" 67d11'7"
Kerri Kathleen Wolf
Cas 0h9'14" 60d34'40"
Kerri Lauren Esbaum
Sgr 18h7'15" -18d34'27"
Kerri Lee
Uma 9h53'40" 67d34'57"
Kerri Leigh Shuman
And 0h58'30" 36d11'16"
Kerri Lynn Pohl
Lyn 7h20'45" 57d33'45"
Kerri Lynn's Star
Cnc 8h49'49" 27d9'22"
Kerri Michelle
Cyg 19h43'19" 38d54'18"
Kerri & Mike 11/13/2004
Cyg 19h30'48" 52d39'46"
Kerri Ohmes
Uma 10h34'58" 61d42'50"
Kerri S
Lib 15h4'34" -23d27'27"
Kerri ' with an i '
Aqr 21h30'59" 1d10'42"
Kerriann Kelley
Cap 20h32'41" -12d16'35"
Kerri-Anne Aitken
Lib 15h25'49" -20d21'1"
Kerrick Watson
Sgr 17h54'18" -29d22'3"
Kerrie
Ori 5h36'43" 3d55'45"
Kerrie
Ori 6h12'0" 17d50'37"
Kerrie Anne
Cyg 20h59'29" 46d15'3"
Kerrie Beth Pfahl
Ari 2h59'26" 21d30'31"
Kerrie Jo's Beautiful Star
Gem 6h17'2" 21d31'3"
Kerrie Lawton
Ari 2h33'38" 14d6'49"
Kerrie Lee Cupples
Cru 12h37'13" -58d57'1"
Kerrie Mays
Tau 3h46'1" 2d34'31"
Kerrie -The Guiding Light-
Vassallo
Vir 11h55'36" -0d57'33"
Kerrie's Place
Sco 17h26'3" -30d14'15"
Kerrilyn Christine
And 2h16'28" 48d27'6"
Kerri-lynn
And 0h25'44" 27d55'30"
Kerrin
Cam 3h57'24" 69d54'57"
Kerris Megan Heather
Randell
And 1h23'28" 41d2'28"
Kerri's Star
Vir 14h16'3" -1d15'41"
Kerrol
And 1h44'8" 46d6'22"

Kerr's Quest: 9/14/1944
Uma 12h43'23" 55d42'12"
Kerr's Star
Her 16h21'35" 8d50'49"
Kerry
Psc 22h57'30" 6d16'30"
Kerry
Psc 0h56'53" 25d50'35"
Kerry
Leo 11h37'30" 20d1'11"
Kerry
And 1h59'45" 42d21'53"
Kerry
Umi 15h24'30" 72d53'3"
Kerry
Cap 21h29'17" -23d7'7"
Kerry Alexandra Chavoya
Uma 10h54'18" 43d37'45"
Kerry and James Strong
Cyg 21h53'46" 48d41'55"
Kerry and Warwick
Cra 18h52'17" -38d7'28"
Kerry Ann
Sco 16h18'7" -12d31'4"
Kerry Ann
Gem 6h44'45" 17d40'54"
Kerry Ann Fullington
Per 2h23'29" 56d10'13"
Kerry Ann Newton-Mehling
Cnc 9h9'55" 7d35'48"
Kerry Ann Walker *My true love*
And 1h41'43" 38d28'39"
Kerry Ann Wilson
Cam 6h37'22" 63d24'37"
Kerry Ann Wren
Ari 3h17'6" 12d8'4"
Kerry Anne Davies
Lmi 10h49'22" 38d27'22"
Kerry Anne Schultz
Lyr 18h49'44" 32d54'47"
Kerry Barber
Aur 5h45'20" 29d4'30"
KERRY BLOMBERG
Cyg 20h15'57" 52d39'52"
Kerry Burns of Collierville, TN
Aql 20h7'54" 3d22'46"
Kerry Carroll
Cyg 20h39'26" 36d19'4"
Kerry Claire Burke
Cap 21h33'27" -17d20'46"
Kerry Claire Gallagher 17.6.1980
And 1h29'23" 34d12'14"
Kerry Colleen
Psc 0h32'57" 16d27'27"
Kerry Colleen Sullivan
Leo 10h55'41" 19d11'48"
Kerry Cutillo Sullivan
Ari 3h13'21" 28d46'40"
Kerry & Donna Krueger
Cru 12h21'22" -60d36'4"
Kerry Douglas
Sco 16h10'14" -12d36'32"
Kerry Elizabeth
And 23h13'23" 47d36'55"
Kerry Elizabeth Chemnitz
Ari 3h23'33" 27d58'45"
Kerry Ellen Virga
Cap 20h10'15" -10d30'36"
Kerry Eugene Gray
Gem 6h50'50" 18d40'59"
Kerry Finney
Cma 7h0'1" -23d22'9"
Kerry - Forever Shining Bright
And 2h22'14" 50d2'58"
Kerry Generose Smith
Ori 5h43'28" 2d8'29"
Kerry Germon
And 0h54'55" 44d15'19"
Kerry Graham
And 2h36'37" 44d55'31"
Kerry Hawkins
Cru 12h42'52" -57d59'54"
Kerry J. Frasier
Gem 7h5'25" 22d26'28"
Kerry Jane (The Otter)
Aqr 20h54'30" -10d40'1"
Kerry & Janis Viksne
Sco 17h29'20" -40d57'49"
Kerry Jean Joseph
Cas 3h15'23" 64d15'47"
Kerry Jean Wydick
Lib 14h53'18" -18d54'23"
Kerry Jefferson Raab
Lyn 9h7'54" 44d35'36"
Kerry Kay Jacobs The Star of Wonder
Vir 12h28'21" -4d3'57"
Kerry Kirkwood
Cra 18h46'39" -40d31'59"
Kerry Koch
Uma 10h42'10" 67d40'53"
Kerry "KT" Turner
Tau 5h54'0" 27d13'3"
Kerry Lee Wilen
Crb 15h31'59" 28d19'54"
Kerry Leigh
Cap 20h36'1" -19d47'37"
Kerry Lynn
And 23h43'36" 42d53'15"
Kerry Lyon
Cnc 8h44'48" 7d25'29"
Kerry Marie
Ari 2h7'13" 23d7'0"
Kerry Marie Castaldi
Lib 14h40'56" -16d2'59"
Kerry Marie Moore
And 0h41'40" 26d13'40"
Kerry & Mark's Silver Star
Cyg 20h9'7" 41d16'47"
Kerry McKay
And 2h30'49" 44d33'57"
Kerry McNicoll
Cyg 19h9'47" 50d39'56"
Kerry "MoonStar"
Vir 12h34'26" 11d19'57"
Kerry Morell
Aqr 22h28'58" -9d46'8"
Kerry Moriarty
Leo 9h22'25" 25d36'44"
Kerry My Lucky Star
Ari 2h54'14" 10d42'10"
Kerry my southern princess
Ori 6h12'23" -1d29'14"
Kerry O'Keefe
Cru 12h39'5" -59d44'17"
Kerry O'Reilly
And 1h40'5" 41d29'30"
Kerry "Pete" Milton Cummins, Jr.
Aql 19h23'28" 4d10'31"
Kerry "Pusan" Carlson
Aqr 23h30'10" -21d4'40"
Kerry Regina Souris
Gem 6h54'3" 21d32'52"
Kerry Schneider
Vir 14h24'42" -0d37'36"
Kerry & Steve's Eternal Light
Ori 5h34'52" -2d51'41"
KERRY STONE
Mon 6h53'26" 9d39'55"
Kerry & Stuart's Wedding Star
Cyg 20h42'27" 34d34'49"
Kerry Suglia
Aur 5h22'13" 33d34'2"
Kerry & Victoria
Cyg 20h52'19" 31d47'36"
Kerry Yelle
Aql 19h27'10" 7d24'12"
Kerry, The Star in my constellation
Uma 10h20'50" 40d27'41"
Kerry-Ann
Uma 12h41'12" 58d9'7"
Kerryn "Kezmeister" Tyrrell
Tau 3h48'24" 10d17'55"
Kerry's Heaven
Aqr 21h45'48" -3d39'12"
Kerry's Hideaway
Ari 2h8'58" 24d12'36"
Kerry's Mustang
Peg 22h22'29" 29d15'31"
Kerry's Sparkly Star
And 23h55'56" 47d10'3"
Kerry's Star
And 0h36'31" 33d45'6"
Kerry...will you marry me?
And 0h8'13" 34d52'24"
Kersey Bisono
Cnc 9h9'6" 25d30'25"
Kersia Ward
Ori 6h13'15" 11d59'22"
Kersteli mis Stärnli
Sge 19h52'5" 18d11'59"
Kersten Suzanne Reid
Vir 14h17'22" 0d1'24"
Kerstin
Vir 13h38'46" -15d24'28"
Kerstin
Uma 8h54'47" 64d22'37"
Kerstin
Uma 10h11'50" 60d47'48"
Kerstin
Uma 9h10'6" 65d43'29"
Kerstin and Nate's Star
Uma 11h24'0" 57d28'16"
Kerstin Binder
Uma 10h24'14" 40d55'36"
Kerstin Denker 29/10 HDL
And 23h36'19" 47d44'17"
Kerstin & Frank
Uma 9h29'29" 44d50'2"
Kerstin Gabrielle Woolfolk
Gem 6h3'52" 27d20'40"
Kerstin Holmenäs Seaton
Tau 3h56'27" 18d11'14"
Kerstin Leimbrink
Cas 0h54'20" 63d23'15"
Kerstin schatz 21
Uma 10h17'27" 57d11'6"
Kery Rodriguez
And 0h8'7" 47d57'3"
Keryie Vickers
And 23h50'10" 46d5'59"
Keryl Michelle Cryer
Cas 23h59'2" 61d7'59"
Keryn Jane Stronach
And 23h5'30" 41d30'55"
Kesamot
Uma 9h50'49" 57d13'44"
kesh
Leo 9h42'20" 31d23'39"
Kesha Lewis
Uma 11h31'29" 41d53'41"
KeshanPaul
And 23h5'9" 41d56'18"
Keshia
Uma 10h58'50" 66d46'9"
Keshia Khokar
Lyn 6h56'29" 53d2'12"
Keshia Stephens
And 0h41'23" 35d10'51"
Keshius L. Williams
Lib 15h28'8" -6d18'1"
Keshonna & Jerry
Cyg 21h40'54" 49d41'17"
Kesi
Lmi 10h10'42" 33d35'43"
Kesi
Cap 20h32'8" -14d30'23"
KESRAS-my favorite girls
Cma 6h30'11" -15d37'11"
Kessa and Corey
Crb 16h12'37" 31d31'31"
Kessler 1981
Vir 13h32'33" 7d34'37"
Kessler Klara Cook
And 1h36'51" 38d36'54"
Kessler Yvonne "Mis Wichtigscht"
Uma 10h47'48" 42d40'54"
Kessler, Dieter
Ori 6h16'10" 11d5'19"
Kesslers' Star
Cyg 19h23'39" 53d21'53"
Kestrel Nest
Uma 13h20'39" 59d56'50"
ket
Uma 8h37'52" 63d2'36"
Ket Saelieo
Uma 10h21'40" 42d23'44"
Ketan Dineshchandra Patel
Lib 15h32'56" -8d27'49"
Ketch
Ori 5h56'10" 11d40'49"
Kete - der hellste Stern am Himmel
Ori 6h15'40" 11d13'40"
Kethryveris and Mooneleafs Hargrave
Ori 5h56'16" 12d41'18"
Ketric & Lawrence
Leo 11h2'25" 22d14'40"
Ketta & Kobie
Gem 6h53'38" 24d55'37"
Ketterling
Uma 10h0'39" 52d54'25"
Ketthip Withiprod
And 0h24'52" 27d9'34"
Keturah
Lyr 19h3'29" 31d29'46"
Keturah Grace
Uma 10h17'14" 68d58'39"
Keturah Maurey Sloan
Lmi 10h27'33" 38d6'6"
Kety SHMILY Star
Aur 5h48'57" 47d55'37"
KEV
Gem 6h16'31" 21d57'59"
Kev Cooper's Cosmic Crash Pad
Per 4h10'45" 41d1'47"
Kev&Kel
Gem 6h21'5" 18d29'36"
Kev & Lyssie 5 Year Anniversary
Cyg 21h51'46" 40d16'43"
Kevalextails
Cam 4h27'47" 64d4'32"
Kevan Stoddard MBE
Cyg 20h2'40" 33d55'10"
Kevan's Star
Ori 5h50'26" 11d30'37"
Kevcolious
Aur 5h53'5" 41d46'57"
Keven Darrel Widener
Sgr 20h6'49" -37d30'17"
Keven L
Aqr 21h46'19" -1d31'57"
Keven Michele Taylor
Cnc 8h17'21" 17d50'16"
Keven, my eternal love forever!
Leo 10h16'59" 16d34'10"
Kevi Christopher Walsh
Cam 4h5'53" 67d37'58"
Keviana
Lyn 6h54'56" 50d58'34"
Kevie Carr
Aqr 23h20'7" -11d17'40"
Kevin
Cap 21h51'42" -10d39'2"
Kevin
Dra 17h28'0" 62d16'20"
Kevin
Dra 18h36'57" 62d39'50"
Kevin
Cyg 20h23'48" 58d51'32"
Kevin
Cap 21h2'32" -26d49'34"
Kevin
Aur 5h45'48" 49d8'34"
Kevin
Per 4h50'16" 45d38'54"
Kevin
Per 3h1'58" 54d30'24"
Kevin
Uma 11h52'45" 50d39'23"
Kevin
And 0h17'20" 44d35'52"
Kevin
Uma 10h58'25" 35d6'53"
Kevin
Ari 3h0'51" 29d28'59"
Kevin
Tau 4h14'1" 26d26'28"
Kevin
Her 16h46'53" 5d35'30"
Kevin
Ori 5h58'34" 4d26'47"
Kevin
Ori 5h22'13" 4d51'25"
Kevin A. Bridges
Dra 16h49'19" 67d50'57"
Kevin A. Dorsey Sr.
Sco 17h58'29" -38d56'28"
Kevin A. Fabiano
Cep 22h25'4" 63d1'23"
Kevin Alan Polcovich
Cyg 21h55'8" 45d4'46"
Kevin Allen Burlison
Lmi 10h37'44" 28d8'37"
Kevin Allen Luberger
Her 16h32'6" 46d44'46"
Kevin Allen Peridore
Uma 12h12'20" 61d18'31"
Kevin Alvestal
Uma 13h5'13" 52d40'2"
Kevin & Amanda's Wish on True Love
Per 2h35'51" 51d14'1"
Kevin & Amy Abian - 4th March 2006
Ara 17h5'42" -56d24'24"
Kevin & Amy Abian 4th March 2006
Ara 17h11'7" -57d5'43"
Kevin and Aimée
Gem 6h45'32" 25d43'23"
Kevin and Amelia Ruebenstahl
Cyg 21h54'48" 48d47'8"
Kevin and Angie "buddies for life"
Her 18h42'53" 19d41'38"
Kevin and Anne Burns
Uma 9h28'1" 46d0'7"
Kevin and Annemarie Hannigan
Cyg 19h40'49" 32d32'19"
Kevin and Ari
Leo 9h42'16" 30d55'31"
Kevin and Barbara Goodman
Cnc 8h21'55" 13d45'32"
Kevin and Holly Forever
Uma 11h31'24" 33d21'20"
Kevin and Jennifers Nana
Psc 0h9'23" 2d42'39"
Kevin and Jill Stress
Her 17h15'54" 34d3'44"
Kevin and Jonathon's Starry Night
Cnc 8h34'2" 22d0'9"
Kevin and June Burke
Cyg 20h38'30" 37d30'5"
Kevin and Katherine: Eternal Love
Lyn 8h55'48" 39d32'59"
Kevin and Kathleen Shafer
Sge 19h51'0" 18d32'35"
Kevin and Kaydean Baalmann
Cyg 20h18'14" 46d23'17"
Kevin and Liz Star
Per 3h26'52" 44d45'34"
Kevin and Lorraine Forever
Tau 4h30'12" 10d58'4"
Kevin and Mary Henebry
Lac 22h27'21" 50d59'38"
Kevin and Melanie Shelton
Ori 5h26'55" 2d9'34"
Kevin and Sarah Strickler
And 1h21'12" 43d47'5"
Kevin and Sherri
Umi 13h13'31" 86d17'44"
Kevin and Sheryl
Her 18h20'49" 18d15'13"
Kevin and Tessa Wayman
Cyg 21h35'47" 53d25'30"
Kevin and Vickie Forever
Ori 5h19'12" -8d9'47"
Kevin Anderson
Per 2h47'29" 54d2'35"
Kevin Andrew Fischer
Leo 11h26'32" 15d34'51"
Kevin Andrew Thompson
Per 3h9'2" 51d38'55"
Kevin Andrew Wolfe
Uma 12h23'56" 59d30'27"
Kevin & Angelo, Father & Son
Per 3h34'14" 52d47'56"
Kevin Ansley Johnson
Her 17h47'46" 47d0'15"
Kevin Anthony Christopher
Cap 21h0'47" -17d22'9"
Kevin Anthony Mushorn
Per 2h37'59" 55d13'1"
Kevin Anthony Staley
Cap 21h41'46" -8d49'56"
Kevin Anthony Stevens
Vir 14h14'53" 4d33'40"
Kevin Arnold
Tau 4h32'46" 19d56'34"
Kevin B.
Tau 3h54'35" 28d1'54"
Kevin B. McMillen
Psc 1h23'37" 32d16'26"
Kevin Badeaux
Sco 16h54'18" -34d10'43"
Kevin Barr
Ori 5h39'52" -0d49'5"
Kevin Batchelor
Cyg 19h59'24" 33d23'48"
Kevin Beukema
Leo 9h41'41" 10d54'41"
Kevin Bingman
Ori 6h3'35" 10d58'40"
Kevin Borg
Ori 5h27'23" -0d49'50"
KEVIN BOURASSA
Her 16h45'41" 44d47'7"
Kevin & Brandi Cloward
Peg 21h29'35" 15d50'15"
Kevin Brandon Cortez
Aql 18h44'43" -0d6'33"
Kevin Brandon Ludwig
Cam 3h53'48" 69d13'57"
Kevin Brewton
Her 18h40'9" 25d2'8"
Kevin Brian Leavy
Ori 5h58'58" -0d21'15"
Kevin Brown
Uma 11h25'41" 66d29'55"
Kevin Brown
Vir 12h44'53" 5d32'6"
Kevin Bruce Chin
Lib 14h54'28" -1d41'56"
Kevin Bruns
Gem 7h30'6" 19d33'55"
Kevin Brzycki
Her 18h19'18" 17d26'59"
Kevin Burke
Per 4h10'36" 50d50'50"
Kevin Burns
Ari 2h25'39" 26d16'45"
Kevin C. Daly
Ori 5h19'6" 16d2'2"
Kevin C. Kaplan
Uma 11h17'48" 39d5'1"
Kevin & Calley Moss
Lib 15h21'58" -15d30'57"
Kevin & Camille's "Xaaliyah"
Cyg 20h50'56" 43d28'24"
Kevin Carl H. and Jennifer Casey J.
Cyg 20h11'50" 33d1'4"
Kevin Carter Wood Mole Productions
Umi 9h19'12" 86d40'34"
Kevin Cash
Mon 7h16'57" -0d4'52"
Kevin Chao
Uma 11h32'30" 47d16'49"
Kevin Charles Coxwell
Aql 19h54'57" 14d28'47"
Kevin Charles King
Cep 22h35'18" 73d37'28"
Kevin Charles Urquhart Junior
Ori 6h12'8" 17d47'13"
Kevin Charles Vrba
Umi 14h54'5" 81d57'37"
Kevin & Chelsea
Ari 2h49'21" 27d36'16"
Kevin Chianta
Lyn 8h22'19" 57d26'34"
Kevin Chicken Monkey Malloy
Dra 18h47'14" 53d7'40"
Kevin Christopher
Cap 21h2'56" -19d30'29"
Kevin Christopher Coughlin
Leo 9h37'45" 26d27'21"
Kevin Christopher Huang
Uma 11h42'31" 42d37'43"
Kevin Christopher Hughes
Umi 15h13'50" 72d25'45"
Kevin Christopher Landeck
Uma 11h47'46" 50d45'39"
Kevin Christopher Whiskeyman
Cep 3h39'6" 81d0'8"
Kevin Clewer
Ari 2h42'17" 27d53'56"
Kevin Clyde Anderson Wilson
Boo 14h44'42" 14d44'23"
Kevin Colleran
Cnc 8h30'59" 26d0'7"
Kevin Comden
Cyg 20h47'26" 47d42'30"
Kevin Corneille
Uma 8h38'59" 60d44'57"
Kevin Cottrell
Umi 15h1'9" 73d17'27"
Kevin Covell
Per 2h49'10" 54d3'50"
Kevin Crawford's Point Of Light
Sco 16h57'8" -43d15'57"
Kevin Cupp
Uma 9h16'9" 53d39'28"
Kevin Curran
Her 16h55'18" 34d58'3"
Kevin D. King — SM up in Heaven
Cap 20h23'50" -12d9'28"
Kevin D Long
Aqr 20h51'26" 2d4'1"
Kevin Dale Roberts
Pho 0h19'12" -42d38'55"
Kevin Daley's Destiny
Her 17h29'29" 26d43'10"
Kevin Dante' Brown
Cnc 9h5'59" 31d5'31"
Kevin David Allsop
Per 4h10'41" 33d57'52"
Kevin David Michael Fletcher
Umi 14h29'51" 74d4'42"
Kevin David Sullivan
Psc 1h8'54" 14d44'0"
Kevin Davis
Uma 10h28'56" 67d18'20"
Kevin & Desirae
Per 2h51'18" 52d51'31"
Kevin DeWitt
Peg 23h16'23" 33d29'51"
Kevin Dick Krause
Her 16h32'6" 21d19'38"
Kevin Don Knight
Her 17h17'39" 26d32'6"
Kevin & Dorie Harrison
Her 17h7'30" 46d4'36"
Kevin & Dot
Cnc 8h37'33" 14d13'44"
Kevin Douglas Smith
Her 17h45'11" 25d25'0"
Kevin Duffy
Uma 8h17'49" 61d5'59"
Kevin E. Flynn
Ari 2h27'21" 19d46'8"
Kevin Earl Ferrell
Lib 15h6'25" -22d42'1"
Kevin Eccleston
Leo 9h31'5" 30d25'22"
Kevin Edward Lawrence
Pho 1h19'44" -43d32'14"
Kevin Edward Skelly
Leo 11h31'42" 7d21'30"
Kevin Edwin
Lyn 8h2'43" 35d30'9"
Kevin Elmer
Ori 5h28'25" 2d56'18"
Kevin & Emily's Starglow
Tau 3h59'31" 9d9'2"
Kevin Enos
Sgr 19h37'41" -12d15'42"
Kevin Eric Hagen
Sgr 18h24'13" -33d1'25"
Kevin & Erica's Galaxy
Sco 17h19'17" -36d42'34"
Kevin & Erin
Cyg 21h22'40" 50d38'12"
Kevin Ethan Kratch
Per 3h21'19" 46d24'20"
Kevin Fahy
Per 4h35'23" 49d41'9"
Kevin Floro
Cep 23h8'20" 86d11'36"
Kevin Foster
Cyg 19h50'28" 57d38'40"
Kevin Fred Estes
Sgr 18h26'11" -32d44'15"
Kevin Frisch
Cnc 8h35'3" 23d17'7"
Kevin G. Anderson
Gem 6h48'34" 17d17'3"
Kevin G. Brogan
Gem 7h0'19" 26d48'35"
Kevin G. Campbell
Cnc 8h52'15" 13d32'8"
Kevin G. Campbell
Her 16h43'29" 46d8'7"
Kevin G. Hinds
Aur 5h30'16" 45d54'40"
Kevin G. Sorgi, Jr.
Crb 16h9'37" 36d47'3"
Kevin G. Toth
Ori 6h5'20" 18d28'30"
Kevin & Gail Mattimore
Cyg 20h16'47" 48d37'48"
Kevin Gamblin
Aql 19h27'18" 12d0'0"
KEVIN GEORGE KLEIN
Cnc 8h47'41" 16d10'54"
Kevin George Lundrigan
Sco 16h16'14" -32d14'31"
Kevin George Walton
Del 20h24'27" 5d15'21"
Kevin Gerard Kniese
Per 2h6'7" 57d18'5"
Kevin Gerard Luczon Ibera
Sco 17h50'41" -32d43'17"
Kevin Gerard Patrick Reagan
Cnc 8h52'35" 31d27'10"
Kevin Gibbon
Uma 12h4'2" 38d22'59"
Kevin Glenn Pledger, Jr.
Sgr 19h51'33" -11d58'43"
Kevin Godfrey Brayshaw
Per 4h33'32" 31d11'0"
Kevin Grew
Sco 17h8'32" -43d14'18"
Kevin Griffin
Cnv 12h52'15" 37d0'30"
Kevin Grissom
Cnc 8h35'54" 8d9'23"
Kevin Gubser
Peg 21h16'40" 18d17'17"
Kevin Guerra
Her 18h35'47" 21d16'30"
Kevin Guieb
Cap 20h50'26" -25d41'57"
Kevin Guillory
Ori 5h11'43" 3d10'30"
Kevin Guthrie
Uma 10h48'38" 66d21'43"
Kevin Hampton
Tau 3h53'50" 1d33'45"
Kevin Healy
Her 17h34'36" 37d0'10"
Kevin Heidrich
Her 17h36'17" 36d54'12"
Kevin Helton
Her 17h42'13" 26d13'53"
Kevin Heumann
Tau 4h39'7" 22d18'4"
Kevin Horst Kessler July 22, 1970
Ori 5h37'31" -0d8'15"
Kevin Howard LeBeouf
Sgr 18h48'31" -32d42'14"
Kevin Hursh
Aqr 22h44'39" -21d54'34"
Kevin J. Baran Jr.
Aur 5h48'31" 39d26'9"
Kevin J. Beyrer
Lib 15h48'36" -19d53'8"
Kevin J. Culligan ~ 9-11-54
Vir 14h0'37" -13d25'13"
Kevin J. & Erin J. Ellis
Cyg 19h46'3" 33d43'41"
Kevin J Hunter (Ailsa)
Ori 5h50'17" 3d19'28"
Kevin J. Kaplan
Lac 22h19'6" 45d30'33"
Kevin J. Wert
Cas 1h43'0" 64d37'40"
Kevin Jack Whetzel
Sco 16h12'36" -10d55'55"
Kevin Jacobi
Cap 21h48'31" -22d19'32"
Kevin Jacy Weiß
Uma 10h9'30" 42d10'32"
Kevin James
Uma 9h40'7" 52d30'5"
Kevin James Brasil Caetano
Ari 3h22'15" 22d8'57"
Kevin James Dumont
Tau 3h33'30" 8d24'22"
Kevin James Farrell
Lyr 18h29'19" 36d49'5"
Kevin James Ford
Tau 3h44'27" 27d18'54"
Kevin James Harvey
Uma 11h50'44" 32d20'3"
Kevin James Jersey
Her 16h34'36" 38d18'1"
Kevin James McGuiness
Psc 1h38'47" 19d49'31"
Kevin James McGuinness
Per 4h43'33" 44d59'1"
Kevin James Michaels
Cap 21h5'32" -18d1'27"
Kevin James O'Neill
Uma 9h24'19" 54d15'17"
Kevin James O'Neill
Uma 9h38'13" 59d42'22"
Kevin James Provost
Cnc 8h15'25" 10d23'53"
Kevin James Young
Dra 18h20'54" 53d58'8"
Kevin & Jamie Forever
Umi 14h52'9" 72d34'43"
Kevin Jay Dana
Lyn 8h3'45" 41d14'0"
Kevin Jay Wolford
Leo 10h55'47" 19d51'57"
Kevin Jeffery Koeing
Sco 17h18'23" -38d7'25"
Kevin Jermaine Nichols
Leo 11h58'8" 26d38'23"
Kevin & Jessica
Ari 2h17'6" 27d16'6"
Kevin John
Her 16h39'45" 27d2'12"
Kevin John Corrigan II
Tri 2h21'34" 32d0'55"
Kevin John Fay
Per 2h33'38" 51d12'23"
Kevin John Furner
Dor 4h53'37" -68d27'46"
Kevin John McCarthy
Lib 15h44'24" -17d20'49"
Kevin John Testa
Cnc 9h5'6" 31d31'2"
Kevin John Woodcheke
Cep 22h8'3" 60d41'0"
Kevin Johnson, my love and my light
Cep 0h0'57" 82d4'13"
Kevin Jones
Her 17h53'51" 49d42'24"
Kevin Jones
Ori 5h46'32" 2d13'2"
Kevin Joseph Grassett
Uma 9h16'39" 68d47'28"
Kevin Joseph Hartranft
Dra 18h31'9" 63d26'47"
Kevin Joseph Mackey
Cap 20h40'49" -22d11'54"

Kiarna Simone Hume
Sco 16h22'22" -29d41'37"
Kiarugy
Cnc 8h51'45" 22d6'15"
Kia's Kosmo
Aqr 23h21'27" -20d55'46"
Kia's Star
Vir 14h18'52" -19d34'3"
Kiatara Mhari McAlpine
Cmi 7h35'29" 4d27'45"
Kiauna Jade
Umi 16h53'52" 77d58'13"
Kiaundra
Uma 9h39'51" 56d32'22"
Kiavash Memari
Aqr 20h47'10" 0d21'48"
Kibbe50
Per 2h44'48" 52d51'49"
Kibbie
Psc 0h46'17" 12d32'56"
Kibibi
Del 20h38'10" 17d6'2"
Kicca
Per 3h25'26" 41d33'5"
Kiccatat
Cas 0h25'15" 62d23'6"
Kickin' It Gloria
Uma 11h16'55" 36d52'20"
Kicsi Beja
Uma 9h43'53" 41d49'26"
Kid
And 23h14'17" 52d10'5"
Kidd, Anthony Bruce Christian
Ori 5h20'7" -8d32'37"
Kidder Meade
Umi 14h37'20" 74d46'38"
Kiddo
Vir 14h32'55" 6d8'49"
Kiddo and Oldman
Dra 16h39'45" 62d53'53"
kiddo and sparky's star
Psc 1h33'53" 25d17'7"
Kiddo's Star
Leo 9h24'43" 11d37'55"
Kiddy Family
Uma 11h56'29" 56d20'12"
Kidene Shanelle McCarthy
And 0h59'13" 45d47'38"
Kidgusto
Lmi 10h4'26" 39d15'43"
Kido's Star
Gem 7h2'16" 18d46'21"
Kids
Cas 1h0'25" 63d44'0"
Kidult Carole
Cas 0h57'3" 59d41'6"
Kieckhöfel, Uwe
Ori 5h7'42" 7d38'10"
Kiefel, Cordula
Ori 6h15'46" 10d40'1"
Kiefer
Uma 9h21'20" 66d27'42"
Kiefer C Eller
Psc 1h33'0" 22d48'26"
Kiefer James Christie
Gem 7h20'59" 31d56'0"
Kiefer James Welch
Her 17h21'35" 45d49'48"
Kiehn, Karl-Heinz
Uma 11h29'5" 32d6'0"
Kiehn, Thomas
Sgr 18h5'32" -27d54'37"
Kiel Brasier
Cap 20h32'36" -21d17'16"
Kieley Humrichouse Major Dworkin
Cap 20h7'23" -9d17'16"
Kieliger Renato (OTTI) 3.10.1974
Her 16h21'47" 13d46'22"
Kienast, Andreas
Uma 9h54'41" 53d45'59"
Kiep, Frank
Ori 5h19'18" -4d44'31"
Kiera
And 22h59'52" 51d22'42"
Kiera
And 23h22'24" 51d14'3"
Kiera
And 1h19'35" 47d13'12"
Kiera Amber Woolley
And 23h15'44" 41d45'20"
Kiera Ashley Ramirez
Aqr 23h7'25" -10d22'14"
Kiera Bridget Flanigan
Lib 15h44'46" -13d0'19"
Kiera Groom
And 1h43'30" 42d53'36"
Kiera James Garry
Lyn 7h53'46" 36d34'43"
Kiera Jane Hackett
Sgr 18h9'57" -31d31'13"
Kiera Lindmark
And 0h42'43" 40d34'46"
Kiera Louise Broughton
Uma 10h6'0" 41d39'28"
Kiera Louise Chitty
Sco 16h56'13" -37d37'46"
Kiera Lynn
And 2h21'6" 48d54'18"
Kiera Miglani
Gem 6h16'0" 25d14'45"
Kiera Miller
Leo 11h12'8" 21d51'4"
Kiera Nicole Chance
Cnc 8h37'38" 19d31'7"
Kiera Rose
Ari 3h16'56" 29d34'36"
Kiera Tilley
And 1h20'15" 50d36'56"
Kieran
Cnc 9h5'27" 20d30'32"
Kieran
Leo 11h16'12" 19d7'36"
Kieran
Ari 2h33'29" 12d52'8"
Kieran
Cam 4h19'9" 61d33'13"
Kieran Barr
Ori 5h27'10" -0d48'55"
Kieran Collins Burke
Leo 11h25'1" 14d32'46"
Kieran & Daniel - Bliss 06
Cru 12h5'35" -64d33'20"
Kieran David Sorenson
Vir 14h53'15" 3d18'56"
Kieran J. Brothers
Lac 22h54'35" 51d37'59"
Kieran James
Per 3h45'40" 48d54'37"
Kieran Joseph Kissane - 23/8/01
Cep 20h40'19" 55d21'51"
Kieran K
Her 18h48'7" 15d18'40"
Kieran Mc Carey
Cep 23h24'8" 83d0'53"
Kieran Michael Taylor
Cap 21h30'27" -20d23'59"
Kieran Nevton Dunn
Crb 16h23'38" 39d9'48"
Kieran O'Brien
Umi 15h24'37" 70d55'40"
Kieran Peter Lee
Cep 22h28'12" 63d47'52"
Kieran Quinn
Umi 14h40'37" 71d2'41"
Kieran Sequoia
Peg 22h3'56" 6d53'55"
Kieran Sleeth
Cra 18h53'54" -41d2'12"
Kieran Thomas Beecham
Col 6h9'18" -32d59'34"
Kieran Wood
Cep 21h19'32" 62d37'14"
Kiera-R-Nibiru-3750-60-AD-C-L
Cnc 7h57'12" 15d23'20"
Kiera's Angel in the Sky
And 2h32'57" 41d1'10"
Kiera's Star
Cam 6h15'36" 84d5'34"
Kierenveer
Gem 7h10'9" 14d58'44"
Kierin Allen Mcfall Perry
Oph 17h11'52" -0d4'10"
Kierk
Ari 2h34'55" 25d24'52"
Kiernan Sullivan Rushford
Vir 14h31'14" 5d57'11"
Kieron Diffenderfer
Del 20h39'51" 14d59'41"
Kieron & Karen O'Sullivan
Uma 14h6'5" 53d44'13"
Kieron Stone
Cep 21h4'56" 57d34'44"
Kierra and Fabio love you babe
Sco 16h10'40" -9d29'41"
Kierra Rose Williams
Sco 17h23'16" -31d10'37"
Kiersa
Mon 6h43'37" 7d16'53"
Kiersten
And 23h26'14" 47d0'1"
Kiersten Andréa
Uma 10h29'3" 58d44'37"
Kiersten E. Page
Gem 6h30'1" 12d46'1"
Kiersten I Albers
Lyn 7h16'22" 49d36'6"
Kiersten "K-10" Lipkin
Leo 10h46'20" 14d46'52"
Kiersten Loren
Tau 5h50'24" 25d56'56"
Kiersten Petesch
Psc 1h25'37" 28d26'59"
Kiersten Rae Folkman
Lib 15h36'21" -26d1'12"
Kierstin Anne Barnett
Sgr 19h27'46" -14d24'49"
Kierstin Trebour - Nini's Star
And 0h18'44" 29d15'57"
Kieryndrea
Leo 11h42'21" 24d17'2"
Kiesha D. Gilmer
Ori 5h24'10" 2d20'19"
Kiesha Officer
Tau 4h17'8" 30d9'48"
Kiet
Tra 16h33'37" -60d45'50"
Kiet Do
Cyg 21h28'17" 38d43'24"
kieth's star
Ori 6h13'51" 20d57'12"
Kiette Leonard
Umi 14h36'43" 79d24'17"
Kihci Awasis
Sco 17h58'5" -37d21'25"
Kii Jennafer Graham
Sco 16h8'26" -29d41'32"
Kijkeow
Lyn 8h22'40" 51d16'39"
Kijoo Nam
Vir 14h8'52" 3d25'38"
Kik And Kak happy forever
Uma 12h7'23" 51d55'26"
Kika Deniell DeGruchy
Gem 6h42'5" 12d44'6"
Kike Pedro's Star
Sge 20h9'8" 17d11'46"
Kiker
And 23h0'54" 48d19'25"
Kiki
Cyg 21h49'29" 46d34'0"
Kiki
Lyn 6h56'34" 51d32'32"
Kiki
And 23h30'49" 38d36'39"
Ki-Ki
Ari 2h50'26" 30d26'29"
Kiki
Psc 1h1'48" 9d28'50"
KiKi
Cnc 8h11'20" 15d33'52"
Kiki
Cnc 9h1'58" 17d53'28"
Kiki
Leo 9h36'30" 27d23'5"
Kiki
Tau 3h52'43" 26d6'21"
Kiki
Lyn 7h7'57" 59d6'54"
"Kiki"
Cam 5h6'54" 63d31'41"
KiKi Fornuto
Lib 14h33'24" -18d30'19"
Kiki Ilic
Lib 15h51'0" -12d40'9"
Kiki Louise Killingback
And 0h13'45" 47d24'46"
Kiki Marinakis
Lib 14h55'43" -1d2'6"
KikiBen
Leo 10h54'28" 10d38'3"
KikiLoveQuincyCole
Gem 6h20'24" 18d33'12"
Kikipania Keo's Wehilani
Del 20h17'31" 16d7'53"
Kiki's Sweet 16 Cheer Star
Cnc 8h14'15" 20d13'45"
Kikitini
Lyn 8h35'58" 44d40'39"
Kikka 30
Sco 17h16'23" -43d42'6"
Kikkola
Aur 6h45'0" 35d28'8"
Kiko
Vir 13h55'14" 5d27'33"
Kikou 01
Cas 3h28'37" 69d56'54"
Kikue Motohara
And 2h23'7" 46d7'2"
Kila Ciss Fagan McDermott
Peg 22h47'16" 25d18'51"
Kila Massett
Peg 22h29'4" 24d40'48"
Kilana M. Gaddis (Kittie Moon)
Tau 4h32'53" 18d44'51"
Kildastar
And 2h13'10" 41d26'1"
Kilee Marie
Aqr 23h30'55" -20d13'5"
Kiley
Aqr 22h2'16" -2d25'50"
Kiley
Cam 4h18'31" 78d26'40"
Kiley A Bueler
Umi 14h49'41" 72d56'10"
Kiley Ann Pollard
Sco 17h36'6" -44d34'12"
Kiley Blake
And 0h18'9" 30d36'10"
Kiley Brynn
Cyg 21h15'33" 38d25'4"
Kiley Doane
Cam 4h46'50" 64d36'25"
Kiley Katherine Southall
Crb 16h8'28" 37d22'20"
Kiley Louise Simonof
Tau 5h9'32" 19d51'47"
Kiley Marie Sutton
And 23h19'46" 41d58'2"
Kiley Munger
And 1h52'30" 41d35'31"
Kiley Noel Preusch
Dra 17h11'38" 61d0'12"
Kiley Pearson
Sco 16h48'12" -30d56'37"
Kiley Perry
Ari 2h10'37" 26d6'8"
Kiley Rae
Vir 13h54'17" -19d49'26"
Kiley Rea Whetzel
Ari 3h8'4" 28d40'47"
Kiley Rose
And 0h30'13" 32d3'42"
Kilgore, Jr. Family Star
And 0h3'45" 45d21'0"
Kilian
Uma 11h24'15" 38d32'49"
Kilian Ronan Weber
Peg 23h3'27" 26d20'44"
Kilian, Klaus
Ori 6h17'9" 19d10'21"
Kilikina
Cnc 8h1'53" 20d30'53"
Kilikina
And 2h20'28" 50d36'51"
Kilikina
Lyn 8h26'29" 55d45'42"
Kilinc, Aslan
Ori 6h21'48" 16d40'33"
Kill Pixie
Cru 12h19'39" -57d40'3"
Killa Jo
Cyg 20h30'33" 47d32'8"
Killah Des
Psc 2h3'21" 5d51'55"
Killer Star of Dave
Uma 9h34'2" 67d37'45"
Killey-Rich 1999
Cas 0h10'23" 56d19'8"
Killian Andrew Smith
Her 18h29'56" 16d35'6"
Killian Birusingh Durick
Psc 1h8'3" 21d1'53"
Killian Gibson
Cnc 8h51'55" 31d32'36"
Killian James Soltero Walker
Tau 5h15'5" 19d4'52"
Killian Lee
Sco 16h15'4" -12d43'59"
Kilowac
Lib 15h37'51" -28d59'41"
Kilrush
Umi 13h10'59" 74d27'4"
Kiltman
Gem 6h23'39" 20d53'52"
Kim
Cnc 8h38'9" 16d34'28"
Kim
And 0h25'57" 33d29'58"
Kim
Cnc 8h45'54" 9d41'39"
Kim
Cnc 8h27'55" 7d27'36"
Kim
Hya 9h22'49" 5d18'27"
Kim
Ori 5h23'35" 7d34'30"
Kim
Lyn 8h26'16" 37d14'51"
Kim
And 0h49'0" 34d20'6"
Kim
And 2h37'48" 44d15'10"
Kim
And 1h44'37" 42d31'5"
Kim
And 0h42'47" 39d0'31"
Kim
And 23h14'38" 47d4'26"
Kim
Cyg 20h53'5" 46d47'9"
Kim
Lyn 8h19'17" 48d29'0"
Kim
Uma 8h47'56" 69d41'37"
Kim
Uma 12h8'2" 60d18'23"
KIM
Uma 11h35'39" 52d31'39"
Kim
Uma 13h16'30" 56d10'11"
Kim
Cas 0h25'15" 61d3'16"
Kim
Aqr 20h45'0" -11d24'6"
Kim
Cap 21h36'45" -10d3'8"
Kim
Aqr 22h22'48" -18d4'18"
Kim
Cap 21h0'28" -18d31'36"
Kim
Sgr 19h10'8" -17d0'5"
Kim A. Jordan
Gem 7h30'49" 30d52'46"
Kim A. Markland
Cas 0h58'39" 60d47'13"
KIM AKA SLF
Cap 21h53'9" -18d19'14"
Kim Alexandra
Vul 21h5'14" 27d40'27"
Kim Always and Forever
Uma 11h12'47" 34d32'24"
Kim & Amanda's Comet! Year of 05
And 1h5'35" 36d29'11"
Kim Amaya Eastwood
Aql 19h10'7" 2d7'27"
Kim and Dan Peat's Star
Ori 5h20'27" -8d43'40"
Kim and David
Ari 2h48'12" 22d8'30"
Kim and James Herrmann
Leo 11h22'2" 17d0'22"
Kim and Kerstin
And 23h46'54" 38d48'9"
Kim and Kris 21-10-2006
Cru 12h23'2" -59d1'35"
Kim and Ralph
Her 17h59'53" 49d2'19"
Kim Andrea Davison
Lib 14h48'56" -12d47'26"
Kim Anh Oanh Tran
Psc 0h5'14" 6d46'18"
Kim Ann Kross
And 1h26'5" 44d43'26"
Kim Annette
Eri 4h25'17" -0d57'22"
Kim Arland Bancroft
Per 3h21'56" 41d54'21"
Kim & Armandos Star
Cyg 21h17'7" 42d28'2"
Kim Ayers
Sco 17h35'34" -33d20'54"
Kim Baker
Mon 7h1'41" -9d11'21"
Kim Balistreri
Leo 9h26'57" 28d54'51"
Kim Balthaz
Psc 23h40'3" 6d55'35"
Kim Beaulieu
Mon 7h7'48" -0d31'36"
Kim Becker
Ari 2h33'36" 23d39'39"
Kim Belly
Leo 11h21'15" 15d35'33"
Kim Birch
Sco 16h23'46" -28d9'8"
Kim Boharsik
Crb 15h54'31" 28d56'51"
Kim Bosse
Sco 16h20'14" -10d37'43"
Kim Bowman, because your you.
And 0h59'15" 35d21'2"
Kim & Brandon's Crabdipper
Vir 14h12'41" -6d24'17"
Kim Buchanan Tilley
Uma 13h56'34" 55d32'45"
Kim Butler
And 0h22'47" 34d10'28"
Kim Byrne
And 1h40'44" 49d12'45"
Kim Campbell
Aql 19h9'19" 5d5'33"
Kim Canepa
And 23h48'43" 45d2'37"
Kim Cansler, CMP-GaMPI Shining Star
Lmi 10h25'1" 37d55'9"
Kim Carney
Cas 23h53'51" 53d38'4"
Kim Chapman
Psc 0h44'1" 9d23'8"
Kim Charlene
Ori 6h10'33" 15d16'18"
Kim Chi
Psc 0h27'44" 17d37'20"
Kim Christina Bulloch
Aqr 22h58'48" -8d20'37"
Kim Christy
Ari 2h2'57" 21d50'27"
Kim Colegrove
Crb 15h34'50" 26d51'16"
Kim Colligan
Cam 5h16'46" 61d46'46"
Kim Collins
Cas 1h15'3" 59d1'38"
Kim Collis
Aql 19h51'59" -0d17'5"
KIM CONNELL
And 23h25'5" 43d14'21"
Kim Cook
Cru 12h28'30" -61d10'9"
Kim Craig Katzer
Uma 10h9'39" 69d19'27"
Kim Crowell
Mon 6h32'18" 5d36'25"
Kim Curry
Uma 11h43'20" 44d12'26"
Kim & Cyril Lesniak
And 0h19'44" 25d36'36"
Kim Daly
And 2h28'6" 49d41'47"
Kim Dangerous
Ser 15h45'27" 21d55'43"
Kim Darnell 102756
Cas 1h24'33" 53d35'12"
Kim & Darren's First Anniversary
Her 17h24'55" 43d22'48"
Kim Denise Kedeshian
Eri 3h41'41" -8d56'23"
Kim DeSerio
Uma 11h22'3" 29d9'24"
Kim E. Perron
And 23h59'44" 42d26'50"
Kim E. Yandow
Cas 0h12'5" 51d38'38"
Kim Eggimann
Cas 1h58'13" 60d37'0"
Kim EJ & Kwon YJ
Cyg 19h34'40" 53d38'40"
Kim Elaine Craig
Cas 23h49'46" 53d10'43"
Kim Elizabeth Wright-Arnold
Cyg 21h33'21" 47d22'15"
Kim Ellis
Uma 9h5'23" 54d13'48"
Kim Eloise
Cru 12h43'4" -64d29'5"
Kim Elworthy
Com 12h32'58" 26d10'0"
Kim Eng Hendrix
Leo 11h16'6" 7d6'10"
Kim & Eric a.k.a. Spanky & krunK
And 0h19'32" 31d55'47"
Kim Erkan
Psc 0h54'52" 29d18'37"
Kim Ewing
Cam 5h16'46" 57d58'8"
Kim Finnerty
Uma 9h56'50" 53d56'59"
Kim Fitzgerald-Lindley
Cap 21h8'55" -26d45'27"
Kim Franklin
And 23h34'30" 41d25'42"
Kim Fumiko Manago
Ori 5h30'16" -4d52'59"
Kim Furzer
Sco 16h12'56" -11d17'13"
Kim Gahlla
Aql 19h13'54" 7d44'36"
Kim Gardner
Aqr 21h14'25" -11d12'50"
Kim George
Peg 23h11'31" 26d41'3"
Kim Gilman
Gem 7h6'47" 31d35'58"
Kim Golightly
Her 18h40'58" 19d29'5"
Kim Griffiths
Cnc 8h37'46" 22d23'37"
Kim H. Gadjo
Uma 8h24'54" 62d21'10"
Kim Haney
Crb 16h7'44" 30d44'38"
Kim Haney
Uma 10h12'45" 45d51'19"
Kim Hanson Prince
Crb 16h10'5" 37d12'8"
Kim Harmon Settle
Sco 17h5'38" -37d18'11"
Kim (Hiner) Miller
Uma 10h55'18" 52d49'16"
Kim Hyo Hyung
Cyg 19h39'35" 47d19'5"
Kim Hyun Taek
Dra 18h25'16" 76d20'2"
Kim Irene
Cap 21h16'23" -16d34'1"
Kim J. Sveska
Uma 10h33'46" 56d18'54"
Kim "Jade Wolf"
Tau 4h33'35" 12d41'8"
Kim & James
Sco 17h47'41" -37d12'45"
Kim Jean Bordelon
Peg 23h8'47" 17d3'49"
Kim Jennings
Cas 0h40'43" 48d48'30"
Kim Jenny Do
Leo 10h28'1" 13d44'41"
Kim & Jim
Cyg 20h27'57" 57d28'41"
Kim Jones
Umi 16h27'54" 77d29'4"
Kim Joy Hughes
Crb 16h16'18" 38d15'28"
Kim K 1
Lyn 6h52'26" 50d27'10"
Kim "Kiki" Renick
Uma 13h10'41" 53d45'12"
Kim Kregloski
Uma 10h58'34" 50d49'35"
Kim la sirène
Del 20h46'20" 5d17'47"
Kim Lanae
Cyg 21h32'49" 48d58'34"
Kim Landreth
Lyn 9h21'49" 33d12'22"
Kim Le
Cap 21h46'17" -8d29'0"
Kim Lemmer
Tau 5h7'20" 17d57'19"
Kim Levin
Gem 6h46'57" 24d57'45"
Kim Lights Up the Sky
Tau 3h55'41" 25d34'8"
Kim Littwin
Cnc 8h23'42" 18d2'31"
Kim Loan Thi
And 2h31'26" 47d34'57"
Kim Loan Tran
And 0h33'21" 27d36'53"
Kim Lorraine Dunham Sequoia
Lib 15h9'27" -5d4'2"
Kim Louise Moy - Always & Forever
Cru 12h25'6" -62d3'8"
Kim Loveridge
Cas 0h8'45" 57d1'34"
Kim Loves Willard C Upchurch STF
Uma 11h21'37" 34d38'16"
Kim Maceira-Brown
Com 12h33'36" 29d11'19"
Kim Marchal
Peg 21h32'46" 23d34'8"
Kim Maree Stehbens
Col 5h59'37" -38d0'4"
Kim Margaret Pope
Sgr 19h27'20" -43d21'19"
Kim Maria Baker
Mon 6h53'18" -0d4'50"
Kim Marie
Lib 15h5'21" -6d9'59"
Kim Marie
Lyn 6h45'47" 54d48'57"
Kim Marie Ciarfella
Ari 2h0'21" 14d24'56"
Kim Marie Cole
Cam 5h40'48" 68d22'6"
Kim Marie Corder
Vir 13h18'39" 12d52'32"
Kim Marie Dilts
Ari 2h39'17" 12d54'23"
Kim Marie Hartsen
Sco 16h8'49" -9d57'21"
Kim Marie Kefalas
And 2h1'18" 45d0'3"
Kim Marie Tassinari
Ari 2h3'37" 27d29'58"
Kim Marie Zacher
Uma 8h28'1" 61d30'4"
Kim Marlow
Ori 6h14'48" 12d31'46"
Kim Martin
Psc 1h7'6" 9d52'27"
Kim Martin
Leo 9h52'26" 21d14'10"
Kim Mason
Sco 17h43'4" -37d41'38"
Kim McKerrell - Padushka
Leo 11h14'55" -3d37'33"
Kim McSwiggan
Leo 10h6'35" 24d50'53"
Kim Mellen
Cyg 19h58'24" 56d32'3"
Kim Morgan
Uma 9h43'29" 47d24'0"
Kim Morris
Per 3h14'48" 42d13'22"
Kim Morrissette - Love of My Life
And 0h45'16" 29d10'37"
Kim Moses Chase
Per 4h45'28" 46d43'3"
Kim Murray
Lib 15h3'50" -6d16'50"
Kim & Myckie
Lyn 7h3'58" 53d56'48"
Kim Nguyen
Lib 15h54'15" -10d3'3"
Kim Nobert - Baby Girl
Gem 7h35'37" 26d1'47"
Kim Noemi
Ori 6h13'12" 10d30'21"
Kim Norris
Lyn 8h9'49" 46d30'6"
Kim O'Brien
Uma 13h0'17" 58d0'13"
Kim Osterberg
Psc 0h44'21" 6d50'33"
Kim Owens Tilman
Aqr 22h47'32" -8d46'18"
Kim Paletta
Aqr 23h23'49" -14d33'52"
Kim Parker
Cnc 9h5'6" 27d54'44"
Kim Patricia Stierli
Cam 6h0'59" 56d28'3"
Kim Pepe, Our Shining Star
Cap 21h31'5" -20d43'42"
Kim Perkins
Dra 14h46'50" 59d40'27"
Kim Peterson
Uma 11h30'28" 63d0'17"
Kim & PJ Forever
Cyg 19h36'13" 30d14'16"
Kim "Princess" Reinstra
Lyn 8h29'49" 56d29'54"
Kim Purdy
Cra 19h5'42" -38d58'3"
Kim Quimby
Uma 9h33'4" 47d34'17"
Kim Rafaela von Mühlenen
Lac 22h42'2" 52d49'20"
Kim Riberal
Cap 21h35'34" -9d14'33"
Kim Riley
Cyg 19h40'50" 38d26'37"
Kim Roach
Uma 11h27'34" 60d43'48"
Kim Robin
Umi 16h21'43" 84d0'51"
Kim Robison
Aqr 22h43'18" -9d10'32"
Kim Robison
Aqr 22h46'49" -9d54'24"
Kim Rome
Her 16h39'29" 25d9'15"
Kim Rose
And 0h37'48" 33d12'47"
Kim Rose
Aql 19h6'14" -0d52'50"
Kim Rummel
Cap 20h22'29" -12d49'15"
Kim S. Rawlings
Ari 2h22'54" 19d32'36"
Kim Sachiyo Anderson
Cma 7h7'20" -27d19'34"
Kim Sarah
Per 4h11'12" 52d20'34"
Kim Sarina Ramsauer
Cyg 21h46'16" 42d20'20"
Kim Saurer
Lyr 18h31'41" 31d52'49"

Kim Schoenholtz & Elvira Pratsch
And 23h7'19" 39d57'27"
Kim & Seaby T.L.A.
Uma 12h12'29" 55d10'37"
Kim Setnicky
Leo 11h56'4" 22d58'52"
Kim (Snugglebug) Hall
Peg 22h51'59" 20d45'0"
Kim & Steve
Cyg 19h52'1" 31d23'30"
Kim & Steve
Cap 21h39'38" -17d38'25"
Kim Swallow
Peg 22h18'59" 7d11'8"
Kim Talay
Vir 15h2'58" 4d11'30"
Kim The Dolphin
Del 20h42'22" 15d23'6"
Kim the Tattooed Crafter
Aqr 23h0'10" -5d26'2"
Kim Theodora
Gem 7h13'27" 26d52'47"
Kim Theresa Benson
And 1h59'38" 46d22'24"
Kim Thuy
Cam 7h56'32" 69d59'21"
Kim Tomlinson 381
Cyg 19h37'32" 31d5'1"
Kim & Tony Fabela
Lib 15h2'49" -2d44'50"
Kim Towns
Cas 0h53'42" 59d14'1"
Kim Tracy
Ari 2h43'44" 25d37'20"
Kim Tschumper
Cap 20h34'40" -10d1'18"
Kim Tuyen
Lyn 8h4'53" 35d31'39"
Kim Tyler
Uma 11h33'45" 32d2'9"
Kim und Elina
Cyg 21h17'50" 30d3'28"
Kim Vahey
Vir 13h57'13" -22d9'20"
Kim Van Geluwe Bakane
Uma 9h43'23" 67d19'28"
Kim Verna
And 0h33'49" 40d19'13"
Kim Vicary "Hey Y'all"
And 23h22'12" 51d19'40"
Kim Walsh
Dra 17h34'50" 67d22'27"
Kim Walsh
Cru 12h49'32" -58d19'56"
Kim White
Cas 1h10'15" 70d3'13"
Kim Williams
Lyn 7h25'56" 52d49'42"
Kim Yoko Kreamer
Vir 13h25'30" -6d46'37"
Kim Yoonjin
Per 4h26'48" 40d18'40"
Kim Yu Mee
Ari 1h50'16" 24d7'37"
Kim Yusun
Cnc 9h3'12" 15d18'39"
Kim Zimmerman
And 23h11'59" 48d3'58"
Kim, the TWINKLE in my eyes !!!
Lyn 8h17'35" 59d8'7"
Kim,William and James
Her 17h38'32" 32d29'44"
Kimako, Ed.D
Cap 21h4'16" -14d58'50"
KIMANY
Uma 10h16'25" 66d1'47"
Kimaquarius67
Aqr 21h53'9" -7d58'23"
KIMARIE
Psc 23h13'22" 3d35'29"
Kima-To
Mon 6h40'36" 6d25'2"
Kimba
Ari 3h16'36" 16d2'5"
Kimba
Uma 10h35'58" 45d10'37"
KIMBA
Lib 15h48'0" -18d11'34"
KIMBA 81
Uma 8h30'56" 65d19'33"
Kimbalina
Uma 9h2'36" 62d14'36"
Kimball M. Crofts "Nsoromma"
Cep 21h54'36" 66d37'2"
Kimber
Tau 5h51'13" 16d53'56"
kimber
Vir 13h41'3" 2d47'3"
Kimber
Her 18h9'56" 18d9'55"
Kimber 16
And 0h15'1" 30d40'41"
Kimber and Brian
Cam 4h9'57" 67d41'24"
Kimber Ann Fox
And 2h22'6" 48d43'27"
Kimber Deanne Muse
Cyg 19h50'35" 35d43'24"
Kimber Fewell
Del 20h36'35" 15d7'5"
Kimber Jean Griffin
And 0h17'19" 31d24'31"
Kimber Sue Frances Christie
Ari 2h6'42" 18d23'2"
Kimberella
Ari 3h12'30" 22d38'59"
Kimberely "Heaven" Fredericksen
Dra 16h23'23" 62d29'54"
Kimberlee
Psc 0h2'8" 7d10'23"
Kimberlee
Lib 15h8'51" -24d17'34"
Kimberlee
Leo 11h21'2" 11d39'8"
Kimberlee Anne McKee
Ari 2h43'43" 25d23'44"
Kimberlee Anne Pericoli
Ari 2h18'46" 27d10'25"
Kimberlee Cristina Avila Libricz
Vir 14h10'27" -0d20'51"
Kimberlee Joy
And 0h42'21" 37d44'28"
Kimberlee Lurana Cowan
And 1h6'59" 36d20'49"
Kimberlee Michele Gaylord
Ori 5h52'56" 21d5'56"
Kimberlee Renee
And 23h9'34" 35d13'15"
Kimberlee the Singer
Lyr 18h46'4" 41d1'41"
Kimberlei
Mon 6h32'8" 11d35'56"
Kimberlei Key
Crb 15h46'37" 32d49'19"
Kimberleigh Ann Johnson
Sco 16h13'17" -16d52'25"
Kimberley
Lib 15h40'21" -19d57'15"
Kimberley
Cas 0h23'24" 61d52'47"
Kimberley
Cyg 21h14'36" 47d6'47"
Kimberley
Vir 13h17'45" 6d11'10"
Kimberley Ann
Vir 13h2'41" -12d56'39"
Kimberley Ann Iaccarino
Dra 15h52'41" 56d4'20"
Kimberley Anne
Cas 0h55'18" 57d13'49"
Kimberley Bosso
Sco 17h56'31" -32d7'40"
Kimberley Bromberg
And 0h32'34" 45d16'10"
Kimberley Elizabeth Maggs
Cas 23h35'51" 54d35'10"
Kimberley Emmett 21
Cyg 21h48'47" 33d52'41"
Kimberley Ford
Cap 20h16'15" -10d0'26"
Kimberley Hawling
Cas 23h25'55" 52d10'43"
Kimberley Jane Lewis
Lib 15h25'58" -25d2'29"
Kimberley Jansen - Kim's Star
And 0h45'20" 30d45'54"
Kimberley Jo Coulter
Cnc 8h47'27" 15d33'15"
Kimberley Kaye
Tau 4h28'33" 3d54'44"
Kimberley Louise
And 23h13'47" 47d48'5"
Kimberley Marie
Ari 2h12'4" 22d23'48"
Kimberley N. Escue
Uma 11h27'46" 34d15'1"
Kimberley - Our Princess
And 2h19'20" 41d56'32"
Kimberley P. Richmond
Uma 8h51'24" 54d43'5"
Kimberley Pearl Oswick
Uma 11h27'41" 60d18'31"
Kimberley Rose
Sco 16h15'11" -29d31'41"
Kimberley Ruth Gayed
Cas 1h57'36" 66d35'34"
Kimberley Stockton
Cas 23h17'53" 55d55'51"
Kimberley Timlock
Lyn 6h35'24" 59d1'24"
Kimberley Van Houten
Ari 3h16'52" 20d25'7"
Kimberley Warnett
And 1h29'31" 34d59'47"
Kimberley Westfall
Aqr 22h52'33" -11d57'52"
Kimberley-Chadney-Paige
Lib 14h57'9" -24d56'32"
Kimberley-Rose
Cru 12h0'15" -60d36'19"
Kimberley's Shining Star
Gem 7h10'29" 22d1'34"
Kimberli
Mon 6h29'53" -1d35'45"
Kimberli DeAnna
Cap 20h38'5" -10d51'33"
Kimberli Erin Cress
Crb 16h9'16" 29d56'41"
Kimberli Marie
Cyg 21h49'20" 47d8'53"
Kimberli Smiech
Uma 13h54'33" 55d46'31"
Kimberlie D. Cobb
Leo 9h58'21" 16d31'12"
Kimberlun Noble
Lyn 8h24'44" 38d39'8"
Kimberly
Lyn 7h40'48" 38d1'6"
Kimberly
And 2h37'26" 43d25'11"
Kimberly
Lyn 8h48'22" 40d12'31"
Kimberly
And 0h35'43" 39d7'25"
Kimberly
And 0h47'12" 42d44'14"
Kimberly
And 2h15'57" 38d7'49"
Kimberly
And 1h48'4" 38d21'13"
Kimberly
And 1h47'59" 39d4'39"
Kimberly
Tri 2h22'54" 34d0'0"
Kimberly
Lyn 7h55'35" 36d2'39"
Kimberly
Cnc 8h51'57" 32d23'57"
kimberly
Uma 11h34'49" 35d16'28"
Kimberly
Lac 22h54'51" 46d27'58"
Kimberly
Uma 10h44'31" 48d27'54"
Kimberly
And 0h51'28" 45d50'20"
Kimberly
And 23h15'43" 40d58'37"
Kimberly
And 1h33'7" 45d8'30"
Kimberly
Peg 21h46'6" 21d40'18"
Kimberly
And 0h13'43" 33d33'32"
Kimberly
Lyr 19h19'55" 29d22'41"
Kimberly
Leo 9h33'58" 25d27'8"
Kimberly
Lmi 10h47'41" 26d5'39"
Kimberly
Gem 7h33'59" 22d35'57"
KIMBERLY
Gem 6h11'20" 23d13'42"
Kimberly
Tau 5h54'11" 25d0'41"
Kimberly
Tau 4h30'42" 17d28'14"
Kimberly
Leo 10h41'49" 8d52'55"
Kimberly
Vir 12h7'53" 12d36'47"
Kimberly
Del 20h30'23" 7d23'2"
Kimberly
Aqr 21h10'22" 1d28'50"
Kimberly
Cas 23h37'3" 57d2'3"
Kimberly
Cas 1h51'8" 65d2'26"
Kimberly
Sco 16h11'7" -13d10'54"
Kimberly
Lib 14h48'57" -10d8'30"
Kimberly
Sgr 19h11'34" -18d34'20"
Kimberly
Sgr 19h9'2" -21d2'13"
Kimberly
Psc 0h0'51" 4d9'48"
Kimberly
Sgr 18h4'4" -28d51'59"
Kimberly 1
Sco 16h54'54" -38d25'17"
Kimberly A. Boyko
Lyn 7h31'6" 57d56'22"
Kimberly A. Cushing
And 1h21'0" 40d15'49"
Kimberly A. DeStefano
Cas 23h1'59" 58d15'24"
Kimberly A. Ellenberger
Lib 14h33'54" -10d43'28"
Kimberly A. Hamilton
Lib 15h26'1" -10d27'3"
Kimberly A. Haynes
Cas 1h31'44" 65d35'39"
Kimberly A. Jones
Cas 23h46'26" 55d45'16"
Kimberly A. Meglio
Uma 8h43'27" 47d48'44"
Kimberly A. Sondy
Aql 19h17'48" 7d38'53"
Kimberly A. Walls
Uma 12h15'55" 58d4'52"
Kimberly A. Welch
Uma 10h0'7" 52d26'4"
Kimberly Alexis
Tau 5h35'23" 23d58'13"
Kimberly Alida Roden
Ori 5h37'36" 12d15'53"
Kimberly Allison Lajoie
Cap 21h17'6" -15d18'36"
Kimberly Alyssa Harris
And 1h18'18" 42d11'6"
Kimberly and Michael Forever
Uma 9h40'1" 54d49'44"
KIMBERLY AND MICHAEL HESS
Lib 15h23'18" -22d3'22"
Kimberly and Stevie
Leo 11h38'47" 18d2'4"
Kimberly and Westley
Lib 15h54'7" -20d7'56"
Kimberly Angelina Caccese
Dra 18h26'6" 77d7'25"
Kimberly Ann
Cas 1h31'34" 62d9'33"
Kimberly Ann
Peg 22h38'6" 21d0'15"
Kimberly Ann
Ari 2h30'14" 25d32'8"
Kimberly Ann
Lyr 19h10'24" 26d17'2"
Kimberly Ann
And 1h40'32" 42d24'56"
Kimberly Ann Amerson
Mon 7h9'7" -0d35'50"
Kimberly Ann Bailey
Vir 14h8'42" -20d24'55"
Kimberly Ann Briscoe
And 23h23'42" 47d44'15"
Kimberly Ann Broderick
Crb 15h30'32" 29d35'5"
Kimberly Ann Broten
Umi 14h47'37" 73d45'47"
kimberly ann bundy
Sgr 19h56'20" -18d20'58"
Kimberly Ann Butler
Cap 20h10'19" -19d37'29"
Kimberly Ann Conroy
Per 3h14'2" 44d35'18"
Kimberly Ann Costa
Psc 1h8'36" 10d5'18"
Kimberly Ann Ebent
Sco 17h50'18" -39d10'2"
Kimberly Ann Fry
And 0h41'29" 26d42'12"
Kimberly Ann Greene
Uma 11h56'28" 62d35'2"
Kimberly Ann Harris
Cam 4h30'16" 57d24'17"
Kimberly Ann Junker
Aqr 21h8'44" -0d52'42"
Kimberly Ann Kilroy Procopio
Eri 4h30'54" -0d28'28"
Kimberly Ann - L41
Sgr 18h34'33" -22d3'28"
Kimberly Ann Lawyer
Lib 15h13'48" -9d28'6"
Kimberly Ann Lindsay Nicole
Cam 4h40'21" 59d37'3"
Kimberly Ann Means Rosenbaum
Ari 2h24'24" 17d37'4"
Kimberly Ann Miller
Vir 14h43'51" -0d51'23"
Kimberly Ann Montesano
Cap 21h35'12" -8d44'9"
Kimberly Ann Morse
Cas 0h14'47" 53d15'57"
Kimberly Ann Norwood
Crb 16h15'17" 35d54'21"
Kimberly Ann Picard
Sco 16h44'18" -29d20'10"
Kimberly Ann Prazenica
Peg 22h32'56" 30d42'34"
Kimberly Ann Pupa 102271
Sco 16h16'41" -9d47'1"
Kimberly Ann Shafer
Aqr 22h0'50" -16d42'46"
Kimberly Ann Shakinis
Tau 3h49'35" 13d54'55"
Kimberly Ann Smith
Gem 7h48'26" 27d0'38"
Kimberly Ann Smith
And 2h2'41" 42d30'18"
Kimberly Ann Szuma
Ari 2h24'58" 18d46'10"
Kimberly Ann Troyer
Lyn 8h57'52" 46d25'32"
Kimberly Ann Young
Vir 13h9'33" 7d8'29"
Kimberly Anna
Tau 5h37'36" 26d10'1"
Kimberly Anne
Leo 11h46'10" 24d50'4"
Kimberly Anne
Cnc 9h12'48" 10d8'33"
Kimberly Anne
Her 17h9'12" 45d23'1"
Kimberly Anne
Lib 15h9'29" -16d4'53"
Kimberly Anne
Umi 14h30'47" 78d59'37"
Kimberly Anne
Lyn 6h35'41" 61d20'30"
Kimberly Anne
Cap 21h26'21" -23d2'0"
Kimberly Anne Borden
Sgr 18h44'4" -28d47'56"
Kimberly Anne Copeland
Crb 15h18'35" 30d40'49"
Kimberly Anne DaRosa
Lyn 7h41'2" 48d34'5"
Kimberly Anne Figueroa
And 23h28'52" 47d57'54"
Kimberly Anne Loiselle
Cap 20h25'56" -27d27'27"
Kimberly Anne Osborne
Crb 15h45'11" 35d51'37"
Kimberly Anne Tardalo
Lib 14h34'49" -14d54'23"
Kimberly Anne Usher
Cnc 9h11'7" 17d11'35"
Kimberly Anne Zebrowski
Cnc 8h30'12" 28d56'9"
Kimberly Annette
Cas 1h0'22" 56d55'51"
Kimberly Ann's Star
Cam 4h32'52" 59d36'36"
Kimberly Arlene
Lib 15h33'35" -17d59'29"
Kimberly Arnold
Uma 11h27'14" 60d9'29"
Kimberly Asher Ericson
And 1h42'46" 37d52'37"
Kimberly Ashley
Cru 12h48'56" -57d15'27"
Kimberly Aultman Day
Uma 11h32'37" 33d0'44"
Kimberly Austin
Aqr 22h22'9" 1d50'23"
Kimberly Ayers
Lyn 6h58'45" 50d58'47"
Kimberly B.
Cas 1h44'22" 61d51'44"
Kimberly Bain
Psc 1h41'55" 5d1'59"
Kimberly Baker Alten
Per 4h6'25" 41d52'17"
Kimberly Barbers Eyes
Cma 6h44'30" -14d35'3"
Kimberly Baum
Uma 11h49'7" 44d27'49"
Kimberly Bea Edwards-Lewis
Cas 0h32'19" 62d44'10"
Kimberly Beeden
Leo 10h47'23" 19d53'28"
Kimberly Beth
Gem 6h43'44" 28d21'51"
Kimberly Beth Jernagan
Lmi 10h7'42" 39d28'49"
Kimberly Beth Kramer Bridges
Ari 1h57'9" 22d32'34"
Kimberly Bevers
And 0h20'59" 29d48'45"
Kimberly Bianchi
Vir 13h18'2" -10d49'37"
Kimberly Bolte
Cap 21h11'52" -19d54'6"
Kimberly Boss
And 2h26'6" 49d47'25"
Kimberly Boydston
Psc 1h6'3" 14d51'41"
Kimberly Brimberry
Cas 1h24'54" 63d44'49"
Kimberly Broadway
Cnc 9h5'45" 32d42'11"
Kimberly Brock Armacost
Sgr 19h12'17" -14d21'48"
kimberly bryant
Leo 9h27'15" 26d9'19"
Kimberly Bryant
And 0h21'19" 32d56'2"
Kimberly Burridge"Angel in the Sky"
Cyg 20h2'30" 57d26'36"
Kimberly Burrow
Oph 16h34'11" -0d14'26"
Kimberly C. Kelley
Tau 3h43'42" 17d58'22"
Kimberly Carpenter
Sgr 18h39'31" -35d43'25"
Kimberly Cess Dela Cruz Valdivieso
And 0h49'56" 41d14'1"
Kimberly Charette
And 23h35'17" 41d52'48"
Kimberly Chen Lei
And 1h15'47" 37d36'47"
Kimberly Christine Dodd
Psc 0h53'35" 9d25'40"
Kimberly Christine Jones
Psc 0h29'31" 15d10'55"
Kimberly Collins Doyle
Sgr 18h59'2" -35d10'28"
Kimberly Coo
Sco 17h49'35" -39d23'26"
Kimberly Cornette
And 23h36'33" 41d50'52"
Kimberly Cosgrove
Ari 3h15'26" 28d32'22"
Kimberly Costa & Robert Dean
Pho 0h30'54" -50d42'24"
Kimberly Crocker
And 0h20'4" 32d16'30"
Kimberly Cronin Francis Finberg
Leo 10h48'23" 17d13'51"
Kimberly D. OSborne
Psc 0h58'28" 13d38'0"
Kimberly D. Sickler
Lyr 18h31'8" 28d17'9"
Kimberly D. Stout
Aqr 21h24'1" -0d50'8"
Kimberly Dailey
Lib 15h49'51" -6d52'45"
Kimberly & Dan's Star
Leo 10h15'21" 16d14'52"
Kimberly Davies
Cas 1h14'30" 65d25'43"
Kimberly Davis
Cam 13h13'18" 78d13'48"
Kimberly Davis Richards
Gem 6h49'34" 16d52'37"
Kimberly Dawn
Ori 5h58'20" 18d25'36"
Kimberly Dawn
Leo 11h17'44" 17d15'28"
Kimberly Dawn
Lyn 8h25'46" 38d20'29"
Kimberly Dawn Dunlap
Psc 2h1'39" 6d25'12"
Kimberly Dawn Everett
Sgr 17h55'40" -29d57'56"
Kimberly Dawn Ewasiuk
Dra 12h18'10" 71d37'49"
Kimberly Dawn Harmon
Vir 12h16'4" -0d57'29"
Kimberly Dawn Kent
Cap 21h15'41" -19d22'28"
Kimberly Dawn Lake
Cap 20h59'16" -21d22'23"
Kimberly Dawn Taylor
And 0h21'36" 42d9'15"
Kimberly Dawn Watson
Aql 19h7'33" 8d59'14"
Kimberly Dawson
Sco 17h5'44" -38d7'32"
Kimberly Defrancesco
Lyn 6h42'36" 56d51'14"
Kimberly Denise Dukette
Lib 15h54'20" -8d23'36"
Kimberly Diana
Ari 3h8'12" 26d25'1"
Kimberly Diana Banda
And 23h55'56" 47d22'58"
Kimberly Diane Banzhoff
And 22h58'18" 40d41'47"
Kimberly Dillon
Ari 3h13'13" 25d1'31"
Kimberly Dixon
Aqr 22h9'32" -0d59'36"
Kimberly Dorsey
Crb 15h51'35" 28d5'3"
Kimberly E. Pennington
Dra 18h50'14" 51d17'20"
Kimberly Ebel
Gem 7h55'5" 16d45'52"
Kimberly Eileen Moggio
Lib 15h37'8" -20d14'0"
Kimberly Eileen Smith
Ori 5h50'8" 6d14'40"
Kimberly Elaine Waller
Sco 16h12'43" -14d59'52"
Kimberly Elizabeth Carnes
Vir 11h47'54" -0d27'57"
Kimberly Elizabeth Cefaratti
Uma 11h41'38" 36d25'45"
Kimberly Elizabeth Cotton
Cnc 8h57'34" 12d59'9"
Kimberly England Haines
Cas 1h3'12" 63d43'1"
Kimberly Erin Birmingham
Uma 11h17'3" 49d33'6"
Kimberly Erin Brewer
Vir 13h14'33" -2d17'4"
Kimberly Ewan
Uma 10h44'38" 51d43'5"
Kimberly Fern Josephine Castaneda
Cap 21h45'39" -16d53'4"
Kimberly Fierst
Lib 15h10'26" -27d28'52"
Kimberly Flores
And 1h58'47" 43d44'58"
Kimberly Flores
Leo 10h10'4" 22d27'9"
Kimberly Fontenot
Lib 15h59'25" -20d2'8"
Kimberly Forgach
Tau 4h51'36" 17d53'10"
Kimberly Frey
And 0h35'19" 34d22'3"
Kimberly G. Garvey
And 0h44'24" 28d7'20"
Kimberly Galan
I.W.M.Y.O.D.I.G.P.
Leo 11h54'6" 26d10'38"
Kimberly Gayle
Nor 16h8'42" -57d47'38"
Kimberly Gayle Bookout
Lib 15h34'31" -6d0'41"
Kimberly Gayle McClellan
Psc 1h45'38" 23d41'42"
Kimberly Giustizia
Cyg 19h35'19" 32d2'8"
Kimberly Glassman "Kimberlina"
Uma 11h45'15" 32d28'22"
Kimberly Glitter and Stitches
Lib 15h25'56" -8d19'11"
Kimberly Grace
Ari 2h22'49" 24d14'55"
Kimberly Graham Shepard
And 0h21'21" 25d58'24"
Kimberly Greene Oliver
Ari 2h10'38" 27d50'20"
Kimberly Grover
Cyg 21h40'39" 49d29'58"
Kimberly Grubianis
Crb 15h47'27" 30d4'57"
Kimberly Gwen
Mon 6h45'48" 3d19'42"
Kimberly H. Quackenbush
Leo 11h18'10" 15d41'14"
Kimberly Han
Lac 22h22'16" 52d24'42"
Kimberly Harmon Ligon
Lep 5h11'52" -12d38'37"
Kimberly Harris
Crb 15h35'58" 33d47'19"
Kimberly Harvanchik
Aqr 22h14'7" -0d16'7"
Kimberly Hasko Sotelo Belknapp
Tau 3h42'23" 18d18'14"
Kimberly Heller
And 2h19'44" 45d1'51"
Kimberly Hiigel
Gem 7h6'56" 27d0'18"
Kimberly Hites
Leo 9h29'7" 25d41'34"
Kimberly Holmer
Lib 14h52'20" -4d28'35"
Kimberly Holston
Ari 3h23'33" 22d44'58"
Kimberly Hope Stroman
Sco 16h14'12" -18d49'23"
Kimberly Hunny Bunny
Uma 10h52'28" 41d4'23"
Kimberly Ivette Davila
Vir 12h38'31" 2d37'57"
Kimberly J. Brown
Sco 16h6'56" -14d56'53"
Kimberly J. Bupp
Cas 1h59'14" 60d52'47"
Kimberly J Carlino
Tau 3h41'6" 9d47'58"
Kimberly J. Hord
Lib 15h12'8" -7d35'0"
Kimberly Jacobs
Uma 10h58'0" 52d57'35"
Kimberly & James Decker
Cyg 21h33'35" 44d39'36"
Kimberly Jane
Leo 11h26'3" -0d58'7"
Kimberly Jane Roberts
Leo 9h29'22" 29d8'3"
Kimberly Jane Sikora
Umi 14h18'25" 76d45'15"
Kimberly Jane Worthman
Lmi 10h40'45" 35d11'43"
Kimberly Jean
And 23h17'31" 48d15'16"
Kimberly Jean
Cas 23h44'43" 53d21'37"
Kimberly Jean Adler
Sgr 19h9'30" -20d59'3"
Kimberly Jean Hordies
Sco 16h43'38" -31d30'9"
Kimberly Jean Keating
Ari 3h0'14" 27d18'35"
Kimberly Jean Serignese
Cap 20h22'9" -9d44'38"
Kimberly Jean Smeals
Uma 10h33'31" 65d50'31"
Kimberly Jean Tyra
Ari 3h27'52" 26d2'42"
Kimberly Jelouane Henderson
And 1h44'54" 45d30'55"
Kimberly Jett Bowlin, A Great Mom
Sgr 18h3'16" -32d37'39"
Kimberly Jezowski
Del 20h34'51" 6d40'10"
Kimberly Jo
Peg 22h23'30" 10d6'10"
Kimberly Joann Smith Morningstar
Cas 1h46'3" 61d23'30"
Kimberly JoAnne Ellis
Gem 6h19'58" 22d52'35"
Kimberly Johnston
Sco 16h38'5" -30d37'47"
Kimberly Jonny
Lyn 7h45'28" 38d57'13"
Kimberly Jordan
Sgr 18h2'51" -30d24'1"
Kimberly & Joseph Reif
Sco 17h51'31" -40d29'25"
Kimberly Joy
Psc 0h24'51" 7d47'14"
Kimberly Joy Hitchcock
Ori 4h48'53" 10d37'34"
Kimberly Joy Redder
Crb 15h36'46" 29d38'9"
Kimberly Joyce
And 23h24'13" 52d13'58"
Kimberly Joyce Furtado
And 23h9'31" 48d22'32"
Kimberly Joyce Steward
Sco 17h45'46" -35d39'40"
Kimberly & Justin Barden
Peg 23h26'22" 33d42'48"
Kimberly Justine Crow
Psc 1h25'36" 15d53'48"
Kimberly K. Carda
Leo 10h18'43" 16d47'52"
Kimberly K. McMorrow
Cnc 8h11'20" 20d57'36"
Kimberly K. Proxmire
Her 17h11'22" 19d26'29"
Kimberly Kay Bolton
Vir 12h23'36" 4d30'44"

Kimberly Kay Hyde
Aqr 22h39'40" -15d34'3"
Kimberly Kay Parr
Leo 10h58'27" 24d47'32"
Kimberly Kelly
Ari 3h18'14" 29d14'33"
Kimberly / Kelsey / Nolan
Aqr 21h24'38" 1d56'56"
Kimberly Key
And 0h34'42" 23d3'55"
Kimberly Kibbe
And 0h47'37" 40d23'10"
Kimberly Kim
Leo 9h30'54" 16d35'47"
Kimberly Kiser
And 23h9'9" 42d27'19"
Kimberly Klein
And 0h18'7" 25d44'34"
Kimberly Kramer
Cas 1h13'38" 55d1'7"
Kimberly Kressel
Cyg 19h38'1" 28d17'9"
Kimberly Krise
Cap 21h4'17" -16d54'51"
Kimberly Kucharski
Aql 19h39'26" -10d33'15"
KIMBERLY KUWAE
Lyn 8h24'18" 56d11'18"
Kimberly L. Roberts
Com 12h21'9" 26d12'38"
Kimberly L. Sojka
Sco 17h20'13" -32d44'0"
Kimberly Lacy
Her 16h19'9" 19d16'11"
Kimberly Lam
Sco 17h54'56" -39d5'0"
Kimberly Lashaun Baker
Ari 2h39'35" 29d4'32"
Kimberly Lauren
Tau 5h46'59" 15d59'50"
Kimberly Layne Maryman
Ari 3h11'22" 27d24'15"
Kimberly Link
Cyg 21h11'8" 53d53'10"
Kimberly Lisa
Psc 0h53'55" 31d8'47"
Kimberly Lorraine Basquez-Chacon
And 1h59'6" 38d58'28"
Kimberly Louise Sehm
Psc 0h30'21" 17d41'4"
Kimberly Lynn
And 0h27'36" 26d51'43"
Kimberly Lynn
Uma 10h1'38" 52d14'10"
Kimberly Lynn
Lyn 7h0'37" 49d27'25"
Kimberly Lynn Fenoff
Vul 19h32'59" 26d15'23"
Kimberly Lynn Miller
Lyn 8h40'27" 35d30'2"
Kimberly Lynn Same
Cap 21h57'4" -18d0'28"
Kimberly Lynn Therkelsen
Vir 14h7'33" -15d9'19"
Kimberly Lynne Eneks
Aqr 20h44'53" 1d40'57"
Kimberly M Belnap
Cyg 21h11'56" 40d24'0"
Kimberly M. Gill, Esq.
Lyn 8h38'10" 33d12'31"
Kimberly M. Osborne
Lyn 7h3'2" 47d57'11"
Kimberly Mae
Gem 7h27'3" 32d31'24"
Kimberly Mae Hinton
Ari 3h5'20" 11d49'33"
Kimberly - MAH
Cnc 8h34'40" 9d38'52"
Kimberly Majewski
Sco 17h48'20" -42d12'23"
Kimberly Malarchuk
Ari 3h5'39" 25d11'29"
Kimberly Maloney
Uma 10h13'16" 62d47'0"
Kimberly Mantiply
Sgr 18h55'53" -35d26'18"
Kimberly Marie
Sgr 18h0'35" -27d15'54"
Kimberly Marie
Cas 23h49'15" 57d46'56"
Kimberly Marie
Psc 1h20'7" 18d59'46"
Kimberly Marie
Cas 1h24'29" 55d54'3"
Kimberly Marie Arnold
And 0h13'7" 35d10'6"
Kimberly Marie Conover
And 0h41'25" 34d51'18"
Kimberly Marie Cook
Gem 6h46'59" 15d22'29"
Kimberly Marie Gilliam
Sco 17h1'5" -33d12'9"
Kimberly Marie Handrahan
Vir 14h34'58" 3d54'22"
Kimberly Marie Paesler
Cnc 8h29'41" 21d34'20"
Kimberly Marie Rose Hunt
Psc 1h38'24" 7d53'16"
Kimberly Mariko Suzuki
Leo 11h28'34" 3d11'53"
Kimberly Marmolejo
And 1h46'56" 42d4'24"
Kimberly Massarelli
Lib 15h10'2" -24d17'50"
Kimberly Maureen Collins
Vir 13h36'28" 4d50'20"
Kimberly McGowan
Crb 15h35'4" 26d41'29"
Kimberly McKee
Cas 1h37'51" 65d32'41"
Kimberly Melvin Dukester
Lyn 7h15'28" 52d19'23"
Kimberly Michele Kayto
Vir 13h49'0" -7d11'42"
Kimberly Michele Williams
Cnc 9h21'59" 27d11'5"
Kimberly Michelle
And 0h22'14" 29d27'55"
Kimberly Michelle
Aqr 22h7'31" -0d8'58"
Kimberly Michelle
Sco 17h4'15" -42d10'37"
Kimberly Michelle Brooks
Aqr 20h41'46" -2d11'31"
Kimberly Michelle Brown Draper
Leo 9h52'36" 6d26'59"
Kimberly Michelle Dorey
Cas 0h30'58" 50d0'59"
Kimberly Michelle Fuller
And 2h22'48" 41d19'0"
Kimberly Michelle Kozak
Lib 14h51'20" -4d55'27"
Kimberly Michelle Levy
Lib 15h42'26" -8d34'22"
Kimberly Michelle Morrison
And 0h8'56" 31d55'15"
Kimberly Michelle Spain
Cap 21h11'5" -19d22'36"
Kimberly Miller
Uma 9h17'49" 67d14'4"
Kimberly Miller
Tau 4h2'16" 17d2'4"
Kimberly M.Kelly
Cap 20h31'7" -21d59'38"
Kimberly Monique
Lyn 7h37'3" 38d15'9"
Kimberly "Mooky" Cotten
And 0h30'58" 32d8'58"
Kimberly Morgan
Cyg 21h56'51" 45d19'57"
Kimberly Morgan Washam
Ari 2h59'39" 26d52'28"
Kimberly Morrell
Cnc 8h11'52" 16d38'19"
Kimberly Morse
Uma 9h27'28" 68d15'53"
Kimberly My Angel
Per 4h18'7" 42d59'48"
Kimberly My Love
Col 5h53'15" -27d49'4"
Kimberly N. Anderson
Cnc 9h12'22" 19d13'29"
Kimberly N. Dirksen
Aql 20h15'59" 0d28'44"
Kimberly Negrette
Gem 6h44'46" 23d22'34"
Kimberly Nelson
Lyn 8h0'41" 38d3'39"
Kimberly New
Leo 9h28'31" 24d12'42"
Kimberly Nicole
Lib 14h55'45" -6d46'23"
Kimberly Nicole Beck
Lib 15h43'56" -28d54'53"
Kimberly Nicole Horner
Aqr 22h37'15" -0d45'16"
Kimberly Noelle McIntyre
Leo 11h44'20" 26d17'32"
Kimberly Oliver
Mon 6h44'19" -0d20'4"
Kimberly Olsen
Cyg 20h14'31" 34d5'22"
Kimberly Oyen
Tau 3h58'53" 24d39'10"
Kimberly Palmer
Leo 11h35'8" 27d12'46"
Kimberly Patricia Schneider
And 0h34'56" 41d5'34"
Kimberly Payne
And 0h15'40" 35d36'22"
Kimberly & Peter
Uma 9h36'6" 52d27'29"
Kimberly & Peter Hecht
Lyr 18h51'53" 42d31'44"
Kimberly Philson
And 23h11'31" 42d38'17"
Kimberly Pickens
And 23h39'25" 41d41'2"
Kimberly "PK Kline"
Tau 5h1'14" 22d3'35"
Kimberly R. Clark
And 0h15'1" 23d7'48"
Kimberly R. Dishmon
Sco 16h8'2" -16d18'7"
Kimberly R. May
Cas 0h31'42" 63d10'7"
Kimberly R. McHenry
Vir 13h25'2" -12d42'27"
Kimberly Rae Porcase
Ari 3h14'4" 15d37'36"
Kimberly Rae Smith
Mon 7h1'14" 5d26'32"
Kimberly "Red Wolf" Winson
Uma 11h34'45" 37d39'3"
Kimberly Renee
Crb 15h36'15" 35d35'10"
Kimberly Renee
Vir 13h21'44" 12d50'10"
Kimberly Renee Heywood
Lib 15h7'57" -3d41'25"
Kimberly Ritz Miller
Vir 12h45'44" -5d39'1"
Kimberly Rockman
Sgr 19h13'4" -15d18'51"
Kimberly Rosanne Painter
Ari 2h30'58" 24d36'54"
Kimberly Rose
Vul 19h49'13" 21d21'52"
Kimberly Rose
Gem 7h42'2" 31d56'45"
Kimberly Rose
And 1h20'56" 47d31'1"
Kimberly Rose
Cap 21h18'30" -17d56'1"
Kimberly Rose Arlene Veltre
Sgr 19h51'43" -14d6'54"
Kimberly Rose Bowcut
Cyg 20h50'27" 44d42'13"
Kimberly Ruppel-Slick
Cam 5h22'30" 64d39'4"
Kimberly S. Passint
Aur 5h59'20" 42d35'5"
Kimberly Samples
Lyn 7h24'58" 55d11'14"
Kimberly Sanders - Eachus' Star
And 2h35'9" 43d24'38"
Kimberly Sarah Dickens
Lib 14h53'3" -7d40'6"
Kimberly Sayles
Leo 9h35'54" 27d43'19"
Kimberly Scalio&RussellMyersJr ILUA
Psc 23h21'10" 2d31'30"
Kimberly Schenck
And 0h38'20" 31d49'50"
Kimberly Schutt
Aqr 22h11'47" 1d13'11"
Kimberly Schwartz
Gem 7h4'21" 21d59'26"
Kimberly Sellers
Tau 5h31'24" 18d35'38"
Kimberly Shanahan
Sco 16h13'23" -17d29'54"
Kimberly Shannon Atkinson
Aqr 23h39'31" -15d10'37"
Kimberly Shewmaker
Ori 5h29'15" -4d42'47"
Kimberly Shizuko Machida
Sgr 18h6'56" -28d26'31"
Kimberly "Shuggy" Butt
Crb 16h10'29" 37d58'14"
Kimberly Soker
Sco 16h10'0" -11d41'41"
Kimberly Sorvillo
Gem 7h17'31" 30d46'50"
Kimberly Spaeth
Cas 0h54'37" 56d10'26"
Kimberly Stack & Christopher Perry
Ori 5h57'46" 7d6'9"
Kimberly Star
Sco 17h22'13" -42d18'21"
Kimberly Stephanie DeMille
Tau 5h45'22" 22d28'10"
Kimberly Stephanie Smith
And 0h47'16" 46d4'46"
Kimberly Stevens
Cas 23h36'19" 53d26'6"
Kimberly Sue
Cyg 19h58'23" 32d24'35"
Kimberly Sue Adler
Ori 5h32'54" 1d40'11"
Kimberly Sue Tiberio
Boo 15h6'52" 34d24'21"
Kimberly Summer Perry
Aqr 22h45'53" -12d28'42"
Kimberly Susan Van Gorp
Sco 17h11'7" -41d28'38"
Kimberly T B Saunders
And 23h21'48" 47d55'6"
Kimberly Tarnowski
Aqr 22h29'5" 0d19'7"
Kimberly Taylor
Cnc 8h31'33" 19d59'58"
Kimberly Taylor Bost
Cyg 20h45'51" 35d42'36"
Kimberly Taylor Thomas
Cyg 20h17'4" 38d22'13"
Kimberly & Thomas Kenney
Mon 7h21'10" -0d24'35"
Kimberly Thorne
And 23h9'46" 40d43'6"
Kimberly Thu Ngo
Cam 5h55'22" 60d28'5"
Kimberly Titus
Ari 3h4'35" 28d34'11"
Kimberly Trentalange
Cas 0h14'25" 63d22'14"
Kimberly Tringali
Leo 11h21'36" 27d10'0"
Kimberly Troutman
Gem 7h24'26" 15d22'0"
Kimberly Vernice
Cnc 8h38'7" 31d11'42"
Kimberly Vincent
Cas 0h20'37" 47d24'35"
Kimberly W
Lib 14h48'7" -16d21'28"
Kimberly Walsh
Umi 15h13'7" 71d6'3"
Kimberly Williams
Ari 2h6'33" 26d10'5"
Kimberly Williams Paisley
Cas 23h3'14" 57d5'31"
Kimberly Wolf
Crb 16h11'1" 33d9'26"
Kimberly Wooten - Balerinarius
Cap 20h14'15" -10d29'52"
Kimberly Wright
Ari 2h25'58" 23d10'13"
Kimberly Yudi Carter
Cyg 19h48'23" 33d18'57"
kimberly1022
Her 18h23'3" 12d18'2"
Kimberly-88
Uma 8h46'49" 71d46'15"
Kimberly-Clark Marketing Services
And 23h23'38" 47d56'40"
Kimberly-Clark Trading (M) Sdn Bhd
Ori 5h39'23" 3d37'2"
KimberlyDavidNeff
Sco 17h58'18" -31d25'32"
KimberlyMay
Leo 9h58'0" 18d5'24"
Kimberly-My Mother My Inspiration
Cas 0h38'39" 58d32'21"
Kimberlynn R.
Ari 3h7'3" 27d24'45"
Kimberly's Eyes
And 23h4'15" 47d3'42"
kimberly's eyes
Lib 15h35'14" -24d17'7"
Kimberly's Morning Star
Mon 7h7'8" -6d38'23"
Kimberly's Star
Psc 0h41'36" 7d50'43"
Kimberly's Star of the Dragonfly
Sgr 19h22'18" -35d28'15"
Kimber's Superstar
Lmi 9h45'7" 39d24'26"
Kimble Webb
Her 16h51'57" 37d44'22"
Kimbo
Sco 16h43'44" -29d39'29"
KIMBO
Uma 9h46'20" 70d50'58"
Kimbo 40
Cam 4h22'50" 68d49'46"
Kimbo's star
Cap 21h17'31" -26d47'16"
Kimbra
Aql 18h53'14" 8d37'10"
Kimburleigh Joy
Lyn 7h33'7" 50d24'48"
Kimburrkay
Sco 17h28'8" -42d12'7"
KimDaveSarsaKbearEmBeastSameDawl
Umi 16h6'11" 70d32'19"
Kim-David
Lyn 8h38'38" 36d4'35"
KimEkiM
Sgr 18h46'33" -29d6'20"
Kimera Martha Jane
Tau 4h22'37" 13d13'22"
Kimerlee Lyn Maison Born 11/3/1978
Uma 10h10'42" 71d19'58"
Kimi
And 0h18'38" 25d51'14"
Kimi
Lyn 8h3'21" 43d46'9"
Kimi Alicia Golonka
Vir 11h55'10" 8d1'19"
Kimi- love u as long as this shines
Ser 15h54'26" -0d8'2"
Kimi Lynn Landry
And 1h4'50" 37d27'55"
Kimi Pappas
Psc 1h29'20" 17d59'54"
Kimi420
Ari 3h13'31" 29d38'18"
Kimia
Lyn 7h8'58" 59d2'22"
Kimi-Ann
Cas 1h43'8" 60d56'46"
Kimico
Mon 6h48'22" -0d46'54"
Kimie Kae
Psc 0h21'20" 18d32'39"
KimieRoo
Uma 11h20'47" 40d6'10"
KimiKi
Cap 21h25'53" -22d37'14"
Kimiko Bullion
Psc 1h21'15" 14d6'19"
Kimiko Kiseki Yanaura
And 1h41'4" 47d34'13"
KIMIKO LAHAELA
Uma 11h43'57" 35d29'27"
Kimiko Treitler
Gem 7h39'0" 23d36'25"
KimikoFranco
Sgr 18h54'45" -34d10'25"
Kimille Dean
Cnc 9h8'43" 11d41'26"
Kimiqo
Sco 16h39'1" -36d30'45"
KimiStar
And 2h14'20" 42d4'9"
Kimiya Sabzghabaian
Cap 21h4'14" -27d7'35"
Kimizee
Aql 19h5'6" -0d3'10"
Kimizzle's Star
Tau 5h30'55" 18d50'31"
KimJack
Sgr 18h45'7" -27d8'42"
Kim-Luise
Uma 12h54'30" 61d53'31"
Kimly
Mon 6h31'21" 7d36'33"
Kimmer
Uma 9h19'29" 62d0'43"
Kimmer
Cas 1h19'14" 64d26'52"
Kimmers
Vir 13h41'57" -11d8'9"
Kimmi
And 0h19'19" 45d53'7"
Kimmi Gayle
Pho 1h20'3" -43d9'10"
Kimmie
Lyn 8h53'59" 37d37'14"
kimmie
Lyr 19h17'28" 29d26'5"
Kimmie Baby Girl H.
Cap 20h22'43" -9d43'49"
Kimmie & Justin
Cyg 19h44'30" 55d6'15"
Kimmie Nguyen
Gem 7h34'42" 29d47'47"
Kimmie Nguyen Freytag
And 23h15'47" 41d57'46"
Kimmie Townsend Sinex
Cyg 20h59'38" 44d39'12"
Kimmieita
Cyg 19h54'17" 38d30'53"
Kimmie's Smith
Mon 7h33'51" -0d47'10"
Kimmo Ronimus 60V
Vir 13h5'46" 11d0'36"
Kimmy
Psc 23h50'23" 4d21'0"
Kimmy
Uma 11h31'18" 47d29'25"
Kimmy
And 0h31'42" 41d28'46"
Kimmy
Cyg 19h37'32" 31d50'58"
Kimmy
Sgr 19h33'47" -14d45'42"
Kimmy
Sgr 18h31'25" -17d8'49"
Kimmy
Sgr 18h16'18" -29d19'58"
~Kimmy And Chris~
Tau 4h32'27" 29d58'57"
Kimmy Angel Baby
Mon 6h43'41" -0d19'41"
Kimmy Chelle's Eternal Hope
Uma 13h56'58" 52d48'6"
Kimmy Dugan
Lyr 19h12'1" 39d2'41"
Kimmy Girl
Uma 11h13'29" 55d37'58"
Kimmy K
Psc 0h12'3" 11d36'44"
Kimmy Kat
Tau 4h5'23" 8d1'49"
"Kimmy" Michelle O'Neal
And 23h13'21" 50d58'34"
Kimmy Shipyan
And 23h47'15" 46d10'30"
Kimmy Stukel
Vir 14h53'5" 2d21'32"
Kimmy Sue
Ori 5h31'51" 6d18'26"
Kimmy V
Sco 16h3'6" -13d25'29"
Kimmydawn
Umi 14h23'25" 66d43'5"
KIMMYPANTS
Umi 14h42'50" 79d43'49"
Kimmyrae
Sgr 18h2'48" -26d55'35"
KimmyRobbins
Cnv 12h25'55" 33d26'51"
KIMMY'S
Dra 16h43'27" 51d34'54"
Kimmy's Shining Star
Cas 1h35'30" 60d13'25"
Kimmy's Star
And 23h41'27" 42d27'30"
Kimmys Star
Crb 15h53'23" 29d2'0"
Kimness
And 23h18'46" 47d19'56"
Kim-N-Kelly
Cyg 20h41'46" 50d48'17"
Kimo
Aql 19h39'8" 11d48'53"
Kimo Assi
Apu 14h52'3" -74d8'21"
Kimommy
Lyn 7h45'38" 55d46'0"
Kimorra Nevaeh Buggs
Cyg 21h23'17" 51d46'37"
Kimosavee
And 0h43'2" 40d4'42"
Kimrita
Aqr 22h21'48" -23d25'36"
KIMROCK
Lib 15h24'28" -26d4'28"
Kim's Drop Of Sunshine
Leo 11h13'36" 5d10'19"
Kim's Everlasting Star
And 1h10'35" 43d5'3"
Kim's Heaven
Ori 6h9'4" 14d39'11"
Kim's Nursery- star growers
Her 17h31'59" 18d11'29"
Kim's Star
Leo 11h4'55" 10d33'1"
Kim's Star
Tau 4h26'23" 22d23'15"
Kim's Star
Uma 11h40'15" 42d33'30"
Kim's Star
And 23h12'45" 51d35'15"
Kim's Star
Cru 12h28'43" -59d40'47"
Kim's Star
Dra 18h19'27" 78d46'1"
Kim's Star of Love
Dra 16h24'31" 54d52'18"
Kim's Starr
Sco 17h23'12" -41d26'34"
Kim's Sweet Angel
Cnc 8h50'53" 31d8'17"
Kimsey
Uma 11h9'58" 67d39'1"
Kimspiration Glitteralus
Cru 12h42'26" -57d59'6"
KimStar
Uma 10h24'41" 48d19'44"
Kimstuhr
Ori 5h18'19" -4d39'59"
Kimthong Moore
Cas 1h30'44" 60d47'35"
Kimy
And 1h23'1" 45d53'57"
Kimz Hope
Cnc 9h16'59" 26d18'45"
Kimzie
Leo 11h17'29" 13d32'5"
Kincaid Paul Wylie
Ari 2h50'14" 14d44'14"
Kinch
Her 17h28'50" 45d21'24"
kindah Ahmed Sais
Cnc 9h16'19" 16d51'16"
Kindal
Her 18h50'31" 23d9'16"
Kindel
Ori 6h25'19" 9d57'34"
Kindergarten All Star
Umi 15h39'5" 77d16'52"
Kindergarten Lippmanngasse Wien
Uma 10h28'42" 65d49'30"
Kindest Lawyer in CA - Mr. Ed Smith
Aur 6h15'40" 29d14'17"
Kindi Kay Scartaccini
And 0h50'49" 40d45'34"
Kindle Kay Higgins
Uma 8h34'55" 52d24'57"
Kindlein, Hildegard und Klaus
Uma 11h47'43" 36d6'11"
Kindra
And 23h22'26" 36d34'25"
Kindra
Ari 2h57'14" 28d18'21"
Kindra
Psc 23h0'31" 7d52'57"
Kindra Powers
Cyg 19h52'40" 38d33'26"
Kindred McGuire Hare
Umi 16h29'1" 75d45'0"
Kindred Spirit
Col 5h23'47" -35d30'46"
Kindred Spirits
Cas 2h10'16" 65d37'43"
Kindred Spirits
And 0h41'42" 25d15'48"
Kind's All-star
Boo 13h48'48" 24d38'41"
Kindy Eileen
Cas 0h17'11" 51d26'39"
Kine Lise Andersen
Lyn 7h43'17" 48d22'0"
KINEMI
Vir 13h20'15" 6d11'52"
Kinemor
Aqr 22h34'44" -0d25'4"
King
Ari 2h32'31" 10d57'10"
King and Protector
Leo 11h11'14" 20d53'54"
King Arthur
Umi 15h13'9" 70d33'32"
King Arthur 09/04/33
Vir 12h18'15" -5d8'43"
King Azie
Cep 22h20'51" 59d16'25"
KING BJ
Sgr 19h29'9" -18d49'56"
King Bob
Cru 12h28'15" -58d33'3"
King Botachalli
Gem 7h6'10" 20d56'24"
King Brainard
Cep 21h39'16" 61d52'12"
King Christopher
Cap 20h35'33" -26d23'25"
King Cole Haughbrook
Cen 12h51'20" -46d57'13"
King Dana M. Thrash "qs"
Cep 22h6'2" 63d55'19"
King Daniel Devoe McNair
Cep 0h7'1" 75d24'36"
King Daniel & Queen Jade
Lyr 18h48'8" 32d30'53"
King Dannie
Vir 13h24'52" -10d21'39"
King David
Leo 11h50'41" 21d13'28"
King David - My Dutchy
Psc 0h5'16" -2d28'42"
King Diller
Cep 22h31'9" 74d44'4"
King Dustin Hubbell
Cyg 21h29'0" 32d37'2"
King Edward's Star
Ari 2h1'21" 24d5'34"
King Fisher
Lib 14h53'29" -3d13'49"
King Fuller
Aqr 20h39'20" -13d40'35"
King George Harvey Strait
Tau 3h30'14" 24d26'9"
King Ghidorah
Boo 14h21'29" 51d41'38"
King Gunner
Sgr 18h39'2" -18d4'24"
King & Hime - Collier Family Star
Vir 13h58'50" -9d31'30"
King James
Dra 17h49'8" 54d35'17"
"King James" Eugene Wardrobe Jr.
Leo 10h12'36" 12d47'3"
King Jeffrey
Ori 5h35'36" 2d54'46"
King Joseph My Beloved
Umi 13h28'33" 71d45'49"
King Kevin
Per 3h47'7" 42d20'30"
King Kevin F. Kast
Cen 13h20'42" -33d18'28"
King Liam
Her 16h49'44" 26d49'0"
King Luis
Leo 11h5'56" 8d25'37"
King Malcolm
Cep 21h45'7" 64d46'9"
King Manuel
Boo 14h35'37" 41d23'17"
King Midge
And 0h26'2" 41d37'12"
King Monkey
Cep 21h49'23" 62d20'38"
King of Whistles
Gem 7h1'33" 20d38'3"
King Pa - Pete Jones
Cep 23h16'11" 77d8'48"
King PePe
Cap 20h58'3" -25d19'39"
King Purse
Cep 21h16'53" 66d43'34"
King & Queen Nubb
And 0h22'18" 46d21'46"
King Ross
Cep 21h3'41" 57d28'14"
King Snogmeister
Cru 12h8'16" -61d54'30"
'King' Stephen Kennedy
Sco 17h6'26" -35d45'8"
King Tadashi
Ari 2h57'39" 26d52'26"
King Terry Scott
Her 17h17'59" 35d35'34"
King Tomas Calderon
Ari 2h5'28" 23d25'32"
King Tracy
Cep 21h25'8" 63d25'18"
King Tutankhamen & Ankhesenamun
Ori 5h29'50" 1d56'19"
'King Vilayvanh'
Ori 5h36'29" 9d4'3"
King/WilsonSumrell'sJo
Sco 17h56'14" -31d44'54"
Kinga and Tim Manders
Lmi 10h7'10" 37d59'22"
Kinga és Dávid Csillaga
Uma 13h13'8" 56d59'51"
Kinga Ogonowska
Leo 11h19'24" 5d5'38"
Kingdom of Tonga
Dra 19h41'31" 73d18'40"
Kingjade
Vir 12h35'47" 6d46'11"
King's Caterers
Uma 11h28'54" 58d52'43"
Kingsbury's Lone Star
Boo 14h33'24" 14d45'42"
Kingsley Leung
Cep 21h49'52" 58d30'27"
Kingston Light
And 0h46'7" 37d12'6"
kinimod
Cyg 21h17'23" 45d54'51"

Kinipela
Cap 21h40'26" -17d49'47"
Kinjia
Del 20h37'44" 7d9'13"
Kinkie
Lib 15h22'9" -17d27'44"
Kinky Coconut
Vir 11h59'53" 9d30'12"
Kinlee Blackner
Leo 10h34'47" 21d38'52"
Kinley Marie Sullivan
Vir 14h0'36" -16d9'15"
Kinnely Joy Wilson
Cam 4h3'2" 57d51'24"
Kinnin Payson
Uma 10h20'30" 63d5'13"
Kinnison Prince
Aqr 22h13'35" 0d1'36"
Kins
Uma 11h6'23" 34d21'57"
Kinsers
Cam 4h50'7" 65d32'10"
Kinsey Parker
Psc 22h52'30" -0d1'55"
Kinsley Faith
Umi 14h23'50" 73d6'55"
Kinsley Mae Jones
And 0h36'50" 30d22'9"
Kinslie
Lmi 10h49'21" 33d21'30"
Kinstar
Uma 9h46'54" 58d50'47"
Kintzel, Tom
Ori 6h10'35" 15d42'53"
Kinumi-Baba
Ari 3h6'31" 12d39'35"
Kinyona Viola Tucker
Sco 17h55'57" -34d34'51"
Kinzee
Gem 6h43'40" 16d38'3"
Kinzey Nicole Powell
And 0h14'39" 44d9'33"
Kiona
Lib 14h38'12" -10d26'54"
Kiora Diane Batts
Gem 7h33'22" 23d6'40"
KIP
Ori 6h12'47" 15d43'10"
Kip Farrar
Ori 5h51'39" 2d14'22"
Kip & Heidi
Sge 19h38'2" 18d49'5"
Kip James Ray
Psc 1h33'18" 26d3'57"
Kip & Kelly Kauffman
Sgr 18h12'48" -31d56'54"
Kip L. Donley
Psc 23h10'0" -0d25'12"
Kip Phillips
Tau 4h9'25" 8d38'22"
Kipling Adulian Mattis
Lyn 8h26'44" 35d51'8"
Kiplynne
Cnc 8h33'14" 15d52'50"
Kipp Ramsey
Ori 5h41'1" 0d22'1"
Kipp, Heidi
Uma 9h44'57" 49d36'50"
kipper
Leo 10h12'14" 23d52'11"
KippyBeth
Gem 7h37'6" 34d43'46"
Kira
Gem 7h25'20" 31d19'12"
Kira
Lmi 10h31'13" 35d14'34"
Kira
Ori 5h30'27" 13d40'28"
Kira
Aqr 21h52'59" 0d50'20"
Kira
Cap 20h17'43" -14d57'11"
Kira 82903
Uma 8h41'14" 65d52'21"
Kira Abigail
Lib 15h53'7" -7d58'31"
Kira Akulova (Mouse)
Cam 5h9'48" 63d7'43"
Kira & Alex
Uma 13h42'10" 62d18'40"
Kira Anastasia Tsamas
Ari 2h24'38" 27d27'48"
Kira and Jon
Gem 6h47'23" 17d43'50"
Kira and Trey
Cap 21h42'3" -12d34'8"
Kira Ann
Com 12h36'6" 23d22'0"
Kira Ann Sellers
Sgr 18h46'1" -31d54'37"
Kira Anne McKiernan
Cnc 8h41'42" 30d29'19"
Kira Charlotte Doyle - Kira's Star
And 23h0'51" 50d25'45"
Kira Cindle Ashley
Cap 21h44'52" -12d36'45"
Kira Elizabeth
Vir 13h6'19" 10d59'28"
Kira Ellestad
Cyg 21h9'47" 48d18'5"
Kira Elyse
Psc 1h30'11" 12d18'51"
Kira Eve Browen
And 23h42'34" 48d14'34"
Kira Faith Pilger
Ori 6h1'31" 19d55'37"
Kira Goolsby
And 2h12'3" 43d34'28"
Kira Griswold
Lyr 18h16'7" 39d19'57"
Kira Kira Majo
Sco 17h24'31" -41d22'57"
Kira Leigh Bible
Gem 7h5'49" 21d14'52"
Kira & Luke ~ Forever in Love
Cyg 20h14'42" 48d54'51"
Kira Lunn
Cru 11h57'21" -63d6'41"
Kira Lynn Bosch
Aql 19h39'12" 0d9'26"
Kira Madison
And 23h13'58" 47d47'31"
Kira Manzo
Lib 14h36'32" -23d28'11"
Kira Marie Raynor
Cnc 8h12'34" 15d53'58"
Kira Nicole Gangi
Cas 0h26'30" 65d30'42"
Kira Papamihalis
Psc 0h40'30" 16d12'40"
Kira Phillips
Sgr 18h33'59" -27d8'25"
Kira Ratliff
Sgr 19h12'45" -16d53'42"
Kira Raven Worrell
And 1h39'28" 42d20'8"
Kiralee Brooke Knotts
Ori 6h12'5" 6d36'0"
Kiraly Rosebud Maxwell
Sco 17h54'7" -41d30'0"
Kirameku
And 2h15'55" 38d51'22"
Kiran
And 0h8'30" 44d32'19"
Kiran
And 2h33'0" 39d13'5"
Kiran Kaur Purewal
Cyg 21h41'0" 50d24'3"
Kiran Kaur Sampley
Ari 2h44'6" 25d58'18"
Kira-Nana's Bright Beautiful Star
Lyn 7h34'6" 53d31'3"
Kiranjit
Vir 11h45'40" 2d53'9"
Kiranmayi
Tau 4h33'35" 16d50'37"
Kira's Bean
Psc 23h39'14" 2d9'2"
Kira's Shining Star
Lib 15h20'37" -15d50'10"
Kirasten Dasha
Uma 13h3'14" 54d11'5"
Kirbie
Cap 20h47'3" -18d48'46"
Kirbie Leigh Head
Tau 4h24'47" 15d55'59"
Kirbie Lynne
Gem 7h36'2" 23d30'28"
Kirby
Vir 11h44'35" -3d25'9"
Kirby
Cap 20h36'18" -24d20'37"
Kirby and Anna Jambon
Umi 16h19'19" 74d0'3"
Kirby and Brandon
Cyg 20h38'45" 46d4'11"
Kirby and Grace
Per 3h6'40" 43d21'53"
Kirby and Terra
Cyg 19h44'22" 31d57'0"
Kirby Beth
Psc 1h11'10" 21d41'54"
Kirby es Chulo
Boo 14h47'43" 45d55'47"
Kirby Keith Duncan
Sgr 18h42'30" -33d28'55"
Kirby Schoepke
Uma 9h0'36" 46d41'11"
Kirby & Scottie's Love Star
Cyg 21h23'48" 46d17'12"
Kirby & Susan
Pho 0h7'13" -42d48'22"
KirbyJ21
And 23h27'58" 38d28'3"
Kirch, Jochen Hans
Ori 5h12'50" -8d16'44"
Kirchbaumer, Carmen
Uma 10h13'10" 55d36'40"
Kirchhoff
Uma 14h2'24" 49d43'1"
Kire
Sge 19h50'6" 17d12'34"
Kiren Srinivasan
Uma 13h33'51" 56d58'58"
Kiri
Sct 18h54'12" -12d31'51"
Kiri Michelle
And 1h42'54" 40d30'35"
Kiriakoula "Kyra" Patuelli Mom&Wife
Cas 0h18'28" 61d40'27"
Kirill et Stéphanie
Dra 19h47'45" 62d50'26"
Kirill Umov Loves Julia Romalis
Tau 4h22'1" 28d5'12"
Kirin Taylor Smith
Crb 15h51'14" 34d37'57"
Kirk
Her 18h20'12" 18d36'55"
Kirk
Psc 0h59'15" 11d56'29"
Kirk
Cap 21h1'54" -16d20'26"
Kirk Addison Evans
Ori 5h49'55" 5d36'8"
Kirk Alan Miller
Aql 19h45'18" -0d50'22"
Kirk and Kristin Forever
And 0h25'13" 38d29'16"
Kirk and Leigh
Sge 19h49'38" 18d44'27"
Kirk Anthony De Jesus
Cep 22h35'11" 75d57'7"
Kirk Auston
Her 16h55'6" 14d19'39"
Kirk & Beth
Lyr 19h20'25" 29d37'1"
Kirk David
Per 3h37'41" 34d5'0"
Kirk E Bengtzen
Dra 19h0'35" 74d1'33"
Kirk Gregory Nazarian
Aqr 22h17'31" 0d23'45"
Kirk L. Thomas
Per 4h19'19" 51d38'39"
Kirk Lyon
Uma 12h38'24" 62d33'35"
Kirk & Mair..Best Thing .....
Gem 7h5'9" 23d38'37"
Kirk McKerrow
Cas 1h23'36" 52d45'41"
Kirk Michael & Stephanie Heather
Tau 3h48'33" 28d45'32"
Kirk -N- Kelly *Always -N- Forever
Cap 21h55'45" -8d54'2"
Kirk Powles
Ari 2h13'8" 25d22'43"
Kirk Richard Phillips
Vir 13h8'27" 13d11'4"
Kirk Terron
Per 4h37'18" 40d50'30"
Kirk Voclain
Per 3h5'40" 47d32'15"
Kirk Warren Imlay Bensemann
Cru 12h10'23" -57d31'31"
Kirk William Pagoota
Cep 20h36'40" 65d10'44"
Kirkja
Aqr 21h12'47" -11d43'41"
Kirkwood & Khrysten 1st V-Day 2004
Cnv 13h46'24" 30d45'3"
Kirobochie
Ari 3h2'3" 29d30'37"
Kiros Papaloizou
Ori 5h16'34" -9d18'54"
Kirra Chesterfield
Cru 12h43'44" -56d50'29"
Kirra Jade - 4 November 1999
Sco 17h2'57" -38d12'38"
Kirra Jade Britton - 13.06.2006
Cru 12h32'21" -64d12'18"
Kirra Quinn Baker
Per 4h17'44" 39d59'18"
Kirren
Cyg 19h37'26" 31d50'59"
Kirsi Anniina Jokinen
Leo 11h2'21" 4d29'20"
Kirst, Jürgen
Ori 4h59'29" 15d36'31"
Kirsta "Kaye" Peterson
Vir 13h21'6" 2d16'58"
Kirsteen
Lib 15h33'48" -25d40'10"
Kirsten
Vir 13h23'6" -13d25'52"
Kirsten
Uma 12h19'52" 57d58'27"
Kirsten
Aqr 22h34'36" 0d27'14"
Kirsten
Tau 4h25'57" 18d26'13"
Kirsten
Cyg 19h52'37" 32d39'27"
Kirsten
Crb 15h49'41" 38d15'29"
Kirsten A. Pernich
Ori 6h1'47" 19d41'1"
Kirsten Alyson Buckle
Lyn 8h4'2" 42d17'21"
Kirsten Anita Joy Firth - 04.03.1977
Psc 23h23'56" -3d1'38"
Kirsten Ann Downing
Gem 7h16'43" 20d0'9"
Kirsten Anna Westerland
Lyn 9h5'52" 39d9'26"
Kirsten Anne Wrisley
Uma 14h15'12" 57d41'54"
Kirsten Annette Mooso
Psc 0h48'54" 18d40'27"
Kirsten Ball
Psc 0h56'45" 4d21'18"
Kirsten Cheyenne Ranoco
Leo 11h29'10" 25d39'34"
Kirsten Chyler (KC) Agravante
And 0h28'50" 38d20'50"
Kirsten Dickens
And 1h15'28" 37d33'55"
Kirsten Elaine
Sgr 18h55'34" -29d20'36"
Kirsten Elyse Ferrigan
Vir 12h14'16" -4d58'21"
Kirsten Faith Pesola
And 0h20'36" 41d34'51"
Kirsten Gabriella
Uma 10h31'25" 55d47'38"
Kirsten Garber
Vir 11h55'37" 5d0'34"
Kirsten Gielnik
Uma 11h38'55" 60d31'52"
Kirsten Gorman
Psc 1h4'49" 29d30'9"
Kirsten J Dunn
And 1h12'43" 44d58'43"
Kirsten James Wheeler
Umi 16h39'51" 82d54'40"
Kirsten Jayne
Uma 8h18'18" 69d16'0"
Kirsten Jensen
Aqr 21h7'10" 0d59'30"
Kirsten Kelso
Lyn 7h51'19" 51d54'13"
Kirsten Lambert
Cnc 8h28'8" 26d3'51"
Kirsten Leah
Gem 7h42'24" 33d6'36"
Kirsten Loves You
Gem 6h52'19" 27d7'47"
Kirsten Maki
Aqr 22h27'7" -2d38'20"
Kirsten Marie
Peg 21h52'26" 12d21'28"
Kirsten Muir Brown
And 1h21'26" 47d28'34"
Kirsten My Love
Cnc 8h18'8" 23d36'37"
Kirsten Nigro
Vul 19h46'20" 24d37'53"
Kirsten Renae Groth
Tau 3h47'36" 28d22'19"
Kirsten & Ryan
Cyg 21h28'26" 51d57'0"
Kirsten Schomacker
Leo 9h30'36" 10d28'16"
Kirsten "Serendipity" - Love Mum x
Cas 1h0'50" 49d21'35"
Kirsten Victoria Deavours
Leo 10h8'58" 15d25'21"
Kirsten & Volker
Ori 6h19'55" 19d20'11"
Kirsten Warner
Sco 16h50'20" -27d2'19"
Kirsten's Lucky Star
Cnc 8h21'40" 14d22'1"
Kirsten's Wishing Star
Cap 20h58'53" -25d6'9"
Kirstin Elizabeth Morin
Cap 21h10'30" -21d23'38"
Kirstie
Pav 18h10'12" -58d45'11"
Kirstie Ann
And 23h38'6" 47d44'37"
Kirstie Bower
And 23h1'50" 51d19'5"
Kirstie Durrant-Cormier
Lyn 6h43'46" 61d10'41"
Kirstie Hall
And 2h15'41" 46d37'9"
Kirstie Joanne Enefer
Cyg 20h30'15" 33d44'21"
Kirstie Lee
Cru 12h26'43" -60d52'31"
Kirstie Leigh
Ori 6h0'11" 16d58'8"
Kirstie Reichers
Ori 5h52'28" -2d15'12"
Kirstin
Uma 9h33'25" 55d45'12"
Kirstin
Ori 6h18'13" 18d55'35"
Kirstin Ashford
Cas 1h57'57" 64d17'3"
Kirstin Elizabeth Karfonta
Tau 4h27'25" 12d47'11"
Kirstin Fischl
Uma 13h27'54" 61d35'33"
Kirstin Johns
Leo 10h22'6" 18d42'10"
Kirstin Kate Renfrow
Aur 6h58'37" 37d42'1"
Kirstin Leigh Distler
Ari 2h39'16" 25d29'28"
Kirstin Pettry
Lyn 9h4'13" 38d45'19"
Kirstin Young
Cnc 8h24'5" 26d37'4"
Kirstn's Reflexion
And 23h7'19" 49d41'3"
Kirsty
And 1h11'53" 38d32'29"
Kirsty
Ari 2h47'5" 12d53'58"
Kirsty and Nathan Forever
Cen 11h47'52" -39d45'30"
Kirsty and Steve forever in love
Dra 17h34'59" 55d15'41"
Kirsty Ann Wharton
And 0h13'22" 46d57'48"
Kirsty Douglas
And 1h4'52" 46d1'49"
Kirsty Ellen Whelan
And 2h10'15" 41d31'19"
Kirsty Emma
Uma 11h47'35" 33d49'8"
Kirsty Green
Tau 3h40'21" 28d27'32"
Kirsty Hawkins sparklies!
Cyg 19h56'32" 40d49'35"
Kirsty Isla Robertson 7
And 1h15'57" 45d32'38"
Kirsty & Jeff
Lib 15h24'18" -21d53'14"
Kirsty Leanne
Lmi 10h18'12" 37d19'18"
Kirsty Leanne Williams
And 1h28'10" 48d29'33"
Kirsty Louise Bain
Cas 1h50'46" 61d40'28"
Kirsty Louise Hand
And 2h21'52" 46d19'15"
Kirsty Louise Nesbitt
Cyg 20h53'10" 31d13'59"
Kirsty Louise Norton Twiselton
And 1h19'40" 43d35'20"
Kirsty MacDonald
And 2h7'48" 37d54'35"
Kirsty Mae
Cnc 8h12'43" 15d33'59"
Kirsty Marie Thomson
Cyg 21h47'52" 38d4'4"
Kirsty McClean
Cyg 21h47'38" 54d43'4"
Kirsty McPhee
Psc 1h33'27" 13d14'11"
Kirsty Mia
Vel 9h49'35" -40d22'47"
Kirsty Michelle Youde
And 1h48'43" 35d42'7"
Kirsty Nickless 1986
And 0h34'12" 27d11'13"
Kirsty Siddons
And 23h39'49" 42d54'22"
Kirstyn Blake Garcia
Lyn 6h58'31" 54d10'28"
Kirstyn Kori McNutt
Cyg 21h14'48" 52d8'27"
Kirsty's Star
Sco 17h10'10" -44d2'51"
Kirt & Abby
Cyg 20h32'7" 37d15'50"
Kirt and Michele
Ari 3h15'46" 26d56'29"
Kirt Foster Reeves
Sco 17h8'32" -45d39'57"
Kirtina's Girl Scout Star
Mon 7h19'25" -0d49'40"
Kirtis Jay Blizzard
Aql 19h36'22" 0d51'9"
Kirven and Charlie Powers
Per 3h46'18" 33d57'44"
Kiryan Paige Cooper Allen
Tau 4h25'10" 21d13'2"
Kirysten
Gem 6h22'41" 18d39'30"
kis baby bogar cica
Cnc 9h9'53" 10d50'53"
Kis Herceg
Leo 9h48'43" 20d14'42"
Ki's Star
Uma 11h59'24" 40d9'18"
KISA
Tri 2h35'26" 34d11'14"
Kisa Lew
Aqr 22h56'22" -22d57'51"
Kisboszi
Sgr 18h58'25" -16d8'28"
Kiscsillag
Lyr 18h48'57" 37d51'47"
Kisha
Lyn 7h11'34" 57d26'37"
Kisha Marie Hesse
And 23h19'19" 49d17'6"
Kishama
Tau 5h37'53" 20d3'50"
Kishta
Vir 11h37'49" -3d55'29"
Kishwar Sarwar xx
Cyg 20h32'3" 40d50'59"
Kislányomnak Sárának szeretettel
Cas 0h47'28" 50d14'54"
Kismajom csillaga Dinótól
Cnc 8h10'1" 30d50'22"
kismat
Tau 3h31'12" 20d6'44"
Kismegike-Szalontay Orsolya
Uma 10h5'56" 58d26'56"
Kismet
Uma 10h55'5" 59d53'49"
Kismet
Cyg 21h18'57" 52d57'25"
Kismet
Uma 13h40'23" 58d22'25"
Kismet
Uma 9h40'27" 64d16'30"
Kismet
Ori 5h11'31" -5d50'24"
Kismet
Cma 7h14'46" -17d59'10"
Kismet
Ori 6h11'54" 17d52'29"
Kismet
Cyg 21h32'31" 52d16'36"
Kismet 9
Crb 16h4'50" 32d33'18"
Kismet Rebecca
Crb 15h41'5" 25d43'40"
Kisner's Everlasting Life
Umi 16h49'39" 75d5'39"
KISS
Leo 9h26'23" 14d29'52"
Kiss About It
Lib 15h45'20" -28d58'8"
Kiss Andrea
Uma 11h59'5" 47d37'49"
Kiss Attila, a legjobb férj
Psc 23h11'12" 3d35'27"
Kiss Bence Dávid
Uma 9h28'34" 59d10'40"
Kiss Cris
Crb 16h18'31" 38d36'58"
Kiss & Hug
Uma 11h8'27" 52d46'7"
Kiss 'n' Lala Powell
Cap 21h16'18" -16d5'22"
KISS UPON A STAR
Sgr 19h1'28" -30d16'5"
kiss8506
Sco 17h31'6" -41d36'13"
Kissdipity
Cap 20h30'50" -12d30'50"
Kisses For Cassie
Gem 6h55'30" 22d13'57"
Kisses from Heaven
Dra 19h25'59" 72d54'19"
Kisses Of Ceria
Leo 9h50'42" 28d20'38"
Kisses till then xx
Cnc 9h16'34" 26d50'24"
Kissie Monster
Dra 10h6'36" 74d47'25"
Kissy
Lyn 7h55'44" 53d19'40"
Kissy May Miller
Cap 21h31'34" -24d15'8"
KisTigris 690205
Uma 9h42'1" 43d16'54"
KISTNA
And 0h3'12" 40d52'29"
Kistybown
Ari 2h51'16" 17d58'35"
Kit
Sco 17h34'6" -34d58'41"
Kit and Hannah Webster
Cyg 21h19'17" 53d59'32"
Kit and Howie
Cyg 21h51'46" 54d4'22"
Kit and Monk
Cyg 21h37'10" 39d8'38"
Kit Benedict
Umi 16h17'8" 75d52'49"
Kit Bryan Horton
Dra 17h14'27" 65d23'2"
Kit Ginn
Lmi 10h13'59" 39d39'54"
Kit Martin
Crb 15h31'6" 28d44'56"
Kit Zinser
Lib 15h6'59" -22d46'34"
Kitamu Kareema
Cap 20h30'59" -10d48'56"
Kitaygorodskaya
Vir 13h0'22" -3d53'10"
Kitcat & Barry 18605 QM
Cas 0h13'39" 53d51'40"
Kitcyushka
Uma 12h11'54" 57d44'25"
KITEK
Lyn 8h37'49" 40d58'10"
Kith mi Roomio
Sgr 18h19'24" -32d1'59"
Kitiera
Ori 6h17'58" 11d14'59"
Kitison
And 23h32'17" 39d7'36"
Kititas
Lyn 7h37'21" 53d10'49"
kitkatbutterfly
Ori 5h42'23" 8d7'29"
KiTo Naldi
Cas 0h38'50" 61d4'9"
Ki'Ton
Lib 15h36'27" -18d1'43"
Kit's Eternal Guardian
Cru 12h41'57" -62d37'51"
Kitsune's Love
Ari 2h46'39" 27d17'52"
Kitt
Lyn 6h42'41" 60d2'28"
Kitt Ashely Westpfahl
Gem 7h7'58" 33d45'21"
Kitt Star
Umi 16h21'23" 70d3'39"
Kitten
Cap 21h32'45" -9d19'14"
Kitten
Sco 17h52'26" -36d21'29"
Kitten
Cnc 8h25'39" 9d54'18"
Kitten and Cougar
Lyn 9h8'36" 38d40'16"
Kitten ~ Precious Love
And 2h20'14" 39d47'47"
Kitten4Evr
And 0h53'17" 41d28'3"
Kittens
Cyg 19h43'17" 52d17'2"
Kittie "Sweetie" Lai
Cru 12h18'17" -56d57'17"
Kittiya Jaravijit
Ori 6h11'38" 7d19'12"
Kittlesen
Leo 10h28'55" 18d52'8"
Kitttie
Uma 13h12'24" 61d16'43"
Kitty
Sgr 18h1'50" -17d35'56"
Kitty
Lib 15h39'17" -28d12'38"
Kitty
Ori 5h28'40" 9d58'4"
Kitty
Ori 6h20'13" 8d15'22"
Kitty
Lyn 7h8'49" 48d28'17"
Kitty
And 0h57'10" 39d58'37"
Kitty
Lyn 9h13'50" 34d53'30"
Kitty Almon
Ori 5h43'12" 7d38'17"
Kitty B (Cathryn Bernice)
Cnc 7h57'32" 15d23'6"
Kitty Belle
Lyn 8h9'20" 49d26'21"
Kitty & Bob Makley
Cyg 20h17'47" 34d32'2"
Kitty Boo
And 1h11'33" 45d5'26"
Kitty Bussell
Vir 12h35'35" 4d55'28"
Kitty Chi
Cap 20h34'45" -21d28'52"
Kitty Daniel
Cnc 8h26'10" 13d39'43"
Kitty Delicata's Guiding Star
Lmi 9h43'3" 40d41'5"
Kitty Dillewyn
Lyn 7h29'46" 49d40'26"
Kitty Efa Davies
Lyn 8h6'13" 59d3'16"
Kitty Gillis
Leo 11h22'2" 12d35'24"
Kitty H
Cet 2h21'46" -4d47'25"
Kitty Kat
Boo 14h31'54" 52d9'59"
Kitty Kat
Lyn 9h3'48" 43d48'49"
Kitty Kay
Lyr 18h16'59" 34d24'17"
Kitty Lee
And 23h56'32" 40d48'51"
Kitty Liczkowski
Aqr 22h25'52" -1d35'32"
Kitty Lou
Leo 10h57'2" 6d9'16"
Kitty loves her Cubby
Sco 16h48'53" -28d1'59"
Kitty n Poodle
Cam 4h21'34" 65d49'12"
Kitty Nebula
Aqr 23h3'28" -6d38'22"
Kitty Nickels
Uma 13h58'52" 54d54'7"
Kitty St. John
Cas 1h35'59" 71d21'53"
Kitty Star
Lyn 7h32'38" 41d44'46"
Kitty Teo
Vir 11h53'14" -4d32'27"
Kitty Wakefield
Uma 12h2'32" 29d50'42"
Kitty623
Cnc 8h29'25" 14d21'57"
Kittybone7
Gem 7h45'33" 34d29'55"
KittyKat
Cap 21h12'38" -24d58'21"
KittyNick
Her 18h36'15" 25d57'41"
Kitty's Genesis
Lyn 7h34'43" 49d39'9"
Kitty's Nightlight
Uma 10h24'39" 61d4'38"
KitWhits
Ori 6h9'4" 8d59'31"
Kitzileo
Ori 5h54'38" 20d49'8"
Kiva
Cma 7h3'2" -29d6'4"
Kiva Harris
Lyn 8h16'15" 39d49'10"
Kiva Marie Kaibob
Per 3h23'33" 45d36'17"
kiwi
Cam 4h8'42" 66d53'51"
Kiwi
Cap 20h38'39" -18d48'30"
Kiwi 250487
Cyg 19h42'29" 40d37'22"
Kiwi Betancourt
Uma 10h26'3" 56d55'39"
Kiwi Reyes
Aql 19h47'3" -0d0'25"
Kiwini Lika Kuonoono Hali
Cyg 19h59'40" 35d12'47"

Kiya 'Maguire' Brown
And 0h49'24" 46d7'36"
Kiya Michelle West
Cnc 8h29'22" 18d48'22"
Kiyah Love Always N Forever
Uma 8h14'53" 65d26'26"
Kiyan 1
Umi 16h20'16" 75d24'25"
Kiyanta Alexander
Leo 10h9'28" 25d17'22"
Kiyara
Cen 11h43'13" -59d36'10"
Kiyomasa Kuwana
Uma 13h30'35" 56d31'5"
Kiyoshi & Erina
Lib 15h5'27" -17d18'17"
Kiyota & Chieno
Aqr 23h7'49" -23d49'27"
Kizzified For Aunt Kizzyz Boyz
Cnc 9h4'5" 12d40'0"
Kizzin
Dra 18h45'6" 70d45'58"
Kizzy
Gem 7h14'51" 27d47'14"
Kizzy#7
Crb 16h0'15" 35d9'46"
Kizzy Serento Hollis
Ori 6h18'47" 14d46'44"
Kizzy Will You Marry Me
Leo 9h56'44" 17d37'21"
KJ
Leo 9h32'54" 27d41'12"
KJ
Leo 10h49'14" 14d26'0"
KJ
Psc 1h3'13" 31d23'34"
KJ 53
Cnv 13h19'25" 43d1'0"
KJ (Kathleen Audrey Johnson)
Per 2h26'25" 54d38'58"
KJ shining love
Lyr 18h54'35" 33d27'10"
KJ, My Love
Cas 1h48'50" 67d34'50"
kjac
Her 17h45'40" 47d26'1"
K.J.C.
Dra 12h20'3" 71d41'18"
KJDarnell
And 2h23'37" 49d1'8"
Kjell Wicksell
Leo 10h6'20" 26d56'9"
Kjersten Jane Turpen's Lucky Star
Umi 15h45'24" 76d40'1"
Kjerstin's Shining Smile
Uma 11h56'25" 62d59'17"
KJJP2006
Sge 19h24'15" 16d56'12"
KJN/R 6/10/60
Gem 6h43'57" 21d56'16"
KJR Star for Mom
Uma 12h5'57" 47d10'50"
KJ's "SSS" Star
Peg 23h21'1" 15d58'37"
Kjujik #1 Gosiaczek
Gem 7h4'17" 22d21'23"
KK
Tau 4h26'1" 26d46'41"
kk
Cas 0h1'1" 53d44'11"
KK
Lyn 8h41'43" 44d32'3"
K.K. Holiday
Uma 11h16'38" 52d47'3"
kk Rainbow
Ori 4h56'26" 3d18'33"
kk111mdm
Lmi 9h54'58" 38d22'3"
KK-AW
Leo 9h49'53" 15d39'49"
K-K-K-K-Kyle
Leo 11h14'0" 17d55'58"
KKMAB&DCB ad infinitum
Sge 19h52'54" 17d29'9"
KKRKPJ Three Hearts
Tau 4h28'46" 16d41'39"
KL 9-17-2004
Uma 10h18'0" 43d28'46"
KL - Alyssa M. Mazzoli
Cnc 8h50'40" 14d8'54"
K.L Reis
Aur 5h48'0" 50d6'39"
KL Wilburn
Gem 6h44'41" 24d3'11"
KL143
Uma 10h13'16" 72d22'13"
Klaassen-Rudman
Uma 8h23'6" 60d48'15"
Kladbritin
Cyg 20h41'38" 36d34'44"
Klagges, Torsten
Uma 11h20'24" 36d25'2"
Klairpôl
Cyg 20h58'49" 32d21'55"
Klamachpin Witscheman's Reign
Uma 12h20'55" 56d6'13"
Klancy Adeline Mickan
Cap 21h53'4" -23d10'13"
Klappstein, Dieter
Ari 2h21'50" 19d34'44"
Klara
Cam 7h45'57" 67d20'6"
Klara
Uma 10h16'9" 68d55'22"
Klara Huncar
Lib 15h29'26" -4d23'48"
Klara Murbach
Umi 15h39'40" 79d18'6"
Klara Salidzik
Uma 9h15'20" 47d35'10"
Klara Smidova Christ Dalmares
Psc 1h24'44" 22d42'35"
Klara Zimmermann-Müller
Boo 13h49'39" 23d28'1"
Klárika emlékére
Sgr 18h55'1" -35d22'32"
Kläritt
Umi 14h22'35" 67d15'37"
Klassen, Harald
Ari 2h35'37" 18d4'2"
Klaudia Bella
Sco 17h3'28" -39d24'4"
Klaudia Katrina Lipford
Psc 1h10'48" 27d34'58"
Klaudia Leszyk, The Greatest Friend
Cnc 8h7'5" 22d34'25"
Klaudia Novak
Lyn 7h51'25" 39d52'29"
KLAUS
Cyg 20h31'27" 43d4'23"
Klaus Albers
Ori 5h57'12" 18d11'53"
Klaus Baldini
Uma 11h2'43" 37d53'3"
Klaus Baumgartner
Uma 13h26'44" 62d59'49"
Klaus Berkel
Ori 5h58'27" 7d17'27"
Klaus Bertram & Manuela Christoph
Uma 13h18'42" 63d13'31"
Klaus Bogus
Uma 9h56'38" 69d41'19"
Klaus Buhr
Uma 9h58'10" 48d35'47"
Klaus Capra
And 1h16'11" 42d29'38"
Klaus D. Machowsky
Uma 9h20'28" 67d46'33"
Klaus Dubberstein
Uma 8h42'44" 64d40'10"
Klaus Eisermann
Uma 8h54'28" 56d41'25"
Klaus Fritsche
Uma 12h1'48" 35d57'26"
Klaus Funke
Uma 9h19'8" 55d32'54"
Klaus George Roy
Aqr 20h43'31" 0d43'6"
Klaus Görner
Uma 11h14'51" 30d6'21"
Klaus Hallmanns
Uma 11h12'30" 42d53'48"
Klaus Heiner Kost
Uma 9h39'14" 47d31'30"
Klaus Huber
Uma 12h33'45" 52d32'10"
Klaus Keiner
Uma 10h30'42" 51d11'24"
Klaus Knauth
Uma 11h41'59" 60d17'26"
Klaus Langenstrassen
Uma 9h33'27" 58d47'8"
Klaus Leciejewski
Sco 17h36'39" -40d49'26"
Klaus Ludwig Geraldy
Ori 6h19'20" -1d7'28"
Klaus Nölle
Ori 5h54'13" 13d6'1"
Klaus Pankow
Uma 8h42'54" 51d59'0"
Klaus Petersen
Uma 8h17'49" 67d42'33"
Klaus Stirmlinger
Ori 6h6'17" 21d20'46"
Klaus Stratmann
Uma 9h19'17" 64d13'48"
Klaus Waldschmidt
Umi 15h32'28" 74d17'23"
Klaus Weckop
Ori 6h18'32" 9d12'51"
Klaus Werner Jähnel
Ori 5h55'42" 12d5'53"
Klaus Wiegard
Uma 9h12'17" 57d47'1"
Klaus Wybranietz
Ori 6h22'31" 15d28'50"
Klaus (Yeti) Tennhardt
Uma 14h8'27" 61d3'29"
Klaus-D. Hielscher
Uma 9h14'12" 66d18'51"
Klauser, Artur
Leo 10h25'12" 19d22'13"
Klaus-Gustav Noelle
Ori 5h56'4" -2d20'43"
Klavdia Petrovna Sklarova
Aqr 20h54'48" -7d22'14"
Klayten Perreault
Ori 5h27'15" 2d27'24"
KLB14344
Her 17h38'1" 28d49'29"
KLD Eagles Wings
Aql 19h37'31" 3d58'38"
KLE21
Lyr 18h59'29" 26d2'47"
Klea
Uma 10h29'11" 43d8'23"
Kleanthe C. Caruso
Gem 7h22'29" 17d43'59"
K-Lee Elentári
Dra 19h19'47" 65d36'19"
K-Lee's Kosmos
Psc 0h18'52" 15d12'28"
Kleigh
Ori 6h24'56" 10d3'24"
Klein, Anne
Uma 8h44'10" 47d21'30"
Klein, Christian
Uma 13h28'44" 59d44'29"
Klein, Julius
Uma 10h31'58" 65d45'28"
Klein, Klaus Jakob
Ari 3h16'32" 18d56'36"
Klein, Roman
Cnc 8h19'49" 19d39'1"
kleine süße Andi
Ori 6h19'12" 19d34'53"
Kleiner
Uma 10h38'36" 42d21'0"
kleiner Entestern
Aql 19h33'45" 4d4'34"
Kleiner Leo
Uma 9h13'35" 49d53'51"
"kleiner Stern"
Umi 16h10'59" 74d59'12"
Kleines Mädchen
Uma 11h4'2" 55d6'8"
Kleingon
Pyx 8h39'40" -34d7'12"
Kleinridders, Horst
Uma 8h45'32" 49d49'40"
Kleintje Cookie
Umi 15h49'31" 77d8'54"
Klemens Behner
Ori 6h17'46" 3d22'41"
Klemens Jungeblodt
Uma 10h18'11" 49d51'54"
KlemensundHertha-Spitzlei
Uma 9h3'58" 56d6'49"
Klengen Tiger
Oph 16h37'30" -4d47'42"
Klenke, Fritz
Uma 10h38'42" 46d27'48"
K-Lepru
Tau 4h25'24" 17d20'17"
klfmd-5
Uma 11h39'29" 29d34'49"
KLG
Cas 0h32'22" 53d48'4"
K.Licona
Aqr 22h29'10" -3d50'7"
Kline Specialties
Uma 10h9'49" 53d14'23"
Klingebiel, Johannes-Georg
Uma 10h17'31" 50d1'17"
Kling-N-Starr
Cyg 20h20'53" 54d25'1"
Klink, Wolf
Aqr 21h54'33" 0d28'24"
Klinky
Cnc 8h16'35" 23d17'45"
Klinny
Leo 11h48'57" 26d2'30"
Klippel Gabriella, örök Szerelmem
Uma 8h27'27" 60d51'47"
Klippenstein, Olga
Uma 8h43'49" 53d56'14"
Klivinyi Zoltán
Sgr 18h59'19" -33d49'32"
KLK111889
Uma 11h22'5" 31d56'15"
KLMMEW
Cyg 21h31'56" 45d39'34"
Klock Family Star
Aql 20h14'5" -7d57'8"
Kloie Halmoni Kim
And 23h55'4" 46d7'57"
Klonar
Tau 4h13'35" 15d13'8"
Kloo & Christine
Uma 11h25'56" 30d24'36"
Klose, Peter
Uma 11h39'10" 38d44'12"
Klotz, Carsten
Uma 8h41'13" 57d51'47"
KLS 03-10-70 Big Daddy
Psc 23h31'28" 3d57'38"
KLS-1
And 23h29'36" 47d47'51"
Kluiber
Ori 6h10'30" 20d53'46"
KLY120
Cnv 13h44'2" 31d46'25"
Klytia Noel Dutton
And 2h6'48" 41d2'25"
KM 4Ever 5683
Vir 12h50'9" 5d12'49"
KM71
And 0h15'47" 27d57'53"
Kmac07
Per 3h3'6" 53d16'33"
KMAG7
Uma 12h21'24" 59d12'50"
K-Man
Psc 1h25'42" 18d42'18"
KMB - Tinkerbell
Sco 16h6'47" -11d47'25"
KME JMS 2/6/04
Aqr 22h54'53" -6d13'21"
Kmeczó János
Uma 9h4'23" 71d17'32"
KMiller551
Uma 9h52'18" 50d5'16"
K.M.J.C.
Srp 18h25'10" -0d11'19"
KML
Cnc 8h29'30" 26d4'39"
KML2004
Aur 5h13'5" 42d50'5"
KMP - 1/23/1980
Ori 5h30'28" 1d49'5"
KMR and KJE 4EVER
And 1h33'0" 49d13'11"
KMR Star - my lady love for always
Aql 19h25'22" -0d9'45"
KMS1
Umi 16h26'11" 74d14'40"
KMStopka 112006
Ori 5h56'16" 20d40'6"
kmtmdy12-07-2005
Uma 9h52'28" 51d37'13"
kmw22872
Psc 1h50'26" 7d42'42"
KN 831 always
Psc 1h19'9" 32d54'53"
K.N.77 Max Zero5 M. Melchor
Dra 17h0'39" 56d5'43"
Knapp Class of 2005
Cma 6h48'38" -14d8'13"
Knapp, Peter
Uma 9h57'19" 60d17'15"
Knatestahl
Tau 4h18'31" 8d24'56"
Knatisha Nelson
Cas 0h33'22" 55d5'8"
Knauby
Uma 9h28'34" 58d10'7"
Knecht
Umi 15h46'1" 85d8'32"
Kneifl, Lothar
Ori 5h36'39" -2d8'49"
Knellex
Her 17h23'41" 29d43'38"
Knetsch, Nicole
Uma 12h3'57" 39d18'8"
KNHCLH
And 0h20'15" 41d17'0"
KNIGHT IN SHINNING ARMOR
Eri 3h56'31" -0d5'28"
Knight of Stars
Sco 16h9'3" -15d18'26"
Knight Warrior
Cyg 20h56'49" 46d47'39"
KnightinShiningArmor
Dra 17h47'4" 54d40'14"
Knightkat
Ari 2h49'17" 28d44'57"
KnightStar
Dra 18h0'19" 78d51'46"
Knighttour
Psc 0h19'49" 19d14'45"
Knigit
Lyn 8h25'9" 56d3'50"
KnnyChsny's baby
And 23h23'46" 38d38'26"
Knoch, Siegfried
Sco 17h15'40" -32d27'28"
Knolli 04.07.1969
Ori 6h12'28" 14d48'32"
Knopf, Thomas
Uma 8h54'21" 47d46'51"
Knoppie
Ari 2h32'40" 26d27'15"
Knot to be undone. Foreverinv.
Aql 20h9'11" 11d58'46"
Knotted Passion
Mon 6h53'57" 3d15'17"
Knox
Aql 19h9'10" 2d34'20"
Knox John Maxwell Heath
Cep 0h22'51" 75d55'11"
KNRBS - 14/8/2003 - Forever -
Tau 3h43'16" 27d11'53"
Knubby
Uma 10h39'41" 66d14'27"
Knucklehead
Uma 11h4'55" 64d50'24"
Knud and Adda
Her 16h47'40" 30d10'52"
KNUDDEL
Ori 5h6'51" 9d15'13"
Knuddel
Uma 9h13'12" 58d52'31"
Knuddelbabe
Umi 15h48'25" 82d13'55"
Knuddeltigerli
Cep 22h12'22" 69d21'11"
Knuddly (Me and you, you and me...)
Aur 6h39'16" 37d18'32"
Knuffel
Boo 15h5'14" 13d37'13"
Knuffi & Sünneli forever
Lac 22h52'29" 50d25'35"
Knut Dresel
Uma 11h3'43" 44d43'35"
Knut-Patrick Olsen
Uma 8h41'35" 66d10'56"
Knutson
Uma 10h7'19" 64d50'26"
KO OLINA
Cam 4h36'45" 66d27'44"
Koa Buddy
Ori 6h13'39" 18d45'8"
Koa Orion Lefebvre
Ori 5h47'5" 2d57'58"
Koala
Ari 2h49'11" 15d41'27"
Koala
Tau 3h42'58" 22d53'20"
Koala
Lyr 19h27'59" 37d58'19"
Koala Bear
Gem 6h42'51" 14d6'9"
Koala Bear
Uma 10h5'17" 69d8'43"
KoalaBaby
Aqr 22h31'59" -3d38'2"
Koaletto
And 1h36'14" 47d44'16"
KOAM Howard A. Stern
Uma 9h35'30" 59d58'38"
Koarbin Thomas Edward Anderson
Ari 3h26'5" 19d56'23"
KOB
Vir 12h48'38" 3d42'31"
KOB FAMILY
Gem 7h39'46" 22d13'35"
Koba's Star
Cep 22h26'5" 82d4'0"
Kobbe's Star
Cep 20h39'20" 58d35'33"
Kobe
Cma 6h44'56" -14d51'52"
Kobe Gold's Ever Rising Star
Leo 10h15'23" 17d33'15"
Kobe Lili-Matise Maher - Sparkling Star
Cru 11h58'32" -55d48'35"
Kobes, Rosemarie
Ori 5h31'41" 10d43'7"
Kobetz, Sonja
Uma 12h30'52" 62d17'35"
Kobi Anne Finley
Cap 20h43'19" -21d10'4"
Kobi : Friend and Master Scientist
Cnv 13h51'49" 33d8'41"
Kobi Stephen Leary - Our Little Star
Cru 12h39'44" -60d24'42"
Kobi-Love's-Ginger
Vir 14h33'53" 3d11'14"
Kobi's Mother
Psc 1h30'22" 20d45'8"
Koby
Cnc 9h9'14" 27d30'45"
Koby Michael McGill
Umi 14h18'15" 74d21'24"
Koby Robert Srijemac
Boo 15h38'53" 41d12'26"
KocaSzajkó
Cnc 9h9'47" 9d36'45"
Koch, Carmen
Aqr 21h24'40" 1d31'43"
Koch, Klaus-Dieter
Uma 10h17'33" 39d25'22"
Koch, Sven
Uma 11h35'41" 30d49'40"
Koch, Wolfgang
Uma 9h58'54" 54d2'16"
Kocham Was Teresa & Zdzislaw
Uma 9h29'29" 53d13'16"
Kochan and Mika Ai no Hoshi
Lib 15h21'56" -24d0'31"
Kochana Helenka
Peg 21h31'23" 18d15'41"
Kochana Mari
Uma 8h19'6" 62d58'41"
Kochana Monisia - Monika Gruziel
And 1h58'20" 37d36'43"
KOCHANIE
Uma 12h45'1" 58d1'17"
Kochav Aloni
Uma 11h31'49" 45d0'57"
Kochav Yair Noam Ha'Levi
Sco 17h5'42" -35d32'8"
Kocieda
Uma 9h48'9" 48d58'44"
Kocikoci
Ori 6h10'32" 5d45'6"
Koco Trpkovski
Lyr 18h45'34" 32d37'40"
Kocsmárszki Kitti Hercegno 1987
Sgr 19h17'25" -24d57'2"
Kodi
Vir 14h22'27" 0d3'30"
Kodi Bear
Uma 13h49'47" 51d59'51"
Kodi Landon Hancock
Uma 13h18'6" 58d9'2"
Kodi We Love You
Cmi 7h58'38" 7d22'48"
Kodiak
Cmi 7h35'53" 5d22'29"
Kodiak
Cap 21h14'26" -21d29'58"
Kodie
Umi 15h14'16" 72d49'32"
Kodman
Uma 11h28'2" 31d50'59"
Kody Alexander Kotimko
Her 17h40'7" 44d20'48"
Kody Bassler
Mon 6h46'51" -0d7'10"
Kody Herman
Lyr 18h53'9" 32d53'17"
Kody Martin's Star
Cap 20h24'3" -12d2'59"
Kody Michael Chapman
Psc 0h19'26" 3d28'22"
Kody Thomas Cimaglio
Leo 10h5'47" 7d57'21"
Kody-Aleczander Ayze
Uma 10h3'34" 70d49'55"
Koenig
Ori 5h28'37" 3d20'33"
Koerdt, Klaus
Sgr 18h1'39" -27d33'6"
Koesmanjalie
Leo 11h9'27" -0d23'17"
Kofi 54
Boo 14h59'20" 13d49'10"
Kofi and Nane Annan
Cyg 21h35'31" 52d35'20"
KOGAR
Boo 14h15'11" 17d43'44"
Koggi
Ori 6h19'44" 10d20'43"
Kohl Tyler
Vir 12h19'2" -0d54'14"
Kohlenberg Cotoc Maria
Cep 21h18'48" 63d53'29"
Köhler, Torsten
Ori 5h11'36" 15d36'15"
Kohlie Ole'
Lib 15h21'4" -22d11'9"
Kohn, Klaus-Dieter
Uma 11h43'1" 28d34'34"
Kohn, Sabrina
Uma 11h7'19" 56d33'50"
Kohpeiß, Jürgen
Uma 8h12'6" 63d56'10"
Kohpeiß, Jürgen
Uma 8h16'56" 64d11'20"
Koi Kimo
Lib 15h35'34" -7d48'41"
Koi to seki to wa kakusarenu
Cyg 21h37'15" 54d9'15"
Koibito
Cra 18h41'42" -45d17'55"
Koichiro Sugimori
Cep 23h53'40" 74d40'30"
Kojima-Cohen
Crb 16h17'6" 27d56'27"
Kokie Swanger
Uma 9h43'14" 54d32'59"
Kokka
Umi 15h39'56" 84d22'53"
Kokkonitsa
Cyg 19h36'55" 30d44'28"
Koko
Her 18h49'3" 22d14'0"
Ko-Ko #One
Uma 10h26'41" 62d27'30"
Koko Salib
Uma 11h8'0" 43d32'22"
Kokobee
Cnc 8h45'2" 21d55'53"
Kokomo
Cap 20h21'56" -13d7'42"
KOKOPELLI
Gem 6h51'26" 22d23'14"
Koko's Akieee
Cam 3h59'36" 65d8'40"
Kolajas Focus
Uma 12h48'38" 57d31'33"
Kolbi
Lmi 10h36'2" 37d48'28"
Kolby Bo
Dra 19h35'41" 74d16'26"
Kolby Brett Moyer
Gem 6h57'28" 27d19'2"
Kolby & Colin
Psc 1h23'13" 24d48'27"
Koldoon
Cap 21h43'16" -13d46'19"
Kole and Carter
Aqr 22h25'31" 0d58'36"
Kolecke-Dwyer
Crb 15h35'29" 27d40'18"
Koli kiscsillaga
Psc 23h30'56" 7d21'29"
Kolibori Titi
Apu 13h58'26" -74d26'41"
Kolie
Leo 9h30'10" 10d48'28"
Koliya Mac Star
Cra 19h5'56" -37d37'47"
Kolja
Cap 21h50'38" -14d47'52"
Kolk, Christl
Uma 8h53'1" 57d57'29"
Kolleen Bartley
And 0h55'7" 37d4'13"
Kollene Mayer Magari
And 23h27'34" 46d56'16"
Kollinslight
And 0h25'48" 29d4'10"
Kollyn Scott Reide Cozens
Cnc 8h59'42" 16d39'58"
Kolob
Uma 10h18'15" 48d28'13"
Kolokea
And 23h14'10" 42d13'4"
Kolton Alexander Alday
Cnc 9h12'39" 20d17'0"
Kolton Marc
Ori 5h29'42" 5d24'25"
Kolton Thomas
Aqr 22h58'19" -8d38'25"
Kolyn Thomas Quinn
Aqr 22h12'30" 1d59'1"
Kolyn Thomas Quinn
Per 3h11'20" 53d32'13"
Koma, Ko'u Mau Loa Aloha
Gem 6h51'23" 31d10'16"
Komal Dave
Psc 23h45'36" 5d48'30"
Komal Karnik
Dra 15h35'17" 63d13'5"
Komal Kaur Dhillon: "Soft Princess"
Cyg 21h17'34" 47d41'43"
Komasim
Cyg 20h47'28" 47d42'15"
Komet Bear
Sco 16h11'8" -12d4'39"
KomJankóm/Komjáti János égicsillaga
Leo 9h42'25" 11d46'8"
Komninos Gus Karellas
Per 4h20'15" 44d7'59"
Komorek, Dieter
Uma 11h17'50" 63d27'47"
Kona
Leo 9h57'41" 15d19'1"
Kona
Cnc 8h43'13" 18d1'39"
Kona 2007
Cnc 9h5'51" 12d19'16"
Kona Hogge
Pup 7h8'32" -38d34'39"
Kona Java Plantation
Cnc 8h8'41" 11d23'52"
Koncseg Zoltan
Dra 18h38'12" 59d47'29"
Koncsik Margit
Ari 2h34'40" 22d31'28"
Konemann
Lyn 7h20'25" 52d57'47"
Koni Anne Dixon
Lib 15h54'3" -17d24'16"
Konietzko Brothers
Her 16h49'39" 33d24'44"
Konitos
Dra 16h51'11" 67d11'23"
Konner
Cam 5h36'10" 76d21'32"
Konner
Lyn 8h48'31" 40d12'37"
Konner Matthew Valen
Ori 5h28'26" 0d55'14"
Konnor Joseph - The Shining Son
Sco 17h33'50" -45d4'51"
Konnor Macalister Perrin
Lyn 6h38'2" 57d15'39"
Konrad
Ori 6h19'15" 15d8'16"
Konrad Douglas Schnabel
Sco 16h14'16" -13d6'28"
Konrad Markwardt
Her 16h48'25" 14d40'52"
Konrad Paul Liessmann
Cma 6h18'41" -32d40'37"
Konrad Slettedahl
Cep 7h17'48" 85d42'48"
Konsey
Gem 7h3'21" 26d33'17"
Konstance Evagelia Koutoulakis
Cas 23h28'35" 52d59'59"
Konstantin
Umi 15h13'57" 69d54'35"
Konstantin
Peg 21h49'36" 25d10'33"
Konstantin
Per 4h20'37" 38d34'41"
Konstantin Brovot
Uma 11h17'6" 41d13'47"
Konstantin Sadin
Per 3h18'0" 52d55'18"
Konstantine
And 1h56'25" 39d8'44"
Konstantine
Gem 6h38'25" 20d53'19"
Konstantine
Uma 9h34'5" 60d4'38"
Konstantine
Umi 14h40'4" 78d21'37"
Konstantine
Vir 12h26'13" -6d51'58"
Konstantine
Ori 6h5'5" -0d47'26"
Konstantine
Ori 5h20'52" -8d31'2"
Konstantine
Aqr 22h4'42" -2d56'39"
Konstantine
Cap 20h22'28" -11d7'5"
Konstantine
Lib 15h12'33" -13d58'16"

Konstantine Vir 13h36'2" -16d24'45"
Konstantine Aqr 23h29'16" -15d44'17"
Konstantine Pho 23h40'24" -57d39'47"
Konstantinos (Dean) Psilopoulos Mon 6h51'55" -0d56'25"
Konstantinos Haritsis Uma 13h37'46" 52d26'17"
Konstantinos Kandilis Uma 11h16'8" 29d16'25"
Konstantinos Korletis Uma 10h57'31" 35d38'2"
Konstantinos Macheras Uma 10h21'4" 64d2'59"
Konstantinos Ntoukas 15-10-1984 Lib 14h56'53" -16d18'27"
Konstantinos Tzembelikos Gem 6h53'5" 25d5'6"
Konstantinou Cru 12h38'35" -64d10'32"
Konstantins Melnikovs Sco 17h20'10" -45d3'37"
Konstanty Henryk Hordynski Vul 19h40'45" 25d31'20"
Konstanze Henschel Uma 12h2'11" 42d31'3"
Konstatine Ari 3h16'56" 22d24'15"
Kontos Her 17h41'41" 15d24'23"
Kony Lyr 18h43'46" 34d32'42"
Koofkie Uma 13h35'45" 56d14'46"
KooKiedaisch Mon 7h15'25" -0d55'33"
Kookla Uma 8h39'5" 67d19'44"
Kookoobabes Dra 17h26'22" 58d57'19"
Kool Kris Uma 11h23'18" 57d13'43"
Kool-Aid Tau 4h41'9" 22d55'53"
Koon-Kor Ori 6h14'50" 8d54'22"
Koontz Tra 16h26'4" -60d48'18"
Kooper K Reed Cep 21h17'30" 64d49'15"
Koos Cra 18h0'6" -37d1'27"
Koos Terblanche se eie reenster Ori 5h37'29" -1d9'54"
Koosh Ari 2h50'9" 22d19'19"
Kooshka And 1h10'51" 42d50'52"
Kopal Madhu Sgr 18h31'1" -25d3'23"
Kopecki, Klaus Uma 11h34'36" 28d25'41"
Koppa's Zvezda Boo 13h49'3" 17d7'17"
Kora Louise Houselander And 23h20'20" 47d27'1"
Koracey Umi 14h48'41" 73d6'38"
Koran Lib 15h16'0" -23d47'52"
Korapat Aqr 22h19'9" -17d7'32"
Korb-A-Naro! Umi 15h32'53" 70d55'33"
Korbel San Felasco 1/15/03 Lmi 10h38'21" 29d48'11"
Korben's Eternal Love for Mummy Cas 0h36'0" 66d17'51"
Körber, Peter Uma 9h36'55" 42d49'26"
Korbin George Aqr 22h4'37" -13d57'44"
Korbin Matthew Hightower Her 16h55'2" 16d10'22"
Kord Cyg 21h20'4" 38d48'12"
Kordbarlag, Sven Uma 11h29'41" 29d45'0"
Kordula Wals Uma 9h30'21" 50d10'36"
Kore Gem 7h33'18" 24d7'22"
Koreen Cyg 19h34'9" 28d46'56"
Korena Lyn Payne Gem 6h54'28" 33d48'5"
Korey Gem 6h51'59" 32d23'35"
Korey Alexander Ford 04-15-1992 Ari 2h12'20" 12d46'43"
Korey and Kevin Lee Woodin Sgr 18h2'13" -21d23'34"
Korey Maurell Edwards II Vir 11h49'38" 3d4'2"
Korey Roati Gem 7h7'44" 15d5'0"
Korey Ryan Feeley Aql 19h42'1" -0d19'38"
Korgaki Crv 12h26'32" -14d24'22"
Kori Leo 9h58'17" 17d18'35"
Kori and Mike Sge 19h50'59" 17d41'26"
Kori Ann Kilass Cap 20h46'24" -20d49'32"
Kori Belanger Gem 7h22'51" 26d27'25"
Kori Joy And 1h39'6" 39d48'10"
Kori Lynn Hickrod Lyn 8h12'37" 55d19'29"
Kori Lynn Martin Lyn 7h22'38" 44d51'49"
Kori McCall Black Cap 21h45'8" -24d2'3"
Kori Paige Cnc 8h36'23" 20d45'11"
Kori Rose Gem 6h28'30" 12d52'30"
Kori Tolbert 6/4/79 6:40 a.m. Lyn 8h2'8" 40d23'28"
Korie Cnc 9h15'54" 12d4'26"
Korie Divine Crb 15h49'18" 37d5'30"
Korie Lynn Dvorak Ori 5h58'37" 8d30'5"
Korin Michelle Taylor Lyr 19h19'5" 35d35'22"
Korina Aqr 21h43'49" -0d9'9"
Korina Dolic Gem 7h29'37" 32d23'5"
Korincifer Kolumbuz Ari 3h25'16" 26d13'50"
Korine Garcia Ari 3h2'30" 15d11'12"
Köring, Hans-Henning Uma 12h53'16" 52d25'0"
Korinna Amelie Dunn And 23h39'9" 47d4'25"
Korinna Pedd 08.07.1975 Uma 11h9'34" 40d10'44"
Korinne Marie Robertson Aqr 22h53'5" -5d44'59"
Kori's Light from Heaven Uma 11h15'44" 64d30'24"
Korissa Dra 18h54'11" 77d21'4"
Korissa Grace Cyg 21h48'41" 49d52'11"
Korissa Grace Diomar And 23h49'35" 38d45'4"
Korn, Peter Uma 11h41'15" 32d46'23"
Korneffel, Lothar Uma 8h52'15" 69d14'55"
Kornél Cas 23h19'40" 59d32'27"
Korogy Cyg 20h59'22" 30d46'25"
Koron Dwayne Davis Cas 1h35'59" 64d24'20"
KorrinHeilman Ari 2h45'57" 11d50'15"
Kortney Ori 5h38'36" 4d13'38"
Kortney And 0h23'36" 41d9'5"
Kortney Kay Hull Uma 13h19'28" 59d33'58"
Kortney Leigh Palomba And 1h45'32" 41d44'11"
Kortney N. Loosli Aqr 21h51'42" 1d26'24"
Kortney Nicole Com 12h35'16" 16d36'47"
Kortney's Shining Star Love Cody Cnc 9h17'32" 31d4'5"
Kortni Tau 4h31'33" 17d23'20"
Kortni Umi 13h10'20" 69d24'18"
Kortni Elizabeth Sloan 17 And 2h35'55" 46d57'49"
Koru Tewanee Manaia Joseph Uma 12h32'32" 61d5'23"
Korus Ori 5h58'1" -0d43'5"
Kory B. Littlefield Per 3h10'55" 48d28'55"
Kory "G-Dogg" Garrison Aur 6h12'56" 33d12'22"
Kory Janelle Cyg 20h53'52" 31d13'22"
Kory "Rev. Wigm" Bishop Lib 15h20'16" -26d54'18"
Kory Robert Grahl Uma 14h3'21" 56d57'54"
Kory Schramm Cep 22h1'59" 70d42'1"
Koryn Joel Gatta Vir 14h15'12" -20d6'46"
Ko's Star Burst Vir 13h7'38" 12d5'3"
KOS-5/11/1970 Tau 4h12'51" 25d45'34"
Koshencia Lyn 7h23'8" 55d50'4"
Kosko Pho 23h55'56" -48d4'11"
Kosky And 1h46'39" 41d34'28"
Kosma Gianginis Cyg 19h45'24" 31d32'50"
Kosmala, Bernd Uma 11h5'21" 71d32'18"
Kosmo Kotlar Ari 2h3'19" 21d28'43"
Kosmos Aster De Kalanchoe Aql 19h17'29" -9d32'30"
Kossert, Joachim Uma 8h12'52" 63d38'49"
Kossert, Joachim Uma 8h17'26" 64d43'56"
Kosslitz, Steffen Ori 6h3'43" 7d24'33"
Kosta Uma 8h37'22" 61d7'41"
Kosta Zimaras Cnc 8h29'19" 19d28'33"
Kostas Giannikos Uma 9h10'27" 61d12'15"
Kostas Hazifotis Psc 1h3'33" 25d34'18"
Kostka Anniversary Star Cap 21h6'28" -25d50'33"
Kota Gem 6h49'9" 21d30'38"
Kote Lib 15h0'8" -6d8'38"
Kotik-Katerinushka Lyn 7h26'36" 57d13'28"
Kotilee Uma 10h9'9" 68d27'54"
Kototh Prime Dra 17h47'43" 69d51'41"
Kotton Uma 12h18'54" 60d58'45"
Kottusch, Gesine Sco 16h58'3" -38d43'32"
Koty Per 3h24'32" 45d16'34"
Kotya # 2 And 23h25'26" 48d51'9"
KOUKLI Crb 16h21'4" 34d36'48"
kOurTle Gem 6h51'45" 17d43'2"
Kourtney Lyr 18h46'40" 32d48'58"
Kourtney Peg 22h51'11" 31d18'38"
Kourtney 1965 Aqr 23h1'13" -6d5'7"
Kourtney Burnett And 1h14'21" 39d22'1"
Kourtney Eryn Rambeau Tau 5h8'26" 24d30'56"
Kourtney Lauren Morrissey And 2h32'24" 44d56'40"
Kourtney Lynn Sco 16h10'42" -13d52'45"
Kourtney Marquis Leo 9h30'9" 16d6'18"
Kourtney Michele Slagle And 0h36'19" 36d35'58"
Kourtney Padgett Ussery Uma 11h1'25" 60d28'11"
Kourtney's Serendipity Cnc 8h40'19" 30d52'53"
Kourtnie Hertlein Uma 9h51'59" 52d17'14"
Kovács István Uma 9h36'9" 44d57'34"
Kovács László, a Bramac csillaga Cap 20h33'27" -11d17'0"
Kovács Monika Niké és Dávid József csillaga Uma 9h24'40" 44d28'37"
Kovács, Georg Ori 6h16'30" 11d13'10"
Kovágó Károly Cnc 8h8'51" 16d50'3"
Kovári Zsolt Csillaga Ari 3h4'40" 23d56'29"
Kowalik, Manfred Ori 5h13'36" -8d9'35"
Kowal's Anvil Uma 10h58'57" 47d6'21"
Kowalski, Hans-Günter Uma 10h55'7" 43d7'56"
Kowana Ragland CMP - GaMPI Star Lyr 18h53'12" 37d7'9"
koy tsun Umi 15h49'19" 71d59'18"
Koyas and Rima Cyg 21h18'2" 40d55'27"
Koyata Sdn Bhd Ori 5h34'5" 4d11'59"
Koytchev, Rossen Ori 6h15'32" 10d11'18"
Kozlik Danny Psc 0h59'52" 13d45'33"
Kozmo Uma 9h23'42" 45d34'31"
Kozy One Leo 11h30'25" 1d28'52"
KP Lmi 9h52'0" 35d23'13"
KP And 1h16'29" 39d9'24"
KP & Kate - February 19, 2005 Cru 12h13'16" -61d52'45"
KP Loves KH, Happy Valentine's Day Boo 14h39'12" 15d54'52"
KPAK "9-5-1997" Uma 8h51'4" 71d47'39"
K-PAX Psc 0h38'28" 9d19'23"
kpc6899 Gem 7h47'38" 21d26'54"
K-Pop Ari 2h52'51" 26d45'27"
Kpoxa Vir 13h55'22" 1d57'40"
KPSTAR 040886 Ari 3h9'15" 30d41'29"
KR + JG Siempre Cnc 8h16'42" 19d3'14"
Krae David Martin Lmi 10h33'2" 37d44'44"
K-Rae-Z Ru-May-Z Lyn 6h37'4" 55d10'30"
Krafty Awww Bernard Crb 16h0'47" 25d59'20"
Krähe, Klaus Uma 8h42'15" 72d38'5"
Krähenbühl Erich Lyr 18h50'57" 26d15'54"
Kraig J. Kolomyski Uma 9h31'18" 71d19'36"
Krain, Ramona Uma 9h46'28" 67d59'5"
Kraite Cru 12h31'8" -61d53'58"
Krake, Ulla Uma 9h2'0" 54d57'50"
Krakower Haven Uma 9h43'3" 64d15'46"
Kralj Steve Stojic Cep 0h14'0" 67d53'3"
Kramenna Suntazilljeech Uma 9h36'54" 64d29'48"
Kramer Leo 11h20'47" 19d32'39"
kramer Lyn 7h5'38" 50d10'13"
Kranti Madhusudan Athale Aql 19h54'39" 4d53'32"
KrashyMac Leo 11h44'37" 26d20'32"
Krate Boo 13h37'53" 21d19'44"
Kratz, Hans-Jürgen Uma 10h14'16" 72d46'56"
Kraus Ori 6h4'4" 10d3'54"
Krause, Gerd Uma 9h27'3" 42d4'50"
Krautkrämer, Erika Ori 5h53'34" 21d46'21"
K-Raye Lib 15h30'56" -7d26'42"
Krayer - Atkinson Cas 23h13'54" 55d57'17"
KrazeeStar Umi 14h41'3" 69d57'12"
Kreena Vaghela And 22h57'27" 51d26'17"
Kregg (LaFong) Thornburg Per 4h25'58" 43d13'14"
Krelwa Cnc 9h8'25" 32d47'6"
Kremena Kovacheva Ari 2h37'54" 19d26'21"
Krén Judit (Jucóka, Hippsy) Tau 3h31'7" 17d35'19"
Krepelka's diamond Gem 7h34'41" 15d34'23"
Kresge And 2h10'44" 38d31'56"
Kresl the Great Lyn 8h8'51" 46d27'54"
Kressley Uma 12h28'43" 59d5'24"
Kretschmann, Leon Noah Uma 8h38'21" 64d51'38"
Kreuz, Dieter Uma 8h18'23" 64d38'19"
Kri Her 16h21'25" 4d25'31"
KRIBBAGE Cnc 8h18'58" 14d26'27"
Krickett Del 20h48'6" 12d46'23"
Kriddy1 Tau 3h42'7" 26d8'56"
Kriders Shining Star Umi 18h12'3" 87d58'28"
Kriegel, Andrea Uma 10h9'48" 65d1'20"
Krieger Sgr 18h4'22" -27d22'53"
Krieger Lac 22h26'1" 43d11'34"
Krieger, Rolf Uma 8h50'41" 59d41'17"
Kriener, Heinrich Uma 9h32'26" 65d40'42"
Krigelondo Ori 6h0'59" 18d18'45"
Krika Gem 6h52'8" 12d9'32"
Krikit Baksh Aql 19h11'22" 4d56'47"
Krimar Ori 5h56'16" 21d2'2"
Kris Sge 20h3'22" 17d26'17"
Kris Psc 1h36'46" 13d53'1"
Kris Cyg 20h19'50" 38d39'39"
Kris Cas 2h2'26" 59d41'14"
Kris Lyn 7h59'38" 43d38'23"
Kris Lib 14h39'52" -18d18'58"
Kris Sco 17h21'10" -42d39'41"
Kris<3Megan Cyg 20h20'35" 43d37'28"
Kris 613 Gem 7h22'9" 14d16'41"
Kris and Cheryl Ayres Cyg 19h32'33" 29d28'28"
Kris and Jacqui's Eternal Love Star Eri 3h50'20" -0d19'32"
Kris and Jared FOREVER Vir 13h18'47" 11d38'40"
Kris and Jen Gem 7h25'31" 31d4'19"
Kris and Suzie Perry Lyn 7h56'55" 37d52'10"
Kris 'Babe' Newman Lib 15h7'13" -8d14'9"
Kris Barreras Aqr 22h35'49" -1d29'8"
Kris Beuning Cas 0h44'45" 57d57'5"
Kris Bradley Nilsson Lyr 18h47'34" 38d15'42"
Kris Costello Uma 11h42'47" 30d32'23"
Kris' Eyes Ari 1h49'50" 22d58'19"
Kris Field Crb 15h27'14" 26d48'54"
Kris GrMM Uma 9h27'55" 59d22'17"
Kris Hall Vir 12h22'22" 9d6'34"
Kris Hill Uma 11h1'22" 65d31'45"
Kris Johnstone Pho 0h39'58" -50d18'41"
Kris K. Tanaka Leo 11h11'38" 0d0'39"
Kris Kelly Her 18h0'9" 18d11'33"
Kris L. Gifford Sco 16h6'37" -18d23'14"
Kris Melag Gem 7h23'4" 18d28'27"
Kris Nicol Umi 13h56'8" 78d53'57"
Kris Radlinski Uma 10h16'30" 68d17'31"
Kris Rhodewalt Sgr 19h15'43" -27d31'42"
Kris Ritchie Sgr 18h53'31" -16d3'26"
Kris Ruberto Cap 20h14'10" -13d22'16"
Kris & Sharon's Heaven Col 6h4'12" -30d48'58"
Kris Shugrue Cas 0h44'56" 47d49'11"
Kris' Star Ori 5h36'22" 1d12'26"
Kris T Cru 12h3'4" -63d15'51"
Kris Tanaka Lyn 7h57'4" 39d52'58"
Kris The Brightest Star Cas 0h45'52" 50d24'6"
Kris Van Treese Sgr 18h36'56" -17d24'48"
Kris W. Oanes Psc 0h26'39" 19d35'33"
Krisaleigh Rae Cyg 20h17'14" 39d27'53"
Krisandsue Uma 9h20'11" 57d30'32"
Krisanne (Forte) Barbara Cas 0h44'37" 60d58'38"
Krischelle Cap 21h32'48" -13d32'58"
KrisDan Umi 15h20'27" 73d16'28"
Krisha Lyn 8h12'48" 50d18'4"
Krisha Kamlesh Gandhi Ori 5h23'35" 14d58'10"
Krisha & Mark Aqr 22h11'7" 1d0'2"
krishauna Lib 15h43'9" -5d40'9"
Krishawna Belle Morris-Escarra Tau 5h33'18" 22d53'59"
Krishna "KK" Konakondla Sgr 18h34'10" -29d36'5"
Krishna Pal Singh Lib 15h11'22" -28d15'16"
Krishna W. Lawrence Ari 2h50'26" 27d49'12"
Krisidia Ann Hinnant Car 7h15'1" -52d28'50"
Krisleigh Lewis Hoermann Cyg 19h34'25" 28d9'19"
Krislen Dawn Owens Uma 11h45'33" 49d9'6"
Krislock Cnv 12h36'35" 45d21'54"
Krisniky Aqr 21h14'33" -8d27'34"
Krispy Supernova Nebula Tau 4h1'1" 19d46'9"
KrisRobinBobbyKristin Her 18h54'10" 23d28'12"
Kris's Shining Star And 0h13'7" 45d44'58"
Krissi Lyn 8h40'48" 33d56'21"
Krissi D'Elia Lyn 7h37'48" 40d14'51"
Krissi Light of My Life Psc 23h21'35" 4d7'44"
Krissy Tau 3h49'56" 28d55'21"
Krissy And 2h37'42" 40d43'35"
Krissy And 0h41'59" 36d18'50"
Krissy Cap 21h57'13" -21d45'30"
Krissy Crowley Tau 5h46'55" 22d19'55"
Krissy G. Sco 17h53'38" -35d50'54"
Krissy Leicht Sgr 17h47'32" -25d21'53"
Krissy Loves Me Star And 0h25'59" 27d0'5"
Krissy Noblett Lyn 8h19'57" 55d53'42"
Krissy & Robert's Lib 15h5'48" -8d51'13"
Krissy Rynn Vir 12h15'22" 7d29'28"
Krissy Vinegar Weinstein Cnc 9h10'5" 31d0'24"
Krissy-Cameron Lyn 7h10'32" 46d33'55"
Krissy's star Uma 9h40'14" 62d40'38"
Kris-t Psc 1h10'11" 4d7'53"
Krista Lyn 8h20'30" 56d10'18"
Krista Uma 8h54'55" 58d56'11"
Krista Cap 20h26'9" -12d17'37"
Krista Cap 21h28'50" -16d57'35"
Krista Mon 6h46'54" -0d4'20"
Krista Lyn 6h59'5" 47d22'21"
Krista And 2h9'29" 45d1'4"
Krista And 23h13'39" 51d32'8"
Krista And 0h20'53" 38d19'12"
Krista And 0h8'4" 40d37'55"
Krista Psc 1h1'38" 10d1'3"
Krista Tau 4h46'12" 19d12'48"
Krista Leo 11h13'34" 13d42'7"
Krista Gem 6h38'17" 19d38'58"
Krista 5/30/04 Gillian Psc 1h10'27" 11d30'41"
Krista & Anatoly Wedding Day Sge 19h51'18" 18d13'26"
Krista and Brett 3/7 Tau 5h23'57" 26d45'24"
Krista Ann Aschbrenner Crb 16h7'49" 37d27'40"
Krista Ann Bentrup Cyg 19h51'22" 37d41'58"
Krista Ann Hepfler Ari 3h8'33" 28d29'57"
Krista Ann McKenna Gem 6h34'21" 20d26'42"
Krista Ann Z Cap 20h41'16" -18d19'3"
Krista Bear Leo 11h57'21" 23d28'6"
Krista Bennett Cas 0h44'34" 50d15'24"
Krista Blair Konuk Pho 0h49'24" -51d51'26"
Krista & Blake Murray Cyg 20h31'23" 50d10'53"
Krista Blazejewski Born: Mar 22/1993 Uma 11h31'34" 56d31'20"
Krista Brianne Blair Cap 20h28'10" -25d28'33"
Krista Burkholder Cyg 20h56'35" 31d21'0"
Krista Carey Lyn 7h19'27" 47d25'38"
Krista Cecil Sco 17h55'26" -42d54'42"
Krista Costa Hartman Dra 19h17'23" 70d44'27"
Krista D. Trammel Psc 0h30'27" 7d14'22"
Krista Darr Tau 4h18'11" 26d40'27"
Krista De Anne And 0h50'50" 37d28'54"
Krista Denali Naujoks Leo 11h35'47" 27d11'42"
Krista Domiquez Leo 11h44'42" 25d47'41"
Krista Elizabeth Billingsley Aqr 22h28'4" -24d46'44"
Krista Engstrom Dreamy Star Sco 16h15'23" -9d10'42"
Krista Faye Bacon Ari 2h32'38" 24d55'59"
Krista G. McCampbell Ori 5h53'35" 21d34'7"
Krista Grace Leo 9h46'35" 6d45'30"
Krista Hayley Allen And 23h24'21" 47d20'53"
Krista Jean And 23h31'33" 48d40'20"
Krista Jean Sgr 17h55'27" -17d12'8"
Krista Jean Lib 15h7'59" -5d13'4"
Krista Johnson Ari 2h56'38" 29d5'24"
Krista Kapalko Psc 0h7'35" 7d36'41"
Krista Kelly Uma 14h15'6" 60d45'8"
Krista L. Freeman Crb 16h5'58" 38d8'7"
Krista Lee Crb 16h18'28" 33d57'47"
Krista Lee-LaFogg Ricchi Ari 2h34'50" 22d17'46"
Krista Leigh Leo 11h7'32" 2d44'40"
Krista Leigh Cas 0h49'12" 61d12'23"
Krista Lynn Cantwell Aql 19h32'22" 6d44'32"
Krista Lynn Daniels-Meyers Cnc 8h55'35" 6d43'19"
Krista Lynn (I love you) Tau 5h28'47" 24d38'16"
Krista Lynn Williams Lib 15h23'45" -4d37'30"
Krista M. Jasso Cnv 13h47'19" 37d4'41"
Krista Marie Tri 2h23'20" 33d10'40"
Krista Marie Cnc 8h10'10" 11d48'52"
Krista Marie Faith Com 12h44'38" 16d6'40"
Krista Marie Fawcett And 23h7'48" 51d32'0"
Krista May Ari 2h32'30" 25d55'23"
Krista Michelea Filion Her 16h41'34" 32d40'25"
Krista my lil' Princess And 1h47'26" 39d10'31"
Krista My True Friend Lyn 7h55'4" 56d53'20"
Krista Nicole And 0h50'1" 39d42'59"
Krista Nicole Ari 2h41'15" 14d2'42"
Krista Nicole (Buggy) Fidanza Peg 22h51'14" 16d1'15"
Krista Nicole Dooley Cam 5h46'42" 57d25'18"
Krista Noel Sparkle Butt Donoven Aqr 23h3'46" -8d26'46"
Krista Park Dra 16h4'6" 60d24'9"
Krista "Peaches" Zetusky Ari 2h48'10" 26d59'44"
Krista Rae Car 8h2'26" -54d19'17"
Krista Rae Bauer Sco 16h6'9" -11d53'49"
Krista Rae Deroo Tau 5h14'16" 19d37'5"
Krista Rae Hendricks And 0h45'20" 37d15'44"
Krista Rhode ~ August 18, 1988 Leo 11h13'44" 22d39'2"
Krista Schroeder Ori 6h16'34" 10d13'10"
Krista Schwartz Cas 2h9'18" 66d23'43"

Krista Shiel
Ari 2h34'8" 13d29'28"
Krista Shirret Waldrep
Gem 6h23'0" 21d6'4"
Krista St. Aubin
Her 17h34'57" 27d4'22"
Krista "Star" LaFever
And 0h31'6" 30d54'39"
Krista & Steve April 22, 2004
Cam 4h9'30" 70d19'12"
Krista Stierwalt 07/23/77-02/05/05
Lyr 18h38'2" 40d10'24"
Krista Superstar
And 0h43'5" 26d10'59"
Krista Winklepleck
Lib 15h6'53" -6d55'3"
Kristabell
And 23h48'34" 41d15'50"
Krista-Brian
Cyg 20h9'37" 33d7'56"
Kristal
And 2h37'52" 38d35'47"
Kristal
Tau 4h42'2" 18d32'46"
Kristal Ann Bonito
Sco 16h9'12" -18d19'51"
Kristal Beharry and Sarah Fernandez
Cap 20h28'23" -14d10'16"
Kristal Brennan
And 0h40'37" 35d34'36"
Kristal Carlisle
And 23h9'4" 47d16'27"
Kristal "chattykdt"
Cas 1h38'3" 65d38'10"
Kristal Greene-Christie
Crb 15h44'25" 28d12'25"
Kristal Joy
Psc 1h23'20" 13d30'32"
Kristal Kayleen
Cam 3h39'47" 64d48'14"
Kristal Marie Miduski
Cyg 21h44'16" 36d52'50"
Kristal Marie Spallone
Vir 14h47'43" 4d25'47"
Kristal McKinney
Vir 11h48'21" -3d7'1"
Kristal Oreon Ellis
Ori 5h52'11" 6d17'24"
Kristalan Brown Gal
Lib 14h55'27" -7d0'1"
Kristali
Tri 2h21'58" 33d3'12"
KristaMyEveningStar
Umi 23h7'10" 88d48'18"
Kristan Bienek
Tau 4h35'26" 19d47'17"
Kristan Gemma Leary
Cyg 20h6'24" 39d35'16"
Kristan & Rachel
Cnc 8h17'4" 6d54'10"
Kristan's Le Star
Tau 4h5'20" 26d28'44"
KristAnton Partusch
Crb 15h34'12" 27d28'57"
Krista's Diamond
Mon 6h50'17" -0d9'39"
Krista's star
Lyr 18h56'15" 33d21'53"
Kristee
Cet 1h8'35" 0d33'57"
Kristeen
Ori 5h42'6" 7d50'8"
Kristeen Ann Rowley
Vir 12h51'55" 11d34'5"
Kristeenager (Doo-Doo, Ba-Ba!)
Uma 9h45'24" 55d52'50"
Kristel Dyan MQR²
And 23h27'31" 43d1'32"
Kristel L. Kalender
Lib 15h2'10" -4d17'47"
Kristel Zarovska
Cap 21h21'52" -22d12'13"
Kristell
And 0h3'42" 33d12'37"
Kristen
Ari 2h35'27" 25d0'28"
Kristen
Cnc 8h38'31" 17d29'12"
Kristen
Tau 5h49'43" 14d39'10"
Kristen
Tau 4h28'13" 18d27'39"
Kristen
Aqr 22h12'0" 1d29'53"
Kristen
And 23h54'35" 48d11'32"
Kristen
Cyg 19h31'36" 32d22'59"
Kristen
Crb 16h7'55" 36d6'11"
Kristen
Lmi 10h31'10" 37d23'45"
Kristen
And 1h6'8" 44d27'8"
Kristen
Ori 5h23'47" -0d57'56"
Kristen
Lyn 6h36'52" 57d33'4"
Kristen
Uma 10h8'0" 58d34'6"
Kristen
Cas 1h2'2" 63d43'59"
Kristen
Uma 11h16'52" 65d37'16"
Kristen
Uma 10h26'21" 66d31'59"
Kristen
Cam 6h17'31" 69d52'18"
Kristen
Sco 17h43'11" -40d28'10"
Kristen
Sco 17h26'46" -40d46'37"
Kristen
Sco 16h59'28" -38d24'51"
Kristen
Sgr 19h4'11" -32d29'15"
Kristen
Sco 16h46'17" -32d29'5"
Kristen
Lib 15h39'42" -24d6'42"
Kristen
Sco 16h25'56" -26d24'9"
Kristen
Eri 4h12'35" -23d11'1"
Kristen 06/29
Cyg 20h41'59" 39d39'32"
Kristen 0711
Umi 15h18'41" 68d42'43"
Kristen 71999
Lib 15h23'57" -22d20'35"
Kristen A. Zgoda, Love of my life
Vir 12h10'4" 0d33'17"
Kristen Adams
Cet 1h9'15" -19d55'0"
Kristen Alexandra Kij
Lmi 10h50'45" 27d2'28"
Kristen Alisha
And 1h15'23" 41d32'14"
Kristen Alyse Doyle
Vir 13h41'36" -18d34'15"
Kristen and Aaron Forever
Lmi 10h15'18" 30d47'16"
Kristen and Brent
Cyg 19h50'33" 33d38'53"
Kristen and Clay Forever
And 1h58'4" 38d35'0"
Kristen and Jeff Smarse
Lib 15h48'5" -18d28'51"
Kristen and Jeff's Shining Star
Cyg 20h24'30" 34d34'25"
Kristen and Justin
Cyg 20h49'44" 45d5'15"
Kristen and Keith 2p's forever
Lyr 18h53'3" 36d23'20"
Kristen Anderson
Psc 1h29'29" 13d39'56"
Kristen Anne Barennes
Cas 2h45'21" 57d57'51"
Kristen Anne Birtell
Ori 6h5'11" 13d41'7"
Kristen Anne Marie
Psc 1h12'1" 28d28'15"
Kristen Annelise Anderson
And 23h17'51" 51d33'3"
Kristen Ashley Vega
Tau 4h24'59" 17d54'37"
Kristen Ashley Yoder
Tau 4h13'14" 26d18'0"
Kristen Baby Bash Seay
And 2h20'37" 37d34'22"
Kristen Baker
Crb 16h7'14" 38d12'49"
Kristen Baresic
Sco 17h1'20" -39d7'30"
Kristen Bedard
Sco 16h55'8" -42d11'12"
Kristen Begley
Vir 13h19'46" -20d37'45"
Kristen Bethell
Cnc 8h56'51" 9d45'37"
Kristen Bleecker
Lyn 7h32'36" 57d36'51"
Kristen Bobo
Com 12h49'30" 30d15'4"
Kristen & Brenda
Cyg 19h54'43" 33d41'5"
Kristen & Brian
Lib 15h9'55" -12d59'29"
Kristen Brissing
Sgr 18h7'59" -25d44'17"
Kristen Brooke Machen Thigpen
Vir 13h20'35" 5d3'35"
Kristen Brown
Ari 2h27'25" 18d7'21"
Kristen Brulee'
Sgr 18h15'25" -21d41'57"
Kristen C. Autero
Leo 9h41'39" 30d43'59"
Kristen Capolino
Sgr 18h8'34" -31d33'11"
Kristen Carmody
Gem 7h12'30" 21d58'14"
Kristen Casaubon
Psc 1h12'16" 28d15'17"
Kristen Cecilia Churnside
Cra 18h46'15" -39d33'16"
Kristen & Charles
Boo 14h22'48" 53d52'36"
Kristen Cirillo
Sco 17h51'55" -36d22'36"
Kristen Coelho
Lyn 8h13'47" 35d52'23"
Kristen Craft
Sgr 18h47'13" -31d1'55"
Kristen Curria
Leo 11h41'41" 18d10'31"
Kristen D. Kerkhoff
Tau 4h4'34" 23d30'20"
Kristen D Meiklejohn
Cnc 8h15'45" 31d46'25"
Kristen D Walker
Uma 13h14'40" 55d50'46"
Kristen Danielle Collins
Vir 13h14'1" 1d38'44"
Kristen Danielle Knotts
Cap 21h52'51" -10d16'14"
Kristen Danielle Lomax
And 0h40'14" 41d50'37"
Kristen Davey
Cap 20h37'32" -9d34'50"
Kristen DeVito
Lib 14h53'22" -3d7'26"
Kristen Di Giacomo
Leo 11h6'17" 7d55'44"
Kristen Dixion
Cyg 19h47'39" 42d56'4"
Kristen Doe
Uma 11h40'55" 31d56'55"
Kristen Dory Gerrish
And 1h13'16" 45d5'20"
Kristen Elizabeth
And 1h2'49" 41d5'19"
Kristen Elizabeth Brody
Gem 7h24'17" 14d3'7"
Kristen Elizabeth Morgan
Dra 18h8'28" 76d2'10"
Kristen Elizebeth Hilliard
And 0h23'36" 32d8'2"
Kristen Fox
Ori 6h4'39" 17d10'28"
Kristen Frenchy Doak
Tau 4h19'38" 12d37'27"
Kristen Furnare
Uma 10h15'46" 51d45'2"
Kristen G
Gem 7h17'5" 22d34'42"
Kristen Gail
Vir 12h41'13" 1d37'41"
Kristen GLus
Dra 19h9'32" 69d4'33"
Kristen Golder
Cyg 20h28'29" 43d48'32"
Kristen Hall
Leo 11h8'42" 2d55'1"
Kristen Harrell
Com 12h9'28" 18d57'18"
Kristen Hartman
Ori 6h14'27" 15d42'13"
Kristen Henry
Tau 4h11'13" 10d35'52"
Kristen Higgins
Gem 7h2'57" 17d9'16"
Kristen Hodgson
Lyn 7h17'35" 56d33'37"
Kristen Hummel
Cas 1h40'49" 65d24'31"
Kristen Jados
And 23h17'16" 47d58'2"
Kristen Jannuzzi
And 0h46'35" 30d42'28"
Kristen & Jason Mahoney
Cma 6h45'19" -13d27'22"
Kristen Jennette
Sco 16h12'45" -9d29'23"
Kristen Johnson
Tau 5h16'30" 27d5'30"
Kristen & Jonathon Forever
Aqr 20h41'35" -7d32'54"
Kristen Jones
Sco 16h47'55" -27d11'46"
Kristen & Joseph Kleinberg
Cnc 8h31'33" 30d54'33"
Kristen & Josh
Gem 6h32'5" 16d35'57"
kristen juliano
Aqr 23h3'43" -8d25'43"
Kristen Juline Elizabeth Kurth
Sco 17h45'3" -40d54'6"
Kristen Karen Paul
Aqr 21h33'19" -0d0'58"
Kristen Kay Anderson
Crb 15h55'59" 27d13'30"
Kristen Kay Griffith
Cap 21h8'53" -24d33'32"
Kristen Kaye Harvey
And 1h18'30" 45d22'25"
Kristen Keeran
Ori 5h55'31" 17d3'54"
Kristen Keller
Aqr 22h35'27" -0d22'10"
Kristen Kelley
Psc 1h23'16" 15d32'14"
kristen_KRM
And 0h46'22" 39d0'57"
Kristen L. Addison
Mon 6h30'12" 0d31'18"
Kristen L. Horne
Sgr 17h58'18" -16d7'40"
Kristen L. Miele
Sgr 19h0'54" -13d54'26"
Kristen L. Smarse
Leo 11h33'9" 7d36'4"
Kristen L. Wright
Cnc 9h15'30" 30d30'3"
Kristen Lally
Leo 9h39'25" 31d30'5"
Kristen Langan
Ori 5h59'45" 17d22'32"
Kristen Langford
Tau 4h47'17" 21d42'7"
Kristen Larsen
And 0h15'29" 36d37'54"
Kristen Laura Nunley
Sco 17h15'11" -32d36'58"
Kristen Laurie Beriloff
Vir 13h26'49" 12d45'43"
Kristen Layton
Uma 8h37'12" 49d2'7"
Kristen Lea Hartwigsen
Per 4h42'17" 50d29'55"
Kristen Leah Snyder
And 2h9'54" 44d6'20"
Kristen Lee Adishian
Uma 10h51'37" 66d8'4"
Kristen Lee Hansen
Lyr 19h26'0" 40d1'6"
Kristen Lee Kleinschmit
And 1h15'26" 49d8'55"
Kristen Lee Victoria Mead
And 2h28'49" 47d28'34"
Kristen Leigh
Psc 1h46'21" 8d2'15"
Kristen Leigh Biondi
Cnc 8h45'17" 31d24'16"
Kristen Leigh Forever
Sgr 18h9'54" -31d9'2"
Kristen Leigh Hinds
Psc 1h21'42" 24d8'52"
Kristen Leigh Kennedy
Uma 11h8'59" 60d23'20"
Kristen Leilani Woisard
Sco 17h38'4" -41d18'25"
Kristen Lena Epifane
And 1h42'39" 41d52'5"
Kristen Lendl
Cap 21h10'52" -21d16'22"
Kristen Lisa
Ari 2h5'36" 22d49'34"
Kristen Lobotsky
Mon 6h40'44" -0d26'37"
Kristen Lomas
Psc 23h20'55" 4d46'10"
Kristen Lorenc
Sco 16h8'50" -12d54'26"
Kristen Luman
And 1h41'25" 46d35'38"
Kristen Lynette Cronon
Cap 20h26'40" -16d10'36"
Kristen Lynn Boscia
And 0h44'23" 42d39'21"
Kristen Lynn Mraz
Lib 15h24'9" -10d40'13"
Kristen M. Binder
And 0h46'48" 43d37'24"
Kristen M Bressi
Dra 16h54'20" 60d11'19"
Kristen M Cooney
Vir 12h30'46" 4d56'58"
Kristen M. Fling
And 2h25'3" 45d5'36"
Kristen M. Kolankowski
Sco 17h25'32" -37d55'11"
Kristen M West, My Princess
Tau 4h9'2" 7d49'28"
Kristen M Wilson
Ari 3h19'38" 26d0'59"
Kristen Mae Narlow
Cnc 8h36'33" 23d26'16"
Kristen Marie
Ari 1h55'4" 20d11'52"
Kristen Marie
Cas 1h20'25" 59d10'6"
Kristen Marie
Cas 23h24'56" 54d57'52"
KRISTEN MARIE
Ser 15h16'56" -0d52'37"
Kristen Marie Bain
Mon 6h30'31" 8d11'54"
Kristen Marie Brouillard
And 0h52'16" 43d18'13"
Kristen Marie DeLorenzo
Aqr 21h29'57" 1d36'45"
Kristen Marie Dobis
Aqr 22h39'13" 1d41'18"
Kristen Marie Ford
Vir 12h14'27" 10d54'8"
Kristen Marie Johnson
Cyg 21h35'17" 34d2'44"
Kristen Marie Johnson
Cap 20h18'17" -13d48'18"
Kristen Marie Livermore
Ori 5h27'28" 7d2'9"
Kristen Marie Morro
Psc 0h52'0" 27d26'28"
Kristen Marie Moyer
Psc 0h40'18" 8d52'15"
Kristen Marie Mucci
Cap 21h26'51" -20d56'47"
Kristen Marie Munzipapa
Aqr 21h16'43" -7d42'41"
Kristen Marie Murphy
Cap 20h36'40" -20d50'29"
Kristen Marie Seielstad
Lyn 6h57'41" 45d22'4"
Kristen Marie Straut
Tau 5h45'52" 21d17'18"
Kristen Martin
Ari 3h8'5" 24d22'55"
Kristen Mary Boltrushek
Cet 1h21'13" -0d8'32"
Kristen Mary Faust
Sgr 18h57'55" -24d31'10"
Kristen McKenna
Psc 1h7'36" 7d49'3"
Kristen Mendoza
And 23h23'18" 38d55'20"
Kristen Meredith
Uma 12h10'52" 48d17'42"
Kristen Michele Magnus
Lib 14h56'46" -1d46'47"
Kristen Michelle
Cas 23h28'11" 57d17'26"
Kristen Michelle Moore
Vir 13h45'25" -19d27'17"
Kristen Miller
Cam 4h16'38" 56d42'32"
Kristen Moyer
Aqr 21h39'7" -2d21'14"
Kristen My Love
Lyn 8h19'5" 38d39'5"
Kristen N. Ackman
And 22h58'12" 48d8'38"
Kristen Natalie Palmieri
Crb 16h4'35" 38d36'39"
Kristen Nicholson
Leo 9h29'35" 13d24'41"
Kristen Nicole
Tau 5h46'32" 22d46'20"
Kristen Nicole
Cyg 20h57'26" 33d11'49"
Kristen Nicole Dezmal
And 23h20'48" 42d31'51"
Kristen Nicole Farley
And 23h51'15" 37d52'17"
Kristen Nicole Fristedt
Ori 5h53'32" 22d4'26"
Kristen Nicole Moehrle
Gem 7h13'45" 17d39'55"
Kristen Nicole Philpotts
Per 4h23'15" 43d21'51"
Kristen Nicole Taft
And 0h40'42" 36d30'4"
Kristen Niedecken
Lib 15h22'11" -27d57'58"
Kristen Noel Alexander
Cyg 21h5'1" 46d44'51"
Kristen Noel Jenkins
Dra 15h43'28" 56d15'1"
Kristen Noelle Zadina
Tau 4h4'2" 27d21'35"
Kristen Ortman
Tau 4h19'55" 17d49'23"
Kristen P Birdsill
Psc 23h6'32" 0d10'20"
Kristen Partipilo's Star
Ari 2h10'25" 26d4'12"
Kristen Penner
Cas 0h47'37" 61d51'19"
Kristen Perone
Sco 16h10'42" -10d56'15"
Kristen Pomeroy
And 0h2'4" 46d11'30"
Kristen Rae Fomby
And 1h1'23" 39d23'29"
Kristen Rafajko's Star
Peg 21h53'46" 26d58'23"
Kristen Robyn Gairdner
Boo 15h36'29" 45d25'30"
Kristen Rochelle
Cas 1h12'22" 52d36'19"
Kristen Rosauer
Psc 1h17'30" 13d46'9"
Kristen Rose Lee
Umi 15h20'47" 69d46'12"
Kristen Ryan Muller
Uma 11h17'37" 36d9'0"
Kristen Sauvigne
Uma 13h29'18" 57d43'18"
Kristen Schumer
And 2h35'40" 43d40'28"
Kristen "Scooter" Woodbury
Psc 0h29'50" 13d51'36"
Kristen & Sean
Cnc 8h47'11" 23d42'40"
Kristen Sholtis
Gem 7h3'1" 25d23'15"
Kristen Simms
Lyn 9h0'47" 36d2'42"
Kristen Spinola
Leo 9h45'35" 6d56'58"
Kristen Starzyk
Tau 4h15'30" 9d22'0"
Kristen Steglich
Psc 1h20'11" 27d3'16"
Kristen Stephanie Castillo
Tau 5h36'10" 22d32'25"
Kristen Stone
Mon 6h54'48" 8d19'8"
Kristen Sweet
And 23h3'30" 36d15'22"
Kristen Taylor
Ori 5h28'20" 14d57'1"
Kristen the Shining Star
And 23h24'42" 51d23'47"
Kristen Trimble
Cas 2h31'11" 66d23'23"
Kristen Waggoner
And 0h47'31" 42d54'28"
Kristen Wenning
And 23h50'30" 45d12'12"
Kristen Wilson AJ
Cas 1h17'40" 59d52'4"
Kristen Woods
Sco 17h10'19" -41d4'47"
Kristen Wudyka and Nick Terech
Her 17h6'8" 30d58'40"
Kristen, Amanda, Blake, & Sydney
Lyn 8h47'23" 34d31'40"
Kristen, Will You Marry Me?
And 23h43'11" 46d56'36"
Kristen111588
Sco 16h9'32" -9d37'30"
KristenLanette
Crb 15h55'28" 39d13'31"
KristenMarieAnnDriscoll
Lib 15h19'49" -11d46'33"
kristennbillalwaysandforever110901
Vir 12h42'8" 2d56'53"
Kristen's Beauty
Ori 4h53'34" 12d10'56"
Kristen's Firefly
Cyg 19h46'48" 30d41'40"
Kristen's Star
Ori 5h56'23" 11d42'15"
Kristen's Star
And 0h13'56" 29d29'2"
Kristen's Twinkle
Crb 15h22'44" 28d17'47"
Kristeph
Lyr 19h20'1" 33d13'41"
Kristeve Hines
Cap 21h50'31" -13d12'38"
KRISTEY
Cap 21h37'11" -11d4'57"
Kristey Hann
Sco 17h51'23" -31d13'53"
Kristi
Sco 16h8'9" -9d33'55"
Kristi
Vir 12h8'30" -5d14'23"
Kristi
Lyn 8h10'25" 57d43'41"
Kristi
And 2h34'34" 49d26'44"
Kristi
Uma 11h4'53" 51d46'6"
Kristi
Leo 9h24'56" 26d36'31"
Kristi
Ori 5h9'34" 6d28'19"
Kristi
Cnc 8h5'17" 14d55'32"
Kristi
Ari 2h34'46" 17d53'33"
Kristi 09-22-77
Vir 11h45'22" -0d44'40"
Kristi A. Serrano
And 23h14'3" 49d24'53"
Kristi A. Smith - Williams
Aqr 21h40'52" 0d52'59"
Kristi Amerson
Mon 6h49'13" -0d35'55"
Kristi and Andy "Your So Cool"
Cyg 20h18'9" 48d6'1"
Kristi and Mark Forever!!!
Cnc 8h52'46" 26d39'11"
Kristi and Steve
And 0h26'22" 39d40'4"
Kristi Anderson
Cas 0h25'36" 64d28'47"
Kristi Ann
Sco 16h44'41" -33d15'9"
Kristi Ann
Pho 0h42'20" -43d22'36"
Kristi Ann Godshall
Cas 23h30'26" 56d55'35"
Kristi Anne Hobbs
Gem 6h5'58" 26d24'36"
Kristi Barton
And 23h59'9" 42d38'45"
Kristi Beaty
Ari 2h9'36" 24d12'38"
Kristi & Brian's Family Star
Umi 13h37'58" 73d7'5"
Kristi Burr
Lyr 18h33'13" 28d50'21"
Kristi Bush
And 22h58'30" 48d19'12"
Kristi Colossi
And 22h58'43" 40d20'1"
Kristi D Moore
Sco 16h9'1" -13d26'2"
Kristi Dawn Gibson
Cyg 21h45'59" 35d41'11"
Kristi Eastwell
Cyg 21h7'56" 52d55'14"
Kristi Elizabeth Raney
Cam 4h42'20" 54d47'48"
Kristi Faris # 23
Mon 6h50'23" -0d32'26"
Kristi G721
Cap 21h7'11" -19d38'58"
Kristi Garner
Cas 1h13'37" 59d27'18"
Kristi Girdharry
Lyn 7h46'53" 41d31'58"
Kristi Groom
Ori 5h40'46" 5d58'55"
Kristi Hesser
And 0h42'18" 43d37'32"
Kristi Hobbs
Psc 1h14'22" 5d55'37"
Kristi Ingles
Lmi 10h41'10" 38d3'41"
Kristi J.
Leo 11h23'44" 15d31'22"
Kristi Jean White
Psc 1h0'48" 19d40'18"
Kristi Jo Burns
Tau 4h1'46" 10d28'44"
Kristi Klatt
Uma 13h1'23" 54d9'0"
Kristi Kruthaup
Mon 6h29'32" 11d11'34"
Kristi L. Shook
Sgr 18h8'17" -35d28'37"
Kristi Lee
And 2h10'23" 46d11'29"
Kristi Lee Garieri
Cas 1h40'52" 60d7'16"
Kristi Leigh
Peg 21h53'51" 12d10'46"
Kristi Leigh Elkins Henry
And 2h34'42" 46d3'36"
Kristi Lizzie Wizzie
Aqr 22h21'14" -1d18'36"
Kristi Lorraine
Ari 2h11'14" 25d48'39"
Kristi Lynn Hatten
Vir 15h2'29" 4d52'35"
Kristi Lynn Johnson
Cas 0h36'43" 60d41'18"
Kristi Lynn Zimmerman
Sco 16h7'23" -16d26'36"
Kristi Lynne
And 0h34'29" 41d15'1"
Kristi Lynne Peterson
Gem 6h34'45" 25d1'53"
Kristi Mae
Crb 15h50'24" 28d37'12"
Kristi Marcos and Ryan Lane
Aql 19h29'41" 12d40'14"
Kristi Marie Barry
Cnc 8h21'59" 10d43'8"
Kristi Marie Vasquez
And 1h7'32" 41d45'40"
Kristi Marie's North
Umi 14h35'0" 81d45'4"
Kristi Michele Day
Ori 5h17'2" -8d28'36"
Kristi Michelle
Ori 5h33'1" 7d44'25"
Kristi Michelle
Gem 6h22'35" 22d1'59"
Kristi Nichole
Crb 16h11'6" 37d14'38"
Kristi Plasencia "Ultimate Star"
And 2h19'10" 49d24'59"
Kristi Reyes AKA Peaches
Lib 15h2'18" -23d16'56"
Kristi Schlewitz
Vir 14h39'51" -3d41'18"
Kristi Sharon Hendrix
Ari 2h22'2" 26d5'9"
Kristi & Shaun McCulley
Vir 13h16'17" 6d44'29"
Kristi Snapp
Vir 12h40'49" 5d9'19"
Kristi Warwick
Del 20h20'31" 13d51'50"
Kristi & Zak 05 to Forever
Cyg 19h43'47" 37d45'54"
Kristiaan M. Lewis
Gem 7h47'48" 31d32'29"
Kristian and Amanda Bearman
Cra 18h36'48" -41d32'16"
Kristian & Bernadette's Star
Ori 5h40'20" -2d23'22"
Kristian Diana
Leo 11h0'42" 3d17'55"
Kristian Erik Keski Kastari
Del 20h49'6" 17d6'33"
Kristian Itty Bitty Nelson
Umi 14h26'27" 76d6'1"
Kristian Matthijs Dick Beuker
Aqr 22h14'15" -14d44'45"
Kristian og Aga
Umi 15h10'30" 81d23'1"
Kristian Ørud
Cep 21h14'47" 66d4'25"
Kristian Paul Ramos
Ori 5h22'11" 3d22'49"
Kristian S. Burkland
Leo 9h29'17" 9d4'30"
Kristian Xavier Conn
Leo 11h56'29" 22d37'43"
Kristiana Ashley Ulmer
Sgr 20h13'30" -37d7'5"
Kristiana Becker
And 2h34'40" 43d46'55"
Kristiana Leigh
And 22h58'25" 46d59'10"
KRISTIANA M. TENORIO
Lyr 19h13'17" 34d22'7"
Kristiana Marie Brown
And 0h14'9" 45d49'23"
Kristianna Alexis Gavin
Cnc 8h7'49" 12d36'26"
Kristianna DeShon Byrd-Nobles
Crb 15h44'17" 35d32'16"
Kristi-Anne Clarke
Cra 19h4'3" -39d10'27"

Kristian's Star
Cnc 8h28'1" 28d33'14"
Kristie
Gem 6h12'33" 24d46'12"
Kristie
Cnc 9h15'14" 13d32'28"
Kristie
Vir 11h49'44" 8d3'22"
Kristie
Ari 2h28'35" 21d51'55"
Kristie
Mon 7h16'11" 0d38'2"
Kristie
Crb 16h2'41" 32d56'52"
Kristie
Umi 15h31'27" 71d48'36"
Kristie
Boo 14h27'22" 52d52'3"
Kristie A. Anderson
Lib 15h42'40" -17d27'32"
Kristie Alyssa
Uma 11h21'36" 35d17'33"
Kristie and Andrew's anniversary
Vir 12h41'17" -8d7'22"
Kristie Ann Ruiz
Gem 7h11'24" 34d29'59"
Kristie Avila
Ari 3h2'46" 26d48'7"
Kristie Bobolis, M.D.
Leo 10h23'29" 23d14'22"
Kristie & Dave Forever
Ori 5h31'4" 0d12'22"
Kristie Denice Ray
Per 3h42'3" 41d53'37"
Kristie Emma Ellington
Peg 21h58'28" 18d42'48"
Kristie Hosch
Lyn 6h53'33" 60d37'17"
Kristie Keith Harmon
Sgr 18h54'58" -36d10'12"
Kristie L. Weiss
Cas 23h29'27" 53d1'4"
Kristie Lee Sporton 22 October 1984
Cru 12h34'18" -58d15'5"
Kristie Lynn "Fatz" Spain
And 0h47'11" 40d22'52"
Kristie Lynn Swanson
And 2h6'33" 38d36'20"
Kristie Lynns Very Own Star
Sgr 20h3'2" -42d16'12"
Kristie Mae Barnes
Del 20h42'30" 8d41'46"
Kristie Marie
Lyn 6h32'55" 59d42'19"
Kristie Marie Jones
Sct 18h43'2" -7d39'34"
Kristie Marie McCort
Aqr 22h26'21" -7d25'51"
Kristie Morgan Dembinski's Star
Crb 15h35'26" 27d1'35"
Kristie - My "Lady in Red"
And 23h39'18" 50d37'46"
Kristie ooOgli Luu
Leo 11h3'6" 22d18'35"
Kristie Renee
Uma 11h33'6" 43d19'35"
Kristie Rogers
Col 5h59'58" -34d51'35"
Kristie Rose
Ori 5h53'12" 6d55'52"
Kristie Russell
Uma 13h43'59" 52d38'41"
Kristie S. Branick
Cas 1h56'43" 64d0'54"
Kristie's Touchpoint
Aqr 22h17'51" -8d54'1"
KrisTim 1023
Cnc 7h56'38" 11d6'7"
Kristin
Cnc 9h19'48" 8d43'38"
Kristin
Tau 4h11'12" 16d29'24"
Kristin
Ori 5h37'57" 3d6'15"
Kristin
Tau 4h23'43" 9d48'40"
Kristin
Aqr 20h40'13" 0d13'45"
Kristin
Gem 6h34'57" 21d9'14"
Kristin
Com 12h19'3" 23d31'57"
Kristin
Leo 11h13'37" 23d9'50"
Kristin
Gem 7h12'24" 24d2'51"
Kristin
Gem 7h36'4" 27d58'21"
Kristin
Lmi 10h30'55" 38d45'40"
Kristin
And 0h40'11" 38d16'10"
Kristin
And 0h46'22" 35d44'57"
Kristin
Psc 1h26'54" 32d31'13"
Kristin
Cas 0h17'17" 55d5'4"
Kristin
Cam 4h3'20" 55d17'38"
Kristin
And 23h27'33" 41d39'9"
Kristin
Aqr 22h40'14" -7d50'3"
Kristin
Cyg 21h11'38" 55d16'38"
Kristin
Cas 1h29'54" 65d35'9"
Kristin
Uma 13h7'8" 53d55'24"
Kristin
Psc 0h42'57" 7d15'42"
Kristin 15.1.78
Uma 11h57'2" 50d56'11"
Kristin 21
Uma 12h35'59" 58d3'49"
Kristin Adcock
Cap 20h10'52" -25d53'59"
Kristin Aitken
And 23h57'35" 48d26'9"
Kristin Alana Lieber
Vir 12h38'29" 0d39'34"
Kristin Alison Terranova
Cas 0h45'8" 61d56'25"
Kristin And Justin
Lyn 7h38'9" 44d30'47"
Kristin And Kalei
Gem 6h12'59" 23d12'53"
Kristin and Owen 2-19-06
Cyg 20h31'39" 40d49'39"
Kristin Angell Lankford
Tau 4h12'44" 12d58'47"
Kristin Ann Jernsted Crossland
Psc 0h45'41" 9d10'25"
Kristin Ann Kinnaman
Uma 12h2'29" 55d46'34"
Kristin Ann & Larrisa LeAnn Murphy
Psc 23h5'51" -1d16'47"
Kristin Anne
Ari 3h10'24" 29d51'58"
Kristin Ashley
Vir 12h13'59" 11d39'41"
Kristin Aviles
Cas 0h52'46" 63d42'35"
Kristin Babist
And 22h59'14" 45d27'10"
Kristin " Baby Girl "
Lib 15h6'17" -1d34'31"
Kristin Baum
Aqr 22h51'41" -9d1'19"
Kristin Beth Kintigh
Lib 14h39'21" -10d41'58"
Kristin "Biscuit" Edwards
Cap 20h30'58" -13d25'47"
Kristin Boo
Cas 1h37'10" 61d0'8"
Kristin Bowman
Psc 1h4'6" 3d15'37"
Kristin & Brian's Star
Lyr 18h34'36" 31d58'47"
Kristin Calder
Tau 3h47'46" 19d31'28"
Kristin Carges
Gem 7h24'56" 32d55'6"
Kristin Catherine Matherne Sample
And 23h13'57" 47d10'10"
Kristin Cavolick
And 1h28'16" 41d30'22"
Kristin Cervantes
Lib 15h32'37" -8d13'1"
Kristin Childers
Gem 7h43'36" 31d50'46"
Kristin Conklin
Ori 6h5'20" 17d10'49"
kristin cuff
Her 17h17'51" 32d20'7"
Kristin D Carlstedt
Psc 2h5'9" 6d43'40"
Kristin Dawn-Alyea Powers
Mon 6h47'48" -0d5'28"
Kristin Doughtie
Lyn 9h12'14" 37d27'43"
Kristin Dunn
Lib 15h26'34" -6d13'21"
Kristin Dunreath Williams
Sgr 17h48'22" -17d26'1"
Kristin E. Honeysuckle
Uma 11h11'26" 36d23'56"
Kristin Elizabeth
Gem 7h32'22" 33d46'31"
Kristin Elizabeth
Cas 0h2'57" 56d1'50"
Kristin Elizabeth
Vir 13h39'25" 2d59'42"
Kristin Elizabeth
Lib 15h5'11" -3d49'46"
Kristin Elizabeth Firmani
Uma 10h46'0" 65d57'22"
Kristin Elizabeth King-Fournier
Ori 5h42'36" -0d0'37"
Kristin Elizabeth McNulty
Lib 15h9'47" -7d15'53"
Kristin Elizabeth McNulty
Lib 15h23'17" -7d6'36"
Kristin Elizabeth Zuerblis
And 2h20'13" 47d9'30"
Kristin Ellen
Lmi 10h28'11" 37d43'50"
Kristin Farinacci
Sgr 19h9'21" -13d19'47"
Kristin Fellner
Crb 15h36'28" 26d39'3"
Kristin Ferris
Tau 3h43'41" 27d15'5"
Kristin Finn
Cyg 21h55'22" 41d27'54"
Kristin Fittante
Peg 21h53'31" 26d13'28"
Kristin Fortune
Gem 6h27'28" 24d49'4"
Kristin Foy
Uma 10h31'25" 52d35'22"
Kristin Funk
Tau 5h24'48" 19d51'55"
Kristin Gay
Ari 2h49'54" 29d24'48"
Kristin Goat
And 0h27'30" 30d2'49"
Kristin Goodwin
Dra 19h34'48" 61d8'43"
Kristin Gotthardt
Mon 6h21'41" -3d28'19"
Kristin Grace Murphy
Uma 10h16'18" 68d19'30"
Kristin Grace White
Lyr 18h45'12" 36d35'5"
Kristin Gunther
And 1h5'28" 45d42'34"
Kristin Hantla
Ori 5h50'52" 19d6'14"
Kristin Hess
Tau 4h45'5" 20d18'38"
Kristin Horton
Cas 23h7'25" 58d7'34"
Kristin Hough & David Rohner
Cyg 21h38'30" 50d19'38"
Kristin + James, Always and Forever
Cyg 20h21'30" 55d10'58"
Kristin Jeanine Ewish
Sco 17h26'0" -42d47'22"
Kristin Johnson
Tau 5h48'43" 14d53'3"
Kristin Jureczki
Lyn 6h58'45" 48d14'43"
Kristin Karen Hawkes
Cyg 19h39'40" 29d58'7"
Kristin Kay Jones
Lyn 8h3'47" 50d49'26"
Kristin Kelli Brown Dec. 1988
Cap 21h20'52" -19d39'34"
Kristin Kelly Najarian
Eri 4h42'33" -8d6'16"
Kristin Kelm
Cap 21h33'0" -11d36'31"
Kristin Kimery
Leo 11h7'52" 8d43'6"
Kristin Kleinjans
Tau 3h43'33" 27d39'12"
Kristin L. McKay
Cyg 21h25'6" 30d10'42"
Kristin Lee
Tau 3h53'40" 11d51'14"
Kristin Lee Drace
Sco 17h10'16" -38d5'42"
Kristin Lee Sims
Aqr 20h51'20" 2d19'6"
Kristin Leigh
Psc 1h7'14" 14d2'8"
Kristin Leigh
Vir 14h42'34" 0d29'17"
Kristin Leigh
And 0h40'27" 22d58'7"
Kristin Leigh Bailey
Psc 23h43'25" 6d52'9"
Kristin Leigh Good
Uma 10h11'43" 65d41'3"
Kristin Leigh Hurst
Cyg 21h23'18" 33d0'51"
Kristin Leigh Norman
Ori 5h31'44" -8d37'58"
Kristin Leigh Rose Flower Krieg
Uma 9h34'39" 63d16'6"
Kristin Leona Wolfe
Cnc 8h23'23" 15d59'16"
Kristin Liana
Tau 5h6'48" 25d25'0"
Kristin Liane Wood
Leo 10h8'18" 15d59'25"
Kristin Lloyd
Tau 5h48'34" 26d28'7"
Kristin Lorraine Curry
Cyg 19h52'17" 36d17'32"
Kristin Louise Stankovic
Psc 2h4'5" 4d49'5"
Kristin Luli Ammar
Com 12h37'52" 17d57'10"
Kristin Lyn Kaspary
And 1h4'44" 38d36'21"
Kristin Lyn Occhionero
Leo 10h1'40" 20d15'10"
Kristin Lynne Lewis
Ari 2h50'31" 30d56'17"
Kristin M. Darrah
And 0h20'25" 37d52'33"
Kristin M. Heilmann
Aqr 23h40'53" -24d34'23"
Kristin M. Huvala
Vir 13h11'38" 5d47'8"
Kristin Manka
Uma 12h34'50" 60d46'29"
Kristin Marie 12
And 1h20'59" 44d30'25"
Kristin Marie Aust
Sco 17h8'50" -43d28'55"
Kristin Marie Deck
Cas 1h5'56" 54d13'30"
Kristin Marie Ettinger
Cap 21h23'31" -23d39'18"
Kristin Marie Evenson
Sco 17h53'51" -35d56'55"
Kristin Marie Haight
Cap 21h46'19" -12d48'19"
Kristin Marie Mayer
Sgr 20h10'32" -32d12'33"
Kristin Mary McMahon
Cnc 9h13'8" 10d33'8"
Kristin Matthews Medeiros
Cnc 9h6'40" 32d38'35"
Kristin Mazur
And 2h22'0" 47d25'29"
Kristin McMahon Westol
Crb 15h46'2" 28d3'8"
Kristin Melissa
And 23h0'49" 42d30'1"
Kristin Michele Jannone
Sgr 19h57'6" -30d15'56"
Kristin Michelle Erickson
Cyg 21h36'52" 52d0'37"
Kristin Michelle Fontaine
Cas 0h43'31" 61d48'10"
Kristin Michelle Gargano
Leo 11h32'15" 8d10'24"
Kristin Michelle Munro
And 23h28'36" 46d15'31"
Kristin Michelle Payne Heaven Sent
Mon 6h50'43" -0d8'11"
Kristin & Mike Sept 10 2005
Lyn 6h23'45" 56d30'30"
Kristin Moncayo
Cnc 8h40'22" 32d15'11"
Kristin Mundie Bauer
Ari 2h55'38" 26d24'50"
Kristin N. Muehlbauer
Sgr 17h56'9" -17d26'12"
Kristin N. Raudonis
Cas 1h40'39" 63d14'54"
Kristin Neafsey
Cyg 20h54'30" 42d3'13"
Kristin Nichole
Cas 0h20'14" 50d31'4"
Kristin Nichole
And 0h21'37" 32d6'49"
Kristin Nichole
Lib 15h5'55" -7d54'52"
Kristin Nicole
Lib 15h7'43" -9d28'48"
Kristin Nicole
Ari 3h9'54" 26d49'7"
Kristin Nicole Barta
And 0h50'36" 34d54'59"
Kristin Nicole Kauffman
Cap 21h42'8" -23d49'7"
Kristin Nicole McLean
Psc 1h7'52" 17d42'17"
Kristin Nicole Wilson
Gem 7h11'22" 28d23'19"
Kristin Noelle Blair
Pic 5h38'47" -43d29'35"
Kristin P. Kiessling
Per 2h25'32" 55d22'10"
Kristin Paige
Lmi 10h6'1" 33d27'0"
Kristin Perrotta
Crb 15h19'53" 26d9'46"
Kristin Poulos
Mon 7h49'50" -0d28'31"
Kristin Reisinger
And 2h23'15" 37d29'1"
Kristin Rose Syverson
Mon 7h35'39" -0d29'23"
Kristin Ruth Wong
Lyr 19h27'11" 38d22'16"
Kristin Sanders
Sco 17h49'19" -38d50'37"
Kristin Sanderson
Aqr 23h2'8" -8d0'53"
Kristin Sarrels
And 1h3'43" 46d44'39"
Kristin Scott
Psc 1h20'48" 18d31'41"
Kristin Shantz
Sgr 18h2'56" -26d59'20"
Kristin Shawne Scott
Aqr 22h15'46" -5d50'15"
Kristin Simpson
Vir 14h1'40" -11d3'22"
Kristin Smith
Uma 11h39'37" 54d36'26"
Kristin "Special K" Ann McCrary
Uma 9h21'45" 56d19'38"
Kristin Squires - Mi Lux Solis
Ori 6h0'6" 5d47'16"
Kristin Stockman
Uma 11h21'17" 40d45'24"
Kristin Sue
Vir 11h46'2" 6d11'10"
Kristin Taylor Lai
Gem 7h5'32" 27d37'17"
Kristin Taylor Matteson
Lmi 10h46'12" 27d18'33"
Kristin & Tony Medved
Aur 5h22'12" 44d55'9"
Kristin & Ty Miller
Cyg 20h23'2" 45d44'20"
Kristin V. Roehling
Ari 2h42'7" 14d27'7"
Kristin Van Kampen
And 0h13'26" 43d25'33"
Kristin Van Oudenaren Robinson
Cas 23h22'24" 55d2'7"
Kristin Viktoria, 29.11.2006
Lyn 7h26'34" 52d37'40"
Kristin Wailgum
Lyn 8h11'21" 56d37'56"
Kristin Wotring
Crb 16h10'5" 37d30'52"
Kristin Wright
Ari 3h7'38" 19d47'22"
Kristina
Cnc 9h20'16" 16d2'35"
Kristina
And 0h35'5" 25d45'57"
Kristina
Psc 1h11'37" 27d18'2"
Kristina
Ari 2h5'0" 24d29'28"
Kristina
Lyr 18h52'1" 42d58'0"
Kristina
Lyn 6h43'42" 52d19'59"
Kristina
Cyg 20h40'29" 45d43'17"
Kristina
And 23h9'38" 52d2'49"
Kristina
And 0h37'36" 42d37'13"
Kristina
Cyg 21h34'55" 34d56'7"
Kristina
Uma 11h51'34" 34d55'41"
Kristina
Crb 15h22'17" 31d10'23"
Kristina
Cyg 19h43'21" 36d57'45"
Kristina
Lib 15h44'40" -9d27'28"
Kristina
Sco 16h6'34" -19d25'2"
Kristina
Vir 13h9'38" -21d35'3"
Kristina
Sgr 18h45'48" -16d43'43"
Kristina
Sco 16h52'14" -38d53'41"
Kristina
Pho 23h55'59" -45d43'0"
Kristina A. Pretsch
Umi 14h47'16" 73d11'55"
Kristina Allemand
Cnc 8h32'25" 19d15'32"
Kristina Allene Houser
Cas 0h42'54" 61d45'43"
Kristina Amira Norman
Uma 10h22'1" 44d19'8"
Kristina and TJ
Cyg 21h39'18" 43d20'3"
Kristina Angel Star
Cnc 8h40'44" 13d20'56"
Kristina Ann Dayley
Cas 1h42'45" 63d20'53"
Kristina Ann Osmun
Ori 5h32'39" 4d21'47"
Kristina Ann Soggee
Ori 5h8'57" 3d45'4"
Kristina Ann Soggee
Ori 5h55'15" 9d38'4"
Kristina Anna Sandmann
Umi 17h8'23" 79d54'45"
Kristina Anne
Ori 6h5'59" 10d45'13"
Kristina Anne Hughes
Lib 14h49'33" -1d52'43"
Kristina Anne (Tina)
Crb 16h5'12" 26d30'19"
Kristina Ann-Marie Mok
Psc 1h9'21" 25d47'7"
Kristina Argabright
Crb 16h22'49" 37d40'26"
Kristina Ashley
Ori 5h12'24" 15d42'38"
Kristina Ashley Grish
Tau 4h31'45" 29d35'13"
Kristina Aston
Ari 2h48'8" 14d37'14"
Kristina Aufiero
Uma 12h0'7" 53d56'33"
Kristina Baker
Sgr 19h11'35" -17d41'40"
Kristina Bingham & M Akra
Vir 13h15'51" -22d36'14"
Kristina Calisto
Lyn 7h51'19" 40d16'57"
Kristina Carbaugh's Star
Leo 11h43'46" 26d19'17"
Kristina Clancy
And 2h14'7" 45d54'25"
Kristina Conner
And 0h37'7" 33d52'38"
Kristina Copp 17
Lyn 8h21'35" 55d3'27"
Kristina Coppolino
Cnc 8h6'44" 27d28'15"
Kristina Dawn Nelson
Ari 2h8'49" 23d35'53"
Kristina Elizabeth Sablan Bermudes
Psc 0h9'40" -1d44'27"
Kristina Frolova
And 0h41'21" 43d37'7"
Kristina G. Buchak
And 0h43'27" 36d36'25"
Kristina G. Shelor
Sco 16h5'32" -19d8'54"
Kristina Galiotto
Cnc 8h16'26" 23d17'37"
Kristina Geleziunaite
Leo 11h12'56" -6d11'33"
Kristina Gevara
And 1h51'26" 38d10'19"
Kristina Grace Gram
Sgr 18h35'46" -29d6'56"
Kristina Graf
Uma 9h36'33" 58d21'32"
Kristina Guerrero
Cas 0h21'1" 63d51'10"
Kristina Higginbothan
Ori 5h51'34" -0d59'2"
Kristina Hiller
Psc 1h45'4" 14d42'56"
Kristina Jane Ruedin
And 2h16'10" 37d37'20"
Kristina Jean Marvin
Cas 0h24'32" 47d1'26"
Kristina Jenica Prince
Lyn 7h41'46" 37d28'6"
Kristina Joy Reh * June 1-2005
Dra 16h36'10" 60d54'12"
kristina juarez
Lyn 7h44'51" 37d47'49"
Kristina Kay Williams-Woods
Cnc 9h17'47" 12d55'57"
Kristina Kaye Gehrke
Lyn 7h20'16" 49d24'58"
Kristina Kaye Miller
Gem 6h57'40" 26d25'6"
Kristina Khorraminia
Vir 11h55'32" -0d20'38"
Kristina King
Cas 3h23'36" 74d23'26"
Kristina Kristianna
Sco 16h6'31" -33d5'20"
Kristina Laine
Tau 5h13'16" 26d26'29"
Kristina Lee
Uma 8h45'30" 50d50'39"
Kristina Leigh Brody
Vir 14h21'56" -0d14'35"
Kristina Leigh Winn
Lib 15h19'2" -12d11'7"
Kristina Leszczak
Leo 11h11'58" 20d33'48"
Kristina Lori
Lib 14h53'53" -2d5'6"
Kristina Lukhmanova
Aqr 22h8'6" -2d28'40"
Kristina Lyn
And 0h17'41" 33d9'31"
Kristina Lynn
Cap 21h15'29" -27d16'31"
Kristina Lynn Redabaugh
Aqr 22h26'54" -2d29'29"
Kristina M Migut
And 1h58'30" 41d22'18"
Kristina M. Pollack
Crb 15h39'18" 35d40'45"
Kristina M. Ward
Uma 9h50'25" 56d34'38"
Kristina Macedonian pride
Leo 11h7'21" 0d18'1"
Kristina Margaret Baril
Tau 5h30'7" 22d14'12"
Kristina Maria Peters
Oph 16h44'25" -0d53'46"
Kristina Marie
Psc 1h41'40" 5d10'57"
Kristina Marie
Lmi 10h22'5" 30d40'20"
Kristina Marie Biffar
Dra 19h40'6" 63d24'44"
Kristina Marie Campbell
Cap 21h31'1" -16d28'0"
Kristina Marie DeHarts Star
Cap 21h15'34" -19d35'49"
Kristina Marie Dittmer
And 1h44'8" 42d26'28"
Kristina Marie Eckman
Gem 6h28'26" 26d53'54"
Kristina Marie Haines
Aqr 22h3'51" 1d18'55"
Kristina Marie Hernandez
And 1h17'46" 35d46'28"
Kristina Marie Hysell
Cam 3h52'21" 53d23'51"
Kristina Marie Matthews
Gem 6h3'13" 25d35'30"
Kristina Marie Mota
Cam 4h33'38" 58d54'16"
Kristina Marie Ozkayan
And 2h9'23" 45d36'3"
Kristina Marie Re
Gem 7h16'23" 26d57'36"
Kristina Marie (Vixen)
Lib 16h1'27" -8d20'9"
Kristina Marija Narkeviciute
Lyr 18h46'39" 44d26'3"
Kristina Marin
Cas 0h3'23" 53d24'15"
Kristina - Marmeduke - Space Odyssey
Ori 6h14'27" -1d38'17"
Kristina Maureen Grogan
Sgr 18h58'52" -33d54'47"
Kristina May's Christening Star
And 23h13'15" 52d53'38"
Kristina McHugh
Cap 21h32'56" -13d35'39"
Kristina Michelle
Ori 6h20'31" 16d18'46"
Kristina Minerva
Cnc 9h7'48" 17d9'51"
Kristina Morris
Vir 12h15'15" 3d45'57"
Kristina Mountain - 06-06-06.
Cyg 21h38'26" 53d9'14"
Kristina Mumtaz
Leo 9h35'17" 7d5'59"
Kristina (My Baby Girl)
Uma 11h30'47" 44d41'23"
Kristina N. Miller
Lib 15h21'25" -16d5'35"
Kristina Nicole Martin
Cas 1h42'36" 64d11'5"
Kristina Nicole Ram
Mon 7h0'28" 5d17'53"
Kristina Nobles
Mon 7h19'16" -0d47'48"
Kristina Noel
Lib 15h16'29" -18d50'45"
Kristina Noel Murty
Uma 10h53'45" 42d32'53"
Kristina Noga Star
Cas 0h45'21" 63d58'6"
Kristina Palmieri
Gem 7h39'17" 31d56'22"
Kristina Pauline Soares
Cnc 8h33'15" 14d44'51"
Kristina Perry
Leo 10h3'34" 6d39'5"
Kristina Phelan
Cap 20h35'5" -24d58'24"
Kristina R Whaley
Tau 4h42'47" 27d36'6"
Kristina Rehna Carnes
Uma 10h15'8" 57d0'23"
Kristina Reising
And 1h32'36" 42d27'27"
Kristina Renee Barton
Ari 2h37'28" 26d45'32"
Kristina Ro
Mon 6h33'53" 9d14'21"
Kristina Rose Lents
Her 18h33'19" 20d7'20"
Kristina Rose Threlkeld
Cnc 8h56'4" 10d10'47"
Kristina & Ryan's Star
Leo 10h46'54" 19d13'25"
Kristina Sagittarii
Sgr 18h40'49" -17d4'39"
Kristina Sanderson
Lyn 8h41'18" 34d1'38"
Kristina Schmoldt
Cam 12h30'57" 78d55'4"
Kristina Shining Star
Cas 0h27'40" 60d14'15"
Kristina Shortell
Leo 11h3'44" -2d36'17"
Kristina Sinagra
Ori 6h17'17" 15d53'27"
Kristina - Sofia
Lyn 7h15'38" 47d12'54"
Kristina Spivac
And 0h54'25" 43d36'49"
Kristina Starr
Vir 12h17'28" 12d7'40"
Kristina Susan Keil
Mon 6h48'47" 0d14'29"
Kristina T. Donahay-Williams
Vir 15h2'21" 5d4'0"
Kristina Teressa Verona Streeter
And 0h54'20" 45d6'32"
Kristina The Beautiful
Uma 9h20'58" 61d2'3"
Kristina Thompson
Lmi 10h28'3" 36d33'26"
Kristina Thorn
Tau 4h8'47" 5d2'26"
Kristina Torres
Sgr 19h21'19" -12d48'12"
Kristina Underthun
Sco 16h1'16" -20d33'31"
Kristina Valdes-Pages
Cnc 8h3'5" 12d58'10"
Kristina Vaughn
Sgr 19h44'47" -14d15'8"
Kristina Wolfe
And 0h43'13" 39d33'10"
Kristina-Lyn Pauline Calandro
Cas 1h30'49" 62d36'28"
KristinaMDC20
Leo 11h26'23" -3d56'0"
Kristina's and Christopher's Star
Uma 10h29'48" 56d18'48"
Kristina's Hope
Leo 11h17'30" 15d39'33"
Kristina's Love
Sgr 19h42'39" -13d29'22"

Kristina's Star
Uma 9h51'36" 64d35'8"
Kristina's Star
Cnc 8h12'53" 25d34'12"
Kristina's Star "Luz de mi vida"
And 23h22'17" 37d50'32"
Kristine
Cyg 21h45'21" 44d7'44"
Kristine
And 0h13'51" 45d39'10"
Kristine
And 23h44'16" 41d45'13"
Kristine
Cyg 21h17'41" 42d30'17"
Kristine
Per 3h6'4" 46d26'41"
Kristine
And 23h17'27" 49d14'19"
Kristine
Cam 3h51'41" 56d32'52"
Kristine
And 0h54'6" 40d49'7"
Kristine
And 0h57'27" 39d7'29"
Kristine
Ori 6h23'51" 15d14'42"
Kristine
Vul 21h20'31" 21d24'10"
Kristine
Cam 7h35'44" 67d38'23"
Kristine
Cas 1h59'15" 60d31'8"
Kristine
Cyg 20h14'27" 53d7'4"
Kristine
Lib 15h24'1" -9d36'32"
Kristine Adams
Tau 4h24'27" 25d34'50"
Kristine and Jim's 1st Anniversary
And 2h37'54" 49d8'39"
Kristine Armstrong Pompa
Psc 0h5'12" 2d37'2"
Kristine B. Albert
Sgr 18h35'13" -17d47'44"
Kristine Bennett
Cas 1h24'9" 63d10'37"
Kristine Bortner
Lib 14h52'43" -1d53'0"
Kristine Brundage
Lyn 7h54'24" 51d30'12"
Kristine Cambra
Aqr 21h12'13" -11d41'27"
Kristine Camille Antonio
Tau 4h43'44" 27d35'36"
Kristine Carol Lepore / Mom
Sco 16h42'24" -30d46'52"
Kristine Diane Cook
Aqr 22h8'6" -0d21'57"
Kristine Elaine Bandfield
Cnc 9h19'22" 14d23'5"
Kristine Elizabeth Gariepy
Aqr 22h37'4" 2d22'48"
Kristine Evonne Gautsch (Munchkin)
Vir 13h7'28" -11d7'6"
Kristine Fila
Ari 1h54'23" 24d28'14"
Kristine Fiore
Cnc 8h33'32" 32d20'19"
Kristine Gluck
Cnc 9h6'38" 17d53'28"
Kristine H. Salley
Leo 10h57'56" 10d35'0"
Kristine Hall
Vir 13h19'26" 4d27'41"
Kristine Holm
And 23h18'41" 47d59'36"
Kristine HoneyBear Martinson
Cnc 8h34'53" 8d16'13"
Kristine Hunsinger
Cap 21h10'28" -24d31'35"
Kristine Ion-Rood
Uma 11h39'42" 52d20'52"
Kristine Joy Mallari
Mon 6h50'6" 8d0'36"
Kristine Kingston
Vir 14h37'38" 3d55'12"
Kristine L. Gibson
Lyn 7h28'41" 45d0'29"
Kristine L. Nash
Uma 11h11'8" 54d56'47"
Kristine Laura
Leo 11h33'16" 24d26'49"
Kristine LeAnne Horton
Vir 13h35'48" -8d48'12"
Kristine Lesur
Uma 8h39'38" 69d56'33"
Kristine Lewis
Gem 6h40'59" 33d45'59"
Kristine Lisa Herbert
Ori 5h28'50" -3d13'44"
Kristine Lynette MacNeil
Ari 3h1'48" 12d8'27"
Kristine Lynn
Cyg 19h48'9" 33d36'48"
Kristine Lynn Large
Uma 11h27'41" 59d17'6"
Kristine Marie
Ari 2h34'52" 18d3'43"
Kristine Marie / Christopher Thomas
Cap 20h47'9" -25d26'27"
Kristine Marie Scanga
Lib 15h9'18" -4d18'38"
Kristine Martin
And 1h7'11" 33d49'44"
Kristine Mary Halvorson
Cas 2h31'39" 66d0'48"
Kristine My Baby
Cas 0h13'7" 59d28'10"
Kristine Nicole
Uma 8h14'11" 59d48'47"
Kristine Nicole Mathisen
And 0h33'24" 45d18'44"
Kristine Pamela Waters
Vir 13h20'40" 7d18'46"
Kristine S. Wong
And 23h18'11" 48d42'30"
Kristine Sawula
Psc 1h12'53" 22d34'10"
Kristine Schwandner's 40th BD Star
Cas 1h34'27" 62d30'26"
Kristine Star
And 23h38'32" 33d21'24"
Kristine Summers
Uma 9h22'26" 66d32'37"
Kristine Vandelac
And 1h47'2" 36d4'55"
Kristine Victoria Kemmer
Crb 16h3'45" 28d50'38"
Kristine Wallin
And 1h31'56" 45d47'52"
Kristine Welsh
Uma 10h20'24" 43d41'32"
Kristine Westin
Cnc 8h49'35" 12d10'39"
Kristine Whisman
And 1h44'24" 39d30'41"
Kristine Yapp
And 0h26'57" 28d16'49"
Kristine Yarussi
Lyn 8h5'2" 34d45'13"
Kristine Yvonne Hillyer-Mohn
Vir 12h33'20" 0d18'47"
Kristine122269
Mon 6h46'48" -0d8'35"
Kristine's Aura
Cas 23h47'20" 59d38'4"
Kristinius Maximus RawR
Uma 13h19'48" 60d51'1"
Kristin-Kiley
Dra 18h34'34" 79d19'18"
KristinLynn
Ori 5h5'22" 6d42'31"
Kristin-n-Patti
Uma 9h17'37" 52d13'50"
Kristin's Angel
Tau 5h15'27" 21d53'32"
Kristin's Guiding Light
Sco 16h3'1" -26d7'2"
Kristin's & Jon's wishing star
Ari 2h55'15" 14d23'0"
Kristin's Light
Ori 5h28'40" 3d18'19"
Kristin's Mystical Light
And 0h32'36" 32d9'43"
Kristin's Shining Star
Cas 1h40'7" 61d49'24"
Kristin's Star
Uma 11h58'53" 39d12'21"
Kristin's Star of Hope
Vir 12h53'26" 7d56'38"
Kristin's stjärna
Psc 1h5'32" 32d45'42"
Kristi's Figment of Imagination
Uma 13h39'16" 58d25'44"
Kristi's Gary
Pho 0h12'46" -41d4'47"
Kristi's Love
And 2h19'10" 41d32'25"
Kristi's Wish Light
Leo 11h1'1" 4d20'16"
Krististar - 31
And 2h11'30" 42d19'11"
Kristita
Ori 6h15'25" 8d56'40"
Kristl
Gem 7h41'7" 19d49'12"
Kristofer
Sgr 18h58'37" -28d27'30"
KRISTOFER DOBRANIC
Ori 6h7'47" 8d48'36"
Kristofer E
Uma 11h37'38" 42d53'29"
Kristofer Erik Marshall
Her 17h45'2" 23d17'36"
Kristofer Holmes Light of Our Lives
Ori 6h0'53" 17d2'11"
Kristofer Sullivan
Her 17h6'33" 47d52'36"
Kristoffer Johan Sundoey
Cep 4h18'51" 82d58'33"
kristophe
Gem 7h43'36" 25d30'22"
Kristopher Adam Ballingall
Uma 10h16'24" 49d24'7"
Kristopher Berlingieri
Sgr 19h28'38" -39d29'31"
Kristopher Brian Galletta
Gem 7h39'1" 34d51'10"
Kristopher Britton
Cep 22h23'4" 62d19'50"
Kristopher Carrington
Cap 21h12'12" -15d18'25"
Kristopher Daniel Scott Grey
Her 17h13'48" 33d31'53"
Kristopher Fabian' Wilson
Cap 21h20'26" -14d53'22"
Kristopher Gabriel Stull
Ari 2h15'56" 14d16'18"
Kristopher Gene Rogers
Her 16h54'5" 29d52'57"
Kristopher Glenn Wojtunik
Per 4h12'55" 52d33'19"
Kristopher John Methodius Hartman
Per 3h9'53" 54d42'5"
Kristopher Lee Brown
Leo 11h23'55" 18d4'9"
Kristopher M. Matz
Uma 11h28'40" 36d13'22"
Kristopher Marc Rosengrant
Psc 1h10'7" 12d51'27"
Kristopher Micheal Shay Kitts
Her 16h25'28" 9d58'40"
Kristopher Norman Ray Vanderpool
Gem 7h10'53" 26d49'17"
Kristopher Peter Lyman
Per 2h21'20" 55d33'4"
Kristopher R. Harrison
Cep 21h17'49" 62d57'42"
Kristopher Schaer
Ori 6h6'33" 5d27'52"
Kristopher T. Larsen
Uma 10h17'12" 43d21'29"
Kristopher Timothy Fox
Aql 19h31'25" 8d6'26"
Kristopher Tyler
Gem 6h27'31" 17d41'33"
Kristopher William Schoonmaker
Tau 5h38'34" 25d23'30"
Kristy
Cnc 8h27'50" 25d21'29"
Kristy
Leo 9h22'51" 27d45'4"
Kristy
Psc 23h1'0" 7d26'27"
Kristy
And 1h55'13" 40d39'57"
Kristy
Cas 1h24'44" 57d0'2"
Kristy 3
Gem 6h53'53" 16d20'6"
Kristy Abel
Uma 8h41'54" 54d27'25"
Kristy Ada
Her 17h4'35" 14d13'47"
Kristy Allen
Psc 1h26'58" 17d44'42"
Kristy and Tony
Cap 21h12'37" -19d39'31"
Kristy Ann
Psc 0h56'10" 5d3'1"
Kristy Ann Dodd
Cyg 21h22'15" 39d27'55"
Kristy Ann Kobylarz
Cas 1h19'13" 63d52'56"
Kristy Blake
Tau 4h6'1" 9d4'55"
Kristy C. Hurst
Tau 3h36'17" 16d15'41"
Kristy Callahan
Aqr 22h25'8" -0d9'15"
Kristy Conlon
Uma 9h47'53" 66d36'22"
Kristy Cook,CMP, GaMPI Shining Star
And 2h27'58" 42d17'13"
Kristy Currier
Leo 11h19'13" 15d54'46"
Kristy Dawn Overstreet
And 2h31'46" 46d1'17"
Kristy Dupont
Lyn 6h52'18" 50d52'5"
Kristy Germano
Vir 12h50'51" -0d43'25"
Kristy Grieco
And 2h17'33" 50d14'34"
Kristy Irene Grover
Tau 3h42'33" 16d56'31"
Kristy Jean McDaniel
Lyr 18h34'47" 36d4'55"
Kristy Jo
Cas 23h0'45" 57d51'40"
Kristy Jo McBride
Leo 11h8'59" 21d36'46"
kristy jones
Sgr 17h55'31" -27d4'28"
Kristy Kennamore
Sco 17h13'19" -44d14'42"
Kristy Knight A.K.A. Pudding Cup
Tau 5h4'39" 23d44'11"
Kristy L. Bittle
Cyg 21h8'27" 47d56'0"
Kristy L. Mulcahy
Gem 6h53'6" 14d31'12"
Kristy L. Waldron
Aqr 22h29'16" -6d34'40"
Kristy Lasch
Peg 22h8'15" 11d7'0"
Kristy Leaver
Uma 8h43'30" 56d55'48"
Kristy Lee
Psc 23h17'32" 7d38'36"
Kristy Lee Kell
Lib 15h14'38" -10d33'14"
Kristy Lee Winterberg
Leo 11h46'30" 27d17'47"
Kristy Leigh
Cap 21h34'56" -17d32'13"
Kristy Leighbear's Light in the Sky
Lib 15h29'24" -10d50'29"
Kristy Lorraine Wilhelm
Lib 14h58'7" -3d47'54"
Kristy Louisa Morse
Cru 12h52'15" -63d2'41"
Kristy "Lovebug" Lawson
Uma 8h58'17" 55d16'0"
Kristy Loza
Com 12h0'15" 20d56'37"
Kristy Luddy Bathgate
And 1h14'44" 46d4'29"
Kristy Lynn
And 2h12'2" 42d47'12"
Kristy Lynn
Uma 9h23'48" 60d44'30"
Kristy Lynn
Cap 21h30'38" -20d14'39"
Kristy Lynn Blackstone
Cas 23h2'45" 54d32'46"
Kristy Lynn Johnson
Uma 11h22'26" 71d52'21"
Kristy Lynn July 25, 1979
Crb 16h10'41" 37d33'0"
Kristy Lynn Nielsen
Uma 12h37'53" 59d23'18"
Kristy Lynn Saunders
Gem 6h48'1" 28d7'24"
Kristy M. Chatterton
Leo 9h23'59" 28d33'33"
Kristy M. Pettit
Umi 21h14'7" 88d41'13"
Kristy MacCarthy
And 22h59'46" 51d17'33"
Kristy Maree Schweinberger
Tau 3h26'25" 16d58'42"
Kristy Marie
Lib 15h3'49" -3d0'35"
Kristy Marie Kah
Psa 22h49'45" -25d51'14"
Kristy Marie McCleary
Cnc 8h38'34" 31d17'34"
Kristy Marshall
Sco 16h14'33" -22d6'40"
Kristy May
Tau 5h42'45" 24d32'28"
Kristy May, happy 21st birthday
Gem 6h51'24" 12d48'24"
Kristy Morgan
Leo 9h33'10" 28d29'24"
Kristy Murray
Lep 6h7'31" -19d17'20"
Kristy Naranjo
Per 2h17'16" 51d27'15"
Kristy & Nathan -Happily Ever After
Cru 12h38'3" -57d53'50"
Kristy Nicole
Lib 14h44'8" -15d16'46"
Kristy R. Harrington
Leo 9h57'34" 9d34'2"
Kristy Rodrigues
Tau 5h36'40" 26d1'5"
Kristy Rose McCormick
Cap 20h24'6" -9d48'19"
Kristy Sprague
Cyg 20h33'17" 41d36'56"
Kristy Thompson
Cas 1h49'38" 63d56'31"
Kristy Vandervalk
Crb 15h39'16" 30d5'59"
KrIsTy ViCtOrIa NoRmAn
Aqr 21h45'18" -1d5'13"
Kristy Waldron
Sco 16h10'14" -24d46'36"
Kristy Whitney Womack
Dra 17h23'40" 63d24'52"
Kristy Woodward Lampson
Mon 6h46'1" -0d10'12"
Kristye Grayson
Mon 7h3'20" -7d3'35"
Kristyn
Vir 14h17'20" -12d40'34"
Kristyn *21*
Cnc 9h5'27" 22d54'47"
Kristyn April Rudnet
Lac 22h25'4" 48d43'49"
Kristyn Gieptner
Cam 4h48'15" 55d57'57"
Kristyn Ida Tredinnick
Lyn 7h25'12" 51d3'36"
Kristyn Marie
Ori 6h7'47" 9d13'39"
Kristyn Michele
Com 13h5'5" 17d15'3"
Kristyn Shearer Lindley Bridges
Ari 2h19'35" 13d2'3"
Kristyn Stem
Lib 14h58'8" -2d42'6"
Kristyn Winger
Lyn 8h46'17" 33d47'5"
Kristyna DeBaca
Sco 17h46'41" -35d20'40"
Kristyna Jazz Mayfield
Gem 7h45'51" 31d16'2"
Kristyne Vickery Born: Thu July 1/1965
Uma 11h31'0" 57d39'12"
Kristy's Astellas Star
Vir 13h22'18" 11d49'13"
Kristy's Shining Light
And 1h3'54" 42d23'31"
Kristy's Sparkle
Aqr 21h5'43" -11d17'26"
Kristy's Star - a token of our love
Psc 23h26'11" -3d6'48"
Kristy's Zeta
Leo 9h23'21" 10d35'55"
Krisy ELrod
Aqr 23h24'5" -7d25'22"
Kriszta csillaga, Jade
Uma 13h57'17" 54d3'2"
Kriszta és Peti szerelemc-sillaga
Uma 9h55'23" 45d37'38"
Krisztina
Gem 6h54'56" 35d3'4"
Krisztina Füredi
Ori 6h2'20" 10d7'2"
Krit
And 0h28'21" 36d46'41"
Kritikal
Sgr 19h49'14" -15d8'20"
k-rito
Uma 9h52'41" 57d38'3"
Kritty Cakes
Cam 7h30'24" 60d2'39"
Kriz Bugaoan
Lib 14h30'7" -22d44'56"
Krizel Ailana Kok
Psc 1h21'10" 17d42'33"
Krizia Pereira
Gem 7h53'1" 29d57'34"
Krizia S. Ty
Sgr 18h58'32" -34d53'51"
KRIZZO.
Lmi 10h29'21" 38d14'29"
KRJ Varda1
And 0h35'33" 41d32'21"
KRL21-10/06/04
Lib 14h49'6" -2d20'49"
Kroath
Psc 1h16'1" 15d21'31"
Kröger, Claus
Sco 17h16'58" -32d35'7"
Krol Triboulot
Ari 2h10'41" 22d14'17"
Krôlewna Lisa K.
Uma 11h42'17" 56d5'4"
Krolstar
Lyn 7h44'13" 53d4'49"
Kromhoff 80
Uma 9h54'39" 58d57'15"
Krompet
And 0h52'50" 37d38'0"
Kron, Hans
Uma 10h48'18" 63d25'9"
Kronberg, Ilse Anna
Uma 8h55'36" 68d53'45"
Krooglik
Cru 12h10'18" -62d24'27"
KROZUT
Uma 9h53'5" 51d48'28"
KRSForever
And 0h38'0" 43d53'46"
Krueger
Uma 8h23'20" 63d53'44"
Krüger, Alfred
Uma 8h21'15" 64d8'43"
Krüger, Frank
Uma 12h48'41" 54d47'49"
Krüger, Thomas
Uma 11h37'13" 50d19'38"
Krüger, Wilfried
Uma 9h10'10" 51d16'11"
Krull the Warrior King
Tri 2h12'41" 35d20'17"
Krümel
Uma 10h0'37" 53d25'39"
Krümel
Cas 3h34'4" 69d52'18"
Krümel forever
Boo 14h36'41" 17d14'21"
Krümelchen
Uma 9h51'46" 50d42'40"
Krumrey, Silvia
Uma 11h11'24" 62d59'24"
Krupesh Kumar Parikh
Umi 16h10'54" 86d12'35"
Krupsta
Uma 10h40'35" 43d32'25"
Kruschke, Manfred
Uma 12h58'12" 53d43'25"
Kruse, Hans Jakob
Sco 16h50'2" -38d44'20"
Kruti & Nimesh
Cyg 20h24'38" 34d44'17"
Krybecca Rock
Gem 6h48'38" 32d15'40"
Kryco
Sco 16h17'24" -18d33'9"
Kryestyn's piece of heaven
Cap 21h21'21" -18d50'2"
Kryn
Vir 12h21'51" 11d27'20"
Krypto
Ari 3h23'16" 16d5'55"
Krypton Bland
Per 3h58'17" 46d45'35"
Krypton Psys
Mon 6h45'29" 7d25'52"
Kryptonite
Uma 10h24'36" 61d32'40"
Kryptonite Star (for Matthew Reed)
Uma 10h25'53" 63d4'3"
Krysia
Gem 6h27'6" 20d18'38"
Krysia Baran
Gem 6h30'30" 25d28'37"
Krysia Gery-Morawski
Leo 11h15'32" 1d16'53"
Krysiunia
Cnc 8h18'42" 20d34'26"
Krysta
Cyg 21h33'59" 37d3'44"
Krysta Brianne
Cas 23h56'2" 59d40'27"
Krysta Brooks
Cas 0h37'14" 48d48'36"
Krysta Dawn Barkus
Cam 3h26'25" 67d20'5"
Krysta Gazdowicz
Cas 1h14'42" 63d3'40"
Krysta Robyn Blackwell
And 2h14'49" 46d46'27"
Krystal
And 23h8'49" 40d32'10"
Krystal
Uma 10h8'18" 44d38'47"
Krystal
And 0h23'31" 34d36'45"
KRYSTAL
Cnc 8h15'58" 22d52'48"
Krystal
Leo 9h33'55" 29d29'55"
Krystal
Crb 15h52'41" 27d22'27"
Krystal
Psc 23h47'47" 5d25'34"
Krystal
Leo 9h25'15" 11d18'56"
Krystal
Sco 16h11'10" -10d38'7"
Krystal
Cap 20h23'48" -26d8'57"
Krystal Alysee
And 1h27'41" 39d27'36"
Krystal and Jacob's Star
Ari 3h12'16" 28d4'2"
Krystal and Jeremy 9 17 05
Gem 7h6'50" 18d49'15"
Krystal Ann Kroom
Psc 0h46'24" 6d15'35"
Krystal Beauty
Lyn 7h34'7" 41d37'3"
Krystal Beierbach
And 2h35'44" 40d22'38"
Krystal Bonsall
Vir 12h45'18" 3d5'9"
Krystal & Brian
Vir 13h18'39" -7d53'34"
Krystal Caroline Gunter
Gem 7h2'55" 18d5'5"
Krystal Cheyenne Shostrand
Her 17h8'43" 46d12'7"
Krystal Claire Rushford
Aqr 21h41'37" 0d50'55"
Krystal D Ericson
Leo 11h2'10" -1d28'56"
Krystal Dawn
Uma 11h19'25" 42d16'34"
Krystal Dennis
And 23h45'7" 42d12'7"
Krystal Ellielle M
Ori 5h2'5" 10d38'2"
Krystal Gibbens
And 1h15'13" 40d54'35"
Krystal Hutchinson
Vir 11h51'30" 4d35'6"
Krystal Irene
Gem 7h29'44" 20d33'48"
Krystal & Isaac Grauer
Cyg 20h12'4" 40d15'9"
Krystal Jade
Ori 5h57'16" 16d22'42"
Krystal Janell
And 0h44'5" 36d44'44"
Krystal & John Staudt
Cyg 20h42'6" 40d56'41"
Krystal Kathleen Hooper
Cam 5h41'13" 59d52'57"
Krystal Koch
Uma 8h10'54" 62d4'8"
Krystal Kris
Ari 2h43'44" 25d14'5"
Krystal Leah Henderson
Uma 11h19'50" 42d47'4"
Krystal Lee
And 1h26'12" 45d5'27"
Krystal Lee Abbott
And 0h13'49" 26d36'35"
Krystal Lee Hartley
Sco 17h53'33" -35d42'41"
Krystal Lynch
Uma 8h48'43" 47d36'52"
Krystal Lynn
And 1h24'45" 45d30'33"
Krystal Lynn
And 0h58'34" 37d8'57"
Krystal Lynn Best
And 23h12'56" 44d13'19"
Krystal Lynn Garcia Mariposa
Vir 12h47'11" 12d48'9"
Krystal Lynn Swinney
Crb 15h44'46" 35d12'53"
Krystal M. Pride
Aqr 22h55'59" -8d37'2"
Krystal Marie Almaguer
Leo 10h10'11" 25d28'14"
Krystal Marie Chuck
Cnc 8h38'20" 32d20'28"
Krystal Marie Dickson
Aql 18h59'39" -0d2'11"
Krystal Marie Flores
And 0h38'23" 37d49'14"
Krystal Marie Rector
Pho 1h38'52" -46d25'22"
Krystal Marie Roberge
Sgr 19h50'53" -14d15'3"
Krystal Masako
Uma 11h59'30" 28d24'19"
Krystal Michelle Akes
Per 3h26'18" 48d42'36"
Krystal Michelle E.
And 0h43'24" 28d52'0"
Krystal N. Larkin
Lyn 7h26'28" 56d16'58"
Krystal Nicole Bridge
Uma 10h27'24" 72d7'5"
Krystal Noell
Lyn 7h33'27" 48d57'41"
Krystal noelle
Sgr 18h0'54" -26d50'6"
Krystal Rene
Cyg 20h0'52" 48d15'55"
Krystal Riddle The Beautiful
Lyn 6h45'9" 57d55'17"
Krystal Rose
Psc 0h8'40" 7d23'53"
Krystal Rose Ambriz
Sco 17h19'18" -30d32'8"
Krystal Rose Stakes
Psc 0h16'10" 12d16'45"
Krystal Schmitt
Peg 23h14'7" 16d54'19"
Krystal Sharee Chapman
Ori 5h35'59" 1d9'31"
Krystal Sherrie Richards
Mon 6h52'44" -0d11'18"
Krystal Skye Schlichting
Lyn 7h37'27" 40d41'8"
Krystal Smith
And 22h59'40" 44d13'28"
Krystal Sweetman, aka: Yee Yee
Sgr 18h32'50" -27d45'40"
Krystal Tait and Ben Isenberg
Aqr 22h32'16" -13d49'1"
Krystal, Glover & Austin
Uma 12h37'33" 58d42'58"
Krystalia
And 0h24'6" 43d24'20"
Krystalin13
Ori 5h44'30" 4d30'22"
Krystal's Dream
And 2h27'50" 50d22'30"
Krystal's Glimmering Light of Hope
Aqr 22h32'22" -1d40'50"
Krystal's Light
Ari 2h4'27" 19d34'20"
Krysteena
And 0h42'35" 41d20'43"
Krystel
Ori 4h51'5" 4d30'2"
Krysten
Tau 5h46'0" 17d0'59"
Krysten
Sco 17h46'45" -36d42'43"
Krysten Ashley
Crb 15h49'44" 27d40'59"
Krysten Raymond
Cyg 19h58'35" 45d47'33"
Krysten Somer
Her 16h26'30" 10d59'52"
Krysti Lynn Waller
And 0h14'51" 29d57'1"
KRYSTIAN CHODON
Psc 1h15'53" 15d36'43"
Krystie Lee
And 2h35'6" 50d36'15"
Krystin
And 1h51'35" 46d36'31"
Krystin and Steve
Cyg 21h40'54" 45d53'54"
Krystin delCueto
Leo 11h51'11" 18d58'57"
Krystin Jean Shank
Gem 6h42'6" 14d28'32"
Krystin Mari
Uma 10h3'17" 58d44'1"
Krystin Marie Cassidy
And 23h22'11" 48d18'20"
Krystin Michele
And 23h17'11" 41d1'21"

Krystina And 23h3'58" 39d34'58"
Krystina Lyn 7h55'11" 59d38'7"
Krystina Cap 20h38'51" -22d12'36"
Krystina Adel Davis Lib 15h20'53" -5d54'2"
Krystina Chickerillo Leo 11h18'9" 23d0'29"
Krystina Lee Koepp Gem 6h59'14" 30d6'35"
Krystina Marie Beauchemin Lyn 7h5'19" 59d31'13"
Krystina Ritchey Cnc 8h22'27" 11d13'45"
Krystina's Smile Psc 2h3'57" 10d17'32"
Krystine Lyn True Uma 10h58'22" 51d44'53"
Krystle Cam 3h39'48" 58d0'17"
Krystle Lyn 8h27'22" 48d43'1"
Krystle Ori 6h7'3" 5d54'37"
Krystle Mon 6h49'0" 7d47'46"
Krystle Psc 1h23'0" 15d43'32"
Krystle Ann Holderfield "My Only" Vir 14h30'49" 7d1'19"
Krystle Berryman Cas 1h24'27" 62d28'53"
Krystle Dawn Aql 19h32'47" 9d27'27"
Krystle Dawn And 2h13'1" 45d58'30"
krystle Ion Umi 11h54'51" 87d47'44"
Krystle Ireland Dra 18h6'28" 73d37'57"
Krystle Jones Ori 5h38'9" -8d58'31"
Krystle Lynn Dubord Psc 1h19'17" 31d59'51"
Krystle Lynn Wells Ori 5h5'49" 4d28'36"
Krystle Marie Cyg 20h45'12" 32d57'32"
Krystle McFadden Dra 17h15'10" 54d16'10"
Krystle Rae Pagaduan Jovillar Ori 5h7'30" 12d8'3"
Krystle Rose Matthews And 1h15'46" 36d43'55"
Krystle Stanfield and Adam Stary And 1h11'8" 45d46'37"
Krystle T Gullace 12.12.84-08.04.99 Car 10h2'47" -63d5'11"
Krystle Thornton And 1h27'13" 39d50'33"
Krystle429 Crb 15h43'6" 36d7'56"
Krystles Star Lyr 18h26'56" 33d37'19"
Krystyl Dawn Wood Cap 20h54'15" -19d18'36"
Krystyn Evrley Cas 1h42'32" 62d10'39"
Krystyna Ori 5h2'42" 15d2'38"
Krystyna and Yanusz Morgenstern Lyr 18h41'56" 40d10'58"
Krystyna & Brian Forever Cas 0h29'54" 52d5'20"
Krystyna Danielle Yvonne Kopacz And 22h58'56" 45d17'19"
Krystyna INDY 77 Rutkowski Sco 16h8'15" -13d57'33"
Krystyna Zduniak Tau 3h40'55" 27d53'11"
Kryswharn Crb 16h20'9" 36d44'53"
Krzysiu Ori 6h8'50" 15d15'14"
Krzyston's Eternal Love Her 18h46'20" 21d3'28"
Krzysztof and Sonja Sge 19h44'0" 18d13'44"
Krzysztof Kaminski Uma 10h8'45" 55d25'53"
K's Star And 2h15'40" 49d50'33"
KsAs1Anvrsry Aqr 22h11'9" -3d4'6"
Ksenia Cas 0h55'38" 61d44'30"
Ksenia And 23h19'9" 51d4'29"
Ksenia Luka Leo 9h23'56" 22d0'54"
Ksenia Olga Vandalov Com 12h7'58" 19d45'15"
Ksenia S. Kuznetsova Tau 6h0'16" 25d10'53"
Kseniya Gulyamova Cnc 8h5'36" 21d13'59"
Ksenyia Zhuzha Cap 21h35'45" -16d23'41"
Kseronus Lyn 7h56'2" 39d32'12"
K.S.F.- My Konstantine Tau 4h6'21" 9d58'42"
Kshama Uma 8h24'43" 65d11'36"
K-Shizzle Uma 10h36'8" 40d11'0"
Ksiusha Sco 16h22'0" -26d49'4"
KSM WRM 6934 Cnc 8h35'56" 32d32'41"
KSON Gem 6h46'56" 16d6'33"
KStuMit Cyg 19h40'42" 32d33'18"
Ksyusha my little star Cap 20h40'17" -18d52'11"
KSZ Mazzitelli Peg 21h58'36" 25d39'2"
KT Ori 6h7'57" 19d47'16"
KT Sco 16h6'23" -15d13'3"
Kt and Jo-Jo Her 17h58'56" 29d54'2"
KT Graham's Illumination Aqr 23h6'2" -16d22'27"
K.T. ONE Ari 2h12'23" 10d49'11"
K.T. Richmond Cap 21h16'48" -18d48'16"
K.T. Sullivan Cam 5h44'45" 59d16'54"
KT With All My Love & Daisy Bo Lee Vir 14h7'59" -17d9'53"
KT04261980 Tau 4h33'2" 17d42'26"
KtBoo Psc 0h45'35" 9d3'44"
KTBUG Psc 1h9'0" 32d24'59"
KTdid And 1h13'44" 45d26'58"
KTJ Sgr 19h27'49" -16d7'8"
K.T.K Uma 10h26'48" 66d40'6"
K.T.McGann 007J.T.Central H.S. Cap 20h26'20" -12d9'43"
KTM-Heini Cap 20h13'26" -10d33'46"
KTO1 Vir 12h34'37" -2d16'37"
Kt's Star Lmi 10h41'29" 38d32'23"
KTWJMLQA Uma 10h15'32" 68d1'18"
Kty Gauthier Cas 23h50'17" 52d37'56"
KU Dan Uma 10h25'29" 62d12'40"
Ku' u Hoaloha' Oi Uma 10h44'40" 63d39'15"
kuani bay cleveland Tau 5h42'18" 22d16'32"
Kubala Sisters And 1h23'6" 44d19'15"
Kubiaks Knightlight Uma 13h44'51" 56d17'52"
Kubiessa, Bernhard Ori 4h56'7" 10d43'35"
Kübler Leonhardt Mon 7h43'35" -1d53'59"
Kubra - Behbood Del 20h28'48" 16d8'1"
Kubzig, Ralf Uma 9h34'45" 49d40'45"
Kucsan-Hood Lyn 8h42'34" 34d28'40"
Kudzu David Vir 13h16'55" 7d29'21"
Kueb-O-Wen Gem 6h43'13" 33d5'13"
Kuei Yun Sgr 19h28'39" -30d48'47"
Kuenke E. Per 3h10'16" 47d29'52"
Kuenz "Love Star" Cas 1h1'3" 57d17'30"
Kuertchen Leo 11h0'20" -5d9'5"
Kugler, Helene Ori 5h18'7" 3d36'31"
Kuhl BlueSkies And 0h23'45" 39d27'1"
Kuhlman McCool Lib 15h20'11" -6d34'49"
Kuhn, Annika Madeleine Uma 8h20'39" 63d47'10"
Kühn, Manfred Erwin Uma 10h57'10" 48d56'50"
Kühnemund, Rudolf Uma 9h55'8" 61d57'25"
Kühner, Stefan Ori 6h15'1" -3d42'56"
Kuhnle Lyr 18h40'5" 26d13'11"
Kuhnus Major Five "12-25-00" Lyn 7h48'51" 49d6'28"
Kuikawa Hoku (Special Star) Vir 13h49'19" -5d33'31"
Kuilee Uma 12h2'0" 32d55'21"
Kuisses Ari 1h50'15" 18d46'8"
KUKLA, FRANZ Ori 5h50'2" 7d13'58"
KULEMA-KULEMA Ori 6h1'23" 18d3'16"
Kulit Mon 7h30'21" -0d28'40"
Kulla Ori 5h14'44" 12d16'1"
Kully Nonohiaulu Kekaula Uma 10h53'11" 41d59'2"
KULSUM KASSAM Tau 4h50'26" 22d24'51"
Kulveen Virdee Cyg 20h27'35" 46d5'29"
Kulvinder Kaur Ahluwalia Cas 0h51'14" 51d38'26"
Kuma Uma 14h22'47" 56d49'9"
Kuma Bear Ori 5h26'25" 3d8'3"
Kumar Uma 13h26'45" 57d34'34"
Kumar Parakala Cru 12h8'3" -60d45'11"
Kuma's Shining Love Aqr 22h14'57" 0d12'12"
Kumi Wauthier Cas 1h8'25" 60d22'35"
Kumquat Umi 14h5'54" 75d19'13"
Kumquat Star Cas 2h50'37" 60d25'39"
Kumu Casey Kono Uma 10h27'35" 45d14'34"
Kumud Shah Uma 8h56'1" 56d48'38"
Kunaja Dra 9h29'57" 75d15'3"
Kunal Ori 6h8'51" 20d49'18"
Kunal Mohindroo Uma 11h45'52" 51d47'17"
Kunal Mohindroo Uma 9h11'12" 69d33'44"
Kung Sing 2/9/71 Cam 4h30'21" 67d3'6"
Kung Yi Chien Lyn 8h12'50" 37d4'1"
Kunta 150 Tau 3h48'40" 27d23'2"
Kunzewitsch, Willi Uma 10h35'33" 72d40'42"
Kuok, Wai Chi Uma 10h37'12" 47d28'0"
Kupabear's Star Uma 12h41'4" 57d8'2"
Kupfer, Mario Uma 10h56'9" 43d11'7"
Kupkova Alena Umi 17h1'1" 79d18'14"
Kupono Momi Pyx 8h42'18" -22d15'36"
Küpper, Peter Uma 12h15'48" 61d10'55"
Kuprian Bernhard Uma 9h11'59" 59d41'17"
Kuraia Libre Leo 10h16'8" 26d31'47"
Kurama Uma 12h0'7" 33d12'16"
Kürbis 19 Ori 5h15'31" 3d19'10"
Kurcik And 0h8'52" 34d15'52"
Kurcsics Rita - El Arwen Uma 11h34'41" 35d43'24"
Kurleen's Star Cam 4h24'0" 67d20'22"
Kurrae Umi 16h24'46" 76d31'31"
Kurrashalynn Miyosha Caver Cyg 20h46'58" 32d57'20"
Kurrie Wells Lyn 8h33'45" 34d52'24"
Kurrle, Angela Ori 6h12'13" 13d49'33"
Kurstin Diane Leo 11h49'51" 23d8'17"
Kurt Leo 9h34'31" 20d51'2"
Kurt Aur 6h27'15" 34d28'34"
Kurt Uma 9h59'8" 42d50'40"
Kurt A. Petry "1963-2005" Aqr 23h29'37" -14d39'33"
Kurt and Mom's Star Uma 9h58'30" 41d53'52"
Kurt and Stephanie Forever Cyg 20h17'25" 54d39'35"
Kurt Arney Per 2h58'10" 45d12'23"
Kurt Bösch Lmi 10h30'3" 31d11'18"
Kurt "Buckwheat" Mueller Lyn 6h52'4" 51d36'24"
Kurt Chen Selko Cep 21h40'23" 56d6'53"
Kurt Douglas Herman Aql 19h7'14" 4d16'43"
Kurt Edward Heinzman Her 16h45'55" 36d52'40"
Kurt Evan Jerrett the Boldest Star Gem 7h47'17" 34d8'29"
Kurt Friederich Tau 5h44'23" 17d12'5"
Kurt Gerald Anderson Her 17h40'27" 36d55'33"
Kurt Hackspiel Uma 11h52'5" 62d26'32"
Kurt Happy 1st Anniversary Love Em Cyg 19h42'9" 32d45'44"
Kurt Heims Psc 1h16'58" 15d47'19"
Kurt Hickam Uma 13h54'48" 47d55'15"
Kurt Hirschberg Uma 8h31'39" 59d43'44"
Kurt J. Giometti Uma 11h35'47" 56d36'24"
Kurt J. Schoeller Her 16h53'18" 37d28'35"
Kurt Johanning 02.05.33 - 28.03.06 Uma 12h17'1" 52d39'33"
Kurt John Wyberanec And 0h21'42" 37d18'2"
Kurt Joseph Obrecht Ori 6h10'55" 5d42'56"
Kurt Karon Centennial Uma 13h22'20" 54d19'29"
Kurt Kohout Psc 0h46'9" 16d46'51"
Kurt & LindaSue Wells Love our Moms Psc 1h28'42" 17d29'42"
Kurt Louis Veitch Sgr 19h17'45" -34d49'51"
Kurt Maass Uma 10h29'35" 65d10'43"
Kurt Manson Cru 11h57'17" -62d16'45"
Kurt Mathew Hunsanger Ori 6h6'44" 8d47'3"
Kurt Matthew Rowland Her 17h59'38" 21d26'38"
Kurt Michael Freund Cap 20h13'50" -18d35'22"
Kurt Patrick Dardis Dra 17h6'44" 58d42'44"
Kurt Pedersen Aql 19h55'7" 11d16'49"
Kurt R. Wimer Aur 5h23'51" 41d5'9"
Kurt Rey, 13.03.1998 Her 17h41'2" 38d52'36"
Kurt Sahli Der himmlische Angelibeck And 2h24'20" 47d28'3"
Kurt Stahl Sco 17h46'41" -33d42'4"
Kurt Stallings Dra 17h54'32" 51d45'27"
Kurt Stirling Hampe Cyg 20h34'30" 57d48'28"
Kurt W. Ineichen Cyg 19h47'12" 35d22'19"
Kurt Walter Ori 5h56'47" 7d14'33"
Kurt Zello Smith Uma 11h50'36" 62d14'38"
Kurtis Aqr 22h37'10" -1d52'48"
Kurtis Per 2h46'44" 53d38'16"
KURTIS BANCHERO Uma 14h0'10" 60d38'27"
Kurtis Blankenship Per 2h44'31" 53d31'41"
Kurtis Fisher Umi 13h52'49" 75d52'10"
Kurtis James Ori 6h7'18" 18d14'1"
Kurtis Letch Cra 18h57'55" -39d3'19"
Kurtis Levi McConnell Vir 13h18'22" -2d15'46"
Kurtis Popp Per 3h26'3" 41d15'8"
Kurt's Cleopatra Eri 4h8'53" -32d36'53"
Kurty's Star Cap 20h28'52" -26d28'6"
Kurtze, Christa-Maria Uma 10h25'30" 66d4'35"
küschall And 0h44'9" 41d50'57"
Kuschelbär Mike Uma 10h14'48" 70d15'4"
Kuschelbärli Crb 15h47'20" 26d20'1"
Kuschelmus Andrea Cam 14h10'40" 83d24'30"
Kushi & Schubbi Tau 3h59'31" 1d52'51"
Kusibärli Umi 14h33'22" 79d10'37"
Kustomaniak Livy'Oz Cas 0h54'36" 55d2'2"
Kusum Malati Topé Ari 3h28'4" 26d56'35"
Kutallee Vir 12h26'4" -5d11'9"
Kutie Uma 11h17'31" 42d58'54"
Kutik Sco 17h57'8" -30d13'47"
Kuti's Light Aqr 22h19'0" -1d50'13"
Kutsal Ori 6h21'6" 15d36'17"
Kutschke, Ben Uma 8h44'55" 64d48'59"
Küttel Andrea Cas 1h25'12" 52d4'58"
Kutzner, Margarete Uma 9h53'29" 72d33'25"
Ku'u Hoku Liko Sge 19h41'12" 19d5'1"
Ku'u lei Per 3h20'32" 42d4'52"
Ku`ualoha Ho`omanawanui Ori 5h59'58" 22d37'20"
Kuuiplet Uma 9h12'38" 48d59'47"
Kuuipo Lib 14h36'42" -18d49'25"
Kuuipo Lekili Chiyomi Sgr 19h9'53" -20d59'58"
Ku'uipo Lilikoi Gem 6h58'14" 29d36'4"
Ku'uipookalani Sco 16h43'6" -29d51'49"
Ku'uli'aleinani Leo 9h35'45" 6d57'37"
Kuzminca "The Princess" Sco 16h11'7" -12d46'58"
Kuzmirek Christmas 2005 Cyg 19h58'1" 31d57'35"
K.V. and Dorothy Nelson Vir 13h45'34" 1d30'15"
KVane Lib 14h50'17" -2d7'52"
KVIETKA Sgr 19h22'14" -35d16'40"
KW 2006 Cma 7h14'15" -23d51'5"
KW1327 Uma 10h39'35" 70d4'21"
KW8511KS Lyn 8h13'32" 47d32'21"
Kwan Nga (Eva) Leung Cnc 8h10'29" 6d56'39"
KWC's Star Psc 0h1'41" 3d45'13"
Kweon II Gem 6h41'40" 18d23'23"
KWIATEK Carmen Masely Gem 7h54'19" 21d23'35"
Kwinn T'Essense Cma 6h37'41" -13d12'3"
Kwitowski's Her 15h56'12" 42d39'42"
Kwon SooYoung Ori 5h36'9" -2d41'12"
KWR3557 Psc 0h57'21" 31d47'29"
KXP Uma 10h9'9" 64d51'32"
Ky Ori 5h23'43" 0d42'58"
Kya Lynn Hann And 1h16'23" 46d28'11"
Kya Michelle Cato Sgr 18h9'21" -19d15'42"
kyafri Vir 12h56'58" 11d23'48"
Kyah Kranz Cap 21h53'58" -13d36'37"
Kyan Michael Alltop Aql 19h46'59" 6d10'40"
Kyan's Guiding Star Pyx 9h24'43" -28d40'18"
Kyauna La'chaun Skinner Cyg 21h56'41" 37d21'29"
KyD Cam 4h38'45" 69d57'31"
Kydee Lib 15h47'54" -18d13'49"
Kye Harley Brown Tau 4h21'40" 2d21'27"
Kye Robert Smith Ari 2h13'56" 14d57'16"
Kyen Blake Gross Leo 10h11'52" 14d17'5"
Kyeng - Hee Neas Per 3h22'18" 44d22'33"
Kyeson Col 6h7'26" -31d4'4"
Ky-Girl Tau 4h9'50" 23d39'13"
Kyhla Topjian Leo 11h39'57" 18d1'6"
KYKY (Cristina Elena Bozocea) Lmi 10h3'25" 39d36'29"
Kyla Sco 16h12'56" -13d36'50"
"Kyla" Ori 5h35'4" -9d56'6"
Kyla Aqr 22h13'55" -2d48'1"
Kyla Uma 9h4'10" 60d45'23"
Kyla Dra 16h13'6" 54d44'48"
Kyla Alvarado Gem 7h13'8" 16d14'46"
Kyla Amoré Pratt Lmi 10h4'41" 36d39'31"
Kyla Ann Ori 6h2'4" 17d37'4"
Kyla E. Alsip Lyr 18h48'6" 36d55'15"
Kyla Elizabeth Uma 11h9'30" 48d51'14"
Kyla Emaline Groller Ari 2h50'36" 30d52'30"
Kyla Evelyn Gitthens Gem 6h47'43" 27d21'48"
Kyla Faith Uma 10h20'14" 46d9'0"
Kyla Gavin Psc 1h56'21" 5d29'11"
Kyla Grace Lyn 6h25'5" 56d37'38"
Kyla Jeanette Powers Gem 7h19'56" 29d53'55"
Kyla & Jeremy 8/15/1999 Del 20h40'15" 15d55'25"
Kyla Knepp Uma 11h58'45" 39d53'27"
Kyla Linn Caprarella And 0h11'25" 34d2'51"
Kyla Mae Wagner And 0h19'55" 45d53'47"
Kyla Margaret Hartigan Cap 21h38'17" -13d24'50"
Kyla Marie Mattioli Leo 11h10'53" -2d38'2"
Kyla Munkberg Vir 12h52'58" -5d49'57"
Kyla Niedermaier Lyn 8h21'22" 47d5'24"
Kyla Price And 23h39'58" 47d15'48"
Kyla RaeAnn Timm Gem 6h59'57" 14d27'31"
Kyla Schmidt Psc 0h42'49" 9d28'14"
Kyla Yohey And 2h37'29" 43d21'10"
Kylah Ori 5h46'47" 10d39'50"
Kylah McCain Umi 16h14'53" 73d1'48"
Kylah Violet Riddell Cru 12h38'20" -59d58'50"
Kyla-Louise Haughton Vir 11h48'16" -0d7'35"
Kylan James Huffman Her 16h50'45" 29d48'11"
Kyla's Star And 23h28'37" 48d36'45"
Kyle Per 2h55'52" 54d48'25"
Kyle Aur 5h12'18" 42d31'11"
Kyle Her 17h47'24" 31d36'47"
Kyle Tau 4h21'20" 24d1'3"
Kyle Psc 23h33'50" 1d12'39"
Kyle Aqr 21h41'17" 2d50'7"
Kyle Del 20h35'26" 14d6'22"
Kyle Aqr 20h48'22" -12d42'58"
Kyle Cap 20h7'34" -14d28'48"
Kyle Sgr 19h40'57" -12d48'19"
Kyle Lib 15h45'9" -20d56'8"
Kyle Dra 20h12'17" 71d23'36"
Kyle Dra 19h27'41" 65d7'52"
Kyle Lyn 6h33'36" 56d34'33"
Kyle Uma 9h0'22" 54d2'27"
Kyle Abrahamsen Per 3h35'49" 42d3'34"
kyle adam Her 16h20'41" 16d52'12"
Kyle Adam Chambers - Kyle's Star Gem 6h29'57" 16d26'27"
Kyle Adam Viguerie Dra 15h46'12" 61d14'51"
Kyle Addison Hitchcock Tau 4h30'2" 15d4'18"
Kyle Alan Sims Dra 17h0'58" 60d24'37"
Kyle Albert Connor Per 3h40'7" 48d27'26"
Kyle Aldinger Ori 6h0'56" 18d28'14"
Kyle Alexander Caudill Cam 3h51'8" 67d30'17"
Kyle Alexander Dubin Aqr 22h13'52" 1d27'2"
Kyle Alexander O'Connell Uma 11h21'19" 53d12'33"
Kyle Allen Tolbert Uma 9h13'25" 66d55'9"
Kyle and Angela McCown Ori 5h39'48" 5d42'56"
Kyle and Annette's Star Sge 19h13'44" 19d40'51"
Kyle and Dawn Per 3h42'38" 49d28'42"
Kyle and EB's Star Cyg 19h52'2" 33d28'46"
Kyle and Elizabeth's Forever Star Cyg 20h11'31" 44d22'31"
Kyle and Felicia Forever Uma 9h34'50" 69d16'42"
Kyle and Jana Biggs Cyg 21h19'49" 41d52'16"
Kyle and Katie Her 16h31'59" 46d43'34"
Kyle and Kirsty - Eternity Col 6h34'30" -35d31'22"
Kyle and Mallory's Star And 0h31'44" 42d20'31"
Kyle and Marcia's Mulberry Creek Sgr 19h38'54" -23d10'21"
Kyle and Rebecca Forever Ori 5h0'2" 15d10'49"
Kyle and Shannon Forever Lib 15h6'18" -16d46'51"
Kyle and Tesa, Amelia Island, FL Ori 5h0'0" -0d14'52"
Kyle and Tiffany Forever Her 17h57'15" 24d27'6"
Kyle Andrew 2806 Leo 10h7'42" 10d56'42"
Kyle Andrew Buck Per 4h39'56" 40d32'58"
Kyle Andrew Kegley Aur 5h55'15" 45d20'22"
Kyle Andrew Lyman Her 17h31'59" 36d43'45"
Kyle Andrew Mangin Lib 15h58'23" -16d47'53"
Kyle Andrew Onufer Cam 5h29'35" 66d24'15"
Kyle Andrew Stevens Vir 12h46'36" 5d17'49"
Kyle & Angela xoxo Cyg 20h35'48" 38d17'33"
Kyle Ann Byers Cyg 20h53'11" 38d23'9"
Kyle & Annaka Magill And 23h22'5" 47d29'8"
Kyle Ansley Cnc 8h4'35" 13d8'19"
Kyle Anthony Donohoe Sgr 18h22'17" -20d54'15"
Kyle Anthony McNally Leo 11h39'31" 25d23'59"
Kyle Anthony Redmond Boo 14h10'3" 27d54'16"
Kyle Armitage Per 2h38'7" 54d3'6"
Kyle Aubrey Nix Per 4h13'27" 43d0'8"
Kyle Auer Per 4h4'8" 34d14'51"
Kyle Aymond Ari 2h7'25" 11d51'25"
Kyle B. Hanks Ori 5h17'52" -0d55'37"
Kyle Battinieri Leo 11h41'36" 16d54'26"
Kyle Benjamin Morris Dra 18h39'17" 60d17'53"
Kyle Bento Gem 7h42'36" 21d18'47"
Kyle Besant-Jones Umi 16h5'47" 77d41'0"
Kyle Blake Hendee Lib 15h11'4" -5d43'48"
Kyle Bollman & Kelsey Hracho Uma 9h33'48" 66d45'43"
Kyle Bradford Doyle Tau 5h32'0" 27d56'46"
Kyle Bradley Lib 15h44'29" -12d13'19"
Kyle Bradley Breckenridge And 2h13'50" 49d22'56"
Kyle Bradley Rucker 7/8/82-11/11/02 Cnc 8h39'49" 17d52'1"
Kyle Brandon Hensley Cnc 8h45'26" 18d42'49"
Kyle Brent Howard-Campbell Uma 11h59'3" 37d30'9"
Kyle Briggs Avirom Tau 4h26'33" 12d44'7"
Kyle Burton Furness March 26 1984 Ori 5h21'49" 6d16'36"
Kyle C. Heideman Ori 5h33'36" 9d3'4"
Kyle Carson Stafford Lyn 7h11'44" 58d13'44"
Kyle Chancelor King Tau 3h54'57" 21d27'41"
Kyle Charles Brennan Umi 14h46'38" 76d4'29"
Kyle Charles/Faith Renee Leo 11h55'52" 23d56'12"

Kyle Chase
Cnc 8h28'41" 28d7'11"
Kyle & Chelsi
Cyg 21h32'15" 38d39'26"
Kyle Christian
Aql 19h45'29" 2d59'11"
Kyle Christian DeMatteo
Aql 19h42'46" 14d27'13"
Kyle & Christina Bordelon
Always
And 1h47'34" 41d53'22"
Kyle Christopher Bryson
Aqr 23h5'12" -10d26'3"
Kyle Christopher Coble
Tau 5h37'20" 25d21'16"
Kyle Christopher Walker
Vir 14h16'43" 0d53'39"
Kyle Claire
Uma 9h40'28" 50d56'57"
Kyle Condie, and Kelly
Masters
Cap 20h21'20" -11d12'52"
Kyle Cossette's Star
Uma 10h46'42" 55d4'51"
Kyle Cresanto - Little Man
Lib 14h23'41" -17d25'12"
Kyle Cullen Hitchens
Uma 13h40'44" 51d22'26"
Kyle Curtis Ruggles
Gem 6h47'37" 28d55'14"
Kyle Dalton Kleckner
Dra 17h35'10" 53d25'40"
Kyle Danger David Burgess
Ari 3h19'13" 29d4'5"
Kyle Daniel Perry
Her 17h34'19" 36d6'8"
Kyle David Fitzgerald
Cnc 9h8'29" 32d50'57"
Kyle David Grant
Eddington
Gem 7h22'19" 23d41'28"
Kyle David Harris
Aql 19h18'2" -0d14'39"
Kyle David Konrad
And 23h35'58" 44d52'14"
Kyle David Page
Lib 15h9'22" -12d31'8"
Kyle David Scott
Aur 5h39'53" 45d14'26"
Kyle Dean Jorgensen
Aqr 22h43'31" 1d4'39"
Kyle DeMonte
Uma 8h32'55" 68d44'44"
Kyle DeSimone
Aur 5h34'10" 44d35'1"
Kyle Dominic Ingels
Tau 3h53'2" 7d10'51"
Kyle Douglas Weichman
Leo 9h53'11" 8d7'30"
Kyle Douglas Whipple
Lib 15h12'44" -21d54'49"
Kyle Dowler
Uma 13h44'34" 55d43'42"
Kyle Duane Fowler
Psc 0h10'18" 2d36'24"
Kyle Dunnigan
Boo 14h49'19" 51d0'22"
Kyle Dustin Grossman
Per 2h47'47" 48d7'9"
Kyle Dylan Tomlinson
Umi 14h54'8" 70d44'38"
Kyle Edward Gonzales
Aql 18h55'46" -0d0'44"
Kyle Edward Riddle- A.K.A
"Koder"
Cap 20h24'36" -9d17'47"
Kyle Elliott
Vir 14h15'8" -13d55'38"
Kyle Emily Rogers
Cep 0h31'10" 78d57'2"
Kyle Engel
Pho 0h47'27" -48d14'14"
Kyle Eric
Sgr 18h41'59" -27d32'37"
Kyle Evan Danielson
Boo 14h20'58" 17d9'56"
Kyle Evan Zamajtuk
Vir 12h42'23" -6d20'41"
Kyle Evans Church
Gem 7h20'10" 24d32'22"
Kyle F. Driscoll
Lyn 8h55'48" 33d32'41"
Kyle Ferrell & Sarah Cappa
Aqr 22h14'44" -3d13'9"
* Kyle Fielding *
Dra 19h39'20" 64d6'2"
Kyle Figgins
Cep 23h20'28" 87d11'45"
Kyle Franzen
Psc 1h19'23" 31d4'31"
Kyle Fritz Valle
Uma 9h24'1" 49d23'3"
Kyle G.
Her 17h29'52" 18d52'20"
Kyle George Laird McKay
Per 2h33'2" 57d2'6"
Kyle Germanton
Aqr 21h18'4" 1d58'59"
Kyle Glenn Holcombe
Uma 12h40'0" 56d52'19"
Kyle Gordon Hollis
Leo 11h9'56" -0d48'57"
Kyle Grady Stringer
Ori 5h24'38" 5d17'39"
Kyle Greenlee
Cap 20h25'13" -10d50'35"
Kyle Haas-Second Star To
The Right!
Lib 15h4'30" -7d21'34"
Kyle Hastings Eagan
Ari 2h34'43" 25d54'17"
Kyle Hayes
Dra 17h23'21" 53d24'9"
Kyle Hemion
Her 17h30'24" 46d38'12"
Kyle Hiebert
Vir 14h42'58" 4d22'41"
Kyle Higgins
Sco 16h7'3" -11d21'5"
Kyle Howe
Her 17h43'53" 39d20'25"
Kyle J. Bergstrom
Cnc 8h49'13" 26d36'2"
Kyle J. Magdich
Uma 13h21'40" 62d8'18"
Kyle J. Miller
Uma 9h53'45" 43d24'42"
Kyle J. Perciak
Gem 6h59'7" 15d45'42"
Kyle Jacob Brooks
Lyn 8h11'45" 55d43'6"
Kyle Jacob Krutilek
Lyn 7h29'5" 49d27'46"
Kyle James
Cep 21h25'17" 62d47'35"
Kyle James Fidler
Psc 1h39'12" 20d49'19"
Kyle James Raleigh
Uma 10h53'38" 52d13'2"
Kyle James Rossiter
Per 3h24'24" 41d27'13"
Kyle James Sinclair
Lmi 9h59'59" 36d22'58"
Kyle Janice
Cap 20h34'19" -24d23'51"
Kyle Jeffrey
Per 3h10'30" 52d31'29"
Kyle Jessie Kirkpatrick
Cnc 8h14'13" 23d35'44"
Kyle Jeter's Great Ball of
Fire
Ori 4h52'34" 4d19'20"
Kyle Jo
Sgr 19h39'5" -15d39'47"
Kyle John
Cma 6h43'1" -15d50'36"
Kyle John O'Connor
Uma 11h12'20" 49d31'58"
Kyle John Seymour
Cnc 8h44'32" 27d51'19"
Kyle Johnson 2004
Uma 11h21'10" 30d34'59"
Kyle Joseph
Vir 13h22'10" 12d13'53"
Kyle Joseph Hoffpauir
Ori 5h25'46" 2d54'43"
Kyle Joseph Rhode
Sgr 19h34'14" -13d44'59"
Kyle Joseph Ryder
Her 16h59'21" 33d12'10"
Kyle Joseph Suozzo
Cap 21h22'1" -24d54'51"
Kyle Joseph Theroux
Uma 10h23'10" 47d9'10"
Kyle Joseph Theroux
Cyg 19h55'40" 37d40'47"
Kyle Joseph Van Loon
Her 16h14'17" 47d51'58"
Kyle Joseph Yarusites
Lmi 10h34'6" 38d49'24"
Kyle & Justin
Wojciechowski
Lib 15h2'37" -0d51'51"
Kyle Keawe Pabo
Cnc 8h41'24" 23d11'57"
Kyle Kendrick McCarthy
Leo 11h39'50" 21d44'20"
Kyle Kenney
Uma 9h56'11" 52d53'23"
Kyle Kevin Herrick
Her 16h2'30" 16d33'18"
Kyle & Krystal
Cyg 21h13'58" 43d12'6"
Kyle Kyle Crocodiles Star
Leo 9h27'8" 15d44'45"
Kyle "Kylie" Gensel
Uma 8h44'21" 67d52'59"
Kyle Lardner
Gem 7h14'23" 33d8'18"
Kyle Laurel Dorothy Nolan
Gem 7h43'15" 34d16'37"
Kyle & Lauren
Psc 1h5'39" 26d20'37"
Kyle + Lauren
Aqr 23h13'41" -8d25'31"
Kyle Lawrence
Her 17h32'33" 44d59'54"
Kyle Lawrence Olszak
Her 16h14'28" 48d1'11"
Kyle Lazarus
Ari 2h37'51" 27d9'31"
Kyle Lindsey
Cam 4h3'50" 68d21'16"
Kyle Livingston
Sgr 19h36'19" -13d41'17"
Kyle Logan
Ari 2h55'19" 18d1'1"
Kyle & Louise's Star
Cru 12h25'56" -56d12'17"
Kyle Lynn Scott
Sgr 17h56'30" -23d49'23"
Kyle M Vail
Cap 21h4'50" -21d48'58"
Kyle Mann
Aql 19h30'0" 2d30'13"
Kyle Marcus Green
Gem 7h3'10" 18d35'5"
Kyle Marquardt
Her 17h13'22" 47d9'44"
Kyle Marshall
Lib 15h15'32" -15d28'17"
Kyle Massey
Uma 11h8'5" 53d1'15"
Kyle Matthew Cook
Her 16h46'40" 37d27'10"
Kyle Matthew Krogstad
Cyg 21h32'57" 41d37'54"
Kyle Matthew Tyler
Pyx 8h39'42" -21d29'17"
Kyle McDiarmid Cameron
Gem 6h25'27" 22d8'38"
Kyle Michael Jacobs
Aur 5h54'28" 42d54'39"
Kyle Michael Sung
Leo 11h46'25" 18d7'31"
Kyle Molidor
Aql 19h51'54" 10d38'52"
Kyle Moon Boy
Uma 11h19'51" 43d49'33"
Kyle M.Young
Lib 15h9'40" -13d37'8"
Kyle n Emma 4ever
Cyg 19h48'34" 37d10'18"
Kyle Neil Thompson
Aqr 22h34'42" -20d52'58"
Kyle Nigel Dell
Aql 19h33'46" 12d4'40"
Kyle Noel
Psc 23h52'7" 0d31'9"
Kyle P. Kenney
Cnc 8h42'22" 31d52'23"
Kyle Paisley
Her 16h48'52" 24d24'41"
Kyle Patrick Cummings
Sco 17h51'44" -35d42'2"
Kyle Patrick Kohmann
Ori 5h15'29" 0d21'44"
Kyle Patrick Margerum
Cnc 8h48'57" 27d13'28"
Kyle Pound
Cas 1h1'24" 61d8'38"
Kyle Principe
Uma 13h55'38" 56d20'49"
Kyle R. Swanson
Her 17h28'13" 32d53'34"
Kyle Rabe
Mon 6h52'52" -0d50'11"
Kyle & Rae Lynn Forever
Cyg 19h21'45" 29d14'25"
Kyle Rau, loved husband of
Jessica
Cep 1h51'18" 85d38'16"
Kyle Ray McGaha
Leo 9h45'49" 29d47'29"
Kyle Raymond Sack
Leo 9h24'43" 13d15'47"
Kyle Reder
Lac 22h41'48" 51d13'24"
Kyle Regan Price
Vir 13h18'15" 3d11'2"
Kyle Richard Rose Ellison
Cap 20h22'9" -14d0'23"
Kyle Richard Waidelich
Aqr 20h55'47" -3d35'17"
Kyle Rizzo
Psc 1h19'53" 21d43'50"
Kyle Robert
Dra 16h0'53" 59d38'17"
Kyle Robert McCall
Cep 3h20'3" 82d40'53"
Kyle Robert Musser
Ari 2h4'56" 24d52'7"
Kyle Robert Tagg
Cnc 9h7'6" 25d41'25"
Kyle Rudig
Ori 5h45'12" 0d35'33"
Kyle Ruey
Ari 3h18'22" 19d24'5"
Kyle Russell Guilford
Lyn 9h7'2" 35d28'2"
Kyle Russell Matz
Eri 4h33'36" -0d11'22"
Kyle S. Hanley
Ari 2h7'30" 12d55'57"
Kyle S. Root
Vir 13h35'47" -4d21'8"
Kyle Schlanger
Aqr 22h9'55" -1d21'55"
Kyle Schofield
Cnc 9h4'45" 31d30'59"
Kyle Scott Morgan
Cru 12h42'45" -57d55'34"
Kyle Shifflett
Her 17h54'8" 49d55'39"
Kyle Shigeo Frommer
Umi 16h21'25" 75d55'57"
Kyle Smith
Lmi 10h5'15" 38d2'15"
Kyle Stephen Adamson
Uma 8h56'52" 48d18'4"
Kyle Steven Pflanz
Her 17h18'58" 36d28'50"
Kyle T. Roberts 5/1/01
Umi 14h21'20" 77d55'29"
Kyle "The Hero of Our
Skies" Dani
Cas 23h10'15" 54d22'53"
Kyle Thomas Christian
Umi 14h18'37" 77d19'1"
Kyle Thomas Himel
Ori 5h8'52" 6d17'32"
Kyle Troy Faille
Equ 21h22'49" 3d49'38"
Kyle Vargas
Umi 15h15'48" 72d47'32"
Kyle W. Baker
Psc 0h7'58" 7d55'53"
Kyle Wade Hankins
Uma 9h45'40" 65d35'4"
Kyle Walter Copija
Sgr 18h7'15" -33d44'24"
Kyle Wesley
Her 18h41'1" 22d8'27"
Kyle William Herman
Ori 5h34'25" 2d54'7"
Kyle William Nowadnick
Cyg 20h9'11" 32d33'54"
Kyle William Wade
Her 17h7'22" 25d11'22"
Kyle Wilson
Tau 5h3'38" 24d42'7"
Kyle Woodiel and Sarah
Silvas
Psc 1h15'7" 27d3'40"
Kyle Zigmund McEneany
Per 3h39'49" 49d4'40"
Kyle, Alyssa & Ethan
Hutton
Mon 6h52'14" 7d49'17"
Kyle, the Bright One
Uma 11h26'23" 70d46'8"
Kylea Corr
Cas 0h14'7" 51d33'29"
Kyleanna
Lmi 10h20'13" 37d37'34"
Kylee
Psc 2h0'40" 3d29'15"
Kylee
Psc 1h34'32" 22d37'59"
Kylee Ann
Cyg 19h50'35" 35d49'40"
Kylee Ann
Psc 0h50'2" 3d20'59"
Kylee Ann Fruin
And 23h32'42" 38d53'51"
Ky-lee Anne Owen
Cru 12h2'30" -63d6'15"
Kylee Hawkins
Per 3h8'46" 43d10'2"
Kylee Joy
Sco 16h39'49" -29d35'6"
Kylee Lynn Sellers
Aqr 22h16'41" 1d3'20"
Kylee Mckenzie Ray
Cnc 9h7'48" 22d17'56"
Kylee Megan Trayer
Cyg 21h33'6" 34d39'42"
Kylee Morgan Clinton
And 1h18'2" 41d32'54"
Kylee Noelle Walters
And 0h37'39" 41d32'7"
Kylee Raines
And 2h15'9" 38d20'18"
Kylee Rose Breedlove
Cyg 21h31'5" 42d14'43"
Kylee Shane Johnson
Lyr 18h47'52" 39d11'40"
Kylee Sue
Sgr 18h26'55" -18d20'26"
Kylee Victoria Vargas
Aqr 22h23'53" -5d32'59"
Kylee Wray Parks
Tau 5h49'23" 17d8'53"
KyleEB120103
Crb 16h16'50" 34d40'51"
Kyleen Michelle
Sco 16h3'45" -13d4'20"
Kylei Brandell Opie
Uma 13h23'21" 57d18'40"
Kylei Lacie Conti
Vir 12h25'55" 11d12'5"
Kylei Lauren Cutts
Cap 21h5'37" -15d25'44"
Kyleigh Elizabeth Vansky
Ari 3h18'21" 27d16'13"
Kyleigh & London Evertsen
Cyg 21h14'10" 33d14'16"
Kyleigh Marcelle Hallam
Psc 1h16'37" 27d3'8"
Kyleigh Yvonne
Ari 2h39'31" 18d41'3"
KyleLeeMichieli
Uma 13h51'18" 52d58'50"
Kylen K. Kamaka
Lep 5h40'22" -21d51'4"
Kylene Dorothy Wright
And 2h24'47" 41d41'58"
Kyler Branden Parker
Uma 11h34'24" 45d48'3"
Kyler Brown
Cap 20h23'25" -13d33'39"
Kyler Connelly
Sgr 18h17'41" -30d11'11"
Kyler D. Vancamp
Aql 20h7'36" 8d12'21"
Kyler Glenn Valentine
Aql 19h47'57" 4d38'54"
Kyler J Edwards
Cyg 20h25'18" 34d47'43"
Kyler Jacob
Dra 18h41'17" 58d28'14"
Kyler John Marutzky
Leo 9h29'12" 26d28'30"
Kyler Martin Del Valle
Uma 10h5'29" 54d45'17"
Kyler Raine
Umi 15h25'59" 71d45'19"
Kyle's Angel
Sco 16h7'0" -12d38'42"
Kyle's Heart
Uma 11h27'41" 62d14'4"
Kyles Keep
Tau 5h9'4" 17d57'51"
Kyle's Satellite
Leo 11h2'43" 8d27'45"
Kyle's Shining Star
Cnc 9h20'40" 27d17'37"
Kyle's Shooter
Vir 13h44'58" 3d9'1"
Kyle's Smile
Uma 11h7'31" 63d53'45"
Kyle's Star
Uma 11h2'7" 63d26'21"
Kylessie's Light
Lyn 8h31'51" 33d13'0"
Kylestar
Ari 3h15'31" 27d31'12"
Kyley
Aqr 21h28'7" 1d33'14"
Kyley Kay-Marie Cyr
9/16/03
And 23h37'55" 47d31'21"
Kyley Tharp
Ori 5h21'25" -3d37'29"
Kyleysaurus Rex
Leo 11h12'55" -3d38'53"
Kyli Leah
Sge 19h42'39" 17d21'1"
Kylian
Tau 5h54'23" 25d6'28"
Kylie
Tau 4h22'35" 23d24'16"
Kylie
And 0h31'43" 32d42'21"
Kylie
Cnc 9h20'39" 16d0'36"
Kylie
And 2h18'48" 48d31'6"
Kylie
Cnv 12h35'55" 40d21'14"
Kylie
Sco 16h30'53" -25d31'42"
Kylie A. Bechtol
And 1h5'41" 34d51'16"
Kylie Adrianna Campbell
And 23h46'0" 45d8'11"
Kylie Alexandra Gangi
And 1h12'4" 36d22'54"
Kylie Ann Jentz
Psc 1h3'53" 31d28'24"
Kylie Anne-Louies Setchell
Psc 0h20'47" 7d27'10"
kylie bear
Ori 5h41'10" 1d21'28"
Kylie Belle
Ari 3h27'6" 21d10'48"
Kylie Bender
Cnc 7h58'32" 11d40'46"
Kylie Bree
Sco 17h38'12" -41d8'21"
Kylie C. Barsch
Vir 12h51'56" 12d41'56"
Kylie Cheyanne Hope
Dosko
Umi 16h38'9" 83d1'18"
Kylie & Dermot - A New
Beginning
Cru 12h39'16" -58d55'31"
Kylie Dianna Wolfe
And 23h57'48" 42d22'52"
Kylie Duval
Dra 16h55'38" 64d48'29"
Kylie Edwards my love my
everything
Gem 6h24'31" 18d1'23"
Kylie Elizabeth
Aql 19h35'0" -11d10'30"
Kylie Elizabeth Bennett
Del 20h49'2" 2d51'40"
Kylie Elizabeth Casey
Ori 5h23'20" 7d12'4"
Kylie Elizabeth Escalante
Cnc 9h4'11" 15d45'56"
Kylie Elizabeth Park
Lib 14h32'29" -13d32'19"
Kylie Elizabeth Smith
Aqr 22h31'52" -11d33'15"
Kylie Elle McLain
Cap 20h32'35" -9d28'54"
Kylie Frances Roy
Vir 13h0'59" 10d40'39"
Kylie Gillgower
Ori 5h35'53" -3d34'20"
Kylie Grace Junkin
Gem 6h4'15" 26d8'15"
Kylie Grace Sebastian
Cnc 8h39'31" 21d27'0"
Kylie Gruber
Uma 10h40'7" 65d57'38"
Kylie Guest
Aur 6h37'38" 47d48'42"
Kylie Hebisen
Umi 14h44'10" 73d42'47"
Kylie Isabel Montoya
Cnc 8h38'22" 16d7'47"
Kylie Isabelle & Chloe
Elizabeth
Umi 14h36'24" 75d10'34"
Kylie Isabelle Wareham
Psc 1h44'3" 21d47'42"
kylie jade
Cnc 8h11'14" 9d4'7"
Kylie Jade McKee
Sgr 18h51'17" -17d37'41"
Kylie Jane Long
Tau 4h47'54" 18d47'17"
Kylie Jo Radtke
Cnc 8h39'53" 6d46'41"
Kylie & Kierra Lindeman
And 0h21'0" 29d32'28"
Kylie Lehman
Gem 6h45'41" 34d42'14"
Kylie L'Eplattenier
Peg 22h16'39" 27d58'22"
Kylie Lorenz
Leo 10h13'44" 25d5'9"
Kylie Louise
Ari 3h19'49" 16d48'9"
Kylie Lyn Rubino
And 0h40'6" 33d44'51"
Kylie Lynn Kidd
Lib 15h45'57" -21d52'16"
Kylie Makenna Marschall
Tau 4h14'8" 26d27'13"
Kylie Marie Brown
Her 16h26'6" 47d38'43"
Kylie Marie Veronie
Cap 21h12'54" -16d56'28"
Kylie Mary
Lib 14h50'28" -5d26'17"
Kylie Matheson Law
Cas 2h26'59" 73d35'13"
Kylie McKenna
Leo 11h51'54" 20d10'53"
Kylie Michelle
And 23h17'23" 48d1'1"
Kylie Michelle
Cap 21h55'46" -23d4'26"
Kylie Miracle O'Gorman
Sgr 18h33'28" -16d52'48"
Kylie Mussay
Cyg 20h8'31" 32d26'11"
Kylie N Curtis
And 1h6'16" 39d6'31"
Kylie Olivia McCoy
Uma 11h36'12" 36d38'25"
Kylie Rae Finley Squeque
Cnc 8h53'4" 30d56'51"
Kylie Rea
Ori 5h20'17" 1d41'31"
Kylie Renae Hall 21
"Queen"
Tau 4h38'0" 1d51'41"
Kylie Rose
Psc 1h16'41" 26d36'24"
Kylie Rose Alves
Leo 10h23'40" 12d37'54"
Kylie Rose Coleman
Ari 2h1'59" 24d1'15"
Kylie Rose Gerrans
And 0h37'45" 36d19'54"
Kylie Rose Stokes
Tau 3h58'53" 21d0'38"
Kylie Samantha Emery
Del 20h45'45" 12d33'47"
Kylie Saul
Ori 5h42'44" -1d31'50"
Kylie Shianne
Cyg 19h33'49" 31d54'54"
Kylie Sioux Congdon
And 1h22'16" 49d42'47"
Kylie Sue McNichols
Cnc 8h44'38" 30d37'52"
Kylie Thoman
Cam 3h55'58" 58d24'5"
Kylie Wastell
Cru 11h57'4" -62d55'51"
Kylie Young
Lyn 7h26'34" 45d17'41"
Kylie-Kathleen Edmond
Hillman
Sco 17h19'16" -31d59'13"
Kylie's Kiss
Cru 12h28'59" -61d5'52"
Kylie's Mira 13
Cnc 9h14'15" 8d10'49"
Kylie's Star
Pho 0h53'25" -43d53'1"
Kylin D. Slade
Sgr 19h12'2" -12d39'12"
Kylin Warbelow
Psc 0h52'43" 31d41'0"
K'ylinda's Freedom
Gem 6h30'11" 12d23'22"
Kylion
Lyn 8h18'41" 50d38'27"
Kylo 02192003
Uma 9h10'4" 49d14'16"
KYLO'S APOLLO
Uma 10h59'13" 52d10'31"
Kyly Zakheim
Umi 14h2'32" 77d53'55"
Kylynn Ivy Stovall
Lyr 19h20'44" 37d39'58"
Kym
Cnc 8h17'28" 12d33'27"
KYM
Uma 10h50'46" 61d53'51"
Kym and David
Cyg 21h18'10" 35d0'11"
Kym J. Blackburn
Ari 1h51'24" 20d58'28"
Kym & Pete 09-07-05
Cru 12h57'17" -59d18'19"
Kym Rahnai Blevins
Tau 5h34'15" 17d31'46"
Kymbah
Peg 21h33'21" 21d28'20"
Kymber Rae Bonham
Aql 20h4'8" 3d49'36"
Kymberlee
Leo 11h6'23" 17d28'53"
Kymberli A. Anthony
Cyg 20h8'48" 32d38'37"
Kymberli Nicole Simken
Mon 6h48'9" -0d6'25"
Kymberlie Alta Eileen Alves
Cap 21h12'8" -19d37'59"
Kymberlie Renee
Aqr 22h24'58" -2d10'34"
Kymberly
And 0h56'49" 38d11'28"
Kymberly Diane Byrd
And 0h49'52" 34d14'17"
Kymberly Rose Lund
Lyr 18h38'49" 31d36'38"
Kymberly's Light
Tau 4h39'22" 9d46'29"
Kymests' Starry Starry
Light
And 1h22'56" 46d25'28"
Kymin Total Eclipse (aka
Floyd)
Cma 6h55'31" -15d5'15"
Kymisaurus
Cam 3h57'46" 66d1'44"
Kymmig
Mon 6h47'42" 6d59'36"
Kymmy Z
Gem 7h26'10" 34d17'46"
Kyna 15 October 1992
"Bugzy"
Lib 14h51'56" -15d53'56"
Kyndall Nikole Durden
And 0h16'46" 34d40'15"
Kyndall Nikole Durden
And 0h34'33" 43d30'11"
Kyndall's Star
Gem 6h29'13" 21d34'18"
Kyndle Faulkner
Lib 15h7'16" -5d28'57"
Kyndra Joyce Springer
And 0h36'53" 41d57'14"
Kynnee' Golder
Cyg 21h9'6" 43d49'37"
Kynou
Aqr 23h43'55" -14d30'4"
Kynzi Rea
Uma 9h38'17" 44d16'21"
Kyodesi
Ori 6h7'43" 14d5'56"
Kyoli
Gem 6h49'5" 29d20'55"
Kyongmi Odell
And 23h3'19" 48d12'6"
KyoSung Lambert
Gem 7h9'36" 25d48'0"
KYOUKO & HI-KO STORY
Ari 3h6'44" 19d46'41"
Kyphi
Cnc 8h27'54" 17d52'32"
Kyp's twinkle in the sky
Cyg 19h40'15" 35d39'39"
Kyra
Lyr 18h35'14" 33d14'57"
Kyra
Uma 11h21'50" 38d20'38"
Kyra
Lyr 18h31'37" 28d25'42"
Kyra
Ari 2h59'56" 18d40'46"
kyra
Vir 14h39'21" 2d11'37"
Kyra
Lib 15h8'1" -1d5'18"
Kyra
Uma 8h20'21" 64d8'25"
Kyra
Lib 15h38'0" -27d26'42"
Kyra Alexis Kavouridis
And 23h24'47" 46d25'44"
Kyra Angel
And 1h46'14" 46d18'21"
Kyra Evangeline
Cyg 20h5'0" 35d55'59"
Kyra Fox
Cas 1h18'50" 59d23'17"
Kyra & Jayde Wicks - Twin
Angels
And 23h34'7" 45d12'48"
Kyra Jedrykowski
Aqr 22h20'58" -18d7'36"
Kyra Linden Tannehill
Ari 2h28'11" 26d0'20"
Kyra Louise Kaplan
Peg 22h43'5" 4d13'4"
Kyra Lucile Khaos
Mon 6h47'27" 8d8'28"
Kyra Lynne Price
Lib 15h22'33" -14d41'6"
Kyra Mae Cerny
Mon 6h47'53" 6d56'23"

LaDonn M. Usitalo AKA D.D.G. Lyn 7h17'58" 44d53'1"
LaDonna Cyg 20h52'26" 36d2'21"
Ladonna Gem 6h37'5" 13d18'50"
LaDonna Camille McCall Brixey Uma 11h18'3" 72d17'50"
LaDonna Deeanne Leo 11h52'32" 23d33'28"
LaDonna Latress Sgr 18h14'21" -19d28'49"
LaDonna Ruth Lmi 10h48'16" 27d6'24"
LaDonna Shelaine Daniel Cap 21h53'58" -22d13'35"
LaDonna Slattery And 1h21'30" 44d16'29"
Ladore Vir 13h52'36" -20d28'18"
LaDonna Stephens Vir 13h25'39" 11d53'48"
LADstar887 Crb 15h29'26" 31d9'41"
Ladwig, Ehrenfried Hermann Uma 13h57'28" 55d0'23"
Lady Cas 23h34'14" 56d28'52"
Lady Uma 10h35'36" 66d50'0"
Lady Uma 11h38'49" 42d39'18"
Lady Dra 17h47'57" 51d11'56"
Lady Tau 4h12'29" 27d19'45"
Lady Ari 3h21'5" 24d44'14"
Lady A And 1h32'4" 39d38'3"
Lady Abdullah Crb 15h46'58" 27d6'26"
Lady Alexandria Gem 7h2'27" 22d14'58"
Lady Ali - Star of my life & love. Sco 17h56'9" -38d54'33"
Lady Alyssa Uma 11h24'36" 30d13'45"
Lady Amanda Salutatorian 2005 Sco 16h5'49" -17d44'29"
(Lady) Ann Cyg 20h51'58" 54d48'49"
Lady Anna Beloved Mama Susan Lac 22h28'23" 42d42'29"
Lady Annie Ari 3h16'7" 21d23'59"
Lady Aurialis Tau 5h25'0" 28d1'59"
Lady Azita Forever And 2h16'46" 39d11'4"
Lady Bear 1058 Uma 13h41'47" 56d26'16"
Lady Bette-Helena Sgr 18h21'23" -24d30'48"
Lady Brenda Fenwick Giammanco of NY Cas 3h24'39" 70d55'14"
Lady Brock Lmi 10h12'38" 40d17'15"
Lady Bug Vir 13h18'42" 12d6'35"
Lady C Leo 9h29'50" 30d7'11"
Lady Cain And 1h19'52" 38d54'2"
Lady Caitlin Uma 8h37'43" 66d5'12"
Lady Cathrine Elizabeth Leo 11h35'34" 25d30'14"
Lady Cheyenne Ari 2h37'42" 28d1'54"
Lady Clare and Colonel Ed Cyg 21h42'1" 44d11'22"
Lady Clista Lyn 7h45'28" 42d13'43"
Lady Cynthia Cap 20h29'28" -9d44'27"
Lady Dale And 22h58'0" 48d4'36"
Lady Denza Ori 5h19'39" 6d55'28"
Lady Di Ori 5h59'54" 18d43'17"
Lady Di And 23h14'7" 47d5'53"
Lady Di Uma 11h1'10" 38d1'31"
Lady Di Uma 13h50'19" 58d50'24"
Lady Di Cas 1h24'44" 62d22'25"
Lady Di Just Joy Art Star Stone Sco 17h40'24" -42d27'5"
Lady Di S.R. Freburg Cnc 8h50'34" 11d39'57"
Lady Diane And 0h19'45" 41d45'42"
Lady Diane Lyn 7h54'31" 54d36'18"
Lady Diane Mon 7h35'25" -3d52'45"
Lady Diane Marie Cas 1h47'34" 64d49'37"
Lady Dianne 10-11-6 Ari 3h15'21" 25d57'49"
Lady Donna And 1h8'52" 39d38'45"
Lady Donna Lee Vir 13h17'0" -2d10'19"
Lady Durie DMCO Cas 0h47'58" 48d28'0"
Lady Eleana Aqr 23h52'13" -11d56'54"
Lady Eleanor Christine Cas 0h17'41" 52d33'11"
Lady Elena Bunina Leo 11h57'59" 26d13'26"
Lady Elizabeth Leo 10h26'32" 21d52'12"
Lady Erica Tanacs Leo 9h27'21" 26d16'3"
Lady Esther Valdez Cas 1h32'30" 67d16'50"
Lady Fourty Ori 5h52'40" 7d29'42"
Lady Gators"Heart and Hustle" 06-07 Uma 8h51'15" 51d24'37"
Lady Gela Vir 13h26'10" -14d43'6"
LADY GRAY STAR Tau 5h43'18" 27d11'20"
Lady Gwen of Modon Lyn 7h2'57" 52d46'42"
Lady H Vir 14h3'53" 3d57'28"
Lady Hannah Brooke Cru 12h44'38" -61d34'23"
Lady Hawk Ari 2h21'25" 20d5'35"
Lady Hawke And 0h20'47" 43d13'45"
Lady Hawker Psc 1h4'5" 30d17'0"
Lady Heather Cas 0h55'11" 59d8'42"
Lady Heidi And 23h21'47" 41d26'39"
Lady in Red - Barbara Clarke. Cas 0h57'19" 47d20'8"
Lady In Red Ruth Ari 2h13'14" 26d8'33"
Lady Inayah Cap 20h14'36" -10d47'56"
Lady Indigo-MCDC Lib 15h27'16" -4d37'56"
Lady J Lyn 7h50'29" 57d19'56"
Lady J Sgr 18h55'45" -34d10'15"
Lady Jane Lib 15h47'9" -17d58'32"
"Lady" Jane Goldenberg And 2h18'36" 46d13'2"
Lady Jean Cas 0h1'9" 56d5'54"
Lady Jessica Sgr 18h12'47" -26d2'54"
Lady Joanne And 23h4'29" 45d56'48"
Lady Josephine Ari 3h10'43" 27d6'7"
Lady Josephine Guidance by Strength And 0h18'58" 31d37'11"
Lady Katharine Dale Bohlmann Sgr 18h3'42" -26d21'48"
Lady KiSS Aqr 22h38'14" 0d17'3"
Lady Klára Printer Women on the Sky Cap 21h40'13" -14d16'3"
Lady L Cas 23h43'7" 54d12'16"
Lady Layce Cnc 8h27'56" 30d49'13"
Lady Leah Cyg 20h52'48" 38d51'51"
Lady Leah Mackenzie Kessinger Her 17h55'21" 22d4'10"
Lady Leo Leo 11h34'36" 19d32'18"
Lady Lin Cas 1h15'1" 55d7'7"
Lady Loretta of the Angels Cap 21h34'43" -10d52'52"
LADY LORRAINE MY LOVE Sgr 18h25'57" -16d30'1"
Lady Lou Cas 0h21'44" 58d50'21"
Lady Lucie Lapointe Ori 5h3'52" 10d18'42"
Lady Lulu Buttercup And 1h33'58" 40d43'47"
Lady M. Cyg 20h48'49" 32d21'40"
Lady Mackenzie And 0h37'19" 23d35'14"
Lady Margaret Cas 1h21'10" 53d48'21"
Lady Marie Srp 18h8'51" -0d13'14"
Lady Marion Gem 7h28'20" 25d3'12"
Lady Michelle 22 NOV 66 Sgr 18h4'35" -27d2'51"
Lady Moondance Uma 8h48'54" 72d47'58"
Lady Nadine Allcock of Dunnottar Cas 1h10'13" 51d2'48"
Lady Natalie G. Retamoza's Star And 1h22'45" 35d36'59"
Lady Nordstern Uma 8h54'21" 54d28'8"
Lady Odette Ori 5h36'29" -0d32'25"
Lady of Grace Aqr 22h28'14" -1d50'9"
Lady of Solitude Leo 10h19'44" 13d39'49"
Lady Oyindamola Pho 23h47'7" -50d27'58"
Lady Prince Leo 9h36'38" 31d39'20"
Lady Raven Aql 19h37'11" -0d21'32"
Lady Robin Ari 2h14'58" 24d9'59"
Lady Sadie of Spring Cma 6h36'40" -15d29'36"
Lady Samberdee's Alexandria And 2h24'23" 48d53'11"
Lady Sarah Joy Cas 1h45'1" 63d27'16"
Lady Shelby And 23h26'42" 48d18'52"
Lady Sondra Cas 0h40'6" 51d21'7"
Lady Susan Crb 15h31'54" 36d20'56"
Lady Sylvia Sco 16h11'13" -20d4'17"
Lady T, Of The Carvalho Gem 6h53'46" 21d13'36"
lady tamara And 23h13'26" 48d43'49"
Lady Thieryn Ori 5h21'31" -5d11'2"
Lady Verdandi Gormsdottir Cap 21h10'19" -25d13'10"
Lady Victoria Burwell Gem 6h3'7" 24d47'7"
Lady Victoria Hannah King Tau 5h58'47" 23d20'25"
LADY VICTORY Psc 1h51'23" 9d0'7"
Lady Wanda Rosario Lib 14h50'50" -1d9'42"
Lady Wendy Cas 23h29'30" 58d57'12"
Lady Whitney Ori 5h8'9" 12d8'4"
Lady Wolfen Mist Shining Love Light Cas 1h34'47" 69d10'47"
ladyBug Umi 14h22'34" 66d20'48"
LadyBug Uma 9h28'25" 57d13'56"
Ladybug Aqr 23h12'43" -9d28'20"
Ladybug Abby Cnv 13h48'2" 31d3'5"
Ladybug D And 0h38'46" 28d40'8"
Ladybug Jessica Wright Per 3h15'34" 48d58'45"
Ladybug-Abby's gentle guiding star And 1h12'15" 37d15'49"
Ladybug's Lucky Star Sgr 19h37'40" -43d0'37"
LadyDiButterfli Aqr 22h20'23" -8d20'10"
Ladyhead's Chickenman Uma 8h22'48" 69d6'22"
Ladypamelasue Psc 1h8'51" 28d39'31"
Ladyra Lyn 7h27'53" 49d42'51"
Lady's Lantern Aqr 23h3'21" -9d10'57"
Lady's Star Leo 10h13'32" 22d23'37"
La.Dy.St Del 20h39'18" 14d39'10"
LADYSWEETLING Cap 20h31'49" -10d13'22"
Lael Eri 3h48'12" -0d46'31"
Laela Nicole Karimi Leo 9h48'4" 26d38'55"
Laela Sage Broad Ori 5h22'47" -0d29'15"
LAEllen333 Lyr 19h10'31" 46d39'58"
Laena Marie Jagoe Cyg 21h53'47" 41d16'34"
Laeti... Cariño mio Cas 3h25'39" 70d21'27"
Laeticia Tau 4h29'25" 29d41'14"
Laeticia et Johnny Col 5h31'20" -42d25'3"
Laetitia Cas 0h2'20" 58d30'48"
Laëtitia Béal Uma 10h1'33" 61d29'8"
Laetitia Bernardini Lyn 7h29'20" 48d52'50"
Laetitia Brierre Umi 16h4'56" 73d53'5"
Laëtitia Dalloz Del 20h50'56" 13d47'57"
Laetitia Haas Sgr 18h54'0" -15d57'13"
Laetitia "Mon Etoile" Leon And 0h45'24" 25d45'58"
Laetitia Monereau Peg 22h37'8" 23d44'43"
Laetitia.H Del 20h34'9" 14d48'29"
Laetitialiana Umi 10h53'0" 87d59'16"
Laetizia & Mickaël Psc 1h45'25" 12d48'33"
Lafern Peg 22h20'52" 7d11'51"
Lafitte Omega 12 V1 Sgr 20h17'18" -38d2'58"
Lafong Lib 15h23'42" -19d38'23"
LAG Gem 7h40'14" 27d24'9"
LaGerald Jawuan Crawford Her 16h51'30" 47d36'26"
Lagniappe Sco 16h37'56" -31d25'47"
Lago De Paz Uma 10h11'25" 56d25'1"
Lagova Leo 10h59'12" 7d25'58"
Lagrada Medicine Crow Cas 23h5'51" 57d39'31"
LAHAKIME And 0h11'7" 42d55'22"
Lahnee - Heavenly Sky - U'Ren Tau 4h27'44" 21d10'20"
Lahnstein, Norbert Uma 8h16'33" 63d9'55"
LaHoma Colley Cnc 8h42'21" 16d56'36"
Lai Sco 17h21'15" -32d53'50"
Lai Kim Tong Sco 16h24'38" -32d44'39"
Lai Tee's Star Lyn 6h34'38" 55d18'42"
LAI, Siempre Mi Hermosa Estrella Uma 11h51'26" 45d7'50"
Laia - Xènia Aur 5h1'56" 40d12'10"
Laiah Abigail Lyn 8h42'35" 39d48'18"
Laicey Woolford Cas 0h20'33" 61d42'4"
Laicsyde Soul Sister (Lucy) Uma 9h8'33" 52d59'25"
Laihla Skye Fairbairn And 0h59'9" 45d7'4"
Laika Vir 13h22'47" -12d56'39"
Laiken Robbins, My "Laiky" Peg 22h9'1" 12d55'13"
Laila Per 3h46'2" 41d44'12"
Laila Tau 4h12'52" 30d33'36"
Laila Alisha Lmi 10h46'22" 34d44'6"
Laila Dagher Cnc 8h32'18" 24d33'22"
Laila Grace Owen Vul 20h33'1" 21d5'35"
Laila Laine Uma 11h41'24" 42d32'34"
Laila Miroku Tau 3h40'51" 27d58'46"
Laila Rose Mon 6h37'57" 6d28'48"
Laila Simone Pickett Lep 5h52'5" -11d40'18"
Laila Star Umi 18h52'59" 88d53'53"
Laila Tenille Murphy And 1h21'59" 36d11'35"
Laila V Lyn 8h6'12" 46d48'1"
Lailas ledestjerne Lib 14h44'8" -9d7'2"
Lailee Ana Hollander Lyn 6h22'41" 57d36'20"
Laina Cap 21h3'25" -24d26'4"
Laina Lyr 18h49'21" 31d48'29"
Laina Jean Sco 17h4'8" -37d50'22"
Laina Nicole Schneider Vir 13h3'44" 10d57'27"
Laina's Star Aqr 22h14'15" 1d24'3"
Laine Anthony Moser Sgr 18h35'38" -22d29'31"
Laine Jewel Uma 8h44'39" 71d2'25"
Laine Lisbeth Maurer Ari 2h12'45" 24d4'56"
Laine Schad And 1h20'5" 39d34'9"
Laine Seal Cyg 19h34'43" 29d42'34"
Laine The Goddess Of Love Vir 13h31'0" -9d23'42"
Lainey Cnc 8h50'43" 27d1'0"
Lainey*16 Uma 10h54'44" 33d55'18"
Lainey Kay Miller Ari 1h49'45" 22d14'43"
Lainey Miller Ori 5h33'0" 13d42'49"
Lainey-Bugg Ori 5h53'51" 18d22'28"
Lainey's Star Mon 8h7'20" -1d7'16"
Laini Marie Lumpp Ari 2h42'4" 27d0'10"
Lainibob Cmi 8h1'57" 0d9'2"
Lainie B. Moore Lyn 7h16'7" 56d39'15"
Lainie Britt 2007 Tau 4h29'11" 21d55'55"
Lainie Erin Hebert Cyg 19h58'17" 38d33'50"
Lainy DeMae Trampush Uma 12h34'45" 57d59'24"
Lair Bear Kircher Sco 17h11'5" -43d23'36"
Laird Uma 13h22'58" 54d36'37"
Laird & Amy Miller Sco 16h15'26" -21d10'18"
Laird Vernon Jones Umi 14h29'38" 86d39'35"
Laird's Light Cnc 9h3'42" 25d6'54"
Lairia's Hearth Ori 5h30'21" -0d53'2"
Laith Cnc 8h29'50" 23d21'23"
Laith Tawfik Frites Vir 14h2'14" -9d39'17"
Lajit Cyg 21h30'27" 42d19'54"
Lajmar D. Anderson Cam 4h3'48" 69d58'57"
Lajos Horvath Uma 11h16'25" 55d45'45"
Lajos L. Nemeth Sco 16h12'2" -10d19'4"
Lajos Polya Crb 15h40'11" 35d43'48"
LaJuanna Sancho Leo 11h16'2" 24d53'52"
LaJune Storgaard Means Gem 7h17'44" 25d58'19"
Lakatos Zoltán Cas 0h0'27" 54d56'21"
LaKaylah Star And 23h12'12" 41d29'59"
Lake Elliott Rabenold Uma 11h38'22" 43d14'52"
Lake of Begude And 1h24'2" 35d34'13"
Lakeesha Monique Jefferson Aqr 23h19'31" -18d9'1"
LaKeisha McGlothin Apu 15h30'50" -71d17'33"
Lakeisha Mon-Que Johnson And 0h22'34" 23d59'40"
Lakeisha S. Munn Cap 21h55'15" -13d34'21"
LaKeisha Sanderlin Lyn 6h34'33" 60d56'28"
Laken N. Buning Gem 7h11'54" 34d37'39"
Laken Noelle Fritz Umi 17h42'34" 81d30'50"
LAKER UNITAS Cae 4h42'50" -31d4'20"
Lake's Aql 19h48'16" 11d39'12"
LaKesha Brooks Cas 1h24'41" 67d43'37"
Laketa Lefevre, Salcido Leo 9h36'23" 12d9'16"
Lakeville Para-Professional Cyg 21h36'51" 38d31'38"
Lakiesha Mo chara Hulgraine And 23h18'6" 39d4'45"
Lakin Robert Glessner Aur 6h37'16" 34d0'47"
Lakin Smith Cap 20h24'45" -12d46'31"
Lakita And 0h57'26" 36d53'32"
LAKNDAB Uma 11h40'2" 52d13'3"
Lakota Leigh Tau 5h47'33" 26d1'48"
Lakshmi Sundaram Tau 4h46'30" 24d10'1"
Lakshmipriya Raghavan And 2h33'28" 43d26'27"
Laksmi's heart Aql 19h46'6" -0d30'15"
LaLa Cam 5h36'9" 72d54'53"
LaLa Uma 10h43'18" 67d0'29"
LaLa Cas 1h7'16" 60d45'47"
LALA Cnc 8h24'18" 27d37'36"
Lala Ori 6h13'46" 15d56'4"
LaLa Aqr 22h16'42" 1d59'21"
Lala Matthews Psc 1h28'5" 6d34'31"
Lala Nazary Leo 10h9'23" 19d13'9"
LALA & WALTER Tau 5h11'6" 24d9'8"
Lala,Theodore,Novena,Isabel,Annica Uma 11h8'0" 57d37'42"
Lalaine Aqr 23h32'21" -15d48'42"
lalaith - elenion ancalima Uma 12h30'0" 61d48'5"
Lalana V Pav 19h18'18" -63d11'37"
LaLa-n-DaNi Sgr 18h53'44" -28d42'39"
Lalany Star of Peace Peg 21h56'39" 10d54'21"
Lalaynia Umi 14h47'22" 69d3'19"
Lale Zaman Lib 14h28'54" -10d55'13"
Laleisha_Miché Aqr 22h5'16" -0d39'27"
LALEX Leo 11h19'42" 17d26'55"
Lali túnáit Cas 23h14'35" 59d39'16"
Lalie - Aurélie Lebreton Lmi 10h4'38" 30d59'26"
Lalla mia Sct 18h54'2" -12d26'11"
LallaManfre2005 Uma 8h40'17" 55d0'16"
Lallan*8 Tau 4h7'13" 5d20'59"
Lalli Lyn 7h45'58" 56d22'55"
Lally - My Girl Tau 5h38'45" 24d30'43"
Lalman Aqr 21h14'29" 1d53'55"
LALO Her 16h58'4" 30d53'14"
Lalo Cyg 21h13'36" 52d30'47"
Lalo Hernandez Lyn 9h4'12" 37d46'48"
La-Love's Light Psc 0h50'30" 16d19'3"
LaLumia-Snyder's Luminous Star Ori 5h30'6" -5d30'11"
Lam 03/30/90 Ori 5h30'21" 1d6'37"
Lam Chapman Dra 17h54'17" 79d13'55"
L.A.M + J.W.B Gem 7h8'19" 17d31'49"
LAM+MVC POOKIE STAR Cas 0h56'27" 59d52'43"
Lam Soon Edible Oils Sdn Bhd Ori 5h39'8" 3d23'10"
Lam Wai Man, Gavin Ari 1h56'40" 18d51'34"
LAM YEUNG Leo 9h51'29" 21d31'41"
Lama Uma 9h38'44" 42d16'10"
Lama R. Mehio Lmi 10h19'28" 28d23'29"
Lamar "Butch" Krohn Per 4h14'39" 41d1'28"
Lamar D. White Uma 12h9'19" 49d28'29"
Lamar Hewett And 2h20'15" 49d30'45"
Lamar Johnson Vir 12h57'7" -11d2'36"
Lamar Marquise Campbell (Mar Mar) Leo 11h2'18" -3d27'22"
Lamart Clay Per 4h29'26" 45d50'11"
LamarValrae Hamblin Cnc 8h3'14" 6d46'24"
Lamb Chop Aqr 22h41'31" -1d6'6"
Lambda Theta Alpha - Beta Sigma Umi 14h42'23" 76d4'46"
Lambda Zeta's Hope Leo 11h7'46" 10d40'21"
Lambertonus Gem 6h29'28" 13d52'19"
Lamberts' Love Cyg 20h45'45" 43d29'18"
Lambrecht, Walter Uma 8h49'21" 58d26'38"
Lambs slice of heaven Cap 20h8'14" -10d17'54"
L.A.M.E. Yeargain Star Crb 15h35'37" 27d50'32"
Lameese Jean Lyn 8h38'59" 38d6'8"
L'amerò per sempre Peg 22h22'15" 6d29'24"
L'amiciza bella Cnc 8h43'39" 24d48'7"
Lamioosh Cas 2h15'31" 69d28'10"
L'amitie Profonde Sgr 18h59'39" -35d26'2"
Lammers Cyg 20h43'51" 40d5'3"
Lammers Family Boo 15h44'2" 40d22'22"
L'amo Kailah! Cyg 20h15'41" 52d27'48"
Lamoes And 1h36'0" 46d45'29"
Lamont Per 3h17'7" 46d47'10"
L'amore Sco 16h51'9" -33d28'48"
L'amore di MOSF Cyg 20h33'43" 59d18'26"
L'Amore luminoso di Niki e Dan Per 3h12'10" 41d25'52"
L'amore trovera sempre un senso Uma 10h47'19" 67d23'40"
Lamount Cranston Juliani Johnson Tau 4h14'6" 26d24'52"
LAMOUR Lac 22h27'28" 43d12'31"
L'amour d'Alice Lmi 10h13'18" 33d21'2"
L'amour de Shahatra Cyg 19h31'40" 53d15'1"
L'amour Della Mia Vita Cap 21h4'43" -25d6'50"
L'amour du Trip et du John Umi 13h55'1" 69d37'55"
L'amour éternel Ori 6h9'57" 8d18'19"
L'Amour Toujours And 23h20'52" 42d16'14"
LaMoyne Cas 0h24'25" 55d24'59"
Lampoviola Ori 6h24'22" 13d18'7"
Lamprey's 25th Oasis Uma 11h13'3" 57d10'51"
Lan Leo 10h27'30" 17d25'10"
Lan Anh Sgr 18h28'39" -18d24'28"
Lan "Bunny" Nguyen Uma 10h6'5" 45d47'1"
Lan Nguyen Wasko Psc 1h16'35" 15d5'59"
Lan Phuong Hoang Ari 2h12'20" 25d10'14"
Lana Leo 9h27'4" 17d48'4"
Lana Ari 3h3'27" 28d23'8"
Lana Leo 11h40'12" 23d31'20"
Lana Ori 5h7'24" 9d15'29"
Lana Vir 11h38'20" 1d40'19"
Lana Crb 16h3'51" 36d58'19"
Lana And 0h33'37" 41d37'21"
Lana Ori 5h5'46" -0d31'17"
Lana Dra 19h11'34" 73d0'33"
Lana Cam 3h46'19" 65d44'14"
Lana 02/05 Cnc 8h40'1" 18d53'5"
Lana Abdulla And 1h20'10" 36d13'53"
Lana B. Lyr 18h45'9" 32d23'27"
Lana Babiy Uma 11h9'57" 68d40'57"
Lana Bell Com 12h34'20" 17d17'13"
Lana Brooke Cnc 7h56'41" 17d28'28"
Lana Carol Smith 5 lb. 11 oz. 18" Gem 7h15'34" 16d50'22"
Lana Daisy Turner And 23h52'37" 42d39'5"
LANA DEANNE PHILLIPS E.S.F.W. Cyg 19h42'59" 32d34'15"
Lana Elise Jones (Baby White Wolf) Cma 6h44'19" -16d36'52"

Lana Elizabeth
And 0h47'31" 24d39'21"
Lana Evelyn's Twinkle
And 0h43'44" 43d1'23"
Lana Gayle
Psc 0h40'21" 17d35'6"
Lana Gene Ryland
Ori 6h5'46" 9d56'5"
Lana Grace Kempson
Peg 21h42'0" 23d39'1"
Lana Jean Bougger
And 1h7'56" 34d14'46"
Lana Jill Murphy
Ari 2h12'11" 22d28'39"
Lana Knaapen
Tau 4h4'58" 5d59'5"
Lana Louise Rice
Psc 1h17'53" 17d35'0"
Lana Luth
And 1h18'47" 47d13'0"
Lana Maria ANDREY
Cam 6h9'49" 65d59'26"
Lana - Maria Blanco Chkaiban
And 23h56'19" 48d1'29"
Lana May
And 0h1'40" 45d53'26"
Lana Mercadante
Ari 3h13'40" 10d57'1"
Lana Noé Riesen
Cas 23h37'5" 55d29'1"
Lana Rachele Moore
Ori 5h27'50" -5d1'35"
Lana Rain Hafley
Lyn 7h40'51" 36d44'52"
Lana Rene Noel
Leo 10h17'2" 11d14'58"
Lana Rios
Her 17h56'35" 29d45'18"
Lana Robinson
Aqr 23h13'29" -9d42'25"
Lana Rose Buckley
Lib 15h38'11" -9d32'27"
Lana Rosone
Uma 8h56'50" 53d52'35"
Lana S. Sarieddine
Lib 15h2'43" -1d45'18"
Lana Sears
Cas 0h2'11" 55d41'6"
Lana Selthofer
Ori 6h24'34" 14d4'1"
Lana Shehadeh
Leo 10h38'10" 22d44'50"
Lana Silverman
And 23h23'31" 41d16'41"
Lana Star
Vir 13h31'58" -22d35'48"
Lana the Writer
Her 16h24'35" 49d18'8"
Lana, Phil & Family
Leo 11h51'48" 23d27'49"
Lana, The Queen Of Peach
And 1h32'31" 48d35'30"
LANA1131
Gem 6h49'22" 17d10'14"
LanaArt
Peg 23h56'33" 20d55'34"
Lanacre
Uma 11h20'51" 55d25'44"
Lanae
Ari 1h47'37" 19d17'46"
Lana-Ed
Sco 16h34'11" -37d57'18"
Lanaette Mi Tesoro
Gem 6h56'12" 12d23'10"
Lanai Kathleen O'Brian
Cas 1h25'1" 53d39'6"
Lanai Shepherd
And 23h9'45" 42d49'15"
Lana-Lee "Leafy" Feeney
Tau 3h41'5" 25d48'21"
Lanamar Blanco
Sge 20h5'47" 20d31'52"
Lana's 21st Valentine Star
Ori 6h11'35" 16d24'28"
Lana's Dream
Lyr 18h49'20" 39d24'18"
Lana's Star
Uma 9h2'19" 50d51'37"
Lana's Star
Lyn 7h8'37" 53d31'5"
Lanay Marie
Cyg 21h55'44" 53d26'7"
Lanayah
And 0h42'30" 43d32'30"
Lancaster Mennonite Class of 2007
Uma 11h45'21" 62d6'30"
Lancaster-Williams, Janie 2
Aqr 22h45'53" -23d26'54"
Lance
Aql 18h53'7" -0d1'59"
Lance
Cap 20h15'43" -17d37'22"
Lance
Leo 11h16'42" 18d3'7"
Lance Adrian Bremer
Lib 14h47'21" -4d24'6"
Lance Agurrie
Her 17h7'35" 17d6'45"
Lance Alan Shaw
Cnv 12h38'31" 43d23'27"
Lance and Erinn's Star of Love
Umi 15h12'40" 69d10'49"
Lance and Jeannette Boekenoogen
And 0h52'57" 39d23'50"
Lance and Kristina
Sge 19h50'36" 18d57'45"
Lance and Lynette Forever
Umi 12h37'6" 88d15'38"
Lance and Wanda star
Tau 4h9'40" 11d45'25"
Lance & Annie
Aqr 22h21'0" -8d59'11"
Lance Buchanan
Aur 6h47'38" 44d49'22"
Lance Copperman
Cap 21h36'6" -14d33'31"
Lance Corporal Eric D. Hillenburg
Gem 6h54'1" 13d16'34"
Lance Corporal Gregory Paul Rund
Aql 19h33'56" 13d13'59"
Lance Corporal Jason C. Redifer
Equ 21h9'13" 9d11'12"
Lance Corporal John Martin Holmason
Cap 21h46'29" -12d58'59"
Lance Corporal Robert Rogers III
Vir 12h38'8" -8d8'47"
Lance Cpl. Brian P. Montgomery
Uma 9h25'31" 43d8'7"
Lance Cpl. Daniel Bubb
Per 2h44'35" 54d51'42"
Lance Cpl. Daniel T. Morris
Per 2h43'42" 56d36'5"
Lance Cpl. Phillip G. West, USMC
Per 3h25'30" 50d5'34"
Lance Cpl. Sean Michael Langley
Per 3h10'44" 44d17'13"
Lance E. Phillips
Her 17h19'26" 47d31'54"
Lance Edward
Uma 11h23'54" 33d2'0"
Lance Furman
Aql 18h48'33" 7d53'42"
Lance H. Eller
Boo 14h33'1" 26d47'20"
Lance Hayden-Kump
Cyg 21h45'8" 48d49'17"
Lance Joseph Adams
Cnc 8h43'1" 19d32'6"
Lance Letho
Her 17h50'46" 44d33'16"
Lance Lorenz
Psc 1h4'54" 22d29'12"
Lance Loves Maryann.....forever
Uma 11h12'23" 50d1'37"
Lance M. Hobbs
Aqr 21h0'18" 1d34'59"
Lance M. Lorenz
Psc 0h56'42" 31d48'21"
Lance M. Shaw
Cnc 8h19'34" 18d22'26"
Lance McNeill Scott
Uma 13h48'4" 52d11'53"
Lance Michaels
Uma 13h49'8" 52d35'54"
Lance Nelkin
Aqr 21h29'53" 2d2'1"
Lance Preston Storz
Lib 14h29'56" -13d59'42"
Lance Russell
Ori 5h50'32" -0d16'34"
Lance Sage
Cnc 8h41'51" 31d45'28"
Lance Steffey
Lac 22h22'0" 52d29'19"
Lance Thompson and Stephanie Downey
Leo 11h22'56" -4d10'7"
Lance Threet
Tau 5h27'24" 27d13'56"
Lance W. Wright
Per 3h4'37" 48d58'50"
Lancer
Her 17h15'9" 26d4'31"
Lance's Guiding Light
Per 3h42'18" 41d21'33"
Land of the Loving Jewish Princess
Ari 2h51'26" 18d19'19"
Landa Coduto
Lyr 18h49'51" 43d18'2"
Landa Joanna Martin
Uma 11h50'18" 51d18'35"
Landee Jaye Lynch
Uma 9h11'49" 71d47'17"
Landen Asher
Ori 5h55'43" -0d20'8"
Landen Christy
And 0h53'25" 34d48'40"
Landen Cole Finley
Lib 14h49'4" -4d14'40"
Landen Matthew Hayden
Her 18h15'32" 26d57'31"
Landen Michael Kolle
Uma 11h56'26" 57d37'28"
Landen Monroe Spangler
Aur 6h24'36" 32d30'22"
Landen Ray Johnson
Crb 15h56'27" 36d40'32"
Landi Star Hough
And 23h1'16" 51d54'42"
Landis estella bateman
Uma 11h42'9" 33d14'30"
Landi's Star
Uma 10h54'25" 49d31'39"
Lando Stracner
Cap 20h25'59" -11d30'2"
Landon
Dra 16h34'37" 59d8'39"
Landon
Cep 22h10'47" 60d24'9"
Landon Aden Kendrick
Cas 0h45'19" 53d13'34"
Landon Alexander Price
Gem 7h18'7" 25d35'42"
Landon Alexander Trask
Per 4h27'6" 35d32'8"
Landon Alexanderia Mullis
Lyn 8h35'52" 33d40'22"
Landon Andrew Buechler
Cnc 8h37'6" 11d12'51"
Landon Andrew Carsey
Cyg 21h49'40" 50d22'40"
Landon Austin Rehg
Uma 13h16'6" 58d38'24"
Landon Bradford White
Uma 12h3'52" 33d37'30"
Landon Bradley
Ori 5h48'10" 2d16'59"
Landon Bryant Boal
Uma 14h25'41" 58d43'46"
Landon C. Hughes
Lib 15h7'18" -9d0'55"
Landon C V Julian
Her 17h18'43" 32d47'47"
Landon Charles Davis
Lyn 7h28'37" 53d39'41"
Landon Daniel
Tau 4h50'6" 21d59'24"
Landon Eric Oswalt
Sco 16h46'8" -34d46'25"
Landon Eugene Graham
Her 17h57'36" 23d43'17"
Landon George Browning III
Aqr 20h53'37" -9d39'25"
Landon Gustave Hansen
Ari 2h4'28" 13d59'13"
Landon Hayes Kopacz
Lmi 9h57'26" 39d57'18"
Landon Hogue
Psc 1h25'30" 6d45'25"
Landon Hugo
Ori 4h49'26" 11d23'3"
Landon James Holland
Per 3h43'40" 48d33'28"
Landon James McCandless
Umi 16h16'15" 76d11'3"
Landon Jason West
Tau 3h48'49" 9d21'42"
Landon John Luis Conaty
Vir 12h37'38" -1d45'0"
Landon Kelly May
Vir 12h16'17" 12d32'46"
Landon Lawrence Bow Ravago
Lib 15h56'58" -7d21'35"
Landon Lee Alexander
Dra 18h43'49" 51d47'2"
Landon Lee Courselle
Uma 11h45'33" 42d11'40"
Landon Meade Henkel
Umi 15h37'46" 73d22'31"
Landon Michael Loreman
Cap 20h15'28" -23d19'53"
Landon Nakata
Boo 14h38'53" 12d30'10"
Landon Nicholas Tash
Uma 8h37'47" 52d13'3"
Landon Nydegger
Psc 0h53'39" 6d26'44"
Landon Parker Harris
Uma 10h29'1" 52d10'1"
Landon Paul
Vir 14h0'43" -2d31'52"
Landon Reese Smith
Cnc 8h1'11" 11d58'47"
Landon Reid Peterson
Her 17h58'30" 17d54'37"
Landon Reid Peterson
Per 3h50'55" 48d20'58"
Landon Robert Heuman
Ori 5h25'36" 3d37'31"
Landon Travis Chappel
Umi 16h23'44" 77d8'56"
Landon Ty Hui
Aqr 21h47'15" -2d4'52"
Landon Tyler Currie April 12, 2005
And 23h27'38" 45d11'16"
Landon Wade Britton
Per 3h3'30" 53d44'47"
Landon Wayne Hilton Anderson
Vir 13h6'0" 13d10'41"
Landon William Coates
Dra 16h35'14" 62d14'51"
Landon Zakheim
Uma 8h56'10" 51d8'46"
LandoStar
Dra 18h53'51" 73d17'39"
Landrey Elizabeth Cincurak
Sco 16h13'44" -10d53'6"
Landrey Fulmer
Uma 13h50'56" 54d48'40"
Landria
Srp 18h8'11" -0d4'20"
Landrie Ryann Perry
Vir 12h59'38" -18d35'25"
Landrigan
Ori 5h28'51" 6d35'44"
Landry Mindungu
Dra 12h49'26" 76d12'12"
Landry Thomas Gardner
Ori 5h22'37" 3d30'28"
Landry's Love
Ori 5h37'0" -2d13'59"
Land's End Nutmeg of Turtle Creek
Cma 7h24'16" -29d30'51"
Landyn Brown
Aqr 23h5'9" -7d59'14"
Landyn Jacob Hossler
Cap 21h38'14" -23d51'29"
Landyn Morgan Evans
Per 3h22'47" 48d10'10"
Landynn Andrews
Cnc 8h40'58" 9d47'33"
Lane
Gem 6h40'6" 17d43'39"
Lane
Uma 9h12'15" 71d1'24"
Lane Aldridge
Sgr 17h52'52" -29d41'16"
Lane Arthur Vitt
Sgr 18h14'44" -22d42'5"
Lane Arthur Vitt
Ori 5h23'36" 13d48'49"
Lane Bennett
And 1h19'53" 42d14'15"
Lane Christopher Cowan
Dra 17h53'41" 58d23'50"
Lane & Lauren
Cyg 20h3'2" 32d31'54"
Lane Luminary of Love
Cas 0h35'20" 51d20'23"
Lane Lyon
Aqr 22h19'33" -6d6'32"
Lane Marshall Dare
Uma 11h16'45" 52d47'40"
Lane Newman
Uma 10h5'20" 43d16'35"
Lane Nicholas Neal
Her 17h35'7" 19d44'43"
Lane Olivia Dreslin
Gem 7h17'15" 20d1'2"
Lane Paul Miller
Cam 4h37'52" 65d13'54"
Lane R. Ludwig 217
Umi 16h42'39" 77d17'43"
Lane Ruotsala
Cap 20h57'41" -16d34'37"
Lane Steven Taege
Her 16h34'33" 34d21'45"
Lane Susan McKenna
Aqr 21h31'48" 2d2'47"
Lane Tennison
Psc 22h53'46" 2d10'37"
Lane "The Bear" Hockman
Tau 3h42'56" 15d52'39"
Lane Thomas Wilson
Sco 17h36'3" -42d57'19"
Lane-Bennett
Ori 6h0'49" 18d37'9"
Lanee Nicole Swiech
Sgr 18h3'19" -21d19'16"
Lanee's Star
Psc 23h12'50" -0d6'42"
Lane-Hanson
Uma 11h8'27" 52d49'10"
Lane's Matthew 6:19
Uma 11h4'42" 54d36'33"
LaNessa
Cnc 8h42'3" 14d56'40"
Lanestar
Psc 0h28'53" 11d17'0"
Lanet Gill
Lyn 6h58'21" 46d37'41"
Lanetri
Cam 5h49'42" 72d56'4"
Laney
Cap 20h50'18" -16d16'32"
Laney
Cam 5h52'26" 57d40'13"
Laney Camille Bates
And 0h10'53" 29d12'4"
Laney Cate 1217
Sgr 17h54'25" -28d43'32"
Laney Fralin
And 1h35'0" 49d55'1"
Laney Jo Lowrimore
And 0h37'45" 39d49'5"
Laney Lane
Vir 12h54'11" -6d42'2"
Laney Marie
Cam 6h8'35" 60d27'14"
Laney Marino
Uma 12h40'28" 61d6'55"
Laney Michelle
Sco 16h11'30" -20d48'4"
Laney Skylar Duvall
Cra 18h36'44" -38d38'51"
Lang DeCoudres
Gem 7h1'17" 18d52'52"
Lang, Hans-Jörg
Ori 5h5'8" 6d29'52"
Lang, Manfred
Sco 17h23'17" -31d59'40"
Lang, Viktor
Uma 9h20'21" 47d48'34"
Langdon P. Horne
Cep 22h15'29" 73d32'2"
L'Ange
Col 5h14'39" -31d10'7"
l'ange de l'or
Aur 6h34'25" 34d31'38"
L'ange de Timothé: Sandra
And 23h25'37" 43d32'46"
Lange, Fritz
Ori 5h24'56" 14d19'32"
Lange, Herbert A.
Uma 9h5'53" 54d11'2"
Lange, Peter
Uma 10h14'20" 52d29'10"
Lange, Wolfgang
Uma 10h50'45" 62d48'30"
l'angelo custode
Uma 12h19'27" 57d15'58"
Langford
Cyg 19h55'28" 37d55'13"
Langley Alvin Shuker
Psc 0h3'26" 3d52'52"
Lan-Hsin
Uma 9h43'47" 53d3'56"
LANI
Sco 16h12'41" -17d4'6"
Lani
Ori 6h9'13" 21d1'13"
Lani Ann
Uma 11h42'48" 50d59'57"
Lani Waters
Crb 15h46'51" 37d36'31"
Lanie
Lyr 18h42'27" 39d11'49"
Lanie
Lib 15h9'16" -12d46'46"
Lanie
Sco 17h28'28" -38d34'15"
Lanie
Sgr 19h16'7" -26d53'11"
Lanier C. Twyman
Her 16h58'6" 36d28'51"
Lanier, C.L.
Ari 2h47'12" 13d57'35"
Lanig, Alisia Verena
Ori 6h3'11" 6d52'47"
Lanigan Matthew McCulty
Vir 12h1'19" 3d24'47"
Lani's Wishing Star
Umi 16h25'31" 70d21'40"
Lanitta Fonville
Cam 4h37'26" 56d27'17"
Lank John Kendrick Sr.
Her 18h23'45" 12d26'38"
Lanlon
Uma 11h41'7" 63d13'5"
Lanner's Love
Peg 22h52'11" 29d40'3"
Lannette R. Huffman
Sco 17h40'21" -45d32'38"
Lannie Delery
Uma 11h47'5" 37d55'57"
Lanny
Leo 11h13'36" -4d30'57"
Lanny A. Musslewhite
Aql 19h37'44" 1d48'20"
Lanny Reinish
Uma 10h8'29" 62d36'20"
Lanora
Crb 16h10'36" 25d49'34"
LanShina Cooper
Uma 11h29'46" 38d40'20"
Lantsnik, Frieda
Uma 9h0'52" 61d14'53"
LANUSHA
Vir 12h32'28" 4d9'49"
Lanza Valiente
Uma 11h15'3" 36d5'1"
Laoise Saorla Leahy
Cyg 19h56'1" 42d26'24"
Laoise Varian O'Mahoney
Vul 20h10'14" 26d49'4"
Laourik Jan Krasivaya
Leo 11h17'36" 16d17'15"
Lapage S&A
Tau 4h42'47" 14d36'31"
Lapifany Renee Collins (Mabane)
Tau 4h26'48" 17d4'16"
Lapin
Vul 21h1'17" 23d41'55"
Lapin
Dra 17h45'27" 54d43'46"
lapinette et lapinot
Tau 5h6'53" 27d23'44"
LaPrade
And 23h56'2" 42d37'41"
Lapril
Uma 8h44'6" 47d37'37"
Laquina Williams
And 2h38'41" 37d48'43"
LaQuincy & Jea - Your Shining Star
Sco 17h20'6" -33d3'46"
LaQuita Joy
Gem 6h28'1" 25d46'9"
LAR099
Lac 22h23'56" 46d14'6"
Lara
Cas 1h42'20" 54d41'25"
LARA
Cyg 21h10'28" 47d3'49"
Lara
And 23h25'17" 42d3'34"
Lara
Cyg 20h40'37" 36d2'49"
Lara
Uma 11h28'30" 32d31'18"
Lara
Lyr 19h5'52" 32d53'22"
Lara
Cnc 8h16'36" 25d34'24"
Lara
Gem 6h6'8" 22d39'33"
Lara
Com 13h9'23" 26d37'33"
Lara
Sge 20h17'35" 17d54'56"
Lara
Ori 5h35'48" 10d3'57"
Lara
Uma 8h25'28" 65d46'3"
Lara Aebli
Lyr 18h29'47" 28d53'58"
Lara Aerni
Her 18h28'56" 12d52'22"
Lara and Filipe Forever
Peg 22h25'59" 7d0'29"
Lara and Massimo's Island
Umi 16h45'54" 81d56'39"
Lara Ann Rose Walker
And 23h13'14" 43d30'51"
Lara Ashfield Gibbs
Gem 6h56'17" 25d33'49"
Lara Bella St.James
Ari 2h40'58" 27d29'11"
Lara Berman "The Brightest Star"
Cas 1h1'7" 63d6'46"
Lara Brody
Cap 20h20'4" -10d2'22"
Lara Brophy
And 1h36'52" 41d45'5"
Lara Camille Pilant
Uma 13h8'29" 52d27'5"
Lara Chow
Lib 15h28'10" -3d52'47"
Lara Cunningham
And 2h35'23" 49d12'17"
Lara De Palatis
Ori 6h14'35" 11d14'15"
Lara El-Khatib
Cas 23h8'57" 59d8'32"
Lara Fabian
Cap 21h27'7" -19d29'6"
Lara Grace Partrick
Lyr 18h42'49" 30d40'31"
Lara Grace Tsomik
And 1h7'43" 42d10'28"
Lara i Milos
Cyg 20h41'19" 42d28'9"
Lara Jan
Ari 2h41'56" 25d14'56"
Lara Jaramillo Savage - #1 Mom
Vir 12h47'16" 4d4'45"
Lara Jean's Dreams
And 23h51'58" 39d26'14"
Lara & Jeff's Eternal Star
Mon 6h36'6" 7d35'8"
Lara Joy
Uma 10h55'45" 44d18'32"
Lara & Karim
Crb 16h10'36" 31d49'46"
Lara Kate Weinberger
Crb 15h46'32" 32d27'59"
Lara Katherine Berry
Cnc 8h17'37" 32d58'19"
Lara Kostun
Leo 11h15'59" 16d59'30"
LARA LaPoe
Uma 11h24'32" 38d37'57"
Lara Little
Lmi 10h48'9" 25d45'23"
Lara & Livia Guttieres
Uma 9h19'55" 49d2'54"
Lara Louise Enright - 30 May 2005
Cru 12h28'50" -61d57'59"
Lara Mae Carandang
And 23h46'55" 35d16'9"
LARA Maria
Her 18h13'1" 18d25'15"
Lara Meyer
Vir 13h26'44" -12d46'58"
Lara Michèle
Gem 7h49'47" 21d5'14"
Lara Michele
Aqr 21h0'12" 1d16'54"
Lara Murphy
Uma 10h4'58" 55d47'49"
Lara N Lloyd
Aqr 21h25'26" -0d44'47"
Lara Nicole Saxe
Ari 2h12'36" 23d32'5"
Lara Noelle Wise
Aqr 22h0'1" -3d36'21"
Lara Olivia Monroe
Ori 5h52'23" -0d0'59"
Lara Sophie
Ori 5h35'16" 3d24'29"
Lara Star
Cnc 8h28'13" 24d52'32"
Lara Trinity Robinson
And 1h17'40" 41d41'9"
LaRae
And 2h16'45" 47d45'55"
LaRae Diane Didier
Leo 9h51'13" 12d58'28"
Larae Townsend
Cyg 21h44'25" 35d15'49"
LaRae's Star
And 0h51'32" 35d22'35"
Larahmine
And 2h24'2" 42d26'46"
Larain Louise Jorgensen
Vir 12h16'6" 5d57'22"
Larain Louise Jorgensen
Vir 13h53'16" -8d42'6"
Laraine 65
Tau 3h54'31" 19d31'44"
Laraine Diana Turk
Tau 4h11'16" 21d59'8"
Laraine Harrison
Lyr 18h50'40" 36d34'35"
Laraine M. Selivonchik
Lyn 7h24'46" 45d24'36"
Laralu
Cru 12h22'6" -57d57'16"
Laramie Paul Gonzales
Com 13h5'58" 23d59'34"
Larank
Uma 11h32'21" 63d28'41"
Lara's Little Place In The Sky
Uma 11h34'47" 62d17'53"
Lara's Star
Pho 1h5'25" -47d36'12"
Larashunda Spann
Umi 15h19'17" 78d45'16"
Lara-Stern-22-03-1990
And 23h59'11" 44d57'34"
LARCOM
Uma 13h46'48" 54d32'59"
Lare
Dra 17h22'12" 52d59'59"
Laree Ann Price
Crb 15h44'19" 33d52'1"
Laree Dawn
Crb 15h29'37" 29d56'57"
LaRee's Star
Uma 11h23'4" 31d40'26"
Larehn Marianna Quinn
And 2h37'26" 37d45'43"
Lareina Ann Galaviz
Cnc 8h28'25" 13d24'13"
Larell - Daddy's Angel
And 0h19'8" 33d26'51"
Laren
Aqr 22h13'33" 2d28'59"
LaRene R. Ward
Lyr 19h21'23" 37d31'53"
larew
Lyn 8h39'20" 33d47'26"
LaReyna Jean Davenport Nowviock
Leo 11h42'46" 21d15'37"
Larf Darf
Uma 10h59'50" 47d48'0"
Lari
Psc 1h23'59" 6d55'17"
Lari Lei Blanco
And 0h44'37" 37d48'49"
Larilou
Ari 2h9'56" 25d5'36"
LaRin J. Rangel
And 0h25'4" 42d9'39"
Lario
Per 4h6'19" 34d17'13"
Larion's Friend
Umi 15h35'14" 81d49'14"
Larisa
Vir 13h49'53" -0d17'54"
Larisa
Psc 1h10'1" 26d33'45"
Larisa Cary
Ari 3h15'40" 29d15'13"
Larisa J. Poplawski
Cyg 21h11'49" 47d16'15"
Larisa "Karabosova"
Mon 6h46'3" -0d54'58"
Larisa Nikolaevna Drozdovskaya
Uma 10h45'9" 42d4'59"
Larisa Rahman
Cap 21h27'52" -11d7'56"
Larisa Vitaliti
Cap 20h32'39" -10d21'31"
Larissa
Sgr 17h48'29" -19d28'8"
Larissa
Uma 9h8'3" 65d46'4"
LARISSA
Cep 22h20'15" 66d4'54"
Larissa
And 0h11'59" 42d12'59"
Larissa
And 0h45'40" 41d52'25"
Larissa
And 1h1'2" 39d57'32"
Larissa
And 1h15'12" 43d27'20"
Larissa
Gem 7h5'55" 33d53'6"
Larissa
Cas 0h28'15" 59d49'35"
Larissa
Her 16h31'59" 14d6'56"
Larissa A Wohl
And 23h18'24" 40d39'42"

Larissa Cascio
Lib 15h48'26" -5d41'54"
Larissa Cristina
Mon 7h37'17" -1d3'3"
Larissa Grace McNeil
And 1h39'20" 47d28'42"
Larissa Hauser
Her 18h22'52" 12d7'11"
Larissa Heinrich
Ari 2h24'46" 26d10'39"
Larissa Johnson
Psc 0h51'54" 13d10'51"
Larissa K. Karas
Sco 17h19'50" -44d49'12"
Larissa Kemmett
Cnc 8h36'33" 9d38'19"
Larissa Kerbel
Vir 15h2'20" 5d38'48"
LARISSA LIDIA TRACZ-PADUCHAK
Sco 17h48'43" -34d10'21"
Larissa Marie Anderson
And 1h48'51" 44d40'58"
Larissa Otypka
Uma 13h15'16" 57d40'9"
Larissa Popp
And 1h14'48" 49d31'13"
Larissa Rae
Uma 9h29'28" 56d25'13"
Larissa Sheryl Rosa
And 0h44'17" 44d35'34"
Larissa Spichiger
Uma 10h41'26" 47d53'35"
Larissa Vila
Lyn 6h43'47" 57d36'22"
Larissa XOX
Cnc 9h18'32" 9d7'43"
Larissa's Little Star 042775
Ori 6h4'5" 17d40'4"
Larizzy
Cyg 21h29'33" 31d50'35"
Lark
Leo 9h24'27" 15d45'51"
Lark Elizabeth Yoder-Hall
Lyn 8h47'23" 44d35'55"
LARKIN
Leo 10h12'8" 14d44'14"
Larkin Elizabeth Sanders
And 1h36'44" 43d23'36"
Larkin "Lucky" Marie Smith
And 0h29'22" 34d6'38"
Larkin Mia Loftin Jacks
Crb 15h30'7" 27d31'55"
Larkin Patricia Mielke
Psc 1h12'12" 10d23'3"
Larkin Stringer
Tau 4h35'29" 19d29'59"
Larks
Gem 7h26'53" 26d39'40"
LarLar SteveStar
Uma 12h1'43" 52d20'21"
LarLea's Love
Lyr 18h33'19" 37d28'39"
Larna Kay
And 1h38'47" 42d50'31"
LAROBY
Cas 0h26'16" 62d27'5"
Larold
Sco 16h3'13" -16d27'29"
LaRona McDaniel
Crb 15h44'0" 26d3'13"
LaRoye C. Vincent Crew
Uma 8h10'15" 60d30'26"
Larra Lopez
Crb 15h42'17" 32d42'37"
Larraine Bodley - Princess
Uma 10h58'7" 35d49'18"
Larraine Kay
Cru 12h13'12" -62d18'3"
Larran
Uma 10h56'53" 34d47'46"
Larrere Pierre
Cnc 9h7'54" 15d52'11"
Larri Alyn Peterson
Cyg 20h35'21" 31d12'47"
Larrissa (Lar Lar)
And 23h0'13" 37d7'8"
Larrp
Vir 14h38'13" 3d59'35"
Larr's Luminary
Ori 6h0'39" 10d16'14"
Larry
Leo 11h23'16" 14d46'49"
Larry
Per 3h38'1" 43d2'49"
Larry
Per 4h7'5" 49d23'59"
Larry
Umi 16h40'41" 87d17'37"
Larry Alan Glaspell
Leo 9h44'38" 22d30'56"
Larry Albin
Cep 21h51'28" 69d10'2"
Larry and Julie Pitts
Cyg 20h5'39" 31d58'10"
Larry and Karen
Cyg 21h20'34" 38d29'43"
Larry and Kay Gordon
Gem 6h46'28" 19d10'4"
Larry and Linda Tucker
Sge 20h4'52" 16d36'25"
Larry and Lynn
Tau 5h52'3" 25d14'13"
Larry and Marie
Cyg 20h29'11" 38d9'20"
Larry and Marlene
Gebhard's Star
Aql 19h23'51" 5d28'24"
Larry and Maxine
Cyg 19h54'20" 32d56'39"
Larry and Paulette
Cyg 20h11'41" 31d58'28"
Larry And Phyllis Shulz
Crb 15h48'53" 27d15'52"
Larry and Rosemary
Lenzmeier
Gem 6h54'13" 30d55'14"
Larry and Venessa's Star
Uma 11h9'18" 48d18'8"
Larry Andrew Goodman
Dra 16h19'4" 59d51'22"
Larry Anthony Prince
Lyn 8h35'47" 33d38'14"
Larry B
Lib 15h0'50" -19d15'40"
Larry Bachewski
Aqr 23h25'4" -19d11'0"
Larry Banish
Per 1h43'43" 50d48'10"
Larry & Barbara
And 1h5'6" 42d27'49"
Larry Barnaby a.k.a. Harley
Sco 17h17'24" -33d13'1"
Larry Bauman
Lib 15h3'10" -1d45'2"
Larry Bimbo Mitchell
Sco 17h27'58" -38d33'59"
Larry Brown
Uma 10h44'33" 50d27'7"
Larry Bruce Goad
Ari 2h41'0" 24d57'13"
Larry Buffaloe's Southern
Star
Tau 4h58'40" 24d11'41"
Larry Bye
Leo 10h8'37" 6d32'34"
Larry C. Flowers
Per 3h38'46" 45d47'5"
Larry C. Gillett
Cap 20h24'43" -12d27'47"
Larry C. Westlake
Aur 6h25'54" 40d58'51"
Larry C. Wittenbach
Crb 16h16'11" 28d19'57"
Larry Cameron
Ari 2h33'33" 25d20'58"
Larry Carl Shelton
Ori 5h35'48" -1d39'34"
Larry Carp
Ori 6h10'14" -0d41'23"
Larry Chance
Cep 0h36'14" 79d46'21"
Larry Charles Clark
Ori 6h7'29" 14d10'35"
Larry & Charlotte
Cyg 19h45'55" 33d55'21"
Larry Chmiel
Equ 21h18'20" 8d34'38"
Larry Christiansen
Sgr 18h33'46" -23d55'7"
Larry Costigan
Dra 18h55'8" 55d16'26"
Larry D. Gibson
Aql 19h46'43" -0d30'2"
Larry Dale Brewer
Psc 1h8'54" 10d39'57"
Larry Dale Dick
Cnc 8h39'46" 27d31'50"
Larry Dalo, Jr.
Leo 11h13'56" 24d31'52"
Larry Dean Buck
Her 17h1'30" 18d10'19"
Larry Dean Davis
Cnc 9h5'23" 27d44'43"
Larry Digiacomo
Cep 23h45'6" 75d10'13"
Larry E. Belger
Her 16h46'35" 38d45'35"
Larry E. Cole, Jr.
And 2h5'27" 42d51'57"
Larry E. McDowell, Sr.
Cnc 9h11'49" 24d30'34"
Larry E Mellor
Uma 9h5'14" 48d39'39"
Larry Edward Arnholt Ph.D.
Cyg 21h28'11" 45d57'32"
Larry & Eileen King
Cyg 20h6'39" 40d28'53"
Larry Eugene Miller
Ori 6h1'7" 17d52'24"
Larry Evie & Jonathan
Woods Star
Cam 4h13'17" 66d48'23"
Larry F. Strate
Cep 20h39'31" 63d51'7"
Larry Fankhauser
Uma 13h33'24" 57d52'59"
Larry Franklin Johnson
Sgr 18h42'40" -16d13'25"
Larry FRENCHIE Weinstein
Gem 7h39'7" 30d47'32"
Larry G. Freed
Psc 1h12'40" 27d30'51"
Larry Gallagher
Vir 14h15'46" 2d27'40"
Larry Gallahue
Aqr 20h49'11" 2d21'28"
Larry Geisinger
Aqr 21h59'51" -15d53'8"
Larry Gene Fletcher
Ari 2h25'10" 26d16'51"
Larry Gerwit
Uma 11h52'12" 51d29'19"
Larry Gordon Cover
Uma 9h57'6" 43d44'55"
Larry Greenwood
Per 3h15'45" 54d55'45"
Larry Herbert Brown Phd
Aur 5h9'37" 47d0'32"
Larry Hernandez
Ori 5h26'52" 6d46'55"
Larry Higgins
Lyn 8h25'39" 39d55'24"
Larry Hinnen
Sco 16h34'45" -33d57'41"
Larry Hite
Ari 3h20'12" 27d32'37"
Larry Holliday
Ori 5h25'19" 3d19'54"
Larry Hoover
Boo 14h36'26" 51d34'17"
Larry Hyde and Michelle
Boudreaux
Sge 20h1'12" 17d32'7"
Larry J. Cioffi TIC 04/01/72
Ari 2h33'40" 26d22'50"
Larry J. Hale
Cyg 20h47'44" 49d10'0"
Larry J. Roberson
Uma 8h51'32" 56d4'53"
Larry J. Rupert - A
Teaching "Star"
Ori 5h21'58" 9d15'20"
Larry Jeffcoate
Cep 1h12'40" 80d30'47"
Larry Jones-Washington
Aqr 21h32'39" 0d50'13"
Larry K.
Cap 21h29'14" -19d45'14"
Larry Kaluzna
Aql 19h12'11" 8d55'5"
Larry & Kate Forever Star
Cyg 20h16'25" 55d39'44"
Larry Keller
Ori 5h44'14" 6d49'35"
Larry Kennepohl
Boo 14h25'30" 35d19'26"
Larry Kindle
Uma 8h36'30" 53d4'52"
Larry King
Sgr 19h22'25" -27d18'22"
Larry King
Her 18h37'18" 17d34'4"
Larry Klein
Ori 5h39'31" -1d57'5"
Larry Knewbow Star
Ori 5h32'31" -1d43'48"
Larry Koral
And 1h41'35" 50d12'53"
Larry L. Facemire Sr.
Ori 6h19'58" 9d25'23"
Larry L. Stange
Uma 11h21'15" 29d33'11"
Larry Lake
Vir 12h7'58" 6d7'0"
Larry Lamore, Sr.
Uma 11h39'17" 60d32'58"
Larry Laskiewicz
Her 18h22'4" 29d13'5"
Larry Lawrence
Cnc 8h22'26" 13d31'28"
Larry Lawson
Ori 5h3'50" 5d32'54"
Larry Lee Love
Sgr 18h32'8" -21d37'19"
Larry Leichtman
Cnc 8h21'4" 16d48'34"
Larry Levy #1
Uma 9h2'33" 46d45'18"
Larry Lima, Jr.
Per 3h8'2" 39d48'2"
Larry Linda
Per 3h46'42" 44d59'50"
Larry & Linda
Uma 9h37'16" 54d44'56"
Larry Linder
Ori 5h46'36" 7d1'18"
Larry Liska
Umi 13h40'17" 71d20'14"
Larry & Liz - Destined
Pho 23h48'53" -41d27'6"
Larry & Lou Berg
Uma 12h40'17" 56d24'10"
Larry Lynn Hodges
Sgr 18h57'42" -33d58'5"
Larry Lynn Sandlin
Ari 2h40'2" 14d41'20"
Larry & Maj Hagman Go
Gold
Cyg 19h19'6" 52d53'20"
Larry Mangiaracino
Ari 3h21'6" 25d45'57"
Larry Mann
Uma 11h25'11" 39d6'53"
Larry Maracle
Cep 20h40'52" 64d27'59"
Larry & Mary Smith, &
Bubba too
Aql 19h21'3" 5d2'39"
Larry Masten
Aql 19h51'23" -0d45'49"
Larry Max Driver
Psc 1h21'7" 24d55'2"
Larry Maxell November
28th, 1951
Uma 8h44'2" 72d43'42"
Larry McComb
Uma 10h47'35" 40d2'35"
Larry Melvin Reynolds
Cep 21h47'0" 63d9'1"
Larry Michael Lon Smith
Per 3h35'0" 33d55'9"
Larry Middleton
Her 16h16'45" 41d36'33"
Larry Mola
Ori 5h55'31" 21d2'52"
Larry Motley
Boo 14h22'58" 14d18'28"
Larry Murdock - We Are
Eri 3h3'31" -23d51'52"
Larry "My Honey" Jacobs
Per 3h20'27" 43d46'21"
Larry My Love
Ori 5h33'20" 10d40'15"
Larry Nelson Carroll
Lyn 7h48'52" 38d34'7"
Larry Owen Cook's ètoile
Gem 6h13'27" 22d42'14"
Larry Patton
Per 3h39'22" 35d11'49"
Larry Paulette
Uma 14h0'47" 59d42'19"
Larry & Penny New 4-ever
Ori 6h0'41" 10d4'11"
Larry Perez
Gem 6h52'20" 21d15'20"
Larry Peterson
Per 2h41'3" 54d0'35"
Larry "Pops" Rotondi
Uma 9h54'45" 42d47'13"
Larry Pratt
Aql 19h13'23" 8d59'42"
Larry R. Hernandez
Per 3h30'56" 51d5'44"
Larry R. Turner
Cep 22h26'36" 62d46'10"
Larry Raveling
Ori 5h34'31" 3d52'47"
Larry Read
Uma 10h5'44" 48d15'22"
Larry Remble's Piece of the
Sky
Ori 6h12'32" 15d25'36"
Larry Robertson
And 1h43'11" 45d11'39"
Larry S. Hewlett Jr.
Aqr 23h1'37" -6d33'48"
Larry Santangelo
Sco 16h52'43" -41d7'47"
Larry Schiefer
Lib 14h34'45" -16d58'23"
Larry Schneider
Cep 22h1'55" 56d27'22"
Larry Scott Macklin
Gem 7h41'35" 19d43'24"
Larry Scott Miller
Her 16h42'36" 46d0'24"
Larry Scott Wilson
Lyr 18h47'57" 44d38'10"
Larry Seigel
Tau 5h8'37" 22d38'36"
Larry Sexton
Her 16h38'46" 23d7'52"
Larry & Shirley Powell
Cyg 20h27'18" 33d8'31"
Larry Simpson
Boo 14h48'27" 18d17'32"
Larry Sprouse
Ari 3h20'43" 29d32'28"
Larry Stephen Roy
Dra 17h19'43" 54d30'45"
Larry Sterina Jr.
Cnc 8h37'42" 14d27'1"
Larry Steven Godden
"Greatest Dad"
Per 1h46'15" 51d15'54"
Larry Stoddard
Sgr 17h46'52" -21d49'49"
Larry & Sue Edwards
Uma 9h8'13" 54d17'30"
Larry Sullivan
Psc 1h44'37" 5d10'31"
Larry & Suzi Dickey
Lyr 18h52'14" 41d11'39"
Larry T. Burgess
Her 18h54'3" 23d35'37"
Larry The Gentle Giant
Gem 7h4'1" 21d40'27"
Larry "The Star" Stevens
Dra 20h7'25" 64d49'38"
Larry Thomas
Uma 9h18'43" 64d50'27"
Larry Trujillo
Uma 13h39'42" 55d44'58"
Larry Truman Olson (O.D.)
Uma 11h28'46" 31d29'33"
Larry Turner
Per 2h18'47" 55d38'30"
Larry "Umpa" Pond
Dra 17h32'36" 62d0'49"
Larry W. Reed
Leo 11h25'36" 14d10'47"
Larry W. Troesch
Cap 20h33'56" -27d9'37"
Larry W Wild
Ori 4h48'42" -0d9'5"
Larry Warren
Per 4h9'15" 37d35'46"
Larry Wayne Cornelison
Cap 21h27'16" -23d27'16"
Larry Wayne Dean
Cnc 8h42'39" 18d37'58"
Larry Wayne Evans
Per 3h4'32" 55d33'14"
Larry Wayne Mankins
Gem 6h46'50" 16d7'13"
Larry Wayne Wick
Gem 7h6'23" 24d35'14"
Larry Winegardner
Cnc 8h23'39" 8d35'20"
Larry Worth-Jones
Uma 9h24'21" 46d51'3"
Larry-Caroline
Cyg 21h46'49" 46d26'6"
Larry-Ian-Marni-Fathers
Day
Uma 9h21'24" 43d23'1"
Larry's Famous Star
Uma 12h43'51" 53d0'51"
Larry's Legacy
Tau 4h32'2" 16d36'23"
Larry's Lucky Star
Vir 13h32'59" -15d8'44"
Larry's Wish
Sco 17h36'17" -33d16'51"
LarrySaladbar
Uma 13h13'14" 56d25'38"
Larrysia Ford
Mon 6h59'36" -0d9'29"
Larry—The Bug of Love
Cnc 8h49'34" 32d6'24"
Lars
Boo 15h27'48" 34d1'31"
Lars
Uma 9h58'59" 50d28'3"
Lars
Sco 17h33'42" -44d51'14"
Lars Antonio Oosterhof
Peg 21h27'51" 2d46'14"
Lars Aristodemos
Her 17h53'50" 49d27'51"
Lars Buchholz
Uma 8h12'8" 62d21'44"
Lars From Mars
Uma 8h30'21" 64d1'35"
Lars III
Dra 18h37'1" 60d47'39"
Lars Jenni
Cam 6h49'36" 69d18'19"
Lars Michael Edeen
Umi 13h32'29" 72d22'7"
Lars Ottiger
Uma 11h15'24" 46d50'3"
Lars Schüler
Uma 8h45'13" 54d48'10"
Lars Senftleben
Ori 5h54'14" 3d36'47"
Larsa Landis
Umi 16h13'59" 74d32'23"
Larsen
And 1h1'59" 39d28'36"
Larsen Ori Riise
Aql 19h49'45" 13d16'3"
LarsenLundgren
Cma 6h48'41" -15d12'48"
Larson Garrett Brock
Dra 19h27'36" 73d9'16"
Larson Levenson
Tau 3h38'38" 24d40'11"
Larson W. B. Knoflick
Sco 16h54'10" -41d40'3"
Larson Wade Davick 9-22-88
Per 2h26'14" 51d7'17"
Larsony Fisher
Vir 14h6'14" 3d37'56"
Larsson Brunsvold
Tau 4h45'48" 21d44'54"
LarStar
Vir 12h5'10" 11d16'28"
laryll
Cyg 20h17'7" 40d48'35"
Larysa Rissmann
Uma 9h41'25" 48d15'36"
Laryssa Marcella Spengler
Tau 5h0'21" 23d52'13"
Las Luces Guias
Ori 6h6'33" 14d30'31"
LAS Star
And 0h48'13" 37d7'10"
Las Trece Aguilas de San
Antonio
Her 16h38'36" 16d57'27"
Lasaw
Her 15h58'16" 44d40'15"
LashaLaine
Cas 0h54'24" 59d59'30"
Lashann Williams
Uma 11h20'37" 47d34'2"
LaShann Williams
Uma 11h55'42" 52d17'4"
Lashann Williams
Umi 15h29'41" 69d31'46"
LaShaunte
Aqr 22h16'46" -2d31'28"
Lashawne Lefotu
Peg 21h28'37" 16d18'45"
Lashonda R. Holt
Ari 3h5'55" 23d14'35"
Lashonda Renee Smith
Lyn 7h53'32" 37d38'19"
Lashone
Lyn 7h48'41" 43d53'41"
Lasius John James
Houston
Uma 11h43'18" 54d41'40"
Laska 82
Cas 0h30'59" 55d23'13"
Laskawy, Stefanie
Ori 6h20'8" 19d1'31"
Laskodi
Cma 6h45'48" -14d36'54"
Lason
Umi 15h11'26" 71d35'14"
LaSonda Grace Baxter
Ari 2h26'5" 19d44'43"
LaSpina
Uma 10h19'12" 63d16'19"
Lassie and Maria Nastou
Aur 6h8'47" 29d4'6"
Lassielyn
Sco 16h39'5" -29d35'31"
last dream
Ari 3h25'7" 27d9'24"
Last Stop Before Heaven
Uma 12h3'43" 31d54'15"
Last Wish
Ori 6h17'48" 13d32'57"
laStellaAgile
Her 17h55'33" 18d1'31"
L'asteroide B 612
Uma 14h1'54" 58d30'37"
Lasting Love Tom & Bern
Sullivan 35
Cyg 19h46'21" 55d18'16"
Lastochka Jenny
And 1h40'48" 50d37'35"
L'Astralia, le 6e continent
Uma 12h10'39" 57d53'44"
Laszlo Hollo
Aur 5h10'3" 42d5'54"
Laszlo Laszlo Istvan
Lib 15h21'51" -28d26'42"
Laszlo Peter Nyary
Vir 13h12'57" 11d34'12"
László Roland
Uma 10h57'31" 37d44'56"
Latania M. "Boo Boo"
Lyn 6h33'37" 55d38'51"
Latanya Ann Stackhouse
Cnc 8h20'28" 17d59'57"
Latanya L. Baines "Boo"
Cas 1h15'47" 55d55'9"
Latarra Marliese
Mon 6h28'28" 9d11'21"
Latarshia M. Charland
Birthday Star
Sco 16h9'14" -12d58'44"
Latasha Ann Loveday
Lib 15h49'27" -14d7'37"
Latasha Cherry
Cap 21h41'10" -13d13'56"
Latasha D. McGowan
Leo 11h26'8" 6d33'56"
Latasha Enderby
Cas 1h23'31" 57d6'28"
Latasha Michelle Mitchell
Tau 5h5'57" 25d21'34"
LaTasha's Shining Star
Vir 14h20'3" 4d14'32"
LaTasi
Ari 2h25'25" 26d9'18"
LaTawnja
Uma 11h38'22" 65d22'14"
Latayne
Ori 4h50'21" 11d40'44"
Laté
And 23h48'50" 37d9'57"
Late Vitolina
Lyr 19h19'43" 33d44'35"
Latena Teree Carter
Aqr 20h42'6" -9d33'0"
Latham's Amor Astralis
Aqr 20h41'41" -2d38'56"
Latham's Peanut
Lyr 18h51'36" 45d6'39"
Lathrop Preston
Her 16h25'11" 7d9'9"
LaTia
Psc 1h22'29" 16d2'15"
Latifa
Uma 11h34'14" 59d43'28"
Latifa Bouras
Del 20h45'35" 15d17'42"
Latifa Mohammed Al Ajmi
Lib 14h43'33" -19d0'22"
Latin's Learners 2004-2005
And 23h22'5" 51d36'19"
Latisha C. Ruffin
Sco 17h45'28" -40d8'44"
Latisha Marie Dobbs
Ari 3h3'36" 18d6'58"
LaTisha Michelle
Grandstaff
And 1h26'0" 43d13'31"
Latitti
Peg 22h33'43" 9d34'55"
Latke
Uma 10h29'45" 66d18'9"
Latoma Gill
Crb 15h29'26" 28d27'50"
LaTona Anne "Stroud"
Paulson
Cyg 19h34'7" 29d12'18"
LATONIA
Lib 15h38'39" -24d53'43"
LaTonya Rene
Sgr 18h28'46" -27d1'56"
Latosha
Cas 0h47'29" 53d28'2"
LaToya L. Price
Leo 11h9'9" -0d15'57"
Latoya S. Washington
Aqr 21h37'29" 1d16'13"
Latraia
Cam 7h25'2" 75d24'29"
Latrice
Sgr 20h5'24" -25d33'40"
La-trice
Cas 0h58'42" 53d42'44"
Latriece Casey West
Sgr 18h31'49" -27d8'57"
Latrisha Holdsworth,
Loved!
Cyg 20h33'13" 51d49'6"
Latta 8
Ori 5h44'45" 5d24'21"
Latte Hogge
Pup 6h42'8" -44d43'10"
Latterman Family
Tau 3h41'54" 7d34'3"
Lattice
Uma 9h21'49" 63d34'14"
Latto Diamond
Ori 5h39'42" -3d16'21"
Lau Hoi Yi
Aqr 21h50'18" -1d19'49"
Lau On Ki, Fion
Gem 7h19'37" 20d17'18"
Laua e
And 1h16'17" 41d50'38"
Lauchlan Connel Callaghan
Per 3h4'31" 45d16'44"
Lauda
Her 16h42'0" 22d36'8"
Laudelina P. de Simas "A
Morena"
Ari 2h47'12" 17d13'45"
Laueri Schätzeli (Kurt
Emch)
Boo 15h29'26" 37d52'4"
Lauf-Coach Klaus Herweck
Uma 12h4'51" 44d21'11"
Laugh Master 9000
And 1h53'32" 40d5'41"
Laughing Star
And 1h37'6" 41d56'46"
Laughlin Lewis Whiteley
Peg 21h48'27" 26d11'37"
Laughs with Love in the
Sky
Cyg 19h37'12" 28d49'17"
Laulita Cardoso Valadao
Cnc 8h30'3" 21d47'22"
Launa
Psc 1h10'31" 29d43'57"
Launa and Whitney Anne
Minson Star
Lmi 10h28'45" 36d59'32"
Launa Jean
Crb 16h10'36" 38d17'59"
Laupichler, Rudolf
Ori 5h52'20" 7d18'8"
Laur
Cnc 8h16'0" 19d11'31"
Laur Bear
Tau 4h30'51" 22d9'52"
Laura
Tau 4h26'3" 16d13'12"
Laura
Peg 21h25'47" 13d4'7"
Laura
Psc 1h16'43" 18d2'4"
Laura
Leo 10h28'23" 13d5'45"
Laura
Ori 5h16'37" 0d35'49"
Laura
Vir 15h7'13" 4d48'48"
Laura
Psc 0h9'24" 8d52'42"
Laura
Leo 11h20'50" 18d16'1"
Laura
Leo 10h43'54" 15d16'43"
Laura
Ori 5h57'58" 16d54'11"
Laura
Psc 0h53'32" 28d1'46"
Laura
Ari 2h28'50" 27d43'9"
Laura
And 0h33'10" 26d21'43"
LAURA
Peg 22h30'59" 20d40'35"
Laura
Her 18h34'51" 17d38'23"
Laura
Peg 21h51'38" 24d33'43"
Laura
Crb 15h37'38" 26d56'12"
Laura
Leo 11h22'54" 23d9'36"
Laura
Aur 6h16'32" 29d12'35"
Laura
Tau 5h51'59" 26d28'5"
Laura
And 1h42'25" 45d46'9"
Laura
And 1h36'43" 45d25'14"
Laura
And 1h20'50" 49d0'25"

Laura
Lyn 7h22'8" 51d19'56"
Laura
Lyn 8h39'27" 46d21'44"
Laura
And 23h25'39" 46d42'27"
Laura
Crb 16h10'54" 35d15'24"
Laura
Lyn 8h37'10" 36d53'53"
Laura
Lyn 7h43'31" 37d12'6"
Laura
And 2h21'56" 38d11'29"
Laura
Cyg 20h58'37" 35d58'36"
Laura
Lyn 8h46'13" 41d45'47"
Laura
Ori 5h19'32" -8d32'38"
Laura
Vir 14h2'15" -11d20'58"
Laura
Vir 13h45'46" -2d24'55"
Laura
Mon 6h54'34" -0d35'25"
Laura
Mon 6h41'17" -0d19'57"
Laura
Mon 7h2'46" -6d7'16"
Laura
Umi 17h44'20" 80d22'54"
Laura
Umi 15h19'46" 83d44'54"
Laura
Umi 17h35'3" 84d39'50"
Laura
Sgr 19h21'49" -15d30'37"
Laura
Aqr 22h44'9" -15d4'9"
Laura
Aqr 21h16'13" -14d26'15"
Laura
Cas 2h26'27" 65d55'43"
Laura
Uma 9h24'45" 68d40'2"
Laura
Uma 13h1'17" 54d14'40"
Laura
Lyn 6h27'57" 54d4'35"
Laura
Lyn 7h42'0" 57d7'3"
Laura
Dra 16h26'52" 57d32'32"
Laura
Cas 23h35'54" 58d32'1"
Laura
Sco 17h48'56" -33d28'18"
Laura
Sco 17h53'3" -33d30'14"
Laura
Ori 5h53'52" -8d17'15"
Laura - 05 04 07
Cas 1h48'20" 56d4'45"
Laura 100
Lyn 7h53'36" 54d0'26"
Laura 143
Gem 6h43'59" 26d46'6"
Laura 18
Psc 1h14'55" 28d49'35"
Laura 2004
Leo 11h6'6" 20d10'23"
Laura 414
Lyn 7h9'35" 57d11'42"
Laura 9-28-01
Vir 14h42'10" 4d28'52"
Laura A Altman
Ari 1h52'23" 20d58'44"
Laura A. Cruze
And 0h33'49" 39d5'59"
Laura A J Jones
And 1h6'20" 35d40'52"
Laura A Kristof
Sco 16h15'53" -9d49'30"
Laura A Louviere
Cas 1h46'29" 67d47'12"
Laura A. Sills
Psc 0h8'18" 7d33'6"
Laura (A Star is Born) Shendelman
Aqr 21h54'4" 1d38'40"
Laura A. Wise
Aqr 22h16'55" -0d31'9"
Laura Above Us
Gem 7h16'32" 32d23'11"
Laura Adams
And 1h25'56" 37d41'6"
Laura Adele Palm
Psc 23h58'13" 0d58'21"
Laura Adele Stewart
And 1h43'36" 43d17'42"
Laura Adora
Cas 2h16'3" 66d19'33"
Laura Alayon
Gem 7h34'19" 20d34'4"
Laura Alese Rzepka
Sco 16h12'12" -17d5'56"
Laura Alexandra Milne
Cas 23h45'19" 53d51'22"
Laura Alexandra Weiss
Sgr 17h56'34" -25d49'29"
Laura Alexis
Tau 4h15'7" 28d1'49"
Laura Alice Kleinberg
And 1h5'39" 44d45'32"
LAURA ALISSA
Dra 10h29'20" 79d28'50"
Laura Allison Williams
Lib 15h31'16" -7d53'1"
Laura Alma Brookes
And 1h19'47" 50d35'13"
Laura Alma Foster Irving
And 23h23'1" 46d34'14"
Laura Aloisi Dowd
Her 16h45'46" 45d46'24"
Laura Altamirano Rayo
Vir 12h8'7" -2d55'50"
Laura Alvarado
And 2h15'33" 46d7'59"
Laura Amador
Com 12h18'37" 27d27'38"
Laura Amélie Wade
And 0h13'4" 46d47'49"
Laura and Betty Smith
Uma 11h32'47" 34d54'42"
Laura and Brad Forever
Lib 14h59'32" -13d24'23"
Laura and David
Peg 23h13'21" 34d2'2"
Laura and Jeff Hunsicker
Gem 7h43'13" 30d49'54"
Laura and John Forever
Aqr 23h1'56" -3d22'52"
Laura and John Hammond
Aql 20h13'20" 7d48'54"
Laura and Paul Higday
Ari 2h7'22" 23d51'46"
Laura and Sal's Star
Per 3h28'4" 35d34'22"
Laura and Scott Forever
And 1h25'9" 47d27'45"
Laura and Vin in Love Forever
Cap 20h32'16" -11d3'7"
Laura Andrae Price
Cnc 8h52'49" 13d57'24"
Laura Andrews
Sco 16h45'6" -31d27'38"
Laura Angel 50
Lyn 7h44'50" 51d8'23"
Laura Ann
Crb 16h1'18" 38d48'14"
Laura Ann
Cas 0h9'42" 57d36'5"
Laura Ann
Aqr 22h42'48" 0d37'32"
Laura Ann
Gem 7h28'33" 19d59'15"
Laura Ann
Leo 10h41'40" 15d15'54"
Laura Ann
Tau 3h59'3" 29d13'28"
Laura Ann
Cnc 8h37'5" 23d1'58"
Laura Ann
Cnc 8h14'34" 24d35'29"
Laura Ann
Lib 15h5'47" -0d32'31"
Laura Ann
Umi 15h45'45" 85d30'1"
Laura Ann Bent
Cas 0h15'54" 58d46'43"
Laura Ann Bright Star
Ari 2h3'14" 22d6'32"
Laura Ann "Cookie Dough" Brown
Peg 22h42'12" 13d53'40"
Laura Ann D'Agostino
Cap 20h46'3" -20d48'53"
Laura Ann Hannett
Sgr 18h49'47" -19d28'21"
Laura Ann Heidtman
Cas 0h31'5" 60d7'17"
Laura Ann Huzar
Cas 0h9'15" 60d58'44"
Laura Ann Iancu
Cas 1h25'36" 51d36'42"
Laura Ann Oliver
Sgr 18h18'22" -32d23'24"
Laura Ann Passante
Vir 13h7'31" 12d3'53"
Laura Ann Quarles
Sco 16h14'34" -17d34'2"
Laura Ann Rain Thomas
Vir 12h51'28" 4d37'40"
Laura Ann Tomasiello
Vir 11h50'33" -4d20'13"
Laura Ann Turner aka "Laurabell"
And 0h37'30" 29d31'56"
Laura Ann "Tweety" Sirratt
Cas 1h15'12" 55d59'6"
Laura Anne
Lyn 7h23'56" 49d25'5"
Laura Anne
And 0h51'59" 40d24'44"
Laura Anne Aikman
Aqr 23h2'35" -19d1'4"
Laura Anne "Anamchara"
Cnc 9h0'7" 8d23'52"
Laura Anne Cover
Aqr 22h40'11" -23d3'36"
Laura Anne Drumm
And 23h6'26" 46d22'10"
Laura Anne Gaskill The Beautiful 25
Lyn 8h29'40" 38d7'38"
Laura Anne Hickey
Leo 9h55'59" 26d28'10"
Laura Anne Pollitt
Crb 15h59'31" 34d29'58"
Laura Anne Schantz
And 23h17'34" 41d28'22"
Laura Annette
Ari 3h13'7" 26d25'15"
Laura Ann's Light
And 1h8'0" 42d36'52"
Laura Anthony Joshua Marc Haynes
Tau 3h44'23" 24d39'5"
Laura Appleby
And 23h33'9" 38d42'58"
Laura Arlinghaus
Sco 17h38'0" -40d26'18"
Laura Armstrong
Vir 12h30'58" 11d55'2"
Laura Arnold
And 23h19'9" 47d27'20"
Laura Arsena 73
Lac 22h31'50" 41d22'2"
Laura Ashleigh Davidson
Cyg 20h7'8" 40d28'42"
Laura Ashley
Lyr 19h13'35" 46d8'34"
Laura Ashley
Lyr 19h9'10" 26d36'6"
Laura Ashley
Lib 15h31'13" -7d14'50"
Laura Ashley Fry
Psc 0h36'58" 8d7'52"
Laura Ashley Lane
Cma 6h37'1" -22d51'4"
Laura Ashley Mintz
Ori 6h16'33" 14d8'55"
Laura Ashley Spry
And 0h16'11" 37d54'37"
Laura Ashley Sypek
And 23h17'32" 48d37'59"
Laura Ashley Wemple
Lib 15h58'39" -5d23'31"
Laura Baby
Ori 5h22'23" 1d0'58"
Laura Balbás
Lmi 10h52'5" 27d2'15"
Laura Barnes
Lyn 7h18'12" 59d41'3"
Laura Barron Sullivan
Uma 10h22'30" 47d48'27"
Laura Bastedo
Cas 1h56'21" 73d11'0"
Laura Battaile Schavrda
Mon 6h28'11" 8d20'10"
Laura Bean
Gem 7h21'25" 32d18'48"
Laura Belle
Sgr 19h41'19" -16d27'15"
Laura Belle Grisevich
Aqr 22h8'57" -1d42'58"
Laura Bello
Vir 13h45'42" 3d10'59"
Laura Bertschinger
Cas 23h59'24" 49d14'23"
Laura Beth
Her 17h21'26" 34d9'53"
Laura Beth
Vir 13h19'39" 5d13'2"
Laura Beth
Cnc 9h3'34" 24d22'23"
Laura Beth
Uma 10h39'16" 54d48'44"
Laura Beth Hancock's Start
Leo 11h13'59" 17d26'36"
Laura Beth Ledford
Per 4h49'16" 45d34'24"
Laura Beth Maclean
Ori 5h0'29" 8d47'30"
Laura Beth Mastrianni
Dra 19h47'57" 67d18'13"
Laura Beth Nickels
And 1h28'23" 44d24'26"
Laura Bettina Schouten
And 1h52'15" 46d31'53"
Laura Bieri
Ori 5h24'6" -3d49'48"
Laura Blazejewski Born: June 7/1990
Uma 11h30'34" 57d14'37"
Laura Boobear Flanagan
Cnc 8h49'22" 11d25'34"
Laura Bottazzi
Peg 22h3'10" 14d46'23"
Laura Bowerman
Cnc 8h53'11" 14d45'49"
LAURA BOYCE
Ari 3h26'44" 27d55'40"
Laura & Brad
Cnc 9h8'50" 16d23'10"
Laura Brady
Uma 9h47'44" 58d27'29"
Laura Braggion
Ori 5h16'48" 7d28'1"
Laura Brandt
Sco 16h6'54" -12d39'54"
Laura Branigan
Cas 0h16'40" 58d51'14"
Laura Brennan
Lib 15h9'27" -27d43'59"
Laura Brines
Uma 9h37'2" 63d28'43"
Laura Bryan
Cas 0h25'13" 61d35'33"
Laura & Bryan
Cyg 19h47'43" 38d12'3"
Laura Bullock 2005 Love, Mom
Uma 10h23'12" 50d27'42"
Laura Burkett
Peg 23h55'44" 25d49'25"
Laura Burton
Ari 3h23'18" 27d38'50"
Laura Butler
And 0h1'24" 44d7'10"
Laura C Springer
Lyn 8h29'25" 50d13'52"
Laura Cairns
Umi 15h27'9" 67d43'52"
Laura Caitlin Moss
Cap 21h58'0" -19d54'49"
Laura Cannon's Star
And 23h25'1" 52d55'35"
Laura Capitelli Rossi
Cam 3h42'9" 55d9'8"
Laura Carmela Gennarelli
Lib 14h43'55" -11d11'36"
Laura Castellucci
Peg 22h17'13" 17d16'14"
Laura Catherine
Psc 1h21'29" 20d35'38"
Laura Catherine Flannery
Cnc 8h2'52" 26d37'26"
Laura Catherine Gimbarti
Cap 21h43'48" -19d52'6"
Laura Catherine Tellefsen
Cap 21h47'31" -20d55'16"
Laura Cecile Farinas
Gem 6h23'7" 26d5'32"
Laura Celdran
Cnc 8h39'29" 30d54'57"
Laura Cho Lambert
Vir 13h10'31" -2d56'13"
Laura & Chris' Anniversary Star
Uma 11h41'46" 48d51'33"
Laura & Chris Graham
Cyg 19h41'48" 37d12'18"
Laura Christine
And 1h22'6" 37d10'55"
Laura Christine
Cas 2h44'35" 57d35'53"
Laura Christine
And 23h46'29" 47d55'48"
Laura Christine Conrad
Gem 6h42'6" 17d11'39"
Laura Christine Henkaline
And 23h43'2" 41d32'45"
Laura Cipres
Psc 23h56'41" 9d10'38"
Laura Claire Bryant
Gem 7h38'18" 21d32'7"
Laura Cobos Alonso
Cas 1h28'42" 56d21'6"
Laura Cochrane
Peg 22h36'23" 9d7'41"
Laura Colozza
Aur 6h37'43" 31d49'5"
Laura Connolly
And 23h46'13" 44d47'4"
Laura Couturier
Ori 6h0'8" 4d4'36"
Laura (Cowgirl) Durfee
Leo 10h5'6" 25d30'49"
Laura Cunningham and Julie Varner
Cyg 21h59'3" 51d16'50"
Laura D Crouse
Aqr 21h23'57" 1d33'56"
Laura Daddi
Umi 17h39'49" 80d31'32"
Laura Danielle
Sco 16h12'14" -14d12'27"
Laura Danielle
And 23h54'18" 41d19'58"
Laura Danielle Catlaw
Cas 1h42'34" 62d12'22"
Laura Danielle McCain
Sgr 19h46'58" -22d22'58"
Laura Darlingiubus McLeod
Sgr 18h47'26" -34d18'33"
Laura & David
Mon 6h52'47" -0d17'57"
Laura & David - Eternal Love
Cyg 21h33'55" 34d13'10"
Laura Dawson
Cas 23h48'57" 53d59'52"
Laura Deborah Baker
Cyg 20h17'42" 56d9'59"
Laura Denise
Sco 17h23'42" -37d53'31"
Laura Diana
Sco 17h24'59" -38d21'56"
Laura Diane Nelson
Leo 10h21'10" 20d6'36"
Laura Dianna
And 23h43'39" 46d34'25"
Laura Dobrovolskyte Khalil
Aqr 22h4'32" -8d10'39"
Laura Doran
Lmi 10h26'33" 34d1'42"
Laura Dorothee Elisabeth
Vir 13h5'41" -0d43'21"
Laura Downhour
Cas 1h21'4" 67d25'48"
Laura "Dunes" Dail
And 1h41'8" 43d28'16"
Laura Durrett (The Love of My Life)
Psc 1h10'33" 28d59'9"
Laura E. Faulds
Leo 11h5'27" 1d26'36"
Laura E Feller
Cyg 20h4'43" 50d50'39"
Laura E Kunzman
Psc 1h9'26" 4d31'43"
Laura E. Rhea
And 1h12'18" 43d0'44"
Laura E Schrodt
Lyn 8h56'33" 38d36'20"
Laura E. Swisher
Lyn 7h48'9" 49d53'21"
Laura Eddie Mattmann
Ori 6h14'40" 15d23'50"
Laura Elena Santos
And 0h12'18" 29d11'48"
Laura Eliana Castaño
Vir 12h39'17" 6d12'16"
Laura Elieen Flanagan
Aqr 21h12'19" -10d28'44"
Laura Elisha Boronkay
Aqr 22h27'37" -1d5'37"
Laura Elizabeth
Aqr 22h28'27" -21d58'7"
Laura Elizabeth
Psc 0h51'51" 16d54'2"
Laura Elizabeth Barr
Cas 1h12'14" 55d18'19"
Laura Elizabeth Berman
And 0h4'9" 47d36'13"
Laura Elizabeth Campbell
Sco 16h49'21" -34d24'59"
Laura Elizabeth Claire Hulbert
Cyg 20h31'25" 42d23'39"
Laura Elizabeth Finch Padden
Vir 13h18'27" 13d36'25"
Laura Elizabeth Heber
Uma 9h31'59" 42d29'19"
Laura Elizabeth Howard
Cas 1h14'2" 64d16'20"
Laura Elizabeth Howse
And 0h39'37" 43d15'38"
Laura Elizabeth Kristof Standard
Uma 11h27'4" 33d48'18"
Laura Elizabeth Lauterbach
Cas 0h41'44" 60d23'34"
Laura Elizabeth [l.e.s.]
Tau 4h6'37" 4d51'17"
Laura Elizabeth May
Cas 0h34'49" 52d29'11"
Laura Elizabeth Perkins
Cap 21h5'39" -19d34'5"
Laura Elizabeth Remick
Vir 12h30'52" 11d58'33"
Laura Ellen Nicolson
And 2h36'20" 50d30'13"
Laura Ellen Wertz, The Yellow Rose
Mon 6h49'39" 8d15'4"
Laura Erli
Cas 0h21'56" 57d9'37"
Laura Etneo
And 23h2'1" 50d50'59"
Laura Eva Welch
Lyn 7h20'28" 48d39'24"
Laura Evelyn
And 0h57'22" 38d17'31"
Laura Fabienne Birkigt
Uma 11h18'10" 43d1'21"
Laura Fagherazzi
Cyg 20h1'2" 33d47'48"
Laura Fair
Uma 10h45'58" 67d42'43"
Laura Faith Goetz
And 0h51'57" 35d55'58"
Laura Faith Peluso
Cnc 8h36'58" 22d38'10"
Laura Fekken
And 0h58'5" 35d53'43"
Laura Felmore
Lyn 7h38'2" 52d50'15"
Laura Fireball!
Lmi 10h41'15" 32d40'40"
Laura - Forever In My Heart - Jeff
Vir 13h27'0" -0d2'39"
Laura Frances Winterroth
Sco 17h51'38" -36d48'7"
Laura Francine
Crb 16h12'2" 34d52'39"
Laura Franco
And 23h24'43" 42d3'28"
Laura Franz 1125
Psc 0h32'21" 7d12'7"
Laura Fullerton
Lmi 10h48'43" 38d31'2"
Laura G. Franciscus
Psc 0h15'16" -1d0'34"
Laura G. Saunders
Leo 10h58'6" 5d49'41"
Laura Gayle
Tau 3h31'8" 18d37'44"
Laura Geary Dunson 5-31-1991
Gem 7h18'11" 25d52'19"
Laura Gerrior
Cas 0h59'32" 57d9'31"
Laura Gonzalez
Lyn 7h35'31" 38d11'15"
Laura Gregory
Uma 11h45'52" 44d8'15"
Laura Grimes
And 23h15'45" 49d17'25"
Laura Grimsley's Shining Star
Lib 15h26'2" -9d29'11"
Laura Grueber
Lyn 7h29'55" 49d54'55"
Laura Guastalla
And 1h46'12" 50d35'8"
Laura Guthridge
Aqr 22h29'12" -0d33'2"
Laura Haessler
Cas 1h33'48" 63d26'40"
Laura Hagyard
Uma 8h40'15" 51d47'6"
Laura Hamilton Stellar Midwife
Gem 7h16'5" 25d50'3"
Laura Hampson
Cas 0h10'15" 54d0'48"
Laura Harrison
Lyn 8h19'38" 43d51'32"
Laura Hayden
Aql 20h1'7" 3d6'18"
Laura Heathy Hansen
Gem 6h57'11" 14d36'20"
Laura Heinemann
Gem 7h23'49" 28d59'30"
Laura Helen Moss
Cas 23h16'59" 54d59'2"
Laura Herman
And 1h40'32" 46d17'48"
Laura Heseltine
And 0h47'48" 45d6'16"
Laura Heydinger Ryan
Lib 15h11'27" -11d35'9"
Laura Heyworth
Aql 19h42'44" -3d56'30"
Laura Hickman
And 23h16'5" 48d32'25"
Laura Hillary Schuman
Cas 0h31'48" 63d52'45"
Laura Hoffman
Sco 17h20'6" -31d19'23"
Laura H.P.
And 1h25'19" 49d16'22"
Laura Hummer
Gem 6h48'39" 34d15'33"
Laura Humphrey
Cam 4h28'9" 59d32'45"
Laura Hunt, my shining star
And 0h13'43" 45d44'9"
Laura Ingalls
Cam 6h23'59" 69d46'9"
Laura Ippolito
Cyg 19h36'1" 31d38'52"
Laura Isabel Quintero Bolivar
Uma 11h16'37" 36d57'49"
Laura Isela Lombera
Uma 12h0'22" 46d1'42"
Laura J. Beard
Lep 6h4'35" -15d8'56"
Laura J. Beyer
Cas 23h13'1" 56d3'43"
Laura J. Durham
Uma 11h31'40" 38d14'16"
Laura J. Gorsuch-Smith"DOGGY STAR"
Cyg 19h43'4" 34d23'55"
Laura J. Patterson
Cmi 7h35'45" 6d41'11"
Laura J. Pinzon
Del 20h37'5" 12d34'53"
Laura J Smith
Cas 23h46'29" 59d14'11"
Laura J. Tarleton
And 1h32'7" 48d41'43"
Laura Jackson
And 23h15'58" 45d27'14"
Laura Jane
Sco 17h12'58" -43d42'46"
Laura Jane (Angel From Above) TLY
Uma 9h6'18" 54d39'49"
Laura Jane Clerehugh
Cas 23h39'36" 54d33'26"
Laura Jane McKiernan
And 0h7'13" 29d54'49"
Laura Jane Molloy
And 23h23'13" 46d56'45"
Laura Jane Scott
Cru 12h30'20" -59d52'7"
Laura Jane Strutt
And 23h39'34" 42d20'16"
Laura Jane Williams
Uma 9h57'10" 69d28'0"
Laura Janese
Cyg 19h25'24" 28d38'27"
Laura "Janie" Moore
Gem 7h11'40" 17d6'13"
Laura & Jared Baker - 04.03.2006
Cru 12h23'31" -58d16'21"
Laura Jayne
Cyg 21h40'33" 52d51'47"
Laura Jean
Cas 23h21'36" 59d54'8"
Laura Jean
Dra 16h50'0" 68d5'13"
Laura Jean
Gem 7h9'39" 27d5'55"
Laura Jean
Ari 2h7'18" 21d31'56"
Laura Jean Adhikari
Vir 14h52'53" 4d1'29"
Laura Jean Bankston Money
Umi 14h33'45" 68d30'17"
Laura Jean Elizabeth Satkowski
Cyg 20h14'47" 35d22'36"
Laura Jean Gala
And 1h1'42" 36d17'30"
Laura Jean Marie
Psc 1h8'10" 4d52'9"
Laura Jean Paine
And 2h3'28" 42d34'42"
Laura Jean Reidy
Tau 5h22'24" 23d37'51"
Laura Jean Shedden
Psc 23h37'27" -2d12'29"
Laura Jean Theresa
Aqr 22h9'56" -2d39'48"
Laura Jeane
And 1h6'17" 37d4'32"
Laura Jeanette Alvarez-Palomino
Gem 6h27'14" 26d51'55"
Laura Jean's Admirinufromafar
Ori 4h47'52" 12d46'45"
Laura Jensen
Ori 6h16'2" 15d26'52"
Laura Jensen
Crb 16h12'26" 33d53'30"
Laura Jindra van den Berg-Sekac
Lyn 7h25'7" 50d40'15"
Laura Jo Miller
Aqr 21h17'30" -1d42'1"
Laura Joanne
And 23h14'30" 43d51'55"
Laura & Joe
Lib 15h19'34" -5d10'45"
Laura & Joe Vahle
And 22h59'56" 47d19'31"
Laura & Joel Where the prayers met
Cyg 20h2'50" 48d54'46"
Laura Johanna Morrison DeGree
And 23h28'30" 45d9'26"
Laura & John John
Lyr 18h53'18" 32d1'44"
Laura Jones
Psc 0h11'52" 9d11'6"
Laura Josephine Mickens Edwards
Ari 2h8'9" 23d51'53"
Laura Joy Hale
And 2h23'26" 49d43'41"
Laura Jukes
And 1h26'53" 49d42'33"
Laura Julia Tapiero Fleischmann
And 0h5'53" 46d3'34"
Laura - Julio, Rocco, Joseph & Ray
Per 4h18'21" 40d45'54"
Laura June
Gem 6h43'26" 16d41'44"
Laura June Sawyers
Uma 11h28'29" 33d6'11"
Laura K. Abramovitz
And 23h12'5" 40d59'2"
Laura K. Grunas
And 0h11'33" 47d15'9"
Laura Kane
Gem 6h47'51" 22d19'51"
Laura & Karen Sisters Forever
Gem 7h22'12" 33d18'5"
Laura Kate
Del 20h39'4" 15d26'58"
Laura Kate Allen - 24th November 2004
And 23h21'51" 41d21'3"
Laura Katherine Simonyan
Gem 7h10'40" 22d14'55"
Laura Kathleen
Uma 9h45'18" 59d19'58"
Laura Kathleen Becker
Uma 11h4'4" 46d8'49"
Laura Kathleen McClung
Sco 16h9'35" -15d24'51"
Laura Kathleen Sallee
Leo 11h36'29" 24d48'52"
Laura Kay Payton
Lyn 8h33'32" 34d38'2"
LAURA KELLER
Cyg 20h55'23" 30d43'28"
Laura Kelly Harper
Uma 11h27'3" 44d1'14"
Laura Kessler
Aqr 22h28'32" -9d20'12"
Laura Keyser
Sco 16h16'3" -8d40'27"
Laura Kimberly Hoover
Tau 3h49'37" 4d53'28"
Laura Kirksey
Lyn 9h36'49" 39d36'12"
Laura Klemin
Cyg 20h3'26" 44d10'5"
Laura Kliebenstein
And 2h13'48" 50d10'43"
Laura Kniolek & Lance Hagerman
Cyg 20h21'59" 55d38'55"

Laura Knowlton - 2004
Lyr 18h16'6" 33d34'39"
Laura L. Klimecki
And 1h16'58" 41d59'4"
Laura L Richardson
Boo 15h24'29" 44d57'34"
Laura Lachok
Sco 17h56'56" -30d55'18"
Laura "La-La" Marie D'Arcy
Lib 15h48'21" -17d54'27"
Laura Lamson Humphreys
Uma 9h24'9" 67d51'57"
Laura Land
And 2h35'3" 45d26'31"
Laura Larissa Sophie
Uma 11h57'11" 37d31'32"
Laura Larkin Moscato
And 1h2'46" 45d24'17"
Laura "Laurabug" Basilio
Lib 14h27'49" -24d40'42"
Laura Lawson
Gem 7h47'7" 24d3'29"
Laura Lea Woods
Sgr 18h44'1" -17d57'58"
Laura Leanne
And 23h13'29" 42d55'25"
Laura Lee
Lyr 18h33'52" 38d57'35"
Laura Lee
And 23h20'42" 51d32'29"
Laura Lee
Psc 23h49'4" 5d52'51"
Laura Lee
Vir 12h59'28" 12d28'28"
Laura Lee
Sgr 18h31'57" -22d45'14"
Laura Lee Anagnos
Psc 1h14'0" 18d45'53"
Laura Lee Bazerjian
And 0h28'37" 46d29'7"
Laura Lee Byrd Olander
Cas 0h39'31" 51d32'23"
Laura Lee Crowley
Gem 7h39'44" 21d5'32"
Laura Lee Lynch
Peg 22h32'55" 10d22'5"
Laura Lee Patrick
Ori 5h52'24" 22d34'58"
Laura Leeann
Lyn 6h45'38" 59d28'0"
Laura Lee's Darkside
Ori 4h48'43" 11d29'2"
Laura Lefarth
Psc 1h11'45" 15d59'39"
Laura Leigh
And 23h1'14" 51d26'50"
Laura Leigh Bain
Sgr 19h43'25" -38d21'38"
Laura Leigh Carroll
And 1h22'41" 38d59'38"
Laura Leigh Hennesy
Uma 10h24'52" 66d32'24"
Laura Leigh Levins
Leo 9h28'49" 24d59'17"
Laura Leigh McGowen
Aqr 23h2'32" -9d4'5"
Laura Leone Special
Cam 3h57'45" 53d24'59"
Laura Leroux
Sgr 17h58'39" -16d55'57"
Laura Liliana
Lyr 19h12'13" 34d45'31"
Laura Lilise
Lmi 11h0'23" 25d4'58"
Laura Loi
Cyg 19h51'47" 38d33'44"
Laura Lomando
Cnc 9h12'7" 30d57'41"
Laura Lombardi
Cas 23h30'57" 53d8'11"
Laura Lonn Tuss
Leo 11h39'7" 20d50'1"
Laura Louise
Lyr 18h30'17" 28d2'29"
Laura Louise Brown
Cas 23h31'24" 57d42'27"
Laura Louise Collins
And 23h49'52" 35d1'59"
Laura Louise Nunn
And 1h0'14" 45d37'45"
Laura Love A Lot
Cyg 19h51'3" 33d42'23"
Laura Loves Gerrad
Tau 4h17'0" 7d24'42"
Laura Loyola
Lib 15h34'51" -19d32'26"
Laura Lu
Vir 13h1'17" 11d34'49"
Laura Lu Lu
Cas 1h37'34" 66d34'38"
Laura Luann Heaton
Aqr 22h18'26" -4d8'39"
Laura Lucero
Vir 13h29'55" 10d32'20"
Laura Lucille Browning Sevy
Psc 0h52'25" 3d22'34"
Laura Lundeen
Uma 10h0'55" 63d19'36"
Laura Lundell
Lib 15h21'49" -7d30'33"
Laura Lyn Evnin
Cap 20h50'54" -19d14'13"
Laura Lynn
Leo 11h28'19" 9d58'15"
Laura Lynn
Tau 3h57'57" 21d50'0"
Laura Lynn Alexson
Com 13h8'43" 28d32'8"
Laura Lynn Cooney
Dra 17h21'19" 51d52'41"
Laura Lynn Finlay
Tau 5h42'32" 27d57'22"
Laura Lynn High
Gem 6h38'30" 22d50'7"
Laura Lynn "Lulu" Susice-Rader
Ari 2h21'0" 18d32'55"
Laura Lynne
And 23h19'33" 42d51'38"
Laura Lysogorski
Ori 6h8'30" 6d4'45"
Laura M. Campbell
And 23h9'17" 40d7'23"
Laura M. Dunne
Sco 17h54'41" -41d34'53"
Laura M. Stoute
Aqr 22h18'0" -0d46'15"
Laura Madaline
Sgr 20h25'21" -42d10'5"
Laura Mae
And 23h8'46" 42d26'6"
Laura Maire Long
And 2h25'40" 48d5'16"
Laura Mandell Danforth
Crb 15h34'5" 38d28'34"
Laura Maness
And 23h6'31" 40d26'19"
Laura Maria
Lyn 8h1'48" 51d5'40"
Laura Maria
Sgr 18h37'49" -33d37'12"
Laura Maria Lauzon
Ari 2h34'33" 26d52'40"
Laura Maria Lin
Cas 1h22'11" 62d28'40"
Laura Maria Macedo
Uma 8h53'23" 47d30'39"
Laura Marie
Gem 7h31'43" 21d34'49"
Laura Marie
Uma 11h54'52" 57d51'41"
Laura Marie
Ori 5h21'25" -5d32'43"
Laura Marie
Cap 21h30'58" -10d20'34"
Laura Marie Begue
Ari 2h46'34" 18d24'13"
Laura Marie (Belanger) Garlough
Cas 23h25'25" 58d37'12"
Laura Marie DeSilva
Psc 1h31'55" 4d4'45"
Laura Marie Higgins
Lmi 10h50'47" 38d42'40"
Laura Marie Leitz
Gem 6h45'9" 32d13'34"
Laura Marie Piotrowicz
Uma 9h30'4" 71d6'57"
Laura Marie Poeta
Aqr 21h45'45" -0d25'38"
Laura Marie Poole
Psc 0h50'40" 7d45'7"
Laura Marie Stapler
Leo 11h15'18" 17d57'24"
Laura Marie Utter
Cap 21h43'57" -12d5'9"
Laura Marks
Gem 6h8'22" 26d40'51"
Laura Mary Svendsen
Psc 0h30'52" 8d44'23"
Laura Mascolini
Cyg 21h5'40" 47d57'12"
Laura Massey
And 23h59'18" 39d48'3"
Laura Massol petite chérie
Sgr 19h23'2" -20d16'28"
Laura May Barbour
Cas 1h20'0" 63d45'27"
Laura May Shorthouse
And 0h46'29" 31d26'26"
Laura Maynes My Angel
Cas 23h9'21" 55d27'38"
Laura McAuley
And 0h47'28" 25d34'57"
Laura McNeill
Lib 15h30'38" -16d30'19"
Laura Medina
Gem 7h33'18" 23d17'35"
Laura Mercer
Uma 13h25'46" 58d19'38"
Laura Mi Amor
Gem 6h30'48" 15d27'47"
Laura Michèle Myriam Clement
Ari 2h36'59" 30d18'11"
Laura Michelle Ginsburg
Del 20h20'52" 19d7'14"
Laura Michelle Melvin
Ori 4h53'41" 13d51'3"
Laura Michelle Paradise
Mon 6h29'49" -0d32'41"
Laura Michelle Wiederhoeft
Lyr 18h34'21" 32d7'27"
Laura Miller
Apu 13h54'4" -71d27'3"
Laura Mills
Psc 22h56'24" 0d1'16"
Laura Mineo
Sco 17h20'26" -38d56'55"
Laura Miressi
Ari 1h58'59" 14d30'59"
Laura Mitrow
Gem 7h43'49" 34d32'23"
Laura (Mommy) Wilson
Gem 7h11'31" 33d32'53"
Laura Montgomery (In memory of)
Psc 1h28'6" 17d16'33"
Laura Mooberry
Ari 3h1'11" 12d32'34"
Laura (Moondancer) DeLaurentis
Cas 1h41'28" 68d57'1"
Laura Morrell Pritchard
Lyn 8h16'6" 38d12'35"
Laura Müller Deutschland 2006
Leo 10h25'29" 23d8'1"
Laura Murphy
Ari 3h15'49" 20d18'22"
Laura Musselman
Vir 13h27'31" 10d50'53"
Laura "My Babydoll" Suminski
Gem 6h54'17" 14d15'4"
Laura "My Lovely Wife"
And 0h20'7" 33d13'51"
Laura - My Princess - 1000 Days
Umi 13h29'48" 75d52'21"
Laura my sweet
Cnc 8h16'21" 20d9'34"
Laura Nadia Trevor Lucho de Castro
Mon 6h49'20" -3d41'54"
Laura Nagelhout
Per 4h15'4" 51d13'25"
Laura Nardelli
Cap 21h34'8" -13d57'49"
Laura Nedved
Ant 9h37'25" -26d35'2"
Laura & Neil
Cyg 19h43'4" 28d28'52"
Laura Nell Rodenmeyer
Vir 13h28'41" 5d14'42"
Laura & Nibs
Cyg 21h40'36" 40d14'14"
Laura Nicole Ginga
Tau 3h37'44" 28d35'19"
Laura Nicole Trahan
Vir 14h5'37" -0d37'21"
Laura Nicole Warren, Star of Stars
Gem 6h41'58" 13d31'54"
Laura Ning
Leo 9h32'32" 27d49'19"
Laura Nolin
Mon 6h49'31" 7d51'39"
Laura Olive
Cas 23h4'14" 56d9'22"
Laura Olivia Gutierrez-Robinson
Vir 12h38'2" 3d54'12"
Laura Olivieri
Cam 3h59'54" 56d35'41"
Laura O'Malley - 18
And 23h37'7" 41d22'1"
Laura Oriani
Vir 13h15'49" 3d46'55"
Laura Osborne
Del 20h38'51" 15d20'33"
Laura Ottinger
And 0h19'10" 31d36'57"
Laura our light in the sky
Lib 15h7'49" -28d9'23"
Laura Ozuna
Com 13h5'23" 16d22'42"
Laura P.
Boo 15h24'56" 48d20'39"
Laura P. Heikes
Cap 20h18'33" -16d13'51"
Laura Padegs Zamurs
Psc 0h30'57" 14d13'58"
Laura Page
Leo 11h38'29" 20d31'40"
Laura Palmer
And 0h35'12" 36d40'5"
Laura Pearl
Sco 16h58'40" -38d50'19"
Laura Peky Perez-Palma
Cas 1h34'5" 64d5'17"
Laura Pene Tara
And 0h17'14" 44d40'57"
Laura Perry
And 23h39'30" 47d24'18"
Laura Pertl
Cnc 8h19'20" 9d26'22"
Laura Pesce
Lyr 18h34'46" 32d33'49"
Laura Pink
And 0h14'50" 48d22'12"
Laura Porvaznik
Lib 15h42'28" -29d11'8"
Laura Postor
Uma 9h37'10" 49d57'54"
Laura Powers
Sgr 19h11'46" -23d53'12"
Laura Precopio +grandchildren
Lib 15h10'1" -12d57'24"
Laura 'Princess' Pearce
And 1h30'34" 40d49'11"
Laura Prusko
Sgr 18h28'41" -16d29'33"
Laura Qualey
And 0h46'38" 37d15'5"
Laura Quasney
Sco 16h51'42" -38d25'3"
Laura Que Escribe
Aur 5h59'10" 42d47'40"
Laura Quincy
Lib 14h50'6" -1d9'6"
Laura Rackette
Lib 14h25'15" -9d32'3"
Laura Radtke
Aqr 21h43'23" -3d0'18"
Laura Rae Bodenstab
Aqr 22h23'27" -5d16'38"
Laura Rae Rosenthal ~ Oct 30, 1988
And 0h29'3" 46d31'22"
Laura Ramsundar
Mon 6h44'13" -0d3'34"
Laura re Romanis Tat
Ori 5h0'9" -0d53'1"
Laura Reese
Uma 8h57'31" 72d6'8"
Laura Renae Owens
Lib 15h25'57" -14d26'31"
Laura Renee
Leo 10h50'25" 18d55'6"
Laura Renee Stewart
Uma 8h53'20" 56d54'40"
Laura Richelle
Aqr 21h21'41" -7d56'26"
Laura Rios
Ori 5h42'8" 6d49'6"
Laura Robinson
Leo 10h46'18" 10d37'31"
Laura Rogers - Lucile Horstman
Lyr 18h47'55" 37d43'47"
Laura Rose
Lib 14h48'41" -3d59'15"
Laura Rose Clayton
Lyn 8h4'34" 37d8'3"
Laura Rose Davis
Uma 11h42'42" 60d59'25"
Laura Rossi
Umi 14h49'18" 69d19'1"
Laura Rouleau
Her 17h35'59" 44d42'21"
Laura Rowley
Ari 2h19'52" 23d24'44"
Laura Russell
Uma 8h40'43" 51d31'53"
Laura Rutkowski
Her 16h48'40" 37d12'51"
Laura Salvel
Leo 10h58'43" 5d45'4"
Laura Sánchez Berrocal
Col 6h20'38" -39d57'28"
Laura Sanders
Sco 17h35'56" -39d10'27"
Laura Savannah Crawford Carlson
Ari 2h2'9" 18d6'31"
Laura Scala
Cap 20h28'6" -19d21'42"
Laura Scassera
Uma 8h49'45" 64d24'47"
Laura & Scott Murray
Uma 11h21'50" 62d59'22"
Laura & Sean
Del 20h39'50" 4d51'24"
Laura Selinsky
Lyn 6h31'27" 56d38'44"
Laura Semper Bellus
Vir 12h59'19" 2d51'58"
Laura Serafini aeterna
Psc 1h20'46" 16d57'17"
Laura Sergi
Cnv 13h32'23" 30d46'49"
Laura Sermonti
Ori 5h53'10" 12d43'33"
Laura Serrano
Cam 4h0'19" 59d13'36"
Laura Sexauer
Cyg 20h13'15" 47d36'16"
Laura Shannon Foster
And 0h12'29" 28d49'32"
Laura Sharman
And 0h19'26" 25d1'36"
Laura Sharon Puett
Lib 15h3'33" -15d14'17"
Laura Shaw
Leo 10h23'21" 16d4'12"
Laura Shorty Chacon
Vir 14h0'39" -16d29'32"
Laura Simenson Vittengl
And 23h15'21" 42d37'12"
Laura SL
Ari 3h27'26" 27d0'9"
Laura "SmallFrie" Manfield
Gem 7h23'15" 20d7'10"
Laura Smith
Ori 6h7'27" 9d45'52"
Laura Snider
Uma 9h39'47" 55d20'8"
Laura Soler Alcántara
Mon 6h38'54" 5d40'20"
Laura Sourant
Tau 4h47'19" 26d38'55"
Laura: Star of Baku
And 2h36'52" 43d13'3"
Laura Starr Moore
Cam 3h47'35" 59d7'59"
Laura Steibel
Gem 8h2'40" 31d51'31"
Laura & Stephen
Cyg 20h16'3" 53d4'8"
Laura Strasser
And 2h31'54" 45d42'52"
Laura Strauß
Uma 11h44'14" 38d6'51"
Laura Strzalkowski
Uma 8h42'25" 52d54'13"
Laura Substalae Christy
Ari 2h36'11" 22d51'12"
Laura Sue
Cam 6h39'4" 69d48'3"
Laura Sue Lary
Crb 15h35'22" 28d6'50"
Laura Suzanne MacLeod
And 23h27'7" 49d57'36"
Laura Suzanne Reed
Vir 12h55'14" 6d18'10"
Laura T. Coviello
Cyg 21h53'22" 48d48'6"
Laura Tavella
Cam 6h41'46" 64d8'56"
Laura Temporin
And 2h13'21" 41d17'53"
Laura Tesch's Little Star
Tau 3h30'44" 12d56'14"
Laura Tesoro
Tau 5h33'20" 19d39'39"
Laura the Beautiful
Lyn 7h3'57" 61d14'26"
Laura Todaro
Sco 17h24'41" -41d31'55"
Laura Tom
Aqr 21h26'22" -8d15'58"
Laura Tonnison
Dra 19h29'50" 71d24'13"
Laura und Gernot
Aqr 21h24'9" 1d24'49"
Laura Uphold
Sco 16h55'49" -42d23'5"
Laura Uribe
And 23h52'2" 35d21'2"
Laura V - Forever My Star -
Cru 12h26'1" -60d25'13"
Laura Vanessa Soldner
Her 18h2'31" 17d56'4"
Laura Veglia
Cyg 19h36'29" 28d1'58"
Laura Vera
Sgr 19h17'21" -16d44'47"
Laura Veronica
Sco 16h13'18" -10d53'4"
Laura Victoria
Sco 16h48'30" -34d31'29"
Laura Vitola Pastina
Tau 3h40'10" 22d15'51"
Laura Vivian Santiago
Lib 15h9'39" -4d17'31"
Laura Vosbikian
Gem 6h50'10" 16d55'9"
Laura Waclawsky's Star
Sco 16h13'20" -11d42'55"
Laura Ward
Tau 5h45'24" 16d22'31"
Laura Warren
And 1h14'36" 43d2'58"
Laura Watson
Leo 9h25'28" 15d39'52"
Laura Wedge - Superstar
And 22h58'35" 52d42'51"
Laura White Swanson Star
Ari 2h42'33" 14d50'48"
Laura White, our Brilliant Mum!
Cyg 20h2'38" 31d13'50"
Laura Whitehouse
Uma 9h10'48" 56d44'59"
Laura Whitney Alden
And 2h25'58" 39d37'56"
Laura Wild - Laureatus Bellus Astrum
And 0h46'0" 26d2'41"
Laura Wilkinson
Sco 16h52'8" -37d31'2"
Laura Williams
Aqr 23h33'57" -19d34'28"
Laura Williams
Lib 16h1'6" -19d5'0"
Laura Wilson
And 1h21'25" 39d54'29"
Laura Withrow
And 1h36'2" 40d4'34"
Laura Wolf
Uma 9h14'40" 55d55'5"
Laura Woodbury
Leo 11h0'30" -1d25'4"
Laura - World's Greatest Mom
Ari 2h30'31" 25d4'9"
Laura Wren Tarlton Stiles
Sco 17h57'32" -37d1'45"
Laura Yagliyan
Ari 2h32'22" 18d28'36"
Laura Yakubov
Aqr 22h57'10" -10d34'53"
Laura Yam Bonnenfant 7/2/16-11/6/04
And 2h23'0" 49d16'31"
Laura Yvonne Rodriguez
Lib 15h18'58" -10d19'53"
Laura Zajdel
Leo 11h41'45" 22d1'8"
Laura Zubani
Cam 4h9'56" 56d45'58"
Laura Zwanziger
Ari 1h57'5" 15d44'54"
Laura, My Everlasting Love
Aql 19h4'37" -0d18'30"
Laura, Our Shining Star
Psc 0h39'13" 15d47'33"
Laura, Paul E.
Uma 10h47'45" 41d27'38"
Laura, this star speaks of our love
Cru 12h38'52" -59d42'58"
Laura121104
And 23h21'29" 43d12'4"
Laura-Anne Bell
Lyn 8h26'43" 42d6'34"
Laura-anne Jane Morse
And 0h17'45" 45d17'2"
Laurabee
Cnc 9h14'55" 12d25'55"
Lauraberth
Cas 0h23'47" 56d31'18"
Laurabeth Mixon
Sco 16h11'14" -18d40'0"
Laura-Calli
Gem 7h4'41" 16d42'50"
Laura-chérie
Per 2h56'36" 46d48'27"
Lauradawn
Ari 2h24'42" 23d50'9"
Lauraffy
Crb 15h58'2" 35d48'7"
Lauragina
Lyn 7h14'31" 52d22'1"
Lauraine
Ari 1h48'6" 15d20'49"
Lauraine Everson
Cas 2h38'57" 66d55'47"
LauraJames - Love Eternally Bright
Car 10h6'22" -62d44'46"
LauraJane
And 1h19'5" 47d38'22"
Laura-Jay
And 2h29'57" 43d1'5"
Laurakins
Lyr 19h9'46" 27d6'28"
Laural
Aqr 21h12'34" -3d44'42"
Laural Francis Zubal
Sgr 19h8'34" -31d15'52"
LauraLee
Psc 23h57'50" -0d6'55"
LauraLee
Lib 15h40'54" -17d47'1"
LauraLee
Tau 4h33'30" 16d42'1"
Lauralee Ann Mora
Uma 9h40'58" 42d37'3"
Lauralee Marie ti amo
Uma 10h17'48" 71d11'15"
Laural's Light
Gem 8h5'57" 28d7'11"
Lauralynn Madison Gravely
Lyn 7h45'38" 38d7'30"
LauraMatthewBartley
Crb 15h31'58" 36d4'57"
Laurana Helen Charron
Cyg 21h25'40" 35d55'49"
Laura'N'Craig
Boo 14h41'51" 47d32'44"
Laurang
Leo 11h31'41" 27d17'3"
Laurangel
Cyg 21h19'53" 52d30'40"
Laurann Fink
Lyr 18h39'14" 30d30'25"
Laura's Angel
Vir 13h22'23" -18d11'21"
Laura's Beam
Tau 4h16'1" 18d43'19"
Laura's Beauty Eternal
And 2h6'35" 39d34'53"
Laura's Blazing Love for Brendon
Uma 9h55'32" 49d7'24"
Laura's Light
And 23h24'51" 47d9'49"
Laura's Light
Crb 15h42'34" 26d19'23"
Laura's light
Aqr 22h36'8" -0d19'23"
Laura's Little Pink Star
Tau 3h50'58" 27d24'5"
Laura's Shining Rose
And 23h25'20" 51d18'37"
Laura's Shining Star
And 0h24'8" 25d18'35"
Laura's Shining Star
Cap 21h33'15" -19d43'28"
Laura's Shining Star of Serenity
Cas 23h22'56" 55d56'41"
Laura's Sparkling Star
Tau 4h33'55" 4d24'31"
Laura's special star
And 23h18'17" 52d37'45"
Laura's Star
Cas 23h39'11" 53d32'7"
Laura's Star
Cas 23h52'59" 57d26'50"
Laura's Star!
Cam 5h14'42" 61d15'7"
Laura's Star
Lib 14h52'40" -8d0'21"
Laura's Star
Cap 21h37'5" -8d40'11"
Laura's Star
Sco 16h51'7" -27d19'52"
Laura's Star
Mon 6h48'21" 7d39'0"
Laura's star
Gem 6h18'50" 26d23'26"
Laura's Star
Leo 11h20'53" 23d39'52"
Laura's Star
Lyr 19h11'57" 27d1'52"
Laura's Star
Cas 0h23'16" 54d16'46"
Laura's Star
And 23h0'45" 51d18'22"
Laura's Star
Uma 9h33'16" 48d30'22"
Laura's Star
And 23h17'29" 41d48'49"
Laura's Star
And 2h18'48" 42d19'8"
Laura's Star
And 0h12'43" 40d50'32"
Laura's Star
And 0h22'22" 41d51'45"
Laura's Star: Shining for Eternity
Uma 14h12'58" 58d44'34"
Laura's Stella
Uma 11h47'14" 59d55'34"
Lauras Stern
Uma 11h10'40" 67d25'8"
Lauras Stern
Uma 9h18'49" 72d36'52"
Laura's "Super" Star
Gem 6h59'20" 26d32'2"
Laura's Tango
Mon 6h54'48" -0d52'8"
Laura's Wish
Aql 19h14'46" -7d46'12"
Laura's Wish
Ari 2h24'45" 25d44'47"
LauRay
Cyg 20h41'26" 31d18'36"
Laurcelia71406
Lyn 8h41'38" 41d43'20"
Laure
Psc 1h24'19" 23d29'33"
Laure Bonin
Cmi 7h33'50" 2d45'13"
Laure Cabirol
Mon 8h8'20" -3d1'22"
Laure Jean-Daniel
Ari 2h19'23" 10d57'0"
Laure Marchal
Cnc 9h2'23" 14d51'6"
Laure Milleville
Del 20h52'43" 12d24'59"
Laure Traoré
Cas 0h5'28" 55d58'58"
Laure-Anne Luckystar Steverlinck
And 23h2'15" 42d51'56"
Laureate 12/24/46
Cap 20h26'0" -10d42'3"
Laureen
Cas 1h57'4" 63d56'6"
Laureen
Psc 1h2'13" 4d28'24"
Laureen Ballard
Cas 23h39'21" 57d50'13"
Laureen Bouchard
Cas 0h34'36" 63d47'18"
Laureen Hazel
Crb 15h47'44" 38d2'23"
Laureen Patricia Flynn Baccaro
Cap 20h50'51" -21d59'48"
Lauregis
Cam 7h20'45" 73d55'13"
Laureid
Cyg 21h44'25" 50d4'9"
Laurel
Lyn 8h7'12" 47d12'3"
Laurel
Uma 11h17'2" 42d4'52"
Laurel
Vir 13h18'19" 8d58'52"
Laurel
Tau 4h9'38" 14d24'53"
Laurel
Ari 3h28'37" 22d50'20"
Laurel
Boo 14h10'27" 29d23'32"
Laurel A. Latt
And 23h32'17" 42d53'13"
Laurel Ann
And 2h6'43" 43d57'36"
Laurel Ann Rogers
Gem 7h16'19" 19d7'36"
Laurel Ann Seneca
Leo 9h43'5" 17d33'26"
Laurel Anne Burrows
Gem 7h2'12" 28d56'59"
Laurel Betty Glew
Cru 12h5'23" -59d16'27"
Laurel "Blue Eyes" Shiver
Ori 5h32'8" 12d7'15"
Laurel (Boo)
Ori 6h19'13" 7d20'0"
Laurel & Chip
Cyg 20h0'58" 36d16'13"
Laurel Dauer's Star
Vir 12h10'51" 3d51'38"

Laurel Eileen Escoll
And 23h30'45" 45d29'59"
Laurel Glaze 10/05/80
Lib 14h59'15" -3d44'0"
Laurel June Fleming
Uma 11h40'7" 31d7'34"
Laurel Kim
Cnc 8h18'38" 14d19'20"
Laurel Kloth Myers
And 0h38'51" 32d46'58"
Laurel Lanae Milton
Cnc 8h41'1" 21d57'34"
Laurel Leaves
Psc 1h11'44" 7d13'44"
Laurel Marie Livingston
And 1h14'28" 35d55'37"
Laurel Priest
Vir 14h37'56" -6d29'50"
Laurel Reanier
Lmi 9h26'37" 36d6'5"
Laurel Richmond
Cas 0h15'34" 59d58'24"
Laurel & Rink In Love
Cyg 20h48'33" 47d48'58"
Laurel Vana
Ori 5h35'23" 9d53'19"
Laurel Wehler
Her 17h50'55" 23d1'19"
Laurelay
And 2h37'16" 50d12'57"
Laurelbug McCall Littler
Pho 1h4'38" -48d34'36"
Laurelea Rose
Ori 5h54'15" 20d59'34"
Laurelei
Lyn 7h6'44" 48d14'0"
LaurelEStewartInSilenceWalkInLight
Cep 22h47'39" 66d6'32"
Laureli
Pho 0h56'41" -57d47'40"
Laurelito
Lyn 8h32'2" 34d36'19"
Lauren
Gem 7h51'48" 32d35'33"
Lauren
Cyg 20h1'26" 31d55'57"
Lauren
And 1h49'43" 43d26'0"
Lauren
And 1h27'54" 43d28'38"
Lauren
And 2h27'33" 44d27'53"
Lauren
Per 3h18'54" 44d47'20"
Lauren
Lyn 8h12'43" 44d49'45"
Lauren
Cas 1h12'4" 49d24'37"
Lauren
And 1h7'36" 45d51'28"
Lauren
And 1h51'44" 45d28'36"
Lauren
And 23h32'7" 42d32'1"
Lauren
Cas 0h27'51" 52d22'39"
Lauren
And 23h25'42" 40d55'5"
Lauren
Crb 16h6'37" 38d39'11"
Lauren
Cyg 19h55'22" 43d5'52"
Lauren
Cam 4h35'18" 53d20'1"
Lauren
Gem 7h25'58" 16d31'11"
Lauren
Psc 1h20'10" 24d26'40"
Lauren
Ari 2h45'31" 28d39'55"
Lauren
Crb 16h1'32" 25d54'18"
Lauren
Leo 9h32'34" 27d7'16"
Lauren
Tau 3h32'50" 29d41'54"
Lauren
Tau 4h40'49" 26d29'24"
Lauren
Aqr 21h41'32" 1d58'10"
Lauren
Psc 0h10'4" 10d42'32"
Lauren
Cmi 7h41'31" 1d4'9"
Lauren
Mon 6h44'37" 0d54'18"
Lauren
Ori 6h0'57" 0d11'19"
Lauren
Cnc 8h50'59" 7d36'50"
Lauren
Cnc 8h17'12" 7d36'16"
Lauren
Psc 0h30'22" 16d11'54"
Lauren
Psc 1h0'49" 17d43'22"
Lauren
Cru 12h22'1" -56d4'4"
Lauren
Psc 0h33'25" 6d40'42"
Lauren
Sgr 18h23'42" -22d43'25"
Lauren
Dra 19h49'28" 69d43'5"
Lauren
Uma 11h38'33" 62d19'5"
Lauren
Cas 23h48'44" 57d1'26"
Lauren
Cas 23h19'30" 55d47'33"
Lauren
Uma 11h6'40" 58d51'14"
Lauren
Lyn 7h12'32" 58d16'49"
Lauren
Lib 14h52'39" -3d54'12"
Lauren
Lib 14h53'48" -6d44'3"
Lauren
Lib 15h55'9" -6d56'42"
Lauren
Lib 15h6'55" -2d46'30"
Lauren
Lib 15h0'40" -6d49'44"
Lauren
Aqr 22h21'15" -0d3'29"
Lauren
Lep 5h50'44" -11d16'6"
Lauren
Cap 20h54'14" -19d22'9"
Lauren
Vir 13h50'15" -17d29'9"
Lauren
Cap 20h49'30" -14d59'1"
Lauren
Sco 16h15'37" -9d6'42"
Lauren
Lib 15h22'27" -11d44'38"
Lauren "12-25-02"
And 2h30'9" 37d33'38"
Lauren 143
Vir 13h14'28" -5d35'59"
Lauren 434
Ori 6h21'15" 15d52'5"
Lauren A. Hepplewhite
And 23h11'56" 51d1'48"
Lauren A. Shoalmire
Cas 0h58'47" 55d34'24"
Lauren A. Warner
Uma 10h10'7" 49d9'6"
Lauren A. Weisbrodt
Lib 14h26'3" -19d16'8"
Lauren A. Williams
Sco 16h31'18" -25d54'58"
Lauren A. Williams
Tau 3h49'58" 22d50'50"
Lauren Adele Hosford
Gem 6h52'8" 14d50'5"
Lauren Adelle Colon
Vir 13h19'26" 8d43'37"
Lauren & AJ Forever and ever, babe
Cnc 7h58'31" 16d49'24"
Lauren aka Purple Haze
Cap 21h55'33" -11d57'31"
Lauren Alanna Reid
Cap 20h32'58" -20d31'56"
Lauren Albright
Uma 9h47'22" 62d44'18"
Lauren Aletha Mos
And 1h39'51" 41d6'29"
Lauren Alexa Bernstein
Ari 3h18'27" 29d42'7"
Lauren Alexa Farris
And 1h25'6" 43d32'35"
Lauren Alexandra
Cap 20h12'30" -8d51'6"
Lauren Alexandra Ford
Dra 18h50'37" 58d46'23"
Lauren Alexandra Read
Cap 21h40'5" -10d23'4"
Lauren Alexandra Schekeloff 8.12.80
Cru 12h38'57" -58d4'31"
Lauren Alexandra Valvo
Her 17h9'15" 18d7'53"
Lauren Alice Chambers
Cas 0h40'42" 53d16'52"
Lauren Alicia Holt
Uma 12h51'52" 57d58'20"
Lauren Alise Forkner
And 1h30'36" 45d12'26"
Lauren Alissa Watthuber
Cyg 21h31'38" 39d47'7"
Lauren Allen
Vir 13h4'4" -17d51'33"
Lauren Alyse's Light
Cru 12h32'47" -59d31'20"
Lauren Amber Lewakowski
Gem 7h48'17" 31d50'59"
Lauren Anastasia Stabinski
Psc 22h54'44" 7d15'50"
Lauren and Bob
Per 3h13'50" 31d22'20"
Lauren and Dan's (Laurdan's) Star
Cap 20h9'41" -22d16'32"
Lauren and Jonathan
Cnv 13h46'6" 32d29'8"
Lauren and Josh
Cyg 19h51'35" 31d27'25"
Lauren and Kevin - Only You
Vir 12h50'17" 3d30'58"
Lauren and Kylie's Star
Uma 11h27'7" 66d28'41"
Lauren and Matt's Star
Vir 14h16'31" 0d49'27"
Lauren and Michael's Star
Cnc 9h7'57" 28d56'7"
Lauren and Scott's
Cyg 20h24'36" 45d13'53"
Lauren Anderson
Crb 16h19'7" 36d6'20"
Lauren Anderson
Aqr 22h54'50" -16d53'44"
Lauren Angel
Ari 3h21'46" 29d0'58"
Lauren Ann
Gem 6h58'53" 18d46'9"
Lauren Ann
And 1h25'24" 39d33'29"
Lauren Ann Baker
Del 20h36'16" 15d9'44"
Lauren Ann Lester
Ari 3h15'52" 28d9'47"
Lauren Ann McConnell
Uma 9h21'27" 57d51'2"
Lauren Ann Miller
Del 20h27'14" 18d51'53"
Lauren Ann Richmond
Leo 11h10'37" 21d21'15"
Lauren Ann Sinacore
Sco 16h55'31" -44d54'8"
Lauren Ann Veronica Reich
Cas 0h19'50" 61d23'10"
Lauren Anna
Crb 16h9'48" 39d0'59"
Lauren Anne
Lyr 18h15'36" 32d1'59"
Lauren Anne Alcantar
Aqr 21h46'44" 0d13'55"
Lauren Anne Garber
Leo 11h47'57" 17d22'49"
Lauren Anne My Love
And 23h19'54" 47d29'48"
Lauren Ashlee
Cap 21h38'15" -13d46'16"
Lauren Ashlee Smith
Aqr 22h1'55" -14d44'9"
Lauren Ashley
Ori 5h40'52" 0d4'48"
Lauren Ashley Borick
Cas 23h43'48" 54d57'7"
Lauren Ashley Bowen
Gem 7h44'2" 34d53'11"
Lauren Ashley Fehl
Umi 14h16'12" 68d31'57"
Lauren Ashley Fleming
And 2h26'22" 50d30'8"
Lauren Ashley Halstead
Uma 10h32'51" 51d22'2"
Lauren Ashley Hefner
Cas 0h46'8" 64d29'54"
Lauren Ashley the magnificent
Psc 23h23'33" 7d41'54"
Lauren Ashley VanGorder
Boo 14h38'7" 35d49'41"
Lauren Ashley Wilson
And 0h18'5" 23d37'26"
Lauren Ashley Wood
Ori 5h28'14" 5d6'54"
Lauren Atkins
Crb 15h50'50" 26d29'55"
Lauren Atwood
And 0h38'52" 27d18'4"
Lauren Audrey Braman
Tau 4h25'13" 24d6'37"
Lauren Avery Mitchell
And 0h25'31" 30d38'58"
Lauren B
And 1h18'22" 40d41'5"
Lauren B. Pinto
Lep 5h6'43" -11d14'3"
Lauren B. Wyma
And 0h59'40" 42d49'53"
Lauren Baby
Cnc 9h4'37" 24d59'11"
Lauren Bailey
Vir 14h27'46" 4d34'5"
Lauren Bailey Cutshall
And 1h3'51" 42d20'44"
Lauren Baker
Leo 10h34'44" 13d15'18"
Lauren Barrett
Sgr 18h21'26" -31d36'5"
Lauren Bastardo
Lyn 7h8'52" 58d28'35"
Lauren Bean Hayward Brunt
Cas 2h22'50" 72d42'3"
Lauren Bell
And 1h3'52" 44d2'23"
Lauren Bene
And 1h53'35" 36d22'28"
Lauren Bennett Varner
Aqr 21h42'46" 0d15'18"
Lauren Beth Rawa
Cap 20h57'34" -22d10'4"
Lauren Bethany Doucette
Uma 12h45'37" 57d51'25"
Lauren Betty Mathieson
And 23h12'54" 46d49'57"
Lauren Blaire Tonniges
Vir 14h10'4" -5d46'55"
Lauren Bonenfant
Lyn 7h20'0" 53d51'54"
Lauren (Boo-Delicious) Haubach
Cnc 8h9'16" 14d47'3"
Lauren Brenna Holly
Leo 9h38'2" 29d51'4"
Lauren Brooke
Ari 2h2'6" 22d40'34"
Lauren Brooke Gallanthen
Umi 15h16'33" 84d27'27"
Lauren Brooke McGlohon
Tau 5h14'25" 18d39'53"
Lauren Brown
Cap 20h32'44" -14d19'13"
Lauren Burroughs Cook
Per 3h1'17" 54d32'50"
Lauren Butler
Uma 11h44'36" 43d46'12"
Lauren C. Iles
Cas 2h21'13" 73d36'24"
Lauren C. Maheady
Vir 12h19'16" 1d56'55"
Lauren Caldas
Psc 1h14'17" 21d48'14"
Lauren Capotorto
Sco 17h19'5" -39d40'39"
Lauren & Carl
Psc 1h38'20" 15d9'55"
Lauren Carlson
Vir 13h6'52" 4d53'34"
Lauren Carter
Uma 9h10'3" 59d5'58"
Lauren Castro
Cap 20h21'44" -11d13'58"
Lauren Catherine White
Cru 12h37'20" -56d46'19"
Lauren Cathrine Campbell
Peg 0h11'42" 16d34'23"
Lauren Cayla Wittlin
And 1h24'39" 47d23'43"
Lauren Charee Cutright
Uma 10h34'52" 48d35'40"
Lauren Cheung's Bit Of Heaven
Ari 2h21'54" 13d1'52"
Lauren Chilcote
Dra 15h27'36" 60d31'46"
Lauren Chloe "Arnaciiraar" Smith
Umi 13h28'44" 73d20'41"
Lauren Christie 7
Aqr 22h12'12" 1d35'32"
Lauren Christina-Marie Reyes
Dra 15h30'1" 56d26'22"
Lauren Christine Zimmerman
Sco 15h53'51" -20d48'7"
Lauren Cierra Demko
Lmi 10h21'14" 36d2'55"
Lauren Cipperly
And 23h8'57" 35d19'8"
Lauren Clem Davis
And 0h42'14" 30d10'36"
Lauren & Cleve - Forever One
Cru 12h38'20" -58d49'48"
Lauren Cohen
Lyn 8h21'57" 33d13'6"
Lauren Colleen Piccolini
Her 17h35'42" 33d11'38"
Lauren Collier
Crb 15h38'13" 39d20'52"
Lauren Conrad
And 2h34'43" 46d42'8"
Lauren Copeland Ragsdale
And 0h47'43" 33d52'2"
Lauren Corbett
Lyr 19h18'52" 28d47'7"
Lauren Cornelius
Uma 10h37'19" 43d35'16"
Lauren Coulston
Leo 11h43'18" 14d14'10"
Lauren Cristine Lahna
And 23h19'3" 49d38'38"
Lauren Czap
Cyg 19h27'50" 35d41'17"
Lauren Dahl Foley
Her 16h46'6" 30d46'57"
Lauren Dalecki
And 0h3'25" 37d55'43"
Lauren Daly
Psc 0h56'40" 2d46'58"
Lauren Danielle Barr
Uma 8h57'5" 49d17'17"
Lauren Danielle Blanchette
Umi 15h12'1" 70d2'44"
Lauren Danyelle Stewart
Psc 1h31'5" 15d57'10"
Lauren Daria Pollack
Leo 11h3'15" 20d48'41"
Lauren & David Mines
Cyg 21h30'41" 35d44'26"
Lauren Day
Cap 21h36'19" -24d41'41"
Lauren DeBritton
Per 3h50'8" 32d37'26"
Lauren Dee Meshel
And 0h54'51" 38d56'9"
Lauren DeJoy
Lib 15h59'54" -9d41'39"
Lauren Dewind
Aqr 23h16'49" -16d56'43"
Lauren Diane Irving Forever
And 0h31'50" 32d8'43"
Lauren Diane Mills
Cnc 8h8'14" 32d22'53"
Lauren Dianne
Gem 6h47'36" 17d4'24"
Lauren DiGregorio
Umi 16h18'36" 73d11'44"
Lauren Dinapoli
Umi 15h12'57" 68d50'34"
Lauren & Doug
Cap 20h57'23" -20d23'8"
Lauren Dulcinea Pawlika
Vir 11h49'12" -4d6'34"
Lauren Dyer
Lib 15h26'13" -7d31'18"
Lauren Dyer Nicholson
Psc 1h39'46" 3d42'16"
Lauren Dyson
And 1h22'59" 43d38'32"
Lauren E. Litscher
And 2h35'12" 43d34'31"
Lauren E. Muehlheuser
Ari 2h12'29" 11d9'10"
Lauren Eberhart
Lyn 8h14'16" 52d2'27"
Lauren & Eddie
Cyg 19h45'14" 35d12'59"
Lauren Eighmy
Cnc 8h48'18" 30d32'1"
Lauren Eighmy
And 23h12'54" 50d57'23"
Lauren Elaine
Sco 16h39'1" -38d29'4"
Lauren Elaine "The Dancing Star"
Psc 1h31'43" 22d17'29"
Lauren Elisabeth Anne Bosco
And 2h17'48" 46d20'3"
Lauren Elise Childress
And 1h44'12" 39d45'16"
Lauren Elise Willie
Umi 15h24'43" 72d2'3"
Lauren Elizabeth
Lib 15h35'20" -10d8'43"
Lauren Elizabeth
Sco 16h14'25" -8d36'50"
Lauren Elizabeth
Sco 16h14'27" -11d47'43"
Lauren Elizabeth
Psc 1h9'31" 7d29'33"
Lauren Elizabeth
And 23h58'50" 34d32'53"
Lauren Elizabeth
Lyr 18h28'37" 30d28'2"
Lauren Elizabeth
Tau 4h16'48" 1d21'47"
Lauren Elizabeth
And 0h34'22" 25d36'19"
Lauren Elizabeth
And 0h34'34" 29d17'17"
Lauren Elizabeth
Cnc 8h24'10" 24d3'39"
Lauren Elizabeth
Tau 3h47'36" 28d45'33"
Lauren Elizabeth
Tau 3h43'21" 26d43'56"
Lauren Elizabeth Bragg
Psc 1h27'45" 25d56'23"
Lauren Elizabeth Burianek
Lmi 9h57'0" 33d54'38"
Lauren Elizabeth Clemons
And 23h36'39" 49d13'59"
Lauren Elizabeth Collier
Cam 4h11'33" 68d8'58"
Lauren Elizabeth Correll
Cas 23h53'7" 57d58'17"
Lauren Elizabeth Cox
Dra 9h45'37" 79d17'40"
Lauren Elizabeth Crovato
Sco 17h18'7" -40d1'13"
Lauren Elizabeth DeShazo
Aqr 23h8'44" -8d55'40"
Lauren Elizabeth Estes
Cnc 8h47'11" 18d16'23"
Lauren Elizabeth Fernberg
Cnc 9h12'10" 15d20'9"
Lauren Elizabeth Foulkes
Mon 6h52'47" -0d54'55"
Lauren Elizabeth Green
Sgr 20h20'4" -36d20'59"
Lauren Elizabeth Harvey
Peg 21h28'36" 15d58'15"
Lauren Elizabeth Henrikson
Cam 5h13'57" 66d34'10"
Lauren Elizabeth Hubbard
Cas 1h50'5" 63d47'7"
Lauren Elizabeth Hughes
Umi 14h11'10" 77d20'5"
Lauren Elizabeth Irwin
Cap 20h20'23" -16d48'55"
Lauren Elizabeth Janis
And 1h44'10" 45d39'47"
Lauren Elizabeth Lampkin
Tau 3h49'49" 25d26'52"
Lauren Elizabeth Lobue
And 0h20'49" 32d34'27"
Lauren Elizabeth MacIntosh
And 1h24'15" 45d42'1"
Lauren Elizabeth Martin
Cam 13h18'28" 77d59'54"
Lauren Elizabeth Mathews
And 23h6'35" 51d40'10"
Lauren Elizabeth McCarthy
Uma 10h37'9" 61d14'51"
Lauren Elizabeth McDonnal
Lib 15h7'36" -6d6'45"
Lauren Elizabeth Miegel
Ari 2h32'56" 25d47'3"
Lauren Elizabeth Molz
Tau 4h37'24" 20d13'21"
Lauren Elizabeth Moore
Gem 6h46'51" 20d59'20"
Lauren Elizabeth - My Shining Star
And 1h15'15" 41d40'9"
Lauren Elizabeth Ottinger
Uma 10h53'3" 55d38'7"
Lauren Elizabeth Patricia Parsons
Vir 14h6'19" -13d41'19"
Lauren Elizabeth Poe
And 23h17'0" 44d0'42"
Lauren Elizabeth Pope
Cas 0h2'54" 59d58'52"
Lauren Elizabeth Rainwater
Tau 5h11'48" 18d28'2"
Lauren Elizabeth Riley (My Stella)
Lyn 7h55'9" 42d57'31"
Lauren Elizabeth Rollette
And 0h7'51" 47d42'37"
Lauren Elizabeth Sklena
Per 1h38'47" 54d32'29"
Lauren Elizabeth Sprengel
Per 3h6'3" 54d55'6"
Lauren Elizabeth Suty
And 0h6'40" 44d18'54"
Lauren Elizabeth Tinnin
And 1h34'7" 46d7'19"
Lauren Elizabeth Wells
And 23h7'40" 51d37'23"
Lauren Elizabeth Wheeler
Uma 11h3'32" 36d55'17"
Lauren Elizabeth White
Aqr 22h8'7" -2d8'22"
LAUREN ELLIS COLEMAN
Psc 0h41'1" 14d36'53"
Lauren Emiko
Ari 2h43'52" 12d43'34"
Lauren Emily Matoushek
Sgr 18h55'22" -32d55'22"
Lauren Enkerud
Sco 16h9'17" -18d18'5"
Lauren & Evan June 11 2006
Sge 19h42'7" 17d51'32"
Lauren Falk
Lyn 8h56'41" 39d52'8"
Lauren Farber
Cyg 20h19'21" 52d58'50"
Lauren Ferguson
Cap 21h24'52" -20d14'15"
Lauren "Fifle" Rayle
Leo 10h45'33" 14d36'52"
Lauren Fitting
And 1h10'38" 37d5'17"
Lauren " Flea" Hough
Tau 4h26'10" 12d30'38"
Lauren Ford
Vir 12h13'35" -0d57'14"
Lauren Foreman
Sgr 18h23'52" -32d37'50"
Lauren Fowler
Gem 7h38'59" 27d10'7"
Lauren Frances Glatstein
Del 20h55'38" 8d53'11"
Lauren Francesca Guccione
Lib 15h10'48" -5d20'30"
Lauren Frye
Cam 7h35'45" 63d35'26"
Lauren Fugate
Lyn 7h51'5" 53d17'59"
Lauren Gabrielle Knaffo
Leo 9h49'53" 31d40'42"
Lauren Gabrielle Williams
Lyr 18h35'29" 33d43'2"
Lauren Gail
And 23h23'38" 48d26'52"
Lauren Garner
Uma 8h26'25" 64d29'53"
Lauren Gasaway
Sco 16h4'56" -18d19'31"
Lauren Gerchow
Umi 15h35'38" 72d15'48"
Lauren Grace Croxford
And 23h19'29" 48d34'30"
Lauren Grace Edmondson
And 1h46'9" 41d15'48"
Lauren Grace Hight
And 2h22'36" 47d40'6"
Lauren Grace Pasquale
And 1h43'44" 39d18'12"
Lauren Grace Rasmussen
Lib 14h56'16" -2d4'15"
Lauren Grace Smith
Lib 15h43'3" -16d9'35"
Lauren Graham
Aur 5h41'38" 40d20'13"
Lauren & Grant Nicholls
Cru 12h41'12" -57d43'25"
Lauren Grillo
Vir 12h36'39" 1d53'13"
Lauren Haag
Leo 9h30'13" 11d41'24"
Lauren Harms
Leo 10h33'45" 25d32'36"
Lauren Hausmann
Tau 5h57'23" 23d43'15"
Lauren Hawley
And 0h30'19" 28d53'4"
Lauren Hayes Parry
Crb 16h18'56" 30d56'12"
Lauren Heflin- Angel from GOD
Cyg 19h37'48" 27d56'40"
Lauren Hennrich
Uma 8h43'56" 47d41'44"
Lauren Hess
Cas 1h30'14" 64d31'24"
Lauren Hill
Psc 2h4'45" 6d59'16"
Lauren Hoffman
Aqr 22h45'3" -0d56'56"
Lauren Holloway-Drown
Cyg 21h16'34" 44d43'10"
Lauren Honey Robison
Lib 15h49'18" -9d54'7"
Lauren Hope Bottom
Gem 7h24'27" 31d4'30"
Lauren Hope Stabiner
Cnc 8h55'44" 23d10'25"
Lauren Hough
Ari 2h1'12" 13d57'38"
Lauren Irene Snow
Sco 16h52'3" -38d51'14"
Lauren Ivy Hasenour
Tau 3h54'36" 26d0'56"
Lauren J. Grupper
Uma 11h59'13" 29d35'25"
Lauren Jacqueline DiSilvestro
Uma 13h39'2" 58d33'18"
Lauren Jamieson
Tau 3h36'8" 24d44'41"
Lauren Jane Holmes - Lauren's Star
And 0h1'34" 35d40'49"
Lauren Jane Innes
And 1h23'34" 43d50'40"
Lauren Jayne Fear
And 23h49'13" 45d58'29"
Lauren Jean
Gem 6h53'53" 26d32'5"
Lauren Jeane Bakos
Sco 16h15'36" -17d23'43"
Lauren Jenna Burke
And 23h45'12" 39d4'24"
Lauren & Jeremy
Cyg 20h5'3" 57d32'7"
Lauren Jessica Bender
Leo 9h37'33" 31d16'51"
Lauren Joanne
Cas 1h52'11" 64d12'0"
Lauren/Joel Connors-Ephesians 5:31
Psc 23h11'23" 5d58'16"
Lauren Johnson
Lyr 19h27'23" 42d48'15"
Lauren & Jonas Svoboda
Pup 7h33'58" -35d26'44"
Lauren Jordan
And 23h56'22" 43d16'23"
Lauren Joy
Aqr 23h39'14" -14d2'19"
Lauren Justina Sarofim
And 0h20'19" 35d50'5"
Lauren K. Grebb
And 0h22'5" 37d48'54"
Lauren K. Hunter
Cap 20h26'17" -23d7'10"
Lauren K Lafey
Gem 6h46'54" 18d55'9"
Lauren K. McCaully
Lyr 19h3'17" 40d13'50"
Lauren K. Trainor
Leo 10h51'30" 14d57'28"
Lauren Kane
Lib 15h50'57" -6d22'57"
Lauren Kathleen
Aqr 21h57'0" 0d22'49"
Lauren Kay
And 0h49'57" 36d2'25"
Lauren Kay Myers
Ori 5h58'11" 21d0'10"
Lauren Kayla Nolan
Aqr 20h42'4" -8d51'8"
Lauren Kelly Stokes
Gem 7h8'9" 34d2'29"
Lauren Kelsey Talbot
Cnc 8h27'53" 7d6'11"
Lauren Kendall Brown
Ori 6h5'26" 10d7'15"
Lauren Kenison
Cyg 20h41'12" 34d41'51"
Lauren & Kevin
Lyr 19h13'18" 26d22'3"
Lauren Khair
Sco 17h18'42" -38d54'54"
Lauren Kimberly Hammock
Cap 21h4'48" -27d26'3"
Lauren Kirchmeier
Psc 1h28'12" 27d24'38"
Lauren Kosecki
Umi 14h33'32" 75d45'27"
Lauren Krystyna Call
Uma 14h3'24" 57d44'6"
Lauren L. Vasquez
Aqr 22h46'0" -11d52'44"
Lauren LaGuerre
Aqr 22h27'10" -23d19'31"
Lauren Lane
Leo 11h49'56" 26d48'16"
Lauren LaPlante
Uma 11h48'58" 59d29'11"
Lauren Leah Reason - 25.12.2004
Col 5h49'15" -29d2'43"

Lauren Lee
Sgr 19h48'57" -27d21'13"
Lauren Lee
Sgr 19h9'16" -17d9'14"
Lauren Lee
Ori 6h8'29" 11d36'55"
Lauren Lee
Cyg 20h18'52" 39d24'27"
Lauren Lee Clifford
Cas 23h23'38" 58d44'13"
Lauren Lee Jimolka
Ori 5h44'53" 6d39'19"
Lauren Leigh Ouellette
Vir 14h9'53" -5d42'38"
Lauren Leviton
And 1h38'56" 44d18'32"
Lauren Lightcap
Lib 15h14'31" -22d53'39"
Lauren Lisa Schunemann
And 23h12'11" 52d2'2"
Lauren Lochetto
Psc 0h27'29" 10d28'3"
Lauren 'LolaStar' Reiser
Lyn 7h34'53" 35d37'43"
Lauren Louise
Gem 6h26'37" 25d45'56"
Lauren Louise Castaldi
Tau 4h52'31" 27d1'19"
Lauren Loves Chris
Sco 16h17'57" -8d43'53"
Lauren loves Kevin
Aqr 21h2'17" -11d35'59"
Lauren Lowry
Uma 11h11'6" 38d11'23"
Lauren Lucko
Cam 6h59'36" 65d26'51"
Lauren Lyla
Lyr 18h43'4" 33d45'31"
Lauren Lynch McCullagh
Leo 11h25'44" 18d7'8"
Lauren Lynn Lewis
Gem 7h36'34" 26d26'42"
Lauren M. Grigolon
Cas 1h43'0" 68d4'28"
Lauren M Heinonen
Cyg 19h36'2" 29d43'27"
Lauren M. Wallace
And 1h9'58" 41d58'51"
Lauren Machin
Uma 9h29'26" 47d40'25"
Lauren Mackenzie Cox
Lib 14h57'33" -2d12'45"
Lauren Mackenzie Gill
Cnc 8h49'1" 29d47'52"
Lauren Madison
Aqr 22h59'49" -7d54'37"
Lauren Maree Dorsa
Col 5h8'14" -33d38'3"
Lauren Margaret Wisdom
Lyn 6h50'33" 57d46'11"
Lauren Marie
Vir 13h3'54" -19d26'52"
Lauren Marie
Ari 3h12'4" 28d56'22"
Lauren Marie
Vul 19h54'9" 22d47'49"
Lauren Marie
Leo 11h14'22" 16d30'19"
Lauren Marie
Leo 10h7'14" 18d51'52"
Lauren Marie
Mon 6h42'13" 7d20'39"
Lauren Marie
And 23h50'49" 36d40'25"
Lauren Marie
Cnc 8h19'54" 30d28'32"
Lauren Marie
Lyn 8h41'5" 35d1'34"
Lauren Marie Anastasia Palenski
And 23h11'11" 47d25'32"
Lauren Marie Festa
Sco 16h54'32" -38d37'26"
Lauren Marie Floyd
Tau 4h50'13" 20d40'34"
Lauren Marie Glagavs
Sgr 19h0'16" -11d59'30"
Lauren Marie Hobson
And 0h43'50" 46d36'33"
Lauren Marie Kosovec
And 1h58'13" 37d44'25"
Lauren Marie Kostak
Leo 10h54'58" 10d8'7"
Lauren Marie Pendergast
Ari 2h25'43" 25d47'5"
Lauren Marie Peterson
Mon 6h49'14" -4d26'2"
Lauren Marie Pinske
And 1h37'33" 40d42'41"
Lauren Marie Shallow
Vir 12h46'56" 11d32'20"
Lauren Marie Vaknine
Psc 1h42'19" 14d53'48"
Lauren Marie Vetere
Cyg 21h27'5" 51d53'5"
Lauren Marie Zile
Uma 8h38'46" 69d38'39"
Lauren Marie Zwick
And 1h22'2" 40d1'39"
Lauren Marie's Star
Boo 14h42'0" 34d46'52"
Lauren Marissa DiGiovanni
Del 20h34'59" 19d29'11"
Lauren Marshall
Sgr 19h3'20" -20d15'57"
Lauren Mary
Psc 1h15'32" 31d3'14"
Lauren Mary Licciardi
And 23h13'40" 47d21'23"
Lauren Mary-Elizabeth Brown
Sgr 18h59'16" -26d37'11"
Lauren Mason - Sweet 16
And 23h15'59" 35d58'18"
Lauren Mathews
Cam 3h29'59" 54d13'45"
Lauren Maxwell
Umi 15h21'25" 75d30'55"
Lauren Mayfield
Mon 7h8'44" -0d59'44"
Lauren McArdie
And 2h4'51" 41d20'19"
Lauren Mcguire
Cnc 8h39'32" 17d17'1"
Lauren Mcwalter
And 1h35'59" 49d57'11"
Lauren MeChele
Lib 14h53'55" -3d50'6"
Lauren Megan Beasley
And 1h58'25" 37d44'55"
Lauren Megan Stryjewski
Cyg 19h38'26" 31d52'35"
Lauren Meier
Cap 20h20'21" -10d54'41"
Lauren Melissa Kortsha
And 0h42'19" 36d52'31"
Lauren Meyer
Cnc 8h28'52" 18d39'29"
Lauren Michael Byrd
And 1h57'24" 41d47'30"
Lauren Michele
Peg 21h29'48" 14d26'40"
Lauren Michele Stephens
Lyn 7h56'0" 54d41'10"
Lauren Michele, The Star
Lib 15h27'56" -9d25'44"
Lauren Michelle
Lib 14h50'54" -1d27'6"
Lauren Michelle
Oph 16h24'46" -0d3'27"
Lauren Michelle
Tau 5h17'53" 19d13'13"
Lauren Michelle
Leo 10h35'30" 14d50'25"
Lauren Michelle
Ori 6h12'13" 2d28'43"
Lauren Michelle Anderson
And 1h56'50" 46d35'11"
Lauren Michelle Costello
Sgr 18h56'51" -17d35'53"
Lauren Michelle Edwards
Cyg 20h59'4" 37d44'15"
Lauren Michelle Flannery
And 0h12'33" 46d27'21"
Lauren Michelle Harkins
Ori 5h36'13" -1d44'7"
Lauren Michelle Kelley
Cyg 19h39'20" 33d50'59"
Lauren Michelle Popkey
Sco 16h12'32" -11d30'19"
Lauren Michelle Potter
Sgr 19h48'9" -28d58'21"
Lauren Michelle Taveira
Leo 10h52'10" -2d24'36"
Lauren Michelle Yoggy
Psc 0h39'40" 8d56'33"
Lauren & Mike
Cnc 8h29'46" 24d30'47"
Lauren Milana
Cas 1h12'37" 54d35'57"
Lauren M.M. Hypes
And 0h17'11" 27d17'48"
Lauren Moseng
Com 12h17'38" 31d20'11"
Lauren Mudd
Lyn 7h32'35" 48d22'16"
Lauren Muehlheuser
And 2h12'4" 46d46'20"
Lauren Muehlheuser
Uma 10h51'31" 44d49'36"
Lauren Murdock
Aql 19h48'31" 14d52'58"
Lauren Museus
Cas 23h33'53" 58d20'5"
Lauren Mustapick
Gem 6h46'24" 16d18'40"
Lauren My Pandora
And 1h13'37" 46d47'8"
Lauren Mynhier
Ari 2h12'23" 25d34'36"
Lauren N. McGary
Tau 3h38'34" 29d42'8"
Lauren Nance
Vir 12h4'3" -7d35'23"
Lauren Nancy
Vir 12h20'3" 11d12'52"
Lauren Naomi Kajiura
Ari 3h27'25" 28d2'30"
Lauren NB7
Sgr 17h59'37" -22d47'6"
Lauren Necciai
Aqr 20h49'15" -8d44'59"
Lauren Nichole McLeod
Leo 10h26'19" 13d53'55"
Lauren Nicole
Gem 6h53'35" 12d37'4"
Lauren Nicole
Gem 6h17'33" 25d10'34"
Lauren Nicole
Cas 0h59'13" 62d13'50"
Lauren Nicole
Uma 8h42'3" 53d14'11"
LAUREN NICOLE BROWN
And 0h50'32" 40d34'14"
Lauren Nicole Cincurak
Sco 16h5'0" -10d57'31"
Lauren Nicole Cochrane
Ari 1h53'7" 18d45'2"
Lauren Nicole Crouch
Cas 0h52'17" 53d18'54"
Lauren Nicole Dengle
Peg 21h54'15" 34d7'47"
Lauren Nicole DiMaria
Lyn 9h0'19" 34d23'52"
Lauren Nicole Edmonds
Dra 17h50'27" 52d9'29"
Lauren Nicole Hudson
Lyn 8h49'23" 35d17'25"
Lauren Nicole Jacoway
Lyn 7h32'35" 57d29'31"
Lauren Nicole Puckett
Psc 1h11'11" 10d0'13"
Lauren Nicole Stander
Cas 0h26'55" 54d59'53"
Lauren Nicole Thompson
And 0h4'24" 47d50'48"
Lauren Noel
Cap 21h34'39" -10d23'23"
Lauren Noelle Littlefield
Cap 20h12'56" -10d2'8"
Lauren Noelle Yukiko Regan
Cmi 7h39'53" -0d3'27"
Lauren Norris
Sco 17h31'1" -38d30'35"
Lauren Nuckolls
Cnc 8h18'16" 18d30'19"
Lauren O
Uma 10h34'20" 43d40'6"
Lauren O'Keeffe's Star
Uma 13h56'58" 54d16'47"
Lauren Olivia Lemoi
And 1h42'40" 38d53'24"
Lauren P Candler
Psc 1h25'44" 32d21'25"
Lauren Paige Barber
And 0h24'12" 32d10'46"
Lauren Paige Bradley
Tau 3h45'1" 17d43'19"
Lauren Paige Roland
Vir 14h15'4" 5d16'48"
Lauren Paige Williams
Aqr 22h35'31" -3d1'17"
Lauren Patterson
Crb 15h30'12" 28d51'8"
Lauren Patty
And 0h32'5" 28d56'30"
Lauren "Peanut" L Scavitto 10/22/03
Uma 13h50'36" 56d43'6"
Lauren Peiffer's Lilting Laughter
Lmi 10h7'6" 34d12'1"
Lauren Penner
Vir 12h47'5" -6d20'55"
Lauren & Pete Jackson - Stars 2005
Her 17h52'50" 22d44'34"
Lauren Phillips
Cap 21h18'20" -15d54'2"
Lauren Polinsky
Ori 6h4'2" 14d11'7"
Lauren Pratt
Tau 5h23'42" 25d31'56"
Lauren Rachel Fink
Col 5h46'59" -33d7'44"
Lauren Rachel Marquardt
And 1h43'3" 41d32'18"
Lauren Rachel Will
Ori 5h51'10" 6d59'23"
Lauren Radicchi
Ari 2h15'17" 24d23'12"
Lauren Rae
And 23h53'0" 47d14'19"
Lauren Rae
Cas 0h10'52" 53d14'41"
Lauren Rae Harvill
Her 16h24'57" 7d7'13"
Lauren Rautanen
Uma 11h36'55" 64d14'53"
Lauren RBD Brennan
Psc 1h9'48" 31d37'51"
Lauren Reed
Cam 5h11'7" 68d36'54"
Lauren Rees
Ari 2h13'34" 26d3'7"
Lauren Reese
Cas 0h11'42" 50d1'21"
Lauren Regina Kreitzman
Ari 2h45'21" 18d59'57"
Lauren & Reid's Eternal Light
And 2h37'29" 39d25'56"
Lauren Renee
Uma 10h42'38" 48d51'29"
Lauren Renee' Biddy
Lyn 6h45'41" 57d22'50"
Lauren Renee Giambrone
Psc 23h15'36" 1d7'25"
Lauren Revis
Sco 16h9'36" -15d44'16"
Lauren Riho
Sco 16h53'13" -38d9'7"
Lauren Rivers 2/14/03
Lyr 18h35'51" 36d42'22"
Lauren Roberts
Cam 3h28'30" 66d48'13"
Lauren Rodgers and Justin Nekoloff
Cyg 20h54'45" 47d37'37"
Lauren Rose Desiderio
Aqr 23h3'45" -12d58'38"
Lauren Rose Mesnekoff
And 23h47'56" 46d12'4"
Lauren Rose Polomsky
Cap 20h57'7" -25d22'54"
Lauren Rose Woods
Uma 10h51'37" 54d48'0"
Lauren Routhier
Sco 17h41'59" -42d57'44"
Lauren Rudman
Cas 1h19'16" 51d50'52"
Lauren Rzoski
Cap 21h33'48" -14d29'42"
Lauren Santin - Beautiful Princess
Cru 12h56'39" -58d38'12"
Lauren Scalise
Lib 15h32'10" -6d38'17"
Lauren Schmitt
Aqr 21h22'58" -0d46'8"
Lauren Schulze
Ari 1h53'19" 18d22'6"
Lauren Shampine's Star
Ari 3h26'50" 23d17'21"
Lauren Sienna
And 23h3'13" 40d41'1"
Lauren Sierra Hildebrandt 3/11/03
Psc 1h27'59" 5d34'29"
Lauren Skylar + Matthew Tristan
Aqr 22h2'32" -16d37'30"
Lauren Smith's Star
Crb 16h16'1" 32d21'7"
Lauren "Smoren" Callum
Lib 14h39'45" -24d30'7"
Lauren "Snuggle Bunny" Franco
Lyr 19h6'7" 28d42'11"
Lauren Sommerfield
Cap 21h26'17" -17d10'15"
Lauren Spada
Leo 11h19'22" -0d14'5"
Lauren Speers
Umi 16h55'34" 79d18'32"
Lauren Spencer
And 0h43'45" 30d8'46"
Lauren Spicher
Psc 1h5'10" 3d13'54"
Lauren Star
Lib 14h26'8" -20d3'39"
Lauren Stein
And 0h46'59" 32d25'53"
Lauren Stephanie
Uma 12h2'11" 28d43'55"
Lauren Stockli
Aqr 22h11'26" 1d55'14"
Lauren Stretton's Star
Sge 20h9'14" 17d18'15"
Lauren Suzanna Ina Carroll
And 1h54'40" 40d54'15"
Lauren T. 2/18/81
And 23h14'27" 43d31'24"
Lauren Taylor
Aql 20h10'21" 8d10'34"
Lauren Taylor
Lmi 10h49'53" 27d27'51"
Lauren Taylor Smith
Cnc 8h14'33" 15d42'48"
Lauren Tesoroni
Her 17h36'22" 15d32'54"
Lauren T.H. Wasson
Cap 21h35'58" -17d44'47"
Lauren Theobald
Ari 1h58'4" 19d30'57"
Lauren Therese Baumann
Leo 10h29'39" 18d36'32"
Lauren Thie Grace
And 1h13'30" 45d24'47"
Lauren Thompson
Psc 1h21'57" 7d56'27"
Lauren Tiffany's Birthday Star
Sgr 18h7'42" -25d6'14"
Lauren Tuvell
Pho 1h36'10" -43d1'46"
Lauren & Val Moro Together Forever
Lyn 7h19'27" 54d8'21"
Lauren Valerie Wilson
Gem 7h49'7" 32d18'54"
Lauren Venustus
Aqr 23h47'52" -17d23'47"
Lauren Victoria Kensington
And 2h2'11" 39d33'13"
Lauren Walker
Per 2h23'38" 56d10'24"
Lauren Walker
Sgr 19h19'6" -21d16'22"
Lauren Walker
Mon 7h0'10" -6d8'19"
Lauren Williamson
And 23h12'10" 39d53'13"
Lauren Woodrick
And 0h19'25" 37d7'53"
Lauren Wynters
Gem 7h46'17" 33d18'57"
Lauren Zibens
Sco 17h45'34" -40d33'45"
Lauren, beauty beyond this world
Cyg 21h57'27" 48d34'2"
Lauren, Brightest Star in the Sky
Umi 14h34'53" 72d25'31"
Lauren, Jeffery Colan & Family
Uma 11h47'14" 53d7'50"
Lauren, My Beautiful Baby
And 23h2'29" 52d7'47"
Lauren, My Special Love
Vir 13h34'12" 3d28'2"
Laurenalis
Ori 4h57'44" 10d54'9"
Laurenaria 1972
Lyn 7h43'2" 41d29'52"
Laurence
Uma 11h39'59" 41d33'46"
Laurence
And 0h5'40" 45d5'19"
Laurence
Psc 0h22'59" 12d19'59"
Laurence
Psc 0h15'40" 18d56'21"
Laurence
Umi 14h50'23" 76d9'24"
Laurence A. Woolfson
Sco 16h42'1" -31d40'5"
Laurence Carmé
And 2h30'36" 45d36'32"
Laurence Courbin
Cep 21h37'10" 55d56'12"
Laurence D. la passionnée
Psa 22h11'7" -32d34'32"
Laurence De Ridder
Sco 17h53'1" -34d39'58"
Laurence Gainsborough
Per 3h21'32" 38d53'10"
Laurence H. Mass
Leo 10h29'29" 12d45'58"
Laurence Henriet 20121962
Sgr 18h51'55" -15d54'21"
Laurence J. Silvi, Sr
Uma 11h4'1" 35d16'29"
Laurence & June Orne - 9/24/44
Uma 9h43'59" 52d27'34"
Laurence la Chanteuse
Cep 21h43'0" 63d45'52"
Laurence M Gillen
Leo 9h46'5" 27d44'16"
Laurence Marie 29-09-60
Lib 15h1'42" -24d45'13"
Laurence Nouvel-Puillandre
Sgr 18h13'56" -19d19'22"
Laurence 'Papillon' Duhamel
Cep 23h19'14" 75d7'21"
Laurence Peugeot
Sgr 18h18'27" -27d53'19"
Laurence Ralph LoPresti
Aqr 20h56'37" -12d47'10"
Laurence Samuel Field
Cru 12h42'13" -58d5'36"
Laurence "Teddy" Bohannan's Star
Uma 9h44'32" 50d39'17"
Laurence Villarin
Her 18h32'20" 19d2'49"
Laurence Wilcox
Uma 11h39'21" 53d14'45"
Laurence Wilson Sharpe
Leo 11h37'16" 16d52'38"
Laurence-Lyne Petit Coeur
Sco 16h11'19" -23d43'6"
Laurene Puzo
And 0h48'42" 37d54'50"
Laurenelle
And 23h0'57" 50d37'32"
Laurenius
Sco 16h17'53" -17d35'55"
Lauren-Kailee-Cameron-Martinez
Uma 11h19'13" 68d13'45"
LaurenRoliglovesherfamily &friends7
Sge 20h6'26" 16d43'36"
Lauren's Destinie
Aqr 22h10'6" 0d4'6"
Lauren's Dream
Lib 15h29'1" -27d32'19"
Lauren's Dream Come True
And 0h45'45" 37d7'18"
Lauren's Drop Of Jupiter
Aqr 22h30'10" -1d50'23"
Lauren's Light
Sgr 18h2'26" -18d41'51"
Lauren's Light
Cru 12h27'6" -63d14'2"
Lauren's Little Star
And 2h11'45" 46d52'20"
Lauren's Lullaby
And 2h37'38" 46d11'36"
"Lauren's Lunar Light"
And 23h48'37" 35d8'29"
Laurens' Smile
Crb 15h41'37" 25d54'37"
Lauren's Special Sparkle
Pho 2h10'39" -42d1'48"
Lauren's Star
Lib 15h9'53" -6d37'16"
Lauren's Star
Lyn 7h37'48" 54d19'25"
Lauren's Star
Psc 1h25'27" 27d56'47"
Lauren's Star
Gem 7h11'2" 21d58'7"
Lauren's Star
Ori 5h49'36" 2d31'34"
Lauren's Star
Leo 11h22'1" 10d6'46"
Lauren's Star
Leo 11h26'1" 11d30'41"
Lauren's Star
Cnc 8h21'39" 12d44'7"
Lauren's Star
Aur 7h3'1" 39d9'39"
Lauren's Star
Lmi 10h43'30" 31d38'8"
Lauren's Star: Always Shining
Sco 16h43'16" -33d30'35"
Lauren's Star LKD
Aqr 20h53'52" -8d34'54"
Lauren's Starlight
Uma 10h39'4" 41d8'2"
Lauren's Wishing Star
Uma 14h1'30" 53d27'25"
Lauren's Yellow Roses
Ari 2h11'30" 25d16'5"
Laurent
Vul 20h25'1" 26d32'6"
Laurent
Umi 14h22'26" 71d15'25"
Laurent
Cru 12h1'6" -62d3'20"
Laurent Agache
Lib 15h34'56" -27d26'45"
Laurent Agache
Lib 15h34'50" -16d17'5"
Laurent Badonnel
Ori 5h7'14" 5d47'30"
Laurent Bel 26/03/1979
Ari 3h3'49" 25d59'14"
Laurent Bourgnon
Ari 2h34'21" 27d27'13"
Laurent Brunetti
Leo 9h55'52" 23d10'31"
Laurent Chandler Rose
Gem 6h41'46" 17d59'53"
Laurent Cliet
Umi 16h9'18" 75d28'48"
Laurent Colomer
Gem 6h23'18" 19d53'7"
Laurent Davenas
Leo 10h29'47" 21d59'35"
Laurent et Sandrine
Leo 11h12'30" -1d33'13"
Laurent Gras
Cas 1h21'25" 55d44'18"
Laurent Krakowski
Vir 13h32'24" -18d10'15"
Laurent Plumat
Del 20h39'20" 10d33'53"
Laurent Schaiblé
Gem 6h48'42" 16d43'10"
Laurent, lumière et feu
Uma 8h41'17" 71d36'53"
Lauretta DiBenedetto-Minnick
Gem 7h29'46" 20d12'27"
Lauretta Margarite G. Gast
Cas 0h48'10" 64d55'37"
Laurette
Her 17h50'17" 29d50'17"
Laurette
Tau 4h26'47" 21d9'59"
Laurette Deschamps
Uma 10h51'1" 60d35'19"
Laurette Gauthier
Lyr 18h50'16" 43d41'48"
Laurette & Ray
Cyg 19h44'31" 28d14'35"
Laurhidil
Boo 14h41'58" 23d47'59"
Lauri
And 0h52'38" 40d30'36"
Lauri Jeantheau Lambeth
And 1h8'33" 44d18'38"
Lauri L. Terrell - My Buttercup
Vir 13h21'22" -12d31'47"
Lauri Sipila
Pho 0h34'58" -45d5'53"
Lauriana Marie Portoles
Vir 12h40'29" -6d34'57"
Laurice Morgan
Uma 10h26'57" 49d0'28"
Laurie
And 23h48'22" 47d29'59"
Laurie
Lyr 19h0'1" 42d25'13"
Laurie
And 1h27'5" 48d25'3"
Laurie
And 23h4'40" 37d22'21"
Laurie
Cnc 8h51'53" 31d34'48"
Laurie
Lyn 8h3'9" 36d43'22"
Laurie
Gem 6h7'25" 24d19'46"
Laurie
Mon 6h59'2" 8d59'54"
Laurie
Cmi 8h0'26" -0d12'49"
Laurie
Psc 0h49'21" 6d25'10"
Laurie
Sgr 19h25'0" -27d56'21"
Laurie A. Apple
Tau 3h41'1" 4d59'8"
Laurie A. Badea
Cas 1h13'18" 49d59'38"
Laurie A McNeill
Mon 8h2'35" -0d25'11"
Laurie A.Barbour
Aqr 21h55'34" -8d11'26"
Laurie Ann
Psc 0h7'35" 1d13'58"
Laurie Ann (Butterfield) Lecrone
Cap 21h56'9" -17d31'14"
Laurie Ann Green
Ari 3h22'56" 29d23'0"
Laurie Ann Kania
Leo 11h16'43" -1d10'30"
Laurie Ann Lathouse
And 0h43'26" 32d44'6"
Laurie Anne
Crb 16h3'2" 39d2'30"
Laurie Anne
Vir 14h18'14" -1d6'26"
Laurie Anne Bruton
Cas 1h35'24" 71d40'36"
Laurie Anne Harrison
Cnc 8h17'33" 10d48'49"
Laurie Anne Pero
Cam 5h12'59" 68d1'38"
Laurie Axworthy
Uma 11h55'14" 33d45'26"
Laurie B. Anderson
Uma 11h57'51" 35d51'9"
Laurie Baker
And 1h4'31" 41d7'49"
Laurie "Ballz" Dunn
Sgr 19h44'4" -19d5'50"
Laurie Belle
Leo 10h45'15" 14d2'44"
Laurie Black
Cyg 19h32'7" 28d35'50"
Laurie Brevick
Vir 13h12'44" 3d32'29"
Laurie Brockie Ackerman Rosa
Cas 1h42'10" 64d24'1"
Laurie Caple
Peg 22h20'49" 4d7'3"
Laurie Chabra
Her 16h44'58" 48d35'50"
Laurie Christine Collier
Uma 12h25'26" 59d47'34"
Laurie Cruddas
Gem 6h55'4" 22d21'6"
Laurie Cummins
And 2h30'14" 43d44'42"
Laurie Davis
Cas 0h54'54" 62d9'0"
Laurie Davis
Sco 17h51'36" -36d26'13"
Laurie DiPalma
Ari 1h58'23" 22d55'15"
Laurie Doll's Wish Star
Uma 8h32'53" 64d17'26"
Laurie & Doug
Cyg 20h56'44" 47d58'57"
Laurie E. Stevens
Lib 14h26'37" -18d25'26"
Laurie Eternal
And 2h15'2" 45d20'23"
Laurie Faye
Lyn 7h32'15" 52d59'15"
Laurie Figaniak
Cas 1h27'8" 68d42'48"
Laurie Folse
Lyr 18h45'24" 32d45'26"
Laurie Gamache
Cnv 13h53'33" 32d0'48"
Laurie Givens
Vir 13h10'41" 13d26'45"
Laurie Gonzalez
Lyn 6h44'26" 56d26'3"
Laurie Hanson's Star
Tri 2h6'6" 34d7'34"
Laurie Her
Vir 14h28'3" 1d23'23"
Laurie Hill McQuillan
Per 4h14'41" 49d20'11"
Laurie Hodgkins
Ori 5h15'3" 8d0'50"
Laurie Hunt Elliott
Leo 10h47'44" 17d46'14"
Laurie Jean
Dra 18h42'19" 67d0'4"
Laurie Jean The Star in my Life
And 1h50'6" 43d3'46"
Laurie Jeanne My Guiding Light
Gem 7h27'54" 32d10'59"
Laurie & Jim Lauria
Sge 19h53'48" 18d46'17"
Laurie Jo Lang
Cas 0h11'59" 51d49'34"
Laurie John Eldon Campbell
Cep 0h16'44" 80d32'1"
Laurie June
Cap 21h32'58" -16d13'29"
Laurie K.
Leo 10h4'4" 23d43'14"

Laurie Kay
Mon 6h54'9" 0d29'34"
Laurie Kay James
Psc 0h25'43" 8d5'20"
Laurie Kiler
Cyg 19h44'11" 29d58'54"
Laurie Kramer
Cnc 9h20'32" 31d18'53"
Laurie L. Cunningham
And 0h11'16" 31d59'25"
Laurie L Smith
Lib 15h31'44" -18d52'8"
Laurie & Lee
Her 18h1'6" 36d59'0"
Laurie Lee Lourie
Psc 0h52'46" 17d20'57"
Laurie Lee Raybuck
And 23h9'12" 45d32'17"
Laurie Lee Watson
Aqr 23h4'19" -9d6'38"
Laurie "Loli"
And 1h3'23" 45d41'37"
Laurie "Lollydab"
Uma 9h37'39" 65d5'6"
Laurie Louise ANP Major
Cyg 20h38'2" 36d0'50"
Laurie Love
And 0h39'9" 35d24'57"
Laurie Lucas
Vir 11h48'57" 7d41'52"
Laurie Lynn
Per 3h29'58" 33d13'27"
Laurie Lynn
Uma 11h24'46" 41d30'30"
Laurie Lynn
Cyg 20h1'29" 43d47'55"
Laurie Lynn
Cam 4h32'9" 58d55'50"
Laurie Lynn Jessen
And 0h47'27" 43d47'22"
Laurie Lynn Veda Bay
Cyg 20h19'31" 30d4'10"
Laurie Lynne Dyson
Cas 23h30'8" 52d58'22"
Laurie Lynne Garrett
Cas 0h49'21" 59d0'43"
Laurie M.
Leo 11h38'38" 20d25'40"
Laurie M Kramer
Sgr 19h6'44" -13d52'33"
Laurie Marie Montano
Ari 2h46'21" 28d1'57"
Laurie Mariea
Lyn 6h40'47" 54d31'46"
Laurie McCauley
Uma 9h15'46" 57d9'39"
Laurie Mcvey
Sco 16h6'25" -18d12'49"
Laurie Motyka
Cap 21h31'26" -10d53'42"
Laurie "My Shining Star" Rivera
Lyn 6h51'34" 51d3'52"
Laurie O'Mara
Mon 6h54'19" -0d59'43"
Laurie Peel
Uma 14h26'37" 60d21'19"
Laurie Pena
Sgr 18h45'8" -23d52'43"
Laurie Piquette
And 1h29'13" 44d16'2"
Laurie Reisinger
Cas 23h21'21" 58d33'19"
Laurie Renee~K.L.
Uma 10h34'51" 65d44'48"
Laurie S Cuccurullo
Cas 1h6'57" 69d25'32"
Laurie S. Huze
Uma 11h35'28" 55d41'54"
Laurie Schlekeway-Burkhardt
Per 4h44'6" 37d12'16"
Laurie Shea Watsabaugh
Cas 23h33'17" 58d47'41"
Laurie Stapleton Lanza
Tau 4h18'8" 27d7'41"
Laurie Susan Walloch
Per 3h12'52" 55d37'35"
Laurie Taylor Perrault
Sgr 18h31'17" -16d22'58"
Laurie Tuchrelo
And 0h13'22" 46d26'53"
Laurie Walker
And 1h6'56" 44d8'47"
Laurie Ward
Lyn 6h22'19" 56d22'27"
Laurie Winkless
Peg 21h46'3" 24d5'31"
Lauriearoarus
Aqr 20h41'25" -10d46'49"
Laurielle
Vir 13h26'39" -20d9'31"
Laurien Corey
Cnc 8h22'11" 14d51'32"
Laurie-Riley McDaniel
Vir 12h23'38" 3d46'12"
Laurie's Blue Diamond
Cmi 7h20'5" 1d0'58"
Laurie's Light
Crb 15h48'48" 34d21'29"
Laurie's Star
Boo 14h35'49" 37d31'28"
Laurie's Star
Peg 21h58'10" 14d29'58"
Laurie's Star
Crb 15h38'47" 26d19'8"
Laurie's star
Aqr 21h13'28" -3d29'26"
Laurie's Star
Uma 8h28'53" 61d7'24"
Laurie's Star-Earth's Greatest Mom
Aqr 22h37'26" 0d50'58"
Laurieus Ann Birdus
And 0h40'35" 24d50'10"
Laurie-Zoe Schuper Star
Gem 6h45'59" 21d2'56"
LAURIN
Per 4h32'48" 45d44'38"
Laurin Michelle Maier
Uma 10h50'59" 50d55'8"
Laurin Miles
And 1h40'45" 47d7'10"
LAURIN RAMON
Cyg 19h37'19" 31d0'36"
Laurin Rochelle Kelly-Seeley
Cnc 8h51'25" 12d40'43"
Laurine June
Cru 12h43'45" -55d48'4"
Laurinette
Del 20h34'32" 6d50'4"
Laurinity
Lyn 7h4'46" 51d3'49"
Laurino
Uma 12h29'12" 56d8'53"
Lauris L. Antoine
Dra 18h39'20" 52d23'33"
Lauri's Paradise
Uma 10h21'29" 71d58'5"
Laurisa
Cnc 8h54'15" 25d0'38"
Laurita
Vir 13h2'44" -17d51'39"
LaurJey
Psc 1h0'19" 5d28'16"
Laurla
Ari 3h14'52" 29d47'10"
Lauro Chavira III
Per 4h16'19" 51d25'30"
Laurose
Cas 0h26'58" 61d20'20"
Laurrie Co
Cmi 7h20'30" -0d13'30"
laurs scion of eternity
Dra 19h25'47" 69d24'37"
Laursenium
And 0h13'48" 34d21'39"
Laurus
Ori 5h13'11" 15d21'52"
Laury
Uma 10h18'9" 65d37'27"
Lauryn
Aqr 23h10'51" -7d43'37"
Lauryn
Cnc 9h8'45" 16d18'51"
Lauryn
Uma 10h23'12" 45d58'48"
Lauryn Ashley
Psc 1h5'21" 14d20'59"
Lauryn Elizabeth Lounsbury
Vir 12h40'14" 11d56'12"
Lauryn Feltham
Cru 11h58'50" -59d17'45"
Lauryn Jacqueline Dunn
Cyg 20h6'54" 35d45'5"
Lauryn Jessica
Cyg 20h6'58" 33d47'51"
Lauryn Joyanne Hess
Lyn 8h20'13" 51d42'11"
Lauryn Kassidy O'Hagan
Cas 23h20'54" 57d48'3"
Lauryn Kate Abbott - Guardian Angel
Cru 12h31'12" -64d17'14"
Lauryn Marie Hynson
Lib 15h33'34" -7d11'45"
Lauryn Mary DeCrescenzo
And 2h18'54" 41d54'4"
Lauryn Monique Morris
Lyr 18h48'38" 31d49'22"
Lauryn Neff
Cas 0h8'49" 54d48'26"
Lauryn Nicole
Peg 22h40'9" 2d56'31"
Lauryn Rose Marks
And 23h28'49" 49d28'1"
Lauryn Sydney Brown's Birthday Star
Gem 6h55'26" 26d58'42"
Lauryn Taylor - LoLo
And 0h11'0" 32d31'48"
Lausebub
Ori 5h36'45" 5d52'4"
Lautari & Killian's Star
Cyg 19h53'36" 31d50'41"
Lauter, Joe
Ori 6h14'5" 16d9'36"
Lava Star
Pho 1h0'2" -40d13'43"
LaVae Beach Mastrangelo
Leo 10h43'14" 14d21'52"
Lavana
Cap 21h1'11" -24d52'58"
Lavana "Shyanne" Tanner
Uma 11h46'3" 58d47'29"
Lavanya
Gem 6h27'6" 26d7'54"
Lavar & Clair's Eternal Flame
Sgr 19h19'29" -14d59'49"
Lavasia April-Lia Williams
And 0h20'1" 38d4'19"
Lavell
Uma 12h24'27" 56d54'33"
Lavelle Mousset
Uma 12h44'58" 58d20'39"
Lavena
Ari 2h27'50" 27d28'24"
LaVera D. Moss
Cyg 19h57'32" 34d21'49"
LaVera Fay Hawk
Sgr 19h21'57" -17d40'2"
LaVera Irene Purdy Huff
Ori 6h17'13" 1d55'31"
Laveranues
Aqr 22h25'13" -21d19'21"
Lavern Ralph Halzle
Crb 15h46'42" 34d55'56"
LaVerna with Passion
Cas 23h26'5" 57d37'35"
LaVerne
Cap 20h17'3" -20d51'45"
Laverne
Cas 0h6'35" 53d40'12"
LaVerne
Tau 4h1'48" 26d13'4"
LaVerne A. Witmyer
Peg 22h49'58" 19d24'22"
Laverne & Jimmy Franco
Cnv 12h33'55" 44d6'32"
LaVerne Margaret Csik
Leo 11h35'29" 26d11'20"
LaVerne Marie and James Sturgeon
Ari 2h36'11" 24d42'49"
Laverne Surgala
Cas 0h22'57" 60d58'46"
Laverne V. Young
Psc 0h44'13" 7d12'1"
LaVerne W. Effinger
Crb 15h48'53" 39d35'52"
LaVerne W. Effinger
Crb 16h14'10" 38d3'23"
Laverne-My Sister in Sobriety
Psc 1h20'41" 31d38'18"
Laverne's Smile
Tau 4h17'46" 9d21'24"
Laveta
Leo 9h25'33" 13d17'53"
Lavina's Love
Uma 9h31'29" 49d13'18"
Lavinia
Uma 11h12'28" 33d7'47"
Lavinia Carol Lazzaretti
And 1h7'22" 46d50'34"
Lavinia Chiara
Cam 4h17'57" 56d36'44"
Lavinia Gardella
Ori 6h19'59" 9d54'29"
Lavinia Jean Lenhart
Umi 15h28'58" 73d30'28"
Lavinia Klein
Uma 9h35'20" 65d38'33"
Lavinia Scarfone
Cru 12h20'29" -58d40'56"
Lavon Lacey,GaMPI Shining Star 8-06
Uma 9h5'25" 50d6'44"
LaVonne
Cas 1h14'28" 60d0'40"
Lavonne
Aqr 22h29'51" -1d42'25"
Lavonne Jacobs Ellsworth
Gem 6h38'11" 20d44'13"
LaVonne M. Badzik
Lyn 7h56'26" 48d8'37"
LaVonne's Angel
Lyr 18h45'1" 41d9'8"
L.A.W.
Lyn 7h7'15" 51d57'45"
Law
Cyg 19h57'39" 30d28'25"
Lawahna " Wanie " Thurston
Lyn 7h46'49" 58d12'39"
Lawana C Hatch
Sco 17h43'57" -35d49'46"
Lawarence A. Lanzi
Sgr 18h10'44" -29d53'35"
Lawease BIG DAWG Parker
Cnc 9h4'48" 15d13'0"
Lawra
Psc 0h51'37" 29d4'28"
Lawren
Uma 10h38'35" 47d48'14"
Lawren Askinosie
Uma 11h0'8" 58d0'1"
Lawren Bocock
Cnc 8h47'54" 19d36'20"
Lawrence
Her 17h44'40" 48d43'54"
Lawrence
Cap 21h37'10" -16d19'20"
Lawrence A. Reumann
Per 2h33'59" 55d14'9"
Lawrence A. Shirley III
Dra 18h45'53" 72d0'23"
Lawrence Albright
Tau 4h34'53" 6d11'46"
Lawrence and Elizabeth Fuccella
Lyn 7h13'22" 50d55'56"
Lawrence and Lois
Uma 8h50'58" 66d11'38"
Lawrence Andrew Ferguson
Aur 5h53'42" 53d23'5"
Lawrence Arthur Russell III
Her 17h21'47" 14d26'40"
Lawrence at Fifty
Cep 21h33'7" 69d43'43"
Lawrence Austin Waits
Cyg 21h24'33" 39d42'59"
Lawrence B. Barraza
Her 17h8'53" 32d44'12"
Lawrence B. Marcus
Ori 6h7'5" 17d36'22"
Lawrence Benjamin Denny
Her 16h25'29" 47d40'34"
Lawrence Brooks Fogdall
Cyg 20h25'58" 43d31'35"
Lawrence Calamia Jr.
Uma 11h41'54" 40d50'15"
Lawrence Charles Erdmann
Uma 10h12'7" 42d54'14"
Lawrence Charles Wadsworth
Gem 6h47'51" 33d8'56"
Lawrence Christopher Cook
Psc 1h33'15" 4d45'17"
Lawrence Collon White
Lib 14h24'8" -18d10'2"
Lawrence Craven Holdsworth V
Aqr 22h7'31" -8d44'20"
Lawrence D. Carter
Her 17h7'50" 32d19'44"
Lawrence D. Foote
Ari 2h16'1" 26d19'1"
Lawrence D Porter
Ori 6h17'54" 9d29'59"
Lawrence Dean Stone
Leo 11h4'29" 0d16'53"
Lawrence Doherty
Cep 21h56'49" 64d51'1"
Lawrence Dorion
Cap 20h29'43" -9d29'35"
Lawrence Drexel Flagg
Uma 10h35'42" 45d7'28"
Lawrence E. Brandenburg
Per 3h19'2" 42d45'22"
Lawrence E. Kiehart Sr.
Her 16h50'7" 31d0'52"
Lawrence Edward Dudek Circa 1951
Per 3h27'14" 35d36'20"
Lawrence Edward Gemmill
Sgr 17h55'56" -29d16'30"
Lawrence Edward Lang
Per 2h39'55" 54d21'37"
Lawrence Edward Starkey
Cap 20h51'23" -22d22'50"
Lawrence Edward Townsend
Sco 17h49'30" -34d59'10"
Lawrence Eric Harris
Sgr 18h15'21" -31d51'24"
Lawrence Evan Roseberry, Jr.
Uma 9h57'47" 45d1'0"
Lawrence Family Legacy
Sgr 18h18'9" -32d35'49"
Lawrence Fields
Sco 16h11'0" -26d5'6"
Lawrence Floyd Moshier
Lyn 8h29'57" 43d57'46"
Lawrence Franklin Thompson "LFT"
Uma 11h19'50" 50d14'1"
Lawrence G. Pugh, III
Per 2h52'0" 45d59'17"
Lawrence G. Thomas
Dra 19h27'1" 69d24'40"
Lawrence G. Zagardo
Dra 15h1'7" 62d34'31"
Lawrence & Genevieve Genovesi
Cyg 19h48'51" 46d45'52"
Lawrence Gordon
Lyn 7h36'36" 47d51'59"
Lawrence H Cuccurullo
Per 4h28'50" 42d0'9"
Lawrence H. Helm
Uma 11h27'3" 33d0'2"
Lawrence Hamlett
Ori 5h51'45" 6d51'25"
Lawrence Harvey Leigh
Umi 17h24'22" 82d19'59"
Lawrence Henry Schwartz
Ori 5h45'58" 5d17'16"
Lawrence Hirsch
Leo 9h32'21" 10d48'16"
Lawrence I. Rodriguez
Uma 9h42'17" 61d1'45"
Lawrence I."Larry" George
Dra 20h7'49" 69d24'39"
Lawrence J. Costantini
Ori 5h36'24" -0d11'54"
Lawrence J. Freaso "Larry Fallon"
Per 3h48'49" 46d12'54"
Lawrence J. Hill
Per 4h5'41" 37d14'53"
Lawrence J. Howard III
Psc 1h23'4" 25d31'43"
Lawrence J. Raduazzo
Mon 6h44'40" -0d9'13"
Lawrence Jay Gagen
Sco 15h52'26" -20d53'9"
Lawrence Jean Pierre
Per 2h49'20" 54d19'27"
Lawrence "Joe" Wayne Roland 3/25/42
Ori 5h34'23" -1d21'41"
Lawrence Joseph Bassett
Uma 14h7'19" 61d52'10"
Lawrence Joseph Kaifesh
Uma 11h30'6" 49d22'30"
Lawrence Joseph Labate
Cyg 19h38'22" 39d49'50"
Lawrence Joyce JR.
Per 3h9'8" 47d43'54"
Lawrence K. Heck
Cas 2h37'51" 66d58'6"
Lawrence K. Novick
Cas 1h21'49" 59d55'32"
Lawrence L. Pilon
Aur 5h55'28" 37d4'49"
Lawrence Lavendol Jr.
Lyn 7h36'14" 48d55'32"
Lawrence Lee Lorenzo Jr.
Gem 7h13'15" 34d16'20"
Lawrence Lee Nevitt
Per 3h52'7" 40d30'50"
Lawrence M. Goldberg
Cep 0h22'40" 76d59'22"
Lawrence M Kaile
Uma 14h15'55" 58d31'59"
Lawrence Malone
Cyg 19h31'49" 53d44'4"
Lawrence Marvin Jones
Per 4h3'43" 40d28'41"
Lawrence Matthew Claeys
Her 18h5'12" 18d7'22"
Lawrence & Meiko Dukes
Sge 20h8'54" 21d5'7"
Lawrence Merrill Kraay
Psc 1h33'41" 11d34'24"
Lawrence Moore
Ari 2h6'59" 19d54'41"
Lawrence Munizza
Her 17h20'13" 20d8'27"
Lawrence & Norma Struck
Lmi 10h29'52" 30d22'48"
Lawrence Paul Tomasiello "Paul"
Aqr 22h6'59" -9d1'5"
Lawrence R. Anderson
Ori 5h33'26" 3d4'57"
Lawrence R Cavalieri
Lyn 8h10'52" 56d54'5"
Lawrence R. James
Ser 18h18'30" -14d2'4"
Lawrence R. KinCannon
Sgr 18h24'38" -23d3'13"
Lawrence R. Tiernan
Leo 10h22'10" 8d58'24"
Lawrence Richard Weber
Per 2h24'0" 57d13'40"
Lawrence Robert Briscoe
Ari 2h4'58" 21d41'56"
Lawrence Roberts
Psc 1h36'49" 24d33'44"
Lawrence Roy Bell
Lib 15h45'38" -17d4'17"
Lawrence Ruzzo
Cep 22h45'24" 65d54'0"
Lawrence Shawe
Sge 19h16'54" 16d43'7"
Lawrence Spurrell
Umi 5h30'44" 88d35'18"
Lawrence the Great
Ori 6h0'53" 18d9'55"
Lawrence Thomas Gregan
Psc 1h9'3" 26d53'40"
Lawrence Tobias "Baboo"
Ori 5h7'59" 13d6'29"
Lawrence Trent Lane
Cap 20h42'6" -14d49'21"
Lawrence Vande Vyvere
Her 16h43'43" 25d28'23"
Lawrence Vierheilig
Boo 14h51'29" 28d25'18"
Lawrence W. Abel
Tau 5h26'50" 17d15'26"
Lawrence W. Campbell
Lib 15h54'19" -13d29'43"
Lawrence W. Cooper
Sco 16h11'26" -9d9'33"
Lawrence W. Kelley
Gem 6h13'34" 25d12'19"
Lawrence W. Muckelroy, Sr.
Aqr 22h39'48" -3d7'27"
Lawrence Wade Smith
Ori 5h6'20" 3d46'23"
Lawrence Watson AKA Bugs Bunny
Per 4h37'13" 41d6'15"
Lawrence Wilfert
Dra 15h55'43" 58d32'22"
Lawrence Winslow
Leo 11h5'25" 1d52'15"
Lawrence X. Lopez
Gem 6h42'23" 24d23'20"
Lawrence-Rebecca
Uma 12h39'30" 62d5'23"
Lawrie-Anne
And 23h33'41" 42d55'53"
Lawson Alexander Cox, II
Cap 20h34'18" -11d16'53"
Lawson James Lovell
Cnv 12h45'1" 38d8'16"
Lawson Lee Haveman
Leo 10h41'2" 18d49'32"
Lawton W Johnson
Psc 1h16'1" 6d56'35"
Lay
Lib 15h15'46" -8d20'51"
Laya Reeves
Psc 0h32'20" 9d8'53"
Layah, my light, now and forever.
Cap 20h24'6" -26d52'45"
Layanne Wright
Ori 5h0'14" 9d47'47"
Layer, Harald
Uma 8h38'10" 59d42'38"
Layla
Cas 23h27'50" 57d47'31"
Layla
Sgr 19h41'45" -14d15'0"
Layla
Cap 20h31'14" -11d8'41"
Layla
Ori 5h17'58" -1d1'36"
Layla
Ari 2h49'40" 17d38'28"
Layla
Gem 7h15'9" 24d17'50"
Layla
And 1h28'54" 41d11'45"
Layla
And 2h34'14" 50d33'51"
Layla Abigail Morgan Bickley
And 23h37'27" 46d57'50"
Layla Alane
Gem 7h17'20" 17d57'30"
Layla Alsharifi
Tau 4h26'42" 23d18'20"
Layla Andrews
And 1h21'35" 44d9'51"
Layla Ariana Eastep
Sgr 19h14'59" -22d6'57"
Layla Barkawie
Cyg 21h32'17" 46d29'10"
Layla Caramel Lund
Cru 12h32'50" -60d20'45"
Layla Elizabeth Mihal
Vir 13h14'24" -20d16'51"
Layla Farraj
Lyr 18h32'20" 36d53'12"
Layla Faye Handsaker
Leo 10h12'51" 15d28'28"
Layla Hernandez Lopez
Eri 3h54'35" -2d51'35"
Layla Isobel
And 2h34'5" 46d31'45"
Layla Kearney
Oph 16h36'39" -0d57'28"
Layla Love
Sgr 19h29'0" -18d26'45"
Layla Mae
And 1h21'3" 33d53'56"
Layla Marie Carlson
And 0h26'55" 29d3'40"
Layla Michelle Kelly
Cam 4h37'42" 55d37'45"
Layla Moody
Cap 20h37'0" -11d24'25"
Layla Nicole Mason
Crb 15h27'17" 26d16'59"
Layla Roberts
Lyn 8h22'55" 49d42'12"
Layla Rose
And 2h24'49" 48d52'48"
LayLay
Aqr 20h43'26" -9d9'56"
Layli
Ori 6h14'24" 15d37'3"
Layluhbee
Lyn 9h5'44" 34d25'46"
Layna Zinich - My Boo
Lyn 7h33'16" 49d32'39"
Layne Antonio Grajeda
Tau 4h27'58" 21d49'15"
Layne Cole West-Hale
Uma 11h39'44" 32d39'35"
Layne Joseph Miller
Psc 0h36'57" 5d52'12"
Layne Ledding
Cam 3h36'44" 59d16'22"
Layne McGee
Cap 20h31'40" -14d42'23"
Layne Reve
Her 18h26'31" 27d16'57"
Layne Scoles
Leo 11h10'31" -5d55'27"
Layne Sigman
Gem 7h18'36" 19d59'56"
Layne Tade
Sco 17h38'24" -41d15'20"
Layne William
Sgr 18h46'13" -29d40'23"
Layney Joe Henson
And 1h58'52" 46d21'42"
Laynla Randersfyniki
Uma 10h31'11" 65d31'30"
Layshka
Uma 10h46'13" 61d31'30"
Laytam
Uma 8h39'43" 48d34'19"
Layton
And 0h44'31" 38d54'4"
Layton
Sgr 19h44'19" -12d44'10"
Layton James Parker
Lyn 7h32'2" 48d21'59"
Layton & Marilyn's Christmas Star
Lyn 8h14'57" 46d46'47"
Layton Robert Maybury
Cyg 21h38'39" 51d43'38"
Layton Victoria
Lyr 18h27'8" 39d24'33"
Layzell Lotus Lazerbeam 33
And 23h33'7" 42d32'12"
laz
Peg 22h16'55" 7d44'9"
Lazanowski
Ori 5h20'9" 1d41'58"
Lazar
Lib 15h45'36" -8d44'35"
Lázár Bence Csillaga
Ari 2h31'24" 25d30'57"
Lazara
Vir 12h47'9" 4d25'5"
Lazara Eberkys Viera
Sco 17h11'50" -37d42'45"
Lazaro Collera 2:53pm
Sco 17h33'0" -38d41'23"
Lazarus
Vir 12h46'38" 2d5'3"
Lazarus Henton
Uma 8h36'19" 64d55'10"
Lazhar's Star
Uma 13h44'53" 53d29'49"
Lazie J 2-10-1979
Uma 11h21'44" 46d12'35"
Lazo-Bradley
Cas 23h14'25" 56d9'4"
Laz's Brumbettes
Cru 12h32'19" -55d48'15"
Lazyboy's Trampoline
Aur 5h45'43" 48d8'12"
Lazzarini Livio 12/10/2005
Cam 3h22'8" 64d33'6"
LB 20-07-2002
Aur 5h38'44" 32d44'52"
LB Always Forever
Uma 11h41'45" 35d4'24"
L.B. Holcombe
Boo 14h59'58" 54d39'58"
L.B. Kilman
Aur 6h35'28" 34d21'30"
LB2
Ari 3h15'24" 11d17'56"
LBBHTHDBJB
Crb 16h22'32" 33d25'52"
L.B.G. Ritchie
Mon 6h46'33" 10d55'43"
LBM
Gem 7h3'9" 22d33'14"
LBolshaw65
Per 2h40'11" 56d26'42"
LBS
Gem 7h46'44" 34d1'31"
LBS Amor Para Siempre
Lyr 18h26'19" 38d44'24"
LC
Ari 2h12'29" 24d58'18"
LC Star
Cnv 12h36'48" 45d5'47"
LCD Senior
Sgr 17h55'6" -29d39'36"
LCDR John Douglas Trask
Vir 14h15'12" -7d40'2"
LCH#41
Aqr 21h44'18" -7d39'30"
Ich Liebe Dich!
Her 17h18'50" 47d14'11"
L.Charlene Smith
Sgr 19h0'23" -13d38'5"
LCL Storm Bridger
Uma 11h20'20" 36d57'28"
Lcon
Lyn 8h14'40" 56d56'16"
LCPL Aaron Boyles
Per 3h23'48" 48d28'21"
LCpl Branden P. Ramey
Per 4h12'8" 47d9'51"
LCpl. Christopher Mayo Pruitt
Aql 20h35'27" -7d6'54"
Lcpl David Toms
Leo 9h43'36" 24d59'40"
LD "Grandpa" Jurrens
Aur 5h51'8" 51d33'27"
LD Rae 9-6-1984
Vir 12h51'56" 5d8'34"
LDA29
Lyn 9h9'46" 38d59'26"
L'e
Cam 5h17'33" 61d38'16"
Le Beau Larme Diamant
Cam 4h11'6" 56d14'42"
Le Beau Realt
And 1h30'53" 49d17'50"
Le beau violoncelliste
Psc 0h36'29" 9d34'10"
Le centre de mon univers.
Leo 9h53'33" 9d53'8"

Le Coeur de Kona
Pho 0h39'15" -43d3'56"

Le coeur de Pierric
Uma 9h8'24" 48d2'1"

"LE CORBUSIER"
Cyg 21h43'15" 40d56'7"

Le Deesse, Laura
Tau 5h45'41" 15d43'49"

Le Donna
Cas 2h9'9" 73d36'40"

L.E. "Gene" Hughs
Her 17h22'26" 17d26'18"

LE Grand International
Ori 5h38'13" 3d24'0"

Le Habryle
Lib 15h46'16" -4d54'22"

LE Hickey (Toby) Dad
Cra 18h8'58" -37d19'24"

Le Hien Allen
Ori 4h51'55" 1d11'6"

Le Jardin Secret D'Ysa
Cas 2h20'1" 68d47'14"

Le Le Wood
Lmi 10h34'57" 34d49'51"

Lé LucyMae Purpureus
Cyg 20h20'24" 54d28'43"

le monde d'F
Tau 5h59'4" 25d20'55"

Le Mont, Le Petit Chien (Monty)
Uma 9h24'11" 53d50'17"

le mouton Chloé
Tau 3h54'39" 8d54'51"

le petit prince
Cnv 13h25'33" 30d6'30"

Le Petit Prince
Cep 21h59'37" 65d51'28"

Le Petit Prince Robin
Cas 23h29'58" 56d34'42"

Le Petit Prince William
Umi 14h7'29" 76d1'55"

Le Phare
Vir 13h6'27" 8d31'44"

Le Poèt Charles Nestoret
Ori 6h14'15" 16d7'52"

Le point sublime
Crb 15h59'31" 31d24'33"

Le Prof Bessette
Aur 7h29'4" 35d16'15"

L.E. Pyle
Uma 14h23'9" 59d10'14"

Le Roch
Uma 8h24'36" 60d36'58"

Le Stelle Si Sono Allineate Per Noi
Lyr 19h10'45" 38d16'6"

Le Thi Ngoc Phuc (Camille Le)
Lib 15h26'38" -7d19'50"

Lê Thi Riêng
Cra 18h42'0" -38d56'47"

Le Thi Thu Lieu
Tau 4h0'29" 26d52'40"

Le Thuy Hong
Uma 13h47'2" 48d1'58"

LEA
Cyg 21h12'46" 45d21'58"

Lea
Cnc 8h4'18" 19d19'5"

Lea
Cap 21h35'32" -22d45'14"

Léa
Leo 9h59'28" 15d56'6"

Lea Aileen Ford
Vir 12h32'44" 6d48'22"

Lea and Jack Winn
Dra 15h12'58" 63d11'1"

Lea Ann
Ori 6h18'40" 14d21'29"

Lea Ann Little A.K.A. Sugarbear
And 2h15'29" 50d13'27"

Lea Aoun
And 2h30'21" 50d9'14"

Lea Behanna
Gem 6h35'4" 15d5'22"

Lea Butkiewicz
And 0h46'18" 36d51'38"

Lea Fargano
Lyn 8h54'22" 37d37'34"

Léa Forelli
Aqr 22h27'21" 1d57'16"

Lea Gebrezghier
Umi 14h59'37" 69d32'14"

Lea & Gene Klein
Ori 5h40'33" 4d54'37"

Léa Griffoul—Coustet
Aqr 23h23'55" -18d24'22"

Léa Hottier
Ori 5h9'30" 7d47'38"

Lea Howie (L.A.P.)
Sco 16h15'39" -10d42'50"

Lea & John
Com 12h15'50" 20d23'11"

Léa la Libellule
Lib 15h18'43" -16d56'40"

Lea - Lifting Everyone's Affection
Vir 13h43'34" -13d17'41"

Lea Loves Mark
Boo 15h36'57" 47d44'45"

Lea Marie
Uma 10h0'49" 50d7'49"

Lea Marie Martin
Umi 15h25'16" 85d19'29"

Lea McDowell
Ari 3h24'46" 23d38'56"

Lea McManus - Our Guiding Star
Cru 12h7'44" -62d40'32"

Lea & Mikey
Cyg 19h39'41" 29d55'26"

Lea Musiol
Leo 9h22'13" 15d47'34"

Lea Nicole
Uma 12h44'43" 62d39'30"

Léa Obligisecotieresauzayjumel
Uma 9h38'57" 42d11'14"

Lea Renee Petcheny
Leo 11h45'18" 25d28'4"

Lea Rose Leavell
Ori 5h54'17" 18d24'19"

Lea Sarah Florian
Cas 23h0'55" 57d48'34"

Lea Sophia
Aur 6h38'43" 38d47'35"

Lea - Sophie Steffen
Uma 10h31'58" 48d59'58"

Lea Spitser
Psc 1h28'37" 16d25'35"

Lea und Toby in Love
Cas 23h53'26" 50d10'16"

LeaAnn
And 2h6'38" 41d57'0"

Leadership
Cep 21h58'33" 64d8'43"

Leadfon Garment Sdn Bhd
Ori 5h38'19" 3d44'12"

Leading Lady
Cas 23h53'20" 50d2'31"

Leah
And 23h9'33" 45d23'16"

Leah
And 23h5'14" 48d22'24"

Leah
And 23h40'43" 40d52'25"

Leah
Lyr 18h39'40" 34d43'23"

Leah
Leo 11h19'35" 2d27'22"

Leah
Ori 5h47'47" 7d36'39"

Leah
Ari 3h18'52" 18d59'22"

Leah
Vir 13h23'26" 8d14'1"

Leah
Aql 19h10'59" 8d20'57"

Leah
And 0h11'33" 26d38'3"

Leah
And 0h30'52" 27d27'23"

Leah
Ari 2h33'5" 24d38'35"

Leah
Cam 5h5'54" 66d49'45"

Leah
Cas 0h13'39" 62d28'56"

Leah
Lyn 8h33'39" 55d19'36"

Leah
Uma 12h8'33" 58d47'45"

Leah
Vir 12h21'12" -5d2'37"

Leah
Mon 6h47'57" -4d5'32"

Leah
Psa 22h53'56" -32d13'29"

Leah
Cap 21h21'45" -23d17'0"

Leah
Sgr 18h25'56" -36d17'15"

Leah
Sco 16h43'50" -33d2'24"

Leah 1991********Ric and Lucy 20th
Crb 15h48'50" 27d15'9"

Leah A. Hines (A.K.A. - L)
And 23h0'34" 47d30'43"

Leah Allison Stokes
Aur 6h24'1" 46d8'53"

Leah and Charlie Forever
Ari 2h35'39" 21d36'55"

Leah and Justin Forever
Ari 2h27'34" 12d17'21"

Leah and Ryan - Forever Love
Aqr 21h40'29" 2d52'41"

Leah Angelina Barna
And 0h16'53" 34d52'7"

Leah Ann
Ari 3h6'57" 28d42'7"

Leah Ann Murtha
Lyr 18h34'12" 27d8'26"

Leah Annabelle Rose Sullivan
Cyg 21h38'47" 48d34'39"

Leah Arielle
Psc 1h25'10" 23d18'10"

Leah Audrey Skarbnik
Her 16h50'31" 36d52'9"

Leah Audrey Wood
Leo 11h36'31" 9d38'37"

Leah Baldwin
Lyn 8h14'4" 54d46'5"

Leah Bartu's Star
Uma 11h26'50" 44d7'22"

Leah Battaglia
Cap 21h47'23" -11d8'26"

Leah Baurley
Leo 10h49'52" 15d53'28"

Leah Beth Radow
And 0h44'24" 35d55'30"

Leah Bo Yang O'Neill
And 1h9'58" 43d34'53"

Leah Bradford
Crb 15h17'20" 27d9'4"

Leah Broyles
Lib 14h45'56" -18d53'16"

Leah Burns Watt
Umi 17h52'6" 87d19'0"

Leah Catherine Potter
And 0h10'31" 44d58'43"

Leah Clark
Uma 14h16'46" 58d34'9"

Leah Coleman
And 0h25'21" 25d42'29"

Leah Crank
Cap 20h25'34" -26d19'58"

Leah D. Hazuda
Psc 0h11'8" -4d10'44"

Leah D. Savicki
Lyn 7h41'47" 43d32'12"

Leah Danielle Austin
Vir 12h57'57" 2d21'29"

Leah Danielle Grant (Beautiful)
Sge 20h14'18" 17d47'17"

Leah de Castro
Mic 20h32'6" -29d50'24"

Leah Dianne Mooney
Leo 10h52'9" 24d4'28"

Leah Dieter
Gem 7h20'25" 31d9'14"

Leah " DJ Keiyra " Stanbro
Umi 14h48'59" 70d49'10"

Leah Dolph Stafford
Mon 6h47'27" -0d26'43"

Leah Dorothy Anderson
And 2h32'43" 50d2'36"

Leah Dzhindzhikashvili
Ori 6h19'10" 9d45'21"

Leah Elisabeth James Happy Heart
And 2h27'59" 46d50'41"

Leah Elizabeth
Ori 5h37'1" 9d49'5"

Leah Elizabeth Jones
Cru 12h31'6" -60d15'59"

Leah Elizabeth Rouscher
Cas 0h19'58" 58d42'10"

Leah Ellen
Umi 16h8'32" 81d14'56"

Leah Erin Madsen
Lmi 10h6'23" 38d24'42"

Leah Farber
Lib 14h52'43" -1d2'17"

Leah Fay Naylor
Psc 1h6'56" 3d45'8"

Leah Faye
Sco 16h13'33" -23d17'41"

Leah Fox
Cas 23h45'47" 57d48'44"

Leah Fuller Hofer
Tau 3h47'51" 27d51'12"

Leah Glass
Tau 4h5'48" 7d24'12"

Leah Grace Lunsford
And 1h52'31" 38d26'12"

Leah Grace Mawhinney
And 1h36'14" 44d35'20"

Leah Hanoud
Cnc 8h43'35" 30d46'8"

Leah Humphrey
And 23h31'51" 47d49'26"

Leah J.A.N.
Lyn 6h55'45" 51d35'32"

Leah Jayne
Lib 14h55'32" -17d54'22"

Leah Jeanette's Star
Sco 17h33'40" -43d59'20"

Leah Joy
Psc 0h37'18" 8d39'45"

Leah Joy
Leo 10h38'4" 17d31'23"

Leah Joy Zonis
Cnc 7h56'53" 15d0'41"

Leah K. Pursel
Cap 20h7'1" -9d54'29"

Leah Kaitlyn Graham
Lyr 18h45'8" 39d46'22"

Leah Kando
Lyn 7h45'22" 49d1'12"

Leah Katherine Gillette
Cap 21h39'23" -14d30'6"

Leah Kay
And 0h19'53" 32d38'59"

Leah Kay Brown
Gem 6h27'15" 24d57'12"

Leah Kelly
Pyx 8h36'44" -20d8'0"

Leah Kinney Polacek
Cap 20h17'17" -10d52'13"

Leah "Le Le"
Lyn 7h11'4" 57d30'16"

Leah LeBow
Tau 3h43'53" 29d1'54"

Leah Lin Preston
Vir 12h35'38" 3d24'29"

Leah Lubuff Senavsky
Ari 2h57'28" 27d36'54"

Leah Lynn
Gem 7h28'21" 28d38'53"

Leah & Lynn
Cyg 20h5'0" 51d19'15"

Leah Lynn Gaeta
And 1h48'45" 39d6'36"

Leah M.
Cyg 21h59'13" 50d46'17"

Leah M. Freeman
Ari 2h14'0" 26d44'41"

Leah M. Hudson
Psc 0h54'18" 18d9'36"

Leah M. Register
Vir 12h28'59" -5d49'2"

Leah M. Schessel
And 1h5'28" 46d29'20"

Leah Madeleine Welch
Tau 5h35'31" 24d16'7"

Leah Madeline Levitan
Tau 4h20'49" 29d13'27"

Leah Malberg
Tau 5h16'41" 21d57'1"

Leah Marie
Tau 4h26'48" 11d28'31"

Leah Marie
And 23h1'33" 49d10'14"

Leah Marie
Uma 9h53'30" 47d26'26"

Leah Marie Barton
And 23h57'35" 45d36'55"

Leah Marie Delcourt
Uma 8h59'36" 71d42'35"

Leah Marie Good
Cas 1h46'56" 63d27'34"

Leah Marie Snavely
Sgr 18h28'48" -21d31'52"

Leah Marie Snelgrove
And 0h41'19" 38d2'52"

Leah Marie Visingardi
Uma 10h27'19" 41d46'8"

Leah Mariel
Ari 2h12'36" 24d38'56"

Leah Martinez
Tau 3h38'28" 4d17'9"

Leah Mays
Cas 1h6'58" 61d40'50"

Leah McKenzie Thomas
Sco 16h4'51" -11d37'40"

Leah Meinhart
Cap 20h19'25" -22d59'30"

Leah Melina Gillespie Donnelly
Cyg 21h11'40" 51d43'59"

Leah Michele Jacobs
Psc 1h10'45" 32d30'25"

Leah Michelle
Leo 9h47'20" 31d30'42"

Leah Michelle
Psc 23h41'27" 6d25'10"

Leah Michelle Riley
Sgr 19h2'35" -15d42'56"

Leah Michelle Ross
Lib 15h45'6" -18d46'31"

Leah Michelle Schauer
Vir 12h31'22" -9d1'11"

Leah Michelle Zimmerli
Cyg 19h50'23" 31d38'43"

Leah & Mike Forever
Cru 12h38'28" -58d5'46"

Leah Minshull
And 23h11'55" 38d33'57"

Leah Moeller
Vir 13h23'15" 7d21'24"

Leah - Mommy
Peg 23h33'11" 12d32'41"

Leah Monet
Crb 15h40'13" 28d55'28"

Leah Moonlight - You light up my Life
And 23h43'56" 34d40'56"

LEAH "MY ANGEL"
Psc 1h35'22" 9d5'44"

Leah "My Love" Lindsey
Leo 11h23'53" 0d25'36"

Leah Nicole
Cra 18h45'50" -42d24'19"

Leah Nicole Darnaby
Cnc 8h24'24" 13d44'15"

Leah Nicole Downes
And 1h12'35" 46d2'6"

Leah Nicole Hauff
And 23h42'36" 37d1'50"

Leah Nicole Zucca
Lmi 10h29'41" 29d18'47"

Leah Nusia Wojtaszek
Tau 4h53'12" 25d11'14"

Leah Olivia Cramblit
And 1h49'18" 41d47'7"

Leah Parks
Mon 7h2'56" -6d51'28"

Leah Pearl
Sco 17h56'37" -31d21'52"

Leah Perez
And 1h21'18" 49d6'44"

Leah Peteet - GaMPI Shining Star
Lyn 7h45'50" 38d8'26"

Leah Porter
And 0h53'24" 43d19'3"

Leah Queen of My Heart
Cas 2h50'41" 63d54'35"

Leah R. Ferguson
And 1h50'5" 42d42'6"

Leah Rae
And 0h54'31" 33d41'28"

Leah Rae
Leo 9h56'53" 7d37'46"

Leah Ray Heynsbroek
And 1h23'30" 49d37'40"

Leah Raye Bean
Vir 13h20'30" 12d18'8"

Leah Rebecca
And 1h53'51" 40d23'34"

Leah Reichert
Sco 16h27'23" -32d16'50"

Leah Renee Chaney
And 1h33'50" 43d35'15"

Leah Renee Leighton
Leo 11h14'50" -3d5'22"

Leah Reneé Sherwood
Lib 15h17'17" -6d31'11"

Leah Rich
Crb 15h40'46" 33d43'29"

Leah Riley's Star
Sgr 18h17'41" -32d17'23"

Leah Rose Hasmorh
Lib 15h45'24" -6d7'53"

Leah Rospierski
Uma 11h19'37" 53d18'11"

Leah Russell
And 0h22'25" 43d18'38"

Leah Santeene Jenkins
And 23h45'32" 37d17'3"

Leah Silcock
Tri 2h37'2" 31d35'52"

Leah Suzana Pallares Feingold
Lmi 10h37'6" 30d31'55"

Leah Suzanne Kendall
And 1h40'56" 46d42'9"

Leah Szarek
Leo 11h14'46" -2d7'25"

Leah Taylor Croom
Sgr 18h7'7" -29d50'24"

Leah Taylor Garland
Aqr 22h38'6" -2d7'34"

Leah Teresa Sichenzio
And 2h35'42" 49d0'29"

Leah Tetzlaff
And 1h13'1" 45d11'34"

Leah Tortilla
Aqr 23h45'52" -23d43'53"

Leah Tracy Fleming Hand
Cyg 20h3'43" 32d28'40"

Leah Trammell
Lyn 8h18'59" 41d57'15"

Leah Tringali
Mon 6h23'56" 7d6'4"

Leah Vega
Lib 15h20'12" -21d20'7"

Leah Victoria
Peg 21h56'16" 33d50'15"

Leah Vuillemot
Leo 11h33'29" 20d35'19"

Leah Westfield
Cyg 20h46'52" 35d43'20"

Leah & Will
Cyg 20h46'56" 36d41'51"

Leah Wyar
Equ 21h16'5" 9d8'5"

Leah, Laura, Morgan - Stars Forever
Cyg 20h52'42" 46d13'6"

Leah, Light of My Life
Dra 16h12'16" 57d2'9"

Leah, Sky & Elisha's Star
Cru 12h10'8" -62d28'30"

LeahAndJoe10152006
Cyg 21h19'24" 43d25'2"

Leahanna & Brandon's Star
Cnc 9h8'22" 17d2'51"

Leah-Ayden-Derek
Gem 6h50'26" 23d40'53"

Leahbear
Dra 20h31'46" 69d8'26"

LeahLorene
Sco 16h10'41" -23d10'12"

Leahpar
Leo 11h28'46" 10d35'44"

Leahra
Cru 12h48'13" -57d35'26"

Leah's Bright Star
And 23h17'33" 47d36'37"

Leah's Lite
Uma 11h52'21" 35d8'25"

Leah's Sparkle - Baker 1984
And 0h50'57" 44d16'16"

Leah's Star
Cap 21h40'14" -9d53'6"

Leah's Star - celebrating her 21st
Cru 12h27'12" -60d29'37"

LEA-JAY WILSON (SUPERWOMAN)
Psc 1h17'39" 32d58'47"

Léakim
Uma 12h41'30" 60d20'24"

Leala B
Vir 12h26'42" -9d13'45"

Léan Noa
Cas 1h56'57" 61d45'38"

Lean Sally Jean
Ori 5h27'7" -9d3'50"

Leana
And 1h50'16" 46d41'19"

Leana
Cnc 8h38'58" 23d58'9"

Leana Angel Girl
Del 20h35'27" 17d46'23"

Leana Laba
And 1h22'44" 35d44'39"

Leana Liberatas 724
Lyn 8h18'11" 40d15'27"

Leana Marie Semoes
Uma 9h57'44" 56d32'54"

Leana Massardi
Cyg 19h47'48" 34d15'32"

Leana Patricia's Wishing Star
Uma 9h29'18" 52d6'59"

Leanda MaryAndrea
Leo 11h49'8" 21d46'43"

Leandra
Cyg 21h30'44" 46d35'49"

Leandra
Cas 2h1'52" 59d39'26"

Leandra Catherine Preece
Cas 2h5'28" 62d52'40"

Leandra Kern
Lib 15h48'15" -16d53'21"

Leandro
Lyr 18h28'45" 27d53'20"

Leandro & Gwyndelon Sanz
Vir 12h24'56" -6d43'36"

Leandro Stöcklin
Umi 16h3'50" 73d2'13"

Leandro Taveras
Cnc 8h38'2" 12d54'12"

Leani
Ori 5h55'11" 22d12'49"

LeAnn
Vir 12h48'33" 4d53'37"

LeAnn
Lyn 6h38'18" 59d9'37"

LeAnn
Cap 21h52'56" -19d42'43"

Leann Ashley Bickford
Cnc 8h27'30" 23d10'58"

LeAnn Feldhaus
Lmi 10h34'7" 36d33'0"

Leann Lynn Gervin
Uma 8h48'21" 48d31'20"

LeAnn M Pearce
Cnc 9h18'2" 15d32'38"

Leann Mené Frost
Lyn 8h34'17" 33d35'12"

Leann Michelle Tillman
Lib 15h34'13" -11d26'58"

Le'Ann of Lamesa
Lib 14h52'30" -1d14'52"

Leann Pomponio
Aqr 22h40'0" -4d57'8"

Leann Rachel Bach
Vir 14h6'21" -17d7'42"

Leann Spurlin
Lyr 18h53'36" 34d50'14"

Leann The Bright and Beautiful
Ori 6h11'41" 16d37'1"

Leann Veronica
Ari 3h1'21" 11d55'23"

Leann Virginis
Vir 12h8'11" -10d57'57"

Leann Yi DellaMonica
Sco 16h7'1" -16d19'38"

LeAnna
Psc 0h50'27" 15d42'25"

LeAnna
Lyn 8h36'31" 33d6'53"

Leanna Alicee Shulterbrandt
Cyg 19h50'49" 33d22'12"

Leanna Dianne
And 23h31'33" 41d25'11"

Leanna Georgina Zuniga Cutler
Ari 2h46'15" 25d47'42"

Leanna Hiroko Tamura
Tau 4h43'19" 27d27'5"

Leanna Inez Rybacki
Cyg 20h59'0" 42d18'15"

Leanna Marie
Sco 16h34'9" -43d3'21"

Leanna Marie Inman
Psc 0h15'47" 6d3'24"

Leanna Murray
Cap 20h27'7" -18d37'49"

Leanna Sultan
Cas 1h46'49" 60d0'46"

Leanna The Heart Of Life
Psc 0h11'18" 11d14'55"

Leanna Victoria Brown
Lyr 18h58'35" 27d28'50"

Leannah Louise
Cas 23h49'41" 54d17'14"

LeannaRachaelGabbyAbramBenJason2004
And 2h22'26" 45d17'49"

Leanna's Star
Cra 18h4'37" -37d17'47"

LeannDave
Cyg 21h12'40" 34d20'53"

Leanndra
Lyn 7h44'7" 46d19'26"

Leanne
Cam 4h40'23" 59d15'17"

Leanne
And 23h15'7" 35d54'1"

Leanne
Uma 9h26'59" 41d29'41"

Leanne
Psc 23h30'26" 6d46'45"

Leanne
Psc 0h41'43" 18d11'24"

Leanne
Ari 1h59'10" 20d52'0"

Leanne
Ori 5h35'13" -6d19'24"

Leanne Adele Duke
Leo 11h21'38" 17d30'48"

Leanne Bakken
Crb 15h39'36" 28d58'29"

Leanne Brant Star
Cnc 8h59'6" 22d35'33"

Leanne Cahill
Cru 12h25'18" -59d1'23"

Leanne Carol Williams
Ori 5h37'8" 0d28'15"

Leanne Christine Morgan
Tau 4h36'23" 9d53'14"

Leanne Clare Speight
Tau 5h10'4" 18d32'25"

Leanne Clement
Uma 9h54'39" 42d7'4"

Leanne Dawn McPherson
And 23h2'6" 41d38'7"

Leanne Edmonds
Cru 12h8'7" -62d0'12"

Leanne Ehnat
Umi 10h6'0" 70d4'43"

Leanne Elizabeth Fox
Vir 12h39'12" 6d6'8"

Leanne Garduno
And 0h23'32" 27d50'50"

Leanne Geering
And 1h57'20" 42d16'21"

Leanne Larkin
Boo 13h53'55" 12d5'35"

Leanne Latondress
Psc 1h52'12" 5d48'36"

Leanne Lee
Aqr 22h33'43" -0d21'32"

Leanne Leele
And 0h46'32" 40d32'50"

Leanne Marcoux
Aur 7h11'1" 41d49'37"

LeAnne Marie Beck
And 2h24'29" 49d11'2"

LeAnne Marie Gustafson
And 1h20'18" 41d10'14"

Leanne Marie Schmid
Psc 1h48'12" 2d55'30"

Leanne Marie Smith
And 2h30'12" 41d3'46"

Leanne Marie Wenjing Chance
Ori 5h34'21" -4d18'38"

Leanne Marie Wheatley
Ari 2h58'35" 24d43'17"

Léanne Mathilde
Uma 8h51'17" 51d46'24"

Leanne McGee
Lmi 10h28'21" 35d20'35"

Leanne Minges
Cnc 8h3'55" 11d11'47"

Leanne Morris
And 23h14'47" 47d22'41"

Leanne Murray 1/1/86
Cap 20h59'53" -20d26'1"

Leanne Paton
Pho 0h44'42" -43d48'4"

Leanne "Pebbles" Leacock
And 1h54'20" 35d46'43"

Leanne Renee Casher
Ari 3h7'56" 27d32'49"

Leanne & Rick at "Triple H"
Cas 23h13'12" 54d25'55"

Leanne Schmall
And 1h10'54" 45d26'20"

Leanne Shtayyeh
Lib 15h40'28" -29d4'32"

Leanne Smith of Blackpool England
Tau 4h15'13" 9d42'10"

Leanne & Thomas Fiscoe
Dra 17h38'44" 58d27'27"

Leanne Tucker
And 0h16'57" 28d7'9"

Leanne 'Tweetie' Stockham
And 1h18'34" 38d8'17"

Leanne's Dream
Aur 5h51'34" 52d5'55"

Leannie Giralt
Del 20h46'1" 15d6'24"

Leanochka Farber
Cnc 8h49'43" 28d22'37"

LeaNora Kordova
Cas 23h28'53" 53d26'43"

Leanore Becker
Lyn 9h19'11" 33d46'37"

Leap of Faith 2-29-1996
Cyg 20h5'27" 34d18'48"

Leaphy
Tau 4h48'0" 18d19'39"

Leapold
Leo 11h56'12" 27d57'37"

Leara Alina
Uma 8h56'32" 47d8'39"

Learning To Fly / Liz
Lyr 18h50'13" 39d6'34"

Lea's Determination
Uma 9h26'55" 41d56'32"

Leatrice
Vul 20h42'14" 25d11'49"

Leatrice Salky
Lyr 18h41'12" 36d33'58"

Leawicious
Dra 18h31'1" 62d39'59"

Lebenslang
Cap 21h39'49" -18d30'31"
Leberer, Markus
Uma 9h59'5" 54d32'20"
Lebetkin Family Star
Per 3h46'14" 46d21'54"
Lebron-Lagoa
Lib 15h31'53" -26d46'2"
Lecardonnel Guillaume
Ori 6h19'15" 7d3'11"
Lech W. Losiowski
Sgr 18h44'11" -28d53'17"
LECHTKEN
Leo 11h10'40" -2d23'19"
Le'Cinda Joi
Leo 10h59'3" 11d49'57"
Leco1
Sco 16h14'50" -14d32'40"
LED 7/11/45
Cnc 9h6'19" 32d14'31"
LED4D112004SAV
Umi 15h24'27" 69d14'12"
Leda
Aqr 23h49'50" -14d53'13"
Leda
Per 3h0'20" 49d21'38"
Leda Arituzhe
Cyg 20h5'55" 34d30'9"
Leda Rose
Psc 23h50'19" 5d38'8"
Leddy Stuart
Uma 10h20'49" 53d48'31"
L.E.D.II
Uma 11h45'19" 65d10'56"
Ledroit
Her 16h33'44" 30d54'28"
LedUsDownOurBrokenRoads2eachOther
Cyg 21h23'23" 39d18'34"
Ledys
Ori 5h46'8" 12d23'4"
Lee
Ori 5h27'5" 13d7'32"
Lee
Per 3h24'42" 47d6'21"
Lee
Cyg 21h40'43" 49d13'33"
Lee
Her 17h9'5" 33d31'54"
Lee
Uma 9h34'45" 60d53'15"
Lee
Lyn 7h46'53" 57d57'28"
Lee
Aqr 21h8'7" -6d29'18"
Lee A. Dunn
Psc 1h9'21" 14d55'57"
Lee A. Gholson
Ori 5h36'36" -0d22'0"
Lee A Shafford
Aql 20h12'20" 13d23'30"
Lee A. Zagar
Sgr 18h26'4" -26d46'39"
Lee Aimee Walton
Uma 13h0'44" 56d28'8"
Lee Alan McCuish
Cep 20h40'28" 63d15'48"
Lee & Alexandra Hesketh
Cyg 20h57'20" 30d56'55"
Lee Amos
Aur 5h52'28" 53d34'46"
Lee and Alysia's Star
Umi 15h30'39" 73d29'14"
Lee and Annabel
Cru 12h32'38" -55d54'59"
Lee and Chris
Uma 11h23'40" 48d24'10"
Lee and Harry
Psc 0h53'26" 17d20'52"
Lee and Holly Snelling
Uma 9h12'39" 67d18'37"
Lee and Kristy
Cap 21h38'32" -18d34'43"
Lee and Leighann
Cas 0h56'1" 58d33'3"
Lee and Michelle Forever
Cyg 20h14'48" 34d40'59"
Lee and Sheila Kation
Cyg 20h53'52" 35d3'10"
Lee and Trevor
Ari 1h59'44" 21d37'6"
Lee Andrew Gasper
Aqr 23h35'48" -16d27'15"
Lee Ann
And 2h20'35" 49d17'8"
Lee Ann and Matt
Cyg 21h25'35" 33d55'59"
Lee Ann Crosby/Nguyen
The The
Lyn 7h21'4" 48d45'17"
Lee Ann Davis
Lmi 10h32'8" 36d58'2"
Lee Ann Hall
Psc 0h8'36" -2d45'11"
Lee Ann L'Ecuyer
Psc 1h0'45" 7d34'11"
Lee Ann LOVES Monty
Ari 2h58'22" 24d42'33"
Lee Ann Olinger
Peg 21h35'28" 22d1'15"
Lee Ann Smart-Koch
Cap 21h0'42" -25d27'16"
Lee Ann Sue Zeman
Psc 0h54'51" 30d38'17"

Lee Anna Danna's Planet Splendorama
And 0h45'43" 41d12'46"
Lee Anne
Cas 1h4'6" 69d39'0"
Lee Avery's Star
Sco 16h25'35" -32d30'24"
Lee Bennett Bradford
Gem 7h14'28" 18d10'20"
Lee Boehmer
Aql 18h52'45" -0d42'2"
Lee Broderius
And 0h5'1" 46d12'18"
Lee Brophy 21
Uma 9h47'1" 56d36'6"
Lee Champagne
Her 16h44'5" 6d41'13"
Lee Clark Jr.
Ori 5h55'10" 20d38'18"
Lee Coulthard
Del 20h35'39" 4d25'13"
Lee "Cowboy" Patten
Cnc 9h13'39" 32d46'17"
Lee "Dani" Nelson
Aqr 21h5'28" -10d2'25"
Lee David Stephens
Ori 5h8'8" 4d31'5"
Lee David Stratmeyer
Her 17h14'26" 48d30'10"
Lee David Wiseman
Cha 13h26'31" -79d47'38"
Lee Deflorio
Sco 16h42'15" -44d27'32"
Lee & Denise Meling
Her 17h55'34" 22d27'24"
Lee DeZuzio
Uma 10h3'35" 70d20'37"
Lee Dillion
Leo 10h3'48" 24d16'0"
Lee Doll
Uma 11h15'48" 54d21'49"
Lee Edward Brun
Lib 15h48'39" -7d41'19"
Lee & Emma
Cyg 21h36'54" 41d47'37"
Lee Etnyre Keiser
Gem 6h58'41" 14d11'44"
Lee F. Cresta
Leo 10h13'34" 25d19'52"
Lee Fleming
Cnc 8h35'11" 27d38'23"
Lee Freeman
Aql 19h10'18" -0d57'38"
Lee Fuller Best Mom Ever!
Psc 1h14'16" 18d6'0"
LEE "GATOR" BUCHER
Sgr 18h56'41" -30d4'47"
Lee Gauld
Per 3h45'52" 32d17'3"
Lee & Gisela Phelan
Cyg 21h55'36" 54d40'58"
Lee Grossnickle
And 23h40'3" 42d50'28"
Lee "Hallie" Albright
Cnv 12h44'54" 42d8'2"
Lee Halligan
Aqr 22h16'19" -8d27'19"
Lee Hamilton
Her 17h26'48" 27d3'42"
Lee Harris Nash
Leo 11h25'23" 16d56'10"
Lee Harrison Scott
Sgr 19h40'1" -12d29'29"
Lee & Helena's Everlasting LoveStar
Cas 23h20'53" 54d55'17"
Lee Henry
Umi 15h19'48" 81d29'42"
Lee Herberholz
Gem 7h14'31" 20d8'29"
Lee Hiu Ham (Ada)
Cap 20h28'16" -12d0'49"
Lee III
Umi 13h32'58" 69d52'4"
Lee J. Whiteley 9/9/39
Uma 10h38'17" 40d33'56"
Lee James Eastbourne
Tau 5h36'47" 26d57'34"
Lee James Principe
Per 2h42'38" 56d14'5"
Lee & Jenni 040504
Uma 13h15'50" 59d44'8"
Lee Jordyn Danielsen
Sco 16h6'29" -13d2'10"
Lee Jorgensen
Uma 11h51'5" 40d0'23"
Lee Joseph Grande
Sco 16h16'7" -8d49'29"
Lee Jung-Ok
Uma 11h37'48" 41d35'4"
Lee K. Newsom
Uma 10h40'13" 66d34'22"
Lee Kaufman
Vir 13h37'27" -8d44'51"
Lee Kinder's Perfect Present
Sgr 18h44'51" -32d22'9"
Lee Knapp
Umi 15h22'45" 79d29'2"
Lee Laine 10-23-55
Uma 10h50'51" 46d24'25"
Lee Laughinghouse
Per 3h19'44" 46d8'25"
Lee Lee
Cnc 9h7'13" 25d4'29"

Lee Lee
And 0h40'25" 22d11'48"
Lee Lee
Cnc 8h44'31" 14d1'35"
LEE LEE LMF
Sco 16h34'32" -28d49'58"
Lee Lee & Scooby Forever
Uma 9h26'27" 53d22'2"
Lee Logs Barrs
Leo 10h53'16" 12d25'45"
Lee London
Aql 19h19'45" 0d45'12"
Lee Loush
Cnc 8h43'1" 14d16'13"
Lee Lu
Cnc 8h6'5" 21d52'55"
Lee Lunardi - My True Love for Life
Cep 22h57'43" 71d37'0"
Lee M. Barry 007
Leo 10h3'52" 22d7'36"
Lee Makepeace
Per 2h53'42" 42d13'7"
Lee Martin
Uma 14h23'47" 57d37'18"
Lee & Mary Edwards
Cyg 19h57'55" 46d37'59"
Lee Masangkay
Cap 21h40'30" -21d51'35"
Lee McCloskey the love of my life
Pup 7h44'6" -22d20'55"
Lee McKenna
Cnc 8h9'0" 24d53'37"
Lee & Meika Manghan
Per 4h1'4" 43d10'59"
Lee Merklinger endless dreams & possibilities
Uma 10h19'50" 63d34'54"
Lee Michael Tetreault
Per 2h52'43" 44d8'13"
Lee Michelle Kelpy
Psc 0h57'41" 9d19'17"
Lee Milton Machemer III
Gem 7h15'24" 32d48'30"
Lee "Minitree" Weisbrod Memorial
Cnc 8h26'51" 10d4'44"
Lee "Miracle Mom" Berlin
Ari 2h55'53" 13d11'45"
Lee Mobley Shining Star
Uma 10h44'3" 54d9'33"
Lee (My Baby)
Lib 14h54'48" -24d30'32"
Lee & Nancy Harmon
Cyg 21h14'7" 32d17'30"
Lee Nordstrom
Boo 14h45'21" 52d2'54"
Lee O Webb
Leo 10h19'6" 25d17'40"
Lee Patrick Neenan
Del 20h37'2" 14d53'3"
Lee & Patty
Cyg 20h1'24" 49d43'15"
Lee R. Snyder a.k.a. Pop
Sco 16h7'23" -13d15'50"
Lee Redfearn Shubert, Amicus Carus
Aqr 22h2'56" -0d45'42"
Lee & Rikard
Cyg 20h2'12" 49d44'20"
Lee Robert Gibson
Uma 8h46'33" 50d59'15"
Lee Robert Ottulich
Aql 19h0'14" -7d37'27"
Lee Roy Boehme Sr.
Cep 21h18'40" 58d24'41"
Lee Roy William Hylton
Cap 20h10'32" -26d8'52"
Lee RYAN
Cyg 21h50'32" 44d59'41"
Lee' s Birthday Star
Cyg 19h37'58" 31d54'6"
Lee Santiago
Aqr 22h45'34" -4d41'48"
Lee Sisselsky
Tau 5h5'11" 20d59'23"
Lee Steven Ford
Umi 16h36'22" 76d3'26"
Lee Stolzenhein
Eri 4h17'35" -34d4'28"
Lee Sutherland Burrows
Lyn 8h59'47" 39d37'16"
Lee' Tanya Janel Adams
Mon 6h34'22" 7d39'11"
Lee The Most Wonderful Star
Cnc 9h16'37" 7d42'23"
Lee Theaker
Per 3h6'6" 37d31'45"
Lee Thomasson
Lyn 7h48'0" 44d1'3"
Lee Triumph ELP Orc Coulbeck
Dra 15h51'2" 52d31'55"
Lee Tsukie Hirata
Tau 4h25'6" 23d50'45"
Lee Ulrich
Leo 10h5'11" 22d34'14"
Lee Van Omen
Mon 7h17'5" -0d38'1"
Lee Vue
Leo 11h42'23" 17d10'45"
Lee W. Brandes
Per 3h25'49" 33d30'1"

Lee Wiley Harned
Per 3h48'53" 35d33'30"
Lee Yu Chen
And 0h8'34" 29d11'50"
Lee, 21 Years Shining Bright-lee
Aqr 21h8'58" -8d54'51"
Lee,Brenda,Sam,Abby,and Ben Fritsch
Her 18h37'39" 20d39'16"
Leea Lamminaho
Her 16h43'20" 6d52'3"
Leea Marie
Ori 6h1'51" 8d12'16"
Leeann
Leo 11h42'54" 17d7'49"
Lee-Ann
And 2h24'53" 50d25'14"
LEEANN
Cap 20h28'42" -22d56'50"
LeeAnn Carol
Psc 1h22'32" 7d49'5"
LeeAnn Elizabeth Davis
Cnc 8h17'15" 29d46'7"
Leeann Elizabeth Turner
Aqr 23h1'8" -16d16'37"
LeeAnn Hutchinson
Uma 11h6'56" 54d44'55"
LeeAnn Marie Kirschner
Peg 23h37'11" 13d55'36"
LeeAnn Michelle Kuiper
Cam 7h45'22" 76d23'13"
Leeann Shumway
Cnc 9h10'17" 13d37'10"
Leeann:Summer's Creator
Aqr 21h9'10" -9d40'15"
LeeAnna Elizabeth Nohs
Cnc 8h32'29" 10d27'27"
LeeAnna Meche McGee (Maw-Maw)
Aqr 22h28'29" -0d39'5"
LeeAnne
Cas 23h58'5" 56d45'36"
Leeanne
Vul 19h4'43" 23d43'25"
Leeanne Blake
Vir 13h20'49" -17d6'7"
LeeAnne Code
Uma 11h25'24" 59d55'15"
Lee-Anne's Little Piece of Heaven
Gem 6h21'51" 19d35'16"
LeeAnn's Everlasting Star
Gem 7h34'27" 24d7'1"
Leebo
Ori 6h11'51" 15d23'53"
Leela
Uma 11h11'13" 67d51'58"
LeeLa
Pho 0h56'19" -44d42'6"
Leela Elizabeth
Psc 1h13'57" 26d59'55"
LeeLee
Leo 11h58'7" 20d1'37"
LeeLee
Uma 9h38'25" 56d25'29"
LeeLee R Washaleski
Sco 17h28'37" -38d18'12"
Leelee's "Golden" North Star
Cas 1h24'55" 61d45'28"
Leelo
Aqr 22h45'9" 0d1'16"
Leeloo
Cyg 19h59'0" 45d48'34"
LEELOO
Per 3h19'27" 41d53'21"
Leelou
Uma 9h19'19" 48d33'24"
Leen
Cas 1h33'18" 57d50'59"
LEENA
Lyn 7h32'26" 37d13'59"
Leena
Ari 3h8'19" 12d53'35"
Leena Kohli
Ari 3h15'40" 28d37'8"
Le'na Rose Michaels
Cyg 19h58'5" 59d47'8"
leena spiritus cupiditas
Sgr 19h19'34" -21d36'28"
Leena99
Aqr 22h33'56" -17d33'7"
Leena's Heart
Cru 12h26'16" -58d34'35"
Leena's Serenity
Ari 1h58'53" 21d28'29"
Leendert
Her 16h43'53" 6d19'5"
Leeney
Sgr 19h35'29" -14d18'27"
LEEN'S Shining Star
Lib 14h35'45" -14d56'42"
Leeona Carstairs - 02/02/1960
Aqr 23h2'43" -18d49'33"
Lee's Glory
Ori 6h15'57" 15d22'55"
Lee's Love Ray 12/25/00
Lyn 8h5'1" 34d34'24"
Lee's Shooting Star
Pho 1h55'23" -43d33'1"
Lee's Star
Uma 10h42'49" 50d13'54"

Lee's Star
Ari 2h43'37" 13d10'22"
Lee's Star
Her 17h55'11" 23d30'7"
Lee's Star of Mostad
Sco 16h48'5" -27d6'38"
Leesa Andrea
Cru 12h24'8" -58d8'34"
Leesa Csolak, The Starmaker
Cas 0h20'4" 65d13'41"
Leesa Love
Cyg 20h51'34" 33d6'54"
Leesa Marie
Cap 20h53'23" -17d2'33"
Leesah Uterus
Uma 11h17'35" 51d27'24"
Leesh
Sgr 18h23'38" -31d58'58"
Leesie
Lmi 10h45'35" 36d13'10"
LeeStar
Tau 4h38'36" 0d19'1"
Leeta Mook
Aql 20h5'36" 5d7'57"
Leetah Higgins
Sgr 18h10'35" -16d46'3"
Lee'Tayna Janel Adams
Crb 15h49'24" 38d39'19"
LeeTi
Uma 9h35'34" 58d56'9"
Leevi
Aql 19h42'28" 7d51'30"
Leevon Sarah
Eri 4h26'13" -2d52'45"
Leeza Bock
Psc 0h28'6" 16d42'0"
Leeza Palmer
Lyn 8h51'42" 41d16'29"
LeFaucheur-Cullinane
Cyg 19h46'41" 39d56'59"
Lefkie Sophia Germanides
Ari 3h26'2" 26d43'45"
Left Coast Pizza
Uma 9h38'7" 41d30'13"
"Legacy"
Uma 12h1'11" 56d19'10"
Legame Vita
Lyn 8h37'37" 38d54'42"
Legan Family
Uma 10h19'11" 51d12'1"
legas6580
Cen 11h19'49" -44d4'9"
Legdrágább Kincsem
Tau 5h34'33" 24d3'59"
Legend
Uma 9h22'54" 66d58'23"
Legend Ruel
Uma 10h26'4" 64d4'50"
Legends Never Die Paul
Lyn 7h14'33" 52d49'11"
Legény Margit
Cyg 21h14'52" 46d59'4"
Legionnaire
Tau 4h40'5" 23d28'17"
Legna
Lup 15h46'36" -32d22'14"
Legna Anaid
Psc 1h34'11" 8d21'28"
Lego
Sco 17h20'42" -32d34'21"
Legolas
Cep 3h42'52" 81d23'20"
Legolas
Ari 3h17'5" 21d51'9"
LeGrys
Uma 10h24'20" 46d0'50"
LEH
Cap 20h23'3" -9d20'17"
Lehmann, Heinz
Uma 9h44'30" 63d26'29"
Lehmann, Wolf-Dieter
Uma 10h12'0" 55d40'5"
Lehmarvellous
Cru 12h43'14" -58d10'12"
Lehoczki Marcsi
Sco 16h23'31" -32d24'3"
Lei Le Ani
Cnc 8h37'14" 24d50'14"
Leia
And 23h13'14" 46d45'45"
Leia
Gem 7h1'53" 34d29'16"
Leia
Sco 16h10'23" -12d6'42"
Leia Adelle Harper
And 23h25'45" 42d43'6"
Leia Danielle Lem
Aqr 22h36'6" -1d54'59"
LEIA FREESE
Lyn 9h7'50" 37d40'34"
Leia Hayes - Princess of the Sky
Vir 13h5'35" 2d31'42"
Leia Jaiden Star
Gem 7h27'38" 32d51'38"
Leia Rebecca
Cyg 20h25'36" 33d24'5"
Leia Ryan
Uma 11h50'8" 37d3'44"
leia vellner
Cnc 8h17'45" 20d14'45"
Leialoha
Lyn 9h1'23" 39d31'52"

LEIALOHA
Aqr 21h37'31" -0d13'9"
Leidi Collins
Uma 12h36'46" 57d53'2"
Leif & Ann Marie Eltoft
Cyg 20h40'15" 40d2'54"
Leif DeWayne Biffle
Tau 5h38'22" 25d49'32"
Leif Erik Larsson
Vir 14h26'11" 6d46'35"
Leigh
Cas 0h57'55" 60d38'21"
Leigh
Lep 5h48'3" -12d26'27"
Leigh
Sco 16h13'7" -8d51'33"
Leigh 0324
Uma 10h33'56" 62d14'37"
Leigh A. Belden-GaMPI Shining Star
Crb 15h45'20" 31d30'43"
Leigh Adan Dolan
Lyr 18h31'25" 35d49'35"
Leigh Amey
Gem 6h53'32" 24d58'59"
Leigh and Mike
And 0h39'0" 26d3'2"
Leigh and Rick Wolany
Cnc 8h2'2" 10d30'52"
Leigh and Stacey De Col
Cyg 20h4'8" 31d16'55"
Leigh Ann
Gem 7h19'55" 32d48'46"
Leigh Ann
Vir 12h52'11" 8d23'0"
Leigh Ann
Dra 16h56'56" 53d12'14"
Leigh Ann and Nathan
Ori 5h52'25" 11d45'32"
Leigh Ann Bonnette
Psc 0h21'0" 21d30'2"
Leigh Ann Daffern
And 2h27'3" 43d50'10"
Leigh Ann Dzvonick
Cas 1h12'30" 63d50'56"
Leigh Ann Gray
Ari 2h46'17" 26d6'11"
Leigh Ann Howard. Love of my life.
Ori 6h4'29" 17d26'31"
Leigh Ann Kauffman
Uma 10h0'8" 66d42'3"
Leigh Ann Lara
Cap 20h58'47" -15d46'6"
Leigh Ann Lewis
Cyg 20h20'30" 48d38'22"
Leigh Ann Padgett
Sco 16h38'23" -30d9'30"
Leigh Anna Theresa Wetherington
Gem 7h29'4" 29d17'54"
Leigh Anne Rogers
Eri 4h24'0" -0d8'10"
Leigh Blaine
And 23h19'45" 40d5'5"
Leigh Bowman
Aur 7h17'53" 42d5'46"
Leigh & Brian
Boo 14h45'33" 36d37'16"
Leigh Cecelia Evans
Aqr 21h38'57" 0d41'56"
Leigh Claire & Taylor Brianna Cryer
Lyr 18h27'39" 28d26'15"
Leigh Diane (Ward) Linebarger
Aqr 21h21'24" -7d28'15"
Leigh Ella
Uma 9h32'13" 44d51'45"
Leigh Farrell
Per 4h22'57" 35d2'3"
Leigh Ford Kane
Tau 5h57'5" 26d9'20"
Leigh J. Barreto
Crb 15h51'7" 37d31'7"
Leigh James Felstead - 22.05.1987
Aur 6h32'15" 34d43'3"
Leigh Jones Orthodontics
Pyx 8h30'21" -28d43'3"
Leigh Joyce
Cap 20h18'20" -13d10'14"
Leigh K Kern
And 2h16'12" 49d28'12"
Leigh Lei
Pho 0h33'59" -42d5'15"
Leigh Letson
Ori 6h3'59" 12d45'44"
Leigh & Manuel's Wedding Star
Cyg 19h50'45" 51d22'15"
Leigh Margaret Anderson
And 23h50'7" 36d8'20"
Leigh & Michael Demarco
Cnc 8h42'49" 31d44'10"
Leigh Milton Roberts
Uma 9h9'46" 66d22'30"
Leigh Percival Frankel
Uma 10h2'22" 60d17'8"
Leigh Pritchett
Vul 20h27'22" 29d7'4"
Leigh Reilly
Lyn 8h6'10" 42d0'36"
Leigh Vazquez
Aqr 21h37'2" 0d49'17"

Leigha
Uma 10h24'40" 69d17'21"
leigha
Lib 14h48'19" -9d0'31"
Leigha Ann Rybnick
Sgr 18h32'42" -19d19'12"
Leigha Brooke Gallagher
Cas 0h46'45" 66d36'35"
Leigha Christine Stankewich
Ori 5h42'41" -1d35'12"
Leigha Marie
Vir 14h40'34" -1d11'46"
Leighalaine Morgan Muse
And 0h25'22" 42d44'51"
Leighaltys
Sco 17h3'40" -32d59'13"
LeighAndy
Lib 14h24'1" -12d21'45"
Leighann
Cap 20h27'31" -12d19'30"
Leigh-Ann
And 1h26'13" 44d22'55"
Leighann
Tau 3h47'51" 7d58'34"
LeighAnn
Vir 12h44'9" 8d23'25"
Leighann and Stephen Forever
Gem 7h11'20" 26d59'13"
Leighann Etheridge
Tau 5h3'22" 24d29'24"
Leigh-Ann K. Anders Stripling
Sgr 18h33'38" -29d32'16"
Leighann Koets
And 1h24'34" 49d16'4"
Leigh-Ann *Leiash* Helen Foster
Gem 7h15'5" 30d11'11"
Leigh-Ann Viramontes
Crb 16h18'9" 30d52'28"
LeighAnne Manwiller
Gem 7h23'44" 30d22'41"
Leighism
Crb 15h47'5" 26d33'8"
Leigh's Love
And 0h2'21" 44d42'38"
Leigh's star
Vir 14h19'45" 5d35'49"
Leigh's Star
Cam 5h23'8" 68d39'48"
Leigh's Star 50
Psc 1h34'14" 16d21'11"
Leigh's Wish
Gem 6h38'10" 12d43'14"
Leighton
Sgr 19h51'27" -15d26'15"
Leighton Bo Ewart
Her 16h40'19" 17d45'41"
Leighton Grace Heard
Crb 15h46'56" 25d52'4"
Leighton Lambert
Per 4h8'15" 33d20'38"
Leighton Michael MacDonald
Cep 20h56'4" 69d48'32"
Leighton Olivia Farrar
Tri 2h21'32" 32d14'47"
Leighton Richards
Cru 12h37'21" -58d12'31"
Leighton's Star
Per 2h16'13" 57d23'31"
Leiko & Robert Beck
Cnc 8h24'22" 9d49'50"
Leila
Gem 6h59'42" 14d41'29"
Leila
Leo 11h42'41" 25d33'22"
Leila
Cam 3h53'17" 59d54'8"
Leila
And 23h39'13" 47d13'30"
Leila
Lyr 19h11'12" 39d4'59"
Leila
And 0h2'15" 40d49'29"
Leila
Sco 17h36'18" -37d53'59"
Leila
Cyg 20h22'37" 55d54'43"
Leila
Vir 11h56'45" -1d17'35"
Leila Allen
Lyr 18h49'54" 35d42'11"
Leila Ashmahan (Renee Beers)
Cas 3h22'1" 70d5'11"
Leila Belal
Umi 16h7'6" 73d15'26"
Leila Dorothy
Cra 18h12'14" -45d28'8"
Leila e Paolo
Peg 21h30'13" 23d30'54"
Leila Elizabeth Robinson
And 2h28'8" 42d26'18"
Leila Faye Coons
Cas 0h42'14" 67d14'31"
Leila Grace Weck
And 0h1'45" 44d42'34"
Leila Jean Kavanagh
Lep 5h52'57" -17d33'53"
Leila Joan
Cra 18h1'10" -37d35'57"

Leila Mafoud
Mon 6h48'56" -0d10'3"
Leila Marie Coulliette (Bug-a-boo)
Cas 0h30'23" 62d6'17"
Leila Masinaei
Ari 2h5'26" 14d17'42"
Leila Michelle Nouri
Peg 23h21'8" 32d13'9"
Leila Oswald
Cyg 20h40'52" 51d11'15"
Leila Perry
Sco 16h12'42" -13d18'55"
Leila Rose Rivera
Psc 0h22'30" 17d7'51"
Leila Sian Matthews
And 2h16'32" 42d15'58"
Leila Simone Hinton
Gem 7h5'24" 20d43'22"
Leila Sleet
And 1h25'40" 40d8'26"
Leila Straffi
And 23h22'23" 42d35'56"
Leila Tahira Haselden
And 2h25'25" 46d12'51"
Leila Zanjani
And 0h9'15" 28d42'25"
Leilani
Ari 2h16'36" 26d13'50"
Leilani
Cnc 8h40'34" 19d38'1"
Leilani
And 1h6'31" 42d14'50"
Leilani
Lyn 8h15'21" 42d26'40"
Leilani
Lib 15h6'47" -11d20'53"
Leilani
Sgr 18h27'38" -16d45'2"
Leilani and Jamison Veitch
Ari 2h44'57" 21d53'46"
Leilani Esther Cochran
Psc 1h13'48" 7d1'24"
Leilani Hill
Peg 22h41'15" 19d38'58"
Leilani Kiyoko Muneno
Peg 23h4'39" 13d9'55"
Leilani "Little Flower from Heaven"
And 0h41'42" 23d20'15"
Leilani May Will
And 1h51'45" 45d10'44"
Leilani Monet
Cnc 8h52'19" 30d22'57"
Leilani Raz
Crb 15h47'13" 31d23'38"
Leilani Renee Gaucin
And 0h52'34" 41d42'23"
Leilani Rose
Per 4h34'34" 31d58'19"
Leilani Winkfield
Psc 0h33'57" 7d16'6"
Leilani, Divine Flower
And 2h38'41" 45d55'34"
Leilanie Pulido Gaon
Tau 4h23'41" 21d15'59"
Leilani's Oasis
Umi 14h55'31" 77d15'25"
Leilani's Star
Sco 17h50'48" -30d12'6"
Leila's Dragon
Dra 17h30'22" 67d8'44"
Leïlatoon Shahshahani
Umi 13h45'12" 73d4'44"
LeiLei
And 0h7'17" 43d27'32"
Leili Z.
Leo 10h19'7" 17d15'21"
Leimomi Kahailiaioha
Ari 3h6'4" 11d53'58"
Leina Rae Hutaff
Tau 3h43'1" 13d57'51"
Leiola Haunga
Uma 10h14'37" 60d7'42"
Leipert, Joachim
Uma 11h35'49" 30d41'46"
LEISA
Gem 7h29'17" 20d13'52"
Leisa Ann Bishop
Sgr 18h17'14" -30d41'7"
Leisa J. Onzo
Mon 7h31'13" -0d40'13"
Leisa Jayne Durrett
Ari 2h54'37" 21d21'6"
Leisa Moore
And 2h28'41" 42d14'42"
Leisenberg, Peter
Ori 6h17'26" 16d41'35"
Leisha
Cru 12h40'13" -56d17'14"
Leisha Aliberti
And 23h8'34" 48d1'38"
Leisi...My Love
Cap 20h41'26" -14d49'20"
Leita & Robert Mum & Dad Forever
Cru 12h28'27" -61d25'6"
Leitet
Leo 11h12'12" -1d53'22"
Leith Ellen Louise Turner
Sgr 19h37'55" -41d46'22"
Leith Elliott
Ari 3h17'6" 12d28'52"
Leitha
Crb 16h20'54" 39d31'8"
Leiting the Way! Lincoln K4 2005
Uma 11h53'25" 58d41'42"
Leiu Justine Blais
Crb 16h11'41" 37d37'9"
Leizel Miranda
Aqr 21h2'16" -7d20'3"
Lejla Nicevic
Ari 3h10'0" 29d38'22"
LEK
Sgr 18h13'16" -16d45'44"
Leka
Aqr 22h54'30" -15d22'0"
Leka
Uma 10h30'49" 54d37'29"
Léka
Peg 22h35'13" 20d10'59"
Leke
Per 3h20'22" 48d9'25"
Le-Khanh Thi Pham
Psc 0h28'33" 18d33'44"
LEL1984
Lyn 8h13'44" 38d45'22"
Lela Ann Barker
Leo 10h30'19" 20d58'51"
Lela Attilio
Umi 14h57'26" 79d4'45"
Lela E
Sco 16h43'6" -29d39'3"
Lela Gillispie
And 23h23'33" 48d27'11"
Lela Harkness Edwards
Ari 2h21'37" 27d42'55"
Lela Iline Ploussard
Cyg 20h27'30" 40d1'11"
Lela Lenora Ripamonti
Ori 5h56'1" 9d20'9"
Lela Marie
Cnc 9h4'5" 20d22'0"
Leland Earl "Grandaddy" Biggs
Cep 22h27'23" 81d7'57"
Leland Gage Wibel
Uma 11h18'13" 62d57'2"
Leland James Kahn
Aur 5h44'43" 55d40'16"
Leland Julian Dix
Her 17h13'3" 34d2'14"
LELAND L. 'LEE' KATZ JR.
Uma 9h2'27" 68d33'9"
Leland Luttmer
Uma 11h57'2" 62d51'28"
Leland Pants
Aqr 21h40'28" -3d23'52"
Leland Ray (The Wood Star)
Per 3h5'56" 41d57'4"
Leland Robinson
And 2h27'52" 44d7'20"
Leland Sodapop Wiackley
Uma 13h9'43" 56d30'28"
Leland Steven Lehnus
Uma 8h39'49" 51d16'50"
Leland Taylor Stewart
Uma 10h8'49" 67d54'45"
LELE
Uma 11h34'50" 65d6'38"
Lele
And 1h10'50" 46d10'38"
Lelia Mae
Leo 11h7'6" 16d29'3"
Lelio Marino
Her 16h42'15" 33d15'13"
L'Elisa + JV
Tau 4h26'0" 21d21'59"
Lella
Cas 2h43'35" 58d3'51"
Lella
Lyn 6h20'30" 56d13'56"
Lella 1965
And 23h16'46" 50d52'50"
Lelli, Antonio
Ori 6h11'39" 13d51'3"
Lellie
Per 3h47'49" 43d45'28"
Lelliott's Law
Sgr 19h21'28" -17d52'40"
Lelly Fornuto
Psc 23h42'5" 7d0'6"
Lelu's Little Stellina
Ari 2h31'52" 22d3'23"
Lely Hunkie Hang
Gem 6h49'11" 17d12'33"
Lemac
Dra 19h31'44" 73d36'19"
Lemaia Timanica
Cru 12h23'24" -58d35'54"
LeMay
Gem 7h34'9" 16d46'40"
"LeMetric" Elline Surianello
Leo 10h17'43" 25d17'54"
Lemniscate 8
Uma 11h49'26" 44d37'44"
lemo
Cap 20h27'7" -8d38'24"
Lemondrop
Cnc 8h23'47" 31d59'20"
Lemonsa
Cas 0h2'46" 56d14'38"
LeMoyne X
Aql 19h47'55" -0d28'52"
Lem's Star
Ori 5h20'35" 6d9'7"
Lemur
Cyg 21h43'35" 33d10'37"
Lemusu T. Tafunai
Leo 10h16'14" 20d37'5"
Len
Ari 2h9'22" 16d37'40"
Len and Janine Harrison
Cyg 19h37'15" 32d25'29"
Len & Joan "Our 2 Northern Lights"
Cyg 21h43'36" 48d53'56"
Len&June Shine Together 4ever As 1
Ari 2h10'5" 17d55'8"
Len & Phyl
Eri 3h51'12" -7d41'35"
Len R. Tatore
Sco 17h48'7" -41d30'16"
Len & Selma Fisch 50th
Cyg 20h37'5" 45d54'33"
Lena
Ari 2h28'26" 20d52'0"
Lena
Leo 11h23'50" 4d47'26"
Lena
Leo 10h35'19" 18d50'21"
Lena
Ori 5h58'2" 21d23'29"
Lena
Crb 16h24'54" 28d45'54"
Lena
Psc 1h31'23" 2d56'39"
Lena
Lib 14h51'45" -0d44'25"
Lena
Aql 18h51'59" -0d23'40"
Lena
Mon 6h43'6" -0d0'14"
Lena
Dra 9h25'4" 74d7'36"
Lena
Uma 10h17'6" 65d8'29"
Lena
Uma 9h5'36" 54d13'38"
Lena
Cas 23h28'2" 55d35'31"
Lena 417
And 1h13'6" 44d32'24"
Lena Albazie
And 23h25'38" 37d15'36"
Lena Alexander
Cap 20h46'27" -20d52'48"
Lena Anne Beul
Ari 2h17'29" 17d5'0"
Lena Anthony Campanella
Leo 10h59'1" 16d2'56"
Lena Artemis
And 1h12'49" 47d17'24"
Lena Beutler
Cas 0h59'6" 60d38'28"
Lena Blue Eyes
Tau 3h29'24" 24d46'48"
Lena Christine Karas
Sco 17h23'57" -33d1'23"
Lena Cowan Jones
Uma 11h9'18" 68d47'0"
Lena Emely
Her 17h59'36" 16d56'36"
Lena Etienne
Cas 0h53'37" 61d43'12"
Lena Finnerty
Lyr 18h25'49" 35d17'16"
Lena Grace
Lmi 10h37'28" 37d46'35"
Lena Jill Austria-Cayamanda
Vir 14h57'50" 0d36'43"
Lena Joon
Aqr 22h59'3" -13d50'31"
Lena Kay
Cnc 8h54'47" 25d59'13"
Lena Koydl
Aqr 22h24'53" -6d23'59"
Lena L. Ruffatto
Peg 23h10'25" 26d42'43"
Lena Langton *Nanna*
Cas 1h17'45" 50d43'44"
Lena Li
Crb 16h5'13" 30d38'0"
Lena Mae Bolton
And 23h44'25" 43d44'53"
Lena "Mama" Herpin
Uma 8h44'58" 53d47'47"
Lena Marie *14.04.2004
Uma 10h43'17" 41d0'25"
Lena Marie Flynn-Fuller
Vir 13h56'11" -4d38'29"
Lena Marie Reyes Patrick
Pho 23h45'19" -45d32'19"
Lena Marie Weyers
Com 12h43'15" 27d42'14"
Lena Masiello
Crb 15h44'42" 31d33'44"
Lena Nalbone
Lib 15h47'24" -5d24'18"
Lena "Nanny" Leone
Cas 0h3'56" 53d42'12"
Lena Nila
Uma 11h57'56" 31d37'16"
LENA NOELLE
Boo 13h48'43" 23d23'18"
Lena / Prue
Aqr 22h43'18" -24d5'6"
Lena Rafaela
Lmi 10h33'31" 37d50'12"
Lena Rain Wilson
Lyn 7h9'41" 51d6'44"
Lena Razo
Uma 14h26'28" 59d25'10"
Lena Rhodes
Ori 6h11'5" 15d10'24"
Lena Rose
Cas 1h24'28" 72d37'48"
Lena Sotelo
Ori 5h0'46" 5d35'56"
Lena Wanda Ciolkosz - 100 Yrs.
Uma 10h47'9" 61d46'39"
Lenarcik, Emilia
Uma 9h15'2" 68d6'30"
Lena's Pixie Dust 32
And 0h46'26" 30d14'18"
LenBet
Uma 11h15'44" 52d15'47"
LenDoreen
Uma 11h23'36" 57d40'0"
Lene
Lib 15h54'47" -5d45'48"
Lene Sivertsen
Lyn 7h0'18" 53d44'34"
Lene Susanne
And 1h21'57" 43d22'44"
Lene Vinding-Rasmussen
Cap 20h11'42" -15d47'47"
Lenea
Gem 6h49'51" 27d26'53"
leni
Uma 11h47'44" 45d28'50"
Leni
Uma 11h12'53" 61d33'1"
Lenica Kwiatek
Psc 0h28'46" 18d11'13"
Lenicka
Ari 1h56'8" 18d43'33"
Lenida Olaes Ventus
Tau 5h29'27" 25d8'39"
LENIO
Aqr 21h41'24" -0d25'3"
Lenise
Uma 10h31'41" 47d14'35"
Lenise C. Plummer
Aqr 22h34'15" -1d25'31"
Lenisima
Lyn 7h32'16" 52d20'13"
Lenja Rahel
Cyg 21h15'10" 45d29'34"
Lenka
Lyr 19h5'54" 36d47'7"
Lenka
Peg 22h52'34" 13d20'24"
Lenka Hermanova & Antoine Avignon
Ari 2h48'56" 28d38'26"
Lenka Ledvinka
Cnc 8h28'27" 6d58'23"
Lenka Menyhartova
Lyn 7h48'48" 47d25'3"
Lenka Ravaszova
Cyg 21h31'24" 39d27'25"
Lenka - the most beautiful...
Ori 5h20'58" 2d28'19"
Lenlen
And 23h1'3" 48d24'38"
Len-Linn Star of the future
Aql 19h51'14" 3d35'2"
Lennard Benedikt Böhm
Uma 8h18'30" 67d2'48"
Lennard du Toit
Ori 6h17'19" 14d26'32"
Lennart de Vet (Bkyt)
Her 17h49'32" 45d51'25"
Lennart Simonsson
And 1h0'44" 34d32'40"
LENNART STAR
Uma 11h51'4" 47d51'26"
Lenna's Christmas Star
Sgr 19h8'29" -17d42'20"
Lenna's Star
Lyn 8h12'49" 57d43'10"
Lenné - In My Heart Forever
Psa 22h35'32" -26d6'7"
Lenné, Heinz
Sco 17h16'7" -32d57'43"
LENNEA
Tau 5h6'58" 24d49'55"
Lennell H. Dunbar
Vir 14h7'40" -13d28'39"
LeNnie
Cap 20h52'14" -20d35'20"
Lennie Allen
Uma 10h1'34" 54d16'38"
Lennie and Danielle Gill
Cyg 21h57'57" 53d4'34"
Lennie and Lesli
Her 16h21'31" 20d4'56"
Lennie Lee Copeland Mills
Uma 8h32'6" 63d29'14"
Lennon Alfredo
Cap 21h27'40" -15d11'57"
Lennon James
Per 2h40'0" 52d52'49"
Lennon James Lee Warwick
Her 16h41'55" 12d58'45"
Lennon Vincent Campo
Lib 15h25'0" -6d55'11"
Lennox Thomas McDermott
Lmi 10h6'8" 35d9'21"
Lenny
Per 3h52'25" 34d36'24"
Lenny
And 2h17'23" 38d24'50"
Lenny
Lmi 9h37'6" 39d9'40"
Lenny
Vir 12h45'8" 11d45'1"
Lenny
Umi 16h12'55" 77d16'28"
Lenny and Helen Santoro
Cyg 20h16'8" 33d12'21"
Lenny Bissonnette
Ori 5h41'24" 1d43'49"
Lenny C. Frahme
Sco 16h12'41" -9d14'11"
Lenny F. Bragg
Uma 11h40'8" 41d52'28"
Lenny Grandchamp
Cyg 20h2'49" 40d18'11"
Lenny Lhérault
Ori 6h15'52" 13d46'37"
Lenny & Marian's "Lucky Star at 75"
Dra 16h0'6" 57d57'58"
Lenny miin Star
Com 13h2'56" 18d14'17"
Lenny R. Yocum
Vir 12h36'16" 3d39'41"
Lenny St. Gerard
Her 16h24'3" 11d3'7"
Lenny V
Lib 14h45'10" -17d33'39"
Lennys star
Sgr 19h0'9" -31d49'32"
Lenoa K.
Leo 11h57'42" 24d43'47"
Lenochka Koroeva
Cyg 19h39'57" 29d50'19"
Lenok
Uma 11h29'11" 30d28'32"
Lenok
Aqr 22h42'1" -4d51'19"
LENOK 8
Uma 13h45'51" 56d6'6"
Lenora
Cam 4h48'55" 69d12'25"
Lenora
Cyg 21h21'14" 45d56'22"
Lenora
Leo 9h49'55" 22d36'50"
Lenora
Cnc 8h28'38" 20d12'27"
Lenora
Ori 6h17'41" 7d5'4"
Lenora D'Agostino Elliot
Cas 0h29'53" 63d48'26"
Lenora F. Chandler
Uma 13h9'31" 58d0'52"
Lenora Floyd Shining Brightly
Cas 0h13'41" 55d2'30"
Lenora McCarty
Eri 4h33'39" -0d41'29"
Lenora V.E.V. Cox (peanut)
Sgr 18h12'49" -24d22'28"
Lenora Williams
Cas 23h32'57" 51d10'49"
Lenora's Star
Sgr 19h18'34" -27d0'13"
Lenore
Tau 4h49'34" 16d58'49"
Lenore
Ari 3h24'17" 27d58'45"
Lenore 121052
Umi 14h13'52" 76d50'45"
Lenore Conuniello
Cas 23h47'14" 52d45'0"
Lenore E Churgay
Lyn 7h58'21" 38d45'53"
Lenore Janis
Psc 1h3'13" 25d7'9"
Lenore (Libby) Gould
Psc 1h23'23" 24d43'23"
Lenore M. Casteel
Gem 7h43'31" 24d24'32"
Lenore My Love
Crb 15h48'45" 26d21'3"
Lenore Namm
Cyg 21h51'18" 49d6'22"
Lenore Olson
Aqr 22h25'24" -12d51'38"
Lenore Shaw Bjurberg
Crb 15h21'37" 29d26'46"
Lenore Stark Topinka
Sgr 17h53'3" -29d45'29"
Lenroe Beicker
Aql 19h16'56" 7d36'8"
Len's Star from Baby Doll
Ori 6h18'11" 5d52'32"
Lenta Mae Kump
Cyg 19h44'11" 37d14'54"
Lenton Odgon Bush Jr.
Uma 9h9'28" 58d27'21"
Lentzner
Uma 12h3'5" 41d36'42"
Lenusya Mityakova
Aqr 22h11'2" -2d36'39"
Leny Adams
Uma 8h11'5" 67d18'42"
Lenz, Norbert
Uma 12h20'32" 61d49'47"
Lenzen, Marc Levin
Uma 9h42'32" 68d4'32"
Leo
Dra 18h56'57" 74d52'46"
Leo
Lac 22h7'11" 52d44'44"
Leo
Ori 5h30'34" -8d39'17"
Leo
Sco 17h54'51" -31d33'22"
Leo
And 1h44'50" 38d43'0"
LEO
And 23h17'29" 47d44'16"
Leo
Psc 1h36'53" 16d37'11"
Leo
Boo 14h42'17" 24d5'0"
Leo
Her 17h29'1" 27d35'0"
Leo
Leo 10h28'48" 21d0'42"
Leo
Leo 11h2'37" 22d13'10"
Leo Abaunza
Cnc 8h56'10" 7d21'38"
Leo and Anita Dufresne
And 1h26'25" 41d20'31"
Leo and Diane
Cnv 12h43'36" 39d58'47"
Leo and Jillian Forever
Sgr 18h6'13" -17d3'36"
Leo and Travis Senior
Dra 16h9'29" 66d33'2"
Leo & Anna Chernyakhovsky
Aql 19h39'49" 12d38'32"
Leo & Anne Fernandez Eternal Luv
Cas 0h47'58" 60d50'41"
Leo Anthony Giacometto
Tau 4h31'20" 19d33'44"
Leo Asher Fein
Sco 17h19'37" -33d9'9"
Leo Bearman, Jr.
Ori 6h6'31" 18d44'29"
Leo Bell
Boo 14h43'2" 21d13'33"
Leo Bertuzzi
Per 4h20'30" 41d54'45"
Léo Botton Bors
Cnc 8h0'47" 23d48'19"
Leo Bowman
Aur 6h32'46" 39d3'44"
Leo Buurman
Cru 12h33'45" -60d38'45"
Leo Cantagalli
Leo 9h37'32" 28d37'13"
Leo Cassidy Giacometto
Cnc 8h34'13" 29d12'23"
Leo Charles Grenier
Per 3h14'23" 51d4'16"
Leo "Chief" Langlois
Per 2h23'5" 52d40'36"
Léo Collet
Dra 19h10'32" 56d59'24"
Leo Craig
Lmi 10h0'20" 35d53'54"
Leo Daughtrey
Lmi 9h45'46" 34d29'26"
Leo & Dawn Cole
Cep 21h39'29" 60d39'23"
Leo E. McDevitt (09-05-59)
Vir 11h59'17" 9d59'44"
Leo Edelstein
Cyg 20h17'59" 38d14'10"
Leo Eric
Uma 10h18'33" 42d17'31"
Leo Galvan
Per 4h14'37" 35d9'34"
Leo Geoffrey
Leo 10h33'5" 11d44'13"
Leo Goddard
Leo 11h9'2" 6d59'34"
Leo "Guitar Hero" Gunderman
Tau 4h13'39" 20d34'4"
Leo H. Diekhans
Uma 10h25'39" 47d31'35"
Leo Henry Hijduk
Lib 16h1'43" -15d22'7"
Leo Hindery 2005 Le Mans Champion
Uma 10h27'3" 43d53'37"
Léo Huot
Umi 16h11'8" 76d14'29"
Leo J. Breirather
Her 17h7'58" 34d47'7"
Leo J. Nosal Sr.
Cep 21h57'39" 65d44'53"
Leo James Fagan
Cep 22h47'45" 67d55'9"
Leo James Oehlschlager
Cap 21h48'6" -18d51'37"
Leo Kelly
Lmi 10h43'51" 28d9'53"
Leo Lamb Brown
Her 17h53'27" 23d36'58"
Leo Leonard Heisinger
Uma 8h48'53" 48d36'24"
Leo Lewandowski
Per 3h13'9" 52d40'39"
Leo Lombardi
Cyg 21h55'7" 41d36'3"
Leo Major
Cnc 8h16'47" 19d55'29"
Leo Mattison Moore
Lyr 18h28'43" 36d43'57"
Leo Michael Wisniewski
Ari 2h46'32" 30d46'3"
Leo Mullen
Uma 11h12'38" 31d46'33"
Leo & Nancy 1314
Cyg 19h40'13" 41d6'9"
Leo Piechowicz
Uma 12h22'36" 53d10'44"
"Leo" Praxis Dr.Hussel & v.d.Lühe
Uma 8h26'47" 68d55'22"
Leo Ralph McIntyre
Vir 13h18'33" 4d44'21"
Leo Raymond McCracken
Cep 22h32'58" 65d9'22"
Leo Robinson-Hurst
Leo 10h45'40" 12d46'34"
Leo Rossi
Ori 5h5'58" 13d20'4"
Leo S. Gallo
Lyr 19h27'1" 37d36'58"
Leo Sadala
Sco 17h2'33" -43d15'54"
Leo Star Jean Hunton
Leo 10h53'55" 12d10'50"
Leo Steinberg
Her 16h27'11" 49d21'3"
Leo Stember
Sco 16h11'36" -14d9'34"
Leo Stockwell
Her 16h46'15" 37d17'40"
Leo & Tatiana
Leo 11h58'22" 14d58'10"
Leo the Gentle Teacher
Per 3h39'38" 44d42'39"
Leo Tsoir
Tau 5h33'38" 20d55'26"
Leo W. Ziemlak
Aqr 21h7'29" -9d8'15"
Leo Ward
Leo 11h54'5" 20d43'49"
Leo Wedow
Psc 1h7'25" 29d51'52"
Leo Wheeler
Ari 2h23'33" 22d5'12"
Leo William Dietlin "Who is Loved"
Leo 10h9'37" 12d13'51"
Leo William Weber
Lmi 10h57'53" 26d10'9"
Leo Wolfe
Lmi 10h17'32" 35d8'42"
Léo Yohan Joshua Wurpillot
Leo 9h27'32" 24d38'10"
LEO ZAC
And 0h6'43" 44d2'31"
Leo Zelvis
Lmi 10h18'27" 36d38'19"
Leo Zieman
Lmi 9h30'30" 38d6'17"
Leo, Lisa, Sandy & Baby Whipps
Uma 13h58'56" 58d59'49"
Leo28
Leo 9h29'12" 15d8'17"
LEOASH
Lmi 10h32'6" 36d6'11"
Leobardo Ramirez
Cap 21h40'33" -12d43'12"
LeoDeGario Lopez Sr.
Cyg 21h8'1" 51d29'28"
Leojunepiter
Lmi 10h58'41" 32d47'17"
Leokadia Gougeon
Mon 6h46'29" -0d4'17"
Leola
Cap 20h20'35" -11d6'38"
Leola Carolyn Weltzin
Uma 9h9'59" 57d31'20"
Leola L. Shaw
Uma 11h22'14" 60d11'7"
Leola "Lee" Brown
Peg 21h49'14" 18d35'47"
Leola Odem Satberry
Per 3h50'1" 33d17'41"
Leola Zech
Lyr 18h49'38" 38d0'39"
Leon
Her 17h52'42" 39d8'42"
Leon and Kelli
Eri 4h20'6" -31d51'55"
Léon André KIENER
Cap 21h56'45" -20d55'37"
Leon Andrew Jeffers
Her 15h53'48" 42d36'25"
Leon Andy Cooper
Peg 23h42'30" 21d27'32"
Leon Artin
Per 3h22'35" 48d36'10"
Leon Bartz
Uma 11h48'0" 43d15'43"
Leon Cooper "Daddy"
Per 3h11'12" 38d23'43"
Leon Dale Kalbfleisch
Lib 15h17'54" -4d18'50"
Leon Easlon
Vir 12h51'47" 4d29'54"
Leon Fulcher of Manakau & Tuai
Cru 12h46'34" -62d10'51"
Leon George Innes
Umi 11h51'10" 86d32'34"
Leon Giles
Ari 3h15'6" 27d59'46"

Leon Goldstein
Leo 9h27'7" 24d51'49"
Leon Gomes
Cep 21h39'35" 66d5'50"
Leon "Habibi" Francis
Cas 23h21'53" 57d57'42"
Leon Hartnett & Tania Johnson
Cru 12h25'13" -58d12'7"
Leon J Carroll
Sco 17h58'27" -40d23'9"
Leon James Fournier
Uma 10h57'56" 54d39'46"
Leon Joseph Guerette Sr.
Cap 21h46'49" -10d5'3"
Leon Joseph King
Cep 22h49'7" 71d3'30"
Leon K. Reyniers
Ori 5h58'55" 17d13'43"
Leon & Lillian Berard
Aur 5h11'8" 47d36'21"
Leon loves Mijung
Cnc 8h7'48" 18d10'15"
Leon M Zakrzewski
Ari 3h20'55" 29d45'1"
Leon & Maria Mills
Ori 5h37'47" 2d41'6"
Leon Morrow
Sco 17h9'55" -40d33'6"
Leon P. Sanders
Cnc 8h49'32" 32d48'37"
Léon Philippe
Leo 10h22'6" 10d30'0"
Leon Preston Robinson
Psc 1h22'20" 13d38'35"
Leon Richard Bishop
Cru 12h36'50" -57d43'25"
Leon Robin Netzer
Her 18h9'50" 18d38'48"
Leon Sheridan
Her 16h36'0" 7d38'34"
Leon T. McKeon
Leo 11h22'45" 17d25'30"
Leon Vick
Ori 6h8'3" 20d44'17"
Leon Wayne Hoyt
Uma 13h20'13" 56d58'9"
Leon Wilson
Uma 8h43'11" 72d15'4"
Leon´s Glücksstern
Uma 9h55'18" 54d34'43"
Leona Angeline Klepek
Uma 13h43'42" 55d38'16"
Leona Berkowitz
Per 2h40'13" 53d6'18"
Leona & Bill
Umi 15h35'50" 74d55'58"
Leona Cornelius
Lyr 18h47'40" 38d9'51"
Leona Jean Hull
Crb 16h17'16" 31d30'24"
Leona Krueger
And 0h43'8" 40d8'36"
Leona Mae 5/22/1948
Lyn 8h35'0" 59d12'44"
Leona Margaret
Ari 3h15'50" 28d0'40"
Leona Marie Kocher
Ari 3h19'50" 27d43'4"
Leona Markl Moonstar
And 1h10'6" 41d38'10"
Leona Morgan Stracqualursi
Crb 15h39'48" 38d11'49"
Leona Quilici
Sgr 17h51'49" -25d39'46"
Leona Raines Pettigrew
Her 16h49'8" 29d31'40"
Leona Sobierajski
Cas 2h7'24" 69d57'59"
Leonard
Her 16h59'39" 27d55'58"
Leonard
Gem 7h20'50" 21d17'29"
Leonard
Peg 22h46'26" 8d32'5"
Leonard A. Weber, Sr.
Uma 8h57'38" 56d14'47"
Leonard Alan
Gem 6h47'15" 28d8'8"
Leonard Aloyious Carolan
Sgr 18h13'39" -19d55'19"
Leonard Anthony Mosley
Cep 21h49'13" 58d52'0"
Leonard Assante
Cap 20h20'8" -10d9'28"
Leonard Baldonado
Aqr 22h8'54" -17d6'25"
Leonard Beaumont
Uma 9h9'51" 48d14'59"
Leonard Bell
Per 2h57'59" 47d8'46"
Leonard Bell's Star
Cap 21h8'33" -23d57'36"
Leonard C. Christofferson
Crb 16h16'4" 30d59'9"
Leonard Calvert Wood III
Lmi 10h47'51" 35d11'7"
Leonard Campbell
Ari 2h44'46" 28d57'40"
Leonard Clifford
Cyg 20h47'8" 32d33'26"
Leonard E. Harvey My Forever Star
Uma 10h8'53" 48d17'44"
Leonard E. Woods
Psc 1h29'45" 7d23'26"
Leonard Earl Carreathers
Lmi 10h47'30" 24d53'35"
Leonard Francis de Geus
Lib 15h14'45" -6d49'1"
Leonard Franklin Sutton
Cma 6h14'1" -22d28'44"
Leonard G. Larsen IV
Cas 1h9'5" 58d57'9"
Leonard Georg
Lmi 10h6'4" 35d28'24"
Leonard & Gloria Coulson
Sco 17h32'7" -41d49'46"
Leonard Grover Galloway
Her 17h26'27" 35d7'51"
Leonard H. Townsend
Uma 13h28'21" 55d17'59"
Leonard Hamilton
Uma 11h7'59" 36d3'43"
Leonard Harvey Klekner, JR.
Cnc 9h7'32" 32d18'44"
Leonard J. Imperial
Leo 10h39'37" 18d50'2"
Leonard John Adams
Ori 6h3'15" 11d6'4"
Leonard John Gibely
Ari 1h57'25" 20d27'17"
Leonard Joseph Rossi
Cnc 8h51'39" 31d38'48"
Leonard Joseph Walker
Aql 18h57'37" -0d30'53"
Leonard Kauffman
Uma 12h55'6" 52d28'31"
Leonard L Redding, Jr
Tau 5h16'58" 19d51'29"
Leonard Lafleur 08.08.1925
Cas 23h29'39" 54d56'35"
Leonard Lee James
Uma 8h26'54" 69d18'11"
Leonard 'Lenny Grandad' Greenwood
Cep 22h50'8" 65d9'44"
Leonard Leo Giustino
Per 3h7'8" 53d56'47"
Leonard Lewis
Her 17h46'28" 44d14'47"
Leonard Loba
Aur 5h39'46" 54d59'7"
Leonard Lopez, Jr.
Gem 7h6'49" 25d36'25"
Leonard Louis Poenaru
Vir 12h45'51" -1d26'35"
Leonard Malech
Pup 8h4'23" -35d3'15"
Leonard Mardian
Her 17h11'10" 17d38'27"
Leonard Michael Giannone
Her 17h21'43" 37d13'31"
Leonard Mielke
Uma 9h32'45" 65d50'29"
Leonard N. DeMartino
Cep 22h10'4" 61d33'25"
Leonard Newman "Light Of My Life"
Cep 5h54'19" 85d42'15"
Leonard Nimoy
Ari 2h7'21" 22d14'32"
Leonard P Vella
Cyg 19h31'57" 28d48'28"
Leonard Parisi
Her 16h46'27" 30d2'48"
Leonard Peter Cookingham
Dra 19h17'38" 56d28'25"
Leonard & Phyllis Starr
Lyr 18h49'29" 37d44'8"
Leonard R Putnam
Dra 17h34'28" 54d40'13"
Leonard Rodney Felts
Leo 11h35'54" 26d24'59"
Leonard S. Krusinski "05/07/1926"
Tau 3h34'9" 30d38'57"
Leonard S Marrella
Cnc 9h5'36" 30d53'44"
Leonard Sasso
Per 4h12'15" 50d17'25"
Leonard Scimemi, Jr.
Boo 14h36'31" 52d9'45"
Leonard & Shirley Robison
Cyg 20h7'16" 37d54'2"
Leonard Smug
Sco 16h2'28" -12d13'22"
Leonard T. Eagan
Sgr 19h41'41" -12d44'34"
Leonard T. & Irene M. Corcoran
Per 3h19'28" 42d12'48"
Leonard Terman
Cnc 9h7'2" 17d13'35"
Leonard Thomas Kelly
Cas 0h46'10" 61d44'35"
Leonard Tippens
Lib 15h57'47" -5d38'35"
Leonard Weisenberg
Aqr 22h29'5" 1d42'48"
Leonard Woolf "Star Extraordinaire"
Psc 1h37'36" 9d33'40"
Leonard, Lydia, Frank, & Mary LeDuc
Uma 10h35'50" 60d1'7"
Leonardo
Cam 7h1'35" 68d59'0"
Leonardo B. DeGuzman, Jr.
Per 3h5'47" 48d27'4"
Leonardo Carvajal
Dra 19h34'43" 69d11'38"
Leonardo Chiari
Ori 4h49'38" 4d37'16"
Leonardo David
Sgr 17h52'46" -28d34'31"
Leonardo Di Lorenzo
Cam 4h47'21" 59d35'56"
Leonardo Gallo
Uma 11h28'7" 59d29'7"
Leonardo John Frisoli
Cyg 19h21'26" 54d46'3"
Leonardo Love Star
Per 4h9'10" 34d29'44"
Leonardo Palmieri
Uma 9h47'31" 56d14'15"
Leonardo Ramacci
Umi 13h12'39" 75d38'47"
Leonardo & Rosalina Colegio
Sgr 18h36'14" -24d2'0"
Leonardo Seibel
Cep 20h55'25" 58d16'25"
Leonardo Tribelli
Vir 12h40'49" 11d46'6"
Leonardo Volontieri
Uma 9h40'5" 62d48'17"
Leonard's Ladys
Cas 1h16'11" 63d8'11"
Leonastan
Ori 6h12'23" -0d4'38"
Leonato 320
Psc 0h55'29" 25d43'56"
Leone
Lmi 10h27'5" 36d35'42"
Leone
Cap 20h21'20" -9d21'13"
Leone B. Thayer
Uma 10h16'39" 70d57'27"
Leonel De La Torre Gonzalez
Boo 14h31'24" 34d1'28"
Leonel Ortega
Lib 15h7'56" -29d15'37"
Leonel Quintana
Per 2h56'48" 45d11'3"
Leonel Siliezar
Uma 10h43'27" 46d18'52"
Leonela F. Martinez
Aqr 22h45'54" -10d1'55"
Leonetta Esposito
Cap 21h38'14" -10d47'16"
Leonic-Elliott
Del 20h35'5" 16d24'14"
Leonid
Leo 11h39'13" 23d43'56"
Leonid Brodskiy
Gem 6h51'33" 27d58'16"
Leonid Rabinovich
Leo 11h10'54" 12d14'58"
Leonid Yalomkovich
Cyg 21h12'23" 36d33'10"
Leonidas Driskill
Aur 5h17'21" 42d51'17"
Leonidas Vrettakos
Uma 9h14'46" 60d19'13"
Leonides Reyes
Lyn 7h38'4" 39d8'58"
Leonie and David Apps
Cap 21h45'24" -16d23'25"
Léonie Elsaesser
Cas 0h12'45" 51d57'36"
Leonie Geschray
Her 18h7'55" 28d41'25"
Leonie Goedknegt
Leo 9h56'27" 20d4'53"
Leonie Heather Scott
Cru 12h6'9" -58d26'50"
Leonie Herz
Sgr 18h6'12" -22d43'9"
Leonie Lana
Ori 6h9'50" 7d35'39"
Leonie Margaret Preston
Ori 5h43'23" -7d57'29"
Leonie Niendorf
Ori 5h27'52" -8d33'57"
Leonila Bernal Liwanag
Uma 12h46'15" 58d38'31"
LeonNicholas04
Cnc 8h19'47" 13d7'53"
Leonor Esther Ramirez
Cyg 19h47'9" 35d23'8"
Leonor Greiner
Tau 5h30'35" 25d24'18"
Leonora
Sco 17h28'42" -43d21'26"
Leonora Anne Ceresa
Ari 2h22'14" 25d27'46"
Leonora Biagi Galvao
Vir 12h0'50" -0d11'11"
Leonora Marie Theresa Moran
Cas 0h42'24" 65d20'18"
Leonora Stella Piombino
Ori 5h52'56" -10d14'6"
Leonora YeoEun Yi
Lyn 7h43'40" 38d37'0"
Leonore Applebaum
Cas 0h28'39" 64d27'12"
Leonore Marguerite Cortazar Biever
Psc 0h39'19" 5d6'45"
Leon's Light
Cru 12h28'47" -59d9'15"
Leon's Light
Uma 9h30'8" 71d36'30"
Leon's Light
Sco 16h13'36" -15d36'56"
Leonsito
Her 17h27'34" 18d56'36"
Leony Concepcion
Lmi 9h53'27" 35d2'13"
Leony Mandavy
Cru 12h36'30" -58d21'37"
Leopold
Uma 9h18'59" 72d55'28"
Leopold
Cep 0h22'41" 75d48'57"
Leopold Cannon Steiner
Leo 9h25'28" 11d18'25"
Leopoldo Mazzaresi
Uma 11h33'16" 47d34'31"
Leora
Cas 23h54'40" 53d43'12"
LEORA
Sgr 19h12'36" -14d27'4"
Leora Bryan
Umi 14h20'11" 67d36'27"
Leora Pearl
Lmi 10h5'42" 38d15'31"
Leo's Light
Cnc 8h16'41" 9d28'52"
Leo's Love
Cyg 19h49'44" 37d21'18"
Leo's Star
Cru 12h18'17" -57d52'33"
Léost Jean-Pierre
Del 20h17'50" 15d8'36"
Leovi
Tau 4h32'40" 4d43'27"
Lepekah
Aur 5h40'20" 48d53'24"
Leppy_91 Patrick McVey
And 23h45'46" 48d5'4"
Lepresle
Her 18h50'39" 25d41'42"
Leprotto
Cnv 12h37'16" 43d48'23"
Lera's Star
Cyg 20h16'15" 38d55'7"
Lericann
Ori 6h5'10" 10d49'19"
Leriotis George 29/9/58
Umi 13h28'14" 74d48'9"
Lerma
And 1h3'33" 45d19'20"
Lerone3
Lyn 7h40'6" 48d56'48"
LeRoy
Boo 14h25'41" 47d24'18"
Leroy
Ari 2h18'25" 23d22'41"
LeRoy A Olson
Her 17h22'14" 34d8'11"
Leroy Albert Spurlock
Gem 6h34'1" 23d7'25"
LeRoy August Heuler
Uma 13h42'42" 57d42'48"
LeRoy C. Brown
Lib 15h12'56" -25d26'21"
LeRoy D Spencer Memorial Star
Sco 17h40'3" -40d12'40"
Leroy David Trenholm
Lyn 8h17'17" 58d53'13"
Leroy Donley
Vir 11h51'0" 7d1'31"
Leroy E. St. Clair
Aur 5h49'53" 44d37'17"
LeRoy Ellis
Lib 15h44'0" -17d17'12"
LeRoy F. Willner, Sr.
Cnc 8h48'50" 31d38'41"
LeRoy Francis Jr.
Umi 14h46'44" 88d47'11"
Leroy Goff & Stacy Goff
Dra 18h37'46" 63d50'36"
Leroy Gutierrez Jr.
Uma 9h14'38" 69d27'40"
Leroy & Helen Dodge
Umi 17h19'41" 81d38'11"
Leroy Irvin Pease
Aql 19h55'57" 11d2'8"
LeRoy J. Pearson
Ori 5h8'21" 0d34'8"
LeRoy "JaJa" Wieczorek
Cas 0h4'23" 57d49'32"
Leroy James Gallegos
Leo 9h53'7" 16d40'40"
LeRoy James Glass
Aqr 20h45'18" -13d25'4"
Leroy James Handford
Uma 11h29'57" 58d24'40"
Leroy Joel Norlander
Gem 7h12'4" 16d35'2"
Leroy M. Calhoun
Uma 8h35'20" 62d20'24"
Leroy Prielipp
Lmi 10h28'36" 33d55'12"
Leroy Rehli
Per 4h18'31" 44d13'19"
LeRoy Robert Anderson
Ori 6h18'54" 14d47'32"
Leroy Steven Hurt
Cyg 21h46'20" 54d11'39"
Leroy Tangradi
Uma 13h27'47" 56d29'16"
LeRoy Taylor
Cep 20h40'7" 59d30'19"
Leroy Valentine Schrumpf
Gem 6h21'49" 17d29'58"
LeRoyce
Cam 5h52'45" 73d3'6"
Leroy's P38
Cnv 12h23'54" 33d23'34"
Lerrick and Oliver
Lyr 18h36'59" 27d17'19"
LER's California Star
Sgr 18h19'33" -22d46'10"
Leryndika
Uma 8h36'20" 55d14'29"
Les
Lib 15h6'3" -8d57'8"
Les
Cep 0h28'43" 82d53'31"
LES
Lyn 7h30'58" 45d11'53"
les Ailes de Cheetah
Ori 5h10'25" -1d51'15"
Les Amoureuses
Lib 15h34'55" -12d42'0"
Les Anton
Uma 9h17'23" 64d46'2"
Les Blakeman, Sr.- Electric Light
Leo 10h5'55" 20d31'53"
Les Boucans
Crb 16h20'57" 29d20'4"
Les Bowman
Psc 0h43'56" 15d40'18"
Les ebats de prophetes et de saints
Per 3h15'47" 47d0'3"
Les Frères Meunier
Uma 9h54'27" 46d59'6"
Les Gueux
Cyg 20h46'10" 47d36'57"
Les Hayden
Uma 12h4'52" 44d21'44"
Les' JIAB
Uma 9h44'0" 70d22'37"
Les Li'l Guy Glass
And 23h8'43" 46d24'46"
Les & Mary Spinner
Cyg 21h37'28" 53d50'58"
Les Osborne
Tau 4h8'8" 7d22'8"
Les & Pat Becker
Cyg 19h44'31" 39d17'33"
Les Ross
Uma 10h44'47" 62d18'42"
Les Schreck
Her 17h29'29" 26d34'34"
Les Stone
Per 4h26'35" 40d51'27"
Les & Theresa - Infinity and Beyond
Cru 12h23'14" -58d34'31"
Les Tomac
Her 17h36'41" 44d16'35"
Les Wagner
Dra 15h39'55" 62d31'6"
Lesa
Leo 10h22'4" 23d58'50"
Lesa Ann
Psc 1h0'11" 6d5'13"
Lesa Ann Ritchie
Cma 6h26'48" -16d32'53"
Lesa Caryl
Crb 15h29'28" 26d53'52"
Lesa G. Addington
Leo 10h25'7" 22d4'15"
Lesa Halford
Ari 2h49'21" 22d29'23"
Les-a-lie
Per 1h59'53" 53d52'22"
LesaLisa Butterfly
Crb 16h13'3" 38d21'53"
Lesephanie
Her 18h52'2" 23d9'55"
Lesia Brockell
Sco 17h27'7" -42d26'25"
Leslee Ann Fagan
Cap 21h53'36" -13d31'25"
Leslee Ann Picone
Gem 6h49'20" 34d30'1"
Leslee Anne Mathis
Dra 18h20'49" 51d59'4"
Lesleh G. Bodine
Cap 20h53'19" -24d44'57"
Lesleigh's Beauty
Dra 10h59'29" 75d36'46"
Lesley
Cyg 20h57'41" 36d29'34"
Lesley A. Underwood
Tra 16h34'53" -62d26'35"
Lesley Ana
Gem 6h53'22" 24d11'11"
Lesley and James
Crb 16h10'48" 37d41'10"
Lesley Ann
Lib 15h45'21" -7d22'37"
Lesley Ann Sheppard
Vir 12h4'33" 6d52'0"
Lesley Anne
Crt 11h29'55" -15d13'14"
Lesley Autumn Willard
Lyn 8h40'17" 40d17'5"
Lesley Bishop
Uma 8h45'52" 55d54'23"
Lesley C. Dinwiddie
Uma 10h31'54" 44d37'6"
Lesley Collingwood
Cma 6h27'11" -15d30'25"
Lesley (Coton) Sheldon
Cas 1h2'40" 48d54'36"
Lesley DeFelice - "Luv"
Sco 16h17'36" -13d11'22"
Lesley Delaney
Cru 12h28'36" -60d6'26"
Lesley Elizabeth Pitfield - Mum's Star
Cas 0h56'55" 63d45'30"
Lesley Ferguson
Cas 0h17'18" 51d12'5"
Lesley Foote
Cas 23h56'37" 56d25'15"
Lesley Francis Bowler
Cas 23h46'28" 55d33'26"
Lesley Gale Chamberlain
Cyg 19h55'24" 37d29'0"
Lesley Hewer
Cas 23h53'49" 57d51'22"
Lesley Hollaway
And 0h39'23" 36d31'13"
Lesley Jane
Leo 10h55'8" 11d15'52"
Lesley Jaydon
Uma 9h55'14" 58d2'5"
Lesley Jayne Thomas 22nd July 1961
Umi 16h49'55" 82d58'8"
Lesley Leann Merritt
Gem 6h8'26" 27d29'50"
Lesley "Lel" Catherine
Cas 1h26'24" 57d56'14"
Lesley Mad Golf Lady
Cas 23h41'34" 54d11'5"
Lesley Marchand Beers
Lyr 18h54'25" 32d50'3"
Lesley Marie Ho Sang
Ori 6h19'23" 16d14'52"
Lesley Mijatovic 012273
Aqr 22h36'31" 2d1'59"
Lesley Nicole Beers
Cap 20h54'34" -14d40'20"
Lesley Nicole Lozano
Cnc 8h43'23" 31d28'10"
Lesley Nixon
Cyg 21h59'38" 45d59'59"
Lesley Nordahl
Uma 12h24'5" 55d47'36"
Lesley Perry
Sco 17h20'24" -41d48'35"
Lesley Rawlings
Aur 6h34'36" 35d29'28"
Lesley Romanoff
Pic 5h51'48" -47d26'38"
Lesley & Russel Martin
Her 16h23'52" 22d19'12"
Lesley Russo
Cyg 21h16'25" 46d36'18"
Lesley Sammons
Cas 0h56'32" 63d43'41"
Lesley Sandra 24.11.1944
Cru 12h45'38" -61d39'30"
Lesley Tan
Ori 4h58'54" 13d55'1"
Lesley Thomas
Peg 21h31'36" 11d23'19"
Lesley Warren
Uma 10h5'3" 56d22'34"
Lesley-ann Wilson
Cru 12h47'5" -56d53'57"
Lesley's Child
Gem 7h32'24" 25d48'0"
Lesli Brown
Vir 13h16'34" 5d16'39"
Lesli R Fisher
Ari 2h52'59" 19d3'46"
Lesli Summerhill
Lyn 6h38'46" 57d31'23"
Lesli Wybiral, Esq.
Dra 17h34'39" 57d41'17"
Lesliann Sarah Higgins
And 2h19'44" 47d17'6"
Leslie
Cas 0h41'28" 51d23'14"
Leslie
Lmi 10h7'43" 35d40'43"
Leslie
Leo 9h38'3" 11d23'19"
Leslie
Com 12h51'40" 21d10'7"
Leslie
Lyn 8h23'37" 56d36'57"
Leslie
Cep 21h50'54" 64d11'33"
Leslie
Mon 8h3'38" -0d53'50"
Leslie
Oph 17h33'0" -0d23'25"
Leslie A. Beer
Uma 10h21'26" 64d30'36"
Leslie A. Bonfield
Cas 1h17'45" 64d32'9"
Leslie A. Gorenc Laipple
Gem 7h10'3" 16d33'19"
Leslie A. Macias
Lyr 19h4'29" 31d53'25"
Leslie Alan Matthews
Mon 6h46'35" -0d9'35"
Leslie Allen & Vienna Arlene Harju
Leo 10h59'41" -0d16'43"
Leslie and Andrew Person UNION Star
Cyg 20h13'22" 30d3'28"
Leslie and David Roche
Uma 11h26'41" 58d59'28"
Leslie and Dorothy Forbes
And 0h55'48" 36d40'18"
Leslie and Georgia Rogers
Uma 11h28'56" 50d30'43"
Leslie and Marcus
Sgr 19h49'57" -12d37'46"
Leslie Andrea
Leo 9h44'16" 32d33'57"
Leslie Anise
Del 20h35'57" 3d19'6"
Leslie Ann Burns
Leo 9h47'52" 20d58'4"
Leslie Ann Canada
Gem 6h31'1" 23d50'35"
Leslie Ann Eastin Southwick
Pho 0h13'44" -43d1'10"
Leslie Ann Erdman
Uma 11h47'24" 42d19'14"
Leslie Ann Fraser Loewner Bell
Leo 9h39'39" 27d1'39"
Leslie Ann Hackett
Her 17h1'36" 27d1'40"
Leslie Ann Hardy
And 0h13'48" 43d56'54"
Leslie Ann Mudgett-Kutz
And 1h31'31" 49d31'10"
Leslie Ann Neff
Lib 14h41'14" -16d53'14"
Leslie Ann Noble
Sgr 19h38'28" -25d16'1"
Leslie Ann Palmer
And 2h11'54" 38d50'26"
Leslie Ann Rudny
Aqr 22h14'18" -3d54'55"
Leslie Ann Wilsey
Uma 11h44'10" 29d37'32"
Leslie Anne Cary
Vir 12h43'23" -7d32'58"
Leslie Anne Cruse
And 2h9'27" 45d16'15"
Leslie Anne Davis
Aqr 23h11'23" -5d33'38"
Leslie Anne Hendrickson
Cas 2h17'37" 62d40'11"
Leslie Anne Menichini
Leo 11h24'54" -5d8'1"
Leslie Anne Metcalf
Cnc 8h31'48" 31d32'44"
Leslie Anthony Ward
Cnc 8h51'24" 25d38'6"
Leslie Aubineau
Lyr 19h1'1" 28d9'33"
Leslie B.
Cyg 20h43'8" 40d18'43"
Leslie Beth
Leo 9h28'38" 25d37'17"
Leslie Bitz
Cma 7h3'31" -31d22'38"
Leslie Bland
And 1h4'0" 34d33'50"
Leslie Breest
Lyr 18h36'21" 30d37'6"
Leslie C. Lewis
Uma 10h25'17" 66d33'0"
Leslie Campanelli
Gem 7h29'14" 29d51'37"
Leslie Carol Brown
Mon 7h37'12" -0d51'3"
Leslie Carter
Cep 22h19'4" 72d3'44"
Leslie Casto
And 0h41'17" 39d49'27"
Leslie Chyten
And 1h9'51" 44d28'45"
Leslie Collins
Cyg 20h33'56" 46d38'55"
Leslie Cooper
Uma 10h32'39" 51d39'40"
Leslie Couchot
Cyg 20h52'37" 31d9'23"
Leslie Cruz
Aqr 22h35'29" -0d17'4"
Leslie D. Mulkey, Jr.
Lyn 7h39'40" 53d36'22"
Leslie Daley
Sco 16h51'42" -42d12'2"
Leslie Dann
Lib 15h9'21" -6d41'16"
Leslie David Knox
Ori 5h39'9" 1d12'5"
Leslie Davis
Mon 6h46'45" 6d35'9"
Leslie Dawn Corman
Cyg 21h55'13" 40d27'26"
Leslie Denise's Syzygy
Aqr 22h55'50" -10d9'18"
Leslie Denny
Umi 15h11'40" 69d31'17"
Leslie Diane
Cap 21h7'29" -25d2'54"
Leslie Diane Kitson
And 0h42'55" 23d48'19"
Leslie Diane Martin
Lyr 18h34'28" 39d27'56"
Leslie Duncan
Vir 13h16'29" -21d34'2"
Leslie E. Andersen
Crb 15h35'4" 33d49'34"

Leslie Elaine Francis Smith
Cap 21h37'26" -14d49'51"
Leslie Elizabeth Merritt
Aql 20h20'57" 5d11'24"
Leslie Ellen Harrison
Ari 2h37'41" 25d24'27"
Leslie Elliott
Lib 15h32'37" -25d43'6"
Leslie F Scott
Leo 9h59'21" 20d50'30"
Leslie Fairchild
Mon 6h53'53" -0d32'20"
Leslie Fay Brown
Uma 12h17'19" 57d30'50"
Leslie Fleck
Ari 3h24'26" 27d13'16"
Leslie & Frances
Mon 6h50'9" 8d8'2"
Leslie Fuchs
Cap 20h14'2" -10d19'49"
Leslie "Fungus Butt" LaVictoire
Aqr 21h36'52" -0d38'52"
Leslie George Wigginton
Ari 2h29'18" 25d35'24"
Leslie Gertrude Long
Cnc 9h4'52" 31d49'56"
Leslie Goodlin
And 23h41'16" 42d27'2"
Leslie Grace Brooks
Lyr 18h32'27" 36d0'9"
Leslie Gray
Lyr 18h57'2" 45d50'56"
Leslie Gray
Ori 5h37'32" 8d44'48"
Leslie Groulx
Gem 7h44'27" 32d54'27"
Leslie Haruyasu Ono
Mon 6h47'11" 7d55'35"
Leslie Hill
Leo 10h12'37" 7d23'12"
Leslie Hope
Tau 5h46'12" 22d7'9"
Leslie Howard Billington
Peg 22h29'25" 30d39'14"
Leslie Howard Singer
Sco 17h29'52" -42d17'8"
Leslie Hrovatic
Ori 5h40'1" 3d54'24"
Leslie J. Barreto
Cyg 21h24'56" 46d55'59"
Leslie Jane Johnson-The Curious One
Leo 9h36'13" 22d51'4"
Leslie Jean
Lyn 8h1'59" 40d59'18"
Leslie Jenkins
Uma 11h10'22" 54d19'24"
Leslie Jill Cira Klaus
Dra 14h51'34" 55d15'55"
Leslie Jo
Gem 7h30'12" 19d3'50"
Leslie Jo Coy
Lib 15h5'7" -7d9'34"
Leslie John Muff
Lmi 11h4'53" 26d50'47"
Leslie Jordan Magee
Sco 15h54'15" -22d15'10"
Leslie K. Arslan
Sco 16h45'24" -38d53'40"
Leslie K. Epstein
Cyg 19h31'51" 28d16'15"
Leslie K Marsh
Sgr 18h29'29" -20d58'1"
Leslie K. Ring
Umi 16h21'23" 73d52'37"
Leslie Keller's Star
Cas 23h45'2" 51d27'5"
Leslie Kern
Cas 1h13'37" 55d55'12"
Leslie Kroschel
Uma 9h38'59" 69d4'11"
Leslie Kubas
Aqr 22h34'16" -1d19'3"
Leslie L. Mildenhall
Aur 6h33'42" 34d11'22"
Leslie Lamar Perez
Leo 11h14'56" 11d17'13"
Leslie Lancaster-Smith, CCC-SLP
Cyg 20h27'9" 30d21'33"
Leslie Leah Jones
And 0h34'27" 29d8'39"
Leslie Lee Woods
Cas 1h36'26" 61d44'32"
Leslie Leon Cooper
Dra 18h57'40" 54d29'45"
Leslie Lewit Milner
Leo 11h10'17" -2d15'4"
Leslie Liberman
Gem 7h5'25" 20d37'54"
Leslie Lohmeier
Sgr 18h31'34" -22d29'12"
Leslie Looney
Leo 10h44'40" 22d12'34"
Leslie Louise
Cam 4h28'54" 70d17'51"
Leslie Mae Roach
Hya 10h20'29" -16d22'56"
Leslie Marie
Vir 12h5'53" 11d53'50"
Leslie Marie Carbiener
Aqr 22h33'40" -0d15'2"
Leslie Marie Paoluccio
Tau 5h5'59" 24d46'0"
Leslie Michelle Carnarius
And 1h22'34" 34d45'57"
Leslie Mobley
Sco 16h49'28" -45d17'8"
Leslie Mueller
Eri 3h25'45" -20d42'23"
Leslie: My Wife, My Life, My Love
Cas 1h24'4" 63d33'56"
Leslie Nicole
Lib 15h11'55" -14d6'27"
Leslie Nicole
Ari 2h51'23" 27d8'36"
Leslie Nicole Peacock
Lib 14h54'27" -5d33'46"
Leslie Oates Erdman
Aqr 22h38'22" -23d1'3"
Leslie Olson Crincoli
Lyn 8h13'57" 40d25'39"
Leslie P Clark
Gem 7h13'6" 22d57'31"
Leslie Parker Shropshire II
Per 2h50'14" 53d50'28"
Leslie Paulette Roberts
Mon 7h38'14" -0d45'24"
Leslie Perrico
Vir 11h47'47" -2d55'19"
Leslie Peters - So Very Special
Cyg 20h10'5" 31d51'36"
Leslie Pockets
Psc 0h41'25" 18d41'55"
Leslie Pulliam
Crb 15h49'10" 37d11'51"
Leslie Rena Aley
And 1h17'30" 34d2'11"
Leslie Renee Adams
Mon 6h45'26" -0d4'59"
Leslie & Robin - Forever Friends
Uma 9h22'46" 65d58'33"
Leslie Ruth Crist
And 1h10'13" 39d22'56"
Leslie & Ryan
Gem 6h39'17" 14d35'35"
Leslie Sandoval
Crb 15h41'40" 36d23'21"
Leslie Saul Ann
Uma 10h37'9" 52d28'13"
Leslie "Scooby Doo" Hall
Cyg 21h18'31" 52d40'7"
Leslie Seabury
Vir 14h16'13" -6d43'20"
Leslie Sellman
Sco 17h24'34" -38d32'26"
Leslie Shadle
Per 3h30'54" 45d21'3"
Leslie Singer
Cas 0h51'29" 53d54'30"
Leslie Soldier
Cap 21h1'58" -26d3'3"
Leslie Star crumpler
Lyn 7h5'31" 61d36'31"
Leslie "Star Gazer" Meyers
Psc 23h0'18" 5d23'33"
Leslie Steitz
Cyg 21h31'17" 45d50'11"
Leslie Susana Vasquez
Cas 0h23'56" 54d30'55"
Leslie Tallis
Uma 9h28'51" 68d48'44"
"Leslie" The Mother Star
Cas 1h38'36" 61d41'20"
Leslie Tomko, Asst Director of CCS
Ari 2h41'9" 23d49'14"
Leslie Trekie and Stargazer
Her 16h35'49" 37d59'1"
Leslie Truax
Psc 1h12'57" 28d28'52"
Leslie V. McClellan, Princess Star
And 22h58'8" 49d29'39"
Leslie Valentine
And 0h38'46" 42d16'8"
Leslie W. Drake
Ari 2h43'59" 12d50'3"
Leslie Wayne Davis
Psc 0h38'48" 7d29'49"
Leslie Wilkes
And 0h33'11" 41d14'28"
Leslie Woodcock
Per 1h31'25" 52d8'32"
Leslie-Ann Scipio-Humes
Gem 6h21'31" 21d42'41"
LeslieAWarren
Leo 11h19'13" -4d47'59"
Leslie's 20 Years of Success
Uma 11h38'12" 53d25'3"
Leslie's Diamond in the Sky
Tau 3h51'51" 25d4'23"
Leslie's Forever Love
Lyn 9h4'46" 39d24'24"
Leslie's Light
Ori 4h54'6" 13d55'7"
Leslie's Luminary
Sgr 18h48'27" -24d47'3"
Leslie's Star
Uma 14h25'8" 55d20'16"
Leslie's Star
Uma 9h7'34" 72d23'11"
Leslie's Star
Psc 1h8'57" 26d12'44"
Leslie's Star
Lmi 10h16'16" 31d40'10"
Leslie's Tigger Lou and Zen Man
Lib 15h28'24" -6d34'50"
Leslie's Wish
Cas 1h32'26" 63d8'14"
LeslinaLuvsKallie
Lib 15h25'2" -25d30'35"
Leslita Linda
Cnc 8h29'40" 12d3'58"
Lesly
Lyr 18h30'47" 32d29'15"
Lesly Leone
Psc 1h17'59" 12d1'37"
Lesly Markie Myles Morency
Uma 8h19'15" 61d5'50"
Leslye Ann Boyd
And 0h38'13" 38d6'7"
Lesnie Orehi
And 0h15'14" 48d39'13"
Lesnik
Crb 16h11'8" 32d22'56"
Lesnik, Ferdinand
Uma 11h33'58" 39d16'36"
L'Esprit de Notre Mme Wilson
Tau 4h20'27" 23d25'17"
Lessa Amicus
And 1h43'25" 43d4'36"
Lessard
Ori 5h57'9" 13d11'19"
Lesser Trog a.k.a Mike Sheffield
Pho 0h39'55" -47d41'33"
Lesson
Her 18h50'45" 24d9'46"
Lesta Thompson
Uma 8h34'17" 61d9'59"
Lestat
Sco 16h6'49" -12d21'40"
Lester and Angeline Campbell
Cyg 21h41'33" 43d0'38"
Lester and Lillian Johnsen
Per 3h20'31" 43d27'24"
Lester Berry
Her 16h33'22" 44d31'31"
Lester C. Parker
Per 3h12'31" 52d25'33"
Lester Eugene Neill "06/23/1962"
Cnc 8h28'33" 14d52'3"
Lester & Gertrude Siegel
Cyg 19h35'41" 34d12'25"
Lester John
Tau 3h51'49" 25d4'6"
Lester Joseph Knispel
Gem 6h48'48" 23d36'29"
Lester Lee Wingate
Ari 3h17'10" 29d5'20"
Lester Michael Berkeley
Dra 20h27'26" 67d26'2"
Lester Paul Allison
Uma 13h45'55" 57d4'40"
Lester Paul Mathieson III
Ori 5h41'50" -0d41'9"
Lester Pearson
Cma 6h44'26" -15d49'20"
Lester "Skip" McClintock
Sgr 18h14'9" -19d22'14"
Lester Valentine Thierwechter Jr.
Her 18h46'13" 19d53'46"
Lester William Schulz
Ori 6h5'16" 18d50'23"
Lestersmith Kairos
Leo 10h27'25" 19d24'56"
Leszek Perth
Uma 8h57'27" 48d11'36"
Let It Shine
Scl 0h38'52" -28d16'11"
Let Your Little Light Shine
Lyn 7h34'17" 58d17'34"
Let Your Love So Shine
Cyg 21h44'40" 49d12'28"
Leta Corrine Turner Wallace
And 23h23'52" 52d18'1"
Leta Jo Ford
Uma 10h45'39" 52d48'50"
Leta M. Banther
Oph 17h11'19" -0d46'38"
LeTeMeKeMiDa
Peg 22h11'31" 7d3'53"
L'eternamente Bel Victoria
Uma 10h17'44" 60d38'44"
L'éternel scintillement Laurie
Psc 1h13'19" 28d27'41"
LeTerra Marie
Sgr 17h53'56" -29d21'38"
Letha Ann Wismer
And 1h0'17" 37d49'0"
Letha Bradshaw
Cyg 20h4'10" 55d19'6"
Letha Hoover
Cas 1h21'48" 67d37'51"
Letha-Jimmy
Cyg 20h25'45" 52d12'49"
Lethie Lou
Leo 9h39'21" 27d42'23"
Leti
Cap 21h20'6" -17d44'3"
Leti Dinsmore
Mon 6h28'20" 5d17'27"
Letia Lynn Schimpf
Cap 21h58'58" -23d56'50"
Leticia
Sgr 20h24'43" -34d12'17"
Leticia
Cnc 8h17'33" 9d5'40"
Leticia
Gem 6h28'4" 14d44'7"
Leticia
Ari 3h14'30" 22d9'13"
LeTiCiA
Cyg 19h56'51" 35d51'55"
Leticia
Uma 9h25'32" 44d22'32"
Leticia Angelic
Aql 19h20'18" 4d9'3"
Leticia Arballo
Her 17h10'59" 36d1'38"
Leticia Bravo
Leo 11h52'41" 20d28'49"
Leticia Castaneda Palos
Gem 6h45'35" 26d24'57"
Leticia Charmed
Mon 7h21'48" -0d50'39"
Leticia Citlalic
Cam 3h52'2" 55d8'12"
Leticia Dueñas Flores de Alvarez
Per 3h20'37" 41d21'59"
Leticia Duncan Morquecho
Lyn 8h11'19" 38d19'48"
Leticia Gutierrez
Lyr 18h24'2" 39d29'25"
Leticia Hernandez
Lib 15h24'46" -6d23'39"
Leticia M. Saenz
Psc 23h2'33" 1d15'14"
Leticia Maria Rodriguez
Gem 6h22'56" 21d16'48"
Leticia Michelle Smith
Leo 11h12'47" 2d58'38"
Leticia Quinones
Sco 16h37'0" -35d16'33"
LETICIA RENEE MONTE-CINO
Gem 7h31'58" 16d23'42"
Leticia Rubalcaba
Cnc 8h54'22" 17d43'17"
Leticia Salas (Stellar)
Cnc 8h44'32" 9d17'44"
Leticia Soto
Psc 1h34'51" 18d1'18"
Letisha, Princess of the Stars
Psc 1h18'6" 16d23'40"
Letitia Dean
And 23h43'9" 42d12'4"
Letitia Lycke
Cas 2h2'34" 58d59'54"
Letitia's Quiet Place
Cas 0h19'45" 61d26'33"
Letizia
Uma 10h37'13" 56d55'8"
Letizia
Aql 19h44'48" -0d34'59"
Letizia
Aql 20h9'7" -0d3'37"
Letizia Airos
Leo 9h50'54" 29d49'24"
Letizia Alía
Tau 4h14'50" 22d26'42"
Letizia Hobbs
Ori 5h10'4" 0d1'34"
Letizia Principato
Lyn 7h24'7" 50d28'32"
Letizia Remoto
Uma 8h17'37" 69d14'17"
leto efloresco ab solis
Ori 5h50'10" 12d10'6"
L'Etoil d'Aventure
Uma 9h47'14" 46d37'51"
L'étoile Chez Cate
And 1h32'58" 48d38'40"
L'étoile d'Alison
Ori 6h2'28" 13d22'47"
l'étoile d'Amour
Vir 14h13'18" 2d30'15"
L'étoile d'Amour Fameux de David
Tau 5h16'28" 22d9'14"
L'Etoile de Beau Josee Routhier
Per 4h30'34" 45d0'20"
L'étoile de Césarine
Peg 21h57'46" 21d18'25"
L'etoile de Charlotte
And 1h55'35" 46d30'55"
L'étoile de Clochette
Mon 6h53'42" -1d52'23"
L'étoile de Dédé
Ori 5h54'33" -0d42'13"
L'étoile de Franck
Uma 9h53'26" 48d53'9"
L'etoile de Kay
Cnc 9h19'31" 32d22'51"
L'etoile de La Fleur
And 0h11'15" 45d42'9"
L'Etoile de l'Amour Eternel
Sco 16h52'40" -43d45'23"
L'Etoile de Papi
Ori 6h5'49" 13d52'44"
L'étoile de Pterri
Leo 11h42'7" 24d42'24"
L'étoile de Torres
Hya 10h24'4" -13d42'50"
L'étoile Deena
Tau 5h21'45" 17d9'20"
L'étoile d'Emilie
Cas 0h9'40" 54d1'42"
L'étoile des
Leo 11h32'45" 19d25'9"
L'étoile des amoureux...
Uma 9h54'3" 69d6'56"
L'étoile des amoureux...
Uma 13h55'54" 58d1'42"
L'étoile des Crocodiles
Ori 6h20'30" 6d40'27"
L'étoile des Tatous
Vir 11h39'17" 9d30'8"
L'étoile d'Ofé
Leo 10h22'14" 16d35'23"
l'étoile Erin
Cap 21h22'23" -15d25'31"
L'Etoile étincelant de Jennifer
Uma 13h56'12" 50d32'34"
L'Etoile McFaul
Umi 15h22'39" 74d8'48"
L'Etoile Patou
Ori 5h40'51" -2d21'0"
L'étoile Seta & Jacques
Uma 9h26'42" 59d43'53"
L'Etoile Sherwood
Aqr 21h33'32" -0d46'15"
L'étoile Solitaire
Peg 22h42'35" 29d3'25"
Let's Get Ready To Rumble
Per 3h41'8" 43d56'21"
Let's Roll!!
Per 4h39'54" 39d25'28"
Letteriello
Tau 4h55'36" 28d8'19"
Letti und Tom
Ori 5h21'12" -1d3'0"
Lettie Marie Shifflette
Umi 14h14'42" 70d53'13"
Lettie Merle
Per 3h11'43" 42d40'9"
Lettie's Presence
And 23h25'42" 51d45'56"
Lettini's Eternity + 1 day
Cyg 20h59'58" 46d24'9"
Lettner, Alois
Uma 10h31'13" 68d54'13"
Letty
Eri 4h32'34" -36d10'43"
Letty Fowell
Umi 14h50'36" 75d14'25"
Letty M
Uma 11h55'0" 35d40'8"
Letty Ortiz
Peg 22h21'23" 9d25'36"
Letty Sustrin
Lib 14h50'5" -2d14'25"
Lety Mota
Aqr 22h11'15" 0d13'42"
Letycia's Light
Cas 0h20'51" 62d0'49"
Leuchtende Sabine
Her 16h40'36" 6d25'0"
Leung Sai Mui
Leo 11h42'25" 26d24'55"
LEV
Lib 15h5'22" -7d1'7"
Lev A. Zhivotovsky
Aur 7h19'22" 42d8'29"
Lev Blinov
Psc 1h19'13" 32d53'17"
Lev Friedman
Gem 7h27'24" 24d24'37"
Lev Jacob Kaplan
Lmi 10h14'19" 30d2'24"
Lev Meas
Gem 6h57'7" 24d52'33"
Lev Mints
Psc 23h20'25" 4d4'9"
Lev Zak
Aql 19h49'56" -0d13'25"
Lévai István és Lévai Laci csillaga
Uma 8h38'35" 46d44'3"
Leveque 50
Lmi 10h33'42" 37d36'11"
LeVerne Engh
Psc 1h11'58" 25d9'55"
Levette Bosley
Uma 10h38'36" 56d1'12"
Levi
Psc 1h30'55" 13d14'48"
Levi
Uma 11h49'50" 44d20'47"
Levi Andrew Raymond Latif
Tau 4h4'49" 4d16'12"
Levi Benjamin Dunn
Her 18h54'5" 25d14'13"
Levi Coby Morris
Uma 10h0'12" 47d38'29"
Levi Cornelius Doerr
Cap 20h28'41" -12d27'16"
Levi David Judd
Boo 14h42'49" 43d46'48"
Levi Feltes
Per 3h34'18" 44d3'46"
Levi Garrett Windle
Lib 15h23'50" -11d46'51"
Levi Hunter Major
Cnv 12h41'54" 42d37'15"
Levi Lovejoy
Peg 23h26'58" 19d40'1"
Levi Mark McGlinchey
Dra 10h57'26" 76d6'15"
Levi Michael Funk
Aqr 22h19'27" 2d5'38"
Levi Richard Steuck's Star
Cnv 12h37'57" 46d14'19"
Levi Shawn Hendrix
Per 3h7'26" 47d49'50"
Levi Tofino Miller
Gem 6h40'8" 27d32'57"
Levi Trent McCleskey
Leo 11h44'47" 20d48'59"
Levi Xavier
Ara 16h52'43" -59d47'38"
Leviathan's Tear
Ori 6h15'11" 17d9'59"
L'évidence de nous
Uma 13h15'4" 58d14'48"
Levin
Her 18h15'10" 17d47'16"
Levin Hunkeler
Aur 6h28'3" 34d23'4"
LEVIN SCHMID
Umi 13h9'16" 88d36'42"
Levi's Eagle
Aql 19h39'7" 6d5'20"
Levi-Sara
Ari 2h7'52" 21d13'51"
Levon Self
Ori 6h9'56" 9d14'40"
Levon Ter-Petrosyan
Her 18h37'7" 19d57'44"
Levona
Cyg 21h44'39" 47d12'36"
LeVonda Ritchie
Lib 14h59'27" -14d13'51"
Levoris J Harmon 3rd
Vir 12h47'39" -1d4'23"
Levyn
Cnc 8h23'55" 14d53'20"
LEW
Uma 10h21'19" 59d56'37"
Lew and Hermine Horwitz 50th Anniv
Uma 11h20'32" 55d36'48"
Lew and Lois Chiappelli
Cyg 19h47'41" 31d43'7"
Lew and Melissa Brunhaver
Crb 15h56'13" 30d12'8"
LEWIE-LEWIE
And 23h8'37" 48d37'33"
Lewin Aurel
Per 4h11'10" 34d55'40"
Lewis
Per 4h35'17" 47d17'15"
Lewis
Lyn 7h57'58" 48d3'40"
Lewis
Ori 5h56'30" 12d29'43"
Lewis
Uma 9h5'22" 58d24'55"
Lewis
Umi 14h59'55" 74d49'58"
Lewis
Sco 17h52'46" -35d45'17"
Lewis "18"
Per 3h1'54" 47d47'52"
Lewis Alexander MacDonald
Ori 5h32'46" 0d35'39"
Lewis Alfie Whitnell
Per 4h12'31" 35d44'11"
Lewis and Clark
Her 16h34'23" 37d37'59"
Lewis and Marienel Kaufman
Cyg 21h39'50" 43d25'0"
Lewis and Martha Thomas Forever
Cyg 20h3'50" 47d19'7"
Lewis and Shannon: Ad Infinitum
Cir 14h37'7" -65d2'21"
Lewis Anthony Farinella
Umi 15h14'0" 70d59'4"
Lewis Anthony Pugh - "Dink's Star"
Cep 21h56'45" 81d55'19"
Lewis Anthony Wiles
Per 3h3'11" 47d14'48"
Lewis Arthur Jones
Per 2h52'52" 46d41'14"
Lewis Ashley Brindley
Ori 5h54'7" 17d37'25"
Lewis Ashmore
Boo 15h15'58" 25d49'47"
Lewis Bearden III
Ori 6h15'42" 10d32'54"
Lewis Christopher Moses
Cru 12h47'35" -60d31'51"
Lewis Clay Robertson IV
Cnc 8h19'18" 19d50'27"
Lewis D. Zinkin
Boo 15h34'15" 49d2'42"
Lewis Daniel Meyers - 17.07.2000
Cru 12h40'44" -57d52'10"
Lewis "Dragon Master" McCann
Dra 9h31'25" 74d20'35"
Lewis Dromey
Aqr 21h17'34" -11d20'39"
Lewis E. Daidone
Lyn 6h34'1" 59d6'45"
Lewis Edward Bronfeld
Boo 14h43'5" 37d22'26"
Lewis Frankie Golby
Ori 6h21'27" 16d23'21"
Lewis Gilmore Steele
Ori 5h19'28" 6d22'36"
Lewis Glyn Jones
Cyg 20h22'34" 51d37'4"
Lewis Graham-Giddins
Cep 1h58'45" 82d25'10"
Lewis H. Green
Aqr 21h43'18" -1d20'26"
Lewis Harold Patten
Per 4h7'42" 45d2'14"
Lewis Hepburn
Peg 21h56'26" 26d8'59"
Lewis Herbert Swartwout
Leo 11h23'5" -2d59'2"
Lewis is a star.
Per 2h19'14" 52d20'0"
Lewis J. Neelands
Gem 7h6'0" 33d23'46"
Lewis Jacoby
Cru 12h55'34" -58d5'32"
Lewis James Brooks - 24 May 1988
Gem 7h26'54" 16d9'36"
Lewis James Cass
Cen 13h43'4" -42d13'41"
Lewis James Hall
Umi 17h20'22" 86d18'18"
Lewis James Tattersall
Uma 14h3'16" 61d3'14"
Lewis John Best
Lmi 9h44'14" 40d15'8"
Lewis John Obi
Uma 12h2'30" 37d38'13"
Lewis Leventhal
Leo 9h45'20" 27d46'11"
Lewis (Lew-Lew) Rothery
Umi 17h9'34" 75d42'35"
Lewis M.
Uma 9h32'5" 47d11'11"
Lewis McMahen
Uma 10h37'40" 56d52'45"
Lewis Michael
Ori 5h56'42" 11d29'59"
Lewis Oliver Paul Hickey
Her 18h2'42" 35d44'32"
Lewis Quinn
Lmi 9h31'4" 33d34'42"
Lewis R. Caffo
Psc 0h40'23" 7d44'7"
Lewis R. Loebe, Jr.
Psc 1h28'44" 21d48'23"
Lewis Ray Harris Jr.
Psc 1h4'39" 3d34'35"
Lewis Robert Thomas Barnsley
Uma 9h0'12" 68d9'1"
Lewis S. Bookwalter
Umi 15h57'41" 83d15'22"
Lewis & Sheryl Miles, III
Cyg 20h51'58" 46d10'1"
"Lewis Skywalker" - Lewis Peter Fenney
Ori 5h55'39" 11d26'35"
Lewis Stephen McKean
Cep 0h9'57" 82d14'1"
Lewis Stephens
Aqr 22h41'13" -1d51'53"
Lewis Stine
Per 3h34'8" 34d52'27"
Lewis T. Hochevar Aurora Stardust
Ori 5h26'38" 5d30'38"
Lewis Taylor Claiborne
Her 16h43'59" 26d49'17"
Lewis The Gunner
Vir 12h29'25" -7d16'9"
Lewis Wayne Lytle, Jr.
Dra 17h38'45" 52d29'43"
Lewis Webb Cary
Cyg 21h40'59" 41d27'50"
Lewis Zafer
Aql 19h13'28" 14d48'9"
Lewis Zane Christie
Psa 22h31'29" -34d34'54"
Lewis-Mary Ruby
Ori 6h19'0" 2d31'38"
LEX
Leo 11h27'53" -5d15'57"
Lex
Cap 21h50'25" -15d44'3"
Lex and Mark Ozmen-Mr and Mrs Goodoo
And 23h46'14" 35d24'45"
Lex and Sarah
And 0h18'9" 36d8'8"
Lexa
Uma 11h4'28" 67d14'57"
Lexarkana
Sco 16h20'1" -18d57'23"
Lexa's Lucky Star
Vir 12h12'57" -0d55'33"
Lexcea Valerio Corpuz
Tri 2h1'24" 31d54'1"
Lexen, Olaf
Ori 5h7'52" 15d37'18"
Lexi
Leo 10h14'40" 14d33'41"
Lexi
Gem 7h19'24" 20d11'48"

Lexi
Lmi 10h40'13" 33d1'49"
Lexi
And 0h44'42" 44d11'17"
Lexi
And 2h13'48" 50d25'18"
Lexi
Umi 16h32'46" 76d33'13"
Lexi
Lib 15h18'57" -4d39'53"
Lexi
Uma 10h35'48" 54d31'12"
Lexi
Cas 1h39'41" 62d25'1"
Lexi Anne Southall
Lyr 18h41'18" 41d7'27"
Lexi Brianne Anderson
Uma 10h57'14" 49d34'52"
Lexi George
Lyn 7h46'0" 53d30'6"
Lexi Hensley
Lyn 6h30'31" 57d31'53"
Lexi Jasmine Lopez
Vir 13h20'42" 3d45'46"
Lexi Kaye
Aqr 23h0'27" -24d12'0"
Lexi Lee Oetterer
Lyn 6h55'28" 59d59'2"
Lexi loves Jimmy!!
Cyg 19h47'25" 42d35'43"
Lexi Lynn (143)D.T.
And 1h31'55" 45d33'0"
Lexi MacDonald
Sgr 18h5'18" -32d8'40"
Lexi Nicole
Lyr 19h19'2" 29d29'3"
Lexi Rae Lopriore
Psc 23h0'1" -0d30'58"
Lexi Stein
Umi 16h9'8" 72d17'6"
Lexi Williams
Cap 20h28'39" -11d29'40"
Lexi, Hallie & Mikenzi Carlo
And 1h5'21" 45d1'31"
Lexia
And 0h15'13" 43d51'12"
Lexiana Childress
And 0h11'36" 33d31'33"
Lexi-Carmel
And 1h12'46" 45d59'50"
Lexie
Crb 16h7'34" 37d49'31"
Lexie
Umi 17h11'24" 75d24'35"
Lexie
Cru 12h12'33" -62d18'21"
Lexie and Brandon
Aqr 22h19'59" -0d52'1"
Lexie Chambers
Tau 5h51'17" 25d39'22"
Lexie Fulmer
Uma 13h37'48" 58d40'53"
Lexie Kimble
Ori 6h8'12" 8d28'33"
Lexie Kunz
Cas 1h42'42" 69d9'7"
Lexie Lauren
Ori 6h10'21" -0d50'45"
Lexie Lyne Taborda
Cyg 20h0'24" 30d44'29"
Lexie Van Maanen
Cru 12h5'3" -61d21'41"
Lexie Volakis
Cyg 21h33'33" 30d41'31"
Lexie-Becks
Cas 0h16'0" 53d37'40"
Lexington Weiss
Uma 8h27'42" 70d29'56"
Lexi's Hope
Cyg 20h16'22" 34d11'15"
Lexi's Light
Ori 6h13'32" 16d36'25"
Lexis Marie
And 2h17'11" 47d0'27"
Lexi's Mimi
Leo 10h25'26" 18d5'17"
Lexis Nunez
Sgr 18h32'51" -27d15'42"
Lexi's Star
Lib 15h41'32" -26d3'58"
Lexi's Star
Sco 16h55'40" -34d36'50"
Lexi's Star
Gem 6h52'45" 28d46'1"
Lexiss Rose Antle
Leo 9h27'50" 8d24'30"
Lexlee Alderman
Lyn 8h48'19" 41d20'55"
Lexus Elise Fritz
Umi 14h24'25" 78d48'59"
Lexus Monroe
Cnc 8h39'15" 10d39'56"
Lexus-Marine
Lyn 8h13'32" 44d4'16"
Lexy
And 0h14'8" 42d36'55"
Lexy
Tau 5h48'10" 22d7'13"
Lexy
Mon 6h54'30" -0d40'40"
Lexy Bryans
Cam 3h58'18" 63d14'25"
Lexy - Carla Jo
And 1h17'14" 41d51'44"
Lexy Poozer
And 22h58'11" 47d15'15"
Lexy, A Star in the Heavens
Cnc 8h42'42" 16d11'11"
Lexye Gail Wright
And 0h37'58" 25d46'1"
Leya Katerina Scheel
Cyg 21h1'37" 36d32'46"
Leya Mathew
Uma 14h4'15" 55d9'27"
Leya Wilson
And 1h24'30" 43d20'33"
Leyawiin & Alayu
Tau 3h35'51" 29d21'56"
Leyha Marie
Tau 5h2'50" 26d2'18"
Leyla Enver
Uma 11h18'58" 32d29'13"
Leyla Marie
Leo 11h38'35" 22d26'49"
Leyna Ellen Stover
Psc 1h22'35" 21d1'56"
Leyna Taguchi
Uma 13h42'14" 56d56'49"
Ley's Light
Lmi 10h57'39" 30d11'25"
Leysa Katlyn Miner
Sgr 19h47'43" -13d18'59"
Leyton Briol
Cap 20h54'44" -15d25'10"
Leza Geovich
Cyg 19h54'14" 33d25'39"
Leza Szweras Goldlist
Uma 12h48'27" 61d43'59"
Lezleigh Y Cook
Cnc 9h21'15" 11d20'24"
Lezlie Brooke Rogers
Vir 14h52'45" 3d12'24"
Lezlie R. Johnson
Tri 1h42'37" 29d27'35"
Lezlie Rae Bourque
Lib 15h24'8" -12d20'46"
lfbudzinski
Cnc 8h16'33" 18d56'42"
LFM & GMN
Umi 16h8'25" 75d5'10"
L.F.S. Forever
Cnc 8h44'39" 6d44'20"
L.G.C.C.
Uma 11h46'10" 62d29'40"
LGFB Forever
Psc 1h5'12" 31d3'58"
LHHP
Leo 10h51'4" 18d24'42"
L-hub
Boo 14h40'58" 46d1'17"
Li amerò per mai
Cyg 21h48'30" 51d54'1"
Li Ding Ni
Lyn 6h49'4" 56d47'52"
Li Gai the Omniscient
Leo 9h52'41" 16d3'55"
Li Jingjing
Lib 14h57'19" -10d28'31"
Li Li
Aql 19h5'49" -0d4'16"
Li Ma
Sco 17h36'25" -42d13'56"
Li Xiao Lin and Jim Murphy
Cyg 21h43'11" 38d17'30"
Li Xigang
Crb 16h4'23" 38d15'24"
LI, Xutong and WANG, Quan
Sco 16h3'1" -24d45'36"
LIA
Sgr 18h19'47" -19d26'3"
Lia
Lib 15h17'57" -16d30'33"
Lia
Cas 3h13'16" 69d58'40"
Lia
Ori 5h56'31" 20d47'55"
Lia
Mon 7h1'46" 7d37'33"
Lia Anna
Uma 10h9'25" 67d31'0"
Lia & Casey
Mon 6h45'21" -0d52'25"
Lia Claire Cerracchio
Cap 21h46'2" -13d0'50"
Lia D. Coyne
Sco 16h28'7" -25d32'6"
Lia Daniéle
Lyn 7h49'53" 53d5'26"
Lia De Paoli
Aqr 21h57'16" 0d17'38"
Lia Delucchi
Lyr 18h53'43" 34d53'58"
Lia - Elisabeth
And 23h56'25" 48d33'47"
Lia Elizabeth
Gem 6h53'19" 18d57'5"
Lia Grace
And 2h13'35" 48d44'23"
Lia Grace Harrington
Cas 0h0'19" 53d25'58"
Lia Isono
Leo 10h18'55" 25d44'4"
Lia Marie Carroll
Cnc 8h39'33" 29d31'22"
Lia Marie Rowe
Lib 15h47'25" -10d53'41"
Lia Nicole Tepker
Tau 4h15'5" 26d26'49"
Lia Stites
Per 3h34'29" 51d59'41"
Lia Valeria
Ori 6h3'25" 5d58'1"
Liam
Cnc 8h31'16" 25d29'51"
Liam
Uma 11h15'0" 34d9'2"
Liam
Psc 1h7'23" 31d58'38"
Liam
Lyn 9h27'51" 40d14'36"
Liam
Umi 16h23'5" 72d13'49"
Liam
Cru 12h25'45" -60d47'56"
Liam
Cru 12h32'20" -58d52'28"
Liam A Brady
Cnc 8h12'20" 11d9'20"
Liam A. Clifford
Cyg 21h20'23" 34d20'46"
Liam Adrian Mabey
Umi 14h52'22" 74d53'52"
Liam Aiden McDermott
Uma 10h8'46" 53d50'38"
Liam and Mommy Doyle
Leo 10h59'33" 22d42'56"
Liam Andrew Gill
Umi 16h30'23" 77d58'45"
Liam Anthony Neville - Liam's Star
Her 18h49'43" 17d20'11"
Liam Austin Wellard
Umi 16h38'9" 81d48'59"
Liam Barron
Umi 14h55'25" 68d1'58"
Liam Boland
Sgr 18h48'21" -29d33'5"
Liam & Brendan Thompson Star
Ori 5h41'28" 0d53'11"
Liam Briggs
Umi 14h17'20" 75d59'30"
Liam Brooks McConnell
Umi 16h9'9" 85d45'23"
Liam Charles Heady
Cyg 19h55'41" 32d25'10"
Liam Chicoine
Boo 14h50'6" 52d58'36"
Liam Connor
Sco 17h21'9" -32d32'22"
Liam Connor
Her 18h50'59" 23d8'18"
Liam Cooper Welsh
Vir 12h30'45" 4d38'17"
Liam Crabbe
Aqr 21h51'29" -1d22'46"
Liam Curtis
Leo 9h35'23" 23d54'47"
Liam Danger Ornelas
Ari 3h3'56" 13d11'13"
Liam David Newton's Birthday Star
Cyg 19h33'45" 29d36'24"
Liam Dominic
Srp 18h10'23" -0d14'5"
Liam Douglas Ferretti
Cyg 20h25'19" 56d1'55"
Liam Douglas Smith
Gem 6h25'51" 26d24'34"
Liam Eames
Umi 14h25'0" 69d39'42"
Liam Emerson
Cap 21h45'50" -12d45'48"
Liam Emmet Roberts
Leo 10h43'3" 15d8'46"
Liam Everett Grau
Aqr 21h12'3" -11d3'22"
Liam Forest Stewart
Tau 4h20'57" 29d10'44"
Liam Frank McGrath
Cep 0h5'1" 78d12'15"
Liam Gabriel Gallagher
Sco 16h7'7" -10d48'42"
Liam Gilbert Jackson
Ori 5h59'5" 18d37'24"
Liam. H. Gillies
Sgr 18h27'46" -35d5'20"
Liam Harold Phair
Ara 17h19'53" -50d43'41"
Liam Hawke Moore
Aqr 22h19'59" -1d23'0"
Liam Hearne
Vir 12h40'41" -8d52'0"
Liam Henry
And 0h42'32" 29d26'17"
Liam I'll love you always, Jesi
Her 16h59'5" 30d6'38"
Liam Jack Wilkinson
Her 17h35'34" 18d38'55"
Liam Jahanbigloo
And 0h7'31" 41d57'41"
Liam John Obey Pitt
Cep 0h0'56" 84d40'25"
Liam Jonathan Koralek Catlin
Uma 11h57'15" 52d30'26"
Liam Jordan
Uma 13h34'34" 58d34'52"
Liam Joseph Glessner
Boo 14h51'4" 20d4'13"
Liam Joseph Kilpatrick
Her 17h31'19" 32d18'53"
Liam Joseph Mihalko
Vir 12h6'18" -9d54'29"
Liam Joseph Nulty
Aqr 21h44'3" -1d41'16"
Liam Joshua Renden - Baby Angel
Ori 5h51'28" 6d53'13"
Liam Judson Quinn
Per 4h37'8" 40d32'25"
Liam Kenneth O'Brien
Vir 12h8'24" 7d26'16"
Liam Lights Our Lives Forever
Cru 12h46'31" -64d17'24"
Liam & Lily Sheeran
Cyg 19h37'33" 29d41'2"
Liam Lynch
Uma 11h8'50" 70d15'2"
Liam Matthew Cunningham
Lyn 7h24'10" 44d53'55"
Liam Matthew Doyle
Leo 9h29'46" 11d42'56"
Liam Matthew Macomber
Uma 8h55'48" 58d39'32"
Liam McNamara
Dra 18h52'27" 59d22'59"
Liam Michael Shea
Her 16h49'48" 49d55'51"
Liam & Michelle Simpson - Forever
Col 6h19'1" -42d3'55"
Liam Miles Cunningham
Aqr 21h41'44" 2d10'44"
Liam Miles Greene
Aql 19h35'24" 8d17'51"
Liam Mulcahy
Boo 14h56'7" 19d6'49"
Liam Murray
Umi 13h58'59" 73d47'13"
Liam O Dochartaigh
Cru 11h57'8" -58d20'12"
Liam Oliver Schlosser
Dra 19h22'6" 66d27'13"
Liam O'Malley
Per 3h11'12" 56d18'35"
Liam P. Barr
Ori 5h28'12" -2d26'25"
Liam Patrick Baggott
Cnc 8h23'22" 26d20'24"
Liam Patrick Gallo
Gem 7h43'19" 33d42'57"
Liam Patrick Kelso Age 2
Her 17h20'20" 30d42'1"
Liam Paul Motherway
Cru 11h58'2" -59d30'0"
Liam Philip
Lib 15h44'40" -10d8'18"
Liam R J Power's Star!
Lib 15h17'1" -14d8'24"
Liam Rajeev Connors
Aqr 21h40'59" 1d43'23"
Liam Rankl McGrath
Sgr 19h9'51" -13d49'17"
Liam Richard
Per 3h9'31" 52d53'50"
Liam Richard Catron
Per 3h8'45" 50d51'28"
Liam Robert James Costello
Leo 11h55'50" 23d16'12"
Liam Robert Schwartz
Tau 4h25'20" 15d36'8"
Liam Robert Tullio McCarthy
Uma 11h30'43" 33d51'43"
Liam Russel Paulsen
Sgr 19h24'10" -16d11'46"
Liam Smith's Arse Star
Cen 13h38'43" -41d19'51"
Liam**The Dark Knight
Aqr 21h15'45" -0d44'49"
Liam Thomas Roback
Lib 15h54'43" -12d46'14"
Liam Tomas Cunningham 08/09/2004
Ori 6h0'56" 11d25'44"
Liam Walker Baldwin
Cap 20h56'6" -18d25'54"
Liam Walsh McShane
Ari 2h20'57" 18d36'10"
Liam Wolfe
Her 17h28'19" 31d30'21"
Liam Wood Parker Star
Cnc 8h35'4" 9d55'25"
Liama's Light
Nor 15h50'50" -54d11'15"
Liam's Blessing
Tri 2h9'25" 33d28'39"
Liam's Star
Crb 16h23'35" 34d41'25"
Lian
Sgr 19h46'17" -13d2'54"
Lian Jean Ulrey Adoption Star
Uma 12h6'24" 48d53'44"
Lian Lynn
Lep 5h42'36" -21d34'0"
Lian Thuy Bach
Mon 7h56'33" -0d56'7"
Liana
Cam 6h57'48" 66d2'42"
Liana
Sco 17h25'37" -38d36'21"
Liana
Lib 15h9'1" -23d47'30"
Liana
Psc 1h21'57" 20d46'42"
Liana
Gem 6h20'53" 21d40'28"
Liana Alexa Gordon-Jimenez
Vir 12h15'21" -4d17'24"
Liana and Claude Belair
Uma 9h10'30" 63d16'12"
Liana Bonavita
Ari 2h7'40" 21d0'44"
Liana Joy
Cyg 21h40'23" 45d44'13"
Liana Mi Amor
Lmi 10h52'55" 27d21'37"
Liana Moussaeva
Uma 13h35'50" 55d19'19"
Liana Quill Camper-Barry
Aqr 23h42'26" -7d29'35"
Liana Quill Camper-Barry
Aqr 22h10'7" -16d14'7"
Liana R. Sarzoza
Sgr 18h27'11" -17d19'40"
Liana Wendy Sarapas
Aqr 21h41'35" 1d11'52"
Liana's Light
Cyg 20h41'24" 52d9'2"
Liane Alison Jones
Tau 4h6'36" 7d23'18"
Liane Cyrene
And 0h25'26" 41d34'10"
Liane Kiyomi Calamayan
Sco 16h49'35" -27d17'51"
Liane Sottile
Uma 11h28'27" 33d10'48"
Liane Yomogida
Lyr 18h33'37" 38d51'23"
Liane's Cool Bean
And 0h46'10" 36d45'29"
Liang Yu Heng - 18.04.1964
Ari 2h58'18" 25d30'14"
Liani
Aqr 21h37'9" 2d25'29"
Li-ann Pagès
Umi 13h23'16" 70d44'32"
Lianna
And 0h19'26" 40d32'58"
Lianne
Ari 3h10'43" 29d46'50"
LIANNE BOUCK FOREVER LOVED BY ALL
And 1h18'1" 39d6'24"
Lianne Zara Thompson
Umi 14h30'38" 74d38'40"
Lianne's Jewel
Crb 15h37'43" 29d20'6"
Lia's Light of Eternity
Ori 6h7'54" 6d47'29"
Lias, Juste de l'autre côté du chemim
Uma 12h28'8" 61d9'28"
Liba Ronit
Her 18h52'35" 23d49'33"
Libardo
Uma 10h58'9" 48d41'16"
Libbey Michele Nastri
Tau 4h39'31" 27d43'20"
Libbey Zetzer
Sco 16h15'28" -15d43'23"
Libbie Jay Hand
And 2h29'24" 37d33'34"
Libbie Jude
And 0h31'18" 26d49'28"
Libbie Lea Black
Lyr 18h49'49" 43d47'20"
Libbie May
And 1h8'48" 35d43'50"
Libbie & Oli
Sge 19h55'34" 16d51'26"
Libby
Cnc 8h23'24" 19d37'23"
Libby
Gem 6h52'5" 25d6'3"
Libby
Ori 6h15'48" 9d36'16"
Libby
Leo 11h32'8" 14d19'40"
Libby
And 1h13'48" 36d22'40"
Libby
Aur 6h29'33" 35d6'57"
Libby
And 1h49'43" 39d27'7"
Libby
And 2h5'12" 42d49'6"
Libby
Lyn 7h20'34" 48d55'18"
Libby
Lib 15h24'12" -7d43'46"
Libby
Lib 15h6'29" -6d34'53"
Libby
Umi 14h19'25" 67d14'38"
Libby 11
Gem 6h42'18" 26d38'4"
Libby Bradshaw
And 1h27'46" 50d27'32"
Libby Daniel
And 1h39'53" 42d11'54"
Libby Ellen Parker
And 23h20'52" 52d58'22"
Libby Francesca Wilson
And 23h31'33" 48d27'11"
Libby Gaoa
Gem 6h43'13" 18d40'50"
Libby Georgia Froome
And 23h34'45" 46d42'21"
Libby Green
Cnc 9h11'12" 23d30'3"
Libby Hinote
Psc 23h13'39" 0d47'49"
Libby & Irwin Stein
Uma 13h48'4" 56d48'27"
Libby Lea Anderson
Gem 6h29'1" 25d55'41"
Libby Leah
Psc 1h8'55" 31d39'46"
Libby Louise Barton
And 23h57'29" 45d43'3"
Libby Lynn
Sco 16h28'15" -35d17'35"
Libby Maria Irving - Libby's Star
And 1h20'5" 49d40'56"
Libby Marie
And 23h37'5" 49d45'14"
Libby Mitchell
And 23h32'28" 44d1'32"
Libby My Love
Vir 12h24'2" 6d11'9"
Libby Pearl
Leo 9h33'42" 29d27'17"
Libby Rakes/ Pahnke
Lyr 18h50'51" 31d57'57"
Libby Rose Hall
Leo 11h9'34" 27d7'16"
Libby Senese
Uma 9h33'59" 63d28'8"
Libby Swain - Libby's Star
And 22h58'34" 43d37'11"
Libby Swanson
Sgr 18h43'39" -22d1'31"
Libby Wills
Sco 17h35'49" -30d5'32"
Libby's Dancing Star
Uma 10h26'34" 61d5'51"
Libby's Gate
Leo 9h37'30" 29d12'9"
"Libby's Light"
Tau 3h53'30" 22d49'16"
Libby's little twinkle Star
Cyg 21h10'23" 48d9'18"
Libby's Star
And 0h58'57" 45d52'3"
Libby's Star
Sgr 19h22'26" -31d54'9"
Libby's Star 1954
Vir 12h13'55" 8d25'32"
Libera Pensa
Uma 12h4'6" 56d17'49"
Libero
Lib 14h52'9" -4d6'9"
Libertad
Cap 21h27'4" -8d43'10"
Liberté
Aqr 23h26'20" -18d42'35"
Liberti Salvatore
Her 17h15'9" 27d18'14"
Liberto Navarro
Mon 6h24'38" -4d41'43"
Liberty
Lyn 8h29'46" 54d4'20"
LIBERTY
Per 4h32'38" 41d5'15"
Liberty
Leo 9h30'15" 26d30'28"
Liberty
Cnc 8h13'48" 24d28'35"
Liberty
Ari 3h8'4" 20d31'17"
Liberty
Her 16h43'11" 48d12'55"
Liberty
And 23h24'59" 51d29'43"
Liberty
Cnc 9h22'26" 31d25'42"
Liberty Abigail
And 0h40'32" 36d56'7"
Liberty Bell
Crb 15h48'24" 35d39'6"
Liberty E. Rivera
Tau 5h54'12" 26d3'2"
Liberty Faye
Lib 14h50'49" -4d23'36"
Liberty Grace Jewson
And 23h48'23" 42d58'4"
Liberty Grace Liddle
And 1h18'24" 49d55'20"
Liberty Jane
Umi 13h59'49" 75d8'14"
Liberty Katie
And 23h3'4" 49d3'59"
*Liberty*Nathan Dean Johnson*
Cma 6h58'43" -22d55'33"
Liberty Rose Above
Uma 12h37'23" 54d7'23"
Libor Husnik, Ph.D.
Uma 11h30'11" 63d37'42"
Libor Tamasi
Psc 0h0'46" -2d3'48"
Libra Karma 926
Lib 15h48'33" -11d14'27"
Libra's Gem
Lib 14h55'42" -17d15'56"
Liccy
Lyr 18h57'17" 41d11'16"
Licenciada Bitxore Petralanda
Uma 9h34'59" 66d15'2"
Lices Pieces for Lisa Ulshoeffer
Ari 2h31'18" 24d36'18"
Licette
Psc 1h9'5" 27d41'36"
Lichita
Gem 6h35'51" 22d21'40"
Licht der Liebe von Andrea & Frank
Vir 12h8'59" 10d38'14"
Licht Kinami
Vir 14h16'54" -19d6'38"
Lichterfeld, Ralph-Robert
Uma 9h37'56" 48d56'21"
Licia
Cas 0h27'52" 63d52'55"
Licia Scaccia
Sct 18h36'9" -14d34'46"
Licuada Me Ato
Cyg 19h49'4" 32d24'49"
Lida
Ori 5h30'18" 3d25'56"
Lida Blossom Caples
Lyn 7h25'53" 54d17'12"
Lida Conley
And 0h23'9" 38d9'10"
Lida Virginia & Harold Vincent KING
Cep 2h23'45" 82d34'26"
LidaChris
Cas 0h53'52" 59d27'39"
Lidaki
Uma 10h20'11" 69d47'26"
Lidden Biddie
Ari 2h54'7" 11d12'16"
Liddia's Love
Tau 4h21'47" 18d12'9"
Liddicappidee
Ori 5h33'11" 5d22'33"
Liddlefoot
Lib 15h17'36" -20d32'3"
Lidia
Lep 5h33'0" -16d45'53"
Lidia
Cas 0h26'29" 61d3'28"
Lidia
Eri 4h31'13" -35d39'24"
Lidia
Cnc 9h14'26" 22d36'52"
Lidia
Crb 15h35'45" 38d40'25"
Lidia Angella Aymar
Uma 10h29'34" 47d1'26"
Lidia Antonia Maltez
And 0h1'36" 32d27'10"
Lidia Di Pietro
Dra 19h43'1" 62d11'3"
Lidia & Jerzy Zyskowski
Aqr 21h49'6" -6d7'27"
Lidia Macia
Sgr 17h52'22" -28d38'29"
Lidia Sánchez Vega
Crb 15h21'33" 25d39'43"
Lidia Z Tsiporenko
Leo 11h12'38" -4d51'47"
Lidia Zylowska, M.D.
Crb 15h59'31" 38d41'30"
Lidia, la meva nena petitoneta
Cas 1h2'3" 49d49'35"
Lidia, la mia adorata Micetta :)))
Per 2h21'59" 51d47'15"
Lidianna Salt
Leo 9h22'55" 9d28'26"
Lidija
Cap 21h7'4" -16d17'24"
Lidija BV 1961
Uma 10h28'43" 55d25'52"
Lidor
Her 17h7'14" 14d13'19"
Lidwina Buettner
And 23h18'22" 48d23'8"
L'Idylle Choubamour 031206
Uma 9h35'15" 44d26'28"
LIEA
Cnc 8h0'8" 17d2'21"
Liebe
Lmi 9h36'6" 38d55'12"
Liebe
And 23h3'47" 52d52'42"
Liebe Kruse
Cap 21h44'16" -12d47'36"
Liebe & Vertrauen
Lac 22h37'54" 38d30'42"
Liebe zu meinen Batz
Ori 6h24'20" 16d52'37"
Liebe, Leonie
Uma 11h42'10" 35d55'30"
Liebeanfang
Dra 16h59'19" 62d38'53"
Lieben Sie ewig
Crb 16h11'35" 30d21'38"
LIEBES STERN
Cyg 20h41'37" 46d15'42"
Liebespfand
Cma 7h15'29" -30d7'47"
Liebesstern
Lyr 18h32'0" 36d12'50"

Liebesstern MADA
Cam 7h53'54" 72d53'42"
Liebesstern: Urs - Madeleine
Tau 5h5'20" 24d42'56"
Liebesstern von Alex u. Katharina
Ori 5h1'29" 5d34'33"
Liebevoll Aushalten
Lib 14h24'9" -10d36'35"
liebling
Peg 22h56'56" 25d56'50"
Liebold, Andre
Ori 6h16'36" 10d13'17"
Liedtke, Ditta
Uma 9h4'46" 53d41'47"
Liefjes
Crb 16h3'41" 27d57'57"
Liehr, Günter
Ori 6h4'8" 9d43'13"
Lieke en Maarten
Lyn 7h8'6" 58d1'6"
Liela & Larry Hanson
Cyg 20h14'25" 39d20'9"
Lielah Leighton
Cnc 9h4'55" 30d53'51"
Lien P. Tran
Sco 16h45'39" -44d50'30"
Lien Thi Huynh
Leo 10h13'18" 13d52'56"
Lies (s beste mami für immer)
Uma 8h57'55" 65d6'15"
Liesa
Cnc 8h13'38" 15d27'16"
Liese, Aldona
Uma 8h20'38" 62d32'4"
Liesel
Leo 11h45'9" 12d38'16"
Liesel
Ori 6h3'15" 6d30'49"
Liesel Nicole Powell
Psc 1h38'55" 9d33'22"
Liesel's Star
Ori 5h59'22" 20d59'23"
LIESJE
Sgr 18h50'24" -29d20'37"
Lieske Tigchelaar
Cru 12h34'57" -62d19'15"
Liesl
Lmi 10h16'19" 34d16'27"
Liesl
Per 2h59'43" 55d8'45"
Liesl and Anthony Forever 08201999
Leo 11h8'44" -0d24'22"
Liesl Epperson
Cam 3h56'25" 65d16'15"
Lieutenant Colonel John Moring III
Lib 14h53'55" -2d23'36"
Lieutenant Colonel M.L. Schiller
Lib 14h34'59" -9d1'9"
Lieutenant Malcolm Reed
Aql 19h48'27" 14d34'50"
Lieutenant Marc Stephen Olson, SPD
Tau 4h27'16" 7d44'45"
Lieutenant's Blue Knight
Cap 21h25'44" -24d20'36"
Lieve Konijn
Lep 5h35'35" -25d58'39"
Liezel
Crb 16h1'18" 34d58'26"
Life
Cam 4h29'35" 69d2'30"
Life
Uma 12h20'57" 53d28'20"
Life is Good CT+SP
Uma 11h49'31" 51d26'31"
Life is Love
Aqr 21h35'53" 0d58'45"
Life long memories
Vir 14h2'35" -4d55'45"
LIFE ~ LoveInfinityForeverEternity
Gem 7h15'2" 16d24'18"
Life Point Church Night MOPS
Cyg 21h49'55" 45d38'50"
Life star go with SiyuanDu&JohnShen
Sgr 18h12'3" -35d50'49"
Lifeboat Eric Klien
Ori 5h49'21" 8d58'57"
Lifen "Flora" Huo
Crb 15h46'6" 35d2'42"
Life's A Beach, Grammy
Cap 21h44'7" -11d50'11"
Life's little coincidences
Gem 7h19'32" 25d27'44"
Liga 09.09.06
Ori 5h12'36" 8d35'25"
Ligall Abitbol
Cyg 19h44'53" 35d5'38"
LIGEFEDE
Her 18h19'46" 28d52'57"
Ligeia Rose
Gem 6h59'41" 19d54'20"
Light and Love of My Life
Cyg 21h12'46" 37d18'56"
Light Fantastic - Sylvan R. Shemitz
Uma 8h28'55" 62d59'3"
Light Like Frank and Kara
Sge 19h48'28" 18d57'7"
Light & Love in a Luminous Embrace
Crb 16h0'53" 31d58'59"
Light My Way_Leslie Mefford
And 23h29'24" 47d55'59"
Light of Alice
And 1h1'25" 37d54'39"
Light Of Amen
Sco 17h39'2" -40d29'12"
Light of Andrew's heart - Janine
Cru 12h14'14" -64d5'58"
Light of Barbara
Cas 1h43'51" 66d17'8"
Light of BeRich
Cyg 20h25'13" 43d57'56"
Light of Brooke Lynn
Crb 15h53'1" 26d17'45"
Light of Cassandra's Eyes
Lyn 7h13'11" 50d29'44"
Light of Cathy Gibbons
Ari 2h35'17" 12d45'9"
Light of Christ for Bruce and Beth
Cyg 22h0'56" 51d4'4"
Light of Dirty Jenny
Leo 11h32'40" 5d13'41"
Light of Eternal Devotion, LJ & MJ
Cyg 21h6'34" 35d12'35"
Light of eternal Hope and Faith
Cnc 8h39'21" 26d18'38"
Light of Hopes and Dreams
Cru 12h37'33" -58d1'49"
Light Of Joy
Mon 6h44'47" -0d5'34"
Light of Judy
Cru 12h14'0" -64d24'34"
Light of Lauren
Mon 6h25'12" -4d23'54"
Light of Lee
Lib 15h4'28" -12d22'41"
Light of Linda
Lmi 10h19'3" 30d37'17"
Light of Linda
Lmi 10h27'32" 35d1'14"
Light of LoPicolo
Lib 14h45'29" -9d36'16"
Light of My Heart
Vir 13h48'50" -3d46'54"
Light of my heart Ronny
Tri 2h18'30" 36d50'48"
Light of My Life
And 2h28'28" 43d41'33"
Light Of My Life
Gem 6h49'11" 27d25'17"
Light of My Life
And 0h22'18" 28d48'26"
Light of my life
Aql 19h2'56" 1d36'41"
Light of My Life
Ori 5h31'30" 2d53'52"
Light of my Life
Cap 20h12'6" -10d41'36"
Light of my life
Aqr 22h56'18" -9d19'49"
Light Of My Life - Brilliant Pamela
And 0h40'51" 33d37'40"
Light of Naomi
Aqr 22h55'28" -10d38'2"
Light of Nativita
Leo 11h29'16" 28d6'17"
Light of Our Love
Cas 1h26'27" 66d42'40"
Light of Peter's Life
Per 2h37'15" 55d49'58"
Light Of Ronni
Sco 17h56'46" -35d12'20"
Light of Sheila
Psc 0h11'3" 8d7'44"
Light of Sonja
Lyn 8h29'16" 45d57'18"
Light of the Eagle
Ori 5h32'17" -2d21'14"
Light of the IB
Aqr 23h11'24" -7d50'42"
Light on Linzi
Cas 23h38'47" 53d58'41"
Light the Universe
Uma 12h12'49" 52d54'12"
Lightman
Cap 20h14'35" -18d21'47"
Lightning Kitty
Peg 22h15'2" 12d21'0"
Lightning Larry Sapp Of The Cosmos
Aqr 21h17'31" 2d4'42"
Lightning McQueen
Uma 8h56'40" 55d6'5"
Lights of all Lights
Per 3h27'48" 45d35'27"
lights will guide you home
Uma 12h35'17" 57d21'11"
Lights Will Guide You Home
Uma 13h39'15" 56d4'48"
Ligia Corina
Ori 6h9'38" 15d16'40"
Ligia Corredor Ryan "The Greatest!"
Leo 10h51'39" 18d30'57"
Ligia "Mashi" Diana
Cas 1h15'33" 56d16'36"
Lihnida Josevski
Ari 3h17'32" 15d23'52"
Lihy
Lyn 8h53'10" 37d18'51"
Liiis
Aqr 21h9'53" -8d50'55"
Liisa Mooney
Cas 1h55'10" 69d16'10"
Liisi
Peg 21h43'26" 8d51'18"
Lija Your Eternal Light
Cyg 21h46'19" 51d28'7"
Lika Malia
Vir 12h44'43" 2d4'8"
Lika-Kami
Ari 3h16'46" 19d25'0"
Like a Rose - Robin Elizabeth
Ara 17h37'49" -47d12'25"
Like Crazy
Uma 11h6'55" 60d21'35"
Like Father Like Son
Gem 7h50'9" 30d43'56"
Like The Daisy Loves The Sun
Lyn 8h33'26" 33d50'52"
Like You
Sgr 19h28'27" -15d24'25"
Likey 22
Tau 5h53'13" 26d8'55"
Liko Hewett
Gem 7h5'16" 14d51'40"
Li'l 1
Umi 16h27'50" 77d43'56"
Lil' Andy Roo—Andrew Jackson Hines
Lmi 10h49'43" 26d24'49"
lil angel
Cnv 12h49'47" 37d0'53"
Lil Angel Kyle
Sco 16h10'51" -31d41'53"
Lil Baby Boo & Sprinklez
And 1h28'49" 40d32'32"
Lil' baby Ethen Tyler Wilson
Psc 1h15'34" 26d7'7"
Lil Baby Girl
Del 20h46'29" 6d44'17"
Lil' Bear
Uma 8h56'58" 54d50'49"
Lil Bec Bec
Leo 11h12'17" -6d2'48"
"LIL BEST"
And 0h10'19" 31d13'3"
Lil' BigHead
Umi 14h20'13" 72d28'35"
Lil' Bit
Cnc 8h59'14" 8d59'30"
Lil Bit Bair
Dra 18h13'34" 76d46'53"
Lil' Bit of Sunshine
Sco 16h38'23" -34d37'1"
Lil Bits
Umi 16h37'4" 77d7'15"
Lil' Blond Chick
Cas 23h40'5" 55d4'7"
Lil' Booger
Psc 1h46'1" 22d40'14"
Lil' boy
And 1h49'30" 42d32'38"
Lil' Buddy
Uma 9h20'15" 43d22'13"
Lil C
Ari 2h32'16" 25d40'20"
Lil Chris Zubia
Her 17h31'25" 35d17'53"
Lil' Cookie
Cyg 21h52'32" 50d21'2"
Lil Darlings Missy Kiki Heather
Peg 22h18'38" 17d0'39"
Lil Day
Aqr 22h34'44" -2d47'17"
lil dee wright
Cyg 21h8'55" 32d16'18"
Lil Doc
Aqr 21h48'45" -7d3'21"
Lil Dorse
Her 17h7'13" 46d42'52"
Lil Drof
Leo 10h45'45" 14d49'36"
LiL Ducky
Her 16h57'6" 31d17'16"
Lil' E
And 0h13'2" 46d5'23"
L'il Essie
Aql 19h43'35" -0d1'17"
Lil G
Leo 10h32'52" 20d20'44"
L'il Gary Whybark McCune 4 22 1995
Crb 15h45'20" 31d7'42"
Lil Girl
Ori 5h45'3" 5d59'1"
Lil Glenn
Uma 11h18'34" 29d42'39"
Lil Hulkster
Uma 11h2'8" 38d39'51"
Lil' Jase
Cam 4h11'28" 71d28'40"
Lil Jesse
Ari 2h16'31" 23d13'37"
LiL Jessica Kiriwas
Tau 3h58'27" 19d39'2"
Li'l Jill
And 23h20'52" 42d18'22"
Lil Jones
Lyn 8h45'13" 46d17'42"
Lil' K
Tau 3h49'42" 16d34'46"
Lil Keebs
Leo 10h28'34" 16d52'48"
Lil' Krissy
Sgr 17h44'54" -21d23'29"
LiL K's Hero
Ori 6h15'7" 8d23'46"
lil Lady
Tau 5h28'4" 26d49'57"
Lil Lady Sonja
Col 5h10'32" -34d41'3"
Lil Lin-Z Lou
Vir 11h47'9" 9d47'36"
lil lulu
Vir 13h5'18" -12d24'53"
L'il Mac
Lyr 18h54'11" 43d38'1"
Lil' Magic
Peg 22h24'26" 9d48'38"
lil mama
Ari 2h37'43" 26d47'33"
L'il Margie
Psc 1h4'17" 8d47'34"
Lil & Max "Pappy" Mandel
Cyg 19h30'52" 53d11'52"
LiL me
Vir 14h7'31" -14d46'48"
LiL Meat & Filet
Umi 16h10'57" 75d24'14"
lil monkey
Gem 7h10'44" 17d40'50"
Lil' Niffer
Lib 15h17'45" -5d13'1"
Lil Ninja Dad- Hero In The Universe
Vir 14h34'3" -3d55'32"
Lil' Pamapaja
Lmi 10h24'16" 32d26'52"
Lil' Paradise
Ori 6h7'58" 6d52'30"
Lil' Plum
Cyg 20h48'2" 30d34'45"
Lil' Poppet
Pup 7h42'42" -18d19'5"
Lil Princess Stopar
Sgr 18h31'37" -24d32'19"
Lil Pup
Mon 7h33'11" -0d36'9"
Lil QT
Umi 15h39'56" 70d39'0"
Lil Red
Cyg 20h43'35" 34d52'28"
Lil' Red
Per 2h45'12" 51d38'55"
Lil Reggie Chelle
Umi 14h35'30" 86d42'24"
Lil' Ricky
Ori 5h37'32" 12d29'10"
Lil' RJL
Sgr 19h53'51" -24d10'31"
Lil' Ry Ry
Ori 6h10'47" 6d7'49"
Lil' Sessa (Vanessa Braun)
Uma 9h44'23" 51d51'56"
Li'l Sir Alex of The Big Big Heart
Sco 16h51'8" -27d14'39"
Lil Spoo
Vir 13h52'23" -7d49'43"
Lil' Stacey
And 23h55'37" 41d56'17"
Lil Star
Lib 15h9'58" -1d29'33"
Lil Star Cynthia
Del 20h39'1" 19d22'2"
Lil Steph
Aqr 22h28'43" -0d38'36"
Lil Sunshine
Lyn 8h26'32" 38d15'52"
Lil' Teeny
Leo 11h54'34" 26d14'37"
Lil Thumper's Tabbyland
Lib 15h9'0" -5d55'55"
Lil Tomo
Sgr 18h19'21" -25d4'16"
Lil' Trooper
And 0h43'27" 35d53'51"
Lil Val
Uma 12h40'40" 62d11'26"
Lil Walla Walla
Aur 5h19'59" 43d39'27"
Lil Willie Dippel
Peg 22h15'2" 11d45'26"
Lil Yellow Flower
Uma 8h57'43" 46d54'43"
Lila
Boo 14h49'5" 23d1'34"
Lila
Cnc 9h8'56" 15d3'12"
Lila
Uma 11h32'46" 56d29'47"
Lila
Cap 21h7'18" -26d41'39"
Lila Belle
Umi 16h43'39" 75d14'53"
Lila Elizabeth Goldberg 1/3/2004
Lyn 9h15'31" 46d7'26"
Lila Frances
Aqr 22h36'12" -1d48'40"
Lila Gatlin
Ori 5h32'49" -0d49'52"
Lila Gonzalez
Cam 9h17'39" 79d39'29"
Lila Jane Constance Murray
And 23h55'52" 42d11'57"
Lila June Wittman
Boo 15h19'10" 40d48'30"
Lila Kassem
And 1h46'25" 42d17'48"
Lila & Rafael
Cyg 19h46'39" 51d21'45"
Lila Roberts
Leo 11h7'5" 4d24'35"
Lila Rockwell
Sco 16h14'0" -16d22'6"
Lila Rose Diamond
And 2h19'1" 41d27'36"
Lila Vivian Rosenberg
And 23h27'13" 46d50'56"
Lila Wattenberg
Leo 11h14'42" 13d44'51"
Lilac Lady
Crb 16h6'15" 38d30'17"
Lilah Gordon Fienberg
Lyr 18h43'14" 34d55'25"
Lilah Juliette Filtranti Pacifico
Cnc 8h28'29" 31d0'34"
Lilah Mae Colbert
Ari 3h18'35" 29d3'9"
Lilah Rose
And 1h21'19" 43d59'2"
Lilakade
Psc 1h17'4" 22d33'23"
Li-Lan
Sco 17h40'39" -39d51'42"
Lilani
And 23h45'27" 34d45'8"
Lila-Rose
And 0h17'24" 40d36'44"
Lila's Estrella
Leo 10h45'31" 17d27'14"
Lila's L3 Legacy
And 1h29'31" 48d47'51"
Lila's Star
Sgr 18h33'38" -21d9'11"
LiLbit Duke
Lyn 8h12'29" 57d0'35"
lilbleustar
Lib 15h29'26" -6d20'14"
Lilcelt Amy
Lib 15h3'32" -18d36'39"
Lilea
And 23h25'11" 39d11'28"
Li-lei
Leo 11h11'14" 16d8'56"
Lilene
Her 16h53'28" 27d34'34"
LilHart
Uma 10h22'12" 47d44'0"
Lili
Vir 13h1'5" 10d57'20"
Lili
Psc 0h31'25" 9d19'59"
Lili
Sgr 18h50'53" -16d32'13"
"Lili"
Cap 21h2'7" -17d47'30"
Lili
Vir 13h2'52" -6d38'9"
Lili
Mon 7h16'2" -4d45'3"
Lili
Cas 1h2'35" 63d42'54"
LILI
Mus 13h44'47" -73d31'50"
LiLi
Psc 1h4'4" 5d4'48"
LILI
Lib 15h28'9" -25d33'7"
Lili
Her 16h36'27" 36d35'34"
Lili Bedoya
Vir 13h12'44" -22d4'53"
Lili Elizabeth Neumann
Gem 6h47'19" 26d31'10"
Lili Fröse
Uma 11h37'30" 37d13'39"
Lili Jenkins
Lmi 10h30'45" 34d28'44"
Lilia
Tri 2h35'6" 35d8'56"
Lilia
Per 3h24'21" 41d36'38"
Lilia
Lep 5h51'39" -11d47'11"
Lilia
Uma 10h7'30" 65d42'26"
Lilia Andrade
And 1h22'9" 39d9'12"
Lilia Georgina Ondarza Cisneros
Ari 2h53'25" 19d2'35"
Lilia Ines
And 1h15'48" 35d15'37"
Lilia & Leonid Mamut
Cnc 8h38'51" 32d39'18"
Lilia- Lita- Moreno
Cas 1h16'27" 66d14'59"
Lilia Makena Underwood
Dor 4h41'44" -53d58'53"
Lilia Martinez Hall
Sco 16h53'20" -38d51'16"
Lilia P. Garcia
Gem 7h10'26" 16d31'50"
Lilia Reyann Gomez
Cnc 9h1'58" 12d46'26"
Lilia Shirin
And 0h25'17" 42d36'6"
Lilia Vardanyan
Psc 1h23'8" 16d0'54"
Lilia Yanes
Ari 3h10'39" 22d20'56"
Lilian
Ori 5h36'51" 8d18'37"
Lilian
Crb 16h20'10" 37d47'54"
Lilian
Cas 23h16'6" 58d9'27"
Lilian
Cyg 20h29'20" 57d42'13"
Lilian Anshumala Dayal
Ari 2h16'34" 27d46'52"
Lilian Asquith
Ari 2h5'47" 24d50'46"
Lilian Blackwood
Cas 3h34'8" 69d23'33"
Lilian Chevalley
Cas 23h43'41" 56d7'59"
Lilian Corcuff
Sco 17h52'32" -36d27'46"
Lilian Espinoza Cerna
Uma 10h48'7" 49d31'41"
Lilian Finnemore 70th Birthday Star
Cas 3h13'0" 59d45'44"
Lilian Fowler
Leo 10h36'45" 22d49'37"
Lilian Garcia
Leo 11h14'54" -5d11'0"
Lilian Grace
Ari 2h27'29" 21d36'11"
Lilian Hautemulle
And 0h56'48" 37d11'12"
Lilian Mabel Winder
Cas 22h57'59" 56d24'55"
Lilian Manger 1923-2003
Uma 9h18'50" 48d58'37"
Lilian Mary + John William Milner
Cyg 22h0'40" 52d35'47"
Lilian Maud Meechan - Lily's Star
Cas 1h51'26" 64d57'6"
Lilian Pacheco
And 0h58'23" 46d15'57"
Lilian Rios
Cma 7h26'34" -18d46'12"
Lilian Riot
Psc 0h47'8" 19d50'38"
Lilian Von Villon Garcia
And 0h34'12" 28d22'39"
Liliana
Psc 0h52'57" 11d35'36"
Liliana
Uma 10h10'38" 49d15'24"
Liliana
Cam 7h3'29" 78d5'56"
Liliana
Cma 6h57'28" -31d37'30"
Liliana
Cma 6h57'28" 31d37'30"
Liliana Aceves
Ant 10h9'56" -29d37'42"
Liliana "Anjinha" Oh
Ari 2h45'42" 14d20'51"
Liliana Beer
Cas 0h32'33" 63d4'1"
Liliana Bonilla
Umi 16h23'24" 78d16'16"
Liliana De Los Angelas
Lib 15h37'16" -17d7'35"
Liliana Elzbieta
Uma 11h8'21" 55d1'35"
Liliana Falcone
Aqr 22h34'30" 2d24'54"
Liliana & Gianna Stella
Umi 14h3'26" 75d3'52"
Liliana & Gino I Love You Always
Vir 13h3'6" 3d25'4"
Liliana Grace
And 2h12'11" 42d22'38"
Liliana Grace Fox
Umi 15h2'53" 72d36'30"
Liliana Kaye Meissner
Cnc 9h6'2" 31d31'42"
Liliana Longo
And 0h19'36" 43d48'4"
Liliana M. Mendieta Psi-Aquarii
Aqr 21h38'27" 1d7'57"
Liliana + Marco Moretti
Cam 6h16'39" 66d12'16"
Liliana Maria Battistella
Tau 5h25'38" 24d50'56"
Liliana Mares Evans
And 0h46'58" 40d22'3"
Liliana Martina
Cnc 9h0'52" 30d6'26"
Liliana Mejia
Gem 6h23'5" 26d54'51"
Liliana Putra
Del 20h33'38" 17d19'28"
Liliana Sangre De Mi Vida
Umi 16h56'42" 77d35'46"
Liliana Tarcea Fitch
Uma 8h9'19" 60d35'25"
Liliane Eberle
Umi 17h5'3" 82d41'43"
Liliane Massara
Aqr 22h54'10" -19d59'51"
Liliane Rose Morley Cottrell
And 0h21'26" 26d24'17"
Liliane Schneider
And 2h37'16" 48d2'25"
Lilianna Alexandria
Sco 16h36'25" -35d11'0"
Lilianna Koledin, BFA
Ari 3h17'28" 28d5'34"
Lilianna Marisol Kehnle
Lib 15h25'25" -19d22'50"
Lilian's Wish Come True: Davey
Cyg 21h56'23" 50d22'3"
Lilias Larkin
Uma 9h13'2" 61d47'26"
Lilibeth Camargo
Sco 16h7'43" -20d58'29"
Lilie Cin Peng Lowe Walshe Angel
Cra 18h35'56" -43d58'26"
Lilie Queen
Vir 12h52'36" -11d32'7"
Lilija Juzeniene
Uma 10h56'59" 71d41'13"
Lilinoe Ka'ihikapumahana T. Kehano
Gem 7h30'9" 30d55'43"
Lilipuce
Sgr 18h57'16" -36d0'14"
liliquoi
Cnc 8h41'17" 16d22'12"
Lili's little piece of the heavens
Uma 10h52'4" 44d39'41"
Lili's Miracle
Ori 5h37'3" -1d11'57"
Lilish
Cap 21h28'15" -23d30'2"
lilit oganisyan
Uma 10h38'33" 43d29'44"
Lilit Sahagian
Tau 4h22'3" 9d40'25"
Lilit Topchian
Cyg 21h42'17" 48d5'21"
Lilita
Cnc 8h49'2" 8d47'3"
Lilith
Aqr 23h11'11" -8d16'7"
Lilith Pandora Rayne Gibson
And 0h43'44" 45d54'36"
Liliya
Vir 13h19'13" 12d19'15"
Liliya Bezmen
Crb 15h20'35" 26d55'37"
Liliya Prinos
Leo 10h23'0" 24d15'33"
Lilja Laatikainen
Cap 20h50'50" -23d16'38"
LILL 1
Umi 16h21'32" 77d21'32"
Lilla
Lib 14h48'34" -5d43'52"
Lilla G. 21859
Dra 19h49'53" 69d4'13"
Lilla Maryn
Cam 3h29'45" 55d15'40"
Lille
Uma 8h35'0" 59d35'59"
Lille West
Sco 17h6'9" -30d25'8"
Lillegren
Ari 2h11'11" 27d7'36"
lillemus
Ori 6h9'16" 13d25'45"
Lilli
Her 16h18'36" 6d41'37"
Lilli Bean
Umi 16h18'50" 80d27'39"
Lilli Harrianne Gardner
And 0h53'40" 43d3'10"
Lilli Machon
Lyn 6h44'14" 50d19'55"
Lilli Rayne
Cru 12h25'9" -61d15'32"
Lillia Anne Tarman
Uma 12h36'17" 59d52'46"
Lillia Hannah Agadjanyan
Lib 14h43'9" -17d20'43"
Lillia Shân Reilly
Cas 23h11'25" 59d32'41"
Lilliam Marie Cruz
Psc 1h52'24" 9d8'43"
Lillian
Vir 13h14'51" 2d9'38"
Lillian
Crb 15h53'3" 33d17'39"
Lillian
Cas 23h40'6" 55d25'41"
Lillian A Kelly & Spurgeon A Kelly
Cyg 20h14'10" 49d46'19"
Lillian A Milaef
Cyg 20h48'47" 39d44'28"

Lillian A. Notopoulos
Peg 22h24'16" 18d36'46"
Lillian Abrams' Star
Lyn 8h48'7" 40d11'13"
Lillian Alexandra
Umi 13h55'13" 71d13'48"
Lillian Alice Morton
Uma 9h37'39" 70d13'57"
Lillian and Chloe's Smile
Gem 7h34'7" 29d26'2"
Lillian and Murray
Cyg 19h56'37" 32d39'52"
Lillian Ann
Uma 9h0'20" 65d6'41"
Lillian Anna Blouin
Ari 1h59'27" 13d11'56"
Lillian Antoinette Guerin Ranalla
And 0h9'5" 44d21'44"
Lillian Bailey Simpson
And 0h22'17" 30d14'26"
Lillian Beesch Carden
Cet 0h54'20" -0d7'4"
Lillian Belle Tonole
Cas 2h23'32" 64d44'53"
Lillian Boni
Cas 2h15'13" 64d3'39"
Lillian Bronte Schad Burwell
And 0h30'27" 41d19'39"
Lillian Bruce
Cas 0h43'30" 60d43'49"
Lillian Camille DeLellis LiPuma
Sgr 17h57'53" -17d57'28"
Lillian Castillo de Strong
Ari 2h57'11" 13d35'55"
Lillian Christine
Lyr 18h42'39" 41d24'48"
Lillian Claire Goldberg
Crb 16h20'13" 38d36'57"
Lillian Clare
Aql 19h45'48" 6d11'12"
Lillian Corinna
Psc 1h26'27" 18d29'9"
Lillian & Dale Watson Forever
Ori 6h22'1" 13d19'13"
Lillian Downs Powell-Worrilow
Cnc 9h5'5" 31d39'59"
Lillian Dumas
Cyg 21h13'43" 43d32'29"
Lillian Dumas
Lyn 7h6'56" 52d21'56"
Lillian Excog
Peg 21h31'57" 16d15'36"
Lillian "Fabulous Moolah" Ellison
Cnc 9h3'30" 6d43'58"
Lillian Fay Penprase
Uma 9h22'40" 62d51'14"
Lillian Fiegl
Cas 23h23'2" 53d14'15"
Lillian Flora
Cap 21h15'16" -18d7'30"
Lillian G. Mazza (alpha MIMIonus)
Ori 5h2'56" 9d42'47"
Lillian Giana
And 0h36'18" 30d23'38"
Lillian Gipson
And 0h53'8" 41d26'9"
Lillian Goldberg
Mon 7h20'14" -5d10'25"
Lillian Grace
Sgr 18h37'51" -23d11'42"
Lillian Grace
Gem 6h50'29" 14d56'41"
Lillian Grace Cavanagh
Leo 9h51'7" 18d49'22"
Lillian Grace Harhay
Lyn 8h1'38" 58d4'18"
Lillian Grace Levee Lambert
And 0h37'55" 33d10'14"
Lillian Gusmano
Uma 9h56'48" 54d3'18"
Lillian Heyworth
Boo 15h35'37" 41d58'35"
Lillian Hope Anderson
Lib 15h10'15" -11d57'25"
Lillian Isabella Gottlieb
And 2h7'32" 45d2'49"
Lillian Ivette Santiago
And 2h34'24" 46d48'17"
Lillian Jane
Aqr 22h28'22" -0d37'45"
Lillian Jean Noriega
Cnc 8h29'12" 12d18'19"
Lillian Josefchak
Leo 10h39'28" 13d24'2"
Lillian Kari
Cam 3h21'48" 55d53'39"
Lillian Kelly Ahlers
And 1h0'40" 41d9'37"
Lillian Kelsey Chenault
Lyn 8h42'55" 34d42'41"
Lillian Kupferman
Aqr 22h29'8" -4d18'15"
Lillian L. Rose
Gem 6h49'29" 33d29'17"
Lillian Layne Hanmer
Leo 10h19'37" 11d36'31"
Lillian Lea Spar
Sco 16h37'7" -30d24'43"
Lillian Lekas
Cyg 20h47'48" 47d27'33"
Lillian Leonia
Tau 5h36'48" 23d36'4"
Lillian (Lilly) Hamnett
Aqr 23h5'14" -10d21'7"
Lillian Louis
Uma 9h58'44" 68d21'52"
Lillian M. Burton
Aql 19h39'22" 0d4'55"
Lillian M. Steinberg
Uma 9h42'9" 55d39'19"
Lillian Mae Vazquez
Tau 5h49'54" 27d41'53"
Lillian Malm
Ari 3h14'19" 22d19'12"
Lillian Margaret
Oph 17h4'43" -0d54'25"
Lillian Marie
Sco 16h7'38" -9d39'39"
Lillian Marie Lincoln
Uma 11h18'57" 59d47'12"
Lillian Marie Roosa
Lyn 8h27'16" 37d14'58"
Lillian Marie Spaid
And 1h47'12" 38d46'5"
Lillian Mattos
Cnc 9h14'56" 10d45'46"
Lillian May
And 1h24'53" 42d51'35"
Lillian May Pachico-Burovac 7/13/34
Umi 15h27'23" 71d31'52"
Lillian McCormick
Cnc 9h5'4" 18d42'8"
Lillian Messner
Ari 2h42'38" 29d17'13"
Lillian Moss Bleecker
And 23h37'52" 35d38'11"
Lillian Mutuku Nthenge
Her 16h6'12" 47d48'3"
Lillian Neva
Mon 6h59'2" 8d43'16"
Lillian O'Brien Soucie
And 0h13'0" 39d33'26"
Lillian Olivia-Jean Mauro
Crb 16h8'29" 37d41'58"
Lillian Parks
Pho 23h57'42" -56d18'47"
Lillian Pattinson Todd - "Lillian's Star"
Cas 0h53'42" 67d0'2"
Lillian Paz Esteban
Aqr 21h3'5" -7d6'0"
Lillian Petra Bickman
And 23h36'15" 48d41'49"
Lillian Phelps
Leo 10h46'30" 15d20'33"
Lillian R. Kelly
Cnc 8h18'52" 12d33'40"
Lillian Raiola
Ori 5h54'55" 18d37'5"
Lillian Rene Pickart
Ori 5h26'6" -1d13'33"
Lillian Rose
Ari 2h37'0" 26d36'41"
Lillian Rose McCarthy-Cole
Cen 12h24'14" -49d8'50"
Lillian Rose Ryan Karges
Vir 12h16'45" 11d7'52"
Lillian Rose Thorne
Cas 1h4'18" 60d46'23"
Lillian Ruth
Cas 1h24'1" 63d21'28"
Lillian Ruth
Del 20h33'16" 8d57'52"
Lillian S. Steele
Cas 0h29'31" 61d44'38"
Lillian Sandra Laiks
And 1h33'14" 47d30'3"
Lillian Sherry
Sgr 19h14'26" -19d34'0"
Lillian Sky Hall
Cru 12h3'1" -62d23'23"
Lillian Sloan Whatman
And 0h18'41" 41d13'2"
Lillian & Stanley
Umi 15h29'18" 75d10'49"
Lillian Star
And 2h17'14" 41d35'34"
Lillian Tanner
And 2h16'29" 46d33'57"
Lillian Taylor-Brock
Cru 12h37'36" -60d39'11"
Lillian Theaker - 30.03.1938
Cas 1h36'8" 61d25'2"
Lillian Trkula
Cas 0h35'45" 60d49'57"
Lillian V. Anthony
Cap 21h32'50" -22d51'25"
Lillian V & John E Leposky
Cyg 21h4'10" 38d4'31"
Lillian V Roberts
Aql 19h42'36" 3d23'48"
Lillian Velez Crawford
Cyg 19h53'30" 36d59'25"
Lillian (Vicki Bear) Sloan
Ari 2h29'33" 24d49'45"
Lillian Violet Lockwood
Cnc 8h13'40" 22d9'28"
Lillian Virginia
And 2h11'1" 42d3'58"
Lillian Walker
Lyr 18h55'45" 32d34'16"
Lillian Wilcox
Uma 11h33'57" 53d19'7"
Lillian Wilson
Sgr 18h11'53" -20d0'51"
Lillian York (Matey)
Srp 18h11'6" -0d9'3"
Lillian Yvonne West
Lyn 6h33'30" 54d0'59"
Lilliana
Uma 10h19'2" 50d16'59"
Lilliana DeFinizio~Our Newest Star
Lyr 18h57'10" 40d45'12"
Lillianna Elizabeth Andreoli
Uma 9h28'16" 42d24'8"
LilliannaGomes 9-54-57
Mon 6h45'6" 8d46'59"
Lillians Angel
And 1h10'57" 42d55'51"
Lillian's Eastern Star
Crb 16h3'49" 35d49'36"
Lillians Love Shines Forever
Aqr 22h32'25" -2d38'2"
Lillian's Shining Star
And 0h45'1" 39d28'28"
Lillie and Ellis Jordon
Lyn 7h54'0" 58d11'6"
Lillie Angel
Sco 17h34'12" -42d0'35"
Lillie Ann
Aqr 22h56'51" -9d30'23"
Lillie Ann Johnson
Ari 2h10'6" 24d44'55"
Lillie Ann Johnson
Ari 2h42'42" 25d20'30"
Lillie Anna
Ari 2h14'36" 23d11'29"
Lillie Bean
And 23h17'13" 47d28'2"
Lillie Bear
Umi 14h25'18" 73d4'14"
Lillie Bell
Tau 5h26'0" 17d52'11"
Lillie Catherine Kruger
Uma 13h55'55" 61d18'34"
Lillie Francis
Peg 21h52'3" 24d38'2"
Lillie George Mills
And 23h31'17" 37d39'43"
Lillie I. Gillen
Psc 0h49'38" 18d37'19"
Lillie J. Tulli
Cyg 19h58'9" 38d37'14"
Lillie Jewell Grasser
Hya 8h43'6" 4d19'21"
Lillie Lamar
Cas 1h3'43" 60d51'19"
Lillie Mae & Cliffords Star
Cma 7h22'29" -12d30'23"
Lillie Mae Tatom
Crb 15h45'24" 25d55'36"
Lillie May Wilkinson
And 23h6'27" 40d21'23"
Lillie Michelle Allen
And 23h42'11" 37d7'24"
Lillie's Star
And 1h0'42" 45d51'11"
Lilli's Christmas Star
Uma 11h21'44" 34d13'48"
Lillith Olivia Radford
And 2h16'10" 39d12'22"
Lilly
And 0h57'35" 44d32'52"
Lilly
Cnv 14h1'49" 35d24'2"
LILLY
Gem 7h5'58" 34d19'7"
Lilly
Cyg 21h53'17" 46d4'40"
Lilly
Leo 10h7'35" 24d7'42"
Lilly
Peg 22h9'50" 4d55'46"
Lilly
Aqr 22h13'9" -0d50'45"
Lilly
Lyn 6h54'31" 58d31'7"
Lilly
Lyn 8h18'40" 56d37'53"
Lilly
Cra 19h1'6" -39d53'34"
Lilly and Anthony Randazzo
Uma 10h57'29" 46d43'38"
Lilly and Carlie
Cap 20h56'7" -21d15'32"
Lilly Ann Owens
Vir 13h10'2" -4d23'13"
Lilly Bean Oyler
Uma 8h34'44" 62d34'16"
Lilly Blossom Davies
Del 20h36'12" 15d19'6"
Lilly Christine Jackson
Vir 12h33'39" -0d40'44"
Lilly Chu
Dra 18h59'20" 52d59'34"
Lilly & Daddy's Star
Aqr 22h58'9" -8d42'56"
Lilly "Danielle" Gillespie
Lyn 7h46'31" 47d6'16"
Lilly Ela
Cyg 20h56'30" 54d43'40"
Lilly Elizabeth Broder
Cyg 19h46'33" 34d34'20"
Lilly Grace Bowman
And 2h21'27" 46d55'38"
Lilly Grace Feeney
And 0h46'50" 40d0'35"
Lilly Grace McKinley
Ori 6h1'54" 16d56'40"
Lilly Grace McQuade
And 2h2'9" 47d24'17"
Lilly Grace Spence
Lib 15h43'41" -8d15'47"
Lilly Joy Gamel
Aqr 22h35'47" -1d59'0"
Lilly Kathleen Gordon
Uma 11h50'58" 61d23'42"
Lilly Kathleen King
And 2h38'0" 45d36'49"
Lilly Kathryn
And 2h37'21" 45d37'11"
"Lilly" Klähnhammer
Uma 9h4'30" 72d55'15"
Lilly Kliman
Lyn 7h19'7" 53d37'26"
Lilly Le
Cap 21h42'13" -10d5'10"
Lilly Loeding
Ori 6h18'38" 15d22'43"
Lilly Lou's 21st Birthday Star
Sgr 18h1'38" -26d45'56"
Lilly Mae
Crb 15h34'4" 30d23'30"
Lilly Mae Parrish
And 23h36'17" 48d41'58"
Lilly Maree Lesley Nutley
Cru 12h47'43" -62d20'8"
Lilly Marie and Leah Elizabeth
Ori 5h38'36" 3d1'27"
Lilly MaryAnne Compton
Cnc 9h9'39" 28d14'43"
Lilly Mathew Abraham
Per 4h22'14" 52d38'10"
Lilly May Mann
Cas 0h22'56" 55d25'19"
Lilly O'Brien
And 2h0'53" 42d6'31"
Lilly Quay
Lib 15h24'15" -8d21'15"
Lilly Santana Bartz
Uma 9h20'35" 69d29'43"
Lilly The Dosser
Cep 23h13'24" 86d41'31"
Lilly the Dosser
Cep 21h30'34" 83d1'46"
Lilly the Lover
Ari 1h56'18" 24d46'24"
Lilly Victoria Gagen - Lilly's Star
And 23h53'39" 36d57'59"
Lilly Vie Elmstrom
Ari 3h23'35" 28d29'45"
Lillyanna Blu Ramirez
Crb 15h48'17" 35d12'41"
Lilly-Ella Hall
And 23h44'49" 39d7'44"
Lilly-Faye
Umi 16h23'35" 79d0'56"
Lilly's Dream
Psc 1h50'31" 7d17'48"
Lilly's light
Cnc 8h46'37" 20d1'32"
Lilly's Star
Aqr 20h51'42" -12d12'26"
Lilly's Star - our Angel Bear
Umi 16h49'45" 84d19'19"
Lilly's Twinkle Star
Umi 13h4'48" 73d36'4"
LillyStar
Vir 12h48'58" 5d38'39"
LilMark41975
Umi 10h26'3" 86d49'24"
LiLo
Uma 11h34'44" 64d37'55"
Lilo
Sco 16h57'43" -31d52'23"
Lilo
Tau 4h48'47" 22d23'0"
Lilo
Tri 1h43'42" 29d29'52"
Lilo and Adis
Ara 17h27'6" -45d59'4"
Lilo Ann
Sge 20h6'25" 18d42'25"
Lilo Light and Love
Uma 13h54'20" 54d25'48"
Lilo's Perfect Moment
Sco 17h58'39" -40d8'17"
Lilou
Cen 13h42'29" -41d56'45"
Lilou
Cas 23h36'0" 58d15'21"
Lilou
Ori 6h20'11" 10d46'41"
Lilou
Del 20h35'29" 6d30'45"
Lilredz
Cyg 21h24'59" 47d5'46"
LilRenee32
Uma 11h35'13" 38d12'39"
Lil's Light
Cnc 9h2'58" 8d40'7"
Lilù
Her 17h11'33" 49d27'56"
Lilwenn
Cas 0h20'33" 58d54'0"
Lily
And 23h45'31" 47d51'8"
Lily
And 2h11'39" 50d34'56"
Lily
And 0h16'56" 43d35'14"
Lily
Crb 16h18'29" 30d1'25"
Lily
Lmi 10h33'48" 32d10'2"
Lily
Cnc 8h30'37" 13d25'34"
Lily
Cnc 9h2'45" 20d44'39"
Lily
Cnc 8h55'28" 15d26'44"
Lily
Ari 3h3'6" 26d36'55"
Lily
Cap 21h55'41" -23d32'47"
Lily
Cas 23h38'36" 58d7'47"
Lily
Lyn 7h52'38" 54d0'22"
Lily
Aqr 20h39'54" -8d43'22"
Lily
Aqr 23h49'8" -14d13'28"
Lily 3
Aqr 23h13'49" -9d34'17"
Lily A. Maier
Lib 15h20'10" -20d12'15"
Lily Aldridge
And 0h21'9" 29d54'19"
Lily Alexandria Notthoff
Psc 1h22'4" 11d42'33"
Lily A'marie David-Fortune
Sgr 18h21'52" -35d20'19"
Lily and Ruby we love you
Cyg 21h30'36" 39d16'9"
Lily Ania's Special Star
And 23h34'47" 38d8'5"
Lily Ann Cashman
And 23h1'50" 35d14'56"
Lily Ann Corpuz
Uma 9h8'59" 51d32'1"
Lily Ann Martin
Leo 9h38'46" 23d50'30"
Lily Anne Bradley
And 0h28'14" 27d22'8"
Lily Anne Dykema
Gem 6h43'25" 15d15'20"
Lily Ann-Marie Zucks
Aql 20h17'27" 3d15'0"
Lily Ann's Star
Uma 9h7'20" 68d42'15"
Lily Arianne Lipman
Uma 10h11'12" 63d37'54"
Lily Atkins Hunt
Cap 20h37'29" -9d30'27"
Lily B Taylor
Gem 7h40'32" 34d5'6"
Lily Bane's Night Light
And 0h28'7" 35d3'50"
Lily Beatrice
And 2h37'55" 45d6'16"
Lily Beatrice Ann Buckler
And 2h24'8" 41d52'25"
Lily Beth Errington
And 23h39'41" 41d27'58"
Lily Campbell
Del 20h45'17" 4d40'59"
Lily Caroline Canter
And 23h44'33" 44d52'13"
Lily Caroline Hatch
Uma 9h30'21" 43d59'18"
Lily Cate Weisskopf
Cyg 20h11'33" 44d6'49"
Lily Catherine Anouk Bretos
Peg 22h25'59" 29d54'44"
Lily Charlotte
Ari 3h23'45" 18d53'47"
Lily Claire Ecoff
Tau 4h33'50" 16d21'18"
Lily Clara
Crb 16h5'48" 39d29'6"
Lily Claro. Dancing with the Angels
Lib 15h53'52" -14d6'22"
Lily Cornachio
Dra 17h52'52" 52d3'34"
Lily Davis
Uma 11h3'29" 63d58'40"
Lily Dean Skeldon
Tau 4h4'38" 17d7'27"
Lily Dell-Siden
Cas 23h1'22" 53d37'16"
Lily Dickson - "Forever Shine"
Cas 1h55'35" 61d50'30"
Lily E. Pearce
And 23h22'52" 50d57'39"
Lily Elizabeth Duffy
Cas 22h59'58" 54d55'56"
Lily Elizabeth Molnar
And 23h18'16" 51d51'40"
Lily Ella Rolfe
And 1h20'1" 44d19'6"
Lily Ellen Hillyer
And 0h41'16" 43d36'4"
Lily Eve Evans
And 2h25'35" 44d36'45"
Lily Evelyn Grace Slade
Cas 23h23'24" 58d45'16"
Lily Faith Sterner
And 23h49'32" 47d57'9"
Lily Faith Williamson
Uma 9h48'6" 58d53'33"
Lily Fangman
Gem 7h25'4" 20d16'4"
Lily Fast
And 0h37'37" 37d56'16"
Lily Fay Griffiths
And 1h30'52" 45d11'30"
Lily Florence Fleat
And 23h7'27" 41d4'1"
Lily Flower
Uma 9h2'5" 51d46'17"
Lily Forever In our Hearts
Cnv 13h48'40" 39d25'25"
Lily Frances Bittle
And 23h15'19" 36d44'5"
Lily Franciskovic-Goss & John Goss
Lib 15h17'45" -9d44'28"
Lily Grace
Cyg 21h46'9" 54d29'51"
Lily Grace
Cru 12h25'2" -60d52'28"
Lily Grace
And 1h31'15" 39d56'42"
Lily Grace
Lyr 18h48'58" 33d52'23"
Lily Grace
Cnc 8h39'48" 10d27'25"
Lily Grace Padden
And 23h26'17" 48d10'13"
Lily Grace Peters
Sco 16h16'50" -16d30'53"
Lily Grace Stokes
And 2h1'30" 46d23'10"
Lily Grace Yancey
Ori 5h22'39" 13d55'12"
Lily Guile Smith
Ori 5h33'29" 9d36'9"
Lily Hee Kyung
Cas 1h45'16" 64d54'55"
Lily Holme
And 2h22'32" 37d41'59"
Lily Hope Denney
Ori 6h15'8" 15d5'22"
Lily Hoven
And 0h26'18" 45d18'35"
Lily Huer
Cas 1h6'28" 57d0'38"
Lily Iona McCormick
And 23h13'43" 40d34'13"
Lily is 1 Today, Our Little Star.
Cam 4h10'31" 52d57'35"
Lily Jacqueline
Cas 2h16'24" 62d42'49"
Lily Jane Sladen
Aqr 22h7'12" -18d40'2"
Lily Jean
Psc 1h12'6" 5d28'24"
Lily Jo
And 0h8'51" 40d39'3"
Lily Kae Littrell
Uma 11h6'59" 33d48'32"
Lily Kate
Psc 1h27'17" 19d4'34"
Lily Katilin Moyer
Tau 5h9'40" 18d41'7"
Lily Kourkounis
Leo 10h14'27" 9d59'54"
Lily Lai
Mon 6h47'53" 4d43'50"
Lily Louise
Ari 2h44'32" 11d38'27"
Lily*Loves*Alex*Forever
Cyg 20h27'47" 50d13'20"
Lily Luna
And 2h3'54" 37d48'23"
Lily M. W. King
Lyn 6h47'50" 58d33'49"
Lily Mackenzie Wilde
Uma 11h30'17" 57d12'41"
Lily Mae
Cas 23h3'58" 56d19'39"
Lily Mae
And 23h10'45" 41d40'14"
LILY MAE
Vir 13h1'16" 12d9'8"
Lily Mae Dehaan
And 2h11'22" 50d33'2"
Lily Mae Moore
And 22h59'13" 51d18'15"
Lily Mae Winter Grimes
Cru 12h14'48" -60d46'37"
Lily Margaret Nowicki
And 23h12'41" 52d43'33"
Lily Marianne
Cas 0h28'15" 55d25'13"
Lily Marie
And 0h9'11" 29d21'2"
Lily Marie J
Cyg 20h11'46" 58d29'21"
Lily Marie Kershner
And 23h50'13" 39d30'59"
Lily Marie Reinold
And 2h22'32" 48d30'7"
Lily Marin Wesp
Lib 14h55'38" -18d33'26"
Lily Maxine
Lyr 18h27'22" 29d24'37"
Lily May
And 23h12'8" 39d39'39"
Lily May
Uma 11h13'38" 50d22'23"
Lily May
And 2h38'44" 37d42'30"
Lily May Miles
Cas 0h14'38" 54d5'0"
Lily May Smith
And 1h53'32" 41d19'44"
Lily Mia
And 23h54'11" 42d24'0"
Lily Michelle
And 23h23'23" 44d1'13"
Lily Morales
Ari 2h36'56" 24d59'26"
Lily My Princess
Dor 5h49'59" -63d2'38"
Lily Neff
And 23h21'10" 41d43'21"
Lily Nicole
And 23h51'6" 48d27'52"
Lily Nicole Bosh
Cyg 19h37'58" 39d27'39"
Lily Nicole Haimes
Vul 21h24'38" 26d18'0"
Lily Noelle Boyle
And 23h16'37" 47d28'0"
Lily of the Heavens
And 1h42'28" 46d22'31"
Lily Paige Gleason
Crb 15h58'5" 31d33'9"
Lily Parisa Tajiki
And 2h32'37" 50d3'2"
Lily Patricia Andersen
Lyr 18h43'21" 34d12'3"
Lily Payne
And 2h25'3" 41d36'22"
Lily Rae-Ann
And 23h27'34" 41d58'17"
Lily Raquel Burling
Cnc 8h47'14" 11d1'54"
Lily Rebecca Jones
And 0h33'15" 46d1'46"
Lily Rose
And 23h59'47" 46d30'12"
Lily Rose
Vir 14h19'49" -5d8'11"
Lily Rose
Cas 23h35'30" 55d35'43"
Lily Rose
Cas 1h3'1" 61d44'7"
Lily Rose Anderson
And 0h23'25" 39d7'2"
Lily Russenberger-Rosica
Cnv 12h45'25" 44d3'42"
Lily Ruth Asbee
Gem 6h44'4" 33d46'53"
Lily Sammaciccia
And 23h23'19" 45d58'46"
Lily Sherman
Cnc 8h2'0" 11d14'17"
Lily Simone Kosminsky
Sco 16h10'46" -10d22'50"
Lily Spadafora
Cas 0h36'18" 65d40'53"
Lily Starr Fabbrini
Peg 22h40'24" 10d42'58"
Lily Sue Fletcher
Gem 6h59'43" 15d45'5"
Lily Thomson Harris
And 0h45'32" 36d48'40"
Lily Tran
Lyr 19h3'4" 32d3'39"
Lily und Andi - Come what may
Uma 8h30'21" 60d51'43"
Lily Vaughan
Cas 0h28'15" 58d51'29"
Lily Weidman
Uma 10h24'30" 66d2'35"
Lily With An Angel Face
Cap 20h39'10" -15d47'30"
Lily Xuan Thanh Tran
Crb 15h25'2" 31d29'23"
Lily Yan
Lyr 18h48'47" 43d7'34"
Lilyan
Cnc 8h18'55" 32d13'43"
Lilyan Dawn
Cap 21h11'29" -24d25'29"
Lilyana
And 0h8'58" 33d9'47"
Lilyana Mirelis Mojica
Cnc 8h36'21" 23d52'22"
Lilyana Soto
Cnc 8h52'37" 30d46'55"
Lilyann Dina Soots
Gem 6h46'46" 16d33'40"
Lily-Ann May
And 0h55'30" 44d6'40"
Lilyanna Grace Wagstaff
Umi 16h39'57" 82d39'4"
LilyAva
Lmi 9h56'24" 35d58'17"
Lilybeth Craw-Seaman
Cas 0h6'47" 55d35'24"
Lily-Bug
Peg 22h37'12" 20d42'14"
Lily-Ella
Umi 14h39'24" 81d15'44"
Lily-Ella Aldworth
And 23h38'26" 41d16'12"
Lily-Ella Beatrice
And 23h22'30" 43d35'36"

Lily-Jade Russell
And 1h0'17" 45d28'51"
Lily-Mae Roisin Cooney
And 2h29'17" 40d29'58"
Lily-Mae Star
And 23h56'38" 46d25'29"
Lilypad
Sco 17h35'49" -36d22'9"
LilyRenea
Dra 19h46'49" 62d26'36"
Lily-Rose
And 23h15'43" 39d50'35"
Lily's Acer
Ori 5h40'15" 15d7'49"
Lily's Angel
And 2h13'59" 50d10'45"
Lily's Diamond in the Sky
Sco 16h10'27" -11d49'40"
Lily's Hero
Aqr 22h34'33" -19d36'56"
Lily's Love
Umi 16h27'51" 81d20'29"
Lily's Love Light
Cnc 8h56'36" 12d35'25"
Lily's Star
Lyn 8h3'22" 57d21'21"
Lily's Star for Erica
Lib 15h34'19" -10d6'45"
Limar's star
Ari 3h9'53" 29d53'20"
Limbostarr
Cnc 9h16'53" 27d25'22"
LIME Financial
Per 4h17'18" 33d3'13"
LIMIGAL
Sco 16h54'36" -35d56'46"
Limited
Uma 11h27'15" 29d19'21"
Limor Rostamian
Uma 11h47'9" 38d36'2"
Lin
Cas 1h6'20" 70d49'48"
lin#25love
Cyg 20h12'38" 55d27'53"
Lin Bothe
Cyg 21h35'18" 34d53'27"
Lin DeVaun
Cap 20h49'36" -19d26'58"
Lin Haldeman
Sgr 19h8'31" -18d38'28"
Lin & Jim
Cyg 20h47'29" 30d31'39"
Lin Jiyue
Uma 11h21'27" 46d12'46"
Lin Ren
And 23h11'34" 40d0'19"
Lin Shih-Chieh
Sco 16h19'23" -10d1'15"
Lin Star
Sgr 17h55'11" -29d14'31"
Lina
Pyx 8h37'23" -22d30'41"
Lina
Uma 14h24'34" 56d24'46"
Lina
Uma 9h59'36" 45d39'1"
Lina
Uma 9h45'54" 45d13'39"
Lina Alba
Gem 6h46'32" 33d23'25"
Lina & Bill Lento
Lyn 7h17'39" 58d42'30"
Lina & David
Cyg 21h43'52" 38d3'13"
Lina Eskander
Her 18h55'8" 18d46'17"
Lina Flurina
Her 17h53'35" 32d45'52"
Lina Haraké Khansa
Cyg 19h50'8" 32d51'58"
Lina Immerz
Uma 10h5'38" 48d35'19"
Lina Mae Lee
Srp 18h17'35" -0d9'11"
Lina Majed
Boo 15h22'30" 37d50'16"
Lina Marcela Garcia
Crb 15h39'12" 26d54'27"
Lina Maria
Uma 10h21'49" 45d53'55"
Lina Perrotti
Cas 23h19'41" 54d41'40"
Lina Petit Ange
Ari 2h51'30" 22d16'2"
Lina & Sven
Cyg 21h21'5" 51d49'52"
Lina Tamm
Her 17h53'33" 26d54'45"
Lina Valencia
Psc 1h37'55" 17d51'20"
Lina Velez
Cmi 7h26'30" -0d8'28"
Lina-Amélie
Uma 9h54'27" 62d24'40"
Linake Ku'u Makakehau
Sgr 18h33'3" -28d7'54"
Linal Harris
Uma 11h36'55" 52d9'57"
Linald Star
Cap 20h26'50" -10d16'51"
Linas and Sam
Tau 4h17'55" 23d19'47"
Linas — Lion
Lmi 10h12'29" 29d3'20"
Lina's Universe
Leo 11h51'58" 26d59'32"
Linbabes
Cmi 7h8'15" 10d17'49"
Lincoln B. Snyder
Cnc 8h41'37" 32d29'41"
Lincoln Brydon Zachary Burke's Star
Sco 17h32'45" -42d24'50"
Lincoln Douglas Durey
Tri 2h11'42" 32d56'36"
Lincoln James Feuerstein
Uma 10h16'55" 42d29'9"
Lincoln & Kay Wray
Mon 7h9'8" -0d0'54"
Lincoln Logger
Uma 9h35'21" 44d12'28"
Lincoln Robert Alger
Uma 13h44'57" 52d34'33"
Lincoln Robert Morin
Uma 11h42'2" 39d26'12"
Lincoln-Star Mary Ann Erdmann
And 0h32'36" 42d22'53"
Linda
Lyn 8h7'38" 44d11'37"
Linda
Lyn 7h31'38" 38d9'25"
Linda
Cas 1h10'37" 57d30'0"
Linda
And 23h27'8" 41d48'22"
Linda
And 2h11'8" 49d17'29"
Linda
Ari 2h2'13" 18d6'9"
Linda
Ori 4h52'56" 10d40'59"
Linda
Vir 13h32'32" 5d27'5"
Linda
Crb 15h24'43" 28d28'51"
Linda
Psc 1h18'0" 27d16'31"
Linda
Ori 6h0'8" 22d29'21"
Linda
Leo 10h53'37" 21d9'1"
Linda
Cam 3h48'12" 71d55'49"
Linda
Uma 9h43'0" 69d39'56"
Linda
Mon 7h38'5" -7d19'50"
Linda
Lib 15h6'1" -2d20'33"
Linda
Aql 19h59'47" -5d43'50"
Linda
Lib 15h18'36" -20d40'5"
Linda
Sco 16h6'45" -19d10'15"
Linda
Umi 14h12'35" 78d49'49"
Linda
Psc 1h43'10" 6d47'11"
Linda 143
Leo 10h52'50" 16d41'55"
Linda 143
Gem 6h11'23" 26d39'12"
Linda (3 year anniversary)
Ari 2h9'14" 27d3'12"
Linda 52550
Gem 6h28'49" 24d55'50"
Linda 7-17-51
Ori 5h3'57" 13d28'8"
Linda A. Coulombe
Lib 15h5'35" -2d2'25"
Linda A. Curcio
Gem 6h44'17" 13d24'3"
Linda A. McLaughlin
Crb 15h48'16" 25d54'36"
Linda A. Stopford Shines Forever
Leo 10h38'16" 18d37'53"
Linda A Wegrzyn
And 2h5'1" 43d29'57"
Linda Abrahamsen
And 2h26'18" 45d1'35"
Linda Addington
Peg 23h22'31" 32d51'45"
LINDA AKINS
Leo 11h36'20" 15d33'10"
Linda Alison Pymont 23-03-2007
Ari 1h57'57" 20d41'9"
Linda Allen
Cnc 8h55'52" 18d22'57"
Linda - An Eternal Shining Light
And 23h41'23" 47d29'22"
Linda and Brian
Sgr 17h51'53" -29d49'30"
Linda and Celia Velez
Cyg 21h16'14" 48d25'11"
Linda and her Daughter
Cyg 20h17'46" 47d4'9"
Linda and John Maderious
Lyr 18h38'54" 38d16'48"
Linda and John Zacchio
Cnv 13h35'15" 33d41'24"
Linda and Lai Forever in Love
Cnc 8h37'41" 31d24'3"
Linda and Nathan Wang
Umi 16h3'48" 79d22'32"
Linda and Rich Calabrese
Crb 15h33'31" 32d23'56"
Linda and Robert Quattrone
Dra 16h50'1" 58d4'49"
Linda Ann
Cas 0h39'7" 61d26'37"
Linda Ann
Psc 23h54'2" -0d13'28"
Linda Ann Cameron
Gem 7h35'6" 16d54'20"
Linda Ann Carol Brown
And 0h5'11" 45d43'43"
Linda Ann Eckhardt
Gem 6h52'27" 29d46'27"
Linda Ann McCright
Lyn 7h30'34" 53d27'1"
Linda Ann Skitt
Cas 23h44'2" 55d4'50"
Linda Ann Stafford
Cap 21h32'22" -8d59'10"
Linda Ann Symos
Cas 2h17'47" 63d39'34"
Linda Ann Wods
Leo 9h36'14" 20d22'35"
Linda Anne
Mon 6h47'1" 3d7'6"
Linda Anne
Cap 21h42'0" -11d9'58"
Linda Anne Marousek
Cyg 20h39'14" 46d48'48"
Linda Antonette Johnson
Cap 20h52'56" -25d42'47"
Linda April
And 2h35'14" 48d14'46"
Linda Arcangeli -GaMPI Shining Star
And 0h34'40" 36d38'34"
Linda Ascendant
Tau 4h32'54" 14d11'56"
Linda Atkinson's honored place
Cyg 19h58'42" 55d19'57"
Linda Aydostian
Cap 20h22'8" -25d1'8"
Linda B
Cam 7h31'31" 60d37'8"
Linda B. Goldsmith
Lyr 19h19'33" 29d44'44"
Linda B Star
Ari 3h14'58" 20d26'20"
Linda B. Suydam Lawnik
Uma 8h13'16" 70d43'32"
Linda B & Sydney E
Cnc 8h35'39" 15d51'40"
Linda B. Weinstein
Tau 4h2'31" 15d53'51"
Linda Baker's Ruby Celebration
And 23h42'19" 37d7'32"
Linda Ball Scovill
Lyr 18h49'7" 34d14'16"
Linda Banaszek
And 0h18'21" 43d25'7"
Linda Barca
Cas 2h12'46" 65d4'32"
Linda Bast
Cnc 8h35'13" 24d22'47"
Linda Beatrice
Srp 18h4'39" -0d31'29"
Linda Bernstein
Lib 15h29'47" -9d44'33"
Linda Betty Barlow
Lyn 7h52'45" 37d5'17"
Linda Bewley
Uma 8h46'2" 64d53'28"
Linda Blooms' Personal Wishing Star
Cyg 20h47'23" 51d40'44"
Linda Bordeaux Star of Courage
Vul 19h29'27" 23d13'17"
Linda Bouliane
Lyn 6h45'55" 53d50'41"
Linda Bradley-Star mom
Cas 0h32'59" 55d47'15"
LINDA BRENNAN
Lmi 10h8'40" 38d57'59"
Linda & Brian Bishop
Cnc 8h20'36" 14d29'54"
Linda Bridges
Crb 15h33'5" 37d4'24"
Linda Brittain Atwell
Ari 3h4'54" 27d33'6"
Linda Brogle
Sco 17h56'54" -31d13'37"
Linda Buck
Tri 2h7'35" 34d26'25"
Linda Bui
Ori 6h12'7" -0d52'5"
Linda Burke
And 0h52'53" 37d16'57"
Linda Butler
Cas 2h12'50" 72d54'35"
Linda C. Farina
Lyn 7h39'42" 41d19'39"
Linda C. Garcia
Gem 7h30'50" 29d50'37"
Linda C Goodwin
Uma 8h20'45" 62d41'16"
~ Linda C. Kirschke 12-26-1956 ~
Cap 21h48'48" -21d9'37"
Linda C Russell
Tau 4h8'25" 26d35'2"
Linda C Silva
Tau 4h36'25" 20d53'26"
Linda C. Taylor
And 23h46'58" 41d25'25"
Linda C. Werbrock
Sco 17h54'41" -36d16'4"
Linda Cacace
Cas 1h31'30" 62d1'33"
Linda Cafarelli Love Amy and Eddy
Cyg 21h4'51" 36d33'50"
Linda Caltha Belle Astrum
Leo 9h34'25" 28d54'42"
Linda Campanelli
Ari 2h1'32" 22d1'6"
Linda Cantrell
Vir 13h39'26" -9d29'6"
Linda Carlotta Osborn
And 1h12'0" 37d39'56"
Linda Carlson Gunn
Leo 9h50'54" 7d48'21"
Linda Carol
Peg 23h42'45" 10d10'47"
Linda Carol
Crb 16h10'10" 29d58'32"
Linda Carol Gomme
Ari 2h28'6" 25d45'55"
Linda Carol Yeater Winings
Aql 19h4'6" -0d55'43"
Linda Carol Yerg Dickinson "Ribbit"
Uma 10h28'8" 64d35'3"
Linda Caroline - December 12, 1997
Sgr 18h19'39" -32d35'22"
Linda Caston
Psc 23h24'38" 6d0'58"
Linda Catherine Dupre
Tau 3h57'13" 27d22'15"
Linda Cecilia Payne Bryson
Lib 14h49'23" -18d9'45"
Linda Chant. My Wife...
Cyg 19h36'3" 35d7'24"
Linda Charleston
Uma 10h35'25" 44d39'44"
Linda Charlotte Stone Johnson
Gem 6h50'6" 20d58'51"
Linda Cheng - Rising Star
And 1h43'3" 42d5'6"
Linda Chisholm
Cyg 21h47'57" 50d9'49"
Linda Christine Mathisen
Col 5h54'54" -32d59'10"
Linda & Chuck Buie
Uma 12h41'20" 56d49'2"
Linda Claire Babl
Uma 11h25'52" 71d17'1"
Linda Clark
Cam 3h55'43" 59d57'48"
Linda Clarke
And 2h17'37" 38d6'2"
Linda Compton
Cam 5h10'8" 65d31'39"
Linda Constas
And 1h24'6" 50d30'37"
Linda Contessa
Gem 7h4'55" 25d37'55"
Linda Corbett
Lyn 7h55'17" 41d53'59"
Linda Corporon Arnold
Tau 4h22'28" 25d56'0"
Linda Coulter
Cas 0h13'38" 52d45'6"
Linda Couture
And 22h57'33" 51d51'46"
Linda Cracknell Rau
Uma 11h32'51" 36d18'26"
Linda D. O'Neal
Cas 0h29'23" 56d51'30"
Linda D. Ruscito
Lib 14h49'12" -2d31'9"
Linda D'Andrea
Uma 10h10'50" 59d25'28"
Linda Daniel
Uma 13h43'28" 53d41'22"
Linda Daniel
Uma 14h4'50" 50d21'2"
Linda Danielle Bowen
Sgr 18h59'24" -14d6'8"
Linda Darlene Krahn
Ari 3h27'12" 23d1'44"
Linda Darlene Miller "We love you"
Uma 11h54'19" 65d3'2"
Linda Darlene Ostrander
Uma 10h9'32" 51d4'44"
Linda Davey
Cyg 21h16'12" 44d17'51"
Linda Dawn
Leo 9h43'32" 31d36'33"
Linda Deann
Cas 0h54'45" 57d12'49"
Linda Deborah Lowell Oeth
Vir 13h9'51" 5d56'9"
Linda Dee Dee Hershey
Vir 13h59'0" -14d57'38"
Linda Dell Saxe
Sco 17h15'31" -32d8'58"
Linda Dettmann
Cas 0h46'46" 60d53'46"
Linda DeVita 1955
Aqr 21h41'9" 2d14'53"
Linda Diana Hill
And 0h27'31" 42d47'44"
Linda "Diane" Doyle
Uma 10h3'15" 42d10'26"
Linda Diane Ryan
Lyn 8h7'0" 33d57'17"
Linda Diane Sartor Latzoni
And 22h58'33" 40d9'34"
Linda Diane The Star of Angels
Umi 16h1'8" 82d5'27"
Linda & Don Wendler
And 1h35'55" 47d49'50"
Linda Donaldson
And 23h59'17" 45d3'41"
Linda Dow
Tau 5h12'9" 24d38'51"
Linda E. Shipp
Col 5h52'18" -35d48'23"
Linda Elisabeth Schneider
Aqr 22h45'44" -3d59'24"
Linda Elizabeth Arciniega (Eya)
Lib 15h41'22" -28d49'42"
Linda Elizabeth Twyman, Mommy
Vir 11h42'16" 7d35'32"
Linda Ellen Vey
Vir 14h38'32" 3d44'14"
Linda Emily Hughes
Psc 0h17'29" 12d8'12"
Linda Esquivel
And 0h44'4" 40d24'42"
Linda et Yvan
Uma 13h21'25" 54d18'10"
Linda F. Burritt
And 1h25'57" 47d45'45"
Linda F. Jeter
Cyg 19h35'29" 30d40'49"
Linda F. Willinger
Cam 13h28'9" 79d6'18"
Linda Face
Psc 23h6'17" 7d47'24"
Linda Fagin
Lmi 10h26'5" 34d51'53"
Linda Fairstein
Tau 5h48'9" 27d12'44"
Linda Faith
And 0h47'29" 39d58'12"
Linda Farley Fisher
Cas 0h14'5" 51d32'51"
Linda Faye
Cap 20h40'29" -15d1'19"
Linda Faye Lemon
Ari 2h12'58" 26d14'3"
Linda Finkelstein
Ari 2h56'31" 26d3'24"
Linda Fire Hawk
Pho 23h34'39" -39d42'46"
Linda Fisher
Gem 7h28'32" 33d13'41"
Linda Fornasier
And 2h36'16" 49d4'5"
Linda Fortunato
Ari 1h59'38" 14d50'58"
LINDA FOTI
Uma 10h10'56" 42d15'12"
Linda Frances
Crb 15h38'14" 26d37'23"
Linda Frances Eneks
Per 3h14'46" 31d53'56"
Linda Francine Sebolt
And 0h49'38" 45d36'50"
Linda G. Brooks
Tau 4h38'28" 22d54'0"
Linda G. Christian
Cam 5h34'24" 57d45'34"
Linda G. Davis
And 1h47'43" 49d22'52"
Linda G. Ketner
Tau 3h39'48" 15d49'50"
Linda G. Peltz "Blondie"
Leo 9h34'52" 26d28'18"
Linda Gabrielsen Loe
Ari 2h50'4" 16d49'40"
Linda Gail
Ori 4h54'8" 10d7'29"
Linda Gail
Mon 7h17'29" -0d25'1"
Linda Gail Childers Milam
And 0h11'36" 32d53'43"
Linda Gail Novak
And 2h5'50" 45d1'33"
Linda Gale
Lib 14h50'12" -1d45'43"
Linda & Gary
Lyn 7h58'6" 35d4'11"
"Linda Gay"
Lib 15h31'49" -6d3'17"
Linda Gayle Hallead
Cas 0h19'41" 58d3'59"
Linda & Gene Young
Leo 11h4'26" 21d48'12"
Linda George
Gem 6h45'58" 17d25'37"
Linda Gilbert Wyble, Jan. 1, 1957
Cma 7h25'25" -24d31'9"
Linda Gill
Del 20h41'23" 15d6'5"
Linda Gillis
Tau 3h39'12" 30d2'10"
Linda Gionet
And 0h37'35" 41d26'57"
Linda Gomez
Ari 3h8'39" 25d29'30"
Linda Goodwin
And 2h21'40" 38d1'8"
Linda Grace Clark
Psc 23h40'48" 7d16'23"
Linda Group
Uma 10h58'10" 33d42'54"
Linda Gullo Rodriguez
Mon 6h31'46" 8d26'38"
Linda Guss
Cas 1h19'30" 52d20'37"
Linda H. Richardson
Leo 11h7'20" 9d33'55"
Linda Hall
Aqr 21h31'20" -0d14'41"
Linda Hamm
Ori 6h0'29" 9d36'52"
Linda Hammer
And 1h55'29" 40d48'40"
Linda Hardin
Psc 23h15'18" 5d0'26"
Linda Haremza
Aqr 20h40'44" 1d34'1"
Linda Hargan
Vir 13h45'41" 3d24'0"
Linda Hayes
Uma 13h35'28" 60d44'15"
Linda Hazlett
Cas 1h37'21" 65d30'50"
Linda Heldt
And 0h14'41" 42d41'48"
Linda Helen
And 23h11'22" 48d41'5"
Linda Helen Gisborne
Cas 1h33'17" 64d31'4"
Linda Herman
Aqr 21h41'5" 0d49'39"
Linda Hickey
Cyg 20h51'39" 34d58'32"
Linda Hiroko Gibbons von Studnitz
Ari 3h18'34" 22d42'54"
Linda Hirsch Wilson
Lib 14h54'18" -11d10'8"
Linda Hoak
Crb 15h33'20" 38d19'25"
Linda Hook
And 0h14'19" 42d8'45"
Linda Hope
And 0h12'55" 31d10'22"
Linda Hough
Sgr 18h3'0" -27d55'59"
Linda Hughes
Uma 10h15'51" 55d57'50"
Linda Huynh Nguyen
Lib 15h24'24" -9d29'12"
Linda Ilana
Crb 15h41'4" 31d51'46"
Linda Irene Wood
Ori 5h37'42" -2d29'8"
Linda Irvine
Aqr 21h10'16" -8d44'55"
Linda Isabell Hunter-Olthoff
Peg 21h54'37" 7d41'35"
Linda Isabell Hunter-Olthoff
Peg 21h55'38" 7d38'7"
Linda Isabella May
Cas 23h45'7" 55d14'27"
Linda J. Bilton
Lyn 6h28'38" 57d56'41"
Linda J. DeSantis
Lib 15h46'24" -6d29'11"
Linda J. Jones
Lyn 7h40'40" 37d33'25"
Linda J. Layton
Gem 6h26'42" 17d33'59"
Linda J Weaver
And 0h53'43" 40d40'15"
Linda Jade
Sco 16h48'47" -34d35'0"
Linda Jane
Uma 11h23'22" 32d51'30"
Linda Jane
And 23h0'32" 41d1'12"
Linda Jane - 22/07/1967
Cru 12h41'17" -57d55'56"
Linda Jane Collins
Cam 7h12'42" 76d4'22"
Linda Jane Mackie
Cyg 20h28'15" 41d27'39"
Linda Jane Miller
Crb 15h19'19" 26d47'46"
Linda Jean
And 1h3'23" 34d54'33"
Linda Jean Drake
Mon 7h33'39" -0d36'42"
Linda Jean Forsyth
Sgr 20h6'49" -30d12'45"
Linda Jean Janson
Cas 0h17'28" 53d24'58"
Linda Jean Kelley
Cap 21h49'12" -14d51'28"
Linda Jean Kerpa
Sgr 18h45'39" -31d8'22"
Linda Jean McAllister
Ari 2h8'22" 23d12'13"
Linda Jean Miles
Cas 1h56'47" 62d0'31"
Linda Jean Wagner Moore
Eri 4h24'42" -0d14'23"
Linda Jean Walker
Cyg 21h4'7" 46d4'48"
Linda Jeane Chubb
Sco 16h10'35" -11d23'42"
Linda Jeanine Karwatka
Sco 16h15'19" -42d16'49"
Linda Jewell
Cyg 20h3'34" 31d3'49"
Linda Jill
Aqr 22h37'40" -0d34'47"
Linda Jo
Cas 2h11'25" 64d56'49"
Linda Jo
And 23h43'27" 38d51'15"
Linda Jo Gallaway
Cas 23h13'9" 55d40'50"
Linda Jo Laymon
Tau 5h33'2" 21d59'54"
Linda Joan
Cas 1h27'28" 63d6'34"
Linda Joan Bloom
Lyn 9h6'0" 33d8'7"
Linda Joann Russell
Aql 19h16'33" 4d33'41"
Linda Joanne Banthrall
Tau 5h9'31" 24d15'34"
Linda Joanne Powell
Sco 17h19'26" -43d19'24"
Linda & Joe BarKatt *39th*
Cyg 20h48'10" 46d12'45"
Linda&Joe Yeager "9-1-83"
Ori 6h22'19" 14d31'53"
Linda & John Bain
Uma 13h46'56" 49d1'19"
Linda & John's Star
Umi 11h29'5" 87d46'1"
Linda Joines
Psc 0h39'55" 11d59'59"
Linda "JoJo" Garber
Cas 0h33'0" 53d48'38"
Linda&Jonas
Gem 7h50'10" 31d33'24"
Linda Joy
Cet 1h26'33" -0d2'56"
Linda Joyce Eccleshall
Cnv 12h43'25" 39d55'19"
Linda Joyce Faxon
Lyr 18h47'17" 39d38'20"
Linda June
Sco 16h42'22" -40d57'58"
Linda K. Booth
And 0h55'32" 39d19'11"
Linda K Coley
Com 12h16'19" 24d27'51"
Linda K. Hill
Uma 10h34'13" 71d16'33"
Linda K. La Plante
Gem 6h51'48" 12d46'12"
Linda K. Phan
Cnc 8h23'14" 30d47'15"
Linda K. Richardson
And 0h28'29" 37d10'17"
Linda Kahn
Tau 4h34'55" 28d46'0"
Linda Kalaj
Umi 13h19'48" 74d33'16"
Linda Katherine Wyatt
Cas 1h56'22" 61d4'18"
Linda Kathleen Lechowicz
And 1h40'11" 47d23'37"
Linda Kay
Psc 1h15'53" 26d56'32"
Linda Kay
Mon 6h44'47" 7d28'40"
Linda Kay Dodd
Cas 2h50'36" 60d0'13"
Linda Kay Flowers
Cap 20h20'14" -14d31'13"
Linda Kay Graves
Lyn 7h48'20" 42d8'56"
Linda Kay Larson
Lyn 8h30'41" 38d7'48"
Linda Kay Porter
Psc 1h14'2" 14d39'32"
Linda Kay Schreiber
Vir 13h28'59" 6d5'54"
Linda Kay Sharp
Cas 0h46'58" 56d24'24"
Linda Kaye Lillehoff Acker Elliott
Ari 3h7'31" 25d55'44"
Linda Kempf
Cas 0h36'28" 64d20'4"
Linda Kenny
Gem 7h36'45" 27d1'32"
Linda Kindberg
Vir 13h10'23" -3d54'42"
Linda King
Cas 0h19'53" 57d48'22"
Linda Kinney
Cas 0h12'14" 53d23'37"
Linda Kitch
Ori 5h52'45" 6d17'39"
Linda Kovanitz
Uma 10h12'21" 52d49'36"
Linda Kucukistipanoglu
Sco 16h4'31" -11d26'49"
Linda L. Dulmovitz
Tau 5h10'38" 27d53'48"
Linda L. Greene
Uma 10h1'34" 66d38'26"
Linda L. Hanke "The Dancer"
Sco 16h2'25" -12d58'37"
Linda L. Hudson
And 0h46'33" 31d43'50"
Linda L. Nogrady
And 0h28'50" 37d41'19"

Linda L. Reed
Vir 13h17'24" 5d39'4"
Linda L. Rios
Cnc 8h10'12" 7d35'7"
Linda L. Robinson
Lyn 8h23'41" 56d40'1"
Linda L. Zagar
And 23h40'30" 46d51'23"
Linda Lam
Tau 5h49'20" 26d6'10"
Linda Langlais
Aql 20h3'0" 15d24'4"
Linda Langley
Ori 5h34'8" -0d45'6"
Linda Lanham
Cyg 20h13'21" 36d39'22"
Linda Laski Halibrand
Lyn 8h7'54" 34d22'23"
Linda LaVitola
Leo 11h12'47" 9d45'22"
Linda Lea Heimbuck
And 1h43'8" 42d49'1"
Linda LeBlanc
Sco 16h43'21" -44d34'43"
Linda Lee
Cas 0h44'0" 60d18'35"
Linda Lee
And 0h57'48" 39d54'4"
Linda Lee
Ari 2h24'31" 18d10'43"
Linda Lee Bina
Uma 13h55'37" 48d45'41"
Linda Lee Duffy
Leo 11h41'58" 25d43'1"
Linda Lee Korn
Leo 11h19'38" 1d59'35"
Linda Lee Lemak
Lyn 7h23'0" 57d16'48"
Linda Lee Lipowski 38
Cap 21h18'34" -15d53'24"
Linda Lee Lovenburg
Vir 13h16'35" 12d9'58"
Linda Lee Luedeman
Ori 6h1'22" 20d7'35"
Linda Lee Marchio
Aqr 22h21'46" -0d44'44"
Linda Lee Marchio
Aqr 22h13'16" -4d2'29"
Linda Lee McQueary
Vir 13h26'58" 8d23'45"
Linda Lee Osmon
Lib 15h49'44" -11d26'49"
Linda Lee Over
Ori 6h20'24" 11d3'19"
Linda Lee Payne
Cyg 19h41'16" 55d50'38"
Linda Leigh Gilliland
Lac 22h26'24" 47d5'52"
Linda Leonardelli
Cyg 20h0'52" 45d21'22"
Linda Liang
And 23h21'47" 50d52'44"
Linda (Ling dai) Eva Perry
Umi 13h12'4" 70d53'10"
Linda Lokos
Uma 11h47'26" 42d26'5"
Linda Lomonaco
Sgr 17h57'24" -17d59'14"
Linda Lopez
Sco 16h29'12" -37d48'48"
Linda Lou
Sgr 18h49'5" -24d50'56"
Linda Lou
And 23h14'19" 46d54'34"
Linda Lou
Vir 13h18'50" 11d8'5"
Linda Lou Evans
Sgr 17h53'46" -17d59'21"
Linda Lou Hightower
Cap 20h9'55" -9d15'48"
Linda Lou McGhee, Beautiful Linda
Lyn 7h52'22" 39d15'51"
Linda Lou Spencer
Dra 17h40'30" 51d28'6"
Linda Lou Stephens
Tau 3h39'20" 13d2'48"
Linda Louden
Uma 8h36'50" 69d23'28"
Linda Louise
Sco 16h15'41" -11d42'15"
Linda Louise
Lib 15h6'21" -12d17'22"
Linda Louise Butcher
Cap 20h9'53" -13d52'0"
Linda Louise Janota
Tau 3h32'28" 25d39'56"
Linda Louise Lockwood
Ari 1h55'18" 18d43'35"
Linda Lue Mauck
Lib 15h46'56" -20d16'2"
Linda Lukey
Crb 15h48'28" 26d39'30"
Linda Luysterborg
And 2h7'1" 40d34'38"
Linda Lynn Rogers #46
Psc 0h49'54" 19d20'0"
Linda Lyon
Cas 2h46'9" 57d34'2"
Linda M
Aqr 21h45'19" -7d21'36"
Linda M. Chapman, Rock Solid
Uma 12h56'16" 55d39'35"

Linda M. McKay
Ari 2h20'12" 27d15'39"
Linda M Parker 711946
Cyg 20h53'58" 31d15'21"
Linda M. Shaw
Aqr 22h22'43" -4d9'14"
Linda M. Skinner
Vul 19h51'51" 23d22'22"
Linda M. Wasyln
Aqr 22h52'7" -8d7'22"
Linda M. Whiteford
Lmi 10h36'10" 32d58'9"
Linda Mac 3152
Aur 5h49'46" 53d48'31"
Linda Mae
Gem 6h48'56" 33d18'1"
Linda Mae
Lib 15h47'8" -25d19'46"
Linda Maines
Ori 6h7'8" 6d32'34"
Linda Mak
Uma 11h33'17" 40d2'42"
Linda Manning
And 0h43'4" 31d35'0"
Linda Margaret Johnson
Crb 15h50'53" 27d56'18"
Linda Maria Cross
Cru 12h27'41" -58d31'16"
Linda Marian Perry
Aqr 22h12'34" -0d45'25"
Linda Marie
Lyn 6h21'58" 57d57'59"
Linda Marie
Cam 4h14'55" 68d58'7"
Linda Marie
Cas 1h54'54" 63d57'30"
Linda Marie
Col 5h48'59" -34d55'36"
Linda Marie
Leo 10h13'1" 22d56'31"
Linda Marie
And 0h19'25" 28d0'51"
Linda Marie
Cas 0h13'58" 52d58'36"
Linda Marie
Lyn 6h59'5" 45d19'5"
Linda Marie Angelillo "My Angel"
Gem 7h45'10" 25d33'45"
Linda Marie Bendl
Lyn 7h6'49" 58d44'5"
Linda Marie Centofante
Sgr 19h1'25" -24d49'43"
Linda Marie Collins
Aqr 22h33'58" -3d34'7"
Linda Marie Deakins
Ori 5h2'5" 15d6'59"
Linda Marie DelMasto
And 23h14'3" 43d55'33"
Linda Marie Diers-Johnson
Psc 23h20'10" 3d26'18"
Linda Marie Griffith
Aql 19h25'20" 7d52'26"
Linda Marie Henderson
Cas 0h53'41" 55d6'6"
Linda Marie Knosalla
Dra 19h48'2" 71d55'33"
Linda Marie Knox Cardona
Cyg 20h41'58" 51d14'58"
Linda Marie Mitchell
Psc 1h34'45" 19d29'11"
Linda Marie Nehrich
Umi 13h54'18" 78d6'35"
Linda Marie O'Neil
And 1h9'15" 34d29'45"
Linda Marie Onesy 061105
Psc 0h20'13" -4d6'18"
Linda Marie Posch
Cnc 8h15'8" 7d19'40"
Linda Marie Traxinger
Cas 0h2'33" 53d55'21"
Linda Marie Van Riper
Cnc 8h16'14" 23d4'16"
Linda Marie Wolfbrandt
Tau 4h24'28" 14d32'53"
Linda Marino
Lib 15h34'3" -10d27'15"
Linda Marsh
Pho 1h13'35" -47d59'17"
Linda Martin
Vir 13h52'32" -16d8'58"
Linda Martin
Uma 11h4'36" 52d38'57"
Linda Martin, la Mia Anima, Ti Amo.
And 23h50'45" 34d42'13"
Linda Mary Lou Vangelof
Cas 1h36'42" 62d7'35"
Linda May
Cap 21h10'20" -15d42'32"
Linda May
Her 17h49'31" 39d1'26"
Linda May
Her 16h33'22" 19d26'10"
Linda May
Gem 7h31'12" 16d55'14"
Linda May Leaser
And 23h38'52" 47d43'21"
Linda Maynes, My Angle
Cas 2h23'24" 67d9'23"
Linda Mays
Cyg 19h59'33" 35d13'2"
Linda Mazade
And 0h16'32" 43d56'45"

Linda McCleave
Aur 6h37'47" 30d7'54"
Linda Medina #1 Wife, Mom, Grandma
Peg 0h5'16" 17d44'36"
Linda Mendelsohn
And 0h30'48" 34d23'26"
Linda Mercadante
Tau 3h55'36" 26d32'17"
Linda Mignosa
Cam 4h5'16" 54d36'12"
Linda & Mike's Little Bit of Heaven
Cru 12h22'25" -57d35'22"
Linda (Mom Mom)
And 1h28'4" 46d27'25"
Linda "Mom" Putnam
Uma 9h6'20" 58d43'47"
Linda Mooney
Gem 7h23'55" 24d43'53"
Linda Moran Evans
Peg 22h45'54" 21d35'53"
Linda Moran Evans
Uma 10h31'40" 57d11'54"
Linda Morettie
Crb 15h36'39" 31d32'28"
Linda Morgan
And 0h12'14" 46d27'44"
Linda Moses Sam
Uma 14h17'28" 56d46'20"
Linda Mouacheupao
Cas 1h21'57" 66d2'41"
Linda Murphy
Uma 9h43'33" 71d54'39"
Linda Murtaugh
Lmi 9h23'8" 33d10'35"
Linda my love
Lmi 10h23'22" 32d5'32"
Linda 'My Love' Carlsen
Lib 14h46'6" -15d53'51"
Linda "My Love Star"
Lyr 18h51'4" 46d1'40"
Linda Myrick
Mon 6h31'46" 6d37'19"
Linda Nelle Edmond
Sgr 18h21'17" -24d11'46"
Linda Neukum-Kozak
Lyn 6h48'53" 50d15'56"
Linda Nga Loflin
Tau 4h23'48" 22d48'9"
Linda Nguyen
Sgr 19h12'17" -19d44'29"
Linda O'Keefe Lamkin
Vir 15h8'49" 3d55'36"
Linda Orrender
And 1h42'31" 42d57'32"
Linda Ortiz
And 0h46'12" 38d49'0"
Linda Ota
Ari 2h1'48" 10d34'31"
Linda our shining light
Ari 2h47'48" 11d6'40"
Linda P 1118
Lac 22h23'38" 46d25'15"
Linda P. BeMiller
Uma 11h24'17" 46d10'11"
Linda Palczewski Goltz
Cas 0h48'55" 57d1'32"
Linda Paloma
Per 1h42'21" 50d45'3"
Linda Pane
Uma 9h4'9" 50d28'0"
Linda Papianni
Leo 10h52'46" 0d42'0"
Linda Paquet
Lyr 18h45'2" 43d15'12"
Linda Partak
Sgr 18h14'27" -18d8'32"
Linda Pastore
Ori 5h54'55" 18d17'56"
Linda Patricia Ann Harrison
And 1h56'24" 43d26'23"
Linda Payette
Leo 11h32'9" 5d6'8"
Linda Pejka
Psa 22h37'34" -32d0'56"
Linda Perkin
Peg 23h43'36" 12d7'0"
Linda Pettis
Leo 9h58'37" 18d13'40"
Linda Phipps
Peg 23h42'37" 16d59'28"
Linda Pickard
Lyn 9h14'39" 39d0'41"
Linda Pillot
Vir 13h34'21" 3d14'13"
Linda (Pink Princess) Spiecker
Lib 14h46'14" -12d39'2"
Linda Pollot
Vir 13h28'39" -19d31'43"
Linda Psomas
And 0h17'51" 44d36'30"
Linda R. Bauer
Cyg 21h39'32" 35d24'15"
Linda Rae
And 0h43'36" 39d22'5"
Linda Rae Chapman
And 0h19'17" 33d31'3"
Linda Ragsdale
Crb 15h24'22" 31d9'0"
Linda Raper
And 1h16'42" 37d46'25"
Linda Ray Miller-Phillips
Vir 11h39'49" 4d25'9"

Linda Rees 17/12/1974
Cra 17h59'38" -37d20'38"
Linda Renee
And 0h44'35" 40d52'49"
Linda Reppucci
And 23h16'22" 49d35'33"
Linda Reusch
Sco 17h50'25" -34d37'22"
Linda & Richard Battista
Cyg 19h16'54" 51d23'21"
Linda Rita Stysiack
Uma 8h10'6" 64d10'29"
Linda Roberts
Vir 14h5'12" 4d36'7"
Linda Robinson's Personal Light
Aqr 22h44'46" -18d37'55"
Linda Rose
Cas 23h5'30" 59d8'16"
Linda Rose Castanada
Cas 23h34'25" 52d22'9"
Linda Rose Mendolia Pepe
Aqr 23h18'30" -18d26'44"
Linda Roush Dean
Uma 9h46'55" 56d54'20"
Linda Ruth Cornell
Leo 11h26'46" 22d6'0"
Linda S. Bizier
Per 3h20'8" 33d0'23"
Linda S. Bostian
Cap 20h59'29" -21d16'57"
Linda S. Hewitt
Lyn 8h49'36" 34d57'0"
Linda S. J. Goytia
Aqr 23h15'34" -7d31'28"
Linda S. Kipp
Psc 1h1'57" 14d0'2"
Linda S. Nelson's Shining Star
Vir 12h54'50" 1d58'28"
Linda S. Wanaselja
Ori 5h26'37" -4d48'20"
Linda S. Wolf
Leo 9h52'18" 17d36'57"
Linda Saenz
And 1h35'34" 47d3'42"
Linda Schneider Wilke
Cas 1h31'24" 63d21'22"
Linda Schultz
Uma 9h8'18" 58d49'59"
Linda Schultz
And 0h30'3" 38d41'51"
Linda Schulz
And 0h28'15" 37d9'29"
Linda Scott Keenan
Leo 10h54'51" 6d39'24"
Linda Scully
Aqr 22h15'57" 1d25'6"
Linda Shans-Tony, Kris & Erics Star
Cyg 19h51'37" 31d1'9"
Linda Sharp
Cnc 8h35'20" 18d7'10"
Linda Shehee
Lib 15h40'36" -19d52'5"
Linda Sheppard
Vir 12h19'1" -0d57'12"
Linda Shirey
Per 3h53'28" 49d47'35"
Linda Shultz
Cyg 20h18'54" 39d7'36"
Linda Simons
Cas 1h16'34" 55d33'48"
Linda Smith
And 1h1'30" 46d31'39"
Linda Smith
Lyr 18h43'29" 34d3'5"
Linda Smith
Cas 2h22'27" 66d18'4"
Linda S.N. Chang
Cap 21h8'43" -21d37'29"
Linda Snow
Umi 14h26'25" 75d22'25"
Linda Spradley Dunn
Aqr 23h48'50" -23d57'44"
Linda St. Coeur
Cyg 21h30'31" 42d59'16"
Linda. Star and Sparkle Of My Life
Cyg 19h46'48" 37d27'37"
Linda Stavrakis
Lep 5h33'55" -11d11'31"
Linda Stewart
Cas 1h14'9" 55d5'8"
Linda Stone
Aqr 21h13'56" -9d4'43"
Linda Strange
Psc 0h47'57" 20d19'51"
Linda Stuart
Psc 1h33'5" 26d31'19"
Linda Szelecki
Lib 15h49'44" -3d49'46"
Linda Sue Comer Lindsey
And 0h19'0" 30d46'58"
Linda Sue Davis
Vir 13h30'10" -3d50'49"
Linda Sue Dreier
Uma 9h38'27" 72d8'56"
Linda Sue Erickson
Cas 0h29'12" 57d6'22"
Linda Sue Ferrel
Cap 20h12'37" -26d9'4"
Linda Sue Lebbs
Uma 9h52'43" 48d4'58"

Linda Sue McDaniel
Sco 16h12'23" -8d50'22"
Linda Sue Morris
Sgr 18h37'6" -21d20'52"
Linda Sue & Ronald John Rzepecki
Cyg 21h41'38" 52d19'10"
Linda Sue Roney
Vir 13h16'54" 8d32'39"
Linda Sue Spence
Leo 10h35'23" 7d47'10"
Linda Sue Sturm
Ari 3h18'16" 21d29'53"
Linda Sue Widmann
Cas 1h44'0" 61d33'38"
Linda Sunness Lessin1
Sgr 18h14'30" -19d23'40"
Linda Susan
Cas 0h12'11" 53d47'52"
Linda Susan Buglass "Little Nana"
Cas 23h2'35" 57d36'45"
Linda Susan Gibson
Uma 9h48'41" 43d26'1"
Linda Susan Gilbert
Lyn 7h10'20" 52d49'27"
Linda Susan Hamm
Lyr 18h46'8" 36d36'38"
Linda Susan Kluber
Cas 1h1'0" 48d43'40"
Linda Susan Whitson
Peg 22h21'15" 4d36'0"
Linda Suzanne Batewell(Sweet Baboo)
Cnc 8h45'3" 16d16'34"
Linda Taylor *Light Of My Life*
Sco 16h17'19" -21d18'23"
Linda Teel
Gem 7h27'43" 16d16'58"
Linda Teresa Dilks
Sgr 18h34'0" -34d46'34"
Linda Thompson
Lyn 7h37'59" 38d15'22"
Linda Thoms
Sgr 17h56'27" -17d21'43"
Linda Tichy
Leo 9h28'34" 11d45'18"
Linda Tiefenthal
And 1h38'45" 37d39'47"
Linda Titus
Cap 21h43'33" -10d10'36"
Linda Tobler
Uma 8h37'8" 60d22'30"
Linda Tolodxi
Psc 1h18'23" 17d22'34"
Linda & Tom S F M H 898 Forever
Sco 17h37'14" -40d30'1"
Linda Trevena
And 23h58'48" 46d31'22"
Linda "Turtle" Ferguson
Leo 10h26'10" 24d58'22"
Linda Ulatoski
Lyn 8h15'15" 34d57'25"
Linda Underhill
Cas 0h24'28" 56d13'16"
Linda Van Noy Stamp
Lib 15h19'29" -8d12'58"
Linda Vaughn
And 0h50'23" 38d45'59"
Linda Vercillo
Psc 23h24'2" 5d22'24"
Linda Voirin
Cnc 9h17'35" 30d26'4"
Linda Volz
Cas 0h23'58" 53d13'24"
Linda Vottero
Cap 21h12'15" -19d22'7"
Linda W. Roehrig
Vir 12h49'51" 8d34'12"
Linda Walleen Will
Vir 13h3'37" 12d57'27"
Linda Walter
Cnc 8h37'12" 32d54'48"
Linda Wasserstein
Vir 13h27'51" 13d27'6"
Linda Waterman
Uma 9h33'9" 51d9'20"
Linda Weber Daywalt
Cyg 19h59'22" 45d11'26"
Linda Wegenaar
Uma 9h12'21" 61d37'44"
Linda Weimer comet
And 23h12'37" 41d32'29"
Linda Wheelock Gruno
And 23h37'56" 36d43'32"
Linda Wildey Place
Aql 19h18'39" -10d46'40"
Linda Wilson
And 1h11'49" 42d22'57"
Linda Wilson
And 0h9'36" 45d32'4"
Linda Woloshansky
Cap 21h15'20" -24d50'16"
Linda Wood
Gem 7h6'2" 15d4'28"
Linda & Woody
Lyr 19h5'42" 41d50'29"
Linda Wright
Umi 18h3'15" 88d17'15"
Linda Wyss
And 0h33'58" 34d36'24"
Linda Yanez
And 23h55'25" 41d15'10"

Linda Yost - Sweetheart
Sgr 18h46'28" -30d48'37"
Linda Yosua
And 0h16'53" 26d58'6"
Linda, James, and Andrew
Sco 16h51'24" -25d10'8"
Linda, Jonathon, Nathan
Ori 5h14'17" -7d31'50"
Linda, My Princess
Ara 17h32'13" -47d59'47"
Linda88
Sco 16h8'11" -10d30'36"
LindaDanielSpitz
Cas 2h27'49" 72d2'4"
Linda-Dawn
Cru 12h45'10" -60d20'36"
LindaHallem2aGr8Mother
Sco 17h21'30" -41d53'53"
LinDale JoNathaniel RobinsonThyne
Gem 6h41'21" 23d28'6"
Linda-Marie Singer
Uma 9h47'18" 47d32'47"
Lindamood-Bell
And 0h26'30" 32d49'21"
lindao
Tra 16h30'47" -60d55'43"
Linda's Angel
Leo 10h53'2" -3d10'59"
Linda's Birthday Star
Ari 2h5'18" 23d19'44"
Linda's Chooch
Cas 23h32'33" 55d14'14"
Linda's Colin
Del 20h44'28" 19d13'38"
Linda's Dawn
Lyn 7h25'30" 54d7'58"
Linda's Diamond
Aql 19h14'41" 14d23'30"
Linda's Dimples
And 0h19'57" 31d28'57"
Linda's Dream
Vol 8h49'28" -72d55'25"
Linda's Dream Star LLL
And 2h28'24" 41d56'14"
Linda's Eternal Love 6/2/1946
Gem 6h45'2" 17d47'50"
Linda's Gaurdian Angel
Lyn 8h34'29" 56d51'41"
Linda's Light
Uma 10h47'3" 62d38'40"
Linda's Light
And 0h23'45" 32d12'7"
Linda's Little Piece of Heaven
Cap 20h27'50" -14d6'48"
Linda's Retreat
Uma 9h41'23" 67d18'8"
Linda's Shining Light
Cas 23h42'14" 58d30'31"
Linda's Star
Ari 3h24'47" 29d58'18"
Linda's star
Tau 4h17'23" 27d6'57"
Linda's Star
Cnc 8h18'37" 15d39'4"
Linda's Star
Cnc 8h40'13" 21d28'20"
Linda's star
Uma 11h42'30" 38d18'54"
Linda's Star
Cyg 20h41'51" 46d51'37"
Linda's Star for Angel Mira
Vir 14h4'50" -7d8'0"
Linda-Schäfchen
Uma 11h40'15" 31d44'34"
Lindel Hawkins Marsh
Vir 12h36'42" 7d33'49"
Lindell
Dra 15h24'42" 60d24'8"
Lindell Brinser Smith
Uma 11h49'50" 62d53'32"
Lindelou 5555
Uma 9h54'24" 43d12'55"
Lindemann, Waltraud
Uma 8h56'42" 50d27'30"
Linden D.
Umi 14h35'4" 73d44'28"
Linden I Adams
Uma 9h34'28" 45d33'6"
Linden Vogelhut
Tau 4h37'53" 18d27'56"
Linden, Brigitte
Uma 11h13'33" 64d14'31"
Lindenhurst Family Practice
Uma 11h33'31" 58d24'38"
Linder
Uma 13h4'18" 54d31'13"
Lindi
Cru 12h25'28" -62d3'41"
Lindina y Lluiset
And 1h35'45" 42d5'39"
Lindi's Angel
Sgr 17h55'31" -29d35'18"
Lindley Dauphine
Cap 20h16'28" -11d5'0"
Lindo
Uma 10h31'22" 58d51'59"
Lindsay
Cas 1h45'11" 67d1'40"
Lindsay
Cap 20h38'52" -10d13'52"
Lindsay
Sgr 18h56'3" -18d48'4"

Lindsay
Lib 15h11'6" -4d42'46"
Lindsay
Sco 17h15'59" -32d4'27"
Lindsay
Sgr 19h14'17" -35d18'4"
Lindsay
And 1h36'28" 41d41'29"
Lindsay
And 0h54'55" 40d46'22"
Lindsay
Boo 15h23'22" 32d44'39"
Lindsay
And 23h54'58" 45d26'43"
Lindsay
Lyn 6h51'46" 50d36'2"
Lindsay
Ari 3h20'7" 16d5'10"
Lindsay
Leo 10h43'28" 14d51'31"
Lindsay
Ori 5h8'11" 1d14'35"
Lindsay
And 0h19'37" 28d31'58"
Lindsay
Gem 7h50'51" 29d31'59"
Lindsay 03-13-88
Psc 23h56'0" -0d1'41"
Lindsay 21
Leo 10h48'10" 7d31'12"
Lindsay * 817
Leo 11h39'56" 20d55'26"
Lindsay Adams
Cyg 21h53'56" 41d29'31"
Lindsay Alexus Wrather
Crb 15h38'53" 35d46'17"
Lindsay Alice Pogemiller
Gem 7h15'49" 31d23'29"
Lindsay and Alan Griffiths
Uma 9h21'9" 47d59'8"
Lindsay and Colin
Uma 9h24'6" 60d48'4"
Lindsay and Shoop's Star
Ari 2h45'47" 27d46'22"
Lindsay Ann
Ori 5h27'24" 3d21'36"
Lindsay Ann
Uma 13h48'25" 50d14'50"
Lindsay Ann
Sgr 18h50'1" -20d11'42"
Lindsay Ann Allen
Uma 8h36'32" 56d32'11"
Lindsay Ann Bucholtz
Vir 13h43'39" 12d17'39"
Lindsay Ann Elizabeth Libby
Aql 19h35'2" -10d6'28"
Lindsay Ann Heller
Cas 0h14'44" 60d42'12"
Lindsay Ann Massey Fackrell
Lyn 7h45'27" 43d29'37"
Lindsay Ann Shorenn Coolie Agro
Psc 0h15'5" 1d10'13"
Lindsay Ann Super
Lyn 8h54'58" 39d45'35"
Lindsay Anne Byrne
Gem 7h16'20" 32d46'9"
Lindsay Anne Galloway
And 0h39'33" 36d52'38"
Lindsay Anne Kennedy
Aqr 23h2'18" -20d28'4"
Lindsay Anne Shaul
Sco 16h50'30" -38d34'24"
Lindsay Anne Smith
Gem 6h54'27" 24d21'40"
Lindsay Austin
Cas 0h55'45" 47d28'3"
Lindsay B. Lawson
Sco 16h15'22" -9d40'0"
Lindsay Bacon
Lep 5h51'9" -12d10'36"
Lindsay Bird
Sgr 18h6'11" -27d3'50"
Lindsay Bitonti
Cyg 20h8'56" 41d50'30"
Lindsay Blair
Sge 20h9'14" 18d53'22"
Lindsay Brooke Boen
Cap 21h13'50" -15d52'2"
Lindsay Brooke Grosfeld
Leo 10h23'46" 23d41'52"
Lindsay Brooke Simcoe
Sgr 18h23'32" -22d44'16"
Lindsay Brooke Ybarrondo
Cam 5h39'57" 65d43'50"
Lindsay Burke
And 1h25'30" 49d26'58"
Lindsay C McGhan
Cmi 7h32'17" 5d52'31"
Lindsay Calvert's Star ;) :) :-0
Leo 10h45'32" 14d9'0"
Lindsay Carter
And 23h41'51" 47d39'22"
Lindsay "Chowder Boo" Jones
Sgr 19h46'3" -15d52'2"
Lindsay & Chris Ryalls
Cyg 21h36'19" 50d6'0"
Lindsay Clatterbuck
Psc 0h26'11" -1d48'47"
Lindsay Clinton
And 23h43'9" 48d34'42"

Lindsay Coleman Lyn 7h36'28" 52d36'24"
Lindsay Cooper Cnc 8h0'48" 11d53'46"
Lindsay Cumm Cas 23h52'58" 53d33'1"
Lindsay D Schultz Gem 6h31'51" 14d59'59"
Lindsay David Anna Edwards Uma 11h7'33" 43d38'19"
Lindsay Dayle Forbis Sgr 18h25'23" -33d5'15"
Lindsay DeBonis And 23h14'41" 48d4'46"
Lindsay Denise Siminiski Crb 16h13'20" 39d26'8"
Lindsay Densmore Umi 15h23'23" 73d40'0"
Lindsay Diann Ruoff And 0h35'16" 46d23'18"
Lindsay Dirlam Lyn 7h58'27" 44d20'22"
Lindsay Doughty And 23h18'49" 48d46'4"
Lindsay Dupie Lib 14h59'29" -19d2'6"
Lindsay E. Magil Lib 14h47'4" -11d41'5"
Lindsay Elaine And 0h18'40" 37d29'36"
Lindsay Elaine Blick Vir 13h22'11" 0d55'38"
Lindsay Elisabeth Merker Uma 13h59'47" 60d17'54"
Lindsay Elise Novak And 0h25'36" 40d26'8"
Lindsay Elizabeth Aqr 22h37'28" -0d11'1"
Lindsay Elizabeth Sco 16h58'40" -32d4'39"
Lindsay Elizabeth Nielander Tau 3h42'3" 21d31'12"
Lindsay Elizabeth Thornton Leo 11h12'47" -2d22'44"
Lindsay Ellen Oph 16h50'30" 0d56'0"
Lindsay & Emma Mon 6h30'21" 6d17'50"
Lindsay Erin Martin And 2h29'39" 46d31'40"
Lindsay Ezsol and David Pfefferles Uma 9h35'5" 42d5'29"
Lindsay Fernandes Psc 1h22'58" 28d22'9"
Lindsay Fisher Aqr 22h13'14" -13d15'57"
Lindsay Fisherback Aql 20h5'2" 9d9'39"
Lindsay Floyd Lyn 7h36'27" 45d2'51"
Lindsay Franz And 0h15'20" 40d13'26"
Lindsay Frazier Uma 11h57'47" 60d41'41"
Lindsay G. Ari 2h45'13" 25d13'49"
Lindsay Gagliano Ari 2h41'13" 31d6'2"
Lindsay Gail Ari 3h24'11" 26d15'0"
Lindsay Gale Tau 5h34'46" 18d27'23"
Lindsay Gehrls Tau 5h44'3" 23d26'2"
Lindsay Giolzetti Cnc 8h27'52" 18d13'24"
Lindsay Goodrow Leo 9h37'28" 27d57'51"
Lindsay H Aur 6h27'21" 32d47'12"
Lindsay Hart Crb 15h35'51" 26d14'22"
Lindsay Hartung Cnc 8h52'51" 31d10'17"
Lindsay Heavens One and Only Angel Lmi 10h6'18" 39d20'55"
Lindsay Hillyer Sgr 18h31'38" -18d37'22"
Lindsay ingen Columb Peg 21h59'47" 29d17'48"
Lindsay Irene Clutter Ari 3h9'2" 24d20'19"
Lindsay Irrthum Ari 3h16'13" 28d14'50"
Lindsay Ives Vir 11h52'39" 8d40'30"
Lindsay J. Martin Peg 0h0'5" 21d49'36"
Lindsay Jane Hainzinger Leo 11h18'32" 3d0'51"
Lindsay & Jason Cyg 20h6'32" 34d38'37"
Lindsay Jean Uma 12h10'31" 51d11'46"
Lindsay Joan Taylor And 1h40'52" 45d17'3"
Lindsay Jones Sco 17h34'1" -44d28'3"
Lindsay Julia King Tau 5h19'25" 18d34'2"
Lindsay K. Wilbert Aqr 21h11'59" 0d44'30"
Lindsay Kate McAlpine And 2h9'46" 46d13'8"
Lindsay Katherine Malkiewicz Lib 14h48'13" -8d8'47"
Lindsay Kesten Sgr 18h55'46" -23d17'25"
Lindsay Kye And 2h11'39" 38d51'9"
Lindsay & Kyle's Everlasting Love Umi 15h22'1" 74d33'18"
Lindsay Laird And 1h49'54" 38d59'47"
Lindsay Land Sgr 18h19'19" -32d56'3"
Lindsay Lea Ori 6h15'18" 15d4'19"
Lindsay Lewis And 23h29'40" 41d24'48"
Lindsay Littler Lyn 9h2'15" 44d30'1"
Lindsay Llaca Lib 15h7'44" -19d3'24"
Lindsay Louise Green - Lindsay's Star Aur 4h49'13" 35d44'11"
Lindsay Love Crb 16h10'41" 36d58'16"
Lindsay Love of My Life Foreman And 23h6'57" 47d24'24"
Lindsay (Lu) Rosso 9-19-1989 Vir 13h8'53" 0d25'10"
Lindsay Luijs -My Starlight- Ari 2h21'55" 14d51'31"
Lindsay LuLu Neese Cas 0h35'36" 54d32'2"
Lindsay Lynne Megara Tri 2h27'3" 36d43'19"
Lindsay M. Ackerman Vir 11h54'51" 8d0'10"
Lindsay M. Bilski Lib 14h53'23" -18d26'15"
Lindsay M. Dettwiler Vir 13h22'47" 6d44'40"
Lindsay "mama's sunshine" Rae Lib 15h6'24" -5d33'18"
Lindsay Marie Cas 1h17'46" 58d35'1"
Lindsay Marie And 0h7'25" 46d59'20"
Lindsay Marie Black Sco 16h37'32" -36d5'37"
Lindsay Marie Brant Lib 15h6'2" -8d32'6"
Lindsay Marie David Leo 9h46'16" 26d2'0"
Lindsay Marie Flick Cnc 7h59'8" 18d15'46"
Lindsay Marie Hawkins Leo 11h35'28" 12d45'0"
Lindsay Marie LaPointe Leo 10h50'5" 18d29'31"
Lindsay Marie Moffitt Lib 15h5'18" -0d39'13"
Lindsay Marie O'Dell Cam 7h52'46" 73d32'59"
Lindsay Marie Schlemmer Lib 15h48'49" -4d8'56"
Lindsay Marie Steffen Lep 5h17'3" -26d12'4"
Lindsay Marie Tippett Sco 16h16'11" -14d22'20"
Lindsay Marisa And 1h38'38" 40d10'53"
Lindsay Massaro Uma 9h0'47" 50d7'9"
Lindsay McLean And 23h38'40" 37d32'34"
Lindsay McNeil Peg 22h13'50" 30d13'50"
Lindsay Megan Arduino Sco 16h7'14" -11d21'22"
Lindsay Meyer Cnc 9h4'18" 30d26'4"
Lindsay Michelle Phillips Uma 8h39'20" 62d21'19"
Lindsay Miller-Domer Cam 7h56'20" 60d31'19"
Lindsay Mowbray And 0h4'19" 44d10'52"
Lindsay N. Goff Leo 11h10'25" -1d50'22"
Lindsay N. Payne Lib 14h33'52" -23d0'29"
Lindsay Nicole Umi 15h31'14" 70d26'52"
Lindsay Nicole Uma 11h21'51" 53d39'36"
Lindsay Nicole Myers Cas 1h26'3" 68d11'35"
Lindsay Nicole Schmersahl Cas 0h1'17" 57d2'56"
Lindsay Noel And 0h38'50" 36d26'22"
Lindsay Norment Lmi 10h43'2" 27d23'0"
Lindsay Owen Cox Psc 23h13'19" -2d20'18"
Lindsay Paulina Dodd Leo 9h55'48" 14d38'15"
Lindsay Payne And 23h18'48" 46d46'21"
Lindsay - Precious Grandaughter Lyn 7h10'8" 46d50'42"
Lindsay Rae Gem 7h21'9" 16d56'46"
Lindsay Raquel Heron And 0h16'3" 43d36'14"
Lindsay Ray Ori 5h26'24" -10d52'34"
Lindsay Reday Shining Sco 16h6'44" -19d48'11"
Lindsay Rene' Grayson Mon 7h22'6" -0d52'14"
Lindsay Renee Cumm Vir 14h55'48" 6d36'21"
Lindsay Resis Sgr 19h13'6" -22d13'59"
Lindsay Rieman Ori 5h56'14" 11d54'54"
Lindsay Rivas Sgr 18h9'47" -34d55'25"
Lindsay Rocks Cnc 9h19'22" 13d4'10"
Lindsay & Roger Watson Cyg 19h41'50" 27d56'48"
Lindsay Rose And 1h1'47" 36d10'17"
Lindsay Rose Lyn 6h42'59" 56d52'43"
Lindsay Rose Hall And 23h34'3" 44d38'42"
Lindsay Ryan Lyr 18h49'46" 37d49'12"
Lindsay Ryan Machin Uma 10h18'33" 64d0'28"
Lindsay 's Heart And 23h11'18" 46d54'1"
Lindsay Sarah Willemse Lyn 8h3'3" 43d34'4"
Lindsay Schmidt Cnc 8h37'39" 31d56'55"
Lindsay Shankman And 2h11'22" 45d8'27"
Lindsay Shea Walker Psc 0h56'37" 31d20'8"
Lindsay Slade Pillsworth Mon 6h53'50" -0d46'5"
Lindsay Smithberger Cyg 19h54'49" 32d34'12"
Lindsay Sue Noel Gerhardt And 1h7'3" 43d47'59"
Lindsay Sullivan Gem 7h27'14" 31d25'59"
Lindsay Taylor Weickert Cap 20h33'31" -16d33'43"
Lindsay Tolley Sco 16h18'10" -11d59'19"
Lindsay Vega Tau 3h56'10" 25d56'27"
Lindsay Ventura Psc 1h4'14" 3d1'37"
Lindsay Vlachkis Sco 16h30'21" -37d28'54"
Lindsay Wallace Cas 23h28'8" 53d0'2"
Lindsay Warren Byers Cam 3h47'2" 55d59'28"
Lindsay Wheeler And 0h5'6" 41d12'7"
Lindsay Williams And 23h38'47" 41d20'27"
Lindsay Wilson - "Will" Aqr 20h59'33" 1d45'19"
Lindsay (Zaylin) Nicole Buck Gem 6h54'45" 14d31'33"
Lindsay2100 Gem 6h6'59" 26d54'43"
LindsayHeartJones Gem 6h47'28" 16d22'49"
Lindsayloo And 1h59'2" 43d0'14"
LindsayRyan05 Ori 6h11'19" 7d28'33"
Lindsay's Eye Ari 2h22'41" 26d2'57"
Lindsay's Guiding Light Peg 22h9'40" 10d29'26"
Lindsay's Own Dawn Sgr 19h11'12" -24d22'26"
Lindsay's Shining Star Ori 6h12'17" 15d16'42"
Lindsay's Special Wishing Star Aqr 22h4'37" -14d29'39"
Lindsay's Star Lib 15h48'31" -19d59'49"
Lindsay's Star Aqr 23h6'37" -4d51'0"
Lindsay's Star Lib 15h21'26" -6d25'22"
Lindsay's Star Sco 17h22'50" -36d30'18"
Lindsay's Star Ori 5h6'9" 15d41'39"
Lindsay's Star Gem 6h39'42" 24d11'47"
Lindsay's Star - 27th September 1989 Cru 12h12'11" -62d4'44"
Lindsay's Wish Sco 17h17'38" -41d28'56"
Lindsee Michelle McAlister Peg 21h52'39" 17d26'14"
Lindsey Leo 10h14'7" 22d13'5"
Lindsey Com 12h0'16" 21d34'57"
Lindsey Gem 6h9'49" 23d57'44"
Lindsey Tau 4h26'4" 23d42'51"
Lindsey Uma 11h19'5" 29d10'54"
Lindsey Psc 0h52'14" 15d51'34"
Lindsey Cnc 8h53'56" 13d52'54"
Lindsey Vir 13h16'53" 7d31'0"
Lindsey And 1h44'7" 44d53'22"
Lindsey Per 2h46'7" 44d59'10"
Lindsey Crb 16h7'39" 33d47'59"
Lindsey And 23h51'14" 40d28'25"
Lindsey And 1h21'40" 46d22'48"
Lindsey Per 3h9'13" 45d50'21"
Lindsey Per 4h12'27" 51d5'25"
Lindsey Vir 14h34'33" -6d16'5"
Lindsey Mon 6h48'9" -0d22'26"
Lindsey Leo 11h15'46" -1d20'29"
Lindsey Lyn 6h25'56" 54d34'54"
Lindsey A. Greer Cap 20h35'36" -23d48'4"
Lindsey A.K.A. Craigs Angel Baby And 0h24'18" 40d54'18"
Lindsey America Ryan Lyn 6h59'37" 52d19'58"
Lindsey and Adam - Forever Uma 11h31'9" 57d19'57"
Lindsey and Jay - Anniversary Star Cyg 21h31'42" 50d23'57"
Lindsey and John's 1st Christmas Uma 10h31'0" 47d31'12"
Lindsey and Michael Beard Cyg 21h14'25" 44d56'5"
Lindsey and Shane Uma 8h26'33" 63d2'58"
Lindsey and Shane Miller Gem 6h45'34" 26d34'50"
Lindsey Ann Lmi 10h35'54" 30d41'3"
Lindsey Ann Loyd Lib 14h52'7" -3d8'36"
Lindsey Ann Medine Sgr 20h11'13" -38d40'35"
Lindsey Ann Monti Tau 3h51'1" 16d25'18"
Lindsey Anne Hanebury Uma 9h27'19" 45d19'52"
Lindsey Anne Hanebury Lib 15h50'4" -11d13'5"
Lindsey Anne Miles Umi 16h59'17" 81d51'4"
Lindsey Aronson Aqr 20h57'47" 0d43'34"
Lindsey Aurora Per 1h44'55" 54d29'56"
Lindsey "Babe" Pool Vir 12h35'27" -0d1'25"
Lindsey Baby And 23h43'43" 35d48'1"
Lindsey "Babygirl" Parsons Lmi 10h11'53" 34d25'51"
Lindsey Bauer Lyn 8h18'51" 38d25'9"
Lindsey Bergo Lib 15h46'20" -7d2'4"
Lindsey Blair Lyn 7h42'15" 37d1'18"
Lindsey Blake Boyce Sgr 20h7'37" -31d31'6"
Lindsey Blume-Lowery And 2h26'23" 47d4'33"
Lindsey Bonistall Cas 0h39'47" 62d23'36"
Lindsey Brea Koehler And 23h15'31" 48d22'50"
Lindsey Brooke And 1h2'46" 47d12'6"
Lindsey Butor Uma 9h25'25" 46d18'47"
Lindsey Callaway Lib 15h28'58" -22d42'14"
Lindsey Carol Scardino Cas 23h40'1" 52d30'52"
Lindsey Caroline Lilly Psc 1h23'50" 25d40'9"
Lindsey Catherine Cam 6h4'2" 62d59'48"
Lindsey "Celestial" Vickers Cnc 9h9'52" 22d29'14"
Lindsey Charlotte Mintz Ori 5h23'32" -4d10'35"
Lindsey Charlotte Rose Lyn 8h34'35" 45d38'30"
Lindsey Connelly And 0h17'4" 44d35'9"
Lindsey Danielle Rivers And 0h38'18" 32d35'40"
Lindsey Denise Crb 15h52'0" 26d56'20"
Lindsey Dennison And 23h24'14" 48d0'16"
Lindsey Denson Scharer Lyn 8h1'16" 35d5'17"
Lindsey "Dirty" Evans Dra 19h40'29" 61d24'22"
Lindsey E. Cook Cam 4h51'54" 56d9'31"
Lindsey E. Johnson Cas 2h8'36" 59d50'46"
Lindsey Elaine Farris And 1h18'42" 50d31'6"
Lindsey eleven Leo 10h18'51" 26d32'41"
Lindsey Elianna 4637 Ari 2h30'58" 24d59'13"
Lindsey Elise Cap 20h9'15" -8d36'47"
Lindsey Elizabeth Uma 10h22'20" 51d50'54"
Lindsey Elizabeth Barrett Cas 1h0'0" 60d39'14"
Lindsey Elizabeth Floyd And 0h36'41" 29d26'24"
Lindsey Elizabeth Gerzabek And 0h8'14" 44d54'9"
Lindsey Elizabeth Heister Per 3h22'47" 33d30'21"
Lindsey Erin Cap 21h4'30" -15d52'9"
Lindsey Erin Glick Lyn 9h0'20" 35d28'43"
Lindsey Faith Miller Lib 14h28'55" -18d40'0"
Lindsey Ferris Uma 8h50'16" 59d14'26"
Lindsey Gall Cam 4h43'35" 55d30'24"
Lindsey Gates Sloan Claire Bushong Vir 11h48'21" -3d31'54"
Lindsey Gayle Psc 1h12'16" 10d37'46"
Lindsey Geneva LG Cas 1h26'20" 68d37'16"
Lindsey Grace Boyd Vir 14h55'17" 5d37'31"
Lindsey Hall Per 2h49'52" 37d50'11"
Lindsey Hancock's Dreams Aql 19h42'6" -0d2'22"
Lindsey Helen Podpora Sgr 18h12'19" -19d1'42"
Lindsey Henderson And 0h22'38" 28d20'42"
Lindsey Hendrick Cnc 8h46'26" 26d17'49"
Lindsey Hill Sco 16h8'13" -11d10'59"
Lindsey Horan Lmi 10h22'1" 38d19'19"
Lindsey Howard Gem 6h29'58" 13d38'56"
Lindsey Hull Lib 15h9'48" -12d52'45"
LINDSEY I Ari 2h15'35" 26d15'44"
Lindsey J. Cooper Leo 9h59'11" 9d0'6"
Lindsey Jane McDonald Leo 9h36'10" 28d10'50"
Lindsey Jeffs Cas 23h57'55" 56d44'33"
Lindsey Joan Gutierrez And 0h57'5" 34d14'40"
Lindsey Joe Deardorff Gem 7h20'46" 14d10'52"
Lindsey Joy Pedacchio And 1h58'11" 39d17'9"
Lindsey & Juan Ramos Cas 1h17'59" 62d23'19"
Lindsey Julia 12.13.94 Sgr 18h40'19" -23d19'30"
Lindsey June Shelton Tau 4h23'21" 26d8'14"
Lindsey K. Cane Cnv 12h44'35" 39d15'22"
Lindsey K. Iacofano Psc 0h33'1" 14d52'9"
Lindsey K. Oliver Ori 6h21'0" 15d39'19"
Lindsey Kathleen Pendergrass And 1h24'25" 49d6'16"
Lindsey Kay Johnson Sgr 19h33'20" -25d21'56"
Lindsey Kim Preston Hanks Per 3h20'8" 41d20'42"
Lindsey Kimball Cas 1h20'36" 61d46'40"
Lindsey Kristin Voigtlander Ori 5h23'43" 3d33'53"
Lindsey Lamp Uma 11h29'38" 32d41'12"
Lindsey Land Ari 2h17'8" 12d20'32"
Lindsey Law Cyg 19h58'8" 58d51'50"
Lindsey Lee Park Lyn 8h25'11" 51d7'45"
Lindsey Leigh Hobbs Uma 10h57'59" 50d37'43"
Lindsey Loreno Gem 7h18'3" 25d51'44"
Lindsey Lou Ori 5h31'7" 3d25'50"
Lindsey Lou Lyn 6h54'16" 59d59'49"
Lindsey Lou Lib 15h35'7" -20d33'27"
Lindsey Lou Epperson Cap 20h37'3" -20d42'44"
Lindsey Louise Holdridge Lib 15h59'5" -6d59'57"
Lindsey Love Lac 22h27'45" 48d23'26"
Lindsey - Love & Star of My Life Cas 0h19'22" 56d34'50"
Lindsey M. Ash Gem 6h20'10" 23d59'16"
Lindsey M. Caron Cnc 8h51'49" 14d59'37"
Lindsey M. Petree Sgr 20h20'18" -34d16'10"
Lindsey M & Shannon B Trichler And 0h24'37" 32d55'13"
Lindsey M. Truett Per 4h4'41" 44d55'20"
Lindsey Mae Cap 20h20'42" -16d16'23"
Lindsey Mae Bogaczyk Psc 0h38'55" 19d36'2"
Lindsey Margaret St.George Psc 1h51'17" 6d35'31"
Lindsey Marie Tau 3h29'39" 9d20'33"
Lindsey Marie And 1h10'37" 35d8'31"
Lindsey Marie Bonistall And 1h37'44" 48d16'19"
Lindsey Marie Bonistall Cas 1h26'43" 65d50'10"
Lindsey Marie Borg Cyg 21h55'12" 53d53'34"
Lindsey Marie Claar Gem 7h6'47" 32d24'22"
Lindsey Marie Felton Gem 7h49'3" 17d26'38"
Lindsey Marie Frederickson And 0h24'23" 32d27'33"
Lindsey Marie Greening Lib 14h37'4" -12d48'48"
Lindsey Marie Korn Aql 20h0'25" 5d13'4"
Lindsey Marie Kreye Lyn 7h28'54" 57d3'10"
Lindsey Marie Partridge Leo 9h49'48" 28d47'26"
Lindsey Marie Peecher Cnc 8h51'43" 31d41'49"
Lindsey Marie Shanahan And 1h9'14" 37d14'2"
Lindsey Marie Slaton 5/26/82 Gem 6h42'26" 15d57'21"
Lindsey Marie Szczygiel Crb 15h41'3" 36d48'34"
Lindsey Marie Wheeler Lyr 18h41'13" 27d26'28"
Lindsey Marija Tau 4h28'47" 29d37'17"
Lindsey Marlene Cas 23h15'9" 55d48'16"
Lindsey Martine Seales "SuperStar" Lib 15h40'48" -23d18'7"
Lindsey Merline Michels Uma 13h13'58" 60d57'44"
Lindsey Michele And 0h58'20" 45d10'6"
Lindsey Michelle And 0h24'37" 43d3'8"
Lindsey Michelle Adair Per 3h27'13" 32d46'4"
Lindsey Michelle Ellis Ori 5h34'49" -0d37'17"
Lindsey Michelle Hiser Her 16h45'54" 33d1'2"
Lindsey Michelle Nichols Leo 11h22'45" 17d39'7"
Lindsey Miller's Valentine Star Sgr 19h11'42" -27d33'28"
Lindsey Miranda Ori 6h5'56" 17d7'32"
Lindsey Morgan Nolan Vir 14h19'41" 1d18'39"
Lindsey Morgan's Star Cam 4h32'30" 67d9'30"
Lindsey My Cambria Pyx 8h46'23" -30d4'54"
Lindsey my love. Cnc 8h4'54" 12d37'8"
Lindsey N. La Chiana Leo 10h44'1" 11d22'35"
Lindsey Nahodil Cnc 8h14'46" 8d24'2"
Lindsey Nichole Gray Psc 1h24'45" 14d53'26"
Lindsey Nicole Cnc 8h32'57" 28d46'24"
Lindsey Nicole Hess And 23h40'21" 45d16'40"
Lindsey Nicole Kearns Star Ori 4h51'45" 13d48'36"
Lindsey Nicole Nixon Uma 12h14'45" 59d35'20"
Lindsey Nicole Salmons Lmi 10h49'31" 29d18'12"
Lindsey Niemeier, love of my life! Vir 12h30'18" 2d49'43"
Lindsey Noel Deitz Simpson Cnc 8h36'1" 8d33'20"
Lindsey Noel Hemsley Her 17h28'59" 32d13'43"
Lindsey O'Neil Pyx 8h35'22" -36d50'26"
Lindsey Pacita Marie Heffron Del 20h31'57" 12d23'33"
Lindsey Pai Uma 8h33'15" 71d8'27"
Lindsey Paige Velez Psc 1h10'33" 31d4'7"
Lindsey Parker And 1h40'30" 42d25'12"
Lindsey Patterson Boo 14h19'20" 52d0'45"
Lindsey Paul Seagroves Her 18h22'41" 21d6'29"
Lindsey Pea Farina Sco 17h31'40" -45d12'37"
Lindsey Peters Crb 15h31'23" 32d7'13"
Lindsey "Pretty Lady" Janies And 0h29'56" 28d3'11"
Lindsey Princess Leo 10h12'40" 15d36'3"
Lindsey R. Mallek And 0h26'44" 33d19'52"
Lindsey Rae And 0h49'14" 26d55'53"
Lindsey Rae Walker Reeder Tau 4h24'31" 17d57'59"
Lindsey Reese Cam 7h25'9" 62d49'57"
Lindsey Renee And 23h42'17" 48d10'30"
Lindsey Renee Heim Leo 11h36'4" 13d51'59"
Lindsey Roberts And 0h28'9" 27d48'48"
Lindsey Roberts Uma 10h43'36" 45d15'17"
Lindsey Rochelle Pasley Vir 13h2'29" 6d50'19"
Lindsey Rondou Vir 13h13'47" 8d14'12"
Lindsey Rose Garrison And 0h45'50" 36d0'54"
Lindsey Rose Mitchell Lib 14h49'10" -2d37'28"
Lindsey Rose Schultz And 23h17'33" 49d2'11"
Lindsey Rounder happy 3 years And 1h41'21" 46d25'53"
Lindsey Samantha Broedel And 2h27'7" 45d6'27"
Lindsey Santacross Uma 13h12'21" 53d26'46"
Lindsey Sarah Becker Ori 4h49'49" 12d46'37"
Lindsey Shay Petrik Aql 19h29'9" 8d16'41"
Lindsey Suzanne Baker Vir 13h47'55" -16d37'55"
Lindsey Taylor Lyn 6h26'48" 57d21'18"
Lindsey Taylor DelColletti 11/13/02 Cyg 21h54'6" 50d9'23"
Lindsey Taylor LaCasse Cnc 8h51'53" 31d55'39"
Lindsey Tennyson Uma 8h55'39" 54d42'28"
Lindsey Vennard Ari 2h50'30" 14d0'39"
Lindsey Vicary Cas 2h36'34" 67d59'50"
Lindsey Vickers Cnc 8h59'29" 11d26'13"
Lindsey Victoria Zaretsky Sco 17h52'39" -32d36'20"
Lindsey Volpe And 23h19'10" 48d14'40"
Lindsey Wenzen Ori 5h42'14" -3d4'3"
LindseyAnn Tau 4h35'50" 17d8'28"
Lindseybean Ari 3h8'38" 27d49'58"
lindsey-fatty Crb 15h28'12" 29d14'13"

Lindsey-lou
Cas 23h55'55" 53d22'56"
Lindsey's Light
Lib 15h8'20" -13d11'30"
Lindsey's Midnight Star
Sgr 18h55'18" -33d54'30"
Lindsey's Only One
And 1h43'1" 45d24'26"
Lindsey's Sparkle in the Sky
And 23h20'47" 43d36'42"
Lindsey's Star "P.S.S.K.I."
Uma 8h35'5" 46d49'49"
Lindsey's Starlight
Vir 12h46'25" 3d53'45"
Lindseystar
And 23h18'53" 51d0'40"
Lindsi Elizabeth
Ori 5h39'1" 0d29'11"
Lindsy
And 2h18'41" 49d13'4"
Lindsy Alisa McCorgary
Ari 3h7'2" 11d53'39"
Lindsy My Lobster
Uma 11h23'20" 31d10'5"
Lindy
Lyr 18h35'27" 38d5'35"
Lindy
Ori 5h48'22" 11d47'34"
Lindy
Sgr 19h12'42" -16d11'20"
Lindy
Uma 10h26'57" 67d23'17"
Lindy Breest
Lyr 18h57'42" 35d45'10"
Lindy Brooks' Star
And 0h32'23" 26d1'46"
Lindy Danielle
Crb 16h15'49" 37d33'27"
Lindy Gardner
Cam 4h40'11" 54d54'59"
Lindy Hayes
Dra 10h44'23" 76d42'1"
Lindy Kay Carmichael
Sgr 18h36'55" -18d25'2"
Lindy Lou
Lac 22h24'58" 45d13'24"
Lindy Lou 'Discovered'
Gem 6h25'17" 22d6'54"
Lindy M. Dul
Vul 19h30'26" 24d35'39"
Lindy Michelle Lindsey
Ari 2h21'8" 25d34'33"
Lindy Musick
Gem 7h23'40" 33d20'5"
Lindy R. Brandon
Psc 23h20'46" 4d46'39"
Lindy Star
Uma 12h37'2" 61d23'27"
Lindy's Light
Vir 12h47'40" 0d46'33"
Lindy's Magic Source
Her 17h58'4" 21d59'49"
Lindz Boo
Pho 1h20'20" -42d25'1"
Lindzey LaShay Sandefur
Lib 15h47'49" -18d22'30"
Lindzster
Tau 3h51'55" 24d12'50"
Line Adelheid Stang
Cap 20h29'30" -19d7'26"
Line Osmundsen
Cam 7h48'17" 83d24'12"
Line Robert
Cas 1h41'32" 60d38'53"
Line Tømmerholt
Cnc 8h38'8" 19d59'16"
Linear Concept Sdn Bhd
Ori 5h39'36" 3d8'47"
Linell's Freedom
And 23h8'46" 47d17'22"
Line's Light
Tau 4h42'58" 19d29'4"
Linetta Marie
Ori 5h47'15" 3d4'15"
Linette Ann Babiarz
Sco 16h15'0" -9d42'5"
Linette M. Rivas
Mon 6h40'35" 4d2'42"
Ling
Srp 18h4'37" -0d12'38"
Ling Chang
Sco 17h46'22" -33d31'2"
Ling Chantal Taylor
Ori 6h5'28" 13d54'55"
ling ling
Aqr 21h11'12" -2d2'3"
Ling Ling Zhu
Lyn 8h44'54" 33d46'4"
Lingdar
Vir 12h23'29" 3d24'42"
Ling-Ling
Lib 15h37'22" -10d54'33"
Linh Bui
Lyn 7h24'35" 53d55'52"
Linh Trieuly in My Sky Forever
Cyg 19h44'46" 30d39'44"
Linh Vescovini
Boo 15h18'35" 42d31'21"
Link
Dra 17h22'54" 62d52'27"
Link
Uma 10h32'6" 63d34'31"
LINK 12394
Lyn 9h15'3" 38d8'31"
Link Alexander Matozzo
Cap 21h19'7" -27d22'20"
Link To Meg's Heart
Cnc 8h56'30" 28d46'3"
Link, Herbert
Uma 10h56'38" 41d34'0"
Link, Kirsten
Uma 9h25'53" 68d53'3"
Link, Nicole
Uma 10h19'4" 53d32'1"
Linki
Psc 1h21'24" 25d24'53"
LinMark
Cyg 20h47'20" 33d41'45"
Linmike
Sco 16h54'42" -12d4'16"
Linna Hauser
Sgr 19h12'44" -18d8'13"
Linnea
Uma 11h38'22" 62d46'2"
Linnea Grace Dreslin
Sgr 18h57'10" -22d58'20"
Linnea Johanna
Uma 10h58'27" 44d46'20"
Linnea Marie Larson
Cnc 8h5'23" 14d24'6"
Linnea Rochelle Samuels
Aqr 22h4'50" -2d39'48"
Linnie
And 2h20'20" 49d50'3"
Linnie Barlow my every heartbeat
Uma 12h10'36" 46d29'13"
Linnie Roberts
Cas 0h55'30" 62d5'47"
Linnie's Star
Lyn 6h17'46" 54d26'37"
Linni's Honey Joe
Mon 6h53'3" -0d28'5"
Linny
Lyn 8h3'37" 41d16'35"
Linny B
Aqr 21h15'38" -3d55'29"
Linnys' Star
Cas 23h24'36" 53d14'26"
LinnyStar514
Cyg 21h24'49" 35d1'41"
Lino
And 23h58'12" 41d45'4"
Lino Kios Emilian
Uma 11h56'37" 46d28'10"
Lino Lourenco
Ori 6h4'13" 21d10'57"
Lino S.82
Per 2h27'30" 55d17'37"
linoemy
Uma 12h26'35" 54d51'55"
Linos' Amanda
Lyn 6h29'4" 61d35'0"
Linos Kolb
Uma 8h51'36" 61d17'14"
Lin's Star
Boo 15h41'12" 40d8'7"
Linsay
And 0h38'35" 42d18'55"
Linsey
And 2h26'58" 48d18'22"
Linsey
Vir 13h15'3" -20d59'13"
Linsey Ann Plummer
Umi 13h48'49" 69d26'53"
Linsey Faith
Leo 10h35'19" 14d44'58"
Linsey Jo
Aqr 22h11'1" -1d19'51"
Linsey Kaye Churchill
Psc 0h24'23" 8d29'4"
Linsey Michele Altavater
And 0h31'35" 38d22'28"
Linsey Renee
Uma 10h25'47" 65d37'22"
Linsey Talley
Lyn 9h40'4" 40d26'22"
Linsey Terese Bared
Lib 15h15'47" -12d29'1"
Linsey Thomas Earl Sr.
Per 4h9'26" 46d41'26"
linseyrenae05
Vir 13h15'10" 6d51'44"
Linsey's Light
Lyn 8h25'59" 42d43'10"
LinStar
Vir 13h5'45" -12d36'20"
Linus
Lib 15h49'27" -17d42'35"
Linus
And 23h4'1" 51d3'29"
Linus
Cmi 7h39'23" 0d30'57"
Linus Christopher
Cap 20h16'38" -22d20'5"
Linus Lowery
Lyr 18h49'59" 43d48'36"
Linus - Mutz - forever
Umi 14h23'28" 68d50'59"
Linus Star
Umi 16h33'28" 75d12'9"
Linwood
Uma 10h14'0" 55d49'53"
Linwood
Her 17h22'54" 37d48'11"
Linz 21
Dor 5h0'5" -68d59'27"
Linz & Steve: Side-by-side Forever
Cyg 19h58'3" 33d45'31"
Linzey Wilson
And 1h10'56" 48d24'34"
Linzi
And 0h48'3" 21d55'35"
Linzi Alayna & Ryan Michael
Cyg 19h28'55" 52d18'12"
Linzi Gilliam
Dra 9h59'39" 73d16'55"
Linzi Jo Jasso
Aqr 21h5'54" -13d10'5"
Linzi & Seth
Cyg 20h21'16" 34d6'50"
Linzie Gail Derrick
Cyg 19h50'3" 37d11'32"
Linzie Jo
Umi 16h22'40" 73d32'11"
Linzy Baby
Cas 0h22'29" 54d15'21"
Linzy Zorn
Dra 18h19'37" 57d41'58"
Lion Heart
Gem 6h8'47" 24d23'46"
Lion Noel
Uma 12h19'23" 59d58'56"
Lion Sparkes
Lib 15h36'49" -23d32'58"
Lionel Boucher 14/11/2003
Uma 11h31'11" 57d1'31"
Lionel Bühler
Uma 9h45'19" 67d34'32"
Lionel C
Cnc 8h54'51" 7d52'27"
Lionel Douglas Lueck
Dra 20h8'48" 74d18'41"
Lionel James Dawson
Sco 16h16'6" -19d3'49"
Lionel Kooper Superstar
Tau 3h55'33" 5d51'57"
Lionel Micah Riley 3/31/57
Ari 1h54'29" 22d16'28"
Lionel Olive
Cep 22h35'6" 58d4'24"
Lionel Ovadia
Ari 2h37'25" 25d22'27"
Lionel Pasco
Cap 20h59'3" -22d32'32"
Lionel Preston Joyner
Leo 9h34'37" 28d2'33"
Lionel Sanchez
Gem 7h13'58" 27d58'31"
Lionel Tim
And 23h23'14" 43d9'7"
Lionel's Shining Star of Peace
Cap 21h33'44" -13d43'46"
lioness
Lmi 10h5'49" 37d38'46"
LionGirl
Ari 2h59'12" 25d17'33"
Lionheart
Ori 5h8'57" 15d11'12"
Lion's Flower
Cnc 8h0'0" 16d53'4"
Lior Shav
Psc 1h9'15" 10d24'42"
Liora
Peg 23h8'11" 14d32'19"
Lioubov P
Cru 12h18'3" -57d29'51"
Lipari 2006
Lyn 8h57'23" 35d29'43"
Lipika Mohanty
Mon 7h35'54" -0d39'37"
Lipimoni
And 0h21'27" 32d53'46"
LIPS OF AN ANGEL
Leo 9h28'2" 9d44'0"
Lips of an Angel
Pho 0h38'11" -49d26'17"
Lira
Gem 6h42'42" 24d44'53"
Lireca
Dra 16h26'40" 57d23'5"
Liridona
Cam 4h20'10" 61d25'35"
Liridona Hajredini
Aql 19h43'50" 8d19'25"
Lis and Pauls Love Star
Lyn 8h0'35" 44d19'32"
Li's Star
And 0h29'36" 28d32'33"
Lisa
Ori 5h56'1" 21d11'27"
Lisa
Cnc 8h3'5" 27d4'24"
Lisa
Gem 7h34'10" 26d16'31"
Lisa
Gem 7h27'16" 28d13'4"
Lisa
Tau 3h58'23" 25d9'40"
Lisa
Vul 20h2'35" 22d37'38"
Lisa
Leo 9h53'26" 29d56'52"
Lisa
Leo 11h40'54" 24d36'50"
Lisa
Aql 19h29'6" 7d59'6"
L.I.S.A
Psc 0h51'3" 17d13'52"
Lisa
Aur 6h40'9" 44d1'31"
Lisa
And 2h3'10" 38d57'9"
Lisa
And 1h12'11" 37d58'2"
Lisa
And 1h14'9" 39d10'20"
Lisa
And 1h19'12" 35d24'47"
Lisa
And 23h27'47" 42d10'26"
Lisa
And 0h22'20" 46d28'54"
Lisa
And 23h24'52" 37d36'59"
Lisa
And 23h14'3" 41d54'10"
Lisa
Cnv 12h49'42" 38d50'1"
Lisa
Cas 0h9'7" 53d31'11"
Lisa
And 23h12'55" 48d18'5"
Lisa
Cas 0h19'52" 54d30'34"
Lisa
Uma 9h46'49" 51d23'22"
Lisa
Uma 10h15'26" 71d15'15"
Lisa
Umi 14h12'55" 70d58'44"
Lisa
Cas 0h43'38" 64d10'48"
Lisa
Cas 0h37'40" 64d27'12"
Lisa
Uma 9h4'3" 59d1'23"
Lisa
Vir 12h34'44" -0d24'5"
Lisa
Eri 4h27'39" -3d26'2"
Lisa
Cap 20h34'23" -20d51'36"
Lisa
Dor 5h13'22" -58d59'30"
Lisa
Aqr 23h46'11" -24d37'13"
Lisa
Cma 6h41'49" -25d16'26"
Lisa
Sgr 18h53'41" -22d34'42"
Lisa
Sgr 17h53'49" -29d49'52"
Lisa
Umi 17h43'21" 81d41'21"
Lisa A D Quattrochi
And 1h30'55" 42d26'34"
Lisa A. Parker
Cnc 8h59'33" 15d9'54"
Lisa A. Webb
Srp 18h8'36" -0d21'10"
Lisa Abbott
Del 20h24'40" 16d42'26"
Lisa Acho
Crb 15h47'50" 34d55'58"
Lisa Adam
Cap 20h21'11" -10d59'4"
Lisa & Adam's Star
Cyg 20h12'13" 38d22'38"
Lisa Addington Moore
And 0h31'10" 32d44'44"
Lisa Aileen Reinsdorff
And 0h25'52" 41d55'3"
Lisa aka Bubbles
Sgr 18h4'59" -22d5'3"
Lisa Albaugh
And 1h17'32" 40d0'22"
Lisa Almanza
Gem 7h30'44" 21d48'33"
Lisa and Adam
Ori 6h19'4" 16d43'38"
Lisa and Aiden Joseph Laroche
Com 12h44'8" 15d20'0"
Lisa and Brendan Burba
Cnc 8h51'17" 27d40'30"
Lisa and Chris
Lib 15h40'58" -19d36'34"
Lisa and Chris 4 ever
Lib 14h42'55" -13d16'27"
Lisa and Clay
Umi 15h15'9" 75d35'37"
Lisa and Ed's Engagement Star
Cyg 20h31'29" 59d44'46"
Lisa and Foofey's Star
Her 17h50'27" 20d29'19"
Lisa and George
Ori 5h42'39" 8d13'39"
Lisa and Jim Johnson
Uma 9h35'45" 71d6'49"
Lisa and Jim Mailler
Gem 7h32'39" 28d45'0"
Lisa and Lisa
Cyg 21h46'53" 46d11'21"
Lisa and Nicki One Year Anniversary
Cyg 19h39'53" 31d3'4"
Lisa and Rocky Concepcion
Uma 12h3'29" 29d9'57"
Lisa and Solo's Eternal Place
And 2h17'35" 37d50'14"
Lisa and Stephan's 1st Anniversary
Cyg 20h46'25" 41d39'22"
Lisa and Tucker's Day Star
Tau 4h29'28" 26d39'35"
Lisa Anderson
Lmi 10h34'27" 38d1'10"
Lisa André Milam
And 1h9'50" 36d20'46"
Lisa Angela Dorozio
Cas 2h11'12" 72d39'5"
Lisa Angela Marie Bilancia
Tau 5h53'30" 27d5'43"
Lisa Angelina Lera
Cas 23h1'57" 59d30'43"
Lisa Angelique Koller
Lmi 10h5'0" 40d5'58"
Lisa Ann
And 0h49'42" 44d30'30"
Lisa Ann
Tau 3h45'43" 30d7'40"
Lisa Ann
Crb 15h41'1" 34d6'58"
Lisa Ann
And 1h42'51" 45d13'4"
Lisa Ann
Cas 0h8'24" 56d50'30"
Lisa Ann
Aqr 21h52'10" 0d11'24"
Lisa Ann
Ori 6h15'29" 15d3'25"
Lisa Ann
Lyn 7h15'23" 56d18'50"
Lisa Ann
Aqr 23h10'46" -9d35'22"
Lisa Ann 5671
And 1h13'7" 33d47'5"
Lisa Ann Allor
Cnc 8h40'28" 6d34'1"
Lisa Ann Cole
And 23h22'44" 36d47'12"
Lisa Ann Dever-Cruz
Uma 10h26'29" 68d21'15"
Lisa Ann Edwards
Cas 0h7'56" 54d35'7"
Lisa Ann Gordon
Tau 4h20'46" 26d14'54"
Lisa Ann Griffith
Crb 15h51'34" 39d33'51"
Lisa Ann Guthrie
Sco 16h12'37" -11d56'34"
Lisa Ann Hamner
Srp 18h19'16" -0d8'32"
Lisa Ann Healey
Cyg 20h0'30" 46d46'25"
Lisa Ann Karr
Tau 3h37'49" 3d51'32"
Lisa Ann Knick
Pho 0h30'31" -41d49'18"
Lisa Ann Kringel's Star
Lyn 7h12'50" 44d52'19"
Lisa Ann Nelson
And 1h33'57" 36d37'16"
Lisa Ann Perillo
Ari 3h23'22" 27d46'58"
Lisa Ann Rich
Aql 19h5'35" 2d49'2"
Lisa Ann Roper
And 0h10'15" 45d25'3"
Lisa Ann Smith
Per 4h49'56" 45d46'54"
Lisa Ann Steele
Lyn 8h3'31" 56d51'24"
Lisa Ann Tam
Aql 18h58'31" 15d25'9"
Lisa Ann Usserman
Uma 13h9'14" 56d33'49"
Lisa Ann Walters
Her 18h44'38" 20d7'53"
Lisa Anna-Mae Burwell
Cas 1h12'41" 57d57'1"
Lisa Anne
And 22h58'26" 48d32'12"
Lisa Anne
Uma 9h37'20" 47d49'19"
Lisa Anne Bender
Dra 18h22'55" 73d53'28"
Lisa Anne Brady
Sgr 19h29'42" -15d32'3"
Lisa Anne Chouinard
And 0h13'54" 25d9'36"
Lisa Anne Harper
Mon 6h53'49" -3d37'28"
Lisa Anne James Gardner
Vir 13h3'9" 5d33'43"
Lisa Anne Phillips
And 23h9'55" 40d22'35"
Lisa Anne Taylor Villeneuve
Cas 0h29'30" 58d41'36"
Lisa Annie
Ari 2h20'16" 13d27'7"
Lisa & Anthony - "Tosaliny"
Cyg 21h56'25" 49d26'15"
Lisa Antoinette Savastano
Psc 0h35'41" 6d48'40"
Lisa Antonelli
And 23h46'22" 47d46'25"
Lisa Antonuccio
And 1h49'48" 41d30'31"
Lisa Arancibia
Vir 14h54'33" 0d6'9"
Lisa Arbach
Leo 11h16'10" 16d20'27"
Lisa Armstrong
Cam 3h50'27" 59d47'5"
Lisa Arnold
Uma 11h38'58" 62d51'30"
Lisa Augustine
Cap 20h46'33" -15d0'30"
Lisa Austin
Uma 9h26'15" 56d33'13"
Lisa B
And 2h16'33" 39d7'23"
Lisa B.
Psc 23h46'46" 7d32'32"
Lisa Baby Girls Bowers' Star
And 1h20'7" 46d29'42"
Lisa Balisa Shining Star
And 22h58'35" 45d41'41"
Lisa Barbara
Aqr 23h35'2" -8d38'5"
Lisa Barrie
Psc 0h57'41" 10d5'33"
Lisa Baruzzi
Cnc 8h0'16" 18d4'41"
Lisa Beavan
And 1h40'4" 50d4'7"
Lisa "Bee" Brosnan
Cnc 8h4'26" 20d0'36"
Lisa Begley
And 0h55'19" 40d49'53"
Lisa Belcher Flatt
Leo 10h17'23" 27d13'36"
Lisa Bell
And 23h4'53" 42d37'47"
Lisa & Benny Thornhill For Eternity
Cru 12h28'52" -61d47'26"
Lisa Benzak M&M
Cnc 8h14'19" 16d22'24"
Lisa Betar
Sgr 18h54'26" -33d47'24"
Lisa Beth DePedro
Cnc 8h44'30" 26d50'53"
Lisa Beth Neidich
Cas 2h34'22" 65d18'49"
Lisa Birse's Angel Above
Aqr 21h7'44" -9d9'40"
Lisa Blanchard
Cas 0h25'37" 65d3'11"
Lisa Blinman - Daughter of the King
Cru 12h32'0" -59d26'11"
Lisa Boeglin
And 0h12'18" 26d49'49"
Lisa Bolen
And 23h32'22" 47d44'3"
Lisa Boni
Aur 4h57'6" 33d10'52"
Lisa Bowles
Lyn 7h39'30" 52d52'54"
Lisa Boyens
Uma 10h22'25" 57d59'34"
Lisa Bradley
Lmi 10h49'18" 26d43'12"
Lisa Bramucci
Uma 11h48'30" 29d22'16"
LiSA BroWn
Psc 1h12'7" 26d46'3"
Lisa Brown
Lib 14h56'31" -0d56'45"
Lisa Browne
Vul 19h5'47" 23d39'50"
Lisa Bruce
Ori 5h32'51" 1d43'20"
Lisa Brunner
Uma 8h53'49" 59d25'4"
Lisa Bui
Ari 3h7'58" 29d50'46"
Lisa Bulich
Cas 1h19'16" 61d44'23"
Lisa Burton, I will love you forever
Del 20h35'54" 18d54'5"
Lisa Byrne
Crb 16h1'51" 32d57'32"
Lisa Carleen Adrie
Ari 3h12'9" 21d0'31"
Lisa Carme Pimentel
Umi 15h59'20" 72d51'34"
Lisa Carr
Cyg 19h37'34" 54d54'40"
Lisa Carretta
Sco 17h16'24" -32d27'30"
Lisa Carter
Lmi 10h29'44" 32d32'11"
Lisa Cavalli
Cyg 19h55'26" 38d26'36"
Lisa Cawley
And 0h42'19" 25d4'9"
Lisa Chambers
Aqr 22h46'10" -21d17'32"
Lisa Christin Allison
Cnc 9h13'6" 29d19'10"
Lisa Christine Anthony
Tau 4h5'35" 15d32'19"
Lisa Christine Campbell
Uma 11h0'53" 47d0'41"
Lisa & Chuck Harrison
Uma 9h28'23" 56d48'12"
Lisa Clayton
Cas 3h12'56" 67d51'12"
Lisa Coane
Ari 2h37'47" 19d31'25"
Lisa Cockey
And 2h37'1" 43d20'15"
Lisa Combs 1
Ori 6h20'56" 16d36'36"
Lisa Conama
Aql 19h26'20" -0d39'19"
Lisa Cooper
Ari 2h16'31" 22d36'27"
Lisa Cortney
Crb 15h57'15" 37d22'10"
Lisa Costello - Forever in our Hearts
Umi 16h49'42" 82d48'7"
Lisa Cuba
Cas 0h2'40" 53d39'3"
Lisa Cummins Ramirez
Lyn 8h31'37" 48d22'31"
Lisa Custer
Dra 16h18'1" 62d57'44"
Lisa D.
Crb 16h5'16" 25d50'38"
Lisa D. Colapietro
Gem 7h41'54" 32d32'21"
Lisa D. Weathersby
And 23h34'57" 38d28'36"
Lisa & Daniel McMahan
Uma 11h18'33" 61d25'54"
Lisa Danielle Conger
Ori 6h4'48" 7d12'29"
Lisa ~ Danny & Caleb Cyr
Peg 22h42'56" 26d56'6"
Lisa Darlene Mitchell
Cas 1h27'41" 75d48'35"
Lisa & David always&forever&beyond
Cas 0h46'34" 61d6'21"
Lisa & David's Wedding Star!
Cyg 21h18'35" 46d47'58"
Lisa Davison
And 2h25'9" 47d50'56"
Lisa Dawn
Uma 13h3'54" 53d43'28"
Lisa Dawn Carnes
Cnc 8h3'39" 20d35'14"
Lisa Dawn Daniels
And 0h25'8" 29d4'23"
Lisa Dawn Esq.
And 23h19'24" 40d42'25"
Lisa De Silvestri
Aql 20h8'30" 8d11'9"
Lisa Degenhardt
Tau 4h15'32" 23d34'2"
Lisa Delphine Vargas
And 0h26'37" 33d23'24"
Lisa DelPriore Gray
And 0h23'2" 27d19'23"
Lisa DeLynn Hathcock
Vir 14h22'51" 4d44'55"
Lisa Deming Ladd
Lib 15h23'47" -10d54'49"
Lisa Denise
Peg 21h54'51" 12d36'0"
Lisa & Des
Pav 18h43'53" -61d36'23"
Lisa Deuley
Cam 4h27'35" 64d33'31"
lisa di maio
Ari 3h8'44" 26d59'0"
Lisa Diamond in the Sky
Uma 11h31'5" 58d12'56"
Lisa Diane
And 23h4'28" 52d57'33"
Lisa Diane Ragan
Cas 0h23'24" 53d13'15"
Lisa Diane Reed
And 0h54'54" 38d44'33"
Lisa Diane Rizzo
And 23h57'38" 41d11'15"
Lisa "Dino" May O'Dell
Gem 6h38'13" 21d8'26"
lisa doyle
Aqr 23h49'43" -19d42'12"
Lisa Drew
Uma 9h49'9" 58d6'54"
Lisa Duncan
Cas 0h18'16" 62d4'44"
Lisa Durant
And 0h40'35" 21d46'24"
Lisa Dyan DeGarmo
Uma 9h6'48" 55d4'5"
Lisa E. Hemp
Sco 16h46'13" -32d21'30"
Lisa e Matteo
Peg 21h56'1" 12d14'54"
Lisa E. Seibert
And 1h28'48" 34d52'34"
Lisa Eap
Aqr 22h33'0" 0d22'58"
Lisa Elaine Kay
Ori 5h23'44" 6d39'9"
Lisa Eliat Fiss
And 1h38'2" 37d21'36"
Lisa Elizabeth
And 0h32'47" 34d12'44"
Lisa Ellen
Leo 11h8'49" -0d42'44"
Lisa Erhardt
Vir 13h44'42" -0d3'41"
Lisa & Eric
Sgr 18h33'2" -30d7'29"
Lisa & Erica
Cyg 20h2'42" 51d3'33"
Lisa Erin 010501
Dra 15h53'14" 60d8'25"
Lisa Esposito
Psc 1h20'13" 31d16'13"

Lisa Estell Arnold
Cap 21h5'34" -22d15'58"
Lisa Eufemia Carlson
And 1h6'29" 37d46'24"
Lisa Evans
And 0h20'19" 33d9'35"
Lisa Evans Poczatek
Tau 5h52'9" 26d38'9"
Lisa F.
Psc 1h24'9" 17d29'47"
Lisa Face
Eri 4h0'40" -14d56'45"
Lisa Fassi
Cnc 8h29'52" 31d25'34"
Lisa Fay
Lyn 6h26'27" 58d7'11"
Lisa Faye Hooper 32
Vir 13h28'47" 3d42'11"
Lisa Ferranti
Cas 1h43'47" 65d16'41"
Lisa Ferrone
Uma 11h9'11" 53d7'47"
Lisa Fiorita
Lyn 7h45'42" 49d44'5"
Lisa Fleming
Psc 1h12'46" 27d17'30"
Lisa Forever
Uma 8h45'8" 63d56'46"
Lisa Francis
Cap 20h53'36" -24d45'37"
Lisa Free
Crb 15h50'18" 30d4'40"
Lisa Freeman
Leo 9h29'3" 22d54'14"
Lisa Freyenhagen
And 2h23'37" 47d14'28"
Lisa from Nibans
Leo 11h15'56" 21d51'47"
Lisa Gae Morris
Cas 1h19'18" 58d5'9"
Lisa Gail
Lib 15h25'52" -15d12'31"
Lisa Galatio
Lib 15h9'11" -10d12'48"
Lisa Garman's Dream
Vir 12h33'48" -0d16'46"
Lisa Gayle
Tau 5h47'51" 17d2'40"
Lisa Gerace
Tau 4h40'29" 18d52'23"
Lisa Gerling Serenity Love Infinity
Tau 3h46'37" 24d20'37"
Lisa Giannotti
And 1h28'6" 44d46'36"
Lisa Gillian
Peg 21h35'2" 24d46'28"
Lisa "Gnome" Elliott
Cru 12h21'54" -57d36'30"
Lisa Goodman
And 2h11'30" 38d33'31"
Lisa Goodwin
Cyg 20h2'15" 32d9'49"
Lisa Graessle
And 1h38'43" 42d29'2"
Lisa & Graham - First Anniversary
Cyg 20h45'30" 41d29'4"
Lisa Grebe
Vir 14h13'28" -8d59'17"
Lisa Groh
Ori 5h34'30" -0d20'34"
Lisa Guagliardo
And 23h28'48" 47d40'19"
Lisa Gullatt
Gem 7h10'14" 28d50'18"
Lisa Gwin
And 0h27'23" 46d41'2"
Lisa H Richardson
Oph 17h20'24" -0d35'26"
Lisa Hacker
Lyn 7h30'36" 37d58'31"
Lisa Hancock - Beautiful Nursey
And 2h29'26" 50d10'26"
Lisa Hardy, Chelsie and Lachlan
Cyg 20h59'25" 45d48'38"
Lisa Harrison
Sco 16h57'22" -37d23'10"
Lisa Hart's Star
Leo 10h27'2" 12d48'12"
Lisa Hayes ~Always Watching Over Us
Uma 10h39'2" 44d50'50"
Lisa Hegyi
Aur 5h51'21" 36d0'58"
Lisa Helene
And 0h50'16" 40d24'21"
Lisa & Henry Rooney
Cyg 19h45'52" 33d6'8"
Lisa Hess Lynch
Cap 21h41'5" -12d34'35"
Lisa Hickernell
Mon 8h9'22" -0d36'14"
Lisa "Honey Bunny"
Tau 5h29'8" 21d14'16"
Lisa Honeybabe
Cru 12h16'44" -57d33'15"
Lisa Hotsky Lehotsky
Cap 20h39'37" -25d42'26"
Lisa Howard
Sco 16h40'38" -29d21'56"
Lisa & Ian's Wedding Star
Cyg 20h17'4" 45d53'27"
Lisa J Aguon
And 23h22'49" 41d24'58"
Lisa J Cooke
Ori 6h19'47" 18d50'56"
Lisa J. Fikany
Uma 13h29'14" 57d57'59"
Lisa J. Yancey
Cas 0h39'4" 58d18'31"
Lisa Jacobs
Cas 1h39'34" 63d33'51"
Lisa Jade Smith - My Little Lady
Cen 13h57'9" -60d33'26"
Lisa Jane
And 2h28'37" 44d48'17"
Lisa Jane
Gem 7h28'47" 29d30'44"
Lisa Jane
Aql 20h3'46" 5d3'14"
LISA JANE
Tau 4h16'36" 9d19'8"
Lisa Jane Curtis
Cas 1h20'6" 64d10'1"
Lisa Jane Evans Lye
Leo 9h43'42" 16d29'58"
Lisa Jane Rance
Cas 2h51'20" 60d36'0"
Lisa Jane Salas-Hamilton
Cnc 8h50'47" 21d54'5"
Lisa Jane Sheppard
And 1h42'23" 35d49'27"
Lisa Janine Offerman
Umi 14h30'51" 74d11'36"
Lisa Jasso
Vir 14h29'15" 2d30'6"
Lisa & Jay UTEOT
Ori 5h38'40" -3d37'21"
Lisa Jayne Corkhill
And 0h26'18" 23d25'25"
Lisa Jayne Jackson
Cyg 20h8'16" 43d33'10"
Lisa Jean
Lyr 18h38'16" 41d33'4"
Lisa Jean
Cyg 19h44'11" 29d23'43"
Lisa Jean Baus "Lisa's Luminary"
Umi 16h12'11" 70d10'41"
Lisa Jean Sirois
Ari 2h42'58" 27d45'47"
Lisa&Jeff JorgensenTrueLoveTenYears
Ari 3h17'55" 26d17'39"
Lisa Jeffries
And 1h25'34" 44d25'47"
Lisa Jennifer Walters
Gem 6h17'37" 22d4'31"
Lisa Jewell
Cam 5h59'25" 61d14'21"
Lisa & Jimmy Ziats' Shining Star
Cyg 21h54'17" 55d17'50"
Lisa Jo
Gem 7h53'21" 29d53'58"
Lisa Jo
Cas 1h22'2" 59d2'27"
Lisa Jo Corneil 1963
Uma 8h20'29" 62d52'10"
Lisa Joanne Gossage
And 23h3'28" 41d47'47"
Lisa Jones
Cap 21h31'31" -8d45'4"
Lisa Joy
Aqr 22h56'1" -16d54'59"
Lisa Joy
Cap 20h23'1" -19d5'16"
Lisa Joy
Gem 8h2'11" 29d52'52"
Lisa Joy
Vir 13h23'52" 3d37'57"
Lisa & JP Mitchell
Dra 11h8'55" 77d38'41"
Lisa K. Crafton
Leo 10h56'33" 8d3'57"
Lisa K. Garretson
And 1h17'38" 37d58'28"
Lisa K. Whitmire & Gregory S. King
And 1h17'47" 41d1'29"
Lisa "Kat" Hockman
Cnv 13h49'21" 35d31'42"
Lisa Kate Turley
Cru 12h32'7" -61d5'3"
Lisa Kathleen Campbell
Vir 13h17'14" -0d4'49"
Lisa Kautz
Leo 10h57'57" 6d43'55"
Lisa Kay
And 1h11'32" 45d5'51"
Lisa Kay
Cas 0h22'7" 61d59'4"
Lisa Kay Bowman
Com 12h46'11" 15d39'52"
Lisa Kay Lockwood
Cas 3h12'7" 67d10'9"
Lisa Kay Rose
Crb 15h27'24" 28d23'18"
Lisa Kaye
Ari 2h42'12" 25d44'55"
Lisa Kaye
Lib 15h37'40" -27d11'32"
Lisa Kaye Busman
Tau 3h53'29" 20d54'27"
Lisa Kaye Prescott
And 0h12'52" 44d0'18"
Lisa Kelly
Lib 15h30'54" -24d43'25"
Lisa Kelly Burke
Cas 1h10'37" 64d18'1"
Lisa & Kenny
Ori 5h15'7" -0d16'28"
Lisa Kilpatrick
And 23h12'14" 47d0'45"
Lisa Kilpatrick Broussard
Mon 6h45'9" 9d11'1"
Lisa Kim Benton
Ari 2h14'9" 15d51'6"
LISA KIM DAVIS
Tau 5h11'41" 18d5'58"
Lisa Kimberly Smith
Psc 0h49'48" 16d20'18"
Lisa Kirkpatrick
Tau 3h49'26" 22d36'15"
Lisa Kitchens
Aqr 22h37'10" -15d42'16"
Lisa Kravchenko
Cnc 8h14'29" 9d40'46"
Lisa Krawczyk
And 23h38'46" 46d24'13"
Lisa Kristine
Sgr 18h5'28" -29d40'5"
Lisa Kulakowsky
Cas 0h24'45" 56d39'37"
Lisa L. Di Loreto
Aqr 22h28'28" 0d45'18"
Lisa L. Diehl
Cyg 21h15'42" 38d23'52"
Lisa L. Moody
And 23h15'32" 52d22'25"
Lisa Lafond
Cap 20h41'38" -21d12'52"
Lisa Laghezza
Cas 0h56'56" 57d5'2"
Lisa Lassandrello
Lib 15h2'24" -25d41'41"
Lisa Lavery & Mark McBride
Cyg 20h50'20" 47d5'25"
Lisa Lavoro
Cas 0h38'52" 63d59'4"
Lisa Le
Psc 0h41'21" 4d1'9"
Lisa Lea Race
Cyg 19h11'26" 52d5'46"
Lisa Leah
And 23h3'19" 43d46'2"
Lisa Lee
Aqr 22h38'2" 0d52'31"
Lisa Lilac
Uma 9h16'35" 67d45'19"
Lisa (Ling Ling) Pham
Per 3h43'3" 54d42'45"
Lisa Lipman
Cas 0h23'54" 61d22'13"
Lisa Little
And 2h36'18" 45d11'32"
Lisa Little (Wisdom)
And 0h17'53" 32d31'7"
Lisa Littleton McClain
Gem 7h29'8" 19d56'7"
Lisa Lorraine Hemphill
Aqr 22h56'45" -10d47'54"
Lisa Love
Lib 15h48'30" -15d40'50"
Lisa Lucy Collins(my northern star)
Umi 16h37'16" 83d50'26"
Lisa Lyn Reigelsperger
Cas 1h28'49" 62d37'9"
Lisa Lynn
Cas 0h13'14" 53d32'59"
Lisa Lynn
Uma 11h37'9" 34d11'33"
Lisa Lynn Carman
Uma 11h19'47" 31d55'53"
Lisa Lynn Hanson
Cas 1h16'47" 53d16'50"
Lisa Lynn Loves David Lynn More!!
Sge 19h53'17" 18d43'40"
Lisa Lynn Olivo
Aqr 22h3'1" 1d59'36"
Lisa Lynne Whitney
Gem 7h16'36" 15d3'2"
Lisa M
Ori 5h49'47" 11d18'19"
Lisa M.
Lyn 7h50'7" 41d55'29"
Lisa M Alderfer
Lyn 8h8'4" 37d37'18"
Lisa M. Alfaro
Ari 2h31'43" 20d0'11"
Lisa M. Bell
Lyn 8h13'26" 33d47'47"
Lisa M Byars
Cnc 9h12'55" 10d56'10"
Lisa M Calhoun
Lib 15h2'51" -2d10'4"
Lisa M Carr
Cas 0h7'20" 58d38'51"
Lisa M. Darling (Beautiful Lady)
Cas 0h22'54" 52d45'37"
Lisa M Kendall
Leo 10h22'17" 26d40'53"
Lisa M. Pena
Cnc 8h18'45" 14d29'33"
Lisa M. Smith
Ari 2h30'47" 20d29'13"
Lisa M Virell
Psc 1h11'19" 13d16'26"
Lisa MacDonald
And 1h52'37" 46d9'48"
Lisa Mae Lefebvre
Sgr 18h9'1" -18d26'55"
Lisa Mae Smith
And 0h16'29" 46d1'35"
Lisa Mallory Brew
Cas 23h6'39" 53d53'19"
Lisa Manganiello
Aql 19h22'31" 16d19'52"
Lisa Marani Ryan
Mon 6h45'35" -7d21'18"
LISA MAREE
Psc 23h1'37" 5d49'56"
Lisa Marguerite Franzel
Aqr 23h29'46" -7d20'51"
Lisa Maria
Cas 0h46'31" 57d6'27"
Lisa Maria
Lyn 7h58'19" 38d44'49"
Lisa Maria Berquist
Lyn 7h50'10" 44d17'17"
Lisa Maria Kumar
Aqr 22h27'14" -5d18'50"
Lisa Maria Mariano, March 8, 1985
Uma 11h32'52" 58d10'42"
Lisa Maria Medeiros
Cyg 21h36'3" 39d21'26"
Lisa Marian Drew
Vul 19h56'14" 23d28'18"
Lisa Marie
Leo 10h5'30" 27d11'21"
Lisa Marie
Leo 9h35'31" 23d35'53"
Lisa Marie
Ari 2h20'1" 26d2'57"
Lisa Marie
Psc 0h42'58" 15d36'58"
Lisa Marie
Cas 0h9'7" 50d26'51"
Lisa Marie
And 2h17'53" 48d10'9"
Lisa Marie
Cas 1h10'31" 58d8'10"
Lisa Marie
Uma 9h3'41" 48d3'35"
Lisa Marie
Gem 7h40'18" 33d31'2"
Lisa Marie
And 1h20'45" 37d1'52"
Lisa Marie
Cas 23h59'11" 57d31'30"
Lisa Marie
Uma 10h53'35" 67d7'32"
Lisa Marie
Mon 7h35'0" -1d14'55"
Lisa Marie
Mon 7h21'41" -7d8'8"
Lisa Marie
Sgr 17h53'42" -17d47'38"
Lisa Marie
Vir 13h57'53" -20d48'44"
Lisa Marie
Sgr 18h1'34" -27d56'6"
"Lisa Marie"
Sco 17h12'42" -34d32'17"
Lisa Marie
Sco 16h45'23" -32d44'36"
Lisa Marie 09221976
Cyg 19h59'1" 30d35'44"
Lisa Marie 10/24/1985
Sco 17h49'23" -31d14'20"
Lisa Marie 4/14/1963
Uma 11h9'21" 64d59'6"
Lisa Marie Antoinette
Lyn 6h20'46" 57d3'19"
Lisa Marie Arnone-Jackson
Gem 7h12'52" 19d53'24"
Lisa Marie Baurle
Ori 4h50'22" 4d9'33"
Lisa Marie Belano Vigliotta
Tau 4h52'6" 21d8'10"
Lisa Marie Belles
Leo 11h8'32" 15d36'7"
Lisa Marie Bellmore
Ori 6h25'8" 10d36'4"
Lisa Marie Berg
Ari 2h23'15" 26d37'23"
Lisa Marie Berotti
Uma 10h46'32" 66d40'8"
Lisa Marie Bieri
And 1h18'55" 49d54'23"
Lisa Marie Boneberg
Aur 5h45'51" 48d8'53"
Lisa Marie Bringhurst
Aqr 22h28'1" -2d34'4"
Lisa Marie Browne
And 0h48'12" 46d17'6"
Lisa Marie Caboot
And 1h31'35" 45d17'10"
Lisa Marie Cannariato (Lisa's Star)
Leo 10h36'0" 14d19'4"
Lisa Marie Crawley
Leo 10h43'53" 14d17'5"
Lisa Marie - Daughter
Ari 3h7'53" 10d51'22"
Lisa Marie DeVore
Gem 6h51'18" 26d39'21"
Lisa Marie Diaz
Psc 1h23'55" 15d55'40"
Lisa Marie Flood
And 0h27'30" 38d39'55"
Lisa Marie Gagliardi
And 2h32'24" 49d25'41"
Lisa Marie Guerra
Sgr 19h34'36" -12d25'3"
Lisa Marie Hanley
Gem 6h45'5" 33d52'35"
Lisa Marie Harte
Per 4h49'39" 43d24'50"
Lisa Marie Hatcher
Cyg 20h51'15" 31d43'18"
Lisa Marie Iannarelli
Mon 7h19'9" -0d51'38"
Lisa Marie Irvine
Col 6h6'36" -29d7'32"
Lisa Marie Isaac
Psc 1h24'11" 16d49'41"
Lisa Marie Lyons
And 0h45'0" 25d36'50"
Lisa Marie " Maggie " Geary
Uma 11h37'52" 40d53'49"
Lisa Marie Marino *30th*
Vir 14h14'52" 5d9'11"
Lisa Marie Marshall
Ari 2h29'16" 24d52'39"
Lisa Marie Martinez
Cap 21h33'12" -21d40'21"
Lisa Marie Mettert
Sco 16h39'28" -37d18'54"
Lisa Marie Mosel Harrington
Lyn 6h17'18" 57d0'49"
Lisa Marie Muench
Ori 6h19'8" 16d22'47"
Lisa Marie Nichols
Hya 8h55'44" 4d23'38"
Lisa Marie Nowell
Vir 14h49'6" 2d0'58"
Lisa Marie Pallick
Leo 9h38'25" 28d37'39"
Lisa Marie Perez
Aqr 22h37'22" -0d49'28"
Lisa Marie Piper-Jackson
And 2h36'4" 43d38'55"
Lisa Marie Pirkel
And 22h59'35" 43d57'47"
Lisa Marie Plumhoff
Aqr 20h40'38" -2d13'53"
Lisa Marie Roberts
Cas 1h39'26" 60d34'19"
Lisa Marie Rochelle
Gem 6h41'27" 21d27'3"
Lisa Marie Roden
And 0h53'34" 40d37'30"
Lisa Marie Rubino
And 23h21'33" 44d38'23"
Lisa Marie Schmidt
Ari 2h15'27" 25d24'29"
Lisa Marie Sharp
Cnc 8h37'48" 7d19'39"
Lisa Marie Shore
Sco 17h34'3" -45d4'57"
Lisa Marie Sinclair Johnson
Uma 11h49'24" 33d17'26"
Lisa Marie Smith
Crb 15h35'0" 37d20'46"
Lisa Marie Stewart
Vir 12h19'22" 5d22'35"
Lisa Marie Unruh
Uma 9h58'48" 46d0'54"
Lisa Marie Vega
Cra 18h35'31" -38d46'48"
Lisa Marie Velez
Lyn 9h29'10" 40d0'35"
Lisa Marie Viera
Lyn 8h6'6" 41d12'7"
Lisa Marie Waraksa
Psc 0h18'47" 9d34'21"
Lisa Marie Wise - BaByCaKeS
Sgr 17h58'24" -17d14'7"
Lisa Marie Woods
Psc 23h30'3" 6d24'37"
Lisa Marie Woody Frilot
Uma 11h41'52" 43d30'18"
Lisa Marie Wyckoff
Leo 9h54'0" 27d41'9"
Lisa Marie Zerrlaut
Psc 1h25'44" 21d33'42"
Lisa Marie Zhou
Ari 2h52'5" 31d6'35"
Lisa Marie, My Munchkin
Aqr 22h32'50" -19d57'30"
Lisa Marie's Star
And 1h33'12" 46d39'23"
Lisa Marino
And 0h25'19" 42d19'46"
Lisa & Mark Null
Umi 16h28'2" 77d42'43"
Lisa Marmon and Lori Patzer
Gem 6h20'34" 21d37'53"
Lisa Martin
Vir 13h10'58" 8d4'58"
Lisa Martin
Cru 12h2'53" -56d26'53"
Lisa "Matchmaker" Church
Eri 3h50'29" -29d16'52"
Lisa & Matthew
Uma 9h3'57" 68d38'55"
Lisa Mattson
And 0h50'12" 40d51'20"
Lisa&Matty
Sco 17h38'57" -37d57'25"
Lisa May
And 2h36'3" 50d11'1"
Lisa McClary
Cas 1h4'31" 52d59'40"
Lisa McDonald
Umi 14h24'48" 73d22'23"
Lisa McFaul 230678
Cnc 8h47'21" 32d14'0"
Lisa McGill
Ori 6h14'17" 15d29'4"
Lisa McKnight
And 1h53'56" 46d12'29"
Lisa McTheny
Sco 16h15'30" -12d31'38"
Lisa Medeiros
Cas 0h0'42" 54d18'25"
Lisa Meinen 17.06.1935
Uma 14h10'51" 61d52'30"
Lisa Meisel
And 0h55'10" 40d24'44"
Lisa Mellows
Gem 6h11'41" 25d27'2"
Lisa Melnichuk
And 1h48'58" 46d42'18"
LISA - mi Stärnli am Himmel
Cep 20h39'54" 58d59'56"
Lisa Michele Chester
Cnc 8h45'58" 30d39'52"
Lisa Michelle
And 1h14'27" 38d1'1"
Lisa Michelle
Uma 9h32'43" 43d3'22"
Lisa Michelle
Cap 21h31'12" -17d12'19"
Lisa Michelle Burnham July 26, 1974
Leo 11h43'57" 24d52'29"
Lisa Michelle Croley
Cap 20h11'38" -9d11'50"
Lisa Michelle Gillette
Lyn 7h51'9" 57d11'42"
Lisa Michelle Martino-Roberts
Leo 10h20'16" 17d12'12"
Lisa & Mike
Lyr 19h17'33" 34d28'32"
Lisa & Mike Gaertner 3-9-91
Aql 20h4'27" -0d4'2"
Lisa Miller
Psc 0h23'12" -4d26'7"
Lisa Monique Tchume
Ari 2h21'54" 12d59'28"
Lisa Moore
Cas 0h42'32" 61d13'29"
Lisa -Murfee's Mom-Andrews
And 23h50'51" 37d12'8"
Lisa Muriel Avery
Vir 12h39'33" -10d10'9"
Lisa Murphree
Sgr 18h47'5" -18d39'56"
Lisa Murphy
Cyg 20h56'5" 30d34'10"
Lisa Murray
And 0h20'8" 32d6'53"
Lisa My Love
Sco 16h57'41" -42d38'39"
Lisa My One and Only
Gem 7h38'25" 20d37'26"
Lisa My You
And 23h27'41" 41d38'56"
Lisa N. Moffo
Tau 4h41'12" 27d34'5"
Lisa 'n' Robbie Dale's daughters
Uma 9h31'5" 53d48'24"
Lisa Najibe
Lyr 19h7'26" 26d25'2"
Lisa Naomi Levine
Vir 12h54'46" -5d20'52"
Lisa Neef
Peg 23h27'49" 19d45'25"
Lisa Newbold
And 23h33'46" 41d46'35"
Lisa Nicole
Crb 15h34'28" 29d33'9"
Lisa Nicole
Peg 21h49'52" 11d12'55"
Lisa Nicole
Uma 9h36'40" 53d50'57"
Lisa Nicole Brink
Cas 23h34'24" 51d51'51"
Lisa Nicole Cizek
And 23h25'43" 44d24'37"
Lisa Nicole Cooper
Lib 15h37'3" -29d42'43"
Lisa Nicole Henderson
Tau 5h56'41" 26d58'17"
Lisa Nicole Szucs
Sgr 18h16'25" -17d2'50"
Lisa Nininger Hale
Cnc 9h5'1" 17d0'1"
Lisa: November 20, 1966
Cas 23h52'32" 56d22'6"
Lisa Novick
And 23h11'53" 43d59'40"
Lisa O'Connor
Crb 15h39'36" 38d57'1"
Lisa Owen
And 2h13'45" 49d11'11"
Lisa Owens
Cyg 19h58'41" 30d35'59"
Lisa Oxenham
And 2h9'21" 40d25'39"
Lisa Palm-Joy
And 1h4'16" 34d45'29"
Lisa & Paris King
Cyg 20h22'49" 33d10'19"
Lisa & Peter Pookalooka
Per 2h22'35" 52d21'38"
Lisa Phelps
Psc 0h17'16" 12d15'2"
Lisa & Phil Martorana
Cyg 20h21'2" 46d40'2"
Lisa Pontorno
Vir 15h0'24" 3d57'16"
Lisa 'Poochie' Leong
Cap 20h31'57" -9d41'37"
Lisa Pooh Steward
Uma 8h35'18" 61d15'54"
Lisa Pougnet
Sgr 18h7'18" -24d4'10"
Lisa Powell
And 23h39'59" 34d36'52"
Lisa Prechel
Leo 9h43'38" 30d16'14"
Lisa(Precious)Sedwick
Cas 0h18'21" 52d14'41"
Lisa Prophet
And 23h25'3" 47d50'44"
Lisa R. Courtney
Vir 14h16'50" -2d42'38"
Lisa R. Larson
Cam 7h34'48" 75d8'4"
Lisa R. Portier, My Favorite Star
Cnc 8h49'54" 9d7'8"
Lisa R. Stevenson
Tau 4h19'20" 14d19'0"
Lisa Raiola
Ori 5h35'8" 6d59'0"
Lisa Ramsaroop
Leo 11h37'20" 18d45'31"
Lisa Rasmussen a.k.a. "Mrs. R."
Uma 10h45'43" 65d11'1"
Lisa Reid
Cru 12h14'16" -61d29'22"
Lisa Reinert
Uma 10h27'3" 49d7'47"
Lisa Rena Tisdale-Gonzalez
Cnc 8h43'41" 20d43'46"
Lisa Renae
Cas 2h1'32" 58d41'22"
Lisa Rene
Sco 17h43'11" -42d9'7"
Lisa Rene Hohn
Del 20h37'48" 15d56'5"
Lisa Renee
And 0h52'45" 41d34'13"
Lisa Rice
Gem 7h6'16" 27d53'13"
Lisa Richardson+Jeff = Love 4ever
Ori 6h13'43" 2d24'17"
Lisa Richmond
And 23h8'49" 47d47'16"
Lisa "ROCK" Barbosa
Lyn 8h34'58" 42d15'1"
Lisa Rodriguez
And 2h18'18" 49d24'39"
Lisa Rooke
Cru 12h28'22" -62d17'40"
Lisa Rooney
Cas 0h22'20" 63d49'13"
Lisa Rose
Psc 23h7'49" 0d14'11"
Lisa Rose
And 0h36'32" 24d13'19"
Lisa Rose Cyr Desrosiers
Uma 8h42'20" 58d39'13"
Lisa Rotter
Psc 1h45'49" 5d54'28"
Lisa Ruiz
Tau 4h38'10" 17d24'25"
Lisa S. Cataldo
And 2h19'6" 50d13'20"
Lisa S. Rahal
Sco 16h28'8" -28d26'34"
Lisa Sabatini
Uma 8h46'53" 51d51'38"
Lisa Sanders
And 23h8'58" 36d7'13"
Lisa Sanford
Leo 11h18'49" 0d37'10"
Lisa Schlatter
Leo 11h31'42" 12d47'27"
Lisa Schoyen
Psc 0h43'43" 17d0'25"
Lisa Schram
Vir 14h40'28" 4d7'1"
Lisa Schuster
Ari 2h19'17" 25d36'34"
Lisa Schwartz
Cam 3h51'52" 57d41'40"
Lisa Seaton
Gem 6h23'52" 26d46'21"
Lisa Shaffer
Lyr 18h41'21" 38d1'14"
Lisa Shala
Psc 1h20'33" 6d1'14"

"Lisa" Shannon Marie Melton Cato
Uma 10h9'9" 71d8'30"
Lisa Shaw
And 1h33'56" 43d31'9"
Lisa Sheff
Sco 16h14'21" -21d21'10"
Lisa & Shelby Gority
Vir 14h26'24" 3d17'35"
Lisa Shissler
And 2h23'47" 50d36'44"
Lisa Simms
Uma 9h45'35" 54d49'1"
Lisa Simone Cleveland
Sco 16h15'47" -9d32'25"
Lisa Slocum
Gem 7h28'3" 21d41'20"
Lisa Smith
Ari 3h12'53" 27d45'20"
Lisa Smith-McDonald
Gem 6h20'13" 22d26'14"
Lisa Solis Brown
Gem 6h4'3" 21d45'6"
Lisa & Sonia Gercie
Aqr 22h20'47" -15d54'54"
Lisa Sophia Shoemaker
Psc 23h13'52" 7d40'45"
Lisa Sparber
Cam 5h32'0" 74d21'13"
Lisa Spaventa
Cam 5h46'57" 61d36'40"
Lisa Spector
Vul 19h32'12" 25d5'4"
Lisa "Squiggly" Hayes McQueen
Cap 21h6'59" -15d51'47"
Lisa Star
Tau 3h39'16" 17d38'42"
Lisa Star Mommy Rebello
Crb 16h13'11" 32d12'41"
Lisa Star Rees
Cyg 21h13'19" 31d24'29"
Lisa "Stareyes" Depaola
Sco 17h41'24" -39d46'35"
Lisa "Starpie" Quirk
Uma 10h44'1" 72d32'37"
Lisa Stauffer "My loving wife"
Lyn 7h3'59" 46d22'27"
Lisa & Steve
Tau 5h47'39" 17d26'45"
Lisa Stewart
Crb 15h40'46" 29d32'19"
Lisa Stiver
Cas 0h58'54" 53d35'38"
Lisa Stuart
Tau 4h19'11" 26d17'7"
Lisa Sumlin
Lyn 7h15'36" 50d30'9"
Lisa Susan McLeod-Simmons
Gem 7h32'22" 26d0'11"
Lisa Susan Rachbind
And 0h14'26" 24d35'49"
Lisa Sutton
Cyg 19h44'42" 41d5'1"
Lisa Swann - Mummy
Cyg 19h43'25" 35d32'38"
Lisa T Yeh
Tau 4h50'55" 20d22'16"
Lisa Tabita
Uma 13h20'28" 61d12'54"
Lisa Taub
Tau 5h46'33" 26d48'58"
Lisa Taylor
Peg 22h42'12" 23d27'4"
Lisa the Angel of the Night
Ori 5h35'7" 10d37'26"
Lisa Thorpe
Lyn 8h39'30" 39d41'30"
Lisa Tilton
And 2h22'55" 42d45'4"
Lisa Tomkinson
Sgr 18h46'31" -17d54'19"
Lisa Tomlinson
Aqr 22h35'16" -1d29'15"
Lisa Tope
Cyg 21h20'25" 33d7'23"
Lisa Torre
Ori 5h50'22" 9d17'39"
Lisa Tuffnell
Her 17h1'28" 13d36'56"
Lisa Tuttle
Gem 7h20'3" 32d30'17"
Lisa "Tyger" Band
Cyg 20h50'27" 37d14'12"
Lisa Unwin
And 23h12'24" 43d28'11"
Lisa V.
Lmi 10h42'33" 23d45'33"
Lisa V. Sherratt
Cnc 9h15'44" 23d8'58"
Lisa V. Talareck
And 1h46'14" 44d13'9"
Lisa Vachovetz, Now and Forever
Cnc 8h47'56" 30d49'59"
Lisa Vasquez
Aqr 23h53'54" -14d15'0"
Lisa - Veritas Dea
Sgr 18h44'52" -22d57'52"
Lisa Vernacchio
Uma 10h28'50" 52d38'58"
Lisa Vestal
And 0h37'33" 42d8'38"

Lisa Victoria Cano
And 0h25'35" 43d54'54"
Lisa Victoria Tomasiello
Cnc 9h9'33" 19d27'24"
Lisa Vinayak
Vir 12h46'6" 5d52'35"
Lisa Wagner
Aqr 22h12'42" 0d37'50"
Lisa Walker
Aqr 22h11'32" -22d47'16"
Lisa Walls
Cap 20h20'45" -16d3'13"
Lisa Watts
Cas 1h8'24" 66d57'32"
Lisa "Wave Dancer"
Sco 16h36'5" -44d40'24"
Lisa Wentworth
And 0h31'42" 27d23'38"
Lisa Wilkinson Baker
Cam 4h15'46" 67d35'13"
Lisa Willbarger
Cas 1h43'18" 64d8'25"
Lisa Wilshaw
And 23h13'41" 41d29'51"
Lisa Winters
And 22h57'35" 51d15'22"
Lisa Witney
Leo 9h56'0" 21d8'56"
Lisa Yasi Seigel
Cet 2h22'31" -2d48'13"
Lisa Yasmin
Uma 10h0'5" 72d11'17"
Lisa Yvonne Clark
Cas 1h12'53" 53d46'23"
Lisa Zweber
Aqr 22h43'59" 1d44'41"
Lisa, Alex, Josh and Kyle
Umi 15h0'21" 69d18'0"
Lisa, Alex, Josh and Kyle
Uma 11h6'50" 62d30'46"
Lisa, Angel of the Heavens
Cas 23h11'19" 59d16'37"
Lisa, Peter, Callie, Griffin Hughes
Umi 14h13'38" 78d8'41"
Lisa, Sister of Dolly
And 0h36'1" 31d56'50"
LISA08132004
Cas 0h14'34" 50d33'36"
Lisa1113
Sco 16h46'15" -30d22'14"
Lisa-Ann
Cnc 9h11'26" 29d0'50"
Lisa-Ann Barajas
Ari 2h58'34" 24d55'10"
Lisa-Ann Ellen Rudd
Gem 6h41'42" 31d39'24"
LisaB
Gem 6h53'17" 22d20'47"
Lisabeth
Uma 9h31'42" 45d29'10"
LisaBrown4me4ever
Uma 8h44'16" 70d18'40"
Lisa-Chanel
Uma 11h42'50" 34d18'40"
Lisa-Chayla-Katelynn
Cyg 20h17'24" 46d37'17"
Lisa"Chicky" McTighe
And 2h23'51" 42d54'11"
LisaJ
Ori 5h40'37" 3d12'0"
LisaJanelle0202
Aqr 22h16'43" -14d39'15"
Lisa-Jayne Maguire
Vir 13h32'45" 12d15'50"
Lisala
Sgr 18h32'0" -16d1'7"
LisaLisa
Crb 15h43'24" 32d53'44"
Lisa-Louise Crisafulli
Aql 19h25'27" 3d48'14"
Lisa-Lover of Beauty
Uma 12h3'46" 59d45'41"
Lisa-Marie
Dra 18h33'38" 58d59'33"
LISA-MARIE
Uma 8h54'52" 67d41'22"
Lisamarie
Sgr 18h13'46" -34d29'22"
Lisamarie
Vir 13h50'11" 3d59'2"
LisaMarie
Ari 2h0'26" 18d4'55"
Lisamarie Ulman
Aqr 22h0'12" 0d41'3"
Lisandra
Lyr 18h50'19" 35d13'28"
Lisandro Depaula
Cap 21h41'11" -14d19'12"
Lisanne
Psc 1h21'35" 16d4'58"
Lisannette M. Ramirez
Tau 4h7'29" 5d51'37"
Lisanntenacity
And 0h42'22" 30d9'41"
LisaPeffley
Cas 1h44'6" 61d31'21"
Lisa..Perfect..Beautiful..Mine
Sgr 19h27'28" -21d42'57"
Lisarita
Ari 3h22'15" 28d59'23"
Lisa's Eyes
Cma 6h54'28" -31d4'23"

Lisa's Giraffe
Sgr 19h51'57" -30d51'23"
Lisa's Giraffe
Ari 2h44'46" 14d14'17"
Lisa's Glittering Snuggle Bug
Cap 20h57'29" -15d39'38"
Lisa's Guiding Light
Vir 12h32'8" -9d44'34"
Lisa's Joy
And 2h16'31" 46d38'40"
Lisa's Krümmelchen
Tau 5h3'24" 25d33'46"
Lisa's Last Unicorn Star
Mon 6h40'43" 7d28'38"
Lisa's Leap Of Faith
Umi 15h3'37" 79d13'42"
Lisa's Light
And 23h25'0" 51d56'57"
Lisa's Light 17-11-1970
Cru 12h39'44" -60d35'42"
Lisa's Love Forever
Tri 1h53'17" 30d48'7"
Lisa's Love Lucent
Cyg 19h48'40" 53d16'40"
Lisa's Mai
Uma 10h3'15" 72d3'1"
Lisa's Mom
Ari 2h49'11" 28d27'1"
Lisa's Olivia
And 0h22'5" 40d21'1"
Lisa's Ray's
Cas 23h32'13" 53d12'3"
Lisa's Shining Star
Aqr 22h10'19" 1d37'48"
Lisa's Smile
Cyg 19h11'35" 54d8'7"
Lisa's Smile
Aqr 21h17'33" -10d12'7"
Lisa's Soldier Boy
Boo 14h15'52" 49d19'32"
Lisa's Sparkle
And 23h53'48" 35d3'8"
Lisa's Star
And 1h54'48" 43d47'13"
Lisa's Star
Leo 9h39'9" 31d19'21"
Lisa's Star
And 22h59'38" 50d15'23"
Lisa's Star
And 0h35'45" 45d4'45"
Lisa's Star
Cyg 21h47'2" 41d19'52"
Lisa's Star
Peg 21h37'43" 13d11'46"
Lisa's Star
Leo 11h49'30" 20d50'42"
Lisa's Star
Ari 1h56'19" 24d8'42"
Lisa's Star
Uma 8h46'56" 61d7'15"
Lisa's Star
Sco 17h45'13" -37d50'50"
Lisa's Star
Sco 17h57'27" -39d45'0"
Lisa's Star
Sco 16h23'39" -26d45'27"
Lisa's Star - Forever With Us
Cru 12h26'24" -56d45'19"
Lisa's Star. Love Forever, Stu xxx.
Cyg 21h40'31" 44d4'33"
Lisa's Stardust
Ari 2h55'34" 25d27'10"
Lisa's Starr
Sgr 19h34'42" -24d46'51"
Lisa's White Light
Sct 18h46'27" -5d31'32"
Lisa-Seesi
And 23h43'30" 34d17'29"
LisaStar 8151971
Leo 9h32'23" 28d57'26"
LisaSteve
Cra 17h59'40" -37d28'47"
Lisbet Kline
Uma 9h21'6" 45d5'47"
Lisbeth
Leo 9h31'6" 25d56'54"
Lisbeth Armentano
Umi 10h50'0" 87d52'54"
Lisbeth "Gummie Bear" Garcia
Lib 14h52'42" -10d49'57"
Lisbeth Höhener
Uma 10h40'42" 39d34'13"
Lisbeth My Lady Carrera
Cas 1h8'1" 51d12'9"
Lisbeth Ocondi
Cap 21h53'24" -11d5'14"
Lisbeth Studer
Cas 0h22'46" 55d13'58"
Liscea
And 0h40'40" 26d16'24"
Lise'
Aql 19h51'48" -0d58'2"
Lise E
Psc 0h16'39" 19d49'31"
Lise Joyal
Psc 23h25'27" 2d30'59"
Lise L. Langley
Aql 19h58'58" -0d25'14"
Lise Lemaire
Cyg 21h44'23" 43d57'59"

Lise & Léo Huber
Uma 9h44'7" 47d40'55"
Lise Lippé
Cas 3h28'22" 68d34'9"
Lise Lucienne Ouellette Jarvis
Vir 11h41'54" 8d50'3"
Lise Raspail
Leo 11h36'4" -4d21'36"
Lise Stern
Mon 6h36'5" 10d47'47"
Lise Yvonne Melnyk
Lyn 8h37'16" 37d59'27"
Lise, the one love for matty
And 0h8'57" 38d28'49"
Lisee
Vir 12h27'5" -1d45'0"
LiseF030304
Tau 3h40'21" 16d0'3"
Liselotte and Peter Friedman
Cyg 20h30'16" 57d21'44"
Liselotte Zadrawez, 07.08.1947
Cam 5h27'21" 69d33'23"
Lisett
Uma 13h44'54" 52d39'46"
Lisetta
Cnv 12h39'5" 39d52'16"
Lisette
Umi 15h56'0" 71d57'51"
Lisette and Rafael Crespo
Cyg 20h23'57" 46d9'11"
Lisette Estrada
Aqr 22h17'26" -0d46'47"
Lisette Santana Archondo
Crb 16h8'25" 38d9'23"
Lisey Broom
And 1h49'10" 46d45'26"
Lish
Uma 12h8'38" 55d49'0"
Lish Family Star
Lyn 6h48'34" 54d16'15"
Lisha (Kay Kay) & Jon (Yeung Yeung)
Sco 16h1'7" -20d48'35"
Lishan Olivia Brasile
Lyn 7h25'31" 44d48'38"
Lisi
Sco 17h49'50" -36d19'35"
LisiMM
And 2h37'9" 44d3'15"
LiskanDaniel
Eri 4h35'49" -15d44'30"
Lisle's Star
Aqr 21h31'52" 0d25'41"
Lisnik Berger
Uma 12h6'13" 46d42'20"
L'isola che non c'è dei bimbimeden
Per 3h3'43" 51d39'12"
Lison Léna Leroux Mailly
Aqr 22h11'55" -7d30'6"
Lispetharjam
Uma 12h5'43" 38d1'53"
Lissa
And 23h37'17" 45d7'34"
~Lissa
Ari 2h46'4" 15d17'47"
Lissa
Cnc 8h41'16" 27d45'7"
Lissa
Lib 15h39'34" -19d11'30"
Lissa and Ray James
Cyg 20h20'49" 53d29'44"
Lissa Lain Coffman
Uma 10h14'38" 56d56'39"
Lissaelle
Her 17h14'45" 26d47'19"
Lissandra Marie Salinas
Cyg 20h12'8" 35d58'21"
Lissa's Light
Gem 6h39'45" 27d8'3"
Lissbeth Rumbaut
And 0h18'2" 30d8'18"
Lisse in the sky with diamonds
Uma 11h27'10" 48d45'8"
Lisset Garza
And 23h32'1" 41d22'5"
Lissett Vargas
Sco 17h15'34" -32d18'43"
Lissette Martinez
And 0h48'40" 40d46'25"
Lissette's Star
Cas 0h39'13" 51d12'25"
Lissia
Aqr 23h24'32" -18d36'2"
Lissie
Ari 3h17'25" 28d29'41"
Lissithan
Mon 7h32'2" -0d57'44"
Lißmann, Uwe
Lib 15h45'21" -8d12'44"
Lissy Poo
Sgr 19h20'56" -26d20'51"
Listelle Ignacio
Uma 9h38'27" 44d54'58"
Listori
Her 17h27'23" 27d32'1"
Lita Alvarez
Uma 11h16'44" 45d24'14"
Lita Cohen
Cas 2h38'36" 67d24'8"

Lita & Franko Forever
Cyg 19h47'49" 46d47'40"
Lita's Star
Tau 5h37'20" 24d53'38"
Lite Of God Susan Hoover
Eri 4h31'14" -22d54'27"
Lit'le Unicorn
Psc 1h10'2" 17d33'41"
Lit'lin
Ori 5h35'45" 0d34'57"
Little
Umi 14h18'5" 73d6'17"
Little
Umi 16h21'21" 75d43'55"
Little A and Little B
Umi 15h1'31" 86d10'4"
Little Ang
Cnc 9h6'51" 25d59'52"
Little Angel
Lyr 18h40'6" 37d53'38"
Little Angel
Lyr 18h48'12" 37d6'16"
Little Angel
Umi 15h18'50" 73d29'36"
Little Angel Autumn Taylor
Lyr 18h49'8" 32d57'5"
Little Angel Kaitlyn Nicole Turner
Leo 11h53'11" 25d18'5"
Little Angel Orth
Cyg 20h32'9" 39d9'7"
Little Angel Riley
Cnc 8h27'31" 26d40'48"
Little Angel Robin
Uma 11h32'4" 47d54'38"
Little Angel, Georgia Elise Barrett
And 2h35'38" 49d6'27"
Little Angels
Umi 13h10'37" 76d19'42"
Little Anne
Leo 10h13'50" 27d29'17"
Little Apple
Cap 20h40'17" -18d3'36"
Little Baby Carter
And 0h36'8" 33d40'25"
Little Baby Cutter
Umi 15h23'51" 85d51'21"
Little Ball & Little Lips Love
Sgr 18h46'57" -25d35'15"
Little Bambina
Cmi 7h30'23" -0d4'8"
Little Bay Hotel
Dra 16h21'29" 65d4'26"
Little Bear
Umi 15h57'46" 72d26'58"
Little Bear
Umi 15h16'16" 70d17'33"
Little Bear
Sco 16h2'32" -11d12'17"
Little Bear Brandon
Umi 15h35'45" 74d52'52"
Little Bear - Jennifer Anne Clark
Umi 17h7'33" 76d15'0"
Little Beebs
Dra 12h16'42" 70d28'10"
Little Ben
Umi 13h5'34" 72d16'22"
Little Bethy
Aql 19h2'29" 5d33'59"
little BIG
Umi 14h23'34" 73d23'14"
Little Big Man
Umi 13h39'26" 74d8'0"
Little Bill
Dra 17h52'41" 67d6'14"
Little Billy
Umi 16h8'30" 74d9'28"
Little Birdie Jenny
Crb 15h38'25" 28d58'1"
Little Birdy
Cap 21h37'35" -16d20'53"
Little Bit
Sco 16h5'56" -9d46'6"
Little Bit
Umi 14h31'19" 74d57'11"
Little Bit
Umi 13h58'11" 71d24'26"
Little Bit
Vul 19h23'55" 25d59'39"
Little Bit
Psc 1h22'33" 27d47'46"
Little Bitty
Umi 16h8'59" 73d16'52"
Little Bitty
Cap 21h52'14" -18d31'13"
Little Black Eagle
Umi 10h17'24" 89d17'54"
Little Bobby Golas
Vir 13h21'7" 8d13'30"
Little Bochicho
Cnc 8h35'40" 7d45'6"
Little Boo Jeanna
Lmi 10h6'57" 34d41'56"
Little Boomba
Vir 13h5'32" -16d12'9"
Little Boy
Ori 6h5'29" 16d33'6"
Little Brians Star of Courage
Ori 6h11'38" -1d28'15"
Little Brick
Aql 19h13'6" -0d0'18"

Little Britches
Lib 15h8'6" -14d37'7"
Little Bro
Aqr 22h31'13" -2d7'43"
Little Bro
Uma 11h2'34" 38d53'19"
Little Brother Flynn
Aqr 23h5'35" -6d22'4"
Little brother Nero
Cep 23h19'40" 84d0'19"
Little Brother - Phillip
Leo 10h52'11" 10d1'36"
Little Brother with Big Dreams
Her 17h25'37" 39d9'49"
Little Bubba
Leo 10h48'34" 9d17'17"
Little Bud
Tau 5h16'46" 24d59'11"
Little Buddy
Her 16h44'19" 24d45'42"
Little Buddy
Leo 9h25'37" 18d32'9"
Little Buddy Ty
Her 18h41'12" 15d56'14"
Little Buffalo
Dra 16h54'24" 57d26'57"
Little Bug
Uma 9h28'25" 67d3'43"
Little Bug
Vir 11h55'2" -1d22'47"
Little Butch
Ori 5h50'23" 9d48'17"
Little Cal
Umi 14h27'40" 70d22'21"
Little C.C.
Umi 14h44'49" 71d30'40"
Little Champion
Her 17h12'16" 15d48'12"
Little Charlie
Sgr 18h7'38" -32d3'0"
Little Chicken Pleasure Saro
Sco 16h15'43" -10d32'16"
Little Chim Chim
Uma 10h15'5" 44d18'46"
Little China Doll
Her 18h47'8" 17d49'14"
Little Cindi Star
Aqr 21h48'57" -1d20'8"
Little Corey
Umi 15h35'15" 80d28'54"
Little Curly Finlay's Cradle
Cru 12h54'26" -57d45'47"
Little Currant
Uma 12h3'36" 57d0'34"
Little Cyn 58
Lib 14h44'28" -17d18'9"
Little Danielle
Uma 11h21'0" 40d26'51"
Little Darlin
Her 16h57'4" 36d14'35"
Little Darling
Umi 14h33'39" 77d46'2"
Little Dave
Her 16h23'56" 11d13'39"
Little David
Umi 14h24'56" 71d46'7"
Little Debbie
Vul 19h35'24" 26d20'1"
Little Debbie
Cas 0h10'38" 58d59'21"
Little Debbie Cakes
Umi 15h24'15" 75d38'11"
LITTLE DEER FOOT
Boo 14h47'18" 35d40'11"
Little Dibber
Ori 6h3'29" 6d48'50"
Little Dreamer
Umi 16h2'6" 73d58'40"
Little Dreamer Chance
Ari 3h7'34" 14d20'46"
LITTLE DUCK ALWAYS
Lyr 19h3'25" 34d21'37"
little e
Leo 10h30'21" 15d11'32"
Little "E"
Umi 15h18'42" 72d18'49"
Little E
Sgr 19h15'48" -20d41'16"
Little Edie
Cas 2h49'29" 60d8'4"
Little Ellen
Cam 7h38'58" 64d49'28"
Little Emilie
And 22h57'54" 35d49'20"
Little Ern
Cam 9h31'35" 82d33'5"
Little Ernie
Aqr 20h38'57" 1d52'14"
Little Eskimo
Tau 5h12'32" 20d5'57"
Little Fighter
Lib 15h30'25" -7d4'32"
Little Finn
Boo 14h31'29" 34d43'19"
Little Fish
Aqr 22h31'36" 0d57'42"
little flower
Crb 15h25'49" 28d10'30"
Little Flower School Reif '06
Cma 6h34'35" -17d0'15"
Little Foot
Ori 6h18'55" 15d2'28"

Little Foot and Pookey Shining High
Lib 14h25'54" -13d19'3"
Little Fox
Vul 20h15'6" 22d54'38"
Little Frannie Goonie
Cas 23h28'1" 57d28'13"
Little Fur Seal
Umi 14h27'27" 74d46'34"
Little G
Umi 15h59'17" 70d13'12"
Little G
Umi 15h19'30" 69d28'10"
Little Gecko (Matt's Star)
Leo 9h28'18" 26d14'25"
Little Gem
And 0h22'1" 27d9'42"
Little Gem
Gem 6h56'50" 14d38'8"
Little Gem
Dra 17h51'24" 68d4'43"
little germangirl Doreen
Cas 1h43'52" 67d59'13"
Little Gina
Dra 15h30'10" 56d58'2"
Little Ginger Renee Smith
Cnc 8h53'46" 25d21'42"
Little Ginny (Virginia)
Aql 19h7'12" 14d40'41"
Little Girl
Peg 21h38'1" 25d5'39"
Little Girl
Uma 12h2'59" 61d29'48"
Little Glenny
Uma 8h13'34" 65d9'2"
Little Glowworm
Uma 10h55'53" 53d22'25"
Little Good Harbour
Umi 14h20'56" 74d50'17"
Little Grandpa
Uma 11h40'10" 57d34'48"
Little Grrrr Boy
Ari 2h11'4" 25d44'48"
Little Hal
Ori 5h41'6" -0d21'58"
Little Harry
Uma 8h13'12" 72d23'30"
Little Heater
Ori 5h35'58" 1d11'52"
Little Heaven
Cnc 8h19'56" 29d55'25"
Little Hershey Kiss
Gem 6h43'50" 26d16'23"
Little Hottie
Sgr 18h6'24" -26d40'42"
Little Imi
And 23h16'14" 52d6'30"
Little j- Jackson Carter Boers
Aqr 20h40'44" -11d40'45"
Little Jalei Grace
Sco 16h34'33" -34d1'36"
Little James
Psc 1h39'46" 16d0'35"
Little JB
Vir 12h42'13" -2d16'13"
Little Jess
Vir 12h24'57" 8d36'29"
Little Jimmy
Uma 9h23'29" 52d33'45"
Little Jimmy Johnson
Cnc 8h26'44" 22d33'11"
Little Jo
And 1h20'34" 43d38'5"
Little Joe
Cyg 20h10'39" 56d23'37"
Little Joey
Vir 13h50'47" -5d11'29"
Little John's Angel Star
Cyg 22h1'52" 45d14'5"
Little Jose
Equ 21h18'45" 9d12'48"
Little Juapaleno Cakes
Aur 5h14'38" 40d42'7"
Little Katie
Umi 14h21'2" 74d7'25"
Little Kelin Star
Aqr 22h37'43" -2d10'3"
Little Kiki
And 0h50'27" 41d24'37"
Little King
Vir 11h53'59" 6d13'49"
Little King and the Mermaid
Gem 7h37'26" 19d41'44"
Little Krazi
Aql 19h43'33" -0d0'15"
Little Kyle Sunshine 2
Cap 21h29'42" -9d27'8"
Little Lady
And 0h31'14" 36d47'59"
Little Lady Bug Allison
And 1h15'44" 37d3'48"
Little Lady Roxanne
Col 5h46'49" -32d34'45"
Little Laura
Sgr 19h18'4" -23d27'49"
Little Laura
Aql 19h25'19" 5d51'20"
Little Lee
Uma 10h18'27" 49d53'56"
Little Leesey
Ari 2h37'13" 17d53'4"
Little Legs
Umi 15h48'58" 70d13'40"

Little Leionna's Light
And 23h49'18" 42d44'30"
Little Lenny
Umi 14h15'18" 69d8'58"
Little Leo
Ari 2h16'10" 24d58'57"
Little Light Paquette
Lyn 7h4'45" 50d19'27"
Little Lil and G.T.
Cyg 21h15'39" 46d14'53"
Little Lilly Putin
Ari 3h5'29" 18d45'44"
Little Linda's Star
Leo 9h30'33" 27d35'11"
Little Lindsey Loo Who
Aqr 21h50'35" 2d1'32"
Little Lindz
Psc 1h21'50" 32d22'33"
LITTLE LION
Boo 13h46'4" 23d26'17"
Little Lisa
And 1h57'52" 46d31'42"
Little Lisa
Sgr 18h11'50" -18d9'50"
Little Lisa in the Sky
Cap 20h30'9" -10d53'15"
Little Little Princess
Vir 12h6'11" 6d41'59"
Little Lizzy
Vir 14h22'8" 4d56'7"
Little Llama
Umi 16h44'48" 77d17'3"
Little Lu
Cyg 20h32'45" 59d13'42"
little lucKy
Psc 1h11'25" 28d4'26"
Little LuLu
Tau 5h33'9" 23d43'39"
Little Mac
Leo 10h19'9" 26d13'7"
Little Mac
And 2h37'24" 43d30'39"
Little Mae
Col 6h10'22" -33d41'28"
Little Magic
Cas 1h54'12" 68d46'12"
Little Mama
Umi 14h14'42" 75d19'13"
Little MAMA
Aqr 22h32'18" 2d29'22"
Little Man
Her 17h54'12" 26d22'12"
Little Man
Lmi 10h5'35" 35d2'49"
Little Man
Cap 20h48'31" -16d28'28"
Little Man
Lyn 6h54'44" 57d1'55"
Little Man - Connagh Jay
Uma 9h36'12" 46d58'14"
Little Man Dan
And 23h6'0" 51d1'33"
Little Man Sebastian
Umi 14h13'11" 72d17'33"
Little Man's Star
Psc 1h13'15" 15d12'43"
Little Maria
And 0h6'45" 34d21'22"
Little Mark
Umi 21h6'3" 88d49'49"
Little Master Chang Ka
Ming Dennis
Leo 11h29'48" 4d30'36"
Little Max
Aur 6h53'57" 37d41'35"
Little Maya
Sco 16h12'36" -17d47'23"
Little Me
Cmi 7h30'16" 4d32'56"
Little Meg
Cas 23h37'6" 52d8'54"
Little Mia
Sgr 19h22'3" -24d30'48"
Little Miho
Tau 4h26'1" 10d12'50"
Little Miss
And 23h11'1" 42d57'52"
Little Miss Anna Sanzone
Sco 17h57'59" -38d40'51"
Little Miss Bianca
Col 6h11'53" -34d14'7"
Little Miss Blue Eyes
Marlene Duhon
Cnc 8h18'58" 13d53'52"
Little Miss Christina Citrola
And 0h23'17" 30d19'46"
Little Miss Cuccolo
Lib 15h29'18" -6d25'55"
Little Miss DeSoto
And 1h31'26" 45d28'55"
Little Miss Emma
Sco 16h48'19" -30d53'16"
Little Miss Holly
Cap 21h30'11" -13d32'19"
Little Miss Sassy Strokker
Uma 11h23'27" 42d12'42"
Little Miss Sunshine
Aqr 22h29'56" 2d6'51"
Little Missy
And 23h35'21" 41d15'6"
Little Molly Thompson
Cyg 21h18'40" 43d17'56"
Little Moments
Crb 15h43'18" 31d58'46"
Little Moments
Cyg 19h45'23" 31d54'23"
Little Monkey - Curt Agee
Uma 9h45'31" 54d3'35"
Little Mrs Fiddy Lisa Muro
Cnc 8h49'5" 31d38'35"
Little Muffin
Lib 15h3'17" -5d49'51"
Little Muse
And 23h22'35" 35d36'43"
Little Nana Gleason
Cyg 21h32'36" 40d55'46"
Little Nancy
Cap 20h43'39" -22d5'53"
Little Nemo
Crb 16h16'51" 38d10'5"
Little Nick
Aqr 22h23'47" -6d7'18"
Little Nicky
Aqr 22h32'5" -0d31'1"
Little Old Winemaker
Per 3h1'41" 53d51'30"
Little One
Her 17h51'6" 42d20'23"
Little One
Ori 6h2'36" 4d34'39"
Little One
Ari 2h2'36" 17d31'11"
Little One
Tau 5h53'42" 26d36'3"
Little One
Umi 13h12'50" 86d32'42"
Little One
Lib 15h7'6" -19d11'1"
Little One
Umi 14h4'20" 69d49'46"
Little One
Umi 15h16'10" 72d22'8"
Little One
Uma 8h54'57" 63d38'43"
Little One Mary
Sco 16h46'33" -40d49'26"
Little Pam Cobalt
Aqr 22h33'1" 1d37'27"
Little Patriots Day School
Umi 15h31'5" 76d59'17"
Little Patty
Lib 15h32'43" -11d27'44"
Little Peach
And 2h3'36" 38d7'30"
Little Peanut
Umi 14h15'53" 67d43'41"
Little Peet Star
Aqr 22h9'21" 0d31'25"
Little Penguin
Umi 14h41'5" 74d1'48"
Little Pepper
Ori 6h2'19" 5d9'27"
Little Phil
Uma 8h19'22" 67d31'19"
Little Phillip Lark from
Chelsea St
Cru 12h31'1" -63d11'13"
Little Pie
Cnc 7h59'7" 19d38'48"
Little Piggy in the Sky
Lyn 6h51'31" 53d13'30"
Little Prince
Psc 0h56'7" 31d28'35"
Little Prince Cooper
Leo 11h32'41" -0d23'30"
Little Prince of Costas &
Magdalene
Cep 21h23'45" 60d59'3"
little Prince & Princess
Dra 9h44'6" 74d21'0"
Little Prince Star
Vir 12h50'10" 4d57'13"
Little Princess
Del 20h24'7" 10d9'39"
Little Princess
Tau 5h34'45" 21d47'37"
Little Princess
And 23h56'10" 35d13'52"
Little Princess
Uma 10h41'35" 66d36'36"
Little princess ANI
Aqr 21h42'22" 1d59'13"
"Little Princess" BLP ~ May
11, 2004
And 0h9'53" 35d9'38"
Little Princess Buski
Leo 11h29'59" 8d41'26"
Little Princess Jennifer
And 0h11'16" 46d0'46"
Little Raj
Ori 6h12'4" 17d47'57"
Little Rand
Cap 20h22'31" -11d48'2"
Little Rascal & The Queen
(KPB/NMG)
Mon 7h32'6" -0d44'36"
Little Red
Lmi 9h59'32" 36d18'26"
Little Red Dragon
Vir 14h14'49" -2d47'29"
"Little Reed"
Lyn 7h55'50" 39d12'49"
"Little Rew"
Lyn 6h59'4" 50d30'42"
Little Ricky
Uma 10h13'32" 49d22'24"
Little Ricky
Cep 23h2'38" 70d18'21"
Little Rinka
Psc 1h18'56" 25d41'5"
Little Rita
Umi 15h32'50" 71d13'20"
Little Roo
Her 17h22'21" 48d8'17"
Little Roo
Lmi 9h57'5" 39d21'7"
Little Rose
Pho 23h53'43" -45d29'5"
Little Ryan
Lmi 10h22'5" 35d22'10"
Little Saint
Uma 12h8'7" 62d26'53"
Little Samantha Star
Ari 3h16'3" 20d23'21"
Little Sampson
Her 18h33'53" 20d13'22"
Little Saoirse
And 1h46'49" 42d58'38"
Little Sara
Cas 23h48'41" 52d51'9"
Little Sarah Jayne
Sco 17h55'6" -37d5'17"
Little Shine Brite For
Corine&David
Umi 16h18'6" 71d30'25"
Little Shnoo / Little Shnee
Leo 9h55'59" 18d11'30"
Little Sis Jayna
Gem 6h51'7" 20d9'11"
Little Sister
Lib 15h5'57" -9d40'9"
little sister star
Cap 20h22'28" -11d0'1"
Little Slick
Dra 16h38'14" 63d53'11"
Little Snowy Palko
Cam 7h4'15" 77d13'27"
Little Soft Motz
Tau 5h45'28" 16d44'12"
Little Spink
Umi 16h18'2" 71d2'33"
Little Spittle
Uma 14h22'35" 59d37'46"
Little Splott (never forgot-
ten)
Umi 13h13'16" 69d27'57"
Little Spooky
Cyg 19h51'17" 30d36'22"
Little Spoon
Per 3h7'48" 52d28'23"
Little Spoon 23
Uma 11h16'13" 57d49'29"
Little Star aka "Squirt"
Umi 15h15'39" 69d27'50"
Little Stevie's Wonder
Aqr 23h24'50" -19d30'45"
Little Su
And 0h24'58" 38d33'22"
Little Sue
Uma 10h33'49" 48d16'45"
Little Superstar Laird
And 0h47'4" 37d5'35"
Little Susie
Tau 5h34'33" 21d40'17"
Little T my shining star
Tau 5h47'12" 22d28'7"
Little Teddy Poe
Aqr 22h7'20" -16d48'25"
Little Tierney Grant
Umi 15h18'9" 71d36'42"
Little Tiger
Ori 5h30'12" -0d28'35"
Little Tigrus
Leo 10h11'46" 12d49'43"
Little Timmy
Cap 21h6'8" -16d1'11"
Little Tin
Cmi 8h1'34" -0d4'13"
Little Tinker
Lib 15h51'49" -9d31'17"
Little Treva
Leo 11h53'8" 20d43'11"
Little Tucker Boy
Umi 13h39'32" 69d43'29"
Little Turtle
Aqr 22h40'41" 1d58'58"
Little Tyler
Vir 12h44'1" 2d35'54"
Little Vesuvio
Cyg 21h53'13" 38d26'3"
Little Whiskey Girl
Aqr 22h13'43" -13d3'5"
Little Wing
And 0h22'38" 45d40'53"
Little Wing
Gem 6h24'41" 17d29'3"
Little Woody
Boo 14h37'28" 18d41'27"
Little Yolushka
Vir 13h22'7" 13d34'36"
Little Zebbie
Tau 4h15'32" 27d36'9"
Little Zen
Umi 13h34'12" 70d8'32"
Littledipper
Umi 14h21'18" 67d6'39"
Littlee
Umi 15h0'16" 78d34'1"
littlehempstead
Cap 20h58'49" -21d52'40"
Littlekatpaw
Ari 2h12'4" 25d33'4"
Littlelittlestar Shirley
Sco 16h56'14" -41d54'3"
Littleone's Constellation
Dra 17h44'6" 55d48'3"
Littlest Angel
Lyr 18h53'14" 35d52'51"
Littlestar Mandolin
Aqr 22h41'36" -11d5'27"
Littleton
Cyg 20h14'58" 59d25'52"
LittleZoca
Cas 23h48'4" 51d55'48"
Littlun
Sco 16h13'52" -13d47'38"
Litton Burns
Col 6h1'36" -34d49'42"
LITWILER
Aur 5h55'12" 42d4'17"
Litxia Dianna Miranda
Gem 6h46'52" 25d52'29"
Liu Dai
Cep 22h58'57" 72d14'58"
Liu Dan Qing
Cap 20h30'27" -16d44'27"
Liu Ling
Cnc 8h35'29" 7d57'50"
Liu Tao and Song Jia Yang
Aqr 21h5'41" -8d47'30"
Liu Xuanming
Uma 11h37'42" 47d23'57"
Liu Yan
And 1h13'55" 45d2'8"
LIUBOMIRA
And 23h18'47" 47d23'31"
Liusa
Cyg 20h7'55" 36d9'38"
Liv
Cnc 7h57'37" 13d4'51"
Liv Carrington Eidsness
Gem 7h47'18" 32d20'12"
Liv Edna Spellerberg
Boo 14h51'51" 37d31'53"
Liv Gaard
Aur 5h21'2" 44d51'50"
Livange
Ori 6h6'8" 11d30'40"
Live And Let Die
Tau 5h45'10" 16d45'25"
Live in Love Forever
Cyg 20h30'3" 47d51'3"
Live Laugh Love, Roy +
Sandy Bamber
Cap 21h40'41" -15d37'51"
live the passion
Cap 21h1'29" -16d28'11"
Live Your Dreams
And 2h8'10" 41d44'19"
Live, Laugh, Love
Vir 11h47'58" 5d27'43"
live.laugh.love
Cra 18h55'38" -39d59'38"
Liverbones
Ori 6h2'26" 14d15'27"
Livers
Her 16h15'34" 25d20'4"
Livestrong Mikey
Cnc 8h55'18" 22d34'14"
Livi
Tau 3h25'31" 17d0'22"
Livi Browne
Sgr 19h16'22" -15d7'48"
Livia
Mon 6h17'33" -4d12'15"
Livia
Lyn 6h43'33" 53d46'24"
Livia
Tau 5h48'37" 16d53'35"
Livia
Cas 1h10'4" 51d19'19"
Livia and Andy's
Lib 14h37'54" -10d56'37"
Livia Bianca
Uma 11h19'20" 28d50'2"
Livia Caliopi
Peg 21h31'59" 15d52'16"
Livia Del Bue
Crb 15h21'30" 26d23'48"
Livia Elsa Linnea
Ari 2h39'51" 12d50'50"
Livia Estelle Fields
Lmi 10h49'33" 26d58'56"
Livia Hurrell
And 23h59'7" 45d11'14"
Livia Jana
Uma 9h20'50" 51d58'58"
Livia "Muchtuk" Kohn
Psc 1h37'51" 20d55'38"
Livia Müller
Cep 22h12'33" 68d11'14"
Livia Trimiño
Cas 0h31'48" 57d35'10"
Liviette Lomboy
And 1h20'21" 37d56'38"
Living Memory
Psa 21h34'12" -36d9'51"
Living Stones
And 1h27'12" 36d53'51"
Living the dream
Cyg 21h0'47" 54d47'54"
Living The Dream
Lib 14h52'3" -3d33'19"
Livingstar
Ari 2h39'12" 20d11'20"
Livingston
Psc 1h17'2" 16d31'6"
Livingston
Per 3h40'47" 43d51'17"
Livingston's Ebony Lady
Uma 11h3'26" 52d13'14"
Livio
Ori 6h6'6" 6d45'58"
Livio Pasquini
Uma 10h9'11" 60d48'44"
Livi's Star
Psc 1h24'40" 14d42'28"
LiWi
Uma 11h29'9" 38d46'16"
Lixa Maria
And 0h35'0" 32d22'35"
Li-Ya Chan
Lyr 19h14'37" 26d28'32"
Liyola
Tau 4h34'56" 26d26'17"
Liz
Tau 4h37'5" 28d43'32"
LIZ
Leo 9h26'54" 25d54'33"
Liz
Gem 6h39'23" 20d45'2"
Liz
Psc 23h46'46" 6d16'18"
Liz
Uma 10h40'56" 42d54'6"
Liz
And 1h18'31" 39d19'11"
Liz
Cyg 19h57'3" 35d34'9"
Liz
And 23h11'38" 47d37'45"
Liz
Lyn 7h11'34" 45d41'55"
" Liz "
Cyg 19h47'50" 39d11'17"
Liz
Aqr 22h32'18" -0d27'44"
Liz
Lib 15h34'14" -10d2'14"
Liz
Lib 15h51'51" -17d36'7"
Liz
Cap 20h36'51" -26d2'26"
Liz A. Wilcox
And 1h30'55" 42d35'37"
Liz and Brian
Per 3h9'32" 54d6'1"
Liz and Lou - Oct 30, 2004
Lyr 18h30'14" 36d6'45"
Liz and Mike
Gem 7h6'33" 22d0'22"
Liz Athey
Uma 11h23'29" 33d47'32"
Liz B. Norman
And 2h6'45" 43d4'27"
Liz Bartram
Uma 14h18'55" 59d20'54"
Liz Bat
Peg 21h46'31" 25d48'51"
Liz Buttimer.
Crb 15h31'26" 36d37'33"
Liz Cottrell
Cyg 19h53'50" 33d20'50"
Liz Crick
Cas 0h4'21" 52d59'5"
Liz Davis
Uma 11h12'47" 62d11'7"
Liz DeLaLuz
Cas 23h13'23" 59d17'0"
Liz Ellis
Sgr 18h17'37" -16d19'47"
Liz & Eric's Love Star
Cyg 20h26'20" 57d43'35"
Liz Gallaher
Ari 2h7'54" 12d44'50"
Liz Geoghegan
Vir 13h10'50" 9d44'45"
Liz Hatfield
Cap 20h36'33" -18d10'32"
Liz Hill
Tau 5h26'20" 28d23'6"
Liz Holland
Cnc 9h15'8" 31d2'56"
Liz Iannone
Cas 23h38'16" 56d13'49"
Liz & Jonas Forever
Leo 11h28'21" 4d42'36"
Liz Konzen
Uma 10h48'43" 44d49'26"
Liz Koppelman
Lyn 8h2'27" 42d47'50"
Liz Kowalski
Lyn 6h29'49" 54d42'46"
Liz Kreutzer
Sco 17h55'8" -32d32'45"
Liz ~Lizziebeth~ Inglis
Cra 18h14'4" -40d54'51"
Liz Lord
Cas 0h27'29" 58d43'1"
Liz Loves Shane Always
Cyg 19h32'35" 52d4'27"
Liz M.B.
Cam 4h43'55" 72d40'56"
Liz McGowen
Mon 7h34'49" -4d43'5"
Liz&Mikey
Tau 5h28'1" 27d19'55"
Liz Nickerson
Aqr 23h45'42" -10d53'43"
Liz Paez
Lib 15h18'21" -5d43'50"
Liz Pickett
And 1h51'40" 37d5'24"
Liz Putton Koch an immac-
ulate heart
Gem 6h10'20" 23d17'3"
Liz Reed's Star
Cas 1h30'37" 63d22'13"
Liz Richmond
Aqr 21h21'21" 2d21'24"
Liz Roque
Aql 19h6'59" 3d51'1"
Liz & Rowan Hurley
Cyg 21h35'48" 49d20'46"
Liz Shaughnessy
Cas 0h14'4" 59d26'29"
Liz Shimmering
Leo 11h5'42" 11d19'31"
Liz Smith
Cas 1h23'6" 61d51'24"
Liz Spencer
Ori 5h59'56" 17d58'48"
Liz Vaiana-Cavanagh's
"Artsy" Star
Sgr 19h38'25" -13d1'21"
Liz Vinca
Cas 23h0'44" 57d36'40"
Liz Wiz
Lyn 8h9'13" 54d33'11"
Liz y Ouadi
Aql 19h33'37" 11d46'7"
Liz & Yura
Cyg 19h47'16" 54d11'5"
Liz "Zil" Stein
Cnc 8h4'40" 22d9'46"
Liza
Gem 6h59'53" 26d37'27"
Liza
Tau 4h39'36" 22d4'48"
Liza
Lib 15h12'25" -9d41'44"
Liza Alons Star
Sco 16h48'42" -27d8'9"
Liza Ann Tuttle
Ari 3h5'5" 29d34'43"
Liza B.
Gem 7h39'48" 33d31'32"
Liza Barbarello
Lmi 9h31'47" 38d50'53"
Liza Beck
And 23h35'10" 36d46'35"
Liza Chandler
Uma 11h54'36" 59d54'22"
Liza & David Gest
Eri 4h22'43" -13d39'19"
Liza Flor Mendoza San
Diego
Lyn 6h47'8" 51d19'43"
Liza Gillette
Cyg 19h49'24" 47d49'25"
Liza Jane
Umi 13h38'14" 72d30'1"
Liza K. Miles
Lyn 8h12'54" 33d51'34"
Liza Leigh Ann Srock
Psc 1h18'57" 25d48'23"
Liza Liberty Toto
Aqr 21h43'13" 0d26'54"
Liza Maeve Nelligan
Mon 7h19'6" -0d50'8"
Liza Masters
Ari 2h5'44" 24d14'3"
Liza McNamara
Cas 1h27'29" 55d47'15"
Liza Minor
Vir 12h34'44" -10d23'48"
Liza Morales
Leo 9h29'58" 11d21'34"
Liza Mott Ultima
Liberfemina Stella
Sgr 19h41'42" -39d31'30"
Liza Mutis "My Kitty"
Lyn 7h58'59" 53d48'31"
Liza Suela Benedito
Leo 11h29'16" 14d24'39"
Lizabear
Umi 14h11'48" 74d52'32"
Lizabeth
Vir 11h41'17" 6d7'53"
Lizah Good 0701
Cnc 8h19'10" 16d22'21"
LizamaFamilyStar
Sco 17h26'33" -38d21'35"
Lizana
Leo 11h17'48" 3d51'59"
Lizanne Alvarez
Her 17h3'26" 39d14'34"
Lizar 01242004
Uma 11h46'46" 49d47'29"
Lizard
Cam 7h59'37" 61d5'7"
Lizard's Dream Space
Lac 22h21'10" 54d20'10"
Liza's Beauty
Lib 15h20'7" -23d33'11"
Liza's Star
Lmi 10h43'33" 30d45'39"
Lizbeth
Vir 13h9'2" -7d18'28"
Lizbeth Ann
Uma 9h5'2" 71d37'18"
Lizbeth Vargas
And 0h51'48" 41d50'43"
Lizbo 24
Crb 15h18'3" 26d35'25"
Lizcakes
Sco 16h7'31" -13d47'26"
Lizet Star Santos
And 0h25'3" 27d27'44"
Lizeth De Carmen Osuna
Virgen
Lyn 8h8'44" 54d5'19"
Lizeth's Star
Uma 13h41'11" 56d5'5"
Lizett Alvarez
Psc 0h57'53" 19d47'11"
Lizette
Vir 13h13'16" -0d11'2"
Lizette Amin Ortega
Caudillo
Ori 6h13'3" 9d7'22"
Lizette+Antonio
Vir 12h35'37" 3d2'4"
Lizette Gonzalez
Cnc 8h52'41" 26d23'47"
Lizette Marie Yrarragorry
Umi 14h55'45" 69d16'0"
Lizette Ramirez
Lib 15h12'35" -10d59'23"
Lizetti Acurio
And 0h48'7" 22d37'8"
LizFrank
Lyn 8h18'23" 48d55'43"
Lizi Westcoast
And 0h34'12" 36d36'6"
Lizie Sweeney
Cas 0h17'6" 50d10'25"
Lizrom
Cas 0h27'15" 59d40'56"
Liz's Guardian Angel
Lyr 18h32'38" 37d28'57"
Liz's Legacy of Laughter &
Love
Cas 2h46'26" 57d40'4"
Liz's Star
Cas 23h19'49" 59d43'59"
-*- Liz's Star for Dad -*-
Col 5h19'16" -34d20'44"
Liz's Valentine's Star
And 0h57'37" 36d11'24"
Lizzapyjaba
Umi 14h14'21" 75d16'31"
Lizzard's Star: 40 Years &
Glowing
Lib 14h49'46" -3d22'49"
Lizza's Hot Tamale
Ari 1h57'24" 18d36'30"
Lizze
Lac 22h28'10" 41d39'21"
Lizzete Hartz
And 1h5'22" 35d29'23"
Lizzeth Silva
Lmi 9h50'54" 40d5'23"
Lizzette Muñoz
Cap 21h9'53" -19d40'43"
Lizzi Smith
And 23h4'33" 37d1'8"
Lizzia Angelis
And 23h42'10" 44d44'16"
Lizzie
Cnv 12h46'40" 40d24'32"
Lizzie
Gem 7h49'10" 31d10'34"
Lizzie
Tau 5h24'5" 21d49'48"
Lizzie
Vir 12h41'52" 3d54'35"
Lizzie
Com 12h51'27" 20d30'17"
Lizzie
Cma 6h14'4" -14d44'11"
Lizzie
Sco 17h57'55" -36d55'27"
Lizzie and Artie
Cyg 21h46'10" 40d49'53"
Lizzie Dee
And 23h23'40" 41d54'59"
Lizzie DeGaetano
Lib 15h11'30" -23d4'58"
Lizzie "Hoops" Lee
Cyg 19h47'48" 43d29'34"
Lizzie Noon's Wonderstar!
Sge 20h9'28" 16d38'49"
Lizzie Sylvain
Cas 23h46'41" 51d11'30"
Lizzie The Star
Lyn 8h56'36" 45d42'58"
LizzieB
Sco 16h3'53" -26d1'59"
Lizzie-loo
Dra 17h46'16" 59d0'22"
Lizzies Light
Cas 2h5'47" 70d0'48"
Lizzie's Little Nebula
Sco 17h15'24" -33d8'3"
Lizzy
Cru 12h35'57" -57d51'47"
Lizzy
Psc 0h39'26" 6d44'3"
Lizzy
Cas 1h25'32" 63d33'19"
Lizzy
And 1h37'34" 42d43'10"
Lizzy
Tau 4h11'23" 28d58'56"
Lizzy
Ori 5h22'22" 2d46'52"
Lizzy
Psc 0h38'15" 14d45'32"

Lizzy
Ari 2h8'10" 14d47'43"
Lizzy and Kate's BFF star
Umi 13h42'4" 70d31'44"
Lizzy Byrant
And 0h21'21" 41d32'9"
Lizzy DelRey
Pho 0h55'50" -46d9'8"
Lizzy Mac
And 1h17'33" 50d10'50"
Lizzy Major
Sgr 17h52'45" -28d57'25"
Lizzy Selimi
Gem 6h43'53" 16d58'21"
Lizzy Tish Konis
Crb 16h7'15" 28d25'59"
Lizzy V
Uma 12h58'43" 55d29'22"
Lizzy, My Love For a Lifetime
Ari 2h22'46" 25d4'27"
Lizzy719
Uma 10h26'26" 64d52'6"
LizzyBeth
Cyg 20h35'8" 43d43'12"
Lizzybethannie
Uma 9h6'57" 49d5'8"
LizzyM11805
Cap 20h57'7" -20d15'35"
Lizzy's Destiny
Mon 6h43'27" 7d19'36"
Lizzy's Light - Shining Forever
Cru 12h39'28" -60d27'22"
Lizzy's Little Star.... Forever.
Cyg 20h3'27" 40d58'43"
Lizzystarz
Tau 4h55'57" 20d42'53"
Lizzytish
Cyg 19h46'55" 51d59'15"
LJ Foun10
Lyn 6h20'47" 54d1'52"
L.J. Hartig
Uma 10h45'48" 45d26'28"
LJ - ONE
Lib 15h25'28" -6d26'55"
LJ Villa
Boo 14h29'40" 13d51'17"
L.J., Angie and Jace
Aql 19h45'4" 12d35'32"
LJG Shines Forever In My Eyes
Umi 15h2'15" 68d15'43"
LJH1982
And 2h8'35" 40d26'27"
Ljiljana Dragojevic
Cyg 20h54'29" 30d2'53"
Ljiljana Panin
Cas 0h15'13" 57d16'57"
LJ-Jada
Cam 3h52'32" 71d50'19"
L.J.L.'s Star
Uma 11h23'18" 37d48'18"
LJM16
Tau 5h23'24" 25d2'37"
Ljoki, Mustafije
Aqr 20h39'35" -7d0'42"
L.J.P.H.C.
Uma 11h52'28" 47d53'18"
LJ's Rock Star
Lyr 18h49'33" 33d38'57"
Ljubilionov Prasilion
Cas 1h53'0" 64d11'16"
L.J.W.8.2.
Gem 6h48'6" 31d28'54"
LK
Lyn 9h13'23" 39d15'21"
LKL09142005
Vir 13h15'21" 12d27'42"
L.L. PITT
Ari 2h9'25" 16d36'10"
L.L. - SUEBEE
Mon 6h49'58" 7d5'0"
Lladnar the Destroyer (RW Hunter)
Ari 2h7'35" 20d39'6"
Llama
Vir 14h34'13" 5d5'56"
Llama
Cru 12h16'26" -58d58'8"
LLD
Tau 5h43'9" 24d18'2"
Lletel Noel
Cap 20h23'34" -20d12'56"
Llewel Nigel Mancebo
Lib 15h11'15" -4d40'28"
Llewellyn
Ori 5h23'53" 2d54'10"
Lleyton Perry
Uma 14h13'40" 60d57'12"
LLF Margrithli
Cas 0h10'32" 52d46'30"
llissa and ciso
Lmi 10h25'56" 36d0'42"
L.L.Keeling
Cnc 8h15'15" 7d25'36"
lLogan Patrick
Her 18h47'22" 24d29'36"
Llorenç Llobera Silva
Sgr 17h55'18" -26d49'13"
LLOYD
Ari 2h27'14" 27d43'37"
Lloyd
Leo 10h11'51" 21d4'15"
Lloyd and Lynn Leger
Crb 16h14'31" 25d41'45"
Lloyd and Ruth Hayward
Ori 5h33'30" -4d42'12"
Lloyd Anthony Kieffer
Boo 15h27'59" 42d26'41"
Lloyd Avant
Ori 5h6'26" 4d6'42"
Lloyd Brooker's Piece of Heaven
Ori 5h40'48" -0d50'21"
Lloyd Charles Townsing
Nor 16h25'8" -51d23'40"
Lloyd Duane Shedden
Ori 6h3'1" 19d48'33"
Lloyd E. Trotter
Lmi 11h5'10" 29d36'13"
Lloyd Forever
Aqr 22h6'7" -1d30'4"
Lloyd George Pickering
Leo 11h17'13" 14d47'53"
Lloyd Gepfer
Per 3h8'5" 54d12'12"
Lloyd J. Forman
Leo 10h50'33" 12d31'7"
Lloyd J. Montgomery
Her 17h19'15" 36d23'46"
Lloyd J. Sullivan, Sr.
Gem 6h54'21" 24d3'27"
Lloyd James Prytherch
Per 3h11'12" 46d16'38"
Lloyd Paul Bowlby
Uma 11h5'55" 57d41'11"
Lloyd Robert Goulette
Leo 9h46'4" 28d51'21"
Lloyd Roland Robinson, Jr.
Lyn 8h50'51" 41d2'0"
Lloyd & Ruth Deem
Cyg 20h54'32" 40d45'44"
Lloyd Sergent
Aql 19h13'55" 14d23'57"
Lloyd Simpson
Per 2h11'34" 54d36'16"
Lloyd Steward 9/1/1952-2002
Aql 19h55'7" -0d58'9"
Lloyd Stough Jr.
Uma 11h16'59" 55d7'52"
Lloyd Stough Sr.
Uma 11h34'13" 54d2'27"
Lloyd the Norwegian
Cep 4h18'27" 81d38'7"
Lloyd Wayne Deering
Cnc 8h44'17" 7d19'49"
Lloydie B
Uma 9h28'36" 56d2'52"
Lloydy's Buckle
Dra 14h42'47" 55d44'30"
Llúcia Casanovas Coma
Cnc 8h16'10" 31d29'14"
Lluis Alba
Cap 20h14'7" -16d31'33"
Lluis Garcia i Lapuente
Ori 6h17'30" 14d39'23"
Lluliana Gutierrez
Vir 13h26'15" 5d24'1"
Lluvia Karina Sotelo
Aqr 21h38'35" 1d15'12"
Llywelyn C. Graeme
And 0h37'55" 37d30'1"
LM
Vir 13h32'2" -11d12'56"
"LMCYS" - Doris and Rod Sutton
Cyg 20h22'58" 52d24'30"
LMD 8/9/1956
Leo 10h15'33" 20d53'51"
LMFBeautifulButterflySpag hettiHead
Uma 11h18'5" 38d8'49"
LMJ
Uma 12h51'38" 55d24'36"
LMN
Lyn 7h28'31" 47d55'14"
LMNO
Sco 16h10'31" -13d53'34"
L.M.PeTuNiA 16
Aql 19h47'11" 10d52'55"
LN
Peg 22h42'40" 3d21'1"
LnC Infinity
Cnc 8h46'4" 32d31'13"
LnG 2gether 4ever
Uma 10h40'40" 52d55'23"
LnJb95
Lyn 9h22'28" 41d27'23"
Lo
And 0h45'1" 37d13'0"
Lo
Cas 1h36'27" 58d32'32"
Lo
Tau 4h41'44" 0d42'12"
Lo
Leo 10h39'19" 17d23'21"
Lo + Ve = Lofe Always..
Cyg 21h31'43" 51d32'13"
Loa
Cas 23h42'17" 53d17'10"
loAlon
Cas 0h34'8" 54d24'46"
Loan
Cnc 8h59'40" 9d38'51"
Loann Nguyen
Vir 13h15'27" -16d7'32"
Loanna Tran Ramirez
Cnc 8h2'57" 24d49'7"
Lobna
And 1h30'52" 48d41'19"
Lobo
Tau 4h14'45" 2d38'41"
Lobsang-Nima-Jordhen
Umi 15h58'20" 79d45'34"
Lobster & Bunny
Uma 10h1'13" 43d51'17"
Lobster Pumpkin
Lyn 7h58'33" 42d50'12"
Local Beauty
Uma 10h29'7" 68d1'4"
Locherbie
Cyg 20h27'21" 41d25'2"
Lochinvar's Angel
Sco 16h30'7" -26d58'9"
Lochlain T
Umi 15h29'1" 71d43'32"
Lochlan Angelo Hewitt
Per 3h41'28" 42d30'45"
Lochlan James Patterson
Cyg 20h5'8" 46d55'34"
Lochlan Luke Bocking - 2005-Our Star
Cru 12h43'20" -56d15'29"
Lochlund
Uma 10h7'32" 58d43'50"
Lockard Landing
Uma 11h18'11" 55d29'26"
Locksmith USA, Inc.
Tau 4h27'25" 16d10'50"
Lockwood Guiding Light
Ori 5h38'6" -2d18'59"
Lockwood & Stephanie - Eternal Love
Cnc 8h28'1" 14d7'10"
Locky
Uma 11h39'50" 40d27'10"
Loddonvale Gustav
Cmi 7h34'48" 5d3'20"
Loddonvale Louis
Cmi 7h19'14" 1d35'20"
Lode & Jennifer Holtslag
Cyg 20h32'56" 43d7'58"
Loebert, Lothar
Ori 5h12'0" 15d51'58"
Loel (Light of Eternal Love)
Ari 2h43'3" 27d4'14"
Loeric
Uma 11h51'34" 60d0'14"
Loesch, Wilhelm
Uma 9h45'27" 65d37'49"
Loesje
Cas 0h19'34" 58d14'46"
Löffler, Willibald
Ori 6h16'45" 15d33'20"
Loftis
Lyr 18h38'22" 31d18'25"
Lofton Benjamin Nall
Uma 10h39'23" 52d54'17"
Lofton E. Stewart
Uma 11h36'54" 47d12'54"
Logadyl
Cas 0h16'18" 53d51'7"
Logan
Uma 9h58'50" 50d57'49"
Logan
Aur 6h25'26" 51d37'29"
Logan
Per 2h56'45" 33d41'38"
Logan
Psc 23h25'12" 6d41'51"
Logan
Umi 16h16'46" 78d13'15"
Logan
Cap 21h27'56" -20d25'48"
Logan
Lib 15h35'25" -10d3'50"
Logan 04 2005
Ari 2h59'19" 19d10'18"
Logan 1
Umi 15h4'16" 72d23'21"
Logan Alan Proctor
Ori 5h25'41" -8d35'32"
Logan Alexander Kenney
Mon 6h46'33" -3d46'39"
Logan Alexander Merkt
Aqr 22h9'58" -14d8'20"
Logan Alexander Tucker
Psc 23h10'52" 5d14'55"
Logan Alexandria McCarrell
Per 3h16'5" 47d4'49"
Logan Alvstad
Umi 14h24'10" 75d46'4"
Logan Aniceto
Sgr 19h19'34" -20d52'33"
Logan Antonio Anaya
Aql 20h8'1" 10d37'31"
Logan Armstrong Hall
Her 18h7'53" 38d43'16"
Logan Avedis Watson
Peg 23h24'4" 12d44'6"
Logan Bahr
Ori 5h51'35" 6d57'24"
Logan Blake
Ori 6h10'38" -0d2'45"
Logan Blake Dionisio
Uma 13h36'12" 49d16'40"
Logan "Bugg" Taylor
Lib 15h4'25" -23d52'0"
Logan Cade Tinsley
Aql 19h23'17" 2d58'29"
Logan Castillo
Cap 21h42'31" -13d22'10"
Logan Charboneau
Gem 7h25'2" 27d19'27"
Logan Charles Metelits
Ori 6h7'18" 14d7'17"
Logan Chase Dreamer
Cam 4h15'25" 61d54'39"
Logan Christian
Ori 6h4'6" 16d12'48"
Logan Christopher Vratny
Her 17h36'26" 43d16'7"
Logan Clay Trimble
Cap 20h32'16" -25d54'48"
Logan Cole LeMay
Vir 13h23'16" -20d32'32"
Logan Daniel Filidei
Lyn 7h32'42" 38d22'11"
Logan Daniel Lazarus
Uma 11h49'7" 43d10'40"
Logan Daniel Roberts
Aqr 21h22'31" -8d34'12"
Logan David Cecil
Her 17h29'11" 35d29'18"
Logan David D'Arezzo
Gem 6h34'23" 12d27'6"
Logan David Stephen Carr
Dra 10h55'20" 74d38'49"
Logan Dean
Boo 14h41'58" 38d40'19"
Logan Dewain Yarborough
Ari 2h47'46" 14d29'32"
Logan Drew Klenk
Sgr 18h17'11" -23d20'4"
Logan Elias Feller
Uma 10h41'41" 71d37'31"
Logan Elizabeth
Psc 0h45'56" 20d34'33"
Logan Emmet Paulsen
Ari 2h47'30" 26d18'18"
Logan Forsyth
Per 3h58'12" 31d6'32"
Logan Goodwin
Cyg 20h33'46" 31d31'50"
Logan Gregory Binder
Uma 14h2'40" 62d0'55"
Logan Jack
Uma 10h11'16" 53d45'19"
Logan James
Cnv 12h33'21" 39d15'1"
Logan James DeSimone
Her 17h30'26" 43d35'14"
Logan James Iversen
Dra 16h42'28" 68d39'2"
Logan James Lustre
Sgr 18h13'37" -24d5'22"
Logan James Mathias
Psc 1h0'5" 31d31'15"
Logan James Roy Soutar
Per 2h58'26" 41d46'11"
Logan James Shaffer
Aqr 22h6'59" -3d54'25"
Logan James Walker
Ari 3h27'15" 19d26'37"
Logan Jean
Crb 15h39'7" 27d52'22"
Logan Jeffrey Weaver
Lib 14h52'46" -1d2'20"
Logan "Jellyfish" Myers
Psc 0h53'11" 27d6'50"
Logan John Hollis
Cru 12h24'42" -62d15'56"
Logan Jude Vaughan
Cnc 8h54'34" 31d26'58"
Logan Jurkas
Sgr 18h23'42" -32d53'17"
Logan Keith
Cep 23h1'52" 77d10'48"
Logan Kristopher Welch
Ori 5h52'1" 18d21'37"
Logan Leavitt
And 0h38'7" 37d59'57"
Logan Lee
Ori 5h59'24" 17d44'17"
Logan Lopez
Ori 5h29'34" 1d3'0"
Logan Louis Corvisiero
Leo 10h1'42" 16d25'54"
Logan M. Blunt
Lib 14h48'53" -18d30'0"
Logan M Kane
Dra 18h32'32" 54d25'2"
Logan Maria Jordan
Lyn 7h43'25" 52d0'49"
Logan Martin
Dor 5h2'25" -67d7'21"
Logan Mason Greene
Tau 5h46'0" 18d13'53"
Logan Matthew Boffi
Uma 10h17'44" 54d59'4"
Logan McKinney
Aur 5h39'39" 45d4'25"
Logan McMillan
Aql 18h51'25" -1d25'32"
Logan Medenica
Sco 16h16'59" -15d31'53"
Logan Michael Dion Seymour
Cap 20h48'27" -16d20'27"
Logan Michael Dolch
Uma 11h0'34" 48d20'34"
Logan Michael Fuhs
Cap 20h48'5" -18d40'17"
Logan Michael Grant
Lib 14h35'3" -14d20'9"
Logan Michael Moore
Psc 1h23'42" 18d27'32"
Logan Michael Spear
Sgr 18h51'43" -21d7'26"
Logan Michael Touve
Psc 1h1'6" 9d54'39"
Logan Michael Xanders
Her 16h21'24" 18d23'46"
Logan Mitcheltree
Ori 6h8'33" 8d43'49"
Logan - Mommy's Lil' Leprechaun
Psc 0h11'38" 0d48'35"
Logan Moore Hines, II
Sgr 19h44'10" -15d14'58"
Logan Moskunas Cookerly
Aql 19h55'38" 16d3'58"
Logan Nathaniel
Vir 14h16'7" -9d0'36"
Logan Odell
Umi 15h14'35" 69d9'56"
Logan Oscar Schmidt
Sco 16h12'7" -15d21'27"
Logan Owen Pagluiso
Ari 2h34'0" 28d19'31"
Logan P. Flanagan
Gem 6h52'5" 34d52'41"
Logan Patrick DeVos
Uma 10h6'50" 46d24'45"
Logan Patrick Doyle
Lmi 10h51'16" 35d37'55"
Logan Patrick McGough
Lac 22h23'47" 51d40'59"
Logan Patrick Stevens
Ari 2h50'31" 29d17'0"
Logan Paul McHugh
Per 4h41'33" 46d30'15"
Logan Paul Miller
Psc 1h25'30" 23d9'6"
Logan Paul Schultz
Sgr 18h21'4" -28d21'5"
Logan Phillip Herger
Psc 1h11'40" 21d2'46"
Logan Radney
Gem 6h23'28" 25d17'3"
Logan Raeann
Lib 15h43'25" -15d40'23"
Logan Randolph
Ari 3h12'51" 26d8'52"
Logan Reece Stephens
Ori 5h53'7" 21d37'49"
Logan Rhea Miller
Leo 9h37'46" 28d26'48"
Logan Richard Sattell
Dra 17h50'36" 54d18'28"
Logan Robert Montgomery
Cap 21h18'26" -22d32'8"
Logan Sarah Reeves
Lmi 10h12'23" 30d16'28"
Logan Scott Zelen
Uma 10h48'38" 70d9'34"
Logan Sean Snider
Her 16h51'19" 32d22'2"
Logan Selby (Yogi Bear)
Psc 23h24'53" 6d33'30"
Logan Shaun and Hunter Chase
Gem 7h26'56" 26d34'20"
Logan Shea Gane
Cra 19h5'7" -38d22'55"
Logan Stoner
Uma 10h25'21" 41d19'25"
Logan the Terror
Sgr 18h20'58" -32d26'13"
Logan Thomas Altomare
Ori 5h1'56" 2d44'27"
Logan Thomas Turner
Psc 0h50'9" 8d26'57"
Logan Thomas Woodall
Gem 7h13'20" 31d7'1"
Logan VanCuren (Light Of My Life)
Crb 15h36'9" 37d39'47"
Logan Vincent Reed
Lib 15h52'8" -5d33'50"
Logan Waldeck
Uma 9h45'39" 62d41'19"
Logan Walker Adams
Umi 16h19'19" 75d57'38"
Logan Walker Gabler
Uma 10h17'26" 64d16'38"
Logan Walzak
Per 4h17'16" 52d18'1"
Logan William Cowan
Gem 6h23'50" 24d21'32"
Logan William Finley
Aql 19h57'49" 5d54'38"
Logan William Tracia
Aur 5h49'15" 49d57'55"
Logan X-1
Cyg 20h57'14" 46d5'24"
Logan Xaivier
Cyg 20h13'38" 54d24'21"
Logan Xavier Kellogg
Lib 15h29'50" -17d34'5"
Logan, Collin, Sheldon
Cyg 21h14'31" 47d0'34"
Logan, Danielle & Ron
Umi 15h27'3" 73d43'59"
Logan, Nathan, & Alyssa Bohannon
Tri 2h13'49" 35d30'31"
Logan, Our Star
Her 16h37'51" 7d53'28"
Logan, Star of the Universe
Psc 0h17'39" 15d28'17"
Logan's 1st Birthday Star
Gem 7h23'21" 27d14'5"
Logan's Eternal Light
Uma 11h23'37" 61d12'39"
Logan's First Christmas
Leo 10h54'11" 5d51'40"
Logan's Light
Lmi 10h35'11" 33d9'59"
Logan's Light
Cam 3h58'15" 68d28'26"
Logan's Piece of Heaven
Ori 5h47'22" 6d20'23"
Logan's Star
Lyn 8h39'15" 46d24'42"
logans star
Aql 19h40'30" -3d58'58"
Logan's Wish
Ori 5h42'10" 2d22'26"
Logan's World
Sco 17h33'6" -33d53'25"
Loganus Major
Uma 11h12'12" 35d35'53"
Logen Bryce Krapf
And 2h36'43" 39d45'57"
Logen Carlogero
Cap 21h25'36" -27d10'51"
logie
Sgr 18h12'26" -17d46'47"
Logix 1.0
Cma 6h58'46" -19d51'17"
Logyn Alexandra Caveza DeBaca
Cap 20h55'59" -27d15'5"
Lohan Khamsada Tuan Dung Chevy
Ori 6h16'2" 9d4'58"
Löhr, Karl-Heinz
Aqr 21h34'53" 3d1'4"
Lohse
Per 3h28'31" 48d54'55"
Loi Duong
Aqr 21h42'4" -2d51'12"
Loi Liang Tok
Ori 5h37'50" -3d37'33"
Loïc Hagens
Ori 6h0'39" 20d57'35"
Loïc Hamel
Her 17h43'36" 38d47'17"
Loïc Laforet
Boo 14h24'43" 46d13'14"
Loïc Lorin
Lmi 9h25'31" 33d3'44"
Loic Muller
Cyg 21h26'18" 33d6'3"
Loïc Romuald Robert - Australia 06
Ori 5h40'13" -1d4'13"
Loïc Serge Bouchard
Tau 5h44'25" 15d3'29"
Loida Graham
Ari 3h8'21" 29d53'24"
Loida Peña Ortiz
Cap 20h9'24" -10d41'20"
Loie Hayes - Julie Ogletree Wedding
Lib 14h23'3" -9d27'1"
Loïk Pagès
Ori 4h59'16" -3d23'9"
Lois
Aqr 23h55'23" -7d16'8"
Lois
And 22h59'6" 52d13'18"
Lois #1
Cap 21h26'30" -27d9'3"
Lois 80
Cep 21h42'22" 60d5'15"
Lois A. Drumheller
Peg 23h25'19" 17d3'48"
Lois A. Schlussler
Cyg 20h17'12" 40d13'12"
Lois A. Traverso
And 1h47'29" 50d32'13"
Lois and Allen Van Cleve
Cyg 19h59'38" 34d43'53"
Lois and Jerry's Fire Light
Tau 3h57'19" 30d50'16"
Lois and Paul Guynes
Uma 12h1'21" 45d33'42"
Lois Anderson
Cam 5h35'15" 62d32'1"
Lois Angela "Blueylo" Tucci
Ari 3h2'55" 27d11'12"
Lois Ann 03-22-43
Ari 1h47'5" 23d8'1"
Lois Ann Anderson Larry
Ori 5h31'42" 3d39'12"
Lois Ann Fetters
Uma 9h40'45" 54d53'49"
Lois Ann Jenness Studley
Uma 10h29'14" 57d45'41"
Lois Ann Palonis
Cnc 8h22'53" 11d19'33"
Lois Ann Saddler's Star of Faith
Gem 7h30'24" 20d32'34"
Lois Anne Berger
Cyg 21h44'53" 33d5'19"
Lois Anne Cudworth Diakoff
Vir 14h43'20" 3d19'27"
Lois Audrey Cameron
Uma 11h32'22" 34d51'5"
Lois Bailey 2/17/32 - 8/24/99
Sge 19h51'18" 18d36'43"
Lois Beina Rosenzweig Always a Star
Cas 23h57'36" 57d33'12"
Lois Boyd
Aqr 21h50'57" -0d42'12"
Lois C. A. Smith
Peg 22h22'23" 24d8'28"
Lois Campbell Pearson
Uma 10h43'56" 53d10'13"
Lois Carolyn Homberg
Lyn 9h40'46" 40d58'26"
Lois Carroll
Aqr 21h19'58" -13d39'13"
Lois Casteel
Cas 0h2'22" 54d9'18"
Lois Cathrine Kaye
Cru 12h37'16" -59d48'51"
Lois Clark McCoy
Aql 19h36'56" 11d17'4"
Lois Clarke
Cap 20h51'41" -25d27'46"
Lois Comfort
Her 17h29'40" 32d51'24"
Lois Corvisiero
Cap 20h17'22" -23d18'40"
Lois Davenport
Cas 1h20'41" 64d51'7"
Lois Demi Centuri
Cen 15h0'9" -38d30'11"
Lois E. Carter
Cas 1h37'27" 61d18'17"
Lois E. Neville
Aql 19h52'57" 11d33'40"
Lois Elise Eldred
And 2h27'26" 42d41'47"
Lois English
Gem 6h42'18" 33d45'58"
Lois F. Wetherell
Ori 6h0'8" 18d14'5"
Lois Farra
Dra 9h31'30" 76d58'58"
Lois Fay
And 0h58'48" 46d12'18"
Lois Frye a.k.a. Lois Lane
Vir 14h14'36" -17d21'34"
Lois Grieco
Leo 10h7'50" 26d0'8"
Lois Grotewold
Cnc 8h11'1" 8d53'5"
Lois Horner
Peg 23h15'23" 32d55'24"
Lois I
Lyr 18h43'32" 35d3'30"
Lois Irene Cheatum
Cyg 21h19'2" 40d8'9"
Lois J
Psc 0h47'48" 10d24'13"
Lois J. Allen
Cas 23h35'7" 51d3'34"
Lois & Jack, Our Angels in Heaven
Uma 9h56'26" 41d37'26"
Lois Jean
And 0h33'11" 41d19'26"
Lois Jean Mack
Cnc 8h53'37" 30d30'58"
Lois Johnson
Lyr 18h15'19" 30d47'44"
Lois Joy Johnson
Mon 6h43'36" -1d3'39"
Lois Joy Johnson
Cam 7h7'58" 72d47'55"
Lois Kate Ridgway
And 1h7'11" 44d57'32"
Lois Lane
Cas 0h6'42" 59d14'18"
Lois Lea Robson - Lois' Star
And 23h23'28" 52d58'59"
Lois Lieverdje
And 23h32'51" 37d38'20"
Lois Lockhart
Cnc 9h21'21" 12d24'42"
Lois M. Bryant
Leo 10h22'56" 14d1'21"
Lois M. Carlsen Worthy Matron 2004
Cas 23h22'1" 58d27'28"
Lois M Knowles
Crb 15h58'34" 31d43'43"
Lois Mae
And 23h26'49" 44d22'13"
Lois Mari
Cas 0h19'22" 57d47'56"
Lois Marie
Sgr 18h13'5" -29d57'55"
Lois Marie Austin
Lyn 7h49'2" 38d29'40"
Lois Marie Smead
Mon 7h15'5" -0d7'59"
Lois May Straw
Crb 15h39'25" 35d53'20"
Lois McWilliams Galbraith
Vir 12h39'48" 2d37'43"
Lois Mindi Almond
Del 20h40'17" 15d39'18"
Lois Morehead
And 2h28'51" 41d33'33"
Lois Nadeen Rayburn
Sgr 19h28'6" -41d59'40"
Lois Neff
Uma 9h31'54" 61d37'15"

Lois Parker Glick
Cas 1h19'53" 67d25'11"
Lois Potts
Cas 0h25'4" 56d47'59"
Lois Raye Lonergan
Dra 17h49'25" 52d31'39"
Lois Reisinger
Vir 14h47'46" 4d24'3"
Lois Rottersmann
And 0h35'23" 42d14'36"
Lois Ruisi
Uma 12h9'41" 53d25'26"
Lois Sidney
Leo 11h25'47" 16d4'50"
Lois & Stanley Kramer
Cyg 21h19'32" 46d48'54"
Lois Stout
And 23h18'49" 44d30'29"
Lois Stryson
Psc 1h22'53" 15d59'23"
LOIS THERESA EVANS
Lib 15h20'10" -7d12'53"
Lois Voth
Vir 12h43'48" 12d45'17"
Lois Webster
Umi 14h37'18" 72d35'45"
Lois "Your Angel in the Sky"
Lyn 8h13'37" 35d32'8"
Lois, Don & Adam Family Star
Cyg 19h36'30" 34d27'4"
Loisi Rauter vlg. Messner
Cas 0h29'6" 58d44'4"
Lois-Mae
And 0h5'27" 35d46'52"
Loizos Kolokotronis
Uma 8h35'8" 71d39'13"
Loken
Peg 22h23'13" 34d16'21"
loker crystal
Leo 10h39'32" 15d33'17"
Loki
Psc 0h24'38" 19d50'26"
Loki
Ori 5h19'30" -9d16'52"
"Loki Ravenstone Alexander"
Cma 7h25'1" -20d20'8"
Loki Rhythm
Uma 11h31'24" 61d49'27"
Loki Valentine Kandel
Cma 6h18'51" -29d38'47"
Lokiedog Star
Cap 21h10'6" -15d7'15"
Lola
Sgr 19h12'10" -14d15'12"
Lola
Uma 11h27'8" 63d21'41"
Lola
Uma 8h49'20" 57d25'16"
Lola
Lyn 7h45'27" 57d0'42"
Lola
Cap 20h11'14" -23d48'5"
Lola
Sco 17h1'2" -41d16'18"
Lola
Psc 23h54'41" 6d16'57"
Lola
Aqr 22h34'42" 2d21'49"
Lola
And 0h56'42" 39d31'22"
Lo-La
Uma 11h42'16" 30d25'42"
Lola
Per 4h10'18" 47d14'42"
Lola Ala Nia Reynolds
Uma 8h39'59" 51d41'50"
Lola Armani Rowland
And 1h5'59" 45d25'16"
Lola & Baky
Sgr 18h19'57" -34d22'46"
Lola Bean
Aqr 22h15'55" 1d50'11"
Lola Bryant
And 0h33'29" 27d55'38"
Lola Bunny
Lep 5h14'15" -11d16'13"
"Lola" Catherine A. Bliss
Cyg 19h18'38" 54d47'4"
Lola Catherine Bradley
Lib 14h53'23" -6d56'36"
Lola Creel
Com 13h9'17" 28d30'55"
Lola D.
Cam 7h22'20" 79d57'50"
Lola Elise
And 2h12'5" 41d52'30"
Lola Elise Elwell
And 1h1'5" 42d16'42"
Lola Elizabeth Lee
And 23h14'56" 40d35'39"
Lola Elizabeth May Harrison
Umi 14h48'46" 74d2'42"
Lola Fering Lascano
Cas 0h11'10" 53d58'23"
Lola Gabrielle Sophia
Uma 9h38'15" 65d10'8"
Lola Hafenstein
Lyr 18h34'2" 32d24'16"
Lola i Mladen 17.02.1979.
Per 2h12'6" 52d7'43"
Lola June
Peg 21h46'2" 27d50'17"
Lola Kathleen Rifley
Cyg 19h38'31" 37d36'34"
Lola Kaye Morse
Cyg 20h52'28" 53d45'49"
Lola L. Shelton
Cas 23h36'23" 55d6'37"
Lola Loraine Liddell
Crb 16h13'0" 38d2'11"
Lola Maisie Eynon
And 22h59'24" 43d21'47"
Lola Marie Swint "1966"
Gem 6h2'11" 26d14'33"
Lola Marie Terrell
And 1h35'25" 47d12'41"
Lola Mathewson #1 Grand Daughter
Ori 5h1'3" 14d13'23"
Lola May Butler
Sgr 18h36'50" -29d45'10"
Lola Millie Wakefield
And 22h58'50" 46d50'36"
Lola Nadja Weller
And 23h14'28" 47d22'2"
Lola Nibbi Jose
Uma 11h42'16" 43d22'9"
Lola Ren
Cam 5h34'56" 63d32'6"
Lola Rose
And 1h26'49" 46d53'51"
Lola Ruben
Cma 6h44'54" -12d33'52"
Lola Shines On
Ari 3h23'46" 21d34'35"
Lola Snow
Cyg 21h37'19" 43d6'13"
Lola Sonnenschein
Aqr 22h38'36" -0d7'58"
Lola Vegas Halloum
And 1h16'30" 34d49'20"
Lola-Gary
Gem 6h40'12" 17d23'26"
LolaK
Cas 1h58'24" 67d8'6"
Lola's Star
Psc 1h31'37" 23d38'53"
loLFDBve
Lyn 8h10'27" 33d58'18"
Loli
Cnc 8h20'31" 30d27'21"
Loliacw
Ari 2h45'53" 28d26'13"
Lolita
Psc 1h18'13" 16d39'36"
Lolita Tatiana Sosa
Sco 17h44'51" -37d56'47"
Lollie
Lyn 7h15'45" 50d30'46"
lollipop
Lib 15h37'18" -23d12'44"
LOLLY
Lyn 7h1'31" 60d49'34"
Lolly
Cyg 19h43'26" 52d45'17"
Lolly
Crb 15h49'49" 33d44'46"
Lolly
Cyg 19h55'12" 37d22'20"
Lolly Leigh
Cas 23h56'17" 53d42'45"
Lolly Mahnaka
Sgr 19h26'44" -28d24'19"
Lolly Pop
Crb 15h45'14" 32d41'57"
LoLo
Cyg 20h57'46" 42d19'13"
Lolo
Cam 4h21'31" 56d17'2"
Lolo
Tau 5h49'45" 17d1'6"
Lolo Sunshine
Ari 2h0'23" 23d30'15"
Lolonyo
Cam 3h54'35" 66d12'8"
Lolotea-gift from God
Cnc 8h39'48" 7d22'24"
Lolotte
Lyn 9h15'5" 39d1'45"
Lolotte
Uma 9h47'27" 64d56'1"
Lolurenzocia
Mon 6h48'28" 7d41'38"
Lom
Leo 10h4'45" 13d38'7"
Loma
Dra 16h26'28" 51d49'12"
Loma AlTalibi
Gem 7h25'54" 15d56'25"
Lomas' One Hand Clapping
Uma 9h21'56" 44d44'45"
Lomeenieimadookin
Mon 7h33'51" -0d30'53"
Lomeron-K
Crb 15h45'0" 35d18'32"
Lomitoywilly
Sgr 19h12'31" -29d11'22"
L.O.M.J.
Cam 4h6'43" 52d58'57"
LOML
Her 16h46'12" 37d11'42"
LOML
Lyr 18h50'0" 33d51'2"
loml
Gem 6h29'56" 16d52'17"
LoMl
Ari 2h9'46" 23d35'44"
"LOML Eternal" by JRichard & Regina
Cap 21h48'11" -20d13'23"
Lomls Gallegos
Uma 11h30'48" 44d17'54"
loml's star
Gem 7h47'23" 33d4'40"
LOMLuscious
Uma 9h31'16" 44d29'23"
Lomney
Uma 12h20'45" 59d17'56"
Lona Brooklyn Peppers
Gem 7h8'46" 21d57'34"
London
Ari 2h36'17" 23d55'32"
London Elizabeth Remley
Sgr 19h15'58" -19d39'12"
London Parker Jones
Gem 7h23'56" 21d12'38"
London Sydney Clark Adams
Lib 14h23'8" -21d8'40"
Londyn Douglas Weisman
Uma 10h4'34" 63d49'15"
Londyn Willow
And 2h21'37" 45d4'21"
Lone Bennedsen
Lyn 7h14'2" 57d31'53"
Lone Pine
Sco 17h26'7" -39d34'42"
Lone Star Blue
Vir 12h19'1" 1d22'49"
Lone Star Matthew
Boo 14h16'22" 17d47'6"
Lonea Rand
Tau 5h34'49" 22d48'59"
Lonestar
Boo 14h19'48" 19d13'10"
Long Live The Billy Goat Curse!
Lyn 7h38'41" 48d43'16"
Long Phan: wueng yün
Cep 22h32'57" 62d44'3"
Long Tall Woman
Cyg 19h45'41" 56d9'12"
Longhorn Architect
Uma 9h26'42" 62d36'44"
Longo Emanuela
Per 3h5'35" 44d9'11"
Long's House
Uma 9h5'36" 50d22'29"
Longshot
Eri 3h52'38" -0d49'51"
Longy
Lyn 8h58'18" 34d32'10"
Loni & Eric
And 1h28'6" 49d21'9"
Loni Kayleen Gonzalez
Cnc 9h10'24" 30d41'17"
Loni Krystal Ostermaier
Aqr 23h55'52" -23d27'12"
Loni Nagwani
Leo 10h18'28" 15d0'58"
Loni Sue
Uma 8h54'15" 54d17'41"
Loni Toca Murphy
Cnc 9h2'26" 16d9'12"
Lonie
Aqr 22h16'1" -2d33'11"
Lonie/David #1 Toute Les Jours
Per 3h21'51" 51d22'58"
Lonna
Crb 16h21'48" 32d5'32"
Lonna Jeanne Lysne
Leo 11h17'15" 24d55'18"
Lonna little flower
Cas 22h59'25" 56d42'11"
Lonna Montrose
Psc 0h29'47" 14d2'26"
Lonnie and Dona Cruse
Crb 15h30'33" 27d18'58"
Lonnie and Staci's Star
Tri 1h51'20" 28d56'59"
Lonnie Bellview
Cam 4h39'34" 66d11'19"
Lonnie C. Freels Jr.
Lyn 7h28'8" 47d19'57"
Lonnie Cameron
Lyn 6h38'11" 57d9'43"
Lonnie Claude Vess
Per 4h23'43" 40d42'39"
Lonnie Dean DeClue
Uma 8h52'16" 49d20'4"
Lonnie Ellis Benson II
Cnc 8h29'7" 26d23'48"
Lonnie Gene David
Leo 9h49'24" 32d12'31"
Lonnie Gene Mullins
Uma 12h2'36" 39d49'59"
Lonnie Hudkins
And 0h12'51" 39d28'2"
Lonnie L. Latham, Jr.
Aql 19h56'45" -0d10'35"
Lonnie Lee Allen
Leo 9h28'36" 11d0'3"
Lonnie Paul Cefalu
Uma 11h47'18" 43d52'28"
Lonnie Ray Henderson
Men 5h24'2" -70d55'45"
Lonnie Rivera
Uma 9h26'22" 44d25'8"
Lonnie Wade
Dra 18h54'41" 60d7'37"
Lonnie Werner
Lyn 8h45'48" 33d41'39"
Lonny Anderson June 29, 1980
Aql 19h44'52" 12d46'2"
Lonny L Gwynn
Aql 19h56'59" 8d56'2"
Lonny Snow Super Sponser
Uma 13h35'20" 49d49'11"
Lons Star
Psc 23h46'33" 5d10'43"
LoNvE
Uma 10h17'55" 43d14'31"
Looby
Per 3h13'1" 47d32'3"
Looby Loo
And 2h35'24" 39d49'2"
Looby Star
Cam 8h57'52" 78d31'21"
Look above & we are together 4ever
Cyg 20h25'49" 54d54'37"
Look After You
Uma 10h48'21" 60d28'12"
Look at me! I'm your star, Becky!
Sco 17h51'13" -36d37'34"
Look How They Shine~Dan & Alicia
Uma 11h0'52" 48d26'47"
Look up and I will be there
And 2h38'40" 40d2'42"
Look up & see my love shine for you
Cap 21h35'58" -10d12'9"
Look Who's Here, Star Sudar-Hezer
Umi 16h0'14" 71d18'55"
Looking After You
Ori 5h12'8" -5d56'35"
Looking at the stars
Cyg 19h44'5" 32d54'33"
LookUpMaileCPollingtonPs 148
Sco 17h53'25" -39d17'46"
LooLou
Vir 13h16'17" 6d29'57"
Loomanator
Col 6h7'25" -29d17'0"
Loona DeBora
Dra 19h1'56" 70d46'7"
Loony Moon Maiden
Mon 7h17'24" -0d3'59"
Looper's Muse
Gem 7h28'18" 20d14'19"
Looper's Wife
Cas 1h59'47" 66d22'3"
Loopstar
Aqr 22h42'20" -21d7'7"
Loose, Corinne
Uma 8h43'43" 57d34'16"
Loosey Goose
Uma 9h6'43" 58d35'27"
Loosie Lou
Gem 7h45'12" 34d55'52"
Lopaka Poha Laimana
Tau 5h10'18" 23d2'44"
Lopez Bryan
Uma 8h51'31" 57d31'45"
Lor Lynn's Twinkling Little star
And 0h37'17" 26d10'13"
Lora
Umi 16h6'37" 70d20'28"
Lora
Dra 20h12'52" 65d3'14"
Lora
Cas 2h18'16" 72d52'55"
Lora Ann
Leo 9h25'41" 10d36'7"
Lora Ann Grace Metz Sonnick
Cyg 21h23'30" 35d6'26"
Lora Beth I Love You
Vir 11h58'51" 2d37'2"
Lora D. Caton
Cas 0h12'18" 54d48'37"
Lora Davis Lyle
Cas 0h20'5" 57d12'12"
Lora Diane Dasero
Peg 21h46'29" 27d55'48"
Lora Lee
Ori 5h59'34" -0d3'18"
Lora LeeAnn Peppin
Vir 13h19'38" 11d16'19"
Lora Lei Gorospe
Ori 6h11'31" -1d49'28"
Lora Lilavati Stice
Crb 15h44'43" 36d50'53"
Lora Lynn Birkett
Uma 11h7'23" 62d10'5"
Lora M.
Cnc 8h43'58" 21d52'57"
Lora & Michael Mogilewski
Cyg 20h47'56" 31d39'29"
Lora Nelson
Gem 7h14'20" 33d17'41"
Lora Screen
Boo 14h17'6" 47d6'1"
Lora Timberlake
Sco 17h15'46" -44d5'36"
Lora Todorova Hadjimiteva
Aqr 22h16'27" -9d24'33"
Lora Vallijo
Lmi 10h16'59" 29d5'38"
Lora Virginia Keleher
And 0h38'49" 39d55'25"
Lora Viti
And 0h18'54" 45d48'9"
Lorabill
Cyg 19h42'50" 54d56'2"
Loraine Abegaile G. Pinon
Gem 6h35'44" 18d41'27"
Loraine and Charles Keligian
Cyg 21h41'52" 43d29'8"
Loraine O
Sgr 18h57'44" -34d36'52"
Loraine Rodgers Greene
Aqr 22h27'32" 0d15'20"
Loraine Ruth Steele
Cas 1h3'15" 54d3'6"
Loraine S. Nevill
And 0h50'36" 45d54'15"
Loraine's Star
Ori 6h11'24" 8d20'49"
Lorainne & Andre
Cyg 19h45'32" 32d30'46"
Lorainne Constant
Aqr 23h27'15" -22d51'31"
LoraJean Fyfe the Love of My Life
Ari 3h3'42" 24d55'41"
LoraJoe
Uma 9h26'19" 66d1'38"
Loralee
Lib 14h55'26" -4d57'8"
Loralee Lynn
Uma 9h3'54" 55d23'23"
LoraLeiRineySparkie!
Lyn 7h52'36" 56d37'43"
Loralie & Joseph
Psc 0h44'3" 15d16'24"
Loralou
Tau 4h49'55" 17d11'31"
Loralyn Jean Delich
Lyr 18h48'31" 43d8'3"
Lorana Monagas
Lyn 7h47'27" 40d6'35"
loranmax-mildredelizabeth
Uma 10h25'54" 61d54'45"
LORANN
Ari 3h8'55" 19d31'14"
LorBear
Psc 1h20'37" 7d13'48"
Lorcan
Aqr 22h37'19" -3d0'2"
Lorcan Augustus McMahon
Per 3h49'21" 48d47'54"
Lorckavinney
Uma 13h35'51" 57d23'58"
Lord Bill
Ori 6h5'1" 13d4'56"
Lord Bobus Maximus
Aql 19h50'54" -0d29'10"
Lord Christopher
Tau 4h13'46" 27d34'28"
Lord Dain Miller
Sco 16h15'8" -16d29'33"
Lord David Wayne Steyer
Cap 21h6'0" -16d12'57"
Lord Fong
Ori 5h35'37" -6d53'45"
Lord Harold & Lady Joyce Whitfield
Cas 0h46'21" 67d5'7"
Lord Murphy's Company
Aur 5h22'18" 32d37'58"
Lord Nathan
Leo 9h41'32" 27d29'4"
Lord Neil Clive Sinclair
Cep 22h48'55" 79d30'5"
Lord of the Rings - Glenn Stumpff
Crb 15h59'11" 39d7'1"
Lord Shiva: Babaji
Umi 4h23'6" 88d31'40"
LordJensen2005
Umi 15h47'34" 77d19'41"
LORDY
Lac 22h17'47" 37d34'32"
Lore
Tau 3h56'25" 29d40'52"
Lore' Elizabeth Beach
Peg 21h33'49" 13d11'59"
Lore Lore's Star
Cap 21h16'13" -21d13'23"
Lore von Rymon Lipinski
Uma 8h41'27" 54d39'59"
Lorea
Col 5h54'19" -27d53'35"
L'Oreal Malaysia Sdn Bhd
Ori 5h38'44" 3d32'9"
Loreal Moreno
Uma 11h14'10" 70d13'33"
Loreal's Heavenly Light
Cyg 20h15'41" 56d38'25"
Loredana
Cas 2h18'59" 73d31'30"
Loredana
Lyn 6h50'43" 60d45'48"
Loredana
Cas 1h18'51" 54d13'48"
Loredana
Cam 3h52'12" 54d29'19"
Loredana
Gem 7h44'7" 33d30'6"
Loredana
Cyg 19h46'52" 31d21'45"
Loredana Aprile
Cap 20h18'22" -15d32'32"
Loredana Colombo
Boo 14h39'40" 10d44'22"
Loredana Di Scisciola
Leo 11h54'6" 21d45'9"
Loredana + Remo
Umi 14h25'22" 76d14'33"
Loree and Rad Payne
Tau 4h27'56" 20d34'13"
Loreen
And 1h0'49" 44d12'11"
Loreen Amaro Heart Star
Psc 0h56'9" 11d14'27"
Lorelai
Per 4h29'11" 36d0'17"
Lorelai Cadence
And 0h17'10" 29d34'13"
Lorelei
And 0h51'37" 21d43'48"
Lorelei
Uma 10h22'51" 40d14'46"
Lorelei Elise Morgan
Uma 8h54'41" 66d43'24"
Lorelei Elizabeth Sandberg
And 23h8'30" 42d44'13"
Lorelei J. Borland
Lyr 18h49'54" 31d49'16"
Lorelei May Sherman
Psc 1h23'21" 5d7'44"
Lorelei Starr Merritt
Del 20h38'43" 15d56'28"
Lorelei Walker
Lyr 19h15'41" 30d0'30"
Lorelei's Pinky
Ari 2h14'26" 23d7'5"
Lorelei's Wishing Star
Peg 22h28'3" 11d48'44"
Loreli
Vir 12h15'11" -3d26'29"
Lorell Gifford
And 23h49'3" 45d20'26"
Lorella
Ari 2h2'0" 10d43'20"
Lorella Barbi
Boo 14h47'43" 27d34'27"
Lorella Roberts
And 2h32'53" 50d22'30"
Lorelle Nicole White
And 0h44'7" 44d30'54"
Lorelli Susan Stone
Cas 23h18'20" 55d2'49"
Lorellin
Lyn 6h49'34" 52d24'57"
Loren
Psc 1h8'46" 32d2'45"
Loren
Lib 14h50'34" -5d30'16"
Loren A. Crowder
Ari 3h23'23" 28d30'3"
Loren Alexander Dunn
Sco 16h15'32" -25d30'31"
Loren Alexandra
Lib 15h44'54" -9d22'17"
Loren Alexis Osborne
Cyg 19h40'52" 39d48'3"
Loren and Donna Appleby
Umi 14h50'13" 75d49'17"
Loren and Jennings
Cyg 20h21'5" 45d58'17"
Loren Bosso DiFranco
Per 3h36'39" 47d40'54"
Loren Carrillo King
Cep 22h14'23" 67d20'14"
Loren Cate
Cap 21h35'27" -16d2'42"
LOREN CHARLOTTE
Sco 17h51'45" -36d27'56"
Loren David Adams
Ari 3h27'14" 19d29'30"
Loren & Donna Gunn
Sco 17h35'42" -36d10'3"
Loren Edgcomb Hale
Uma 9h51'4" 47d10'4"
Loren Houckham
Aqr 23h37'18" -15d8'2"
Loren & Kellie LaPlant
Umi 15h24'19" 73d21'52"
Loren Macchiaroli
Psc 1h20'12" 17d2'15"
Loren Mian
Cam 3h59'49" 54d37'24"
Loren Peterson
Uma 12h32'4" 52d38'9"
Loren Reilly
Umi 15h44'5" 72d56'15"
Loren Rochelle the Friendship Star
Ari 2h13'23" 25d0'42"
Loren Sten Tunforss
Gem 7h30'30" 26d38'49"
Loren William
Boo 14h49'12" 17d38'56"
Loren Woods
Umi 15h56'7" 72d50'29"
Loren y Claudia
Aqr 23h4'13" -17d20'13"
Lorena
Aqr 22h41'10" -23d42'36"
Lorena
And 0h16'2" 25d29'57"
Lorena
Gem 7h38'28" 23d12'54"
Lorena
And 0h57'47" 34d38'9"
Lorena Ayup
Cnc 8h37'8" 15d52'47"
Lorena Casadei
Cas 0h38'41" 60d14'41"
Lorena D' Angelo
Her 18h1'46" 18d3'53"
Lorena Grasso
Ori 6h13'4" 6d21'17"
Lorena H Bishop
Ori 5h24'47" 9d38'39"
Lorena Huizar
And 2h16'39" 39d11'45"
Lorena Idelmar Del Prado
Cam 5h11'58" 66d36'19"
Lorena Jervasi
And 0h8'26" 28d47'23"
Lorena Lynn Sardella
Lib 15h6'16" -3d54'32"
Lorena prime
Sco 17h41'1" -34d28'28"
Lorena Proietti
Cyg 20h11'38" 35d49'37"
Lorena Salazar
Cam 4h4'15" 68d50'32"
Lorena Torralbo
Com 13h33'46" 27d38'33"
Lorena Viera
Gem 7h28'1" 26d34'2"
Lorenay
Vir 12h44'0" -9d20'37"
Lorenda Jardey
Vir 12h50'22" 12d46'24"
Lorene A. Martin
Peg 23h9'52" 19d5'57"
Lorene Frye Burtner 100
Cas 1h20'16" 54d40'12"
Lorene Wayne Smith
Cnc 7h58'11" 12d10'41"
Lorenna R. Freeman "Pooky"
Sgr 17h52'42" -18d8'56"
Lorenz Ferdinand
Per 3h22'33" 36d45'18"
Lorenz Momo
Leo 10h51'27" 18d37'10"
Lorenz Theo
Cas 0h30'24" 60d12'17"
Lorenza
Cyg 19h37'52" 31d41'39"
Lorenza
Cas 1h13'39" 52d49'58"
Lorenza e Saverio
Per 4h17'57" 45d24'56"
Lorenza Torlo
Peg 22h4'0" 20d32'11"
Lorenzen, Barbara
Ari 2h45'9" 10d58'1"
Lorenzo
Per 3h4'5" 52d8'58"
Lorenzo
Her 18h11'34" 42d16'23"
Lorenzo
Cam 3h40'40" 55d14'19"
Lorenzo
Aur 5h57'38" 53d54'24"
Lorenzo
Her 17h14'57" 30d42'20"
Lorenzo
Crb 16h1'15" 31d31'3"
Lorenzo
Uma 11h41'59" 31d39'1"
Lorenzo
Aur 5h41'5" 32d41'10"
Lorenzo
Cnc 8h50'27" 31d14'25"
Lorenzo
Per 3h24'25" 41d31'58"
Lorenzo
Uma 9h11'24" 54d50'41"
Lorenzo
Uma 14h5'44" 62d17'47"
Lorenzo
Cma 7h19'0" -12d47'9"
Lorenzo
Sgr 18h18'7" -34d59'31"
Lorenzo Accomasso
Her 17h51'42" 16d56'7"
Lorenzo Barber
Psc 0h20'42" 8d28'15"
Lorenzo Barsacchi
Psc 1h33'17" 6d53'3"
Lorenzo Bastianello
Cyg 20h14'56" 36d55'22"
Lorenzo Breazile
Uma 11h35'16" 39d50'23"
Lorenzo Cammilleri
Cet 2h25'25" -14d50'25"
Lorenzo Dolfi
Per 2h49'54" 50d23'36"
Lorenzo Gregory Scavio
Cap 21h47'4" -17d14'24"
Lorenzo & Jeanne Wyche
Her 17h5'4" 31d57'59"
"Lorenzo" Lawrence Peter Fairfield
Cnc 9h10'51" 25d21'18"
Lorenzo Nicola Venturini
Ori 5h43'10" 1d55'14"
Lorenzo Oldani
Uma 9h9'35" 54d53'27"

Lorenzo Pagni
Boo 15h18'12" 42d31'50"
Lorenzo Perfetti
Crb 16h0'21" 35d32'30"
Lorenzo Piliego
Umi 14h9'2" 66d41'2"
Lorenzo T. (Ben) Mozley
Uma 11h2'9" 53d30'35"
Lorenzo & Tara For All Eternity
Uma 10h32'48" 67d47'22"
Lorenzo Terzo
Per 3h17'41" 43d2'13"
Lorenzo Tollari
Lyr 19h19'30" 33d7'54"
Lorenzo V. DeLaGarza Jr.
Gem 6h25'27" 21d38'31"
Lorenzo V. Picinic
Lib 14h50'44" -3d22'54"
Lorenzo W. Curry
Gem 6h52'55" 24d46'1"
LorenzoGKidd
Uma 10h32'10" 52d25'32"
Loreta IR Sigitas 02-14-2005
Crb 16h13'21" 27d45'32"
Loreto Carmine Gobbi
Cam 4h13'51" 74d36'6"
Loretta
Uma 8h54'48" 69d24'26"
Loretta
Cas 23h43'29" 55d36'35"
Loretta
Crb 15h27'9" 27d54'50"
Loretta
And 0h16'21" 27d37'37"
Loretta
Aqr 20h44'59" 1d42'8"
Loretta
Cnv 12h34'35" 46d20'49"
Loretta
Cyg 19h59'43" 41d12'55"
Loretta A. Dokas-Beier
Sgr 19h28'27" -40d19'36"
Loretta Ann Grijalva
Sgr 18h7'27" -26d35'34"
Loretta Balfour
Tau 5h25'5" 21d0'19"
Loretta Bambrick
Leo 11h37'56" 20d29'15"
Loretta Beckwith
Uma 10h28'48" 45d39'43"
Loretta Bleier
Cyg 21h55'54" 46d29'15"
Loretta Canaday
Cyg 19h45'3" 35d43'41"
Loretta Caterino
Ori 5h52'5" 22d7'36"
Loretta Childs
Ori 6h8'51" 6d14'4"
Loretta Corinne
Crb 15h34'17" 36d6'1"
Loretta Elaine Long 1936
Gem 7h48'26" 32d17'55"
Loretta Gene Horn
Vir 13h24'21" 8d11'2"
Loretta Genevieve Maas
Ari 2h29'17" 22d28'59"
Loretta Glynn
Psc 1h23'54" 24d55'43"
Loretta Gnocchi Paternollo
Her 18h22'26" 21d49'35"
Loretta Green
Sco 16h17'6" -15d11'35"
Loretta Irene Veltri
Uma 12h10'18" 48d31'34"
Loretta J
Psc 1h8'58" 10d44'45"
Loretta J. Manning
Aqr 23h0'9" -4d50'17"
Loretta Johncola
Uma 10h46'49" 62d46'44"
Loretta (Jones) Coffman
Cap 20h21'28" -20d59'14"
Loretta Joy Duncan
Umi 13h20'55" 71d3'58"
Loretta "Kay" Grable
And 1h25'4" 43d19'5"
Loretta Kayleen Schultz
And 0h16'17" 28d10'38"
Loretta Lazeski Arnold
Sco 16h56'46" -43d13'54"
Loretta Lee Binkley
Vir 12h52'12" 2d54'53"
Loretta LePeak
Cam 8h10'17" 68d19'48"
Loretta Loudin Parker
Lib 15h42'20" -5d38'29"
Loretta Lynn Lindemann
And 23h58'52" 34d30'20"
Loretta M. Smalley
Cas 1h4'37" 59d17'9"
Loretta Mae "Asbury" Hurley "Mom"
Uma 9h16'12" 64d24'7"
Loretta Mae Sanchez Jonkers
Uma 8h38'53" 49d31'16"
Loretta Margaret Herman O'Brien
Vir 11h52'8" 3d33'12"
Loretta Marie Hoog
Cyg 20h17'22" 34d41'36"
Loretta Mary Rita Emanuele
Vir 13h40'34" -5d32'16"
Loretta McMillan
And 0h37'37" 26d49'59"
Loretta Meredith
Vir 13h31'4" 11d38'22"
Loretta Merritt-Superstar Educator
Cas 1h37'43" 64d11'44"
Loretta Meus Decorus Matris Raftis
Cru 12h51'5" -64d17'51"
Loretta "Mom" Country Grandma
And 23h30'22" 41d54'43"
Loretta Morgan
Sco 17h27'31" -41d4'53"
Loretta Natter
Cas 0h37'22" 57d10'26"
Loretta Nester
Ori 5h38'41" 0d6'41"
Loretta O'Brien
Del 20h41'38" 16d14'8"
Loretta Palmer
Lyn 6h50'55" 56d48'48"
Loretta Peter
Psc 1h23'48" 23d54'54"
Loretta Reiter
Cas 0h58'21" 50d18'22"
Loretta Ruth Cruz
Ari 3h9'59" 14d42'48"
Loretta Santa Lucia
Cnc 8h23'15" 12d51'29"
Loretta Schatz Rozankowski
Leo 10h52'20" 23d3'36"
Loretta Skittone Giles Martin White
Lmi 10h50'44" 36d52'38"
Loretta Star
Lyn 7h19'9" 53d35'0"
Loretta "Suave" Dodge
Lib 15h56'39" -17d24'4"
Loretta Sue Mercer
Crb 15h45'19" 30d37'58"
Loretta Thomas
Cnc 8h38'45" 16d16'0"
Loretta Todisco
Uma 11h8'49" 47d7'5"
Loretta Winter
Uma 9h52'46" 71d14'11"
Loretta Wittich
Tau 5h52'9" 15d40'3"
Loretta Woulfe
Cas 1h43'26" 61d19'23"
Loretta, A Wonderful Wife
Ori 5h21'29" 1d37'42"
Loretto
Cyg 19h41'35" 30d37'15"
Lori
And 0h30'1" 40d33'40"
Lori
And 1h53'59" 43d52'21"
Lori
Uma 11h50'28" 47d57'53"
Lori
Uma 9h46'8" 47d44'41"
Lori
Lyn 7h17'31" 50d16'7"
Lori
Cas 0h34'48" 47d18'55"
Lori
Tau 4h35'0" 15d41'52"
Lori
Leo 9h39'1" 27d30'2"
Lori
Uma 10h6'49" 55d21'52"
Lori
Uma 8h33'0" 69d12'29"
Lori
Cas 2h27'53" 65d41'54"
Lori
Uma 11h2'28" 61d36'52"
Lori
Aqr 23h5'2" -5d19'5"
Lori 12-06-60
Cas 0h56'35" 58d53'35"
Lori 8-27-1977
Vir 13h47'16" -0d32'8"
Lori A. Brown
Lyn 7h2'27" 51d1'44"
Lori A. Cruz
Gem 6h37'55" 17d29'49"
Lori A. Meacham, DVM
Lyn 8h4'48" 34d47'9"
Lori Alcala
Cas 23h58'20" 56d32'35"
Lori Alison
And 2h5'56" 40d54'0"
Lori and Bobby
Psc 1h21'45" 24d39'59"
Lori and Caryn's Devotion
Gem 7h3'45" 26d6'10"
Lori and Christopher Stearman
And 23h37'44" 43d6'8"
Lori and Jonathan
Lyr 18h44'16" 39d28'43"
Lori and Kevin
Ori 6h18'45" 13d27'46"
Lori and Steve
Sge 19h41'44" 18d6'54"
Lori Anderson Mundy
Tau 5h14'40" 18d28'52"
Lori Ann
Ari 3h15'16" 18d47'56"
Lori Ann
Gem 6h30'55" 21d46'32"
Lori Ann
Cnc 8h34'53" 19d16'16"
Lori Ann
Uma 12h40'59" 63d1'5"
Lori Ann
Uma 9h13'45" 62d56'39"
Lori Ann
Lib 15h4'44" -7d16'45"
Lori Ann
Sgr 18h19'25" -32d25'2"
Lori Ann Abbruzzese
Lyn 8h0'39" 47d7'6"
Lori Ann Ball
Uma 11h1'32" 69d21'25"
Lori Ann Binns
And 0h31'39" 32d26'37"
Lori Ann Edmonds
Lyn 9h13'0" 37d36'30"
Lori Ann Ergott
Ori 5h35'50" 4d49'50"
Lori Ann Ferreira
And 0h0'11" 44d35'18"
Lori Ann Forman
Cap 21h58'18" -21d23'39"
Lori Ann Halbert
Cas 0h7'37" 53d17'15"
Lori Ann Hart
Lib 15h36'25" -9d43'56"
Lori Ann Lehman
Vir 13h47'6" -11d14'31"
Lori Ann Maffei, Sugerfoot
And 1h6'52" 42d31'44"
Lori Ann Rainwater
Crb 16h13'56" 28d56'51"
Lori Ann Rollings
Leo 9h24'4" 27d18'20"
Lori Ann Sapp
Lmi 10h47'34" 23d30'40"
Lori Ann Saroli
Cas 23h24'37" 57d54'59"
Lori Ann Yares
Psc 0h36'55" 14d19'11"
Lori Anne
Ari 3h5'45" 28d27'33"
Lori Anne
Psc 1h18'12" 30d28'4"
Lori Anne
Cas 0h12'5" 54d52'45"
Lori Anne Fisher
Uma 8h13'1" 67d59'52"
Lori Anne McDonald
Ori 6h17'40" 14d40'52"
Lori Anne Swearingen (Titi)
Psc 0h33'49" 15d49'35"
Lori Avanzato Arbitell
Tau 4h15'22" 21d52'59"
Lori B.
Lib 14h30'48" -9d34'4"
Lori B. McIntire
Lyr 19h11'25" 26d31'41"
Lori B Minassian
Gem 6h38'24" 17d36'53"
Lori Bay
Gem 7h9'30" 24d59'22"
Lori Bea
Crb 16h4'12" 26d52'5"
Lori Beth
Leo 10h15'34" 25d48'24"
Lori Beth
And 0h45'55" 40d42'27"
Lori Beth Hunt Sutphen
Uma 8h36'51" 52d44'35"
Lori Biassoni
And 2h20'42" 44d9'5"
Lori Billingsley
Aqr 22h50'25" -10d17'42"
Lori Blair-Graves
Leo 11h39'57" 18d38'32"
Lori & Bradley Mahon
Sco 17h55'21" -37d2'38"
Lori Brandon
Dra 19h45'29" 70d13'19"
Lori Brown
Uma 8h50'28" 73d0'17"
Lori Brudzisz
Cyg 19h48'37" 31d19'5"
Lori Burns Fredrickson
Ori 6h24'49" 16d58'44"
Lori Burr
Ari 2h12'47" 25d58'6"
Lori Camp
Uma 9h21'4" 51d55'22"
Lori Campbell
Aqr 23h3'53" -10d23'10"
Lori Carrin Young
Sco 16h53'13" -42d34'39"
Lori Chelius
And 0h32'35" 31d52'32"
Lori Crownover
Aqr 22h4'17" -21d40'24"
Lori D. Ross
Peg 21h50'12" 16d27'35"
Lori & Dane Martens
Umi 15h13'39" 74d54'37"
Lori Danielle Gould
Lmi 10h41'17" 27d15'0"
Lori Darlin
Col 5h33'48" -34d15'1"
Lori Dawn
Psc 1h8'7" 3d17'44"
Lori Dawn McClelland
Sex 9h44'17" -0d35'19"
Lori Deanna
Leo 9h23'26" 11d47'26"
Lori Dell
Mon 6h43'27" -0d19'43"
Lori Delores (Dee Dee) Cleesen
Aqr 22h50'7" -8d4'3"
Lori DeMonte
And 1h36'50" 46d37'40"
Lori Denise Blalock
Sgr 18h21'42" -21d29'15"
Lori Diesbach
Vir 11h48'3" -2d40'22"
Lori Dittus
Aqr 22h44'47" -3d54'11"
Lori D'Onofrio
Tau 5h48'50" 22d14'41"
Lori Dumas
Peg 23h47'43" 27d39'46"
Lori Durbin
Tau 5h47'23" 18d57'11"
Lori E. Lazenby
And 2h32'53" 38d30'50"
Lori Ellen Earle
Cas 0h37'53" 64d19'16"
Lori Ellen Peterson
And 23h25'18" 48d22'26"
Lori Ellen Silva
Lyn 7h40'13" 56d42'39"
Lori Eschagary
Psc 0h20'13" 16d50'15"
Lori Evans
Ori 4h46'22" -2d15'40"
Lori Foehr
Lmi 10h8'20" 31d40'42"
Lori Gearhart Peters
Lib 15h50'30" -3d46'38"
Lori Gene Gilliam
Ori 5h2'55" -0d0'28"
Lori Gentry
Gem 7h5'13" 18d20'21"
Lori Ginger Snap Rudolph
Aqr 20h57'46" -7d10'57"
Lori Hanko Mendoza
Uma 10h22'44" 64d47'25"
Lori Haviland-Foley (Mommy)
Cas 0h43'56" 63d49'33"
Lori Hohn
Sgr 18h13'19" -29d43'37"
Lori Holdheide
Aqr 21h24'28" -7d38'35"
Lori Hynes
Ori 6h13'14" 19d42'37"
Lori Ivey
And 23h3'37" 48d17'33"
Lori Ivey
And 23h2'47" 47d44'45"
Lori J Popodi
Cas 1h42'12" 66d12'58"
Lori Jean Dombrowski
Cnc 8h16'35" 19d24'46"
Lori Jean Elkins
Cas 2h13'34" 67d1'11"
Lori Jean Orzulak
Gem 7h5'10" 23d7'13"
Lori Jeanne Bopp
Psc 1h5'29" 32d34'34"
Lori Jewkes Fenton
Uma 11h7'9" 56d59'17"
Lori Jo
Psc 1h13'1" 17d34'16"
Lori Jo Martin
Tau 4h16'53" 12d38'1"
Lori 'Jon Hearon
Mon 6h49'34" -0d8'56"
Lori Jon Rubin
Her 16h39'3" 45d17'41"
Lori June Henderson
Lyr 19h14'8" 27d2'7"
Lori K. Craig
Cas 0h46'36" 64d50'38"
Lori Katherine Couch Star
And 0h45'26" 35d14'7"
Lori Kay Hanson
And 23h27'54" 47d37'16"
Lori & Ken 2005
Cyg 20h46'23" 39d54'51"
Lori Kennedy Brittan
And 1h15'59" 42d16'22"
Lori Kilpatrick
Cas 23h49'52" 55d44'45"
Lori Kimberly Stamps
Sgr 19h22'41" -14d24'50"
Lori Lambchop Pedemonti
Tau 5h9'7" 22d3'32"
Lori Lee
Vir 13h19'47" -0d39'31"
Lori Lee Lewis Star
Sgr 19h45'22" -30d24'33"
Lori Lenay Eickleberry Patterson
Peg 21h55'52" 35d17'31"
Lori Lewis
Lib 15h41'35" -13d55'21"
Lori Lilly Preston
Cnc 8h48'55" 32d13'58"
Lori Lynn
Aqr 21h23'41" 0d23'30"
Lori Lynn
Com 12h46'47" 28d16'55"
Lori Lynn
Tau 5h32'17" 26d8'32"
Lori Lynn
Cam 3h27'40" 68d12'7"
Lori Lynn Ballinger
Lib 14h53'20" -4d42'51"
Lori Lynn Bauer
Gem 6h46'35" 17d8'9"
Lori Lynn Eilo
Cam 4h6'24" 62d42'13"
Lori Lynn Garbison
Vir 13h20'22" 6d31'34"
Lori Lynn Lamprecht
Uma 11h28'39" 32d20'24"
Lori Lynn Little
And 2h30'12" 43d59'46"
Lori Lynn Sykas
Cnc 8h40'11" 7d35'42"
Lori M. Calderas
Per 3h15'45" 46d1'8"
Lori M Smith
Gem 6h33'40" 18d23'7"
Lori Malina George
Sgr 19h41'8" -15d16'14"
Lori Marie Frain "06/04/1988"
Umi 13h9'16" 76d2'56"
Lori Marquez
Tau 5h49'22" 22d20'1"
Lori Martelli
Leo 11h23'53" 14d5'35"
Lori Matley
Crb 15h36'4" 37d9'7"
Lori Mazrimas
Cnc 9h5'49" 9d22'53"
Lori McCleary
Gem 6h49'45" 33d49'28"
Lori McLaughlin
Her 16h44'6" 41d39'54"
Lori McLellan
Cas 0h14'44" 50d46'43"
Lori Michelle Roberts
Cas 1h19'14" 54d25'37"
Lori Minton my love
Dra 15h38'22" 64d33'37"
Lori Mock Robinson
Leo 11h10'47" -1d12'59"
Lori Mockridge
Cas 1h33'51" 66d41'9"
Lori Moretti
Ari 3h20'21" 27d14'36"
Lori Murray
And 1h36'19" 46d21'36"
Lori my Angel
And 1h25'38" 48d15'32"
Lori (My Angel) Faubert
Uma 10h51'36" 52d49'39"
Lori - my best friend
Cnc 8h35'17" 8d16'9"
Lori My Love
And 1h1'59" 39d50'53"
Lori My One & Only Mom
Tau 5h26'34" 21d57'32"
Lori Myers
And 0h32'16" 42d38'37"
Lori Nolen
Cnc 9h4'26" 31d28'22"
Lori O'Brien
Ari 3h18'46" 20d11'44"
Lori Phelps
Cnc 8h35'25" 22d24'27"
Lori Pisaneschi
Vir 13h4'45" -3d18'54"
Lori Pollard
Cnc 8h12'48" 32d11'41"
Lori Preston
Psc 1h0'55" 12d3'16"
Lori Princess Bride
And 23h35'27" 45d43'50"
Lori Pushckor
Lyn 6h36'46" 57d6'33"
Lori Rae
Cas 1h34'50" 62d51'28"
Lori Rae Martinson
Tri 2h30'36" 31d7'49"
Lori Rae Stephens
Ari 2h12'13" 25d17'27"
Lori Rene Brown
And 2h8'34" 47d7'37"
Lori Rhinesmith
Aqr 22h39'21" -20d0'8"
Lori Ricci
Cas 23h14'44" 58d30'58"
Lori Rios
Mon 6h24'45" 9d13'37"
Lori Rospierski
Uma 11h36'34" 54d0'34"
Lori Ryerson
Ari 2h27'44" 27d42'56"
Lori Sands
Cyg 21h52'24" 38d39'45"
Lori Saylor loves Brad Curtis
Cyg 21h19'11" 51d43'2"
Lori Schlachter
Cas 23h28'50" 52d19'40"
Lori Schmidt
Sgr 18h23'21" -32d12'32"
Lori Shannon
Cnc 9h4'55" 22d42'7"
Lori Sherer
Tau 5h29'24" 25d39'47"
Lori Shiffler (Mommy)
Psc 0h17'9" 11d35'20"
Lori Skretta
Aql 19h33'38" 15d59'51"
Lori Spinks
Aqr 22h7'23" -16d6'53"
Lori Star
Leo 9h22'34" 10d8'38"
Lori & Steve
Uma 11h30'10" 61d4'19"
Lori " Sweet Pea "
Uma 12h57'44" 54d0'22"
Lori T. Nichols
Gem 6h46'35" 26d56'58"
Lori Taylor
Tau 3h56'32" 29d23'10"
Lori the Leo's 35th Birthday Star
Leo 10h54'16" 14d45'34"
Lori & Tyler Gear
Uma 11h40'45" 39d1'22"
Lori Van Duzer
Cas 0h56'23" 65d20'45"
Lori Veerkamp
Lyn 9h30'5" 40d21'59"
Lori Wallace
Cnc 8h33'50" 17d55'0"
Lori Watson
Cnc 8h51'48" 29d32'55"
Lori Werner
Gem 7h12'48" 32d24'25"
Lori Wisz
And 2h15'56" 50d33'0"
Lori, 10.08.1995
Umi 16h1'16" 73d7'6"
Loriana Aviles
Gem 7h14'6" 34d7'26"
Loriann
Lyn 8h41'42" 46d32'46"
Lori-Ann Lambert (An Angel)
And 1h30'13" 48d55'52"
Lorianne Gambrell
Cnc 8h38'55" 29d37'2"
Loriboo
Aqr 22h57'4" -7d12'41"
Lorick's Gift
Sco 17h23'57" -33d52'22"
Lorie
Aqr 22h52'2" -6d19'36"
Lorie Bliss Nichols
Lib 14h31'57" -22d40'50"
Lorie Lindahl & Jacquelyn Hightower
Crb 15h49'39" 34d48'45"
Lorie McCaulley
Lib 15h11'30" -10d32'15"
Lorie Sherman
Lyn 7h27'24" 44d37'7"
Lorie Villanueva
Cnc 8h40'37" 14d2'4"
Lorien Mace Foux
Del 20h49'51" 14d30'36"
Lorilee Elizabeth Toole
Cas 2h23'11" 74d38'31"
Lorilee Goodall
Mon 7h47'35" -0d44'32"
Loril's love
Uma 10h47'35" 55d48'22"
LoriM
Cap 21h27'36" -9d30'9"
Lorimik
Gem 7h27'41" 16d27'6"
Lorin - 05/11/2006
Sco 17h33'39" -35d51'30"
Lorin Lynn
And 2h29'48" 50d9'50"
Lorin Samuel Holmes
Per 3h23'28" 51d42'52"
Lorina L. Aguilar
Lyr 19h17'22" 28d27'21"
Lorina Salinas
Ori 5h20'22" 3d53'26"
Lorinda
Ori 5h25'56" 5d54'27"
Lorinda
Cyg 20h0'19" 40d6'12"
Lorinda
Uma 11h19'50" 51d28'7"
Lorinda
Cap 20h10'5" -12d53'0"
Lorinda M. Barker
Leo 9h42'44" 16d21'29"
Lorin's Star
Tau 4h8'39" 27d48'31"
Loris
Lyr 19h11'26" 37d47'11"
Loris
Crb 16h22'35" 38d29'41"
Loris
Cas 23h26'15" 58d10'55"
Loris Angel
Her 18h44'14" 22d0'28"
Lori's Angel
Gem 7h0'40" 14d27'23"
Lori's Baubles and Beads
Aqr 21h47'54" -2d32'21"
Lori's Light
And 0h32'40" 45d15'47"
Lori's Light
And 0h18'37" 37d30'56"
Lori's Paradise
Aqr 22h35'58" -3d2'33"
Loris Soldato
Cyg 21h10'31" 31d49'41"
Lori's Star
Cyg 21h40'21" 38d42'58"
Lori's Star
Cam 4h45'54" 53d56'3"
Lori's Star
Vir 13h4'47" 11d26'17"
Lori's Star
Aql 19h37'12" 3d54'9"
Lori's star
Leo 11h43'49" 19d50'42"
Lori's Star
Uma 9h6'27" 65d52'35"
Lori's Star
Sco 17h45'42" -41d18'25"
Lori's Wishing Star
And 23h25'32" 41d41'34"
Lori's Zahira Zabana Suha
Uma 12h36'7" 55d38'47"
Lorisa Bellus
Sgr 18h54'27" -31d46'9"
Lorissa
Ari 2h33'29" 13d38'3"
Lori-Volino-Bizzarro
Leo 11h10'23" -1d10'19"
Lorjas
Cep 21h26'56" 63d9'32"
Lorna
Umi 13h27'40" 72d51'40"
Lorna
Cas 23h4'43" 55d38'32"
Lorna
Leo 11h7'26" -2d37'53"
Lorna Ann Bourke
Cas 23h52'52" 56d37'9"
Lorna Ashworth Wenzel
Lib 15h54'6" -13d9'58"
Lorna Banzon
Peg 21h58'25" 21d24'7"
Lorna Betteridge
Tau 4h10'52" 3d29'11"
Lorna Doone
Crb 15h47'36" 34d25'28"
Lorna Dyson Centenary Star
Cru 12h16'15" -61d48'48"
Lorna Fernbacher
Aqr 22h2'31" -17d12'8"
Lorna Fox
Vir 12h38'11" -0d5'13"
Lorna "Gram-Gram" Collar
Uma 13h38'36" 56d9'3"
Lorna Jacqueline
Cas 23h22'52" 54d58'11"
Lorna Jean
And 23h13'58" 42d40'39"
Lorna Jean Dodge
Lib 15h43'27" -29d12'2"
Lorna Jean Dykstra
Sgr 19h4'58" -20d45'41"
Lorna Jean Heuer
And 0h29'57" 45d19'39"
Lorna Joan Styles 15th January 1927
Cru 12h28'58" -61d8'15"
Lorna Johnson
Cas 0h28'49" 65d43'8"
Lorna Joseph
Lyn 9h5'22" 36d7'40"
Lorna Joyce Cobbin
Lib 15h27'8" -9d27'21"
Lorna Jury & David Gladstone
Cyg 21h31'11" 39d56'37"
LORNA L. Pryor
Lyn 9h18'10" 38d2'25"
Lorna M. West
Cas 2h12'34" 59d51'33"
Lorna & Marvin ~ Anniversary Star
Cyg 20h11'7" 35d52'59"
Lorna McQuillan (nee Marshall)
Cas 23h13'48" 58d26'17"
Lorna Murray
Uma 8h55'52" 50d53'19"
Lorna Rae
Lib 14h48'51" -6d49'22"
Lorna Sargent
Cas 0h35'28" 51d48'1"
Lorna Wright
And 0h31'24" 25d56'42"
Lornah D. Schulte
Uma 8h48'7" 69d46'8"
Lorna's Evening Sparkling Light
Gem 7h27'44" 14d44'37"
Lorna's Love
And 0h3'10" 44d9'31"
Lorne McCrae
And 1h47'48" 44d31'33"
Lorne Poppa Congdon
Uma 8h17'29" 59d52'29"
Lorne & Wendy Abony
Cyg 20h39'53" 45d2'50"
Loron
Oph 17h45'19" 2d44'8"
Lorraine
Leo 9h25'28" 11d33'19"
Lorraine
Cas 0h10'20" 54d0'36"
Lorraine
And 23h3'51" 41d35'56"
Lorraine
Lyr 19h27'27" 37d40'13"
Lorraine
Gem 8h6'4" 30d27'53"
Lorraine
Lmi 10h15'14" 30d50'25"

Lorraine
Uma 8h48'47" 56d59'25"
Lorraine
Cas 0h29'1" 61d46'41"
Lorraine
Cam 5h54'29" 66d14'6"
Lorraine
Vir 13h7'20" -15d53'44"
Lorraine
Ret 3h44'39" -53d10'11"
Lorraine 11201
Uma 8h34'42" 67d4'46"
Lorraine A. Batina
Srp 18h36'20" -0d27'10"
Lorraine Adele Schneider
Uma 9h57'35" 45d41'25"
Lorraine and James Calderazzo
Cyg 19h38'12" 34d22'59"
Lorraine and Michael Nolan
Cyg 20h11'25" 33d17'54"
Lorraine Ann Sobieck VanHale Wagner
Aqr 22h35'55" -9d9'52"
Lorraine Ann Wisse
Uma 9h33'1" 72d41'25"
Lorraine Anne
Gem 7h6'30" 17d43'46"
Lorraine B. Blaikie
Uma 10h58'13" 70d49'7"
Lorraine Bernice Henry
Lmi 10h8'10" 37d38'50"
Lorraine "Best Mom " White
Ari 2h48'57" 29d10'58"
Lorraine Beth
Cap 20h29'15" -11d52'32"
lorraine boswell
Uma 11h46'20" 30d3'28"
Lorraine C. Jessen-8/11/1923
Leo 11h4'55" 20d32'13"
Lorraine Caraballo
Cas 0h47'45" 65d46'37"
Lorraine Coppola
Cnc 8h48'3" 17d48'32"
Lorraine Couture
Sgr 19h15'17" -21d31'44"
Lorraine Cullina
Vir 13h20'57" -11d4'2"
Lorraine Dallas
Del 20h36'36" 15d4'3"
Lorraine Debenport
Cap 21h22'18" -19d22'14"
Lorraine Dickenson
Crb 15h44'20" 29d32'49"
Lorraine Dorsey
Oph 16h47'22" -0d12'2"
Lorraine Elizabeth Little Sweetie
Sgr 19h29'59" -32d24'30"
Lorraine Ethel Beno
Leo 11h34'26" 18d54'27"
Lorraine Evans
Cas 0h59'23" 57d18'3"
Lorraine Ewell Dingman
Vir 13h19'2" 10d41'39"
Lorraine Fairish
Lib 14h42'12" -10d14'37"
Lorraine Farrell
Ori 6h16'54" 16d2'18"
Lorraine Favaro
Peg 23h14'47" 25d53'26"
Lorraine Faye Grant
And 23h37'21" 47d15'44"
Lorraine Fota
Ori 5h1'33" 5d25'9"
Lorraine Frances (Brothers) O'Neil
Cnc 9h5'31" 31d50'23"
Lorraine Franzi
Uma 10h44'15" 62d42'55"
Lorraine Gilbert
And 0h18'26" 33d38'50"
Lorraine Gresl
Peg 23h56'16" 24d54'17"
Lorraine H. Whitten Gilley
Sgr 19h6'15" -30d53'56"
Lorraine Harty
Sgr 18h43'40" -19d29'35"
Lorraine Haruye Aoyagi
Cyg 20h14'19" 43d29'46"
Lorraine Hoku of Ikaika
Cyg 19h49'37" 29d53'19"
Lorraine Ilene Rosado
Uma 10h41'24" 46d53'25"
Lorraine Johnson
Lib 15h13'42" -11d37'28"
Lorraine Joss
Vir 13h43'20" -7d1'2"
Lorraine Kirk
And 23h31'18" 48d9'46"
Lorraine Lapetina
Vir 13h26'57" -19d16'5"
Lorraine Lazaro
Cam 4h4'41" 55d46'41"
Lorraine Lombardo
Leo 11h48'16" 27d43'53"
Lorraine: LT&LL 2204
Leo 10h57'9" 21d32'21"
Lorraine Lux Lucis
Tau 4h29'11" 26d11'47"
Lorraine M. Neri
Sgr 20h9'7" -28d32'47"
Lorraine M. Olmsted
Lyn 6h36'26" 61d20'21"
Lorraine M. Pereira
Leo 11h32'0" 15d28'41"
Lorraine Manfredi
And 0h35'32" 43d49'44"
Lorraine Maynard
Lib 14h58'22" -11d16'46"
Lorraine Medina ~ Mona Lisa
And 0h35'41" 25d16'4"
Lorraine Melia
Gem 6h23'48" 21d3'25"
Lorraine Mitchell
Leo 10h31'20" 13d21'36"
Lorraine Monnot
Lib 14h39'4" -13d14'8"
Lorraine My Light of The Morning Star
Uma 10h18'13" 61d55'51"
Lorraine Neilson-Blauvelt
Lyr 18h26'40" 31d0'12"
Lorraine Nicole Smith
Leo 11h9'55" 0d43'31"
Lorraine Noble
Cas 0h2'18" 63d40'23"
Lorraine Ostojic
Cru 12h38'4" -58d41'43"
Lorraine P. Rosati
Uma 11h17'38" 32d27'44"
Lorraine Patricia Plummer (Mom-Gma)
Cas 1h29'42" 61d55'2"
Lorraine 'Raine' Kalk
Lyr 18h50'41" 34d17'19"
Lorraine Rodriguez
Lyn 8h57'37" 37d20'45"
Lorraine Ruderer
Lib 15h4'35" -12d50'26"
Lorraine Ruth Miller
Cap 21h45'6" -9d45'53"
Lorraine Savoie mon amour
Cyg 20h52'26" 32d19'56"
LORRAINE SCHRAMKE
And 0h49'29" 37d7'20"
Lorraine Smith
Aql 19h11'44" -2d8'8"
Lorraine Smith Brown
Tau 5h34'56" 26d42'36"
Lorraine Sparveri
Cam 4h14'57" 56d59'0"
Lorraine Summers
Cap 21h19'1" -24d58'23"
Lorraine T. Rosica
Cnc 8h50'21" 32d6'9"
Lorraine Theresa Verstraete
Aqr 22h38'8" 1d32'29"
Lorraine Thompson
Lyr 19h17'37" 29d38'45"
Lorraine Wiedemann
Gem 7h20'43" 26d59'37"
Lorraine Willard
Leo 10h38'59" 20d33'16"
Lorraine Wilson
Uma 11h30'30" 36d37'22"
Lorraine Wilson, DCGS-N PM
Aqr 22h8'34" -15d43'28"
Lorraine Yeager
Cas 0h51'22" 61d8'42"
Lorraines Beauty
And 2h8'11" 42d0'59"
Lorraine's Gagster
Aqr 20h46'56" -8d49'30"
Lorraine's Nightlight de Amor
Leo 10h40'57" 12d45'36"
Lorraino
Cas 2h41'25" 57d35'48"
LORRA'S LIGHT
Ari 3h15'39" 27d33'55"
Lorren Michelle Conroy
Leo 11h6'49" 10d34'8"
Lorri
Uma 9h41'20" 44d18'10"
Lorri Ann Tagliaferro
Lyn 6h19'1" 57d48'29"
LORRI AZAR
Psc 0h45'1" 5d12'21"
Lorri B
Cap 21h19'35" -15d1'30"
Lorri M. Booth
Cam 7h40'1" 61d16'50"
Lorri Maria Dursi
Crb 15h40'59" 29d28'21"
Lorri Stanislav
Ori 5h58'40" -0d10'30"
Lorriane Partaine Freeman
Uma 9h32'51" 61d40'13"
Lorrie
Aqr 23h4'53" -6d59'26"
Lorrie Ann Besemer
Cyg 20h45'46" 34d30'17"
Lorrie Anne Ruff Gray 3-7-1963
Mon 6h26'13" 9d11'21"
Lorrie Austin
Lyr 19h12'9" 34d30'15"
Lorrie Beth
Ari 2h39'47" 25d55'53"
Lorrie Beth's Dream Star
Psc 23h44'30" 8d4'20"
Lorrie Cynowa
Cas 23h0'51" 56d13'8"
Lorrie & Dean
Cas 1h13'31" 62d44'49"
Lorrie Deysher & Randall Jarvis
Cyg 20h37'39" 36d26'47"
Lorrie Ellen Hickman
Leo 11h3'23" 10d50'30"
Lorrie Green
Uma 11h43'57" 64d43'24"
Lorrie Hope 50
Vir 12h17'34" 3d43'46"
Lorrie Lee Koahou
Tau 5h1'51" 15d42'19"
Lorrie Lew
And 23h25'28" 43d1'3"
Lorrie - Loving Wife and Mother
Cnc 9h14'23" 7d35'31"
Lorrie Macha
Crb 15h32'54" 28d51'44"
Lorrie Melton
And 23h39'46" 49d13'4"
Lorrie & Steven Young
Lyr 19h25'57" 35d59'22"
Lorrin
Dra 16h17'20" 61d56'21"
Loruana Rodriguez
Del 20h32'33" 18d50'24"
Lory
Cas 0h8'56" 56d21'34"
LORY 1947
Uma 13h28'5" 53d41'14"
Lory et Stéph, unies pour la vie
Cas 23h1'35" 57d29'19"
Lory Masters
Lib 15h47'22" -17d8'33"
Lory McGuire
Lmi 10h27'35" 37d17'33"
Lory & Rick Ankiel
Cyg 19h42'25" 37d31'45"
Loryn S Wright
Gem 7h44'45" 25d32'56"
Los and Lori Love Star
Tau 4h20'15" 15d39'13"
Los Angeles Urban League
Aql 19h46'21" -0d12'9"
Los hermanos Gandarillas
Gem 7h1'42" 23d47'6"
los pingüinos de la plata del vient
Ari 2h16'22" 24d44'42"
Los Quetos
Uma 8h18'11" 61d20'57"
Los Tres Vigos
Dra 17h15'57" 65d19'54"
Löscher, Stefan
Uma 12h31'29" 54d32'2"
Losert, Felix
Uma 11h35'20" 30d30'40"
Losey
Cma 6h35'53" -18d29'31"
LOSOCH AJA
Leo 11h5'2" 20d3'6"
Losonczi Ottó
Aqr 20h41'18" 0d49'42"
Lost Love
Ori 5h36'53" 10d42'19"
Lote & Aline & Mary Jane
Uma 10h43'5" 45d14'52"
Lothar Gebhardt
Aqr 22h19'56" -8d24'33"
Lothar Güth
Uma 10h1'33" 58d44'5"
Lothar Maier
Uma 11h41'25" 65d39'54"
Lothar Schramm
Uma 10h5'49" 41d29'15"
Lothar Struckmeier
Uma 9h30'40" 42d17'33"
Lots and Lots
Sgr 19h33'59" -13d52'36"
Lot's of love (Jes & Gene)
Tri 2h24'29" 36d33'54"
Lots of Pride
Uma 12h32'3" 58d50'58"
Lotsie Webster
Cam 7h22'16" 68d56'1"
Lotta Andersson
Uma 10h45'39" 58d4'7"
Lotta Sonja Carolina Simberg 280589
Cas 23h28'22" 56d13'57"
Lottchen
Aqr 20h39'15" -1d52'56"
Lotte "Beste Oma Der Welt" Woerz
Crb 15h49'27" 26d9'30"
Lotte Blumenbach
Uma 12h16'39" 53d13'31"
Lotte Hansen
Dra 16h22'46" 62d8'37"
Lotte Lena Frank
Uma 9h53'44" 65d0'24"
Lotti Adora
Cyg 21h6'1" 47d31'45"
Lotti Ganz
Uma 11h7'46" 30d1'23"
Lotti Shimkowski
Gem 7h49'5" 17d59'50"
Lottie
And 23h35'51" 47d11'51"
Lottie
And 23h44'41" 38d25'5"
Lottie Allene Glyn
Uma 10h22'20" 61d44'53"
Lottie (amma) Martin
Peg 23h22'39" 22d41'13"
Lottie Bell Lee
Ant 9h39'31" -26d9'51"
Lottie Jean Elliott
Uma 9h23'50" 72d55'19"
Lottie Skipka
Cyg 21h42'19" 47d58'0"
Lottie Tucker and James De Leston
Leo 9h52'40" 28d1'52"
Lottie's Star
And 1h8'16" 45d47'8"
Lotty
Cap 20h51'17" -16d2'11"
Lotus Maree Grant - 29 April 2005
Tau 5h3'38" 17d19'41"
Lou
Vir 12h42'13" 13d16'43"
Lou
Uma 13h45'47" 50d38'8"
Lou and Lee Freedman
Uma 8h51'28" 69d51'6"
Lou Ann Brooks
Ori 6h6'48" 19d27'29"
Lou Ann: Mother of Mine
Cnc 8h3'54" 11d31'40"
Lou Ann Urso
Cap 21h41'5" -10d52'16"
Lou Ann Williamson
And 2h36'23" 43d23'18"
Lou Anne
Umi 15h5'40" 70d51'36"
Lou Anne Bouty
Uma 11h11'14" 30d15'17"
Lou Avallone, Super Star
Tau 4h22'33" 7d33'8"
Lou Berner, Jr.
Ori 5h32'45" 2d41'37"
Lou Buddy's 60th
Lib 15h57'46" -10d36'58"
Lou Cicardo
Psc 0h3'54" 1d18'49"
Lou & Colette Forever
Lyn 7h34'48" 41d58'19"
Lou Doelling
Her 16h38'27" 16d53'56"
Lou & Doll Riddle
Cas 0h16'43" 55d27'43"
Lou Dondero
Aql 19h44'46" -0d2'22"
Lou Ella
Vir 11h42'29" 4d24'59"
Lou & Emma Jean Bader
Psc 1h15'0" 13d37'38"
Lou Gatto #60
Her 17h9'8" 30d52'4"
Lou Hopper Jr.
Leo 9h47'48" 31d48'50"
Lou King
Cyg 21h57'19" 38d11'30"
Lou Kovanda
Boo 14h39'5" 54d47'10"
Lou Ku Dé Ta 4ever in my heart Trev
Cnc 9h0'7" 13d4'30"
Lou Lou
Psc 0h30'33" 4d28'1"
Lou Lou Belle
Sco 16h55'32" -38d39'59"
Lou Lou Belle Pugh
Uma 10h6'22" 51d17'53"
Lou Lou Lou Lou Lovely
Lou Lou
Cas 0h12'1" 51d47'15"
Lou Lou's Monstar
And 1h20'39" 49d49'7"
Lou Love4ever
Lib 15h4'14" -19d0'21"
Lou & Lucille Pelletier (Lou-Lou)
Mon 7h15'24" -0d41'54"
Lou Lukemeyer
Leo 11h29'3" 17d55'19"
Lou Manos
Per 2h36'53" 56d9'10"
Lou & Marianne Allred
Cyg 19h38'41" 37d47'51"
Lou Marie Hertz
Vir 13h21'37" 13d14'36"
Lou Micaël
Uma 12h42'31" 59d28'16"
Lou Michael Reyes
Lyn 7h49'57" 51d44'42"
Lou Mick
Lyr 18h47'19" 39d27'57"
Lou Parasimo
Per 4h0'44" 32d14'4"
Lou Russo
Uma 12h50'50" 58d4'28"
Lou Ruvo
And 1h55'11" 37d59'52"
Lou Schuettenberg
Gem 7h48'10" 31d56'22"
Lou Seijido
Uma 13h46'25" 55d27'0"
Lou Snowball Ghrammanski
Her 17h12'17" 16d22'44"
Lou "The Dude" Milano
Lib 14h52'4" -3d41'10"
Louane
Gem 6h1'49" 23d24'56"
Louane Hamilton
Crb 16h17'33" 33d9'9"
Louane Nozeret
Del 20h33'14" 4d44'52"
LouAnn
Cam 3h35'47" 57d50'9"
Louann and Vernon Semper Fi
Ari 2h9'49" 26d2'45"
LouAnn Tramonte Caruthers
Ari 3h21'22" 27d49'34"
Louanne Grace Atkinson
Vir 14h44'32" 6d27'41"
Louanne Hamilton
Crb 16h2'10" 33d42'54"
Louanne Lalande
Cas 0h47'14" 67d29'50"
LouAnne Taylor Dale
Aur 5h59'16" 46d3'0"
Louburn Anselze Brown
Uma 11h38'32" 64d0'17"
Louby - Samantha Louise Sykes
Cyg 21h20'12" 32d31'51"
Loucille D'Or
Lyn 8h31'48" 46d35'30"
Loucita 16/01/1976 chiquita
Gem 7h19'1" 20d15'56"
Louee and Daniel
Dra 17h24'37" 57d4'42"
Louella M. Shipman
Lyr 19h3'36" 39d39'16"
louellafloatjohnson
Cap 20h52'37" -15d25'21"
Louella's Star
Cas 0h16'25" 52d15'19"
Louen
Cyg 20h27'19" 54d20'13"
Loui Carrington
Her 16h8'2" 20d55'8"
Loui Kaytsa
Cep 21h35'46" 68d9'1"
Louie
Lib 15h12'19" -19d20'27"
Louie
Vir 13h45'10" -4d29'57"
Louie
Cru 12h38'12" -59d56'19"
Louie
Ari 1h49'55" 25d9'34"
Louie
Cmi 7h40'43" 0d14'35"
Louie
Aqr 20h56'36" 0d39'45"
Louie!?, What J!
Nor 16h34'55" -59d55'15"
Louie A. Maier V
Ari 3h22'8" 27d29'37"
Louie Anderson
Cas 23h3'5" 56d33'9"
Louie & Angie Borg
Cru 12h29'16" -57d53'8"
Louie Barragan
Aur 5h38'52" 33d9'48"
Louie Beddingfield Morris
Cnc 8h50'11" 32d22'55"
Louie Blu Harmer
Umi 13h53'7" 78d25'9"
Louie Blue Eyes
Leo 10h34'12" 13d24'47"
Louie Cerveny
Sco 17h19'23" -33d22'59"
Louie Cordero
Cap 21h33'10" -9d30'30"
Louie Edgi
Uma 10h33'18" 46d5'51"
Louie Ellis Wright
Per 3h2'24" 43d4'38"
Louie "Elnino" Rivera
Dra 15h4'59" 64d22'50"
Louie Falcone
Tau 4h12'56" 27d5'12"
Louie Gustov
Aql 20h31'11" -7d10'53"
LOUIE JOSEPH LEVEY
Aql 19h51'23" -0d14'34"
Louie L. Leos
Uma 10h46'49" 72d15'10"
Louie & Lori
Cyg 20h40'10" 35d59'50"
Louie Perna
Cas 0h59'18" 58d23'9"
Louie Pilkington - Louie's Star
Her 17h47'5" 20d58'6"
Louie Teincuff
Crb 15h43'32" 33d10'51"
Louie Tirio
Her 16h31'42" 39d7'57"
Louis
Boo 15h37'23" 41d29'48"
Louis
Per 4h8'35" 46d32'57"
Louis
Lyn 7h5'54" 44d55'43"
Louis
Aql 19h1'19" -7d16'30"
Louis A. Grande
Sco 16h44'53" -39d18'29"
Louis A. Papaleo
Ori 6h4'7" 18d52'22"
Louis Albert Rossano
Uma 8h20'44" 61d38'13"
Louis Albert Valier
Uma 11h22'35" 37d56'40"
Louis Alex Rodriguez
Ari 2h56'53" 30d17'39"
Louis Alexander Pettit-Birch
Her 17h43'6" 39d57'1"
Louis and Alyssa
Cyg 20h19'41" 54d29'6"
Louis and Dolores
Cyg 21h13'19" 42d35'16"
Louis and Gwen
Uma 11h49'1" 36d11'1"
Louis and Jessica
Cam 6h44'5" 82d10'6"
Louis and Jill's 20th anniversary
Sge 20h2'29" 17d52'5"
Louis and June Nelson
Cyg 19h47'8" 46d45'27"
Louis and Mattie Setters
Sco 17h21'32" -42d23'6"
Louis & Angela Rutigliano
Cyg 21h16'51" 42d26'30"
Louis Aubert
Boo 14h35'59" 42d5'0"
Louis B. Robin
Gem 6h33'31" 25d29'5"
Louis (Babe) Licata
Leo 11h40'41" 20d13'27"
Louis Bargellini
Ori 5h32'24" -0d53'22"
Louis "Bear" Bauer, III
Psc 0h30'8" 8d14'12"
Louis & Becky
Dra 16h34'14" 54d23'31"
Louis Biggerstaff
Her 17h23'43" 38d53'50"
Louis Borgogno, 19.12.1955
Uma 13h56'21" 49d55'21"
Louis Burgener
And 0h47'11" 31d10'43"
Louis Buscio
Psc 0h56'13" 9d49'51"
Louis "Butch" Beni
Lib 15h34'19" -15d23'2"
Louis Camael
Umi 15h57'7" 87d5'57"
Louis Castelli The Great
Her 18h2'49" 16d2'45"
Louis Charles Germain
Her 16h15'26" 44d15'22"
Louis Christian Shirley
Cep 21h45'39" 65d10'17"
Louis Cipriano
Dra 14h35'17" 64d23'17"
Louis Crosetti
Lyr 18h47'53" 37d55'43"
Louis D' Ambrosio
Leo 11h51'28" 23d33'14"
Louis D. Perani
Psc 0h47'12" 8d7'33"
Louis Danto
Tau 5h35'30" 27d37'20"
Louis De Rosa
Umi 13h52'21" 69d48'3"
Louis Delagarde
Sgr 19h11'49" -33d30'43"
Louis Dellis
And 23h43'21" 34d54'11"
Louis & Dianna Roy
Ori 5h54'48" -0d41'19"
Louis E. Westercamp
Uma 10h10'38" 48d46'38"
Louis Edward
Leo 11h53'32" 16d38'0"
Louis Enrique Cento
Dra 17h19'58" 51d17'56"
Louis Eunson
Uma 11h1'27" 66d35'2"
Louis Evan Markowitz
Aqr 20h46'58" -12d1'30"
Louis F. "Mugzy" Riccelli, Jr
Cyg 19h40'31" 32d56'43"
Louis F. Provenzano, Jr.
Uma 10h44'13" 44d27'29"
Louis F. Saunders III
Psc 1h24'54" 18d34'55"
Louis Ferenc Taylor
Cep 22h46'9" 60d21'43"
Louis Fernand Danty
Uma 11h4'44" 43d38'3"
Louis Feyt
Cru 12h46'36" -61d55'35"
Louis Fleury Damerval
Ser 16h15'13" -0d2'13"
Louis Flowers
Aur 7h27'28" 35d24'53"
Louis Francis Haddon
Cnc 8h20'49" 9d41'17"
Louis Fried
Uma 8h44'0" 66d1'25"
Louis Fuscaldo
Sco 16h42'6" -40d46'10"
Louis G. Lussier
Ori 5h52'28" 12d3'29"
Louis George Chiavola
Uma 11h0'26" 67d10'14"
Louis George Pescatore
Psc 1h9'6" 23d57'4"
Louis Gervasio
Lyn 7h50'20" 56d47'59"
Louis Gonzalez
Uma 11h29'56" 64d49'55"
Louis Grassi
Sex 10h5'48" 1d58'10"
Louis Guislain 14/05/2006
Lyn 6h44'36" 56d59'38"
Louis "Gum Boy" Galiano
Dra 16h39'24" 65d47'30"
Louis H Anderson (Pops)
Per 4h42'52" 49d11'4"
Louis Harmond Machado
Leo 11h18'57" -4d34'17"
Louis & Ila Mae Edgett
Cyg 20h32'42" 47d36'18"
Louis J. Bath
Aql 19h12'38" 2d43'1"
Louis J. Chunko
Leo 11h22'18" 21d19'39"
Louis J. Fanfa Jr.
Ori 6h16'23" 10d6'57"
Louis J. Hansen
Her 16h52'10" 27d10'9"
Louis J. Notaro Sr.
Cep 21h43'7" 61d39'51"
Louis J Richt
Cep 21h55'31" 60d49'19"
Louis J. Tatarko - 100th Birthday
Uma 13h45'21" 61d30'36"
Louis J Tiburzi Jr.
Aqr 21h41'35" 1d30'50"
Louis James Alfano Jr.
Aur 5h25'0" 54d24'47"
Louis James Benjamin Kebell
Her 18h15'55" 19d57'29"
Louis James Boyd
Vir 14h34'21" -0d4'36"
Louis & Jenifer Forever
Uma 9h32'59" 65d1'53"
Louis John Cosentino Jr.
Psc 1h21'6" 32d49'4"
Louis John Czanko
Psc 0h51'3" 8d41'15"
Louis John Parenteau
Uma 10h10'19" 54d32'26"
Louis Jordaan III
Sgr 19h10'25" -23d48'21"
Louis Joseph Glagovich
Cep 20h48'36" 61d21'43"
Louis Joseph Shaheen, Sr.
Lib 15h34'38" -9d24'43"
Louis Jr.
Her 16h15'20" 48d42'8"
Louis L. Orr IV and Sheri M. Cramer
Psc 1h28'36" 18d27'28"
Louis LaMariana & Rachele Goldman
Uma 11h25'43" 42d23'58"
Louis Long Leung
Her 17h5'55" 27d16'36"
Louis Lynn Grube
Ori 5h53'45" 21d37'59"
Louis M Bridgwaters
Per 3h26'48" 48d6'28"
Louis M. Gettes
Cep 22h33'18" 62d13'4"
Louis M Michaels you are my Heart31
Her 17h13'2" 17d36'45"
Louis Marius Benigni
Sgr 18h49'37" -15d54'7"
Louis Merlin
Gem 7h12'27" 18d48'50"
Louis Michael Criscuolo
Tau 3h40'33" 16d57'36"
Louis Michael Sonnenberg
Psc 1h21'19" 33d4'22"
Louis Monaghan - 13.12.2004
Cru 12h12'28" -61d15'11"
Louis Moran
Cnc 8h47'1" 13d49'52"
Louis Moreau
Psc 0h43'4" 16d9'36"
Louis Moreno Brown
Ori 5h23'7" 2d50'17"
Louis Nicholas Esposito III
Peg 23h52'18" 29d27'33"
Louis Nocks Pandolfo
Cnc 8h43'0" 15d36'53"
Louis Otto Bohling
Her 16h57'31" 34d4'23"
Louis P. Vitti, Esq.
Uma 9h3'23" 54d12'40"
Louis P. Young, Sr.
Lib 15h2'57" -3d49'46"
Louis Perritt
Vir 12h43'54" 3d0'44"
Louis Peter Baratta, Jr.
Ari 2h11'15" 23d34'51"
Louis Philippe
Her 16h36'39" 46d36'16"
Louis R. Foster
Uma 14h11'15" 58d2'14"
Louis Respet
Umi 15h44'54" 72d5'9"
Louis Riccelli
Her 17h48'35" 43d48'17"
Louis Richard Pena
Per 3h38'3" 47d10'34"

Louis Richard Salvatore
Lib 15h29'23" -15d54'27"
Louis Richardson
Uma 9h19'28" 52d36'24"
Louis Robert Kelly
Ori 5h36'28" -0d51'45"
Louis & Roberta Lombardo
Cyg 20h12'5" 46d8'39"
Louis Rosenberg
Uma 13h19'33" 56d12'5"
Louis Scripa
Sgr 19h47'0" -26d54'7"
Louis Shuster "Star Designer"
Cnc 7h57'17" 18d1'42"
Louis Sicilian
Her 17h24'19" 16d51'33"
Louis T. "Grandma O" O'Byrne
Cap 20h29'14" -23d2'38"
Louis Takkos
Aql 20h6'9" 11d3'1"
Louis Tejeda
Per 3h15'28" 54d43'20"
Louis "The Only Toga" Augis Jr
Ari 2h51'16" 28d11'40"
Louis Thomas Hemming
Lmi 9h23'0" 35d47'37"
Louis Tomei
Lyr 18h47'26" 31d48'42"
Louis Toscano
Her 16h11'6" 49d19'3"
Louis & Trish Starita
Cyg 19h32'6" 52d42'39"
Louis Valentino Martinez
Aqr 21h45'21" -1d42'12"
Louis Van den Rijse
Lmi 10h6'27" 35d7'0"
Louis Villiarola Jr.
Cyg 21h39'48" 41d16'33"
Louis Vincent Antonino
Uma 10h25'54" 50d26'28"
Louis W. BUZZY Boisvert, Jr.
Cru 12h3'6" -56d9'20"
Louis W. Camp III
Uma 11h7'3" 35d55'11"
Louis Wesley Williamson
Per 3h19'32" 44d35'58"
Louis William Berklich, Jr.
And 1h8'34" 45d42'11"
Louis Zammit
Sco 17h19'33" -43d24'36"
Louis Zucaro
Psc 0h53'32" 33d16'20"
Louis, Frank
Uma 10h29'42" 40d35'36"
Louis, René Duperray
Lib 15h31'39" -5d11'14"
Louisa
Sco 16h50'21" -42d20'18"
Louisa
Lib 15h42'37" -17d13'11"
Louisa
Tau 4h39'2" 23d3'6"
Louisa Angela Santos Morea
Ari 2h55'1" 26d39'12"
Louisa Cornacchio-Jacobi
Cnc 8h1'42" 11d27'23"
Louisa Dorothy
Cas 1h38'53" 60d16'12"
Louisa Elizabeth Wilkes
Cas 23h4'3" 56d16'19"
Louisa Hirsch
Uma 9h53'1" 68d41'8"
Louisa & Jonathan
Sco 17h43'19" -33d21'32"
Louisa M. Misemer
And 23h1'54" 45d52'33"
Louisa Maye
Cnc 8h45'40" 16d7'53"
Louisa Messenger's Light
Dra 17h9'34" 57d22'26"
Louisa My Love
Lyn 8h22'4" 38d37'30"
Louisa Rose
Tau 4h52'38" 17d57'18"
LOUISA RUBINSTEIN
Sgr 18h58'55" -23d29'8"
Louis-André Villeneuve
Boo 14h19'0" 34d1'32"
Louisanne Claich
Gem 6h56'58" 16d6'21"
Louise
Psc 1h16'56" 25d17'1"
Louise
Lmi 9h57'56" 36d37'40"
Louise
Psc 1h7'40" 31d3'48"
Louise
Per 2h54'59" 33d30'16"
Louise
Tau 4h36'33" 30d30'13"
Louise
Lyn 8h6'7" 42d3'13"
Louise
Cas 0h21'42" 52d47'59"
Louise
And 23h18'1" 41d28'34"
Louise
Lib 15h28'15" -26d13'32"
Louise
Cap 21h1'54" -24d37'11"
Louise
Col 5h45'26" -31d8'46"
Louise
Aqr 22h38'13" -23d24'43"
Louise
Cas 1h46'45" 63d36'44"
Louise
Uma 9h51'53" 56d48'3"
Louise
Uma 8h32'1" 72d25'8"
Louise Adrianne James
Cyg 20h58'50" 46d10'59"
Louise Alice Mathews
Cas 23h53'38" 53d16'35"
Louise and Karl
Aql 19h51'12" 16d11'5"
Louise and Loveless
Sge 19h43'36" 19d2'31"
Louise and Mark Strycharske
Gem 7h23'56" 31d54'9"
Louise and Raymond Severance
Crb 15h41'13" 35d34'0"
Louise and Roger
Cyg 21h27'4" 30d50'10"
Louise Ann Lee
Cnc 8h49'51" 25d16'51"
Louise Ann McClosky
Uma 11h50'30" 61d55'41"
Louise Ann Patricia
And 0h49'26" 31d11'50"
Louise Anna Hulsey
Uma 11h12'44" 60d36'40"
Louise B 1965
Tau 5h10'55" 18d20'13"
Louise B. O'Donnell
Cnc 8h47'13" 28d27'59"
Louise "Babe" Atkinson
Uma 9h25'52" 48d18'49"
Louise Baigneres
Del 20h23'16" 20d2'14"
Louise Baker MacDonald
Uma 9h8'23" 54d43'59"
Louise Banner Welch
Lyn 8h28'44" 55d41'46"
Louise Bates-Terrey
Tau 5h20'54" 18d26'52"
Louise Bianka Anne
Pho 0h42'0" -44d59'32"
Louise Boland
Cru 11h57'50" -62d41'9"
Louise Bowles Ballew
Lmi 10h18'8" 37d8'52"
Louise Brockett
Aql 20h24'32" 1d20'30"
Louise Buch Gurkoff Grass
Cas 23h10'16" 59d41'24"
Louise C Kratoville
Uma 9h22'22" 49d58'56"
Louise Cara Ferguson
Gem 6h55'7" 13d11'25"
Louise Carstairs
Cyg 19h58'19" 39d44'52"
Louise Chantelle Emmett
Aur 5h12'32" 42d28'10"
Louise Chorley
Peg 22h19'25" 17d19'52"
Louise Claire - 4.10.2004
Col 6h6'43" -42d30'10"
Louise Clare French
Cas 23h2'34" 55d2'3"
Louise Cochran Abraham Born 8/23/21
Vir 13h13'19" 8d57'48"
Louise Comolli
Umi 16h8'18" 80d23'29"
Louise Coulombe
Her 18h6'36" 43d2'32"
Louise Creswell
Uma 9h56'48" 65d8'39"
Louise Daem
Cyg 21h37'5" 52d18'38"
Louise Darragh
Cyg 20h14'12" 35d27'29"
Louise De Asis Cruz
Aqr 23h19'49" -17d47'16"
Louise DiAbundo Sarah Cassidy Quirk
Cyg 21h36'50" 38d47'31"
Louise Dorothea Mitchell
Cru 12h0'18" -61d11'10"
Louise Duhamel
Gem 7h49'25" 29d57'55"
Louise Elaine Hansen Bakken
And 0h20'46" 46d30'11"
Louise Elizabeth Skaggs 11/11/1914
Sco 16h6'23" -15d15'38"
Louise Elizabeth Speranza Schoyen
Gem 6h47'43" 32d5'5"
Louise Estelle Eleanor Vitello
Aqr 22h22'56" -6d58'5"
Louise Esther Maynard
Cas 0h30'41" 61d57'52"
Louise & Felix
Uma 8h23'5" 61d54'16"
Louise Frances
Ori 5h28'42" 7d24'10"
Louise Frances Carlson
Sco 16h46'42" -33d5'47"
Louise Gabrielle Renaud
Gem 7h45'21" 34d21'22"
Louise Garcia
Cra 18h49'51" -41d16'55"
Louise & Gerald's Shining Love Star
Gem 7h5'37" 34d10'25"
Louise Gibson
Ari 2h37'28" 21d4'46"
Louise Girl, with minimum Curl
Cas 0h43'56" 64d18'4"
Louise Godin
Peg 21h42'42" 19d53'48"
Louise Gunn
And 0h9'23" 29d21'2"
Louise Hale 2025
And 23h19'26" 48d43'34"
Louise Hooley
And 1h46'4" 39d16'11"
Louise Hopkins
Gem 6h33'2" 13d7'22"
Louise Howley
Del 20h23'3" 9d52'47"
Louise J Trollope - 11 Feb 1991
Cru 12h40'9" -58d16'48"
Louise Jane
Cas 0h12'53" 53d25'2"
Louise Joel
And 0h35'45" 45d29'43"
Louise Johnson
Vir 13h23'48" 12d57'28"
Louise Jones Willis
Uma 13h56'21" 58d6'23"
Louise Katie Mckeever
Cas 23h28'1" 55d28'0"
Louise Kissane-Lynch
Cas 0h19'49" 54d41'49"
Louise Klein
Lmi 10h45'13" 25d46'57"
Louise Kramer
Gem 6h18'19" 22d3'1"
Louise Krongrad
And 1h59'52" 46d50'6"
Louise Lee (Princess)
And 23h2'10" 51d27'12"
Louise Legault
Uma 13h34'51" 60d18'6"
Louise Leonard-Taylor
Aql 20h30'46" -6d49'15"
Louise Llewellyn-Jones
And 23h59'23" 45d33'18"
Louise Long
Psc 1h51'2" 7d13'10"
Louise - "Lupe"
And 1h25'43" 44d8'24"
Louise Lynn Carter
And 0h44'14" 39d10'1"
Louise M Buker
Ari 2h56'54" 26d18'26"
Louise M. Scavitto Feb. 20, 1908
Uma 13h18'39" 57d28'35"
Louise M Sklar
Leo 11h40'19" 14d39'46"
Louise Manley - So Beautiful
Sgr 18h49'1" -16d47'28"
Louise Manning
Dra 12h44'19" 70d31'44"
Louise Marchand
Cas 1h38'40" 63d57'7"
Louise Margaret Duncan
And 2h34'24" 42d19'13"
Louise Marie Dalton
Peg 21h21'1" 24d0'57"
Louise Marie Martinez
Uma 9h3'23" 57d39'36"
Louise Marie Pullen
And 1h11'35" 46d43'13"
Louise Marion Stahl
Cas 1h26'47" 62d1'0"
Louise Marjorie Barnacle
And 1h15'33" 42d16'33"
Louise Marshall
Leo 10h40'2" 7d49'39"
Louise Martel Vance
And 1h35'2" 48d17'48"
Louise McCall
Mon 6h53'19" 8d35'21"
Louise McKinney Willis
Lyr 18h50'38" 35d49'4"
Louise Mellor
Cas 0h22'54" 53d50'52"
Louise Menard Maman
Uma 14h2'30" 56d29'34"
Louise Mimms
Crb 15h28'37" 27d58'19"
Louise "Mom" Holiday
Cas 0h39'28" 53d1'51"
Louise Monteverde
Uma 11h23'18" 43d34'33"
Louise Morgan - World's Best Mom
Cnc 8h5'52" 11d50'16"
Louise Muse
Sco 17h23'40" -34d6'28"
Louise Nesbitt Gordon
And 23h36'45" 48d40'16"
Louise Nixon
Sgr 19h7'29" -20d44'59"
Louise O'Donnell
Lib 15h27'53" -16d17'6"
Louise Rachel Smith
Cas 0h4'54" 57d15'36"
Louise Reynolds
Cmi 7h18'11" 12d14'10"
Louise Ribotta
Lyn 8h14'29" 33d44'20"
Louise & Riley's Guiding Light
Cyg 20h3'36" 33d32'40"
Louise Rose Creutz
Cas 1h31'39" 58d50'34"
Louise Ryan Anderson
Vir 11h47'10" 3d2'30"
Louise S R Waterhouse 50 Star
Cru 12h31'6" -61d8'52"
Louise Sadler
Vir 14h2'46" -0d56'6"
Louise Sarandria
Cap 21h7'40" -20d36'18"
Louise Sargent Akins
Ari 2h41'48" 21d5'29"
Louise Sebald & Lucille Geisler
Cyg 20h36'34" 42d32'50"
Louise Squeeze Joynson
Ari 2h22'16" 12d57'28"
Louise Stafford
Cas 1h34'44" 73d6'40"
Louise Sweet Jackson
Sgr 19h36'33" -13d21'18"
Louise T. Joseph
Aur 5h27'9" 46d53'17"
Louise Thomas a.k.a "Marshmallow"
Cyg 21h13'35" 42d49'46"
Louise Thomas (NANA)
Cas 0h38'35" 47d44'29"
Louise Tomlinson, RN
Uma 13h10'20" 56d6'44"
Louise Tronchet
Col 5h56'25" -37d41'37"
Louise Trottier Kavanagh
Uma 13h21'10" 55d47'46"
Louise Tyler
Cyg 19h44'24" 27d54'14"
Louise Umeki
Tau 5h33'13" 25d56'14"
Louise Urbanovich
Lyn 7h52'9" 57d28'17"
Louise V. Larimer
Lmi 10h5'53" 31d46'46"
Louise Victoria's Christmas Star
Sco 17h24'44" -42d10'15"
Louise Virginia
Com 12h41'8" 31d2'10"
Louise Wall
And 1h22'1" 35d26'31"
Louise Watson
Boo 14h13'50" 34d54'33"
Louise Winslow Yeatman
Lib 15h8'25" -1d1'41"
Louise Winstanley
Cas 0h5'50" 59d17'15"
Louise York Rose
Tau 5h56'10" 23d9'51"
Louise Zeppetelli
Sgr 19h25'55" -27d3'14"
Louisemarie
Ari 2h6'18" 1d40'24"
Louise's Heavenly Light
Aqr 23h29'12" -12d25'30"
Louise's Luminous Light
Pyx 8h58'47" -26d32'12"
Louise's Star
Tau 5h56'27" 25d16'59"
Louis-Frédéric Decam
Dra 14h31'6" 58d53'27"
Louisiann
Sco 17h55'19" -39d1'23"
Louis-Philippe Poitras
Uma 9h7'45" 53d48'38"
Louita Ann
Cas 2h39'35" 68d40'28"
Louix Tziortzis-Foskett
Dra 18h26'47" 51d29'56"
Louiza Tsarolastra
Vol 8h25'52" -67d19'37"
Loujibna
Col 5h24'5" -28d38'5"
Loukas Ktistakis
Uma 11h52'10" 43d49'7"
LoukasMerve
Cyg 19h55'9" 41d44'45"
Louke
Uma 11h40'43" 45d11'10"
Loula Isabel Simmons
Gem 7h59'3" 32d5'40"
Loulette
Her 17h43'53" 37d46'6"
Loulla (We Miss You So Much)
Tau 4h17'1" 2d9'29"
Loulou
Psc 23h3'23" 8d1'30"
Loulou
Umi 13h54'27" 74d35'17"
Loulou
Uma 10h53'36" 57d52'40"
Loulou
Uma 13h45'23" 54d52'41"
Loulou & Zion
Peg 21h55'31" 9d0'6"
Loum's Luminous Superstar Pulsar
Oph 16h35'11" -0d27'57"
Loup Bram dit Saint Amand
Cas 23h28'18" 56d3'19"
Loup's Light
Cep 20h55'2" 61d33'34"
Louraine Victorio-Cudworth
Vir 14h53'24" 5d44'12"
Lourdes
Peg 22h14'59" 11d1'5"
Lourdes
Leo 9h31'18" 12d55'45"
Lourdes
Dra 18h59'56" 52d29'34"
Lourdes
Cap 21h8'31" -19d15'45"
Lourdes Amparo Delgado
Cam 6h36'1" 74d20'29"
Lourdes and Robert Forever
Ari 2h42'28" 17d7'1"
Lourdes Bosch-Taylor
Lib 15h53'59" -11d41'9"
Lourdes Caroll Kastner Ilacad
And 23h43'52" 39d11'24"
Lourdes Evans
Uma 11h59'53" 49d6'2"
Lourdes LaPuente
Sco 16h47'2" -32d0'51"
Lourdes Lozada
Vir 11h44'1" 8d4'11"
Lourdes Maria Tirado
Ari 3h14'16" 29d29'57"
Lourdes Menedez Prieto
Sco 17h1'32" -37d52'10"
Lourdes My Love
Uma 11h25'37" 62d46'28"
Lourdes Philamena Krogmeier
Cap 20h45'38" -15d59'15"
Lourdes & Rory
Aqr 21h36'10" 0d53'29"
Lourdes (shia)
Psc 0h8'20" 5d40'7"
Lourdes Star
Sco 16h7'40" -11d20'29"
Lourdes Stevens & David Stevens
Uma 14h24'5" 57d11'53"
Lourdes Torres
And 0h41'28" 24d25'14"
Lourissa
Cap 21h13'7" -15d43'25"
Lou's Angel
And 2h35'42" 49d10'36"
Lou's Gardian Angel
Tau 5h45'35" 14d16'21"
Lou's Star of Stefani
Cyg 20h16'49" 36d46'52"
Lou-Sam21
Uma 9h41'6" 43d14'7"
Lousaria-keharti
Ori 5h11'21" -10d45'39"
Lousie "Weezy"
Uma 12h2'27" 40d13'41"
Lousine Kalajian
Cnc 8h18'49" 26d14'39"
Louvane
Cru 12h23'36" -62d17'33"
Louvenia
Cas 0h7'46" 59d56'33"
Louzieh Doces
Gem 6h49'20" 32d49'45"
LOVaughan
Lib 14h47'11" -0d50'10"
Lovdore
Uma 11h51'52" 59d35'38"
Love
Uma 9h32'45" 52d55'53"
Love
Cyg 20h30'19" 52d32'15"
Love
Lib 14h49'56" -1d19'8"
Love
Aql 19h43'58" -0d6'4"
Love
Ori 5h34'30" -1d4'50"
Love
Aqr 21h4'59" -10d14'51"
Love
Gem 7h24'14" 35d12'6"
Love
Cyg 19h53'21" 31d5'27"
LOVE
Uma 10h2'4" 43d38'24"
Love
Cyg 21h3'2" 46d20'42"
Love
And 0h17'8" 46d14'49"
Love
Cyg 21h9'48" 42d57'15"
Love
Leo 11h1'25" 20d54'5"
Love
Ori 6h14'13" 2d12'48"
Love Actually
Umi 15h19'50" 71d12'1"
Love Adeldave D.
Uma 10h21'25" 62d35'14"
Love Always
Vir 12h41'12" 11d53'8"
Love Always
Her 17h30'46" 44d4'33"
Love Always
Lyn 7h13'36" 45d58'35"
Love Always and Forever
Umi 15h42'40" 78d6'12"
Love Always Kelly Ann Hausker
Uma 12h16'41" 54d51'39"
Love and Peace, 50
Aur 6h18'40" 28d58'13"
Love & Appreciation: the Star of Sue & Barry
Cyg 19h41'22" 38d38'26"
Love At First Sight
Ori 5h30'15" 3d37'3"
Love at First Sight
Cir 15h21'47" -56d45'59"
Love Baka
Lmi 10h48'27" 36d15'47"
Love Beyond Words
Uma 11h6'3" 50d9'52"
Love Birds
Cyg 20h25'11" 35d16'27"
"Love" Brian Nairn & Chantel Panton
Cyg 20h17'4" 58d9'3"
Love Bug
Vir 13h6'59" 10d24'31"
Love Bug
Tau 4h49'31" 22d19'30"
Love Bug
Gem 6h45'6" 26d59'5"
Love Bugg
Cnc 9h10'51" 12d51'42"
Love Derek
Psc 0h11'43" 8d34'15"
Love Divine
Ori 6h0'57" 19d41'6"
Love Does Exist
Dra 19h16'25" 65d8'29"
Love Dottie
Cas 23h32'17" 52d28'17"
love dumplin
Mon 8h3'4" -0d33'9"
Love E.M.
Uma 9h46'3" 66d8'19"
Love Everlasting
Sgr 18h36'28" -23d15'8"
Love Everlasting
Cnc 9h4'13" 32d32'15"
Love for all eternity! Brad & Anna
Aqr 22h12'52" 0d43'1"
Love for Blake
Uma 11h28'19" 34d2'49"
Love for eternity(Sarah & Chris)
And 23h29'40" 43d11'28"
Love for Mary
Ari 3h3'33" 12d53'46"
Love For Pcyeta
Ari 2h31'55" 26d43'57"
Love for Sonya
Cnc 8h25'58" 28d36'21"
Love for Stephen Scruggs
Her 17h28'41" 44d39'30"
Love Forever
Cnc 8h52'31" 31d48'1"
Love Forever
Aql 19h21'34" -0d8'29"
Love forever Roland&Corinne
Lyr 19h5'40" 42d52'26"
Love Forever - Theresa & Niklas
Uma 10h16'18" 45d13'15"
Love Forever Yves & Brenda
Cyg 21h32'23" 39d22'37"
Love & Friendship - Andrew & Kerrie
Cru 12h24'27" -62d27'56"
Love Goddess
Oph 16h31'56" -0d2'26"
Love Goes Toward Love
Uma 11h51'28" 41d42'27"
Love Grandma & Grandpa Oct 7th 1945
Cap 20h25'34" -12d58'45"
Love Grows
Lmi 9h53'39" 37d27'59"
Love & Heart
Sco 16h50'36" -38d40'45"
Love & Hope
Lyn 7h26'7" 59d42'5"
Love & Hope
Psc 22h54'54" 4d49'26"
Love in Coney Island
Cyg 21h45'46" 43d56'54"
Love Is...
Cyg 21h18'12" 47d7'16"
Love is...
Lyr 18h52'37" 35d44'24"
Love is a moment that lasts forever
Cyg 20h48'46" 43d17'21"
Love is Forever Holly
Ori 6h19'6" 14d38'59"
Love Is In The Sky
Vir 14h12'57" 5d0'27"
Love Is Just, You And Me
Tau 5h38'34" 26d10'9"
Love is Life, Adele and Bob
And 0h51'39" 38d18'9"
Love is patient.
Ori 6h3'54" 18d51'33"
Love is Patient . . .
Tau 4h17'47" 15d59'33"
love IS the answer
Leo 11h25'14" -2d22'14"
Love is....Damien and Jen
Ara 17h29'44" -54d31'20"
Love Is...Our One Year Versary
Cyg 21h30'50" 36d54'14"
Love is...Our Wishes Coming True
Cyg 20h42'54" 48d41'32"
Love Joce!
Sgr 18h40'23" -28d6'48"
Love Jones
Peg 22h19'41" 7d5'37"
love kenjirou 2004
Lib 15h22'42" -28d9'28"
Love Knot
Uma 10h9'49" 61d12'52"
Love & Laughter
Vel 9h37'19" -45d1'15"
"Love & Laughter" Susie's Star
Cas 23h59'37" 52d37'19"
Love Light
Leo 11h13'18" 23d43'4"
Love Like First Kisses
Pho 2h0'54" -41d26'30"
LOVE (Lisas Obvious Vibrant Energy)
Leo 11h26'19" 11d19'17"
Love Love
And 0h57'46" 39d52'1"
Love Love Love
Umi 16h12'42" 76d30'10"
Love makes life a para-dise...
Cyg 20h38'44" 34d9'21"
Love Me Tender
Ari 2h55'42" 25d46'58"
Love Monkey
Ori 6h18'30" 9d41'25"
Love My Tony
Lyn 6h53'18" 57d5'9"
Love Never Fails
Cyg 19h59'52" 46d58'55"
Love Never Lost
Uma 12h51'18" 56d48'6"
Love Nugget
Aqr 21h48'19" -3d3'45"
Love of a Lifetime- Amber & Aaron
Her 16h39'57" 35d9'18"
Love of a Son
Lyn 6h47'34" 55d13'1"
Love of Ashley Mrozowski
Sco 17h27'1" -41d19'57"
Love of Destiny
Umi 16h47'2" 81d3'38"
Love of Jagiela
Uma 9h12'12" 61d39'36"
Love of Lisa
Umi 15h23'21" 71d9'26"
Love of my Dreams
Psa 22h7'14" -28d53'16"
Love of My Heart
Col 6h36'27" -40d33'19"
Love Of My Life
Apu 17h20'24" -69d32'29"
Love of My Life
Aqr 23h9'57" -3d26'28"
Love of My Life
Aqr 21h0'57" -12d56'18"
Love Of My Life
Cnc 8h48'38" 32d34'13"
Love of my Life
Uma 10h6'10" 42d7'14"
Love Of My Life
Per 3h29'5" 45d16'3"
Love Of My Life
Ari 2h9'42" 23d31'52"
Love of my Life
Tau 5h27'52" 26d7'39"
Love of my Life
Ari 3h14'48" 28d51'47"
Love of my life Ashley
Tau 3h27'15" 4d17'50"
Love Of My Life Brad William Gierke
Her 16h37'52" 38d45'54"
Love of My life - Katherine
Uma 11h11'43" 32d48'23"
Love of my Life Petra + Thomi
Cas 1h23'47" 71d28'45"
Love Of My Lifetime - MPLavoie
Lib 15h43'59" -11d59'34"
Love of Ruth
Aqr 20h49'13" 2d4'44"
Love of Sandra
Leo 11h40'13" 18d44'56"
Love of Teresa
Cyg 20h30'41" 40d59'1"
Love on Fire
Tri 2h16'37" 33d37'44"
love & peace for wendelin
Cas 23h56'34" 53d24'32"
Love & Peace Taraneh Karim
Aqr 22h27'41" -2d7'6"
Love Penny 55
Ari 2h11'33" 26d1'50"

Love Petal
Lyr 18h49'36" 35d53'11"
Love~ Ron Ungerman ~Love
Psc 1h16'58" 10d15'25"
Love Song
Sgr 18h55'5" -27d53'43"
Love star
Sgr 18h19'51" -32d36'59"
Love Star
Cas 1h16'19" 69d22'24"
Love Star
Cnc 8h2'13" 16d28'22"
Love Star
Leo 9h49'58" 29d38'56"
Love Star of Atsusi & Mihoko
Psc 0h44'12" 4d27'2"
Love Star Sheldon and Shirley Terry
Sge 19h44'16" 18d19'43"
Love & Strength. Together forever.
Her 17h32'47" 16d22'33"
Love Thy Self
Sgr 18h23'22" -31d46'25"
Love Ting
Sgr 18h26'34" -32d24'1"
Love to David Schwed ~ Love Mom
Her 17h8'12" 29d24'38"
Love to Infinity-Anthony and Vicky
Per 2h53'0" 50d9'13"
Love To Maizie!
Lib 14h57'17" -9d41'17"
Love to Schienny Forever
Lyn 7h44'33" 48d21'33"
Love To The "Oldyweds" 08-18-1956
Cas 0h6'22" 50d7'9"
Love Trasa
Vir 11h43'45" 8d41'6"
Love Turtle 17
Umi 16h14'31" 82d35'27"
Love U Mom Andrew, Brandon & Kayla
Cas 1h19'32" 65d57'53"
Love Under A Watertower
Cyg 19h36'12" 51d50'5"
Love United Forever
Cyg 21h27'54" 39d51'59"
Love will keep us alive
Boo 14h19'37" 17d38'40"
Love Written in a Star
Ori 6h17'20" 15d54'20"
Love Ya Big
Her 16h44'36" 43d50'42"
Love You A Little Bit
Her 18h53'53" 24d0'33"
Love You Always Big G & Baby G
Gem 7h0'28" 27d17'38"
Love you Andy
Ori 5h39'30" -0d6'40"
love you BIG
Leo 10h36'31" 9d47'44"
Love you Dad - Leonard's Star
Cep 22h57'48" 79d27'36"
LOVE YOU DAD! Love, Meghan!
Lmi 10h35'47" 36d28'21"
Love You Dad, You Light Up My Life
Per 2h15'57" 55d35'55"
Love You Forever
Aur 5h51'0" 53d26'6"
Love You Forever
Cyg 19h52'11" 48d7'49"
Love You Forever
Lib 14h43'27" -12d17'40"
Love You Forever...Mommy
Crb 16h2'8" 32d11'28"
Love You Locks & Locks Mom's - Star
Cyg 20h8'53" 45d29'15"
Love You Megan
And 2h13'47" 46d1'17"
Love You More
Tau 4h30'58" 26d22'52"
Love You Pumpkin
Leo 11h26'17" 11d48'36"
Love You Stephanie
Aqr 20h48'13" -1d54'24"
Love You Ter
Cyg 19h37'5" 37d31'36"
Love you to the stars and back Dave
Cyg 21h27'42" 44d20'8"
Love You Whole Big Bunches
Pic 5h9'4" -49d15'9"
Love You...Love You More
Cyg 20h24'51" 54d45'59"
Love, Always and Forever
Psc 0h44'6" 11d5'41"
Love, Fate and Destiny
Ori 5h41'37" -2d34'40"
Love, Hope, Honesty = Christina
Psc 1h15'37" 16d11'31"
Love, Laughter, Happily Ever After
Cyg 20h56'3" 36d39'51"
Love, Peace & Happiness
Aqr 21h16'11" -14d13'55"
Love, Unity and Trust
Ari 1h59'0" 11d27'5"
Love4Jen10Forever
Tau 4h8'0" 7d50'36"
Love-A-Nikki-Today
Uma 8h16'7" 61d16'45"
Lovebird
Cyg 20h47'24" 44d5'10"
Lovebud
Mon 6h46'45" -0d40'22"
Lovebug
Sco 16h15'10" -15d57'17"
LoveBug
Umi 15h50'15" 74d15'54"
Lovebug
Boo 14h56'18" 52d42'27"
LoveBug
Cap 20h28'56" -26d35'54"
LOVEBUG
Per 3h8'28" 52d2'54"
Lovebug
Cas 1h42'49" 54d40'13"
Lovebug
Cnc 8h38'43" 18d39'54"
Lovebug Jones
Uma 8h56'38" 68d14'45"
Lovecat
Cma 6h42'40" -12d23'46"
Loved Forever
Uma 10h6'39" 67d24'32"
Loved Forever - Bruce Gilbert
Col 5h25'28" -40d40'16"
Loved Friend Kara Elizabeth Latham
Cap 21h38'7" -19d25'27"
Lovedrops Fairley
Crb 15h42'53" 26d21'1"
LoveEverlastingSteven&KarenSink
Lmi 10h44'30" 27d3'31"
Love...Forever & Always
Cyg 19h52'48" 57d49'54"
Love.Forever.True
Cyg 20h20'50" 55d44'57"
Love-Lanza
And 2h12'32" 49d57'28"
loveleen
Cyg 21h16'29" 41d6'25"
Loveleigh
Cap 21h40'20" -8d58'9"
Loveless Peace
Cap 20h19'30" -9d47'12"
LoveLight
Ari 3h16'49" 24d3'39"
Loveline
Uma 10h40'42" 61d36'15"
Lovell, Richard
Uma 11h43'24" 32d37'53"
lovelove (Gideon & Lauren)
Pho 0h38'7" -50d12'17"
Lovely
Sco 16h12'6" -16d5'12"
Lovely
Lmi 10h12'54" 29d32'14"
Lovely Alison
And 23h0'54" 51d19'20"
Lovely Allison
Cnc 8h36'47" 28d21'5"
Lovely Andra
Uma 13h27'40" 58d34'19"
Lovely Anna
And 1h30'30" 45d19'6"
Lovely Ashlee Lopez
Vir 13h14'43" -18d32'51"
Lovely Baby Star Faby
Psc 0h44'43" 10d45'15"
Lovely Caroline
Aqr 20h42'8" -2d30'51"
Lovely Cassandra
Vir 11h42'17" -3d2'41"
Lovely Chelsea
And 0h32'47" 33d53'22"
Lovely Dianne
Umi 3h30'5" 88d58'22"
Lovely Erika
Mon 8h10'31" -0d24'49"
Lovely Fiona
Cas 2h4'41" 74d25'28"
Lovely Gary
Per 4h29'33" 42d31'18"
Lovely Heather
Ori 4h57'28" 15d22'43"
Lovely Inez Jordan Powell
Ara 16h37'16" -47d28'14"
Lovely Joanne
Gem 7h9'16" 34d9'14"
Lovely Kiran
Aqr 23h21'22" -18d53'10"
Lovely Ladies Liturgical League
Lyr 18h47'23" 28d12'22"
Lovely Lady
Vir 11h41'56" 4d52'53"
Lovely Lady Larkin
And 0h16'3" 30d15'1"
Lovely Laura's light
Cas 3h12'33" 57d56'19"
Lovely Lauren
And 2h34'52" 45d1'33"
Lovely Lauren
Cap 20h43'24" -21d5'1"
Lovely Leo Starshine
Leo 10h39'40" 17d18'36"
Lovely Lese
Gem 7h4'52" 10d27'24"
Lovely Leslie
Cap 20h53'34" -14d52'37"
Lovely Lexi
Aqr 21h57'38" -1d35'15"
Lovely Lida
Vir 12h36'13" 0d28'35"
Lovely Lily
Ori 6h8'20" 20d45'28"
Lovely Linda
Tau 5h43'12" 22d45'16"
Lovely Linda
Aqr 20h43'4" -8d6'56"
Lovely Lindsay
And 23h42'17" 38d22'16"
Lovely Lisa
Uma 13h36'50" 49d40'0"
Lovely Little Jess
Vir 12h52'48" 7d56'14"
Lovely Liz
Leo 10h31'7" 17d59'55"
Lovely Lori Jo
And 2h21'59" 48d27'30"
Lovely Luane
Per 3h29'44" 49d38'33"
Lovely Lucy
Cas 0h50'26" 56d30'10"
Lovely Lyn Michelle Tiller
Cam 6h5'22" 62d2'13"
Lovely Marlen
Cyg 19h43'46" 34d39'2"
Lovely Maus
Lib 15h38'30" -23d54'52"
Lovely MeowP00 and her MrFluffy
And 23h23'5" 51d58'19"
Lovely Ms Molli
Sgr 18h26'0" -26d33'51"
Lovely Pamela Jeanne
And 23h18'30" 35d27'52"
Lovely Princess Lindsay
Lmi 9h58'59" 36d42'12"
Lovely Renee
Cas 1h49'47" 66d42'44"
Lovely Sarah
Lib 15h23'40" -12d33'56"
Lovely Susanne J. Harrelson
Gem 6h51'58" 25d52'41"
Lovely sweet Aurora
Ori 5h54'17" 22d38'53"
Lovely Sweet Lisa
Vir 12h55'5" -5d50'53"
"Lovelyness"
Lyr 18h25'8" 32d47'47"
Lovemuffin/The Sauce/My Sunshine
Sco 16h6'59" -10d20'26"
Love.Peace.Happiness.Strength. B&L.
Aqr 22h36'42" 2d1'42"
Lover
Psc 1h32'28" 23d29'39"
Lover
Cyg 19h57'24" 39d14'45"
.lover.
Cyg 20h34'5" 57d47'27"
Lover
Uma 9h41'27" 57d41'26"
Lover Bean
Uma 12h2'40" 62d43'35"
Lover Bear
Leo 9h51'45" 22d9'55"
Lover Bunnies
Cas 0h5'51" 56d25'43"
Lovers
And 23h42'24" 42d4'22"
Lovers
Per 2h48'1" 37d28'19"
Lovers
Lib 15h47'21" -17d29'5"
Lover's Destiney
And 2h25'12" 50d28'19"
Lovers Forever
Umi 16h4'36" 71d23'1"
Lovers Forever Edna and Sam
Lyn 8h27'38" 35d6'50"
Lovers & Friends
Cap 21h34'17" -23d53'29"
Lover's Nest
Cyg 20h12'14" 37d6'25"
Lover's Star. Jim & Jane Facchini
Gem 6h48'54" 15d32'34"
Lovers Union
Aqr 21h41'48" 0d23'53"
Loves
Cnc 9h9'41" 31d26'32"
Loves
Cyg 20h49'37" 42d6'18"
Love's Beginning
Vir 13h1'13" -21d48'56"
Love's Light
And 0h38'58" 36d23'44"
Love's Light
Mon 6h40'40" 7d42'2"
"Love's Light" - Marc & Erica
Ori 6h20'38" 2d10'36"
Love's Little Star
Cyg 20h42'43" 47d33'44"
Love's Poet
Cnc 8h2'37" 24d2'45"
Loves Shining Star
Ara 17h48'9" -48d26'15"
Love's Strength
Uma 12h1'42" 34d24'3"
Loves True Course (Mary and Mike)
Cyg 21h38'25" 42d44'56"
LovesEye
Per 3h38'8" 33d59'53"
Lovesome Laura
Cnc 9h4'0" 21d54'23"
LoveStar
Ori 6h10'39" 11d12'58"
Lovestar
Vir 11h47'53" -0d45'5"
Lovestar for Isabella und Max
Uma 10h50'17" 41d33'53"
Lovestar Myléne & Jenny
Uma 10h5'55" 66d17'30"
Lovestar Ralf + Myléne
Uma 8h36'31" 58d30'45"
Lovestar Viviane Kurath
Her 16h15'2" 44d48'20"
LoveStar100805
Cyg 21h39'17" 35d51'42"
Lovetten
Vir 13h31'42" 7d12'44"
LoveU
Ori 6h5'40" 18d13'50"
LoveULongTime
Peg 21h31'52" 26d0'21"
Lovey
Lmi 10h11'9" 28d20'24"
Lovey
Crb 15h44'24" 26d56'42"
Lovey
Uma 12h29'10" 54d0'44"
Lovey and Ali Bruner
Leo 9h31'23" 9d59'33"
Lovey Dovey
Cyg 21h35'13" 46d58'25"
Lovey World
Ori 6h11'51" 8d52'35"
Lovey's Star
Ori 6h15'19" 15d37'18"
Lovie
Cnc 8h36'42" 21d54'44"
Lovie
Cap 21h9'53" -15d28'32"
Lovie & Erv Kowalski Oct 24, 1953
Cyg 21h40'2" 41d22'14"
Lovie's Star
Vir 12h43'59" 11d7'16"
Lovin Bugger- Cait
Lyn 8h18'48" 49d16'22"
Loving Amy
Ori 5h22'36" -8d49'27"
Loving Dad, Joseph Siegel
Cep 22h14'20" 61d40'36"
Loving Dad,Mate,Friend: John Kopec
Her 18h22'35" 29d1'25"
Loving Grandmother Victoria Wyhopen
Sco 16h44'47" -39d0'7"
Loving Heart
Lib 15h8'45" -1d32'0"
Loving Husband & Father - Todd Bean
Umi 14h32'2" 79d0'41"
Loving Lucille Hopwood
Lyn 9h11'14" 38d13'47"
Loving Marissa Marie
Cnc 8h13'54" 16d48'31"
Loving Miranda
And 2h13'38" 37d37'27"
Loving Mom to Three
Lib 14h51'10" -3d9'22"
Loving Mother
Cas 1h47'9" 61d18'43"
Loving Mother
Cas 0h50'35" 60d11'43"
Loving Mother and Father
Cyg 21h41'46" 47d19'47"
Loving Mother Marsha Wyhopen Garvey
Ari 2h52'5" 22d4'52"
Loving Mother & Wife-Tracy Hammack
Cap 20h24'50" -13d47'48"
Loving Rita Gamache Sparkling Star
Cas 1h15'12" 58d10'40"
Loving True: Samantha and Me
Sco 17h56'50" -37d5'22"
Loving U
Ari 1h51'35" 19d1'18"
Loving VRV
Cas 0h29'32" 62d25'54"
Loving Wife B. Diane Holder
Sco 17h25'37" -43d43'11"
Loving You Always
Umi 15h5'23" 68d30'9"
Loving you always
Cyg 21h53'58" 40d52'29"
Loving you whether...
Ori 6h0'58" 21d18'22"
lovingdannyyes
Uma 8h43'31" 57d1'37"
lovishMC
Cyg 20h44'45" 53d53'25"
Lovona
Lyr 19h13'5" 27d11'13"
LovróJózsef1
Aqr 21h39'11" -0d21'26"
Low Shu Fen Sophia
Aql 18h51'49" -0d3'26"
Löwe Stefcio der erste
Uma 9h8'4" 55d28'22"
Lowell
Aqr 20h56'31" -7d4'43"
Lowell
Sgr 19h48'57" -12d19'42"
Lowell
Psc 0h53'13" 30d40'17"
Lowell "Brent" Jones
Uma 11h31'41" 60d25'33"
Lowell Duane Dahl
Ori 5h32'28" 9d2'40"
Lowell M Ward
Leo 10h11'21" 20d42'16"
Lowell Thomas Cline
Her 16h46'58" 28d57'1"
Lowell's Soul Shine
Uma 8h37'10" 59d40'55"
Lowrey Sisters-Frances Bettye Ann
Tau 3h47'20" 25d8'36"
Lowri Mia Huxley
And 23h15'18" 52d28'25"
Lox and Curry
Cyg 21h33'25" 52d9'59"
loxl (nedaisy coltini)
Per 3h17'0" 41d21'24"
Loxley
Cas 0h28'23" 55d37'43"
Loyal Stark's Eastern Star
Eri 4h11'22" -14d4'31"
Loyalty
Gem 6h37'48" 18d2'41"
Loyce Huff (Our Queen Angel)
Cyg 21h37'29" 32d1'39"
Loyd Lightfoot
Boo 14h40'20" 25d55'25"
Loyd & Maxine McGowen 50th
Sge 20h3'26" 18d34'4"
Loye Lee Free
Cyg 20h15'24" 33d46'9"
loysa zenoble
Cyg 21h53'43" 51d36'52"
Loza Lu
Cep 23h1'43" 73d0'24"
lozd Infantry Regiment (Iron Grays)
Uma 10h7'36" 71d48'14"
Lozon Jr.
Cas 1h21'24" 57d11'38"
Lozzie
Cap 20h28'50" -19d59'54"
Lozzie Stardust
Cas 0h21'32" 50d12'11"
lozzi's destiny
Sgr 17h54'28" -29d43'59"
LP
Aqr 23h3'16" -9d20'21"
LP+HP
Crb 15h35'5" 30d27'39"
LP22173
Lmi 10h0'44" 36d5'49"
LPG
Aqr 23h56'18" -15d33'32"
LR Harrison
Leo 10h15'29" 17d5'10"
L-R Ramsey
And 1h26'32" 43d33'31"
LRG
Cnc 9h11'55" 14d4'46"
LRJ, Sr.
Uma 12h4'21" 52d59'14"
L.R.J.S.
Cam 4h3'30" 66d25'59"
LRMBDV - 3/9/1996
Ori 5h35'16" 7d19'29"
LS
Uma 8h49'14" 68d9'42"
LS
Umi 14h23'27" 82d13'13"
LS71
Aur 5h56'36" 32d56'0"
L.S.B. 1
Her 16h39'43" 36d25'40"
LSchneider and CGiarraputo
Sgr 18h19'47" -22d56'7"
LSD The Beautiful
Sgr 19h0'13" -21d22'58"
LSGPV4
Uma 11h40'28" 53d1'52"
L.S.M.- The Brightest Star At VB
Uma 11h8'27" 46d45'32"
LT & Bubba's Love Star
Ori 4h45'37" 4d7'58"
Lt. Cdr. Andrew John Griffiths R.N.
Cet 1h5'33" -6d59'49"
Lt Col Dale Christensen,USAF (Ret)
Aql 19h44'30" 7d25'7"
Lt Col EMG J-Ph Gaudin/Cdt Bat Cyc 1
Crb 15h34'20" 26d1'34"
Lt. Col. Frederick H.G. Sahl
Uma 9h35'19" 56d40'28"
Lt. Col. Gregory J. Wick, US Army
Uma 11h17'1" 57d10'39"
Lt. Col. Jack J. Watts
Aql 19h13'5" 14d7'21"
Lt. Col. John J. Zentner
Aql 19h22'3" -0d36'25"
Lt. Col. Kenneth I. Brown
Lyn 7h0'38" 48d35'16"
Lt. Col Marshall Kirk Bolyard Ret.
Uma 13h22'30" 60d28'7"
Lt Col Mike Clark, USAF (Ret)
Aql 19h46'55" 7d43'58"
Lt. Colonel Walter Morgan
Per 3h44'37" 42d46'14"
Lt. Commander Howard L. Randall
Uma 11h32'40" 58d21'9"
Lt. D. Schmokle
Uma 10h26'3" 48d51'54"
Lt. Dan
Leo 9h41'27" 31d31'48"
Lt. Dick Criswell, SCFD
Vir 13h12'42" -14d57'36"
Lt. Gen. Brian Arnold
Per 3h17'5" 49d35'28"
Lt. General George S. Boylan, Jr.
Ori 6h4'40" 10d37'38"
Lt. Gregory Floyd Medlin
Umi 16h7'42" 76d46'52"
Lt. Ivan and Laura Loock
Cyg 19h36'29" 52d25'30"
LT. Jessica Ellen Hill
Sco 17h28'5" -36d51'30"
LT Matthew Echo Myers
Lyn 6h39'30" 58d54'17"
Lt. Officer Bobby Cabral
Per 3h41'3" 40d5'8"
L.T. Price
Eri 4h40'54" -0d37'12"
LTB Kiddo
Ori 4h48'26" -3d20'32"
LT.Colonel Robert Alvarez
Ori 5h13'57" 6d11'46"
LTG 56
Her 18h36'50" 18d4'29"
Ltjg Jeffrey Shoup
Sgr 19h1'27" -35d27'20"
LTP-SM Forever
Cnc 9h16'34" 32d13'7"
Lu
Sgr 19h19'38" -21d44'4"
Lu
Cap 21h0'25" -17d0'26"
Lu Ann Kinney
Ari 2h41'59" 12d35'4"
Lu Lu Bellew
Aqr 22h36'42" -1d52'58"
Lu Lu's North Star
Leo 10h51'21" 18d23'13"
Lu, Anfang
Cyg 20h13'17" 44d53'4"
Lua Daniella
Psc 0h35'40" 9d1'21"
Luahiwa
Mon 6h50'21" 7d8'45"
Luaire407
And 23h10'19" 40d6'2"
Luan Giannone
Cyg 19h46'5" 38d7'19"
Luana Cavaldoro
Cas 23h30'26" 57d52'44"
Luana Maria Mullins Honeybunny
Gem 7h32'53" 26d24'24"
Luana Monia Hell
Uma 10h9'9" 49d3'13"
Luane Query
Ari 3h20'25" 29d33'44"
Luann Carol
Uma 10h0'19" 55d49'40"
Luann D. Fogg
Aqr 21h28'55" -5d19'43"
Luann Dail
Uma 13h47'53" 50d7'0"
LuAnn Fardel-Linker
Aqr 21h15'28" 1d53'10"
LuAnn Fee
Cas 23h30'18" 51d36'24"
Luann Loves Paul More!!!
Tri 2h9'35" 34d16'6"
Luanna
Vir 12h11'42" 4d59'17"
Luanne
Cas 0h25'35" 61d34'33"
LuAnne
Lib 15h2'38" -28d23'44"
Luanne Burkholder
Psc 22h55'30" -0d1'26"
Luanne R Citrin
Dra 19h15'38" 76d50'33"
Luany
Cnc 9h6'30" 31d0'48"
Luap Oksuk
Ori 5h17'14" -7d31'8"
Luba
Sco 16h9'38" -30d28'14"
Luba
Psc 1h45'21" 17d12'18"
Luba and Gena Belitsky
And 0h16'3" 44d14'27"
Luba From Ludington
Uma 11h23'49" 58d3'48"
Luba Lillian
Tau 4h10'17" 5d16'48"
Luba Pala
Umi 14h39'40" 74d58'54"
Lubby
Her 17h17'26" 17d55'51"
Lübke, Jens
Uma 12h3'44" 60d18'57"
Lubna
Lyn 6h22'15" 56d50'59"
Lubomir Kassadjikov
Uma 9h40'39" 69d20'50"
Lubomir Lamplot
Aqr 21h58'45" 1d29'52"
Lubomira Landers
Dra 19h54'43" 75d14'45"
Lubov Sorokina
And 23h13'17" 41d6'56"
Lubulé
Uma 11h22'19" 68d13'54"
Luc Alexander Roberts
Ori 5h7'20" -0d27'49"
Luc and Louise
Cyg 20h25'15" 45d34'31"
Luc&BillieJean Past,Present &Future
Her 16h41'1" 13d25'40"
Luc Charlebois
Ori 5h19'15" 16d5'49"
Luc David Wolff-Merovick
Sgr 19h31'59" -13d29'18"
Luc Dubois
Psc 1h15'59" 4d34'30"
Luc Hardyn
Lyn 9h7'55" 33d42'44"
Luc "My Love" Bellemare
Cas 23h28'53" 58d21'43"
Luc Rey Drymer Graham
Lib 15h49'42" -17d18'33"
Luc Rey Drymer Graham
Boo 14h36'26" 44d28'54"
Luc Sacha Gutbub
Aql 20h7'40" -0d13'15"
Luc William Wastiaux
Her 18h22'28" 24d52'31"
Luc, mon amour de papa
Sco 17h31'36" -32d18'39"
Luca
Cas 0h23'54" 61d39'47"
Luca
Uma 14h23'4" 56d50'5"
Luca
Dra 18h5'32" 72d45'3"
LUCA
Aql 19h37'35" 8d47'55"
Luca
Cas 1h9'7" 48d54'39"
Luca Alessandro's Little Star
Cru 12h50'33" -58d2'37"
Luca Andrew DeLisi
Umi 20h59'32" 89d7'30"
Luca Antonio Tassone
And 23h9'11" 40d27'17"
Luca Benni
Ori 5h55'11" 12d43'17"
Luca Bernardini
Ori 6h8'42" 16d4'36"
Luca Carmine Tomasso
Ari 2h41'51" 27d19'19"
Luca Cerasani
Cam 4h5'56" 57d12'48"
Luca D'Arelli
Uma 8h47'59" 68d29'30"
Luca Frigerio
Sco 16h4'20" -20d21'18"
Luca Frost - Luca's Star
Per 4h8'22" 43d44'47"
Luca Gennaro Phillipy
Uma 12h30'50" 56d59'42"
Luca Gordini
Uma 10h54'15" 59d21'8"
Luca James Homer
Lmi 10h0'11" 34d53'26"
Luca James Meikle - Luca's Star
Ori 5h53'9" 7d2'18"
Luca Janik
Uma 9h8'39" 64d11'4"
Luca Jimmy Garas
Ori 5h22'5" 3d37'43"
Luca John
Per 2h16'9" 58d8'20"
Luca Johnson
Sgr 19h5'33" -24d16'15"
Luca Joseph Aidala
Cnc 8h7'58" 17d22'17"
Luca la mas estrella
Cas 0h27'21" 62d23'13"
Luca - la stella della mia vita -
Ori 6h24'18" 13d9'56"
luca laube
And 23h59'16" 44d1'7"
Luca Limon Sperka
Uma 9h39'0" 66d39'9"
Luca Marciano
Cru 12h35'26" -58d25'51"
Luca "Moose" Ettore
Cyg 20h12'27" 33d25'30"

Luca Noah Patrick Purdy Gem 6h4'53" 23d37'49"
Luca Pelliccia Lyr 19h27'6" 40d23'32"
LUCA PISU TI AMO Dra 10h4'5" 74d31'13"
Luca Russell van Eck Uma 9h8'0" 65d26'45"
Luca Saade Cas 0h0'1" 58d23'18"
Luca Scatena Leo 11h25'54" 21d20'22"
Luca Sebastian Molinari Uma 9h40'9" 50d19'6"
Luca & Silvia Gem 6h18'24" 24d57'51"
Luca Sunshine Vir 13h19'25" -13d26'47"
Luca Thomas Hamilton Her 16h35'16" 47d35'15"
Luca Tusche Uma 10h20'58" 46d37'53"
Luca, a csillag Psc 0h52'44" 10d45'20"
Lucadia Leo 11h10'45" 6d57'0"
Lucania Cnc 8h51'30" 11d35'11"
Lucanus David Juist Lac 22h19'7" 48d15'25"
Lucas Her 16h31'26" 14d8'2"
Lucas Peg 22h40'19" 14d8'40"
Lucas Cma 6h38'10" -20d2'23"
Lucas Cap 20h31'12" -10d41'44"
Lucas Cru 12h4'15" -64d23'12"
Lucas Sgr 18h38'22" -30d18'49"
Lucas 1 Uma 12h52'19" 59d56'4"
Lucas A. Jones Ari 2h15'51" 23d28'2"
Lucas A. R. O'Keefe Dra 15h13'0" 59d23'16"
Lucas Alexander Leavell Ori 5h34'54" 7d25'11"
Lucas and Lauren Ori 6h4'53" 13d51'59"
Lucas and Teresa Zapata Family Cyg 20h43'57" 54d52'2"
Lucas & Andrew Gem 6h54'39" 26d15'14"
Lucas Andrew Gem 7h48'55" 32d12'5"
Lucas Andrew Flowers Ari 2h40'7" 14d48'7"
Lucas Anthony Monteiro Aqr 23h5'42" -21d41'14"
Lucas Arthur Cooney Carrillo Aur 6h44'19" 35d28'48"
Lucas Benjimin Buxton Rowley Her 18h45'44" 17d43'35"
Lucas Bougriot Sgr 19h35'51" -32d14'30"
Lucas Carl Michael Geary Uma 11h40'38" 51d4'28"
Lucas Carson Sanders Per 3h25'46" 47d22'55"
Lucas Chavaudra Her 16h51'12" 12d40'5"
Lucas Christopher Pennington Gem 6h52'56" 34d16'9"
Lucas Cieri Uma 11h13'36" 30d49'0"
Lucas Cristobal Mendoza Lib 14h52'35" -1d46'54"
Lucas De La Paz Her 16h46'39" 43d4'46"
Lucas Dejmek's Baptism Star Tau 4h5'57" 27d8'11"
Lucas Denton Uma 11h33'6" 41d48'32"
Lucas Duane Carlson Ori 5h11'26" 9d9'32"
Lucas E. D. Reid Uma 11h27'49" 60d26'4"
Lucas Earl Lane Lib 14h58'42" -4d14'31"
Lucas Edward Rossi Vir 13h47'42" -5d10'2"
Lucas Farrand Tierney Her 17h10'51" 39d40'26"
Lucas Fernando Benitez Sgr 19h47'11" -44d50'47"
Lucas Frank Feller Uma 10h51'36" 52d34'51"
Lucas Fraser Aqr 23h44'23" -22d32'49"
Lucas Gaukler Lyn 7h27'44" 52d38'36"
Lucas Gerhart Minken Per 4h28'39" 34d2'7"
Lucas Glen Sharp Ori 5h30'9" 10d37'45"
Lucas Glenn Santise Dra 17h31'59" 67d23'22"
Lucas Gordon Andrjeski Dra 14h36'52" 56d21'41"
Lucas Greenlee Her 17h26'55" 35d20'10"
Lucas Gustave Hansen Psc 23h38'14" 4d32'16"
Lucas Henry McGuire Her 17h11'19" 35d36'13"
Lucas Henrygue Sgr 19h4'36" -16d13'36"
Lucas Hobbs & Katie Schelzel Uma 8h52'25" 68d57'58"
Lucas Hubbard Her 16h46'37" 16d48'7"
Luca's Independence Uma 13h28'53" 53d15'57"
Lucas J Umi 14h35'36" 75d10'12"
Lucas Jamal Sterling Ori 6h8'35" 13d26'25"
Lucas James Delgado Umi 16h51'38" 79d50'3"
Lucas James Duzik Uma 13h55'14" 51d24'19"
Lucas James Ellenberger Psc 1h7'46" 4d41'58"
Lucas James Vittoria Gem 6h45'21" 18d4'59"
Lucas Jeffrey Burdick Ori 5h35'38" 1d20'21"
Lucas John Adam Peg 22h49'1" 21d27'47"
Lucas John Umberto Valente Umi 14h13'50" 70d56'55"
Lucas Joseph Eckhardt Her 17h22'23" 38d12'5"
Lucas Joseph Reichert Her 17h46'11" 33d31'45"
Lucas Kellen McBrearty Dra 19h21'41" 73d56'48"
Lucas Kevin Jones Her 18h48'5" 22d27'11"
Lucas Lane Vines Per 3h26'41" 43d41'16"
Lucas (Luke) James Floren Uma 10h0'4" 71d31'57"
Lucas 'Luke' John Saathoff Sco 17h20'57" -40d25'1"
Lucas Major Lib 15h42'40" -24d6'5"
Lucas Mario Catuogno Tau 3h52'23" 27d35'41"
Lucas Matthew Young Dra 18h32'20" 66d27'10"
Lucas Maximus Gregory Rohde Lib 14h51'39" -2d40'32"
Lucas Michael Gabriel Reid Umi 14h7'10" 86d38'1"
Lucas Michael Malmos Ori 5h21'11" 6d55'4"
Lucas Michael Sell Boo 14h55'54" 24d56'41"
Lucas Miller Lyr 18h15'49" 31d12'32"
Lucas Nikolai Sco 17h10'35" -45d12'54"
Lucas Nolan Gurney Lyn 8h18'24" 56d13'0"
Lucas Nolan Pastor Ari 3h18'55" 23d32'26"
Lucas Owen Blankenship Gem 6h7'32" 23d51'36"
Lucas Patrick Havens Psc 0h11'10" 10d2'55"
Lucas Patrick Sullivan Ori 6h4'39" 13d13'36"
Lucas Paul Tait, Born 01-22-1976 Aqr 22h23'15" 1d17'58"
Lucas Payton Bland Leo 11h57'12" 22d2'58"
Lucas Peter Pflanz Per 3h34'35" 45d43'31"
Lucas Pine Cnc 9h17'25" 24d59'17"
Lucas Quinn Balentine Boo 14h32'40" 27d26'15"
Lucas Quintin Payette Per 3h45'28" 48d6'36"
Lucas R. Saul - Our Star And 0h57'48" 38d54'50"
Lucas Ray Cruz, Jr. Her 17h28'14" 43d55'42"
Lucas Rex Katz Uma 11h39'53" 37d43'0"
Lucas Reyes Lyn 7h39'0" 36d16'42"
Lucas Robert Boyle Per 3h28'22" 34d48'53"
Lucas Robert Khowaylo Cep 22h14'39" 65d22'39"
Lucas Robert Walters - Squirrel 07 Ori 5h58'2" -1d45'21"
Lucas Salt Aur 5h50'43" 31d55'34"
Lucas Scott Psc 1h13'43" 13d56'50"
Lucas Star Her 18h55'36" 24d20'34"
Lucas Steven Cathey Ori 5h31'59" 2d11'7"
Lucas Tanner Malmsten Uma 12h39'13" 53d9'56"
Lucas Thayer Dreamer Cam 4h46'31" 61d33'34"
Lucas Tsegaye Burke Umi 14h17'20" 72d1'44"
Lucas W. Darling (Son Star) Ori 4h53'43" 12d34'44"
Lucas W. LaBeau Ori 5h5'59" 4d4'22"
Lucas Walter Schuh Her 16h36'56" 17d38'24"
Lucas Walters Cru 12h2'44" -62d13'36"
Lucas William Henningsen Ori 4h50'11" 11d24'33"
Lucas Zettle (Love, Sky) Aur 5h25'12" 40d15'30"
Lucas, Hannah, & Nathan Hensler Uma 11h23'44" 41d11'8"
LUCAS25 Lac 22h28'22" 41d25'58"
Lucca Ori 5h31'29" 3d15'54"
Lucciola Psc 1h13'14" 28d30'15"
Luce de Lichere Uma 9h44'51" 48d16'46"
Luce Del Brigon Per Il Mundo Psc 1h8'17" 10d3'14"
Luce Del Cielo Leo 11h18'15" 14d55'34"
Luce del Sole Psc 0h1'47" 8d17'14"
Luce del Sole Pho 0h15'50" -39d27'24"
Luce dell' Amore Gem 7h37'6" 14d9'10"
Luce Della Madre Vir 14h34'55" 6d4'19"
Luce di Hannah Crb 16h9'52" 37d58'50"
Luce di Sara Cyg 19h57'6" 30d13'51"
Luce di Sara Vir 15h1'43" 4d55'32"
Luce My Little Light And 0h45'41" 27d13'12"
Luce Suzanka Leo 9h42'57" 28d25'16"
Lucel Ori 5h12'45" 2d19'32"
Lucelie Hovsepian Kerr Aql 19h4'15" -0d38'53"
Lucelle Margaret Wallerstedt Ori 5h39'49" -7d50'15"
Lucem Lyr 19h12'15" 27d13'12"
Lucero Del 20h46'49" 15d5'50"
Lucero Cnc 8h40'18" 17d7'20"
Lucero Ori 5h14'49" 7d14'54"
Lucero Ramirez Uma 10h14'17" 44d16'7"
LuceroJr Cep 21h54'42" 65d59'46"
Lucero's Seancy 01/25/1990 Dra 17h40'48" 52d36'12"
Lucesita Mae Lyn 6h40'20" 53d14'40"
Lucette & Robert Lmi 10h12'29" 37d3'14"
LUCHAR Gem 6h55'3" 14d8'25"
Luche Ari 1h53'47" 20d3'8"
Luchia Shapiro Aqr 22h57'51" -20d1'18"
Luci Cyg 20h53'26" 48d9'7"
Luci 22.09.1944 Uma 11h7'39" 63d29'40"
Luci Ann Mon 6h44'29" -0d7'18"
Luci Carol Stubbs Cas 0h28'40" 53d40'17"
Luci Goldberg Lyn 9h3'6" 38d2'13"
Lucia Aur 5h16'33" 41d54'37"
Lucia Cnc 9h0'24" 14d1'58"
Lucia Vir 12h46'15" 2d28'10"
Lucia Ori 5h53'58" 16d52'43"
Lucia Lep 5h16'54" -11d22'43"
Lucia Cap 20h11'14" -21d49'35"
Lucia Aqr 22h13'43" -10d26'18"
Lucia Cas 1h59'11" 67d54'42"
Lucia Cyg 20h39'13" 54d30'54"
Lucia Cas 23h43'43" 58d55'10"
Lucia Aqr 23h23'48" -17d44'25"
LUCIA ALARCIA DEL OLMO Crb 16h4'42" 38d39'51"
Lucia Alessandra Maffeo Mon 6h26'58" 8d18'24"
Lucia- Bella Gem 6h18'37" 27d6'24"
Lucia C Cas 0h15'16" 52d10'26"
Lucia C. Roderique Cas 3h27'10" 74d23'7"
Lucia Carla Gasparinho Cas 23h13'49" 59d38'25"
Lucia Cerbo Psc 0h52'25" 8d46'5"
Lucia Chirinos Pereira Sgr 18h58'50" -34d36'19"
Lucia Christina Scotece And 23h17'1" 51d11'48"
Lucia Cicorella Uma 8h23'5" 62d10'29"
Lucia Cucci Ori 5h59'26" 22d42'37"
Lucia Davila Per 3h27'22" 43d52'58"
Lucia Della Principessa Cnc 8h40'23" 23d32'22"
Lucia D.G. Tau 5h34'48" 18d11'27"
Lucia Di Marzo Cap 21h34'53" -20d36'12"
Lucia e Andrea Lyr 18h41'55" 46d25'53"
Lucia e Gianbattista Umi 15h35'28" 69d1'34"
Lucia Edlund Cam 5h48'20" 64d2'22"
Lucia Fernandez Leo 11h37'40" 15d39'58"
Lucia Florence Mahoney Sgr 18h11'26" -28d9'30"
Lucia Gabriella Riotto Lib 14h49'2" -1d25'24"
Lucia Giovanna Cuffari Tau 4h39'42" 28d37'45"
Lucia Hallett Sgr 19h22'11" -18d51'26"
Lucia Hannah Cannon And 1h51'28" 36d5'0"
Lucia Iman Leo 11h14'25" 14d50'17"
Lucia Jane Franco Vir 13h44'33" 2d41'6"
Lucia Lengua Nebula Cas 23h43'25" 53d14'31"
Lucia Lobraico And 1h29'0" 48d36'34"
Lucia Luccarelli Lac 22h17'53" 53d40'29"
Lucia Luna You Are My Star Uma 8h27'8" 64d55'35"
Lucia Maria Chiavola Cas 0h26'46" 54d9'55"
Lucia Marie Daubel Cas 1h45'53" 62d31'12"
Lucia Martarano Tau 3h50'59" 24d40'51"
LUCIA - Mia Amorevole Mamma Uma 11h18'8" 62d39'14"
Lucia Palazzo Cep 22h11'31" 71d16'43"
Lucia Papa And 2h32'19" 50d1'23"
Lucia & Pierre Uma 9h50'51" 58d26'41"
Lucia Rose Walsh Pochynok Cas 0h41'44" 72d51'29"
Lucia Salce Uma 9h20'38" 44d17'19"
Lucia Saveria Romano-Bright Cyg 19h58'20" 39d37'20"
Lucia Six Ari 2h51'1" 29d15'41"
Lucia Smith - My Shining Star Cru 12h36'46" -61d13'32"
Lucia Suzanne Cody Baker And 0h33'56" 27d40'41"
Lucia V. Moffa-Mills Cas 1h43'44" 61d9'9"
Lucia Valle Cyg 21h11'45" 36d19'39"
Lucia Versteeg Sgr 18h35'20" -22d37'36"
Lucian Lyn 8h19'34" 34d58'55"
Lucian Alexander John Plack Per 2h49'40" 53d27'21"
Lucian Anthony Lennon Cyg 19h49'33" 35d59'15"
Lucian Hollingsworth Cousins Sr Gem 6h41'49" 21d30'45"
Lucian Mar Vaughan Per 3h34'18" 36d10'47"
Lucian Mark Trudeau Her 17h49'21" 47d57'58"
Lucian White Lyr 18h47'5" 34d56'34"
Luciana 060660 Gem 7h30'16" 20d25'47"
Luciana Bilotta Lyn 7h27'34" 57d10'22"
Luciana & David Umi 15h43'38" 75d4'34"
Luciana DeMarchi Cas 22h58'58" 59d35'39"
Luciana Major 1 Ori 5h7'15" 4d44'13"
Luciana Ortolani Umi 15h51'17" 78d6'3"
Luciana Rhodes' Love Cap 21h22'10" -15d15'54"
Luciana Spedaliere Cyg 20h2'42" 53d38'11"
Luciana Valentina Gregorio And 0h45'2" 42d44'28"
Luciani's Hope 2005 Leo 9h55'16" 13d11'6"
LucianMichaelNievesWalker Gem 7h21'36" 16d48'29"
LuciAnn And 1h44'40" 45d51'34"
Lucianna Orion Cole Ori 4h47'31" 0d28'32"
Lucianne McTaggart And 1h11'47" 42d28'31"
Luciano Alu 1936-1998 Uma 11h6'55" 34d3'25"
Luciano Mario Prosperi Lib 15h24'31" -21d34'45"
Luciano Ramos Vir 13h13'13" -18d27'23"
Luciano Sgherri Aur 6h17'54" 29d3'42"
Luciano Steven Vivino Lib 14h48'33" -13d12'50"
Luciano Weingartner Uma 9h30'7" 57d57'5"
Lucia's "MTN ROAD in the Sky" Uma 11h42'9" 44d0'32"
Lucia-Skye Dougall Hart And 1h11'58" 46d15'17"
Lucibello Lyn 7h51'11" 42d11'16"
Lucid Dreamer Lmi 10h37'25" 26d50'32"
Lucid Lynn Gem 7h21'0" 32d2'0"
Lucie Lyn 7h48'0" 36d14'55"
Lucie And 2h36'49" 46d7'26"
Lucie Peg 23h11'24" 31d11'23"
Lucie Alice Glover And 23h1'7" 52d12'10"
Lucie and Rachel's Uncle David Dra 17h27'21" 57d13'49"
Lucie Angèle Amado And 0h44'48" 30d32'58"
Lucie Brantenaar Psc 0h9'5" 8d29'59"
Lucie Deroche aishiteru by F.C Gem 6h59'38" 22d50'50"
Lucie et Benoît Leo 10h45'39" 10d59'55"
Lucie et Julien Peg 22h35'28" 20d57'37"
Lucie G Cas 23h58'16" 61d16'49"
Lucie Gallova Uma 10h15'18" 42d32'5"
Lucie Gibson And 23h50'28" 44d55'24"
Lucie Honor Birney Cas 23h38'25" 56d10'30"
Lucie Isabella Williams And 22h59'55" 51d53'15"
Lucie Lillian Michaloski Cas 23h18'40" 54d40'41"
Lucie Louise Dodd Del 20h38'11" 15d23'40"
Lucie M. Wynne And 23h43'32" 40d37'46"
Lucie Marie And 23h28'48" 37d43'17"
Lucie May Sco 16h58'7" -42d44'52"
Lucie Moore And 23h0'28" 45d31'3"
Lucie Waddell And 23h7'14" 50d38'29"
Lucien Camilleri Cru 12h24'56" -55d51'6"
Lucienne Cas 1h16'21" 67d44'46"
Lucienne Lyn 7h18'14" 48d4'43"
Lucienne Psc 0h58'55" 19d43'58"
Lucienne D'allenger Kershaw Cep 1h3'33" 80d1'9"
Lucienne Dorais Delorme Umi 7h9'32" 88d14'11"
Lucienne Lachance Ori 5h33'46" 3d40'17"
Lucie-Rose Robinson Cnc 9h0'37" 19d54'4"
Lucie's Little Light And 2h8'29" 43d47'47"
Lucie's Love Cyg 20h39'54" 35d5'52"
Lucifer Her 16h55'4" 32d12'54"
Lucija Koledic Lib 15h44'53" -8d15'40"
Lucila Fasano Psc 23h27'16" -3d15'34"
Lucila Lapus Lopez Mon 7h22'14" -0d18'27"
Lucile et Benjamin Umi 15h40'41" 77d40'9"
Lucile May Uma 13h36'30" 55d49'55"
Lucile with One L Cam 4h29'46" 73d43'32"
Lucilla Berni Peg 22h18'49" 24d0'24"
Lucille Aqr 21h2'10" 0d59'17"
Lucille Leo 11h35'53" 13d55'32"
Lucille Crb 16h6'6" 36d26'18"
Lucille And 2h19'58" 39d2'43"
Lucille Lyn 7h28'27" 49d27'8"
Lucille And 1h9'10" 47d54'53"
Lucille Cas 1h21'40" 57d52'25"
Lucille Uma 12h36'51" 53d6'29"
Lucille A. Petrarca And 0h37'4" 32d2'34"
Lucille Anne Gem 7h26'29" 33d0'19"
Lucille B Copp And 0h6'24" 44d17'26"
Lucille Bishop Per 2h52'20" 42d27'33"
Lucille Bourguignon Gem 6h10'4" 27d50'54"
Lucille Brown Leo 10h22'54" 14d14'29"
Lucille Burch Lib 15h4'12" -14d10'51"
Lucille Butler Ori 6h22'52" 10d2'5"
Lucille Campbell Gem 7h21'38" 28d9'53"
Lucille Charlene Halzle Crb 15h51'40" 27d52'35"
Lucille Conti Oates Ari 2h47'11" 29d34'49"
Lucille Crocilla And 0h8'57" 45d33'45"
Lucille Croteau Lyn 8h15'0" 37d24'28"
Lucille E. Hyder Ari 3h15'39" 28d31'35"
Lucille E. Silvernail Vir 15h1'59" 2d39'5"
Lucille Foyt Lmi 10h28'37" 35d47'34"
Lucille Frances Lorang Crb 15h40'30" 37d11'2"
Lucille Gertrude Casagrande Mon 6h53'58" -0d2'10"
Lucille Irene O'Dea Helmer Cas 1h19'44" 58d14'7"
Lucille Joy Lyn 7h12'35" 47d37'40"
Lucille M. Baxter Lyr 18h54'35" 32d20'35"
Lucille M. Deeb Cap 20h26'3" -22d31'25"
Lucille M. Hass Lib 15h41'25" -19d52'57"
Lucille M. Osborne Leo 10h36'17" 18d1'25"
Lucille Marie And 0h0'47" 46d20'15"
Lucille Marjolaine Uma 8h34'4" 62d41'17"
Lucille Martinez Ramirez Vir 12h49'17" 2d57'13"
Lucille Moore Cas 2h5'53" 72d0'44"
Lucille Nadeau Cas 1h25'9" 53d3'12"
Lucille Rock Lyn 7h56'28" 33d55'10"
Lucille S. Rinaldi Cam 7h38'0" 65d47'34"
Lucille Schultz And 1h14'19" 38d19'30"
Lucille T. Menard Tau 4h13'19" 7d3'6"
Lucille Tevis McClure Cas 1h19'14" 59d7'10"
Lucille The Twinkle In My Eyes Tau 5h49'55" 16d18'30"
Lucille Thomas Psc 1h18'58" 31d13'48"
Lucille Tockstein And 1h5'31" 45d6'13"
Lucille Untch Cas 0h31'28" 52d6'43"
Lucille Vernice Sullivan Psc 23h53'20" 6d33'12"
Lucille (Villani) Legnini Psc 1h17'19" 31d36'50"
Lucille Virginia Gibbs Cap 20h28'40" -13d52'58"
Lucille Vitelli Uma 10h57'47" 38d54'5"
Lucille Zanghi Ari 2h51'1" 26d14'18"
Lucille's Soul Cap 21h42'37" -9d25'46"
Lucille's Star Sco 16h6'53" -20d39'13"
LuciMar 06 Ori 4h49'9" 13d16'35"
Lucina Elena Uma 9h0'33" 61d12'34"
Lucinda Uma 8h45'11" 69d16'47"
Lucinda Ari 3h4'58" 25d26'58"
Lucinda Gem 7h26'36" 33d54'9"
Lucinda Baker Cnc 9h9'1" 15d1'35"
Lucinda (Cindy Korff Rattray) Cap 21h16'21" -24d45'48"
Lucinda D'Angelico Lyn 8h47'26" 42d10'21"
Lucinda Dauterman Lyn 8h16'31" 34d58'28"
Lucinda Jane Carlile Cyg 20h4'58" 31d18'36"
Lucinda K. Bourn Her 16h33'52" 43d43'21"
Lucinda Kay Kisinger Holland Cyg 20h0'4" 33d25'9"
Lucinda Marie Smith Birthday Star Sco 16h21'57" -28d14'23"
Lucinda Nell Scott And 0h34'18" 25d8'22"
Lucinda Pearl And 1h25'0" 44d58'57"
Lucinda Rangel Gem 7h15'57" 30d42'30"
Lucinda Williams Lyn 7h48'56" 42d30'6"
Lucinda's Light And 23h45'8" 37d3'57"
Lucio Uma 12h6'55" 44d40'4"
Lucio Ferrarini Uma 9h5'58" 51d31'19"
Lucio Magini Aur 5h39'54" 32d34'3"
Lucio Satar Leo 10h10'51" 26d13'33"
Lucio Sicignano Vir 13h19'45" -20d24'39"
Luciole Tau 5h46'39" 22d57'46"
Luciole (Lucille Rancourt) Aqr 21h17'19" -3d12'10"
Lucious Leonora Sgr 18h58'26" -23d43'24"
Lucious Lorna Gregory Pup 7h44'37" -19d46'22"
LUCISTE Crb 16h8'39" 26d38'43"
Luciu Vir 13h23'16" 11d38'14"
Lucius Uma 11h4'19" 36d4'7"
Lucius & Manu Uma 10h18'27" 62d55'8"
Luck Be My Lady Tonight Ari 2h26'30" 11d12'53"
Luckenbach Crb 15h41'42" 31d25'30"
Lücker, Andreas Uma 8h11'31" 68d47'1"
Lücker, Georg Cap 20h36'42" -23d3'59"
Luckey&Skittles Psc 0h53'37" 17d22'55"
LuckieLeo Lmi 10h17'55" 28d34'51"
Luckina Hviezdicka *33* Cas 23h11'23" 54d11'14"
Lucky Uma 11h59'7" 56d10'54"
Lucky Uma 8h12'55" 69d1'6"
Lucky Umi 16h13'56" 74d58'28"
Lucky Umi 15h40'57" 68d52'29"
Lucky Uma 11h32'5" 63d16'33"
Lucky Lib 14h47'31" -7d37'51"
Lucky Cma 6h40'48" -29d56'37"
Lucky Cma 7h10'2" -22d55'0"
Lucky Ari 2h50'43" 24d22'58"
Lucky Her 17h11'51" 16d42'55"

Lucky
Leo 11h52'4" 20d22'55"

Lucky
Psc 1h22'37" 16d23'53"

Lucky
Vir 13h11'8" 7d49'19"

Lucky
Ori 6h19'58" 10d28'10"

Lucky
Uma 11h46'21" 50d6'47"

Lucky
Per 2h37'58" 55d23'52"

Lucky
And 23h17'2" 39d21'4"

Lucky 13
Umi 16h59'43" 81d44'5"

lucky 7
Uma 10h29'50" 65d25'9"

Lucky Ahlfeldt
Cyg 20h32'48" 52d34'48"

Lucky Amor
Gem 6h49'4" 23d5'51"

Lucky Bastard "9-16-1986"
Uma 9h39'7" 54d27'18"

Lucky Boy
Crb 16h19'2" 29d35'42"

Lucky Charm
Ori 6h13'51" 20d52'13"

Lucky Charm
Aqr 20h44'38" 0d2'40"

Lucky Clover JSM&SJM
Uma 11h9'44" 35d27'16"

Lucky Couple
Dra 15h15'51" 60d44'8"

Lucky Cy-Cy
Leo 10h26'22" 12d5'46"

Lucky Day
Uma 11h16'24" 42d20'4"

Lucky Day "13th July 1979"
Uma 9h2'28" 64d5'22"

Lucky Girl
Vir 14h41'27" 6d9'4"

Lucky Gold
Her 17h46'47" 25d58'36"

Lucky JG
Uma 9h55'20" 51d37'1"

Lucky Laura
And 23h31'16" 43d3'19"

Lucky Linda
Ari 2h26'3" 22d21'58"

Lucky Louie / Zaira Diaz
Cyg 21h49'51" 52d15'25"

Lucky Luca
Ori 5h28'52" 14d3'18"

Lucky Lucas
Aqr 21h30'28" 1d51'12"

Lucky Lucas
Uma 13h46'13" 53d27'27"

Lucky Lucia Maria
And 23h51'4" 47d26'39"

Lucky Luckow
And 23h57'56" 33d34'8"

Lucky Man
Leo 10h54'2" 2d17'38"

Lucky Marie
Cmi 7h26'27" 1d52'33"

Lucky Noah
Ori 5h29'35" 2d11'38"

Lucky One
Ori 6h9'48" 15d16'53"

Lucky One
Leo 10h50'26" 19d50'55"

Lucky Penny
Cnc 8h20'55" 8d39'7"

Lucky Poljak
Cyg 21h42'16" 32d32'9"

Lucky Porter
Cam 3h58'36" 66d23'43"

Lucky Rae
Uma 9h19'27" 72d34'44"

Lucky Seven
Dra 18h7'6" 71d26'46"

Lucky Smart
Ori 5h25'45" 0d31'6"

Lucky Sophie
Vir 12h48'21" 3d40'49"

Lucky Star
Ari 2h9'39" 24d21'27"

Lucky Star
Umi 16h41'43" 83d50'58"

Lucky Star
Cet 1h4'6" -0d8'36"

Lucky Star
Cap 21h29'45" -8d33'55"

Lucky Star of Andrey Preobrazhensky
Sgr 18h32'22" -16d31'8"

Lucky Star-13
Uma 11h34'17" 57d26'14"

Lucky Strike
Leo 9h54'0" 28d42'42"

Lucky Thirteen
Cen 12h4'17" -47d58'13"

Lucky Winner Sample
Per 3h4'46" 45d59'38"

Lucky Woods
Sgr 18h52'43" -29d49'42"

LuckyBug
Uma 10h30'31" 64d49'29"

LuckyLady 248
Aqr 22h57'23" -7d23'14"

Luckymans' Angel
Cnc 8h37'50" 27d40'21"

Lucky's Charm
Uma 8h55'53" 49d38'35"

LuckyTink83
Ari 2h12'54" 15d27'12"

LuckyWe
Lib 15h59'15" -11d52'28"

Luco
Her 16h52'0" 27d49'8"

Lucrecia
Cas 1h48'4" 64d46'56"

LucreciaZ1
Lyn 9h30'22" 40d46'7"

Lucretia Mercer
Tau 4h27'26" 22d24'44"

Lucretia's Perfect Gift
Lyn 7h39'33" 46d55'25"

Lucrezia
Umi 13h10'25" 73d33'23"

Lucrezia Bosco
Lib 15h10'38" -23d24'35"

Lucrezia Magini
Cam 5h33'52" 71d29'51"

Lucrezia Petrella-Mitterhoffer
Lyr 18h35'50" 37d39'32"

Lucus McFucus
Ari 3h17'39" 26d38'51"

Lucy
Her 17h46'43" 17d20'9"

Lucy
Leo 10h35'5" 17d39'17"

Lucy
Cnc 9h11'50" 16d17'20"

Lucy
Psc 0h45'38" 19d33'27"

Lucy
Ori 6h3'19" 13d46'50"

Lucy
Aqr 22h27'42" 1d2'48"

Lucy
And 0h56'31" 45d14'47"

Lucy
Cas 0h51'40" 49d0'16"

Lucy
Uma 10h8'24" 45d35'33"

Lucy
And 2h15'27" 44d38'4"

Lucy
And 0h35'38" 39d2'56"

Lucy
Lib 15h21'4" -24d44'26"

Lucy
Cas 22h58'12" 55d4'5"

Lucy
Cas 0h31'2" 62d5'4"

Lucy
Lib 15h53'57" -9d33'49"

Lucy
Cet 2h10'34" -21d19'24"

Lucy
Aqr 23h18'43" -14d24'20"

Lucy
Cap 21h36'15" -16d38'16"

Lucy
Psc 23h9'47" -1d23'23"

Lucy
Psc 22h57'43" -0d4'9"

Lucy
Lib 15h10'34" -6d19'54"

Lucy 2005
Gem 7h36'43" 33d38'37"

Lucy 211187
And 0h39'55" 44d19'7"

Lucy A. Maier
Aqr 22h38'55" -15d1'20"

Lucy Alexandra
And 23h46'45" 42d47'27"

Lucy Alicia
Tau 5h50'25" 14d3'21"

Lucy Alleyne
And 23h19'37" 42d33'11"

Lucy and Leo Valen
Uma 12h20'15" 59d22'27"

Lucy and Tom Adamo
Cyg 20h49'5" 46d13'59"

Lucy Anderson
Vir 12h48'39" 6d28'7"

Lucy Ann
Lib 15h20'57" -13d0'46"

Lucy Ann 1949
Cnc 8h4'26" 26d53'40"

Lucy Ann Collins
Cyg 21h2'30" 43d22'55"

Lucy Ann Farrelly
Cas 23h59'6" 65d13'41"

Lucy Ann Funnell
And 0h10'30" 33d14'9"

Lucy Ann Neville
Cas 23h50'55" 53d7'24"

Lucy & Annette Forever Friends
Uma 9h55'25" 70d37'28"

Lucy Apperson Hanes-Mamalou
Cas 0h4'6" 59d44'18"

Lucy Arksey
And 23h48'30" 37d0'48"

Lucy Ava Tanner
Cnc 8h11'39" 32d27'34"

Lucy Bailey
And 1h52'0" 38d35'6"

Lucy Bain
Tau 3h45'37" 19d21'2"

Lucy Beatrice
And 0h15'22" 40d2'16"

Lucy Beatrice Petreus
Psc 0h5'16" -0d3'17"

Lucy Bednarek
Gem 6h57'28" 14d45'11"

Lucy Bensley
Cas 23h41'11" 58d34'56"

Lucy Boyd
Ori 5h16'14" -0d9'2"

Lucy Cannamela
Cas 0h43'19" 65d5'35"

Lucy Chavez
And 0h35'21" 35d45'34"

Lucy Claire Mills
Vir 13h22'50" 12d32'37"

Lucy Cohen
Per 4h36'15" 49d38'44"

Lucy Colendich
Lib 14h59'43" -10d11'15"

Lucy Daile Wilcox
And 2h36'7" 44d3'19"

Lucy Davies' Star
Cyg 21h22'9" 34d9'37"

Lucy Davis
Cas 23h35'8" 54d17'18"

Lucy del Rosario
And 23h20'36" 43d4'25"

Lucy DePaola
Sgr 18h3'26" -24d6'26"

Lucy Dew White
Ari 2h35'19" 14d55'14"

Lucy Diamond
Crb 16h11'39" 33d19'5"

Lucy Diamonds McDunlap
Ori 5h19'15" 15d2'51"

Lucy Diponzio
Psc 0h4'9" 4d11'52"

Lucy Doodle
Aqr 20h43'41" -8d43'20"

Lucy Downing
Ari 2h21'28" 18d8'8"

Lucy Elizabeth
And 0h50'28" 23d30'49"

Lucy Elizabeth
Uma 11h49'25" 39d55'31"

Lucy Elizabeth Eron
Lmi 10h52'8" 26d35'33"

Lucy Elizabeth Falconer
And 0h21'16" 25d57'56"

Lucy Elizabeth Gleason-Geise
Gem 7h57'31" 19d41'29"

Lucy Elsie Pick
And 0h47'51" 32d43'34"

Lucy Emily Magenta Chambers
Cas 0h45'40" 54d5'49"

Lucy Eva Pearl
Sco 17h48'56" -42d33'55"

Lucy for Linds
Uma 11h9'50" 63d40'31"

Lucy Francis Hunt
And 2h14'11" 45d50'2"

Lucy Francome Studdert
Cyg 20h0'36" 30d2'5"

Lucy & Frank DeFrenza
Cyg 19h17'18" 53d55'53"

Lucy Gardner White
And 23h46'55" 34d29'55"

Lucy Georgina Diamond
Umi 13h44'31" 77d51'27"

Lucy Gould Butterbauch
Uma 10h55'23" 62d0'19"

Lucy Grace Chisholm
And 2h32'45" 44d7'28"

Lucy Grace Green
And 23h37'24" 48d38'11"

Lucy Grace Jewett
Sgr 18h31'46" -27d54'13"

Lucy Grace Radley
Sge 19h47'33" 17d6'31"

Lucy Grace Sansom - Jamil
And 1h12'51" 46d17'42"

Lucy Graham - "Lucy's Star"
And 2h11'10" 47d13'14"

Lucy Grogan
Lmi 10h37'51" 37d39'25"

Lucy H. Johnson
And 23h20'28" 43d36'31"

Lucy Hailson
Cas 2h21'10" 63d3'22"

Lucy Helen McAuliffe - 1/03/57
Cap 20h40'45" -22d21'35"

Lucy Hernandez
Mon 8h3'56" -0d43'23"

Lucy Hopkins
Uma 9h16'8" 62d35'29"

Lucy Hughes Abell
Sgr 18h17'30" -24d2'40"

Lucy Hunter
Leo 10h25'58" 19d20'31"

Lucy- In Hopes Of A Speedy Recovery
And 23h43'13" 36d59'47"

Lucy in the sky
And 0h56'40" 45d16'23"

Lucy In The Sky
Cnc 8h47'46" 19d8'28"

Lucy in the Sky with Angels
Uma 9h53'0" 47d26'31"

Lucy in the Sky With Diamonds
Cyg 19h41'13" 46d40'12"

Lucy in the Sky with Diamonds
And 2h36'47" 37d43'52"

Lucy in the Sky With Diamonds
Leo 10h51'56" 11d2'25"

Lucy in the Sky with Diamonds
Cen 12h38'51" -49d47'59"

lucy in the sky with diamonds
Lep 5h46'7" -20d59'3"

Lucy - J
Lyr 18h47'18" 39d13'3"

Lucy J. Pomales
Crb 15h47'36" 33d22'24"

Lucy Jackson Eade
Sco 16h14'40" -16d33'39"

Lucy Jane Michna
And 1h13'7" 45d46'18"

Lucy Jane Pollard - Lucy's Star
And 0h1'10" 32d17'45"

Lucy Jane Rudd Moloney - Lucy's Star
And 23h9'5" 51d46'30"

Lucy Jane Scriven x
And 2h37'13" 50d19'50"

Lucy Jayne Grainger
Cyg 21h15'41" 46d58'28"

Lucy Jean
Lmi 10h36'33" 34d20'7"

Lucy Jean Quayle - Lucy's Star
And 23h56'50" 41d42'56"

Lucy Jeanne
And 1h10'46" 43d51'35"

Lucy & John
Cyg 21h37'6" 35d14'40"

Lucy Joy Miller
Leo 10h27'13" 11d57'14"

Lucy K
Uma 10h20'42" 51d33'1"

Lucy Kelly
And 1h56'38" 42d13'28"

Lucy Kirk Jennings
Gem 6h43'54" 15d33'18"

Lucy Lang
And 1h17'16" 35d49'54"

Lucy Lee
Lyn 8h16'45" 52d20'3"

Lucy Locket
Cru 12h11'42" -60d9'51"

Lucy Louise Belhomme
And 2h22'33" 42d32'36"

Lucy Louise Jane Woodhouse
And 0h13'20" 25d28'31"

Lucy Lucinda
Cas 0h41'43" 65d21'45"

Lucy Maria June Muir
And 2h8'3" 37d22'39"

Lucy Marie Smith
Uma 11h45'22" 63d40'43"

Lucy & Mark's Wedding Star
Cyg 19h48'37" 38d52'30"

Lucy Mars
Cas 0h12'32" 50d34'20"

Lucy Marzocca
Cyg 21h33'34" 31d59'16"

Lucy May Currie
And 2h0'31" 47d19'58"

Lucy May Gooding-Williams
And 1h10'18" 45d21'12"

Lucy May Lucas
Vir 13h13'40" -14d54'38"

Lucy May Wilkins
Sgr 18h25'39" -28d4'46"

Lucy Meehan
Cyg 19h47'46" 31d55'45"

Lucy Michelle
And 1h40'53" 38d53'12"

Lucy Mines
And 2h22'28" 50d44'2"

Lucy Minor
Cap 21h29'29" -8d29'7"

Lucy Mooney
Leo 10h22'41" 25d32'3"

Lucy Murphy in the sky w/diamonds
Psc 22h51'34" 2d37'28"

Lucy Novak
Vir 12h11'58" 7d59'3"

Lucy & Nuno
Tau 5h33'10" 27d28'59"

Lucy Opeña
Vir 14h25'18" -6d55'25"

Lucy Orozco
Lib 14h34'4" -17d31'37"

Lucy Piraneo Cervone
Crb 16h9'31" 27d43'8"

Lucy Rae Bland
Cyg 20h42'48" 32d43'47"

Lucy Ramos
And 0h40'53" 27d33'55"

Lucy Rebecca Gemmell
And 1h25'47" 49d45'1"

Lucy Rohn Worthington
Sgr 18h45'24" -25d53'13"

Lucy Rose
And 0h52'9" 46d23'56"

Lucy Rose Adkin
Mon 7h8'57" -8d21'15"

Lucy Rose Moore
Tau 3h52'19" 17d9'41"

Lucy Serrani
Sco 17h51'49" -36d35'14"

Lucy Stanton Doan
Cas 2h45'6" 65d13'44"

Lucy Star (Lucille Maleno)
Tau 4h48'6" 21d42'44"

Lucy Star Zalayet
Cas 0h40'29" 62d46'25"

Lucy Susan Green
Cas 0h12'58" 56d59'55"

Lucy V.
Mon 6h51'1" -0d10'31"

Lucy VanTine
Leo 11h17'53" 16d4'10"

Lucy Victoria Wood - Lucy's Star
And 22h58'7" 40d8'43"

Lucy Virginia Christening Star
And 23h33'13" 42d22'41"

Lucy Whyte Agnew - Lucy's Star
And 2h25'20" 39d11'1"

Lucy Williams "Our Shining Star"
Psc 1h45'56" 17d38'29"

Lucy Zeimet
Psc 1h10'0" 28d33'43"

Lucya Alexandrovna Guselnikova 4846
Ari 2h38'1" 17d41'9"

Lucyandy
Mon 6h47'4" -0d0'31"

LUCYANNE
Sco 16h45'15" -31d51'32"

Lucy-Kate Bestford
Col 5h47'35" -35d41'26"

Lucyna
Cas 23h39'16" 55d12'5"

Lucyna Kugel
Del 20h36'42" 15d23'12"

Lucyna Przywara
Vir 12h5'33" 6d31'28"

Lucyna Webb
And 1h9'43" 35d9'3"

LUCY'S ETERNAL LOVE
Uma 11h20'37" 63d31'11"

Lucy's Fire
Mon 6h53'35" -0d11'2"

Lucy's Heart
And 0h39'31" 32d33'5"

Lucy's Lamp
Tau 3h40'25" 14d30'1"

Lucy's Wish
And 23h25'38" 47d17'5"

Luczak-Herod
Mon 7h47'21" -8d41'14"

Lud & Jen
Vir 14h55'2" 4d28'18"

Luden
Ori 6h13'51" 18d46'55"

Lüders, Klaus
Ori 6h6'29" 14d1'37"

Ludger Bernhard Evers
Uma 10h14'58" 65d50'16"

Lüdicke, Horst
Ori 5h35'6" 10d0'20"

Ludivine McGrew
Sgr 19h4'26" -35d9'50"

Lüdke, Leonhard
Uma 11h56'58" 51d42'3"

Ludmila
Cyg 19h59'53" 51d5'59"

Ludmila
Tau 4h8'48" 29d20'23"

Ludmila Kirmusova
Cep 21h23'32" 64d30'19"

Ludmila & Sergei Yushiny
Cyg 20h30'59" 34d35'7"

Ludmila Sirbu
Per 4h14'0" 36d50'58"

Ludmila Zalesskaya
Ori 5h32'58" -2d0'3"

Ludo
Cnc 9h5'43" 23d39'4"

Ludo&Manu
Psc 1h14'41" 26d4'5"

Ludovic Gérard
Col 5h14'42" -28d26'47"

Ludovic Pimenta
Tau 4h18'19" 29d50'22"

Ludovic Prédal
Vir 14h11'25" -8d12'28"

Ludovica
And 2h24'49" 43d54'53"

Ludovica Acanfora
Dra 17h49'42" 66d7'1"

Ludovica Chiacchiera
And 0h19'31" 45d2'57"

Ludovica Lucarelli
Per 3h7'5" 51d38'43"

Ludovica Palatiello
Lyr 19h10'52" 37d49'6"

Ludovico Attilio
Cas 23h34'24" 53d16'41"

Ludovico De Mezzo
Lyn 8h46'36" 44d21'46"

Ludovico & Marla Iezzo
Cyg 20h11'34" 57d14'2"

Ludovik
And 1h23'57" 34d52'11"

Ludoxia
Uma 10h25'56" 59d16'49"

Ludwig Ray
Vir 11h40'50" 2d36'7"

Ludwig, Uwe
Ori 5h52'24" 21d45'0"

Luella Coffey
And 2h38'6" 50d26'47"

Luella "Our Dolly"
Aqr 22h49'37" -6d16'23"

LuellaRainn Alice
Tau 5h34'52" 26d11'48"

Luena
Cyg 21h56'1" 41d32'47"

Luetyville
Cyg 21h34'41" 54d33'11"

LUF2 Bolton
Aql 19h44'47" 8d2'52"

Lufae Ash Ash
Sgr 19h41'22" -25d48'8"

Luftgräfin Ingrid, Hinz
Uma 12h36'45" 61d46'45"

Lugnut
Cap 20h29'10" -14d28'45"

Luguangshi
Lib 14h23'28" -12d8'58"

Luhmannescent
Vir 13h44'49" -18d53'27"

Lui
Cma 7h8'34" -31d23'22"

Luigi
Cap 20h48'17" -26d23'7"

Luigi
Cam 6h5'17" 60d44'10"

LUIGI
Uma 9h10'59" 52d10'0"

Luigi
Uma 10h23'23" 41d19'42"

Luigi Bucefalo
Uma 10h34'6" 71d39'17"

Luigi Danesi
Cam 7h6'43" 79d48'51"

Luigi et Tatiana
Uma 13h46'9" 52d12'35"

Luigi Fosco
Cap 21h44'39" -14d19'7"

Luigi Gregory Scolieri
Per 2h37'14" 54d56'25"

Luigi Langella
Cas 0h27'21" 62d23'20"

Luigi Piva
Her 18h3'47" 17d26'1"

Luigi Sesso
Cam 3h26'31" 56d34'57"

Luigi Sr/Jr Colubriale (MVCMR)
Umi 16h35'53" 82d1'7"

Luigi Vita
Cyg 21h29'19" 37d0'16"

Luigina Schincariol
Lyr 18h49'18" 31d38'46"

Luigy Ceron
Uma 9h32'57" 58d2'20"

Luikart's Fire
Aur 5h25'47" 31d16'31"

Luis
Gem 6h40'42" 20d22'47"

Luis
Equ 21h5'5" 8d54'21"

Luis
Ori 5h24'7" 1d12'56"

Luis
Vir 12h45'30" -7d14'54"

Luis A. Alverio
Psc 0h58'47" 29d10'21"

Luis Alberto Cueva
Aqr 23h36'41" -13d33'3"

Luis Alcala
Psc 1h33'32" 18d16'19"

Luis Alejandro Nuñez Barron
Cas 0h37'31" 54d25'31"

Luis Almanza
Per 4h4'18" 47d31'49"

Luis Alonzo Valdez
Boo 14h23'25" 17d59'34"

Luis Alvarado
Aql 19h21'23" 1d48'41"

Luis and Anto's
Per 3h20'10" 48d25'49"

Luis and Fransisca Charqueno
Uma 10h26'23" 56d37'8"

Luis and Jenny Forever and always
Sge 19h57'15" 16d48'38"

Luis and Kristina
Crb 16h19'3" 29d59'8"

Luis Andrea Hürlimann
And 1h34'39" 46d22'44"

Luis Angel Flores
Cnc 9h1'18" 23d47'54"

Luis Anthony - James Robinson
Per 3h2'24" 47d13'25"

Luis Antonio Ramirez
Lyn 8h29'42" 47d43'26"

Luis Ariaza Rodriguez
Sco 16h5'10" -15d51'28"

Luis Bustos
Per 4h12'59" 48d38'51"

Luis Calvo
Dra 18h30'46" 78d54'30"

Luis Campos
Cnc 8h34'30" 23d27'46"

Luis Carlos Rodriguez
Tau 5h31'42" 18d28'21"

Luis Cervantes
Leo 9h41'36" 21d3'26"

Luis Cornejo
Gem 6h31'49" 24d48'56"

Luis Coronado Blanco
Uma 9h9'50" 68d19'12"

Luis David Johnson
Ori 6h13'6" 21d7'4"

Luis Diaz-Rousselot
Lib 15h12'18" -9d42'49"

Luis E. Campos (Estrellita)
Crb 16h9'38" 31d1'10"

Luis Fernando Peraza
Per 4h26'19" 43d25'52"

Luis Fernando Ruiz Gutierrez
Her 16h50'38" 25d48'20"

Luis G. Moreno
Aqr 22h57'26" -14d42'54"

Luis Garcia
Cma 7h26'38" -20d11'35"

Luis Gonzalez
Vir 12h8'5" 12d15'59"

Luis Guillermo F. Gonzalez Esteves
Ori 5h52'49" 22d46'1"

Luis Hoffer
Boo 14h30'20" 26d12'25"

Luis Iglesias Velazquez
Cap 21h52'12" -9d35'29"

Luis Ignacio Velutini Morasso
Lib 15h28'57" -25d31'8"

Luis & Jackie, Linked in the Stars
Vir 14h17'1" -20d51'54"

Luis&Jen Forever
Cyg 20h50'42" 46d26'6"

Luis & Josefa Hernandez
Uma 11h0'13" 49d49'36"

Luis "Lou Lou" Jimenez
Leo 9h54'11" 24d55'56"

Luis M. Alvarado, M.D., F.A.C.P.
Sco 17h52'26" -36d18'29"

Luis M. "Bambalante" Molina Monroig
Aql 20h10'37" -6d34'46"

Luis Manuel Perez Cruz
Cap 21h43'56" -13d34'10"

Luis Mario Romero
Aql 19h56'7" 13d45'58"

Luis "Melo" Diaz
Leo 9h43'36" 31d33'31"

Luis Miranda Izquierdo
Aql 19h49'20" 3d56'54"

Luis Monterroso
Uma 11h7'33" 46d21'7"

Luis Muñoz Jr.
Her 16h43'28" 29d17'9"

Luis "My Superman" Silva
Her 16h14'55" 47d45'46"

Luis N Karina Ducky
Cas 1h25'43" 61d54'44"

Luis N Kriti
And 0h38'21" 42d55'17"

Luis Palomero López de Armentia
Sco 17h50'3" -39d38'32"

Luis Pato Gonzalez
Ori 6h6'44" 21d27'20"

Luis R. Orozco
And 2h31'46" 49d20'57"

Luis Rafael Andrade Leon
Aqr 22h37'57" -2d49'22"

Luis Ramirez
Sco 17h51'34" -36d24'52"

Luis Rey Herrera
Sgr 19h22'54" -22d8'42"

Luis Reynaldo Brand
Aql 19h10'11" 2d56'43"

Luis Rivera
And 0h23'54" 44d41'1"

Luis Tarrina Niko
Lib 14h24'53" -10d57'13"

Luis Vidal Soria
Equ 21h14'16" 8d2'57"

Luis y Marga
Her 18h22'53" 16d55'51"

Luis Y Rebeca
Ret 3h42'36" -55d0'2"

Luisa
Col 6h15'34" -35d6'7"

luisa
Ori 5h15'54" -0d22'34"

Luisa
Peg 23h55'16" 27d6'33"

Luisa
Lyn 9h9'31" 40d5'30"

Luisa A. Da Silva
Cas 0h11'19" 51d30'28"

Luisa Adelfia Cruz
Cas 0h29'1" 55d22'46"

Luisa Aru
Aur 6h21'57" 29d22'57"

Luisa Cipelletti 26 settembre 1958
Per 4h15'24" 33d2'46"

Luisa Da Re
Aqr 22h49'21" -13d39'12"

Luisa Di Falco
And 2h31'53" 40d3'1"

Luisa Gerola
Umi 17h53'29" 87d15'45"

Luisa Guida
Cas 1h53'37" 61d49'41"

Luisa Joanne Carroll
Cyg 19h43'59" 27d49'42"
Luisa May Mulvihill
And 0h5'27" 39d25'38"
Luisa Rainone
Lyn 7h47'58" 58d26'33"
Luisa Santucci
Uma 10h28'44" 44d9'18"
Luisa Simoni
Gem 7h46'22" 29d11'9"
Luisa191052
Lib 15h6'53" -3d47'40"
Luisana Mayerlin Wills
Ari 2h15'42" 24d33'0"
LUISE LUNA - Sprüche 8,17
Uma 8h12'42" 62d43'27"
Luise S. Johnston
Uma 10h26'48" 61d4'56"
Luise Sophie Gradl
Uma 13h18'50" 61d38'20"
Luisella
Her 16h38'15" 6d16'3"
Luisiana & David Kommers 11/7
Ori 6h17'44" 14d54'28"
luisycynthia
Ari 2h51'39" 27d53'9"
Luitgard
Uma 9h37'52" 49d14'37"
Luiz Anthony
Peg 22h9'35" 26d45'52"
LUIZ ANTONIO LIMA DE ANDRADE
Col 5h39'1" -27d36'16"
Luiz de Souza Junior
Cra 18h39'32" -37d51'42"
Luiza Reingatch
Leo 10h50'3" 16d57'11"
Luiza Ruger Mussnich
Cen 14h32'45" -36d16'16"
Luiza Vieira Biaggi
Lib 15h3'39" -21d49'25"
Lujan
Ori 6h10'13" 6d58'22"
Lujan Angel de Dios
Psc 0h0'39" 5d42'50"
LuJuana Debra Tedford
Com 12h0'51" 21d33'50"
Luka
Cma 6h44'28" -26d58'42"
Luka Guy Bracey
Aql 18h59'37" -8d36'12"
Luka Jack Frangos
Ori 6h5'3" 21d29'17"
Luka Kolomejec
Cyg 21h36'52" 50d48'27"
Luka Moon Zachary Greenwood
Umi 15h35'50" 79d26'36"
Luka Soracolli
Leo 10h51'7" 22d39'58"
Luka Vjestica
Umi 14h21'26" 76d45'10"
Lukács Barnabás Csillaga
Cnc 8h13'39" 16d1'1"
Lukah Kahlil Lierman
Uma 11h13'29" 52d48'48"
Lukaka nau ko'u aloha mau loa Keli
Uma 12h35'33" 56d16'9"
Lukas
Cep 22h57'8" 62d42'10"
Lukas
Psc 1h42'23" 20d2'49"
Lukas Allen Becker
Cnc 8h10'21" 9d13'49"
Lukas David Weiler
Per 3h11'0" 46d48'40"
Lukas Delcourt
Cap 21h57'59" -10d34'29"
Lukas Eberle
Cas 2h1'6" 62d25'30"
Lukas Ellingsen Janout
Ori 6h20'48" 6d1'12"
Lukas Hayes Bednar
Umi 15h26'17" 70d20'17"
Lukas Holzer
Uma 10h28'43" 62d4'14"
Lukas James Burress
Psc 1h1'13" 26d45'44"
Lukas Julian
Uma 13h50'45" 62d11'42"
Lukas Koch
Per 4h15'45" 44d6'23"
Lukas Laurent Bouchard
Sco 17h49'14" -36d37'41"
Lukas Skye
Leo 9h46'40" 30d55'39"
Lukas Stavros
Cap 21h44'50" -16d52'45"
Lukas Stürmlinger
Her 18h22'57" 24d34'19"
Lukas Vytautas Narkevicius
Peg 23h9'56" 34d14'46"
Lukas Wayne Rust
Per 2h57'9" 51d8'2"
Lukas William Lazorishchak
Agr 21h41'0" 0d50'26"
Lukatie
Dra 15h59'59" 60d46'12"
Luke
Dra 19h19'6" 65d0'6"
Luke
Uma 10h45'47" 65d12'10"
Luke
Umi 14h46'17" 74d51'13"
Luke
Lyn 6h35'48" 54d26'10"
Luke
Cma 6h22'3" -27d46'33"
Luke
Sco 17h46'11" -38d26'59"
Luke
Cyg 20h14'36" 39d58'49"
Luke
Cnc 8h57'12" 6d42'58"
Luke
Her 16h59'52" 13d17'38"
Luke
Vir 12h35'56" 11d30'3"
Luke
Tau 3h38'8" 17d5'13"
Luke
Leo 9h51'28" 24d11'27"
Luke
Tri 1h52'29" 29d21'26"
Luke
Her 18h22'39" 17d57'24"
Luke
Cnc 9h3'22" 19d9'16"
Luke
Per 3h36'2" 47d56'35"
Luke
Lmi 10h56'34" 32d55'46"
Luke 4305
Cmi 7h17'37" 9d34'48"
Luke A Harthan
Cnc 8h40'48" 23d23'34"
Luke A627
And 22h58'55" 41d26'32"
Luke Abbate
Cap 21h25'38" -17d50'42"
Luke Adam Ford's Lucky Star
Cru 12h9'52" -62d49'9"
Luke Adam Prentice
Cap 21h50'27" -18d23'53"
Luke Adam Stalker
Per 3h0'21" 41d11'51"
Luke Addison Fox
Gem 7h38'41" 33d20'32"
Luke Alden Fuller
Per 3h45'13" 49d3'5"
Luke Alex Standing
Sco 17h1'24" -39d36'54"
Luke Alexander Fletcher
Per 1h38'40" 51d34'58"
Luke Alexander Junod
Her 16h57'34" 15d35'26"
Luke Alexander Scot
Cyg 20h1'39" 30d46'18"
Luke Allan Larson
Lib 15h29'52" -28d1'3"
Luke & Allie
Sge 20h5'44" 16d50'55"
Luke Allsopp
Gem 7h49'56" 33d11'34"
Luke and Angela's Star
Cyg 20h34'43" 47d12'14"
Luke and Krissy's Love Light
Cyg 21h24'9" 50d40'36"
Luke and Kristina
Lyn 7h7'25" 51d46'6"
Luke and Maria
Vir 14h2'4" -0d9'17"
Luke and Monika Tilmans
Peg 21h12'17" 19d55'42"
Luke Anderson Jackson
Lyn 7h25'22" 52d14'14"
Luke Andrew Robert Raymond Burgess
Dra 17h1'36" 53d33'41"
Luke Anthony
Sco 16h8'18" -13d20'16"
Luke Anthony
Per 4h45'52" 47d18'57"
Luke Anthony Baca
Ori 6h8'32" 7d34'40"
Luke Anthony Clarke
Her 17h53'47" 40d24'11"
luke anthony harper hewwitt
Lyn 7h31'16" 38d14'28"
Luke Anthony Sielicki
Cnc 8h21'12" 22d57'17"
LUKE ANTOINE COSTANZO
Ori 5h40'40" -0d14'55"
Luke Aram Anderson Wilson
Aql 19h50'54" 11d35'5"
Luke Ashby Gouldthorpe
Ori 6h19'17" 14d3'50"
Luke Atkinson
Tau 4h5'7" 21d20'51"
Luke B. Waller & Jessica
Ori 5h56'34" 20d50'53"
Luke Barber
Cyg 20h40'56" 40d24'12"
Luke Benjamin Laukaitis
Tau 5h36'15" 26d18'23"
Luke Benjamin Norman Docherty
Lmi 10h39'27" 34d8'9"
Luke Bishoy Sarofim
Sgr 18h31'35" -16d10'6"
Luke Bowler
Umi 14h25'2" 67d35'17"
Luke Bridger - Luke's Star
Uma 9h39'56" 50d21'54"
Luke Bryan Thurston
Lyn 7h15'25" 53d29'11"
Luke Caruthers
Lyn 6h52'10" 54d7'54"
Luke Charles Hlavaz
Per 4h30'12" 39d44'45"
Luke Christopher Brennan
Per 3h57'0" 38d46'7"
Luke Christopher Brooks
Psc 0h14'45" 11d16'11"
Luke Christopher Ryan
Ori 6h2'37" 18d26'2"
Luke D. Jaffe White Tiger Guppie
Ari 2h7'29" 24d6'0"
Luke Daniel Bird
Tau 4h21'23" 19d49'24"
Luke Daniel Bojko
Cru 12h43'9" -64d31'17"
Luke Daniel Sturley
Cnc 8h26'5" 14d5'41"
Luke David Bonnes
Ori 5h35'27" -0d32'51"
Luke Davis
Lyn 7h46'22" 36d37'44"
Luke E. Dobie
Cnc 8h51'22" 6d59'38"
Luke Edward Coble
Cap 21h32'50" -18d2'14"
Luke Edward Falco
Gem 6h51'49" 16d56'31"
Luke Edward Martin
Her 18h23'5" 14d24'0"
Luke Edward Sparshott
Umi 14h41'59" 77d56'14"
Luke Edward Vanderpool 5/19/04
Tau 4h12'42" 17d59'48"
Luke Edwin Stults
Sco 17h15'27" -44d7'49"
Luke Ellis Robinson
Cnc 8h53'44" 32d15'48"
Luke & Emily
Uma 9h44'50" 47d35'5"
Luke Ethan Saunders 1st Birthday
Umi 16h47'22" 77d0'2"
Luke - Eyes of an Angel
Per 4h48'19" 45d34'44"
Luke Finley-Gillis
Uma 11h17'23" 45d20'37"
Luke - forever my star - 26 Jan 06
Cru 12h38'11" -58d49'46"
Luke Francis
Ori 4h53'35" 12d20'23"
Luke Franklin
Aqr 21h27'30" 2d6'37"
Luke Fuller Stokes
Gem 6h46'32" 16d39'1"
Luke George Carr
Umi 17h26'47" 83d33'30"
Luke George Mackin
Cap 21h30'33" -17d48'49"
Luke Glynn
Aqr 21h26'25" -9d54'43"
Luke Goodwin
Ori 5h51'1" 19d39'7"
Luke Gruosso
Her 17h58'33" 29d23'37"
Luke Guerriero
Ori 6h0'58" -0d42'46"
Luke H. Fields Jr.
Cma 6h30'42" -16d36'19"
Luke Hampton Webster
Ori 5h27'34" 0d22'20"
Luke Harrison Baker
Cap 20h19'35" -14d34'2"
Luke Harrison Smith
Dra 16h14'24" 55d24'42"
Luke Henry
Lyn 8h14'50" 51d50'46"
Luke Herbert Mills
Dra 16h24'42" 66d29'19"
Luke Hudson Lemonds
Her 18h1'36" 24d53'36"
Luke Hughes Franklin
Umi 15h13'3" 79d33'33"
Luke Jacob
Per 4h11'51" 50d36'20"
Luke James
Her 18h1'3" 21d50'27"
Luke James Bollinger
Boo 15h17'8" 38d34'31"
Luke James Connor "The Chubster"
Umi 14h6'14" 68d36'51"
Luke James MacDonald
Peg 21h31'22" 25d59'25"
Luke James Slater - Luke's Star
Her 17h26'31" 18d3'50"
Luke Jason Hudson
Aqr 22h5'48" -11d3'44"
Luke Jefferson
Aqr 22h52'52" -9d29'49"
Luke Joseph
Ori 5h53'14" 21d49'48"
Luke Joseph
Aur 7h7'52" 39d4'3"
Luke Joseph Coleman
Sgr 18h41'27" -27d14'6"
Luke Joseph Mangan
Aqr 20h54'48" -8d38'57"
Luke Joseph Robert Daly
Cep 22h20'46" 62d8'27"
Luke Joseph Schreder
Leo 11h14'51" 21d26'14"
Luke Joseph Villaluz
Ori 6h19'39" 10d36'38"
Luke & Joshua McMahon
Gem 6h54'39" 18d21'37"
Luke Kristian O'Flynn
Peg 22h49'59" 4d12'50"
Luke Lattig
Per 3h27'49" 41d13'20"
Luke Lee Armstrong 4/23/1983
Tau 5h48'52" 14d13'7"
LUKE LINDSAY WOLF
Ori 5h27'43" 0d36'43"
Luke Lourdes David Titus
Lib 14h30'55" -10d45'40"
Luke Luv
Tau 4h9'21" 28d36'30"
Luke Mallaro
Uma 12h5'8" 46d41'35"
Luke & Marie - In Love We're United
Ori 5h39'2" -3d6'33"
Luke Marinaro
Cam 4h37'20" 62d17'9"
Luke Markus Herdade
Ori 5h54'21" 17d58'30"
Luke McCormick Albu
Ori 6h3'56" 14d14'1"
Luke McCoy Whitsell
Her 17h35'15" 26d34'4"
Luke- Memory of A Shining Star
Uma 11h32'51" 42d53'24"
Luke Merchant
Ori 6h11'8" 7d11'9"
Luke Michael Haggart
Leo 11h33'35" 17d40'44"
Luke Michael Mueller 1/4/82-4/20/06
Umi 14h26'35" 75d56'31"
Luke Michael Rowntree
Dra 14h53'41" 59d28'47"
Luke Michael Whiting
Ori 5h42'33" 1d38'42"
Luke Milton Gregory
Her 18h45'24" 14d51'38"
Luke Molzen
Tau 4h4'28" 26d35'26"
Luke Myles
Psc 1h52'33" 7d25'9"
Luke Naheem Shah - Luke's Star
Per 4h35'33" 47d37'51"
Luke Nelson Simpson
Cyg 19h24'13" 48d35'44"
Luke Nordman
Lyn 6h45'32" 51d56'34"
Luke Orion LaFemina
Ori 6h10'29" -0d48'26"
Luke - Our Guiding Light
Cru 12h25'3" -58d53'38"
Luke - Our Love Is Forever
Cru 12h45'25" -64d38'26"
Luke Patrick Boulos
Boo 14h7'10" 22d43'8"
Luke Patrick Carter-Schelp
Uma 9h14'20" 72d9'1"
Luke Patrick Vaccaro
Sgr 18h58'41" -26d12'5"
Luke Peter Ashmore
Aqr 22h25'54" -24d20'2"
Luke Pierce Freeman
Cnc 8h26'14" 29d17'43"
Luke R. Jacoby
Ori 5h37'40" -0d38'26"
Luke Randolph
Tau 4h33'43" 19d45'31"
Luke Raynor
Psc 1h18'48" 30d41'13"
Luke Robert Attard - 4 September 2006
Cru 12h27'29" -58d38'26"
Luke Robert Fletcher
Uma 8h26'36" 68d56'48"
Luke Robert Howard
Boo 15h33'45" 48d52'35"
Luke Robert Levy
Cnc 8h9'55" 18d5'8"
Luke Robertson Bennett
Per 4h4'48" 46d36'24"
Luke Robinson Griggs
Uma 11h52'1" 37d11'9"
Luke Roma Hughes
Sco 17h54'57" -38d41'6"
Luke Ronald Heckler
Umi 14h41'56" 76d13'34"
Luke Russell Weber
Her 18h36'40" 20d28'56"
Luke Ryan Valentas
Her 16h37'16" 49d16'42"
Luke Sample
Sgr 19h41'17" -13d2'8"
Luke Schultz
Tau 4h10'50" 2d33'30"
Luke Sohns' Star
Ari 2h42'20" 15d47'14"
Luke Steven Thomson
Per 2h16'15" 57d22'8"
Luke Telljohann
Aql 20h2'32" 8d54'9"
Luke "The Candyman" Jackson
Cep 21h35'10" 59d9'32"
Luke the "Duke"
Aqr 22h32'21" 1d31'25"
Luke Thomas Albert Strong
Her 17h44'6" 38d53'54"
Luke Thomas Briggeman
Lmi 10h8'47" 34d50'39"
Luke Thomas Nicholson
Cyg 20h48'33" 31d10'27"
Luke Thomas Peeples
Ari 2h51'13" 14d52'3"
Luke Thomas Ralston
Her 16h31'2" 25d6'6"
Luke Thomas Rumage
Her 17h36'14" 40d47'53"
Luke Thomas Waverka
Dra 17h43'37" 58d59'15"
Luke Thomas Z
Uma 10h43'53" 46d12'46"
Luke Timothy Hoffman
Sgr 19h25'22" -20d6'4"
Luke Tindill
Aql 18h51'29" 11d12'27"
Luke Tristin
Umi 15h21'23" 72d25'54"
Luke Turozci
Umi 14h20'25" 75d36'45"
Luke Vernon Riddle
Aqr 22h51'5" -9d19'57"
Luke Vigna
Ari 3h26'30" 28d18'55"
Luke Vincent Riccio
Boo 15h33'17" 44d19'49"
Luke Warren
Her 17h7'46" 46d35'14"
Luke Webb
Her 16h47'12" 34d5'27"
Luke Weston Brewer
Tau 5h44'22" 23d6'23"
Luke White
Psc 0h51'14" 11d39'29"
Luke William Eaton
Agr 23h28'59" -21d39'45"
Luke William Harrington
Per 2h53'52" 56d33'37"
Luke William Justin Harford
Gem 6h47'1" 12d40'57"
Luke William Mangini
Dra 20h24'13" 69d26'54"
Luke William McPake
Per 3h48'10" 49d37'34"
Luke William Nilsen
Per 4h13'10" 37d10'13"
Luke Zammit
Umi 14h33'22" 76d21'52"
Luke, Holly and Tom
Psa 23h1'57" -33d6'49"
Luke, Star of Light.
Umi 13h34'10" 70d52'16"
Luke0127 Joseph O'Shaughnessy
Cnc 8h10'30" 17d56'50"
Luker
Her 18h55'14" 24d0'24"
Luke's Guiding Light
Cnc 8h14'13" 12d10'35"
Luke's Mum - Aileen
Cyg 19h52'52" 37d25'37"
Luke's Shining Star
Her 17h13'10" 35d3'14"
Luke's Star
Cra 18h6'59" -37d4'58"
Lukestar
Cnc 8h19'57" 22d22'21"
Lukestar Morris
Cru 11h59'20" -62d12'42"
Lukey Daniel Resta *King Minga*
Cru 12h25'27" -58d50'54"
Lukey's Angel
Cap 21h12'8" -16d13'12"
Lukey's Dream Girl Skye 18.02.2005
Cru 12h31'46" -60d2'54"
Lukey's Star
Cap 21h41'37" -14d39'55"
Lukey's Star
Her 15h56'49" 42d54'0"
Lukie
Cam 6h42'20" 63d35'27"
Lukky Charm
Aqr 21h26'58" -0d13'2"
Luko
Ari 2h33'1" 27d2'11"
LUKUS DANIEL CROSS-MAN
Ori 6h11'53" 15d33'51"
Luky
Her 17h41'17" 20d24'23"
Lula
Gem 7h29'11" 26d37'27"
Lula
Psc 1h32'30" 18d39'57"
Lula
Psc 1h27'7" 10d31'51"
Lula
Cas 1h22'18" 52d5'35"
Lula B. Allison "Dear"
Vir 14h41'41" 6d4'9"
Lula Mae
Lib 15h36'36" -5d18'51"
Lula Sledge
Psc 1h19'26" 23d20'2"
LULABELLA 49
Psc 1h10'11" 14d35'11"
Luli McLain
Cnc 8h50'1" 12d30'26"
Lullaby Star
Lyr 19h11'2" 37d25'3"
Lult
Col 5h25'32" -35d45'33"
Lulu
Col 6h11'16" -28d17'55"
Lulu
Sgr 18h50'9" -30d27'35"
Lulu
Lib 15h0'49" -14d0'46"
LuLu
Uma 14h5'1" 53d15'25"
Lulù
Cas 0h29'32" 62d25'54"
LuLu
Lyr 18h51'48" 38d16'0"
LuLu
Psc 1h25'23" 17d56'3"
LULU
Vir 13h45'29" 2d3'29"
Lulu
Tau 4h21'30" 3d41'41"
LuLu
Leo 10h43'39" 15d8'8"
Lulu
Vul 20h22'7" 23d45'51"
LuLu "05/28/1954-05/18/2005"
Gem 7h38'26" 21d24'37"
Lulu "07/05/1997-07/14/2001"
And 0h4'1" 43d4'51"
Lulu and Gismo
Ori 4h57'41" 10d5'0"
Lulu Cream
Cnc 8h36'28" 31d34'32"
LuLu & Grandpa Jim
Uma 11h24'32" 70d17'30"
"Lulu" Lauren Amanda
Lib 15h34'50" -6d57'41"
Lulu My Love
And 2h5'33" 45d8'37"
Lulu notre tite étoile à nous
And 0h15'35" 46d8'39"
Lu-Lu Seaton
And 2h20'26" 41d51'11"
LuLu Star Luke
Cas 2h40'20" 57d51'31"
LuLuBelle
Crb 16h2'39" 29d21'57"
Lulue
Sgr 18h15'44" -21d34'52"
Lulu's Star
Sco 17h58'29" -39d24'50"
LULYNCH5.0
Cam 3h53'7" 54d52'12"
Luma Hawk
Umi 15h32'28" 72d5'1"
Lumar
Uma 10h23'5" 46d38'15"
Lumber Liquidators - Store 999
Uma 12h21'38" 55d37'33"
Lumen Daniel Reuben
Umi 14h54'5" 71d1'6"
Lumen Fidei
Crb 16h17'21" 32d47'5"
Lumeninis
Lyn 8h13'7" 40d41'6"
Lumi
Lyr 18h47'21" 45d4'36"
lumière de l'amour
Pyx 8h41'41" -22d12'47"
Lumiere de l'amour (light of love)
Lyr 18h46'56" 37d40'30"
Lumie're de Suzanne
Tau 4h14'9" 1d19'33"
Lumiere Etoile
Crb 16h15'30" 36d40'40"
Lumiere Lexa
Cyg 21h44'18" 43d53'6"
Luminaria Rogeri
Uma 9h54'53" 49d32'53"
Lumination of Strawberry Heaven
Uma 11h9'21" 50d31'32"
Luminesce
Ori 5h33'11" -0d57'32"
Luminescent Lucas
Uma 10h30'57" 39d49'14"
Luminosa
Psa 22h40'28" -25d43'29"
Luminoso Emilios
Cnc 8h38'38" 10d4'47"
Luminous 1
Crb 15h36'50" 27d7'16"
Luminous C Chandler Gerrard
Sco 16h6'28" -12d44'48"
Luminous Grandma Lu
Cap 20h51'22" -15d9'13"
Luminous Laura
And 23h13'23" 47d39'46"
Luminous Laura
Lyr 18h36'23" 41d24'30"
Luminous Natalie Kane
Cas 1h0'4" 67d22'58"
Luminous Renee
Uma 8h53'36" 57d30'14"
Lumpa
Lyn 6h45'8" 54d49'21"
Lumpi Laypon Feickert
Cas 23h21'59" 59d16'42"
LUMTL
Sge 19h26'37" 18d26'17"
L.U.M.U.
And 0h16'24" 47d45'50"
lumu
Umi 14h8'8" 72d46'39"
lumu
Cam 6h6'45" 63d8'14"
Lumumba Mosquera
Cap 20h23'54" -8d51'56"
Lumus Lamphear
Uma 10h56'7" 36d48'29"
L'Un Che Brilla
Vir 13h9'13" -5d26'27"
Luna
Crb 16h15'22" 37d1'32"
Luna
And 22h58'16" 37d57'35"
Luna
Ari 2h53'35" 22d27'33"
Luna
Ori 5h58'3" 3d11'48"
Luna Ailani Castillo Camacho
Cam 4h42'59" 64d51'16"
Luna de miel
Cyg 21h11'23" 51d46'50"
Luna (Holly & Oliver's Star)
Cyg 19h57'5" 54d7'3"
Luna - Hope
Eri 3h51'13" -0d20'19"
Luna Lake
Lyr 18h49'23" 34d31'28"
Luna Llena
Dra 14h46'5" 60d53'43"
Luna Louise
Vul 20h17'0" 23d6'33"
Luna Marie 2006
Uma 10h48'30" 42d46'19"
Luna Mileen
Uma 8h27'38" 69d14'26"
Luna Munday-Ruiz
Sco 17h49'39" -35d5'27"
Luna Roberta
Cma 7h19'21" -30d8'47"
Luna Star
Ori 6h21'20" 10d38'5"
Lunabear
Cam 4h11'4" 72d20'34"
Lunar Ali Lloyd
Sgr 18h40'55" -22d21'13"
Lunar "Nelson" Luettotarian
Cap 21h58'0" -10d55'17"
Lunar Orange
Umi 14h5'25" 67d44'31"
LunaStar
Dra 15h48'51" 64d3'59"
Lunchbox
Ari 2h15'41" 20d33'15"
Lundquist's United We Stand
Per 4h27'20" 43d19'35"
Lundy Ray Loosli Loved Forever
Tau 3h58'35" 25d23'19"
L'Unicorno
Mon 6h32'1" 3d42'26"
Lunkewitz, Stephanie
Uma 14h24'2" 59d5'29"
Lunnatuk
Aql 19h55'36" 1d38'50"
Luo Xiaowen
And 2h23'41" 45d24'56"
LUPE
Ari 2h8'0" 23d33'7"
Lupe Armijo
Umi 14h12'17" 77d26'6"
Lupe Dibernardo
Lyn 8h18'52" 40d44'59"
Lupe Freeman
Uma 9h3'58" 48d36'53"
Lupe Hall
Lyn 8h29'21" 34d8'35"
Lupe & Neomi Garza
Psc 0h8'37" 7d43'32"
Lupe Talamantez
Cyg 20h26'5" 32d16'24"
Lupe the Jellybean
Leo 10h24'50" 21d33'33"
Lupe the love of my life
Leo 10h56'49" 8d17'9"
Lupecolefe
Lyn 7h59'26" 42d21'0"
lupemylove
Leo 11h55'17" 21d58'53"
Lupita
Psc 1h42'22" 12d43'6"
Lupita Mascarenas
Psc 23h50'27" 6d24'17"
Lupita Moreno
Lib 15h0'8" -22d55'33"
Lupus
Col 6h4'20" -28d30'58"
Lupus
Lyn 7h48'1" 46d38'56"
Lura
Tau 4h30'14" 16d12'6"
Lura
Uma 9h16'40" 65d28'36"

Lura Jean Armas Blair
Tau 5h32'19" 27d14'32"
Lura McKelva Morris
Vir 12h54'12" -5d45'26"
Lurmia
Leo 10h21'0" 26d44'54"
Lurve
Cyg 20h45'20" 47d36'59"
Luscious Love Omelet
And 0h52'57" 36d26'2"
Luscious Lucius
And 23h44'2" 41d59'27"
LUSH
Uma 10h11'49" 48d36'57"
Lush Bucket
Lyn 7h21'18" 47d19'16"
Lushka Bainbridge
Cma 7h21'30" -20d23'11"
Lushus
And 23h7'36" 47d10'45"
Lusi
Aur 5h15'3" 39d31'10"
Lusi I. Flerova
Aqr 21h8'49" -2d56'31"
Lusi & Viken Happy Silver Anniv
Cyg 20h33'2" 48d25'8"
Lusia
And 23h20'30" 47d39'41"
Lusiana Gimenez Caruso
Lyr 18h48'38" 40d26'38"
Lusina Funk
Sco 17h5'59" -36d38'10"
Lusine Balikyan
And 1h46'51" 37d51'5"
Lusine Balikyan
Ari 2h11'34" 22d49'25"
LuSi-Stern
Uma 12h54'17" 62d59'44"
Lustgarten
Vir 13h48'30" -12d23'58"
Lustin Hartley
Cnc 8h26'57" 32d47'53"
Luszczewski
Lyn 8h4'46" 41d35'15"
lut - lut
Lyn 7h43'38" 47d53'25"
Lutheian
Vir 12h14'58" 12d1'44"
Luther Hayward Frost IV
Cyg 19h47'48" 42d22'1"
Luther Oliver Buchert
Her 17h24'39" 16d14'29"
Luther Tucker
Her 17h22'12" 26d50'26"
Luthien 05
Cyg 19h56'2" 35d52'49"
Lutvija Hrnjic
Ari 2h52'58" 25d44'41"
Lutz Behrens
Ori 6h23'2" 16d8'28"
Lutz Bittner
Uma 11h17'36" 42d14'59"
Lutz Funke
Ori 5h29'35" 10d31'14"
Lutz' Heaven's Eleven
And 2h20'21" 46d39'34"
Lutzler Pierre
Uma 10h21'28" 63d38'27"
Luucee
Psc 2h2'5" 6d23'4"
Luule Suozzi
Dra 16h1'28" 58d1'15"
Luv
Uma 8h47'54" 70d19'13"
Luv Le
Sco 16h13'27" -10d4'35"
Luv U Viv Elcock
Cyg 21h43'5" 48d13'52"
Luv4Tyrsis
Uma 11h29'22" 51d25'52"
Luv-a-bug
Psc 23h0'43" 7d19'13"
LuvButton
Vir 12h8'3" 10d53'55"
Luvencups
Cyg 19h57'14" 50d3'45"
Luvert
Leo 11h55'56" 22d8'2"
Luvfertanya
Ari 2h16'39" 22d30'54"
Luvie
Cap 20h41'54" -19d58'47"
Luvius Huckleberrius
Lyn 7h37'44" 53d22'25"
Luvpug - Gregg Steven Pobanz
Lib 15h49'10" -18d19'31"
Luvunook
Sco 15h53'46" -20d37'32"
Luvy Ibarra
Tau 5h51'26" 23d44'9"
Lux Aeternus
Tri 2h34'8" 34d3'52"
Lux et Caduceus
Cnc 8h2'3" 27d14'43"
Lux Liberum
Uma 10h18'13" 48d7'36"
Lux Lucis
Dra 18h46'19" 50d59'18"
LUX MEA
Ori 5h27'34" -7d53'3"
Lux Mundi (The Light of the World)
Gem 6h30'0" 24d45'48"
Lux Mundi, ALH
Ari 2h47'59" 25d18'10"
Lux, Hans-Dieter
Uma 9h1'42" 52d16'41"
Luxia
Aur 5h3'27" 45d37'56"
Luxor
Lyn 6h45'4" 61d12'11"
Luxtorius
Leo 9h25'20" 14d23'16"
LUXY
Peg 22h28'23" 29d10'37"
Luxy Star Litzner
And 0h29'42" 32d31'47"
Luz
Oph 17h33'38" -18d42'24"
Luz Adriana
Umi 15h51'59" 77d37'2"
Luz and Luna's Love Eternal
Cnc 8h8'54" 29d30'22"
Luz Berenice
Lib 15h8'26" -6d32'18"
Luz Caridad Freed
Lib 14h57'23" -1d59'16"
Luz de Fulya
Vir 12h27'6" 11d36'21"
Luz de Krista
Ori 6h5'22" -0d33'56"
Luz De Liz
Cap 20h54'43" -15d2'3"
Luz de los Angeles
Gem 6h33'15" 26d45'23"
Luz dos Olhos Meus
Cas 0h32'7" 50d3'51"
Luz Erenia Price
Tau 4h46'26" 29d10'49"
Luz Gonzalez - Rising Star
Lyr 18h24'3" 45d11'34"
Luz (Grandma Lucy) Padilla
Uma 12h13'53" 53d34'45"
Luz Janet
Vir 13h16'10" 7d4'1"
Luz Maria
Lyr 19h14'38" 37d19'25"
Luz Maria Alvarez Montes de Oca
Sgr 18h3'51" -32d48'26"
Luz Marina
Psc 0h2'3" 3d58'24"
Luz Marina
Lyn 7h34'40" 36d54'58"
Luz Marina Arias Barquet
Cas 1h7'1" 63d7'17"
Luz Nereida
And 0h32'36" 42d2'49"
Luz Quirina y Ernesto
Tau 5h41'42" 23d37'52"
Luz Sanchez
Aqr 21h43'20" -0d18'20"
Luz Stella Garces
Ori 5h36'31" -1d25'59"
Luz Tacoronte
Cas 1h37'58" 64d7'35"
Luz Zaragoza Chavez Leister
Sco 17h17'27" -32d24'9"
Luzandhenrysstar
Uma 11h14'38" 34d54'30"
Luzdella Deloatch
Lyr 18h25'13" 34d8'5"
LUZIA
Umi 16h10'41" 87d25'54"
Luzia Kilias
Uma 8h59'6" 63d16'51"
Luzia Weston
Sgr 18h5'50" -27d14'55"
luzstephanie
Leo 11h5'15" 11d44'5"
LV2
Lib 15h38'4" -21d16'48"
LVB III
Peg 21h30'59" 15d53'16"
LVCR Banquet Rock Stars
Crb 15h51'34" 32d12'19"
LVK-052415
Gem 7h16'20" 15d21'32"
LVVS & CSC
Sco 16h47'53" -27d15'37"
L.V.W. Prime
Uma 13h36'36" 51d1'28"
LVY
Dra 18h53'21" 57d50'12"
lwb6749
Umi 19h7'24" 87d21'53"
Lwrece
Gem 6h44'5" 26d17'36"
"LY" IN LOVE FOR EVER
Umi 16h22'49" 78d59'23"
Lya Adina
And 2h9'50" 41d54'38"
Lya je t'aime ! Fred
Cap 20h13'22" -15d43'27"
Lyal (Bubba) Albert Witchey III
Cap 21h23'53" -26d12'29"
Lyam Matthew Belleville
Gem 6h18'8" 22d19'9"
Lyana "AQ" Othman Star
Leo 10h19'45" 17d26'47"
Lyanh Nguyen
And 0h51'12" 36d37'17"
Ly-Ann Fortin-Savaria
Umi 13h8'7" 72d26'11"
Lyann Nishihira
Cyg 19h48'49" 37d20'6"
Lyazzat Daurenbekova
Sco 16h51'43" -38d36'45"
Lycette
Lib 15h39'13" -4d3'36"
Lyda Sutherland
Vir 12h14'33" -4d45'20"
LydeeLove (aka HaneeBunch's star)
Leo 11h10'10" 11d23'4"
Lydia
Del 20h37'7" 14d4'15"
Lydia
Ori 6h7'28" 12d58'25"
Lydia
Peg 21h46'21" 21d17'55"
Lydia
And 0h52'45" 43d54'39"
Lydia
Gem 7h49'14" 31d17'24"
Lydia
And 23h5'15" 50d42'59"
Lydia
And 2h31'28" 50d23'25"
Lydia
Sco 16h20'37" -30d22'52"
Lydia
Tau 3h38'23" 18d25'6"
Lydia
Cas 1h51'46" 73d47'10"
Lydia A. Flier
And 23h27'46" 48d10'10"
Lydia A. Henry
Tau 4h21'15" 16d59'47"
Lydia A Magyar
And 23h8'0" 50d47'52"
Lydia Ann
Cyg 19h42'57" 32d35'17"
Lydia Annette Lopez
Uma 11h14'38" 69d57'50"
Lydia Aprile Masi
Cyg 19h42'59" 54d54'54"
Lydia Asbury
And 1h34'32" 39d38'5"
Lydia Ashley Mooney
Cyg 20h59'16" 34d35'21"
Lydia Berry
Cap 20h24'58" -21d17'46"
Lydia Booker
Uma 11h31'34" 44d0'22"
Lydia Catlyn
Lyn 8h47'56" 40d0'25"
Lydia Claire
Psc 1h27'16" 26d35'28"
Lydia Dawn
Gem 7h35'5" 14d24'11"
Lydia Degen Hutchins Wife,Mom,Nana
Vir 13h49'35" -10d48'47"
Lydia Elizabeth Bailey
Cas 1h53'36" 61d24'6"
Lydia Elizavyeta Vlasov
Gem 7h49'45" 31d5'6"
Lydia Florence Beebe
Psc 1h14'3" 24d11'41"
Lydia Furter
Cep 2h52'52" 82d41'2"
Lydia Gail Hall
Cas 0h35'35" 57d32'52"
Lydia Gerwig
And 0h17'24" 44d42'54"
Lydia Govinden
Ori 6h10'11" 12d25'6"
Lydia Grace
And 0h17'3" 44d13'49"
Lydia Grace Martin
Cnc 8h39'30" 7d23'53"
Lydia Helene Bordfeld (Little Sis)
And 0h37'10" 40d33'20"
Lydia Hope
Uma 10h5'13" 43d1'16"
Lydia Imboden-Pont
Per 1h33'53" 54d23'46"
Lydia Jean Jacob
Per 2h40'20" 40d4'22"
Lydia Kay
And 1h26'57" 47d19'54"
Lydia Kerr Binanay
Ori 6h9'21" 2d57'51"
Lydia King
Uma 14h0'25" 54d0'40"
Lydia Koury Hornyik
Sco 16h46'2" -44d16'27"
Lydia L. and James R. Graham
Cyg 20h17'8" 38d59'29"
Lydia LaPlante
Pho 23h46'22" -47d48'49"
Lydia Lauren Damron
Aqr 22h50'40" -7d23'33"
Lydia & Lauri Forever
Aqr 22h56'5" -24d11'24"
Lydia Loveday Willow Saccomani
Cyg 20h54'13" 47d15'31"
Lydia Marie
Cas 1h2'59" 57d12'52"
Lydia Marie Moell
Vir 12h50'6" -4d8'28"
Lydia Marsha Folckomer
And 1h20'46" 37d3'25"
Lydia Marston Albanesius
Sgr 19h30'34" -14d1'57"
Lydia McDaniel
Cas 0h2'3" 56d25'19"
Lydia Melissa Rose Summons
And 1h5'36" 46d18'57"
Lydia Michele Hopson
And 23h44'27" 36d48'9"
Lydia Michelle Peer
Psc 1h7'0" 17d31'49"
Lydia Moralde
Vul 20h37'47" 26d55'18"
Lydia Morene
Lib 15h15'1" -11d50'23"
Lydia Pan
Aqr 22h26'14" -2d2'8"
Lydia Pereira Solis
And 0h55'1" 33d44'56"
Lydia Perez
Vir 13h30'47" -17d57'45"
Lydia (Pie) Jones
Vir 12h57'52" 11d5'46"
Lydia Polgar
Uma 14h15'15" 57d4'17"
Lydia R. Chubb
Cap 21h37'45" -10d40'49"
Lydia Rachel Lowe
Crb 16h3'29" 38d49'1"
Lydia Renee
Cas 0h18'1" 50d32'34"
Lydia Renee
Vir 14h36'10" -2d27'14"
Lydia Ritter
Tau 5h44'56" 16d47'36"
Lydia Rose
And 1h1'59" 41d33'19"
Lydia Ryu, The Most Beautiful!
Aqr 22h29'8" -9d30'8"
Lydia Soletti
Sco 16h2'2" -20d42'46"
Lydia Troyce
Ori 6h5'47" 20d49'49"
Lydia Veryl Sturch Sell
Crb 16h15'57" 31d52'32"
Lydia Walker
Aqr 20h51'37" -10d50'44"
Lydia Wallace-Hancock
Uma 12h5'5" 60d19'44"
Lydia's Light
Cnc 9h20'15" 14d59'55"
Lydia's Shining Star
Vir 13h4'40" -4d29'1"
Lydia's wish
Per 3h0'5" 48d57'38"
Lydie Durieux
Col 6h19'41" -34d41'27"
Lye Vera Giliath
Sco 17h18'46" -33d11'14"
LYF1
Cap 21h20'50" -18d56'42"
Lyfe
Cam 3h55'20" 54d51'1"
Lyia Harris
And 1h48'36" 36d20'8"
Lykaon
Her 17h15'39" 32d25'45"
Lyl Sinclair
Dra 18h40'18" 59d40'24"
Lyla Beda Priebe Lloyd
Lib 15h21'1" -4d15'15"
Lyla Kathleen Rehill
Lyn 7h21'25" 56d44'21"
Lyla Louise Rassi
Ari 3h8'58" 15d24'11"
Lyla Marie
Uma 8h36'36" 61d13'7"
Lyla Walsh
Cyg 19h37'50" 36d19'23"
Lylace Dawn
And 23h22'49" 48d28'32"
Lylarex
Lyr 19h7'40" 36d39'14"
Lyla's eyes
Lyn 6h45'47" 58d26'19"
Lyle
Cap 21h49'32" -14d48'48"
Lyle 08 05 1925
Uma 10h40'26" 67d31'8"
Lyle and Sandy Millard, Mendon MI
Lyn 8h7'31" 49d2'28"
Lyle Blaine
And 2h16'30" 50d36'14"
Lyle Charles Tom
Vir 12h24'5" 1d11'22"
Lyle Eager
Her 18h14'36" 26d59'4"
Lyle Eugene
Ari 2h27'13" 18d36'48"
Lyle Firnhaber
Uma 10h40'52" 66d53'8"
Lyle Hunter
Umi 14h3'25" 71d2'18"
Lyle & Jane Feller
Uma 10h30'52" 64d21'56"
Lyle K. St Mary
Lyn 8h2'24" 34d46'48"
Lyle Kasprick
Cyg 21h50'16" 37d50'27"
Lyle Lange
Cap 20h24'9" -13d12'56"
Lyle Lewis
Gem 7h11'8" 15d23'51"
Lyle Mark Olson
Uma 8h56'4" 67d41'36"
Lyle Nelson
Per 4h5'21" 33d49'9"
Lylia Benyelles
Cyg 19h51'15" 30d13'51"
Lylian Jo Bush
Ari 3h7'53" 14d0'34"
Lylle Breier
Uma 11h23'41" 46d4'29"
L.Y.M.
Cyg 19h55'47" 55d54'25"
Lyman C. Dye
Aql 18h55'1" -0d2'19"
Lyman J. Scripter,MD
Aql 18h47'14" 8d21'9"
Lyman W. Packard
And 23h50'9" 40d28'7"
LYMKAT
Crb 15h44'49" 36d21'53"
LYMTYLTT Dottie Wheeler
Uma 10h57'50" 52d44'2"
Lyn
Lep 5h43'38" -14d6'11"
Lyn
And 2h37'44" 44d56'50"
Lyn
Uma 11h47'50" 38d42'44"
Lyn
Crb 15h34'43" 26d9'9"
Lyn Allam
Vir 13h35'4" -14d48'26"
Lyn and Nigel
And 2h19'16" 38d10'12"
Lyn Brown
Tau 5h4'13" 24d46'25"
Lyn Dillies
Lmi 10h8'10" 38d23'25"
Lyn & Jay 4Ever
Cap 21h40'15" -10d37'36"
Lyn & Jim Levitt
Cyg 21h40'35" 53d25'13"
Lyn K. Berlin
Ori 5h53'23" 21d6'55"
Lyn Le Fevre
Cru 12h24'26" -61d5'20"
Lyn M. Price
Tau 4h26'59" 26d40'16"
Lyn Müller
Cas 1h8'58" 67d2'51"
Lyn Rogers
Vir 11h38'13" -4d44'24"
Lyn Scherry
Ari 3h17'38" 29d20'29"
Lyn & William McPheat
Gem 7h26'34" 34d45'28"
Lynae Elizabeth Renz
And 0h11'19" 37d9'52"
Lynann Jolene Butkovich-Blake
Cap 21h33'26" -14d35'11"
Lynare & Dave
Dra 18h32'39" 57d38'55"
Lynch Family
Her 17h10'30" 21d33'39"
Lynch Star
Cam 6h6'38" 63d17'31"
LynchMob5 Forever
Her 17h47'7" 14d44'13"
Lynda
Ori 4h54'43" 2d15'26"
Lynda
Cnc 8h11'50" 15d8'57"
Lynda
Cyg 20h29'44" 37d24'49"
Lynda
Cyg 21h31'28" 38d53'14"
Lynda
Per 2h58'6" 48d24'1"
Lynda
Cap 20h37'25" -13d51'31"
Lynda
Aqr 22h38'38" -16d54'50"
Lynda
Cap 20h48'35" -21d44'56"
Lynda
Cap 20h17'55" -18d57'17"
Lynda
Psc 1h24'20" 5d2'17"
Lynda A. Dean
Cyg 19h55'19" 52d10'52"
Lynda Ann Brown - Happy 50th Birthday
Ori 5h47'25" 12d20'24"
Lynda Ann Buckley
Tau 5h51'52" 15d4'58"
Lynda Apple Love at first sight
Cnc 8h18'6" 20d15'58"
Lynda Avellino Abbott
Psc 0h26'24" 17d8'19"
Lynda B. Horenstein
Ari 2h52'30" 27d50'15"
Lynda "beautiful" Magarity
Leo 9h25'49" 18d31'2"
Lynda Beth Pinki Ausnus Riess
Ori 5h24'17" 4d4'47"
Lynda Bradley
Aqr 22h14'14" 0d46'50"
Lynda Brophy
Cas 0h1'48" 54d11'31"
Lynda Carol Assur
Sgr 18h41'25" -25d29'11"
Lynda Cheryl Millar
Vir 12h59'13" 6d42'11"
Lynda Conder
Lyn 8h27'11" 47d26'20"
Lynda Connolly
Leo 10h4'23" 19d25'36"
Lynda Cook
Leo 11h12'58" -1d57'56"
Lynda Daniel
Sco 16h4'47" -11d25'13"
Lynda Diana Legros
Psc 1h18'52" 3d4'59"
Lynda Emery Cope
Cap 21h0'20" -17d2'7"
Lynda Erlene Davis
Aqr 22h1'30" -13d28'25"
Lynda Frautnick Shining Star
Cnc 9h7'13" 22d38'33"
Lynda Gail Wachob
Boo 14h16'43" 49d26'57"
Lynda Green
Uma 11h31'24" 62d18'12"
Lynda Hodge
Cyg 20h31'8" 53d16'19"
Lynda Hunter
Uma 11h2'16" 50d3'27"
Lynda J 2007
Aqr 22h42'26" -1d37'39"
Lynda J. Barreto
And 1h6'36" 44d36'24"
Lynda Jane
Cas 23h24'21" 59d29'27"
Lynda Jean Supplee
Lib 14h51'17" -2d40'57"
Lynda Joy
Uma 11h8'3" 49d28'45"
Lynda K. Jones
Uma 9h44'6" 49d15'17"
Lynda Kay Trullinger
And 0h12'26" 45d49'38"
Lynda Lee Lindsey
Sco 16h23'42" -33d24'0"
Lynda Leigh "My Shining Star"
Lyn 7h15'58" 59d33'19"
Lynda Leone
Ari 1h53'51" 21d23'11"
Lynda Ling - Lynda's Star
Cyg 20h47'53" 47d13'38"
Lynda Lvu1422
Psc 0h37'54" 11d31'25"
Lynda Marie
Peg 21h40'52" 23d46'46"
Lynda Marie
Cap 21h1'31" -20d8'57"
Lynda Marie
Psc 1h2'47" 6d53'34"
Lynda & Mark Written in the Stars
Cyg 20h4'10" 31d7'29"
Lynda May Smith
Ori 6h23'4" 15d35'36"
Lynda McKinney
And 0h39'46" 39d15'28"
Lynda Megarry
Cas 0h46'18" 62d40'3"
Lynda Munro
Vir 14h44'28" 5d14'44"
Lynda Palmeri
Cap 21h36'0" -16d0'29"
Lynda Pauline
Ori 6h11'36" 19d18'47"
Lynda Rosenthal: Star Polisher
Per 2h47'48" 54d47'7"
Lynda Ryan Baker
Peg 23h39'42" 11d14'25"
Lynda Samuels
Lyn 7h6'57" 47d36'5"
Lynda Stafford
Sco 17h19'52" -44d56'7"
Lynda Susan
Uma 9h57'39" 71d40'1"
Lynda T
Aqr 23h47'1" -4d46'43"
Lynda Tseng
Ori 5h51'26" -0d7'47"
Lyndall's Starlight
Cra 18h3'58" -40d25'3"
Lynda's "Big Tipper"
Cap 21h16'42" -27d9'5"
Lynda's "Little Tipper"
Cap 21h20'3" -27d26'10"
Lynda's Loving Light
Crb 15h31'57" 27d47'1"
Lynde Teresa
Uma 9h39'54" 54d46'47"
Lyndee
Sgr 18h11'44" -24d45'26"
Lyndi
Cnc 8h36'28" 7d33'31"
Lyndi
Gem 7h45'42" 33d28'1"
Lyndi Dawn Cances
Tau 4h7'17" 11d52'26"
Lyndi Lynn
Cnc 8h20'47" 11d34'14"
Lyndon A. Smith
Ori 5h24'14" 3d12'45"
Lyndon John Jarrett
Ori 6h19'55" -1d27'52"
Lyndon Travers
Cet 2h47'3" 8d14'13"
Lyndon Zvonek
Boo 14h32'16" 22d12'33"
Lyndsay
Ari 2h57'21" 19d47'31"
Lyndsay
Cas 2h35'28" 65d10'9"
Lyndsay Albarado
And 23h43'49" 43d56'37"
Lyndsay Ann Long
Uma 10h50'3" 45d55'53"
Lyndsay Danae South
And 0h10'14" 43d31'55"
Lyndsay Elizabeth Hayne
Col 5h45'40" -35d24'4"
Lyndsay Gates
Cyg 21h20'40" 32d25'36"
Lyndsay Griffin
And 1h44'8" 50d24'38"
Lyndsay Jane Kelley
And 23h21'35" 40d11'28"
Lyndsay Joycelyn Orton
Vir 13h3'45" -21d12'17"
Lyndsay L. Tilton
Psc 0h47'25" 15d54'57"
Lyndsay Marie
Uma 11h29'38" 58d48'27"
Lyndsay Susan Gasser
Cnv 13h42'41" 39d27'34"
Lyndsay Tryba
And 0h14'15" 27d37'34"
Lyndsay's Papaw
Ari 1h47'23" 24d9'52"
Lyndsee and Brian
Cyg 20h48'25" 31d27'32"
Lyndse's Wish
And 23h15'3" 42d36'50"
Lyndsey
Ari 2h7'10" 23d13'11"
Lyndsey
Vul 19h48'40" 21d33'55"
Lyndsey Aitken
And 23h7'0" 50d48'32"
Lyndsey and Tim's Star to Wish On!
Gem 6h37'41" 15d14'50"
Lyndsey Ann Marie
Leo 9h47'29" 17d21'13"
Lyndsey Barge
Oph 16h42'20" -0d49'23"
Lyndsey Beshears
Leo 9h39'6" 22d6'53"
Lyndsey "Bunnie" Reynolds
Psc 1h22'5" 29d42'34"
Lyndsey Dianne Shaffer
And 2h24'15" 42d24'16"
Lyndsey Ellen Kaiser
Lyn 7h50'11" 56d18'39"
Lyndsey Grace Carter
Lib 15h26'47" -21d1'23"
Lyndsey Hart
Cas 0h25'36" 55d8'56"
Lyndsey & Hussein
Gem 6h56'16" 20d13'28"
Lyndsey Ione
Cap 20h8'35" -8d38'31"
Lyndsey J. Gauthier
And 0h40'36" 42d58'14"
Lyndsey Jane - Flexible Flame
And 2h37'32" 37d30'4"
Lyndsey Morgan Richmond
Aqr 21h53'32" -1d52'3"
Lyndsey Nicole Ricker
Sgr 17h56'58" -28d41'47"
Lyndsey Richardson
And 22h59'52" 51d1'53"
Lyndsey Shinn
Ori 6h8'21" 15d48'47"
Lyndsey-Renee
Lmi 10h26'43" 35d45'24"
Lyndsey's Angel
Lmi 10h15'22" 31d34'30"
Lyndsie Nicole Cobb (strawberry)
Cyg 20h18'11" 33d16'19"
Lyndsy A. Listenberger
Cas 1h27'42" 57d28'58"
Lyndy
Ari 2h17'23" 20d6'23"
Lyndy Shephard
Vir 12h57'9" -18d19'38"
Lyndy & Wade
Cyg 21h36'39" 33d34'4"
Lyndzi
Ari 2h51'25" 18d51'49"
Lyne Chevrette
Uma 10h16'5" 49d27'4"
Lyneah and Anthony's Star
Ori 4h52'44" 13d31'54"
Lynelle
Uma 10h41'13" 59d2'56"
Lynelle Mitschele
Crb 15h28'25" 27d11'5"
Lynerd
Sgr 18h44'47" -32d13'24"
Lynese N David
Cyg 20h15'58" 50d1'27"
Lynette
And 0h12'59" 46d11'50"
Lynette
Lyn 8h34'21" 39d45'30"
Lynette
Del 21h6'42" 14d36'21"
Lynette
Cam 7h35'8" 60d30'7"

Lynette Ann Cristianna Monaghan
Ari 2h42'21" 14d40'8"
Lynette Anne
Sgr 19h39'48" -28d49'2"
Lynette Figueroa
Aqr 20h39'12" -9d23'9"
Lynette Jeanne Gagnon
Umi 15h10'44" 79d11'22"
Lynette Jenkins
Cyg 19h34'5" 32d14'49"
Lynette Lebron, My Love Forever
Aqr 22h40'24" -3d14'14"
Lynette Louise Massie
And 1h32'24" 41d24'16"
Lynette Marbley
And 0h17'50" 45d14'35"
Lynette Marie Johnson
Cyg 21h0'53" 46d6'40"
Lynette Marie Lu-Hsueh
Ori 5h29'28" -4d49'21"
Lynette Marie Nicolas
And 23h18'9" 47d14'58"
Lynette Marie Nocito
Lyn 8h46'5" 40d46'55"
Lynette May - 24.03.1979
Col 5h58'5" -34d48'59"
Lynette Miller
Cas 1h38'36" 62d56'33"
Lynette Rowe's Star
Lmi 10h34'32" 35d52'0"
Lynette Santa Massaro
Aqr 21h59'52" -7d43'49"
Lynette Sherrie Evans
Crb 15h48'35" 36d13'56"
Lynette. Simply the Best
Cas 0h52'22" 47d35'44"
Lynette Smith, My Baby Boo
And 1h9'58" 38d50'47"
Lynette Tanaka
Gem 7h7'34" 25d25'54"
Lynette Tara Vick
And 2h16'55" 48d51'28"
Lynham1942
Lyr 18h38'46" 35d4'29"
Lynhen 05-30-1965
Umi 15h1'36" 69d46'13"
Lynie
Lib 15h6'53" -6d9'11"
Lynlee Maree Creef
Ori 5h36'6" -9d1'22"
Lynley's Star
Ori 6h19'28" 14d58'37"
Lynn
Gem 6h44'38" 26d28'47"
Lynn
Cnc 8h33'33" 24d7'12"
Lynn
Lyr 19h14'52" 27d3'2"
Lynn
Crb 16h0'25" 28d12'32"
Lynn
Leo 10h17'30" 16d37'57"
Lynn
Uma 8h47'39" 50d6'59"
Lynn
Aqr 22h44'35" -12d18'16"
Lynn
Uma 11h30'7" 65d40'7"
Lynn
Cap 21h44'42" -24d21'20"
Lynn
Sco 17h19'42" -43d8'15"
Lynn
Psc 1h43'39" 6d54'57"
Lynn A. Hopper
Lyn 7h3'0" 54d56'14"
Lynn A. Hoth
Vul 20h14'12" 23d2'55"
Lynn A. Vanderboegh
Lmi 10h41'57" 30d43'34"
Lynn Anczarski
Psc 0h11'23" 8d36'31"
Lynn and Brian Hughes
Umi 15h16'25" 73d40'48"
Lynn and Charles
Cyg 21h9'33" 47d23'49"
Lynn and Lee
And 0h54'24" 44d38'54"
Lynn and Shadow
Aqr 22h23'24" -23d1'58"
Lynn and Terri White
Pho 0h59'13" -48d1'40"
Lynn Andreen Bowen
Ari 2h15'25" 23d15'55"
Lynn Ann
Psc 1h32'31" 9d38'16"
Lynn Ann
Crb 15h58'32" 34d28'31"
Lynn Ann
And 2h16'48" 48d14'1"
Lynn Ann 'Babydoll' Larson
Ori 6h14'41" 6d2'26"
Lynn Ann Hindman
Ari 3h18'11" 27d31'14"
Lynn Ann Holland/Hess
Cnc 8h36'40" 16d54'23"
Lynn Ann Keiser
And 23h21'42" 42d19'47"
Lynn Anne Cedrone
Uma 10h55'9" 42d47'33"
Lynn Anne Studards
Leo 11h10'44" 17d29'55"
Lynn Ann's Star
And 0h31'0" 45d2'54"
Lynn Audrell Mims
Psc 1h1'27" 10d23'37"
Lynn B. Hall
Uma 9h56'15" 64d55'17"
Lynn Barbara Pilchak
Aqr 23h9'40" -12d50'13"
Lynn Beauregard
Vir 14h26'19" -0d26'20"
Lynn Benet
Cnc 8h2'49" 20d37'38"
Lynn Brennan
Cas 1h16'18" 53d41'36"
Lynn Briere
Ori 6h3'22" 21d25'38"
Lynn Brown
Vir 14h49'10" 3d41'59"
Lynn Burgess Moriarty
Uma 10h58'46" 40d7'45"
Lynn C. Kinney
Vir 13h14'33" -0d17'1"
Lynn Charland
Uma 11h24'15" 59d23'12"
Lynn & Charlotte's Star
Cyg 21h45'36" 44d9'9"
Lynn Connolly
Sgr 18h33'5" -16d24'37"
Lynn Constantin Hecker
Lmi 9h59'25" 34d14'42"
Lynn Crum
Tau 4h18'12" 19d36'42"
Lynn Cruz
Cas 1h8'18" 63d52'11"
Lynn D Knauss
Ari 2h39'5" 27d41'21"
Lynn D. Ritter
Cyg 21h55'14" 54d22'28"
Lynn Danielle
Uma 11h2'4" 47d16'0"
Lynn David
Tau 4h8'9" 28d23'29"
Lynn Davilla Shields
Per 3h33'2" 45d39'51"
Lynn Diehl Dudley
Sco 16h10'25" -34d4'32"
Lynn Donahue Simpkins
And 0h19'31" 43d43'34"
Lynn E. Berding
Cyg 21h2'17" 50d39'29"
Lynn Elder Schirm
Uma 11h9'35" 51d38'15"
Lynn Elizabeth Bernhardt
Cas 0h57'10" 62d14'43"
Lynn Ellen Shinn
Vir 12h35'46" 5d10'56"
Lynn Elsdon (the Ruby Twinkle)
Ori 6h14'51" 15d59'35"
Lynn F. Ramsdell
Lyn 8h13'17" 48d51'37"
Lynn Fairchild
Lib 15h43'58" -18d23'44"
Lynn Farris
Tau 3h50'46" 29d7'53"
Lynn Feiffer
Lyn 8h15'16" 40d53'33"
Lynn Fisher
Cam 3h58'34" 67d35'14"
Lynn Ganz
Uma 12h15'2" 63d3'15"
Lynn Goodrow
Crb 16h15'28" 38d7'32"
Lynn Greetis
Ari 2h7'55" 25d32'15"
Lynn Grystar
Cap 20h24'30" -14d16'4"
Lynn Guiliani's "Hiazena"
Lyn 8h33'56" 57d54'57"
Lynn Hamilton
Uma 9h7'11" 69d43'13"
Lynn Hamilton
Tri 2h31'19" 30d25'15"
Lynn Haynes
Lyr 19h13'47" 26d25'50"
Lynn Hedges, My Best Friend
Psc 0h50'47" 13d51'13"
Lynn & Howard Reiss
Leo 11h34'36" 15d42'20"
Lynn Hudson
Del 21h7'37" 19d29'17"
Lynn & Ian Chappell
Cyg 20h37'29" 52d17'0"
Lynn J. Ingros
Uma 12h2'52" 59d25'55"
Lynn & Jeff Kornblau
Lyr 19h0'59" 33d50'28"
Lynn Jenny
Ori 6h1'8" 7d3'6"
Lynn & Jeremy
Uma 9h56'6" 51d41'56"
Lynn Johnson
Uma 10h4'8" 58d18'38"
Lynn & Jonathon's Guiding Light
Cyg 19h59'46" 40d36'35"
Lynn Joy Hansen
Cnc 8h48'45" 18d31'25"
Lynn Kari Dean
Sco 16h41'40" -32d14'48"
Lynn Keeling
Lmi 10h48'7" 27d35'55"
Lynn & Kelly Liddington
Ari 3h17'1" 28d39'50"
Lynn L. Boudon
Cas 1h23'22" 68d35'48"
Lynn L. Griffith
Lmi 10h45'41" 26d5'28"
Lynn Leann Towers
And 2h27'0" 45d32'27"
Lynn & Lee Tirrell
Cyg 20h15'40" 32d42'20"
Lynn Ling Zhou
Vir 14h42'25" 5d27'0"
Lynn Love
Lib 14h56'36" -0d30'41"
Lynn Lundquist
Ari 2h35'5" 24d40'28"
Lynn M Keenan
Cas 1h12'24" 50d38'2"
Lynn M. Lynch
Uma 11h45'12" 43d38'34"
Lynn M. Russell
And 0h45'58" 39d5'24"
Lynn M Sanford
Leo 11h57'42" 20d47'8"
Lynn M. Sussina
And 1h32'57" 48d9'45"
Lynn Marie
Ari 2h40'50" 27d48'13"
Lynn Marie
Cnc 9h12'22" 6d48'53"
Lynn Marie
Cnc 8h38'15" 14d22'52"
Lynn Marie
Aql 19h31'41" -10d1'52"
Lynn Marie Davidson
And 0h58'45" 40d1'19"
Lynn Marie Eunisius
Leo 11h3'24" 6d17'42"
Lynn Marie Jackson
Gem 7h25'21" 28d10'5"
Lynn Marie Marion
Tau 4h42'17" 22d28'55"
Lynn Marie Martyn
Vir 13h26'39" 12d8'13"
Lynn Marie Mcintosh
Ari 2h53'4" 25d5'14"
Lynn Marie Miller
Gem 7h32'12" 28d7'4"
Lynn Marie Minicozzi
Cyg 19h48'42" 33d25'49"
Lynn Marie Overstreet
Per 3h5'0" 46d22'7"
Lynn Marie Reedy
Lyn 8h11'30" 57d22'39"
Lynn Marie Staffeldt
Cnc 7h58'22" 15d12'44"
Lynn Marie's wish
Ori 5h21'38" -1d18'19"
Lynn Mason
Lyn 7h10'11" 48d15'15"
Lynn Merwin Tiefenbrun
And 2h9'26" 43d4'5"
Lynn Michael
Cas 1h34'29" 65d26'47"
Lynn Michele Kelly
Cas 0h3'13" 62d36'18"
Lynn Michelle Reed
Lib 15h0'11" -10d28'57"
Lynn Miletto
Gem 7h6'12" 34d26'1"
Lynn "Mo Cuishle"
And 2h28'54" 42d20'24"
Lynn Moran
Cnc 8h47'9" 32d32'17"
Lynn My Angel
Leo 10h11'6" 27d17'42"
Lynn Nash
And 23h13'36" 48d16'2"
Lynn Nutt
Uma 11h46'24" 45d48'22"
Lynn Owens
Aur 5h49'30" 49d53'16"
Lynn Parker Bellin
Uma 11h33'55" 62d56'28"
Lynn Parsons
Cyg 20h39'13" 45d3'3"
Lynn Poss' Diamond
Psc 23h55'9" 1d0'32"
Lynn Rainey
Lyr 19h8'31" 42d46'21"
Lynn Raymond Griebahn
Lyn 9h8'29" 36d32'32"
Lynn Reid Perkins
Gem 7h19'56" 33d27'40"
Lynn Reine
Vir 14h41'19" -2d6'56"
Lynn Rembert
Sgr 18h35'22" -20d32'28"
Lynn Rita
Vir 12h6'37" 4d14'38"
Lynn Roe
Cas 0h26'32" 57d3'10"
Lynn Roger
Lib 14h51'54" -0d50'53"
Lynn Rolfe
Cyg 19h42'40" 30d29'28"
Lynn Rosenblum
Cas 0h7'57" 60d31'39"
Lynn Rubenstein Nicholson
Gem 6h33'37" 20d42'14"
Lynn Schultz
Uma 12h4'54" 56d59'10"
Lynn Seitzinger
Cas 0h20'49" 48d0'28"
Lynn Simonson
Lyn 7h44'43" 38d42'19"
Lynn Skipper
And 1h13'12" 45d49'2"
Lynn Smith
Cru 12h37'18" -57d17'2"
Lynn Soulby
Del 20h35'38" 16d37'52"
Lynn Spahr
Gem 7h10'58" 34d20'38"
Lynn Star
Uma 14h3'44" 61d3'1"
Lynn Steblecki
Ori 5h20'11" 9d5'13"
Lynn Sterling
And 0h48'16" 36d11'42"
Lynn Stevens - Lilly's Star
Cas 23h23'46" 55d16'5"
Lynn Stewart
Cyg 21h24'43" 54d28'27"
Lynn Stought
Cmi 7h36'31" 3d48'21"
Lynn Suzanne
Cyg 19h57'40" 41d34'9"
Lynn Tanguay "Love Star"
Lyr 19h4'25" 42d0'34"
Lynn - The Star of My Life
And 1h3'49" 36d5'44"
Lynn Tosh
Sgr 19h10'18" -13d7'48"
Lynn Tschirky
Cyg 19h46'47" 34d37'32"
Lynn Usher
Sgr 18h14'43" -23d24'58"
Lynn Vecqueray
Tau 3h35'23" 4d9'25"
Lynn & Vito
Gem 7h19'6" 20d43'57"
Lynn Weldon Steele
Ori 5h53'39" 3d14'53"
Lynn Wilhelm "Mom"
Sco 17h54'41" -43d0'14"
Lynn Willis
Crb 15h38'11" 31d7'6"
Lynn Worcester
Cnc 7h57'52" 16d3'26"
Lynn Z Richardson
Aqr 23h39'5" -7d16'9"
Lynnae
Cnc 8h46'53" 30d54'23"
Lynnae Willette
Dra 16h1'13" 53d33'7"
Lynnai and Charlie Kernan
Ori 6h10'45" 19d33'13"
lynnandtaro
Cyg 20h51'18" 32d13'10"
Lynnard
Peg 23h31'7" 26d2'1"
LYNN....Be Mine
Ari 2h45'28" 20d29'50"
LynnD
Uma 11h50'18" 59d30'44"
Lynnda Lee Kratovil
And 23h18'27" 47d24'39"
LynnDale
Ori 6h13'49" 15d4'34"
Lynndsey C.A. Smith Birthday Star
Vir 14h53'33" 4d24'21"
Lynne
Aqr 21h39'34" 1d11'51"
Lynne
Vir 12h21'48" 12d31'3"
Lynne
Leo 9h55'56" 9d27'48"
Lynne
Ari 2h22'18" 26d1'30"
Lynne
Psc 1h13'57" 26d3'59"
Lynne
And 0h53'16" 40d19'23"
Lynne
And 1h55'46" 37d12'27"
Lynne and Lynn
Cyg 20h10'47" 54d8'22"
Lynne and Rob Barrowman
And 1h17'40" 46d4'9"
Lynne Arrowsmith's Everlasting Star
Cru 12h22'44" -56d37'29"
Lynne B. & Peter Adam Krell Star
Cyg 19h51'22" 33d20'56"
Lynne Brick
Cnc 8h20'11" 18d5'27"
Lynne Bush Ponte - My magical mom<3
Cam 5h3'6" 63d8'7"
Lynne Byrwa
And 1h4'50" 42d13'16"
Lynne CHAMP Manderano - # 2
Her 17h32'46" 33d48'35"
Lynne (Duckiebear) Duc-Leung
Ari 2h29'45" 11d43'56"
Lynne Elaine
Cyg 20h55'28" 48d15'49"
Lynne Garden
And 1h1'11" 43d7'28"
Lynne Hawkes
Cas 2h0'48" 62d37'50"
Lynne Herman
Crb 15h50'22" 27d22'4"
Lynne Hope
Cnc 8h10'29" 16d26'47"
Lynne & Jim Kozak
Leo 10h32'6" 11d23'47"
Lynne Kuglar
And 0h44'45" 35d15'14"
Lynne L. Cardinal
Cap 20h53'59" -18d35'38"
Lynne Margaret MacDonald
Cas 2h47'10" 65d36'2"
Lynne Marie
Aql 19h17'23" 4d59'8"
Lynne Marie Ellis
Cas 23h35'24" 56d39'48"
Lynne Marie Leslie
Sco 16h41'41" -31d21'12"
Lynne Marie O'Hara
Aqr 22h18'5" 1d9'22"
Lynne Marie Rousseau
And 1h47'25" 36d25'2"
Lynne Marie Russo
Leo 11h14'26" 13d53'19"
Lynne Marie Teeter Burgeson
Ori 4h49'12" 5d31'52"
Lynne McCabe
Gem 7h5'33" 18d12'46"
Lynne McVeigh
Uma 13h33'12" 53d41'28"
Lynne Myers
Cap 21h39'37" -24d37'30"
Lynne P. Leonard
Ori 6h15'26" 16d47'38"
Lynne Pascale Walker
And 1h39'55" 47d27'50"
Lynne Pinkerton Platten
Aqr 20h42'15" 0d8'59"
Lynne Tetreault
Uma 10h39'58" 40d2'55"
Lynne Votta, Mom of JC,Katie&Beth
Uma 12h5'51" 29d47'44"
Lynne777
Cas 1h20'23" 55d38'42"
Lynnea
Uma 10h45'53" 51d7'58"
Lynnea
Uma 10h6'38" 62d13'34"
Lynnea Marie Harabedian
Cyg 20h17'28" 54d47'43"
Lynnerama
Cnc 8h46'59" 30d47'49"
Lynne's "Happy Golden Dancing" Star
Cap 21h51'14" -11d9'59"
Lynne's Star
Uma 11h38'58" 49d35'4"
Lynne's Star of Friendship
Pho 0h48'32" -40d10'23"
Lynnet Castaldi
Sco 17h18'24" -44d49'53"
Lynnette
Lyn 7h12'3" 58d41'26"
Lynnette
Leo 10h43'44" 10d52'36"
Lynnette Carol Certa
Leo 9h31'1" 22d56'14"
Lynnette Dawn
Cap 20h9'1" -8d39'41"
Lynnette & Jon
And 0h19'9" 26d49'35"
Lynnette Margarette Hodgins
Cas 0h43'31" 73d32'38"
Lynnette Perke Zillessen
Mon 6h44'34" 11d23'56"
LynnGossett
Lyn 7h59'47" 41d52'7"
Lynniam Star
Ari 2h19'0" 21d58'35"
Lynnie
Psc 1h53'42" 9d34'13"
Lynnie
Per 3h11'14" 55d33'58"
Lynnie 4/27/55 - 2/19/99
Tau 4h34'43" 18d18'7"
Lynnie and Graham
Uma 11h51'39" 33d43'14"
Lynnie Lu Lu Tschappler
And 0h52'10" 36d10'11"
Lynniemac, Luminary
Psc 1h31'50" 20d40'46"
Lynnlee and Matthew
Sge 19h52'27" 17d54'49"
LynnLew
Cyg 20h40'45" 55d14'15"
Lynnmare 1
And 2h27'44" 44d44'19"
Lynnmarie DeBolt
Her 17h35'22" 29d34'14"
LynnMarie Kilmister
Cap 20h25'15" -26d23'7"
LynnMarie MacDougall
Lyn 8h30'25" 37d13'36"
LynnMarie Panzarella
Lmi 10h51'57" 35d8'56"
Lynnmarie-Hope Has A Voice
Cru 12h23'40" -56d16'33"
Lynn'nroger
Tau 4h21'43" 23d53'23"
Lynnowen
Uma 8h59'38" 56d11'1"
Lynn's Bayou Dew
Sco 16h40'34" -31d17'9"
Lynn's Butterfly
Sgr 19h26'22" -32d52'57"
Lynn's Dream
Lib 15h46'24" -18d28'52"
Lynn's Heart of Mine
Cap 20h53'34" -17d14'24"
Lynn's Place in Space
Cnc 8h39'52" 17d58'26"
Lynn's Star
Ori 5h32'48" 0d47'40"
Lynn's Star
Cyg 20h18'39" 38d11'27"
"Lynn's Star" - L.S.B. - 1947
Lyn 9h5'53" 34d57'28"
Lynnsey Butler
Lyn 9h2'30" 41d47'5"
Lynnwood Hollins Tucker
Tau 5h47'56" 22d15'12"
LYNNYPOOKLES
Leo 11h26'0" 17d36'50"
Lyn's Heavenly Body
Aqr 23h23'11" -11d19'0"
Lyn's Tear
Eri 4h34'12" -16d22'17"
Lynsay Sobon
And 0h14'57" 43d55'0"
Lynse
And 23h33'2" 47d34'56"
Lynsey and Jim's LoveStar
Cap 21h0'59" -15d49'15"
Lynsey Angel 1
Com 12h30'54" 13d44'40"
Lynsey Belle
And 2h30'0" 50d34'16"
Lynsey Bruce
Psc 1h7'27" 24d23'6"
Lynsey Bulloch
Cas 1h24'7" 68d58'53"
Lynsey Dredge
Sco 17h29'35" -43d2'0"
"Lynsey Lou" Monin
Peg 22h40'11" 24d49'18"
Lynsey Loves Andrew
Cyg 19h32'30" 28d54'51"
Lynsey Marie Frye
Umi 13h38'56" 72d48'43"
Lynsey McKenzie
Psc 1h22'31" 30d45'8"
Lynsey Patricia Roy
Cas 0h30'22" 66d1'45"
Lynsey Peterson - Shining Star
Cam 6h3'14" 60d0'41"
Lynsey Rose
Cas 0h1'29" 56d3'39"
Lynsey's Star
And 23h3'51" 50d50'31"
Lynsey's Star
Lyn 6h39'12" 61d19'12"
Lynsi Jordan
Peg 22h41'2" 10d20'25"
Lynsie
Umi 14h11'31" 86d39'29"
Lynsie Crespo
And 23h22'27" 48d22'32"
Lynsie Elise
Psc 1h41'55" 6d13'15"
Lynstar
Peg 21h10'44" 19d35'26"
Lynton Hunt III
Sgr 19h40'32" -27d47'29"
Lynwood Ray
Umi 16h16'38" 70d52'56"
Lynwood Wayne Kelch
Ori 6h17'11" 14d52'19"
Lynx
Leo 11h46'1" 21d14'40"
Lynx Soleil
Lyn 6h38'15" 57d44'7"
Lynyrd Skynyrd Erickson
Cma 6h46'24" -13d30'6"
Lynzee Mo
Ari 3h22'28" 16d2'48"
Lynzi, light of the night!
Uma 12h18'41" 63d2'57"
Lynzi's Angel in the Sky
And 0h25'25" 44d14'48"
Lyon Skylab Case
Cnc 8h31'49" 17d55'11"
Lyona M. Conner
Vir 11h51'7" 2d40'59"
Lyonshall
Vir 12h58'29" -15d52'45"
Lyonskazee
Ari 2h53'47" 26d51'47"
Lyora- Leigh Rachel Gromadzyn
Sco 16h5'59" -9d45'44"
Lyra Brooke Cassell
And 0h17'0" 42d14'49"
Lyra Fumiko
Vir 13h8'43" -4d41'53"
Lyra G. Tabornal
Lyr 18h31'0" 27d47'4"
Lyra / Scott Bolin
Her 16h37'46" 6d31'37"
Lyra Swan Hara
Lib 15h9'47" -12d54'15"
Lyric Marie Garcia
Lyr 18h31'10" 37d36'1"
LYRIC RAIN
Vir 12h18'42" 4d28'24"
Lyris A. Ceja
Lyn 7h41'11" 55d48'33"
LYRITH
Cas 0h23'23" 47d5'13"
LY's India Marie
Aqr 21h9'27" -3d6'37"
Lysa
Sco 17h16'24" -33d12'33"
Lysa
Gem 6h50'58" 22d27'6"
Lysa Ansaldi
Lyn 7h56'56" 39d40'1"
Lysa M. DiDonato "With A Y"
Lyr 18h35'45" 38d7'46"
Lysabad
Psc 1h3'52" 23d53'57"
Lysandra
And 0h59'16" 37d22'38"
Lysandra
Lib 15h24'36" -6d53'56"
Lyse
Cnc 8h6'59" 25d7'41"
Lyseth
Lmi 10h10'28" 27d58'7"
Lysette Aimee DeBoard
Ari 2h41'27" 26d57'43"
Lysiane mon Amour
Ori 6h20'2" 16d51'22"
Lysianne Joanette
Umi 13h12'56" 71d14'44"
Lyssa
Ori 6h14'58" 16d22'17"
Lyssa Bear
Aqr 21h48'36" 2d18'7"
Lyssa Bergman
Crb 15h37'54" 26d2'4"
Lyssah Kell
Aqr 22h3'5" -1d49'30"
Lyssy
Leo 11h36'29" 9d34'26"
Lyssy Lou
Vir 12h45'12" 2d46'39"
Lytha Naasko
Uma 9h2'39" 67d57'12"
Lyuda & Anton
Eri 4h13'42" -0d11'30"
Lyudmila
And 0h19'27" 25d25'21"
Lyudmila Grivko
Cyg 21h14'10" 36d13'25"
Lyudmila Nikolaevna
Sgr 18h3'54" -21d22'33"
Lyvonne Rogers
Cas 0h21'28" 53d40'39"
Lyyli Mallula
Lib 15h39'36" -26d3'25"
Lyz
Cap 20h21'2" -11d43'12"
Lyz's Star
Ari 2h34'38" 19d51'50"
LZT
Cnc 9h7'39" 16d29'4"
M*
Her 18h23'8" 15d0'27"
M
Ari 2h7'7" 22d43'18"
M
Leo 10h9'4" 26d7'44"
M
Tau 4h36'19" 28d8'47"
M
Her 17h49'51" 38d46'31"
M.
Umi 13h44'2" 73d26'16"
m 06 noel
Sgr 19h15'19" -16d39'25"
M & A
Leo 11h42'9" 20d18'57"
M & A: corazones dulces
Ori 5h41'43" 8d36'24"
M A K Y C I D A
Umi 15h56'15" 78d40'4"
M A Korves
Tau 4h18'35" 26d42'59"
M A N U L A
And 1h13'16" 44d52'20"
M A T W
Cru 12h17'22" -57d35'37"
« M A Y A »
Cyg 21h44'32" 44d4'9"
M Adam Wethington & Amanda E Thomas
Cyg 21h26'12" 46d24'12"
M and M
Ari 2h50'26" 27d56'30"
M. and R. Leap Year Leprechaun
Uma 10h45'17" 52d39'6"
M B T - Lost - 08.12.05.
Umi 13h45'4" 70d55'18"
M&C 9-Oct-2004
Uma 9h19'36" 55d15'54"
M. Carla Wambach
Cam 5h53'18" 59d57'25"
M & D Forever Our Guiding Light
Cru 12h37'11" -58d25'8"
M & D Rhodes
Cyg 20h11'1" 46d2'31"
M E A Z
Uma 9h21'39" 56d29'41"
M. E. (Betty) Pickard
And 23h44'25" 34d37'21"
M E H - C H I - 10/22/06 - 6:16:24
Cas 23h22'37" 54d46'49"
M e l l i e
Ori 6h19'39" 10d30'31"

M. Eleanor Spaulding
Cas 23h27'29" 52d56'31"
M & Em 140
Per 2h59'31" 50d51'16"
(M & F)
Aur 6h19'17" 47d20'38"
M. Foehr
Ori 5h57'43" 18d10'37"
M. G. and Helen Y. Mason
Lyn 6h57'37" 46d5'28"
M&G Bleichner's 50th Anniversary
Uma 8h30'14" 62d36'48"
M. Grayden D. Orlando
Ori 5h55'23" 21d37'59"
m&hsalazar1289
Cyg 21h22'27" 36d21'24"
M & J
Lib 15h2'6" -0d57'55"
M J
Cru 12h15'18" -58d47'2"
M J Miksch
Uma 8h22'32" 64d56'53"
M J R - Infinity
Ara 17h22'5" -51d58'2"
M J Tibbs-Scout Master For Life
Uma 8h56'58" 59d4'9"
M. Jane Betzendahl
Cas 1h17'7" 62d43'24"
M. Jane Sumera
Lyn 7h14'53" 58d59'33"
M. Jill Wehmer
Boo 15h25'9" 47d41'11"
M. JoAnn Baker
Gem 7h34'51" 15d19'28"
M/J's Love
Ori 5h29'36" 1d57'21"
M. Judith Stattmiller
Vel 9h24'59" -42d36'55"
M + K 4 Ever
Umi 15h33'54" 74d2'29"
M & K Du'Cole
Sco 17h30'20" -43d17'36"
M K P Tanja
Uma 10h37'40" 42d24'9"
M K Rasmussen
Ari 2h56'49" 18d55'0"
M. Kim Mynatt
Ari 2h31'16" 26d14'46"
M L Palomo 523
Uma 8h40'48" 61d25'18"
M. LARENZ CHAPMAN
Eri 3h47'48" -0d46'59"
M. Lee Murray Star
Her 17h31'36" 34d42'12"
M. Lee Thompson
Sco 17h17'38" -32d59'15"
M. Leydet
Lib 14h32'50" -10d54'44"
M Lorraine Mancuso
Lib 15h3'52" -3d25'59"
M. Lo's Wishing Star
Psc 0h45'29" 5d16'3"
M. Louise Horn
Vir 13h28'57" 12d51'23"
M Loves A Forever Times Infinity
Cyg 19h37'56" 28d43'46"
M & M
And 0h21'25" 31d32'15"
M&M
Ori 6h24'34" 17d2'40"
M & M
Ori 6h2'45" 19d38'59"
M & M
Cnc 8h35'37" 18d37'2"
M & M
Cnc 9h7'6" 10d37'35"
M&M
Ori 6h18'21" 15d42'2"
M&M
Gem 7h21'57" 32d32'11"
M & M
Uma 11h13'5" 44d5'43"
M & M
Cyg 20h21'45" 52d13'39"
M&M
Lib 15h5'45" -6d24'25"
M & M
Aql 20h13'18" -3d25'37"
M & M 10th Anniversary star
Cyg 20h44'41" 37d23'16"
M&M 4Ever
Cyg 20h18'59" 54d14'20"
M&M/Bichara
Gem 6h50'36" 15d57'35"
M & M Eagle
Lib 15h15'21" -12d35'37"
M/M Ecstasy
Leo 11h16'43" -1d57'59"
M&M Felkay 001
Lyn 7h14'59" 57d24'19"
M & M Fishman
Cyg 21h38'56" 30d57'46"
M & M Forever
Cyg 20h3'35" 43d59'25"
M & M Forever and For Always
Cyg 20h50'18" 54d18'46"
M & M Grinnan
Lyn 6h26'19" 60d56'50"
M&M (Melody & Morley)
Del 20h39'1" 16d16'28"
M & M Mills
Aqr 22h46'46" -14d17'34"
M&M Nuptia
Cas 1h16'12" 51d33'58"
M & M "To Our Love"
Uma 13h38'24" 50d3'43"
M&M Vest
Vir 14h38'53" 3d50'22"
M & M, 10/28/89 Almost Paradise
Lyr 18h50'48" 34d0'12"
M. Marie Phetteplace
Uma 14h12'9" 56d44'6"
"M" - Marilyn
Sco 16h15'5" -12d57'31"
M. Martin Lacelle
Uma 11h23'28" 57d32'22"
M & Ms and Wine
Sco 17h19'35" -30d36'23"
M & M's Evening Star
Del 20h30'59" 12d55'1"
M & M's Light
Cru 12h29'22" -60d36'27"
M & M's Silver Anniversary Star
Vir 13h10'22" -22d39'10"
M&M'S STAR
And 0h6'21" 35d54'39"
M & M's Treinta Perlas
Cas 0h20'6" 55d44'5"
M. Munsur K. Ahmed
Uma 10h31'44" 53d17'40"
M N P B *Doc*
Cyg 21h55'57" 49d33'50"
M. Norman Buchanan born March 11, 1925
Cep 22h54'32" 70d48'10"
M. O. S. Shining Knight
Lib 15h1'58" -17d8'37"
M. Patricia Duggan
And 0h6'41" 45d15'5"
M & R
Vol 7h36'19" -70d53'53"
M. R. (Bubba) Hawk
Aql 19h42'20" 4d31'43"
M & R Shooting Star
Per 3h5'46" 56d44'21"
M&RLUV4EVER
Umi 16h19'40" 73d10'2"
M. Roxanne Veliz
Lmi 9h30'29" 34d14'5"
M!/S!
Ari 2h37'40" 22d21'40"
M&S
Cen 11h31'9" -52d33'32"
M S Cat One
And 0h25'38" 35d54'32"
M S Matheis 040404
Ori 6h5'44" 20d38'4"
M. Schmidt
Uma 11h0'37" 51d44'45"
M/Sgt. John "Jack" W. Wood
Cnc 8h0'57" 21d57'40"
M Shannon Siobhan
Ori 5h15'45" -1d27'45"
M Shaunessy
Uma 9h22'3" 48d58'39"
M Shimadazizi
Lib 14h26'18" -14d33'56"
M Squared
And 23h3'32" 47d29'5"
M. Susan Duffy
Cyg 20h42'21" 32d50'45"
M. Suzanne a.k.a Cakes
Cam 5h24'41" 58d2'10"
M. Suzanne Rubincan
Leo 10h21'16" 22d11'8"
M. Theresa Jessome
Cas 1h0'38" 57d1'26"
M Toomey 6103
Tau 4h23'37" 18d27'45"
M. Townsend 1944
Cep 20h40'30" 64d23'20"
M Y M I
Lyr 18h43'55" 33d57'31"
M1 2nd Grade Class - 2004
Ori 5h35'33" -1d57'26"
M2
Gem 6h51'14" 31d51'39"
M2
Ori 6h11'46" 17d45'56"
M2MKelly
Lyn 7h42'26" 49d37'30"
M2RZ
Cyg 20h48'25" 44d1'23"
MA
Cyg 19h57'8" 42d3'1"
Ma
Cas 23h23'16" 53d14'0"
MA 4 8 30
Uma 11h44'50" 63d39'7"
Ma and Pa 3/24/56
Uma 12h23'5" 61d47'32"
Ma and Pop Ralls
Crb 15h51'57" 26d46'42"
Ma (Barbara Palumbo)
And 1h5'40" 46d38'36"
Ma Beleneverinarig V 2 Andrenecia M
Cru 11h58'55" -59d46'57"
Ma Belle Amoureuse
Equ 21h6'7" 12d18'3"
Ma Belle Michelle
Leo 11h16'37" 23d4'19"
Ma Belle Nadou
Gem 7h36'53" 26d15'44"
Ma Boubou San
Col 5h5'2" -33d13'19"
Mª Carmen Llantada
Vir 14h9'13" -9d5'2"
Ma Chère Grand-mère
Cap 20h29'24" -15d43'0"
Ma & Dad Tomeo Star
Uma 12h2'35" 30d20'7"
Ma Dame
Cyg 21h34'6" 49d4'22"
Ma Dawny d'à moaaa
Cas 1h19'52" 59d23'49"
Mª del Aguila Cordero Bulnes
And 2h11'47" 37d58'7"
Mª del Rocio de la Yglesia Hidalgo
And 0h18'15" 37d8'25"
Ma Didi
Sgr 18h58'19" -25d56'7"
Ma Douce Roma (ERB)
Cas 0h56'41" 63d44'59"
Ma Doudinette
Del 20h47'10" 10d41'29"
MA DOUDOU CELESTE
Sct 18h24'38" -15d23'15"
Mª Elena S. Orviz
Cyg 19h31'0" 28d39'12"
Ma Foley
Uma 8h15'38" 61d1'44"
Ma. Isabelita Argana Pineda
Sgr 18h26'24" -17d7'52"
Ma Jaya Sati Bhagwati
Lyn 7h44'55" 35d59'15"
Ma Joie
Dra 17h6'40" 62d47'2"
Ma Ka Li, Kelly
Ser 18h30'16" -0d34'57"
Ma&Kia
Boo 14h35'23" 54d52'33"
Ma Lipo
Cas 1h5'19" 64d52'57"
MA MA HANCOCK
And 0h56'13" 35d32'49"
Ma Maman 'Manag' Huguette Eva
Ori 6h8'29" 13d36'44"
Ma Maman, mon Etoile
Uma 9h48'30" 43d13'21"
Ma - Marie Souther Cox
Sco 16h6'29" -13d26'21"
Ma Mere
Sco 16h4'55" -16d56'0"
Ma Nath...
Uma 9h1'52" 52d8'20"
ma Patricia Dondi Escudero Cabigon
Cap 21h39'26" -15d27'24"
Ma Petite Amie, Mon Amour
Lib 14h50'11" -7d23'35"
Ma Petite Belle
Uma 11h45'2" 42d15'7"
Ma petite Florence Gillot
Uma 9h40'40" 50d9'49"
Mª Reyes Cordero Bulnes
Ori 5h43'23" 9d14'23"
MA Shining Star '76
Gem 6h53'39" 30d52'50"
Ma Shuu Rozsas
Cas 1h29'37" 66d5'46"
Ma Soeur Kari
And 0h12'46" 44d19'5"
Ma Steph
Leo 10h55'33" 5d7'58"
M.A. Summers (Nanny)
Cas 0h0'6" 55d14'4"
Ma Sylphide Sandrine
Vir 14h41'23" 6d23'33"
Ma tête de Turc
Lib 15h39'37" -8d12'36"
Ma. Theresa Uy
And 23h3'56" 48d42'49"
Ma und Pa, 23.10.1967
Cam 5h40'6" 70d50'55"
Ma Violette
Cmi 7h24'47" 2d30'3"
M.A. Worthington
Uma 9h14'15" 68d25'16"
Ma Xiao Jun
Sgr 18h36'12" -23d56'52"
MAA120106SDB
Sgr 18h14'40" -23d37'7"
MAAGIE
Cnc 8h49'43" 27d39'32"
Maamaw
Sco 16h34'28" -37d47'0"
Maarten - Lover Friend Companion - Ro
Vir 12h49'17" -9d16'55"
Maartje & Eelco voor altijd
Uma 10h1'29" 56d48'59"
maasan & Tyami
Cap 20h31'31" -27d0'13"
Maayan (gozland)
Tau 3h49'53" 9d29'37"
MAB 4201933
Lyr 18h41'40" 39d10'39"
Mabel
Per 2h54'30" 45d30'37"
Mabel Laclau Miro
Ari 2h35'30" 21d53'33"
Mabel Leialoha Kekauoha Barros
Peg 21h51'48" 10d2'58"
Mabel Mildred Otis
Uma 8h56'8" 48d27'31"
Mabel & Ray
Leo 11h24'42" -0d16'26"
Mabel Rose Baughman
Cyg 20h50'14" 46d10'16"
Mabel Ruth Eblin Timberlake
Uma 11h1'54" 38d11'26"
Mabel S. Maurer
Psc 0h58'32" 31d39'5"
Mabel Vieta
Ari 2h44'3" 16d3'31"
Mabelle "Bonnie" Huttig
Psc 1h13'28" 22d9'9"
MABELY
Leo 10h55'47" -0d7'4"
Mable & Joanne-Sent With Love, Bev
Uma 11h27'35" 63d39'11"
Mable Rae
Leo 11h22'17" 15d43'31"
Mabnah
Aql 20h9'14" 12d20'58"
M.A.C.
Psc 1h18'43" 14d54'7"
Mac
Leo 9h44'14" 25d25'1"
Mac
Aur 5h55'17" 37d3'35"
MAC
Uma 11h47'7" 34d53'46"
Mac
Uma 11h6'52" 57d27'12"
Mac
Cma 6h30'42" -23d53'12"
Mac 1927
Psc 1h12'16" 25d19'7"
Mac 2005
And 23h30'28" 41d23'27"
Mac and Norma
Cap 21h42'12" -22d44'13"
Mac and Sophie McCormack
Cyg 20h45'2" 32d31'48"
Mac & B
Sgr 19h43'9" -12d58'52"
Mac Black
Cyg 21h46'5" 43d55'12"
Mac & Danielle
Cyg 21h13'40" 36d50'33"
Mac Homer
Dra 17h30'28" 51d10'17"
Mac & Mim
Ori 5h56'39" 21d57'58"
Mac Mountain
Tau 4h24'28" 23d12'7"
Mac n Ish
Sge 19h51'19" 17d36'32"
Mac & Prov
Cyg 19h38'19" 54d59'33"
Mac Rubble
Uma 8h55'47" 69d33'0"
Mac Say
Sco 17h27'50" -36d40'44"
Mac Seekford
Leo 11h37'5" 23d14'50"
Mac Seisear
Ori 4h53'52" 14d1'3"
Mac Wright Porter
Umi 10h38'21" 88d35'43"
Mac & Zoe
Uma 9h11'29" 60d48'50"
Maca
And 0h28'43" 36d45'49"
Maca Honorato
Cnc 8h20'6" 7d27'26"
Macaleb Fitzsimmons
Uma 11h34'35" 63d10'47"
Macanas
Eri 4h18'22" -34d40'6"
Macanmel
Cra 18h29'4" -40d48'28"
Macaque
Umi 16h42'22" 77d33'25"
Macaroni Herrad
Uma 10h34'7" 67d1'6"
Macatt 4-2-2007
Ari 2h16'16" 21d54'51"
MacBain
Crb 15h40'33" 28d12'30"
MACC
Ori 5h37'46" -5d50'55"
Maccrea
Sgr 18h7'55" -28d3'39"
MacDonald All Stars
Uma 11h20'1" 62d38'47"
Mace
And 23h36'39" 43d2'30"
Mace Barrett
Cap 21h51'52" -10d29'35"
Mace Hitchens
Leo 10h28'48" 12d47'20"
MacEachern
Cep 21h24'36" 59d53'38"
Macee Joanne Lanier
Tau 4h29'29" 13d37'46"
Macen Colby Mitchell
And 0h34'54" 33d24'50"
Macen Patrick Sanantonio
Leo 11h17'52" 12d7'29"
Macey
And 23h16'0" 42d17'38"
Macey
Vir 13h12'57" -2d51'33"
Macey Furstenau
Lmi 10h5'55" 37d31'31"
Macey "Gentle Giant"
Boo 14h32'41" 14d18'2"
Macey Marie Wallace
Uma 11h37'37" 53d28'22"
Macey Nichola Sexton
And 2h34'0" 41d7'57"
Macey Rose Parker
Vir 11h42'8" 2d34'58"
Macey S Henderson - Macey's Star
And 1h22'55" 41d21'25"
Macey's Star
And 1h5'0" 45d30'43"
Macey's Star
Sco 17h28'30" -39d12'47"
Macey's Traveling Star
Sgr 19h4'33" -27d13'5"
MacGregor
Uma 11h48'48" 60d30'30"
MacGYVER
And 23h8'29" 51d18'55"
Machaela Jane Munoz
Aqr 23h9'52" -6d52'49"
Machauer, Rudolf
Ori 6h15'57" 13d17'34"
Machelle Adcock Shoptaw
Cap 21h31'25" -16d13'50"
Machelle Miller/Baby Girl
Sgr 18h2'15" -27d38'12"
Machmad Chemdah
And 0h18'49" 44d36'33"
Maci Hicks
Umi 15h59'4" 78d42'9"
Maci Jo Eisenhauer
Cyg 21h51'31" 52d4'41"
Maci Lynn
And 0h16'54" 27d3'4"
Macian
Cep 22h45'7" 66d44'56"
Macie Ann
And 23h3'58" 51d27'38"
Macie Ann McCloud
And 2h20'23" 38d51'40"
Macie Clark
Cas 0h16'14" 51d22'21"
Macie Leona Mace
Cyg 19h57'6" 40d19'44"
Maciek Matt Trella
Cep 22h2'51" 67d55'6"
Macifiú (Dr. Skrabák Péter)
Vir 14h29'18" -1d22'24"
Macikámnak, Sipos Lászlónak
Uma 10h38'27" 63d1'10"
Mack
Boo 15h37'3" 43d7'12"
Mack 14
Leo 11h37'9" 6d7'59"
Mack Castleman Fuqua, PhD
Ori 6h11'16" 15d53'36"
Mack J. Smolek
Uma 11h4'40" 58d10'53"
Mack Macksimus
Tau 4h22'25" 18d53'39"
Mack1er
Leo 9h26'54" 8d13'47"
MacKensey Hope
Dra 17h49'11" 52d32'51"
MacKensey's Star
Gem 7h27'55" 33d39'31"
Mackensi Grace
Uma 9h23'21" 51d4'21"
Mackenzie
Cas 0h37'42" 54d7'19"
Mackenzie
Lyn 7h26'34" 50d19'53"
Mackenzie
Lmi 10h9'24" 32d7'1"
Mackenzie
Lyr 18h39'19" 30d14'10"
Mackenzie
Psc 23h42'1" 1d48'30"
Mackenzie
Uma 11h57'47" 55d59'46"
Mackenzie
Aqr 22h35'47" -1d7'0"
Mackenzie A. Barth
Cnc 9h11'44" 32d19'27"
Mackenzie Alex McHale
Ori 5h56'3" 21d50'1"
Mackenzie Ann Battle
Cam 4h14'52" 60d31'22"
Mackenzie Ann LaLena
Tau 5h7'9" 25d11'17"
MacKenzie Ann Suppes
And 23h39'36" 47d25'1"
Mackenzie Autumn
Tau 4h34'4" 30d9'54"
Mackenzie Briana Maddox
Psc 0h38'23" 18d19'40"
Mackenzie Brooke Brown
Cnc 9h10'50" 20d55'21"
Mackenzie Brooke Jimerson
Mon 6h32'38" 8d34'2"
Mackenzie Brooks
Uma 8h58'1" 57d45'32"
MacKenzie Catherine Wales
Psc 1h26'52" 13d5'5"
Mackenzie Charlotte Cornes
And 0h21'44" 45d35'19"
Mackenzie Christine
Ori 5h37'18" 1d44'33"
Mackenzie Claire Washington
Uma 9h33'42" 47d46'55"
Mackenzie Doris
Uma 11h10'11" 30d10'17"
Mackenzie Dougherty Chapman
Uma 11h55'46" 38d29'53"
Mackenzie E. Campbell
Leo 11h23'44" -6d38'46"
Mackenzie Elizabeth
Cap 20h37'47" -16d35'50"
Mackenzie Elizabeth Foy
Tau 3h56'53" 4d7'45"
Mackenzie Elizabeth Taylor
And 1h18'18" 46d6'11"
Mackenzie Erin Oster
And 23h33'3" 43d7'17"
MacKenzie Faith Deshae Cooper
Sgr 18h3'56" -27d36'4"
Mackenzie Faith Hagedorn 7
Uma 9h33'54" 68d17'43"
Mackenzie G
Uma 10h28'40" 64d10'25"
Mackenzie Grace
And 23h0'46" 51d28'55"
MacKenzie Grace Jones
Lib 15h29'20" -24d37'32"
Mackenzie Grace Young
And 23h35'3" 43d1'12"
Mackenzie Gray
Peg 22h38'56" 11d58'21"
Mackenzie Hoover
Gem 7h12'40" 29d3'52"
Mackenzie J Henderson - Mackenzie's Star
Ori 6h13'37" -2d16'48"
Mackenzie Jake Toal
Per 4h4'58" 48d31'47"
Mackenzie Jane Lanning
And 0h31'49" 28d17'59"
MacKenzie Jayde Zwillman
And 0h19'44" 36d48'23"
Mackenzie John Pengelly
Uma 10h5'47" 60d58'40"
Mackenzie Jordan
Crb 15h42'21" 38d29'27"
Mackenzie Joy Dolle
Cas 23h41'14" 57d58'12"
Mackenzie Joyce
Per 4h17'45" 41d51'38"
Mackenzie Kaidith
Uma 10h12'19" 46d28'30"
Mackenzie Katz
Lib 15h9'40" -16d22'20"
Mackenzie Lee
Cnc 8h24'0" 30d47'52"
Mackenzie Leigh
And 1h1'3" 42d33'24"
MacKenzie Leigh
Cap 20h49'8" -16d4'15"
Mackenzie Lorraine Scrimgeour
Aqr 21h39'56" -2d15'14"
Mackenzie Louise Dann
Cru 12h23'11" -57d41'31"
Mackenzie Lucius
And 0h53'59" 39d22'17"
Mackenzie Lynn
And 2h28'53" 44d38'53"
MacKenzie Lynn Russo
Aqr 22h24'12" -18d44'32"
Mackenzie Lynne
Uma 8h44'54" 48d0'8"
Mackenzie Mae Gerrans
And 0h38'23" 36d15'11"
Mackenzie Mae Pegher
And 0h5'25" 35d30'5"
Mackenzie Marie
And 0h20'43" 29d48'6"
Mackenzie Marie
Boo 14h37'43" 29d58'13"
Mackenzie Marie
Tau 4h38'5" 19d56'19"
Mackenzie Marie Cassidy
Lib 14h52'30" -0d56'52"
Mackenzie Megan Knode
Ant 9h28'6" -29d24'42"
Mackenzie MiChele
Sgr 19h5'46" -27d31'4"
MacKenzie Morgan
And 0h24'47" 33d15'54"
Mackenzie R. Branyon
Lib 15h3'50" -17d21'43"
Mackenzie Rae
Umi 15h32'2" 73d23'27"
Mackenzie Rae Zara
Psc 0h38'53" 6d59'21"
Mackenzie Rawe
Ari 3h21'53" 26d32'3"
Mackenzie Rea Pepper
And 0h19'43" 37d44'36"
Mackenzie Reis Moyer
Ari 2h49'20" 24d23'11"
Mackenzie Renae
Lyn 8h40'14" 42d5'13"
Mackenzie Rene
Lib 14h49'4" -8d22'58"
Mackenzie Ridgway
Uma 13h56'30" 54d19'35"
Mackenzie Rose
And 1h55'45" 37d11'53"
Mackenzie Ryan
Ori 6h2'53" 17d22'34"
Mackenzie Starr Johnson
Cam 7h25'18" 70d32'30"
Mackenzie Swan
Cyg 19h40'31" 36d50'53"
Mackenzie Taylor Chizmadia
Umi 15h26'38" 72d54'33"
Mackenzie Taylor Gilliard
And 0h53'5" 42d3'38"
Mackenzie Warren-Brownhill
Umi 18h21'14" 87d23'46"
MacKenzie Yeager
Peg 22h22'51" 19d47'16"
Mackenzie, Madison, & Mia Meyer
Ori 4h48'1" 12d42'18"
Mackenzie's Magic
Sgr 19h31'50" -13d30'6"
Mackenzie's Noni
Crb 16h9'16" 33d34'3"
Mackenzie's Rae
Cam 4h0'12" 68d10'58"
Mackenzie's Star
Dra 19h2'40" 73d24'36"
Mackenzie's Star
Umi 15h31'25" 75d5'22"
MacKenziopeia
Peg 21h45'20" 8d56'3"
Mackers Silva
Uma 9h16'5" 50d24'31"
Mackey
Uma 11h44'18" 45d6'1"
Mackie
Lib 15h45'7" -13d34'49"
MacKinley
Cma 6h40'54" -14d34'16"
Mackinley Erin Minor
Mon 6h48'37" 8d36'29"
Mackintosh
Sco 16h12'48" -14d11'18"
MackLyla
Umi 15h26'48" 80d43'26"
Mackó csillag
Uma 9h40'51" 44d14'2"
Mackovski
Per 3h0'8" 40d59'26"
Mack's Dreams
Ori 5h41'4" -2d9'2"
Mack's Little Piece Of Heaven
Cnv 13h54'42" 32d36'47"
Mack's Star
Tau 5h1'25" 24d5'18"
Macky
Ari 3h16'35" 26d5'30"
MacLaren
Vir 13h20'45" 9d39'52"
MacLean's Ceiligh
Uma 12h0'24" 58d33'56"
Macnessa
Uma 12h42'47" 59d13'15"
Macó & Denis
Lyr 18h32'31" 35d53'3"
Macoemejen Galea
Cra 18h9'45" -44d51'28"
"Macondo" Graham - Paige 1928
Cru 12h51'15" -58d5'22"
Macrae Denice Wood
Lyn 6h54'32" 59d17'29"
Macri
Sco 16h6'57" -12d18'25"
Mac's Ree
Ari 2h13'39" 25d21'11"
MAC's Shining Star
Uma 9h12'26" 65d23'23"
Macskacicó
Aqr 22h26'51" -21d11'45"
MACTANGENT
Aql 19h42'44" 15d24'13"
Mactastic P Flash
Tri 2h38'55" 35d21'53"
MacTavish's Komet
Cma 7h8'45" -25d15'48"
Macushla
Sco 16h49'53" -39d50'33"
Macushla
Uma 11h51'57" 48d3'59"
Macy
Tau 3h38'57" 10d59'47"
Macy and Marco Enciso
Cyg 20h17'10" 35d9'54"
Macy Ann Huntley
And 1h41'0" 45d11'5"
Macy Bear
Umi 15h42'26" 78d23'55"
Macy Brooke Page
Ari 2h56'29" 18d40'27"
Macy Claire Gartland
Cap 20h45'3" -17d37'37"
Macy Eve Gray
Umi 10h49'30" 88d53'6"

Macy Fouracre
And 23h14'33" 52d15'24"
Macy Irene Pearce
And 23h13'58" 48d17'56"
Macy Isabella Cook
Dra 10h5'59" 75d56'15"
Macy Jane
Aql 20h16'22" 2d20'12"
Macy Jane Lindstrom
And 0h20'55" 21d51'56"
Macy Kenan Trego
Eri 4h10'19" -0d13'4"
Macy L. Dobbins 'n Joshua R. James
Cnc 8h34'54" 13d44'34"
Macy Lynn Gross
Tau 5h24'48" 21d32'58"
Macy Mae - Too beautiful for Earth
Cru 12h30'27" -61d57'32"
Macy Rae and Vinny
Ori 4h52'45" 8d35'32"
Macy Shae & Reese Emerson Cohn
Cnc 8h42'20" 28d57'46"
Macy Yip
Aqr 21h28'15" -7d39'3"
Macyn Lee Motichka
Umi 16h28'35" 76d31'52"
Maczey, Dieter
Uma 8h15'37" 64d49'50"
Maczkó Havaska
Gem 7h2'20" 29d37'0"
Mad At Pink Hair
Gem 7h39'5" 33d14'37"
MAD CREW
Uma 9h26'20" 46d19'18"
Mad Dog
Sco 17h16'58" -35d51'23"
Mad Mac
And 2h31'37" 50d32'16"
Mad Man Matthew
Aqr 21h43'22" -3d21'55"
Mada Nagrom
Ori 6h0'38" 13d24'17"
Madahoochi
Cma 6h40'8" -15d21'19"
Madai Cuevas
Uma 14h14'18" 58d18'13"
Madalein
Per 3h12'25" 54d12'6"
Madaleine Jade Clark
Dra 19h8'34" 71d48'33"
Madalena Jael A.R. Malpique
And 0h5'2" 40d25'36"
Madalena pour l' éternité
Dra 18h44'39" 54d8'43"
Madalina
And 0h26'57" 38d43'37"
Madaline J. Sassaman
Gem 7h13'26" 34d25'57"
Madaline Rose Linhardt
Cap 20h26'20" -10d39'8"
Madalyn Jane Neice
And 0h4'12" 34d20'36"
Madalyn Lewand
Aqr 21h38'38" 1d47'27"
Madalyn Olivia Rife
Tau 4h19'43" 27d18'57"
Madalynn Grace
Lyr 19h18'54" 29d26'16"
Madalynn Lane
Crb 15h51'16" 33d45'28"
Madalynne Reneé Rizzo
Psc 23h45'10" 7d31'0"
Madalyn's Juliet
Sco 17h56'52" -39d25'23"
Madam Di
Cas 23h2'38" 53d19'46"
Madama
Ori 5h29'26" -4d5'59"
Madame Annette Kennedy
Cas 1h2'51" 62d24'7"
Madame Fry Professeur Formidable
Cas 0h34'22" 55d26'16"
Madame Jessi Lyn Wetzel Joly
Uma 12h5'6" 42d33'0"
Madanydee
Cas 0h26'8" 54d4'2"
Madasyn Marie
Sco 17h53'4" -41d28'51"
Maday
Sco 17h36'32" -33d14'14"
Maday
Vir 12h36'29" 2d5'17"
Madchen Alexandra
Peg 22h14'23" 11d27'26"
Maddalena
Peg 22h16'52" 10d37'31"
Maddalena
Uma 13h35'38" 59d36'1"
Maddalena Alvarez
Crb 15h54'7" 28d21'42"
Maddalena and Franco Amarini
Cyg 20h28'6" 51d52'24"
Maddalena Cialdella
And 23h57'20" 36d54'15"
Maddalena Cinzia
Her 18h37'21" 25d47'45"
Maddax Cole
Sco 16h13'37" -10d31'42"

Maddelyn Whittaker
Tau 4h26'24" 17d32'36"
Madden Robert Orozco
Ari 2h50'44" 14d26'52"
Madden Stephen Browen
Uma 11h26'14" 61d13'27"
"Maddernicus" Two Hearts - One Love
Pho 1h32'31" -39d58'0"
Maddi and Daddy
Uma 13h35'25" 61d8'17"
Maddi and Daddy
Aqr 22h19'46" -2d29'11"
Maddi Hayes
Cmi 7h20'43" 7d55'25"
Maddi Mae
Umi 14h44'39" 82d0'49"
Maddie
Uma 10h11'32" 63d8'54"
Maddie
And 2h31'13" 43d20'21"
Maddie and Matt
Cyg 21h59'2" 49d13'42"
Maddie Arbuckle
Vir 14h17'51" -20d38'43"
Maddie Jayne
Gem 6h29'1" 21d19'57"
Maddie Jayne Young
Uma 8h58'39" 51d2'26"
Maddie Jones
Psc 1h33'27" 25d24'45"
Maddie Love
Peg 22h18'11" 16d29'39"
Maddie Mae
And 1h33'31" 42d52'39"
Maddie Mae Pavlico
Vir 14h16'48" -13d0'15"
MADDIE MOYSE
Sgr 19h42'31" -15d47'1"
Maddie Pimenta
Gem 6h30'2" 21d5'48"
Maddie Thelander
Uma 11h37'53" 59d11'48"
Maddie-Ella
And 0h50'6" 41d24'54"
MaddieJ
Leo 10h19'50" 25d57'7"
MaddieK
Aqr 23h11'35" -7d35'4"
Maddies Daddy
Uma 10h41'28" 44d10'33"
Maddie's Star
Peg 22h34'4" 26d1'13"
Maddie's Star
Tau 4h24'42" 7d6'29"
Maddie's Wishing Star
Psc 1h1'27" 6d34'45"
Maddiestar 1107
Uma 10h14'6" 46d22'35"
MaddiLovesNana
Mon 6h41'27" -1d49'11"
Maddi's Star
Tau 3h45'15" 22d41'44"
Maddison
Ari 2h31'29" 25d57'23"
Maddison Antonia
And 2h25'38" 38d57'9"
Maddison Dawn
Cnc 8h12'34" 16d19'59"
Maddison Ellen
And 1h26'20" 47d31'25"
Maddison Grace Fisher
And 1h22'29" 43d49'40"
Maddison Jade Todd - Maddie's Jewel
And 1h21'56" 43d58'23"
Maddison Logue
Cru 12h56'46" -60d2'3"
Maddison Lylie Stanford
Sco 17h54'3" -35d45'31"
Maddison Nancy Allen
And 0h18'0" 32d12'6"
Maddison Olivia Lax
And 2h14'25" 49d53'31"
Maddison Olivia Owen
Cap 20h53'21" -15d10'36"
Maddison Paige Compton
Uma 9h20'52" 61d41'5"
Maddison Paige Lloyd
Sgr 18h5'19" -30d52'17"
Maddison Rowan Corrall-Jackson
Ori 5h58'50" 18d43'46"
Maddox
Boo 14h19'43" 17d25'6"
Maddox
Tau 4h9'56" 27d23'33"
Maddox Daniel Sherman
Cap 20h21'46" -17d19'59"
Maddox Isaiah Olenick
Uma 13h59'50" 52d7'43"
Maddox Joseph Carter Gomes
Cnc 8h26'12" 26d6'2"
Maddox Joseph Gomes
Cnc 8h3'17" 12d19'38"
Maddox Kade DeLong
Sco 16h12'40" -12d26'24"
Maddox Ray Von Gunten
Ari 2h7'42" 23d47'59"
Maddox Riley Ludlum
Psc 1h26'46" 16d46'57"
Maddox Wells
Sgr 18h2'20" -28d17'3"

Maddux Cade
Uma 8h46'46" 47d48'44"
Maddux Gordon
Her 17h35'17" 34d7'7"
Maddy
Uma 10h11'10" 49d45'26"
Maddy
Mon 6h46'31" -0d24'44"
Maddy
Ori 5h22'11" -1d30'20"
Maddy Jane
Tau 3h54'44" 20d55'38"
Maddy Moo
And 0h0'37" 43d40'36"
Maddy Rawlings
Cas 1h7'31" 61d46'47"
Maddy S. Mc Bride
Cam 4h10'34" 66d47'13"
MaddyHarry
Crb 15h53'10" 26d37'48"
MaddyMona
Cyg 20h28'40" 53d22'15"
Maddy's Hope
Uma 11h32'40" 52d57'1"
Maddy's Hope
Uma 8h54'24" 49d58'36"
Maddy's Light
Lyr 18h32'30" 37d9'31"
Maddy's Star
And 1h39'24" 49d33'40"
Maddy's Star
Umi 15h22'55" 72d45'20"
Maddy's Star
Uma 9h8'19" 68d29'57"
Made for Each Other
Cyg 20h31'19" 47d8'43"
Madeha
And 1h22'0" 35d53'16"
Madeira Shammas
Cyg 20h28'41" 51d12'11"
Madel Angeles
Vir 13h7'49" -3d12'16"
Madel Martinez
Tau 4h17'19" 8d58'10"
Madelain Zonis Land
Sgr 17h56'17" -27d28'40"
Madelaine
Ori 5h9'18" 15d18'46"
Madelaine Emelie Ludwig
Uma 10h54'34" 47d52'18"
Madelaine Lankford
Cnc 9h0'2" 27d20'6"
Madelaine Lucy Taylor
Cas 1h32'42" 61d27'46"
Madelaine Notaro
Lyn 7h55'35" 49d41'46"
Madelaine Ruby Marsden
And 23h5'8" 40d28'5"
Madeleine
Aur 5h19'12" 31d30'59"
Madeleine
Cnv 13h25'57" 30d56'50"
Madeleine
Tau 5h42'53" 27d0'5"
Madeleine
And 0h9'35" 32d39'46"
Madeleine
Leo 11h2'40" 20d23'52"
Madeleine
Sgr 18h58'19" -24d54'17"
Madeleine
Sco 16h14'41" -27d56'4"
Madeleine
Uma 9h24'3" 62d3'46"
Madeleine Adelaide Haskell
Vir 15h7'30" 4d39'44"
Madeleine Amick-Kehoe
And 1h0'5" 44d0'26"
Madeleine and Kevin Mahnken
Cyg 19h29'34" 51d58'47"
Madeleine Angles
And 2h17'39" 41d52'52"
Madeleine Beayni
And 1h22'37" 49d59'58"
Madeleine Deavers
Uma 14h27'24" 58d51'13"
Madeleine Elise
Tau 4h12'5" 12d59'5"
Madeleine Elizabeth Quinn
And 2h36'14" 46d16'42"
Madeleine Elouise Scarlett Benson
And 1h30'59" 40d16'54"
Madeleine Grace Carrier
And 1h2'32" 38d30'16"
Madeleine Grace & Lorien Elizabeth
Cas 1h37'54" 73d4'24"
Madeleine Grace Volz
And 0h47'48" 38d8'44"
Madeleine Grace Zacks
Cas 0h52'9" 63d59'19"
Madeleine Greder
Ori 5h53'1" 6d32'57"
Madeleine Grenier et Emery Gagné
Del 20h14'59" 15d3'6"
Madeleine Halacy
Psc 1h28'12" 12d10'14"
Madeleine Hali
Ari 3h17'36" 19d58'5"
Madeleine Holly Delforge
Crb 15h51'9" 34d2'50"

Madeleine Jane Ipsen
Crb 16h6'16" 36d9'7"
Madeleine Jane Salkeld
And 0h49'24" 31d27'17"
Madeleine Jeanne Ware
Uma 14h5'6" 54d31'27"
Madeleine Jude Brown
Uma 9h50'2" 54d53'50"
Madeleine Kate Hayes
And 2h34'15" 44d57'31"
Madeleine Lee Benham
And 2h26'31" 46d15'14"
Madeleine Mary Dossett
And 23h50'54" 34d34'29"
Madeleine & Maurice Gravelle
Tau 5h10'40" 21d16'40"
Madeleine May
Aql 19h21'11" 0d50'35"
Madeleine Müller
Cas 1h5'25" 62d27'46"
Madeleine Nielson
Ori 5h19'8" 9d18'2"
Madeleine Roberta Chalmers Meikle
And 0h16'26" 48d4'53"
Madeleine Rose
Uma 11h33'17" 34d17'20"
Madeleine Rose Kirkup
And 0h49'25" 22d19'20"
Madeleine Sasha Ann Stevens
And 2h13'20" 49d27'11"
Madeleine & Stefan
Lyr 18h46'8" 46d53'1"
Madeleine Stratton
Uma 8h53'19" 66d14'9"
Madeleine "Sweet Pea" Hess
Aqr 22h36'31" -0d51'56"
Madeleine Victoria
Cas 23h58'3" 60d3'29"
Madeleine's Wish
And 22h58'37" 52d2'3"
Madelena
Cnc 9h5'19" 9d1'12"
Madelene
Cnv 13h31'20" 42d30'14"
Madelene's Light
Cru 12h46'26" -57d35'7"
Madelin Cox
Ori 5h50'44" 3d22'42"
Madelin Skye Doyle - 22.08.2005
Cru 12h28'55" -58d44'22"
Madeline
Sgr 19h14'46" -26d35'3"
Madeline
Cas 1h31'48" 68d52'4"
Madeline
Cas 23h48'37" 54d37'4"
Madeline
Ari 3h5'13" 13d10'46"
Madeline
Her 16h50'50" 17d55'54"
Madeline
Crb 15h35'25" 26d36'39"
Madeline
Cnc 8h38'35" 24d51'41"
Madeline
Lyn 7h2'22" 50d42'49"
Madeline
Cyg 20h14'42" 49d50'40"
Madeline 3/8 Shannon 3/27
And 1h58'17" 38d55'44"
Madeline Alaine
Cas 23h47'43" 53d34'47"
Madeline Alice Nevis
And 0h13'58" 43d10'16"
Madeline and David's Star
Vir 14h24'49" 1d3'21"
Madeline & Andy Martinez
Uma 13h37'56" 57d49'51"
Madeline Anelia Blandini
And 0h21'10" 28d34'20"
Madeline Anne
Gem 7h7'48" 28d9'56"
Madeline Ariana
And 2h23'37" 50d1'56"
Madeline Axelle Knauss
Ari 2h46'31" 30d30'20"
Madeline B. Brann
Leo 10h39'5" 13d50'55"
Madeline Barbara McDougal
Uma 11h20'49" 50d3'47"
Madeline Bell Rubicam
Lmi 10h34'5" 36d52'10"
Madeline Benoff
Tau 4h40'49" 11d37'32"
Madeline Berry Moore
Cyg 21h53'8" 38d48'27"
Madeline Beth
Tau 5h40'57" 26d35'18"
Madeline Bozyk Connaughton
Umi 13h58'59" 70d13'1"
Madeline Britta
Crb 15h51'24" 34d54'36"
Madeline Brooke Young
Aqr 23h9'50" -6d14'5"
Madeline Bunny Star
Uma 9h57'36" 42d56'41"
Madeline Celestina McCoy
Vir 13h4'1" -19d48'57"

Madeline Chase
Crb 15h34'0" 37d34'28"
Madeline Chiarenza
Tau 5h59'38" 23d31'29"
Madeline Christine Waldock
And 1h13'28" 45d34'2"
Madeline Claire
Uma 8h46'56" 70d50'56"
Madeline Cortesi Bernardi
Cas 0h24'23" 64d42'41"
Madeline D. Herth
Cam 6h23'27" 65d5'1"
Madeline Doyle
Cas 2h24'47" 74d53'31"
Madeline Elaine Gray June 8, 1999
Gem 6h56'4" 17d23'40"
Madeline Elizabeth Frisina
And 2h22'8" 46d49'45"
Madeline Elizabeth McArthur
Cas 0h15'2" 57d9'32"
Madeline Emiley
Uma 11h36'48" 36d13'46"
Madeline Eve Pregulman
Sco 17h53'40" -35d58'22"
Madeline Felicia Tarabola Sumner
Uma 8h53'19" 53d28'53"
Madeline Frances
Peg 22h28'36" 33d45'36"
Madeline Gennette DeLuca
Aql 19h48'51" 10d52'27"
Madeline Goodwin
Cas 1h0'6" 49d30'1"
Madeline Goss
Mon 6h51'4" -0d5'41"
Madeline Grace
And 0h11'29" 33d18'55"
Madeline Grace Campagnolo
Aqr 23h6'44" -19d56'39"
Madeline Grace Holmes
Lyn 7h13'17" 48d27'25"
Madeline Grace Moore
Cas 23h37'51" 55d4'47"
Madeline Grace Rumble
Ari 2h8'52" 13d2'25"
Madeline Grace Zeiter
Tau 3h34'57" 21d35'33"
Madeline Irene Phillips
Lyn 8h55'23" 41d39'29"
Madeline & James Goff
Lyr 18h58'25" 27d42'7"
Madeline Jane
Lmi 10h50'30" 22d58'11"
Madeline Jean Polk
Ari 2h4'6" 24d34'25"
Madeline Jeanette Snow
Cas 1h19'54" 62d44'16"
Madeline Jo Bish
Cas 1h31'54" 64d49'29"
Madeline Josie
And 0h59'2" 44d34'27"
Madeline Kos
Uma 10h7'59" 62d30'20"
Madeline Lee Cobb
Lmi 10h43'27" 26d12'28"
Madeline Lee Wills
And 1h42'42" 39d13'37"
Madeline Lena Getty
Psc 1h18'20" 32d21'42"
Madeline Lily
Ori 6h3'14" 19d28'13"
Madeline Lisy
Ari 2h36'50" 20d10'31"
Madeline Lucille Villmer
Cap 20h32'5" -13d56'46"
Madeline Margaret Setzer
Vir 13h21'44" -2d47'33"
Madeline Marie
Aqr 22h29'32" -0d45'53"
Madeline McGinnis
And 1h58'54" 39d10'0"
Madeline McMillion
Cap 20h28'32" -13d33'51"
Madeline Meidl
And 1h12'8" 44d43'18"
Madeline&Michael
Ari 2h42'43" 21d16'1"
Madeline Michael Tafoya
And 1h12'36" 46d24'16"
Madeline Michelle-Leigh Manriquez
Cas 1h35'57" 58d5'40"
Madeline Miller
Cap 20h37'12" -14d41'22"
Madeline Monica
Peg 22h16'28" 7d9'14"
Madeline Murphy
Aqr 22h0'7" -13d48'21"
Madeline Pamela Colantuono
Lyn 7h58'49" 57d5'9"
Madeline Paszkiewicz
Sco 16h11'33" -9d15'34"
Madeline R. McNamara
Uma 12h50'22" 62d46'31"
Madeline & Ralph Bova
Cyg 19h58'13" 36d51'28"
Madeline Reese
Crb 16h13'56" 35d58'8"
Madeline Reese Hills
Uma 9h59'12" 56d36'31"

Madeline Reese Meehan
Lib 15h6'41" -8d22'38"
Madeline Reimer
And 0h18'26" 43d18'13"
Madeline Renee' Kelley
Cnc 8h44'32" 16d38'36"
Madeline Rogers
Vir 13h16'35" 3d32'39"
Madeline Rose
And 1h42'14" 40d32'26"
Madeline Rose
Sco 16h30'16" -25d35'35"
Madeline Rose Cox
Sgr 19h17'20" -31d19'3"
Madeline Rose Holub
Cnc 9h10'46" 24d36'6"
Madeline Rose Jarman
Vir 13h22'1" 1d30'58"
Madeline Rose Vandervegt
And 23h11'23" 51d9'57"
Madeline Ryan Connelly
Cas 1h23'14" 64d51'41"
Madeline S. Carlson
And 23h15'3" 47d25'5"
Madeline S. Monroe
Tri 2h31'23" 31d43'3"
Madeline Sarah
And 2h16'48" 47d24'6"
Madeline Senese
Her 17h52'47" 29d19'9"
Madeline Sky Wensel
Tau 4h23'6" 19d9'24"
Madeline Sloan Way
Lyn 8h37'22" 45d56'6"
Madeline Smerin
Mon 6h49'21" 7d2'3"
Madeline Smolenski
Lyn 8h4'5" 33d38'39"
Madeline The Beautiful
Cas 1h28'41" 56d37'52"
Madeline Theresa Mader
Cas 0h49'36" 56d35'13"
Madeline Veyrenc
Ari 2h51'47" 31d9'24"
Madeline Victoria
Sgr 19h16'19" -13d5'56"
Madeline Virginia Sellers
And 1h23'40" 50d26'30"
Madeline Wissmann
Crb 16h23'34" 30d11'0"
Madeline136
And 2h7'21" 45d38'19"
Madeline56
Cnc 9h9'13" 7d15'7"
Madeline's Sparkle
Cap 21h27'3" -10d35'23"
Madelyn
And 0h33'25" 41d43'23"
Madelyn Ann Yeager
Cas 0h43'3" 66d56'26"
Madelyn Elizabeth Lipke Miner
Psc 0h16'15" 15d24'24"
Madelyn Elizabeth Wiseman
Aqr 21h18'35" -0d25'28"
Madelyn Elyse
Cas 0h6'22" 57d17'27"
Madelyn Falcone
Cap 20h39'50" -24d37'33"
Madelyn Gails Downs
Cas 1h8'35" 49d33'2"
Madelyn Grace
Cap 21h43'0" -18d18'29"
Madelyn Gray Pearson
And 1h39'57" 41d45'9"
Madelyn Harn
Mon 6h31'52" 8d22'14"
Madelyn Jean Tess Dubman
Peg 22h18'5" 17d22'18"
Madelyn Joyce Smith
Ori 5h37'49" 3d38'11"
Madelyn K. Brunner
Ari 2h13'57" 22d47'5"
Madelyn Kate Scaringe
Vir 13h29'40" -7d3'11"
Madelyn Kent
And 2h19'40" 47d21'1"
Madelyn Larsen Miles
Cas 1h47'13" 63d32'3"
Madelyn Lee Wright
And 0h27'35" 44d57'46"
Madelyn Ly-Lan Vo
Leo 11h44'19" 18d1'29"
Madelyn (Lynn) Kahana
Vir 14h5'40" -20d47'20"
Madelyn ("Madge") DiCocco
Vir 12h0'19" 5d26'27"
Madelyn Marie Massanelli
Vir 12h1'32" -5d52'16"
Madelyn Marie Wade
And 23h36'20" 44d12'17"
Madelyn Mary
Cas 1h15'43" 64d31'28"
Madelyn McKay Moore
Cap 20h47'41" -17d49'26"
Madelyn Nichols Fleck
Aqr 22h7'26" -10d42'43"
Madelyn Nicole Fernandez
Umi 14h18'55" 76d33'48"
Madelyn Nicole Wilson
Lib 15h43'24" -17d44'35"

Madelyn Okun
And 23h36'25" 47d3'58"
Madelyn R. Williams
Uma 10h51'53" 41d58'56"
Madelyn Rose Dodd
Peg 21h43'46" 20d50'40"
Madelyn Ryan Berault
Peg 22h21'22" 29d27'41"
Madelyn Sanchez
And 0h59'14" 40d20'42"
Madelyn Taylor Murphy
Crb 15h51'2" 33d56'57"
Madelyn10/04/04
Cas 23h7'10" 59d22'24"
MadelynAnnStewich041819 98
Ari 3h9'30" 29d2'58"
Madelyne
Cnc 7h58'53" 14d59'37"
Madelyne Jo Liebler
Leo 11h20'36" 18d2'48"
Madelyne Nicole Page
Ari 3h14'40" 19d0'38"
Madelyne Richards
Uma 13h51'38" 57d5'22"
Madelynn
Vir 13h29'50" -7d0'32"
Madelynn Grace
Uma 9h7'10" 57d1'42"
Madelynn M Graham
Tau 5h39'38" 27d3'26"
Madelynn Raeann Fenton
Leo 10h38'43" 14d26'59"
Madelynn Sol-Luna
Cma 6h43'55" -22d10'14"
Madelyn's Angel
And 0h16'4" 26d4'9"
Madelyn's Star
Lyr 18h37'27" 31d26'4"
Madelyn's Star (Sloan)
Cas 0h44'8" 49d35'53"
Mademoiselle Amrika
Gem 6h8'34" 27d18'9"
Mademoiselle Ashli
Lib 14h52'31" -3d33'12"
Mademoiselle Pétrone
Uma 9h30'25" 55d41'18"
"Madge" Alice Margaret Ryan
Cru 12h25'5" -55d42'2"
Madge & George Wilkinson
Cyg 20h30'29" 39d16'30"
Madge Hunt Tanner
Vir 12h28'56" 4d33'17"
Madge & Steve King
Umi 15h34'23" 73d26'36"
Madge's Star
Cas 0h56'20" 55d7'11"
Madhatter
Ari 3h23'50" 26d1'22"
Madhavi Deepak
Lib 15h9'28" -17d16'41"
Madhavi Manyam
Crb 15h59'21" 36d7'53"
Madhu
Ori 5h35'17" 5d50'22"
Madhuri Pinnamaneni
Psc 1h10'27" 27d26'21"
Madhusmita and Saurav
Cyg 21h27'17" 45d38'43"
Madi
Tau 4h27'52" 27d4'13"
MADI
Her 17h29'57" 26d40'28"
Madi - Will You Marry Me - Jason
Leo 10h4'57" 20d21'30"
Madi14
Mon 7h2'42" -8d42'53"
Madie
Cyg 20h27'48" 49d3'6"
Madie Levin
Leo 10h16'2" 17d46'36"
Madie Ventavoli Steiner
Umi 13h55'13" 78d9'53"
Madie's Star
Cnc 8h7'6" 20d27'14"
MadiEssieAna
Lyn 8h20'28" 52d28'33"
Madilyn
And 2h23'13" 46d28'8"
Madilyn Adele Hosford
Gem 7h10'9" 20d45'25"
Madilyn Ava
Umi 15h34'5" 85d10'32"
Madilyn Christine Simpson
Umi 15h25'51" 81d23'1"
Madilyn E Thomas
And 0h54'47" 39d3'18"
Madilyn Grace Morgan
Cyg 20h54'37" 33d21'14"
Madilyn Margaret Haskett
Sgr 18h16'56" -16d41'41"
Madilyn Salo
And 0h13'21" 34d27'35"
Madilynn Sierra Morris
Peg 22h3'16" 6d14'41"
Madimac
Leo 10h20'18" 14d31'49"
Madina Bugulova
And 0h44'58" 44d30'44"
Madisen
Ari 3h25'46" 28d6'17"
Madisen Elizabeth Oey
Tau 4h6'44" 14d58'0"

Madisen Keane Baldwin
Ori 5h23'8" 9d9'31"
Madisen Keely Murawski
Mon 6h51'47" -0d3'17"
Madisen Rose Mazzio
And 1h19'7" 42d28'56"
Madison
Uma 10h6'21" 43d20'15"
Madison
Lyn 9h6'59" 39d0'23"
Madison
Lyn 8h4'42" 40d42'58"
Madison
Lyn 7h35'25" 41d25'14"
Madison
And 0h51'35" 37d25'1"
Madison
Per 2h10'16" 50d36'25"
Madison
Lyn 7h22'46" 48d26'22"
Madison
Ari 3h6'58" 13d29'34"
Madison
Vir 14h34'35" 3d12'18"
Madison
Leo 10h48'0" 10d19'29"
Madison
Tau 4h13'27" 25d33'4"
Madison
Tau 4h0'12" 27d6'3"
Madison
And 0h39'38" 23d51'18"
Madison
Vir 11h49'26" -0d29'29"
Madison
Sco 16h8'22" -18d6'12"
Madison
Sco 16h15'10" -15d39'46"
Madison
Lib 15h23'9" -16d40'45"
Madison
Cam 4h7'46" 70d22'9"
Madison
Uma 10h17'3" 54d58'46"
Madison
Uma 9h58'0" 59d22'41"
Madison
Cen 11h47'27" -52d8'44"
Madison
Cap 21h9'44" -25d29'8"
Madison Abigail Evans
Uma 11h41'12" 28d51'12"
Madison Alexandra Smith
Sgr 18h20'25" -22d32'8"
Madison Alexandra
Sorenson
Sco 16h5'11" -19d32'52"
Madison Alexis
And 0h38'33" 25d19'35"
Madison Alyn Dorman
Uma 8h53'55" 52d24'2"
Madison Alyssa Cupelli
Tau 5h42'40" 25d43'11"
Madison and Andy
Ori 4h51'28" 13d37'51"
Madison and Kylie Barrick
Sge 19h41'7" 18d45'10"
Madison Andre Wright
Del 20h40'13" 13d37'15"
Madison Angela Lange
Lyr 18h46'12" 39d11'21"
Madison Ann DeMattia
Uma 11h6'47" 61d55'34"
Madison Ann Peon
Vir 13h16'9" 6d41'16"
Madison Anne Walker
Her 18h6'44" 17d50'49"
Madison Anne Wraga
Cap 21h39'43" -13d44'32"
Madison Anon Shelton
Tau 3h48'22" 27d8'51"
Madison Austin
Psc 1h15'14" 16d26'22"
Madison Avenue Salon and
Day Spa
Crb 15h20'28" 32d29'46"
Madison "Baby Love"
Hughes Twinkles
Sgr 19h15'4" -17d34'25"
Madison Blankenship
And 23h46'30" 41d57'11"
Madison Blevins
Leo 11h55'50" 19d10'39"
Madison Bonawitz
Sco 16h45'27" -32d12'6"
Madison Brannigan
And 2h35'59" 50d0'13"
Madison Brianna Black
And 0h13'49" 33d34'22"
Madison Bridgewater
Peg 21h51'28" 18d13'52"
Madison Bright
And 0h28'15" 33d15'23"
Madison Brooke
Aql 19h26'52" 8d4'45"
Madison Brooke Alexander
Lyr 19h9'9" 42d14'13"
Madison Brooke Blanton
And 0h51'11" 41d21'48"
Madison Brooke Francis
And 2h19'1" 37d38'40"
Madison Brooke Truess
And 1h4'22" 38d52'11"
Madison Camille
Cnc 8h35'13" 22d59'1"
Madison Candace
Montgomery
Umi 16h45'45" 79d46'33"
Madison Carole Maywalt
And 0h34'52" 29d50'10"
Madison Catherine Graham
Her 17h59'31" 21d22'4"
Madison Catherine Verhulst
Lyn 7h47'54" 36d4'13"
Madison Chelsea Leander
Cyg 21h22'54" 50d40'7"
Madison Chiolan
Peg 21h40'9" 25d34'45"
Madison & Chris
Psc 0h45'56" 15d52'20"
Madison Christine
And 1h8'31" 41d8'42"
Madison Christine Halloran
Gem 7h6'48" 10d46'31"
Madison Claire Schabacker
Crb 15h40'22" 29d5'11"
Madison Cramer
Tau 4h19'9" 11d4'30"
Madison Dae
Ori 6h6'15" 16d58'24"
Madison Dale Schon
Vir 12h43'17" 4d52'42"
Madison Danielle Rolfe
Ari 3h6'1" 21d8'44"
Madison Dawn
Sgr 19h53'55" -17d57'16"
Madison Demi
Lyn 7h45'26" 53d55'18"
Madison & Doug's Star
Cap 21h44'23" -14d39'42"
Madison Dreyer
And 2h13'21" 45d23'42"
Madison Duda
Aqr 22h54'49" -5d16'47"
Madison Eliece Petschk
And 23h38'27" 47d12'0"
Madison Elise
And 0h37'33" 27d48'18"
Madison Elise Jordan
And 23h46'32" 42d24'44"
Madison Elise Morton -
05.05.05
Cra 18h3'27" -45d3'16"
Madison Elizabeth
Lib 14h58'58" -16d19'35"
Madison Elizabeth Brown
Sgr 19h11'40" -21d2'57"
Madison Elizabeth Caron
And 0h13'43" 26d6'10"
Madison Elizabeth Cesari
Uma 12h31'15" 58d31'12"
Madison Elizabeth Noguez
Psc 1h10'49" 25d54'50"
Madison Elizabeth Osborne
Vir 12h55'2" -2d24'36"
Madison Elizabeth Schafer
Psc 1h39'54" 21d58'12"
Madison Eller
Uma 8h50'54" 57d57'34"
Madison Eloise Torres
And 23h42'28" 38d19'56"
Madison Emma Fuller
And 2h26'46" 41d31'30"
Madison Emma Perrin
Cap 20h37'36" -10d24'17"
Madison Faith
And 1h53'19" 41d47'19"
Madison Faith Findlay
And 23h41'7" 48d24'12"
Madison Faith Schaefer
Uma 12h18'14" 60d6'26"
Madison Fei Rose
Cap 21h11'34" -22d47'39"
Madison Fia Louise
Banting
Umi 16h26'28" 76d42'17"
Madison Finney
Peg 21h28'45" 12d4'17"
Madison Fionnait of
Mulberry Creek
Uma 10h20'2" 53d27'52"
Madison Fluharty
Cmi 7h20'3" 0d1'10"
Madison Friedsam
And 23h57'16" 39d53'48"
Madison Fucci
And 1h40'49" 46d5'56"
Madison Gabriella Ryan
Psc 1h16'47" 14d57'59"
Madison Gabrielle
Widdecombe
Tau 5h38'44" 26d37'53"
Madison Grace
Ori 5h52'32" 11d27'36"
Madison Grace
Lib 15h21'37" -21d31'11"
Madison Grace Becker
Cap 21h15'56" -15d19'12"
Madison Grace Binding
And 0h22'37" 39d48'49"
Madison Grace Fernandez
Tau 4h35'49" 18d36'32"
Madison Grace Meyers
Lyn 9h12'0" 43d16'33"
Madison Grace Sepe
And 23h10'4" 49d29'59"
Madison Grace Taylor
And 0h16'6" 46d59'38"
Madison Grace Wilkey
Bundren
Vul 19h40'45" 19d35'42"
Madison Greene
Sco 16h36'34" -43d59'39"
Madison Grundler
Lmi 10h34'25" 33d35'39"
Madison Guzman
Crb 16h6'26" 27d39'30"
Madison Gwenn Rosier
Sgr 17h52'47" -17d23'2"
Madison Harris
Gem 6h48'46" 16d53'1"
Madison Hart
Mon 7h14'50" -0d53'29"
Madison Hartford
Cas 23h45'31" 54d39'8"
Madison Hertzog
And 23h24'55" 51d2'14"
Madison Hope
Lyn 7h54'3" 42d21'34"
Madison Hurdt
Mon 6h25'46" -10d12'13"
Madison Isabella Bullock
And 2h29'56" 38d29'4"
Madison Jane Ramirez
Cap 20h11'57" -9d40'18"
Madison Jane Rogers
And 1h47'40" 46d32'54"
Madison Jean Davis
Cap 20h29'20" -19d30'4"
Madison Jewel
Cas 23h10'0" 59d13'7"
Madison Jewel Galetka
Uma 11h49'31" 54d22'45"
Madison Jones
Umi 15h43'17" 77d57'25"
Madison Josephine Scott
Cyg 20h27'3" 37d5'51"
Madison Joy Cecilia
Ari 2h46'12" 30d9'7"
Madison Joyce Seymour
Tau 5h32'8" 24d56'28"
Madison Julia Scardino-
Rettig
Psc 1h37'58" 27d37'34"
Madison Justice Wright
And 0h30'12" 25d22'54"
Madison Kate Bocking-
2003-Our Star
Cru 12h51'39" -57d40'11"
Madison Katherine Silcox
Uma 11h27'42" 43d27'4"
Madison Kathleen
Lyn 6h58'41" 44d55'27"
Madison Kay
Leo 11h54'15" 19d3'10"
Madison Kay Cockrell
Aqr 23h26'35" -8d20'32"
Madison Kaye Mills
And 2h35'26" 39d40'25"
Madison Kaylee Waltemire
And 2h35'1" 44d14'51"
Madison Kelly
Lmi 9h49'22" 38d31'51"
Madison Kim Bird
Cmi 7h24'23" -0d0'44"
Madison Kneeland
Vir 12h58'36" -0d9'51"
Madison Koepke
Ari 3h5'18" 19d16'30"
Madison Kunkel
Sco 16h50'8" -35d39'2"
Madison Langley
Uma 10h26'15" 64d43'39"
Madison Lanter
Cnc 8h30'50" 19d31'11"
Madison Laura Moore
Lib 15h51'38" -17d29'42"
Madison Lauren Hart
Gem 7h28'17" 29d25'5"
Madison Lee
Vir 11h58'36" -0d36'54"
Madison Lee Lewis
And 2h18'58" 49d10'9"
Madison Lee Wisdom
Tau 5h21'18" 22d24'57"
Madison Lee Wolfe
Cap 21h13'29" -19d19'49"
Madison Leean Allen
And 0h39'27" 35d41'16"
Madison LeeAnn Meyer
And 0h35'40" 36d42'34"
Madison LeeAnne Morrow
Cas 23h6'19" 53d27'11"
Madison Leffler
Lyn 8h47'46" 41d21'32"
Madison Leigh Tierney
Sgr 17h54'21" -29d4'7"
Madison Lena Hall
Gem 7h22'42" 29d12'26"
Madison Lewis
Tau 5h48'44" 17d8'17"
Madison Lindsay
Lib 15h1'29" -9d51'22"
Madison Louise Tablas
Psc 1h27'18" 11d11'39"
Madison Lovera
And 2h38'42" 46d32'15"
Madison Lynn
And 1h3'50" 41d27'5"
Madison Lynn Nott
And 1h9'52" 36d6'38"
Madison Lynn O'Steen
Aqr 21h24'48" -2d49'3"
Madison Lynne Kalkstein
Cas 1h24'21" 65d28'56"
Madison Mabel Wynne:
Grandma's Girl
Uma 11h23'20" 42d18'51"
Madison "Maddi" Grace
And 23h8'23" 39d32'6"
Madison Mae Schreiber
Sco 16h58'8" -42d32'24"
Madison Mae Zygadlo
And 2h37'33" 46d14'28"
Madison Mansi
And 1h17'15" 45d38'23"
Madison Maria Muise
And 0h56'23" 39d50'11"
Madison Marie
Crb 15h36'22" 28d15'24"
Madison Marie
Leo 11h31'27" -5d7'23"
Madison Marie Adkins
And 1h43'13" 45d58'10"
Madison Marie Dessert
Crb 16h4'29" 26d0'13"
Madison Marie Hays
Ari 2h50'13" 29d38'12"
Madison Marie Tropiano
Leo 11h28'54" 24d40'13"
Madison Marie Weigand
Cap 21h52'51" -17d40'12"
Madison Matthews
Crb 16h7'13" 33d28'34"
Madison May Westerfield
And 0h34'5" 32d21'24"
Madison McGeorge 1
Lyn 8h33'47" 57d34'17"
Madison McGeorge #2
Gem 6h17'52" 22d52'14"
Madison McKenzie Morgan
And 23h28'42" 50d4'46"
Madison McRay
Sco 17h43'47" -39d6'15"
Madison McSparran
Ori 5h9'34" 6d7'15"
Madison Michelle Beach
09/02/04
And 0h17'12" 26d7'11"
Madison Michelle Kemp
Sco 16h41'50" -28d39'52"
Madison Michelle Llewellyn
And 23h12'49" 41d48'51"
Madison - Mom & Dad's
Precious Star
Cyg 20h4'37" 34d13'49"
Madison Neomi Beard
Cap 21h47'14" -10d46'17"
Madison Nichole Keller
"9/01/1996"
Vir 13h17'40" 11d37'32"
Madison Nicole
Lyn 8h16'44" 41d6'27"
Madison Nicole
Cunningham
Leo 9h55'49" 21d34'4"
Madison Nicole Davis
Lep 5h22'9" -11d57'31"
Madison Nicole Hall
Ori 6h6'34" 14d22'22"
Madison Nicole Merrick
Psc 1h9'25" 7d16'48"
Madison Nicole Schaller
Vir 14h17'16" -12d14'26"
Madison Noel
Cam 3h27'49" 61d28'59"
Madison Olivia Forrest
And 0h49'27" 37d12'0"
Madison P Mccauley
Psc 0h8'33" 2d32'14"
Madison P. Riley
Peg 22h44'11" 11d23'26"
Madison Paige
Uma 8h37'13" 64d59'34"
Madison Paige Barris
Lyn 7h46'31" 58d4'24"
Madison Paige Delgado
Lyn 7h4'55" 58d13'10"
Madison Paige Pearson
And 1h32'56" 39d47'36"
Madison Paulle
And 1h46'21" 45d49'5"
Madison Ponte
Uma 10h54'40" 53d33'24"
Madison Rae Kline
Lyn 6h58'24" 46d59'1"
Madison Rae Lowden
Crb 15h43'36" 37d21'9"
Madison Rae Taylor
Sco 16h16'31" -9d8'27"
Madison Renate Smith
And 0h28'9" 27d14'5"
Madison Renee Foster
Crb 15h39'58" 34d8'18"
Madison Riley Hape
And 2h38'10" 38d35'59"
Madison Riley Thomas
Peg 23h52'49" 27d55'25"
Madison Ronda
Sgr 19h38'5" -15d29'6"
Madison Rose
Uma 11h22'51" 62d4'33"
Madison Rose
And 0h56'33" 22d10'29"
Madison Rose Bockol
And 0h50'51" 41d13'43"
Madison Rosenblitt
Vir 12h30'38" 6d3'50"
Madison Ross
Uma 8h48'47" 63d14'54"
Madison Ruby (8:14 P.M.)
Sco 17h16'3" -33d4'43"
Madison Sara
Lowenschuss
And 23h5'42" 40d35'46"
Madison Sarah Murphy
Cru 12h24'2" -56d39'44"
Madison Shay Viger
Uma 8h27'13" 62d28'38"
Madison Shelbie Jones
And 2h13'54" 38d25'30"
Madison Skye Arroyo
Cap 20h17'55" -14d45'1"
Madison Slagowski
Sgr 18h46'52" -23d10'1"
Madison Stanko Blue
Lyr 18h49'34" 38d57'8"
Madison Starr Chickey
Del 20h48'54" 17d40'28"
Madison Stephanie Blake-
Pharr
And 23h24'32" 48d9'17"
Madison Stetler
And 23h16'17" 39d42'10"
Madison Sue Feuerbach
And 0h14'37" 36d0'47"
Madison T. Page
Cnc 8h40'9" 32d49'37"
Madison Tate Freeman
Vir 12h11'32" 12d2'11"
Madison Taylor
Sco 16h9'6" -17d50'55"
Madison Taylor Farley
Crb 16h18'7" 38d6'29"
Madison Taylor Grinnell
Hya 9h32'50" -0d52'11"
Madison Taylor McKoin
Gem 7h17'20" 27d6'48"
Madison Taylor Wolski
Tau 3h46'57" 28d42'6"
Madison Theresa Langley
And 2h29'16" 49d47'39"
Madison Therese Medina
Sco 17h20'27" -32d41'27"
Madison Thorn
Uma 8h33'29" 65d11'35"
Madison Virginia Ponte
Uma 11h54'36" 59d23'59"
Madison Whitfield Wilson
Peg 23h44'22" 17d39'24"
Madison-Belle
Gem 6h19'12" 27d35'12"
MadisonForeverlovesScoob
yDad&Lauren
Lyr 18h43'28" 33d47'5"
MadisonLynne12042002
And 0h24'24" 43d1'1"
Madison's Delight
And 0h26'49" 41d49'9"
Madison's Fairy
Leo 9h39'45" 27d11'15"
Madison's Light
Sgr 18h46'47" -23d39'17"
Madison's Miracle
Lyn 8h3'29" 57d51'7"
Madison's Star, My Moon
Ari 2h41'37" 24d53'56"
Madison's Starr
Leo 9h33'29" 29d33'2"
MadisonsPlace
Peg 21h31'58" 15d25'34"
Madisyn
And 2h26'50" 37d43'46"
Madisyn
And 0h54'26" 38d46'38"
Madisyn
Uma 11h12'11" 33d31'49"
Madisyn Elizabeth Kamilee
Medeiros
Sco 16h10'49" -22d35'28"
Madisyn Marie Banis
Mon 6h45'48" -5d59'8"
Madisyn Marie Gibson
And 0h13'7" 30d7'21"
Madisyn Nicole Lunner
Aqr 22h19'49" -5d39'26"
Madisyn & Pascal
And 0h18'11" 25d56'20"
Madisyn Rose Kafer
Vir 11h49'25" 6d1'2"
Madisynn Lee
Gem 6h47'18" 24d26'11"
Madivia
Her 16h26'3" 48d57'28"
Madjie Lee Thornton
Uma 8h51'51" 52d53'12"
MADJIMBA
Boo 15h30'43" 46d41'21"
Madlaina
Peg 21h30'39" 21d40'29"
Madlaina
Cam 6h20'28" 66d50'34"
Madlen Bremer
Uma 8h53'19" 68d28'19"
Madlyne Bolton
Vir 11h50'47" -0d39'17"
Madolyn Francis Sweaney
Uma 10h46'55" 59d10'4"
Madolyn's Dream
Lib 15h40'59" -17d15'39"
Madona Jade Casini
Vir 12h35'46" -0d28'58"
Madonna and Tom Forever
And 23h44'58" 33d46'30"
Madonna Ann Casella
Cyg 20h21'35" 59d51'24"
Madonna C. Jeffries
Umi 13h39'7" 72d26'34"
Madonna Dwenger
Crb 16h12'5" 37d17'13"
Madori Woo
Lib 15h16'37" -21d20'44"
Madre Alla Figlia
Umi 15h6'46" 79d56'59"
Madre Catherine
Uma 9h5'20" 64d51'28"
Madre di Curante
Cyg 20h22'53" 31d57'46"
madre 'n padre Ti Amo
Psc 1h21'25" 32d49'12"
Madrichat Justine
Lyn 7h59'50" 40d10'13"
Madrid21/05/2006
Uma 14h2'54" 53d45'25"
Madsen
Sco 16h13'14" -12d20'53"
Madsion M Carter
Sgr 17h49'23" -28d9'0"
Mady
Aqr 22h32'3" -22d42'43"
Madylin Lee Harris
Peg 21h58'39" 12d32'36"
Madysen Rose Isenhour
And 0h49'17" 40d7'26"
Madyson Kate Weis
Sco 17h50'10" -44d10'10"
Madzia
Leo 10h37'44" 7d57'17"
Madzos
Psc 1h13'51" 17d55'22"
Mae
And 0h47'0" 40d48'13"
Mae
Cru 12h49'47" -64d23'42"
Mae.....
Sgr 18h43'39" -22d9'55"
Mae
Aqr 22h22'50" -7d15'5"
Mae
Vir 12h53'58" 8d12'24"
Mae Alice Burke
Uma 11h1'36" 43d22'25"
Mae & Alvin Arterbury 12-
26-1951
Lyr 19h10'5" 27d21'31"
Mae Belle Boyles
Per 3h13'49" 48d39'22"
Mae DelPrincipe
Cam 7h59'22" 70d14'57"
Mae (Dominica) Spada
Lyn 8h45'2" 38d48'44"
Mae Esther beaster
Sco 16h40'40" -33d25'44"
Mae Estoy
Cap 21h0'42" -20d50'21"
Mae Hoeper Schaumburger
And 0h45'3" 34d32'25"
Mae Lee Thompson
Lyn 8h40'10" 33d47'5"
Mae- Ling Pacin
Cma 7h8'1" -30d40'42"
Mae Louise Haines
Umi 16h25'5" 70d57'52"
Mae Mae
Lmi 10h20'31" 36d51'38"
Mae Mae Lomeli
Leo 10h10'41" 26d54'44"
Maé Petit
Umi 14h52'42" 76d15'38"
Mae Shurter 100 Years
Cnc 8h42'45" 23d23'28"
Mae & Steve DiPietro:
Love&Remember
Sco 16h2'38" -17d43'27"
Maea Hamby
And 1h54'5" 42d55'1"
Maebelle Helena Johnson
(Simonson)
Uma 9h58'7" 54d43'11"
Máebh Ann Grady -
Máebh's Star
And 1h24'30" 49d20'55"
Maedave
Ori 5h19'43" 15d25'29"
Maeg
Ori 5h33'5" -0d54'4"
Maegan
Uma 9h32'28" 52d56'57"
Maegan
And 23h50'35" 41d57'12"
Maegan
And 0h31'37" 41d24'18"
Maegan Ann Jordan
Gem 6h38'53" 12d33'9"
Maegan Cecelia Harpool
And 2h31'26" 37d22'22"
Maegan Elizabeth Smith
Lib 15h1'51" -8d5'40"
Maegan Kananimauloa
Brandt
Gem 7h14'48" 24d14'5"
Maegan (Noah's Ark)
Aqr 22h27'6" -1d28'22"
Maegan Rose
And 0h18'23" 26d22'34"
Maegan Shull
Lyn 7h25'11" 54d5'36"
Maegan Warner
Cap 20h32'38" -16d45'56"
Maegann and Cameron
Sco 16h4'3" -16d56'31"
Maegan's Wish
Uma 10h10'36" 48d25'30"
Maeghan Grace Eldridge
Cru 12h40'49" -64d16'28"
Maegn Patin
And 1h44'19" 50d29'34"
Mae-Joe Honigman
Mon 6h40'44" 7d21'56"
Maël ton étoile du bonheur
Leo 10h56'4" 7d36'23"
Maëlie
Umi 14h21'30" 79d7'22"
Maelin
And 0h23'46" 38d9'25"
Maelin Marie Koester
And 0h8'17" 47d18'21"
Maelle Enzo Noah
Maâyane
Uma 8h44'24" 71d28'25"
Maëlys Quillevere
Leo 9h29'47" 23d15'3"
Mae-Mae
Aqr 20h53'56" 2d13'45"
Maemarie Kristovich
Vir 12h41'11" 1d6'29"
Maerie
Uma 13h42'13" 48d30'39"
Mae's Star
Leo 11h44'6" 22d3'16"
Maestar
Ori 5h53'33" -0d44'32"
Maestra Karen Dietz (NM)
Per 4h25'57" 39d28'12"
maestracissyv
Cyg 20h6'33" 34d25'25"
Maestro
Pho 0h56'3" -40d35'6"
Maestro and Mrs. Kurt
Masur
Cnc 8h53'51" 27d18'5"
Maestro Gogi Chichinadze
Aqr 23h5'17" -9d21'40"
Maestro Henry J Piermattei
Psc 0h1'32" 9d40'28"
Maestro Herbert S.
Gardner
Uma 9h3'0" 66d42'11"
Maestro Thomas E.
Newby, Jr.
Lyr 18h48'4" 44d32'53"
MAETINEE
Psc 2h1'32" 6d42'22"
Maeva Sauvé
Lib 15h37'56" -19d49'41"
Maeva Steible
Aqr 22h45'27" -22d44'30"
Maeve
Psc 0h47'8" 2d58'43"
Maeve
Aqr 21h46'24" -6d23'50"
Maeve
And 0h29'3" 41d38'19"
Maeve Amelia Shramko
Aqr 22h59'12" -9d5'26"
Maeve Dannon
And 23h35'25" 50d8'3"
Maeve Donovan Bambery
Psc 1h30'23" 27d57'9"
Maeve Eliza Parsons
Aqr 22h52'59" -15d40'54"
Maeve Elizabeth
Cas 0h19'33" 61d3'59"
Maeve Elizabeth Bennett
Gem 7h16'24" 30d39'15"
Maeve Elizabeth Ruane
Umi 14h56'20" 74d25'59"
Maeve Kelly McBride
Gem 7h19'16" 32d55'2"
Maeve Margaret Sheehan
Cap 21h40'4" -12d37'4"
Maeve Schallert
Umi 15h40'56" 84d49'9"
Maeve Skye Ramsey
Lyn 7h52'6" 38d35'58"
Maeve Teresa Dunne Tilley
Aqr 21h6'42" -8d37'6"
Maeve's Faith
Lib 14h50'51" -2d22'26"
Maevyn Aurora Becker
Sgr 19h35'46" -16d13'18"
MAFALDA
Psc 1h45'29" 13d24'37"
Mafalda Lopez Migliozzi
Dra 19h18'36" 59d37'25"
Mafalto2
Cas 0h12'48" 50d54'53"
Maffer Arango
Cyg 20h59'34" 31d27'21"
Maffre Melyssa
Cas 1h27'47" 59d14'25"
Mag
Umi 15h59'23" 74d3'8"
Maga
Lyr 19h2'14" 33d15'47"
Maga Joans
Cnc 8h49'46" 9d27'49"
Magali
Vul 19h50'20" 21d35'59"
Magali
Per 3h17'33" 47d43'3"

Magali Batel
Umi 13h45'36" 76d49'41"
Magali Dardenne
Ori 6h15'21" 11d5'2"
Magali et Etienne
Lib 14h30'49" -11d46'39"
Magali Grosset
Psc 0h47'9" 15d15'45"
Magali & Nicolas
And 23h20'56" 38d4'9"
Magali petit caillou
Uma 11h51'24" 33d28'15"
Magali40
Ori 5h4'32" 10d24'44"
Magalie
Umi 13h19'18" 74d11'56"
Magalie et Christophe
Crb 16h8'46" 30d39'38"
Magalis Vargas
Cas 23h45'25" 54d39'57"
Magaly
Lib 15h47'27" -16d46'8"
Magaly
Sgr 18h35'42" -34d47'28"
Magaly Mitstifer
Cam 3h55'50" 67d34'6"
Magan Elizabeth
Vul 20h27'9" 23d11'32"
maganda
Aqr 21h55'34" 1d16'47"
Maganda Marikit
Dra 18h41'30" 63d4'13"
Magay Mozsár Eszter
Uma 11h46'11" 62d45'24"
Magcol Dancer
Eri 4h14'42" -24d25'32"
Magda
Lyr 18h21'34" 41d28'31"
Magda 1931
Cnc 9h2'35" 22d37'16"
Magda Dimakopoulou
Uma 11h46'17" 59d56'10"
Magda Fichter
Cyg 20h52'47" 48d54'20"
Magda Harmala
Psc 1h1'9" 19d42'8"
Magda Karina Joy
Sgr 18h37'22" -23d16'2"
Magda Kavallieratos
Sgr 19h14'19" -28d23'55"
Magda Levatich Martin
And 0h45'18" 46d9'3"
Magda Maria de Melo
Sgr 18h5'44" -24d52'32"
Magda Mendel
Aqr 20h49'51" -9d25'55"
Magda Morales
Uma 10h42'33" 43d3'42"
Magda my one true love, forever
Leo 11h37'27" 25d28'27"
Magda Pasecka
Vir 12h29'44" -5d26'12"
Magda Pitera
Lyr 19h9'26" 42d51'41"
Magda Sylvia
Cas 23h45'58" 52d49'55"
Magda Von Gruen
And 1h44'28" 36d12'54"
Magdalana Szmydt
Lyn 8h54'24" 34d15'55"
Magdalena
Uma 11h55'50" 43d24'49"
Magdalena
Uma 9h51'3" 50d23'39"
Magdalena
And 23h6'4" 48d2'18"
Magdalena
Gem 7h34'5" 26d18'13"
Magdalena
Lib 15h4'56" -6d7'38"
Magdalena
Cap 20h31'3" -10d5'55"
Magdalena Ann Alexander
And 1h12'36" 41d34'35"
Magdalena Anna Podstawek
Lyn 8h37'1" 38d40'47"
Magdalena Berniker
Sco 17h22'45" -36d51'40"
Magdalena Broutin
Uma 10h46'48" 40d12'33"
Magdalena Fernandez
Cnc 8h47'2" 31d0'32"
Magdalena Gaujour Grzyb
Cas 23h9'54" 55d36'6"
Magdalena Josefa Tirado
Uma 13h16'9" 57d40'4"
Magdalena Lopez
Uma 12h29'49" 59d49'50"
Magdalena Luminita
Psc 1h6'27" 4d7'59"
Magdalena Montemayor, DTMMBR
Uma 13h24'59" 54d56'3"
Magdalena "Muce"
Tau 5h58'23" 23d6'36"
Magdalena Nowak
And 2h37'2" 40d38'40"
Magdalena O'Dell
And 0h40'11" 44d5'7"
Magdalena S Schuler
And 23h3'53" 47d5'9"
Magdalena Sitkowska
Ari 2h58'39" 24d46'37"

Magdalena Star
Aqr 21h13'45" -3d33'59"
Magdalena Szeliga
Cas 0h7'22" 56d56'13"
Magdalena Taya
Psc 1h5'11" 31d25'27"
Magdalena-1992
Col 5h42'10" -37d43'8"
Magdalena's Light
And 23h20'45" 51d20'10"
Magdalena's Lucky Star, Love Peter x
Dra 19h40'57" 73d35'0"
Magdalene Chatigny
Psc 1h5'7" 4d57'33"
Magdalene Leah Carlsson
And 1h46'18" 44d37'59"
Magdalene O'Dell
Oph 16h41'6" -0d40'55"
Magdalene Therese McCallum 26.02.86
Cru 12h44'35" -64d16'21"
Magdalene & Wsewolod Popov
Cyg 20h43'46" 36d46'49"
Magdaline Johnson
And 0h30'15" 44d25'28"
Magdealena Clark
Cas 1h50'0" 60d52'3"
Magdelena
And 2h12'39" 49d55'11"
Magdelena Mercedes Martinez
Aqr 22h6'23" -13d22'26"
Magdeline DeAngelis
Crb 15h43'43" 26d46'10"
Magdklena Valero Lamprecht
Lyn 6h46'17" 60d25'48"
Maged
Vir 13h6'19" -14d20'41"
Magee Funk McDonald
Cyg 21h11'3" 52d0'44"
Mageelam30
Crb 16h24'25" 32d41'35"
Magen Brittany Essey
Tau 4h43'26" 2d17'48"
Magen Michelle Routier
Gem 6h39'35" 26d50'16"
Magen Nicole Cabezas
Sco 16h9'34" -15d55'1"
Magen Riley
Cas 23h21'27" 58d8'56"
Magenta
Del 20h38'56" 12d58'38"
Magenta
Cnc 8h36'34" 32d8'46"
Maggan
Mon 6h46'26" 10d36'50"
Maggie
Vir 13h12'46" 13d7'58"
Maggie
Mon 6h56'11" 2d28'46"
Maggie
Crb 16h2'43" 26d9'8"
Maggie
Ari 2h30'30" 25d40'49"
Maggie
And 0h33'11" 27d27'34"
Maggie
Lyr 18h33'56" 31d24'44"
Maggie
Cyg 19h37'40" 30d2'11"
Maggie
Lyn 9h21'55" 41d18'18"
Maggie
And 23h36'41" 48d0'8"
Maggie
Cas 1h18'17" 55d55'20"
Maggie
And 2h21'9" 47d7'33"
Maggie
And 2h28'42" 47d58'3"
Maggie
Cas 1h5'7" 63d41'19"
Maggie
Uma 9h44'8" 55d11'38"
Maggie
Uma 13h16'28" 61d31'21"
Maggie
Uma 10h20'21" 68d32'43"
Maggie
Uma 8h39'57" 68d8'19"
Maggie
Sgr 19h43'10" -16d40'54"
Maggie
Cma 7h24'56" -18d29'56"
Maggie
Cap 20h26'49" -10d38'29"
Maggie
Aqr 20h51'26" -6d22'23"
Maggie
Sco 16h50'33" -30d33'35"
Maggie
Sgr 18h25'23" -28d10'44"
Maggie
Ori 5h44'50" 1d52'17"
MAGGIE 1
Sgr 19h28'48" -25d31'53"
Maggie 2004
Umi 16h15'20" 79d47'13"
Maggie and Ben forever
Cyg 20h34'53" 45d59'21"
Maggie and Jon
Ara 17h19'19" -50d45'50"

Maggie and Papa
Uma 8h21'15" 64d12'28"
Maggie and Tate
Her 16h48'15" 44d44'30"
Maggie B
Gem 6h54'39" 26d27'53"
Maggie Baby
And 22h58'56" 44d20'52"
Maggie Baker
Cas 1h50'58" 61d23'3"
Maggie Bartlett
Ori 5h55'33" 22d33'13"
Maggie Bennett
Cnv 12h49'33" 39d35'40"
Maggie Bochnak
Sgr 19h29'50" -32d16'4"
Maggie Brink
Cas 1h30'2" 59d42'20"
Maggie Bullock
Mon 6h59'41" -7d3'57"
Maggie Capalbo
Dra 18h39'57" 50d52'47"
Maggie Claire
And 0h39'22" 42d8'17"
Maggie Cocca
And 23h14'18" 48d20'22"
Maggie Colgan's Shining Star
Cam 3h56'24" 57d28'53"
Maggie Danielsen
Aqr 22h9'6" -8d5'13"
Maggie Dimas
Cas 0h44'8" 49d4'29"
Maggie Dooley
And 23h16'1" 48d29'58"
Maggie Doonan
Cas 1h29'44" 66d5'58"
Maggie Elizabeth Murphy
Aqr 23h10'33" -11d5'49"
Maggie Elizabeth Roiger
And 2h34'19" 44d16'13"
Maggie Elizondo
Sgr 18h34'12" -16d46'27"
Maggie Erin Allen
Aqr 23h38'40" -22d34'12"
Maggie Ferris
Leo 11h42'2" 25d5'43"
Maggie Francis Lorentzen
Lib 15h28'28" -17d14'2"
Maggie Fraser
Ori 5h35'3" -2d19'53"
Maggie G.
Cnc 8h1'9" 24d30'19"
Maggie Germain
Cas 2h31'0" 65d51'8"
Maggie Grey Watson
Gem 6h50'19" 14d13'53"
Maggie Haill
Uma 11h18'8" 32d31'30"
MAGGIE HALTER
Lyn 7h55'11" 42d51'50"
Maggie Hamilton
Cap 21h32'58" -16d4'1"
Maggie J. Smith
Aqr 22h2'4" -16d36'49"
Maggie Janeen Massaro
And 2h17'15" 47d27'24"
Maggie Johanna Croal - Precious One
Cra 17h59'48" -37d17'42"
Maggie Kondrath
And 1h57'9" 44d3'8"
Maggie L Klumpp Happy 94th Birthday
Cas 1h59'46" 63d9'14"
Maggie Leah Blankenship
Cas 0h41'18" 61d42'13"
Maggie Loomis
Cam 5h59'58" 56d41'47"
Maggie Lou Nelson
Cyg 21h8'28" 52d0'11"
Maggie Louise Frye
And 1h11'40" 36d35'58"
Maggie Lucille Vyskocil
Psc 1h19'59" 7d0'3"
Maggie Lynn
Gem 7h8'34" 27d20'11"
Maggie Lynn Eike
Leo 9h51'3" 23d54'11"
Maggie Lynne
And 2h28'54" 38d54'9"
Maggie Mae
Lib 15h37'12" -10d37'52"
Maggie Mae Barone
And 23h15'31" 49d16'23"
Maggie Mae Rizzi
Leo 9h48'36" 30d37'49"
Maggie Maloney
Uma 10h16'18" 63d44'8"
Maggie Margaret Wilkerson
Aqr 20h46'38" -8d9'37"
Maggie&Matthew The Little Pinguino
Uma 13h54'0" 52d28'17"
Maggie May
Cnc 8h54'49" 24d18'15"
Maggie May (Ann Addino)
Tau 4h5'52" 6d39'21"
Maggie May Bachak
Leo 9h31'48" 29d46'11"
Maggie May Monday
Leo 10h23'19" 12d36'6"
Maggie May Mook
Sco 17h46'52" -31d4'26"

Maggie May Sadlon
Cma 6h47'37" -16d5'53"
Maggie May Volkman
And 23h31'23" 41d17'9"
Maggie McCain
Cnc 9h8'42" 15d11'0"
Maggie McClure
And 0h48'58" 43d26'11"
Maggie McGeorge 1
Tau 4h12'12" 15d14'52"
Maggie McGeorge 2
Cyg 19h39'45" 37d52'51"
Maggie McGeorge #3
Tau 5h42'1" 24d27'28"
Maggie Mealy
Cas 23h59'17" 57d3'24"
Maggie Merrick Worthington
And 1h44'11" 45d22'53"
Maggie Miller
Lyn 6h46'0" 51d44'37"
Maggie Moran
Cnc 8h44'26" 10d57'20"
Maggie My unique & rare shiningstar
Aqr 23h41'9" -14d33'22"
Maggie N' Lil
Uma 13h0'23" 58d13'50"
Maggie Nicole Deuser
Sgr 18h57'29" -28d35'48"
Maggie Ostroff
And 2h35'10" 45d41'50"
Maggie Peck
Aqr 21h39'6" 0d35'0"
Maggie Rae Desmond
And 1h29'55" 49d24'36"
Maggie Rianne Tuttle
Leo 11h19'7" 13d18'41"
Maggie Rivera
Cam 5h41'32" 56d23'2"
Maggie Rogalski
Vir 12h47'21" -10d23'47"
Maggie Ro's Star
Sco 17h36'28" -33d4'25"
Maggie Rose
Mon 6h50'2" -0d1'24"
Maggie Rose House
And 2h19'59" 46d55'38"
Maggie Rose; My Sunshine
And 2h26'24" 49d19'32"
Maggie S
Ari 3h12'57" 28d17'42"
Maggie Sansone
Cnc 8h8'27" 17d6'4"
Maggie Scott
Cas 1h6'59" 52d58'29"
Maggie Scott
Sgr 19h48'44" -15d58'22"
Maggie Shairs/ Wood
Leo 9h41'33" 31d45'54"
Maggie Sheehan
Col 5h58'36" -32d43'22"
Maggie Tanner
Lyn 7h28'22" 45d15'53"
Maggie Thursam
And 0h3'1" 44d56'10"
Maggie Weinert
Tau 4h8'16" 14d33'6"
Maggie Wick
Uma 12h5'2" 41d59'9"
Maggie Williams
Uma 11h8'43" 63d33'32"
Maggie Woodman
Gem 7h1'23" 16d55'27"
Maggie Wrba
Sco 17h54'4" -35d18'36"
Maggie Wright
Sco 16h48'42" -38d1'59"
Maggie Wright
Sco 17h53'2" -37d47'49"
Maggie Xifo
Sgr 18h55'34" -34d22'47"
Maggie83
Tau 4h6'12" 25d18'59"
Maggie-Home Is Where Your Heart Is
Aqr 22h8'52" -23d3'3"
MaggieLuv88
Tau 5h31'22" 25d48'47"
Maggie's Majesty
Leo 10h7'58" 17d55'2"
Maggie's Shining Star
Aqr 20h55'12" -11d32'5"
Maggie's Star
And 2h34'44" 38d13'27"
Maggie's Wish Upon a Star
And 0h45'24" 38d47'17"
Maggy Carrera
Cap 21h52'43" -10d30'45"
Maggy Lee Swyers-Oppelt
And 0h39'4" 29d18'24"
Maggy Rose
And 1h35'23" 45d37'53"
Maggypie
Vul 19h31'28" 24d36'32"
Maghan Kathryn Rea
Uma 9h11'11" 56d45'40"
Maghan Pufahl
Cap 20h33'42" -23d22'9"
Maghen LeAnn Shipley
Vir 12h21'0" -1d38'36"
Mägi & Iwan
Dra 10h0'7" 77d45'44"
Mägi & Sepp
Vul 19h54'3" 22d40'31"

Magic
Ari 2h13'23" 24d52'14"
Magic
Leo 11h3'3" 17d3'44"
Magic
Lyr 19h11'2" 39d21'59"
Magic
Cyg 21h14'37" 31d56'14"
MAGIC
Cma 6h45'24" -20d59'33"
Magic
Uma 9h15'54" 55d5'21"
Magic at Eight
Eri 4h14'12" -11d3'45"
Magic Browse
Uma 9h52'14" 42d18'1"
Magic Crystal
Cyg 20h52'35" 31d14'9"
Magic Evalina 1965
Sgr 18h16'16" -26d29'43"
~magic happens: *believe*
Dra 18h53'12" 62d8'43"
Magic Love
Cyg 21h39'15" 39d14'20"
Magic Man Vazquez
Tau 4h19'35" 4d13'55"
Magic Mark
Psc 1h11'4" 16d48'44"
Magic Marty
Gem 6h47'40" 32d37'48"
Magic Michel
Uma 11h13'3" 29d20'1"
Magic Moley
Gem 7h43'6" 33d19'32"
Magic Smith Mattioli
Cma 6h35'44" -16d57'21"
Magical
Cyg 21h50'27" 48d52'27"
Magical Mackenzie
Crb 15h43'16" 28d28'4"
Magical Maggie
And 1h50'36" 46d37'26"
Magical Mia
Crb 15h32'41" 28d27'36"
Magical Mikah
Sgr 19h2'37" -18d40'22"
Magical Wong Yi Kwan
And 0h9'10" 39d38'53"
Magic's Star
Cap 20h23'3" -10d38'0"
magicvenetianstarlove09-23-03-TJ
Uma 9h30'31" 45d23'17"
Magida
Psc 0h55'19" 26d31'32"
Magie
Ari 2h42'9" 24d41'25"
Magik coin coin
Uma 11h54'57" 33d20'26"
Magin & Karine
Lac 22h20'21" 47d45'11"
Magina
Cnc 8h15'59" 7d58'37"
Magister Francesco Perrulli
Per 2h43'2" 54d15'44"
Magistra Elke Führer
Uma 8h55'45" 68d18'9"
Magistra Terrestrialis
Leo 9h42'9" 7d29'3"
Maglyn "12-1-01"
Uma 12h44'48" 58d2'16"
Magnestar
Umi 15h1'58" 76d22'56"
Magnificant Morgan
Her 16h50'36" 34d53'22"
Magnificent Ashley Shaw
And 0h50'11" 41d55'50"
Magnificent Manda
And 23h20'38" 48d3'38"
Magnificent Mason 2005
Uma 8h52'0" 69d50'20"
Magnificent Millicent
Lyn 7h4'25" 51d14'34"
MagnificentMaddie12/11
And 0h28'22" 32d37'58"
Magnificus Flammeus
Uma 8h54'38" 57d48'46"
Magnificus Triplus
Cnc 8h51'59" 11d46'53"
Magnolia Gordon
And 22h57'50" 52d25'25"
Magnolia Noriega
Vir 13h20'49" -10d2'9"
Magnolia's Angel
Cyg 20h7'54" 38d13'28"
Magnum
Crb 15h19'44" 27d24'6"
Magnum Ten Bear
Cnv 13h48'21" 39d27'38"
MAGNUS AMDAM
Her 16h16'21" 43d4'0"
Magnus Candidus
Uma 11h30'46" 41d30'49"
Magnus Generalis Riemer
Uma 11h22'7" 56d55'10"
Magnus George Moore Kelly
Her 16h35'51" 7d14'59"
Magnus Josef Matter
Ori 6h19'14" 6d6'8"
Magnus Paternus
Sgr 17h45'41" -24d20'57"
Magnus Reuben Jacobsen
Cru 11h57'22" -59d38'12"

Magnus van Duijse
Uma 11h57'17" 65d0'24"
magnuson-connolly pan-asian hoshi
Aqr 22h55'30" -22d55'30"
Magoooo
Sco 17h14'53" -40d48'37"
Magorganite
Her 16h36'40" 47d3'8"
Magrathea
Cyg 21h48'30" 38d45'56"
MagrathM
Gem 6h37'5" 27d39'37"
Magrette Cogdell Young
Oph 17h33'43" -0d39'19"
Magruder J. Fick, Sr.
Her 16h44'13" 47d18'8"
Maguerite "Mom-Mom with the Beads!"
And 23h16'39" 47d11'41"
Maguire 2000
Lac 22h49'22" 41d42'23"
Maguire - 8
Uma 9h0'45" 59d32'21"
Maguire Carl Sullivan
Per 3h13'57" 45d29'19"
Maguire Davis Gilner
Dra 18h42'16" 70d43'41"
Maguire Zachary Gentz
Per 3h26'25" 45d34'21"
Magurf
Sgr 18h46'48" -28d56'35"
Maguzula
Sgr 18h56'31" -27d38'2"
M.A.H. & R.J. Martiensen
Cen 11h29'4" -48d33'31"
Maha
Uma 11h11'26" 45d25'39"
Maha
Cnc 8h27'49" 25d13'48"
Maha Al-Baadi
Sco 16h17'0" -9d30'16"
Mahaffey
Gem 6h51'34" 14d39'16"
Mahal
Uma 9h19'10" 50d2'3"
Mahal
Cyg 19h52'8" 38d28'39"
Mahal
Umi 15h35'59" 85d41'47"
Mahal Kita
Lyn 8h23'50" 51d6'51"
Mahal Kita Boobey
Cma 7h6'5" -24d27'1"
Mahal Kita Kailanman
Ari 2h47'38" 27d53'44"
Mahal Lucinda Ma Mathai
Sco 16h13'18" -14d16'54"
Mahal Two
Eri 3h50'36" -0d31'20"
Mahala
Com 11h59'33" 28d6'53"
Mahala's Mom
Peg 22h25'21" 4d58'44"
Mahaley Thomas
Leo 9h48'1" 26d34'40"
Mahalia Eleanor Flanagan
Cap 21h40'49" -20d5'4"
Mahalia's Light
Cas 0h45'25" 64d32'48"
MAHAR
Uma 8h29'54" 67d19'59"
Maharani Nan
Cas 23h58'39" 53d47'48"
Maharlika Amper Hutts
Psc 1h40'12" 23d45'27"
Mahatma Indira Velázquez Solorzáno
Cam 6h57'6" 65d10'30"
Mahaut
Boo 14h36'35" 22d25'56"
Mahayeto
Ind 21h29'12" -53d17'35"
Mahdi
Uma 10h52'44" 50d12'54"
Maheen
Cap 21h47'33" -11d2'11"
Mahesh & Madhu Chander
Uma 9h40'41" 51d30'24"
Mahfoud Zbir
Uma 10h28'47" 63d39'27"
Mahfouz Juliette Doss
Crb 16h3'43" 29d21'44"
Mahina Dawn Crowley
Tau 4h30'50" 19d17'4"
Mahinahoku
Tau 5h48'0" 19d40'27"
Mahinda & Indrani
Eri 3h38'9" -5d52'23"
Mahira Hussain
Sgr 19h8'13" -22d49'46"
Mahita
Cnc 9h5'30" 31d2'53"
Mahler Family
Uma 8h20'49" 71d31'38"
Mahmood Fassih-Khoshgard
Vir 11h54'1" 5d5'55"
Mahmoud Hasanat
Aur 6h23'25" 41d45'21"
Mahmoud Ibrahim Mansour
Aql 19h15'31" 11d47'24"
Mahmuda Sultana (Tina)
Cyg 21h41'32" 52d26'18"

Mahogany... Always All Ways
Uma 11h53'47" 29d22'38"
Mahon
Cyg 19h40'36" 29d8'7"
Mahoneys Star
Aqr 23h3'1" -9d32'55"
Mahrukh Ahmed
And 1h6'1" 34d30'25"
Mahryah Lynn Lovett
And 0h51'12" 38d48'6"
Mahshid
Sgr 18h16'11" -18d24'15"
Mahshid Rajaee
Sgr 18h23'1" -21d11'25"
Mahvelicious
Per 3h22'2" 48d20'44"
Mai
Ori 5h22'48" 3d26'9"
Mai
Sgr 20h4'17" -40d8'57"
Mai Ashraf Salama
Vir 14h59'10" -0d13'45"
Mai Chang "GORGEOUS" Vang
Sco 17h24'12" -43d13'1"
Mai da Solo
Uma 11h19'52" 60d23'22"
Mai Iizuka
Gem 7h48'3" 30d2'52"
Mai Ko Emu
Aql 20h6'10" -0d17'37"
Mai Koishii Shannon
Cnc 8h33'51" 24d2'32"
Mai Le
Cas 0h32'44" 50d15'58"
Mai Lee Doan
Aqr 22h11'42" 0d21'26"
Mai Nguyen Abelson
Ari 3h3'23" 27d54'59"
Mai Non Concludere Luce Luminosa
Aur 5h29'7" 40d26'9"
Mai Pham
Uma 8h11'4" 67d18'14"
Mai poina
Cyg 21h21'30" 37d50'9"
Mai Star
Uma 9h37'51" 68d56'17"
Mai Tai
Uma 11h25'52" 45d11'39"
Mai Tong Vang
Cnc 7h58'1" 12d6'42"
Mai Yang Xiong
And 0h7'41" 44d34'38"
Mai Yia Yang
Crb 15h34'43" 28d42'24"
Maia
Psc 1h16'46" 12d12'46"
Maia
Uma 11h46'41" 44d28'48"
Maia
Vir 13h36'10" -13d33'25"
Maia
Cma 6h36'19" -12d50'20"
Maia
Aqr 22h44'3" -21d43'43"
Maia 1
And 0h30'28" 26d8'53"
Maia Alise Anderson Cabusao
Uma 8h55'37" 72d42'59"
Maia Browning
Uma 13h39'42" 54d41'44"
Maia Hvatum Sweeney
And 1h16'35" 50d20'51"
Maia Isabella Swasey-Stephenson
Cnc 8h4'57" 20d42'37"
Maia Jordan Wong
Ari 2h59'52" 22d16'49"
Maia Krist Johnson
Sgr 19h36'35" -24d4'42"
Maia Leigh Curtis
And 23h48'2" 42d13'9"
Maia Mahala
And 1h9'15" 41d30'17"
Maia Pauline
And 0h20'34" 24d0'52"
Maia Rain Cariadus
Gem 6h44'54" 30d16'57"
Maia Saige Hartman
Tau 4h28'9" 1d31'22"
Maia the Infinite
Uma 12h45'14" 56d54'23"
Maiah Lucas
Gem 7h18'37" 29d6'31"
Maica
Cap 20h12'25" -22d4'45"
Maicee Marie McKee
Lib 14h42'32" -11d15'41"
Maid Marianne
Gem 7h18'50" 19d37'6"
Maiden of Beauty, Snare of My Heart
Cas 1h39'59" 63d7'1"
Maiden Star Of Jen
Vir 13h48'20" 4d54'39"
Maier Runyon - Tupperware
Boo 14h43'1" 16d54'36"
Maier, Bernd
Lib 14h38'49" -10d54'35"
Maier, Johannes
Cap 21h29'1" -14d22'15"

Maifredi Angelo
Lyr 18h30'45" 32d33'10"
Maigheo
Sco 16h50'3" -32d33'18"
Mai-Gunn 1975
And 23h4'32" 47d27'47"
Maihlee Tiffany Vang
Tau 4h29'56" 22d16'46"
Maija D. Lazenby
Crb 16h10'45" 30d35'11"
Maija Star
Gem 6h40'35" 19d0'35"
Maik
And 0h18'54" 36d56'49"
Maik
Ori 5h40'30" -10d14'11"
Maik Karbautzki
Ori 5h53'51" 11d41'47"
Maik Martin
Aql 20h16'27" -3d13'12"
Maik, I love you forever!
Nicole
Uma 9h15'55" 64d35'18"
Maika'i Po Elikapeka
Cas 23h1'56" 57d41'58"
Maikalima O Makua
Vir 12h44'39" 8d14'30"
Maike
Vir 13h9'30" 8d16'5"
Maike
Uma 13h51'23" 48d37'13"
Maike July-Grolman
Uma 11h9'37" 42d14'36"
Maiko & Katsuaki
Ari 2h36'52" 18d32'23"
Maiko St.Hilaire
And 1h42'3" 48d36'28"
Maila Rose
And 1h3'30" 45d0'13"
Mailani
Aqr 22h52'30" -8d5'50"
Maile
Aql 19h20'42" -8d48'35"
Maile
Uma 11h9'58" 54d40'23"
Maile
Tau 4h39'39" 15d56'45"
Maile Kapuniai
Crb 15h36'41" 26d6'2"
Maile Lu Heikka
Sgr 18h44'38" -32d26'42"
Maile Michelle Hoover
Ari 2h15'7" 26d40'3"
Maile William Fatafehi Latu
Uma 11h48'2" 46d22'59"
Maile1
Uma 11h9'10" 46d15'52"
Mailee Saraphun
Cap 20h22'17" -9d52'48"
Maillor
Crb 16h23'28" 35d29'20"
Maily Acosta
Cas 0h13'15" 59d56'55"
Maily Do
Aql 19h30'36" 9d41'29"
Maily Weirbach
Uma 11h57'45" 39d9'58"
Maïlys
Cas 0h57'20" 55d52'7"
Main Four 06-07 Loves
Mrs. Haskins
Uma 11h59'20" 45d25'15"
Maïna
Dra 15h0'55" 60d33'4"
Maine Darragh Lewis
Sco 16h34'0" -26d2'1"
Maine Munchkin Head
And 0h8'46" 45d0'55"
Maintenant et Toujours
Uma 9h51'42" 70d3'51"
Mainuee
Sco 17h34'31" -45d28'43"
Maiou, Maria
Uma 11h48'6" 34d52'15"
Maira Cabelto Adolfo
Tau 3h49'24" 29d7'45"
Maira Molina
Dra 18h35'5" 56d37'53"
Maira Sanguino
Cam 5h22'38" 56d46'34"
Mairah
Tau 5h49'52" 27d21'55"
Mairaman
Sgr 18h2'23" -18d12'12"
Maire Bresnan
Ari 2h36'27" 27d53'23"
Maire DeFago
Cnc 8h51'9" 13d37'15"
Maire Rose Chubb
Leo 9h28'33" 16d31'58"
Mairead Garbhie
Per 2h42'12" 41d5'34"
Mairead Grogan
And 23h8'58" 39d30'53"
Mairhead Alison Bruce
Cas 23h25'8" 55d3'41"
Mairi Orsolya Hazel
Sherriffs
Cas 0h54'56" 57d16'22"
Mairion and Ronnie Walsh
Cyg 19h40'7" 38d44'45"
Mai's Heart
Aqr 21h59'24" -16d22'10"
Maisa
And 23h25'8" 48d25'3"
Maisey Jane Greenhow
And 23h56'13" 46d31'57"
Maisi (Ma)
Lyn 7h41'4" 45d50'0"
Maisie
And 23h13'59" 43d6'54"
Maisie
And 0h13'35" 45d50'45"
Maisie Allen
And 0h18'59" 39d29'30"
Maisie Ann: The Milky Way
Superstar
Cyg 20h11'50" 54d52'21"
Maisie Anne Wilmer
Cru 12h2'46" -57d2'19"
Maisie Catherine Lambert
Cmi 7h34'16" 4d10'20"
Maisie Dobson
And 23h21'21" 51d32'32"
Maisie Dolly
And 23h17'56" 51d31'17"
Maisie Eliza Emily Davis
And 23h18'42" 48d26'48"
Maisie Elizabeth
And 0h22'18" 25d38'2"
Maisie Ella Dresser
And 23h59'13" 40d46'44"
Maisie Florence Dyson
And 23h40'42" 50d5'24"
Maisie George Fuller
And 0h2'10" 32d51'26"
Maisie Gregg
And 2h33'5" 42d51'33"
Maisie Jane Edge
And 23h44'53" 35d20'29"
Maisie Rose
And 23h46'18" 33d49'39"
Maisie Valerie Simeoni
And 0h59'23" 42d58'49"
Maisie Williams - Maisie's
Star
And 23h22'54" 42d57'9"
Maisies' Limitless Future
Peg 22h39'57" 22d40'44"
Maisie's & Marena's Star of
Light
Sgr 18h26'29" -27d11'20"
Maison James Pike
Umi 16h25'2" 73d22'50"
Maison Zachariah
Wilkinson
Ari 2h30'19" 23d35'51"
Maissane Ramla
Ori 5h39'4" 15d31'4"
Maisy Bo
And 23h10'14" 52d56'40"
Maisy Dean
And 23h58'42" 38d41'2"
Maisy Higginson
And 0h4'46" 37d45'32"
Maisy Lynn Wennerberg
And 23h46'21" 37d51'26"
Maisy Smith
Uma 13h10'56" 53d21'39"
Maital Stern, I Love You
More!
Cyg 21h1'44" 46d10'24"
Maite Duarte
Aqr 22h34'41" -9d35'42"
Maite' Michelle Bee
Sgr 18h16'0" -24d10'15"
Maiteli
Aql 19h34'43" 5d54'26"
Maitha BuHumaid
Aqr 21h52'27" 0d21'59"
Maitiú Michael Lavin
Peg 22h20'58" 27d55'4"
Maitland Abbey Meade
Lyn 9h11'42" 37d46'9"
Maitscher Andrea Masüger
Psc 1h28'57" 29d50'50"
Maitzie - Star of the Stan
Family
Crb 16h14'11" 29d51'6"
Maiya
Col 6h11'31" -35d35'15"
Maiya
Cap 20h24'38" -14d53'12"
Maiya Arlene McNamara
Psc 0h7'29" 4d54'20"
MaiyaBaum
Cas 1h15'55" 64d25'15"
Maize Diane Neice
And 0h29'2" 26d57'19"
Maize Justice Carter
Leo 11h16'36" 24d5'35"
Maizie Lu Hamm
Lyr 18h36'5" 31d39'20"
Maizie Mary Newton
And 0h3'58" 44d57'2"
Maizie Suzanne
And 0h44'5" 40d24'59"
Maizy
Uma 13h56'18" 54d10'44"
Maizy
Uma 9h35'49" 60d26'1"
Maizy Ann Criss
Mon 7h3'27" -8d34'0"
MAJ David Michael
Braddock
Uma 12h46'5" 55d12'57"
Maj. Dwight A. Holland
PhD., MD
Men 5h22'18" -72d42'13"
Maj. Gen. Brian Arnold
Per 3h11'56" 48d26'0"
Maj Harilela
Cyg 19h35'4" 36d4'43"
Maj - Tündér
Cas 23h49'9" 56d16'24"
MaJa
Uma 12h44'32" 60d12'47"
maja
Lib 14h47'53" -1d46'17"
Maja
Per 2h14'19" 55d28'25"
MAJA
Aqr 22h20'39" 1d34'48"
Maja
Tau 4h6'6" 6d33'33"
Maja
Ori 6h11'40" 15d24'0"
Maja D Jagodic Serbia
3.7.1984.
Cyg 21h5'36" 55d17'20"
Maja i Deja
Peg 23h53'2" 24d51'18"
Maja Isabelle
Uma 10h21'31" 43d37'55"
Maja moj MED
Cyg 19h50'21" 32d11'16"
Maja og Frands
Her 16h46'58" 42d45'4"
Maja Rose Farner
And 2h14'42" 45d14'38"
Maja Tanja Thompson
And 2h4'36" 47d24'23"
Maja's Star
Uma 9h56'14" 43d26'20"
Maja-Sofie
Ori 6h7'16" 6d20'45"
Maj-Britt Vogl
Umi 16h7'4" 76d19'3"
Majda
Aur 5h24'1" 36d22'1"
Majdán Ági
Sgr 18h58'50" -34d27'14"
Majedian
Cam 4h52'42" 71d51'28"
Majeed
Cap 20h19'59" -12d38'56"
Majella
Lyn 7h42'31" 41d40'15"
Majerka
Sgr 18h48'48" -19d39'48"
Majestic
Uma 10h40'5" 45d9'12"
Majestic Buddy
Peg 21h27'39" 15d58'58"
Majestic Gran
Ori 5h33'38" 1d49'30"
Majestic Marian
Cas 2h10'36" 59d44'11"
Majestic Maureen
Mon 6h45'30" -0d45'30"
Majestic Princess
Tau 5h25'46" 26d59'33"
Majid
Aqr 23h30'32" -8d51'26"
Majie
Tau 5h41'26" 19d43'56"
Majiel Mary Doris Noonan
Ori 4h49'26" 0d40'47"
Majik Camel
Uma 9h47'28" 46d43'46"
MaJö
Ori 5h18'12" 7d29'32"
Majo
Aqr 23h32'1" -24d47'0"
Majomi
Cyg 19h30'57" 33d5'42"
Major Bode
Her 16h48'47" 33d24'51"
Major Cecil S. Shelton
"Tazs"
Aql 19h29'20" -7d9'19"
Major Charles "Rob"
Soltes, Jr.
Aql 19h4'16" 14d23'18"
Major Daniel S. Keenan
Cnc 8h22'19" 10d16'2"
Major Daniel S. Keenan
Ori 5h56'9" 22d11'9"
Major Daniel S. Keenan
Sgr 20h3'37" -28d28'34"
Major David Ridgway
Aqr 22h25'57" 0d16'55"
Major Dean A Freytag
Per 3h28'46" 32d48'48"
Major Donald Burn Bell
Uma 10h57'25" 43d51'57"
Major Francis Cornelius
Lawler Jr.
Sco 16h12'24" -18d14'12"
Major Golden
Aql 19h46'33" 5d39'7"
Major Jason Luke Amerine
Tau 5h30'4" 17d41'51"
Major Ken Bole #28
Per 2h55'4" 54d53'12"
Major Megan
Uma 10h15'26" 66d35'27"
Major Nick Carter
Ori 5h37'47" 6d46'15"
Major Pa
Cru 12h7'37" -61d41'54"
Major Richard John
Gannon II
Per 2h51'48" 44d51'16"
Major Robert Kavanaugh
Breazeale
Per 4h11'29" 44d34'38"
Major Rózsa
Gem 6h53'9" 15d31'2"
Major Smitty
Per 3h19'9" 41d33'15"
Majorie Victor
And 0h30'11" 42d10'28"
Majtényi Zsolt és Majtényi
Zsoltné
Uma 9h34'22" 43d20'18"
Mak
Her 18h2'41" 26d55'59"
MAK JR
Uma 11h9'56" 49d18'22"
MAK-10
Uma 10h45'4" 45d40'32"
MAKA
Sgr 17h56'42" -18d10'36"
Makaela Grace McComas
Eri 3h48'37" -0d49'28"
Makahla
Cru 12h50'50" -58d33'30"
Makahla E. McCarty
Gem 6h45'59" 26d50'45"
Makai & Abuelita Chita
Sge 20h3'5" 18d30'43"
Makai Millay
Leo 9h32'3" 10d56'3"
Makaila
Leo 9h35'13" 16d10'28"
Makaila Brie Juarez
Wormington
Sco 17h11'45" -38d42'0"
Makaila Nicole Stierlin
Psc 2h1'28" 7d24'59"
Makaio Daniel Groos Van
Tassel
Dra 16h16'18" 53d28'2"
Makai's Star
Vir 13h5'42" 12d55'6"
MaKala-Poppy
Peg 22h47'13" 22d21'47"
Makana Kamaha'o
Leo 11h11'7" -4d4'46"
Makanani
Gem 7h17'14" 24d20'53"
Makay Christensen Morris
Hya 9h16'53" -0d19'37"
Makayla
Uma 13h59'58" 55d49'5"
Makayla
Uma 13h21'27" 54d31'42"
Makayla
Uma 11h31'49" 43d29'54"
Makayla Amanda
Ori 5h57'21" -0d53'41"
Makayla Ann
Cnc 9h7'9" 32d9'55"
Makayla Ariel Perez
And 1h1'14" 45d22'20"
Makayla Ashton
Gem 6h20'47" 18d24'27"
Makayla Audrey
Cap 20h35'21" -18d54'42"
Makayla Brooklin Baca
Sco 16h55'8" -34d59'57"
Makayla Chesnick
Gem 6h45'5" 20d4'31"
Makayla Helen Yeatman
Ori 5h14'22" 6d27'39"
Makayla Irene Gravatt
Lyn 8h3'54" 41d56'45"
Makayla Jade Vernalls 05-
01-2006
Cru 12h36'50" -60d41'18"
Makayla Kathryn Dilg
And 2h15'7" 40d51'35"
Makayla Marie Merrick
And 0h34'58" 34d57'32"
Makayla Nicole Dupuis
And 23h1'22" 42d17'54"
Makayla's Star
Tau 5h8'58" 17d53'31"
Makayla's Star
Sco 16h10'46" -17d54'10"
Make A Wish Foundation of
Erie, PA
Uma 11h27'16" 58d33'51"
Make A Wish, The Sky Is
The Limit
Uma 11h38'24" 58d26'28"
Make Way for Ducklings
Per 4h13'11" 31d14'18"
Make-A-Wish Foundation
of Australia
Cru 12h36'51" -60d57'31"
Make-A-Wish Kids of WNY
Uma 11h37'59" 56d30'19"
Makeda's Herdemsaé
And 1h44'50" 47d18'30"
Makena Caroline Hanson
Cas 2h1'41" 69d30'46"
Makena Lee
Com 12h34'1" 26d22'17"
Makena Morgan Courtright
Tri 1h49'51" 31d15'31"
Makena Rita Cynthia
Anglin
And 1h13'50" 44d35'9"
Makenna
Tau 5h46'35" 26d30'48"
MaKenna
Del 20h35'4" 17d4'57"
Makenna Ashlyn Taylor
And 0h31'10" 33d33'18"
Makenna Colby
Umi 13h12'16" 72d21'44"
Makenna Grace Hartley
Mon 6h27'37" 4d50'4"
Makenna Hattie Rowe
Aqr 22h39'31" -1d6'57"
Makenna Kaylynn Stone
Cas 1h17'48" 69d0'40"
Makenna Lindsay
Uma 10h13'44" 69d32'37"
Makenna Nicole Sentz
Aqr 22h28'33" -10d16'6"
Makenna Paige
Lazorishchak
Gem 6h47'35" 33d54'25"
Makenna Samuel
Cyg 20h57'38" 41d58'57"
Makenna, Elizabeth,
Samuel Schuyler
Umi 14h5'26" 72d12'7"
Makenna's Birthday Star
Cnc 8h42'44" 28d13'8"
Makenzi Dawn Bower
Sgr 19h26'46" -16d51'38"
Makenzi Marie Baker
Ari 3h11'57" 28d30'23"
Makenzie
Lyr 19h8'7" 26d33'8"
Makenzie
Tau 4h28'31" 11d56'45"
MAKENZIE
Uma 10h0'51" 48d19'9"
Makenzie Anne - Our
Penny
Ari 3h20'29" 29d19'48"
Makenzie Arlynn Kelsey
Ari 2h27'52" 21d32'3"
Makenzie Elizabeth
Uma 12h8'4" 56d8'33"
Makenzie Elizabeth Devine
And 0h43'36" 44d19'16"
Makenzie Gerry
Psc 1h20'21" 26d3'25"
Makenzie Grace
Uma 8h53'23" 60d11'3"
MaKenzie Grace Allmond
Aqr 23h46'36" -19d28'40"
Makenzie Hope Massey
Aqr 22h37'56" -9d25'16"
Makenzie Kaylyn Spencer
And 0h27'49" 29d6'32"
Makenzie Marie
Leo 11h4'52" 9d37'57"
Makenzie Marie Smith
And 1h5'16" 45d25'47"
Makenzie Mary Bardgett
Aqr 21h15'25" -12d37'42"
Makenzie Oster Douglas
Patch
Aqr 22h0'8" -21d4'58"
Makenzie Paige Sharp
And 1h5'7" 45d55'26"
Makenzie, Jake, and Mimi
Roberts
Aql 19h51'20" -0d59'55"
Makenzy Lou
And 2h32'45" 49d14'1"
Maker of Treasure Chest
Umi 13h50'14" 77d21'10"
Makesha Lynn West
Cap 21h0'44" -16d11'21"
Makhi Stanley Gilman
Tau 5h50'18" 27d18'45"
Maki
Sgr 18h17'6" -18d28'29"
Maki and Eiichiro Takada
Uma 11h48'45" 56d1'20"
Maki Tamura
Vir 13h18'41" 5d3'18"
Makia Patrice
Tau 5h36'2" 26d58'29"
Makiko & Hajime
Per 4h16'8" 45d14'38"
Makila Lee Rose Tryon 02-
17-2002
And 0h41'58" 35d56'59"
Making Memories Of Us
Sge 19h13'34" 19d37'31"
Makinlee Jayde Seay
Gem 6h29'21" 16d28'46"
Maki's Heaven
Sco 17h49'44" -40d9'3"
Makis Theodorou
Uma 10h46'36" 39d41'33"
Makison
Vir 14h52'19" 5d1'56"
Mako's Star
Ori 5h35'58" -8d29'23"
Makoto
Ori 5h58'34" 18d11'56"
Makoto Ikebe
Leo 11h11'4" 12d59'6"
Makoto ni Utsukushi
Del 20h35'30" 12d4'19"
makteto
Per 3h33'16" 34d23'11"
Maktub - Steve & Shona
Rickard
Cyg 19h21'16" 46d59'40"
Makua Kane
Crb 15h55'39" 26d3'9"
MAKUSUN
Boo 14h52'20" 15d2'48"
MaKuzzin
Leo 11h39'56" 27d2'47"
Makyndra Alison Starr
Coder
Cnc 8h49'18" 31d20'26"
Makynzee R. Salazar
And 0h20'10" 30d54'34"
MAL 74ever
Uma 8h46'27" 65d23'12"
Mal Mal
Psc 1h18'21" 16d17'54"
Mal Rivkins
Aqr 22h0'43" -1d10'8"
Mala
Leo 11h43'10" 26d42'22"
MaLaBar
Uma 12h49'4" 60d29'38"
Malabar
Cru 12h22'25" -58d23'18"
Malacarne Beatrix
Dra 17h59'3" 79d33'14"
Malacha
Tau 4h26'35" 23d18'25"
Malachi
Umi 13h7'47" 88d25'51"
Malachi Christopher
Tau 5h48'27" 26d17'59"
Malachi Data Gruenhagen
Cam 4h38'8" 66d22'55"
Malachy Bryce
Uma 11h26'12" 60d18'37"
Malaclyttic
Sgr 18h2'51" -27d32'44"
Maladon
Umi 4h8'37" 88d41'9"
Malaël
Com 12h42'54" 14d54'46"
Malaia
Sgr 19h39'48" -18d28'32"
Malaika
Uma 10h39'31" 57d6'49"
Malaina cookieface Jane
Fondriest
Sgr 18h0'58" -25d42'42"
Malaisha Kharee' Allen
Pettway
Uma 10h35'58" 47d36'12"
MALAK A RIZK
Psc 1h51'49" 6d53'48"
Malaka M. Russell
Crb 15h39'20" 36d27'8"
Malakai Andrei Cirencione
Psc 0h10'49" 9d56'39"
Malakai Hector Gustilo-
Rios
Her 17h57'53" 22d16'55"
Malakai Xavhier Williams
Cet 3h20'30" -0d10'27"
Malary Ann
Ori 5h43'4" 2d14'53"
MALARZ
Ori 5h49'13" 6d2'26"
Malathy Srinivasan
Uma 12h43'6" 58d11'55"
Malaynee
Cnc 8h58'44" 6d45'18"
Malayni Wilkinson
And 2h28'51" 44d1'52"
Malca Rad
Uma 12h35'49" 55d45'15"
Malchia Olshan
Uma 8h36'24" 64d50'54"
Malcolm
Vir 11h46'28" -4d24'57"
Malcolm<\@>50
Col 5h35'20" -39d19'20"
Malcolm A. Haffie
Lib 15h18'50" -9d59'59"
Malcolm Antony Stanley's
Star
Psc 23h4'59" 5d11'27"
Malcolm Cain
Cap 21h26'17" -16d49'13"
Malcolm David Greene
Cnc 8h6'48" 27d26'16"
Malcolm Douglas Whiting
III
Her 17h1'12" 31d15'18"
Malcolm Drew Meadows
Lmi 11h6'40" 26d57'41"
Malcolm Dyson
Uma 11h29'21" 37d18'40"
Malcolm Farrow
Uma 11h4'56" 35d9'21"
Malcolm - "Forever Yours"
Col 6h0'54" -36d32'35"
Malcolm Forrest
Her 16h41'4" 4d13'53"
Malcolm Hantien Hsu
Cam 3h37'59" 52d56'52"
Malcolm Hector Holt
Ori 5h19'19" 6d30'37"
Malcolm J Riddlestone
And 2h32'13" 41d18'35"
Malcolm James Bage
Psc 0h49'2" 5d4'4"
Malcolm Keith Cottrill
Per 3h18'12" 56d42'20"
Malcolm L. Knell
Cap 20h8'49" -25d0'16"
Malcolm Looi
Cru 11h57'17" -55d53'13"
Malcolm Louis Stumpf
Her 17h51'0" 34d28'13"
Malcolm Lowe
Cep 21h23'41" 57d20'41"
Malcolm N. Keehner
Sco 17h39'0" -39d33'16"
Malcolm Newell & Iris
Martinello
Cyg 21h37'52" 50d8'32"
Malcolm (of Love and
Peace)
Cnv 13h11'3" 48d36'27"
Malcolm Patrick Macdonald
Ari 3h15'18" 29d6'38"
Malcolm Peatfield
Uma 12h1'57" 35d14'19"
Malcolm Pepper Heard
Aqr 22h11'52" -9d12'37"
Malcolm Richard Sarahs
Per 4h30'49" 49d5'52"
Malcolm Ryan King
Her 16h45'58" 33d18'37"
Malcolm Tyler Ewing Star
Uma 9h10'57" 53d44'45"
Malcolm Walker
Cyg 19h40'52" 39d54'37"
Malcolm West White
Ori 5h38'40" -0d29'10"
Malcolm's Star - M. Donne
1972-2001
Cru 12h28'16" -62d24'44"
Malconian
Gem 6h2'12" 23d50'19"
Malcum Elvis Beaner
Tesnow
Gem 7h16'59" 13d47'17"
Maldan<\@>1
Aqr 21h42'18" -3d35'48"
Maldwyn L Treharne
Per 2h55'1" 52d23'15"
Malea and Bryan
Per 3h25'19" 40d11'52"
Maleah
Tau 4h14'24" 27d21'6"
Malealea
Peg 22h29'25" 30d52'16"
MALEE
Sco 16h10'50" -15d36'42"
Maleiah's Star
And 0h53'46" 39d43'18"
malema
Tau 4h15'50" 10d17'57"
Malen Ferguson
Aur 5h32'31" 41d50'54"
Malena
Umi 15h0'46" 71d23'51"
Malena Morales "M & M"
Lyr 18h42'48" 34d9'6"
MALERIE
Ari 3h6'55" 17d55'27"
Malgorzata A. Szczotka
Sgr 19h11'40" -17d55'21"
Malgorzata Goddess of
Beauty
Lyn 6h31'30" 57d10'24"
Malgorzata Oster
Psc 23h27'59" 1d0'52"
Malgosia
Leo 11h28'57" 10d10'31"
Malia
And 1h23'41" 38d17'26"
Malia
Lyn 7h11'7" 46d4'2"
Malia
Sco 17h48'26" -38d10'48"
Malia Brooke
Lyr 19h13'48" 27d16'29"
Malia Dale
Mon 6h35'6" 1d51'34"
Malia Dawn Tumulak
Umi 15h17'11" 73d41'57"
Malia Lotus Willow Jenner
Gem 7h27'0" 33d46'25"
Malia R. Layne
Tau 5h5'38" 20d39'43"
Malia Redzinski
Vir 13h22'20" -0d49'50"
malia star cleveland
Vir 13h0'44" -21d47'8"
Malia2623
Tau 4h46'33" 18d58'52"
Malia's Wish
Tau 4h31'4" 7d42'14"
Malia'sMom
Gem 6h13'58" 26d3'28"
Malibooty
Ari 3h12'22" 28d56'35"
Malibu
And 0h18'46" 30d29'54"
Malibu
Cnc 9h2'44" 15d46'8"
Malibu
Sco 17h17'16" -32d59'58"
Malicity
Uma 11h41'29" 42d48'46"
Malif812
Leo 10h56'16" -0d6'16"
Malik
Dra 19h45'1" 66d25'46"
Malik
Uma 8h36'35" 60d52'37"
Malik Ernest Smith
Lyn 6h31'43" 57d46'20"
Malik Thomas John Rocha
Leo 9h32'53" 17d21'5"
Malika
Ori 6h19'45" 14d32'0"

Malin Ramona
Aqr 22h42'13" -3d5'55"
Malin Stalfors
Psc 1h1'46" 10d40'39"
Malina A. Mansell
And 2h36'45" 40d33'30"
Malina Pearl Ruiz
And 0h25'48" 31d53'45"
Malinda
Tau 4h41'21" 19d15'34"
Malinda and Ronnie
Lib 15h59'46" -10d58'35"
Malinda Jean Maginley
And 1h26'54" 49d13'44"
Malinda K Corder
Lyn 7h21'15" 53d3'45"
Malinda Leigh Hogan
Umi 14h3'12" 70d32'51"
Malinda & Robert For Eternity
Pho 1h21'11" -41d30'9"
Malinda Sliger
Oph 17h39'7" -0d19'22"
Maling-Topham
Psc 1h46'11" 13d43'14"
malioh
Leo 11h14'45" -1d50'54"
Malisa Chaophrasy
Sgr 17h46'30" -22d43'9"
Malishka
Tau 4h20'35" 17d2'13"
Malison
Uma 10h25'33" 44d46'40"
Malissa
Ori 5h47'37" 11d25'8"
Malissa Crawford
Aql 19h20'19" 4d1'54"
Malissa Lea Sines
Sco 17h44'1" -41d29'13"
Malissa Mae
Lib 14h29'58" -16d55'51"
Malissa's Eternal Light
And 0h42'18" 42d46'24"
Malith & Chandani
Cap 21h32'1" -19d21'37"
Maliusinkaja Olia
Cas 1h34'10" 69d23'21"
Malivérich
Ori 5h53'6" 5d39'33"
Malka
Leo 11h21'4" -3d12'1"
Malka
Lib 14h53'8" -2d27'3"
Malka Kassem Gellani
Dra 16h15'46" 55d21'37"
Malka Rus
Tau 4h37'24" 19d57'35"
Malkat Ha-Regina
Psc 0h53'28" 14d41'49"
Malkin Rosette
Uma 11h23'8" 42d28'59"
Malklavious
Cnc 7h56'35" 17d11'33"
malldoon
Uma 10h28'25" 66d57'15"
Mallely Ledezma
Umi 16h22'29" 81d4'38"
Mallery And Chris
Cyg 21h20'57" 38d42'32"
Malley Driscoll-Luttringer
And 0h22'32" 43d16'46"
Mallie
Cnc 8h43'44" 19d4'33"
Mallie Lanier
Lyn 8h14'7" 45d42'25"
Mallora
Cas 23h26'39" 58d0'15"
Mallori Downs
Cnc 8h20'48" 10d17'0"
Mallori McArthur
Vir 13h57'28" 2d30'26"
Mallorie Lee Ferguson
Cam 4h37'55" 72d49'57"
Mallory
Lyn 8h18'47" 55d33'3"
Mallory
Uma 12h16'33" 59d27'21"
Mallory
Lep 5h33'44" -20d4'43"
Mallory
Psc 0h38'22" 9d11'56"
Mallory
Ori 6h12'49" 9d12'50"
Mallory
Her 18h44'5" 20d28'52"
Mallory
Psc 1h44'5" 22d41'28"
Mallory
And 0h13'21" 46d36'36"
Mallory
Crb 16h6'56" 38d44'22"
Mallory
Crb 15h51'1" 36d11'55"
Mallory Amber Burkett
And 0h26'10" 33d27'59"
Mallory and Brandon Forever
Per 3h41'54" 39d13'17"
Mallory Barr Wilbourn
Aqr 22h58'22" -6d46'14"
Mallory Brooks
Leo 11h7'13" 11d21'31"
Mallory Camille Webster
Lyr 18h45'57" 32d8'19"
Mallory D. Willman
Leo 10h37'40" 18d1'2"
Mallory Elise
Cnc 9h15'48" 12d4'28"
Mallory Elizabeth Hinderer
Cap 20h45'43" -15d20'26"
Mallory Elizabeth Lawrence
Lib 15h19'28" -14d31'45"
Mallory Ellen Smith
And 0h43'18" 29d6'23"
Mallory Faith Hicks
Pho 1h17'1" -40d52'11"
Mallory Funkhouser
Ori 5h59'43" 15d22'13"
Mallory Gayden Ewing Star
And 0h28'6" 44d53'16"
Mallory Houston
Uma 11h1'59" 68d15'28"
Mallory "...how they shine for you"
Ari 3h16'32" 29d8'52"
Mallory Jo Ellis
Ari 2h34'54" 14d14'31"
Mallory Jo Rowley
And 23h16'29" 40d25'37"
Mallory Johnson
Cas 0h53'57" 54d20'59"
Mallory Kay
Sco 17h22'22" -42d29'48"
Mallory Lynn Rossi
Cap 20h15'54" -26d12'18"
Mallory & Maggy's Twilight Twinkle
And 0h35'37" 33d21'36"
Mallory McQuay
And 0h49'38" 39d55'48"
Mallory Monroe
Uma 9h56'55" 50d35'59"
Mallory Nicole Hicks
Psc 23h39'18" -2d5'29"
Mallory Nicole Kopp
Gem 7h1'16" 22d41'55"
Mallory Nicole Vincent
And 0h22'56" 33d11'47"
Mallory O'Day
Ori 5h42'11" 2d59'28"
Mallory Oross
Peg 21h29'22" 14d7'34"
Mallory Oross
Cyg 20h1'38" 34d52'15"
Mallory Paige Rich
Lyn 8h16'32" 41d45'47"
Mallory Paige Wortham
Ori 5h48'48" 5d42'56"
Mallory Payton Rujawitz 1-14-01
Cap 20h22'24" -19d37'26"
Mallory Pendleton
Lib 15h26'20" -13d21'19"
Mallory Pivron
Crb 15h19'50" 25d40'46"
Mallory Rose Wurzburg
Vir 12h2'57" 9d31'53"
Mallory Salmons
And 0h12'38" 44d5'30"
Mallory shea Zumbach
Cyg 20h33'51" 55d8'41"
Mallory Siegfried
And 2h18'31" 44d43'27"
Mallory Smith
Uma 13h34'11" 55d58'36"
Mallory Star
Uma 12h5'14" 42d23'33"
Mallory Willson
Peg 22h22'31" 20d23'50"
Mallory & Zac Smale
Umi 15h4'46" 80d55'52"
Mallory's Radiance
Vir 13h27'37" 8d9'20"
Mally
Uma 11h6'3" 45d57'3"
'Mally' (Paul Alexander Mallaghan)
Per 3h1'5" 46d29'24"
Mally Sainty
Eri 2h25'26" -45d16'16"
Malmal Dorothy
Cnc 8h40'11" 7d16'25"
MalMal & RoRo Love Forever 8/23/80
Cyg 19h22'33" 48d35'35"
Malmquist 061905
Uma 12h30'35" 58d34'14"
Malo Dramatic
Uma 10h12'15" 44d20'24"
Malone Galster
Uma 8h57'59" 49d0'34"
Malori Kaye Flanagan
Leo 11h17'9" 25d10'20"
Malori & Matt's Starlit Love
Cyg 21h26'14" 45d22'22"
Malorie Elise Hamby
Ori 4h49'32" 8d40'36"
Malorie LeeAnn York
And 0h59'23" 40d50'36"
Malorie Lynn
Sgr 18h26'30" -32d45'57"
Malory
Aqr 22h40'17" -1d21'15"
Malory
Her 17h23'29" 35d45'51"
MALORY LAEMMLE
Leo 10h28'21" 12d17'0"
Malou Buffagni
Aql 20h14'55" 1d4'59"
Malou Shanti Marie
Lyr 18h43'17" 35d52'53"
"Malta" Anthony Sammut
Cnc 8h40'38" 30d47'9"
Malte
Uma 9h37'24" 57d41'36"
Malte Leesemann
Ori 5h19'30" 15d57'27"
Malte Saß
Uma 14h7'28" 57d2'58"
Maluaka Makena
Lmi 10h47'43" 32d44'40"
MaLucy
Aqr 23h28'37" -10d38'52"
MALUKE
Uma 11h23'59" 52d38'0"
Malus
Lyn 7h27'1" 52d28'14"
Maly Anialek
Cap 21h54'56" -19d5'1"
Maly, Hartwig
Uma 8h25'32" 64d51'35"
Malyce Bielawski
Aqr 22h5'37" -0d49'39"
Malyiah Haney
Leo 11h9'9" 23d0'52"
Malynn Devers
Psc 1h22'57" 16d34'18"
Mam
Leo 11h5'39" 12d24'52"
MAM
Cyg 21h40'11" 44d19'49"
MAM
Vir 13h13'5" -19d58'22"
Mam our star that shines so bright.
Cep 22h32'36" 65d26'48"
M.a.m.a
Uma 13h42'12" 59d51'52"
Mama
Cas 23h53'22" 53d31'8"
Mama
Vir 12h34'19" -4d52'29"
Mama
Cap 21h34'45" -22d41'9"
Mama
Cas 0h53'31" 57d29'46"
mama
Cas 0h34'41" 53d28'35"
Mama
Lmi 10h34'30" 37d17'51"
Mama
Crb 15h55'56" 31d6'15"
Mama
Per 4h27'42" 43d32'36"
Mama
Tau 4h31'6" 19d8'2"
"Mama"
Ori 5h39'14" 11d50'59"
Mama and Pammy
Leo 11h52'32" 25d59'44"
Mama and Papa
Vir 14h40'22" 4d59'29"
Mama and Papa Mahaven
Her 16h43'11" 32d39'42"
Mama Babe
Sgr 19h13'37" -28d1'17"
MAMA BEAR
Uma 13h13'32" 56d6'59"
Mama Bear
Uma 8h34'39" 63d49'27"
Mama bear De Leon
And 1h12'54" 33d54'48"
Mama Bear Loves Papa Bear
Com 13h22'46" 14d32'5"
Mama Beez Neez
Sco 17h33'31" -41d46'8"
Mama Betty Lee Arnold
Cap 20h13'51" -9d39'41"
Mama Bev With Love, Bobbie
Cas 2h1'36" 62d56'8"
Mama & Bianca's Star
Cyg 21h55'0" 53d21'28"
Mama BooBoo
Aqr 22h16'43" -17d54'12"
MAMA Chatz
Ori 5h31'39" 10d3'47"
"MaMa Choo Choo" Shirley Lane
Cas 0h24'30" 52d14'19"
Mama Christine
Boo 13h48'31" 12d15'42"
Mama Chunk's Bling in Da Sky CRK79
Dra 20h24'38" 75d16'12"
Mama D
Aqr 22h45'40" 2d8'19"
Mama Day
Per 3h11'26" 55d53'34"
Mama Dee
Cnc 8h43'41" 23d39'41"
Mama Doll
Aqr 22h37'16" -1d59'19"
Mama Doris
Cas 0h21'54" 52d0'17"
Mama Dorothy Elizabeth Margaret
Cas 23h8'23" 58d45'13"
Mama Dukes
Per 4h7'32" 45d18'35"
Mama egyletlen csillaga
Uma 14h15'36" 57d9'1"
Mama G
Psc 1h14'55" 30d49'12"
Mama Gladys
Cas 0h23'25" 61d35'33"
Mama Hill
Crb 15h56'24" 30d2'26"
Mama Hill
Crb 15h43'22" 28d20'56"
Mama Isia
Lib 15h11'52" -7d55'45"
Mama Jo
Aqr 21h41'12" 2d20'37"
Mama Jo Russell
Ori 5h26'48" 14d47'8"
Mama Karlea M.A. Allyn
Uma 11h39'28" 50d36'51"
Mama Kelly
Gem 7h48'0" 24d51'54"
Mama Lee Ann Star
Del 20h41'53" 16d47'0"
Mama Lee Pagitt
Cnc 8h50'55" 26d34'15"
Mama Leoncia
And 1h31'8" 48d52'20"
Mama Liz
Gem 6h46'34" 33d42'9"
Mama Ludmila
Vir 12h33'22" -4d52'34"
MaMa Luk
Psc 23h44'48" 5d46'49"
Mama Magdolna Juhasz
Umi 16h20'22" 76d55'51"
Mama Mary
Cas 2h22'44" 74d46'6"
Mama Mary Scro "Mother Goose"
Psc 1h18'38" 30d35'6"
Mama Megan
Cas 0h44'15" 51d57'50"
Mama Mia
Cnc 8h51'59" 27d45'0"
Mama Michele
Leo 10h14'24" 17d17'0"
Mama Neek
Cnc 8h42'49" 18d51'8"
Mama Nikki
Ari 3h8'48" 29d55'12"
Mama "O"
Lyn 8h30'24" 46d32'11"
Mama & Papa Williams 13.05
Dra 14h38'44" 55d7'9"
Mama & Pup McCullough
Uma 12h6'2" 46d33'10"
Mama Renea
Vir 14h10'48" -12d48'53"
Mama Robin
Sgr 19h57'19" -35d23'15"
Mama Rose Trbovich
And 0h6'34" 43d54'38"
Mama Ruth Aliene Green
Lib 15h52'26" -7d6'1"
MaMa Ruthie 1919-2006
Sgr 18h17'16" -28d38'5"
Mama Squeege
Sgr 19h45'16" -12d19'0"
Mama T
Cyg 20h29'56" 58d22'2"
Mama Vas's Star
And 23h38'40" 38d6'59"
Mama Wanda
Cnc 8h20'9" 28d31'26"
Mama2798
Lyn 7h52'56" 46d19'15"
Mamacita Ault - Glenna Maxine Ault
Cyg 19h51'4" 35d48'27"
Mamacita Smootessa
Gem 6h29'54" 22d58'39"
MamaCita's Star
Cap 20h48'18" -15d13'4"
Mamadoodi
Leo 10h12'12" 20d31'42"
Mamaita
Peg 21h50'23" 26d35'15"
Mamakat
Cnc 8h43'26" 31d53'20"
MamaLena
Lac 22h20'14" 45d33'46"
Maman Nambi
Vir 12h30'25" 11d46'55"
Maman Sabah
Uma 11h19'21" 43d43'29"
Maman Tatayet
Aqr 23h5'51" -8d54'59"
Mama's Angles
Pho 23h49'31" -50d27'3"
Mama's Big Boy
Ari 2h49'2" 28d36'49"
Mama's Boy
Aqr 20h58'7" -12d55'59"
Mama's Love (Elnora Wheeler)
Mon 7h36'3" -0d54'46"
Mama's Migaloosh
Gem 6h43'26" 17d16'9"
Mama's Star
Leo 9h22'50" 27d23'54"
Mama's Star
And 2h34'58" 43d58'10"
Mama's Star
Sgr 17h53'35" -29d47'22"
Mamasita And Poo Dee Head
Cap 20h24'35" -13d8'38"
Mamasita's
Cas 1h15'23" 64d8'21"
Mamaw
Cyg 21h35'41" 38d19'38"
Mamaw
Vir 12h43'45" 4d16'59"
Mamaw and Papaw
And 0h50'12" 36d43'15"
Mamaw Betty Ann
Vir 14h6'9" -16d43'7"
Mamaw Major
And 0h46'40" 43d54'22"
Mamaw's Becky
Cap 20h54'30" -20d49'41"
Mamaw's Magic
Ari 2h56'37" 29d3'57"
Mamber Monkey
Her 16h19'0" 5d46'22"
Mambo Augustine Frances Hunter, MD
Cyg 19h9'11" 48d55'40"
Mambo Tonya
Her 16h30'44" 34d50'57"
Mamede & Silvia
Cyg 19h38'44" 35d15'10"
Mamere
Aql 20h1'52" 9d1'56"
Mamére's Clayton Christopher Hall
Umi 16h59'3" 75d54'29"
Mamére's Taylor Madison Hall
Umi 16h58'50" 78d22'47"
Mami
Del 20h45'57" 13d15'4"
MAMI
Her 16h43'1" 6d9'58"
Mami
Uma 11h6'25" 41d3'58"
Mami and Lee
Gem 7h19'55" 24d1'52"
Mami Berthi
Boo 14h51'45" 15d3'3"
Mami Doris
Umi 16h2'29" 78d7'48"
Mami Kawamura
Vir 13h27'8" -6d28'56"
Mami Lotti & Papi Alfred
And 2h24'16" 39d13'45"
MAMI NR. 1
Cep 22h17'10" 67d26'9"
Mami Silvi
Aql 19h37'19" 6d55'2"
Mami Takeshita
Ari 2h29'20" 15d7'54"
Mami + Thomi
Uma 9h31'26" 43d24'53"
Mami & Toshihiko
Leo 9h41'34" 8d52'31"
Mami und Papi mir händ euch gern - Sternli
Mon 6h35'54" 5d45'26"
Mami y Papi Babes
Sge 20h2'59" 17d35'52"
Mami, 27.04.1939
Crb 15h34'19" 25d37'10"
Mamidream
Uma 9h40'8" 63d37'48"
Mamie
Lyn 8h6'47" 59d30'49"
Mamie
Cyg 19h59'37" 46d38'26"
Mamie Almon Rogers
Leo 10h4'56" 21d15'14"
Mamie Howard Golladay's Star
Cas 1h18'37" 56d46'10"
Mamie Kathryn Duggar
Sgr 19h31'37" -21d22'17"
Mamie Lynn Maxwell
And 0h52'21" 39d41'2"
Mamie Mammours
Cas 23h15'9" 56d6'8"
Mamie Prouty
Lyn 7h56'45" 44d30'27"
Mamie Webb-Hixon
Ari 2h52'17" 17d33'43"
Mamily
Cas 23h14'18" 55d31'18"
Mamima Nicole Meyer
Uma 10h37'2" 41d9'12"
Maminka
Cnc 8h24'53" 8d58'29"
Maminou Shay
Dra 19h34'10" 74d47'11"
Mamish
And 0h24'58" 32d33'5"
Mamita
Tau 4h10'39" 7d13'27"
Mamitu
Cyg 19h35'9" 28d31'2"
Mam-"Looking Down and Smiling"
Uma 11h55'46" 65d42'47"
Mamma
Uma 9h43'57" 54d45'31"
MAMMA
Cap 20h51'33" -19d45'41"
Mamma Daniela Frisco
Umi 16h49'23" 78d22'42"
Mamma Giulia
And 2h36'3" 50d32'57"
Mamma Liliana
Cap 20h28'19" -14d4'38"
Mamma Marjorie Mining
Cas 1h18'18" 68d17'45"
Mamma Miko's Star
Sco 17h40'53" -32d43'24"
Mamma Rose G.
Aqr 22h15'20" -2d30'54"
Mammie's Engeltjies
Gem 6h46'3" 34d1'1"
Mammolo
And 1h14'55" 42d42'6"
Mammy/Nico
Lyn 7h54'9" 53d9'49"
Mamo
Peg 21h45'43" 9d30'33"
Mamon Aminov
Tau 4h46'51" 26d52'24"
Mamouchnaka d'Amour
Uma 10h9'5" 57d32'35"
Mamoun
Uma 11h1'6" 65d29'23"
Mamour
Umi 17h6'35" 80d26'48"
Mamour
Del 20h20'37" 10d9'59"
Mamour
Tau 5h51'53" 17d53'50"
Mamour
Her 17h56'0" 49d59'21"
Mamourelodie
Sgr 19h14'56" -22d12'49"
Mamta Kalra
And 1h12'22" 46d2'11"
Mamuchis
Hor 3h13'48" -48d5'10"
Mamucika védocsillaga
Cap 20h35'34" -20d52'17"
'MaMummaBub' Stacy V Gorman 2004
Tau 5h40'19" 22d2'12"
Mamy Doreau
Peg 21h56'34" 13d27'31"
Man of Hope
Per 3h12'52" 53d25'54"
Man of Lego
Per 3h52'11" 41d42'1"
Mana Banana
Lib 15h25'23" -5d28'19"
Mana Namura
Lyn 7h18'0" 53d24'31"
Manacca
Tau 4h26'4" 21d37'59"
Manaika'ekolu
Sco 16h33'42" -25d55'51"
Manaka Fujioka
Leo 9h34'2" 8d11'54"
Manal
Sco 16h7'14" -13d22'47"
Manal Al Kareem
Per 4h39'35" 39d46'17"
Manal Al-Hasawi
Psc 1h14'30" 10d12'39"
Manal Franklin - Tupperware
Ori 5h19'45" 5d45'42"
Manal M. El-Shimy
Lib 15h9'18" -13d45'8"
Manal Peracha
Uma 10h32'25" 64d7'53"
Manami
Ari 2h33'31" 29d32'10"
Manan "Manu" Minasyan
Del 20h45'34" 12d56'17"
Manarda
Uma 9h37'6" 48d51'38"
Manav
Per 2h24'27" 57d28'26"
Mance
Uma 9h35'15" 46d46'41"
Manda
And 1h12'35" 45d41'53"
Manda
Psc 0h18'59" 15d36'10"
Manda
Cas 0h22'55" 61d9'53"
Manda
Psa 22h8'8" -25d43'5"
Manda Belle
And 23h3'13" 47d3'1"
Manda Faye
Umi 15h8'48" 81d33'45"
Manda Jo
Lyn 7h54'15" 33d58'57"
Manda Lynn
Tau 5h23'48" 17d59'11"
Manda Panda
Cam 4h3'8" 55d40'44"
Manda Prus, Honors Graduate
Cyg 19h51'32" 32d16'59"
Manda21805
Aqr 21h13'5" -9d57'10"
Mandabear
Uma 8h26'10" 61d25'54"
Mandalee
Peg 22h12'47" 7d3'10"
Mandalynn
Cyg 21h56'16" 47d41'27"
Mandalynn
And 1h12'48" 45d9'52"
Mandana Jones
Psc 23h51'19" -3d5'37"
Mande L. Lewis
Uma 10h3'32" 51d13'3"
Mandee
Vir 13h18'2" 4d0'45"
Mandee The Beautiful
Cas 0h14'16" 58d24'14"
Mandeep Janda
Cas 0h50'17" 70d8'36"
Mandelbezium
Psc 23h1'13" 7d38'26"
Mandelyn Belle
And 1h8'49" 38d8'13"
Mandelyn Belle Babcock
And 23h30'31" 35d55'18"
Mander Babe
Cap 21h22'39" -19d57'56"
Manders
Vir 12h32'21" -1d53'33"
Manders
Gem 6h42'29" 12d11'24"
MANDERS Q
Uma 10h26'43" 47d37'42"
Mandi
Ori 4h51'53" 8d35'6"
Mandi
Ori 5h26'55" -5d14'55"
Mandi
Lyn 6h42'4" 55d8'9"
Mandi
Sco 17h39'33" -36d38'1"
Mandi and John Forever
Cyg 20h1'40" 31d10'12"
Mandi and Noah
Ari 2h59'31" 24d58'16"
Mandi Erin
Lib 15h31'57" -12d23'55"
Mandi & Gary Christensen - Our Star
Cru 12h26'50" -61d11'20"
MANDI & JAY
Vir 11h40'50" -3d14'51"
Mandi Jo Veronica Wilkens
Uma 12h22'43" 61d54'35"
Mandi K. Vennard
Vir 13h48'52" -7d30'0"
Mandi Konst
And 23h27'42" 49d14'45"
Mandi Louise Shawley
Cas 23h33'13" 56d12'54"
Mandi Lynn Lewis
Gem 6h30'27" 24d32'42"
"Mandi Pie"
Ori 4h59'8" 14d32'12"
Mandi Rea Wolford
Her 16h59'2" 32d45'33"
Mandi Rochelle
Leo 11h32'24" 24d4'47"
Mandi Skylar
Cas 0h27'12" 50d28'0"
Mandi the Little Witch
Ari 2h46'7" 27d26'35"
Mandi Wright
Cap 20h56'31" -24d19'21"
Mandi, My Guiding Light, I Love You
And 0h7'59" 46d34'3"
MandiCase
And 2h20'41" 46d25'8"
Mandie
Sco 17h4'40" -39d58'35"
Mandie
Cam 6h54'22" 63d13'14"
Mandie Jane
And 1h55'59" 37d40'49"
Mandie Lee Martinolich
Ari 2h48'54" 28d59'22"
Mandie Marie Lee
Tau 4h2'25" 17d12'38"
Mandie Pierce
Lib 14h50'43" -7d1'39"
Mandie Renee'
Sco 16h14'5" -15d58'0"
Mandi's Angel Light
Col 5h47'52" -34d1'12"
Mandi's BFS
Cnc 7h59'50" 18d6'27"
Mandi's Star
And 23h13'59" 48d21'1"
Mandi's Star
Lib 14h51'12" -3d23'8"
Mandi's Wish
Umi 5h5'36" 88d45'54"
Mandi's Wish
Cas 1h42'6" 60d16'27"
MANDIYSTEVENS
Lib 15h4'44" -10d34'20"
Mandlebum
Pyx 8h38'17" -22d21'8"
Mandola
Uma 10h43'1" 61d52'24"
Mandu
Leo 11h12'46" -2d57'14"
Mandy
Vir 13h51'14" -9d39'59"
Mandy
Uma 12h20'2" 63d13'50"
Mandy
Uma 8h44'32" 52d35'50"
Mandy
Lac 22h55'19" 46d41'48"
Mandy
Uma 9h21'55" 51d21'1"
Mandy
And 2h35'46" 45d30'20"
Mandy
And 2h14'34" 47d29'37"
Mandy
Lyn 8h11'10" 51d13'27"

Mandy
Uma 8h59'37" 49d22'53"
Mandy
Cas 0h43'9" 50d58'30"
Mandy
Her 16h46'23" 38d47'9"
Mandy
And 0h25'18" 41d47'51"
Mandy
And 0h37'39" 38d54'34"
Mandy
Aur 5h57'26" 42d36'32"
Mandy
And 2h31'32" 44d40'39"
Mandy
Vir 12h29'21" 11d21'27"
Mandy
Leo 9h29'53" 10d8'3"
Mandy 27
Leo 10h47'47" 17d32'22"
Mandy and Allen
Crb 15h35'27" 32d37'18"
Mandy and Chris - Together Forever
Vir 12h2'25" -0d7'27"
Mandy and Eric
Boo 14h33'7" 19d52'1"
Mandy and Fish 9/30/05
Lib 15h3'40" -6d50'16"
Mandy B Pavlovec
And 0h54'44" 37d35'30"
Mandy "Baby Girl" Deol
Gem 7h2'44" 18d18'25"
Mandy Beagles "Star of Angel's"
Cas 0h33'14" 52d48'27"
Mandy Blevins
Tau 4h29'10" 6d56'8"
Mandy Brown
Tau 4h33'29" 28d16'37"
Mandy Carr
And 23h48'16" 45d15'32"
Mandy Caso
Sco 17h4'51" -37d13'56"
Mandy Catoe
Lep 5h41'24" -21d4'41"
Mandy & Chris Forever
Lyn 7h48'19" 36d1'43"
Mandy Crouch
And 0h2'12" 45d38'2"
Mandy Danyel Severson
And 0h31'21" 27d7'24"
Mandy Dare Gauger
Crb 16h9'12" 36d33'54"
Mandy & Darren
Aql 19h2'4" 17d6'56"
Mandy Estrada
Lyn 7h25'0" 45d23'19"
Mandy Fenton Tyson
Sco 17h57'22" -37d18'7"
Mandy Flanders
Cnv 13h41'14" 34d5'16"
Mandy Gene Mooring
Sco 16h16'4" -13d36'20"
Mandy Glover
Psc 1h33'30" 11d53'51"
Mandy Gonzalez
Cap 21h40'50" -14d35'27"
Mandy Gonzalez
Aqr 20h45'54" -3d29'34"
Mandy Gonzolaz
Aqr 21h14'40" -7d25'45"
Mandy Hammond
And 0h44'15" 43d28'24"
Mandy Jean Healy
Ari 2h38'58" 18d32'41"
Mandy Joann Buffone
Sco 16h18'12" -9d34'51"
Mandy Joy Werbil
Sgr 18h14'38" -16d10'12"
Mandy Kae
And 0h58'7" 40d33'25"
Mandy Kate Jacobs
Psc 1h21'11" 26d30'4"
Mandy Kay
Gem 7h6'4" 34d11'52"
Mandy Keith
Tau 5h56'1" 25d0'26"
Mandy L. Dusseau
Sco 16h12'22" -15d9'7"
Mandy Lee Pensiero
Umi 16h32'58" 75d18'6"
Mandy Lisa Boothe
Uma 11h52'19" 32d45'29"
Mandy Loftis
And 0h22'17" 42d8'44"
Mandy Louise Blore
Umi 15h27'18" 79d57'4"
Mandy Louise Sturrock
Cru 12h35'3" -56d2'29"
Mandy Lynn
Crb 16h6'5" 36d54'40"
Mandy Lynn
Gem 7h4'6" 32d20'56"
Mandy Lynn
And 0h44'33" 28d31'37"
Mandy M. Kirk
And 23h50'7" 44d13'23"
Mandy M. Vallinas
And 23h14'15" 46d52'51"
Mandy Mae Darling
Aqr 20h51'39" -8d59'55"
Mandy Marie
Psc 1h18'58" 18d12'56"
Mandy Marie Erwin
And 1h26'29" 39d48'0"
Mandy Marie Martin
Sco 16h15'21" -18d25'7"
Mandy Maurer
And 0h45'43" 37d40'23"
Mandy Moo
Lib 15h5'34" -1d48'55"
Mandy Mooney
Tau 4h43'44" 17d15'5"
Mandy Moquin
Cyg 19h42'45" 31d31'6"
Mandy "my angel" Hughes
Vir 13h32'7" -0d3'16"
Mandy My Heart
And 0h40'31" 28d58'25"
Mandy Nation
Lyn 8h50'41" 42d18'37"
MANDY NGA
Aqr 20h47'35" 2d21'2"
Mandy Nicole Altman
Dra 17h36'36" 59d0'56"
Mandy Pamela Taylor's Star
And 2h36'40" 37d36'29"
Mandy Powell
Ori 4h57'48" -2d1'23"
Mandy Rae
Aur 5h34'28" 45d19'45"
Mandy & René
Uma 9h56'17" 53d45'43"
Mandy Rimer
Aql 19h4'39" 15d19'17"
Mandy "Rockstar Barbie" Stone
Lib 14h30'1" -16d35'20"
Mandy Shines
Aqr 22h39'10" -16d27'12"
Mandy Skinner
Leo 11h54'38" 21d50'17"
Mandy Starr
Uma 11h49'22" 62d35'3"
Mandy & Sue
Cyg 20h1'12" 37d5'38"
Mandy Taylor
And 2h29'37" 49d38'57"
Mandy Thistle-Natalie
Ari 2h34'30" 24d38'32"
Mandy Ungar
Ser 15h39'2" 21d53'14"
Mandy Vandegrift
Cap 21h33'19" -17d34'17"
Mandy Warlick
Aqr 21h36'46" 1d20'19"
Mandy Wilkins
Ori 4h50'51" 12d41'0"
Mandy's First Star of the Night
Ari 2h17'24" 20d0'7"
Mandy's Light
And 23h40'16" 42d3'20"
Mandy's Love
Ari 2h20'18" 23d5'32"
Mandy's Shining Star
And 23h46'24" 48d18'9"
Mandy's Star
And 2h35'23" 50d56'48"
Mandy's Star
Uma 11h28'31" 44d55'39"
Mandy's Star
Crb 15h38'34" 31d4'8"
Mandy's Star
Leo 10h30'4" 23d5'25"
Mandys Star
Cap 20h39'17" -24d59'54"
Mandy's Star ARC+DES
Gem 6h45'41" 20d1'35"
Mandy's Wish
And 1h6'24" 34d1'58"
Mandy's Wishing Star
Dra 17h24'45" 60d34'2"
Maneka Shewa
Mon 6h46'16" -0d18'33"
Manet
Aqr 22h1'8" -16d42'20"
Manette Mancuso
Sco 17h56'25" -40d4'18"
Manette Marie Chadwick
Leo 11h47'27" 21d48'6"
Manfred
Psc 23h13'12" -2d10'50"
Manfred
Uma 8h49'7" 56d29'0"
Manfred Bahr
Uma 12h59'57" 56d26'37"
Manfred Bylicki & Petra Ortmann
Ori 5h3'17" 15d21'6"
Manfred Carlsen
Uma 10h3'52" 61d19'25"
Manfred F. Reinold
Lmi 10h51'11" 29d50'6"
Manfred Ganser
Uma 10h22'0" 54d52'13"
Manfred Gröbel
Uma 11h10'35" 48d54'39"
Manfred Hannich
Uma 10h18'51" 44d13'34"
Manfred J. Wirfs
Uma 11h14'53" 50d23'16"
Manfred Kaiser
Uma 9h57'10" 64d58'40"
Manfred Kellenberger
Uma 8h37'3" 57d23'47"
Manfred Kropp
Ori 5h54'22" 12d51'44"
Manfred Lengies
Ori 5h57'41" 2d18'40"
Manfred Maus
Psc 23h6'44" 3d11'27"
Manfred P. Drost
Cnv 13h39'12" 30d15'19"
Manfred Pleasant "Ron" Morgan
Aqr 21h15'20" -5d21'31"
Manfred Ramlow
Uma 10h32'32" 40d4'18"
Manfred Schiebel
Vir 14h39'14" 2d52'51"
Manfred Schuster
Uma 10h6'49" 41d58'4"
Manfred Wagner
Uma 8h46'4" 68d2'47"
Manfred Waller
Uma 9h26'15" 41d48'53"
Manfred Werner
Ori 5h53'28" 3d10'37"
Manfred Wilhelms
Uma 10h16'16" 47d57'44"
Manfred Züchner
Uma 12h56'9" 56d15'47"
Manfredi Claudia
Lyn 7h59'27" 52d31'30"
Manfredi Monteduro
Sco 17h28'30" -38d39'56"
Manfredo Hammerschlag
Lmi 10h23'6" 28d27'39"
Manfriend "8-25-71"
Per 3h3'10" 55d14'46"
Mang, Claus
Uma 10h55'7" 57d24'25"
Manga
Cyg 21h52'13" 39d20'14"
Manga Kryddar
Her 16h34'15" 27d37'22"
Mangela
Tau 5h16'17" 18d29'45"
Mangherini Alessia
Her 17h16'53" 26d44'47"
Mango
Cnc 9h12'27" 32d17'42"
Mango Bay Hotel & Beach Club
Peg 21h54'0" 34d20'31"
Mango Maddie
Umi 14h49'42" 86d2'23"
Mangus
Crb 16h14'2" 29d13'46"
Mani
Cas 23h55'45" 50d28'30"
Mani Subramanian
Uma 8h44'16" 68d17'50"
Mani Trihima
Gem 6h39'1" 14d4'52"
Manideep Kamineni
Psc 0h26'54" 10d38'14"
Manijeh & Fray
Cyg 21h58'33" 50d55'0"
MANIKANDAN
Crb 15h46'16" 27d55'24"
Manimala
Aqr 20h40'36" -2d53'23"
ManiNav
Cnc 8h42'19" 11d1'6"
Mani's Magic
Ori 5h36'37" -1d31'2"
Manjari Mardithaya
Uma 8h16'32" 65d29'52"
Manjinder
Mon 6h53'20" 2d15'10"
Manjula
Aqr 22h9'38" -13d30'55"
Manko's Legacy
Her 17h25'36" 37d48'43"
Manley
Aql 19h48'18" 11d31'59"
Manlio Cardona
Aql 19h44'16" 16d17'2"
Manna
Cyg 21h46'3" 39d25'54"
Manna
Cam 3h58'19" 66d35'12"
Manne
Ari 3h7'49" 15d21'33"
Mannherz, Heinz
Ori 6h15'53" 11d11'38"
Manning
Uma 10h43'57" 56d35'3"
Manning
Cru 12h19'51" -58d14'57"
Manning "Punk" Jenkins
Tau 4h4'54" 7d12'0"
Manning's Cujo
Uma 11h31'38" 59d41'22"
Mann's Joy
Cnc 8h51'41" 13d4'3"
Manny Adams
Cap 20h59'17" -19d15'23"
Manny and Monica Valdez
Cnc 8h2'13" 19d34'41"
Manny Byron
Gem 7h31'18" 22d22'47"
Manny Miller
Psc 1h45'45" 6d2'15"
Manny Moreira Jr.
Tau 4h49'36" 21d30'52"
Manny & Scott
And 23h8'2" 48d25'20"
MaNNY-LoVES-DENNiS
Eri 2h12'53" -53d1'58"
Manny's Charity
Aqr 22h11'48" -23d40'39"
Manny's Girls
Cnc 8h38'53" 23d28'17"
Mano
Umi 11h15'38" 86d47'32"
Mano Ich liebe Dich! Dein Stern
Ori 6h19'36" 11d12'40"
Mano (Mina)
And 0h9'9" 46d25'33"
Manoj Astrey Naresh Shivani Parwani
Umi 14h58'16" 70d43'34"
Manókám, Csurgó István
Gem 7h10'46" 25d14'50"
Manola
Cam 3h56'33" 55d47'12"
Manoli
Gem 6h44'36" 23d48'16"
manoli
Cap 21h35'15" -11d24'4"
Manoli Martin Estrella
Aur 6h24'33" 29d26'47"
Manolis Apostolou
Uma 10h47'7" 40d56'50"
Manolis - Our brightest star
Cru 12h55'0" -59d51'56"
Manolis Th. Karvounis
Ori 6h4'3" 14d32'45"
Manolo Maccioni
Uma 11h26'8" 35d18'29"
Manon
Leo 11h30'53" 28d6'34"
Manon Alice Maya G—G
Uma 12h5'45" 59d30'32"
Manon Lamontagne
Lyn 8h15'50" 40d15'36"
Manon Prevost
Ori 5h15'4" 5d54'7"
Manon Vranderick
Cas 23h21'10" 55d20'52"
Manonbisha
Lib 15h36'21" -27d13'41"
Manong Jay
Aqr 22h3'10" -0d52'42"
Manos Lupassakis
Uma 12h26'58" 55d21'26"
Manou
Uma 9h57'34" 67d15'32"
Manou
Cyg 21h19'50" 41d26'28"
Manou - Denis Feuilloley
Ori 6h0'35" 6d24'11"
Manou et Manounette
Leo 10h42'4" 7d35'23"
Manouchehr Khalilian
Sco 17h2'40" -31d13'51"
Manouk & Shakeh
Cnc 8h48'23" 23d10'51"
Manoula
Leo 10h42'48" 16d27'27"
Man-Pui
Gem 7h27'3" 33d19'10"
Manrico
Per 3h0'14" 51d39'29"
Mansa Kwame-Turé 3/6/2000
Cyg 20h4'50" 36d14'43"
mansan459tteot
Cyg 21h10'38" 43d56'6"
mansan459tteot
Cyg 21h27'56" 45d52'59"
Mansfield
Uma 11h36'29" 44d49'18"
MANSI
Eri 4h34'5" -15d38'21"
Mansi Jailwala
Sco 16h9'45" -13d39'57"
Mansour
Ori 5h27'22" 2d54'34"
Mansur Alam
Aql 20h5'21" -0d58'27"
Mantea
Pho 1h15'19" -44d55'11"
Mantra
Lyn 7h12'51" 49d8'14"
"MANU"
Dra 18h21'58" 49d6'12"
Manu
And 0h5'25" 40d58'55"
Manu & Flo
Uma 9h50'16" 66d41'2"
manu for ever love
Cep 22h10'14" 56d57'34"
Manu Manu
Ori 5h56'2" 18d37'42"
Manu & Michi
Equ 21h5'5" 9d20'49"
Manu the thirties
Lyr 19h13'59" 28d56'31"
Manu, 13.12.1975
Cyg 19h38'39" 31d52'57"
Manue
Cnc 9h0'10" 7d49'3"
Manuel
Ori 6h15'52" 13d12'35"
Manuel
Ori 6h17'42" 7d8'17"
Manuel
Uma 14h2'53" 50d54'44"
Manuel
Lyn 8h32'13" 50d22'0"
Manuel
Uma 9h37'43" 62d49'52"
Manuel
Uma 9h20'38" 67d44'9"
Manuel
Cam 4h42'42" 67d43'43"
Manuel
Dra 20h26'15" 62d56'0"
Manuel
Sgr 19h38'49" -17d16'39"
Manuel Alarcón Almazán
Gem 6h43'45" 14d59'37"
Manuel Andres Penalver
Aql 20h2'4" 8d41'30"
Manuel Angel Pardo Gutiérrez
And 0h49'0" 35d59'33"
Manuel&Anna
Umi 17h36'8" 80d6'45"
Manuel Antonio Castro Crespo
Psc 0h49'24" 3d28'43"
Manuel Brandon Monsibais III
Lib 15h18'16" -21d17'43"
Manuel Coelho Aguiar
Per 2h26'53" 52d45'6"
Manuel Cunha
Ori 5h57'48" -0d47'35"
Manuel De Jesus Corral
Oct 21h24'21" -75d22'11"
Manuel Debus
Uma 8h56'3" 49d49'48"
Manuel Delgado Jr
Boo 14h36'59" 30d12'37"
Manuel Dente
Cyg 19h55'6" 48d46'0"
Manuel Dima
Ori 6h19'58" 5d56'57"
Manuel Frattini
Uma 9h13'19" 50d45'43"
Manuel Garcia del Rey
Mon 6h59'27" 8d38'16"
Manuel (Grandpa) Fontes
Aqr 22h4'10" -1d11'46"
Manuel Illescas Taboada
Uma 13h2'28" 58d36'5"
Manuel Joaquim SOUSA
Mon 7h44'3" -11d16'13"
Manuel L'Ecuyer
Cep 22h22'20" 60d11'9"
Manuel Lee Chip Oliver, Jr.
Umi 13h17'16" 74d52'32"
Manuel Lo Martire
Dra 19h2'52" 71d40'43"
Manuel Lopez Vidal
Uma 9h35'44" 63d30'58"
Manuel Mannolo Cepeda
Ori 5h26'19" 0d30'48"
Manuel "Manny" Mayenco
Cru 12h22'17" -56d48'37"
Manuel Maria
Aur 6h6'5" 54d5'6"
Manuel My Love
Sco 17h23'47" -32d28'45"
Manuel N. Garza
Com 12h36'39" 30d11'15"
Manuel Pacheco
Lib 15h54'20" -7d6'22"
Manuel "Pedro" Silva Rodrigues
Cap 21h46'26" -12d15'34"
Manuel Perez
Cyg 19h52'58" 54d58'19"
Manuel Peter McPolin
Boo 14h58'50" 54d15'9"
Manuel Planas Estrada
Cnc 9h0'55" 31d23'24"
Manuel R. Curielmillan
Her 16h43'51" 38d26'13"
Manuel Recuero Manzanal (Paspas)
Tau 4h12'9" 12d34'42"
Manuel Ricardo Salas Garita
Cap 21h16'8" -19d5'37"
Manuel Rubio
Uma 10h37'8" 41d26'37"
Manuel Silva
Tau 4h45'31" 24d12'31"
Manuel Sylvest
Cyg 20h11'11" 35d39'2"
Manuel & Tanja
Dra 20h11'33" 73d12'20"
Manuel Torres Jr.
Cep 21h33'57" 68d37'9"
Manuel Valentine
Per 3h33'33" 51d20'12"
Manuel y Norma Coto
Aql 19h39'8" -0d4'6"
Manuel Zürcher
Psc 23h56'14" 5d2'30"
Manuel, Tracy, Leeann
Tri 2h31'15" 35d41'2"
Manuela
Cyg 20h45'48" 32d4'20"
Manuela
Per 2h59'47" 40d54'7"
Manuela
Lyr 19h17'10" 37d49'17"
Manuela
Lyr 18h44'57" 40d3'26"
Manuela
Cam 4h6'4" 55d38'51"
Manuela
Cyg 20h39'39" 45d56'38"
Manuela
Ori 4h58'44" 14d38'18"
Manuela
Ori 5h58'2" 2d40'31"
Manuela
Umi 15h45'20" 77d32'10"
MANUELA
Sct 18h29'40" -15d29'49"
Manuela
Uma 9h52'22" 72d30'14"
Manuela
Uma 14h23'53" 57d38'5"
Manuela
Uma 13h42'54" 53d59'23"
Manuela
Lyn 8h7'38" 44d11'37"
Manuela Alessandrini
Cas 0h30'19" 61d33'18"
Manuela Annis
Cas 0h44'21" 55d58'12"
Manuela Barragan
Cyg 20h0'42" 38d35'20"
Manuela Boss
Com 13h5'10" 15d50'9"
Manuela Carolina Marin
Gem 7h21'15" 15d59'52"
Manuela Cuomo
Vir 13h51'8" -7d24'31"
Manuela de la Hislop
Tau 3h41'41" 27d54'36"
Manuela e Michele
Her 17h44'45" 16d23'45"
Manuela Elisabeth Johanna
Uma 10h38'54" 63d40'27"
Manuela Eva
Cyg 21h51'26" 46d44'5"
Manuela Fasano
Per 2h52'39" 50d51'44"
Manuela Frère
Cas 0h58'40" 54d1'26"
Manuela Frischknecht
Boo 14h8'56" 29d1'33"
Manuela Gehrig
Tri 2h22'20" 37d9'37"
Manuela Gübeli
Cas 23h14'27" 56d11'34"
Manuela Hoefling
Uma 11h33'3" 59d13'11"
Manuela Leal Fernandes - Fofinha
Lib 15h27'57" -11d21'31"
Manuela Maria Ferger
Ori 6h21'18" 10d37'27"
Manuela Mike Pascale
Umi 19h23'39" 88d12'13"
Manuela Ramirez
Cas 23h59'13" 50d31'13"
Manuela Regis Rodrigues de Oliveira
Sco 17h50'35" -44d5'26"
Manuela Rosania
And 2h18'55" 39d57'26"
Manuela Soldi
Boo 14h45'45" 15d3'12"
Manuela Tati Ferrari
Uma 10h40'48" 71d34'56"
Manuela Tomic
Peg 21h36'2" 11d23'43"
Manuela & Tommy forever
Umi 15h32'26" 72d36'56"
Manuela VANNI
Cnc 9h6'43" 27d14'17"
Manuela Vettor
Mon 6h19'25" -5d19'32"
Manuela Weinhardt
Uma 8h34'11" 68d18'43"
Manueli
Uma 11h9'5" 37d34'9"
Manuelito
And 1h18'37" 45d1'48"
ManuellynK.
Ori 5h53'55" 17d45'15"
Manuokalani P C
Leo 11h19'15" 21d51'55"
Manvir Singh Kandola - our Shining Star
Uma 12h24'8" 54d44'4"
Many e Ivonne
Ari 2h43'42" 25d41'34"
Manya Kuzemchenko (muf-fin)
Leo 10h38'44" 18d22'59"
Manye
And 2h22'47" 48d52'30"
Manzelli
Cyg 20h52'15" 34d31'15"
Mao
Cam 3h34'20" 62d23'20"
Mao Armando Belloni
Umi 14h58'39" 80d10'18"
Maou
Psc 1h31'28" 27d44'22"
(Map) Sonya Felicia Walker
Leo 10h10'49" 26d7'16"
Mapi
Per 3h28'28" 41d38'17"
Mapridechotinbell, "M.P. Tink"
Ari 3h9'14" 25d6'41"
Mapstar 010701
Cru 12h37'44" -58d28'53"
Mar
Cas 23h1'24" 56d43'27"
Mar
Leo 9h42'59" 23d24'45"
Mar
Uma 11h7'45" 42d13'48"
Mar Mar
Per 3h35'0" 45d17'39"
Mar Mar
Ari 2h34'25" 24d55'35"
Mar Mar's Star
Tau 5h47'59" 18d15'28"
Mar Mer
Aur 5h44'3" 49d38'12"
Mar San
Ori 5h32'17" 13d1'33"
Mar Skylar
Ari 2h35'22" 13d1'22"
MAR-10.05.1949-Cakehead-Knee
Uma 11h30'30" 34d56'2"
Mar3zuri Alo
Peg 21h29'42" 14d36'53"
Mar5
Lyn 7h38'21" 39d13'5"
Mara
Cyg 21h31'38" 47d29'36"
Mara
Uma 9h2'27" 52d14'4"
Mara
Ori 6h4'15" 19d21'10"
Mara
Uma 10h1'0" 55d52'44"
Mara
Ori 5h44'51" -2d16'45"
Mara
Aql 20h15'34" -0d6'2"
Mara
Vir 12h59'38" -19d10'31"
Mara
And 0h2'3" 37d36'50"
Mara Ann Star Estrada-Surratt
Leo 10h20'37" 22d6'20"
Mara Carinn Thompson
Uma 9h48'21" 61d42'46"
Mara De Bernardo
And 23h8'6" 37d6'4"
Mara Filogamo
Umi 16h17'17" 73d39'21"
Mara Fuffa
Aur 5h52'50" 32d43'28"
Mara Gabrielle Leake
Mon 6h41'57" 7d4'44"
Mara Galeazzo
Cas 0h15'26" 52d11'8"
Mara Gleda
Aqr 21h46'5" -6d31'1"
Mara Jaymes-Marie
Leo 10h17'5" 24d15'50"
Mara Julia
Cas 1h3'36" 60d31'36"
Mara Kaylin Dawe
Uma 8h42'11" 60d53'38"
Mara Kaylin Dawe
Uma 8h24'37" 60d33'34"
Mara La Rocca
Uma 9h23'31" 51d14'47"
Mara Lynn Maxwell
Crb 16h2'44" 36d14'34"
Mara Risa
Uma 10h40'10" 55d41'8"
Mara Rose Zigler
And 0h21'56" 43d14'19"
Mara Star
Umi 13h51'43" 69d36'23"
Mara Star
Sco 16h15'9" -15d57'40"
Mara Tekach
Uma 10h28'5" 66d3'40"
Mara ~ Woofes Cannon Forever
Uma 11h46'14" 43d57'33"
Maráczy Endre
Uma 8h53'15" 70d29'39"
Maradu Paereya Sathasivam
Aql 20h18'17" -7d59'33"
Maral Avetian
Uma 10h27'0" 61d15'54"
Maral Frendjian
Ori 5h19'9" 9d1'33"
Maral Mouchmouchian
And 2h36'11" 43d29'1"
Maral - Princess Mimo Melhem
Cas 0h25'49" 48d9'29"
Mara"la Patrizia"
Peg 22h41'11" 27d22'5"
Maralee Ruth Parker
Ari 2h1'55" 22d17'17"
Maralyn Gilbert Howe "Gwendolyn"
Cas 0h45'21" 58d0'31"
Maranda Davis
Lib 15h45'28" -11d8'15"
Maranda Harris
Psc 0h22'59" 15d22'29"
Maranda Jane
Cyg 21h10'49" 48d23'44"
Maranda Row
And 0h34'47" 37d20'2"
Maranesi Anna
Ori 5h56'3" 12d1'59"
Marania
Ari 2h12'30" 17d15'55"

Mara's wish
Per 4h10'1" 49d42'5"
Marash Gjolaj
Tri 2h13'55" 34d15'54"
Maravilla
Uma 13h29'14" 57d8'12"
MaRaya
Her 16h19'25" 6d30'59"
Maraya Lorraine Lee Lucio
Crb 16h0'53" 38d55'45"
Marbella
Ori 6h15'57" 14d46'17"
Marbles
Uma 13h41'16" 54d8'55"
Marby, Per
Uma 8h56'4" 54d36'54"
Marc
Sco 17h53'39" -35d54'35"
Marc
Uma 9h34'20" 48d46'56"
Marc & Mollie
Sgr 18h45'58" -28d57'5"
Marc 18.5.03
Boo 15h33'53" 46d48'18"
Marc A. Calabrese
Sgr 18h53'7" -18d59'52"
Marc A Sikorski
Aql 20h9'56" 8d27'1"
Marc Aidan Molina
Sco 16h15'2" -12d14'43"
Marc Alan Holliday
Tau 4h27'8" 19d34'20"
Marc Allan Patla
Per 2h53'12" 45d38'55"
Marc Allen Finkhouse
Cas 0h58'34" 60d51'6"
Marc and Aimee Vignocchi
Uma 10h37'47" 57d16'6"
Marc and Annaline Forever
Mon 6h45'44" -0d49'31"
Marc and Charlotte
Cap 21h40'16" -10d23'12"
Marc and Sherry
Aqr 22h35'52" -2d58'54"
Marc and Tonya Wislen
Cyg 20h21'36" 45d41'11"
Marc André Bachofen
And 22h59'51" 38d24'20"
Marc Anthony & Amber's Magical Star
Peg 21h45'34" 20d47'19"
Marc Antonelli
Aql 19h58'5" 8d54'51"
Marc Arlen Nelson
Tau 4h43'23" 29d13'26"
Marc Binggeli
Uma 10h29'36" 54d35'54"
Marc Blesteau
Aqr 22h48'41" -22d42'33"
Marc Blewitt - Marc's Guiding Star
Ori 6h18'51" 14d20'1"
Marc Bourdages
Sco 17h43'46" -39d56'54"
Marc C. Keller
Uma 10h50'47" 45d40'48"
Marc Cairoli
Per 2h41'53" 56d6'16"
Marc Charles Weir
Umi 17h1'20" 80d11'19"
Marc Chavanis
Com 13h4'39" 16d5'26"
Marc & Claudia Player
And 2h17'11" 50d26'58"
Marc & Cris
Her 17h59'59" 25d50'22"
Marc Cruz
Cyg 19h39'14" 32d50'37"
Marc D. Howard, Jr.
Vir 12h38'52" 2d34'18"
Marc D'Angelo
Per 3h48'29" 50d55'9"
Marc Daniel
Her 16h32'21" 48d58'31"
Marc Davis
Her 16h19'1" 46d57'23"
Marc Dellenbach
Crb 16h9'19" 36d9'14"
Marc Dominik
Lyr 18h43'1" 40d0'36"
Marc E. Johnson II
Dra 18h38'35" 56d37'59"
Marc' Ele
Cam 5h35'19" 71d27'16"
Marc Enver Torres "Spen"
Tau 4h6'0" 6d1'32"
Marc Eric Jensen
Lib 15h41'20" -12d43'24"
Marc Estévez Arnau 01/02/1983
Lyn 8h15'36" 50d25'7"
Marc et Caroline
Psc 22h57'18" 6d18'43"
Marc F Peterson
Uma 11h46'23" 55d0'55"
Marc Fetscherin
Lyr 19h11'52" 34d23'3"
Marc Fielmann
Uma 8h24'53" 61d21'8"
Marc Franco
Cep 22h29'23" 71d39'45"
Marc G Bélanger
Cyg 20h50'52" 31d22'0"
Marc Gittelman
Vir 14h1'33" -0d52'31"
Marc Glatz
Uma 9h48'56" 48d14'25"
Marc Guillaume
Cma 7h5'28" -30d38'43"
Marc Heany
Lyn 7h7'57" 58d49'43"
Marc Henzi, 16.11.2003
Ori 6h5'53" -3d57'22"
Marc Hopkins
Tau 3h45'5" 9d35'30"
Marc I Brodlieb
Aqr 22h33'8" -0d56'9"
Marc J Dyke
Dra 20h17'56" 69d0'48"
Marc J. P. Archambault
Uma 10h27'37" 49d40'44"
Marc Jackson Shelton
Her 17h51'47" 23d26'5"
Marc Jacob Duquella
Lib 15h5'18" -28d35'0"
Marc James Muswell
Cep 22h48'41" 70d20'1"
Marc & Jenny Pekny's 1st Christmas
Aql 19h14'5" -0d36'26"
Marc John Davies
Per 4h2'57" 45d23'42"
Marc & Katy Normington
Her 17h41'20" 37d31'11"
Marc Koller
Her 16h28'33" 14d30'2"
Marc L. Marshall
Per 2h52'42" 55d59'44"
Marc L. White
Psc 0h50'8" 27d49'3"
Marc Lamothe
Psc 1h54'50" 7d7'18"
Marc Landtwing
Per 2h52'48" 54d49'21"
Marc Lane Robbins
Tau 3h44'19" 29d33'27"
Marc Latham
Sgr 19h8'19" -24d33'16"
Marc & Laura Gallant
Cyg 20h19'32" 52d44'51"
Marc Lavett
Leo 10h2'23" 6d45'1"
Marc Lenny Cecchet
Ori 6h9'44" 9d57'3"
Marc Leon Julian Pachler
Cnc 9h7'6" 11d49'35"
Marc Lowenberg
Psc 0h19'40" 3d20'32"
Marc Martinez
Per 4h25'25" 50d44'35"
Marc May
Her 17h22'41" 14d37'55"
Marc Mertz
Crb 15h47'23" 36d27'56"
Marc Michael Beda
Lac 22h44'40" 36d23'6"
Marc Mora Ventura
Tau 5h34'57" 17d38'52"
Marc & Nica
Cyg 19h45'25" 44d48'31"
Marc Nicholas Jennings
Ori 4h46'49" -0d26'50"
Marc Oldham
Per 3h7'5" 37d48'54"
Marc P. Clements
Ori 6h16'13" 14d59'34"
MARC P. GRUENBERG
Leo 11h0'20" 4d31'0"
Marc P. Pearson
Uma 8h55'35" 66d50'56"
Marc P. Pfeffer
Uma 10h21'7" 44d28'46"
Marc Pankin
Psc 1h28'41" 26d9'58"
Marc Pattee
Her 17h46'46" 35d26'49"
Marc Paustian
Gem 6h46'35" 33d32'5"
Marc Pfüpf
Aql 19h32'44" 9d33'19"
Marc Poizot
Psc 1h31'53" 16d19'15"
Marc R. J.
Her 18h46'37" 19d59'55"
MARC R. WELTON
Psc 1h43'8" 27d45'2"
Marc Riccio
Cmi 7h27'43" 8d29'3"
Marc Richards Birthday Star
Uma 11h0'30" 61d27'35"
Marc Robert Achermann
Vul 19h53'28" 23d56'57"
Marc & Romy
Ori 5h8'15" 7d52'19"
Marc Saghbini
Sco 16h54'48" -31d44'46"
Marc & Sarah's Australien
Ori 5h38'26" -3d14'13"
Marc Scott Indelicato
Gem 7h0'56" 16d20'12"
Marc & Sean Always
Tau 3h49'33" 28d28'52"
Marc Simmons
Cma 6h22'46" -30d30'58"
Marc Sorgman
Dra 18h33'37" 76d37'26"
Marc Steingrand
Cra 18h12'17" -36d51'37"
Marc Stephen Dowdy
Uma 10h57'5" 44d20'31"
Marc Szymanik
Psc 1h31'44" 13d2'18"
Marc & Tammy's Anniversary Star
Cyg 21h39'2" 54d55'46"
Marc Terenzi
Per 3h9'37" 54d19'47"
Marc Thomas Robinson
Lmi 10h16'0" 30d50'41"
Marc "Tiko" Murin
Vir 12h12'10" 8d56'53"
Marc Timothy Smith
Her 17h52'22" 50d0'30"
Marc Todd Abrams, PhD
Lib 14h42'34" -9d56'58"
Marc Toquebiau
Umi 13h51'56" 78d16'49"
Marc Torras Piulachs
Gem 7h17'19" 19d40'54"
Marc Wächter
Cyg 19h42'13" 29d2'5"
MARC WASHINGTON
Vir 12h50'53" -1d2'31"
Marc William Hutchison
Uma 9h12'58" 55d0'56"
Marc Winslow
Uma 10h7'51" 56d59'11"
Marc Younker
Ari 2h10'37" 24d58'52"
Marc, 03.03.1956
Her 16h12'6" 22d19'0"
Marca
Cnv 13h20'39" 45d38'59"
Marc-André Daunais
Cnc 8h32'57" 7d18'40"
Marc-Antoine
Col 5h10'4" -28d20'0"
Marc-Antoine et Scaphandre
Boo 14h30'6" 19d57'37"
MarCasie
And 23h16'30" 48d6'16"
Marcco and Candice
Cyg 20h5'57" 54d33'25"
MARCEL
Ori 6h18'48" 10d53'12"
Marcel Alden Moore "Mar"
Lyr 18h47'4" 30d36'12"
Marcel and Joyce Pons
Cas 1h8'0" 63d38'51"
Marcel André Pérez Graulau
Tau 4h22'39" 4d20'9"
Marcel Bruno Butcher
Tau 4h26'35" 19d50'14"
Marcel Büchi
Uma 12h59'10" 55d12'53"
Marcel Dagher
Cra 18h15'9" -40d6'54"
Marcel Desrosiers 09-07-1945
Cep 21h10'27" 67d4'46"
Marcel Fortin
Cep 21h42'31" 72d2'4"
Marcel Haasis
Cas 23h6'30" 55d46'52"
Marcel Harmann
Cas 1h34'17" 60d43'3"
Marcel "Hasi" Jounais
Uma 10h21'22" 55d43'19"
Marcel Herman Ludwich Stomp
Boo 14h51'50" 29d54'3"
Marcel Howald, 15.01.1981
Ori 5h57'31" 22d14'58"
Marcel Huppé
Gem 6h48'35" 18d24'18"
Marcel Isaac Russell Adams
Aqr 20h53'8" -9d47'49"
Marcel Kappeler
Cnc 8h38'30" 30d44'46"
Marcel Levy
Ori 5h5'34" 9d41'29"
Marcel Mensink
Aur 5h23'22" 44d56'17"
Marcel Mischler
Uma 13h56'4" 49d27'22"
Marcel Nguyen
Lyr 19h7'13" 33d44'59"
Marcel Q. Coronel
Ori 6h8'57" 11d11'2"
Marcel & Robin Sim
Lyr 19h15'32" 40d53'53"
Marcel Schmidberger
Vul 19h49'58" 22d46'3"
Marcel Theriault
Sgr 18h58'37" -35d15'11"
Marcel und Andrea
Lyr 18h56'30" 33d1'42"
Marcel Wessels
Sco 16h10'51" -10d55'55"
Marcela
And 1h11'11" 43d19'6"
Marcela
Her 18h5'23" 37d34'53"
Marcela
Cyg 21h59'30" 48d14'15"
Marcela Aragon
Ori 6h21'21" 14d44'53"
Marcela B. Flores
Leo 11h24'56" 9d1'36"
Marcela Bracamonte Laguna
Ari 1h51'59" 12d41'44"
Marcela Eusse
Cas 0h31'31" 61d36'36"
Marcela Stephanie Amaya
Gem 6h42'57" 20d34'49"
Marcela "The Star of Brasil"
Ari 2h34'9" 28d1'47"
Marcela's Fire
Aqr 22h56'19" -6d30'3"
Marcelene Williams Painter Plemmons
And 1h58'24" 42d44'41"
Marcelia: The Love of My Life.
Ari 2h27'58" 26d58'4"
Marcelina Alexandra Samuel
Cnc 8h33'57" 7d35'57"
Marcelina Lindo-Smith
Cap 21h30'10" -19d52'33"
Marceline Strain
Crb 16h14'59" 37d42'8"
MARCELINO
And 0h46'42" 23d34'55"
Marcella
Leo 11h27'26" 21d5'35"
Marcella
Her 18h54'4" 23d28'46"
Marcella
And 1h12'10" 44d1'38"
Marcella
Cap 21h2'43" -17d3'37"
Marcella
Cas 23h3'8" 55d15'31"
Marcella
Sco 17h22'3" -33d1'29"
Marcella A. Meehan
Cep 23h26'22" 71d41'30"
Marcella Albano 1
Uma 9h46'8" 47d43'14"
Marcella Alexia Montesinos
Mon 6h51'22" -0d33'25"
Marcella and Patrick
And 23h9'31" 35d24'10"
Marcella Carlock
Gem 6h45'46" 26d33'14"
Marcella di Prospero
Umi 17h26'10" 86d18'53"
Marcella Gill
Uma 11h46'42" 50d20'43"
Marcella & Ingo
Uma 9h38'5" 49d23'7"
Marcella J. Smith
Peg 21h40'46" 23d44'54"
Marcella L. McEniry
Cap 21h29'19" -18d40'55"
Marcella Mae Boyle Hoover
Vir 13h40'8" -17d41'40"
Marcella Marie Bee
Cas 0h40'51" 51d2'48"
Marcella Mary Tovar
Cap 21h30'36" -8d59'42"
Marcella Sabine (Hasenpfötchen)
Uma 11h28'12" 28d55'47"
Marcella Scire
Sgr 18h24'25" -22d31'57"
Marcella Simona Franzese
And 2h13'24" 49d31'8"
Marcella V. Russell
Gem 6h29'54" 17d42'52"
Marcella Veronica Duba
Uma 12h28'24" 54d11'11"
Marcelle
Uma 10h28'1" 57d58'3"
Marcelle Léontine Juliette
Vir 13h10'52" -19d14'27"
Marcelle Marie Major
Cnc 8h48'41" 31d35'53"
Marcellino
Uma 9h35'10" 56d2'37"
Marcello
Ori 6h18'33" 12d38'40"
Marcello
Eri 1h51'1" -52d36'7"
Marcello Coscia
Umi 15h57'47" 85d44'30"
Marcello Deon Flowers
Psc 1h24'9" 11d20'40"
Marcello e Paola
Lep 5h54'35" -19d12'50"
Marcello Facco 16/1/1931
Peg 22h25'54" 16d53'15"
Marcello Ficaloro
Cap 21h40'59" -16d6'34"
Marcello Luis
Boo 14h53'19" 52d29'5"
Marcello Verrocchi 15th March 1966
Psc 1h7'4" 12d17'36"
Marcellus Dee Mecham
Per 2h24'59" 54d53'39"
Marcel-Marie LeBel
Ori 6h14'3" 13d58'58"
Marcelo Fonseca
Cru 12h14'50" -56d51'12"
Marcelo Tello Campodonico
Lyn 7h46'40" 41d22'27"
Marcelo Veras The Great
Her 18h47'13" 18d51'38"
Marcelo y Aurora
Cam 5h52'33" 56d16'50"
Marceny-Penelo
Cru 12h56'4" -59d39'48"
Marcey
Lyn 6h52'4" 53d5'37"
Marcey Virginia Pettus Lemmond
And 1h5'17" 45d47'48"
March Andrew Maher
Lib 15h11'23" -2d22'59"
MarChar
Cap 20h20'23" -22d16'27"
Marchar
Aur 6h30'9" 36d26'23"
Marc-Henry & Catharine Anne David
Cam 4h11'24" 66d17'56"
Marchesini Annamaria
Umi 15h2'18" 73d5'21"
Marci
Umi 13h22'28" 73d54'37"
marci
Cas 23h30'1" 53d0'57"
Marci
And 1h21'57" 43d22'28"
Marci and Mabs
Cap 21h40'29" -14d17'31"
Marci Bee Dayton *Pickle Puss Jr.*
And 1h10'59" 45d51'6"
Marci Beth Cabrera
Vir 14h4'7" -9d46'6"
Marci "Cat" Kaczynski
Lmi 10h21'59" 35d54'18"
Marci Clark
Cnc 8h50'55" 11d17'35"
Marci Donovan
Sgr 18h50'41" -31d55'27"
Marci Gauthreaux
Sco 16h8'23" -18d7'54"
Marci Jean Nichols
Com 12h45'37" 28d38'41"
Marci Lynn
And 0h6'6" 44d4'11"
Marci Morrison
Lyn 8h2'13" 47d30'48"
Marci Nicole
Gem 7h26'9" 26d22'43"
Marci Renee Paul
Tau 3h53'5" 19d49'22"
Marci Robyn Hayden
Ari 2h0'53" 18d46'36"
Marci Seltzer
Ori 6h14'52" 15d35'17"
Marci Susan Kornblut
Cyg 19h55'4" 56d45'27"
Marci Taub
Cyg 19h54'39" 30d24'11"
Marcia
Cyg 20h0'51" 37d22'13"
Marcia
Uma 9h39'43" 44d5'32"
Marcia
And 0h9'40" 45d18'33"
Marcia
Tau 3h34'38" 14d39'48"
Marcia
Cnc 8h50'29" 22d35'14"
Marcia
Tau 5h53'53" 28d18'32"
Marcia
Tau 3h42'23" 28d43'41"
Marcia
Dra 18h58'5" 66d3'48"
Marcia 52
Cnc 8h33'17" 15d53'52"
Marcia A. Stephans
Lib 15h20'44" -18d9'26"
Marcia Ann
Vir 11h55'21" -4d22'51"
Marcia Berg
Gem 7h14'54" 16d45'34"
Marcia "Booba" Gambale
Aqr 21h33'16" 1d50'33"
Marcia C. "Sunshine" Hendrick
Cas 0h18'57" 55d40'48"
Marcia Cassidy
Cyg 20h3'22" 34d7'36"
Marcia Dos Santos Costa
Uma 11h15'47" 41d37'38"
Marcia Elizabeth Hook
And 1h41'42" 42d16'30"
Marcia Feinstein
Sgr 18h10'33" -18d45'11"
Marcia Ferstle
Tau 5h52'27" 15d36'37"
Marcia Gail
Cas 0h19'57" 61d55'24"
Marcia Geary
Cyg 19h39'9" 34d40'33"
Marcia Hagedorn
And 2h35'49" 43d15'24"
Marcia Harrison
Vir 12h10'27" 7d11'18"
Marcia Hazard
Uma 11h53'11" 58d34'15"
Marcia & Howard
Lyr 19h6'42" 42d33'14"
Marcia J Miller The Star of My Life
Cas 23h40'37" 54d59'23"
Marcia J. Misiorski
Cnc 8h53'36" 31d58'32"
Marcia Jane
Cyg 20h52'53" 31d10'5"
Marcia Jane Kerr
Psc 1h17'8" 18d30'37"
Marcia Jean Moses
Per 4h49'22" 46d24'27"
Marcia K. Mitchener
And 1h44'26" 43d43'53"
Marcia K. Wharton Thrush Kasper
Vir 14h4'7" 3d17'16"
Marcia Kay McKenna
Uma 12h7'4" 51d52'17"
Marcia Kay Page Altshuler
Cyg 20h55'3" 47d10'58"
Marcia Kay Radelet
Per 3h46'56" 33d20'29"
Marcia Kay Smith
Sco 16h18'51" -10d39'22"
Marcia Koch
Vir 13h28'10" -6d11'6"
Marcia Kosin
And 1h27'41" 47d21'53"
Marcia Kyburz
And 0h29'41" 37d39'48"
Marcia L. Arnold
Lmi 10h33'38" 33d50'46"
Marcia Lee Leonard
Ori 5h59'11" -0d6'29"
Marcia Loves Kathy Beyond Infinity
Cyg 20h50'52" 32d37'23"
Marcia Loves Will
And 0h20'31" 42d57'57"
Marcia Lussier
Tau 4h34'51" 13d32'29"
Marcia Lynn Laporte
Lib 15h58'12" -17d23'24"
Marcia Mandi
Cas 23h39'38" 62d40'1"
Marcia Marie Shuchat
Lyr 18h40'19" 31d47'0"
Marcia Mattice - Rising Star
Cam 4h34'2" 73d32'35"
Marcia McMurtrey's Star
Cyg 21h18'27" 34d41'19"
Marcia Menard
Tri 2h30'30" 30d48'31"
Marcia My Love
Cas 0h25'57" 61d7'24"
Marcia (puce russe)
Dra 18h8'16" 79d53'58"
Marcia Regina
Sgr 19h8'13" -18d19'22"
Marcia Resnick
And 2h38'3" 43d37'56"
Marcia Richman
Cas 1h32'8" 62d37'4"
Marcia & Robert Skelly
Cyg 20h7'14" 54d16'33"
Marcia Rosati Rainbows Butterflies
Uma 14h5'46" 53d11'17"
Marcia S Boothe & Tina L Parks
Ori 6h9'31" 13d55'35"
Marcia Shelite
And 0h40'19" 45d41'56"
Marcia T. DeWitt - Botanical Artist
Lyr 18h50'12" 43d51'48"
Marcia & Terry Leek
Cap 20h8'44" -8d37'25"
Marcia Traer
Mon 6h45'46" -0d2'44"
Marcia Yoko French
Cas 0h25'25" 48d38'19"
MarciaandDickWalker
Cyg 20h47'10" 42d51'18"
MarciaBonita
Ori 4h51'17" 6d43'16"
Marciadee
Cyg 21h20'1" 36d31'17"
Marcial Mateo
Cnc 8h30'22" 26d43'24"
Marciann,Timothee,Brendyn and Loren
Ori 5h8'16" 15d2'24"
Marcianne
Per 3h33'46" 40d29'8"
Marcia's Gift
Vir 14h1'45" -17d39'20"
Marcia's Millenium Marble
Ori 6h13'23" -0d50'31"
Marciastar
Cyg 21h42'49" 48d9'21"
Marcie
Vir 12h39'8" 3d29'46"
Marcie
Umi 8h48'43" 86d55'58"
Marcie
Pic 5h10'26" -56d12'20"
Marcie Olive Burnett
Gem 6h26'48" 25d51'30"
Marcie Segura
Ori 6h4'51" 19d48'25"
Marcie "Squirrel"
Del 20h33'20" 13d45'52"
Marcie-Mu
Cyg 20h32'6" 33d33'54"
Marcie's Light
Crb 16h15'18" 32d57'43"
Marcie's Sun
Umi 14h53'37" 78d59'44"
Marcifer
Pho 0h33'18" -43d25'34"
Marcille Travis Hill
Lmi 10h41'9" 30d47'31"
Marcin & Ewelina
Cyg 21h44'2" 49d18'55"
Marcin Kniaz
Cap 20h50'10" -24d32'16"
Marcin Kolodynski
Uma 10h5'34" 62d18'16"
Marcin M.
Uma 11h37'48" 53d8'26"
Marcin Maslowski
Lib 15h0'5" -24d56'23"
Marcinek Magnus
Uma 12h59'3" 53d34'23"
Marcio & Valerie
Vir 13h30'39" -21d37'30"
Marcirena
Uma 9h40'55" 48d35'35"
Marci's Jewel
Crb 16h10'5" 39d14'22"
Marci's Star
Ari 3h24'48" 26d1'12"
Marcky 30
Aur 5h12'30" 42d8'8"
Marco
Aur 4h53'43" 39d40'26"
Marco
Per 3h27'57" 34d54'33"
Marco
Uma 11h12'37" 46d5'33"
Marco
Cyg 21h13'59" 46d25'2"
marco
Cas 1h1'11" 56d32'48"
Marco
Ori 5h58'57" 21d28'59"
marco
Boo 13h44'39" 18d20'6"
Marco
Umi 17h4'38" 80d12'44"
Marco
Lib 15h7'57" -6d48'50"
Marco
Cas 23h27'54" 57d52'28"
marco
Uma 10h3'8" 61d22'31"
Marco
Cep 20h55'11" 63d42'3"
Marco
Umi 14h55'30" 68d54'28"
Marco & *Désirée*
Uma 9h38'56" 49d33'25"
Marco '05
Umi 16h8'35" 79d32'32"
Marco 55 "9-13-50"
Vir 11h57'36" 8d4'5"
Marco A. Baltazar, Jr.
Her 17h40'25" 38d15'8"
Marco A. Sanchez
Tau 4h21'44" 10d12'36"
Marco Albergo
Uma 9h37'46" 57d45'41"
Marco Alexandre
Peg 22h22'27" 6d26'50"
Marco Andre Rua Cruz
Cnc 9h12'40" 26d47'34"
Marco Antonio Daniel
And 2h31'32" 44d18'27"
Marco Antonio Hernandez
Hya 9h26'44" -10d58'49"
Marco Antonio Soto
Crb 15h27'37" 26d17'13"
Marco Babbi
Ori 6h12'58" 5d58'22"
MARCO BARRETTO
Lyn 7h50'48" 58d57'1"
Marco Bartolomeo
Uma 8h39'32" 71d7'9"
Marco Beddoni
Cyg 19h49'19" 38d48'10"
Marco Benso
Tri 2h30'53" 31d15'55"
Marco Bertolini
Her 18h4'33" 36d18'49"
Marco Chimilio
Psc 1h7'43" 18d54'59"
Marco Cifuentes
Uma 10h28'54" 45d33'23"
Marco Dörig
Peg 23h22'43" 32d4'6"
Marco e Ambra
Her 18h33'55" 18d55'15"
Marco e Samanta
Crb 15h57'39" 26d31'54"
Marco et Kathia
Del 20h34'26" 3d37'24"
Marco Ferreira
And 0h30'51" 33d41'13"
Marco Fischer-Stocker
And 2h27'38" 43d4'55"
Marco Furrer
Cas 1h7'9" 49d43'9"
Marco Gachnang
Cyg 20h35'12" 42d14'48"
Marco Geboers
Uma 11h20'28" 31d24'31"
Marco Giambattista
Cep 22h39'54" 66d1'37"
Marco Gian Novaro
Her 17h36'37" 40d27'44"
Marco Giorgi (Cathy's Shining Star)
Lyn 7h14'53" 58d14'24"
Marco Gottardello
And 1h49'39" 42d35'28"

Marco Hardmeier
Cas 23h32'7" 59d56'2"
Marco Indelicato
Lyn 7h50'21" 57d1'31"
Marco + Ines
Leo 11h2'4" 20d20'15"
Marco Krebs
Uma 10h22'37" 43d0'16"
Marco Kuhnen
Uma 10h2'23" 47d52'47"
Marco I hope too
Uma 9h29'59" 44d25'31"
Marco Laubscher, 05.05.1966
Cas 0h16'52" 54d24'2"
Marco Ledri
Gem 7h6'20" 18d32'54"
Marco Liberati 14/04/1979
Peg 21h44'48" 21d14'29"
Marco & Luce's Monkey Wedding Star
Cyg 19h47'52" 34d49'22"
Marco Maltempi
Cam 5h14'1" 65d10'33"
Marco Mandolfo
Umi 15h12'9" 82d37'49"
Marco Manuel
Uma 13h35'14" 53d20'49"
Marco*Maurizi
And 2h16'11" 42d29'30"
Marco Meli, 10.08.1969
Cep 0h28'21" 77d52'33"
Marco Muzzioli
Per 2h52'30" 43d55'27"
Marco Nardelli
And 1h42'6" 45d35'13"
Marco & Pamela
Cam 7h23'54" 64d27'4"
Marco Piellucci
Cam 7h59'9" 70d32'5"
Marco Potestio
Ori 4h53'59" -0d54'48"
Marco Reese
Uma 9h0'25" 54d14'57"
Marco Rodolfi
Cas 23h38'28" 53d49'42"
Marco Samuel Scarmardo
Uma 10h29'41" 47d19'49"
Marco Santagata
Dra 17h55'19" 52d13'22"
Marco Santoro
Leo 9h25'17" 6d51'42"
Marco Schröder
Uma 10h7'12" 72d31'24"
Marco Sciarrillo
Umi 16h24'50" 71d18'34"
Marco Solo
Uma 11h19'1" 60d22'22"
Marco & Sonja
Umi 14h13'15" 65d26'9"
MARCO SPEZIALE
Cyg 20h40'58" 45d38'29"
Marco Su
Lib 15h11'21" -10d46'13"
Marco the Star
Sgr 18h32'56" -28d7'34"
Marco Truckenbrod-Fontana
Per 4h17'26" 34d36'8"
Marco & Valeria
Mon 7h9'28" -0d46'8"
Marco Venturini
Uma 8h39'6" 56d51'49"
Marco Villa
Cyg 20h25'21" 33d28'19"
Marco Virgara
Uma 9h15'7" 65d53'46"
Marco Wyatt
Her 17h50'58" 15d41'31"
Marco Zampieri
And 23h51'37" 43d44'34"
Marco Zanetti
Aql 19h18'3" 5d5'3"
Marco Zingg
Dra 20h25'41" 67d35'23"
Marco, Marty, Miguel Aexel
Dra 16h38'21" 68d6'34"
Marcolis
Ori 6h6'33" 10d49'54"
Marc-Olivier
Her 15h50'51" 40d10'11"
Marcos
Lib 14h24'6" -14d38'42"
Marcos Aguado
Lyn 7h29'19" 53d47'15"
Marcos and Teresinha Hoff
Sgr 18h22'39" -33d1'33"
Marcos del Sol
Cam 6h33'45" 66d23'9"
Marcos Diaz-Huerra
Vir 12h20'2" 10d1'1"
Marcos Geraldo
Psc 1h6'51" 26d8'49"
Marcos Gorman Yanes
Lib 15h4'46" -12d15'31"
Marcos Mario Garcia
Her 18h14'44" 22d42'27"
Marco's Miracle
Cnc 8h45'25" 14d49'42"
Marcos Ortega
Ari 3h2'56" 26d50'14"
Marcos Pilot Jato
Sco 16h12'12" -13d29'47"
Marcos Rocha III
Uma 10h41'13" 67d33'29"
Marcos Villalobos Love You Grandma
Umi 14h52'49" 74d17'53"
Marcos y Isabel
Cyg 19h53'37" 36d4'47"
MarcoSilvia
Lyn 6h27'39" 61d4'46"
Marcquesé Paisley
Lyn 6h49'29" 55d46'56"
Marcrista
Vel 9h53'22" -51d14'48"
Marc's Eye
Sgr 19h32'16" -37d59'29"
Marcsi és Balázs Boldogságcsillaga
Uma 11h8'50" 69d36'52"
MarcTomko
Ori 5h46'31" -2d2'3"
Marcus
Cas 0h52'56" 60d11'45"
Marcus
Per 2h43'43" 55d47'59"
Marcus
Her 16h47'53" 23d9'12"
Marcus
Leo 9h35'56" 13d12'36"
Marcus
Aql 19h46'47" 7d31'52"
Marcus
Vir 14h19'29" 5d50'10"
Marcus 06-18-06
Lyr 18h51'36" 41d6'52"
Marcus Adam Brischetto - 14.11.2005
Sco 16h42'15" -42d11'51"
Marcus Alan Reyna
Ori 6h13'50" 8d50'32"
Marcus and Amanda's Star
Her 17h11'50" 19d32'2"
Marcus and Nicole Brown
Psc 1h21'53" 14d45'35"
Marcus and Tanya's Star of Love
Cyg 20h0'53" 51d48'39"
Marcus and Taylor Williams
Cyg 21h26'22" 53d20'31"
Marcus Andreas Krydsby - Haug
Cap 21h39'49" -19d24'52"
Marcus Andrew Golczynski
Her 16h17'14" 43d51'32"
Marcus Andrew Hernesman
Dra 10h1'44" 75d16'39"
Marcus Andrew Morris
Gem 6h31'3" 19d30'6"
Marcus Andru Figueroa
Cyg 20h59'9" 39d13'35"
Marcus Anthony Fisher
Lib 15h18'44" -19d22'44"
Marcus Anthony Lovingood
Uma 9h9'15" 60d54'37"
Marcus Antwone Simmons
Aqr 23h5'56" -8d35'31"
Marcus Arthur Elrod
Sco 16h12'0" -13d22'3"
Marcus Aureus
Ori 5h20'28" 5d48'11"
Marcus Bolton
Per 2h43'56" 53d19'23"
Marcus Bright
Per 3h10'18" 54d23'56"
Marcus Crudus
Per 2h19'16" 55d29'34"
Marcus Cutts
Hya 10h12'0" -13d54'23"
Marcus D. Fink
Lib 14h51'53" -20d23'54"
Marcus D Miller
Psc 0h26'58" 7d36'37"
Marcus D. Moore
Cep 21h25'34" 77d31'38"
Marcus Damion Anderson
Cnc 8h57'16" 7d50'38"
Marcus DeAngelis
Per 2h49'8" 54d8'54"
Marcus Dominique Johnson 3/2006
Ari 2h33'53" 24d7'54"
Marcus E. Guzman
Lib 15h7'11" -16d26'56"
Marcus E. Yandle - A Song at Twilight
Cmi 7h28'47" -0d0'51"
Marcus Earl
Boo 15h20'11" 42d23'8"
Marcus Elric Lane
Dra 18h54'41" 69d41'2"
Marcus Elwood Gumm
Lib 15h10'50" -6d0'45"
Marcus Estrella
Cep 23h39'42" 74d36'43"
Marcus Everett Cole
Umi 13h58'59" 75d0'48"
Marcus Ezekiel
Her 16h44'23" 22d44'28"
Marcus Haycook-O'Hara
Uma 10h12'26" 55d8'46"
Marcus Horner
Uma 11h0'3" 35d34'50"
Marcus Jai Signorelli
Sco 17h29'9" -35d33'12"
Marcus John Morgan Jinks
Umi 16h40'52" 76d48'57"
Marcus Joseph Nistico
Psc 0h9'48" -1d57'17"
Marcus Knox
Her 18h19'1" 14d36'21"
Marcus Lee Breaux
Aql 20h35'36" -0d44'49"
Marcus McDonald
Umi 14h47'24" 72d29'25"
Marcus Miguel Broster
Her 16h18'28" 3d43'56"
Marcus Mitchell Cush
Gem 7h3'2" 9d51'26"
Marcus Moody The Mighty Warrior
Uma 11h58'41" 44d44'58"
Marcus N. Nisar
Per 3h32'22" 37d22'47"
Marcus Neal McCalister
Tau 5h58'16" 23d31'59"
Marcus Patrick Moody 'Buster Boo'
Per 3h22'41" 39d2'10"
Marcus Paul Donkersley
Per 3h22'23" 39d33'40"
Marcus Pullen
Umi 15h58'44" 80d1'5"
Marcus Randall Haycook-O'Hara
Per 4h14'22" 36d29'41"
Marcus Robert Heygster
Tau 5h35'25" 27d26'38"
Marcus Roger Mabey
Dra 17h24'25" 56d23'32"
Marcus & Sara
Col 6h3'46" -28d15'26"
Marcus Schwiening
Uma 10h1'25" 41d28'56"
Marcus & Sharon Roden
Cyg 21h7'35" 46d22'49"
Marcus' Shining Star
Boo 13h56'7" 23d27'15"
Marcus Solomon
Cep 22h26'23" 73d9'19"
Marcus' Star
Dra 19h9'47" 67d17'48"
Marcus Stevens
Cep 4h34'52" 80d33'51"
Marcus Terence Coleman
Ari 2h48'52" 22d15'54"
Marcus Tyler Rhodes
Aqr 22h33'9" -1d19'38"
Marcus V. Ray Sr.
Lmi 11h0'26" 26d41'24"
Marcus Volle
Tri 2h26'14" 36d58'3"
Marcus Walter Schoenberg
Ari 3h13'48" 27d6'13"
Marcus William McCarty
Lyn 6h36'59" 61d51'30"
Marcus Woods
Per 3h28'31" 51d26'52"
Marcus2
Aur 5h18'31" 43d1'27"
Marcusoni
Aur 5h22'5" 30d20'24"
Marcus's Star
Cas 23h50'35" 52d40'23"
Marcy
Dra 16h30'19" 53d41'41"
Marc"y"
Umi 13h42'1" 72d57'39"
Marcy
Psc 0h33'6" 8d57'1"
Marcy and Bill
Lyn 6h59'54" 49d26'7"
Marcy Crouser
Lib 15h4'28" -12d6'54"
Marcy Cusimano
Ari 2h7'40" 23d9'54"
Marcy Fancher Green
Com 13h0'28" 19d34'29"
Marcy Grundner
Cas 23h54'2" 56d33'50"
Marcy Jacomet
Vir 13h41'35" 3d33'23"
Marcy Jane Lyle Christian
Psc 1h22'37" 17d42'46"
Marcy Jeannine Smith
Sco 16h10'35" -15d35'30"
Marcy Kay
Psc 1h20'31" 9d48'40"
Marcy Kupiec
Ari 2h39'45" 29d38'8"
Marcy L. Kassie M. Devon L. Ziegler
And 23h25'24" 50d43'37"
Marcy L Rhoades
Sco 17h57'18" -32d22'59"
Marcy Lynn Davis
Sgr 18h16'32" -27d57'5"
Marcy Malay
Psc 0h59'25" 7d55'27"
Marcy Maxwell
Sco 16h14'3" -12d26'7"
Marcy McConaughey
Psc 0h17'46" 11d40'38"
Marcy Nadel
Aqr 22h34'12" -1d23'36"
Marcy Ramirez
Cas 23h29'11" 52d17'57"
Marcy Rose Dion
Lyn 7h30'49" 35d56'10"
Marcy Stevens
Cyg 20h12'38" 44d7'34"
Marcy Stoddard
Cep 21h49'41" 68d45'10"
Marcy Studzinski
Del 20h47'27" 14d38'30"
Marcy Winsch
Vir 13h41'40" 2d6'33"
Marcy, Karl & Nolan
Lyn 8h41'3" 33d54'38"
Marcy-Caitlyn-Andrew Mother's Day
Cam 7h57'35" 69d43'9"
Marcyne
Sco 17h22'40" -32d33'59"
Marcy's Ray, Illume My World
Crb 15h50'22" 27d26'56"
Marcy's Very Own Big Bright Star
Vir 14h40'34" 2d25'21"
Mardel Bertha
Lib 15h25'0" -16d9'12"
Mardell H. Kisner
Vir 12h49'48" 1d41'27"
Mardella Henrie
Sco 17h44'11" -35d11'42"
Mardelle Feenstra
Psc 0h24'23" 18d32'57"
Mardelle Louise Vonderohe 2004
Uma 10h48'29" 62d26'41"
Mardene Willis
And 2h36'20" 40d59'28"
Mardobug
Uma 9h57'5" 46d4'32"
Mare
Ari 2h40'30" 15d37'26"
Mare
Uma 9h43'12" 67d1'8"
Mare
Dra 17h17'44" 53d12'8"
Mare and Bear
Psc 1h9'35" 26d44'42"
Mare Baer
Sco 16h13'18" -15d47'57"
mare sweetness
Ari 2h26'3" 22d28'19"
Mare Tranquil
Sco 16h39'15" -40d6'14"
Marea S. Thompson-Hunter
Gem 6h13'17" 24d41'19"
Maree C. Dick - Our Great Nan
Sgr 19h13'46" -23d26'50"
Maree Finkelstein's Starlight
Pyx 8h59'42" -34d0'18"
Mareen Saskia
Uma 10h22'31" 39d34'13"
Mareika
Uma 12h51'1" 57d45'2"
Marek
Uma 11h31'6" 53d32'49"
Marek
Vir 13h20'55" -1d2'11"
Marek
Leo 9h43'4" 28d44'52"
Marek Jankowski
Lib 15h9'45" -22d34'22"
Marek Juliusz Madej
Uma 11h7'4" 45d49'20"
Marek Kluj
Uma 11h56'39" 31d14'9"
Marek Rawicki
Aur 6h0'17" 36d33'25"
Marek, Henry
Ari 2h45'43" 12d12'36"
Marek-Jan
Cap 20h49'6" -20d26'45"
Marel Thomson
Uma 11h31'55" 38d5'47"
Marelle and Jim
Cyg 20h23'21" 45d36'42"
Marelli Manuel
Per 3h3'5" 51d39'29"
Marelva
Uma 10h31'43" 55d59'54"
Maren
Crb 16h1'20" 36d6'17"
Maren & Dillon Richards
Cyg 21h33'52" 46d48'17"
Maren Elizabeth Cohen
Tau 5h52'23" 26d53'40"
Maren Rose Dufour
Uma 8h15'58" 63d12'51"
Maren Sophie
Uma 12h5'44" 35d22'25"
Maren Sweet
Crb 15h56'40" 36d50'12"
Maren Sweetiepie Jensen
Aqr 21h49'51" 0d16'24"
Marena
Aql 19h2'3" -0d1'10"
Marenda Gail
Gem 6h3'0" 22d26'11"
Maren's Star
Tri 2h21'23" 33d39'56"
Mareska
Uma 9h0'20" 56d13'3"
Maressa Vincenza Ciccone
Tau 4h31'19" 19d27'23"
Marett
Vir 14h4'40" 3d51'38"
Mareve Corbett
Per 2h54'39" 56d2'54"
MARF
Aqr 21h50'13" -0d29'29"
Marfilius, Gaby
Uma 10h31'35" 57d40'35"
Marg
Sgr 19h34'47" -29d5'21"
Marg - The Guiding Star
Pyx 8h32'38" -25d44'21"
MARGA AMOUR
Cas 1h26'48" 72d13'43"
Marga & Jürgen Koch
Uma 9h12'45" 54d20'54"
Marga Meier
Uma 9h8'12" 62d10'17"
Marga Rosalie
Sco 16h37'32" -42d48'11"
Margalit Sheli
Vir 12h40'54" 12d16'10"
Margaret
Leo 10h40'45" 8d33'26"
Margaret
Tau 4h39'22" 19d35'30"
Margaret
Ari 2h42'16" 16d27'49"
Margaret
Leo 11h55'4" 22d8'31"
Margaret
Psc 1h43'16" 23d9'0"
Margaret
And 0h33'58" 25d54'23"
Margaret
Cnc 8h28'52" 23d54'47"
Margaret
Ari 3h29'2" 27d51'15"
Margaret
Lyn 7h16'7" 48d32'14"
Margaret
And 2h17'21" 49d31'40"
Margaret
Gem 7h11'25" 34d18'28"
Margaret
And 2h11'24" 40d57'57"
Margaret
And 2h33'56" 43d1'48"
Margaret
Psc 1h4'13" 4d7'54"
Margaret
Cam 6h57'10" 62d29'13"
Margaret
Cas 0h30'38" 64d33'38"
Margaret 12
Crb 15h35'38" 38d34'22"
Margaret A. Ditto
Uma 9h56'23" 55d44'7"
Margaret A Hunter (Marg's Heart)
Tra 16h32'50" -60d44'23"
Margaret A. Lewis
Crb 15h48'31" 27d34'42"
Margaret A. Milton
Lyn 7h41'8" 49d58'37"
Margaret A. Raven / Angel
Aqr 23h33'7" -7d19'49"
Margaret A510
Vir 11h41'31" -3d9'30"
Margaret Aguilar
Psc 0h9'49" 5d38'19"
Margaret Allman
Cap 20h41'55" -24d25'13"
Margaret (Ana) Giovingo
Aqr 22h9'32" -13d38'8"
Margaret and Al Parisian
Per 4h5'11" 43d41'56"
Margaret and Alan
Uma 9h51'26" 69d30'54"
Margaret and Jack Cooper
Cyg 21h42'2" 54d44'8"
Margaret and Paul Celillo
Cnv 14h5'47" 37d7'20"
Margaret and Robert Pettigrew
Lib 15h51'9" -17d35'48"
Margaret and Thomas Bastable
Her 17h3'40" 13d30'4"
Margaret and William Whelan
Psc 1h33'21" 22d57'3"
Margaret Andersen
Cas 2h1'4" 70d6'18"
Margaret Ann
Vir 13h30'46" -8d56'0"
Margaret Ann
Psc 1h4'31" 11d15'58"
Margaret Ann Banks
Crb 15h50'19" 26d50'1"
Margaret Ann Castillo
Vir 13h46'59" -17d31'0"
Margaret Ann "Cookie" Zengou
Pho 2h23'54" -46d0'45"
Margaret Ann Klansnic
Cnc 9h2'56" 10d23'13"
Margaret Ann LaVoy Schoenborn
Lyr 18h51'28" 36d33'32"
Margaret Ann MacGibney
Del 20h28'33" 20d41'59"
Margaret Ann Mather
Leo 10h26'1" 18d2'48"
Margaret Ann Murphy
Tau 3h50'8" 20d48'50"
Margaret Ann "Peggy" Marsh
Per 4h14'32" 42d49'33"
Margaret Ann Pritchard
Lyr 19h8'56" 26d28'45"
Margaret Ann Ragle
Aqr 22h45'11" -4d36'3"
Margaret Ann Resce Milkint
Ori 5h48'29" 9d9'31"
Margaret Ann Vaughan
Cas 1h36'53" 64d2'19"
Margaret Ann Wallace Collins
Tau 4h25'18" 24d26'23"
Margaret Ann Ward
Vir 14h22'57" 5d16'29"
Margaret Anne
And 0h29'45" 27d35'38"
Margaret Anne
Cyg 21h27'19" 44d19'31"
Margaret Anne
Sco 17h30'3" -37d9'57"
Margaret Anne Moniz
Psc 1h14'9" 3d31'17"
Margaret Anne Peterson
Crb 16h17'28" 31d1'7"
Margaret Anne Short
And 0h7'37" 44d20'11"
Margaret Anne Williams-Clayton
Cen 13h3'1" -50d14'43"
Margaret Annette Sendak
Cnc 8h34'20" 23d7'35"
Margaret Annette Smith
Sgr 18h40'19" -28d16'59"
Margaret "Annie" Lang
Cas 1h17'22" 65d11'40"
Margaret "Aunty M"
Cas 0h31'17" 53d34'40"
Margaret B Callahan
Cap 20h12'20" -25d14'14"
Margaret B. Harbison
Lyn 6h39'12" 54d21'17"
Margaret B. Mercer
And 1h47'47" 39d57'51"
Margaret B. Self
Per 2h48'8" 40d53'58"
Margaret Baba
And 0h50'16" 36d20'36"
Margaret & Barney Lewellyn
Cyg 21h41'35" 54d59'35"
Margaret Barton
Vir 12h21'33" -6d40'9"
Margaret (Batus) Gutzmer
Uma 10h41'33" 70d2'52"
Margaret Berge-Mallison
Cyg 20h35'10" 41d46'38"
Margaret Bergner-Penniman
Sco 17h41'52" -39d12'46"
Margaret Beth Avila
Uma 11h10'1" 58d6'44"
Margaret Bieterman
Tau 4h27'45" 22d50'38"
Margaret Bombaski
Cyg 20h21'50" 51d27'54"
Margaret Bourdeau
Uma 11h25'59" 60d0'17"
Margaret Bowen McElreath
Cnc 8h34'49" 23d7'35"
Margaret Bradley
Cas 1h27'32" 57d48'41"
Margaret Bradley
Lep 5h39'31" -13d29'12"
Margaret Brauer
Cap 21h40'16" -17d34'54"
Margaret Brazier
Cas 1h0'23" 49d36'33"
Margaret Buck
Cas 0h3'34" 50d34'43"
Margaret Buhman
Uma 10h44'16" 55d7'0"
Margaret Bunker
Cas 1h36'37" 61d25'6"
Margaret Byron McDonnold Pannitto
Gem 6h57'32" 22d13'37"
Margaret C. Beck
Sgr 19h19'5" -13d11'22"
Margaret C. Folse, an Angel's Mommy
Leo 10h24'29" 23d58'9"
Margaret C. Lamb
And 2h2'34" 42d13'43"
Margaret Calvey
Vir 11h48'14" -3d50'44"
Margaret Camille Newtown
Lyn 8h39'7" 35d44'17"
Margaret & Carl Plack
Cyg 19h54'39" 39d17'20"
Margaret Caroline
Crb 15h20'42" 31d50'54"
Margaret Carroll Huizenga
Cas 23h52'38" 57d9'1"
Margaret Catherine Elizabeth Roberts
Cas 1h27'2" 50d55'56"
Margaret Catherine McPherson
Cas 0h45'6" 60d36'26"
Margaret Champlin Lloyd
Vir 12h43'39" -8d41'55"
Margaret Clare
Vir 13h1'43" -7d39'32"
Margaret Clarice Peake
Crb 15h51'3" 35d35'58"
Margaret Cleo Hodges
Cas 0h37'8" 56d59'22"
Margaret Cochran Ault
Sgr 18h32'50" -27d14'18"
Margaret Connors
And 22h59'17" 52d24'52"
Margaret Craft
Crb 15h57'37" 38d9'43"
Margaret Crosby Trimble
Ari 2h38'50" 21d31'36"
Margaret Custodio
Psc 0h44'28" 9d6'0"
Margaret D.
Tau 4h6'27" 27d46'59"
Margaret Darling Herman
Ori 6h22'41" 14d40'37"
Margaret Davies
Com 12h43'45" 15d6'54"
Margaret Dawn Galloway
Uma 11h27'4" 68d32'35"
Margaret Day Miller
Sgr 19h22'4" -25d21'38"
Margaret Day Serbin
Cas 0h28'50" 56d31'38"
Margaret Deck
Uma 10h35'27" 52d27'23"
Margaret Delia
Sgr 18h45'4" -16d24'53"
Margaret DeMarzi
Lyr 18h26'50" 40d50'30"
Margaret Dempsey RSM
Tau 5h27'7" 26d26'30"
Margaret Denise McSherry
Psc 0h14'6" 7d24'4"
Margaret Doris Farace
Ari 2h4'29" 14d29'32"
Margaret Drew
Cnv 12h36'17" 44d52'5"
Margaret Dubuque Scott
Uma 9h6'56" 64d29'44"
Margaret E. Bennett
And 1h3'22" 45d36'49"
Margaret E. Macke
Lyr 19h14'16" 28d2'2"
Margaret E. Rex
Cas 1h6'3" 63d6'43"
Margaret Edith Sanchez
Leo 9h54'25" 14d31'17"
Margaret Elaine
Aqr 23h20'38" -7d13'52"
Margaret Elder Neuhaus
Tau 4h35'45" 19d47'42"
Margaret Eleanor
Tau 5h44'48" 25d35'53"
Margaret Eleanor
Leo 10h25'20" 22d28'33"
Margaret Elizabeth
Cas 1h55'0" 61d1'58"
Margaret Elizabeth
Psc 0h34'38" 5d32'48"
Margaret Elizabeth Blair
Cru 11h56'55" -63d49'55"
Margaret Elizabeth Burton
Cas 23h41'44" 57d35'48"
Margaret Elizabeth (Jarvis) Southam
Cas 2h35'36" 63d59'23"
Margaret Elizabeth Messina
Aql 19h53'52" 10d13'11"
Margaret Elizabeth Tull
Peg 22h37'6" 17d55'2"
Margaret Elizabeth Wright
Cas 1h13'34" 50d45'27"
Margaret Ellen
Psc 1h8'59" 17d34'57"
Margaret Ellen Fuller
Uma 11h54'16" 57d50'44"
Margaret Elmore
Aql 19h47'14" 2d51'9"
Margaret Emelia Lee
Uma 9h47'5" 44d41'3"
Margaret Emily Munro
Uma 9h12'4" 47d9'45"
Margaret Emma Daphne Conti
Cas 0h52'15" 53d34'35"
Margaret Ernestine Harrell
Her 17h38'29" 26d24'5"
Margaret Estelle Aiesi
Cas 23h17'22" 56d8'35"
Margaret Estep
Cas 0h44'33" 54d47'11"
Margaret Ewing
Uma 9h56'55" 63d52'54"
Margaret Fawlk
Tau 5h57'38" 24d55'31"
Margaret Fernandez
Lyr 19h12'59" 37d24'5"
Margaret Fishbein
Cap 20h38'10" -10d24'2"
Margaret Fitzgerald Reiling
Gem 7h33'54" 32d8'34"
Margaret Francais
Aqr 21h11'19" 0d55'18"
Margaret Gainsborough
Cas 2h15'50" 70d46'47"
Margaret Garcia Valenzuela
Cnc 8h48'20" 22d0'12"
Margaret Gardiner (Bellus Domina)
And 1h9'14" 38d58'58"
Margaret Geddes
Aqr 23h32'58" -17d34'18"

Margaret Ginger Sperling, Beloved
And 0h14'46" 31d57'51"
Margaret Glasgow
Psc 0h23'55" 10d58'6"
Margaret Golden
Cnc 8h52'11" 30d33'18"
Margaret Gowan Webster
Sco 17h46'36" -41d32'6"
Margaret Grace Rowlands
Lyn 8h2'47" 34d53'34"
Margaret "Grammy" Helen Pracht
Ari 2h13'2" 23d8'38"
Margaret Gray
Leo 9h44'14" 8d26'3"
Margaret Green
Gem 7h39'31" 33d16'41"
margaret green
Cap 21h49'21" -10d53'32"
Margaret Gwenneth James
And 23h1'2" 48d2'15"
Margaret H Shaw
And 0h46'5" 35d51'47"
Margaret Hayden
Ari 2h40'34" 21d38'4"
Margaret Helen Marie
Vir 13h41'6" 3d22'16"
Margaret Herrington SanAntonio Star
Ori 6h19'16" 13d15'21"
Margaret Hodgson Hegel
Lib 14h49'35" -1d10'34"
Margaret Hood
Crb 15h47'26" 36d4'30"
Margaret & Horrie - always together
Cru 12h46'49" -64d19'5"
Margaret Howell
Lmi 10h40'59" 31d55'11"
Margaret Imms
Umi 14h47'16" 68d53'12"
Margaret Irene Gandee
Cas 1h19'44" 57d44'27"
Margaret Irene Greeno
Mon 6h54'5" -0d6'43"
Margaret Irovando
Peg 22h36'25" 7d59'24"
Margaret Ivy Walker
Vul 20h42'28" 26d25'32"
Margaret J. Cummings
Cyg 19h49'20" 32d12'26"
Margaret J. Gruzensky
Gem 7h47'24" 31d16'49"
Margaret J. "Peggy" Villont
Psc 1h23'15" 16d27'35"
Margaret Jane Lein Gormley
Gem 7h7'8" 33d31'2"
Margaret Jane Norton
And 1h41'1" 40d24'45"
Margaret Jane Zemla-Saglam
Sco 17h50'38" -30d13'39"
Margaret Jean
Ari 2h48'8" 29d41'14"
Margaret Jean Dodge (Pretty Star)
Lyr 18h45'56" 35d45'31"
Margaret Jean Sullivan "Peg"
Lyr 18h16'26" 34d32'15"
Margaret Jeannine Beard
Aur 5h25'42" 42d1'22"
Margaret Jenkins
Cas 23h4'54" 55d19'18"
Margaret Jerry
Ari 3h15'56" 26d58'48"
Margaret Jessey
Uma 10h54'3" 71d25'55"
Margaret Jester
Cas 1h24'20" 57d43'14"
Margaret Jones
Lyr 18h40'45" 31d21'28"
Margaret Jones
Cas 0h56'58" 62d49'8"
Margaret Joy Hughes
Ori 5h16'42" 15d22'4"
Margaret Joye
Gem 6h43'43" 17d8'0"
Margaret Juanita Northcutt
Uma 8h59'28" 56d24'47"
Margaret & Jules Equuleus
Cas 2h1'22" 62d30'50"
Margaret K. Crossman
Sgr 18h42'51" -36d19'17"
Margaret Kate Gibson
And 2h31'27" 49d27'0"
Margaret Kay "Peggy" Clayman
Uma 9h6'26" 59d34'4"
Margaret Keay
Cas 0h14'0" 59d10'20"
Margaret Kibler
Cas 1h33'29" 65d22'2"
Margaret Krauss
And 0h34'12" 37d27'21"
Margaret Krebs
Cas 1h17'20" 63d31'53"
Margaret Kulp
Cas 1h4'51" 54d10'38"
Margaret L. Joseph
Cap 20h49'6" -15d30'49"
Margaret L Wynn
Lyn 7h55'12" 52d56'27"
Margaret LaDean Griffin-Green
Ori 5h8'34" 15d44'52"
Margaret LaRue
Lib 14h35'20" -17d29'6"
Margaret Lawson
Cyg 19h53'15" 36d41'55"
Margaret Lee
Gem 6h54'19" 31d5'3"
Margaret Lembo
Cnc 8h49'19" 22d4'57"
Margaret Lena Jensen
Sco 16h9'6" -10d29'58"
Margaret Liegel
Her 17h36'5" 32d39'24"
Margaret Lillian Barth Myhrer
Sco 17h40'35" -39d19'24"
Margaret Link
Sco 17h51'22" -35d56'44"
Margaret Lisle
And 0h22'40" 32d28'34"
Margaret Lois Troutman
And 1h9'3" 44d12'43"
Margaret Lord
Psc 1h8'10" 27d49'45"
Margaret Lorriane Barber
Cas 0h34'23" 54d10'55"
Margaret & Louis Sofo
Leo 11h7'53" -0d38'57"
Margaret Louise Gogolin Wykoff
Cap 21h1'15" -19d10'13"
Margaret Louise Hurll-Dyment
Cas 1h24'21" 52d48'9"
Margaret Louise O'Neill Gillespie
Ori 5h31'29" 0d13'13"
Margaret Louise Schwarz Bray
Cas 0h43'3" 52d10'22"
Margaret Lucius
Psa 22h44'26" -25d30'31"
Margaret Lumia
Uma 10h22'37" 51d45'30"
Margaret Lutterloah
Cas 1h7'27" 55d50'33"
Margaret Lynn Hoisington
Gem 6h4'56" 26d9'51"
Margaret M. D'Ambrogio
Uma 11h45'16" 49d8'8"
Margaret M. McLaughlin
Lib 14h53'54" -14d45'2"
Margaret M. Morales-Zuniga
Cas 1h43'21" 64d35'2"
Margaret M. Murphy
Psc 1h15'21" 23d12'23"
Margaret M. Willoughby
Ori 6h17'37" 14d59'25"
Margaret M. Zelasko
Vir 13h25'6" -5d41'20"
Margaret Mabugu
Vir 12h7'36" 10d46'50"
Margaret Macfarlane Gordon Fraser
Lyn 9h19'53" 34d51'14"
Margaret Macko Sosoka
Cas 2h13'10" 66d19'20"
Margaret "Maggie" Landers
Cas 1h43'56" 61d53'42"
Margaret Magnarelli
Cas 0h37'26" 61d45'33"
Margaret Maguire
Cas 1h26'50" 65d13'22"
Margaret "Marge" Bossard 4/3/22
Cmi 7h32'5" 2d58'59"
Margaret (Margie) Cloughessy
Cas 0h54'58" 61d24'26"
Margaret Marie
Crb 15h55'17" 38d30'46"
Margaret Marie Weiper Parker
Pho 23h38'24" -40d39'39"
Margaret Mariko
Psc 1h24'24" 25d37'18"
Margaret Mary
Cas 1h16'21" 62d42'28"
Margaret Mary Booth
Umi 15h25'53" 71d47'17"
Margaret Mary Bothe
Uma 13h8'44" 59d12'6"
Margaret Mary Caulfield
Cas 0h26'8" 62d52'18"
Margaret Mary DeSellier
Cap 21h36'28" -9d53'25"
Margaret Mary Jenkins
Vir 14h7'36" -9d29'23"
Margaret Mary Linnell
Cas 0h10'53" 56d25'17"
Margaret Mary Macnie
Cnc 9h6'9" 17d40'2"
Margaret Mary Moriarty Lawrence
Vir 14h15'48" 1d16'25"
Margaret Mary Mucci
Lib 15h59'12" -14d46'12"
Margaret Mary Murphy
Cap 20h35'54" -9d54'17"
Margaret Mary Parmese
Vir 13h21'26" -13d43'12"
Margaret Mary Ramsey-Davis
Cnc 8h18'56" 8d20'20"
Margaret Mary Robinson
And 0h41'44" 41d41'44"
Margaret Mary Sobey
Lyn 8h57'43" 36d32'27"
Margaret Mary Stanton
Crb 16h11'4" 36d44'13"
Margaret Mary Taylor
Aqr 21h51'12" -7d57'15"
Margaret Mary Valiquette Donnelly
Uma 11h46'42" 44d42'45"
Margaret Mary Wells Salcewicz
Cas 1h39'34" 64d23'1"
Margaret Mates
And 2h22'38" 42d41'53"
Margaret McBride Allison
Ari 3h16'44" 29d56'17"
Margaret McCartney
Cnc 8h42'49" 10d46'31"
Margaret McEldowney Bracken
Cas 0h40'51" 50d47'33"
Margaret McKenna
Uma 8h22'38" 65d2'23"
Margaret McKittrick
Cma 6h39'53" -17d0'37"
Margaret Melia
Cas 23h32'56" 53d14'8"
Margaret - Missy - Sheehan
Ari 3h18'24" 19d43'33"
Margaret Mitchell Star
And 1h48'54" 42d53'47"
Margaret Mogul
Oph 17h5'51" -29d41'28"
Margaret (Morningstar)Schreiber
Psc 0h11'49" 8d18'26"
Margaret Mullaney
Cas 23h41'39" 54d41'53"
Margaret Murray McCaffrey
And 1h22'4" 44d33'6"
Margaret Nancy Rehr
Sgr 18h0'30" -27d56'9"
Margaret & Nick
Cyg 20h8'20" 41d30'41"
Margaret Oden Hamaker
Crb 16h13'9" 30d49'38"
Margaret "Ol' Grammy" Plikaitis
Cas 0h32'33" 60d35'15"
Margaret O'Leary
Vir 12h59'41" 12d20'11"
Margaret "OMa" Herrler 102nd B'day
Cas 22h58'43" 55d43'0"
Margaret O'Malia Snyder
Cap 20h21'53" -13d24'9"
Margaret Osner "Organist"
Uma 11h17'6" 42d43'32"
Margaret Ozols
Uma 11h35'47" 46d17'56"
Margaret P. Martz
Ari 2h24'30" 25d45'52"
Margaret Pakulska
Psc 1h19'10" 18d21'47"
Margaret Paradee
Vir 12h31'48" 3d0'36"
Margaret Patricia McCoy
Aqr 22h36'52" -4d57'21"
Margaret Pauline & Lawrence G. Gay
Cyg 19h33'34" 52d49'38"
Margaret Pavicich
Uma 13h45'54" 55d20'38"
Margaret Peddel
Cas 23h20'31" 58d12'12"
Margaret (Peg) A. Keane
Psc 1h8'25" 11d7'5"
Margaret "Peg" Sohst
And 2h12'17" 45d52'7"
Margaret (Peggy) Adams
Cas 0h11'34" 54d11'4"
Margaret Peggy Baker-Hope, Beloved
And 0h21'53" 33d38'50"
Margaret (Peggy) Benedict
Lyr 19h10'9" 36d34'12"
Margaret "Peggy" Haynes
Cas 0h57'32" 48d43'12"
Margaret "Peggy" Johnson
Leo 9h51'57" 16d56'17"
Margaret Peggy Morgan 180322-191105
Cas 23h22'34" 59d8'52"
Margaret Penzkofer
Cam 4h31'22" 54d6'35"
Margaret Perricone-Mom Grandma
Cas 1h41'28" 62d50'16"
Margaret Philbrook
Aur 5h22'6" 43d30'6"
Margaret Porter O'Herron
Crb 15h53'18" 28d17'24"
Margaret Price
Aqr 20h41'27" -12d33'54"
Margaret Priscilla Boatman
Cyg 20h22'35" 48d54'19"
Margaret Quigley
Ari 2h41'49" 21d31'36"
Margaret R. Cole
Lmi 10h7'51" 34d18'55"
Margaret Radoff
Uma 9h8'21" 57d38'54"
Margaret ReBell
Leo 11h35'15" 22d1'24"
Margaret Reuter
And 2h16'58" 45d5'45"
Margaret Rieger 100 Year Star
Cas 1h24'36" 62d25'17"
Margaret Rito
Cas 0h58'47" 70d21'14"
Margaret Rohrs Higgins
Leo 11h15'49" 14d34'48"
Margaret Rooney
And 0h26'33" 40d48'5"
Margaret Rosciano Cerniglia
Aqr 21h44'40" 1d13'2"
Margaret Rose
Lib 14h54'44" -6d15'27"
Margaret Rose Briody Pigott
Vir 14h37'39" 3d9'13"
Margaret Rose Fisler
Sgr 18h33'35" -27d41'35"
Margaret Rose Rogers
Vir 13h57'10" -19d20'15"
Margaret Rossini
Cnc 8h48'32" 32d21'41"
Margaret S. Ryan
Sgr 18h25'15" -17d23'48"
Margaret Saunders
Lyr 18h49'12" 43d47'54"
Margaret Schatzle
Cas 1h36'47" 60d34'14"
Margaret Scherrer Mohatt
Cap 21h34'0" -16d52'45"
Margaret Schlitt Davis
Leo 10h20'56" 12d44'57"
Margaret Schuler
Cas 1h19'0" 72d53'0"
Margaret Schwerin
Cnc 8h49'31" 14d14'24"
Margaret Scios
Cas 1h41'5" 62d19'30"
MARGARET S.D.1
Uma 13h55'34" 57d12'33"
Margaret Sears
Crb 15h27'41" 26d57'20"
Margaret Shelters
And 0h35'16" 38d41'7"
Margaret Sim Rodrique
Cas 0h43'31" 62d39'56"
Margaret Simon
Lyr 18h41'25" 41d52'54"
Margaret Sinkhorn
And 23h39'50" 41d0'28"
Margaret & Skip Adams
Sge 19h40'28" 17d58'3"
Margaret Squires Prince 11/25/04
And 0h29'31" 38d6'52"
Margaret Stalker Hill
And 0h44'38" 39d57'44"
Margaret Steele Taylor
Sgr 18h48'13" -24d41'55"
Margaret Stoddard
Mon 7h49'10" -0d56'24"
Margaret Stoneris Majoris
Ari 2h53'34" 24d48'50"
Margaret Stoutenburg
Gem 7h41'12" 25d3'34"
Margaret Susan Longhurst
Cas 1h29'46" 69d14'13"
Margaret Suzanne Fairman
Uma 10h14'48" 54d12'22"
Margaret Sweeney
And 0h49'6" 39d32'58"
Margaret T. Kantar
Ori 5h35'40" 14d13'59"
Margaret Tashiro Caccamo
Lib 15h28'34" -24d29'59"
Margaret TenElshof
Cyg 21h17'54" 30d21'58"
Margaret Thatcher Unger
And 1h42'1" 44d31'10"
Margaret Thompson
Per 3h48'44" 49d36'51"
Margaret Tognacci
Cas 0h17'57" 63d0'31"
Margaret Turner
Cyg 20h6'20" 41d9'20"
Margaret Valerie Ann
And 0h33'3" 26d9'13"
Margaret Van Valen
Lmi 10h43'28" 25d40'27"
Margaret V-Angel Holt
Psc 1h20'26" 29d42'52"
Margaret Vanilla Bean
And 23h10'35" 51d27'29"
Margaret Vera Harris
Psc 23h12'22" 6d45'43"
Margaret Vonbehren
Psc 0h31'7" 18d58'37"
Margaret W. Joyce
Crb 15h28'55" 29d57'25"
Margaret Wagner
Aqr 22h59'25" -12d18'9"
Margaret Westlund
Cap 21h43'28" -14d47'28"
Margaret Wien Hines
Ori 5h33'6" 3d26'30"
Margaret, Mum, Nan
Cas 23h38'54" 57d32'3"
Margareta Hartmann
Cnv 13h24'51" 31d59'36"
Margaret-Anne Sytsma
And 23h56'31" 35d38'53"
MargaretBob
Uma 11h28'29" 60d0'23"
Margarete and Eric Timmerman
Uma 10h15'56" 59d41'48"
Margarete Messinger
Cyg 20h6'37" 48d57'39"
Margarete Trinkl
Uma 12h10'39" 47d53'51"
Margareth
And 2h18'58" 47d59'57"
Margaretha Uniar Chandra
Sco 17h41'39" -41d33'46"
Margaretium Blumus
Lib 15h28'56" -10d6'35"
Margaret-Maude Bickhart
Lyn 8h32'33" 37d49'51"
Margaret-May
Cas 3h12'33" 63d46'37"
Margaret-N-George
Sgr 18h37'39" -16d23'26"
Margaret-Our Unity Star
Gem 7h32'49" 19d42'56"
Margaret's Angel
Ari 1h49'16" 20d32'49"
Margaret's Christmas Star
Ori 5h30'6" 3d10'22"
Margaret's Desire
Leo 9h33'14" 27d19'2"
MARGARET'S KE-KA-KJ-KL
Cyg 21h20'21" 53d27'33"
Margaret's Red 7+13
Cam 4h28'12" 56d18'9"
Margaret's Shining Light
Cyg 19h34'47" 54d42'23"
Margaret's Star
Cam 5h20'28" 60d3'10"
Margaret's Treasure
Uma 8h18'17" 71d31'42"
Margarette Thomas
Lmi 10h25'19" 33d48'55"
Margarida Vila-Nova
Gem 6h3'22" 27d31'9"
Margarita
Ori 5h20'39" 6d55'26"
Margarita
Aqr 21h39'45" 2d21'26"
Margarita
Ori 6h9'32" 7d38'19"
MARGARITA
Cnc 8h37'2" 31d43'15"
Margarita
And 2h17'47" 42d27'11"
Margarita
And 23h18'35" 47d48'4"
Margarita
Lib 15h26'39" -9d32'15"
Margarita
Sco 16h15'31" -12d56'49"
Margarita
Vir 14h9'40" -5d21'57"
Margarita Andrea
Uma 9h33'11" 46d51'39"
Margarita Arriagada
Peg 23h41'36" 28d54'3"
Margarita Brillantes
Tau 3h37'11" 11d55'34"
Margarita Comas
Lyn 8h29'5" 35d0'20"
Margarita Crispina Moyano
Ari 2h16'57" 20d21'1"
Margarita Delbrook Villa
Mon 6h49'37" -0d3'12"
Margarita Esperanza
Mon 6h53'9" -0d54'44"
Margarita Ferrer
Peg 22h18'10" 12d10'36"
Margarita Hunt-Tejada
Cas 1h3'25" 63d34'24"
Margarita Kazakova
Tau 4h5'43" 27d19'41"
Margarita Krasnova
Lib 14h50'52" -4d8'36"
Margarita L. Barreto
Cyg 21h52'53" 48d22'16"
Margarita Milla
And 22h59'4" 35d39'2"
Margarita Novikova Goskirk
Psc 1h19'28" 11d54'26"
Margarita R. Colon
Tau 3h33'30" 8d51'21"
Margarita R. Santiago
And 23h5'8" 48d20'53"
Margarita Rodriguez
Crb 15h40'25" 29d38'50"
Margarita Roman
Uma 10h14'39" 43d0'25"
Margarita Veronica
Ori 5h20'49" 4d14'53"
Margarita Wuthrich
Uma 11h37'50" 60d49'42"
Margaritha
Lyn 7h10'28" 48d45'48"
Margaux
Peg 22h23'15" 21d19'27"
Margaux
Sgr 19h35'27" -17d46'46"
Margaux
Lib 15h6'33" -28d46'17"
Margaux
Uma 11h39'49" 45d59'17"
Margaux Béguin
Vir 14h12'38" -0d42'44"
Margaux et Alexis
Leo 11h36'22" 20d43'19"
Margaux Jacqueline Hamel
Vir 13h16'36" -22d13'27"
Margaux Jo Elliott
Cas 1h9'45" 66d51'17"
Margaux Luttenbacher
Cnc 8h4'54" 22d41'10"
Margaux Miller
Lib 15h11'57" -22d5'39"
Margaux Praplan
Uma 9h14'58" 51d21'41"
Margaux Ryan Hackett
And 23h18'50" 41d4'44"
Marge Bell
Ori 5h37'30" 1d27'10"
Marge Donnelly
Cyg 19h49'40" 36d36'14"
Marge E. Stickel
Leo 9h35'41" 28d2'43"
Marge Fromi
Lyn 9h10'5" 40d9'33"
Marge Gray
Leo 9h57'15" 12d55'25"
Marge Groene
Uma 11h33'0" 55d40'21"
Marge & Harold
Uma 11h31'59" 46d44'18"
Marge Hartranft
Lyn 7h0'4" 56d42'34"
Marge Hirdler
Leo 11h26'52" 25d46'17"
Marge Lybarger Gardner
Her 16h58'8" 29d54'43"
Marge Marissa Olivia JonJon Nick
Uma 14h12'33" 62d6'58"
Marge Rahu
Aqr 22h29'3" -20d10'11"
Marge Weber
Cyg 21h6'44" 46d12'25"
MarGee
Ari 2h1'59" 20d2'52"
Margee Iddings
Aql 19h27'6" 11d33'17"
Margee's Boys
Cnc 8h22'34" 18d24'54"
Margeit, Alfred
Uma 9h2'37" 52d12'49"
Margeret Ann Lubeck
Psc 1h33'9" 21d29'43"
Margeret Ridley
Lyn 7h0'29" 55d30'43"
Margery A. Gruber
Aqr 22h50'38" -15d45'12"
Margery Ann Ferch
Cyg 20h2'29" 38d41'29"
Margery Ann Steinhouse
Uma 11h26'56" 53d30'57"
Margery Ann Wyatt
Crb 16h23'11" 33d42'16"
Margery Anne Anderson
Leo 10h35'43" 7d31'44"
Margery Arlene Kinser
Lyn 7h24'50" 57d6'21"
Margery Christina Locke
Cyg 21h43'3" 51d42'8"
Margery Hosman
Sgr 19h18'45" -11d59'48"
Margery Mary Schepis
Cas 1h26'2" 55d46'13"
Margery Prior
Uma 9h5'51" 62d55'9"
Margery Rose Friend
Uma 10h11'9" 63d48'22"
Margery Schepis
Uma 9h34'40" 42d50'2"
Marge's Star
Tau 4h35'54" 19d27'46"
Margey Ann Williams Alger
Psc 23h34'38" -2d1'49"
Margherita
Cyg 21h7'20" 49d7'31"
Margherita
Cyg 21h3'33" 50d28'5"
Margherita
Cma 6h54'29" -30d36'29"
Margherita
Peg 21h25'40" 21d40'0"
Margherita
Aur 5h14'42" 41d23'54"
Margherita
Cyg 21h3'33" 50d28'5"
Margherita
Cyg 21h7'20" 49d7'31"
Margherita
Aur 6h8'44" 50d31'18"
Margherita Di Crescienzo
Cyg 20h56'32" 31d36'38"
Margherita Praticò
Cas 23h0'13" 53d47'24"
Margherita Yasmina Novic - 26.04.1977
Tau 4h29'14" 20d53'50"
Margi Halloran
Sco 16h17'43" -17d47'52"
Margi K. Patel
Sgr 19h28'58" -17d9'57"
Margiana
Aqr 21h46'38" -8d13'43"
Margie
Aqr 22h20'3" -15d53'20"
Margie
Ari 2h46'38" 19d2'51"
Margie
Psc 1h43'9" 22d18'56"
Margie
Ori 5h34'26" 2d59'25"
Margie
Vir 12h41'22" 0d50'14"
Margie
And 2h2'40" 37d15'1"
Margie
Her 16h14'22" 44d40'52"
Margie
Cyg 22h0'27" 51d20'38"
Margie A. Warner
Cam 5h27'59" 62d32'2"
Margie and Joe's 50th 1956-2006
Cyg 19h28'14" 52d6'7"
Margie and Leo Keys
Sco 16h3'19" -21d45'23"
Margie Bactad
And 2h38'39" 46d5'5"
Margie Blum
Lep 5h7'13" -11d6'24"
Margie Carter
Sgr 20h1'11" -30d27'3"
Margie Chelini
Com 13h8'13" 26d29'3"
Margie " Dancer" Scheller
Cap 20h29'30" -18d29'19"
Margie Del Pozzo
Lyr 19h7'28" 37d22'32"
Margie E. Venner
Cas 0h45'11" 66d40'40"
Margie Ervin
Lyr 18h51'5" 31d38'36"
Margie Fay Martino
Cyg 21h6'38" 46d4'32"
Margie Girl
Cas 23h31'44" 53d36'18"
Margie Hall
Aqr 20h51'44" -9d14'52"
Margie Helen Floyd
Uma 10h16'10" 45d39'58"
Margie & Henry Weaver
Vir 11h50'30" 6d35'31"
Margie K Reinholz
Cyg 21h50'18" 54d35'18"
Margie/Kikay- The Light of Laughter
Umi 14h3'45" 68d26'45"
Margie Kramer
Cap 21h7'38" -16d48'12"
Margie Lou
Tau 3h29'38" 24d30'38"
Margie Marie
Leo 10h39'53" 18d5'26"
Margie Mater Mia Filia
And 0h17'33" 22d26'35"
Margie Mittleman
Cas 23h45'10" 57d2'51"
Margie Murray
Sco 17h53'18" -36d17'47"
Margie Our Star
Cru 12h38'13" -59d33'27"
Margie Walker Horsch
Cap 20h17'18" -19d18'9"
Margie Youngblood
Lmi 9h54'21" 36d20'40"
Margie, My Shining Star
Lyn 8h1'43" 55d43'43"
Margie's 60th
Gem 6h54'24" 31d37'14"
Margie's Apollo
Mon 7h5'46" -7d56'59"
Margie's Family
Leo 10h54'55" 10d42'48"
Margie's Star
Aqr 21h37'5" 1d26'10"
Margiotta, Guarino & Cohen
Uma 13h43'58" 57d53'32"
Margit
Lib 14h53'25" -16d9'12"
Margit Christel
Ari 2h50'10" 20d25'49"
Margit und Marion Aicher
Uma 8h44'18" 59d10'16"
Margit Wagner
Cas 0h58'7" 58d51'37"
Margita Elizabeth Satterlee
Uma 11h44'57" 54d35'14"
Margitta Marquardt
Uma 11h4'42" 55d39'24"
Margitta - mein Glücksstern
Uma 10h12'33" 46d44'14"
Margo
Cas 0h56'38" 54d8'21"
Margo
Leo 11h7'25" 3d0'1"
Margo
Cap 20h16'27" -14d38'15"
Margo
Sgr 18h36'6" -35d7'58"
Margo
Sco 17h25'50" -42d16'23"
Margo Catherine
And 23h32'14" 45d17'6"
Margo Colman
Sco 16h12'42" -16d24'40"

Margo Fox
Leo 10h12'26" 13d31'14"
Margo Haugum
Mon 7h11'42" -9d11'16"
Margo Hintz Ragone
Cas 0h19'37" 58d54'7"
Margo Kustar
Psc 1h56'56" 5d9'32"
Margo Lamb
Psc 0h55'19" 31d44'57"
Margo Layton
Per 3h1'55" 54d31'10"
Margo Luminessa
Vir 14h4'7" -9d46'6"
Margo Lynn
Psc 1h3'20" 4d49'53"
Margo Lynn Harkness Wilkinson
Ari 2h46'18" 27d20'20"
Margo Lynn Pusateri Smith
Leo 10h9'44" 7d3'34"
Margo McKinley Heintz
Cas 1h27'53" 60d30'23"
Margo Oakes
Per 3h45'58" 49d0'45"
Margo Patricia Hudson Star 27101954
Cru 12h46'56" -60d48'11"
Margo Potvin
Sco 16h11'22" -17d12'10"
Margo Trudel
Cas 1h15'40" 51d6'41"
Margo W. Williams
Lyn 6h27'48" 54d15'39"
Margolinka
Cnc 8h37'51" 17d56'43"
Margorie Baker
Lyn 6h52'46" 53d30'25"
Margorie Clarkin
Lyn 8h29'0" 33d22'10"
Margorie Goforth
Uma 13h56'35" 51d26'29"
Margo'sTrouble
Cnv 12h51'2" 43d29'5"
Margot
And 1h35'43" 47d17'18"
Margot
Uma 12h3'16" 33d53'6"
Margot
Lyn 6h34'59" 56d54'52"
Margot Albertini
Umi 15h38'33" 69d34'9"
Margot Ann Cerone
Ori 5h18'11" 0d38'6"
Margot Cavallini
And 22h58'22" 46d47'53"
Margot Chapelain L'Officier
Leo 9h52'47" 8d57'28"
Margot_Fielden_O'Brien
Gem 7h43'14" 34d0'49"
Margot & Martin Kramer
Lyr 18h53'42" 43d7'56"
Margot Miglior
Cap 20h19'12" -26d50'25"
Margot Müller
Ori 6h14'39" 8d24'37"
Margot Romanski
Sgr 18h28'36" -25d37'0"
Margot Wombacher Our Guiding Star
Cet 1h7'19" -0d19'1"
Margot's First Light
And 1h39'46" 42d5'18"
Margot's Star
Sco 16h4'26" -8d46'2"
Margot-Theodore Kleisinger
Uma 8h21'33" 65d5'35"
Margoverduin.04.23.56
Dra 18h36'44" 51d28'7"
Margret Ann Crone
Cap 21h37'23" -12d51'55"
Margret Elain Schneider
And 23h25'18" 47d3'4"
Margret Howells - Our Star
Cas 0h17'0" 50d13'18"
Margret Joan White
Ari 2h49'11" 21d33'13"
Margret "Lucky Lumae" Mix
And 0h18'49" 46d31'33"
Margret Marion Edwards
Tau 3h37'38" 25d28'44"
Margret Mary
Cam 3h21'59" 55d32'50"
Margret W. Raymond
Crb 16h4'42" 35d38'20"
Margrethe Quach
Psc 0h50'36" 15d39'6"
Margrett Fogelsonger
Uma 14h6'10" 50d25'54"
Margrett Killpack
Pho 0h36'47" -40d24'51"
Margrette
Aqr 20h39'19" -12d20'3"
Margrit
Cep 22h8'20" 53d26'27"
Margrit Ruf
And 23h32'54" 50d38'43"
Margrit und Eugen Koch
Lyr 18h29'6" 37d5'39"
Margrit Wakefield
Vir 13h45'55" 1d10'28"
Margrit & Xavier Perle d' Amour
Lyr 18h57'37" 33d36'6"
Marguerita
Vir 13h19'7" 4d50'54"
Marguerita
Peg 22h48'15" 19d21'32"
Marguerita Iniesta Castellano
Crb 15h55'25" 31d53'3"
Marguerite
Cas 0h5'44" 54d16'36"
Marguerite
Cas 0h3'29" 54d25'16"
Marguerite
Psc 1h22'9" 29d27'4"
Marguerite
Crb 15h34'57" 25d52'49"
Marguerite
Crb 16h6'57" 26d36'44"
Marguerite
Lib 15h53'5" -7d5'43"
Marguerite C. Zambito-Ferrante
Ari 2h27'25" 22d28'36"
Marguerite Carter
And 2h32'33" 45d35'39"
Marguerite Charles
Cap 20h26'2" -19d36'33"
Marguerite Currie
Aqr 22h2'40" -21d25'41"
Marguerite de Chivré de Color
Cap 21h13'35" -19d19'24"
Marguerite Demerest Borman
And 23h8'46" 51d21'12"
Marguerite Duffy Degen
Uma 11h26'59" 32d37'22"
Marguerite Elizabeth
Leo 11h20'51" 14d52'54"
Marguerite Epei
Ori 5h39'26" 5d18'50"
Marguerite Gallagher
Cas 2h22'14" 65d46'42"
Marguerite Gauthier
Tau 4h8'2" 26d34'16"
Marguerite Gore
Pav 20h45'23" -60d16'41"
Marguerite Halpin
Aqr 22h27'59" 1d57'32"
Marguerite Janisse
Lyr 18h41'0" 34d46'27"
Marguerite Jenkins Clark
And 23h56'15" 45d33'17"
Marguerite & Joseph DiRenzo
Cyg 20h19'14" 58d46'27"
Marguerite Le Messurier Cartledge
Cas 1h3'25" 60d57'25"
Marguerite Mackey
And 1h42'52" 46d41'59"
Marguerite Marjorie Thomas
Vir 13h26'41" -0d14'7"
Marguerite McArthur
And 0h46'23" 26d38'37"
Marguerite Mom and Nanny Garvin
Vir 13h41'9" -6d53'45"
Marguerite Orr
Vir 13h48'0" -5d50'7"
Marguerite "Peggy" Rosecky
Aur 5h43'48" 40d6'43"
Marguerite Richter
Umi 16h42'21" 77d32'45"
Marguerite Ross
Cas 23h24'9" 56d2'50"
Marguerite Ryan
Cas 23h37'2" 56d35'50"
Marguerite Sexton
Ari 2h29'42" 11d36'56"
Marguerite Soler
Sgr 18h10'49" -26d43'35"
Marguerite Vizien Simard "Cent Ans"
Uma 9h15'6" 60d27'25"
Marguerite Wilkinson
Lyr 18h19'44" 38d55'45"
Marguerite Wurtz
Cam 5h10'3" 65d38'28"
Margueritte Kibel
Lyr 19h7'7" 27d9'3"
Margurite Mary Lewandowski Gagnon
Cyg 19h45'4" 35d20'3"
Margus Merilo
Lyn 7h50'6" 53d6'13"
Marg-Vic
Cyg 20h48'18" 30d28'30"
MargWay Star of Love
Cyg 20h13'12" 48d52'47"
Marharyta Perelman
Leo 10h41'35" 6d50'35"
Mari
Psc 0h43'23" 13d30'52"
Mari
Aqr 22h24'33" -1d0'22"
Mari Alexander Holland
And 1h1'26" 39d18'9"
Mari and Mike Mariano
Col 5h40'32" -36d24'56"
Mari Annelise
And 0h47'42" 24d12'37"
Mari Boswell
Leo 11h20'38" -4d44'57"
Mari Crystal Sanchez (Mariposa)
Lyn 6h43'48" 52d17'13"
Mari e Tino
Aur 6h45'58" 35d36'19"
Mari Giuliana
Uma 11h6'42" 44d11'54"
Mari Hirsch Meyer
Uma 12h13'54" 52d24'41"
Mari Homma
Aqr 21h9'19" -12d3'49"
Mari Isis
Lib 15h32'33" -8d35'0"
Mari Jo
Gem 7h9'28" 25d51'30"
Mari Karen
Gem 6h43'36" 25d5'21"
Mari Kathleen Watson
Tau 4h26'43" 26d38'9"
Mari Keith
Crb 15h42'15" 33d58'51"
Mari Lekarczyk
Tau 4h12'59" 5d40'46"
Mari Martinez
Ori 5h36'49" -1d23'4"
Mari Mater O'Neill
Psc 1h7'26" 28d23'25"
Mari n Ray
Cyg 20h6'23" 58d26'36"
Mari Peri
Peg 22h53'51" 32d22'43"
Mari Rocks
Leo 11h56'3" 20d55'40"
Mari Vicente Garcia
Ori 6h0'43" 14d18'37"
Maria
Cnc 8h55'2" 9d42'59"
Maria
Ari 1h56'29" 20d42'48"
Maria
Psc 23h1'54" 1d43'38"
Maria
Ari 2h21'0" 12d26'35"
Maria
Ari 2h34'12" 12d57'41"
Maria
Ori 5h57'37" 7d39'50"
Maria
Cnc 8h22'27" 16d23'56"
Maria
And 0h20'13" 27d38'53"
Maria
Gem 6h43'4" 27d38'19"
Maria
Gem 6h49'8" 29d25'11"
Maria
Cyg 19h54'5" 32d3'50"
Maria
Lyn 7h37'29" 35d42'51"
Maria
Lyn 8h53'47" 37d8'8"
Maria
And 0h52'59" 36d10'59"
Maria
And 2h19'34" 41d29'11"
Maria
Cas 0h12'32" 53d40'0"
Maria
And 1h1'32" 45d49'31"
Maria
And 0h29'57" 45d59'34"
Maria
And 23h50'56" 39d24'54"
Maria
Lyn 7h50'51" 45d4'12"
Maria
Sgr 19h17'47" -31d17'42"
maria
Dra 16h8'4" 54d52'7"
Maria
Cas 23h32'55" 56d48'56"
Maria
Cas 1h36'24" 60d56'3"
Maria
Uma 12h10'41" 59d26'44"
Maria
Cam 7h34'48" 63d14'59"
Maria
Uma 8h26'20" 60d24'20"
Maria
Cas 2h25'2" 67d4'22"
Maria
Cas 2h39'22" 63d36'10"
Maria
Cas 1h58'39" 65d58'2"
Maria
Uma 9h5'53" 64d53'10"
Maria
Uma 11h23'5" 60d7'43"
Maria
Umi 14h29'59" 71d7'20"
Maria
Uma 8h56'1" 69d37'0"
Maria
Vir 11h55'27" -0d4'17"
Maria
Umi 17h35'49" 80d46'0"
Maria
Umi 17h11'8" 80d24'37"
Maria
Lib 14h50'29" -4d25'48"
Maria 09061985
Dra 16h42'46" 63d37'40"
Maria 213 ChibiLouLou
Aqr 22h35'41" -1d51'44"
Maria A. Yuricevic
Aqr 22h33'52" 0d26'32"
Maria Ad. Dimopoulos
Uma 13h51'37" 52d36'54"
Maria Adelaida
Crb 15h48'19" 32d25'31"
Maria Adriana
Vir 12h48'15" 4d10'43"
Maria Agranovskaia
Gem 7h15'8" 32d56'53"
Maria Agustina Lizarraga
Cas 2h25'50" 67d5'50"
Maria Ahrens für 30 Jahre
Ori 5h2'39" 5d51'15"
Maria Aissa
And 0h55'47" 40d46'44"
Maria Alarcon
And 22h58'15" 51d16'16"
Maria Alba
Aqr 22h12'33" -24d5'44"
Maria "Ale" Chinchilla
Umi 15h50'41" 78d43'36"
Maria Alejandra
Gem 6h52'42" 23d24'2"
Maria Alejandra Muniz
Lmi 10h52'27" 27d24'27"
Maria (Ali) Politi
And 23h3'25" 51d59'10"
Maria Alicia Gomez
Ari 2h25'37" 11d48'52"
Maria Allegra Wright
Lib 15h32'46" -5d2'15"
Maria Alondra
And 0h41'19" 25d10'1"
Maria Amalie
Cnc 8h55'43" 29d53'40"
Maria Anastasiou
And 23h22'31" 49d45'32"
Maria Anastasiou
Vir 14h24'15" -4d38'52"
Maria And Francis's Star
Dra 18h4'43" 70d30'21"
Maria and George
Cyg 20h7'35" 38d32'45"
Maria and Selena Vaccarella
Lyn 7h46'1" 55d6'26"
Maria Andrea Salas Bernal
Vir 13h29'41" -12d7'39"
Maria Andrea San Andres
Cap 20h15'17" -17d1'37"
Maria Angela
Cyg 21h13'50" 52d43'8"
Maria Angela Piazza
Cnc 8h19'12" 13d59'28"
Maria Angeles Arroyo
Gem 7h46'57" 22d47'28"
Maria Angeles Jimenez Lopez
Cas 0h55'54" 52d27'14"
Maria Angelina
Ori 5h27'5" 13d17'10"
Maria Ankele
Sco 16h6'50" -21d9'10"
Maria Ann
Ari 2h0'31" 23d48'27"
Maria Ann Perna
Leo 9h36'29" 30d59'20"
Maria Ann Wood
Psc 0h40'35" 7d41'44"
Maria Anna Foglia
Lyr 19h21'22" 30d57'5"
Maria Anna Howell
Lyn 7h21'34" 57d56'44"
Maria Antoinette Pugliese
Aqr 20h43'58" -2d5'44"
Maria Antonette Maruzza
Leo 9h45'5" 27d40'4"
Maria Antonietta Coronado
Lyn 6h41'8" 60d44'26"
Maria Apessos
Sgr 18h57'26" -26d41'50"
Maria Aphrodite Patramanis
Sco 17h18'42" -44d26'38"
Maria Arbelo
Lyn 8h25'7" 36d23'32"
Maria Argyros
Uma 11h19'12" 68d16'10"
Maria Arranz Martinez
Sco 16h14'14" -14d24'46"
Maria Aurora Juanillo Bolivar
Leo 9h43'15" 11d12'49"
Maria Aurora Marzoa
Vir 12h19'35" 6d23'34"
Maria Avila
And 2h22'54" 50d9'49"
Maria B. Van de Zilver
And 23h23'39" 47d25'32"
Maria Baldwin
Uma 9h18'34" 59d24'31"
Maria Benites
Peg 21h40'23" 21d34'27"
Maria Bernadette
Sgr 19h32'2" -16d31'3"
Maria Bernardita Patrocinio Vidal
Aqr 22h34'47" -11d57'38"
Maria Betancourt
Cap 21h29'48" -17d44'32"
Maria Bitsch
Lmi 10h16'3" 29d16'23"
maria bj
Lyn 8h56'19" 39d49'58"
Maria Bollhalder
Cyg 19h33'34" 30d12'16"
Maria Bon Bon
Crb 15h34'33" 28d33'16"
Maria Bonilla
Ori 5h46'43" 3d5'30"
Maria - Boubou Samara
Gem 7h50'29" 34d39'42"
Maria Bourne
Cas 0h12'54" 59d41'58"
Maria Bova
Ori 6h5'8" 6d49'48"
Maria Bruna Perlina
Cas 0h32'54" 62d24'0"
Maria Bukovinsky
Psc 1h33'34" 19d15'50"
Maria Bunnell
And 0h37'17" 28d23'58"
Maria Buontempo
Sgr 19h17'31" -14d52'32"
Maria C. Sepulveda
Sgr 18h35'54" -27d28'32"
Maria C. Vega "Magy"
Lib 14h45'19" -19d22'37"
Maria Cambreri
And 0h9'31" 30d23'36"
Maria Camila Acosta
Lib 15h2'59" -0d36'48"
Maria Campbell, DVM
Lmi 9h56'41" 35d14'56"
Maria Capella Torres
Del 20h33'50" 14d27'56"
Maria Caraveo
Uma 9h47'42" 46d11'19"
Maria Carmela Rosa
Tau 4h32'40" 25d33'58"
maria carmen alfonso
Lyn 7h4'17" 55d53'3"
Maria Castillo Allen Classique
Leo 11h34'54" 19d21'14"
Maria Catalina Posada
Leo 9h45'29" 22d7'45"
Maria Cecilia
Psa 22h13'15" -31d58'10"
Maria Cecilia Almonina
Vir 13h6'2" -0d17'59"
Maria Cecilia Byrd
Aqr 23h18'25" -18d20'23"
MARIA CERISSA YSA NAGUI
Vir 12h47'25" 6d37'45"
Maria Chiara Concas
Uma 11h33'22" 52d7'14"
Maria Chiara Tigani
Per 3h41'9" 45d26'30"
Maria Christina Duffy
Her 16h32'20" 13d30'39"
Maria Christina Lopez
Leo 9h35'47" 28d32'4"
Maria Christina Rios
Cma 7h22'31" -17d39'17"
Maria Christina Sorbara
Cas 2h11'43" 64d1'52"
Maria Christine Benevento
Cap 21h11'4" -20d43'6"
Maria Clara
Cas 1h28'22" 60d12'44"
Maria Claudia Perez
And 23h20'28" 48d49'40"
Maria Coatl Rios "My Luv"
Lyn 8h30'48" 33d13'2"
Maria Coco Lopez
Lyn 8h10'44" 45d21'35"
Maria Colon-Droz
Cap 20h10'18" -22d40'18"
Maria Concetta Monte
Peg 21h38'15" 14d36'1"
Maria Contreras
Gem 7h42'40" 31d50'31"
Maria Corrales Hernández
Mon 6h52'34" -0d37'59"
Maria Cristina
Uma 11h32'2" 36d41'53"
Maria Cristina 15122004
And 23h19'16" 42d10'18"
Maria Cristina Marchand
Sco 16h11'20" -11d52'2"
Maria Cruz Huerta
Gem 6h21'42" 19d7'5"
Maria "Cuddlebug" Toth
Cas 0h31'52" 58d50'33"
Maria D. Armendariz
Cap 21h14'46" -25d42'22"
Maria D. Rodriguez
Tau 5h57'58" 24d56'36"
Maria Dalla Costa
Aur 6h26'16" 33d27'25"
Maria Davis
Gem 7h44'38" 34d4'27"
Maria Davis
And 23h37'16" 33d14'9"
Maria de Fatima Tarrinha
Uma 10h26'10" 62d19'30"
Maria de Jesus Alvarez de Rigoli
Del 20h53'4" 7d58'56"
maria de la luz ticknor
Sgr 19h17'29" -26d25'20"
Maria de la Paz Vides
Pho 2h17'23" -46d28'55"
Maria De Los Angeles Aleman
Tau 4h14'44" 11d0'5"
Maria de Los Angeles Bellido
Cas 2h2'20" 58d8'17"
Maria De Los Angeles Chirino
Sgr 17h57'0" -29d10'6"
Maria de Lourdes Cunha
Cap 20h38'21" -24d44'37"
Maria De Lourdes Tankersley
Cas 1h21'15" 58d26'15"
Maria de Mater Gelpi
Aqr 23h25'1" -11d29'12"
Maria Deane
Gem 7h3'20" 21d16'5"
Maria deJesus Becerra Reyes
Uma 14h13'27" 58d9'43"
Maria Del Carmen
Vir 13h19'58" 11d15'32"
Maria Del Carmen Acosta
And 0h11'34" 35d20'31"
Maria del Carmen Placencia
Leo 11h37'45" 14d8'20"
Maria Del Carmen Rivera
And 1h12'56" 45d29'46"
Maria del Carmen Toussaint Gallardo
Com 12h11'29" 32d17'26"
Maria del Consuelo Salinas Stephens
Sco 16h25'56" -31d15'51"
Maria Del Mar
Lib 14h39'39" -17d5'9"
Maria del Mar Obispo Casado
Cas 1h1'47" 61d6'21"
Maria Del Pilar
And 0h50'9" 22d23'48"
Maria del Pilar Madera
Lib 15h34'17" -27d26'51"
Maria Del Refugio Maldonado Cossio
Cnc 8h29'51" 31d35'39"
Maria Del Rosario Curiel
Gem 7h11'27" 26d25'34"
Maria Del Rosario Yanez Forgash
Lib 16h0'3" -8d42'52"
Maria Delgado
Ori 5h29'6" -0d45'13"
Maria Delia Gaytan
Gem 6h39'57" 14d18'51"
Maria Denise
Cap 20h29'42" -24d42'45"
Maria Denise Miller
Aql 19h47'6" -3d22'58"
Maria Deramus
Crb 15h45'29" 26d7'45"
Maria Di Tommaso
Cap 20h47'44" -23d41'52"
Maria DiBartolo, My Brightest Star
And 23h20'41" 38d7'11"
Maria Diffley
And 23h6'11" 42d28'41"
Maria Difonzo
Psc 1h2'27" 26d15'53"
Maria do Céu Pereira Barros Alves
Cas 0h10'33" 59d29'53"
Maria Dolores Roman
Cas 0h34'40" 50d21'27"
Maria Dolors B.B
Cas 0h23'16" 59d32'47"
Maria Donicia
Vir 14h30'29" -3d31'21"
Maria D'Onofrio
Cas 0h50'12" 60d4'4"
Maria Dorothea Zerola
Ari 3h14'45" 27d44'37"
Maria Dower
Leo 10h30'5" 18d23'48"
Maria Draganovic
Cen 11h49'50" -52d3'50"
Maria Drakopoulos
Psc 1h6'49" 26d45'54"
Maria Dubitsky
Tau 4h18'17" 7d23'50"
Maria Dulce deLemos Williamson
Uma 9h30'49" 66d32'44"
Maria Dulce DeSousa
Uma 10h49'1" 70d43'15"
Maria Duran Garcia
Uma 8h42'16" 71d0'12"
Maria E Delpapa
Cas 23h24'33" 57d58'54"
Maria e Gianni
Cas 0h7'2" 53d38'12"
Maria E. Jahja
Dra 16h29'18" 56d26'22"
Maria e Marco
Cam 7h15'56" 67d36'44"
Maria E. Rodriguez
Lib 15h5'23" -24d7'49"
Maria E Torres
Cyg 21h36'55" 36d56'55"
Maria een stjerne
Cas 23h22'38" 55d30'14"
Maria Elaina
Lyn 8h58'11" 46d8'3"
Maria Elaine - "My Spanish Beauty"
And 23h8'27" 41d56'19"
Maria Elena
And 0h16'16" 45d54'41"
Maria Elena
Lyn 7h18'46" 46d46'9"
Maria Elena
Crb 15h49'20" 32d8'16"
Maria Elena
Vir 12h43'42" 6d4'56"
Maria Elena
Leo 11h24'0" 2d41'30"
Maria Elena
Lyr 18h28'32" 28d13'36"
Maria Elena Aguilar
Aqr 21h37'20" 1d10'19"
Maria Elena Becerra
Ori 5h8'39" 1d23'32"
Maria Elena Fabello
Lib 15h20'49" -4d24'44"
Maria Elena Garcia
Uma 9h32'41" 42d21'9"
Maria Elena Gomez
Uma 10h52'44" 58d53'0"
Maria Elena Madariaga
And 23h7'48" 48d14'19"
Maria Elena Orellana
Vir 11h57'2" -4d17'49"
Maria Elena Oyanarte de Villanueva
Ori 5h24'5" 10d18'15"
Maria Elena Perez
Crb 15h50'11" 25d51'31"
Maria Elena Perez - Teacher
And 0h11'47" 39d32'26"
Maria Elena Sanchez
And 0h45'55" 38d34'12"
Maria Elena Tostado
Cyg 19h34'25" 28d18'24"
Maria Elena Vargas
Uma 9h25'42" 44d35'24"
Maria Elena, My Love
Psc 0h19'53" 8d17'50"
Maria Elisa
And 1h7'40" 45d11'52"
Maria Elisa Grimstad
Aqr 22h33'5" -1d44'56"
Maria Elizabete Ferreira
Cap 20h34'44" -12d37'2"
Maria Elizabeth Davis
Umi 15h5'55" 73d53'29"
Maria Elizabeth Smith
Cas 1h44'24" 60d21'6"
Maria Emmanuela Tsitsoulas
And 23h10'49" 44d52'30"
Maria Enid Serrano
Equ 21h12'8" 10d52'24"
Maria Enidina Jordan
Ori 5h36'53" 2d35'44"
Maria Enriqueta Ongaiz
Lyn 8h21'30" 57d15'17"
Maria & Ernst
Her 17h56'7" 41d39'20"
Maria et Jean-Charles
Cyg 19h55'54" 39d48'46"
Maria Eucharistica Aguilina
Ori 5h57'38" 18d31'56"
Maria Eugenia e Manuel Lima
Uma 9h50'26" 66d29'54"
Maria Eugenia Ferreira Malugani
Leo 10h57'32" 22d26'3"
Maria Eugenia Velasquez Mendoza
Lib 15h5'40" -4d54'12"
Maria Eunis Canon
Tau 5h27'25" 22d26'6"
Maria Felicita Quiles Ramos
Cas 0h48'15" 58d37'45"
Maria Fernanda Reyes Ayala
Tau 4h3'52" 5d3'49"
Maria Ferris
Tau 4h10'46" 29d24'52"
Maria Fiandaca
Tau 4h8'54" 15d58'14"
Maria Fine
Leo 11h8'28" 15d14'42"
Maria Flor De Lis Wilma South
Sex 10h45'40" -0d24'4"
Maria Flurry
Lyr 18h58'21" 27d52'47"
Maria Fornari
Umi 16h19'31" 85d6'43"
Maria Fox
And 1h12'43" 37d44'16"
Maria G. Morlando
Psc 23h56'2" -0d59'20"
Maria G. Wise
Vir 11h48'55" -0d25'8"
Maria Gagliardi
Lyn 7h34'48" 46d28'20"
Maria Galley
Mon 6h48'11" 8d36'11"
Maria Gantman
Leo 11h25'40" 25d57'12"
Maria Garcia
Umi 14h4'20" 78d1'39"

Maria Gatta
Tau 5h38'26" 26d31'34"
Maria Gavaris
Uma 9h29'25" 60d32'43"
Maria Gema Ayuso Galan
Lib 15h44'24" -28d38'26"
Maria Gessica Pecana
Crb 16h12'58" 38d46'35"
Maria Giovanna
And 1h0'50" 47d56'15"
Maria Giovanniello
Psc 23h9'55" 2d21'51"
Maria Glidewell Beloved Wife & Mom
Cap 20h22'32" -11d34'20"
Maria GOC
Ari 2h20'7" 19d8'4"
Maria Gomez McKinney
Ori 5h41'46" 7d41'21"
Maria Graciela y Carlos Nevarez
And 0h15'6" 28d20'39"
Maria Gramble
Lyn 7h45'35" 44d29'40"
Maria "Grandmama" Cesena
Cap 21h30'52" -22d49'48"
Maria Grazia
Cyg 21h26'51" 37d0'16"
Maria Grazia
And 23h41'59" 38d29'43"
Maria Grazia Albareda
Cru 12h10'45" -59d38'19"
Maria Grazia Caravaggio
Aql 20h9'9" -0d18'43"
Maria Grazia Lucchini
Peg 22h19'7" 17d17'4"
Maria Grazia, 06.12.1971
Ori 5h57'58" 14d26'6"
Maria Guadalupe Bautista
And 1h4'39" 34d2'2"
Maria Guadalupe Cunha
Cas 1h35'26" 64d28'36"
Maria Guadalupe Macias
Tau 5h41'21" 24d29'2"
Maria Guerriero
And 1h36'55" 46d30'36"
Maria & Guido
Cyg 19h46'35" 35d1'23"
Maria H. Salomé Uranga
Sco 17h40'9" -39d14'9"
Maria Halaby
Cyg 21h2'20" 52d21'31"
Maria Harte
Umi 14h53'26" 69d19'54"
Maria Herlinda Rooney
Crb 15h51'35" 27d1'25"
Maria Herrin
Tau 4h41'6" 18d56'30"
Maria Hildner-Hessling
Uma 11h45'54" 38d3'43"
Maria Hudson
Ori 5h52'9" 11d49'13"
Maria I. Escobedo
Tau 4h37'3" 29d40'57"
Maria I. Monteiro
Cyg 20h7'59" 33d12'26"
Maria Ignacia Prado de Galilea
Cru 12h24'37" -60d11'24"
Maria Inés García López
Mon 7h26'51" -0d49'56"
Maria Ines Madrigal De Leon
And 23h0'57" 51d38'42"
Maria Ines Pellecer
Ori 5h0'11" -1d25'27"
Maria Irma Rosa
Psc 0h8'42" -2d8'43"
Maria Isabel
Cap 20h32'1" -11d58'46"
Maria Isabel Jiménez Gentil
Her 18h28'28" 23d51'0"
Maria Isabel Mudry
Vir 14h21'57" -0d55'37"
Maria Isabel "Ria"
Leo 11h19'45" 7d20'3"
Maria J. Rodriguez
Ori 6h5'8" 13d59'33"
Maria Jeannette Morris
Crb 15h47'35" 32d38'26"
Maria Jedrych
Uma 9h41'10" 42d46'35"
Maria Jensen
And 2h11'18" 42d26'18"
Maria Jesus
Aqr 21h35'6" 0d30'28"
Maria Jesus
Lib 15h46'33" -28d59'59"
María Jesús de Santos Mas
Lyn 8h20'28" 50d25'29"
María Jesús Ruiz
Lmi 11h2'30" 29d26'18"
Maria & John
Ori 6h17'29" 16d29'47"
Maria Jose
Ari 2h35'42" 21d58'53"
Maria Jose Acevedo
Uma 12h39'40" 55d0'36"
María José García López
Ori 5h58'4" 17d59'40"

Maria Jose Girones Osequeda
Ori 6h13'40" 7d13'9"
Maria & José Guevara
Sge 20h14'9" 17d29'38"
Maria Jose Vega
Cnc 8h59'46" 19d40'50"
Maria *Juicy* Hendricks
Ari 2h10'16" 22d57'32"
Maria Kamari
And 0h20'21" 26d44'53"
Maria Kathleen Linton
Leo 11h11'39" 1d55'2"
Maria Katina Hadjiyanis
Lib 15h18'3" -5d17'20"
Maria Kavli
Vir 13h47'0" -6d16'20"
MARIA KKALLI
Tau 4h14'59" 13d22'38"
Maria Koutsakis
Cnc 8h9'40" 17d2'29"
Maria Kristina Dossey
And 0h35'6" 35d20'39"
Maria Kubis
Vir 12h35'28" -2d23'34"
Maria Kyriakou
Cnc 9h14'28" 21d2'45"
Maria Kyrmegalou
And 23h16'16" 48d16'26"
Maria L. Esparza
Cas 23h30'2" 52d16'28"
Maria L. Polanco
Aqr 23h4'10" -6d14'43"
Maria Lago Vigil Morris
Cas 0h14'28" 52d15'52"
Maria Laina
Cas 23h4'51" 53d29'44"
Maria Leonor Garcia
Lib 14h55'5" -3d42'6"
Maria Leonora Teresa
Sgr 18h28'35" -16d52'5"
Maria L.F.I.V
Pho 1h6'11" -48d32'38"
Maria Linda 06/05/1987
Gem 7h30'2" 16d33'27"
Maria Linda Galante
Aqr 22h39'23" -0d14'30"
Maria Lisa
Cyg 20h51'43" 46d54'33"
Maria Livingston
Cas 0h50'20" 69d24'43"
Maria Loizou
And 0h47'1" 26d23'0"
Maria Lopez
And 1h5'36" 38d16'56"
Maria Louisa Angelica Galang
And 23h19'52" 42d27'18"
Maria Louisa Powers
Uma 10h36'57" 59d26'38"
Maria Louise
Psc 1h2'26" 7d22'3"
Maria Lozano Cantu
Sgr 19h42'59" -20d59'40"
Maria Lucy
And 1h0'35" 41d40'10"
Maria Luisa
Gem 7h26'21" 30d48'57"
Maria Luisa
Com 13h6'17" 29d3'47"
Maria Luisa Feola
And 23h28'5" 36d44'56"
Maria Luisa Piraquive
Ori 5h57'32" 17d47'29"
Maria Luise
Her 18h3'16" 18d21'46"
Maria Lupita Regalado
Aur 5h48'26" 45d33'48"
Maria "Lupo" Ducote
Mon 6h15'32" -5d42'42"
Maria Luz Afable Campbell
Sco 17h24'55" -42d32'28"
Maria Luz Jaramillo
Sgr 19h23'44" -38d49'12"
Maria Lynn
Umi 13h30'24" 69d43'32"
Maria Lynn Clovis Seta
Sgr 18h8'20" -34d50'9"
Maria Lynn Ellen Ayers
Cas 0h22'35" 54d28'29"
Maria M. DeFilippis
Uma 10h20'13" 41d6'36"
Maria M. Macon
Uma 11h26'26" 58d32'0"
Maria M. Waizenegger
Aqr 22h38'41" -0d28'26"
Maria Magana
Mon 7h16'4" -0d2'2"
Maria Magana - huerta
Ari 2h43'22" 27d42'44"
Maria Magdalena de Moujan
Vir 14h4'57" -1d46'37"
Maria Magdalena Hinojosa
Cnc 8h54'49" 22d43'29"
Maria Magdalena Rangel
Uma 11h24'38" 62d54'59"
Maria - Major
Uma 11h2'42" 66d51'36"
Maria Manuela - "Ella"
Crb 16h17'31" 31d13'57"
Maria Margarida Martins
And 23h43'55" 46d35'1"
Maria Marissa Mia
Leo 11h53'43" 25d21'9"

Maria Markoulli
And 1h25'55" 49d4'57"
Maria "Mar-Vada" Smith-Moskwa
Crt 11h24'41" -15d41'1"
Maria März
Ori 6h17'4" 16d51'10"
Maria Matthews
Cyg 21h37'11" 32d56'46"
Maria 'Maz' Connelly
And 22h58'1" 37d3'52"
Maria Medeiros & Sophia Olivia York
Sco 17h19'51" -40d52'6"
Maria Melissa-Louisa Pimm
Cyg 19h54'26" 37d5'30"
Maria Mendez Thomas
Mon 7h9'52" -1d21'52"
Maria Mercedes
Uma 9h13'44" 58d29'44"
Maria Mercedes Luque
And 1h50'8" 39d20'59"
Maria Mercedes Vargas
Leo 11h51'34" 23d42'26"
Maria "Meshi" Rizzuti
Crb 15h45'56" 31d42'43"
MARIA MICAELA ALVARADO
Uma 9h51'19" 46d46'31"
Maria Michalak
Uma 9h34'44" 41d49'48"
Maria Milazzo
Lyn 8h0'10" 50d25'3"
Maria Mirkovich
Lyn 7h57'29" 57d29'40"
Maria & Mitch's Little Angels
Uma 8h47'50" 64d57'39"
Maria Moares
Aqr 20h49'4" -7d56'32"
Maria Molina de la Cueva
And 23h20'15" 48d11'52"
Maria Monaco & Associates
Cru 12h3'8" -57d46'6"
Maria Monica Esguerra
Sgr 17h53'56" -29d43'45"
Maria "Monk" Cerrato
Lib 14h51'53" -1d26'48"
Maria Moreno
Sco 16h14'23" -14d48'15"
Maria Morningstar Jaime
Sco 16h4'38" -16d43'32"
Maria Morreale
Leo 11h15'43" 6d39'47"
Maria Morritt
And 2h30'15" 46d22'28"
Maria Muckey
Cap 20h30'57" -26d45'1"
Maria Murabito
And 23h14'30" 49d40'39"
Maria Murphy
Cnc 8h27'35" 18d31'43"
Maria - My Skooliki
Psc 0h58'48" 9d35'10"
Maria N. Little
Tau 5h42'35" 25d34'4"
Maria N. Onorato "The Goddess"
Cas 0h37'34" 47d14'34"
Maria Nau
Peg 21h39'46" 21d28'9"
Maria Nichole
And 23h29'47" 45d53'37"
Maria & Nikolai Zaborskiye
Cyg 20h38'16" 36d15'17"
Maria Noel Martin
And 0h52'35" 45d20'5"
Maria Noelle Camerota
Her 17h6'49" 34d55'40"
Maria & None
Cnc 8h21'32" 23d30'4"
Maria Nora Jimenez
Ari 2h37'3" 21d4'23"
Maria O'Connor
And 1h0'50" 38d26'59"
Maria October 21,2004
Uma 11h17'7" 43d49'56"
Maria Olive May Scandinaro
Psc 22h58'6" 5d3'8"
Maria (Oma)
Lyn 8h34'17" 43d27'18"
MARIA OQUIÑENA
Lyr 18h25'22" 31d4'19"
Maria Otero
Cap 21h32'43" -15d36'1"
Maria P
Uma 10h51'56" 62d51'2"
Maria P. Roman
Psc 0h45'24" 13d45'55"
MARIA - PANTELIS
Her 17h27'59" 26d5'22"
Maria Paschal
Cas 23h51'50" 57d54'40"
Maria Paz Clark
Ari 2h0'19" 19d48'41"
Maria Pecci 18
Ori 5h24'40" -8d14'29"
Maria & Pedro
Umi 16h22'32" 76d21'50"
Maria Pellatt-Morley (Sveeeetie)
Uma 9h24'19" 53d12'21"

Maria Pelo
Cam 3h50'37" 72d48'58"
Maria Pelo
Lyr 18h45'58" 37d13'56"
Maria Perandones Alarcon
Ari 2h32'9" 11d46'10"
Maria Perez
Cyg 20h16'46" 50d6'19"
Maria & Peter Sauter
Dra 16h48'2" 56d12'28"
Maria Phasmatis
Cas 1h58'17" 64d28'12"
Maria Piedad Do Nascimiento
Vir 12h44'43" 0d38'58"
Maria Pietro
Vir 13h28'11" 5d19'5"
Maria Pizzarusso
Her 18h57'10" 16d41'11"
Maria Plankensteiner
And 0h39'23" 31d13'44"
Maria Popi Giatis
Aqr 21h38'14" -2d47'44"
Maria Primrose Howeson
Aqr 22h5'26" -14d58'24"
Maria Quarato
Gem 7h11'21" 34d17'39"
Maria Quezada,Educator
Cas 0h38'41" 55d7'57"
Maria R. Fiore
Cas 23h8'19" 56d37'14"
Maria Raiche
Psc 23h18'16" 6d35'9"
Maria Ramos
Mon 6h32'22" 8d2'21"
Maria Refujio Lopez
Mon 6h34'32" 0d53'10"
Maria Regina Lim (Toinky)
Sco 16h4'45" -8d31'33"
Maria Rena Gomez Aliga
Vir 14h13'8" -2d5'8"
Maria Reyes Mi Reina Mi Estrella
Cyg 21h31'37" 48d34'1"
Maria Ribolini
Gem 6h29'57" 13d35'53"
Maria Riedel
Uma 8h21'39" 67d0'22"
Maria Rinde
Gem 7h17'20" 26d37'2"
Maria Rita
Tau 3h31'53" 25d4'11"
Maria Rita
Her 16h29'45" 30d50'8"
Maria Rita - Lona Branca
Lib 14h50'49" -1d56'25"
Maria Rita Paciullo
Aur 5h48'7" 29d47'29"
Maria Riza Padilla
Aqr 23h27'36" -14d37'49"
Maria Rizza Llorada
Tau 4h18'25" 24d18'19"
Maria Romano
Cyg 21h40'38" 48d0'37"
Maria Rosa Veneziano
Aur 5h53'5" 36d6'50"
Maria Rosaria
Peg 22h41'6" 27d24'13"
Maria Rosaria Auricchio
Umi 16h53'43" 78d16'59"
Maria Rosaria Ciaravolo
Uma 9h30'25" 53d25'27"
Maria Rosaria Coccalalla
Cas 23h59'36" 60d47'33"
Maria Rosaria Maietta
Cyg 20h13'18" 36d31'31"
Maria Rosaria Napola
And 2h25'16" 41d35'20"
Maria Rose
And 2h15'36" 43d29'27"
Maria Rose
Lib 15h50'42" -17d31'43"
Maria Rose MacAllister
Umi 16h1'23" 70d23'56"
Maria Rovinelli
Uma 10h39'47" 40d43'1"
Maria Rucsandra Dobre
Leo 9h42'26" 26d57'48"
Maria Rueda
Ori 5h55'53" 7d0'22"
Maria Ruth Baxter MD
Uma 11h42'36" 36d22'0"
Maria RV54CR54
Lyn 6h44'46" 56d44'58"
Maria S. Gietzen
Vir 13h46'34" -5d11'32"
Maria Salcido
Cyg 19h43'12" 37d25'1"
Maria Samira Maroufi
Cas 0h52'8" 53d24'30"
Maria Sanchez
Cnc 8h44'52" 23d23'44"
Maria Saozinha Pagano
Psc 1h29'50" 21d13'12"
Maria "Shorty" Lopez
Leo 11h2'2" 6d5'51"
Maria Siannas
Gem 6h52'0" 19d24'41"
Maria Silva (Nina)
Leo 11h25'32" -1d23'56"
Maria Silvia Giannoni
Lib 15h31'5" -5d7'53"
Maria Simon
Uma 9h25'38" 47d8'15"

Maria Sloan DuPont
Cap 20h50'27" -25d47'18"
Maria Smith
And 2h19'47" 42d35'48"
Maria Socorro Leon-Andrade
Uma 12h59'13" 53d19'35"
Maria Sofia Mura
And 23h58'35" 38d44'28"
Maria Soledad López
Cas 1h39'30" 61d31'57"
Maria Soledad Rodriguez
Sco 16h15'41" -11d6'32"
Maria Sonia J. Oli
Aqr 21h11'54" -3d41'50"
Maria Soonsup
Cas 23h10'10" 59d40'48"
Maria Sophia
And 23h10'30" 42d58'36"
Maria Souli Dioyenis
Uma 13h0'17" 55d34'9"
Maria Sprouse
Psc 1h25'33" 16d47'35"
Maria & Sven
Ori 5h52'4" 22d4'14"
Maria T Johnston
Psc 0h27'20" 20d12'57"
Maria Taguas
Cas 23h2'43" 59d16'32"
Maria Teata
Psc 0h14'56" 9d2'4"
Maria Tedesco
Cyg 21h11'4" 47d14'7"
Maria Teesch
Crb 15h43'27" 25d54'28"
Maria Teresa 04
Aqr 23h5'51" -19d1'25"
Maria Teresa Almaraz
And 1h11'47" 37d1'44"
Maria Teresa Buhajla
Uma 11h26'54" 58d54'20"
Maria Teresa Castello Bertrand
Cas 1h28'44" 64d0'45"
Maria Teresa Henriques Salgado
Ari 2h43'29" 25d9'46"
Maria Teresa "Maite" Arguelles
Lib 15h1'25" -4d30'0"
Maria Terezia Ferencz
Crv 12h5'9" -14d25'10"
Maria The Beautiful Angel
Aqr 22h42'16" 0d44'15"
Maria - The Brightest Star
And 2h25'49" 38d7'16"
Maria Theresa
And 2h7'19" 43d20'32"
Maria Theresa Catanzaro
Sco 16h6'53" -15d47'55"
Maria Theresa DeVera
Cam 4h1'4" 68d21'31"
Maria Theresa Milstrey
Cam 9h11'14" 77d52'24"
Maria Theresa Van Egmond
Cma 7h24'8" -31d53'14"
Maria Thérèse Horton
And 23h15'28" 51d10'33"
Maria & Thomas Rooney
Cyg 21h47'16" 44d5'28"
Maria Tigre
And 23h19'28" 47d30'15"
Maria Tina Isbella Waters
Cas 23h22'35" 59d35'6"
Maria Tobin
Crb 15h19'9" 26d58'33"
Maria Tommelise Hovden Knudsen
Aqr 22h35'44" 0d7'13"
Maria Torralva
Vir 14h0'43" -14d34'9"
Maria Trifilio
Vir 13h3'43" -4d50'25"
Maria Trikilis, Angel Of The Light
Vir 13h35'55" -4d49'1"
Maria Trosdal Sparkman
Cnc 8h15'3" 12d31'54"
Maria Tveritina
Ori 6h10'35" -2d10'46"
Maria Tweety Julian
And 23h15'12" 52d56'58"
Maria U. Aviles
And 0h29'14" 42d0'4"
Maria V McGuire
Tri 2h2'13" 32d59'54"
Maria Vacca's Shining Light
Aqr 23h5'6" -7d34'41"
Maria Varela Rusnak "Mimi"
Cnc 8h51'12" 28d43'13"
Maria Velasquez
Lyn 8h6'20" 33d12'4"
Maria Ventura
Leo 10h48'0" 22d35'49"
Maria Vernetti
Cas 0h47'4" 53d38'33"
Maria Vicenta
Uma 8h50'32" 67d26'25"
Maria Victoria Matheu-Delgado
And 23h49'10" 45d41'3"

Maria Victoria Munoz
Lib 14h55'8" -19d12'11"
Maria Victoria Nanny Won
Leo 10h24'54" 18d23'28"
Maria Vittoria Novi
And 2h27'22" 50d27'15"
Maria Voit
Lep 5h8'51" -11d30'54"
Maria Watt
Cas 23h34'53" 52d53'43"
Maria Wendy Tremillo Kim
And 22h58'14" 45d42'41"
Maria woman of grace
Tau 4h58'15" 19d30'9"
Maria Woodhouse - Mummy's Star
And 2h20'20" 44d32'14"
Maria xx
Oph 17h32'2" -3d0'47"
Maria Yepes
Aqr 21h5'44" -11d49'33"
Maria Zuleta
Ari 3h23'29" 24d19'34"
Maria, Aaron and Katie
Uma 13h49'23" 49d21'50"
Maria, Chiara Corsaro
Her 17h52'45" 29d28'0"
Maria, mi mama
Sgr 18h32'2" -24d20'20"
Maria05071965
And 0h13'46" 46d20'11"
Maria07
Ori 6h23'30" 14d16'38"
Maria-Anna
Cam 6h22'21" 66d2'18"
Maria"Bebe"Ortiz Camacho Mi Sunchi
Mon 6h52'38" -7d27'6"
MariaBogri
Ari 3h21'33" 17d51'1"
Mariachristina
Cyg 19h37'59" 37d37'13"
Maria-Christina Pattichis
Vir 12h4'23" 3d11'12"
Mariacristina Annarita Abbruzzese
And 1h2'11" 42d6'39"
Mariacristina e Mario
Lyr 18h26'20" 34d42'59"
mariadelpilar
Lib 15h9'31" -0d30'51"
Maria-Dolores Zamora
Lmi 10h44'42" 27d51'6"
MariaElena Bauer
Cas 0h36'46" 55d0'43"
MariaElena Del Pizzo
And 23h56'5" 34d35'38"
Mariagrazia
Her 17h48'33" 42d33'54"
Mariagrazia e Gianfranco
Cyg 21h48'21" 54d6'37"
Mariagrazia Galasso
Uma 12h9'47" 57d0'25"
Mariagrazia Trasente
Uma 11h23'38" 59d25'57"
Mariah
Uma 13h43'33" 54d3'49"
Mariah
Vir 12h17'32" -8d46'26"
Mariah 15
Peg 22h15'57" 7d39'37"
Mariah Acton
Cyg 19h53'36" 32d5'11"
Mariah Ariel Clemons
Psc 0h47'23" 17d11'33"
Mariah Carolina Bohnstehn
Cap 21h16'23" -16d37'10"
Mariah Christine 12/25/02
And 23h49'3" 45d40'1"
Mariah Cortez
Psc 0h18'2" 6d39'33"
Mariah Dawn
Aqr 23h30'20" -12d37'16"
Mariah Dawn Smith
Cyg 21h38'52" 48d36'18"
Mariah De Aspen Y Manzanita
Dra 18h48'4" 50d49'45"
Mariah Grace Olson
And 1h11'49" 44d43'10"
Mariah Hope
Gem 6h44'32" 13d25'6"
Mariah J. Riddle
Ari 2h31'22" 24d31'52"
Mariah Kaylee
Cas 2h26'46" 65d22'16"
Mariah L Riney
Cas 0h22'11" 59d12'52"
Mariah Lynn
And 2h18'24" 49d29'47"
Mariah M. Folk
Lyn 6h36'13" 57d20'24"
Mariah Manley * Brytney Goathe
Ori 4h56'2" 2d59'24"
Mariah Marie Rogers
Cap 21h30'0" -10d33'44"
Mariah McCulloch
Aqr 22h54'33" -9d19'54"
Mariah N. Mandy
Leo 10h46'30" 14d45'13"
Mariah Pilar Jones
And 0h31'23" 30d30'4"
Mariah Vega
Aqr 21h31'25" 1d6'33"

Mariah Victoria Robinson
Crb 15h52'2" 27d20'6"
mariahbabygirl
Tau 5h3'58" 24d47'44"
Mariah's Star
Ori 5h39'30" -1d58'25"
Marialessandra
Ori 5h45'59" 1d56'15"
Marialexia Alfano
And 23h0'26" 36d21'23"
MariaLuana
Umi 13h6'3" 72d30'54"
Marialuce Barbuzzi
Umi 13h45'15" 71d18'11"
Marialuisa
Cyg 21h27'14" 36d59'55"
Mariam Afshar
Psc 23h59'25" 10d30'23"
Mariam Al Zain
Sco 17h57'42" -39d45'7"
Mariam Elizabeth Bacon
Sgr 17h55'23" -28d33'8"
Mariam Entezari
Mon 6h30'39" 7d38'59"
Mariam Jamali
Leo 11h26'1" 1d13'55"
Mariam "Mary" Keshishyan
Ari 2h12'57" 26d37'15"
Mariam "Mermaid"
Sgr 19h58'26" -40d30'1"
Mariam Mikati
Uma 12h17'4" 61d43'31"
Mariam Zehra Rizvi
And 1h37'48" 42d17'17"
Marian
Cnc 8h14'28" 22d46'52"
Marian
Ari 3h17'25" 15d45'13"
Marian
Ari 2h42'54" 16d18'28"
Marian
Mon 6h52'49" -0d42'38"
Marian
Umi 16h22'21" 77d21'41"
Marian A. Harris
Cas 0h35'28" 59d24'1"
Marian and David
Cap 20h19'13" -11d28'52"
Marian and David
Cap 20h29'55" -18d26'27"
Marian Annette Kinkopf
Lib 15h54'6" -7d5'43"
Marian B. Kelson
Uma 10h39'3" 59d8'9"
Marian Bacher
Uma 10h21'9" 42d22'8"
Marian Curras Garcia
Psc 0h16'29" 2d46'38"
Marian D. Ess, The Love Of My Life
Aql 19h56'48" 7d22'29"
Marian Dolgoff
Uma 10h34'5" 68d37'4"
Marian Dorothy
Psc 0h15'35" 6d9'6"
Marian E. Jackson
Umi 17h4'42" 86d20'52"
Marian E. Lindberg
Sco 16h9'57" -11d3'44"
Marian Elizabeth Daniels
Vir 12h44'37" 4d51'22"
Marian Elizabeth Hollingsworth
Com 12h47'51" 27d45'7"
Marian Eloise Rissler
Psc 1h32'14" 18d57'0"
Marian Elsaesser
Ori 4h58'52" 10d36'31"
Marian Ferine
Gem 6h46'11" 12d16'6"
Marian Full of Grace
Cas 0h30'43" 53d13'30"
Marian Grace
And 1h7'0" 41d57'36"
Marian Herr
Cyg 19h36'25" 28d43'57"
Marian & Jack Bleakley
Uma 11h3'21" 68d12'8"
Marian Jessica Dayhuff
And 0h33'5" 35d31'7"
Marian Judd
Sgr 18h21'34" -33d1'19"
Marian Kohn
Uma 10h55'39" 63d20'59"
Marian Loeffler
And 0h52'52" 46d29'38"
Marian Main
Cap 21h26'44" -16d3'12"
Marian Marcolini
Lyn 7h4'47" 48d37'44"
Marian Marie McNabb
Cnc 8h9'25" 24d38'12"
Marian McCall
Cap 20h38'55" -13d59'51"
Marian McCormick
Ari 2h40'55" 13d34'57"
Marian Misinco
Ari 2h47'1" 26d1'20"
Marian My One True Love My Star
Gem 7h28'52" 29d29'14"
Marian O'Connor Franklin
Cap 21h0'10" -20d22'35"
Marian Partee
Lyn 7h11'49" 45d30'36"

Marian Powell
Lyn 6h23'46" 55d15'50"
Marian Pressler
Per 3h38'0" 47d36'57"
Marian R. North
Uma 9h40'33" 70d17'41"
Marian Rose
And 2h32'47" 45d13'38"
Marian Ross
Ori 5h52'37" 21d8'43"
Marian Ruth Stone
Cru 12h57'21" -59d15'38"
Marian Ryan Tuttle Loved Art Nature
Lib 14h54'54" -3d31'53"
Marian Schlieper
Psc 0h56'29" 33d18'2"
Marian Spencer CMP - GaMPI Star
Umi 14h57'4" 68d56'53"
Marian V. Dodson
Sgr 19h37'11" -16d8'4"
Marian & Winford Rawlins
Ori 5h27'2" 6d38'14"
Marian, Beloved Little Flower
Peg 23h15'54" 19d25'8"
Mariana
And 0h3'31" 44d49'9"
MARIANA
And 23h10'49" 41d56'36"
Mariana
Vir 14h14'27" -8d47'28"
Mariana
Ori 4h46'19" -2d52'53"
Mariana
Lyn 7h0'51" 52d40'24"
Mariana 69 Torres
Cnc 8h26'51" 14d22'44"
Mariana Angelina Perricone
Dra 16h3'52" 52d41'55"
Mariana & Armando's Whole New World
Cnc 9h13'22" 17d13'32"
Mariana Bruni
And 0h18'54" 34d53'44"
Mariana Camilla Lopez
Ori 6h20'49" 7d22'5"
Mariana Cantrell
Sgr 19h8'43" -19d16'44"
Mariana Cardenas
Cas 1h31'30" 65d3'46"
Mariana Garzezi Peres Bernauer
Leo 11h22'52" 7d3'43"
Mariana Gomez
And 0h36'24" 35d22'38"
Mariana Isabel Hernández Gutierrez
Lib 15h25'16" -11d45'38"
Mariana Luna
Lyr 18h23'56" 39d44'43"
Mariana Nedelea aka Mama Tzitzi
Cas 23h54'26" 54d18'14"
Mariana Noriega
Cap 21h9'0" -15d52'36"
Mariana Ramos
Lep 5h36'13" -15d36'35"
Mariana Rosario
Cas 23h56'50" 56d59'2"
Mariana Sebastian
Dra 18h50'39" 66d34'6"
Mariana y Cédric estrella de amor
Cyg 21h10'35" 44d23'27"
Mariana Zabaleta
Crb 15h47'3" 27d55'34"
Mariana's Star
Aqr 23h40'44" -9d38'38"
Mariane Potts
Cap 20h29'37" -9d23'39"
marianella1111
Sco 16h9'20" -17d5'30"
Mariangela
And 0h27'18" 36d8'6"
Mariangela Gurney
Lib 15h19'42" -7d49'43"
Mariangela Soriano
Cyg 21h59'5" 46d39'16"
Mariangela (tvttb ombra)
Cas 23h4'54" 59d9'15"
Mariann April Happney
Ari 2h44'59" 14d31'46"
Mariann Elvie Bayliss
Umi 13h45'31" 71d33'53"
Mariann & István
Uma 9h23'43" 43d4'5"
Mariann Leandrez Collado
Vir 13h56'50" -10d21'9"
Marianna
Aqr 21h56'22" -2d16'12"
Marianna
Mon 6h46'54" -0d6'24"
Marianna
Umi 13h32'19" 72d48'28"
Marianna
Uma 9h7'44" 69d12'21"
Marianna
Uma 8h53'54" 60d31'25"
marianna
Uma 14h26'43" 55d35'55"
Marianna
Uma 12h18'39" 56d37'40"
Marianna
Uma 12h42'3" 53d30'43"
Marianna
Sct 18h54'35" -12d32'59"
Marianna
Com 12h48'55" 27d9'22"
Marianna
Leo 9h41'47" 26d38'50"
Marianna Augelli
And 1h14'41" 46d1'14"
Marianna Calabrese
Ori 5h4'32" -0d5'1"
Marianna Christine
Sgr 19h21'13" -40d11'51"
Marianna Elizabeth Vazquez
Lyn 7h1'25" 55d18'13"
Marianna Kleemann
Dra 15h14'51" 62d23'47"
Marianna "Nutellina" Leardini
Lyn 8h1'20" 53d44'46"
Marianna Vetere
Ori 5h55'27" 18d5'53"
Marianna Visnadul
And 0h54'43" 41d15'28"
Marianna99
Psc 0h56'8" 32d35'43"
Marianne
Psc 1h31'9" 32d39'18"
Marianne
And 0h57'12" 41d58'29"
Marianne
And 1h28'15" 42d10'40"
"Marianne"
And 0h38'17" 40d40'45"
Marianne
And 2h17'16" 45d2'36"
Marianne
Cas 0h17'34" 51d55'56"
Marianne
And 0h12'36" 45d38'29"
Marianne
Cyg 20h19'59" 42d18'39"
Marianne
And 23h45'28" 48d31'54"
Marianne
Ari 2h9'37" 14d43'57"
Marianne
Ori 6h7'26" 0d38'34"
Marianne
Ori 5h30'58" 3d21'32"
Marianne
Lyn 6h58'32" 55d57'54"
Marianne
Uma 9h12'29" 55d25'55"
Marianne
Cas 0h32'44" 62d40'39"
Marianne
Uma 9h56'9" 61d7'15"
Marianne
Uma 9h15'2" 68d23'47"
Marianne
Ori 5h43'26" -2d9'12"
Marianne and Fred Forever
Eri 3h40'20" -41d32'51"
Marianne Baumgartner
Crb 16h11'11" 26d35'7"
Marianne Becker
Cas 23h39'19" 55d39'47"
Marianne/Bob Our Hearts Together
Cyg 20h56'33" 44d36'23"
Marianne Bringeman
Cas 0h55'34" 60d24'51"
Marianne Catherine Johnston
Leo 11h13'3" -1d54'12"
Marianne Celeste Bachman
Cnc 8h46'16" 24d36'43"
Marianne Charlotte
Cas 0h36'17" 47d4'42"
Marianne D Swann - Jewel of the Sky
Ori 5h34'21" 2d22'30"
Marianne Decher
Lib 14h51'17" -2d11'24"
Marianne E. Larkin
Sco 16h33'30" -25d59'10"
Marianne Eisenschmidt
Uma 8h55'19" 48d14'48"
Marianne Elizabeth Wolfe
Tau 5h15'31" 24d56'3"
Marianne/Felicia
Cyg 19h42'24" 31d19'58"
Marianne Fennell
Cap 21h27'16" -13d58'11"
Marianne Friolet
Ori 5h44'43" -2d41'58"
Marianne Funke
Uma 10h0'6" 64d6'36"
Marianne Gadbois
Ori 5h33'5" -1d16'26"
Marianne & Georg Meier-Schäfli
Per 4h50'15" 49d23'11"
Marianne Grashoff
Lib 15h4'57" -7d3'22"
Marianne Harding
Cas 0h19'46" 51d18'32"
Marianne Hawwa
Cyg 19h30'24" 30d57'21"
Marianne Hazel & Emma Andrea Dunn
Gem 6h39'33" 21d25'59"
Marianne Jecmen
Ori 4h48'48" 12d57'14"
Marianne Kathleen Minwegen
Cas 0h38'9" 57d49'11"
Marianne Katrina McDermott
Cyg 20h21'1" 48d7'22"
Marianne Kosmider
Cas 0h33'58" 66d46'48"
Marianne L. Firestone
Crb 16h12'9" 36d5'16"
Marianne M. Filali
Lyn 7h44'43" 55d48'13"
Marianne "Momma" Behr
Crb 15h40'22" 27d22'7"
Marianne Montgomery
Tau 5h6'3" 19d52'10"
Marianne Moon Thomas
Lyr 18h52'35" 40d46'48"
Marianne Myland
Cyg 21h35'7" 46d1'35"
Marianne Orschel
And 23h2'27" 50d56'8"
Marianne Palmer
Sco 17h32'28" -38d4'47"
Marianne Peffall
Uma 13h26'46" 57d38'39"
Marianne Pendlebury (FireFly)
Cnc 8h37'58" 19d55'23"
Marianne petit Coeur
Sgr 18h26'15" -36d46'22"
Marianne Preker
Sgr 18h42'31" -23d10'25"
Marianne Reith Hatfield
Tau 5h44'52" 22d18'30"
Marianne Robinson
Lyn 8h5'31" 40d37'38"
Marianne Schoenmann
Ari 2h11'14" 18d27'58"
Marianne Spain
And 23h52'20" 42d17'20"
Marianne Sprecher
Cas 1h5'28" 63d49'29"
Marianne Strotz-Graf, August Strotz
Cap 20h28'53" -12d29'12"
Marianne Truglio
Cas 0h19'36" 51d26'20"
Marianne Woszczynski
Lib 15h25'26" -9d51'7"
Marianne, 15.06.46
Ori 6h14'50" 20d42'9"
Marianne-Brian
Cnc 8h8'26" 15d14'39"
Marianne's Baby
Tau 5h9'16" 25d26'34"
Marianne's Eye
Ari 2h21'50" 25d32'35"
Mariano
Cyg 21h46'34" 45d57'4"
Mariano Alexander
Uma 9h29'1" 68d48'30"
Marianovella
Cam 4h1'33" 57d17'32"
Marian's Dream
Mon 6h35'39" 7d14'49"
Marian's Star
And 23h50'58" 44d42'6"
Maria's 21st Birthday
Ari 1h58'32" 20d2'16"
Maria's Angel
Cas 0h7'51" 57d26'20"
Maria's Angel
Umi 14h24'50" 78d22'28"
Maria's Family Star
Tau 4h30'17" 0d58'49"
Maria's Forever
Cam 4h52'58" 52d54'21"
Maria's Star
And 23h53'59" 40d29'9"
Maria's Star
Ori 5h46'53" 9d6'20"
Maria's Star of Love
And 0h40'41" 31d44'46"
Mariasole Csermely
Ori 6h18'33" 12d38'40"
Mariateresa Novello
Dra 15h54'9" 56d18'19"
Mariavittoria Laschena
Uma 10h48'52" 59d22'9"
Maribel
Psc 23h34'54" 2d27'27"
Maribel
Ori 5h56'34" 18d1'35"
Maribel
Gem 7h47'55" 26d46'6"
Maribel
Tau 3h48'20" 31d2'54"
Maribel
Lmi 10h18'59" 38d55'14"
Maribel
And 0h18'4" 43d16'58"
Maribel Beckwith
Psc 1h20'47" 18d24'21"
Maribel Diaz-Garcia
Cnc 8h17'39" 11d19'54"
Maribel G. Rodriguez
And 22h59'58" 47d53'38"
Maribel Juarez Howard
Sco 16h51'38" -30d21'48"
Maribel Lopez
And 2h8'4" 45d21'14"
Maribel Rios
Ari 2h49'5" 13d26'10"
Maribel Sanchez
Cap 21h48'44" -9d33'25"
Maribel the mother of three stars
Cnc 7h59'6" 12d5'31"
Maribel, mi sol, mi luna, mi amor
Psc 1h27'15" 32d0'7"
Maribella2
Mon 6h49'6" -3d59'0"
Maribeth
Lyn 6h59'35" 44d20'56"
Maribeth Browne
Cas 0h15'5" 52d5'15"
Maribito
Her 17h27'9" 27d0'16"
Mariby
Pho 0h33'51" -44d5'30"
Marica "Glee's Star"
And 0h59'34" 40d4'27"
Maricar Cruz
Lyn 9h10'15" 37d2'54"
Maricarmen
Ari 1h56'48" 22d18'32"
Maricarmen
Aql 19h44'8" -0d6'40"
Maricarmen Herrera
Cnc 8h26'10" 13d9'55"
Marice
Cap 21h46'8" -9d42'13"
Maricel Balala
Cyg 20h23'29" 35d54'27"
Maricel "Sweet" Bolos
Cnc 8h6'14" 13d56'6"
Maricela A. Vasquez
And 2h17'57" 41d37'35"
Maricela Janeth
Dra 16h57'3" 55d24'9"
Maricela Jazmin
Uma 9h37'18" 41d59'19"
Mariceli Venegas
Cet 3h10'54" -0d9'30"
Maricky
Cnc 9h4'41" 9d16'4"
MARICRUZ CERDA
Vir 13h27'7" -4d47'59"
Maricruz Martinez
Cas 23h31'29" 55d42'17"
MariDan
Mon 7h20'12" -0d18'35"
Marideth Pasamba-De Leon
Uma 8h37'55" 70d7'23"
Maridys
Psc 0h52'43" 6d16'13"
Maridyth & Charles Nardone
Uma 11h25'26" 53d32'14"
Marie
Uma 11h9'17" 56d27'21"
Marie
Cas 0h48'31" 61d56'2"
Marie
Cas 1h26'52" 67d37'7"
Marie
Uma 13h12'57" 62d15'26"
Marie
Aqr 21h52'34" -7d54'46"
Marie
Sco 17h9'5" -35d59'31"
Marie
Sco 17h29'7" -31d57'52"
Marie
Leo 9h54'57" 11d15'59"
Marie
Ori 5h35'32" 11d8'0"
Marie
Ori 5h18'10" 6d49'4"
Marie
Cnc 8h43'40" 16d27'45"
Marie
Lyn 7h46'5" 42d26'42"
Marie
And 0h45'5" 41d17'32"
Marie
Lyn 9h20'33" 33d46'36"
Marie
And 0h18'37" 35d8'45"
Marie
Crb 16h15'47" 38d13'35"
Marie
And 1h41'35" 50d4'12"
Marie
Cas 1h22'9" 53d32'59"
Marie
Uma 9h20'16" 46d48'12"
Marie
Uma 10h40'24" 51d28'1"
Marie 15/05
Tau 5h9'56" 17d55'28"
Marie A. Leonick
Lib 14h48'58" -0d50'53"
Marie A. Riegel
Cap 21h55'25" -17d30'25"
Marie A Sikora
Umi 15h19'32" 72d29'29"
Marie A. Watson
Peg 22h38'5" 5d21'37"
Marie Abbey Blair
Cap 21h3'59" -24d55'37"
Marie Abell
Cas 23h33'37" 52d51'13"
Marie Agnes Grundy
Cas 0h13'42" 58d11'6"
Marie Agop
Ori 5h21'22" -4d6'26"
Marie Ahern
Cas 0h11'8" 63d25'21"
Marie Alice Mitchell Reid
Gem 6h46'33" 26d32'14"
Marie and Frank Lupi
Ori 5h52'42" 20d43'17"
Marie and Robert Cloud
Ori 5h51'18" 2d35'27"
Marie Angeli "Ganda" A. Hilario
Cnc 8h23'38" 11d15'35"
Marie Ann
Umi 16h49'25" 79d0'59"
Marie Ann Owen
And 1h25'3" 36d25'4"
Marie Anna Vogt Reed
Cap 21h35'46" -13d40'37"
Marie Anne
Leo 11h56'31" 20d58'34"
Marie Anne Loncher
Uma 9h42'22" 53d51'35"
Marie Antionette
Lyn 8h35'8" 46d19'10"
Marie Aurelia
Cep 21h36'27" 60d52'30"
Marie B. Worthy (Nana)
Vir 13h41'9" -14d31'10"
Marie & Barbara Hoeing
And 1h23'11" 34d15'26"
Marie Bartolone
Cnc 9h17'36" 22d36'49"
Marie Benhard
Ori 5h25'36" -9d21'44"
Marie Bergeron / Sultrie Sorceress
Tau 5h55'20" 24d28'13"
Marie Bernier
Uma 10h24'59" 68d7'6"
Marie (Betty) Bergantino
Cap 20h13'22" -25d0'8"
Marie Boucavalas
Peg 23h46'48" 23d29'5"
Marie Braaksma
Peg 21h57'3" 29d24'42"
Marie Brennan WMD ALA '06 - '07
And 23h26'39" 45d5'6"
Marie Broderick
Aqr 22h25'20" -14d26'56"
Marie Bunny Scotto
Sco 16h38'49" -27d35'29"
Marie Busuttil
Cas 0h5'42" 50d18'12"
Marie C. Albers
Leo 10h28'54" 12d38'53"
Marie Cant - The Eternal Time Stepper
Sgr 19h33'20" -19d22'59"
Marie Catherine
Leo 9h38'59" 27d55'42"
Marie Cécile et Bruno
Vir 12h10'27" -0d37'33"
Marie Cécile Saunier
Cas 23h58'27" 59d53'19"
Marie Chriqui Mellman
And 0h42'57" 42d29'10"
Marie Christine
Umi 14h47'48" 75d3'35"
Marie Christine
Col 5h55'18" -39d5'9"
Marie Christine Carey Horman
Cas 0h44'14" 67d22'37"
Marie Christine Whitmore Anstine
Mon 7h16'12" -0d44'6"
Marie + Christophe
Sge 20h18'57" 17d35'53"
Marie & Chuck Murphy
Del 20h48'30" 6d35'30"
Marie Ciochetti
Cas 22h59'55" 57d47'54"
Marie Claire Aubry
And 2h34'42" 45d9'38"
Marie Claude Soupizon
Leo 10h56'10" 7d11'34"
Marie Conn
Cas 0h21'5" 57d10'35"
Marie Covinsky
Sco 17h51'28" -36d24'34"
Marie Crimi Roser
And 23h27'29" 44d39'28"
Marie D. Lemire
Cas 1h1'23" 61d12'5"
Marie & David
Cra 18h8'19" -37d8'6"
Marie Deeds
Uma 11h38'57" 55d11'58"
Marie & Denis
Del 20h17'7" 15d1'19"
Marie Denison
Cnc 9h7'4" 17d53'44"
Marie DeVitto
And 2h1'1" 38d13'49"
Marie Digaetano
Lyn 8h11'24" 52d40'54"
Marie Donnadille
Uma 10h7'46" 52d34'35"
Marie Donovan
Cas 23h26'39" 52d58'12"
Marie Dorothy LaLonde Mayhew
Leo 11h9'12" 13d9'13"
Marie Dreßbach
Uma 9h41'57" 51d5'46"
Marie E. (Betty) Stilwell
Uma 9h23'31" 71d20'27"
Marie E. Combs
Cas 1h37'25" 63d44'0"
Marie E. Fucci
Crb 15h39'40" 26d44'12"
Marie E. Matonovich
Crb 16h7'17" 37d17'3"
Marie E. Wahl
Ori 5h54'57" 20d34'0"
Marie Edwards Webb
Vir 12h50'24" 7d6'19"
Marie Elisabeth
Cas 1h7'50" 50d45'58"
Marie Elizabeth
Cas 2h10'46" 59d15'14"
Marie Elizabeth
Gem 7h32'52" 33d51'21"
Marie Elizabeth
Cas 0h20'48" 65d34'30"
Marie Elizabeth Powers
Cap 20h7'46" -19d24'57"
Marie Elizabeth Ries
Cyg 20h39'26" 37d41'16"
Marie Elizabeth Rueter
Vir 13h27'38" 0d26'54"
Marie Elizabeth Theresa Notte
Tau 5h56'2" 23d59'26"
Marie Ellen Long
Uma 8h48'17" 70d24'27"
Marie Ellen Sager
Ret 4h10'51" -58d13'19"
Marie Ellen Watson
Vir 13h21'54" -15d20'4"
Marie Embleton
Cas 23h51'41" 52d55'2"
Marie Emilie Bouchard
Peg 22h11'22" 12d58'48"
Marie Esther
Lyn 6h39'42" 61d6'13"
Marie Eternité
Aqr 21h39'6" -0d33'11"
Marie Eve
Gem 7h46'4" 34d53'45"
Marie F Benvin
Crb 15h40'50" 25d43'51"
Marie Felipe
Lyn 7h53'40" 41d16'38"
Marie Ficher
Peg 22h27'41" 10d5'27"
Marie Flaherty
Cnc 8h47'42" 12d39'40"
Marie Frances Dildine Costantino
Her 18h56'12" 15d48'32"
Marie Frances Sullivan
Cap 21h39'33" -16d48'34"
Marie & Frank
Aqr 22h56'39" -8d30'56"
Marie Fristachi
Crb 15h32'44" 37d50'31"
Marie Fromont
Per 2h52'43" 53d44'31"
Marie G. Rodriguez Mimi & Sidney
Cyg 20h44'14" 45d50'30"
Marie Gemma Smith
And 23h58'4" 42d6'13"
Marie Gilbert
Cas 23h16'23" 55d26'19"
Marie Glennen
Tau 5h9'43" 18d15'15"
Marie Guinta
Ari 3h26'34" 19d51'12"
Marie H. Krekeler
And 0h38'3" 34d37'1"
Marie Halt
Cas 0h44'2" 64d45'2"
Marie Hargreaves
Aur 5h25'32" 47d20'23"
Marie Harvey
Cap 21h34'48" -10d23'19"
Marie Helen Rose
Uma 8h53'54" 64d21'27"
Marie Hellmann
Uma 9h24'59" 64d16'5"
Marie Henshaw McCain
And 2h4'51" 39d48'26"
Marie Herbert West
Cap 20h45'24" -19d49'11"
Marie Hewitt
Leo 10h51'53" 5d4'44"
Marie Hodge
Ari 3h1'2" 28d3'57"
Marie Irene Love
Dra 16h13'18" 52d4'5"
Marie J. Allard
Cas 1h39'25" 64d5'17"
Marie J Moreno
Cas 1h45'15" 60d56'27"
Marie J. Panico
Cas 0h18'49" 61d5'35"
Marie Jeanne Haenni
Uma 9h34'25" 42d58'45"
Marie Jesus Da Silva
Ari 2h2'42" 17d39'54"
Marie Joe Germano, My MJ
Cnc 9h4'23" 16d50'12"
Marie Jose Florvil Marlene
Lib 14h27'14" -18d23'37"
Marie Josephine Cimino
Psc 1h11'43" 26d20'51"
Marie Josephine O'Connor
And 23h24'56" 49d4'3"
Marie Keating
Sgr 19h49'41" -28d23'33"
Marie Keita
And 23h14'58" 51d16'25"
Marie Kiely
Uma 12h2'57" 63d8'42"
Marie King Caumartin
Aql 19h50'33" -0d57'18"
Marie la Belle
Tau 4h40'12" 23d29'20"
Marie la Maya - Marie-Louise Jung
Cyg 20h29'0" 37d0'50"
Marie Lavaud
Sco 17h51'4" -41d11'37"
Marie Leatham
Lyr 19h7'59" 27d9'9"
Marie Leffler Henry
Uma 10h2'0" 47d31'37"
Marie & LeRon
Cyg 19h51'1" 39d58'0"
Marie Louise
Cas 23h16'56" 59d44'19"
Marie Louise Beardslee
Tau 5h28'24" 18d45'20"
Marie Louise Grenier
Lib 15h9'30" -14d43'41"
Marie Louise Leno
Uma 9h47'2" 49d6'8"
Marie Louise Wong
Tau 5h22'11" 21d28'43"
Marie Lourdes Jimenez
Peg 23h6'59" 31d25'34"
Marie Maddaloni Addis
Ari 2h51'42" 27d50'2"
Marie Madeleine Puissant
Per 2h57'57" 45d45'2"
Marie Maggioli
Psc 0h50'20" 29d16'20"
Marie Mahoney
Crb 16h17'19" 30d15'48"
Marie Malaea
Vir 13h23'12" -17d17'10"
Marie Mariano
Aqr 23h30'38" -19d57'26"
Marie Marlene
Crb 16h11'38" 37d13'51"
Marie Martha
Leo 11h6'7" 3d29'56"
Marie & Martine
Crb 15h55'5" 31d45'19"
Marie Masayo Garcia
Uma 12h27'12" 60d28'51"
Marie Matuska
Crb 16h10'16" 36d31'55"
Marie Maureen
Del 20h48'17" 14d46'16"
Marie Mauroit
Peg 23h11'11" 31d4'29"
Marie May Jeanotte
Cas 1h29'11" 62d59'9"
Marie Mc Carron
Tau 3h35'3" 2d39'39"
Marie McDevitt
Sgr 18h25'53" -26d52'51"
Marie McIlrath
Aur 6h11'6" 29d4'35"
Marie Mesnooh
Cas 1h25'12" 52d11'37"
Marie "Mimi" McCarthy
Aqr 22h41'30" -8d32'35"
Marie "Mimi" Sidwell
Lyr 18h45'49" 39d5'20"
Marie Miracle
Lyr 19h2'11" 32d0'54"
Marie Monastero
Ori 6h6'20" 19d13'40"
Marie Monson
Uma 10h15'13" 68d44'24"
Marie Morgan Yandle
Tau 4h26'29" 24d6'22"
Marie Muller
Cyg 21h21'48" 38d59'55"
Marie N. Garza
Cam 5h51'58" 63d58'41"
Marie Nathalie Estelle Drouin
Umi 14h41'26" 69d29'55"
Marie "NEO"
Sco 16h22'35" -26d59'21"
Marie Nesius Schweidler
Aqr 21h4'59" 1d52'23"
Marie Noelle
Aql 19h42'4" 8d31'34"
Marie & Norm Peterson
Cyg 19h59'1" 58d24'24"
Marie Odile Batt
Umi 17h12'41" 83d46'12"
Marie Olive Headley March 16, 1938
Psc 1h18'24" 5d56'33"
Marie Olmstead "Babea"
Psc 0h1'6" 4d7'13"
Marie Ottomeyer
Aqr 22h37'31" -1d4'47"
Marie P Carr
Cas 0h21'31" 60d36'57"
Marie P. Gallip
And 1h49'34" 42d51'15"

Marie Papaleo My Mom, My Star
Cas 1h23'52" 62d31'20"
Marie Parent
Del 20h49'54" 15d3'30"
Marie Pascal
Tau 4h9'50" 30d1'57"
Marie Patricia Barry
Uma 10h29'42" 56d36'27"
Marie Pelletant Garcia
Gem 6h54'55" 17d21'29"
Marie Pelletier
Sco 17h57'19" -30d30'49"
Marie Poulain
Gem 6h21'35" 23d2'33"
Marie Protat
Leo 9h53'31" 8d45'22"
Marie R. Angelicchio
Cas 1h40'51" 68d0'11"
Marie Reale
Uma 11h9'12" 29d25'55"
Marie Reed
Aqr 21h9'37" -9d2'27"
Marie Renee Emery
Tau 3h56'1" 3d16'50"
Marie Rinell
Ori 4h51'40" 13d28'38"
Marie Rita Buckridge Kelland
Ari 2h12'55" 24d59'24"
Marie Roberta Rose
Cnc 9h6'2" 26d6'32"
Marie & Rocco Bellantoni
Uma 10h12'34" 43d24'38"
Marie Roland Ramsdell
Sco 16h15'17" -9d49'31"
Marie Romaine Croughn
Crb 16h13'4" 36d52'4"
Marie Roppolo
Cnc 8h46'47" 32d45'46"
Marie Rose
And 0h52'11" 40d25'36"
Marie Rose
Cas 0h54'22" 56d29'49"
Marie Rose
Vir 13h53'28" -17d8'6"
Marie Rosse
Cap 20h30'42" -13d4'50"
Marie Roudaire
Ari 2h29'46" 25d18'25"
Marie Ruby Kluge Simpson
Gem 7h14'0" 32d15'34"
Marie Rufus Taglialatela
Tau 4h35'22" 28d6'37"
Marie Ruth Huc
Tau 3h29'13" 2d13'13"
Marie S. Fausset
Leo 10h39'5" 14d32'35"
Marie Samuels
And 0h27'32" 43d5'28"
Marie Santos Boeker
Cnc 8h20'17" 29d47'46"
Marie S.B.
Sgr 20h11'53" -39d3'37"
Marie Schmitt 25th H.D. Hudson Star
And 0h14'44" 28d34'14"
Marie Shaughnessy Suriano
Uma 12h0'29" 60d51'18"
Marie Shawn Armstrong
Aqr 21h34'51" 2d23'57"
Marie Silveira
Eri 3h23'28" -12d17'4"
Marie Sky Flower
Ori 5h26'46" -0d55'53"
Marie Slevin
Cas 0h45'33" 53d45'28"
Marie Spanbauer
And 23h8'24" 46d54'28"
Marie Spence-Bisset
Cas 23h2'3" 55d56'39"
Marie Spevak
Ori 5h36'0" 13d32'51"
Marie Stetina of Chicago, Illinois
Aqr 21h50'22" 1d13'55"
Marie Stone
Crb 15h55'1" 32d53'23"
Marie Straub
And 1h27'6" 47d24'37"
Marie- T
Lyn 8h44'57" 35d18'59"
Marie T. Dehoff
And 1h9'12" 39d39'34"
Marie T. Ferro
Uma 10h2'17" 44d40'0"
Marie T. Slaney
Gem 6h39'22" 21d14'54"
Marie Tavolacci
Tau 5h37'59" 18d58'35"
Marie Teresa Bundy
Dra 14h41'50" 56d33'11"
Marie the Love of my Life
Cyg 21h35'1" 35d11'26"
Marie The Miracle Maker
Psc 1h9'23" 27d17'29"
Marie Theresa Napoliello Codella
Lyn 7h46'39" 36d57'49"
Marie Thérèse et Jean
Crb 15h17'19" 29d1'2"
Marie Valmai Morgan
Cru 12h37'12" -60d14'23"
Marie Vanouse
Lib 15h32'0" -5d54'15"
Marie Veronica Bernadette Rose
Leo 9h23'21" 24d51'4"
Marie Véronique Mathilde Bouchard
Cas 0h4'51" 53d20'20"
Marie Villeneuve
Uma 11h27'29" 45d19'1"
Marie Vozzo Allbritton
Lyn 7h49'38" 57d47'35"
Marie Wehmeyer
Cas 1h42'41" 63d38'41"
Marie Winifred Ulmer Weekes
Uma 11h2'18" 54d46'35"
Marie Wright
And 0h54'8" 40d38'48"
Marie Zanca
Psc 0h14'0" 12d15'53"
Marie, Love Forever Always, Larry
Cyg 19h56'26" 50d51'25"
Marie, Love U more than anybody can
Cyg 19h34'2" 52d16'56"
Mariea & Destiny
Ori 5h30'53" 3d30'45"
Marie-Afining La Belle
Cap 21h35'26" -18d48'28"
Marie-Andrée
Crb 15h57'38" 35d50'48"
Marie-Andrée Eternelle
Umi 5h19'11" 89d34'8"
Marie-Andrée Leduc
Cas 0h33'32" 61d41'26"
Marie-Ange
Crb 16h0'13" 28d39'26"
MarieAngela
Uma 9h6'17" 61d11'4"
Marieanni
Gem 7h19'46" 32d42'51"
Marie-Beth
Lib 15h52'10" -6d41'20"
Marie-Chantal Toupin
Cnc 9h4'41" 7d53'53"
Marie-Chantal Vinrech
Lib 15h21'54" -16d18'0"
Marie-Christine Casals
Aql 19h56'46" 1d6'6"
Marie-Christine Ferraye
And 0h18'42" 22d15'56"
Marie-Christine Mossu-Marchal
Cas 2h40'8" 57d53'52"
Marie-christine Parisot A
Aqr 23h40'15" -10d11'37"
MarieClaire Philippe2005
Crb 16h20'46" 36d17'30"
Marie-Claire's Star
And 23h29'33" 52d33'43"
Marie-Claude Lapierre
Umi 13h38'32" 87d56'35"
Marie-Douce
Ori 6h5'32" 10d9'29"
Marieelena
Lyn 8h30'52" 46d49'21"
MARIE-ELENA
Ret 3h41'8" -53d19'14"
Marie-Eve Pineault
Cas 2h28'51" 68d21'9"
Marie-Fleur Stevenson-Deane
Gem 7h27'20" 33d14'41"
Marie-France
Peg 21h53'27" 13d27'54"
Marie-France
Crb 15h56'50" 28d49'5"
Marie-France Boyle
Cas 0h57'46" 50d30'52"
Marie-France Lachance
Cas 0h46'1" 73d24'59"
Marie-France Taquet
Cnc 8h39'56" 7d45'39"
Marie-Hélène Bordellier
Gem 7h7'13" 13d25'3"
Marie-Hélène Gauthier 10.09.05
Aql 19h36'29" 5d53'45"
Marie-Hélène Gendron
Del 20h30'35" 16d36'40"
Marie-Isabel
Leo 10h3'59" 22d16'33"
Marie-José Arias
Ori 5h38'29" 12d20'14"
Marie-Josée Bisson & Martine Lahaie
Cyg 20h11'57" 33d16'31"
Marie-Josée Vivier
Cyg 21h41'39" 43d51'35"
MarieKat
Lyn 8h15'36" 35d49'2"
Marieke
Boo 14h32'37" 18d4'20"
Marieke Perchik
Cas 1h39'11" 64d3'20"
Mariel
Uma 11h37'49" 58d46'27"
Mariel
Ori 5h40'53" -1d59'12"
Mariel
Gem 6h5'34" 27d54'3"
Mariel
Crb 15h49'37" 34d6'47"
Mariel Del Valle
Uma 11h21'34" 41d44'17"
Mariel Del Villar
Gem 7h9'7" 14d28'42"
Mariel Elizabeth Gunther
And 0h24'11" 32d27'38"
Mariel Esquerra
Lyr 18h37'55" 35d24'15"
Mariel J. Bernhardt
Gem 7h20'20" 23d3'31"
Mariel Lee
Her 17h57'22" 49d23'57"
Mariel " My Angel "
Vul 20h58'52" 27d7'38"
Mariel Vannel Scott
Sco 17h16'56" -32d29'39"
Mariela
Sco 16h11'29" -22d21'24"
Mariela A. Moyano-Trochez
Crb 15h19'42" 27d26'22"
Mariela D Constantine
Sgr 18h42'58" -17d44'7"
Mariela: "Mi amor de mi Vida"
Pho 0h40'15" -41d40'31"
Mariela Monge
And 1h9'5" 46d15'2"
Mariela Sanchez Victoria
Crb 15h45'53" 26d30'0"
Mariela Theresa Fasnacht
And 23h44'50" 41d18'27"
Marie-Laure et André
Dra 19h30'19" 64d3'30"
Marieleedoodles
And 2h31'54" 49d47'41"
Marielena
And 23h47'11" 47d20'14"
Mari-Elena Baldwin
Crb 15h50'20" 26d52'44"
Mariella
Her 18h35'32" 17d31'45"
Mariella Louise Frisina
Leo 11h57'58" 23d31'46"
Marielle
Peg 22h30'0" 8d52'50"
Marielle
Lyn 7h35'12" 37d14'8"
Marielle D. Marne
Cam 6h12'12" 61d46'53"
Marielle Faith
Her 17h0'5" 32d43'27"
Marielle Ghanem
Cyg 20h0'54" 42d36'52"
Mariellen Elizabeth Day
Aqr 22h12'52" 1d6'39"
Mariellyn Lawson McCrystal
Sco 16h16'52" -11d1'49"
Marie-Lou
Cas 0h59'1" 58d51'55"
Marie-Louise Burness
Cmi 7h26'25" 3d40'7"
Marie-Louise Michaelsen
And 1h36'26" 40d43'44"
Marie-Lyne
And 1h8'51" 38d30'12"
Marie-Madeleine Eicher
Tau 3h37'41" 8d53'3"
Mariena Nevada
Sco 15h51'50" -22d38'19"
Marienne
Cas 23h42'8" 57d13'14"
Marie-Odile Maso 08/07/1943
Cnc 8h12'43" 12d25'21"
MARIEOLS
Ori 5h25'40" 0d23'3"
Marie-Paule Vincensini
Col 6h26'56" -39d5'33"
Marie-Pier Lemay
Cap 21h50'31" -20d37'57"
Marie-Pier Simard
Ari 2h4'29" 18d47'41"
Marie-Pierre Ronchetto
Uma 12h1'32" 29d24'33"
Marie-Rose Kohler
Uma 12h46'14" 55d5'38"
Marie's Celebration of Life
Cas 0h41'50" 61d28'44"
Marie's Dream
Aqr 22h0'56" -15d10'28"
Marie's love; Joseph and Jacob
Lib 15h25'53" -5d45'56"
Marie's Magical Star
And 2h36'7" 44d26'58"
Marie's Star
Sgr 17h59'11" -17d6'11"
Marie's Star
Sco 16h59'31" -39d48'24"
Marieta
Cas 0h38'48" 61d12'53"
Marie-Thérèse
Psc 1h30'40" 16d35'4"
Marie-Thérèse Cloutier
Uma 13h28'53" 62d2'51"
Marie-Thérèse Duvieilh
Gem 7h27'59" 20d45'54"
Marie-Thérèse Girod
Cep 0h14'6" 70d2'24"
Marietje Maree
Ori 5h36'55" -0d11'0"
Marietta
Tau 5h37'4" 25d31'27"
Marietta Beougher
Peg 22h52'59" 19d14'45"
Marietta Collins
Cas 1h33'54" 68d4'39"
Marietta L Julienne, Miss Margarita
Cra 18h29'13" -38d35'59"
Marietta Marinoni
And 1h9'7" 37d8'45"
Marietta Miller
Gem 6h34'4" 23d28'13"
Marietta Myers "Blue Eyed Angel"
Cas 1h23'57" 55d52'35"
Marietta Rodgers
Her 16h49'49" 38d14'5"
Marietta Weist Ulbrich
Ant 9h58'40" -39d8'3"
Marietta Wewee
Lib 15h30'59" -27d44'20"
Mariette
Cyg 19h10'33" 52d23'14"
Mariette L. Lawrence
Gem 7h8'48" 34d39'55"
Mariette's Eternal Beauty
Cyg 19h41'5" 29d26'48"
MarieTwo - MJ Talbot & MC Lévesque
Ari 2h58'25" 22d50'16"
Marie-Urs
Lyr 18h31'42" 36d49'35"
Marifaust
Eri 3h26'19" -22d32'20"
Marife Ann Importante Ortega
Tau 4h45'37" 18d5'21"
Marifields
Leo 11h14'41" -1d0'39"
Mariflor DeSousa (The Mar Star)
Boo 13h51'24" 11d23'5"
Marigrace DiNello
Com 12h1'13" 21d43'57"
Mari-Healer of the Universe
Oph 17h26'32" 8d32'3"
Marija Bozic
Peg 21h25'43" 15d49'36"
Marija Fitness
Cas 1h27'9" 51d17'0"
Marija Lina Weingarten
And 0h25'59" 32d1'52"
Marija - Macina najsjajnija zvezda
Peg 22h50'35" 26d47'54"
Marija Serifovic
Per 4h19'37" 45d32'47"
Marija Steblaj
Cas 0h48'23" 53d8'51"
Marijean Moulton
Cnc 8h46'33" 32d11'38"
Marika
Cyg 19h35'32" 30d10'36"
Marika
Crb 15h38'12" 31d39'7"
Marika Schulz
Ori 6h15'36" 8d53'45"
Mari-Katherine
Oph 16h35'33" -0d37'57"
Mariko Katayama
Aql 19h1'13" -0d42'21"
Mariko Lavirginia Hart
Cyg 21h10'44" 44d23'51"
Mariko Naso
Lyn 8h32'42" 47d30'42"
Mariko's Babies' Christmas Star
Ari 3h12'14" 27d6'57"
Marikris Abigail Gatchalian
Gem 6h45'43" 32d20'37"
Marilayn Hannan
Lib 15h3'5" -18d13'59"
MARILCE
Ari 3h28'2" 26d29'47"
Marilee
Tau 4h13'32" 27d48'46"
Marilee Catherine Joyce
Lib 14h22'14" -23d55'20"
Marilee Joan Minard Stephens
Tau 4h7'9" 1d8'15"
Marilee Robin
And 1h32'7" 47d4'0"
Marilee's Star
Uma 10h36'34" 52d58'20"
Marilen
Umi 16h2'13" 86d32'25"
Marilena
And 2h22'0" 45d8'56"
Marilena
Lyr 18h52'13" 33d15'52"
Marilena Acquati
Umi 13h10'25" 73d33'23"
Marilena Segreto
And 1h57'9" 40d16'52"
Marilene Mimi Woms Marquis
Leo 11h14'40" 25d7'34"
Marilex
Sge 19h33'28" 18d56'40"
Marili Libke Benjamin
Umi 13h50'18" 75d6'2"
Marilia
Leo 11h5'26" 22d5'21"
Marilia
Tau 3h35'27" 1d15'48"
Marilia da Silva
Sgr 18h14'58" -33d36'12"
Marilinda Cunanan
Cam 3h18'42" 64d38'13"
Mariline Sciortino
Cyg 21h45'40" 44d47'7"
Marilisa
Her 18h14'25" 18d16'29"
Marillyn Lou Nicholson
Sco 16h52'15" -28d1'49"
Marilou
Aqr 22h26'47" -1d27'31"
Marilou Blanco Dizon
And 1h58'45" 36d12'46"
Marilou Mozdzen
Ari 2h48'25" 28d43'48"
Marilou Pilman
Lyn 7h39'57" 37d27'42"
Marilou Velasco
Psc 22h55'56" 6d54'36"
Mari-Louise Donovan
Lib 15h55'22" -17d24'27"
Marilu Henner
Lyr 18h48'39" 37d12'35"
Marilu Otero-Rivera
Sco 16h13'35" -24d47'3"
Marilú y Pablo
Cyg 20h0'34" 33d28'32"
Marilu's Star
Aql 19h16'18" -7d28'5"
Marilyn
Umi 13h38'3" 74d58'31"
Marilyn
Uma 12h30'34" 58d7'10"
Marilyn
Cas 1h20'44" 63d57'13"
Marilyn
Sgr 19h4'33" -32d42'32"
Marilyn
And 1h48'56" 38d3'59"
Marilyn
Lyr 18h50'52" 36d25'49"
Marilyn
Crb 16h24'38" 31d20'14"
Marilyn
Peg 22h23'53" 5d41'52"
Marilyn
Mon 6h41'37" 7d27'29"
Marilyn
Ari 2h30'57" 20d47'48"
MARILYN
Leo 10h14'9" 24d24'20"
Marilyn
Lyr 19h4'15" 27d11'27"
Marilyn
Cnc 7h59'4" 19d11'57"
Marilyn
Gem 6h28'32" 17d49'51"
Marilyn 1 Stella Pectorius Magnus
Psc 0h33'27" 7d34'55"
Marilyn 5
Cap 21h2'0" -17d5'26"
Marilyn A Gallo
Aql 19h32'54" 6d14'45"
Marilyn and Bill Carhart
Lyn 7h23'52" 44d37'50"
Marilyn and Doug 1974
And 1h11'12" 46d32'15"
Marilyn and Jason
Cyg 21h18'37" 32d45'7"
Marilyn and John Dudley 4ever 92504
Uma 10h15'26" 48d54'7"
Marilyn and Michael Faricellia
Cyg 20h13'43" 53d54'7"
Marilyn Anderson
Lib 15h37'44" -6d4'14"
Marilyn Anderson Chase
Uma 11h1'0" 33d20'33"
Marilyn Angela Casassa Vancini
Ari 3h9'23" 17d40'58"
Marilyn Ann
Cas 2h2'18" 58d42'1"
Marilyn Ann Dillon
Cnc 8h39'42" 8d45'35"
Marilyn Ann Fleming
Cnc 9h4'21" 11d13'16"
Marilyn Ann Simpson
Cas 0h22'10" 59d10'49"
Marilyn Ann Whitten
Lyn 6h49'51" 54d44'36"
Marilyn Arlene Sieber
Tau 3h47'42" 28d22'40"
Marilyn & Augustine Fuimefreddo
Cyg 21h10'56" 53d47'22"
Marilyn B Pangadakis
Crb 16h4'27" 31d17'12"
Marilyn Barc
Cas 1h54'28" 61d41'34"
Marilyn Bellinger
Uma 8h46'53" 52d56'47"
Marilyn Beth
Sco 16h45'15" -39d59'3"
Marilyn Beverly Horowitz
Uma 13h54'57" 48d42'27"
Marilyn Black
Peg 22h56'0" 26d0'40"
Marilyn Borroto
Del 20h46'30" 14d18'6"
Marilyn Bundschuh
Tau 4h12'38" 15d16'47"
Marilyn Burnell S. Maddox
Gem 6h10'40" 23d28'2"
Marilyn & Carl Goldman
Cyg 21h35'19" 52d39'46"
Marilyn Carollo
Cas 1h29'57" 57d48'57"
Marilyn Castro Torreguitar
Cap 20h17'49" -14d47'56"
Marilyn Charleen Leader
Lyn 9h1'32" 34d33'22"
Marilyn Cogliati Walag
Lib 15h7'51" -21d51'25"
Marilyn Coons
Psc 1h45'27" 21d33'6"
Marilyn DeVore Troppmann
Vir 13h46'28" 4d36'19"
Marilyn Divine
Tau 5h42'29" 24d52'34"
Marilyn Dunstan
Aqr 22h15'38" -3d10'13"
Marilyn E Grimm
Cas 1h39'39" 61d14'25"
Marilyn E Kurman
Psc 1h25'20" 32d25'16"
Marilyn E. Sova
Uma 10h21'13" 44d33'55"
Marilyn E. Woolson
Cyg 20h35'52" 40d52'0"
Marilyn Elaine Kaut
And 23h28'23" 38d40'50"
Marilyn Elizabeth
Aqr 20h41'24" -9d13'31"
Marilyn Elizabeth Cutright
Ari 1h59'48" 17d43'54"
Marilyn F. Thomas
Umi 17h17'48" 83d56'53"
Marilyn Ferguson's Star
Uma 9h45'32" 49d41'19"
Marilyn Figueredo
Crb 15h30'42" 27d41'47"
Marilyn Ford
Gem 7h34'48" 33d26'1"
Marilyn Gardner
Gem 7h38'1" 15d33'50"
Marilyn Gardner Constine
Sex 10h31'47" -0d6'56"
Marilyn Garofalo
Gem 6h52'41" 22d49'8"
Marilyn Gayle
Dra 19h35'26" 73d15'49"
Marilyn Gray; Stellar Teacher
Cas 1h20'30" 62d41'37"
Marilyn Henriques
Vir 14h14'5" -8d35'31"
Marilyn Herbison
Crb 15h47'23" 32d48'10"
Marilyn Ingraham Grande
Cas 0h7'4" 57d4'22"
Marilyn Ivison
Cmi 7h36'16" -0d6'50"
Marilyn J. Altenbach
Cam 6h4'46" 61d52'58"
Marilyn J. Brinkley-Cotter
Crb 16h5'28" 25d41'30"
Marilyn J. Callahan
Uma 14h20'48" 59d51'44"
Marilyn J. M. Daley
Uma 11h16'10" 46d50'46"
Marilyn Jane Lister
Cas 0h54'2" 56d34'44"
Marilyn Jane Menefee Zajac
Lyn 8h9'33" 50d8'50"
Marilyn Jean Cholin
And 1h24'53" 47d42'52"
Marilyn Jeanne Fogarty
Aqr 21h38'48" 0d17'40"
Marilyn & Jim Jenkins - Hopper Glen
Cyg 19h35'24" 31d57'42"
Marilyn Joan LaFroscia
Vir 12h25'33" -0d39'29"
Marilyn Joan Markle Sachs
And 23h27'45" 48d23'9"
Marilyn Joy Staines
Cru 12h39'45" -58d32'36"
Marilyn Joyce Bell
Crb 16h9'38" 35d59'52"
Marilyn Joyce Soderquist
Cyg 19h42'50" 29d0'48"
Marilyn & Justin
Lyn 8h53'42" 40d13'8"
Marilyn K
Mon 6h46'3" -0d22'16"
Marilyn K Brand
And 1h0'58" 44d44'3"
Marilyn Kathleen Smith
Cap 21h44'0" -24d44'6"
Marilyn Kay
Ori 5h30'20" -0d42'3"
Marilyn Kay
Uma 13h38'37" 57d37'11"
Marilyn Kay Clark
Uma 11h32'50" 48d1'29"
Marilyn Kay Croston
Crb 15h53'28" 26d21'38"
Marilyn Kay Sisco Prox
Gem 7h3'20" 24d2'11"
Marilyn Kepler Bird
Uma 9h19'31" 62d22'54"
Marilyn Knehans
Crb 16h15'12" 32d4'31"
Marilyn Kozinski
Cas 0h2'23" 53d29'15"
Marilyn L. Ballard
Psc 23h0'20" 2d26'18"
Marilyn Lane Little
Vir 12h1'49" -1d58'25"
Marilyn Last - VAVA's
Umi 14h36'37" 74d36'23"
Marilyn Leona French
And 23h47'54" 38d23'55"
Marilyn Lewis Tsi-Da-Da-Chlutch
Cas 0h37'50" 75d13'43"
Marilyn Lightstone
And 0h21'4" 26d1'9"
Marilyn Linhares
Lib 14h33'7" -12d24'6"
Marilyn Loechner Deiter
Gem 7h20'48" 34d14'18"
Marilyn Logsdon Mennello
Lmi 10h23'3" 28d35'46"
Marilyn Lopera
Gem 6h18'20" 22d10'27"
Marilyn Lou
And 2h25'40" 49d58'52"
Marilyn Louise Maltby
And 0h45'20" 35d49'13"
Marilyn M. Bruneau
Crb 15h48'1" 37d23'0"
Marilyn M. Holloway Heckler
Gem 6h56'41" 13d22'23"
Marilyn M Priebe Parker
Cap 20h20'56" -12d47'17"
Marilyn Mae
Ori 5h32'49" -2d47'26"
Marilyn Martell
Cnc 9h4'23" 25d28'16"
Marilyn Matthews (Pearce) Precourt
Aqr 22h29'39" -3d22'57"
Marilyn McDonald
Sgr 18h59'8" -24d28'17"
Marilyn McDowell
Uma 8h55'39" 59d14'49"
Marilyn McIntyre
Aqr 22h13'11" -9d50'23"
Marilyn McPherson Mother/Doc
Gem 7h2'59" 25d26'5"
Marilyn Mills Nitz
Lib 15h53'53" -17d30'20"
Marilyn "Mimi" Hartnett
Per 2h53'1" 42d30'57"
Marilyn Monroe
Tau 5h33'17" 25d58'35"
Marilyn Moore
Cyg 19h57'42" 39d56'33"
Marilyn Murdoch Peklo
Mon 6h40'13" -6d26'19"
Marilyn Noda Swartz
Com 13h8'20" 28d34'14"
Marilyn of the Arts
Uma 11h9'46" 29d45'24"
Marilyn Papa Genovese
And 0h20'27" 30d0'57"
Marilyn Pekter
Cas 0h42'16" 64d28'49"
Marilyn Pierini
Tau 5h31'7" 24d34'59"
Marilyn Pomeroy
Lyn 8h42'24" 33d32'26"
Marilyn Poppino
Sco 16h40'18" -30d52'1"
Marilyn "Princess" Rawlingson
Lyn 6h42'19" 60d20'8"
Marilyn & Ray Rivell
Cyg 21h11'33" 46d50'51"
Marilyn Raymond
Cas 1h26'47" 69d14'49"
Marilyn Rena Stevens Carter
Cnc 9h5'53" 8d43'26"
Marilyn Rinaldo
Cas 0h41'26" 59d34'1"
Marilyn Rodriguez
Sco 16h7'40" -11d46'40"
Marilyn Rose Brandfass
Pic 5h49'3" -54d40'42"
Marilyn Rose Hayes
Umi 13h11'25" 72d8'21"
Marilyn Rothman
Leo 10h22'30" 16d3'56"
Marilyn Russell
Uma 8h45'11" 55d33'19"
Marilyn Ruth (Daniell) Redden
Tau 4h16'50" 0d52'36"
Marilyn Ruth Stromberg
Cas 2h8'22" 59d40'42"
Marilyn S. Allen
And 0h15'20" 36d57'56"
Marilyn Sandra Panting
Tau 4h32'40" 18d12'54"
Marilyn Shute
Cas 0h48'23" 60d58'32"
Marilyn Siempre Cerca De Mi Corazon
Lmi 10h46'55" 29d2'20"
Marilyn Spivey
Lyr 18h52'53" 31d7'8"
Marilyn Stonestreet
Cas 23h21'53" 57d42'11"
Marilyn Sue
Aql 19h35'31" 14d44'24"

Marilyn Sue Bolinger
Lmi 10h36'46" 28d55'31"
Marilyn Sue Vaught
And 2h33'41" 39d16'16"
Marilyn T. Corrow
Cas 1h29'53" 61d4'19"
Marilyn T. Shore
Cas 1h33'42" 62d13'22"
Marilyn Tarantino
Uma 10h31'27" 49d40'0"
Marilyn Tarlow
Cas 1h13'49" 59d58'44"
Marilyn the Beautiful Mommom
Sco 17h56'24" -31d6'57"
Marilyn the Magnificent
Lib 15h50'50" -13d52'49"
Marilyn Therese
And 0h15'21" 29d40'23"
Marilyn Timmons (Grammie)
Vir 12h17'48" 3d56'49"
Marilyn Vasquez
Psc 23h25'34" 4d8'19"
Marilyn Villano
Lyn 9h42'13" 40d6'8"
Marilyn Villegas
Gem 6h49'35" 33d49'47"
Marilyn Viola Wacholz
Psc 1h19'31" 3d12'52"
Marilyn Vogt
Cap 20h10'33" -15d22'8"
Marilyn Wiles
Cas 1h31'28" 61d25'16"
Marilyn-mjcat12<\@>
Per 2h29'12" 55d58'47"
Marilynn
Sgr 17h53'7" -18d4'23"
Marilynn Hendrickson
Sgr 17h53'22" -28d26'51"
Marilynn Jean Jones
Dra 19h49'36" 69d50'1"
Marilynn Joyce (Anacker) Weber
And 0h21'8" 41d43'18"
Marilyn-Pete-Zoee-Welch
Lib 14h41'25" -25d0'17"
Marilyn's Eternal Love
Aqr 22h44'6" -20d47'17"
Marilyns lil star
Cyg 20h8'41" 44d36'34"
Marilyn's Majesty
And 0h58'36" 39d21'22"
Marilyn's Michael - Never Forget
And 23h43'13" 35d5'17"
Marilyn's World
Aql 19h8'16" 4d5'25"
Marimel
Dra 17h24'36" 51d0'12"
Marimer Fraumont "9-25-1952"
Lyn 7h41'43" 35d40'52"
Marimuthu
Sco 16h30'37" -26d21'54"
Marin
Aql 19h23'35" 13d35'12"
Marin 561979
Ori 5h41'58" -0d13'28"
Marin Brighid Moore
Uma 10h16'44" 49d29'28"
Marin Mahoney
Gem 7h16'59" 34d15'11"
Marin Makenna Butler
And 1h20'54" 37d46'11"
Marin Pierre Aznarez
Cnc 8h52'47" 19d3'9"
Marina
Psc 1h11'46" 26d25'35"
Marina
Her 18h4'56" 17d5'12"
Marina
Gem 6h49'44" 12d46'10"
Marina
Aqr 21h0'37" 0d17'25"
Marina
Psc 0h59'17" 14d22'24"
Marina
Psc 1h7'40" 11d11'18"
Marina
Vir 12h34'17" 6d45'4"
Marina
And 1h6'54" 44d51'49"
Marina
And 2h5'14" 37d37'58"
Marina
And 0h36'4' 43d32'55"
Marina
And 0h3'45" 44d12'26"
Marina
And 0h55'8" 41d47'26"
Marina
Uma 11h25'27" 39d45'26"
Marina
And 2h29'30" 43d39'40"
Marina
Cyg 21h49'36" 47d46'45"
Marina
Her 17h41'56" 41d29'55"
Marina
Lib 15h4'36" -2d10'50"
Marina
Cap 20h52'51" -21d18'24"
Marina
Sgr 19h43'38" -16d11'30"
Marina
Lib 15h30'33" -10d14'56"
Marina
Uma 8h52'15" 70d26'24"
Marina
Cas 2h50'55" 60d24'41"
Marina
Sgr 18h0'48" -27d37'15"
Marina 13
Psc 1h6'45" 12d2'4"
Marina Alexandra Ark
Lib 15h8'2" -6d22'24"
Marina and Jarred
Uma 11h17'30" 34d38'20"
Marina Armstrong
And 1h3'51" 40d27'5"
Marina Artamonova
Cam 7h24'44" 76d43'37"
Marina Bellus
Del 20h31'34" 16d12'29"
Marina Boltyanskaya
Cap 20h31'46" -19d1'51"
Marina Campbell Turpen's Lucky Star
Umi 15h56'3" 77d19'36"
Marina Clausen
Cas 1h59'0" 65d38'32"
Marina D. Korotaeva
Cnc 8h25'56" 27d40'45"
Marina Daligault
Ori 5h1'19" -1d30'38"
Marina Diedrich
Uma 11h57'52" 61d59'28"
Marina Elise Dingman
Cnc 8h17'45" 20d35'25"
Marina Gallo
Cap 20h40'45" -14d39'31"
Marina Gawlinski
Cas 0h12'34" 59d57'32"
Marina Goldman
Lyr 18h45'20" 36d25'41"
Marina Gomez
Tau 5h6'50" 22d57'53"
Marina Grace Giovannetti
And 2h22'23" 48d54'20"
Marina i Buda
Per 4h20'50" 46d50'24"
Marina (i love u)
Del 20h39'19" 14d4'54"
Marina (i love u)
Del 20h38'56" 14d37'57"
Marina & Joe
Aqr 20h48'4" -10d48'16"
Marina & Joe
Eri 3h41'58" -0d9'5"
Marina Jozinovic
Sco 17h12'58" -43d42'46"
Marina K.
Aqr 22h19'25" -9d59'21"
Marina Katherine Pattilio
Ari 3h13'39" 27d40'23"
Marina Kiseleva
And 2h35'9" 50d8'23"
Marina Kostromkina
And 0h52'59" 41d4'24"
Marina Legeley Demas
Cas 23h42'16" 53d9'38"
Marina Libro
Psc 1h6'39" 14d48'49"
Marina Lorie Hodgini
Ari 2h10'57" 26d22'37"
Marina Luz Valadez Georgallides
Cap 21h51'39" -17d20'34"
Marina Messina Mamalis
And 1h22'16" 37d48'47"
Marina Michele Cushing
Tau 4h46'41" 22d38'35"
Marina Mogollon
Leo 11h56'9" 24d5'21"
Marina Mountain
Uma 11h36'50" 45d30'11"
Marina Noelle Pardee
Cap 20h29'50" -13d21'48"
Marina P.
Per 3h37'1" 45d18'17"
Marina Parolari
Uma 10h2'35" 42d29'30"
Marina Patti "KIKA - KEA"
Lyr 18h27'21" 32d22'52"
Marina Peredo
And 23h3'27" 50d42'35"
Marina Perks
Cas 23h57'34" 56d46'3"
Marina Petra Gross
Uma 14h18'54" 56d54'39"
Marina R. Alfonso
Ari 3h13'57" 27d13'8"
Marina Ramos
Lyn 6h42'9" 50d5'43"
Marina "Rina Star"
Dra 18h28'7" 74d1'21"
Marina Rodriguez Cabeza
Uma 9h22'39" 63d43'4"
Marina Rosa
Dra 16h11'52" 56d50'9"
Marina Rose Berger
Sgr 17h59'57" -25d8'40"
Marina Sachi
Uma 12h1'24" 58d45'42"
Marina Sagalova
Uma 9h3'56" 58d3'26"
Marina Scheglov "Your My Star"
And 0h6'52" 44d3'22"
Marina Shaforost
Cyg 21h36'4" 49d22'41"
Marina Shelton
Sgr 18h25'45" -21d15'54"
Marina Silvia
Lyn 7h13'13" 52d55'59"
Marina Taufer
Umi 16h14'56" 71d16'32"
Marina The Dancer
Tau 4h14'23" 5d12'16"
Marina Wilson
And 23h24'25" 51d7'8"
Marina y Jairo por siempre
Vir 11h38'41" -2d0'23"
Marina Zaiats
Sgr 18h37'5" -31d24'41"
Marina, 06.03.1990
And 23h56'58" 35d34'16"
Marina, I love you
Aur 5h25'47" 41d15'0"
Marina, Your Days Be Always Bright
Vir 12h0'29" -0d7'25"
MarinaAlejandra
Gem 7h16'30" 31d43'34"
Marina's Everlasting Inspiration
Cap 20h33'58" -23d4'54"
Marinbakh
Tau 4h35'8" 27d41'39"
Marinda*
Cap 21h13'45" -17d34'4"
Marinda Pezzetti
Vir 12h50'57" 7d57'38"
Marinda Ruth
Leo 10h9'12" 15d0'18"
Marine
Aur 5h13'26" 37d27'28"
Marine
Cnc 8h9'22" 22d7'43"
Marine Gaetan
Mon 6h18'4" -4d9'28"
Marine Pistre
Mon 6h49'26" -0d19'17"
Marine Sgt. Byron Wayne Norwood
Per 3h51'40" 33d51'45"
Marine Wife
Uma 12h9'37" 60d41'50"
Marinella amore per sempre
Her 16h12'57" 5d41'42"
Marinero Valiente
Cam 4h17'7" 67d47'28"
Marinna Reina
And 1h8'0" 46d40'42"
Marino Fanelli
Uma 9h25'22" 71d33'52"
Marino Fontana
Uma 8h49'46" 61d19'11"
Marino & Nancy Curra
Eri 4h32'38" -10d13'37"
Marino Sandoval
Ari 3h20'48" 28d25'18"
Marino Scipioni
Per 4h46'15" 48d59'19"
Marinou
Uma 10h38'56" 62d40'23"
Marinus
Cen 11h59'30" -52d26'21"
Mario
Cyg 20h2'37" 53d8'24"
Mario
Per 4h19'56" 44d2'7"
Mario
Peg 23h46'3" 11d46'44"
Mario A. Falu
Leo 11h54'31" 26d3'38"
Mario A. Iavicoli
Cep 21h20'44" 66d1'32"
Mario A. Scolaro
Lib 15h14'11" -19d59'48"
Mario Alberto Martinez Jr.
Per 3h1'47" 41d53'44"
Mario Alberto Perez
Sgr 19h9'16" -35d37'11"
Mario Alberto Rangel
Ori 5h54'27" 21d2'38"
Mario and Denise
Cas 0h21'28" 53d25'44"
Mario B. Del Conte
Aur 5h29'31" 35d37'48"
Mario & Beth's Love!!
Cyg 21h13'7" 55d4'21"
Mario Bischof
And 23h24'44" 52d53'39"
Mario C. Hodge
Ori 5h30'8" 8d23'19"
Mario C. Veilleux - Our Buddy
Per 3h8'0" 53d48'14"
Mario Caballero
Lib 15h32'46" -11d41'1"
Mario Cantando
Per 3h11'56" 54d56'52"
Mario Carratelli
Uma 8h48'4" 58d7'35"
Mario Castro M.D.
Vir 13h22'5" -10d34'19"
Mario Cherrio
Cep 21h13'5" 55d47'52"
Mario & Christine
Lyn 7h55'45" 55d15'16"
Mario Cosme Dornelas Costa
Ori 5h22'35" 1d55'25"
Mario & Daniela
Uma 12h9'14" 53d26'46"
Mario Decarolis
Uma 8h20'3" 65d9'6"
Mario Enrique Estevez
Lyn 8h1'24" 56d38'35"
Mario Ficaloro
Cap 21h27'46" -15d38'29"
Mario Forcellati
Her 17h22'42" 39d37'37"
Mario G. Nargi (« Hope »)
Ori 4h45'16" 5d18'8"
Mario Gabel
Uma 13h4'11" 52d44'5"
Mario Gabriel Hernandez III
Ari 2h17'5" 20d41'53"
Mario Garza
Lyn 8h45'13" 33d3'23"
Mario Garza Cano
Her 17h13'50" 31d50'9"
Mario Gaudino
Uma 8h52'58" 49d27'37"
Mario Giannella
Ari 3h7'55" 25d32'14"
Mario Giuliani
Per 3h26'26" 41d31'45"
Mario Gonzalez
Per 3h17'9" 46d49'55"
Mario Hernandez
Ari 3h7'48" 22d26'32"
Mario "Hi Ho" Rocco
Her 18h16'33" 23d34'40"
Mario J. Lario
Uma 8h46'33" 63d37'26"
Mario J. Lewin
Psc 1h21'23" 16d6'12"
Mario Jaile, Jr.
Her 17h29'7" 38d4'0"
Mario Joao DaEira
Her 18h41'39" 18d53'14"
Mario Joseph & John Daniel
Gem 6h9'46" 24d5'19"
Mario Kohring
Uma 8h13'21" 72d21'2"
Mario & Lata's Little Star Santana
Lyn 7h20'19" 50d29'37"
Mario Louis Clemente
Her 17h28'1" 46d2'51"
Mario Maccarone
Ori 6h2'28" 15d49'26"
Mario Maggia I Will Always Be Yours
Cyg 19h29'59" 51d27'39"
Mario Man Handsome Face
Ori 6h4'6" 18d31'54"
Mario Mariani
Aql 20h12'4" 13d20'31"
Mario Meli
Tri 1h45'48" 31d13'44"
Mario Mendoza - Love is Forever
Gem 6h36'56" 19d49'43"
Mario Micucci
Cep 22h57'6" 77d16'19"
Mario Mikeal Maglieri Jr
Uma 11h9'35" 62d27'11"
Mario My Love
Ori 5h36'2" 1d44'19"
Mario my protector
Her 17h26'12" 32d50'30"
Mario & Odessa
Sco 17h49'58" -42d52'6"
Mario Oriolo
Crb 16h14'23" 26d32'28"
Mario P. Chiarello's Star
Cep 22h19'23" 76d53'19"
Mario R. Gonzalez
Tau 4h26'16" 3d12'17"
Mario Renee Vega
Leo 9h28'28" 26d23'30"
Mario S. Sneen
Umi 16h26'52" 78d54'40"
Mario Schmidt
Uma 9h51'25" 42d19'51"
Mario Sili
Ori 5h37'1" 12d50'45"
Mario Susie Jr.
Uma 14h2'24" 56d59'46"
Mario Szymik
Ori 6h8'47" 7d42'52"
Mario Täuber
Uma 9h12'51" 69d14'19"
Mario Zwyssig, 23.02.2004
And 23h22'59" 35d30'44"
Mario, Peggy, Eva, Joseph and Meyer
Uma 11h29'48" 35d51'27"
Mariola Szafranski
Ori 5h24'6" 2d3'51"
Mariola Torres Hernandez
Col 5h31'59" -28d48'2"
Marion
Ari 1h54'8" 23d1'4"
Marion
Umi 13h16'37" 72d17'30"
Marion
Umi 16h11'0" 77d8'34"
Marion
Ori 5h55'8" 12d30'32"
Marion
Crb 15h38'58" 30d32'5"
Marion
Leo 9h47'35" 30d47'3"
Marion A. Doele-Johnson
Uma 9h29'12" 48d46'21"
Marion A. Hill
Ari 2h14'59" 24d8'48"
Marion Achinger
Ori 6h15'28" 9d58'3"
Marion Ahalt Gaydos
And 0h48'2" 36d37'59"
Marion Ahern
Umi 14h36'43" 77d23'29"
Marion Alexander Palmer
Aql 19h42'49" 11d52'35"
Marion Alice Snekenberg
Ari 2h44'27" 30d7'7"
Marion and Nadinia
Cyg 20h2'10" 33d23'22"
Marion Barraclough
Ari 2h12'12" 16d42'4"
Marion Benedict
Cam 3h21'43" 61d20'56"
Marion C. Tavares
Ari 1h56'59" 13d12'5"
Marion Chan
Cas 0h17'29" 56d40'7"
Marion & Charles Osgood
Cyg 19h47'23" 32d56'45"
Marion Chatot
Psc 23h38'9" 6d2'41"
Marion & Christian
Uma 9h11'4" 63d1'5"
Marion Coleman Pitts Stephens
Cap 21h31'49" -18d57'1"
Marion Coyne
And 1h22'45" 47d15'56"
Marion & Donna Pawlowski
Her 17h54'24" 23d47'15"
Marion E Gwynn
Aql 19h59'0" 8d42'6"
Marion E Roark
Cnc 8h39'29" 14d44'20"
Marion E. Strozik
Psc 1h16'55" 32d24'35"
Marion Estelle Rogers
Gem 7h3'29" 10d46'45"
Marion & Frank Ehrhardt
Cyg 21h49'49" 49d54'28"
Marion Galante
Cas 2h41'11" 57d55'40"
Marion & Gene Wetzel (Star of Love)
Uma 8h59'56" 60d8'24"
Marion Genevieve Clark Bowen
Crb 16h16'52" 34d21'3"
Marion Grace
Ari 2h59'26" 12d53'55"
Marion Green
Peg 22h22'7" 26d46'35"
Marion Hagmann
Cap 20h27'39" -18d25'47"
Marion Hendricks Lonsdorf
And 0h38'29" 39d20'14"
Marion Hoare
Uma 11h12'0" 38d31'50"
Marion K.
Cam 4h21'33" 54d48'40"
Marion Kassab
Uma 10h35'50" 54d2'45"
Marion Kathleen Kolb Kammerling
Ori 5h35'10" 13d21'38"
Marion Kay Johnson
Lyn 7h40'23" 37d18'5"
Marion Kelley
Uma 9h17'15" 47d39'34"
Marion Kovach
Tau 4h22'33" 27d43'39"
Marion L
Mon 7h33'20" -0d29'18"
Marion L. Kroha
Cas 1h29'40" 61d19'41"
Marion La Blonde
Crb 15h56'56" 28d39'42"
Marion Lang
Umi 16h6'34" 70d14'21"
Marion Leo 79 81 75
Uma 10h29'5" 63d22'15"
Marion & Liddie Johnson Family Star
Cyg 20h14'43" 48d53'48"
Marion Logie
Vir 12h51'12" 3d35'52"
Marion Louis Yarger Middleton
Cnc 8h10'20" 7d58'46"
Marion Louise Chamblee
Uma 11h6'15" 39d25'43"
Marion Lucille Dokken
Del 20h40'10" 15d0'57"
Marion "Mary" J. Egan
Uma 13h28'58" 53d28'55"
Marion McRae Holloway
Tau 5h49'24" 21d18'7"
Marion Mennona
Lmi 9h52'25" 36d35'14"
Marion Nakken
Umi 15h14'58" 67d46'56"
Marion North - Mother of 7 Star
Cru 12h28'47" -58d58'19"
Marion & O'Pierre Harcourt
Umi 14h5'35" 76d29'39"
Marion "Our Star" Chang
Cas 1h50'23" 63d53'11"
Marion Our Star, Born 10/13/1908
Uma 9h11'23" 56d55'56"
Marion P. Church
Cas 23h29'3" 54d45'13"
Marion P. Flanders
Cap 20h10'22" -26d9'52"
Marion Patricia DeYoung
Cnc 8h49'48" 31d55'10"
Marion R. Schmitz 8-31-1933
Vir 13h23'40" 11d58'23"
Marion Regina
Ori 5h8'41" 9d13'36"
Marion Rollie Sims
Psc 0h47'53" 18d20'40"
Marion Rose Welchert-Garcia
Ari 2h31'46" 24d8'16"
Marion S. Harrison
Lyn 7h30'3" 49d38'4"
Marion S. Osborne
Uma 10h33'30" 41d42'22"
Marion Schenher Guagliardo
Peg 22h35'56" 8d47'3"
Marion Sue Stewich
Cas 0h26'35" 58d53'41"
Marion T. Hovanec
Aqr 21h44'28" 2d4'34"
Marion Theresa
Leo 11h37'16" 18d59'49"
Marion Victoria
Gem 7h46'28" 33d8'43"
Marion W. Cleary
Cnc 8h9'14" 16d47'53"
Marion Walters Lawrence
Vir 14h38'28" 7d7'7"
Marion Welsh Van Winkle
Vir 12h17'0" 11d13'5"
Marion Widger
Uma 9h21'3" 66d12'38"
Marion Wigfall Bradshaw
Psc 0h59'31" 15d22'0"
Marion Wilkie
Cas 2h50'38" 60d43'26"
Marion Wilson
Cyg 21h58'34" 45d33'55"
Marion, Alf, Sam & Dorothy Sage
Ori 4h50'28" -3d17'39"
Mariona Font Zuazua
And 0h11'47" 25d28'10"
Marion's Passion
Vir 12h46'40" 2d2'7"
Marion-Wintersonne 7Dec1961
Aur 5h15'15" 43d25'36"
Mariopaulette
Uma 9h28'20" 68d28'13"
Mario's 4th of July Star
Aql 19h41'34" 13d16'48"
Mario's Jewel
Vir 12h16'8" 11d5'1"
Maripepagarciabootello
Ori 5h45'18" 11d30'28"
Maripier Alexandrine Prudhomme
Uma 12h4'27" 38d30'1"
Mariposa
Uma 11h28'31" 54d37'44"
Mariposa
Lib 14h47'51" -8d8'44"
Mariposa púrpura de Debbi
Psc 1h0'11" 22d59'9"
Maris
Lib 15h35'4" -28d9'4"
Maris Diluculo
Ori 5h12'10" 8d58'26"
Maris Jade Branyan
Psc 1h37'24" 7d35'47"
Mari's Piece of Heaven
Cru 12h25'19" -58d54'53"
Mari's Stellar Sky Diamond
Aqr 22h43'38" 0d25'53"
Marisa
Gem 7h5'21" 14d34'45"
Marisa
Ori 6h21'9" 11d38'38"
Marisa
Peg 21h51'36" 11d13'46"
Marisa
And 1h19'49" 42d0'26"
Marisa
Gem 6h46'19" 34d41'34"
Marisa
Lyn 9h13'37" 33d1'25"
Marisa
Cyg 20h31'15" 40d33'23"
Marisa
Sgr 19h18'17" -14d45'21"
Marisa
Sgr 19h47'36" -20d23'20"
Marisa
Vir 13h42'30" -5d42'20"
Marisa
Cas 23h28'21" 53d30'29"
Marisa
Cas 23h32'49" 57d59'13"
Marisa
Dra 18h36'35" 71d22'58"
Marisa "Always & Forever"
And 0h21'46" 28d21'55"
Marisa Amy Shea
Tri 2h29'3" 36d35'16"
Marisa and Cecelyn
Gem 6h54'36" 30d52'46"
Marisa and Maria's Star
Umi 13h25'43" 73d59'0"
Marisa Anne
Aqr 21h11'40" -7d59'4"
Marisa Anne
Cnc 8h37'57" 18d2'53"
Marisa Araceli Livingston
Tau 5h51'9" 26d35'11"
Marisa Barrera
Cap 20h32'3" -12d7'12"
Marisa Bennett
Uma 12h3'5" 43d8'3"
Marisa C. Karabin
Aql 19h54'19" 15d40'50"
Marisa Christina Ratliff
Lyn 8h30'59" 46d32'28"
Marisa Christine Penkauskas
And 23h40'31" 37d26'58"
Marisa D'Amario
Leo 11h43'20" 25d35'32"
Marisa Escalante
Boo 14h15'54" 27d43'2"
Marisa Fuchs
Cas 23h53'37" 58d57'58"
Marisa Giori
And 23h43'43" 43d19'10"
Marisa Graziano
Cam 4h57'2" 59d7'33"
Marisa (il mio amore eterno)Johnson
Her 17h16'46" 26d42'42"
Marisa Isabella
And 0h50'37" 41d17'22"
Marisa Jo Samantha Palma Everage
Vir 13h8'39" 11d47'32"
Marisa Kaufman
Psc 2h5'40" 9d23'41"
Marisa L. Perez
Cap 21h7'9" -15d39'48"
Marisa La Corte
Cyg 19h55'41" 33d7'42"
Marisa Lynn Craft
Ari 2h32'36" 18d51'16"
Marisa Lynn Seander
Vir 12h48'38" -8d30'3"
Marisa M Frenandez
Sco 16h47'36" -33d5'34"
Marisa Maggiore "Guiding Light"
Cas 1h26'4" 60d57'44"
Marisa Maloney
Vir 13h29'3" 12d18'47"
Marisa Maloney
And 0h24'58" 41d8'5"
Marisa Marie Buol
Cam 5h49'13" 65d27'15"
Marisa Minute
Lib 15h44'46" -29d37'46"
Marisa Ottinger
Lib 14h54'3" -2d55'26"
Marisa Phelps
Cap 21h39'59" -8d40'58"
Marisa Raas
Cas 1h18'20" 51d24'50"
Marisa Rae
Lyn 8h45'30" 35d16'34"
Marisa Raquel Osorio
Uma 10h51'54" 46d44'18"
Marisa Riley
Cnc 9h14'5" 8d38'50"
Marisa Rose Valentina Cheley
Ori 5h38'55" 8d6'15"
Marisa Simona
Cas 1h54'42" 59d26'33"
Marisa (Snuggles) Ingraffia
Tau 4h27'58" 12d14'36"
Marisa Spicer
Sgr 20h17'5" -43d35'5"
Marisa Stella
Tau 3h53'30" 24d53'55"
Marisa Varino
Uma 10h16'28" 47d34'11"
Marisa Victoria
Ari 2h0'48" 21d30'21"
Marisa Wintrow
Leo 11h32'11" 27d35'4"
Marisa, Baby Lamb
Cap 21h35'26" -11d46'1"
Marisa, My Brown Eyed girl
Her 18h43'45" 23d22'49"
Marisa, my everlasting star =:)
Vir 12h26'8" -6d0'54"
MarisaAndreaCeciliaRaymondPeteMaria
Uma 9h7'17" 48d37'56"
Marisabel
And 0h42'56" 26d9'32"
Marisabel Gomez Diaz
Tau 3h33'57" 1d27'26"
MarisaChris06
Her 18h11'57" 15d0'49"
Marisaluna
Peg 21h53'11" 13d5'13"

Marisa's Other Shining Light
Ori 5h30'6" -2d40'18"
Marisa's Rose
Vir 13h8'54" -11d22'59"
Marisa-Shnookums & Ben-Snuggle Bear
Cyg 20h17'13" 48d14'18"
Mariscotas
Lyr 18h39'18" 34d12'23"
marisela
Lmi 10h36'55" 36d21'13"
Marisela
And 23h17'4" 50d40'22"
Marisela
Psc 0h22'44" 18d6'40"
Marisela
Com 12h28'20" 18d31'54"
Marisin Elida
Lyr 18h49'56" 31d43'23"
Mariska Hogger
Aqr 23h28'40" -7d48'1"
Mariska van Vuuren
And 23h19'0" 52d10'13"
Marisol Arunee
Cnc 8h13'50" 9d56'48"
Marisol Castillo Guivas
Lib 14h49'12" -3d34'22"
Marisol Elena Sámano-Hopper
Crb 16h5'46" 38d52'33"
Marisol Gallardo
Cap 21h55'57" -19d14'12"
Marisol Gonzalez
Lyn 9h16'43" 34d30'53"
Marisol Laureano Mercado
Peg 22h22'29" 4d37'13"
Marisol Michaud
Cas 23h56'7" 50d20'55"
Marisol Q. Castillo
Psc 1h53'4" 9d47'51"
Marisol Reyes Bargo
Uma 13h1'55" 62d59'29"
Marisol Roman
Ori 5h54'8" 17d54'58"
Marisol's Star
Cap 21h54'25" -24d28'47"
Marissa
Sgr 19h30'37" -32d59'49"
Marissa
Cam 6h9'4" 64d8'6"
Marissa
Umi 13h30'38" 69d45'48"
Marissa
Cyg 20h11'4" 54d45'39"
MARISSA
Cap 20h20'15" -12d28'48"
Marissa
Aqr 22h31'47" -1d25'54"
Marissa
Umi 15h55'33" 75d8'9"
Marissa
Tau 4h35'26" 28d33'24"
Marissa
Gem 7h39'26" 23d50'11"
Marissa
Cnc 8h40'5" 12d4'7"
Marissa
Tau 4h34'14" 18d55'34"
Marissa
And 1h0'52" 47d33'32"
Marissa
Cnc 9h12'19" 30d35'29"
Marissa
And 1h0'20" 40d44'58"
Marissa ~ 07-05-05-11:16
And 1h30'11" 43d43'19"
Marissa Alexandra
Lyn 7h26'33" 49d15'31"
Marissa Angeletti
Uma 11h21'56" 64d35'49"
Marissa Ann
Tau 4h36'14" 29d12'58"
Marissa Ann Donnelly
Ari 2h14'10" 21d6'53"
Marissa Ann Napier
Cnc 8h14'3" 9d0'25"
Marissa Ann Olivieri
Sco 17h39'4" -45d10'8"
Marissa Ann Pirkey
And 2h26'22" 43d5'44"
Marissa Ann Timoner
Uma 11h20'10" 29d50'23"
Marissa Audrey
And 0h42'4" 44d0'51"
Marissa Bari Lewis
Ari 2h40'44" 12d31'16"
Marissa Barnes
Tau 4h42'13" 20d27'29"
Marissa Basile
And 23h15'29" 48d1'46"
Marissa Bitty
Uma 9h29'38" 62d32'39"
Marissa C. Bentley
Sco 16h56'36" -40d33'46"
Marissa C. Rosenthal
Uma 13h48'45" 61d20'11"
Marissa C. Woodward
Tau 3h52'29" 23d55'48"
Marissa Christina Robinette
And 1h27'35" 49d8'26"
Marissa Demma
Her 16h42'33" 7d20'8"
Marissa Denua Smith
Crb 15h52'13" 37d24'15"
Marissa DiStefano
Ari 2h52'59" 29d19'49"
Marissa Dozier
Ori 5h59'27" 21d29'31"
Marissa Elizabeth Holland Marqusee
Ori 6h20'6" 7d26'1"
Marissa Elizabeth Spungin
And 0h24'6" 42d31'33"
Marissa Elizabeth Tymm
Umi 15h33'46" 73d41'25"
Marissa Heiken
Psc 1h21'43" 6d16'52"
Marissa Helt
And 23h21'49" 49d4'58"
Marissa J. Tait
Vir 14h8'24" 6d59'55"
Marissa Jade
Cas 1h33'27" 66d58'14"
Marissa Jane
And 23h11'7" 47d49'54"
Marissa Jayne
Cap 21h16'12" -23d4'41"
Marissa "Joy"
Cap 21h29'42" -9d23'37"
Marissa Joy
Lyn 9h5'41" 36d47'42"
Marissa Kane Breton
Vir 13h22'31" 7d24'12"
Marissa L. Vanderkin
Cas 2h24'51" 64d46'16"
Marissa L. Woodley
Umi 14h24'49" 76d15'8"
Marissa LaBelle
Sco 17h49'43" -39d21'40"
Marissa Lauren
Lib 14h39'2" -22d55'0"
Marissa LeeAnn Raley
And 0h33'24" 32d38'13"
Marissa Lynn
Cyg 20h45'17" 44d14'45"
Marissa Lynn German
And 0h36'11" 38d44'50"
Marissa Lynn Gomez
Lib 14h54'54" -7d6'28"
Marissa Lynn Hovious
Vir 14h6'7" -15d14'34"
Marissa Lynn Pellicane
Ari 3h14'42" 29d19'7"
Marissa Lynn Zucker Star
Vul 20h51'16" 22d22'10"
Marissa Maggio
Tau 5h49'7" 21d44'14"
Marissa Mahealani
Gem 7h6'52" 17d41'49"
Marissa Margaret Fidler
Psc 0h54'22" 24d39'34"
Marissa Maria Montalban
Umi 13h31'55" 72d48'49"
Marissa Marie McKiernan
Aqr 22h55'33" -8d6'11"
Marissa Marino
Ari 2h14'34" 24d33'45"
Marissa Mayer
Oph 17h53'18" -0d59'29"
Marissa Metcalf
And 0h24'50" 44d49'34"
Marissa Michelle Hageman
And 2h8'29" 45d42'37"
Marissa Moran
Cnc 8h43'3" 12d12'30"
Marissa Nicole Buck
Mon 6h30'4" 8d25'8"
Marissa Nicole Hight
And 2h13'40" 44d22'22"
Marissa Nicole Vallette
Leo 11h37'57" 26d45'48"
Marissa Noll
Crb 15h51'31" 34d13'11"
Marissa Olson
And 2h17'57" 46d42'59"
Marissa Osato
Cap 21h57'54" -18d37'1"
Marissa Page Young\Derum 3-12-94
Psc 1h8'59" 32d38'3"
Marissa Peña
Aqr 22h5'46" -14d13'31"
Marissa Ramey
And 2h16'11" 50d13'33"
Marissa Ren Johnson
Aql 18h59'48" 9d16'35"
Marissa Rodriguez
Lyn 6h23'11" 54d54'23"
Marissa Rojas
Leo 11h8'39" 19d58'46"
Marissa & Ryo
Ori 6h6'9" 6d1'0"
Marissa Sonja
Uma 10h32'56" 50d59'4"
Marissa Suzie
Sco 16h5'55" -14d30'35"
Marissa " The Dancing Queen" Salemi
Psc 0h54'5" 30d49'54"
Marissa Tosh
Vul 21h17'2" 21d53'7"
Marissa Valenti
Lyr 18h52'16" 43d44'37"
Marissa Valenti
Cas 1h36'3" 63d54'33"
Marissa Victoria Mortillaro
Aqr 22h53'45" -8d1'26"
Marissa Yvonne
Sgr 19h58'32" -41d48'45"
MarissaBean
And 23h25'30" 48d47'57"
MarissaJo
Leo 10h42'29" 18d53'30"
Marissa's Shining Star
Tau 4h24'0" 23d40'17"
Maristela Marinelli
Cru 12h53'19" -56d16'24"
Maristella Marinelli
Cru 12h7'45" -63d25'53"
Marit Rebecca Thomas
Uma 12h9'41" 51d28'27"
Marit Renate Midthjell - My Viking
Ori 5h39'22" -1d5'46"
Marit Sørensen
Cnv 13h45'21" 38d1'45"
Marit Strand
And 0h43'49" 43d43'13"
Marita
Uma 13h53'29" 59d0'47"
Marita Aguilar
Aql 19h47'53" -0d11'45"
Marita "BabyGurl" Ryle
Vir 12h49'2" 7d12'18"
Marita Foot - Star Princess
Col 6h25'47" -34d2'36"
Marita mi
Ori 5h42'40" 9d18'29"
Marita & Norbert´s Love Star
Uma 10h18'54" 47d31'36"
Marita Wilhelmy
Uma 11h9'37" 28d37'1"
Maritere
Ori 5h53'14" -0d39'22"
Marites Cochran
Crb 15h18'1" 31d6'27"
Marithea Anne Carisella
Cam 7h41'9" 64d0'31"
Maritime Magic
Umi 14h0'48" 74d59'22"
Marito
Lib 14h34'8" -18d17'12"
Maritom
Sco 17h47'38" -33d57'12"
Maritza
Uma 9h22'34" 59d8'50"
Maritza Acosta
Cam 6h33'39" 74d56'57"
Maritza Alvarenga
Lyn 7h51'9" 44d35'54"
Maritza Kelesis
Lib 15h49'36" -8d1'35"
Maritza Leong
Vir 12h40'30" -9d0'26"
Maritza Martinez
And 2h34'9" 45d12'14"
Maritza Nazario
Cnc 8h52'56" 27d23'38"
Maritza Valentin
Uma 8h27'12" 65d49'55"
Maritzabel
Dra 17h31'26" 58d10'39"
MARIUS
Cas 23h40'57" 55d17'22"
Marius
Ori 5h2'38" -0d23'8"
Marius
Leo 11h14'57" -2d22'22"
Marius
Lib 15h1'8" -11d30'49"
Marius Adam 2001
Her 16h32'32" 47d30'14"
Marius Alfred
Per 2h31'55" 52d25'21"
Marius Garleng
Cnv 13h52'14" 39d59'34"
Marius Stropus
Aur 7h13'53" 42d42'34"
MARIUSZ B
Cyg 21h40'8" 49d29'30"
Marivania ~ Eternamente ~
Uma 11h26'11" 55d18'41"
Marivi
Leo 11h1'5" -5d26'7"
Marivic Espejo
Per 4h29'4" 35d6'15"
Mariwil
And 0h55'27" 41d35'7"
Mariya
Gem 6h52'3" 23d19'52"
Mariya Lynn Rau
Leo 11h49'7" 21d4'47"
Mariya Panfilova
Leo 11h7'40" 2d47'31"
Mariya Pima Chakarova
And 0h27'57" 26d16'51"
Mariya, Raisa & Valentina Grygorash
Umi 14h34'11" 72d42'53"
Mariyah Hope
Aqr 22h3'21" 1d31'43"
Mariza & Andrew
Cyg 20h30'59" 57d43'4"
Marja Birgit Ahlqvist
Leo 9h24'7" 25d51'27"
Marjan
Psc 1h42'43" 13d22'57"
Marjane Belayachi
Col 6h22'44" -35d6'24"
Marje Sandford Luce
Ori 6h19'34" 7d40'41"
Marj-Geri
Cas 1h24'27" 68d56'51"
Marji
Ori 4h48'32" 11d35'52"
Marjie
Cap 21h27'35" -9d4'22"
Marjie (Mag) King
Lib 15h1'45" -17d11'34"
MarJo
Lib 15h20'42" -9d29'26"
Marjolein
Aqr 22h18'44" -6d9'1"
Marjolein
Ari 2h1'33" 18d41'40"
Marjolein Elgersma
Peg 23h41'2" 29d26'11"
Marjolie15
Ari 2h32'21" 25d18'48"
Marjolyn
Tau 4h7'23" 26d48'42"
Marjon
Sge 20h2'46" 18d27'41"
Marjori Russo
And 23h13'50" 47d49'4"
Marjorie
Cas 0h16'16" 54d17'15"
Marjorie
Cas 0h42'6" 54d31'28"
Marjorie
Dra 18h30'45" 51d31'14"
Marjorie
And 0h10'5" 41d57'56"
Marjorie
Ari 2h37'43" 27d28'38"
Marjorie
Tau 4h18'45" 15d42'43"
Marjorie A Coulombe "1924-2003"
Vir 14h3'21" -16d0'18"
Marjorie A. Fulmer
Uma 11h15'51" 30d35'19"
Marjorie A. Johns
Srp 18h12'51" -0d2'25"
Marjorie A. Stock
Scl 0h37'50" -28d17'7"
Marjorie Abagnalo
Sco 16h8'17" -9d4'28"
Marjorie Amelia
And 1h18'27" 42d26'38"
Marjorie Ann
And 23h7'2" 50d54'6"
Marjorie Ann
Cap 20h27'57" -14d43'31"
Marjorie Ann
Cas 1h10'17" 60d25'20"
Marjorie & Ann Friends Since 1949
Tau 3h51'43" 24d33'6"
Marjorie Ann Lyne Stickelman
Sco 16h13'46" -12d44'19"
Marjorie Ann Saunders
Gem 6h4'15" 25d25'45"
Marjorie Anne Cronin
Aqr 22h27'11" -7d18'7"
Marjorie Anne Thomas
And 1h0'23" 37d3'6"
Marjorie Anthony
And 2h0'6" 42d40'10"
Marjorie Arlene Riddler "Nana"
Psc 1h15'51" 28d34'42"
Marjorie Cole
Uma 10h5'49" 47d38'43"
Marjorie Cuenca, MC
Ari 2h4'19" 20d14'45"
Marjorie Dates Fishbain
Ari 2h13'44" 10d34'21"
Marjorie Dehou
Cas 1h29'56" 68d17'5"
Marjorie Diane
Uma 10h58'52" 50d34'42"
Marjorie Diaz
Tau 4h50'45" 22d12'37"
Marjorie Dona
Sgr 18h12'4" -31d31'40"
Marjorie E. Durre
Cas 0h58'9" 54d26'27"
Marjorie Ellen Mello
Cyg 20h18'47" 44d57'53"
Marjorie Elnora Pearson (Boze)
And 23h33'39" 37d44'9"
Marjorie Faye
Cas 23h43'50" 53d40'11"
Marjorie Ferguson
Ari 3h22'29" 22d11'40"
Marjorie Fields
Vir 13h46'14" -6d2'48"
Marjorie Fior
Cas 0h19'50" 51d48'42"
Marjorie Goetz
Lyr 18h45'2" 38d34'1"
Marjorie Greninger
Tau 5h9'27" 27d13'2"
Marjorie Hall
Uma 11h19'31" 30d50'0"
Marjorie Hicks
Lyn 7h59'11" 54d22'44"
Marjorie Howorth
Cas 23h35'50" 55d1'31"
Marjorie J. Vaughan
Tau 4h35'21" 12d51'34"
Marjorie Joan Pinkham
Lyn 7h31'6" 36d44'32"
Marjorie Johnson Hood
Cep 23h36'48" 70d34'15"
Marjorie Joyce Dance
Vir 12h58'42" 11d53'39"
Marjorie K. Sims
Ori 6h5'51" 11d23'26"
Marjorie Krommenhoek
Psc 1h11'35" 9d5'46"
Marjorie Lacson
Cas 0h1'26" 53d17'12"
Marjorie Lee Matlock
Uma 8h17'53" 69d52'21"
Marjorie Lilac
Crb 16h18'57" 37d20'11"
Marjorie Lorraine Chaffin
Crb 16h17'19" 32d30'34"
Marjorie Lorraine Graves Yamamoto
Vir 13h3'41" -2d24'18"
Marjorie Lynn Sutton
Uma 9h22'37" 48d44'19"
Marjorie Lynn Walters
Lib 14h29'33" -14d19'17"
Marjorie M. Arsht
Uma 8h29'2" 64d1'28"
Marjorie M. Collins
Uma 9h15'44" 61d46'43"
Marjorie M. Olsen
And 2h31'54" 43d41'17"
Marjorie Mae Neal
Cas 0h24'41" 58d19'3"
Marjorie(Marj) Ellen Mahoney Dusay
Cas 23h46'10" 55d57'36"
Marjorie Meftah
Uma 8h45'47" 69d5'51"
Marjorie Mell
Uma 11h47'46" 31d31'54"
Marjorie Moonbeam Wiley
Ori 6h6'16" 11d47'9"
Marjorie Morgan Weatherford
Sgr 17h47'59" -28d8'9"
Marjorie 'Morningstar' Lamb
Ari 3h27'39" 28d7'26"
Marjorie Nell Maxwell
Crb 15h35'10" 26d54'1"
Marjorie Osborn Shook 85
Cas 0h36'24" 56d55'11"
Marjorie Provost Morea
Crb 15h53'40" 27d36'2"
Marjorie Redford-Glatt
Umi 20h59'30" 89d5'43"
Marjorie Rose
Psc 1h53'11" 5d50'40"
Marjorie Rosen
Uma 11h25'38" 65d14'30"
Marjorie Schleppenbach Family Star
Sco 17h33'25" -33d51'13"
Marjorie Steiner
Sgr 17h53'43" -29d9'4"
Marjorie Street Dandridge
Leo 10h48'6" 14d24'55"
Marjorie Sublett (MOM)
Cas 0h40'29" 57d59'46"
Marjorie Tarplee
Tau 4h16'44" 17d15'50"
Marjorie Vail McCone
Cyg 20h29'51" 33d52'51"
Marjorie Virgnia Stillwell
Gem 7h41'52" 22d19'21"
Marjorie Weiss
Cas 23h32'50" 54d49'5"
Marjorie Welham
Lyn 7h25'15" 53d30'17"
Marjorie Whitney Sadler
Uma 11h36'58" 55d9'44"
Marjorie, Always and Forever
And 0h11'35" 45d10'23"
Marjorie's Knight
Mon 6h45'50" -0d6'51"
Marjorie's Star
Sgr 18h15'54" -19d3'4"
Marjorie's Star
Uma 8h48'24" 61d47'27"
Mark
Cep 20h46'9" 62d13'9"
Mark
Cep 21h20'46" 62d11'18"
Mark
Uma 10h22'59" 71d47'15"
Mark
Vir 11h50'59" -0d57'8"
Mark
Aqr 22h37'14" -2d49'37"
Mark
Cyg 20h58'9" 32d10'45"
Mark
Uma 11h56'44" 33d56'6"
Mark
Cnc 8h54'57" 16d23'59"
Mark
Her 17h16'43" 17d15'5"
Mark 2 Linda 05
Cyg 21h35'24" 50d21'28"
Mark A Beehn
Cnc 8h47'7" 26d16'32"
Mark A. Benedict
Vir 13h20'32" 11d0'46"
Mark A. Biedell
Per 3h54'12" 38d5'32"
Mark A Garant
Her 17h16'25" 45d36'34"
Mark A. Glackin
Lyr 18h26'1" 31d11'55"
Mark A. Little
Gem 6h47'24" 26d22'43"
Mark A. Mehalick
Sgr 18h3'7" -27d56'16"
Mark A. Michelis
Cnc 8h42'4" 16d56'28"
Mark A. & Michelle Poperowitz
Gem 6h52'25" 14d54'43"
Mark A. Moore 10/01/00
Aur 5h50'45" 37d49'28"
Mark A. Prost, Jr.
Her 17h39'47" 39d17'5"
Mark A. Prost,Jr.Esq.
Per 3h2'5" 41d6'52"
Mark A. Sommer, Jr
Her 17h26'42" 37d11'17"
Mark A. Vichas
Leo 11h36'18" 25d32'45"
Mark A Wright SuperStar
Lib 14h46'16" -17d8'37"
Mark Abate
Cnc 8h52'26" 11d41'28"
Mark Adam Ackerman
Cyg 20h51'47" 54d49'52"
Mark Adam Boden
Uma 9h55'48" 51d9'2"
Mark Alan Andrews
Lyr 19h8'13" 27d10'4"
Mark Alan DeWaters
Cep 22h18'24" 62d40'6"
Mark Alan Montgomery
Leo 11h29'12" 0d40'4"
Mark Alan Murray
Uma 10h47'58" 61d14'28"
Mark Alan Wells II
Gem 6h49'21" 13d57'38"
Mark Albert Mellor
Vir 11h59'48" 7d59'27"
Mark Alexander French
Cen 13h14'11" -44d44'30"
Mark Alexander Smith
Per 3h11'4" 51d29'27"
Mark Allan Lehar
Ori 6h9'36" 9d22'27"
Mark Allan Townsend
Peg 21h35'43" 26d41'30"
Mark Allen Donahue Loving Son & Dad
Sco 17h23'32" -30d13'11"
Mark Allen Hahn
Per 3h33'25" 35d25'29"
Mark Allen Kamin
Per 3h8'12" 53d28'26"
Mark Allen Koehring
Umi 14h41'35" 72d32'46"
Mark Allen Lambert
Sgr 19h41'4" -17d20'37"
Mark Allen Lubben
Lyn 6h52'14" 60d30'10"
Mark Allen Richard
Lyr 18h32'50" 33d21'45"
Mark Allen Rushing
Uma 10h25'54" 41d43'39"
Mark Allen Taylor
Gem 6h7'14" 24d48'38"
Mark Allen Todd
Cap 21h15'32" -15d57'52"
Mark Allen Ulm
Psc 0h28'43" 9d11'7"
Mark Allen Wilson
Peg 22h21'47" 26d13'38"
Mark Allison Goodwin
Cnv 14h0'56" 37d5'52"
Mark Allum
Cyg 20h46'54" 45d54'57"
Mark Allyn Pullen
Cap 21h29'47" -13d40'15"
Mark Ambrosino
Psc 23h13'16" -1d30'51"
Mark & Amy! 11/17/02 Until Forever!
Cap 21h39'7" -14d10'21"
Mark & Amy 'Our love shines bright'
Ori 5h38'48" -2d27'14"
Mark and Anita
Uma 11h15'35" 29d28'35"
Mark and Becky: One year of Bliss!
Ari 1h48'4" 22d46'30"
Mark and Brandy
Cnc 8h35'50" 23d31'16"
Mark and Christine
Psc 0h38'39" 11d32'53"
Mark and Christine
Cyg 21h22'30" 55d2'48"
Mark and Cora
Cyg 21h29'16" 37d30'40"
Mark and Crissy Tibbitts
Aql 19h35'42" -0d3'53"
Mark and Dorothy Fleming
Cyg 21h11'12" 44d32'34"
Mark and Elizabeth Tonn
Sge 19h30'29" 17d38'7"
Mark And Erin
Cyg 20h41'30" 53d57'59"
Mark and Eugene
Sge 19h51'12" 18d12'51"
Mark and Holly
Ori 6h15'20" 5d56'1"
Mark and Jamison
Per 3h16'43" 46d10'49"
Mark and Jessica's Christmas Star
Eri 3h41'49" -0d28'27"
Mark and Kendall, The Best Parents
Uma 9h34'42" 61d22'34"
Mark and Laura
Cyg 20h15'10" 52d27'49"
Mark and Lauren
Cyg 20h10'53" 47d6'30"
Mark and Lisa Workens
Vir 14h28'9" -4d40'20"
Mark and Lori
Cnc 9h4'35" 30d37'36"
Mark and Mabel Rodrigue
Cyg 21h54'58" 45d58'35"
Mark and Mary
And 23h25'29" 48d0'32"
Mark and Maureen
Her 16h41'33" 46d53'16"
Mark and Maureen LeRoy
Per 4h27'8" 49d9'39"
Mark and Mitchell's star
Uma 10h57'36" 47d17'41"
Mark and Rachel Searle
Cyg 21h10'3" 46d28'9"
Mark and Robyn's Silver Anniversary
Cyg 19h18'51" 51d31'33"
Mark and Sally Smogor
Uma 11h59'34" 55d26'34"
Mark and Sherry's Star
Uma 11h39'52" 49d48'34"
Mark and Steffanie Cline
Uma 9h55'53" 64d10'27"
Mark and Whitney
Vir 14h30'2" -6d31'0"
Mark & Andrea's Anniversary Star
Cru 12h52'31" -57d54'14"
Mark Andrew
Cyg 20h23'59" 48d45'55"
Mark Andrew
Cas 0h10'14" 56d29'34"
Mark Andrew Caramore
Sgr 19h38'9" -14d5'53"
Mark Andrew Dynes
Del 20h36'17" 14d59'50"
Mark Andrew Lyons
Uma 10h43'42" 72d48'18"
Mark Andrew Mercante
Umi 15h32'35" 72d3'47"
Mark Andrew Murphy
Gem 7h27'5" 31d9'53"
Mark Andrew Nathan
Umi 14h38'32" 71d5'0"
Mark Andrew Salter
Lib 15h21'3" -16d4'18"
Mark Anthony
Crb 15h42'34" 26d31'22"
Mark Anthony Callahan II
Per 3h11'18" 43d2'52"
Mark Anthony Croce "Carino"
Ori 6h4'46" -0d19'6"
Mark Anthony Daniels
Lib 15h43'6" -17d18'47"
Mark Anthony Donovan
Gem 7h6'33" 33d51'10"
Mark Anthony Dreher
Lyr 19h19'46" 29d45'45"
Mark Anthony Grifo
Cep 23h16'20" 71d4'52"
Mark Anthony Hornbuckle
Uma 11h16'16" 52d56'24"
Mark Anthony Ingram
Gem 7h28'45" 20d36'17"
Mark Anthony Jacobi
Ari 1h47'25" 22d0'57"
Mark Anthony Mestaz My Love Bear
Uma 10h15'40" 56d16'35"
Mark Anthony Murray
Per 3h8'9" 53d53'24"
Mark Anthony Nixon
Per 1h36'57" 51d49'40"
Mark Anthony Salwoski
Aql 19h51'56" 12d9'11"
Mark Anthony Sandvig
Ori 6h3'48" 19d38'46"
MARK ANTHONY SHAND
Umi 14h0'29" 73d50'29"
Mark Anthony Smith
Her 17h15'49" 30d54'17"
Mark Anthony Stevenson
Per 4h44'4" 49d9'59"
Mark Anthony Toresdahl
Leo 11h22'52" 2d49'40"
Mark Anthony Whitby
Cen 13h13'23" -42d44'9"
Mark Antony Anyfandakis
Uma 8h47'46" 70d30'31"
Mark Antony Haycraft
Cep 21h19'40" 60d13'21"
Mark Arnold Dagys
Sgr 18h36'5" -30d6'48"
Mark Ashley Jones
Per 2h54'31" 40d35'24"
Mark Atkins Mitchell
Leo 9h34'24" 27d47'2"
Mark Baaden
Gem 7h14'39" 34d37'56"
Mark Backlund
Cep 20h37'3" 61d23'18"

Mark Baker
Aql 19h35'28" 15d2'2"
Mark Ballard
Cap 20h16'27" -18d50'35"
Mark Bartup
Cep 2h18'12" 80d37'32"
Mark Baruch Hazelbaker
Vir 12h45'32" -1d56'29"
Mark Benson
Her 16h53'58" 41d5'9"
Mark Bertsche
Her 17h50'35" 39d37'2"
Mark Besand
Her 18h50'51" 23d43'4"
Mark & Betti Mastrippolito Forever
Cru 12h36'35" -58d55'19"
Mark Bird
Cyg 20h6'37" 33d37'45"
Mark Birnbaum
Cnc 8h49'45" 26d41'53"
Mark Blair
Ori 5h32'19" 3d11'20"
Mark Blazejewski Born: May 17/1964
Uma 11h31'1" 56d55'26"
Mark Bowden
Per 4h22'53" 49d22'11"
Mark Bridges
Uma 8h27'26" 69d47'35"
Mark "Brother Bear" Cascella
Umi 15h12'57" 73d43'24"
Mark Burlington's Star
Lib 14h54'12" -6d30'32"
Mark Buttner
Per 3h27'35" 46d45'12"
Mark Byelick
Sgr 18h25'53" -32d8'4"
Mark Byrnes
Tau 3h27'9" 10d51'10"
Mark C. Del Guercio
Aur 6h30'20" 34d30'47"
Mark C. Jimenez
Ori 5h30'57" 3d23'10"
Mark C Johnson
Sgr 18h51'21" -20d19'15"
Mark C. Kelley
Uma 14h1'47" 58d16'51"
Mark C. Kris, Zach & Jason Bertoni
Leo 11h27'48" 25d20'33"
Mark C Obert Jr.
Lib 15h24'46" -7d10'55"
Mark C. Sullivan
Tau 4h14'59" 16d57'15"
Mark Calhoun
And 22h58'5" 36d57'56"
Mark Calloway's Star
Ori 6h1'17" 5d51'27"
Mark & Carmela Dovale
Cep 20h40'47" 65d36'21"
Mark & Carol
Vir 14h16'47" -4d29'44"
Mark Caudill
Aqr 22h11'50" 1d14'2"
Mark Champion
Dra 9h53'18" 77d27'27"
Mark Chan
Her 18h5'40" 35d36'22"
Mark Chapman
Cap 20h59'21" -22d12'34"
Mark Chipping
Cyg 20h35'23" 48d51'20"
Mark Christopher Averitt
Sgr 17h53'24" -29d36'44"
Mark Christopher George Dickins
Dra 18h52'44" 68d32'10"
Mark Christopher Johnson
Vir 13h25'27" -4d8'26"
Mark Christopher McHugh
Aur 5h11'33" 47d28'23"
Mark Clarricoats
Uma 8h35'12" 62d27'21"
Mark Clay
Tau 4h34'13" 28d29'1"
Mark Collins
Cep 22h6'53" 53d31'51"
Mark Conner Wilfong Sr.
Cnc 8h45'59" 24d30'7"
Mark Coulstock
Aqr 22h33'55" -1d12'51"
Mark & Courtney Basile
Aqr 22h17'30" -1d13'17"
Mark Cox
Lib 15h5'49" -8d27'40"
Mark Cusenbary
Uma 12h30'57" 56d17'28"
Mark Cushman
Uma 10h44'26" 41d51'40"
Mark D. Avart
Aur 6h24'10" 41d43'13"
Mark D. Brown M.D., PhD.
Uma 9h44'13" 71d43'14"
Mark D. Coffino
Uma 11h25'40" 29d26'23"
Mark D. Flaten's Star
Psc 0h58'25" 19d39'9"
Mark D. Kobart "Dad of McKinnon"
Uma 11h32'24" 58d30'34"
Mark D. Marek
Tau 3h53'32" 26d37'0"

Mark D Tropeano
Uma 10h47'6" 58d28'34"
Mark D. Verry
Aqr 23h30'5" -23d26'40"
Mark "D" Wildstein
Cap 21h26'23" -26d26'3"
Mark D. Wright
Sco 17h42'14" -39d33'13"
Mark D. Young
Cap 21h27'29" -16d1'43"
Mark Daniel
Ori 5h25'40" 0d59'41"
Mark "Dark" L. Vranderic
Her 17h35'55" 34d0'42"
Mark Daszkiewycz
Per 3h13'7" 46d17'19"
Mark David and Patricia Ann
Dra 18h58'33" 52d13'2"
Mark David Arteta
Leo 9h25'18" 24d34'19"
Mark David Bassner
Per 4h49'51" 45d40'49"
Mark David Hamilton
Lmi 10h37'5" 32d3'34"
Mark David Herbert
Her 17h27'46" 17d27'33"
Mark David McVicker
Sco 16h6'0" -10d17'26"
Mark David Williams
Vir 13h19'2" 3d7'58"
Mark David Zolner
Ori 5h52'24" 5d22'20"
Mark Davidson
Her 17h0'29" 46d40'14"
Mark Davis
Sco 17h57'56" -41d15'3"
Mark Dean Russell
Her 16h29'7" 46d59'26"
Mark DeGeorge
Tau 4h6'57" 27d29'9"
Mark & Deina
Lmi 10h43'13" 24d39'23"
Mark DeLaCruz
Aql 19h40'47" -0d5'1"
Mark Demarr Solomon
Gem 7h16'42" 20d0'45"
Mark - Destiny's Love Eternal
Psc 23h47'14" 4d48'0"
Mark Doerbeck
Uma 10h50'35" 67d9'30"
Mark Dominic Firmani
Tau 5h3'7" 15d42'17"
Mark & Dominique
Cru 12h22'9" -57d10'2"
Mark Douglas Dill
Vir 11h53'33" -3d33'25"
MARK DOVIDAS
Leo 10h16'15" 15d2'38"
Mark Duane McClendon
Ori 6h16'13" 10d16'53"
Mark Dubois
Lib 14h51'25" -4d17'9"
Mark Dvorak
Lyn 7h22'0" 53d8'51"
Mark E. Bakken
Sco 16h55'33" -42d15'20"
Mark E. Gurnow
Aql 19h36'8" 12d40'3"
Mark E. Hammitte
Sco 16h11'46" -16d32'59"
Mark E. Kennedy
Her 17h40'16" 15d5'27"
Mark E. Koenig
Gem 7h23'6" 32d3'34"
Mark E. Mackerman II
Ori 5h29'40" -0d11'45"
Mark E. Riggs
Ori 5h23'28" 2d32'57"
Mark E. Smith
Her 17h38'52" 48d23'57"
Mark Earnest's Star
Gem 6h47'0" 23d35'20"
Mark Eastaff
Cep 21h55'9" 65d31'10"
Mark Easton Taylor
Psc 1h46'5" 6d38'45"
Mark Edgemon
Uma 8h40'39" 59d6'21"
Mark Edward Aramino
Ori 5h55'55" 17d25'26"
Mark Edward Hosko
Sgr 18h11'30" -23d7'54"
Mark Edward Lichvar & Lydia Tejada Lichvar
Uma 9h21'38" 46d48'17"
Mark Edward Little's Star
Aql 19h45'58" -0d3'48"
Mark Edward Lucas
Aqr 21h40'59" 2d5'6"
Mark Edward Sandersk Jr.
Her 17h3'59" 34d17'15"
Mark Edward Thomas
Lib 15h3'54" -4d13'33"
Mark Edward Wiener
Leo 9h22'51" 10d5'51"
Mark Edward Williamson
Vir 11h49'47" 4d58'7"
Mark Edwards
Cep 4h8'21" 84d30'1"
Mark Eliot Wheeler
Gem 7h9'19" 16d7'49"

Mark Emerson Lassman's Star
Tau 3h49'15" 29d6'20"
Mark Emile DuPont
Sco 17h1'20" -40d3'43"
Mark + Emma Parker Wedding 14.07.07
Cyg 21h14'20" 47d43'12"
Mark Eric Avery
Ori 5h19'53" -0d22'0"
Mark Eric Bise
Leo 11h39'7" 16d55'56"
Mark Eric McKinney
Per 3h38'54" 43d23'11"
Mark Eric Romich
Sco 16h11'5" -11d7'12"
Mark Ervin Walcheske
Cas 23h49'7" 61d9'54"
Mark Eugene Stephens
Leo 10h55'7" 15d32'47"
Mark F. Herbert
Boo 15h2'58" 41d31'19"
Mark F. Karl
Per 4h30'0" 40d38'30"
Mark F. Mason
Per 4h48'29" 49d10'23"
Mark & Farah "Forever Shining"
Cyg 19h21'26" 47d6'43"
Mark Fink
Aur 5h10'15" 29d8'4"
Mark Finkeldei
Her 17h6'2" 32d39'14"
Mark Fleser
Her 17h36'56" 16d7'10"
Mark Foncannon
Cap 20h48'49" -16d14'8"
Mark Foster
Leo 11h54'19" 20d7'30"
Mark Francis Palos
Vir 13h27'41" -0d47'43"
Mark Frato
Aql 19h24'23" 7d42'13"
Mark Fricker
Her 17h50'16" 48d26'6"
Mark Fuller
Lib 14h50'28" -2d3'54"
Mark "Fuzzy" Andrew Edwards
Per 3h38'14" 45d55'46"
Mark G. Heming
Per 2h21'25" 51d40'12"
Mark Gabriel Wolfe
Per 4h17'14" 49d23'31"
Mark Gambrill
Cep 3h44'22" 78d31'4"
Mark Garcia
Uma 11h23'30" 65d16'16"
Mark Gary Rodrigues
Per 3h47'16" 41d5'20"
Mark Geisseler
And 23h20'43" 44d40'4"
Mark George Kuzyk
Ori 5h54'21" 22d46'21"
Mark Gerald Johnson
Sgr 19h21'41" -14d17'31"
Mark Gerard Artall
Aqr 21h49'40" -2d21'17"
Mark Giaconia
Sgr 18h4'36" -30d5'13"
Mark Gillen
Lac 22h28'45" 54d41'39"
Mark Goodman M.D.
Per 4h21'7" 33d17'50"
Mark Goodwin
Gem 6h51'4" 20d39'46"
Mark Gutierrez
Sgr 19h17'23" -28d9'52"
Mark H. Buckingham
Vir 15h8'59" 4d21'52"
Mark H. Splettstoeser
Per 3h30'38" 39d41'40"
Mark Haines 02.07
Lyn 8h51'53" 38d56'57"
Mark Halliwell
Ori 5h35'26" -5d48'49"
Mark Hallowell Jr.
Cyg 19h30'21" 28d2'29"
Mark Hamilton Warfe
Leo 9h32'57" 19d21'44"
Mark Hankins
Uma 9h55'49" 53d36'48"
Mark Happy Birthday
Psc 1h9'49" 11d48'20"
Mark Harold Kalagher
Aqr 23h55'42" -3d45'31"
Mark Harris
Cep 0h19'22" 67d44'39"
Mark Harrison
Cep 23h9'51" 71d2'41"
Mark Harter
And 0h42'42" 45d6'13"
Mark Haspel & Cory Hall
Tau 4h19'42" 13d43'50"
Mark Hassel
Uma 9h8'9" 51d12'32"
Mark Havens
Aur 5h17'15" 41d46'6"
Mark Hayduk
Leo 10h20'57" 25d21'53"
MARK HEFNER
Ori 6h18'23" 10d37'45"
Mark Heinz
Aur 5h20'52" 31d14'37"

Mark Hellis-Tatum
Per 3h20'3" 42d8'22"
Mark Herron (Babes)
Uma 9h44'45" 43d55'44"
Mark Hespenheide
Tau 4h46'11" 26d59'2"
Mark Hoffmann
Her 17h19'17" 24d45'56"
Mark Holloway
Uma 11h45'35" 48d57'50"
Mark Holt
Her 17h51'2" 46d2'11"
Mark Hudec
Lib 15h36'31" -27d52'18"
Mark Hunter Holsman
Ori 5h52'20" 3d37'26"
Mark Hurst
Cyg 20h47'33" 31d14'6"
Mark I. Berk
Cap 21h39'53" -8d49'45"
Mark Ian Drysdale
Uma 12h59'20" 59d14'49"
Mark Iapaolo
Per 2h39'30" 54d55'25"
Mark Ira Carlin
Aqr 22h16'5" 2d12'45"
Mark J
Leo 11h14'56" -2d5'57"
Mark J. Constantino "Tino's Star"
Cnc 9h14'1" 14d40'54"
Mark J Dinnes - Mark's Star
Sgr 17h59'20" -17d35'1"
Mark J Gerencser
Gem 7h11'22" 26d30'47"
Mark J. Harrison
Per 3h14'29" 53d53'6"
Mark J. Kendrick (Go Cubs)
Leo 9h24'41" 9d50'59"
Mark J. Smith
Ori 5h33'30" 9d53'15"
Mark J. Stephens
Gem 7h43'52" 24d14'6"
Mark J. Weitz
Uma 10h38'3" 52d7'33"
Mark James Coldwell
Dra 19h53'42" 75d44'17"
Mark James McAlpine
Lib 15h14'54" -7d11'5"
Mark James Olson
Leo 10h24'3" 7d1'43"
Mark James Pius X Ashton
Sgr 19h40'19" -15d33'6"
Mark Jamison
Cas 1h41'9" 62d2'51"
Mark & Janet Holecek
Cyg 19h59'9" 34d35'43"
Mark & Janna's Star
Ori 6h15'49" 15d10'3"
Mark Jarvis Biddinger
Uma 11h26'56" 59d5'28"
Mark Jason
Ori 5h23'48" 3d18'36"
Mark Jeffrey Bouyer
Per 3h26'21" 52d7'8"
mark jenn chris-best friends 4ever
Uma 9h9'34" 52d13'29"
Mark & Jody's Silver Star
Cyg 20h25'24" 56d11'39"
Mark Joe Brandon Brent Brianna
Lyn 6h28'9" 57d35'23"
Mark John Hanna
Psc 23h59'0" -3d43'37"
Mark John Meyer
Her 17h38'29" 37d29'32"
Mark Johsla
Cap 20h21'31" -15d9'31"
Mark Jonathan Mancini
Cet 2h38'59" 7d27'17"
Mark Jones
Cyg 19h40'24" 38d9'56"
Mark Jordan Snedden
Leo 9h42'1" 32d42'52"
Mark Joseph & Alison Marie
Cyg 19h32'16" 28d55'3"
Mark Joseph Hernandez
Cap 20h29'57" -12d7'40"
Mark Joseph - My Evening Star!
Cap 20h11'52" -10d18'8"
Mark Joseph Perez
Lmi 9h38'12" 35d21'48"
Mark Joseph Salopek
Psc 0h5'42" 4d50'5"
Mark Juliano
Cep 22h36'31" 72d1'23"
Mark June
Umi 14h0'44" 71d17'59"
Mark K. and Judy G. Love Star
Cyg 20h44'22" 54d41'23"
Mark K. Lundquist
Cyg 19h40'32" 31d29'13"
Mark Kalash
Leo 11h55'31" 25d28'33"
Mark Karolich
Cet 2h13'52" 7d10'2"
Mark & Katrina Anderson
Uma 11h0'23" 46d44'4"

Mark & Katrina Sutherland
Cru 12h17'44" -62d12'56"
Mark Keeling "Dad"
Per 2h55'45" 53d51'29"
Mark & Kelly Conner
And 0h51'29" 38d23'17"
Mark & Kelsie
Cyg 21h41'20" 31d54'37"
Mark Kennedy Rider
Ori 5h31'23" 10d23'36"
Mark Kenneth Kreps
Umi 16h19'1" 72d58'38"
Mark Kent
Tau 4h4'56" 26d40'29"
Mark Kerekes
Lyn 8h30'31" 42d4'52"
Mark Kofalt
Leo 9h24'13" 10d33'44"
Mark Koppe
Cyg 19h43'10" 38d55'52"
Mark Koppe
Sco 16h11'26" -9d56'5"
Mark Kulvinskas
Leo 9h29'2" 17d44'56"
Mark Kupferberg
Her 18h26'43" 27d42'53"
Mark L. Gray "Dee O Cee"
Cnc 9h6'44" 10d10'31"
Mark L Stanley of Boone Grove IN
Dra 17h6'46" 62d29'54"
Mark L. Tustin
Uma 12h14'19" 56d52'6"
Mark L Williams
Dra 19h5'39" 60d28'57"
Mark L. Woolfson & Diane M. O'Leary
Sge 19h50'33" 18d11'6"
Mark Langrehr
Her 17h11'53" 47d49'17"
Mark & Lauri Pettengill 4ever
Uma 10h25'46" 66d57'0"
Mark Lavallee
Uma 11h55'24" 49d15'33"
Mark Lee Sanders
Gem 6h14'48" 27d15'13"
Mark Leo Sanders
Uma 11h35'21" 47d43'56"
Mark "Leon de Corazon" Maras
Her 17h51'11" 46d33'57"
Mark Lewis Brierley
Cep 22h11'26" 53d59'40"
Mark & Libby's Wedding Star
Cyg 20h32'37" 38d27'7"
Mark Lindsey Coleman
Uma 10h25'53" 41d27'53"
Mark & Lisa Brettschneider
Cru 12h24'28" -58d26'36"
Mark & Lorevell Okuda
Cyg 21h26'10" 43d57'44"
Mark Louis Rose
Uma 10h9'18" 53d26'33"
Mark "Love" Carman
Cru 12h18'22" -57d21'17"
Mark loves Heather
Gem 6h49'16" 34d27'44"
Mark Loves Jessica
Cyg 21h29'50" 39d8'11"
Mark Loves Lisa
Lib 15h6'24" -25d13'13"
Mark loves Sandy - forever!
Tau 4h26'2" 18d9'58"
Mark Lucero
Her 17h36'50" 26d57'52"
Mark M. Cislo
Vir 13h7'23" -14d29'5"
Mark Mallord
Cep 3h21'35" 80d47'51"
Mark Malone
Her 17h22'4" 33d31'28"
Mark & Margie - Our Special Place
Leo 9h27'55" 18d53'37"
Mark Mariano
Uma 9h20'58" 44d47'4"
Mark & Marilyn Richmond
Uma 9h19'49" 69d46'42"
Mark/Marlene
Cyg 19h52'49" 32d22'13"
Mark Marske
Boo 14h58'52" 15d33'23"
Mark & Mary Forever
Psc 1h5'34" 20d37'31"
Mark & Mayra's Star
Uma 11h36'5" 42d28'48"
Mark MC
Tau 4h35'3" 28d2'28"
Mark McKenna
Cnc 8h52'12" 29d29'28"
Mark McNulty 1976-1995Cherished Son
Lib 14h47'42" -2d12'49"
Mark & Meghan
Gem 7h25'4" 33d18'40"
Mark Meister
Sco 16h8'5" -10d0'34"
Mark & Melody Wedlock
Cyg 21h30'6" 46d10'31"
Mark Meredith Jones
Per 2h26'31" 51d32'59"

Mark & Michele Daniels' 25th Anniv.
Cyg 20h20'40" 41d20'56"
Mark & Michelle's Engagement Star
Cyg 19h51'47" 30d46'43"
Mark Miemietz
Cnc 8h17'36" 19d25'55"
Mark Miller
Umi 16h49'35" 82d40'40"
Mark Milotay Mordechai ben Avraham
Cas 0h46'30" 52d21'16"
Mark Mirl Burton
Ori 5h27'58" 1d52'10"
Mark Misenheimer
Cnc 9h13'47" 26d22'32"
Mark Moloney "Baby"
Umi 14h50'37" 74d54'7"
Mark Montoya, Jr.
Her 16h41'39" 33d15'39"
Mark Moran Ph.D.
Per 3h12'34" 51d42'22"
Mark Morris
Cru 12h35'50" -60d30'46"
Mark Mullane
Psc 1h0'56" 33d17'51"
Mark My Love
Boo 14h18'39" 46d6'6"
Mark (na-na) Remsik
Leo 9h53'8" 23d15'22"
Mark & Natalie
Lib 15h34'7" -20d3'28"
Mark Navin *7/9/82-9/8/05*
Per 4h19'59" 36d29'19"
Mark Neveu
Boo 14h56'30" 46d44'34"
Mark Nhu
Cma 6h59'57" -18d20'59"
Mark & Nicci Ryan
Leo 10h28'30" 16d41'9"
Mark Nicholas
Leo 11h24'38" 5d33'19"
Mark Nicholas Corbett
Leo 10h28'39" 6d29'9"
Mark Nicholas Esposito
Peg 23h49'20" 26d36'52"
Mark Nunn
Per 2h24'19" 57d14'6"
Mark Ocondi
Sco 16h6'14" -9d1'36"
Mark of Helen
Cyg 21h31'42" 36d33'46"
Mark Orze
Ori 5h29'2" 4d45'16"
Mark Osburn
Her 18h10'27" 28d17'9"
Mark O'Shaughnessy
Per 3h46'6" 32d45'44"
Mark Ottenstein
Uma 8h28'26" 61d40'21"
Mark Owen Eason
Ori 5h32'12" 0d18'37"
Mark P. Madias
Sco 16h8'44" -10d3'38"
Mark P McNeil
Cep 23h15'30" 79d54'55"
Mark Pacheco
Uma 11h0'11" 52d39'6"
Mark (Papa Bear) James Weiland, Sr.
Lib 15h12'5" -13d8'35"
Mark Patrick Chisholm
Per 3h38'27" 47d20'4"
Mark Patrick Conboy
Gem 6h25'13" 25d34'26"
Mark Patterson
Sgr 19h34'59" -14d41'37"
Mark Patterson II
Cyg 20h26'25" 38d48'20"
Mark Pearce
Cyg 19h45'24" 33d35'8"
Mark Pendleton
Uma 9h15'36" 47d5'36"
Mark Pepin
Uma 14h3'13" 52d1'43"
Mark Perez
Aql 20h14'14" 5d21'21"
Mark Philip Hall
Ari 2h3'18" 24d1'59"
Mark Philip & Neil Stuart Higgins
Uma 12h7'29" 58d55'49"
Mark Phillip Haaser
Uma 9h18'44" 43d22'36"
Mark Pierson
Her 17h10'41" 29d31'23"
Mark Pietig
Per 3h33'13" 36d25'29"
Mark Pimentel
Cyg 20h2'34" 33d6'49"
Mark Polino
Aur 5h40'38" 41d43'39"
Mark Potocnjak
Aql 18h52'33" -0d39'22"
Mark Prokes
Sco 16h7'26" -10d45'26"
Mark Provancher
Cnv 13h39'57" 39d58'55"
Mark "Q" Quam
Leo 10h19'33" 23d45'42"
Mark R. Burkett
Sco 17h57'15" -30d24'57"
Mark R Grant
Lyn 8h21'34" 52d6'15"

Mark R. V. Willard
Ori 5h36'42" -1d2'1"
Mark Rae Mills
Cep 20h42'47" 60d14'35"
Mark Ralph Jung
Lyn 8h12'40" 45d51'31"
Mark Ramon Nail
Her 17h4'46" 33d3'6"
Mark Randal Sims, We love you!
Sgr 17h50'42" -17d25'5"
Mark & Randi DeGennaro
Umi 16h30'39" 75d24'50"
Mark Ratkovic
Uma 11h38'36" 34d37'8"
Mark Raymond Carbone
Cnc 9h10'17" 32d29'47"
Mark Raymond Florkowski
Her 16h33'45" 7d49'58"
Mark Raymond Zorick
Her 16h48'23" 27d7'20"
Mark Reid Tucker
Per 4h9'37" 34d30'55"
Mark Reiman
Uma 13h6'50" 56d49'51"
Mark Richard Astin
Pho 0h47'8" -52d14'26"
Mark Richard Connell
Tau 5h53'56" 23d47'47"
Mark Richard Koppe
Aur 5h52'40" 51d11'26"
Mark Richard Koppe <3
Per 4h9'16" 51d13'32"
Mark Richard Meili
Uma 12h26'56" 61d38'2"
Mark Richard Peterson
Aur 5h23'45" 31d18'46"
Mark Riggenbach
Ari 2h16'49" 24d14'30"
Mark Rinaldo - Chief Of Police
Per 2h41'36" 52d0'48"
Mark RJ Dupuis
Per 3h41'4" 44d4'5"
Mark Robert Cahill
Per 3h42'17" 39d9'33"
Mark Robert Hottovy
Leo 11h32'2" -6d40'47"
Mark Roberts
Boo 14h31'25" 14d50'57"
Mark "Rockstar" Kraner
Lyr 18h52'0" 33d23'13"
Mark Roland Dixon
Ori 6h1'2" 18d7'24"
Mark Roper
Aur 6h38'34" 37d53'33"
Mark Rosen
Leo 11h39'38" 16d33'9"
Mark Rosen
Cep 22h58'8" 76d53'1"
Mark Rosengarten Gadjitfreek
Gem 7h9'29" 33d50'22"
Mark Roy Mittelman
Aqr 22h55'56" -24d48'41"
Mark Rudolph
Ari 2h12'6" 12d3'33"
Mark Russell & Nina Terese Swanson
Gem 7h30'6" 20d31'2"
Mark Ryan Struble
Her 16h32'47" 33d13'1"
Mark Ryberg
Her 16h53'4" 34d20'54"
Mark S. Bailine
Sgr 18h5'38" -21d21'22"
Mark S. Phillips
Cyg 21h54'8" 39d28'10"
Mark S. Pierno
Leo 9h44'47" 26d33'29"
Mark S. Samson
Gem 6h51'28" 24d34'42"
Mark S. Wood
Ori 6h13'51" 6d9'46"
Mark Saliba
Ori 5h7'3" 3d54'13"
Mark & Sara Quaife
Cyg 21h36'7" 47d52'7"
Mark Sarnoff
Boo 14h56'8" 17d27'15"
Mark Schoeller Porter
Ari 2h4'12" 11d26'13"
Mark Schreiber
Cnc 9h16'28" 8d42'51"
Mark Schultze
Vul 20h15'43" 23d32'3"
Mark Scott Agnew
Ori 5h53'24" 0d15'37"
Mark Scriven
Uma 13h19'10" 62d53'49"
Mark Shannon Luke
Per 4h43'15" 49d29'2"
Mark Sheridan Star
Ori 5h47'3" 5d55'4"
Mark Sherry
Uma 13h53'23" 53d30'48"
Mark Shorr
Her 17h35'48" 31d54'53"
Mark Short
Her 18h34'37" 22d2'33"
Mark Shubatt
Sco 16h19'14" -25d34'23"
Mark Sias
Psc 0h38'2" 14d19'43"

Mark Sickles
Dra 15h30'22" 57d7'16"

Mark Simmons
Boo 15h15'17" 47d42'34"

Mark Simpson
Cnc 8h55'2" 9d31'59"

Mark & Siobhan Levy
Cyg 20h37'23" 34d54'15"

Mark Slorach
Per 3h58'33" 45d7'37"

Mark Smelcer
Leo 9h36'2" 29d47'49"

Mark Smith
Per 3h9'52" 41d9'45"

Mark Snyder
Cnc 8h46'38" 32d36'31"

Mark & Sonia Lukow
Cyg 20h10'18" 34d35'46"

Mark Spicola
Cnc 8h59'54" 8d48'4"

Mark Stamback
Vir 13h53'41" 1d14'36"

Mark Stanson 50
Psc 1h20'31" 19d5'55"

Mark Staub
Aqr 21h53'13" 0d47'23"

Mark & Stephanie Bass
Cma 6h58'30" -22d49'6"

Mark Stephen Adams
Umi 16h53'41" 75d24'0"

Mark Stephen Cleary
Cru 12h24'22" -62d45'54"

Mark Stephen Clelland
Uma 8h39'7" 70d1'44"

Mark Stephen Fletcher
Per 3h44'15" 44d54'20"

Mark Stephen Harvey
Cas 0h42'10" 51d37'34"

Mark Stephen Pisch
Uma 14h26'34" 56d38'8"

Mark Sterba
Ser 18h17'15" -14d1'9"

Mark Steve Ian Taladua
Cmi 7h7'9" 6d17'23"

Mark Steven Conrad
Ori 5h57'32" 18d19'53"

Mark Steven Cummings
Ari 2h51'0" 27d12'51"

Mark Steven Edwards
Her 18h36'35" 19d54'22"

Mark Steven Mungle
Cnc 8h33'57" 30d54'55"

Mark Steven Zangara
Per 3h27'52" 51d22'20"

Mark Steven Zenoni
Cep 21h26'48" 61d4'2"

Mark Stier
Tau 4h25'54" 24d18'2"

Mark (Stretch) Johnson
Cep 0h42'11" 82d38'16"

Mark Stuhr
Uma 9h24'1" 68d59'35"

Mark Sullivan
Cyg 19h59'34" 31d41'50"

Mark & Sydney
Aqr 22h2'24" -21d18'8"

Mark Sykes, Missed by us All.
Per 2h27'38" 54d29'28"

Mark T. Leonard, Super Star
Tau 5h41'3" 21d6'17"

Mark T. Manetta
Lyr 18h51'51" 44d16'36"

Mark T. McKernan
Uma 10h24'37" 42d44'21"

Mark Tabaka
Aql 19h24'52" 8d47'6"

Mark & Teresa Lazarov (Our Star)
Leo 11h53'4" 23d37'23"

Mark Terry
Cap 20h28'50" -13d52'40"

Mark Teshoian
Aql 19h3'36" 18d16'21"

Mark "The Shark" Allen
Ori 5h15'53" 7d36'7"

Mark Thomas Alvino
Cnc 8h51'13" 31d3'34"

Mark Thomas Blackman
Dra 17h24'23" 62d26'45"

Mark Thomas Heimiller
Her 18h2'3" 26d54'41"

Mark Thomas LaCross
Ori 6h5'58" 18d50'37"

Mark Thomas Pennebaker
Gem 6h37'10" 20d40'25"

Mark Thomas Terveer
Aql 19h22'8" 6d23'58"

Mark Thomas Wilfahrt
Boo 14h51'41" 52d43'15"

Mark Tigner
Aql 18h46'34" 8d27'13"

Mark Tillar — The Best Father
Per 3h16'55" 43d38'49"

Mark Timothy Macmanus
Ori 6h4'56" -0d19'39"

Mark Tonelli
Ori 5h40'54" -1d30'28"

Mark & Tracy - 4.3.2005
Cyg 19h35'50" 28d36'56"

Mark Tullio Zarantonello
Aqr 23h29'12" -23d16'21"

Mark Turnbough
Ori 5h25'43" 14d44'18"

Mark Twain McClary
Aur 6h25'34" 32d56'43"

Mark V. D'Agostino
Dra 18h29'45" 55d22'45"

Mark V. Matthews
Cnc 8h17'50" 9d33'32"

Mark van Eeuwen
Aql 19h27'33" 3d20'21"

Mark Varidin
Umi 15h43'59" 75d55'15"

Mark Verne Reed
Ori 5h17'48" 0d50'37"

Mark Vincent Sudbeck
Cnc 9h20'40" 15d18'23"

Mark Visconti
Her 17h14'13" 27d3'6"

Mark VonBorstel
Aqr 22h35'39" -18d56'9"

Mark W. Cash
And 23h22'2" 52d29'58"

Mark W. Cobb
Dra 16h36'37" 68d24'26"

Mark W. Foreman
Boo 15h9'35" 33d10'29"

Mark W. Jackson
Her 18h37'25" 21d59'24"

Mark W. Lowe
Dra 15h29'2" 62d58'35"

Mark W. Wood
Tau 3h45'25" 5d24'12"

Mark Wagner
Boo 14h34'58" 29d43'59"

Mark Waldo
Uma 14h2'43" 60d56'56"

Mark Wallace
Ori 6h16'35" 16d3'39"

Mark Wallis
Per 2h52'38" 41d22'16"

Mark Walter Breede
Ori 5h8'38" 1d15'52"

Mark Walter Breede
Pav 18h26'13" -59d17'50"

Mark Warren Dean
Cnc 9h16'27" 15d14'45"

Mark Warren Goddard
Tau 4h46'49" 19d27'1"

Mark Warren Jenney
Her 17h9'20" 33d34'17"

Mark Wasylik
Boo 14h16'59" 45d29'19"

Mark Wattenberg
Leo 9h56'24" 29d53'17"

Mark Wayne Fitts
Per 2h24'14" 51d8'45"

Mark Wayne Guy
Cap 21h14'44" -17d10'29"

Mark White
Tau 4h0'30" 25d19'34"

Mark Wigton
Psc 23h22'23" 3d43'16"

Mark William Croshier III
Her 17h19'41" 46d1'36"

Mark William Evans Jr.
Per 2h59'10" 54d30'26"

Mark William Goik 33
Per 3h27'29" 39d42'45"

Mark William Nelson
Per 2h36'48" 51d42'23"

Mark William Nelson
Leo 9h41'14" 23d40'53"

Mark William Steward
Per 1h45'22" 51d24'38"

Mark Williams
And 23h46'50" 47d40'41"

Mark Winterfeld
Uma 8h52'16" 68d54'19"

Mark Withers
Cnc 8h14'22" 12d5'18"

Mark Workman
Ori 4h52'40" 9d2'56"

Mark Wyman
Per 3h23'52" 55d25'41"

Mark Yost
Her 18h44'16" 20d11'34"

Mark Z. Johnson
Her 18h34'19" 20d10'16"

Mark Zeveckas
Tau 4h23'4" 18d10'3"

Mark Zika's Star
Aql 19h38'33" 12d23'5"

Mark Zimmerman
Ori 5h55'57" 12d58'53"

Mark Zout
Cet 2h26'53" 8d48'57"

Mark, Annemarie & Lindsay Dantuono
Tri 2h21'47" 35d17'51"

Mark, Barb, Tim & Becky Vogel
Uma 9h59'47" 63d46'21"

Mark, Caroline, Flennigan Stoeklen
Ori 6h14'27" -0d24'3"

Mark, Francesca, Ian & Ethan Stieg
Ori 5h23'51" -0d29'13"

Mark, Helen, & Megan Caranci
Uma 11h7'59" 35d27'47"

Mark, Linda, Jen + Tim Dillehay
Peg 22h29'49" 9d16'30"

Mark, Mo anam cara, ta gra agam ort
Ori 5h34'17" 9d13'47"

Mark, The Love Of My Life
Cap 21h13'49" -19d43'48"

Mark, you light up my life, Carly
Sgr 19h22'50" -20d26'27"

Mark2005
Cas 0h45'5" 69d24'17"

Markalan
Ari 2h3'50" 14d44'21"

markanjesslila
Crb 15h47'0" 34d58'43"

MarkAnthonyThomasMummaMartinBell
Her 17h23'9" 35d15'18"

Mark-A-Saurus
Per 3h45'36" 46d3'5"

MarkASpencer
Per 3h21'32" 45d44'14"

Markayla Kamille
Ari 2h0'18" 23d38'11"

MarkE
Aqr 22h39'23" 2d12'14"

MarKe
Cap 21h51'41" -17d46'12"

Marke Levene
Cyg 19h39'51" 35d32'44"

Markee Marie Stowell
Psc 0h3'30" 8d2'54"

Markeen Otis Beam
Aqr 21h10'22" -9d12'41"

Markeli, Bernhard
Sgr 18h3'14" -27d44'55"

Markell
Uma 10h18'7" 47d36'38"

Markelle Rae Frei
Uma 10h26'36" 43d16'10"

Markenna
Per 4h5'0" 47d29'43"

Märki
And 2h12'34" 41d35'41"

Markie
Per 3h12'44" 43d51'35"

Markie
Cap 20h16'53" -10d11'38"

Markie Falzone Jr.
Umi 16h35'12" 76d3'42"

Markie & Nerrisa Forever and a Day
Sco 17h49'3" -41d12'0"

Markie Sanderson
Aqr 22h31'15" -21d59'10"

Markie Starr Crutcher
Lyn 6h57'20" 59d9'25"

Markins Corp. Sdn. Bhd.
Ori 5h37'37" 3d38'47"

Markio
Cyg 21h19'31" 41d22'13"

Markku J. Rajaniemi
Umi 14h24'44" 74d19'11"

Markle Farkle
Cnc 8h10'32" 21d36'43"

markley
Cyg 19h55'28" 32d17'29"

Marko
Tau 5h35'27" 25d37'27"

Marko
Ori 5h41'52" 0d16'39"

Marko
Vir 14h11'35" -11d8'49"

Marko Gotovcevic
Crb 16h9'32" 33d36'24"

Marko Loves Lotz
Aqr 21h52'33" -0d35'55"

Marko Muzic Polomcic
Psc 0h57'59" 31d12'28"

Marko Vranjes
Ari 2h13'6" 24d17'55"

Markos
Lyn 7h40'32" 57d9'7"

Markou Mary
Tau 4h18'45" 26d39'10"

Markowsky
Per 3h23'57" 49d15'9"

Markree
Cyg 21h33'59" 31d0'7"

Mark's Angel
Tau 5h44'8" 22d51'17"

Mark's Astellas Star
Ari 2h43'54" 26d49'24"

Mark's Cakie forever
Crb 15h34'34" 27d54'58"

Mark's Caro 3296
Sgr 19h33'28" -21d6'43"

Mark's Causeyland
Uma 9h25'49" 50d25'44"

Mark's Chrystal
Aur 6h0'51" 47d22'30"

Mark's Converse All Star
Psc 1h18'29" 18d26'8"

Mark's Eternal Light
Srp 17h58'11" -0d23'48"

Mark's eternal star, love Kelly
Aur 5h19'5" 42d39'49"

Mark's Girls
And 0h26'15" 31d47'49"

Mark's Journey
Aql 20h6'19" 0d25'44"

Mark's Levatio Olympus 7
Uma 13h46'1" 56d1'53"

Mark's Luz de las Estrellas
Sco 17h29'43" -41d44'41"

Mark's Maestro
Uma 13h58'10" 58d26'24"

Mark's Music
Gem 6h51'52" 16d17'31"

Mark's Spark
Tau 4h26'52" 17d27'45"

Mark's Star
Aur 7h14'39" 42d24'12"

Mark's Star
Uma 10h42'30" 52d51'11"

Mark's Star
Cep 21h6'15" 61d27'11"

Mark's Star
Aql 19h30'55" -0d3'49"

Mark's Tucker
Boo 14h29'7" 16d51'26"

MARKUS
Tri 2h27'55" 29d16'11"

Markus
Vul 18h59'58" 24d12'17"

Markus
Lyr 19h16'8" 34d26'23"

Markus
Cyg 20h33'9" 42d13'20"

Markus
Cyg 20h41'34" 46d11'17"

Markus
Cyg 21h24'0" 49d8'31"

Markus
Uma 10h42'30" 57d21'36"

Markus
Cas 0h52'46" 62d19'7"

Markus Alexander Gilgen
Tri 1h44'1" 30d56'2"

Markus Altstätter, 30.12.1973
Crb 15h50'28" 27d51'3"

Markus - amici a vita
Uma 8h41'0" 49d49'9"

Markus Andreas Grill
Gem 6h57'43" 31d4'29"

Markus Esser
Uma 11h4'31" 37d50'57"

Markus Gedon
Uma 8h54'7" 60d42'1"

Markus Graf
Lyr 19h15'36" 35d10'9"

Markus Greiner
Ori 6h5'10" 0d15'19"

Markus Grill
Vir 13h4'12" 3d3'40"

Markus Hampel
Boo 14h27'9" 24d57'50"

Markus Hilty
Lac 22h41'35" 49d44'11"

Markus Hodge
Ori 5h43'8" 1d35'49"

Markus Joseph Perrotta
Gem 6h40'52" 21d27'25"

Markus & Kimberly
And 1h1'2" 39d22'58"

Markus Kroner
Uma 10h26'33" 51d14'29"

Markus Lukas
Eri 4h31'10" -14d1'30"

Markus & Marc
Uma 9h37'20" 48d41'21"

Markus Meier
Uma 10h32'52" 42d40'9"

Markus Monnig
Ori 5h44'55" 0d46'9"

Markus Oberle
Lyr 18h59'26" 33d2'7"

Markus Onni Elias Malmio
Uma 8h39'54" 54d41'20"

Markus Reto Gutschi
Cyg 20h45'2" 52d8'4"

Markus Schmoll
Her 16h19'30" 12d45'14"

Markus Schwarz
Ori 6h17'28" 18d59'12"

Markus Stemper
Uma 9h20'40" 72d51'52"

Markus Stephan Feuchter
Uma 9h3'24" 69d29'39"

Markus Veneer-us
Uma 11h29'39" 41d0'45"

Markus Vetter
Ori 5h55'24" 13d2'3"

markus waldvogel
Ori 5h59'58" 21d49'44"

Markus Widmer
Ori 5h8'26" 8d58'58"

Markus Worm
Uma 11h19'14" 44d5'1"

Markus y Graciela Kitzmüller
Gem 7h47'48" 14d44'47"

Markus Zenk
Uma 11h17'36" 36d39'46"

Markusberg
Ori 5h44'20" 6d54'43"

MarkuStar
Uma 9h15'0" 61d14'20"

Markwith's Midnight Sun
Ori 5h34'28" -0d30'1"

MarkWong.W.K
Sco 17h56'21" -39d1'18"

Marky
Pho 0h28'57" -41d53'11"

Marky
Ari 3h15'45" 26d28'51"

Marky
Per 2h14'44" 53d51'44"

Marky Mark Austin
Leo 11h57'2" 21d6'52"

Marky's Star (Marky Ramone)
Tri 1h59'26" 33d52'49"

Marla
Cas 0h57'54" 54d12'4"

Marla
Psc 1h10'49" 24d56'0"

Marla
Cas 23h7'20" 56d5'22"

Marla B. Turner
Lyn 7h20'41" 45d52'19"

Marla Balas
Uma 8h15'13" 62d0'7"

Marla Eileen Carver
Sco 16h8'34" -18d19'12"

Marla G. Wells
Oph 17h7'8" -0d48'26"

Marla Gowans
Tau 5h10'11" 20d57'26"

Marla Hosie
Col 5h7'57" -32d37'56"

Marla Jean Gregory
Uma 9h34'20" 52d37'30"

Marla Kaye Mouton
Gem 6h57'11" 13d21'28"

Marla M Downs
Cas 0h28'20" 61d26'30"

Marla Marie Musso
And 23h49'3" 44d58'24"

Marla Peggy Hudkins
Cnc 8h43'12" 30d41'52"

Marla Porrazzo
Gem 7h15'1" 28d5'30"

Marla Read
Sco 16h50'49" -36d21'9"

Marla Sherill Macoubrie
Com 13h3'31" 19d3'43"

Marla Smith, My Princess
Sco 17h10'20" -36d58'33"

Marla Stewart
Com 12h33'22" 17d50'57"

Marla Virginia Myers (i love you)
Psc 1h55'55" 5d47'50"

Marlaena
Ori 5h53'22" 18d12'3"

Marlana
Lyn 7h7'24" 50d44'52"

Marlane
Lep 5h47'2" -11d47'2"

Marla's glistening smile
Lmi 10h23'52" 35d27'0"

Marla's Star
And 23h29'38" 41d35'20"

Marla's Star
And 0h31'8" 28d17'41"

Marla's star
Vir 13h12'19" -2d26'21"

MarlaSophiaAnnaBrianMuniFamilyStar
Cas 0h26'40" 50d15'13"

Marlea Ruth Heizer Anderson
Ori 4h46'59" 7d47'52"

Marlee
And 0h11'55" 34d27'23"

Marlee Criaco
Leo 11h16'16" -1d8'47"

Marlee Elizabeth Mayer
And 0h18'5" 42d20'47"

Marlee Johnston
Cyg 20h15'34" 31d57'12"

Marlee Kiet
And 23h20'11" 45d48'17"

Marlee Roberta Morning
Ari 2h18'40" 25d9'21"

Marlee Sophia Schuesler
Leo 11h45'17" 19d39'2"

Marlee Vincenza
Dra 16h35'32" 57d50'24"

Marleen
Sco 16h4'31" -9d20'36"

Marleigh Agner
Lyn 8h41'2" 41d23'22"

Marleine Marie Spetz
Ari 2h1'30" 13d31'28"

Marlella E. Gantt
Uma 12h2'34" 50d45'8"

Marlem
Lib 15h10'25" -9d36'28"

Marlen Bueno
Leo 11h40'1" 20d0'18"

Marlena
Lyn 7h34'38" 37d15'51"

Marlena Dawn Powers
And 1h45'25" 46d8'45"

Marlena Nisbet
Aqr 21h52'50" -2d30'38"

Marlena Palmer McMurchie
Ari 3h8'2" 26d25'38"

Marlena Raye
And 1h30'34" 45d44'21"

Marlena Spina
Del 20h53'5" 7d39'57"

Marlena's light
Her 16h19'4" 23d4'9"

Marlene
And 0h31'3" 29d48'2"

Marlene
Del 20h57'26" 15d7'33"

Marlene
Uma 11h55'9" 51d51'0"

Marlene
Crb 15h43'23" 35d19'57"

Marlene
And 1h10'1" 43d7'20"

Marlene
Mon 7h18'9" -0d19'36"

Marlene
Sco 16h8'58" -9d24'26"

Marlène
Oph 17h34'48" -17d20'8"

Marlene
Cas 0h53'34" 61d33'51"

Marlène
Uma 11h48'32" 59d12'56"

Marlène 21.04.1956
Uma 10h32'31" 60d22'18"

Marlene A. Bryden
Lyn 9h6'5" 38d3'15"

Marlene Agnes (Kelly) Cochrane
Cas 22h59'54" 56d14'30"

Marlene Alexis Sullivan
Cas 0h59'4" 54d11'2"

Marlene B. Mendez
Sgr 18h0'8" -28d30'5"

Marlene Beatriz Rodriguez
Aqr 23h24'32" -18d28'48"

Marlene Capaldo
And 0h29'20" 39d37'43"

Marlene Concepcion
Psc 1h7'47" 29d35'1"

Marlene Connor
Uma 8h15'36" 60d7'15"

Marlene Czajkowski
Sco 17h43'28" -35d41'31"

Marlene & Dave Falconer
Cyg 21h8'22" 37d25'26"

Marlene Dorothea
And 23h25'0" 48d41'37"

Marlène Dreuilhe
Cnv 12h43'19" 43d57'23"

Marlene E. Burks
Her 16h20'25" 44d17'29"

Marlene Elise
Ari 2h35'42" 26d3'3"

Marlène Floc'h
Ari 2h17'30" 21d24'34"

Marlene Foster
Cas 0h50'39" 62d40'50"

Marlene Garberding
And 2h33'19" 50d12'21"

Marlene Gearey
And 23h25'11" 52d6'5"

Marlene Griswold
And 1h47'16" 49d31'29"

Marlene H Elgersma
Ari 2h3'58" 13d6'51"

Marlene Herrera
Sco 17h18'56" -32d24'52"

Marlene J Wadhams-Knoeller
Gem 7h18'13" 25d39'49"

Marlene Jeanne
Cnc 8h52'13" 14d22'8"

Marlene Jule
Uma 10h22'50" 58d36'41"

Marlene - Light of my Life
Cas 1h28'7" 51d20'20"

Marlene Linda Tanner
Cas 1h28'26" 60d1'33"

Marlene Luitwieler
And 0h24'37" 31d43'35"

Marlene M. Garcia
Uma 11h45'53" 29d2'54"

Marlene M. Mihojevich
Tau 5h39'29" 24d29'38"

Marlene Machut
Lyr 18h28'51" 31d53'1"

Marlene Maria
And 0h37'48" 33d43'22"

Marlene Marie
And 0h23'25" 36d12'46"

Marlene Mendosa
Cnc 9h10'34" 25d33'15"

Marlene Mikhail Brikho
Tau 3h41'52" 26d51'24"

Marlene O'
And 1h56'7" 39d11'35"

Marlene ONeil
Cas 23h38'47" 54d41'57"

Marlene Pegel
Cnc 8h54'55" 29d3'48"

Marlene R. Marne
Cas 0h47'40" 58d4'42"

Marlene Rocks
Sco 17h45'34" -32d57'58"

Marlene Rose Mazza
Col 5h36'52" -30d44'32"

Marlene Sarich
Lib 14h23'22" -18d37'38"

Marlene Selby
Lyn 6h31'40" 57d56'46"

Marlene Shingleton
Uma 9h12'20" 67d57'24"

Marlene Speece-Clifford
Aqr 22h58'5" -21d4'22"

Marlene T. C.
And 0h29'47" 31d25'38"

Marlene T. Robinson
Vir 13h12'34" -20d6'38"

Marlene Thank You GOD For Blessings
Uma 11h18'53" 50d55'15"

Marlène & Thomas
Aqr 22h46'11" -16d19'55"

Marlène Tiberi
Sco 16h48'19" -26d35'19"

Marlene Valeria
Uma 11h12'47" 67d6'37"

Marlene Velecky - 21/12/1951
Cru 12h28'14" -59d29'24"

Marlene Wenger
And 23h16'16" 41d48'10"

Marlene Wilson
Del 21h8'41" 19d27'26"

Marlene's Inner Light
Leo 11h29'50" -2d10'56"

Marleny
Cam 7h22'22" 61d33'38"

Marleny Torres
Aqr 22h37'17" -18d34'41"

Marleon Walsh-Watson
Ori 5h12'52" 5d34'36"

Marlette's Star
Ori 6h13'14" 17d30'28"

Marley
Ori 5h54'53" 18d33'52"

Marley
Uma 10h8'33" 47d59'25"

Marley
Lmi 10h26'29" 38d41'19"

MARLEY ALEXIS
Tau 3h43'37" 27d29'41"

Marley Ann Shuford
And 2h5'58" 38d9'3"

Marley Blu Mayo
Leo 11h7'43" 14d36'20"

Marley Brooke
Psc 0h55'57" 8d11'52"

Marley Dale Martin
Sco 16h17'6" -11d30'14"

Marley Danielle Cross
Cru 12h46'8" -56d41'47"

Marley Dias
Cma 7h23'56" -12d42'2"

Marley Elizabeth
Lib 15h4'50" -10d18'8"

Marley Elizabeth Leard
Lyr 18h52'52" 35d32'11"

Marley Eve
Aqr 21h46'42" -0d28'41"

Marley Jane Winningham
And 0h28'51" 45d27'36"

Marley Louise
Psc 23h58'21" 1d46'25"

Marley Payan
Cnc 8h1'26" 26d25'44"

Marley Williamson
Uma 9h21'46" 49d6'7"

Marley Winn Bryden
Uma 11h51'20" 35d28'3"

Marli
Ari 2h8'50" 23d38'8"

Marli Bee - A Star Is Born
Ori 5h35'21" -1d48'10"

Marlicia
Vir 14h29'26" 3d14'25"

Marlie Viola Shrier
Psc 1h32'0" 28d7'43"

Marlies
Umi 15h48'24" 86d25'15"

Marlies Hammann
Uma 9h0'47" 51d27'28"

Marlies J (Fritz) Schneider
Aqr 23h28'15" -14d46'37"

Marlies & Mario together in LOVE
Cep 22h3'26" 68d58'30"

Marlies Meinhold
Mon 6h50'54" 7d27'12"

Marlies Single
Aqr 22h59'35" -7d2'23"

Marlin
Cyg 20h33'51" 43d30'28"

Marlin Christin
Uma 10h7'12" 45d36'45"

Marlin & Gina Mann
Uma 11h7'50" 49d13'35"

Marlin & Hazel Wynn
Lyn 8h59'33" 33d59'32"

Marlin Hydal
Peg 21h44'56" 13d12'9"

Marlin Lee Jolly
Sco 16h16'39" -10d35'15"

Marlin Oliver Mehring 14.06.1994
Uma 9h17'46" 67d49'17"

Marlinda
Her 16h34'56" 44d51'29"

Marline et Dominique
Tau 4h27'30" 15d12'57"

Marlis Gerda Röschli
Aur 5h54'38" 42d32'8"

Marlis June
Psc 1h3'19" 10d3'9"

Marlis Marie Mangan
Uma 8h54'1" 65d57'28"

Marlo Marie Quezada
Cyg 19h39'59" 32d3'36"

Marlo Musco
Psc 1h9'8" 31d57'35"

Marlon Gsell
Cas 1h45'9" 68d13'22"

Marlon Lasa
Sgr 19h6'24" -28d40'26"

Marlon Lening Villagra Rodriguez
Aqr 23h25'33" -12d4'38"

Marlon Panjo
Lmi 9h26'25" 34d37'7"
Marlon Spearman
Dra 18h33'15" 52d5'58"
Marlon Taylor
Vir 14h3'0" -13d34'35"
Marlon TeRon McGee
Sgr 19h44'58" -16d19'43"
Marlon's Star In The Starry Heavens
Umi 13h46'38" 77d23'24"
Marlou D. Davis
Uma 11h26'37" 53d31'28"
Marlou Elaine
Cas 1h42'46" 65d4'54"
Marloupa
Mon 6h30'56" 9d5'14"
Marlow Hicks
Uma 11h36'26" 64d25'19"
Marlowe Grace's Crawler Omega
Leo 11h38'47" 25d15'13"
Marlowe Whispering Paws
Pho 0h42'49" -42d29'37"
Marly
Gem 7h20'12" 14d19'3"
Marly
Ori 4h54'24" 10d7'8"
Marly Lima Freitas
Lyn 7h1'21" 59d7'17"
Marlyn
Lmi 10h47'28" 26d3'32"
MARLYN
Cyg 20h1'5" 41d3'6"
Marlyn Kilpatrick
Mon 7h44'48" -0d45'58"
Marlyn Lazo
Cyg 19h42'15" 34d20'26"
Marlyn Sepulveda
Umi 14h58'34" 81d17'45"
Marlyne Gina McLeod
Cam 5h53'8" 61d36'50"
Marlys
Uma 10h40'11" 68d59'42"
Marlys Cleon
Uma 13h21'53" 56d6'30"
Marlys Merna
Aqr 22h21'38" -7d29'31"
Marlyse
Cnv 13h32'29" 30d52'44"
MarMad6656
Lyn 6h18'37" 57d27'37"
Marmalade (Marmy) Schork
Uma 12h39'58" 56d54'20"
MarMar
Uma 11h48'58" 44d12'53"
Marmar Superstar - Marlies Staples
Sgr 19h49'39" -12d31'24"
Marmee
Crb 15h50'3" 37d41'53"
Marmie
Mus 12h19'46" -64d52'55"
Marmie Cole-Provence
Peg 22h7'30" 30d49'22"
Mar-Morr
Ori 5h55'31" 7d4'51"
Marn
Lib 15h32'49" -9d56'23"
Marnè M. Boen
Uma 11h50'6" 39d38'32"
Marnee J Green
Lib 15h59'47" -14d24'57"
Marnell's Twinkle
And 0h27'18" 38d37'45"
Marner & Ruby Smith
Uma 11h25'12" 48d28'16"
Marni Alyson
Sco 16h57'44" -40d42'31"
Marni(Mommy) & Ashley
Cnc 8h55'49" 16d3'58"
Marnie
Cas 23h28'12" 57d54'32"
Marnie Anne Terry O'Neill
Cas 23h33'34" 55d29'1"
Marnie Michelle
Gem 7h17'13" 25d18'30"
Marniela
Uma 13h2'42" 62d48'25"
Marnie-Meredith
Cnc 8h10'29" 15d26'22"
Marnie's Dream Star
Cru 12h23'48" -55d58'13"
Marni's Star
Ari 2h38'29" 18d12'59"
Marni's Star
And 23h6'6" 35d33'15"
Marny & Mia Ridling
And 1h58'58" 37d38'27"
Maron Thomas Spohn
Aur 7h30'10" 44d10'5"
MAROSPORTS
Per 4h27'24" 44d31'10"
Marót Viki
Cas 1h20'13" 68d44'25"
Maroth
Umi 10h5'18" 87d0'55"
Marotta
Cap 21h3'27" -17d39'46"
Maroun E. Khater Cute King
Aqr 22h52'34" -8d29'27"
Maroyan
Crb 15h34'54" 26d46'51"

Marphi-Tatadri
Cru 12h18'35" -56d46'49"
Marq Edward Hollingsworth
Lmi 9h49'31" 34d19'36"
Marquardt, Dieter
Uma 8h32'42" 72d45'0"
Marquee Cairns
Cmi 7h43'32" 1d19'11"
Marques Noble Gary
Umi 14h18'24" 69d34'26"
Marquesa Michael Wilson
Umi 16h4'6" 76d19'50"
Marquis LeeAnn Mart
Uma 9h13'23" 58d0'38"
Marquisa
Gem 6h49'58" 26d46'35"
Marquita
Cnc 9h4'2" 10d40'15"
Marra
Her 17h51'24" 22d39'42"
Marra Karamanoli
Sge 19h52'18" 17d10'55"
Marrah C. Peebles
Vir 14h25'3" 5d10'14"
Marraine Chouchou
Vir 14h5'1" -18d49'8"
Marri A Wallace
Ari 2h24'0" 25d13'19"
Marri Romina
Ori 5h54'37" 22d42'39"
Marriage Star - Yvette & Nick Kelly
Col 5h28'2" -35d26'51"
Marriana Catherine
Cas 0h36'43" 64d22'48"
Marrianne Rose Backous
And 1h40'34" 43d50'32"
Marrinan
Aqr 20h52'53" 1d48'19"
Marrionme
Lyr 18h47'59" 31d55'54"
Marrisa S. Bishop
Lib 15h10'11" -14d10'34"
Marrisa Zophie Contreras
And 0h48'33" 42d37'11"
"Marroquin" Family Star
Sge 20h3'22" 16d43'1"
Marry Ellen O'Donnell
Tau 3h46'49" 16d53'23"
Marry me again: Happy 10th Miriam
Cnc 8h33'14" 24d46'3"
Marry Me Erin
Aqr 22h39'28" -1d17'1"
Marry Me Hillary
Cyg 19h52'27" 39d10'15"
Marry Me *Kristen*
And 0h54'2" 37d6'53"
Marry Me Sue
Vir 13h9'35" 11d38'3"
Marry Me, Nicole
Cap 21h39'11" -20d51'10"
M.A.R.S.
Umi 16h3'51" 76d38'48"
M.A.R.S.
Uma 9h6'32" 70d5'27"
Mars
Ori 6h2'46" 9d42'33"
Mars' OB-ERN 2b
Uma 11h40'54" 52d19'7"
mars pixie
Gem 7h25'12" 31d13'13"
Marsal
Uma 9h35'38" 70d32'0"
Marsdiener-Martina
Uma 9h35'17" 51d29'11"
Marsha
Cam 3h51'34" 56d15'19"
Marsha
Ari 3h0'21" 19d43'58"
Marsha
Cas 23h11'1" 55d9'44"
Marsha and Olivia's World
Tau 4h21'46" 5d45'41"
Marsha Ann Demaske
Sco 17h16'6" -39d41'43"
Marsha Ann Smith
Aqr 21h7'29" 2d19'12"
Marsha Ann Yarnell
Peg 21h44'18" 11d14'53"
Marsha B. Elliott
Cyg 19h52'33" 33d32'58"
Marsha Bullock
And 23h3'51" 41d35'56"
Marsha C. Avery
And 2h30'42" 49d59'56"
Marsha C. Ferebee
And 1h18'59" 44d46'15"
Marsha Cohen
Vir 14h42'20" 0d50'1"
Marsha Coleman
Crb 15h53'46" 27d10'53"
Marsha D. Johnson
Ari 2h49'8" 26d18'15"
Marsha D'Arrigo
And 23h22'27" 47d16'47"
Marsha Deeana White
Psc 1h39'14" 6d46'20"
Marsha Foreman
Cas 0h48'52" 49d39'16"
Marsha Gaynor Lewis
Cam 3h22'14" 57d23'39"
Marsha Gaynor Lewis
And 23h41'35" 36d21'54"

Marsha Gentry
Crb 15h44'31" 37d27'43"
Marsha Goldman
Psc 0h40'16" 7d2'5"
Marsha Gravesen
Cyg 19h45'22" 36d43'45"
Marsha Hammer
Cyg 20h41'19" 45d55'38"
Marsha Hendrickson
Lyn 8h4'53" 46d38'10"
Marsha Hotkowski
And 1h43'10" 36d46'16"
Marsha J. Tuttle
And 0h36'35" 42d42'0"
Marsha Jean Rogers
Uma 12h54'50" 58d57'43"
Marsha Joy
Per 3h19'15" 44d42'46"
Marsha L Blowers
Lmi 9h29'38" 36d36'48"
Marsha L. Shreve
Uma 8h44'35" 62d57'30"
Marsha Lynn Nygren Star #1
Vir 12h41'21" 2d12'44"
Marsha M. Limbaugh
Lyn 8h28'16" 55d4'27"
Marsha Marsha Marsha Ainsworth
Cam 5h41'33" 57d24'2"
Marsha Michelle 05.15.82
And 1h9'18" 36d51'41"
Marsha Mother of Jack and Emma
Cnc 8h52'54" 25d32'12"
Marsha Mudd
Uma 10h59'59" 44d15'5"
Marsha (My Sweetie)
And 0h38'58" 36d16'21"
Marsha Our Bright Angel In The Sky
Vir 14h7'3" -15d53'29"
Marsha Rand
Uma 11h54'41" 57d42'36"
Marsha Ray - Tupperware
Peg 23h16'32" 32d53'30"
Marsha Reiniche
Cmi 7h29'48" 9d13'22"
Marsha Shyanne O'Dell-Montroy
Vir 13h33'59" 10d35'29"
Marsha Tansey Four
Aqr 21h42'40" 2d46'28"
Marsha Taylor
Sgr 18h4'28" -29d38'52"
Marsha Taylor Fisher
Uma 12h12'12" 52d51'37"
Marsha V. Brown
Lib 15h33'8" -15d13'50"
Marsha Wagner
Cam 5h8'47" 66d26'34"
Marsha Weikle
Lyn 7h59'42" 42d3'45"
Marsha, The Star Of My Life
Psc 0h42'57" 7d48'46"
Marsha.L. 05
Sgr 17h47'36" -23d24'40"
Marshal E. Jones
Ari 2h30'28" 21d32'23"
Marshall
Ori 5h5'50" 14d27'22"
Marshall
Cnc 8h4'19" 26d21'22"
Marshall
Cep 0h12'40" 78d22'36"
Marshall A. Kapp
Ori 6h18'50" 10d34'5"
Marshall Abram Deuel
Vir 11h45'30" 9d46'45"
Marshall and Becky in Love.
Uma 13h18'15" 62d1'52"
Marshall and Betty Bonds
Gem 7h40'3" 23d1'27"
Marshall Ascendant
Pho 2h18'36" -45d36'38"
Marshall & Bunny Field
Eri 4h41'7" -0d53'29"
Marshall C. Darrow
Cep 21h37'52" 68d4'13"
Marshall Christopher Corazza
Gem 6h58'3" 16d12'23"
Marshall Dean Wylie
Ari 2h44'24" 13d32'5"
Marshall "Donald" Tooker
Cep 21h45'45" 62d35'14"
Marshall Edward Lewis
Psc 1h33'10" 12d0'6"
Marshall Edwin Phillips
Sgr 18h0'29" -26d41'40"
Marshall G. Geisser
Ori 5h30'21" 3d11'20"
Marshall Gregory Thomas Cline
Cnc 8h26'34" 11d50'46"
Marshall J
Uma 9h17'30" 72d53'48"
Marshall James Ives
Gem 6h32'23" 14d56'50"
Marshall James Quearry III
Ori 5h38'20" 0d22'35"
Marshall Lee Cain, Jr.
Ari 2h8'21" 10d57'56"

Marshall Lucifero
Cep 20h25'29" 60d9'18"
Marshall Manning
Uma 12h41'8" 56d5'40"
Marshall Miller
Cyg 19h59'40" 33d15'56"
Marshall Montgomery
Ari 2h52'27" 28d35'30"
Marshall Morgan Clark
Aql 19h22'43" 4d21'2"
Marshall R. Kistner
Cyg 20h37'57" 34d46'17"
Marshall "Rocks"
Ori 5h25'33" 8d59'56"
Marshall & Square
Uma 12h12'45" 55d24'58"
Marshall Sylver
Uma 13h12'26" 54d4'37"
Marshall W. Diemert
Lib 15h2'56" -0d43'29"
Marshall Wilson Dawson, Jr.
Uma 9h23'59" 47d44'21"
Marshall Z. Ray
Uma 9h36'3" 44d41'14"
Marsha-Lovely Eagle When She Flies
Aql 19h11'3" 11d39'32"
Marsha's Shining Wonder
Lyn 6h47'10" 50d36'51"
Marshele Walker
Crb 15h34'25" 34d43'55"
Marsheri Elaine Woodard
Sco 16h14'15" -18d11'19"
Marshmallow with love from Dunbi
Psa 22h0'11" -28d14'10"
Marsie
Vir 11h38'55" -4d12'29"
M.A.R.S.-Millie Alva Rosenbaum Star
Dra 17h55'17" 67d21'31"
Marssoft
Lyn 8h40'23" 40d52'33"
MARSTAN
Boo 14h17'42" 40d40'0"
Marstan - Fifty/Fifty
Vir 15h7'21" 2d6'52"
Marsteller
Tau 5h57'54" 25d6'15"
Mart
Umi 16h18'6" 77d17'17"
Marta
Dra 16h43'9" 61d18'17"
Marta
Uma 9h49'48" 68d19'9"
Marta
Cap 20h19'42" -25d7'26"
Marta
Lib 15h3'56" -23d33'42"
Marta
Sct 18h54'28" -12d22'34"
Marta
Del 20h48'43" 17d47'32"
Marta
Ari 2h20'10" 26d3'37"
Marta
Ori 5h29'1" 11d23'22"
Marta
Ori 6h23'20" 13d46'55"
Marta
Crb 16h7'11" 36d55'44"
Marta A Macias
Tau 4h38'8" 30d9'23"
Marta Abad Boveda
Lyn 7h54'41" 34d25'36"
Marta & Alexandria McNaughton
And 0h28'36" 26d47'38"
Marta and Julianna
Sgr 19h28'48" -14d40'56"
Marta Carrera
Cnc 9h14'38" 30d3'29"
Marta Cruz Moscoso
Lib 14h51'55" -1d4'6"
Marta Curti
Cyg 20h10'32" 36d43'18"
Marta Diez Parrado AMM
Ori 5h23'38" 14d39'14"
Marta do Nascimento
Lib 15h43'16" -17d49'2"
Marta Estela Ramos Wells
Psc 1h20'43" 22d44'8"
Marta & Fabian
Her 17h55'5" 14d34'28"
Marta Fioravanti
Sco 16h11'21" -23d47'57"
Marta Flis & Tom Welna Love Ukotku
Cyg 21h47'5" 39d7'18"
Marta Forcolin
Cyg 21h0'31" 40d12'31"
Marta Garces Ingles
Her 16h34'16" 23d38'19"
Marta Gioia
Per 2h24'26" 52d30'58"
Marta Granat
Gem 7h22'29" 32d42'29"
Marta Isabel (Gigi) Bonilla
Sco 17h15'45" -32d36'36"
Marta Lau
Mon 6h53'41" 8d14'41"
Marta Maria Ferreira Almeida Duarte
Uma 13h3'46" 53d49'8"

MARTA MECINSKA
Psc 1h17'39" 16d25'43"
Marta & Mercedes Aguiar
Uma 10h7'25" 52d40'20"
Marta Messori
Lac 22h14'51" 49d16'46"
Marta Moskowitz
Psc 0h57'40" 9d56'50"
Marta Murillo
Cyg 19h44'27" 27d52'29"
Marta och Bengt
Lyn 7h33'32" 48d36'1"
Marta para siempre
Uma 12h4'33" 35d39'7"
Marta Peso Vilella
Cas 0h17'38" 52d14'44"
Marta Ramponi
Umi 15h12'37" 70d57'38"
Marta Santambrogio
Cas 0h54'51" 56d52'18"
Marta Schifone
And 22h59'30" 52d19'38"
Marta Talayero
Ser 17h33'30" -14d3'8"
Marta Teresa Fiol
Ari 2h19'23" 18d37'4"
Marta Wadowiec
And 23h10'32" 52d10'8"
Marta y Kike
Uma 9h53'22" 69d27'48"
Marta Yelverton
Sgr 19h41'4" -29d30'20"
Marta-Maria
Cas 23h29'30" 57d44'36"
Martandrea
Cam 9h25'36" 82d4'19"
Marta's love
Cas 0h5'7" 63d34'37"
Marta's Loving Dedication to Jared
Del 20h49'44" 18d57'11"
Marte <3 Kristian - for alltid!
Tau 4h24'38" 23d13'39"
Marte Asheim
Ori 6h5'2" 12d34'24"
Martella Valentino
Per 4h2'27" 49d8'52"
Martellaro's Family Star
Uma 12h42'4" 59d26'26"
Martes Maria Chaves
Cap 21h28'57" -16d37'55"
Martha
Vir 14h39'0" -0d52'19"
Martha
Lib 14h54'3" -1d42'31"
Martha
Psc 0h21'33" 6d18'39"
Martha
Sco 17h20'50" -37d36'4"
Martha
Leo 10h34'13" 14d7'57"
Martha
Gem 7h29'12" 26d16'50"
Martha
Gem 6h19'12" 21d27'45"
Martha A. Fricks "Semper Fi"
Cas 23h29'2" 56d10'28"
Martha A Lewis
Cap 20h43'40" -20d46'33"
Martha A. Lopez
Lib 15h31'17" -10d38'38"
Martha Aguilar
Ari 2h0'29" 14d49'0"
Martha and Frank Puletti
Ori 5h30'27" -5d16'48"
Martha and Laurence Kusek
Cnc 9h18'44" 32d19'12"
Martha Andrade " Gasolina "
Uma 11h36'4" 40d21'22"
Martha Ann
Aqr 22h34'9" 0d36'14"
Martha Ann Kirkpatrick
And 0h40'2" 35d36'5"
Martha Anne
Leo 10h42'55" 18d52'52"
Martha Anne Borowski
Leo 9h47'18" 21d35'40"
Martha Avgita
Gem 6h35'48" 13d22'20"
Martha B. Hendrix Shelton
Uma 8h34'17" 69d55'23"
Martha Baker
Uma 8h44'47" 51d3'25"
Martha Barberena
And 23h50'59" 48d5'54"
Martha Berzosa
Per 3h2'4" 46d27'7"
Martha Biagi
Cas 0h56'28" 57d18'33"
Martha Binetti
Sgr 18h31'12" -27d57'42"
Martha Bischof Ritter
And 2h29'45" 42d50'30"
Martha Bodenheimer Elder
Vir 13h56'4" -20d45'25"
Martha C. P. Christensen
Ori 5h21'36" -5d17'41"
Martha Cao
Leo 11h12'0" 17d18'16"
Martha Carol Kimmons
Mon 6h49'0" -0d12'54"

Martha Carr
Sco 17h54'28" -29d54'39"
Martha Catalina
Sgr 17h58'4" -17d14'41"
Martha Christiansen
Crb 15h44'27" 38d18'9"
Martha Claire Rhodes
And 0h37'59" 36d54'32"
Martha Colunga
Lyn 6h48'28" 51d4'14"
Martha Cora Gilbertson Rogstad
Uma 12h1'54" 33d50'26"
Martha Crystal Abigale Mariott
Vir 14h18'7" 2d42'36"
Martha D. Neal
Her 17h2'18" 29d1'47"
Martha D. Sasser
Leo 9h34'18" 14d8'49"
Martha Darlene LaPointe
And 0h52'30" 45d13'43"
Martha de Leija
Col 5h26'36" -42d20'18"
Martha & Donna Stark
Cas 23h39'48" 51d11'30"
Martha Douglas
Cam 12h27'9" 77d19'31"
Martha E. & Frank J. Speicher
Cyg 19h46'50" 49d3'53"
Martha E. Kempf (Coffman)
Tau 4h17'54" 4d17'45"
Martha Elizabeth
Aqr 20h55'30" 0d32'17"
Martha Ellen Cannon
Cas 1h9'20" 64d11'6"
Martha Evelyn Lopez
Aqr 21h47'2" -2d43'57"
Martha F. Miller
Cas 1h22'17" 64d7'48"
Martha Fearns
Ari 3h7'18" 28d23'34"
Martha Featherman
And 0h36'50" 37d53'9"
Martha Fitzgerald
Cyg 19h35'47" 29d37'19"
Martha Frias
Psc 0h25'0" -2d59'1"
Martha Gail
Cap 21h9'33" -20d54'7"
Martha Gibbons
Cas 23h32'12" 58d1'26"
Martha Gibbs Meeks
Lyr 19h19'10" 39d23'13"
Martha H Bobbys
Uma 12h33'40" 54d42'27"
Martha Haarloev
Aqr 22h55'22" -17d10'4"
Martha Hansen
Cmi 7h35'23" -0d0'36"
Martha Hattie Eiden
Dra 17h32'49" 54d3'14"
Martha Hoyle Hardee Tattersall
Leo 10h30'1" 17d49'24"
Martha I. Arvizu
Ori 5h22'31" 1d5'16"
Martha Imelda Benavides
Cma 7h20'24" -12d39'34"
Martha J. Maddox "M.J.M."
Uma 11h43'18" 35d6'18"
Martha J Reedy
Tau 4h20'6" 25d9'46"
Martha Jane Doggett
Leo 9h31'17" 10d27'36"
Martha Jane Stansbery
Cam 4h13'6" 67d25'50"
Martha Jane Wilhoite
Aqr 22h19'38" -0d29'59"
Martha Jane Wills Garza
Peg 21h52'35" 13d3'3"
Martha Jean
Lyr 18h25'23" 32d6'43"
Martha Jean Bennett
Gem 6h43'54" 15d25'43"
Martha Jean Cummings
Mon 6h38'38" -1d31'10"
Martha Jean's Super Star
Ori 6h19'36" 15d24'46"
Martha Kondroski's Wish 11.02.06
Sco 16h1'27" -28d9'24"
Martha L. Lester, Esq.
Cas 23h30'49" 56d23'3"
Martha L. Maness
And 0h48'56" 41d59'15"
Martha L. Valencia
Lyn 7h44'7" 35d21'30"
Martha L. Valencia
Tau 4h7'7" 10d6'52"
Martha Lanier Graham
Tau 5h44'30" 17d53'46"
Martha Lee Sullins
Vir 12h36'33" 4d49'48"
Martha Lily Trethaway
And 23h48'11" 36d52'43"
Martha Lisa
Lyn 7h44'55" 55d59'30"
Martha Lisa Holman
Ari 2h18'49" 26d6'12"
Martha Logan Thomas
Uma 11h39'15" 34d5'16"
Martha Lou-Bob
Peg 22h21'30" 22d30'34"

Martha Louise Karbler
Sco 17h48'59" -38d28'41"
Martha Lucia
And 0h37'47" 23d57'10"
Martha Lynne
Ori 5h48'29" 9d12'10"
Martha M. Hutchins
Cap 20h57'55" -17d18'33"
Martha Maria Alberto
Cas 0h33'58" 54d52'14"
Martha Maria Suzy Corona
And 1h38'53" 39d22'57"
Martha "Martie" Anne Carlson Rudd
Del 20h40'42" 14d12'11"
Martha Martin
Dra 18h29'22" 55d11'31"
Martha Matienzo
Umi 12h8'55" 89d16'50"
Martha McCormick Estin
Dra 15h40'38" 59d24'18"
Martha McDowell
Gem 6h44'55" 16d39'26"
Martha McSweeney
Ari 2h30'3" 11d5'26"
Martha Melhorn Wilhoite
Uma 12h52'15" 61d57'54"
Martha "Mema" Thompson
Uma 9h52'59" 66d12'48"
Martha Miller
Uma 10h35'47" 57d29'26"
Martha Minton
Cam 7h20'52" 63d51'30"
Martha Noemi
Cnc 8h19'28" 8d26'23"
Martha Nonnie Alpha
Gem 6h33'40" 21d32'34"
Martha Noreen Beloved Star Mother
Vir 13h24'50" 12d17'9"
Martha Owens
Cyg 21h26'5" 37d24'20"
Martha Polen
Sco 16h10'27" -12d50'48"
Martha Prime
Psc 1h10'5" 12d0'3"
Martha Raney Gaddy
And 1h0'27" 35d48'51"
Martha Rebecca Little
Aur 5h28'52" 45d41'8"
Martha Richardson Sisco
Psc 1h8'42" 26d42'11"
Martha Rose Johnson
And 2h38'9" 38d49'8"
Martha Ruth McDermott
And 1h1'26" 45d59'19"
Martha Sanclemente
Lib 15h55'10" -8d3'50"
Martha Silverman
Cap 20h57'5" -15d53'18"
Martha Sloan
Aql 19h57'55" -0d26'13"
Martha Snow
Sgr 17h54'50" -29d45'5"
Martha Star
Sco 17h51'13" -36d46'55"
Martha Szczerba
Cnc 9h11'41" 21d14'7"
Martha VanDusen
Uma 9h26'2" 51d26'0"
Martha W. Gowder
Uma 10h59'17" 67d18'56"
Martha & Warren Campbell Eternity Star
Cyg 21h31'9" 36d28'39"
Martha Yanez
Lyn 6h46'59" 49d57'27"
Martha Yelinek Kish
And 0h19'53" 44d1'2"
Martha & yiayia
Cru 12h1'2" -60d42'35"
Martha "You Are A Shining Star"
Uma 9h41'44" 62d46'19"
Martha Zeltner
Psc 0h36'38" 8d42'35"
Marthabelle Lencioni
Cam 6h13'47" 61d16'57"
MarthaDandridgeCustisWeaverBolton
Ari 3h18'44" 26d9'36"
Martha-Linda's Kept-My-Promise Star
Uma 13h20'39" 56d36'12"
MarthaMongeau
Lib 15h23'15" -9d10'14"
MarthaRose
Cas 0h43'49" 64d6'40"
Marthe "Alix" Wilderman
Lib 15h22'1" -6d6'10"
Marther The Menacer
Lib 15h9'21" -21d24'57"
Marthostar
Sco 16h36'43" -37d46'32"
Marti Ann Mammarelli
Vir 12h35'40" -0d59'48"
Marti Deighan
Vir 12h34'1" -1d10'18"
Marti Fitz Cook;11/12/64-11/15/2006
Sco 17h50'19" -39d0'34"
Marti Kondzielaski
Sgr 18h9'48" -16d59'47"
Marti Steindl 1
Aur 6h26'14" 41d17'41"

Marti T. Miller
Psc 1h18'51" 32d54'13"
Martica
Mon 7h9'1" -0d22'41"
Martie
Uma 11h17'6" 65d2'54"
Martie
Ori 6h4'13" 17d26'52"
Martien, Seema, Marco & Denise
Lib 15h59'8" -11d13'23"
Martijn Breed
Her 18h1'23" 18d12'15"
Martijn de Vries
Uma 9h23'33" 46d20'2"
Martika
Sgr 18h22'3" -20d33'8"
Martin
Ori 5h13'48" -8d23'30"
Martin
Dra 18h9'9" 72d13'46"
Martin
Cep 22h56'1" 57d13'18"
Martin
Lmi 10h40'15" 36d5'33"
Martin
Her 17h35'46" 32d39'2"
Martin
Uma 11h27'47" 43d25'8"
Martin
Her 18h43'57" 18d57'37"
Martin
Peg 22h37'25" 27d22'11"
Martin
Gem 7h37'48" 23d23'27"
Martin
Ori 6h22'52" 9d56'35"
Martin "70" Richman
Cep 2h1'27" 82d27'20"
Martin A. Wilke
Her 18h38'10" 23d39'5"
Martin Abrahamyan
Sco 16h37'26" -26d51'36"
Martin - Alessia
Umi 14h36'56" 84d53'37"
Martin and Joy
Cyg 21h19'22" 41d57'1"
Martin and Louise - 28 May 2005
Crb 16h6'10" 33d14'15"
Martin & Anita Wynne
Cyg 19h58'39" 35d14'23"
Martin Arnold
Uma 9h37'8" 61d13'5"
Martin Baetz
Ori 6h9'36" 8d6'59"
Martin Barbosa
Uma 10h46'32" 56d58'1"
Martin Baron
Sgr 18h9'43" -18d19'0"
Martin Barrett FDNY Ladder 79
Per 3h25'48" 48d33'50"
Martin Benteli
Ori 4h50'20" 10d27'53"
Martin Birch
Uma 13h58'52" 58d19'51"
Martin Birrane - Martin's Star
Ori 6h15'11" 7d8'27"
Martin Blank
Uma 9h42'6" 52d1'9"
Martin Bradley Smith
Aql 19h35'19" 14d36'55"
Martin & Caz Ford
Crb 16h5'36" 36d34'8"
Martin Chamberlain
Uma 9h55'0" 48d34'24"
Martin Christian Giger
Dra 18h14'50" 71d16'40"
Martin Christopher
Her 16h24'25" 22d9'17"
Martin Collas
Gem 6h20'9" 26d40'36"
Martin Craft
Per 3h8'52" 45d40'52"
Martin D. Braver
Lib 15h26'6" -19d50'37"
Martin & Daniela, 21.10.1997
Cas 1h16'7" 59d9'34"
Martin David Bookallil
Ori 5h36'42" -1d35'43"
Martin David Davidson *Marty*
Lib 15h2'0" -22d27'30"
Martin Demierre
Peg 22h21'38" 21d30'48"
Martin Derek Ley
Umi 13h15'47" 72d10'10"
Martin Disler
Ori 6h16'32" 18d57'29"
Martin Dobmann
Uma 8h15'24" 60d8'8"
Martin (Du bisch mi Stärn)
Peg 22h10'43" 18d5'0"
Martin E Bay CMP-GaMPI Shining Star
Ori 6h11'36" 13d44'9"
Martin E. R. Nuttall
Lib 15h24'35" -28d49'32"
Martin E. Ross
Boo 15h8'56" 40d7'38"
Martin Edward Pickett
Cap 21h26'0" -16d54'42"
Martin Eichner 06/28/30
Ori 6h22'14" 14d23'38"
Martin F. Mulgannon
Sco 16h17'7" -17d2'32"
Martin "Fajha" Wheeler
Gem 6h43'13" 25d13'55"
Martin Foresome
Aqr 21h55'54" -7d33'56"
Martin Fry
Lyn 6h53'39" 56d21'55"
Martin G Hansen
Lmi 10h48'22" 25d52'6"
Martin Gafner 24.03.1986
Uma 13h0'38" 55d3'26"
Martin George Duff Seward
Aur 5h53'27" 37d2'9"
Martin Gottlieb
Uma 8h29'22" 69d45'16"
Martin Gracie
Cam 6h49'10" 63d35'30"
Martin Greene Light of the Galaxy
Lib 15h27'47" -12d56'9"
Martin Grossbach
Gem 7h30'25" 34d13'1"
Martin Henry Arslan Jr.
Lib 15h22'1" -20d19'51"
Martin Hernandez
Ari 2h10'25" 20d32'4"
Martin Holst
Vir 12h10'47" 12d5'8"
Martin Howe
Cas 23h5'2" 59d30'49"
Martin Huser
Vul 20h24'17" 26d19'25"
Martin J. Blumenthal
Per 3h4'9" 40d24'37"
Martin J Salberg
Per 3h14'15" 53d6'9"
Martin J. Terry
Aqr 22h36'31" -15d25'36"
Martin Jacob Koenig
Ori 6h19'30" 14d48'31"
Martin Jacobsen - January 26, 1996
Per 4h27'9" 37d13'59"
Martin Jaehn
Tau 4h40'22" 27d56'8"
Martin James Fay
Umi 15h49'8" 85d18'37"
Martin James McGowan
Uma 11h30'42" 53d14'20"
Martin Javier Rosso
Ari 2h2'30" 11d23'34"
Martin Jean
Uma 13h22'15" 62d53'37"
Martin Jeffrey Norman
Psc 0h52'54" 5d55'3"
Martin John Bacon
Per 4h9'45" 36d50'8"
Martin John Pavitt
Cra 18h39'17" -41d24'27"
Martin Joseph Keane
Gem 6h41'29" 34d40'19"
Martin Joseph Mullee
Uma 8h16'59" 72d6'11"
Martin Jourdy
Col 6h24'14" -34d26'42"
Martin Jude Lopez
Sco 16h16'57" -14d26'5"
Martin Karkosiak
Lyn 7h21'40" 46d10'49"
Martin Kehl "my sweet and great Love forever"
Crb 15h32'26" 36d11'26"
Martin L. Huber
Ori 5h41'22" 0d18'41"
Martin Leff
Vir 13h9'29" 12d44'58"
Martin Leo
Aqr 22h25'35" -5d16'15"
Martin Liam Donnelly
Per 3h0'5" 45d43'18"
Martin Louis Espinosa
Sco 17h28'19" -38d51'49"
Martin Love
Cyg 19h37'15" 54d47'31"
Martin M. Garcia
Ari 2h20'30" 13d54'12"
Martin Maag
Ori 4h55'31" 3d59'24"
Martin Malkowski
Boo 15h31'48" 40d14'32"
Martin Manuel Martinez
Vir 14h31'27" -4d56'52"
Martin Marshal Pollak
Lmi 10h5'46" 38d26'29"
Martin "Marty" Reisch
Sgr 18h11'32" -20d15'21"
Martin (Marty) Richard Schwartz
Sgr 17h52'40" -24d42'26"
Martin Mayers
Peg 22h25'4" 6d34'24"
Martin McKeever
Lib 15h38'5" -28d7'43"
Martin McMahon
Uma 11h14'10" 53d28'2"
Martin Melrose Smith
Lmi 10h35'49" 29d32'48"
Martin Mézière
Leo 10h23'17" 12d1'25"
Martin Michael Medina
Her 18h32'4" 19d1'48"
Martin & Michelle Sutton
Crb 16h19'38" 30d2'40"
Martin "Moose" Stanovich
Her 17h25'12" 22d52'7"
Martin Morris
Cep 21h47'56" 67d42'9"
Martin N Wolding
Vir 14h31'38" -4d8'45"
Martin Obertrifter
Uma 9h30'42" 47d26'38"
Martin (Onid) Gautschi
Her 17h0'25" 13d10'21"
Martin Owen
Cnc 9h16'36" 21d13'40"
Martin P.
Cep 22h45'10" 76d20'13"
Martin P Nolan
Boo 14h48'46" 24d26'29"
Martin Palik
Sgr 18h35'26" -17d28'56"
Martin Pankau
Uma 8h46'24" 53d50'4"
Martin Paul Gunther
Uma 9h2'41" 48d8'48"
Martin Pflugi
Ari 2h8'49" 27d3'41"
Martin Phillips
Per 3h9'3" 40d25'29"
Martin Raymond Darby
Per 3h25'36" 40d52'21"
martin rays' perfect star
Tau 5h15'40" 24d11'46"
Martin Reginald Rozelle
Ori 5h57'36" 12d19'9"
Martin Reichard
Uma 9h35'17" 45d23'38"
Martin Rivera
And 23h19'18" 48d42'7"
Martin Rocha
Her 16h50'25" 33d27'40"
Martin "Rocketman" Orona
Leo 10h53'6" 19d30'22"
Martin Rumpold
Peg 22h40'25" 24d10'47"
Martin S Mazur
Aqr 21h19'50" -8d56'50"
Martin Schatz du bisch mini Melodie
Uma 13h33'28" 53d13'0"
Martin Schermond
Uma 8h43'29" 71d22'31"
Martin Schindler
Ori 5h29'2" -9d13'14"
Martin & Sharon Together Forever x
Cyg 20h12'38" 35d33'20"
Martin Simon
Uma 10h23'18" 41d58'47"
martin sonderegger löwe 06.08.1951
And 23h11'19" 41d45'51"
Martin Spornraft
Uma 11h7'50" 29d35'24"
Martin Stone
Per 4h10'31" 50d10'16"
Martin Stralendorf
Uma 12h45'41" 53d58'7"
Martin & Sue
Her 17h49'14" 48d23'7"
Martin T. Crowe, C.Ss.R.
Uma 10h15'38" 45d54'9"
Martin Th. Bartholet
Lyr 19h9'17" 45d11'9"
Martin "The Duke" Alterman
Lyn 7h53'8" 57d48'20"
Martin Thomas Reed
Cnc 9h8'40" 31d46'9"
Martin Thomas Theodore Muckerheide
Per 4h18'14" 51d44'57"
Martin Towsey
Per 4h24'24" 47d11'37"
Martin Valtwies
Uma 11h0'37" 71d56'25"
Martin & Vicky, Mauritius 2007
Cyg 20h50'50" 54d3'2"
Martin W. Oberle
Tau 3h50'51" 9d43'57"
Martin Witkoff
Umi 14h6'43" 72d35'23"
Martin Wittwer
Crb 15h16'39" 26d11'51"
Martin Wlecke
Uma 10h14'33" 43d54'34"
Martin Wolfe
Uma 12h42'16" 60d17'38"
Martin Woolford
Cep 20h53'3" 59d24'30"
Martin, 11.12.1967
Uma 11h57'35" 53d54'57"
Martin, Jennifer & Sierra Cuillerier
Per 3h35'34" 36d28'40"
MARTIN,ARI,ISAIAH,ARIA H
Uma 11h38'13" 45d15'41"
Martina
Boo 15h29'4" 43d32'18"
Martina
Cyg 20h10'42" 36d41'33"
Martina
Com 12h16'12" 24d49'36"
Martina
Peg 21h29'35" 23d32'51"
Martina
Psc 23h39'59" 7d10'3"
Martina
Mon 6h31'56" 6d32'50"
Martina
Uma 9h35'46" 53d56'52"
Martina
Cas 23h13'19" 56d2'23"
Martina
Cam 5h33'52" 71d29'51"
Martina
Umi 17h24'28" 79d30'28"
Martina
Sco 16h59'29" -39d0'27"
Martina - Aaron
Cas 1h2'13" 57d13'42"
Martina Ashford
Uma 11h28'21" 61d14'9"
MARTINA BERNASCONI
Aur 5h21'56" 30d4'42"
Martina Brown
Vir 13h3'28" -7d57'20"
Martina Cappelletti
Per 1h43'0" 51d15'2"
Martina Cecilia
Lyr 18h24'17" 30d19'31"
Martina Elizabeth
Ari 3h12'57" 24d3'0"
Martina et Thierry Scherz
Uma 14h16'44" 57d11'41"
Martina Fey
Uma 10h29'22" 47d54'3"
Martina Hofbauer - ILY
Cas 0h10'16" 54d53'28"
Martina Isler
Vul 21h8'16" 27d27'50"
Martina Jungmann
Ori 6h1'24" 21d29'27"
Martina Kubis
Ori 5h57'34" 21d54'29"
Martina Lucantoni
Umi 15h54'35" 76d30'42"
Martina & Marco Friends For Ever
Uma 9h36'40" 48d37'59"
Martina Maria Thompson
Cas 0h29'27" 75d24'8"
Martina Marie Plafcan
Leo 11h47'25" 26d0'9"
Martina Milada
Leo 11h15'5" 23d54'26"
Martina Miller
And 23h17'9" 52d21'34"
Martina Morgen
Uma 11h4'56" 42d22'47"
Martina Olsen *Irish Princess 143*
Lib 15h49'9" -11d49'31"
Martina Reville
And 23h43'2" 48d31'47"
Martina Stebler
Uma 12h6'18" 47d18'20"
martina steiner
Cam 7h46'0" 62d30'0"
Martina stella nerazzurra
Her 15h59'1" 44d43'1"
Martina Sterling
Sco 17h56'13" -41d30'52"
Martina Sweets
Vir 12h14'24" -4d59'30"
Martina Tabozzi
Cam 5h15'7" 62d12'35"
Martina—> the best friend
Mon 7h56'27" -4d4'44"
Martina und Bernd
Uma 10h22'1" 63d7'1"
Martina und Thomas
Her 18h19'51" 16d8'14"
Martina-maus!
Ori 6h12'5" 16d35'40"
MartinaMayerova
Lib 15h27'20" -10d34'3"
Martina's Glow
And 0h9'53" 33d27'40"
Martina's Smile
Cnc 8h5'41" 20d35'21"
Martine 2006
Cyg 20h12'41" 31d49'41"
Martine et Moza
Cap 21h44'22" -9d31'42"
Martine Leray
Ori 6h24'36" 14d9'39"
Martine Sabban-Obadia
Umi 13h26'0" 88d15'0"
Martine' The Pirate Star
Leo 10h32'43" 13d57'10"
Martine Véjux Rolland Hamitou
Lib 15h49'38" -9d45'55"
Martine-Hélène Wicker
Cas 0h11'4" 57d23'58"
Martinenza
Uma 11h27'31" 69d27'14"
Martiney
Dra 17h47'33" 53d47'6"
Martinez 25
Uma 10h28'58" 65d2'41"
Martinha
Gem 7h22'0" 24d24'6"
Martinha de Bahia
Peg 22h22'57" 3d39'5"
Martini & Rossi II
Vir 12h32'26" -1d37'53"
Martini's Miracle
Lib 14h53'36" -2d41'22"
Martinita Kubesova
Sco 17h55'4" -31d43'36"
Martinka
Psc 1h3'0" 11d32'43"
Martinnia
Tau 3h56'19" 8d14'42"
Martino
Ori 5h9'24" 15d20'33"
Martino Canalini
Uma 10h55'7" 49d58'10"
Martinous 14 Mai 1983
Tau 3h26'41" 19d21'23"
Martinovic, Katarina
Uma 9h54'28" 54d54'18"
Martins
Per 3h35'31" 45d12'54"
Martins + Carolines "Lovestar"
Cam 7h50'17" 63d2'28"
Martins Heart
Crb 15h30'26" 26d25'22"
Martin's Jubilee
Lyr 18h43'42" 40d3'0"
Martin's Ra
Vir 13h29'29" -8d11'21"
Martin's Shining Mark in the Sky
Uma 11h26'57" 37d30'15"
Martinstern - Andrea
Uma 8h19'18" 64d16'30"
Martinus
Uma 11h35'28" 39d36'47"
Martinus McEachern
Cep 22h28'39" 57d31'27"
MartinWilliamBatesMyMarty 459
Gem 7h4'5" 17d27'51"
Martire Luca
Lmi 10h15'52" 37d33'53"
Marti-Roger
Uma 13h27'28" 57d45'53"
Martis
Tri 1h51'34" 34d43'21"
Marti's Shining Star
Ari 2h4'16" 20d1'47"
MartMart100
Cap 21h27'25" -18d56'11"
Marto's Rock And Roll Star
Cru 12h39'12" -57d11'45"
Martuni
Vir 14h37'24" -3d7'41"
Martunia
Del 20h41'5" 13d9'34"
Martusia
Uma 9h45'49" 65d20'17"
Martuska
And 1h14'15" 38d32'8"
Marty
Her 17h33'14" 37d39'53"
Marty
Ari 2h35'52" 22d11'23"
Marty
Boo 14h30'15" 14d25'58"
Marty
Vir 15h6'31" 3d52'23"
Marty
Cnc 8h25'14" 28d11'1"
MARTY
Dra 16h0'39" 62d8'41"
Marty
Umi 14h3'0" 74d11'43"
Marty
Umi 14h24'42" 68d57'23"
Marty
Uma 11h33'57" 58d6'7"
Marty
Ser 16h18'7" -0d57'59"
Marty
Aql 18h58'53" -0d1'25"
Marty
Umi 15h57'47" 85d44'30"
Marty
Sco 17h33'53" -32d47'59"
Marty
Sgr 18h22'0" -32d51'39"
Marty Abrams
Ori 5h16'3" 7d5'31"
Marty Ackerman
Uma 10h26'23" 61d44'13"
Marty and John Feagin Jr.
Cyg 21h33'30" 47d8'13"
Marty and Peg
Cap 21h43'18" -9d11'21"
Marty Anderson
Cap 20h38'46" -19d25'52"
Marty Andrew Black
Uma 10h49'12" 66d6'44"
Marty Arndt
Sgr 18h39'12" -25d1'13"
Marty Arno
Psc 0h53'53" 7d27'5"
Marty Bayer
Uma 14h5'13" 57d6'9"
Marty Belanger
Per 3h35'13" 45d39'50"
Marty Bevilacqua
Per 2h46'34" 53d28'20"
Marty Bishop
Uma 10h27'23" 58d13'53"
Marty Britt
Uma 11h29'45" 54d10'14"
Marty Buchan
Ori 5h54'50" -0d18'21"
Marty C
Cap 20h40'2" -20d49'50"
Marty Davis
Uma 9h3'42" 55d36'43"
Marty Davis
Gem 7h0'59" 18d11'27"
Marty Domovich
Her 17h35'38" 30d48'20"
Marty Drummerman
Dra 16h30'25" 60d4'32"
Marty Edelston's Bottom Line Star
Dra 16h29'5" 52d56'44"
Marty Fraga 2006
Pic 5h37'38" -51d8'56"
Marty Garrett
Per 3h49'19" 43d53'47"
Marty Goldberg
Lmi 10h25'37" 33d48'44"
Marty is 40
Her 16h15'44" 6d3'51"
Marty Larson
Psc 23h20'38" 6d36'24"
Marty Loves Sandy
And 23h20'0" 51d33'50"
Marty Marion McKinnie
Lib 15h14'47" -8d33'30"
Marty: Mom, Gma, Friend
Aqr 22h17'25" 1d30'9"
Marty Murphy
Aqr 22h33'42" 1d23'12"
Marty O'Connell
Gem 6h47'50" 26d47'59"
Marty Papai
Leo 9h32'29" 24d55'50"
Marty Peters
Uma 9h37'33" 56d27'59"
Marty Salisbury
Boo 14h27'11" 25d34'54"
Marty Schofield
Tau 3h54'12" 16d1'47"
Marty Sheehan
Cep 21h40'39" 70d38'54"
Marty Susan
Cap 21h47'12" -11d38'7"
Marty The Light Of My Life
Psc 23h57'47" -4d34'3"
Marty Weiner
Tau 4h41'47" 1d23'31"
Marty Weisman
Her 18h50'3" 25d25'56"
Marty Whisel
Her 16h27'59" 23d43'52"
Marty Wolf
Vir 13h49'6" 6d1'58"
Marty Worrall -* Shooting Star*-
Aqr 22h29'55" -7d13'33"
Marty, Aidan and Owen Nix
Aqr 21h52'12" 1d10'40"
Marty-Angie
Cyg 19h53'52" 42d26'36"
Martyn and Maggie Warner
Cyg 19h35'33" 29d31'44"
Martyn Cole
Aur 7h16'35" 41d55'22"
Martyn & Michelle
Cyg 20h39'39" 46d1'47"
Martyn Owen
Cep 21h51'37" 60d18'37"
Martyn Thomas
Per 3h17'28" 42d45'32"
Martyna Halsey
Del 20h33'41" 14d11'12"
Marty's Favorite
Vir 13h41'26" -15d26'59"
Marty's Light
Sco 17h35'34" -32d34'23"
Marty's Star
Uma 11h32'18" 46d39'36"
MartysGreg
Lib 15h41'36" -6d24'52"
Maru
Ari 2h55'21" 30d36'24"
Maru Barguera
Cnc 8h35'29" 15d8'57"
Maruhi 3804
Sgr 18h43'16" -19d39'45"
Marukosu Always
Ari 3h7'3" 14d33'19"
Marv My Love
Cep 22h32'17" 77d4'6"
Marv Our Star
Lac 22h10'52" 49d28'46"
Marva J Lewis
Gem 6h25'6" 17d34'25"
Marva Scott-Starks
And 0h26'4" 37d3'9"
Marvasco
Aqr 22h16'45" -8d11'54"
MarvEdna
Vir 12h36'57" -8d17'18"
Marvel
Lib 15h9'42" -1d0'6"
Marvel
Vir 12h19'47" 4d54'17"
Marvella Villegas
Cap 20h12'29" -15d7'21"
Marvellous Mary Anne - 30/07/1957
Leo 9h56'19" 14d7'49"
Marvelous
Uma 10h5'35" 53d38'30"
Marvelous Mabel - Guys Other Half
Cnc 8h46'26" 7d17'54"
Marvelous Mom Lisa
Sgr 18h16'30" -20d45'26"
Marvin
Vir 13h24'5" -12d31'4"
Marvin
Per 2h24'3" 54d18'45"
Marvin and Amy Klavans
Leo 11h40'48" 19d31'15"
Marvin and Marguerite Honeycutt
Uma 11h38'17" 47d36'32"
Marvin and Ruth's 59th
Lmi 10h7'49" 37d9'8"
Marvin Bernard Jones
Ori 5h1'20" 5d31'59"
Marvin & Billie Rogers
And 2h5'32" 46d44'47"
Marvin Blaine Owensby
Psc 1h29'22" 17d45'33"
Marvin Blumberg
Lib 15h14'27" -20d20'34"
Marvin Bruce Meijerink
Cnv 13h24'20" 31d30'3"
Marvin & Donna Edwards
Per 3h20'10" 42d47'15"
Marvin F. Clarke
Boo 14h43'33" 15d56'53"
Marvin Fleisher
Ori 5h57'0" 21d8'48"
Marvin Flicker
Aqr 23h4'40" -10d29'58"
Marvin Ginsberg's 75th. Birthday
Vir 14h49'12" 0d59'25"
Marvin Gloven
Ori 4h59'11" 15d6'27"
Marvin I. Enderlin
Uma 10h32'47" 68d58'31"
Marvin J. Cohen
Dra 17h25'11" 69d55'41"
Marvin K. & M. Denise Walls
Tau 4h18'46" 12d54'12"
Marvin & Kendra
Lyr 19h11'28" 26d23'31"
Marvin L. Gelsinger Jr.
Aur 5h27'29" 40d57'22"
Marvin Leon Sokolow
Leo 9h42'26" 7d22'48"
Marvin Lichtman - Best Dad
And 0h50'27" 38d28'18"
Marvin M. Henkins
Sgr 19h13'42" -16d17'59"
Marvin M. Whatley
Cas 23h21'39" 55d2'7"
Marvin Masely
Vir 11h49'5" 3d12'20"
Marvin McLaughlin
Cas 1h38'1" 60d18'5"
Marvin Murray Sims
Uma 8h12'27" 69d32'50"
Marvin Myers
Gem 7h0'49" 26d44'49"
Marvin Pierson
Cyg 21h1'50" 43d16'37"
Marvin Rappoport
Aql 19h26'8" 7d59'3"
Marvin Ray Persall
Lyn 7h34'5" 47d50'23"
Marvin Shields, Jr. (MSJR)
Vir 13h59'21" 1d46'9"
Marvin Stuart Village
Her 18h29'10" 23d56'33"
Marvin & Susie Forever
Vir 13h22'10" 6d27'32"
Marvin "The Marshin" Tornaden
And 23h21'12" 49d16'31"
Marvin W. Ferguson
Ori 5h29'13" 3d17'21"
Marvin Wittwer
Cas 1h18'9" 51d19'49"
MarvSel-estial
Crb 15h43'58" 31d19'25"
Marwa
Eri 3h46'28" -11d51'14"
Marwan Angelface Pharaon
Uma 10h5'48" 45d19'59"
Marwan Kishek
Cap 21h53'51" -14d10'24"
Marwood
Uma 10h11'40" 60d15'6"
Marwyn Waddell
Cep 0h29'27" 83d32'38"
Mary
Umi 21h42'34" 89d13'47"
Mary
Lib 16h0'27" -6d22'48"
Mary
Sgr 18h14'54" -16d41'50"
Mary
Uma 10h5'55" 67d24'9"
Mary
Cas 0h44'7" 64d25'45"
Mary
Cas 23h1'33" 58d31'7"
Mary
Cas 23h59'59" 56d16'27"
Mary
Uma 10h5'17" 54d54'12"

Mary
Lyn 7h37'32" 54d9'46"
Mary
Sco 17h25'2" -32d53'26"
Mary
Peg 23h27'12" 12d21'7"
Mary
Uma 10h33'52" 50d2'25"
Mary
Cam 4h27'20" 56d58'20"
Mary
Crb 15h47'36" 32d19'40"
Mary
Her 17h9'56" 31d28'55"
Mary
And 1h14'28" 37d9'20"
Mary
And 2h10'19" 38d35'3"
Mary
Cyg 21h33'16" 36d28'50"
Mary
And 0h21'39" 38d11'40"
Mary
Tau 4h11'46" 3d32'1"
Mary
Vir 13h16'48" 3d21'27"
Mary
Aql 19h22'50" 6d37'35"
Mary
Aql 19h45'31" 1d19'24"
Mary
Psc 0h57'13" 10d25'38"
Mary
Leo 10h17'19" 12d54'21"
Mary
Lyr 19h15'43" 28d22'42"
Mary
And 0h41'32" 26d32'43"
Mary 101
Cyg 21h57'51" 47d13'43"
Mary #2
Uma 8h27'6" 72d6'26"
Mary 51185
Com 12h45'44" 30d15'43"
Mary 53 from 1953
Cap 21h28'10" -10d8'4"
Mary 70
And 23h12'29" 48d15'11"
Mary A. Bidey Berry
Lyr 18h16'46" 34d14'41"
Mary A. Burrows
Mon 7h35'55" -0d54'55"
Mary A. Crawford
Lib 14h53'4" -9d23'5"
Mary A. Haegert 5/23/45 - 2/2/02
Uma 10h41'20" 40d55'29"
Mary A. Harlan
Cnc 8h38'44" 24d11'13"
Mary A. Leiker
Psc 1h21'1" 18d2'27"
Mary A. Pacifico
Cas 1h3'22" 56d31'58"
Mary A. Patrick
Aql 19h1'57" 16d54'54"
Mary A. Sherman
Crb 16h5'36" 39d32'31"
Mary A. Zappia
Sco 16h16'44" -15d8'45"
Mary A. Ziarno
And 1h18'24" 43d4'43"
Mary Agnes Coughlin Yates
Cma 7h19'36" -31d5'47"
Mary Agnes O'Donnell
Lib 15h28'13" -4d44'15"
Mary Ahlberg - The Day Camp Diva!
And 23h20'48" 43d18'42"
Mary Alessandra Noone
And 1h40'4" 50d29'26"
Mary Alexis Shope
Aqr 22h29'3" -7d0'33"
Mary Alice Bond
Umi 15h55'35" 84d59'46"
Mary Alice Delaphine
Tau 4h6'37" 6d38'34"
Mary Alice Fletcher
Psc 1h34'58" 23d21'38"
Mary Alice Hubbard Price Werner
Vir 13h2'32" -21d42'43"
Mary Alice O'Cain
Aqr 23h24'58" -20d4'27"
Mary Alice Wiles
Mon 7h8'36" -0d26'0"
Mary Alicia Ferguson
And 1h39'24" 41d47'23"
Mary & Allen
Sgr 19h16'41" -18d29'7"
Mary Allison Clarke
Aqr 21h45'22" -1d26'42"
Mary Alma Elizabeth Recendez
And 1h39'17" 41d11'2"
Mary Alter
Tau 3h57'29" 6d29'41"
Mary Alyssa Murray
Leo 10h38'7" 8d26'24"
Mary Amanda
Cas 2h27'28" 67d55'17"
Mary and Clive Watson
Umi 16h53'52" 79d50'25"
Mary and Dana
Uma 9h17'6" 56d57'15"
Mary and Daniel Roche
Per 2h17'17" 51d14'20"
Mary and Dick Plude
Cae 4h20'14" -47d52'30"
Mary and Doug Bowman
Uma 11h4'1" 70d59'25"
Mary and Edward
Cyg 21h10'47" 45d46'37"
Mary and E.T.
Aqr 21h50'44" -0d14'10"
Mary and Hannah
Uma 8h57'14" 54d17'46"
Mary and Jerry's
Aqr 23h42'26" -16d18'34"
Mary and Joshua
Umi 11h29'54" 89d29'41"
Mary and Richard Heath
Cyg 21h48'29" 52d28'27"
Mary and Robert
Cas 1h27'22" 57d8'48"
Mary and Steve Matthews
Cyg 21h49'19" 39d7'36"
Mary and Tom O'Dwyer
Uma 10h13'1" 58d30'33"
Mary and Tony's Rocco
Cap 20h31'11" -12d10'53"
Mary and Vic Williams
Cyg 19h43'54" 55d26'8"
Mary Anders and Patrick Keating
Uma 9h3'29" 68d8'41"
Mary Andrews
Cas 0h2'18" 62d33'6"
Mary Andrews
Aqr 22h28'12" -23d42'29"
Mary Andrews DeMeo
Ori 6h16'4" 14d7'41"
Mary Angel Augustine
Ari 3h5'54" 29d22'0"
Mary - Angel watching over her son
Pup 8h2'42" -35d11'45"
MARY ANGELA TTF
And 0h3'8" 45d2'21"
Mary Anita Graham
Ara 17h36'46" -46d1'51"
Mary Ann
Sgr 20h27'4" -42d27'52"
Mary Ann
Uma 13h25'39" 53d29'44"
Mary Ann
Uma 8h48'9" 69d50'39"
Mary Ann
Vir 12h22'19" -5d37'55"
Mary Ann
Peg 21h40'1" 26d56'5"
Mary Ann
Leo 11h18'41" 18d4'57"
Mary Ann and Matthew Oszurek
Cyg 20h42'25" 37d16'13"
~ Mary Ann Anderson ~
Com 12h29'7" 22d14'48"
Mary Ann Bartosch
Per 4h33'14" 40d35'10"
Mary Ann Beeres
Vir 12h51'55" 13d1'49"
Mary Ann Berry Palumbo
Uma 11h25'49" 47d10'12"
Mary Ann Biagetti
Cap 21h25'14" -26d13'57"
Mary Ann Biscieglia
Crb 15h52'33" 37d15'53"
Mary Ann Bkrtstx
Cnc 8h19'20" 29d25'39"
Mary Ann Blaine
Vir 15h5'4" 0d31'4"
Mary Ann Bowen
Psc 0h29'37" 5d5'59"
Mary Ann Brown
And 23h25'16" 51d26'5"
Mary Ann Buehlman
Crb 15h37'0" 28d58'37"
Mary Ann (Bunny) Saxman
Lep 5h30'40" -12d44'4"
Mary Ann Byrd Bobbitt
Lyn 9h9'9" 34d23'35"
Mary Ann Canino
Uma 12h58'5" 52d22'50"
Mary Ann Corts
Aqr 23h54'16" -10d23'57"
Mary Ann Crista Nagorr
Her 17h14'53" 29d28'54"
Mary Ann D'Ambrosio
Aqr 20h39'25" -11d16'31"
Mary Ann David
Sgr 19h52'35" -28d25'5"
Mary Ann Elekes
Leo 11h7'32" 3d27'43"
Mary Ann Farraye
Psc 1h13'44" 4d1'7"
Mary Ann Garver
Lyr 19h9'47" 39d29'26"
Mary Ann Gilson
Leo 9h40'42" 28d46'44"
Mary Ann Greuel, Pres, Cuore Arte
Cas 23h33'4" 58d27'45"
Mary Ann Hagstrom
Cas 1h10'34" 66d1'0"
Mary Ann Hallal
Lyr 18h48'8" 41d31'7"
Mary Ann Hannum
Cas 1h21'54" 66d46'6"
Mary Ann Howard
Sgr 18h31'25" -23d9'37"
Mary Ann Ingria
Cas 0h36'37" 61d39'36"
Mary Ann Johnson Burghgrave
Cap 21h1'16" -17d40'30"
Mary Ann Jones Francois' 60th B'Day
And 0h15'17" 38d15'28"
Mary Ann Jozwiak
Sgr 18h36'23" -23d43'35"
Mary Ann Knee
Lib 14h58'4" -9d48'41"
Mary Ann Lambert
Cnc 8h41'10" 16d5'16"
Mary Ann Langguth
Cyg 21h57'32" 53d40'4"
Mary Ann Lubchuk
Sgr 19h46'36" -14d38'17"
Mary Ann MacKenzie Farrell
Umi 13h28'23" 74d27'12"
Mary Ann Maloney Place "10/17/1919"
Mon 7h17'48" -0d34'9"
Mary Ann "Max" Bernardi
Vir 15h4'40" 5d35'1"
Mary Ann McBride Harrington
Vir 13h22'43" 11d19'36"
Mary Ann Meils
Sco 17h52'29" -36d19'27"
Mary Ann Michaels-Youngblood
Lib 16h0'28" -10d9'22"
Mary Ann Moore
And 0h26'40" 37d38'20"
Mary Ann Nevera Rice
Cas 2h35'5" 65d53'31"
Mary Ann Pelc DeSantis
Per 3h0'59" 46d23'27"
Mary Ann Pelster Henderson
Aqr 22h35'37" -0d17'6"
Mary Ann Pieshala
Uma 11h26'7" 42d52'33"
Mary Ann Prejza
Ori 5h37'23" -1d20'8"
Mary Ann Roma
Cyg 19h38'52" 51d45'51"
Mary Ann "Roo" Voorhees
Vir 12h43'0" 10d36'43"
Mary Ann Schaffer
Cam 4h25'50" 62d54'1"
Mary Ann Shenep
Cam 4h8'25" 58d6'37"
Mary Ann Signore
Cnc 8h24'51" 8d34'55"
Mary Ann Starkey
And 1h14'45" 38d51'9"
Mary Ann Stiles
Sco 16h9'21" -28d30'20"
Mary Ann Sweas
And 1h43'41" 49d22'44"
Mary Ann Thompson
Cnc 9h2'23" 13d27'6"
Mary Ann Trione Knorr
Aqr 22h27'52" 1d13'14"
Mary Ann Tucker
Uma 11h16'45" 52d50'45"
Mary Ann Vasquez
Leo 11h43'49" 10d26'50"
Mary Ann Walters
Crb 15h49'35" 38d30'30"
Mary Ann Yochim
Uma 13h56'20" 53d16'44"
Mary Ann, Eternally Heavens' Best
Cas 1h26'32" 52d22'27"
Mary Anna Hawkins
And 0h14'27" 39d16'10"
Mary Anna King
Cas 1h34'16" 62d6'11"
Mary Anna Revay
Cyg 21h44'12" 44d22'15"
Mary Anna's Guiding Star of Wisdom
Sgr 18h4'47" -32d36'3"
Mary Anne
Crb 15h27'49" 32d16'50"
Mary Anne Allyn
Vir 12h58'29" -13d8'26"
Mary Anne Bradley
Sco 16h9'19" -13d47'21"
Mary Anne Burns-Duffy
Vir 13h8'29" -1d3'26"
Mary Anne Cerneka Davis
And 1h20'39" 45d55'36"
Mary Anne Collard
And 1h36'47" 49d59'20"
Mary Anne Damiano
Sco 17h25'6" -42d41'11"
Mary Anne Fisher
Lyn 8h34'4" 38d55'49"
Mary Anne Frances Hunter
Cam 7h57'33" 69d50'5"
Mary Anne & Haley
Uma 13h31'29" 57d8'23"
Mary Anne Hart
Cnc 9h8'37" 30d50'20"
Mary Anne Loftus
Cap 20h31'26" -18d57'15"
Mary Anne Mazanec
Gem 7h2'26" 18d10'57"
Mary Anne Rawson
Psc 1h22'4" 31d12'30"
Mary Anne Reiman
Uma 13h20'3" 60d31'43"
Mary Anne Wentink
Uma 9h17'55" 53d57'17"
Mary Anne Wolling
Uma 10h53'19" 44d6'23"
Mary Anne's Kiss
Sco 15h53'15" -23d54'3"
Mary Anterhaus
Crb 16h11'59" 34d12'20"
Mary & Anthony Stokes
Cyg 20h5'9" 52d10'49"
Mary Antonia Dusek
Cas 23h22'56" 57d42'54"
Mary Arnethia Elliott 3.15.1973
Psc 0h59'58" 9d29'18"
Mary Ashbaugh
Cas 0h55'37" 63d34'13"
Mary Atwell
Uma 10h26'19" 71d57'53"
Mary Ava
Psc 1h22'58" 17d1'2"
mary avis
Vir 12h34'57" 7d27'27"
Mary Avis Pacior Olon
Lib 15h14'12" -27d0'33"
Mary B
Sgr 18h40'51" -28d46'22"
Mary B. Deeb
Cap 21h47'54" -14d43'0"
Mary B. Ellis
Uma 11h8'43" 33d6'55"
Mary B. Krisch
Cas 0h39'14" 64d10'23"
Mary B. Poole
Pho 0h31'1" -39d58'6"
Mary B Somerville
Crb 16h9'26" 38d48'8"
Mary B. Zuro
Cyg 20h48'52" 34d49'25"
Mary Bachelder Sproul
Cyg 20h1'48" 33d13'12"
Mary Baldridge
Uma 14h25'20" 61d53'51"
Mary Balogh "UG"
Ari 3h13'29" 27d46'53"
Mary Barager Hughes
Lyn 8h26'57" 41d46'52"
Mary Barbara Bitter
And 1h21'23" 40d45'10"
Mary Barbara Chilcott
Cyg 19h44'0" 28d48'23"
Mary Barrett
Lyn 7h32'15" 39d20'38"
Mary Barry
Cas 1h37'4" 63d5'7"
Mary Barsic
Dra 18h40'39" 51d6'7"
Mary Baxley Shuler
And 0h39'14" 41d10'27"
Mary Bebus Tortorici
Cap 21h44'27" -19d21'45"
Mary Beisser
Cyg 19h39'25" 34d24'5"
Mary Belinda Beard
Peg 23h53'11" 11d52'14"
Mary Bella Davis Custardo
Ari 2h38'42" 12d36'42"
Mary Belle Dixon 1926
Gem 7h34'26" 33d54'57"
Mary Bennett Gilliam
Peg 22h46'27" 15d38'34"
Mary Bergamini
Lyr 19h18'43" 32d24'6"
Mary Bernice
Lmi 10h6'14" 30d11'1"
Mary Beth
And 0h52'8" 40d8'20"
Mary Beth
And 2h35'24" 50d18'6"
Mary Beth
Leo 11h54'2" 21d50'55"
Mary Beth
Mon 6h42'43" -0d22'56"
Mary Beth
Cam 4h26'51" 66d57'38"
Mary Beth
Umi 14h11'32" 68d5'12"
Mary Beth and Glen Gratz
Lyn 7h33'55" 48d56'13"
Mary Beth Biggs
Tau 4h45'29" 22d55'31"
Mary Beth Brennan
Lyn 8h6'21" 51d8'13"
Mary Beth Burt
And 1h9'54" 46d50'21"
Mary Beth Cates
Vir 13h1'1" 10d50'40"
Mary Beth Coleman
Lyn 6h42'4" 49d58'59"
Mary Beth Conti 11/20/67
Sco 17h58'44" -42d37'38"
Mary Beth Devault
Ori 5h38'25" -0d29'13"
Mary Beth Dutton
Per 4h36'34" 39d9'45"
Mary Beth Garber
Aqr 23h56'21" -11d4'10"
Mary Beth Grater
Leo 10h47'29" 19d14'30"
Mary Beth Hewitt
Leo 10h20'45" 6d23'57"
Mary Beth Honey Baby Doll Brinkman
Cnc 8h0'31" 14d41'11"
Mary Beth Jackson 12/10/88
Uma 10h48'0" 68d51'14"
Mary Beth Kipila
Lyn 7h6'22" 47d8'58"
Mary Beth Malcolm
And 1h12'47" 38d15'45"
Mary Beth Massi
Psc 23h46'44" 6d15'17"
Mary Beth O'Dea
Her 16h45'6" 43d58'15"
Mary Beth Sattler
Cap 20h40'34" -22d20'54"
Mary Beth Skelley
And 23h34'39" 36d33'2"
Mary Beth Strong
Leo 10h17'48" 25d2'2"
Mary Beth Trotter
Aqr 20h43'59" -0d43'33"
Mary Beth Walsh
Psc 23h58'27" -4d40'44"
Mary Beth Wilson
Sgr 19h35'9" -16d53'17"
Mary Beth, will you marry Rob?
Psc 1h34'14" 11d10'21"
Mary Beth's Heart
Cas 23h9'59" 58d18'43"
Mary Beth's Rose
Uma 11h51'2" 57d30'22"
Mary Beth's Star
And 0h15'58" 31d23'26"
Mary BIGBIRD Baker
Lib 15h19'54" -5d9'31"
Mary & Bill McGarrigle
Cyg 20h21'37" 55d18'22"
Mary Billingsley
Crb 15h50'30" 36d16'48"
Mary Billot Cobb
Lyn 8h38'58" 44d12'29"
Mary & Bill's Angel Star
Uma 9h59'27" 58d13'58"
Mary Bissett
Vir 12h1'37" 8d2'34"
Mary Bladow
Ori 5h24'29" 7d24'27"
Mary Blumberg
Cas 0h30'55" 50d4'13"
Mary Bob Welborn
Aqr 22h23'46" -9d44'39"
Mary & Bob Witowski
Uma 8h51'41" 54d38'40"
Mary Bolling
Vir 12h27'23" -10d32'22"
Mary Bonita Bednar
And 23h6'25" 49d15'16"
Mary Bordeaux
Uma 10h19'58" 60d0'59"
Mary Bottone
Uma 10h12'56" 42d4'26"
Mary Branham
And 0h56'20" 39d37'58"
Mary Brechin Cumming
And 1h31'25" 42d19'48"
Mary & Brendan McBrearty
Cyg 19h48'42" 55d51'8"
Mary Bridget 8880
And 1h18'12" 36d32'50"
Mary Britt
Aql 19h47'4" -7d15'43"
Mary Bryan Kemmerer
And 23h33'41" 36d42'7"
Mary Buonfiglio
Cas 1h27'46" 51d44'33"
Mary Buonfiglio
Cas 0h32'23" 60d8'40"
Mary Burdue
Aqr 21h4'4" -10d15'30"
Mary Burns of Collierville, TN
Cam 13h10'43" 77d38'49"
Mary Busam
Cas 23h22'52" 55d19'54"
Mary "C"
Sgr 18h12'55" -30d36'48"
Mary C. Barber
Gem 7h27'26" 33d48'41"
Mary C. Bence
Cap 20h13'43" -16d41'4"
Mary C. George
Cap 21h57'25" -18d51'12"
Mary C. Hendershot
Cam 4h50'24" 55d41'26"
Mary C. Maglione
Uma 8h23'36" 64d34'44"
Mary C. Myers
Lyr 18h25'9" 35d33'9"
Mary C. Nordberg
Uma 9h44'45" 65d50'38"
Mary C. Ross
And 23h0'6" 42d32'36"
Mary C. Thorsten
Cap 21h4'27" -16d46'51"
Mary Candusio
Uma 10h42'40" 71d42'15"
MARY CANNON
Cas 0h8'21" 56d17'32"
Mary Cannon Smith
Crb 15h51'18" 28d45'28"
Mary Carmen's Star
Aqr 23h47'9" -18d14'47"
Mary Carol
Cas 1h13'16" 55d38'54"
Mary Carol
Crb 16h21'23" 34d48'13"
Mary Caroline
Lyr 19h4'49" 44d52'18"
Mary Caroline
And 0h25'30" 32d31'48"
Mary Caroline Litchman
Uma 12h0'55" 60d54'35"
Mary Carolyn Johnson Doak
Lib 15h53'41" -10d7'1"
Mary Carroll Groff
Umi 15h34'28" 74d48'47"
Mary Carroll the Love of my Life
Ori 6h17'43" 14d23'59"
Mary Carter Bennett
Sco 16h23'38" -40d56'19"
Mary Casella
Uma 9h55'55" 59d52'12"
Mary Casey
Cas 0h41'36" 52d2'46"
Mary Casper Richey
Cap 21h18'43" -16d24'1"
Mary Castrogiovanni
Cas 0h22'40" 57d39'30"
Mary Catherine
Cas 0h23'28" 51d28'1"
Mary Catherine
Ori 6h22'55" 11d14'39"
Mary Catherine
Cas 22h59'31" 53d27'8"
Mary Catherine Amerine Our Jewel
And 1h37'1" 49d17'14"
Mary Catherine Bowman
Her 17h19'2" 45d47'3"
Mary Catherine Burtis
Vir 12h8'54" 12d9'17"
Mary Catherine Cohn
Mon 6h48'18" -0d19'57"
Mary Catherine Crist
Ori 6h3'32" -0d54'49"
Mary Catherine Fassler
Tri 2h5'7" 32d6'10"
Mary Catherine Faygo
Uma 10h49'18" 47d37'58"
Mary Catherine Gianaris
Aqr 23h38'26" -7d6'17"
Mary Catherine Habeeb
Aqr 23h4'7" -9d51'29"
Mary Catherine Holguin
Aql 19h39'51" 5d9'53"
Mary Catherine Hopkins
Cnc 9h17'23" 14d9'0"
Mary Catherine (Kay) MacLean Young
Sgr 18h10'1" -20d6'8"
Mary Catherine Kelly
Cas 0h1'28" 56d39'51"
Mary Catherine Letton
Cyg 20h47'57" 52d32'10"
Mary Catherine MacGoy
Per 2h17'27" 51d30'21"
Mary Catherine Madel Davis
Crb 15h37'21" 31d26'27"
Mary Catherine Mileto
Ari 2h56'19" 30d15'7"
Mary Catherine Moola
Tau 5h53'55" 23d58'17"
Mary Catherine Pastore
And 1h7'21" 39d1'1"
Mary Catherine Sharp
And 0h28'49" 28d49'15"
Mary Catherine Skelton
Sgr 18h59'47" -11d52'4"
Mary Catherine Therese
Tau 3h38'55" 16d23'49"
Mary Catheryn Kratochvil
And 2h11'44" 43d11'25"
Mary Cauley Hodges
And 1h24'57" 44d19'41"
Mary Cay & Brad Sargent 26 Years
And 0h21'20" 44d13'37"
Mary Cecelia
Uma 11h31'19" 45d57'58"
Mary Cecelia Boss
Leo 10h24'24" 11d3'59"
Mary Cecelia Rose Kapnis
Lyn 7h36'30" 38d2'11"
Mary Cecilia Harker
Cas 1h28'47" 67d22'34"
Mary Cecilia Parker
Dra 20h18'38" 62d24'43"
Mary Chalmers
Lyr 18h49'1" 36d30'26"
Mary & Charles Marsh
Cyg 20h17'59" 46d9'53"
Mary Christina Elizabeth Bradbury
And 23h24'9" 47d1'54"
Mary Christine Bird
Leo 11h45'30" 19d24'1"
Mary Christine Poyner
Ari 3h16'43" 24d18'31"
Mary Christine von Drachenfels
Crb 15h57'55" 33d51'25"
Mary Christine Williams
Uma 10h39'31" 55d36'53"
Mary Christmas Morales
Cap 21h7'5" -16d12'38"
Mary Clairey
And 23h24'59" 52d20'43"
Mary Clancy
Sco 16h13'37" -14d6'21"
Mary Clapp Woelper
Uma 10h31'51" 62d44'27"
Mary Clara Eichel Madden
Mon 6h31'47" 9d2'0"
Mary Clare
Cas 1h18'52" 63d9'1"
Mary Clementine Peters
Lyn 9h13'2" 37d6'26"
Mary Condon
Sgr 18h48'51" -20d27'43"
Mary Connolly
Cas 1h30'23" 66d8'2"
Mary Cooper Pitts
Cnc 8h49'22" 15d44'43"
Mary Cornelius
Uma 11h32'50" 38d6'39"
Mary Coussan
Sgr 17h52'21" -18d5'39"
Mary Criscuolo
Lib 15h47'18" -9d50'54"
Mary Cutshall
Uma 8h36'31" 47d59'53"
Mary Cynthia Wagoner
Ari 3h6'57" 25d12'33"
mary d
Her 16h47'45" 41d44'37"
Mary D. Mitchell Elementary School
Ori 6h9'59" 14d55'33"
Mary D. Scalio
Uma 11h34'23" 56d43'53"
Mary Dailey
Dra 16h29'37" 57d1'12"
Mary Damato
Gem 7h7'50" 32d42'24"
Mary Daul
Cas 0h34'36" 69d29'33"
Mary & David's Star
Cyg 21h42'6" 47d49'1"
Mary Davin
Leo 11h6'34" 22d8'15"
Mary Deborah - my beacon
Cyg 20h17'41" 47d56'23"
Mary Del Rosario
Tau 5h23'28" 25d6'25"
Mary De-Leston-Bailey
Lyr 18h30'16" 32d51'13"
Mary Della
Uma 10h5'42" 68d27'30"
Mary Denise Ervin
Gem 6h20'7" 21d59'49"
Mary Derra
Lib 14h52'22" -3d14'28"
Mary Desper
Uma 10h55'29" 55d41'53"
Mary Dews
Aql 19h48'29" 12d13'59"
Mary Dixon
Leo 9h23'22" 27d3'32"
Mary Dixon Chapman
Cas 0h30'30" 55d53'22"
Mary Dolores Bayless
Tau 5h52'46" 16d48'24"
Mary Dolores Quinones
Psc 1h32'56" 11d16'10"
Mary Dombrosky
Tau 4h35'8" 22d33'12"
Mary Doran
Uma 10h52'12" 55d29'38"
Mary Dorothy Henry - Dorie's Star
Cas 3h31'58" 69d30'10"
Mary Duffy
Uma 10h9'53" 44d8'34"
Mary Duncan Carney
Gem 7h28'1" 17d36'47"
Mary E. Barr
Ori 5h34'25" -2d55'36"
Mary E. Cicero
Cas 23h31'49" 52d59'30"
Mary E Cole
Aqr 22h32'7" -2d27'33"
Mary E. Cummings
Sco 16h4'42" -14d41'36"
Mary E. Davenport
Vir 14h26'2" 0d58'29"
Mary E. Davis 12/20/1924
Sgr 18h3'41" -26d38'34"
Mary E. Groman
Uma 10h58'25" 72d15'42"
Mary E. Haynes Phillips Jorgenson
And 1h13'13" 38d3'46"
Mary E. Hiscock
Lyr 18h41'34" 39d37'47"
Mary E. Holland Amour de ma vie
Cas 0h10'57" 57d3'14"
Mary E. Kemmesies
Ari 3h29'4" 23d22'22"
Mary E. Mills
And 23h28'57" 42d19'14"
Mary E. Olsen
Cap 20h32'29" -13d35'48"
Mary E. O'Malley
Cnc 8h19'6" 30d23'11"
Mary E. Peifer
Ori 6h14'52" 8d17'51"

Mary E. Rose Sgr 18h47'18" -25d11'51"
Mary E. Searle Cas 0h55'41" 50d33'17"
Mary E. Smith - Star of the North Umi 16h17'19" 72d7'6"
Mary E. Speer Cep 23h55'38" 74d23'10"
Mary E Trent Sgr 18h15'12" -19d15'34"
Mary E. Walter Cap 21h36'15" -10d12'52"
Mary E Whittington Metts Cap 21h44'55" -17d48'51"
Mary E. (Wood) Duquette Sgr 19h11'54" -27d17'18"
Mary Edwards Cas 0h47'51" 56d34'1"
Mary Egan Lyn 7h36'48" 47d49'41"
Mary Eileen Lyn 6h53'25" 51d18'21"
Mary Eileen 06231962 Lyn 7h47'53" 48d45'6"
Mary Eileen Gisler Crb 15h45'50" 32d9'19"
Mary Eileen Hartnett Crb 15h30'35" 30d28'12"
Mary Elaine (Ferguson) Backous Cas 2h35'50" 65d45'49"
Mary Eleanor Larson Lib 14h49'45" -3d42'19"
Mary Eleni Righos And 0h44'30" 38d32'28"
Mary Elinor Moehlenkamp Lyn 6h50'13" 52d5'7"
Mary Elizabeth And 23h20'19" 37d29'49"
Mary Elizabeth Cnc 8h35'52" 23d42'11"
Mary Elizabeth Com 12h47'26" 27d8'36"
Mary Elizabeth Lyr 19h17'18" 28d27'13"
Mary Elizabeth Leo 9h23'22" 17d0'18"
Mary Elizabeth Ori 5h15'49" -4d59'9"
Mary Elizabeth Uma 8h11'34" 64d48'20"
Mary Elizabeth Uma 13h7'34" 58d7'32"
Mary Elizabeth Adams And 23h34'30" 44d39'10"
Mary Elizabeth Anderson Lib 14h26'12" -11d54'16"
Mary Elizabeth Ann Lorence Blessing Gem 6h48'38" 23d7'25"
Mary Elizabeth Anne Her 18h33'13" 15d45'10"
Mary Elizabeth Aspinall 1989 Cas 1h8'29" 51d9'57"
Mary Elizabeth Benner And 0h57'29" 36d39'3"
Mary Elizabeth Billing Crb 16h8'21" 35d46'51"
Mary Elizabeth Brooks Sgr 17h57'13" -28d59'13"
Mary Elizabeth Bullock Psc 0h15'37" 10d1'16"
Mary Elizabeth Campbell And 0h17'1" 21d42'5"
Mary Elizabeth Carter Ari 2h4'21" 22d49'29"
Mary Elizabeth Cauble Walker Leo 9h59'18" 27d47'11"
Mary Elizabeth Christmas Lep 5h13'43" -12d45'41"
Mary Elizabeth Clark Cas 0h38'42" 48d28'0"
Mary Elizabeth Cole Uma 10h48'38" 47d48'54"
Mary Elizabeth Cosgrove Miller Lyr 19h4'57" 41d41'28"
Mary Elizabeth Crawford Uma 11h48'7" 52d19'53"
Mary Elizabeth Danielson Cap 21h32'22" -13d59'48"
Mary Elizabeth Donegan Exner Leo 11h13'49" 17d4'43"
Mary Elizabeth Ference Cas 0h28'5" 61d6'13"
Mary Elizabeth Flayer Gem 7h20'28" 33d2'57"
Mary Elizabeth Gumbel Uma 10h44'11" 48d28'21"
Mary Elizabeth Hart Nardo Cam 5h55'24" 59d51'26"
Mary Elizabeth Henry Cas 23h34'27" 52d43'0"
Mary Elizabeth Higgins Sco 16h15'7" -41d59'37"
Mary Elizabeth Holland Psc 23h48'28" 6d23'59"
Mary Elizabeth Hoskins Ahls Lyr 18h53'19" 33d6'30"
Mary Elizabeth "Judy" Marut Psc 1h48'8" 5d6'35"
Mary Elizabeth Kromling Cam 7h38'51" 64d50'53"
Mary Elizabeth Lister Sgr 18h5'8" -25d17'16"
Mary Elizabeth Lyons Sco 17h29'7" -40d27'42"
Mary Elizabeth Mathes Leo 10h15'24" 26d45'51"
Mary Elizabeth McCrae Per 1h51'10" 51d18'14"
Mary Elizabeth McQuaid And 0h14'10" 45d0'20"
Mary Elizabeth Moorhead And 23h57'6" 39d30'9"
Mary Elizabeth (Nana & Mom) Sco 17h53'37" -38d16'55"
Mary Elizabeth Naughton And 2h36'39" 44d21'20"
Mary Elizabeth Oliver Cas 1h25'37" 65d32'24"
Mary Elizabeth O'Neill And 2h10'24" 43d46'28"
Mary Elizabeth Renner Vir 13h39'11" 5d1'48"
Mary Elizabeth Ross Hickman Cyg 21h55'22" 49d12'37"
Mary Elizabeth Rowland 6/13/76 Gem 7h31'10" 33d55'27"
Mary Elizabeth Scarola Vul 20h20'28" 23d46'10"
Mary Elizabeth Scarpetta Ori 5h34'33" 4d25'3"
Mary Elizabeth Sheppard And 1h39'43" 36d5'58"
Mary Elizabeth - Star of our Hearts Apu 16h4'57" -75d36'30"
Mary Elizabeth Thompson 1919 Del 20h42'35" 15d27'12"
Mary Elizabeth Titus Sco 16h14'25" -17d0'24"
Mary Elizabeth Wenzel Cyg 19h40'13" 33d11'14"
Mary Elizabeth Whitaker Gem 7h24'51" 26d39'47"
Mary Elizabeth Williams Cas 0h19'50" 57d59'6"
Mary Ellars Sgr 18h9'52" -34d31'37"
Mary Ellen Sgr 19h11'32" -17d22'13"
Mary Ellen Cyg 19h53'35" 37d47'47"
Mary Ellen And 0h36'23" 42d1'31"
Mary Ellen Tau 5h38'7" 26d9'0"
Mary Ellen Leo 11h20'58" 22d18'2"
Mary Ellen Anne Scerra Dachel And 2h16'31" 49d25'58"
Mary Ellen Beaty Crb 15h40'44" 26d32'27"
Mary Ellen & Betty Jane Uma 9h49'51" 52d32'26"
Mary Ellen Bibb Uma 13h39'51" 61d25'39"
Mary Ellen Boan Ari 3h25'46" 27d35'44"
Mary Ellen Bradshaw Parsons Cra 18h12'4" -40d30'23"
Mary Ellen Burnett Cap 20h20'9" -12d30'9"
Mary Ellen Butler Cas 0h59'1" 57d6'42"
Mary Ellen Campbell Vir 13h19'54" 12d3'58"
Mary Ellen Cavallaro Leo 11h47'47" 19d49'46"
Mary Ellen Cotton "Mamaw" Psc 1h1'36" 19d3'39"
Mary Ellen DAmato Ori 6h5'20" 13d37'20"
Mary Ellen Donnelly Cyg 20h9'54" 41d14'36"
Mary Ellen Fry Vir 13h23'2" -14d28'54"
Mary Ellen "Grammy" Gelinas Gem 6h35'3" 20d38'4"
Mary Ellen Johnston And 23h9'39" 51d27'9"
Mary Ellen Kuta Ari 3h27'41" 27d40'19"
Mary Ellen Larkin Cap 21h15'32" -15d52'26"
Mary Ellen Levin Vir 12h46'43" 10d37'34"
Mary Ellen Mahoney Vir 12h26'57" 2d36'0"
Mary Ellen "McDonald" Donner Cyg 21h31'8" 46d48'35"
Mary Ellen Murphy And 0h32'27" 42d21'19"
Mary Ellen Nevin Keating Tau 5h29'27" 27d24'47"
Mary Ellen Norris Ari 1h53'19" 19d32'16"
Mary Ellen Overdier Jackson Cyg 20h35'37" 60d38'23"
Mary Ellen Payne Uma 9h40'30" 58d3'40"
Mary Ellen Payne Aur 6h58'35" 38d15'57"
Mary Ellen Rhein Cap 20h24'31" -12d41'31"
Mary Ellen Ryan And 1h31'41" 42d46'15"
Mary Ellen Scott Cas 23h1'12" 56d8'12"
Mary Ellen Seiger Cas 2h21'56" 66d5'25"
Mary Ellen Slusser Sgr 18h16'20" -27d19'52"
Mary Ellen Smith Oph 16h52'2" -0d3'9"
mary ellen stankeys star Per 3h15'27" 42d18'46"
Mary Ellen Sullivan Sco 16h3'9" -29d14'29"
Mary Ellen Wallar-Panzano Tau 4h27'54" 15d37'42"
Mary Ellen Williams Lyn 7h43'17" 36d31'52"
Mary Ellen's Rose And 23h20'21" 51d47'25"
Mary Elliott Tau 4h57'5" 21d31'44"
Mary E.M.B. Stitt Cas 23h27'32" 56d4'27"
Mary Emily Greenwood And 1h58'45" 40d10'2"
Mary Emma Nott Zumwalde And 0h44'8" 42d52'18"
Mary Ercolano Per 4h26'58" 42d47'56"
Mary Ernestine Watts-Short Crb 15h44'57" 36d51'17"
Mary Ertel Cas 1h36'48" 62d2'49"
Mary Estelle Theresa Josephine Uma 10h47'27" 49d27'49"
Mary Esther Baxter-Hernandez And 0h43'46" 37d29'32"
Mary Esther Filosi Cas 23h21'28" 53d35'5"
Mary Esther Johnson Lyn 7h34'48" 38d38'42"
Mary Etta McClarren Cas 0h17'22" 57d51'40"
Mary Etta Watson Boo 14h11'14" 32d19'48"
Mary Eva Cyg 20h6'39" 57d1'16"
Mary Evelyn Crb 15h52'11" 33d14'1"
Mary Evelyn Church Ari 3h24'35" 26d58'35"
Mary Evelyn Rose Mennecke 1-24-1915 Eri 4h9'8" -0d45'46"
Mary F. 419 Cam 7h36'51" 66d10'12"
Mary F. Adams Sgr 19h11'26" -27d50'2"
Mary F. Battenhausen Sco 16h56'46" -44d1'7"
Mary F. Douglas Uma 14h20'44" 57d24'16"
Mary F. Haughton Per 3h11'19" 40d14'41"
Mary F. Lyons Cas 0h3'35" 56d44'53"
Mary F. Messina Cas 2h51'19" 61d15'25"
Mary F. & Niko S. Forever Friends Leo 10h4'20" 25d46'1"
Mary F. Pinter Vir 13h7'59" 11d49'49"
Mary F. Puckett Cyg 19h55'39" 36d19'50"
Mary F Russell Ari 2h50'17" 25d8'6"
Mary Fairy Ari 2h49'19" 27d50'10"
Mary Faith Cas 0h55'44" 59d47'12"
Mary Faith Gallant Gem 7h29'47" 31d53'43"
Mary Falco Aqr 21h8'53" -14d27'36"
Mary Fangman Tau 5h5'20" 23d55'7"
Mary Farnan Vir 13h19'47" 10d26'5"
Mary Fatima Sgr 19h13'6" -23d29'59"
Mary Faye Sears Dodson Vir 13h8'0" -1d3'21"
Mary Federle Ray Lyr 18h51'7" 36d5'20"
Mary "Feet" Smith Crb 15h42'52" 31d22'39"
Mary Fefy Crb 16h20'27" 38d29'17"
Mary Feiring - A True Treasure Ari 2h55'0" 27d36'59"
Mary Fellner Cas 0h16'44" 64d46'13"
Mary Ferrara Leo 9h31'41" 28d56'18"
Mary Fleming And 23h37'11" 45d43'2"
Mary Florance Lyr 19h2'58" 33d38'56"
Mary Foster Cyg 20h42'45" 52d11'41"
Mary Fox Bartley Vul 20h41'56" 25d17'49"
Mary Frame Gallagher Vir 15h7'45" 5d10'36"
Mary Frampton Cap 20h42'5" -15d42'18"
Mary Frances Barron And 1h9'51" 37d43'40"
Mary Frances Brown And 0h17'45" 26d22'41"
Mary Frances Calocino Sgr 19h38'6" -26d25'57"
Mary Frances Hegarty Cas 2h49'8" 60d20'36"
Mary Frances Jones And 22h58'55" 47d31'45"
Mary Frances Marchbank Cap 20h12'6" -15d1'22"
Mary Frances Smalling Napier Uma 11h13'41" 29d6'51"
Mary Frances Thein Uma 11h4'23" 51d30'9"
Mary (Frances) White Furino Ari 2h2'21" 19d24'3"
Mary Francis Lib 14h37'9" -16d27'54"
Mary Francis Anderson Chambers Vir 12h41'0" 4d12'24"
Mary Francis Bryan Willett - TWT Mon 6h47'22" -0d10'30"
Mary Francis Zagrodnik Cma 7h13'41" -30d20'16"
Mary Frank Artist/Fine Friend Cas 0h31'13" 54d37'41"
Mary Frazier Lyn 7h13'14" 55d18'36"
Mary Frenette Wife, Mom, & Grandma Cas 23h7'47" 56d17'24"
Mary Frost "Tootsie Pot" Cap 20h7'48" -11d13'54"
Mary Furman Waldron Lib 15h56'53" -12d54'0"
Mary Fusilier Uma 11h28'19" 33d31'29"
Mary G. Barricks Cyg 20h42'48" 45d53'50"
Mary G. Virgilio Per 2h42'11" 53d55'40"
Mary G. Zabierek Leo 9h48'24" 6d32'3"
Mary Gallahue Sco 17h5'15" -43d19'45"
Mary Garce Poleo Tau 3h49'12" 3d41'46"
Mary Garcia Psc 1h15'40" 24d58'14"
Mary Garzone (Loving Gram & Nanny) Cas 1h21'11" 63d30'36"
Mary Gelormino Leo 10h4'34" 23d29'34"
Mary Gemma Lib 14h51'24" -0d38'49"
Mary Gersema's Morning Star Cas 0h53'32" 56d52'56"
Mary Gertrude Gray And 1h31'15" 33d42'31"
Mary Gethings Cas 2h0'45" 62d58'7"
Mary Gibbons Beloved Mother Lib 15h25'16" -6d8'54"
Mary Gibson Jeanmard And 23h22'59" 51d37'26"
Mary Gillanders Cas 0h10'45" 53d57'32"
Mary Glackin Vir 12h10'5" 11d44'29"
Mary Glassmaker Dickes Eri 4h24'33" -0d19'14"
Mary Glenna Dolly Cep 22h22'7" 73d57'18"
Mary Glenney, Mom Extraordinaire Cyg 19h32'33" 30d50'45"
Mary Golden Aqr 22h26'23" -7d57'33"
Mary Grace And 22h59'9" 41d7'17"
Mary Grace And 1h47'37" 48d50'48"
Mary Grace Ori 4h48'11" 12d6'22"
Mary Grace Cummings And 1h9'10" 48d6'35"
Mary Grace Lemier Dra 17h11'20" 65d24'45"
Mary Grace Milne And 1h14'31" 34d55'17"
Mary Grace Rawlings And 0h44'9" 41d55'27"
Mary Grace Servello Cas 0h18'14" 57d50'41"
Mary Grace's e-Smart Star Lib 14h45'12" -18d5'52"
Mary Grammy Wagner Aqr 22h28'13" -0d38'19"
Mary "Gramster" Flood Lyr 18h43'0" 31d34'39"
Mary Grasso Uma 14h22'56" 56d39'4"
Mary Graves And 2h35'53" 49d42'12"
Mary Gresick Cas 0h29'45" 55d7'54"
Mary Grosvenor Vir 11h42'24" 10d15'51"
Mary G's Crb 15h37'55" 32d26'5"
Mary Gustafson Cnc 8h54'36" 25d40'28"
Mary Guzick Crb 15h36'33" 39d9'48"
Mary Gwendolyn Lax Phillips Srp 18h17'47" -0d57'20"
Mary H. Costlow 6-26-19/10-03-2004 Cma 7h2'38" -23d58'16"
Mary H Donald Psc 0h19'38" 4d42'34"
Mary Hammett Leo 9h38'19" 32d8'32"
Mary Hanna Uma 11h32'1" 58d11'17"
Mary Hannah Rzasa Peg 21h37'41" 18d17'44"
Mary Harlan Crb 16h8'57" 35d19'30"
Mary Harvey Vay Cas 2h29'1" 68d14'41"
Mary & Hayden Morgan Dra 14h48'14" 55d18'2"
Mary Heckwolf Staylor Ari 2h25'31" 21d36'38"
Mary Hedrix Cas 0h2'20" 59d40'9"
Mary Heil Grimm D.M.D. Lib 14h47'55" -24d31'27"
Mary Helen Cap 20h49'26" -20d19'22"
Mary Helen And 1h43'19" 42d33'0"
Mary Helen and Nip Uma 11h12'55" 45d31'44"
Mary Helen and Nip Uma 9h40'16" 58d31'50"
Mary Helen Hernandez Gem 7h19'8" 15d28'52"
Mary Helen Organ (Gommie) Aqr 22h36'15" -5d36'6"
Mary Helen Simon Crb 15h23'49" 30d52'26"
Mary Helena Lemos Brum Uma 8h38'17" 65d40'3"
Mary Hensley Evans Leo 10h18'16" 26d2'53"
Mary & Herman Beyersdoerfer Umi 15h29'30" 72d32'46"
Mary Hernandez Cnc 9h6'26" 23d19'8"
Mary Hernandez Magdaleno Cas 0h42'35" 50d17'1"
Mary Hiswa, Bob & Grace Davidson Tri 2h4'27" 32d41'57"
Mary Hoch Snieckus Cas 0h51'41" 55d52'51"
Mary Holmes Uma 9h19'19" 51d9'39"
Mary & Howard Langdon Cyg 21h27'16" 31d21'39"
Mary Huffman Lyn 8h16'28" 47d19'16"
Mary Hutton Crb 16h19'2" 31d56'11"
Mary Imelda Scanlon Kye Leo 9h59'10" 14d47'6"
Mary Ingalls, Loving Mother Cas 23h45'48" 58d15'26"
Mary Innes Bowman Cap 20h8'9" -9d54'27"
Mary Ireland Colucci Vir 13h20'30" 8d12'15"
Mary Isabella Agostinacchio Cas 1h34'58" 62d57'26"
Mary J Bridges Uma 9h31'23" 57d39'58"
Mary J. Citron Cas 23h58'0" 56d53'35"
Mary J. Concardi Tau 5h40'57" 26d39'36"
Mary J Lyons Uma 9h32'9" 67d29'29"
Mary J Peirson Uma 11h8'43" 47d2'34"
Mary J. Townley Crb 16h4'35" 37d9'25"
Mary J. Weiss Gem 7h32'36" 26d25'1"
Mary J. Westphal 6/25/26-12/15/95 Ari 1h47'12" 20d11'4"
Mary Jane Psc 23h42'14" 7d53'52"
Mary Jane Cas 0h56'15" 75d1'4"
Mary Jane Cas 23h21'45" 55d54'53"
Mary Jane Lyn 8h3'46" 58d59'22"
Mary Jane Leo 11h12'2" -3d44'3"
Mary Jane Aqr 23h39'36" -16d41'26"
Mary Jane Sco 17h51'56" -36d16'11"
Mary Jane & Alfred Good Uma 13h47'48" 62d8'5"
Mary Jane and Kip Reuter Cas 0h44'54" 60d42'48"
"Mary Jane and Spiderman" Cyg 21h14'18" 43d59'0"
Mary Jane Barber Uma 10h27'16" 43d56'47"
Mary Jane Baumgartner Psc 0h5'22" 2d57'26"
Mary Jane Blackmore Lib 14h50'35" -2d12'20"
Mary Jane Bryan Aqr 22h18'26" -22d50'51"
Mary Jane Cochran & Roy Marshall Gem 6h31'26" 16d30'3"
Mary Jane Combs Del 20h19'46" 9d47'0"
Mary Jane Crawford And 2h9'38" 41d50'22"
Mary Jane Dubin Cnc 9h12'44" 16d39'51"
Mary Jane DuCoin Sgr 18h38'2" -36d2'50"
Mary Jane Duncan~Robert Lee Duncan Cyg 20h1'50" 39d47'3"
Mary Jane Elizabeth Hall Cas 0h21'24" 53d33'15"
Mary Jane Faienza Uma 12h52'40" 55d5'38"
Mary Jane "Grandma Spice" John Gem 7h37'7" 32d35'9"
Mary Jane Heims Gem 6h14'59" 26d4'11"
Mary Jane Henly Wechter Her 17h54'16" 50d1'2"
Mary Jane Hite Ari 3h5'28" 26d1'47"
Mary Jane & Kristina Jane Uma 10h34'41" 40d10'14"
Mary Jane Lakey Leo 9h46'11" 13d9'28"
Mary Jane Laneve's Star Cas 1h21'36" 61d52'47"
Mary Jane Lawler Cas 2h14'3" 67d0'0"
Mary Jane Macy Lyn 8h12'47" 41d40'13"
Mary Jane Martin And 0h7'15" 36d49'25"
Mary Jane Nelson Aqr 22h20'3" -3d48'53"
Mary Jane Nolan Per 3h41'41" 45d25'3"
Mary Jane Palmer And 23h49'31" 47d2'47"
Mary Jane Palmer Lyn 8h0'35" 37d43'15"
Mary Jane Palmer And 1h34'42" 41d21'49"
Mary Jane Pierce Lyr 18h54'59" 32d55'3"
Mary Jane Polsgrove Gem 7h0'35" 22d53'36"
Mary Jane Rorrer And 0h42'52" 37d44'28"
Mary Jane Smith Woodbury Uma 10h55'56" 71d7'20"
Mary Jane Stewart Bright Star And 0h29'14" 40d48'32"
Mary Jane Strycharski Cap 21h3'21" -21d33'13"
Mary Jane Taylor Cam 5h33'42" 69d14'32"
Mary Jane Wickenkamp Cas 0h34'24" 61d41'9"
Mary Jane's Grandma Star Uma 11h28'14" 28d38'5"
Mary Janet Trigg Cas 1h16'59" 54d24'35"
Mary Janine Nottingham Tau 4h24'48" 26d43'16"
Mary Jarvis Uma 10h39'56" 61d41'53"
Mary Jean Cra 18h9'29" -44d39'23"
Mary Jean Dewire Cas 1h0'43" 60d35'0"
Mary Jean Martin Cas 1h24'17" 57d3'40"
Mary Jean Oros And 2h19'11" 48d14'1"
Mary Jean Shea Cam 3h46'43" 59d2'13"
Mary Jean Wells Cnc 8h38'46" 23d52'45"
Mary Jeanne Uma 10h26'21" 63d57'38"
Mary Jeanne Aqr 23h7'28" -11d3'44"
Mary "Jeanne" Gardner Crb 16h18'4" 30d51'50"
Mary & Jerry Vogel Uma 11h45'49" 57d5'38"
Mary Jessica Von Guilleaume Uma 11h48'31" 43d10'24"
Mary Jessie Gibson Gem 7h20'35" 13d48'12"
Mary Jirauch and Dwight Bowen Cas 0h27'28" 61d43'16"
Mary Jo Lib 14h26'29" -17d56'15"
Mary Jo Cnc 8h50'28" 26d9'39"
Mary Jo Lyn 7h4'7" 51d29'57"
Mary Jo Altman Vir 12h37'10" -2d41'42"
Mary Jo Callahan Cas 23h31'57" 53d37'43"
Mary Jo Christine Thomas Sco 16h13'3" -15d19'25"
Mary Jo Clark Leo 11h56'51" 22d1'48"
Mary Jo Davis Cnc 8h0'0" 18d29'24"
Mary Jo DiPardo Sco 16h28'3" -25d2'47"
Mary Jo Goesse And 0h54'8" 38d13'50"
Mary Jo Gould Cas 2h11'44" 59d41'18"
Mary Jo Guthrie Uma 11h3'25" 67d45'24"
Mary Jo Henger Uma 11h22'11" 70d26'23"
Mary Jo Maggiano And 2h14'46" 40d54'21"
Mary Jo Matekovich Gem 6h3'4" 24d57'6"
Mary Jo McManus - Light of my Life Uma 11h3'18" 41d27'1"
Mary Jo Ross Lib 15h7'47" -7d5'45"
Mary Jo Russell Vir 13h24'54" -2d47'44"
Mary Jo Stella Cam 3h17'27" 60d30'33"
Mary Jo Tennant Lyn 8h40'10" 36d47'14"
Mary Jo Urbaniak Cas 0h40'6" 60d57'0"
Mary Joan Lib 15h25'29" -26d45'40"
Mary (Jodi) Hicks And 0h19'9" 29d20'50"
Mary Joe Williams Uma 9h4'7" 57d42'25"
Mary & Joel Lyr 19h19'38" 35d19'24"
Mary Johnston And 0h4'17" 45d6'41"
Mary Jones And 1h28'45" 39d41'13"
Mary Jordan Lmi 10h43'40" 38d47'10"
Mary Jo's Little Lamb Aqr 22h28'51" -20d24'35"
Mary Jo's Twinkle Crb 15h28'35" 28d55'11"
~Mary Joyce Racki~ Cas 23h38'26" 58d13'38"
Mary Julia "Jerry" McClurkin Cas 0h31'14" 50d49'48"
Mary Julia Michalski - 2007 Leo 10h22'14" 17d6'4"
Mary June Manske Gem 7h26'11" 32d33'40"
Mary "June" Matta Gem 6h19'15" 21d37'6"
Mary K Lib 14h22'15" -17d33'19"
Mary K Davis Sgr 18h37'13" -17d57'12"
Mary K Fries Cap 20h22'28" -11d43'18"
Mary K Langston Uma 10h17'59" 63d37'9"
Mary K. Meyer Cas 0h5'53" 53d51'19"
Mary K. Pogorzelski Uma 9h44'59" 41d30'26"
Mary K. Radeke Ari 3h28'21" 27d37'11"
Mary K Tucker And 0h44'59" 36d6'5"

Mary K Vaughn - Resident of Heaven
Cap 20h17'51" -16d43'39"

Mary K. Willis
Uma 12h10'47" 61d45'29"

Mary K. Yochum
Lyr 18h31'12" 28d30'56"

Mary Ka
Umi 17h8'14" 86d25'2"

Mary Kalin Puent
Peg 21h19'31" 19d6'59"

Mary Kane
Ori 5h11'54" -0d54'13"

Mary Karin
Sco 16h5'43" -12d27'4"

Mary Kate
Cap 21h21'17" -22d31'42"

Mary Kate Denault
And 1h41'52" 45d30'5"

Mary Kate Dunn
Sco 16h5'50" -10d50'24"

Mary Kate Hatcher Weber
Gem 7h10'1" 27d56'22"

Mary Kate Marlow
And 2h19'36" 45d39'46"

Mary Kate McKenzie
And 22h58'35" 40d15'23"

Mary Katelynn Atkinson
And 2h24'31" 40d15'24"

Mary Kate's Star
Cyg 20h52'16" 38d37'21"

Mary Katherine
Cnc 8h50'12" 13d58'20"

Mary Katherine
Vir 11h44'21" -2d50'13"

Mary Katherine Harper
Lib 15h23'3" -17d21'40"

Mary Katherine Mason
Psc 0h25'18" 17d39'2"

Mary Katherine Moniak
Cap 20h30'31" -25d6'33"

Mary Katherine Newcomb
And 23h25'6" 47d59'13"

Mary Katherine Scheeler
Crb 16h12'0" 35d41'56"

Mary Kathleen
Aqr 22h40'15" -3d29'54"

Mary Kathleen
Lyn 6h56'26" 52d37'15"

Mary Kathleen Buckley
Cas 0h51'2" 58d6'25"

Mary Kathleen Gianutsos
Psc 1h28'42" 6d31'54"

Mary Kathleen Groch
Uma 8h43'2" 56d39'22"

Mary Kathleen Jacobs
Sco 16h11'8" -20d42'19"

Mary Kathleen Nicholson McKenzie
Ari 2h21'48" 26d28'43"

Mary Kathleen Stone
Cas 23h35'7" 59d3'10"

Mary Kathryn Jones
Cas 0h21'3" 56d9'6"

Mary Katie Andrews
Aqr 22h18'25" -23d12'19"

Mary Kay
Sgr 19h20'42" -22d30'55"

Mary Kay
Her 16h37'30" 13d7'39"

Mary Kay Addington
Uma 10h49'25" 50d20'12"

Mary Kay and John Taylor
Cyg 21h22'59" 50d26'28"

Mary Kay and Patrick Estep
Per 4h32'9" 41d21'51"

Mary Kay & Bruce Wilson
Cyg 21h56'59" 52d54'44"

Mary Kay Byley-Hardy
Cyg 20h44'59" 31d15'52"

Mary Kay Jeynes
And 23h36'8" 49d12'24"

Mary Kay Jones
Cyg 19h27'30" 46d54'56"

Mary Kay Mooney
Lyr 19h25'9" 37d35'58"

Mary Kay Nations
Sco 17h38'13" -34d17'4"

Mary Kay Schiel Donahue
Psc 1h25'50" 18d38'21"

Mary Kay Slowikowski
Cam 4h2'58" 69d16'22"

Mary Kay Tasker
Tau 5h45'59" 18d1'49"

Mary Kaye Malik
Uma 11h9'40" 64d59'0"

Mary Kay's Light
Lyn 7h22'32" 50d22'19"

Mary - Kelly Ashley Valentine Fry
Ori 6h2'21" 12d9'19"

Mary Kennedy
Cas 23h42'10" 56d30'38"

Mary Kennelly McCoy Swan aka Kelly
Cyg 19h47'56" 29d47'1"

Mary & Kenneth Kidd Family Star
Lyn 7h9'19" 59d6'52"

Mary Kent
And 1h4'45" 33d46'2"

Mary Kephart Star
Leo 10h58'34" -3d8'59"

Mary Kern
Lyn 8h18'37" 35d50'29"

Mary Kettenacker
Sco 16h11'21" -11d32'32"

Mary Keyzer Miller
Leo 9h37'18" 10d46'44"

Mary Kiker Davis
Vir 13h2'27" -17d29'25"

Mary Kilcoyne
Aqr 22h1'36" -9d0'4"

Mary Kim Partridge
Aqr 23h18'14" -21d22'24"

Mary Kimbrell
Umi 13h21'24" 72d6'15"

Mary Kirby
Oph 17h44'59" -0d33'43"

Mary Klueg
Cas 1h32'23" 62d45'0"

Mary Knoebel
Cas 1h25'1" 60d56'12"

Mary Knoll
Cap 20h25'19" -12d57'7"

Mary (Kobles) Pribulsky
And 23h24'23" 48d25'0"

Mary Konig
Peg 22h4'58" 36d8'34"

Mary Kraft
Lib 15h40'52" -4d45'11"

Mary Krasnansky Purcell
Tri 1h53'4" 32d22'14"

Mary Kris
Gem 8h6'51" 29d12'14"

Mary K's Star
Vir 13h22'37" -13d3'56"

Mary L. Arias
Cyg 21h50'14" 46d13'55"

Mary L. Becker
Cas 0h11'23" 52d23'7"

Mary L. Caruso
Cas 0h24'10" 58d31'19"

Mary L. Franey
Leo 10h50'25" 18d34'10"

Mary L. Gasiewski
Cas 0h36'7" 60d54'5"

Mary L. Martin
And 1h33'58" 47d14'7"

Mary L. Mooter
Ori 5h56'33" 11d58'47"

Mary L Reecamper
Lyn 7h55'26" 47d38'37"

Mary L. Salmon
Lyn 7h39'13" 59d30'4"

Mary L. Schweitzer
Uma 8h14'8" 67d6'32"

Mary Lalor
Leo 10h29'37" 17d38'8"

Mary Lamar
Ori 5h57'24" 22d40'56"

Mary Lampariello Scios
Vir 12h28'37" 11d17'44"

Mary LaSalle
Mon 6h44'46" -0d5'0"

Mary Lasley Simpson
Lyn 6h25'18" 60d14'34"

Mary Laura Minor
Umi 15h20'47" 75d21'18"

Mary Lauren Tinnon
Leo 9h46'39" 21d41'50"

Mary Laurette Kirk
Peg 23h52'57" 23d42'34"

Mary Laurilla Nash
Leo 10h25'23" 27d14'5"

Mary Leali
And 0h19'3" 34d47'44"

Mary Leatherman
And 1h48'43" 38d49'10"

Mary Lee
And 1h19'2" 46d26'4"

Mary Lee
Leo 10h12'14" 12d16'37"

Mary Lee Anderson
Cas 1h13'52" 58d15'29"

Mary Lee Byra Jewel
Cas 1h59'31" 62d9'42"

Mary Lee Covington
Ori 5h35'48" 1d59'9"

Mary Lee Cunill
Vir 13h36'2" 2d14'23"

Mary Lee Dolvin Bagwell
Sco 17h30'21" -30d18'25"

Mary Lee Gill
Cap 21h7'5" -19d32'23"

Mary Lee Gravning Always25
Lib 14h51'49" -7d57'32"

Mary Lee Houser
Gem 6h41'11" 27d13'51"

Mary Lee Moran Our Brighest Star
Ari 2h23'50" 19d45'58"

Mary Lemuel
Lmi 10h41'56" 32d39'29"

Mary Leone
Uma 10h45'43" 54d19'11"

Mary Leonice Jenks Scalli
Gem 6h55'26" 24d5'3"

Mary Liese
Lyn 7h32'47" 38d26'6"

Mary Linda
Sgr 18h17'33" -19d2'4"

Mary Linda Watson
Vir 14h42'19" 5d11'27"

Mary Lindholm
Umi 15h15'40" 70d1'31"

Mary Lindsay Bennett
Cas 1h33'11" 60d57'47"

Mary Lindsay Wilkinson
Tau 5h49'24" 17d50'23"

Mary Lions Ochsner
Lmi 10h36'16" 31d27'0"

Mary Lisa
Cap 21h3'39" -21d26'53"

Mary Littman
Psc 1h19'14" 24d54'32"

Mary Livengood
Cap 21h58'22" -11d38'5"

Mary Liza Case
Lib 14h52'34" -12d26'34"

Mary Lohndorf Deacetis
Crb 15h30'13" 27d28'3"

Mary Lord
And 2h24'23" 46d46'48"

Mary Loretta Glenn Gromowsky
Cas 1h58'29" 63d14'36"

Mary Lorraine C. Ibasco
Vir 13h59'24" 0d16'52"

Mary Lorraine Hiatt
And 0h11'36" 30d30'40"

Mary Lorraine Kramm
Umi 13h44'19" 75d1'59"

Mary Lou
Mon 6h48'33" -0d34'14"

MARY LOU
Cap 20h35'6" -17d38'40"

Mary Lou
Cyg 20h39'40" 53d37'7"

Mary Lou
Cas 1h43'54" 64d28'27"

Mary Lou
Uma 9h30'43" 66d29'24"

Mary Lou
Tau 3h33'53" 27d38'12"

Mary Lou
Tau 4h27'41" 23d25'13"

Mary Lou
Mon 6h45'11" 7d2'16"

Mary Lou
Ari 2h32'11" 13d4'30"

Mary Lou
Cnc 8h37'2" 9d56'31"

Mary Lou Abrams
Com 12h42'15" 30d55'28"

Mary Lou Babe Swartz
Sgr 18h59'35" -35d25'52"

Mary Lou Beisser 1955-2005
Tau 3h57'8" 18d33'14"

Mary Lou Carbajal
Sgr 18h8'40" -27d0'3"

Mary Lou Catherine Reichel "Nana"
Cas 23h30'38" 58d49'49"

Mary Lou Crowley
Leo 10h5'45" 24d33'25"

Mary Lou & Dave
Uma 8h20'17" 67d55'50"

Mary Lou Ellis
Dor 5h23'24" -68d31'46"

Mary Lou Eskew
Cap 21h44'37" -12d39'41"

Mary Lou Hartinger
Uma 13h48'52" 58d56'57"

Mary Lou Hipple
Cas 0h22'6" 61d57'33"

Mary Lou Hood
Vir 11h39'51" 3d41'33"

Mary Lou LaRoche
Psc 23h14'58" 1d0'31"

Mary Lou Liebau
Sco 16h34'3" -29d5'15"

Mary Lou Miller
Lep 5h11'42" -11d30'21"

Mary Lou Murphy
Sgr 18h0'22" -28d41'46"

Mary Lou Nerone
Crb 16h9'45" 35d16'56"

Mary Lou Ohrman
Vir 12h28'26" 5d51'2"

Mary Lou Peluso
Cas 1h25'11" 62d1'43"

Mary Lou Puckett Kauffman
Vir 12h47'45" 2d34'50"

Mary Lou Sanchez Lettig
Aqr 22h33'27" 1d12'14"

Mary Lou Shonk House Wright
Cnc 9h4'2" 27d39'46"

Mary Lou Stallings
Uma 11h59'42" 33d2'38"

Mary Lou Swiatek
Lyn 7h3'15" 53d57'21"

Mary Lou Timerding
Cas 0h55'21" 61d34'57"

Mary Lou Whitmer
Ari 3h16'19" 12d24'44"

Mary Lou, The Navigator
Cyg 20h1'8" 37d5'16"

Mary Louise
Cas 0h30'48" 55d25'24"

Mary Louise
Mon 6h44'16" -0d9'51"

Mary Louise Chapman
Lyn 6h33'20" 56d15'59"

Mary Louise Claxton Smith
Cyg 20h56'14" 46d2'36"

Mary Louise Cox
Aql 19h7'19" 2d46'50"

Mary Louise & David Amerine 20th.
Cyg 20h53'35" 47d55'55"

Mary Louise Henke
Her 16h39'19" 7d13'27"

Mary Louise Jones
Crb 15h36'7" 37d18'36"

Mary Louise Kelly
Uma 9h48'26" 45d25'53"

Mary Louise Kim
Cas 23h30'10" 51d38'37"

Mary Louise Kline
Psc 1h20'50" 3d50'55"

Mary Louise MacKenzie Meriwether
Mon 6h46'8" -0d5'34"

Mary Louise Mele
Cas 1h37'17" 63d3'7"

Mary Louise Radleigh Kilwein
Cas 1h26'52" 57d37'55"

Mary Louise Rothacker
Vir 12h39'8" -0d4'34"

Mary Louise VanThournout
Uma 13h15'22" 55d43'47"

Mary Lou's Diamond In The Sky
Uma 9h39'13" 51d45'40"

Mary Lou's Star
Sgr 18h8'16" -26d21'18"

Mary Love
Uma 11h56'47" 38d45'42"

Mary Lu and Vince
Sco 16h47'28" -34d2'42"

Mary Lu Shull
Cas 2h27'41" 65d31'21"

Mary Lucero
Cap 21h28'18" -14d38'38"

Mary Lucile Ward James
Ari 2h52'47" 22d19'40"

Mary Lucinda McKeever Sanchez
Leo 9h30'44" 27d23'26"

Mary Lupi
Cma 6h50'25" -22d20'23"

Mary Lynda (Piccoli Dolci) McCann
Leo 9h40'11" 7d24'38"

Mary Lynda Spraggs
Tau 4h21'53" 22d59'7"

Mary Lynn
And 1h32'34" 39d31'18"

Mary Lynn
And 0h43'35" 43d11'3"

Mary Lynn Beale
Mon 7h15'18" -0d44'4"

Mary Lynn Bohannon
Lyn 8h36'2" 35d49'33"

Mary Lynn Cala
Cas 2h50'57" 60d44'27"

Mary Lynn Manns
Aql 19h55'55" 12d15'3"

Mary Lynn Morici
Cas 1h30'4" 61d59'20"

Mary Lynn Porter
Per 4h8'6" 47d13'23"

Mary Lynn Shelor-Cason
Cap 20h28'8" -18d33'46"

Mary Lynn Tate Hereford
And 0h38'23" 43d21'3"

Mary Lynn Teeling
And 23h27'11" 37d43'18"

Mary Lyse Crellin
Gem 6h37'23" 20d51'59"

Mary M 2.3
Lyn 8h52'2" 40d40'53"

Mary M. Benedict Hendrix
Aqr 20h51'36" -13d4'32"

Mary M. Bolish
Uma 11h42'2" 54d19'37"

Mary M - Cherished Wife Mom Grandma
And 23h57'45" 41d47'20"

Mary M. Chuhinka
Tau 5h20'14" 27d25'5"

Mary M. Frare
Lib 15h33'57" -19d29'57"

Mary M. Mitchell
Lyn 6h59'20" 56d45'15"

Mary M. Murtaugh
Lib 15h29'44" -5d18'37"

Mary M. Riffel
Lib 14h52'42" -4d26'14"

Mary M. Wright
Crb 16h8'45" 35d20'13"

Mary Mac
Lyr 18h50'17" 32d31'49"

Mary Macchione
Aqr 22h5'52" 0d43'12"

Mary MacLean
Lyr 18h40'39" 30d47'12"

Mary Madeline
Cas 0h35'13" 54d23'30"

Mary Madeline
And 23h40'18" 45d37'21"

Mary Madeline
Cas 1h23'53" 62d26'37"

Mary Mae Jordan (Mamaw)
Uma 10h18'16" 45d39'29"

Mary (Mae) Layden
Psc 1h24'38" 32d32'34"

Mary Maegan Brundage
Mon 6h51'47" 7d19'51"

Mary Magaret
Tau 4h44'49" 29d14'50"

Mary Mailly
And 0h7'13" 42d55'57"

Mary Maloney
Ari 2h2'45" 20d30'18"

Mary Maravelia
Uma 9h18'32" 42d13'18"

Mary Margaret
Cyg 21h21'56" 34d8'33"

Mary Margaret
Uma 10h6'15" 47d20'37"

Mary Margaret
And 23h10'18" 41d13'13"

Mary Margaret
Peg 21h31'1" 12d25'44"

Mary Margaret Cross
Uma 11h45'9" 53d22'4"

Mary Margaret (Currie) Starkey
Cap 20h28'53" -14d37'15"

Mary Margaret Heisler
Sgr 18h55'5" -22d0'41"

Mary Margaret Kinkele
Aql 19h39'56" -0d14'17"

Mary Margaret Kroupa-Casey
Ari 2h22'15" 25d23'52"

Mary Margaret Nagle
And 1h36'6" 48d24'5"

Mary Margaret Nolan
Psc 0h24'14" 18d17'55"

Mary Margaret Pinkerton
Ari 2h7'1" 20d32'58"

Mary Margaret Rathburn
Psc 1h24'47" 31d34'19"

Mary Marine Dzheyranyan
Cnc 8h38'58" 8d32'5"

Mary Marissa
Mon 6h47'25" 6d43'48"

Mary Marshall
Aqr 22h33'27" -0d47'31"

Mary Martin *Mom-Grandma*
Tau 3h46'34" 15d24'38"

Mary Mary
Sgr 19h35'25" -12d58'16"

Mary Mary
Uma 9h29'59" 64d26'14"

Mary Masuhr
And 23h29'48" 48d12'13"

Mary Matilda Amo
Cyg 21h16'36" 31d49'36"

Mary Maxine
Uma 10h17'27" 45d11'13"

Mary McAlister
Vir 13h18'42" 4d59'3"

Mary McClean
Cas 0h31'3" 51d4'35"

Mary McCollum -Mama
Sco 16h11'48" -15d36'45"

Mary McCune
Cas 1h39'29" 60d58'15"

Mary McDonough
Cas 0h2'56" 53d24'54"

Mary McFarland
Del 20h36'35" 19d11'28"

Mary McGee Lailhengue Durel
Tau 5h7'40" 18d21'33"

Mary McKenna
Uma 12h3'57" 49d36'28"

Mary McKenna
And 0h28'51" 40d13'18"

Mary McLean Carpluk
Crb 15h40'10" 36d38'7"

Mary McPherson - 6 February 1949
Cru 12h25'59" -61d50'26"

Mary Meadows
Cyg 20h3'37" 33d30'1"

Mary Meissel
Cnc 8h51'16" 26d46'35"

Mary Mesagno
Mon 7h21'1" -0d16'56"

Mary Messina
Lib 15h3'2" -2d34'56"

Mary Metzger
Vir 14h1'39" -16d12'9"

Mary Michele St. Germain
Lyn 7h45'49" 39d6'28"

Mary Michelina
Cas 0h51'25" 56d52'26"

Mary Milano
Leo 11h5'44" 20d27'51"

Mary Mockridge
Aql 19h40'37" 14d42'21"

Mary Monica Lynne
Vir 12h46'59" 1d20'54"

Mary Monica Scott
Aqr 22h17'36" 0d59'59"

Mary Montalbano
And 0h9'19" 32d15'17"

Mary Moore Whistler
Sgr 19h22'25" -16d54'25"

Mary Mulcahy Mattoni
Sco 17h24'22" -39d55'2"

Mary Mullery
Uma 10h57'43" 53d59'56"

Mary Mulvihill
Peg 23h57'56" 26d39'39"

Mary Munroe
Lib 15h3'46" -23d50'52"

Mary My Love
Uma 8h31'43" 63d24'35"

MARY MY PERFECT ANGEL
And 0h16'34" 46d31'31"

Mary (my sweetheart for life)
Aqr 23h1'29" -10d13'56"

Mary N. Porter
Cas 22h59'49" 57d27'48"

Mary Nader
Vir 13h43'49" -5d19'49"

Mary Nadine G. Giberson
Cap 20h25'58" -10d37'22"

Mary (Nana) Pica
Crb 15h26'53" 27d35'44"

Mary Napoli
Mon 7h20'45" -0d18'56"

Mary Nichole Francis
Vir 14h45'3" 1d14'28"

Mary Nicholson
Ori 5h28'32" -4d33'33"

Mary Noakes
Gem 6h50'54" 22d4'0"

Mary & Noble
Ari 2h16'40" 25d38'26"

Mary & Nora Richardson
Gem 7h40'17" 34d5'43"

Mary Novosedliak
Uma 8h38'21" 58d2'32"

Mary O
Uma 9h51'20" 72d49'19"

Mary O'Connor
Mon 6h24'13" 8d28'32"

Mary Olds
Aqr 22h42'2" 0d11'14"

Mary Olson
And 23h48'8" 42d44'15"

Mary Osborne
Cas 0h21'5" 54d40'29"

Mary Our English Goddess Star
Eri 4h29'53" -10d4'37"

Mary Owen Hebert
Sgr 19h4'49" -14d14'16"

Mary Owings Shriver Pierrepont
Lyn 7h17'8" 52d11'5"

Mary P. Potter mother of Stephen
Uma 11h38'57" 47d44'43"

Mary Pack
Cas 2h7'23" 62d12'20"

Mary Paige
And 1h9'24" 38d17'30"

Mary Paquette
Crb 16h8'21" 38d0'7"

Mary Pat
Cas 0h39'55" 57d53'3"

Mary Pat
And 1h2'40" 44d9'24"

Mary & Pat Fusco
Umi 14h21'42" 78d56'59"

Mary Pat Garry
Cyg 19h58'39" 40d5'13"

Mary Patricia Aker
Sco 17h29'57" -44d24'23"

Mary Patricia Caruso
Cyg 19h30'42" 28d13'38"

Mary Patricia Fallon
Cnc 8h58'41" 14d49'58"

Mary Peace
Crb 15h40'55" 28d58'8"

Mary Pehr Rubin
Cnc 8h52'46" 14d32'13"

Mary Perrin
Cas 1h31'19" 66d56'51"

Mary & Peter Drake (Mum & Dad)
Cyg 21h31'42" 31d40'2"

Mary Petersen
Uma 10h7'45" 56d27'7"

Mary Petras Hink
Cas 23h57'0" 52d8'5"

Mary Peyton and Adam
Sge 20h7'13" 20d50'11"

Mary Phelan
Lmi 10h6'29" 41d1'25"

Mary Pillow Colmore
Equ 21h14'8" 8d37'48"

Mary Piotto - Forever a Star
Ari 2h36'15" 14d38'45"

Mary Pishny
Peg 23h26'26" 32d0'26"

Mary Pleich-Santich
Tau 4h21'35" 23d52'32"

Mary Plum
Lyr 18h54'8" 39d29'7"

Mary Polcer
Sco 16h6'19" -10d19'31"

Mary Poppins
Lib 15h4'21" -3d15'35"

Mary Potter
Crb 16h22'46" 38d36'7"

Mary Presswood's Star
Crb 15h46'1" 31d20'21"

Mary Price
Crb 15h36'22" 30d29'19"

Mary Prothero
Uma 13h30'56" 53d51'54"

Mary Pyrzynski
Lyn 8h37'15" 40d34'38"

Mary R. Boyer
Leo 10h32'24" 13d46'6"

Mary R. Evans
Uma 10h18'45" 52d51'43"

Mary R. Falica-Lewis
Cas 23h30'42" 53d7'1"

Mary R. Molinaro-Hamada
Cas 0h44'47" 64d14'42"

Mary R. Van Pay
Lib 15h52'46" -12d32'40"

Mary Rachel Burchwell
Psc 1h19'38" 21d42'7"

Mary Rachel Carpinelli
Peg 23h19'14" 32d59'18"

Mary Rachel Sheeran
Cas 0h22'38" 52d34'22"

Mary Radebaugh Bowles
Srp 18h17'31" -0d13'52"

Mary & Ralph Shirk
Ori 5h19'59" 0d25'13"

Mary Randolph
Sgr 18h57'13" -29d12'33"

Mary Rebecca
Cyg 20h58'35" 47d13'20"

Mary Rebecca Callahan
Ori 5h10'37" -1d38'49"

Mary Rebecca Pearson
Vir 12h38'14" 3d25'33"

Mary Rebecca Streble
Psc 1h18'53" 17d11'6"

Mary Regina Pham
Sco 16h47'53" -32d39'39"

Mary Reuter's Angel
Mon 6h45'55" -0d2'47"

Mary & Richard Henszey
Del 20h46'39" 12d41'57"

Mary Rita Murnane
Uma 9h27'45" 44d49'43"

Mary Rizzo
Lyr 19h13'38" 37d42'13"

Mary Roberts
Lyr 18h33'34" 38d14'31"

Mary Roberts
Cas 1h29'29" 67d48'11"

Mary Robinson
Umi 16h55'46" 79d18'45"

Mary Rodriguez
Crb 15h53'17" 35d56'36"

Mary Roland David James
Lyn 6h29'4" 60d48'45"

Mary Rosalie
Sco 16h47'18" -44d1'25"

Mary Rose
Leo 9h53'17" 31d3'21"

Mary Rose Cascaes
Leo 11h13'1" 23d54'22"

Mary Rose Pinter
Cyg 20h40'40" 52d20'33"

Mary Rose Reitano
Sgr 18h3'39" -30d16'42"

Mary Ross Bowman
Cap 21h31'18" -20d28'56"

Mary Ross Thomas
Aqr 23h2'22" -6d2'10"

Mary Rowley
Mon 6h49'2" -0d8'31"

Mary Rudnicki
Sgr 19h21'26" -36d15'38"

Mary Ruh
Psc 1h49'43" 5d3'59"

Mary Russell
Cas 23h4'23" 54d43'4"

Mary Ruth
Cyg 20h59'32" 34d57'41"

Mary Ruth Briggs
Vir 12h18'6" 12d35'31"

Mary Ruth Gray
Cnc 8h57'39" 14d8'48"

Mary Ruth Sexton
Lib 15h21'13" -13d8'3"

Mary Ruth & Vivian Stringfield
Ori 4h50'20" 7d55'6"

Mary S. Harris
Her 16h17'45" 6d18'14"

Mary S. Jones (diva for life)
Sco 16h46'4" -29d11'36"

Mary S. Shainberg
Umi 14h53'15" 69d13'24"

Mary Sandra Ryan
Gem 7h39'24" 25d7'54"

Mary Sarbello
Cam 4h6'11" 68d17'21"

Mary Savoy Queen
Vir 11h40'30" -2d47'39"

Mary Scharf 12/18/30-8/1/03
Sgr 19h13'1" -22d3'20"

Mary Scheer
Sgr 18h15'56" -34d21'12"

Mary Schelich
Cas 0h16'38" 51d24'25"

Mary Schempf
Cas 1h41'51" 64d5'37"

Mary Scherff
Cas 1h28'4" 52d14'11"

Mary Schultz
Psc 1h29'59" 33d6'11"

Mary Sciarrino Estruch
Ari 2h0'55" 22d2'23"

Mary Scrivines
Cas 0h19'38" 51d32'57"

Mary Selena
Gem 6h33'31" 23d16'59"

Mary Seneschal
Cyg 21h51'8" 47d51'25"

Mary Settles
Gem 6h52'43" 30d26'37"

Mary Shaw
Cas 2h59'15" 57d35'4"
Mary Shaw
Sco 17h19'37" -33d9'57"
Mary Shields
Tau 4h25'55" 6d48'34"
Mary Smith
Cru 12h6'34" -62d25'52"
Mary Sneed
Aqr 21h59'47" -12d27'16"
Mary Sneed Pharris
Uma 12h12'54" 63d13'10"
Mary Soon Ae Henkin
Crb 16h10'21" 38d37'11"
Mary Stadler
Leo 11h41'17" 17d37'15"
Mary Standish Higdon
Lyr 18h58'30" 26d26'51"
Mary & Stanley Swerdoski
Uma 8h29'1" 69d24'49"
Mary Stanzione
Cas 0h10'0" 53d18'42"
Mary Star of the Sea
Uma 9h48'46" 46d18'44"
Mary Stella Glowacki Meizis
Cas 1h27'30" 68d49'38"
Mary Stephanie Songco
Cap 21h2'22" -21d28'14"
Mary Stephens
Sco 16h41'11" -43d58'55"
Mary Stepp
Her 16h54'45" 17d22'30"
Mary Stimus
Vir 13h0'46" 4d28'8"
Mary Stockard
Cas 2h22'51" 67d50'40"
Mary Stokesberry
Cas 0h58'28" 49d56'1"
Mary Stranghoener
Sgr 19h22'52" -17d43'12"
Mary Strong
Gem 6h54'45" 25d10'2"
Mary Sue
Tau 3h49'17" 26d54'52"
Mary Sue Badami
Sco 16h25'20" -27d5'48"
Mary Sue Busch
Vir 14h17'12" 1d44'34"
Mary Sue Elder
Ari 2h22'20" 14d33'15"
Mary Sue Erickson
Com 13h10'5" 28d43'1"
Mary Sue Hunt
Umi 21h19'45" 89d34'0"
Mary Sue McGee
Cap 20h51'56" -16d35'54"
Mary Sue Pallerino
Lmi 10h28'58" 35d44'30"
Mary Sue Smith
Ari 2h37'49" 22d0'44"
Mary Sue Welch
Uma 9h58'25" 63d56'18"
Mary Sue's Wish
Vir 13h56'31" -11d35'55"
Mary & Sully
Cyg 19h40'43" 29d17'54"
Mary Summerlin
And 0h12'12" 46d10'12"
Mary Susan Jane Richards
Crb 15h52'49" 26d21'53"
Mary Suzanne Harrison
Cap 21h56'52" -21d20'46"
Mary Szabo
Uma 9h12'59" 57d30'57"
Mary T.
Uma 10h44'4" 63d37'28"
Mary T
Lib 15h54'54" -6d9'2"
Mary T. Batchelor "6-23-96"
Crb 15h46'1" 34d42'39"
Mary T. Brassard
Cas 1h37'41" 63d11'59"
Mary T Foster
Cap 21h39'5" -15d41'58"
Mary Taylor
Cep 22h45'31" 63d46'29"
Mary Taylor - Mom & Sis
Cas 0h53'37" 60d44'46"
Mary Taylor Zimbalist
And 23h43'51" 44d7'9"
Mary Tennien
Cas 1h3'14" 61d20'2"
MARY TERESA JACOBY FARO
Leo 10h48'15" 19d12'43"
Mary Teresa Morgan
Cap 20h35'55" -17d15'30"
Mary The Brite Star
And 0h37'55" 29d17'57"
Mary Thelma
Leo 10h23'54" 24d58'46"
Mary Theresa Bucknavage
Cyg 20h3'22" 35d42'18"
Mary Theresa Goulding
And 23h6'5" 47d26'9"
Mary Theresa Klein
Uma 10h59'51" 56d52'2"
Mary Theresa Suarez Colon
Ori 6h6'32" 4d21'25"
Mary Therese
Cas 1h45'53" 61d11'56"
Mary Thrasher
Crb 15h51'14" 36d38'23"
Mary Tiara Cirnigliaro Lamb
Uma 13h13'53" 59d15'26"
Mary Todd Marr
Vir 13h12'22" 7d28'8"
Mary Toder
Cnc 8h11'29" 8d14'33"
Mary & Tony 50
Cnc 8h40'46" 30d42'37"
Mary Tootsie Mullenbach
Lyn 8h14'34" 59d35'10"
Mary Travis
Crb 15h38'25" 37d47'32"
Mary Travis Haisten Doepner
Leo 10h45'39" 16d11'52"
Mary Tscheulin Bagwell
Gem 6h48'57" 26d0'51"
Mary "Twinkle" Geray
Leo 10h43'4" 18d26'53"
Mary V. Kinahan
Sgr 18h54'38" -21d30'0"
Mary Valentine-Terranova
Cas 1h24'39" 62d7'59"
Mary Vasquez
And 0h16'38" 28d18'7"
Mary Vieth
Psc 0h39'18" 17d37'33"
Mary Vine
Psc 0h37'10" 8d7'21"
Mary Violet Collins
Uma 8h49'20" 68d40'57"
Mary Virginia Carnell - Christening
Cyg 21h10'45" 43d27'50"
Mary Virginia Foley
Uma 10h54'50" 71d56'23"
Mary Virginia Housel
Uma 10h4'24" 49d18'18"
Mary Virginia Jerome
Cas 23h56'30" 54d13'41"
Mary Virginia Schmauss
Cyg 21h30'54" 36d18'23"
Mary Vollmer
And 0h21'4" 38d45'47"
Mary Volpe
Crb 15h36'40" 25d56'56"
Mary Walker
Cas 0h53'9" 50d7'4"
Mary Wark Biear
Cas 1h23'1" 58d41'43"
Mary Warshal
Uma 8h13'10" 62d43'55"
Mary Wells
And 1h20'38" 40d57'27"
Mary Weyker
Cas 0h13'6" 56d41'54"
Mary Whalen's Eastern Star
Cet 1h40'14" -0d30'55"
Mary Wicklund
Cas 23h7'16" 58d55'4"
Mary & William Myers
Uma 10h36'4" 60d6'44"
Mary Williams
Lyn 7h27'24" 56d30'24"
Mary Wissen-Neumann
Uma 11h44'52" 38d7'14"
Mary Woychuk Cocchiarella
Ari 2h48'20" 26d55'29"
Mary Wright
Cyg 20h39'23" 48d41'26"
Mary "Yia Yia" Lambros
Gem 6h43'41" 14d40'3"
Mary Zavaglia
Col 5h59'45" -34d51'53"
Mary Zemola
Uma 9h16'58" 58d29'44"
Mary Zimmerman
Cyg 19h44'51" 53d44'18"
Mary Zuella
Crb 15h43'29" 33d2'31"
Mary, Jeffrey, Kris & Katie Shapiro
Lyr 19h8'36" 46d41'22"
Mary, Star of Trabia.
Tau 5h27'37" 21d55'16"
Mary, the Most Amazing Mum
Cyg 21h42'21" 45d50'51"
Marya Isabelle
Vir 13h30'6" 11d55'38"
Marya Swenson
Aqr 22h30'30" -2d8'45"
MaryAlexander Myers
Cas 3h13'3" 64d43'58"
Maryalice Multari
Gem 6h43'54" 34d12'50"
Maryam Eftekhari
Lib 15h16'42" -15d49'59"
Maryam & Sameer
Gem 7h38'45" 32d37'54"
Maryam's Soul
Aqr 23h36'38" -7d4'10"
Maryams Wish
Cyg 20h58'1" 31d11'32"
Maryan Rijkhoek
And 23h47'20" 39d26'41"
Maryana Panayiotou
Crb 16h18'59" 38d37'41"
Mary-Angel
Ori 5h0'10" 15d5'9"
MARYANN
Aql 19h10'44" 3d31'22"
Maryann
And 2h15'5" 45d53'35"
Maryann
Cyg 20h21'32" 46d24'31"
Maryann
And 0h45'56" 39d33'28"
Maryann
Cnc 8h19'22" 31d2'32"
MaryAnn
Lyn 8h23'49" 35d20'6"
Maryann
Aqr 22h54'5" -9d29'59"
Maryann
Cas 3h30'47" 69d34'14"
Maryann
Uma 13h46'10" 57d23'27"
Maryann
Lyn 7h14'2" 56d54'26"
MaryAnn 1
Uma 10h53'45" 51d22'7"
Maryann 102
Lyn 7h34'20" 35d40'42"
Mary-Ann Abbott
And 1h13'21" 46d35'35"
Maryann - Ang
Aql 19h42'49" -0d1'35"
MaryAnn Brown
Sgr 17h59'23" -28d33'31"
Maryann Catherine Bland
And 23h46'10" 32d58'22"
MaryAnn Clark
Crb 15h43'0" 31d42'12"
MaryAnn D'albis
Cyg 20h44'46" 35d19'28"
Maryann Davis (Always Shining)
Ari 2h40'10" 21d9'16"
MaryAnn Fara
Aql 19h26'14" -11d15'15"
Maryann (Fluffy) & Paul
Lib 14h52'12" -13d30'44"
MaryAnn Fox
Psc 1h25'27" 25d46'8"
MaryAnn Grecca Kelly
Ori 6h13'5" -3d53'9"
Maryann Harrington
Cnc 8h48'25" 6d51'32"
Maryann Lillian
Cas 1h32'57" 69d10'2"
Maryann Macintosh
Vir 14h16'11" 3d35'51"
MaryAnn Malzone
And 23h13'47" 50d43'55"
MaryAnn Minervino
Cas 1h22'50" 56d36'10"
MaryAnn "Momma Vita" DeToro
And 1h7'36" 42d25'55"
MaryAnn Murphy
Ari 2h18'5" 23d56'33"
Maryann My Love
Sgr 18h56'50" -29d12'32"
Maryann Nordengreen
Lib 15h30'4" -3d48'18"
Maryann Pantano Davis
Ori 5h29'51" -0d24'46"
MaryAnn Pranulis
Aqr 22h5'59" -21d38'22"
Maryann Preziosi
Cyg 21h26'22" 35d6'53"
Maryann Prouty
Cap 20h44'35" -22d29'50"
Maryann Raffa
Leo 10h48'28" 9d35'37"
Maryann Richmond Luther
Cas 0h15'50" 64d27'23"
MaryAnn Rose Walker
Cas 3h3'12" 65d3'37"
Maryann Russo
Sco 16h48'44" -42d35'21"
Maryann Salafia
And 1h43'32" 45d7'5"
Maryann Shurtleff Coons
Aqr 22h34'45" -16d12'16"
Maryann Splendora
Cas 2h18'53" 67d57'37"
MaryAnn Stanford
Cnc 8h20'52" 6d58'24"
Maryann Sylvander
Uma 8h29'11" 61d29'10"
Maryann Walton
Cas 23h25'54" 57d57'42"
Maryanna
Cap 21h20'16" -14d59'58"
Maryanna Martinez
Psc 0h2'11" -2d59'43"
Maryanna Sophie
Leo 9h33'36" 15d34'46"
MaryAnne
Cnc 8h56'14" 18d7'47"
MaryAnne
And 23h8'19" 50d38'41"
Maryanne
Cnv 13h8'46" 46d25'10"
Maryanne Alves
Sco 16h8'43" -10d19'54"
MaryAnne Bonetti
Peg 0h1'14" 30d17'4"
Maryanne Casey
And 1h48'1" 46d35'27"
Mary-Anne Celine
Cas 0h57'13" 53d57'30"
Maryanne Cestaro
Tau 3h55'22" 24d45'14"
Maryanne Ford Friend & Confidante
Sco 16h7'10" -11d10'33"
Maryanne Fotter
And 23h16'50" 48d30'18"
Maryanne Kosis
Cyg 19h47'20" 33d36'28"
Maryanne Mata
Gem 7h44'50" 26d32'31"
Maryanne Shea
Leo 11h0'4" 19d41'4"
Maryanne Shula
Cnc 8h57'52" 8d8'12"
Maryanne Stein
Mon 6h51'35" 3d7'34"
Maryanne's Reflection
And 0h17'23" 45d4'4"
Maryann's Star
Uma 9h12'38" 56d32'47"
Maryapril
Ari 2h44'31" 23d29'58"
MaryArden Jones
Vir 13h59'48" 4d32'53"
Maryasha Lurye-Soybelman
And 1h15'51" 37d27'21"
MaryAshleyDanaBrianJudy NathanJoshJr
Cyg 19h54'31" 55d19'21"
MaryB
Uma 13h54'0" 61d26'17"
Marybeall
And 0h40'17" 39d51'21"
MaryBear
Lib 15h2'59" -1d47'12"
Marybelle Gregory
Ari 2h52'39" 11d10'41"
MaryBeth
Sgr 18h49'34" -19d8'14"
Mary-Beth "BooBear" Evans
Psc 0h53'10" 8d40'51"
Mary-Beth Danko
Umi 15h31'40" 70d7'54"
Marybeth Francis
Cas 1h28'34" 57d25'29"
MaryBeth Lucius
Crb 16h1'45" 35d59'30"
Marybeth Merrill
Lyn 9h16'39" 37d56'49"
MaryBeth Musial
And 2h10'29" 41d28'37"
MaryBeth N. Doyle
Gem 7h37'14" 27d5'35"
Marybeth Seligmann
Aql 19h32'34" 4d20'20"
MaryBeth Shannon
Gem 7h6'35" 32d47'47"
Marybeth Vanko
Ari 3h28'27" 27d29'29"
MaryBeth23
Gem 7h6'10" 22d33'18"
MaryBeth's Forever Love
Vir 12h15'59" 11d27'48"
MaryBob1955
Lyr 18h37'13" 29d43'45"
MaryBoyd01
Crb 15h35'34" 32d43'22"
MaryBud
Ori 6h20'32" 10d32'41"
Maryca Chesterman
Uma 9h24'36" 71d35'36"
MaryCarmen Rodriguez
Sgr 18h26'20" -18d18'15"
Mary-Carter Darlington Kniffen
Cnc 8h6'27" 24d36'29"
Marychris pour la vie
And 1h1'54" 46d2'41"
Mary-Christine
Gem 6h30'18" 22d21'34"
Maryclaude
Uma 9h15'47" 58d33'55"
Marydele Turtle
Uma 13h23'16" 58d36'40"
MaryEdu
Del 20h32'14" 17d16'54"
MaryEllen 11-01-82
Crb 15h52'18" 37d23'15"
Maryellen Celinski
Lyn 8h26'3" 34d53'34"
Maryellen Elizabeth Rush
And 0h2'50" 44d30'35"
Maryellen Kline
And 0h37'53" 42d52'24"
MaryEllen McLeod
Crb 16h11'42" 36d49'49"
MaryEllen Mulrenan - Cardaci
Dra 19h46'15" 76d36'59"
Maryellen the Magnificent
Gem 6h34'5" 14d2'21"
MaryEllen Tremendously Great Tucker
Gem 7h28'39" 26d54'35"
Maryellen's Star of Love
Tau 5h11'32" 18d24'20"
Mary-Eve D'Amours
Umi 14h13'27" 75d22'33"
MaryFran Hopper Deachin
Aqr 21h5'42" 1d33'22"
MaryFrances (Marilyn Mineo Paladin)
Aqr 22h10'8" -1d57'18"
Mary-Grace Marie Moore
And 23h19'51" 49d7'52"
Maryhallixus Major 01-09-1987
Aql 20h12'43" 6d7'40"
Maryiri Alexandra Tejeda Mateo
Peg 22h21'12" 20d13'15"
Maryjane
Tau 3h41'13" 26d11'12"
Maryjane
Crb 15h43'3" 31d5'24"
Maryjane Drajewicz is Loved!
And 23h43'33" 38d38'17"
MaryJo Audiss
Lyr 19h1'58" 26d1'53"
Mary-Jo Louise
Cas 23h13'18" 55d49'2"
Mary-Jo Lyman
Ori 6h16'55" 14d52'44"
MaryJo Lynn Lindquist
Cas 1h37'30" 65d33'58"
MaryJo Muller
Dra 15h37'14" 60d57'4"
Mary-John Godfrey-Dalrymple
Leo 10h53'27" 19d31'13"
Maryjo's Sparkle
Cap 21h46'54" -12d40'29"
Mary-Kate Longyear
And 2h22'46" 45d31'16"
Marykate Smith
Crb 16h7'44" 36d5'3"
Mary-Kathryn
Lib 15h13'44" -15d37'59"
Marykay
And 1h53'12" 42d59'18"
Maryland's Itty Bitty Writing Light
Umi 15h15'45" 78d38'12"
Marylee
Lyn 6h27'37" 57d57'38"
Marylee Ann Abreu
Cap 21h4'8" -16d58'6"
Marylène Baud
Lac 22h49'59" 49d17'40"
Maryline
Per 1h43'50" 54d23'22"
Maryline & Sylvain
Crb 16h23'33" 31d4'57"
Marylise
Aqr 23h35'22" -9d43'3"
Marylise, together in eternity
Leo 9h32'51" 10d49'59"
Marylou and Anthony 3/4/05
Lyn 8h57'5" 40d22'36"
Marylou Descoteaux née 26-04-02
Umi 13h5'44" 69d32'13"
Marylou England
Sgr 19h50'46" -18d45'58"
MaryLou Hannon
Ori 5h39'9" -5d27'8"
Marylou J. Amor
And 0h15'0" 44d36'0"
MaryLou & Jerry
Cyg 19h23'37" 53d18'56"
Mary-Lou Llull
And 0h35'31" 42d25'18"
MaryLou M. Newman
Lmi 10h25'47" 32d18'11"
Mary-Lou Mackie-Tomlin
Uma 9h2'10" 50d50'20"
MaryLou Vanzini
Cyg 19h51'32" 33d41'28"
MaryLouise Chavoyo
Uma 10h24'16" 65d54'55"
Marylynn Rowe Miles
Lyr 18h44'16" 38d56'46"
MaryMorganSuperStarRed book2006
Cyg 21h26'27" 39d13'33"
Maryn
And 2h16'35" 45d28'36"
Maryn Asher
Uma 11h9'59" 60d57'32"
Maryn Eve Simonof
Tau 4h41'57" 22d36'15"
Maryna Synytska Wagner
Cyg 19h43'58" 35d26'10"
Maryocéane Claude Célestine Guy
Ari 2h6'21" 23d6'55"
Maryoma
Cma 7h0'22" -26d32'22"
Mary-Rachael
Uma 9h54'50" 57d7'17"
Maryrose Barber
Psc 1h19'12" 15d21'45"
Maryrose Leeann Ratchko-Maggio
Tau 3h45'4" 27d29'56"
Mary's Angel
Uma 9h21'38" 51d56'18"
Mary's Angel Eyes
Lyn 7h55'20" 54d29'37"
Mary's Diamond
Crb 16h7'47" 38d33'29"
Mary's Diamond
And 0h41'8" 34d53'0"
Mary's Heart
Cnc 8h5'34" 26d8'11"
Mary's Inner Light
Cap 20h30'40" -20d7'19"
Mary's Lucky Star
Vir 13h13'37" 4d32'18"
Mary's Morrin
Gem 6h23'28" 21d18'44"
Mary's North Star
Dra 19h12'51" 78d13'18"
Mary's Shining Star
Eri 4h23'41" -0d58'49"
Mary's Smile
Cas 23h41'24" 55d23'12"
Mary's Sparkle
Her 16h18'31" 19d8'40"
Mary's Star
Ori 5h26'3" 9d55'17"
Mary's Star
Cnc 8h52'41" 12d46'8"
Mary's Star
Cnc 8h38'48" 31d46'14"
Mary's Star
Leo 9h43'59" 30d45'25"
Mary's Star
And 23h38'3" 47d50'23"
Mary's Star
Cas 1h52'17" 65d43'18"
Maryse
Aqr 23h1'26" -8d33'35"
Maryse Thomas
Cnc 8h4'4" 11d31'44"
Maryshell
Sco 16h17'38" -19d34'10"
Marysol Villegas
Cam 7h18'9" 65d45'42"
MarySue e Domenico
Uma 12h10'18" 62d30'18"
Mary-Sue Mc Nary
Vir 13h43'4" -15d6'50"
MaryTom
Lyn 6h48'28" 51d52'54"
Mary-Walker Watson
Gem 6h46'27" 18d22'34"
Marz: Mi Amor, Mi Estrella
Gem 7h44'59" 18d52'57"
Marzena
Aqr 21h51'22" -4d12'9"
Marzenka Kubis
Umi 14h48'40" 85d32'35"
Marzia
Aql 20h12'20" -0d43'48"
Marzia 4374
Mon 6h30'34" 7d26'10"
Marzia Moretti
Crb 16h24'1" 29d2'58"
Marzio Arditi
Umi 13h10'31" 70d3'28"
Marzio Pietro Luca
And 0h59'26" 47d39'10"
mas 4 pls
Cap 21h29'36" -24d49'54"
Ma's LiL One's
Cas 0h30'25" 50d14'28"
Ma's Loving Heart
Cas 0h32'53" 53d18'51"
Ma's Star-Forever Brightest 3/28/39
Ari 3h28'31" 21d27'57"
Måså
Her 17h53'2" 39d32'28"
Masa Djordjevic - 29.06.1984.
Cas 23h40'6" 52d58'26"
Masaki Oya
Leo 9h35'58" 21d43'9"
Masako
Psc 1h7'47" 23d35'32"
Masako Hama
Cen 13h4'29" -48d34'57"
Masalu
And 1h56'8" 38d48'56"
Masami
Lib 15h6'13" -10d49'43"
Masanao Yamashita
Cru 12h27'50" -61d48'32"
Masashi & Hiromi Love Forever
Psc 1h18'59" 5d34'12"
Masataka
Leo 10h21'45" 11d54'3"
"Masel" Marcel Bertschi
Psc 0h48'31" 15d38'53"
Masella's Magic
Cet 2h30'31" 8d45'28"
Maset Fabio
Her 18h49'21" 21d58'1"
Masey Ann
And 0h36'52" 36d11'1"
MASH
Sco 16h2'46" -17d9'38"
Mash
Uma 9h20'30" 54d42'34"
MASH Choir-2005
Uma 10h47'57" 59d21'47"
M.A.S.H. - Madly in Love, Forever
Lib 14h27'57" -15d44'43"
Masha
Boo 14h33'11" 54d54'55"
Masha
Uma 10h30'58" 68d26'9"
Masha
Tri 2h37'24" 34d44'23"
Masha
Tau 4h38'11" 19d51'18"
Masha
Vir 13h21'44" 12d12'14"
Masha Fishman
And 2h27'19" 49d18'44"
Masha Galeb
Aqr 21h46'54" -1d45'47"
Masha Perelman
Lib 14h39'33" -9d9'47"
Masha Silja Bock
Umi 13h13'52" 74d51'44"
MASHapella-2005
Uma 12h10'42" 59d50'26"
Masha's Star II
Ori 6h6'52" 19d21'6"
Mashee
Uma 9h45'39" 53d57'31"
Mashenka
Sco 16h15'35" -8d59'17"
Mashenka V.
Psc 1h25'0" 32d22'24"
Mashha
Aqr 21h7'45" -7d26'50"
Mashhoor
Psc 2h1'26" 4d22'51"
Mashley
Cam 6h31'29" 68d23'4"
Mashly
Vir 14h34'57" 1d45'9"
Mashona
Crb 15h44'47" 27d15'47"
Mashoo
Sgr 19h42'0" -19d52'52"
Masie Emma Thomas
And 0h13'57" 26d18'36"
Mäsis Angostura Bar
Umi 14h42'24" 76d51'9"
Maskowitz
Gem 6h37'28" 22d18'32"
MASKRAMYAK
Aur 5h19'56" 41d57'25"
Mason
Leo 9h36'1" 17d31'59"
Mason
Aqr 22h18'27" 1d32'50"
Mason
Ori 5h50'6" -4d17'57"
Mason
Dra 18h38'44" 67d48'39"
Mason
Uma 11h37'28" 58d22'26"
Mason 1 today. Love Grampy and Nana
Umi 16h22'22" 70d26'48"
Mason Adam Pierle
Her 17h42'42" 26d34'16"
Mason Alexander Campbell
Umi 13h42'44" 74d46'43"
Mason Allen Miller
Uma 11h25'15" 32d31'26"
Mason Andrew Davenport
Cnc 9h17'12" 10d55'9"
Mason Angelo Comaites
Leo 11h20'6" -4d23'13"
Mason Anthony Lee
Aur 5h35'12" 33d38'37"
Mason Bauer Tallman
Psc 1h4'30" 21d9'56"
Mason Borum's Smile
Per 4h1'56" 47d11'26"
Mason Bosch-Bird
Sco 16h17'57" -12d35'46"
Mason Bowling Stanley
Sgr 18h4'26" -26d43'59"
Mason Brade
Uma 11h40'29" 40d19'31"
Mason Broc Duncan
Lib 15h17'20" -6d4'30"
Mason Bryce Moxley
Aur 6h31'31" 43d24'0"
Mason Chamberlain Turner
Umi 15h20'33" 73d49'33"
Mason Charles Morosky
Umi 14h18'54" 76d51'37"
Mason Charles Swanger
Hya 9h23'11" -3d32'51"
Mason Christopher Dale
Umi 14h21'10" 68d44'5"
Mason Colby
Ari 3h9'3" 29d52'54"
Mason Cole Beardmore
Dra 18h59'6" 63d19'31"
Mason Cole Heer
Her 18h50'6" 22d34'15"
Mason Cole Smith
Lib 15h15'59" -22d53'54"
Mason Corey Weldon
Her 16h33'13" 13d42'38"
Mason Crist Heller
Umi 15h33'5" 71d12'16"
Mason Crowther-Sullivan
Umi 16h45'28" 80d10'27"
Mason D
Cam 4h18'36" 73d50'8"
Mason Dane Stark-Nichols
Cap 21h25'10" -17d25'22"
Mason Daniel David Cline
Cnc 8h50'41" 12d17'0"
Mason Daniel Freddie Pivarro-Monaghan
Her 17h2'54" 19d38'26"
Mason Danielle Blalock
Vir 14h38'2" 2d14'8"
Mason David
Uma 11h4'34" 53d33'18"

Mason Dean Sidle
Leo 11h27'1" 8d50'25"
Mason Diego
Her 17h10'11" 32d43'36"
Mason E. Etman
Aqr 23h50'11" -19d43'7"
Mason Edward Payne
Tau 5h42'57" 22d58'4"
Mason Emilee Dixon
Psc 0h58'30" 28d19'12"
Mason Engelke
Tau 5h46'47" 19d52'52"
Mason Eugene Wise
Vir 13h24'31" -0d51'52"
Mason Evan Bard
And 23h41'39" 38d45'16"
Mason Evelyn Madden
Peg 22h8'28" 12d58'58"
Mason Fredrick Sanders
Umi 15h17'58" 72d31'40"
Mason Hennigan
Uma 11h41'13" 41d10'52"
Mason Isaiah Krueger
Cap 20h31'7" -11d6'5"
Mason Isiah Wood
Vir 11h49'34" 3d44'2"
Mason J. Swanson
Aur 5h41'45" 48d28'54"
Mason Jack Longley
Umi 14h51'51" 68d54'28"
Mason Jack Swift
Per 2h13'29" 51d13'59"
Mason James
Gem 6h43'22" 19d44'32"
Mason James Carpenter
Cyg 20h56'48" 46d49'23"
Mason James Hallchurch
Dra 17h30'52" 67d36'40"
Mason James Leppard
Sgr 18h7'51" -30d29'3"
Mason James Minitti
Ari 3h10'4" 11d45'6"
Mason James Redmond
Sco 16h51'31" -42d6'40"
Mason James Sardella
Psc 1h25'33" 31d21'59"
Mason James Yeager
Umi 16h12'49" 76d1'26"
Mason Jared Levine
Cnc 8h42'20" 31d7'20"
Mason Jay Matthews
Tau 4h4'13" 23d39'48"
Mason Jefferson Alexander
Ori 4h52'11" 2d47'38"
Mason Jeffrey
Tellinghuisen
Aur 5h9'14" 40d16'46"
Mason J.J. Kusek
Her 16h10'30" 24d37'32"
Mason Joseph Moore
Tri 2h28'26" 31d25'51"
Mason Kerr
Her 17h19'23" 38d29'59"
Mason Lane
Boo 14h35'3" 33d4'37"
Mason - Love / Love -
Mason
Per 3h20'39" 51d5'33"
Mason Mabee
Dra 12h12'9" 68d43'30"
Mason Maclean
Her 17h44'30" 22d44'29"
Mason Matthew Bowen
Sco 17h50'2" -30d8'17"
Mason Maybe
Dra 17h5'19" 63d55'19"
Mason Michael Opie
Her 16h40'42" 48d47'35"
Mason Moppin's Star
Uma 9h31'59" 45d43'41"
Mason Navon Wood
Leo 11h54'11" 20d43'2"
Mason Olivia Gossip
Uma 9h1'48" 68d10'2"
Mason Owen Whitten, May
14, 2007
Tau 5h34'18" 22d51'22"
Mason Patrick Feliciotto
Ari 2h38'56" 22d4'35"
Mason Patrick Feliciotto
4/18/07
Uma 9h14'30" 62d30'55"
Mason Ralph Lemont
Lmi 10h48'58" 35d48'4"
Mason Randall King
Uma 11h6'57" 48d15'25"
Mason Riley Hosek
Sgr 18h33'30" -22d43'12"
Mason Robert Fitts
Boo 15h28'29" 34d27'34"
Mason Robert Snider
Ori 5h22'43" 3d28'39"
Mason Rory Weiss
Vir 13h25'48" 13d12'20"
Mason Scott Knocke
Her 18h43'25" 20d35'40"
Mason Shawn
Sgr 19h27'37" -41d20'31"
Mason Sheffer
Cnc 8h44'14" 18d54'58"
Mason Stachowski
Aqr 22h7'37" -19d37'22"
Mason Thomas Kemp
Leo 9h23'28" 12d24'40"
Mason Todd Wagner
Boo 14h15'36" 48d0'46"
Mason Tyler McGraw
Srp 18h31'3" -0d38'28"
Mason Tyler Welchko
Sco 16h42'33" -34d42'22"
Mason Ward Graham
Dra 18h59'0" 53d58'0"
Mason Warren
Cyg 20h18'6" 49d29'23"
Mason William Finke
Cnc 8h18'7" 31d3'12"
Mason William Krneta
Houghton
Uma 8h45'0" 47d12'41"
Mason William Pascoe
Lib 14h53'34" -12d5'31"
Mason William Peterson
Uma 11h32'2" 59d3'43"
Mason Xavier Dutton
Leo 9h51'22" 19d0'37"
Mason's Magic
Cep 0h10'6" 70d21'1"
Masons Mate
Cra 18h8'28" -44d46'14"
Mason's Star
Umi 14h20'34" 65d33'31"
Mason's star
Her 17h15'3" 27d10'52"
Mason's Star
Cyg 20h17'43" 35d31'30"
Massa Joe Fuscaldo
Sco 17h21'2" -44d48'26"
Masseure Isabelle Asselin
Cas 1h20'8" 55d53'24"
Massiel
Aqr 23h55'16" -11d25'0"
Massimiliano
Lyr 19h15'17" 38d40'33"
Massimiliano
Her 16h47'30" 13d4'30"
Massimiliano
Ori 5h57'14" 5d39'23"
Massimiliano Bianchi
Tri 2h39'17" 35d4'27"
Massimiliano Citarelli
18/10/1983
Cyg 19h58'46" 59d16'17"
Massimiliano Fantozzi
Her 16h42'6" 22d52'9"
Massimiliano Il Saggio
Uma 11h56'9" 31d58'30"
Massimiliano Patrizio
Cibelli
Cap 20h43'27" -24d39'54"
Massimiliano Secci
Ori 5h1'7" -0d56'26"
Massimiliano Teia
Cnv 12h27'6" 37d24'14"
Massimo
Her 16h32'48" 23d11'47"
Massimo
Umi 14h43'8" 76d16'0"
Massimo Arrighi
Umi 14h7'38" 75d25'59"
Massimo Barducci
Cyg 19h43'59" 34d53'42"
Massimo Califiore's
Wishing Star
Cep 2h54'31" 83d45'12"
Massimo Carpinteri
Leo 10h22'30" 14d4'54"
Massimo Conti
Ori 5h53'43" 22d38'57"
Massimo Fornasari
Tau 4h20'45" 13d17'16"
Massimo Franceschiello
Cas 0h50'40" 58d20'13"
Massimo Marrosu
And 2h10'0" 38d36'48"
Massimo*Orlando
Uma 10h50'40" 59d0'19"
Massimo Pasini
Per 4h21'53" 39d16'34"
Massimo Stevani
Per 4h11'33" 42d47'49"
Massoni
Ori 5h24'18" 2d14'43"
Master
Lyr 19h2'58" 42d51'43"
Master
Dra 19h39'45" 70d0'44"
Master Aiden Porton
Her 17h26'10" 31d55'59"
Master Alfie Wallace
Her 18h9'57" 21d23'47"
Master Andre Cameron
Gem 7h36'26" 29d36'47"
Master Andrew James
Cutcher
Sco 16h6'32" -13d28'57"
Master Augusto Ferraiuolo
Lib 15h37'6" -10d21'15"
Master Brat
Psc 1h22'38" 32d59'50"
Master Bruce's Star
Ori 5h44'13" 8d12'7"
Master Callum James
Phillips
Umi 14h34'18" 67d45'51"
Master Camden Tyler
Hamel
Sco 17h52'16" -40d37'46"
Master Chase
Sco 15h51'12" -23d52'16"
Master Chief
Ori 6h4'23" 11d0'43"
Master Chief C.E.Evans
Her 16h24'9" 22d14'43"
Master Clark
Sgr 19h21'23" -31d1'37"
Master Coppins
Lib 15h6'57" -27d27'54"
Master David
Peg 22h46'49" 31d29'6"
Master Elizabeth Filleti
Peg 23h14'45" 26d33'8"
Master Ernie Temple
Vir 12h10'22" 11d0'18"
Master J W Dally
Cas 0h33'13" 63d9'43"
Master Jack Watson -
"Grandad Jack"
Ori 6h6'5" 0d53'10"
Master James
Psc 1h8'33" 28d48'40"
Master Jason Landau
Cnv 12h56'54" 41d26'31"
Master John Robert
Schwalke
Gem 7h24'46" 17d28'20"
Master Max Grunau
Sco 17h51'19" -35d16'22"
Master Mom
Psc 1h10'31" 9d38'45"
Master Nicholas Tyler
Callaway
Her 18h2'41" 36d25'15"
Master of all that is Good
Uma 9h51'48" 64d3'55"
Master of the Sky
Aql 19h5'4" -0d42'21"
Master Reese Adonis
Winborn
Boo 14h41'30" 14d49'12"
Master Rick Hoon Lee's
Magical Star
Per 3h50'13" 32d26'56"
Master Rig
Gem 6h58'44" 20d17'16"
Master Sargent Gerd N.
Hoffmann
Her 16h48'2" 9d18'32"
Master Sergeant Grant
Leigh
Per 3h28'24" 54d21'40"
Master Sergeant Loren J.
Zimmer
Dra 19h21'7" 74d42'19"
Master Seth Samuel
Weissman
Cnc 9h4'37" 18d32'15"
Master Simon Thompson -
Simon's Star
Per 3h17'10" 40d55'2"
Master Taylor
Lyn 8h39'59" 33d33'23"
Master Tobias of
Weatherford
Ori 5h30'55" 1d30'22"
Master Vincent Chun
Umi 14h20'3" 73d3'52"
Master William Houston
Howie
Umi 15h21'32" 72d11'18"
Master/Woman 12-15-05
Aqr 22h21'29" -12d30'4"
MasterChief Christopher
Grantham
Pho 0h41'2" -41d33'54"
Masters
Aql 19h43'8" 6d29'4"
Master's Fire Angel
Sco 16h9'18" -10d44'21"
Master's jewel
Umi 14h23'21" 81d57'51"
Mastroianni
Uma 11h21'22" 62d41'56"
Masugana
Ori 5h36'41" 6d13'17"
Ma-Suzanne
Cas 1h14'53" 54d23'5"
Masy
Cas 23h48'16" 56d11'13"
Mat Mat
And 23h5'9" 47d59'12"
Matalena Lynn Sciulli
Uma 9h53'14" 66d33'18"
Matalyn Rose
And 0h49'29" 35d54'55"
Matan H. Freedman
Vir 13h19'52" 12d46'39"
Mataya
Cap 20h24'20" -13d1'12"
Matchbox Man
Sgr 19h41'31" -13d56'11"
Matchy
Lib 15h34'31" -13d25'36"
Mate 17
Lmi 10h39'30" 38d49'25"
Matej Brezina
Cmi 7h23'55" 4d24'17"
Matelyn Carrin Carrico
Uma 8h41'20" 52d29'10"
Mateo
Leo 10h16'9" 9d1'47"
Mateo Aidan
Tri 2h9'11" 34d20'14"
Mateo Joaquin Baiza
Aur 6h10'51" 52d11'10"
Matéo Mouton
Col 6h1'6" -28d15'42"
Mateo Natividad Cantu
Lib 15h54'14" -19d54'19"
Mateo's Forever Birthday
Star
Cap 20h50'51" -26d7'16"
Mater
Leo 11h33'35" 6d26'55"
Mater Avis
Uma 9h28'13" 49d38'1"
Mater Carty
Cas 1h26'50" 52d5'15"
Mater Lux Prima
Sgr 19h24'32" -28d57'57"
Mater Matris
Cas 1h19'48" 57d30'5"
mater matris
Leo 11h19'58" 15d31'34"
Materika
Leo 11h26'33" 21d45'29"
Mates
Psc 1h0'48" 19d21'21"
Mateusz Borowiecki 2005
Lmi 10h8'16" 37d31'6"
Mathair Alainn
And 2h20'17" 42d4'20"
Mathelkeo Gale
Cyg 19h53'46" 33d41'36"
Mathéo
Gem 6h45'58" 18d55'27"
Matheson
Crb 16h13'40" 30d3'17"
Matheson Clyde
Gem 7h34'40" 24d29'10"
Mathew
Ori 5h20'48" 6d24'49"
Mathew
Tau 4h25'51" 5d42'30"
Mathew Adam Montanari
Ori 6h13'43" -1d12'42"
Mathew Alistair Bookallil
Ori 5h35'21" -1d30'55"
Mathew Allen Sweatland
Uma 9h57'59" 59d37'14"
Mathew and Laura
Aql 19h13'32" 7d25'8"
Mathew Andes
Psc 1h24'55" 21d39'32"
Mathew B. Baker
Uma 11h5'7" 34d28'17"
Mathew Bradley Heil
Uma 11h15'38" 36d12'38"
Mathew Casaubon
Sco 17h28'26" -37d58'51"
Mathew Charles Gambler
Her 16h36'33" 37d8'46"
Mathew D. Divito
Per 3h44'51" 38d20'8"
Mathew D. Negri
Uma 12h3'23" 54d25'15"
Mathew David Breach -
Mattie's Star
Aql 19h53'8" 14d53'21"
Mathew Davis Puckett
Lib 14h52'28" -14d26'4"
Mathew Demopoulos
Per 4h16'58" 47d37'19"
Mathew Drue Livingston
Cap 21h22'53" -15d33'2"
Mathew Galan
Vir 13h26'51" 7d41'43"
Mathew Garner
Boo 14h40'21" 14d37'24"
Mathew Gentry
Her 18h25'36" 14d55'3"
Mathew J Bronson
Uma 10h4'26" 56d48'15"
Mathew & Jayson Martin
Lyn 8h55'41" 46d14'20"
Mathew Lawson Wilson
Lib 15h30'11" -7d58'37"
Mathew Lianne Chloe Rilee
Vir 12h51'47" 8d0'42"
Mathew Luke Brink
Col 5h53'31" -35d2'10"
Mathew - Matty - O'Brien
Cru 12h38'57" -58d55'42"
Mathew Robert McKelvey
Sgr 18h4'26" -27d19'0"
Mathew Ryan Baisley
Her 16h40'33" 48d5'31"
Mathew Ullo
Uma 10h23'57" 60d51'54"
Mathew's Star January 26,
1979
Aqr 22h36'58" -21d43'25"
Mathia O'Neal Harris-
Jackson
Aur 5h44'15" 41d27'55"
mathias
Uma 11h23'37" 39d27'29"
Mathias
Peg 22h26'6" 12d42'2"
Mathias
Tau 3h40'48" 9d55'23"
Mathias & Brigitte
Ori 6h12'20" 7d9'52"
Mathias Domes
Uma 13h21'26" 62d41'13"
Mathias Eichmann
Uma 9h48'45" 42d2'12"
Mathias Jessup Bartels
Uma 10h7'52" 47d26'39"
Mathias Jöstl
Gem 6h29'48" 26d12'24"
Mathias Lehnen
Ori 5h22'56" 8d18'17"
Mathias Linnemann
Uma 8h58'22" 52d46'53"
Mathias mon homme pour
la vie
Lac 22h43'10" 43d57'3"
Mathias Podolski geb.
21.06.1981
Uma 10h0'5" 62d32'8"
Mathias Schädel
Uma 10h57'22" 57d53'0"
Mathice
Cnc 8h16'31" 19d30'45"
Mathieu Boutin Lobel
Uma 11h23'28" 57d53'11"
Mathieu Courteault
Tau 3h59'5" 23d28'43"
Mathieu Gril
Cam 3h57'56" 56d8'9"
Mathieu Hogue
And 2h14'6" 48d51'17"
Mathieu James Konarski
Cas 2h22'14" 68d52'3"
Mathieu Johann
Vir 14h13'9" -3d47'33"
Mathieu Johann
Col 5h55'30" -37d5'48"
Mathieu Pierre Long
Cep 21h23'19" 60d23'29"
Mathilda Rose Chapman
Cru 12h27'46" -61d44'32"
Mathilde
And 23h19'56" 50d44'49"
Mathilde Constensoux
Umi 16h7'28" 81d58'40"
Mathilde Ferra Marie
Hamacher
Cnc 8h7'33" 22d10'1"
Mathilde Laffon
Sco 17h53'12" -32d14'1"
Mathilde Niggli-Aerni
Her 18h26'48" 12d59'11"
Mathis Deckmyn
Gem 6h10'2" 22d59'10"
Mathis Dupont (Regarde je
suis là)
Umi 16h32'59" 83d40'55"
Mathis Ryan Martin
Dra 16h19'57" 57d27'56"
Mathy Jr
Per 3h12'50" 54d35'40"
Mati Recolons
Ori 6h22'33" 14d42'26"
Matias
Ori 5h43'43" -2d36'33"
Matie-Elaine Thibert
Ari 2h25'58" 20d27'5"
MATIK DE TE
Psc 23h13'30" 1d19'2"
Matika
Cyg 20h2'45" 39d14'48"
Matika - A Match Made in
Heaven
Aqr 22h7'14" -0d17'8"
Matilda
And 0h53'58" 39d45'9"
Matilda
Cyg 19h51'13" 32d21'42"
Matilda B. Clipner
And 0h24'51" 42d8'7"
Matilda Betsy Burch
Cas 23h39'26" 53d1'48"
Matilda Chase
Lib 15h15'7" -26d59'14"
Matilda Constance Jean
Owen
And 1h41'8" 49d2'44"
Matilda & Elmer Lerner
Cyg 19h30'31" 51d28'20"
Matilda Estelle's
Christening Star
Aqr 22h59'17" -10d10'0"
Matilda Grace Mathieson
Boo 15h35'53" 45d17'0"
Matilda Harris Lowell
Aqr 22h51'6" -4d22'41"
Matilda Ivona Bennett West
And 2h25'27" 45d39'37"
Matilda J. Klipsch
Cas 0h23'40" 52d2'25"
Matilda Josephine
Uma 11h20'49" 67d4'19"
Matilda Rose
Sgr 19h29'30" -39d25'7"
Matilda Ruby Avci-Pollard
And 2h27'21" 39d45'8"
Matilda (Tillie) Petroni
Aqr 21h59'42" -14d9'20"
Matilda Zenobia
Sgr 18h42'3" -31d59'53"
Matilde
Her 17h11'21" 24d47'43"
matilde
Lac 22h44'41" 37d5'48"
Matilde Carrara
Lyr 18h35'40" 38d50'13"
Matilde Formica
Cyg 20h47'10" 36d56'23"
Matilde "Nonna Bella"
Uma 11h57'31" 31d3'49"
Matilde Pavesi
And 23h5'33" 52d6'9"
Matilde Rodriguez
Sco 16h15'10" -9d59'8"
Matinicus Maggie
Lib 15h39'43" -4d43'50"
Matisse Zen Shepherd
Leo 10h11'24" 19d11'58"
MatLis
Lyn 7h12'42" 51d8'23"
MATLOCK 7
Vir 13h16'30" 5d34'27"
Matnat
Aql 19h58'1" 10d29'11"
Matochi Kala...Miye
Ogliglewakan
Umi 15h21'16" 73d22'23"
Matopia
Ori 5h49'55" 18d22'52"
Matota
Cra 18h45'55" -37d8'25"
Matrimonium
Mon 7h35'39" -0d17'50"
Matris Amie
And 2h17'1" 46d34'57"
Matriseternus Dilingo
Uma 10h50'10" 42d8'15"
MatsandKanya
Leo 10h26'45" 21d30'11"
Matschke, Heinz
Uma 9h20'20" 64d6'10"
Matsue Shimabukuro
Cyg 20h55'20" 49d24'29"
Matsumoto.T.S.2006.11.11
Sco 17h29'45" -32d1'40"
MATSUNAGA
Eri 3h30'26" -22d2'55"
Matsunaga-Ellison
Leo 11h7'50" 21d48'12"
Matt
Uma 10h7'45" 47d21'39"
Matt
Cyg 20h10'16" 37d33'28"
Matt
Gem 6h49'36" 33d40'23"
Matt
Cma 6h50'30" -17d18'34"
Matt
Cap 20h34'29" -21d5'31"
Matt
Uma 9h2'14" 69d1'55"
Matt 40
Ori 5h41'44" 7d1'15"
Matt A Schommer
Cyg 20h52'59" 47d54'40"
Matt Aman's Star of Angel
Puffs
Cyg 20h13'36" 41d32'21"
Matt and Adrien's Star
Peg 23h53'34" 19d25'44"
Matt and Amanda-Forever
Lyr 18h46'44" 39d39'37"
Matt and Ashlee
Uma 10h16'57" 55d8'50"
Matt and Becky's
Gem 6h56'31" 16d8'11"
Matt And Char
Tau 4h5'24" 22d14'39"
Matt and Christine
Cyg 20h50'18" 45d42'43"
Matt and Connie Always
and Forever
Vir 14h38'26" 2d43'37"
Matt and Dana Schmidt
And 0h45'55" 40d9'7"
Matt and Danelle
Aqr 22h47'2" -24d38'23"
Matt and Emily's Star
Lyr 19h7'37" 45d15'22"
Matt and Erin, Forever in
Love
Cnc 8h49'5" 28d48'47"
Matt and Jaime 2005
And 2h0'22" 41d26'38"
Matt and Julie Boomsma
Lib 15h47'25" -12d7'10"
Matt and Kasia
Cyg 20h42'22" 35d56'41"
Matt and Kristin Forever
Tau 4h9'48" 27d51'2"
Matt and Lani's Star
Cyg 19h49'30" 33d21'12"
Matt and Leah Bingham
Cyg 21h39'6" 33d27'33"
Matt and Leigha
Lib 14h31'3" -24d17'43"
Matt and Lindsey
Sco 16h10'55" -17d19'36"
Matt and Lindsey
Uma 11h14'56" 49d41'58"
Matt and Lindsy Anderson
Aql 19h43'30" 9d32'55"
Matt and Martha 021407
Sgr 18h24'0" -31d17'59"
Matt and Nancy 2-11-2006
Cyg 21h39'43" 41d59'2"
Matt and Nicole
Psc 1h8'46" 4d31'37"
Matt and Stacy Rende
Cyg 20h25'58" 52d8'41"
Matt ~ and ~ Tonya
Sge 19h41'45" 18d7'37"
Matt & Anna Lillis
Cas 1h1'47" 63d43'5"
Matt & Audrey Goolsby
Lyr 19h3'46" 32d12'41"
Matt Barton
Vir 13h19'54" 6d16'58"
Matt "Beefy" Hopper
Her 16h35'14" 37d47'15"
Matt Behoff
Gem 6h44'14" 22d56'49"
Matt Berube
Lyn 7h12'4" 44d56'16"
Matt Breest
Lyr 18h18'26" 30d59'18"
Matt Brown
Her 17h32'45" 40d15'44"
Matt Brown "Lovebug"
Ori 5h28'22" 2d0'42"
Matt "Bulldog" Baugh
Sco 17h57'34" -30d9'18"
Matt Canberg
Tau 4h12'52" 17d42'55"
Matt & Cari
Lac 22h25'2" 50d58'43"
MATT CARMODY
Uma 11h4'34" 49d37'27"
Matt Caster
Her 18h15'57" 27d31'55"
Matt & Chelsea
Cyg 19h55'48" 33d34'8"
Matt Church
Cru 12h28'37" -60d27'5"
Matt Codding-Forever in
our hearts
Leo 11h54'10" 23d34'39"
Matt Cooper
Vir 14h5'27" -16d6'17"
Matt Corvin
Uma 12h5'1" 33d5'51"
Matt Cowdrey
Ori 6h4'8" 4d9'15"
Matt Daley's Star
Tau 4h8'59" 6d52'13"
Matt & Danielle's Star
Leo 11h50'25" 25d55'42"
Matt DiCamillo
Her 17h14'11" 34d20'59"
Matt Dooley
Her 18h11'12" 19d38'0"
Matt & Echo Groff
Uma 10h33'27" 67d15'11"
Matt Edwards
Dra 17h25'43" 51d23'19"
Matt Erazo Hayden
Umi 15h54'33" 79d31'17"
Matt & Erin
Cyg 20h14'53" 49d35'1"
Matt & Erin
Lyr 18h53'28" 35d57'25"
Matt Erwin
Lyn 6h54'57" 52d24'35"
matt - eternal light of my
heart
Per 3h22'6" 39d53'15"
Matt & Eva
Cyg 21h18'28" 52d15'46"
Matt - Foreva My Shining
Light - Jas
Cru 12h25'50" -60d40'57"
Matt Funk and Aurora Ford
Umi 17h7'19" 80d52'42"
Matt Goike
Uma 10h16'26" 59d54'34"
Matt Gracey, Sr.
Sgr 18h53'53" -27d42'29"
Matt Grebe
Per 3h11'58" 42d15'27"
Matt Hardy
And 23h30'41" 37d4'11"
Matt & Heather
Cyg 21h20'59" 51d59'30"
Matt Hilson
Boo 14h18'52" 28d17'43"
Matt J. Demoore
Vir 13h42'45" -0d48'16"
Matt J. Dube
Umi 14h41'44" 74d51'27"
Matt & Jackie Barkley
Lyn 7h45'16" 35d26'27"
Matt & Jen
Ori 5h27'54" 3d21'25"
Matt & Jenny
Sgr 18h34'4" -28d38'19"
Matt Joyce
Ori 4h48'18" 4d32'39"
Matt & Karen
Cyg 20h27'48" 54d41'57"
Matt & Kelly
Uma 9h57'12" 53d59'35"
Matt & Kelly Ferguson
Lyn 7h37'1" 36d59'3"
Matt & Kelly, True Love is
Forever!
Cyg 20h50'21" 33d36'39"
Matt King
Vir 13h42'10" -5d28'40"
Matt LaPrade
Ori 5h29'36" 2d48'20"
Matt & Latoya Patterson
Cap 20h45'22" -15d36'26"
Matt Leans
Sco 16h58'12" -37d33'26"
Matt & Lori Petrunyak
Ind 22h24'47" -70d30'0"
Matt Lynott & Kate Ciocca
Sco 18h16'14" -10d59'19"
Matt M. May 18, 1969
Cyg 19h59'2" 39d41'32"

Matt Maldonado
Aqr 22h6'54" -20d7'44"
Matt Malone Family Star
Aql 19h28'51" -0d37'36"
Matt Man
Vir 13h47'46" -2d48'7"
Matt Maroun
Cnc 8h56'54" 16d55'28"
Matt McCarthy
Her 18h57'0" 15d44'11"
Matt McQuarrie
Psc 1h31'54" 18d22'15"
Matt McVey
Per 3h17'43" 54d57'37"
Matt Milad
Per 4h30'31" 47d24'43"
Matt Mitchell
Per 3h27'20" 48d31'49"
Matt Mitchell
Cep 23h3'24" 79d52'58"
Matt & Monica Anniversary Star
Vir 13h11'51" 11d17'23"
Matt "Mookie" Cunningham
Tau 5h55'38" 24d55'26"
Matt: My Forever Love
Gem 7h13'19" 25d58'35"
Matt: My One and Only Love
Her 18h7'51" 37d8'4"
Matt -n- Ashton's Foreverus
Cyg 20h14'5" 50d27'8"
Matt & Nikki's Twins All Our Love
Gem 7h22'19" 25d54'15"
Matt & Noelia
Sge 20h3'18" 18d15'37"
Matt Ouimet
Per 3h11'24" 53d58'18"
Matt Overton
Her 17h39'46" 33d12'45"
Matt Palermo
Per 3h43'52" 45d52'37"
Matt Penoyer
Ori 5h29'30" 3d47'0"
Matt Posluszny
Uma 10h42'29" 50d36'20"
Matt Reed
Leo 11h32'41" 14d50'27"
Matt Roberson (Wulfe)
Per 3h16'15" 44d19'0"
Matt Rocks!!!
Aqr 22h6'16" -13d33'13"
Matt S Hall
Per 4h48'36" 42d33'4"
Matt Sagan
Uma 11h23'34" 59d7'47"
Matt&Sam
Cyg 20h1'59" 45d54'34"
Matt & Sarah Lampka, Nov. 25, 2006
Lyr 18h29'52" 33d8'13"
Matt & Shannon Thompson
Uma 10h29'52" 64d53'30"
Matt & Shelly Burks 10/22/05
Ori 5h33'5" -7d58'48"
Matt & Shelly Forever
Per 4h22'41" 48d5'52"
Matt Shining Bright
Tau 5h35'6" 27d33'20"
Matt Smith
Cru 12h51'55" -58d0'58"
Matt Sneaky Saba
Boo 15h15'34" 48d12'39"
Matt Sostaric
Cam 5h11'21" 76d26'54"
Matt "Spooky Cat" Miquelon
Lyn 7h44'33" 49d10'45"
Matt Star
Uma 10h3'43" 48d19'52"
Matt Stockard
Her 16h25'56" 10d29'12"
Matt & Tamara
Vir 12h39'44" -8d56'54"
Matt Thain
Cru 12h46'40" -64d25'26"
Matt & Toni
Cyg 19h42'7" 34d13'58"
Matt Vander Molen
Uma 10h38'24" 60d15'37"
Matt Verdeflor
Her 18h53'24" 22d31'6"
Matt Vine
Per 4h48'10" 47d37'35"
Matt W. Redington
Gem 6h31'57" 23d59'3"
Matt Walling
Ori 5h39'52" 1d28'50"
Matt Watts
Dra 17h46'38" 58d56'45"
Matt Welty
Uma 11h43'18" 45d10'2"
Matt Wensauer
Ori 5h42'8" 2d11'30"
Matt Wieland
Cep 22h40'15" 69d22'3"
Matt Wilkerson
Sco 17h21'0" -42d54'18"
Matt Young
Cam 4h27'42" 67d2'35"
Matt, Tam & Cooper Forever - 2005
Cru 12h12'28" -61d59'15"
MattaDub
Uma 9h25'54" 48d52'10"
Mattanré
Ori 5h35'4" -2d17'24"
Mattasun
Cma 6h45'14" -15d34'41"
Mattchu
Sco 16h15'41" -13d57'6"
Matteo
Cnc 8h52'16" 6d54'55"
Matteo Balocchi
Uma 11h22'34" 55d3'32"
Matteo Balzano
Gem 7h11'35" 27d39'16"
Matteo Calandrelli
Lyr 19h20'27" 34d6'53"
Matteo Ciavatta
Her 16h52'12" 27d53'15"
Matteo Coradin
Her 18h22'51" 13d31'50"
Matteo De Filippo
Cap 21h32'50" -8d45'14"
Matteo de Luca
Uma 11h20'9" 71d52'1"
Matteo Ferrantino
Lib 15h37'35" -17d14'35"
Matteo Fini Chioccioli
Lyr 18h59'29" 26d7'9"
Matteo Gianluca Meyer
Cyg 20h50'56" 47d1'37"
Matteo Joseph Perrotta
Ari 3h9'58" 29d34'16"
Mattéo KEMPFF
Psc 0h48'38" 4d11'14"
Matteo Olcelli
Umi 13h21'36" 69d48'38"
Matteo & Steph = True Love Forever
Cyg 21h30'2" 45d32'8"
Matteo Vidotto
Her 18h11'23" 18d35'37"
Matteo Zuaboni
Ori 6h16'28" 9d48'31"
Matthaios
Tau 5h45'32" 22d15'24"
Matthanda
And 2h36'53" 39d38'30"
Matthew
Her 17h14'31" 34d41'25"
Matthew
Her 17h21'7" 34d6'59"
Matthew
Boo 14h42'32" 50d17'27"
Matthew
Her 16h48'43" 47d52'30"
Matthew
Aur 5h32'9" 46d6'55"
Matthew
Ari 2h24'41" 21d35'19"
Matthew
Cnc 8h2'0" 12d49'26"
Matthew
Ari 2h28'19" 12d19'0"
Matthew
Tau 3h42'27" 27d36'8"
Matthew
Gem 6h12'58" 25d13'57"
Matthew
Crb 15h27'16" 25d41'3"
Matthew
Leo 11h30'30" 25d46'28"
Matthew
Leo 11h13'15" 24d34'16"
Matthew
Leo 9h39'42" 27d20'32"
Matthew
Cep 20h46'51" 64d55'16"
Matthew
Uma 11h30'46" 57d48'26"
Matthew
Uma 12h3'11" 58d20'8"
Matthew
Dra 16h19'54" 57d50'45"
Matthew
Cas 23h8'37" 55d48'25"
Matthew
Sgr 17h59'20" -17d19'30"
Matthew
Lib 15h0'50" -13d37'49"
Matthew
Lib 15h11'7" -7d37'25"
Matthew
Aqr 21h38'17" -0d47'21"
Matthew
Vir 12h37'56" -0d20'36"
Matthew 40
Cep 22h39'55" 74d0'3"
Matthew 7 Neves
Her 17h1'17" 33d21'37"
Matthew 7-13-1984
Ori 5h15'46" -5d4'41"
Matthew A. Boykan
Cyg 21h44'38" 39d24'51"
Matthew A. Cunningham, Jr.
Dra 17h32'26" 54d34'36"
Matthew A. Henson
Vir 13h29'50" -7d4'31"
Matthew A. Jeffcoat
Aqr 21h43'19" -3d25'44"
Matthew A Lye - Our Three Wise Men
Cru 12h0'25" -58d45'0"
Matthew A. Tascione
Umi 14h35'2" 76d59'46"
Matthew Aaron
Cnc 8h4'6" 26d42'10"
Matthew Aaron
Aqr 22h40'26" 2d31'11"
Matthew Aaron Balogh
Per 3h13'59" 47d0'39"
Matthew Aaron Glenn
Her 17h46'26" 24d55'25"
Matthew Aaron Nicely
Her 17h46'12" 36d39'47"
Matthew Aaron Stark
Sgr 19h53'7" -28d45'42"
Matthew Aaron Thurmond
Per 4h5'5" 51d3'59"
Matthew Adamczyk's Star
Uma 11h36'49" 57d39'56"
Matthew Adams Elliott
Vir 13h31'36" -17d7'43"
Matthew Addison Richard
Aqr 22h22'45" 0d5'51"
Matthew aka Lumpy
Sgr 19h36'23" -14d17'33"
Matthew Alan Bean
Psc 1h14'59" 26d34'11"
Matthew Alan Croft Mi Amor
Sgr 19h9'36" -25d16'21"
Matthew Alan Cunningham, Jr.
Vir 13h13'49" 11d40'53"
Matthew Alan Jones
Uma 10h13'16" 65d31'5"
Matthew Alan MacDonald
Per 3h53'8" 32d53'39"
Matthew Alan Schommer
Ori 5h27'8" 2d47'7"
Matthew Alan Zinkand
Dra 18h39'58" 56d45'32"
Matthew Alexander Heman
Ori 5h28'37" 4d49'17"
Matthew Alexander Jackson
Vir 13h35'35" 2d44'23"
Matthew Alexander Stoddart
Vir 13h16'7" 11d7'56"
Matthew Alexander Stuart
Sco 16h3'32" -24d1'18"
Matthew Allen
Ori 6h4'19" 12d55'45"
Matthew Allen Beebe
Her 16h17'44" 13d1'53"
Matthew Allen Henderson
Uma 9h13'3" 47d25'45"
Matthew Allen Levi Pierce
Per 3h10'42" 42d54'34"
Matthew Allen Loser
Sco 16h9'2" -11d3'41"
Matthew Allen McIntyre
Aql 18h56'38" -0d57'29"
Matthew Allen Reynolds
Dra 19h21'17" 62d26'28"
Matthew Allen Yates
Uma 11h3'40" 70d5'59"
Matthew & Ally Niblett 2/27/03
Sge 19h13'59" 19d46'49"
Matthew Alpha
Oph 17h19'49" -22d33'18"
Matthew and Aimee Kuhn
Cyg 20h29'11" 55d40'29"
Matthew and Allison Fournier
Cyg 19h45'48" 46d45'27"
Matthew and Amanda
Cyg 21h19'16" 40d58'7"
Matthew and Amanda
Cyg 19h24'44" 53d44'6"
Matthew and Christina 11.15.06 x5e.
Ori 5h33'18" 0d15'2"
Matthew and Crystal
And 1h49'13" 42d6'7"
Matthew and Elizabeth White
Cyg 20h58'9" 46d20'27"
Matthew and Eryl's Moon Star
Uma 9h22'6" 56d21'22"
Matthew and Hannah
Lmi 10h6'40" 34d46'39"
Matthew and Hayley
Her 16h53'18" 36d43'30"
Matthew And Jennifer
Cyg 20h45'40" 54d11'39"
Matthew and Karen Bridgeman
Cyg 19h31'31" 28d56'42"
Matthew and Karen Neave
Cyg 21h12'27" 46d14'42"
Matthew and Megan's Star
Cyg 20h43'44" 51d36'30"
Matthew and Nicole Brotherton
Sge 19h43'11" 18d56'34"
Matthew and Paula
Leo 10h59'6" 1d10'55"
Matthew and Rachel's Love Star
Cyg 21h13'13" 53d39'58"
Matthew and Samantha 06172006
Cyg 19h36'16" 47d12'59"
Matthew and Wendy Cycenas
Cyg 19h37'25" 35d40'51"
Matthew Andrew Kokodynski
Uma 9h59'18" 58d55'54"
Matthew Andrew Pelley
Dra 14h45'5" 56d26'25"
matthew andrew taylor
Uma 8h34'40" 66d57'22"
Matthew Andrew Thor Rollin
Eri 4h23'34" -8d30'9"
Matthew Angel Perez
Sco 16h3'7" -15d45'27"
Matthew Angelillo
And 0h19'37" 43d2'0"
Matthew Anthony Chenault
Uma 11h42'7" 46d21'42"
Matthew Anthony Rice
Aur 5h49'20" 51d6'13"
Matthew Ari Parsky
Lib 15h7'20" -12d26'21"
Matthew Arron Sweet
Cma 6h41'40" -14d59'26"
Matthew Artingstall
Her 17h16'43" 34d17'23"
Matthew Austin Jordan
Lib 15h33'50" -12d1'14"
Matthew Austin Shephard-Lupo Star
Per 2h19'11" 53d26'43"
Matthew "Avarus" Baltes
Her 17h6'52" 31d45'52"
Matthew B. Loomis
Boo 15h12'9" 45d43'37"
Matthew "Babe" Riccelli
Umi 13h57'32" 73d54'35"
Matthew Baby
Per 4h47'35" 49d23'20"
Matthew Baker
Uma 10h1'50" 47d22'48"
Matthew Baker
Aur 5h52'14" 43d40'32"
Matthew "Balz" Borowski
Cep 22h45'7" 63d31'6"
Matthew Baron
Cnc 8h15'38" 31d1'11"
Matthew Barron - My Eternal Love
Cru 12h15'22" -62d3'54"
Matthew Baumhardt
Ori 6h12'28" 16d28'46"
Matthew Beattie
Cnc 8h19'32" 18d12'47"
Matthew Benjamin Jaffe
Tau 4h38'19" 20d43'48"
Matthew Bennett Welling
Sco 16h24'29" -37d37'4"
Matthew Bennett Wood
Per 3h30'42" 49d3'3"
Matthew Bernard Schaer
Her 17h26'52" 35d27'35"
Matthew Bing Rogers
Uma 11h55'23" 64d2'17"
Matthew Blaine Jones
Leo 10h33'38" 18d55'9"
Matthew Blake Ball
Aqr 21h13'3" -13d15'19"
Matthew Bly Cowdrey
Vir 13h18'18" -1d57'19"
Matthew Bolar
Per 3h0'27" 46d14'22"
Matthew Borrett
Cap 21h32'59" -21d32'7"
Matthew Bradley Pierce
Psc 23h48'46" 6d57'46"
Matthew Braedon Starnes
Her 17h42'37" 29d18'38"
Matthew Brandon Chill
Her 18h27'32" 21d20'50"
Matthew Brenton Farwell
Leo 11h42'39" 14d15'14"
Matthew Brian Bulow
Uma 11h25'20" 48d9'50"
Matthew Brian George Dulski
Ori 6h7'51" 21d25'40"
Matthew Brian Grinchell
Cet 1h25'27" -0d45'48"
Matthew Brian Ramirez
Her 17h56'30" 28d49'12"
Matthew & Brittany
Cyg 20h13'12" 52d41'47"
Matthew Brown
Cyg 20h0'58" 57d26'9"
Matthew Brown
Uma 10h16'55" 66d23'15"
Matthew Brown
Crb 15h44'36" 34d27'12"
Matthew Browning
Her 17h43'47" 31d54'55"
Matthew Budway Morycz
Psc 0h56'49" 17d41'13"
Matthew Bugbee
Lib 14h28'23" -20d54'29"
Matthew Burgoyne
Cep 21h41'52" 61d18'26"
Matthew Burton
Uma 9h21'29" 57d31'45"
Matthew C. Brockway
Uma 9h13'14" 55d10'29"
Matthew C. Brown, II.
Umi 14h24'42" 75d55'25"
Matthew C. Cummins
Tau 4h7'57" 4d44'39"
Matthew C. Hensinger
Leo 11h38'0" 24d31'17"
Matthew C. Limtiaco
Cnc 8h55'42" 30d36'18"
Matthew C. Mattingly
Ori 4h50'34" 3d34'3"
Matthew C. McDermott
Uma 11h32'37" 63d7'11"
Matthew C. Noone II
Her 18h39'30" 24d30'22"
Matthew Capbarat
Her 18h19'16" 25d56'27"
Matthew Capritto
Lmi 10h17'16" 29d42'18"
Matthew Carl Behling
Tau 5h9'59" 22d6'44"
Matthew Carroll Smith
Uma 10h12'24" 65d18'59"
Matthew Carson Pechan
Lmi 10h46'40" 33d20'36"
Matthew Cartwright
Dor 4h17'2" -55d57'56"
Matthew Cassidy
Cep 22h2'41" 72d3'33"
Matthew Catherine Werner
Ari 2h9'11" 13d43'51"
Matthew Caulfield Martinez
Dra 19h15'13" 71d54'1"
Matthew Charles Hood
Per 3h6'51" 44d41'35"
Matthew Charles Iberger
Aqr 22h17'44" -15d33'19"
Matthew Charles Mullen
Cnv 12h41'1" 40d33'14"
Matthew Charles - Our Angel In Heaven
Vir 12h15'13" -0d54'29"
Matthew Chase Griffin
Boo 14h53'8" 18d56'25"
Matthew Chase Meserole
Tau 3h29'26" 17d23'56"
Matthew/Chelsea
Peg 23h59'29" 11d23'13"
Matthew Chiel Brooks
Umi 13h40'34" 73d29'42"
Matthew Christian Vigilante
Cnc 8h34'59" 24d53'34"
Matthew Christman
Uma 9h44'9" 46d28'1"
Matthew Christopher
Cnc 7h56'22" 16d1'0"
Matthew Christopher Angel Infinatum
Aql 19h44'47" -0d4'43"
Matthew Christopher Dahl
Lyr 18h49'34" 31d58'33"
Matthew Christopher Johnson
Gem 6h48'23" 26d13'39"
Matthew Christopher Kennelly
Aqr 22h11'28" -0d44'50"
Matthew Christopher Rogers
Aur 5h46'16" 30d17'45"
Matthew Christopher Tyo
Ori 5h30'34" 1d14'52"
Matthew Christopher Wilhelm
Tau 5h47'39" 25d58'35"
Matthew ChristopherSmith God's gift
Uma 12h23'14" 56d37'36"
Matthew Chulla
Uma 9h11'11" 50d56'41"
Matthew Ciaran Moore
Her 17h48'10" 34d58'22"
Matthew Cimini
Ori 6h20'15" 18d53'12"
Matthew Cirincione
Tau 3h39'50" 28d39'2"
Matthew C.J. Cummins
Gem 6h40'52" 21d47'10"
Matthew Clark Shields
Her 16h50'32" 33d18'20"
Matthew Clay Bettinghouse 9/19/1983
Vir 13h9'25" 3d35'14"
Matthew Colin Michael McFarlane
Per 3h22'39" 39d20'2"
Matthew Collier Richardson
Aqr 21h38'58" 1d15'18"
Matthew Collins Chandler
Cap 21h36'15" -14d29'7"
Matthew Connor Lombardo
Cep 21h40'41" 61d57'50"
Matthew Corrigan
Her 18h29'20" 18d23'28"
Matthew Coté Norton
Uma 9h23'28" 64d31'31"
Matthew Cramer
Her 17h42'8" 37d39'35"
Matthew Curran
Per 3h49'23" 48d52'22"
Matthew D. Chenoweth
Her 18h7'37" 49d6'54"
Matthew D. Dollar D.D.S.
Uma 9h25'54" 45d42'18"
Matthew D. Ediger
Sgr 19h11'13" -24d6'22"
Matthew D. Hunter III
Umi 14h36'12" 75d49'38"
Matthew D. Killen
Aql 19h10'50" -0d11'59"
Matthew D. Morland
Psc 1h16'6" 16d49'51"
Matthew D. Semple
Vir 13h10'26" -22d10'25"
Matthew D. Swanson
Cep 22h42'16" 73d5'8"
Matthew Dale Soular 2003-2004
Leo 10h23'48" 13d48'0"
Matthew Dalton Price
Gem 6h30'36" 26d2'3"
Matthew Damien Carreiro
Aql 19h3'52" -0d33'4"
Matthew & Dana Brown
Leo 10h14'59" 26d44'5"
Matthew & Dana Fox
Vir 12h17'37" -2d27'43"
Matthew David
Uma 12h23'38" 54d14'22"
Matthew David Allen Claar
Uma 9h9'6" 48d20'11"
Matthew David Bennett
Tau 4h27'44" 19d3'43"
Matthew David DeMatteo
Aur 5h14'41" 31d58'23"
Matthew David Guthrie
Uma 10h8'29" 63d44'24"
Matthew David Hackshaw
Crb 15h35'25" 31d33'49"
Matthew David Hollins
Aqr 23h4'17" -8d55'38"
Matthew David Jahn
Her 17h9'20" 37d33'8"
Matthew David Jones
Tau 4h7'18" 7d42'27"
Matthew David Jones
Uma 9h3'53" 62d6'5"
Matthew David Latham
Tau 3h43'37" 24d13'57"
Matthew David Mariano
Leo 9h44'7" 31d26'2"
Matthew David McKenna
Aql 20h12'9" 14d39'9"
Matthew David Moir - 29/08/1984
Pho 2h19'44" -43d15'10"
Matthew David Paczkowski
Ari 3h21'54" 19d48'10"
Matthew David Roberts
Cep 2h11'18" 80d43'42"
Matthew David Ross
Leo 11h15'59" 23d51'48"
Matthew David Schommer
Tau 4h35'52" 19d17'29"
Matthew David Semon (1/23/04)
Umi 16h57'55" 87d9'30"
Matthew David Weitort
Lmi 9h48'36" 35d3'26"
Matthew David Ybarra
Aqr 20h42'44" 0d49'2"
Matthew Dean
Lyr 18h35'19" 39d53'47"
Matthew DeFrancis
Leo 11h29'14" -4d41'52"
Matthew Delamar
Umi 15h32'4" 78d3'24"
Matthew Dennis Carey
Per 3h22'13" 52d21'40"
Matthew Dlugos Gift of God
Aur 5h33'44" 42d43'37"
Matthew Dominick Rosso
Lib 15h39'32" -10d51'23"
Matthew Donald Kyffin
Aqr 20h55'57" -12d56'33"
Matthew Donovan Hubner
Ori 5h50'7" 3d10'30"
Matthew Douglas Johnson
Uma 10h43'15" 57d25'0"
Matthew Douglas Poorman
Sco 16h8'46" -15d25'14"
Matthew Durrigan
Gem 7h44'37" 14d37'21"
Matthew Dylan Upton
Ori 5h50'14" 1d59'3"
Matthew E. Anderson
Dra 16h10'15" 61d32'59"
Matthew E Eggers
Ori 6h8'37" 20d44'30"
Matthew E. Wilson
Her 18h16'47" 18d55'50"
Matthew Earl Lancaster
Leo 10h12'58" 22d40'42"
Matthew Edward Duncan
Ori 5h12'17" 12d52'21"
Matthew Edward Galatioto
Aqr 23h27'16" -8d16'22"
Matthew Edward Homan
Gem 6h2'40" 27d37'40"
Matthew Edward Kraft
Per 4h10'31" 51d21'0"
Matthew Edward Otsuki
Tau 3h32'3" 22d21'24"
Matthew Edward Simon
Tau 3h51'33" 3d43'37"
Matthew Edward Webb
Aqr 22h14'31" 1d30'30"
Matthew Edward Zick
Ori 4h54'27" 2d21'22"
Matthew Eldridge
Her 18h32'25" 15d25'9"
Matthew Eli Bogoyevac
Uma 9h24'47" 49d57'26"
Matthew Elijah Fox
Gem 7h39'12" 33d27'28"
Matthew Elijah Merz
Leo 11h24'59" 9d35'19"
Matthew Ellis
Ori 5h57'39" 20d50'45"
Matthew Elmer Clark
Ari 2h12'26" 24d52'5"
Matthew Eric Serna
Uma 8h54'21" 48d49'28"
Matthew Erick Bill
Umi 14h37'15" 68d3'52"
Matthew Exton Harmon
Her 17h33'1" 47d20'6"
Matthew Faires
Her 17h16'57" 32d31'30"
Matthew Farley
Vir 11h37'43" -0d40'53"
Matthew Farnum
Ari 3h26'40" 26d23'23"
Matthew Ficaloro
Cap 21h33'9" -15d49'51"
Matthew Finley Senogles-Ball
Umi 14h7'30" 73d6'40"
Matthew Finnerty
Boo 15h9'0" 35d6'30"
Matthew Flesch - June 9, 1975
Gem 6h39'37" 19d10'11"
Matthew Flynn
Cnc 8h46'22" 16d41'8"
Matthew Forney
Her 16h50'52" 42d47'27"
Matthew Foye
Per 2h56'19" 48d2'2"
Matthew Freeman
Leo 11h44'27" 18d18'55"
Matthew Fries
Psc 1h18'59" 30d50'13"
Matthew Fuller
Tau 5h10'13" 27d42'5"
Matthew G. DiSalle
Uma 11h51'28" 58d29'38"
Matthew G. Nielsen
Aql 19h5'14" 3d4'53"
Matthew Gabriel
Lup 15h35'2" -40d51'41"
Matthew Galea
Tau 4h17'41" 2d58'30"
Matthew Gary Fogg
Vir 13h16'40" 6d11'13"
Matthew Gates Ringquist
Ori 5h48'48" 5d35'16"
Matthew Gauvin
Cyg 19h46'28" 31d30'45"
Matthew George Rea Phelps
Per 2h15'39" 52d28'31"
Matthew Gerard Camillery
Uma 11h5'55" 62d20'44"
Matthew Gerard Guinta Jr.
Her 16h7'19" 48d17'27"
Matthew Gibran Ghazali
Sco 17h56'6" -38d34'40"
Matthew & Gina-Love's special stars
Cyg 19h39'51" 33d29'27"
Matthew Glenn Cammack
Vir 12h37'59" 6d4'50"
Matthew Glenn Williams, Disciple
Ori 4h49'36" 5d52'13"
Matthew Golden
Ori 5h27'15" 2d8'50"
Matthew Gordon Browne
Per 3h50'32" 49d30'55"
Matthew Grant
Ori 4h46'39" 3d54'26"
Matthew & Grant -Forever and Always
Sgr 18h15'51" -25d24'49"
Matthew Gregory Morris
Lib 15h42'17" -7d57'51"
Matthew Gregory Stom
Uma 13h5'56" 53d39'34"
Matthew Grey Malmsten
Uma 11h39'39" 29d22'29"
Matthew Guastamacchia
Cnc 8h51'2" 31d21'36"
Matthew Gustaitis
Lib 14h50'8" -10d3'12"
Matthew Gutauskas
Vir 15h3'40" 1d39'3"
Matthew H. Hale
Cep 21h32'38" 67d43'36"
Matthew H. Knight
Uma 9h31'7" 42d7'43"
Matthew H. Lee
Ari 2h47'18" 24d31'5"
Matthew Haines
Gem 7h12'55" 20d26'46"
Matthew Haines
Col 5h52'39" -38d56'46"
Matthew Harrington Hale
Eri 3h56'40" -29d40'36"
Matthew Harris
Peg 21h30'40" 11d1'29"
Matthew Harrison
Gem 6h42'51" 16d21'25"
Matthew Harvey
Ori 4h51'23" -0d44'23"

Matthew Hauptman
Aqr 23h35'50" -21d0'53"
Matthew & Helen's Star
Lyr 18h40'16" 30d56'19"
Matthew Hewlett
Per 2h59'25" 51d21'3"
Matthew Hite Voss
Her 18h23'16" 14d19'2"
Matthew Hoang
Her 17h23'23" 44d11'54"
Matthew Hocking - Hocking Family
Her 17h45'4" 47d37'15"
Matthew Hoeltje
Ori 4h56'34" -2d1'37"
Matthew Holland
Ori 5h56'27" 18d2'15"
Matthew Honeycutt
Uma 10h37'31" 47d58'20"
Matthew Houser's Star
Aqr 22h40'25" -0d28'53"
Matthew Huerta
Sco 16h4'3" -16d49'1"
Matthew Hunter
Sco 17h24'9" -38d17'31"
Matthew Hunter Gray
Lib 15h4'6" -28d54'49"
Matthew Hyner "Love E. M. C."
Lyn 6h17'4" 57d2'2"
Matthew - I Love You - Sarah
Cru 12h30'18" -58d34'15"
Matthew Ian Klein
Uma 14h25'43" 59d48'54"
Matthew Ian Weiss
Leo 9h35'37" 26d15'34"
Matthew J. Banks
Aur 7h14'13" 41d28'40"
Matthew J. Bond
Lyn 7h8'41" 58d25'14"
Matthew J. Braun
Cyg 21h59'42" 49d5'53"
Matthew J Buckley
Sgr 19h25'37" -17d21'26"
Matthew J. Budde
Ori 5h36'0" 2d40'41"
Matthew J Cicotta
Her 18h29'55" 21d44'8"
Matthew J. Driscoll
Ori 5h34'4" 11d5'19"
Matthew J Goodwin
Aql 20h6'37" 5d50'9"
Matthew J. Holley
Lib 14h37'1" -9d21'48"
Matthew J. Magdic
Cap 20h35'30" -22d18'48"
Matthew J. Markle
Aqr 22h51'36" -6d48'28"
Matthew J. Melchionda
Her 16h50'43" 29d22'11"
Matthew J. O'Donnell
Sgr 19h19'50" -14d33'16"
Matthew J P (Phillips) Pasquerillo
Psc 0h27'37" 7d35'49"
Matthew J. Preziosi
Cyg 20h4'32" 35d39'23"
Matthew J. Spedale
Cap 20h32'36" -13d55'3"
Matthew J Vondeling
Cap 21h10'56" -23d24'15"
Matthew J. Wood
Ori 6h23'46" 14d46'24"
Matthew Jack Legg
Gem 6h31'42" 24d42'43"
Matthew Jackson
Tau 4h9'44" 0d45'41"
Matthew Jackson
Uma 11h10'42" 31d50'26"
Matthew Jacob Slover
And 0h23'51" 42d15'50"
Matthew James Adams
Ori 6h8'17" 19d5'5"
Matthew James Arciniega
Vir 13h47'44" -15d35'55"
Matthew James Beason
Cma 6h25'27" -16d25'45"
Matthew James Bradley
Cep 22h48'28" 70d10'37"
Matthew James Brown
Vir 14h18'40" -21d4'33"
Matthew James Cole
Ori 6h1'15" 18d22'10"
Matthew James Cooney
Gem 7h24'46" 29d9'30"
Matthew James Di Giovanni
Sco 17h15'39" -41d32'37"
Matthew James DiStefano
Aqr 22h17'20" -4d36'8"
Matthew James Frey - L.T.D.
Per 3h10'35" 49d31'20"
Matthew James Garito
And 0h44'43" 37d5'13"
Matthew James Gerold
Psc 22h59'17" -0d4'23"
Matthew James Grzymalski
Per 3h12'0" 51d33'41"
Matthew James Heller
Per 4h0'21" 40d59'3"
Matthew James Howard
Cyg 19h38'20" 31d30'9"
Matthew James Hunt
Cep 0h18'36" 76d19'15"
Matthew James Hurd
Her 18h19'11" 23d4'58"
Matthew James LaLonde
Per 4h3'57" 32d56'24"
Matthew James Lee
Per 4h26'50" 33d22'20"
Matthew James Martin
Cap 21h45'18" -13d32'8"
Matthew James Marzocca
Ari 3h18'59" 29d34'56"
Matthew James Matherne
Tau 3h55'3" 26d29'2"
Matthew James McManus
Cnc 9h9'39" 28d52'9"
Matthew James Orange
Cep 22h22'49" 74d18'57"
Matthew James Poitras
Uma 8h54'35" 47d17'44"
Matthew James Redfield
Per 3h8'55" 57d14'50"
Matthew James Shorthouse
Uma 9h35'54" 65d41'37"
Matthew James Star
Per 3h1'51" 45d35'3"
Matthew James Stonebridge
Umi 15h0'24" 75d45'24"
Matthew James Sumner
Sco 17h24'46" -31d26'56"
Matthew James Tomlinson
Cam 5h38'57" 56d32'53"
Matthew James Williams
Ori 5h46'33" 5d29'27"
Matthew Jamie Monaghan
Umi 13h6'44" 86d10'44"
Matthew Jason
Uma 9h23'48" 45d28'11"
Matthew Jay Hooks
Ori 5h28'7" 8d29'31"
Matthew Jay Savage
Per 2h20'13" 55d25'12"
Matthew Jeffrey
Leo 9h22'31" 15d13'25"
Matthew Jeffrey Kemmer
Ori 5h12'25" 11d58'18"
Matthew & Jennifer Galloway
Vir 12h21'44" 8d47'36"
Matthew Jerome
Cap 21h43'56" -12d59'56"
Matthew Jimmy
Cyg 19h59'27" 51d35'45"
Matthew & Joanna's Wedding star
Cyg 20h40'44" 45d53'7"
Matthew Joeseph Randall
Ari 1h57'4" 20d33'17"
Matthew John
Ari 2h41'19" 29d3'2"
Matthew John Bannister
Umi 17h26'51" 83d42'9"
Matthew John Bunch
Umi 16h1'32" 72d32'1"
Matthew John DeCerbo
Psc 1h19'37" 31d40'44"
Matthew John Hemmick
Psc 1h31'25" 18d39'7"
Matthew John Holley
Her 17h27'36" 36d33'24"
Matthew John Jeakle
Aql 18h58'59" -0d31'3"
Matthew John Kenneth King
Lep 6h7'14" -24d37'11"
Matthew John Lindsell
Cru 12h36'39" -64d23'10"
Matthew John McCarthy
Sco 16h51'20" -27d10'35"
Matthew John Morgan
Crb 15h33'11" 27d31'9"
Matthew John Profita
Uma 9h46'37" 63d51'8"
Matthew John Quinlan
Her 17h32'14" 39d16'7"
Matthew John Rodgers
Lyn 7h41'26" 45d28'38"
Matthew John Whitcomb
And 2h17'0" 46d39'15"
Matthew John Williams
Per 2h21'38" 52d58'28"
Matthew Johnson
Gem 7h18'42" 14d31'23"
Matthew Jon Edgerton Purdie
Cep 20h38'35" 66d47'57"
Matthew Jonathan Belmont
Ori 5h57'29" 17d43'49"
Matthew Jonathan Dawe
Uma 11h23'23" 58d35'2"
Matthew Joseph 5402
Her 16h50'20" 34d14'10"
Matthew Joseph Corena
Uma 9h34'12" 47d49'22"
Matthew Joseph DeBlauw
Lmi 10h24'51" 35d29'50"
Matthew Joseph Espinosa
Dra 17h20'28" 68d26'37"
Matthew Joseph Flores
Per 4h47'18" 41d22'35"
Matthew Joseph Galea
Ari 3h25'44" 22d41'8"
Matthew Joseph Goers
Leo 11h21'53" 16d38'56"
Matthew Joseph Hiles
Lib 14h59'5" -3d40'49"
Matthew Joseph Jenifer Rae
And 1h34'46" 45d35'1"
Matthew Joseph Konz
Leo 10h11'49" 14d37'7"
Matthew Joseph LoPiccolo
Gem 7h42'3" 32d54'16"
Matthew Joseph Randall
Uma 10h47'20" 47d26'1"
Matthew Joseph Schulze
Lib 14h55'35" -0d33'14"
Matthew Joseph Vetter
Uma 10h29'11" 45d30'0"
Matthew Joseph Vodon
Ori 5h41'47" 1d41'38"
Matthew Joseph Vodon
Tau 3h54'27" 4d41'47"
Matthew Jr. and Macy Ames
Sgr 19h28'28" -15d46'25"
Matthew Judge
Uma 10h47'58" 53d16'19"
Matthew Justin Gaines
Lib 15h0'21" -20d27'5"
Matthew Justin White
Cap 20h42'24" -16d52'45"
Matthew Karl Biesiada
Uma 14h12'27" 58d52'4"
Matthew Kauranen
Boo 15h23'54" 34d18'43"
Matthew Kendzior
Tau 5h10'24" 22d23'50"
Matthew Kenneth Walter
Per 3h17'4" 52d1'7"
Matthew Kenney
Per 3h28'49" 46d20'41"
Matthew Kent O'Hara I Love You
Lib 14h58'14" -11d30'15"
Matthew Kester
Cas 22h58'43" 57d16'18"
Matthew Ketner's Star
And 2h35'7" 48d50'20"
Matthew Kevin Alban
Ori 5h40'19" 6d7'33"
Matthew Kevin Loughney
Uma 13h40'11" 58d1'26"
Matthew Kliewer
Per 3h33'8" 44d19'29"
Matthew Knapp
Sco 17h6'48" -36d58'11"
Matthew Krebs
Lib 15h50'29" -12d24'28"
Matthew Kroon McEvoy
Vir 14h2'42" -16d22'23"
Matthew Kuhn
Tau 3h38'16" 19d48'40"
Matthew L. Hoeckelman
Per 3h5'8" 48d59'35"
Matthew L. Knight
Her 16h58'28" 16d20'42"
Matthew L. Meyer
Ori 5h31'51" -1d33'19"
Matthew Lackett
Cas 22h58'30" 57d19'54"
Matthew Lael Carr
Cap 20h38'52" -17d6'7"
Matthew Lamb
Her 17h45'52" 44d40'55"
Matthew & Laura Laverdiere
Cyg 20h35'45" 30d51'14"
Matthew Laurent Bishop
Mon 7h25'19" -3d7'56"
Matthew Lawrence Peck
Psc 0h57'51" 7d34'49"
Matthew Lawrence Scharf (Sunshine)
Cnc 9h11'51" 32d40'4"
Matthew Lee
Leo 11h35'35" 24d50'40"
Matthew Lewis Corrall
Ori 6h4'37" 20d42'59"
Matthew Lincul
Uma 14h6'8" 61d38'3"
Matthew & Lisa Pratt
Ori 6h5'28" 21d26'21"
Matthew Locke
Cru 11h56'19" -58d38'19"
Matthew Louis Schneider
Lyn 7h47'28" 37d12'20"
Matthew Louis Van Eykeren
Uma 9h58'30" 49d0'59"
Matthew Love
Ori 5h20'26" -5d8'22"
Matthew Lovell
Her 17h59'22" 28d44'35"
Matthew loves Carla
Cru 12h25'4" -60d6'21"
Matthew Lowell Watkins # 3 1/2
Per 3h29'6" 44d51'18"
Matthew Luigi Fedele
Lmi 9h26'37" 36d13'54"
Matthew Luke Fadden
Sgr 18h34'40" -26d43'10"
Matthew Luke Ferrara
Per 3h28'35" 45d28'26"
Matthew Lusty 17.10
Uma 14h26'39" 56d39'44"
Matthew Lynn Baltzell
Ori 5h42'25" 0d29'7"
Matthew Lyons
Gem 7h31'35" 27d30'25"
Matthew M. and Thaniya S.
Leo 10h19'4" 21d22'43"
Matthew M Leffingwell
Cap 20h41'45" -16d59'52"
Matthew M. LiVigni
Vir 13h6'55" 5d43'28"
Matthew M. Vahimian
Uma 8h25'12" 68d3'25"
Matthew Mackrill Buzzle's Light
Cru 12h31'48" -59d12'47"
Matthew Mario Caruso
Cyg 19h41'38" 30d31'57"
Matthew Mark
Ori 5h25'58" 2d39'25"
Matthew Mark
Sgr 19h11'35" -12d37'1"
Matthew Mark Wadsworth
Sco 17h27'0" -41d6'5"
Matthew Marshall
Her 17h38'36" 33d8'25"
Matthew Matera
Sgr 19h13'35" -24d18'4"
Matthew "Mattman" Ellis
Tau 5h55'39" 25d25'44"
Matthew McCray
Her 16h12'39" 4d55'44"
Matthew McGrath
Lmi 9h47'4" 37d56'19"
Matthew McMullen - "Matthew's Star"
Umi 13h22'54" 70d30'52"
Matthew Meeh
Uma 8h17'52" 70d13'30"
Matthew Meininger
Ori 6h17'25" 14d58'5"
Matthew & Melanie - 30 October 2004
Vir 11h56'30" 7d33'39"
Matthew & Melissa
Leo 11h28'45" 15d35'15"
Matthew Michael
Sco 16h11'51" -17d9'48"
Matthew Michael Honiker
Per 3h26'55" 41d10'16"
Matthew Michael James Conrad
Aqr 20h57'55" -11d45'19"
Matthew Michael Jose
Leo 11h38'55" 23d9'47"
Matthew & Michael Marciano
Her 18h3'14" 17d51'33"
Matthew Michael Miulli
Sgr 20h4'45" -23d24'6"
Matthew Michael Wagner
Lep 5h10'49" -12d40'56"
Matthew Michaels, Jr.
Cep 22h10'21" 71d37'16"
Matthew Micheal Ruegsegger
Her 16h48'21" 23d20'27"
Matthew Mikeal Gagnon
Crb 16h6'9" 36d54'1"
Matthew Miller
Uma 13h35'52" 53d31'24"
Matthew Morgan
Leo 10h11'2" 11d31'46"
Matthew Moriarty
Her 16h36'35" 35d51'26"
Matthew N. Mudd
Ori 5h30'50" 13d19'10"
MATTHEW NAIMAN
Cnc 8h50'36" 31d0'40"
Matthew Nash
Aqr 22h9'0" -2d33'11"
Matthew Nathan Taylor
Ori 5h31'9" 1d24'11"
Matthew Neiger
Psc 0h44'36" 15d25'15"
Matthew Nello Spires
Dra 17h4'22" 59d30'46"
Matthew Newman
Ari 2h31'4" 21d23'39"
Matthew & Nicole "Love Star"
Umi 14h40'37" 72d40'30"
Matthew Nieman
Her 18h19'1" 23d34'48"
Matthew Noel McMillan
Sco 16h12'53" -14d57'47"
Matthew Nolan
Dra 19h45'51" 66d25'33"
Matthew Oelschlaeger
Per 2h24'59" 51d18'12"
Matthew Oliva
Her 17h47'43" 47d31'30"
Matthew Oliver
Cyg 21h35'29" 38d7'31"
Matthew O'Neal Thomas
Ori 5h31'23" 3d35'12"
Matthew Opsahl
Uma 11h8'52" 56d51'37"
Matthew Orin Thurber
Aqr 22h33'50" -3d13'59"
Matthew P. Campione, M.D.
Psc 1h38'34" 26d56'18"
Matthew P. Kennedy
Oph 17h18'32" -22d34'50"
Matthew P. Tallmadge
Lib 14h51'30" -3d11'20"
Matthew P. Wamsley
Uma 11h2'48" 48d26'8"
Matthew Pane
Her 18h8'39" 27d34'51"
Matthew Patrick Berry
Cap 21h4'54" -16d56'59"
Matthew Patrick Young
Ari 2h53'14" 29d30'36"
Matthew Paul Allen
Ari 2h8'4" 22d57'31"
Matthew Paul Ashmore
Her 16h32'57" 17d17'32"
Matthew Paul Austria
Sgr 18h26'0" -27d58'42"
Matthew Paul Boucher
Cep 23h2'57" 70d48'4"
Matthew Paul Dixon
Her 17h39'21" 30d24'27"
Matthew Paul Looser
Lib 15h28'58" -7d10'42"
Matthew Paul Seaman
Sco 17h40'57" -35d38'47"
Matthew Paulo Moreira
Vir 12h40'51" 12d45'46"
Matthew Pearson - 27.01.1977
Aqr 22h54'45" -10d50'32"
Matthew Pearson Bishop
Ari 2h36'11" 26d1'14"
Matthew Peter Harris
Boo 14h27'37" 41d15'28"
Matthew Peter Lanzer
Uma 10h50'4" 65d56'6"
Matthew Peter Reaney
Per 4h11'28" 46d4'27"
Matthew Phelps Poulin
Lyr 18h34'1" 39d50'4"
Matthew Philip Carrella
Cap 21h16'1" -19d44'46"
Matthew Phillip Adams
Her 17h6'25" 31d4'48"
Matthew Phillips
Her 17h27'4" 25d33'39"
Matthew Prekupec
Ori 5h44'11" 9d8'17"
Matthew Price shines bright forever
Lib 15h12'1" -21d6'2"
Matthew Proman
Gem 6h56'22" 16d40'10"
Matthew Quay Ammon
Cen 12h17'37" -38d24'10"
Matthew R. Hodgson - my shining star
Ori 6h13'58" -1d14'11"
Matthew R. Lesinski
Per 3h44'4" 49d45'24"
Matthew R Taylor
Lyn 7h31'32" 58d6'36"
Matthew Radford Morgan
Ori 5h49'59" -0d55'14"
Matthew Rakoski
Psc 0h48'2" 8d17'48"
Matthew Rand
Uma 11h42'15" 55d24'27"
Matthew Ravner
Per 2h24'44" 57d7'58"
Matthew Raymond King
Cap 21h12'43" -20d44'12"
Matthew Raymond Robertson
Sco 16h4'41" -24d46'37"
Matthew Rea
Per 3h6'5" 49d44'1"
Matthew Reif
Aqr 23h28'52" -17d33'37"
Matthew Remsen
Ori 5h29'4" 1d26'37"
Matthew Richard
Umi 16h44'16" 76d54'48"
Matthew Richard Luton *3:37am*
Vir 12h37'57" -3d15'59"
Matthew Richard McKee
Her 16h51'9" 36d20'23"
Matthew Richard Perello
Ori 6h10'3" 5d51'22"
Matthew Richard Shirk
Cnc 8h53'46" 22d44'58"
Matthew Richard Swift
Umi 14h58'4" 72d33'43"
Matthew Robert
Her 16h51'21" 35d42'11"
Matthew Robert Bagliore
Her 17h51'53" 44d21'34"
Matthew Robert Dadswell
Uma 11h18'38" 31d34'42"
Matthew Robert Duguay
Sgr 18h43'34" -31d7'15"
Matthew Robert Gardner
Cnc 8h20'2" 10d22'25"
Matthew Robert Hunter
Per 4h27'14" 33d58'46"
Matthew Robert Keates
Per 2h56'4" 51d53'52"
Matthew Robert Laforge
Uma 11h8'10" 55d31'37"
Matthew Robert Marco
Psc 1h41'8" 22d54'28"
Matthew Robert Polland
Cep 21h13'0" 56d6'2"
Matthew Robert Quish
Aql 20h1'59" -7d0'34"
Matthew Robert Sims
Aqr 23h50'35" -11d3'47"
Matthew Robert Walker
Her 16h14'26" 25d20'29"
Matthew Roberts
Crb 15h35'45" 28d36'31"
Matthew Roy Woodall
Ori 6h23'59" 10d16'25"
Matthew Ryan
Ori 4h52'14" 11d2'52"
Matthew Ryan
Lyn 9h1'29" 33d26'22"
Matthew Ryan Buttross
Her 18h41'47" 23d47'26"
Matthew Ryan Curley
Dra 12h58'15" 71d12'40"
Matthew Ryan Ferrari
Lib 15h9'24" -14d50'37"
Matthew Ryan Liedtka
Ori 6h7'39" 15d16'49"
Matthew Ryan Spalding
Aqr 21h36'23" 0d2'17"
Matthew Ryan Villanueva
Uma 11h9'44" 33d22'2"
Matthew S. Daigle
Leo 9h35'22" 11d0'57"
Matthew S. O'Keefe
Dra 16h22'26" 62d19'55"
Matthew S. Perron
Umi 14h17'47" 74d25'37"
Matthew S. R. Purcell
Her 17h19'43" 32d21'53"
Matthew Samuel Lahar
Ari 2h54'51" 29d26'2"
Matthew Sandman Robenhymer
Ori 6h3'57" 19d6'10"
Matthew Sanguinetti
Uma 9h39'41" 44d21'51"
Matthew Santino Ignatius McCluskey
Umi 12h6'18" 89d2'10"
Matthew & Sara Grogan
Uma 10h7'33" 54d57'31"
Matthew Schapiro
Ori 5h46'49" 4d53'11"
Matthew Scott
Sgr 20h13'59" -35d24'37"
Matthew Scott Keith
Ori 6h8'12" 15d35'3"
Matthew Seaman
Psc 0h13'44" 7d49'44"
Matthew Sean Sedberry
Sco 17h12'35" -30d13'9"
Matthew Shane Renko
Aqr 23h27'34" -8d31'8"
Matthew Shannon Hernandezium
Ori 6h0'25" 19d29'10"
Matthew Sierra Richardson
Uma 8h33'21" 61d54'53"
Matthew Sigafoos
Uma 13h43'0" 55d17'50"
Matthew Silverman
Cnc 8h19'59" 20d8'57"
Matthew Simon Farmer
Uma 9h4'47" 49d41'17"
Matthew Simon Running Martinez
Cyg 21h43'40" 54d14'51"
Matthew Simon & Sophie Marie
Umi 16h19'47" 72d58'21"
Matthew Siver
Gem 7h26'24" 19d9'42"
Matthew Sloim
Uma 13h23'59" 58d8'2"
Matthew Smith
Vir 13h54'57" -10d57'58"
Matthew Smith
Her 17h38'11" 37d14'52"
Matthew Sobkowski
Psc 0h41'50" 19d37'58"
Matthew - Son of Danielle & Bruce
Cnc 8h27'1" 16d35'25"
Matthew & Sophie Watts - 26/02/2004
Psa 21h31'28" -36d6'45"
Matthew & Sophie's Star
Vir 13h25'5" -0d54'12"
Matthew Soto
Psc 1h14'24" 26d19'11"
Matthew 'Sparky' Edler
Aqr 22h33'11" -3d8'59"
Matthew Spencer Hill
Gem 6h50'49" 17d12'15"
Matthew Staropoli
Vir 13h23'14" 13d26'37"
Matthew Stefhenson
Tau 4h33'15" 21d44'24"
Matthew Stephen Billingham
Umi 16h20'28" 75d18'42"
Matthew Stephen Kenney
Aur 5h57'51" 52d41'5"
Matthew Stephen Luongo
Sgr 18h18'0" -28d52'33"
Matthew Stephen Prettyman
Tau 5h28'2" 19d43'14"
Matthew Stephen Quinn
Per 3h15'14" 52d24'53"
Matthew Stephen Rockwell
Her 18h22'49" 23d59'21"
Matthew Stephen Wells
Gem 7h28'8" 28d13'2"
Matthew Stephen-Terr Winer
Uma 11h13'40" 69d22'55"
Matthew Steven Goldstein
Per 4h38'30" 40d20'59"
Matthew Steven Keenum
Cyg 21h49'53" 54d35'22"
Matthew Steven Larmon
Her 17h38'2" 36d48'53"
Matthew Steven Pellegal
Ori 4h53'44" 1d32'13"
Matthew Stone Nir
Lac 22h19'41" 54d15'37"
Matthew "Sunshine" Siruchek
Uma 11h48'28" 42d35'0"
Matthew Survance
Aur 5h39'41" 49d56'7"
Matthew Swasty
Per 3h10'39" 41d56'48"
Matthew T. Spencer
Her 16h55'50" 33d49'17"
Matthew T. Yates
Leo 10h31'2" 13d48'58"
Matthew Taylor Franklin
Uma 9h33'11" 54d36'7"
Matthew "The Mop" Peterson
Uma 10h59'9" 45d9'49"
Matthew Thomas Booth
Uma 12h34'21" 53d31'33"
Matthew Thomas Capra
Vir 12h35'44" 12d25'48"
Matthew Thomas Coleman
Ari 2h37'16" 20d26'41"
Matthew Thomas Dunn
Vir 12h34'20" 8d11'53"
Matthew Thomas Greer
Umi 16h14'7" 75d55'3"
Matthew Thomas Lightfoot
Uma 9h39'4" 67d18'5"
Matthew Thomas Maher
Cyg 19h44'36" 29d1'16"
Matthew Thomas Osowski
And 23h18'44" 43d22'27"
Matthew Thomas Rotchford
Vir 13h16'44" 12d22'48"
Matthew Thomas Salz
Uma 12h32'11" 63d10'32"
Matthew Thomas Scaglione
Cnc 8h11'28" 16d21'55"
Matthew Thomas Thom
Leo 11h26'29" 9d19'39"
Matthew Thomas Waligora
Uma 10h25'57" 53d25'21"
Matthew Thomas Weddle
Lyn 6h52'16" 55d40'33"
Matthew Thomas Wilkinson
Uma 9h32'14" 50d53'34"
Matthew Timothy Coffey
Cap 21h47'21" -8d42'23"
Matthew Todd Becker
Uma 9h42'36" 57d9'55"
Matthew Todd Berman
Uma 10h30'29" 61d27'24"
Matthew Todd Henderson
Vir 11h40'37" -2d49'41"
Matthew Todd Howarth
Aqr 22h48'2" -9d25'32"
Matthew Toole
Psc 0h37'15" 5d49'12"
Matthew Torre
Lac 22h49'17" 52d50'58"
Matthew & Tracy McKenna
Cru 12h18'12" -57d32'39"
Matthew Treanor Coleman
Lib 15h27'2" -28d29'29"
Matthew Treanor Coleman
Per 3h15'20" 53d5'59"
Matthew Troy Benedito
Per 2h48'34" 53d42'11"
Matthew Tyler Grimm
Uma 10h22'59" 52d57'22"
Matthew Tytla
Per 2h45'54" 53d38'0"
Matthew und Nicole Vanderboegh
Gem 7h39'54" 31d37'6"
Matthew Vecchione
Cep 22h8'34" 56d29'3"
Matthew Vernon Bredehorn
Aqr 21h28'52" 2d4'32"
Matthew Victor Clewley
Dra 18h42'29" 58d34'30"
Matthew Vincent Persico
Psc 0h52'19" 32d12'36"
Matthew W. Fetzer
Gem 7h51'51" 16d54'47"
Matthew W. Flynn - No Expression
Sgr 18h10'2" -19d58'44"
Matthew W. Montgomery's Dwarf Star
Ori 6h13'52" 8d15'10"
Matthew Wade Avery
Aur 5h57'21" 54d2'1"
Matthew Wade Hunt
Cyg 19h37'56" 35d30'28"
Matthew Walter Santa
Cen 11h48'58" -52d2'44"
Matthew Wayne
Leo 9h29'54" 29d24'34"

Matthew Webb
Her 18h3'58" 24d6'7"
Matthew Weber
Sco 16h25'27" -36d9'2"
Matthew Weitzel
Per 3h28'53" 39d25'11"
Matthew Wellner
Ari 3h6'16" 22d47'4"
Matthew Wester
Sgr 19h26'17" -16d16'30"
Matthew Wilbur
And 0h48'10" 30d31'43"
Matthew William Compton
Vir 12h46'3" 0d8'10"
Matthew William Devany
Umi 14h20'42" 78d2'6"
Matthew William Fox
Umi 15h34'9" 79d36'31"
Matthew Willshire
Psc 1h28'47" 31d11'12"
Matthew Witteman
Uma 10h4'49" 47d25'46"
Matthew Wyatt Sunday
Leo 10h53'38" -0d11'23"
Matthew XXX Yost
Vir 13h5'48" -12d54'31"
Matthew Zane Akerson
Cmi 7h23'17" -0d13'6"
Matthew Zilch
Lmi 11h4'26" 27d36'53"
Matthew, my love, always
Her 17h25'17" 14d14'30"
Matthew, Nathan & Carissa Meeks
Aql 18h58'4" 7d14'39"
Matthew, Stephen, Neefe, and Kwirk
Ari 2h48'7" 27d58'7"
Matthew1027
Sco 16h10'33" -26d10'19"
Matthew-61285
Ori 5h7'27" -0d32'42"
MatthewJoy
Ari 3h6'20" 28d40'43"
MatthewJRodwick021768
Aqr 22h34'36" 1d42'30"
Matthew's Angel
Cap 20h59'10" -24d43'3"
Matthew's Angel
Cen 13h25'25" -46d39'50"
Matthew's Baptismal Light
Umi 14h33'51" 77d5'43"
Matthew's Destiny
Aql 19h1'33" -0d53'51"
Matthew's Gabrielle
Lyr 18h50'49" 31d58'28"
Matthew's Haven
Tau 5h57'54" 25d21'36"
Matthew's Heart
Cap 21h1'42" -25d35'42"
Matthew's Inspiration
Cep 21h35'54" 60d59'12"
Matthew's Light
Leo 11h37'52" 24d53'45"
Matthew's Little Star
Per 4h5'33" 47d45'54"
Matthew's Lucky Christmas Star
Umi 14h57'38" 78d15'14"
Matthew's Milky Way
Uma 12h3'1" 33d16'31"
Matthew's Night Light
Dra 17h46'26" 53d39'13"
Matthew's Star
Cyg 19h50'44" 53d58'8"
Matthew's Star
Uma 8h25'30" 65d33'30"
Matthew's Star
Sgr 17h54'9" -29d40'28"
Matthew's Star
Uma 11h46'26" 36d54'11"
Matthew's Star
Ori 5h46'52" 8d43'47"
Matthews Star
Psc 1h26'18" 13d3'44"
Matthew's Star : Matthew Kay 131205
Ori 5h47'47" -5d28'12"
Matthias
Uma 12h36'0" 58d4'43"
Matthias
Ori 5h16'57" 3d14'58"
Matthias
Uma 10h20'33" 40d26'19"
Matthias Aenishänslin
Vir 13h20'32" 2d38'6"
Matthias Fugmann
Leo 10h36'0" 24d42'59"
Matthias Heintke
Uma 11h4'51" 58d25'42"
Matthias + Jantina
Uma 14h19'46" 58d16'49"
Matthias Kaul
Ori 4h50'12" 0d57'52"
Matthias Knöbel
Uma 11h31'21" 59d11'35"
Matthias Konstantin Tschach
Psc 23h30'27" 4d44'40"
Matthias Lebert
Uma 9h1'53" 64d33'21"
Matthias Meyer (mit Liebe)
Sgr 18h8'22" -35d46'11"
Matthias Michael Meixner "Hasi"
Uma 12h0'50" 51d14'57"
Matthias Möhl
Ori 5h48'5" -2d4'25"
Matthias Müller
Uma 9h41'47" 68d43'55"
Matthias Oberholzer
Cas 1h13'45" 72d33'29"
Matthias Schenkel
Uma 10h20'20" 49d4'30"
Matthias Southwick
Aur 6h27'24" 34d13'31"
Matthias Spohr
Ori 5h42'18" 9d19'15"
Matthias Weber
Uma 10h27'53" 47d47'48"
Matthias.14.04.1969
And 1h29'4" 39d27'42"
Matthieu
Umi 15h5'24" 72d0'21"
Matthieu
Uma 13h59'40" 55d19'50"
Matthieu
Sco 17h22'43" -37d37'26"
Matthieu Barre
Sgr 19h54'18" -29d22'5"
Matthieu Brisebois Rioux 22.01.2003
Uma 9h43'7" 47d55'15"
Matthieu et Thomas Sabattier
Gem 7h11'25" 23d30'41"
Matti
Uma 10h20'37" 48d47'4"
Matti
And 23h49'49" 43d23'24"
Matti Gottesfeld Sciss
Mon 7h35'48" -0d33'40"
Matti Max
Aql 19h59'58" 0d5'3"
Mattia
Cas 1h1'13" 54d17'59"
Mattia
Vir 12h11'26" -1d41'10"
Mattia
Lac 22h22'44" 54d20'34"
Mattia Di Gangi
Peg 22h57'53" 31d54'29"
Mattia Marzulli
Ori 6h7'11" 5d48'51"
Mattia Piero De Santis
Her 18h45'52" 13d50'17"
Mattia Silvan
Uma 8h42'7" 62d27'25"
Mattias Lorenz
Uma 11h0'16" 37d50'31"
Mattiasbecker14.04.2005
Umi 13h12'4" 73d18'35"
Mattie
Lyn 7h12'9" 53d38'25"
Mattie
Ori 5h33'34" -0d33'0"
Mattie
Lib 14h28'13" -17d41'3"
Mattie
Lmi 9h28'29" 38d41'12"
Mattie
And 0h25'42" 40d26'41"
Mattie
Uma 11h34'30" 46d8'58"
Mattie
Cnc 8h4'59" 21d55'29"
Mattie Blakely
Ori 5h26'18" -8d14'0"
Mattie Boy
Umi 4h39'26" 89d1'3"
Mattie Brianne Allison
Vir 12h23'6" 3d16'18"
Mattie E. King
Gem 6h55'44" 31d7'36"
Mattie Elizabeth Jordan
And 0h25'40" 29d3'21"
Mattie Grace Guoxian Spoelker
Uma 11h52'37" 61d31'53"
Mattie Griffin
Ori 6h19'58" -0d58'0"
Mattie Kathleen
Uma 8h38'6" 68d34'50"
Mattie & Lloyd Burgess
Sge 20h4'16" 21d6'53"
Mattie Marie Piazzi
Leo 10h59'23" 11d51'34"
Mattie McCracken Sharpe
Crb 16h8'24" 28d45'11"
Mattie Ruth Hicks
Leo 11h58'14" 21d50'59"
Mattie Terrell
Cas 0h44'47" 51d13'24"
Mattie Whitmire
Umi 16h21'20" 76d49'18"
Mattie Yates
And 0h15'11" 47d10'6"
Mattise (Matthew & Elise)
Dra 18h22'52" 48d51'30"
Mattison Ryan
Uma 9h57'49" 41d37'6"
Mattninsu
Cnc 8h44'48" 28d47'48"
Matt-N-Missys Peanut Butter-N-Jelly
Cyg 19h38'13" 38d30'54"
Matto
Per 3h48'35" 47d20'38"
"Mattozzetto"
Uma 10h20'39" 45d48'12"
Matt's 35th
Pho 0h29'59" -40d20'8"
Matt's Astellas Star
Cnc 8h36'12" 27d47'10"
Matt's California Girl
Psc 1h21'29" 5d11'40"
Matt's Heart
Cnc 8h38'13" 19d46'31"
Matt's Hooch - 2006
Umi 16h11'0" 79d45'10"
Matt's Icarus
Lmi 10h33'28" 28d2'30"
Matt's Lucky Star
Uma 11h15'8" 41d28'34"
Matt's Millennium Planet
Lib 15h25'43" -6d13'37"
Matt's My Star
Vir 12h10'1" -11d4'53"
Matt's Star
Umi 15h22'11" 72d14'28"
Matt's Star
Uma 9h35'54" 49d36'35"
Matt's Star ~ I Love You
Psc 0h24'37" -3d57'54"
Matt's Star January 10th, 1990
Cap 20h21'47" -11d25'2"
Matt's Super Star
Uma 11h33'3" 36d22'59"
Matt's White Lightning
Her 17h6'55" 33d4'26"
Mattshani "6-6-99"
Lyn 7h12'2" 50d51'6"
Mattula
Uma 9h35'57" 53d34'18"
Matty
Lib 14h46'19" -12d32'21"
Matty
Vir 12h16'54" -0d57'14"
Matty
Cma 7h18'26" -14d28'6"
Matty
Her 17h44'12" 38d5'54"
Matty
Uma 9h52'4" 44d24'33"
Matty
Uma 10h32'16" 42d5'21"
Matty
Gem 6h36'5" 14d20'43"
Matty and Holly's star
Gem 7h23'31" 32d48'39"
Matty And Lola
Ori 6h2'13" 18d6'25"
Matty and Melissa
Aqr 22h43'24" -0d48'15"
Matty and Sammy
Umi 16h40'28" 77d38'52"
Matty Bear
Her 18h25'39" 22d13'3"
Matty Daniel Melling
Per 4h21'6" 46d35'15"
Matty Dubuc
Umi 14h43'58" 73d13'33"
Matty & Haley
Cru 12h38'42" -60d7'22"
Matty is Our Shining Star
Cnc 8h44'10" 15d44'3"
Matty Kylene Morgan
Umi 15h19'22" 71d52'46"
Matty loves Asha
Ori 6h0'19" 18d49'3"
Matty mathmagician
Cas 0h37'57" 54d21'32"
'Matty' my true love and my friend
Psc 0h18'59" -3d9'34"
Matty Of Lara's Heart
Sgr 19h45'57" -17d5'8"
Matty, my Love
Per 4h23'17" 43d17'7"
MattyCriswellLovedCherishedLVU ^^~*
Umi 16h18'36" 76d45'5"
Mattyhallbaby
Dra 10h20'40" 75d30'36"
MattyMatt
Tau 4h13'17" 29d42'28"
Matty's Bright Star
Cap 20h12'33" -18d25'14"
Matty's Flame
Cru 12h0'37" -60d35'38"
Matty's Star
Uma 8h20'4" 65d34'47"
Matuszak, Patrik
Uma 11h10'7" 70d28'40"
Matvey Vladimirovich Sirotkin
Lyn 7h59'11" 34d17'0"
Maty Ndir
Cas 23h56'7" 55d13'48"
Matya Avidan
Gem 6h59'20" 32d38'21"
Matylin Nicole Barrow
Psc 0h40'4" 8d56'35"
Matza
Cma 6h42'3" -12d58'1"
Mau
Ori 5h50'11" 7d16'22"
Mau loa, Ka wa pau 'ole, ola
Cnc 9h7'23" 14d22'34"
Mau, Manuel
Uma 14h3'5" 56d5'50"
Mauareen Bisogano Neumann
Lib 15h8'28" -3d17'14"
Maud
Ori 6h8'35" 17d8'35"
Maud
Sco 17h57'57" -43d24'53"
Maud
Cnc 8h1'25" 24d47'30"
Maud
Uma 11h43'40" 45d42'53"
Maud Heurtier
Uma 9h42'20" 46d25'45"
Maud la douce
Sco 17h49'38" -39d55'13"
Maude Ellen LaReau
Cas 0h39'20" 64d13'49"
Maudie Rob
Crb 16h9'35" 39d2'51"
Maudry Mae Smith
Cas 1h50'17" 60d46'54"
Maudy
Lib 14h31'23" -17d5'47"
Maugie & Papa
Tau 5h37'16" 25d16'1"
Maui
Umi 15h45'7" 80d52'56"
Mauiwowie
Uma 9h28'29" 62d1'1"
Maulshree
Lib 15h36'19" -12d2'40"
Maura
Umi 13h45'49" 73d43'12"
Maura
Cnc 8h48'57" 26d3'12"
Maura
And 0h35'1" 32d23'24"
Maura A. Sullivan
Uma 10h1'45" 41d49'56"
Maura Bowen
Aqr 21h5'58" -13d0'4"
Maura C.
Ser 18h18'56" -14d11'32"
Maura E. Sweeney
Leo 11h24'22" 12d43'0"
Maura Jeanne McDevitt
Cap 21h56'40" -18d7'23"
Maura Lynch
Cap 21h58'44" -21d24'35"
Maura Lynch
Mon 8h7'42" -2d54'46"
Maura Martinez
Uma 8h29'30" 71d56'43"
Maura "Mo" Spellman
And 23h6'56" 43d47'3"
Maura Prunty
Leo 11h33'23" 22d24'6"
Maura Rail
Crb 15h44'23" 27d32'13"
Maura Wolowski
Vir 14h58'45" 7d18'3"
Maura's Eyes
Didmyheartlovetillnow?
Boo 14h15'13" 46d57'44"
MAUREA
Vir 13h55'0" -17d32'6"
Maurea Lee Walsh
Vir 12h40'26" 6d17'18"
Maureanna Beth Bruce-Parkes
Psc 0h45'54" 15d55'42"
Maureen
Psc 0h40'52" 10d25'50"
Maureen
Cyg 20h26'50" 45d43'5"
Maureen
Uma 10h1'1" 46d36'7"
Maureen
And 2h38'29" 50d10'2"
Maureen
And 1h35'13" 46d39'34"
Maureen
And 0h52'27" 39d48'1"
Maureen
And 0h42'55" 35d38'43"
Maureen
Aqr 21h30'36" -1d45'55"
Maureen
Cas 0h40'28" 64d9'59"
Maureen A. Bigelow
Ori 5h52'55" -2d5'56"
Maureen A. Dumouchel
Cas 1h18'0" 59d8'4"
Maureen A. Temple
Mon 6h52'0" 8d41'42"
Maureen Addis
Cyg 19h55'11" 32d28'8"
Maureen Albright
Cas 1h24'34" 59d45'23"
Maureen - Always Mine
Cnc 8h44'52" 12d51'32"
Maureen and Don Jackson
And 23h13'16" 48d1'33"
Maureen Ann Carroll
Cas 0h21'23" 54d9'59"
Maureen Ann Dulmer
Uma 8h43'52" 49d15'51"
Maureen Ann Howells
And 1h45'19" 40d31'14"
Maureen Ann Lowis "1970-2005"
Cas 0h35'8" 66d10'24"
Maureen Ann Mercier
Uma 14h8'3" 62d15'22"
Maureen Anne Claiborne
Psc 1h38'25" 8d2'38"
Maureen Anne McManus
Sco 17h12'5" -43d14'17"
Maureen Aquino
Cnv 12h44'11" 39d41'39"
Maureen Armstrong
And 0h47'51" 40d14'36"
Maureen B. Sunshine
Lmi 10h18'59" 31d21'15"
Maureen Bambury - My Beloved Mum
Car 10h1'37" -63d2'46"
Maureen Belford
And 1h14'12" 42d19'44"
Maureen & Bill
Cyg 19h34'52" 54d3'19"
Maureen Boyd
Crb 15h45'5" 34d50'26"
Maureen Branca
Crb 16h10'7" 37d17'9"
Maureen Brigid Dwyer
Uma 8h40'32" 47d49'49"
Maureen C Hemingway
Uma 10h33'27" 67d5'16"
Maureen C Veltman
Gem 6h39'41" 27d39'53"
Maureen Cahill
Uma 11h15'40" 34d46'27"
Maureen Carlson
Cas 1h33'24" 64d43'24"
Maureen Cartwright
Cas 1h27'52" 50d50'57"
Maureen Cecile Finn
Sco 17h45'19" -37d43'49"
Maureen & Christopher Desmarais
Cyg 19h41'47" 52d5'11"
Maureen Ciesluk
Psc 1h16'56" 32d51'53"
Maureen Clark
Cas 23h38'1" 55d9'58"
Maureen Cordwell
Col 6h25'6" -35d0'59"
Maureen Costello
Lib 14h27'29" -17d41'44"
Maureen Davey
Cas 23h32'49" 52d59'1"
Maureen De Camp
Cas 23h55'22" 50d3'55"
Maureen Dowd - VAVA's
Umi 14h43'19" 72d49'30"
Maureen Elizabeth Byrnes
Cas 1h28'10" 64d20'59"
Maureen Elizabeth Horan
And 0h51'42" 39d38'42"
Maureen Elizabeth Jenkins
Cap 21h11'32" -23d15'18"
Maureen Elizabeth Payne Allen
Com 13h12'37" 18d1'41"
Maureen Elizabeth Stinson
Cnc 9h18'4" 14d59'57"
Maureen Erin Nagel Delaney
Vir 12h37'27" 10d39'25"
Maureen F. Ferrante
Sco 16h17'3" -23d57'55"
Maureen F. Remie
Boo 14h35'48" 52d36'31"
Maureen Farrell Edlund
Cas 23h31'16" 52d13'13"
Maureen Fortin-Mentor Extraordinair
Cas 1h39'8" 63d36'20"
Maureen Frances Pritchard
Cas 22h58'27" 56d4'35"
Maureen Fuller
Peg 21h58'36" 13d0'53"
Maureen G. McDonnell
Tau 4h8'43" 15d29'48"
Maureen Greeley
Lyn 7h3'54" 49d45'31"
Maureen H
Uma 9h20'42" 56d13'34"
Maureen Halligan
And 0h50'5" 35d36'34"
Maureen Hendricks
Leo 11h5'10" -1d33'17"
Maureen Hook
Cap 21h43'48" -19d31'10"
Maureen Horner
Cas 1h15'30" 57d56'6"
Maureen Hughes
Cam 4h9'35" 56d13'42"
Maureen J. Lynah
Cas 1h42'0" 68d29'43"
Maureen & Jim Murphy
Sge 19h26'38" 17d19'34"
Maureen Joel
Sco 17h43'3" -43d6'6"
Maureen Josephine - 1936
And 1h56'2" 37d30'31"
Maureen Joyce
Lib 14h51'55" -21d44'19"
Maureen Karen Gatta
Cyg 20h57'25" 46d29'20"
Maureen Kerne
Lyn 8h17'52" 38d28'36"
Maureen Klee
Lmi 10h47'44" 37d51'10"
Maureen Lynne Ridgway
Cyg 21h19'14" 42d28'46"
Maureen Malay
Cas 1h36'0" 66d28'14"
Maureen Mancini
Lib 15h3'28" -6d15'27"
Maureen Marie
Tau 5h44'50" 26d1'25"
Maureen Martin - Ma
Cru 12h39'50" -60d13'35"
Maureen McInerney
And 0h21'37" 45d57'32"
Maureen McManus
Lmi 10h7'42" 33d54'33"
Maureen 'Mo' Gannon
And 23h38'50" 38d19'4"
Maureen Monroe
Lib 14h51'33" -2d23'38"
Maureen Mooney
Cnc 9h0'22" 28d10'5"
Maureen Mullally
Uma 10h12'37" 53d48'9"
Maureen. Mum's Star, Nan's Star.
Sgr 18h29'15" -31d39'10"
Maureen Murphy Roller
Psc 0h54'13" 19d36'27"
Maureen Nass
Uma 8h48'31" 47d37'17"
Maureen O'Bryan
Tau 5h14'27" 19d50'31"
Maureen O'Day
Cnc 8h27'0" 14d39'58"
Maureen Page
Uma 12h28'18" 56d59'27"
Maureen (Penny) Dittmar
Cnc 8h47'14" 25d7'55"
Maureen Petrucelli
Uma 10h25'7" 62d59'46"
Maureen Pettersen
Cas 1h29'34" 68d39'16"
Maureen Pheeney Theriault
Gem 7h29'13" 32d35'23"
Maureen Pierce
Ari 2h4'51" 24d43'34"
Maureen Porter
Gem 6h35'18" 21d49'32"
Maureen Rae De Bose
Crb 15h50'56" 35d0'47"
Maureen Rebekah Tan
Sco 16h9'22" -17d6'6"
Maureen REENI O'Donnell
Tau 4h33'7" 25d5'17"
Maureen Regan
Leo 10h40'25" 13d39'58"
Maureen Renee Ayna
Ari 2h14'50" 24d58'30"
Maureen & Riley Dooly
Ori 5h19'29" -0d38'41"
Maureen & Roger Murray
Cyg 20h59'8" 35d9'54"
Maureen Ryan
Cap 20h27'52" -12d31'57"
Maureen Sawh
Ari 2h27'54" 21d54'33"
Maureen Senick
Gem 7h3'33" 18d47'37"
Maureen Sinacore
Umi 15h4'34" 73d19'3"
Maureen Stansfield
Cyg 21h59'34" 46d34'4"
Maureen 'Sunshine' Iannucci
Psc 23h41'3" 7d20'24"
Maureen T-1
And 2h7'5" 38d22'45"
Maureen Y. Taylor
And 23h44'54" 44d31'9"
Maureen Zirbel "Reenie's Star"
Psc 23h12'31" 7d23'34"
Maureen-1
Cap 21h41'8" -12d58'14"
Maureen's Light Still Shines Bright
Aqr 22h4'10" -11d1'57"
MaureenSiobhan
Leo 11h32'4" 11d56'14"
Maurer, Ernst
Uma 8h19'28" 64d35'53"
Maurer, Ingo
Ori 6h19'58" 15d22'7"
Mauri
Her 17h58'8" 49d51'2"
Mauri
Lyn 7h16'44" 44d57'41"
Mauri
Umi 14h46'58" 80d59'0"
Maurice
Dra 17h45'38" 51d2'50"
Maurice
Psc 1h22'20" 18d42'20"
Maurice Aidan Dawkins
Ori 5h5'39" 15d35'22"
Maurice and Caitiriona
Umi 15h22'58" 72d43'59"
Maurice Antonio Van Pelt
Cnc 8h13'29" 28d59'36"
Maurice Boulaud
Ari 2h42'5" 25d6'58"
Maurice (Buck) Cummings
Uma 11h31'20" 47d37'27"
Maurice Burke Jr.
Uma 8h39'15" 59d29'48"
Maurice Chavez
Vir 13h41'33" 3d14'32"
Maurice Choukrane
Vir 13h18'0" 10d59'26"
Maurice Crockerham, My Baby
Sgr 18h21'16" -24d16'41"
Maurice Davidson
Ori 6h19'47" 15d3'25"
Maurice Elias
Per 3h31'14" 47d14'11"
Maurice Fortoul
Vir 12h41'12" -6d8'25"
Maurice Hall
Per 3h9'25" 54d13'11"
Maurice Harri Angel Star
Uma 9h14'6" 52d31'21"
Maurice Henry Craft
Uma 8h25'49" 69d41'17"
Maurice Isaiah Stewart
Vir 13h19'52" -22d18'46"
Maurice Mitts, Esq. of MMS
Crb 15h40'37" 37d37'41"
Maurice "Mossie" Twomey
Uma 10h51'6" 58d48'27"
Maurice Mursky Gowen
Gem 7h22'41" 14d29'48"
Maurice N. Mason
Vir 11h57'37" 9d29'43"
Maurice Nuzzo
And 22h58'41" 39d54'20"
Maurice Owens
Uma 14h20'14" 58d20'2"
Maurice Paul Hassan
Cnc 8h45'33" 10d37'13"
Maurice Pelletier
Cyg 21h47'37" 44d12'53"
Maurice Peter Walsh
Cnc 8h16'42" 20d0'54"
Maurice "Reese" Carr
Sgr 18h51'42" -24d15'14"
Maurice Stan Yow
Per 4h14'48" 50d30'50"
Maurice "The Superman" Star
Cyg 20h8'16" 45d43'22"
Maurice Towney
Her 16h49'31" 37d51'38"
Maurice V. Sexton
Equ 21h17'31" 8d53'6"
Maurice W. Gerry
Uma 10h51'13" 40d58'39"
Maurice W. McRae
Psc 1h9'28" 27d32'56"
Maurice Wilder
Psc 0h27'6" 9d16'16"
Maurice,Julien, Etienne
Peg 22h42'46" 25d15'5"
MauriceBes&SimoneStelleman NL140707
Uma 13h6'42" 54d47'27"
Mauricio
Cnc 8h52'54" 14d47'56"
Maurille
Aql 19h51'13" 16d11'46"
Maurine Hunsaker
Pho 0h7'45" -42d44'46"
Maurine Onat
Uma 9h0'9" 47d18'10"
Maurissa Meyers
Cas 23h28'17" 54d54'31"
Maurita Malini Prasad
Leo 11h24'54" -4d58'5"
Maurizio
Ori 6h14'48" -3d58'41"
Maurizio
Uma 10h39'35" 71d33'38"
Maurizio
And 1h22'34" 33d58'26"
Maurizio
Peg 21h45'55" 27d50'44"
Maurizio Andrea Orena
Ori 5h34'40" 3d9'46"
Maurizio & Barbara
Ori 5h32'49" -4d8'2"
Maurizio Cutrona
Cam 5h56'11" 59d30'53"
Maurizio Di Fava
Her 17h52'45" 20d56'15"
Maurizio Massarin
Uma 13h43'21" 53d20'19"
Maurizio Perseu
Her 18h14'4" 18d43'47"
Maurizio Rivolta
Her 18h8'46" 15d51'47"
Maurizio Testa
Cas 23h1'39" 53d47'16"
Mauro Casiraghi
Lac 22h46'49" 53d59'46"
Mauro Faleri
Aql 19h58'36" -11d9'10"
Mauro Fogli
Uma 11h33'21" 44d52'28"
Mauro Luciani
And 2h23'55" 43d57'24"
Mauro Meli
And 2h21'42" 42d6'12"
Mauro Parella
Her 17h49'26" 39d2'57"
Mauro Pennacchia, Jr.
Cyg 19h57'19" 43d29'46"
Mauro Realty
Uma 8h26'1" 62d28'13"
Mauro Simonazzi
Crb 16h2'21" 31d58'22"
Mauro Tedeschi
Per 3h49'58" 47d5'25"
Mauro Vendittelli
Uma 10h38'54" 71d33'1"

Mauro..son, father,artist, & poet
Uma 11h2'15" 54d15'55"
Maury*95
Cyg 20h23'3" 48d26'31"
Maury Friedburg
Per 3h48'32" 47d51'0"
Maury M. Nichols, esq.
Cnc 8h40'27" 20d32'3"
Maus
Aql 19h30'44" 4d45'54"
Maus
Uma 8h41'48" 57d48'31"
Maus & Babe 10.06.1995
Uma 8h35'58" 58d3'5"
Mau's Shine
Uma 11h17'33" 51d16'18"
Mäuschen Julia
Tau 4h35'37" 16d29'53"
Mausefrau Tanja & Mausemann Torsten
Uma 11h37'26" 49d51'3"
Mausi
Sge 20h16'38" 19d45'29"
Mausi
Cas 1h7'21" 61d43'54"
MAUSI
Dra 20h29'31" 67d32'25"
Mausibärle Yvonne
Uma 8h23'33" 63d8'57"
Mausi's Stern
Uma 12h10'25" 60d34'21"
Mauz
Leo 9h39'2" 27d16'14"
MauziPuh-Knuddy
Uma 9h59'29" 56d41'33"
MAVANA
Vir 11h46'11" 3d47'58"
Maverick
Tau 4h11'10" 17d58'28"
Maverick
Uma 11h17'32" 29d25'9"
Maverick
Psc 1h18'27" 32d26'53"
Maverick
Uma 9h49'54" 58d37'9"
Maverick
Uma 13h45'51" 54d45'26"
Maverick Diaz
Gem 7h47'28" 32d57'50"
Maverick John MacKenzie
Cnv 12h30'26" 32d10'59"
Maverick Patrick Baer
Lib 15h3'25" -9d48'30"
Maverick Perry Major
Cnv 12h32'48" 43d3'28"
Maverick "Son" Carman
Cru 12h18'25" -56d29'59"
Maverick Thomas Hauser
Boo 14h25'9" 14d34'39"
Maverick Wayne Board
Her 18h48'55" 17d20'34"
Mave's Star - Jacque's Mommy
Gem 6h45'53" 32d43'40"
Mavie Dell Chancellor
Ari 3h4'47" 15d34'37"
Mavis
Aur 7h24'2" 40d22'3"
Mavis
Lyn 7h0'25" 49d18'27"
Mavis
Sgr 19h16'28" -14d57'21"
Mavis
Cas 0h23'9" 61d59'17"
Mavis Austen
Cas 0h15'33" 56d58'30"
Mavis B. Prinie
Cnc 8h46'12" 18d54'45"
Mavis Firth
Ari 1h59'9" 24d25'5"
Mavis Gallagher
Cas 0h50'57" 60d47'30"
Mavis Hale
Uma 10h55'1" 47d37'55"
Mavis Lily Davis - My Mum xx
Umi 16h32'16" 75d6'25"
Mavis Tregoning Morgan
Cas 23h39'13" 54d33'44"
Mavis Tsai
Lib 15h25'14" -11d1'49"
M.A.V.P. 42: Adoration
Aqr 22h44'46" -13d51'57"
Mavrik Esteban Boucher Therrien
Cep 21h44'43" 64d38'32"
Mavroidi Georgia
And 23h23'25" 52d16'52"
Mavsy's Star
Uma 8h38'25" 62d27'23"
Maw Maw-Gran Daddy Star
Uma 8h54'17" 50d50'40"
Maw Maw's Star
Sgr 18h54'56" -29d57'29"
Mawa
Cas 0h42'53" 64d52'57"
mawe
Ori 5h27'11" 13d39'32"
Mawka The Beautiful
And 23h8'14" 41d43'1"
Mawmaw Millicent Johnston
Uma 11h18'47" 29d53'23"
Mawusi
Vir 13h23'41" 13d33'20"
Max
Aql 19h42'22" 1d59'58"
Max
Tau 5h55'8" 26d10'16"
Max
Her 17h13'34" 37d37'22"
Max
Boo 15h38'39" 43d29'28"
max
Cyg 21h33'5" 50d9'6"
MAX
Uma 11h29'58" 43d58'22"
Max
Per 3h39'46" 33d58'3"
MAX
Tri 2h18'51" 32d10'28"
Max
Dra 14h59'58" 59d12'46"
Max
Uma 10h51'20" 69d41'33"
Max
Cma 6h39'12" -16d11'54"
Max
Cma 7h22'28" -15d40'0"
Max
Cam 5h35'42" 78d0'18"
MAX
Cep 23h8'40" 79d43'20"
Max
Lib 14h46'28" -0d47'50"
Max "2-2-92"
Cma 6h28'8" -25d57'59"
Max A Million
Cap 21h6'12" -24d44'7"
Max Aaron
Boo 14h46'18" 36d58'19"
Max Aaron Fink
Cnc 8h10'1" 6d48'13"
Max Aaron Tapper
Uma 8h52'40" 61d55'42"
Max Alexander
Sco 16h9'53" -22d34'43"
Max Alexander Carboni
Cyg 20h40'5" 31d48'31"
Max Alexander Mariasch
Psc 0h8'38" -0d44'7"
Max Allsopp
Gem 7h34'49" 33d50'59"
Max and Jenny Ahlvers
Cyg 21h26'31" 47d56'50"
Max Anthony Nicholas Smith
Umi 15h13'2" 78d44'25"
Max Assoulin
Per 2h40'33" 56d34'40"
Max Bech
Ori 6h18'47" 5d46'47"
Max Beck
Ari 3h20'19" 28d48'4"
Max Briley
Lib 14h55'41" -6d12'4"
Max Brown
Psc 1h40'13" 7d47'28"
Max Byles
Cep 22h7'0" 61d41'51"
Max C. Belcher
Ori 5h26'6" 0d2'51"
Max Cannon
Cma 6h42'48" -14d38'29"
Max Charles Whittington - Max's Star
Her 16h20'5" 12d16'52"
Max & Christie
Uma 9h25'31" 60d59'9"
Max Clements
Per 4h5'56" 32d30'36"
Max Connor Eitelberg
Uma 13h0'46" 53d14'28"
Max Cowan
Per 4h36'49" 46d53'27"
Max Daigle
Uma 9h40'18" 47d48'49"
Max Davis O'Guinn, Jr.
Per 1h48'38" 50d55'41"
Max Dugan
Uma 13h45'20" 58d11'42"
Max F. Barresi (MAX1126)
Sgr 19h47'1" -14d29'56"
Max Feldman
Uma 9h21'56" 42d18'33"
Max Ferroni
Cnc 8h40'14" 24d32'14"
Max Frailey-Capitol Cement
Ori 5h32'10" 0d6'47"
Max Gallegos
Cet 3h20'5" -0d45'24"
Max Garry
Umi 23h53'4" 89d27'54"
Max George
Umi 12h25'22" 88d0'4"
Max Gregor Verveckken
Uma 10h6'46" 68d9'49"
Max Haigney
Sgr 19h5'50" -26d53'10"
Max Hall
Lib 16h1'18" -8d34'44"
Max Hamish Badger
Uma 12h16'23" 61d54'32"
Max Harry David Ebdon
Per 4h42'40" 43d21'19"
Max Hawksley
Cas 1h24'32" 59d46'28"
Max Herman
Aur 5h53'43" 46d39'20"
Max Hickey
Ori 6h0'41" 10d21'28"
Max Höcht
Ori 6h12'40" 7d58'44"
Max Holmes
Dra 16h31'42" 58d34'28"
Max Horrocks
Umi 13h40'36" 79d10'19"
Max Horwitz
Leo 10h3'16" 17d49'53"
Max Hurlimann
Psc 0h42'59" 15d45'4"
Max In A Million
Uma 14h7'7" 58d48'38"
Max & Irene Winkler
Cyg 19h31'10" 47d9'48"
Max Jackson's Dream
Per 4h42'22" 43d45'6"
Max Jacob Miller
Lib 15h25'8" -27d44'16"
Max James Allison
Cru 12h2'32" -62d26'22"
Max James Edwards
Cam 4h37'36" 59d18'45"
Max James Fordyce
Per 4h44'19" 37d54'41"
Max James Leech
Umi 7h53'2" 88d42'29"
Max James - Our Godson
Umi 15h36'58" 81d33'39"
Max & Jean Neuman
Cyg 20h51'53" 33d19'19"
Max Jerry Dodson
Aqr 21h15'17" -9d14'59"
Max John Cherry
Umi 16h25'2" 72d14'8"
Max John Hetherington
Her 17h43'43" 19d44'27"
Max Josef Carr McCarron
Cen 13h27'1" -40d58'12"
Max Joseph Baczynski
Leo 10h45'29" 10d39'40"
Max Kapnis
Cnc 8h16'27" 18d50'30"
Max Kenneth Timko
Her 16h32'15" 45d43'25"
Max King
Ori 5h41'43" 1d16'16"
Max King Courtier-Dutton
Gem 6h31'13" 20d13'25"
Max Kurt Mislow
Umi 14h36'4" 73d34'53"
Max Lewis
Umi 13h29'17" 75d17'9"
Max Lloyd Stephen Pritchard 4.7.05
Cnc 9h6'50" 9d2'20"
Max Louis Nardella 'The Great'
Ori 5h48'50" -3d3'42"
Max Louis Reinhardt
Leo 11h9'2" 3d0'8"
Max Love
Sge 20h4'35" 20d29'28"
Max Loves Me
Cyg 21h10'5" 42d26'35"
Max LX
Dra 12h57'28" 75d32'2"
Max M.
Psc 0h41'8" 21d0'7"
Max Markarian PM
Lyr 19h10'6" 37d15'49"
Max Markham Grinnell
Lyn 8h41'32" 36d2'43"
Max (Marshall B)
Peg 0h32'20" 18d8'48"
Max McDonald
Pyx 8h46'24" -27d49'1"
Max McKelvey
Uma 11h50'47" 34d36'21"
Max Michael Howard
Her 18h14'24" 15d57'44"
Max Mills
Uma 8h36'57" 57d21'1"
Max Montgomery Jones
Leo 10h16'11" 26d7'0"
Max Mutchnick
Aql 18h45'30" -0d45'33"
Max Nuki
Aql 19h27'32" -0d10'29"
Max of a Million Dreams
Vir 13h20'54" 11d0'28"
Max Okesha Star
Cnv 13h57'14" 32d47'44"
Max Olivas
Leo 10h53'7" -4d10'30"
Max Owen Fitzpatrick
Tau 3h39'3" 28d21'15"
Max Park-Beach
Umi 14h35'8" 75d2'23"
Max & Patricia Seeds
Psc 1h8'10" 24d21'58"
Max Pfeiffer
Uma 10h44'17" 41d12'13"
Max Philip Burton - Max's Star
Per 4h6'39" 32d55'40"
Max 'Philo' Philips 1969-2004 xxejj
Cru 12h52'27" -57d24'48"
Max R. Peterson
Cyg 20h53'7" 32d30'7"
Max Riley Faass
Umi 14h29'38" 68d21'57"
Max Robert Allen
Ori 6h4'19" 6d38'21"
Max Roger Adams
Uma 11h36'49" 58d9'44"
Max Ryan Carse
Lmi 10h34'16" 38d5'29"
Max Samuel Groves
Gem 6h16'2" 25d41'18"
Max Samuel Mathews
Cen 13h51'30" -42d36'57"
Max Seel "Summer Wind"
Uma 8h48'41" 71d12'34"
Max selvaggia Vale
Ori 5h50'38" 6d51'56"
Max Shapiro
Uma 11h28'37" 44d37'54"
Max Simon
Cnv 12h43'13" 39d2'1"
Max Steel
Ori 5h26'32" 3d40'41"
Max Stephen Pepper
Leo 10h25'6" 13d13'5"
Max Sullivan
Per 4h33'17" 43d43'7"
Max Taplin
Aqr 23h3'38" -16d37'41"
Max & Tee
Lyr 19h23'52" 30d39'54"
Max Thoman
Boo 14h47'36" 52d4'20"
Max Valerio - 19th September 2005
Col 5h42'59" -29d56'8"
Max W Berger Universal Best Brother
Leo 9h26'2" 12d54'52"
Max Walsh
Peg 21h39'35" 27d4'14"
Max Wheeler
Her 18h14'33" 29d19'16"
Max William Helm
Ori 5h59'51" 21d50'7"
Max William Reason - 9.12.2006
Col 5h49'30" -30d16'53"
Max William Stewich
Cep 22h44'22" 66d42'6"
Max Wilson Martin
Cyg 20h4'41" 34d32'45"
Max Winter
Uma 10h58'52" 64d33'31"
Max Yorke
Ori 5h41'48" -1d28'4"
Maxamaflo
Lac 21h59'11" 41d2'33"
Maxamarjac
Uma 9h48'1" 42d16'59"
"MAX-A-MILLION 007"
Gem 7h26'25" 32d18'28"
MaxBe
Her 17h19'9" 36d24'3"
Maxelle 20 Février 1986
Psc 1h39'28" 20d30'30"
MAXEMA
Her 16h47'16" 30d4'21"
Maxence
Del 20h41'48" 10d0'49"
Maxence Kemy Maggioli
Vir 13h3'16" 0d22'54"
Maxence Rousselet
Umi 14h13'53" 65d45'51"
Maxene Bates
Cru 12h26'57" -59d2'12"
Maxfield Joseph Kline
Cep 21h43'55" 56d59'59"
MAXH
Lib 15h23'31" -5d58'51"
Maxi
Cmi 7h28'39" 10d44'22"
MaXi
Ori 5h24'14" 2d14'49"
Maxi
Uma 10h24'42" 50d16'3"
Maxi Hohl
Aql 19h4'34" 9d16'13"
Maxi Million Masluk
Ori 6h17'41" 15d37'44"
Maxianne Allsopp
Lib 15h5'59" -0d29'36"
Maxie
Umi 16h40'5" 75d43'0"
Maxie Cantwell
Vir 13h32'10" -9d57'54"
Maxie Morrone
Cnc 9h14'34" 30d8'26"
Maxim Agrest
Ari 3h24'57" 26d57'40"
Maxim Jayden Moya
Leo 10h17'33" 9d47'17"
Maxim John Taylor
Cep 20h57'26" 65d20'43"
Maxim Picard
Uma 10h44'30" 66d30'15"
Maxim Wezenbeek
Tau 5h11'27" 22d25'33"
Maxim William Burrows
Cep 0h7'51" 67d36'27"
maximamassimiliano
Per 2h57'11" 44d21'37"
Maxime
And 1h10'36" 46d52'40"
Maxime
Psc 23h31'38" 0d0'14"
Maxime Boileau 29jul04
Umi 16h51'29" 79d0'51"
Maxime Eizaguirre
Leo 10h7'21" 14d21'19"
Maxime et Emma
Cnc 8h51'56" 11d9'32"
Maxime et Eva
Cyg 20h55'0" 48d29'20"
Maxime Giethlen
Sgr 18h14'35" -29d34'29"
Maxime Heisei Alien Lebrun
Dra 16h53'12" 66d28'34"
Maxime je t'aime
Leo 11h26'51" 6d44'7"
Maxime Launay - Bussière
Del 20h28'44" 16d34'41"
Maxime Roxanne Leger
Per 2h20'30" 54d49'12"
Maximelena
Ori 6h13'43" 8d58'13"
Maximile Perricone
Cnc 8h31'45" 22d47'17"
Maximilian
Ori 6h0'59" 11d9'4"
Maximilian
Uma 10h1'19" 72d44'3"
Maximilian
Cma 6h14'29" -13d47'30"
Maximilian
Cap 21h11'58" -23d9'36"
Maximilian Alan Altenhofen
Gem 7h31'8" 34d9'39"
Maximilian Brunner
Uma 10h44'57" 63d32'15"
Maximilian Frederick
Lac 22h52'23" 37d31'28"
Maximilian Gartner
Sco 16h16'53" -16d47'46"
Maximilian Nyman
Cnc 8h42'15" 7d49'59"
Maximilian&Suntka
Uma 9h39'34" 49d26'52"
Maximilian W Lempriere
Per 2h37'26" 53d0'40"
Maximilian-Johannes Oder
Uma 13h53'21" 57d4'45"
Maximilian's Lucky Christmas Star
Umi 17h1'8" 76d51'7"
Maximillian
Lib 15h37'48" -15d15'58"
Maximillian
Cma 6h14'29" 13d47'30"
Maximillian Cole
Dra 19h28'27" 75d10'17"
Maximillian Du-pre
Cru 12h34'49" -63d46'24"
Maximillian Fernandez
Cep 21h44'7" 55d32'29"
Maximillian G. Yarabek 1st Birthday
Uma 10h46'47" 52d52'16"
Maximillian James
Leo 11h3'52" 21d42'24"
Maximillian R.F. King
Cep 22h44'0" 72d42'58"
Maximillion
Gem 7h20'17" 25d44'50"
Maximillion Halle
Leo 11h36'5" 7d9'59"
Maximo Paul Tamayo-Bullivant
Cnc 8h19'1" 16d2'3"
Maximus
Sge 19h42'32" 19d3'5"
Maximus
Leo 11h56'33" 24d45'14"
Maximus Abraham Taylor - Max's Star
Per 2h38'37" 51d57'42"
Maximus and Mariah
Gem 7h40'41" 33d40'8"
MAXIMUS ANTONIOUS (ANTHONY LINDSEY)
Aqr 22h2'50" -8d40'29"
Maximus Gregory Battistella
Gem 7h19'43" 30d58'37"
Maximus Joseph Morton
Cnc 8h48'47" 31d26'53"
Maximus Lee Cannon
Sgr 18h50'38" -16d3'5"
Maximus Makrounis
Per 4h41'10" 42d58'1"
Maximus Neilius Stiegelstar
Sco 17h18'10" -32d27'54"
Maximus Xavier Petitte
Uma 9h22'27" 51d9'21"
Maxine
Lyr 18h30'19" 28d17'35"
Maxine
Gem 6h53'20" 27d35'17"
Maxine
Cas 1h19'32" 63d53'52"
Maxine 1928
Cas 0h51'48" 57d18'40"
Maxine Alisa Ballin
Cas 1h34'9" 61d43'33"
Maxine Ann Baker
Dra 16h41'41" 66d58'11"
Maxine Ann Turner
Sco 16h55'57" -42d18'58"
Maxine Bass
Cas 1h44'5" 61d33'45"
Maxine Bergfeld Halm Baenziger
Lyn 7h28'34" 58d27'35"
Maxine "Buttercup" Jones
And 0h45'26" 43d2'42"
Maxine E. Mann
Cas 1h32'10" 62d37'7"
Maxine E Mielczarek (1915-2005)
Sco 16h8'29" -14d15'56"
Maxine E. Sloan
Lyn 9h15'38" 44d53'46"
Maxine Eddings
Cyg 20h36'56" 36d52'3"
Maxine Elenore Willming
Lyn 8h27'3" 39d8'32"
Maxine Elizabeth Pauley
And 1h25'23" 39d30'25"
Maxine Elizabeth Walker
And 0h39'31" 40d27'52"
Maxine Ellis Adkins
Gem 6h31'47" 16d43'10"
Maxine Feildsend
Cas 0h4'0" 54d7'48"
Maxine G. Haneman
Vir 12h47'44" -10d53'19"
Maxine Gibson
And 2h24'20" 42d29'1"
Maxine Grey Puffenbarger
And 1h37'51" 44d12'0"
Maxine Hilton
And 0h22'15" 43d1'56"
Maxine Lewis
And 0h33'40" 37d38'10"
Maxine Louise Schroder
Cas 0h47'42" 57d17'22"
Maxine Ma Baker
Uma 10h58'14" 54d7'41"
Maxine (Ma) Gray
Cas 2h22'29" 67d28'29"
Maxine Madrid
Cnc 7h56'40" 9d50'50"
Maxine "Mod" Goodell
Psc 1h12'50" 16d21'29"
Maxine P. Chadwick
Mon 7h16'16" -0d8'22"
Maxine & Samuel Schlyen
Umi 15h51'20" 76d51'10"
Maxine Sandbeck O'Riley
And 0h33'36" 42d32'21"
Maxine & Sergio Shining Together
Cru 12h15'1" -62d1'7"
Maxine Shakesby
And 0h47'10" 26d30'2"
Maxine Turk
Cnc 8h51'50" 11d33'19"
Maxine Walloch
Leo 10h55'2" 7d51'46"
Maxine Wilder
Uma 9h55'26" 48d45'37"
Maxine's Babe
And 0h1'57" 44d31'36"
Maxine's Star
Lmi 10h14'19" 29d44'23"
Maxland
Crb 16h21'29" 31d34'35"
Maxmillian Anthony Stratton
Cnc 9h6'46" 30d55'58"
Max's Halo
Ari 3h13'53" 28d14'34"
Max's Midnight Star
Cep 21h50'23" 60d6'10"
Max's Passion
Gem 7h21'0" 23d40'57"
Max's Shining Star
Uma 10h11'16" 65d30'26"
Max's Star
Aqr 22h59'55" -5d27'1"
Maxton Gig Beesley
Uma 9h12'0" 55d31'16"
Maxwell
Leo 9h26'43" 30d49'28"
Maxwell 1
Vir 14h19'26" -21d43'44"
Maxwell 9
Leo 9h25'0" 10d52'9"
Maxwell A. KinCannon
Sgr 19h19'18" -27d52'8"
Maxwell Alexander St Ledger McCarthy
Ori 6h18'22" 9d10'5"
Maxwell and Megan
Cyg 19h40'30" 34d22'45"
Maxwell Arnold Heth
Sgr 18h21'24" -32d6'38"
Maxwell Austin Ransom
Ori 5h24'34" 3d15'33"
Maxwell B Thomas - Flying Forever
Cen 15h2'7" -39d16'19"
Maxwell & Barbara Schmidt KBJJCM
Nor 15h50'6" -42d59'23"
Maxwell Caiden Pearce
Lmi 10h41'9" 29d37'47"
Maxwell Cameron Braunstein
Sco 17h11'28" -30d25'33"
Maxwell Chase Matthews
Per 3h22'47" 37d40'19"
Maxwell Christopher Erdley
Cnc 8h31'34" 7d10'57"
Maxwell Clark
Lmi 10h44'4" 25d35'21"
Maxwell Connelly Anderson
Lib 14h34'5" -16d13'18"
Maxwell Conner Eitelberg
Cnc 8h56'50" 16d37'10"
Maxwell Cordeiro Camara
Cap 20h14'27" -23d12'37"
Maxwell D. Bullard
Umi 14h26'25" 74d39'8"
Maxwell David Hasenour
Ari 2h44'13" 16d39'24"
Maxwell Douglas Kramer
Aqr 22h4'20" -21d39'36"
Maxwell Ethan
Gem 7h42'43" 32d26'46"
Maxwell Evan Weeks
Cnc 9h1'26" 6d43'2"
Maxwell Forrest
Cep 21h44'22" 85d53'12"
Maxwell Frank Tafoya
Sct 18h47'52" -11d28'2"
Maxwell Gerard Farina
Psc 1h19'10" 25d27'52"
Maxwell Glen Kohler
Ori 5h40'42" 12d30'7"
Maxwell Glenn Danser - 02/23/95
Her 18h41'27" 20d36'52"
Maxwell Harrison Vitucci
Ori 5h15'5" 16d8'9"
Maxwell James Konik
Her 16h17'14" 12d45'5"
Maxwell James Puscheck
Per 3h20'34" 46d30'14"
Maxwell James Raymond McNally
Ori 6h20'0" 10d42'47"
Maxwell John Pearson
Ori 6h20'26" 15d50'19"
Maxwell John Rodewald
Aql 20h20'17" -0d49'51"
Maxwell Joseph Merritt
Per 3h11'46" 53d38'50"
Maxwell Joseph Simon
Cyg 21h46'36" 37d23'52"
Maxwell LaValley
Cnc 9h10'3" 11d57'28"
Maxwell Lindauer Miller 2004
Sgr 18h34'12" -25d36'33"
Maxwell M. Mitchell
Her 17h47'20" 46d19'26"
Maxwell & Mary
Lyn 8h43'52" 33d8'3"
Maxwell McLeod Brown
Dra 17h4'35" 61d13'12"
Maxwell Michael Merriman
Cnc 8h35'47" 8d59'13"
Maxwell & Oscar Taplin
Pyx 8h27'15" -32d2'18"
Maxwell Owen Colley
Tau 4h39'5" 23d18'38"
Maxwell P Quinn
Ori 4h51'10" 3d55'8"
Maxwell Peter Allen
Umi 15h42'48" 80d39'22"
Maxwell Prizant
Sco 16h15'57" -14d31'29"
Maxwell Reily Lipshutz
Ori 5h46'20" 8d26'10"
Maxwell Silverhammer
Cma 6h40'9" -15d12'12"
Maxwell Sterling Schroeder
Uma 10h12'8" 57d3'6"
Maxwell Swanson
Tau 4h4'27" 2d3'11"
Maxwell Theodore Townsend
Cnc 9h16'41" 26d39'11"
Maxwell Thomas Garner
Ari 2h8'4" 15d54'42"
Maxwell Thomas Greenhalgh
Umi 15h52'24" 73d34'21"
Maxwell Thomas Kemp
Her 17h59'22" 17d19'0"
Maxwell Thomas Peters
Gem 7h29'9" 33d48'14"
Maxwell Ward Alfond
Aur 5h50'16" 44d25'6"
Maxwell Welch Landrey
Umi 16h23'51" 77d50'29"
Maxwelll Covington Cott III
Her 17h7'7" 32d55'40"
Maxwell's Dream
Umi 16h23'40" 83d23'39"
Maxwell's Star
Cep 23h7'46" 70d0'30"
Maxwell-Young Starr
Gem 6h32'38" 21d23'31"
Maxx
Uma 9h42'22" 69d2'26"
Maxx Stone
Ori 5h34'59" 0d47'27"
Maxxi Lee Page
Cam 3h57'0" 59d55'16"
Maxxy Kolterer
And 1h13'50" 45d23'35"
Maxy Boy
Aqr 21h16'5" -0d18'5"
Maxyne Miller
Lmi 9h38'31" 33d50'26"

May
Crb 15h37'24" 28d0'42"
May 05'
Uma 11h32'59" 63d19'2"
May Apacible
Psc 0h23'10" 8d21'29"
May Badr El Deen
Lmi 10h19'33" 28d50'1"
May Bellringer
Cas 0h25'40" 57d31'47"
May Cheung Gee
Gem 6h55'58" 16d19'14"
May Chidiac
Uma 10h57'27" 68d12'17"
May Chong
Uma 10h46'43" 51d40'36"
May Cioci
Vir 13h43'29" -6d22'59"
May Duce
Cnc 8h23'54" 14d32'52"
May East
And 0h8'40" 29d13'14"
May Evelyn O'Connor
Sgr 18h42'53" -32d24'15"
May Florence Coon
Cas 2h40'21" 57d49'7"
May Juliet Valero Plater
Tau 3h55'48" 17d18'48"
May Le
Cam 6h41'54" 65d11'36"
May Night
Cyg 20h56'52" 45d14'51"
May Noureddine
Uma 10h36'2" 62d6'6"
May Our Love Shine Eternally
Aqr 21h48'4" -2d44'54"
May our years out number the stars.
And 0h34'36" 27d27'23"
May R. Woods
Ari 2h46'3" 29d36'3"
May Ruby Peter "Sex N' Heaven"
Crb 16h19'42" 31d4'5"
MAYA
Crb 15h52'27" 33d17'14"
Maya
Cnv 13h58'41" 35d31'58"
MAYA
Lac 22h26'38" 48d1'40"
Maya
And 0h5'44" 45d15'10"
Maya
Gem 6h41'57" 23d7'32"
Maya
Mon 6h45'38" 8d48'57"
Maya
Vir 13h15'47" 11d16'26"
Maya
Cap 21h33'0" -11d3'58"
Maya
Lyn 8h17'34" 55d1'24"
Maya
Psc 0h16'8" 1d0'38"
Maya Abigail Kalfon
Peg 23h50'30" 29d36'25"
Maya Alexandria
Crb 15h27'12" 28d12'23"
Maya Alina
Com 12h22'9" 27d25'41"
Maya and Charlie The Best, Always!
Cyg 20h9'42" 41d22'11"
Maya Angelica Wong Orr
Lyn 7h0'40" 57d4'14"
Maya Ann
Cyg 21h10'39" 31d41'25"
Maya Ashley
Aqr 21h45'15" -0d28'44"
Maya Barsky
Tau 5h28'22" 18d17'14"
Maya Brazal Goostree
Crb 15h26'29" 25d52'53"
Maya Carmen Camba De Ala
Gem 6h57'50" 25d30'55"
Maya Charlotte Scott
And 1h13'8" 41d29'15"
Maya Claire MacCorquodale
And 2h3'37" 42d25'39"
Maya Curtis Farwell
Aqr 21h38'26" 2d55'50"
Maya Dedhia
Sco 16h8'6" -15d42'23"
Maya Dürst
Cas 1h32'10" 72d33'5"
Maya Elle
Leo 11h49'46" 26d17'36"
Maya Escovedo
Cap 20h36'48" -19d52'45"
Maya F. Cookus
Lyn 8h34'43" 33d42'35"
Maya Forbes-Rolling
Cru 12h22'7" -62d36'45"
Maya G Grantis
Umi 16h9'58" 75d22'17"
Maya Gabrielle
Aqr 22h10'47" -8d40'35"
Maya Gabrielle Shurte
And 1h16'18" 46d23'47"
Maya Ghannoum
Aqr 23h43'42" -14d41'22"
Maya Gonzalez
Umi 14h53'43" 68d7'26"
Maya Grace Zeissig
Uma 11h5'25" 53d56'37"
Maya Gracie Murray
Crb 15h45'58" 28d17'2"
Maya Hannah
Uma 9h21'19" 66d45'44"
Maya Helen Brosnick
Tau 4h26'40" 11d31'19"
Maya Indigo Pangelinan
Lyr 18h41'20" 39d33'4"
Maya Isabel
Psc 1h19'17" 15d8'14"
Maya Janelle Mason
And 0h8'35" 34d22'50"
Maya Julie-Ann Sangara
Leo 11h10'25" 6d19'57"
Maya Kay Rogers
Cap 21h32'40" -12d4'18"
Maya Kelly Cravy
Ari 2h18'44" 18d40'7"
Maya & Kerys Dodds
Lmi 10h36'43" 28d55'58"
Maya Killer
Her 18h6'47" 18d39'58"
Maya Liana
Uma 9h33'9" 62d1'44"
Maya Lily Levi
And 23h7'10" 41d31'16"
Maya Lily Renae Strong 18.02.2006
Aqr 22h51'29" -11d10'32"
Maya Loise Neely
And 0h45'29" 46d14'50"
Maya M.
Uma 11h13'45" 54d51'12"
Maya Madrid Thompson
Lyn 7h45'28" 39d21'8"
Maya Marchese-Reilly
Cas 1h26'31" 62d2'33"
Maya Michele Jones
Crb 15h31'38" 27d47'34"
Maya Michelle Blackman "Sweet Pea"
Lyr 18h41'42" 38d52'59"
Maya Michelle Yesalonis
Lib 15h28'49" -6d25'4"
Maya Monet
Mon 6h49'8" 6d55'34"
Maya Nicole Allen
And 0h22'27" 42d34'53"
Maya Nicole Holloman
Psc 1h14'0" 25d19'46"
Maya Olson
Crb 15h52'43" 25d45'6"
Maya R. Clark
Gem 6h40'51" 29d49'1"
Maya Rae Cracchiolo
And 23h19'30" 41d27'6"
Maya Rain Purcell
And 1h43'20" 45d27'33"
Maya Ratna Majumdar
Lib 15h35'49" -27d55'17"
Maya Rebecca Dunlap
Cap 21h51'36" -24d24'40"
Maya Rhee
Sco 17h44'22" -41d39'1"
Maya Rose Servedio
Aqr 21h40'19" -5d47'34"
Maya Sally and Jake Forever
Sco 17h56'14" -31d46'21"
Maya Shandra Capuano
Per 3h42'59" 43d46'42"
Maya Sofia
Lyn 7h25'54" 57d51'28"
Maya - Star of a Princess
Leo 9h57'11" 13d37'19"
Maya Stojan
Dra 16h3'58" 62d12'13"
Maya Strebel
Ori 6h13'14" 13d45'25"
Maya Suzan
Tau 5h38'24" 27d41'34"
Maya Teresa Sindos
Uma 8h57'9" 54d9'15"
Maya the Watergirl
Ori 5h56'2" 22d50'40"
Maya Toula Longo
Cam 4h36'21" 67d50'4"
Mayani
Uma 11h20'36" 65d16'39"
Mayank & Divyaa's Dee-vine Energy
Her 17h58'4" 22d15'28"
Mayank & Punita Patel
Cyg 21h14'41" 47d36'16"
Maya's Angelo Home In India
Cra 19h7'55" -38d30'19"
Maya's Destiny
Per 3h21'35" 48d55'17"
MayaSher
Umi 16h31'58" 83d54'3"
Maya's-star
Ari 1h57'17" 21d42'41"
MAYBE
Psc 1h7'30" 3d29'45"
maybe...definately
Gem 6h36'49" 23d6'40"
Maybelle
Pup 7h38'42" -13d40'46"
Maybelline Ann Chan
Mon 6h43'59" -0d17'46"
Mayberry
Ari 2h46'17" 13d20'43"
May-Britt
Uma 9h42'8" 62d43'9"
Maycee Jo Conrad
Uma 10h27'38" 59d30'11"
Mayda
Psc 0h54'17" 14d46'7"
Mayde
Ori 4h44'7" 4d1'58"
Maye Machnouk
Lyr 18h49'16" 30d0'14"
Mayela Nieves Alicea
Vir 13h38'15" -15d12'51"
mayelei
And 0h29'22" 41d13'56"
Mayer Kagan
Cap 21h34'28" -19d41'49"
Mayer, Hartmut
Uma 9h12'55" 50d54'44"
Mayerocchhelle
Lib 15h39'52" -15d56'10"
Mayfield School
Col 6h22'57" -33d54'23"
Mayleena Christine Padre
Psc 1h8'21" 26d10'48"
Maylen Danielle Hightower
Lib 15h42'30" -18d39'56"
MayLena
Cap 20h38'45" -21d13'46"
Maylis Launoy
Peg 22h41'52" 24d4'41"
Maylyn Michelle Lugones
Cap 21h3'42" -19d39'0"
Maymer
Cnc 8h49'29" 13d59'37"
Maynard Blosser D
Her 16h34'40" 33d38'56"
Maynard & Dolores Stratton
Cyg 20h17'5" 33d9'23"
Maynard Syvilla Parental Love Star
Pho 1h10'28" -44d41'51"
Maynard's Star
Uma 14h5'21" 52d28'43"
Maynou Koua-Tong Her
Vir 15h8'7" 2d25'53"
Mayo
Cap 20h18'49" -9d41'11"
Mayo Bendiciendo
Lyr 18h39'18" 38d23'46"
Mayo Fujii
Sgr 19h36'1" -37d31'55"
Mayo Lyles
Uma 13h55'4" 49d40'6"
Mayodi
Cas 1h18'11" 69d8'4"
Mayoucha
Vul 20h15'56" 22d58'11"
Mayouf Mourad
Tau 3h42'44" 18d23'33"
Mayra
Sgr 18h11'17" -31d23'51"
Mayra
Umi 14h48'24" 70d31'24"
Mayra
Gem 6h50'9" 22d59'31"
Mayra
Ori 5h47'33" 7d48'29"
Mayra
Ari 3h16'57" 19d7'31"
Mayra Alejandra Ramirez
Vir 13h20'25" 7d24'41"
Mayra Ayala
Cnc 8h36'53" 15d46'29"
Mayra C Alcazar
Cnc 8h2'37" 15d14'52"
Mayra Cristina
Sco 17h34'5" -41d28'56"
Mayra Jazmin Mirabelli
Cen 13h39'44" -41d21'19"
Mayra Jocabed
Vir 14h14'13" 3d31'32"
Mayra Magana
Lib 15h12'59" -9d51'49"
Mayra "Mamas" Saldaña
Leo 11h23'44" 16d6'27"
Mayra Mi Corazon
Sco 16h11'40" -15d12'59"
Mayra *My Angel* Zuniga
Leo 11h40'43" 15d47'9"
Mayra My Love
Lyn 7h44'29" 38d49'19"
Mayra "My Rainbow" Garcia
Cnc 9h3'32" 7d37'23"
Mayra Ortega
Mon 6h20'4" -9d8'19"
Mayra siempre seras mi estrella tqm
Sco 17h33'53" -40d2'25"
Mayra Susana
Tau 3h32'15" 4d49'8"
Mayra Yvette
Crb 15h35'44" 32d41'29"
Mayra's Smile
Ari 3h10'59" 11d49'14"
May's Star
Leo 10h33'3" 8d8'2"
Mayssa (my little leo) and Khalil
Cyg 19h33'56" 30d5'9"
Mayte
Ari 3h7'3" 22d47'33"
Mayte Blanco
Uma 11h12'57" 42d32'58"
Mayte dela Caridad Suarez
Ori 6h2'24" 21d2'42"
Mayte Nazareth Ramos Gallegos
Ori 5h24'41" 4d14'25"
Maytee and Natty Cuervo
And 2h12'2" 50d6'32"
Maytinee
Gem 6h43'36" 25d41'47"
Mayu since 2004
Cap 20h12'59" -16d22'16"
Mayukh
Vir 13h19'19" 5d27'48"
Mayuko Suzuki
Sgr 18h18'15" -32d22'19"
Mayumi I. Titus
Uma 8h35'29" 58d9'15"
Mayuri Patel
Leo 10h21'0" 22d39'27"
Mayvin Jeanelle Toki "Mayve"
Uma 9h32'42" 44d13'44"
May-Yay Gerzer 92085
Uma 8h14'9" 65d46'4"
Mayzie
Uma 11h53'7" 35d1'22"
MAZ-97
Uma 9h30'19" 63d43'59"
M.A.Z.A.
And 23h7'54" 51d3'42"
Mazali Omri
Lyn 9h9'28" 34d21'36"
Maze's Star
Sgr 17h59'34" -16d58'18"
Mazette Bel
Uma 11h2'50" 67d24'6"
Mazie J. Stefanovic
Her 16h43'49" 27d13'55"
Maziel
Uma 11h49'43" 58d57'24"
Mazing
And 0h41'22" 31d33'13"
Mazz
Uma 11h54'50" 59d26'19"
Mazza
Sco 16h58'50" -35d13'26"
Mazza
Her 17h10'16" 23d59'18"
Mazza Marie Bowles
Cas 1h31'21" 69d13'34"
Mazzaccheri Flavia
Lyr 19h21'50" 31d17'51"
Mazzi
Cmi 7h29'30" 10d15'33"
Mazzone
Cyg 20h22'46" 38d27'15"
Mazzy
Tau 3h42'2" 24d34'3"
Mazzy Star Jones
Psc 1h1'22" 4d55'57"
MB
Tau 4h28'12" 18d4'38"
MB Love
Gem 6h50'10" 26d32'7"
MB n ET 12/23/2005
Leo 10h0'10" 29d54'57"
MB77
Sgr 18h31'33" -29d34'33"
MBDA 1
Aqr 23h27'58" -18d25'11"
MBDA 1
Hya 9h56'33" -25d37'10"
MBDA 1
Sco 17h56'53" -31d11'27"
MBDA 1
Col 6h26'38" -35d9'12"
MBDA 1
Cap 21h55'35" -18d41'48"
MBDA 1
Vir 14h58'20" -0d25'19"
MBDA 1
Umi 16h55'49" 75d45'29"
MBDA 1
Her 18h33'59" 19d58'27"
MBDA 1
Peg 22h22'38" 21d24'26"
MBDA 1
Cnc 8h40'2" 7d56'14"
MBDA 1
Ori 6h12'55" 3d18'24"
MBDA 2
Ari 2h53'41" 18d1'56"
MBDA 2
And 1h46'12" 49d0'56"
MBDA 2
Cyg 20h14'39" 49d19'54"
MBDA 2
Umi 19h32'52" 86d34'42"
MBDA 2
Ori 5h39'40" -2d25'47"
MBDA 2
Sgr 18h50'49" -18d49'28"
MBDA 2
Lib 15h53'53" -17d16'29"
MBDA 2
Cap 20h28'38" -11d2'31"
MBDA 3
Sco 17h51'23" -35d54'25"
MBDA 3
Tau 5h45'31" 18d2'32"
MBDA 3
Vir 11h40'26" 3d17'46"
MBDA 3
Gem 6h17'51" 21d39'34"
MBG Becky's Wish
Psc 0h44'2" 21d4'45"
M.C.
Cyg 19h59'15" 40d32'38"
MC
Lib 15h25'22" -20d42'48"
M.C. Lonergan
Uma 12h32'10" 57d14'26"
MC6275283052PH4107231376EE
Ori 5h41'18" 8d24'36"
McAllister Family Star
Uma 13h17'7" 56d17'39"
McBooObieBear
Vir 14h28'54" 0d17'37"
McBrayer
Cap 20h55'39" -22d51'49"
McBride First Anniversary
Cyg 20h39'40" 42d21'44"
McCall
Umi 15h16'36" 72d6'25"
McCann Nugget
Leo 10h33'37" 21d58'59"
McCarley Grace Castillo
Lyr 18h43'43" 33d48'54"
McCarrick
Uma 11h1'49" 59d54'48"
McCarthy Q Devine
Vir 13h37'22" -13d25'46"
McClain *Our Brightest Star*
Per 4h15'26" 48d31'41"
McClatchy's Beacon
Tau 3h37'45" 19d0'38"
McClintock-Ray-Lane-Castle-Scruggs
Aqr 23h51'22" -23d46'18"
McCloud's Volantis
And 2h33'51" 42d57'57"
McCollum 5
Ori 5h35'30" 8d34'20"
McCormick Star of Saint Patrick
Per 3h18'50" 53d44'21"
McCosmo
Cma 6h35'37" -16d6'26"
McDaniel
Ori 5h28'9" 3d35'17"
McDizzle
Uma 8h36'7" 54d52'34"
MCDSSCPS2005/SB, DG, LL, PP, SP, CR
Cyg 19h58'59" 39d46'1"
McECLES MAJOR
Tau 3h50'43" 20d2'50"
MCEHGRHIASN
Cap 20h57'36" -18d46'56"
McFarren Clan
Ori 5h57'36" -0d45'9"
MC-Gaucho
Lyr 18h58'13" 33d42'12"
McGowan Insurance 75th Anniversary
Cyg 21h42'38" 45d11'39"
McGowan's Destination
Uma 9h42'38" 41d41'20"
McGraw
Uma 11h59'2" 45d48'40"
McGraw Family Star
Cam 3h56'36" 59d43'14"
McGregor Davis
Cma 6h57'32" -12d5'12"
McGregor Ponticelli
Umi 16h47'24" 81d22'7"
McGuigan Believer
Uma 8h38'26" 59d6'0"
McHenry Clan
Lmi 10h55'8" 26d4'43"
McIlg
Sgr 19h7'18" -20d26'26"
McIllwrick's Light
Her 18h25'8" 23d59'24"
McIntosh-Velarde
Vir 14h11'11" 2d40'4"
McIvor's Star
Boo 14h32'36" 19d25'39"
MCJA9798
Umi 16h11'51" 80d38'50"
McJemjcarn
Lyn 9h41'24" 40d3'4"
M.C.K
Uma 9h41'16" 57d45'3"
McKaila Paige Bailey
Cam 4h24'9" 55d17'11"
McKaila Steffes
And 1h23'43" 48d35'0"
McKayla
Ori 5h11'48" 7d16'26"
McKayla Geisinger
Sgr 17h52'39" -29d2'47"
McKayla - Jennifer
Cas 0h54'2" 57d31'52"
McKayla Marie Fogarty
Lyn 6h49'32" 51d24'16"
McKayla PEANUT Burtnett
Leo 10h27'21" 23d31'28"
McKayla's 1st Christmas Star
Leo 11h7'33" 16d6'46"
McKenna
Del 20h14'53" 9d54'27"
McKenna #1
Uma 12h38'41" 53d28'25"
McKenna Brynn
Tau 3h47'52" 28d54'4"
McKenna Carissa Coleman
And 0h29'31" 37d9'4"
Mckenna Carmond
Vir 14h33'52" -3d12'33"
McKenna Delaney Klenk
Lyn 7h45'34" 46d50'10"
McKenna Della-Vedova
Cam 3h52'33" 56d6'30"
McKenna Jean Chandler
And 0h31'55" 28d12'36"
McKenna Law
Umi 14h32'18" 76d33'23"
McKenna Livia Hollosi
Sco 16h12'35" -13d17'33"
McKenna Marie Ecker
Sco 16h40'21" -39d55'23"
McKenna Matthews
Lib 14h51'59" -0d41'37"
McKenna Morgan Tennell
Leo 10h29'54" 18d21'32"
McKenna Murphy Clary
Aqr 21h37'55" 0d51'31"
McKenna Noel
Cas 1h16'9" 68d59'3"
McKenna Reign 052298
Ori 5h26'41" -0d32'21"
McKenna Rose Mabry A Precious Gift
Gem 6h46'50" 26d9'12"
McKenna Ryleigh Miller
And 1h6'8" 41d58'37"
McKenna Toonen
Sco 17h57'55" -43d1'20"
McKenna Tracy Aubin
Sgr 18h51'51" -27d18'22"
McKenna Williams
Mon 6h49'5" 9d47'2"
McKenna-Kay
Cnc 8h37'20" 14d29'27"
MCKENNAPAIGE
Lmi 10h39'3" 31d0'20"
McKenna's Heaven
And 0h29'7" 32d15'29"
McKenna's Stars Class of 2005
Aqr 21h55'43" 0d22'0"
McKensie Kate
And 0h21'47" 28d34'7"
McKenzey Lynn
Ori 6h1'58" 17d16'55"
McKenzie
Cyg 21h40'53" 50d26'31"
McKenzie
Lyn 8h7'18" 46d15'0"
McKenzie
Sgr 18h2'35" -22d19'41"
McKenzie Allen
Cap 21h50'0" -21d35'13"
Mckenzie Cheek
Cas 1h50'34" 67d41'53"
Mckenzie Clippinger
Sgr 19h6'55" -25d38'19"
McKenzie - Decker - Clover
Cnc 7h57'44" 16d24'15"
McKenzie Elizabeth Tiara Cohs
And 0h21'29" 44d22'16"
Mckenzie Gaglianese-Sale Woody
Cnc 8h36'36" 31d10'11"
Mckenzie Grace Coffman
Uma 10h45'18" 54d0'43"
McKenzie Kay
Cap 21h36'40" -15d52'39"
McKenzie Lakin Wotring
And 0h35'54" 37d42'38"
McKenzie Lynn
And 0h17'53" 43d27'58"
McKenzie Mae-Marie Magnus
Vir 13h1'20" 1d37'11"
McKenzie McBee
Crb 16h7'0" 29d31'27"
McKenzie Narissa Munger
Ari 3h18'26" 27d24'30"
McKenzie Nichols and Brian Jeffrey
Uma 8h56'39" 63d35'34"
McKenzie Rae Scutt
And 0h35'33" 21d58'5"
McKenzie Rayelynn Fessenden
Psc 23h30'22" 4d34'8"
McKenzie Rhiann Wilke
Cnc 8h48'29" 28d2'56"
McKenzie Sabrina Stevens
Crb 15h54'18" 32d14'9"
McKenzie Templer
Cnc 8h53'18" 12d26'26"
McKenzie Trantham
Lyn 7h21'29" 46d0'14"
Mckenzie-May
And 1h52'57" 46d14'25"
McKenzie's Celestial Guiding Star
Pyx 8h47'39" -30d11'49"
McKinley
Uma 8h48'33" 60d37'54"
McKinley Gardner
Umi 15h34'47" 84d7'19"
McKinley Kate Yelton
Cap 21h30'46" -9d25'6"
McKinley's Gaze
Ori 5h9'44" -0d12'47"
McKinna
Lyn 7h19'37" 59d36'13"
McKinney's Outlaw
Peg 23h43'38" 24d51'40"
McKnight
Dra 17h37'43" 54d42'31"
McKynna Fisher
Aql 19h20'57" -7d49'49"
McLaren 525
Her 16h46'23" 22d1'39"
McLaren K. Gould
Umi 14h57'13" 82d4'8"
McLaughlin
And 23h40'34" 47d39'46"
McLaughlin Clan-Billy-Duncan-Blaise
Her 17h5'10" 30d59'27"
McLemore
Uma 9h51'22" 60d30'11"
McLorne Ashby Love
Sco 16h42'39" -27d42'51"
McManus
Sco 17h53'5" -36d14'38"
McMechan Schneider
Psc 1h26'47" 16d26'26"
McMeisl
Cyg 21h30'50" 47d16'55"
McMurray
Aqr 22h17'52" -10d21'37"
McMurray Family
Eri 3h28'25" -4d44'42"
McMurtrie
Uma 10h13'49" 64d26'36"
McNeil
Umi 15h25'50" 70d40'12"
McNeil III
Cep 22h56'49" 73d50'30"
McNulty98
Uma 10h37'0" 52d3'41"
mcp
Tau 5h22'33" 27d32'30"
McParland
And 1h44'38" 46d5'27"
McPartlands
Her 17h18'41" 45d39'55"
McPherson.315
Cas 1h7'19" 63d39'25"
McPlumley
Vir 13h8'44" 13d24'8"
McRae
Uma 13h58'16" 50d0'18"
McReynolds
Her 17h38'0" 32d4'29"
MCris
Uma 11h36'41" 62d39'14"
Mc's Dreams
Psc 1h51'25" 8d41'39"
McShane
Uma 12h1'20" 41d40'43"
McSkittles
Gem 7h20'41" 24d46'5"
McSorley Star
Cyg 20h11'21" 38d13'47"
McWilliams , Ballard
Lyn 7h42'29" 44d22'13"
MD Engel
Her 16h25'34" 11d1'48"
MD Grimes 070707
Crb 15h38'31" 28d19'18"
Md. Stiffer
Uma 8h17'18" 70d40'30"
MD Zehr
Vir 14h26'19" -0d28'57"
MDCCI MMV
Uma 10h3'9" 45d37'51"
MDC's passion, Belinda Willoughby
Uma 11h33'10" 34d27'4"
M.D.G.Greemore
Cas 0h7'11" 56d14'12"
M.DiGi
And 0h39'33" 36d2'20"
MDN-0103060016
Uma 9h26'56" 57d2'22"
MDP 42404 SRW ILY
Uma 13h31'57" 56d43'40"
M.D.S.-Bubbles
Umi 14h45'0" 68d10'27"
MDugas19Jan1969
Aql 19h48'13" 2d23'1"
Me 2
Her 17h36'34" 20d10'59"
Me and My Babes Star
Sge 19h55'7" 18d16'23"
Me and You
Psc 1h10'57" 24d47'4"
Me and You
Lyn 7h53'49" 56d25'7"
Me & Dad's Place
Uma 12h12'9" 60d20'33"
Me Ke Aloha Pumehana
And 0h51'48" 40d48'57"
ME(L)
Umi 9h20'25" 86d23'22"
Me Obie of Levittown, Pa.
Uma 10h33'15" 64d12'31"
M.E. S. #1
Uma 8h27'59" 62d57'58"
- Me Sidus - ggs:25-12-51
Psc 1h54'42" 7d9'22"
ME Smith
Her 17h37'30" 16d9'47"

Me&Sp Forever and Beyond
Cir 15h8'24" -56d58'2"
Me to You
Ori 5h18'13" 0d14'15"
Me to You-Joanne & Phil's Love Star
Cru 12h37'26" -64d24'7"
Me Too
Her 17h27'36" 36d24'6"
Mea
Uma 11h32'28" 34d3'38"
Mea 2.8 Grams
Lyn 9h4'58" 44d21'3"
Mea Calista Amata
Equ 21h17'18" 9d5'59"
Mea Columba
Aqr 22h28'39" -7d15'13"
Mea Kardia
And 0h34'44" 38d14'12"
Mea Peterson
Ori 5h33'39" 1d42'58"
Mea Stella
Cas 0h51'4" 67d6'33"
Meabh Aine Mc Polin
And 0h23'11" 46d10'55"
Meachco's Star
Crb 16h3'56" 26d51'11"
Mead Z. Mier
Leo 10h35'15" 13d12'19"
Meade R. Siever, Jr.
Psc 1h4'22" 3d41'9"
Meadhbh
Dra 19h31'49" 68d1'56"
Meadlins Dream
Uma 11h41'14" 44d30'31"
Meadow
Uma 12h46'59" 58d51'29"
Meadow Marie
Cnc 8h52'15" 30d56'6"
Meadow Of Mischief
Cru 12h18'57" -56d59'34"
Meadowlark
Uma 11h57'50" 46d54'40"
Meag & Andrew
Cap 21h51'53" -9d56'0"
Meagan
Oph 17h13'6" -0d57'52"
Meagan
Cas 23h41'46" 62d4'35"
Meagan
Cnc 8h18'27" 20d13'40"
Meagan Ann Mendoza
Cyg 21h51'56" 54d2'55"
Meagan Ann Yuchas
And 0h57'43" 39d48'43"
Meagan Anne
Cam 3h36'46" 57d8'39"
Meagan Anne Heckers
Aqr 20h44'52" -7d35'4"
Meagan Arnheiter
And 0h17'9" 44d39'8"
Meagan Catherine Fisher
Sco 16h17'58" -14d36'46"
MEAGAN CHRISTINE FREDRICK
Gem 7h10'18" 27d20'36"
Meagan Crockett
Mon 7h1'37" 7d58'31"
Meagan Curry
Cas 0h51'38" 57d3'26"
Meagan D.
Ori 4h47'1" 1d15'43"
Meagan Dawn Reuter
Leo 11h1'35" 20d37'57"
Meagan Diane McIlvoy
Lib 15h43'9" -29d31'14"
Meagan E. Lorenz
Lib 15h16'51" -17d32'4"
Meagan Elaina Sullivan
Aqr 22h37'42" -3d37'42"
Meagan Elaine Kronke
Sco 16h7'37" -10d1'27"
Meagan Elizabeth Falk
Ari 3h1'46" 27d24'50"
Meagan Emily Wilks
Sco 16h54'44" -41d55'9"
Meagan Emory Douple
Ori 4h49'19" 11d41'35"
Meagan Eternal
Gem 6h59'2" 17d44'4"
Meagan Griggs
Aql 20h24'3" 4d5'15"
Meagan Hadley
Psc 1h38'34" 24d40'16"
Meagan Hannah
Cyg 19h35'50" 36d26'22"
Meagan In My Heart Forever LoveTrev
Vir 13h58'4" -13d40'53"
Meagan J. Byrne
Cnc 8h25'48" 17d34'5"
Meagan Key
Lyn 7h57'9" 38d16'17"
Meagan & Kyle
Aqr 21h51'16" -1d58'19"
Meagan Leah Cecilia Lambros
Tau 4h11'7" 19d49'39"
Meagan LeAnn Johnson
Ari 2h56'38" 21d33'13"
Meagan Lorene LaBarre Jones
Umi 14h20'33" 74d32'34"
Meagan Louise Mercieca
Col 5h57'20" -28d27'21"
Meagan Lyn
Aqr 20h41'47" -11d27'48"
Meagan Marie
Cap 20h47'54" -14d50'59"
Meagan Marie
Vir 14h38'8" 2d19'26"
Meagan McDowell
Leo 10h32'28" 26d17'46"
Meagan Meadows
Sco 17h16'54" -32d26'4"
Meagan Michelle LeBlanc
Cnc 8h9'10" 29d52'7"
Meagan Moore
Psc 1h12'3" 7d22'25"
Meagan Moye
Sco 17h52'12" -36d35'36"
Meagan Nichole Womack
And 23h2'48" 48d24'11"
Meagan Nicole Marr
Uma 9h37'42" 47d35'26"
Meagan Noell Oglesby
And 0h21'24" 29d15'55"
Meagan O'Connor Hood
And 0h5'29" 43d38'40"
Meagan Renee Melton - Granddaughter
Lib 15h18'36" -10d54'2"
Meagan Renee Ragland
Psc 1h24'59" 21d46'19"
Meagan Rose
And 23h31'29" 46d45'11"
Meagan Royal
Tau 3h25'20" 15d30'3"
Meagan Schroeder
Mon 6h31'48" 0d41'44"
Meagan Sharissa
Per 3h5'27" 45d26'31"
Meagan "Star Light of Brian's Life"
Cyg 20h47'31" 31d26'56"
meagan stenzel
Sco 16h7'30" -25d5'11"
Meagan Woodall
And 0h2'27" 47d53'42"
Meagan's
And 1h7'59" 37d39'59"
Meagans Heaven
Leo 11h6'10" 15d32'43"
Meagan's Light
And 0h39'56" 31d2'25"
Meagan's Star
Sco 16h14'38" -12d29'43"
Meagen, Rachel, & Melody Hatfield
Uma 11h51'46" 38d0'54"
Meagermuffin
And 23h48'11" 43d1'21"
Meaghan 16
And 0h24'55" 40d40'6"
Meaghan Aileen Ince
And 0h25'53" 42d42'40"
Meaghan and Carl
Cyg 20h18'8" 33d15'21"
Meaghan & Anthony Forever
Dra 19h33'18" 67d34'56"
Meaghan Bailey Schlossmacher
And 1h29'53" 44d21'10"
Meaghan Catherine Tobin
Cnc 8h38'30" 16d26'59"
Meaghan Collins
Sgr 18h50'41" -19d43'51"
Meaghan Diane Livingston
Leo 11h43'37" 10d18'42"
Meaghan Elizabeth Mooney
Gem 7h10'37" 23d43'1"
Meaghan Farris
Aqr 21h45'36" 2d18'6"
Meaghan Hamilton
Cas 23h6'17" 57d0'10"
Meaghan Hohman
Ori 5h39'35" -5d57'2"
Meaghan L. Moody
Ari 2h36'51" 29d22'49"
Meaghan Lea Heffelfinger
And 2h22'53" 45d21'25"
Meaghan Leigh Malone
Crb 16h2'39" 36d57'37"
Meaghan Marie
Cnc 9h3'42" 24d30'37"
Meaghan Marie Cole
Lyn 7h23'4" 46d2'14"
Meaghan My Love
And 23h18'56" 47d26'4"
Meaghan Nicole
And 0h21'59" 42d5'44"
Meaghan O'Keefe Marsala
Vir 12h59'45" -10d35'53"
Meaghan Patricia Goodwin
Gem 7h14'26" 16d26'15"
Meaghan Patricia Holmes
And 0h47'48" 39d57'21"
Meaghan "Piranhar" O'Hare
Lyn 7h13'14" 58d11'8"
Meaghan Rae DuTart
Sco 17h8'2" -34d3'21"
Meaghan S. Adams
Cas 1h2'41" 63d35'29"
Meaghan Schwarting
Mon 6h55'15" -6d34'50"
Meaghan Stetson
Oph 16h57'55" -0d22'34"
Meaghan's Majesty
And 1h43'18" 44d50'10"
Meaghan's Star
Sco 17h40'26" -45d16'28"
Meagon Lynnae Gentles
And 0h40'7" 36d25'36"
Meaita Nenn
Lyn 6h46'8" 56d30'37"
Meakam
Ori 5h58'2" 17d44'11"
Mealeva Tam
Sco 17h13'41" -31d57'5"
Mean Old Man
Per 3h13'35" 44d19'57"
Meanest Girl's Daddy
Per 3h1'38" 44d18'15"
Meangycle
Gem 7h36'12" 32d50'56"
Meaningful Simplicity
Lmi 10h32'36" 32d19'44"
Meanor Juice
Del 20h31'33" 12d51'51"
Meant to Be
And 2h17'1" 43d4'40"
Meant to Be
Uma 10h42'22" 54d22'2"
Meant To Be: Greg & Mairead Reardon
Gem 7h5'50" 29d32'21"
Meant To Be In Love
Cyg 19h45'40" 53d17'21"
Meant to Be ~ T.C. & M.D.
Cyg 21h32'7" 36d49'5"
Meaphiter
Lib 14h49'48" -2d1'8"
Meara Orem
Cnc 8h6'49" 8d50'50"
Mearl Faye Bennett
Ori 5h29'52" 1d52'3"
Measuremarker
Lmi 10h28'14" 35d25'16"
Meatball Head Loves Sticky Head
Vir 13h19'7" -20d45'46"
Meats + Meats = Love
Cyg 20h54'37" 39d14'18"
Meatums Gibley
Her 18h35'25" 13d10'39"
Meave Mae Macfarlane
And 1h55'59" 42d11'47"
MEB
Cru 12h33'5" -56d11'24"
MEB Jr. AKA Mikie
Umi 14h14'36" 74d54'54"
mebet's Prince Charming
And 1h15'1" 42d9'49"
MEB's
Cyg 21h31'29" 46d37'39"
Mecagni Family Star
Tau 5h15'23" 21d22'42"
Mechel Kay Green 06/18/1966
Lyn 8h40'9" 44d21'27"
Mechele's Angel
Cap 20h59'5" -21d13'34"
Mechelle Kimberly Dattilo
Cnc 9h0'11" 32d35'20"
Mechelle White
And 2h19'33" 47d48'6"
Mechelle's Everlasting Starlight
Sgr 18h10'16" -30d6'2"
Mechelles' Star
Aqr 22h44'42" -17d26'50"
MECHIA MAE
Psc 1h7'9" 18d59'20"
Mechil's Star
Umi 13h58'23" 72d41'37"
Mechthild Hack
Uma 9h7'52" 68d29'28"
mecp2000
Tri 2h8'38" 35d10'1"
med loves jcrc
Tau 3h36'2" 1d22'16"
Meda Hutchins
And 0h20'51" 41d22'55"
Medad & Memom-Heavenly Love-April 6
Uma 8h28'6" 70d7'42"
Medardo Canelas Agurcia
Psa 22h40'52" -27d10'59"
Medardo Chita Eisma
Uma 10h25'58" 47d0'50"
Medco
Lyr 19h25'12" 37d32'48"
Medea
Aqr 20h45'21" -7d12'23"
Medea
Cep 2h30'33" 83d45'37"
"Medel" Estrella de Amor Eterno
Vir 13h27'17" -8d43'53"
Medelyn Sibal
Cam 5h30'11" 63d7'23"
Mederer, Iris
Uma 14h13'28" 58d56'34"
Media Alliance
Sex 9h49'12" -0d30'47"
Médiane
And 1h42'6" 49d48'6"
Mediaspec. F.C.
And 0h43'49" 43d33'43"
Mediavision Lifestyle
Cep 3h54'43" 82d32'51"
Medicine River Drum
Eri 4h12'12" -2d15'58"
Medics Decree Bev Hills Retirement
Dra 18h57'40" 52d11'19"
MeDida
Gem 7h41'32" 22d24'56"
Medina Law Firm
Her 17h1'16" 31d52'45"
Medina S. Rhew
Ari 2h24'53" 25d48'49"
Medinas
Lyn 8h42'38" 40d19'15"
MediStar-BRD817
Leo 10h17'30" 16d17'23"
Medora Evangelista
Cyg 19h33'8" 28d14'45"
Medvidek David Kuzela
Vir 13h45'54" 2d15'2"
Medwyn
Crb 15h27'19" 25d59'28"
M.E.E
Uma 12h35'41" 54d41'19"
Mee Jung's Star
Aqr 22h1'38" -9d49'52"
MEE120606
Ori 6h2'28" 20d58'33"
Meech
Tau 3h53'44" 19d1'50"
Meechie Amato
Ari 2h14'0" 14d45'28"
Meedo Elvir
Dra 18h22'45" 71d25'42"
Meegan
Aqr 21h56'30" 1d42'38"
Meegan's Star
Cru 12h22'50" -59d19'6"
Meehan
Cma 6h47'39" -14d37'17"
Meehan Falkenthal 2005
Lyn 8h32'17" 45d4'1"
Meejeong Park
Aqr 22h45'6" -3d56'9"
Meek
Lib 15h5'12" -2d23'54"
Meeka Monsta Star
Her 16h44'57" 43d2'34"
Meeker
Ari 2h17'10" 25d32'9"
"Meeko"
Uma 10h52'41" 43d37'24"
MeeMa Shirley
Ari 2h59'17" 18d34'34"
Meemes
Lib 15h29'15" -6d9'36"
Meenakshi "Minnie" Krishnamurthy
Uma 10h40'53" 47d3'20"
Meenatchi
Sco 16h22'55" -25d49'26"
Meep
Lyr 18h17'18" 45d13'19"
Meera
Cap 21h9'59" -16d15'44"
Meera's Star
And 23h59'56" 46d21'47"
Meerkats Danny-Lisa
Gem 6h29'46" 20d3'26"
MeeRyoung Chung
Psc 1h12'48" 19d30'26"
Mees Gussinklo
Gem 7h22'50" 13d49'44"
Meese
And 2h32'29" 44d58'8"
Meeshell
Cap 20h30'36" -13d17'29"
Meeshi
Tau 5h7'20" 22d30'55"
Meeso Messimer
Cnc 8h36'44" 29d12'33"
Meeta Masi
Ari 3h21'48" 21d35'51"
Meeta Mimi Savatdy
Ari 3h8'1" 27d25'49"
Meeting Place
Sco 16h59'34" -44d43'56"
Meetz, Emma
Uma 9h15'18" 64d52'12"
MeeYun
Sgr 19h13'7" -13d37'40"
Meg
Vir 14h1'33" -13d7'27"
Meg
Aqr 22h32'24" -3d50'50"
Meg
Crb 15h55'48" 27d38'57"
Meg
Lmi 10h23'53" 36d38'9"
Meg Ab Imo Pectore
Ori 6h1'10" 17d13'34"
Meg Allnatt
Psc 1h6'9" 33d26'52"
Meg and Tony
Her 16h59'26" 29d22'53"
Meg and Ty Always
Gem 6h26'55" 18d24'17"
Meg Cheever Lifetime Achievement
Uma 9h17'25" 48d54'14"
Meg Dorothy Olson
Aur 5h58'4" 41d45'32"
Meg Elizabeth
And 0h52'17" 43d50'27"
Meg & Grandma's Star "Rainbow"
Sgr 18h59'18" -34d21'38"
Meg Hinton
Lib 15h18'36" -9d37'7"
Meg Hood
Cnc 8h6'24" 26d25'39"
Meg Isabel Clarke
And 0h49'12" 46d13'25"
Meg Louise
And 0h36'21" 31d37'34"
Meg Mahoney
Ari 2h4'10" 24d43'44"
Meg Mary Ellyn Rygiel
Cas 1h25'42" 65d49'5"
Meg Messmer 6366
Crb 15h45'18" 28d13'8"
Meg Norton
Vir 14h32'4" -0d17'12"
Meg Olivia Barkman
Lib 14h37'18" -9d46'31"
Meg P
And 0h26'45" 41d44'39"
Meg Stevens
Sco 16h57'47" -42d21'33"
Meg Vigar
Peg 21h31'41" 15d53'45"
Meg Wanser Yule *Mom-Wife*
And 0h58'6" 38d45'28"
Meg1VoB
Tau 4h12'7" 4d7'4"
Meg-2004-04-16-Say 'Oun'
Uma 9h55'29" 71d22'39"
Mega Boss Etem Postacioglu
Uma 9h46'54" 50d43'22"
[Mega]losaurus
Uma 9h15'10" 56d45'16"
Mega Woman
Cyg 20h9'10" 41d14'34"
Megahlee Alpha
Ari 2h16'41" 25d41'10"
MegaK
Uma 10h16'51" 50d38'29"
Megalicious
Sco 16h51'29" -39d12'50"
MEGALITE
Sgr 17h55'44" -29d57'53"
Megalmw
Sge 19h43'12" 17d17'45"
Megamite Bright
And 0h21'33" 27d57'36"
Megan
And 0h21'18" 25d54'16"
Megan
And 0h14'50" 29d41'30"
Megan
Ori 6h9'11" 17d18'59"
Megan
Gem 7h30'51" 19d54'45"
Megan
Leo 10h51'10" 15d6'15"
Megan
Aql 20h11'10" 4d54'27"
Megan
Psc 0h23'33" 11d52'35"
Megan
Tau 4h18'27" 21d25'0"
Megan
And 0h40'3" 22d9'43"
Megan
Cnc 7h59'32" 12d38'11"
Megan
Ori 6h15'26" 14d44'13"
Megan
And 0h2'3" 45d7'35"
Megan
And 0h9'34" 46d17'52"
Megan
And 1h43'51" 46d26'23"
Megan
And 1h32'32" 46d12'30"
Megan
And 0h40'51" 41d13'45"
Megan
And 1h38'38" 43d18'28"
Megan
And 0h30'59" 42d50'47"
"Megan"
And 0h30'46" 42d29'14"
Megan
And 0h20'59" 38d19'49"
Megan
Lyn 7h19'51" 44d43'59"
Megan
Uma 11h52'3" 41d15'0"
Megan
Lyn 9h10'44" 44d37'46"
Megan
Psc 1h9'59" 32d22'6"
Megan
Cnc 8h45'1" 32d17'22"
Megan
Sco 16h42'9" -28d51'15"
Megan
Lib 14h27'38" -23d50'44"
Megan
Aqr 22h28'31" -24d19'34"
Megan
Sgr 19h6'44" -31d11'39"
Megan
Sgr 19h33'31" -37d55'50"
Megan
Dra 16h33'43" 58d13'52"
Megan
Cas 0h50'49" 69d54'32"
Megan
Sco 16h8'31" -14d18'37"
Megan
Aqr 23h20'52" -8d0'33"
Megan
Cap 20h39'21" -20d1'51"
Megan
Vir 13h45'20" -12d48'12"
Megan
Aqr 23h1'17" -6d35'6"
Megan 041985
Cas 1h0'41" 61d47'4"
Megan 2004
Uma 9h19'9" 49d40'18"
Megan 4
Cap 21h26'22" -23d52'5"
Megan A. Mullins & Chris J. Edwards
Lyn 7h28'47" 44d44'5"
Megan Adamczyk's Graduation 2007
Lyn 6h40'52" 59d1'44"
Megan Adina Townend
Tau 4h28'34" 7d33'11"
Megan Ailene Hornbuckle
Uma 11h44'54" 43d18'31"
Megan Alana San Pietro
Aql 19h48'8" -0d18'13"
Megan Alanea
Vir 12h39'54" 11d12'38"
Megan Aleece
Ari 2h47'39" 26d39'33"
Megan Alexandra Harvey
And 1h18'50" 42d12'35"
Megan Alexandra Shakespeare
Uma 11h30'4" 55d56'41"
Megan Alexis DeMattia
Uma 10h50'30" 60d36'10"
Megan Alice Hattier - Perfection
And 0h20'15" 30d26'57"
Megan Alicia
And 0h30'11" 39d4'35"
Megan Alida Johnson
Cas 23h26'28" 51d46'36"
Megan Allen
Cas 3h27'34" 68d37'21"
Megan Allisha Graham
Lib 15h42'32" -20d7'47"
Megan Alyssa Brown
Sco 16h44'1" -32d28'59"
Megan and Brian Forever
Cyg 19h56'12" 44d51'6"
Megan and Erin Hilker
Gem 6h46'4" 20d35'36"
Megan and Melanie
Aqr 22h36'2" -2d16'35"
Megan and Nanas Star of Life
Leo 11h24'0" 16d9'34"
Megan and Nathan
Sco 16h7'6" -19d31'28"
Megan and Tim
Leo 11h39'17" 17d38'57"
Megan "Andy" Nelson
Lyn 8h7'13" 35d31'48"
Megan Angela Suchowski
Lyn 6h41'13" 56d51'35"
Megan Ann
Cyg 20h55'7" 34d57'56"
Megan Ann
Uma 11h45'24" 49d37'11"
Megan Ann
Ari 2h51'13" 16d59'30"
Megan Ann
Ori 5h50'1" 6d2'13"
Megan Ann Cajigas
Uma 10h54'33" 51d8'41"
Megan Ann Motto
And 2h29'57" 48d55'56"
Megan Ann Stanley
Per 4h32'7" 31d49'44"
Megan Anne Dickinson
Leo 10h21'48" 6d29'24"
Megan Annie's Everlasting Star
Leo 10h15'23" 26d11'6"
Megan Anthony Cottrell
Tau 4h17'12" 17d36'54"
Megan Antonia
And 23h12'2" 41d33'34"
Megan Arruda
Cas 2h19'16" 71d35'15"
Megan Ashley
Gem 7h7'32" 27d11'30"
Megan Ashley Nitschke
Uma 8h26'46" 60d49'32"
Megan Ashley Puente
And 1h20'1" 43d49'44"
Megan Ayres Mangione
Aqr 22h20'24" -14d19'48"
Megan B. Copley
Ari 2h5'2" 24d48'33"
Megan B. Myers
And 0h40'55" 43d18'14"
Megan Baby Girl
Cnc 8h53'48" 19d49'2"
Megan Baby Girl Garber
Cnc 8h36'55" 10d22'14"
Megan Bartelmie
Ser 15h24'35" -0d18'15"
Megan Becca Keller
Lyn 7h29'40" 52d43'33"
Megan Berio
Lyn 8h43'12" 34d28'37"
Megan Beth
And 23h40'33" 47d5'38"
Megan Beth
Cyg 21h58'37" 49d7'33"
Megan Black
Cas 1h24'4" 52d27'29"
Megan Blair Farinholt
Lib 15h20'42" -6d28'8"
Megan Bracy
And 23h11'46" 42d6'50"
Megan Brady RN
Crb 15h19'40" 26d46'11"
Megan Brooke
Psc 1h2'45" 23d17'53"
Megan Brooke
And 0h26'11" 39d54'14"
Megan Brooke Poston
Lyn 7h50'20" 41d47'25"
Megan Brooke Ward
And 23h8'9" 48d25'9"
Megan Brunell
Tau 3h43'52" 27d27'3"
Megan Bryant
Sco 16h8'23" -15d24'31"
Megan Buckley
Per 4h31'44" 40d11'59"
Megan Burkle
And 0h24'45" 31d10'15"
Megan Butt
Tau 4h10'13" 10d55'37"
Megan C. McCormick
And 0h6'30" 45d49'53"
Megan C Wasnieski
Sco 16h11'42" -9d49'17"
Megan Candice Carter
Lyn 8h0'30" 40d2'8"
Megan Caris
And 0h20'31" 26d51'9"
Megan Catherine Gahan
Vir 14h24'57" -6d53'51"
Megan Cavazuti, a best friend - Liv
Peg 23h51'49" 12d5'23"
Megan Celeste Hayes
Lib 15h31'26" -6d34'42"
Megan Chamberlain
Uma 9h35'4" 56d53'36"
Megan Charlotte Perkins
And 1h34'36" 40d8'14"
Megan Charneco
Cnc 8h55'8" 11d45'38"
Megan Chidester
And 0h12'19" 26d42'35"
Megan Chloe Anne Earnshaw
And 0h5'43" 39d20'11"
Megan Christiansen
Uma 11h36'36" 38d59'43"
Megan Christine
Gem 7h45'27" 33d19'23"
Megan Christyna Wilson
Cyg 19h29'2" 30d52'44"
Megan Claire Crutchfield
Cap 20h23'33" -12d19'34"
Megan Claire Prusia
Ari 3h13'49" 20d56'40"
Megan Clare Leon McMahon
Psc 1h50'24" 5d16'0"
Megan Clare Taylor
And 0h45'59" 30d48'38"
Megan Colleen Reas
Ari 3h10'16" 29d4'24"
Megan Conley
And 23h50'6" 40d3'42"
Megan Convey
And 2h25'3" 42d0'47"
Megan Corey
Leo 11h38'33" 14d26'31"
Megan D. Harrah
Cnc 8h8'27" 16d16'58"
Megan Daisy Ransford
And 2h36'2" 50d6'31"
Megan Dalgewicz
Uma 8h57'33" 62d25'21"
Megan Danielle Blevins
Del 20h38'4" 15d7'32"
Megan Dannielle
Gem 7h18'12" 22d56'17"
Megan Dawson
And 0h29'4" 45d44'24"
Megan De Cort
Ori 5h28'47" 12d50'13"
Megan Deem
Cas 23h48'34" 65d3'23"
Megan Dees Sermoneta Friedman
Ari 2h11'18" 24d12'50"
Megan Detweiler
Cas 0h24'30" 52d38'49"
Megan Diane Carter
Srp 17h57'59" -0d21'58"
Megan Diane Gibson
And 23h19'54" 47d13'1"
Megan Diaz
Her 17h28'48" 27d17'52"
Megan Dinnes
Lyn 8h32'15" 38d40'12"
Megan + Dougs Everlasting Love Star
Pup 7h12'53" -38d9'43"

Megan E. Boyer
And 1h51'56" 46d16'54"
Megan E. Farley
Tau 4h19'53" 17d29'30"
Megan E. Larkin
Sgr 18h56'34" -25d51'57"
Megan E. Limpert
Cnc 8h56'23" 7d19'44"
Megan E. Lister
Cas 1h28'19" 57d44'12"
Megan E. with a Z
Leo 10h51'51" 8d52'6"
Megan Einhart
Leo 11h50'52" 27d3'32"
Megan Elaine Stout
Uma 10h30'47" 42d41'45"
Megan Eleri Jones
And 23h48'45" 42d25'14"
Megan Eliza Gathers
Ori 5h11'25" 7d28'55"
Megan Elizabeth
Cnc 8h55'42" 31d19'23"
Megan Elizabeth
Cap 21h16'0" -25d16'36"
Megan Elizabeth
Lib 15h31'9" -16d43'21"
Megan Elizabeth
Lyn 7h18'20" 58d4'42"
Megan Elizabeth Bourque
Lyn 7h19'59" 45d29'34"
Megan Elizabeth Brockington
Sgr 18h18'57" -18d1'6"
Megan Elizabeth Cichowski
Lib 15h6'41" -2d9'17"
Megan Elizabeth Cowling
Crb 15h49'43" 28d0'28"
Megan Elizabeth Dickerson
Tau 4h29'58" 20d14'40"
Megan Elizabeth Edwards
Sgr 19h47'11" -14d11'7"
Megan Elizabeth Feeney
Cma 6h57'0" -15d36'33"
Megan Elizabeth Hall
Cnc 9h9'6" 13d43'43"
Megan Elizabeth Hammonds
Vul 19h53'1" 23d15'3"
Megan Elizabeth Hlavaty
Leo 9h59'13" 22d0'35"
Megan Elizabeth Jones
Uma 11h23'20" 45d43'29"
Megan Elizabeth Kupferman
And 1h24'38" 46d25'41"
Megan Elizabeth Lowery
And 0h37'44" 44d18'40"
Megan Elizabeth McLaughlin
Leo 11h8'25" 11d34'36"
Megan Elizabeth Parfitt
And 23h10'42" 38d34'34"
Megan Elizabeth Payne
And 23h36'21" 46d52'46"
Megan Elizabeth Rothmund
Lyn 7h39'4" 56d15'54"
Megan Elizabeth Schneider
Gem 6h49'11" 24d23'4"
Megan Elizabeth Shaffer
Aql 18h58'48" -0d0'26"
Megan Elliott
Sco 17h48'25" -40d42'58"
Megan Elyse Lindsey
Gem 7h9'10" 27d45'21"
Megan Erskine
Gem 6h31'20" 18d15'52"
Megan es la estrella más hermosa.
And 23h13'16" 52d58'17"
Megan Estelle
And 1h22'45" 37d7'43"
Megan Eve Marchant
Cnc 8h34'52" 31d5'21"
Megan F. Levkovich
Cas 1h18'47" 58d6'6"
Megan Faye
Lib 15h48'45" -19d10'56"
Megan Felter's Star
Psc 1h12'27" 23d45'53"
Megan Fessak
Tau 4h26'57" 28d12'41"
Megan Fielding
Lib 14h26'25" -17d51'59"
Megan Finn
Leo 11h46'43" 27d3'37"
Megan Florence
And 1h23'27" 43d35'22"
Megan Fox
And 0h6'36" 43d51'43"
Megan Frances Dugan
Leo 11h22'39" 14d14'24"
Megan Francine McLaughlin
Ori 5h35'49" -6d16'41"
Megan Francisco
Vir 12h51'12" 12d33'39"
Megan Fron
Ari 2h58'57" 26d49'17"
Megan Georgia
And 23h28'52" 42d27'17"
Megan Gibson-Waby
Crb 15h38'54" 35d30'15"
Megan Gloria O'Brien
And 0h9'16" 39d23'40"
Megan Goldblatt
Peg 22h46'35" 22d10'34"
Megan Gomez
Cap 21h23'2" -15d59'7"
Megan Goshorn
Uma 11h53'19" 35d24'17"
Megan Grace
And 1h22'41" 50d26'57"
Megan Grace Buddemeier
And 2h17'27" 46d29'3"
Megan Grace Thomas
And 22h59'59" 40d1'32"
Megan Greer
Sco 16h32'51" -30d51'50"
Megan Gregory
And 2h11'34" 37d57'36"
Megan Gurine
Ori 5h43'12" 7d44'33"
Megan Guza
Crv 12h36'8" -12d16'5"
Megan Hana's Star
Mon 6h48'33" 8d25'58"
Megan Harrell
Cap 21h21'12" -19d34'43"
Megan Hassell's Star
Ori 6h2'51" 21d21'51"
Megan Heather Carter
Mon 7h19'35" -0d30'22"
Megan Hillary Whiting
Cam 5h39'9" 76d39'24"
MEGAN HINRICHS
Sco 16h8'21" -22d14'53"
Megan Hoifeldt's Star
Cap 21h40'3" -13d14'3"
Megan Hoisington
Sgr 18h1'40" -18d14'33"
Megan Holly Alcorn
Tri 2h29'41" 31d49'19"
Megan "Honey Bunny" McClain
Ser 15h29'13" 5d57'36"
Megan Hudman
And 0h25'58" 40d31'27"
Megan Humphrey
Crb 16h8'2" 31d25'42"
Megan Hunt
Ari 2h40'2" 25d38'59"
Megan Infantino
And 0h15'29" 41d24'51"
Megan Isaacson
Cyg 19h36'21" 46d3'59"
Megan Isabella
And 23h5'39" 51d19'5"
Megan & Isaiah's First Anniversary
Cyg 20h43'0" 38d30'26"
Megan J.
Vir 12h42'6" 8d12'34"
Megan J. Hatswell
Psc 1h42'26" 12d53'35"
Megan Jacqueline Winters
Aqr 22h46'23" -5d20'23"
Megan Jade Bowling
And 0h23'32" 41d3'49"
Megan Jae
Gem 7h35'49" 35d11'37"
Megan Jane
Tau 5h33'36" 26d18'56"
Megan Jane Babunovic
Lmi 10h39'26" 33d29'37"
Megan Jane Hyland
And 0h33'32" 46d11'9"
Megan Jane Kaspar
Cap 20h24'13" -10d20'18"
Megan Jane Miller
Leo 10h13'39" 25d25'3"
Megan Jane Peaty
And 2h32'44" 50d17'4"
Megan Jane Radcliffe
Cnc 8h5'45" 17d39'4"
Megan Jean McClennan-16th Birthday
Cyg 19h42'55" 41d28'26"
Megan Jeanie
Cap 21h8'19" -20d59'33"
Megan Jeannine
Gem 6h20'49" 21d19'10"
Megan & Jeffrey Barefoot - true love
Lyr 18h40'10" 31d10'6"
Megan Jo Sullivan
And 0h49'33" 36d13'28"
Megan Jones
Cyg 19h36'7" 39d20'27"
Megan Jones
Leo 10h23'35" 10d22'37"
Megan Josephine McCaffrey
Gem 6h47'15" 33d3'24"
Megan Kandace-Alice Kublank
Lyn 7h58'35" 57d38'12"
Megan Kane
Cas 2h27'23" 67d46'44"
Megan Katherine Keys
And 23h0'10" 36d45'18"
Megan Katherine Laforge
Uma 10h48'18" 60d52'35"
Megan Katherine Smith, child of God
Uma 11h51'16" 60d3'51"
Megan Kathleen
Uma 9h8'24" 57d1'29"
Megan Kathleen Bowles
Psc 0h40'39" 12d12'50"
Megan Kathleen Gallagher
Sgr 19h43'5" -12d51'51"
Megan Kathleen O'Keefe
Ari 2h46'42" 22d56'19"
Megan Kathleen Wagoner
Leo 10h15'45" 7d12'59"
Megan Kayleen
Cnc 9h8'25" 31d57'14"
Megan Keely
And 1h23'49" 37d20'17"
Megan Keely Wright
Cnc 8h51'2" 31d55'0"
Megan Key
Cas 0h32'11" 50d16'40"
Megan Kim Carey
Uma 10h31'33" 57d14'54"
Megan King
Cam 3h38'6" 60d8'15"
Megan Kitten Schmeggles Diorio
And 0h39'11" 25d52'59"
Megan Knies
Aql 19h47'13" 12d10'13"
Megan Kristin Riggs
And 1h17'31" 36d22'1"
Megan Kroening
And 0h13'7" 43d30'55"
Megan & Kyle
Leo 11h29'21" 18d11'21"
Megan L Lhota-Calderaro
Mon 7h20'43" -0d28'0"
Megan L. Temples
Tau 3h41'39" 25d47'29"
Megan "La Brujita" Hamilton
Ori 5h32'17" 1d47'44"
Megan Lacole
And 0h38'31" 26d42'20"
Megan Lea
Cru 12h38'8" -63d6'43"
Megan LeCastre
Ari 2h44'47" 27d42'50"
Megan Lee
Tau 4h4'58" 6d19'14"
Megan Lee
Cnc 8h24'28" 13d19'43"
Megan Lee Campbell
Gem 7h10'49" 27d13'20"
Megan Lee Isaman
Cnc 8h33'13" 30d51'44"
Megan Leese
Mon 6h40'22" -1d4'39"
Megan Leigh Dehn "Megsie"
Cru 12h41'3" -60d19'53"
Megan Leigh Johnston
Uma 9h35'41" 57d53'29"
Megan Leigh Martin
Cas 1h17'11" 55d55'26"
Megan Leigh Perks
And 1h52'32" 46d3'32"
Megan Leigh-Ann Laws
Sco 17h33'54" -36d44'32"
Megan Leonard's Urban Star
Ari 2h15'48" 26d36'35"
Megan Leticia Vogt
Leo 10h12'46" 14d43'54"
Megan Lily
And 2h0'2" 46d21'22"
Megan Lily Towell
And 23h33'18" 42d50'43"
Megan Linahan
Psc 23h18'48" 5d9'54"
Megan Linsley Howell
And 1h55'14" 37d52'51"
Megan Lockard
Uma 11h29'31" 39d40'14"
Megan Logan
Cnc 9h1'26" 15d22'45"
Megan London
Leo 11h13'48" 23d0'53"
Megan Loughman
And 1h57'38" 40d18'7"
Megan Louisa Wyborn
And 0h7'29" 33d31'19"
Megan Louise
Cra 18h41'36" -36d48'52"
Megan Louise Hodson - Megan's Star
And 23h43'42" 38d5'40"
Megan Louise Simpson
And 23h16'52" 43d28'14"
Megan Louise Smith
And 2h33'54" 48d47'25"
Megan Ludwig
Ari 2h47'28" 15d32'51"
Megan Lyn Schneberg
Dra 17h5'13" 60d17'20"
Megan Lyn Urbain
Lib 15h12'50" -7d5'5"
Megan Lyndsay
Ari 3h17'0" 16d38'36"
Megan Lynn
Leo 9h41'21" 29d26'35"
Megan Lynn
Psc 1h42'14" 23d50'17"
Megan Lynn Holzwarth
Ori 5h58'34" -0d24'49"
Megan Lynn Mackin
Lyn 7h11'34" 53d24'17"
Megan Lynn Moritz
Cnc 8h27'7" 24d5'32"
Megan Lynn Newman
Cap 20h24'16" -10d4'43"
Megan Lynn Raymond
And 23h43'31" 44d43'30"
Megan Lynne
And 2h9'0" 45d35'57"
Megan Lynne McCroskey
Gem 6h59'29" 25d37'1"
Megan MacKenzie Saywell
Uma 11h0'25" 71d41'4"
Megan Mae Knipe
Uma 9h42'13" 42d44'3"
Megan Mae Kriska
And 2h19'2" 49d10'18"
Megan Mailani
Ori 5h10'22" 7d56'4"
Megan Mansir Davis
Crb 15h41'55" 28d25'57"
Megan Margaret McLaughlin
Ari 2h11'21" 21d50'58"
Megan Marguerite
Leo 11h0'48" -4d21'11"
Megan Marie
Lib 14h52'50" -3d16'9"
Megan Marie
Ori 4h52'45" 4d32'34"
Megan Marie
Leo 9h33'54" 29d5'22"
Megan Marie
Cam 4h44'52" 54d42'21"
Megan Marie
And 0h39'44" 42d51'15"
Megan Marie Abernathy
Aqr 21h50'20" -6d0'29"
Megan Marie Allen
Leo 10h14'6" 12d5'22"
Megan Marie Birge
Del 20h35'24" 16d29'20"
Megan Marie Blackledge
And 1h45'25" 37d6'32"
Megan Marie Chandler Misale
And 0h18'21" 28d27'24"
Megan Marie Corkins (Sweety Pie)
And 0h54'48" 36d20'52"
Megan Marie Cox
Lib 15h40'16" -12d35'3"
Megan Marie Deal
Lyn 7h21'8" 45d2'17"
Megan Marie Ellis
Uma 11h33'1" 38d39'59"
Megan Marie Fields
And 0h15'53" 39d48'15"
Megan Marie Gallagher
And 0h52'10" 46d9'13"
Megan Marie Hennessey
Ari 2h48'41" 13d45'3"
Megan Marie Hertig
And 2h21'22" 49d4'43"
Megan Marie Hill
Cnc 9h4'53" 26d59'42"
Megan Marie Janis
Ari 2h4'11" 14d50'18"
Megan Marie Jones
Lib 14h53'43" -5d28'22"
Megan Marie Kirkman
Aur 6h1'44" 30d40'43"
Megan Marie Martin
Tau 4h20'26" 26d38'32"
Megan Marie Milano
Gem 7h43'38" 21d52'1"
Megan Marie Nelson
Aqr 22h31'32" -6d30'25"
Megan Marie Nepshinsky
Psc 0h42'46" 8d13'57"
Megan Marie O'Donnell
Cap 20h46'18" -21d23'39"
Megan Marie Pagel
Cap 21h29'42" -8d28'7"
Megan Marie Powers
Ori 6h15'36" 15d6'48"
Megan Marie Richardson 7-22-88
Cas 0h34'58" 75d2'56"
Megan Marie Roach
Uma 13h27'41" 61d9'55"
Megan Marie Rossi
Psc 23h2'13" 7d54'32"
Megan Marie Rounds
Psc 0h36'39" 15d8'20"
Megan Marie Seals
Vir 12h40'7" 2d27'24"
Megan Marie Shelby
Uma 9h18'22" 61d36'43"
Megan Marie Shoemaker
Cas 1h10'50" 65d54'14"
Megan Marie Smith
Gem 7h20'41" 27d30'26"
Megan Marie Stookey
Cyg 20h53'19" 38d20'53"
Megan Marie White
Cyg 20h4'33" 52d48'30"
Megan Marie Zerby
And 1h41'32" 39d2'52"
Megan Martin
Tau 3h32'20" 18d5'19"
Megan Mary
And 1h31'40" 42d17'2"
Megan Mary Riley
Vul 21h10'15" 26d0'42"
Megan Matherne
Cas 1h13'4" 54d37'10"
Megan May
And 23h26'48" 36d53'6"
Megan May
Vir 14h5'27" -16d40'56"
Megan Mayme
Sco 17h51'50" -36d17'29"
Megan McIntyre
Del 20h48'23" 8d6'52"
Megan McKenna Caputo
Vir 13h13'23" -13d4'17"
Megan McKim Hurley
Peg 21h26'43" 9d31'32"
Megan "Meggers" Snyder
Psc 1h26'56" 27d24'7"
Megan "Megs" Winkeljohn
Lyn 7h35'18" 36d1'38"
Megan Michelle Bridges
And 0h23'54" 37d12'28"
Megan Michelle Merritt
Aqr 22h40'40" -2d14'40"
Megan Michelle Smalling
And 0h24'13" 29d24'15"
Megan & Miguel's Super Star
Uma 11h33'29" 33d51'51"
Megan Miller
Lyr 19h25'1" 38d29'12"
Megan Mitchell
Her 18h31'31" 12d13'46"
Megan & Molly Egr
Cmi 7h27'27" 10d41'41"
Megan (Moo Moo)
Lup 15h48'30" -30d31'21"
Megan Morack
Vir 13h55'29" -9d23'15"
Megan Moreaux
Gem 7h15'48" 15d21'37"
Megan Morgan
Cap 20h7'23" -18d16'19"
Megan Mowbray, Will You Marry Me?
Tau 4h48'48" 20d24'52"
Megan Moy
Peg 22h25'38" 22d44'57"
Megan Murphy
Vir 13h49'12" -0d5'1"
Megan Murphy
Uma 11h2'24" 69d20'48"
Megan : My Diamond In The Sky
Ori 5h15'37" 6d48'1"
Megan Naylor
Tau 5h55'58" 26d14'58"
Megan Nichole
Tau 4h23'30" 9d54'30"
Megan Nichole Denney
Gem 6h14'18" 25d56'6"
Megan Nicole Ayers
And 0h11'50" 27d54'6"
Megan Nicole Bailey
And 1h20'47" 39d55'29"
Megan Nicole Bentley
Uma 9h54'36" 43d9'46"
Megan Nicole Brown
And 1h38'34" 38d43'10"
Megan Nicole Bryant
Oph 16h50'19" -0d38'17"
Megan Nicole Conley
Cas 1h19'40" 60d42'52"
Megan Nicole DeVaughn
Tau 3h43'11" 25d51'18"
Megan Nicole Harper
Tau 4h25'46" 23d32'46"
Megan Nicole Jensen
Ari 2h15'50" 24d30'58"
Megan Nicole Ripley
Aqr 22h1'27" -10d3'39"
Megan Nicole Truskoski
Del 20h52'47" 5d27'27"
Megan Nielsen (Sparkle)
Sco 17h8'54" -30d25'1"
Megan Noble
Cas 1h12'8" 50d31'35"
Megan Norris
Com 13h11'52" 16d3'5"
Megan Odell Flowers
Tau 3h40'45" 15d56'32"
Megan O'Dowd
And 0h10'19" 45d30'58"
Megan O'Grady
Gem 7h1'5" 30d49'3"
Megan Olivia King - Megan's Star
And 23h36'0" 48d5'50"
Megan Olivia's Star
And 23h38'40" 37d30'35"
Megan O'Rourke
Cap 21h57'32" -13d56'47"
Megan P Armstrong
Lib 15h43'14" -18d45'32"
Megan Paige Boyer Rayzor
Cnc 8h55'9" 22d46'16"
Megan Patricia
Tau 4h10'21" 7d21'50"
Megan Pazdersky
Cnc 8h47'53" 15d35'48"
Megan Peavy
Ori 5h9'57" 8d24'22"
Megan Pee Wee Lyon-Mankowski
Cas 1h15'41" 68d35'19"
Megan Pensyl's diamond in the sky
Lyn 7h48'3" 49d31'7"
Megan Petter
Ori 5h49'0" 6d5'58"
Megan Pitcavage
And 2h37'12" 42d6'0"
Megan R.
Cap 21h44'40" -12d34'51"
Megan R. Hoeft
Lyr 18h16'39" 30d6'19"
Megan Rachel Chichester
Crb 15h49'47" 34d26'16"
Megan Rae Hannon
Sgr 18h2'8" -28d16'10"
Megan Raymond
Mon 6h48'15" -0d3'16"
Megan Re
Lyn 8h3'38" 56d5'32"
Megan Rebecca Richerts
Tau 4h32'18" 8d45'31"
Megan Reinhart
Leo 9h31'28" 27d9'14"
Megan Rene Costello
And 0h9'8" 46d32'48"
Megan Rene Mose
Cnc 8h16'37" 12d45'16"
Megan Rene Porter
And 1h5'21" 41d32'48"
Megan Renee Brandon
Ari 2h39'20" 13d7'12"
Megan Renee Duggleby
Cep 21h42'16" 61d15'25"
Megan Renee Schooley
Psc 0h14'53" -1d8'18"
Megan Renee Tuggle
Psc 23h48'58" 6d26'53"
Megan Reynolds
Cyg 19h44'25" 38d56'3"
Megan Rhea Brown
Ari 2h1'24" 13d18'40"
Megan Richards
Ari 2h45'40" 13d34'54"
Megan Riley Novak
Cam 3h30'54" 59d31'7"
Megan Ritter
And 2h0'49" 42d46'12"
Megan Rose
Ori 5h29'47" 3d39'41"
Megan Rose Byrnes
Ori 5h37'26" -1d30'6"
Megan Rose Ferguson
Cyg 21h9'32" 31d40'12"
Megan Rose Hodgkiss
And 23h18'44" 52d21'52"
Megan Rose Landis
Del 20h47'50" 4d1'52"
Megan Rose Miller
Cap 21h30'52" -23d31'6"
Megan Rose Rivera
And 23h45'12" 34d35'45"
Megan Rose Wessel
Gem 6h1'8" 26d18'2"
Megan Rowan Taylor
Uma 8h22'1" 68d4'43"
Megan Ruth Weber
Lmi 10h30'52" 38d7'18"
Megan Ryan O'Leary
Her 16h5'39" 46d56'37"
Megan S. Anderson
Cam 9h15'25" 76d20'20"
Megan 's star
Cyg 21h10'1" 47d28'37"
Megan Sandra
Uma 11h8'3" 69d18'4"
Megan Sara Alyssa Winkler Sisters
Uma 12h43'7" 57d16'32"
Megan Scheffer
And 1h48'15" 38d30'39"
Megan Schwarb Star
Lyn 7h2'42" 58d44'26"
Megan Scott My SweetHeart
Vir 15h6'34" 3d34'23"
Megan Scott Wacker
Vir 13h33'54" 6d44'56"
Megan Searle
And 1h22'17" 47d15'20"
Megan Self
Cap 20h56'12" -20d2'41"
Megan Sharee
And 0h19'37" 45d12'53"
Megan Shavahon Sasser
Sco 17h40'5" -35d39'8"
Megan Simrall Burgener
Dra 18h35'25" 58d28'44"
Megan Sironen
Aqr 22h25'45" 1d52'50"
Megan Sophia Eaton
And 1h20'20" 44d1'11"
Megan Sophianna Kusznier
Vir 13h20'18" 13d24'17"
Megan Southcott
Lib 15h38'26" -28d10'3"
Megan Standifer
Vir 12h18'29" 3d33'34"
Megan Steffy
Aqr 23h9'30" -8d10'9"
Megan Steimer
Leo 10h10'15" 10d25'23"
Megan Sullivan
Psc 23h16'22" 3d35'4"
Megan Swider
Gem 7h15'5" 20d5'20"
Megan T. Valentine
Uma 10h2'31" 62d11'13"
Megan Talla Higgins
Tau 4h37'25" 23d41'24"
Megan Taylor Meier
Uma 10h51'37" 61d52'14"
Megan Teresa Golwitzer
Cnc 8h45'17" 21d51'51"
Megan Teresa Golwitzer
And 23h41'33" 44d18'6"
Megan Texada
Tau 4h31'27" 17d6'25"
Megan The Angelic Soprano
Umi 15h8'58" 70d29'12"
Megan the Shining Star
Tau 3h59'36" 6d7'6"
Megan the Sloff!
Aqr 21h42'28" 0d38'23"
Megan Therese Monroe
And 2h8'16" 37d38'17"
Megan Thoensen
Aql 19h45'13" 16d25'51"
Megan Thomas
Peg 21h47'51" 22d50'7"
Megan Thomson
And 23h30'11" 39d13'27"
Megan Thurk
Uma 12h45'57" 58d14'29"
Megan & Tony's Marriage Star
Psc 0h25'14" 4d39'17"
Megan Tower
Lib 15h33'14" -16d48'54"
Megan Travers
Ari 3h10'48" 11d28'11"
Megan Trouble Rose
Dra 20h25'14" 63d51'14"
Megan Turcotte
Cnc 9h9'22" 32d19'22"
Megan Tyler
Psc 23h44'58" -1d50'7"
Megan Valenti
And 2h11'32" 38d13'31"
Megan Veronica LaSala
Lyn 7h21'49" 44d26'23"
Megan Victoria Beck
Vir 12h4'5" -9d55'43"
Megan Victoria Chacon
Ari 2h43'4" 16d11'42"
Megan Victoria Gentry
And 23h20'6" 51d37'53"
Megan Warczynski
Uma 10h8'25" 43d25'31"
Megan Wattie
Cas 2h22'28" 74d18'12"
Megan Weber
Uma 11h54'32" 34d2'33"
Megan Wieland
Cas 1h23'38" 62d1'34"
Megan Williams
Her 17h15'45" 48d28'10"
Megan Williams
Lyn 6h59'59" 51d34'17"
Megan Willingham
Aqr 22h19'30" -2d56'22"
Megan Winkler
Leo 9h47'38" 27d41'51"
Megan Wright
And 23h37'50" 45d17'17"
Megan & Zach Christmas 2005
Psc 0h54'41" 19d19'42"
Megan, I will always be there
Leo 11h38'0" 10d38'56"
Megan, Jayden and Craig Forever
Ori 6h15'26" 14d3'41"
Megan070587: The Sparkle in My Life
Tau 4h27'14" 21d40'8"
Megan10892
And 23h40'14" 45d56'42"
Megan19
Cam 5h30'47" 63d29'27"
MeganAmanda
And 0h28'38" 39d36'47"
Meganbrentsebold
Lyn 6h55'19" 59d20'53"
Megane Collard
Gem 7h18'1" 30d41'14"
Meganjoe
Gem 7h32'11" 25d26'50"
MeganMarie
Tau 4h17'49" 19d20'30"
Méganne Chartier
Uma 11h31'59" 57d56'21"
Megano
Sgr 18h10'3" -21d4'52"
Meganopolis21
Psc 0h32'41" 6d28'18"
Megan's 18th Birthday Star
And 0h50'6" 44d49'52"
Megan's Beautiful Birthday Star
Sco 17h52'45" -35d52'3"
Megan's Destiny
Cyg 21h46'39" 52d13'26"
Megan's Evening Star
Cru 12h45'4" -61d18'53"
Megan's Grandma Millie special star
Lmi 9h54'30" 36d57'9"
Megan's Guardian Dragon
Dra 19h44'18" 74d34'26"
Megan's Guiding Light
Vir 13h49'10" -15d13'48"

Megan's Heaven
And 1h21'24" 49d17'27"

Megan's Incubus
Vir 13h21'41" 5d17'45"

Megan's Legacy
Sgr 18h1'9" -28d50'30"

Megan's Light
Lib 14h52'2" -7d23'44"

Megan's Light
Cas 0h11'15" 59d38'31"

Megan's Muse
Mon 6h41'2" -0d24'16"

Megan's own wishing star
Vir 13h52'21" 4d6'19"

Megan's Schneckenburger Star
Gem 7h10'18" 27d24'11"

Megan's Star
Ari 3h9'50" 27d36'52"

Megan's Star
Leo 11h14'32" 23d33'12"

Megan's star
Ori 4h53'36" 9d29'0"

Megan's Star
Leo 11h16'0" 11d19'46"

Megan's Star
And 23h3'10" 40d8'32"

Megan's Star
Lyn 7h59'3" 33d19'15"

Megan's Star
Ori 5h37'16" -1d34'29"

Megan's Star
Uma 11h13'3" 56d28'42"

Megan's Star
Cru 12h1'8" -62d40'45"

Megan's Starr
Uma 9h50'30" 51d42'53"

Megan's Super Star
Cnc 8h54'39" 19d54'18"

Megan's Tinkerbell 1
And 0h36'30" 23d26'34"

Megan's Wish
Vir 12h53'45" -7d46'36"

Megara
Her 16h24'8" 20d26'7"

Megarooney's Light
Cnv 13h18'25" 42d51'14"

megaskuon
Lyr 18h52'52" 34d4'25"

Meg-a-star
And 23h57'31" 39d53'47"

Megasztár
Sgr 17h59'22" -18d8'12"

Megasztár 2.
Tau 3h37'17" 5d24'35"

Megatraining
Gem 7h26'37" 34d23'22"

Megawattpwr
Psc 1h30'6" 3d29'19"

Megean Jones
And 0h20'25" 27d36'22"

Megen Christina Barger
Aqr 20h55'29" -7d30'15"

Meggan
Aqr 21h17'36" -7d22'45"

Meggan Holly
And 0h45'47" 44d24'30"

Meggan Montuori
Tau 3h25'36" 7d44'11"

Meggan Wilkes
And 0h20'4" 29d34'49"

Meggan's Riviera
Lyn 7h36'9" 40d15'44"

Meggers the Wonderfull
Ori 6h3'50" 20d47'13"

Meggie
And 0h29'48" 37d52'39"

Meggie
Lyn 8h50'2" 45d0'16"

Meggie
Lyn 6h42'26" 58d35'44"

Meggie Jo Lungstrom
And 0h36'27" 26d28'48"

Meggie Moose
Leo 10h13'52" 26d6'33"

Meggie Poo
Lib 15h25'47" -6d11'52"

Meggie Weggs
Ari 3h22'10" 27d55'23"

Meggie's Star
Gem 6h39'39" 26d23'39"

Meggin Taylor Hobbs
Psc 0h22'6" 19d46'25"

Meggy
Ari 2h57'44" 26d16'14"

Meggy MooMoo
And 0h36'23" 35d20'38"

Megha Ladha
Gem 6h26'39" 17d21'49"

Meghan
Psc 1h2'38" 16d40'59"

Meghan
Psc 1h4'48" 15d50'12"

Meghan
Tau 4h29'46" 19d51'11"

Meghan
Tau 4h44'56" 19d3'2"

Meghan
Vir 13h14'43" 13d21'33"

Meghan
And 1h7'13" 35d50'49"

Meghan
And 23h49'25" 48d3'0"

Meghan
Cap 21h49'27" -12d26'49"

Meghan
Sco 17h57'22" -31d4'39"

Meghan and Bill - Forever in Love
Ari 2h35'46" 29d56'27"

Meghan and John's Shining Star
Psc 1h35'34" 8d15'34"

Meghan and Papaw
Uma 8h52'46" 58d18'35"

Meghan and Ryan's
Cyg 21h18'33" 46d32'17"

Meghan and Torie
Sgr 18h57'59" -30d7'56"

Meghan Ann Albrecht
And 1h34'27" 43d25'19"

Meghan Ann Marie
Gem 7h22'22" 33d10'25"

Meghan Ann McDonald
Cas 0h28'55" 56d21'33"

Meghan Anne Gallagher
Gem 6h20'15" 26d10'13"

Meghan Anne McCormick
And 0h39'28" 29d27'40"

Meghan Ashley Hill
Tau 4h48'21" 28d18'35"

Meghan B
Per 3h4'23" 55d9'6"

Meghan Barennes
Ori 5h24'5" -5d10'2"

Meghan Barry
And 23h3'44" 38d14'12"

Meghan (BCMB) O'Rourke
And 0h59'22" 40d0'32"

Meghan Bell
Crb 16h10'42" 38d39'41"

Meghan Boyd
Lib 14h50'22" -1d32'33"

Meghan Breen Hartigan
Tau 4h46'49" 23d7'2"

Meghan Burroughs
And 0h19'40" 45d1'0"

Meghan & Chris
Vir 14h53'56" 6d17'51"

Meghan Colleen Burke
And 23h14'57" 47d56'49"

Meghan Danielle O'Malley
Ari 2h52'24" 24d47'21"

Meghan Danis Estes
Ari 3h27'47" 19d38'47"

Meghan Deborah
Psc 1h2'22" 10d33'10"

Meghan Ducky Overby
Sgr 18h55'0" -28d50'22"

Meghan E. Scotti
Crt 11h30'20" -22d29'30"

Meghan Elise Aase
Lyn 7h54'36" 53d29'3"

Meghan Elizabeth
Ori 5h49'25" 7d31'56"

Meghan Elizabeth
Cnc 8h18'22" 13d10'8"

Meghan Elizabeth
Cyg 21h10'54" 47d17'27"

Meghan Elizabeth
And 0h14'58" 45d5'42"

Meghan Elizabeth Ahern
And 1h41'57" 44d29'4"

Meghan Elizabeth Auer
Cyg 20h0'45" 42d31'11"

Meghan Elizabeth Collins
Lmi 10h31'41" 37d50'28"

Meghan Elizabeth Dye
Sco 16h32'23" -28d15'57"

Meghan Elizabeth Lauder
Aqr 22h42'5" 2d21'10"

Meghan Elizabeth Lee Chrissakis
Leo 10h37'46" 14d28'27"

Meghan Elizabeth O'Malley
Ori 5h38'12" -7d22'23"

Meghan Elizabeth Schafer
Uma 9h17'13" 57d10'56"

Meghan Elizabeth Shirling
Uma 14h17'22" 57d37'48"

Meghan Elizabeth Stacy
And 1h15'42" 45d59'35"

Meghan Fecteau
Leo 11h29'1" 9d11'36"

Meghan Flynn Jude Zajac
Leo 11h57'43" 20d30'31"

Meghan Foley
Uma 13h49'0" 56d59'36"

Meghan Folsom Derick
And 1h9'12" 37d48'41"

Meghan Gallagher
And 2h36'6" 44d43'33"

Meghan Gillette
Ari 3h24'42" 28d57'57"

Meghan Grace Emanuel
Aqr 22h30'45" -0d26'53"

Meghan Griffin Norton-Shining Star
And 1h15'11" 39d30'30"

Meghan Hepworth
Cra 19h0'42" -39d34'20"

Meghan Hunter Skiba
Cap 20h42'19" -25d23'25"

Meghan Irene
Per 3h12'16" 51d41'18"

Meghan Janelle Hutton
Cnc 9h5'26" 28d50'20"

Meghan Jeanne
Psc 1h34'43" 24d36'23"

Meghan Kathleen Currier
Uma 11h45'1" 64d47'38"

Meghan Kathleen Downs
Leo 10h23'47" 22d21'21"

Meghan Leigh
Psc 0h43'15" 9d50'58"

Meghan Leigh Jackson
Sco 16h59'1" -42d30'19"

Meghan Leigh Kerwood
Sgr 19h39'41" -15d35'36"

Meghan Leighanna Speaks
And 0h42'34" 22d26'22"

Meghan Liane
Ori 5h27'37" 4d36'11"

Meghan Lilone
Tau 4h51'27" 20d59'10"

Meghan Lily
Uma 10h47'57" 51d48'5"

Meghan Louise
Cnc 8h35'12" 19d51'1"

Meghan Lucille Burch
Sgr 19h17'43" -12d40'54"

Meghan Lynn Cleary
Lyr 18h47'37" 44d15'22"

Meghan M. Gibson
Ori 6h18'34" 10d19'30"

Meghan Malone
Peg 22h4'56" 31d24'55"

Meghan Maloney
Cyg 20h25'1" 38d14'41"

Meghan Mann
Ori 5h42'50" 1d26'9"

Meghan Marie
And 1h34'30" 43d51'48"

Meghan Marie
Crb 16h16'1" 37d19'0"

Meghan Marie<234
Sco 16h4'44" -9d4'22"

Meghan Marie Passey
Per 3h12'57" 54d33'41"

Meghan Marie Sweeney
Ori 5h8'22" 6d19'16"

Meghan McDonald
Tri 2h25'40" 35d56'56"

Meghan McDonald
Peg 23h17'36" 30d58'7"

Meghan McNamara, Mother
Uma 12h35'46" 56d57'10"

Meghan "Muma" Smith
Cap 21h32'18" -24d45'35"

Meghan My Love
Cas 1h7'49" 60d47'7"

Meghan Naylor
Tau 3h36'48" 20d43'10"

Meghan Nicole
And 0h17'25" 42d50'39"

Meghan Noel
Ori 5h51'53" 6d32'42"

Meghan Opal Whipple
Umi 15h13'46" 68d9'17"

Meghan Paige
Lyn 8h27'38" 45d10'8"

Meghan & Patrick Star
Gem 6h19'3" 22d25'49"

Meghan Rae
And 0h31'27" 41d53'17"

Meghan Rae Jenkins 22779
And 0h17'9" 38d43'51"

Meghan Rose
Cnc 8h37'4" 31d38'24"

Meghan Rose Krank-Burke
Aqr 22h12'11" 0d54'20"

Meghan Ryan
And 2h27'10" 42d1'33"

Meghan Ryan Foley
Uma 11h29'30" 42d33'9"

Meghan Scharbauer
Ari 3h20'52" 24d6'9"

Meghan Shaw Kennedy "The Princess"
Aql 19h9'14" -0d24'1"

Meghan Shayne
Crb 16h11'58" 38d42'45"

Meghan Shines Forever
And 2h23'28" 45d26'42"

Meghan Smith
Vir 13h30'12" 3d4'30"

Meghan St.Pierre
And 0h19'54" 32d38'26"

Meghan Suzanne
Uma 11h33'37" 46d4'41"

Meghan Suzanne
Lib 15h25'42" -5d28'4"

Meghan Trizil
Leo 11h44'42" 23d5'3"

Meghan & Vance
Sge 20h3'58" 18d7'1"

Meghan Victoria Andersen
Crb 16h9'10" 33d31'13"

Meghan Violet Holzknecht
Lyn 7h47'32" 58d19'19"

Meghan Williams
Cnv 12h52'19" 43d35'21"

Meghan Winn
Lib 14h51'41" -6d43'38"

Meghan, Will you marry me ?
Cyg 19h37'19" 31d37'17"

Meghana Aggarwal
Lib 14h59'10" -3d19'48"

Meghann Taylor
Peg 22h20'17" 7d10'50"

Meghan's Love
Aqr 21h44'6" -1d46'33"

Meghan's Star
Vir 14h6'8" -21d53'15"

Meghan's Star
Pho 0h35'20" -50d22'39"

Meghan's Star
And 23h30'16" 48d28'7"

Meghan's Turtle
Cnc 8h49'34" 32d10'6"

Meghan's Way
Sgr 18h9'47" -22d45'55"

Meghedie Seza Der-Sarkissian
Ori 5h13'7" 7d10'38"

Meghlos
Uma 10h30'2" 68d13'51"

Meghna
Vir 13h1'32" -4d58'24"

Meghna Menon
Gem 7h21'4" 24d56'46"

Meghna Viral Parekh
Aqr 23h33'55" -8d19'25"

Meghnath Krishnan
Tau 4h17'55" 2d16'32"

MEGHU
Crb 15h34'45" 26d3'24"

Megi Rassette
Sgr 19h21'19" -41d26'31"

Megin Suzanne Fanz
Sco 16h8'58" -13d40'12"

Meglepkék.hu
Cas 23h59'44" 56d30'3"

M.E.G.M.
Aqr 22h25'39" -9d35'17"

Mego Lien
Umi 13h14'58" 74d42'53"

Megron
Cyg 20h22'21" 48d15'22"

Megs
Lyn 8h14'16" 35d33'21"

MEGS
Cnc 8h21'24" 15d39'34"

MegS
Cas 0h44'3" 61d29'23"

Meg's Gemini Twin
Gem 7h4'32" 15d2'20"

Megs N Rick
Leo 11h13'23" 12d44'17"

Meg's Star
Cyg 21h43'43" 37d31'4"

Megsi
Cnc 9h8'29" 29d16'32"

Megstar
Lyn 8h20'34" 58d27'22"

Megumi & John
Cyg 20h41'23" 33d55'5"

Megz
Lyn 8h26'28" 42d26'39"

Megz Star
Lib 15h37'39" -10d38'47"

Megzy's Dreams
Cru 12h48'48" -57d30'55"

Mehagan Family Star
Ari 2h11'13" 24d19'22"

Mehdi
Cap 21h42'54" -8d47'55"

Mehdi Benali
Cas 0h24'25" 53d55'57"

Mehdi Doukari
Vir 14h20'15" 3d45'51"

Mehdi Vlamynck
Cap 21h35'6" -24d56'6"

Mehek
Cnc 8h52'8" 30d34'3"

Meher Jabran Khan
Uma 8h35'6" 69d46'43"

Mehgan Meeks
Aql 20h4'2" 8d14'5"

Mehgels
Mon 6h29'43" 11d1'39"

Mehkia McDowell
Peg 22h27'7" 9d32'52"

Mehl
Per 2h53'16" 33d28'26"

Mehlia
Lyn 7h43'7" 37d11'43"

Mehmet Gundogdu Jr.
Psc 0h36'38" 21d0'6"

Mehmet Hersli
Cru 12h42'36" -56d53'46"

Mehmet Taha Erbil
And 0h43'42" 23d28'35"

Mehrad Loves Jodi
And 0h31'56" 25d53'40"

Mehrangiz "Mary" Rabizadeh
Psc 1h22'39" 23d40'49"

Mehrdad Barkhordar
Per 3h5'16" 39d14'8"

Mehret Fisseha
Cyg 21h38'12" 50d34'24"

Mehrnaz Farrokhi
Cap 20h53'45" -23d24'30"

Mehron
Aur 5h55'24" 53d9'52"

Mehroo Wadia
Cas 0h58'56" 57d7'43"

Mehrunissa
Sgr 17h55'34" -26d39'55"

Mehtal Epoh Nosiddam
Eri 3h37'12" -11d50'57"

Mei Amor
Aqr 23h52'58" -10d51'25"

Mei Amor Semper, Olivia Anne...
Ori 5h26'7" 2d12'35"

Mei Carus Uxor Leslie Ann Cornell
Psc 0h20'40" 16d24'5"

Mei Ling Wong
Cnc 8h25'16" 9d24'24"

Mei Mirus Amor Victoria
Lep 5h12'31" -11d16'12"

Mei Yen Cheng
Vir 13h3'7" 5d11'26"

Meier, Anja
Sco 17h23'15" -33d7'22"

Meighan Fives
Psc 1h0'39" 6d9'46"

Meighan Jayne Silvers
Lib 15h7'10" -17d58'43"

Meike & Christian
Uma 10h25'40" 57d13'45"

Meike210977
Ori 6h15'30" 11d11'12"

Meil Lyne
Dra 17h57'41" 64d16'7"

Meilene U Cook
Psc 0h22'35" 6d47'53"

Meili
Uma 9h33'43" 68d55'59"

Meilyn Rae
Leo 9h22'27" 25d59'57"

Meimei - Wo ai ni till end of time
Cyg 21h55'13" 43d55'16"

MEIN EGNEL ANITA
Uma 8h48'0" 66d35'19"

Mein Engel
Uma 10h9'18" 47d17'56"

Mein Engelchen Stephanie Rühl
Uma 13h7'15" 52d27'28"

Mein Freund Simon
Uma 10h38'19" 63d18'28"

Mein Hase Daniela
Gem 6h22'19" 21d8'0"

Mein Held
Uma 9h27'0" 63d58'13"

Mein Herz
Uma 14h27'53" 58d17'13"

Mein Herzi - Michael Gerle
Ori 6h15'21" 16d30'17"

MEIN KONIGSBLAUER STERN
Cep 22h53'20" 69d29'57"

Mein Liebchen
And 0h17'27" 45d54'51"

mein Liebling Steffen
Cnc 8h8'12" 11d34'34"

Mein Prinz Helge Scholz
Ori 6h4'42" 10d10'6"

Mein Schatz
Psc 0h10'42" 3d10'32"

Mein Schönes
Tau 3h26'36" 5d39'24"

Mein Stern
Sco 17h20'53" -45d17'24"

Mein Stern Thomas
Uma 9h7'35" 60d52'41"

Mein Stern Viktor
Ori 6h18'14" -0d24'25"

"Mein Sternchen Anna"
Ori 5h52'38" 21d58'37"

Mein Traumliebhaber
Cyg 19h51'40" 56d6'35"

mein tumhay pyar karta hun Sarah
Cyg 21h3'53" 36d50'51"

Meindl, Franz Xaver
Uma 8h35'45" 59d15'31"

Meine Einzige Liebe, Valerie
And 2h2'43" 46d13'25"

Meine Ewige Liebe
Uma 9h26'54" 72d13'39"

Meine große Liebe , Susi
Uma 10h16'1" 59d42'45"

Meine grosse Liebe Rudolf
Uma 13h32'22" 54d11'40"

Meine grosse Liebe Silke
Uma 9h21'26" 65d9'54"

Meine kleine "Nicole"
Uma 9h41'12" 58d20'43"

Meine Liebchen
And 2h37'9" 45d14'35"

Meine Süße - Monika - Ma Douce
Ori 6h14'11" 16d50'40"

Meineken
And 1h2'5" 36d54'38"

Meinem Stern Thomas Schäfer
Aqr 20h50'6" -11d39'17"

MeinemSchatzMichaelz.1.Hochzeitstag
Uma 8h55'46" 48d46'1"

Meinhard Holger
Uma 9h56'12" 64d30'56"

Meinhardt, Ehrenfried
Uma 9h41'11" 66d34'55"

Meinnier Bernard
Ori 6h2'45" 6d32'7"

MeinStern 111203 Paphos/CY FZ
Ori 5h17'37" -8d1'1"

Meira Owen
Cas 23h43'6" 53d19'57"

Meira Rivka
Lib 14h50'35" -1d8'8"

Meiramanda
Lyn 7h19'9" 58d13'12"

Meisaa Salloum
Cap 21h6'9" -16d38'28"

Meisel
Tri 2h22'0" 33d30'29"

Meish
Tau 5h12'47" 23d27'51"

Meishla's Star
Dra 16h15'50" 60d46'7"

Meissiana
Lup 15h42'52" -31d38'43"

Meißner, Heinz-Hermann
Ori 6h17'51" 19d33'25"

Meister, Karl-Heinz
Uma 8h55'40" 72d51'40"

Meitzner, Thomas
Uma 11h34'28" 28d35'3"

Meiwa Renée Alyssa D'Archangelis
And 23h50'1" 40d8'38"

MEJ II
Lib 15h42'44" -8d25'50"

Meja
Tau 4h2'6" 24d51'13"

Meka and John
Uma 10h55'22" 61d43'59"

Mekana Mino'aka
Mon 7h0'44" -7d36'41"

Meka's Fascination
Ari 2h57'18" 28d15'5"

Mekhi Lewis Coburn
Tau 4h20'37" 19d45'31"

Mekhi Noah Simeon Crichlow
Cyg 19h45'0" 33d14'51"

Meknewreet
And 22h59'1" 51d15'34"

Mel
Lyn 7h33'54" 39d23'16"

Mel
Lyn 7h53'14" 38d47'20"

Mel
Lmi 10h39'1" 25d15'31"

MEL
Sgr 18h12'59" -19d2'9"

Mel
Sgr 18h19'9" -26d22'48"

Mel
Aqr 23h52'16" -15d5'17"

Mel 1 (Melitta)
Aur 6h37'51" 39d13'1"

Mel & Ali's Star - Best Buds Always
Pyx 8h53'5" -31d50'20"

Mel and Dorothy Funfsinn
Cyg 21h40'0" 38d40'16"

Mel and J
Lyn 9h5'36" 39d19'3"

Mel And Shirley Seffinger
Lyr 19h15'20" 26d15'6"

Mel Chisholm
Cyg 19h36'14" 29d15'27"

Mel Fleming
Lyr 18h51'26" 34d28'11"

Mel Freda
Sgr 18h39'46" -22d57'11"

Mel Harper 17-1-77
Uma 10h3'58" 60d58'28"

Mel Kee Way
Uma 12h41'45" 58d14'13"

Mel "Lillem" Howard
Lib 15h23'7" -4d15'1"

Mel Mel
Peg 23h12'37" 26d44'43"

Mel Noble
Cyg 19h45'31" 56d1'2"

Mel Owens
And 23h8'53" 43d14'12"

Mel Paulsen
Ori 6h10'59" 15d5'58"

Mel&Rev91104
Vir 14h8'15" -16d7'38"

Mel Simmons
Leo 9h36'25" 26d35'58"

Mel "The Star" Kammerer
Uma 8h31'7" 68d33'33"

Mel Weinberg #1 Star
Vir 13h24'51" 6d19'51"

Mel Yackovich
Uma 8h20'42" 62d42'52"

Mela
Cas 1h19'37" 72d32'8"

Mela
Mon 6h55'28" -0d59'45"

M.E.L.A.
Cru 12h28'56" -59d30'55"

Mela Ile Maeglin
Uma 12h34'52" 60d58'44"

Méla1303
Sco 17h45'53" -33d44'0"

Melaina
Sgr 19h40'28" -14d12'8"

Melaine
And 2h28'38" 46d22'25"

Melaine Ann Bletzer
Per 3h37'35" 46d59'5"

Melalee
Ori 5h39'57" -5d25'59"

Melamine
Uma 9h2'25" 48d15'39"

Melancholia Mater Lynn
Ori 5h57'1" 8d5'21"

Melanee Wyatt
Vir 12h16'26" 4d14'9"

Melani Cainto
Lib 14h55'9" -16d22'24"

Melani Schoening
Psc 23h7'24" 7d33'44"

Melania
Crb 16h20'56" 30d50'27"

Melania
Sco 16h11'23" -18d5'2"

Melania
Umi 16h21'16" 71d45'24"

Melania Dark Beauty of the Universe
Cas 1h1'4" 54d0'5"

Melania Janelle DeVane
And 23h32'12" 37d55'4"

Melanie
Lyn 8h20'24" 46d33'58"

Melanie
And 2h26'1" 45d50'31"

Melanie
And 2h33'58" 50d8'54"

Melanie
Cas 0h12'55" 56d54'13"

Melanie
And 22h59'22" 50d44'24"

Melanie
Cyg 19h38'19" 46d47'49"

Melanie
Uma 11h42'21" 37d8'30"

Melanie
Uma 11h23'53" 35d2'28"

Melanie
And 0h46'33" 36d24'27"

Melanie
And 1h16'7" 35d16'56"

Melanie
Psc 1h2'29" 32d30'29"

Melanie
Psc 1h8'26" 32d29'48"

Melanie
Leo 9h22'41" 31d10'28"

Melanie
Lyn 8h51'40" 41d6'32"

Melanie
Per 4h23'54" 41d1'54"

Melanie
Peg 23h32'57" 10d38'10"

Melanie
Tau 5h0'11" 21d53'23"

Melanie
Cnc 8h54'5" 13d21'8"

Melanie
Ori 6h15'20" 10d56'26"

Mélanie
Aql 19h1'52" 3d17'35"

Melanie
Tau 4h30'41" 3d0'6"

Melanie
Her 18h1'31" 25d15'44"

Melanie
Tau 4h42'36" 23d25'9"

Melanie
Tau 5h30'44" 26d29'38"

Melanie
Ori 6h1'16" 21d9'47"

Melanie
Ari 2h22'13" 23d59'25"

Mélanie
Uma 8h33'26" 71d57'58"

Melanie
Dra 15h46'38" 63d30'14"

Melanie
Lib 15h40'37" -19d55'59"

Melanie
Lib 15h18'43" -11d11'15"

Melanie
Cap 21h55'49" -13d35'2"

Melanie
Sco 17h53'7" -35d52'23"

Melanie
Sgr 18h18'47" -32d0'50"

Mélanie
Sco 17h35'30" -41d7'57"

Mélanie
Leo 11h10'35" 26d36'5"

Melanie 21.09.1971
Ori 6h17'32" 10d38'3"

Melanie 92
And 0h39'4" 42d36'30"

Melanie A. Knight
Cas 1h45'1" 64d31'41"

Melanie A. Salafia
Ari 3h13'38" 27d8'11"

Melanie Ablalos Lopez Webb
Ari 3h10'18" 25d7'14"

Melanie Agaliotis
Cnc 8h40'20" 31d49'48"

Melanie Alexander
Lib 14h33'41" -9d47'31"

Melanie and Steve's Star
And 1h3'42" 46d2'51"

Melanie Andrews
Cru 11h57'29" -58d0'20"

Melanie Angela DeMartinis
Vir 14h34'59" -2d8'37"

Melanie Ann
Leo 11h17'9" 18d21'24"

Melanie Ann Lopez
And 23h19'20" 49d29'54"

Melanie Ann Morgan
Gem 7h1'52" 16d31'40"

Melanie Ann Ross
Cap 21h45'40" -9d36'55"
Melanie Ann Worley
And 23h10'24" 44d55'17"
Melanie Anne Jackson
Vir 12h21'38" 12d10'16"
Melanie Annette Dolce
Lyn 7h28'53" 53d42'44"
Mélanie Arbona
Sco 17h57'17" -30d49'2"
Melanie Arzt
Uma 9h45'29" 46d39'33"
Melanie Austria
Umi 7h2'27" 88d56'16"
Melanie (Babe) Hadzima
Gem 7h32'42" 28d18'6"
Melanie Bauso
Lib 15h53'17" -17d11'51"
Melanie "Baybee" Roberts
And 2h29'24" 49d50'21"
Melanie Beckenhauer Heller - #10
Crb 16h11'36" 34d6'49"
Melanie Bell Weaver
Aqr 22h54'18" -7d36'30"
Melanie Best
And 23h51'37" 46d13'44"
Mélanie Boidin 27/07/1982
Leo 11h43'5" 16d28'29"
Melanie Bosshard
Her 18h11'3" 27d56'57"
Melanie Bowie
Cnc 9h6'3" 12d1'21"
MELANIE BROOKE MCLAIN
Gem 7h20'36" 14d43'55"
Melanie Burns
Ari 2h23'37" 19d29'49"
Melanie Büsser
Per 4h44'26" 46d48'33"
Melanie C Schumacher BSc (Hons)
And 1h15'13" 45d46'26"
Melanie Carole
Psc 0h39'27" 7d53'32"
Melanie Christina Martin
And 2h36'46" 49d10'34"
Melanie Clemente
Cyg 19h57'30" 40d2'23"
Mélanie Corcoran
Sco 16h18'58" -18d3'11"
Melanie Crow - Our Shining Star
And 0h38'11" 33d43'22"
Melanie Darlene Ward
Cap 20h29'37" -13d52'24"
Mélanie Darrigade
Her 18h35'42" 19d13'9"
Melanie Davenport
Uma 11h6'20" 59d57'17"
Melanie Davenport
Uma 11h6'1" 62d30'6"
Melanie Del Pozo
And 0h26'20" 29d17'39"
Melanie & Dennis
Ori 5h13'46" 5d42'55"
Melanie Diane Edwards
Sco 16h45'37" -31d0'9"
Melanie Diane Lytle
Ori 4h52'26" 13d33'21"
Melanie Elizabeth
Leo 11h16'45" 13d16'23"
MELANIE ELIZABETH
Vir 13h39'9" -8d23'18"
Melanie Elizabeth Bauer
Cas 23h13'21" 59d30'54"
Melanie Elizabeth Hanlon
Gem 6h45'30" 23d9'8"
Melanie Elizabeth Irene
Cas 23h36'42" 58d39'28"
Melanie Ellen Mitchell
Uma 14h6'50" 59d46'35"
Melanie Elyssa
Cas 0h47'49" 57d58'28"
Melanie Esquivel
Aur 5h14'46" 37d1'29"
Melanie F Clifford
Sgr 19h29'10" -44d57'2"
Mélanie Fairburn Red Summer 95
Cyg 21h45'50" 38d30'0"
Mélanie Fais
Leo 10h20'44" 22d27'30"
Melanie Fawn
And 0h15'15" 44d36'22"
Melanie Forster
And 1h28'57" 34d4'51"
Melanie Frances Davis
Sco 17h40'21" -32d8'12"
Melanie G. Lively
Vir 12h49'24" 3d39'20"
Melanie Germino
Sco 16h36'8" -31d45'58"
Melanie Gill
Vir 12h27'17" 3d54'10"
Melanie Gonzalez
Lyr 19h3'55" 41d59'53"
Melanie Grace DiFabio
Vir 11h39'3" 3d8'34"
Melanie Grace Laime
Lib 15h2'1" -23d41'17"
Melanie Grace Lindmark
Vir 13h40'33" 6d15'47"
Melanie Guarino
Psc 1h5'48" 28d14'26"

Melanie Hae Eun Park
Cam 6h41'38" 65d20'23"
Melanie Heathfield
Tau 5h0'34" 16d26'11"
Mélanie Hélène Aurélie Peyrin
Umi 4h42'9" 89d19'14"
Melanie Henderson
Cnc 8h33'42" 8d5'39"
Mélanie Hernoult
Ari 3h5'29" 22d41'36"
Melanie Hooie
Aqr 21h41'53" -1d18'29"
Melanie Hope Witte
Cas 23h4'54" 59d15'40"
Melanie Jaclyn Downey
Lyn 7h25'47" 50d29'42"
Melanie Jane
Cyg 19h40'11" 35d16'55"
Melanie Jane
Vir 13h7'29" 8d35'51"
Melanie Jane
Sco 16h50'45" -41d55'20"
Melanie Jane Aycock
Cas 0h45'50" 69d43'50"
Melanie Janette Phillips
And 2h27'35" 46d4'57"
Melanie Jean
Ari 2h20'27" 24d12'20"
Melanie Jean Eakin
And 1h3'54" 36d0'48"
Melanie Jennifer Madison Scot Krimm
Lmi 10h7'57" 36d13'38"
Melanie Jessica Steven Erin & Steve
And 2h38'38" 42d51'40"
Melanie & John
Mon 6h54'54" -0d12'48"
Melanie Johnson
Cas 0h15'18" 57d25'39"
Melanie Joy
Aqr 22h51'43" -7d21'56"
Melanie Junean Wheeler
Cas 1h26'33" 55d12'59"
Melanie Kate Oliveiro
Tau 5h11'27" 17d56'16"
Melanie Kaye Smith
And 1h35'7" 43d8'44"
Melanie Keller
Cyg 20h51'3" 48d28'54"
Melanie Kogelbauer
And 2h30'59" 49d30'34"
Melanie Kolifrath
Ori 6h21'58" 16d30'57"
Melanie Koontz-DeMarchi
Cyg 20h33'52" 39d59'8"
Mélanie L. Perna
Sco 17h42'55" -42d42'49"
Melanie Lane
Cra 18h7'49" -40d35'55"
Melanie Lauren Hendel
Cap 21h21'11" -22d29'45"
Melanie Lauren Kay
Lib 14h47'1" -9d17'42"
Melanie Lee Denno
Cas 0h15'15" 51d0'37"
Melanie Lisk
Lib 15h39'51" -17d18'20"
Melanie Logan
Uma 8h35'41" 52d54'28"
Melanie Louise Green
Aqr 21h19'2" -8d42'23"
Melanie Louise Stevens
And 23h20'55" 41d33'14"
Mélanie Loutan
Cas 23h24'47" 58d17'31"
Melanie Luise Hartwick
Uma 9h30'28" 42d33'1"
Mélanie Luraschi
Cnc 8h58'29" 7d36'35"
Melanie Lynn Bridwell
Leo 10h29'0" 23d12'39"
Melanie Lynn Campbell
And 0h22'26" 39d47'16"
Melanie Lynn DeVaughn
Tau 3h39'39" 24d9'40"
Melanie Lynn Tompkins Jell
Uma 11h34'40" 51d44'48"
Melanie M. Ward "My Shining Star"
Cas 0h25'42" 48d1'49"
Melanie Mae
Psa 21h40'14" -31d42'17"
Melanie Mango Lyons
Leo 10h57'36" 15d44'52"
Melanie & Marco
Uma 10h12'1" 68d43'21"
Melanie Marie Baker
Ari 3h21'52" 28d28'16"
Melanie Marie Crowson "MEL-15"
Ori 5h15'1" 5d56'18"
Melanie Marie Dano
Cas 0h20'49" 58d12'52"
Melanie Marie Flores
Cap 20h30'49" -13d55'36"
Melanie Marie Godwin
And 0h41'27" 25d7'37"
Melanie "Me Me"
Crb 15h44'52" 29d3'31"
Melanie Mehr Tessendorf Khan
Lyn 9h13'13" 41d23'23"

Melanie Mills
Cyg 19h51'59" 55d35'35"
Melanie "Mimi" Ohl
Gem 6h43'58" 18d13'18"
Melanie My Love!
Col 5h28'55" -33d4'28"
Melanie & Natalie Sharp
Lyr 19h19'59" 35d17'59"
Melanie & Nicholas Schumacher
Leo 10h0'9" 21d29'45"
Mélanie Olivier
Ari 1h56'57" 14d32'1"
Melanie Ortiz Ibarra
Leo 11h18'53" -1d34'17"
Melanie Our Shining Star
Vir 14h31'46" 4d10'53"
Melanie Paternoster
Sgr 19h18'20" -25d18'48"
Melanie Perez
Sgr 17h58'18" -29d25'24"
Melanie Perez
Aqr 23h26'23" -16d25'29"
Melanie Pero
Ori 5h8'38" 7d54'55"
Melanie Perry
Tau 4h24'7" 21d47'23"
Melanie R. Hundsdorfer
Ori 6h7'46" 13d32'58"
Melanie R. Weickert
Cas 1h43'57" 63d42'10"
Melanie Rae Macomber
Leo 11h23'25" 14d50'1"
Melanie & Ralph Sudan
Peg 22h22'41" 2d42'5"
Melanie Rath
And 0h15'46" 27d48'15"
Melanie Rempe
Cyg 20h39'49" 46d17'18"
Melanie Reyes Kraftchak
Tau 5h45'29" 23d27'37"
Melanie Rhoads Gravelle
Uma 8h53'55" 48d22'14"
Melanie Richards
Cas 23h47'56" 57d56'0"
Melanie Rose
Sgr 17h52'29" -28d36'43"
Melanie Rose Coronado
And 0h38'0" 40d34'4"
Melanie Rose Novikov
Uma 12h1'24" 64d48'35"
Mélanie Roy
Crb 15h59'30" 35d16'19"
Melanie Russo
And 1h31'19" 48d35'36"
Melanie Ruth Rudnick
Eri 3h50'45" -0d34'40"
Melanie Sanborn
Umi 15h57'17" 73d26'47"
Melanie Sandra Wiedrich
And 1h17'34" 34d35'38"
Melanie Schleicher
Lib 14h51'24" -3d52'7"
Melanie Schneider
Cep 23h0'48" 86d39'40"
Melanie Schugardt
Uma 9h32'17" 68d5'12"
Melanie Shae Spinelli
Aqr 23h8'19" -15d24'35"
Melanie Sue Purdy 9-19-77
Cyg 21h4'29" 36d37'56"
Melanie "Sunshine Face" Farrar
Leo 9h22'5" 7d0'2"
Melanie Therrien
Aqr 22h36'11" 1d49'49"
Melanie Tiffany B
Leo 11h44'27" 25d2'6"
Melanie & Tihomir
Uma 10h54'30" 48d16'48"
Melanie Treff Stiekman
Cas 1h15'31" 63d53'2"
Melanie Vesco
Uma 10h48'43" 40d8'58"
Melanie Waller
Mon 6h37'52" 9d9'14"
Melanie Webb Smith
Leo 10h9'36" 23d47'20"
Melanie Whelan
Leo 10h28'9" 18d44'14"
Mélanie & Yanick pour l'Éternité!
Crb 16h0'9" 27d7'31"
Melanie Young
Tau 3h37'53" 3d1'54"
Melanie Zebrowski
Uma 10h59'23" 55d33'57"
Melanie101178
Sco 17h15'7" -33d8'13"
Melanie1437
Boo 14h24'57" 13d36'34"
Melanie-Pearl of Great Price Star
Eri 4h30'1" -0d59'34"
Melanie's Dark Star
Cnc 8h23'4" 9d33'43"
Melanie's Love
Cyg 21h14'54" 33d27'50"
Melanie's Milestone
Ari 2h42'36" 15d51'28"
Melanie's Night Light
And 1h28'59" 47d42'38"
Melanie's Place
Cas 23h2'20" 56d45'37"

Melanie's Precious Rock
Peg 23h55'1" 8d25'21"
Melanie's Smile
Crb 16h24'25" 31d38'33"
Melanie's Smile
Cru 12h41'53" -57d0'54"
Melanie's Star
Cet 1h9'10" -0d27'20"
Melanie's Star
Lmi 10h8'43" 35d14'32"
Melanie's star
Lyr 18h40'40" 31d40'20"
Melanie's Star
Cnc 8h12'45" 31d21'42"
Melanie's Star
And 1h2'12" 45d14'8"
Melanie's Wishes
Cam 3h36'19" 59d5'56"
Melanio Gomez
Umi 17h29'21" 85d21'46"
Melannie Najera
Gem 6h30'1" 25d52'19"
Melany is Love
And 23h35'40" 48d18'14"
Melany Kristine Economakos
Com 12h50'41" 27d53'29"
Melany Yerimias u r my star sayang
Cru 12h27'5" -61d56'24"
Melari 1
Cas 0h54'9" 47d35'7"
Melaya Mitsy
Umi 15h42'14" 71d25'24"
Melba
Aqr 20h45'34" 2d8'13"
Melba 1
Per 3h32'54" 46d54'45"
Melba Blanzaco
Psc 0h8'37" 5d33'51"
Melba Ditz Dahl
Lib 14h47'3" -8d7'19"
Melba Palys
Aql 18h52'1" 8d52'0"
Melba Shelby
Lyr 19h7'47" 31d28'51"
Melba Simpson-Johnson
Cam 5h42'46" 63d8'11"
Melba's Magic
Vir 12h54'30" 6d19'45"
Melbee
Peg 22h38'9" 24d15'8"
MELBIL
And 1h44'22" 45d10'57"
MelChad
Cyg 19h51'41" 46d33'27"
Melchizedek Andrada
Cap 20h31'51" -8d48'24"
Melda Louise
Leo 9h58'31" 6d29'13"
Meldan 1027
Sgr 19h7'5" -17d35'36"
Melde Eruvande
Lib 15h28'58" -9d49'29"
Meldee Ruth Love
Cas 1h39'46" 61d17'11"
Melea Dyan
And 2h9'32" 42d21'7"
MelearohnKiser3
Cnc 8h29'41" 28d10'1"
Meledie A. Bruneau "Angel"
And 23h38'56" 36d41'44"
Melek
Ori 4h48'31" 12d57'4"
MELEM
Uma 10h16'22" 47d12'0"
Melenda Love
Gem 7h30'34" 20d22'38"
Melendez Love
Lib 15h52'17" -6d24'37"
Meleo
Pyx 9h17'17" -30d50'42"
Melessa&RobAmaya
Sge 19h51'0" 18d41'52"
Meleth Galad Aaron
Sco 16h6'8" -13d1'21"
Meletta Eulane Mellon
Lib 15h16'24" -24d11'38"
Melfi Stefko, Dana
Vir 13h23'45" -14d2'38"
MelGrub
Cyg 19h50'18" 37d31'2"
Meli
Tau 4h14'20" 18d22'26"
Meli the Little Mapiet
And 23h28'57" 37d41'30"
Melia
And 1h45'23" 49d57'23"
Melia
Lib 14h59'30" -9d55'30"
Melia Bills
Uma 9h9'8" 67d7'51"
Melia Hopton
And 23h15'36" 42d15'30"
Meliah Marie Cook
Ari 3h27'34" 26d11'27"
Meliana Liong
Psc 1h17'48" 27d11'28"
Melicia Dianne Brewer
Dra 18h33'14" 54d44'13"
Melida
Psc 22h59'47" 7d44'23"
Melida Barton
Aql 19h46'30" 16d10'4"

MELIKA
And 23h19'44" 38d17'36"
Melika
Cap 21h34'32" -14d29'3"
Melika Kahealani
And 23h33'51" 42d37'26"
Melike
Cma 6h37'21" -16d11'36"
Melike Berksu
Ari 2h32'36" 22d18'49"
Melike Gumus
Uma 10h29'32" 55d3'4"
Melina
Uma 11h53'44" 54d48'24"
Melina
Uma 12h5'42" 61d43'36"
Melina
Cap 20h17'31" -15d27'3"
Melina
Dra 19h30'5" 76d19'44"
Melina
Psc 0h30'25" 17d21'51"
Melina
Leo 11h53'8" 10d27'22"
Melina
Gem 7h38'38" 22d52'50"
Melina
Uma 13h59'4" 48d33'0"
Melina
And 23h31'57" 48d22'36"
Mélina
Lyr 18h25'15" 34d22'58"
Melina Carrillo
Aql 20h19'40" 4d12'36"
Melina Doyle's Star of Strength
Uma 9h8'49" 48d57'3"
Melina Johnson
And 1h43'19" 46d37'51"
Melina Katarina Zoolakis
Umi 18h21'8" 87d9'39"
Melina Kate Keller
Del 20h43'57" 16d25'12"
Melina Lambrakos
And 0h49'57" 40d33'20"
Melina Sawell A Hayn
Lyr 18h34'39" 38d11'23"
Melina Smirnoff
Aqr 22h20'23" -6d52'19"
Mélinda
Uma 10h9'22" 70d48'37"
Melinda
Lyn 7h21'29" 58d32'19"
Melinda
Cam 5h59'55" 58d12'41"
Melinda
Cas 0h13'25" 63d30'49"
Melinda
And 2h29'39" 50d17'46"
Melinda
And 2h16'26" 49d39'1"
Melinda
Dra 17h57'48" 52d19'18"
Melinda
And 0h45'3" 39d57'22"
Melinda
Lyn 8h34'45" 39d7'7"
Melinda
Peg 22h36'2" 25d4'41"
Melinda
Leo 11h6'10" 11d22'0"
Melinda
Peg 21h48'10" 10d33'37"
Melinda and Craig
Cyg 19h57'21" 30d42'57"
Melinda and Jeffery
Cyg 19h43'32" 33d26'7"
Melinda and John
Vir 14h1'32" -21d35'42"
Melinda Ann Gates
Cyg 20h35'1" 47d10'34"
Melinda Anne Manaut
Gem 7h7'7" 27d21'19"
Melinda Beth
Tri 2h26'19" 31d23'27"
Melinda Blansett
And 0h19'45" 43d53'11"
Melinda Bowe
Leo 10h33'18" 23d3'57"
Melinda Castillo
Mon 7h48'9" -0d59'57"
Melinda Chicken Tarbell
Vir 13h10'31" 5d41'21"
Melinda & Chris McRae
Cyg 19h42'16" 35d44'53"
Melinda Dale
Cas 1h26'56" 67d16'19"
Melinda Dawn
Lmi 10h49'58" 27d26'23"
Melinda Dawn Andreoli
Leo 11h3'28" -0d10'27"
Mélinda Delaunay
Del 20h51'45" 2d40'21"
Melinda Dumplin Baggett
And 1h52'42" 45d21'19"
Melinda Elizabeth Hall
Uma 10h43'47" 44d35'21"
Melinda Elliott
And 0h42'26" 39d49'7"
Melinda Emily Rushton
Cyg 20h32'45" 55d7'14"
Mélinda et Guillaume
Cas 0h50'56" 56d11'27"
Melinda H. Morris
Ara 17h9'45" -48d55'19"

Melinda Hartford
And 0h43'38" 41d46'57"
Melinda Hays-Clawson
Uma 8h10'0" 61d18'53"
Melinda & Heather's Eternal Flame
Sge 20h1'43" 18d5'10"
Melinda Hudson
Cas 1h19'19" 63d13'5"
Melinda J. Miller
Sgr 19h14'58" -22d3'51"
Melinda & James Coenen-Eyre
Gem 6h51'40" 19d4'11"
Melinda Kate Loesch
Lyn 7h53'22" 38d51'31"
Melinda Kathryn
Uma 11h34'10" 53d49'23"
Melinda Kay Sander
Sgr 19h9'14" -17d59'49"
Melinda Knauss
Aqr 22h56'55" -7d16'43"
Melinda L. Biehler
And 2h7'11" 43d22'58"
Melinda Lamm
Psc 1h5'45" 29d55'50"
Melinda Leigh
Umi 15h28'42" 73d2'23"
Melinda L'Etoile
Cap 21h39'0" -8d39'52"
Melinda Lynn Holladay
Psc 23h59'26" 1d15'31"
Melinda Mae Gamble
Cam 4h4'50" 65d42'36"
Melinda Marie
Cas 0h9'57" 62d43'37"
Melinda Mary Miller-Pfeufer
Lyr 18h52'24" 43d37'1"
Melinda May
Cyg 21h57'39" 50d47'21"
Melinda Meagher
Vir 13h28'48" -4d36'27"
Melinda Melton
Leo 10h32'25" 21d51'27"
Melinda Munn
Tau 4h22'40" 14d25'5"
Melinda ~ My Sunshine
And 1h16'59" 38d41'42"
Melinda Myers
Lyn 7h18'24" 57d5'37"
Melinda Nelson
And 0h59'59" 37d11'15"
Melinda Nessen
Lmi 10h51'47" 32d49'50"
Melinda Night-art
Ori 6h12'17" 9d21'2"
Melinda Raine
Cas 0h28'17" 55d9'39"
Melinda Ramjass
And 0h21'47" 26d15'39"
Melinda+Rodney Eternal Love18/09/04
Ara 17h38'17" -47d5'24"
Melinda Sue Bascos
Ori 5h43'7" 9d0'41"
Melinda Sue Drummond
Uma 9h31'27" 46d42'4"
Melinda Sue Myers
Cyg 19h36'56" 28d49'13"
Melinda Tiny Redd
Uma 11h30'11" 36d24'33"
Melinda und Manfred Moser
Cyg 20h53'38" 47d28'28"
MelindaJoy
Psc 0h46'29" 11d28'30"
Melinda's Alex
And 0h30'56" 39d20'39"
Melinda's Shining Star
Cyg 20h45'51" 42d47'31"
Melinda's Shining Star
Sgr 18h15'2" -17d16'52"
Melinda's Star
Gem 6h35'36" 21d7'20"
Melinh
Ari 2h0'27" 18d3'21"
Melinka Thompson-Godoy
Cap 21h45'34" -12d43'7"
Melis
Lib 15h40'43" -17d17'50"
Melis
Ari 2h4'45" 21d54'8"
Melis & Ali Deha Otmar
Cyg 21h57'4" 53d48'9"
Melisa
Lyn 7h54'52" 54d54'55"
Melisa
Ori 5h22'55" 13d26'30"
Melisa A. Odzic
Psc 0h49'8" 20d51'58"
Melisa Ann
Cas 0h7'7" 54d14'17"
Melisa D. Salandy
Peg 22h16'43" 33d43'41"
Melisa Garay
Lib 15h4'48" -6d42'26"
Melisa Hansson
Cam 4h29'18" 54d43'24"
Melisa Jean
Psc 0h36'8" 8d22'13"
Melisa Jernigan Garay
Tri 2h38'40" 34d3'13"
Melisa Marie *MaryJane*
Umi 14h10'9" 72d41'55"

Melisa Martin-Cook, the Honeybee
Cnc 8h17'0" 10d1'53"
Melisa My Love
Ari 2h50'51" 27d7'2"
Melisa Newman
Cyg 20h38'43" 35d50'30"
Melisa Reilly - The Star of Melisa
Peg 22h32'30" 20d51'57"
Melisa Rose
Uma 11h54'25" 36d11'0"
Melisa Sambath
Ori 5h27'43" 14d5'14"
Melisa, My Love, My Sweetheart
Lib 15h6'41" -1d44'52"
Melisa, my one and only
Sgr 18h47'19" -27d25'59"
Melisean 5-8-68
Vul 19h2'6" 23d17'12"
Meliss
Aqr 21h39'16" -1d18'39"
Meliss "Love Always"
Aqr 22h31'55" -1d53'31"
Melissa
Aqr 22h29'4" -0d5'54"
Melissa
Vir 14h8'56" -3d53'56"
Melissa
Aqr 23h4'5" -17d4'13"
Melissa
Cap 21h4'37" -17d58'45"
Melissa
Lyn 6h38'51" 56d52'48"
Melissa
Uma 10h54'13" 57d9'2"
Melissa
Uma 12h32'44" 52d53'55"
Melissa
Uma 13h52'30" 56d29'56"
Melissa
Cas 0h37'41" 61d59'22"
Melissa
Cas 0h42'46" 65d9'34"
Melissa
Cas 1h45'35" 63d38'4"
Melissa
Sco 17h32'43" -33d13'58"
Melissa
Cru 12h28'17" -59d34'28"
Melissa
Sgr 19h14'28" -33d51'32"
Mélissa
Peg 21h36'57" 27d17'7"
Melissa
And 0h30'25" 32d52'54"
Melissa
Lmi 10h39'4" 25d16'2"
Melissa
Leo 10h8'13" 27d36'2"
Melissa
Leo 10h8'58" 24d12'9"
Melissa
Cnc 9h21'3" 26d58'48"
Melissa
Tau 5h28'58" 26d21'33"
Melissa
Tau 6h0'23" 23d20'21"
Melissa
And 0h19'57" 26d33'39"
Melissa
Ari 2h42'45" 26d36'29"
Melissa
Ari 2h27'35" 27d4'53"
Melissa
And 0h50'53" 24d5'54"
Melissa
Psc 1h14'56" 26d43'53"
Melissa
Gem 6h24'59" 21d32'24"
Melissa
Leo 10h10'46" 15d50'6"
Melissa
Cnc 9h8'44" 15d22'17"
Melissa
Ari 3h19'11" 11d57'31"
Melissa
Aqr 22h32'52" 0d16'2"
Mélissa
Peg 22h42'38" 3d36'15"
Melissa
Aqr 21h12'7" 1d23'0"
Melissa
Aqr 21h46'36" 0d45'0"
Melissa
Cnc 8h19'49" 14d6'38"
Melissa
Leo 10h34'48" 14d34'58"
Melissa
Leo 9h38'20" 11d29'50"
Melissa
Psc 1h6'17" 16d3'19"
Melissa
Ari 1h58'11" 22d13'10"
Melissa
Ari 2h42'16" 20d19'19"
$Melissa$
Tau 4h15'47" 16d3'24"
Melissa
Tau 4h30'48" 21d2'57"
Melissa
Tau 5h7'25" 17d20'14"
Melissa
Ori 6h10'11" 16d29'22"

Melissa
Her 17h38'26" 36d2'14"
Melissa
And 2h6'44" 35d50'46"
Melissa
Per 3h37'4" 33d54'2"
Melissa
And 1h3'30" 33d46'46"
Melissa
And 0h28'24" 37d16'4"
Melissa
Gem 7h49'39" 33d1'56"
Melissa
Gem 7h41'57" 33d41'53"
Melissa
Cnc 9h12'48" 31d56'29"
Melissa
And 0h22'37" 37d56'34"
Melissa
And 0h46'32" 40d46'9"
Melissa
And 2h29'15" 44d39'17"
Melissa
And 2h31'12" 43d19'41"
Melissa
And 2h31'57" 44d11'50"
Melissa
Uma 10h40'55" 41d8'42"
Melissa
Cam 3h52'1" 57d7'10"
Melissa
Cas 1h31'34" 57d51'45"
Melissa
And 23h19'43" 47d27'48"
Melissa
And 23h18'14" 48d26'27"
Melissa
Cyg 21h13'31" 50d50'13"
Melissa
Crb 15h47'2" 38d39'0"
Melissa
Crb 16h0'0" 39d9'44"
Melissa
Uma 8h38'38" 47d15'55"
Melissa
Lyn 7h21'46" 45d48'41"
Melissa 15
Vir 12h40'13" 2d55'7"
Melissa 28
Cas 0h31'23" 50d26'53"
Melissa 42
Pho 0h41'26" -51d11'43"
Melissa A. Benson
Sgr 18h50'50" -31d59'49"
Melissa A Bortmess
Lib 14h49'1" -0d38'52"
Melissa A. Garrity
And 0h26'9" 44d4'56"
Melissa A. Labadie
Sgr 19h49'46" -14d8'56"
Melissa A Milton
Lib 14h51'18" -3d41'57"
Melissa A. Owens
Cas 2h17'44" 67d45'49"
Melissa A. Shirley
Leo 9h29'42" 19d55'51"
Melissa Acosta
Leo 9h31'9" 16d59'26"
Melissa Agnes Ruppart
Vir 11h56'56" 8d37'21"
Melissa Alexandria
Ari 1h59'17" 12d45'54"
Melissa & Alexis
And 1h31'12" 44d35'43"
Melissa Allison Alpha
Omega One
Ari 1h54'38" 17d51'23"
Melissa Alpers
And 23h8'14" 48d40'57"
Melissa Amber Cabrera
Ori 6h16'13" 7d10'22"
Melissa Amber Cecil
Vir 12h15'46" 12d58'4"
Melissa Amber Ortiz
Uma 11h33'19" 54d21'31"
Melissa Amin
Sgr 18h31'26" -16d15'54"
Melissa and Andrea
Forever
Cnc 8h17'57" 11d2'10"
Melissa and Bobby
Sgr 17h57'37" -17d53'52"
Melissa and Brendan's
Per 4h47'17" 45d44'44"
Melissa and Charles
Cyg 21h23'1" 46d8'37"
Melissa and Chris 2006 -
Forever
Uma 10h9'25" 46d49'59"
Melissa and Christian
Ulanch
Sge 19h40'24" 17d43'9"
Melissa and Doug
Mon 6h46'15" 6d30'34"
Melissa and Elvin Valmores
Cas 0h12'48" 56d45'45"
Melissa and Gary's
Cyg 20h51'15" 38d8'9"
Melissa and Greg
Cyg 20h11'44" 34d44'35"
Melissa and Joey
Lyn 7h59'3" 40d25'42"
Melissa And Kevin's Star
Sco 16h57'46" -38d37'22"

Melissa and Martin
Together Forever
Ori 5h25'54" 14d52'24"
Melissa and Michael Morelli
Aqr 22h58'30" -10d24'0"
Melissa and Neal's Shining
Devotion
Uma 8h45'14" 64d17'21"
Melissa and Ted
Uma 11h24'38" 71d48'40"
Melissa and Vince
Gem 7h23'5" 29d27'49"
Melissa Anderson
Vir 13h1'27" -19d57'54"
Melissa Andriano(Missa)
Cap 20h27'51" -26d24'58"
Melissa Ann
Leo 11h24'7" -1d4'23"
Melissa Ann
Uma 9h12'57" 67d5'59"
MELISSA ANN
Tau 3h59'44" 28d41'37"
Melissa Ann
Ari 2h43'46" 14d52'52"
Melissa Ann
Lyn 9h34'34" 40d33'27"
Melissa Ann
Cnc 9h20'11" 31d6'16"
Melissa Ann
And 23h8'13" 50d45'5"
Melissa Ann
Cam 4h37'27" 59d10'50"
Melissa Ann Bain
Mon 7h4'7" -5d17'55"
Melissa Ann Bohli
And 1h35'11" 45d51'20"
Melissa Ann Borrego
Leo 10h38'33" 18d0'47"
Melissa Ann Elizabeth
Zahralban
Cyg 19h35'10" 29d50'56"
Melissa Ann Elmer
Gem 6h50'41" 13d33'43"
Melissa Ann Garcia
Leo 9h45'44" 12d49'18"
Melissa Ann Gianelle
Psc 1h8'26" 4d14'33"
Melissa Ann Gillette
And 0h33'24" 27d57'55"
Melissa Ann Guariglia
And 1h15'54" 38d17'22"
Melissa Ann Homiski
Cap 21h45'52" -14d17'12"
Melissa Ann Hutcheson
Cap 20h33'29" -11d2'27"
Melissa Ann Lingardo
Psc 1h18'11" 10d9'32"
Melissa Ann Lopes
Crb 15h51'12" 36d43'26"
Melissa Ann Lynn Ringstaff
Uma 13h5'6" 59d37'41"
Melissa Ann Manglaviti
Cas 1h36'51" 62d2'25"
Melissa Ann Manriquez
Cas 1h20'38" 55d52'53"
Melissa Ann Martin
Leo 9h45'40" 29d22'58"
Melissa Ann McGee
Lib 15h12'49" -9d58'3"
Melissa Ann Medina
Lyn 7h57'33" 53d58'26"
Melissa Ann Mills
Sgr 19h34'31" -16d33'10"
Melissa Ann Mohler
Lib 15h12'48" -26d35'17"
MELISSA ANN MORONEY
Dra 18h57'2" 55d30'28"
Melissa Ann My Sunflower
Vir 12h44'53" -4d53'51"
Melissa Ann Newland
Sco 17h48'37" -40d13'6"
Melissa Ann Renshaw
Ori 6h14'28" 15d3'27"
Melissa Ann Rybarczyk
Sgr 19h14'4" -26d4'11"
Melissa Ann Schaffter
Gem 7h19'34" 14d34'20"
Melissa Ann Taylor
Eri 1h48'37" -53d36'53"
Melissa Ann Thorbahn
And 0h41'13" 36d6'30"
Melissa Ann Velasquez
Cyg 20h24'58" 47d36'44"
Melissa Ann Voeltz Rowlett
Sco 16h10'8" -16d18'47"
Melissa Ann Williams
And 0h19'28" 29d43'7"
Melissa Anne
Cyg 20h19'0" 33d49'27"
Melissa Anne
Eri 4h30'4" -0d48'14"
Melissa Anne
Uma 8h58'57" 56d51'8"
Melissa Anne "05/17/1974"
Tau 5h36'23" 26d28'16"
Melissa Anne Arsenault
Sco 16h35'21" -36d0'5"
Melissa Anne Biederman
Cnc 9h6'57" 10d41'59"
Melissa Anne Kraynak
Lib 15h13'27" -6d45'22"
Melissa Anne Meador
And 0h46'10" 44d26'45"
Melissa Anne Moffat
Vir 14h27'28" -0d8'28"

Melissa Anne Morgan
Cnc 8h3'41" 22d32'10"
Melissa Anne Riley
Tau 3h37'9" 18d2'33"
Melissa Anne Sasaoka
Cas 23h57'14" 57d34'5"
Melissa Anne Silva
Mon 7h16'28" -0d45'0"
Melissa Anne Solomon
Crb 15h36'15" 38d39'43"
Melissa Anne Thies
And 23h23'50" 48d40'15"
Melissa Anne Tooker
And 23h20'54" 46d18'39"
Melissa Ann's Star
Cap 20h43'54" -26d29'24"
Melissa Armas Arciniega
Cap 21h37'43" -16d3'11"
Melissa Ashley Wilson
Mon 6h44'57" -0d23'22"
Melissa Ashleyanne
Stewart
Lyn 6h57'52" 60d26'58"
Melissa ATCLMJ
Gem 6h44'15" 33d15'4"
Melissa B Mahoney
Vir 13h9'45" -1d11'31"
Melissa "Babedoll"
Vir 14h13'22" 3d4'14"
Melissa "Baby Girl" Renda
Ari 3h28'29" 27d9'16"
Melissa Baby King
Cap 21h13'54" -18d23'11"
Melissa Barabasch
Mon 6h49'43" -0d57'9"
Melissa Barnickel
Cyg 19h21'4" 29d59'33"
Melissa Bell
Psc 0h44'23" 14d7'6"
Melissa "Bella" Valera
Cap 20h22'28" -12d17'18"
Melissa & Ben Taylor
Cyg 20h41'43" 34d33'51"
Melissa Benner
Lyn 8h16'17" 34d45'50"
Melissa Bennett
Ari 3h18'57" 21d4'30"
Melissa Beth
And 0h49'21" 35d31'51"
Melissa Bickel
Mon 6h48'24" -0d31'28"
Melissa Billings
And 2h13'33" 38d14'5"
Melissa Bioy
Uma 10h0'42" 53d53'29"
Melissa Bohn
Lib 14h50'43" -0d52'34"
Melissa Born to Ed & Aleta
Tau 4h35'45" 28d11'5"
Melissa Boroski
Cnc 8h44'40" 22d11'43"
Melissa Bradley
Cam 3h16'28" 60d9'3"
Melissa Bradly
Vir 13h14'51" 11d14'58"
Melissa Bruns
Ari 3h29'13" 27d25'44"
Melissa Burras
Cnc 9h4'40" 30d48'19"
Melissa C.
Cas 0h9'59" 56d46'7"
Melissa C. Castro
Dra 17h14'17" 58d44'44"
Melissa Cairo
Ari 3h19'42" 26d22'22"
Melissa Camacho
Aqr 21h8'1" 0d46'31"
Melissa Carmen Vanessa
Hopps
Vir 13h32'7" -4d42'49"
Melissa Carol
Hya 8h58'14" -2d53'20"
Melissa Caroline Lockwood
Cam 4h51'31" 52d48'10"
Melissa Carro
Uma 11h8'4" 41d3'17"
Melissa Cartiglia
And 23h18'21" 46d53'36"
Melissa Cascia
Aqr 22h13'6" 0d52'57"
Melissa Casey
Cap 21h47'44" -12d48'36"
Melissa Charnisky
Crb 16h15'28" 39d33'31"
Melissa Chavarria
Cas 1h47'55" 64d57'2"
Melissa Christine
Lib 14h52'37" -8d46'53"
Melissa Christine Wall
Uma 11h42'58" 35d40'9"
Melissa Cioppa
Vir 12h5'8" -4d32'35"
Melissa Clark
Sco 16h4'18" -9d59'35"
Melissa Collins
Aqr 22h36'12" 0d21'34"
Melissa Cona
And 1h34'15" 49d54'40"
Melissa Cone
Psc 1h4'36" 10d26'33"
Melissa Conner
Mon 6h42'26" 5d56'10"
Melissa Constantiner
Vir 13h37'29" -12d33'56"

Melissa Conway
Ori 6h14'35" 15d7'47"
Melissa Cooper
And 2h25'39" 39d20'46"
Melissa Corazon Sevilla
Vir 13h25'34" 8d20'58"
Melissa Cordelia Hull
Leo 10h9'7" 12d59'12"
Melissa Corinne omnia
vincit amor
Cru 12h16'27" -57d6'53"
melissa cornelia
Lyr 19h11'9" 34d36'48"
Melissa Cortes
And 0h35'40" 43d49'37"
Melissa Cortes
Cas 2h13'27" 62d20'48"
Melissa Costello
And 1h5'9" 41d33'6"
Mélissa Cournot
Crb 16h1'24" 28d5'19"
Melissa Cristine Stayton
Aqr 22h15'18" 1d32'26"
Melissa Crulli
Cap 20h15'14" -22d4'35"
Melissa Cruz
Sco 16h23'58" -41d16'52"
Melissa Crystal Leaman-
Hentgen
Psc 23h40'4" 2d11'1"
Melissa Cuneo
Cas 0h53'56" 61d28'56"
Melissa D. Graske
Ari 2h48'22" 15d14'49"
Melissa D. Haro
Ori 5h58'45" 20d58'58"
Melissa D. Vantrease
Aqr 22h36'23" -0d30'17"
Melissa DaLuz
Uma 11h36'24" 35d17'5"
Melissa & Daniel
Lyr 18h34'37" 35d49'58"
Melissa & Daniel Berger
Lyr 18h22'4" 39d20'19"
Melissa Danielle McLeod
Aqr 23h38'50" -6d55'7"
Melissa Daniels
And 0h45'33" 37d40'59"
Melissa & David
Cap 21h2'21" -27d21'27"
Melissa & David Mason
10th. Ann.
Cyg 22h1'44" 51d15'39"
Melissa Davidson
Sgr 17h54'2" -29d54'40"
Melissa Dawn
Uma 11h29'41" 63d22'31"
Melissa Dawn Austin
Sgr 18h3'54" -20d50'33"
Melissa Dawn Hayes
Vir 14h5'52" -15d59'47"
Melissa Dawn Martin
Vir 13h54'50" -11d13'9"
Melissa Dawn Seaver
Sco 16h13'3" -11d6'12"
Melissa DeAnn Davis-Doll
Tau 5h9'46" 17d59'23"
Melissa Deanne Gamez
Williams
Gem 7h29'6" 14d7'10"
Melissa DeBernardo
Cyg 21h8'19" 54d8'20"
Melissa DeJesus
Cas 1h41'25" 61d14'21"
Melissa delaMerced
And 1h16'14" 44d52'28"
Melissa Delgrande
Per 3h16'25" 42d46'10"
Melissa Denise Ruhberg
Mic 21h8'50" -39d44'4"
Melissa Denton
Crb 15h22'24" 26d49'50"
Melissa Desantos
And 23h6'58" 51d17'15"
Melissa Diane
Lyn 7h41'56" 39d57'11"
Melissa Diane
Sgr 19h11'33" -22d15'37"
Melissa Diane Herald
And 1h11'1" 39d58'3"
Melissa Diane Suarez
Lyn 7h21'22" 53d12'2"
Melissa Dills
Lyr 19h7'32" 27d11'7"
Melissa & Donny's Love
Umi 17h24'7" 78d57'38"
Melissa Dooley
Lib 15h8'4" -16d25'10"
Melissa Dorrell
Ori 5h38'18" 0d59'54"
Melissa & Doug
And 1h1'52" 46d38'58"
Melissa E Smith
And 0h45'28" 22d40'25"
Melissa Elaschuk
Umi 16h10'36" 72d52'43"
Melissa Ellen Johnson
Lyn 6h58'29" 47d47'24"
Melissa Elliott
Lyn 7h20'18" 52d32'48"
Melissa Ellison
And 1h13'4" 45d34'14"
Melissa Ellsworth
Cyg 21h41'33" 43d0'39"

Melissa Emerita Suder
Uma 8h20'37" 60d9'43"
Melissa Emily Bellere
Sco 16h13'41" -10d37'40"
Melissa Emily Jennings
Lib 14h40'33" -11d32'49"
Melissa & Erik
Sgr 18h5'19" -30d0'14"
Melissa Estoesta Mallorca
Sgr 19h59'45" -25d23'30"
Melissa Estrada
Ori 6h15'41" 14d59'2"
Melissa Eva Barroso
Ori 5h5'35" -1d18'19"
Melissa F. Mirzaian
And 23h15'51" 42d11'20"
Melissa Faughnan
Lyn 9h9'6" 44d58'23"
Melissa Faye Bear
And 0h41'42" 40d11'21"
Melissa Fernandez's Star
of Aquarius
Aqr 20h39'38" -12d26'10"
Melissa Ferraioli
And 0h8'30" 41d13'56"
Melissa Fisher 26-10-1970
Cru 12h18'9" -58d40'5"
Melissa Fitta
Crb 16h8'18" 39d2'11"
Melissa Fowner
Ari 2h58'35" 14d0'11"
Melissa Fragen
And 0h9'11" 44d0'3"
Melissa Frances James
And 0h14'52" 28d53'14"
Melissa Frances' Way
And 0h45'14" 31d32'16"
Melissa & Franklin
Tau 4h1'51" 26d1'14"
Melissa Free
Cas 23h22'46" 59d56'27"
Melissa Fruge' Marquez
Cyg 19h46'30" 37d15'28"
Melissa G
Tau 4h39'37" 22d15'54"
Melissa G. Hardcastle
Lyr 18h48'3" 30d10'19"
Melissa Gail
Uma 11h53'29" 42d38'39"
Melissa Gail Hunt
Ari 2h58'0" 29d43'2"
Melissa Gaiser
Peg 22h59'32" 14d13'40"
Melissa George
Cyg 19h31'56" 29d16'32"
Melissa Godshaw
Sco 16h51'31" -42d40'59"
Melissa Goldstein
Vul 20h33'9" 24d19'20"
Melissa Grace
Ari 2h50'7" 25d45'25"
Melissa Grace
Lyn 6h22'45" 57d41'27"
Melissa Green " Wissa
Sue"
Tau 4h15'38" 7d40'34"
Melissa Guerrero "Boo"
Aqr 22h10'49" -13d21'0"
Melissa Guerreso
Cyg 21h55'25" 50d18'59"
Melissa Halvorsen
Uma 14h6'26" 55d20'37"
Melissa Harrington
Aqr 20h45'42" -10d8'35"
Melissa Harshaw
Lyn 7h23'22" 52d8'36"
Melissa Heather Sobel
Psc 0h39'2" 19d24'31"
Melissa Heinzel
Tau 4h28'19" 15d12'24"
Melissa Herbert Mills
And 23h2'23" 46d21'27"
Melissa Hermida
Sco 16h46'25" -33d38'6"
Melissa Hess
And 0h45'51" 31d12'14"
Melissa Hielman
Tau 5h45'25" 22d6'55"
Melissa Holman - GaMPI
Shining Star
And 0h55'15" 36d7'2"
Melissa - Howie
Vir 12h12'46" -0d58'11"
Melissa Hubner-Augie
Boehm
Sco 16h12'43" -13d47'15"
Melissa I. Smallwood
Cam 4h21'45" 68d21'31"
Melissa Ille
Vir 13h19'3" 5d55'22"
Melissa Irene Gonzalez
Tau 4h19'2" 14d45'53"
Melissa J Casterton Sweet
16
And 23h41'38" 48d6'47"
Melissa J. Gerlach
Oph 16h41'28" -0d22'7"
Melissa J. Gilfoyle
Ori 5h36'11" 6d31'15"
Melissa J. Heeg
Sco 17h18'2" -41d22'9"
Melissa J.A. Golab
Lyr 19h9'21" 26d32'4"
Melissa Jameson
Cas 0h22'52" 56d25'34"

Melissa Jane
Vir 12h0'58" -3d33'31"
Melissa Jane Falco
Sgr 19h1'18" -20d46'36"
Melissa Jane Kelly
And 22h59'26" 51d27'30"
Melissa Jane Kopec
Lyn 7h50'52" 58d7'17"
Melissa Jane Pisani
Cru 12h13'56" -62d2'1"
Melissa Jane Thorp (Missa)
Dra 20h22'52" 68d4'42"
Melissa Jean
Uma 11h7'11" 67d3'53"
Melissa Jean
Gem 7h17'29" 24d1'41"
Melissa Jean Farlow
Lib 15h1'23" -15d47'53"
Melissa Jean Hopwood
Cru 12h19'25" -56d45'57"
Melissa Jean Thomas
Mon 7h39'34" -0d41'59"
Melissa Jean Walsh
Cap 21h14'35" -20d10'23"
Melissa Jeanette Kramer
Leo 11h12'35" 15d26'56"
Melissa/Jenny
And 0h13'21" 42d44'16"
Melissa Jett
Vir 14h19'10" -6d54'19"
Melissa Jo Howard
And 0h39'52" 38d16'26"
Melissa Jo Smith
And 1h10'21" 38d34'5"
Melissa Jo Wyeth
Ori 5h53'22" -0d46'14"
Melissa & Joël
Cyg 19h36'11" 27d55'2"
Melissa & John
Psc 1h0'56" 4d2'15"
Melissa Johnson
Uma 11h20'52" 59d41'41"
Melissa Joshua Nathanel
Laliberte
Peg 21h57'46" 15d51'58"
Melissa Joy Detwiler
Leo 11h33'34" 19d10'8"
Melissa K Aarskaug
Aqr 22h46'36" 2d38'55"
Melissa K Aarskaug
Aqr 22h52'58" -7d33'40"
Melissa K. Heard
Cas 0h26'18" 53d25'28"
Melissa K. Hendrix Olken
Aqr 22h37'34" 0d30'24"
Melissa K Lovett
Gem 7h4'42" 17d8'14"
Melissa K. Morris
Leo 9h51'17" 32d35'13"
Melissa K Vaught
Psc 0h9'17" 7d51'21"
Melissa Karen Jagdeo
Sco 16h9'13" -21d2'13"
Melissa Karpecki
Ari 2h53'45" 17d49'14"
Melissa Katharine Dafoe
Aqr 22h41'28" -19d29'5"
Melissa Kathryn Campbell
Cap 21h30'45" -15d6'3"
Melissa Kay Schulze
Cas 0h43'31" 60d48'35"
Melissa Kaye Harr
Del 20h31'21" 6d7'47"
Melissa Kazimier
Cnc 8h18'3" 11d5'58"
Melissa Kedaitis
Sgr 19h25'17" -15d38'57"
Melissa Kendall
Sco 16h48'42" -27d23'22"
Melissa Kennedy
Cnc 8h50'14" 26d50'37"
Melissa Kettler
And 0h38'34" 26d26'48"
Melissa Kiernan
Gem 7h59'16" 30d7'13"
Melissa Kile
Cas 0h45'46" 57d9'51"
Melissa Kimberly Dye
Uma 11h55'5" 53d16'52"
Melissa King
Dra 17h52'47" 50d55'53"
Melissa Kirk-Singer
Peg 23h17'47" 34d6'24"
Melissa Kment
Cyg 21h16'48" 42d11'42"
Melissa Knowlson
Ori 6h11'51" 9d19'55"
Melissa Kochsiek
Cnc 8h20'1" 25d19'2"
Melissa Kraft
And 2h8'6" 45d56'3"
Melissa L. Davis
Leo 10h29'48" 8d9'49"
Melissa L. Mann
Leo 10h25'55" 26d0'53"
Melissa L. "Missy"
Cas 0h1'42" 56d59'2"
Melissa L Watson
Cyg 21h6'59" 53d11'35"
Melissa Lai
Gem 7h9'58" 34d2'41"
Melissa Lanzaro
Lyn 7h28'15" 50d13'57"
Melissa Laura Wierzbowski
Sgr 19h34'17" -31d35'32"

Melissa Lauren Smith
Ori 5h29'57" 4d9'54"
Melissa Lauren Verges
And 0h46'4" 37d21'6"
Melissa LaVonne Malone
Mon 8h8'24" -2d58'8"
Melissa & Lawrence
And 1h14'56" 45d35'58"
Melissa Lee
Ori 5h29'54" 10d5'45"
Melissa Lee
Ari 2h31'15" 24d5'16"
Melissa Lee
Aqr 20h43'39" -2d20'21"
Melissa Lee Chalfant
Cas 0h33'20" 52d48'58"
Melissa Lee Jameson
Leo 10h31'42" 14d40'11"
Melissa Lee Kading
Gem 7h17'37" 25d49'4"
Melissa Lee Krull
Cnc 8h5'45" 19d41'22"
Melissa Lee Loydall
Aqr 23h11'6" -16d3'59"
Melissa Lee Neece
Srp 18h11'15" -0d27'51"
Melissa Leigh
Cas 1h3'12" 60d57'9"
Melissa Leigh Honeycutt
Cyg 20h36'38" 41d8'12"
Melissa Leigh Morrison
Hatter
And 23h44'52" 34d12'5"
Melissa Leilia Curtis
Umi 14h46'53" 69d29'59"
Melissa Lieng
Lib 15h53'6" -4d12'13"
Melissa Lin
Psc 1h26'13" 27d52'6"
Melissa Loder
Vul 20h31'59" 27d57'22"
Melissa Long
Cyg 20h4'31" 38d41'22"
Melissa Lopez
Mon 7h40'34" -8d58'54"
Melissa Lopez
Lib 15h41'58" -12d50'3"
Melissa Loprete
Cas 1h13'49" 64d46'25"
Melissa Loves Adam
Lyn 7h33'12" 40d25'27"
Melissa Loves Jeff
Sge 20h3'50" 16d59'4"
Melissa Lucy
And 23h20'34" 44d27'38"
Melissa LueAnn
And 1h2'7" 42d15'53"
Melissa Lynn
Cnc 8h44'14" 32d19'0"
Melissa Lynn
And 1h47'14" 50d8'59"
Melissa Lynn
Tau 4h5'26" 7d0'20"
Melissa Lynn
Ori 4h49'56" 4d58'47"
Melissa Lynn
Sco 16h10'32" -16d1'5"
Melissa Lynn Chacón
Lib 14h58'45" -2d11'47"
Melissa Lynn Marron
Tau 5h41'18" 26d11'6"
Melissa Lynn Mumbach
Cap 20h33'37" -11d1'32"
Melissa Lynn O'Neal
Gem 7h14'36" 26d10'42"
Melissa Lynn Smith
Gem 6h53'32" 14d55'5"
Melissa Lynn Weathers
Leo 10h6'19" 21d49'7"
Melissa Lynne Jacoby
Aqr 21h52'14" -4d47'54"
Melissa Lynne Metcalfe
Vir 13h38'59" 2d38'39"
Melissa M. Gattuso
Gem 6h59'49" 31d23'26"
Melissa M. Harvey
Crb 16h8'43" 26d57'57"
Melissa M. Roome
Ori 5h16'15" 15d57'28"
Melissa M. Therien
Lyn 9h30'34" 40d37'38"
Melissa Madeline Mader
Cyg 20h34'32" 35d41'14"
Melissa Mae
Gem 6h42'54" 20d59'16"
Melissa Mae Matthews
And 1h14'22" 36d31'12"
Melissa Mae Thomason
Ori 5h16'49" -9d15'59"
Mélissa Maës
Uma 9h48'27" 43d14'28"
Melissa Maivor
Cap 20h32'41" -19d18'1"
Melissa Maree Lawrence
Psc 23h12'12" 0d26'49"
Melissa Maria
Cyg 21h23'7" 55d1'1"
Melissa Marie
Aqr 21h44'6" -1d56'35"
Melissa Marie
Del 20h47'42" 18d59'57"
Melissa Marie
Leo 11h23'18" 25d20'32"
Melissa Marie
Lmi 9h47'7" 39d1'7"

Melissa Marie
And 0h36'54" 41d21'35"
Melissa Marie
Crb 15h42'34" 35d20'40"
Melissa Marie Bracken
Gem 6h38'34" 19d59'50"
Melissa Marie Duchsherer
Cas 0h42'32" 49d7'6"
Melissa Marie James
Lib 15h19'54" -10d13'33"
Melissa Marie Knopp
Lyn 7h35'4" 49d18'1"
Melissa Marie Lewis
Sco 15h55'9" -20d27'7"
Melissa Marie Lopez
Psc 0h39'49" 17d44'31"
Melissa Marie McDonald
And 1h48'37" 47d31'45"
Melissa Marie Neveroski
Uma 10h22'45" 52d21'45"
Melissa Marie O'Connor
And 1h10'56" 46d12'16"
Melissa Marie Pluta "My Baby"
Psc 0h47'14" 12d17'30"
Melissa Marie Roggenkamp
Cam 4h37'10" 58d2'37"
Melissa Marie Roman
Leo 10h54'42" 15d33'47"
Melissa Marie Smith
Dra 18h34'2" 72d33'52"
Melissa Marie Sunnenberg
And 0h14'42" 38d29'19"
Melissa Marii Griffin
Tau 5h49'12" 25d40'51"
Melissa Mariko Kinoshita
Tau 5h40'4" 24d34'33"
Melissa Marin-Hunter
Ori 5h47'45" 6d19'48"
Melissa Marko
Cru 12h49'40" -64d24'24"
Melissa Martin ~ My Best Friend !
And 23h16'13" 45d4'51"
Melissa Mary Espinosa
And 23h25'35" 47d2'48"
Melissa Maticka
Cnv 13h38'15" 37d2'47"
Melissa Matthews
Lyr 18h19'40" 31d13'20"
Melissa Maurine Gilbert
And 0h43'4" 44d22'29"
Melissa May
Crb 16h10'15" 36d27'54"
MELISSA MAY
Ori 5h33'26" 4d20'13"
Melissa May Marie
And 2h4'25" 40d23'25"
Melissa McCarthy and Matthew Leo
Cyg 20h21'13" 55d33'27"
Melissa McClintock and Robb Myrtle
Uma 11h1'25" 62d21'20"
Melissa McDermott
Cas 1h41'6" 62d42'44"
Melissa McGrath
Uma 11h15'31" 63d29'32"
Melissa Mckellar
Her 17h53'54" 27d20'59"
Melissa McKim
Crb 15h42'39" 29d32'57"
Melissa Meisje (Infinity)
And 0h1'44" 44d7'18"
Melissa Melini Trexler
Cap 21h2'12" -20d57'21"
Melissa Mello
Tau 5h14'48" 16d45'7"
Melissa "Melly" Alston
And 1h45'42" 45d4'26"
Melissa Mendoza
Vir 12h45'15" 4d30'6"
Melissa Mica
Dra 17h14'23" 67d46'49"
Melissa Michalek
Aqr 23h24'50" -12d34'5"
Melissa (Missa)
Ori 5h46'35" 7d53'35"
Melissa "Missy" Marie Hlavacik
Cas 23h25'14" 58d16'40"
Melissa Mitchel
Uma 13h17'51" 58d10'22"
Melissa Mitchell
Cnc 8h24'31" 25d18'5"
Melissa "Mo" O'Brien 1950-2006
Vir 12h22'56" -10d39'41"
Melissa Mo-Mo McAuliffe
Psc 0h2'54" 5d36'0"
Melissa Monique Morataya
Sgr 19h18'45" -23d59'1"
Melissa Moo Nolan
Leo 11h22'3" 15d17'25"
Melissa Morgan
Cas 1h21'5" 58d36'53"
Melissa Morgan
And 1h24'58" 43d58'24"
Melissa Motta
Sco 16h7'23" -14d47'31"
Melissa Mussmann
Lyn 8h0'1" 33d54'11"
Melissa My Heart
Ari 3h19'5" 28d45'4"
Melissa My Love
Ari 1h55'44" 20d40'14"
Melissa "My Shining Star"
And 23h24'31" 38d28'49"
Melissa " My Tinkerbell" Poche
Gem 7h36'49" 27d43'30"
Melissa Myers
Cas 1h46'19" 67d56'32"
Melissa N. Bolton
Vir 13h35'57" -20d35'33"
Melissa "Nana" Thorsby
Cas 23h22'44" 59d33'27"
Melissa Nash
And 1h26'34" 45d2'40"
Melissa Nelson
And 23h51'1" 41d21'50"
Melissa Nguyen
Dra 16h13'16" 59d26'0"
Melissa Nguyen
Vir 13h17'19" -19d4'50"
Melissa Nichol
And 0h40'34" 45d27'30"
Melissa Nicholson
And 1h47'46" 46d47'19"
Melissa Nicole
Psc 23h50'27" 1d0'8"
Melissa Nicole Elizabeth Provenzano
Gem 6h56'34" 22d52'24"
Melissa Nicole, my one and only
Cnv 13h45'56" 30d10'59"
Melissa Niemeyer
Ori 5h53'30" -0d46'11"
Melissa Noe Sterling
Psc 23h47'58" 5d7'23"
Melissa Noriega
And 0h45'3" 42d4'34"
Melissa Odette
Cru 12h43'35" -64d6'3"
Melissa Ondrey
Cap 20h44'22" -20d32'0"
Melissa P. Mchenry Pirolli
Cnc 8h42'19" 15d53'30"
Melissa P. Rae
Leo 11h19'0" -1d1'42"
Melissa Pappas
And 0h38'36" 31d24'35"
Melissa Paredes
And 2h11'23" 41d18'43"
Melissa Pascoe
Cas 0h31'37" 52d59'51"
Melissa Paul
Lyn 7h3'30" 51d11'28"
Melissa Pawson
Gem 7h32'55" 24d40'48"
Melissa PC Greenwood
Lib 15h7'19" -1d46'52"
Melissa Pease
Aqr 20h39'16" -8d33'1"
Melissa Perkins
Sco 16h40'0" -29d5'56"
Melissa Perrotti
And 23h16'18" 47d26'11"
Melissa Petrie
Aqr 20h55'0" -11d54'55"
Melissa Pierce
Gem 7h48'51" 29d34'59"
Melissa Pineda
Gem 7h17'13" 19d6'35"
Melissa Pinger
Lyn 9h6'23" 35d54'40"
Melissa Poghossian
Tau 4h8'41" 8d1'24"
Melissa Pohlmann
Crb 16h11'27" 33d59'48"
Melissa Polonsky
Lib 15h57'18" -6d44'21"
Melissa Pordon
Vir 13h56'40" -1d37'46"
Melissa "Princess" Rodriguez
Sco 16h22'33" -16d51'7"
Melissa "Pumpkin" Keith
And 0h22'2" 31d25'43"
Melissa R. Crain
Cas 23h52'21" 58d47'36"
Melissa R. Heckmanestan
Gem 7h44'59" 34d6'47"
Melissa R. Sparks
And 0h41'14" 42d58'11"
Melissa Rabaglia - Rising Star
And 2h16'31" 42d36'6"
Melissa Rach
Sgr 18h7'45" -16d23'27"
Melissa Rae
Tau 5h59'20" 22d54'47"
Melissa Rae Finnerty
Uma 10h20'59" 47d18'7"
Melissa Rae Ludlum
Ari 2h37'46" 27d54'20"
Melissa Rae Nevitt
Vir 15h7'30" 5d13'43"
Melissa Raes
Cas 1h47'2" 61d27'2"
Melissa Regina Schmidt
Sgr 19h2'19" -18d13'11"
Melissa Renee'
Cas 1h36'52" 66d4'9"
Melissa Renee
Cap 21h32'11" -22d47'33"
Melissa Renee Golding
Lyn 9h13'5" 38d7'53"
Melissa Renee Kruse
Cyg 20h58'57" 45d48'36"
Melissa Renee Ortiz
And 23h15'14" 40d34'33"
Melissa Renee Patrick
Gem 6h38'43" 14d6'50"
Melissa Reynolds
Sgr 18h11'24" -26d36'15"
Melissa Ricci
Cnc 8h17'35" 19d52'50"
Melissa & Ricky
Sge 20h6'4" 21d3'2"
Melissa Robin Warren
Ari 1h47'50" 23d45'56"
Melissa Roque
Psc 1h9'23" 12d19'54"
Melissa Rose
Leo 11h5'23" 3d13'8"
Melissa Rose
And 1h19'12" 41d32'0"
Melissa Rouchell
Uma 9h11'14" 46d34'28"
Melissa Rummell
Cnc 8h48'37" 30d35'2"
Melissa Ruth
Psc 23h6'31" 5d53'40"
Melissa Ruth Bennett
And 0h47'31" 38d2'30"
Melissa Ryan's Radiance
Leo 11h18'3" 19d11'29"
Melissa S Catala
Uma 9h10'54" 59d2'56"
Melissa S Hand
Sco 17h56'52" -31d2'36"
Melissa S. Marsh
Ari 2h6'32" 25d54'16"
Melissa Sabin
Ori 6h5'51" 10d2'44"
Melissa Sanchez
Crb 16h13'10" 31d48'16"
Melissa Sanders
Oph 17h53'49" -0d27'32"
Melissa Sandra
Lib 14h51'16" -4d5'25"
Melissa Sarah
Tau 4h42'16" 28d33'12"
Melissa Schaffer
Vir 14h13'24" -13d39'4"
Melissa Scoville Corcilius
Ari 3h12'26" 28d54'25"
Melissa & Sebastian Wedding Star
Ara 17h0'23" -58d14'36"
Melissa Shannon Beadling
Lib 15h28'2" -22d30'34"
Melissa Smith
Sco 16h13'44" -10d59'42"
Melissa Smith
Leo 9h50'35" 30d18'33"
Melissa Smits
Cap 20h25'55" -11d12'20"
Melissa Solomon
Tau 3h49'45" 15d48'57"
Melissa Sommer
Ori 6h5'56" 0d59'46"
Melissa Spear
And 1h27'8" 44d31'22"
Melissa Spencer
Tau 4h18'55" 29d32'6"
Melissa Stahl
Gem 6h55'21" 13d32'36"
Melissa Star - 4Ever Shall It Shine
Lyn 8h27'12" 55d18'2"
Melissa STAR of my Life xxgg
Uma 13h57'14" 62d10'29"
Melissa Starr
Sgr 18h42'35" -32d37'7"
Melissa Stenson
Sco 17h13'3" -33d50'0"
Melissa Sterling 5/20/1989
Uma 11h46'52" 54d47'3"
Melissa Stone
Sgr 19h1'20" -28d8'29"
Melissa Stone
Lyr 18h29'39" 36d50'48"
Melissa Strader
Sco 16h40'30" -32d35'57"
Melissa Sue
Tau 4h37'14" 19d5'58"
Melissa Sue Alkema
Cyg 19h50'50" 54d20'39"
Melissa Sue-Mylissa
Psc 23h42'36" 5d49'9"
Melissa (Sunahsah) Mackechnie
Sco 17h26'54" -38d52'45"
Melissa Suyapa Valencia
Vir 14h15'12" 4d37'10"
Melissa Suzanne Snead
Cnc 8h48'23" 32d14'13"
Melissa T. Lespier
Aqr 23h6'21" -8d55'57"
Melissa Tarin
Lyr 19h8'51" 38d3'46"
Melissa Tesoro
Ari 2h10'37" 14d4'14"
Melissa the Bee
Ori 5h52'19" 8d56'22"
Melissa - the one slays me
And 0h42'5" 26d13'39"
Melissa Thompson
Tau 4h13'47" 2d30'19"
Melissa Torres
And 23h18'1" 51d26'57"
Melissa V Ward
And 0h46'36" 31d15'25"
Melissa Vaknin
Vir 12h9'39" 1d46'0"
Melissa Valdez-Griffin
Vir 13h4'50" 0d35'44"
Melissa Ward
And 1h26'38" 37d4'17"
Melissa Welch
And 0h44'6" 27d37'53"
Melissa Wendy Rainville
Lib 15h37'12" -12d2'3"
Melissa White
And 2h19'22" 49d26'13"
Melissa Wilshire
Gem 7h15'12" 19d7'9"
Melissa Wissen
Lyr 18h55'1" 33d33'19"
Melissa Wolfe
Cnc 8h51'53" 11d48'21"
Melissa Y. Arakelian
Ori 5h32'12" -0d29'12"
Melissa y Estefan
Vir 14h43'42" -0d31'55"
Melissa Zaffin and Mike's Star
Sco 17h54'13" -42d40'23"
Melissa "Zoe" Star
Sco 16h6'15" -11d43'55"
Melissa, ESQ
And 2h3'45" 38d15'32"
Melissa25
Lib 14h50'25" -1d52'44"
Melissa3811
Cnc 8h16'26" 30d40'31"
Melissa72
Ari 3h27'18" 20d54'7"
MelissaAulisio
Psc 1h40'55" 23d44'32"
Melissa-Brianna-Ethan-Peter
And 0h14'21" 25d52'53"
Melissah
Tau 3h40'11" 25d54'0"
Melissa..........Little Star
Gem 7h14'37" 14d3'40"
MelissaMeyers&JohnSillery I LoveYou
Tau 4h13'50" 2d56'47"
Melissa-N-Kevin
Mon 7h21'38" -0d29'8"
Melissa's Big Ball of Gas
Dra 18h33'23" 52d12'3"
Melissa's Black Zero 29
Aqr 22h41'35" 1d11'10"
Melissa's Dream
Dra 17h19'14" 62d17'51"
Melissa's Dream Come True
Peg 22h56'26" 26d30'36"
Melissa's Esperanza
Psc 1h6'56" 22d48'1"
Melissa's Estes
Psc 1h32'24" 24d53'14"
Melissa's eye
Cyg 21h17'5" 47d13'16"
Melissa's eyes
Lyn 7h9'35" 55d32'51"
Melissa's L.A.M.B.
Cap 21h36'31" -8d32'34"
Melissa's Love
Ori 6h7'54" 11d49'46"
Melissa's Lucky Star
Lib 15h50'14" -17d53'45"
Melissa's Muse
Cam 4h22'27" 53d48'21"
Melissa's Sean
Lmi 10h5'16" 38d53'39"
Melissa's Shining Star
And 2h26'46" 40d38'40"
Melissa's Shining Star
Ari 3h6'54" 29d36'40"
Melissa's Shining Star
Lib 15h37'4" -27d14'48"
Melissa's Smile
Ori 6h9'35" 3d26'17"
Melissa's Smile
Psc 1h19'33" 32d5'8"
Melissa's Sparkle
Lib 14h51'11" -6d12'57"
Melissa's Star
Uma 8h55'27" 62d21'47"
Melissa's Star
Lib 15h8'37" -28d25'39"
Melissa's Star
Crb 15h51'44" 35d4'15"
Melissa's Star
Crb 15h45'10" 38d1'16"
Melissa's Star
Vir 12h44'14" 2d6'24"
Melissa's Star
Ari 3h20'33" 27d23'30"
Melissa's Wish
And 2h27'15" 44d38'54"
Melissa's Wish
Lyn 6h44'54" 53d47'0"
Melissa's Wishing Star
Lib 15h33'19" -7d17'58"
MelissaScott
Ori 6h16'19" 14d42'27"
MelissaSean
Tri 1h46'18" 32d6'53"
Melissa-Warrior for the Under Dog
Psc 1h35'20" 7d42'22"
Melissa Sue Franz
Tau 5h15'6" 26d29'52"
Melita C. Valandra
Vir 13h16'51" 6d43'20"
Melita Jane Simmonds
Cru 12h42'36" -55d54'0"
Melita's Legacy
Cas 0h11'2" 62d34'28"
Melitta
Sco 16h12'37" -9d11'20"
Meliza
Gem 6h30'54" 22d26'9"
Meliza
Lyr 18h40'7" 37d23'13"
Mella
Sco 16h17'59" -22d23'25"
Mellannie Leanne Brown - Superstar!
Ari 2h36'12" 26d13'47"
melleri
Cap 20h28'46" -10d45'34"
Melli&Sascha
Uma 10h12'3" 48d22'43"
Mellicious Sweets
And 2h32'22" 40d6'58"
Mellie Joy
Cyg 20h54'51" 42d23'15"
Mellie's Star
And 0h43'53" 37d9'5"
Mellie's Star
Cas 2h15'53" 62d34'31"
Mellisa
Lyr 19h9'7" 27d18'13"
Mellisa Ann Watson
Leo 11h5'7" 20d7'7"
Mellisa Blair
Psc 23h45'12" 3d9'7"
Mellisa Leon-Guerrero
Psc 23h12'20" -2d28'46"
Mellisa "Milly" Boytim
Sgr 19h21'4" -23d27'40"
Mellisha Sherise Culpepper
Leo 11h21'47" -4d20'39"
Mellissa Bear
Umi 17h43'53" 80d8'33"
Mellissa Lynn Eby
Lyn 8h38'17" 41d33'9"
Mellissa Simon Elliott Banks
And 1h28'52" 41d54'15"
Mellissa Suzann
Leo 9h22'59" 11d32'53"
Mellon
Vul 19h53'8" 23d0'25"
Mellow
Lyn 8h26'30" 40d58'43"
Mellow Yellow
Tri 2h21'14" 31d54'15"
Melly
Ari 3h6'43" 28d41'17"
Melly Leal
And 0h25'6" 41d33'45"
Melly Lee
Ari 2h47'20" 24d37'47"
Mellybell76
Cnc 8h32'26" 15d19'54"
Mellymel30
And 2h35'35" 48d19'50"
Mellynogs
Aql 19h26'37" 6d52'2"
Melmatt
Cyg 20h19'36" 58d28'3"
Mel-Mel
Ori 5h22'21" 2d56'18"
Melnyk
Ori 6h18'27" 18d49'33"
Melo
Del 20h28'59" 16d42'17"
Melo
Dra 18h42'24" 79d45'41"
Melo722
Lib 15h32'5" -25d46'1"
Melodee A. Fletcher
And 2h25'20" 49d32'16"
Melodee Mae Miller
Aqr 21h33'35" 0d37'55"
Melodie April Spiller
Tau 3h27'47" 9d31'34"
Melodie Denise Jenkins
Cas 1h14'14" 58d0'56"
Melodie Ehret
And 2h30'34" 44d9'2"
Melodie Moncy
Cas 1h42'43" 66d23'59"
Melodie's Queen of the Nile Elim 76
Cnc 8h6'28" 19d57'53"
Melody
Cnc 8h55'58" 15d27'48"
Melody
Leo 10h33'11" 17d10'21"
Melody
Leo 11h8'5" 21d33'5"
Melody
Ari 3h11'57" 29d35'36"
MELODY
Gem 6h10'32" 24d6'28"
Melody
Ori 5h20'9" 4d8'14"
Melody
Leo 9h32'14" 11d24'27"
Melody
Ari 3h11'24" 21d24'27"
Melody
And 0h38'8" 41d44'48"
Melody
And 0h55'56" 39d56'28"
Melody
Crb 15h49'20" 34d19'51"
Melody
Cas 1h13'9" 55d24'39"
Melody
And 23h14'14" 42d37'34"
Melody
And 23h27'36" 42d48'12"
Melody
Umi 14h42'38" 73d16'51"
Melody
Uma 11h54'13" 59d35'10"
Melody "9-27-1987"
Lyn 7h22'25" 44d47'35"
Melody A. Bush
Cap 20h24'10" -15d49'26"
Melody Anderson
Sgr 19h10'16" -14d39'14"
Melody Ann
Cas 0h32'0" 61d28'52"
Melody Ann Floyd
Ari 3h8'5" 18d45'38"
Melody Anne
Leo 11h12'17" 1d19'5"
Melody Anne Hodges Walker
Lyr 18h37'29" 35d4'50"
Melody Anne Naifeh Thomas
Tau 4h3'8" 6d37'59"
Melody Black
Uma 8h55'56" 70d16'38"
Melody Borris
Vir 13h16'37" 4d52'42"
Melody Carter-Wright
And 1h11'38" 38d41'12"
Melody Casna
Crb 16h12'49" 37d35'27"
Melody & Chris's Star
Aqr 23h25'54" -18d42'16"
Melody Christine Scott
And 0h42'9" 26d18'32"
Melody Crookston
Cap 21h15'34" -15d4'59"
Melody Ferne Mhyre Spotts
Gem 7h39'53" 33d9'37"
Melody Janel Kay Ehrlich
Vir 13h16'59" -0d25'45"
Melody Jeneé
Sco 17h51'16" -36d17'35"
Melody Jo
Lmi 10h36'39" 35d3'10"
Melody Johnson
Cyg 21h21'41" 34d1'41"
Melody Johnson
Cyg 19h54'14" 35d57'47"
Melody Joy Pruitt
Leo 11h43'59" 14d41'53"
Melody K. Farnham
Ari 2h12'40" 20d1'55"
Melody L
Cas 2h13'5" 64d45'13"
Melody L. Martin
Per 3h29'33" 47d36'9"
Melody Lamb
Cyg 19h56'43" 51d57'23"
Melody Lee Kilcrease
And 23h46'45" 44d8'52"
Melody Lee Leblanc
Uma 11h32'30" 56d8'44"
Melody Lewis
Ori 6h10'37" 6d1'15"
Melody Liu
And 1h54'48" 40d37'19"
Melody Marie Gritz
Ari 3h13'59" 29d17'23"
Melody Nikki Craff
Leo 9h25'49" 11d34'0"
Melody of dreams
Cas 0h20'33" 52d2'52"
Melody Orta
Leo 11h11'5" 23d49'18"
Melody Reed
Cnc 8h19'21" 19d48'8"
Melody Rose
And 1h38'3" 47d9'14"
Melody S. Wright
Aqr 22h0'38" -15d30'46"
Melody Sky
And 23h38'46" 42d25'11"
Melody Stahl
And 1h16'16" 42d7'26"
Melody Stalbaum
Leo 10h16'43" 16d29'51"
Melody Terrazola
Lyn 9h2'2" 36d31'28"
Melody the Beautiful
Cas 23h21'30" 56d3'8"
Melody "Turtledove" Barroga
Ari 2h42'28" 26d43'22"
Melody Whitlock Boylston
Gem 7h20'42" 25d45'21"
Melodye L. Dunn-O'Neal
Uma 9h57'19" 49d15'37"
Melody's Radiance
Crb 15h52'26" 35d39'5"
Melody-Shayenne
Lyr 19h2'34" 28d0'54"
Melondy Spears
Tau 4h42'40" 19d50'52"
Melonie
Ari 3h12'27" 19d18'54"
Melonie Bailey Nelson
Ari 1h49'39" 12d10'30"
Melonie Balonie-Land
Per 3h25'8" 32d33'48"
Melonie's Special Place 7-25-04
And 0h17'52" 44d48'16"
Melony
And 2h33'56" 46d10'12"
Melony and Randy
Cyg 20h28'56" 55d41'0"
Melony Elaine Kuchta
Vir 12h25'35" 6d43'37"
Melony My Love
Cyg 19h20'6" 48d29'32"
Melora
And 23h56'54" 40d29'59"
Melora Zenitha Cornelia A Horvat
Leo 9h38'0" 31d12'17"
Melori Justine Gube Lee Mabunay
Sco 16h15'12" -15d46'44"
Melosch Ewiger Glücksstern 100 Jahr
Uma 11h5'19" 49d26'42"
Me-Lost?
Cap 20h40'34" -14d39'14"
melpomenesorrow rachel
Cyg 20h36'30" 36d18'59"
Melquethia
Lyn 6h57'40" 48d48'15"
MelRay
Lyn 7h43'44" 52d34'43"
Melrose Beach Apts.
Ori 6h8'38" 13d48'47"
Mel's 40th Birthday Star
And 23h6'5" 40d33'32"
Mel's Glod Star
Leo 10h7'44" 10d55'42"
Mel's Princess Star
Gem 6h42'50" 18d25'37"
Mel's Star
Gem 6h50'28" 16d53'17"
Mel's Star
Vir 13h32'6" -11d12'8"
Melski
And 1h52'18" 44d36'7"
Melstar
Cnc 9h16'35" 9d28'31"
Melstar
Ori 5h30'20" -3d55'45"
Meltem Akhan
Leo 11h44'34" 19d1'55"
Meltem Akol
Aqr 23h30'3" -6d46'7"
Meltem'cim, "The Little Monster"
Sco 17h52'0" -42d51'1"
Meltini Stamatopoulou
Uma 11h14'42" 28d57'55"
Melton
Uma 11h24'38" 36d17'23"
Melton
Cam 4h4'11" 66d17'20"
Melva Miller Pate
And 1h16'52" 43d7'0"
Melva Myre
Cas 1h25'17" 61d25'14"
Melvatean
Vir 12h7'39" 7d15'42"
Melvin Anthony Peyerk, Jr.
Cnc 9h3'2" 32d34'19"
Melvin Charles Shryer
Lyr 18h41'50" 30d32'17"
Melvin Curtis Weibel
Lyn 6h51'40" 57d59'10"
Melvin Earl Hale
Per 4h14'12" 51d26'35"
Melvin Hughes
Per 3h51'33" 32d5'13"
MELVIN J. Matanoski
Aqr 22h49'8" -6d27'34"
Melvin Jackson Sr.
Uma 8h40'32" 58d45'22"
Melvin & Joyce Back
Gem 6h47'3" 25d59'58"
Melvin Keith Nielson
Uma 10h30'54" 53d40'39"
Melvin L. Hilliard, Jr.
Gem 6h54'44" 18d34'8"
Melvin & Marilyn Shelton
And 23h0'46" 47d12'21"
Melvin *Mass* Remster Jr.
Umi 15h38'3" 69d44'53"
Melvin Michael Siler
Her 16h46'10" 33d26'47"
Melvin Orenstein
Cep 22h20'8" 62d10'14"
Melvin Robinson
Per 2h48'25" 38d37'56"
Melvin Seidl
Ori 5h51'50" 22d50'33"
Melvin Vaughn
Ari 2h20'59" 12d58'38"
Melvin Wood III
Aqr 21h26'54" -8d1'42"
Melvina Yeardie
Ori 5h38'51" 1d16'19"
Melvyn and Jackie Roost
Uma 10h15'36" 63d50'8"

Melvyn J Hudson Psc 23h58'42" -3d28'39"
Melvyn Ray Budke Aql 19h29'40" -4d11'55"
Melvyn Sutherland Uma 9h28'46" 69d47'4"
Melvyn with a Y Ori 5h41'57" 12d55'24"
mely my love forever Ori 5h58'0" 6d37'55"
Melynda J. Frohlich Cap 21h44'2" -19d53'4"
Melynda J. Morace Uma 8h45'4" 56d18'9"
Melynda Sue Sco 16h7'58" -14d28'52"
Melyndalea Hess Cyg 20h35'3" 53d15'42"
Melynn Elizabeth Cobaugh Sco 16h10'13" -16d55'5"
Melyssa Ann Cowles Crb 16h5'45" 34d14'50"
Melyssa & Morgan "Daddy Loves You" Lyn 8h26'44" 33d32'56"
Melyssa My Sweetheart Her 16h27'37" 45d26'19"
Melyssa Rose Ortega Cnc 8h44'32" 11d22'18"
Melyssa Syll the Chicklet Aqr 20h48'46" 2d6'55"
Melyssa's Dreaming Star Cnc 8h43'28" 31d56'14"
Melza Peg 23h11'8" 16d24'42"
Melzy's Star Cap 21h36'9" -9d18'53"
Mem & Pep Gendron Cyg 20h19'58" 47d1'55"
Mema Per 3h46'59" 31d50'6"
Mema Leo 10h37'17" 17d2'17"
MeMa Gem 6h47'16" 23d0'55"
Mema Vir 12h9'56" 5d24'43"
Mema' Cam 3h55'43" 69d2'47"
Mema Bethel Peg 23h4'28" 29d31'25"
Mema Froglet Ari 2h25'22" 26d43'25"
Mema & Seaboss Uma 11h48'17" 51d57'32"
Mema Violet Clark Per 3h31'12" 34d21'58"
MeMa's Blessing Cas 1h12'56" 59d29'12"
Mema's Shining Star-We Love You '04 Umi 11h55'2" 88d25'34"
Memaw Lib 14h54'45" -5d54'54"
MeMaw's Love Cnc 8h9'55" 22d27'17"
Memaw's Sunshine Tau 4h13'12" 14d19'49"
Members of St Mary's Parish Uma 10h7'6" 58d2'22"
MEMCJF Uma 8h24'55" 66d52'23"
Meme Sgr 18h11'24" -29d48'50"
Meme Tau 4h9'54" 29d8'25"
Meme And 0h13'21" 43d14'15"
MeMe and PopPop Psc 1h8'51" 29d2'7"
Meme+Boo Dra 19h1'20" 62d10'7"
Me-Me we love you! Leo 11h55'35" 26d10'1"
MEME, Georgina, Elena, Alex. Lyn 7h41'50" 52d56'19"
Memere Del 20h37'6" 16d34'6"
Memere Bea Ori 6h12'54" 15d16'17"
MeMere Ketchie Aqr 22h52'9" -16d22'35"
Meme's Crush Leo 9h30'48" 7d23'32"
Meme's Love Leo 11h52'44" 21d56'15"
Meme's Star Aqr 20h51'3" -11d52'14"
Meme's Star Cas 1h32'26" 60d6'3"
Memi Tau 4h35'54" 20d32'54"
Memom Cap 20h51'28" -24d22'31"
Memoria Psc 0h43'10" 12d37'28"
Memoria in aeterna Cas 2h39'9" 64d56'22"
Memorial Star Fumio & Keiko Leo 11h13'31" 12d50'46"
memoriam Grandpa John Love Sally Lib 14h48'47" -10d54'47"
Memoriam Norman Wolff 1929-1998 Ari 2h35'30" 18d52'35"
Memories Uma 9h19'47" 59d43'52"
Memories of Granite Cnv 13h42'3" 33d10'36"
Memories Of Us Ari 3h12'43" 26d51'24"
Memories of William Leo 10h2'56" 21d35'13"
Memory Uma 11h29'9" 39d24'38"
Memory Cyg 21h30'44" 45d4'52"
Memory Lib 15h22'29" -11d17'40"
Memory Mus 13h12'50" -65d23'24"
Memory of Alice Pettengill 6/13/91 Uma 12h15'22" 56d6'57"
Memory of Sgt. Mjr. William Butcher Peg 22h4'53" 7d31'6"
Memory Star Umi 16h35'22" 81d59'37"
Memory Star John J. Duncan "Bear" Uma 10h37'46" 64d18'7"
Memosa Vir 13h17'39" 7d42'28"
Memphis Lmi 10h27'51" 34d29'1"
Memphis Nolan Cnc 8h31'15" 12d19'3"
Memphis Rapture Crb 16h8'48" 37d58'9"
Memre Lyn 6h57'25" 53d10'17"
Memsor Kamarake Uma 11h48'31" 30d33'4"
Men & Women Asso 626th Engrs WW II Uma 11h32'19" 58d37'33"
Mena Alexia Xayavong Crb 15h38'1" 31d59'14"
Menaggio Tau 4h47'16" 18d32'12"
Menauhant Uma 12h34'0" 57d37'12"
Mendelsurf Lib 15h39'24" -16d8'41"
Mendi Dra 19h43'31" 60d44'0"
Mendigito Uma 9h34'39" 56d25'40"
Mendy Rhoades Sgr 20h2'40" -27d9'28"
Mendy Sue Uma 10h37'29" 55d19'13"
Menegay Aqr 21h40'39" -6d2'28"
Menelaos Valianatos Cep 21h45'6" 64d14'32"
Menelik Steiner Ori 6h2'6" 19d55'3"
Menemur Uma 9h18'16" 43d36'46"
MENINA e LUIGI Aur 6h6'59" 38d20'51"
Mennito Gabriele Ori 6h5'59" -2d9'30"
Menráth Nikolett Uma 13h38'54" 49d22'28"
Mentore Ed Amico Uma 8h43'16" 47d3'38"
MEnUnI Ori 6h18'32" 13d15'22"
Menyhert "Met" Szabo Gem 6h33'22" 23d17'43"
Menzel, René Uma 9h45'44" 43d3'54"
Menzello Aur 5h23'12" 36d1'7"
Meo Thit n Dam Tac Cnc 9h4'49" 27d37'9"
Meow Ari 3h22'49" 27d58'52"
meow Lyn 8h21'0" 46d22'54"
MeowFace Cap 20h33'43" -16d38'18"
Méphilalie Cnc 8h24'24" 30d3'2"
Mephisto Psc 23h21'58" 3d33'6"
MEP's Neon Cap 21h21'16" -16d36'14"
Mer Umi 13h7'57" 75d46'21"
Mer Mer Uma 12h3'55" 50d18'25"
Mera Uma 12h15'22" 55d41'0"
Mera Barbara Bouharoun Lyr 18h44'30" 37d39'54"
Mera Renee Uma 11h58'21" 64d44'11"
Meral Peg 21h45'21" 8d31'29"
Meral Sarper Tau 5h49'27" 27d0'28"
Meranda Hyman Sgr 17h52'33" -17d0'10"
Merari Leo 9h34'57" 22d36'13"
Merav Psc 1h9'11" 29d41'45"
Merav Adel Srp 18h22'20" -0d18'19"
Meravigliosa Manuela Uma 9h36'58" 55d1'47"
Meray El-Ahdab Cap 21h19'59" -18d48'11"
Mercadees Our Twinkle Star Per 3h52'50" 35d44'21"
Mercadell Lewis Tau 5h28'11" 23d22'22"
Merce Colomines Vel 10h26'25" -41d47'12"
Mercedeh Aqr 22h34'53" -5d22'20"
Mercedes Umi 15h34'6" 84d49'22"
Mercedes Lib 15h46'37" -19d7'16"
MERCEDES Sgr 18h0'54" -27d24'45"
Mercedes Tau 3h41'13" 27d16'13"
Mercedes Mon 6h52'41" 8d35'1"
Mercedes Cnc 8h22'7" 11d19'59"
Mercedes Uma 11h9'30" 34d57'33"
Mercedes Uma 11h2'41" 33d43'48"
Mercedes And 1h8'16" 42d46'52"
Mercedes Arrielle Carver Cma 6h46'45" -19d32'22"
Mercedes Ashley Gem 6h48'38" 19d2'11"
Mercedes Burgos Mon 6h36'42" 8d47'4"
Mercedes Casanova And 1h8'19" 45d46'17"
Mercedes Cyntia Yazmin Carozzi Lib 14h40'32" -18d5'30"
Mercedes David Sco 16h48'30" -27d13'35"
Mercedes E Mercep Gem 6h56'42" 14d42'12"
Mercedes Elizabeth Campbell And 0h42'14" 35d50'24"
Mercedes Elle Lyn 7h8'31" 50d43'25"
Mercedes Fabro Rivera Uma 10h28'8" 64d37'52"
Mercedes Faith Lib 15h42'44" -12d7'25"
Mercedes Hernandez Leo 10h30'59" 17d15'38"
Mercedes Jamenson Cap 20h12'21" -10d45'5"
Mercedes Jo Sco 16h48'38" -27d25'9"
Mercedes Judith Barrios Aur 4h59'28" 36d27'45"
Mercedes Kay-Lynn Fillebrown Leo 10h54'54" 6d6'3"
Mercedes Kelly Uma 9h46'57" 61d49'11"
Mercedes Kottka Cas 1h5'47" 61d50'36"
Mercedes L Jones Precious Star Del 20h31'52" 16d45'41"
Mercedes Limon Mon 6h52'37" 7d46'54"
Mercedes Madison Leo 10h17'1" 22d16'27"
Mercedes McMurray And 1h8'30" 39d37'18"
Mercedes Padilla Boo 14h52'7" 23d15'41"
Mercedes Scotti Ser 15h26'32" -0d28'36"
Mercedes Sloan Del 20h18'27" 9d42'23"
Mercedes Violet Sgr 19h53'44" -36d30'9"
Mercedes Wheeler Crb 16h18'22" 33d38'18"
Mercedes, Octavio and Tristan Cyg 19h35'57" 36d17'2"
Mercedez Alvarado, Mi Angelita Lib 15h42'45" -29d41'22"
Mercedita Evangelista Cam 5h59'49" 61d0'7"
Mercel Brillante Lyr 19h14'28" 38d42'34"
Mercelle Smith Cen 14h28'20" -31d42'28"
Mercer Marie Chamblee Lyr 18h49'36" 38d22'9"
MERCHE And 2h11'31" 41d26'5"
Merche A.C. Lib 16h1'50" -17d52'52"
Merchita (21-10-02) Del 20h45'53" 18d7'44"
Merci Ori 5h28'36" 8d30'31"
Merci A2 Quad ! Aur 6h16'35" 49d25'11"
Merci Angela Townsend2.7.47-25.4.04 Cru 12h50'19" -57d15'14"
Merci Beaucoup Tau 5h58'50" 25d54'28"
MERCLB Sgr 18h59'52" -27d55'57"
Mercury Magic 'Shane' Cmi 7h32'15" 9d56'13"
Mercuzio Leon Vir 13h2'5" -2d32'41"
Mercy Leo 10h25'21" 13d58'11"
Mercy Leelanee's Majestic Sea Sco 17h53'16" -35d47'5"
Mercy Rice (Belle) Uma 11h23'30" 36d25'46"
Mercy & Scott Lyr 18h52'2" 31d40'54"
Merdith McGaw And 23h15'34" 51d56'28"
Mère Leo 11h10'9" 2d22'33"
MERE Sco 17h39'43" -38d23'44"
Mère Uma 8h14'7" 66d36'38"
Mère Charmante Aqr 23h6'46" -5d54'57"
Mère de L'amour éternel Sgr 18h15'41" -18d4'50"
Meredith Sgr 18h13'16" -20d3'10"
Meredith Cap 20h44'33" -18d23'1"
Meredith Uma 10h44'59" 65d27'31"
Meredith Sgr 19h23'11" -30d51'19"
Meredith Lib 15h33'33" -25d2'41"
Meredith Psc 1h23'17" 26d55'38"
Meredith And 23h37'54" 48d19'33"
MEREDITH And 23h17'15" 47d38'18"
Meredith And 23h34'6" 45d38'16"
Meredith Per 4h48'1" 46d50'16"
Meredith And 1h0'47" 46d21'56"
Meredith And 2h24'48" 46d53'22"
Meredith And 23h21'58" 42d10'33"
Meredith Cnc 8h13'22" 30d43'21"
Meredith Lyn 7h34'59" 36d8'17"
Meredith Adele And 0h46'3" 38d26'18"
Meredith Adella Logan Sgr 19h17'20" -21d31'28"
Meredith Albert Tau 5h23'41" 28d36'4"
Meredith Allen Cap 21h11'3" -15d59'35"
Meredith and Maggie Ara 17h27'35" -47d28'31"
Meredith Ann Jez Leo 9h52'25" 29d27'58"
Meredith Ann McGowan Cap 21h7'35" -24d15'59"
Meredith Ann Moody And 2h33'1" 41d37'17"
Meredith Ann Reifschneider Lmi 10h49'11" 26d12'46"
Meredith Anne Liedy Crb 15h31'44" 27d34'2"
Meredith Ashley Semon Uma 11h26'21" 67d56'19"
Meredith Ballard Lynden And 0h30'30" 39d11'47"
Meredith Besser Tau 5h31'57" 19d47'59"
Meredith Black Per 3h35'4" 45d23'33"
Meredith Blackwood Lib 15h0'23" -9d27'13"
Meredith Brewster Uma 9h56'35" 45d59'20"
Meredith Bryan Gem 6h29'7" 21d36'34"
Meredith Carson Adele Whitted Aql 20h10'43" 1d30'16"
Meredith Causey Bruner Uma 10h38'54" 46d29'25"
Meredith Chiaro Gem 6h32'44" 15d16'9"
Meredith D. Nix Crb 16h16'20" 39d22'28"
Meredith DeLynn And 23h23'32" 47d15'28"
Meredith E. Werden Cyg 21h7'11" 40d17'10"
Meredith Elizabeth Hudson Uma 10h26'46" 42d45'14"
Meredith Elsie Cooper And 23h58'57" 44d42'13"
Meredith Fowlers' Star And 1h32'8" 42d26'56"
Meredith Frances Volz Peg 22h19'5" 13d55'6"
Meredith Freedhoff Aqr 22h31'51" -4d13'16"
Meredith Gautrau Vir 13h47'48" -7d14'17"
Meredith Grace Baham Uma 12h37'48" 54d54'24"
Meredith Grace Oster Aqr 22h24'32" -5d48'59"
Meredith Grube Tufts Sco 16h11'4" -13d12'31"
Meredith Halle Moch Crb 16h12'48" 36d18'29"
Meredith Haygood Cam 5h54'48" 61d27'50"
Meredith Helen Ari 2h55'13" 26d59'42"
Meredith Hope Cnc 8h47'43" 14d44'6"
Meredith & Irene Sprunger's Star Lyn 6h57'59" 47d57'51"
Meredith Jenne Nicole Cnc 8h27'51" 20d56'53"
Meredith Jo Umi 14h21'34" 69d6'59"
Meredith Jones Ori 5h1'41" -0d20'2"
Meredith Keith Psc 23h49'49" 5d39'24"
Meredith L. Craig Cas 0h49'37" 60d23'18"
Meredith Laurence Tweed Sgr 18h27'42" -25d59'15"
Meredith Lee Hathon Psc 0h42'20" 7d48'0"
Meredith Leigh Brown And 23h10'10" 45d7'6"
Meredith Leigh Severance And 0h38'21" 39d10'49"
Meredith Lyn Reyna And 23h50'33" 46d1'23"
Meredith Lyndsey Jackson Tau 3h41'42" 16d22'58"
Meredith Lynleigh Leo 10h40'1" 8d23'43"
Meredith M. Passey Psc 1h15'13" 6d13'8"
Meredith Margratha Sielke Leo 11h26'29" 11d18'54"
Meredith Martini GaMPI Shining Star Lyn 6h53'22" 52d10'6"
Meredith Michele Miller Sgr 19h36'37" -40d57'46"
Meredith Nicole Brown Lib 14h38'1" -9d44'10"
Meredith Paige Denney Gem 7h6'40" 16d17'28"
Meredith Rayne Brooks And 0h19'40" 36d31'19"
Meredith Rene Her 17h37'58" 44d40'42"
Meredith Renfree And Raymond Magill Lib 14h52'0" -2d22'33"
Meredith Richard Aqr 22h32'5" 1d21'53"
Meredith Ruther Cyg 20h0'25" 59d27'38"
Meredith Ryan (Beeb) Sgr 17h57'16" -18d50'55"
Meredith S. Kenneff Leo 11h30'50" 16d48'12"
Meredith Sarah Kessler Leo 9h30'30" 6d47'38"
Meredith Seton Morenz Cnc 9h2'16" 14d33'6"
Meredith T. Moore And 0h6'31" 44d36'23"
Meredith Weatherly Gross Leo 10h25'46" 25d9'20"
Meredith Williams Tau 5h29'34" 23d23'37"
Meredith Wood Juengel Aur 5h21'0" 48d5'52"
Meredith Wright And 23h28'56" 47d50'13"
Meredith Wroblewski And 0h44'17" 43d59'50"
meredithlee Ari 2h20'25" 25d53'5"
Meredith's Angel And 1h47'35" 50d0'19"
Meredith's Diamond Aqr 22h12'56" 1d6'7"
Meredith's Light Aqr 23h2'20" -7d33'5"
Mereika Denee Lelanie Umi 15h0'36" 68d52'28"
Merel Allen Lucas Uma 11h55'19" 34d10'53"
Mereland Lyn 7h55'31" 57d30'55"
Merere Cap 21h23'11" -19d5'45"
Meresi Cap 20h49'49" -24d51'48"
Mergern Gern Leo 10h16'19" 16d5'11"
Mergim Sahin Tri 1h53'31" 33d55'45"
Merholz, Erika Ori 6h20'27" 14d7'27"
meri Leo 11h33'37" 14d14'38"
Meri Leo 10h29'37" 14d35'20"
Meri and Ivan Cyg 20h59'19" 47d22'57"
Meri De Bruin Uma 8h16'29" 66d11'30"
Meri Ervin Cap 21h30'44" -10d6'36"
Meri Farre And 0h19'21" 34d46'56"
Meri Ghazaryan Lyn 8h37'1" 39d35'32"
Meri mama csillaga Cnc 8h10'16" 21d7'28"
Meriah Rose McCash Cnc 8h10'35" 26d58'34"
Meriam Cnc 8h11'7" 25d12'0"
Meriam Lmi 10h38'19" 31d6'6"
Merian Crt 11h25'50" -10d51'55"
Meriana Sco 17h25'54" -37d44'2"
Meribeth Privett Per 3h36'15" 46d54'31"
Meric Lyn 7h56'6" 34d22'10"
Meridee Leo 10h24'37" 26d56'37"
Meridee Anna The Dusty Star Leo 10h19'0" 10d14'28"
Meriden Fire Department 11-23-1961 Lyn 7h41'46" 36d21'55"
Merideth Petz And 1h19'54" 40d56'47"
Meridith Mathers Rohrbaugh Uma 8h19'42" 72d40'57"
MeridithandMichael Lib 15h39'5" -15d44'59"
Meridith's Meteor Peg 21h38'47" 27d34'11"
Meridy Dawn Carpenter Long Crb 15h33'43" 31d33'19"
Meridyth Lib 15h57'58" -10d15'29"
Meriel Umi 14h11'35" 77d21'40"
Merifluff, Protector of the Sea Aql 19h45'54" -0d1'35"
Merilee Ah Lan Sco 16h10'43" -14d3'18"
Merili & Sonny McCoy Crb 16h5'16" 26d31'51"
Merilu Moreno-Smith Mon 7h11'26" -2d47'37"
Merilyn Bodden Milam Leo 10h56'52" 12d6'54"
Merin Jeanine Dunkle Ari 3h10'50" 19d23'40"
Merin Maldonado Legacy And 0h7'4" 34d52'21"
Merina Joy Cardwell Vir 12h41'8" 5d40'31"
Merina Kelsey Miner Gem 7h36'56" 34d32'39"
Merina Lippis Sgr 18h0'22" -18d20'17"
Merissa A. Rudkin Psc 0h17'54" 7d19'23"
Merissa and Gabriel Peg 22h54'47" 15d58'58"
Merit Sgr 18h22'44" -31d58'51"
MeRita Vir 11h47'21" 2d49'28"
Meritage (Taj) Umi 16h16'34" 76d13'35"
Meriweather Mon 7h16'57" -3d33'56"
Merkki Lyn 7h23'46" 56d55'5"
Merle Cap 21h27'7" -14d27'1"
Merle Ori 5h17'42" 3d33'46"
Merle D Lauritsen Cam 4h5'3" 77d6'22"
Merle Elspeth Fitchett Cas 0h21'38" 54d5'7"
Merle L. Brenner Cas 0h46'51" 60d17'46"
Merle LaVerne Monroe Uma 13h50'27" 61d25'15"
Merle Morrison Crb 16h19'50" 38d22'52"
Merlene Byler Uma 8h38'27" 53d15'44"
Merletta Rose Leo 10h47'50" 14d21'26"
Merlin Lmi 10h25'14" 29d54'24"
Merlin Cyg 20h41'13" 46d12'58"
Merlin Uma 9h20'26" 42d54'1"
Merlin Uma 9h25'36" 61d30'39"
Merlin 3/26/02 Uma 10h32'40" 52d10'56"
Merlin Hope Aql 19h40'48" 15d1'33"
Merlin Magic Paws Leo 9h43'31" 7d7'7"
Merlin Reyna Deleon Sgr 20h0'21" -25d59'45"
Merlin Seibel Aqr 23h2'22" -11d30'13"
Merlin the King My Friend Crb 15h37'49" 29d18'26"
Merlin V. Terrill Ori 6h13'25" 15d57'46"
Merlin's Magick Diamond Cas 1h4'50" 64d13'15"
Merlis Yvette Peg 22h24'30" 3d2'38"
Merlo Cap 20h33'49" -13d9'27"
Merlon Shane Woodard Uma 8h50'43" 63d44'55"
Merlyn Cas 23h48'42" 51d31'44"
MERMAIDSLAIR Uma 10h56'54" 45d50'56"
Mermer "Mickey" Gilbert Umi 15h33'20" 79d55'5"
Merna Marie Aqr 21h39'34" -7d43'59"
Mernie Psc 1h9'14" 10d32'25"
Mernis Crb 16h20'30" 37d7'44"
Mer'o CsillagPéter Sco 17h38'35" -42d38'19"
Meropi Umi 14h37'37" 69d41'59"
Meropi And 1h59'26" 38d18'24"
Merovitch - Alioto Lib 14h30'7" -10d22'50"
Merran And 22h59'9" 36d8'27"
Merranda Louise And 2h21'58" 47d57'0"
Merri Lawler Aql 20h9'27" 15d57'10"
Merrick Edward Slade Cep 21h57'44" 63d4'8"
Merrick Pinterich Uma 9h30'4" 44d23'22"
Merrick Rae Avery Gem 6h42'39" 17d23'58"
Merrick Vincent McClain Cep 20h33'44" 60d2'52"
Merrideth Cantwell Lyn 8h9'7" 56d53'54"
Merriee Ori 5h26'43" 13d22'29"
MERRIL Umi 15h1'29" 76d20'10"
Merrilee Hill Gem 7h17'20" 22d33'52"
Merrill and Edna Lundgren Uma 9h22'40" 53d37'34"
Merrill Anne Peg 22h34'40" 29d47'58"
Merrill Dunbar Umi 14h45'57" 69d18'36"
Merrill Francis Redden Uma 10h32'2" 47d30'58"
Merrill Heather and Madison Alexis Tau 4h12'59" 4d45'22"
Merrill-Anita Forever and Always Cyg 20h34'22" 38d59'16"
Merrilyn Hurd Sco 16h14'2" -17d16'18"
Merrilyn Kramer Lyr 19h16'14" 35d14'12"
Merrin Uma 10h15'10" 42d53'18"
Merritt Cyg 20h17'19" 54d36'46"
Merritt Louise Wille And 2h33'56" 45d19'7"
MerrittAndKasey Ori 5h26'54" 4d58'1"
Merry Lib 14h33'59" -18d18'27"
Merry 0856 Leo 11h38'14" 18d23'53"
Merry Chika Ari 1h57'30" 19d54'18"
Merry Christmas Uma 10h33'41" 43d12'1"
Merry Christmas Lori 2005 Lyn 6h20'54" 59d53'7"
Merry Christmas Momma, I Love You!! Gem 7h21'13" 18d11'54"
Merry Christmas To A Star-Sherri Lmi 10h26'19" 36d46'4"

Merry E. Kraemer
Lyn 7h39'24" 44d35'35"
Merry Mary Lou
Uma 11h40'56" 45d19'12"
Merry Upshaw Irwin
Ari 2h51'31" 19d46'28"
Merry, Chris, Ali and Jack Wills
Cyg 21h40'23" 36d32'6"
MerryBear
Cyg 19h55'35" 56d19'36"
Merryl 30
Crb 16h11'26" 30d29'12"
Merryn
Uma 11h8'4" 32d49'29"
Merryn Louise Macdonald
Cru 12h55'17" -57d27'25"
Mersades Jynx
And 1h9'30" 47d11'46"
Merse
Uma 11h24'48" 46d14'14"
Mersini Feda
Cas 0h22'59" 56d22'56"
Mertz, Sonja
Uma 9h32'48" 59d30'12"
Merve & Memo
Cap 21h4'52" -20d30'8"
Merveilleuse Brigitte
Umi 14h8'21" 68d55'26"
Merveilleuse Mimi
Uma 13h40'55" 52d33'45"
Merville-Anne
Umi 15h43'17" 80d1'56"
Mervyn & Charlotte Forever
Cyg 21h13'17" 42d19'57"
Mervyn & Gwladys Roberts
Cyg 19h44'19" 34d1'26"
Mervyn Hirsch
Cru 12h27'4" -60d18'32"
Mervyn Peskin MD Scholar/Gentleman
Lyr 19h27'31" 43d5'17"
Mervyn Robert Rivett
Uma 11h58'6" 65d23'55"
Mervyn Skinner
Her 16h40'12" 6d38'30"
mery
Dra 14h35'51" 58d18'40"
Mery Frey
Uma 12h41'53" 57d32'50"
Mery Paz
Aql 19h32'1" 8d29'21"
Mery Tony
Peg 21h37'17" 25d4'24"
Meryl
Uma 9h43'45" 61d16'43"
Meryl & David(for u to wish upon)
Gem 6h19'50" 23d11'46"
Meryl E. Yasmer
Uma 10h35'53" 65d30'39"
Meryl Geller
Uma 11h10'54" 36d34'13"
Meryl & Johnathan
Cap 21h52'3" -18d12'29"
Meryl Lee
And 1h28'13" 43d38'1"
Meryl Markowitz
Gem 7h48'53" 22d19'35"
Meryl Virga
And 1h42'12" 49d57'51"
Meryl Virga
Cas 0h2'53" 56d24'7"
MERYLINDA
Leo 10h51'52" 15d16'22"
Meryll Jane
Cas 0h34'58" 57d18'31"
MERYSAL
Peg 22h43'28" 27d27'3"
MES
Uma 9h42'40" 57d54'6"
Mes Anges
Cyg 19h38'19" 29d1'24"
Mes Belles Filles Dometria&Monique
And 1h8'47" 45d0'19"
Mesa
Sco 17h57'9" -30d13'37"
Meschelle Elizabeth Ann Atwood
And 23h2'12" 50d50'44"
Meschenmoser, Sigrun und Wilfried
Uma 13h37'0" 61d49'37"
Meschke, Peter
Ori 6h18'55" 16d45'57"
MeSheBow
Her 17h56'14" 14d53'44"
Meshel
Gem 6h3'21" 24d40'28"
Meshell Lynn Smith
Lyr 18h48'46" 35d7'11"
Meshia
Tau 4h18'53" 12d26'28"
Meshia
Sgr 18h25'12" -27d37'48"
Meshkin
Per 3h2'42" 51d32'24"
Mesmerising Miya
Umi 16h48'11" 83d16'14"
Mesmerize
Sgr 19h2'4" -24d26'40"
MESMERIZE, The Future is Bright
Lib 15h10'28" -5d3'54"
Meso & Mezo
Ari 2h24'7" 18d54'34"
Messan
Sge 20h15'15" 18d22'21"
Messer, Gerhard
Ori 6h7'3" 10d58'33"
Messias Couto The Great
Uma 10h30'36" 64d50'38"
Messy Jessie
Gem 7h12'44" 32d49'17"
Mestayer
Aqr 22h36'37" -0d9'12"
Mester, Klaus-Peter
Sco 17h18'12" -33d11'25"
Mészáros Dóra és Orsolya
Uma 11h46'22" 39d48'21"
Mészáros Márti csillaga
Uma 10h55'18" 35d17'21"
Met in a Dream
Tau 5h6'52" 23d41'6"
Meta Goodwin
Cap 20h28'57" -12d11'34"
Meta Marie Kolbjornsen
Lyn 7h55'39" 55d11'11"
Meta Osda Adanvdo
Uma 11h14'7" 29d43'16"
Metcalfe
And 2h12'17" 38d39'20"
Metejihka
Cnc 9h13'54" 31d31'45"
Meteor Mike
Gem 6h11'24" 27d37'17"
Meteoric Monte
Ari 3h8'4" 28d4'44"
Metis 60
Uma 9h7'27" 58d51'35"
Met-Ori Billings
Uma 13h48'48" 54d51'54"
Metra
Psa 22h31'26" -32d23'3"
Metro Elite
Umi 13h30'52" 70d34'3"
Metro Tech High School Class 2005
Uma 11h22'9" 44d18'11"
Metta Harms
Cam 4h21'33" 75d46'49"
Mette 09-03-1979
Psc 1h6'21" 29d6'14"
Mette´s Star
Umi 16h9'3" 84d39'4"
Mette-Louise
Vir 13h25'3" -19d50'39"
mettos-reesa
Cap 20h23'15" -12d25'58"
Meu Amor
Tau 5h17'8" 19d17'51"
Meu Coração
Cnv 14h2'10" 35d14'6"
Meu Crido
Aqr 22h6'51" -5d22'32"
Meum Astrum Lucidum E-Shing Sheu
Uma 8h15'49" 62d26'49"
meum mater
Uma 13h42'20" 52d44'38"
Meurice
Cas 0h8'49" 56d27'50"
Meus Angelus Meus Amor
Cyg 21h26'2" 34d0'10"
Meus Diligo Stacey
And 0h2'2" 33d23'33"
meus patronus
Per 2h15'40" 52d48'20"
Meus Uxor Eternus
Lyr 18h36'29" 36d55'40"
Meusser, Hans
Uma 11h39'32" 31d33'6"
MeusStellaLucidus0209
Aqr 20h40'25" -11d30'58"
MEvE
Uma 10h52'47" 49d45'58"
Me-We
Cas 23h48'10" 59d36'55"
Mewten
Dra 15h23'51" 57d9'14"
Mexican Radio
Lyn 7h20'56" 45d47'7"
Meyer
Uma 9h2'36" 46d57'43"
Meyer Feldberg
Psc 1h34'46" 10d8'33"
Meyer, Gertrud
Uma 8h13'24" 64d46'39"
Meyer, Ingo
Uma 10h28'38" 40d38'8"
Meyer, Lutz
Uma 9h51'9" 63d31'36"
Meyjohn
Aql 18h53'39" 7d51'50"
MeyMeyV
Vir 13h22'34" -4d56'42"
Meyou
Tau 5h11'0" 26d24'25"
Mezclaloca
Cnc 8h34'55" 7d36'17"
Mezpha
Lyr 19h10'11" 26d30'13"
Mezzena Andrea
Cam 4h5'47" 56d58'56"
MF 010405
Ori 6h5'8" 21d5'9"
MF-2001-2006-NG
Uma 11h44'39" 49d6'54"
MFDEV
Mon 7h17'39" -10d26'57"
M.F.E.O. 2269
Sgr 19h18'42" -27d9'58"
M.G.
Sco 17h42'34" -39d49'42"
MG AP 4EVER
Lyn 7h22'17" 52d52'55"
MG Candidus Caelestis
Cyg 22h1'39" 50d43'45"
MG Lloyd J. Austin III, CENTCOM COS
Dra 20h21'11" 63d29'12"
MGM
Uma 11h11'16" 64d51'31"
Mgp/81894 (Melanie's Star)
Ori 5h14'2" 3d50'59"
MGP. BSA. BANTAM
Aur 6h35'52" 35d4'23"
MGS-03-12-05
Cam 4h12'49" 66d34'14"
MGunns K.L. Fitzgerald-Case
Uma 12h10'23" 56d52'40"
M-H K D "Mary Angel"
Uma 10h27'14" 64d33'8"
MH Kaswen
Uma 10h47'26" 50d5'39"
Mhaa-liq Robertson
Cap 20h59'4" -15d5'23"
MHADTL
Her 16h55'50" 17d19'34"
Mhairi MacLeod
Uma 12h5'56" 53d29'0"
Mhairi Milne
Ori 5h8'11" 5d54'38"
Mhari
Leo 10h8'24" 21d56'36"
MHD-50
Vir 15h10'12" 0d11'1"
MHH
Uma 10h47'43" 49d53'21"
Mhiy Rachael
Uma 13h46'20" 58d25'15"
MHJT 2320
Cma 6h46'6" -12d18'9"
Mhokah
Tau 4h43'17" 2d34'9"
Mhychele
Lyr 18h47'33" 37d53'29"
Mi
Aql 19h5'52" 12d50'44"
mi
Uma 9h41'0" 61d40'17"
"Mi Amigo" Baron Bishop
Uma 10h57'36" 38d4'16"
Mi Amigo-Mi Amor-Mi Para Siempre
Umi 16h42'24" 76d13'6"
Mi Amor
Psc 23h14'0" -2d5'31"
Mi Amor
Cas 1h29'59" 69d59'12"
Mi amor
Cyg 21h36'22" 52d51'45"
Mi Amor
Tau 4h37'25" 25d24'32"
mi amor dulce
Gem 6h47'15" 14d54'7"
Mi Amor Fatima
Sge 20h4'53" 19d22'12"
Mi Amor Para Jeanette
Gem 6h37'23" 15d38'55"
Mi amor, John P. Hernandez
Umi 14h50'50" 79d2'50"
Mi Amore
Cyg 20h27'49" 36d9'24"
Mi Amore Mi Principessa per Eterno
Sgr 19h9'26" -12d34'58"
Mi Amour
Sgr 18h58'37" -24d57'22"
Mi amour
Leo 11h8'37" 22d54'32"
Mi Angel
Lyr 18h49'47" 38d56'45"
Mi Angel, "Elian"
Cap 21h58'26" -10d25'50"
Mi Angel, Mi Amore
Ori 5h57'50" 16d58'10"
Mi Angelita
Vir 12h41'33" 3d3'4"
mi angelito Cary
Lac 22h43'21" 49d10'51"
Mi Angelito Nubia
Sco 16h27'58" -27d44'1"
Mi Antojito
Cyg 21h22'22" 36d2'41"
mi Augestärn
And 2h35'5" 44d32'4"
Mi Bella
And 2h19'54" 47d31'16"
Mi Bella Madre, Ginger
Ari 2h16'21" 23d2'51"
Mi Bellisima Carla
Cnc 9h18'58" 11d52'54"
Mi Cariña
Uma 9h35'54" 66d31'46"
Mi Chelle Biggs
Leo 10h8'11" 15d58'3"
Mi Chiquita Muchacha
Col 6h24'39" -37d23'34"
Mi Chumbi
Aqr 21h44'56" -5d39'57"
Mi Cielo
Lyn 7h23'51" 46d1'56"
Mi Cielo
Cyg 21h51'48" 48d30'37"
Mi Corazón
Ori 4h53'37" 11d49'8"
Mi Corazon
Umi 14h38'58" 78d6'9"
Mi Corazon, Mi Vida
Crb 16h9'48" 33d48'1"
Mi corizan, my squishy, my Kenn
And 1h23'51" 42d37'12"
Mi Dana Dulce Para siempre
Ori 5h32'37" 10d25'20"
MI & DB Friends Throu Thick-N-Thin
Ari 2h17'58" 26d25'53"
Mi Dolores
Cyg 19h35'57" 35d5'57"
Mi Dulce Amor
Cyg 20h46'17" 35d38'14"
Mi Esposa Claudia
And 1h47'6" 39d9'16"
Mi Estrella Bella
Leo 9h43'55" 27d47'0"
Mi Estrella Lelita
Pho 23h38'18" -40d43'16"
Mi Estrella, Silvia
Aqr 21h55'24" 0d20'44"
Mi Estrellita Jou
Psc 0h36'52" 20d58'2"
Mi "Estrellita"Rojas
Cam 3h28'2" 64d23'42"
Mi Flor Elegante de Cielo
Tau 5h8'34" 24d57'7"
Mi Lampara
Dra 19h46'29" 64d29'12"
Mi Laurena Bella
Vir 12h19'48" 6d40'26"
Mi Luv
Leo 11h44'38" 25d4'2"
mi madre sally
Uma 8h18'20" 69d13'7"
Mi Mami Yolanda
Sgr 18h52'53" -21d46'0"
Mi Maria
Gem 7h37'37" 32d8'49"
Mi Ming Min
Crb 16h7'49" 34d17'6"
mi muneca con ojos azul
Tri 1h53'52" 30d45'21"
Mi Nefer
Her 18h49'9" 22d22'2"
Mi negri
Cnc 8h24'44" 6d51'0"
Mi niña
Gem 7h25'56" 32d48'52"
Mi niño hermoso JC
Uma 11h42'51" 43d19'55"
Mi Papi
Gem 7h49'53" 29d50'46"
Mi Papi Francisco
Cnc 8h49'53" 20d0'12"
Mi Princesa Nellita
And 1h3'34" 45d56'39"
Mi Prometida
And 1h42'56" 42d36'47"
Mi Pulgosa
Peg 22h37'25" 11d45'14"
Mi Reina Bonita
Cnc 8h42'8" 23d26'54"
"Mi Rey" Armando Rodriguez
Cam 4h20'20" 70d49'10"
Mi Rona Bellissima
Psc 0h53'59" 17d39'19"
Mi Ryang Kim
Tau 4h35'6" 10d35'47"
Mi sueño
Del 20h48'53" 4d35'21"
Mi Sueno,Cynthia Stevens
Uma 13h39'47" 56d26'23"
Mi Tesoro
Cnc 8h50'45" 29d36'5"
Mi Todo Hermoso
Sco 17h10'2" -44d28'1"
"Mi Unico y Verdadero Amor"
Cyg 21h34'37" 40d57'38"
Mia
Cyg 20h8'13" 47d18'44"
Mia
And 0h4'58" 39d50'2"
Mia
Lyn 9h7'41" 41d0'25"
MIA
Aur 5h43'9" 41d37'20"
Mia
Crb 16h3'32" 36d22'39"
Mia
Tau 4h29'54" 27d37'49"
Mia
And 0h17'6" 32d54'48"
Mia
Lyr 18h34'30" 27d57'41"
Mia
Sco 17h57'21" -41d33'33"
Mia
Cas 0h51'59" 63d23'29"
Mia
Uma 8h31'2" 67d30'57"
Mia
Uma 13h35'5" 61d58'12"
Mia
Lib 14h43'23" -17d35'47"
Mia
Aqr 23h7'46" -7d40'27"
Mia
Vir 13h34'52" -3d54'44"
Mia Aida Barilleaux
Sco 17h50'47" -32d10'14"
Mia Alexander Oehler
Leo 9h53'53" 13d45'23"
Mia Alicia Prout
And 2h25'37" 44d49'27"
Mia Alise
And 1h15'35" 45d7'12"
Mia Amato
Cas 1h42'7" 64d42'24"
Mia and Dung Nguyen
Cae 4h49'22" -36d43'20"
Mia and Niall
Dra 19h16'10" 72d50'12"
Mia and Zachary Chadwick
Gem 7h37'9" 15d45'18"
Mia Angelina
And 0h30'6" 30d52'51"
Mia Anne Bland
Ara 17h5'48" -58d28'50"
Mia Anne Hartie
Psc 1h34'9" 24d58'30"
Mia Anne Thiessen
Umi 18h45'46" 87d44'47"
Mia Ashtyn Malpedo
Vir 13h46'58" -6d35'39"
Mia B.
Cnc 8h13'47" 15d7'42"
Mia B
Cnc 9h14'36" 12d26'25"
Mia B B
Umi 15h12'34" 78d8'28"
Mia Beattie
Dra 14h39'56" 55d45'30"
Mia Bella
Her 17h25'11" 36d10'26"
Mia Bella Amor
Cam 4h16'37" 68d7'51"
Mia Bella Laudien
Uma 8h22'3" 60d43'47"
Mia Bella Mealissa Searing
Cnc 8h48'44" 16d4'47"
Mia Bella Scocozzo
Vir 15h7'24" 4d31'12"
Mia Bella Stella
Aqr 22h35'10" -2d44'4"
MIA BERNAUER
Lac 22h24'42" 41d24'52"
Mia Beth Zaretsky
Crb 15h38'2" 39d9'49"
Mia Boyle
Ari 2h48'7" 26d57'37"
Mia Brandon
And 2h31'38" 49d1'41"
Mia Celeste
Aqr 22h31'25" -5d58'41"
Mia Cesarello Zeikowitz
Cap 20h18'27" -27d5'19"
Mia Charlotte Merritt
Ori 6h17'11" -0d12'0"
Mia Christina Volpe
Cap 20h13'29" -27d14'12"
Mia Coppin
And 0h0'11" 33d22'33"
Mia Coste
Her 17h21'49" 29d25'45"
Mia Daniela
Lmi 10h39'30" 28d58'27"
Mia Dawn Markiet
And 23h2'23" 36d34'26"
Mia de Lasson Johansen
Sco 16h51'52" -41d51'42"
Mia Deanna
Gem 7h40'18" 34d36'31"
mia dominga willkomm
Aqr 20h39'39" 0d42'19"
Mia Edge
Lyr 18h43'1" 38d15'17"
Mia Elisa (Tuo G.)
Cam 5h57'8" 61d22'23"
Mia Elise
Lyn 9h1'54" 33d21'14"
Mia Elizabeth
And 1h47'24" 47d32'56"
Mia Elizabeth "Moomie" 1/19/1991
Cas 1h16'12" 55d24'36"
Mia Elizabeth Reid
Sgr 17h59'25" -17d35'27"
Mia Elizabeth Ross
Cmi 7h35'47" -0d8'15"
Mia Elizabeth Santana
Lyn 6h54'18" 51d38'37"
Mia Elizabeth Sumner Reineke
Uma 9h53'23" 45d58'1"
Mia Ellen Dunning
And 1h33'29" 36d45'58"
Mia Elspeth
And 23h42'44" 35d48'6"
Mia Estrella Sartori
And 23h21'32" 48d57'42"
Mia Evgenia Tsirekas 2003
Psa 22h8'48" -32d50'4"
Mia Fadhel
Leo 10h6'58" 15d34'7"
Mia Fernanda Velasquez
And 1h20'37" 44d1'49"
Mia figlia Andrea (femmina)
Uma 10h31'11" 51d38'52"
Mia G Mouron-Adams
And 23h32'47" 48d43'25"
Mia Gabrial
Umi 15h7'28" 69d42'53"
Mia Georgia
And 23h12'55" 41d45'4"
Mia Grace
And 0h15'30" 45d36'19"
Mia Grace Bonomolo
And 1h34'24" 46d38'40"
Mia Grace Evans
Psc 0h10'47" 12d9'4"
Mia Helen Otoski
Uma 8h25'27" 72d6'52"
Mia Helene Thomas
Tau 4h37'51" 19d29'29"
Mia Hunt (Beautiful)
Mon 6h51'40" -0d20'11"
Mia Ilys Staveley
And 0h58'42" 34d46'56"
Mia India Mae Jones - Mia's Star
And 0h43'42" 30d17'6"
Mia Isabella
And 0h31'18" 32d21'26"
Mia Isabella Bacci
Cnc 8h50'51" 22d39'55"
Mia Isabelle
Cap 21h18'46" -19d11'12"
Mia Isabelle Gandini
And 23h12'6" 40d30'34"
Mia Izabella Mastrov
Mon 6h53'35" -0d7'17"
Mia Jones
Lyn 8h4'21" 41d14'35"
Mia Jordana Bensusan
Crb 16h17'46" 38d46'16"
Mia Josephine Pickett
Leo 11h19'44" 16d24'2"
Mia Joy Meadows
Lib 14h32'25" -9d36'57"
Mia Joy Poncia
And 23h28'24" 48d11'44"
Mia Julie Nam
Cap 21h39'23" -15d3'38"
Mia Kate Recine
Cam 4h10'50" 56d23'48"
Mia Katherine
Cnc 8h54'46" 32d37'43"
Mia Kathryn
Aqr 23h2'22" -7d54'40"
Mia Kineta Elliot
Cas 23h6'11" 58d48'16"
Mia Leckey - Mia's Star
Cru 12h8'36" -58d41'4"
Mia Lily Smyth
Cas 23h50'39" 52d42'29"
Mia Linda Scavitto
Uma 9h30'12" 65d56'7"
Mia Lisa
Lib 15h56'11" -10d4'17"
Mia Louise
And 1h0'25" 43d34'27"
Mia Louise Harding - Mia's Star
And 23h21'6" 42d53'44"
Mia Louise Kinley - Mia's Star
And 23h7'1" 51d38'25"
Mia Louise Marilyn Crutcher
And 2h6'59" 40d46'1"
Mia Madre, Bella & Brillando
Tau 5h7'1" 20d30'7"
Mia Michelle Tomaro
Aqr 22h8'26" -19d52'51"
Mia Moreing Russell
Ori 5h26'37" -5d8'52"
Mia Natalie
Cas 1h13'30" 68d16'8"
Mia Nevaeh Corazo
And 2h21'4" 46d13'24"
Mia Nicole
Tau 5h33'2" 25d23'14"
Mia Nicole Gonzalez
Uma 9h41'5" 60d49'40"
Mia Nicole Ruiz
Uma 11h25'40" 39d56'10"
Mia Noorzai
Ori 5h49'46" 9d55'36"
Mia Olive Noble
And 23h16'50" 48d4'43"
Mia Olvia Valter
Cas 23h51'16" 57d16'13"
Mia "Our Gift"
And 1h21'58" 50d29'39"
Mia Payne
Cas 0h37'1" 60d40'53"
Mia Pivirotto
Lyr 18h47'13" 44d38'7"
Mia Propato
Lyn 8h27'42" 55d28'50"
Mia Rachel Robinson
Sco 17h15'39" -33d3'50"
Mia Riley
Psc 0h33'16" 9d15'3"
Mia Rose
Lyn 7h3'52" 44d46'13"
Mia Rose
And 0h44'3" 30d48'43"
Mia Rose
Cas 0h43'40" 67d29'24"
Mia Rose Pellegrino
Ari 3h2'10" 25d48'49"
Mia Rose Shakeshaft
And 0h0'2" 45d14'58"
Mia Rose Spinozzi
Lib 14h28'59" -20d22'12"
Mia Rose's Rising Star
Vir 13h33'7" 5d7'21"
Mia RuZhen Casparian
And 0h35'40" 43d0'25"
Mia Sky
Ori 5h44'11" -4d11'59"
Mia Sofia Campos
Uma 8h11'28" 62d43'49"
Mia Soledad de la Fuente
Crb 16h15'24" 33d22'44"
Mia Soleil
Vir 11h52'38" 4d49'25"
Mia Sophie Leu
Her 16h14'49" 5d47'9"
MIA STELLA RALF
Aur 5h28'19" 40d45'21"
Mia Stokes
Cyg 19h55'29" 40d52'57"
Mia Tarricone
Ori 5h46'35" 0d8'59"
Mia The Cat
Sgr 17h52'17" -17d0'5"
Mia Theresa Boger
Uma 10h47'54" 50d3'23"
Mia Victoria
Psc 1h21'32" 26d2'17"
Mia Victoria Bowring
And 23h45'34" 36d30'1"
Mia Violet
And 22h59'58" 39d43'54"
Mia Yara Wottka
Ori 5h6'33" 3d59'18"
Mia12192006
Cyg 19h27'54" 53d13'14"
Miabobia
Crb 15h43'57" 26d16'35"
Miah
Umi 14h23'41" 77d1'33"
Miamagia
Gem 6h1'44" 26d51'45"
MiaMatt
Leo 11h2'1" 20d8'10"
Miana R. Stewart
Sgr 19h34'31" -43d4'27"
Miao-Zhen Forever
Leo 9h43'31" 12d54'23"
MiaPettey
Lyn 8h31'3" 50d6'56"
Miariail
Lyn 8h57'56" 41d49'21"
Mia's Eternal Flame
Peg 22h56'25" 30d19'15"
Mia's Magical Moment In Time
Sco 16h17'32" -17d22'8"
Mia's Own Special Star (for Wishes)
And 0h19'49" 29d25'51"
Mia's Shining Sparkler
And 1h12'30" 43d4'2"
Mia's Star
And 0h57'31" 43d26'53"
Mia's Star
And 0h1'1" 40d45'45"
Mia's Stella Cadente
Cam 6h18'32" 68d40'55"
Miavirgotexas
Vir 13h17'8" -18d39'59"
Mica
Uma 8h38'35" 72d51'3"
MICA
Leo 9h47'21" 32d36'39"
Mica
Dra 17h50'56" 51d16'7"
Mica Anthony LaBiche
Psc 0h56'46" 13d20'56"
Mica Baby
Peg 22h35'4" 14d56'33"
Mica Love
Tau 3h43'56" 27d27'23"
Micaela
Peg 21h30'47" 23d31'32"
Micaela
Crb 15h38'59" 38d24'43"
Micaela
Cyg 19h45'7" 46d34'47"
MICAELA ADRIANA MASCETTI
Lyr 18h57'22" 26d49'4"
Micaela Ann
Peg 22h32'42" 17d59'32"
Micaela D. Woodley
Umi 14h46'38" 74d49'0"
Micaela Florence Lipman
Uma 10h33'27" 62d32'49"
Micaela Grace Adame-Rocha
Gem 6h43'22" 13d22'11"
Micaela Josephine Scimone
Tau 4h28'40" 17d41'40"
Micaela Julia Gaither
Lyr 18h46'41" 34d16'33"
Micaela Manteo
Cnc 8h46'28" 32d51'42"
Micaela Marie Lazarra
Cyg 21h44'6" 37d21'9"

Micaela Ora Elizabeth Barber
Cyg 19h54'12" 36d55'36"
Micaela Wells
Uma 8h28'40" 61d58'29"
Micah
Uma 13h35'59" 58d14'26"
Micah
Sco 16h4'29" -9d30'26"
Micah
Her 17h35'26" 36d50'51"
Micah
Leo 9h39'3" 27d21'38"
Micah
Cnc 9h0'20" 23d43'8"
Micah And Donnie
Aqr 22h26'53" -2d36'25"
Micah Caleb
Ari 2h54'43" 25d44'33"
Micah (Carebear)
Sco 16h11'59" -19d6'52"
Micah David Rigg
Cru 12h20'57" -57d12'53"
Micah DeLeon
Cap 21h6'48" -20d25'14"
Micah Deon Washington"Millionaire"
Ori 6h17'28" 14d30'42"
Micah Earl Pierson
Cap 20h17'10" -20d4'21"
Micah James Herber
Aqr 20h43'32" 0d4'23"
Micah James Miles
Uma 11h41'27" 31d28'28"
Micah Jon Crabtree
Her 17h17'26" 22d20'29"
Micah & Kathy Wright
Umi 15h35'48" 73d27'2"
Micah Kealiiokahale Byung Kong
Dra 17h31'42" 62d27'5"
Micah Kenyon Sullivan
Umi 14h51'15" 80d44'34"
MIcah & Krissi Lenox
Lib 14h54'26" -16d59'29"
Micah Lands
Ori 5h55'37" 13d44'10"
Micah Lawrence Tath-Slezak
Cap 21h53'39" -9d21'11"
Micah Lazenby
Lyr 18h28'1" 36d27'35"
Micah Lonnie Pilger
Lyn 8h11'54" 50d6'36"
Micah Paul Lester
Cen 13h50'42" -43d6'37"
Micah Scott Ives
Tau 4h15'39" 3d30'12"
Micah Sean
Lib 15h7'20" -24d40'43"
Micah (The Gift of GOD)
Cyg 19h57'53" 33d25'39"
Micah Warren Lawrence
Aqr 23h4'57" -10d13'59"
Micah Woodley
And 1h26'38" 44d0'59"
Micah Alfen
Cap 20h35'21" -21d29'47"
Micah's
Umi 14h59'51" 81d15'54"
Micah's Star
Uma 9h2'3" 63d30'58"
Micaile Elizabeth Lopez
Vir 13h46'11" 5d9'13"
mical
Sgr 19h43'11" -13d15'32"
Micala
Lyn 9h1'32" 44d49'9"
MicaLovesTomForever
Ori 6h24'43" 16d59'5"
Micariñitis
Aqr 22h10'33" -16d39'54"
Micayla Bree
And 0h15'5" 42d34'15"
Micayla Elise Black
Crb 15h36'24" 28d8'42"
Micayla Linn Walling
Lib 15h32'11" -13d1'48"
Micayla Lynn Farber
Gem 7h16'15" 15d26'52"
MicBec X
Cru 12h34'54" -64d3'26"
MicBeth
Cyg 21h53'15" 49d39'55"
Miccio Family
Dra 20h44'50" 80d35'3"
Micè
Cyg 20h47'13" 32d10'49"
Mich and Arons Star
Cyg 20h31'19" 48d22'12"
Micha
And 0h49'14" 42d33'34"
Micha
Crb 15h32'52" 35d42'57"
Micha
Ori 5h3'39" 7d10'39"
Micha Ariel Soble
Tau 5h34'22" 27d0'12"
Micha Renee Dalton
Tau 4h43'7" 7d56'36"
Micha Shuman
Umi 13h54'32" 89d19'4"
Micha Toni Flückiger
Umi 14h36'43" 69d9'47"
Michael
Cep 21h21'45" 62d9'31"
Michael
Cas 1h19'17" 66d12'32"
Michael
Uma 14h19'21" 60d39'17"
Michael
Uma 10h17'15" 63d29'31"
Michael
Uma 9h2'10" 57d19'9"
Michael
Uma 10h1'23" 53d44'20"
Michael
Lyn 6h23'0" 54d34'53"
Michael
Lyn 7h31'18" 58d37'12"
Michael
Uma 13h33'36" 54d39'43"
Michael
Umi 14h35'35" 84d24'51"
Michael
Ori 5h33'19" -0d36'14"
Michael
Vir 12h1'33" -0d40'47"
Michael
Lib 14h54'15" -1d53'34"
Michael
Lib 15h14'57" -9d54'24"
MICHAEL
Sgr 19h7'9" -12d0'34"
Michael
Cma 7h25'47" -19d48'0"
Michael
Sgr 18h48'40" -20d5'35"
Michael
Lib 15h53'9" -20d7'29"
Michael
Sgr 19h28'31" -29d4'4"
Michael
Ori 6h3'7" 10d7'16"
Michael
Ari 2h26'10" 11d1'26"
Michael
Aql 19h24'28" 3d14'46"
Michael
Psc 23h26'23" 4d33'9"
Michael
Leo 9h29'38" 10d44'39"
Michael
Tau 5h14'21" 22d28'48"
Michael
Aql 19h47'2" 13d8'7"
Michael
Psc 1h9'17" 15d59'50"
Michael
Ari 2h4'21" 22d0'38"
Michael
Boo 14h34'32" 23d4'35"
Michael
Com 12h45'46" 28d45'57"
Michael
Uma 11h17'21" 29d41'1"
MICHAEL
Leo 9h45'17" 27d46'54"
Michael
Leo 11h18'33" 24d8'38"
Michael
Her 16h30'3" 28d39'11"
Michael
Leo 11h20'57" 16d52'39"
Michael
Leo 11h44'7" 20d45'42"
Michael
Leo 9h40'32" 21d41'31"
Michael
Ari 2h34'14" 25d59'6"
Michael
Uma 12h0'56" 33d12'26"
Michael
Gem 6h44'21" 33d53'30"
Michael
Gem 6h56'48" 32d4'58"
Michael
Cyg 20h4'13" 32d10'27"
Michael
Uma 9h26'21" 42d21'58"
Michael
Per 3h40'6" 39d7'49"
Michael
Cyg 21h21'25" 46d14'23"
Michael
Boo 15h29'43" 46d11'39"
Michael
Uma 11h44'33" 47d4'32"
Michael
Uma 9h45'43" 45d43'36"
Michael
Uma 8h41'3" 46d41'7"
~ Michael ~
Per 3h30'1" 51d5'20"
Michael
Aur 5h48'6" 49d40'14"
Michael
Aur 5h48'18" 50d23'44"
Michael
Per 4h8'43" 51d16'51"
Michael
And 1h45'31" 49d48'38"
Michael
Lac 22h49'6" 40d55'43"
Michael 17 June 1935
Uma 11h27'39" 58d36'12"
Michael A. Ausiello (Little Pookie)
Cnc 8h41'55" 14d16'47"
Michael A. Battaglini
Aqr 23h11'26" -9d18'23"
Michael A. Berard
Gem 6h36'12" 12d34'58"
Michael A. Blanchard
Cap 20h29'24" -18d37'19"
Michael A. Brandes
Ori 6h12'57" 13d44'28"
Michael A. Brandt
Leo 10h32'3" 9d54'14"
Michael A. Cancel
Her 17h46'5" 32d41'14"
Michael A. Cappuccio
Ari 3h18'43" 29d17'24"
Michael A Castellani
Per 2h48'35" 53d30'54"
Michael A. Ciccarelli
Ori 6h3'25" 11d3'1"
Michael A. Cilento
Tau 4h30'42" 17d0'34"
Michael A. Dickinson II
Umi 15h18'11" 71d1'32"
Michael A Edwards
Her 16h17'35" 19d40'10"
Michael A. Eldridge
Ori 5h41'33" -0d15'54"
Michael A. Fuoch
Tau 5h31'1" 20d35'40"
Michael A. Gajewski
Tau 4h32'23" 17d27'53"
Michael A Gonzales
Sgr 18h33'49" -17d39'58"
Michael A. KinCannon
Tau 4h13'44" 9d18'46"
Michael A. Koplen
Her 17h57'28" 16d19'51"
Michael A. Koveleski
Vir 12h27'8" 4d58'4"
Michael A. Landolfa
Sco 17h0'56" -39d18'27"
Michael A. Lindstrom
Uma 11h15'7" 62d31'23"
Michael A. Mencarelli
Sco 17h19'12" -31d57'36"
Michael A. Missel
Lyn 7h58'1" 59d28'5"
Michael A. Moberly
Her 18h20'41" 17d44'56"
Michael A. Montalbano
Her 18h15'5" 21d35'31"
Michael A. Montano
Boo 15h19'50" 49d28'2"
Michael A. Morris
Psc 1h6'47" 10d56'3"
Michael A. Nachman
Crb 15h45'9" 26d34'30"
Michael A. Page
Cap 21h47'52" -16d58'1"
Michael A. Peck
Cnc 8h47'12" 32d45'17"
Michael A. Petitti
Cyg 20h40'7" 51d14'57"
Michael A. Ponzi
Vir 13h38'12" -6d8'8"
Michael A. Popson
Ori 6h18'37" 14d53'31"
Michael A. Raftery
Ori 5h27'15" -1d36'6"
Michael A. Ruth
Cap 21h7'38" -18d21'17"
Michael A. Sacco
Cep 22h44'29" 66d41'45"
Michael A. Scholl
Vir 12h36'11" -7d48'10"
Michael A. Schori
Cep 21h55'49" 71d41'38"
Michael A. Schwadron
Aqr 22h39'28" 0d19'36"
Michael A. Squillini, Jr.
Vir 12h58'48" -5d26'22"
Michael A. Sturman
Dra 18h41'43" 73d38'21"
Michael A. Tozzi
Sgr 19h1'56" -14d32'7"
Michael A. Wilson
Cep 22h18'25" 62d20'8"
Michael A Wilson
And 23h7'25" 42d37'29"
Michael A. Zonfrillo III
Cnc 8h24'40" 32d9'46"
Michael Aaron Timmons
Lyn 8h9'59" 33d59'20"
Michael Aaron Wechsler
Psc 2h4'38" 5d43'8"
Michael AbouEzzi
Boo 14h34'58" 18d40'58"
Michael Abraham Akaka Wong
Sco 16h11'47" -25d56'28"
Michael Abraldes
Uma 11h26'47" 47d58'28"
Michael Abrams loves Natalie Gunn
Psc 1h50'52" 5d32'31"
Michael Adam
Aqr 22h16'8" 1d37'15"
Michael Adam McDermott
Her 16h59'39" 32d23'54"
Michael Alan Ackerman
Lyn 8h40'9" 43d26'11"
Michael Alan Bouchard
Tau 5h51'58" 16d49'32"
Michael Alan Collins
Lyn 7h53'11" 34d48'45"
Michael Alan Grasmeyer
Cep 1h47'8" 82d59'37"
Michael Alan Price Waterhouse
Ori 6h18'45" 10d34'32"
Michael Alan Schumann
Her 17h36'29" 36d49'40"
Michael Alan Smeltzer
Ori 5h59'52" 21d52'48"
Michael Albert Santarelli
Cas 1h43'46" 67d19'15"
Michael Alen Collier
Cnc 8h58'39" 29d50'23"
Michael Alex Wheeler
Ori 5h3'52" 5d18'8"
Michael Alexander
Per 4h13'56" 45d3'46"
Michael Alexander
Cyg 20h12'58" 55d54'17"
Michael Alexander Copley
Col 5h13'36" -28d28'28"
Michael Alexander Miller
Hya 9h41'21" -21d59'15"
Michael Alexander Olazabal
Ori 5h30'36" 0d59'52"
Michael Alexander Uriati
Dra 19h49'56" 61d57'0"
Michael Alger
Tau 3h59'18" 9d12'54"
Michael Aligata
Cyg 21h30'36" 51d36'29"
Michael Allan Conliffe
Ari 3h24'38" 26d25'28"
Michael & Alleen Borys
And 0h13'16" 28d30'28"
Michael Allen
Aur 5h18'29" 49d47'48"
Michael Allen Bonner
Her 16h28'17" 14d1'34"
Michael Allen Gentilini
Leo 11h39'44" 25d21'7"
Michael Allen Grismore
Her 17h28'38" 36d36'36"
Michael Allen Howe
Aql 19h45'48" -0d56'16"
Michael Allen Kallok
Sco 16h5'16" -11d58'4"
Michael Allen Lochner
Cnc 9h12'48" 27d49'29"
Michael Allen Lollar
Cam 3h28'8" 61d48'7"
Michael Allen McDonald
Psc 1h56'17" 7d2'36"
Michael Allen Napier Jr.
Her 17h49'45" 26d8'55"
Michael Allen Ritter
Sgr 19h23'13" -24d52'16"
Michael Allen Williams
Lib 15h5'13" -5d8'9"
Michael Alley Strong
Aur 7h16'32" 42d17'30"
Michael Allsberry
Aur 6h44'31" 44d13'59"
Michael Aloisio Nastali
Sgr 19h1'5" -17d38'0"
Michael Alvin Bukis
Per 4h6'19" 34d16'14"
Michael & Ana Paigo
Ori 5h59'44" 21d43'26"
Michael Ancona Randy
Lib 15h40'37" -12d48'42"
Michael and Adrina
Cyg 21h41'53" 43d55'43"
Michael and Amy Murphy
And 0h5'42" 39d37'16"
Michael and Anne Marie
Cyg 20h38'38" 45d38'55"
Michael and Beverly Silkey DVM
Lyn 8h11'2" 40d11'18"
Michael and Bobbi Allen
Cyg 20h8'33" 47d7'43"
Michael and Brandi Miller
Sex 9h50'50" -0d38'7"
Michael and Carol Fournier
Sco 17h18'54" -40d15'56"
Michael and Christina Improta
Cyg 20h35'55" 48d55'29"
Michael and Christy Nielsen
Cyg 21h46'30" 46d30'7"
Michael and Cinnamon Reiter
Cyg 19h49'23" 31d11'43"
Michael and Deanna Klafka
Lib 14h45'40" -10d39'56"
Michael and Dolores Hughes
Cyg 19h52'0" 32d52'45"
Michael and Dylan
Aqr 20h49'59" 0d50'44"
Michael and Elizabeth's Star
Uma 13h58'29" 55d48'41"
Michael and Francesca Hensley
Aur 6h12'17" 31d8'38"
Michael and Grace Terranova
Lyr 19h10'24" 27d15'30"
Michael and His Own 1
Ori 5h35'16" 1d50'12"
Michael and His Own 2
Ori 5h42'18" 0d29'15"
Michael and His Own 3
Ori 5h31'51" 0d3'21"
Michael And Holli Cutler Forever
Cyg 20h3'46" 38d55'7"
Michael and Janelle
Per 3h25'35" 46d41'33"
Michael and Jeannette
Cyg 21h26'8" 34d23'5"
Michael and Jeff Lilly
Lyr 18h52'27" 36d52'13"
Michael And Jessica
Uma 13h42'33" 54d34'59"
Michael and Joly
Lyr 18h53'30" 32d26'44"
Michael and Julianne Panikkou
Cyg 20h7'0" 37d57'11"
Michael and Julie
Cyg 21h13'7" 47d12'19"
Michael and Kara
Srp 18h4'29" -0d47'19"
Michael and Kristi Gaffery's
Cyg 20h34'27" 30d43'37"
Michael and Kristina
Aqr 22h57'3" -9d59'1"
Michael and Kristy Hedler
Cyg 21h10'49" 47d43'57"
Michael and Laura Lee, Love Forever
Her 17h50'43" 48d50'33"
Michael and Lauren
Cap 20h24'58" -22d57'0"
Michael and Lauren's Star
Cyg 21h17'6" 51d26'30"
Michael and Linda Gold 07/16/1972
Cyg 20h34'19" 47d11'3"
Michael and Lindsey
Cyg 19h20'49" 51d53'5"
Michael and Lisa Eastman
Uma 9h59'21" 49d25'20"
Michael and Martha McKinnon
Cyg 21h8'5" 47d56'50"
Michael and Martha Rumney
Uma 9h24'38" 41d59'48"
Michael and Mary Ritz
Gem 6h27'27" 20d48'52"
Michael and Mel's Star
Cnc 8h23'33" 31d14'39"
Michael and Paige Fuller
Vir 12h50'55" 12d31'4"
Michael and Paula Heffner
Uma 10h30'38" 40d57'20"
Michael and Sarah
Lyr 18h30'59" 36d46'42"
Michael and Shannon's Star
Leo 11h14'47" 23d6'31"
Michael and Sherry Cook
Psc 23h49'29" 6d6'35"
Michael and Susan Forcellina
Uma 13h42'6" 53d29'45"
Michael and Suzanne Kearns 2004
Peg 22h6'49" 30d53'57"
Michael and Tinamarie
Per 4h26'19" 50d40'14"
Michael Anders Gulow
Ari 3h17'40" 29d39'12"
Michael Anderson-Rockson
Peg 22h8'48" 14d19'16"
Michael Andre Cammarata
Ari 3h11'44" 28d9'52"
Michael Andrew
Per 2h41'44" 56d27'12"
Michael Andrew Butala
Cep 3h4'55" 82d12'2"
Michael Andrew Cardinalli
Gem 7h18'51" 31d35'5"
Michael Andrew Coudoures
Gem 7h17'17" 27d13'16"
Michael Andrew Devney
Lyn 7h33'20" 48d1'56"
Michael Andrew Howard
Ori 5h42'52" 9d4'54"
Michael Andrew Laidlaw
Vir 13h27'24" -4d36'32"
Michael Andrew Lovda
Uma 10h29'48" 61d24'52"
Michael Andrew Milinovich
Aqr 22h8'8" 1d39'20"
Michael Andrew Sanchez
Cnc 8h18'1" 16d41'16"
Michael Andrew Seiter Jr.
Cap 21h32'26" -24d34'27"
Michael Andrews
Vir 12h27'45" 2d12'50"
Michael Andrew's Mercury
Ori 6h3'57" 12d53'3"
Michael Angel Zamora
Leo 11h50'10" 23d26'42"
Michael Angelakos
Tau 4h19'53" 29d7'10"
Michael Angelo
Tau 5h19'41" 20d58'13"
Michael Angelo Bayer
Tau 3h32'17" 8d51'31"
Michael Angelo Colleli
Sco 16h24'42" -18d54'15"
Michael Angelo Greenwald
Gem 6h42'17" 14d42'43"
Michael Angelo Prudenti
Cep 21h51'18" 61d26'30"
Michael Angelo Rosato
Gem 7h3'9" 27d12'46"
Michael Angelo Wright
Lib 16h0'49" -18d4'26"
Michael Angelo, Star of My Heart
Per 3h25'37" 33d5'3"
Michael/Angelo's Magic
Cyg 19h36'39" 52d18'50"
Michael & Ania's Star
Eri 1h36'51" -54d46'6"
Michael & Anita
Cru 12h2'33" -62d47'16"
Michael Ann "Mikey"
Cas 1h36'24" 66d39'40"
Michael Anthony
Lib 15h35'17" -18d46'33"
Michael Anthony
Her 17h40'9" 47d54'55"
Michael Anthony
Uma 10h44'14" 48d33'40"
Michael Anthony
Per 2h21'26" 57d16'56"
Michael Anthony
Psc 0h45'44" 21d0'59"
Michael Anthony Anderson
Her 16h52'46" 37d14'16"
Michael Anthony Ashcraft
Gem 6h36'55" 24d42'4"
Michael Anthony Baldesarra
Tau 4h11'42" 30d45'42"
Michael Anthony Calaro
Gem 6h57'26" 14d31'42"
Michael Anthony Cartwright
Uma 12h3'40" 64d5'56"
Michael Anthony Castillo
Sco 16h17'13" -19d35'44"
Michael Anthony Coleman
Ari 2h15'43" 25d36'31"
Michael Anthony Croft
Cep 5h3'20" 84d33'12"
Michael Anthony Dupree
Tau 5h38'32" 24d58'31"
Michael Anthony Edwin McCarthy
Gem 7h11'20" 27d38'37"
Michael Anthony Flowers
Tau 5h38'23" 26d20'8"
Michael Anthony Georgeo
Uma 13h18'9" 56d50'50"
Michael Anthony Gureghian
Vir 13h18'27" -21d39'6"
Michael Anthony Handel Jr.
Vir 12h31'52" 7d20'36"
Michael Anthony Hunsinger
Psc 23h32'4" 0d2'57"
Michael Anthony Joniec
Uma 10h20'12" 45d5'28"
Michael Anthony LaJoice
Aqr 21h52'13" 2d4'43"
Michael Anthony LaMas
Aqr 21h33'50" 1d43'50"
Michael Anthony Lobello
Uma 13h59'17" 53d42'20"
Michael Anthony Magill
Vir 13h31'49" 3d28'27"
Michael Anthony Mason "Dad"
Boo 15h33'11" 48d40'57"
Michael Anthony McNew
Boo 15h19'17" 38d48'15"
Michael Anthony McTarsney, Jr.
Uma 11h43'30" 63d43'54"
Michael Anthony Mincey
Uma 9h49'4" 70d53'1"
Michael Anthony Mincey
Leo 10h13'1" 17d27'20"
Michael Anthony Minor, Jr.
Leo 10h36'39" 15d10'49"
Michael Anthony Nixon
Sco 17h25'22" -45d14'45"
Michael Anthony Nunno Jr.
Sgr 19h10'18" -21d46'35"
Michael Anthony Packett
Cru 12h19'34" -56d56'30"
Michael Anthony Palermo: My Light
Lib 15h13'32" -20d47'36"
Michael Anthony Parker
Her 16h35'51" 34d59'7"
Michael Anthony Pascarella
And 23h40'8" 44d57'39"
Michael Anthony Pattilio, Jr.
Leo 11h41'37" 25d52'47"
Michael Anthony Penzo
Uma 11h3'43" 36d31'49"
Michael Anthony Perez
Leo 11h51'12" 20d7'10"
Michael Anthony Potenza
Her 16h27'8" 44d20'25"
Michael Anthony Rivera 9/20/84
Ori 5h57'15" 17d52'48"
Michael Anthony Rodarte Jr.
Sco 16h53'56" -32d40'42"
Michael Anthony Simms, II
Psc 1h14'52" 7d6'8"
Michael Anthony Simon
Aqr 23h4'20" -17d8'27"
Michael Anthony Thoma (Dad)
Ari 3h6'29" 22d42'58"
Michael Anthony TTIML Mizzone
Leo 11h33'45" -5d22'27"
Michael Anthony Turner
Uma 11h13'22" 70d41'43"
Michael Anthony White
Lib 14h52'6" -4d36'23"
Michael Arena
Cru 12h19'30" -56d53'54"
Michael Arra
Cnc 8h49'53" 9d13'44"
Michael Arrom
Ori 5h0'38" 14d20'13"
Michael Arthur Barber
Ori 5h28'41" -9d0'49"
Michael Arthur Harris
Cap 20h28'28" -22d41'18"
Michael Arthur Janzen 1969
Gem 7h14'57" 19d51'22"
Michael Artress
Aur 5h23'45" 41d15'34"
Michael & Ashlea's Forever Star
Cnc 8h13'56" 16d10'41"
Michael Astor
Sgr 17h58'56" -26d24'0"
Michael Aubert
Lib 15h42'47" -5d39'52"
Michael August Kemmer
Her 17h35'56" 33d9'0"
Michael Augustine
Per 4h34'33" 50d50'13"
Michael Augustus Shaw
Uma 9h46'54" 56d32'22"
Michaël Aurélie pour toujours
Del 20h34'20" 16d51'30"
Michael Austin
Uma 10h55'30" 61d5'3"
Michael Austin Vickers
Sgr 18h45'30" -17d57'49"
Michael B. Baranowski
Dra 19h50'51" 72d6'23"
Michael B. Matney
Aur 6h33'18" 35d4'14"
Michael B. Somwar
Cyg 20h59'10" 32d16'4"
Michael Babin
Gem 7h14'0" 17d19'20"
Michael Baker
Uma 9h42'0" 56d44'27"
Michael & Barbara Southcott
Dra 18h11'50" 55d9'51"
Michael Barilli
Her 17h8'42" 24d2'25"
Michael Barker
Psc 0h18'32" 16d50'56"
Michael Barnett
Ori 6h19'27" 9d4'47"
Michael Bartilucci
Lib 15h4'57" -27d25'18"
Michael Bayer
Uma 9h53'39" 42d10'33"
Michael Bayroff
Lac 22h12'32" 49d45'39"
Michael (BB) Filteau McNeilly
Psc 0h19'23" 17d1'23"
Michael Beavens
Sco 16h11'11" -10d27'40"
Michael Beaver Tarasoff
Gem 7h33'29" 15d21'35"
Michael Becker -Spinni-
Ori 6h23'41" 10d35'4"
Michael Behnke Alias Donkey Boy
Lyn 6h19'12" 61d57'25"
Michael Bell
Aql 19h24'58" 14d48'43"
Michael Bellinghiri
Cyg 19h42'47" 32d49'25"
Michael Belliveau
Aqr 22h31'36" -0d20'53"
Michael Bello
Per 4h43'45" 45d45'48"
Michael Benfante
Cep 20h42'42" 61d43'35"
Michael Benjamin Dudek
Cep 21h50'57" 63d11'47"
Michael Benjamin Matthew Hutchinson
Per 2h47'4" 54d13'27"
Michael Benkovic
Uma 9h32'24" 67d59'6"
Michael Bernard Demarco
Gem 7h2'46" 26d46'12"
Michael Bernard Green III
Psc 1h19'11" 16d43'49"
Michael Betlow
Cep 20h6'48" 60d42'21"
Michael Bin
Sgr 18h30'19" -27d28'15"
Michael Biondolillo, my universe!
Cap 20h19'45" -10d38'25"
Michael Blackwood
Aur 5h45'13" 49d33'34"

Michael Blade Taylor
Sco 17h7'48" -38d30'7"
Michael Blumenfeld
Uma 9h52'43" 52d22'48"
Michael Bolton
Sgr 18h3'44" -29d48'58"
Michael Bonay
Leo 11h12'4" 20d29'46"
Michael Booth The Great
Ori 6h16'56" 11d5'33"
Michael Borukhov & Karina Natanova
Lib 14h49'12" -1d4'29"
Michael Bosse
Ori 5h35'48" -5d10'30"
Michael Bosse
Boo 14h48'9" 24d49'2"
Michael Bowe
Boo 13h49'52" 11d5'33"
Michael Boyd
Aur 5h54'38" 53d48'23"
Michael Boyden
Uma 11h42'4" 50d7'57"
Michael Bozarth "My Starman"
Uma 11h36'3" 29d39'51"
Michael Bozikakes
Uma 11h7'15" 44d41'10"
Michael Bradley Marcon
Dra 19h1'25" 63d50'1"
Michael Braeden McMillan
Umi 15h24'21" 71d56'32"
Michael Brainerd
Sgr 18h30'6" -18d8'47"
Michael Brandon Pfannenstiel "P"
Her 16h55'51" 13d15'48"
Michael Braum
Lib 15h27'51" -6d30'46"
Michael Braun
Ori 6h19'16" 8d13'49"
Michael Braun "50th Birthday"
Psc 23h6'13" 5d19'40"
Michael Brendan Brodeur
Aql 19h59'12" 3d38'53"
Michael Brennan
Sgr 17h53'36" -29d47'45"
Michael Brent Bowlby
Lib 15h36'27" -7d30'4"
Michael Brian Dolan
Psc 0h43'34" 18d57'14"
Michael Brian Oade
Ari 3h26'55" 25d53'46"
Michael Brisky
Her 17h37'59" 36d53'35"
Michael & Brittanie Larsen
Cyg 19h43'46" 30d42'35"
Michael Broche
Uma 11h23'34" 66d1'51"
Michael Broening
Aqr 22h37'56" 0d34'6"
Michael Brothers - Mike's Star
Her 18h25'43" 24d58'7"
Michael Brown
Aql 19h44'11" -11d19'26"
Michael Browning
Gem 6h27'39" 25d9'44"
Michael "Bruce" Armbruster
Cnc 8h11'22" 16d21'2"
Michael Bruce Budd
Cru 12h17'43" -63d47'46"
Michael Bruce Goodman
Per 2h21'7" 55d17'36"
Michael Bruno John Franz England
Cep 22h56'24" 59d5'11"
Michael Bryan Cahill
Per 4h32'9" 39d19'48"
Michael Bryon Zenk
Mon 6h58'51" 0d14'30"
Michael Büchner
Uma 10h33'37" 41d47'39"
Michael Buel
Dra 20h25'15" 66d41'55"
Michael (Bunny) Thompson
Sco 17h1'38" -44d10'8"
Michael Burke
Lyn 7h43'22" 59d3'31"
Michael Bürsing
Uma 10h2'30" 68d18'37"
Michael Burt Moore
Ser 18h17'41" -13d53'26"
Michael Bush
Uma 10h10'44" 51d48'17"
Michael Busse
Her 18h21'44" 23d35'5"
Michael Butterworth
Cnc 8h39'12" 8d5'3"
Michael C. Baker
Ari 3h21'22" 18d28'4"
Michael C Darling
Cyg 20h46'7" 43d42'25"
Michael C. Dedovich
Cap 20h7'43" -10d20'37"
Michael C. Fischer
Aqr 22h22'49" -23d52'26"
Michael C. Giza
Crb 15h42'30" 27d4'29"
Michael C. McGee
Oph 17h37'35" -0d42'51"
Michael C. Morales
Sco 16h10'56" -10d38'5"
Michael C. Morelli
Leo 9h34'22" 31d40'11"
Michael C- My Heart-My Desire
Cap 20h15'13" -10d50'41"
Michael C. Napoleone-Shining Light
Tau 3h37'38" 0d54'36"
Michael C. O'Brien
Uma 11h47'26" 48d28'34"
Michael C. Prihodka
Aur 5h44'1" 50d10'23"
Michael C. Sangee
Boo 14h13'51" 28d20'0"
Michael C. Skinner
Aur 5h49'23" 40d42'52"
Michael C. Stead
Her 17h40'41" 21d3'42"
Michael C Walden Sr
Her 17h36'19" 36d24'12"
Michael Caleb Lassiter
Cru 12h48'28" -57d20'39"
Michael Calomino
Cep 20h44'13" 66d44'11"
Michael Campanella
Aur 6h38'50" 38d55'49"
Michael Cannon Kerwin
Her 17h10'37" 30d45'1"
Michael Capbarat
Aql 19h5'42" 8d13'45"
Michael Caprai
Per 2h40'43" 56d42'33"
Michael Carl Lamm
Sgr 18h49'14" -27d51'27"
Michael Carl Moberg
Dra 15h32'6" 60d57'1"
Michael Carmine Gargano
Aur 5h40'6" 29d9'36"
Michael Cary Reynolds
Her 17h57'43" 49d34'19"
Michael Casey
Uma 9h31'19" 53d27'18"
Michael Castellano
Cep 21h27'6" 78d22'22"
Michael Catizone
Tau 4h11'25" 2d44'30"
Michael Cavanagh
Her 17h3'57" 24d42'53"
Michael Cea, Jr.
Sco 17h4'14" -34d51'18"
Michael Ceslok
Sco 16h12'41" -11d39'0"
Michael Chad Hannigan
Cep 23h40'46" 78d44'34"
Michael Chad Thaysen,Jr.(Bubba)
Her 16h7'44" 23d0'52"
Michael Charles Bee
Per 4h36'24" 38d20'25"
Michael Charles Costanza
Leo 11h20'18" 0d42'34"
Michael Charles Egner
Lyn 9h13'35" 34d39'20"
Michael Charles Kress
Cas 23h50'20" 62d39'34"
Michael Charles McCarty
Tau 5h16'44" 17d15'28"
Michael Charles Napoleone
Cep 22h20'0" 63d40'27"
Michael Charles Reardon
Vir 14h18'19" -19d3'6"
Michael Charles Riedel
Ori 5h16'29" -5d6'58"
Michael Charles Webster "Starboy"
Psa 21h44'55" -30d49'1"
Michael Chepkwony's Guiding Light
Uma 8h17'49" 64d41'33"
Michael Chesley
Uma 10h48'24" 60d22'47"
Michael Chesley Jr.
Uma 10h49'5" 57d25'53"
Michael Chezem Star
Hya 9h40'23" -0d26'35"
Michael Chirco
Aql 18h57'30" 8d2'4"
Michael Chrepta
Aqr 22h22'57" -22d50'44"
Michael Christian
Cnc 8h17'16" 25d12'51"
Michael Christian Craig
Gem 6h30'18" 25d27'38"
Michael & Christine
Lyn 7h48'20" 47d51'18"
Michael & Christine Semb
Sgr 19h1'41" -33d28'7"
Michael Christmas Rito
Umi 15h31'10" 69d57'48"
Michael Christopher Abt Jr.
Peg 22h25'1" 24d3'35"
Michael Christopher Albright
Leo 10h6'6" 23d17'2"
Michael Christopher Benes
Uma 9h40'22" 57d56'17"
Michael Christopher Cox
Lib 14h54'7" -6d2'14"
Michael Christopher Cunningham
Ori 6h20'55" 16d42'49"
Michael Christopher Moore
Uma 10h56'10" 63d2'26"
Michael Christopher Villani
Leo 9h49'6" 28d20'1"
Michael Christopher Johnson
Gem 7h33'7" 26d24'50"
Michael Ciampi IV
Uma 12h4'51" 36d33'27"
Michael Ciarlo
Uma 11h34'51" 37d35'49"
Michael & Cisco Cosgrove
Dra 16h10'50" 53d39'36"
Michael Claridge
Per 3h30'29" 50d10'14"
Michael & Claudia 25.September 1999
Uma 9h11'53" 52d13'32"
Michael Clayton Barnett
Boo 14h25'18" 23d59'58"
Michael Clayton Miracle
Her 17h19'51" 37d15'47"
Michael Clayton Moore
Cnc 7h57'26" 16d59'20"
Michael Clemmons
Ari 1h48'59" 21d42'54"
Michael Clifford Arey
Leo 11h37'48" 14d1'46"
Michael & Cody Hooper
Uma 11h38'1" 62d0'5"
Michael Colagioia Jr.
Her 16h50'45" 34d12'57"
Michael Cole
Per 3h5'34" 47d31'30"
Michael Colin Rogers
Vir 12h46'51" 0d46'52"
Michael & Colleen McCabe
Cyg 21h37'56" 52d24'33"
Michael Collier
Uma 9h54'10" 53d31'39"
Michael Collins Piper
Cnc 8h47'55" 13d1'24"
Michael Colwell Libby aka Uncle Lib
Aqr 22h1'57" -21d0'38"
Michael Conacher
Cnc 8h46'55" 32d48'29"
Michael Conley
Aqr 22h56'57" -5d29'45"
Michael Connor
Sco 17h56'33" -37d59'2"
Michael Conrad
Aqr 21h29'43" 2d12'31"
Michael Conrads
Uma 10h22'41" 45d49'8"
Michael & Constance Wanniski
Aqr 22h46'42" -20d1'46"
Michael Coogan
Uma 10h49'18" 72d32'37"
Michael Cooker
Sco 16h4'47" -19d17'8"
Michael Cooper
Cen 13h30'22" -45d19'39"
Michael Cooper Mandel
Dra 16h0'37" 61d6'0"
Michael Cordell Webb
Ori 6h11'22" -0d20'37"
Michael Costa
Tau 3h47'2" 14d5'18"
Michael Craig Barry
Tau 4h21'16" 28d1'14"
Michael Crawford
Car 7h24'42" -52d21'49"
Michael Crawford
Cru 12h17'44" -56d44'49"
Michael Creech
Ari 3h20'29" 22d24'57"
Michael Creegan
Per 4h13'44" 37d13'24"
Michael Curtis Davito
Cep 21h24'25" 57d33'56"
Michael Curtis Robinson
Her 16h45'41" 27d52'33"
Michael Custer
Lmi 10h54'43" 25d47'55"
Michael/Cynthia/&~James Carbone
Uma 13h35'6" 58d22'19"
Michael D. Brewer
Her 18h0'1" 21d39'51"
Michael D. Brock
Her 17h39'44" 49d28'8"
Michael D Coon
Psc 1h7'35" 25d48'9"
Michael D. Hoffman
Her 17h15'1" 32d20'50"
Michael D. Jones
Cep 23h34'59" 78d8'26"
Michael D. Kirk
Pho 0h43'41" -42d20'53"
Michael D. Lappin
Boo 15h18'54" 43d29'21"
Michael D. McGrail
Psc 0h5'44" 0d18'2"
Michael D. Medich
Aqr 21h43'23" -1d0'47"
Michael D. Miele
Ari 2h21'26" 12d8'15"
Michael D. Morigeau
Her 17h39'7" 47d43'40"
Michael D. Provost
Psc 1h16'29" 28d24'35"
Michael D. Rosen
Her 16h51'10" 33d49'9"
Michael D. Schmidt
Leo 9h50'25" 32d4'52"
Michael D. Spiliotis
Aqr 20h51'29" 2d13'58"
Michael D. Stancheck
Cap 20h28'26" -11d1'12"
Michael D. Unley
Uma 11h34'29" 43d50'3"
Michael D. Waggoner
Psc 0h33'5" 2d52'57"
Michael D Watts
Gem 6h28'17" 15d18'5"
Michael (Dad & Da) Livingstone
Ori 5h28'34" -5d36'59"
Michael "Daddy" Laino
Her 16h36'36" 36d59'55"
Michael & Daisie's Anniversary Star
Cas 1h46'5" 69d2'31"
Michael Dale
Sgr 19h26'41" -13d7'14"
Michael Dale Coleman II
Uma 8h43'50" 66d40'58"
Michael Dale Jackson
Sgr 18h18'33" -24d38'3"
Michael Dallas
Peg 22h49'49" 13d13'47"
Michael Dallas Superstar
Her 16h40'17" 32d18'59"
Michael Dalvano
Aqr 22h33'25" 2d3'55"
Michael Damian Andersson
Cyg 20h17'58" 55d44'9"
Michael Damian Terrance Beekman
Psc 1h23'22" 28d42'33"
Michael Dan
Cep 22h19'49" 58d30'32"
Michael Dan Coker
Her 18h36'6" 19d1'6"
Michael Dana Voss
Per 4h1'26" 47d8'36"
Michael Dane Baldoni
Lyr 18h48'53" 30d42'35"
Michael Daniel
Ori 5h20'21" 1d13'48"
Michael Daniel Fregara
Per 2h39'1" 54d16'15"
Michael Darren Hymel Naquin 1967
Cnc 8h19'14" 29d23'0"
Michael David Bender
Ori 6h13'40" 19d16'20"
Michael David Burmaster
Uma 10h50'14" 64d32'12"
Michael David Chini
Uma 13h46'28" 61d34'16"
Michael David Corr
Uma 9h5'35" 58d35'44"
Michael David Demirgian
Ari 1h52'2" 16d50'31"
Michael David Fleisher
Sgr 19h49'27" -12d0'24"
Michael David Ford
Tau 4h36'10" 25d12'53"
Michael David Francis
Sco 16h8'41" -10d50'15"
Michael David Hall
Psc 1h7'12" 32d34'12"
Michael David Hansford
Uma 11h22'30" 31d4'12"
Michael David Holstine
Uma 8h49'41" 72d19'48"
Michael David Kärcher *MiMi*
Cam 7h55'50" 72d13'39"
Michael David King
Umi 15h46'33" 76d48'25"
Michael David Kralick
Ari 2h7'32" 14d40'12"
Michael David Lynch
Leo 10h54'30" -4d36'10"
Michael David McDowell
Cnc 8h10'13" 32d27'55"
Michael David McGlauflin
And 1h33'39" 42d36'53"
Michael David Morrison
Lyn 7h41'42" 58d33'47"
Michael David Moyer
Lyn 8h9'22" 42d55'45"
Michael David Mueller
Uma 9h16'40" 47d51'7"
Michael David Nelson
Per 4h19'14" 50d44'11"
Michael David Papendick
Cap 21h0'49" -20d11'18"
Michael David Pohndorf
Tau 5h59'11" 25d13'48"
Michael David Rawls
Cap 20h10'2" -10d30'55"
Michael David Shahan
Uma 10h18'13" 46d1'55"
Michael David Shaw
Leo 10h22'25" 26d19'17"
Michael David Snethen
Psc 1h21'29" 12d19'6"
Michael David Soffer
Lyr 18h36'14" 37d37'47"
Michael David Spicer "Frog" ;)
Dra 18h25'24" 75d58'11"
Michael David Walters
Ori 6h16'20" 12d50'56"
Michael David Zaring
Lib 15h18'34" -26d18'7"
Michael Davis
Lyn 7h31'49" 40d24'44"
Michael Davis' Success
Ori 5h20'29" -0d0'11"
Michael De Camp
Her 18h30'51" 14d3'36"
Michael Dean Neal
Dra 17h27'19" 54d22'41"
Michael Dean Parrill Junior
Crb 15h36'38" 29d12'7"
Michael Dean Wells
Her 18h37'51" 21d36'59"
Michael DeBonis IV
Per 2h15'20" 56d24'8"
Michael Dee Markum
Crb 16h11'41" 29d4'33"
Michael Deen Chronister
Vir 11h43'53" -0d43'52"
Michael DeFrancis
Sco 17h27'42" -35d17'20"
Michael DeFranco
Uma 10h55'46" 71d4'53"
Michael DeJoris
Per 3h34'40" 32d1'46"
Michael DeMarco Jennifer Scholts
Cyg 21h20'55" 53d32'17"
Michael Dennis Dearing
Ori 6h13'27" 16d7'13"
Michael Dennis Tortorello
Aqr 23h41'24" -9d53'53"
Michael Dennis Travers My Heart
Psc 1h2'53" 27d45'52"
Michael Dennis Walraven
Lib 15h8'41" -5d40'59"
Michael Dennis Wenner
Cap 21h50'28" -13d18'7"
Michael DeNonno-My Cherished Love M
Aur 5h39'19" 49d53'25"
Michael Dereck Clark "Sunshine"
Per 4h34'21" 48d43'6"
Michael DeRienzo
Uma 11h22'40" 31d45'48"
Michael Desiderato
Uma 11h28'16" 59d48'37"
Michael Devito
Lyn 9h15'41" 37d48'13"
Michael Dewey Gallagher
Umi 14h17'8" 69d0'42"
Michael & Dianna Santeufemia
Cyg 20h41'34" 46d4'8"
Michael DiBari
Sgr 19h1'51" -25d47'4"
Michael DiBisceglie
Sco 16h12'25" -14d24'57"
Michael DiVenere, Jr.
Psc 1h7'8" 5d12'40"
Michael Dobrushin
Psc 0h25'46" 9d49'5"
Michael Dodgson Memorial Star
Sgr 19h6'54" -13d58'52"
Michael Dominish
Sgr 18h51'18" -18d53'36"
Michael Donald Soignet
Uma 11h36'15" 53d49'16"
Michael Donato
Per 3h46'2" 52d9'14"
Michael Donaway
Lmi 10h30'49" 36d53'0"
Michael Doodles
Gem 7h23'23" 32d59'28"
Michael Doran
Boo 14h43'11" 28d4'5"
Michael Doreen
Cap 21h38'38" -14d46'17"
Michael Douglas Deese
Uma 10h42'25" 40d32'5"
Michael Douglas Gore
Tri 2h32'0" 30d47'28"
Michael Douglas Seitz
Tau 5h45'35" 26d58'28"
Michael Douglas Taylor
Sco 16h12'55" -13d12'2"
Michael Dreher
Ori 6h15'39" 18d29'34"
Michael Drew Dominic Coogan 5/19/04
Boo 14h8'33" 50d0'53"
michael drieberg for ever
Uma 13h57'4" 48d0'41"
Michael Duane Adams
Ari 2h32'16" 25d59'34"
Michael Duffett
Her 18h13'28" 18d55'12"
Michael Duggan
Per 2h46'31" 52d20'23"
Michael Duncan
Cap 20h20'50" -12d48'5"
Michael Durham
Uma 8h28'1" 69d5'56"
Michael Dustin
Uma 9h21'46" 49d41'51"
Michael Dwayne Surovik
Aur 5h52'20" 53d5'23"
Michael Dymicki
Cap 20h31'4" -14d40'59"
Michael Dymock
Cep 7h10'58" 86d49'23"
Michael E. Bell
Her 17h47'1" 44d42'39"
Michael E. Byrnes #1 Dad's Star
Uma 11h34'27" 33d52'59"
Michael E Christensen
Umi 14h54'41" 72d36'25"
Michael E. Conrad
Boo 15h15'26" 48d17'30"
Michael E Dew
Her 17h33'5" 33d59'8"
Michael E. Gazzano
Aql 19h52'43" -0d31'42"
Michael E. Hall
Sco 17h34'22" -33d22'34"
Michael E. Henningsen Jr.
Umi 15h44'43" 70d49'18"
Michael E. Jordan
Cyg 21h31'0" 37d20'4"
Michael E. Klehm
Gem 6h18'27" 27d26'56"
Michael E. Most
Cep 0h6'2" 66d54'14"
Michael E. Muta
Lyn 8h31'56" 56d31'35"
Michael E. Parenzan
Cnc 8h20'14" 8d33'23"
Michael E. Solomon
Uma 9h26'16" 49d26'17"
Michael E Wilber
Aql 19h8'33" 10d59'28"
Michael Edgar McGarry
Cep 22h28'8" 62d4'23"
MICHAEL EDWARD
Lib 15h52'9" -5d54'7"
Michael Edward
Per 3h28'43" 50d27'46"
Michael Edward Ball
Ori 5h28'47" 2d18'50"
Michael Edward Bonet
Sgr 18h10'25" -31d26'54"
Michael Edward Burg
Apu 14h25'54" -73d36'38"
Michael Edward Fetkowitz
Vir 12h11'37" 1d51'23"
Michael Edward Goeller
Dra 19h1'36" 71d29'42"
Michael Edward Greenaway
Ori 5h19'2" -0d59'33"
Michael Edward Mattes
Cnc 8h48'31" 17d59'52"
Michael Edward Roberts
Lyn 6h52'25" 61d11'54"
Michael Edward Robledo
Cap 21h46'37" -17d7'21"
Michael Edward Walsh
Per 3h9'48" 54d0'38"
Michael Edwin Ashurst
Sgr 19h32'54" -14d11'51"
Michael "El Rey" Trujillo
Cap 21h41'22" -23d11'22"
Michael Elder
Uma 13h40'30" 50d8'36"
Michael Elder
Per 4h24'56" 43d41'35"
Michael Elias Hayek
Her 16h31'39" 29d46'43"
Michael Elias Tavares Moreira
Uma 11h25'37" 59d5'45"
Michael Ellis
Her 17h23'31" 37d17'9"
Michael Ellis Anastasio Bowman
Ori 5h24'45" 0d16'36"
Michael Engelman's Shining Star
Gem 6h32'51" 20d55'48"
Michael Entrup
Uma 8h35'22" 66d29'4"
Michael Erana
Aqr 23h3'11" -9d20'43"
Michael Eric Reitman
Her 17h22'46" 37d26'55"
Michael Eric Stevens
Sco 16h8'59" -14d1'50"
Michael E.Turner
Vir 13h17'42" 6d14'34"
Michael Eugene Arpin
Sco 17h23'50" -32d18'10"
Michael Eugene Goerndt
Her 17h45'43" 22d2'57"
Michael Eugene Gravier
Her 16h34'17" 31d56'29"
Michael Eugene Henry
Psc 0h11'51" 1d5'40"
Michael Everett Cannedy
Ori 5h56'4" 7d27'13"
Michael Everett Waldemar Gumprecht
Her 17h20'24" 36d57'57"
Michael Evgeny Rubin
Cep 21h40'0" 62d58'17"
Michael Ezechiele Ippoliti
Cyg 20h3'24" 33d50'28"
Michael F. Abbate Sr.
Tau 5h5'37" 18d31'7"
Michael F. DiNardo
Leo 11h41'27" 25d38'52"
Michael F. Gomulka
Her 16h46'39" 16d30'37"
Michael F. Horwitz
Psc 1h18'57" 24d58'35"
Michael F. Lane
Her 17h24'0" 17d19'52"
Michael F. Lynch
Sgr 19h49'28" -21d55'36"
Michael F. McHugh
Vir 11h48'45" 2d36'35"
Michael F Pettis 05-06-1956
Tau 5h25'58" 21d54'42"
Michael F. Zielinski
Cnc 8h3'9" 10d35'35"
Michael Falgares
Per 3h35'1" 45d48'47"
Michael Farland
Ari 2h52'50" 24d35'28"
Michael Farreny
And 2h19'57" 41d45'42"
Michael Fatovic
Lyn 7h33'24" 50d52'18"
Michael Felton
Lac 22h15'42" 49d46'28"
Michael Fields
Sct 18h48'3" -9d31'55"
Michael Finch
Pho 1h31'51" -43d21'36"
Michael Finefter
Per 3h25'4" 43d50'5"
Michael Fisch
Cyg 21h22'28" 39d15'25"
Michael Fitzgerald Ostrander
Uma 10h7'41" 58d14'25"
Michael Fitzpatrick
Lmi 10h42'52" 25d17'17"
Michael Fleischer
Uma 9h37'28" 69d3'52"
Michael for Peter
Cep 21h55'42" 60d36'10"
Michael Forrest Scott
Ori 5h42'1" 2d16'19"
Michael Fortunato Mellow
Cnc 9h11'15" 23d40'45"
Michael Fox
Uma 9h26'9" 49d18'36"
Michael Frame
Cap 21h40'18" -9d6'37"
Michael Francis Daley
Cnc 8h28'56" 23d27'12"
Michael Francis Malone
Peg 21h56'19" 14d18'5"
Michael Francis O'Shea
Her 18h27'22" 18d27'8"
Michael Francis Petricko
Per 3h37'27" 47d49'17"
Michael Francis Reed
Cnc 8h47'37" 22d42'47"
Michael Francis Savino
Her 18h48'38" 17d31'40"
Michael Frank Lombardo
Sgr 18h3'10" -28d3'57"
Michael Friedlich
Ari 3h8'56" 26d44'4"
Michael Fritzsch
Uma 9h39'35" 72d42'36"
Michael Fuller
Ori 5h27'54" 10d9'12"
michael g
Lyr 19h5'50" 27d26'35"
Michael G. and Jeanette F. Langone
Uma 10h25'20" 57d13'4"
Michael G. Bash
Peg 23h2'3" 28d26'13"
Michael G. Cimorosi
Uma 10h49'35" 47d22'54"
Michael G. Conn
Sco 17h18'54" -44d21'47"
Michael G. Fox
Aqr 22h15'37" 0d40'46"
Michael G. Littleton
Ori 6h15'48" 15d18'56"
Michael G. Okash
Cep 22h16'19" 61d52'3"
Michael G Rafferty
Her 16h35'40" 47d20'32"
Michael G. Salfity
Her 17h6'27" 33d36'24"
Michael G. Weisgerber (UB4)
Ori 5h28'7" -0d18'48"
Michael Gabriel Hug
Umi 15h29'44" 78d21'25"
Michael Gale-Mick's Star
Sco 17h58'37" -39d25'10"
Michael Gallagher
Uma 14h27'28" 59d56'51"
Michael Galligher
Her 17h18'25" 32d53'16"
Michael Gary Pittman
Cep 21h13'39" 67d28'31"
Michael Gauthier
Ori 5h9'10" 3d17'11"
Michael Geers
Srp 18h30'26" -0d6'22"
Michael Gengo
Aqr 22h12'4" 0d57'37"
Michael George
Tau 4h53'11" 21d36'42"
Michael George Kissack
Ori 5h56'26" 17d52'44"
Michael George Musso II
Leo 10h20'20" 9d19'11"
Michael George Shea
Per 3h22'12" 47d33'7"
Michael George Smith
Uma 11h12'56" 31d28'6"

Michael Gerald Breslan
Per 3h8'45" 42d37'19"
Michael Gerard O'Connor
Per 1h44'47" 51d12'20"
Michael Gerard Schneider
Lyn 7h28'51" 57d7'43"
Michael Geraty
Vir 13h35'38" -9d1'59"
Michael Gerber
Uma 12h6'44" 47d58'23"
Michael Gerchow
Umi 15h8'46" 72d0'20"
Michael Gerhard Kippnick
Ori 6h16'1" 12d32'21"
Michael Gervasio
Cap 20h32'7" -20d27'7"
Michael Gestrich, Jr.
Per 4h32'43" 39d37'6"
Michael Giammarco
Aur 6h7'37" 30d21'5"
Michael Gianinni
Umi 16h24'19" 73d58'23"
Michael Giffel 20.04.1972
Uma 9h24'27" 70d1'10"
Michael Gillespie
Aqr 21h54'59" -7d54'14"
Michael Gillis
Gem 6h42'50" 21d49'40"
Michael Ginder
Cyg 21h25'14" 34d39'26"
Michael Gladwell
Dra 20h36'32" 75d2'11"
Michael Glenn Ripley
Tau 4h28'30" 21d47'41"
Michael Glenn Stenson
Lyn 7h58'32" 55d14'16"
Michael Gloor
Per 4h35'1" 41d11'14"
Michael Glynn Beck
Boo 14h52'8" 24d51'59"
Michaël Godzik
Aqr 22h37'7" -10d27'59"
Michael Goedken
Lac 22h14'7" 44d11'49"
Michael Goff
Cnc 8h49'53" 27d36'17"
Michael Goings
Her 17h7'36" 15d30'24"
Michael Goldenberg
Aqr 22h53'3" -10d50'36"
Michael Goldklang
Sgr 18h25'24" -20d2'40"
Michael Goldman
Men 6h43'51" -75d30'41"
Michael Golliker
Del 20h37'41" 13d49'33"
Michael Gordon Duncan
Per 2h59'55" 43d28'49"
Michael Gordon
Mastrogany
Sco 16h12'57" -10d44'44"
Michael Gorecki
Uma 10h47'35" 56d6'46"
Michael Gorman Mann
Sco 17h35'20" -33d42'57"
Michael Gorski
Ori 5h56'27" 18d12'25"
Michael Gould
Aql 19h44'6" -11d23'27"
Michael Graham
Leo 11h29'10" 12d37'6"
Michael Grant Kainalu
Sgr 18h18'57" -22d37'28"
Michael Greeley
Cnc 8h8'35" 14d12'35"
Michael Gregory Cordes
Uma 8h35'43" 68d44'22"
Michael Gregory Morgantini
Sgr 19h46'7" -13d7'8"
Michael Gregory Rackliff
Brown
Vir 12h24'11" 3d28'19"
Michael Gros
Per 4h27'34" 35d40'41"
Michael Guajardo
Leo 10h8'24" 17d14'1"
Michael Guccini
Per 2h21'44" 57d12'54"
Michael Gunther
Gerberding
Lib 15h47'30" -13d6'33"
Michael Gus McCauley
Aqr 22h5'48" 1d47'34"
Michael H Blaine
Cnc 9h2'4" 9d43'15"
Michael H Nielsen
Uma 9h39'45" 49d21'12"
Michael H. Wieland
Dra 20h40'47" 68d48'46"
Michael Haas
Uma 14h12'21" 58d9'36"
Michael Habberjam
Cyg 21h15'58" 47d28'17"
Michael Hackett
Cep 21h23'18" 59d26'0"
Michael Hall
Cap 20h27'18" -23d56'2"
Michael Hanly Manetta
Sco 17h51'7" -42d22'5"
Michael Hardy's Star
Tau 3h44'58" 28d50'11"
Michael Hare
Lyr 19h10'15" 46d18'7"
Michael Harlan Smith
Her 16h55'5" 33d22'30"
Michael Harmon
Aql 19h44'35" -0d4'28"
Michael Harrington
Vir 11h37'53" 3d44'34"
Michael Harrison
Gem 6h51'40" 32d58'52"
Michael Harrison
Sco 17h56'5" -43d2'10"
Michael Harry McKenzie
Uma 13h24'32" 57d51'44"
Michael Haug
Her 17h27'56" 39d55'21"
Michael Häussler
Uma 10h3'58" 64d21'29"
Michael Havekost
Gem 6h31'59" 22d43'48"
Michael Haydn Ringwald
Sgr 18h29'8" -16d47'9"
Michael & Hazel
Mulholland 14/5/77
Cru 12h37'26" -59d46'8"
Michael Hazuda
Sco 16h21'37" -17d22'4"
Michael Helseth
Ori 4h57'4" -0d33'37"
Michael Hembra Lim
Lyn 8h57'51" 36d24'37"
Michael Hemingway
Uma 10h30'29" 54d10'1"
Michael Henry Wickson
Ori 6h13'34" 19d0'10"
Michael Hesler
Sgr 18h7'45" -17d52'0"
Michael Hinzmann
Her 18h8'20" 15d45'2"
Michael Hites
Per 3h37'51" 45d34'32"
Michael Hjelmstad
Ori 5h56'11" -0d0'23"
Michael Holden
Del 20h49'59" 13d25'32"
Michael Horvath
Gem 6h42'22" 33d34'3"
Michael Houston
Tau 4h40'41" 23d17'38"
Michael Howarth
Ori 5h53'39" 21d57'58"
Michael Howarth
Aqr 20h49'33" -12d5'59"
Michael Hummel
Uma 13h29'52" 61d3'57"
Michael Humphrey
Boo 15h24'38" 44d35'14"
Michael (Hunny Bun)
Scholl
Her 18h24'52" 14d13'53"
Michael Huot
Aqr 21h22'11" 20d10'38"
Michael Hutchence
Ori 6h16'9" 14d34'44"
Michael Hutchence..shine-
likeitdoes
Per 4h1'43" 43d9'52"
Michael I. Rotcher
Srp 17h58'36" -0d28'32"
Michael Ian Grauerholz
Wise
Uma 12h43'34" 57d38'16"
Michael Idell
Leo 10h29'54" 21d38'8"
Michael Ihler
Uma 12h4'37" 44d51'18"
Michael Isaac Maestas
Uma 12h45'15" 61d52'56"
Michael Ivory "Our Shining
Star"
Tau 4h57'6" 22d1'30"
Michael J. Auger
Lib 14h26'41" -22d57'58"
Michael J. Barnett
Per 2h40'38" 56d20'57"
Michael J. Basile
Ari 3h4'7" 15d24'10"
Michael J Begun Jr
Per 3h7'16" 52d28'23"
Michael J. Bogue
Per 3h3'54" 54d13'24"
Michael J. Bookstaver
Vir 13h30'6" -4d18'39"
Michael J Buccino
Gem 6h6'32" 26d30'16"
Michael J Burke
Sgr 18h7'45" -25d14'40"
Michael J. Buscemi
Tau 3h42'3" 7d55'58"
Michael J. Buttars
Lyn 8h52'50" 33d41'9"
Michael J. Cawley
Cyg 19h59'19" 52d16'41"
Michael J. Coblentz
Psc 0h24'32" 10d37'31"
Michael J Debets
Leo 11h21'27" 16d42'46"
Michael J. DePeel, II
Per 2h18'0" 51d42'49"
Michael J. DiGiovanni
Per 3h19'38" 54d1'4"
Michael J. Durante
Uma 9h55'50" 60d24'5"
Michael J. Elersic
Ori 6h18'59" 13d39'35"
Michael J. Ferguson
Vir 11h38'15" -0d28'43"
Michael J Francoeur,Jr.-
Our Leader
Her 17h19'29" 47d6'33"
Michael J. Freda
Sco 16h13'38" -10d6'37"
Michael J. Frye
Hya 9h36'27" -0d9'29"
Michael J Graham, Jr
Ari 3h11'36" 28d6'35"
Michael J. Hall
Her 18h16'4" 15d48'46"
MICHAEL J. HARRISON
Uma 10h59'44" 68d26'1"
Michael J. Hartman
Lib 15h13'48" -19d50'27"
Michael J. Harwin
Gem 7h27'4" 25d24'58"
Michael J. Hawkins
Vir 12h35'6" 10d35'14"
Michael J Helfrey
Tau 4h56'19" 21d10'30"
Michael J. Herbert
Gem 7h14'18" 21d26'6"
Michael J. Hester
Dra 17h34'1" 63d38'5"
Michael J. Iovine
Cyg 20h32'43" 46d13'49"
Michael J. Kadylak
Cnc 8h51'33" 31d4'0"
Michael J. Kaplan
Uma 11h8'51" 36d57'36"
Michael J. Kennedy
Leo 9h26'40" 11d11'39"
Michael J Killeen III
Uma 10h43'15" 67d15'36"
Michael J. Kort
Aur 5h54'2" 49d39'30"
Michael J. LeRose/Laurie
S. Howell
Gem 7h31'30" 21d1'45"
Michael J. Liptak
Ori 6h17'2" 10d24'50"
Michael J MacKinnon
Uma 9h6'11" 50d30'43"
Michael J. Malone, Sr.
Ori 5h30'19" 1d2'2"
Michael J. Martin
Cma 6h49'38" -14d25'56"
Michael J. McGregor a
wonderful man
Aur 5h39'8" 44d8'39"
Michael J. McIntyre
Sco 17h15'27" -36d25'23"
Michael J. McMahon
Uma 10h44'2" 50d12'33"
Michael J. Medina
Her 17h1'45" 14d1'13"
Michael J. Miller
Sco 17h57'53" -37d42'47"
Michael J. Murawski
Boo 14h35'24" 34d46'47"
Michael J. & Natalie R.
Hanna
Ari 2h14'28" 25d17'10"
Michael J Neureuther II
Cyg 21h50'19" 49d1'51"
Michael J. O'Connell
Cep 23h2'50" 70d18'22"
Michael J. Oehler
Per 3h44'46" 46d50'23"
Michael J O'Grady Sr. 55
Tau 4h19'48" 3d43'33"
Michael J. OLeary
Per 4h34'30" 51d23'17"
Michael J. Ostrosky
Per 3h16'21" 39d3'49"
Michael J. Palladino
Ori 4h53'25" -3d40'37"
Michael J. Palmateer
Aur 5h48'28" 45d10'9"
Michael J. Payne
Cnc 8h40'20" 16d3'57"
Michael J. Perron
Umi 14h47'34" 73d6'4"
Michael J. Piccolo
Ori 6h16'27" 11d2'28"
Michael J. Reddick
Sco 17h56'24" -40d55'34"
Michael J. Reminger
Vir 13h16'34" -1d30'39"
Michael J. Rowan
Cnc 9h7'0" 24d44'33"
Michael J. Russick Sr.
Del 20h40'15" 15d28'1"
Michael J. Ryan
Cnc 8h43'53" 6d38'33"
Michael J. Scuzzese
Lib 14h52'35" -3d45'12"
Michael J. Silverstein
Vir 12h47'52" 6d53'28"
Michael J. Simpson Jr.
Dra 17h19'16" 68d4'25"
Michael J. Snyder
Cep 5h8'24" 82d11'6"
Michael J. Solomon
Psc 1h3'29" 27d37'11"
Michael J. Stabile
Leo 10h12'17" 11d51'56"
Michael J. Sullivan
Uma 11h23'2" 58d30'4"
Michael J. Traficante
Per 2h19'25" 55d3'11"
Michael J. Travers, Jr.
Uma 11h55'9" 61d20'53"
Michael J. Travis, Jr.
Tau 3h27'9" 8d48'29"
Michael J. Tucker, Sr.
Per 3h10'36" 45d59'55"
Michael J. Tylavsky, Jr
Her 17h10'43" 30d56'17"
Michael J Usiak
Aqr 22h23'1" -23d39'13"
Michael J. Ventrice Star
Uma 10h22'27" 61d20'7"
Michael J. W. Ho
Lib 15h8'48" -5d12'54"
Michael J. Weremay
Lib 16h0'9" -6d22'35"
Michael J West
Her 17h24'57" 34d3'2"
Michael J. Wilderspin Dec.
6, 1942
Leo 10h15'54" 25d25'37"
MICHAEL J. WILLIAMS
Her 17h39'42" 21d11'22"
Michael J. Wolsh
Cep 22h34'29" 72d36'33"
Michael Jacob Holford
Uma 11h41'7" 31d36'38"
Michael Jacob Kirchner
Mon 7h24'6" -3d38'11"
Michael & Jacqueline
Lmi 10h1'31" 32d15'16"
Michael Jakoweiczuk's
Light
Cyg 20h5'41" 35d14'16"
Michael James
Cep 22h9'30" 56d17'20"
Michael James
Cap 20h18'8" -24d45'6"
Michael James Aburn
Webster
Cru 12h29'59" -63d3'4"
Michael James Aldarelli
Leo 9h31'29" 9d35'58"
Michael James Bailey
Uma 9h41'53" 46d32'11"
Michael James Ballard
Her 17h43'51" 20d56'8"
Michael James Bolingbroke
Lib 14h57'48" -18d19'25"
Michael James Boston
Aur 5h11'32" 42d35'9"
Michael James Bruno
Cnc 8h48'35" 31d45'35"
Michael James Dowdall II
Lyn 8h34'24" 45d20'6"
Michael James Dudo
"Mike"
Uma 10h29'41" 50d55'50"
Michael James Ellenberg
Ari 2h31'39" 11d30'27"
Michael James Fanning
Cap 21h28'54" -10d54'12"
Michael James Felsinger
Psc 1h18'0" 19d41'47"
Michael James Gillard
Lib 15h17'27" -5d59'23"
Michael James Godwin
Uma 9h19'35" 64d24'51"
Michael James Gropp
Per 4h50'16" 49d21'57"
Michael James Haines
Her 16h8'43" 24d53'51"
Michael James Jacob
O'Hanlon
Uma 8h56'59" 62d41'26"
Michael James Jones
Uma 11h44'48" 50d59'48"
Michael James Keady
Per 3h23'51" 51d24'48"
Michael James Kerwin
Cep 22h53'29" 64d43'5"
Michael James Kuhens
Ari 3h16'37" 29d49'20"
Michael James LaRue
Aur 5h46'38" 54d56'14"
Michael James Lennon
Sgr 18h28'56" -16d13'33"
Michael James Lewis
Vul 20h44'33" 26d4'39"
Michael James Lubin
Uma 10h18'55" 50d16'50"
Michael James Muehler
Vir 13h16'9" -20d34'12"
Michael James Pilato
Uma 11h35'0" 64d46'56"
Michael James Raber
Psc 1h8'41" 25d51'15"
Michael James Roy II
Lyn 8h12'44" 55d12'32"
Michael James Ryder
Umi 16h12'11" 82d4'49"
Michael James Sbandi
Leo 9h26'21" 26d11'4"
Michael James Sears
Uma 11h52'56" 40d15'7"
Michael James Smith
Ori 5h29'0" 8d20'33"
Michael James Sorenson
Cnc 9h10'49" 13d13'14"
Michael James Wardell
And 23h35'6" 37d34'28"
Michael Jansen
Aql 19h33'25" 15d21'36"
Michael Jay Beculheimer
Her 17h36'10" 27d34'41"
Michael Jay Brooks
Per 4h10'8" 35d48'27"
Michael Jay Ermi
Uma 11h39'37" 43d12'9"
Michael Jay McCleod Jr.
Uma 9h33'51" 44d15'26"
Michael Jay Rogers
Gem 6h35'18" 13d5'20"
Michael Jeffs
Cru 12h34'51" -61d22'48"
Michael & Jennifer
Leo 11h2'18" 3d49'32"
Michael & Jennifer
Eastlund
Cas 0h17'41" 57d35'20"
Michael Jennings Berry
Aqr 21h41'46" 2d24'3"
Michael & Jessica
Her 17h1'58" 32d59'0"
Michael & Jessica Cortez
Uma 11h27'10" 47d12'26"
Michael Jiggetts
Boo 15h5'42" 33d26'5"
Michael (Jimmy) Bush
Her 17h18'10" 20d48'23"
Michael & Jo Erickson
Aql 20h0'7" 6d22'25"
Michael Joe Buzzard
Aqr 22h15'35" -0d23'37"
Michael Joe Mueller &
Family
Uma 9h14'16" 58d37'59"
Michael Joel Hedrick
Her 17h16'23" 14d47'24"
Michael Joel Stueve
Thoren
Ori 5h29'49" -0d2'13"
Michael & Joeys Twinkling
Star
Cyg 19h38'46" 30d9'39"
Michael_John
Umi 13h46'20" 75d43'23"
Michael John
Ser 18h20'30" -13d32'12"
Michael John Archer
Aqr 22h6'36" -21d2'46"
Michael John Burk
Lyn 8h0'17" 34d11'0"
Michael John Cordes
Leo 11h46'59" 26d18'2"
Michael John Cotton
Cyg 19h27'50" 36d6'17"
Michael John DiDomenico
Aur 5h38'17" 45d11'31"
Michael John Druzak
Aqr 20h47'37" 2d4'19"
Michael John Felci
Ori 5h57'10" 22d19'4"
Michael John Genna
Sco 17h54'50" -36d41'33"
Michael John Howe
Per 3h24'36" 40d41'27"
Michael John Hutchinson
Umi 16h23'20" 72d2'48"
Michael John Iberger
Psc 0h31'53" 5d26'31"
Michael John Kellogg
Cap 20h27'35" -15d49'58"
Michael John Lalonde
Cyg 21h46'19" 44d33'9"
Michael John Langford
Her 18h6'26" 32d30'36"
Michael John O'Connor
Lyn 7h39'44" 55d32'34"
Michael John Padovano
Cap 21h43'29" -9d30'30"
Michael John Patrissi
Ari 2h13'45" 25d22'16"
Michael John Paul Delaney
Her 18h4'47" 24d8'56"
Michael John Pietroniro
Aqr 20h49'58" -1d14'57"
Michael John Polley
Psc 1h2'17" 11d18'49"
Michael John Richartz
Gem 6h48'56" 22d58'24"
Michael John Simone - The
Seven P's
Cap 20h27'14" -15d32'44"
Michael John Sutcliffe
Leo 9h55'57" 23d43'53"
Michael John Waite -
Michael's Star
Per 4h4'41" 34d11'10"
Michael John Wright
Uma 9h9'32" 58d23'15"
Michael John, Forever
Strong
Her 18h14'22" 28d9'23"
Michael Johnson
Vir 12h47'19" 0d24'17"
Michael Johnson
Aur 5h53'31" 46d14'53"
Michael Johnson & Lisa
Joyce
Umi 15h30'28" 75d7'23"
Michael Jon
Uma 13h42'24" 56d42'15"
Michael Jon Givens
Cyg 20h50'23" 35d56'55"
Michael Jon Skiles
Aqr 22h4'30" 1d41'54"
Michael Jonathan Hotra
Gem 6h34'9" 19d45'24"
Michael Jones
Lac 22h30'26" 48d45'55"
Michael Jones
Cap 21h53'29" -10d25'57"
Michael Jones Allora
Umi 15h20'54" 71d21'45"
Michael Jons Willumat-
Paganoni
Lyr 18h48'14" 40d10'46"
Michael Jordan
And 1h17'19" 38d2'17"
Michael Jordan Lombardi
Lmi 10h25'27" 37d22'31"
Michael Joseph
Her 17h40'18" 43d54'52"
Michael Joseph
Gem 7h3'38" 26d29'34"
Michael Joseph Andrew
Rigg
Tau 4h11'49" 15d26'42"
Michael Joseph Bartolini
Uma 13h50'46" 56d38'30"
Michael Joseph Brugger
Per 4h14'5" 46d27'3"
Michael Joseph Cappiello
Jr.
Sco 17h2'24" -45d1'12"
Michael Joseph Catchot
Ori 4h54'26" 11d4'46"
Michael Joseph DellaValle,
Jr.
Her 17h24'10" 45d0'10"
Michael Joseph
Diedrickson
Ari 2h21'55" 11d56'20"
Michael Joseph Dill Sr.
Cep 22h42'7" 77d17'12"
Michael Joseph Driscoll
Gem 7h28'59" 20d31'58"
Michael Joseph
Edmondson
Ori 5h55'34" 3d12'4"
Michael Joseph Fealy
Her 17h22'25" 45d50'32"
Michael Joseph Freres
Umi 15h27'21" 73d29'41"
Michael Joseph Garrison
Ori 5h26'49" 2d24'34"
Michael Joseph Giles
And 0h35'27" 35d41'33"
Michael Joseph Hayman
Gem 6h2'21" 25d52'12"
Michael Joseph Hofelich
Leo 11h41'27" 16d33'35"
Michael Joseph Jackson
Gem 7h7'41" 32d30'4"
Michael Joseph Jankovic
Lyn 7h8'18" 58d40'20"
Michael Joseph Keenan
Psc 1h6'28" 33d8'38"
Michael Joseph Kelly
Shines Forever
Umi 15h15'21" 75d33'25"
Michael Joseph Killeen
Cma 6h35'54" -15d5'5"
Michael Joseph Kovach
Uma 11h50'40" 57d30'47"
Michael Joseph Legere
Her 16h30'14" 45d43'41"
Michael Joseph Lombardo
Sco 16h14'21" -12d5'33"
Michael Joseph Maragni
Uma 11h25'40" 50d45'45"
Michael Joseph Martinez
Ari 2h48'16" 13d27'49"
Michael Joseph McCluskey
Umi 12h5'50" 88d32'38"
Michael Joseph Medrano
Jr.
And 1h29'14" 44d31'29"
Michael Joseph Minikel
Cas 1h18'53" 57d31'7"
Michael Joseph Mitchell
Cnc 8h51'29" 27d31'14"
Michael Joseph Nipper
Tau 5h28'13" 26d24'21"
Michael Joseph O'Hara
Cep 0h4'0" 67d25'25"
Michael Joseph Quinn
Uma 11h20'36" 66d43'32"
Michael Joseph Rago
Leo 9h48'30" 28d52'20"
Michael Joseph Ranalli
Vir 13h59'52" -16d8'48"
Michael Joseph Richard
Scionti
Lib 15h43'39" -25d59'28"
Michael Joseph Russo
Umi 14h55'9" 76d53'38"
Michael Joseph Sabella
"baby boy"
Her 17h21'21" 47d33'38"
Michael Joseph Scali, Jr.
Tau 5h45'58" 22d6'59"
Michael Joseph Schanderl
Her 17h16'51" 32d20'2"
Michael Joseph Schiumo
8:52 a.m.
Per 2h21'6" 56d52'12"
Michael Joseph Silvia
Aqr 23h5'45" -9d49'13"
Michael Joseph Sluzis
Gem 6h25'48" 23d39'34"
Michael Joseph Sunder
Leo 11h18'3" 6d23'46"
Michael Joseph Vana
Lib 15h45'16" -6d6'3"
Michael Joseph Wills
Per 3h18'32" 53d50'53"
Michael Joseph Wright
Aqr 22h14'34" -0d57'13"
Michael Joshua Alicea
Cnc 8h49'7" 19d41'11"
Michael Jr Savino
Uma 9h5'39" 51d32'58"
Michael Juarbe
Ori 5h28'16" -2d22'18"
Michael Judd
Aur 5h22'6" 31d51'1"
Michael Jude Mariani
Aqr 22h41'41" -17d8'35"
Michael & Julia Roy
Cyg 21h17'20" 45d13'48"
Michael Julian
Ori 4h53'46" 12d49'8"
Michael Julian
Sgr 19h10'24" -21d59'33"
Michael Justiniano
Leo 9h23'40" 10d28'27"
Michael K.
Uma 11h27'31" 40d11'37"
Michael K
Cap 21h39'12" -11d50'40"
Michael K. Aufmann
Lyn 6h22'53" 54d50'32"
Michael K. Bradash
Her 17h31'13" 47d40'3"
Michael K. Harvey
Tau 3h32'49" 25d4'3"
Michael K. Hudson
Uma 10h19'9" 46d22'1"
Michael K. Raggio
Her 18h1'36" 14d42'13"
Michael K. Rose
Leo 9h44'22" 26d34'48"
Michael K. Sawyer
Per 2h53'32" 54d24'43"
Michael K. Theroux
Sgr 18h11'46" -31d39'15"
Michael Kamm
Uma 11h14'9" 64d44'33"
Michael Kaplan
Ori 5h28'39" -3d34'4"
Michael Karen
And 0h28'57" 43d3'29"
Michael & Karri Boyd
Del 20h39'36" 16d25'23"
Michael Kasick
Per 3h29'22" 43d56'56"
Michael Kasino
Cep 22h50'8" 66d24'3"
Michael Kathleen Gnekow
Garcia
Ari 3h7'2" 28d57'2"
Michael Katsevman
Tau 3h58'1" 15d54'7"
Michael Kebabjian, Jr.
Cep 21h20'11" 67d7'16"
Michael Keil
Uma 9h14'25" 50d57'38"
Michael Keith Boyd
Aql 19h28'26" 5d42'57"
Michael Keith Marsh
Lyn 7h45'27" 58d53'53"
Michael Keith Mayes
Cnc 8h25'9" 17d10'31"
Michael Keith McDonald
Ori 5h33'24" -5d21'16"
Michael Kelly
Cas 1h40'17" 60d38'57"
Michael & Kelly James,
With Love
Col 5h46'8" -34d48'43"
Michael Kendrick Finney
Aql 20h37'2" -0d3'43"
Michael Kennett O'Rourke
Tau 3h43'12" 28d26'1"
Michael Kent
Boo 14h44'25" 37d34'24"
Michael Kevin
Uma 8h19'39" 64d30'40"
Michael Kevin Ireland
Cap 20h13'21" -10d30'33"
Michael Kindling
Cnv 13h45'43" 30d17'58"
Michael Kinney
Sgr 18h0'40" -28d4'59"
Michael Kinskey
Per 3h36'39" 41d48'49"
Michael Kirchner
Her 18h0'43" 20d52'5"
Michael Kirkland
Uma 11h27'25" 45d4'56"
Michael Kjentvet
Cap 21h39'41" -19d47'3"
Michael Klatzkin
Sco 16h49'19" -42d10'43"
Michael Klik "My Shining
Star"
Cnc 8h13'20" 32d30'25"
Michael Klimczak - Rising
Star
Dra 17h59'55" 53d52'27"
Michael Knap
Uma 10h47'58" 50d59'29"
Michael Kogut
Cap 21h56'39" -11d17'9"
Michael Kohlmaier
Cep 22h51'34" 61d46'57"
Michael Konstantin
And 2h25'15" 41d26'44"

Michael Korte Walsh
Aqr 22h15'51" 1d9'51"
Michael Koziol
Lyn 7h7'55" 50d42'16"
Michael Krasnow
Her 16h14'46" 43d50'44"
Michael Krause
Her 16h32'1" 47d18'6"
Michael & Kristen
Vir 13h20'32" -6d8'19"
Michael Kuperman
Cyg 20h19'8" 38d34'41"
Michael Kushner
Uma 11h1'19" 41d18'17"
Michael Kwame Addo
Cyg 19h44'21" 28d13'32"
Michael L. Cline 07-29-70/1-21-90
Per 3h1'10" 46d58'48"
Michael L. Cobaugh, Jr.
Sgr 18h32'22" -24d8'48"
Michael L. Connell
Uma 9h41'1" 69d23'10"
Michael L Craviotto
Aur 5h6'58" 35d8'35"
Michael L. Fischer
Tau 4h3'54" 24d47'21"
Michael L. Girtman
Her 16h36'33" 37d23'14"
Michael L Gorden/My Love
Leo 10h20'25" 23d41'5"
Michael L. Healey
Uma 10h24'43" 59d18'37"
Michael L. Horvath
Per 2h40'39" 53d57'9"
Michael L. Lombardi
Her 17h43'59" 48d10'56"
Michael L. Mackie
Per 3h11'11" 56d24'57"
Michael L. Mattingly, Sr
Per 3h38'9" 45d50'29"
Michael L. Myers
Ari 3h22'13" 29d2'16"
Michael L. Pulley
Crb 15h37'54" 32d28'32"
Michael L. Ressegieu "Mikey"
Her 18h47'40" 25d4'31"
Michael L. Weber
Lib 15h55'24" -12d44'16"
Michael Ladd Thompson
Cnc 8h53'36" 28d26'0"
Michael Ladenheim Frankel
Lib 15h5'51" -3d13'4"
Michael LaDue
Her 17h23'21" 15d23'19"
Michael Lajos Szanto
Tau 5h54'2" 25d35'50"
Michael Lanners III
Per 3h43'25" 45d40'21"
Michael LaPollo
Sco 17h24'14" -40d58'4"
Michael Larkin
Cap 20h22'11" -19d44'12"
Michael Laudisio
Sgr 17h57'36" -16d55'7"
Michael & Lauren 11-10-2004
Sco 16h45'15" -26d24'27"
Michael Lauretti
Her 17h18'36" 48d18'22"
Michael Lawrence
Dra 10h32'14" 79d57'25"
Michael Lawrence Pearson
Crb 15h29'58" 30d45'30"
Michael Lawrence Tamburrino
Uma 9h18'51" 58d27'45"
Michael Lawrence Walker
Aqr 23h40'56" -19d11'34"
Michael Lawson
Ari 2h37'57" 26d56'18"
Michael & Leah Sillix
Cyg 20h28'27" 52d31'45"
Michael Leander Perry
Sgr 19h35'55" -16d7'34"
Michael Lee
Vir 12h7'48" 8d19'39"
Michael Lee Berger
Per 4h16'1" 51d9'31"
Michael Lee Fletcher
Per 3h25'10" 44d47'40"
Michael Lee Hopkins
Vir 13h44'23" 5d47'32"
Michael Lee Jefferson
Leo 11h15'38" 22d41'12"
Michael Lee Jorge
Cyg 20h52'26" 41d29'19"
Michael Lee Knowles
Her 17h12'27" 31d1'58"
Michael Lee Orchard
Gem 6h15'52" 25d34'2"
Michael Lee Perry
Per 3h47'34" 43d53'15"
Michael Lee Ruede
Her 18h23'38" 19d2'9"
Michael Lee Smith
Uma 8h25'2" 66d14'59"
Michael Lee (Snuggles) King
Sgr 18h36'46" -22d31'44"
Michael Lee Thomas
Umi 17h37'27" 82d44'58"
Michael Lee Vanlandingham, My Dad
Her 17h38'52" 16d39'33"
Michael Leo Ellis
Aql 19h40'42" 5d14'31"
Michael Leonard Moore (Poppa)
Uma 12h10'11" 51d34'43"
Michael Leo's Inspiration
Aql 19h32'57" 7d9'28"
Michael Leroy Grove
Per 3h20'24" 33d3'29"
Michael Leslie Martin
Ori 5h22'16" -1d22'35"
Michael L'Etoile de Malibu
Her 16h5'41" 17d54'17"
Michael Leva
Cep 23h51'12" 74d19'12"
Michael Lewis Galloway
Cap 21h8'0" -26d57'46"
Michael Leyrer Rompf
Sgr 18h40'24" -36d7'44"
Michael Libro
Per 3h15'17" 52d3'34"
Michael & Lindsay 10 years-forever
Cyg 20h38'13" 40d57'17"
Michael Lis
Uma 9h4'21" 69d22'25"
Michael & Lisy Trueba Jr.
Uma 11h39'47" 59d41'53"
Michael "Little Cappy" 1957-2005
Ori 6h17'44" 16d10'46"
Michael & Lizzy's Wedding Star
Cyg 22h0'25" 53d45'15"
Michael Lloyd
Gem 6h44'54" 23d26'18"
Michael Lopez
Ori 5h30'44" 5d44'55"
Michael Lopez
Cyg 19h58'30" 35d9'55"
MICHAEL LOREN HIGH-FILL
Her 18h30'40" 18d8'48"
Michael Loren Lipton
Cnc 9h9'54" 31d52'33"
Michael Lorenzo
Uma 10h32'5" 53d7'59"
Michael Losavio
Aur 5h26'13" 43d49'0"
Michael Louis Flink
Sgr 19h12'58" -30d35'47"
Michael Louis Kikos
Gem 7h48'9" 28d33'33"
Michael Louis Stoddard
Per 3h20'45" 50d59'20"
Michael Love
Her 17h0'39" 14d25'50"
Michael Lovell
Cyg 20h46'36" 43d34'42"
Michael Loves Jean
Eri 4h24'19" -0d48'39"
Michael loves Steph Weston Forever
Cnc 8h51'31" 27d38'51"
Michael Lund
Uma 9h7'5" 57d27'48"
Michael Lüthi
Crb 16h14'29" 37d27'44"
Michael Lyborg
Lib 15h11'24" -26d16'24"
Michael & Lyn - The Life Partners
Uma 10h55'1" 68d11'48"
Michael Lynn Burnett
Cnc 8h13'13" 11d25'49"
Michael Lynn Klier
Uma 10h36'1" 72d14'55"
Michael Lynn Looney
Uma 13h15'11" 56d33'23"
Michael Lynn Thomas *02/15/1964*
Uma 11h34'56" 57d6'15"
Michael M. Arlesic
Her 17h20'7" 38d33'46"
Michael M. Lewis
Her 17h33'16" 15d54'34"
Michael M. Scherrer 1986
Cnc 9h19'50" 10d2'31"
Michael "Mac" McCabe
Aqr 21h37'42" 1d24'23"
Michael MacInnes Shaw
Uma 11h55'5" 38d3'3"
Michael MacKinnon Whiteoak
Her 17h46'28" 37d40'2"
Michael Madson - My Angel
Lmi 9h24'9" 37d48'43"
MICHAEL MALIK CRUZ HOOK
Her 17h57'35" 27d58'10"
Michael Mangini
Cma 7h26'26" -15d8'16"
Michael Maniaci
Aql 20h23'57" 7d59'32"
Michael Marcel Nester
Cnc 8h8'58" 8d9'32"
Michael Marcello's Kingdom
And 23h23'56" 42d50'34"
Michael Margolies
Uma 11h39'13" 43d41'20"
Michael & Maria
Cyg 21h15'26" 46d20'53"
Michael & Maria's Guiding Star
Cra 19h6'38" -38d19'18"
Michael Marie
Uma 9h54'4" 53d21'11"
Michael & Marina
Cyg 21h31'18" 44d30'34"
Michael Mark Piasecki
Sco 17h30'24" -30d34'21"
Michael Marks
Her 18h25'30" 25d38'57"
Michael Marra
Her 16h39'50" 7d4'34"
Michael Martin
Ari 2h1'15" 22d39'46"
Michael Martin
Leo 11h11'42" 17d7'5"
Michael Martin
Uma 10h29'42" 49d8'37"
Michael Martin
Uma 10h0'16" 63d15'2"
Michael Martin Hook
Uma 11h23'28" 57d48'12"
Michael Martino
Her 16h38'6" 35d4'41"
Michael & Maryellen
Cyg 21h56'9" 54d56'7"
Michael & MaryRose's Star Of Love
Cru 12h27'11" -59d42'49"
Michael Mase
Sco 17h19'3" -44d42'31"
Michael & Masume Moore
Cyg 21h23'46" 53d58'2"
Michael & Mats
Sco 16h21'30" -16d55'59"
Michael Matthew Gordon Perkins
Uma 14h2'30" 58d52'47"
Michael Matthew Miller 9/11/01
Vir 14h33'50" 6d10'35"
Michael Matthew Schneider
Cnc 8h54'25" 27d56'38"
Michael Mattison Jones
Leo 11h2'18" -6d10'25"
Michael Mauro
Per 1h50'1" 50d45'49"
Michael M.C.
Aur 6h19'5" 52d41'43"
Michael McCarron
Her 18h44'0" 20d13'3"
Michael McDonald
Aur 5h37'34" 51d57'20"
Michael McGorisk
Her 17h19'2" 27d15'37"
Michael McKinley Morgan II
Ori 6h10'1" 18d32'50"
Michael McManus
Cnc 8h58'2" 18d13'37"
Michael McQuillen
Tau 4h6'30" 26d6'1"
Michael Medeiros
Gem 6h45'29" 18d27'6"
Michael Meighan - Shining Star
Per 3h53'24" 37d2'48"
Michael Meisenberg
Uma 10h21'41" 55d46'41"
Michael & Melinda Moore -I Love You
Lyn 6h41'25" 54d35'17"
Michael & Melissa
Cru 12h33'54" -57d50'56"
Michael Merrill
Sco 16h13'33" -11d50'4"
Michael Mescolotto
Aur 5h54'49" 46d54'15"
Michael Mezzacappa
Gem 7h17'16" 20d12'34"
Michael Mi Amor
Cnc 8h18'10" 19d51'18"
Michael & Michelle Adler
Cyg 20h20'38" 46d34'24"
Michael & Michelle Farah
Cnc 8h10'14" 25d36'43"
Michael "Mike" C. Viola
Ari 2h32'23" 18d45'34"
Michael "Mikey" Francis Schafer
Leo 11h53'12" 18d23'26"
Michael "Mikey" Giaccone
Aqr 23h34'7" -9d31'21"
Michael Miller
Uma 8h27'49" 66d21'42"
Michael Milstead
Ari 2h48'50" 13d7'17"
Michael Minerd's Freedom
And 0h1'18" 46d10'11"
Michael Minh Peterson
Gem 7h33'49" 30d50'20"
Michael Mink
Ori 5h40'3" -0d36'6"
Michael Mintz
Her 16h56'44" 19d7'45"
Michael Mirtsopoulos
Lib 15h38'18" -11d50'48"
Michael Mobert
Uma 13h55'3" 60d17'14"
Michael Mocarski
Pho 0h11'55" -43d54'38"
Michael Molinelli
Cnc 8h47'42" 31d32'18"
Michael Monhart
Lib 15h3'55" -2d39'0"
Michael & Monica
Cas 1h26'51" 63d4'5"
Michael & Monica Petrucci
Cyg 21h15'58" 31d37'2"
Michael & Monika Wübbens
Umi 14h47'31" 74d46'41"
Michael Montalbano
Ari 2h47'7" 16d59'39"
Michael Montiel
Psc 0h3'45" 8d4'43"
Michael Moore Furfey
Vir 12h38'28" -8d3'13"
Michael Moran Horvath
Dra 18h30'38" 64d48'30"
Michael Morlan
Uma 9h41'17" 51d19'36"
Michael Morrison Lee Schwab
Ari 2h17'34" 21d42'59"
Michael Morway
Uma 11h17'0" 45d36'28"
Michael Morykwas
Cnv 12h39'12" 41d31'11"
Michael Mosca
Uma 9h22'4" 71d57'55"
Michael Mugs
Aur 5h45'22" 43d50'58"
Michael Muldoon
Cnc 8h42'31" 8d45'58"
Michael Muldoon, The Real Batman
Cap 21h44'9" -13d8'12"
Michael Mullen
Psc 0h21'54" 8d0'50"
Michael Müller
Umi 16h20'11" 71d45'20"
Michael Musser
Cep 20h35'10" 61d38'54"
Michael My Angel
Her 16h10'18" 47d52'10"
Michael My Angel Shining Down On Me
Tau 5h17'30" 27d33'49"
Michael My Diamond Setting
Sgr 17h51'31" -27d16'30"
Michael My Guardian Angel
Sco 16h16'28" -10d2'19"
Michael "My Shining Star"
Psc 1h11'40" 26d27'51"
Michael N. Macaronis
Aql 19h7'20" 15d1'3"
Michael N. Spector
Cep 22h10'59" 61d19'54"
Michael&Nadine
Cep 1h46'11" 78d2'19"
Michael (Neta 34) Iannetta
Vir 12h38'10" 7d5'36"
Michael Nettleton
Lib 15h1'53" -5d42'54"
Michael Newby One
Sgr 18h44'29" -29d42'18"
Michael Nicholas Gorman
Per 3h8'16" 53d59'41"
Michael Nicholas Smee
Aur 5h47'12" 54d38'13"
Michael Niedziejko
Cep 23h13'59" 71d2'7"
Michael & Nina
Uma 12h48'25" 55d24'41"
Michael No Doubt Viola
Sco 17h16'36" -44d57'21"
Michael Norman Ruppert
Oct 19h55'16" -83d12'26"
Michael Nowicki
Lib 14h46'49" -7d44'52"
Michael Nugent
Sco 17h28'26" -42d17'7"
Michael Nuzzo 7/31/1970 - 2/22/2004
Leo 11h15'54" 16d5'59"
Michael O'Brien
Lib 15h46'16" -24d15'59"
Michael Odysseos
Per 4h37'9" 43d56'47"
Michael Oksanen
Her 18h0'23" 21d40'55"
Michael O'Malley
Aqr 22h20'43" -7d10'30"
Michael On is loved
Cas 23h0'4" 53d15'7"
Michael Orapello III
Umi 15h22'7" 71d36'15"
Michael O'Ravitz
And 2h21'7" 50d16'12"
Michael Ordaz "The Mad Texan"
Cyg 20h24'7" 30d53'55"
Michael - Our Little Star
Vul 19h32'49" 23d46'6"
Michael Owen
Leo 10h9'48" 22d9'33"
Michael Owen - Y Gwyddon Seren
Uma 9h30'58" 65d25'42"
Michael Owens
Aqr 22h42'16" -16d46'45"
Michael P Andrjeski
Lac 22h25'52" 47d47'40"
Michael P. Capatina
Aql 19h27'52" 14d58'41"
Michael P Carlson
Gem 6h50'25" 32d33'31"
Michael P. Chaiken
Ori 5h35'16" -2d21'8"
Michael P Cook
Cnv 13h53'12" 35d52'38"
Michael P. Hanks
Gem 7h23'25" 14d55'37"
Michael P. Iaquinta
Aqr 21h49'9" -3d14'34"
Michael P Knapp (QB loves BD)
Uma 14h27'8" 55d32'42"
Michael P. Lawson
Cnc 8h25'14" 13d43'22"
Michael P. Manning
Ori 6h6'28" 17d32'38"
Michael P. Miller
Leo 9h47'42" 29d7'33"
Michael P O'Neill
Leo 11h9'44" 4d46'30"
Michael P. Sylvester Family Star
Leo 11h26'12" 17d39'40"
Michael P Van Kampen
Ori 6h16'10" 10d51'6"
Michael P. Wagner
Lib 14h24'18" -9d28'23"
Michael P. Walsh
Leo 11h50'20" 20d12'12"
Michael Page
Vir 13h1'5" 5d19'17"
Michael Palermo
Sgr 19h38'36" -16d23'39"
Michael Palma
Aql 19h20'51" -0d22'2"
Michael Palmo
Aql 20h11'40" 4d9'51"
Michael Panko
Dra 17h3'20" 56d47'16"
Michael Paoletta
Aql 19h38'2" 11d51'20"
Michael "Papa" Collar
Uma 13h52'17" 54d5'35"
Michael Pardee
Uma 9h9'5" 62d44'17"
Michael Parga
Per 2h16'10" 54d6'48"
Michael Parker
Aur 5h47'21" 50d55'37"
Michael Parker, Jr.
Tau 5h30'37" 22d52'25"
Michael Parrotte
Sco 17h52'51" -30d4'47"
Michael Patrick
Per 3h7'57" 51d17'9"
Michael Patrick
And 2h3'55" 38d47'7"
Michael Patrick Bracco
Aql 20h5'25" 4d22'11"
Michael Patrick Breen
Crb 15h34'41" 35d37'40"
Michael Patrick Crimmins
Cyg 21h13'3" 30d9'57"
Michael Patrick Dooley
Tau 5h5'37" 21d19'59"
Michael Patrick Fleming
Cyg 20h10'19" 51d28'35"
Michael Patrick Marron
Tau 5h36'59" 21d17'57"
Michael Patrick McNamara
Uma 9h29'46" 44d40'53"
Michael Patrick Murphy
Per 3h32'47" 47d54'51"
Michael Patrick Murphy
Ari 3h12'59" 29d44'18"
Michael Patrick Nolan
Cnc 8h38'47" 17d0'16"
Michael Patrick O'Connell
Gem 6h34'29" 25d44'57"
Michael Patrick Reese
Uma 8h16'16" 62d32'42"
Michael Patrick Riley
Ari 3h3'16" 13d9'52"
Michael Patrick Sullivan
Dra 16h56'2" 58d19'9"
Michael Patrick van der Kuyp
Cep 22h35'18" 66d58'57"
Michael Patrick Welby
Her 17h16'46" 15d34'59"
Michael Patrick Windham
Vir 12h35'14" -0d7'4"
Michael & Paul
Uma 11h38'2" 34d14'57"
Michael Paul Bazzell
Psc 1h9'10" 10d12'58"
Michael Paul Cole
Per 4h32'14" 31d59'28"
Michael Paul Harris
Tau 4h6'3" 29d3'12"
Michael Paul Langwasser
Cyg 21h0'21" 47d35'3"
Michael Paul LeBlanc
Tau 5h36'0" 25d12'51"
Michael Paul Manz
Per 4h36'3" 41d21'45"
Michael Paul Place
Cnc 8h48'27" 22d25'14"
Michael Paul Rogalin
Ori 5h46'37" 1d48'30"
Michael Paul Rubin
Sco 16h49'6" -26d58'33"
Michael Paul Shenigo
Her 17h56'3" 49d3'58"
Michael Paul Sisti
Uma 12h33'38" 54d59'54"
Michael Pavelchak
Psc 0h16'15" 17d13'50"
Michael Peck
Uma 8h19'32" 65d15'42"
Michael Penney
Her 16h48'24" 17d16'38"
Michael Peragine
Leo 11h43'35" 26d16'4"
Michael Peroni
Psc 1h23'30" 28d42'41"
Michael Perry Webb
Her 16h38'44" 45d56'17"
Michael Peter Carlo
Lac 22h23'2" 52d13'48"
Michael Peter Gambardella
Her 17h1'56" 32d11'4"
Michael Peter Gardner
Uma 13h46'4" 56d44'24"
Michael Peter Hemingway
Leo 9h49'57" 29d24'45"
Michael Peter Katz
Ori 5h31'57" 12d6'23"
Michael Peter Rock
Sco 16h9'40" -11d23'44"
Michael Peter Stevens
Uma 10h46'29" 54d2'52"
Michael Peters
Gem 7h8'0" 26d42'23"
Michael Peters I Love U 4Ever, Elia
Per 3h47'47" 42d49'30"
Michael Pettenato IV
Tau 3h32'36" 22d12'48"
Michael Philip Jayson
Leo 11h58'9" 23d38'8"
Michael Philips
Ori 5h54'51" -0d44'43"
Michael Phillip Bookmyer
Lmi 9h24'27" 39d6'33"
Michael Picarazzi
Cep 22h37'52" 58d4'11"
Michael Pietro
Cyg 20h51'19" 34d42'37"
Michael Pinkiewicz
Per 2h22'35" 55d36'23"
Michael Pinkney A Star 2 B Proud of
Cep 22h3'59" 57d6'11"
Michael Platt
Ori 5h59'49" 2d4'43"
Michael Podesta
Her 18h57'29" 16d2'43"
Michael Polcino
Aur 5h54'45" 52d35'14"
Michael Pommarane
Uma 10h25'27" 48d31'41"
Michael Pontoriero
Sco 16h52'33" -37d24'22"
Michael "Pop" Melfi
Gem 7h31'18" 33d58'31"
Michael Porter Wilson
Uma 10h10'20" 46d24'42"
Michael Poynor Chapman
Uma 9h23'32" 43d51'27"
Michael Pratt Jones
Cnc 8h17'56" 9d17'0"
Michael Preston Maday
Boo 14h45'19" 38d43'52"
Michael Prsa
Gem 6h42'33" 21d4'36"
Michael Putman
Cep 2h9'41" 80d41'45"
Michael (Quimby) Lewis
Sgr 18h47'59" -31d32'53"
Michael Quinn
And 22h59'57" 41d44'0"
Michael R. Cintolo
Per 3h19'1" 51d51'19"
Michael R. Colby
Her 16h35'46" 36d55'8"
Michael R. Connor "The Superstar"
Per 3h16'27" 47d14'17"
Michael R. Crosta
Cru 12h36'2" -56d0'24"
Michael R. Hitchcock 10/4/67
Ori 5h38'6" 11d21'32"
Michael R. Jopinko, Sr. "Poo"
Uma 10h49'44" 69d34'45"
Michael R. Mills
Per 4h42'25" 50d33'41"
Michael R. Myers
Cyg 21h52'3" 43d13'12"
Michael R. Naif
Cep 22h36'24" 57d47'19"
Michael R. Riley
Umi 14h19'23" 74d35'40"
Michael R. Scott
Uma 9h58'6" 53d57'48"
Michael R. Tune
Sgr 17h53'19" -28d6'9"
Michael R. Vaccaro
Her 16h50'37" 41d25'59"
Michael Raffaelle Scocozzo
Sgr 18h53'54" -20d27'32"
Michael Ralph Vivone
Her 17h24'5" 38d53'21"
Michael Ramsdell
Ori 5h38'13" 6d43'23"
Michael Randall Thompson
Sco 17h50'4" -32d4'54"
Michael Randolph Cumings
Boo 14h34'6" 48d58'59"
Michael Rappaport
Uma 8h41'13" 68d24'45"
Michael Raso
Ori 5h35'10" -5d46'33"
Michael Ratcliffe
Cep 21h48'19" 58d57'26"
Michael Ray Bruce
Tau 4h41'35" 5d15'6"
Michael Ray Farley
Her 17h39'11" 33d47'14"
Michael Ray Rathje
Aqr 21h14'49" -3d35'48"
Michael Ray Swanson
Tri 2h8'53" 35d32'13"
Michael Raybom
Ori 5h32'53" 11d0'40"
Michael Raymond Buzard
Uma 8h55'37" 54d40'27"
Michael Raymond Matthews
Uma 10h59'17" 52d26'17"
Michael (Raymonde's Star)
Per 2h56'19" 56d3'53"
Michael Reardon
Uma 9h25'30" 71d57'25"
Michael & Reba Beratan
Uma 11h44'10" 35d3'10"
Michael "Red"
Uma 11h15'52" 60d31'39"
Michael Redmond
Cep 23h36'58" 74d20'30"
Michael Reinhard
Uma 10h0'34" 64d23'17"
Michael Reinhold
Ori 5h59'2" -2d14'4"
Michael & Renate Salerno
Cyg 19h56'59" 59d27'5"
Michael Rene Meza-McDonough
Peg 22h44'36" 8d4'55"
Michael Reshard Clayton
Lib 15h32'46" -8d5'2"
Michael Reul
Uma 8h12'23" 66d6'33"
Michael Rex Thomas
Psc 1h14'4" 31d28'35"
Michael Rey
Per 3h45'30" 39d16'9"
Michael Reza Clark
Uma 11h48'5" 61d24'41"
Michael Rich
Vir 13h24'43" 10d2'47"
Michael Richard
Uma 11h18'42" 54d1'52"
Michael Richard Cohs
Her 17h28'6" 39d0'3"
Michael Richard Klein
Her 17h11'55" 47d36'15"
Michael Richard Lliam Mowdy
Tau 5h31'25" 26d15'42"
Michael Richard McCorkhill
Sge 20h18'11" 18d33'17"
Michael Richard Smith
Per 3h45'46" 44d1'34"
Michael Richard Tuymer
Ari 3h27'1" 27d4'55"
Michael Ricigliano
Uma 10h47'18" 55d53'34"
Michael Rieger
Uma 9h14'16" 69d47'39"
Michael Riley
Cnc 9h8'48" 31d28'14"
Michael Riley Dadswell
Uma 11h47'10" 50d54'51"
Michael Riley Thomas
Gem 6h51'29" 33d57'30"
Michael Rizzo
Uma 8h16'8" 67d18'1"
Michael Robbins
Sgr 18h0'36" -27d57'24"
Michael Roberson
Aqr 21h5'41" -9d30'3"
Michael Robert Allen Hanna
Sco 17h33'57" -34d3'25"
Michael Robert Ferczak
Uma 11h14'19" 52d34'56"
Michael Robert Gebhardt
Uma 14h23'19" 55d49'7"
Michael Robert Gerrie
Cep 23h7'1" 70d55'32"
Michael Robert Hutchison
Her 18h51'24" 23d31'35"
Michael Robert Latham
Vir 13h49'12" -7d12'36"
Michael Robert McGloin
Cep 20h54'2" 66d51'44"
Michael Robert Ochsner "Uncle Mike"
Dra 18h25'35" 51d7'0"
Michael Robert Pertain
Crv 12h26'24" -18d8'54"
Michael Robert Roach
Cap 21h34'26" -10d55'22"
Michael Robert Roche
Aqr 23h55'48" -4d8'1"
Michael Robert Stobbart
Per 4h39'55" 47d4'12"
Michael Robert Zuckerman
Uma 10h37'52" 41d59'23"

Michael Roberts ASU Graduation Star
Lmi 10h27'41" 35d40'6"
Michael & Robin
Cyg 19h35'26" 30d14'46"
Michael "Robin" Smith-Latino
Cep 21h55'0" 59d57'51"
Michael Roger Burkhalter
Psc 0h31'10" 7d19'38"
Michael Roger Corbeil
Aur 5h35'14" 43d22'45"
Michael Rogers
Cnc 8h39'46" 23d36'52"
Michael Rokicsak
Aur 6h31'34" 34d40'43"
Michael Ronan O'Shea
Per 2h22'34" 53d34'2"
Michael Rose
Peg 22h48'27" 16d8'49"
Michael Rose
Aqr 20h52'33" -11d14'2"
Michael & Roseann
Ori 6h7'8" 12d32'44"
Michael Rosenberg
Lyr 18h31'55" 31d41'18"
Michael & Rosie
Sge 19h36'44" 17d34'33"
Michael Rossi
Ori 6h5'1" 13d32'54"
Michael Roy
Sco 16h7'3" -11d8'56"
Michael Rozenberg
Uma 10h40'20" 45d6'36"
Michael Ruben "Angel Face" Valverde
Lyn 6h19'12" 57d30'43"
Michael Rue Johnson
Vir 14h4'22" -12d56'46"
Michael Rusz
Her 16h35'27" 44d57'24"
Michael Rutter - Twinkle's Star
Uma 8h43'10" 48d30'28"
Michael Ryan
Per 3h43'25" 45d19'28"
Michael Ryan Bradshaw
Leo 11h3'39" 20d1'27"
Michael Ryan Graves
Ari 2h10'26" 17d8'18"
Michael Ryan Pickard
Gem 6h18'28" 21d35'44"
Michael Rykaceski
Aqr 22h30'37" 0d26'51"
Michael S. Blackmon, Sr.
Cnc 9h8'2" 30d48'15"
Michael S. Douglas
Tau 5h15'8" 19d59'0"
Michael S. Druschel
Her 18h24'41" 12d10'12"
Michael S. Franford
Lac 22h12'27" 53d0'22"
Michael S J Eichner
Ori 6h12'51" 15d39'3"
Michael S. Janis
Sgr 18h26'11" -15d59'32"
Michael S. Kerkowski
Uma 11h44'45" 35d8'19"
Michael S. Litterio
Sco 16h11'58" -21d14'1"
Michael S. Manoukian
Tau 5h24'37" 26d15'54"
Michael S. Myers
Aur 5h20'8" 31d42'49"
Michael S. Patten
Her 17h27'27" 46d5'37"
Michael S. Pillow
Aqr 21h50'3" -0d28'4"
Michael S. Robinson
Tau 5h10'7" 19d7'3"
Michael S Sheehan
Cyg 20h41'40" 36d18'38"
Michael S. Trent
Sgr 18h35'5" -23d41'49"
Michael S. Wilson
Aqr 22h44'41" 1d46'55"
Michael Salvatore Russo
Vir 14h29'27" 5d32'35"
Michael Salvatore Stanton
Aql 19h31'1" 3d27'9"
Michael Salz
Uma 11h4'56" 42d32'52"
Michael Samuel
Umi 9h45'52" 88d23'54"
Michael Samuel Curatola
Lyn 7h1'45" 50d58'29"
Michael Samuël Philipoom
Sco 16h48'45" -27d20'20"
Michael Samuel Schwart
Sgr 18h38'17" -24d53'31"
Michael Sandford
Cnc 8h12'0" 7d40'9"
Michael & Sandra Kurtyka
Uma 8h34'1" 67d57'48"
Michael & Sandra McCandless
Dra 16h52'1" 53d34'36"
Michael & Sandy Bierly
Per 4h16'15" 45d25'52"
Michael Sansivero
Vir 11h47'35" 8d52'39"
Michael Santino Canales
Lib 14h34'50" -9d13'51"
Michael Santo Emanuelo, Jr.
Ari 2h48'43" 17d54'39"
Michael Santos, Jr.
Her 18h56'57" 13d25'56"
Michael Savidge
Cep 21h45'15" 70d39'2"
Michael Savino
Uma 12h48'25" 59d29'58"
Michael Saxida
Sco 17h25'50" -35d30'55"
Michael Schalli
Uma 8h49'6" 55d4'28"
Michael Schiavone
Lib 15h56'16" -9d42'4"
Michael Schoenmann
Ari 2h13'54" 19d57'30"
Michael Schurick
Per 3h24'5" 52d20'26"
Michael Sciara - The Best Dad Ever
Lyn 8h46'51" 40d39'53"
Michael Scibelli
Aql 19h47'32" 15d30'45"
Michael Scott Adams
Cep 22h27'53" 65d28'8"
Michael Scott Georgen
Cma 6h36'29" -14d30'22"
Michael Scott Johnson
Vir 13h47'56" 2d53'33"
Michael Scott Kagarise Jr.
Cam 5h44'52" 59d39'46"
Michael Scott Lacey
Gem 6h44'14" 29d8'24"
Michael Scott Lentz
Umi 14h26'23" 78d20'8"
Michael Scott Lieberthal
Uma 8h45'6" 63d48'18"
Michael Scott Lopez
Uma 8h23'55" 66d10'26"
Michael Scott North
Uma 9h43'34" 70d33'36"
Michael Scott Palosaari
Uma 8h33'12" 65d5'37"
Michael Scott Peterson
Tau 4h26'24" 13d18'2"
Michael Scott Scheibner
Her 17h7'46" 47d4'22"
Michael Scott Tetro Jr.
Her 17h57'58" 29d41'3"
Michael Scott Wiltshire
Aur 5h19'38" 31d38'33"
Michael Scowen
Dra 19h14'33" 73d40'1"
Michael Sea
Cyg 20h32'16" 46d6'41"
Michael Sean Oblad Mikesell
Cnc 8h27'41" 26d30'5"
Michael Seligman
Her 17h26'33" 45d12'16"
Michael Selvey
Dra 16h5'8" 57d28'57"
Michael Seth Steingart
Lyn 6h22'5" 57d57'46"
Michael Shane Crater
Psc 0h36'31" 8d41'9"
Michael Shane Miller
Aur 5h29'14" 44d17'31"
Michael Shane Taylor
Uma 10h55'55" 55d2'4"
Michael Shannon Seewald
Aur 6h2'17" 50d28'7"
Michael Shapiro
Cnc 8h4'33" 17d17'33"
Michael Shaun Reed
Ari 2h23'56" 27d45'48"
Michael Shawn
Leo 10h17'51" 17d32'9"
Michael Shawn Coppock—8/18/1988
Leo 10h13'28" 14d51'56"
Michael Shawn Null
Gem 6h51'43" 14d8'18"
Michael Sheehan
Cep 21h14'37" 65d49'20"
Michael Sheldrick
Lyn 7h25'13" 57d39'8"
Michael Sherrill
Uma 9h30'5" 44d23'50"
Michael Shook
Her 17h15'42" 34d56'21"
Michael Shrader - 24.03.1981
Ori 5h26'41" -1d41'7"
Michael Simon Blumenthal
Sgr 19h18'54" -15d36'44"
Michael Sing
Lyn 7h46'15" 37d55'38"
Michael skateboarder Leary Jr 1987
Cnv 12h38'51" 43d9'23"
Michael Slemensky
Dra 16h12'0" 54d5'7"
Michael Smith
Tau 4h7'15" 6d21'45"
Michael Smythe
And 1h11'50" 36d51'3"
Michael Sokol
Sco 17h7'13" -43d27'52"
Michael Sokolowski
Dra 19h28'39" 62d36'55"
Michael Solo Wilson
Vir 14h22'27" 1d43'39"
Michael Spae
Uma 9h52'56" 55d41'7"
Michael Spicer
Her 16h48'18" 30d51'53"
Michael SPIDER Rae
Ari 2h15'4" 24d1'57"
Michael Sponzo
Cep 21h26'23" 64d38'46"
Michael Stacey - Light of Peace
Psc 0h18'19" -0d21'49"
Michael Stahr
Uma 9h0'22" 47d44'16"
Michael Stamatopoulos
Cyg 19h58'54" 54d53'6"
Michael Stanley
Uma 10h35'13" 39d51'35"
Michael Staudacher, M1 ILD D M2 F I
Ori 5h17'39" 3d33'49"
Michael & Stefanie
Cyg 20h21'53" 52d36'32"
Michael Steffen
Tri 2h13'13" 32d26'19"
Michael Stempler
Umi 14h39'9" 72d33'49"
Michael "Stephan" Williams
Uma 9h1'9" 72d21'9"
Michael Stephen Euculano
Uma 10h3'51" 43d27'3"
Michael Stephen Leander
Lmi 11h6'37" 33d0'24"
Michael Sterling
Her 17h33'15" 19d0'36"
Michael Steven Everidge
Per 3h39'55" 34d36'52"
Michael Steven Foss Agape Love
Her 18h7'15" 31d55'7"
Michael Steven Gutierrez
Her 16h33'6" 34d49'59"
Michael Steven Kurowicki
Sgr 18h2'3" -28d43'15"
Michael Steven Lennie
Tau 4h18'46" 0d56'37"
Michael Steven Long
Uma 10h26'48" 55d19'14"
Michael Steven Modell
Psc 1h3'50" 12d58'19"
Michael Steven Pakech
Psc 1h18'32" 31d41'29"
Michael Stevens
Aqr 20h39'45" -8d53'26"
Michael Stewart Woods
Lib 15h43'48" -7d3'48"
Michael & Storm - Alight Forever
Sco 16h55'2" -30d15'46"
Michael Streicher
Uma 9h9'20" 64d15'26"
Michael Strickland, MD
Per 2h25'29" 55d13'16"
Michael Strickland, MD
Per 2h21'15" 55d15'19"
Michael Strout
Umi 15h15'40" 71d14'15"
Michael Stubberfield
Cru 11h58'56" -62d44'53"
Michael Suk Woo Brady
Psc 0h43'55" 15d39'24"
Michael Sullivan
Cep 23h19'23" 77d29'27"
Michael Super Daddy Kosinski
Lib 15h14'42" -11d36'35"
Michael & Susan McLemore
Ori 5h36'57" 8d4'22"
Michael & Susanne Troumouliaris
Cyg 21h16'2" 44d47'16"
Michael Swenson
Aql 19h16'1" 11d7'20"
Michael Swerdloff
Uma 10h29'27" 65d42'0"
Michael Szafranski
Lyn 8h47'54" 36d56'4"
Michael Szymankis
Uma 10h22'26" 40d8'23"
Michael T. Browne
Aql 19h14'56" -4d34'3"
Michael T. Butler
Uma 11h52'53" 43d35'44"
Michael T Debies
Dra 17h10'19" 58d3'14"
Michael T. Greene
Ori 4h50'34" 4d19'44"
Michael T. Ivy, Jr.
Her 17h47'18" 37d4'3"
Michael T. Kapinos
Aur 5h40'27" 49d13'12"
Michael T. Katz
Cyg 21h32'2" 39d7'37"
Michael T. Landy
Per 3h40'23" 34d42'51"
Michael T. Pugh
Eri 4h26'19" 0d14'15"
Michael T. Rasely
Her 18h34'54" 15d38'23"
Michael T. Starr
Ori 5h36'14" -2d54'20"
Michael T. Thelen
Psc 0h42'8" 12d16'56"
Michael & Tainya
Cyg 20h28'15" 47d28'27"
Michael & Tami Rittberg
Cra 18h7'36" -36d58'1"
MICHAEL TAROMINA
Her 18h21'58" 18d30'48"
Michael Tatlow
Per 4h6'54" 41d47'5"
Michael Taylor McGee
Ori 5h8'26" 0d21'16"
Michael Teggae
Per 4h18'38" 45d29'54"
Michael Teoli
Dra 18h54'57" 66d48'13"
Michael Terrance Sadleir
Sco 17h52'25" -37d32'47"
Michael Thanos
Her 17h56'24" 49d39'34"
Michael the Borrowed Angel
Psc 0h34'22" 7d14'51"
Michael "THE DRUMMER" Arnold
Sco 17h20'55" -31d24'29"
Michael the Love of my Life
Ori 5h3'10" 5d24'44"
Michael - the star in the east
Aur 5h40'16" 47d36'13"
Michael&Theresa-Starlight Sanctuary
Cyg 19h49'45" 43d12'42"
Michael Thomas
Ori 6h17'16" 14d22'10"
Michael Thomas Berard
Cep 20h38'17" 64d20'56"
Michael Thomas Bock
Lib 14h33'25" -10d27'55"
Michael Thomas Dickson
Tau 4h7'48" 6d14'11"
Michael Thomas Flick
Dra 16h2'31" 54d49'57"
Michael Thomas Furlough
Cnv 12h29'22" 32d59'44"
Michael Thomas Goldberg
Ori 5h4'25" 15d21'36"
Michael Thomas Gordon
Ari 2h29'47" 18d18'9"
Michael Thomas Gray
Ori 6h6'13" -0d0'22"
Michael Thomas Lanza
Sco 16h13'29" -10d16'32"
Michael Thomas Miosek
Ori 5h14'51" -0d14'17"
Michael Thomas Pinette
Uma 8h53'17" 51d41'57"
Michael Thomas Ray
Her 17h20'12" 45d9'29"
Michael Thomas Vacca
Cep 22h35'7" 66d25'13"
Michael Thomas Watson
Sco 16h7'24" -10d14'22"
Michael Thompson
Uma 10h23'52" 46d39'52"
Michael Thornton Ray
Her 16h9'14" 22d26'7"
Michael "Tig" Tignor
Sco 16h50'55" -38d50'24"
Michael Tim Richards
Psc 0h44'39" 9d58'49"
Michael Timothy
Per 4h45'51" 44d27'31"
Michael Tod Sawyier
Cyg 19h42'58" 31d31'7"
Michael Todd Brown
Lyn 7h12'17" 47d44'38"
Michael todd Cordray
Cap 21h31'19" -16d58'1"
Michael Todd Evans
Per 3h26'15" 45d1'26"
Michael Todd McCray
Ori 5h24'0" 3d0'57"
Michael Todd Meyers
Gem 7h23'10" 27d43'56"
Michael Torline
Sco 17h48'30" -41d35'1"
Michael Torre
And 0h46'51" 37d42'43"
Michael Townsend Martin
Aql 19h24'22" 7d58'28"
Michael Toyne's 21st Birthday Star
Cru 12h34'10" -60d24'18"
Michael Tracy
Her 17h44'46" 35d54'12"
Michael Trainor
Crb 15h50'4" 32d55'50"
Michael Tran
Uma 13h0'20" 53d51'40"
Michael Tray Ring
Leo 10h18'57" 17d11'59"
Michael Trevellyan Jeffery - Daddy's Star
Per 3h2'24" 52d54'58"
Michael Trillizio Super Husband/Dad
Cep 21h57'26" 65d48'1"
Michael Trojczak
Cru 12h6'24" -61d54'1"
Michael T.S
Lib 14h49'49" -0d37'34"
Michael Tubbiola
Ari 3h9'34" 19d32'31"
Michael Tuttle
Cep 21h19'2" 62d45'47"
Michael Tyler Buskirk
Ori 6h18'59" 5d41'48"
Michael Tymczyszyn
Uma 11h28'8" 55d29'45"
Michael Uhl
Lyn 8h21'5" 54d37'14"
Michael Urban Laundroche
Uma 9h44'20" 46d39'16"
Michael V. Josefson
Gem 6h58'20" 17d15'11"
Michael V. Muffuletto
Her 17h13'37" 32d17'27"
Michael V. Pugliesi
Aqr 23h2'53" -14d44'38"
Michael Val Heathcock
Her 17h20'23" 35d57'47"
Michael Valdez My Hearts Twin
Sco 17h10'12" -44d16'15"
Michael Valentino
Tau 4h2'20" 24d7'46"
Michael Van Duyne Cowenhoven
Uma 11h20'3" 66d58'45"
Michael Vaughn
Lib 15h16'30" -6d25'17"
Michael Vecchio
Pho 0h11'29" -41d59'31"
MICHAEL VENDITTO JR
Aqr 21h8'9" -9d17'15"
Michael Verrengia
Per 3h1'28" 54d49'48"
Michael Veto
Aqr 21h22'10" -10d17'37"
Michael Vincent
Cap 21h5'31" -20d26'38"
Michael Vincent
Leo 9h46'20" 28d40'18"
Michael Vincent Corradino
Vir 13h1'48" -1d15'59"
Michael Vincent Guzman
Her 17h34'56" 28d39'45"
Michael Vincent Iozzi
Cep 22h32'53" 72d23'15"
Michael Vincent Prestonise
Umi 14h50'56" 70d12'1"
Michael (Vinnie) Thomas
Sgr 18h20'50" -33d47'43"
Michael & Virginia Carl
Cyg 19h59'31" 49d2'6"
Michael Vito Truncali
Cap 20h7'48" -10d53'50"
Michael Volker
Aur 5h44'44" 47d19'25"
Michael Volpe
Lib 14h52'48" -3d29'19"
Michael W. Gordon
Tau 5h32'16" 24d22'22"
Michael W. Heydlauf
Ori 6h10'25" 2d1'40"
Michael W. Ireland
Umi 15h39'22" 70d15'5"
Michael W. Joern
Dra 19h6'13" 73d18'20"
Michael W. Lopez
Leo 10h8'55" 25d47'41"
Michael W. Lubold
Boo 14h46'18" 21d36'28"
Michael W. Metzger
Uma 9h38'27" 71d28'24"
Michael W. Millwood
Gem 7h23'13" 25d39'24"
Michael W. Murray
Aur 5h18'26" 46d13'8"
Michael W. Murtha
Ari 2h31'40" 24d33'41"
Michael W Pearson
Aqr 22h47'42" 2d36'33"
Michael W. Rainville
Sco 16h8'33" -9d55'29"
Michael W. Wilson
Aqr 21h30'23" 1d46'40"
Michael W. Zehner
Her 17h31'43" 35d54'5"
Michael Wade
Boo 15h5'21" 48d29'31"
Michael Wagner
Uma 11h13'49" 34d3'54"
Michael Wagner
Uma 10h21'1" 42d59'7"
Michael Walker
Uma 9h29'34" 49d56'13"
Michael Walker
Umi 14h48'55" 73d20'15"
Michael Wallace Schaerr
Her 17h25'13" 42d31'16"
Michael Walsh
Cep 22h54'45" 70d5'48"
Michael Walter
Uma 8h14'8" 69d12'36"
Michael Walter
Uma 10h1'10" 51d58'59"
Michael Ward
Ori 6h14'53" 21d3'43"
Michael Ward Superstar
Ari 3h8'31" 28d43'28"
Michael Wareing
Cep 5h5'28" 82d32'10"
Michael Warren Kirkner
Sco 16h45'48" -35d2'29"
Michael Warren Pitts
Aqr 22h29'51" -9d24'48"
Michael Warren Schmitz
Per 2h41'46" 54d15'56"
Michael Warren Sousa
Per 2h20'12" 56d22'21"
Michael Washburn
Psc 1h10'6" 21d58'28"
MICHAEL WASHINGTON
Sgr 18h14'31" -19d29'39"
Michael Waterman
Psc 0h48'33" 7d28'41"
Michael Wayne Kitchens
Aqr 23h21'24" -19d47'47"
Michael Wayne Miller
Cnc 8h7'22" 12d47'13"
Michael Wayne Pogany
Uma 11h27'16" 54d28'8"
Michael Wayne Richeson
Lyr 18h50'17" 33d37'50"
Michael W.Burke
Ori 5h45'13" -0d19'19"
Michael Webb - TKW
Per 4h42'49" 50d19'47"
Michael Weiand
Her 17h45'43" 46d45'40"
Michael Weisback
Her 18h54'30" 20d23'38"
Michael Weiss
Her 18h31'13" 12d27'3"
Michael Weiss
Her 16h21'27" 48d4'7"
Michael Weltie Ware
Per 4h33'30" 39d59'55"
Michael W.Gardner
Lib 15h17'5" -10d36'20"
Michael Wharton Mellor
Her 18h4'36" 16d41'36"
Michael Wheeler
Cep 22h11'9" 64d19'19"
Michael Wicke
Boo 14h19'2" 53d34'42"
Michael Wilcox Grace
Ori 6h15'11" 3d33'34"
Michael Wild-A Shining Star Forever
Her 16h31'33" 48d22'51"
Michael Willan
Uma 11h25'40" 60d0'12"
Michael William
Sco 17h3'18" -34d46'9"
Michael William
Her 16h16'57" 7d21'47"
Michael William Bain
Cru 12h13'0" -59d29'44"
Michael William Bressler
Uma 9h57'38" 49d27'2"
Michael William Dean
Boo 14h41'28" 26d31'24"
Michael William Esposito
Uma 11h44'13" 41d0'56"
Michael William Farrington, MF
And 0h45'33" 40d40'52"
Michael William Garcia
Aql 19h10'12" 5d56'49"
Michael William Haderer
Lib 15h4'39" -1d54'10"
Michael William Higgins
Aqr 22h22'1" -21d41'25"
Michael William Jackson
Psc 1h6'17" 26d41'18"
Michael William McArthur
Ori 6h4'29" 17d5'58"
Michael William McArthur
Ori 6h6'33" 13d55'37"
Michael William Schoener
Crb 16h4'25" 36d51'14"
Michael William Soteras
Cru 12h2'5" -62d21'5"
Michael William Tate
Cru 12h40'45" -60d6'18"
Michael William Ucci
Her 16h31'2" 29d14'35"
Michael William Valdez
Sco 16h3'55" -24d32'34"
Michael William Woan
Uma 10h34'31" 70d30'17"
Michael William Yearsley
Ori 5h39'42" -0d41'21"
Michael Williams
Sgr 18h27'46" -32d5'41"
Michael Williams
Tau 4h27'57" 18d15'12"
Michael Williams
Gem 7h34'50" 34d24'9"
Michael Wilson
Uma 11h44'51" 31d38'39"
Michael Wilson
Per 3h44'49" 48d54'5"
Michael Wilson
Boo 14h29'11" 40d11'9"
Michael Wilson
Aql 18h52'43" 8d41'44"
Michael Winebold ~ Precious Son
Cyg 19h47'29" 55d0'52"
Michael Winters...My Guiding Star
Vir 12h47'20" 8d28'17"
Michael Wollschlager
Cep 21h34'12" 59d12'46"
Michael Wood
Aql 19h54'25" -0d59'5"
Michael Woodliff
Uma 13h10'50" 53d53'21"
Michael Woodside
Gem 7h16'40" 19d59'13"
Michael Worsley
Her 17h28'41" 15d0'54"
Michael Wrobel
Uma 8h11'4" 68d55'37"
Michael Wyatt Pressley
Psc 0h17'17" 10d52'53"
Michael Y. Singh
Gem 6h59'32" 34d56'47"
Michael Yabsley
Cru 12h40'3" -60d4'55"
Michael "YAYA" Kategianes
Uma 11h25'24" 44d22'48"
Michael - You are my Shining Star
Aur 5h17'2" 41d55'29"
Michael - You are my Shining Star
Ori 6h4'59" 5d53'59"
Michael Young
Her 18h44'32" 20d9'40"
Michael Zackery Ostrander
Uma 10h45'47" 58d28'26"
Michael & Zandra Forever
And 23h35'15" 41d21'28"
Michael Zanger Tishler
Psc 2h0'32" 4d46'33"
Michael Zapfe
Lyn 6h27'53" 56d41'32"
Michael Zaradich
Uma 11h23'20" 45d24'42"
Michael Zatto
Sco 16h16'45" -24d12'4"
Michael Zettinig
Dra 18h34'59" 59d15'27"
Michael Zimmerman, Love of My Life
Her 17h7'37" 25d32'41"
Michael Zimmermann
Ori 5h39'25" 0d36'22"
Michael Zwiezynski
Ari 2h28'33" 18d46'23"
Michael, Dana, & Ashley Forever!
Tri 1h53'50" 27d51'30"
Michael, der Stern in meinem Herzen!
Ori 5h25'21" -8d15'16"
Michael, Kathy and William Couvion
Umi 16h44'0" 83d26'7"
Michael, Margie & Sarah Khoury
Lyn 7h22'18" 44d35'44"
Michael, Mary and Rebecca Kades
Crb 15h37'1" 35d37'55"
Michael, meus amore
Leo 11h21'11" 10d29'0"
Michael, my shining star
Cep 0h13'10" 75d51'26"
Michael, The Archangel
Cas 1h15'57" 53d25'36"
Michael, the love of my life
Lmi 10h26'51" 33d2'48"
Michael, Two Hearts One Friendship
Per 3h5'31" 56d31'24"
michaela
And 23h12'34" 48d28'50"
Michaela
Lyr 18h46'12" 36d35'45"
Michaela
Lyn 8h47'4" 40d45'25"
Michaela
Cyg 21h20'57" 32d29'46"
Michaela
Vir 12h54'33" 7d47'55"
Michaela
Ori 6h6'42" 15d16'41"
Michaela
Gem 7h9'54" 23d44'12"
Michaela
Uma 8h47'52" 68d22'25"
Michaela
Dra 15h9'59" 64d4'44"
Michaela
Uma 12h11'52" 60d32'14"
Michaela
Uma 9h14'51" 52d33'34"
Michaela
Uma 10h28'36" 52d34'24"
Michaela
Ori 5h35'52" -6d10'1"
Michaela - 18-06-1980
And 1h2'49" 46d49'18"
Michaela & Alfred Rudolf
Cas 1h1'4" 63d27'23"
Michaela Alissa O'Connor
Gem 6h50'25" 33d42'55"
Michaela Bezonek
Uma 10h0'58" 64d9'10"
Michaela Brody
Aqr 21h22'50" -10d50'39"
Michaela Carolyn Darby Donovan
And 0h39'56" 27d15'48"
Michaela Christine
Vir 11h59'8" -0d11'25"
Michaela Criss
Lyn 6h56'39" 50d10'59"
Michaela Dawn
Gem 6h57'28" 14d34'7"
Michaela Dominique Salerno
Ari 2h23'54" 18d49'37"
Michaela Faith
Her 17h18'56" 14d24'46"

Michaela Gelwick Vir 12h4'20" -10d3'39"
Michaela Gennuso Sgr 18h19'2" -22d22'6"
Michaela Harris And 23h17'34" 41d47'13"
Michaela Honor Moore Cam 4h21'35" 67d49'51"
Michaela Horvat And 1h38'38" 46d39'38"
Michaela Isabella Maria Quimby Uma 12h6'27" 61d7'13"
Michaela Kahle Psc 0h0'4" 3d56'47"
Michaela Kanoelani's Star Lib 15h55'26" -4d17'18"
Michaela Larsen Lib 14h30'10" -9d43'16"
Michaela Leider Aqr 22h25'18" 0d3'26"
Michaela LG Lyn 8h24'35" 45d8'58"
Michaela Lillian Ori 5h20'3" 13d8'57"
Michaela Maria D Marinko And 23h3'50" 40d15'27"
Michaela Marie Ari 3h10'27" 28d58'26"
Michaela Marte And 1h47'29" 49d11'51"
Michaela Mc Niffe Gem 7h27'19" 31d51'33"
Michaela McCloskey-Ternes Cas 1h46'32" 63d34'42"
Michaela *Mia* Zingl Uma 9h7'27" 69d30'9"
Michaela Mikusova Uma 11h23'18" 58d38'57"
Michaela O'Malley Cap 20h22'41" -10d37'52"
Michaela Ritz Uma 8h59'39" 69d55'59"
Michaela Skye Mazur Aqr 23h21'1" -14d55'9"
Michaela Solise Tau 3h45'10" 8d58'26"
Michaela Stary Lib 14h57'34" -0d47'27"
Michaela Suzanna And 23h9'42" 44d1'40"
Michaela Torres - Mishka Pie Cam 5h21'12" 64d22'16"
Michaela1304 Uma 8h30'23" 64d8'47"
Michael-a-Abate, New London, CT Ori 6h9'4" 20d1'13"
MichaelAidriana Cyg 21h14'13" 45d45'15"
michael.alissa.d'amato Aql 19h35'30" -7d20'34"
MichaelandElliot Her 17h42'45" 22d14'26"
MichaelAngela Ori 5h44'45" 3d21'21"
Michaelangelo Leo 10h29'47" 12d44'7"
Michaelangelo Agnello Umi 15h43'53" 78d9'11"
Michaelangelo and Mona Lisa Uma 11h21'54" 52d37'27"
Michaela's eigener Stern Uma 14h9'23" 59d17'52"
Michaela's Herz Ori 6h19'51" 9d32'15"
Michaela's Little Star And 23h42'37" 34d7'20"
Michaela's Magical Make A Wish Star Lyn 7h9'40" 58d57'47"
Michaela's star Cas 0h27'8" 54d37'49"
Michaelaura Cyg 19h36'31" 32d0'4"
Michael-Brittany Devlin Ori 6h12'2" 7d25'13"
MichaelDawnForeverFriends Cyg 19h46'53" 58d0'38"
Michaele Cas 1h22'44" 57d44'4"
Michaelé Marie Jarrell Peg 23h12'6" 17d59'43"
Michael....Esther Uma 11h21'44" 33d18'42"
MichaelHowie Ari 2h5'40" 22d12'17"
Michaeli, Dorothea Uma 11h5'30" 49d10'45"
MichaelJosephGraham Ara 17h24'7" -47d41'27"
Michaella Umi 15h57'59" 76d11'1"
Michaella Anne & Mark Donald Wilson Cas 0h39'9" 60d21'30"
Michaella Loveridge And 2h30'51" 45d35'1"
Michaelle And 2h13'34" 37d59'15"
Michael-Marcella Cyg 21h49'50" 53d38'16"
Michael's 1st Star Her 17h37'44" 16d30'38"
Michael's Alli Lyr 18h45'34" 36d24'19"
Michael's Angel Sgr 19h21'24" -13d43'28"
Michael's Ball Of Gas Her 16h15'39" 18d8'8"
Michael's Brooklyn Star Gem 7h40'56" 19d9'39"
Michael's Eternal Light Umi 16h36'21" 78d5'47"
Michael's Girl Gem 6h43'42" 15d58'40"
Michael's Heavenly Star Aur 5h47'2" 46d10'30"
Michael's Limit Sco 16h11'53" -11d34'34"
Michael's Love Aqr 22h37'50" -3d26'56"
Michael's Lucky Christmas Star Umi 13h57'33" 77d38'29"
Michael's Lucky Star Sgr 18h58'28" -30d21'33"
Michael's Marchers for Miracles Uma 11h26'56" 34d56'36"
Michael's Perfect Date Cyg 21h6'53" 46d26'10"
Michael's place up in the stars Psc 1h19'42" 29d7'45"
Michael's Princess Psa 22h46'55" -29d27'55"
Michael's Serenity Pho 1h7'23" -44d41'47"
Michael's Star Psc 0h46'43" 7d11'41"
Michael's Star Dra 19h27'15" 70d36'49"
Michael's Star Her 18h2'2" 14d54'56"
Michael's Star Uma 8h34'41" 47d38'39"
Michael's Star Uma 9h40'57" 44d8'43"
Michael's star 3-25-76 Lyn 7h58'59" 42d25'1"
Michael's Star For All Time Sgr 18h45'27" -29d22'3"
Michael's Star of Hope Cyg 21h3'1" 46d12'9"
Michael's Star-Apud Crucis Cyg 19h45'43" 35d11'48"
Michael's Sunshine Her 16h38'37" 8d58'42"
Michael's Tracie Cap 20h13'43" -13d10'25"
Michael's Twelfth Gem 7h45'28" 32d36'13"
Michael's View Uma 9h36'8" 53d52'20"
Michaelson Cap 20h17'2" -12d43'9"
Michaelu Miller Per 2h21'2" 55d27'47"
Michaelus Bensonus 2006 Uma 8h10'5" 64d51'56"
Michaelus Youngbearus Uma 11h26'29" 41d47'39"
MichaelVito Sgr 17h51'52" -20d17'5"
MichaelVonStar Ser 15h42'33" 22d20'42"
Michaiah Ruth Thompson Cap 21h29'0" -12d21'24"
Michal Cep 3h56'15" 80d48'29"
Michal Beaumont Ori 5h50'52" 6d37'49"
Michal Brusilovsky Lib 15h39'16" -17d38'53"
Michal Caraline Ari 2h14'32" 12d34'29"
Michal Julia and Blake Kyle Lib 15h45'54" -19d49'52"
Michal Klakla Uma 11h39'9" 53d54'47"
Michal Kolda Uma 12h21'26" 55d14'51"
Michal Milan Strajanek - Vascak Her 18h2'52" 15d13'13"
Michal Rose Lee Chrissakis Vir 14h2'2" -15d5'31"
Michal S. Botting Tau 5h42'52" 21d6'18"
Michal Schauer Cas 0h17'53" 52d6'26"
Michalak-Baldacci Her 17h59'43" 21d57'4"
Michale Coffey, "Hero" Leo 9h44'41" 25d55'36"
Mic-hale Wright Gem 7h40'30" 25d24'21"
Michalea Diane King Tau 5h10'57" 24d53'32"
Michalek, Franz-Erhard Uma 11h1'7" 39d13'46"
Michalene Rishelle Hunt Her 18h10'10" 28d16'6"
Michalis Pagidas Uma 11h47'22" 59d31'48"
Michalis Vlazakis Uma 11h11'11" 29d37'32"
Michalyn Cnc 8h53'40" 22d49'9"
Michar Del 20h27'14" 15d5'47"
Michari Pho 0h43'54" -41d39'18"
Michas Cyg 19h21'51" 28d29'50"
Michayla Siemion Lib 15h3'24" -0d35'42"
Miche Oph 16h56'28" -0d17'2"
Miche Cyg 21h17'18" 31d9'42"
Micheal A McCudden - 23rd August 1957 Uma 9h26'50" 68d58'21"
Micheal Alexander Cap 21h38'23" -15d36'24"
Micheal and Stephanie Hamill Tau 5h27'55" 18d28'16"
Micheal & Ann Sabag Uma 9h29'55" 54d29'24"
Micheal Anthony Nader Cru 12h37'12" -59d17'40"
Micheal Best Tau 3h58'49" 20d45'46"
Micheal Childs Aql 20h8'1" 8d23'4"
Micheal D McGuire Dra 15h44'44" 62d14'31"
Micheal "Dork" Payes Lmi 9h48'42" 38d21'39"
Micheal E Shaw Sco 17h51'48" -36d17'46"
Micheal Glenn Lavender Gem 7h12'13" 20d19'28"
Micheal J. Ziembowicz Tau 4h9'24" 5d41'59"
Micheal James McConnell Lib 14h44'55" -12d49'14"
Micheal Kinsman Lac 22h25'11" 50d48'37"
Micheal Louis Ari 3h13'44" 15d24'16"
Micheal Moon Henderson Per 3h47'6" 51d32'13"
Micheal Paul Stein Cyg 20h26'32" 40d26'52"
Micheal Rady A.k.a MikhA Leo 11h16'17" 14d20'42"
Micheal Ray DeWeerd Sco 17h53'24" -35d53'12"
Micheal T. Kilborn Aur 5h53'53" 53d27'55"
Micheala Bennett Sco 17h47'12" -41d46'21"
Micheala Mann And 23h48'11" 46d50'15"
Micheale E. Smith "Mama Schell" Aqr 20h48'33" -11d4'58"
Michel Cep 2h27'55" 85d48'14"
Michel Lib 15h25'33" -29d9'33"
Michel Aqr 23h43'58" -16d56'57"
Michel Per 4h37'28" 31d52'2"
Michel Equ 21h23'32" 7d5'7"
Michel and Christie Cas 0h56'25" 58d27'15"
Michel Bardet - Papa - Paddy Del 20h25'58" 4d3'45"
Michel Bontemps Psc 1h19'33" 5d18'33"
Michel Carrier Tau 5h56'7" 25d27'23"
Michel Chartol Ori 5h41'20" -2d0'12"
Michel Delcenserie Cas 0h0'7" 54d10'44"
MICHEL DPS Crb 16h18'6" 37d11'46"
Michel et Marianne Bonnewitz Dra 18h59'26" 53d50'12"
Michel Francois Denis Coret Cnc 8h25'1" 14d9'31"
Michel & Gail Masotta Del 20h19'55" 9d6'47"
Michel Herrmann Ori 6h9'15" -2d7'49"
Michel Legrand Boo 14h52'31" 27d54'35"
Michel Leo Rousseaux Uma 11h47'55" 37d35'43"
Michel Menghetti Ori 6h20'44" 7d28'34"
Michel Milliat Lib 15h34'44" -24d51'39"
Michel Papou Hubert Uma 10h57'26" 44d29'3"
Michel Pisani Vir 12h45'6" -7d43'39"
Michel Provost Umi 13h31'34" 73d51'8"
Michel Renaud Dra 15h5'15" 55d42'20"
Michel (Sherif) Loiselle Del 20h37'21" 18d8'57"
Michel Sluijs Forever by your side Cru 12h12'56" -61d21'29"
Michel und Manuela Uma 9h21'17" 52d58'58"
Michel Valentin Leo 11h5'9" 3d35'6"
Michel Wafer Uma 11h33'35" 48d46'43"
Michel, Heinz-Joachim Sgr 18h7'20" -26d46'35"
Michela Cas 0h26'47" 62d20'50"
Michela Peg 22h45'3" 33d51'1"
Michela Lyr 19h20'22" 43d24'27"
Michela Leo 9h33'29" 6d38'30"
Michela Crb 16h2'38" 25d52'18"
Michela 30/05/2005 Boo 14h51'27" 28d18'20"
Michela Agazzi Cam 6h33'53" 62d8'57"
Michela Calussi Cam 4h36'0" 58d14'6"
Michela e Simone And 2h36'0" 40d18'46"
Michela Magretti Ari 2h12'18" 22d8'14"
Michela Stefanello Leo 11h9'20" 3d56'13"
Michela Tosatto Cam 9h30'25" 81d58'11"
Michelangelo Crimi Cep 23h6'15" 87d46'44"
Michelangelo Iansito Tau 5h37'37" 26d33'15"
Michelangelo Poletti Uma 11h33'12" 65d46'3"
Michele Uma 9h2'31" 62d29'9"
Michèle Lyn 6h23'59" 56d32'16"
Michele Umi 17h12'39" 85d55'54"
Michele Umi 16h11'52" 76d48'26"
Michele Lib 15h26'55" -7d26'10"
Michele Cap 20h56'20" -15d43'33"
Michele Sgr 17h58'55" -17d33'25"
Michele Uma 8h34'22" 60d13'16"
Michele Sco 16h52'20" -27d10'1"
Michele Sco 17h23'58" -42d36'44"
Michele Sgr 19h56'4" -35d39'52"
Michele Her 17h55'25" 20d46'33"
Michele Cnc 8h9'48" 15d59'33"
Michèle Gem 6h37'38" 18d25'31"
Michele Gem 6h21'36" 22d5'53"
Michele Leo 10h33'6" 17d48'24"
Michele Cnc 9h12'44" 16d3'16"
Michele Vir 12h31'13" 0d0'38"
Mi'chele Psc 1h26'32" 11d58'57"
Michele Per 4h47'5" 41d10'54"
Michele Lyn 9h18'0" 38d5'37"
Michèle Cam 3h43'46" 55d58'42"
Michele Cyg 21h15'5" 46d59'13"
Michele And 23h7'35" 51d14'38"
Michele Cyg 20h59'32" 46d17'50"
Michele Cyg 20h37'20" 41d37'35"
Michele Cas 0h17'56" 51d12'32"
Michele And 1h41'15" 45d12'20"
Michele And 1h6'23" 46d43'53"
Michele +1 Vir 11h58'23" 4d3'15"
Michele A. Cassaro And 23h27'1" 48d0'26"
Michele Adalia Anderson Peg 21h37'1" 24d58'20"
Michele a.k.a. Baby Girl And 23h25'22" 47d27'12"
Michele Albanese Men 5h25'57" -70d48'36"
Michele Alessandro Montalbano Lyn 7h32'28" 56d37'5"
Michele Alicia Arce Cnc 8h59'30" 11d41'4"
Michele Alyse And 0h29'42" 24d37'11"
Michele and Gaspare Cyg 19h36'44" 50d52'47"
Michele and James Sweet Cyg 21h58'8" 54d39'4"
Michele and Jeffrey Kirsh Cnc 9h7'26" 30d38'1"
Michele and Lance Sco 17h10'18" -39d30'25"
Michele and Michael Lazure Cyg 19h58'21" 37d56'27"
Michele Ann Barton AKA Mikki Leo 9h22'58" 27d26'4"
Michele Ann Flores Dra 19h48'3" 62d29'16"
Michele Ann Omholt Psc 1h24'51" 18d53'33"
Michele Ann Rizzatti Cas 23h7'56" 59d37'55"
Michele Annette Psc 1h39'42" 9d40'15"
Michele AnnMarie Reinboldt And 0h44'32" 39d50'46"
Michele Anthony Ari 2h51'41" 27d31'34"
Michele Arla Siravo Sco 16h34'6" -38d32'7"
Michele Ashley Gates Cam 4h58'14" 58d31'52"
Michele "Babydoll" Freeman Gem 6h20'12" 23d40'48"
Michele Barton Cas 0h4'44" 56d58'31"
Michele Baugher Cas 0h16'21" 53d14'4"
Michele Bellesi Cas 0h27'31" 62d26'26"
Michele Bloom A rising star Cap 20h21'22" -16d0'9"
Michele Boyce Cnc 8h49'16" 12d51'34"
Michele Britton Lyn 7h24'4" 52d42'43"
Michèle & Bruno Umi 16h2'31" 78d12'53"
Michele & Bryan Ori 5h54'9" 11d43'18"
Michele Buchanan Ari 2h28'9" 14d17'33"
Michele Burger Leo 11h14'53" -5d17'44"
Michele Burton Matuszewski And 0h29'24" 43d48'28"
Michele Carol Marcus Psc 0h7'16" 4d36'26"
Michele Castaline Cyg 20h55'28" 34d55'37"
Michele Castelli Lyn 7h49'36" 46d0'31"
Michele Celona Ari 2h2'6" 24d21'33"
Michele Conger Ori 5h52'8" 18d28'39"
Michele Conrique Aql 20h12'25" 14d23'44"
Michele Craft Sgr 18h14'22" -32d43'44"
Michele Crowl Ori 5h27'44" 4d58'30"
Michele Dawn Griffith - little beama Sco 16h13'7" -29d44'57"
Michele Della Porta Umi 15h51'8" 80d45'39"
Michele DeNicola Cnc 8h24'0" 10d53'37"
Michele Denise Cnc 9h17'11" 30d27'35"
Michele DeRitis And 0h27'55" 41d7'32"
Michele Diane Lyn 7h28'28" 48d54'49"
Michele Diane Pontier-Kruckenberg Ori 5h8'2" -1d27'31"
Michele DiCarlo Sco 17h16'57" -32d38'15"
Michele Dorothy Fraser - Happy 30th And 1h58'31" 41d53'26"
Michele E. Lowery Cnc 7h59'14" 17d46'11"
Michele Egan Sgr 17h45'11" -20d12'3"
Michèle et François Couillaud Uma 12h15'59" 57d13'8"
Michele Fabiana Kader Vir 13h19'40" 3d52'17"
Michéle Faoro Cas 0h55'41" 55d36'41"
Michele Fitz And 0h28'49" 43d25'28"
Michele Foss Bartholomew And 2h37'3" 50d15'43"
Michele Frances Schreiber Shuler Vir 13h12'43" 5d31'0"
Michele Franzan Ori 5h42'51" 1d49'22"
Michele from Norwich Ari 3h16'3" 19d18'9"
Michele Garcés Com 12h21'55" 25d8'1"
Michele & Gary Peterson Uma 8h47'16" 53d4'11"
Michele Genung Tau 5h22'49" 27d13'19"
Michele Giorgio Filippo Stillavato Cyg 20h11'39" 35d32'23"
Michele Gordon Cyg 19h56'1" 37d52'36"
Michele Gort Leo 9h47'29" 27d42'36"
Michele Goulet Ori 5h54'45" 20d42'54"
Michele Greene & Andrew Hollander Lyr 18h50'23" 44d46'57"
Michele Hailey Tau 5h56'0" 24d37'31"
Michele Hoang And 1h35'24" 48d10'36"
Michele Huffman Ari 2h33'56" 27d42'9"
Michele J. Borugian Tau 5h36'16" 27d13'18"
Michele J. Leonard Leo 9h39'8" 32d34'44"
Michele Jean Anderson Cnc 8h27'23" 30d12'18"
Michele Jeanne Ohannesian Tau 4h26'41" 18d1'53"
Michele Joan Galeski Cap 20h7'58" -22d21'37"
Michele Josephine Tau 4h6'22" 5d37'58"
Michele Jung Our Shining Star Cas 1h20'19" 61d11'21"
Michele & Kathleen Vir 13h15'42" 5d40'54"
Michele Kay Smith Lyn 8h14'33" 55d8'19"
Michele Kinsolving And 23h51'48" 45d4'8"
Michele Koutsoftas Cas 23h5'6" 55d55'29"
Michele L. Beliveau Per 3h25'52" 47d16'25"
Michele L. Giusto Tau 5h9'6" 24d7'41"
Michele Lail Sgr 18h53'8" -16d39'48"
Michele Lauren Hoyt Ori 5h40'11" 8d13'54"
Michele Laveaux Ari 2h44'37" 17d19'57"
Michele Lawson Blackwell And 0h16'39" 45d57'49"
Michele Leah's Neverending Dreams Gem 7h18'2" 13d43'48"
Michele Lee Aqr 22h5'57" -0d39'26"
Michele Lee Catalona Cam 4h19'27" 66d8'27"
Michele Lee Lanam Mon 6h50'52" -0d4'24"
Michele Lee's Star Vir 12h31'3" -0d43'33"
Michele Leggett And 23h19'4" 42d54'32"
Michele Leigh Covell And 23h1'52" 36d37'23"
Michele L.O.M.L. 3 Aqr 22h33'43" -2d23'9"
Michele Lorraine Sco 16h45'31" -32d31'8"
Michele Lott Cmi 7h9'55" 11d9'25"
Michele Loughney Cas 1h37'43" 66d56'9"
Michele Lumiere du Soleil Psc 0h52'52" 6d40'49"
Michele Lynn Cas 1h12'13" 54d55'26"
Michele Lynn Raia And 0h12'34" 28d50'31"
Michele M. Balme Tau 4h46'53" 18d8'20"
Michele Ma Belle 123 Mon 6h21'14" -3d49'47"
Michele Madeleinat Tau 4h40'26" 23d23'14"
Michele Malinoski And 23h52'46" 41d31'29"
Michele Marie Vir 12h25'1" 3d11'29"
Michele Marie Chatsey Ari 2h32'32" 27d41'54"
Michele Marie FOSTER Vir 13h56'49" -15d29'51"
Michele Marie Holt Lmi 10h31'2" 36d24'23"
Michele Marie Trowbridge Leo 9h40'13" 14d52'5"
Michele Marion Lib 14h52'40" -2d16'7"
Michele Maynard & Gavin Bankert Ori 6h0'33" 14d18'2"
Michele Mc Sherry Tau 5h39'14" 27d8'34"
Michele McBride Psc 1h12'4" 12d28'39"
Michele McCann, Daddy's Little Girl Gem 6h2'59" 25d14'6"
Michele McNicol Sco 16h11'31" -38d22'34"
Michele - Miriam Uma 8h10'44" 60d52'52"
Michele "Missy" Annette Sawyer Psc 1h8'28" 29d47'22"
Michele Moir Uma 11h44'42" 44d31'8"
Michele Molner Lib 14h51'27" -3d34'41"
Michele/Mom Psc 1h6'32" 31d8'54"
Michele *Mommy Is My Angel* Vir 12h36'11" 3d49'0"
Michele Monica Supernova Lib 14h33'29" -10d3'14"
Michele Moore Cas 1h12'46" 58d4'53"
Michele Munding And 0h6'36" 39d33'51"
Michele My Belle Uma 13h25'8" 57d18'46"
Michele Nastalczyk And 1h24'14" 48d18'21"
Michele Pisani Her 16h38'26" 24d17'2"
Michele Pringle Lyn 7h44'54" 52d35'20"
Michele R. Wilcox Vir 13h25'35" -16d34'41"
Michele Reina Rodriquez Leo 10h13'6" 14d27'47"
Michele Rene Armijo Cyg 19h36'5" 31d56'24"
Michele Rene Straub Dra 18h37'36" 68d3'19"
Michele Rifkin And 1h21'24" 37d54'12"
Michele Rowein And 1h3'52" 41d48'8"
Michele Rubinetti Cas 0h31'13" 62d20'24"
Michele S. Kopa Cas 23h55'3" 54d19'40"
Michele S. Rice And 0h40'52" 34d49'2"
Michèle Sani Tau 5h33'49" 18d6'51"
Michele SantaMaria Ari 2h13'36" 11d7'35"
Michele Santarella Cam 3h42'41" 68d17'33"
Michele Sarah Roddie Psc 1h21'4" 27d10'25"
Michèle "SCHELMI" Steinfels Cep 3h58'2" 81d28'38"
Michèle Schneider-Schöchlin Uma 10h23'5" 44d47'8"
Michele Set Me Furlong And 0h18'12" 28d29'3"
Michele "Shamel" Martinez Mon 6h54'17" -0d59'46"
Michele Shaw Uma 11h10'47" 55d48'32"
Michele Sheehan Vir 13h17'21" 12d48'36"
Michele Shelbrack Per 4h49'58" 44d35'17"
Michele "Shelly" Maxian Crb 16h24'9" 27d42'33"
Michele Sivori Lyr 18h26'55" 44d16'54"
Michele Smith Col 6h18'5" -34d18'36"
Michele Snacki Uma 10h7'17" 47d37'26"
Michele St. John Vir 14h45'36" -1d49'25"
Michele Stabile Cas 23h3'10" 53d53'48"
Michele Stafford Sco 16h47'5" -32d36'36"
Michele Stefanello Uma 11h33'30" 59d24'42"
Michele Sweet Mon 6h50'54" -0d36'0"
Michele T Cotta Cas 22h59'36" 58d15'51"
Michele Tai Cas 1h33'43" 74d24'7"
Michele Talamo Lyn 6h30'15" 56d45'15"
Michele Taylor "J. T.' S Girls" And 23h27'15" 41d55'49"

Michele "The Nurse" Totaro Wall
Oph 17h49'50" 12d22'0"
Michele Tierney
Del 20h44'5" 19d38'2"
Michele & Tony Jones
Uma 8h49'49" 60d3'26"
Michele Valle
Aqr 22h10'52" -0d33'20"
Michele Verduci
Uma 8h22'18" 65d41'19"
Michele Vitucci
Cyg 20h37'2" 53d32'49"
Michele W
Aql 19h29'32" -2d44'40"
Michele Wade
Lyr 18h48'30" 37d47'14"
Michele Walburger
Cas 1h36'0" 63d18'27"
Michele with 1 "L"
Gem 7h12'9" 28d11'18"
Michele Zappia
Cep 21h35'17" 64d11'19"
Michele Zarrella
Cyg 20h44'54" 45d19'52"
Michele, 16.02.1971
Dra 18h3'34" 71d47'18"
Michele, A Shining Star
Aql 19h6'54" 1d46'58"
Michele, My Special Belle
Psc 1h12'41" 17d38'19"
MicheleB
Leo 10h13'28" 16d10'5"
MICHELEDAY
Uma 11h14'26" 59d0'9"
Michèle-Jeanne
Cas 1h13'15" 65d58'17"
MicheleLisa2005
Lyn 7h23'10" 54d54'38"
michelelori
Uma 11h35'25" 51d58'57"
Michele-Marie Vella
Leo 11h51'24" 24d19'8"
Michelene Staton
Psc 0h1'10" 7d14'3"
Michele's 40th Birthday Star
Cas 23h33'33" 51d1'4"
Michele's Astellas Star
Sco 16h6'3" -11d49'43"
Michele's Beautiful Smile
Ori 5h53'43" 12d55'59"
Michele's Fairy Tale
Cas 1h36'33" 63d49'45"
Michele's Light
Lib 15h19'50" -7d2'19"
Michele's Light
And 2h22'26" 43d4'9"
Michele's Stairway
Ari 2h4'48" 22d44'29"
Michele's Star
And 0h57'34" 40d42'44"
Michele's "Star of Dreams"
And 1h12'43" 43d3'45"
Michele's Star Science Class 05-06
Aql 19h6'19" -7d31'49"
Michele's Wishing Star
Uma 11h29'44" 33d58'21"
MicheleSara
Umi 15h15'28" 77d53'16"
Micheley
Ari 3h22'13" 28d20'37"
Michelin Star
Cru 12h55'23" -58d39'5"
Michelina
Lyn 7h36'57" 57d17'50"
Michelina
Leo 10h13'50" 11d20'12"
Michelina Arcuri-Queri
Lmi 9h48'43" 33d11'16"
Micheline
Cyg 21h9'55" 51d43'0"
Michelino
Boo 14h50'58" 28d32'21"
Michell Marie Rigsby
Leo 11h54'57" 15d30'59"
Michella Bella
Leo 10h8'22" 27d0'4"
Michella Bella
Tau 4h18'25" 14d38'54"
Michelle
Ori 6h6'21" 11d2'53"
Michelle
Psc 23h29'42" 2d58'27"
Michelle
Aql 19h12'35" 1d57'43"
Michelle
Vir 13h36'2" 1d12'11"
Michelle
Vir 12h46'33" 5d2'49"
Michelle
Vir 12h56'31" 0d32'28"
Michelle
Tau 4h38'28" 4d44'59"
Michelle
Leo 10h38'11" 12d34'42"
Michelle
Psc 1h14'40" 20d1'50"
Michelle
Ari 2h59'33" 18d18'13"
Michelle
Leo 10h21'41" 27d25'42"
Michelle
Lmi 10h9'42" 28d59'5"
Michelle
Tau 3h42'56" 25d39'45"
Michelle
Cnc 8h51'16" 28d21'35"
Michelle
Cnc 8h39'56" 27d31'29"
Michelle
Cnc 8h8'53" 21d33'57"
Michelle
Gem 6h46'22" 20d57'49"
Michelle
Gem 6h27'8" 21d30'37"
Michelle
And 0h25'35" 29d4'49"
Michelle
And 0h11'30" 26d25'0"
Michelle
Cam 4h6'49" 57d56'39"
Michelle
Per 2h15'15" 54d26'3"
Michelle
Lyr 18h19'51" 38d2'24"
Michelle
Cyg 21h48'34" 42d53'41"
Michelle
Cas 0h48'32" 50d21'50"
Michelle
And 1h22'43" 50d35'39"
Michelle
And 2h37'11" 49d11'7"
Michelle
And 1h43'45" 49d46'38"
Michelle
And 2h18'6" 46d5'0"
Michelle
Lyn 7h43'22" 48d24'8"
Michelle
Crb 15h37'27" 36d57'0"
Michelle
Lyr 18h45'42" 36d52'16"
Michelle
Leo 9h32'18" 30d46'43"
Michelle
Gem 7h10'49" 33d42'52"
Michelle
And 0h43'12" 43d18'31"
Michelle
And 0h33'26" 42d33'0"
Michelle
Cyg 20h35'3" 31d24'24"
Michelle
Per 3h53'50" 37d37'25"
Michelle
Uma 11h59'53" 43d59'3"
Michelle
Uma 11h31'43" 43d44'18"
Michelle
Uma 11h57'5" 38d7'56"
Michelle
Lyn 7h57'21" 57d10'47"
MICHELLE
Lyn 6h41'10" 54d2'35"
Michelle
Uma 10h42'24" 55d3'49"
Michelle
Cas 1h25'48" 64d46'56"
Michelle
Cas 0h31'3" 63d7'56"
michelle
Cas 23h59'19" 53d7'10"
Michelle
Cas 23h45'10" 56d51'49"
Michelle
Cyg 21h35'36" 52d39'21"
Michelle
Uma 9h9'30" 65d22'18"
Michelle
Umi 15h11'39" 68d20'25"
Michelle
Mon 6h54'35" -0d47'22"
Michelle
Ori 5h59'11" -0d35'3"
Michelle
Vir 13h43'30" -12d53'16"
Michelle
Sgr 19h12'6" -13d27'31"
Michelle
Sco 16h16'30" -13d42'20"
Michelle
Lib 15h23'8" -10d50'38"
Michelle
Aqr 20h41'20" -10d10'16"
Michelle
Aqr 23h5'13" -9d31'2"
Michelle
Vir 13h50'51" -17d32'46"
Michelle
Vir 13h14'57" -16d1'10"
Michelle
Lib 15h49'28" -18d30'47"
Michelle
Aqr 22h0'0" -15d47'40"
Michelle
Psc 0h28'7" 3d33'38"
Michelle
Sco 17h19'31" -40d37'28"
Michelle
Sgr 20h11'55" -38d46'18"
Michelle
Aqr 22h16'54" -22d33'9"
Michelle
Sco 16h38'56" -27d2'31"
Michelle
Sco 16h15'5" -22d42'21"
Michelle 01/01/78
Cap 20h57'30" -16d47'2"
Michelle 16
Cas 23h19'43" 55d18'55"
Michelle 21
And 23h14'25" 39d46'28"
Michelle 21405
Cap 20h27'37" -10d45'1"
Michelle A. Fill 60th Birthday Star
Gem 6h52'38" 34d53'10"
Michelle A. McVeigh
Cyg 20h8'3" 41d48'58"
Michelle A. Norton
Leo 10h6'0" 15d43'49"
Michelle A Phillips
Ori 5h35'37" 2d57'59"
Michelle A. Russo
And 2h37'1" 43d23'40"
Michelle A. White
Cas 1h27'13" 51d55'38"
Michelle & Abe
Uma 13h29'25" 58d6'44"
Michelle Adrianne Sadler
Psc 23h58'13" 8d49'57"
Michelle Aileen
Cyg 21h50'23" 49d59'43"
Michelle Albrecht
Lyn 8h54'56" 35d41'48"
Michelle Alexandra Dittrich
Ari 2h55'49" 22d7'0"
Michelle Allendorf
Uma 11h27'5" 59d25'58"
Michelle and Adam's Lucky Star
Cma 6h35'48" -16d56'15"
Michelle and Andy
Cyg 21h17'14" 47d32'26"
Michelle and Ben's Eternity
Lyn 7h34'2" 35d41'56"
Michelle and Bradley
Leo 11h19'55" 17d29'14"
Michelle and Don Forever
Cyg 20h37'8" 52d26'18"
Michelle and Jackie
Cet 0h15'56" -20d24'58"
Michelle and Jane
Vir 12h30'54" -0d9'25"
Michelle and Joe's Star
Tau 5h7'19" 17d57'7"
Michelle and Matthew Heger
Aqr 22h43'11" 1d42'27"
Michelle and Michael Forever
Lib 14h49'50" -5d5'27"
Michelle and Mike
Lib 14h53'24" -10d42'44"
Michelle and Mikes Star
Leo 11h16'16" 22d57'13"
Michelle and Nate's Cosmo
And 23h22'57" 47d35'37"
Michelle and Todd Calomino
Sgr 20h3'3" -37d30'16"
Michelle and Willie Brett
Cyg 20h35'11" 38d19'9"
Michelle Anderton
Sex 9h50'16" -0d39'53"
Michelle Ang Shi Hui
Ari 2h48'6" 29d17'21"
Michelle Angel
Ori 6h6'6" 21d8'11"
Michelle Angela Archibald
Cas 23h14'23" 55d18'35"
Michelle Ann
Dra 16h13'50" 62d7'19"
Michelle Ann
And 1h0'42" 45d51'8"
Michelle Ann Amstutz
And 1h20'46" 44d8'0"
Michelle Ann Carter
And 23h24'0" 47d35'33"
Michelle Ann Enriquez
Gem 6h20'52" 18d22'42"
Michelle Ann Hickam
Cas 1h19'35" 55d57'1"
Michelle Ann Petersen
Cas 0h7'2" 56d15'34"
Michelle Ann Schartung
Tau 4h29'45" 18d33'5"
Michelle Ann Serviss
Aqr 22h41'37" 0d59'6"
Michelle Ann Yordy Miller
Sco 17h57'20" -31d8'47"
Michelle Anna Hall Kong
Uma 11h28'31" 59d53'43"
Michelle Anne Allard
Sgr 18h37'28" -28d9'52"
Michelle Anne Douglas
Cnc 8h17'21" 6d59'21"
Michelle Anne George
Uma 10h53'18" 44d52'50"
Michelle Anne Gessner
Tau 5h48'35" 14d45'57"
Michelle Anne Sullano Ortiz
And 0h39'53" 24d50'48"
Michelle Apadula
Tau 4h19'51" 2d38'33"
Michelle & Ashley
Leo 10h17'12" 21d19'20"
Michelle Ashley
Dra 18h56'53" 78d3'46"
Michelle Ashley Dickey
Cas 0h38'26" 55d33'18"
Michelle Balthrop
Lyn 8h3'22" 55d6'7"
Michelle Barbour Jacobs
Lyn 8h0'12" 40d53'44"
Michelle Barouk
Dra 20h18'54" 67d43'2"
Michelle "Bella" Dubos
Cnc 7h57'16" 16d16'10"
Michelle Belle
Ari 2h47'5" 27d48'56"
Michelle Belle
And 1h35'27" 39d29'6"
Michelle_Berger2005
Cru 12h38'35" -58d19'27"
Michelle Blue Eyes
Aqr 21h45'52" 1d23'42"
Michelle (Bonnie)
And 1h8'30" 39d41'44"
Michelle Braden
And 0h53'49" 41d3'32"
Michelle Brickner
Ari 2h9'13" 24d42'35"
Michelle Bridier-Duhamel
Cnc 8h9'31" 16d32'37"
Michelle Brooke
Lmi 10h39'48" 32d18'44"
Michelle Brown
Cnc 8h35'41" 13d17'25"
Michelle Bunny Maline
Sco 16h57'26" -38d25'3"
Michelle Burchette
And 2h33'50" 44d59'8"
Michelle Byrum
Cyg 19h47'38" 46d30'8"
Michelle C. Dizon
Cnc 8h8'44" 28d41'49"
Michelle C. Russo
Cnc 8h0'37" 12d4'59"
Michelle Cabrera
Ori 5h21'23" -6d22'3"
Michelle Cammarano
And 1h5'7" 42d11'8"
Michelle Carlton
Cra 18h33'9" -37d27'17"
Michelle Carpenter
Sgr 19h45'0" -12d36'20"
Michelle Carroll
Gem 7h1'32" 25d14'35"
Michelle Cervantes Duarte
Sco 17h45'37" -31d40'4"
Michelle Cheri' Meyer
Aqr 22h57'30" -7d35'2"
Michelle Christ
Uma 11h5'10" 60d27'42"
Michelle & Christian
Cas 0h27'27" 56d21'27"
Michelle Christina
Sco 16h12'53" -12d32'36"
Michelle Christine Padron
Ari 2h46'30" 16d57'5"
Michelle & Christopher Andrews
Cyg 19h45'34" 30d45'48"
Michelle Clarkson
Cru 12h43'53" -58d18'41"
Michelle Clemens
Her 17h36'52" 32d25'9"
Michelle Coffman
Cas 0h5'51" 54d46'43"
Michelle Colin
Tau 4h6'27" 5d55'42"
Michelle Cook
Aqr 20h38'56" -3d16'46"
Michelle & Correy's Star
Lyr 18h47'0" 36d45'6"
Michelle Cox
And 1h31'22" 35d26'50"
Michelle Cox
Ori 6h8'20" 16d29'35"
Michelle Crespin
And 23h58'24" 45d24'45"
Michelle Cristina Nahas
Uma 8h36'52" 46d55'42"
Michelle Critchfield
Aqr 22h39'53" 1d41'42"
Michelle "Cubby" Renee Moore
Lib 15h46'30" -18d28'22"
Michelle CUW 2007 - Love you Misha
Leo 10h18'45" 25d21'28"
Michelle D. DiCicco
Ori 5h23'16" 14d12'34"
Michelle D. Edgar
Psc 1h42'30" 20d18'41"
Michelle D Lassley
Lyn 6h50'13" 51d51'19"
Michelle Daconto Martone
Psc 0h9'24" 4d21'19"
Michelle DaDante
Aqr 23h2'45" -11d6'40"
Michelle Daschel
Leo 9h36'12" 27d52'24"
Michelle Davis
Cyg 21h25'11" 34d47'17"
Michelle Dawn
And 0h29'29" 34d1'4"
Michelle Dawn Macey
Dra 16h13'19" 67d17'58"
Michelle Dawn Williams
Cyg 21h29'4" 31d21'22"
Michelle Deanne Cavanaugh
Crb 16h20'12" 29d0'43"
Michelle Debra Evans
Lyn 7h33'59" 36d32'59"
Michelle Debrocq
Lyn 8h19'10" 50d40'35"
Michelle - Dein Stern leuchtet dir den Weg!
Ori 6h22'34" 10d20'17"
Michelle DeJong
Sgr 17h48'19" -26d5'55"
Michelle Delena
Aqr 20h39'9" -7d21'7"
Michelle Dellaterza
Cap 21h55'20" -13d53'29"
Michelle DellaVecchia
Leo 9h45'11" 25d38'22"
Michelle Denise
Uma 10h33'2" 43d18'27"
Michelle Denise Lucas
Ari 2h18'43" 25d22'16"
Michelle dePass-Lund
Ari 2h8'8" 21d51'34"
Michelle Dianne Traxler
Leo 11h49'28" 20d18'19"
Michelle Dickson
And 0h42'43" 30d24'0"
Michelle DiPaolo
Sco 17h28'36" -42d14'46"
Michelle DLS
Vir 12h40'31" 4d28'30"
Michelle DLT
Sgr 19h27'3" -13d49'27"
Michelle Dolleas Henry Bey
Vir 13h45'34" 0d28'2"
Michelle Domnique Mattson
Lib 15h40'1" -10d23'59"
Michelle Don Carlos
Dra 18h31'57" 54d57'46"
Michelle Doris Malusky
Cam 7h47'43" 78d21'34"
Michelle & Dozer
Dra 19h1'0" 68d40'11"
Michelle Dreamspeaker
Her 17h17'1" 34d59'53"
Michelle Dubord Lagacé de Laroche
Aql 20h19'51" 0d13'37"
Michelle Dujka
Tau 3h27'25" 7d19'41"
Michelle Durdens's Star
Mon 6h30'57" 10d54'24"
Michelle Eannucci
Lyn 9h21'49" 39d9'7"
Mi'Chelle Ebony
Uma 12h25'39" 55d5'41"
Michelle Eichstedt Buffin
Sco 17h19'18" -40d31'33"
Michelle Elaine Goodman
Vul 20h27'16" 27d43'58"
Michelle Elise
Cnc 9h3'43" 31d59'37"
Michelle Elise Reynolds
And 23h3'35" 52d27'43"
Michelle Elizabeth
Lyn 8h50'2" 37d32'37"
Michelle Elizabeth Camacho
Crb 15h54'54" 28d39'55"
Michelle Elizabeth Chaban
Cap 21h40'4" -18d19'25"
Michelle Elizabeth Konzem
Aqr 21h9'12" -14d23'3"
Michelle Elizabeth Landry
And 23h6'19" 42d39'55"
Michelle Elizabeth Pitchford
Boo 14h40'21" 29d39'41"
Michelle Elizabeth Ricci
Gem 7h27'28" 32d59'55"
Michelle Elizabeth Rosenbaum
And 0h26'16" 35d50'8"
Michelle Elizabeth Tafoya
Her 17h15'38" 46d21'35"
Michelle Elizabeth Winkler
Cyg 20h24'3" 44d39'22"
Michelle Ellen Schueller
Psc 1h7'45" 23d6'51"
Michelle Ellis
Cas 23h24'37" 59d48'45"
Michelle Emme Buan
Psc 0h49'2" 6d31'15"
Michelle Englebert
And 0h42'27" 31d20'54"
Michelle Epperly
And 1h19'8" 47d51'2"
Michelle Evans
And 0h35'4" 42d45'23"
Michelle Favreau Dansereau
Cas 0h58'20" 50d22'10"
Michelle F.E.B
Cnc 9h6'7" 10d58'12"
Michelle Federico
Sgr 19h18'10" -21d52'40"
Michelle Field
Psc 1h11'12" 11d46'50"
Michelle Florinda
Vir 12h47'42" 2d15'2"
Michelle Foord
And 0h10'24" 44d32'0"
Michelle for ever
Lyr 19h8'22" 45d8'44"
Michelle Ford
Sgr 18h51'16" -18d24'57"
Michelle Foreman-BridgeStar Q1-2005
And 1h44'37" 42d0'24"
Michelle Forster
Psc 1h10'4" 11d54'49"
Michelle Franzman
Crb 16h7'44" 34d49'14"
Michelle Frechette
Cas 0h8'37" 53d26'26"
Michélle Fuggle
And 23h15'38" 47d52'3"
Michelle Furnare
Dra 14h39'10" 59d42'18"
Michelle Galactic Rose
Crb 15h43'44" 29d6'52"
Michelle & Gary Forever
Aqr 22h40'55" -2d9'13"
Michelle Gaudreau
Umi 15h37'38" 70d23'45"
Michelle Gay
And 23h0'19" 45d33'8"
Michelle Gehr
Cyg 19h36'42" 27d51'51"
Michelle Gill
Cnc 8h44'59" 11d43'9"
Michelle Gillespie
Leo 11h50'53" 20d26'51"
Michelle Gilyeat
Aql 20h12'53" 7d50'39"
Michelle Goble
And 1h16'8" 49d21'3"
Michelle Gonsalves
Sco 16h53'45" -43d43'26"
Michelle Grace Combs
Uma 9h53'34" 50d49'20"
Michelle & Graham Thurston
Cyg 20h40'55" 54d11'10"
Michelle Gray
Cnc 8h28'34" 32d48'40"
Michelle Greg Liam
Ori 6h18'51" 10d27'44"
Michelle Grossman
Cas 23h3'31" 53d48'5"
Michelle Gupta (Babe)
Ari 2h22'23" 23d59'53"
Michelle Guzik
Lyn 7h38'54" 47d43'54"
Michelle H. Coleman
Aqr 23h20'59" -18d40'35"
Michelle Haché is La Calisto
Uma 14h0'41" 49d19'47"
Michelle Halsey
Cnc 8h9'22" 18d9'5"
Michelle Hamilton
Tau 5h11'41" 24d3'11"
Michelle Hamming
Psc 1h13'26" 26d22'58"
Michelle Hart
Lib 15h9'13" -1d31'47"
Michelle Hathaway
Sgr 18h57'8" -32d13'39"
Michelle Hemming
And 1h30'37" 43d42'32"
Michelle Herbert
Vir 13h14'0" 3d34'36"
Michelle Hicks
And 0h4'18" 39d24'40"
Michelle Hodgdon
Sco 17h54'28" -36d37'31"
Michelle Hogeveen
Ari 2h11'1" 24d56'39"
Michelle Hone
Gem 7h3'56" 17d15'4"
Michelle Hope
Her 17h40'2" 21d28'11"
Michelle Horton
Sgr 17h52'21" -29d53'53"
Michelle Hough & Michael Daly
Cyg 19h42'21" 37d36'34"
Michelle "Hougie" Isherwood
Tau 4h4'23" 5d27'27"
Michelle House
Psc 0h58'39" 7d2'44"
Michelle Howe
Cap 21h1'21" -20d36'41"
Michelle Hullett
Dra 16h8'26" 51d55'45"
Michelle Hyde
Cas 0h18'42" 54d50'52"
Michelle I. Gallavan-Orris
Psc 1h10'46" 9d58'41"
Michelle Ianacone
Sco 17h4'8" -43d32'27"
Michelle Irene Nelson
Aqr 22h33'1" -1d34'9"
Michelle Irwin
Ori 6h13'55" 15d5'53"
Michelle Isabel Tamburini
Psc 0h29'49" 15d14'36"
Michelle J Cutler & Vlad Cutler II
And 1h49'11" 43d55'30"
Michelle J Lacarte Born: July 11/1964
Uma 11h32'48" 56d52'58"
Michelle J. Stewart
Cas 23h54'23" 53d52'34"
Michelle Jackson
Leo 10h14'1" 17d31'34"
Michelle Jacquelyn Monroe
Sgr 18h52'13" -31d39'33"
Michelle Jaent
Lyn 8h27'45" 36d12'52"
Michelle Jane - Forever Our Angel
Leo 11h35'51" -1d53'5"
Michelle Janisheck
And 23h19'16" 51d56'3"
Michelle & Jayden Minor
Umi 14h57'24" 76d27'14"
Michelle Jean Westgard 12-11-1964
Sgr 19h20'43" -34d32'44"
Michelle Jennings
Cas 1h17'53" 54d2'54"
Michelle & Jeremy Perera
Tau 5h46'12" 19d4'0"
Michelle JoAnn Fitch
Lyr 19h12'8" 27d22'23"
Michelle Jodi Narson
Cas 23h32'8" 53d52'43"
Michelle Johns
Mon 6h30'47" 8d54'43"
Michelle Jordan Levine
Crb 15h44'42" 31d36'42"
Michelle & Josh 1 year anniversary
Cyg 20h50'28" 31d9'7"
Michelle Joy Freeman
Psc 0h43'59" 16d17'13"
Michelle Joy Morrison
Cru 12h33'20" -58d26'37"
MICHELLE JUDD
Uma 9h44'27" 63d22'25"
Michelle Julia Bonds
And 0h34'32" 33d58'30"
Michelle Julie Luscombe
Cra 18h45'38" -40d12'23"
Michelle Juliet Elezovic
Cap 20h25'17" -16d16'21"
Michelle K
Psc 0h49'59" 15d34'55"
Michelle K. Goetz
Uma 12h4'15" 61d53'19"
Michelle K. Labriak
Lyn 8h42'29" 40d48'21"
Michelle K. Ramsey
And 2h0'9" 43d2'43"
Michelle Karena Richmond
Cnc 8h30'50" 24d13'34"
Michelle Kathleen Prezioso
Lyn 7h32'52" 49d32'47"
Michelle Kay Harmston
And 1h48'11" 38d11'58"
Michelle Kay Horstman
Cas 1h40'31" 64d49'34"
Michelle Kelly
Crb 16h4'6" 27d42'45"
Michelle & Kenny
Tau 4h38'18" 24d44'24"
Michelle Kerr
Lyn 7h35'4" 41d14'37"
Michelle Kerri Trantina
Umi 16h56'32" 76d59'49"
Michelle King
Cas 23h4'51" 56d38'17"
Michelle Kirkby
Crb 15h29'41" 28d9'2"
Michelle Koffel
Cyg 20h25'21" 31d8'57"
Michelle Koller
Ari 3h0'4" 11d21'29"
Michelle Koutnik
Sgr 19h5'42" -23d11'45"
Michelle Koval
And 0h21'1" 46d21'2"
Michelle Kowalske
Cyg 19h45'11" 34d13'1"
Michelle Krcmarik
Mon 8h9'47" -0d36'54"
Michelle Kretin
Sgr 19h14'12" -14d55'28"
Michelle L. Bridenstine
Ari 3h19'17" 27d42'13"
Michelle L. Kerner *40th*
Gem 7h11'37" 27d11'26"
Michelle L. LeGros
And 0h25'32" 33d37'30"
Michelle L. Papnia
Psc 1h18'36" 18d28'38"
Michelle Lacy "Daisystar"
And 0h21'59" 43d35'57"
Michelle Lane Fahl
Lyn 8h36'24" 34d20'27"
Michelle Lane - Forever In My Heart
Gem 7h42'18" 28d47'47"
Michelle Lauren
Leo 9h38'23" 20d15'52"
Michelle Lea Mann
Umi 14h24'23" 83d56'7"
Michelle Leah Gardiner
Dra 16h40'34" 52d55'46"
Michelle Leanne
Lib 14h47'35" -4d13'56"
Michelle Leblanc
Lmi 10h51'58" 35d6'2"
Michelle Lee
Cyg 20h22'43" 39d41'14"
Michelle Lee
And 23h24'29" 47d57'2"
Michelle Lee
Umi 10h18'54" 89d17'47"
Michelle Lee February 10, 1969
Aqr 22h29'43" -22d50'11"

Michelle Lee Gaskill
Lib 15h13'31" -9d4'21"
Michelle Lee Jones
Cyg 19h36'35" 47d9'47"
Michelle Lee Lawings
And 0h15'50" 42d54'8"
Michelle Lee Leininger
And 2h34'22" 37d26'20"
Michelle Lee Needham
Aqr 21h52'1" -0d55'58"
Michelle Lee Shylanski
Aql 19h40'38" -0d5'2"
Michelle Lee Sneddon
Cyg 20h27'27" 40d22'33"
Michelle Leeann
Vir 13h17'43" 7d40'28"
Michelle LeeAnn
Cas 1h35'58" 66d1'14"
Michelle Leidy
Cnc 8h35'15" 30d55'37"
Michelle Leigh
Cnc 8h50'11" 31d14'48"
Michelle Leigh Roy
Ori 5h58'27" -0d36'13"
Michelle Leigh Strickland
Psc 0h59'27" 2d53'56"
Michelle Leigh Vaars/Evans
Lib 15h23'1" -7d31'17"
Michelle Lerner
Vir 12h33'22" 9d23'35"
Michelle Lilly
Cas 0h12'1" 63d0'33"
Michelle Lisa Distance
Sco 16h9'55" -9d20'9"
Michelle Lisa Piazza
Lmi 10h34'6" 32d18'53"
Michelle Little Momma McKee
Vir 13h12'30" 8d31'53"
Michelle Liv Brinch
Tau 3h46'14" 17d40'36"
Michelle Lloyd
Cas 1h18'38" 56d34'2"
Michelle Loizzo
Ori 5h19'57" -8d14'3"
Michelle Lorrie La Spina
Lyr 18h54'54" 33d25'20"
Michelle Louise
Cru 12h55'27" -60d26'55"
Michelle Louise Dolby
Cap 20h20'47" -24d18'14"
Michelle Louise Keck
Uma 14h21'5" 61d46'23"
Michelle Louise Leonard
Lyn 7h54'44" 57d44'42"
Michelle Louise Taylor
And 2h37'1" 50d25'7"
Michelle love from Georgia
Cas 0h22'2" 55d58'4"
Michelle Lucke
Leo 10h30'55" 17d49'56"
Michelle Luna
Ari 3h9'58" 25d40'10"
Michelle Lutz
Lyn 7h54'40" 41d22'51"
Michelle Luvuh Lujan 10152006SNL
Peg 22h42'31" 30d46'46"
Michelle Lyn Rada
Sco 17h57'41" -36d52'55"
Michelle Lyn Saeteurn
Tau 4h12'30" 21d54'6"
Michelle Lynn
Cyg 20h57'47" 33d41'11"
Michelle Lynn
And 0h41'39" 43d48'15"
Michelle Lynn
Psc 1h15'21" 31d31'28"
Michelle Lynn
And 23h18'58" 52d23'51"
Michelle Lynn
Cas 23h30'27" 52d5'50"
Michelle Lynn
And 0h11'45" 45d4'22"
Michelle Lynn
Lib 15h14'1" -29d10'22"
Michelle Lynn
Cas 23h58'0" 56d50'13"
Michelle Lynn
Lib 15h5'29" -1d37'23"
Michelle Lynn
Aqr 22h0'30" -15d56'55"
Michelle Lynn Adams
Aqr 22h57'2" -15d5'22"
Michelle Lynn Baker
Tau 4h15'50" 15d46'44"
Michelle Lynn Bankoske
Cnc 8h40'0" 19d34'40"
Michelle Lynn Carano
And 23h14'53" 48d5'15"
Michelle Lynn Carder
And 0h22'16" 34d58'31"
Michelle Lynn Ciambella
Cnc 8h12'13" 18d27'58"
Michelle Lynn Clark
Mon 6h59'52" -0d53'21"
Michelle Lynn Cox
Cnc 8h49'54" 12d11'9"
Michelle Lynn Culbertson
Cas 23h35'0" 51d56'9"
Michelle Lynn Day
Tau 3h41'13" 29d23'41"
Michelle Lynn Foster
Leo 9h59'7" 13d2'43"
Michelle Lynn Glazier
Cyg 19h34'20" 54d26'19"
Michelle Lynn Gormley
Gem 7h13'35" 23d12'1"
Michelle Lynn Isabel Acree
And 23h41'46" 43d3'19"
Michelle Lynn Knarr Jan. 24, 1983
Crb 15h48'51" 27d25'54"
Michelle Lynn Main
Lyn 8h7'13" 39d56'6"
Michelle Lynn & Matthew Bennett
Cyg 20h13'14" 36d50'8"
Michelle Lynn McDonald
Lib 15h15'0" -13d0'54"
Michelle Lynn Moore Powers
Ari 2h19'59" 21d31'49"
Michelle Lynn Penman
Cas 0h40'38" 65d35'48"
Michelle Lynn Perriman
Ori 6h17'13" 10d8'23"
Michelle Lynn Redden
Vir 12h53'42" -9d59'56"
Michelle Lynn Richardson
Gem 7h16'51" 13d34'5"
Michelle Lynn Schlegel
Gem 7h16'4" 19d42'7"
Michelle Lynn Smith
Cap 21h49'34" -12d25'1"
Michelle Lynn Steele
Ari 2h20'8" 25d42'54"
Michelle Lynn White
And 1h10'19" 38d28'34"
Michelle Lynne Leonard
Ori 6h7'41" 17d57'58"
Michelle M. Atkins
Sgr 18h15'34" -19d58'26"
Michelle M. Garcia
Lyn 7h54'41" 41d24'27"
Michelle M. Hash
Cyg 21h29'54" 33d15'4"
Michelle M. Kernan
Cam 3h42'39" 58d27'35"
Michelle M. Manke
Cyg 20h33'6" 38d24'13"
Michelle M. Mannings
Cnc 8h34'16" 19d46'11"
Michelle M. Monitzer
Sgr 19h52'15" -28d30'37"
Michelle M. Ward
Cas 1h47'37" 64d45'28"
Michelle MacDonald
Aur 6h36'44" 30d17'1"
Michelle Madlener
Cas 23h15'4" 54d33'4"
Michelle Mae
Tri 2h18'20" 31d59'21"
Michelle & Maëva
Dra 19h9'10" 71d23'8"
Michelle Mahal Na Mahal Kita
Aqr 23h6'19" -11d34'11"
Michelle Malek Felker
And 2h35'49" 43d35'53"
Michelle Manesh
Tau 4h11'29" 29d24'29"
Michelle Maria Lopez
Aqr 23h14'21" -23d3'56"
Michelle Mariani
Crb 15h53'58" 26d59'34"
Michelle Marie
Gem 7h24'17" 26d52'2"
Michelle Marie
Cyg 21h33'44" 36d45'39"
Michelle Marie
Cyg 20h6'58" 42d43'15"
Michelle Marie
Ori 5h41'27" -5d15'54"
Michelle Marie
Uma 9h40'44" 63d30'9"
Michelle Marie Borilla
Psc 0h9'18" 4d38'37"
Michelle Marie Braley
Uma 8h50'50" 46d40'21"
Michelle Marie Edwards
Boo 14h37'10" 36d33'40"
Michelle Marie Hale
Cyg 19h35'28" 28d26'44"
Michelle Marie Hopkins
Tau 4h40'25" 8d13'32"
Michelle Marie Lopez
Sgr 18h14'23" -19d42'56"
Michelle Marie Maddox
Uma 9h44'23" 59d0'53"
Michelle Marie Manor
Cyg 20h10'35" 31d52'8"
Michelle Marie Martin
Aql 19h45'29" -0d35'55"
Michelle Marie Medeiros
Uma 11h37'42" 40d31'36"
Michelle Marie Moretti
Uma 9h22'2" 67d50'55"
Michelle Marie Runyan
Sgr 18h32'29" -27d6'6"
Michelle Marie Wynn
Aqr 21h44'24" -7d36'15"
Michelle Marie Young
Uma 11h32'44" 45d15'1"
Michelle Marlene Meester
Cnc 8h39'19" 24d15'46"
Michelle Marrama
Crb 15h36'58" 38d11'37"
Michelle Marrero Girona
And 0h6'15" 45d43'59"
Michelle Marvelle
Cnc 8h14'11" 27d23'36"
Michelle Matsuda
Umi 15h53'40" 70d46'24"
Michelle & Matthew Broadhecker
Cyg 19h56'33" 42d49'1"
Michelle Matthews
Sgr 19h10'4" -20d21'58"
Michelle Maziar
Sgr 18h2'51" -27d39'34"
Michelle McCauley
Cas 1h2'16" 61d18'23"
Michelle McCoy
Mon 6h39'33" 4d11'19"
Michelle McEneaney
Lyn 7h53'15" 36d20'15"
Michelle McGlynn
And 0h9'57" 40d39'55"
Michelle McGoo Starr
Cnc 8h23'2" 27d41'50"
Michelle McGuire
Aqr 21h58'25" 1d20'53"
Michelle McGuirk
Leo 9h58'57" 32d3'59"
Michelle Melissa Malina
Gem 7h32'55" 31d25'8"
Michelle Mercer
Psc 0h51'40" 14d38'24"
Michelle Michalowski (Honeybear)
Tau 4h37'46" 20d22'16"
Michelle Miller
Cas 1h4'6" 48d58'41"
Michelle "Mindi" Louise Rees
Cap 21h15'43" -24d0'30"
Michelle Monika von Gunten
Com 12h15'52" 21d48'36"
Michelle Monteith
Lyn 6h59'30" 45d24'23"
Michelle Montes Mendoza
Lep 5h17'12" -13d53'28"
Michelle Moody
Sco 17h30'59" -39d43'19"
Michelle Moran
Cas 1h46'29" 61d13'22"
Michelle Moran
Uma 10h2'7" 44d30'3"
Michelle Morissette
Uma 13h39'5" 61d35'24"
Michelle Muhart
Lep 5h52'7" -12d23'25"
Michelle Mullins
Crb 16h15'9" 36d41'17"
Michelle "Murph" Clausi-Ferrara
And 23h53'20" 43d15'52"
Michelle "My Angel" Guerra
Cam 7h46'4" 66d4'30"
Michelle My Belle
Cnc 8h35'54" 17d44'18"
Michelle My Guiding Light
Dra 18h49'7" 52d19'21"
Michelle N. Lewis
Mon 7h14'59" -9d49'17"
Michelle Nash
Aqr 22h32'42" -2d1'43"
Michelle Natividad Villanueva
Crt 11h23'59" -17d10'7"
Michelle Neilson
Oph 16h41'18" -0d58'40"
Michelle Nell
Aqr 23h4'31" -21d46'24"
Michelle Nicholas
Cap 20h40'14" -21d20'41"
Michelle & Nick's star
Umi 14h3'34" 77d41'34"
Michelle Nicole
Pho 1h24'52" -43d6'29"
Michelle Nicole Michon (My Savior)
Tau 5h46'7" 19d55'38"
Michelle Nightingale
And 0h23'13" 38d16'57"
Michelle Nimmo - 21
Cru 12h41'22" -58d4'45"
Michelle Noel
Sgr 17h56'0" -25d50'35"
Michelle Norma Hersey
Cnc 9h1'56" 8d8'29"
Michelle Norris - A True Star
Cru 12h38'17" -58d52'23"
Michelle of Instock
Cas 0h38'20" 50d39'59"
Michelle Ogonowski
Lmi 10h9'52" 30d20'20"
Michelle Oles
Lib 14h47'58" -12d30'55"
Michelle Ordway
Sco 16h15'25" -10d47'33"
Michelle Orrock
Cnc 9h17'8" 21d41'2"
Michelle Palmer
And 23h22'38" 48d7'30"
Michelle Pamela
Ari 2h6'36" 17d43'33"
Michelle Park
Sgr 18h47'17" -17d15'3"
Michelle Parker
Cnc 8h34'18" 32d43'56"
Michelle Parks
Cas 1h27'47" 62d52'10"
Michelle Patrice Madrid
And 1h10'42" 41d50'43"
Michelle Paula Hamilton
Tau 5h48'6" 25d37'7"
Michelle Paulinho
Cyg 20h26'27" 47d56'0"
Michelle Pepitone
Cas 0h42'2" 60d6'19"
Michelle Pfeifer
Cap 20h56'21" -18d11'51"
Michelle Phann
Uma 11h49'40" 28d24'14"
Michelle Piechowski
Cnc 9h4'17" 30d42'18"
Michelle Pineda Mogen
Ari 3h11'43" 27d56'15"
Michelle Pobocik
Mon 6h54'58" -0d30'51"
Michelle "Popstar" Laxina
Gem 7h31'43" 31d23'6"
Michelle Prewitt
Sgr 18h0'8" -26d47'29"
Michelle Prokes
Leo 9h48'40" 27d17'44"
Michelle Purring Babeah
Gem 6h40'17" 21d4'43"
Michelle QR Martinez
Cnc 8h33'8" 23d7'10"
Michelle Qualls
Aur 5h52'31" 43d41'9"
Michelle & Quinton
Cru 12h8'19" -59d46'1"
Michelle R. Betchel
And 2h29'34" 40d57'44"
Michelle R. Diehl, Dylan and Tyler
Uma 11h26'53" 46d24'20"
Michelle R. Hiller
And 0h27'50" 45d5'6"
Michelle R Naimola
And 1h20'36" 50d4'27"
Michelle R Tyler
Cnc 8h32'6" 10d40'51"
Michelle Rae Fields
Gem 6h19'13" 21d58'19"
Michelle Rae Schouten
And 0h3'30" 45d11'58"
Michelle Rae Wilson
Uma 9h39'21" 69d21'52"
Michelle Ramirez - Turtle
Psc 0h1'29" 6d31'55"
Michelle Ramoutar
Cas 2h17'41" 73d47'17"
Michelle Ray - My Shining Light
Col 5h43'11" -36d11'51"
Michelle Rebecca
Uma 13h52'10" 54d40'49"
Michelle Reed
And 0h48'42" 45d46'14"
michelle reed
Leo 11h25'40" 21d17'4"
Michelle Remy
Vir 14h18'16" -2d34'45"
Michelle Renaé
And 1h58'41" 39d10'4"
Michelle Rene
Ori 5h36'44" 11d5'38"
Michelle Rene` Bean
Cap 21h55'54" -22d11'45"
Michelle Rene Miranda
And 0h39'56" 36d42'10"
Michelle Renea
Umi 17h34'51" 82d7'0"
Michelle Renee
Crb 16h3'8" 35d37'35"
Michelle Renee'
And 1h6'20" 46d29'7"
Michelle Renee
Cnc 8h39'0" 18d10'45"
Michelle Renee Bettcher
Ori 5h40'12" -3d8'43"
Michelle Renee Carini-Carder
Uma 9h12'25" 63d26'38"
Michelle Renee Dykes
Crb 15h45'0" 26d43'26"
Michelle Renee Marsengill
Tau 4h7'35" 8d10'12"
Michelle Renee Moore
Crv 12h26'7" -17d46'7"
Michelle Renee Orr
Cnc 9h4'46" 13d30'56"
Michelle Renee Richardson
Cap 21h17'25" -16d10'47"
Michelle Renee Roberts
Sco 17h43'28" -41d30'30"
Michelle Renee Thomas
Uma 10h57'33" 57d33'24"
Michelle Renee Thurston
And 1h38'1" 37d15'4"
Michelle Rivera
Lyn 8h43'23" 36d17'30"
Michelle Robinson
Cas 0h22'13" 51d22'35"
Michelle Robyn Birk
Mon 7h16'37" -0d10'12"
Michelle Rocio Ortiz
Psc 0h34'31" 11d11'23"
Michelle Rockefeller
And 23h22'42" 42d14'27"
Michelle Rosales
Sco 17h26'54" -39d57'9"
Michelle Rose
Aqr 21h13'48" -7d4'21"
Michelle Rose
Psc 1h31'52" 15d20'52"
Michelle Rosemary Greywitt
Uma 11h29'39" 35d51'2"
Michelle Roth
Psc 1h31'11" 10d0'29"
Michelle Ryba
And 1h39'43" 45d57'2"
Michelle S. Fremuth
Ori 6h11'29" 7d28'42"
Michelle S. Watson
Cas 0h36'30" 53d34'27"
Michelle Sachiko Kirihara
Tau 4h18'39" 1d42'11"
Michelle Sagert
And 2h13'59" 40d6'40"
Michelle Sally Wantuch
Lyn 6h42'10" 61d14'43"
Michelle Salvetti
And 1h33'46" 37d2'2"
Michelle Sanders
Psc 1h5'3" 31d30'29"
Michelle Sanders
Ari 2h52'18" 21d40'28"
Michelle Sara
Lib 15h36'21" -19d3'16"
Michelle Schimel
Uma 9h50'41" 64d41'6"
Michelle Schrader Erks
Uma 11h27'31" 52d59'7"
Michelle Schy
Ori 6h14'41" 2d47'4"
Michelle Scott
Dra 18h30'59" 73d43'1"
Michelle Seidel
Boo 14h41'29" 54d15'36"
Michelle Senyk
Sco 16h13'48" -12d5'56"
Michelle Shefflin
And 0h24'20" 42d5'9"
Michelle Shelly Arout
Cap 20h40'29" -17d36'19"
Michelle (Shelly) Elaine Watts
Sgr 17h51'26" -16d56'17"
Michelle (Shelly) Marie Robidoux
Cap 20h51'46" -17d46'37"
Michelle "Shortcake" Thurlow
Cyg 19h39'21" 30d50'31"
Michelle Simons
Psc 23h7'41" 8d4'1"
Michelle Skidmore
Tau 5h51'43" 16d55'38"
Michelle Smith
Lyn 8h23'38" 53d3'4"
Michelle Smith Hendry
Mon 7h15'11" -0d5'12"
Michelle Smittenaar
Cap 21h19'17" -16d16'6"
Michelle "SNAPPER" Racine
Aqr 23h23'3" -20d43'19"
Michelle & Sons
Tau 4h57'24" 16d15'19"
Michelle - Soulmate
Leo 9h24'15" 13d3'43"
Michelle Speicher
Cyg 19h44'34" 29d19'44"
Michelle Stacey Birnbaum
Aqr 23h9'52" -17d52'39"
Michelle Star 18
Aqr 22h52'41" -10d42'57"
Michelle Starr
Psc 0h33'37" 16d24'9"
Michelle Stem Cook
Ori 5h54'10" 19d6'37"
Michelle Stevens Star
Aqr 21h49'27" -6d17'54"
Michelle Storm Baffuto
Uma 11h55'56" 44d4'41"
Michelle Stover
And 23h25'8" 47d57'12"
Michelle Sweet Pea Lee
And 23h55'14" 36d47'24"
Michelle Talbot
And 0h41'11" 40d44'9"
Michelle Tanner
And 0h38'11" 24d21'35"
Michelle Tantalla
And 2h29'5" 45d39'26"
Michelle Taylor
Ari 2h55'40" 26d25'28"
Michelle Taylor Asselin
Lyr 19h20'14" 38d21'45"
Michelle Taylor loves Joseph Labib
Psc 1h19'27" 15d20'58"
Michelle Te Quiero Te Adoro
And 23h46'26" 34d34'13"
Michelle Thao
Cas 0h59'16" 54d2'57"
Michelle Theresa Peterson my Star
Tau 4h40'36" 17d22'27"
Michelle Therese
Tau 4h9'41" 8d1'10"
Michelle Therese Marriott
Psc 23h30'49" 5d26'2"
Michelle Thornton
Ori 4h51'38" 11d55'29"
Michelle Vaiana
Uma 9h5'15" 70d8'33"
Michelle Vernali
Cap 20h26'38" -11d11'32"
Michelle Villarin
And 2h13'10" 47d19'47"
Michelle Vue
Sco 17h45'6" -43d42'21"
Michelle Waczkowski
Cyg 19h31'20" 44d36'33"
Michelle Wall
Vir 14h13'54" 4d26'14"
Michelle Wallace
Ser 15h28'18" -0d7'56"
Michelle & Walter
Uma 9h5'19" 70d35'11"
Michelle Wendt's beautiful star
Vir 15h1'39" 0d26'12"
Michelle Whalen
Dra 18h11'6" 79d10'8"
Michelle Whitney Dodson
Lyn 7h2'55" 45d37'15"
Michelle Williams, Angel of the stars
Cru 12h27'38" -59d37'8"
Michelle Willie
Vir 13h34'48" 3d2'20"
Michelle Wolfe
And 1h18'42" 44d0'12"
Michelle Wotring
Cnc 8h10'35" 27d49'28"
Michelle Y. Osuna
Cnc 9h8'50" 27d34'46"
Michelle Y. Siudut
Leo 11h25'52" 19d27'41"
Michelle Yarbrough
And 0h15'33" 34d57'25"
Michelle Yeux Bleus
Aqr 21h52'57" 1d0'32"
Michelle Yoshiko-Tai Yin
Tau 4h33'6" 8d39'21"
Michelle Yotter
Lib 14h44'20" -20d42'49"
Michelle Young
Uma 8h21'5" 64d1'20"
Michelle Zinn
Psc 1h7'37" 2d50'25"
Michelle, Dennis Clarence
Uma 8h54'16" 47d11'18"
Michelle, Finn, and Amber - Forever
Ari 3h26'22" 25d46'16"
Michelle, I will always remember...
And 0h16'24" 33d11'2"
Michelle, Ma Belle
Sgr 18h8'57" -27d32'41"
Michelle, the star in my eyes
Uma 10h17'16" 51d10'27"
Michelle, You Shine Upon Us All
Cas 0h35'3" 53d35'36"
Michelle2005
Tau 4h28'17" 15d19'5"
Michelle28
Uma 8h43'14" 55d29'5"
Michelle4EVR
Peg 21h46'2" 11d11'28"
Michelle-Andres
Sco 16h9'34" -10d12'49"
Michellearias
Gem 7h28'32" 15d45'49"
Michelle-Jack
Sgr 18h49'58" -16d49'46"
Michelleelliott
Cnv 12h35'33" 51d37'31"
MichelleLucindaKing
Cyg 20h27'59" 34d23'14"
Michelle-Marcella
Vir 13h37'59" 1d40'2"
MichelleMarie
Cnc 8h55'34" 29d2'13"
Michelle-Mélanie & Kevin McLean
Cyg 20h39'23" 44d55'13"
Michelle's
Sco 16h10'3" -20d42'6"
Michelles Amour Etoile
And 0h38'56" 41d45'27"
Michelle's Angel
Ori 6h11'34" 16d2'25"
Michelle's Astellas Star
Tau 3h41'48" 28d6'19"
Michelle's Beautiful Heart
Leo 9h58'1" 14d49'32"
Michelles Echo In Eternity
Dra 17h25'27" 67d43'11"
Michelle's Eternal Brightness
And 1h39'11" 44d23'45"
Michelles Grace
Psc 0h35'2" 10d31'22"
Michelle's light
Dra 16h6'37" 51d58'5"
Michelle's Little Angel
Sco 17h25'53" -42d59'12"
Michelle's Little Wing
Aql 19h59'17" 8d32'53"
Michelle's Love
Cnc 8h28'57" 14d41'37"
Michelle's Mine
Cap 20h31'24" -20d1'58"
Michelle's Perfect Diamond
Cnc 8h45'32" 25d30'58"
Michelle's Prayer
Psc 22h56'26" 5d32'27"
Michelle's Rose
Tau 4h11'36" 15d12'4"
Michelle's Runaway with Me Star
Lib 15h5'46" -1d58'8"
Michelle's Shining Star
Lib 14h49'55" -0d31'38"
Michelle's Star
Col 5h56'23" -38d48'6"
Michelle's Star
Sco 15h50'36" -25d46'9"
Michelle's Star
Lib 15h35'26" -25d37'11"
Michelle's Star
Tau 5h11'51" 17d32'57"
Michelle's Star
Ari 1h58'55" 21d36'40"
Michelle's star
Vir 11h53'21" 8d40'11"
Michelle's Star
Ori 4h49'47" 12d11'27"
Michelle's Star
Cas 0h0'5" 54d52'41"
Michelle's Star
Gem 7h45'47" 31d35'16"
Michelle's Star
Crb 15h27'57" 30d4'40"
Michelle's Sweet Light
And 1h46'50" 45d24'42"
Michelle's Wonder
Cas 23h11'7" 55d37'24"
Michellestar
Lyn 7h35'51" 49d20'22"
Michelle-TheOriginalPossum-1.9.1975
Vir 14h7'32" -15d6'10"
MichelleZap
Cap 21h57'57" -19d9'49"
michelly
Sgr 18h33'49" -24d5'18"
Michelly
And 0h43'41" 41d0'22"
Michelsson
Uma 9h42'57" 43d2'39"
Michesf
Vir 13h9'44" 5d42'57"
Michey 09
Uma 10h24'24" 48d11'39"
Michi
Uma 9h33'8" 49d19'13"
Michi
And 1h58'4" 37d47'12"
Michi
Tau 5h27'10" 25d58'56"
Michi
Lib 15h4'1" -0d51'11"
Michi Du bisch min Stern für's Läbe
Boo 15h5'36" 13d35'54"
Michi & Franzi
Sgr 18h5'1" -27d22'38"
Michi Kieber
Cas 0h59'36" 63d28'24"
Michi Manola
Lyr 18h51'37" 42d31'32"
Michiaki & Tomoko
Psc 1h41'31" 17d15'11"
Michie
Ori 5h25'9" 2d12'51"
Michiel Hagens
Umi 16h55'8" 76d8'36"
Michigan State Sally
And 2h33'55" 45d36'10"
Michiko Brown
Ari 2h18'55" 25d31'57"
Michiko Endo Turner
Umi 14h34'8" 75d8'42"
Michi-Mausi
Uma 13h1'45" 62d48'50"
Michiru-Ninamori Tomiko Yoshiki
Ori 6h17'32" 10d38'57"
Michl, Christine & Karl
Uma 10h26'24" 67d52'3"
Micho & Franny
Lib 15h36'17" -24d58'55"
Michong Beisser
Cam 4h32'23" 66d49'56"
Michou
Cas 23h36'25" 55d29'45"
Michrit
Lyn 8h25'3" 45d24'43"
Michsteve
Aqr 23h11'43" -21d0'9"
Michu mi Schnügu
Cep 22h45'4" 66d54'43"
Michu Welch
Uma 9h57'34" 56d0'52"
Micio e Nata
Aur 5h19'30" 28d43'17"
Mick
Uma 9h28'44" 55d8'41"
Mick A. Armoogam
Sco 16h11'19" -17d26'28"
Mick Adam - My Hero
Per 4h26'41" 42d20'15"

Mick and Patty
Cas 1h19'40" 71d58'13"
Mick and Stef's Star
Lyn 6h26'20" 57d19'21"
Mick & Carly's Wedding Star
Cap 21h47'24" -14d27'8"
Mick Dunn Loving You Forever Steph
Cru 12h19'23" -56d26'14"
Mick Heaney
Her 18h56'10" 17d1'31"
Mick Keeble
Tau 5h44'34" 14d38'29"
Mick Leinweber (K7ZFI)
Lyn 6h21'25" 60d35'24"
Mick Logan
Del 20h51'17" 7d9'36"
Mick Mitchell
Ori 5h24'29" 3d18'28"
Mick Reardon
Leo 9h24'34" 11d50'28"
Mick Shima Shima
Ori 5h42'55" 0d3'40"
Mickael Junior Hayek
Uma 10h19'34" 45d0'14"
Mickael Marcant
Tau 4h7'45" 9d57'4"
Mickaela Rose Toscano
Ari 3h14'13" 27d46'22"
Mickale Lorrayne Getts
Lib 15h3'53" -11d11'45"
Mickayla
Uma 11h1'1" 33d28'48"
MicKenzie Lynn Brown
Aql 20h1'38" -0d31'59"
Mickenzie MML Gudenrath
Her 17h3'10" 31d3'47"
Mickenzies' Diamond In The Sky
Uma 11h34'3" 46d58'43"
Mickey
Her 16h45'48" 33d0'50"
Mickey
Uma 11h47'26" 30d29'12"
Mickey
Cnv 13h51'57" 32d49'33"
Mickey
Uma 11h13'8" 43d46'42"
Mickey
Aqr 21h54'3" -8d16'49"
Mickey
Uma 13h23'41" 58d33'19"
Mickey Bell
Leo 9h31'5" 16d47'19"
Mickey Burkholder
Aqr 22h1'16" -3d59'8"
Mickey Caruso's Magic Kingdom
Aur 7h7'18" 39d6'36"
Mickey & Clif Kleager
Apu 16h26'38" -76d41'13"
Mickey Collin McGinty
Umi 0h7'36" 88d48'32"
Mickey Cook
Lyr 18h24'5" 36d15'50"
Mickey E. Snyder
Sgr 19h39'2" -14d47'30"
Mickey Hall Cheeseman
Psc 0h12'21" 8d0'21"
Mickey Inkanish
Ori 4h58'46" 15d9'13"
Mickey Klein
Uma 9h51'30" 65d46'15"
Mickey & Lori
Ari 2h2'8" 24d19'40"
Mickey Lowery
Her 16h35'12" 33d6'9"
Mickey Mabire
Sgr 18h18'41" -32d20'56"
MICKEY MARIE GAPEN
Gem 6h54'1" 13d34'29"
Mickey Marsh... Landing
Her 17h28'43" 15d7'4"
Mickey & Mary
Umi 15h33'50" 76d12'16"
Mickey Merkle
Uma 12h14'45" 56d56'43"
Mickey Moss
Lyn 7h42'53" 39d28'36"
Mickey Mouse's Daddy BJS
Cap 21h36'50" -13d58'4"
Mickey My Knight In Shining Armor
Aqr 22h35'52" -2d41'51"
Mickey Obradovich
Uma 10h14'51" 53d39'50"
Mickey Parrish
Uma 10h24'24" 51d33'17"
Mickey Rankin
Leo 10h17'21" 23d45'44"
Mickey Richard Conroy
Per 2h11'8" 53d50'45"
Mickey Ryan Jurenka
Uma 10h0'50" 63d4'3"
Mickey the periwinkleblue love
Cap 21h8'27" -16d58'40"
Mickey Wortmann
Aqr 20h47'27" -6d40'7"
mickeyg070103
Lib 15h8'37" -28d42'45"
Mickey's Monkey Star
Ori 6h3'17" 21d8'2"
Mickey's Star "8-9-1917"
Dra 16h50'59" 67d46'14"
Micki
Cet 2h19'4" -13d23'46"
Micki
Del 20h46'53" 13d50'12"
Micki Endliss
Tau 4h26'29" 6d21'13"
Micki Hares
Aql 19h27'6" 10d52'23"
Micki Leann Reynolds Star
And 0h39'32" 44d26'10"
Micki Selbert-Smith
Ari 2h46'55" 25d14'47"
Micki Supple's Eye in the Sky
Lyn 6h41'53" 61d48'1"
Micki Wisely
And 0h26'49" 40d51'33"
Mickie
Uma 12h7'56" 50d57'28"
Mickie Gibson
Lmi 10h26'12" 29d29'52"
Mickie Knuckles
Tau 5h58'20" 24d58'28"
Mickie O'Brien, a wonderful woman
Lib 14h49'36" -24d35'6"
Mickie Weissburg
Umi 14h18'54" 70d5'48"
Micki's Angels
Gem 6h32'56" 20d2'25"
Micklewright Magic
Uma 9h30'34" 62d48'35"
Micko, 24.10.1980
Crb 16h10'58" 35d25'12"
Mickol
Umi 16h30'12" 76d16'2"
Mick's Angel
Aur 5h20'55" 40d47'15"
Mick's Star
Cep 20h44'1" 70d28'18"
Mickstar71
And 1h27'32" 46d22'6"
Micky
Lmi 10h0'55" 29d42'41"
Micky
Sex 10h5'27" 3d23'1"
Micky has the Moon
Cnc 9h14'54" 20d20'5"
Micky Parrish Decker
Uma 11h39'13" 54d27'55"
Micky's Star
And 1h46'5" 42d7'56"
Miclara
And 1h28'42" 34d44'20"
MicMouse
Cru 12h39'16" -59d27'32"
Micniferwilpeau
Cyg 20h3'9" 55d41'35"
Mico
Umi 15h23'51" 75d55'50"
Mico
Tau 5h17'44" 20d55'43"
Microsoft Barcelona
Cyg 21h42'53" 44d12'13"
Miczky Zsolt
Uma 8h41'27" 60d18'6"
Middleton-Vroman bfk year 1
Vir 11h47'30" -5d21'45"
Midge Carver
Tau 4h33'29" 23d54'18"
Midge Dragoo-Crizer
Crb 16h8'26" 26d55'9"
Midge Grace
Psc 1h37'52" 27d57'46"
Midge Modge
Aqr 22h4'29" -15d12'20"
Midge Our Angel
Cas 0h7'36" 56d36'3"
Midge Patrick
Lib 14h37'4" -19d7'14"
Midge Rosie
Ari 2h46'57" 11d57'58"
Midge Souza
Ari 3h3'12" 17d42'23"
Midgee
Uma 8h53'41" 65d12'37"
Midge's Night Light
Cru 12h32'33" -55d55'46"
Midge's Shining Star
Cas 0h55'33" 61d43'4"
Midge's Star
Leo 11h41'7" 26d12'52"
Midget 40 143 13
Sco 17h4'14" -39d41'38"
Midget & Fidget Forever
And 1h57'42" 38d32'4"
Midnight Angel
Psc 1h0'18" 13d17'55"
Midnight Cheyenne
Lyr 18h45'37" 39d18'17"
Midnight Mike
Per 3h28'2" 52d11'33"
Midnight Moon
Tau 4h47'57" 21d6'30"
Midnight Star (Ann Raabe)
Ori 5h44'31" 11d46'38"
Midnight Summer Breeze
Uma 8h27'57" 61d37'36"
Midnight's Love
Uma 11h31'18" 61d22'0"
Midnite
Vir 15h1'43" 3d0'48"
Midori
Sgr 19h8'56" -23d14'55"
Midori Nishi
Aql 19h50'19" -0d33'22"
Midway 10/17/02
Umi 15h31'32" 75d4'53"
Mieczyslaw Basara
Uma 11h8'37" 37d25'11"
Mieirs
Uma 10h55'10" 34d55'17"
Miejek
Lib 14h50'32" -6d46'13"
Mieko
Sco 16h16'51" -16d48'36"
Mieko
Cyg 20h7'44" 35d22'47"
MiG Ayesa ~ Our Superstar
Her 16h29'21" 18d51'24"
Migas
Cyg 20h25'21" 45d47'13"
Migdalia
Vul 20h42'6" 25d56'55"
Migdalia Diaz
Uma 10h9'33" 43d54'24"
Migdalia Santos
Uma 8h39'36" 52d38'36"
Migdalia White
Ari 2h4'33" 22d18'57"
Migdonia
Gem 7h15'47" 34d52'1"
MigElona
Ari 2h32'52" 27d36'56"
Miggeli
Tri 2h30'24" 31d19'54"
Mighty Kim
Ori 5h13'52" -10d40'53"
Mighty Max
Dra 15h58'13" 53d2'39"
Mighty Maxwell
Sgr 19h47'36" -41d23'6"
Mighty Millie
And 23h41'56" 40d27'43"
Mighty Minnie
And 23h26'6" 41d43'22"
Mighty Mo
Her 16h18'18" 23d27'4"
Mighty Mo
Lyn 7h12'5" 54d39'26"
Mighty Quinn "08/16/2005"
Leo 9h41'52" 28d39'55"
Mighty Quinn von Haven
Uma 9h42'42" 69d3'9"
Mighty Tundra
Cma 6h43'46" -13d40'32"
Migliorati Cavalier Angelo
Ori 6h1'20" 20d51'26"
Mignon
Psc 1h15'20" 11d14'5"
Mignonne Trempe
Gem 7h34'3" 25d40'3"
Miguel
Psc 0h53'51" 21d35'36"
Miguel
Uma 13h34'52" 52d52'4"
Miguel A Rivera
Her 17h27'37" 36d54'28"
Miguel Amano
And 2h25'51" 48d37'44"
Miguel and Anne
Cap 21h47'33" -10d20'9"
Miguel Angel
Lib 14h54'39" -11d58'31"
Miguel Angel
Aur 6h27'34" 34d31'6"
Miguel Angel Castillo Acero
Ari 2h16'5" 14d15'14"
Miguel Angel Granada
Sgr 20h23'31" -40d43'23"
Miguel Angel Morales Servin
Per 3h32'2" 45d17'2"
Miguel Antonio
Leo 11h42'22" 24d52'1"
Miguel & Atty
Aqr 22h50'15" -20d44'49"
Miguel Crisantes
Lyr 18h51'27" 31d38'27"
Miguel delgado (El Antiacuario)
Uma 11h56'39" 36d57'44"
Miguel Francisco Santana Jr.
Uma 11h9'52" 52d24'11"
Miguel Gil
Aur 5h16'50" 41d47'57"
Miguel Isiah Masferrer
Psc 0h11'20" 8d2'10"
Miguel R. Patterson
Vir 13h22'26" -20d31'15"
Miguel Ramirez, Jr
Leo 11h50'24" 24d16'47"
Miguel Ramon Vazquez
Sgr 19h45'51" -12d24'54"
Miguel Ribeiro
Lmi 10h15'42" 30d24'6"
Miguel Robles-Coles
Cnc 8h25'28" 11d44'19"
Miguel Rodriguez-Munoz
Psc 1h20'38" 24d41'9"
Miguel Siderakis
Col 6h2'9" -30d58'26"
Miguel Tamayo Leal
Sco 17h40'18" -39d55'48"
Miguel Tugas
Vir 12h43'38" 4d31'42"
Miguel Valls Jr.
Gem 7h7'51" 30d43'50"
Miguelito
Ari 2h4'25" 11d54'21"
Miguel's Heavenly Body
Cen 13h45'52" -40d31'24"
Miguel's Nebula
Lmi 9h25'28" 34d11'9"
Mihaela
Lib 15h18'23" -20d27'26"
Mihaela
Dra 19h39'58" 68d47'14"
Mihaela Felicia Macesanu
Lyn 6h53'5" 55d1'48"
Miharo Atua, Kitea Aroha
Pic 5h52'12" -55d29'50"
Mihavecz Petra Szerencsecsillaga
Tau 5h52'34" 25d7'57"
Mihika Kothari
Tau 5h32'54" 19d18'50"
MIHO
Gem 7h22'45" 26d49'25"
Mihoko
Cnc 8h27'12" 24d3'30"
Mihoko MKI
Cnc 8h58'38" 20d23'38"
Mii Love Joe
Uma 11h6'8" 56d40'32"
Miiiin Tiiiine
Umi 14h11'58" 67d9'40"
Miika H Salonius
Uma 13h14'14" 57d49'13"
mija
Sgr 18h20'22" -32d18'29"
Mijanou
Ori 6h14'36" 7d19'50"
Mijini
Aqr 22h40'42" -0d35'0"
Mijn Papa
Uma 8h18'3" 65d15'23"
MiJo
Gem 6h27'1" 24d39'54"
Mik Jet 7
Leo 10h29'50" 9d58'25"
Mika
Lyn 7h58'43" 41d17'12"
Mika
Cma 7h2'2" -30d39'51"
Mika L. Weeks
Psc 1h40'50" 12d45'34"
Mika Michele Lindsay
Cas 23h38'2" 55d32'31"
Mika Mouse
Lyn 7h49'15" 38d20'13"
Mika1002
Her 18h43'19" 20d5'46"
Mikade Pottorff
Mon 8h6'39" -0d42'38"
Mikado
Leo 10h30'26" 11d44'46"
Mikael
Ori 5h5'6" 9d52'23"
Mikael David Levin
Cap 21h26'33" -25d28'26"
Mikael Fontaine
Ori 5h59'18" 1d59'31"
Mikaël Graindorge
Psc 1h22'36" 16d7'29"
Mikael Holmlund
Lyn 7h8'50" 59d31'5"
Mikael Tarning
Crb 15h23'37" 26d56'25"
Mikaela Bradford
Cru 12h35'3" -55d46'32"
Mikaela Dawn Pasquin
Leo 10h42'27" 19d44'15"
Mikaela DLC
Vir 13h29'35" -7d22'26"
Mikaela Jane Blackely
Cru 12h1'5" -62d50'8"
Mikaela Lynn
And 0h57'19" 37d40'0"
Mikaela Madison Thiboutot
And 2h32'23" 39d9'5"
Mikaela Mae - our angel forever
Cru 12h23'11" -57d46'12"
Mikaela Oneto
Leo 10h20'49" 21d32'20"
Mikaela Renee Lynch
Cap 21h19'55" -16d20'43"
Mikaela Sophie
Leo 10h40'56" 21d4'16"
Mikaela Thi
Vir 13h6'12" -20d16'8"
Mikaela Victoria Hawkinson
And 1h16'50" 45d10'4"
Mikaela Ysabelle Advincula Conjuang
Vir 13h24'14" -18d20'32"
Mikaels
Tau 4h54'47" 17d43'32"
Mikaila Ashley Brundick Jorgensen
Psc 23h21'9" 4d51'32"
Mikaila Renae
Cnc 8h18'59" 15d6'3"
Mikal Jon Morton
Tau 5h48'24" 25d36'46"
Mikal N Tayler
Uma 10h38'44" 44d56'54"
Mikal Shawn Evelyn, JR.
Uma 10h8'55" 41d42'46"
Mikal Victor 2005
Uma 9h0'3" 65d50'3"
Mikala
And 23h23'21" 51d56'24"
Mikala
Psc 1h17'23" 12d28'36"
Mikala Babb Ampson
Ari 3h9'29" 12d8'13"
Mikala Kuchera
Psc 1h13'33" 30d16'28"
Mikala Marie Rossi
Ari 3h17'28" 28d8'54"
Mikaliconmahan
Dra 17h54'58" 65d21'37"
Mikayla
Peg 22h19'30" 16d31'38"
Mikayla
And 1h55'0" 37d37'50"
Mikayla
Cam 4h42'43" 56d0'9"
Mikayla and Tyler's Eternity Star
Vir 13h0'56" 5d46'53"
Mikayla Ann Johnston - 6 May 2007
Cru 12h29'3" -59d11'2"
Mikayla Anne Manriquez
Cas 1h41'48" 58d0'45"
Mikayla Ann-Elizabeth Barrier
And 2h10'24" 45d46'15"
Mikayla Coenen
Gem 7h35'3" 31d19'9"
Mikayla Elizabeth O'Sullivan
And 1h0'56" 45d57'3"
Mikayla Fisher
Leo 9h57'53" 15d25'59"
Mikayla Grace Stout
Cap 21h38'41" -14d47'49"
Mikayla Jade Derge
And 23h12'58" 41d33'50"
Mikayla Jolee
Uma 12h3'22" 30d21'50"
Mikayla Keona Bird
Uma 10h54'38" 45d44'6"
Mikayla Lynn
Vir 13h24'22" 11d3'29"
Mikayla Lynn Gladieux
Vir 12h46'38" 2d4'13"
Mikayla Lynn Whiting
And 0h30'25" 42d16'13"
Mikayla Madeline McCoy
Peg 23h55'41" 23d21'4"
Mikayla Marie
Cnc 9h6'15" 14d58'33"
Mikayla Marie
And 0h21'31" 42d15'7"
Mikayla Marie
Aqr 22h7'54" -1d42'53"
Mikayla Marie Argüelles
Leo 9h53'55" 14d26'54"
Mikayla Marie Plafcan
Psc 1h24'28" 31d13'50"
Mikayla Marie Waters
Peg 21h43'54" 21d20'29"
Mikayla Nicole Wills
Peg 23h43'51" 26d6'4"
Mikayla Paige Hert
Cas 3h21'17" 74d25'54"
Mikayla Paige Schoenrock
Umi 11h7'34" 88d35'50"
Mikayla Payne
Her 17h5'8" 36d37'23"
Mikayla Raye Gibson
Sco 17h23'36" -30d30'59"
Mikayla Remley Tourigney
Cas 0h54'25" 54d45'53"
Mikayla Rose
Gem 7h13'48" 34d12'19"
Mikayla Rose Schiller
Peg 22h5'31" 15d39'36"
Mikayla Shining
And 0h14'6" 43d22'20"
Mikayla Simone Nowell
And 0h48'7" 41d59'35"
Mikaylah Jaye Osinga
Sco 17h5'23" -40d29'43"
Mike
Aqr 22h31'36" -1d23'11"
Mike
Aqr 23h1'29" -8d38'25"
Mike
Cyg 21h14'59" 46d23'42"
Mike
Cyg 20h24'58" 41d38'31"
Mike
And 0h59'35" 48d30'19"
Mike
Ari 2h9'45" 24d25'11"
Mike
Tau 4h10'49" 26d52'27"
Mike "04/25/1981-01/02/2004"
Tau 3h42'40" 3d11'45"
Mike A. Sitar
Dra 12h21'53" 72d21'27"
Mike A Thompson-Forgione
Sco 17h29'4" -36d56'59"
Mike Amos
Aql 19h45'38" -0d29'49"
Mike and Amanda Star
Cas 0h56'26" 57d32'8"
Mike and Amy Bell
Cnc 8h19'13" 32d21'2"
Mike and Ann Rizzo
Sco 16h5'53" -24d47'55"
Mike and Ashley
Ari 2h27'44" 18d54'22"
Mike and Barbara
Gem 7h42'17" 34d38'22"
Mike and Cindy Rock the World
Cyg 21h34'57" 39d53'29"
Mike and Clare's Star
Ari 2h12'51" 24d39'7"
Mike and Claudia
Uma 8h39'48" 57d22'30"
Mike and Courtney
Uma 11h6'33" 45d13'15"
Mike and Dana Inskeep
Lib 15h24'31" -8d15'19"
Mike and Debbi...Buddies Forever
Per 3h44'52" 34d0'9"
Mike and Dewi Tamzil
Cnc 8h5'28" 12d46'56"
Mike and Donna Tokarski
Her 18h51'10" 24d10'58"
Mike and Erin's Star of Love
Cas 0h22'56" 58d20'47"
Mike and Holly
Sgr 19h17'56" -17d16'11"
Mike and Jessica
Lib 14h52'20" -2d57'24"
Mike and Julie
Col 5h13'57" -27d43'12"
Mike and Kasey Detwiler
Lib 14h51'3" -1d50'6"
Mike and Kathi Watson
Cyg 19h40'31" 28d25'27"
Mike and Katie's Wedding Star
Cyg 21h22'36" 50d29'20"
Mike and Kay Popper
Cyg 20h18'28" 40d59'21"
Mike and Kelly Plath Star
Cyg 19h39'34" 38d50'22"
Mike and Kim Anderson's Star
Lib 15h4'29" -6d16'54"
Mike and Kristin Gies
Crb 16h7'25" 33d20'41"
Mike and Lauren Speer Enternal Love
Sco 16h15'13" -10d52'26"
Mike and Lauren's Wishing Star
Tau 5h44'18" 17d51'24"
Mike and Lisa
Vir 12h51'13" 4d51'32"
Mike and Liz Forever
Uma 10h34'27" 64d47'25"
Mike and Marguerite
Cyg 19h33'19" 31d59'19"
Mike and Meg
Lyn 8h13'12" 35d12'54"
Mike and Meghan
Cnc 8h45'59" 17d43'4"
Mike and Nadines Star of Love
Uma 13h37'19" 49d20'4"
Mike and Peg Domenick
Cyg 20h36'18" 52d32'33"
Mike and Rachel
Cyg 20h28'31" 40d46'59"
Mike and Roger celebrate 17 years
Aql 19h57'59" 10d42'49"
Mike and Todd
Uma 12h11'48" 53d18'1"
Mike And Vanessa
Her 17h10'43" 18d48'33"
Mike and Vanessa Ray
Cyg 21h30'33" 36d27'36"
Mike Andrews
Cyg 21h57'46" 45d11'4"
Mike Angel
Leo 9h38'53" 16d0'26"
Mike Angeles (Emily's Angel)
Aqr 22h17'38" 0d44'4"
Mike April 27, 1950
Uma 11h25'22" 37d19'57"
Mike & April, In Love Forever
Cap 20h34'20" -16d17'3"
Mike Aronhalt
Lac 22h51'35" 52d20'22"
Mike & Ashley
And 0h46'58" 43d50'43"
Mike & Aubrie's "April Fool"
Sgr 19h33'33" -15d37'44"
Mike Baade
Dra 19h14'39" 71d39'50"
Mike Baker
Uma 13h1'5" 54d24'25"
Mike Banks
Boo 15h2'32" 41d10'40"
Mike & Barbara
Cyg 19h32'51" 53d44'51"
Mike Barkus
Dra 16h1'25" 58d16'17"
Mike Barnetts Jinx
Leo 9h27'25" 30d43'33"
Mike Barrett
Uma 11h33'0" 34d57'12"
Mike Bass
Gem 6h36'4" 20d16'55"
Mike & Becky - 17 Nov 06
Uma 11h24'42" 59d48'46"
Mike Bedau
Ori 5h36'23" 13d42'43"
Mike Bernstein
Per 3h0'15" 46d25'49"
Mike Bess
Ori 4h47'38" 13d6'58"
Mike Big Blue Romano
Cap 20h23'15" -12d25'57"
Mike Blondin Survey Flyer
Cas 0h55'44" 63d42'36"
Mike & Brenda Keough *Forever*
And 2h9'4" 38d16'56"
Mike Briggs
Cep 21h10'42" 67d53'51"
Mike Brotherton
Cep 2h35'40" 80d47'23"
Mike Brunett
Vir 12h48'23" 5d33'25"
Mike Butler
Ori 5h53'28" 12d8'14"
Mike Carey
Per 3h53'52" 42d21'48"
Mike Carringer
Sgr 17h48'49" -16d40'34"
Mike & Cary's Pool Light
Cyg 20h19'39" 55d28'11"
Mike & Christina Proctor 1/20/1996
Aql 19h47'30" -0d12'46"
Mike Cicero
Her 18h27'35" 25d38'4"
Mike & Cindy Hutchcroft Star
Leo 11h19'23" 15d41'30"
Mike&Claudine
Umi 15h38'52" 69d25'21"
Mike Clemens
Cep 23h27'50" 71d33'31"
Mike Crivaro
Ori 5h33'3" -1d22'58"
Mike "Cuchulain" Ramsey
Sco 17h20'50" -36d50'2"
Mike D. Lorenzana
Uma 11h47'28" 31d40'4"
Mike Daino
Crb 15h46'44" 37d58'17"
Mike Danby
Per 2h12'34" 51d6'9"
Mike & Darla Raybuck
Cyg 19h50'30" 38d40'18"
Mike & Deb 12/23/69
Eri 3h54'42" -0d37'52"
Mike & Dee Bernhard
Uma 9h30'19" 57d25'44"
Mike DeLuca
Cep 22h43'13" 64d29'16"
Mike & Diane
Lyr 18h29'50" 36d14'2"
Mike & Diane McGinnity
Umi 14h23'38" 75d38'27"
Mike Downs
Uma 14h4'56" 61d22'6"
Mike Drexler
Umi 14h42'37" 69d4'38"
Mike Easly
Sco 17h48'27" -42d25'40"
Mike Edward Potter
Sgr 20h0'7" -40d43'5"
Mike & Emily Larson
Lyr 18h39'13" 37d59'39"
Mike Flatt
Ori 6h15'13" 15d52'0"
Mike Flenz
Dra 18h50'52" 50d22'35"
Mike & Florence
Psc 1h9'23" 14d4'59"
Mike Florida Flyboy Snyder
Sgr 19h28'1" -24d46'25"
Mike Fumelle
Ari 3h2'43" 25d2'53"
Mike & Gerda Hannas
Sgr 20h4'30" -23d57'46"
Mike Giuliano
Vir 13h8'19" 5d0'2"
Mike Goebel
Leo 10h43'28" 9d52'37"
Mike Grasmick
Sgr 19h12'22" -17d26'10"
Mike Gritz
Psc 1h18'16" 32d33'32"
Mike Gubicza, Sr.
Lib 15h6'58" -21d52'13"
Mike Guzina
Umi 14h39'1" 70d5'28"
Mike Hardie - Shine On
Sgr 18h35'14" -35d59'37"
Mike Haycock
Cep 23h46'14" 75d54'42"
Mike Heinz
Uma 14h24'32" 55d33'49"
Mike Hernandez
Uma 11h21'43" 38d55'23"
Mike Hogan
Per 4h18'0" 40d43'42"
Mike & Holly Allways and Forever
Uma 11h51'9" 57d36'37"
Mike Hosken
Per 2h43'27" 54d46'10"

Mike Hudec
Cam 3h58'54" 70d12'4"
Mike Hyde
Lib 15h18'12" -4d14'35"
Mike Hynson
Aql 19h8'13" 0d18'57"
Mike & Irene Davis
Her 17h12'59" 30d47'54"
Mike & Irene Minetti's Special Star
Uma 11h25'19" 61d47'21"
Mike Iturrey
Sgr 18h10'38" -17d51'42"
Mike & Jackie Blackwell
Cyg 19h51'35" 48d4'1"
Mike & Jaimi Grether
Umi 15h32'18" 70d47'38"
Mike & Jamie
Del 20h37'27" 15d5'29"
Mike & Jamie Forever in Love
And 2h29'25" 41d55'53"
Mike Jankowski
Uma 14h16'27" 58d21'45"
Mike & Jennifer
Lib 15h40'53" -5d8'22"
Mike Jennings
Vir 13h19'1" 6d30'49"
Mike Jess Always and Forever
Uma 9h49'47" 58d18'19"
Mike (Jimmy) Giznik
Cep 22h21'25" 55d55'8"
Mike & Joann Zelenak
Cyg 21h53'58" 50d10'28"
Mike Jones
Sge 20h5'6" 18d27'15"
Mike Jones
Uma 8h28'57" 71d6'55"
Mike Jones "My Loving Angel"
Umi 11h5'14" 89d30'12"
Mike & Julie's Star
Uma 10h35'27" 62d9'10"
Mike & Kara Taggart
Dra 16h58'45" 68d53'49"
Mike Kelley
Her 18h32'5" 18d30'36"
Mike Kelly
Her 16h54'1" 36d35'6"
Mike Kelz Yamamoto
Eri 4h37'20" -14d11'16"
Mike & Kerrie Sullivan
Cyg 20h20'58" 54d36'15"
Mike King
Lib 15h5'40" -27d35'40"
Mike Klayman 1/27/1948
Aqr 21h1'22" -7d30'21"
Mike Kooiker
Eri 3h48'3" -0d23'27"
Mike Kraml
Car 7h23'43" -53d28'57"
Mike & Kris
Sco 16h2'32" -28d13'44"
Mike LaCella
Lyn 6h31'0" 59d22'6"
Mike Lapinsky "My Daddy"
Per 4h26'53" 44d14'56"
Mike L'Hommedieu
Aqr 22h32'37" 0d18'36"
Mike & Linda
Tau 3h43'54" 27d43'27"
Mike & Lisa
Lib 14h51'8" -24d53'53"
Mike Long
Ori 4h55'20" 1d54'7"
Mike Loper
Psc 23h0'8" -1d1'52"
Mike & Lori Conley
Gem 7h24'35" 24d34'56"
Mike Loves Cindy Narvaez
Sco 17h38'5" -40d28'42"
Mike Loves Val
Cyg 19h32'16" 28d34'28"
Mike Lowhogg Hill
Her 17h50'18" 45d22'14"
Mike Lund
Leo 11h21'36" 6d52'30"
Mike Magee
Cyg 21h14'3" 46d19'26"
Mike Magers
Uma 12h58'22" 55d5'16"
Mike & Mal's Shining Star
Tau 5h16'49" 24d31'54"
Mike Mancari
Tau 5h30'41" 22d14'34"
Mike & Mandy's shining star
Aqr 20h54'50" 1d11'38"
Mike Mannes
Uma 11h4'34" 64d44'16"
Mike & Marcia Greenzeig
Cyg 20h10'22" 57d48'42"
Mike Markert
Uma 9h18'52" 51d7'22"
Mike Marlo
Per 3h0'29" 49d14'16"
Mike Martin Loves Katie Caswell
And 1h26'32" 40d55'23"
Mike & MaryAnn Tierney
Her 16h49'10" 36d13'26"
Mike McCann
Uma 12h7'54" 49d33'43"
Mike McDonald
Gem 6h1'6" 27d28'43"
Mike McDonald
Aql 19h4'0" -0d51'43"
Mike McKale
Crb 15h45'8" 34d54'21"
Mike McKenzie
Psc 0h33'49" 6d41'13"
Mike - Michael Reichert
Uma 8h16'42" 72d36'19"
Mike & Michele Viehman - Forever...
Uma 9h44'1" 57d16'9"
Mike Micliz
Cep 21h34'29" 62d1'9"
Mike & Missy Meyers
Aql 19h37'57" -0d4'25"
Mike & Monica Hardman Wedding Star
Uma 11h40'52" 49d10'9"
Mike Moore
Gem 6h50'8" 31d52'26"
Mike Moskovitz
Per 2h46'52" 54d5'23"
Mike "Mumbling Jumba" Bardill
Uma 9h7'5" 48d34'2"
Mike *My Shining Star*
Cnc 9h8'12" 16d17'51"
Mike 'n' Deni
Cyg 19h40'43" 30d30'16"
Mike N. Kalin
Dra 19h46'22" 78d47'45"
Mike & Nancy Lumia
Uma 10h53'28" 56d6'47"
Mike "Nino" Rodriguez
Uma 11h12'32" 68d31'12"
Mike Noonberg
Ori 5h36'25" 13d38'18"
Mike Novy
Dra 18h3'12" 76d47'40"
Mike O'Brien
Per 4h24'58" 34d4'41"
Mike O'Bryan's Diamond In The Sky
Tau 4h10'33" 6d23'32"
Mike Ortega
Lac 22h19'30" 53d7'17"
Mike Pesarchick
Per 2h44'50" 51d36'30"
Mike Pettinari
Per 3h16'37" 45d30'15"
Mike "Phenom" Williamson
Uma 10h22'25" 50d1'25"
Mike Piotrowski's Blaze of Energy
Psc 0h47'18" 10d54'10"
Mike Prebish
Aql 19h7'23" 3d16'27"
Mike Preston McGraw
Psc 0h43'9" 7d27'1"
Mike Raether
Lac 22h32'6" 39d1'7"
Mike Ratner
Uma 9h26'27" 46d48'24"
Mike Reinbold Forever
Uma 8h42'19" 56d17'18"
Mike Renowden
Ari 2h9'28" 24d40'0"
Mike Robertine
Aur 5h24'42" 40d41'14"
Mike Rocco
Ori 6h8'45" 19d45'49"
Mike & Rose Giammichele
Cyg 20h36'21" 35d13'4"
Mike Rowe
Cas 0h37'52" 53d50'0"
Mike Rowell
Cap 20h19'21" -11d55'0"
Mike & Sandy Whalen
Uma 11h25'25" 33d7'29"
Mike Schrec:"Mio Pilota, Mio Amore"
Sco 16h12'23" -10d7'18"
Mike Schulman
Uma 10h58'20" 57d23'1"
Mike Schultz
Per 2h12'17" 56d29'27"
Mike Schutz
Per 3h18'58" 52d13'26"
Mike Scicolone
Per 2h51'8" 53d53'5"
Mike Scott Family Star
Tri 2h8'30" 34d37'29"
Mike S.H. Sin
Lib 15h35'48" -19d51'48"
Mike & Shannon's Star
Lyr 19h20'28" 37d41'20"
Mike & Sharon Brenizer
Cyg 20h5'26" 39d19'36"
Mike & Sheri
Aql 19h1'57" -0d8'15"
Mike Simpson's Star
Uma 11h59'56" 29d24'7"
Mike Smith
Her 17h42'45" 19d7'6"
Mike Snelgrove
Dra 12h7'46" 67d32'32"
Mike Stuttgen
Uma 11h26'29" 41d33'8"
Mike & Tessa Hergenreter
Vir 13h22'8" 11d32'29"
Mike Teta
Umi 13h47'40" 72d8'38"
Mike: The Champion of My Heart
Her 16h39'2" 35d21'35"
Mike the Poolman
Gem 6h20'24" 22d12'33"
Mike Tomlin
Dra 16h7'10" 55d27'44"
Mike Toye
Her 16h41'9" 26d4'12"
Mike Upchurch
Boo 14h13'20" 35d51'36"
Mike & VV's 1st Anniversary Star
Cyg 19h42'57" 33d31'28"
Mike Walsh
Cyg 20h23'7" 54d41'34"
Mike Warkentin
Cep 20h44'48" 61d41'11"
Mike White
Ori 6h19'28" 8d2'5"
Mike "Wild Bill Hickock" Reynvaan
Her 17h37'27" 15d11'51"
Mike Wright
Cyg 20h39'6" 30d22'4"
Mike Yassine (Big Huss)
Dra 19h36'24" 65d10'27"
Mike Young
Cep 22h17'36" 62d7'16"
Mike & Yvonne Docherty - 26.12.1941
Cyg 20h34'48" 44d55'29"
Mike Zelinski
Per 3h8'21" 51d8'20"
Mike Zenoni (#13)
Uma 9h49'4" 57d16'29"
Mike Zilly
Aur 5h54'46" 47d0'18"
Mike, Angel, and Maddy Pennell
Aqr 23h54'58" -12d4'31"
Mike, Kathy, Chris, Scott Williams
Her 17h14'45" 20d34'30"
Mike, Zach + Zoe Dillehay
Tri 1h54'53" 28d32'35"
Mike-Boo-Boo
Her 18h31'8" 16d55'4"
MikeDiana9204
Ori 6h18'38" 15d37'23"
MikeG95
Uma 9h54'16" 61d15'54"
MikeHan
Ori 5h37'7" 4d25'4"
Mikel
Tau 5h51'50" 17d43'5"
Mikel
Ari 2h27'21" 25d43'28"
MikeL
Gem 6h23'40" 22d23'6"
Mikel Alexander Brown
Her 18h47'56" 20d52'20"
Mikel J Velasquez
Uma 10h26'56" 70d27'43"
Mikel K
Gem 6h30'30" 25d49'33"
Mikel Reyna
Cyg 19h34'13" 29d6'3"
Mikel / Shellie's Star
Cyg 19h30'13" 31d55'2"
Mikel "Smiley" Mukai
Vir 12h48'7" 11d41'17"
Mikel Terluk
Sco 17h27'33" -44d46'53"
Mikell Family
Uma 8h12'31" 62d7'57"
mikenalli 09302001
Dra 19h13'2" 60d55'14"
Mikenlo
Lyn 7h23'16" 52d33'23"
Mikenna Garbin
Cyg 21h41'35" 36d23'22"
Mikenna "Shortstack" Lawson
And 2h5'30" 42d54'46"
MiKenzie
Ari 2h47'40" 29d26'38"
Mike's 52
Vir 13h9'7" -4d57'18"
Mike's Diamond In The Sky
Per 3h35'23" 47d15'4"
Mike's Golf Star "50"
Leo 10h32'55" 24d56'26"
Mike's Goodnight Kiss
Leo 11h31'7" 6d10'36"
Mike's Jaguar Star
Psc 23h13'12" -2d19'15"
Mike's Love
Cyg 20h16'42" 55d7'19"
Mike's Mighty Missile
Uma 13h48'12" 51d19'6"
Mike's Star
Per 3h41'11" 41d33'47"
Mike's Star
Psc 1h8'8" 20d55'55"
Mike's Star
Ari 2h13'41" 24d14'14"
Mike's Star
Vir 14h13'26" -10d59'6"
Mike's Sunshine
Uma 13h2'31" 59d19'46"
Mike's World
Leo 11h40'9" 22d23'5"
Mike's World
Her 17h51'42" 46d40'43"
Miketz
Lyn 8h32'11" 33d23'19"
Mikey
Uma 11h18'14" 38d52'21"
Mikey
Lyn 8h31'21" 44d24'10"
Mikey
Cyg 21h21'7" 34d11'42"
Mikey
Per 2h42'39" 55d31'58"
Mikey
Aqr 20h54'21" -7d22'35"
Mikey
Sco 17h50'56" -36d21'51"
Mikey and Ai's Love Star
Eri 3h47'9" -0d51'45"
Mikey and Heather
And 1h10'0" 37d35'8"
Mikey and Lauren
Aql 20h11'36" 14d26'56"
Mikey + Angela = Love Forever
Cyg 20h24'21" 52d4'15"
Mikey Boo
Umi 16h32'20" 76d40'38"
Mikey Boy
Aql 18h52'19" -0d11'11"
Mikey & Dusty
Aqr 21h31'37" 0d12'18"
Mikey Edwards
Cnv 13h47'28" 31d51'19"
Mikey Elhippo Austin
Ari 2h42'13" 25d26'36"
Mikey & Jessie
Cyg 21h17'13" 52d55'40"
Mikey n Gem 4eva
Cyg 20h53'4" 48d32'43"
Mikey Palumbo & Kristin Noble
And 0h37'18" 46d23'2"
Mikey Poo
Cnc 8h58'47" 27d54'35"
Mikey S Mitchell
Ori 5h41'9" 1d18'0"
Mikey's Star
Ori 5h53'44" 21d2'30"
Mikey's Star
Crb 16h13'59" 37d56'50"
Mikey's Star
Ori 5h58'10" -0d34'24"
Mikhael
Sco 16h3'58" -22d43'37"
Mikhail Dzamashvili
Uma 9h27'17" 56d49'10"
Mikhail Kholodenko
Uma 9h51'36" 66d5'57"
Mikhail Korchemkin
Dra 18h52'18" 60d47'10"
Mikhail Latinsky
Umi 14h37'26" 74d13'58"
Miki
Crb 16h15'0" 28d7'54"
Miki Leigh
Cyg 21h58'51" 55d24'36"
Miki & Shuji
Sco 17h27'38" -42d9'16"
Miki U
Ori 6h14'13" 3d11'28"
Mikia Kalina
Gem 6h24'16" 19d3'19"
MikiChris
Leo 9h47'52" 28d9'17"
Mikie
Crb 15h33'11" 31d34'27"
Mikie and Kelly Perry
Lyn 8h16'26" 52d4'55"
MIKIE da FATHER
Aqr 22h16'52" 1d0'48"
Mikie Data
Uma 9h37'17" 59d25'51"
Mikie Williams
Vir 14h16'10" -5d26'2"
Miki-Gina Vayloyan
Her 17h33'7" 45d13'30"
Mikiko Okabe
And 0h32'11" 36d7'57"
Mikinna Lee Sigerud
Gem 6h32'3" 27d53'7"
Mikka
Umi 16h33'48" 82d31'44"
Mikkena Bonselaar
Peg 21h9'3" 14d34'7"
Mikki and Mark's Star
Uma 13h40'4" 51d58'13"
Mikki and Mark's Star
Cyg 20h40'33" 46d4'21"
Mikki Dupree
Mon 6h47'1" -0d8'6"
Mikki In Paradise
Ari 2h49'56" 14d44'5"
Mikki Jang
Mon 6h48'2" -0d49'51"
Mikki Simmons
Uma 9h7'4" 62d54'45"
Mikki Taylor
Umi 16h7'44" 74d37'19"
Mikki Taylor
Vul 19h36'8" 27d38'38"
Mikki Torres
Psc 1h24'39" 16d44'2"
Mikkiala
Gem 7h18'23" 32d20'43"
Miklos B.Szekely
Vir 13h34'15" 3d1'37"
Miklos Kisiday
Uma 11h27'57" 42d11'51"
Miklós & Szabina (Gyenesdiás20050827)
Tau 3h54'30" 1d54'54"
Miko
Per 3h19'40" 52d52'9"
Miko Chua 5
Lyr 18h31'54" 42d6'33"
Miko & "Mike" Gantt
Uma 11h7'34" 60d41'44"
Mikó Timi
Gem 7h1'39" 28d54'44"
Mikou Mariani
Cap 21h54'31" -18d32'19"
mikus wilderness
Uma 13h18'43" 57d37'43"
Miky4Eternity
Lyr 19h14'50" 28d49'3"
MikyOry26-09-2001
Per 3h24'47" 41d31'40"
MILA
Cap 20h33'11" -27d31'7"
Mila Ann Marie
And 1h54'51" 41d44'59"
Mila Deborah Prince
Peg 22h8'50" 8d27'53"
Mila Flores
Cyg 20h4'33" 37d33'5"
Mila & Jacob's Star
Sco 17h0'56" -44d22'47"
Mila Jane
Ari 2h14'0" 24d51'30"
Mila Lukinovic
Cnc 8h25'1" 28d3'41"
Mila Miskovsky
Ari 3h6'24" 25d9'32"
Mila Pita (stella communi-camus)
Cas 0h46'36" 53d55'36"
Mila Skakie 30112005
Uma 14h13'30" 56d56'6"
Mila Volynsky
Cas 1h1'42" 60d35'17"
Milachy December 25
Uma 14h1'36" 58d54'18"
Miladis & Louis Della Cava
Cyg 19h48'29" 57d11'29"
Milagri
Vir 12h45'44" 1d22'52"
Milagros
Uma 12h37'37" 58d39'12"
Milagros Asencio
Leo 10h32'33" 27d4'26"
Milagros Castaneda
Mon 6h52'1" -0d37'44"
Milagros Raquel
Cyg 21h22'56" 44d44'46"
Milagros Reyes
Vir 11h38'50" 0d34'46"
Milagros Stallings
Sco 16h13'1" -33d15'50"
Milah Kaur
Gem 6h55'18" 21d50'32"
Milam
Cen 13h34'25" -52d3'31"
Milan
Cma 6h47'34" -15d21'31"
Milan
Tau 4h46'2" 18d24'10"
Milan Becker
Cru 12h40'55" -56d11'36"
Milan Chevillard
Vir 14h17'27" -4d28'4"
Milan Lauzon
Umi 17h10'49" 82d3'7"
Milan Michael Malina
Gem 7h40'10" 27d58'14"
Milan Perhac
Sco 16h13'44" -9d37'2"
Milan Stephen Cerstvik
Aur 5h40'52" 40d54'50"
Milan Troy Jurovich
Her 17h47'44" 36d58'0"
Milana
Vir 14h39'0" 3d57'2"
Milana Glogovac
Lib 15h2'57" -12d41'47"
Milano
Aql 19h49'12" -0d44'3"
Milda B. Bartis
Uma 12h56'22" 56d5'7"
Mildred
Ari 3h21'9" 27d50'33"
Mildred
Gem 7h25'10" 17d53'29"
Mildred A Bagley
Vir 13h23'4" 12d51'58"
Mildred Ann Johnson
Lyn 8h9'5" 50d25'28"
Mildred B Knapp
Psc 1h20'40" 31d35'42"
Mildred Bakley
Cnc 9h5'30" 32d38'54"
Mildred Beaulieu
Cas 1h27'48" 57d53'49"
Mildred Benson
Vir 13h18'14" 0d17'56"
Mildred Brenner
Ori 5h5'17" 13d50'4"
Mildred Burdorf
Del 20h44'1" 18d4'51"
Mildred Catherine Crosby Reheis
Vir 13h15'52" 11d42'37"
Mildred Chastin
Ari 3h23'33" 29d37'33"
Mildred Cookman Frick
And 2h18'1" 49d29'33"
Mildred Dawn Blosser
Vir 13h24'51" 13d44'16"
Mildred Donakowski
Ari 2h59'4" 17d56'51"
Mildred Dubov
Gem 7h43'36" 24d58'57"
Mildred Duhl Grackin
Mon 7h32'18" -0d48'2"
Mildred Elenore
Umi 15h19'41" 73d15'23"
Mildred Fairfax ("Mittens") Crow
Ori 5h37'26" -6d14'44"
Mildred Friedman
Gem 7h45'49" 21d29'49"
Mildred Frindt
Ari 2h2'53" 23d47'25"
Mildred Gabriela Alvarez
Leo 10h15'9" 12d3'56"
Mildred Georgene Madsen
Cas 2h24'38" 73d56'5"
Mildred Hankey forever a STAR!
Cas 23h21'58" 59d14'37"
Mildred Hill -Beloved Wife & Mother
Uma 12h55'0" 62d3'51"
Mildred Hoag
Vir 12h42'46" 12d15'17"
Mildred Hudson
Tau 4h4'46" 17d7'56"
Mildred Humphrey
Boo 15h27'21" 37d40'8"
Mildred Irene Waiton
Ari 2h15'43" 25d10'40"
Mildred "Jeanette" Rowe Bristow
Cas 0h40'29" 61d38'41"
Mildred Joan Taafe
Uma 10h17'57" 51d19'33"
Mildred Loretta Perry
Crb 15h48'41" 27d10'18"
Mildred Louise Kerr
Mon 6h44'25" 8d47'14"
Mildred Lucille
Lyn 8h29'57" 48d46'48"
Mildred M. Perkins
Crb 15h58'35" 31d29'29"
Mildred MacMichael
Cap 21h13'16" -17d30'55"
Mildred Mae Bickel
Leo 11h54'42" 19d34'6"
Mildred Manis
Cap 21h6'49" -21d19'30"
Mildred Marie Kelly
Vir 12h45'27" 11d58'48"
Mildred Marie Voelske
Crb 15h42'56" 31d52'45"
Mildred McDowell Broussard
Lib 14h57'50" -9d34'16"
Mildred Merlin
Ori 5h49'40" 20d6'52"
Mildred (Micki) Louise Perschbacher
Cas 1h22'16" 59d0'2"
Mildred "mildieo" Oberkotter
Tau 5h6'5" 24d24'0"
Mildred "Nana" Hutchinson
Cap 20h37'54" -14d43'16"
Mildred O. Rodriguez
Cas 1h10'25" 63d11'41"
Mildred Oslica Star
Her 16h8'29" 49d28'34"
Mildred Palmieri
Vir 14h11'24" -1d57'23"
Mildred Pantaleon
Cap 21h21'51" -20d1'57"
Mildred Pearl Zebrowski
Gem 6h33'53" 22d15'38"
Mildred Regina
Sco 16h8'35" -15d35'4"
Mildred Resendorph Lemal
Cyg 21h41'16" 41d30'42"
Mildred Rivers Harvey
Uma 13h35'3" 55d1'53"
Mildred & Robert Rosenthal
Leo 9h59'24" 16d54'53"
Mildred Rubin Klein
Cas 1h36'24" 61d46'21"
Mildred S. Tanzman
Cap 21h43'20" -8d36'19"
Mildred Sauer
Lyr 18h19'13" 46d15'38"
Mildred Scordo
And 23h0'26" 42d5'27"
Mildred Siegel
Uma 10h56'49" 71d51'43"
Mildred Spears
Ari 2h49'28" 14d24'54"
Mildred U.
Psc 1h9'54" 28d24'25"
Mildred Waguespack Hymel
Lib 15h45'48" -20d21'26"
Mildred Wangerin
Lyn 9h33'43" 40d37'49"
Mildred & Wesley Simpson
Crb 15h41'4" 34d52'36"
Mildred Wilma Gambogi
Tau 4h16'33" 27d15'19"
Mildred Wolfe
Leo 9h27'55" 15d2'21"
Mile Kosanovic
Cru 12h29'39" -55d51'23"
milele na milele
Ori 5h47'51" 6d25'45"
Milena
And 1h16'57" 42d41'20"
Milena
Lib 14h38'48" -17d55'19"
Milena
Lib 14h51'40" -8d20'59"
Milena
Umi 13h47'47" 71d49'41"
Milena
Cas 1h46'7" 60d40'18"
MiLENA DiZENFELD
Vir 13h20'41" -16d54'0"
Milena Geymonat
Ari 3h3'58" 27d25'44"
Milena Hamilton
Ari 2h49'51" 17d8'50"
Milena Ljubica
And 1h53'21" 40d12'48"
Milena Maria "January 30, 1983"
Per 2h18'15" 57d14'32"
Milena Marta
Aqr 23h8'16" -8d16'51"
Milena Mauri
Her 16h37'48" 29d39'26"
Milena Mersini
Tau 4h25'58" 22d23'35"
Milena Petrova Kostova
Ari 2h10'50" 25d4'55"
Milena Prota
Aur 6h21'13" 42d53'49"
Milena Spasojevic
Cap 20h24'3" -9d38'48"
Milena Velinova
Lib 15h29'14" -16d7'1"
Milena Verhousek
Cas 23h58'7" 63d49'44"
Milena09021984
Vir 12h8'48" 6d5'39"
Milene Argo
Gem 7h41'43" 21d33'33"
Mileny R. Crews Rivera
Uma 10h37'48" 59d28'5"
Miles A. Alfieri
Aql 19h36'42" 10d39'47"
Miles and Miles Forever
Cru 12h38'7" -60d8'0"
Miles Anthony Haynes
Gem 6h45'43" 21d17'20"
Miles B. Kehoe
Aql 19h56'23" -0d24'54"
Miles Brandon Graut
Ori 6h4'24" 18d1'53"
Miles Carter Reinhart
Her 17h41'12" 32d36'35"
Miles Casey
Lib 15h48'19" -17d17'12"
Miles David Benson
Vir 12h46'1" 0d16'4"
Miles Durkee
Gem 7h8'30" 15d44'39"
Miles E. Lusiak
Tau 4h30'50" 21d10'34"
Miles Felix Kujawa
Aql 19h15'50" 4d58'57"
Miles Hubley
Umi 15h57'53" 74d15'27"
Miles & Julie Gilman
Cyg 21h42'3" 46d14'12"
Miles K. Ettman
Cep 20h43'13" 60d31'51"
Miles Knight III
Aqr 22h50'34" -6d52'59"
Miles Layne Radde
Sgr 18h48'56" -33d51'58"
Miles Lloyd White - Jan. 15 - 1989
Uma 12h43'22" 60d21'8"
Miles Long
Uma 8h43'15" 61d11'9"
Miles McMyron
Lyn 7h31'31" 49d8'10"
Miles Newsome - SHINING STAR
Per 3h29'27" 45d31'40"
Miles of Mine
Her 17h27'26" 47d31'37"
Miles O'Malley
Aur 5h43'45" 44d26'19"
Miles Silverberg Boehler-Young
Aqr 21h49'2" -2d14'59"
Miles to Koho
Uma 12h45'39" 58d32'21"
Miles Vincent Hill
Vir 12h27'51" -9d8'30"
Miles Wallingford
Umi 15h38'0" 81d26'34"
Miles Zen Ra Harwell
Cnv 13h46'45" 36d3'39"
Miles Zurich Wright
Ori 5h50'37" 6d9'33"
Milestone, On your 40th Anniversary
Cyg 20h2'22" 33d43'39"

Milford & Dorothy Beck Dinner
Uma 10h21'54" 72d13'10"
Milford T. and Ruth E. Wilson
Cyg 21h39'32" 39d42'11"
Milfred Abner Hunt
Uma 13h44'26" 51d38'50"
Mili
Aqr 22h34'35" -16d40'8"
Milica
Vir 14h4'58" -14d52'41"
Milica i Damir - 01.01.2001.
Her 18h48'30" 13d33'22"
Milijana
Cam 7h34'2" 68d17'0"
Milinda Delilah
And 23h25'31" 48d18'0"
Milinda DiCello Long
Cap 21h33'59" -8d44'36"
Milinda Jo Giddy
Uma 10h19'12" 42d46'52"
Military Matt
Crb 15h29'45" 31d42'42"
Militza
Leo 10h11'46" 24d8'19"
Milk & Oreo
Cru 12h28'50" -60d6'30"
Milkdud & Moops Forever
Cyg 19h55'34" 37d50'26"
MILKEE
Sgr 19h14'22" -21d20'28"
Milkellyn
Sge 19h54'36" 17d36'24"
Milkster
Lyn 9h30'39" 41d18'35"
"Milky Monstar" Shannon Alexis
Cyg 20h3'20" 30d47'43"
Milky Rays of The Moon
Gem 6h41'50" 18d34'4"
Milkyway Midnight
Leo 11h5'43" -5d4'46"
Milla
Umi 15h29'7" 73d12'15"
Milla
Lmi 10h16'1" 34d26'16"
Milla Eila Rowe Key - 12.04.2006
Cru 12h34'19" -55d55'48"
Milla Juul Jensen
And 1h50'20" 41d51'55"
Milla Kari Christensen
Cru 12h29'53" -61d10'1"
Milla Kutsin
Uma 11h48'23" 52d5'7"
Millard
Aql 19h41'51" -0d6'19"
Millard L. Strait
Uma 11h44'1" 37d46'53"
Millard Pretzfelder, Jr.
Cnc 8h48'56" 30d51'0"
Millar's Saphire
Cyg 21h45'21" 38d38'26"
Milla's Gift
Crb 15h48'6" 32d14'55"
Milleda
Mon 6h29'59" 6d17'8"
Miller
Uma 11h29'43" 38d38'1"
Miller
Sco 16h10'6" -17d21'47"
Miller
Umi 14h12'26" 68d58'55"
Miller Batcheller
Boo 14h28'10" 52d57'45"
Miller Beekman
Per 3h23'18" 47d42'24"
Miller/Boyle Canine Family
Uma 9h42'39" 66d2'16"
Miller Clyde 2005
Crb 15h36'29" 26d54'42"
Miller Jamie Parker
Umi 15h6'0" 76d46'58"
Miller Light
Ari 2h35'57" 31d3'8"
Miller Time
Uma 11h37'57" 58d48'26"
Miller, Dave Miller
Her 17h51'41" 29d8'28"
Millers Love
Cyg 20h41'2" 37d29'33"
Millertime
Uma 8h57'21" 71d14'5"
Millette's Shining Star
Ori 5h28'46" 3d33'35"
Milli Piyango - SAMPLE
Cep 22h47'2" 65d42'58"
Millicent
Cas 1h10'50" 62d25'25"
Millicent
Gem 7h28'51" 31d52'58"
Millicent A. Douglas
Cyg 21h35'48" 37d55'49"
Millicent and Leonard Chaplinski
Sco 16h6'16" -17d52'21"
Millicent B
Vir 13h12'59" -2d4'44"
Millicent Hurn
Cas 0h29'4" 54d49'13"
Millicent Jane Harris
And 2h21'24" 47d5'21"
Millicent Jean Garrison Casey
Ori 5h40'52" -0d42'40"
Millicent McIntyre
Uma 8h28'32" 69d27'9"
Millicent "Millie" Pyka
Gem 7h34'37" 25d51'52"
Millicent Morse
Leo 10h8'30" 9d12'58"
Millicent Renee Garza
Aqr 21h5'7" -10d44'30"
Millicent-Ryan 6-10-2005
Pho 1h10'17" -43d52'37"
Millicus Umbilicus
Per 3h0'11" 55d49'37"
Millie
Cas 1h23'53" 56d26'37"
Millie
Cas 0h39'18" 59d29'12"
Millie
And 22h59'44" 46d2'12"
Millie
Ori 6h9'29" 2d9'50"
Millie
Aql 19h48'30" 16d21'47"
Millie
Psc 0h0'13" -0d54'7"
Millie
Cas 23h4'43" 53d48'20"
Millie - 19.07.1942
Cas 0h8'36" 59d21'45"
Millie Allinson
Lyr 18h41'36" 30d23'50"
MILLIE ANN MARTIN
Peg 21h46'58" 13d16'52"
Millie Anne Butler - Millie's Star
And 0h32'24" 46d27'47"
Millie Aponte
Cnc 8h56'16" 32d9'59"
Millie Beach
Cyg 20h55'43" 46d24'13"
Millie Beau
Umi 14h54'31" 71d23'25"
Millie & Bo Tiehes
Uma 11h4'3" 49d49'17"
Millie Caradine Franklin
And 23h51'14" 41d59'22"
Millie Christine Court
Cyg 19h48'35" 58d58'4"
Millie Eisenhardt
Gem 6h31'50" 24d32'14"
Millie Ellison 1
And 22h59'34" 52d10'19"
Millie Eloise Tyler
And 22h57'36" 51d13'57"
Millie F. Buxton
Gem 7h21'6" 28d34'47"
Millie Fluff
Umi 16h55'7" 78d4'24"
Millie Grace Latham
And 2h28'15" 50d2'42"
Millie Grigor
Cyg 20h26'23" 49d43'55"
Millie Jae Woods
And 1h37'31" 40d22'44"
Millie Jo Vass - Millie's Star
And 22h59'4" 51d4'27"
Millie June Taylor
Gem 7h3'32" 27d29'32"
Millie Lam - 27.03.82
And 1h22'3" 47d24'37"
Millie - Leo Minor
Lmi 10h58'7" 26d35'40"
Millie Louise Hardman
And 2h28'25" 41d4'38"
Millie Margarita Cudworth
Vir 13h36'50" -6d7'6"
Millie McLane
Cyg 21h39'33" 41d15'19"
Millie McNorton
And 23h15'41" 52d10'8"
Millie Milnarevich
Tau 3h59'15" 7d13'16"
Millie Palmer
Tau 3h45'34" 28d50'6"
Millie Parady Mathews
Com 12h40'59" 17d20'38"
Millie Parker
Ari 3h13'25" 11d0'48"
Millie Pavek
Del 20h39'7" 15d29'56"
Millie "Pudge" LaMay: Our Pillar
Uma 11h36'4" 46d58'33"
Millie Sack
Sco 16h55'51" -31d42'57"
Millie Simcox
And 0h51'24" 35d52'27"
Millie Star
Sgr 18h19'10" -32d21'38"
Millie Thompson
And 23h24'52" 47d10'55"
Millie Valentin
Sco 17h29'41" -37d50'13"
Millie Vanillie
Uma 11h16'43" 29d22'34"
Millie Widberg Robson
And 0h26'16" 43d1'10"
Millie Wilson
Mon 7h35'16" -0d29'6"
Millie & Wyman Bankston
Umi 15h28'36" 76d7'22"
Millie-Mae Coco
Psc 0h15'53" 10d18'53"
Millie's Magic Star
Del 20h43'51" 17d7'24"
Millie's Star
Cru 12h27'34" -60d9'20"
Milliken
Uma 13h44'30" 52d48'46"
Million Athena Taylor - Millie's Star
And 23h48'45" 37d51'59"
Millon and Dillon
Umi 15h24'23" 83d56'17"
Mills Albert George Fluker
And 0h3'57" 44d10'28"
Mills Ryan Arnold
Lib 14h51'31" -11d0'0"
Millward Brown
Umi 16h45'0" 78d9'1"
Millward Brown North America
Umi 14h40'44" 79d7'34"
Milly
And 23h52'37" 42d43'4"
Milly
Del 20h30'28" 7d19'33"
Milly Carlucci
Peg 21h58'27" 25d26'54"
Milly Loizides
And 0h27'33" 39d5'29"
Milly Lou
Lmi 9h50'9" 37d21'39"
Milly Maureen McGuigan
And 0h17'37" 25d0'5"
Milly Morrison
Cas 1h30'18" 67d34'46"
Milly Thompson's Star
Equ 21h7'31" 7d27'31"
Milly Tinsley
And 22h59'17" 52d26'29"
Milly Watmore
And 23h19'52" 50d2'48"
Millynolly
Cap 20h16'59" -13d9'22"
Millz
Lib 14h51'44" -3d29'17"
Milman Fania
Uma 8h46'38" 62d30'58"
milnat Kranitz
Leo 11h29'46" 15d27'29"
Milne Cecilia Crean
Ari 2h52'27" 29d20'48"
Milne Star
Her 17h16'51" 48d20'49"
Milo
Per 3h5'28" 35d49'18"
Milo
Per 4h48'25" 43d40'15"
Milo
Tau 5h8'8" 26d20'47"
Milo
Aqr 22h1'34" 2d4'16"
Milo
Cap 20h56'8" -23d3'52"
Milo 2006
Per 2h51'38" 51d0'19"
'Milo' - Celebration of Love
Cru 12h1'51" -62d49'13"
Milo Dane
Umi 15h4'20" 89d5'13"
Milo Harrison McGuffey
Umi 15h0'1" 71d6'34"
Milo J
Her 18h53'39" 24d14'51"
Milo loves Bisto
Per 4h11'6" 35d18'24"
Milo Lyon
Lmi 10h42'14" 27d37'7"
Milo Marius Sean Mehring 22.03.1996
Uma 11h1'45" 33d44'50"
Milo Rhodes Nochera
Leo 11h37'59" 22d17'15"
Milo Stringfield Joseph
Ari 3h6'41" 22d55'33"
Milo Williams
Uma 12h1'26" 65d46'47"
Milorad Mike Maksimovic
Ari 3h7'12" 13d11'48"
Milos K
Peg 23h59'29" 11d0'53"
Milo's Love
Ori 6h9'33" 18d30'9"
MilouCyrilou
Umi 14h11'26" 75d58'23"
Milovan Cvijovic
Per 2h57'14" 52d45'2"
Milski
Cap 21h18'12" -19d25'44"
Miltiadis Varvitsiotis
Uma 12h36'23" 58d4'8"
Milton
Cap 20h45'8" -19d13'8"
Milton A. Spiro
Uma 9h51'58" 61d31'9"
Milton and Margie Camps
Gem 6h45'23" 14d38'14"
Milton & Audrey Ezzo
Vir 13h50'59" -16d10'35"
Milton Barry Price
Her 16h40'12" 14d12'45"
Milton & Brenda Whitehead
Cyg 21h15'36" 45d34'27"
Milton Chamberlain, Jr.
Aqr 22h36'46" -1d20'39"
Milton & Helen Gough
Cnc 9h3'17" 10d8'53"
Milton Holt
Leo 11h37'55" 19d31'31"
Milton & Janet Collins
Aql 19h12'24" -0d49'41"
Milton "Junior" Hunter, Jr
Tau 5h34'20" 25d48'8"
Milton Kindrick
Ori 5h40'4" 3d20'48"
Milton L. Havens, Jr.
Aqr 22h19'20" -3d51'0"
Milton Mutts Adolphson
Ori 5h22'43" 13d10'33"
Milton Pond
Cep 21h53'39" 63d13'32"
Milton Ray Foote
Uma 13h23'49" 54d28'28"
Milton Ronald Grahl Sr.
Cep 20h43'4" 64d16'28"
Milton Spector
Cyg 19h43'24" 38d24'41"
Milton T. Huston's Star
Uma 10h5'43" 45d7'26"
Milton T. Kyle, Jr. & Lana L. Kyle
Tri 1h54'58" 29d47'37"
Miltonia Sabine
Uma 8h30'28" 62d48'6"
Milton's Soul Mate
Lyn 8h13'15" 34d58'4"
Miltos Maroulidis
Uma 10h46'10" 40d18'14"
Miltz, Sven
Aqr 20h38'35" -2d11'55"
Milu
Cyg 19h57'15" 33d34'53"
MiLu
Peg 22h24'6" 19d25'17"
Miluji Te
Sge 19h55'55" 16d48'21"
Miluna
Cnc 8h44'28" 16d21'11"
Miluv
Ori 5h52'49" 9d38'33"
Milwaukee Electric Tool Corporation
Uma 9h29'14" 71d10'54"
Milwyn
Cyg 20h12'21" 44d8'24"
Milya's Magical Star
Umi 16h44'46" 89d10'14"
mim Sunneschiin
Lmi 10h6'31" 32d19'6"
MIM5387
Lyn 8h59'2" 33d21'6"
Mima
Gem 7h18'8" 35d3'31"
Mimakos
Sgr 18h6'11" -27d28'18"
Mimama
Dra 16h46'34" 55d36'15"
MiMar 8
Lyn 8h44'13" 40d32'6"
MIMARI
Tau 5h46'17" 25d8'42"
MiMeliRigo
Tau 3h29'8" 17d5'52"
Mimi
Ari 2h53'59" 21d59'0"
Mimi
Tau 5h23'12" 17d17'5"
Mimi
And 2h5'41" 44d47'38"
Mimi
And 0h29'23" 38d27'34"
Mimi
Cnc 8h43'27" 31d33'51"
Mimi
Crb 15h57'41" 32d33'50"
Mimi
And 0h36'39" 45d54'0"
Mimi
Cyg 20h25'24" 48d3'26"
MIMI
Cyg 19h59'56" 46d15'44"
Mimi
Cas 0h26'39" 61d38'0"
Mimi
Uma 11h25'53" 59d25'11"
Mimi
Uma 10h52'46" 64d57'24"
MiMi
Mon 6h44'49" -0d3'53"
Mimi
Aqr 22h46'47" -4d7'47"
Mimi
Aqr 22h53'57" -16d32'23"
MiMi
Aqr 22h55'55" -19d42'46"
Mimi
Ari 2h8'9" 22d36'40"
Mimi 58
Lyn 8h38'49" 39d52'26"
Mimi 75
Cnc 9h1'25" 30d21'17"
Mimi 8
Srp 18h26'23" -0d2'12"
Mimi and Geordie's Star
And 1h40'35" 49d18'56"
Mimi and Lilly's Fire
And 0h2'22" 45d0'4"
Mimi and Martin Lytje's star
Her 16h43'25" 42d10'58"
Mimi Ashkar
And 23h57'50" 45d26'25"
MIMI BUI
Sgr 18h37'33" -23d18'52"
Mimi Colcord
Tau 4h36'11" 28d12'1"
Mimi Costa (sunshine)
Crb 15h51'40" 37d36'22"
Mimi & Dana Forever
Sco 17h53'22" -35d47'49"
Mimi & Didi
Cnc 9h5'0" 14d32'4"
MIMI ECKER
Cnc 8h46'43" 16d17'5"
Mimi Fields
Sgr 19h38'54" -14d46'32"
Mimi Friedfeld
Per 3h44'2" 38d14'21"
Mimi Glass
And 2h18'53" 44d37'31"
Mimi Manahan
Vir 11h48'28" 5d6'29"
Mimi Marcal
Vir 13h25'2" 8d6'51"
Mimi Marie Mirriam McNamara
Sco 16h9'7" -15d2'39"
Mimi McConnell
Lib 15h45'50" -5d43'29"
Mimi Mitzie Bessie Pauline Stone
Psc 1h46'16" 7d37'30"
Mimi Morse
Lmi 10h5'7" 32d28'47"
Mimi Palladino
And 23h14'20" 48d49'15"
Mimi & Papa
Lyr 19h19'26" 28d30'9"
Mimi&Papa_50yrs6kds11gr ndkds_Boola2
Per 3h56'52" 38d41'44"
Mimi Pedersen
Sco 16h55'9" -34d7'50"
Mimi Rogers
Lyr 19h16'30" 29d19'57"
Mimi Ruh
And 0h51'32" 36d51'47"
Mimi S. Hitchcock UK 9th Birthday Star
Psc 1h36'7" 22d56'48"
Mimi 's Star
Mon 7h48'0" -0d29'45"
Mimi Siu
Mon 6h45'50" -0d18'55"
Mi-Mi... the love of our lives!
Vir 13h21'25" 12d10'51"
Mimi Winstanley
Umi 14h57'19" 71d23'9"
Mimi Young
Uma 9h41'43" 60d43'23"
Mimi, Leanne, Laurel
Psc 1h20'0" 6d59'29"
Mimi-Claudette
Cas 0h36'39" 58d30'27"
Mimie
Umi 16h30'3" 75d50'25"
mimilachance
Umi 19h29'8" 86d29'22"
MimiMartin Rosenberg 60years bliss
Uma 11h26'22" 51d52'4"
Mimi-Mirando
Psc 1h6'23" 5d27'34"
Mimipat
Cyg 20h56'4" 39d14'24"
Mimi's Island
Cas 0h55'21" 56d29'19"
Mimi's Perpetual Patience
Psc 0h39'21" 3d15'42"
Mimi's Star
Sco 16h11'55" -17d40'0"
Mimi's Star
And 23h48'2" 38d10'45"
Mimi's Star
Per 3h23'3" 47d41'53"
MiMi's Star
Gem 6h38'43" 24d31'20"
Mimi's Twinkle
Uma 10h52'0" 44d26'10"
Mimi's Twinkle
Umi 14h16'42" 70d17'55"
Mimistar
Psc 1h20'9" 4d53'30"
MimiTay
Ori 5h12'53" 6d15'50"
Mimma & Mark
Lmi 10h4'35" 40d58'20"
Mimmi
Per 4h2'6" 42d33'7"
"Mimmie" James G. Stockwell
Uma 11h25'41" 63d55'11"
Mimo
And 22h58'8" 42d8'50"
MiMorena
Vir 12h1'3" 1d44'59"
Mim's Place in Heaven
Ori 5h30'32" 6d21'26"
MimsStar You light up my life 15/02
Aqr 21h24'3" -9d34'27"
Mimy
Cas 0h21'27" 47d28'23"
Min & Andrew
Cyg 20h21'41" 52d11'15"
min chline Bär-Markus Siegenthaler
Uma 11h5'48" 47d9'12"
Min Costas
And 23h12'41" 48d3'53"
Min Hee Nam
Psc 1h23'28" 16d14'12"
Min Ho Yoon
And 23h18'13" 52d23'8"
Min Jeong Kwak
Psc 1h4'56" 19d6'36"
Min Lee
Ari 2h1'9" 23d19'22"
Min Moder Stjarna
Cnc 8h50'45" 16d0'35"
Min Stern
Peg 23h23'17" 32d2'16"
min stern..
Ori 5h40'25" -3d36'20"
Min Stern Iwan
Cas 23h1'56" 57d55'14"
Min Un
Gem 6h46'11" 16d38'19"
Mina
Gem 7h29'53" 18d56'15"
Mina
Cnc 8h33'46" 23d40'24"
Mina
Ori 6h5'36" 16d13'15"
Mina
Cam 4h59'47" 56d2'34"
Mina
Ser 18h18'42" -13d52'8"
Mina
Cap 20h19'36" -15d51'56"
Mina Defazio
Uma 10h21'25" 44d8'15"
Mina Hoshina
Psc 0h17'59" 16d40'22"
"mina" kudler-flam
Vir 12h58'15" 2d36'29"
Mina Marie
Aqr 20h39'2" -9d11'50"
Mina Mitidieri
Ara 17h25'0" -47d41'59"
Minä Papi Ruedi
Lyr 18h38'22" 41d16'1"
Mina Queen
Her 16h45'50" 9d5'42"
Mina Sovtic Belgrade 03.08.1995.
And 23h0'0" 37d25'5"
Mina Thabet
Sco 16h55'18" -42d19'27"
Mina & Willie B.
Cyg 20h47'29" 32d44'29"
Minaki Mou
Cnc 8h46'1" 13d1'28"
Minal
Lib 14h23'2" -9d31'56"
Minal Desai
Aql 18h54'24" 8d29'35"
MinaLer
Gem 6h47'45" 27d44'26"
Minanoa
Cas 0h20'21" 51d55'57"
Minari Nicoletta
Cam 3h19'58" 61d48'0"
Minas Tanes
Uma 10h38'30" 64d5'29"
Minas Tarakhchyan
Tau 3h48'13" 5d2'56"
Minavae Arwen
Sco 17h15'9" -32d44'28"
Minaz
Per 2h39'31" 50d38'2"
Mind Spring Laurie Jane
Ari 2h25'48" 27d35'12"
Mind, Body, Soul
Crb 15h50'21" 35d34'50"
Minda Golez
Lib 15h7'6" -20d55'32"
Minda N Jamilo
Cir 15h1'12" -65d28'25"
Mindala
Tau 4h28'8" 5d4'46"
Mindi
Cyg 21h26'36" 33d58'19"
Mindi
And 2h24'43" 45d55'1"
Mindi Dianne Keese
And 2h22'30" 50d42'17"
Mindi Hernandez
Psc 1h1'34" 24d33'43"
Mindi's Star
Sgr 18h13'21" -20d1'7"
Mindi-Sue
Dra 15h17'37" 55d12'40"
Mindseye
Cru 12h9'47" -62d55'37"
Mindy
Pup 6h33'25" -45d18'41"
Mindy
Lyn 7h6'33" 54d16'32"
Mindy
Ori 5h40'31" -0d34'56"
Mindy
Vir 14h33'12" -6d26'37"
Mindy
Lep 5h18'40" -13d18'52"
Mindy
Gem 7h18'49" 19d44'9"
Mindy
Gem 6h52'42" 16d45'28"
Mindy
Ari 2h13'43" 22d23'20"
Mindy
And 1h57'47" 45d53'8"
Mindy
And 1h5'22" 45d13'8"
Mindy
Gem 7h40'34" 32d9'13"
Mindy A
And 23h19'35" 48d51'30"
Mindy Alexis Le Page
Sco 16h14'36" -9d54'25"
Mindy Allison
Cnc 8h44'19" 18d18'14"
MINDY AMIYA LAYLA LEWIS
And 23h22'46" 47d40'32"
Mindy and Chris For Ever
Cyg 21h11'42" 52d44'48"
Mindy Angel
Lyr 18h40'34" 36d35'51"
Mindy Anne Martin
And 1h40'13" 42d16'7"
Mindy Dolan 240180
Peg 22h29'32" 8d11'38"
Mindy Donatelli
Ori 6h18'59" 14d9'49"
Mindy & Emrys
Cyg 20h15'13" 55d35'1"
Mindy Erica Durkee
Cnc 8h15'4" 23d36'4"
Mindy G. Villella
Psc 1h11'5" 10d58'12"
Mindy Gail Harris - 5/19/1986
Umi 14h32'55" 71d24'10"
Mindy Hamman
Cam 6h2'56" 69d58'16"
Mindy Joan McCoy
Vir 14h3'13" -19d20'6"
Mindy & Jon Peters
Per 3h50'47" 43d8'28"
Mindy K
Lyn 7h54'35" 42d27'51"
Mindy K
Lyr 18h40'17" 30d57'5"
Mindy Kay Gray
Sgr 18h54'10" -22d33'25"
Mindy Kay Slubar
Leo 9h27'16" 25d51'52"
Mindy Knight
Psc 1h17'36" 15d33'20"
Mindy Leeds Zankel
Ari 3h16'22" 20d30'57"
Mindy (LOU) Beckwith
And 0h38'0" 37d41'50"
Mindy Lynn Hodge
Leo 11h56'44" 25d4'53"
Mindy Lynne
Ari 2h45'43" 25d50'5"
Mindy Mantz
And 1h37'40" 46d15'38"
Mindy Marie
And 1h51'54" 46d29'34"
Mindy Marie Miller
Psc 1h21'27" 24d49'40"
Mindy Mazie
Cyg 21h53'55" 47d22'27"
Mindy McStroul
Sgr 20h11'31" -37d33'42"
Mindy Michele Martin VanVeldhuizen
Uma 8h52'33" 48d25'10"
Mindy & Nathan - 15 April 2006
Cru 12h21'44" -58d33'37"
Mindy Nicole
Lyn 7h42'10" 37d3'55"
Mindy Nicole Alford
Sco 16h57'8" -43d22'32"
Mindy our Beautiful
And 0h45'52" 36d43'27"
Mindy Paz - Shining Star
And 1h24'5" 46d41'47"
Mindy Rager
And 1h13'29" 36d37'0"
Mindy Rivas
Sgr 18h37'33" -30d45'39"
Mindy Robin Zajchowski
Sco 16h33'46" -31d16'0"
Mindy Rose Noel
Cyg 21h23'27" 50d52'31"
Mindy Shelton
Crb 15h23'14" 26d27'1"
Mindy Smith
Leo 9h57'46" 31d8'0"
Mindy Starr
Psc 1h5'51" 18d44'26"
Mindy Sue Morrow
Leo 11h19'9" 17d19'56"
Mindy Thomas
Cnc 8h38'20" 13d58'54"
Mindy Timmerberg
Lyn 8h25'44" 57d4'12"
Mindy will you marry me?
Crb 15h41'3" 27d14'29"
Mindy, God's Perfect Mom.
Lyn 7h52'26" 58d31'48"
mindyandgreg9-8-2004
Psc 0h24'24" 3d7'18"
MindyE
And 0h19'12" 33d6'14"
MindyJean Bruehl
Ori 6h12'22" 20d53'8"

* Mindy's Star *
Leo 11h15'58" 16d6'18"
Mindy's Star
Uma 10h57'23" 67d14'15"
Mine
Ori 5h41'27" 4d30'4"
Mine Ashley
Cyg 21h33'30" 51d58'14"
"Mine" - Sunny Patel
Cnv 12h58'0" 48d24'36"
MINERVA
Lac 22h21'11" 48d22'46"
Minerva
Leo 9h30'37" 14d47'15"
Minerva Agiss Andres
Vir 12h38'31" 3d16'28"
Minerva Alejandra Olivera
Dra 18h59'31" 51d35'34"
Minerva Kristen Mohabir
Cas 0h3'9" 53d20'55"
Minerva Medica Amy Doneen
Vir 12h34'54" 0d56'2"
Minerva Mercedes Miranda
Uma 8h23'35" 60d31'56"
Minerva Velazquez Vives
Eri 4h17'32" -6d36'53"
Mines Star II
Umi 15h54'57" 72d20'53"
Mines Stern
Uma 8h44'20" 48d51'12"
Minesh Anbreen
Cyg 19h40'41" 28d33'23"
Minette Gomez
Per 3h27'1" 31d59'23"
Miney
Uma 11h42'14" 39d30'20"
Mineyuki & Miyuki 10.10
Lib 15h25'40" -14d6'56"
Mingy
Sco 16h50'37" -42d23'47"
Mingyawna Hall Satterwhite
Cyg 20h48'37" 34d42'16"
Ming-Yi Hwang
Leo 9h22'15" 14d46'44"
Minh Chau Hai Long
Cap 20h28'16" -13d42'51"
Minh Trang Lee
Her 18h2'51" 25d19'54"
Minh's Legacy
Sgr 19h27'48" -15d41'6"
Mini
Umi 14h43'57" 68d5'37"
Mini
Cyg 20h5'37" 42d23'3"
Mini <3's PUNK
Ari 1h54'40" 17d56'56"
Mini C
Umi 14h13'54" 76d49'39"
Mini Chicken Amigo
Sgr 18h35'48" -23d40'57"
Mini Koso
Leo 11h32'44" 7d48'42"
Mini Mac and Moe
Lib 15h6'2" -11d46'19"
Mini Me
Sgr 19h52'20" -12d54'19"
Mini Meng
Lib 15h11'9" -20d59'33"
Mini Muffin
Ori 6h18'9" 14d41'37"
Mini Sole
Umi 15h24'3" 71d47'5"
Mini Strandberg 95
Ori 5h13'57" 0d13'37"
Mini Val
Umi 14h59'14" 68d24'42"
MINIKHW338
Uma 9h23'13" 64d53'4"
Minina
Uma 11h37'26" 52d51'56"
Minipuri Ramesh Singh
Peg 22h54'45" 20d9'27"
Miniqper1
Cam 3h58'14" 63d7'44"
Mini's Star
Vir 12h40'38" 2d31'41"
Miniuk
Ori 5h33'59" 10d43'14"
Minix
Del 20h36'8" 19d36'10"
MinJae
Lib 15h21'9" -9d49'18"
Min-Jeong Shin
Uma 13h33'54" 58d32'56"
Minkey
Aqr 22h32'12" -16d11'51"
Minky
Ari 2h42'21" 20d53'55"
Min-Kyong
Sco 16h45'14" -31d56'51"
Minn
Umi 14h4'53" 66d39'19"
Minna Caputo Pringle
Uma 10h13'5" 63d51'1"
Minna Elina Lehtinen
And 0h19'36" 45d9'50"
Minna Hartmayer
And 2h18'27" 49d43'35"
Minna Star
Tau 5h28'12" 25d58'11"
Minna Wayne
Vir 14h32'21" 3d46'26"
Minnah
Cas 0h25'55" 60d51'59"
Minnelli
Psc 0h30'40" 7d28'38"
Minnie
Cas 23h29'36" 53d28'24"
Minnie
Uma 10h53'30" 69d15'55"
Minnie
Cnc 8h39'25" 17d10'11"
Minnie Alice
Lyr 18h59'11" 26d0'47"
Minnie Alice Wagoner
Cap 21h39'34" -8d43'21"
Minnie Cesta "Happy 75th"
Psc 1h8'26" 17d56'37"
Minnie Davies
Cas 23h44'23" 57d17'9"
Minnie F. LaVigne
Vir 13h14'55" -4d14'57"
Minnie & Mervin
Lyn 6h33'2" 54d35'47"
Minnie & Mickey
Crb 16h23'1" 28d41'53"
Minnie Nova
Tau 3h44'13" 24d37'26"
Minnie P. Sansteby
Sco 16h53'38" -44d0'11"
Minnie Schaeffer
Peg 22h59'55" 21d58'54"
Minnie & Tony Morales
Cnv 12h49'13" 39d15'32"
Minnie Wilhelmina
Cas 1h0'55" 59d15'5"
Mino
Uma 9h44'36" 42d29'47"
Mino
Ori 5h24'32" 2d0'20"
Mino
Uma 11h15'24" 62d20'41"
Minor Alvin
Aqr 22h44'23" -15d48'33"
Minor Galactic
Umi 14h39'53" 75d45'4"
Minos
Her 18h35'16" 22d16'26"
Minstral Andy Lear
Lyr 19h19'43" 28d7'27"
Minty Clapp
Cas 1h14'10" 49d56'14"
Mintzer-Altschuller "MA" Star
Lyr 18h16'52" 39d11'31"
Minx
And 0h15'53" 45d17'11"
Minxy Becky-Sue
Pyx 8h30'32" -28d49'57"
Mio bello piccola
Ori 6h7'31" 16d17'40"
mio cuore, mia vita, mia madre
Cas 1h21'42" 53d46'41"
Mio Destino
Cas 0h56'51" 53d15'7"
Mio Gino
And 0h23'56" 28d21'53"
Mio respiro, cuore ed anima!
Cyg 19h49'8" 30d6'49"
mio tutto
Gem 6h4'9" 27d53'21"
Mio Tyann Dolce
Umi 15h19'9" 71d38'13"
Miodrag Todorovich
Sco 16h11'42" -16d31'26"
Miossotti
Uma 14h3'34" 51d18'42"
MippySueStar
Ari 2h59'11" 10d42'33"
Miqqy & Kiran
Umi 16h18'8" 77d10'27"
Miquel
Her 16h11'22" 25d14'25"
Miquel
Cnc 8h10'3" 15d22'59"
Miquel A. Lattimer
Her 16h59'19" 16d51'53"
Miquel Cruz
Her 18h47'10" 19d3'57"
Miquel Gonfaus
Ari 3h27'49" 26d47'32"
Mir Ali Pourrahmat
Cep 20h55'57" 65d31'23"
Mir zwei gäge de rest vo de wält
Sge 19h57'27" 16d55'13"
Mira
Leo 11h3'53" 6d35'27"
MIRA
Cnc 9h15'48" 12d54'4"
Mira
Umi 15h39'37" 71d3'38"
Mira
Uma 12h19'4" 56d27'19"
Mira
Cas 23h26'49" 53d31'55"
Mira
Dra 17h49'48" 53d48'33"
Mira Adanja - Polak
Her 16h34'37" 18d5'58"
Mira Bush
Tau 3h59'26" 9d11'47"
Mira Harfouche & Remy Costantine
Uma 11h31'56" 39d35'19"
Mira Lisa
Aqr 20h45'1" -13d44'42"
Mira Marchio
Ari 2h25'29" 24d13'20"
Mira Parekh
Cap 20h28'3" -13d38'13"
Mira Stanislawski
Uma 13h17'30" 53d16'43"
Mirabella Lee
Peg 21h39'32" 27d0'22"
Mirabile Adinfinitum
Cnc 8h29'43" 22d36'19"
MiraCarol
Lyr 18h54'50" 35d27'29"
Miracle
Cyg 21h11'35" 43d52'14"
Miracle
Psc 1h7'13" 22d51'11"
Miracle
Cnc 8h53'22" 17d58'2"
Miracle
Uma 9h38'59" 57d12'24"
Miracle Matt McFadden
Her 17h25'4" 24d20'32"
Miracle McClintock
Umi 16h12'11" 78d12'47"
Miracle out of Lancelot
Her 16h6'28" 46d9'6"
Miracle Star of Margaux Buttula
Tau 4h25'32" 21d45'42"
Miracles~Jonathan & Elliott James
Ori 5h46'17" 8d44'3"
Miracles Precious One Hunt
Lyr 18h34'8" 32d20'46"
MiraculeWater, Inc
Lyn 7h56'57" 48d20'1"
Miraculous
Her 18h41'16" 18d53'37"
Miraculous-Lee
Leo 9h47'45" 31d28'44"
Miraculum
Aqr 21h41'11" -0d57'28"
Mirage Morosi-Khalifa
Umi 14h32'42" 89d21'48"
Mirage-Arianna Wynter Forrest
Peg 22h20'36" 25d13'59"
Mira-Jessica
Uma 11h42'41" 29d45'42"
Miramar Diamond
Cyg 19h45'2" 29d22'38"
Miran
Lib 15h27'1" -6d15'54"
Miranda
Vir 12h24'35" -3d18'55"
Miranda
Cap 21h32'16" -13d53'45"
Miranda
Cas 2h26'33" 66d6'9"
Miranda
Cas 1h41'50" 67d3'40"
Miranda
Aqr 22h34'29" 1d44'38"
Miranda
Lyr 18h53'38" 32d7'15"
Miranda
Cyg 19h42'32" 31d55'52"
Miranda
Lyn 7h37'10" 42d52'12"
Miranda
Lyn 7h48'18" 49d39'47"
Miranda
And 2h36'13" 49d5'11"
Miranda
Lyr 18h41'24" 41d12'43"
Miranda
Uma 10h10'12" 45d23'46"
Miranda Amae Pomeroy
Cnc 8h46'59" 29d18'46"
Miranda And Fraser's Star
Sco 17h24'25" -45d31'38"
Miranda Ann Van Atta
Lyn 7h10'40" 57d28'8"
Miranda Baby Miranda
Leo 11h17'6" 5d33'23"
Miranda Baugh
Vir 13h19'50" 2d57'32"
Miranda Cabarga
Lib 15h42'37" -19d58'9"
Miranda Catherine Williams
Peg 22h54'0" 31d6'33"
Miranda C.J Roberts
Lyn 7h14'50" 59d38'46"
Miranda Collette Elam's 1st star
And 1h6'59" 43d7'44"
Miranda Corlieto
Lyr 18h43'28" 39d28'38"
Miranda Cox "Yasmeena"
And 23h34'22" 41d16'20"
Miranda Danielle Grider's Star
Ari 3h12'52" 15d19'53"
Miranda Diane Blauvelt
And 23h46'19" 46d53'6"
Miranda DiMare
Uma 8h41'2" 47d19'22"
Miranda Dorothy Barker
Crb 16h0'57" 27d12'58"
Miranda E. Fegan
Vir 13h14'3" 11d14'35"
Miranda Faith Kunda "1998-2006"
Sco 16h59'15" -40d26'57"
Miranda Grace Our Shining Star
Sgr 17h56'3" -17d57'0"
Miranda Gwendolyn
Crb 16h6'20" 39d3'51"
Miranda Hirsch
Cap 21h53'40" -21d29'7"
Miranda James
Uma 13h19'54" 57d4'31"
Miranda Jane
Cyg 19h59'31" 31d2'43"
Miranda Jane
Tau 5h49'35" 13d59'59"
Miranda Jane Albers
Lib 15h13'0" -20d44'20"
Miranda Jane Penistan
Lib 14h24'37" -10d31'25"
Miranda K. Kilgore
And 0h24'35" 44d19'13"
Miranda Kee
Cam 3h59'57" 58d31'40"
Miranda & Kevin 04-28-01
Lyn 8h53'20" 36d24'12"
Miranda L. O'Keefe
Dra 14h29'24" 63d44'9"
Miranda Langan
Lib 15h2'37" -24d42'19"
Miranda Langdon
Mon 6h36'30" -4d10'41"
Miranda Lay
And 23h47'19" 44d33'31"
Miranda Lee
Uma 11h13'16" 28d25'23"
Miranda Lee Stutler
Dra 18h18'45" 76d0'24"
Miranda Leigh Sterne
And 2h16'24" 47d24'3"
Miranda Lyn
Vir 11h39'52" -0d52'13"
Miranda Lynn
Gem 6h33'26" 24d36'11"
Miranda Lynn Compton
Tau 4h34'6" 23d50'54"
Miranda Lynn Johnson
Cyg 21h47'13" 47d26'12"
Miranda Lynn Navarro
Aqr 23h1'24" -12d52'19"
Miranda Marie Leavitt
Lib 15h39'41" -23d44'5"
Miranda Marie
Aql 19h20'49" 6d35'26"
Miranda Marie Cole
Uma 9h5'28" 57d5'45"
Miranda Marie Grice
Crb 15h47'52" 34d50'12"
Miranda Marie Michael
And 2h17'29" 46d43'39"
Miranda Michelle
Com 13h9'36" 28d54'11"
Miranda Mudrich
Lyn 8h21'39" 45d59'43"
Miranda. my sunshine bar-biedoll.
Gem 7h29'32" 16d0'44"
Miranda Nicole Easton
Ori 6h1'9" 10d53'22"
Miranda Panda
And 23h40'20" 39d49'9"
Miranda Phan
Lyr 19h14'39" 34d32'26"
Miranda Phillips
Lib 15h8'39" -20d51'21"
Miranda R. + Keaton T. = Love
Cap 21h40'15" -14d13'53"
Miranda Rhoda Booth
Umi 9h29'21" 86d21'32"
Miranda Ricks
And 0h46'48" 36d24'27"
Miranda Roo Haramia
And 23h22'12" 43d2'30"
Miranda Rose
Leo 9h55'18" 31d55'22"
Miranda Sagastume
Leo 11h0'2" 9d7'48"
Miranda Smith
And 23h47'59" 34d14'33"
Miranda Steed
Ori 5h57'51" -0d56'25"
Miranda Sue
Sgr 19h46'41" -23d31'39"
Miranda Super Star
And 23h50'27" 40d14'33"
Miranda Taylor
Uma 10h35'20" 39d21'22"
Miranda Theresa
Lib 14h34'16" -20d51'4"
Miranda Turner
And 2h6'57" 41d29'4"
Miranda Winchell
Cas 23h58'8" 57d38'42"
Miranda Wohlers
Ori 5h59'40" 18d23'58"
Miranda, la niña de mis sueños
Cnc 8h14'17" 9d41'24"
MirandaS
Ari 3h23'17" 28d9'8"
Miranda's Dream
Lyn 7h12'15" 59d7'36"
Mirandia's Dream
Crb 16h8'18" 32d58'20"
Mirawen
Ari 3h22'48" 28d13'4"
Mirawr
Uma 8h21'1" 64d44'42"
MIRAX.FIMH/HHOOO3113
Cyg 21h13'7" 46d13'51"
Miray Serra Kaya
Lib 15h19'49" -7d27'4"
Mirc
Lib 14h42'19" -16d54'20"
Mirco Fiaschi
Peg 22h36'46" 27d29'2"
Mirco Lassandro 4-ever
Cas 1h24'8" 56d14'19"
Mirco Meinel
Uma 9h33'26" 51d57'57"
Mirco und Marielle
Lmi 10h15'23" 40d50'33"
Mire i Nil
Ari 2h15'42" 20d45'57"
Mireia Dolçeta
And 2h13'24" 37d33'47"
Mireille
Cyg 21h38'9" 34d19'45"
Mireille
Uma 9h30'24" 61d20'34"
Mireille & Alain
Dra 12h20'10" 73d28'55"
Mireille Forever
Ori 5h54'47" 6d52'24"
Mireille JeBailey
Cyg 21h41'6" 43d54'4"
Mireille Matheron et Léo Asfez
Lyn 7h42'47" 52d53'11"
Mireille Piel
Psc 0h46'6" 15d18'34"
Mireille & Toby Forever
Cra 18h7'20" -38d32'48"
Mireille Verne
Crb 15h53'2" 34d13'54"
Mireille's True Love
Aqr 23h55'13" -17d59'20"
Mirek
Ari 3h28'30" 22d21'51"
Mirel Carolyn
Tau 4h28'31" 12d18'12"
Mirela
Cap 21h34'0" -18d16'41"
Mirelis M. Negron Rentas
Sgr 19h43'51" -24d43'4"
Mirella
Vir 13h15'58" -22d34'9"
Mirella
Ori 5h37'41" -0d5'56"
mirella
Her 17h54'32" 18d33'42"
Mirella
Cyg 21h14'37" 51d41'7"
Mirella e Massimiliano
Her 17h11'33" 49d27'56"
Mirella e Silvano
Peg 23h54'53" 22d22'54"
Mirella the Kitty Princess
Lyn 8h32'2" 55d30'24"
Mirelle Isabella
Gem 6h57'25" 26d6'54"
Miren80
Col 6h36'51" -42d7'37"
Mirenko
Leo 11h27'37" 10d0'13"
Mireya Hernandez
Crb 15h55'41" 34d57'27"
Miri e Gustavo
Peg 22h42'36" 3d41'20"
Miria Lia
Uma 13h58'24" 56d56'50"
Miriah Savannah Charbonneau
And 0h12'2" 42d1'10"
Miriam
And 1h57'9" 42d37'15"
Miriam
Ari 2h57'59" 30d27'18"
Miriam
And 2h35'46" 46d40'57"
Miriam
Vir 13h14'24" 3d53'38"
Miriam
Leo 10h24'54" 22d7'36"
Miriam
Umi 17h45'56" 87d6'47"
Miriam
Umi 14h42'46" 77d47'5"
Miriam
Lib 15h28'31" -10d4'8"
Miriam
Sco 17h55'53" -39d21'33"
Miriam
Cap 20h46'18" -24d5'41"
Miriam A. Luedecke ~ "My Mib"
Cyg 21h34'41" 39d43'18"
Miriam Adams - Mir Mir
And 1h2'56" 35d2'6"
Miriam Adele Kennedy
Cas 0h53'59" 54d44'24"
Miriam Angelica
Aqr 23h6'40" -10d19'11"
Miriam - Ashley's Beautiful Flower
Cru 12h17'7" -56d20'16"
Miriam Barraza
Sco 17h23'1" -38d9'16"
Miriam Bartolome Rodriguez
Vir 12h44'27" -3d45'17"
Miriam Borkert
Uma 10h43'3" 65d37'59"
Miriam Borrero
Apu 16h8'22" -75d37'54"
Miriam Boylan
Cas 23h28'39" 57d46'24"
Miriam Bridget Cecelia
Psc 1h8'37" 20d58'34"
Miriam Dearson, Loving Mother
Crb 15h44'40" 26d7'54"
Miriam Dessureau
Cap 20h51'34" -18d49'53"
Miriam Dorothy Smith
Cnc 8h27'30" 31d13'6"
Miriam Elisa Alejandro
Vir 11h58'58" -3d21'40"
Miriam Elizabeth Medellin
Psc 0h51'16" 25d31'42"
Miriam Esther Fischer Pentelovitch
Aqr 21h51'11" 2d15'19"
Miriam Findlay
And 2h25'24" 37d19'39"
Miriam Franca Dos Santos
Uma 8h45'20" 55d43'6"
Miriam Franziska
Cas 23h4'21" 53d16'24"
Miriam Friedlander
Cas 0h11'11" 53d54'48"
Miriam Galen
Sco 16h55'54" -40d28'26"
Miriam Garcia (mi sister)
Cnc 8h19'14" 30d37'32"
Miriam Gergelova
Cap 20h53'51" -15d23'30"
Miriam Gonzalez
Mon 6h26'52" 8d13'10"
Miriam Hannah Soud
Lmi 9h47'7" 35d13'46"
Miriam & Hermann
Crb 15h52'35" 33d54'49"
Miriam Katharine Quigley
Uma 10h28'51" 69d19'0"
Miriam Katharine Quigley
Uma 8h30'1" 67d31'44"
Miriam Kunz
Com 13h35'44" 25d28'44"
Miriam L. Henry
Aqr 22h41'33" -6d12'35"
Miriam Lancer
Lyr 18h53'29" 39d4'36"
Miriam Levy-our STAR we look up to
And 2h3'40" 45d51'13"
Miriam lovely lion Leary RN 1959
Lmi 10h32'17" 39d0'52"
Miriam Mack
Cas 0h40'56" 60d53'21"
Miriam Maduro
And 23h23'27" 47d57'17"
Miriam "Mariposa" Sequeira
Ari 1h48'3" 22d14'2"
Miriam "Mars" Carstons
Sco 16h5'31" -11d40'37"
Miriam Massi
Vir 13h14'47" -1d4'48"
Miriam Mavia
Cas 23h11'9" 56d3'38"
Miriam & Michael
Uma 14h10'49" 56d43'38"
Miriam "Mickey" Rothenberg Lorber
And 0h9'51" 38d9'43"
Miriam "Mimi" Reyes
Uma 13h30'58" 57d57'58"
Miriam Mini Prinzässin
Cep 22h58'51" 79d26'13"
Miriam Mirbear Magid
Cnc 8h42'25" 28d35'35"
Miriam Modulon
Ori 5h27'48" 11d16'1"
Miriam Monje Bautista 50190236
Peg 22h45'35" 15d47'14"
Miriam Monte Dugas
Tau 4h23'32" 0d59'26"
Miriam My Hart
Lyr 19h14'45" 26d27'9"
Miriam Pollock
Uma 12h17'10" 54d49'20"
Miriam "Princess" Lytle
Sgr 18h22'35" -33d2'9"
Miriam Rachelle Davis
Psc 0h10'3" 4d58'34"
Miriam Tai
Cyg 20h6'4" 41d17'42"
Miriam the Delicious
Per 2h40'18" 41d6'29"
Miriam The Diamond Rose
Gem 7h9'33" 30d41'40"
Miriam Trepicchio
Uma 9h5'38" 53d45'28"
Miriam Tsipora
Lyn 6h52'53" 56d26'49"
Miriam Windsor
Crb 15h48'44" 36d22'48"
Miriam Würsch
And 2h16'32" 39d10'27"
Miriam Yu
Tau 4h20'10" 9d44'51"
Miriam's Delight
Vir 13h12'3" -20d8'14"
Mirian Thompson
Uma 13h20'53" 56d16'23"
Miriel May Estipona
Ori 5h5'32" 15d3'21"
Miri-Miri-Killara
Cru 12h52'50" -56d14'46"
mirimori great love
Cas 23h31'39" 51d19'17"
Mirinda Ann Gillis
Crb 15h57'17" 34d17'5"
Mirj Mus
Crb 16h2'8" 35d7'54"
Mirjam Djula
Dra 17h31'37" 56d3'10"
Mirjam Egeris Karstoft
Ori 6h6'49" 19d26'11"
Mirjam Hofer
Aur 4h52'30" 39d43'34"
Mirjam Hollinger
And 23h37'37" 43d3'26"
Mirjam & Nick
Cas 0h46'25" 53d38'19"
Mirjam und Manuel
Cyg 19h33'33" 32d8'18"
Mirjo
Cam 4h27'41" 64d26'20"
Mirka Morselli
Peg 23h40'32" 19d5'55"
Mirko
Peg 22h35'0" 9d31'29"
Mirko
Umi 13h18'35" 76d10'20"
Mirko
Cru 12h36'38" -62d6'40"
Mirko Bazzichet
Uma 9h29'18" 56d26'59"
MIRKO BISCUSSI
Aur 5h31'33" 33d28'55"
Mirko Herbst, My Heart, Love & Soul
Uma 10h26'35" 45d9'12"
Mirko & Julie Vuksic
Cyg 20h46'9" 52d3'8"
Mirko Kriebel
Uma 9h23'18" 48d28'51"
Mirlok
Per 3h58'23" 35d24'25"
Mirmi
Vir 12h31'30" 6d33'43"
Mirmonkey
Uma 9h49'23" 63d2'32"
MIRNA 4
Cam 3h56'35" 64d17'36"
Mirna Janeth Dickinson
Gem 7h36'37" 27d31'13"
Mirna Khayat
Cyg 20h18'58" 52d32'24"
Mirna's Star
Tau 3h43'14" 25d16'13"
Mirò
Aur 5h17'23" 42d24'36"
Miro
Cyg 21h21'43" 54d23'51"
Miro
Uma 10h21'47" 60d37'43"
Miro Rames
Ori 5h23'15" 13d14'13"
Miroc
Lib 15h38'37" -15d51'47"
Miroslav Katic
Lac 22h29'27" 54d41'1"
Miroslav Lausman's
Aur 5h40'34" 54d33'27"
Miroslav Pragerova
Umi 13h24'54" 70d45'4"
Miroslav Stupar
Per 4h34'41" 39d15'30"
Miroslava Stäheli
Cas 0h59'59" 53d45'14"
Mirra
Lyn 9h5'21" 42d11'55"
Mirra Lee Maheden
Cas 23h34'6" 53d11'48"
Mirranda
Mon 7h41'43" -7d17'43"
Mirras Miracle
Boo 14h46'1" 52d38'58"
"Mirron"
Cyg 20h38'32" 37d45'29"
Mirta Juana Fisher
Lmi 10h25'3" 35d23'29"
Mirta M Iglesias
And 0h16'40" 30d1'0"
Mirth
Cyg 21h37'43" 51d58'59"
Mirtha Cotto
Mon 6h47'50" 8d19'32"
Mirthew, My Handsome Prince
Cam 4h18'46" 69d0'41"
mirty
Uma 9h53'56" 49d45'50"
Miryam Ghodsian Aghai
Ori 5h59'53" -0d47'56"
Miryam Zihlmann
Ori 5h52'45" 22d5'39"
Mirz
Ori 5h59'21" 21d23'6"
Mirza
Lib 15h50'58" -11d20'4"
Mirza and Enisa
Cyg 19h35'25" 30d10'23"
Mirza Chikita Marquez
Uma 9h20'28" 54d29'21"

Mis
Dra 17h8'39" 54d52'32"
mis amores estrellan (Jessie Cain)
Sco 16h11'0" -12d36'19"
Mis liäbe Mami Rita Schibli
Dra 17h12'53" 54d58'8"
Mis Sternli
Ori 5h44'6" 8d43'28"
Misa
And 0h26'6" 31d38'25"
Misa 11.04.2004.
Per 4h17'21" 41d34'34"
Misa i Jelena
Ori 6h1'21" 0d2'15"
Misa Michael Josef
Tau 5h47'43" 16d17'53"
Misa Palace
Uma 10h27'1" 61d31'57"
Misa Rahib Ankar
Ari 2h5'53" 23d41'39"
Misael Gonzalez Jr. "M.J."
Tau 3h47'36" 19d6'16"
Misak Mike Palanghyan
Boo 15h7'45" 33d35'26"
MISAYUJI KOBAYASHI
Uma 13h40'27" 57d45'33"
Mischa
Lyn 8h32'41" 45d32'23"
Mischa
Leo 9h58'43" 16d18'30"
Mischa Brooke Staton
Leo 11h14'17" -5d52'37"
Mischa Malia Korta
Lyr 18h52'33" 37d46'13"
Mischa's Place in The Heavens
Vir 12h43'6" -0d26'17"
Mischka Wolfe
Leo 11h55'5" 21d27'53"
Miscreant Robert
Umi 19h57'54" 88d38'0"
Misenhelter's Star
Aql 19h39'4" 14d34'57"
MISH 1
Psc 1h25'31" 10d47'37"
Mish and Tarz
Sco 17h25'3" -38d46'49"
MISH LOVE
Cyg 21h14'3" 32d53'14"
Mish Mish
Uma 11h25'40" 46d12'49"
Misha
Lyn 8h19'59" 45d26'18"
Misha
Lyn 9h6'13" 42d4'48"
Misha
Lyn 7h54'33" 35d31'0"
Misha
Ori 5h18'7" 8d31'5"
Misha
Crb 15h49'32" 27d42'33"
Misha
Mon 6h49'58" -0d8'47"
Misha
Lib 15h2'47" -1d11'17"
Misha & Aileen
Uma 10h4'56" 63d55'33"
Misha Dawn
Cyg 19h47'52" 33d57'2"
Misha Marie Hawksworth
Uma 11h29'45" 32d41'4"
Misha Midnight Star
Cas 1h0'43" 61d6'42"
Misha Noel Klein
Uma 13h54'45" 55d20'55"
Misha nute-azzarello
Cnc 9h9'46" 24d29'1"
Misha Tomaye Leong
Tau 5h18'55" 27d25'52"
Mishal Zehra Rizvi
Ari 2h15'27" 24d21'54"
Mishaman
Per 3h1'21" 40d26'46"
Mishan Andre
Aqr 22h28'23" -0d32'34"
'Mishca-Moo' Is Our Star Forever
Cru 12h37'4" -60d56'58"
Mishele Megumi Miyake "M3"
Uma 9h57'2" 44d14'3"
Mishi
Cam 4h12'1" 66d9'1"
Mishi Ilikai
Cap 20h19'50" -14d36'10"
MishIris
Lyn 6h50'15" 52d29'59"
Mishka Krichevsky
Cas 0h38'35" 51d11'9"
Mish's One In A Million
Col 6h20'59" -40d42'35"
Mishu
Aqr 22h43'17" 1d19'27"
Mishwiljem
Cra 18h3'15" -43d42'45"
Mishy Lynn
Lyr 18h54'57" 39d24'52"
Misia 85
Sco 16h14'31" -15d37'50"
Miski Chuspi Maduhu
Cyg 21h35'51" 49d8'18"
Misko-anang
And 2h19'23" 45d49'34"
MisMarLin McCounSel, B/F/F
Vir 12h38'6" -8d20'1"
Misrah Naomi
Sco 16h9'47" -12d0'37"
Miss
Tau 4h35'18" 9d51'16"
Miss A
And 0h34'41" 43d46'44"
Miss Abby
Lyr 18h49'21" 44d47'31"
Miss Abigail
Ari 2h44'44" 28d6'20"
Miss Aeris Rose
Gem 6h21'0" 19d1'40"
Miss Aimee Biedermann - 5/2000
And 2h6'4" 42d38'50"
Miss Alabama
Sco 16h45'39" -32d9'52"
Miss Alexander
And 0h13'36" 27d6'26"
Miss Alexi Comer
Uma 11h42'48" 64d24'54"
Miss Alia Riley
And 23h23'27" 51d5'20"
Miss. Alicia
And 0h22'4" 42d11'44"
Miss Alyssa
Vir 12h27'56" 4d22'17"
Miss Amanda Lynn Tyner
Leo 11h25'59" 1d4'7"
Miss Amber B.
Leo 10h53'6" -3d6'5"
Miss America's WETLAND
Psc 1h9'58" 30d36'40"
Miss Amy
And 1h34'11" 44d48'46"
Miss Amy Lou (Amy Mitchell)
Psc 0h32'8" 13d16'12"
Miss Amy Victoria Allsop
And 1h15'4" 45d53'55"
Miss Anne Thrope
Leo 9h55'53" 8d22'37"
Miss Annette Li
Ori 5h33'44" 3d35'54"
Miss April & Danielle Clark-Scott
Sge 19h40'34" 17d28'16"
Miss Arely E. Diaz
Cas 23h1'47" 58d18'45"
Miss Ashley Lane Mullinax
Mon 7h21'58" -0d29'29"
Miss Athena
Vir 14h17'9" -0d59'10"
Miss Babe
And 1h15'43" 44d53'47"
Miss Bailey Harrison
Ari 2h42'38" 27d27'49"
Miss Becca
And 1h17'33" 37d54'56"
Miss Belle
And 1h37'13" 41d28'4"
Miss Biddie
And 0h50'19" 41d32'32"
Miss Bryonna Marie Barnhart
And 23h5'56" 51d23'31"
Miss "Candy" Girl
Cma 6h59'8" -31d45'58"
Miss Carla Maria
And 1h50'12" 44d7'40"
Miss Casey Good
Cmi 7h20'51" 8d10'30"
Miss Catherine
And 0h49'4" 37d23'45"
Miss Charlotte Cripps
And 23h9'13" 42d52'58"
Miss Chris
And 1h53'27" 42d39'56"
Miss Christine's Great Expectations
Cap 21h40'13" -13d57'9"
Miss Cindy, Love 3M Class 2006
Umi 15h21'42" 77d13'32"
Miss Colette, Love 3M Class 2006
Umi 15h58'30" 75d7'32"
Miss Courtney T. Isabella
And 2h19'51" 47d34'28"
Miss Cynthia Janea
Lyn 8h20'13" 34d49'46"
Miss Daisy Mae
Lyr 19h1'58" 44d20'46"
Miss Dana
And 0h21'57" 33d8'44"
Miss Defiant
Dra 10h1'31" 75d12'44"
Miss Denise
Lib 14h23'42" -14d9'22"
Miss Denisha Bond
Lyn 7h12'46" 52d5'8"
Miss Diana
And 0h35'38" 31d47'43"
Miss Dimple - Love from your Zimbo
And 23h29'45" 47d43'13"
Miss Donna
Cas 1h25'18" 53d40'35"
Miss Donna
Cas 1h46'0" 60d29'59"
Miss Doodle
Mon 6h24'8" -4d25'9"
Miss Edna
Gem 7h36'29" 24d43'50"
Miss Elizabeth
Gem 7h12'52" 15d38'25"
Miss Elle-Mima
Psc 0h43'39" 7d37'5"
Miss Ellie
Cnc 8h53'22" 19d20'45"
Miss Ellie
Umi 13h57'49" 70d21'40"
Miss Ellie Catherine Dukelow
And 23h9'3" 52d40'1"
Miss Ellie Kalski
And 0h17'3" 43d13'6"
Miss Ema-Jane
Leo 9h45'53" 8d41'46"
Miss Emily
Cyg 21h52'18" 37d29'14"
Miss Emily
Lyn 7h42'43" 36d33'5"
'Miss Emily'
And 2h19'42" 37d26'1"
Miss Emily Begg
And 1h25'6" 50d17'14"
Miss Emma Barnett's Star
And 0h16'42" 39d57'33"
Miss Emma Grace Andersen
And 2h29'1" 50d17'34"
Miss Emma Margaret Casey
And 23h20'4" 48d42'40"
Miss Erin Elizabeth McGinnis
Vir 12h24'52" 2d3'25"
Miss ETC
Tau 4h11'4" 17d16'15"
Miss Evelyne
Cyg 19h43'27" 53d29'14"
Miss Ezzie Lee
Crb 15h36'21" 35d32'22"
Miss Favreau
And 23h40'55" 46d30'55"
Miss Frances
Peg 21h13'56" 19d47'31"
Miss Gabrielle Kalnins
And 0h54'1" 36d44'49"
Miss Georgia Smith
And 0h2'2" 44d49'33"
Miss Haley
And 0h57'2" 37d19'38"
Miss Hasannah Terry
Lyn 7h30'1" 47d32'41"
Miss Hattie E. Caveney
And 23h16'47" 36d47'50"
Miss Holly
And 23h25'26" 48d22'24"
Miss Holly Janel Woods
And 1h54'7" 37d11'11"
Miss Isabella Milano
Peg 21h36'31" 25d25'38"
Miss J
Cyg 20h58'3" 34d29'53"
Miss J — Star of Bethlehem
Tau 5h56'57" 25d25'11"
Miss Jackie
Uma 9h46'56" 68d7'38"
Miss Jaimy
Uma 8h9'12" 60d51'5"
Miss Jeannie
Lib 15h40'28" -14d37'19"
Miss Jeannie Lee
Psc 1h7'50" 27d30'8"
"Miss Jennifer" Newton
Cnc 8h51'58" 30d52'1"
Miss Jennifer Schneider
Sco 16h16'38" -10d41'35"
Miss Jodi Kay
Leo 10h15'21" 26d7'29"
Miss Jordan
Sgr 19h40'48" -14d27'58"
Miss Josie Lynn
And 23h0'13" 50d13'45"
Miss Julia's Shining Star
Cnc 8h33'32" 21d30'50"
Miss June, Love 3M Class 2006
Umi 15h8'28" 71d39'11"
Miss Karen
Aqr 22h32'8" 1d27'20"
Miss Kate Mary Munhall
Uma 8h17'20" 70d21'56"
Miss Kathy Lee
Cap 20h49'26" -20d50'57"
Miss Katie
And 0h44'23" 45d37'12"
Miss Kerri McCamant
And 0h53'50" 35d40'58"
Miss Kimie
Cap 20h56'31" -26d31'27"
Miss Kitty
Aqr 22h9'29" -6d17'51"
Miss Kitty
Cam 6h53'49" 65d3'57"
Miss Laree
Peg 23h31'18" 17d15'4"
Miss. LaShawn Moody aka.My Star
Gem 6h49'51" 34d10'16"
Miss Laura
And 0h28'47" 42d13'22"
Miss Lippy
Leo 11h15'12" 11d15'35"
Miss Lisa G
Ari 3h0'46" 29d0'22"
Miss Lisa Nuszbaum
Psc 1h5'53" 29d13'21"
Miss Lisa's Dimpled Radiance 627
Ori 5h55'27" 21d40'9"
Miss Lucy
Cnc 9h3'24" 7d40'32"
Miss Madeleine Keeling Clark
Aqr 22h5'52" -0d17'47"
Miss Madison Monroe
Cap 20h18'13" -15d26'53"
Miss Mae
And 2h37'18" 42d57'12"
Miss Mandy
Gem 6h29'35" 14d56'51"
Miss Margaret
Mon 6h45'55" -0d31'20"
Miss Marisa
Aql 19h4'10" -0d7'0"
Miss Mary "Kitten" Kinder
Gem 7h31'52" 26d14'54"
Miss Maryann
Psc 1h35'24" 24d11'13"
Miss Meka
Uma 10h57'36" 42d17'33"
Miss Melanie The Light Of My Life
Sgr 18h56'7" -31d37'8"
Miss Meliss
Cap 21h44'48" -24d21'11"
Miss Michelle Reigan Lynette
Cas 1h11'9" 69d40'49"
Miss Milkyway's 32/11
Cnc 8h42'42" 24d40'16"
Miss Min
Ori 6h24'36" 17d21'9"
Miss Minton
Vir 13h7'50" 11d37'53"
Miss Missie
Ori 5h36'42" -2d36'22"
Miss Molly
Lmi 9h39'20" 34d29'16"
Miss Molly
And 1h34'43" 46d58'18"
Miss Molly Morgan Smith
Aqr 22h37'57" -1d53'27"
Miss Molly's Sweet 16
And 1h13'53" 46d0'53"
Miss Mom
Leo 9h40'28" 25d2'44"
Miss Monkeyface
Tau 5h50'49" 12d43'36"
Miss Mum
Uma 8h50'0" 67d3'7"
Miss Myra
Lyn 6h57'6" 61d19'40"
Miss Nancy Wyatt
Lib 15h18'14" -5d22'0"
Miss Nhi
Psc 1h3'41" 27d25'9"
Miss Nicola Gail Black
And 1h16'0" 34d58'22"
Miss Niksic
And 2h22'33" 43d28'7"
Miss. Norris
Sgr 19h29'11" -32d42'47"
Miss Ooow!
Uma 10h26'5" 47d20'6"
Miss Opel
Cru 12h31'15" -59d59'49"
Miss Pamela
Crb 15h41'20" 31d50'44"
Miss Patti Twinkle Bright
Aqr 20h54'51" 1d56'43"
Miss Pearlie *** Dora
Psc 2h4'45" 6d22'37"
Miss Peggie
Cas 2h49'34" 65d31'19"
Miss Pettit
Ori 5h3'18" 4d31'32"
Miss Pink
Ori 5h16'30" 15d26'58"
Miss Pixie
Peg 22h29'53" 34d17'8"
Miss Poppy Amelia Buckley
And 2h11'39" 49d33'58"
Miss Prancing Vivian 11-24-2002
Peg 22h33'2" 32d4'55"
Miss Princess
Vir 14h1'55" -0d50'12"
Miss Priss
Peg 23h22'20" 33d6'35"
"Miss Rachel" Newton
Aqr 21h38'38" 0d22'47"
Miss Randi's Shining Light
Crb 15h31'39" 30d17'11"
Miss Randy
Vir 12h28'50" 7d37'54"
Miss Reagan Olino
And 0h20'22" 27d10'54"
"Miss Regina" White
Cnc 8h31'38" 32d32'8"
Miss Riley Lynn
Uma 13h37'51" 49d21'49"
Miss Ro Carlton
Per 3h49'53" 35d6'17"
Miss Roe
Aqr 22h24'44" -13d38'29"
Miss Ruby Phoenix
And 2h34'43" 49d28'12"
Miss Sally
Ori 5h36'27" -2d7'15"
Miss Sara
Tau 4h18'3" 26d56'37"
Miss Sara Grace Clark
Uma 8h28'15" 61d19'59"
Miss Sarah Hawkett
And 1h24'0" 35d30'34"
Miss Sarah Palladino
And 23h39'22" 48d50'18"
Miss Savannah Paige
Uma 10h23'10" 41d29'58"
Miss Scarlet O'Hara The 3rd
And 23h36'46" 46d8'23"
Miss Scarlett Cutrell
Sco 17h48'20" -41d44'47"
Miss Scarlett Voclain
And 2h21'58" 39d45'10"
Miss Shaline Ranat
And 0h24'19" 26d19'25"
Miss Sharon
And 2h22'23" 48d15'20"
Miss Shelley Lynne
Sgr 19h46'0" -20d13'45"
Miss Skye & Bonnie Lindholm
Sge 19h51'20" 17d27'15"
Miss Stacey Buckenham
And 23h37'51" 42d39'57"
Miss Stacy's Star
And 1h46'28" 46d42'37"
Miss Star Dog
Uma 13h29'37" 52d26'23"
Miss Starkesha the Super Woman
Vir 13h12'4" 12d24'42"
Miss Stephanie & Haley Powers
Cyg 20h24'25" 34d44'16"
Miss Sugar
Vir 11h53'19" -5d30'58"
Miss Susan
Cha 13h41'20" -76d49'11"
Miss Sweetpea
Aqr 23h2'55" -11d44'21"
Miss T
Lyr 19h20'28" 28d37'47"
Miss T 1002
And 0h30'3" 28d34'53"
Miss Taryn
Sgr 18h51'41" -34d53'1"
Miss Taylor Elizabeth Jones
Uma 9h55'54" 67d28'36"
Miss Teen Lacey
Cas 1h0'2" 60d5'15"
Miss Tess
And 0h42'35" 37d35'43"
Miss Tester
Per 3h30'36" 39d29'37"
Miss Texas
Tau 4h44'19" 18d59'3"
Miss Theresa Condora
Per 3h19'0" 45d47'36"
Miss Trina Marie
Cap 20h48'11" -19d4'56"
Miss Tupper
Mon 7h31'3" -0d19'52"
Miss Tyler Marissa Rabe
And 2h0'0" 40d9'57"
Miss Valerie Saponara
Lmi 10h54'8" 33d16'55"
Miss Vanessa Carney
Psc 1h28'23" 15d25'44"
Miss Vicky
And 1h33'32" 43d16'40"
Miss Vivi
Cap 21h9'44" -21d7'22"
"Miss Vivian" Otto
Uma 9h49'6" 51d52'53"
Miss Weasel
Vir 13h8'55" 7d59'57"
Miss Whitney
Leo 11h48'54" 17d21'16"
Miss Whitney
Uma 13h58'11" 61d59'5"
Miss Wright
Leo 9h49'24" 7d0'37"
Missi Russell
And 23h45'37" 34d12'20"
Missi Sue Wilson
Ori 5h42'7" 0d12'49"
MissIce
Ori 5h41'2" -1d58'58"
MissiCol
Uma 12h19'17" 58d42'52"
Missie & Dirk
Ori 6h5'51" 17d16'56"
Missie Hilemn (UB 9)
Ori 5h50'35" 2d56'37"
Missie Loves Andy
Cyg 19h43'20" 34d6'51"
MISSIE LYN
Psc 1h6'43" 11d50'49"
Missie11
Aqr 21h23'40" -5d3'7"
Missing Chris
Dra 17h11'17" 54d11'56"
Mission Hughey
Uma 10h24'58" 64d19'57"
MissMaddie
Cas 1h53'17" 65d9'58"
MissNewYorkTeenUSACandaceKuykendall
Sgr 19h41'7" -12d47'43"
MissTeenAmerica 2004, Chelsea Fahey
Cas 0h13'32" 53d50'7"
MissTracey
Ori 4h51'32" 11d29'27"
Missue
Uma 11h32'16" 60d17'58"
Missy
Uma 12h46'30" 58d26'32"
Missy
Uma 11h16'14" 56d58'36"
Missy
Sgr 18h13'33" -19d38'25"
Missy
Cma 6h44'53" -14d9'43"
Missy
Vir 14h24'46" -4d29'59"
Missy
Psc 1h44'38" 10d20'31"
Missy
Cnc 8h11'41" 6d48'16"
Missy
Vir 13h4'45" 12d4'16"
Missy
Psc 1h29'26" 27d35'1"
Missy
And 23h17'52" 47d13'9"
Missy
And 23h20'12" 48d10'1"
Missy
And 22h58'31" 51d43'25"
Missy
And 23h2'24" 41d54'36"
MISSY
Leo 9h46'22" 32d29'31"
Missy
Per 3h9'22" 42d49'59"
Missy and Jim *Forever and Always*
Cap 20h38'2" -10d9'23"
Missy Ann Corson
Oph 17h35'13" 6d10'43"
Missy Ann Leitch
Leo 11h25'1" 25d19'13"
Missy Ann Nielsen-Jerome
And 1h17'36" 41d24'29"
Missy Anne
And 0h53'4" 41d6'8"
Missy Barker
Gem 7h35'19" 32d49'54"
Missy Benmark
Sco 16h8'9" -9d19'5"
Missy Berman Graham
Cyg 20h4'15" 43d41'40"
Missy Boiseau
Cas 1h41'7" 64d21'16"
Missy Dale
Ori 5h29'4" 3d10'32"
(Missy) Ella Pauline Mesker
Cnc 7h55'49" 16d22'4"
Missy Girl
Cyg 21h42'24" 33d50'52"
Missy Higgins
Cru 12h8'1" -61d47'55"
Missy " Infinity + 1 "
Sco 17h4'24" -30d29'19"
Missy Jones
And 0h58'57" 43d23'3"
Missy Lee Bray
Cap 20h35'56" -9d45'46"
Missy Lee Hildreth "4-9-1966"
Uma 10h16'57" 46d54'28"
Missy Lynn
Cyg 20h19'7" 38d44'26"
Missy Lynn
And 0h34'49" 35d46'13"
Missy Maat
Cas 1h37'28" 69d19'54"
Missy Macias
Ari 2h37'53" 24d56'10"
missy mafia
Crb 16h13'17" 36d41'49"
Missy Mary Schwartz
Lyn 7h50'10" 50d55'36"
Missy mei Caelestis Angelus
Vir 15h1'52" 1d52'55"
Missy Missy Moo
Col 5h14'31" -35d3'42"
Missy Moo
Cru 12h50'26" -60d30'0"
Missy Moo
Vir 13h16'33" -21d48'27"
Missy New
Cep 22h26'39" 61d13'32"
Missy P
Cnc 8h20'45" 15d34'23"
Missy Peare
And 1h44'1" 43d54'50"
Missy Poltarak
Cnv 12h47'2" 37d32'31"
Missy Prescott
Lyn 9h16'26" 38d43'33"
Missy Renee Segota
Cha 10h3'34" -80d39'30"
Missy Rogers
And 1h46'19" 49d4'1"
Missy Sue
Cas 1h44'50" 68d39'19"
Missy, my Valentine Forever
Ari 2h51'42" 25d56'5"
MissyB
Aqr 23h20'52" -10d52'48"
Missylu
Lyn 6h34'52" 56d19'58"
Missy-Marie
Tau 5h48'7" 14d42'59"
MISSYmffG821
Leo 10h12'29" 13d36'19"
Missy's Love
Aqr 21h8'20" 1d51'11"
Missy's Love
And 23h50'37" 43d49'14"
Missys Lucky Star
Lib 14h54'28" -22d59'52"
Missy's Promise
And 1h41'19" 42d19'11"
Missy's Silver
Crb 15h37'29" 26d31'49"
Missy's Star
And 0h33'49" 26d28'26"
Missy's Star
Peg 22h42'31" 11d33'8"
Missy's Star
Cyg 20h16'54" 33d51'7"
Missy's Star of Life, When it began
Vir 13h9'37" 8d50'55"
Mista Spakle
Her 17h27'58" 31d37'11"
Mistaken Identity
Cru 12h42'18" -61d47'3"
Mista's Jewel
Sgr 19h19'41" -28d49'22"
Mistchevelenle
Peg 23h44'48" 11d38'27"
Mistee
Ori 5h49'49" 11d52'26"
Mister
Cap 20h51'48" -17d43'12"
Mister A
Ori 6h18'41" 10d46'28"
Mister Babe
Uma 9h32'25" 57d28'58"
Mister E.T.
Uma 11h17'36" 64d44'33"
Mister Hottie Doll
Ori 4h49'16" 1d29'2"
Mister Jass
Psc 0h1'55" 8d13'13"
Mister & Little Princess's 1 Year
Ori 6h10'36" 16d30'15"
Mister Lorraine
Umi 16h31'42" 82d13'29"
Mister "Milo"
Cma 6h20'20" -30d26'34"
Mister & Pumpkin
Sex 10h24'42" -6d6'42"
Misti
Gem 7h29'14" 22d21'37"
Misti Pederson
Leo 11h5'20" -0d0'51"
Misticat
Cas 0h14'33" 50d50'14"
Mistress Kaiel
Gem 7h20'59" 31d21'5"
Mistress Lily
Uma 9h13'24" 62d18'21"
Mistress of the wolf
Sgr 18h51'58" -20d27'32"
MistressGoodPain
Ari 2h18'16" 26d13'2"
Mistruvion
Cas 1h29'58" 60d11'6"
Mistrzuniu
Hya 9h15'44" -0d47'56"
Misty
Mon 7h9'44" -0d11'41"
Misty
Vir 12h32'34" -1d14'56"
Misty
Lib 15h39'5" -6d32'47"
Misty
Cma 6h57'51" -11d26'48"
Misty
Sco 16h9'20" -20d4'49"
Misty
Lyn 8h27'13" 52d34'45"
Misty
Uma 8h55'11" 69d18'54"
Misty
Leo 10h6'49" 24d53'47"
Misty
Gem 6h55'43" 28d26'29"
MISTY
Leo 11h21'8" 14d21'57"
Misty
Leo 11h11'18" 11d38'22"
Misty
Cnc 8h43'13" 6d45'25"
Misty
Gem 7h11'44" 32d27'48"
Misty
Crb 16h14'23" 37d7'21"
Misty
Her 16h42'33" 34d11'39"
Misty
And 2h10'52" 38d19'6"
Misty
Uma 13h36'28" 48d23'48"
Misty & Alyssa 4E
Uma 10h9'45" 68d52'52"

Misty (Angel) Perry
Leo 9h34'36" 16d56'2"
Misty Ann Addis
And 0h15'8" 43d32'23"
Misty Anne
Lyn 8h13'11" 37d20'10"
Misty Ann's Star
And 0h36'11" 37d0'40"
Misty Blue
Peg 23h10'55" 19d2'6"
Misty Blue
Ori 5h33'11" -0d58'43"
Misty Brandt
Vir 14h47'2" 5d41'29"
Misty Buggy
Com 12h31'3" 24d25'6"
Misty Cheyenne Cody
Cas 0h39'39" 48d54'20"
Misty Chrosnister
Cmi 7h26'41" -0d7'55"
Misty Davis
And 2h27'41" 45d6'9"
Misty Dawn
Cam 5h31'20" 57d36'13"
Misty Dawn
And 1h18'28" 35d40'6"
Misty Dawn
Her 16h44'4" 25d45'56"
Misty Dawn
Her 18h43'4" 19d13'16"
Misty Dawn
Sco 16h15'36" -13d16'38"
Misty Dawn Bernard
Mon 7h8'9" -0d3'58"
Misty Dawn Iacobucci Sheldon
Sco 16h7'5" -16d11'3"
Misty Dawn Locke
Psc 1h43'6" 5d4'45"
Misty Decker
Ori 5h18'43" 1d6'38"
Misty Elizabeth Hoeflich
Cnc 8h26'22" 29d37'2"
Misty Garrett
Lmi 10h9'29" 38d0'38"
Misty Girl
Her 17h39'56" 48d24'6"
Misty Gwin
Uma 8h55'19" 63d34'42"
Misty Hansen
And 23h17'22" 43d52'32"
Misty Heather Lightner
And 0h39'28" 37d17'24"
Misty J
Sgr 18h40'4" -17d35'9"
Misty Jestina
Cnc 8h41'53" 27d55'59"
Misty Kay Rugg
And 0h41'38" 38d20'6"
Misty L Reddin
Vir 14h22'34" -2d54'21"
Misty L. Roberson
Cnc 9h14'29" 11d37'14"
Misty LaVina
Ari 2h34'14" 30d59'28"
Misty Leah
Cap 20h31'15" -12d12'24"
Misty Lee
Cas 23h2'20" 58d43'47"
Misty Leigh Smith
Uma 10h44'1" 64d53'1"
Misty Luv
Umi 14h0'40" 72d15'38"
Misty Lynn
Umi 14h20'16" 66d42'10"
Misty Lynn
Cas 0h22'20" 55d3'16"
Misty Lynn
Ari 2h38'30" 26d31'40"
Misty Lynn Taylor
Uma 11h31'41" 59d55'46"
Misty Marie
Lyn 7h32'43" 56d49'12"
Misty Marie
Lib 15h30'16" -8d40'49"
Misty Marie
And 1h46'48" 42d57'55"
Misty Marie Minor
Cam 3h18'10" 60d42'51"
Misty Marie Smith
Cas 0h30'38" 67d3'14"
Misty "May 14, 1992"
Tau 4h5'20" 26d29'46"
Misty Michelle
Cyg 20h38'49" 31d46'54"
Misty Michelle Wofford
Sgr 19h27'59" -23d17'32"
Misty & Mikes' Star
Tau 4h19'59" 23d4'36"
Misty Montejano
Gem 6h50'14" 34d54'6"
Misty Moon
And 0h14'40" 32d44'22"
Misty Moore
Ori 6h19'43" 15d46'26"
Misty N. Scaldaferri
Aqr 22h20'54" -0d47'12"
Misty Nicole Baker
Cyg 19h43'33" 38d23'26"
Misty Nies
Umi 14h58'45" 76d30'28"
Misty Noel Hoku
Cas 0h17'56" 60d53'47"
Misty Phelps
Umi 14h18'49" 67d54'16"
Misty Pickens
Sgr 19h35'54" -30d39'31"
Misty Rogers
And 1h48'17" 46d34'18"
Misty Rose Bishop
And 23h40'58" 46d25'45"
Misty S. Hynes
Oph 16h33'47" -0d55'24"
Misty Sloan
Sco 17h53'27" -30d50'29"
Misty Smith
Leo 11h6'19" 12d29'9"
Misty Star
Sgr 18h42'59" -18d10'46"
Misty Tanaya Gomez
Crb 16h13'53" 28d19'43"
Misty Thesen
Vir 12h47'54" 7d33'8"
Misty Toi
Dra 19h41'24" 78d13'0"
Misty Two Shoes
Aqr 22h46'49" -22d36'57"
Misty Van Horn
Cas 0h5'12" 55d58'34"
Misty Walden
Cap 21h1'44" -22d24'51"
Misty Weatherford
Cnv 13h40'35" 41d8'35"
Misty Wolf
Cam 5h16'45" 69d0'25"
Misty, Brownie, Hiawatha Truzzolino
Cir 15h18'38" -58d16'7"
mistybear
Uma 13h23'13" 56d3'58"
Misu
Uma 8h59'2" 70d7'55"
mita
Uma 8h44'42" 57d44'24"
Mitali
Uma 9h28'53" 62d11'48"
Mitch
Cep 20h59'31" 55d50'28"
Mitch
Boo 14h40'56" 22d7'48"
Mitch 1987
Lib 15h45'43" -27d8'28"
Mitch and Bianca Forever
Cru 11h59'5" -63d56'48"
Mitch and Heather Day
Cyg 21h35'20" 48d59'0"
Mitch Blyumin
Psc 0h18'39" 20d16'31"
Mitch & Candace Horton
Uma 8h23'7" 67d24'3"
Mitch Cangelosi
Per 3h9'17" 54d45'55"
"Mitch" - DAD
Cep 22h7'23" 82d44'35"
Mitch & Dan Forever
Per 2h24'5" 57d3'8"
Mitch Erdely
Aql 19h33'31" 5d3'45"
Mitch Forney
Cep 23h3'8" 77d59'31"
Mitch Gossard
Cep 22h38'54" 77d40'35"
Mitch Herbets
Per 4h37'37" 44d15'56"
Mitch Homan Dancing in the Moonlite
And 1h46'41" 42d20'11"
Mitch Huhem
Boo 15h10'35" 48d7'35"
Mitch Kimbrell
Her 18h18'11" 23d53'42"
Mitch Neil Johanni
Aqr 21h50'40" -0d45'32"
Mitch Philip Mondrowski
Lib 15h25'27" -26d52'56"
Mitch Sands
Her 17h54'44" 18d31'9"
Mitch Stallings
Mon 6h40'23" 8d35'52"
Mitch & Tina
Aqr 22h7'12" 0d4'7"
Mitchel Fishman
Psc 1h18'55" 3d34'54"
Mitchel Kraskin
Crb 15h40'31" 38d3'23"
Mitchel L. Chaikin
Gem 6h48'5" 35d18'31"
Mitchel Lee Mullins
Vir 13h17'9" -4d34'56"
Mitchel Lynn Morse
Ori 6h10'19" 6d22'34"
Mitchel1796
Uma 10h34'44" 43d30'53"
Mitchell
And 2h25'32" 43d2'15"
Mitchell
Ori 5h57'0" 6d48'9"
Mitchell
Lyn 7h42'1" 52d56'24"
Mitchell
Cap 20h50'6" -26d10'11"
Mitchell 16.01.2004
Cap 20h23'4" -10d17'56"
Mitchell A Bettinger
Sgr 18h39'38" -31d30'28"
Mitchell A. Cox
Her 17h26'26" 42d29'35"
Mitchell Alan Dahl
Ori 5h56'19" 17d40'1"
Mitchell Allen Goldstein
Uma 8h46'9" 47d59'12"
Mitchell Armster Whitehurst
Psc 1h9'2" 32d14'0"
Mitchell Byrnes
Peg 22h5'26" 7d59'47"
Mitchell Cangelosi
Uma 11h43'25" 28d56'10"
Mitchell Caudill
Uma 13h46'36" 53d40'52"
Mitchell Chamberlain
Tau 4h29'49" 19d58'47"
Mitchell Chase Daniels
Umi 14h24'9" 73d10'53"
Mitchell Cory Cronin
Gem 6h41'9" 19d31'51"
Mitchell & Danielle
Vir 12h24'8" 6d36'27"
Mitchell Dixon
Uma 11h53'22" 64d35'51"
Mitchell Dubois
Aql 20h5'18" 4d46'23"
Mitchell Erickson
Aqr 22h0'33" -16d26'30"
Mitchell Francis Gallant
Uma 13h59'51" 57d35'10"
Mitchell George Osmond
Tau 4h5'39" 24d16'59"
Mitchell Gilbert
Uma 11h29'2" 60d7'13"
Mitchell Gray Tait
Boo 14h51'54" 20d56'36"
Mitchell Henderson's Star
Cas 1h20'12" 56d28'23"
Mitchell James Burns Jackman
Leo 10h49'47" 18d26'38"
Mitchell James Verhulst
Her 18h7'54" 22d56'35"
Mitchell James Wagoner
Leo 10h30'7" 12d11'23"
Mitchell John Rose
Cam 6h38'35" 63d25'57"
Mitchell Keaton Burnett
Cnc 8h57'8" 9d2'38"
Mitchell & Kelly
Lyr 18h40'7" 31d30'19"
Mitchell Kevin Rhodus
Psc 23h46'42" 5d30'58"
Mitchell Kim Cook
Uma 12h27'37" 57d29'25"
Mitchell Kinnaman
Sgr 19h41'47" -16d50'7"
Mitchell Kupinski
Uma 9h16'57" 67d9'20"
Mitchell Kyle Spinnell
Sgr 19h17'14" -16d28'58"
Mitchell L. Garrett
Sco 17h22'46" -38d13'15"
Mitchell L. Marine
Uma 9h2'1" 56d51'8"
Mitchell Laing
Per 3h39'38" 47d6'22"
Mitchell Lee Harris
Ori 5h18'12" 9d2'46"
Mitchell Lee Vance II
Ori 5h41'45" 2d51'47"
Mitchell Lester
Boo 15h35'56" 48d44'5"
Mitchell Levi Daniel Krumm
Ari 3h12'15" 12d13'37"
Mitchell Lewis
And 0h44'29" 42d20'56"
Mitchell Luke Allomes
Cru 12h36'9" -58d13'49"
Mitchell Mabee
Uma 10h41'50" 54d49'55"
Mitchell Malcolm Loizi
Psc 1h38'1" 14d54'59"
Mitchell Malissia Willis
Lyn 7h48'41" 35d55'35"
Mitchell*miracle*Lynch 21022006
Cru 12h31'17" -60d48'56"
Mitchell - Mitch Kandi Zack Tiffany
Peg 23h31'44" 18d34'19"
Mitchell Modell
Cnc 8h47'34" 8d35'58"
Mitchell Owen - Our Shining Star
Leo 11h35'14" 7d30'11"
Mitchell R. Hummel
Uma 12h57'16" 52d25'1"
Mitchell Ray Mielke
Cnc 9h9'16" 31d4'30"
Mitchell Reid Benson
Aql 19h2'14" -10d20'54"
Mitchell Riner 5-3-1997
Tau 5h38'58" 25d53'14"
Mitchell Ryan Springer's Réalta
Col 5h50'5" -38d39'10"
Mitchell S. Holt
Ari 2h37'28" 17d44'46"
Mitchell S. Zingman
Uma 10h30'6" 40d25'3"
Mitchell Scott Carter
Sgr 18h40'28" -36d11'36"
Mitchell T. Isert
Uma 11h56'29" 61d57'44"
Mitchell Thomas O'Rourke
Uma 10h10'17" 66d12'34"
Mitchell Thompson
Cap 20h23'3" -11d25'26"
Mitchell Walsh
Ori 6h3'1" 16d54'12"
Mitchell Wayne Stomner
Per 3h47'17" 41d24'37"
Mitchell West Brooks
Uma 11h43'1" 43d0'32"
Mitchell William Oliver
Psc 1h29'47" 15d53'46"
Mitchells Medwin
Ori 5h2'2" 15d2'11"
Mitchell's Star
Pho 0h58'46" -41d9'6"
Mitchell's Star In Heaven
Cru 12h39'51" -59d2'32"
Mitchell's Starlight Eagle
Aql 19h7'12" -0d49'54"
Mitchell's Teddybear
Cen 13h22'58" -54d44'25"
Mitch—King of all Fishermen
Eri 4h52'21" -22d59'53"
Mitchy Derose - Djiovani Martin
Uma 12h0'56" 42d10'58"
"Mitchy The Kid"
Umi 14h58'50" 70d37'40"
Mither Fither Lane & Tammie Perkins
Uma 13h41'23" 56d17'43"
Mithun - Anjali
Eri 4h18'44" -32d15'23"
Mitis Speciosus Vir Kevin
Ori 5h30'1" 1d26'28"
Mitok
Leo 10h13'31" 25d12'27"
Mitra
Lup 15h29'39" -35d46'52"
Mitra Anatasia
Gem 7h22'7" 24d7'4"
Mitra Hadim
Crb 16h4'36" 34d44'55"
Mitra Leigh Grubb
Cas 1h27'16" 60d45'2"
Mitralex Libriski
Per 3h51'12" 49d45'24"
Mitsue
Cam 4h9'27" 64d42'37"
Mitsuko Fujimoto
Sco 17h23'17" -41d58'6"
Mitsuo Nakashima
Aql 18h52'11" 7d52'31"
Mitten Wedding Star 9-16-2006
Cyg 20h53'1" 36d19'15"
MittenHealy2163
Ari 2h40'48" 18d35'2"
Mittens De Los Santos
Uma 9h25'13" 45d30'43"
Mittens Sochylle Eternal Light
Uma 12h12'12" 58d21'55"
Mittens the Cat
Boo 14h52'56" 51d14'10"
Mitterer, Winfried
Ori 5h27'3" 13d17'19"
Mituniewicz, Wanda
Uma 11h51'6" 48d34'50"
M.I.T.Z.E.K.
Uma 10h16'41" 56d12'0"
MITZELPLIK
Cyg 21h52'20" 39d28'39"
Mitzi
Leo 11h28'11" 9d51'46"
Mitzi Faythe
Leo 9h24'8" 30d7'59"
Mitzi Grayson Wimberly
Cyg 19h46'59" 35d17'28"
Mitzi Jeanne Crawford
Vir 13h16'7" -21d9'41"
Mitzi & Justice
Uma 8h48'31" 46d37'56"
Mitzi Lawson
Lyn 8h40'6" 45d35'52"
Mitzi Lynn
Lep 5h41'30" -14d55'51"
Mitzi Trivett
And 0h36'54" 41d27'34"
MITZI01
Cnc 8h18'50" 6d39'41"
Mitzie Cossrow
Psc 1h24'19" 21d30'28"
Mitzy
Vir 13h4'26" 12d37'14"
MiVal
Cam 4h30'11" 53d50'26"
Mixa, Nicole
Uma 11h59'56" 35d30'33"
Mixie
Ori 6h12'53" 6d5'44"
Miya Louise Cook
And 23h36'14" 49d33'5"
Miyah Ashlyn
Uma 11h2'0" 70d7'36"
Miyako
Lyn 6h57'3" 58d53'43"
Miya's and Sean's PAR-ADISE
Leo 10h33'59" 26d31'23"
Miyoko Strong
Ori 5h37'50" 3d19'8"
Miyoung Shook
Uma 11h44'25" 35d12'19"
Miyuki
Gem 7h25'18" 33d2'7"
Miyuki
Umi 14h38'43" 68d11'37"
Miyuki Buchholz
Vir 13h11'27" 12d24'23"
Miyuki Tsukayama
Ori 5h33'30" 5d49'9"
Miz Mona Mohawk Nina
Sgr 18h14'32" -28d24'55"
Mizu Ran
And 2h9'20" 44d23'20"
Mizzi
Cam 7h10'12" 69d21'39"
Mizzy Mazzella
Cmi 7h34'30" 6d52'45"
M.J.
Ori 5h48'3" 11d46'54"
MJ
Lyr 18h37'17" 41d36'28"
MJ and Stephanie
Lib 15h45'56" -7d9'13"
MJ Champion
Her 17h52'46" 29d11'22"
MJ Cook
Uma 8h41'43" 52d47'31"
MJ DeGarmo
Eri 3h42'40" -0d11'53"
MJ Maynard
Ori 5h28'43" 0d35'29"
MJ Musselman
Gem 6h41'32" 34d0'10"
MJ Smile
Gem 7h16'9" 19d54'1"
M.J. Smits..light of my life
Cyg 21h56'2" 52d54'42"
MJ24
Her 17h29'32" 47d10'31"
MJ2K "1-26-00"
Uma 12h44'38" 58d52'57"
MJ42306
Cyg 19h34'33" 51d26'15"
MJ-5501
Cap 21h53'36" -8d33'3"
mjb & lilc
Ori 6h16'42" 13d21'6"
MJB05252007
Umi 17h6'7" 76d44'7"
MJD
Dra 16h53'5" 61d43'59"
MJD=My brightest star/favorite. Formal?
Psc 0h9'40" 12d17'43"
MJEternal Light Guides Eternal Love
Ori 5h49'53" -2d48'8"
MJF
Lyn 7h22'12" 50d54'48"
MJGJAK11/03/03
Lyn 7h0'19" 54d20'19"
M.J.G.P.-1: Pacchiano
Vir 12h37'42" 12d0'36"
MJI 1&2
Cap 20h30'26" -9d15'34"
MJK
Tri 1h53'5" 31d45'18"
MJKP02
Leo 11h17'51" 23d20'34"
MJL
Ori 6h3'23" 18d13'24"
MJM
And 0h55'29" 40d31'59"
MJP 143
Uma 12h46'41" 61d43'2"
MJP917
Hya 9h15'23" -0d34'5"
Mjr. Andrew Francis Starman 33 Sack
Apu 16h35'25" -74d13'25"
MJR GELD
Crb 16h18'41" 37d34'25"
M.J.S.
Boo 14h23'4" 14d28'33"
MJ's Sparkle
Aqr 21h18'52" -11d8'50"
MJW & CYT
Uma 11h5'10" 50d56'7"
MJW (Michael Jason Wilkey)
Cap 21h46'31" -13d48'37"
mjwolford
Aqr 22h56'28" -9d13'59"
MK
Dra 18h44'23" 64d56'47"
M.K.
Uma 11h41'31" 43d52'32"
MK
Lyn 8h51'25" 41d10'43"
M.K. Baby Hogerheide
Uma 12h28'29" 53d40'52"
MK De Los Reyes
Sco 17h24'5" -42d20'13"
M.K. Mulholland
Uma 13h52'22" 60d3'58"
MK329
Peg 22h50'50" 26d49'12"
M.K.D. STAR
Sco 16h14'5" -10d47'37"
MKH + ZKM
Cyg 21h27'41" 51d43'46"
MKK
Umi 16h30'47" 81d37'38"
MKS67
Cnc 8h51'26" 20d21'3"
MKTM112604
Sco 16h53'29" -42d17'13"
M.L. Furby Star
Uma 11h45'11" 61d42'30"
M.L. & Idell
Umi 15h41'13" 70d54'40"
M.L. Lewis
Aur 5h3'24" 37d18'5"
ML n°1 Marcella
Tau 5h8'42" 21d28'13"
M.L. Nayyar
Mon 7h20'42" -0d26'20"
MLAG'S Peaceful Place
Cap 21h32'1" -24d4'32"
M'Lee S. Harvill
Tau 4h12'53" 9d24'49"
MLEHRJAVA
Uma 11h33'47" 41d49'19"
MLEN8
Leo 9h45'35" 22d38'17"
MLFALKTX
Gem 6h17'39" 27d31'6"
MLH82998
Cyg 19h58'42" 34d14'17"
MLI
Uma 11h32'46" 61d35'31"
M'Liss
Lyn 6h52'55" 51d33'32"
M'Liss R. Miller
Lyr 18h20'26" 32d33'34"
MLJC102604
Vir 12h43'15" -8d51'16"
MLL4EVR
Uma 10h28'49" 66d36'24"
Mlle Ho
Lyn 7h28'24" 50d18'16"
MLMJBS
Umi 17h6'21" 75d48'44"
MLP
Her 16h50'8" 32d17'33"
MLR
Cnv 12h43'52" 39d27'42"
' Mls Katerinaki '
Leo 10h59'15" 14d16'31"
MLV
Uma 10h52'44" 62d53'14"
mm
Sgr 19h34'23" -15d4'4"
MM
Ari 3h9'39" 18d48'52"
M.M.
Lyn 8h59'2" 33d21'6"
MM Giese
Psc 23h13'11" -1d9'41"
Mma Bco
Leo 11h41'54" 23d21'6"
M.Mammitzsch Eternity
Del 20h36'32" 17d9'30"
MMCCS Pricer
Her 17h27'10" 41d49'40"
MMCK
Ori 5h53'14" 18d22'1"
MMD & JFN
Tau 3h36'2" 16d23'35"
MMG
Cas 0h56'38" 59d13'10"
MMLatone
Her 17h49'13" 33d3'5"
M.M.LeVerrier
Aql 19h49'17" -0d40'34"
MMM
Sgr 18h59'33" -33d47'59"
MMM "Candareenie" Yoh
Aqr 22h34'42" -1d39'31"
Mmm Donuts & Vroom Egli-Vincents
Nor 15h49'44" -53d57'38"
Mmmm... Soobute & Big Rubz 4—>
Cas 1h22'35" 62d10'48"
mmw *Our Love* enf
And 23h8'52" 51d24'30"
MN2 Edward Gardner on U.S.S. Scout
Aql 19h44'14" 7d2'29"
MnE4EVR+1
Uma 10h23'6" 71d57'46"
MNS062078
Gem 7h2'49" 27d6'35"
Mo
Com 12h33'57" 19d26'40"
Mo
And 0h47'14" 23d3'12"
Mo
Crb 15h56'16" 36d36'59"
Mo
Aql 20h22'37" -0d51'55"
Mo #1
Ari 2h35'55" 21d37'49"
MO - 14
Mon 7h16'44" -0d12'10"
Mo aingeal Muireann
Vir 12h45'58" 3d49'27"
Mo 'an' Al Steen
Cyg 19h46'44" 34d3'36"
Mo Anam Cara
Cyg 20h0'27" 30d33'37"
Mo Anam Cara Kendra
Cas 0h49'44" 64d4'27"
Mo Armstrong
Sco 17h58'26" -30d31'51"
Mo Chroi
Uma 10h24'35" 47d5'28"
Mo chuisle
Uma 12h6'18" 52d20'4"
Mo Chuisle
Uma 10h17'30" 61d43'42"
mo chuisle
Cap 21h47'2" -19d49'10"
Mo Cuishle
Lyn 6h32'53" 57d56'3"
MO CUISHLE
Cnc 8h42'11" 24d20'38"
Mo Cuishle Stephen
Lyn 8h13'20" 49d42'10"
Mo & David's Anniversary Star
Cyg 21h36'38" 46d6'2"
Mo Fior Gra
Aqr 22h11'56" 0d24'36"
Mo Ho's Star
Uma 9h46'18" 62d26'51"
Mo Leannan
Uma 9h11'48" 46d50'0"
'Mo' My Beautiful Angel
Uma 9h27'18" 52d16'58"
Mo' s Jenny
Gem 6h52'15" 21d28'15"
Moana
Cyg 20h31'35" 41d41'7"
Moana Ipo Nancy L Berry
Uma 11h36'1" 47d58'44"
Moazzam Ahmed
Ori 6h17'4" 10d38'5"
Mobi
Ori 5h56'16" 20d58'4"
Mobie
Aqr 22h28'41" -1d36'5"
Mobile Tech Thom
Uma 9h16'26" 52d48'3"
Mobius
Lyn 7h24'3" 53d37'19"
Möbius, Andrea
Sco 17h23'4" -32d22'55"
Möbius, Berit
Uma 11h56'38" 52d11'39"
mobstar*
Cnc 9h10'0" 15d53'11"
mobu
Cyg 21h19'10" 44d50'46"
Moby
Cap 21h32'35" -15d14'57"
MOC
Vir 13h7'31" -15d13'52"
Moc
Ari 1h49'5" 22d12'8"
Mocha
Ari 2h37'0" 28d4'59"
Mocha
Lib 15h32'17" -19d52'39"
Mocha's Buddy
Cas 1h17'2" 72d39'15"
Mochito
Lac 22h51'5" 49d42'36"
Mocita Celestial Clara Sanchez AHNI
Leo 10h52'57" 19d35'18"
Mocky Castro
Sco 16h11'8" -16d39'51"
MoClair
Cyg 21h28'35" 34d30'17"
Móczár Tibor Attila
Lib 15h47'31" -17d22'30"
Modena Hope Jacques
Tau 5h34'20" 19d41'6"
Modesto Fury 2005 Champs U12 Girls
And 0h50'6" 42d54'22"
Modoc
Aqr 21h8'28" -11d3'56"
MoDoQuRu 2004
Uma 11h25'31" 35d5'26"
Mody Ann Westcott
Crb 16h6'27" 36d21'28"
moe
Lib 15h9'45" -18d7'15"
Moe
Sco 16h36'43" -28d42'1"
Moe Bizari
Uma 9h29'54" 44d46'55"
Moe Joe
Leo 9h34'41" 27d47'39"
Moe & Joe 1976, 30 years
Per 4h25'43" 40d50'16"
Moe Joseph
Boo 14h36'4" 30d28'21"
Moe Moe
Peg 22h6'33" 6d54'35"
Moe Renfro
Vir 13h42'11" 3d57'56"
Moeglow
Umi 14h32'33" 71d8'58"
Moena
Cam 4h33'15" 54d34'13"
Moe's Fight
Her 16h42'40" 37d4'13"
Moesa's Destiny
Lyn 7h18'2" 59d2'26"
Moëurrier
Ori 5h41'44" 6d25'15"
Moey
Psc 1h13'1" 25d50'37"
Moffa/Savallo Family Star
Aql 19h4'27" 15d52'3"
Moford
Ori 5h37'0" 1d28'46"
MOG - 22
Cap 20h51'38" -19d55'54"
Moggel
Uma 10h31'45" 45d31'38"
Moggie
Per 4h28'42" 49d5'24"

Mohamad and Hayat Cyg 20h12'0" 50d46'56"
Mohamed and Zainab Lyn 7h46'31" 57d35'22"
Mohamed Bakri Cru 12h2'33" -63d8'46"
Mohamed Boudaher, my moudi Uma 10h3'37" 47d2'10"
Mohamed Omran Dra 18h41'45" 75d43'25"
Mohamed Saad Tau 4h7'3" 25d42'27"
Mohammad & Nancy Fallah Vir 12h18'12" 9d27'7"
Mohammad Polli Cyg 20h39'23" 45d31'29"
Mohammad Reza Goodarzi Vir 12h45'15" -7d35'52"
Mohammad Salahuddin Chowdhury Per 4h20'39" 51d49'12"
Mohammed Abdrabboh Cnc 8h15'5" 7d3'27"
Mohammed Akram Sco 17h36'47" -41d38'23"
Mohammed Al-Gadhi Al-Juaid Gem 6h44'36" 29d54'22"
Mohammed Arif Bhatti Per 2h51'51" 46d43'32"
Mohammed Ilyas Eri 3h10'7" -22d48'22"
Mohammed Rasol, My Soulmate Her 16h50'34" 15d59'40"
Mohammed Salman Saad Al - Saud Ori 5h56'20" 20d26'15"
Mohammed Sha Jahan And 1h0'35" 46d44'30"
Mohammed Sheikh - the Other Sock Cnc 8h36'19" 10d26'24"
Mohammed Slaibi Leo 11h8'41" 1d5'44"
Mohammed Usama Saleem Sgr 19h7'54" -14d13'27"
Mohammed Z Badal Sgr 19h36'35" -16d1'9"
Moharry Uma 8h35'26" 54d4'16"
MoHawk ClemSha LasNaasNer Ori 5h40'38" 0d6'58"
Mohegan 12.03.2005 Cyg 21h35'36" 47d10'17"
Moheganmoon1 And 23h12'43" 48d16'41"
Mohini Mehta Cep 22h23'37" 61d59'59"
Möhring, Claus Uma 13h9'36" 61d42'22"
Möhrli And 2h11'18" 50d9'48"
Mohsin Quinn Soliman, M.D. Cnc 9h7'9" 15d12'21"
Mohsin Raza Bajwa Dra 17h16'31" 52d24'21"
moi lapochka Vincent Vir 12h15'59" 11d2'15"
Moia Cyg 20h4'18" 35d55'2"
MoiMiliPlusik (Andre Debakhapouve) Tau 4h9'0" 8d20'34"
Moira Cam 4h36'11" 58d13'50"
Moira Cap 20h32'30" -17d37'19"
Moira A.J. Stewart Aql 18h48'29" 7d14'23"
Moira Amore della mia vita Cam 6h17'52" 69d56'26"
Moira Capela Uma 9h59'36" 57d35'33"
Moira Casali Cas 0h27'8" 61d27'30"
Moira Clair Crb 16h4'7" 39d13'15"
Moira Emilie Pho 0h37'11" -44d41'13"
Moira Emma Olberding Vir 12h39'5" -8d43'17"
Moira Erin O'Donnell Per 3h9'36" 42d9'53"
Moira Fasick Dra 19h54'1" 76d57'25"
Moira Flynn Cas 0h46'12" 57d22'33"
Moira Klich And 0h43'48" 43d37'42"
Moira Mattioni Sct 18h54'34" -12d55'19"
Moira McAN-Barrer Lyn 9h13'4" 33d21'47"
Moira Smith Lyn 8h43'11" 39d26'30"
Moira Teresa Ragen Vir 13h52'7" -10d36'44"
Moira (topina), 24.08.1975 Uma 11h4'42" 56d9'1"
Moiragirl Ori 5h29'40" 13d27'47"
Moirai Ori 5h46'35" 3d4'35"
MoiraMaus Uma 10h25'8" 67d50'51"
Moirin McNamara, Daughter Uma 12h47'50" 57d14'56"
MOISES Ari 3h21'0" 25d47'45"
Moises & Ginger Jane Cyg 20h31'36" 44d2'5"
Moisés II Rope Uma 14h11'41" 57d20'44"
Moises Iordanis G. C. Fonseca Alexiades Ori 6h14'7" 14d38'33"
Moissl, Ralf Uma 11h42'42" 36d30'58"
Mój Milosc Vir 14h7'32" -15d25'11"
Moj Misic Nenad Cyg 20h29'38" 56d54'23"
Moja Gwiazda Umi 17h31'31" 80d36'9"
moja jedina sreca B. A. Aur 6h10'54" 39d19'46"
Moja Kota Kochana Pawelek Cnc 8h59'35" 6d52'21"
Moja Maki Cas 1h21'7" 68d59'13"
MOJA MAMA Tri 2h41'27" 34d41'35"
Mojdeh Moayyed And 1h7'17" 45d1'12"
Moje Kochanie Lac 22h3'7" 45d1'35"
Moje Sunce 11.09.1979 Her 17h30'28" 32d53'26"
MojeOlgishte Peg 22h23'8" 22d58'19"
Mojgan Javid Smulewitz Vir 11h45'26" 4d36'3"
Mojo Ori 4h51'57" 5d45'30"
Mojo Ori 6h1'22" 10d42'43"
mojo Her 18h4'40" 25d52'24"
Mojo Her 17h43'37" 22d56'39"
MOJO Gem 6h36'39" 21d45'0"
Mojo And 0h15'2" 23d38'22"
MOJO Cyg 20h48'7" 43d4'31"
Mojo Umi 13h47'28" 73d20'15"
mojo Psc 1h36'42" 5d57'26"
MOJO Forever Uma 9h21'35" 67d24'16"
MoJoe Uma 12h31'51" 55d6'31"
Mok<\@> loca Lyn 8h30'38" 53d28'16"
MOK Meachum Cnc 8h50'0" 11d25'41"
Mokey-Nova Boo 14h27'17" 18d24'58"
Mokhat7 Lyn 6h38'38" 58d47'31"
Mokita Ori 5h4'30" 5d7'14"
Mokros, Gabriele Uma 8h43'18" 58d53'25"
MoK's_susan Boo 14h57'6" 54d8'18"
MoKy Lyr 18h48'56" 31d26'15"
Mokytoja Gelina Ari 2h36'29" 25d13'58"
Molaris Greens at Engel's Star Dra 16h36'39" 51d30'50"
Moldova Steel Works Cap 21h52'10" -13d29'51"
Moley Uma 8h57'10" 61d59'29"
Molica Mam Per 3h28'44" 34d19'12"
Molimau Haunga Uma 11h3'26" 53d36'24"
Molin Tau 5h20'51" 27d44'34"
Molittieri's Star Psc 0h33'8" 7d20'38"
Moll, André Leo 9h28'49" 10d39'54"
Moll, Marianne Ori 6h6'12" 7d28'44"
Mollee Markham Uma 11h10'54" 38d32'11"
Mollee Rae Cam 5h8'13" 67d51'29"
Möller, Claus Uma 14h2'57" 55d19'19"
Möller, Peter Ori 6h20'4" 14d24'57"
Möllers Vir 13h8'57" 13d39'28"
Molley "Mudder" Ross Psc 1h41'50" 5d6'47"
Molli And 0h21'32" 41d55'57"
Molli G Marie Halcomb Uma 9h1'34" 46d53'28"
Molli Maloney (MM) Aqr 21h14'39" -7d38'36"
Mollie Cma 6h22'9" -16d34'17"
Mollie Vir 14h0'40" -15d24'12"
Mollie Cam 7h35'2" 68d41'57"
Mollie And 2h32'27" 45d9'35"
Mollie And 0h3'30" 42d17'36"
Mollie Peg 23h10'51" 13d7'42"
Mollie Adam Sco 16h52'52" -30d51'18"
Mollie Allyn Lyn 7h34'21" 49d44'54"
Mollie Ann Her 15h56'27" 44d58'11"
Mollie Ann Huntsinger Tau 3h42'17" 26d3'4"
Mollie Anne Bruss And 2h10'56" 37d41'40"
Mollie Anne Jones And 23h15'32" 52d22'48"
Mollie Atkin And 1h6'4" 45d42'25"
Mollie Elaine Gross Ori 5h40'18" -0d32'8"
Mollie Elizabeth Tobin And 1h10'16" 46d32'4"
Mollie Elizabeth White Cru 12h43'9" -56d15'12"
Mollie Erin Ruby And 1h10'32" 42d25'45"
Mollie Grace Sheppard And 2h20'8" 49d57'31"
Mollie Hannah Levine And 1h22'15" 48d37'42"
Mollie Harris - Holy Molie Cap 21h21'22" -26d28'32"
Mollie Hayes & Daniel Arrant Tau 3h58'32" 4d15'49"
Mollie is Kari's Miracle, Always And 0h50'18" 37d43'53"
Mollie Jean Ori 5h56'32" 12d20'50"
Mollie Kelly-Hutson Umi 15h23'26" 87d52'43"
Mollie L. K Penrod Lyn 6h50'49" 56d51'31"
Mollie Michelle And 0h4'32" 33d2'29"
Mollie & Rachel Sorensen Uma 12h7'34" 56d37'5"
Mollie Rebekah Gavin And 23h31'27" 46d9'56"
Mollie Rolnick Crb 15h52'49" 27d35'12"
Mollie Sawyer Cru 12h48'45" -56d31'4"
Mollie Steinberg Aqr 23h43'49" -3d46'55"
Mollie the Baby Doll Lyn 8h12'20" 53d14'26"
Mollieana And 23h20'33" 47d59'15"
Mollie's Star Lib 15h40'44" -15d46'9"
Mollie's star Cap 21h39'11" -10d59'39"
Mollie's Super Cool 16th B-day Star And 0h39'8" 41d46'14"
Mollnick Cyg 20h25'59" 54d25'30"
Molloy's Cru 12h43'44" -57d57'50"
Molls Uma 10h31'48" 67d11'54"
Moll's Star And 1h11'44" 39d3'29"
Molly And 0h44'26" 39d33'40"
Molly And 1h21'14" 43d50'37"
Molly Lyn 8h1'2" 34d53'10"
Molly Cyg 21h16'37" 46d58'30"
Molly Cas 1h20'18" 52d9'55"
Molly Lyn 7h58'2" 46d55'27"
Molly Ari 2h2'53" 24d0'10"
Molly Vir 14h40'13" 3d59'54"
Molly Ari 3h5'58" 16d5'6"
Molly Vir 12h38'7" 11d1'59"
Molly Cas 2h1'46" 70d36'19"
Molly Aqr 22h32'0" -8d2'49"
Molly Umi 17h5'2" 84d41'48"
Molly Ori 5h16'51" -7d22'52"
Molly Mon 7h20'57" -6d48'8"
Molly Sco 17h16'17" -39d48'11"
Molly Lup 15h17'36" -42d52'28"
Molly A. Bardong Sgr 18h32'20" -27d59'38"
Molly A. Johnston Cnc 8h49'22" 24d13'5"
Molly A Peck Cnc 8h51'39" 11d50'4"
Molly Alice And 2h24'20" 46d45'37"
Molly Amber Gore And 1h57'47" 41d21'50"
Molly and Doug 5/1/06 Tau 4h49'26" 21d8'5"
Molly and Gabes Endless Love Star Lyr 19h11'34" 26d17'4"
Molly and Gary And 0h23'32" 41d59'59"
Molly and Keith Healy Ori 5h50'21" 5d54'53"
Molly and Nate Per 3h34'6" 44d42'16"
Molly and Tim Uma 10h20'34" 68d30'21"
Molly and Todd's "Joy" Cap 20h13'1" -8d40'49"
Molly Ann Croxall Uma 10h18'1" 45d27'44"
Molly Ann Kroenke Sgr 19h11'34" -29d14'19"
Molly Ann Proffitt Crb 15h51'56" 28d25'18"
Molly Ann Simpson And 1h25'57" 49d15'57"
Molly Annabelle And 1h29'1" 40d27'50"
Molly Anne Shipley Gem 7h4'41" 14d9'33"
Molly B. Brown Lmi 10h21'21" 33d40'52"
Molly Beck Fergustar 1 Ori 6h15'44" 13d49'16"
Molly Belle Sco 17h34'17" -44d37'32"
Molly Byers Ari 3h17'42" 29d22'55"
Molly C. Gray Cap 21h45'51" -11d14'16"
Molly Carlson Cam 6h33'1" 63d49'34"
Molly Cat Lyn 6h40'45" 52d11'4"
Molly Catherine Holt (9-13-2003) And 1h2'30" 42d14'40"
Molly Cathryn Koets Lyn 8h24'36" 55d23'44"
Molly Chase And 0h34'15" 37d43'47"
Molly Claus Cyg 19h47'19" 37d8'45"
Molly Colleen McWilliams Cas 0h25'41" 63d0'29"
Molly Crabb Aql 19h44'21" 14d52'46"
Molly Crawford Cnc 8h19'6" 11d31'27"
Molly Crawford Lyn 7h57'43" 58d7'22"
Molly Crismond Psc 1h7'25" 32d17'34"
Molly Crowther's 90th Birthday Star. Cas 0h54'52" 62d31'13"
Molly Cunningham's Star Lyn 6h52'52" 55d16'35"
Molly Dee Semerteen Cap 21h32'24" -10d59'31"
Molly E. Miller Gem 6h50'50" 32d26'56"
Molly Edythe Balsam And 2h12'13" 43d36'28"
Molly Elizabeth Boek Cyg 21h24'55" 30d51'33"
Molly Elizabeth Cook Sgr 19h1'57" -28d53'16"
Molly Elizabeth Costello 9/02/1999 Vir 13h14'4" 11d20'26"
Molly Elizabeth Gray And 1h7'3" 35d7'54"
Molly Elizabeth Morgan Faber - Molly's Star And 1h42'40" 41d13'58"
Molly Elizabeth Sacksteder Gem 7h3'59" 23d38'42"
Molly Elizabeth Sharp And 22h59'12" 52d27'34"
Molly Erin And 23h19'36" 47d13'33"
Molly Eve Per 3h10'57" 53d10'13"
Molly Eyberg Cnv 13h16'47" 41d17'27"
Molly Fish Ari 2h55'8" 29d47'17"
Molly Fitzgerald And 2h19'10" 48d24'4"
Molly Florence Risley's Star And 1h28'32" 35d6'33"
Molly Frances Fee Cet 1h8'13" -0d58'52"
Molly Fyfe Umi 15h31'18" 75d23'12"
Molly Gaffney Mulligan Lyr 18h58'20" 28d1'22"
Molly Gaspar And 23h44'56" 37d37'17"
Molly Gassner Leo 9h33'12" 28d38'54"
Molly Gottlieb Eri 4h36'24" -0d43'56"
Molly Grace And 23h9'57" 40d0'44"
Molly Grace Balog Mon 6h53'13" -0d29'48"
Molly Grace Carico Leo 9h28'52" 24d15'17"
Molly Grace Glaser "4-29-03" And 1h5'50" 45d5'22"
Molly Grace Robison 1989 Cas 0h10'3" 57d7'17"
Molly Huber's Star Per 3h26'9" 37d0'2"
Molly Irene Gill And 23h17'52" 35d12'37"
Molly J. Halterman Gem 6h53'28" 22d10'15"
Molly Jane Coghan Cnc 8h50'19" 14d3'28"
Molly Jane Sullivan Aqr 22h12'14" -1d48'32"
Molly Jayne Ori 6h2'15" 19d15'48"
Molly Jayne Carey Leo 11h13'34" 23d56'41"
Molly Jean Box And 0h44'6" 25d27'40"
Molly Jean Demaison And 1h41'14" 41d43'50"
Molly Jean Rositzki Uma 12h6'38" 53d48'47"
Molly & Jerry Wood Lib 15h27'54" -8d57'59"
Molly Jessica Stapleton Equ 21h5'49" 9d34'31"
Molly Jo Lmi 10h43'6" 30d47'59"
Molly Jo And 23h24'57" 40d31'3"
Molly Joiner Cap 21h43'9" -18d12'18"
Molly June Terenzi Leo 10h9'56" 26d29'16"
Molly K. Kearns Cas 0h38'54" 57d31'46"
Molly K Martin Vul 19h48'5" 28d5'25"
Molly K. Welter Sco 16h10'45" -29d22'30"
Molly Katherine Uma 9h52'12" 49d3'52"
Molly Katherine Gangelhoff Psc 1h17'9" 21d1'40"
Molly Kathleen Lib 15h13'35" -27d45'18"
Molly Kathleen Dolye St.Angelo Cap 21h31'34" -11d7'5"
Molly Kathleen McGuire Sco 16h16'22" -28d45'55"
Molly Kirsten Lyon Vir 11h55'11" -0d23'38"
Molly Knight Forever Shining Cru 12h4'27" -62d4'44"
Molly Lane Sgr 18h19'51" -16d46'54"
Molly Lee Thomas Leo 10h59'46" 11d40'34"
Molly Lewis Sco 17h59'9" -37d2'59"
Molly Lim Kheng Yan Leo 9h35'40" 23d57'26"
Molly Louise Terrano Tau 5h28'57" 26d10'41"
Molly Luraine Kendrick Lyn 9h7'50" 34d30'22"
Molly Lynn And 2h25'13" 47d55'15"
Molly MacGregor And 23h31'4" 41d32'0"
Molly Magdalen Allison Lyr 18h46'15" 37d1'45"
Molly Magullian Gem 6h24'38" 20d26'44"
Molly Mardell Lyn 6h27'54" 56d32'56"
Molly Marie Psc 23h30'40" 4d8'26"
Molly Marie Evenson Cas 1h12'46" 54d55'43"
Molly Martinez Cas 3h20'50" 74d20'16"
Molly Maureen Cyg 19h44'30" 43d47'47"
Molly May Barnes Vir 11h44'5" 2d37'51"
Molly McGoldrick Cnc 7h56'37" 18d6'48"
Molly Melane Morgan Vir 13h22'31" -15d0'3"
Molly Meredith Umi 14h41'17" 74d18'44"
Molly Moo Lmi 10h24'5" 36d39'35"
Molly Moore Lyn 8h26'48" 42d29'3"
Molly Morag Staddon And 23h13'17" 52d15'56"
Molly & Morgan Lewis Umi 13h55'19" 76d14'41"
Molly Nelson Tau 4h45'19" 17d51'45"
Molly Nover Crb 16h1'23" 28d17'55"
Molly Nover Mon 7h26'26" -7d49'40"
Molly Nownes Ori 5h52'8" 21d17'17"
Molly Olivia Drake And 23h4'12" 45d29'40"
Molly Paris Cnc 8h44'25" 7d10'57"
Molly Parks Cap 20h57'50" -22d20'51"
Molly Pin Li McLaren And 1h20'56" 49d34'28"
Molly Rae Cas 1h1'40" 60d59'0"
Molly & Reece Mitchell Dra 15h5'19" 57d51'34"
Molly Rhea Mann Cyg 21h6'54" 46d17'2"
Molly Roberts "My Star Forever" And 1h17'47" 42d0'5"
Molly Rocks Leo 10h31'18" 9d23'13"
Molly Rose Tau 5h7'20" 21d39'16"
Molly Rose And 0h6'39" 35d7'44"
Molly Rose Danowski Lyr 18h54'43" 32d24'15"
Molly Rose Farrell Crb 15h54'41" 27d43'53"
Molly Rose Johnston And 23h29'56" 41d46'1"
Molly Rose Zastrow Cas 0h15'18" 53d47'20"
Molly Ruby Margaret Nottage And 0h49'20" 27d52'8"
Molly Samantha Spurlock And 0h34'25" 30d52'21"
Molly Sankowski Cyg 20h24'5" 55d21'58"
Molly Snyder Boo 14h38'16" 27d29'33"
Molly Suzanne Reed Leo 10h27'31" 20d18'42"
"Molly T'AKA Alice Brown Deransburg Cas 1h25'7" 67d47'26"
Molly the Dog Cma 6h45'18" -14d20'55"
Molly the Mooch And 0h20'51" 38d58'53"
Molly the Owl Cas 1h30'48" 57d37'31"
Molly the Schmoo Uma 13h25'13" 59d23'28"
Molly Trace Reynolds Uma 13h33'0" 53d23'28"
Molly Victoria Forman And 23h32'33" 47d32'27"
Molly W. Gilbert Cam 4h46'0" 56d5'8"
Molly Walker 1919-2005 Ara 17h4'9" -50d53'25"
Molly Welker Cangiolosi Uma 9h39'33" 55d28'13"
Molly Wiley And 1h17'10" 38d56'53"
Molly "Wolly" Irene Olimb Leo 10h21'18" 8d54'29"
Molly69 Sgr 19h43'27" -13d4'26"
"Mollybol" A Star Forever Peg 23h9'13" 17d54'5"
Mollye Ann Psc 0h10'39" -4d57'52"
MollyEllenMayAlfiePeterGary Umi 16h8'0" 79d36'14"
MollyLance Cyg 19h30'46" 31d56'36"
mollynora Uma 13h59'19" 48d35'12"
Molly's Boris Ann Vir 14h43'6" 4d26'43"
Molly's Eternal Superstar Uma 10h47'21" 67d37'59"
Molly's Promise Uma 11h32'28" 56d2'14"
Molly's Shining Star Cma 6h38'39" -14d42'15"
Molly's Star Lib 15h43'9" -10d54'58"
Molly's Star Aqr 21h5'43" -13d2'44"
Molly's Star And 23h30'24" 47d7'2"
Molly's Star And 2h35'5" 45d4'4"
Molly's Star And 23h52'34" 35d19'20"
Molnár Balázs József Cas 23h41'57" 55d7'21"
Molnár Ferenc "Caramel" Aqr 22h6'25" -9d59'37"
Molnar Sabau Nicoleta And 0h1'46" 44d15'9"
Molo Vir 13h10'21" -4d19'17"
Moloss Uma 10h36'44" 69d8'33"
Molov Per 4h29'18" 48d51'11"
Molove Crb 15h28'37" 32d1'43"
Molter, Ernst-Joachim Ori 5h5'45" 5d38'16"
Molto et Téamo, Forever... Cas 1h7'52" 49d39'9"
Molyneux Cas 0h56'44" 56d2'43"
Mom Cas 0h35'10" 57d34'15"
Mom Cnv 13h41'52" 44d59'20"
Mom Cyg 19h46'18" 34d24'12"
Mom And 0h27'52" 40d53'27"
MOM Lyn 8h14'16" 42d32'24"
Mom Uma 10h51'19" 40d24'43"
Mom Aqr 21h2'38" 0d56'29"
Mom Umi 14h4'20" 71d17'29"
M.O.M Cas 1h44'31" 65d0'49"
Mom Cas 23h55'11" 57d53'35"
Mom Cas 1h0'41" 60d44'25"
Mom Cas 1h24'7" 62d7'29"
Mom Lib 15h48'51" -9d36'49"
Mom Lib 15h3'49" -14d48'59"
Mom Sgr 18h6'17" -20d22'50"
Mom 4 Eternity - MLHBF Ari 2h3'32" 22d15'34"
Mom and Dad Uma 13h33'33" 58d16'50"
Mom and Dad Uma 8h34'11" 71d57'30"
MoM and DaD FOR3V3R Uma 11h11'42" 29d19'21"
Mom and Dad Miller Lyn 7h45'47" 48d49'32"
mom and dad my special stars Cnc 8h50'11" 30d12'57"
Mom and Dad Wiseman Per 3h44'6" 45d50'52"
Mom and Dad, My Guiding Lights Cyg 19h35'20" 29d39'27"
Mom and Dads Eternal Wish Gem 7h28'33" 28d28'16"
Mom and Dad's Piece of Heaven Psc 1h19'0" 24d21'59"
Mom and Dad's Star Uma 12h33'52" 60d57'16"
Mom and Daniel's Star Lyr 18h43'26" 41d12'36"
Mom and Gram Durfor Cas 0h38'34" 60d57'47"
Mom and Mark, My Shining Stars Ori 5h58'57" 5d40'48"
Mom and me Sco 16h54'40" -34d36'55"
Mom Barb Gem 7h21'25" 33d28'9"
Mom / Barbara Hopta Lib 15h2'42" -27d13'10"
Mom(CathyS) Tau 4h33'57" 23d4'42"
Mom & Dad Per 4h48'57" 39d21'5"
Mom & Dad And 23h21'34" 48d18'29"
Mom & Dad Cyg 21h25'1" 52d6'5"
Mom & Dad 10th Aya & Russell Uma 10h49'32" 52d32'22"
Mom & Dad — 1974 Sgr 18h1'21" -26d17'29"
Mom & Dad 50 years & still shining Aqr 20h38'42" 1d43'46"
MOM & DAD FLEEHER Lyn 7h53'35" 52d2'6"

Mom & Dad Gordon-Smith Cyg 19h54'56" 33d20'52"
Mom & Dad Happy Anniversary Star Cyg 19h52'47" 55d51'13"
Mom & Dad Martin Cma 6h51'48" -13d5'17"
Mom & Dad's 40th Anniversary Star Cyg 20h11'20" 40d1'53"
Mom & Dad's Golden Promise Cyg 20h7'32" 40d29'20"
MOM & DADS STAR Lyn 8h22'47" 36d17'10"
Mom Darlene Per 4h41'55" 46d47'53"
Mom - Dillard Aqr 21h40'50" 1d51'12"
Mom "Energizer Bunny" Christine Lib 15h26'5" -12d29'51"
Mom Forever 06 Cyg 20h0'7" 56d41'12"
mom & fretty Lyr 19h17'16" 28d35'19"
MoM G Uma 8h23'35" 67d14'53"
Mom Girl's Twinkler Uma 14h17'19" 55d7'31"
Mom/Grandma/Marie Uma 11h54'16" 45d14'38"
Mom "Grandma T" "Auntie Em" Lyr 18h50'31" 42d26'15"
Mom - Irene Tolmie Cap 20h33'1" -19d8'4"
Mom is our Star Cas 0h56'30" 60d9'21"
Mom Jody's Star Cnc 9h7'48" 22d5'50"
Mom " Larissa " Lyn 7h0'25" 48d40'47"
Mom Louise Uma 11h15'47" 39d19'58"
Mom - Mary Freeland Per 4h40'44" 40d0'25"
Mom Milliken Vir 12h34'21" 11d35'17"
Mom Mom Bonnie Tau 5h19'34" 17d58'0"
Mom Mom Gloria Beheler Uma 9h20'33" 56d54'51"
Mom Mom Nancy Cyg 21h17'20" 47d6'0"
Mom Mom Rita Gem 7h37'10" 26d48'37"
mom mom's star Ari 2h58'46" 19d40'5"
Mom...... My eternal light. Psc 1h36'6" 16d56'54"
~Mom~ My Guardian Angel Cas 1h29'38" 67d57'46"
Mom n Da Lyr 19h20'22" 29d14'34"
Mom & Nama Gurr, worlds best. Aqr 22h15'38" 2d35'15"
Mom Peña Crb 15h37'4" 26d43'29"
Mom & Pop Ari 2h10'26" 20d26'15"
Mom & Pop Umi 4h44'59" 88d57'13"
Mom Rocklin Sgr 18h23'49" -21d15'22"
Mom - Sally Anne Cyg 20h57'12" 37d21'47"
Mom Sam Sco 16h12'35" -10d16'28"
Mom ... Star of Our Eyes Leo 11h14'1" 18d36'41"
Mom Superstar Peggy Hollenbeck Mon 6h51'11" -0d36'32"
Mom & The Old Man Umi 15h24'24" 76d0'43"
Mom Twinkle Star Cas 23h38'6" 55d46'30"
Mom Viv Cas 1h7'7" 61d46'57"
Mom Weezy Benson Ari 2h14'29" 23d33'18"
Mom, Beverly, Nana, Toolie Cep 22h15'25" 64d42'32"
Mom, Florence R. Gartland Uma 11h39'23" 63d16'8"
Mom, Gary, and Lauren Carstetter Uma 11h29'1" 33d25'0"
Mom, Mary and April Uma 11h50'16" 43d53'33"
Mom, My Best Friend Uma 11h14'7" 53d3'13"
Mom, my guiding star Ori 6h12'8" 13d20'5"
Mom, Oma Cyg 20h5'39" 36d57'39"
Mom, You are the love of my life Aqr 22h54'6" -5d56'58"
Moma & Popa Sge 19h50'2" 18d28'21"
Moma Wosheta Psc 1h9'59" 28d0'33"
MomALou Crb 15h52'36" 26d36'41"
MomanTom Cyg 21h52'47" 49d59'19"
Momar Peg 21h52'33" 10d29'27"
Momaw Stella Leo 11h20'47" 18d0'47"
Mome and Papa Soul mate star Lyn 7h36'24" 56d41'14"
Moment Uma 8h12'0" 69d6'41"
Moments in the Rain Cyg 20h22'40" 54d29'56"
Momentum Lyn 7h9'30" 57d40'38"
Mominator Dawn Char-Lou Aqr 21h1'36" -12d46'20"
Mom-Kate Cas 0h36'13" 58d5'49"
Momm Uma 9h16'22" 60d49'13"
Momma Uma 9h35'0" 46d57'48"
Momma Aqr 21h2'2" 0d54'51"
Momma Ori 5h25'15" 14d44'32"
Momma Ari 1h59'12" 14d3'5"
Momma Angie Lib 15h6'14" -7d6'34"
Momma B. Psc 23h9'32" 5d27'4"
Momma Bear Uma 11h24'50" 35d46'28"
Momma Bear Cha 13h50'41" -79d48'59"
Momma Bear's Love Ori 5h53'8" 21d25'8"
Momma Brooks Psc 23h47'37" 6d0'45"
momma d Lib 14h54'4" -6d5'26"
Momma Gin Leo 9h39'35" 27d35'34"
Momma Gwen Lib 15h25'43" -9d31'38"
Momma Joan Tau 4h15'23" 5d22'20"
Momma Joy Harker Cas 1h47'26" 54d49'33"
Momma Kay Ori 4h53'49" 10d56'16"
Momma Korol Leo 9h41'20" 24d57'9"
Momma Linda of the E's Sgr 18h11'18" -17d45'41"
Momma Lorraine Lyn 6h36'29" 58d31'33"
Momma Lynn Cas 23h25'59" 58d4'48"
Momma Maggie Arnold Our Angel Leo 10h43'38" 17d23'54"
Momma Mel Uma 13h43'25" 59d36'20"
Momma & Papa's Cyg 20h25'2" 46d29'3"
Momma / Paulette Harley-Lynch Her 18h23'32" 15d12'59"
Momma Puent's Star Mon 6h42'3" 6d25'12"
Momma Salamma Cas 1h12'10" 56d27'2"
Momma Sonnett Peg 21h33'47" 16d24'55"
Momma Star LJL Gem 6h56'14" 24d32'24"
Momma Sue Leo 11h45'8" 25d32'18"
Momma T-Horse Cas 23h27'54" 56d8'29"
Momma Vetor Cyg 20h16'58" 38d35'57"
Momma Wood Cas 23h7'0" 58d26'54"
Mommae Crb 15h49'47" 35d43'38"
mommae-ross Mon 6h27'42" 6d9'20"
Momma's Boys Cep 22h21'11" 73d46'28"
Momma's Star Cas 0h31'36" 64d7'28"
Momma's Star Uma 8h53'21" 56d9'33"
Momma's Star -Lonnie Lyn 8h19'26" 44d25'31"
Mommaw Lib 15h31'6" -11d8'1"
Mommaz Pup 7h44'53" -23d36'9"
MOMMIE DEAREST Gem 6h48'24" 26d12'3"
Mommie (Melanie Ann) And 0h25'18" 42d59'24"
Mommiegirl Cas 0h24'41" 51d51'26"
Mommie's baby-Zoe Alexandra Aqr 22h38'11" -5d3'35"
Mommies little piece of Heaven And 0h12'43" 44d18'39"
MomMom Sco 16h40'23" -42d25'1"
Mom-Mom and Pop-Pop's Golden Star Uma 12h17'0" 56d53'20"
MomMom Anderson Tau 4h22'20" 20d26'37"
Mom-Mom & Poppie's Christmas Star Uma 9h2'46" 49d3'42"
Mommy Cas 0h12'57" 57d3'35"
Mommy And 0h48'35" 40d32'36"
Mommy Gem 7h22'58" 32d46'11"
Mommy Cas 23h58'7" 54d21'2"
Mommy Uma 10h39'43" 62d8'48"
Mommy Cap 21h43'29" -9d22'11"
Mommy Lib 15h38'3" -19d43'7"
Mommy Col 5h40'26" -31d23'52"
Mommy 1 Lyn 7h12'54" 47d10'47"
Mommy 1956 Cas 1h36'27" 57d56'26"
Mommy aka. Nadia Marie Psc 0h52'56" 31d17'40"
Mommy and Daddy Sge 19h51'44" 17d58'41"
Mommy and Jacob's Star Gem 8h4'8" 32d31'57"
Mommy and Johnny Cnc 9h19'43" 14d25'56"
Mommy and Kylie Forever Cyg 19h55'58" 46d31'21"
Mommy and Miranda's May Star Uma 10h8'25" 52d43'32"
Mommy Angel Gem 6h31'51" 14d0'37"
Mommy Bemis Per 3h46'19" 35d9'9"
Mommy & Best Friend Christine Ladu Leo 10h15'17" 16d39'18"
Mommy Cinderella Cas 1h21'25" 64d41'3"
Mommy D Tau 4h34'21" 21d28'3"
Mommy D & Bill Lighting the Way Lyn 8h8'53" 38d38'38"
Mommy & Daddy's Shooting Star Cma 7h2'15" -21d43'8"
Mommy & Daughter (Brandy & Mia) And 2h17'34" 41d37'29"
Mommy Deb Tau 4h31'4" 13d23'34"
Mommy Jelley Cas 1h36'4" 58d5'42"
Mommy Lorrel Uma 10h2'30" 66d2'48"
Mommy Loves Bradyn Cyg 21h42'31" 53d12'20"
Mommy Loves Mason Forever Leo 11h51'22" 22d36'42"
Mommy Lynne Sco 16h43'9" -29d20'47"
Mommy & Me - Susan and Taylor Umi 14h18'7" 74d6'36"
Mommy Michelle Starr Leo 9h23'27" 16d48'54"
Mommy My Shining Star from Heaven Gem 7h23'42" 29d12'0"
Mommy MZD Vir 12h34'34" -0d23'46"
Mommy N Lauren Cas 1h0'15" 54d3'32"
Mommy Okon Cas 1h12'43" 56d39'47"
Mommy Paula Crb 15h40'13" 27d36'55"
Mommy - Ryan's Brightest Star - Sheralyn Umi 5h18'1" 89d25'20"
Mommy Star Cnc 8h16'4" 7d58'29"
Mommy Stasia Per 3h43'41" 50d41'0"
Mommy T loves Macy Maree Sgr 18h31'27" -27d30'16"
Mommy Theresa's Star Sco 17h10'31" -39d10'26"
Mommy Tina Cas 1h59'35" 61d35'11"
Mommy Vicky Cas 0h47'14" 50d21'16"
Mommy, The Star in My Life Aqr 21h40'43" -7d46'17"
MommyAndMatthewAndKayceeAndSean Lib 16h0'11" -9d32'14"
Mommyfeet Sco 16h15'28" -21d13'25"
Mommyommyommys! Sco 17h12'33" -35d40'30"
MommyRanda Tau 5h46'46" 28d3'49"
Mommy's 40th Vir 13h20'40" 7d11'5"
Mommy's Angel Ori 5h45'31" -0d12'32"
Mommys Angel Baby Maleena Jayne And 23h27'25" 47d39'33"
Mommy's Angel Mason Salvator Her 16h35'45" 24d10'56"
Mommy's Guiding Light - Jill & Anna Cas 1h29'39" 67d9'17"
Mommy's Little Emma Uma 11h22'15" 53d27'48"
Mommy's Little Girl Aur 5h51'19" 35d40'29"
Mommy's Little Princesses Leo 10h16'36" 14d56'17"
Mommy's Mr. Moon Uma 11h24'56" 59d53'32"
Mommy's "Our Everlasting Love Star" Ari 2h18'46" 24d45'38"
Mommy's peep hole Cnc 7h58'50" 16d50'11"
Mommy's Sparkling Touch Cas 1h30'36" 62d21'8"
Mommy's Star Umi 15h25'49" 70d7'14"
Mommy's Star Cas 0h37'35" 54d22'8"
Mom-Nanny's Star Ari 2h44'28" 26d0'26"
Mommy Lim Uma 8h54'21" 55d0'28"
Momo Uma 14h0'50" 54d4'41"
MoMo Tau 3h40'39" 16d6'20"
Momo Uma 14h2'16" 49d35'36"
MoMo Uma 11h42'42" 44d29'46"
Momo Kesen Her 18h3'4" 18d18'58"
Mom's Angels Ari 2h47'53" 27d51'3"
Mom's Birthday Star Cyg 19h43'24" 32d2'6"
Mom's Butterfly Ari 2h28'25" 25d3'14"
Mom's eternal light Ori 6h4'32" 10d42'59"
Mom's Hope Star Cas 22h58'22" 53d21'13"
Mom's Inspiration Vir 12h42'26" 12d4'23"
Mom's Lil Men Uma 10h59'52" 61d6'59"
Mom's Little Getaway Ari 3h14'16" 28d26'56"
Mom's Little Twinkle Sco 16h2'14" -27d54'47"
Mom's Love And 23h43'22" 47d35'22"
Mom's Lovely Growing Garden Cas 0h14'32" 51d42'59"
Mom's Lucky Star Cas 0h23'40" 53d1'34"
Mom's Lucky Star Ari 2h9'4" 24d31'40"
Mom's Miracle Vir 13h22'24" 8d15'56"
Mom's Miracle - Stefran Lyr 18h53'25" 33d35'37"
Moms Miracle, Daddys Girl Shae Psc 1h18'10" 17d7'46"
Mom's Shining Star Cas 1h41'0" 63d50'47"
Mom's specail star Uma 8h18'35" 72d48'7"
Mom's Star Cnc 8h27'21" 10d33'13"
Mom's Star Tau 4h36'50" 28d20'4"
Mom's star Crb 16h20'13" 32d58'42"
Mom's Star Crb 16h3'42" 35d1'58"
Mom's Star (Tanya Hall) Aql 19h5'21" -7d48'5"
Mom's Star - The Best in the Sky Cas 1h21'41" 57d31'33"
Mom's Star-Linda Scott-from Chris Cyg 19h59'15" 36d55'51"
Mom's Superstar Cas 1h23'56" 62d6'41"
Mom's Tinkerbell Lyn 8h0'30" 42d15'43"
Mom's Twinkle Umi 15h8'3" 68d5'17"
Mom's Wish Aqr 22h3'42" 0d34'45"
momstar Cnc 8h44'14" 31d43'47"
momstar Sco 16h18'50" -27d40'51"
Mon 7éme Ciel Ori 5h37'59" -2d41'0"
"Mon Amour" Aql 19h8'28" 2d9'31"
Mon Amour Tau 3h40'4" 7d16'21"
Mon Amour Her 18h56'37" 16d48'50"
Mon Amour Del 20h31'6" 16d18'29"
Mon Amour Absolu Cas 0h27'37" 61d42'21"
Mon Amour de Pauline Cas 22h57'30" 55d51'40"
Mon Amour et Ange ~ Candice Marie Cyg 21h29'40" 32d53'47"
Mon amour et ma lumière Cru 12h19'49" -57d54'46"
Mon amour éternel pour Danielle Lep 5h58'57" -18d12'33"
Mon Amour Je T'aime Cyg 21h47'8" 44d35'12"
Mon Amour pour l'Eternité Uma 9h31'25" 61d32'57"
Mon amour, Niki Sue Uma 11h54'19" 28d21'0"
Mon Ange Psc 23h8'25" -1d23'32"
Mon Ange Uma 8h36'11" 70d43'16"
Mon Ange Aurélie André Leo 10h29'57" 25d33'26"
Mon Ange (Christine BB.) Sco 17h40'8" -39d14'43"
Mon Ange Karine Boutin Tau 5h11'20" 21d12'24"
Mon Ange Sophie et Philippe Umi 14h20'27" 72d1'23"
Mon Autre Tau 4h40'43" 22d38'12"
Mon BB Nico Cap 21h8'47" -20d39'9"
Mon Beau Malcolm Parmenter Col 6h0'4" -42d50'15"
Mon Bel Amour, Un Amour Infini Lyr 18h30'30" 41d52'0"
mon cher et petite choux Cyg 20h22'21" 56d6'55"
Mon Cheri Uma 11h46'18" 32d2'28"
Mon Coeur Gem 6h29'7" 25d18'35"
Mon Coeur Boo 14h43'36" 20d2'24"
Mon Coeur Aqr 21h42'39" -0d38'57"
Mon Coeur Ori 5h32'50" -0d52'56"
Mon Coeur Umi 14h26'37" 85d19'16"
Mon Coeur... And 1h30'48" 46d59'31"
Mon coeur Delphine Durant Sco 16h52'8" -26d27'28"
mon coeur wolfgang Umi 14h56'0" 71d43'52"
"mon diam's" Laetitia Maréchal And 23h24'13" 43d42'9"
Mon Dodo Tau 4h5'36" 13d46'49"
Mon Doudou Del 20h34'38" 7d26'14"
Mon Doudou Cyg 20h44'8" 45d48'2"
mon et seulement Ari 2h39'30" 25d8'54"
Mon Etienne plus lumineux Uma 11h19'24" 47d8'31"
Mon étoile, pour te voir Dra 17h16'5" 57d28'44"
Mon Huer Uma 11h50'50" 50d0'47"
Mon Meilleur Ami Crb 16h12'53" 28d19'29"
Mon Monde Leo 10h17'19" 7d28'9"
Mon Nino Ori 6h6'25" 13d48'12"
Mon nounours Gem 6h40'59" 32d38'27"
Mon Petit Coeur Tau 4h30'27" 15d51'5"
Mon Petit Lumière Ori 6h18'0" 3d24'5"
Mon P'tit Loup - Emmanuel St-Laurent Umi 16h53'21" 78d5'3"
Mon Titou Uma 9h23'14" 52d46'35"
Mon tout Uma 9h6'21" 49d8'39"
Mon trésor bien-aimeé Uma 11h16'20" 30d26'30"
Mona And 0h29'31" 37d8'29"
Mona And 23h23'12" 47d39'23"
Mona Ori 5h51'6" 6d49'25"
Mona Leo 10h47'22" 7d8'50"
Mona Tau 5h7'15" 21d2'8"
Mona Com 12h20'24" 25d50'34"
Mona Ari 2h14'33" 24d33'24"
Mona Cas 1h38'41" 61d47'46"
Mona Uma 9h4'0" 64d28'30"
Mona Bauer Oakes Her 18h41'26" 19d29'13"
Mona Behnammoradi Crb 16h11'54" 32d15'42"
Mona Bregman Borealis Leo 9h38'20" 26d41'40"
Mona Catherine Ori 5h53'39" 8d3'3"
Mona Dawn Ward Sco 16h14'1" -17d1'1"
Mona Elisa Hayel Malawi Cyg 21h40'55" 44d6'58"
Mona Hanf Uma 8h11'18" 68d9'49"
Mona Jean Nicholas Cas 0h52'12" 53d44'3"
Mona Joyce Rust Lyn 8h5'56" 52d40'36"
Mona Kay Cnc 9h4'2" 21d42'47"
Mona Lee Williams Leo 10h15'47" 23d36'54"
Mona Leroy Cnc 8h40'27" 31d24'15"
Mona Likes Tacos Cam 3h27'58" 61d9'50"
Mona Lisa Baker Sco 16h14'43" -19d20'46"
Mona Lisa's Heart Aqr 22h18'15" 0d17'29"
Mona Louise Uma 12h3'10" 58d1'16"
Mona Magness And 2h35'36" 45d37'42"
Mona Mickael Uma 11h1'57" 37d18'37"
Mona Monaghan Vir 12h15'20" 12d45'29"
Mona My Love Cas 0h28'7" 64d2'12"
Mona of Langhorne Uma 10h0'47" 71d59'50"
Mona O'Sullivan Cas 1h1'50" 63d38'47"
Mona Rae Tau 5h52'3" 26d39'29"
Mona Rae Peterson Ari 2h44'19" 13d49'51"
Mona - The Light Of My Life Crb 15h25'55" 29d38'3"
Mona Tichatschke Ori 5h56'36" 18d39'29"
Mona Wiechers Tau 4h12'41" 5d59'43"
Monagan's Magic Night Cyg 22h1'41" 50d59'7"
Monahans High School Drama Lib 14h53'46" -9d16'49"
MonaLisa Leo 10h6'51" 13d31'1"
Monalisa Ari 3h15'50" 21d25'26"
Monalissa Malheiros And 1h18'10" 48d49'32"
Monalou Lyn 7h42'52" 48d43'43"
MonAnge Uma 11h32'37" 55d48'54"
Monarch's Song 05-29-05 Crb 15h17'49" 26d44'6"
Monas Taufstern Umi 16h23'19" 78d4'35"
"Mona's World" Crb 15h52'19" 34d22'48"
Monay And 23h15'32" 42d30'44"
Moncha Ferrell Ari 1h52'48" 19d13'6"
Monchichi Psc 1h17'47" 31d29'24"
Monchichi143 Cen 11h45'22" -59d17'1"
Monday, 4th of August 2003 Cas 3h3'35" 61d45'9"
Monde du Lapin Vir 13h7'13" -7d17'31"
Mondo M Vir 12h40'0" -4d19'17"
Mondragon's Bull Tri 2h24'13" 30d36'0"
moné Uma 8h31'55" 64d14'31"
Monet Peterson Sgr 18h43'55" -17d2'36"
MONEY LOVE Cyg 21h36'2" 34d7'23"
Money Stacks Lyn 6h33'14" 53d59'37"
Money Tree Leo 11h17'20" 18d56'0"
Mong Thanh Cnc 8h44'44" 26d20'58"
Mongania Uma 10h28'44" 61d57'14"
Mongeon Star Umi 15h5'28" 73d2'57"
Moni Uma 14h3'59" 56d18'56"
MONI Sct 18h27'31" -15d36'49"
Moni Tau 5h31'56" 22d45'16"
Moni Tau 5h58'52" 24d41'47"
Moni Gambrel Uma 11h48'28" 36d34'35"
Moni + Gion Her 18h42'31" 20d33'9"
Monia Ori 5h53'43" 21d21'26"
Monia Lyr 18h25'36" 31d28'40"
Monia Uma 13h42'2" 55d56'33"
Monia & Woody's B & B Peg 23h21'30" 29d1'20"
Monic Gutbier Uma 10h59'31" 58d54'20"
Monica Uma 12h21'12" 56d21'54"
Mònica Dra 17h51'25" 64d3'49"
MONICA Uma 14h5'59" 62d4'41"
Monica Cas 2h23'10" 65d57'40"
Monica Cam 5h36'23" 62d21'22"
Monica Cap 21h21'56" -15d58'3"
Monica Lib 14h50'15" -7d37'35"
Monica Sgr 19h22'1" -13d26'31"
Monica Mon 6h49'10" -0d3'42"
Monica Sgr 18h53'56" -32d0'49"
MÓNICA Sco 17h24'41" -42d17'26"
Mónica Eri 4h27'57" -24d55'42"
Monica Cap 20h23'59" -24d9'36"
Monica Lmi 10h51'9" 29d7'33"
Monica Tau 4h33'8" 25d49'6"
Monica Gem 7h23'7" 20d57'51"
Monica Ari 1h51'4" 24d5'37"
Monica And 0h36'25" 29d0'16"
Monica Leo 11h26'5" 12d46'35"
Monica Ori 6h13'57" 6d31'21"
Monica Ori 4h47'55" 10d23'15"
Monica Aqr 21h37'18" 2d18'55"
Monica Crb 16h11'20" 33d23'41"
Monica Crb 15h41'59" 36d23'4"
Monica Aur 4h55'24" 36d49'20"
Monica Lyn 8h50'39" 34d39'47"
Monica And 0h34'41" 41d33'43"
Monica Cyg 19h56'17" 35d23'32"
Monica And 1h37'47" 44d4'38"
Monica Lyn 8h49'59" 39d52'39"
Monica And 23h12'49" 43d20'31"
Monica And 0h1'27" 45d59'32"
Monica Cas 0h2'22" 56d16'23"
Monica 16 Lib 14h59'41" -16d42'35"
Monica 21 Uma 12h17'56" 60d17'59"
Monica 28/02/70 Psc 23h1'38" 0d48'33"
Monica Alana Sex 9h41'47" -0d18'28"
Monica "Alex" Hall Cyg 19h45'47" 29d43'53"

Monica Alison Adolfie
And 2h28'2" 44d36'4"
Monica Alvarez
Sgr 18h50'52" -32d50'33"
Monica Anai Navarro Morales
Peg 23h56'17" 12d18'44"
Monica and Jayden
Gem 7h3'48" 24d6'36"
Monica Angel Blas
Umi 15h5'56" 70d46'30"
Monica - Angel precioso de Michael
Cas 23h24'0" 56d4'58"
Monica Anita Rabe
Ari 3h3'48" 27d24'46"
Monica Ann Carman
Lib 14h54'38" -6d21'10"
Monica Ann September 5, 2002
Cas 1h30'3" 62d59'20"
MONICA ANN TABOR
Cas 0h47'50" 65d8'35"
Monica Arnone
Lib 15h30'45" -21d29'17"
Monica Asha Dookharan
Del 20h26'22" 20d44'52"
Monica Auer-Müller
And 23h18'41" 51d26'51"
Monica Barcati e Omar Camata
Cam 11h14'9" 85d12'17"
Monica Bestelmeyer 22.07.1980
Uma 14h22'5" 55d8'6"
Monica & Bill Winters
Cyg 19h30'49" 51d35'18"
Monica Blair Kreindel
Lyn 8h48'15" 35d27'40"
Monica Boemi
Gem 7h34'12" 16d9'34"
Monica Börjesson (Mandy)
Uma 14h17'26" 60d13'18"
Monica Bruhn
Aqr 21h41'49" -0d11'2"
Monica Bua and Francisco Levine
Ara 17h18'58" -47d21'11"
Monica Buckley Price
Lyn 8h13'4" 49d3'12"
Monica Bysinger
And 0h47'36" 40d2'4"
Monica C. Giddings-Luk
And 23h11'39" 51d21'27"
Monica C. Lal
Cas 0h27'26" 48d24'7"
Monica C. Peterson
Cam 7h32'31" 61d5'19"
Monica Castro
Lyn 8h6'39" 41d47'52"
Monica Christabel Ruth Wauchope
Sco 17h53'35" -36d32'29"
Monica Conti
Cyg 21h36'38" 50d13'9"
Monica Cordeiro Guerra
Tau 4h7'57" 27d52'15"
Monica D. Whatley
Cas 1h33'37" 69d40'18"
Monica Danielle Leach
Lyn 7h1'33" 54d20'49"
Monica Dawn Hibner
Cnc 7h57'26" 13d11'7"
Monica DeLaGarza
Tau 5h11'54" 24d31'3"
Monica Demello
Cam 5h29'35" 68d26'3"
Monica Diane
Uma 10h31'38" 47d55'9"
Monica Diane Acosta
Uma 9h12'11" 53d29'34"
Monica Diane Murphy 102670
And 1h4'57" 47d45'24"
Monica Doreen Muro/Queen
Uma 10h38'33" 54d37'9"
Monica & Duane
Cyg 21h34'11" 38d8'6"
Monica E. Ramos
Cam 5h41'30" 56d35'29"
Monica Enders
And 2h37'3" 45d11'49"
Monica Esteves - The Star of Monica
Pho 23h29'41" -56d2'1"
Monica F. Covert
Ari 3h2'26" 14d33'30"
Monica Foran
And 23h12'43" 49d12'52"
MONICA GERMAN PEREZ
Col 6h17'24" -33d39'14"
Monica Gomez
Cam 3h17'39" 64d52'42"
Monica Gormley
Ori 5h52'27" 18d25'2"
Monica Grace Milkint
Ori 5h33'20" 1d28'57"
Monica Gutierrez
Psc 0h57'12" 4d44'56"
Monica Holland - Birthday Star.
Tau 3h44'25" 29d6'32"
Monica Iglesias Palacio
Ori 5h58'54" 20d43'56"

Monica il Bello
Lmi 10h39'40" 32d27'9"
Monica Imani
Lmi 10h34'13" 32d52'55"
Monica J Oberthaler
Ari 2h48'16" 29d19'56"
Monica Jane Patten
Sco 17h52'15" -32d31'16"
Monica Jean Farwell Weber
Leo 11h21'37" 14d31'57"
Monica Jeanette Maria VanTrieste
Lup 15h47'7" -31d34'5"
Monica Jenkins
And 0h22'45" 41d42'49"
Monica Joy Walsh
Leo 10h40'15" 14d52'56"
Monica Kaltreider
Crb 16h11'19" 36d37'7"
Monica Karam
Sco 17h53'28" -30d41'44"
Monica Karpecki
Sgr 18h26'54" -16d44'17"
Monica Kay
Cap 20h36'13" -13d54'58"
Monica Kim Wright
Cas 23h38'4" 55d50'24"
Monica Kristen McKay 120488
Lyr 18h38'5" 30d40'26"
Monica L. Benavidez
Ori 5h57'48" 22d30'54"
Monica L. Brener
Sco 16h8'45" -16d30'0"
Monica L. G. Forbes
Sco 17h34'13" -38d34'33"
Monica L. Griffith
Mon 6h48'7" -0d41'1"
Monica L. Juarez
Mon 6h45'34" -0d17'34"
Monica L. Larca
Sco 17h3'2" -35d14'27"
Monica Lalwani
Cas 23h2'39" 55d52'3"
Monica Lamay
Aqr 21h2'8" 1d41'22"
Monica Lee
Mon 6h45'33" -0d41'12"
Monica Lee Angella
Cam 5h26'15" 57d20'10"
Monica Lee Foote
Cep 0h13'21" 69d58'29"
Monica Lennon Golden
Cam 3h51'45" 59d13'52"
Monica Leon
Cyg 19h10'9" 52d17'17"
Monica Lor Novak
And 0h46'42" 33d23'4"
Monica Lorraine Williams
Leo 10h30'20" 13d38'39"
Monica Losavio
And 1h57'0" 39d8'51"
Monica Lynette
Aqr 22h44'59" 1d18'36"
Monica Lynn Reed
Vir 12h44'46" 8d12'49"
Monica Lynn Trepiccione
And 1h54'39" 39d3'6"
Monica Lynn Tyson
Psc 1h25'25" 17d6'40"
Monica Lynne Roy
Tau 3h32'1" 4d53'27"
Monica M. Cerros
Tau 3h39'48" 8d12'8"
Monica M. Palko
Cap 20h7'45" -25d14'33"
Monica Mainardi
Aur 7h27'30" 42d31'5"
Monica Malone
Lyn 7h2'30" 50d56'17"
Monica Margaret Stomner
And 1h41'52" 41d22'39"
Monica Maria la mejor, dulce y querida mamá
Uma 9h32'49" 52d51'12"
Monica Maria Neira
Aur 5h21'6" 31d37'49"
Monica Marie
And 1h17'22" 45d23'7"
Monica Marie
Tau 4h40'19" 24d32'35"
Monica Marie Campbell
And 0h46'6" 43d27'18"
Monica Marie Mora
Ari 2h18'57" 24d26'21"
Monica Marie Tate "Lady Girl"
Leo 10h4'14" 8d50'41"
Monica Mariuzzo
Umi 16h18'59" 76d58'43"
Monica Marlenne Valencia Haro
Cap 20h8'51" -22d21'44"
Monica Martorell Edreira
Vir 14h25'40" -1d1'16"
Monica Mary
Tau 5h58'16" 28d12'37"
Monica Mary Josephine Hatfield
Cyg 20h0'18" 30d10'12"
Monica Mary Stella Hewitt
Cas 2h5'12" 73d31'39"
Monica Matlock
Umi 14h28'36" 73d32'50"

Monica Mattie Sharkansky
Lyn 8h29'51" 56d33'18"
Monica McGavran
Uma 9h19'18" 63d0'55"
Monica My Love
Tau 4h31'38" 16d47'25"
Monica Myrdahl
Umi 15h20'20" 71d55'13"
Monica "Nanina" Venturi
Peg 22h37'58" 24d10'19"
Monica Nuno
Sgr 18h47'8" -21d9'15"
Monica Origel
Mon 6h26'53" 0d1'18"
Monica Ortiz
Tau 4h6'45" 5d29'19"
Monica Pauline Buzzanca
And 0h37'58" 30d0'24"
Monica Pauline Kellar
Col 6h32'0" -41d30'8"
Monica Pell
Aqr 22h7'49" -2d45'14"
Monica R. Barnett
Sco 16h53'51" -44d23'22"
Monica Rae
Uma 9h19'23" 52d36'27"
Monica Ramos
Aqr 21h57'4" 0d27'42"
Monica Ratley
Psc 0h7'34" 3d21'10"
Monica Raye Gallegos
Her 16h45'40" 42d57'1"
Monica Reed
Leo 11h1'47" 3d56'16"
Monica Renee Galsterer
Cnc 8h2'9" 12d47'9"
Monica Renee' Morgan
Cyg 21h18'13" 43d58'6"
Monica Riddle
Cyg 20h55'2" 49d36'1"
Monica Robledo
And 2h21'30" 49d8'35"
Monica Rodriguez
And 0h52'53" 40d31'43"
Monica Rosario
Lyn 8h20'57" 38d47'39"
Monica Rose
Lib 15h31'45" -29d49'14"
Monica Rose Ellerbrock "Blondie 2"
Ori 4h46'51" 12d14'9"
Monica Rose Meyers
And 0h45'32" 45d28'7"
Monica Rose Strickefaden
Uma 13h34'58" 56d15'7"
Monica Rosu -"Mona"
Cyg 20h59'42" 47d32'49"
Monica S. Smiley
And 0h49'23" 39d14'25"
Monica Sievert
Cas 0h9'7" 53d16'54"
Monica Silberberg
Tau 3h42'46" 23d27'12"
Monica Soisic Ralfe'
Pho 0h17'35" -39d20'48"
Monica stregabona
Ori 6h20'54" 11d17'28"
Monica Sundberg
Pho 1h36'20" -41d41'28"
Monica Teresa Lewis
Lyr 19h9'15" 43d2'17"
Monica Torneiro
Cep 0h8'7" 72d24'37"
Monica Weyant
Hya 9h24'37" 4d26'9"
Monica Wonzong
Cam 4h55'0" 59d18'0"
Monica Worley
Cas 0h7'30" 57d3'19"
Monica Yacoub David
Sco 17h54'16" -35d44'54"
Monica Young
Uma 10h46'1" 50d22'23"
Monica Youngblood
And 0h53'11" 36d35'49"
Monica Yvonne Choi
And 0h23'31" 27d9'38"
Monica, 26.07.1980
Sct 18h48'59" -11d53'3"
MonicaBeserra
Aqr 22h40'55" 0d4'37"
MonicaPatel
Gem 6h33'50" 17d21'54"
Monica's 40
Cas 1h42'22" 62d29'12"
Monica's Beautiful Smile
Leo 10h16'20" 16d56'20"
Monica's Fire 140207
Ori 6h16'57" -1d14'55"
Monica's Free
Cas 0h41'48" 65d5'57"
Monica's Heart
Aqr 20h48'44" -2d36'33"
Monica's Star
Ori 6h16'50" 10d55'29"
Monicas Wish
Ari 2h10'46" 24d9'10"
Monick
Pho 0h42'53" -41d38'20"
Monicutza
Gem 6h58'51" 25d9'27"
Monieca Yniguez
Aqr 22h4'56" -3d53'38"
Monie-Monie-Hoonie
Psc 0h22'43" 8d50'7"

MONI-in ewiger Liebe-Andi
Uma 9h44'13" 68d18'6"
Monika
Dra 17h47'19" 69d20'20"
Monika
Umi 14h31'45" 79d9'24"
Monika
Lac 22h46'52" 54d1'34"
Monika
Ori 5h8'35" 8d15'11"
MONIKA
Ori 6h3'32" 10d18'26"
Monika
Leo 10h41'48" 7d55'34"
Monika
Psc 0h52'22" 18d10'27"
Monika
Ori 6h2'32" 19d2'38"
Monika
And 1h0'57" 38d29'24"
Monika
Lmi 10h20'17" 37d50'4"
Monika
Cas 0h9'35" 57d7'6"
Monika 1961 we will always love you
Uma 10h50'19" 42d34'8"
Monika 22
And 0h31'42" 40d39'42"
Monika 8871
Lmi 10h4'54" 36d41'40"
Monika Absolonova
Cas 23h44'49" 53d41'29"
Monika Alissa
Uma 10h15'37" 72d7'3"
Monika and Chapman
Crb 16h19'36" 33d51'1"
Monika and Danny
Sgr 19h47'55" -11d57'48"
Monika & Anthony
Cyg 20h5'9" 41d30'1"
Monika Bednarowicz
Sco 16h56'47" -37d13'57"
Monika Blaschke
Uma 9h22'23" 41d39'55"
Monika Dreves
Uma 11h58'37" 31d17'13"
Monika Eichholz
Uma 10h55'23" 71d39'4"
Monika Espinoza and Lauren Pilnick
Vir 13h47'58" -1d5'51"
Monika Fischer
Umi 13h32'54" 72d3'0"
Monika George Stamatiou
Ori 6h22'31" 10d46'7"
Monika Giess
Ori 6h14'43" 8d22'36"
Monika Gorska - MIKA
Sco 17h51'39" -36d28'17"
Monika H-1
Per 3h45'20" 32d28'42"
Monika I Wojtek
Cas 2h21'18" 68d52'32"
Monika Izabela Waller
Uma 11h29'6" 64d42'14"
Monika Johanna
Uma 9h52'48" 63d58'16"
Monika Kälin
Boo 14h10'29" 27d34'0"
Monika Kramer
Tau 4h43'28" 4d23'25"
Monika Krawczyk
And 1h17'59" 43d16'11"
Monika Lenzi
Com 13h4'29" 15d16'4"
Monika Lisa Piotrowska
Dra 17h52'48" 53d37'47"
Monika Luna
Sco 16h42'16" -30d6'2"
Monika Lynn McQuarrie
Sco 17h34'29" -30d52'50"
Monika Lys Helms
Uma 11h4'19" 47d35'33"
Monika Margret Ruhle Poole
Cas 1h24'17" 61d31'10"
Monika "Mika" Morawiec
And 23h38'50" 45d18'42"
Monika Orr
Ari 1h47'8" 18d13'24"
Monika (Penguin)
Vir 13h27'26" 11d50'10"
Monika Perkins
Ori 6h2'12" 6d26'56"
Monika Poza López
Tau 3h53'24" 17d42'35"
Monika & Rami
Eri 3h52'20" -0d58'25"
Monika Razeng
Uma 11h31'57" 40d58'55"
Monika Romani
Ori 5h12'12" -9d33'16"
Monika Stansbury
Aqr 23h12'0" -10d11'24"
Monika*Star
Tri 2h37'9" 34d54'3"
Monika Stern
Uma 9h47'16" 56d58'10"
Monika Suzanne Fennell
Crb 16h24'11" 31d22'14"
Monika - the Star
Leo 11h47'0" 19d43'55"
Monika Therese Farmer
Uma 9h59'3" 41d40'32"

Monika van Bellen
Ori 6h17'14" 8d54'19"
Monika Villanueva
Lyn 7h35'52" 45d56'43"
Monika Weinfurter
Umi 16h15'16" 71d49'36"
Monika's Star
Dra 16h46'11" 51d53'26"
Monique
Ori 6h13'44" 9d17'53"
Monique
Mon 6h28'22" 9d26'50"
Monique
Ori 6h12'0" 13d45'32"
Monique
Leo 11h36'45" 17d36'52"
Monique
Gem 7h13'50" 19d58'34"
Monique
Dra 15h8'34" 64d33'35"
Monique
Sco 16h18'6" -15d10'40"
Monique
Dra 18h3'8" 79d40'52"
Monique and Mike's Star (m&m star)
Ori 5h40'14" -0d5'31"
Monique Ann Williams
Leo 11h21'54" 14d31'29"
Monique B.
Lac 22h50'22" 49d49'32"
Monique/Babies
Gem 7h0'8" 27d29'5"
Monique Beaty
Lyn 8h12'0" 34d31'44"
Monique C Monteallo: The Angel Star
Gem 7h47'42" 19d49'35"
Monique Charisse Tenez ,Cap 21h28'20" -16d22'26"
Monique Chavez "Mocha"
Cas 0h46'38" 65d1'21"
Monique Cheri Montellano
Gem 6h5'32" 25d51'15"
Monique Choquette
Ori 5h52'53" 18d10'22"
Monique Dufour (Floçon de Neige)
Cas 0h15'48" 55d17'19"
Monique Eloise Rossi
Cra 18h40'2" -42d18'41"
Monique Evans
And 23h42'58" 48d12'26"
Monique Finch
Tau 5h46'40" 20d19'26"
Monique Georgette Guindon
Cnc 9h7'30" 32d41'18"
Monique GRMN GRL
Cnc 8h19'49" 11d49'13"
Monique Hernandez
And 1h40'13" 38d40'24"
Monique Huckabee
Leo 11h52'34" 24d30'23"
Monique L J Hafrey
Eri 3h57'4" -0d47'36"
Monique Langlois Kronberger
Vir 13h44'43" -8d46'12"
Monique LaShawn Gaines
Cyg 20h36'55" 52d44'45"
Monique Leilani Otufangavalu
Cas 2h34'38" 67d19'32"
Monique Lyane Berends
And 23h49'38" 46d31'18"
Monique Marconi
Lyn 7h24'56" 54d13'39"
Monique Monasterio
Ori 5h57'21" 21d50'26"
Monique My Love
Aqr 22h8'23" -0d52'39"
Monique Nicole York
Ari 3h14'30" 16d26'34"
Monique Noel
Ari 3h5'22" 23d14'40"
Monique Ramos
Lyn 6h46'39" 51d48'16"
Monique Renee Gauna
Sco 16h41'19" -31d49'1"
Monique Samuel
Dra 19h11'38" 62d34'9"
Monique Sayaka Nakamura
Lib 15h34'44" -8d41'24"
Monique Sebag
Cnc 9h5'3" 25d41'36"
Monique Shanghelayne
And 1h20'8" 36d29'59"
Monique "Squishy" Malone
Cas 0h46'53" 48d21'53"
Monique - Star Of My Eyes
Tau 3h53'27" 27d13'25"
Monique- Treasured & Chosen One
Pic 5h37'36" -45d25'37"
Monique van Buuren
Uma 10h56'20" 34d8'7"
Monique van Zutphen
Uma 14h13'45" 60d48'35"
MoniqueAlexia
Aqr 21h47'34" -2d34'9"
Monira & Mansoor
Cas 2h12'34" 74d42'46"
Monja Maria Wagner
Uma 11h45'56" 60d49'49"

Monjalina
Ori 5h44'43" -8d55'58"
MonJo07281958
Leo 11h20'32" 12d0'49"
Monk
Cas 1h1'37" 57d8'43"
Monk
Uma 13h10'4" 62d49'17"
Monk Hodges
Ori 4h53'48" 2d4'15"
Mönkemöller, Jürgen
Uma 9h38'25" 65d9'2"
Monkey
Uma 11h5'12" 63d0'1"
Monkey
Aqr 22h4'7" -13d9'46"
Monkey
Cnc 8h55'51" 13d15'16"
Monkey
Leo 10h14'4" 25d27'8"
monkey junk
Cnc 8h37'51" 16d0'51"
Monkey Love
Leo 11h2'55" 15d48'17"
Monkey Love
Cyg 19h57'37" 32d48'33"
Monkey Magic
And 0h24'37" 45d56'11"
Monkey Man Peter J. Spezia
Boo 15h18'40" 32d40'19"
Monkey Mandi
Pho 0h22'48" -50d56'49"
Monkey Snugbug
And 1h50'29" 36d23'37"
Monkey Wrangler
Mic 20h54'20" -27d41'44"
Monkey's Star
Dra 15h2'59" 61d8'56"
Monkey's Wish
Peg 21h37'3" 25d16'35"
monkeystern
Per 3h41'32" 45d34'18"
Monny, Daddy, & Matt
Lyn 8h16'10" 46d20'51"
Mono
Tau 3h38'26" 30d2'49"
Mono y Oso
Cap 20h16'56" -8d44'18"
Monocera Muse of Oakhaus
Mon 6h45'44" -0d20'13"
Monoceros Sandra
Tau 3h30'32" 11d53'36"
Monola Family Star
Cra 19h9'35" -43d5'2"
Monooshag / Mae Fundukian Wright
Tau 5h40'47" 26d27'29"
Monroe
Ari 2h20'37" 12d30'3"
Monroe
Crb 16h8'13" 37d2'22"
Monroe
Per 3h33'10" 42d37'12"
Monroe Jose Ratchford, M&P WF
Umi 14h48'43" 69d11'22"
Monroe Schaffer
Sco 16h59'34" -44d37'19"
Monroe Scott Pomelow
Aqr 22h9'46" -2d49'51"
Monroe Sydney Masa
Cen 11h11'9" -49d18'13"
Monroy's Star
Aur 5h55'36" 41d7'29"
monsajem
Aqr 22h17'26" -7d54'39"
Monseigneur
Ori 6h18'59" 5d41'38"
Monsieur Buddy Bonaparte - "Buddy"
Cep 21h28'41" 66d24'34"
MonsigneurPaul
Aur 5h49'31" 50d44'18"
Monsignor John Esseff
Umi 15h51'26" 74d24'57"
Monsignor Timothy P. Harnett
Lib 15h13'6" -6d26'59"
Monsina
Lyn 6h55'54" 58d48'15"
Monsita
Lyn 7h7'56" 44d53'10"
(Monsta) Seth Francis Cummings
Lmi 11h3'23" 28d23'10"
MONSTAR
Pho 1h10'29" -44d53'54"
monster-heads
Cyg 20h22'55" 33d20'31"
Monsy Corbera
Umi 13h54'59" 74d52'43"
Montague Frederick
Cep 21h41'25" 59d13'56"
Montana
Uma 9h35'42" 42d52'7"
Montana
Uma 11h46'39" 44d28'29"
Montana
And 2h19'48" 49d1'50"
Montana
Cmi 7h29'24" 7d32'44"

Montana 7/29/03
MRT*MEB
Sco 16h14'59" -8d39'13"
Montana Alexa Spears
Psc 1h10'40" 16d40'34"
Montana Grace
Uma 8h34'23" 60d32'56"
Montana Helen
Uma 10h15'54" 59d31'57"
Montana Koy Pace
Ori 6h11'24" -3d48'0"
Montana Leigh
Lyn 6h26'39" 61d29'35"
Montana Marie
Cyg 20h31'31" 36d21'15"
Montana Mary Belle Stuart
Aqr 22h23'16" -1d38'12"
Montana Pullen
Umi 17h1'33" 81d41'42"
Montana Rose Powell
Vir 13h24'33" 13d33'23"
Montana Sky
Uma 11h33'7" 62d33'48"
Montana's Golden Light
Cma 6h36'55" -15d46'36"
Montanna Lane
Umi 16h57'52" 89d31'20"
Montanna Michelle
Umi 15h33'52" 73d20'4"
Monte A. Yocum
Srp 18h11'7" -0d18'36"
Monte and Mary Kestell's Star
Tau 5h44'49" 22d27'2"
Monte & Brenda
Cyg 21h38'47" 45d53'40"
Monte Holm's Leading Light
Tau 4h40'22" 6d53'11"
Monte Kemper
Vir 12h40'16" -6d39'15"
Monte Kuhn
Uma 11h46'15" 57d47'40"
Monte Lynn Mayer
Uma 10h11'23" 56d13'41"
Monte-Haak
Cyg 20h10'41" 30d37'26"
Montell Elaine Edgar
Sco 16h17'47" -42d2'49"
Montello Shining
Dra 16h26'26" 68d38'19"
Montemayor
Lyn 8h45'51" 40d39'46"
Monterroza1G6
Lyn 7h24'13" 44d38'43"
Montey Sanford
Cnc 8h40'10" 7d42'47"
Montgomery Armstrong
Cap 20h45'28" -25d1'33"
Montgomery County Astronomy
Uma 10h22'45" 53d33'45"
Montgomery Jefferson LaFond
Lib 14h47'48" -17d39'27"
Montoriah
Sgr 18h50'12" -31d1'17"
Montre Destiny
Aql 19h21'3" 0d28'44"
Montrice
Ari 3h8'43" 26d29'7"
Montrose Gladys Smith
Ari 2h48'13" 14d46'35"
Montse Ayala
Ori 5h45'32" 8d50'13"
Montse Eterna
Ori 6h6'28" 5d24'52"
Montse Jimenez Gonzalez
Her 18h57'20" 13d50'13"
Montse la nena maca
Aur 6h6'21" 38d50'7"
Montse Marcet
Umi 14h56'20" 78d30'57"
Montserrat
Uma 12h18'16" 55d15'39"
Montserrat Sanz
Gem 7h12'36" 18d56'6"
Montserrat-Martinez-Alonso-Spain
Cyg 19h32'40" 32d18'59"
Montsita Junquera
Dra 20h7'32" 73d14'0"
Monty
Her 17h41'7" 37d53'16"
Monty & Dee Dee
Per 2h38'57" 52d11'47"
Monty Moorman
Uma 11h25'7" 49d36'36"
Monty Palmer Wilson (13 July 59/06)
Umi 15h26'29" 71d16'20"
MONTY RAAB
Cma 6h59'29" -21d53'36"
Montycito
Psa 22h12'54" -25d42'25"
Monu
Vir 12h52'51" -1d33'45"
Mony
Lyn 7h35'1" 38d30'15"
Mony
Ori 6h20'47" 6d23'22"
Monz, Christian
Uma 14h0'8" 57d11'24"
Monzella
And 1h26'51" 47d39'7"

Moo
Cyg 20h56'17" 30d48'31"
Moo Cow (for I love you forever)
Ori 5h52'16" 1d53'4"
Moo Ma
Tau 4h38'11" 18d54'57"
Moo Moo
Cyg 21h47'20" 54d10'49"
Moo Moo's light
Cap 21h33'45" -19d28'3"
Moo Shu Sloopaloop
Sgr 18h3'40" -27d59'40"
Moo Star
Sgr 17h55'31" -22d47'39"
Mooch
Vir 12h33'39" -7d41'26"
Mooch
Uma 12h30'45" 56d6'44"
Moocow Littler-Cain 30
Cas 1h18'12" 53d49'9"
Moo-Dawg
Her 18h29'31" 20d43'18"
Moodi
Uma 10h46'3" 67d53'49"
Moody Daniels
Cyg 21h58'58" 55d7'40"
Moody-Rowton
Cnc 8h24'43" 30d35'1"
Moofie 2
Tau 5h24'30" 20d9'34"
Moogie & Mimi
Ori 6h23'52" 10d43'33"
Moogie & Mimi
Ori 5h52'31" 6d44'28"
Moohie
Uma 9h28'21" 66d57'1"
Mook
Tau 4h38'9" 19d12'22"
Mook Mook
Her 16h44'45" 34d31'12"
Mookie
Per 4h21'46" 43d8'55"
Mookie
Cma 6h21'42" -29d3'55"
Mookie - 02.21.1957
Psc 1h36'42" 17d34'24"
Mookie Dog
Cma 6h42'10" -16d41'34"
Mooky
And 1h10'20" 36d5'39"
Moolie Kajoolie's Star
Leo 9h27'34" 16d31'28"
Mooma
Gem 6h45'16" 33d5'30"
MOO-MFH
Leo 10h25'50" 21d27'50"
MooMoo
Vir 13h18'39" 12d10'5"
Moo-Moo
Lib 14h49'47" -1d2'56"
Moon Beam
Aqr 22h34'55" -2d23'9"
Moon Dancer
Sgr 19h19'59" -21d8'50"
Moon Dates
Ori 5h7'18" 9d12'5"
Moon Etifier
Aqr 23h17'18" -17d33'9"
MOON FLOWER
Psc 0h45'5" 16d53'49"
Moon Hawkins
Pup 7h51'49" -19d25'42"
Moon light
Tau 4h33'21" 29d5'41"
Moon Light Rose
Gem 7h21'48" 32d6'50"
"Moon Mommy Goddess Jihnah" & Emily
And 0h20'11" 29d32'18"
Moon Muffin
Lib 14h47'39" -15d43'29"
Moon N Back 2 U
Cyg 21h43'42" 48d35'59"
Moon Shine
Cep 22h34'58" 66d49'27"
Moon Unit <3
Uma 12h15'13" 53d4'34"
Moon Young Oh
Gem 7h29'3" 26d0'57"
Moonchild
Cnc 8h9'14" 16d1'55"
Moonchild
Cas 0h6'3" 56d17'34"
Moonchild
Umi 14h34'35" 77d58'0"
Moondiva
Aqr 21h4'22" -10d51'12"
Moondocker
Uma 10h48'4" 46d48'12"
Moondoggie
Cap 20h26'46" -12d29'40"
MoonDoggie
Uma 10h19'46" 68d27'11"
Moonflower
Aql 19h46'8" -0d19'10"
MoonFriend
Umi 7h53'4" 88d34'41"
Moonglade - Ro & Bo
Aqr 20h49'59" -11d10'2"
Moonie S. A. R.
Leo 9h32'21" 27d27'13"
Mooniverse KLM
And 2h3'21" 36d7'5"
Moonlight
Ori 5h17'28" 11d47'27"
Moonlight
Uma 13h21'3" 58d30'6"
moonlightmemorieswaiting
AKLremember
Lyn 8h43'38" 40d48'20"
Moonlite37
Uma 14h7'32" 59d31'30"
Moo'nPoo40
Pho 1h19'57" -44d43'7"
Moon's Star
Ori 5h31'7" 1d56'24"
MoonShine
Her 18h34'50" 19d0'18"
Moonshine
Uma 10h28'4" 43d30'37"
Moonshine King
Aqr 20h39'22" -11d21'6"
MoonStar Kelly
Tau 5h24'31" 18d34'25"
Moonstruck
Lyr 19h16'24" 33d26'44"
MOOP
Lyr 18h58'16" 27d17'58"
moop143
Ret 3h44'43" -54d35'0"
Moopstar
And 0h12'0" 41d33'57"
Moopy
Cam 3h57'29" 64d19'17"
Moore
Ori 6h11'0" 13d33'57"
Moore 07-22-06 Meet Me at Our Star!
Uma 11h20'10" 62d49'44"
Moore - Vstar
Uma 10h0'25" 71d38'7"
MooreTower Love 8-6-06
Cyg 19h44'39" 38d24'29"
Moorhead and Louisa Kennedy
Peg 21h45'17" 15d44'32"
Moorhoff, Jobst
Vir 12h53'22" 2d18'11"
Moorlang, Dagmar
Ori 5h5'6" 0d56'8"
Moo's Destiny
Gem 6h59'19" 17d21'1"
Moosayy
Lyn 7h8'37" 48d25'59"
Moose
Ori 5h36'44" 5d9'55"
Moose
Uma 11h35'49" 61d39'36"
MOOSE
Cnv 13h14'58" 48d47'20"
Moose & Frog
Vir 13h14'56" -22d22'14"
Moose Tracks
Cnc 8h42'36" 32d10'42"
Moosedog Max
Cma 6h38'27" -16d31'8"
Moosehead
Tau 5h39'23" 25d46'43"
mooserun m-1 kobe
Cnv 12h48'32" 39d6'26"
Moosey
Cam 3h57'20" 65d43'47"
Mooshi
And 0h33'4" 46d15'58"
Mooshoo JJ O'Sullivan
Cyg 20h18'57" 52d44'31"
moosieandlulu
Cap 20h24'29" -19d29'25"
Moosky McGrubbles
Ret 4h13'19" -57d37'0"
Moosler, Thomas
Sgr 17h54'29" -28d40'33"
Moosmeier, Manuela
Ori 6h0'21" 10d48'47"
Mop
Aqr 22h15'3" 0d11'33"
Mopi
Leo 10h13'7" 11d21'3"
Moppie
Uma 12h33'53" 57d40'48"
Mora Liebe Jackson
And 0h40'26" 44d23'1"
Morabito-Kardash Alexandra
Ori 6h20'36" 13d55'0"
Moracco
Sco 17h2'45" -34d44'10"
Morad
Boo 15h30'12" 49d14'59"
Moradel Family
Lyn 7h58'25" 42d29'16"
Morag Boardman
Cas 1h25'33" 55d12'8"
Morag Fiona McInnes
And 2h30'28" 49d56'32"
Morag's Star
Lyr 18h43'38" 31d26'38"
Morahya Shalou
Mon 6h53'20" 8d7'2"
Morail McCaw
Per 4h44'8" 46d47'13"
Morana
Dra 16h52'44" 55d10'32"
Morando
Aql 18h51'1" 1d35'55"
Morayma Amaro
Leo 10h50'0" 13d57'46"
Mordecai
Ori 6h25'23" 10d34'15"
Morden Ash Born: Sun Jan 11/2004
Uma 11h31'20" 56d18'15"
"More"
Sct 18h47'9" -5d26'40"
More than A Love Song - AAJ~JPH
Ori 6h5'1" 15d58'24"
More Than Love
Cyg 20h5'48" 35d48'4"
More Than the Stars
Uma 9h13'50" 55d32'34"
More To Life
Pho 23h57'31" -48d34'9"
MORELLI-CATT
And 1h33'49" 42d0'45"
Morena Marongiu
Vul 20h24'48" 26d23'36"
Morenita
Tau 4h48'31" 19d23'6"
Morenkanou (Julien Bouvier)
Boo 14h58'52" 18d49'15"
Moreno
Uma 13h37'0" 48d30'54"
Moreno-Hayek
Ori 5h7'35" 5d30'58"
Moreno's Taufstern
Cas 23h5'28" 55d36'13"
Morey
Psc 0h1'57" -0d7'43"
Morey Amsterdam
Uma 10h22'36" 48d29'55"
Morey Junod
Ori 5h25'47" 1d54'43"
Morey Star
Uma 8h43'43" 50d42'35"
Morfayan2003
Leo 11h53'42" 22d7'12"
Morg and Ry
Vir 12h41'12" -0d43'55"
Morgan
Umi 16h17'12" 75d46'44"
Morgan
Uma 11h19'4" 56d4'5"
Morgan
Uma 12h6'30" 54d44'33"
Morgan
Uma 10h37'12" 63d46'35"
Morgan
Umi 14h16'58" 65d58'18"
Morgan
Umi 15h16'32" 74d43'6"
Morgan
Gem 7h16'0" 20d0'33"
Morgan
Gem 7h15'35" 26d7'59"
Morgan
Ori 5h37'26" 3d47'27"
Morgan
Psc 0h37'33" 14d21'24"
Morgan
Peg 21h42'2" 14d20'55"
Morgan
Lyr 18h52'19" 44d31'54"
Morgan
Cyg 20h6'3" 45d2'43"
Morgan
And 0h36'48" 41d7'2"
Morgan
Lmi 10h38'49" 38d42'59"
Morgan
Aur 5h35'46" 33d26'7"
Morgan A. Dufault
Uma 11h54'11" 63d40'16"
Morgan A. Langendorf
Cas 0h19'32" 53d50'46"
Morgan Adrianna Washington
Lyr 18h52'15" 37d28'17"
Morgan Aimee Louise Johnson
And 23h27'52" 46d53'54"
Morgan Ainsley Leon
Lmi 10h42'0" 37d49'11"
Morgan Alan Solomon
Sco 17h51'14" -35d56'56"
Morgan Alayne
Ori 5h35'59" 10d40'52"
Morgan Alexander Clements
Uma 11h37'25" 44d17'25"
Morgan Alexander Raskin
Gem 7h40'48" 26d6'53"
Morgan Alexandra
Cas 0h49'26" 57d34'20"
Morgan Alexandra Jointer
And 22h59'42" 36d52'6"
Morgan Alexandrea
Cnc 9h8'13" 28d49'15"
Morgan Alexis McAllister
And 2h21'32" 49d28'37"
Morgan Alison Christian
Aqr 22h33'33" 0d45'54"
Morgan Allen Aanes
Vir 12h11'17" 10d47'42"
Morgan and Lauren White
Peg 23h43'54" 26d4'33"
Morgan Ann
Umi 16h59'32" 79d17'13"
Morgan Ann Formosa
And 1h9'14" 36d53'6"
Morgan Ann Hickman
Crb 16h8'34" 37d20'49"
Morgan Ann Noriega
And 1h33'15" 41d6'13"
Morgan Anne Nunley
Aqr 21h5'43" -14d20'33"
Morgan Arenth
Uma 13h44'19" 54d45'23"
Morgan Arianna Reed's Star
Lib 15h24'4" -9d36'17"
Morgan Arlene
And 23h35'25" 48d16'30"
Morgan Ashleigh Piehl
And 2h35'23" 39d34'38"
Morgan Ashley
Leo 11h11'32" 15d12'33"
Morgan Ashley Carrero
Peg 21h40'41" 23d51'41"
Morgan Ashley Eplin
Cyg 20h0'51" 32d2'37"
Morgan Ashley Thoma
And 23h22'37" 37d58'37"
Morgan astride Aslan
Lib 15h10'36" -6d34'43"
Morgan Avery Dowdell
Vir 14h26'57" -4d19'6"
Morgan Avery Stokes
Mon 6h51'37" 8d7'58"
Morgan B. Hunt
Sgr 18h15'11" -19d21'1"
Morgan Billy Spicer
Gem 7h0'0" 20d36'2"
Morgan Blac
Vir 12h20'30" 8d17'47"
Morgan Boyd
And 23h33'58" 47d5'32"
Morgan Brittani Churchill
Leo 9h44'29" 18d18'0"
Morgan Brooke Elizabeth Murphy
And 1h7'43" 36d42'7"
Morgan Brudvik
Leo 10h44'1" 17d36'38"
Morgan Buchanan
Vir 12h44'33" -9d18'55"
Morgan Buckwalter
Uma 10h32'54" 66d59'59"
Morgan C. Arenth
And 23h9'2" 52d17'27"
Morgan Cade Schoening
Equ 21h14'55" 8d47'46"
Morgan Calderon
Leo 10h31'6" 14d21'12"
Morgan Camille Roberts
Sex 10h43'12" -0d25'13"
Morgan Carey Pavlot
Lyn 7h31'29" 42d8'9"
Morgan CarolAnn
Cnc 8h18'5" 9d51'19"
Morgan Carr
Lyn 8h16'52" 34d17'0"
Morgan Christina Montgomery
Gem 7h42'25" 23d19'0"
Morgan Coirier
Ari 2h57'20" 21d56'30"
Morgan D. Chapman
Uma 12h41'53" 59d4'24"
Morgan D Dobberstein
Aql 18h51'10" 7d36'37"
Morgan Danner
Psc 1h9'51" 14d22'33"
Morgan Dawes Marshall
Ori 5h50'41" 8d31'27"
Morgan Deluce
Lyn 8h34'7" 34d29'23"
Morgan Diane Neal
Com 12h41'31" 18d3'3"
Morgan Dominique
Cam 4h37'5" 69d49'7"
Morgan Dorothy
Cyg 21h16'40" 41d31'4"
Morgan Douglas Hostetler
Uma 11h18'30" 62d19'48"
Morgan Doyle's Star of Courage
Uma 9h6'36" 49d52'23"
Morgan E. M. Horn
Lyn 8h17'42" 56d23'9"
Morgan Elaine
Leo 11h54'48" 20d5'23"
Morgan Elaine Daniels
Ari 2h16'51" 25d53'34"
Morgan Elisabeth
Del 20h26'5" 18d43'8"
Morgan Elise
Uma 11h47'55" 42d37'20"
Morgan Elise Ballard
Aqr 21h49'27" -6d11'51"
Morgan Elise Jackson
Cnc 8h30'22" 25d17'40"
Morgan Elizabeth
Leo 11h36'15" 23d25'33"
Morgan Elizabeth
And 1h5'22" 42d40'14"
Morgan Elizabeth
And 23h34'32" 46d24'32"
Morgan Elizabeth
Lyn 6h40'55" 61d4'37"
Morgan Elizabeth Bowan
Gem 7h49'23" 16d49'33"
Morgan Elizabeth Humphrey-Clay
Leo 10h16'51" 12d37'12"
Morgan Elizabeth Kuminkoski
Uma 9h23'57" 54d28'11"
Morgan Elizabeth Lutz
Sco 17h16'33" -32d6'12"
Morgan Elizabeth Redden
Psc 23h42'49" 2d10'25"
Morgan Elizabeth Sleter
Tau 4h22'16" 26d13'1"
Morgan Elizabeth Snider
Cap 21h35'31" -18d11'16"
Morgan Ellen Hall
Gem 7h35'11" 26d37'52"
Morgan Ellis Hawthorn - Morgan's Star
Umi 15h56'45" 74d56'41"
Morgan Elyse Santos
Lib 15h51'23" -17d55'53"
Morgan Emma Gasior
Aqr 22h37'15" 1d21'22"
Morgan Emma Kate Berry
Lib 15h11'0" -24d22'14"
Morgan Emma Migliore
Psc 1h4'5" 27d58'28"
Morgan Faith Marie Chard
And 0h42'44" 40d21'12"
Morgan Faith Marie Chard
And 1h35'58" 41d53'53"
Morgan Farmer
Ori 6h24'56" 14d49'23"
Morgan Fisher Clarke
Lyn 7h34'29" 39d43'9"
Morgan Forrester's Star
Cru 12h24'34" -58d46'51"
Morgan Frances
Psc 1h29'40" 13d2'22"
Morgan Freya Benjamin
And 0h22'38" 25d52'22"
Morgan G. Graziadei
Lib 15h21'3" -4d39'41"
Morgan Gomes
Cap 21h53'24" -24d51'25"
Morgan Grace
Lmi 10h37'48" 30d38'2"
Morgan Grace Dall
Uma 10h21'14" 56d8'4"
Morgan Green
And 1h38'19" 45d10'2"
Morgan Grimes
Oph 17h3'50" -0d37'54"
Morgan Haggerty
Cas 1h39'51" 62d3'40"
Morgan Harris-Little Angel
Vir 14h11'31" -13d41'7"
Morgan Harrison Cooper
Aql 19h31'38" 6d46'42"
Morgan Hennis
Cnc 8h10'51" 17d47'9"
Morgan Hilsin
And 0h41'37" 37d6'36"
Morgan Hostetter
Sco 17h48'30" -39d12'43"
Morgan Howe
Lib 15h49'36" -18d48'10"
Morgan Hummel
Psc 0h38'42" 5d49'41"
Morgan Hummel "01/01/1986"
Vir 13h27'25" 12d38'13"
Morgan Isaac Allen
Peg 23h26'25" 15d4'52"
Morgan James
Gem 6h4'0" 22d41'35"
Morgan James Bechthold
Sco 16h10'51" -14d12'50"
Morgan James Fitzgerald
Cyg 21h58'24" 47d6'32"
Morgan James Hollingsworth's Twinkling Star
Umi 15h23'3" 79d24'43"
Morgan James Ingram
Psc 22h59'53" 8d8'44"
Morgan Jane Martinez
Sex 10h14'40" -0d49'51"
Morgan Janis McCarthy
Ari 2h39'21" 21d37'33"
Morgan Jasmine
Cas 0h24'14" 51d31'53"
Morgan Jaye
Umi 17h32'49" 81d25'1"
Morgan Jennifer Primrose
Cas 23h51'7" 58d32'46"
Morgan Jet Pack
Ori 6h20'25" 8d58'15"
Morgan & Jill
Ori 6h13'38" 6d16'5"
Morgan Joe Claunch III
Del 20h31'7" 15d58'12"
Morgan John Francis
Per 4h45'25" 41d9'5"
Morgan Johnsson "Loved by Anders"
Sgr 19h46'10" -14d1'34"
Morgan Joseph Long
Ori 6h10'23" 18d42'43"
Morgan Joseph Neeser
Ori 6h3'31" 18d42'44"
Morgan Joyce Oakes
Lyr 18h44'41" 31d5'12"
Morgan Kate McGreevey
Aqr 22h29'12" -0d12'54"
Morgan Kay Geiger
Aqr 23h22'21" -4d46'56"
Morgan Kayleigh
And 23h32'38" 41d9'41"
Morgan Kelly Hamlin
Umi 14h44'40" 74d1'39"
Morgan Kellye Hediger
Gem 7h39'53" 14d43'41"
Morgan Krueger
Psc 0h39'10" 10d56'54"
Morgan Lane
Ori 4h48'11" -3d3'26"
Morgan Leanne Haberle
And 0h38'47" 45d44'28"
Morgan Leanne Stewart
Psc 0h48'58" 15d32'1"
Morgan Leder
Uma 11h6'16" 49d6'22"
Morgan Lee
Tau 4h2'51" 30d26'55"
Morgan Lee DeBoard
Leo 10h54'28" 19d19'7"
Morgan Lee Desjardins
Cyg 20h34'19" 53d40'10"
Morgan Lee York
Lyn 8h0'6" 57d38'56"
Morgan Leigh
Lyn 8h0'36" 46d10'30"
Morgan Leigh Belsole
Aqr 21h17'43" -10d29'8"
Morgan Leigh Gentry
Lyr 18h34'0" 36d28'29"
Morgan Leigh Taylor
Pup 7h43'50" -19d30'8"
Morgan Linda Ringsaker
Uma 11h32'11" 64d21'9"
Morgan Lyndsay Buckwald
Lyn 6h49'16" 61d9'1"
Morgan Lynn
Lib 15h25'40" -20d17'13"
Morgan Lynn
Crb 16h5'51" 27d36'52"
Morgan Lynn
Ari 2h8'40" 18d11'56"
Morgan Lynn
Vir 14h40'35" 2d36'58"
Morgan Lynn Sligar
Uma 10h0'17" 66d1'47"
Morgan Lynne Helleberg
Sco 16h18'24" -18d5'47"
Morgan Lynne Hiza
Lyn 8h21'7" 52d31'7"
Morgan Mann
Cas 0h56'53" 61d26'39"
Morgan Margaret Sousa
And 2h27'5" 38d12'36"
Morgan Marie
Umi 15h1'0" 72d25'15"
Morgan Marie McDougal
Uma 10h26'43" 65d32'43"
Morgan Marie Urban
And 2h25'55" 50d54'28"
Morgan May Cornelius
And 0h15'35" 43d13'29"
Morgan May Saunders
Gem 6h27'44" 21d46'42"
Morgan McIntyre
Ari 2h48'17" 25d39'49"
Morgan McKinley
Umi 16h19'41" 76d5'9"
Morgan Meadows
Peg 22h26'41" 26d8'19"
Morgan Meagher
Cas 1h14'39" 58d21'8"
Morgan Merlanti
Gem 7h32'16" 29d15'19"
Morgan (MoMo) Lindsey Larrouy-Cross
Crb 15h48'56" 36d59'10"
Morgan Morgan
Lib 14h52'56" -1d9'32"
Morgan Mosby 7/23/01-2/5/02
Mon 6h50'53" -0d9'26"
Morgan - My shining star
Ari 3h5'36" 15d58'32"
Morgan Nanice Marquez
Mon 7h11'2" -9d59'33"
Morgan Nicole Brand
And 2h35'12" 50d30'9"
Morgan Noble
Ari 2h38'39" 14d54'39"
Morgan Olivia O'Neal
Lib 14h57'26" -2d40'51"
Morgan Patrick Wilkes
Ori 6h4'33" 21d29'3"
Morgan Paul Baker
Gem 6h21'21" 22d5'42"
Morgan Pawluk
Uma 10h20'58" 40d57'53"
Morgan Peal
Umi 16h24'0" 72d8'25"
Morgan Peterson
Cyg 19h48'35" 30d45'44"
Morgan Petras
Ori 5h55'26" 21d16'32"
Morgan Prescott Paladenic
Sgr 18h5'42" -27d56'21"
Morgan "Princess Petunia" Belitz
Per 2h49'14" 54d16'47"
Morgan R. Lewis
Uma 12h11'58" 58d46'0"
Morgan R. Linville
Tau 3h41'57" 16d56'5"
Morgan Rachael DuPont
Gem 7h17'1" 30d52'40"
Morgan Rae Spencer
Gem 7h14'24" 28d51'38"
Morgan Ray Jones
Sgr 18h12'28" -20d12'46"
Morgan Rebecca Davie
Lyn 9h6'47" 34d52'25"
Morgan Red Deer
Lib 15h56'20" -9d30'49"
Morgan Reddy
Ori 5h55'50" 11d31'7"
Morgan Reese Piel
Sgr 18h29'1" -16d31'17"
Morgan Robert Isobel
Sgr 19h34'4" -28d52'38"
Morgan Rondeau
Sgr 19h47'9" -13d42'23"
Morgan Rose Fitzgerald
And 23h16'52" 49d27'57"
Morgan Rose Trask
And 23h13'31" 43d24'14"
Morgan Scheid
Lyn 7h3'33" 53d24'7"
Morgan Scott Ferguson
Ari 2h47'34" 13d14'33"
Morgan Serpe
Cma 6h44'51" -15d16'58"
Morgan Shaw
Cnc 7h58'46" 11d39'2"
Morgan Silver
Vir 14h22'55" 0d32'58"
Morgan Slavens
Vir 12h0'18" -8d14'7"
Morgan Smith
Cam 3h25'27" 65d35'21"
Morgan Spencer Bond
Aqr 21h45'3" 1d1'29"
Morgan Spicer
Crb 15h34'2" 31d6'28"
Morgan Star
Mon 6h29'53" -2d40'59"
Morgan Star Pravato
Sgr 19h28'56" -31d25'5"
Morgan Strickland
Leo 9h40'26" 27d42'49"
Morgan Taylor Batchelder
Cep 22h25'23" 83d33'16"
Morgan Taylor Corbin
Aqr 21h42'22" -3d14'53"
Morgan Taylor Smith
Lyn 8h10'47" 40d54'57"
Morgan Uhler
Ori 5h32'1" 11d43'33"
Morgan Webb Perkins
Sco 17h25'29" -42d20'19"
Morgan White
Uma 9h21'16" 46d0'45"
Morgan Whitney Fairfax
Uma 10h44'32" 57d6'22"
Morgan William McDuffee
Per 2h30'12" 55d45'49"
Morgan Wilson
Peg 22h58'15" 25d7'3"
Morgan Yonge
Lyn 8h30'15" 45d47'49"
Morgan Zsolt
Uma 11h31'35" 45d11'37"
Morgan Zwimpfer
Uma 11h37'46" 53d49'40"
Morgana
Umi 17h0'46" 80d9'39"
Morgana Batista
Ari 2h56'38" 27d11'32"
Morgane
Cas 0h18'19" 54d53'37"
Morgane
Uma 9h35'23" 55d24'14"
Morgane Delcellier
Cnc 8h26'9" 9d59'39"
Morgane Elisabeth Marie Danan
Vir 13h13'29" 6d7'9"
MorganInFlames
Vir 14h24'4" 5d4'52"
Morganlacey Bullock
Leo 9h29'23" 27d11'42"
MorganLea
Sco 16h40'22" -32d2'40"
Morganne Noelle
Uma 10h30'38" 57d20'51"
Morganne's Star
Uma 11h27'25" 70d28'10"
Morgannis (MEU)
Vir 12h49'25" 6d29'48"
Morgan's Eternal Love
Vir 12h16'45" -1d41'47"
Morgan's Lucky Star
Sgr 19h2'27" -27d3'3"
Morgan's Magical Mystery Star
Sco 17h47'21" -37d53'21"
Morgan's Max and Cujo
Lyn 7h30'14" 55d1'41"
Morgan's Star
Umi 16h11'20" 70d59'33"
Morgan's Star
Aqr 21h55'0" -1d54'25"
Morgan's Star
Lib 15h26'12" -6d21'49"
Morgan's Star
Cru 12h3'4" -63d20'52"
Morgan's Star
And 0h14'52" 32d53'0"
Morgan's Star
Her 16h22'36" 25d6'22"

Morgan's Star
Cyg 20h56'42" 36d10'42"
Morgan's Star From Patricia
Mon 7h31'58" -0d40'2"
Morgen
Uma 11h55'26" 39d57'22"
Morgen Martinez
Aqr 23h5'15" -22d5'38"
Morghenna
Umi 14h43'25" 78d35'48"
Morgs
Cru 12h29'17" -61d6'47"
morgster
Cnc 9h8'40" 32d38'12"
Morgsy
Uma 11h55'30" 65d1'10"
Mori, the love of my life
Lib 14h49'0" -1d11'51"
Moriah
And 0h42'57" 37d54'41"
Moriah
And 0h43'4" 41d24'27"
Moriah Daniel
Umi 15h54'52" 72d39'30"
Moriah Mersky
Uma 11h17'3" 61d32'12"
Morina
Uma 11h28'37" 53d16'51"
Morio Teraoka
Tau 5h27'55" 25d28'2"
Moritz Benedikt * 18.11.2004
Uma 11h11'45" 62d34'15"
Moritz Sudhof
Cap 21h53'15" -12d53'43"
Morlene Chin:Most Wonderful Friend
Cnc 8h36'37" 32d53'2"
Morley Gwirtzman
Ori 6h17'10" 14d41'38"
Mormor - Grandma's Star
Cas 23h4'12" 59d21'40"
Mormor Grethe
Cas 1h23'34" 62d35'18"
Morms
Uma 9h47'4" 47d46'21"
Morning Angel
Leo 9h37'40" 12d15'13"
Morning Glory
Vir 12h58'13" 1d7'21"
Morning Gorgeous - Lisa's Star
Psc 1h41'11" 18d24'18"
Morning Pony
Uma 10h4'18" 61d33'28"
Morning Star
Sgr 19h37'14" -13d44'27"
Morning Star
Aql 19h31'11" 10d35'12"
Morning Star
Aql 20h11'8" 2d29'45"
Morning Star Morgan Farm
Uma 8h28'49" 60d10'28"
Morning Star Noleaf
Ari 2h34'4" 13d31'30"
Morning Sun
Cas 23h57'6" 56d45'41"
Morningstar
Crb 15h19'29" 29d52'58"
Morningstarr* Roses' One and Only
Uma 14h0'9" 52d14'43"
Morphée
Vir 14h3'3" -18d51'5"
Morpheous Bryan
Sco 17h56'54" -38d49'23"
morpheus-hyperion
Cyg 19h54'58" 37d57'16"
Morphoeugenia
Psc 23h6'21" 7d37'16"
Morrea Lynn Yankey
Cam 5h59'10" 57d40'11"
Morrie Kimmel
Cnc 9h21'31" 22d52'7"
Morris & Alvirda Hyman
Aql 19h5'35" -0d51'15"
Morris and Dorris Hansen
Cyg 21h33'41" 44d45'36"
Morris Berenbaum
Sgr 19h1'32" -34d13'36"
Morris Charles Givens Star
Vir 13h6'16" 6d11'20"
Morris Daniel Eaton II "Ginkers"
Cyg 19h58'18" 40d53'22"
Morris Family Star
Cma 7h10'15" -18d24'42"
Morris' Galactic Star 49
Lyn 7h31'59" 38d5'51"
Morris Hennessy
Tau 5h49'52" 16d1'53"
Morris & Hortense ~ 12/30/1944
Leo 10h52'17" 23d54'27"
Morris Lamana
Lib 14h23'53" -19d52'54"
Morris Markowsky
Uma 10h26'19" 49d57'51"
Morris "Marty" Silverman
Per 4h26'58" 34d50'12"
Morris Percival
Uma 12h48'55" 58d37'10"
Morris Richter
Vir 12h46'54" 10d44'44"
Morris Saletan
Cap 20h38'39" -23d23'14"
Morris T. Houck
Boo 14h38'2" 17d48'28"
Morrison and Helena Gisclair
Uma 11h24'4" 28d32'40"
Morrison (Joanne, Parke & Parke Jr)
Ori 6h0'37" 21d13'20"
Morse
Aql 19h8'38" -0d0'20"
Morsy, Ayman & Elli
Uma 10h16'33" 65d11'49"
Mort and Paula Together Forever
Cyg 19h41'10" 31d23'48"
Mort Helland
Uma 11h38'39" 63d29'15"
Mort Kasman - February 24, 1933
Psc 1h6'56" 30d50'12"
Mort Stein "Poppy"
Uma 9h13'57" 58d34'30"
Mort Waimon
Aql 19h27'1" 16d21'46"
Morten
Aur 5h22'39" 43d24'58"
Morten Bundgaard-Kaas
Uma 8h40'8" 48d15'50"
Morten Lie
Uma 8h31'56" 61d56'22"
Morten-Benedikt-Schobert-Stern
Uma 10h15'18" 55d51'6"
Morticia Deal
And 0h19'25" 27d56'26"
Mortimer
Ori 6h14'18" 15d59'39"
Morton and Diane Lupowitz
Cyg 20h14'54" 49d23'31"
Morton Blaugrund
Her 17h22'52" 37d39'42"
Morton E. C. Willever
Equ 21h19'45" 2d38'21"
Morton Family, Love For Eternity
Per 3h45'16" 35d19'4"
Morton Kesler
Gem 6h41'7" 24d40'12"
Morton Klasmer
Lyr 18h36'34" 30d25'6"
Morton Meadow
Gem 7h40'12" 32d5'54"
Morton "Pop Pop" Rosen
Her 17h54'10" 22d2'18"
Morton Ray
Dra 17h20'17" 68d13'47"
Morwenna
Lyr 18h22'43" 32d13'58"
Morzsa (Emike)
Vir 14h32'42" 5d53'54"
MOS
Leo 11h9'9" 3d50'19"
Mo's Mojo
Tau 4h26'39" 27d32'31"
Mo's Star
Lmi 10h14'25" 35d15'34"
Mo's very own star!
Cas 0h52'3" 63d59'54"
Mosby Costens-Bauman
Uma 10h42'52" 69d20'27"
Moscaritolo
Ari 2h29'54" 22d9'34"
Mosch, Victor Adrian
Uma 9h53'41" 52d20'26"
Moscheni Andrea
Cam 6h49'34" 69d37'41"
Moschner, Ekkehard
Uma 10h23'13" 65d30'17"
Moser, Manfred
Uma 9h38'18" 53d12'32"
Moser's Magic
Uma 9h41'25" 53d36'55"
Moses
Uma 10h0'33" 66d48'2"
Moses
Uma 10h57'44" 63d46'41"
Moses Bermudez
Lyr 18h40'36" 31d9'18"
Moses Brach
Ori 5h19'18" -4d20'37"
Moses Dmitri Stojanovic
Her 17h16'41" 25d4'0"
Moses James Armstrong Hookway
Dra 19h39'45" 70d18'21"
Moses Znaimer
Cep 23h52'45" 74d27'35"
Mosey - Roni
Leo 11h1'21" 21d31'18"
Moshe Leabovize
Per 3h44'23" 44d52'16"
Moshe Quinn
Lmi 9h59'6" 33d22'55"
Mosie
Cyg 21h16'47" 46d57'21"
Mosses
Uma 11h32'17" 64d33'17"
Mossy
Cru 12h41'10" -64d5'56"
Mossy's Star
Uma 10h12'51" 64d53'29"
Most Beautiful Girl in the World
Tau 3h43'49" 25d43'47"
Most loved lady
Cyg 20h53'49" 48d24'14"
Mostafa&RoghiehAzariMajd 6Zoroaster
Sct 18h25'3" -15d37'34"
Moster
Pho 0h44'20" -43d17'8"
Mostest
Lyn 7h55'12" 45d25'11"
MoStew
Sco 16h9'9" -13d1'50"
Mostyn
Sgr 20h1'21" -30d26'57"
Mota
Uma 8h44'9" 69d45'38"
Mother
Uma 10h48'31" 64d28'48"
Mother
Cas 23h44'58" 56d0'3"
Mother
Cyg 20h32'6" 58d14'49"
Mother
Lyn 8h31'58" 44d29'45"
Mother Agape
Gem 7h14'44" 27d9'34"
Mother and Son
Sgr 18h40'18" -25d6'27"
Mother and Son Viktoriya and John
Lyn 7h41'25" 48d22'33"
Mother and Wife Katie Lynn Kettler
Aqr 22h25'11" -7d7'38"
Mother Deaton
Cas 0h45'39" 56d58'12"
Mother Elaine
And 2h18'56" 38d28'52"
Mother Ethel Sharelle
Leo 9h40'28" 26d26'23"
Mother Goose
Cnc 8h44'29" 15d20'46"
Mother Goose
Uma 11h37'4" 52d17'53"
Mother Goose
Cas 1h57'44" 61d40'1"
Mother Harriet
Ori 5h58'48" 22d29'42"
Mother Hot Dog
Ari 2h46'56" 26d53'40"
Mother Hubbard
Cas 0h2'22" 54d7'41"
Mother Kaye
Uma 12h22'54" 62d14'27"
Mother Kodiak Bear
Cas 0h49'40" 61d11'12"
Mother Marion
Lib 15h23'26" -16d42'44"
Mother Mary
Aqr 22h25'24" -1d33'14"
Mother Mary Claire
Del 20h42'33" 15d52'18"
Mother (My Guiding Light)
Uma 9h26'20" 51d56'10"
Mother - Nana
Cas 1h27'16" 63d5'26"
Mother of Elijah
Gem 6h53'56" 16d54'28"
Mother Of Faith
Cas 0h21'55" 57d33'48"
Mother of Katie
And 1h8'43" 36d53'35"
Mother & PaPa Jr's Constellation
Leo 9h36'3" 32d32'24"
Mother Perfect
Col 5h43'28" -33d2'20"
Mother Seville
Ori 5h43'41" 7d49'25"
Mother -Shauna Missy Brandon Trevor
Vir 13h28'9" 5d33'7"
Mother Teresa
Cnc 8h38'45" 23d18'15"
Mother Teresa
Sgr 18h33'33" -33d57'14"
Mother Tucker
Aqr 22h35'8" -17d22'0"
Mother, Friend, Yvonne Sagraves
Per 4h1'36" 45d28'28"
Mother-Aunt Libby
Leo 9h48'31" 27d7'45"
Motherbird
Sco 17h51'2" -36d50'12"
Motherly Love
Uma 9h13'50" 47d2'59"
MotherLynns
Aqr 23h37'35" -23d29'39"
Mother's Day Everett Star
Tau 5h12'58" 19d51'26"
Mother's Light
Ari 2h21'10" 24d20'50"
Mothers Love and Unity
Gem 7h34'52" 28d2'39"
Mother's Magic Gift
Lyn 6h36'58" 57d14'57"
Mothers Twinkling Eye
Vir 13h25'24" -14d59'19"
Mother's View - Stephanie A McGlynn
Vir 12h23'50" 11d34'39"
Mothnad H.o.l.l.a.n.d
Lib 15h18'58" -9d14'39"
Motning
Ori 5h58'11" 20d52'59"
Motoko Amelie
Cas 0h19'26" 50d10'34"
Moton Altreen
Tau 5h45'13" 14d23'18"
Motor City Josh
Cap 20h58'41" -15d46'55"
Motorcycle Nightlight
Cap 20h23'56" -14d17'18"
Mottershead I
Per 3h0'41" 45d21'31"
Motts
Psc 0h28'58" 8d6'51"
Motts1950 and AppleMBS
Vel 11h1'4" -41d49'15"
Mou Agapo
Her 16h27'54" 49d14'18"
Mou aporo otoichos
Cyg 19h47'38" 47d19'49"
Moui Saetern
Per 4h19'10" 41d37'23"
Moulin Rouge
Ori 6h15'3" 10d11'6"
Moumou
Cas 0h22'4" 61d39'49"
Moumounette
Uma 9h43'38" 62d41'30"
Moundville North
Her 17h51'46" 23d29'46"
Mouneshwarappa Gadagkar
Cap 20h21'34" -14d16'43"
Mounir
Crb 16h8'5" 32d59'23"
Mount Mark
Cas 0h2'27" 56d23'58"
Mount Shea the Mighty Shiner
Aqr 21h48'27" -7d27'14"
Mountain Black Hawk
Sgr 19h34'45" -13d13'48"
Mountain Man
Cap 20h13'19" -8d49'5"
Mountain Sister Too
Leo 11h33'56" 25d53'31"
Mountjoy, TJ1
And 0h44'42" 40d7'58"
Moura
Cas 2h36'54" 72d15'33"
Mourids Gion
Ori 5h25'57" 13d28'5"
Mouse n' Pup
Ari 2h50'34" 28d35'45"
Mouse Schroeder
Lmi 10h21'7" 37d29'49"
Mouser
Uma 8h21'30" 71d38'29"
Mouse's Bear
Uma 10h14'45" 49d56'17"
Mouse's House 5 February 1962
Uma 11h12'59" 55d21'24"
Mousie
Lyr 18h36'37" 27d17'6"
Mousse Rabin Kanaan
Ori 6h3'2" 17d17'42"
Moussia
Psc 0h28'15" 4d4'16"
Moustapha Niang/ Senegal
Cyg 19h35'30" 31d33'33"
Movie Star
Cma 6h38'52" -14d24'39"
Mowgli
Aqr 23h7'50" -7d29'25"
Mowgli
Per 3h10'23" 41d53'23"
MoxiReiStarDancer
Sgr 19h11'5" -21d42'13"
Moya
Aqr 23h46'55" -13d52'2"
Moya Jean McCabe
Nor 16h9'55" -49d44'32"
Moya Waters
And 0h33'47" 26d10'17"
Moye Wicks III
Ori 5h38'59" 1d14'38"
Moyer
Aql 19h11'43" 4d2'53"
Moyer-Pequeno
Cmi 7h37'21" 3d15'38"
Moyna Snelson
Sco 17h1'29" -44d8'37"
Moyra-joy
And 23h19'19" 52d8'21"
Moyse,III
Leo 11h11'40" 1d11'2"
Mozart and Jennifer Bautista
Aql 19h16'43" 16d5'38"
Mozart Burkholder
And 0h37'41" 37d50'21"
Mozart Horton
Aur 5h53'34" 45d51'42"
Mozelle Brooklyn Turbow
And 23h0'16" 51d53'22"
Mozenrath
Umi 15h55'57" 80d0'15"
Mozie's Star
Uma 11h24'27" 40d14'45"
MP Squared
Uma 9h31'36" 64d50'43"
MP8251981
Vir 12h59'51" -3d3'28"
MPB - Samaum Amplus Vitricus
Per 3h20'25" 39d48'48"
M.P.H.
Per 1h40'59" 54d32'4"
MPI Research
Pyx 8h41'45" -27d44'15"
Mpiaki
Leo 10h59'33" 23d15'28"
MplecoM, Du wirst ewig leuchten!
Uma 14h2'45" 59d57'10"
MPM
Gem 7h27'37" 34d8'24"
M.Pokora
Uma 9h12'6" 62d53'6"
MPWC Gates Sparkly Star
Her 17h58'27" 23d56'0"
MPZ15031990
Uma 11h41'45" 30d2'38"
Mr 147's Little Piece of Heaven
Cru 12h27'16" -60d47'56"
Mr Adrian "Ronnie" Corbett
Lyr 18h35'25" 39d27'41"
Mr. A'Harrah
Per 2h53'31" 55d1'5"
Mr. Amazing
Uma 11h41'38" 37d59'31"
Mr and Mrs Ackroyd
Crb 15h29'21" 25d41'3"
Mr. And Mrs. Adrian J WiseIII
Aqr 22h20'56" 0d47'13"
Mr and Mrs B
Cyg 20h54'25" 38d28'7"
Mr. and Mrs. Blake Alexander Boyd
Cyg 21h23'55" 51d19'3"
Mr. and Mrs. Calvin Seaburn
Umi 15h21'59" 74d3'21"
Mr. and Mrs. Christopher Ledbetter
Cyg 21h25'0" 47d44'32"
Mr. and Mrs. Cook
Ari 2h10'14" 11d32'0"
Mr and Mrs Coster
Cyg 21h53'0" 42d45'48"
Mr. and Mrs. Daniel Arundell
Crb 15h44'59" 36d32'53"
Mr. and Mrs. Daniel Mains
Her 16h48'59" 48d50'30"
Mr. and Mrs. Daniel Vitale
Cyg 21h45'58" 33d18'39"
Mr and Mrs David Jones
Cyg 20h43'19" 55d9'15"
Mr and Mrs David Wass
Cyg 20h54'22" 35d10'6"
Mr and Mrs Elbeik
Cyg 20h40'49" 49d4'18"
Mr. and Mrs. Gerson Reyna
Cyg 19h30'39" 52d15'8"
Mr. and Mrs. Hereford
Cyg 20h45'10" 30d24'3"
Mr and Mrs Hirst
Cyg 21h30'57" 37d5'0"
Mr. and Mrs. James Realbuto
Gem 7h26'4" 16d0'38"
Mr. and Mrs. Jeffrey Snyder
Cyg 20h6'16" 41d50'59"
Mr. and Mrs. Jeffrey Todd Durham
Cyg 21h30'53" 43d31'45"
Mr and Mrs Jon and Carol Travis
Cas 0h45'59" 61d34'6"
Mr and Mrs Justin Brown
Cyg 19h39'6" 53d29'21"
Mr. and Mrs. Kennith E. Branham II
Cnc 7h55'55" 10d20'10"
Mr. and Mrs. Leo and Erica Olbrych
Ori 6h9'15" 18d41'4"
Mr. and Mrs. Long
Gem 6h50'20" 20d7'19"
Mr and Mrs Lunn
And 23h16'29" 49d31'49"
Mr and Mrs Maddison
Cyg 20h1'53" 50d49'39"
Mr. and Mrs. Mann
Cyg 19h38'53" 31d25'57"
Mr and Mrs Manning
Cyg 20h24'52" 33d9'15"
Mr. and Mrs. Mark and Jamie Goebel
Cyg 19h57'3" 32d33'36"
Mr. and Mrs. Mark Steven Vojtko
Cyg 20h33'30" 47d42'20"
Mr. and Mrs. Mcfly
Leo 9h36'15" 28d42'26"
Mr. and Mrs. McQueen
Ori 5h22'52" 7d23'28"
Mr. and Mrs. Michael Mahan
Per 3h10'17" 48d45'50"
Mr and Mrs Morris
Cyg 21h34'12" 35d47'55"
Mr and Mrs Munch
Cyg 21h56'22" 55d5'2"
Mr. and Mrs. Nicholas Sebastian
Lyr 19h7'39" 31d13'55"
Mr and Mrs Pack
Cyg 21h29'57" 37d32'23"
Mr. and Mrs. Patton
Uma 11h29'41" 47d46'54"
Mr. and Mrs. Phothisane
Cyg 21h15'30" 44d3'22"
Mr and Mrs Raymond Cavallo
Gem 7h11'54" 35d1'52"
Mr. and Mrs. Rob Blume
And 0h48'23" 38d15'6"
Mr. and Mrs. Robert Cavallo
Uma 10h33'11" 56d8'27"
Mr. and Mrs. Robert J. Cooperider
Cyg 19h39'53" 36d39'47"
Mr. and Mrs. Schutter
Cyg 20h56'27" 34d55'6"
Mr. and Mrs. Scott Bott
Cyg 20h15'32" 52d17'1"
Mr. and Mrs. Scott Cojei
Umi 15h52'50" 76d10'43"
Mr. and Mrs. Stephen Quick
Leo 11h57'1" 14d44'1"
Mr. and Mrs. Trevor Purdin
Umi 16h29'46" 79d9'28"
Mr. and Mrs. Tyson Young
Aql 19h28'16" -10d45'36"
Mr and Mrs VA "Blackie" Disciglio
Cyg 20h19'12" 54d31'3"
Mr. and Mrs. Wayne L. Keys
Cyg 21h51'8" 44d1'56"
Mr. and Mrs. William L. Plumley
Cyg 21h22'20" 37d15'39"
Mr and Mrs Zach and Megan Wells
Lyn 8h57'31" 34d19'5"
Mr. and Mrs. Zmith
Cyg 20h2'8" 42d9'48"
Mr. and Ms. Whiskey
Mon 6h47'30" -0d57'46"
Mr Arthur, & Mrs Donnamaria Jones
Ari 2h39'13" 14d21'11"
Mr. Austin William Addington-Strapp
Sco 17h24'50" -39d32'12"
Mr. Avery
Sgr 20h23'50" -32d23'42"
Mr. Awesome
Lib 15h9'27" -17d57'36"
Mr. B
Cyg 20h52'53" 38d54'47"
Mr. Barrows
Umi 13h37'3" 78d57'28"
Mr. Bayless
Tau 4h11'49" 27d13'36"
Mr. Bear
Cap 20h26'29" -19d7'46"
Mr. Bernie
Cmi 7h24'29" 9d10'48"
Mr Beseder Love You Always
Tau 5h29'10" 26d38'6"
Mr. Big
Aqr 20h48'20" -8d46'0"
Mr. Bill
Cap 21h4'36" -16d41'39"
Mr. Bill
Aur 5h58'26" 55d26'5"
Mr. Bill, T-RXKV8R
Her 17h32'41" 36d22'14"
Mr. Bob
Umi 14h55'45" 75d26'39"
"Mr. Bob" Duffy
And 2h33'11" 49d48'24"
Mr. Bomo Wilson
Ori 5h39'5" 7d30'28"
Mr. Brandon Holmes
Aur 5h42'19" 49d56'51"
Mr Bunn
Uma 9h26'4" 58d20'46"
Mr. C.
Per 2h36'46" 51d10'45"
Mr. C
Gem 6h39'32" 21d15'11"
Mr. Caleb Rocks Out Loud!
Aql 19h49'5" 14d54'1"
Mr. Canham
Uma 9h16'8" 66d20'30"
Mr. Charles Urner Price
Boo 15h21'58" 48d23'22"
Mr. Charlotte
Per 2h41'53" 56d4'6"
Mr Cheeky Chops Hugh Langham 280805
Vir 14h9'2" -20d20'26"
Mr. Christenson
Tau 3h41'26" 6d8'2"
Mr. Colliton
Per 3h19'17" 51d22'36"
Mr. Cookie Countess Frank Watson
Cyg 20h0'31" 36d13'10"
Mr. Cooper
Gem 7h16'58" 22d23'53"
Mr. Cybulskis
Uma 13h40'16" 57d44'38"
Mr. D
Cas 0h16'19" 62d59'35"
Mr. D
Del 20h30'16" 16d38'55"
Mr. D
Lyn 8h29'1" 37d43'40"
Mr. D
Per 3h8'4" 53d0'54"
Mr. David A Wade
Tau 4h5'3" 23d11'3"
Mr. David Lewis
Sgr 17h52'58" -28d13'36"
Mr. Davis Lee Jahncke
Tau 4h33'58" 19d34'31"
Mr. Dreamy Darryl
Tau 3h30'58" 9d35'15"
Mr Dunbar and Year 7D 2006
Cru 12h42'18" -60d41'39"
Mr. Earl
Cnc 8h24'50" 29d44'23"
Mr. Eric Arlin, Principal & Mentor
Her 18h11'36" 29d5'6"
Mr. Fisher's Academy 2006 - 2007
Uma 9h35'30" 50d50'4"
Mr. Fox
Her 17h37'20" 44d28'54"
Mr. Friendly
Uma 9h41'39" 43d14'21"
Mr. Funnypants
And 23h11'47" 48d27'32"
Mr. G
Aur 5h32'26" 46d0'42"
Mr. G 10.27.26
Sco 17h41'52" -42d12'12"
Mr. Gary
Uma 8h35'30" 61d44'5"
Mr. Gordon L. Woodard
Cep 21h10'16" 66d36'16"
Mr. Handsome
Cap 20h10'19" -14d5'31"
Mr Harris
Ori 5h4'26" 4d55'12"
Mr. Hicks
Per 3h26'9" 47d45'45"
Mr. Hobbs
Cma 7h17'37" -17d56'6"
Mr. Howard
Leo 11h35'44" 16d50'39"
Mr. Huneke
Uma 11h36'31" 40d41'42"
Mr. I. M. Moster
Ari 2h7'24" 23d34'44"
Mr. Iraj Siyani God Bless you Dad
Ori 5h41'40" 4d21'42"
Mr. J. B. Potter
Aql 19h49'35" 9d2'53"
Mr. Jack B
Tri 1h46'27" 33d5'40"
Mr. James H. Luckinbill Jr.
Lib 15h38'1" -19d43'41"
Mr. Jay Jagpal
Lac 22h15'31" 49d0'41"
Mr. Jim
Ori 5h21'52" -0d8'9"
Mr. Jimmy
Lmi 10h5'24" 39d11'38"
Mr. Joe Brekke
Her 18h46'25" 20d17'17"
Mr. John Bayers
Per 3h8'18" 54d57'36"
Mr. John L. Browne
Tau 4h39'1" 18d49'0"
Mr. K
Per 3h38'31" 47d12'28"
Mr K Brett Gen Jade Kylie Hannah 05
Cru 12h9'21" -62d3'41"
Mr. Kent
Lmi 10h27'46" 29d22'6"
Mr. King
Lac 22h20'30" 46d46'30"
MR. L
Lyn 6h53'27" 53d32'27"
Mr. Latimer
Cep 23h34'44" 81d45'45"
Mr. L's Third Grade Class of 2006
Cyg 19h56'59" 30d40'17"
Mr. Lucio M.V.
Ori 5h36'39" 5d21'29"
Mr Mafia
Dra 15h54'24" 64d45'51"
Mr Magu - Callum Billy
Umi 15h9'11" 73d29'19"
Mr Michael Burns
Cep 21h19'50" 81d20'54"
Mr. Michael "Mike C." Cespedes
Psc 0h44'49" 19d43'30"
Mr Michael Patrick Sawford
Per 4h35'15" 48d42'27"
Mr. Mik
Per 4h6'59" 34d10'40"

Mr. Mike
Per 3h8'58" 51d51'6"
Mr. Mike's Diamond 7
Dra 17h41'5" 54d5'40"
Mr Milne
Uma 11h40'3" 31d25'40"
M.R. Mische
Uma 11h59'48" 37d3'48"
Mr. Morse Here
Cap 20h20'35" -23d39'14"
Mr + Mrs
Cyg 19h35'11" 47d27'41"
Mr. & Mrs. Adam & Arissa Pacheco
Ori 5h3'34" 10d15'46"
Mr & Mrs Andrew & Pauline Ford
Cyg 21h23'6" 30d31'52"
Mr. & Mrs. Andrew Peters
Cyg 20h3'1" 55d32'59"
Mr & Mrs - Andrew & Rachel Ferguson
Cyg 19h53'58" 32d26'49"
Mr. & Mrs. Antal Gratzer
Uma 11h50'29" 56d6'1"
Mr & Mrs Anthony & Julee Purificato
Cyg 19h57'10" 31d7'11"
Mr & Mrs Atkinson
Cyg 21h54'53" 49d20'31"
Mr. & Mrs. Audrey & James Oliver
Cyg 21h39'34" 46d46'38"
Mr & Mrs B!
Cyg 20h39'48" 41d23'55"
Mr. & Mrs. Balonis
Uma 13h20'19" 59d17'19"
Mr. & Mrs. Barry and Julie Brown
Lib 15h53'39" -11d21'10"
Mr. & Mrs. Billy K. Richmond
Cyg 21h12'2" 46d59'7"
Mr. & Mrs. Biscoe
Crb 15h41'56" 26d29'6"
Mr. & Mrs. Bitner Browne Spfld Ohio
Cyg 19h49'54" 39d12'22"
Mr. & Mrs. Bradford
Umi 15h22'37" 77d12'9"
Mr & Mrs Brown
Cyg 19h44'25" 30d42'11"
Mr. & Mrs. Bruce Gifford
Lyr 19h12'28" 26d20'27"
Mr. & Mrs. Bubby
Gem 6h33'36" 20d25'2"
Mr. & Mrs. Castro
Del 20h35'34" 15d30'4"
Mr. & Mrs. Cesar & Elizabeth Garfio
Dra 18h44'6" 58d6'58"
Mr. & Mrs. Charles Deon
Cyg 19h11'9" 53d51'43"
Mr. & Mrs. Christian Stokes
Ari 3h18'58" 27d37'37"
Mr.& Mrs. Christina Ratliff
Cnc 8h38'46" 13d47'56"
Mr. & Mrs. Cimafonte
Cyg 19h36'39" 52d18'52"
Mr.& Mrs. Clint West
And 1h10'53" 45d37'20"
Mr & Mrs Cooper
Cyg 21h31'59" 39d23'47"
Mr & Mrs D Bennett, 30/10/2004
Uma 9h36'54" 43d49'48"
Mr & Mrs Dakin
Cyg 21h5'34" 48d21'54"
Mr. & Mrs. David and Lauren Bejbl
Cyg 19h23'20" 54d3'7"
Mr & Mrs Davies
Cyg 21h32'50" 34d5'21"
Mr. & Mrs. D.E. Moreland, Jr.
Cyg 21h14'14" 45d7'4"
Mr. & Mrs. Denny Denbob Zavada
Ori 6h3'42" 17d16'56"
Mr & Mrs Dubberlin
Cru 12h24'53" -59d14'36"
Mr. & Mrs. Frank Borrego
Lyr 18h26'20" 35d18'47"
Mr. & Mrs. Frank & Marie Blanchard
Tau 3h43'2" 22d16'17"
Mr. & Mrs. Gabriel A. Torre, CPA
Ret 3h43'17" -54d2'55"
Mr. & Mrs. Gelinas (Nana & Pepe)
Per 2h39'43" 56d4'52"
Mr & Mrs Gerrard
Cyg 21h5'22" 50d59'46"
Mr. & Mrs. Gregg Vincent Gerelli
Cas 0h44'54" 61d4'48"
Mr. & Mrs. Hoops
Sge 19h37'29" 17d19'34"
Mr. & Mrs. Hoskins
Ari 2h0'57" 22d52'4"
Mr. & Mrs. Hughes
Sge 20h2'47" 17d13'4"
Mr & Mrs J Hammond
Cyg 19h47'30" 32d30'11"
Mr & Mrs J Hutchinson
And 23h14'15" 46d50'36"
Mr & Mrs Jackson
Cyg 21h32'11" 37d10'2"
Mr.& Mrs. James Egbert
Cyg 20h10'52" 41d10'42"
Mr. & Mrs. James R. Jane A. Baldwin
Cyg 19h43'45" 38d39'43"
Mr. & Mrs. Jeffrey Scott Hardy Sr.
Uma 11h48'53" 31d34'21"
Mr. & Mrs. John and Laurel Philibin
Cyg 19h58'16" 42d27'20"
Mr. & Mrs. John Kristoffer Moreau
Gem 7h6'54" 17d4'42"
Mr. & Mrs. K Model Tom Miller
Cyg 21h41'1" 51d47'18"
Mr & Mrs K Robinson
Cyg 20h6'42" 32d58'27"
Mr & Mrs Kealy
Crb 15h52'20" 33d3'19"
Mr & Mrs Keith Gary
Cyg 21h41'51" 52d2'49"
Mr & Mrs King
Cep 22h34'49" 60d21'57"
Mr. & Mrs. Kynon Ingram
Cyg 19h35'26" 32d16'48"
Mr. & Mrs. Larry W. Leverett
Uma 11h3'10" 50d12'22"
Mr. & Mrs. Leonardo
Umi 13h18'3" 71d49'3"
Mr & Mrs Lleyton and Bec Hewitt
Ara 17h2'40" -59d30'24"
Mr. & Mrs. Lowell Leynes
Tau 4h39'21" 17d17'52"
Mr & Mrs M Corcoran - Now & Forever x
Cyg 21h27'2" 34d40'7"
Mr & Mrs M. Crump
Cyg 21h16'46" 41d15'51"
Mr & Mrs MacKenzie
Cyg 20h41'39" 39d15'47"
Mr. & Mrs. Mark Fredriksen
Lyr 18h44'30" 35d6'36"
Mr&Mrs Matthew C & Brenda L Miller
Cyg 21h15'25" 40d25'35"
Mr. & Mrs. Michael Nidiffer
Uma 9h19'58" 57d0'34"
Mr. & Mrs. Michael Vigeant
Uma 8h37'55" 63d11'14"
Mr. & Mrs. Michael Wallace
Cyg 20h49'44" 46d20'27"
Mr. & Mrs. Michael Whyland
Cyg 19h52'55" 56d40'0"
Mr & Mrs Mitchell
Cyg 19h38'12" 31d25'13"
Mr. & Mrs. Monte L. Starr
Cyg 21h45'31" 45d31'38"
Mr & Mrs O'Grady - Wedding Star
Cyg 20h12'17" 30d16'46"
Mr & Mrs Raymond Tangredi
Gem 7h2'42" 33d3'34"
Mr & Mrs Robert Christopher Kozora
Cnc 8h40'37" 30d44'8"
Mr. & Mrs. Robert Kellis Hubbard
Eri 3h55'25" -0d42'10"
Mr & Mrs Robert R. Walker Jr.
Cnc 9h14'37" 15d35'16"
Mr. & Mrs. Robert W. Lefler
Uma 8h56'19" 56d50'26"
Mr & Mrs S. Wallace Wedding Star
Cyg 21h32'16" 50d37'6"
Mr. & Mrs. Scott and Heather Pavich
Cyg 21h1'22" 46d12'48"
Mr & Mrs Sell - A New Journey
Cru 12h4'30" -58d46'54"
Mr. & Mrs. Shaun Bottiglieri
Gem 6h54'59" 32d40'19"
Mr & Mrs Stephen Hill
Cyg 21h23'47" 32d31'56"
Mr&Mrs Stewart's Diamond In The Sky
Cru 12h19'33" -56d30'3"
Mr & Mrs Stronach
And 1h1'18" 41d17'34"
Mr. & Mrs. Thomas and Dana Johnson
Leo 11h18'43" 22d19'45"
Mr. & Mrs. Thomas Carroll Lee III
Cyg 19h33'44" 32d6'31"
Mr & Mrs Trace
Cyg 19h41'17" 28d22'29"
Mr. & Mrs. Trevor and Teresa Perry
Ori 5h27'36" 9d13'58"
Mr & Mrs Vallely - 1 Year Wed
Cyg 20h12'38" 38d7'29"
Mr & Mrs Wayne Hodgson
Per 3h18'14" 41d29'37"
Mr. & Mrs. William James Jones
Cyg 20h0'41" 41d38'13"
Mr & Mrs Willie A. & Lila Bradford
Lyr 19h19'56" 28d58'52"
Mr & Mrs Wilson, Always & Forever x
Cyg 21h51'20" 50d7'20"
Mr&MrsMarkBednarz Infinity & Beyond
Sge 19h52'57" 18d51'37"
MR Music Man
Uma 11h37'4" 40d33'57"
Mr. Muth's Physics Lab Star
Sco 16h7'0" -10d30'20"
Mr. Nevin
Her 17h36'22" 44d32'59"
Mr. O
Cap 20h12'32" -9d20'11"
Mr. Outstanding
Peg 22h45'21" 32d53'16"
Mr. P
Psc 1h44'57" 5d0'11"
Mr. P. #2
And 1h33'52" 48d59'43"
Mr Patrick Roach
Aql 20h8'15" 10d7'36"
Mr. Pau *8-28-04*
Psc 1h21'10" 31d29'58"
Mr. Peoples
Her 16h14'20" 46d53'45"
Mr Peter "Lovebug" Devereux.
Per 4h34'32" 39d2'3"
Mr. Phil
Uma 9h19'15" 66d42'5"
Mr. Phillip Mack
Cyg 20h20'36" 48d13'12"
Mr. Pickles
Gem 6h29'11" 15d44'8"
Mr. Pumpkin Bear
Uma 11h30'27" 65d0'47"
Mr. Purple
Vir 11h38'19" 4d9'46"
Mr. Ray Lesch
Uma 13h4'22" 52d47'20"
Mr. Redman 3
Uma 11h20'43" 52d0'18"
Mr. Richard D. Jackson
Ori 5h19'3" -6d40'55"
Mr. Rick Lankford
Her 17h31'24" 44d30'39"
Mr. Rider
Uma 10h54'43" 52d14'26"
Mr. Rider
Vir 12h11'7" 7d6'5"
Mr. Robert Goins
Her 16h42'17" 46d9'7"
Mr. Roberto's Star
Cep 20h49'44" 59d35'17"
Mr. Ron
Her 17h43'51" 35d44'48"
Mr. Roogies Star of Love
Per 3h53'54" 37d31'31"
Mr Russell & Louiza Briggs
Cyg 20h32'21" 33d8'46"
Mr. Shawn, Love 3M Class 2006
Umi 14h38'39" 75d8'57"
Mr. Shippey
Uma 12h1'6" 60d17'36"
Mr Silk
Cep 22h43'13" 63d20'57"
Mr Simon & Mrs Jane Porombka
Cru 12h42'45" -60d38'2"
Mr. Smart
And 0h55'24" 46d32'21"
Mr. Snitch
Cep 23h10'4" 70d25'11"
Mr. Snuggly Muggly
Sgr 18h26'44" -26d18'40"
Mr. Sprain
Lup 15h47'38" -34d53'31"
Mr. St. Amant
Cas 0h35'3" 55d40'33"
Mr. Sucky Butt
Vir 12h45'3" 3d18'56"
Mr. Tabares
Cnc 8h46'29" 32d5'33"
Mr. Tee-tee Red
Boo 14h39'10" 36d28'36"
Mr .Testesteron
Gem 6h58'59" 16d36'34"
Mr. Thomas Cruciani
Her 16h40'58" 28d35'12"
Mr Thompson and Year 7T 2006
Cru 12h40'31" -60d27'9"
Mr. Tim Albert & Mrs. Jana Reinking
Lyr 18h40'1" 38d30'14"
Mr. Tom Brissette
Lyn 8h4'26" 38d41'58"
Mr. Turtle
Cep 20h24'12" 61d18'44"
Mr. Tyrone Shannon
Psa 21h48'24" -33d38'49"
Mr Villa
Cnc 7h59'8" 18d10'42"
Mr Virgin Real Estate
Aql 18h53'25" 10d54'47"
Mr. Warren Palmeira
Lyn 8h42'4" 39d5'32"
Mr. Wicket
Uma 11h15'53" 48d8'47"
Mr. Wiggles
Lib 14h54'52" -6d24'38"
Mr. Williams' Diamond In The Sky
Ori 5h47'26" 8d17'51"
Mr. Wonderful
Leo 11h7'18" 15d52'40"
Mr. Wonderful
Cap 21h43'46" -14d37'49"
Mr Wonderful Jay Blain
Uma 10h36'13" 60d38'51"
Mr. Wonderful Moser
Lac 22h45'59" 53d52'2"
Mr. Woofman
Uma 11h57'21" 54d38'28"
Mr. WS6 (Mike Maestas)
Sco 16h7'0" -18d17'41"
Mr. X
Ori 5h39'22" 7d25'20"
Mr. Yaldiz
Ori 5h35'46" -0d21'12"
MRA777
Ori 6h19'24" 7d13'50"
Mramor
Tau 4h45'16" 20d29'0"
Mr.Cooper, You Shine! Love Timmy 06
Ara 17h36'16" -49d23'3"
MRGRstar
Uma 13h43'30" 53d36'43"
MRicahardsLPN
Crb 16h12'33" 36d52'38"
Mrinalini Mata
Lyn 8h51'7" 35d38'25"
Mr.Incredible
Lyn 8h46'22" 41d8'8"
Mritheria
Lyn 8h32'17" 34d31'19"
MRM & SBB
Vir 12h24'48" 10d38'25"
Mr.McGinnis~My Sheet Metal Knight
Dra 19h33'37" 69d4'7"
Mr.Onion
Pho 1h44'54" -50d9'25"
Mrozek, Jasmin
Uma 11h19'44" 33d49'51"
MrP Camalot
Lib 14h56'45" -18d38'43"
Mr.Robby "Baboo" Lovisa
Lyn 7h29'21" 57d42'37"
Mr.Ronald and Mrs.Patricia Lowry
Cyg 20h19'7" 38d10'1"
Mrs. A.
Peg 23h46'15" 13d7'46"
Mrs. A. Bond
Uma 11h27'4" 58d49'47"
Mrs. Amanda Grace Basile
Crb 15h38'7" 37d5'37"
Mrs. Ambereen Abbas
Cas 0h13'4" 56d19'29"
Mrs. Anderson
Cas 1h18'10" 65d49'20"
Mrs. Angela Marie Walbey
Cyg 20h32'17" 56d41'21"
Mrs Anne Marie Bannister
Sgr 18h26'12" -21d50'46"
Mrs. April Lukomski
Sgr 19h6'46" -17d17'34"
Mrs. Areej Al Ghanem
Ori 5h16'4" 15d31'11"
Mrs. Ayrian Jayne Goodwin
Cas 1h23'0" 65d20'48"
Mrs. Barbara McElearney
And 0h45'10" 34d44'17"
Mrs. Barbato
Sgr 19h4'51" -21d53'18"
Mrs. Barksdale's Quilting Valhalla
Cas 0h49'10" 61d21'30"
Mrs. Bessie S. Gerst
Aqr 22h51'43" -23d13'19"
Mrs. Boozer
Gem 7h15'18" 22d22'51"
Mrs. Brayton
And 0h33'7" 30d45'56"
Mrs. Brittany
Sco 16h19'0" -27d24'51"
Mrs. Brown's Class
Uma 11h39'16" 64d3'6"
Mrs. Burris
Cnc 8h51'18" 28d18'12"
Mrs. Carlson
Aqr 22h35'32" 1d9'16"
Mrs. Chanda Blay
And 2h34'48" 49d7'44"
Mrs Clair Mackinnon
Lyn 9h25'58" 40d25'21"
Mrs Clare Hunt will love you always
Cyg 20h35'45" 39d3'22"
Mrs. Crystal Bayles
Lyn 7h30'20" 56d59'11"
Mrs. D
Uma 10h15'45" 62d37'23"
Mrs. Debaar's Star
Dra 17h51'51" 52d12'10"
Mrs. Debra A. Evans-Johnson
Cas 23h59'39" 59d59'25"
Mrs. Denise Davis
Uma 13h37'57" 56d37'17"
Mrs Dickinson
Cyg 20h40'46" 36d47'46"
Mrs Dodson - OC Taylor Star Teacher
Cas 0h21'4" 55d44'3"
Mrs. Doherty's Mom
Cas 1h25'52" 57d58'5"
Mrs Ellen Turner
Sge 20h18'38" 17d51'58"
Mrs Emma Jane Begley
Cyg 20h47'57" 33d10'5"
Mrs. Evelyn Kennedy
Cas 23h53'20" 53d33'39"
Mrs. F Pichini & Mrs. A Marini
Uma 11h58'2" 43d24'6"
Mrs. Fat
Cas 0h14'52" 62d33'23"
Mrs. Flora D. Neely
Cas 1h12'53" 52d33'18"
Mrs. Fullmer
Lyn 7h47'34" 48d16'26"
MRS. GINA
Dra 17h34'10" 55d52'33"
Mrs. Greer Callender
Oph 17h4'4" 3d37'24"
Mrs. Ham's 5th Grade Class '04-'05
Umi 15h55'18" 75d36'12"
Mrs. Harahan
And 0h49'34" 37d57'34"
Mrs. Heather Kenny
Leo 10h25'43" 26d39'5"
Mrs. Helen R. McNulty "Mom's Star"
Cas 23h19'6" 54d44'54"
Mrs. Hodits, Kindergarten Teacher
Cas 1h32'7" 60d26'56"
Mrs. Holly Brown
Vir 12h13'13" 5d16'4"
Mrs. J. (MaryAnn Jacobsen)
Leo 10h20'37" 26d39'1"
Mrs. Jacobson's Class gr3 2005/06
Cas 1h24'34" 63d16'54"
Mrs. Janet Kelly Our Golden Heart
Psc 1h20'58" 26d19'32"
Mrs. Jeanne Weiler
Vir 12h41'39" -9d9'47"
Mrs Jennifer Lynn Waid
Psc 0h54'2" 14d35'50"
Mrs. Jennifer & Mr. Brian Bailey
Lib 14h54'48" -3d40'5"
Mrs. Jennybear
Cap 20h18'6" -12d35'36"
Mrs. Jessica & Mr. Jeremie Longley
Gem 6h54'6" 33d15'25"
Mrs. Jessica Renee Harvey
Crb 15h25'33" 30d32'42"
Mrs Josephine Rinaldi-Davis (Bubby)
Aqr 21h54'31" 1d30'1"
Mrs. Joyce
Uma 13h54'2" 59d40'20"
Mrs. Joyce Ann Thomas
Crb 15h50'40" 37d59'35"
Mrs. Joyce Ung
Crb 16h3'13" 36d43'11"
Mrs Julie Andrews
Cyg 20h3'18" 34d2'40"
Mrs. Julie Ohlin
Cnc 8h45'6" 17d21'55"
Mrs. K. Lockyer
Uma 11h27'43" 59d41'1"
Mrs Karen Riding's Star
Cyg 21h16'52" 53d49'3"
Mrs. Karlee Scott
Tau 4h20'31" 17d17'16"
Mrs Katie Marie Lister-Barron
Leo 11h36'40" 0d48'49"
Mrs. Kelly Rich- 1st grade class 05
Aqr 21h56'51" 1d11'41"
Mrs Kelly Russ
Lyn 8h38'15" 35d24'30"
Mrs. Kimmy E. Kato
Psc 23h10'37" 0d7'32"
Mrs. Kulju BA2K5
Crb 16h7'59" 35d50'5"
Mrs. Larson
Leo 11h19'46" 16d30'46"
Mrs Li Jacquet
Uma 8h36'1" 67d51'47"
Mrs. Lillian Marlow
And 0h43'10" 43d7'58"
Mrs. Lillian Stella
Ori 5h41'28" 6d53'17"
Mrs. Lillie B. Booth
Com 12h29'0" 28d13'29"
Mrs. Lisa Hoff
Uma 9h29'35" 65d42'43"
Mrs Lisa Lelliott
Cyg 20h48'30" 47d3'4"
Mrs. Lisa Morris
Ori 5h58'33" 22d23'34"
Mrs. Losser
Uma 10h18'44" 65d41'0"
Mrs. Mac's Star
Uma 10h49'12" 43d5'16"
Mrs. Maria Vella "aka Aunt Chubby"
Uma 12h3'12" 61d0'35"
Mrs Martha Kane
Uma 8h23'48" 60d53'8"
Mrs. Martinelli
Uma 9h6'33" 70d10'22"
Mrs Mary Compton
Cas 23h59'54" 52d48'8"
Mrs. Mary Norton Rhodes
Cas 23h26'46" 56d41'30"
Mrs. McCabe's Bright Student
Lyn 8h9'45" 39d37'48"
Mrs. Melissa Conway Stefanko
Cas 2h4'22" 62d43'30"
Mrs. Melissa Marie Breschuk
Cas 0h8'59" 54d11'49"
Mrs. Michelle Van Nest
Cas 0h50'44" 66d17'25"
Mrs. Moore's 2nd Grade Star
Uma 11h38'14" 56d35'35"
Mrs. M's Looping Class 2003-2005
Crb 15h52'59" 27d10'32"
Mrs. Nak
Lyn 7h31'42" 46d26'36"
Mrs. Nancy Mack
Cyg 20h19'25" 46d17'26"
Mrs Nicola Cullen's Wedding Star
Cyg 20h1'25" 52d45'53"
Mrs Nina Roberts
Del 20h45'22" 13d3'44"
Mrs Phoebe Roberts
Lyn 8h19'31" 35d13'11"
Mrs. Phyllis (Mena) Olko
Cyg 19h43'20" 51d42'43"
Mrs. Pink
Per 3h18'22" 41d4'25"
MRS Puppy
Tau 5h38'58" 26d59'50"
Mrs. Rana Bruce
Lib 15h16'33" -10d25'23"
Mrs. Renee Walter
Uma 10h15'54" 65d12'43"
Mrs. Rose Weiss
Uma 9h59'38" 44d44'19"
Mrs. Rutherford
Uma 13h11'21" 55d20'45"
Mrs. Sarah Holcomb
And 23h23'2" 48d21'55"
Mrs. Sheila Randal
Vir 14h48'34" 2d55'11"
Mrs Sherri Lynn Stacey
Gem 7h45'37" 34d16'53"
Mrs. Sparky
Gem 6h47'19" 13d25'57"
Mrs. Stacy Quate's Wedding Star
Cyg 19h50'29" 32d41'59"
Mrs. Stephanie Armstrong
Ari 3h24'46" 20d15'46"
Mrs. Stephanie Bennett
Vir 13h16'32" -12d30'23"
Mrs Stern
Cas 1h41'53" 63d37'4"
Mrs. Summer
Sgr 18h28'48" -17d1'47"
Mrs. Sunanda Kheraj
Cas 1h13'6" 59d4'2"
Mrs. Susan Carroll
Cas 2h27'54" 62d16'59"
Mrs. Susan Ford
Cas 0h29'4" 57d6'35"
Mrs. Thompson's Star
And 23h18'53" 40d33'2"
Mrs Tina Jane Pritchett
Cyg 21h25'36" 31d25'33"
Mrs Tomaski's Pre-K
Uma 11h42'16" 55d36'1"
Mrs. Tracy Heckenbach IV 07-13-07
And 1h31'2" 48d6'32"
Mrs. Tracy Schwyzer
Umi 15h40'32" 74d45'45"
Mrs Tweet
Cas 1h57'23" 63d38'2"
Mrs. Tyra Lynn Murphy
Lmi 10h22'5" 29d50'41"
Mrs. Watkins~Teacher Extraordinaire
Crb 15h35'14" 27d28'55"
Mrs Wendy Ann Oliver
Cnv 13h38'11" 32d12'14"
Mrs. Whetstone
Cas 23h15'41" 55d44'39"
Mrs. Winifred Lento
Per 2h18'4" 56d12'3"
Mrs. Winkle
Cam 3h17'16" 60d48'23"
Mrs. Yates
Psc 0h18'2" -2d26'22"
MrSchtepsil
Lyn 7h54'59" 39d14'3"
Mrs.Sexy Otter JCP
Sgr 18h52'26" -19d31'31"
MRTEBA 11:11
Dra 12h51'0" 71d56'18"
Mry Shushan
Aql 19h11'54" 7d21'40"
M's
Ori 5h30'14" 4d32'28"
M's
Sgr 18h8'9" -18d33'28"
MS 45 Nova Mini-School
Lmi 10h26'26" 36d27'51"
MS 70
Umi 14h55'10" 67d37'52"
Ms. Ann's Melody
Crb 15h25'9" 30d25'20"
Ms. Ashli Brownsberger
Vir 11h42'16" -2d55'28"
Ms. B - Jo-Ann Bourgeois
Uma 8h35'49" 52d37'40"
M's Beautiful B
Per 3h7'47" 48d18'44"
Ms. Bellisario ~ Our Star (2007)
Uma 13h43'10" 52d44'14"
Ms. Billie
Cnc 8h50'40" 17d31'30"
Ms. Brandice Lauren Ross
Uma 9h53'32" 47d8'54"
Ms Chem
Vir 14h6'26" -3d23'5"
Ms. Debra Ann
Cas 23h24'9" 54d58'7"
Ms. Dora Kahn -100th Birthday
Cam 6h1'30" 60d50'58"
Ms. Drea
Psc 2h4'0" 8d0'53"
Ms. Evelyn
Tau 3h32'32" 17d49'53"
Ms Flayvee Herring
And 1h46'49" 43d22'10"
Ms. Gaylor's S.T.A.R.R.S.
Cnc 8h10'2" 20d54'15"
Ms. Hammann
Lyn 7h4'12" 51d3'29"
Ms. Helen
Lib 15h29'42" -14d5'0"
Ms. Jane Tindall
Gem 7h2'34" 17d12'51"
MS & JM Pouvesle
Cas 23h1'39" 57d53'39"
Ms. Kaleidoscope Eyes
Cyg 20h53'27" 46d51'56"
Ms. Karen
Ari 3h6'4" 28d10'22"
Ms. Kathy Franker
Ari 3h29'17" 24d41'57"
Ms. Kelly Lynn McDevitt
Psc 0h36'4" 7d42'47"
Ms. Kit Woodruff
Uma 11h48'25" 63d41'30"
Ms Knabbe C202 Your First Class
Umi 13h14'7" 87d18'51"
Ms Lau
Lep 6h11'59" -19d4'12"
Ms. Lizzie
Gem 7h31'41" 26d23'25"
Ms. Lorena Jarquin
Tau 5h47'34" 20d13'43"
Ms. Louise Alison Robertson
Ari 1h54'35" 14d27'52"
MS - Love With Abandon
Umi 14h48'8" 79d21'49"
Ms. Madison
Gem 7h4'29" 29d4'21"
Ms. Mariann
Vir 13h43'53" -6d24'20"
Ms. Maryanne Wheeler-Grafmiller
Sgr 18h8'15" -27d14'3"
Ms. Mary's MAC Primary Two
Cas 0h18'46" 52d52'54"
M.S. Matthew
Uma 9h38'42" 66d3'51"
MS Monica 93
And 0h39'23" 23d24'50"
Ms. Monica Miramontez
Peg 23h23'1" 32d57'37"
Ms. Most Beautiful & Mr. In Love
Lib 15h13'21" -6d50'43"
Ms. Muffin
Leo 10h57'18" 20d12'56"
Ms. Nancy
Cam 3h22'47" 60d49'34"
Ms. Olivia Maria Badillo
Sco 15h50'4" -20d42'46"
Ms. Pamela Guelker
Vir 11h48'42" -4d49'21"
Ms. Paula's Class of 2004-2005
Her 17h49'38" 38d44'42"
Ms Peggy
Lyn 8h29'1" 36d2'44"
Ms. Peggy
Psc 2h4'2" 5d50'5"
Ms. Perez 3rd Grade Room 64 2006
Uma 9h26'27" 42d1'15"

Ms Rebecca
Cyg 19h30'40" 28d47'45"
Ms. Reece
And 0h36'25" 41d39'57"
Ms. Reeves
Aqr 23h36'31" -23d20'26"
Ms. Robin
Cnc 8h0'11" 14d19'43"
Ms. Saccardi's 4th Grade Class 2005
Her 17h26'20" 30d46'30"
Ms. Sarah Kassas
Cap 20h10'35" -18d1'58"
Ms. Sheryll McKever
Per 1h50'18" 50d54'47"
Ms. Speed
Crb 15h43'21" 26d12'10"
Ms. T
Crb 15h45'3" 26d38'20"
Ms Tanem Mustafa
Cyg 20h4'40" 50d48'11"
Ms. Thing
Mon 7h17'59" -0d38'23"
Ms. Zami
Del 20h38'46" 12d47'18"
MsCrystal Duplechan
Peg 21h30'7" 17d37'50"
msd11665
Sco 17h52'1" -37d7'16"
MSELLIE Hill
Srp 18h8'39" -0d20'49"
Msgr. Donald Hanson
Uma 13h13'29" 59d43'18"
Msgr. John J. Sempa
Uma 12h35'49" 59d45'22"
Msgr. Robert C Wurtz
Sgr 19h42'19" -35d53'6"
Msgr. Stan Rousseau-A Loving Heart
Lib 15h0'47" -13d22'17"
Msgt Jesse B. Morris, USMC, ret
Ori 6h2'41" -0d58'50"
mshindi
Uma 11h15'37" 59d10'36"
Ms.Keiko Nishida
Lib 15h19'6" -19d26'50"
MSL Camalot
Lib 15h5'8" -12d44'5"
Ms.Mouse2006
Lyn 6h44'9" 56d30'42"
Mssr. Bernard Beal
Psc 0h14'52" 3d37'41"
MStein
Ori 6h17'19" 10d9'39"
Ms.Teresa Spoon
And 1h18'34" 40d35'24"
MSU I Staff -IHS at Somerset Valley
Cnc 8h51'47" 28d40'12"
Ms.U My Favorite Star Love Timmy 06
Cha 10h44'57" -78d10'8"
M.S.Yezarski
Sco 16h54'39" -44d29'49"
M.T. & F.B. Sellers' Stellar Life
Uma 9h0'14" 66d48'30"
M.T. Keyser
Ori 5h8'15" 15d30'7"
Mt. Zion Youth - Seeking The Son
Lyn 8h36'37" 43d30'9"
MT3
Lyn 8h10'11" 55d47'11"
MTB
Pho 0h37'11" -41d58'25"
MTC & JMG's Monkey Paradise
Cyg 20h29'56" 32d37'23"
MTF
Aur 5h32'56" 45d59'8"
MTMWC - 4 Carraig Hill
Lyr 19h1'49" 42d9'20"
Mu
Per 3h18'11" 39d37'2"
Mu
Cap 21h27'55" -13d48'12"
Mu Chong
Ari 2h48'18" 15d31'15"
Muah
Uma 9h23'25" 64d37'1"
Mubarrak E. Morris
Uma 9h42'41" 65d5'36"
muchande'
Cas 23h23'16" 53d12'18"
Muchelito
Uma 10h43'30" 39d41'15"
Muchos Besos
Cnc 8h53'39" 26d57'0"
Muck
Aql 20h8'29" -6d5'25"
Mücke, Gerhard Heinrich Paul
Uma 14h1'21" 61d30'56"
Mucker
Uma 10h17'54" 45d7'23"
Mudcat58
Sco 17h45'40" -43d3'32"
Mudcharee "Poom" Julotok
Umi 16h17'24" 80d57'53"
Muddie
Umi 14h26'53" 73d44'38"
Muddys
Lyn 8h18'43" 47d21'2"
Mudfeets
Uma 12h57'30" 57d14'48"
Mudge
Uma 10h22'47" 59d54'53"
Mudgen
Uma 10h2'38" 70d15'52"
Mudget
And 2h36'4" 50d5'39"
Mudunuri Pullam Raju
Srp 18h40'10" -0d8'13"
Muermeli
Her 18h53'20" 21d6'45"
Mufasa
Leo 11h2'50" 10d34'53"
Muffet
Lib 15h6'2" -8d11'25"
Muffin
Sco 17h17'30" -38d31'16"
Muffin
Sco 16h52'31" -42d5'55"
Muffin
Her 18h55'51" 16d17'53"
Muffin
Cam 3h48'47" 59d2'28"
Muffin
Cyg 20h53'58" 34d44'26"
Muffin 912
Vir 13h42'40" -1d39'48"
Muffin and Tommy
Cyg 20h9'8" 31d56'29"
Muffin Butt
Ari 2h9'22" 27d7'52"
Muffin Carter
And 1h15'42" 38d28'14"
Muffin Elkin
Vir 13h40'37" 2d41'21"
Muffin Love 4 5
Ori 5h2'27" -1d16'32"
Muffin Makenzie Jones
Sgr 19h33'51" -14d25'55"
Muffin Storms
Uma 13h38'50" 53d33'2"
Muffin Thomas Hughes
Lyn 8h1'13" 58d30'14"
MUFFINBRAIN
Her 16h54'8" 34d31'14"
Muffin-Poodle
Pho 0h25'22" -43d27'14"
Muffi's Little Pearl
Cmi 7h35'45" 5d6'12"
Muffkin
Cas 0h55'46" 60d21'18"
Muffy
Uma 11h58'44" 36d28'11"
muffy olinger
Peg 22h44'45" 32d33'58"
Muffy Sepulveda
Ori 5h27'42" 5d53'32"
MUG
Uma 11h41'46" 61d16'8"
Mugatu
Cas 1h36'13" 66d16'34"
Müge Degerliyurt
Uma 9h16'10" 61d22'59"
Mugeli U. Tanneli
Ori 5h58'3" 8d58'19"
Mugg
Mon 7h4'5" 5d29'42"
Muggeli
Her 18h43'23" 19d50'0"
Muggeli Patrick De Bona
Lac 22h41'8" 52d33'5"
Muggzy Esposito
Cyg 21h20'59" 33d47'44"
Mugi
Her 18h11'29" 18d17'59"
Mugsey
Leo 11h2'1" 15d54'3"
Mugsley Lane
Uma 9h29'58" 47d27'27"
Mugsy
And 0h41'47" 38d40'54"
Mugsy
Cnc 9h19'35" 8d56'33"
Mugsy
Lyn 6h21'18" 56d46'30"
Mugsy Stevens"Karol & John's Buddy"
Uma 11h23'14" 58d24'20"
Mugsy Terusa
Cmi 7h20'18" 9d41'59"
Mugwamp
And 0h15'19" 25d28'14"
Muhammad Abdulrahman Abdullah
Leo 9h32'54" 17d46'52"
Muhammad Akeel
Leo 10h36'6" 13d10'34"
Muhammad I Jangda
Ori 6h13'24" 8d19'22"
Muhammad Mommy
Vir 13h39'44" 3d24'10"
Muhand .M. Ganuni
Ari 2h33'29" 22d37'39"
Muhlissa
Dra 17h42'35" 51d5'3"
Muhr Majácska
Gem 6h27'30" 26d12'38"
Múinteoir álainn
Tau 4h53'52" 19d50'43"
Muir
Lib 15h51'57" -11d43'14"
Muireann
Sgr 19h18'25" -16d14'45"
Muj Motylek
And 0h22'37" 26d27'6"
mujaki innocence
Dra 18h21'48" 57d55'54"
MUJI
Uma 8h41'31" 56d56'0"
Mukaddes
Tau 4h49'1" 27d4'21"
MukaRoo
Sco 17h4'43" -35d4'24"
Mukhtiar Begum
Uma 11h56'37" 33d29'52"
Muki
Uma 12h45'41" 61d2'35"
Muksel Sagredo
Ari 2h7'12" 25d58'10"
Mukti
Ori 6h24'12" 10d7'11"
Mukti Soni Gambhir
Cnc 9h13'2" 17d27'14"
Mülbert, Doris & Jürgen
Uma 10h22'36" 70d38'23"
Mule'
Cyg 20h49'36" 35d44'32"
Mulla herzaman için
Dra 20h4'21" 71d50'43"
Mullan Trail 4104
Sgr 19h57'13" -33d25'52"
Müller
Uma 9h46'15" 46d58'37"
Müller Béla
Psc 1h44'50" 9d23'15"
Muller & Tracy for Eternity
Vel 9h29'40" -50d39'34"
Müller, Bernd
Aqr 21h16'22" -2d21'10"
Müller, Bernd
Uma 9h4'32" 52d17'59"
Müller, Erika
Uma 8h55'28" 49d9'57"
Müller, Henri Paul
Uma 9h42'21" 44d15'40"
Müller, Karl-Heinz
Uma 9h10'48" 64d8'51"
Müller, Klaus
Ori 5h19'19" -8d6'30"
Müller, Klaus
Sco 17h18'35" -32d49'52"
Müller, Maria
Uma 9h17'30" 47d51'53"
Müller, Michael
Uma 9h55'11" 55d55'15"
Müller, Thomas
Ori 5h22'1" -4d55'50"
Müller, Werner
Uma 8h41'33" 59d57'23"
Mulley
Lib 15h32'46" -19d45'34"
Mulligan's Monkeys
Mon 6h43'30" 10d25'36"
Mullin
Cyg 20h40'28" 43d18'54"
muluwu serendipity
Uma 8h28'20" 62d37'59"
Muluwunu
Cyg 20h17'15" 48d46'27"
Mum and Dad Kaiser
Cyg 19h51'46" 33d11'29"
Mum and my little star
Aqr 23h21'42" -20d58'55"
Mum chérie
Cap 20h32'41" -18d20'19"
Mum & Dad - 30 Years of Love xxx
Cyg 19h43'55" 35d11'54"
Mum in a Million
Cas 23h35'16" 56d19'34"
Mum in a Million - "Gillian"
Per 3h13'34" 43d1'6"
'Mum' Joyce Clarke
Cyg 21h20'14" 31d30'8"
Mum/Madge's Star
Umi 15h25'57" 70d16'10"
Mum - "Our Guiding Light"
Vir 14h1'28" -13d32'19"
Muma Bear
Uma 12h8'36" 53d45'40"
Mumistern - ewigi Liebi
Uma 10h36'5" 70d4'3"
Mumma
Uma 11h9'41" 41d38'12"
Mumma Annie Pickles
Tau 4h25'24" 21d0'9"
Mummar
Cra 18h7'51" -37d4'32"
Mummer
Lmi 10h22'9" 36d26'57"
Mummie
Per 2h19'29" 51d14'50"
Mummo
Lyr 18h46'49" 30d6'19"
Mummy
Her 18h26'0" 20d0'26"
mummy Jen
Umi 13h4'26" 87d31'12"
Mummy Lady
Psc 23h40'25" 7d12'1"
Mummy - Nichole Sealy
Uma 8h41'12" 63d30'27"
Mummy Treadgold - Best in the World
Cas 1h17'14" 65d40'31"
Mummy's 30th
Peg 22h22'27" 21d36'1"
Mummy's Angel, Emily Rose Bradley
Lib 14h29'17" -15d40'17"
Mummy's Star
Tau 3h49'6" 23d39'16"
Mummy's Star
And 1h9'9" 45d43'20"
Mummy's star from her baby girl.
Cas 1h37'52" 61d48'2"
Mums
Per 3h36'37" 45d50'40"
Mum's Dreaming Star
Cru 12h38'3" -58d46'18"
Mums & Papa
Cyg 21h15'12" 42d21'22"
Mum's Sparkling Mandy
Gem 6h44'53" 25d14'16"
Mum's the best - love you Jack
Cra 18h1'53" -45d24'42"
Mumsie
Cas 0h21'26" 58d45'9"
Mumsy
Ori 6h6'57" 19d23'10"
Mumu
Uma 13h17'23" 62d21'52"
Mumu
Dra 19h24'37" 68d12'46"
Mumzi
Cas 0h58'46" 62d1'13"
Mumzoo
Uma 11h9'12" 57d59'11"
Mun Peikkonen
Uma 10h30'57" 43d10'13"
Mun San Ma
Ori 4h58'49" 15d0'45"
Muna Mashrah
Vir 13h36'42" -8d38'58"
Münch, Kai
Uma 12h30'52" 53d57'20"
Munchakin
Umi 14h31'57" 75d28'9"
Munchie
Umi 16h21'17" 75d53'31"
Munchie
Cyg 21h26'21" 52d35'59"
Munchie
Sgr 18h56'23" -26d30'48"
munchkie
Uma 11h27'18" 43d20'28"
Munchkie
Cyg 21h44'28" 41d21'44"
Munchkin
Tau 5h50'45" 17d26'21"
Munchkin
Cnc 8h20'58" 13d24'13"
munchkin
Tau 4h6'46" 6d25'31"
MUNCHKIN
Uma 9h14'16" 55d22'29"
Munchkin
Uma 9h15'21" 70d14'38"
Munchkin
Umi 16h13'53" 79d44'5"
Munchkin
Cap 21h36'50" -21d45'59"
Munchkin and Mom
Lyn 7h34'49" 35d16'11"
munchkin and princess
Sgr 19h24'46" -16d25'52"
Munchkin Fart
Vir 14h7'24" -4d37'34"
Munchkin of Mine
Leo 11h21'25" 2d39'7"
Munchkinator
Cas 0h21'40" 50d46'24"
Munchkin's Dream
Aqr 22h32'17" -1d43'19"
Munchkins Shining Dream
Gem 6h53'20" 22d27'11"
Munchkyn
Sco 17h43'36" -32d2'42"
Munchworld Marketing Sdn Bhd
Ori 5h38'4" 3d14'56"
Munckin's Sweetheart
Uma 10h1'35" 63d3'57"
Mundle's Brilliant Star
Psa 22h38'20" -32d18'56"
Mundo Gomez
Lyn 8h14'32" 34d39'47"
Mundt Buckey
Psc 0h6'19" 3d56'4"
Muneca
Mon 6h38'0" -4d5'0"
Muñeca Blance
Lmi 11h4'53" 27d11'1"
Muneef Taian
Cnc 8h36'35" 29d50'8"
MUNIBLOND'S Wishing Star - Forever
Ari 2h8'28" 16d28'9"
Munir Afif Sakhleh
Psc 1h4'14" 14d13'56"
Munira
Sgr 19h22'53" -33d37'27"
Munish
Vir 12h31'21" 1d54'46"
"Munkin"
Uma 10h10'46" 47d11'7"
Munkins
Ori 6h7'37" 20d59'57"
Munky N Mermaid Magick
Cru 12h40'47" -58d2'57"
Munnsey U.K.
Uma 9h58'33" 57d15'15"
Munshed Matti
Lyr 18h33'54" 39d18'19"
Muntzer Mughal
Leo 11h55'51" 21d19'54"
Mupet
Eri 3h56'55" -0d8'52"
Mupfel (Peter Josef Ruhrmann)
Uma 8h55'29" 48d12'43"
Muppet
Cap 21h52'15" -23d25'52"
Muppet the Christmas Dogstar
And 0h30'52" 41d50'15"
Muppethead's Star
Uma 14h20'29" 58d30'37"
Muppie 22
Aqr 23h10'37" -9d15'21"
Murad Ringo Chinoy
Cap 21h25'38" -19d56'5"
Murat Yavuz
Cnc 8h0'42" 14d4'19"
Muratori
Aql 19h14'16" 0d21'55"
Murdock Pickens
Cma 6h44'28" -17d55'57"
Mürggeli
Her 16h49'42" 8d53'43"
Muriah
Uma 11h2'19" 46d3'49"
Muriel
Cyg 21h44'35" 46d39'34"
Muriel
Uma 10h42'46" 66d16'55"
Muriel Alice Hammond
And 0h8'55" 39d36'26"
Muriel and Mason
Ori 6h23'1" 14d53'59"
Muriel & Bert Brodsky
Ari 3h20'3" 28d8'38"
Muriel Burum
Peg 22h51'37" 25d34'13"
Muriel : citadelle sensuelle
Uma 13h41'21" 59d18'30"
Muriel Cronkhite
Tau 3h57'57" 3d15'40"
Muriel Dorice Gough
Cas 23h49'47" 52d38'50"
Muriel Freedman
Cnc 8h18'29" 9d17'31"
Muriel Gotthardt
Peg 23h52'14" 26d45'33"
Muriel J. DiVita
Leo 11h22'13" 0d55'4"
Muriel L. Tipton
Mon 6h52'45" 1d46'23"
Muriel Leavitt
Ori 6h18'43" 9d52'36"
Muriel Levy Katsky
Psc 1h16'45" 30d49'56"
Muriel May
And 23h3'40" 47d42'33"
Muriel McIntosh & Edward L Palmer
Uma 11h23'33" 58d48'29"
Muriel Mervis
Gem 7h43'43" 21d35'43"
Muriel Montserrate Sainz
Sgr 18h54'45" -26d2'16"
Muriel Mueller
Uma 9h25'27" 68d58'15"
Muriel Pfeffer
Gem 6h42'27" 26d30'53"
Muriel Pierce Ring
Vir 13h3'33" -3d50'54"
Muriel R. Cohen
Lib 14h53'51" -3d9'23"
Muriel Rose Anne
Ori 5h0'37" -1d25'40"
Muriel Winter
Cas 0h33'40" 51d37'28"
Muriel, Donna, Richard
Uma 12h35'51" 58d44'31"
Muriel-19612006-45637E-454323N
Cap 20h28'12" -12d8'53"
Murielle et Daniel
Cyg 19h33'26" 32d2'45"
Murita Lynn Settles
Sco 16h45'35" -33d54'25"
Murkel
Uma 10h13'49" 61d27'12"
Murli
Uma 10h48'28" 62d11'15"
Murlin Eugene Gregg
Aqr 22h34'23" -4d5'52"
Murmann, Gerd Willi
Uma 9h18'26" 64d42'41"
Murmeli Björn
Sge 19h52'26" 16d57'51"
Murmeli (Sandra)
Cyg 21h27'56" 34d23'41"
Murong
Cnc 8h41'7" 9d45'28"
Murph
Uma 11h2'48" 33d22'32"
MURPH
Uma 11h32'24" 63d40'59"
Murph 36
Uma 9h31'3" 46d49'59"
Murphy
And 23h50'36" 41d47'35"
Murphy
Boo 14h44'51" 31d38'19"
Murphy
Cnc 8h49'3" 31d2'56"
Murphy
Uma 10h42'20" 70d49'10"
Murphy
Uma 9h43'11" 54d30'18"
Murphy
Aqr 22h32'20" -2d39'16"
Murphy
Aqr 21h41'3" -6d35'33"
Murphy
Aqr 23h43'56" -11d37'56"
Murphy
Cma 7h5'48" -25d0'24"
Murphy
Cnv 13h29'24" 41d37'34"
Murphy Ann Marie
Peg 22h19'57" 6d57'5"
Murphy Beim
Crb 15h48'30" 35d20'52"
Murphy & Jake
Gem 7h42'17" 27d22'23"
Murphy Mok
Uma 10h7'28" 45d10'3"
Murphy Mullins
Uma 10h21'16" 66d20'5"
Murphy the Cat McKenna
Lyn 7h32'9" 48d1'30"
Murphy21
Aql 19h9'36" 3d37'1"
Murphy's Bar & Grill, Honolulu
Cyg 21h28'59" 31d58'38"
Murphy's Dawn
Pho 23h56'2" -45d28'26"
Murphy's Light (D. L. Gruntorad)
Crb 16h20'19" 27d16'58"
Murray
Sex 10h27'8" 1d29'34"
Murray
Uma 11h22'31" 31d44'25"
Murray
Her 17h48'40" 42d48'58"
Murray
Uma 10h35'15" 46d5'42"
Murray
Umi 16h35'0" 82d57'18"
Murray A Pearlman
Dra 9h37'37" 78d20'48"
Murray Allan Vallance
Ori 5h59'31" 3d17'27"
Murray Andrew Clark
Per 2h14'55" 52d25'39"
Murray Cohen
Her 16h26'4" 47d55'51"
Murray F. Cheetham
Uma 8h36'7" 66d30'0"
Murray Goldberg
Uma 8h54'48" 61d29'31"
Murray Jack Rubin
Ori 5h22'56" -8d14'43"
Murray Kennedy
Uma 11h10'3" 63d13'9"
Murray L. Gilman
Umi 14h23'50" 78d6'0"
Murray Lubliner
Her 16h39'10" 27d18'25"
Murray Mansion
Uma 9h59'50" 55d26'36"
Murray "Muzza" Kendall
Cru 12h46'51" -62d3'11"
Murray Paul Stevenson
Her 17h41'6" 18d47'24"
Murray Robert Bruce
Sgr 17h56'2" -17d19'0"
Murray Ross, Great Vocalist
Sco 16h12'1" -16d33'49"
Murray Teagan Callahan
Aqr 22h15'53" 0d15'59"
Murray Wilson ~ 14.04.1925
Uma 10h18'23" 49d35'2"
Murray's Guiding Light
Ori 5h43'38" 6d51'57"
Murrell
And 0h47'2" 38d25'16"
Murrell Mae Boughner
And 2h21'15" 48d51'4"
mürrisch
Per 3h2'41" 52d23'35"
Murron Louise Nicholson - Murron's Star
And 2h30'51" 48d57'28"
Murry Connaughton
Uma 8h53'31" 57d39'4"
Musa & Farie
Cyg 19h36'37" 51d54'51"
Musaad Sultan Alhumaidhan
Lyn 8h52'13" 33d55'13"
Musashi 2002-3-10
Psc 0h58'7" 21d45'26"
MUSCARA
Uma 11h19'33" 42d51'16"
Muscatelli Alexis
Umi 19h47'17" 87d2'14"
Muschanow's Hope
Aql 19h45'24" -0d42'19"
Muscles
Sco 16h5'31" -23d46'38"
Muse
Ari 2h17'20" 12d30'32"
Mush
Uma 11h14'27" 28d31'46"
"Mush" Fararre, Dad & (Toods)
Uma 11h27'25" 52d52'23"
Mushkie: Dan and Kristen
Ari 2h37'28" 20d31'57"
Mushy
Cyg 20h15'36" 31d53'16"
MUSHY
Psc 1h12'5" 31d30'12"
Mushy Chipmunk Munchkin
Peg 22h16'15" 6d33'12"
Musical entrapment
Lyr 18h44'42" 37d52'1"
"Musical Soul" Jason Engleman Star
Her 17h37'28" 32d33'0"
Musidlowski, Wolfgang
Uma 12h15'31" 60d47'36"
Musiques Suisses
And 0h43'11" 41d31'0"
Müsli, 15.01.1999
Cam 7h47'25" 62d25'12"
Mussarrat
Crb 15h33'51" 32d13'50"
Mussell
Peg 22h37'4" 27d24'50"
Mussy Vaughan
Gem 6h51'0" 27d46'33"
Mustafa Barzani
Psc 0h22'38" 17d46'41"
Mustafa Lachgar is loved!
Ori 5h56'6" 21d7'8"
Mustafa Mesic
Tau 5h22'55" 26d7'52"
Mustang Mary
Aql 19h16'28" 9d1'31"
Mustang Sally
And 0h30'12" 40d25'53"
Mustapha Shaaban
Mon 6h50'4" -5d38'36"
Musumeci
Lyn 7h31'46" 36d30'8"
Muterhagen
Sgr 18h13'22" -34d30'27"
Mutha Coachie
Umi 16h27'21" 78d16'14"
Muthoni
Cap 20h23'43" -24d31'1"
Muti
Psc 0h58'17" 6d49'18"
Muti
Ori 5h56'2" 21d54'0"
Mutilation Ritual
Ori 5h42'5" 6d53'15"
Mutley
Her 17h19'15" 22d0'7"
Mutschi Stern
Umi 16h15'58" 79d20'44"
Mutschli
Cep 22h21'31" 66d41'11"
Muttertier
Aqr 23h22'33" -20d4'52"
Mutti Rose Heinle
Uma 13h13'16" 58d38'14"
Mutz
Ori 5h10'55" -1d15'42"
Mutzli Charly
Umi 14h30'53" 68d19'36"
Muù & Pà
Umi 17h27'9" 85d13'49"
Muuji
Uma 10h39'27" 58d41'24"
Muus
Dra 14h38'6" 59d39'34"
Muzza
Cnc 8h16'53" 8d17'12"
Muzzi - Star of Heaven
Uma 10h42'3" 59d28'5"
"Muzzy & T"
Cyg 21h34'32" 35d1'14"
MVBF Tami
Uma 8h33'45" 71d9'39"
MVCC 07: KW, JM, PL, TP, DL, LE, HB
Uma 11h35'44" 61d53'3"
MVOM Rose White Barrington
And 1h10'55" 38d12'4"
MVP 30
Eri 4h12'53" -2d1'10"
MVR/JCR
Sgr 18h59'4" -36d34'23"
MVW 2006
Uma 10h28'15" 61d45'32"
M.W. PRINCESS
Psc 1h16'59" 27d18'34"
MW143
Cas 1h51'31" 63d33'29"
MW21
Peg 22h39'43" 13d37'33"
Mwa
Ori 5h40'28" 6d5'40"
MWA Baby Rocko - a star for my love H
Cra 18h33'56" -43d54'24"
MwaTwa
Cnc 8h43'9" 24d9'35"
MWFAN
Uma 11h2'12" 56d0'53"
MWJILWAMH
Psc 23h58'50" 6d20'34"

MWM.TLH
Gem 6h54'50" 22d49'2"
Mwoon-Ya the Bear
Lup 15h29'44" -47d13'3"
MW's Terrific Twenty
Uma 10h35'26" 62d46'17"
MWZ52
Ori 5h3'5" -0d10'16"
My #1 Mommy, Dorotha M Edwards Birr
And 0h24'6" 32d4'59"
My 1 & Only Baseball Star: Lymtats
Per 3h8'37" 51d34'6"
My $100 Balloon Dream Wishing Star
Sco 15h53'31" -22d46'37"
My Adam Boriotti
Sgr 18h11'43" -35d48'23"
my Adam, forever
Lib 15h10'30" -10d57'16"
My Aim
Lyn 8h10'5" 38d38'0"
My Ain True Love
Lyn 7h35'7" 56d24'9"
My Albert
Uma 8h14'9" 59d52'29"
My Aleksandra
Uma 10h42'53" 57d16'28"
My Alessa
Cap 20h23'31" -10d26'53"
My Alexa Amethyst Glow
Psc 1h22'59" 16d38'33"
My Alexxx
Uma 11h9'39" 49d54'20"
My Alysa
Gem 7h27'4" 29d34'27"
My Amazing Dad
Uma 9h9'5" 55d37'22"
My Amazing ONE of a KIND Angel Bear
Her 17h52'46" 29d30'12"
My Amazing Princess
And 2h38'22" 42d5'10"
My American Buddy
Tau 3h38'9" 22d7'4"
MY AMORE
Umi 17h6'54" 84d52'36"
My Anamcara Wagner
Cnc 8h8'12" 26d33'51"
My Angel
Tau 3h39'30" 24d57'59"
My Angel
Peg 21h47'53" 25d57'29"
My Angel
Psc 1h28'19" 25d56'11"
My Angel
Ari 2h5'6" 10d38'30"
My Angel
Lyn 8h18'36" 34d45'47"
My Angel
Lyr 18h52'7" 33d13'6"
My Angel
Lyr 19h0'50" 33d34'18"
My Angel
And 2h15'10" 46d5'10"
My Angel
And 1h47'7" 45d33'0"
My Angel
Umi 16h5'0" 80d20'13"
My Angel
Lib 15h5'44" -7d34'41"
My Angel
Sgr 18h13'28" -18d19'31"
My Angel
Sgr 19h34'38" -17d9'8"
My Angel
Uma 8h58'24" 57d14'3"
My Angel
Uma 12h24'57" 55d56'45"
My Angel
Cas 0h19'14" 61d25'12"
My Angel
Dra 16h57'59" 54d15'3"
My Angel
Uma 10h55'46" 64d6'30"
My Angel
Psc 0h19'17" 6d53'16"
My Angel 143
And 0h40'41" 44d12'30"
My Angel Always
Lyr 18h50'57" 31d51'37"
My Angel Always and Forever
Cnc 8h13'48" 21d47'27"
My Angel And Love Of My Life Jasmin
Lyr 18h53'52" 33d6'47"
My Angel Ann Marie
Leo 9h40'37" 29d26'21"
My Angel Annee
Cru 12h37'12" -59d31'5"
My Angel Arlina
Lyr 18h53'43" 42d26'58"
My Angel Berna
Aqr 22h0'31" -0d50'2"
My Angel Christina Knabe
Peg 22h39'24" 25d11'36"
My Angel Christine
Cas 1h32'9" 57d7'15"
My Angel Cindy Leigh
And 23h14'46" 41d47'54"
My Angel Debbie
And 0h12'53" 46d13'24"
" My Angel " Dee
Sco 16h3'13" -12d42'27"
"My angel Deedee" (Maya Amro)
Uma 8h56'15" 51d0'32"
My Angel Dorothy Durance
Gem 7h33'20" 34d21'52"
My Angel Erin
Uma 10h55'4" 34d47'52"
My Angel Eve
And 23h39'22" 43d55'54"
My Angel From God
Lyr 18h28'18" 33d27'13"
My Angel Gaven
Cru 12h1'33" -62d32'23"
My Angel Genevieve
Ari 3h28'48" 27d59'55"
My Angel I love you always!
Lib 15h26'41" -7d34'35"
My Angel in the sky....LaMarr
And 0h37'15" 40d57'39"
My Angel Jenn
Cap 20h58'18" -17d25'28"
My Angel Jenny
And 1h7'33" 45d40'33"
MY ANGEL JULIE BISHOP
Cyg 20h9'39" 44d19'43"
My Angel Kristine
Ari 2h51'10" 27d5'32"
My Angel Lauren
And 23h18'56" 48d11'55"
My Angel Lori
Lyr 18h43'31" 34d21'29"
My Angel Manny
Ori 6h2'2" 17d25'19"
My Angel Michelle
Psc 1h24'33" 31d56'45"
MY Angel Michelle Brown
Ari 2h55'28" 27d13'24"
My Angel Mother Lynda
Cir 15h17'48" -56d25'32"
My Angel Nancy, My Gift From God
Ori 5h35'2" 7d31'10"
My Angel Natasha
And 1h0'54" 34d42'56"
My Angel Neddy Bear
Psc 0h57'12" 7d49'11"
My Angel Norma
Sgr 18h2'27" -28d0'8"
My Angel - Qiong Qiong Wang
Psc 1h29'2" 5d40'1"
My Angel Queen Darlene Reffel
Cas 1h37'41" 62d31'1"
My Angel Rosie
Tau 5h40'56" 21d31'53"
My Angel Stephanie
Lep 6h5'27" -11d31'37"
My Angel Whitney Walker
Psc 0h12'1" 6d38'43"
My Angel, Dania Karina Borja
Lmi 10h29'55" 38d0'22"
My Angel, My Friend...Barbara
Ori 5h33'3" 3d15'14"
My Angel, My Love
And 23h9'54" 47d54'31"
My angel, my love, Jennifer Waide.
And 0h48'40" 43d6'57"
My Angel, Rachel Leigh Copeland
And 1h25'50" 45d33'28"
My Angel, Victoria Ann Brennan
Aqr 23h30'23" -23d34'26"
My AnGel4EvA
Tau 5h8'19" 22d54'32"
My Angela
Cnc 9h13'7" 15d39'37"
My Angel-Mauz Forever
Uma 11h52'57" 63d2'7"
My Angels
And 1h20'16" 49d17'27"
"My Angels"
Lyr 18h42'26" 38d49'46"
My Angels - Billy & Lois Wood
Sco 16h57'37" -38d23'8"
My Angels Lovestar
Umi 13h41'47" 76d14'46"
My Anna
Cas 0h44'47" 66d35'36"
My Ashley Fleming
Umi 15h14'45" 71d42'3"
My Ashley love ya just as you are
Sco 17h25'46" -43d42'53"
My Aunt & Great Friend, Amy.
Uma 10h56'37" 56d14'46"
My Babby
Ari 1h50'49" 18d0'56"
my babe
Uma 11h39'27" 63d35'1"
My Babe Mi Amore My One Dan
Lib 14h49'6" -10d19'19"
My Babies
Umi 15h29'3" 77d40'15"
My baby
Umi 14h1'0" 76d54'56"
My Baby
Lib 15h11'17" -7d49'23"
My Baby
Cas 1h4'18" 61d2'22"
My Baby
Psc 0h40'56" 5d16'55"
MY BABY
Cap 21h23'36" -24d45'27"
My Baby
Psc 23h55'35" 0d56'43"
My Baby
Aqr 22h12'57" 1d34'41"
My Baby
Cyg 19h21'22" 28d55'11"
My Baby
Uma 10h6'36" 42d50'27"
My Baby Brooke 143
Cyg 20h24'4" 39d11'58"
My Baby Buteaux
Col 5h15'12" -35d55'46"
My Baby Carole
Vir 11h41'8" -2d43'17"
My Baby Catherine
Vir 13h47'3" 2d44'26"
My Baby Chelsea K Coulson
Aqr 23h29'28" -19d43'26"
My Baby Chelsey
Peg 21h26'45" 11d5'23"
My Baby Danielle
Sco 17h27'4" -38d50'22"
My Baby Girl
Sco 17h17'10" -32d35'57"
My Baby Girl
Mon 6h53'3" -0d7'59"
My Baby Girl!
Uma 13h0'24" 62d44'48"
My Baby Girl Claire
And 0h2'19" 32d52'27"
My Baby Girl Laura
Lib 15h37'34" -22d2'43"
My Baby Girl Nicolette's Sweet 16th
And 1h3'1" 43d0'46"
My Baby- Jade Wesley Jeffords
Lib 14h53'54" -15d43'51"
My Baby - Jennifer Lee Watkin
Vir 13h20'4" 4d41'28"
My Baby Jessica Wright
Psc 0h15'49" 2d32'35"
My Baby Kayce
And 23h43'4" 36d42'12"
MY Baby Kristin
Vir 13h44'17" 0d49'37"
My Baby Lisa
Sco 17h52'12" -36d55'40"
My Baby Matthew Wayne Raynor
Her 16h45'1" 30d19'38"
My Baby Mayra
And 0h18'18" 43d15'12"
My Baby Melissa
Sgr 20h9'50" -38d57'33"
My Baby Moot
Uma 10h2'14" 48d22'35"
My Baby Shannon Hurley
Peg 22h40'51" 11d38'47"
My Baby, Carol
And 2h13'42" 39d19'37"
My Baby, Michelle
Gem 7h43'52" 33d12'50"
My Baby, My Life, My Wife...Claudia
Gem 6h32'5" 15d52'47"
My Baby, my Shpirt
Uma 9h32'56" 50d44'55"
My Babyber
Uma 9h18'1" 57d6'3"
My Baby's Got Sauce
Psc 22h53'51" 4d9'56"
my battalion
Vir 12h1'18" 6d4'20"
My Bay Ros
Leo 10h2'50" 15d3'51"
My BB
Umi 18h0'21" 87d6'18"
My Beacon of Life-My Wife-Misty C.
And 23h29'30" 42d27'26"
My Bear Schaefer
Uma 11h44'37" 53d16'45"
My Beast
Gem 7h29'51" 26d27'59"
My Beautiful Delight Stopar
Gem 7h39'41" 25d20'7"
My Beautiful
Cnc 8h50'55" 27d20'46"
My Beautiful
Tau 4h25'15" 22d52'25"
my beautiful
Cyg 19h46'44" 32d17'3"
My Beautiful
Lyn 6h31'39" 54d14'9"
My Beautiful
Cas 0h46'52" 61d46'56"
My Beautiful
Aqr 22h56'25" -11d22'3"
My Beautiful Amanda
Ari 3h24'39" 29d1'56"
My Beautiful Angel
Gem 8h1'43" 28d54'12"
My Beautiful Angel Amanda
Tau 4h32'2" 26d34'21"
My Beautiful Angel Bebe, MKG
Aqr 21h39'21" -1d12'33"
My beautiful baby girl Natalia
Ara 17h0'22" -49d30'29"
My Beautiful Basherta Allie Silber
Cas 2h21'54" 65d24'11"
My Beautiful Belinda
And 23h56'0" 33d7'42"
My Beautiful Bonnie
And 0h13'39" 32d15'45"
My Beautiful Boo
Cyg 19h42'42" 29d10'38"
My Beautiful Bride Aubra Halladay
Crb 15h38'2" 29d0'36"
My Beautiful Brown Eyed Girl
And 0h4'45" 40d4'2"
My Beautiful Cindy
Gem 7h35'59" 26d45'36"
My Beautiful Courtney
Vir 13h16'56" 11d10'49"
MY BEAUTIFUL DAISY
Sco 17h34'6" -41d53'36"
My Beautiful Des
Vir 14h2'44" -7d35'3"
My Beautiful (Erica Wharton)
And 1h16'10" 49d36'0"
My Beautiful Girl Rainy
Psc 23h24'20" 4d10'0"
My Beautiful Girl Taya
Cap 20h37'4" -13d34'20"
My beautiful girl, Katie Delguercio
Peg 21h49'36" 17d48'29"
My Beautiful Girls
Aqr 20h55'13" -9d10'21"
My beautiful Heather
And 0h40'50" 34d48'24"
My Beautiful Honey Bunny
Cyg 20h18'34" 45d41'27"
My Beautiful Jacki Beau
Psc 0h51'29" 9d39'11"
My Beautiful Jeffrey
Per 3h13'18" 36d12'54"
My Beautiful Jennifer
Ori 6h4'59" 21d24'3"
My Beautiful Jennifer
Psc 1h6'44" 3d32'22"
My beautiful Jessica Ashley Gottron
Leo 11h21'3" 16d54'40"
My Beautiful Jessie
Lup 15h45'39" -30d33'0"
My Beautiful Jolene
Cas 23h58'34" 65d52'58"
My Beautiful Kelly Scott
Gem 7h11'5" 34d25'17"
My Beautiful Kerri Baby
Apu 16h28'36" -76d19'47"
My Beautiful Kerry Mary
Cnc 8h13'11" 30d11'52"
My Beautiful Kylie - Forever Beautiful
Cru 12h38'59" -60d42'38"
My Beautiful Lady
Cru 12h18'7" -62d11'33"
My Beautiful Laura
And 0h32'44" 37d19'20"
My Beautiful Lisa
Cru 12h38'53" -58d58'21"
My Beautiful Little Princess Alyce
Vir 13h8'56" 6d51'26"
My Beautiful Loretta
Pho 0h38'52" -40d51'2"
My Beautiful Lucy
Cas 1h39'24" 64d4'50"
My Beautiful Megan
Lib 15h5'38" -10d26'26"
My Beautiful Mercedes
Ori 5h57'4" 12d34'18"
My Beautiful Mother Fleur
Cnc 8h10'22" 32d45'49"
My Beautiful Mother, Gina Kim
And 23h32'23" 41d37'48"
My Beautiful Nicole
Cnc 8h48'49" 26d59'3"
My Beautiful Princess
Sgr 18h3'3" -28d3'58"
My Beautiful Princess Donna
And 23h38'9" 49d8'0"
My Beautiful Princess Nicole
Col 5h54'13" -38d10'29"
My Beautiful Sarah
Ori 5h57'0" 17d55'58"
My Beautiful Shining Light
Lib 14h24'40" -23d26'47"
My beautiful sister Eileen.
Leo 10h24'50" 21d59'0"
My Beautiful Snowflake
Umi 13h13'10" 74d4'43"
My Beautiful Sunshine
Lyr 18h35'52" 33d45'58"
My Beautiful Surprise
Lyn 8h5'47" 42d9'1"
My Beautiful Susie Q
Ori 5h18'38" -6d58'22"
My Beautiful wife Amy
Ari 2h35'32" 27d20'41"
My Beautiful Wife Erin
Cas 0h17'36" 51d41'2"
My Beautiful Wife Jessica Singer
Lyn 7h56'1" 41d8'35"
My Beautiful Wife Lisa Winters
Cap 21h22'31" -25d33'54"
My beautifull Cara
Her 17h45'47" 33d15'17"
My Bebe (Jason Scott Smith)
Leo 9h22'45" 10d58'17"
My Belinda
Vir 13h16'55" -2d42'49"
My Belle
Umi 19h24'56" 88d39'19"
My Beloved
Sgr 18h5'23" -27d15'28"
My Beloved Angel
Leo 11h6'18" 24d32'43"
My Beloved Chanel
Sco 17h44'28" -40d25'55"
My Beloved Courtney
Cyg 20h49'53" 38d8'36"
My Beloved Emily Kroemer
Leo 11h32'58" 14d8'52"
My beloved husband Shaun Jennings
Cyg 21h43'50" 50d44'13"
My Beloved Husband, Gary
Lib 14h30'10" -10d8'21"
My Beloved Iain
Per 2h22'55" 52d34'15"
My Beloved Lori
Lib 15h30'0" -4d20'13"
My Beloved Lori II
Lib 14h51'26" -2d42'28"
My Beloved One
Cyg 21h17'54" 51d50'45"
My Beloved Ricky
Cep 21h23'40" 61d39'11"
My beloved Shannon.
Sgr 19h28'24" -28d55'59"
My Beloved Teresa
And 0h38'17" 36d58'0"
My Beloved Trish
Cnc 8h6'38" 10d36'58"
My Beloved's Birthday Star
Cap 20h22'8" -13d15'57"
My Bernardette
Cnc 9h2'46" 17d56'53"
My Best
Uma 10h19'53" 44d29'45"
My Best Buddy
Gem 6h46'25" 34d15'37"
My Best Buddy
Umi 16h26'38" 80d16'13"
My Best Friend
Ari 3h26'28" 26d44'24"
My Best Friend
Cnc 8h28'10" 13d40'1"
My Best Friend Amy Robbe
Lac 22h26'33" 50d24'55"
My Best Friend Ann Wilkinson
Tau 4h16'41" 18d55'45"
My Best Friend Diem Ngoc Tran
Cap 21h8'33" -17d43'19"
My Best Friend Ellie
Aqr 22h41'45" -20d31'12"
My best friend Nicole
Del 20h37'30" 10d32'34"
My Best Friend, Jeffrey A. Pearson
Psc 1h4'5" 19d17'17"
My best friend, love, and soul mate
Tau 3h49'37" 27d39'58"
My Best Friend, My Love... My Sean
Uma 9h3'39" 57d10'19"
My Best Love Wang Jia Yi
Cyg 19h58'6" 35d53'53"
My Betsey Girl
And 0h47'49" 24d21'57"
My Betty
Cyg 21h38'6" 42d0'13"
My Big Baby
Cyg 19h57'22" 44d28'53"
My Big Boo
Aql 18h42'56" -0d35'28"
My Big Little Stink
Uma 11h10'13" 35d30'9"
My big Love Stephan Gerber
And 1h22'49" 48d53'54"
My Bill
Her 16h41'37" 48d1'22"
MY BILL
Uma 11h18'37" 50d40'18"
My Billy
Cru 12h31'7" -60d1'43"
My Bingus
Lyr 18h34'55" 33d21'16"
"My Birgit" Star of Kindness
Cnc 9h13'10" 15d5'57"
My Blue Eyed Baby
And 1h13'7" 36d37'25"
MY Blue Eyes
Del 20h35'59" 14d17'52"
My Bobby's Shining Star
Her 18h43'8" 20d28'37"
My Bonnie
Sgr 19h48'31" -18d54'23"
My Boo
Ori 5h18'40" -8d45'36"
My Boo
Uma 14h26'48" 56d43'2"
My Boo
Men 5h45'35" -70d30'54"
My Boo
Leo 10h16'4" 15d8'51"
My Boo
Tau 5h8'56" 25d20'25"
My Bow Bay-Bo 1962-05-16
Uma 9h6'17" 49d51'36"
My Boy
Umi 16h37'57" 81d50'14"
"My Boy", Steven
Her 16h28'56" 45d54'50"
My Boys, Jacob, Eli and Luke
Uma 8h56'29" 65d24'59"
My Brazilian Babygirl
Ari 3h28'29" 27d13'20"
My Bride Kathryn Hub
Cas 1h26'54" 57d25'57"
My Brightest Blessing
And 1h18'31" 45d46'1"
My Brightest Star Amy Mae (Grandma)
Uma 10h35'9" 59d21'6"
My Brightest Star Becky
Cyg 20h36'33" 55d58'36"
My brightest star Elise
And 1h43'5" 45d5'7"
My Brightest Star: Eva
Lyn 7h40'27" 36d30'44"
My Brightest Star Forever Chris
Cap 21h30'13" -22d41'26"
My Brightest Star - Trevor
Col 5h41'2" -39d16'17"
My Brightest Star, Mariana
Psc 0h9'43" 8d2'47"
My Bright-Eyed Boy
Per 2h31'27" 55d16'41"
My Brilliant Claudia
And 0h51'30" 22d59'15"
My Brilliant Man Jake
Her 16h24'50" 12d41'1"
My BrittAny 04'
Aur 5h19'42" 42d3'28"
My Brock Grandchildren - K - K & T
Cru 12h48'4" -64d30'18"
My Brother and My Little
Lyn 9h14'14" 36d15'29"
My Brown Eyed Baby
Ori 4h48'40" 11d43'53"
My Brown Eyed Brooklyn Queen Farrah
And 0h15'29" 32d20'56"
My Brown Eyed Girl
Ori 5h43'32" 8d32'51"
My Brown-eyed Girl
Ori 6h19'36" 2d0'32"
My Bubba
Tau 3h54'47" 8d5'39"
My Bubbykinnz Rachel Rose Niemeyer
Cnc 8h35'27" 24d31'34"
My Buddha
Gem 6h32'14" 23d8'45"
My Buddy
Uma 10h50'2" 55d47'7"
My Buddy Arch Darden Bynum III
Aqr 22h36'37" -0d32'17"
My Buddy Mark
Ori 5h32'30" 5d48'25"
My Buddy Nana
Psc 0h31'19" 14d35'4"
My Bug Forever xxx
Ari 2h50'22" 24d30'18"
My Bunkey
Lyn 6h53'18" 51d42'26"
My Bunky - Bryan Shoemaker
Mon 6h55'33" -0d30'0"
My Bunny Crystal Perez
Lep 5h6'25" -12d6'55"
My Butterfly
Tau 4h27'59" 16d15'12"
My Butterfly Molly
Sco 17h22'20" -41d17'41"
My Buttmunch
Her 16h42'56" 7d43'5"
My Candy Girl
And 2h34'21" 48d13'20"
My Capa
Aql 19h46'23" -0d35'54"
My Captain
Per 1h48'54" 50d44'6"
"My Chad" McCullough
Ari 3h24'29" 27d46'44"
My Champion
Umi 15h13'33" 67d50'39"
My Charished Star
Gem 7h26'28" 31d57'34"
My Chellabella
Cap 21h44'17" -17d29'12"
My Chelles
Ari 2h37'44" 26d56'45"
My Chemical Romance
Cyg 21h10'52" 44d15'21"
My Cherished Father
Cnc 8h43'19" 23d45'42"
My Chica 21 today
And 0h31'13" 28d17'15"
" My Chicklet, My Love " 09/01/2000
And 2h14'55" 45d24'9"
My Chicky Mama Heather
Uma 11h35'16" 62d33'16"
My China Doll Nancy
Sgr 18h4'25" -25d30'49"
My Chris
Tau 4h36'28" 25d27'1"
My Chris - Forever & Eternity
Cyg 20h55'3" 31d52'54"
My Christi
Ari 2h46'7" 11d3'11"
My Christine
Psc 0h45'56" 3d20'17"
My Christine My Love My Forever
Cnc 9h5'38" 19d3'10"
My Cin
And 0h53'10" 40d34'2"
My Claire Lucy - always and forever
Cyg 20h37'31" 34d58'53"
My Clare Louise
And 1h6'21" 45d6'31"
My Compass - Peter Maxwell Mallon
Cru 12h28'34" -60d28'59"
My Conquered Prince
Uma 11h41'48" 50d40'2"
My Cookie
And 1h49'30" 37d49'2"
My Cookie
Sco 17h26'39" -38d28'31"
My Cookie
Cma 6h39'54" -14d7'49"
My Corazon Karmin
Tau 4h22'47" 13d0'20"
My Cowboy
Per 3h23'0" 41d10'3"
My Cowboy, Keith Owen Duncan
Uma 10h12'38" 46d44'22"
My Cracker
Lib 15h1'19" -16d36'37"
My Crazy
Per 3h2'52" 57d5'4"
My Cutie - *Kristen Lee O'Connor*
Sgr 18h34'30" -36d12'22"
My Cutiepie Diana
Aqr 22h35'29" -2d45'37"
MY DAD *
Uma 8h53'19" 51d19'31"
My Dad AEJH
Cep 20h46'22" 63d39'55"
My Dad And "Ei" Together Forever
Cap 20h25'30" -11d23'3"
My Dad, Aaron H. Simon, Love, Adam
Aqr 22h34'35" 1d10'13"
My Dad, JB Hunt
Cep 22h16'21" 69d38'28"
My Dad, Michael Dzema, Jr.
Lyn 6h58'56" 45d10'23"
My Dad, My Hero, John Marsh
Gem 6h1'52" 24d37'1"
My Dad, My Star, Stevie's Star
Vir 12h9'50" 2d12'13"
My Daddy
Per 3h45'57" 32d12'43"
My Daddy
Uma 11h0'58" 67d18'38"
My Daddy Christopher J. Cade
Cep 21h25'11" 60d17'59"
My Daddy Ian Edward Davies 31051966
Per 3h2'48" 48d30'58"
My Daddy John
Her 17h21'9" 26d43'52"
My Daddy & Me - Dan & Andrew
Umi 15h51'18" 80d2'52"
My Daddy Neil
Pho 2h13'39" -41d31'53"
"My Daddy" Roger U. Long
Gem 7h35'7" 21d26'20"
My Daddy, Gerald C Riley
Cyg 19h54'55" 37d16'52"
My daddy, Jim Davis
Lib 14h48'2" -12d43'31"
My daddy, Jim Davis
Lib 14h47'32" -15d28'5"
My Daddy, My Hero
Per 2h46'41" 55d11'51"
My Daddy—I Love You
Per 3h21'40" 47d27'20"

My Daddy's Star
Cnc 8h41'13" 32d10'55"
My Dahling H.P.E.B.H.A. Thyssen
Uma 13h38'47" 49d47'34"
My Daisy
Tau 4h9'59" 29d11'24"
My Daniella
Gem 6h38'44" 12d12'4"
My Danielle
Per 3h28'43" 34d57'29"
My Danie's star
Sgr 20h22'26" -42d23'13"
My Dara
Gem 6h26'49" 18d4'30"
My Darlin
Tau 4h9'57" 18d12'27"
My Darling
Cnc 8h43'6" 15d54'3"
My Darling Angel
Peg 22h8'50" 10d8'19"
My Darling Angel Jey
Uma 14h6'33" 51d37'13"
My Darling Carmen Lafferty
Lib 14h32'39" -12d49'28"
My Darling Courtney
Lib 15h53'37" -11d20'16"
My darling daddy spaghetti
Psc 0h23'3" 8d15'54"
My Darling Denise
Ori 5h23'1" 2d15'8"
My Darling Emily - The Brightest Star
And 1h44'24" 48d55'2"
My Darling Eva
Tau 4h53'50" 17d10'34"
My Darling Grace
Tau 5h32'40" 17d41'12"
My Darling Greenie - 03.09.04
Cru 12h31'38" -61d58'46"
My Darling Handsome Husband Michael
Per 3h26'15" 44d41'35"
My Darling Jane
Lib 14h44'20" -9d20'40"
My Darling Jess
Sco 17h0'49" -37d32'40"
My Darling Kristina
And 23h18'15" 39d29'41"
My Darling Nick
Cyg 21h27'23" 54d26'45"
My Darling Sati
And 2h3'48" 42d15'33"
My Darling Stephen
Pho 1h34'56" -43d43'7"
My Darling, My Love, My Everything
Uma 12h36'25" 54d51'10"
My David...
Cap 21h25'52" -14d34'2"
My David Honey
Vir 12h52'46" 5d43'11"
My Daz Angel
Cra 18h5'37" -37d29'32"
My dd
Vir 13h40'50" 1d4'30"
My Dear
Uma 11h42'9" 32d23'26"
My Dear Ali - I'll Love You Forever
Ori 6h17'37" 20d13'3"
My Dear Mother Looking Down On Me
Cas 0h35'34" 64d14'48"
My Dear Ron & Brothers R.E.A.L.M
Cru 12h47'21" -58d4'49"
My Dear Rose
Cap 21h4'8" -19d45'17"
My Dearest Andrea
Ari 2h38'6" 14d13'36"
My Dearest Angel
Ori 5h37'51" 3d37'29"
My Dearest Beautiful Emily
Cnc 8h0'56" 25d13'25"
My Dearest Frances
And 0h0'46" 43d38'42"
My Dearest Jack (N5JAK) Big 50
Ori 5h27'5" 10d20'6"
My Dearest Lisa Rae
Leo 10h11'16" 15d47'2"
My Dearest Lover—Heather
Cas 0h30'7" 61d12'24"
My Debbie
Uma 13h6'54" 60d21'55"
my dede
Aqr 23h40'14" -22d36'50"
My Delight
Tau 4h31'41" 16d24'40"
My Delinda
Cap 21h40'42" -8d48'55"
My Destino
Lib 15h49'9" -10d9'59"
My Destiny
Cnc 8h37'1" 22d36'10"
My Destiny Beholds A Celestial Love
Uma 11h35'51" 62d59'2"
My Dirty Little Secret
Lib 14h51'1" -2d16'44"
My Doll Face Sandrita
And 0h33'17" 34d19'10"
My Dolphin Princess
Lib 15h30'31" -24d31'4"
My Donal
Cep 4h56'47" 83d51'31"
My Dragonfly
Lib 15h26'19" -27d32'17"
My Dream
And 1h19'57" 47d43'46"
My Dulcet Darling
Boo 14h44'11" 38d16'9"
My DuPre Amour
Ari 2h1'18" 12d59'3"
My Dwighty
Aqr 22h29'16" -3d6'33"
My Dziadzik
Vir 11h54'23" 8d48'12"
My Elise's Star
Psc 0h23'52" 15d41'41"
My_Emily
Lmi 10h34'50" 36d48'55"
My Endless Love
Cyg 20h14'1" 32d56'38"
My endless Love
Per 3h17'5" 54d7'39"
"My Endless Love ~ Kip"
And 2h33'34" 47d18'5"
My Eternal Love
Ser 15h57'54" -0d40'33"
My Eternal Love Lori Settle
And 0h41'14" 37d45'36"
My Eternal Love Marilee Silva
And 0h57'32" 38d42'5"
My Eternal Love of Chi
Her 16h16'52" 17d23'24"
My eternal love, Maike
Psc 0h48'13" 14d45'38"
My Eternal Love, Nanette Thompson
Lmi 10h45'3" 27d23'3"
My Eternally Beautiful Carolyn
And 2h33'31" 37d46'36"
My Everlasting Love
Cyg 21h14'43" 44d49'38"
My Everlasting Love
Psc 23h5'55" 6d5'27"
My Everlasting Sunshine
Uma 11h49'2" 50d17'40"
My Everything
Uma 10h11'48" 47d16'23"
my everything
Lyr 18h50'28" 43d39'41"
My Everything
Cnc 8h47'59" 31d22'49"
My Everything
Ori 5h26'54" 9d4'32"
My Everything
Cnc 8h49'21" 6d51'38"
My Everything
Leo 11h34'40" 26d57'40"
My Everything
Cap 21h9'21" -19d17'58"
My Everything
Sgr 18h41'36" -20d20'9"
My Eyes
Uma 11h46'55" 55d34'18"
My Eyes On Constance
Cas 22h58'4" 53d51'34"
My family forever, Elke, Nadine & Tim
And 2h0'34" 42d9'28"
My Father Edward Orovich
And 23h10'58" 49d46'29"
My Father - My Hero
Cep 20h34'14" 63d53'36"
My Father: Nicholas J. Gates
Tau 4h31'26" 12d6'58"
My Favorite
Gem 7h10'47" 31d59'32"
My Favorite
Umi 14h53'58" 69d14'52"
My Favorite Aunt Candy
Aqr 22h40'11" -20d5'43"
My Favorite Goose
Leo 10h28'2" 8d58'51"
My Favorite Person in the World, D.
Eri 3h51'24" -0d14'11"
My Favorite Spic
Uma 10h32'8" 59d24'15"
My Favorite Star
Psc 1h19'33" 16d37'18"
My Favorite Star
Per 3h14'31" 47d26'56"
My Fiona Val
Aqr 21h53'2" 0d42'42"
My Fireman
Per 3h32'19" 47d9'45"
My First Love
Lyr 18h34'42" 43d43'44"
My First Mate April Nicole
And 2h38'5" 43d50'58"
My First & Only Love Joy Jean Blair
And 0h43'28" 35d26'1"
My first star-Takuya Best wishes
Sco 17h22'36" -42d32'35"
My Flower
Gem 6h46'51" 31d57'20"
My Fly's Star
Cep 23h7'13" 71d10'24"
My Forever
Sco 16h34'16" -26d21'13"
My Forever Love
Ara 17h29'59" -47d21'28"
My forever love
Umi 15h24'15" 80d6'43"
My Forever Love
And 0h24'15" 26d53'5"
My Forever Love - Barry
Ori 6h8'19" 15d47'0"
My Forever Love - Jane
Ori 6h2'40" 12d52'50"
My Forever Love Marty
Cnc 8h43'59" 16d45'45"
My Forever Shining Future Crysta Jo
Sco 17h22'41" -38d5'37"
My Forever Shining Star
Cep 22h20'40" 58d10'4"
My Forever Soulmate
Lyr 18h52'16" 33d36'27"
My Forever Star Michael
Lyn 6h24'53" 54d30'55"
My Frankel-Dankel
Psc 1h6'43" 27d34'5"
My Freida
And 0h56'31" 45d54'15"
My Friend Cyndi
Uma 12h4'0" 32d32'8"
My Friend Erica
Peg 22h15'25" 34d29'39"
My Friend Jennifer
Lyr 18h15'25" 32d14'37"
My Friend Julie
Lyn 7h38'5" 37d6'22"
My Friend Little Jellie
Tau 5h6'47" 25d3'27"
My friend, Dick May
Per 3h11'24" 53d24'1"
My Frog Prince - Eddie
Gem 6h37'40" 22d14'32"
"My Fuji"
Cap 21h37'36" -20d33'23"
My Gabriela
Peg 22h24'34" 17d21'2"
My Gail
And 0h34'36" 44d51'50"
My Gail
And 2h0'23" 42d37'59"
My Geneva
Crb 15h16'39" 30d17'7"
My Gillian
Umi 4h24'16" 88d29'41"
My Girl
Peg 22h29'43" 33d57'11"
My Girl
Psc 0h58'19" 18d57'26"
My Girl Cleo
Leo 9h43'31" 7d44'58"
My Girl Melissa McCoy 23/10
Cru 12h32'45" -59d10'29"
My Girl (Michelle)
And 23h6'32" 40d33'12"
My Girl Pearl
Aqr 22h26'39" -6d19'34"
"My Girls" Annemarie 1975 Michelle 1979
Cra 18h49'21" -40d17'27"
My girls-Stacy, Samantha and Jordyn
Gem 7h39'52" 34d9'20"
My Goddess of Light
Ori 5h21'12" 6d55'21"
My Godmother Gloria
Gem 6h59'28" 12d18'55"
My Gorgeous Anna Ferrajina
Sco 16h17'21" -10d28'36"
My Gorgeous Louise
And 23h52'46" 35d1'43"
My Gorgous Girl Lisa
Ari 3h4'31" 19d4'13"
My Grace
Cap 20h21'43" -19d58'8"
My Graeme
Cru 12h23'31" -57d42'19"
My Greg
Ari 1h57'28" 13d13'16"
My Guardian Angel
Tau 4h35'24" 21d54'36"
My Guardian Angel
Sgr 18h11'27" -25d4'36"
My Guardian Angel
Lib 15h48'42" -6d52'37"
My Guardian Angel
Cet 2h48'13" -0d31'40"
My Guardian Angels
Dra 17h34'26" 65d40'27"
My Guiding Light
Cnc 8h53'23" 11d25'6"
My Guiding Light
Psc 0h56'3" 27d43'31"
My Guiding Light
Lyn 6h55'17" 51d21'1"
My Guiding Light - Darling Michael
Cru 12h36'12" -59d27'36"
My guiding light Elaine.
Cnc 7h58'27" 10d35'37"
My Guiding Light Justin Phan 161183
Cru 12h18'38" -56d41'47"
My Guiding Light - Sandra Collister
Cyg 21h17'41" 53d40'29"
My Guiding Lights
Per 3h21'32" 42d36'57"
My Guiding Star - Mo Boyd
Cyg 21h7'23" 53d5'59"
My Guiding Star - Steven J Brown
Cra 18h49'21" -45d20'21"
My Guy's Perpetual Light
Per 3h7'0" 52d14'57"
My H. Nguyen
Sgr 18h18'24" -18d10'57"
My Hannah's Light
Tau 4h16'47" 13d21'59"
My Happiness
Uma 8h31'53" 68d8'48"
My Happiness Mikel
Gem 7h4'24" 11d3'2"
My Heart
Vir 13h40'13" 2d28'48"
My Heart
And 2h19'10" 45d45'47"
My Heart
Cyg 20h25'35" 49d4'24"
My heart
Lib 15h23'10" -14d2'21"
My Heart
Cru 12h1'18" -62d20'59"
My Heart and Soul
Gem 6h59'10" 22d27'41"
My heart belongs 2 Mom Dad & Elyssa
Uma 10h21'21" 62d31'42"
My Heart Cindy
Leo 11h3'3" 15d12'20"
My Heart Goes To Chyanne
Lyr 18h41'36" 39d41'25"
My Heart Karina
Leo 11h1'5" 15d38'10"
"My Heart" Pvt. Peter Frank Bologna
Cep 21h32'26" 61d7'28"
My Heart, My Love, KT -n- AC
Leo 11h39'44" 25d13'55"
My Heart's Keeper
Uma 11h36'12" 61d47'47"
My Heather, My love
Cyg 19h48'41" 56d4'14"
My Heaven Blessed Sweetheart
Psc 0h37'58" 15d58'37"
My Heaven Julie
Lyn 7h28'32" 56d46'40"
My Heaven Sent Angel
Ari 2h51'6" 24d57'4"
My Heavenly Dawnee
Psc 1h35'2" 20d31'53"
My Heavenly Gay
Lmi 10h35'54" 31d56'32"
My Hercules Sal
Her 17h27'25" 25d38'38"
My Hero
Tau 3h42'42" 28d28'41"
My Hero
Her 18h6'44" 32d4'32"
My Hero
Per 3h46'19" 33d55'24"
My Hero
Per 4h16'5" 46d11'16"
My Hero: My Shaya
Uma 11h14'16" 37d18'37"
My Hero R.G.W. m+c 05
Leo 11h3'39" 22d35'19"
My Hero, Bill Baker
Aqr 20h42'17" -2d37'48"
My Hero, My Dad
Uma 11h46'34" 33d6'23"
My Home is where ever Your Heart is
Lyn 9h26'51" 41d14'27"
My Honey
Uma 10h21'14" 50d8'13"
My Honey
Cyg 20h3'19" 53d9'40"
My Honey Bunches of Oats
Lyr 18h41'42" 40d52'31"
My Honey Bunny Tarrah
Tau 4h32'22" 4d21'59"
My Honey Monica's Star
Mon 7h0'45" -0d49'39"
My Honey Pie Pam
Gem 7h38'46" 21d46'35"
My Honey Ron
Her 18h20'38" 17d52'29"
My Honey, Bob
Uma 13h14'27" 52d57'11"
My Hopes and Dreams with Franklin
Her 16h47'5" 38d44'8"
My Hubby
Cyg 19h57'11" 40d37'11"
My Hummy
Crb 15h44'19" 33d12'37"
My HuneyPie's Star
Lib 15h36'17" -27d47'29"
My Hunny Bunny Jose
Sco 17h35'42" -41d4'54"
My Hunny's Star
Cnc 8h35'5" 28d43'50"
my husband Dave Doyle
Cyg 19h32'55" 31d58'42"
My Husband John
Tau 5h46'28" 21d59'2"
My Husband Our Daddy Paul Drummond
Uma 11h32'16" 55d47'22"
My Husband Steve Hearing
Leo 11h18'19" 23d56'43"
My ickle Scamp
Umi 14h48'50" 78d13'12"
My Ima Cynthia Marie Smith(Sarah)
Leo 11h11'57" 15d24'23"
My Immortal
And 2h35'0" 45d2'35"
My Immortal Beloved
Gem 7h36'38" 28d25'16"
My Inga
Vir 13h50'3" -6d58'55"
My Inspiration Kerry Kerr
Sco 17h58'58" -42d31'3"
My Inspiration,My Yellow,My Sylvia
Uma 10h35'41" 44d7'22"
My Jaan My Universe
Uma 11h32'41" 52d15'16"
My Jack
Uma 9h2'6" 63d43'5"
my jackal
Mon 6h51'21" -0d6'38"
My Jacqueline Rose the kiss 3-17-00
Leo 11h18'17" 2d13'27"
My Jan Jan
Aqr 22h45'46" -21d39'56"
My Jane
And 2h10'17" 39d10'17"
My Jannie
Boo 14h20'1" 46d58'59"
"My Jeffio""
Psc 1h12'1" 31d28'6"
My JenLyn
Gem 6h32'59" 12d33'19"
My Jenny
And 1h4'44" 44d29'26"
My Jerbear
Uma 14h2'51" 56d42'16"
My Jessamy
Cru 12h20'10" -61d10'11"
My Jewel of the night Julie Ann
Gem 7h35'31" 33d19'43"
My Jilly Bean Witt
Ari 2h33'53" 24d49'56"
my Jo bear
Leo 11h52'0" 27d48'8"
My Jody, My Valentine
Cnc 8h0'3" 12d19'52"
My Johan
Uma 13h53'16" 60d23'34"
My John <3
Lyr 19h26'41" 40d18'13"
My John, My Love
Vir 14h40'44" -5d47'1"
My Johnny
Umi 14h49'23" 80d14'28"
My Johnny
Her 17h35'20" 33d32'16"
My Jolly Mon DaDaPony
Ori 5h48'3" 2d50'41"
My Jonathan Please
Sco 16h52'29" -27d24'42"
My Jordan
Vir 13h24'7" -8d46'1"
My Joseph
Vir 14h37'54" -5d33'46"
My Judie
And 0h29'8" 39d43'0"
My Karie
And 0h21'39" 45d53'18"
My Katey Curly
Cas 2h46'0" 57d51'58"
My Kathy's Star 72455
Cyg 21h25'37" 32d27'21"
My Katie Elizabeth
Uma 8h49'43" 46d54'29"
My *Kelly* ShinesForAll!
Del 20h29'51" 16d39'37"
MY KERRIE***** MY LOVE *****MY LIFE
And 1h19'48" 46d26'37"
My Kev, My Love
Cap 21h43'8" -12d5'37"
My King
Sco 17h53'30" -35d46'13"
My King
Leo 9h27'55" 22d44'49"
My King Arba
Cas 23h6'13" 53d43'53"
My Klarissa
Tau 4h11'44" 7d9'42"
My Kloud
Uma 12h33'21" 57d39'58"
My Knight in Shining Armour
Her 17h32'39" 47d9'32"
My Knight Of Epicuricus
Her 17h30'39" 36d28'9"
My Koala
Cam 4h7'20" 58d11'15"
My Kris
Sco 16h4'46" -12d13'45"
My Lady
Lib 14h44'55" -18d1'45"
My Lady
Umi 16h32'10" 82d40'14"
My Lady Bug Mary Anne
Sgr 18h58'54" -29d13'41"
My Lady Donna
Sco 16h40'10" -32d12'4"
My Lady Ellen
Cas 0h24'25" 57d21'9"
My Lady Empress Karri Jo
Ori 5h44'54" 7d12'24"
My Lady Jenna
Psc 0h22'46" -4d8'56"
My Lady Josephine's Little Darling
Eri 3h55'31" -0d9'15"
My Lady M. Natasha
Ari 3h23'8" 28d36'50"
My Lady Natalie More
Ari 1h59'58" 18d3'49"
My Lady Tina
Aqr 22h5'13" -16d50'24"
My Lakota Warrior
Tau 4h24'39" 17d57'0"
My Lasha
Sgr 19h42'6" -17d17'1"
My Lattle Girl
Gem 7h19'26" 20d30'8"
My Lazy Suz
Col 5h56'23" -34d42'20"
My Lesli
Ari 3h10'3" 18d15'7"
My Liebling's Star
Cnc 8h43'58" 20d14'15"
My Life, My Love, My Wife, Penni
Psc 1h24'56" 11d17'28"
My Life's Star - Susan Hoffman
And 1h6'46" 39d36'4"
My Light, Brian
Cap 20h52'51" -16d7'34"
My Lil' Angel
Lac 22h25'6" 47d31'28"
My Lil Brit
Uma 10h17'17" 64d20'54"
My Lil' Girl, Doris
Mon 6h42'26" -0d16'29"
My Lil' Lady
Tau 5h58'58" 25d10'14"
My Lil One
Cyg 21h32'36" 36d16'37"
My Lil Piggy
Psc 1h24'48" 6d22'16"
My Lilly
Tau 3h30'27" 12d41'24"
My Lina
Mon 7h39'10" -0d56'55"
My Linda
Cap 20h26'6" -14d2'55"
My Linh
Ari 3h12'30" 29d3'40"
My Lisa
Cas 2h23'42" 65d45'40"
My Little Angel
Tau 5h14'15" 25d38'50"
my little angel
Vir 13h21'22" 2d11'54"
My Little Angel
Ari 2h30'36" 20d30'9"
My Little Angel
Cnc 8h51'11" 30d41'2"
My Little Angel
And 23h0'51" 50d10'41"
My little angel Kirrin
Sgr 18h20'22" -23d6'36"
My Little Annie
Ari 2h21'26" 23d14'22"
My Little Ashteroid
Cru 12h22'30" -57d31'8"
My Little Beezer
Ori 5h28'50" 5d10'19"
My little Bh
Umi 15h33'42" 72d15'35"
My Little Buddy
Aur 5h45'1" 38d55'17"
My Little Cheerleader (Megan)
Lib 14h46'43" -18d19'15"
My Little Country Star
Her 18h22'29" 25d30'41"
My Little Critta - Your Scott
Lib 15h2'38" -2d33'57"
My Little Dumbell
Dra 19h15'36" 65d38'44"
My Little Eskimo
Aqr 23h53'56" -7d7'57"
My Little Fairy
Vir 13h24'16" -0d34'5"
"My Little Fox"
Gem 7h22'13" 32d38'3"
My Little German Boy - Michael Utz
Cnc 8h46'39" 31d3'50"
My Little Girl
Peg 23h37'44" 25d38'12"
My Little Girl
Psc 1h51'4" 6d57'23"
My Little Goose "Michelle"
Sgr 18h33'32" -27d15'48"
My little gum tree Laura
Cra 18h33'21" -40d47'25"
My Little Guy
Psc 1h0'24" 31d6'47"
My Little Heavenly Treasure
Uma 10h29'7" 61d12'3"
my little honey bear
Boo 14h53'24" 34d21'46"
My Little Honeydew Yenya
Cas 0h8'35" 52d42'38"
my little lady hannah
And 1h18'26" 46d17'59"
My Little Lo
Vir 12h40'41" -4d56'31"
My Little Lobster
Ori 6h1'38" 18d20'54"
My Little Missy
Tau 3h45'2" 7d42'10"
My Little Monkey Ian
Lac 22h23'43" 54d45'32"
My Little Monkey (Lisa Fleming)
Sco 17h54'15" -37d29'48"
My Little & My Little One
Gem 6h56'34" 26d24'56"
My little Nafe
Col 5h46'10" -30d44'57"
My Little One
Ori 6h18'7" 14d6'4"
My Little One
And 0h35'50" 45d15'17"
My little one, bubbles
Sco 17h14'29" -32d47'58"
My Little Pal
Psc 1h6'39" 27d37'38"
My Little Peanut
Umi 14h18'18" 74d3'48"
My Little Pea's
Tau 4h19'35" 17d1'52"
My little Platypus with love
And 2h31'43" 46d4'43"
My Little Po
Lmi 10h33'43" 39d7'44"
My little princess
Psc 1h11'48" 25d57'5"
My Little Princess
Cnc 8h30'55" 25d2'59"
My Little Princess
Tau 5h29'52" 23d44'27"
My Little Schnookems
Umi 16h14'15" 78d1'30"
My Little Scruffy, Forever Loved
Cmi 7h34'55" 3d56'35"
My little star
Cas 2h49'4" 60d18'48"
My little star Dylan Paul
Col 6h33'40" -34d24'11"
MY LITTLE STAR RAE-GAN LEIGH
Cnc 8h26'56" 24d16'44"
My Little Star, My Baby... My Josh
Sgr 17h55'11" -28d13'31"
My Little Strawberry
Her 18h25'32" 22d54'15"
My Little Sweetie
Cnc 8h35'34" 8d28'47"
My Little Winky
And 1h37'4" 48d14'9"
My Lizzy 4Ever
Aqr 23h29'8" -11d38'46"
My Lobster
Ori 6h10'36" 9d16'49"
My Lobster - Chris Greenway
Vir 13h23'35" 11d12'13"
My Lobster, Matthew P. Tarkenton
Her 16h39'52" 6d44'32"
My L.O.M.L. Shaina Annette Higgs
Cnc 9h3'2" 28d37'6"
My Louise
Aql 19h11'43" -7d52'22"
My Lou's Sparkling Stargate
Peg 0h11'6" 16d14'7"
My Love
Peg 22h54'47" 11d53'27"
My Love
Ari 2h3'2" 19d28'2"
My Love
Leo 10h14'16" 10d40'29"
My Love
Leo 10h37'52" 14d53'17"
My Love
Cnc 8h14'30" 13d6'56"
MY LOVE!
Ori 5h55'58" 2d53'7"
My Love
Her 17h58'44" 29d11'33"
My Love
Her 16h46'58" 25d46'2"
My Love
Gem 7h35'18" 27d59'43"
My Love
Tau 4h15'30" 29d20'38"
My Love
Ari 2h38'43" 29d21'15"
My Love
Ori 5h52'1" 20d11'9"
My Love
Leo 11h26'4" 15d20'3"
MY LOVE
Cyg 19h42'8" 38d41'39"
My Love
Cyg 20h26'13" 45d49'53"
My love
And 1h17'46" 41d52'18"

My Love
Cyg 19h44'59" 35d14'31"
"My Love"
Cyg 19h47'47" 34d9'34"
My Love
Gem 7h15'39" 33d48'19"
My Love
Lib 15h34'39" -8d13'8"
My Love
Cap 21h52'21" -17d47'1"
My Love
Dra 18h48'24" 79d15'20"
My Love
Vir 14h12'8" -4d35'59"
My Love
Lib 14h50'22" -3d56'12"
My Love
Uma 10h17'4" 60d55'18"
My Love Adrienne Forever
Lib 15h25'53" -15d54'53"
MY LOVE Allison Wronski
And 23h17'5" 49d19'5"
My Love Always
LindsayJean
Cyg 20h55'11" 47d11'7"
My Love Ashley
And 0h14'29" 28d21'29"
My Love Billy Dewayne
Graham
Cap 20h9'30" -11d4'57"
MY LOVE....... (Brandon
C. Mejia)
Sco 16h29'14" -32d19'24"
my love bugs' star
Sco 16h8'35" -14d40'26"
My Love Carroll Eugene
Gray IV
Cas 1h24'6" 62d45'25"
My Love Catherine
Sco 16h8'5" -10d42'31"
My Love Cindy
Her 16h59'31" 13d25'30"
My Love Cortney
And 1h51'39" 41d16'22"
My love Daniela
Tri 2h20'27" 36d37'23"
My Love Darcy Winward
Tau 4h23'10" 15d34'12"
My Love Donna-Rae T.
Homberg
Lib 14h51'30" -8d6'10"
My Love Dorothy Alfaro
Pho 0h1'57" -52d50'8"
My Love Douglas Ray
Ari 2h53'23" 26d31'21"
My Love Dwayne
Aqr 23h31'49" -17d23'23"
My love Elizabeth Glaza
Lyn 7h37'32" 52d33'40"
My love Elizabeth Van
Gelder
And 23h7'1" 51d3'13"
My love for Ahmed
Aqr 22h56'27" -5d54'19"
"My Love for Ashlei"
Lyr 19h17'45" 29d41'54"
My love for Kristie Wilson
Cnc 8h50'31" 31d54'3"
My Love for Laura
Cas 1h29'32" 62d45'7"
My Love For Life
Her 17h7'50" 31d44'36"
my love for you
Ori 4h58'35" 9d31'19"
My Love For You, Across
The Heavens
Tau 4h6'38" 6d50'31"
My Love Forever, Ashley
Whitaker
And 0h33'32" 42d39'44"
My Love Forever, Carrie!
Gem 6h48'59" 22d4'6"
My Love forever, Mythu.
Cap 20h30'3" -26d3'46"
My Love Heather
Cyg 20h25'6" 34d8'9"
My Love "Heather Daniel"
Cas 1h46'25" 61d38'40"
My Love Irene
Vir 12h48'29" 3d23'39"
My Love is Forever
Leo 10h54'55" 19d15'50"
My Love~James D. Shelton
Uma 11h2'52" 66d28'42"
My Love Jennifer
Vir 14h21'54" -3d17'20"
My Love Jodi
Cas 23h54'30" 51d41'0"
My Love Joseph Paul
Bruner
Lib 15h12'43" -7d7'31"
My Love Julie Zyung
Uma 10h44'18" 52d8'4"
My Love Kim Marie Wile
Cas 2h12'15" 64d37'21"
My Love Kim Stern
And 1h2'25" 46d13'1"
My Love Lamson
Cyg 21h43'14" 52d51'11"
MY love Lesa
And 0h19'49" 25d33'57"
My Love Lisa Marie
And 0h54'44" 41d0'20"
My Love Maribeth
Lib 15h57'51" -10d57'30"
My Love - Michael Wulff -
My Light
Leo 11h14'43" 13d3'27"
My Love Michelle
Lyn 6h51'5" 59d15'55"
My Love My Heart
Cap 20h30'18" -14d42'15"
My love/my life 1941 - 2004
Cyg 21h49'49" 49d41'45"
My love My Michelle
Ori 6h8'59" 15d25'47"
My Love Nicki
Vir 12h43'48" -8d28'41"
My love: Nikita Rae
Cas 0h42'9" 64d56'12"
My Love Polly
Cyg 19h38'26" 29d19'59"
My Love Renee
Lyn 8h23'15" 35d7'4"
My Love Rich
Her 16h29'51" 30d38'39"
My love Sandy
Sgr 18h12'5" -21d39'32"
My Love Sara
Cas 1h24'43" 51d40'19"
My Love Scott
Ori 5h35'21" -0d51'2"
My Love Shines For You
Lyn 7h42'17" 55d15'56"
My Love Sonia
Uma 13h30'27" 56d15'29"
My Love Stacey Allen
Suazo
Sgr 19h49'25" -30d28'30"
My Love Stacy
Gem 7h33'1" 18d14'44"
My Love Stephanie
And 0h35'0" 29d35'11"
My Love Steven Wayne
Joy
Her 17h57'20" 45d41'21"
My Love Sven
Her 17h40'21" 20d14'57"
My Love Theresa
Lyn 7h19'9" 58d14'52"
My Love to Joseph reaches
the stars
Lyn 7h22'10" 49d52'37"
My Love... To the End of
Time
Cyg 19h58'27" 33d27'35"
My Love, Adam Carson
Baker
Ori 6h24'41" 10d25'13"
My Love, Amanda Noel
Lyn 7h54'31" 55d32'48"
My Love, Amy Whiting
Sgr 19h9'24" -20d45'23"
My Love, Andres Alvarado
Ari 2h28'12" 18d43'19"
My Love, Deborah Peck
And 0h49'16" 45d24'56"
My Love, Dennis Mark
Beutel
Her 17h28'41" 26d58'23"
My Love, Donna Canham
Dra 17h44'57" 51d43'50"
My Love, Eydie
Lmi 10h48'28" 38d23'50"
My Love, Jayne Cotterhill
Cyg 19h27'59" 29d15'13"
My Love, Jeannie Cho
Leo 10h18'48" 24d34'21"
My Love, Jessy Gibbons
Ari 2h34'22" 19d9'41"
My love, John Vincent
Kloeker
Crb 16h2'26" 32d29'52"
My Love, Justin Michael
Hibberd
Aur 5h14'6" 29d17'26"
My love, Kara
Uma 9h30'32" 45d56'5"
My Love, Kerri Kay
Cnc 9h8'26" 15d1'0"
My Love, Linda Le
Tau 4h39'19" 4d42'30"
My Love, Megan
Boo 14h52'24" 36d3'34"
My love, My Abe
Scl 23h41'33" -27d35'28"
my love, my dorkfish
Ori 5h50'7" -2d52'28"
My Love, My Heart, My
Soul, My Star
Aqr 20h44'18" -0d23'21"
My Love, My Hero
Per 3h3'13" 37d27'45"
My love, My husband, Lacy
loves you
Cep 21h41'51" 64d7'3"
My Love, My Life, My
heart, My Omar
Leo 9h42'40" 23d12'52"
My Love, My Once, Russ
Pollinger
Her 18h55'53" 14d10'15"
My Love, My Prince
Charming
Gem 8h0'37" 27d24'33"
My Love, My Star
Cyg 20h55'55" 35d15'29"
My Love, My Sunshine, Jen
And 2h38'17" 40d44'55"
My Love, My wife, My
Oksana
Aqr 20h55'53" -9d14'20"
My Love..., Neal Ray
Hildebrandt
Ori 5h34'48" 11d7'31"
My Love, Timothy Lee
O'Pry Jr.
Tau 4h13'10" 14d44'30"
My Love, With No
Description...
Cyg 20h46'52" 31d10'51"
My Love, Yajaira
Lib 15h10'51" -8d49'45"
My Love...Jaime Bowerman
Malone
Lyr 19h12'51" 26d22'19"
My Lovely
Uma 8h28'37" 65d32'21"
My Lovely Amer's Star
Per 4h47'54" 45d13'35"
My Lovely Donna
Aql 19h48'26" 12d15'34"
My Lovely Elizabeth
Lib 15h9'51" -14d50'30"
My Lovely Heather
Cas 0h29'3" 61d5'41"
My Lovely Heidi
And 1h39'3" 41d32'50"
My Lovely Jane
Cyg 20h20'38" 51d0'24"
My Lovely Jenna Marie
Vir 12h17'54" 8d43'34"
My Lovely Lady Helen
Col 5h55'22" -29d25'34"
My Lovely Laura
Cnc 8h49'23" 18d49'4"
My Lovely Lisa
And 0h11'9" 29d7'32"
My Lovely Moon Moon
Boo 15h36'15" 41d50'7"
My Lovely Rachel
Ari 3h24'51" 29d6'22"
My Lovely Wife
Leo 11h44'18" 24d7'41"
My Love..My Life..My
All..Fig
And 1h1'1" 46d11'31"
My Love...Now And Always
Ari 2h47'41" 24d47'42"
My Love's Eternal Soldier
Ari 1h50'15" 17d22'11"
My Loves Light
Lyn 9h2'26" 34d14'24"
My loveys Chiara Bermeo
Pineda
Cas 1h38'50" 63d36'25"
"My Lovie"
Cyg 19h41'40" 53d22'30"
My loving brother, Robert
A. Bogle
Cap 21h40'18" -14d12'37"
My Loving Grace
Cnv 12h52'15" 35d41'46"
My Loving Mother, Lydia
Hall
Uma 12h10'14" 47d2'47"
My Loving Wife -
Constance Marie
Mon 6h54'5" -0d18'20"
My LT.
Uma 8h38'13" 72d11'1"
My Lucky Charm
Dra 19h14'14" 60d21'25"
My Lucky Dustin
Uma 10h27'30" 56d37'5"
My Lucky Star
Cnc 9h5'51" 31d0'18"
My Luck-y Star
And 0h52'24" 41d6'48"
My lucky star *Allan Lloyd
Sharman*
Her 17h29'17" 39d45'15"
My "Lucky Star" (Taylor L
Carbone)
Lyn 8h14'37" 55d41'17"
My Lydia
And 0h27'47" 38d49'6"
My Lynette
And 23h55'39" 41d54'19"
My Maggie-Doll
Cnc 8h30'43" 16d21'56"
My Magic Star* MGB&JBB
*My husband
Cyg 20h47'3" 34d0'18"
My Magical Bodyguard
Dra 19h23'17" 67d55'56"
My Malissa
Psc 23h0'26" 5d9'54"
My Mama
Lyn 7h41'40" 51d5'23"
My Mandala
Cyg 20h21'23" 34d21'6"
My Mar - Philippians 1:20-
21
Crb 15h24'39" 31d55'22"
"My Mare" a.k.a. Mary E.
Dock
Cnc 8h23'11" 32d28'20"
My Margaret
Cyg 20h54'35" 46d10'12"
My Maria
Cru 12h1'5" -57d42'47"
My Mariah Julie
Aqr 23h34'32" -14d40'52"
My Marine Your Angel
Gem 7h32'6" 26d14'3"
My Mariya
And 1h26'14" 44d40'55"
My Mark Forever
Lib 15h38'13" -12d5'37"
My Marvelous Mom
Marlene
Ari 2h55'50" 24d45'16"
My Marykins
Cas 0h52'35" 62d55'49"
My Me', Me'. "My Shining
Star"
Gem 7h42'5" 32d52'48"
My Melissa
Lib 15h27'5" -17d27'58"
My Melody for Melody Meg
Gonzalez
Sco 17h10'17" -42d43'26"
My Michael
Umi 14h23'14" 76d21'34"
My Michael
Lyn 6h35'40" 58d5'49"
My Michael
Ari 2h2'7" 23d7'16"
My Michael Scott
Her 17h18'5" 19d13'18"
My Michael Star
Her 16h58'57" 28d59'39"
My Michael, I'll love you
forever!
Her 18h53'40" 23d46'19"
My Michele
Ori 5h11'2" 1d2'16"
My Michelle
Tau 5h38'27" 20d34'50"
My Michelle
Cyg 21h35'44" 33d36'57"
My Michelle
And 0h5'4" 45d15'26"
My Michelle Marie
Vir 12h44'13" 5d14'17"
MY MICK ~ THE MAN OF
MY DREAMS
Cnc 8h52'53" 31d29'37"
My Mignon Jacob, The
Powerful
Ari 3h1'47" 12d15'15"
My Milagros
Uma 9h13'14" 57d51'51"
My Mimo, Mirvat
Cyg 20h40'47" 49d10'6"
My Miracle
Uma 11h17'22" 58d3'35"
My Mitch
Ara 17h17'8" -53d54'17"
MY MOM
Uma 9h55'34" 71d7'15"
My Mom
Ari 3h2'49" 20d14'55"
My Mom
Peg 23h31'13" 20d52'30"
My Mom Marie
Crb 15h39'10" 26d2'50"
My Mom Susie Jones
Lyn 7h28'0" 48d26'3"
My Mom Suzie A True
Shining Star
Gem 6h32'1" 14d51'36"
My Mom - TLC
Sgr 18h31'58" -22d25'20"
My Mom Vi
Sco 17h28'12" -39d40'11"
My Mom, My Angel, Luisa
Napolitano
Cnc 8h59'13" 7d8'12"
My Momma's Star
Aqr 22h22'19" -4d12'22"
My Mommy
Psc 0h36'47" 4d22'51"
My Mommy Normey
Vir 14h27'45" -0d36'7"
My Mommy's Star
Cas 1h57'45" 66d5'15"
My Mommy's Star
Uma 11h25'16" 59d49'26"
My Mom's a Star - Betty
Ann McBride
Vir 13h0'19" -8d5'15"
My Monday Night Guy
Tau 5h46'53" 22d27'53"
My Mother Donna Lea
Kuhlman-Staley
Psc 0h6'17" 3d32'55"
My Mother Eleanor
Boo 15h21'22" 48d38'47"
My Mother Lisa
Cas 0h20'28" 61d38'35"
My Mother Marie
Crb 15h50'5" 27d44'0"
My Mother ~ My Friend
Cas 0h22'54" 61d47'52"
My Mother My Friend My
Protector
Lib 15h9'36" -6d57'55"
My Mother, My Angel
Cnc 8h18'6" 12d48'4"
My Mother, My Best Friend
Cas 1h24'5" 57d10'4"
My mother, my friend
Cyg 20h42'3" 50d13'40"
My Mother, Vicki: Where
Dreams Fly
Per 3h16'53" 48d2'37"
My Mother's Precious Gift
of Love
Cep 22h20'53" 58d6'0"
My Motivation, My Hero,
My Love
Lmi 10h49'43" 25d9'40"
My Munchkin
And 0h56'49" 40d48'11"
My Musara
Uma 11h27'14" 52d39'24"
My Muse ReAnn
Leo 9h33'38" 28d33'48"
My Mutter Silvana
Lombardo
Cap 21h16'20" -15d44'9"
My Mylinda's Beauty
Ari 3h29'1" 26d50'34"
My Nana and My Papa -
Carol & Don
Uma 11h43'50" 54d21'27"
My Nanny
Cas 0h28'58" 53d17'23"
My Natalie
Cas 0h18'36" 53d51'56"
My Neal
Boo 14h33'32" 41d41'21"
My Nessa " The Brightest
Star "
Psc 0h18'23" 4d42'45"
My New Life, Sue
Lyn 8h11'56" 52d49'40"
"My Nicki" Nicholas
Brendon Calus
Ori 6h20'25" 2d52'3"
My Niece Barbara Schiner
Ari 3h5'45" 28d23'51"
My Night in Shining Armor,
Steve
Ori 5h25'40" 4d11'40"
My Night In Shinning
Armour
Gem 7h35'28" 29d23'40"
my nigma
Her 16h47'36" 7d46'50"
My Norwegian Princess
Katie Loraine
Leo 9h36'37" 26d34'15"
My Nothing
Uma 13h45'3" 58d11'37"
My number one, I love you.
always.
Uma 11h35'23" 62d13'53"
My Olive
Crb 15h37'58" 37d5'11"
My Oma
Lib 15h2'8" -0d42'12"
My one and only
Ari 3h21'47" 22d13'36"
My One and Only Elizabeth
Galvan
And 2h30'55" 38d25'48"
My One and Only Flyboy
Lmi 9h49'20" 33d33'7"
My One and Only Light in
the Sky
Cap 20h12'2" -15d42'59"
My One and Only MAB
Tau 4h24'55" 14d33'3"
My One and Only Teekie
Bell
Cas 1h41'55" 64d20'34"
My One and Only, Eric
Lmi 10h7'6" 30d33'49"
My one and only, Heather
And 1h33'15" 49d9'17"
My One And Only...Michael
Keith
Uma 11h2'25" 60d58'36"
My One In A Million
Rhonda
Cas 0h26'1" 53d40'12"
My One Love Semra
Kandas - Hakan K.
Ara 17h5'30" -58d40'11"
My One My Only
Uma 12h26'53" 57d4'57"
My One True - Erik Torres
Cnc 8h27'56" 30d56'8"
My One True Love - "Bill
Bennett"
Per 2h56'40" 41d22'46"
My one true love, Sabrina
And 0h19'43" 35d54'49"
My Only Lie Was, I Didn't
Love You
Uma 11h49'30" 40d1'17"
My Only Love
Ari 1h53'26" 17d51'16"
My only love Candice Hale
Leo 9h33'40" 11d31'29"
My Only One
Sgr 18h14'31" -30d43'29"
My Only One - Jack and
Rose
Cyg 20h14'48" 51d33'53"
My Only Sunshine
Cyg 19h47'39" 39d50'56"
my only true love Rodney
Her 16h32'20" 45d12'21"
My PaCoWoTy Kids
Umi 16h35'5" 82d46'4"
My Pal
Her 18h8'25" 21d5'48"
My Pal Joey
Per 4h8'33" 50d38'9"
My Panda Bear
And 1h33'40" 39d29'52"
My Papa
Ori 6h9'32" 9d10'9"
My Pat
Uma 8h35'3" 59d55'49"
MY Patti Ann Eternal Love
Leo 11h25'48" 16d7'16"
My Peanut
Lib 14h29'53" -18d33'11"
My Pee-Wee
Umi 14h14'14" 74d48'15"
My Penguin
Her 16h23'52" 47d35'24"
My Penguin Matty
Sgr 19h39'13" -13d42'44"
My Penny
Aql 19h28'5" -0d55'54"
"My Penny Boo"
Sco 16h6'13" -11d46'6"
My Penny Catwomun
Per 2h59'3" 55d10'32"
My PePaw
Mon 7h32'10" -0d21'2"
My Perfect Angel-Lane Erin
Stoddard
Lyn 7h36'16" 41d26'5"
My Perfect Mother
Sco 16h7'41" -19d31'41"
My Perfect Star
Uma 13h31'13" 56d10'42"
My Perfect...Schyler
Thomas Gagnon
Sco 16h46'18" -34d55'56"
My Piece
Sco 16h57'27" -38d56'58"
My Pooh Bear - My Love
Aqr 22h20'6" 1d53'48"
My Pooka
Cnc 7h59'13" 16d19'2"
My Pookie, Moni
Vir 14h20'43" -1d20'12"
My Pop John Paulikas
Uma 12h3'38" 32d19'7"
My Precious
Cyg 20h43'52" 52d5'53"
My Precious
Cap 21h4'5" -20d3'52"
My Precious Angel
Cyg 20h46'50" 31d48'55"
My Precious Angel Jillian
Leo 9h46'4" 28d57'26"
My Precious Baby
Uma 11h17'15" 41d23'32"
My Precious Carolyn
Cnc 7h58'6" 16d41'13"
My Precious Cristina
Her 16h50'6" 27d20'3"
My Precious Danielle
Ari 3h11'43" 27d57'46"
My Precious Erica
And 2h15'49" 46d47'28"
My Precious Felicia
Lyn 9h16'30" 36d46'32"
My Precious Leetta
Cas 0h19'52" 58d58'24"
My Precious Niece - Delilah
Rae
And 0h49'39" 43d29'56"
My Precious Pamela
Cam 3h58'6" 66d20'47"
My Precious Taryn
Leo 10h49'50" 19d8'41"
My Precious ....You are
my star!
Tau 4h9'19" 6d11'6"
My Pretty Lady
Lyn 6h45'5" 52d28'43"
My Prince
Leo 10h23'54" 11d59'10"
My Prince Charming
Chadwick
Psc 0h10'55" 5d46'3"
My Prince - Peter Michael
Berrie
Cyg 19h47'13" 34d20'19"
My Princess
And 0h38'10" 40d13'8"
My Princess
And 23h15'53" 41d34'36"
My Princess
And 23h54'40" 45d0'21"
My Princess
Gem 6h24'22" 19d57'48"
My Princess
Cnc 8h9'2" 25d54'41"
My Princess
Dra 19h29'15" 68d43'50"
My Princess
Aqr 22h16'38" -16d54'27"
My Princess
Cap 20h22'15" -10d53'50"
My Princess Amanda
Sco 17h43'35" -39d42'10"
My Princess - Ashley
Nicole Street
And 22h58'57" 47d12'16"
My Princess Bride - Karen
Vorwerk
Psc 0h21'32" -4d58'30"
My Princess Bridget
Tau 4h40'0" 23d58'9"
My Princess Cassie
Uma 9h33'51" 47d22'47"
My Princess Diana
Sco 16h11'13" -14d56'59"
My Princess Jenn
Tau 4h40'10" 19d16'40"
My Princess Karen
Cnc 8h38'29" 15d10'55"
My Princess Kelly Simpson
And 23h14'33" 40d7'38"
My Princess Lena
And 0h48'6" 23d13'16"
My Princess - Maura Lewis
And 23h23'4" 50d34'50"
My Princess Nicole
Kanallakan
Gem 7h27'26" 20d42'55"
My Princess Poopsie
Vir 12h16'9" 4d55'11"
My Princess Rachael
And 0h34'25" 21d54'32"
My Princess Zukee
Lib 14h39'52" -9d17'8"
My Princess, Nicole Laffely
Psc 0h25'57" 12d27'26"
My Prodigy, Kim
Cap 21h0'37" -18d24'19"
My Promise
Psc 23h37'53" 4d23'6"
My Promise
Cnc 8h38'20" 22d59'4"
my promise my stand
Cep 20h44'11" 61d5'55"
My Promise to You
Ori 6h4'46" 17d10'7"
My Protector Chris Since
6/10/1995
Umi 15h27'13" 71d39'38"
My Punchy
Equ 21h8'53" 4d25'15"
My Punk Angel
Psc 1h21'30" 24d58'22"
My Puzzle Piece
Per 2h53'24" 55d20'50"
My Querido Oscar Ribau
Lyn 8h23'15" 55d43'13"
My Rachel
Uma 8h37'28" 58d21'10"
My "Rae" in the sky
Sco 17h23'29" -42d22'5"
My Ray of Sunshine
Cas 1h29'8" 54d8'39"
My Reason
Cyg 20h13'31" 36d33'46"
My Rebecca 06/12/04
Ori 5h35'15" -2d27'1"
My Red
Pho 23h29'54" -51d48'30"
My Reen
Umi 16h22'39" 82d32'2"
My Reminder
Cru 12h26'33" -56d19'1"
My Robert
Cnc 8h20'49" 24d29'25"
My Roosevelt
Uma 11h36'4" 48d44'18"
My Rose
Cyg 21h25'21" 39d30'57"
My Rose Marie
Ari 2h51'21" 27d25'21"
my Rose, my angel, my
Shelby.
Cap 21h11'19" -23d11'19"
My Rosie
Psc 1h28'49" 26d57'35"
My Russ Russ. My love.
Cyg 19h48'20" 42d53'17"
My Sally Forever
And 0h44'34" 40d54'14"
My Sammy
Lib 15h33'32" -20d37'12"
My Samrang
Sco 16h39'58" -30d15'51"
My Sarah Lee
Crb 16h2'2" 38d56'14"
My Sayro
Hya 10h38'12" -27d53'8"
My Scott
Uma 11h57'2" 38d43'36"
My Scotty
Her 17h39'56" 37d17'26"
My Secret
Ori 6h12'49" 17d6'22"
My Secret Lover
Vir 14h50'15" 3d33'4"
My Seraph Kibou
Del 20h49'46" 6d11'1"
My Serendipity...Ron
Patterson
Ori 5h40'28" -0d13'18"
My Sexy Man Bill Hodder
Uma 9h36'51" 56d33'28"
My Sharice
And 0h18'50" 28d26'56"
My Sharona
And 2h37'7" 49d3'13"
My Shelby Star
Leo 11h25'42" 19d10'37"
My Shell
Leo 10h34'32" 18d13'7"
My Shell
Ori 5h15'22" 7d7'18"
My Shell
Ari 3h26'30" 21d58'33"
My Shining Beauty
Cnc 8h24'2" 10d5'10"

My Shining Knight Daniel Stopar
Aqr 23h25'58" -10d36'21"
My Shining Light
And 23h19'14" 48d12'58"
My Shining Perfection
Leo 10h33'52" 25d47'40"
My Shining Sheryl
Peg 21h43'40" 13d20'55"
My Shining Star
Cnc 8h15'21" 8d45'36"
My Shining Star
Leo 9h38'42" 26d43'21"
My Shining Star
Her 17h12'21" 45d55'3"
My Shining Star
Lyr 18h35'43" 33d34'4"
My Shining Star
Cyg 19h54'50" 37d16'33"
my shining star
Cap 20h33'46" -9d55'54"
My Shining Star
Cyg 20h42'2" 55d3'59"
My Shining Star
Uma 10h34'13" 65d30'13"
My Shining Star Aidan McNamara
Cep 22h41'47" 86d51'38"
My Shining Star & Best Friend Nicki
Ara 17h13'42" -53d51'35"
My shining star Charles 5/10/1947
Tau 3h38'8" 16d2'34"
My Shining Star Charles Hall
Lyn 7h53'14" 56d45'20"
My Shining STAR Christopher
Leo 9h27'36" 26d52'43"
My Shining Star - Daisy
Sco 16h14'6" -13d52'3"
My Shining Star Helena
Cas 2h32'22" 63d11'45"
My Shining Star Jeff Dennis
Her 16h46'53" 33d56'7"
my shining star Linus
Peg 22h29'26" 19d53'33"
My Shining Star Mom Dorothy
Lyn 8h50'59" 34d47'6"
My Shining Star Nikita
Vir 13h41'29" -7d3'28"
My Shining Star Robb
Her 17h33'34" 34d21'31"
My shining star Sharon
Cap 20h51'41" -25d46'16"
My Shining Star Timothy
Aqr 22h17'27" -13d3'14"
My Shining Star Tony
Psc 0h55'16" 6d54'28"
My Shining Star, Earle
Sgr 18h36'13" -23d52'46"
My shining star, Joe Shadle
Uma 12h1'33" 57d7'14"
My Shining Star, Moochie
Aqr 22h18'6" -24d10'36"
My Shining Wife, Melene Doney
Uma 9h15'51" 54d25'3"
My Shiraz
Cam 4h9'24" 69d48'23"
My ShMerriah
Leo 10h28'39" 17d31'25"
My Shooting Star Jennifer Baldwin
Gem 7h38'37" 14d18'7"
My Sister
Aqr 22h54'10" -8d49'42"
My sister Anne
Psc 0h32'58" 6d24'34"
My Sister Carol
And 1h48'29" 46d4'43"
My Sister Diane
Tau 4h19'43" 14d43'28"
My Sister Hilda
And 0h44'3" 30d23'8"
My Sister My Friend Marla
Psc 1h27'49" 17d25'42"
My Soldier Ross
Ori 6h15'20" 15d48'38"
My Somnambular
Gem 6h4'22" 27d8'48"
My Son Gareth
Her 18h48'38" 22d4'0"
My Son Justin
Cnc 8h39'35" 28d12'1"
My Son Matt
Leo 9h53'36" 24d5'17"
My Sonal
Cas 0h25'7" 54d10'6"
My Soul Mate
Leo 9h51'53" 27d39'40"
My Soul Mate Hugh
Per 3h16'52" 37d53'25"
My Soul Mate, Jason Wojahn
Per 2h56'23" 55d3'45"
My Soul Mate, Mike Reeves
Her 17h16'47" 27d31'53"
My Soulmate
Cyg 21h55'9" 51d50'22"
My Soulmate David
Cyg 20h47'31" 31d38'11"
My Soulmate Gigi
Uma 8h46'54" 54d24'53"
My soulmate Mark
Cap 20h8'25" -11d19'3"
My Soulmate Patsy Stefan
Psc 0h36'31" 8d14'21"
My Soul's Light Star
Leo 11h36'32" 26d14'54"
My Southern Belle
Cas 2h50'33" 60d29'0"
My Southern Comfort
Psc 0h53'36" 10d43'37"
My Southern Lily
Ari 2h7'36" 24d13'46"
My Sparkling Angel
Ori 5h36'1" 6d31'41"
My Special Angel
Lib 15h49'2" -7d11'58"
My Special Baby
And 0h15'39" 45d47'45"
My Special Dad
Cas 0h57'50" 61d42'21"
My Special Gina
Cyg 19h55'16" 37d58'39"
My Special One
Vir 13h18'16" 13d38'25"
My Squirt's Star (Heather's Star)
Leo 11h38'55" 24d36'2"
My Squish
Her 17h26'18" 48d40'29"
My Star
Uma 11h4'32" 36d54'17"
My Star
Sgr 18h32'24" -17d33'30"
My star and My heart - Daina
Uma 9h40'40" 71d31'10"
my Star Bärbel Wosahlik
Peg 21h42'23" 16d49'22"
My star David
Cyg 21h37'49" 50d5'28"
My star is Hinano
Psc 1h17'28" 5d14'12"
My Star Jay Terry
Uma 9h36'42" 51d46'4"
My Star Karen
Uma 11h32'4" 56d4'31"
My Star Kelly, forever in my heart
Del 20h53'32" 13d19'18"
My Star - Louise
And 22h59'18" 41d4'24"
My Star Marcy
Uma 9h40'43" 63d42'36"
My Star Mum...Best in the Universe x
Cas 23h21'34" 57d3'26"
My Star ~ Pea Pod Oncea
Leo 11h22'43" -2d9'38"
My Star - Simon
Per 3h24'49" 40d40'56"
My Star Soulmate A.Miladys McNulty
Cas 0h59'53" 59d37'30"
My Star TART
Leo 10h28'59" 13d25'55"
My Star, My Debbi
And 0h13'9" 40d33'28"
My Stars - P J D A K R E C J D J
Dra 19h24'26" 61d49'16"
My Star-William C. Brinton Jr.
Per 3h43'37" 47d36'0"
My Stellar Mao
Cep 22h10'2" 60d58'41"
My Stephanie
Gem 7h5'22" 33d23'53"
My Studmuffin 9-15-03
Umi 16h50'20" 75d44'55"
My Suga Buga
Uma 8h56'54" 56d52'11"
"My Sugar"-Anne Tooley!
Vir 12h49'52" -11d12'0"
My Sunny Girl
Vir 13h58'5" -15d44'51"
My Sunshine
Mon 6h45'26" -0d2'50"
My Sunshine
And 2h12'52" 48d52'4"
My Sunshine.
Vir 11h53'26" 2d54'50"
My Sunshine
Her 17h3'19" 29d13'31"
My Sunshine
Her 18h2'20" 25d1'55"
My Sunshine Adri
Aqr 21h39'8" -3d16'53"
My Sunshine at Midnight
Tau 4h40'40" 24d4'28"
My Sunshine Deborah Anne
Sco 17h52'15" -37d13'54"
My Sunshine in the Darkest Night
Cap 21h45'25" -16d55'31"
My sunshine Lisa
Lyr 19h10'11" 27d10'55"
My Sunshine Melanie
Uma 10h48'19" 55d30'48"
My Sunshine- Stacey Marie Vincent
Aqr 21h24'13" -9d7'36"
My Superman Michael
Cnc 8h46'48" 32d6'8"
My Superstar Ellie
Sco 16h20'49" -26d58'33"
My Swear
Psc 0h32'58" 13d54'34"
My Sweet Addiction 'Stuti!'
Tau 5h28'55" 19d34'26"
MY SWEET ALISHA ANN
Lmi 10h49'25" 34d48'41"
My Sweet Angel Baby Faye
And 0h15'7" 32d49'1"
My Sweet Baboo Tiffany Ann
Lyn 7h59'17" 52d48'38"
My sweet baby girl, Nada
And 23h31'29" 48d6'43"
My Sweet Baby Stephie
And 1h0'32" 36d43'35"
My Sweet Baron
Cma 6h56'45" -29d26'1"
My Sweet Bena
Uma 9h3'51" 57d50'39"
My Sweet Bobstar
Ori 5h32'26" 4d46'28"
My Sweet Boy Tony
Gem 7h49'5" 25d39'34"
My Sweet Charlie & Dad 60th B-day
Sco 17h57'1" -38d20'16"
My Sweet Christine
Cyg 20h18'12" 49d33'4"
My Sweet Cindy
Cas 0h56'39" 54d51'53"
My sweet darling Fabienne
Cam 6h53'15" 65d13'2"
My Sweet Dianna
And 1h41'37" 45d6'7"
My Sweet Dionne
Cap 21h52'20" -14d45'1"
My Sweet Gene
Leo 11h34'25" 1d44'34"
My Sweet Hailey Lane
Ari 2h40'14" 25d17'54"
My Sweet Heidi Lynn
Cam 6h42'19" 63d19'9"
My Sweet Husband
Ari 2h16'1" 23d17'52"
My Sweet Ivy
Ori 5h37'44" 3d13'43"
My Sweet Jacqui's Star
Sgr 18h15'55" -17d32'35"
My Sweet Jennie...My Shining Star
Aqr 21h48'31" -2d40'30"
My Sweet Josh
Psc 1h29'45" 18d39'22"
My Sweet Justin
Lib 15h49'42" -10d33'20"
My Sweet Kim
Lyn 7h50'9" 57d21'27"
My Sweet Little Angel Mary Caitlyn
Uma 12h43'45" 52d23'40"
My Sweet Love Frank D. Fredrickson
Vir 13h15'32" 6d3'57"
My Sweet Love, Jimmie Lee Baker Jr.
Ori 5h8'54" 3d1'8"
My Sweet Loving Sarah
Vir 13h30'49" -3d11'41"
My Sweet M
Dra 17h7'40" 69d55'14"
My Sweet Mama
Gem 7h3'26" 16d10'14"
My Sweet Mylinh
Uma 11h43'24" 45d5'26"
My Sweet Owly-Al
Gem 7h24'16" 26d25'42"
My Sweet P
Uma 11h48'22" 43d30'51"
My Sweet Patun
Leo 10h48'8" 13d25'46"
My Sweet Pea
Lyn 7h11'24" 49d21'2"
My Sweet Precious Angel
Tau 3h44'16" 23d10'18"
My Sweet Prince
Sgr 19h13'4" -15d51'33"
My Sweet Prince
Ret 3h43'30" -53d41'32"
My Sweet Sadness
Cru 12h27'48" -58d33'6"
My Sweet Sara Shim
Lyn 7h35'35" 37d37'52"
My Sweet Sugar Vianey
Cyg 19h39'25" 35d6'49"
My Sweet T.J.
Boo 14h49'36" 20d10'51"
My Sweet Tracy
Lyn 9h11'30" 35d23'50"
My Sweet Vickie
And 0h25'4" 31d35'21"
My Sweet Wonderful Som
Leo 11h43'50" 25d45'31"
My Sweet, Sweet Angel Cakes
Cyg 20h40'20" 35d19'16"
My Sweetest Holly
Lib 15h38'33" -23d7'45"
My Sweetheart
Lib 15h25'48" -26d1'35"
my sweetheart
Cas 0h22'59" 55d50'6"
My Sweetheart Bill Gulick
Per 1h41'30" 50d52'37"
My Sweetheart Christina Finamore
Cas 0h42'44" 64d21'59"
My Sweetheart "Daddy"
Cep 21h3'30" 59d3'8"
My Sweetheart Dayna
Vir 13h12'31" 9d25'50"
My Sweetheart Jeff
Lib 14h54'53" -0d33'8"
My Sweetheart Kori
Cap 21h52'50" -18d27'49"
My Sweetheart Melodie
Cyg 21h31'4" 39d16'13"
My Sweetie
Cyg 20h52'9" 53d52'5"
My Sweetie Brittany
Cnc 8h58'34" 16d38'48"
My Sweetie Girl Mony
Cyg 21h56'53" 39d10'55"
My Sweettart Reneé
Uma 8h48'46" 59d44'51"
My Taipan
Aql 20h5'55" 4d19'43"
My Technicolor Dreamcoat
Gem 6h5'40" 22d33'13"
My Teresa / TARS
Aqr 21h42'44" -1d0'4"
My Terry Forever
Ori 5h9'50" 12d57'11"
My Texas Star
Boo 14h13'19" 17d37'19"
My Tiger "Jeannine Apadula"
Lyr 18h35'41" 36d29'9"
My Timbo
Cyg 21h21'33" 51d58'57"
My timeless Love
Vir 12h57'57" -5d30'53"
My Tina Bina
Lib 15h17'53" -20d17'39"
My TinkyWinky
Sgr 19h44'11" -12d33'29"
My Tomass
Uma 10h7'41" 65d30'51"
My Tony Wayne The Sparkle Of My Life
Cru 12h11'34" -62d54'24"
My Trang and Emil
Crb 15h44'12" 32d3'15"
My Trish
Lib 15h26'14" -4d40'42"
My Trouty
Lyn 7h6'44" 51d54'23"
My Troy
Tau 5h48'47" 27d24'58"
My True Booski Bandit
Uma 10h51'46" 46d37'26"
my true companion
Per 3h21'13" 42d33'37"
My True Love
Lyn 7h39'49" 48d19'22"
My True Love Craig Smethers
Cyg 19h45'16" 33d55'57"
My True Love Patty Kinsella 51606
Leo 10h53'22" 22d55'41"
My true love, David P. Contreraz
Ori 5h39'0" 1d18'31"
my true love, Jennifer Knight Haden
Per 3h36'54" 45d41'2"
My True Love, Mitchell J Moore
Dra 15h54'39" 58d12'24"
My True Love, My husband, My Life
Cyg 20h27'40" 57d35'56"
My True Love, Ricardo Novelo
Her 17h7'36" 24d6'35"
My True North
Umi 9h38'14" 89d17'56"
My True North - Adam Bass
Gem 7h37'25" 19d48'38"
My Turkish Girl Merih
Cap 21h29'30" -18d3'59"
My Twin Soul, Kurt
Vir 11h42'6" -3d15'54"
My Twin, Lori Lynn
Uma 11h56'13" 61d51'0"
My Twisted Mister
Leo 11h7'45" 24d41'33"
My Two Angels
Lyr 18h43'16" 39d35'1"
My Two Angels Jo & Isabella
And 1h0'55" 46d11'14"
My Two Front Teeth
Lyn 7h33'42" 45d58'33"
My Two Loves - Brad & Kyle LaCross
Aql 19h27'40" 2d38'31"
My Two Special Ladies Jo-Ra
And 0h21'48" 30d35'25"
My Ultimate Star, Sharon Love Pa & Coz
Cru 12h26'46" -60d45'9"
My Vach
And 0h38'47" 36d47'42"
My Valentine
Uma 10h36'49" 63d44'38"
My Valentine - Melissa Ann Gray
Cas 1h12'39" 48d49'24"
My very own Angel - Jan Reece
Cas 0h25'41" 57d21'40"
My Victorian Princess
And 23h25'8" 52d26'40"
My Vinco
Aur 6h26'1" 41d16'26"
My Warrior
Her 16h45'16" 37d38'55"
My Way
Uma 9h0'19" 68d57'57"
My Wife Kerry Kristen Zielke
Ari 2h45'14" 25d45'24"
My Wife Tracy
And 1h5'55" 38d6'0"
My Wife, My Angel, Jennifer McAbee
Cas 1h27'49" 63d4'35"
My Wifey
And 1h0'29" 44d54'17"
My Wish
Ori 5h56'58" 17d9'26"
My Wish
Cnc 9h21'5" 22d53'31"
My Wish *Spencer*
Ori 5h35'24" 14d31'26"
My Wish... Bradley
Ori 5h40'7" 8d14'32"
My Wish Come True
Tau 4h15'2" 4d9'11"
My Wish Come True - Daniel E. Fox, Jr.
Ori 5h15'10" -0d11'51"
My Wish Star
Cyg 19h41'28" 33d49'54"
My Wish Upon A Star
Aql 19h7'27" 14d55'46"
my wonderful
Crb 15h27'59" 31d17'25"
My Wonderful Husband - John
Cyg 20h44'31" 40d0'27"
My Wonderful Mother
Leo 11h26'54" 9d2'38"
My Wonderful Mother DeAnn
Leo 9h53'29" 22d7'59"
My Wonderful Mother Nancy Barry
Cas 23h49'37" 52d34'39"
My Wonderful Mother, Fran Jones
Cas 1h0'31" 64d50'32"
My Wonderful Mother—I Love You
Her 17h1'1" 36d15'38"
My Wonderful Parents David & Suze
Lyr 18h29'14" 28d11'58"
My wonderful wife, Dawn Mcguire. xxx
And 23h55'19" 45d16'10"
My Wonderwall
Cnc 9h14'29" 30d34'17"
My World
Cep 23h37'46" 84d44'6"
My world revolves around you - Cher
Dra 12h12'6" 72d59'34"
My Zara 2/11/97
Uma 10h14'5" 63d9'52"
MY3-GR8-Inspirations
Tri 2h10'43" 34d1'48"
Mya Angelina Fasula
And 0h32'17" 29d27'49"
Mya Beveridge
Uma 11h50'15" 57d3'48"
Mya Cherrez
Lib 14h53'46" -14d30'8"
Mya Christine Spencer
Vir 13h11'4" 5d9'58"
Mya D'Souza-Le
Lyn 8h5'25" 51d20'55"
Mya Elizabeth Lichtenwalner
And 2h21'53" 48d3'55"
Mya Elizabeth Walp
Cnc 8h32'14" 11d26'24"
Mya Jane Tait
And 23h33'5" 41d55'23"
Mya Lee Carie
Psc 1h26'14" 25d36'53"
Mya Lily Courtney
And 1h4'58" 42d52'48"
Mya Lynn Quinones
Uma 8h55'22" 65d38'45"
Mya Marguerite Carter
And 0h27'21" 43d25'45"
Mya Marie
And 23h36'53" 48d11'16"
Mya Miller
Aqr 21h53'20" -2d8'17"
Mya Pearl Varacalli
Peg 23h53'28" 10d27'46"
Mya Sydney Lowman
And 1h16'35" 42d20'39"
Mya Thompson
Umi 13h42'29" 75d56'28"
Mya V. Machado
Lib 14h51'8" -13d42'38"
Mya Vallentina
And 2h29'2" 43d0'26"
Myaeda
Ari 3h3'22" 18d40'12"
Myah Alexis
Mon 6h45'51" 6d59'56"
Myah Carmen Harris
Aql 19h41'28" 9d12'27"
Myah Elzora Diorio
Ari 3h10'17" 23d0'59"
Myah Helina Wrobel
Leo 10h17'59" 11d18'52"
Myah Rachelle Laws
Uma 9h27'49" 55d47'50"
MyaLeeNicholls
Lib 15h48'38" -19d28'18"
MyAng
Ori 5h15'22" -1d20'30"
MyAngel21
And 0h26'41" 43d15'50"
Myani Milagros Dowell
Pav 18h40'40" -60d42'47"
MyAnne
And 1h27'21" 34d34'52"
Mya's Final Hero
Uma 8h26'19" 63d35'27"
Mybabe
Lmi 10h23'10" 34d19'7"
Mybestgirl
Uma 10h36'16" 67d5'45"
My-BOOOS
Sge 19h51'32" 18d58'25"
MyBrotherBilly*Wm E Gambrell
Lib 14h55'27" -10d32'49"
Mychaela Tickner-Luce
Lib 15h37'25" -5d40'10"
Mychal & Karen Cox
Cyg 20h13'11" 56d0'40"
Mychal Noah
Gem 6h50'26" 15d12'35"
Mychela Valerie Macarz-Jones
And 0h11'24" 33d32'12"
MyChelle Ingram
Cet 2h44'11" -0d29'34"
Mychellebee Nunez
Uma 10h13'47" 71d54'51"
Mychi
Umi 17h10'13" 78d35'36"
Myckala Malu Lani
Uma 10h39'51" 47d55'55"
"Mycko's Baby"——Patty
Psc 0h37'55" 14d44'52"
MYDASRFR
Per 2h46'8" 56d39'4"
MydnightLaelia's Hope
And 23h4'1" 36d6'56"
Myeia STAR Newman
Pho 0h11'36" -47d4'56"
MyEl
Ari 3h22'6" 23d26'4"
MyElla Sweet
And 1h42'14" 49d22'32"
Myers
Cyg 21h55'50" 38d8'50"
myersmojo
Cnc 8h11'49" 17d35'31"
Myers-Wolcott
And 0h38'52" 37d26'33"
Myesha Jameel Saleem
Lyn 8h43'1" 43d39'32"
Myeshia "Miss Molly" Lee
Lyn 8h50'22" 34d57'19"
"Myfanwy" - Delia Robinson
Cru 12h39'51" -60d34'44"
Myfanwy Sunshine
Umi 16h21'28" 82d17'9"
Myga Makrytera
Cam 4h8'15" 65d39'18"
MyGrams
Vir 12h47'19" 6d15'46"
MYHA VAN
Cnc 8h44'54" 19d12'46"
Myhpasiknoh
Her 17h24'46" 43d50'23"
Myisha Ledawn Palmer
Mon 6h54'17" -5d48'1"
Myka Bellisari Always and Forever
Uma 9h28'27" 55d41'15"
Mykale
Leo 10h38'12" 14d24'3"
Myke & Lauren Thom
Lyn 6h27'42" 56d29'28"
Mykhailo Tsymbaljuk
Sco 17h54'45" -36d31'33"
Mykilina
Sgr 18h43'1" -18d7'44"
Mykim Nguyen
Lib 14h48'55" -5d15'59"
Mykolas Banevicius
Cyg 20h54'17" 46d7'3"
Myla Grace
And 23h1'46" 50d52'40"
Myla Jadyn Tralins
Cnc 8h21'10" 16d34'17"
Myla Marie Moody
Cyg 20h0'55" 33d14'5"
Mylah Anne
And 23h43'18" 48d37'53"
Mylan Le
Cyg 20h55'28" 41d17'41"
MyLe
Gem 7h59'32" 19d50'51"
Myleigh
Lib 15h48'25" -17d52'18"
Myleigh Skye
Lib 15h26'44" -16d47'25"
Mylena Cabezas Ujueta
Gem 7h25'15" 28d35'38"
Mylene Batungbacal "OHMY"
And 23h46'35" 44d4'45"
Mylène Beauregard
Cyg 21h18'6" 44d13'45"
Mylène Elizabeth
And 0h32'16" 44d25'19"
Mylène et Jean-Philippe
Psc 1h18'25" 32d8'12"
Mylene Joy Casayuran
Crb 15h30'10" 26d20'33"
Myles
Per 3h35'58" 49d17'24"
Myles and Ciara
Cap 20h53'49" -19d23'44"
Myles and Joan Hyman
Leo 10h54'22" 16d20'56"
Myles Connell
Vir 14h3'29" -19d32'18"
Myles Dalen Glenn
Her 18h16'45" 19d29'48"
Myles Damond
Cyg 19h57'0" 51d28'33"
Myles Haigney
Ori 6h5'45" 14d58'48"
Myles Henry Baruch
Uma 11h26'38" 29d59'6"
Myles "Kilometers" Sansone
Psc 1h30'9" 13d52'2"
Myles Nelson Cox
Leo 10h23'21" 18d58'44"
Myles P. Dunn
Peg 23h20'53" 33d48'42"
Myles Richard Gwyn Davies
Per 4h8'41" 33d48'27"
Myles Sylva
Uma 9h43'17" 42d9'50"
MyLinda
Vir 13h34'44" -20d10'38"
Mylinda A. Black
Tau 4h27'39" 27d37'6"
MyLinda's Piece of Heaven
Mon 6h50'34" -0d4'48"
Mylissa
Tau 4h10'18" 24d11'52"
Mylissa
Uma 13h49'22" 49d48'17"
Mylo Yuteaba Turner
Cam 6h28'57" 72d22'17"
MYlo6069
Uma 11h10'53" 47d48'16"
Myloune
Ori 6h15'50" 13d21'39"
MyloveGeorgette
Ori 5h57'26" -2d28'15"
Mylyn
Sco 16h13'9" -11d26'37"
MYM-06301971
Ori 6h13'45" 13d1'21"
mymommaMary
Cas 23h12'22" 55d47'41"
MYMOMOLGA
Sgr 19h26'40" -12d50'20"
Mymonie
Crb 15h26'47" 26d16'44"
MyMy's
Mon 6h53'36" -0d26'19"
Myndi
Uma 9h20'30" 42d33'26"
Myndi Marie
Cnc 8h5'20" 24d3'10"
Myndi's Light
Uma 10h52'55" 49d1'6"
MYNOR Y VIVIANA
Uma 9h30'49" 61d9'1"
MyNorma
Cas 1h59'42" 65d26'37"
Mynxe
Sco 16h41'39" -31d20'10"
Myo Sook 1956
Ari 3h10'15" 29d24'29"
Myong Cha
Psc 23h1'29" 1d50'49"
Myong S. Choi
Ari 1h52'58" 18d8'26"
Myong-Suk O
Uma 12h39'1" 54d23'23"
Myosotis alpestris
Per 2h59'17" 32d0'25"
Myra
And 0h53'38" 43d8'34"
Myra
Cas 1h33'14" 57d50'53"
Myra
Cnc 9h11'26" 22d53'1"
Myra
Leo 10h9'20" 15d17'4"
Myra
Cas 23h7'39" 59d8'51"
Myra
Cam 5h36'38" 68d55'39"

myra
Sgr 18h20'38" -32d9'27"
Myra
Sco 17h52'52" -31d8'49"
Myra Adams
Cas 0h17'59" 52d18'56"
Myra Aitchison
Ari 3h2'49" 24d37'32"
Myra and Rick Armstrong
Cyg 19h39'3" 32d25'6"
Myra Ashleigh Kirkland
Leo 9h24'0" 12d4'19"
Myra Beth Gasbarro
Cyg 19h52'3" 33d7'11"
Myra & Chance
Sgr 19h40'43" -12d14'54"
Myra Elizabeth Molina
Tau 4h1'15" 23d19'15"
Myra Evangeline Stockburger
Lyr 18h41'13" 34d12'5"
Myra Jean
Aql 19h47'15" -0d12'11"
Myra Jean Robinson
Uma 9h57'29" 67d26'7"
Myra K. Hamin De Leston
Tau 3h48'55" 27d43'56"
Myra Lopez
Ori 5h12'30" 10d36'5"
Myra Mellisa Melendez
Ari 2h14'43" 24d40'2"
Myra Sue Vernon
Gem 7h41'47" 15d44'31"
Myraed Frances David
Gem 7h22'13" 26d19'51"
Myranda Jayde
Ari 2h48'48" 15d24'9"
Myranda Kaye
Lib 15h16'39" -10d29'11"
Myranda Lee Smith
Lib 15h5'52" -1d10'6"
Myranda Lyn Chaddick
Sgr 18h14'35" -20d7'49"
Myranda Nicole Carter
Gem 7h15'44" 30d32'24"
Myranda Payge Gransbury
Tau 3h42'57" 16d50'24"
myrandajean
Cas 0h38'26" 58d19'46"
Myranda's Smile
Uma 11h18'35" 32d3'52"
Myrddrina
Dra 10h13'10" 78d53'2"
MyrEd55
Sgr 18h28'56" -26d53'15"
Myredith S. Gonzales
Aql 19h3'43" 2d39'18"
myreynalda
Mon 6h39'37" 9d23'7"
Myriah Kristine Miller
And 23h45'19" 42d56'44"
Myriah McMillan
Cas 23h46'33" 52d54'8"
Myriam
Dra 15h48'53" 55d22'24"
Myriam
Cru 12h13'37" -58d46'20"
Myriam
Uma 11h42'6" 50d41'43"
Myriam
Peg 22h40'27" 24d18'34"
Myriam
Gem 7h37'46" 16d24'27"
Myriam Abel
Tau 5h11'55" 21d38'7"
Myriam Clémentine
Peg 23h42'29" 27d42'5"
Myriam Del Pozo
Sco 16h26'12" -29d49'57"
Myriam Flores Garza
Vir 13h50'59" -0d6'42"
Myriam & Jacques
Uma 9h6'42" 55d29'42"
Myriam&Pius
Vul 21h7'44" 27d25'52"
Myriam Szablewski
And 0h11'49" 46d34'41"
Myriam Waldvogel
Boo 15h8'22" 49d59'19"
Myriam Zitouni
Lib 14h52'0" -0d33'10"
Myriame Baert
Tau 3h56'52" 26d45'36"
Myrian
Boo 14h35'59" 16d26'49"
"Myr-ion" - Myrtle and Marion
And 23h20'31" 47d12'57"
Myrk Tyva
Uma 11h38'54" 63d17'18"
Myrna and Gordon Spratt
Cyg 20h0'44" 42d48'16"
Myrna "Bunny" Timmons Schoenhardt
Cyg 20h20'32" 43d58'44"
Myrna Caradon
Cyg 21h15'6" 31d52'26"
Myrna D Hendricks Mitchell
Ari 2h2'7" 17d38'6"
Myrna Figueroa
Cyg 19h19'51" 46d55'25"
Myrna L. Hefty
Leo 11h23'24" 15d27'35"
Myrna Lewis
Uma 11h10'29" 56d44'45"
Myrna Lorena
Tau 4h10'59" 15d37'2"
Myrna Martinez
Vir 13h21'50" 13d48'58"
Myrna Mercedes Huebsch
Cas 1h24'51" 56d36'28"
Myrna "Nan" Griffith
Cas 1h8'18" 60d55'18"
Myrna Neff
Sco 17h19'54" -41d14'55"
Myrna Owen "My Star Wife"
Uma 9h56'12" 65d59'4"
Myrna's Brightest Light
Sco 16h42'47" -32d14'54"
Myrna's Twinkling Star
Psc 1h7'27" 24d24'15"
Myron and Marsha Fisher
Ori 5h5'28" 16d7'57"
Myron C. Yocum Jr. 5-22-1926
Gem 6h22'34" 21d14'8"
Myron E. Etienne Jr. "Doc"
Per 4h32'21" 44d31'4"
Myron Eugene Napper
Vir 13h30'30" -12d59'15"
Myron "Mike" Graef
Her 17h52'53" 46d47'44"
Myron P. Ramirez
Cnc 8h29'32" 7d12'3"
Myron Stefon Webb
Ori 6h6'25" 12d1'56"
Myron Thaden
Uma 9h31'14" 63d20'50"
Myroslaw Fedyk
Lyn 8h4'24" 51d56'7"
Myrrhynda's Star
Cam 7h32'21" 67d19'44"
Myrsew
Her 16h21'56" 19d18'16"
Myrt and Marv - Shining on Forever!
Tau 4h26'56" 12d57'31"
Myrt Gipson
Crb 16h13'43" 38d11'53"
Myrta Pauline Reinhart
Cas 1h21'29" 60d42'43"
Myrtha
Lyn 8h28'7" 51d27'58"
Myrtha Hess
Crb 16h21'34" 37d56'38"
Myrtha Hug
Her 17h32'13" 32d47'10"
Myrtie
Ari 2h22'10" 25d3'1"
Myrtle
Gem 6h6'18" 23d41'47"
Myrtle Be'Ach - Miss Adams Morgan
And 23h44'15" 46d12'46"
Myrtle Bernice Stephens Potter
Lib 14h50'41" -1d59'6"
Myrtle Cameron Hall
And 1h46'47" 42d31'56"
Myrtle E. Einhorn
And 23h15'6" 52d45'55"
Myrtle E. Karnell
Lib 15h6'34" -11d11'57"
Myrtle Evelyn Jifkins (Byrne) Pat
Vir 13h51'51" -11d30'52"
Myrtle Irene
Lib 15h59'39" -10d15'17"
Myrtle Kirkwood
Sgr 19h8'11" -29d34'38"
Myrtle Mae
Sgr 19h1'52" -25d26'5"
Myrtle Mae
Ari 2h39'35" 30d45'15"
Myrtle & Marie Forever Sisters
Sco 16h44'35" -33d51'48"
Myrtle Page
Crb 15h47'0" 27d44'41"
Myrtle Rebecca Skop Rutberg
Lyr 18h34'21" 38d34'11"
Myrtle Smith Metcalf 08/20/1915
Aql 20h2'48" -0d22'29"
Myrtle Sue Conley Hicks
And 0h44'36" 37d20'42"
Myrtle West
Uma 13h29'49" 57d0'8"
Myrtle-L
Cnc 8h19'30" 32d27'4"
Myrto
Boo 15h7'49" 49d0'27"
Myrto & Christos
Cyg 19h51'48" 30d38'32"
Myscheil
Peg 21h39'54" 16d58'30"
MySheilaE
Psc 0h32'7" 8d21'51"
MyShell
Vir 12h20'23" 3d21'1"
Myshell
And 1h54'1" 38d35'15"
Myslim na Vas
Psc 1h24'38" 27d58'32"
Myson
Umi 14h41'26" 76d2'51"
MySquishy
Uma 10h2'3" 45d43'21"
Mysteri Elizabeth Gebbia
And 0h48'3" 40d6'0"
Mysterie Ocean
Mon 7h1'8" 9d2'36"
Mysterry
Uma 10h13'8" 72d16'26"
Mystery Hensley
Umi 15h9'26" 84d4'21"
Mystery Star Diamond
Crb 15h47'34" 28d51'29"
Mystery Suska
Ori 6h7'45" 13d1'38"
Mysti Faith
Lyn 7h34'8" 50d25'51"
Mystic Fire
Lib 15h23'47" -28d28'10"
Mystic Unicorn
Ari 2h55'24" 25d37'42"
Mystic2613
Aql 19h27'31" -0d27'17"
Mystica Malone
Cap 20h22'23" -15d32'17"
Mystical
Lyr 18h49'42" 38d8'30"
Mystical Ash
And 0h41'46" 33d31'53"
MystX
Cyg 20h2'17" 33d23'49"
Mysty
Leo 11h15'24" 14d9'56"
MySue18
Uma 10h38'25" 63d44'9"
Myszka
Uma 9h35'33" 57d32'46"
Mythili/Deepa :)
Cen 12h48'59" -43d43'5"
Mythily
Lib 15h11'0" -4d45'11"
MYTHOS
Lyr 18h47'30" 30d12'43"
Mytutu
Lmi 10h28'35" 34d43'52"
MYUH!
Ara 17h26'52" -46d49'11"
Myungho Cho and Yeosun Kim
Vir 13h54'49" -1d5'3"
MyUyen
Sgr 19h4'39" -24d1'36"
Mywave66
Gem 6h53'12" 26d5'19"
MZ
Sco 16h12'58" -10d4'19"
Mz Dodie
Sco 16h14'10" -15d19'7"
MZ71000
Tri 2h25'6" 30d28'35"
MZMOUSE60
Uma 11h25'5" 45d13'33"
911 Director Warren County, Ohio
Per 4h44'14" 47d26'31"
975032029
Psc 23h47'5" 7d17'19"
N A B
Cru 12h32'57" -59d22'35"
N and J Forever
Cyg 21h34'33" 37d27'5"
N and M Mattson: Endless Love
Uma 13h27'12" 52d57'47"
N Hazel Minders
Aur 6h28'1" 40d43'40"
N I B O R
Gem 6h48'22" 20d45'15"
N J M 9-6-43
Ori 5h24'11" -0d5'20"
N. Kathryn E. Knight
Sgr 18h18'40" -17d0'4"
N M Torode Early Years/Theresa
Cyg 22h1'5" 50d3'11"
N o j 30 L a d y B
Leo 10h21'7" 24d11'2"
N&P's Shining Love Forever
Her 17h46'53" 48d39'33"
N1530S
Umi 13h26'5" 71d0'43"
N3NA
And 0h6'8" 44d12'29"
NA831AF
Cnc 8h22'20" 12d44'58"
Naa
Ori 6h8'59" 7d59'50"
Naadir Pookie Cassim
Ari 1h51'7" 18d42'27"
Naalah
Per 1h49'39" 51d0'53"
NA'AMA
Lib 15h41'39" -28d33'54"
Naamat - May 31, 1979
Cas 23h31'16" 53d4'50"
Naannccy
Sco 16h7'16" -17d7'22"
Naarah Deanne Patton
Sgr 19h33'7" -16d20'57"
Naasadore ~ The Eye of My Tonight
Lyn 8h15'59" 55d11'54"
Nabeel Shaukat
Uma 10h55'0" 55d25'58"
Nabi HB
Per 4h26'3" 43d38'12"
Nabih and Jessica
Per 4h26'1" 49d43'30"
Nabil Charbel Arslan
Uma 13h34'14" 57d29'14"
Nabil El Abedin
Lib 15h1'36" -22d8'16"
Nabil et Himenne
Lib 15h35'1" -19d39'15"
Nabil Qaddumi
Cyg 20h33'16" 55d3'33"
Nabil Sawalha
Equ 21h5'15" 9d20'28"
Nabila Chowghule
Tau 5h5'12" 25d16'18"
Nabrielle
Leo 10h24'11" 24d58'11"
NAC 428 <3
Cyg 21h12'51" 44d23'40"
Nace Allen Goldman
Lmi 10h0'46" 32d15'53"
Nacey's Kismet
Lib 15h2'23" -1d12'56"
Nachalo AE
Cyg 21h27'29" 37d20'53"
Nachia Heu
Lyn 7h12'22" 56d32'15"
Nackunstz, Klarina + Olaf
Ori 6h2'6" 14d11'47"
Nacol Catherine Reinmuth
Lyn 7h13'7" 53d53'32"
Nacole Brook Winter
Lib 15h12'28" -18d41'37"
Nacole Love You Always Harvey
And 23h24'53" 42d58'0"
Nacrina Center of the GTA Universe!
Gem 6h56'26" 17d17'10"
Nad2Nad
Per 4h24'34" 43d38'34"
Nada
Leo 9h45'30" 21d0'24"
Nada
Lyr 18h50'9" 29d31'56"
Nada
Umi 15h36'55" 71d18'28"
Nada Alsrayyea
Psc 2h0'24" 5d10'50"
Nada & Issa Naber
Lyr 18h38'54" 30d46'13"
Nada Mehanna
Pho 1h20'49" -45d8'53"
Nada Salamé
Cas 1h49'7" 63d31'55"
Nada Shouhayib
Gem 6h42'4" 23d37'3"
Nada W. Gemayel
Lib 14h24'37" -19d42'30"
Nada Young
Uma 10h13'56" 69d43'38"
Nadati se
Psc 1h12'29" 11d2'3"
Nada...Zelja...Sreca...Ljuba v...
And 0h41'51" 45d54'48"
Naddie M. Garcia
Lyr 18h40'11" 37d31'12"
Nadeem H
Uma 8h20'13" 67d54'46"
Nadeem & Zenib
Cyg 19h36'58" 31d32'10"
Nadège.Dupied.10/06/1982
Uma 13h37'30" 49d36'22"
Nadene (Dene)120191 Our Love Shines
Sco 16h35'58" -44d13'1"
Nadene Sottosanti
Vir 14h3'24" -16d16'46"
Nader
Del 20h49'25" 16d55'37"
Nadeshda
And 1h20'43" 45d41'20"
Nadezchka
Umi 16h10'33" 72d13'52"
Nadezda Abramova
Cap 20h8'35" -17d18'8"
Nadezhda
Cam 3h59'0" 72d27'57"
Nadezhda
Crb 16h5'24" 32d46'57"
Nadezhda
Psc 1h33'32" 24d48'19"
Nadezhda & Anatoliy Muravjov
Ari 2h49'48" 30d10'27"
Nadezhda Nikulina 9 settembre 1977
Per 3h45'40" 38d38'5"
Nadezhda Zhilova Eubanks
Gem 6h47'58" 14d31'13"
Nadi
Gem 7h44'19" 33d46'22"
Nadi El Khoury
Uma 9h52'47" 60d3'58"
Nadia
Uma 10h40'31" 71d38'25"
Nadia
Uma 13h36'24" 56d1'40"
NADIA
Uma 13h6'26" 58d36'29"
Nadia
Lep 5h39'38" -20d40'5"
Nadia
Umi 17h41'19" 80d23'30"
Nadia
Umi 16h51'44" 80d7'24"
nadia
Sgr 19h5'19" -31d38'10"
Nadia
Crb 16h23'45" 31d4'38"
Nadia
Cnv 13h59'37" 36d3'29"
Nadia
Cyg 19h39'46" 30d27'8"
Nadia
Cyg 20h13'56" 35d45'31"
Nadia
And 1h35'58" 46d13'15"
Nadia
And 0h52'55" 22d4'37"
Nadia
Psc 1h28'34" 25d23'24"
Nadia
Gem 7h7'2" 18d46'11"
Nadia
Cnc 7h58'21" 15d13'9"
Nadia
Tau 5h7'16" 26d43'45"
Nadia 30 maggio 2003
Per 3h23'27" 50d11'7"
Nadia Adams
Psc 1h13'39" 31d12'5"
Nadia Ahmed
Umi 16h45'50" 81d54'50"
Nadia and Peter
Cyg 21h33'4" 44d51'14"
Nadia Bergerie
Her 17h52'25" 49d44'12"
Nadia Bibles
Uma 10h26'37" 43d34'34"
Nadia Bulifa
Cas 0h50'6" 64d45'36"
Nadia Cameron
Cru 12h14'44" -62d28'20"
Nadia Comaneci
Her 18h17'56" 22d53'21"
Nadia De Momi
Oph 17h13'24" 4d15'22"
Nadia Fakih
Sgr 17h55'11" -17d24'49"
Nadia Ferandini & Markus Muggli
Cyg 20h40'3" 48d42'45"
Nadia Flavio 270303
Ori 5h16'12" 0d39'23"
Nadia Franciscono
Psc 0h44'34" 15d40'52"
Nadia Fuentes
And 0h30'22" 42d43'49"
Nadia Giovanetti
Uma 11h59'57" 31d49'39"
Nadia Hannah Peinado
Sco 16h18'27" -15d43'39"
Nadia Jamai...Rare & Precious
Gem 6h59'22" 14d43'48"
Nadia Jasmine & Mikko Juhani
Cyg 21h42'20" 44d19'13"
Nadia Kara Hmaidi
Cap 20h29'4" -13d14'10"
Nadia Karys Dickey
Tau 4h19'3" 29d21'21"
Nadia Ladjimi-Thürler
And 23h0'58" 50d59'18"
Nadia Leigh Dropkin
Sco 16h8'19" -9d24'3"
Nadia Luz
Uma 13h58'39" 54d2'46"
Nadia M. Rodriguez
Cas 0h39'38" 66d0'18"
Nadia Mabrouki & Fabrice Bubel
Sco 16h50'18" -36d9'11"
Nadia Marie Pyrdeck
Ari 2h57'51" 30d1'57"
Nadia Maus Schuler
Cep 23h14'0" 79d27'12"
Nadia Milena Smerdka
Sgr 18h9'26" -27d47'49"
Nadia My Lief
Cyg 20h57'43" 34d6'26"
Nadia N Kohler
Peg 22h31'9" 20d46'23"
Nadia N. Lynn
Uma 8h19'11" 66d48'4"
Nadia Nedue
Mon 6h30'33" -10d1'40"
Nadia Nell
Lib 14h49'15" -11d10'8"
Nadia Nizam's Snowflake
Cnc 9h20'56" 23d15'37"
Nadia & Pascal
And 22h58'6" 36d43'44"
Nadia Petris
Per 3h24'4" 41d31'32"
Nadia Principe
Boo 14h38'31" 10d31'34"
Nadia Rana G.P.
Crb 15h36'25" 31d4'2"
Nadia Sarah Landstad
Sco 16h11'33" -12d29'40"
Nadia&Sasha
Cas 2h18'44" 68d51'0"
Nadia Schmitz
Gem 7h4'7" 27d37'3"
Nadia Seemuth
Cyg 20h17'49" 53d6'17"
Nadia Serra
Com 12h21'37" 28d35'26"
Nadia, <<Nimo>>
Cyg 20h34'38" 57d48'18"
Nadia,Jannat,Alisha,Maria m Saeed
Lyr 19h6'35" 32d23'30"
Nadia-jdn-b
Cap 21h0'42" -24d38'0"
Nadia's Angel
Lyn 6h51'28" 54d27'37"
Nadia's Star
Cyg 19h54'31" 33d29'34"
Nadia's Sternchen
Leo 10h59'54" -4d50'47"
Nadiene
Dra 15h57'51" 57d16'24"
Nadieska-C
Dra 18h42'11" 71d3'18"
Nadilynn
Umi 21h10'24" 89d11'42"
Nadim Adriano Titi
Boo 14h36'2" 30d41'26"
Nadim Elias
Tau 5h23'36" 26d13'22"
Nadin & Andreas "the dreamteam..."
Uma 9h49'11" 44d42'41"
Nadine
Uma 10h13'58" 41d37'39"
Nadine
Cyg 19h42'31" 35d49'34"
Nadine
Lyr 18h15'33" 36d20'48"
Nadine
Cas 0h24'2" 55d27'18"
Nadine
Uma 9h52'12" 45d12'50"
Nadine
Cas 1h16'58" 51d20'24"
Nadine
And 1h19'51" 49d9'7"
Nadine
Peg 21h46'34" 21d30'21"
Nadine
Ori 6h24'17" 14d51'15"
NADINE
Equ 20h59'17" 12d7'31"
Nadine
Cnc 8h41'47" 6d33'39"
Nadine
Uma 10h9'30" 70d3'7"
Nadine
Umi 15h13'8" 69d28'32"
Nadine
Cas 2h9'24" 73d11'0"
NADINE
Lyn 7h50'12" 54d28'8"
Nadine
Sco 17h43'50" -32d15'22"
Nadine 2710
Uma 8h47'52" 48d13'52"
Nadine and Mark
And 0h18'50" 46d40'48"
Nadine and Nikolas Together Forever
Uma 10h43'54" 51d43'34"
Nadine and Ramzi
Crb 16h10'24" 31d55'45"
Nadine Baggott
Peg 21h31'18" 25d56'6"
Nadine Bissat Irani
Gem 7h19'15" 24d10'20"
Nadine Catherine
Eri 3h25'59" -42d27'2"
Nadine Chaoui, My angel Nad
Uma 8h56'24" 51d0'17"
Nadine Cowdrey Wilkey
Gem 7h22'39" 26d59'51"
Nadine Davis Boone
Lyr 18h45'26" 28d13'33"
Nadine E. Caratelli
Ori 5h12'53" 15d9'0"
Nadine et Robert Ghougassian
Lib 15h28'35" -9d38'22"
Nadine et Stéphane Puissant
Tau 5h47'1" 19d35'2"
Nadine Gharbi & Louis Vincent Maury
Uma 9h33'44" 44d9'24"
Nadine Guillen-Logan
Aqr 23h1'41" -4d13'22"
Nadine Ingrid Bowen
Lib 14h57'27" -1d56'33"
Nadine Joyce Brown
Leo 10h47'10" 10d45'2"
Nadine Jürgen Katharina
Uma 9h43'32" 69d12'34"
Nadine Keller
Sco 16h15'18" -11d32'54"
Nadine Kirychuk
Vir 13h7'0" 13d20'43"
Nadine Krautscheid
Uma 11h56'7" 55d23'26"
Nadine Krescentia Gazzola
Cap 21h21'25" -22d53'8"
Nadine la Bichette
And 1h17'10" 48d59'11"
Nadine Lee
Cas 1h43'22" 61d36'36"
Nadine Lucille Harris
Sgr 18h21'48" -34d49'2"
Nadine M. Malucci
Oph 17h12'44" -0d46'48"
Nadine Maria Elena
Cnc 8h45'1" 8d25'42"
Nadine Marie Hogan
Aqr 22h11'55" -22d31'46"
Nadine Meyer, 11.12.1977
Crb 16h7'10" 36d44'12"
Nadine "Nana" Bryan
Vir 14h58'31" 4d29'13"
Nadine Newton
Cap 21h37'8" -16d55'16"
Nadine Noodle
And 23h14'40" 44d44'57"
Nadine Pretorius
Her 17h44'21" 46d3'23"
Nadine "Prettiest Mermaid" Gentry!!
Cyg 20h32'7" 33d32'56"
Nadine Renee Price
Tau 4h36'29" 18d9'25"
Nadine Rose Meehan's Guiding Star
Cas 0h26'11" 56d16'46"
Nadine Sophie Rosmus
Psc 0h19'57" 6d59'20"
Nadine & Timo
Uma 12h47'55" 55d34'58"
Nadine & Walid
Vul 20h19'42" 23d19'8"
Nadine Werlen
And 23h22'44" 37d43'41"
Nadine Zacheres
Uma 8h57'49" 47d46'24"
Nadine Zürcher
Per 4h41'55" 48d57'29"
Nadine, 18.05.1979
Mon 6h29'43" -5d15'52"
Nadine, Habib, Aline & M.ali
Cyg 21h47'53" 41d59'13"
Nadine, My Naj
Uma 10h42'43" 62d24'34"
Nadine-Christoph
Uma 8h26'47" 59d58'46"
Nadine's dream
Cas 0h19'6" 54d53'43"
Nadine's Spirit
Sco 17h51'31" -39d18'17"
Nadine's Star 2 Me 2006
Uma 9h42'5" 61d14'10"
Nadine's Star Of Hope
Aur 6h32'27" 38d49'54"
Nadira Ann
Gem 6h49'40" 23d1'8"
Nadira Arreola
Vir 13h4'58" -10d18'15"
Nadiya
Uma 10h43'56" 39d29'3"
Nadiya
Gem 7h8'1" 33d36'22"
Nadja
Uma 11h58'8" 50d26'17"
Nadja
Ori 6h12'46" 13d31'39"
Nadja Andrea
Vir 14h27'53" 0d57'1"
Nadja Ashe
Aqr 22h57'56" -23d52'23"
Nadja Bain
Ari 2h56'30" 24d59'8"
Nadja Cainero I Love you din Tiga
Ori 5h52'48" 22d17'58"
Nadja Hornstrup Mogensen
Lib 15h9'58" -8d11'19"
Nadja "Naddl" Weidner
Ori 6h8'46" 16d5'52"
Nadja Raven
Mon 6h50'58" 7d51'45"
NADJA & RICO
Her 17h45'38" 30d36'38"
Nadja Rohner & Nicola Elsener
Ori 5h55'1" 18d0'43"
Nadja Rusczyk
And 1h56'53" 42d17'4"
Nadja Vincenz
And 2h18'1" 48d27'10"
NadJa2001
Per 4h47'40" 50d37'23"
Nads
Cet 2h29'6" 7d47'1"
Nady & Alex
Ori 6h10'39" 8d11'22"
Nadya Kristine Young
Cnc 7h58'45" 13d48'57"
Nadya Lee
Lyn 6h35'59" 60d22'11"
Nadya Velardo
And 0h31'13" 31d4'13"
NAE42
Uma 10h21'30" 47d51'28"
Naea
Com 12h16'39" 31d2'2"
Naema
Uma 8h11'51" 66d43'46"
Naëmi Fiesolani
Lmi 10h43'4" 28d56'8"
Nafarroa
And 0h49'38" 46d17'25"
Nafesa
Crb 15h23'21" 26d34'44"
Naffy
Uma 10h31'3" 66d55'49"

Nafissa
Cnc 9h18'7" 19d34'8"
Nagasa
Uma 11h51'21" 55d53'27"
Nagel, Ernst
Uma 9h4'45" 54d46'9"
Nagel, Manfred Walter
Ori 5h20'16" -4d44'25"
Nagemelyk
Dra 17h3'31" 58d31'30"
Nagi N. Awad Jr.
Lib 14h45'54" -17d26'1"
Nagley
Crb 15h56'14" 39d8'38"
Nagligivaget Roxanne
Umi 15h36'36" 78d48'8"
Nagore
Ari 3h8'26" 22d38'0"
Nagwe Alsamadisi
Sco 17h50'44" -35d6'22"
Nagy Erika
Uma 11h52'37" 56d4'24"
Nagy Zoltán és Nagyné Koltay Erzsébet
Aqr 20h39'33" -2d50'56"
Nagyapa Angyal~Buppa's Angels
Cyg 20h55'31" 47d17'56"
Nagy-Mihályi Judit (Kiscsillagom)
Uma 10h30'40" 70d20'49"
Nahadennhail
Gem 7h22'16" 15d37'2"
Nahal Chitsazan
Vir 12h40'58" -8d36'13"
Nahal Radfar
Ori 5h36'45" -1d24'33"
Nahama Joy
Cnc 8h27'23" 23d13'21"
Naheed-Ismail-Siddiqui
Her 18h42'11" 20d35'23"
Nahid Jamzadeh
Leo 10h13'14" 22d27'42"
Nahim, Precious and Perfect
Ari 2h24'46" 26d37'18"
Nahm János
Gem 6h30'13" 25d51'12"
Nahtona
Umi 16h33'50" 77d21'12"
Nai
Aql 19h44'39" 16d26'4"
Naia Angel
Peg 22h57'53" 31d54'29"
Naiara
Ori 6h10'30" 7d2'59"
Naica Hoffmann
Uma 13h8'14" 52d47'2"
Naidely
Del 20h33'39" 15d57'6"
Naif
Mon 7h20'35" -1d28'40"
Naiha Ali
Cap 20h46'46" -24d17'8"
Nail 42404
And 0h49'36" 36d26'28"
Naila
And 0h39'45" 39d35'19"
Naila Rivera
Tau 3h38'3" 28d52'32"
Nailah Asha Taylor
Lyn 7h39'35" 46d21'21"
Naima Beatrix Alwi
Uma 9h47'59" 53d44'33"
Naïma mon Amour
Umi 14h53'47" 74d37'12"
Naiman
Ori 5h30'15" 2d33'58"
Naimh Mary Currie
And 1h42'25" 50d21'21"
Nainy
Lib 14h53'22" -10d33'13"
Naiomi Elaine Uncangco
Psc 23h31'34" 4d36'36"
Naira
Lyn 6h36'7" 60d13'24"
Nait Harkes - Always Loved
Aql 19h7'7" 8d57'50"
Naizion
Per 2h15'15" 51d8'58"
Najdraze - Katarinica & Krcka
Cyg 19h47'35" 35d11'53"
Najike
Vir 14h48'38" 5d33'51"
Najla & Karim
Ori 6h6'19" 10d23'27"
Najla "Lita" Saab
Umi 15h19'46" 67d37'37"
Najlaa
Lib 15h3'21" -2d40'24"
Najma Jebari
Psc 0h48'53" 6d4'26"
Najmeh Hannanvash
Tau 4h35'27" 18d26'52"
Najona Michelle Ichimaru
Vir 12h45'19" 0d1'33"
Naju
Vir 14h4'35" -3d34'34"
Najwa
Cap 20h30'44" -23d4'22"
Naka Kazuno
Oph 16h45'38" -0d55'3"
Nakamura Dental Star to Success
Gem 7h7'5" 24d55'17"
NAKED #7
Tau 4h13'31" 17d6'13"
"Naked Angel" - Michael Ciampi, Jr.
Sgr 19h15'31" -15d17'27"
N'akeetula
Uma 12h0'33" 28d49'18"
Nakeshia Ruberg
Aql 20h12'0" 13d37'32"
Naketia loveronish staten
Leo 10h59'35" 1d5'7"
Nakis-Whitney
Lyn 8h31'12" 56d3'11"
Nakita
Leo 9h43'18" 31d59'4"
Nakita Lynn Mathes
Vul 20h32'16" 22d56'36"
Nakoma
Uma 9h17'54" 64d52'14"
Naku Temwa
Ari 2h19'4" 20d34'17"
Nala
Lyn 6h59'40" 44d49'11"
Nala
Cra 18h35'57" -45d5'51"
Nala Bubbles Dzikonski
Uma 10h51'33" 42d31'49"
Nalani & Rylan
Hor 4h5'29" -41d8'33"
Nalee Vwj
Psc 0h50'54" 14d22'23"
Nalei
Uma 12h12'23" 58d51'58"
Nalin
Leo 11h2'13" 8d31'22"
Nalinee most loved mom in the world
Ori 5h55'1" 12d3'20"
Nalini Sridharan
Vir 13h57'54" 2d4'27"
Nalini Yadla
Cap 21h45'30" -11d3'25"
Nallely Mendez
Gem 7h32'53" 16d31'35"
Nallumcm Ecarg Enilorac
Lyr 19h26'40" 38d17'10"
Nalukea o Kena
Per 3h38'15" 34d48'28"
Nam Hee Kim
Vir 13h18'33" 7d42'10"
Nam Mei Amor
And 1h10'33" 45d3'33"
Nam Su Liermann
Lyn 7h39'57" 51d30'44"
NAM VO
Lib 14h25'35" -19d45'30"
NAMA
Leo 9h45'31" 27d13'32"
Namahana's Star Baby
Ari 3h7'7" 11d6'53"
Namaste
Psc 0h37'58" 15d45'59"
Namaste'
Uma 11h11'53" 43d22'36"
Namaste
Sgr 19h29'6" -18d9'17"
Namaste
Aqr 22h4'39" -2d55'20"
Namaste AbbyFloren
Uma 11h31'22" 32d42'24"
Nami No Tara
Gem 6h10'25" 26d27'14"
Nami Nom
Vir 12h22'10" 12d6'48"
Namie and Papa Hindman
Ori 6h1'26" 1d5'40"
Namiki 35th
Ari 1h59'29" 20d53'56"
NAMI-LNW929jb
Lib 16h1'45" -12d0'25"
NAMITA GUJRAL
Gem 7h31'53" 15d24'34"
Namita Sanjay Shah
And 23h46'47" 47d14'36"
Namoo
Per 3h25'21" 31d22'56"
Namour
And 0h14'37" 46d37'39"
n'amour
Psc 1h2'8" 22d39'52"
Namour
Umi 16h30'34" 75d52'55"
Namrata
Lib 14h35'11" -9d30'7"
Nan
Cap 21h1'47" -20d17'0"
Nan
Psc 0h21'58" 9d20'47"
Nan and Grandalf
Uma 11h34'12" 56d34'42"
NAN CHANDLER
Uma 9h9'11" 64d19'57"
nan et margo
Cas 1h16'50" 59d59'14"
Nan & Geoff Williams - Anniversary Star
Cyg 21h45'40" 38d27'4"
Nan Kempner
Uma 10h5'58" 44d7'22"
Nan Martin
Ari 2h34'42" 14d54'49"
Nan McCaskie
Uma 10h15'9" 42d13'49"
Nan Min
Tau 5h52'51" 25d0'56"
Nan N Doug
Boo 14h33'53" 19d49'25"
Nan of Night - Light in our Hearts
And 0h5'21" 42d49'46"
Nan & Oompa, My Shining Stars
Uma 9h10'53" 59d48'53"
Nan & Paul Jones 50 Years Together
Dor 4h17'45" -49d1'4"
Nan Renee Machen
Cnc 8h52'41" 28d19'3"
Nana
Crb 15h53'22" 27d1'5"
Nana
Ori 4h45'53" 13d1'49"
Nana
Ori 5h34'47" 8d9'16"
Nana
Psc 0h30'1" 8d43'55"
Nana
Lyn 8h34'59" 41d39'1"
Nana
Gem 7h27'41" 33d32'3"
NANA
Uma 11h33'39" 30d58'53"
Nana
Lyr 18h53'35" 35d35'40"
Nana
Per 2h43'28" 56d21'23"
Nana
Cam 4h11'30" 57d46'58"
Nana
Uma 10h27'57" 50d42'5"
Nana
Psc 0h39'38" 4d9'13"
Nana
Psc 1h44'14" 6d1'18"
Nana
Cap 20h9'9" -26d10'7"
Nana
Sgr 18h1'53" -28d16'24"
Nana
Uma 10h59'55" 58d50'52"
Nana
Cas 1h31'27" 61d55'17"
Nana
Cas 0h59'53" 65d34'57"
Nana
Cas 23h9'4" 59d16'20"
Nana
Uma 11h34'44" 63d58'24"
Nana
Aqr 23h3'50" -8d10'50"
NANA
Aqr 23h2'51" -4d52'24"
Nana 75
Tau 4h27'4" 13d37'23"
Nana and Pap Oblack
Cyg 21h41'11" 37d1'53"
Nana and Pap Pap Wagner
Crb 15h35'7" 28d13'36"
Nana B
Aqr 22h33'27" -22d28'55"
Nana Bailey
Lyn 7h29'29" 44d56'29"
Nana Banana Whalen
Tau 4h27'28" 16d49'24"
Nana Becker
Ari 1h54'32" 18d21'6"
Nana Bennett
Cma 6h38'23" -17d46'20"
Nana Brady Peacock 104
Cam 4h11'6" 67d33'24"
Nana Bubba
Leo 11h16'4" 15d50'58"
Nana Castle
Aqr 22h0'19" 1d41'13"
Nana Chanin
Cas 2h43'29" 65d14'4"
Nana Chris Triano
Uma 10h34'26" 44d41'9"
Nana Daisy
Cas 0h3'3" 52d43'35"
"Nana" - Deanna Lynn Shepard
Gem 7h41'25" 32d46'34"
Nana Gay-Gay Richey
Her 16h49'10" 20d37'11"
Nana Grace
Lib 14h51'30" -3d35'8"
Nana & Gramps
Pho 23h58'20" -48d59'48"
Nana & Grandy, Mum & Dad, Win & Les
Boo 14h44'29" 44d9'10"
Nana Haynes
Lyr 18h51'3" 33d46'46"
Nana Helen and Grandpy Henry
Uma 10h7'12" 50d16'35"
Nana I
Lyr 18h43'4" 31d12'49"
Nana Iris
Nor 16h8'44" -42d54'54"
Nana Ivy Briggs
Cap 20h27'30" -22d46'8"
Nana Karen
Cas 1h41'30" 63d10'57"
Nana Kaye Zani
Uma 11h24'14" 36d14'13"
Nana Linda
Uma 8h12'50" 60d50'46"
Nana Linda Fay
Aql 19h21'33" 0d43'18"
Nana Lorraine Bernice
Aqr 20h46'49" -11d20'34"
Nana Louise
Cas 0h11'50" 51d40'23"
Nana Louise Hlewicki
And 0h30'16" 27d13'33"
Nana Lynda's Loving Star
And 0h28'21" 40d25'5"
Nana Mabel Ablett 1917
Aqr 21h20'37" -2d31'21"
Nana Margaret's Shining Star
Crb 15h48'43" 27d23'28"
Nana Marie
Cap 21h44'59" -24d44'21"
Nana Marilyn "2-1-40"
Crb 15h38'7" 36d1'29"
Nana Marilyn Sue
Sgr 19h36'49" -15d7'32"
Nana Marjorie
Uma 11h8'51" 47d35'16"
Nana Mary
Sco 16h5'20" -14d0'1"
Nana & Mother
Psc 1h42'50" 14d14'16"
Nana - Our Cross, Our Light
Cru 12h13'42" -63d18'13"
Nana & Pada Shankin
Ori 6h7'52" 7d27'34"
Nana & Papa Sota
Per 3h11'28" 46d12'3"
Nana & Papa Tepe
Gem 7h16'7" 14d7'9"
Nana & Papa Trama
Uma 8h58'55" 57d25'51"
Nana Pellecchia
Tau 4h49'24" 28d11'7"
Nana Poo
Leo 9h50'1" 29d39'30"
Nana & Poppo Barrons
Cyg 20h27'38" 52d35'2"
Nana & Pop-Pop
Uma 13h37'28" 59d2'37"
Nana & Pop-Pop Zerrenner
Uma 11h22'42" 31d43'22"
Nana Rose's Star
Ari 2h10'28" 13d4'49"
Nana S. B.
And 1h22'54" 42d59'32"
Nana (Sam) and Pop (Hank)
Uma 11h55'17" 56d28'0"
Nana Satterfield
Cyg 20h10'11" 36d40'15"
Nana Sue Callon's Star
Cmi 7h32'45" 9d55'14"
Nana Sue star
Sgr 18h58'24" -34d20'24"
"Nana" Susan Avril Hancock
Cas 0h37'59" 67d14'16"
Nana Turns 60
Cas 23h9'8" 56d48'16"
Nana Wickman
Aql 19h57'18" 0d29'0"
Nana Zagerman
Leo 11h20'2" 17d5'8"
Nana, Jeff and Jake's Star of Love
Uma 11h8'11" 32d57'16"
Nana, Poupee D Amour Remillard
Cas 0h13'14" 60d47'8"
Nanabug
Lib 15h39'50" -7d39'39"
Nana-Gail
Gem 7h52'14" 31d16'49"
Nanakumimaru
Tau 5h19'26" 16d59'58"
N'Anamour
Cnc 8h22'36" 14d29'1"
Nanan Hélène
Gem 6h54'5" 26d55'29"
Nanandella Princess
Tau 5h54'34" 26d46'26"
Nana-N-Papa
Uma 9h24'14" 45d34'22"
Nana's Forever Shining Star
Vir 13h26'59" 2d15'26"
Nana's Heart
Peg 23h35'53" 24d39'23"
Nana's Heart
Uma 10h0'27" 48d23'40"
Nana's Hooligan
Gem 6h44'59" 20d56'25"
Nana's Love
Cyg 19h57'41" 34d23'0"
Nana's Monkey
Lyr 18h43'38" 38d35'11"
NANA'S Poppy
Lib 15h20'0" -9d28'50"
Nana's Rana Stella
Lyr 18h40'22" 32d4'47"
Nana's Shining Star
Cyg 21h13'27" 52d51'1"
Nana's Star
Sco 17h56'45" -41d13'54"
Nana's Star
Cas 23h34'41" 51d1'15"
NaNa's Star
Cam 5h36'0" 58d28'13"
Nana's Star
Psc 23h26'53" 7d8'52"
Nana's Star
Vir 12h43'13" 8d2'43"
Nana's Star (Mary Petrilli)
Cap 20h31'40" -19d18'32"
Nanashi
Uma 11h45'51" 40d59'35"
NanaV24
Dra 15h23'41" 60d11'3"
Nanaw Miller
Aqr 20h51'35" -7d30'41"
Nanc
Tau 3h59'55" 27d49'5"
Nanc Colleen Dominguez Adams 2005
Aqr 20h41'13" -13d19'40"
NanC & Tim Zimmerman
Gem 6h5'39" 27d30'32"
Nancaizja Duquaine
And 0h52'52" 37d57'27"
"Nance"
Uma 10h26'17" 62d14'31"
Nance "Namonahan"
Ori 5h35'15" 13d23'13"
Nanci
Psc 0h54'27" 14d49'50"
Nanci Anderson
Uma 8h57'54" 62d46'23"
Nanci Bennett
Cyg 21h30'25" 53d31'44"
Nanci Carys Morgan
And 23h22'50" 51d18'33"
Nanci DiGiaimo
Lyn 9h31'45" 40d31'42"
Nanci Elizabeth Francis
Lyn 8h58'45" 39d30'35"
Nanci J. Roth
Cam 5h24'24" 61d35'26"
Nanci Ruth
And 2h19'10" 47d9'38"
Nanci Stevenson Gergler
Uma 8h56'27" 50d52'55"
Nancie Lynn Morriss
Psc 23h16'0" 6d0'53"
Nancie Skrocki
Lib 15h38'4" -6d28'4"
NanciMiltonFitterman EarthPoet52957
Gem 6h42'12" 14d46'59"
Nanci's Pegasus
Peg 22h28'38" 25d27'50"
Nancita
Umi 15h11'16" 79d17'21"
Nancy
Cet 1h24'7" -0d56'55"
Nancy
Dra 18h22'32" 78d55'59"
Nancy
Lib 15h23'38" -11d26'25"
Nancy
Cas 0h19'34" 64d1'50"
Nancy
Sco 17h56'7" -30d18'9"
Nancy
Peg 23h58'12" 23d45'25"
Nancy
Crb 15h44'2" 28d15'53"
Nancy
Leo 11h42'5" 24d21'41"
Nancy
Gem 7h39'50" 23d6'36"
Nancy
Gem 7h34'11" 26d57'47"
Nancy
Com 13h12'58" 21d7'11"
Nancy
Ari 2h44'43" 26d32'19"
Nancy
Tau 4h25'23" 19d12'39"
Nancy
Ari 3h0'59" 22d8'45"
Nancy
Ari 2h28'56" 19d41'8"
Nancy
Tau 3h48'25" 7d5'57"
Nancy
Cam 5h25'18" 56d54'9"
Nancy
Cyg 19h45'15" 38d11'42"
Nancy
Uma 11h32'19" 41d6'33"
Nancy
And 0h11'14" 43d42'0"
Nancy
And 0h37'12" 43d52'22"
Nancy
And 0h42'56" 41d8'12"
Nancy
And 0h4'15" 40d54'39"
Nancy
Cyg 19h37'49" 34d55'27"
NANCY
Cyg 19h36'35" 31d54'53"
Nancy
Tri 2h0'31" 32d31'58"
Nancy 1951
Cas 2h37'10" 67d11'58"
Nancy '73
Cap 21h51'53" -10d52'24"
Nancy 80
Cas 23h21'0" 56d32'37"
Nancy A. Anderson
Cyg 20h41'24" 35d9'48"
Nancy A. Ferguson " NAF "
Cyg 21h16'46" 45d13'23"
Nancy A. George - Universal Buddy
And 2h20'9" 45d47'6"
Nancy A. Hamel
Leo 9h25'5" 17d10'19"
Nancy A. Heintz
And 23h9'38" 51d19'12"
Nancy A. Moran
Uma 8h59'7" 62d8'18"
Nancy A. "Nan-ny" Freund
Sgr 19h14'51" -16d57'23"
Nancy A Prather
Ari 2h50'39" 26d19'34"
Nancy A. Rand 40th, 2005
Cnc 8h55'1" 13d27'59"
Nancy A. Ranew "1955-2006"
Gem 6h37'57" 19d31'31"
Nancy A. Rodriguez
Gem 6h31'50" 24d41'22"
Nancy A. Rogers
Ori 4h47'2" -0d4'56"
Nancy A Weisner
Cas 0h38'7" 61d40'29"
Nancy Abarza
And 0h43'21" 40d37'25"
Nancy Abellera
Crb 15h51'48" 29d5'58"
Nancy Alario
Lib 15h25'0" -4d54'19"
Nancy Albertson
Cyg 21h58'47" 46d37'39"
Nancy Alexia
Uma 13h43'50" 58d43'45"
Nancy Alvarez
And 0h40'23" 34d35'1"
Nancy Ames Cole
Uma 10h50'11" 57d34'57"
Nancy and Donald Barry
Lmi 10h44'49" 38d10'51"
Nancy and Girls
Uma 12h44'7" 59d16'34"
Nancy and Joseph Armbruster
Cas 23h48'57" 57d40'36"
Nancy and Richard Norling
Cyg 21h20'3" 31d24'37"
Nancy and Stephen Scaturro
Cyg 21h10'9" 48d20'33"
Nancy and Tex
Per 2h25'51" 52d25'43"
Nancy and the Boyz
Lmi 10h16'3" 36d17'33"
Nancy and Wendell Gengler
Cyg 19h46'11" 56d49'57"
Nancy Ann
Lyn 6h19'39" 57d35'55"
Nancy Ann
And 23h2'27" 41d44'15"
Nancy Ann
Ari 2h33'40" 25d47'5"
Nancy Ann Barsocchini
Leo 11h4'40" 8d12'42"
Nancy Ann Burleson
Cyg 21h13'35" 38d22'30"
Nancy Ann Cheek Foss
Sco 17h49'53" -44d4'37"
Nancy Ann Hamel
And 0h1'38" 36d59'40"
Nancy Ann Hill
Lyn 7h13'15" 58d35'46"
Nancy Ann - Joseph (sparky)
Ori 5h47'26" 7d57'52"
Nancy Ann Keane
Uma 10h25'50" 46d15'13"
Nancy Ann Mueller
Cnc 9h16'41" 13d20'12"
Nancy Ann Mullen
Lib 15h11'15" -24d56'16"
Nancy Ann Padden's Guiding Star
Cas 0h32'35" 56d33'28"
Nancy Ann Patricia Romeo
Psc 1h9'25" 26d49'33"
Nancy Ann Scherer
Cyg 19h48'27" 52d35'55"
Nancy Ann Simas
Vir 13h15'26" -3d36'46"
Nancy Ann T. Schulte
Uma 10h54'49" 36d18'41"
Nancy Ann Wilson
Cas 0h40'54" 63d51'53"
Nancy Anne Flynn
Ari 2h7'2" 22d17'19"
Nancy Anne Johnson
Ori 5h12'38" -0d2'0"
Nancy Anne Tannahill's Star
And 23h52'19" 44d59'13"
Nancy Argueta
Dra 18h32'56" 72d38'37"
Nancy Arlene Lutz
Lib 14h35'32" -14d31'35"
Nancy Arlene Martin
Aqr 22h34'42" -18d28'51"
Nancy Arthur
Ari 3h13'26" 20d36'7"
Nancy Ballentine
Cyg 19h22'10" 50d6'12"
Nancy Barbara Cowie
Umi 16h7'57" 70d51'54"
Nancy Benco's Star
And 1h9'0" 38d21'3"
Nancy Bennett
And 23h6'11" 51d27'54"
NANCY BERGEMAN
Crb 16h14'58" 37d45'20"
Nancy Berneking
Uma 11h45'52" 46d6'15"
Nancy Bernice Ulatowski
Cas 1h22'53" 51d40'58"
Nancy Beth Evans
Crb 16h13'47" 32d6'20"
Nancy Beth Wolf
Uma 9h26'36" 69d24'27"
Nancy Bickford
Cas 1h32'20" 64d49'28"
Nancy & Bill's 60th. Ann. Star
Cyg 20h20'24" 41d1'11"
Nancy Blanchard
Lyr 18h35'33" 32d47'2"
Nancy Boydle
Ori 6h5'14" 12d51'50"
Nancy Brewer
Cyg 19h50'15" 33d34'19"
Nancy & Brian together forever
Cyg 20h23'40" 47d0'58"
Nancy Brown
Ari 3h8'45" 21d57'6"
Nancy Bryant
Lyr 18h27'34" 31d34'26"
Nancy Bui
Crb 15h34'15" 26d1'5"
Nancy Bunting Cline "Superstar"
And 0h42'23" 33d57'40"
Nancy Burke
Cnc 9h8'58" 29d56'47"
Nancy Burrows
Cas 2h39'35" 65d23'15"
Nancy Butera
Uma 11h47'43" 29d25'47"
Nancy Byams
Cam 5h42'23" 61d29'37"
Nancy C. Smith
Cyg 20h28'29" 50d37'8"
Nancy Cafferty
Lmi 10h13'49" 41d0'15"
Nancy Capriola
Leo 11h51'23" 23d39'43"
Nancy Carol
Peg 22h12'29" 16d23'56"
Nancy Carol Athanasiadis
Leo 10h19'10" 7d10'36"
Nancy Carol "Boo"
Cas 23h47'1" 54d0'59"
Nancy Carter
Uma 10h3'1" 71d4'49"
Nancy Casseopia Hanson
Cas 23h4'42" 58d44'4"
Nancy Catherine
Ari 2h7'2" 25d31'4"
Nancy Cavett
Del 20h39'24" 15d38'53"
Nancy Charlene
Cas 2h26'23" 72d24'51"
Nancy Charlotte Van Brunt
Cyg 21h45'45" 45d55'34"
Nancy Clare Stern
And 23h56'25" 47d23'41"
Nancy Clark
Leo 11h45'50" 22d31'21"
Nancy Collier
And 1h8'12" 40d59'1"
Nancy Colsa
Cas 0h11'5" 55d50'24"
Nancy Conneely
Cas 0h45'36" 57d18'49"
Nancy Cook Catando
Cas 0h37'35" 60d34'52"
Nancy Coulbourn
Psc 1h12'59" 12d2'7"
Nancy Crotta
Cas 1h26'32" 61d22'13"
Nancy Crotty
Aqr 22h44'41" -0d59'17"
Nancy D. Burt, Our mother and hero
Gem 7h5'17" 17d34'26"
Nancy D. Neuman
And 23h27'58" 42d0'30"
Nancy D. Pless
And 0h42'32" 26d47'5"
Nancy D'Amico
Mon 7h17'33" -0d34'42"
Nancy & David's Star
Cyg 21h27'49" 53d2'17"
Nancy Dawn Gager
Aql 19h39'55" 8d28'31"
Nancy Dee
And 2h20'45" 46d14'1"
Nancy Dee Burt
Gem 6h35'49" 12d9'55"
Nancy Denise Lawler
Leo 11h19'45" 16d59'1"
Nancy Derbyshire (my Einstein)
Gem 6h53'41" 14d57'50"

Nancy Donahue
Gem 7h33'48" 22d25'14"
Nancy Dopman
Umi 16h17'30" 75d35'32"
Nancy Dorsey
Sco 16h5'20" -15d19'55"
Nancy Dorsey
Sco 17h41'43" -39d42'1"
Nancy Doyle's Star
Ari 2h13'40" 24d2'53"
Nancy Driscoll
And 23h24'22" 43d52'14"
Nancy Duncan
Lyn 6h50'53" 56d32'21"
Nancy Dutton Lemos
Sgr 17h52'49" -18d4'54"
Nancy E. Gustafson
Lyn 8h25'5" 57d2'54"
Nancy E. Tomaszewski
Sco 17h53'50" -35d46'7"
Nancy E. Woolger
Lyn 7h9'27" 58d30'12"
Nancy Eberhardt
Cnc 8h17'32" 20d59'55"
Nancy El Ahmar
Mon 6h52'14" -5d43'7"
Nancy Elaine
Tau 4h43'53" 17d34'39"
Nancy Elaine Klann
Lyn 7h55'3" 38d43'41"
Nancy Elaine Patterson
Lib 15h5'12" -29d14'45"
Nancy Elizabeth
Lyr 18h39'27" 34d8'15"
Nancy Elizabeth Hurt
Lyr 19h26'24" 37d32'23"
Nancy Elizabeth Hyatt
And 1h9'5" 35d34'51"
Nancy Elizabeth Robbins
Cnc 8h54'2" 7d10'34"
Nancy Ellen
Lyn 9h5'39" 33d19'48"
Nancy Ellen
And 0h16'4" 43d10'10"
Nancy Ellen
Umi 17h2'57" 80d17'22"
nancy ellen
Uma 9h42'40" 67d35'9"
Nancy Ellen Cota
Cap 20h31'26" -20d52'3"
Nancy Ellen Mark
Cas 0h26'14" 61d16'4"
Nancy Ellen Widfeldt Laeha
Uma 11h12'54" 52d10'37"
Nancy Emmaline Lentz
And 1h33'9" 41d58'26"
Nancy Erin Harper
And 23h33'36" 41d39'51"
Nancy Fallon
Cas 0h33'35" 62d47'46"
Nancy Farina
And 0h40'7" 40d23'53"
Nancy Fernandez
Com 12h41'47" 22d50'27"
Nancy Fife - Shining Star
Umi 15h24'55" 67d33'15"
Nancy Flippen
Sco 16h13'9" -16d2'57"
Nancy Florence Jeske
And 2h36'25" 44d36'7"
Nancy Fodera
Vir 12h56'1" 12d48'53"
Nancy Fox Schweiger
Uma 10h29'58" 41d17'51"
Nancy Franovich Guliuzo
Cap 20h55'51" -25d52'37"
Nancy Frazier Erb
Mon 7h22'27" -0d24'38"
Nancy Freifeld
Leo 9h33'27" 29d27'3"
Nancy G.
Sco 16h38'47" -37d20'34"
Nancy G. Aufenanger
Gem 6h46'42" 17d32'33"
Nancy G. Nazelrod
Lyr 18h27'8" 31d45'18"
Nancy Garcia
And 23h18'43" 52d11'8"
Nancy Gassert
And 2h16'41" 48d32'3"
Nancy Gayle
Sco 17h35'47" -43d58'19"
Nancy George
Cyg 21h29'27" 54d14'18"
Nancy & George Wetherington
Uma 10h32'45" 71d6'11"
Nancy Ghazarian
Ori 5h36'13" 10d48'50"
Nancy Gosselin
Per 3h25'47" 53d34'42"
Nancy Grace
And 0h19'24" 31d4'20"
Nancy Grace
Gem 7h8'26" 27d31'45"
Nancy Grace Welborn
Cap 20h48'0" -25d47'16"
Nancy Grayhek
Lyn 7h53'39" 56d37'1"
Nancy Griffin Hogarth
Sgr 18h11'7" -25d19'25"
Nancy Grubbs Hicks Nelson
And 23h12'14" 36d51'8"
Nancy Guarnieri
Cyg 19h20'46" 54d37'1"
Nancy Guevara
And 23h13'23" 48d19'23"
Nancy Guillen
And 0h20'42" 43d58'15"
Nancy Gurrola-Dixon
And 0h21'29" 34d43'0"
Nancy H.
Aql 19h12'23" 5d16'3"
Nancy H. Force
Peg 23h10'56" 30d44'12"
Nancy H Tran
Cap 20h23'35" -11d9'14"
Nancy Halladay Kaufman
Tau 5h36'57" 25d21'52"
Nancy Havens Geyman
Lib 14h48'36" -6d15'1"
Nancy Healy
Aur 6h28'10" 41d5'27"
Nancy Hebert-Conners
Oph 17h8'38" -0d9'12"
Nancy Helen Morrow-Campbell
Aqr 22h5'50" 2d4'9"
Nancy Hensel Hagy Hinkel
Cnc 8h32'36" 12d58'38"
Nancy Her
Ari 2h58'50" 11d4'8"
Nancy Hernandez
And 0h18'18" 35d35'46"
Nancy Hoffman, S.C.
Lyr 18h49'59" 43d28'26"
Nancy Hogan
Cas 1h24'57" 58d57'6"
Nancy Hopping
Cyg 20h13'45" 50d1'2"
Nancy Horwitz
Gem 6h22'46" 24d21'6"
Nancy Hwang
Peg 23h13'30" 20d59'0"
Nancy Irene LaFont
And 0h25'31" 32d48'17"
Nancy J.
Lib 15h7'23" -27d26'17"
Nancy J. Gavin
Cap 21h47'36" -24d4'31"
Nancy J. Mosca
Tau 5h38'30" 25d52'34"
Nancy J. Nava
Psc 0h42'12" 15d21'25"
Nancy J. Rettinger
Cas 2h29'56" 67d15'9"
Nancy J. Sanger
And 23h21'35" 38d18'2"
Nancy J. Turner
Ari 2h20'29" 13d33'17"
Nancy J. Westmoreland
Psc 0h43'30" 8d10'46"
Nancy James & Hazel Hanson
Per 3h20'50" 48d36'13"
Nancy Jane Alberga
Uma 13h35'27" 59d16'49"
Nancy Jane Auciello {Mom's Star}
Uma 10h34'34" 47d21'0"
Nancy Jane Delicious
Aqr 22h11'32" -0d19'33"
Nancy Jane Lampton
And 0h17'26" 36d11'13"
Nancy Jane McConnell Baldwin
And 1h35'32" 42d40'39"
Nancy Jane Norris
Cas 1h56'13" 61d52'27"
Nancy Jane Stacy
Ari 2h31'32" 25d17'23"
Nancy Jane Vap
Aqr 22h25'21" -7d26'39"
Nancy Jayne Zubaty
Cnc 8h36'5" 24d1'47"
Nancy Jean
Tau 5h29'15" 21d54'14"
Nancy Jean
Cyg 19h44'11" 31d44'58"
Nancy Jean
Vir 11h46'40" -0d18'7"
Nancy Jean
Cap 20h15'1" -17d51'5"
Nancy Jean of Crofton
Sgr 18h13'41" -27d57'10"
Nancy Jean Taplin
Psc 1h33'9" 27d31'49"
Nancy Jean TJ
Lib 15h9'19" -16d30'38"
Nancy Jeanne Fadely
Ori 5h27'13" -0d2'51"
Nancy Jo
Lib 14h53'26" -4d41'52"
Nancy Jo Reynolds
Ari 2h12'5" 23d38'6"
Nancy Jo Smith
And 0h31'8" 33d28'2"
Nancy Jo Wojnar
Gem 6h10'2" 25d31'55"
Nancy Joan Spivey
Ori 4h51'9" 8d16'38"
Nancy & Johnny Together Forever
Sge 19h42'42" 18d0'37"
Nancy Joy
Crb 15h35'42" 37d17'44"
Nancy Joy Cerve
Del 20h48'33" 14d50'40"
Nancy Joy Filippo-FF & P
Vir 14h19'27" 0d12'12"
Nancy Joy Smith
Crb 16h12'2" 32d59'27"
Nancy Joyce Vila
Ori 5h11'44" -1d25'12"
Nancy June King
Crb 15h50'43" 26d16'22"
Nancy June Tkacs Fragala
Cyg 21h10'23" 51d48'9"
Nancy K. Aukeman
Aql 18h51'46" -0d52'39"
Nancy K. Dieterich
Cyg 20h46'13" 31d10'8"
Nancy Karason
Psc 1h22'17" 23d17'6"
Nancy Karcher
And 1h42'2" 38d9'0"
Nancy Karen O'Shanick Lefler
Lib 15h32'35" -6d12'52"
Nancy Karla Lopez
Cyg 20h34'33" 49d26'34"
Nancy Kaufmann Garton
Cas 1h23'33" 65d49'20"
Nancy Kay
Sgr 18h50'13" -15d54'59"
Nancy Kay Gautrey
Cyg 21h58'8" 46d23'0"
Nancy Kay Seifert
Cap 21h35'17" -9d11'51"
Nancy Kelley
Crb 15h47'25" 33d28'41"
Nancy Kesselman
Cap 21h30'11" -9d36'59"
Nancy Kielich
Crb 16h5'28" 29d48'4"
Nancy & Kirk Dyer
And 23h25'45" 51d24'44"
Nancy Klauer
Cnc 8h57'59" 24d28'26"
Nancy Klein
And 23h23'47" 49d6'44"
Nancy L. Auffrey
Eri 4h32'43" -1d18'3"
Nancy L. Clift
Cnc 8h19'24" 28d56'2"
Nancy L. Crocker
Lyn 6h33'30" 61d34'0"
Nancy L. Kreisler "Brightest Star"
Cap 20h57'29" -24d54'46"
Nancy L. Leibold
Aqr 23h18'16" -19d46'18"
Nancy L Tennison
Cap 20h26'58" -18d55'36"
Nancy L. Yocono
Sgr 19h46'28" -12d14'50"
Nancy Lafferty
And 23h48'28" 37d36'49"
NANCY LALONDE
Crb 15h56'2" 34d11'21"
Nancy Laura LaRue
Ori 6h11'0" 7d35'10"
Nancy Lee
Vir 11h52'13" 9d0'26"
Nancy Lee
Cyg 21h45'20" 36d42'4"
Nancy Lee Aka Dog Blu
Sgr 18h47'20" -22d56'13"
Nancy Lee Ball
Lyn 8h41'28" 43d51'55"
Nancy Lee Carroll
Cas 0h47'48" 58d0'15"
Nancy Lee Ewing
Uma 10h18'14" 69d52'13"
Nancy Lee Genzel
Cam 6h11'48" 76d46'17"
Nancy Lee Heiberger
Cyg 21h17'26" 32d18'41"
Nancy Lee Lohr Pyle
Vir 12h15'47" -0d54'58"
Nancy Lee Parsley
Gem 7h41'15" 32d53'0"
Nancy Lewis
Sco 17h47'56" -32d57'34"
Nancy Linn Van Houten - SG
Tau 5h16'41" 24d11'51"
Nancy (Little Red) Arrington
And 1h44'21" 45d26'51"
Nancy Liu
And 2h12'19" 38d12'45"
Nancy Lizette Barrera
Sgr 18h37'42" -35d36'52"
Nancy Locke Smarinsky
Crb 15h43'28" 32d39'38"
Nancy Lou Ross
Sco 17h54'27" -41d47'43"
Nancy Louise Compton
Lib 14h51'38" -3d40'9"
Nancy Louise Kathryn Hart
Lyn 6h41'45" 56d5'26"
Nancy Louise Miller "01/16/1957"
Cap 21h19'48" -19d44'41"
Nancy Lourdes
Cas 1h21'51" 61d21'41"
Nancy Lucadamo
Cas 1h38'14" 66d51'53"
Nancy Lynn
Sco 17h56'35" -40d26'25"
Nancy Lynn
Cas 0h49'25" 56d58'39"
Nancy Lynn Everson
Cas 1h38'33" 62d20'18"
Nancy Lynn Finley
Vir 12h22'50" -10d36'3"
Nancy Lynn Johnson
Ari 2h11'33" 10d52'58"
Nancy Lynn Smith Snyder
Cyg 20h1'59" 39d17'57"
Nancy M
Cas 0h23'28" 57d50'50"
Nancy M Gregorio
Cas 1h2'34" 58d9'15"
Nancy M. O'Toole
Lib 15h17'3" -9d16'39"
Nancy M Pencsak
And 0h18'44" 28d18'28"
Nancy M. Skirchak
Cnc 8h21'32" 19d49'35"
Nancy M Standard
Gem 6h34'18" 27d22'48"
Nancy Mae \ Nonna
Crb 16h9'34" 36d49'55"
Nancy Malanaphy
Cas 1h52'58" 61d41'34"
Nancy Malarney
Ari 2h33'18" 25d11'44"
Nancy Malito
Sco 16h58'4" -44d51'12"
Nancy Mandile
Leo 9h52'58" 28d2'39"
Nancy Marie
Cas 1h42'38" 63d36'32"
Nancy Marie
Vir 13h58'33" -10d46'49"
Nancy Marie Alamo
Cas 23h59'33" 56d24'55"
Nancy Marie Booth Fenton
Cas 0h14'25" 62d52'5"
Nancy Marie Hartman
Aqr 22h11'48" 0d27'39"
Nancy Marie Hartung
And 23h11'20" 43d36'50"
Nancy Marie Unger
Gem 6h2'29" 27d55'44"
Nancy May
And 1h34'2" 50d1'26"
Nancy May
Lyn 8h10'46" 47d6'35"
Nancy May Easton
Sco 16h6'12" -14d31'57"
Nancy May Romani
And 23h45'28" 46d1'15"
Nancy Mayo's Mothers Day Love Jeff
Umi 15h31'43" 76d7'28"
Nancy McBride
Uma 9h36'42" 53d10'26"
Nancy McCullar
Leo 10h7'44" 20d14'2"
Nancy McGonagle August 3, 1951
And 23h38'41" 47d11'42"
Nancy McKay Whyte
Tau 4h38'51" 25d34'17"
Nancy "Midmom" Ballister
And 23h8'38" 42d50'3"
Nancy & Mike Heightchew
Cyg 20h17'19" 38d37'27"
Nancy Miller MacKay Green
Sge 19h51'29" 16d53'35"
Nancy Moore
Ori 5h44'45" -0d46'39"
Nancy Morais
And 1h35'40" 46d20'34"
Nancy Morgan
Cnc 8h45'32" 8d27'47"
Nancy Morrison Ditto
Sco 17h47'54" -33d57'26"
Nancy Mullins
Cnc 8h7'5" 18d49'1"
Nancy " My Mom " Brandon
Cas 1h6'4" 49d35'2"
Nancy N Huang
Ori 5h35'20" 7d59'32"
Nancy Nanh Bui
Leo 10h27'13" 22d10'8"
Nancy "Nannygoat" Djernes
Cas 0h45'24" 52d11'53"
Nancy Nelson Davis
Sgr 18h32'46" -18d30'25"
Nancy Nichols
Lib 14h31'55" -9d45'49"
Nancy Nicoll, Math Teacher
Cas 1h31'46" 62d50'37"
Nancy Noll
Leo 9h50'31" 23d9'24"
Nancy Northrup
Mon 7h38'48" -0d30'7"
Nancy Northway
Uma 8h18'29" 61d30'46"
Nancy Notto
Lib 14h52'55" -4d2'32"
Nancy Nyustem Breton
Lyn 8h21'15" 54d33'44"
Nancy of Norway
Ari 3h28'51" 23d0'56"
Nancy Orozco
Leo 9h29'23" 10d30'33"
Nancy Packard - Worlds Greatest Mom
Tau 3h46'39" 24d1'47"
Nancy Palmieri
Uma 9h32'9" 49d43'39"
Nancy Patterson
Cas 1h28'31" 52d7'45"
Nancy Paula Davis
Sgr 18h2'44" -29d38'43"
Nancy "Peanut" Cusack
Aqr 20h49'56" -13d15'1"
Nancy Petitta
Cas 1h45'54" 61d11'55"
Nancy Phyliss Lyvers
Uma 9h49'31" 47d34'29"
Nancy Pierce
Lyn 8h16'26" 36d24'21"
Nancy Pilkerton
Cnc 9h14'4" 17d1'29"
Nancy Piotrowski
Cnc 8h10'50" 9d7'59"
Nancy Pisaruk
Mon 7h18'42" -4d41'54"
Nancy Poff
Uma 10h40'51" 53d2'54"
Nancy R. Markland
Cap 21h7'3" -27d24'53"
Nancy R. Megas
Uma 9h51'7" 65d46'44"
Nancy R Wygle
Com 13h5'37" 18d9'29"
Nancy Raman
Lyn 6h29'26" 56d46'29"
Nancy Randolph, R.N.
Ari 2h25'15" 25d4'43"
Nancy Reall
Uma 11h25'4" 70d41'3"
Nancy Renell Miller
Uma 11h48'18" 54d40'17"
Nancy Rezachek
Lyn 7h27'59" 53d42'3"
Nancy Richey-Deffendall
Sgr 19h37'3" -37d53'39"
Nancy Rierson
Aqr 20h50'27" 0d13'3"
Nancy Rivas
Per 4h24'49" 42d21'33"
Nancy Robinson
Ari 2h11'43" 22d52'27"
Nancy Rose
Ari 2h51'13" 28d35'37"
Nancy Rose Moore
Aqr 21h21'56" -11d10'15"
Nancy Rose Moroney
Per 3h9'53" 55d3'47"
Nancy Rotundo
Leo 10h59'8" 6d52'5"
Nancy Ruth Seitz
Ori 5h52'19" 8d42'18"
Nancy Shanken
Gem 6h17'3" 26d35'42"
Nancy Shemrah Fallon
Cas 0h6'0" 63d1'30"
Nancy Simmer
Dra 17h29'2" 59d17'57"
Nancy Simms Benson
Cas 0h19'15" 59d23'48"
Nancy Smith
Leo 9h55'9" 8d16'13"
Nancy So Miller
Lyn 8h15'30" 41d41'33"
Nancy Sparrow
Ari 2h2'2" 22d39'58"
Nancy Special
Uma 10h36'39" 40d13'37"
Nancy Stephenson Wettering
Vir 14h19'18" -20d42'26"
Nancy St.John
Vir 12h23'58" 10d37'8"
Nancy Sue Roper born August 8, 1952
And 23h4'51" 42d30'9"
Nancy Sue Stamm
Uma 11h42'28" 39d59'37"
Nancy - Susanne
Uma 9h26'11" 64d37'13"
Nancy Swiderski
Vir 13h5'28" -18d33'44"
Nancy Thagard
Sco 16h11'9" -22d43'12"
Nancy Toby Larrick
Uma 10h57'18" 72d17'58"
Nancy Toledo Salgado
Vir 13h26'44" -10d58'50"
Nancy Tomoko Welling
Uma 10h42'46" 68d10'38"
Nancy Torr Andis
Uma 11h43'1" 54d58'22"
Nancy Torres- Our Shining Star
Sgr 18h2'51" -35d57'12"
Nancy Treadwell Ross
Uma 9h36'32" 54d9'50"
Nancy -Truly, Madly, Deeply
Lib 15h41'1" -12d16'42"
Nancy Ululani Kalani
Mon 7h34'4" -0d23'20"
Nancy Virginia Weber
Cyg 21h11'23" 44d56'52"
Nancy Viscencio Lopez
Cnc 8h32'26" 23d30'40"
Nancy Vue
And 0h45'33" 38d53'2"
Nancy Waters
Psc 1h9'27" 25d37'0"
Nancy Weinberg Simon
Cyg 19h54'43" 38d30'46"
Nancy Wesselmann
And 1h1'27" 48d15'15"
Nancy Wherry
Gem 6h46'29" 34d10'32"
Nancy White
Gem 7h42'8" 31d58'14"
Nancy Wilhelm
And 1h27'21" 45d2'1"
Nancy Williams
Cas 0h40'29" 76d46'7"
Nancy Wills
Lyn 8h0'4" 37d33'5"
Nancy Worrall Boston
Per 3h32'13" 48d13'26"
Nancy Y. Piper
Uma 8h56'32" 51d48'15"
Nancy Yap
Aqr 21h37'36" -0d26'50"
Nancy Young
Aqr 22h44'37" 2d10'49"
Nancy, Forever In My Heart
Vir 13h27'43" -0d37'8"
Nancy8269
Leo 11h13'28" 11d32'16"
NancyAnn
Cnc 9h5'18" 27d35'9"
Nancycjennings
Cyg 20h25'50" 35d4'3"
NancyClaude
Uma 9h11'1" 54d35'22"
NancyLee
Dra 16h56'55" 58d14'10"
NancyLevineArmstrongThe Beautiful
Cap 20h28'42" -22d56'15"
NancyLou
Cyg 20h46'48" 43d47'29"
Nancy-Pants
Psc 1h54'33" 8d51'34"
Nancy's Angels Star
Lyr 18h45'44" 31d36'27"
Nancy's Chariot
Aqr 20h38'40" 0d51'48"
Nancy's Dream
And 0h20'31" 38d11'40"
Nancy's Glitter Wings
Lyr 18h48'28" 38d15'34"
Nancy's Shining Star
Uma 8h52'27" 52d1'35"
Nancy's Shining Star
Del 20h43'7" 15d55'46"
Nancy's Spot
Uma 8h38'20" 47d23'55"
Nancy's Star
Cyg 21h10'17" 49d13'5"
Nancy's Star
Lib 14h47'6" -7d43'37"
Nancys-Angel
Lyn 6h37'4" 60d3'20"
NancyYucius
And 0h40'25" 40d6'8"
Nanda
Cas 2h21'33" 74d58'35"
Nandan Van: A Garden of Heaven
Umi 15h31'40" 75d44'10"
Nandi L Branford
Cyg 21h15'9" 44d50'16"
Nandini Patel
Lib 15h42'45" -18d24'51"
Nando Eileen Jones
Aur 5h20'23" 30d11'50"
Nandranie Persaud
Her 16h55'28" 30d45'41"
Nane
Tau 5h14'7" 23d4'57"
Nanee
Uma 11h50'34" 35d52'35"
Nanethelen
Ori 5h12'52" 0d36'26"
Nanette
Vir 13h13'50" 5d17'27"
Nanette
Cnc 9h12'15" 25d2'5"
Nanette
Sco 17h56'13" -30d21'21"
Nanette Allis
Ori 6h10'39" 17d23'11"
Nanette Anita Burbank
Vir 13h42'5" 2d4'36"
Nanette Barbara Rodney Kelekian
Psc 1h18'30" 32d50'41"
Nanette Elizabeth
Cap 20h31'8" -9d57'56"
Nanette Goldstein
Psc 0h54'13" 32d30'15"
Nanette Kveder
Leo 10h9'27" 24d9'50"
Nanette Lavett
Sco 17h36'53" -42d58'44"
Nanette Lee
Uma 9h15'33" 47d25'36"
Nanette M. Skewis
Cap 20h38'18" -20d55'56"
Nanette McCarter Stanfill
And 0h31'2" 33d47'51"
Nanette Michelle
Lyn 8h49'44" 45d38'25"
Nanette Minahan
Lib 15h11'6" -1d8'54"
Nanette "Newt" Thomas
Sco 17h51'36" -37d50'28"
Nanette P. Benoit
Lib 14h52'8" -1d37'55"
Nanette Rebecca Allen Casteel
Cas 3h14'8" 59d26'54"
Nanette Roxanne McBride
Sgr 18h36'10" -23d52'52"
Nanette Schoeder
Tau 4h19'9" 17d44'51"
Nanette Thomas, ARC
Sgr 20h4'50" -30d15'31"
Nanettie
And 2h28'51" 47d58'51"
(Nang) Somchanh Detvongsa
Umi 14h33'3" 85d34'9"
Nangy
Uma 10h29'31" 69d15'5"
Nanha Tse
Lyn 8h20'12" 33d13'8"
nani
Lmi 10h24'42" 36d16'59"
Nani
Tau 5h44'33" 15d59'50"
Nani
Gem 7h52'48" 18d32'7"
Nani
Psc 0h35'25" 6d56'29"
Nani Naish
Ari 3h5'5" 12d36'58"
Nani & Papa's Little Angels
Cyg 20h23'7" 47d26'18"
Nani-1
Aqr 22h42'47" 2d13'8"
Nanie
Ari 2h50'44" 25d59'15"
nanieyra
Uma 8h32'47" 65d39'25"
Naniloa
Cyg 20h48'25" 33d11'55"
Naninouche
Cam 3h53'8" 56d6'15"
Nanise Malia
Cas 1h26'8" 61d54'42"
Nanita
And 0h11'26" 35d5'50"
NanJan
Lmi 10h7'50" 28d12'46"
Nanky 3631
Psc 0h3'46" 6d50'43"
Nanna Catherine
Lib 15h45'17" -28d46'38"
Nanna Elisabeth Hagard
Tau 4h13'0" 26d50'32"
Nanna - Lucky 19
Cas 23h40'9" 56d10'25"
Nanna Norma Atkins
Nor 15h59'28" -53d57'17"
Nanna & Pa Robinson
Cru 12h38'39" -59d51'35"
Nanna Pearson
Cas 2h21'28" 73d54'35"
Nanna Phyliss Casey
Pho 0h26'23" -46d31'39"
Nanna Vanggaard Hansen
Leo 10h26'34" 23d52'9"
Nanna - you will never dream alone!
Aqr 22h58'20" -10d52'23"
Nannan Lorene
Uma 11h30'40" 56d17'27"
NANNANS
Sgr 18h41'15" -30d48'33"
Nanna's Star (Phyllis Smith)
Cas 0h50'39" 61d23'42"
Nanner
Gem 6h50'5" 26d34'9"
Nannette Darlene Saunders
Lib 14h50'26" -4d47'34"
Nannette -My Shining Star-Nevarez
Aqr 22h52'38" -7d3'29"
Nannette Reeves
And 2h11'35" 38d54'14"
Nanni
Sco 16h7'46" -11d33'29"
"Nanni" ~ 100
Uma 9h6'5" 55d53'52"
Nannie
Lib 15h7'5" -6d31'21"
Nannie
Lyr 18h52'16" 34d50'10"
Nannie
Uma 9h41'38" 51d23'29"
Nannie Dunaway
Uma 8h54'32" 68d52'55"
Nannie Yates
Cas 0h8'10" 53d51'43"
Nannie's Star
Lib 15h10'19" -6d59'4"
Nanny
Ori 5h19'5" -4d41'34"
Nanny
Uma 11h16'41" 71d30'24"
Nanny
Cas 0h48'46" 62d45'13"
Nanny
Lib 15h39'0" -28d35'33"
Nanny
Psc 1h17'5" 3d38'5"
Nanny
Cas 0h39'5" 57d13'39"
Nanny
Uma 13h39'46" 48d19'5"

Nanny
Psc 23h56'15" 7d12'13"
Nanny
Vir 11h58'43" 5d21'10"
Nanny and CC
Crb 16h14'7" 33d54'15"
Nanny and Pappy
Cyg 20h23'40" 54d38'46"
Nanny and Poppy
Crb 15h31'21" 28d25'14"
Nanny Andryshak
Aqr 22h38'55" 0d32'46"
Nanny Ann Thornton
Aqr 23h4'32" -18d53'53"
Nanny Ginny
And 0h37'38" 41d41'49"
Nanny H
Cas 1h27'21" 60d28'32"
Nanny Jill
Cas 3h34'31" 70d25'51"
Nanny Joan
Cas 23h22'39" 54d51'14"
Nanny Jules
Cap 20h51'51" -25d13'49"
Nanny Margaret and Grandad Bill
Ori 5h17'5" -0d14'28"
Nanny Maria's Star
Cas 0h50'46" 73d26'47"
Nanny Marie Guthridge
Cas 23h28'18" 55d36'24"
Nanny Noodle
Leo 9h37'46" 26d45'31"
Nanny Olive and Grandad Bill
Uma 10h48'45" 63d34'6"
Nanny Parry
Cru 12h52'19" -62d3'38"
Nanny Patsy
Cas 22h59'0" 56d1'6"
Nanny & Pop - Pop Dwyer
Uma 11h29'23" 57d31'12"
Nanny Porcelli
Cap 20h11'4" -17d26'3"
Nanny Renee
Lyn 7h35'41" 36d24'28"
Nanny Ringshall's Star
And 1h53'48" 39d45'41"
Nanny Witch
Uma 9h39'8" 50d37'40"
Nanny's Star
Peg 22h44'17" 26d43'27"
Nano
Vir 14h31'49" 1d49'39"
Nano Nana
Sgr 18h4'5" -31d23'37"
Nanon M Williams, in my heart 4ever
Per 3h13'33" 37d43'25"
Nanoo
Tau 4h30'25" 27d48'29"
Nanook
And 1h48'2" 45d3'28"
Nanook Lieberman
Lyn 8h42'29" 39d4'6"
Nanou-Anne
Lib 15h6'27" -26d13'34"
Nan's Devine Light
Dra 19h29'17" 73d27'52"
Nan's Star
Cas 23h32'15" 58d29'20"
Nan's Star - Tina Nicoletta Mezzino
Cru 12h27'7" -60d14'16"
Nansen52
Uma 11h28'46" 32d23'42"
Nanstevia
Tau 5h34'47" 21d40'34"
Nany
Ori 6h15'46" 9d50'39"
Nao
Psc 23h39'37" 6d0'4"
Naocha naoi balluin dearga
Pyx 9h6'16" -34d53'42"
Naoelle
Lyn 7h58'38" 37d59'55"
Naohiro & Tomomi
Cnc 8h4'52" 15d4'20"
Naoko
Her 17h15'21" 16d24'9"
Naoko Abe star
Vir 13h34'42" -12d35'58"
Naoko Kougou
Cnc 8h19'39" 13d45'4"
Naoko Mazany
Uma 10h9'32" 70d4'3"
Naoko Shirane
Uma 11h40'31" 44d57'15"
Naoma Ford
Crb 15h19'16" 27d2'28"
Naoma Lee and Jerry V
Cyg 20h47'3" 43d55'42"
Naomi
And 23h23'14" 39d19'31"
Naomi
Cas 0h1'36" 56d34'30"
Naomi
And 23h4'34" 48d22'23"
Naomi
Aur 6h31'9" 39d3'36"
Naomi
Leo 10h33'56" 8d47'34"
Naomi
Ari 3h12'27" 22d4'43"
Naomi
Uma 9h50'2" 63d20'59"
Naomi
Lyn 6h30'16" 57d55'23"
Naomi
Vir 14h5'11" -14d18'3"
Naomi
Aqr 22h52'33" -5d30'7"
Naomi
Lib 15h27'30" -6d34'34"
Naomi
Sgr 19h33'8" -14d15'49"
Naomi 1
And 1h50'35" 46d0'10"
Naomi and Evan Forever
Gem 6h42'39" 35d8'53"
Naomi and Norihiro Miwa
Cyg 20h22'18" 36d53'57"
Naomi Ann Rhodes
Tau 4h26'21" 21d42'18"
Naomi Baker Starr
Sgr 19h41'29" -14d39'41"
Naomi Bliss & Neil Brooks Stambaugh
Lib 15h31'57" -19d46'47"
Naomi Bogdanovic
Dra 19h31'57" 65d23'59"
Naomi Boyington
Lib 14h57'47" -11d8'21"
Naomi Braswell
Aqr 20h41'20" -11d53'36"
Naomi Catherine Flood
And 23h41'24" 42d30'4"
Naomi Christianson
And 23h3'41" 41d12'5"
Naomi Collins
Cru 12h31'1" -60d21'42"
Naomi Elise Bacon
Psc 1h8'19" 20d58'55"
Naomi Fraser
Cyg 19h47'50" 40d46'55"
Naomi Himmelberger
Lac 22h27'49" 42d16'37"
Naomi Hope
Sco 16h10'50" -17d45'28"
Naomi Jade Morlan
Tau 4h3'42" 21d34'52"
Naomi Jane
Cru 12h32'32" -60d58'21"
Naomi Jasmine Pulido
Cap 21h18'54" -15d58'39"
Naomi Jones
Vir 11h47'16" -0d2'58"
Naomi Joon
And 0h37'9" 39d23'34"
Naomi Leventhal
Peg 22h59'26" 20d9'40"
Naomi & Lily
Aqr 22h40'37" -0d52'38"
Naomi Lou Brodoski, I love you baby
And 1h1'28" 45d33'16"
Naomi Louise Parry
Lyr 18h44'3" 39d51'8"
Naomi Love
Aqr 23h3'30" -21d9'30"
Naomi Lowe
Aqr 22h19'21" -13d26'34"
Naomi Macarthur
And 2h34'28" 44d36'50"
Naomi Marie Mautz
And 0h20'17" 33d19'19"
Naomi McGee
Col 5h59'1" -35d2'44"
Naomi McVay
Pho 0h40'52" -41d8'15"
Naomi Milburn
And 0h10'16" 43d10'34"
Naomi Miranda
Sgr 18h14'14" -32d7'53"
Naomi Miriam
And 0h29'26" 43d54'26"
Naomi (Mummy)
Cas 1h8'38" 51d0'2"
Naomi P Bear
Lib 14h26'31" -19d35'21"
Naomi Rachel
Leo 11h42'10" 18d9'43"
Naomi Robertson
Lib 15h18'32" -6d12'31"
Naomi Rose Ghen-Trachtenberg
Cas 1h22'24" 64d14'11"
Naomi Roxy Bernstein
Sco 16h7'30" -11d24'8"
Naomi Sarah
Cnc 9h13'55" 20d33'50"
Naomi Sarna
Cas 23h59'32" 56d17'50"
Naomi Shaley
Lyr 18h19'50" 44d24'31"
Naomi Soleil
Cas 23h50'14" 52d52'42"
Naomi Soraya van de Vliert
Sgr 18h26'46" -31d50'24"
Naomi Tate
Leo 11h23'30" 11d33'36"
Naomi Wallis
Gem 7h33'25" 16d9'31"
Naomi Wils and George Burella
Cyg 19h29'12" 52d44'3"
Naomi Woloshin
Lyn 8h4'35" 41d7'38"
Naomi Young
Uma 9h36'48" 56d59'21"
Naomi's Everlasting Star
And 2h27'2" 43d46'31"
Naomi's Piece Of Heaven
Leo 9h55'50" 14d2'41"
Naomi's Star
Aur 5h21'45" 30d12'19"
Naomi's Star
Lib 15h42'57" -27d44'40"
Naomi's Twinkle
And 1h15'53" 42d48'8"
Naomistéphane
Gem 6h24'56" 26d23'15"
Napierkowski
Sgr 18h11'18" -28d40'13"
Napierkowski's D Major
Umi 10h40'3" 89d3'55"
Napirai-Stern-01-07-1989
Her 16h43'6" 5d47'36"
Napló
Sgr 17h55'43" -18d8'28"
Napolitano Pack 335 Guiding Star
Uma 8h54'37" 48d4'13"
Nappy
Peg 22h4'0" 20d32'11"
Nap-zthalla
And 23h48'45" 48d25'37"
N.A.R. 80-06
Psc 23h47'56" 5d18'23"
Nara Pegeen Vesely
Lyn 6h59'56" 50d49'18"
Naraa
Peg 22h27'39" 22d50'13"
Narbit
Boo 14h44'55" 22d24'43"
Narciso Saavedra
Sco 17h49'57" -31d38'58"
Narda Rae Mutter
Crb 16h24'16" 35d15'37"
Nardi Leonard L. Swirda
Aqr 21h32'31" 1d53'29"
NARDJIS
Cnc 8h15'18" 16d45'46"
Nareenea Muradian (Nat)
Sco 17h57'43" -44d15'43"
Narelle 111104
Pyx 8h47'59" -24d8'48"
Narelle Anne Drenikow
Cru 12h41'42" -58d4'23"
Narelle 'Relle' Watson 23May1976
Gem 6h45'22" 22d59'3"
Narendra Ganti
Uma 9h47'34" 68d5'38"
Nari
Mon 7h52'2" -3d9'5"
Narina
Leo 11h26'25" 0d12'28"
Narine Gevorkyan
Cap 20h23'45" -13d17'35"
Narisa Naomi Sarintra
Sgr 19h17'39" -27d26'34"
Narishia's Special Star
Tau 3h48'16" 20d34'45"
Naritoshi & Kana
Cap 20h29'48" -27d4'13"
Narjes
Ari 1h54'21" 18d28'39"
Narnia
Lyr 19h17'52" 29d21'32"
Narrow Gate
Her 16h36'54" 10d33'47"
Naruha
Aqr 23h23'44" -16d29'57"
Narzan
Uma 9h19'13" 47d11'55"
Nas & Holly
Cyg 21h42'3" 46d19'45"
NASCAR FAN UNCLE BOB
Leo 10h19'4" 11d0'44"
Naseem
Ori 5h32'52" 14d6'18"
Naseem Torian & Ramin Razavi
Cyg 20h12'27" 42d23'35"
Naser Kraja
Cam 4h18'54" 66d43'5"
Nash
Aur 5h39'11" 46d18'20"
Nash
Psc 0h49'51" 9d55'34"
Nash Houston's Roaring Star
Lmi 10h25'53" 36d28'23"
Nash Lee Villani
Lib 15h34'19" -29d26'54"
Nash Michael Abbott
Cnc 8h10'42" 29d36'16"
Nash Munoz
Aql 20h1'24" 1d46'54"
Nash N. Gerdak
Her 16h25'17" 24d40'23"
Nash Smith
Aur 5h14'18" 47d32'0"
Nash-Ditzel
Cnv 12h42'1" 40d4'14"
Nashili
Lyn 9h13'29" 44d17'5"
Nashla
Ari 2h47'0" 24d46'17"
Nasim
And 0h32'50" 41d0'19"
Nasir Malachi Peoples
And 2h18'6" 48d58'2"
Nasrin Sultana Shampa
Cet 2h33'44" 8d2'25"
Nasrine Dina
Mon 6h52'3" -7d4'12"
Nassar ;*
Cnc 8h6'21" 14d28'59"
Nasse, Walter
Uma 8h35'8" 64d43'2"
Nasser Asses
Sco 17h35'24" -37d0'19"
Nasser - Bubu
Uma 9h12'45" 55d27'55"
Nasser Shahin
Cnc 8h26'44" 31d4'14"
Nassim Abou Fakhar
Cyg 20h5'50" 51d41'16"
Nastasia
And 1h14'25" 43d7'11"
Nastassia Danielle Christen Foster
Lyn 7h51'20" 39d6'33"
Nastusha
Leo 11h11'25" 15d8'8"
Nasty Dog
Cma 6h53'24" -22d23'15"
Nastya
Uma 12h34'33" 53d55'36"
Nastya Morozova
Uma 8h53'31" 72d28'35"
Nasty's Lucent Wink
Dra 17h21'49" 58d56'2"
Nasty's Ultimate 420
Crb 15h51'20" 35d27'14"
NASZ KOCHAC GWIAZDA
Lyn 8h18'28" 55d58'44"
NAT
And 23h3'12" 48d24'46"
Nat
Cnc 8h14'2" 8d47'26"
Nat
Psc 23h55'26" 5d40'21"
Nat and Gladys Swartz
Gem 6h59'18" 31d51'32"
Nat B Eisenberg
Uma 11h24'24" 56d10'10"
Nat & Mary Mistretta
Uma 11h33'41" 32d31'31"
Nat Rat
Her 17h57'39" 23d43'15"
Nat Schell III
Umi 21h19'3" 88d54'47"
Nata
Uma 9h16'13" 49d39'38"
Nata+Erik, Amor for life.
Cyg 20h46'57" 47d45'12"
natacha
Sgr 18h25'16" -16d19'55"
Natacha
Aqr 23h34'41" -21d35'15"
Natacha et David Meyer
Vir 11h46'58" 9d28'5"
Natacha Lucienne Bensoussan
Tau 3h47'25" 23d37'52"
Natacha Minor
Ari 1h53'7" 22d39'51"
Natachka - Anastasia Burlyuk
Sco 17h25'54" -45d31'40"
Natacia's star
Gem 6h57'41" 18d22'33"
Natahlie O'Connor
Lyr 19h1'19" 41d43'35"
Natale Anthony Buono, Jr.
Oph 17h2'6" -0d39'19"
Natale J. Scalzo
Uma 8h26'11" 59d45'33"
Natale Koretskiyi
Mon 6h52'33" 3d28'41"
Natale "Natterz" Stabile
Uma 10h40'27" 69d20'15"
Natalea
Cap 20h51'47" -26d15'20"
Natalee
Cyg 20h30'42" 38d32'29"
Natalee Brooklynn Dunn
Col 5h51'43" -33d49'32"
Natalee Holloway
Lib 14h26'19" -19d22'27"
Natalee Mavis
And 23h16'17" 35d12'54"
Natalee Roseboom
Lyn 8h14'1" 46d39'32"
Natalee's & Lee's serendipity
Ori 5h32'56" -3d14'35"
Natali
Dra 18h5'36" 74d29'31"
Natali
Cyg 20h4'34" 34d26'11"
Natali
Tri 1h50'42" 30d27'17"
Natali Rivas
Aqr 22h29'58" -18d54'10"
Natalia
Leo 11h17'46" -6d0'0"
Natalia
Sgr 19h33'40" -31d55'28"
Natalia
Sgr 17h52'53" -28d34'47"
Natalia
Cnc 9h7'0" 30d43'59"
Natalia
Cyg 21h54'59" 37d42'36"
Natalia
Ori 5h2'7" 15d2'5"
NATALIA
Tau 4h30'7" 25d45'18"
Natalia
Boo 14h39'49" 25d16'34"
Natalia
Leo 10h12'28" 22d56'10"
Natalia 21
And 0h41'20" 38d29'23"
Natalia A. Omelina
Ori 5h26'56" 2d21'7"
Natalia A Perdomo
And 0h39'37" 32d50'13"
Natalia Agudelo
Dra 15h52'25" 55d55'1"
Natalia Akopian
Ari 3h3'23" 27d28'58"
Natalia Anne
Lib 15h14'25" -27d39'52"
Natalia Anne Certa
Lyn 8h56'33" 39d8'29"
Natalia Araneta Morgenstern
Psc 0h39'0" 9d21'21"
Natalia Avendano
Leo 10h27'15" 15d2'11"
Natalia Bolotina
Psc 0h57'9" 18d18'20"
Natalia Castro Botero
Cru 12h20'39" -56d50'16"
Natalia Christine Mohabir
Ori 6h19'51" 11d4'20"
Natalia Chromiak
Lyr 18h32'38" 31d59'15"
Natalia Cortada Garcia
Aqr 22h1'46" -21d4'34"
Natalia Cristina Valdesuso Marrero
Cas 1h35'41" 63d1'32"
Natalia Cristobal
Cnc 8h28'46" 19d0'26"
Natalia Dana Chung
Uma 11h48'33" 59d9'46"
Natalia de Sesma
Lib 15h52'1" -9d35'13"
Natalia E. Perez
And 0h48'58" 37d16'38"
Natalia Elizabeth
And 0h32'22" 29d42'37"
Natalia Enciso
Sgr 19h16'12" -14d14'2"
Natalia Fernandez
And 23h23'6" 52d13'5"
Natalia Gaviria
Tau 4h47'44" 27d14'21"
Natalia Gomez
Ari 2h34'36" 21d15'39"
Natalia Grace Tadlock
Uma 10h39'1" 41d58'25"
Natalia Grinchenko
Cas 1h25'1" 51d7'54"
Natalia Janae Waters
Sgr 18h41'41" -23d28'2"
Natalia Jean Mazzoni
Del 20h38'35" 15d12'52"
Natalia Jerlecki
And 23h22'9" 47d15'2"
Natalia Jimenez Almela
And 23h7'44" 49d57'8"
Natalia Jo Giampa
Lyn 7h8'26" 58d28'15"
Natalia Komasyuk "smart Tash"
Ori 6h7'44" 6d20'55"
Natalia Koneva
Peg 22h35'28" 8d58'18"
Natalia Korzunina
Sgr 18h59'4" -34d16'25"
Natalia la Stella Bella
And 1h22'54" 39d15'26"
Natalia Leukhina
Cyg 19h31'19" 44d2'26"
Natalia Lynn Mauro
Sgr 18h19'36" -27d16'57"
Natalia Marie Shuman
Dra 16h10'20" 61d43'48"
Natalia Mathis
And 1h14'47" 45d22'45"
Natalia Porydsaj
Uma 9h1'9" 57d40'57"
Natalia Prada-Rey
Cas 23h22'59" 55d37'18"
Natalia Radziuk
Vir 12h54'56" -10d57'45"
Natalia Rene
Psc 1h13'54" 17d43'21"
NATALIA ROG
Lib 15h17'18" -8d27'16"
Natalia Rose
Uma 8h58'48" 47d22'49"
Natalia Ruberg
Crb 15h50'12" 30d19'2"
Natalia Sanchez
Vir 14h41'49" 2d56'32"
Natalia Semjonowa
Uma 8h54'7" 63d26'26"
Natalia Sofia Almada
Ari 2h13'41" 14d15'27"
Natalia Starowicz
Aqr 21h42'59" 1d13'51"
Natalia The Diamond
Sgr 17h53'14" -28d26'51"
Natalia Vyorst
And 0h57'51" 37d36'24"
Natalia Wilk
Ari 2h39'15" 27d16'50"
Natalia's Smile
Psc 1h11'18" 18d1'45"
Natalicium
Lib 15h7'45" -6d10'11"
Natalie
Ori 5h28'55" -7d30'42"
Natalie
Aqr 23h52'46" -11d45'46"
Natalie
Vir 14h11'4" -20d41'40"
Natalie
Uma 8h56'54" 66d6'35"
Natalie
Cam 7h59'8" 63d24'11"
Natalie
Cas 2h16'4" 64d0'26"
Natalie
Uma 13h1'3" 63d9'5"
Natalie
Uma 14h14'48" 57d43'8"
Natalie
Uma 10h25'51" 57d24'15"
Natalie
Sgr 18h23'1" -23d2'54"
Natalie
Psc 0h0'1" 6d30'9"
Natalie
Tau 5h50'35" 17d11'22"
Natalie
Leo 10h56'10" 7d58'54"
Natalie
Vir 13h6'16" 11d55'34"
Natalie
Ari 2h16'33" 24d30'21"
Natalie
Ori 5h57'7" 18d5'25"
Natalie
Ari 3h11'33" 27d14'55"
Natalie
Gem 7h31'21" 26d30'17"
Natalie
And 0h42'3" 42d6'2"
Natalie
And 1h28'34" 37d32'6"
Natalie
And 1h42'13" 40d52'45"
Natalie
And 1h44'3" 42d54'30"
Natalie
And 23h46'20" 34d9'24"
Natalie
And 0h34'42" 42d21'30"
Natalie
Crb 16h10'30" 33d36'31"
Natalie
Lmi 9h47'38" 33d14'48"
Natalie
Per 4h10'10" 35d17'39"
Natalie
Lyn 7h39'5" 51d6'40"
Natalie
And 2h17'23" 50d21'27"
Natalie
And 23h9'33" 44d52'25"
Natalie
And 23h9'52" 45d59'26"
Natalie
Uma 11h2'45" 50d44'45"
Natalie A. Inman
Vir 12h9'26" -6d12'3"
Natalie A. Salser
And 23h53'31" 44d7'8"
Natalie Adele Smith
And 23h35'38" 49d58'40"
Natalie & Alfred Hermandorfer
Cyg 20h46'14" 43d51'51"
Natalie Alonso
Uma 8h53'41" 47d48'9"
Natalie Amanda Bailey
And 0h9'41" 44d23'43"
Natalie Amy Jemma Young
And 23h21'0" 47d36'52"
Natalie and Andrea
Lib 15h41'38" -16d0'40"
Natalie and Andy
Col 5h42'43" -31d56'15"
Natalie and Dale's Wish
Ari 3h6'14" 29d51'30"
Natalie and TJ "11-11-04"
Cyg 21h37'24" 41d49'43"
Natalie Anderson Carson
Leo 10h22'2" 17d48'30"
Natalie Andrade
Her 17h57'59" 14d54'39"
Natalie Angel Murphy D.E.F.
Lyr 18h54'45" 32d1'15"
Natalie Ann
Ori 5h5'21" 7d29'15"
Natalie Ann Chambers
Vir 13h22'35" 12d19'50"
Natalie Ann Edge
Lib 15h28'26" -13d0'47"
Natalie Ann Frampton
And 0h15'23" 28d30'50"
Natalie Ann Lobato
Gem 6h31'49" 14d19'37"
Natalie Ann Mosier
Ari 2h36'15" 27d10'45"
Natalie Ann "Princess"
And 0h26'52" 33d46'6"
Natalie Ann Smith
Ari 2h32'4" 23d36'18"
Natalie Ann Wright
Uma 8h47'23" 59d9'44"
Natalie Anne "Bug" Siemek
And 0h55'31" 38d55'37"
Natalie Anne Moore
Tau 4h7'33" 6d26'11"
Natalie Anne Richardson
And 0h39'36" 23d46'36"
Natalie Anne Sole
And 0h25'30" 45d55'20"
Natalie B.
And 1h29'1" 43d9'56"
Natalie Bassil
Com 12h18'34" 24d28'10"
Natalie Batshon
And 2h5'46" 36d26'25"
Natalie Baydon Jackson
Cru 12h41'16" -58d31'49"
Natalie Bernadette Bonello
Psc 23h55'57" 7d27'4"
Natalie Bernstein
Gem 6h34'48" 27d3'48"
Natalie Boyer
And 1h56'44" 37d4'53"
Natalie Britt Champion
And 1h0'9" 45d47'54"
Natalie Britton - Brightly Shining
Cra 18h35'50" -42d47'25"
Natalie C.
Vir 12h8'10" -5d17'54"
Natalie Caroline Delson
Aqr 20h45'7" -10d39'4"
Natalie Carolyn Mathis Hampton
Lyn 7h9'33" 46d31'46"
Natalie Catherine
And 0h30'56" 46d30'1"
Natalie Charchol
Cyg 19h33'23" 47d7'28"
Natalie Christian Belle Farquhar
Vir 13h18'51" 4d19'42"
Natalie Christina Wood
Peg 21h49'35" 31d15'8"
Natalie Christine Beach
Sgr 18h13'12" -24d36'4"
Natalie Christine Giannetti
And 1h42'1" 50d3'53"
Natalie Christine Lewis
Uma 11h31'30" 39d20'34"
Natalie Claire Jurkosky
Cnc 8h39'56" 31d41'59"
Natalie Claire Velehorski
Leo 11h44'56" 23d45'36"
Natalie Colalillo
Cap 20h8'2" -17d45'13"
Natalie Colceriu
Sgr 19h25'6" -31d44'44"
Natalie Cruz
Cru 11h56'25" -61d34'13"
Natalie Cutting
And 1h13'21" 42d26'41"
NATALIE D. NEITZKE
Cyg 20h4'5" 36d3'26"
Natalie & David
Cyg 20h14'55" 50d5'51"
Natalie Dawn DeBassige
Del 20h38'42" 12d53'40"
Natalie Dawn King
Ari 2h24'2" 22d7'44"
Natalie de Reus
Ori 5h51'30" 19d41'24"
Natalie DeAngelo
Mon 7h35'5" -0d47'7"
Natalie Denlinger and Ian Hahn
Cyg 20h27'44" 53d10'47"
Natalie Diane
Lib 16h0'23" -17d43'37"
Natalie Dodaro
Umi 14h43'30" 72d2'25"
Natalie E. Evans 23.05.1982
And 0h22'30" 24d51'6"
Natalie E. Patton
And 23h52'2" 41d24'27"
Natalie Earhart
Vir 14h14'55" -11d14'29"
Natalie & Eddie's Anniversary Star
Vir 12h17'39" -0d53'2"
Natalie Eichler
Psc 1h13'2" 27d3'18"
Natalie Eileen
Cap 20h56'27" -21d49'35"
Natalie Elaine Brown
Dra 18h26'38" 51d6'44"
Natalie Elise Thornton-Webb
Crb 15h42'45" 32d28'26"
Natalie Elizabeth Clair Lusty
And 23h17'22" 41d2'44"
Natalie Elizabeth-Ann Powelko
Lib 14h33'55" -13d23'5"
Natalie Elkinawy
Vir 13h20'32" 11d27'12"
Natalie Emma Robinson
Per 2h57'26" 44d41'38"

Natalie Espinoza
Dra 19h16'41" 72d54'56"
Natalie Estelle Parks
Crb 15h48'1" 30d1'21"
Natalie F. Lewis
And 23h16'48" 39d3'20"
Natalie Fernandez
Sgr 18h17'55" -17d7'49"
Natalie Fowlkes
Mon 7h47'22" -11d10'3"
Natalie Garmany
Sgr 19h59'46" -34d3'56"
Natalie&George A New Life Together
Cyg 19h46'58" 54d55'45"
Natalie Gertrude LaRue Brooks Stout
Cnc 8h43'39" 24d48'8"
Natalie Gilligan
And 0h5'7" 39d24'25"
Natalie Gisele
Sgr 19h17'14" -14d34'48"
Natalie Gomez
Psc 0h54'28" 6d43'8"
Natalie Gonzalez
Uma 11h58'39" 53d33'27"
Natalie Gonzalez
And 0h40'14" 45d55'5"
Natalie Gore My Star Love Stephen
Ari 2h8'32" 17d27'6"
Natalie Grace
Cap 21h55'1" -13d51'0"
Natalie Grace
Lib 15h30'4" -16d27'10"
Natalie Grace
Lib 14h25'23" -19d38'47"
Natalie Graham
Cnc 9h5'34" 7d0'41"
Natalie Haena
Leo 10h18'45" 23d21'36"
Natalie "Hart of the Heavens"
And 2h17'6" 50d28'44"
Natalie Henry
And 0h27'31" 29d23'59"
Natalie Heseltine
Cas 23h20'50" 57d14'40"
Natalie Huryn Seymore
Uma 9h34'38" 56d42'31"
Natalie I love you
Ori 6h17'20" 19d15'56"
Natalie Irene Krentz
Leo 9h26'17" 26d20'28"
Natalie Isabelle
Cnc 8h26'47" 25d28'51"
Natalie Isaeva
Cas 0h33'36" 64d8'23"
Natalie J. Askland(Baby Girl)
Leo 9h43'32" 28d16'46"
Natalie J. Nardone GaMPil Star
And 23h30'42" 41d56'34"
Natalie J Talbot
Gem 6h32'26" 13d8'33"
Natalie Jael
Ori 5h49'48" -5d36'34"
Natalie James Miner
Vir 14h38'19" -5d23'56"
Natalie Jane Garrett
Ari 3h12'48" 27d59'53"
Natalie Jane Johnson
Cru 12h41'12" -56d39'18"
Natalie + Janssen
Ari 2h24'48" 10d32'57"
Natalie Jayne Grace Warren
And 23h48'57" 42d5'9"
Natalie Jayne Panagrosso
Uma 8h34'31" 48d32'57"
Natalie Jean
Peg 21h29'30" 16d14'47"
Natalie Jean Eacrett
Cap 20h21'48" -20d6'29"
Natalie Jessica
Aqr 21h29'41" -0d7'4"
Natalie Jo
Lib 15h49'5" -3d44'55"
Natalie Joi Fluent
Lib 16h1'44" -16d10'4"
Natalie Joy
Sgr 19h4'46" -16d44'3"
Natalie Joy
Umi 14h17'28" 77d23'58"
Natalie Karen Nix
Cas 23h20'43" 56d29'40"
Natalie Kate Rappos
Lyn 9h7'13" 37d36'50"
Natalie Kay
Ari 2h21'53" 14d32'28"
Natalie Kay Moss
And 1h41'43" 50d1'28"
Natalie Kim Taylor-Stroupe
Aql 19h32'26" 8d39'46"
Natalie Koffarnus
Uma 9h59'45" 54d35'33"
Natalie Krohe
Cas 0h22'19" 50d37'23"
Natalie Kumari Wilcox
Leo 11h48'41" 25d45'59"
Natalie L. Gaull
Dra 19h43'18" 61d13'1"
Natalie L Goff
Cyg 21h57'34" 39d23'55"
Natalie Lamar
Psc 1h21'27" 16d5'51"
Natalie Lauren Monari
Tau 4h52'3" 24d26'50"
Natalie Levine
Cas 1h22'39" 61d35'49"
Natalie Libenson
Sgr 19h1'1" -16d58'57"
Natalie Liberte
Lib 14h53'17" -3d33'45"
Natalie Lobo
Cas 0h10'50" 48d42'40"
Natalie Logan Korzuch
Cap 21h37'57" -8d42'43"
Natalie Louise McRanor
Cas 0h13'39" 51d49'38"
Natalie Loves Trystan
Leo 9h26'59" 12d11'57"
Natalie Lynn DeSalvo
Vir 12h13'3" -0d59'20"
Natalie Lynn Lutz
Aqr 22h40'44" -16d3'5"
Natalie Lyon Schlesier
Lyn 6h23'58" 59d37'38"
Natalie M. Kite
Lib 14h58'58" -5d12'6"
Natalie M. Webb
Cap 21h19'33" -19d7'9"
Natalie Macias
And 0h27'26" 26d17'53"
Natalie MacKenna Martin
Cas 1h15'16" 58d45'55"
Natalie MacKenzie Gannon
Cap 20h25'11" -18d5'12"
Natalie Mae
Psc 0h17'33" 3d44'51"
Natalie Mae Barnickel
Lib 15h20'52" -25d33'14"
Natalie Mae Eros
Uma 9h3'2" 65d2'59"
Natalie Mae Eros
Cnc 8h16'20" 21d4'2"
Natalie Mae Lockard
And 0h41'7" 33d55'56"
Natalie Maksimovic
And 0h58'42" 35d3'20"
Natalie Margaret
And 1h23'28" 50d31'56"
Natalie Maria
Uma 10h59'15" 64d5'59"
Natalie Marie
Dra 17h32'12" 52d17'50"
Natalie Marie
Ari 2h35'11" 19d58'20"
Natalie Marie
Tau 4h55'4" 15d58'41"
Natalie Marie Aldridge
Umi 16h23'18" 70d33'11"
Natalie Marie Asghar
Tau 5h25'1" 17d37'53"
Natalie Marie Celano
And 2h13'31" 42d55'47"
Natalie Marie Enciso
Ori 5h33'55" 6d14'51"
Natalie Marie Ferraro
Lmi 10h7'6" 40d47'35"
Natalie Marie Miller
And 0h47'22" 40d43'6"
Natalie Marie Neurauter
Uma 13h7'12" 56d30'19"
Natalie Marie Reeves
And 23h21'15" 47d7'59"
Natalie Marie Seguin
Lib 15h15'20" -23d53'12"
Natalie Marie White
And 23h7'35" 50d50'36"
Natalie Marie Yaeger
Sco 17h57'1" -39d10'5"
Natalie Marlaine Ruby Hernandez
And 1h42'6" 38d13'2"
Natalie Mary Reynolds
Cas 1h19'30" 69d44'24"
Natalie Mary Sparks
Ori 5h15'9" 12d50'50"
Natalie May Jordan
Cap 20h19'34" -19d48'28"
Natalie McDaniel
Lib 15h26'5" -7d22'13"
Natalie McDonald
Uma 8h53'37" 66d55'35"
Natalie McQuade
Cyg 20h31'57" 53d58'33"
Natalie Mei Hebert
Cas 3h34'34" 70d11'24"
Natalie Melcher
Dra 18h43'53" 51d0'58"
Natalie Michelle De La Mora Autran
Crb 16h6'42" 35d52'18"
Natalie Michelle LeMinous
Leo 10h45'35" 17d39'55"
Natalie Minh Chau Pham
Aqr 20h44'5" 1d37'52"
Natalie "Miss Sugar & Spice"
And 0h34'11" 33d54'24"
Natalie Mobbs
And 1h25'25" 50d22'19"
Natalie Munroe Didona
And 2h10'35" 45d5'44"
Natalie - My Angel Forever
Cru 12h7'12" -63d41'18"
Natalie "My Love"
Per 3h46'17" 48d50'52"
Natalie My Princess
Psc 23h42'59" 0d32'57"
Natalie - my slice of heaven
Pyx 9h26'41" -29d8'39"
Natalie Myers
Aqr 21h7'58" -9d20'55"
Natalie N Salse
Cap 20h20'36" -17d10'12"
Natalie N Wood
Ori 5h33'28" 4d11'48"
Natalie Naesheane
Cyg 20h49'26" 35d22'17"
Natalie Neve Bellina
Cnc 8h18'27" 10d29'0"
Natalie Nichole
Tau 3h38'26" 29d22'24"
Natalie Nicole Cuadras
And 0h14'9" 36d55'14"
Natalie Nicole Edwards
Cap 20h45'4" -24d46'3"
Natalie Nicole Garcia
Lib 15h40'42" -19d55'1"
Natalie Nicole Sitton
Vir 13h18'35" 2d44'59"
Natalie Nicole Waltz
Ari 3h5'2" 24d9'32"
Natalie Nienaber
Cyg 21h31'36" 41d38'22"
Natalie P Reid
Tau 5h42'22" 23d7'35"
Natalie Paynter's Heart of Light
Lyn 7h41'35" 41d42'52"
Natalie Penner
Gem 6h30'46" 12d51'13"
Natalie PGH No.7
Ari 3h18'40" 27d53'56"
Natalie Phang
Lib 15h41'54" -19d56'22"
Natalie Philbert
Cap 20h7'19" -10d40'19"
Natalie "Piglet" Brown
Psc 1h7'29" 3d33'29"
Natalie Pittman
Psc 0h49'28" 15d18'59"
Natalie Portia 2812
Umi 15h30'7" 68d39'30"
Natalie - Princess In Pink
And 1h28'35" 40d48'24"
Natalie Quick
Cap 20h44'27" -15d9'5"
Natalie R. Sganga
Lib 14h23'21" -9d39'8"
Natalie R. Soto
Vir 13h16'3" 6d56'28"
Natalie Rabinovich
Cyg 19h37'40" 28d11'38"
Natalie Rene
Sgr 18h30'3" -23d22'59"
Natalie Renee Bingham
And 2h24'55" 42d43'43"
Natalie Renee Housand
Lyr 18h53'54" 32d54'7"
Natalie Renee Spratlin
And 1h53'52" 38d28'9"
Natalie Renee Taylor
Gem 6h46'30" 31d54'17"
Natalie Renee Wood
Sco 17h2'18" -40d58'42"
Natalie Ria Pinnock - Natalie's Star
Pho 1h11'58" -56d47'30"
Natalie Robbins
And 2h31'41" 41d6'43"
Natalie Rochelle
Tau 4h41'58" 18d23'15"
Natalie Rodriguez
Lyn 8h2'42" 43d37'20"
Natalie Roese - Eternal Love
Cru 12h18'57" -64d17'6"
Natalie Rose Ayala
Uma 12h2'40" 52d56'10"
Natalie Rose Burke
Vir 13h17'49" 6d2'13"
Natalie Rose Middleton
And 2h26'44" 50d17'45"
Natalie Rose Ullman
Sco 16h16'56" -29d29'21"
Natalie Rosellen Sisk
Uma 10h44'23" 43d59'4"
Natalie Ruby Mastrantoni
Cnc 8h39'12" 29d45'18"
Natalie Ruggles - Little Sister
And 23h49'58" 37d33'36"
Natalie S. Johnson
Cyg 21h24'38" 38d51'11"
Natalie Salinas
Mon 6h27'51" 5d27'34"
Natalie Salvacruz
Cnc 8h40'38" 18d42'0"
Natalie Sandra
Cyg 21h28'0" 32d38'37"
NATALIE SANTANA HERNANDEZ
And 1h4'9" 37d34'20"
Natalie Shaw
And 0h46'0" 46d24'24"
Natalie Shelly Moran
And 1h0'14" 34d40'53"
Natalie Smith
And 23h30'16" 37d10'28"
Natalie Smith
Lyn 7h0'37" 50d44'22"
Natalie Smith
Crb 16h22'52" 27d7'56"
Natalie Spaulding
Tau 5h58'12" 24d47'7"
Natalie Spencer
Aqr 22h43'40" -0d41'49"
Natalie Stabenow
Ori 5h22'41" 9d37'6"
Natalie Starr Harrison
Gem 6h55'43" 31d25'25"
Natalie Starr Mesusan
And 2h34'52" 45d8'7"
Natalie Strizich
Gem 6h53'17" 17d4'52"
Natalie Sutton
Aqr 21h6'58" -8d40'39"
Natalie Suzanne Rhodes
Sgr 19h11'53" -14d18'38"
Natalie Therese Holland
Cru 12h26'52" -59d5'30"
Natalie Thomas
Lmi 10h56'33" 30d10'1"
Natalie Tran
Ari 2h1'7" 21d35'6"
Natalie Unzueta
And 0h22'3" 39d44'20"
Natalie V Mitchell
And 2h12'51" 39d6'28"
Natalie Valenti
Uma 13h47'53" 53d35'37"
Natalie Vercellotti
Cnc 8h21'40" 22d42'5"
Natalie Victoria
Peg 22h4'16" 13d51'7"
Natalie Wolf
Psc 1h14'56" 20d52'56"
Natalie Yazhary
Psc 1h3'16" 14d43'14"
Natalie Yvonne Mohn
Vir 12h20'51" 0d55'33"
Natalie, My Loving Star From Heaven
Cru 12h18'36" -61d11'35"
Natalie1
And 22h59'27" 40d50'59"
Natalie's 10th Birthday Star
Leo 10h9'8" 12d50'37"
Natalie's Aurora
Mon 6h54'49" -0d32'14"
Natalies' Dream
And 0h13'15" 46d8'43"
Natalie's Enlightened Rainbow Star
Ara 17h12'46" -55d47'29"
Natalie's Heaven
Sgr 19h10'48" -36d43'25"
Natalie's Light
Cru 12h38'7" -60d23'12"
Natalie's Light
Lyn 7h52'58" 49d48'15"
Natalie's Luxlucis
And 1h15'33" 34d53'26"
Natalie's Pop Tart
Ari 2h14'20" 23d46'29"
Natalie's Shining Star Ron Markwood
Tau 3h47'43" 28d37'38"
Natalie's Star
And 0h15'54" 31d3'26"
Natalie's Star
And 0h59'0" 44d26'31"
Natalie's Star
And 0h12'52" 40d40'26"
Natalie's Star
Aqr 21h39'18" -3d48'39"
Natalie's Wish
Mon 7h1'47" -8d14'38"
Natalina
And 22h58'34" 52d26'46"
Natalina Contoreggi
Leo 10h13'56" 26d13'51"
Natalis
Cap 21h23'9" -26d55'3"
Nataliya
Lyn 8h24'34" 33d11'37"
Nataliya Chabanyuk
Aqr 22h20'43" -18d47'30"
Nataliya Dyshuk
And 1h56'25" 46d10'1"
Nataliya Johnston
Uma 12h5'58" 39d38'52"
Nataliya Mayfield
And 0h6'54" 44d16'58"
Natalizia Trebisacce 1983
Sco 17h25'39" -38d9'31"
Natalka Maya
Lib 14h50'42" -3d0'45"
Nataly Attieh
Sgr 19h18'17" -27d50'10"
Nataly Eftalia
Leo 11h50'22" 10d24'35"
Nataly Real
Dra 20h36'6" 69d9'49"
Natalya
And 2h29'6" 44d17'21"
Natalya Dmitrievna Lozovoy
Tau 3h51'38" 26d42'45"
Natalya Kovalova
Psc 1h4'17" 28d34'53"
Natalya Lemberskaya
Leo 10h24'42" 26d57'37"
Natalya Mae
Cas 22h58'10" 58d0'18"
Natalya Ohanesyan
Cnc 8h47'19" 18d11'5"
Natalya Zhilyaeva
Gem 6h30'37" 25d42'21"
Natalya-Alissia
Vir 12h47'41" 5d8'49"
Natalyia
Mon 7h28'7" -3d40'31"
Nataned
Tau 5h44'10" 22d18'34"
Natania Elisheva
Lib 14h56'53" -17d40'19"
Natania Malin Gazek
Psc 0h6'14" 0d5'8"
Nataniel1
Leo 9h26'46" 11d25'30"
Natanja Stadler
Ori 6h13'2" 10d9'26"
Natanya
Tau 5h24'46" 19d39'53"
Natariga
Cru 12h27'11" -58d51'30"
Natarsha Angel Baby
Psc 23h28'14" -3d1'46"
Natas Kincaid
Vir 12h41'12" 7d47'27"
Natasa Markovic
Peg 23h57'58" 21d8'4"
Natasa Vukoje
And 2h14'33" 47d38'57"
Natascha
Ori 6h3'1" 19d28'29"
NATASCHA
Aqr 21h24'34" -0d23'12"
Natascha
Uma 9h18'28" 52d46'22"
Natascha Alexandra
Ori 5h29'29" 15d14'42"
Natascha Migmar
Aql 19h35'47" 11d10'47"
Natascha Tremp
Crb 16h11'5" 36d48'44"
Natascha, mein Glückstern, Völker
Uma 9h44'48" 66d25'29"
Natascha's Destiny
Dra 18h45'57" 62d2'39"
Natascia Anastascia Nol
Lib 15h21'15" -10d12'2"
natascia taverna
Cas 22h58'13" 54d25'16"
Natasha
Cas 1h58'46" 62d4'52"
Natasha
Cru 12h47'8" -57d49'42"
Natasha
Ari 3h4'53" 30d46'53"
Natasha
Lmi 10h8'27" 40d41'34"
Natasha
And 1h7'32" 39d28'40"
Natasha
And 1h53'47" 37d44'55"
Natasha
And 1h52'7" 42d44'9"
Natasha
And 1h32'26" 46d40'36"
Natasha
Dra 16h30'25" 52d5'15"
Natasha
Aql 19h58'6" 9d26'36"
Natasha
Tau 4h21'34" 17d21'26"
NATASHA
Cnc 8h27'45" 11d14'7"
Natasha
Cnc 7h57'15" 14d1'13"
Natasha
Tau 5h50'14" 26d45'17"
Natasha
Cnc 8h52'20" 27d8'22"
Natasha
Vul 20h39'27" 25d57'53"
Natasha Abbie Wakefield
Ori 5h56'52" 17d1'29"
Natasha Aiuto
And 2h16'55" 46d57'31"
Natasha and Andrew's Star
Cep 4h7'6" 83d17'35"
Natasha Ann
Cap 21h38'38" -13d57'37"
Natasha Brodich
Mon 6h37'47" 5d36'3"
Natasha Burson
Cnc 8h18'50" 20d59'21"
Natasha Butenko
Ari 3h13'17" 27d36'43"
Natasha Carroll
Lib 14h53'5" -2d59'55"
Natasha Carter
Lyn 8h59'0" 40d46'35"
Natasha Cekerevac (Naughty)
Cas 0h49'10" 48d33'42"
Natasha Christie Newman
Uma 10h18'28" 46d41'42"
Natasha Clark
Cnc 8h19'15" 10d18'7"
Natasha Corinne
And 23h13'25" 50d51'53"
Natasha D. Earls
And 1h5'22" 48d25'3"
Natasha Ding
Cru 12h2'57" -63d32'41"
Natasha Dubrowskij
Tau 4h19'49" 19d40'4"
Natasha Eggins
Cru 11h56'21" -62d5'32"
Natasha Escareno
And 2h1'14" 37d41'54"
Natasha Fedor
And 1h8'51" 43d55'57"
Natasha Gale Wood
Cas 1h6'35" 53d34'4"
Natasha Hamill
Ari 3h15'54" 29d59'21"
Natasha Helena Jones
And 0h27'16" 40d4'49"
Natasha Hoszowski
Aqr 23h37'47" -22d19'9"
Natasha Hribernik
Cru 11h57'57" -62d42'14"
Natasha J. Hepburn
Ari 3h28'34" 24d0'49"
Natasha & James Drummond
Leo 10h27'10" 25d38'10"
Natasha Jane
And 2h37'37" 49d52'37"
Natasha Jane Ford
And 0h2'39" 40d54'27"
Natasha Jones
And 23h13'7" 42d50'14"
Natasha Kang Mercado
Ori 5h7'47" 7d25'0"
Natasha Katherine Casey-Holowka
Aqr 22h16'5" 2d1'52"
Natasha Katina Washington
Cyg 20h26'34" 35d18'22"
Natasha Knowles
And 22h59'6" 51d52'52"
Natasha Kumari Gatward Burgess
And 0h36'42" 46d15'13"
Natasha L
Vir 12h27'9" 5d52'47"
Natasha L. Binggeli
Ari 1h47'34" 18d26'1"
Natasha Lauren Ascevich
Leo 11h16'25" 6d16'53"
Natasha Leigh Bender
Sgr 18h32'2" -33d30'18"
Natasha Loki Cody Hayter
Crb 16h22'47" 38d3'23"
Natasha Louise Saxton
And 23h17'55" 46d50'59"
Natasha May
Gem 7h27'14" 26d47'18"
Natasha Megan Prior - Tasha's Star
Cas 23h55'11" 59d5'43"
Natasha Meutiara Voisin (eNVy)
Crb 16h5'23" 31d6'48"
Natasha Mills
Ori 5h34'2" -9d5'38"
Natasha Monique Abeyta
Cnc 8h38'20" 31d21'48"
Natasha Nicole Jeter-Finnegan
Sgr 19h14'23" -17d19'25"
Natasha Palmer
Aqr 21h48'35" -5d28'57"
Natasha Perez
Sgr 17h54'18" -17d55'26"
Natasha Raquel Velarde
And 1h3'52" 36d0'22"
Natasha Rosko
Leo 9h38'12" 15d40'37"
Natasha Samarenko
Cas 0h28'31" 50d17'49"
Natasha Sarah & Aubrey Singer
Cyg 21h59'47" 50d52'34"
Natasha Sayani
Psc 0h5'8" -4d25'45"
Natasha Scheepers
Lyn 7h6'23" 47d45'28"
Natasha Streamline, Love you always xx
And 23h34'34" 37d39'38"
Natasha Taglietti
Cas 23h16'42" 58d56'56"
Natasha "Tashi" Kaimeia Southerland
Gem 6h45'41" 22d15'3"
Natasha The Skater
Cap 21h27'26" -15d2'8"
Natasha Tio
Aqr 20h39'32" -12d21'38"
Natasha Watso-Paul
Ari 2h30'54" 22d20'46"
Natasha Winter
Cas 1h27'47" 57d50'17"
Natasha Wood
Psc 0h27'25" 10d50'59"
Natasha050881
Ori 4h53'11" 5d24'23"
Natasha-Raffi
Leo 10h20'41" 22d33'5"
Natasha's Star
Cyg 19h38'30" 28d18'59"
Natasha's Star
And 0h16'14" 48d10'43"
Natasha's Star
And 2h24'11" 42d12'15"
Natasha's Starlight
Peg 0h12'20" 18d54'24"
Natasha-Thilo
Cyg 19h49'9" 33d23'21"
Natashia Chaimouratov
Psc 0h57'53" 25d33'22"
Natashia Juhlin my loving daughter
Cas 0h59'50" 59d19'10"
Natashia Shai
Aqr 22h20'58" -4d8'22"
Natasoula
Gem 6h49'5" 19d59'53"
nata-tasha
Uma 12h11'30" 58d26'44"
Natausha Ramirez
Tau 4h15'51" 5d55'48"
Natausha Ruth Dechant
Lmi 10h33'46" 34d9'46"
Natavia Reynolds
Cap 21h47'42" -16d59'54"
Nate
Sgr 19h23'31" -16d1'19"
Nate
Lyn 8h6'41" 56d5'41"
Nate
Ari 1h54'50" 23d27'3"
Nate and Debby
Cnv 12h44'31" 41d25'34"
Nate and Jax
Uma 11h49'13" 40d18'42"
Nate and Kandice Gilliard
Cyg 20h43'35" 30d32'3"
Nate Bomb
Cru 12h2'36" -58d50'27"
Nate & Courtney Sharpe
Ori 6h19'33" 8d31'34"
Nate Dawg
Gem 7h7'39" 23d35'45"
Nate E. Habert
Leo 10h15'57" 24d50'29"
Nate Forever My Starlight *ILY*
Uma 12h5'56" 32d59'25"
Nate Goulding
Cnc 8h1'9" 26d49'4"
Nate Hydinger
Her 16h41'15" 44d56'57"
Nate Kimball
Lib 15h22'46" -5d54'6"
Nate Kliewer
Lmi 10h30'18" 37d15'7"
Nate Mertens
Her 17h14'15" 32d14'43"
Nate & Shane: Mutual Saviors
Per 3h10'40" 55d9'31"
Nate & Teresa Graham
Cyg 19h44'48" 43d51'3"
Nate Treibitz
Ori 5h32'7" -1d3'8"
Nate Vanek #22
Aqr 22h16'37" 1d47'14"
Nate Waldrop
Ori 5h27'10" 2d37'6"
Nate Weinel, the Love of My Life
Peg 22h9'34" 31d33'20"
Nate Womack + Christina Sears
Aqr 22h36'22" -24d20'49"
NateandJulieAdams
Cyg 21h11'47" 53d28'48"
NATEDGREAT
Aur 5h37'9" 39d56'45"
Nately Boydall
Tau 5h32'40" 26d26'33"
Natercia Cardinale
Uma 8h10'23" 63d7'27"
Natessa
Col 5h25'20" -30d33'47"
Natestar
Aql 20h8'41" 9d9'22"
Natey
Leo 11h44'6" 10d45'4"
Nathadius
Gem 7h25'3" 26d34'19"
Nathaël
And 23h18'31" 52d21'29"
Nathalia
Uma 11h18'1" 70d40'10"
Nathalie
Uma 13h16'51" 62d4'10"
Nathalie
Cap 20h29'15" -10d26'58"
Nathalie
Uma 13h59'23" 51d51'32"
Nathalie
Lyr 18h42'45" 39d36'18"
Nathalie
Her 17h42'12" 30d51'35"
Nathalie
Gem 6h44'19" 23d7'59"
Nathalie
Her 18h46'58" 22d22'9"
Nathalie
Boo 15h10'18" 20d10'26"
Nathalie
Del 20h34'56" 7d38'50"
Nathalie
Psc 1h23'31" 21d3'2"
Nathalie
Ari 2h13'4" 11d58'5"

Nathalie & Alain Cap 20h44'27" -21d42'50"
Nathalie Anna Margaretha Uma 12h58'33" 52d20'2"
Nathalie Antinea Leo 11h6'33" 5d48'28"
Nathalie Attias Cas 0h55'23" 69d55'25"
Nathalie B Sgr 18h7'50" -17d3'37"
Nathalie B Simpkins Ari 2h7'1" 16d27'52"
Nathalie Breton Pagan Umi 16h13'3" 80d5'15"
Nathalie Brown Cyg 19h45'12" 35d28'20"
Nathalie C Gomez Lib 14h35'30" -19d37'11"
Nathalie Caicedo Psc 1h5'19" 30d49'35"
Nathalie Carriat Del 20h48'15" 7d24'22"
Nathalie Cestrone Sge 20h16'2" 20d2'6"
Nathalie Conesa Ori 6h1'5" 16d54'41"
Nathalie Der Sahaguian, "Naty" Cas 23h6'4" 55d57'53"
Nathalie Emma Beauté Uma 11h34'24" 48d31'40"
Nathalie et Wilfried Uma 12h2'23" 59d39'48"
Nathalie Fanelli Leo 9h40'40" 29d21'56"
Nathalie Fournier Uma 11h35'30" 65d29'30"
Nathalie Google Wah Ari 3h0'36" 19d49'4"
Nathalie Grace Torres Vir 13h12'19" -22d20'30"
Nathalie Hoffet Crb 16h3'32" 29d33'35"
Nathalie Kerebel pacific ocean Cas 0h51'52" 63d41'59"
Nathalie Lanta Sco 16h3'42" -22d57'9"
Nathalie Lemay Lyn 7h46'49" 36d30'25"
Nathalie M. Mujica Ser 15h54'11" -0d14'53"
Nathalie + Marie-Aude Umi 16h6'51" 71d44'55"
Nathalie Matta Uma 11h16'36" 30d7'36"
Nathalie Pearl Leach Tau 4h35'53" 29d25'57"
Nathalie Perreault Sco 17h54'0" -36d16'0"
Nathalie & Philipp Boo 13h53'11" 11d37'25"
Nathalie Rosby Uma 11h22'3" 44d9'31"
Nathalie S Barnes Uma 8h47'50" 48d24'20"
Nathalie Seibert Uma 9h33'55" 51d44'45"
Nathalie "Tété" Sebaalani Gem 6h45'33" 22d58'55"
Nathalie Wathelet Leo 11h28'43" -3d38'21"
Nathaly Chicheportiche Aqr 21h38'20" -0d33'8"
Nathaly Moscoso Sco 17h54'3" -38d35'4"
Nathan Umi 16h32'13" 76d45'56"
Nathan Umi 14h53'25" 70d10'17"
Nathan Tau 5h23'8" 22d34'10"
Nathan Ori 6h24'28" 17d17'56"
Nathan Leo 11h18'11" 17d22'23"
Nathan Tau 4h6'42" 2d6'15"
Nathan Uma 11h8'16" 46d4'46"
Nathan A. Ashley Ori 6h5'5" 13d36'18"
Nathan A. Carr Lyn 8h41'0" 33d30'57"
Nathan A. McCutcheon Gem 7h16'29" 16d38'41"
Nathan Aaron Español Her 17h35'1" 26d57'2"
Nathan Alan Jepson Dra 18h37'19" 51d23'6"
Nathan Alexander Uma 9h56'30" 58d3'12"
Nathan Alexander Blades Cnc 8h48'34" 26d20'28"
Nathan Alexander/ NIKA 082804 Uma 11h50'36" 36d47'32"
Nathan Alexander Spinelli Per 3h30'21" 49d8'47"
Nathan and Alyssa Forever Cyg 20h8'18" 34d26'56"
Nathan and Amy Lib 15h1'44" -16d37'0"
Nathan and Catherine Monk Umi 13h53'13" 72d9'24"
Nathan and Denise Lyn 7h47'6" 42d53'2"
Nathan and Jenny Childers Uma 8h48'44" 46d48'31"
Nathan and Marci Cyg 20h3'28" 32d42'48"
Nathan and Rachelle Uma 11h30'19" 63d15'23"
Nathan and Tara Per 1h47'52" 51d3'53"
Nathan and Terri Leo 11h43'3" 25d7'8"
Nathan Andres Quinteros Uma 10h58'39" 67d42'55"
Nathan Andrew Ballentine Aqr 22h43'50" -16d27'44"
Nathan Andrew Riley Suderman Crb 15h21'59" 30d53'21"
Nathan & Annah Forever in Love Sge 19h26'7" 18d22'2"
Nathan B. Benedict Sco 16h9'33" -11d22'0"
Nathan Bell Uma 8h45'48" 68d30'49"
Nathan Benjamin Thompson Uma 10h58'4" 58d53'59"
Nathan Birau Umi 14h57'34" 81d43'4"
Nathan Bloch Ori 6h2'4" 16d53'38"
Nathan "Boo" Powers Umi 15h23'43" 75d40'57"
Nathan Boris Riskas Gem 7h15'5" 24d11'43"
Nathan Boyer Ori 6h8'59" 8d44'34"
Nathan Brackett Cep 4h21'12" 82d54'47"
Nathan Bradley Harpe Cnc 8h2'19" 11d57'56"
Nathan & Breanna Uma 11h53'21" 59d59'17"
Nathan Brogden Her 17h40'4" 36d30'36"
Nathan Brotman Per 3h20'48" 43d31'31"
Nathan C. Mitchell Vir 13h1'1" 0d38'12"
Nathan C Rivas Lyn 7h30'41" 47d49'41"
Nathan C Sylak Cap 21h28'15" -16d12'31"
Nathan Campbell Pyx 8h47'30" -30d10'9"
Nathan Carpenter Her 17h2'27" 13d22'51"
Nathan + Chantal = Forever & Always Umi 17h46'19" 87d21'51"
Nathan Chris Roseman Uma 12h1'58" 56d37'41"
Nathan & Chrissy Dolliver's Star Uma 8h39'44" 60d48'29"
Nathan Christian Reid Ori 6h0'25" 17d17'1"
Nathan Christopher Buckridge Lmi 10h46'8" 29d32'19"
Nathan Christopher Buckridge Cnc 8h1'29" 10d55'38"
Nathan Clark Hamblet Umi 15h49'7" 80d1'6"
Nathan Cole Som Tau 3h39'22" 29d22'37"
Nathan Cortez Leo 10h19'58" 27d28'27"
Nathan Craig Francis Ori 5h39'50" 1d1'16"
Nathan Cross Uma 9h15'22" 50d35'3"
Nathan Dale Brown Cyg 19h52'45" 36d7'46"
Nathan Dale Hughes Uma 11h36'56" 47d17'59"
Nathan Daniel Ezro Her 17h24'46" 34d45'5"
Nathan Daniel Marsh Sgr 19h1'4" -24d26'59"
Nathan Daniel Sligh Uma 11h49'51" 48d40'25"
Nathan Daniel Taylor Ori 4h50'48" 3d4'25"
Nathan David Eastman Uma 10h51'9" 65d13'56"
Nathan David Pleming Per 2h55'51" 45d39'40"
Nathan David Vanderhuff Tau 3h39'27" 16d42'4"
Nathan Davis Cnc 9h3'29" 28d41'19"
Nathan Dean Whicker - I Love You! Sgr 19h46'50" -13d9'21"
Nathan Demers-Gagné Cap 20h18'24" -23d52'52"
Nathan Dettmering "Nate Dogg" Cyg 20h7'12" 58d34'48"
Nathan Dizenfeld Ari 3h16'14" 29d26'32"
Nathan Edward Patterson Cyg 21h14'12" 54d30'57"
Nathan Elliott Kozlica Lyn 6h33'25" 57d50'42"
Nathan Erich Huffman Umi 15h7'58" 75d7'17"
Nathan Estrin Aur 5h35'47" 41d42'35"
Nathan Firedog Nelson Ori 4h55'48" 9d53'44"
Nathan Forrest Nasers Ori 5h32'45" 7d5'43"
Nathan Franco Per 3h26'44" 48d54'4"
Nathan George Per 3h0'17" 48d25'24"
Nathan George Cummings Per 4h29'19" 49d54'56"
Nathan Geribo Levis Vir 14h6'53" -12d38'40"
Nathan Goforth Her 16h42'56" 27d16'15"
Nathan Gordon Fletcher Her 17h19'54" 19d46'57"
Nathan Gregory Hall Sgr 18h2'56" -16d54'16"
Nathan Grossman Ori 5h39'15" 1d4'20"
naThaN h niXs LUD Tau 3h50'45" 21d20'21"
Nathan Hall Uma 11h24'30" 36d24'46"
Nathan Hatfield Uma 9h9'13" 56d30'41"
Nathan Hauxwell's Star Peg 21h37'17" 15d30'38"
Nathan Howard Babcock Leo 11h26'44" 14d13'2"
Nathan I love you Vir 12h5'43" 4d57'51"
Nathan J. Knight Uma 10h45'36" 67d29'17"
Nathan Jade Aqr 21h54'10" -2d24'32"
Nathan James Campbell Uma 11h40'11" 52d22'20"
Nathan James Webdell Uma 12h15'40" 54d46'20"
Nathan Jasiek Tau 5h49'47" 16d41'27"
Nathan Jean Dumais Leo 11h14'9" 14d11'52"
Nathan Jodi Cam 5h21'53" 76d38'31"
Nathan Joel Garcia Lib 14h50'24" -16d56'7"
Nathan John Fields Per 4h26'39" 41d36'37"
Nathan John Gerbig Dra 16h30'12" 57d14'21"
Nathan John Klingensmith Cnc 8h47'59" 14d56'59"
Nathan John Metcalf Her 18h23'51" 17d24'49"
Nathan John Treanor Cep 21h5'19" 56d14'46"
Nathan John VanVlaanderen"Superman" Cru 12h2'53" -62d56'13"
Nathan Jonathan Wells Dra 19h5'15" 63d42'13"
Nathan Jordan Lib 15h14'11" -27d56'45"
Nathan Jorel "Giggle Puss" LaCount Her 18h14'35" 23d56'31"
Nathan Joseph Sco 17h56'54" -39d2'20"
Nathan Joseph Applebee Her 17h12'18" 34d13'5"
Nathan Joseph Kurek Lib 15h39'27" -9d38'20"
Nathan Joseph Pinelli Cnc 8h40'36" 31d42'16"
Nathan & Kate's Anniversary Star Pho 2h5'17" -40d13'10"
Nathan Kerry Crites Her 16h44'46" 30d33'36"
Nathan Kimball & LeAnn Steinbronn Vir 13h44'15" 1d7'25"
Nathan Kirkpatrick Ori 5h42'20" -0d16'49"
Nathan Kyle Simmions Cha 9h53'53" -78d53'5"
Nathan Lane Eri 4h39'33" -0d42'0"
Nathan Lawrence Bavaro Vir 13h18'10" 12d37'33"
Nathan Lawrence Graddon Leo 11h32'19" 15d56'24"
Nathan Lee Per 4h16'31" 51d4'53"
Nathan Lee LaPorte Ori 6h2'7" 11d2'42"
Nathan Lee Miller Gem 7h49'30" 20d17'38"
Nathan Lee Street Cnc 8h41'51" 7d46'54"
Nathan Lewis-Stone Gem 6h49'26" 17d13'54"
Nathan Louis Per 3h7'6" 56d27'58"
Nathan Louis Platt Cyg 19h56'53" 44d10'49"
Nathan Lynn Ford Leo 10h14'3" 17d6'23"
Nathan MacNevin Her 17h39'44" 27d10'10"
Nathan Martin Hargis Gem 6h46'8" 16d2'15"
Nathan & Maylene Lib 15h2'50" -16d44'27"
Nathan McFadden Aur 5h44'27" 53d38'56"
Nathan McNitt Per 3h6'48" 56d21'1"
Nathan Medeiros Her 16h45'46" 48d50'54"
Nathan Michael Sgr 19h27'32" -37d53'9"
Nathan Michael Presley Cap 21h35'18" -18d33'48"
Nathan Michael Sheffer Psc 1h24'54" 33d34'51"
Nathan Monroe Poteet Her 17h27'42" 28d30'58"
Nathan Morgan Herr-Gessell Uma 10h28'49" 42d3'40"
Nathan Morganroth Her 17h36'0" 27d18'15"
Nathan Morris Cen 13h31'17" -37d42'38"
Nathan Murphy Cru 11h58'42" -55d48'39"
Nathan Murray Linden Umi 14h10'29" 66d39'20"
Nathan Myers Ori 5h33'37" -0d23'13"
Nathan - Nate - Clarke - 9.02.1979 Cru 12h43'12" -57d4'11"
Nathan "Nattie" Wayde Owens Ari 2h53'26" 19d44'55"
Nathan (Nicky) Lindy Cyg 19h42'11" 34d53'47"
Nathan Oldham Longabach Her 17h28'43" 37d31'7"
Nathan O'Neill Cru 11h57'1" -61d5'55"
Nathan Orsi Courtas Cas 23h39'44" 51d30'0"
Nathan "Our special little man" Lmi 9h46'43" 34d32'9"
Nathan P. Barr Ori 5h32'11" -4d24'20"
Nathan P. Orlosky Ori 5h49'47" 3d24'5"
Nathan Patrick Leo 10h58'8" 8d16'43"
Nathan Patrick Lappage Umi 16h25'5" 77d11'48"
Nathan Paul Cunningham Aqr 21h15'16" -14d19'18"
Nathan Paul Wiley Per 2h22'5" 54d38'32"
Nathan Peter Myers Psc 1h7'8" 28d56'49"
Nathan Phan Seamans Her 17h48'21" 38d53'49"
Nathan Philip Martin Ogborne Umi 16h19'18" 72d3'39"
Nathan Phillip Davis Cap 20h26'9" -23d14'36"
Nathan Pierre Psc 0h45'22" 15d43'5"
Nathan Plaza Tau 4h21'58" 27d10'47"
Nathan Price Cap 21h30'34" -17d47'31"
Nathan Pye Ori 5h56'18" 20d40'4"
Nathan R. Ari 2h59'41" 17d58'38"
Nathan R. Bogue Psc 0h39'43" 21d29'52"
Nathan R. Cramer Cep 21h15'53" 57d51'15"
Nathan R. Mitchell Cep 22h2'27" 68d15'20"
Nathan R. Stiffler Uma 9h35'54" 68d36'21"
Nathan Ray Mares Cap 20h23'39" -10d3'30"
Nathan Raymond Hemington Ori 5h41'16" -2d56'30"
Nathan Robert Babcock Gem 7h17'43" 14d26'47"
Nathan Robert Kyles Her 17h44'1" 15d17'38"
Nathan Rollins Ori 5h23'32" 1d9'33"
Nathan Ryan Her 16h36'36" 6d4'27"
Nathan Ryan Brown Gem 7h13'56" 25d38'39"
Nathan Satsky Vir 12h29'5" 0d33'30"
Nathan Schaffhouser Uma 10h54'16" 58d52'42"
Nathan Schatz Per 2h57'18" 52d9'13"
Nathan Scott McInnis Her 18h4'3" 17d43'44"
Nathan Scott Siefert Cam 4h17'53" 56d41'29"
Nathan "Sexy Mexi" Munoz Cap 20h46'24" -21d35'59"
Nathan Shawn Paa Psc 23h42'45" 0d26'34"
Nathan Shorter Sco 17h22'48" -43d55'56"
Nathan "Sonny Glenn" Monroe Newman Ori 4h51'9" 5d22'49"
Nathan Soowal Cyg 20h23'16" 52d10'29"
Nathan Stahl Tau 4h54'48" 17d10'4"
Nathan T. Oddy Lyn 8h47'55" 34d11'7"
Nathan the Sweet Ori 6h6'3" 16d0'47"
Nathan Thomas Broadhead Ori 5h47'53" -0d26'25"
Nathan Thomas Detwiler Aur 5h45'40" 44d49'31"
Nathan Thomas Michel Cep 20h42'1" 75d11'7"
Nathan Thomas Zech Uma 9h50'35" 58d42'55"
Nathan Toby Neulander Tau 3h56'36" 21d5'37"
Nathan Turner Her 17h47'39" 25d13'12"
Nathan Vendeland Her 16h44'37" 34d17'3"
Nathan Victor Brudjar Aql 18h57'55" 17d58'7"
Nathan Wayne Cline Aqr 20h39'14" -9d15'44"
Nathan Wayne Stehr Per 2h21'10" 54d36'40"
Nathan Wesley Barling Lyn 7h18'35" 46d36'29"
Nathan Wesley Harris Leo 11h57'31" 26d24'26"
Nathan William Keebler Her 16h22'38" 12d3'29"
Nathan William Wineman Vir 12h13'55" 12d38'33"
Nathan Winter Whitmore Psc 0h45'56" 19d57'19"
Nathan, Karla, & Ariana ~ Friends Uma 11h8'0" 64d51'49"
Nathan, Ryan & Brandon's Boy's Town Peg 23h30'2" 24d4'48"
Nathan, Spencer, & Jarod Malnik Del 20h38'54" 16d5'20"
Nathan05 Cap 21h58'1" -9d4'12"
Nathanael C. Bennett Psc 0h34'44" 3d1'41"
Nathanael Quinten Metke Uma 10h14'2" 70d34'29"
Nathanael T. Gray Ori 5h52'5" 6d41'24"
Nathanartha 10 Aql 19h51'44" 12d18'49"
Nathaneal Jose Uma 12h55'11" 53d3'52"
Nathaneil Seth Laden Vir 13h5'46" -21d25'3"
Nathanial James Kruse Psc 22h52'11" 4d40'3"
Nathaniel Ori 6h13'57" 8d22'8"
Nathaniel Ari 2h17'28" 25d57'28"
Nathaniel Per 3h48'13" 49d10'58"
Nathaniel Umi 14h31'3" 78d59'51"
Nathaniel Abram Pyx 8h53'36" -34d30'32"
Nathaniel Alden Martin Whiting Per 2h49'17" 54d7'17"
Nathaniel Alexander Washam Her 18h56'52" 16d56'48"
Nathaniel and Joanne 28-04-2006 Ori 5h33'27" -5d8'15"
Nathaniel Andrew Elder Sco 16h36'36" -30d48'59"
Nathaniel Anthony Holmes Uma 10h54'34" 50d55'39"
Nathaniel Brody Hyde Barry Her 16h35'56" 20d46'48"
Nathaniel Bruce McClain Dra 18h25'44" 55d28'2"
Nathaniel Caselton Umi 20h2'28" 87d26'48"
Nathaniel Christopher Lindblom Ori 5h44'21" 7d36'40"
Nathaniel Cole Crb 15h41'22" 28d17'16"
Nathaniel D Muirheid Lib 14h22'54" -12d52'16"
Nathaniel D Smith And 0h37'45" 22d17'10"
Nathaniel David Armstrong Lmi 10h20'20" 39d12'29"
Nathaniel David Gloyd Her 17h9'54" 32d13'42"
Nathaniel David McClure Lyn 7h49'32" 48d10'59"
Nathaniel David Shaffer Uma 11h43'51" 52d56'6"
Nathaniel Duncan Carrera Cnc 8h11'54" 27d47'9"
Nathaniel Frederick Jones Her 18h51'43" 23d49'5"
Nathaniel George Topping Umi 14h28'19" 74d2'34"
Nathaniel George Zeiger Leo 11h22'1" 12d3'14"
Nathaniel Glenn Roberts Her 16h50'42" 29d42'28"
Nathaniel Gordon Rippl Umi 14h35'33" 76d3'24"
Nathaniel Hollingsworth Her 16h39'32" 7d38'4"
Nathaniel Hunter Adams Ori 5h25'42" 4d9'21"
Nathaniel J. Eisner Gem 7h5'45" 33d33'44"
Nathaniel Jacob Uma 8h44'57" 50d14'10"
Nathaniel James Hall Psc 0h33'51" 20d57'37"
Nathaniel James Turner Ori 6h19'3" 9d12'31"
Nathaniel Jefferson Belger Leo 11h52'29" 22d39'51"
Nathaniel Jeffrey Durr Ori 5h33'16" 7d30'31"
Nathaniel&Joseph&Samuel GraMa Psc 1h37'30" 27d30'32"
Nathaniel Joshua Carter Ari 3h23'5" 19d17'6"
Nathaniel Kristian Smith Umi 16h31'18" 77d41'17"
Nathaniel Lee Ari 2h14'20" 21d9'9"
Nathaniel Lee Stanley Eri 4h36'58" -0d38'5"
Nathaniel M. Neubauer Ari 2h37'39" 21d41'4"
Nathaniel Martin Hannewald Cnc 8h6'26" 8d22'14"
Nathaniel Mealman Ori 6h5'8" 19d8'54"
Nathaniel Moss Per 4h7'14" 36d48'40"
Nathaniel Norman Flaga Uma 9h48'59" 54d47'50"
Nathaniel of Melody Her 17h41'4" 21d31'6"
Nathaniel P. Pawelczyk Uma 10h17'19" 56d55'11"
Nathaniel Raymond Boo 15h5'44" 33d22'59"
Nathaniel Raymond Marshal Bacon Cap 20h11'27" -9d43'34"
Nathaniel Richard Winters Lmi 10h21'10" 32d57'20"
Nathaniel Robert Hawk Aqr 22h20'31" -15d37'57"
Nathaniel Roland Fiorucci Uma 11h14'23" 33d43'4"
Nathaniel Sancho Kelly Her 18h0'17" 24d44'54"
Nathaniel Steven Garza Uma 9h5'29" 48d18'56"
Nathaniel Thomas Haas Her 16h53'6" 32d59'59"
Nathaniel Van Dette Dra 18h32'43" 60d24'20"
Nathaniel Vincent Nixt Her 16h37'30" 45d17'21"
Nathaniel Weldon Schnader Leo 10h18'42" 17d1'1"
Nathaniel Xavier Pawlowicz Psc 1h0'21" 3d8'20"
Nathaniel-Jacob Uma 11h49'39" 61d26'57"
NathanPeterYoungBeloved Cep 21h28'24" 56d27'11"
Nathan's Celestial Star #120504 Gem 7h31'8" 32d38'58"
Nathan's Fire Crb 16h8'31" 27d7'0"
Nathan's Fire Sco 17h39'9" -34d48'23"
Nathan's Galexy Cnc 8h48'38" 10d49'32"
Nathan's love for Lindsey Cnc 9h17'52" 25d30'20"
Nathan's Nobility Her 17h43'24" 45d27'18"
Nathan's Nugget Ori 6h9'28" 9d0'40"
Nathan's Star Her 18h53'45" 25d40'59"
Nathan's Star Her 16h4'2" 18d38'33"
Nathan's star Per 4h25'43" 32d42'32"
Nathan's Star - 4 August 1971 Cru 12h9'22" -61d32'46"
Nathan's Wish Lib 15h4'49" -17d33'8"
Nathan-William Psc 1h5'49" 31d35'55"
Nathapet Srinivasan Uma 13h51'33" 56d56'7"
Nathasha so so so scandalous Lyn 6h30'58" 57d30'52"
Nathecat Sgr 18h26'30" -16d3'46"
Nathen 16 Lyn 6h55'56" 58d51'41"
Nathen Gunner Watson Aqr 22h38'25" -1d0'57"
Nathi Lyr 19h8'50" 42d29'28"
Nathian & Richelle Myers Sge 19h54'33" 18d24'32"
Nathxine Heart and Soul Cyg 20h31'41" 35d19'46"
Nathy Perez Hya 9h51'0" -11d50'17"
Nati And 1h1'58" 42d48'56"
Nati Ari 3h8'56" 17d41'36"
Nati Ari 2h58'26" 14d10'42"
Nati & Ifati Aloni Aqr 23h33'25" -15d41'20"
Nati Q. Lyn 6h53'36" 53d58'43"
Nati´s Jendrik 05052006 Uma 9h39'37" 47d10'19"
Natia Tateishvili Cam 4h12'10" 60d57'19"
Naticia Simon Lyn 7h29'40" 57d23'43"
Natii - My Star 4 Life Ari 2h26'34" 27d9'14"
National American Miss Umi 15h51'55" 76d46'28"
National Dance Institute Lyr 18h51'28" 34d54'39"
Natisha Virdee And 23h12'48" 41d16'51"
Natisya Item Uma 10h59'28" 67d30'33"
Natividad Jimenez Uma 10h14'20" 68d33'19"
Natividad Vigil Vir 11h38'30" 3d33'36"
Natmat Crb 15h55'6" 28d3'26"
Natnar Dra 17h47'59" 67d19'31"
Nat-n-Pat Cyg 19h41'24" 55d15'46"
Natosha Renee Hill Aqr 23h54'31" -19d17'7"
NATPAT Uma 11h22'47" 66d43'17"
Natree's Loving Light Psc 1h25'12" 11d6'47"
Natsalan Vir 11h46'57" 4d50'25"
Natt & Jay Dra 17h29'1" 51d3'17"
Nattalie Abegail Volcsko Leo 9h33'56" 16d5'14"
Nattalie Mitchell Crb 16h11'10" 27d53'58"
Nattapan Fay Chautavipat Psc 0h33'21" 12d28'45"
Nattie And 1h25'16" 49d27'45"
Nattlebugs Her 16h24'26" 11d1'49"
Natty Bonco Solomon 20.01.1988 Cap 21h44'14" -15d10'30"
Natty & Ricky 3-12-05 Lyn 8h5'9" 46d49'22"
Natty110885 Ori 5h40'57" -2d0'2"
NattyStar Cnc 8h46'27" 27d57'6"
Natulka Sco 17h49'26" -38d15'52"
Natural Aphrodisiac Cas 0h19'41" 62d46'13"
Natural Beauty Gem 6h41'8" 19d8'30"
Nature Cyg 20h43'46" 42d1'12"
Natvar Natha Lac 22h7'33" 35d21'53"
naty Cam 3h58'9" 64d17'14"
NatyRey Col 5h25'51" -29d43'49"
Naü Umi 17h53'29" 87d15'45"
Nau Ko'u Aloha Psc 0h6'28" 9d36'33"
Naughty Paulette Sgr 17h57'55" -17d25'49"
Naughty-talia Lyn 7h18'44" 47d52'20"

Naum & Tamara
Lib 15h46'18" -6d31'32"
Naunus
Umi 15h55'43" 71d57'35"
Naur Coelho
Lyn 9h6'26" 36d18'9"
Nauset Light
Tri 2h8'25" 31d39'39"
Nausicaa
Uma 10h49'14" 59d29'0"
Naut Reality
Cyg 21h16'16" 46d25'19"
Nautilus 40
Ori 6h1'45" 14d0'36"
Nautilus Beach Apts.
Lyr 18h21'18" 45d45'33"
Nava (Baby Joon) Torian
Uma 10h33'56" 60d2'8"
Nava Shir
Ori 6h12'43" 17d45'34"
Navajo
Cma 6h46'22" -17d13'20"
Navajo Shelley
Uma 9h45'22" 45d26'32"
Navarro Anthony Schunke
Uma 12h4'22" 49d24'2"
Navatia
Gem 7h29'49" 19d12'42"
Navaya Riale
Cnv 12h42'23" 37d9'3"
Navdeep Kaur
Uma 10h9'51" 64d36'24"
Naveed Torian
Aur 6h50'37" 41d26'50"
Naveen Jason Punyamurthy
Eri 3h51'36" -0d13'49"
Naveen Moeller
Uma 9h45'42" 63d8'33"
Naveen-Sharon
Ori 6h0'31" 20d6'17"
Naveira
Psc 1h13'14" 3d51'9"
Navi
Aqr 22h22'23" -0d14'46"
Navid
Psa 21h42'15" -30d18'27"
Navid
Gem 6h20'48" 19d3'45"
Navid
Cnc 8h32'11" 28d12'15"
Navid Joon
Lyn 9h4'21" 33d26'2"
Navigator Marty Sauer
Uma 10h29'50" 64d23'24"
Navigo Amo
Cyg 20h20'48" 44d52'2"
Navillus M D
Lyn 7h43'55" 40d51'23"
Navin Dugal
Cap 20h47'59" -21d36'47"
Navita Narinedhat
Aqr 22h29'31" 1d12'39"
Navneet Gill
Lib 16h1'0" -10d43'36"
Navneet Phangureh
Ari 3h16'22" 30d45'13"
Navpreet Kaur Sandhu
And 22h57'29" 51d53'5"
Navy
Lyn 7h38'13" 38d40'12"
Nawaar Binte Farooq
Cap 21h33'8" -12d22'40"
Nawaf & Toosha Forever
Cyg 20h13'46" 44d26'14"
NaWaiLohi
Tau 5h56'28" 25d22'37"
Nawal Chaya
Apu 14h52'58" -73d12'53"
Nawal Elsayed
Psc 0h31'15" 18d10'46"
Nawal Kdary
Lyn 6h34'40" 58d27'28"
Nawang Eden
Aqr 21h31'49" 1d28'36"
Nawar Al Hassan
Cap 20h58'41" -24d4'38"
Nawiel El-Hawari
Uma 10h10'44" 42d30'1"
Nawracaj Family Star
Uma 9h13'43" 62d51'11"
NAY
Cas 0h19'1" 51d33'5"
Nay AliDad
Uma 11h20'43" 72d39'2"
Nay Nay
Aqr 22h28'57" -4d14'28"
Nay Nay
Lib 15h1'13" -6d8'15"
Nay Star
Sgr 18h36'27" -23d51'26"
Nayda Anibel
Psc 0h34'22" 18d50'58"
Nayda Maritza - My Eternal Light
Peg 22h53'5" 19d55'39"
Naydean Rimmer
Lyn 7h43'11" 36d40'32"
Naye Moussa
Crb 16h7'16" 30d23'41"
Nayel
Sco 17h27'16" -34d17'32"
Nayel Jafer Hashem * DESTINY *
Uma 9h32'14" 49d20'53"
Nayeli
Vir 13h45'41" -2d34'2"
Nayeli Cazares
Aqr 22h42'56" -1d33'50"
Nayeli Chiara
Lup 15h17'34" -39d36'4"
Nayendi
Cap 21h0'34" -27d27'11"
Nayisid Cantillo
Gem 7h12'16" 34d39'27"
NaykaNookies
Psc 1h44'50" 5d29'24"
Nayla Nebula
Aqr 23h45'8" -15d10'33"
Naylan
Lmi 10h4'43" 38d39'50"
NayNay's Star
Peg 22h18'56" 9d3'30"
NayNay's Wish Granter
And 0h20'0" 45d15'5"
Nayomie Bonilla
Vir 12h42'50" 12d44'50"
Nayr
Uma 10h38'3" 46d57'49"
Naysa
Sgr 18h7'5" -27d14'40"
Naysa Lovi Meloni
Sgr 18h3'27" -27d27'37"
Naz
Aqr 22h55'41" -8d23'17"
Naz & Daniyal
Cyg 20h5'7" 51d35'11"
Naz & Zuber
And 23h6'23" 51d14'15"
Nazak
Tau 5h46'26" 24d14'27"
Nazanin
Ori 6h2'3" 10d57'2"
Nazanin Jalalian
Lyn 8h35'3" 38d43'44"
Nazanin "MicKey" Khansari
Uma 9h30'1" 52d34'27"
Nazarena Ocon
And 23h16'28" 42d57'42"
Nazaria
Cas 1h28'50" 52d28'31"
Nazario & Julia
Cyg 19h58'17" 40d30'18"
Nazgol
Cnc 8h6'31" 26d51'49"
Nazhada Madani
Ori 5h55'12" 12d30'4"
Nazi Alborz
Aql 19h42'20" -0d1'5"
NAZIAN
Umi 14h10'48" 75d40'26"
Nazie
Cap 20h26'56" -13d3'35"
Nazish Ali
Sco 16h6'9" -16d35'55"
Nazish Iram Dholakia
Boo 14h13'52" 16d52'50"
Nazli's Shining Star "Pishi"
Cam 3h55'47" 62d57'18"
Nazom
Del 20h33'49" 13d51'3"
Nazzari Cristina
Uma 10h1'0" 49d53'28"
NB-11-24-04
Uma 9h38'58" 69d54'26"
NB27
Uma 11h46'15" 54d19'36"
NBB122019
Cap 21h58'9" -24d2'49"
Nbg07062005
And 23h16'55" 51d43'17"
NBM and RAL 04/02/06
Psc 1h41'8" 14d39'20"
NBS-1
Ori 6h8'52" 5d56'0"
NC Cougars 5th Grade 2004/2005
Uma 11h33'53" 57d36'16"
NC3
Agr 23h19'45" -14d56'10"
NCC1701-D
And 2h36'21" 44d6'37"
NCCS' Heavenly Pride
Leo 9h56'59" 21d19'0"
NCG is Loved
Psc 0h45'43" 12d8'28"
Nckad
Lyn 7h41'36" 53d47'30"
NCP
Umi 15h41'17" 77d11'28"
Ndjadi Kingombe
Lyn 9h6'42" 45d12'3"
Ndougou
Aur 5h24'18" 55d24'13"
NDU-AUTUMN
Cam 5h37'9" 66d3'30"
Ne - Ne
Lib 15h56'28" -18d4'49"
Né Né
Ari 3h28'2" 24d16'37"
Ne Sarang
Ari 2h42'11" 22d2'16"
Nea Nastassia Conley
Vir 13h2'52" -11d3'44"
Neal
Sgr 18h11'58" -31d40'42"
Neal Alan Semel / Nachman Ari
Ari 2h8'38" 26d46'47"
Neal and Reba Lame/Long's Star
Sge 19h34'36" 18d47'23"
Neal Anthony Douros
Cas 0h41'15" 48d11'16"
Neal Barry Freuden
Cap 21h32'14" -9d29'24"
Neal Butler
Cnc 8h16'4" 8d49'25"
Neal Clifford King Jr.
Sco 17h55'17" -39d37'23"
Neal D. Apgar Sr
Aql 19h8'11" 6d41'46"
Neal Dominick Steele
Umi 15h24'4" 68d24'14"
Neal Eugene McCormick
Vir 11h57'8" 9d15'12"
Neal James Ross Downs
Sgr 17h45'11" -22d14'13"
Neal Jay Kuipers
Her 18h3'22" 24d41'12"
Neal L. Goldberg "My Shining Star"
Uma 11h53'40" 41d49'17"
Neal Lawrence Sandberg
Ori 5h26'37" 2d45'47"
Neal Lawson
Ori 4h51'34" -0d3'33"
Neal Maciel
Lib 14h31'52" -11d40'52"
Neal & Marie - November 27, 2006
Ori 6h17'34" 9d22'9"
Neal Marshall Nadler
Cep 21h39'35" 67d54'26"
Neal Mc Laughlin
Uma 9h23'53" 61d52'13"
Neal Mc Laughlin
Uma 12h18'57" 55d0'36"
Neal Morgan
Aur 5h44'18" 29d15'47"
Neal & Peggy Phillips
Uma 9h42'18" 49d20'34"
Neal Raper
Sco 16h4'13" -23d26'41"
Neal Schon, The Rock Dawg
Psc 1h18'9" 15d22'9"
Neal "Storage Extraordinaire" Weiss
Boo 14h17'12" 48d6'26"
Neal Witschonke
Vir 13h26'40" 11d29'54"
Neala Bernadette
Tau 3h49'14" 10d41'28"
Neala of Oriole Beach Elementary
Cnc 8h15'23" 26d10'45"
Neale A. Gow
Uma 12h32'44" 60d48'51"
Nealie Anne Williams
Cam 6h8'46" 63d40'58"
Neal's* Everlasting*Astro*Legacy
Ari 2h8'20" 24d39'4"
Nealus Leaningus
Umi 15h30'46" 75d12'9"
Near and Far
Aqr 22h11'0" 1d5'35"
Near, Far, Where Ever You Are
Sgr 18h10'59" -20d7'59"
Nebojsa - 10.6.1964.
Cyg 19h47'54" 56d34'22"
Nebojsa moj Nosorog
Her 16h14'18" 15d43'16"
Nebraska
Cma 6h47'56" -13d25'19"
Nebula Duchess Mason René
Mon 6h53'21" -1d26'55"
Nebula Fankhauser
Dra 19h50'8" 77d27'55"
Nec plus ultra
Cnc 9h8'7" 9d27'32"
Necci
Lyr 18h50'51" 39d25'27"
Nechal&Aidan
Uma 9h48'30" 52d24'24"
Necla Akgül
Cam 7h28'41" 67d34'48"
"NECOLE"
Sgr 19h0'7" -32d28'3"
Nectar
Gem 6h52'29" 24d5'19"
Necy's Krown
Aqr 21h41'21" 1d21'32"
Ned
Psc 1h13'7" 27d57'57"
Ned and Margaret Hess
Uma 9h16'28" 53d57'54"
Ned "Bonfire" Strain
Uma 9h37'58" 65d28'23"
Ned C * 50
Vir 12h47'54" 6d6'34"
Ned & Dorothy - Always
Cyg 19h50'41" 43d6'9"
Ned F. Seeger
Gem 6h21'34" 21d56'7"
ned geueke from cef
Uma 8h15'17" 64d9'57"
Ned Levey
Gem 6h37'21" 15d47'26"
Ned Prince
Lmi 9h40'10" 33d16'20"
Ned Santee
Uma 10h28'48" 44d56'28"
Neda Delgoshaei
Cyg 20h43'43" 30d46'0"
Neda - Meda - Star
Cyg 19h39'39" 35d11'22"
Neda Rad
And 1h25'36" 40d39'32"
Neda's Star
Tau 4h16'6" 24d34'51"
Neddra
Ori 6h5'3" 9d42'51"
Neddy
Vir 12h44'53" 7d47'55"
Nedelka
Leo 11h15'41" 5d57'50"
Nedera
Aqr 23h10'54" -17d15'29"
Nedia
Ori 5h8'29" 8d29'2"
Nedo
Uma 8h20'51" 72d25'24"
NEDRA
Aqr 22h54'5" -10d54'35"
Nedra
Leo 11h19'7" 14d59'13"
Nedra
Cnc 9h18'4" 11d56'41"
Nedra
Lyr 18h50'16" 34d49'20"
Nedra Aileen
And 0h33'11" 31d1'50"
Nedra Giroir
Ari 2h2'48" 21d20'45"
Nedra L. Bell
Sco 17h52'24" -35d38'28"
"Nedster" The Shining Star
Boo 15h38'32" 42d52'29"
Nedulon
Dra 19h33'31" 72d32'13"
Nedwidek, Thilo
Uma 10h25'50" 41d15'7"
Nedzat and Florina Isakovski
Psc 1h17'36" 10d59'30"
Neeche
Cnc 8h24'29" 7d20'41"
Neecy D
Lyn 8h44'47" 45d35'57"
Neel
Peg 22h27'10" 25d27'25"
Neel Mukesh Rathod
Uma 12h24'57" 55d58'58"
Neela
Cas 0h38'16" 50d15'21"
Neela Azaran
Psc 1h14'36" 28d15'49"
Neela Bindu
Crb 15h49'10" 36d20'38"
Neela Isabelle Adrian
Umi 15h28'31" 73d47'27"
Neela K. Ratwani
Peg 22h33'56" 25d12'2"
Neelam
Tau 5h40'52" 20d57'32"
Neelam
Sco 16h2'45" -13d30'54"
Neelam
Pho 0h37'3" -41d0'8"
Neelam Sunder Eternity
Aur 5h10'39" 36d8'18"
Neele Venetz
Ori 6h10'14" 6d44'6"
Neeley
Vir 11h40'8" -5d37'25"
Neelima Singeetham
Gem 7h29'39" 15d37'59"
NeelKamal Agarwal
Uma 11h38'50" 39d5'48"
NEELY
And 0h26'43" 25d14'21"
Neely 655
Uma 8h47'30" 63d37'51"
Neely Ann Wiek
Lib 15h12'16" -4d11'0"
Neely E. Tompkins
And 2h25'12" 46d46'39"
Neelys Legendary #8
Gem 6h50'13" 33d52'44"
Neely's Star
Cap 20h46'24" -14d57'0"
Neemada
Eri 3h46'51" -15d18'30"
Neen
Per 3h28'30" 47d53'15"
Neene
Sgr 18h53'0" -19d58'36"
Neener
Aql 19h12'28" 0d19'34"
Neener's Star
Uma 13h26'19" 55d52'19"
Neenu
Aqr 22h47'35" -6d4'12"
Neer 40
Cyg 20h28'37" 40d37'28"
Neera Bhatnagar-Blankenship
Umi 15h5'37" 81d21'57"
Neesa Morgan
Cmi 7h26'24" 2d58'24"
Neesha
Gem 6h40'19" 26d4'42"
Neesha Mooney
Cma 6h42'17" -12d15'38"
Neesha Stephens
Psc 0h39'3" 17d58'42"
Neet my Angel
Aqr 21h12'32" -9d28'28"
Neeta J Sookhoo
Ari 2h23'1" 11d48'36"
Neeto Begol
And 0h59'26" 34d31'13"
NEETU
Aql 19h23'24" 7d1'13"
Neeyalyn Anngel Ramos Kamelo
Tau 4h6'45" 4d41'22"
Nefeli Papakiriakopoulou
Umi 15h54'10" 81d29'12"
Neftali Cruz, Jr.
Cyg 21h47'38" 41d31'8"
Negar Nooralah
Cnc 8h32'28" 30d52'9"
Negbit
Cnc 7h58'5" 13d2'45"
Negin Shadaram - Shining Star
Dra 16h51'10" 52d23'43"
Negra
Cap 21h57'2" -13d48'49"
negra clara
Uma 10h44'44" 46d13'26"
Negrita
Cnc 8h18'27" 14d3'27"
Negrito Verde
Dra 18h44'6" 81d7'48"
Negro, Marc Dennis
Uma 10h48'59" 71d3'40"
NEHA
Aqr 22h10'3" 0d27'20"
Neha Susan George
And 0h10'34" 40d33'36"
Neha_Taz
And 23h54'19" 40d35'4"
Neha, Love of Justin
Ari 2h22'39" 13d54'55"
Nehad Ebrahim Al-Zayed
Vir 13h54'38" -2d44'4"
Nehemiah Gabriel Bojkovsky
Uma 14h4'0" 56d46'22"
Nehi
Leo 11h7'35" 21d9'16"
Nehmedo Issa
Cam 3h55'22" 67d59'49"
Neianne Cornett
Tau 5h32'35" 21d26'26"
Neidra Beatrice Crocker
Psa 22h18'19" -28d23'15"
Neidy Barrios Pico
Cmi 8h9'29" 0d6'43"
Neiger-Moore
Cyg 20h37'24" 48d43'3"
NE-III-159 Fox
Uma 13h34'48" 58d30'44"
Neikko Antony Mitchell
Gem 7h40'31" 27d38'39"
Neil
Ari 2h33'31" 22d13'35"
Neil
Cnc 8h52'38" 10d42'11"
Neil
Dra 12h29'57" 70d42'55"
Neil A Brooks
Cyg 19h52'38" 39d6'39"
Neil A. Johnson
Uma 11h50'31" 50d35'18"
Neil Aarron
Uma 9h14'16" 54d25'48"
Neil Addis - Dreams can come true
Dra 16h29'28" 71d21'42"
Neil Alexander Mount
Umi 15h1'43" 67d33'54"
Neil Allen Clark
Gem 6h22'9" 17d59'32"
Neil and Lynn, Always and Forever
Mon 7h19'34" -0d49'43"
Neil and Melissa
Her 17h14'12" 39d49'10"
Neil Andrew McCartan
Aur 7h26'2" 40d51'45"
Neil Anthony Santoriello
Vir 13h41'37" -6d7'9"
Neil Artman and Margaret Straub
Her 18h17'26" 29d46'12"
Neil Barclay Dickinson
Ari 2h49'35" 20d18'8"
Neil Buchanan
Cep 22h32'55" 57d58'25"
Neil C. Gibson
Sco 17h56'0" -30d7'23"
Neil Charles Sullivan
Aqr 21h47'44" 1d16'48"
Neil Christopher Young
Cep 22h22'27" 68d24'37"
Neil Curtis Mink
Her 18h23'42" 25d54'26"
Neil & Cyndi Perry
Sge 19h27'39" 18d9'39"
Neil "Dad" Gillespie
Psc 1h7'47" 4d43'42"
Neil David Hines
Per 4h37'7" 49d31'58"
Neil Davies
Aur 5h51'51" 47d7'20"
Neil E. Monello
Lib 15h28'38" -25d30'5"
Neil E. Rife
Lib 14h53'37" -1d4'40"
Neil E. Sorensen
Ari 1h48'7" 24d26'12"
Neil & Emilie Eternally Together
Dra 17h28'46" 69d2'2"
Neil Eric Hirsch
Cap 20h19'52" -13d19'53"
Neil F. Bowser
Cep 22h52'8" 73d3'47"
Neil Fleckenstein
Cnc 8h54'8" 14d22'7"
Neil Flippance
Cru 12h37'46" -57d44'41"
Neil Formisano
Cyg 20h27'52" 50d31'38"
Neil Francis Diedrickson
Aqr 22h31'54" 2d18'47"
Neil Friedman
Ori 5h38'3" 0d38'15"
Neil & Gail's Star
Cyg 21h52'5" 43d25'12"
Neil Geoffrey Bartlett
Cyg 21h57'35" 53d32'30"
Neil Geoffrey Smith
Dra 10h7'13" 75d6'15"
Neil Higgins' 40th Birthday Star
Cep 23h59'22" 74d22'36"
Neil & Howard
Per 3h42'52" 45d59'17"
Neil & Janet
Cyg 20h47'29" 48d6'15"
Neil Juhl Larsen
Cep 21h43'0" 62d33'45"
Neil Kay
Aql 19h17'50" 4d44'49"
Neil & Kerry Wood Together Forever
Vir 12h42'52" -5d9'43"
Neil Kornutick
Uma 11h43'16" 51d41'20"
Neil Kreckler
Cnc 8h43'8" 17d50'4"
Neil Lajeunesse
Her 17h53'15" 20d28'23"
Neil & Lesley Hawtrey
Cru 12h42'58" -62d20'58"
Neil & Lorna Ball - 28.1.95
Cyg 21h33'37" 49d7'20"
Neil - Love You Always
Her 16h51'19" 46d11'14"
Neil Luke - My Shining Light!
Del 20h35'53" 6d43'49"
Neil Martin
Cep 0h1'13" 75d6'4"
Neil Matz
Sgr 19h16'15" -27d19'58"
Neil McCabe
Cyg 21h10'30" 47d19'7"
Neil McPake
Cep 21h35'20" 56d1'37"
Neil Meron
Sco 16h10'26" -16d39'59"
Neil 'Mi Tapa' Sykes
Umi 14h30'20" 68d11'38"
Neil Myers
Tau 4h35'9" 28d25'38"
Neil O'Connor
Her 17h38'5" 43d32'8"
Neil P. Breen
Tau 4h37'42" 3d38'11"
Neil Robertson Duncan
Ari 2h0'17" 14d13'26"
Neil & Rose Mary Ahern
Cyg 19h36'5" 31d30'36"
Neil Ross
Her 18h6'41" 17d31'13"
Neil Savage
Uma 8h57'49" 51d8'8"
Neil Shannon Mascarenhas
Boo 15h16'39" 48d9'18"
Neil Stephen Alpern
Her 18h51'51" 20d52'48"
Neil Stephenson
Cas 0h22'22" 53d33'43"
Neil Sullivan
Tau 4h30'41" 26d10'46"
Neil T. Harkins
Cnc 8h19'55" 28d54'38"
Neil & Terri Soderstrom
Cyg 19h45'38" 32d45'56"
Neil & Tracy
Uma 8h37'12" 69d12'9"
Neil Wallace
Her 17h26'32" 24d24'53"
Neil Wallette
Vir 14h45'40" 4d57'11"
Neil Webley
Uma 14h23'22" 60d46'51"
Neil Weed
Crb 15h27'37" 27d16'14"
Neil Williams
Boo 14h30'8" 24d48'35"
Neil (Zoobie) & Ami (Woobie)
Lyn 9h7'10" 45d19'33"
Neil031104
Umi 15h4'45" 76d56'18"
Neila Louise Lambo Conner
Gem 6h29'57" 22d34'39"
Neila Yara Michiles Bono
Gem 7h4'24" 17d10'46"
Neilana
Cnc 8h42'55" 25d34'39"
Neilana
Cnc 9h5'33" 7d48'43"
Neil-Cole-Ben
Cam 4h30'19" 53d56'25"
Neilee Marie Reuser
Uma 10h24'37" 54d14'55"
Neilium
Boo 14h18'30" 46d43'48"
Neill Gardner Tompkins
Sco 17h44'59" -43d59'54"
neily-bug
Lyn 7h57'27" 56d43'0"
Neisecke, Lieselotte
Ori 5h8'35" 8d53'12"
Neisha Dawn Taylor
Gem 7h2'36" 24d10'39"
Neisha Elizabeth Sargeant
Cra 19h6'46" -39d11'57"
Neishma Luz Lopez
Cam 5h38'29" 74d36'37"
Neita Kekel
Sco 17h32'10" -44d25'14"
Nejla Güryen-Yavuz
Umi 16h8'6" 80d40'52"
* Nejma * * P * Belaïssaoui *
Del 20h34'43" 6d41'13"
Nejra Sahbegovic
Cru 12h48'12" -61d31'12"
Nejume; a bright star among chaos.
Umi 15h50'10" 87d54'22"
Nek Setay One
Lmi 10h31'22" 36d47'53"
Neki Jane NFYTL
Cap 21h29'26" -19d56'26"
Nekia Nodrog
Dra 15h25'40" 61d17'32"
Neko
Peg 21h52'13" 13d33'0"
Nekunk ragyogj!
Vir 13h35'56" -10d36'46"
Nela
Umi 13h37'39" 71d42'33"
Nelda
Tau 4h15'39" 8d35'34"
Nelda Beth
Uma 10h40'47" 53d31'41"
NELDA (FRANCES)
Cas 0h38'6" 52d25'54"
Nelda Lee King Walter
Uma 9h25'35" 43d49'18"
Nelda Sturgeon
Cam 5h51'56" 75d52'11"
Nelda Worthley
Cyg 19h42'6" 52d12'43"
Neldo
Psc 0h28'6" 6d34'52"
Neldon J Bowman
Uma 9h38'51" 67d36'54"
Neldo's Light
Cru 12h20'29" -55d48'41"
Nele De Keyzer
Crb 15h24'49" 25d47'55"
Nele Hannah
Uma 11h58'13" 61d6'51"
Nele Louise Weber
Uma 10h1'34" 43d59'32"
Nele Maus
Uma 8h37'53" 46d59'55"
Neleen Maimone
And 2h19'32" 45d49'51"
Nele-Marie
Uma 10h17'37" 48d57'7"
Nele's Star
Aql 19h28'36" 4d12'8"
Nelia
Umi 16h58'59" 83d20'56"
Nelia
Lib 15h8'21" -17d18'39"
Nelia Hudson Mikes Baby Doll
Gem 7h29'14" 23d35'31"
Nelida and Charles Rodriguez, Sr.
Uma 11h5'6" 58d55'39"
Nelida O. Erneta
Lyr 19h19'57" 29d28'34"
Nelio
Lib 14h38'42" -23d39'16"
Nell
Cas 1h42'11" 62d49'45"
Nell Keyworth
Uma 9h58'37" 51d7'1"
Nell Rose
Cas 0h46'55" 63d56'7"
Nell Valentine Cote
And 0h39'1" 33d54'56"
Nell Yvette
Vir 12h47'45" 11d47'49"
Nella
Aqr 21h47'33" -2d5'49"
Nella Michaell
Sco 17h26'44" -40d25'38"
Nelle Cary Welsh Koepfgen Maharg
Cyg 19h58'17" 30d42'3"

Nelle Lai
Tau 4h13'54" 4d20'29"
Nelle Marie
Aqr 22h39'29" -2d58'56"
NellGus 50
Uma 10h10'25" 55d56'23"
Nelli Ginzburg
Tau 3h43'49" 30d12'52"
NELLI halhatatlan orangyala
Cas 23h49'19" 58d12'1"
Nellie
Boo 14h34'39" 19d8'5"
Nellie B. Hobbs
Mon 6h45'39" -0d27'59"
Nellie Christine Noble
Psc 1h35'44" 16d24'12"
Nellie Cirino
And 0h31'17" 39d1'20"
Nellie Drew
Uma 9h40'12" 45d23'3"
Nellie Jensen
Uma 10h24'33" 67d28'17"
Nellie Lousie Huegel
Vir 13h45'21" 5d40'2"
Nellie Momchilov
Uma 10h6'6" 56d40'5"
Nellie Noble
Psc 1h36'45" 17d59'22"
Nellie Park
Lib 15h9'21" -4d53'14"
Nellie Taylor
Boo 15h29'59" 37d32'35"
Nellie Va
And 23h31'46" 36d39'3"
Nellie Young
Lmi 10h34'54" 32d44'19"
Nellie's Star
Uma 9h29'12" 62d6'58"
Nello
Peg 21h55'39" 12d40'21"
Nellwyn's Mister Maxamillian
Cnv 13h55'58" 34d4'17"
Nelly
Lyr 18h35'16" 32d16'55"
Nelly
Cyg 21h26'9" 33d4'33"
Nelly
Psc 1h29'53" 10d59'26"
Nelly
Oph 17h5'25" -0d53'52"
Nelly & Corsin Camadini
Com 13h33'53" 21d27'17"
Nelly Dath
Uma 8h59'9" 52d11'52"
Nelly Habib
Leo 11h24'37" 2d31'13"
Nelly Hebrard
Sco 15h58'15" -22d55'31"
Nelly Jucker-Wegelin
Cas 1h19'8" 68d21'32"
Nelly M. Molina
And 1h24'29" 47d35'11"
Nelly Maria Cammareri
Cyg 20h0'23" 31d39'1"
Nelly Robles
And 2h12'45" 42d19'44"
Nelly Rodriguez
Cap 21h39'6" -19d14'54"
Nelly Yescas
Tau 4h13'4" 1d54'37"
Nelly's Star
Leo 11h3'26" -4d22'17"
Nellysabel
Sgr 19h30'13" -27d31'18"
Nelsa Mona
Vir 12h8'41" -10d51'28"
Nelson
Per 3h38'10" 46d26'30"
Nelson Ariel Molina
Cap 21h15'8" -22d17'30"
Nelson & Chary Ruiz
Cyg 20h50'55" 37d35'16"
Nelson Edward Ahr
Ari 3h10'16" 24d6'28"
Nelson Elliot Miranda-Rivera
Gem 7h43'4" 32d46'43"
Nelson Ferreira
Sgr 18h41'56" -28d2'51"
Nelson Forrest Small
Uma 12h25'9" 56d2'42"
Nelson Lena Lexane
Aur 5h58'56" 45d31'55"
Nelson McLemore, III, MD
Aur 6h21'47" 41d54'41"
Nelson Omar Juarbe
Ori 5h14'1" -7d29'17"
Nelson Palmer
Vir 13h27'33" 7d48'13"
Nelson Peltz
Tri 1h55'49" 34d38'55"
Nelson Perez & Erica Baessler
Ori 5h38'34" -2d28'48"
Nelson "Shamir" Tyler
Cep 20h40'48" 59d57'28"
Nelson T. Alberto (Papi)
Ari 3h17'6" 26d51'43"
Nelson Woodburys Black Hole
Cnc 8h15'25" 16d13'46"
NeltonZavierNealKimberlyAnnLeathers
Uma 13h18'45" 57d31'10"
Nelva Cristina Centeno Vasquez
And 23h8'52" 36d19'16"
Nemesis (April Sims)
Cyg 21h11'13" 42d2'39"
Németh Attila
Lib 15h47'22" -17d51'34"
Németh Kristóf
Sgr 18h26'23" -16d13'18"
NEMI
Sco 16h12'52" -15d34'26"
Nemik
Umi 15h49'5" 72d43'0"
Nemno and Flunder
Uma 10h51'33" 42d36'52"
NEMO
Cnc 8h43'58" 30d24'2"
Nemo
Per 3h4'48" 50d13'36"
NEMO
Lyn 7h53'37" 46d56'32"
Nemo
Dra 18h40'37" 50d52'9"
Nemo
Psc 1h7'35" 27d7'54"
NEMO
Peg 21h33'29" 20d50'32"
Nemo
Psc 23h29'33" 7d57'10"
Nemo
Sgr 20h1'22" -36d38'17"
Nemo 83-1
Umi 14h49'32" 72d18'57"
Nemo Chamsy
Uma 12h26'0" 52d29'14"
Nemo & Fishey
Del 20h49'8" 17d16'25"
Nemo -Lady D's Love
Gem 6h17'36" 22d26'41"
Nemo Wan
Cnc 9h12'29" 25d23'7"
Nemy
Crb 15h17'20" 25d51'40"
Nena
Gem 7h12'44" 17d11'38"
Néna
Vir 12h59'30" 5d40'7"
Nena
And 2h3'39" 36d33'28"
Nena
Lyn 7h46'15" 42d12'0"
Nena
Cep 21h14'21" 55d36'29"
Nena Barragan
Gem 7h25'5" 26d7'42"
Nena Gerber
Leo 11h27'51" 13d51'23"
Nena Landers
Leo 11h51'38" 17d32'6"
Nena Santos "James' Angel"
Gem 6h29'3" 16d2'51"
Nena Snodgrass
Cnc 8h0'47" 13d3'52"
Nenad
Uma 13h9'46" 56d25'20"
Nenad & Angela
Gem 7h36'11" 16d7'46"
Nenad Angelov - Macedonia
And 1h12'30" 42d9'47"
Nenad's Angel
Tau 3h29'43" -0d29'11"
Nenandro Reyes, Sr.
Uma 11h36'58" 49d38'40"
NENE
And 2h23'40" 50d31'41"
Nene
Mon 7h13'19" 0d17'33"
NENE
Cap 21h43'26" -14d46'56"
Nene Grace
Lyr 18h35'56" 36d20'47"
Nenelandia
Ori 5h13'31" 1d15'51"
Nener
Vir 13h33'16" -3d27'51"
Neni
Aqr 22h22'36" -1d38'26"
Neni - du bist immer bei uns!
Umi 13h13'9" 72d57'1"
Nenita StaAna - VAVA's
Umi 14h35'47" 73d29'46"
Nenito
Gem 7h16'16" 18d58'28"
Nenna
Lyr 18h46'40" 30d36'8"
Nennars Star
Dra 15h0'29" 58d57'49"
Neno K. Gechanov
Uma 9h8'51" 66d58'17"
Neo
Ori 5h56'7" 12d35'27"
NEO
Ori 6h16'6" 8d32'50"
Neo and JJ's Star
Sge 20h4'33" 17d12'46"
Neo Cordingley
Her 17h40'22" 38d41'2"
Neo Eltis
Aqr 22h26'1" -7d25'42"
Neo Irna Ahjs
Del 20h32'45" 17d1'16"
Neo Ross Thomas
Lmi 10h49'1" 30d27'35"
Néo Thonet
Lib 14h34'46" -19d25'15"
Neofitos Economou
Uma 11h46'41" 59d22'24"
Neoma Maynes
Mon 6h52'1" 8d27'48"
Neon
Lyn 7h11'30" 57d7'41"
Neon
Umi 13h5'0" 69d59'56"
Neon Leon
Cap 21h34'11" -21d50'18"
Neon Yellow
Cnc 8h36'26" 21d35'49"
Neons Laughing Eyes
Umi 16h23'24" 72d53'33"
NeoWave
Uma 11h55'34" 38d28'56"
Nephastie - SMF
Cas 0h6'42" 50d23'32"
Nephtaphis
Aql 19h38'13" 15d55'50"
Neptune
Tau 4h32'42" 20d39'17"
Neptune
Uma 8h21'37" 62d35'57"
Neptune's Denn
Cep 22h55'3" 59d58'15"
Neptunis Autumnis Giovannus Jupitus
Cnv 12h29'3" 43d9'18"
N'er One
Cam 3h57'25" 64d35'7"
Nera My Beautiful Angel
Sgr 19h45'57" -16d55'40"
Nera Sarovic
Aur 6h9'30" 29d12'47"
Nerakinanaj
Aqr 22h3'58" 1d34'17"
Neraknllib
Cap 21h38'5" -12d53'50"
Nerdy
Lyn 7h55'3" 55d6'18"
Nereida Melgarejo
Aqr 20h52'14" -1d6'5"
Nereyda Garcigas
And 0h17'51" 40d40'17"
Nergiz
Cyg 21h56'5" 38d40'41"
Neri
Sco 16h15'41" -16d38'12"
Neri "Light of God"
Lyr 18h45'9" 31d20'5"
Neriah Dawn Milis 16-09-2003
Cru 12h49'52" -62d2'2"
Neridah's Bit of Heaven
Ari 2h39'53" 13d2'58"
Nerissa Mae
Lmi 10h29'37" 34d9'13"
Nerissa1
Cyg 20h5'29" 34d20'59"
Nermalicious
Cnv 14h5'16" 36d33'29"
Nermin M. Alanwar
Tri 2h26'0" 30d21'24"
Nermina
Ori 5h31'46" -8d10'44"
nerny-meyer 1217
Lyn 8h0'37" 58d40'16"
Néroliza Quiles
Tau 4h37'11" 19d23'31"
Nero's Star
Cma 6h23'33" -21d30'43"
Nery
Dra 15h34'48" 60d8'38"
Nery
Gem 7h44'24" 34d27'25"
NERY"S HEART
Gem 6h40'49" 33d10'40"
Nesa Michele Londer
Aqr 21h57'38" 1d38'56"
Nesamani Tharmalingam
Ori 5h21'26" 11d19'15"
Neshtemon
Tau 5h15'33" 21d59'31"
ne-si'ka tsil
Uma 11h24'42" 30d36'36"
Nesli Emre - Ceyhan Karacan
Her 17h43'37" 32d55'31"
Neslihan Erten
Cnv 12h44'3" 38d26'52"
Nesrine El Bouhassani
Cap 20h32'59" -26d6'47"
Nesrine Sabrina Hottier
Ori 5h25'54" -3d57'44"
Ness
Cas 23h41'30" 63d3'27"
Ness Marie Poli
Dra 18h32'42" 56d0'50"
Ness Pooh's
Umi 16h8'28" 74d14'1"
"Ness" Steven Robert Collins
Psc 0h32'21" 3d53'28"
Nessa
Gem 6h35'51" 24d51'33"
Nessa Dorphonion
Cra 18h2'5" -37d28'55"
Nessanben
Cnc 8h31'39" 13d41'45"
Nessa's Shining Star
Gem 6h52'1" 19d18'51"
Nessi100780
Uma 11h36'30" 31d35'28"
Nessie '67
Cas 23h20'3" 55d30'10"
Nessie Cymraeg Wood
Cap 20h12'22" -12d39'40"
Nessmann, Josef
Uma 9h8'41" 50d58'41"
Nessy
Tau 5h59'13" 25d9'2"
Nesta
Cma 7h10'50" -25d47'30"
Nestle
Umi 15h32'28" 70d37'2"
Nestle Products Sdn Bhd
Ori 5h39'33" 3d14'55"
Nestor Javier Perez
Uma 9h57'37" 55d13'36"
Nestor Ricardo Catalan
Aqr 22h33'17" -2d17'4"
Nestore
Per 3h7'13" 51d41'54"
Neta
Vir 12h43'42" 7d7'33"
Neta Corley
Lyr 18h35'42" 33d1'59"
NETAT
Cma 6h42'51" -28d47'55"
Neter Naim Cook
Cap 20h35'8" -20d20'26"
Neterka Amun Ra
Tau 4h19'19" 21d52'44"
Netney
And 0h53'13" 44d23'23"
Netomania
Lmi 10h31'53" 37d45'15"
Netta Antonette Rubino Falzone
Sco 17h56'10" -38d13'20"
Netta Sheives
Ori 6h2'7" 6d47'59"
Nettchen
Uma 11h42'54" 28d54'42"
Nette
And 23h25'16" 48d3'23"
Nette
Lib 14h22'32" -11d30'30"
Netti Schechter
And 1h43'56" 43d49'0"
Nettie
Lmi 10h7'44" 33d38'23"
Nettie
Cnc 8h40'43" 10d58'59"
Nettie
Cyg 19h55'27" 55d30'3"
Nettie Faye Umbreit
Cap 21h7'48" -22d21'13"
Nettie Harrington
Dor 5h17'16" -58d59'48"
Nettie Lou McClure Lindsey
And 0h31'6" 42d7'27"
Nettie Lynch
Psc 1h5'33" 30d47'4"
Nettie-My Sunshine, My Love+Friend
Cap 21h46'3" -17d16'43"
Netty
Dra 19h16'33" 63d46'21"
Netty
Uma 11h56'8" 61d11'57"
Netty
Uma 10h7'37" 48d25'3"
Netty
Tau 5h53'54" 25d21'8"
"NETTY" - ich liebe dich
Uma 11h29'33" 29d44'48"
Netty Peddles Joa
Sgr 19h39'21" -12d55'10"
Netty's Star
Umi 15h38'29" 83d46'13"
Netylicious
Ori 5h19'8" 6d29'3"
Netzsch, Helga
Gem 7h17'42" 20d29'22"
Neubauer, Peter
Uma 8h20'56" 64d47'3"
Neufeld Anniversary #1
Vir 14h0'52" -5d52'56"
Neugebauer, Albert
Ari 3h9'28" 19d22'43"
Neuhaus, Norbert
Uma 11h30'8" 36d51'48"
Neuhaus, Ute
Uma 9h25'45" 68d36'37"
Neumann, Mirko
Uma 12h8'51" 60d12'35"
Neumann, Uwe
Ori 5h15'36" -8d45'55"
Neumer, Ulf
Sco 16h51'19" -38d47'54"
Neus. B. E. & Joaquin. A. Z.
Lib 14h26'19" -23d25'50"
Neus Miquel
Del 20h50'8" 4d9'38"
Nev & Bev's little piece of heaven
Ara 17h23'41" -46d35'26"
Neva
Cnc 8h50'48" 13d28'31"
Neva Brown
Uma 12h10'28" 48d26'30"
Neva Christine Austin
Lyn 8h48'10" 34d16'32"
Neva May & Creighton Edward Diener
Ari 3h8'44" 28d42'17"
Neva "Osito Panda"
And 0h25'11" 41d47'12"
Nevada
Ori 5h51'3" 2d36'44"
Nevadah Haley Trent
Uma 11h17'20" 40d36'49"
Nevaeh
Per 3h13'1" 43d34'33"
Nevaeh
Uma 11h44'51" 51d25'39"
Nevaeh
Vir 12h43'41" 4d2'20"
NEVAEH
Tau 4h32'3" 23d37'13"
Nevaeh
Ari 2h18'33" 23d55'24"
Nevaeh
Cru 12h16'43" -63d42'25"
nevaeh
Ori 4h57'15" -0d21'27"
Nevaeh
Cap 21h43'0" -20d34'33"
Nevaeh
Cas 1h23'32" 61d14'1"
Nevaeh Destiny Moore
Psc 0h4'21" -1d53'57"
Nevaeh Grace Onesti
And 1h43'31" 44d30'8"
Nevaeh Jaide
Aqr 22h36'27" -17d18'22"
Nevaeh Jordeyn Rae McLees Bastyr
Uma 10h36'7" 68d50'16"
Nevaeh Joy Feusi
Leo 10h17'11" 25d32'26"
Nevaeh Marie Green
Aqr 22h30'50" 2d16'39"
Nevaeh Mia Riojas
Ori 6h5'18" 19d0'41"
Nevaeh Nickole Crawford
Aqr 23h11'27" -11d42'43"
Nevaeh Ranee Raba
Cam 4h14'5" 61d36'41"
Nevaeh's Heaven
Ari 2h46'42" 27d36'59"
Nevaeh's Little Piece of Heaven
Leo 10h48'39" 10d13'14"
Nevan Cole Jones
Ori 5h42'47" 0d28'32"
Neve Alexandra
Umi 13h57'52" 72d25'44"
Neve Elizabeth Murray - Neve's Star
And 23h24'46" 50d44'42"
Neve Giselle - 15.09.2005
Cru 12h39'34" -61d55'32"
Neveah
Aql 19h46'11" 6d29'56"
Neveah Jaylin
And 2h17'42" 41d17'2"
Neveen Kamal
Lib 15h45'28" -28d26'24"
Neven Kolomejec
Cyg 20h36'49" 48d1'20"
Nevena Slovic
Cyg 20h39'35" 48d22'45"
Nevenka Jurkovic
Vir 12h47'12" 6d59'28"
Never Alone
Cru 12h24'44" -61d16'6"
Never Ending Dawn
Cnc 8h43'1" 26d58'53"
Never Ending Love
Cyg 21h11'57" 47d16'12"
Never Ending Love
Cyg 20h30'22" 45d0'45"
never ending love
Umi 13h42'26" 73d38'24"
Never Ending Love
Sco 16h17'52" -10d51'10"
Never Forget
Umi 15h29'15" 89d22'52"
Never Forget
Crb 16h21'56" 32d44'17"
Never Give Up Hope
Uma 10h28'6" 64d9'0"
"Never Gone"
Per 4h11'38" 39d12'46"
Never Let Go
Dra 18h11'23" 72d34'5"
Never Too Permanent: Sam&Am's Star
Cap 21h5'17" -14d38'31"
Never Touch The Ground
Uma 11h32'43" 50d15'7"
Never-ending Love
Aql 19h51'36" -0d17'4"
NeverForgetMyLoveIsWatchingOverYou
Dra 15h47'28" 59d50'56"
Neverland
Lyn 8h9'22" 55d40'31"
Neverland
Umi 14h47'27" 70d22'18"
Neverland
Dra 20h15'56" 63d54'20"
Neverland
Dra 16h11'36" 61d28'52"
Neverland
Cir 14h53'37" -65d40'33"
Neverland
Uma 10h50'45" 48d56'53"
Neverland
Ori 5h54'38" 6d15'33"
Neverland 14.07.2007
Umi 17h30'16" 81d21'2"
Neverland 9/5/91
Her 16h15'49" 26d5'29"
Neverland for Misfit Children
Peg 22h53'31" 26d33'39"
Nevermind
Her 18h39'54" 15d48'14"
Nevermore
Uma 11h27'7" 67d40'48"
Neville A. Morris
Gem 7h19'21" 32d46'15"
Neville Benvenuti
Cen 13h34'54" -48d51'11"
Neville & Emma Bluckert
Mon 6h30'9" 8d28'46"
Neville H. Cumming
Uma 9h18'43" 42d26'12"
Nevin E. B. Martin
Ari 3h0'17" 11d0'47"
Nevin Jagrup
Per 4h37'12" 42d36'36"
Nevin Minor
Umi 15h16'0" 70d18'30"
Nevin Stoutenburg
Ori 6h11'38" 17d35'28"
Nevio Ugo Mori
Umi 15h53'50" 80d18'41"
Nevio, the shining star forever!
Ori 5h54'44" 12d46'20"
Nevis, West Indies
Vir 12h37'13" 1d7'55"
NEW BALTIJA INC.
Umi 15h11'30" 78d32'14"
New Beginings/Bill & Judy
Uma 8h35'31" 57d28'47"
New Beginning
Pho 0h13'45" -43d54'0"
New Beginning
Cru 12h38'14" -58d12'5"
New Beginnings for Trish
Vir 13h55'53" 1d43'42"
New Bolton Center
Her 17h28'29" 35d23'24"
New Columbia
Uma 10h50'43" 48d0'53"
New Emmeline Mina
Pho 1h15'12" -41d51'37"
New Journey
Uma 9h14'2" 62d1'10"
New Lebanon Flash
Lib 15h4'29" -11d31'18"
New Life
Pho 23h40'22" -48d13'26"
New Love True Love, Pam & Steve
Per 4h17'1" 52d26'19"
New Providence
Uma 8h26'39" 62d57'8"
New Star Nadja - 06.10.2006.
Her 18h54'16" 21d40'0"
New Ventures México
And 1h50'49" 45d46'33"
New York New York
Lib 14h50'48" -7d36'37"
Newberry
Uma 11h27'35" 62d58'39"
Newby 2005 ME of the Year
Uma 9h22'44" 46d22'17"
Newby Hands
Her 18h10'19" 16d33'22"
Newell S. Habermacher
Lib 14h51'29" -6d16'37"
Newell's Star
Cra 18h47'40" -40d53'52"
Newman #14
Ori 6h17'5" 14d25'12"
Newman Riechman
Umi 15h15'36" 68d7'16"
Newman's Star
Aqr 22h32'1" 1d21'17"
Newnsie
Dra 16h20'30" 58d3'48"
Newport Navigator
Lib 15h2'50" -23d22'19"
News at Eleven
Uma 10h13'5" 62d49'8"
Newson's Devotion
Gem 6h3'24" 25d14'9"
NEWSPRINT
Sco 16h5'13" -12d43'28"
Newton
Cma 7h21'45" -21d40'55"
Newton Fox Decker
Ari 3h20'59" 28d57'37"
Newton Peckarsky
Ori 5h39'4" 12d8'21"
Newton Sparkle
Cnc 8h47'7" 14d24'45"
Newton Ward James
Peg 22h32'30" 24d21'11"
Newty
Cnc 8h38'36" 17d17'22"
NexOne
Her 17h23'35" 17d4'16"
Next Portal
Dra 18h45'28" 53d20'55"
Nexus
Leo 9h30'28" 12d18'33"
Neyda Mora
Del 20h43'17" 15d49'34"
Neysa Pineda
Psc 23h59'40" 8d19'23"
Ne-Zhoni Ah-Tad
Leo 9h26'51" 10d53'49"
Nezzar N Selim
Dra 16h15'10" 63d42'50"
NFullerStar
Aqr 22h28'30" -8d51'0"
Ng ShiYing, Vivian
Cnc 8h47'16" 32d12'42"
Ng Tze Yuen & Peggy Teo
Cyg 19h37'53" 31d26'43"
Nga
Lyn 8h27'2" 56d48'26"
Ngaio Helen Felix
And 1h11'34" 42d28'53"
Ngakai. Angie's Angel.
Cru 11h59'24" -62d17'28"
Ngan Kim Do
And 0h7'3" 45d25'28"
Nghi Phung Lu
Gem 7h28'21" 33d53'55"
Ngoc D'Ann Patty Thach
Gem 7h24'22" 15d17'4"
Ngoc K. Le
And 2h21'52" 45d37'10"
Ngozi * Heart of Courage & Strength
Psc 1h24'52" 28d1'1"
Nguyen
Uma 12h38'36" 55d25'44"
Nguyen Family
Lyn 8h41'56" 36d18'55"
Nguyen Hoang Nguyen
Tau 4h4'0" 5d39'44"
Nguyen Quynh Nhu
And 23h19'8" 48d14'38"
Nguyen Thi Khanh Ly
Cnc 9h16'15" 17d7'29"
Nguyen Thi Thu 1
Lyn 8h1'8" 47d22'27"
Nhat Tran
Lyn 7h27'3" 55d36'37"
Nhi Nguyen
Per 3h8'34" 55d23'59"
Nhoj (A.K.A. Planet Johnny Boy)
And 2h33'36" 38d51'53"
Nhu-An Nguyen
Ari 3h17'10" 16d49'6"
Nhu-My
Sco 17h53'12" -36d1'27"
Nhung Ngo "Linda"
Uma 9h15'20" 57d30'26"
Nia
And 0h22'0" 39d27'7"
Nia and John's Star
Cyg 19h40'19" 47d51'37"
Nia Catrin Jones
Ari 2h34'9" 18d48'9"
Nia Elizabeth Etienne
Cap 20h25'1" -18d29'9"
Nia Jones
Cas 23h53'53" 50d5'14"
Nia S
Sco 17h23'24" -32d11'31"
Niagra Falls
Crb 15h36'19" 25d54'58"
Niall Cuinn Gallagher
Per 3h23'38" 47d37'29"
Niall Donohoe 1978
Uma 12h37'6" 52d39'13"
Niall "HT" Masterson
Ori 5h58'17" 7d6'31"
Niall Mullarkey
Tau 4h33'54" 27d18'10"
Niall O'Loingsigh
Sco 16h40'39" -27d38'24"
Niall Patrick Kelly
Ori 5h27'26" 0d16'55"
Niam A. Patel
Lyn 8h35'59" 43d24'56"
Niamh
And 0h5'6" 48d4'21"
Niamh
And 23h32'47" 37d49'26"
Niamh
And 23h2'25" 51d21'43"
Niamh
Leo 9h57'56" 24d58'31"
Niamh and Sarah
Aur 6h26'59" 31d37'55"
Niamh- Angel with butterfly Wings
Gem 7h12'30" 25d55'16"
Niamh Balfe
And 23h2'32" 51d39'36"
Niamh (Bright and Beautiful)
And 0h3'14" 45d54'46"
Niamh Bronagh Lourdes Ryan
And 2h20'45" 42d30'43"
Niamh Caitlin McInerney
Pav 18h18'39" -58d45'0"
Niamh Crowley
Cas 1h57'45" 61d6'40"

Niamh "Cutie" McGoldrick
And 0h29'54" 45d31'29"
Niamh de Warrenne Lee
Peg 22h39'23" 6d1'53"
Niamh Duffy
Cas 23h40'27" 52d41'40"
Niamh Elizabeth
And 2h27'36" 50d13'45"
Niamh Eloise
And 1h27'52" 43d22'23"
Niamh Emma Murphy
And 1h56'32" 35d44'40"
Niamh Erin Wake
And 22h57'36" 51d44'52"
Niamh Fallon
Aqr 22h27'31" -24d49'20"
Niamh Frances Birks
And 23h40'37" 47d46'13"
Niamh Hogan
And 2h4'37" 46d57'45"
Niamh Joanna
And 1h17'26" 43d19'13"
Niamh Kate Garnett
And 0h12'52" 39d56'8"
Niamh Louise Collins
And 23h43'44" 37d58'37"
Niamh Lunny
And 1h23'24" 41d31'12"
Niamh Mary Ettie Stockdale
And 1h42'50" 47d36'20"
Niamh O'Brien
Aur 5h11'1" 42d58'6"
Niamh O'Raw
Uma 8h12'46" 70d12'21"
Niamh Sheehan Wonder Child
And 1h48'23" 39d19'7"
Niamh's Life Star
And 23h38'30" 49d6'18"
Niani Sekai
Sgr 19h11'14" -32d34'6"
Nia's Brilliance
Aqr 22h28'32" -12d5'28"
Nibal & Jennifer
Sgr 18h51'12" -18d23'49"
Nibbit's Heart
Her 18h32'43" 12d15'31"
Nibbles
Lep 5h16'50" -20d9'26"
Nibbs
Aur 5h31'13" 40d32'0"
Niblet
Cyg 19h57'44" 51d57'53"
Niblet is pure bliss
Uma 10h34'51" 50d2'52"
Nic
Crb 16h21'16" 29d22'12"
Nic and Ann Drinkwater
Cyg 19h56'26" 49d32'51"
Nic and Nif Best Friends Forever
Tau 4h22'57" 27d0'7"
Nic + Andreas
Uma 9h54'37" 49d29'8"
Nic B loves Scott P
Cru 12h38'39" -59d41'3"
Nic Bigott
Cnc 8h1'54" 10d53'45"
Nic Francis
Lib 15h39'33" -19d23'31"
Nic Groves
And 2h14'17" 50d26'30"
Nic Hübscher
Equ 21h3'50" 5d12'13"
Nic Lottering Star
Ori 5h38'26" -2d34'29"
Nic 'n' Sil
Psc 0h39'10" 16d14'26"
Nic Radolovich
Tau 4h37'41" 18d16'1"
Nic/Tuck
Aql 19h36'18" 13d32'7"
Nica
And 2h12'35" 41d16'54"
Nica 1918-1987 Siempre Te Recuerdo
Uma 9h34'10" 45d7'4"
Nicä - für immer mi grossi Perle
Lac 22h26'35" 47d32'1"
NicAbbey
Ori 6h14'55" 15d9'54"
NicAli Carabba
Tri 2h21'11" 32d45'17"
Nicci
Umi 16h53'12" 76d30'36"
Nicci and Curtiss' friendship star
Gem 7h6'18" 10d15'43"
Nicci and Don Beeck...loving you
Gem 7h44'22" 32d26'49"
Nicci & Donobin
Sco 16h11'10" -11d32'36"
Nicci Hess
Sgr 18h47'37" -33d37'32"
"Nicci-18" Star of Steve & Denise
And 0h8'41" 28d56'35"
NICCO
Cas 1h23'47" 69d28'4"
Niccole and Robert Moore Forever
Lib 15h9'11" -5d31'28"
Niccole Carmella Mastandrea
And 0h59'49" 44d26'33"
Niccole Christine Clark
Lyn 7h5'10" 56d48'58"
Niccole's Star
Lib 15h5'41" -26d44'40"
Niccolò Bazzini
Per 3h19'23" 31d59'45"
Nice
Cap 20h49'44" -16d16'23"
Nice' - Forever My Bright Star
Cnc 8h14'3" 16d11'37"
NicGarey Amour
Cru 12h19'26" -56d35'44"
Nichelle Marie Oldham
Sco 16h40'45" -30d29'32"
Nichita Yuyen
Cas 1h21'9" 64d18'38"
Nichol Lee Kruger
And 0h41'0" 41d19'26"
Nichol M. Varao
Cyg 20h19'42" 40d14'6"
Nichol Vaino
Per 3h19'54" 50d35'49"
Nichola
Vir 13h17'48" 4d9'29"
Nichola Jayne Bye
And 1h25'9" 42d40'58"
Nichola Julie Stubbins - Tichola's Star
And 1h9'52" 39d24'16"
Nichola Kennedy
Cyg 21h22'53" 50d24'7"
Nichola Shân Jones
Cas 1h19'56" 66d30'14"
Nicholaas Peter Ulrich Leembruggen
Aql 19h13'20" -0d0'29"
Nicholaos S. Kourbelas
Cap 20h51'9" -16d33'36"
Nicholas
Sgr 19h27'54" -16d27'2"
Nicholas
Uma 9h26'3" 65d31'58"
Nicholas
Dra 20h39'31" 68d51'3"
Nicholas
Uma 11h32'36" 57d51'51"
Nicholas
Sgr 18h48'47" -30d9'51"
Nicholas
Cas 0h1'52" 53d3'52"
Nicholas
Per 2h59'46" 52d58'54"
Nicholas
Uma 9h46'58" 49d59'14"
Nicholas
Her 17h49'37" 49d26'49"
Nicholas
Per 4h41'40" 46d51'0"
Nicholas
Her 16h46'25" 43d26'19"
Nicholas
Gem 7h13'12" 31d35'20"
Nicholas
Gem 7h5'14" 35d1'44"
Nicholas
Her 17h11'43" 32d21'35"
Nicholas
Vir 14h39'14" 2d55'9"
Nicholas
Ori 5h47'34" 3d40'39"
Nicholas
Ori 5h30'36" 4d38'40"
Nicholas
Ori 6h22'19" 9d59'34"
Nicholas
Cnc 8h59'42" 8d42'52"
Nicholas
Leo 9h24'7" 12d33'4"
Nicholas
Aql 19h31'13" 12d26'25"
Nicholas
Gem 7h7'37" 20d44'49"
Nicholas
Ori 6h14'32" 18d28'21"
Nicholas
Cnc 9h5'48" 19d11'19"
Nicholas
Boo 14h58'1" 21d51'32"
NICHOLAS
Del 20h40'44" 15d21'10"
Nicholas
Psc 1h13'16" 25d1'14"
Nicholas
Psc 1h11'11" 23d13'54"
Nicholas
Her 18h32'51" 25d26'15"
Nicholas - 10/02/03
Umi 15h55'20" 82d6'42"
Nicholas 51
Sgr 19h1'39" -35d13'17"
Nicholas A. Benton
Uma 10h0'46" 53d27'8"
Nicholas A George
Cam 4h3'57" 68d10'27"
Nicholas A. Greener
Cyg 19h44'46" 30d29'58"
Nicholas A Pruitt
Cap 21h24'37" -23d11'40"
Nicholas A. Ross
Aql 20h11'25" 8d0'4"
Nicholas A. Sisak
Leo 9h31'26" 24d54'43"
Nicholas A. Thom
Aqr 20h53'2" -10d10'27"
Nicholas A. Vaughn
Her 17h39'7" 34d54'47"
Nicholas Aaron Hudson
Boo 14h52'32" 22d18'45"
Nicholas Aaron Ortiz
Sco 17h30'42" -41d18'11"
Nicholas Accordino
Boo 15h14'11" 48d50'43"
Nicholas Adam Berry
Leo 10h14'59" 17d43'19"
Nicholas Adam Bowden
Uma 9h57'58" 57d18'26"
Nicholas Aeneas Betro
Uma 10h2'54" 46d23'1"
Nicholas Albert Yaden
Vir 12h7'37" -3d33'31"
Nicholas Alberto
Her 17h40'23" 38d34'3"
Nicholas Alessi 8/27/1982-3/1/2004
Vir 14h40'43" 2d44'36"
Nicholas Alexander Albrecht
Tau 5h50'2" 15d13'43"
Nicholas Alexander Marin Garcia
Cma 6h55'25" -31d50'4"
Nicholas Alexander Pasquis
Crb 16h22'31" 35d30'2"
Nicholas Alexander Saunders
Ori 6h14'32" 8d18'11"
Nicholas Alexander Stuart
Cet 0h52'42" -0d8'29"
Nicholas Allen Cremeans
Sgr 19h28'35" -17d59'57"
Nicholas Allen Nappier
Cap 20h47'13" -22d17'53"
Nicholas Allen Vance
Peg 22h10'45" 8d3'41"
Nicholas Alsten
Tau 4h12'10" 1d36'15"
Nicholas and Alexandra
Leo 10h8'42" 20d20'33"
Nicholas and Bethany
Ori 6h13'18" 15d38'43"
Nicholas and Elizabeth Tieskoetter
Uma 11h36'37" 36d39'45"
Nicholas and Jennifer Landry
Ori 5h59'58" 21d22'40"
Nicholas and Marguerite DeSantis
Lmi 10h39'44" 38d4'39"
Nicholas and Samantha Tsubota
Gem 6h33'49" 22d16'37"
Nicholas Anderson & Emma McKnight
Gem 6h41'20" 19d13'55"
Nicholas Andrew Borick
Cas 1h8'45" 65d49'58"
Nicholas Andrew DeSimone
Leo 11h29'29" 0d0'31"
Nicholas Andrew Manley
Sgr 19h48'39" -13d32'10"
Nicholas Andrew Pfannenstiel "RED"
Per 2h52'1" 53d1'20"
Nicholas Andrew Ramirez
Uma 13h7'42" 52d29'38"
Nicholas Andrew Scalise
Lyr 18h49'31" 43d20'1"
Nicholas Andrew Schroder
Aur 5h32'0" 43d10'4"
Nicholas Andrew Scott Hauser
Cap 20h34'53" -25d34'8"
Nicholas Andrew Tramontana
Per 2h43'15" 53d57'16"
Nicholas Andrew Vincent Blanda
Lyn 7h18'28" 47d30'35"
Nicholas Angelo DeFilippo
Her 16h42'55" 35d50'7"
Nicholas Angelous
Aql 19h47'28" 5d47'59"
Nicholas Anthony Curol
Lyn 7h31'15" 56d37'44"
Nicholas Anthony Dotro
Cap 21h32'19" -12d57'39"
Nicholas Anthony Giambelluca
Umi 16h20'35" 77d44'26"
Nicholas Anthony Muccia
Lyn 7h28'48" 44d49'38"
Nicholas Anthony Orlando II
Cyg 19h45'34" 39d4'1"
Nicholas Anthony Szotak
Cam 8h49'17" 74d57'40"
Nicholas Anthony Torres
Per 3h12'40" 54d11'58"
Nicholas Archer Peppas Star
Cru 12h43'34" -58d47'49"
Nicholas Arden "12-25-06"
Cep 21h45'17" 71d48'12"
Nicholas Armetta
Lib 14h50'42" -1d14'33"
Nicholas Arthur Bambina
Aqr 22h10'49" -16d20'30"
Nicholas Arthur Ristow
Uma 8h58'51" 68d26'1"
Nicholas Auer
Aur 5h46'23" 44d6'29"
Nicholas Austin Neissa
Aur 5h57'56" 42d34'27"
Nicholas B. Bolerjack
Cnc 8h8'45" 6d46'9"
Nicholas B. Hemby
Leo 9h59'36" 6d46'6"
Nicholas B. VanDeusen
Psc 1h11'58" 28d10'38"
Nicholas Barron
Cep 22h12'26" 58d25'51"
Nicholas Baumann
Uma 8h14'19" 68d39'46"
Nicholas Biagio Vivilecchia
Psc 1h51'35" 5d20'7"
Nicholas Blaine Woodard
Cep 23h55'12" 80d17'15"
Nicholas Boscamp
Lib 14h55'59" -2d22'27"
Nicholas Bradley Licameli (Cole)
Vir 12h34'15" -0d24'18"
Nicholas Braum
Tau 5h35'44" 24d41'2"
Nicholas Brewer
Uma 11h21'3" 72d30'22"
Nicholas Brinkley & Jon Levy
And 0h57'29" 38d26'19"
Nicholas Bruce Rose
Ori 5h36'41" -0d52'33"
Nicholas Bryan Carr
Sgr 18h11'38" -17d5'51"
Nicholas Burke 2007
Cep 21h39'7" 61d43'9"
Nicholas Butters
Cru 11h58'37" -55d43'22"
Nicholas C. Mason
Lib 15h8'20" -15d9'47"
Nicholas Campbell Cavill
Lyn 7h11'59" 59d35'30"
Nicholas Capers & Shadow
Umi 14h17'41" 75d12'18"
Nicholas Carden Bowling
Cep 22h58'11" 72d32'35"
Nicholas Carl Stratton
Gem 6h16'35" 21d48'6"
Nicholas Carl Warren
Hya 10h18'56" -13d14'29"
Nicholas Caselli
Aqr 23h42'22" -12d50'43"
Nicholas & Catherine Neill
Cyg 22h0'10" 52d54'17"
Nicholas Centrone
Boo 15h13'22" 49d8'34"
Nicholas Charles Aaron Huhn
Lyr 18h46'17" 41d22'11"
Nicholas Charles Alliotts
Ori 6h4'58" 9d35'0"
Nicholas Charles Clegg (Daddy)
Ori 6h10'5" 6d5'26"
Nicholas Charles DeSimone
Cyg 20h15'36" 39d13'35"
Nicholas Charles Gasorek
Cep 20h39'37" 55d25'17"
Nicholas Charles Giardella
Uma 10h40'37" 63d10'7"
Nicholas Charles Maidens
Per 3h11'53" 57d25'41"
Nicholas Charles Ryan
Uma 13h29'48" 58d47'25"
Nicholas Charles Sacco
Vir 13h0'12" 5d15'27"
Nicholas Chase Fachko
Cep 22h56'53" 66d34'34"
Nicholas Chi Chen
Cep 21h40'15" 61d49'52"
Nicholas Chilton Brownell
Aql 19h38'37" 13d8'39"
Nicholas Christian Armstrong
Psc 0h22'16" 8d47'7"
Nicholas Christian Giampietro
Uma 11h49'45" 51d33'5"
Nicholas Christian Henebury
Umi 15h20'51" 72d32'42"
Nicholas Christian Thane Weaver
Her 17h42'34" 15d20'8"
Nicholas Christine
Lyn 6h50'12" 54d18'26"
Nicholas Christopher
Uma 9h10'40" 56d27'49"
Nicholas Christopher
Leo 11h30'0" 9d46'33"
Nicholas Christopher Sayles
Umi 14h25'23" 75d32'46"
Nicholas Clark
Uma 10h39'42" 48d46'49"
Nicholas Colby
Her 18h18'23" 15d49'15"
Nicholas Connor Hauffman
Gem 7h49'35" 19d3'35"
Nicholas Conrad Senff
Her 18h16'10" 17d22'56"
Nicholas Conrad Tisdale
Aqr 23h8'42" -13d5'32"
Nicholas Creighton Boles
Dra 16h12'3" 68d44'12"
Nicholas Cuenca Dooley
Cap 20h33'40" -10d0'48"
Nicholas Cyron Betts
Her 16h33'4" 48d34'10"
Nicholas D. Ferrante
Sgr 18h5'32" -23d40'34"
Nicholas D. Lee
Boo 13h37'12" 21d34'9"
Nicholas Dale Kaufman
Sgr 18h41'12" -18d27'41"
Nicholas Dale Smith
Aql 19h51'57" -0d16'50"
Nicholas "Danger" Reed
Sgr 19h33'13" -12d48'9"
Nicholas Daniel Colucci
Per 3h10'17" 56d58'9"
Nicholas Daniel Cox
Aqr 21h37'4" 1d12'23"
Nicholas Daniel O'Hara
Cap 21h3'10" -21d37'19"
Nicholas D'Aurio
Uma 10h54'11" 55d33'54"
Nicholas David Quevedo
Umi 14h16'21" 87d58'27"
Nicholas David Reed
Her 17h33'45" 28d1'51"
Nicholas Deglman
Lmi 10h22'49" 35d46'1"
Nicholas & Denay Sady
Cyg 20h21'15" 47d13'34"
Nicholas Dillingham
Per 3h17'31" 42d43'44"
Nicholas DiNapoli
Per 3h54'37" 31d42'4"
Nicholas Donald Stover
Ari 3h10'14" 11d11'48"
Nicholas Donald Walker
Aur 5h39'29" 44d9'4"
Nicholas Duane Smith
Cyg 21h57'44" 47d56'32"
Nicholas E.
Lyn 8h4'19" 37d15'27"
Nicholas E. Kosanovich Jr.
Per 3h58'21" 33d30'37"
Nicholas Edmund Jakubowski
Aqr 21h19'23" -13d13'28"
Nicholas Edward
Cep 23h32'14" 67d5'25"
Nicholas Edwin Naughton
Dra 18h28'36" 73d12'57"
Nicholas Eller Wilmoth
Cap 21h27'50" -14d48'56"
Nicholas Emma
Cap 20h37'33" -18d38'52"
Nicholas Eric Mohamad
Cyg 20h6'50" 43d32'18"
Nicholas Eric Schutz
Lib 15h41'18" -6d26'14"
Nicholas Ernst Bacci
Uma 11h59'7" 28d49'18"
Nicholas Eroh
Aqr 20h46'32" -6d0'22"
Nicholas Estiva Panganiban
Tau 4h16'20" 17d53'15"
Nicholas & Ethan Brumley
Cru 12h38'38" -60d6'35"
Nicholas Everett Shurley
Lib 14h43'21" -12d56'34"
Nicholas F. Nastal
Vir 13h24'14" -12d55'14"
Nicholas Fan Chang Ilasi III
Cap 21h36'38" -19d18'56"
Nicholas Felix
Dra 12h32'43" 70d59'27"
Nicholas Ferguson
Uma 9h2'27" 65d53'11"
Nicholas Forsgren Poust
Psc 1h13'6" 29d20'58"
Nicholas Frangoulis
Cnc 8h18'10" 32d26'35"
Nicholas Frederick Lavin
Vir 13h15'14" 4d38'16"
Nicholas G. Rodgers
Sco 17h40'45" -41d17'9"
Nicholas Gabriel Becker
Ari 2h40'0" 28d59'15"
Nicholas Gaffney
Sco 16h6'52" -13d37'24"
Nicholas George Espejo Chiotellis
Cnc 9h1'9" 15d51'35"
Nicholas George Garmo
Ari 3h4'56" 26d0'50"
Nicholas George Kostidis
Her 18h54'34" 23d50'10"
Nicholas George "QT Patootie" Bahry
Sgr 19h4'27" -16d29'7"
Nicholas George Ward
Her 17h11'15" 47d12'10"
Nicholas Giannelli
Ari 1h58'39" 21d19'27"
Nicholas Girgolas
Cru 12h41'18" -61d7'42"
Nicholas Glenn
Cep 2h36'50" 87d23'12"
Nicholas Gonzalo Rivero
Leo 9h37'36" 22d40'50"
Nicholas Gregory Chenault
Uma 11h42'21" 48d38'38"
Nicholas Gregory Woodhouse
Umi 15h39'32" 83d13'43"
Nicholas Groene
Uma 11h2'0" 60d0'28"
Nicholas HajiDemetriou
Sco 17h49'7" -35d24'54"
Nicholas Hamann
Per 3h16'50" 52d16'57"
Nicholas Harrie Stuart Oliver
Cru 12h1'41" -62d39'52"
Nicholas Hawrysz
Psc 1h44'55" 8d35'57"
Nicholas & Heather
Tau 5h9'40" 24d14'39"
Nicholas Hennessey
Uma 10h26'26" 66d20'42"
Nicholas Hofmann
Her 17h35'54" 37d49'37"
Nicholas Hripcsak
Psc 1h23'18" 31d47'19"
Nicholas Hunter Swanson
Sco 16h14'12" -17d48'44"
Nicholas Hunter Wilson
Gem 6h40'3" 16d23'49"
Nicholas Ian Justice
Aql 19h20'0" 2d27'11"
Nicholas Iannacone
Her 18h39'22" 17d7'39"
Nicholas Ireland
Cru 12h15'40" -60d47'2"
Nicholas Isaiah Vallejos
Ori 4h52'20" 1d24'46"
Nicholas J. Chamberlain
Sco 17h17'23" -37d55'7"
Nicholas J Cruciani
Cyg 20h39'1" 55d8'55"
Nicholas J. Dirig
Sgr 19h19'32" -18d5'2"
Nicholas J. Fling
Ori 6h5'19" 18d43'7"
Nicholas J. Geissler
Aqr 22h29'33" -19d35'18"
Nicholas J. Holden
Uma 11h49'14" 35d26'24"
Nicholas J. Kline
Cap 21h39'11" -13d11'42"
Nicholas J. Savaiano
Uma 9h55'22" 49d14'46"
Nicholas Jack Sutton
Cap 20h35'16" -10d20'50"
Nicholas Jackson Culling
Her 16h42'41" 37d8'1"
Nicholas James
Lmi 10h32'53" 38d17'36"
Nicholas James
Per 2h20'41" 55d14'16"
Nicholas James
Per 3h36'19" 45d23'2"
Nicholas James
Cnc 8h22'40" 25d52'46"
Nicholas James Altenhofen
Sco 16h10'16" -10d46'16"
Nicholas James Browning
Per 4h20'47" 31d13'2"
Nicholas James Clough
Ori 5h51'25" 4d18'20"
Nicholas James Corrigan
Tri 2h22'40" 32d48'22"
Nicholas James Crocker
Tau 4h47'5" 28d30'20"
Nicholas James Dacey
Cep 22h3'5" 57d13'42"
Nicholas James Deogracia Jannotti
Uma 11h45'18" 34d3'24"
Nicholas James Holt
Uma 12h10'6" 54d51'8"
Nicholas James Junk
Cas 0h18'57" 52d31'41"
Nicholas James Mavilia
Uma 12h2'37" 49d44'12"
Nicholas James Padova
Per 2h36'18" 52d23'16"
Nicholas James Powell
Cep 23h37'40" 74d49'55"
Nicholas James Richard Berry
Umi 15h1'35" 68d33'38"
Nicholas James Shedd
Lib 14h54'19" -2d19'47"
Nicholas James Sikes, My Star
Gem 6h47'59" 20d27'10"
Nicholas James Stamm
Cap 21h39'52" -15d59'33"
Nicholas James Turley
Vir 13h44'53" -4d18'17"
Nicholas Jenkins - Rising Star
Aql 19h47'27" 12d10'43"
Nicholas Jodush
Aqr 21h48'43" -0d0'31"
Nicholas John
Cru 12h24'28" -58d39'24"
Nicholas John
Boo 14h52'33" 19d7'51"
Nicholas John
Gem 8h1'42" 27d57'13"
Nicholas John Abbott
Per 3h5'42" 52d25'58"
Nicholas John Alessi
Vir 14h24'38" 1d10'14"
Nicholas John Djokaj
Ori 6h17'38" 11d23'12"
Nicholas John Krause
Ari 2h21'30" 18d25'42"
Nicholas John Ladan
Aqr 22h15'15" -22d56'42"
Nicholas John Negola
Aur 5h53'19" 44d11'13"
Nicholas John Pregno
Vir 13h12'53" 8d32'5"
Nicholas John Smith
Ari 2h36'50" 23d21'57"
Nicholas John Voikos
Uma 8h47'3" 53d17'59"
Nicholas Johnson
Ori 5h26'0" 1d38'38"
Nicholas Johnson Africano
Vir 11h55'0" -0d45'5"
Nicholas Jorge Monello
Leo 11h29'51" 2d14'18"
Nicholas Joseph Anthony Monticello
Cnc 8h17'22" 6d51'11"
Nicholas Joseph Bernhard
Boo 14h26'14" 37d27'12"
Nicholas Joseph Cendrowski
Ori 5h38'48" 2d0'40"
Nicholas Joseph Cron
Dra 19h45'51" 65d27'56"
Nicholas Joseph Kerekgyarto
Her 18h5'58" 49d35'48"
Nicholas Joseph Maher
Vir 14h2'39" -3d29'35"
Nicholas Joseph Martino-Krueger
Eri 4h27'17" -3d22'3"
Nicholas Joseph Noce
Dra 17h28'9" 66d49'42"
Nicholas Joseph Serrano
Dra 15h16'20" 55d33'24"
Nicholas Joseph Volpe
Per 2h41'10" 54d19'34"
Nicholas & Josephine Caputo
Sgr 18h33'43" -28d5'51"
Nicholas Justin Hazemy
Leo 11h8'31" 11d10'10"
Nicholas Justin Levin
Per 2h10'33" 57d3'9"
Nicholas K Walker
Uma 13h37'11" 57d32'58"
Nicholas Kartes
Aqr 23h39'59" -24d46'52"
Nicholas Kenneth Rizzi
Psc 1h22'17" 32d16'29"
Nicholas Kevin Neier
Sgr 19h14'15" -14d15'56"
Nicholas Kevin Sloan
Gem 6h46'18" 33d40'30"
Nicholas Klinger
Aql 19h5'51" 2d45'17"
Nicholas Kovalcin
Uma 14h19'39" 59d51'39"
Nicholas Kyle Miller - Nick
Cyg 21h21'27" 54d51'20"
Nicholas Kyle Miller-Nickoli
Cyg 19h43'14" 32d5'30"
Nicholas L. Tisinger
Cap 20h24'10" -11d52'4"
Nicholas Lancaster
Her 16h39'34" 37d56'9"
Nicholas Laskowski
Uma 13h52'4" 55d44'26"
Nicholas Lathrop Bold
Cap 20h39'24" -16d48'35"
Nicholas Laurence Walters
Lib 15h8'57" -7d18'35"
Nicholas Lawrence Christofano
Her 16h33'14" 13d29'59"
Nicholas Lee Herz
Umi 16h50'56" 86d0'23"
Nicholas Leever
Uma 9h29'26" 63d14'44"
Nicholas Lembo
Vir 12h30'52" 6d8'0"
Nicholas Leo Milburn
Cap 21h29'32" -16d18'4"
Nicholas Lewis Schnell
Umi 15h26'12" 67d31'59"
Nicholas Loppe
Ori 6h21'26" 13d30'39"
Nicholas Lorenzo Rossi
Cyg 21h8'31" 46d54'7"
Nicholas Loucas Pitaro
Lyn 8h19'48" 37d43'53"
Nicholas' Love
Eri 3h1'33" -48d15'24"
Nicholas - Love Always - Staci
Ori 5h40'56" -2d24'3"
Nicholas Lowe - Graduation Day
Cru 12h48'58" -60d42'44"

Nicholas Lucas
Aqr 22h27'59" -0d12'41"
Nicholas Lyons
Ari 1h53'55" 14d11'41"
Nicholas M. Souza
Uma 11h1'47" 65d14'51"
Nicholas M. Traficante
Ori 5h42'41" 7d33'52"
Nicholas Madison & Keri Ann
Lyr 18h59'45" 33d1'40"
Nicholas Mahnkey
Aur 5h35'19" 44d38'24"
Nicholas Mangini
Aur 5h40'28" 43d13'3"
Nicholas Marcy
Crb 15h49'18" 32d46'19"
Nicholas Marinelli
Lib 14h50'9" -4d46'16"
Nicholas Markatos
Per 3h42'28" 34d10'57"
Nicholas Marks
Aur 5h35'2" 46d41'33"
Nicholas Martin
Per 3h19'14" 33d24'47"
Nicholas Martin
Uma 11h22'58" 58d52'16"
Nicholas Matthew
Cep 22h49'17" 60d39'1"
Nicholas Matthew Woods
Uma 9h4'26" 55d33'4"
Nicholas Melvin Hughes
Sgr 19h14'12" -22d56'41"
Nicholas Michael
Tau 5h30'17" 21d44'40"
Nicholas Michael Asghar
Tau 5h33'21" 19d18'12"
Nicholas Michael Fineo
Cnv 12h29'3" 44d14'3"
Nicholas Michael Iverson
Her 16h48'18" 34d8'48"
Nicholas Michael Kelly
Ori 5h59'9" -0d44'11"
Nicholas Michael Lindo
Aqr 21h18'50" 0d55'27"
Nicholas Michael McCarthy
Her 18h19'21" 21d38'34"
Nicholas Michael Nelsen-Sandler
Psc 1h19'0" 32d5'24"
Nicholas Michael Paradysz
Cnc 8h15'13" 23d8'12"
Nicholas Michael Sage
Umi 16h13'14" 75d52'40"
Nicholas Michael Scarpaci
Her 17h34'20" 40d46'54"
Nicholas My Christmas Angel
Cnv 12h34'40" 43d40'2"
Nicholas N. Stegall
Aql 20h0'52" 14d55'19"
Nicholas "Nick" Andrew Speights
Per 3h17'25" 52d41'19"
Nicholas "Nick" J. Angelos
Uma 11h59'35" 56d59'50"
Nicholas Nickel
And 22h59'10" 48d33'32"
Nicholas & Nicole Nichols
Psc 22h58'26" 3d21'37"
Nicholas' Night Light
Dra 19h6'34" 67d12'45"
Nicholas (Niko) La Fuente
Her 16h47'30" 37d18'15"
Nicholas Nissen
Gem 7h37'36" 27d27'32"
Nicholas Opsahl
Uma 12h7'53" 57d1'51"
Nicholas Opsahl
Uma 12h12'26" 57d28'20"
Nicholas P Manocchio
Tau 4h30'55" 28d45'6"
Nicholas P. Pieroni
Her 16h39'39" 43d0'25"
Nicholas Pankow
Her 18h26'21" 21d7'11"
Nicholas Parma
Tau 5h35'11" 23d51'24"
Nicholas Parsons Tiffany
Cap 20h35'18" -19d42'40"
Nicholas Patrick Bednark
Per 3h49'8" 47d58'21"
Nicholas Patrick Ryan
Lyn 6h28'42" 57d32'34"
Nicholas Patrick Stella
Per 3h36'16" 47d2'23"
Nicholas Patrick Thompson
Ari 2h50'40" 29d35'12"
Nicholas Patrick Tolan
Cep 22h27'47" 62d1'19"
Nicholas Patton
Ori 4h58'16" -0d13'55"
Nicholas Perez
Cnc 9h14'1" 10d27'28"
Nicholas Perrikos
Cnc 8h31'3" 30d47'19"
Nicholas Peter Arkis
Aur 5h9'58" 35d57'55"
Nicholas Peter Conway
Dra 18h30'6" 58d37'32"
Nicholas Peter Kunewalder
Sgr 18h28'10" -25d50'10"
Nicholas Peter Pietkiewicz
Ori 5h54'59" 20d57'14"

Nicholas Peter Theodore
Lib 15h2'1" -0d52'5"
Nicholas Philip Stewart
Psc 1h16'43" 27d26'24"
Nicholas Phillip Seidner
Sco 17h50'4" -33d45'49"
Nicholas Pickolas Grassi
And 0h59'28" 35d53'28"
Nicholas "Pops" Anest
Per 3h49'5" 36d56'42"
Nicholas Porter Hann
Her 17h55'57" 19d35'25"
Nicholas Powell
Dra 16h27'8" 64d26'19"
Nicholas Prince
Cep 22h31'50" 62d1'50"
Nicholas Procaccino
Gem 6h54'40" 25d20'51"
Nicholas Procopiou
Tra 15h50'21" -66d29'39"
Nicholas Quincy Mager
Per 3h56'9" 35d17'39"
Nicholas R Dice
Uma 13h15'4" 53d43'26"
Nicholas R. Enus
Umi 16h19'7" 70d39'57"
Nicholas R. Leach
Ari 3h12'1" 15d2'52"
Nicholas R. Richardson, Jr.
Cnc 8h55'30" 11d44'6"
Nicholas R. Stratton
Ori 6h9'38" 18d33'57"
Nicholas "Raiden Kamiyama" Alonge
Sco 16h14'4" -11d33'41"
Nicholas Ralph Mello
Tau 5h32'23" 17d11'7"
Nicholas Ray
Uma 9h36'27" 58d21'36"
Nicholas Ray McClary SM
Leo 11h19'21" 1d17'8"
Nicholas Ressa
Her 17h8'11" 33d20'58"
Nicholas Richard Demid
Dra 18h31'47" 76d58'18"
Nicholas Richard Stiren
Uma 13h48'26" 53d28'43"
Nicholas Robert Beaudin
Aql 20h8'1" 11d1'10"
Nicholas Robert Dodge Super Star
Lyr 18h43'30" 34d59'45"
Nicholas Robert Fortenbach
Ori 6h17'43" 10d59'40"
Nicholas Robert Kurtzer
Lmi 10h33'32" 31d51'2"
Nicholas Robert Meiborg
Cap 20h28'16" -10d44'52"
Nicholas Robert Messina, Sr.
Cep 22h46'32" 65d56'51"
Nicholas Robert Vucich, Sr.
Dra 18h57'40" 70d21'23"
Nicholas Rocco
Psc 1h0'37" 10d31'25"
Nicholas Rogowski
Uma 8h13'51" 67d18'46"
Nicholas Ryan Cummings
Ori 5h19'16" -7d28'55"
Nicholas Ryan Jackson
Cnc 8h17'26" 19d15'36"
Nicholas Ryan Koerber
Uma 12h54'33" 60d7'57"
Nicholas Ryan Macker
Gem 7h31'11" 20d3'11"
Nicholas Ryan Schofer
Lac 22h22'27" 47d13'2"
Nicholas Ryan Stranges
Ori 5h53'7" 21d6'2"
Nicholas Ryan Welker
Oph 17h39'33" -0d35'1"
Nicholas S.
Uma 10h30'23" 62d16'40"
Nicholas S Hyatt
Uma 8h13'28" 71d3'16"
Nicholas S Lobianco
Lib 15h32'24" -16d37'1"
Nicholas S. White
Per 4h46'21" 45d16'26"
Nicholas Sallee
Uma 8h46'48" 61d43'8"
Nicholas Salvatore
Lyn 9h12'49" 33d41'44"
Nicholas Savino
Aql 20h2'56" 10d3'59"
Nicholas Schackai
Ari 2h19'59" 13d14'37"
Nicholas Schaffer
Lib 15h5'0" -4d12'50"
Nicholas Scott Deubert
Leo 10h46'42" 14d41'2"
Nicholas Scott Heckler
Leo 10h52'8" 15d27'14"
Nicholas Scott Shipe
Uma 10h27'3" 47d27'50"
Nicholas Scott Shollenberger
Cep 2h37'13" 81d18'13"
Nicholas Scott Wilkey
Aqr 22h31'0" -14d24'26"
Nicholas Sean Ware
Ari 3h27'16" 21d47'26"
Nicholas Sebastian Gavin
Ori 4h51'41" 3d50'40"

Nicholas Sebastian Lodwick
Umi 15h27'13" 72d24'22"
Nicholas Semack
Vir 12h42'37" 1d23'22"
Nicholas Sequino
Leo 11h31'42" 25d46'40"
Nicholas Shankar
Her 17h37'47" 16d32'16"
Nicholas Shotkoski
Uma 10h31'10" 54d42'24"
Nicholas Siano
Lyn 7h41'34" 42d10'47"
Nicholas Sidney
Cep 22h41'50" 60d47'22"
Nicholas Simon
Her 18h22'21" 20d33'2"
Nicholas Simon Stephens
Cyg 21h21'17" 32d28'3"
Nicholas Sokolovich
Per 2h27'24" 55d8'21"
Nicholas' Star
Vir 11h54'17" -0d4'10"
Nicholas Stella
Ori 5h34'46" 6d48'23"
Nicholas Stephen Fecteau
Lmi 11h4'38" 28d37'34"
Nicholas Steven Barre
Leo 11h11'45" 11d11'7"
Nicholas Steven Feys
Her 16h23'1" 43d23'24"
Nicholas Stormy Navarro
Sco 17h49'20" -34d51'36"
Nicholas T. Alfano
Sco 17h10'48" -41d49'9"
Nicholas Taylor
Per 3h30'54" 45d6'15"
Nicholas Thomas
Tau 5h33'0" 23d30'25"
Nicholas Thomas Sanzero
Cap 20h33'40" -16d1'45"
Nicholas Thomas Valvo
Her 16h16'27" 13d44'46"
Nicholas Tiedemann
Ara 17h15'21" -50d38'36"
Nicholas & Tina Hruda
Cyg 21h38'25" 49d27'51"
Nicholas Tobin Llanos
Lyn 8h5'28" 37d37'29"
Nicholas Trawczynski
Per 3h43'7" 49d33'16"
Nicholas Tyler Chavez III
Umi 16h18'27" 75d21'37"
Nicholas Vacca
Her 18h17'2" 15d18'28"
Nicholas Valenti
Uma 13h29'33" 56d52'25"
Nicholas Van Jensen
Aql 19h51'4" -0d16'34"
Nicholas Vern Strasburg
Per 3h9'52" 54d13'12"
Nicholas Victor Clescere
Ori 5h26'34" -5d29'11"
Nicholas Vijay Shankar
Equ 21h8'36" 5d56'21"
Nicholas Villareal
Her 16h55'40" 13d42'41"
Nicholas Vincent Desiderio
Uma 10h11'25" 44d41'1"
Nicholas Vincent Grennan
Vir 13h19'1" 6d8'48"
Nicholas Vincent Marzullo
Sgr 18h5'36" -28d12'44"
Nicholas Vincent *Our Shining Star*
Crb 15h39'29" 29d48'8"
Nicholas Vincent Sardi
Sco 17h49'31" -31d28'58"
Nicholas W. Greenhoe
Her 17h13'17" 16d0'41"
Nicholas Watroba
Sgr 17h55'19" -27d58'55"
Nicholas' Way
Lyn 8h20'22" 55d5'0"
Nicholas Webb
Ori 5h21'16" 3d56'39"
Nicholas Welham
Cru 12h39'46" -59d56'1"
Nicholas Whorrall
Her 17h40'26" 19d14'12"
Nicholas William Clark
Per 3h18'43" 51d13'33"
Nicholas William Doreste
Uma 12h59'1" 55d1'15"
Nicholas William Joseph Talbot
Cma 6h37'31" -17d15'53"
Nicholas William Parker
Cru 12h36'57" -58d46'4"
Nicholas William Smith
Her 17h48'10" 17d41'34"
Nicholas Wilson Harsell
Ori 5h37'31" -5d18'13"
Nicholas Yukio Yanagi
Lib 15h8'27" -4d44'4"
Nicholas, My Poney Huney
Cru 12h46'44" -61d31'2"
Nicholas, your my star, I love you
Her 16h48'1" 45d26'45"
nicholasglenn16
Aqr 22h4'14" -0d28'32"
NicholaShirin
Sge 20h5'28" 17d43'32"

Nicholas-My Baby-Our Bro
Her 16h35'51" 21d49'20"
Nicholas's Star
Cma 7h26'28" -19d13'44"
Nichole
Leo 11h15'15" -4d16'59"
Nichole
Cas 23h49'30" 59d55'41"
Nichole
And 0h29'15" 27d20'20"
Nichole
Tau 4h8'19" 27d48'15"
Nichole
Vir 12h41'9" 5d15'7"
Nichole
Vir 14h33'32" 2d21'14"
Nichole
And 0h24'31" 42d8'49"
Nichole 27
Vir 12h10'5" 10d20'27"
Nichole A. Cummins
Tau 4h37'43" 13d41'51"
Nichole Alexis Garcia
Cap 21h3'7" -14d34'23"
Nichole and Jason
Cyg 19h53'50" 32d0'7"
Nichole and Ryan
Ori 6h6'27" 19d15'47"
Nichole Anderson
Gem 7h25'3" 34d50'53"
Nichole Andrea Laggan
And 23h3'10" 36d36'25"
Nichole Ann
And 2h3'3" 45d22'54"
Nichole Ann Otero
Leo 11h17'14" -6d35'15"
Nichole Ashley Adamo
Lib 14h54'53" -1d53'56"
Nichole Briana
And 23h43'18" 37d51'33"
Nichole Brittnay Stanley
Cas 0h39'21" 61d43'8"
Nichole Charise Martin
Cas 0h16'33" 63d26'19"
Nichole Christine Reiland
Leo 9h37'6" 13d50'53"
Nichole Dannielle Flippen
Ari 3h25'14" 27d6'22"
Nichole Danos Hartigan
Aqr 23h2'29" -5d51'59"
Nichole Didier
And 1h9'13" 42d15'23"
Nichole Elizabeth
And 1h5'25" 45d20'54"
Nichole "Elohcin" Hutchins
Sco 16h7'17" -14d35'17"
Nichole Fleischman
Psc 1h24'54" 32d37'52"
Nichole Hammac
Aqr 21h41'7" -1d41'4"
Nichole Helen Antoinette Riley
Tau 4h28'55" 17d33'47"
Nichole Hinson ex adyto cordis
Cyg 21h37'13" 41d32'32"
Nichole J. Honn
Ari 1h59'7" 21d50'39"
Nichole Jane Allan
Vir 14h31'58" -4d24'32"
Nichole Joneson
Cnc 9h17'29" 14d55'26"
Nichole & Karmen
Ori 5h40'8" 8d6'7"
Nichole Kealoha
Cas 23h30'24" 57d51'55"
Nichole Lang
Ari 2h22'30" 20d32'56"
Nichole Lee Jones
Sgr 19h0'16" -33d34'49"
Nichole Louise Elliott
Lyn 7h12'39" 59d3'2"
Nichole Louise Hulse
Lib 15h27'43" -19d5'45"
Nichole Lynn Lucas
Com 12h46'5" 18d48'56"
Nichole Manista
Cap 21h15'23" -25d23'28"
Nichole Marie
Sgr 18h22'54" -32d0'11"
Nichole Marie
Gem 7h52'37" 27d27'24"
Nichole Marie Brown
And 23h39'50" 40d3'45"
Nichole Marie Genevieve McDaniel
And 0h9'52" 44d53'7"
Nichole Marie Kersey
Leo 10h53'40" 6d50'41"
Nichole Marsh
Cnc 7h56'12" 14d37'56"
Nichole Martina Mason
Vir 14h46'18" 2d45'13"
Nichole Meyer
Vir 13h4'13" 12d25'42"
Nichole Michele Kelsey
Vir 14h36'15" 4d8'18"
Nichole my Superstar
Cyg 21h24'16" 30d35'28"
Nichole Owen
Cyg 21h54'23" 51d39'36"
Nichole R. Espinoza
And 0h16'47" 43d36'47"
Nichole Rudny
Tau 4h5'45" 23d28'51"

Nichole<\@>S29A45
Ari 2h48'24" 25d42'19"
Nichole Socha
Aqr 23h9'36" -9d10'21"
Nichole Tenderholt
Mon 6h32'26" 8d7'55"
Nichole Tillison
Ori 6h12'36" 15d0'23"
Nichole Weaver Madison
Dra 12h15'36" 71d4'34"
Nichole Yeakey
Cap 20h11'44" -10d40'17"
NICHOLERIA
Tau 4h41'17" 7d33'57"
Nicholes Piece of the Sky
And 0h35'46" 44d52'41"
Nichole's Star
And 0h37'41" 35d18'14"
Nicholetta Zet
Cas 1h39'16" 62d52'39"
Nicholeus
Cam 5h36'3" 78d39'38"
Nicholi Violette
Uma 8h50'46" 47d2'27"
Nichollas Scott Fawcett
Cru 12h48'27" -60d59'0"
NichoLouise
Lyn 7h57'52" 42d41'8"
Nicholous Rossaert
Crb 16h22'8" 29d24'30"
Nici
Tau 3h40'24" 24d47'43"
Nici
Lib 14h57'51" -6d55'51"
NICI
Uma 8h21'6" 63d44'38"
Nici, 06.03.1987
And 23h30'41" 41d16'35"
Nicio Boltiansky
Cep 21h48'5" 67d22'58"
Nicire Nick-Claire
Crb 16h12'19" 29d54'30"
Nici's Star 20052005
Uma 11h21'10" 31d19'25"
Nick
Her 16h53'53" 28d31'32"
Nick
Her 18h14'2" 25d56'1"
Nick
Her 17h58'18" 23d14'15"
Nick
Boo 14h8'28" 16d58'59"
Nick
Cep 20h23'47" 61d41'6"
Nick
Uma 12h29'0" 57d26'4"
Nick
Eri 3h49'13" -10d42'33"
Nick
Lib 14h32'47" -18d42'30"
Nick A. Caporella
Aql 19h11'40" 5d41'56"
Nick A Henkel "King of the Sky."
Uma 8h27'9" 63d39'49"
Nick Ackerman
Sco 16h7'59" -29d12'38"
Nick & Alicia Welsh
Cnc 8h4'27" 13d59'0"
Nick & Allie's Star
Sco 17h56'18" -30d22'37"
Nick and Abbie
Cyg 21h56'36" 46d7'18"
Nick and Amy 12-11-02
Del 20h48'50" 19d54'55"
Nick and Amy Nabors
Sco 16h55'29" -33d37'44"
Nick and Ashley
Cyg 20h17'0" 51d54'21"
Nick and Ashley
Uma 11h6'55" 50d10'2"
Nick and Betty
Lyn 8h59'34" 36d1'10"
Nick and Bri's Escape!
Cyg 20h22'15" 55d25'21"
Nick and Crystal Burgess
And 23h14'18" 51d51'8"
Nick and Dre Forever and For Always
Dra 18h52'4" 50d26'38"
Nick and Jeni Visser
Cyg 20h7'6" 57d59'4"
nick and kenz
Her 17h13'53" 15d30'23"
Nick and Lauren's Star
And 0h32'56" 45d17'27"
Nick and Mary
Cyg 20h14'13" 50d8'2"
Nick and Megan's star "megnick"
Cyg 20h24'54" 56d1'1"
Nick and Melissa's Anniversary Star
Cyg 19h17'51" 51d40'34"
Nick and Sarah:)
Boo 14h33'19" 23d3'50"
Nick and Tami Tobolski
Crb 15h24'1" 27d36'21"
Nick and Ying Hwei
And 23h27'55" 48d52'19"
Nick & Andrea
Dra 19h38'48" 70d39'10"
Nick & Anna Forever
Cyg 21h17'4" 46d42'44"

Nick Anthony Salimeno
Leo 9h23'22" 14d0'25"
Nick Arkless
Per 3h59'50" 38d1'36"
Nick At Night
Ari 2h47'59" 14d55'36"
Nick Athanasatos
Boo 14h52'31" 26d31'26"
Nick Austin
Per 4h49'35" 46d43'24"
Nick Azzolino
Per 4h4'8" 37d19'47"
Nick Bair
Her 17h54'21" 45d16'22"
Nick & Barb Schreiner
Her 16h16'36" 48d14'48"
Nick Barrett
Sgr 18h12'30" -28d36'50"
Nick Bateson
Per 3h39'55" 41d6'12"
Nick Benton created stars & sired 4
Uma 8h50'58" 46d52'13"
Nick Blais
Gem 6h49'59" 34d3'13"
Nick Brandenburg 1985 - 2006
Cap 20h39'9" -18d24'58"
Nick Bryan
Uma 13h23'21" 53d38'40"
Nick Buxton
Cas 0h21'34" 55d21'43"
Nick Cabot, Teacher Extraordinaire
Her 18h8'45" 49d16'10"
Nick Carnavos
Nor 16h10'43" -46d8'19"
Nick & Cherie Together Forever
Ori 5h22'25" -4d35'20"
Nick Cianci
Del 20h19'2" 9d41'25"
Nick Colmenero
Per 4h15'33" 51d11'9"
Nick Cometa
Per 3h2'21" 47d19'20"
Nick/Dad/PopPop Shines Down on Us
Per 3h22'47" 47d26'6"
Nick DeFinizio "Nick DStar"
Per 2h40'7" 40d12'12"
Nick "Dido" Lozinski
Leo 9h47'9" 28d41'33"
Nick Disla, Jr.
Lib 14h52'2" -2d33'11"
Nick Elder
Crb 15h50'27" 36d15'2"
Nick Facciolla
Cam 4h19'46" 56d47'30"
Nick Flanagan
Cyg 21h39'39" 50d27'54"
Nick Fortunatus
Aqr 21h5'21" 0d5'1"
Nick & Frances Stojanovic
Sgr 19h15'18" -34d51'31"
Nick from Kandersteg
Dra 17h14'47" 66d39'23"
Nick G. & Mayra S.
Sco 16h41'56" -33d45'31"
Nick Gawler's Brilliant Light
Cru 12h41'11" -60d32'40"
Nick Georgiou
Per 3h47'10" 51d23'40"
Nick Girardi
Lib 15h2'24" -15d56'4"
Nick Griepentrog
Her 18h29'20" 24d22'35"
Nick Gross
Aql 19h41'28" -0d1'31"
Nick & Haley's Anniversary Star
Cae 4h51'5" -28d26'56"
Nick Hall
Lmi 9h44'42" 37d50'6"
Nick Hamilton's Night
Leo 11h13'6" 23d9'49"
Nick Hannon
Cnc 8h55'29" 26d6'48"
Nick & Hayley's Love Star
Mon 7h19'27" -0d38'6"
Nick & Heidi
Sco 16h2'31" -18d50'27"
Nick Hinojosa
Uma 11h43'33" 41d12'15"
Nick Hochwender
Lmi 10h24'13" 36d39'52"
Nick is a Doodyhead
Lib 15h37'39" -5d10'24"
Nick J. Elko
Uma 13h56'12" 53d4'6"
Nick Jamons
Cru 12h33'15" -61d15'16"
Nick Janssen
Her 16h48'44" 41d7'42"
Nick & Jessica Holland & Their Kids
Uma 13h51'7" 51d12'9"
Nick John DiFrancesco
Her 16h14'52" 24d23'28"
Nick Jones
Her 16h33'54" 22d43'6"
Nick & Karen Vaughan's Wedding Day Star
Cyg 21h43'20" 46d2'12"

Nick Kirby 9597
Vir 11h38'19" -3d24'25"
Nick Kraus
Ori 5h31'48" 2d18'50"
Nick Latka
Uma 11h39'32" 62d46'27"
Nick & Laura King
Col 5h47'9" -33d20'23"
Nick & LaVonne Mancini
Per 2h10'4" 57d24'18"
Nick Lee Horne
Cam 4h26'41" 70d29'22"
Nick Lerouge
Uma 13h35'52" 52d1'47"
Nick Libertore
Uma 8h29'57" 68d24'59"
Nick Loves Jaime
Ari 2h6'4" 24d35'41"
Nick Loves Krissy
Cyg 20h32'28" 52d41'29"
Nick Lucas
Her 16h30'29" 46d43'49"
Nick Madrid
Vir 12h31'11" 0d36'22"
Nick Maggi
Her 16h44'49" 43d43'11"
Nick Mangle
Psc 23h9'39" -0d18'22"
Nick Marino
Boo 14h40'25" 37d50'22"
Nick Mason
Cap 20h7'28" -8d54'5"
Nick Michael
Per 3h24'44" 40d52'13"
Nick & Michelle Goodness
And 1h44'55" 43d7'30"
Nick Mone $$
Cep 21h0'24" 63d42'39"
Nick Muce
Dra 19h16'36" 64d46'4"
Nick Nack
Leo 9h39'48" 27d20'16"
Nick Nacks
Uma 9h38'53" 46d4'44"
Nick Nicholls
Cep 22h51'56" 61d13'4"
Nick Nilmeyer
Uma 11h14'42" 45d30'47"
Nick Oldham - Daddy's Star - 9.10.06
Cru 12h31'55" -59d37'7"
Nick Panaccione
Vir 13h5'37" 11d58'20"
Nick Parr
Ari 2h40'18" 27d39'0"
Nick Parrish
Gem 7h5'12" 25d3'4"
Nick & Peggy Davidovich
Lyr 18h31'54" 36d20'5"
Nick Petrillo Jr.
Her 18h36'3" 19d9'45"
Nick Poynting Shilo Baker Eternity
Ara 17h38'24" -47d21'34"
Nick Primola
Boo 14h49'16" 37d53'41"
Nick Probert
Leo 10h41'11" 22d48'43"
Nick & Rachael
Cyg 19h38'12" 31d5'45"
Nick Radwanski
Cnc 8h0'47" 17d22'12"
Nick Rasdal
Her 17h35'0" 28d2'3"
Nick Read
Gem 7h9'40" 33d53'57"
Nick Rigg
Uma 12h9'32" 52d10'15"
Nick - R.M.A. - J - F2
Sco 17h26'23" -41d44'33"
- Nick - Rose & Bob Salyha
Uma 11h22'41" 59d19'51"
Nick Salerno
Leo 10h58'7" 21d12'38"
Nick Sampford
Per 4h21'0" 36d24'30"
Nick & Sandy Forever
Umi 15h30'3" 74d21'2"
Nick & Sarah
Cyg 20h17'43" 51d20'13"
Nick Sidoti
Per 3h6'35" 56d37'59"
Nick Sisson
Her 17h22'40" 28d5'53"
Nick Solan
Her 16h11'50" 23d7'46"
Nick Spano
Per 3h7'16" 54d36'0"
Nick & Stacy
Her 18h5'23" 28d16'53"
Nick Sullivan
Lac 22h35'0" 39d21'10"
Nick & Terry
Lib 15h15'24" -11d14'57"
Nick Trainor
Cep 22h52'7" 62d34'56"
Nick Vendetti
Dra 19h21'45" 69d52'59"
Nick Veropoulos
Uma 10h34'13" 57d46'24"
Nick Wallick
Ori 5h55'48" 12d43'36"
Nick Waters
And 0h53'52" 41d0'53"

Nick Webb
Ori 5h57'0" 16d58'23"
Nick, Mike, and Grace's Love
Cam 7h0'42" 80d55'29"
Nick2004
Sgr 18h33'42" -26d22'18"
Nicka
Crb 15h53'20" 31d15'59"
Nickada
Cru 12h47'29" -61d5'1"
NickandKy
Cyg 19h41'21" 30d53'59"
Nickatie
Aur 5h46'55" 49d0'58"
Nick-at-nite
Leo 10h54'56" 10d6'32"
Nickel & Topher -4- Ever
Ori 5h29'31" 3d19'49"
Nickel, Andreas
Psc 1h15'33" 2d47'20"
Nickeshia Tenecia Porter
Psc 1h9'48" 30d21'32"
Nicki
Crb 16h14'17" 36d24'19"
Nicki
And 23h28'47" 36d45'25"
Nicki
Boo 14h15'40" 41d38'2"
Nicki
Lyr 18h49'28" 38d3'55"
Nicki
Dra 18h55'7" 51d54'18"
Nicki
Leo 9h24'11" 18d59'4"
Nicki
Sgr 17h54'33" -29d36'57"
Nicki
Sco 17h30'0" -34d20'3"
Nicki Aitken
And 0h53'41" 45d41'55"
Nicki and Dave
Cyg 20h52'7" 32d2'0"
Nicki "Bebe"
Lyn 8h55'6" 37d12'35"
Nicki Erisman
Sco 16h6'36" -16d6'33"
Nicki Higgins
Lyn 7h46'33" 49d21'33"
Nicki Howell
Cas 0h11'53" 54d41'16"
Nicki Jet Jim Star
Ori 5h35'26" -1d55'23"
Nicki Kinnard
Uma 12h10'21" 50d44'48"
Nicki Lynn Wiens
Aql 19h42'48" 0d24'31"
Nicki M. Korman
Aqr 22h27'52" 0d22'53"
Nicki Marie
Crb 15h40'56" 27d13'21"
Nicki Pecori
Cnc 9h10'20" 28d26'14"
Nicki Pooh
Ori 6h17'43" 14d37'49"
Nicki Rhodes
Aur 5h1'24" 49d58'21"
* Nicki * Rosie * Eepo *
Cmi 7h30'16" 8d31'5"
Nicki Scott
Psc 1h17'19" 20d29'59"
Nicki Smith
And 1h37'2" 50d28'31"
Nickie
Lyn 7h51'4" 46d16'45"
Nickie Astry
And 0h44'8" 37d1'58"
Nickie Bough
Vir 12h22'0" -6d2'48"
Nickie G
Cnc 8h22'34" 31d3'23"
Nickie Harkabusic
Uma 8h50'46" 52d36'0"
Nickie & Jason's Star
Aql 20h28'36" -0d53'1"
Nickie Pocus
And 0h28'20" 42d9'21"
Nickiea's light
And 0h11'48" 45d32'8"
Nick-ie-poo
Uma 10h39'46" 46d59'32"
Nickie's Star
Gem 7h47'1" 33d50'27"
Nicki's Angel
And 1h51'38" 36d11'46"
Nicki's Star
Uma 10h59'7" 52d6'55"
Nicki's Star
Cnc 8h55'18" 26d1'16"
Nicki's Star Tooth - 12-8
Ori 6h10'6" -0d37'19"
Nicki's Wishing Star
And 1h14'47" 44d31'9"
Nicklas Hornstrup Mogensen
Sco 16h17'30" -16d29'10"
Nicklaus and Shami
Her 16h53'55" 31d53'53"
Nicklaus Charles Wiedbusch
Per 3h0'32" 48d47'24"
Nicklaus Picard
Lyn 6h39'7" 56d10'26"
Nicklboni
Lyn 7h42'21" 49d2'46"
Nicklet-Vicklet
Ori 5h29'55" 6d20'7"
Nicko Mercurio
Vir 14h12'56" -13d1'4"
Nickol & Tyson Senalik Forever
Cyg 21h53'26" 42d29'31"
Nickolas Andy Brand
Sco 16h15'35" -21d50'15"
Nickolas Anthony Letizio
Umi 15h36'44" 73d14'16"
Nickolas Channing Langley
Sgr 18h3'29" -29d25'4"
Nickolas Hofmann
Per 3h18'15" 46d9'6"
Nickolas Jaynes
Umi 15h0'50" 78d15'18"
Nickolas John Ash
Aqr 22h4'53" -11d6'17"
Nickolas "My Baby"
Umi 15h9'16" 69d26'6"
Nickolas Smith Rodgers
Lyn 7h49'30" 54d1'30"
Nickolas Willis
Cma 7h26'47" -24d30'12"
NickolasScottPowers03129 2Celestial
Psc 1h29'13" 18d35'39"
Nickolaus A.C. Hahn
Ori 5h38'50" 0d47'14"
Nickole McGurk
Leo 11h26'31" 9d6'9"
Nickole Perkins
Ari 2h58'16" 29d43'59"
Nickolite
Psa 21h38'4" -31d43'6"
Nick's 6th Birthday Star
Umi 16h31'33" 76d55'29"
Nick's Flamethrower 11
Ori 5h6'18" -1d41'49"
Nick's Nifty Nitelite
Uma 13h43'17" 52d54'14"
Nick's Piddidle
Ari 3h9'41" 29d8'50"
Nick's Serenity
And 1h6'33" 40d40'51"
Nickson Rodney Leo
Ari 3h5'22" 17d37'5"
Nickstelauron 1/4/91
Uma 10h52'11" 69d56'52"
NICKTRON - owned by Nicky Irvine
Leo 11h36'6" 15d59'55"
.Nicky.
Ari 2h53'48" 25d19'34"
Nicky
Del 20h39'14" 14d8'45"
Nicky
Peg 21h40'31" 12d52'17"
Nicky
Dra 12h15'31" 68d46'1"
Nicky
Ori 4h56'57" -0d13'15"
Nicky
Umi 17h11'38" 82d4'12"
Nicky
Cma 6h59'37" -23d20'51"
Nicky 06.06.1984 in liäbi din Sacha 28.04.2003
Sge 19h52'19" 17d7'37"
Nicky amore
Her 16h22'16" 6d29'7"
Nicky and Mike's Star
Cap 20h30'34" -11d11'23"
Nicky Balaam
And 1h0'36" 46d35'2"
Nicky Betka
Lyn 9h39'35" 40d6'39"
Nicky Brownlow
Aqr 20h46'9" -7d3'24"
Nicky D
Aql 19h7'10" 1d24'2"
Nicky Dear
Cas 1h14'14" 68d17'18"
Nicky Durbin
Cyg 21h55'31" 52d31'16"
Nicky Galligan 25th Birthday Star
Gem 6h58'54" 21d36'23"
Nicky Graydon
Cnc 8h1'53" 15d48'52"
Nicky Harris
Cyg 21h53'41" 54d40'23"
Nicky Johnson
Lib 15h35'33" -8d0'26"
Nicky Lavoie
Cas 1h48'10" 69d5'22"
Nicky Lea
Cap 20h25'27" -9d18'10"
Nicky Legg
Vir 14h28'48" -0d12'1"
Nicky & Matt's Special Day
Cyg 19h43'39" 34d45'1"
Nicky Mcnair
Leo 11h2'46" 0d54'29"
Nicky (Nicholas Hackney)
Vul 19h6'56" 23d46'56"
Nicky Patel (Goddess of The Stars)
Aqr 21h14'39" 1d17'46"
Nicky Poo
Ari 2h53'21" 26d31'48"
Nicky Rankine
Gem 7h4'28" 11d8'37"
Nicky Rose
Aur 5h33'57" 45d39'49"
Nicky & Sharon's Star
Cyg 21h43'51" 33d31'56"
Nicky Stefenelli
Umi 17h23'13" 86d33'46"
Nicky & Stephanie
Psc 1h15'7" 26d32'20"
Nicky Stevens
And 1h14'31" 45d31'36"
Nicky Tyrrell
And 22h59'23" 51d50'57"
Nicky W.
Uma 10h16'58" 54d13'20"
Nicky, You & Me Against The World
Gem 7h36'44" 33d56'59"
Nicky's Baby
Sgr 20h7'45" -35d44'47"
Nicky's Bright Shining Star
Aqr 22h35'52" -8d59'59"
Nicky's Nastiness
Uma 13h24'18" 58d0'47"
Nicky-Z
Cas 0h56'40" 60d11'50"
Niclas Booz Fischli
Cep 22h12'10" 68d11'1"
NicMic-30
Uma 13h20'19" 59d59'49"
Nico
Uma 12h21'40" 55d22'0"
Nico
Uma 11h52'34" 55d13'34"
Nico
Uma 13h8'33" 61d42'39"
Nico
Uma 12h59'24" 61d28'38"
Nico
Cam 6h16'50" 60d5'34"
Nico
Cra 18h6'26" -37d12'18"
Nico
Cyg 20h44'23" 35d23'35"
Nico
Uma 11h51'53" 39d29'15"
Nico
And 23h36'28" 43d6'29"
Nico
Ari 2h31'55" 27d41'52"
Nico
Leo 10h1'59" 17d25'8"
Nico
Ori 6h14'20" 6d44'14"
Nico Anthony Volpe
Her 17h5'43" 25d17'40"
Nico B
Gem 7h29'23" 16d15'45"
Nico Cianciulli
Ari 2h31'54" 27d3'59"
Nico/Cu
Cam 7h49'56" 75d4'46"
Nico e Ceci
And 0h19'56" 21d57'27"
Nico Fabian
Uma 8h13'58" 64d2'6"
Nico J. Hofmeester
Boo 15h31'59" 46d40'46"
Nico Klauner
Ori 6h13'18" 2d23'59"
Nico Knull
Equ 21h11'22" 11d17'41"
Nico Lorenzo Ferri
Lib 14h23'36" -13d3'16"
Nico Maltarich
Aqr 22h31'30" -2d18'16"
"Nico" - Nicolo Joseph Bahamonde
Leo 9h49'11" 27d11'41"
Nico Panzarea
Cyg 19h40'16" 46d37'54"
Nico Rossi Bernardini
Per 3h26'43" 43d53'47"
Nico Ryan Caprez
Uma 8h36'4" 60d22'20"
Nico Sinigallia
Cyg 21h7'56" 50d37'25"
Nico Sinigallia
Cyg 21h7'56" 50d37'25"
Nico Squishy Mandes
Vir 13h22'5" 12d32'8"
Nico Uttikal Star
Uma 8h27'38" 60d48'39"
Nicokle Kraft
And 0h38'46" 42d13'41"
Nicol Carden Mc Vann Hostettler
Leo 9h31'46" 6d40'17"
Nicol Fortin
Uma 13h20'34" 54d19'47"
Nicol Hall Engagement
Men 5h15'50" -72d37'38"
Nicola
Cas 0h26'25" 62d22'35"
Nicola
Cas 0h41'50" 63d50'14"
Nicola
Uma 9h23'21" 63d41'36"
Nicola
Cas 3h29'22" 70d24'52"
Nicola
Umi 17h31'8" 80d0'23"
Nicola
Cap 20h42'32" -17d12'25"
Nicola
Cyg 19h46'44" 35d17'55"
Nicola
Cnv 13h47'3" 31d48'58"
Nicola
Crb 16h10'55" 32d35'33"
Nicola
And 23h0'26" 50d59'10"
Nicola
Cyg 21h16'1" 42d32'51"
Nicola
And 1h47'17" 50d4'16"
Nicola
Aqr 22h41'26" 1d10'11"
Nicola
And 0h19'29" 27d29'29"
Nicola
Her 18h0'57" 17d59'3"
Nicola Ann Priestley
Tau 5h55'54" 25d5'30"
Nicola Barry
Psc 1h21'39" 17d52'26"
Nicola Bimps Preston
Umi 15h5'24" 82d17'25"
Nicola Brattole
Col 5h44'32" -33d1'35"
Nicola Brennan Ryan
And 23h37'3" 42d47'36"
Nicola Brownlee
Tau 3h31'17" 24d45'36"
Nicola Catherine King
Cep 20h44'51" 61d43'19"
Nicola Etchells
Cas 0h53'8" 52d30'4"
Nicola Gabriele Dallorso
Cyg 20h58'28" 48d17'23"
Nicola Jane
And 0h50'47" 41d50'16"
Nicola Jane Dailly
And 0h57'48" 45d46'43"
Nicola Jane Gifford
Sco 17h55'40" -35d52'46"
Nicola Jane Goode
Vir 14h33'3" -6d31'6"
Nicola Jane Norrie
Cas 23h59'30" 58d48'12"
Nicola Jane Steel B.A.(Hons) "Nikola35"
Cyg 19h49'17" 39d3'21"
Nicola Janette 1969
Ori 6h18'59" 2d26'42"
Nicola Jayne Brown
Ori 5h45'1" -2d24'12"
Nicola Jayne Stubbs
Cas 0h19'42" 57d24'9"
Nicola Kim David Etter
Cas 0h18'42" 51d2'53"
Nicola Kretzschmar
Psc 1h20'31" 7d4'49"
Nicola Kristan
Uma 10h24'7" 39d26'10"
Nicola Leach-Hoobyar
Cas 1h5'35" 61d53'25"
Nicola Louise Johnstone 27/12/85
Cru 12h42'9" -56d1'6"
Nicola Marie Muir
And 2h33'56" 42d12'15"
Nicola Moulton
And 2h5'45" 39d47'9"
Nicola Mulryan
Cyg 20h15'40" 38d41'4"
Nicola Nicoloff
Ori 5h39'29" 8d35'3"
Nicola Parente
Lib 15h55'4" -18d17'31"
Nicola Rose Prior
Gem 6h59'6" 24d20'23"
Nicola Starsky Helwig
Ori 5h40'21" 0d32'31"
Nicola Stephenson (Little Nic)
And 23h23'24" 42d34'14"
Nicola Tasker - Nicola's Star
And 2h4'28" 46d27'29"
Nicola Tillman Live Love Laugh
Cnc 9h6'15" 29d33'37"
Nicola Vernese
Peg 22h16'1" 6d34'57"
Nicola Wilson
Ari 2h24'21" 10d56'43"
Nicola Young
Peg 22h47'15" 2d55'25"
Nicolae Ponici
Ori 6h17'19" 19d17'52"
Nicolai Donald Lee
Cyg 20h6'26" 32d33'18"
Nicolai H. Skridshol
Gem 6h50'43" 26d50'56"
Nicolai Laszlo Nagy
Cnc 8h28'43" 27d36'37"
Nicolai & Nana
Leo 11h22'49" 24d44'41"
Nicolaine
Lyn 7h42'4" 47d33'38"
NicolaMendicinomioAmore amato
Per 3h52'58" 49d52'42"
Nicolaos Dranitsaris
Ari 2h8'12" 26d4'33"
Nicolas
Cnc 9h1'58" 16d22'44"
Nicolas
Ori 6h14'56" 5d52'7"
Nicolas
Cas 0h31'57" 50d20'16"
Nicolas
Uma 11h4'23" 40d18'46"
Nicolas
Uma 13h53'12" 55d2'59"
Nicolas
Uma 12h59'16" 55d59'45"
Nicolas
Uma 9h54'10" 67d54'24"
Nicolas Albert
Uma 10h24'38" 46d43'27"
Nicolas Allan Bowry
Cnc 9h3'41" 31d31'52"
Nicolas Anthony Masciola
Aqr 23h8'39" -8d53'14"
Nicolas Anthony Pellegrino
Cnc 9h3'24" 27d18'17"
Nicolas Anthony Roldan
Gem 7h9'26" 14d44'24"
Nicolas Arden Chase Smith
Gem 6h16'13" 27d43'44"
Nicolas Beunel
Leo 11h10'49" -0d54'13"
Nicolas Billerot mon Tibou
Uma 9h35'33" 70d6'37"
Nicolas Capotosto
Aur 5h49'43" 53d56'23"
Nicolas Charles 6lbs. 2oz.
Aqr 22h22'1" -0d42'1"
Nicolas Daniel Rogers
Boo 14h37'38" 22d24'20"
Nicolas Dion "Ti-Tom"
Per 3h49'24" 32d23'47"
Nicolas Donnez
And 1h43'0" 47d49'34"
Nicolas et Celine !!!
Cas 0h13'10" 53d52'19"
Nicolas et Séverine
Uma 14h0'30" 49d55'19"
Nicolas F. Trujillo "Truji"
Cep 21h35'47" 59d18'17"
Nicolas Franjo Simic
Cru 12h19'11" -62d27'52"
Nicolas Fullen
Umi 15h6'47" 80d17'29"
Nicolás Gallego Sastre
Her 16h30'53" 24d5'36"
Nicolas Giovanni Taboadela
Aqr 22h16'29" -12d31'32"
Nicolas Grlica
Her 17h26'12" 35d20'45"
Nicolas Gurley
Her 18h34'59" 18d44'25"
Nicolas Harry
Aqr 22h25'55" -5d18'22"
Nicolas Herman
Per 4h6'29" 44d12'45"
Nicolas Imirtziadis
Uma 10h39'55" 59d11'55"
Nicolas Jenson Vigilante
Cnc 8h24'16" 25d23'54"
Nicolas John Kalli - Nicolas' Star
Umi 15h16'13" 70d0'37"
Nicolas John Zebrowski
Tau 4h30'32" 30d38'43"
Nicolas Jones
Dra 17h21'50" 55d28'53"
Nicolas K. Marchione
Umi 15h0'37" 74d40'6"
Nicolas LaBrec
Uma 13h34'37" 61d58'16"
Nicolas Leonardo Salinas
Sco 16h58'13" -38d21'3"
Nicolas Leroux June
Col 6h20'52" -34d27'43"
Nicolas Lory
Crb 15h55'3" 39d21'41"
Nicolas Mateo Vogel
Lib 14h49'15" -3d13'26"
Nicolas Michael Madaffari
Cru 12h6'45" -64d1'23"
Nicolas Michael Rhodes
Psc 1h20'40" 17d31'52"
Nicolás "Nico" Gabriel Brenni
Boo 14h33'40" 26d23'50"
Nicolas Novelli
Uma 8h24'0" 70d24'8"
Nicolas Paul Grill
Cap 20h33'49" -19d15'14"
Nicolas Peter Sammond
Tau 3h33'46" 4d58'57"
Nicolas Preguica
Ori 5h32'51" 10d3'51"
Nicolas Rayne Meza
Umi 15h52'59" 76d14'14"
Nicolas Reid Arizon
Umi 14h50'46" 74d22'47"
Nicolas Reza Bordfeld
Her 17h6'58" 30d50'33"
Nicolas Rinella
Vir 12h54'28" 6d42'43"
Nicolas Rioux
Col 5h25'17" -39d18'27"
Nicolas Robert Hyams
Per 3h8'2" 54d54'3"
Nicolas Rouger
Tau 5h45'56" 14d16'50"
Nicolas Ryan Augustine
Sco 16h12'36" -13d45'20"
Nicolas Saleur
Ori 5h15'26" 5d35'3"
Nicolas Seillé
And 23h4'2" 42d10'17"
Nicolas Sosa
Cnc 8h59'57" 11d43'8"
Nicola's Star
Peg 21h13'56" 16d50'56"
Nicola's Star
And 2h31'57" 47d7'42"
Nicolas Tokalatzidis
Aql 19h29'12" 4d28'13"
Nicolas Uttenveiler
Aqr 22h54'32" -23d39'3"
Nicolas William Castellarin
Ari 2h9'7" 24d4'1"
Nicola-Ti-Amo
Cyg 20h58'50" 47d59'43"
Nicolaus V. Jefferes
Uma 8h58'18" 54d58'44"
Nicole
Lyn 7h0'39" 56d13'54"
Nicole
Cyg 21h30'50" 55d9'31"
Nicole
Cyg 20h35'5" 57d47'39"
Nicole
Cyg 21h28'6" 53d19'51"
Nicole
Uma 11h25'15" 61d50'14"
Nicole
Uma 11h27'31" 61d0'52"
Nicole
Uma 8h58'14" 66d41'45"
Nicole
Sco 16h18'5" -11d45'4"
Nicole
Lib 15h47'39" -13d34'54"
Nicole
Cap 20h24'57" -13d13'35"
NICOLE
Cap 20h26'44" -18d49'1"
Nicole
Crt 11h20'16" -15d40'33"
Nicole
Lib 14h47'35" -4d59'31"
Nicole
Lib 15h22'19" -7d7'10"
Nicole
Aqr 21h51'5" -7d3'2"
Nicole
Vir 12h24'58" -8d50'58"
Nicole
Psc 23h57'30" -4d29'47"
Nicole
Sgr 19h21'18" -30d5'30"
Nicole
Sco 17h2'10" -45d33'52"
Nicole
Cra 18h18'55" -39d45'29"
Nicole
Per 3h27'40" 41d34'29"
Nicole
Her 17h58'39" 49d50'30"
Nicole
Uma 9h4'1" 52d9'30"
Nicole
Uma 9h55'46" 50d59'6"
Nicole
Cas 0h21'11" 54d48'53"
Nicole
Cas 0h19'53" 55d36'20"
Nicole
And 23h34'53" 48d29'34"
Nicole
And 22h57'51" 51d52'22"
Nicole
And 2h33'57" 48d48'51"
Nicole
And 2h16'18" 46d37'11"
Nicole
And 1h42'4" 45d15'20"
Nicole
Cas 1h21'42" 52d1'42"
Nicole
And 1h24'43" 47d36'31"
Nicole
Lyr 18h29'27" 37d10'50"
Nicole
And 1h49'43" 36d13'45"
Nicole
And 0h47'55" 30d5'55"
Nicole
Lyn 8h42'55" 36d44'45"
Nicole
Leo 9h42'17" 31d22'54"
Nicole
Lyn 8h35'55" 42d24'28"
Nicole
And 0h57'16" 42d11'34"
Nicole
Cyg 20h19'18" 35d5'22"
Nicole
Ari 2h16'40" 26d43'26"
Nicole
And 0h17'48" 26d53'4"
NICOLE
Ori 6h16'48" 19d29'13"
Nicole
Cnc 8h9'50" 16d25'29"
Nicole
Com 12h54'28" 29d1'8"
Nicole
Crb 15h42'5" 28d17'17"
Nicole
Peg 23h47'25" 23d50'21"
.....Nicole.....
Gem 6h19'37" 23d6'1"
Nicole
Ari 3h22'41" 28d18'33"
Nicole
Tau 5h22'20" 28d33'5"
Nicole
Vir 14h51'32" 6d3'5"
Nicole
Leo 9h32'41" 6d33'20"
NIcole
Vir 12h48'0" 2d11'34"
Nicole
Vir 11h49'50" 3d43'14"
Nicole
Psc 0h2'25" 1d31'17"
Nicole
Cnc 8h1'6" 13d10'48"
Nicole
Vir 13h17'17" 9d17'6"
Nicole
Com 12h7'54" 14d52'34"
Nicole
Vir 11h51'42" 9d7'33"
Nicole
Psc 0h38'44" 18d15'49"
Nicole
Psc 0h53'40" 19d26'39"
Nicole
Tau 5h10'35" 19d49'31"
Nicole 1981
Ori 6h13'41" 16d35'14"
Nicole 24.10.1963
Uma 9h11'14" 58d18'3"
Nicole 7101970
Sco 17h40'27" -42d19'55"
Nicole A. David
Sco 16h12'56" -18d20'22"
Nicole A. Hernandez a.k.a. Nikki
Tau 5h37'35" 23d33'42"
Nicole A. Murphy
Uma 11h9'30" 33d59'42"
Nicole A. Schultz
And 23h18'19" 44d6'25"
Nicole A Williams
Ori 5h34'3" 0d21'58"
Nicole Abdaem
And 1h28'11" 39d31'14"
Nicole Abshire
Cyg 20h38'21" 38d3'10"
Nicole Adams
Cap 20h38'13" -24d13'53"
Nicole Addonizio & Thomas Re
And 2h34'50" 45d9'32"
Nicole Adrianna Ciro
Tau 4h30'16" 23d10'16"
Nicole Alagna
And 2h16'26" 47d37'19"
Nicole Alberta Swan
Cnc 8h43'37" 14d54'40"
Nicole Albrecht
Ori 5h55'53" 17d8'33"
Nicole Aleen Lockhart
Psc 0h31'8" 10d2'7"
Nicole Alexa Carach
Cap 20h23'4" -12d6'48"
Nicole Alexandra
And 1h45'31" 42d22'40"
Nicole Alexandra Stout
Sgr 18h31'37" -22d27'11"
Nicole Alexandra Tropea
Crb 15h39'26" 36d35'42"
Nicole Alexis
Psc 1h10'46" 20d48'41"
Nicole Alfonso
Cnv 13h44'12" 32d23'21"
Nicole Alison Lamanda
Sgr 17h58'58" -16d53'29"
Nicole Al-Jabiri
Ari 2h29'33" 21d30'58"
Nicole Alyse
And 0h47'0" 38d58'58"
Nicole Alyssa
Leo 10h19'53" 23d53'52"
Nicole Alyssa Snyder
And 23h16'55" 48d11'27"
Nicole Anaya
Ori 5h11'0" 15d29'39"
Nicole and Barrett's Star
And 0h55'52" 40d6'51"
Nicole and Bryan
Leo 11h34'28" -6d40'27"
Nicole and Constantinos
Lyr 18h39'46" 30d28'44"
Nicole and James for ever and ever!
Cyg 20h1'20" 33d59'10"
Nicole and Jerry
Leo 10h7'38" 14d59'15"
Nicole and Joe
Col 5h39'32" -32d40'32"
Nicole and Jon Forever
Uma 10h56'36" 50d42'12"
Nicole and Josh's Burning Love
Leo 11h22'53" 11d54'18"
Nicole and Rich
And 1h46'28" 49d59'51"
Nicole and Ryan's Love
Lib 15h33'9" -8d4'54"
Nicole and Shaun
Sco 17h34'57" -30d48'20"

Nicole and Thomas Cyg 20h22'28" 54d45'59"
Nicole and Zakeek Ali Uma 11h38'30" 34d6'40"
Nicole Anderson Aql 19h26'36" 6d43'39"
Nicole Andrea Cnc 8h1'58" 14d14'58"
Nicole Andrea Dominguez Cap 20h17'5" -10d57'2"
Nicole & Andreas Uma 9h51'26" 72d34'21"
Nicole & Andreas Ori 5h11'45" 6d52'9"
Nicole & Andy Peg 23h8'50" 29d27'30"
Nicole Angela Murray Sgr 19h51'54" -23d29'30"
Nicole Angela Petrarca And 0h22'48" 43d0'38"
Nicole Anh Vo Psc 0h26'37" 11d39'25"
Nicole Ann Leo 11h12'43" 9d6'39"
Nicole Ann Armstrong Aql 19h53'38" -0d14'37"
Nicole Ann Berrios And 1h20'38" 49d23'13"
Nicole Ann Hamblet Umi 13h5'51" 70d35'21"
Nicole Ann Kuzma Cap 21h52'20" -14d49'16"
Nicole Ann Martin Uma 11h25'2" 70d55'14"
Nicole Ann Riches Her 16h24'45" 10d32'25"
Nicole Ann Storch Leo 10h6'2" 25d34'11"
Nicole Anne Leo 10h47'47" 10d41'44"
Nicole Anne Southard Ori 6h4'6" 21d7'24"
Nicole Anthony And 22h57'30" 51d0'49"
Nicole Antoinette Mir And 23h30'16" 48d22'52"
Nicole Argyros Cas 23h47'13" 64d43'12"
Nicole Arnold Cas 2h0'56" 70d48'18"
Nicole Ashley Macleod And 23h40'38" 44d45'56"
Nicole Ashley Rodriguez Her 17h19'4" 32d10'24"
Nicole Ashley Salamon Lyn 7h58'1" 35d35'45"
Nicole Ashley White Cas 23h29'14" 52d58'0"
Nicole Avezard Cas 0h15'28" 50d45'26"
Nicole B Cap 21h42'14" -9d42'59"
Nicole B. Kalina Psc 1h20'30" 7d20'54"
Nicole B. Kwetinski Psc 23h44'30" 5d24'16"
Nicole Baillargeon Lib 15h46'8" -18d17'51"
Nicole "Bambi" D'Angelo Cap 20h9'33" -22d11'18"
Nicole Beare & Colin Hluchaniuk Umi 4h45'40" 88d28'26"
Nicole Beautiful Hildebrandt Leo 11h25'30" 15d1'16"
Nicole Benedetto Lyr 18h51'14" 41d32'36"
Nicole Bennett Ari 2h19'59" 17d33'57"
Nicole Bertholet Cas 0h30'24" 51d3'26"
Nicole Beswick Cas 0h17'31" 52d59'8"
Nicole Beth Rowland Lib 16h0'26" -17d5'21"
Nicole Beth Stearns And 2h34'17" 38d56'38"
Nicole Bethany Cox Peg 21h38'10" 7d6'44"
Nicole Biasiello Psc 1h3'35" 32d46'3"
Nicole Blanchard Cyg 20h49'9" 47d35'53"
Nicole Blankenspoors' Star Pho 0h12'8" -40d28'8"
Nicole Boram Lyn 6h28'48" 55d14'52"
Nicole Bouquet Cru 12h33'31" -60d33'6"
Nicole Braaten And 23h35'44" 44d56'1"
Nicole & Brad - For Eternity Cru 12h35'55" -59d7'21"
Nicole Bridges Aqr 22h43'44" -3d46'6"
Nicole Brook Place Sgr 18h57'55" -30d13'38"
Nicole Brown Uma 8h20'6" 67d46'56"
Nicole Bunnell Lib 15h6'44" -8d50'3"
Nicole Burskey's STAR! Cnv 12h53'56" 45d23'40"
Nicole Butterfly Renee Atkinson Crb 15h48'40" 26d15'57"
Nicole C. Glaser Ari 2h13'56" 23d2'43"
Nicole C. Lighthouse Aqr 21h10'48" 1d21'48"
Nicole C. Stroup Cnc 7h57'51" 12d37'55"
Nicole Caitlin And 1h34'56" 37d45'30"
Nicole Campbell Lib 14h36'43" -17d44'23"
Nicole Cannizzaro Cas 1h46'33" 69d4'25"
Nicole Carey Cas 23h49'1" 53d21'52"
Nicole Carnell Umi 15h38'56" 77d41'39"
Nicole Carreon Gem 7h16'11" 32d3'40"
Nicole Castonguay Ori 5h25'22" 14d31'1"
Nicole Catanese Lyn 7h56'43" 59d4'51"
Nicole Catherine And 23h33'13" 47d5'14"
Nicole Catherine and Julia Maxime Umi 15h30'10" 70d2'53"
Nicole Chandler Tau 3h29'48" 18d15'5"
Nicole & Charlie Ori 6h20'57" 10d23'56"
Nicole Chatham Sgr 18h12'18" -19d30'37"
Nicole Cherice Shellito And 0h42'43" 34d44'58"
Nicole Cherie Psa 21h39'25" -31d19'25"
Nicole Christina Lobue And 0h36'11" 31d53'44"
Nicole Christine Karaman Cas 22h57'30" 54d21'23"
Nicole Claire Leo 11h53'58" 18d52'2"
Nicole Clary Mon 8h0'7" -0d26'27"
Nicole Cohen Cas 0h59'31" 67d16'34"
Nicole "Colie" Sunberg And 0h35'54" 43d29'26"
Nicole Colquett Aqr 22h20'28" 1d48'58"
Nicole Cooper Tau 5h25'5" 17d36'28"
Nicole Cori Sundheimer Sco 17h54'9" -42d30'37"
Nicole Corin Wharton Peg 21h51'55" 8d23'12"
Nicole Cortes Ari 2h32'4" 23d46'51"
Nicole Costantino Tau 3h44'15" 28d21'27"
Nicole Coutant Kress Ori 5h30'31" 5d17'38"
Nicole Couvertier Tau 4h41'53" 2d31'37"
Nicole Covington Leo 9h25'15" 30d59'39"
Nicole Crisp Uma 10h49'0" 50d18'14"
Nicole D. Maves Cam 3h17'6" 58d32'26"
Nicole Dalabiras Cen 13h55'53" -59d52'24"
Nicole Daniella Richey And 0h20'9" 32d13'30"
Nicole Danielle Alexanian Tau 4h8'25" 12d5'16"
Nicole Danielle Frammosa And 1h44'53" 36d23'59"
Nicole + Dario Cnv 13h56'11" 36d1'8"
Nicole David Buisson Her 17h54'0" 49d44'15"
Nicole Dawn Kielar Umi 9h58'8" 87d53'16"
Nicole Dell'Angelica Crb 15h43'41" 31d23'55"
Nicole DeLuco And 0h18'37" 44d41'6"
Nicole Denans And 23h49'14" 48d28'32"
Nicole Denielle And 23h30'45" 48d23'49"
Nicole Denise Stranz Sgr 17h54'22" -28d37'32"
Nicole Diane Colbert And 23h24'3" 51d34'38"
Nicole Diane Mims Cas 23h5'37" 57d52'56"
Nicole Dingman Mon 7h5'24" -5d48'15"
Nicole Domenica Loia Cmi 7h22'34" 3d32'53"
Nicole Dominy's Friendship Star Gem 6h45'28" 23d46'43"
Nicole Donna Bode Lyn 8h0'13" 52d46'57"
Nicole Dunithan And 1h44'55" 45d43'46"
Nicole DuPlessis Sgr 19h17'43" -18d13'31"
Nicole E. Fontenault Psc 1h9'19" 25d10'5"
Nicole E Gordon Psc 1h29'32" 9d31'15"
Nicole E. Wills And 1h31'44" 41d37'42"
Nicole East Cru 12h29'46" -59d16'15"
Nicole Elisabeth Steele And 23h30'44" 41d25'7"
Nicole Elise Bailin Lmi 10h6'33" 34d13'54"
Nicole Elizabeth And 1h45'16" 43d55'26"
Nicole Elizabeth Crb 15h51'3" 27d47'42"
Nicole Elizabeth Cap 20h56'35" -20d41'7"
Nicole Elizabeth Blaszak I Love You Sgr 18h10'0" -30d31'8"
Nicole Elizabeth Buonocore Psc 0h15'26" 10d27'11"
Nicole Elizabeth Cook Sgr 19h25'13" -15d18'37"
Nicole Elizabeth & Courtney Sue Uma 11h37'7" 47d47'35"
Nicole Elizabeth Dorey Cas 1h42'56" 67d5'3"
Nicole Elizabeth Dunn Leo 9h39'10" 28d4'55"
Nicole Elizabeth Goodman Cap 20h51'48" -16d29'39"
Nicole Elizabeth Graziano Aqr 22h30'28" -0d51'38"
Nicole Elizabeth Herb Gem 7h39'12" 27d15'0"
Nicole Elizabeth Long And 2h37'52" 50d37'14"
Nicole Elizabeth Schmucker Umi 15h4'20" 75d2'54"
Nicole Elizabeth Sedutto And 23h3'22" 51d18'45"
Nicole Emerson Cas 23h22'11" 55d12'19"
Nicole Erato Ori 6h4'10" 10d31'43"
Nicole Ercole Tau 4h10'10" 20d3'57"
Nicole Eusebio Mon 6h31'5" 7d48'26"
Nicole Evans Aqr 22h59'36" -11d55'56"
Nicole Everett Per 1h40'51" 54d34'11"
Nicole F. Sgr 19h30'8" -18d26'24"
Nicole Faith Currie And 2h20'58" 48d57'24"
Nicole Faith Pierre And 1h46'32" 42d46'59"
Nicole Falduto Sgr 18h52'45" -20d27'39"
Nicole Faye Saputo Leo 9h47'18" 6d36'42"
Nicole Filler Cas 23h46'24" 59d18'15"
Nicole Fisher Vel 9h8'28" -37d31'2"
Nicole Fishkin Cam 6h48'24" 64d18'13"
Nicole Fiumefreddo Cas 23h23'49" 58d1'22"
Nicole Frakes Crb 15h47'27" 33d20'44"
Nicole Francesca King - Nic's Star And 22h59'16" 52d1'11"
Nicole Francine Hall Lib 14h47'34" -24d35'32"
Nicole G. Aqr 21h46'40" -1d21'49"
Nicole G. Eckstorm And 23h37'31" 42d32'40"
Nicole G. Weorriour Ari 2h32'14" 19d25'54"
Nicole Gabrielle Flynn Lyn 8h12'26" 52d17'58"
Nicole Gabrielle Muratore And 2h29'49" 41d34'31"
Nicole Gabrielle Silvestro Ari 1h56'48" 18d49'58"
Nicole Galgano Aqr 23h21'21" -14d1'24"
Nicole Ganz And 23h58'40" 41d10'7"
Nicole Gasser, 20.06.1984 Umi 13h45'51" 79d3'44"
Nicole Gavin Ori 6h8'59" 15d49'10"
Nicole Gayle Vir 13h25'55" 13d30'21"
Nicole Gillet Peg 22h36'13" 19d56'29"
Nicole Gluszek Psc 23h32'6" 7d16'56"
Nicole Goldner Tri 2h19'22" 36d32'53"
Nicole Goodling Cyg 21h33'1" 39d22'28"
Nicole Grace Bjork Lyn 8h45'27" 42d27'17"
Nicole Graham And 23h9'32" 41d25'27"
Nicole Gresiak And 1h40'26" 38d22'14"
Nicole Griffin-Ammons Mon 6h40'40" -0d26'38"
Nicole Gut Umi 14h38'0" 76d45'40"
Nicole Hagan Lib 15h37'13" -16d58'32"
Nicole Hammersmith And 1h38'7" 42d35'33"
Nicole Hammond Psc 1h45'2" 5d48'17"
Nicole Harris Psc 1h16'14" 23d46'13"
Nicole Harrison Gem 6h48'51" 17d44'50"
Nicole Härtl - Nici's Star And 0h2'51" 44d41'29"
Nicole Hays Cas 0h14'20" 61d25'19"
Nicole Hepinstall Sgr 20h2'44" -38d47'15"
Nicole Herald Nox Caelum Umi 14h15'29" 72d54'2"
Nicole Herbert Psc 1h8'30" 12d9'24"
Nicole Hiljus And 23h15'3" 41d42'49"
Nicole Hillier Uma 13h56'39" 49d45'34"
Nicole Hill-Smith Leo 10h16'35" 25d16'4"
Nicole Hobbs - 23 September 1975 Pho 0h41'51" -46d4'33"
Nicole Hope Cnc 8h4'35" 8d2'30"
Nicole Houle Vir 12h59'37" 11d0'32"
Nicole Howley Gem 7h32'43" 20d48'14"
Nicole Huber-Guebeli Per 4h44'10" 41d4'43"
NICOLE ILD FOREVER Lac 22h57'22" 46d5'8"
Nicole Irene Cas 23h24'15" 58d51'13"
Nicole Ivy Goldenberg Vir 13h7'7" 13d44'48"
Nicole J. Senence Ari 2h58'57" 17d34'59"
Nicole Jane Patacca Umi 15h32'59" 67d37'38"
Nicole Janel Cyg 20h38'33" 35d55'57"
Nicole Janssen Uma 10h19'55" 50d45'54"
Nicole & Jason Schwartz - True Love Cyg 19h38'6" 53d13'4"
Nicole Jauaneau And 1h18'44" 42d48'54"
Nicole Jayne Gandy Sgr 19h22'25" -12d34'37"
Nicole Jean Allen And 1h53'34" 45d29'5"
Nicole Jean Bradbury Dra 18h57'8" 50d43'9"
Nicole Jean Briese Lib 15h9'24" -1d3'34"
Nicole Jean Ransden Aqr 22h14'8" 0d36'23"
Nicole Jennifer Ryan Sgr 19h14'3" -16d6'42"
NICOLE JO And 0h48'30" 44d2'59"
Nicole Joan Tau 5h32'52" 22d7'59"
Nicole JoAnn Miller Cas 0h15'55" 58d51'53"
Nicole Jodean Raymond Leo 11h41'28" 26d32'6"
Nicole Jodi Ritchie Vir 15h7'46" 4d48'10"
Nicole Jones Tau 3h29'38" 8d18'0"
Nicole Joyce Sablan Merfalen Lib 14h51'1" -17d13'4"
Nicole Julia Poole Lyn 6h48'41" 50d41'11"
Nicole K. Harris - 21 Aqr 21h34'10" 1d1'32"
Nicole K. Voumvourakis Aqr 21h34'53" -0d33'8"
Nicole Kaplan's Musical Star Uma 9h22'27" 60d33'42"
Nicole Karen Welch Cap 20h16'39" -10d23'36"
Nicole Kastell And 0h29'42" 41d26'38"
Nicole Katharine Bond Crb 16h15'43" 36d48'10"
Nicole Kay Thompson Vir 13h23'28" -9d4'7"
Nicole*Kaylee Sge 19h50'16" 18d29'29"
Nicole Kelley McAlister Cyg 21h16'26" 47d48'27"
Nicole Kennedy Cyg 20h2'14" 35d32'18"
Nicole Kennedy Psc 1h28'29" 10d38'8"
Nicole Kern Leo 10h7'47" 16d49'43"
Nicole & Kevin Forever Vir 13h4'2" 12d58'29"
Nicole Kimberly Taylor Uma 8h22'53" 69d41'25"
Nicole Konarzewski And 0h26'57" 33d13'49"
Nicole Korgeski Lyn 7h38'29" 36d48'58"
Nicole Kristin Dittman And 23h59'22" 39d50'36"
Nicole Kristine Ownby Psc 23h46'17" 5d20'48"
Nicole Kritselis Sgr 19h10'56" -16d50'13"
Nicole Kurth & Robert Levy Her 16h34'28" 15d48'50"
Nicole L. Robinson Cas 0h42'12" 50d45'56"
Nicole L. Robson Cas 23h36'20" 52d9'13"
Nicole La Porte Ari 2h48'46" 26d59'36"
Nicole Lacy Tau 3h58'0" 25d28'41"
NICOLE L'ALLIER Tau 3h31'34" 28d6'30"
Nicole Lamb Leo 10h13'59" 24d50'34"
Nicole & Lance Winterhalder Uma 12h48'39" 62d11'14"
Nicole Langlinais Lib 15h6'49" -8d38'18"
Nicole Langlois Crb 16h11'51" 39d18'49"
Nicole Lauren And 0h26'36" 39d35'24"
Nicole Leandra Morris Cnc 8h2'26" 21d15'53"
Nicole Leanne Smith Lyn 8h23'57" 56d6'13"
Nicole Lee Sco 16h58'2" -38d35'13"
Nicole Lee Cyg 21h27'38" 31d6'9"
Nicole Lee Anne Sadler Sgr 18h17'7" -22d40'24"
Nicole Lee Bletzer Per 3h24'36" 45d27'0"
Nicole Lee Hager And 1h33'41" 37d43'40"
Nicole Lee Heckman Cnc 8h13'32" 15d6'35"
Nicole Lee McCoin And 1h16'2" 38d41'7"
Nicole Lee Prater Her 17h42'8" 33d10'14"
Nicole LeeAnn Estes Cap 21h29'27" -14d51'32"
Nicole Leeann Fisher Tau 4h50'22" 23d40'30"
Nicole Leigh Boyle And 23h24'41" 52d17'35"
Nicole & Leon Cyg 19h47'45" 44d9'40"
Nicole Levay / Frowlow37 Cyg 19h37'54" 32d11'12"
Nicole Lewins And 0h46'46" 40d53'53"
Nicole Liano And 2h25'3" 46d45'53"
Nicole Lin Ross Vir 12h44'33" -9d15'0"
Nicole Lindsay Lib 15h4'56" -6d3'27"
Nicole Lischer Uma 11h7'48" 56d8'46"
Nicole Logan And 2h22'30" 46d48'43"
Nicole Loreen Albert And 23h10'10" 42d41'52"
Nicole LoRene Winters And 23h16'51" 52d16'7"
Nicole Lorraine Earle Umi 15h21'33" 73d29'57"
Nicole Lorraine Hosman Uma 8h14'38" 67d38'23"
Nicole Louise Cyg 20h16'47" 37d7'43"
Nicole Louise Ray Psc 23h40'33" 2d36'34"
Nicole Loves Carl Ari 3h16'38" 15d1'18"
Nicole loves Spence Uma 11h3'50" 53d24'6"
Nicole Lyn Underwood And 1h0'32" 39d46'55"
Nicole Lynae Barker Vir 14h15'36" 0d32'26"
Nicole Lynae Bruice Leo 11h7'28" 23d12'32"
Nicole Lynn Leo 10h22'6" 21d57'36"
Nicole Lynn And 2h33'43" 44d36'34"
Nicole Lynn Chavis And 23h20'43" 48d40'6"
Nicole Lynn Fiske And 2h22'56" 41d25'2"
Nicole Lynn Moran And 2h12'39" 45d58'52"
Nicole Lynn Schluep Cap 21h34'49" -9d52'21"
Nicole Lynn Watson Ori 5h25'8" 6d19'7"
Nicole M. Bala And 0h4'15" 35d26'32"
Nicole M Cedillo Uma 9h26'15" 54d11'49"
Nicole M. Chiudioni Cnc 8h43'29" 16d1'59"
Nicole M. Cram Cnc 8h0'46" 25d50'53"
Nicole M. Cremona Vir 13h29'14" -5d22'11"
Nicole M. Lavin Leo 11h5'29" 10d20'26"
Nicole M. Robidart Uma 8h49'42" 51d6'50"
Nicole M. Swanson Umi 16h18'22" 77d14'41"
Nicole M Wagner Crb 15h46'13" 39d10'51"
Nicole M. Walsh Leo 9h45'34" 27d25'32"
Nicole Macri Cas 23h4'47" 54d57'47"
Nicole Mae Threlkel Sco 16h6'58" -18d29'55"
Nicole Malia Kerrigan And 23h11'31" 48d2'14"
Nicole Maria Cnc 8h12'56" 7d35'59"
Nicole Mariah Molaro Cas 1h13'45" 63d41'29"
Nicole Marie Aqr 21h47'42" -1d43'32"
Nicole Marie Aqr 22h2'54" -0d38'10"
Nicole Marie Cap 21h5'37" -25d9'37"
Nicole Marie Cnc 9h9'48" 11d17'56"
Nicole Marie Ori 5h27'28" 0d37'17"
Nicole Marie Ori 5h42'58" 7d49'35"
Nicole Marie Leo 11h15'45" 23d51'19"
Nicole Marie And 2h26'53" 48d52'56"
Nicole Marie Cnc 9h8'6" 30d59'16"
Nicole Marie And 1h17'40" 42d9'42"
Nicole Marie And 23h41'0" 34d40'13"
Nicole Marie 91 And 2h29'49" 49d0'3"
Nicole Marie Blackburn Sgr 17h59'49" -17d28'7"
Nicole Marie Bradford And 2h13'42" 41d13'40"
Nicole Marie Braun Tau 4h5'25" 18d43'31"
Nicole Marie Busby Ari 2h7'59" 26d9'18"
Nicole Marie Canta Leo 9h44'0" 24d45'39"
Nicole Marie Craft Crb 15h23'24" 27d48'41"
Nicole Marie Cripe Cam 7h28'29" 64d2'39"
Nicole Marie Cubitt And 0h49'55" 46d19'53"
Nicole Marie DiNizo And 23h12'13" 51d59'6"
Nicole Marie Garber Lib 14h23'41" -19d26'39"
Nicole Marie Garza Leo 10h10'55" 22d10'18"
Nicole Marie Guy & Robert Staples Cyg 21h27'58" 46d0'33"
Nicole Marie Hicks Lmi 10h24'34" 39d12'0"
Nicole Marie Hooley Leo 9h38'38" 26d51'56"
Nicole Marie Hurst Cap 20h23'8" -11d4'51"
Nicole Marie Jones And 0h54'44" 35d4'10"
Nicole Marie Jordan Gem 6h51'24" 23d42'0"
Nicole Marie Klimek Tau 3h39'24" 28d4'25"
Nicole Marie Koye Lyn 8h39'47" 39d7'45"
Nicole Marie Kruk Cyg 20h38'25" 52d45'37"
Nicole Marie Leone Psc 1h27'40" 27d23'11"
Nicole Marie Manniti Sco 16h9'8" -11d56'44"
Nicole Marie Mazzaschi Cyg 19h33'31" 28d37'57"
Nicole Marie McQueen Lyn 7h43'41" 47d42'40"
Nicole Marie Miller Her 17h2'7" 29d38'21"
Nicole Marie Milovich Gem 6h30'18" 20d56'19"
Nicole Marie Monica Cas 2h38'59" 66d41'54"
Nicole Marie Perata Tau 3h45'36" 11d15'36"
Nicole Marie Pucci Ari 3h19'27" 29d42'29"
Nicole Marie Puleo Cap 21h48'53" -11d24'14"
Nicole Marie Rasch And 23h17'41" 46d46'57"
Nicole Marie Reed/Gorby Vir 13h23'46" 1d53'43"
Nicole Marie Rivera Tau 4h30'43" 18d49'37"
Nicole Marie Rodriguez Cyg 20h53'9" 31d36'57"
Nicole Marie Rognrud Cas 2h8'49" 62d14'36"
Nicole Marie Romer Psc 0h39'37" 16d54'37"
Nicole Marie Schafroth Gem 6h36'10" 25d2'25"
Nicole Marie Siegel Ori 5h29'11" 4d26'58"
Nicole Marie Smith Tau 4h15'3" 27d26'35"
Nicole Marie St. John Aqr 20h45'10" -8d43'3"
Nicole Marie Stahl Ori 5h54'11" 17d25'50"
Nicole Marie Thomas Sco 17h49'16" -43d3'56"
Nicole Marie Tortoreti Leo 9h31'36" 12d37'36"
Nicole Marie Wilson Cap 20h28'11" -13d10'43"
Nicole Marie Witherell Ori 6h5'35" 13d49'25"
Nicole Marie Wurtele Crb 16h10'36" 26d13'40"
Nicole Marling Gem 7h41'14" 20d13'13"
Nicole Marquez Psc 1h0'20" 18d32'28"
Nicole Martin Uma 10h16'16" 45d30'26"
Nicole Martinson And 23h28'21" 41d42'50"
Nicole Mary Ari 3h6'17" 10d35'49"
Nicole Mary Aqr 21h8'15" -12d14'23"
Nicole & Matt Men 4h26'39" -81d59'43"
Nicole Mattson Tau 4h45'37" 20d50'0"
Nicole Maureen Gertonson Uma 9h3'7" 48d0'4"
Nicole Mayumi Furuta Cnc 8h18'53" 10d0'1"
Nicole McCarthy Cas 2h28'40" 67d47'14"
Nicole Mccollum And 2h19'51" 46d32'9"
Nicole McCullough Uma 11h26'58" 52d24'22"
Nicole McLean Crb 15h35'32" 27d51'5"
Nicole Mele Peg 21h39'5" 17d2'38"
Nicole Mellor And 0h25'57" 25d41'23"
Nicole Michelle Bartz Vir 13h8'15" 6d16'12"
Nicole Michelle Cross And 0h32'52" 33d6'33"
Nicole Miller Crb 16h4'8" 34d20'45"
Nicole Miraglia And 0h21'4" 30d42'15"
Nicole *Mom* Ari 1h56'43" 18d30'46"
Nicole Montero Vir 12h25'56" 6d18'24"
Nicole Morfis And 23h8'15" 46d11'49"
Nicole Mullins Cyg 20h30'32" 37d30'16"
Nicole Murray Cas 2h32'46" 66d24'29"
Nicole - My Forever Star - 31/10/04 Aqr 22h22'21" -7d11'21"
Nicole Nabors Cyg 20h17'1" 47d51'14"
Nicole Nannette Goodson Mon 7h32'46" -0d56'32"
Nicole Nardoux Peg 23h31'0" 18d58'13"
Nicole Natoli And 0h37'30" 37d48'21"
Nicole Nesmith Crb 15h46'9" 27d3'47"
Nicole Nguyen Tau 4h7'21" 7d29'54"
Nicole Noelle Thysens Boo 15h10'4" 40d19'47"
Nicole Norman Lib 15h37'46" -10d22'49"
Nicole Nunn Cam 4h47'11" 58d33'40"
Nicole Oben Aqr 21h15'34" 2d19'31"
Nicole O'Brian And 1h10'41" 42d54'19"
Nicole & Oli Cas 1h18'43" 51d24'4"
Nicole & Oliver Ori 5h56'1" 21d32'40"

Nicole Ora Bershtel
Vir 13h40'23" 7d4'46"
Nicole Osti
Aqr 20h42'38" -8d42'57"
Nicole Overton
Uma 10h55'2" 60d25'49"
Nicole P. Cowell
Crb 15h30'31" 32d14'50"
Nicole Pacylowsky
Sgr 19h11'16" -21d8'9"
Nicole Page Christmas 2005
Cru 12h18'0" -57d34'38"
Nicole Parsons Danner
And 1h45'26" 42d6'11"
Nicole Pashkow
Cas 1h33'59" 65d48'22"
Nicole Patrice Davis
And 0h20'29" 27d48'44"
Nicole Patrice Miller
Uma 9h15'2" 48d24'33"
Nicole & Pedro
Uma 10h44'26" 49d9'10"
Nicole Peterson
Lib 14h29'37" -19d50'28"
Nicole Peyton
Cas 0h15'2" 63d32'58"
Nicole & Pierre
Ori 6h14'6" 20d45'17"
Nicole Piquant
Tau 4h18'46" 14d9'40"
Nicole Pisterzi
Aqr 23h20'59" -9d1'24"
Nicole Planinz
Ori 5h53'28" 15d38'18"
Nicole Pohorenec
Gem 6h5'52" 27d53'24"
Nicole Polivka
Ori 5h25'37" 2d44'49"
Nicole Pourgouridou
Vir 12h13'40" 9d36'51"
Nicole Preste
Ari 3h13'38" 29d29'57"
Nicole Pretet
Cam 6h41'35" 63d42'9"
Nicole Proctor
Tau 4h38'39" 18d48'37"
Nicole Psiaki
Mon 6h59'41" 3d29'56"
Nicole R. Franko's Star
Uma 11h54'8" 34d16'8"
Nicole R. Johnson MY NIKKI
Psc 1h7'9" 29d55'38"
Nicole R Junger
Ari 2h33'19" 18d55'30"
Nicole R. Mckenney
Sco 16h15'32" -10d28'26"
Nicole R. Ralph
Gem 6h23'40" 25d37'13"
Nicole R. Thompson
Sgr 18h25'7" -16d35'14"
Nicole Rachael Evans - Love of Life
Tau 4h16'43" 24d47'15"
Nicole Rachell
Lib 14h27'43" -18d15'12"
Nicole Rae
Sco 16h11'27" -10d11'20"
Nicole Rae Hayes
Aqr 23h17'30" -19d6'29"
Nicole Rae 'Nikki' Chialiva Gillet
Vir 11h39'54" 9d11'34"
Nicole Rae Steiner
Cam 5h20'10" 56d55'27"
Nicole Raphaella Pen
Sgr 17h45'33" -27d54'53"
Nicole Ray Muller
Sgr 19h13'39" -12d22'48"
Nicole Raymond
Psc 1h14'56" 31d56'55"
Nicole Rebecca Dias
And 23h20'41" 39d9'33"
Nicole Reed
Lyn 6h25'57" 55d48'21"
Nicole Regina Leva
And 1h28'38" 40d31'55"
Nicole Renae Howze
Aqr 22h17'47" -1d39'6"
Nicole Rene
And 1h46'41" 49d37'2"
Nicole Renee
Vir 12h52'9" 6d35'55"
Nicole Renee
Cas 23h30'41" 58d6'52"
Nicole Reneé
Dra 16h11'36" 53d30'26"
Nicole Renée Benson
Psc 1h13'26" 16d48'14"
Nicole Renee Ciferni
And 0h34'3" 23d2'30"
Nicole Renee Coffey
Psc 1h20'45" 16d28'53"
Nicole Renee' David
Sco 16h13'1" -12d1'4"
Nicole Renee Gambin
Col 5h32'5" -33d23'4"
Nicole Renee Harsha
Ori 5h30'59" 6d2'17"
Nicole Renee Henry
Psc 0h16'3" 15d10'14"
Nicole Renee Novak
Her 16h42'26" 46d2'20"
Nicole Renee Perrine
And 2h10'13" 38d29'38"
Nicole Renee Pettit
Tau 5h22'50" 28d1'38"
Nicole Renee Richard
Cap 20h29'8" -16d13'32"
Nicole Renee Russo
Tau 4h47'26" 16d2'48"
Nicole Renee Sartori
Lyn 7h57'33" 48d13'22"
Nicole Reyes
Vir 13h54'22" -19d33'34"
Nicole Ricks
Sgr 18h25'10" -27d53'40"
Nicole Rim
Gem 7h44'34" 32d17'48"
Nicole Ringler
Gem 6h20'20" 18d22'57"
Nicole Risinger
Pho 0h53'46" -57d18'10"
Nicole Robinson
And 23h23'56" 41d31'59"
Nicole Rodriguez
Ari 2h22'14" 27d30'22"
Nicole&Rolf
Ori 5h53'54" 17d25'3"
Nicole Rose Bergeron
Lib 15h21'25" -20d4'59"
Nicole Rose Getsie
Lyn 6h34'39" 57d49'39"
Nicole Rose Sanfilippo
Uma 14h19'18" 55d56'21"
Nicole Rotella
Cam 7h5'40" 71d56'29"
Nicole Ruth Marcinkus
Cnc 8h35'13" 24d58'27"
Nicole Rutledge
Ori 6h17'17" 10d6'17"
Nicole S. Hall
Gem 6h4'13" 21d53'7"
Nicole Salis
Uma 9h55'43" 44d13'42"
Nicole Samantha's Star
And 1h27'22" 40d23'53"
Nicole Samone Lobell
And 22h57'53" 47d35'31"
Nicole Sara Brown
Cas 0h26'24" 55d46'34"
Nicole Sarnicola
Cas 2h50'27" 64d2'55"
Nicole Saskia
Cas 0h3'43" 52d6'22"
Nicole Scaiola Chiarelli
Psc 1h1'14" 7d20'58"
Nicole Schempp-Christopher Solomon
Cyg 19h57'37" 47d11'52"
Nicole Schifter
Crb 15h54'51" 34d17'48"
Nicole Schlosser
Cas 0h22'32" 50d7'2"
Nicole Schmid
Lyr 19h8'47" 42d6'57"
Nicole Schmucki
Cyg 20h54'26" 47d28'57"
Nicole Schmutz-Dubey
Lyr 18h30'27" 36d35'25"
Nicole Schneider **Mein Schatz**
And 2h36'5" 45d50'44"
Nicole Schulte
Uma 8h35'30" 57d54'22"
Nicole Serena
Tau 3h59'46" 21d14'8"
Nicole Shababb
Vir 13h15'17" 8d2'48"
Nicole Sharpe
Com 13h13'32" 17d56'4"
Nicole Simmons
And 0h22'29" 41d22'33"
Nicole Simone
Per 4h31'48" 40d59'19"
Nicole Smith
Crb 16h13'12" 39d29'43"
Nicole Smith
Mon 6h51'48" -0d26'30"
Nicole Souris
And 0h14'43" 26d6'49"
Nicole "Starkitten" Martin
Dra 17h10'6" 61d17'16"
Nicole Steward
Leo 11h5'18" 16d20'28"
Nicole Stinker King
Sco 16h3'51" -14d58'5"
Nicole Stumpf
Uma 9h16'20" 50d51'3"
Nicole Supernova Tape
Sgr 19h16'20" -22d33'58"
Nicole Suzanne Osborne
Srp 18h23'37" -0d55'26"
Nicole & Sven Anniversary 1
Gem 7h26'43" 27d21'7"
Nicole "Sweetcakes" Skipper
And 2h17'46" 44d49'40"
Nicole "T" Lynn
Cnc 9h14'13" 22d44'25"
Nicole Taryn Albert
Sgr 18h33'57" -18d8'16"
Nicole Taylor
Cas 2h28'9" 66d18'39"
Nicole Taylor DeSousa
Ori 6h16'36" 3d24'56"
Nicole Teresa Delacruz
Ori 6h1'52" 10d32'21"
Nicole The Great
Psc 1h42'53" 5d0'54"
Nicole " The Lazy Butt" S.
Uma 9h29'7" 61d58'14"
Nicole "The White Dragon"
Cas 0h8'1" 57d19'29"
Nicole Thometz
Sgr 19h19'37" -16d47'18"
Nicole Thum
Sco 16h3'48" -12d6'24"
Nicole Tilley
Uma 13h44'16" 58d26'52"
Nicole Timpone
Ori 5h24'25" -0d51'18"
Nicole "Tink" Saputo
Cyg 20h32'19" 35d7'4"
Nicole & Tony
And 1h29'54" 44d36'54"
Nicole Totzkay
Cnc 8h42'38" 17d17'52"
Nicole Trinity Kratz
And 0h5'56" 44d35'8"
Nicole & Tryston Stone
Sgr 18h56'46" -24d56'0"
Nicole Ullrich
Sgr 18h0'48" -29d46'29"
Nicole und Marcel
Lyr 19h17'58" 35d34'21"
Nicole Unterberger
Dra 18h52'38" 71d50'50"
Nicole Vegas
Cnc 8h1'21" 21d42'16"
Nicole Verbin
And 0h52'39" 45d15'18"
Nicole Verrino
Cap 20h31'49" -14d3'32"
Nicole Vesely's 18th Birthday Star
Aqr 22h38'5" 2d2'50"
Nicole Victoria Mendoza
Gem 6h12'20" 24d45'41"
Nicole Victoria Stanley
Ori 6h11'23" 18d44'51"
Nicole Vitello
Leo 11h41'4" 24d59'0"
Nicole Voney
Lyn 8h29'59" 42d49'32"
Nicole Wagner
Uma 14h25'0" 58d21'38"
Nicole Walker
And 23h24'50" 41d24'44"
Nicole Walker - Teacher & A True Gift
Leo 9h47'21" 12d48'33"
Nicole Walkowiak
And 2h11'56" 39d54'36"
"Nicole" Wang Run Jia
Vir 13h12'33" 6d37'3"
Nicole Wardle
Sco 17h50'31" -41d20'39"
Nicole Wegmann
Boo 14h32'15" 18d56'3"
Nicole Weiland's Lucky Star
Lib 15h1'35" -11d12'40"
Nicole Weinert
Uma 10h53'5" 51d59'37"
Nicole Weismiller
And 1h30'59" 45d30'30"
Nicole Werth
And 23h18'9" 36d54'21"
Nicole White
And 0h55'5" 39d26'17"
Nicole White
And 2h19'24" 47d22'25"
Nicole Wittwer
Del 20h37'8" 6d42'38"
Nicole Young
Her 16h44'43" 9d14'55"
Nicole Yvonne Thorp
Cas 0h38'49" 58d20'6"
Nicole Zanolari
Boo 14h15'40" 40d58'33"
Nicole Zebrowski
Ari 2h30'2" 17d46'54"
Nicole Zürcher, 13.06.1979
Aur 6h35'57" 38d48'11"
Nicole, 01.10.1983
Cas 23h30'12" 59d25'45"
Nicole, Samantha & Andrea
And 0h39'2" 37d32'3"
Nicole1Joel
Psc 23h21'28" 2d29'6"
Nicole25101976
Cnc 8h57'31" 16d33'30"
Nicole-Constantin
Uma 8h28'45" 59d36'50"
NicoleMarie
And 2h19'43" 45d38'33"
NicoleMary1978
Vir 14h27'53" -4d26'52"
NicoleMichelle
Umi 14h17'7" 73d0'6"
Nicole-Michelle Fradette
Lib 14h49'11" -11d56'41"
Nicole-Moj Zivot Moj Svijetla
Cru 12h39'44" -58d35'47"
Nicole-My Starbound Best Friend
Lib 15h26'56" -9d24'16"
Nicolenadora
Cnc 8h37'33" 22d39'0"
Nicolene Dream
Ari 3h28'5" 26d30'53"
Nicole's Aura
And 0h18'28" 46d13'41"
Nicole's Awesome Star
Cas 23h47'47" 54d19'54"
Nicole's Compassion
Cap 20h25'2" -27d22'32"
Nicole's Eternal Light
Cas 1h42'7" 60d1'16"
Nicole's Fire in the Sky
Mon 6h47'17" 7d54'39"
Nicole's Forever And A Day
Ori 6h25'14" 13d48'2"
Nicole's Gift
Lmi 10h34'54" 37d19'52"
Nicole's Guiding Light
Cru 12h43'11" -60d16'30"
Nicole's Light
Ara 18h2'1" -54d1'38"
Nicole's Masterpiece
And 23h45'5" 35d22'5"
Nicoles Name In Lights
And 0h52'20" 38d23'2"
Nicole's Princess Star
Aqr 22h42'51" -6d22'12"
Nicole's Radiance
Ari 3h15'17" 29d33'28"
Nicole's radiant gift
Cru 12h51'6" -64d18'33"
Nicole's Star
Cru 12h40'15" -57d52'33"
Nicole's Star
Uma 13h34'19" 58d45'53"
Nicole's Star
Cnc 8h34'46" 14d44'11"
Nicole's Star
Vir 12h38'59" 10d49'47"
Nicole's Star
Tau 4h30'34" 16d30'34"
Nicole's Star
Psc 23h46'3" 6d57'59"
Nicole's Star
And 1h6'51" 42d15'10"
Nicole's Star
Gem 7h17'32" 31d7'20"
Nicole's Star
Uma 11h28'55" 49d3'21"
Nicole's Star To The Universe
Sco 16h41'7" -25d38'24"
Nicole's Stern der Engel
Sgr 19h29'20" -30d55'48"
Nicole's Super Star
And 0h13'19" 43d26'37"
Nicole's Superstar
Ari 3h10'56" 27d6'27"
Nicole's Twinkle
And 1h16'54" 38d56'19"
Nicole's Wish Upon A Star
And 23h57'16" 41d49'16"
Nicoleta
Cnc 8h44'24" 22d6'10"
Nicoleta
Lac 22h46'29" 53d41'14"
Nicoleta Yasamin
Her 16h41'55" 19d1'49"
Nicolete "New Beginnings"
Peg 21h28'43" 15d53'8"
Nicoletta Candiani
Peg 21h56'41" 12d48'6"
Nicoletta Corso
Umi 14h9'55" 76d42'29"
Nicoletta Costanza
And 23h59'30" 40d23'59"
Nicoletta e Achille Lex
Lyr 19h26'0" 41d37'16"
Nicoletta Evalie Gianopulos
Lib 15h13'54" -4d0'16"
Nicoletta Mafessoni Poma
And 2h27'37" 43d21'58"
Nicoletta Stazi
Lac 22h55'6" 53d28'53"
Nicolette
Leo 11h29'54" -2d19'27"
Nicolette
Lyn 7h52'17" 39d55'38"
Nicolette
Aqr 21h55'8" 1d54'36"
Nicolette Alexandria Bahr
Ari 3h2'40" 27d17'32"
Nicolette Buske
Tau 3h37'38" 27d28'47"
Nicolette Claire Di Donato
Aqr 22h9'14" 1d50'52"
Nicolette Falzone
Her 16h15'10" 48d21'51"
Nicolette Frattellone
Per 3h17'1" 42d16'14"
Nicolette Ha
Cas 23h14'17" 55d38'37"
Nicolette Hawks
Sco 16h9'22" -11d15'40"
Nicolette Jeanne
Crb 16h24'26" 33d10'1"
Nicolette Marie
Gem 6h53'36" 22d53'39"
Nicolette Marie Guillou
Ari 3h14'59" 28d56'39"
Nicolette Matsumoto
Lib 14h32'11" -18d16'9"
Nicolette Podraza
And 0h39'2" 42d4'18"
Nicolette Pretorius
Dor 4h20'1" -55d59'27"
Nicolette Rabadi
Lyn 8h58'57" 45d9'13"
Nicolette Silvestro
Uma 11h30'8" 55d27'11"
Nicolette Suzanne Dunn
Tau 4h3'2" 27d36'7"
Nicolette Suzanne Jones
Gem 7h5'47" 18d28'53"
Nicolibra
Lyn 6h48'11" 51d52'8"
Nicolie
Vir 12h49'23" -10d13'44"
Nicolina
Gem 7h9'15" 16d26'16"
Nicolina
Vir 12h31'6" 11d6'3"
Nicolina Carol Villano Vetter
Cap 21h12'36" -15d56'8"
Nicolina Wall
Ari 3h6'51" 29d53'1"
Nicolitsis
Pho 23h31'22" -46d0'48"
Nicoll Fuller
And 2h21'47" 46d21'28"
Nicolla Constance
Vir 13h46'40" -9d38'45"
* Nicolle Culverwell *
Per 2h27'28" 56d3'4"
Nicolle Nadine Formo
Equ 21h15'52" 7d30'56"
Nicolle (Schnookum)
Aur 5h28'40" 40d25'34"
Nicolle Star
Psc 23h50'35" 6d12'27"
Nicolle Wick
Aqr 23h3'17" -13d25'26"
Nicolò Casino
Cas 0h52'42" 56d39'59"
Nicolò Dall' Olmo
Peg 23h48'21" 28d8'1"
Nicolos Whitopia Prime
Boo 14h46'50" 20d53'17"
NicoNico
Lyr 19h19'29" 28d26'20"
NICORAY
Lyn 7h1'41" 45d3'59"
Nico-Romeo
Cam 12h34'25" 79d9'57"
Nicosia
And 0h36'54" 38d55'42"
Nicosia Riccardo
Uma 10h7'4" 51d25'55"
Nicotitine
Sgr 18h56'46" -21d16'28"
Nicoye
Cap 21h27'41" -8d57'8"
Nicoz
Lib 15h18'56" -27d46'52"
Nicsabluk
Umi 15h33'43" 71d55'47"
Nida
And 1h54'14" 42d1'42"
niddle
Sco 17h27'18" -43d25'23"
Nidhi
Cyg 20h49'21" 31d52'46"
Nidia
Cnc 8h29'8" 22d47'20"
Nidia
Ori 5h30'43" 5d11'7"
Nidia Arellano
Cnc 8h23'4" 9d13'42"
Nidia Garcia
Lmi 10h34'15" 36d34'35"
Nidia Leal
Uma 8h37'54" 57d3'33"
Nidia V. Flores de Brown
Crb 15h49'19" 27d20'37"
Nidschge
Boo 14h22'34" 40d47'27"
Nidya
And 0h49'53" 38d35'38"
Niecy May
Sco 17h33'47" -41d33'27"
Niederheide, Sven
Uma 10h54'40" 43d22'12"
Niedzwiedz Polarny
Uma 11h24'5" 43d10'23"
Niellejay
Vel 9h22'18" -46d20'34"
Niels
Cep 1h19'47" 82d20'45"
Niels Bo Bøggild
Uma 10h12'35" 43d11'44"
Niels Erik
Psc 1h6'33" 30d56'33"
Nielsen-Michaux
Ori 5h19'46" 7d11'34"
Nielsen's Ochlos
Uma 9h25'26" 60d15'22"
Niemann, Steffi
Ori 6h0'22" 14d0'8"
Niemas
Lmi 10h19'36" 36d47'41"
Niemi
Lyn 7h22'19" 49d17'32"
Niemi-Couillard
Her 17h17'21" 26d44'5"
Nienna
Ori 6h18'9" 15d59'43"
Nier Oira Silme
Uma 9h26'8" 46d14'42"
Niesa Brieanne Patton
Aqr 21h1'48" 0d1'16"
Niesha II
Mon 7h30'35" -0d54'59"
Nießing Lara, Manuel
Uma 10h27'2" 69d2'44"
Nieve Elizabeth Greener
And 23h44'0" 38d9'39"
Nieve Jane Horsfield
And 23h3'21" 50d41'53"
Nieves
Uma 11h44'37" 42d6'12"
Nieves
Cep 22h49'11" 68d29'44"
Nieves Collado
Cyg 20h9'18" 42d32'5"
Nieves Maria Cuetara
Lib 15h5'15" -5d28'46"
Nifer
Leo 9h39'7" 25d53'41"
Niff
Aqr 21h5'38" -12d10'23"
Niffer
Gem 7h38'14" 33d2'57"
niffer52
Tau 4h37'47" 27d57'41"
Niflheim
Ori 5h50'13" 7d5'7"
Nigel
Uma 10h20'30" 45d10'36"
Nigel and Glenys
Vul 21h28'42" 26d26'33"
Nigel Bacon
Hya 9h26'21" 4d31'38"
Nigel Beyer-Kay
Leo 11h19'53" 14d40'49"
Nigel & Bridget McBean
Cap 20h9'23" -10d43'47"
Nigel Edward Conder
Cep 21h59'27" 58d53'9"
Nigel & Emily's Wedding Star
Cyg 21h41'27" 55d9'32"
Nigel Eric Bird
Cep 22h33'14" 63d29'46"
Nigel Fox
Uma 10h46'12" 41d53'54"
Nigel Graham Pogson
Cnc 8h20'51" 14d25'7"
Nigel Gray
Cyg 19h56'58" 52d29'35"
Nigel James Kay
Cru 11h57'45" -61d14'23"
Nigel John Hardy
Cap 20h20'30" -13d20'41"
Nigel - lovely lovely shining star
Lib 14h48'27" -9d13'5"
Nigel Mark Safe
Her 18h5'35" 27d37'39"
Nigel "Pooh Bear" Clutton.
Uma 8h45'9" 60d36'45"
Nigel Simon Menezes
Lac 22h46'43" 48d54'3"
Nigel "Tego" Ridgway
Cru 12h37'3" -58d14'7"
Nigel Watson
Uma 8h35'58" 52d54'49"
Nigel Williams
Cep 1h52'52" 80d21'52"
Nigel Wrisdales Hug from an Angel
Peg 22h36'47" 22d27'2"
Nigel's Lori
Ari 2h56'30" 30d5'14"
niger deetz blumpkin
Ari 3h6'16" 29d43'10"
"Niggie"
Gem 7h6'53" 25d43'18"
Night Bliss A.C.
Dra 15h21'14" 60d23'18"
Night Danger
Leo 9h57'52" 20d23'8"
Night Princess - Samantha Wright
Ori 5h28'41" 7d45'45"
Night Shift
Ori 6h9'47" 14d25'46"
Night Walker
Peg 22h5'15" 34d38'17"
Nightingale
Per 4h26'52" 43d27'8"
Nightingale
Psc 0h52'32" 29d15'50"
Nightshine-LMS Class of 2007
Her 18h48'32" 22d1'6"
Nigit'stil Norbert
Cas 1h34'38" 61d19'45"
Nigyar Makmudova
Cep 21h20'41" 64d36'36"
Niha
Dra 18h28'13" 58d43'17"
Niina & Stephen 4.12.2003
Uma 11h21'45" 48d35'49"
Nijma Aljundi
Eri 3h41'50" -0d17'25"
Nik
Tau 5h8'40" 26d26'2"
Nik and Dawna
Cas 1h19'55" 66d20'6"
Nik Colebank "Stargate 17"
Cap 21h54'41" -13d39'15"
Nik & Nol I love You
Gem 7h31'29" 30d58'36"
NIK P.
Ori 5h24'33" 8d33'32"
Nika
Mon 6h43'26" 8d40'19"
Nika
Leo 10h12'56" 17d23'16"
Nika
Per 2h20'40" 54d14'39"
Nika
Sgr 19h48'47" -14d8'29"
Nika Dedvukaj
Uma 13h44'52" 52d47'45"
Nika Hakobyan
Tau 3h42'15" 18d34'5"
Nikaj S. van Wees
Cap 21h43'7" -13d37'54"
NiKayla, Rob's Girl
Uma 9h25'3" 54d45'52"
Nikaz
Cyg 21h56'55" 50d20'5"
Nikcheri
Umi 14h27'31" 74d28'28"
Nike Meade
Cyg 20h44'49" 37d40'50"
Nike Moye
And 1h9'47" 45d53'17"
Nike77
Peg 22h3'30" 19d39'14"
Nikeeta Yadali
Gem 7h2'59" 26d6'0"
Nikell
Lyr 18h32'22" 27d37'56"
Nikhat Azam
Uma 8h57'3" 55d59'43"
Nikhil & Jalpa Fats: Together Shining Forever
Cyg 19h50'44" 30d12'33"
Nikhil Ranjan
Psc 1h17'6" 15d14'56"
Niki
And 1h10'14" 45d30'10"
Niki
And 23h21'5" 42d30'18"
Niki
Vir 14h45'43" -0d13'38"
Niki
Sco 17h4'51" -33d0'13"
Niki 813
Cam 5h31'43" 68d38'18"
Niki and Mike Gillikin
Lib 14h25'10" -18d32'45"
Niki Carlton
Uma 10h14'36" 60d21'12"
Niki Cogliano
And 0h34'38" 36d33'34"
NIKI Doane
Her 16h35'29" 13d41'19"
Niki G
Vir 12h35'40" 11d25'43"
Niki Gates
Umi 14h12'7" 73d3'24"
Niki Jade Leis
Ori 5h40'43" 1d54'3"
Niki L. Savich
Pho 0h53'30" -47d22'26"
Niki Lee
And 0h56'2" 44d0'36"
Niki Lynn
Crb 15h41'7" 37d12'43"
Niki Marie Kontantoulas
Cnc 8h38'42" 22d50'28"
Niki Monbaron
Umi 16h14'28" 73d33'17"
Niki Morris
Psc 0h44'39" 17d24'22"
Niki Pryce
Sgr 18h41'16" -36d35'51"
Niki Rogers Mercy
Uma 10h37'27" 43d5'15"
Niki Sernio
Uma 12h31'20" 56d2'6"
Niki Szathmary
Ari 2h18'10" 11d36'1"
Niki Warner
Cnc 8h15'51" 22d38'16"
Niki Westlund - Corazon de Melocton
Cyg 20h59'54" 39d51'55"
Niki Wheetley
Aqr 23h44'23" -23d59'15"
Niki Zewe Ulbrich
Leo 10h20'3" 22d20'47"
Nikia
Eri 2h28'45" -46d13'48"
Nikia
Sco 16h16'48" -9d39'12"
Nikia Dawn Arvanitis
Aqr 22h18'16" 1d27'46"
Nikia McGlothin
Col 5h31'15" -29d23'20"
Nikias Kounoupas
Cyg 20h6'37" 34d35'30"
nikib
Ori 6h14'50" 8d42'26"
Nikila Jayne Hunter
Leo 9h36'34" 12d1'51"
Niki's Eternal Night Light
And 0h45'24" 36d54'33"
Niki's paradise
Cyg 20h16'50" 59d3'1"
Niki's Star Noster Eternus
Cyg 21h16'32" 52d48'53"
Niki's Wonderland
Umi 16h3'21" 80d53'4"

Nikishka's Dog Star
Cma 6h31'44" -28d33'9"
Nikit Desai
Her 17h30'59" 46d57'21"
Nikita
Lyn 7h40'30" 49d55'23"
Nikita
Ori 5h9'42" 11d39'18"
Nikita
Ori 5h52'36" 21d2'25"
Nikita
Umi 16h55'15" 75d33'3"
Nikita A. Tchaldymov
Cyg 20h33'54" 30d13'47"
Nikita Bragin 4/1/98
Uma 11h27'14" 32d18'41"
Nikita Dyakov
Uma 9h28'10" 46d26'40"
Nikita Maria Kay
Ori 5h34'45" -9d48'35"
Nikita Nicole Markley Hulse
Cnc 8h33'46" 24d49'20"
Nikita Philips
Uma 13h58'21" 53d59'48"
NikJack
Cyg 20h14'43" 55d44'30"
Nikk Wise
Sco 17h44'1" -31d36'24"
Nikka
Lyn 9h4'4" 38d12'8"
Nikkala Martinez
Lib 14h50'37" -2d36'3"
Nikkelly
Sge 19h50'3" 17d8'26"
Nikkethan
Ari 2h44'30" 12d50'44"
Nikki
Ori 6h8'58" 3d38'5"
Nikki
Psc 0h51'37" 29d4'5"
Nikki
Leo 10h27'35" 16d4'8"
Nikki
Gem 8h0'16" 21d0'37"
Nikki
Gem 6h41'24" 24d54'7"
Nikki
Lmi 10h44'12" 23d46'46"
Nikki
Lyn 8h4'36" 40d39'46"
Nikki
Cyg 20h8'51" 35d55'38"
Nikki
And 0h13'58" 39d42'12"
Nikki
Aur 5h57'50" 36d36'14"
Nikki
And 23h27'38" 48d53'28"
Nikki
Cam 4h34'39" 53d56'51"
Nikki
And 1h4'32" 45d1'40"
Nikki
And 1h42'18" 45d41'31"
Nikki
Crb 16h0'55" 39d24'14"
Nikki
Vir 14h1'33" -4d25'52"
Nikki
Lib 15h36'5" -7d5'57"
Nikki
Lib 15h32'39" -6d51'37"
Nikki
Umi 15h24'49" 82d23'0"
NIKKI
Cap 21h52'30" -12d35'17"
Nikki
Sgr 18h46'33" -19d14'55"
Nikki
Aqr 22h0'21" -20d21'41"
Nikki
Uma 14h27'23" 56d41'17"
Nikki
Uma 11h52'25" 53d20'55"
Nikki
Lyn 6h28'31" 58d1'49"
Nikki
Uma 8h20'29" 65d49'10"
Nikki
Cam 7h39'0" 61d44'38"
Nikki
Sco 17h58'22" -30d7'30"
Nikki Aldridge
Sco 17h3'38" -41d45'30"
Nikki and Chad
Sco 16h10'43" -16d45'56"
Nikki and Cody Tucker
Sge 19h27'16" 17d28'53"
Nikki B
Tau 4h11'20" 23d49'54"
Nikki Bacharach
Cnc 8h31'16" 25d9'10"
Nikki Baylor
And 0h16'2" 24d44'13"
Nikki Bear
Sco 16h7'45" -11d56'29"
Nikki Boettcher Beautiful
Ari 2h20'5" 13d25'15"
Nikki Boon
And 2h2'39" 42d22'42"
Nikki Brown
Cas 1h40'30" 60d31'0"
Nikki Bruno
Uma 10h21'49" 72d34'41"
Nikki Byrne
Uma 13h29'58" 55d22'10"
Nikki C. Mass
And 2h13'27" 39d38'17"
Nikki Carter
Lyr 18h49'59" 33d51'19"
Nikki Cinnamon Lucas
Leo 11h36'56" 28d17'9"
Nikki Davis
Cnc 8h17'55" 21d10'0"
Nikki Deason
Lib 14h50'12" -5d43'51"
Nikki Decesaris
Lyn 7h59'14" 40d49'31"
Nikki DiMartino (Bonnie Blue)
Aqr 22h4'0" -19d9'6"
Nikki "Euphrosyne" Thorn
And 1h12'4" 45d18'29"
Nikki Faust
Psc 1h13'42" 7d2'9"
Nikki Galella
Cas 0h23'43" 61d26'33"
Nikki Glo
Vir 13h57'14" 2d25'13"
Nikki Goldstein
Col 5h59'59" -35d12'11"
Nikki & Guido & Cowboy
Her 17h51'41" 22d51'58"
Nikki Gurski
Lib 15h44'31" -24d23'23"
Nikki Helago
Lyr 19h9'0" 45d23'42"
Nikki Hess
Vir 11h39'45" 4d4'34"
Nikki I Love You Forever - Mike
Cyg 19h48'39" 33d33'16"
Nikki I Love You. Love Always Mick
Vir 13h38'36" -20d29'3"
Nikki & Ian Forever
Cyg 19h44'12" 39d16'17"
Nikki J. Allred
Cyg 21h39'26" 53d19'9"
Nikki J. Tillman
Leo 10h47'45" 19d28'23"
Nikki Jade
And 23h50'36" 46d40'47"
Nikki & Jaime Lopez
Tau 5h58'56" 28d13'0"
Nikki Jean Neal
Dra 17h50'45" 56d32'56"
Nikki & Jeff
Tri 1h50'36" 28d26'57"
Nikki Jenkins
Lyn 7h56'7" 57d5'22"
Nikki Jo Ghart
Lyn 7h23'40" 47d38'51"
Nikki & Joey
Sge 19h48'40" 18d32'25"
Nikki Josephine
Leo 10h21'48" 10d15'19"
Nikki June
Lyn 7h53'40" 49d33'36"
Nikki Kemp
And 2h36'18" 39d50'53"
Nikki Kemper
Ori 6h4'8" 18d31'20"
Nikki Koskol
Gem 6h58'4" 14d13'10"
Nikki Kuhn
Umi 16h52'36" 85d0'54"
Nikki Kurland
Cnc 8h51'2" 30d53'0"
Nikki Leanne Loach
And 23h58'6" 35d36'44"
Nikki Leynor (Squirrel)
And 2h31'50" 39d53'42"
Nikki Lore
Cap 20h39'58" -16d34'23"
Nikki Louise Freepons
Cnc 8h11'57" 10d35'0"
Nikki Loves Ross
Cyg 21h54'22" 52d4'44"
Nikki Lynn Adamson
Sco 16h13'20" -10d14'10"
Nikki Lynn Drasal
Sco 16h14'58" -10d57'31"
Nikki Lynn Empson
Cam 3h37'26" 58d7'13"
Nikki Lynn Maes
Umi 14h20'43" 73d24'57"
Nikki Lynn Mathews
Cap 21h39'31" -8d54'44"
Nikki Lyster
Ari 3h24'38" 27d42'44"
Nikki Marie Guetzloff
Ari 2h13'12" 21d35'52"
Nikki Marie Huggan
Psc 1h0'30" 17d17'3"
Nikki Marie Trinkle
And 1h21'33" 35d9'6"
Nikki Martinez
Gem 6h36'10" 13d10'13"
Nikki Meister
Psc 1h29'51" 15d40'4"
Nikki Monique
And 0h40'14" 38d4'26"
Nikki Morgan
Srp 18h9'56" -0d24'4"
Nikki Moscetti
Ari 2h7'19" 26d59'22"
Nikki Myers
Cas 23h3'43" 59d21'56"
Nikki Nielsen
Cap 21h57'7" -18d8'52"
Nikki Novak
Sgr 19h48'1" -25d13'9"
Nikki Nuwar
Lyn 8h25'54" 42d6'31"
Nikki Parker
And 2h10'6" 38d14'19"
Nikki Peebles
Pup 7h57'30" -34d56'32"
Nikki Penelope
Cas 23h37'48" 54d57'59"
Nikki Pepper
Ari 2h33'24" 16d14'56"
Nikki Petrina Gallagher
Aqr 22h12'18" 1d12'29"
Nikki Polesky's Shining Star
Lib 14h52'8" -2d32'19"
Nikki Poteet
And 23h51'59" 39d49'37"
Nikki Prokop
And 0h9'6" 34d57'48"
NIKKI RAE K
Uma 10h32'25" 66d40'57"
Nikki Reiter
Gem 7h4'41" 33d2'6"
Nikki Reiter
Aqr 22h33'27" 1d16'54"
Nikki Renee
And 23h9'1" 51d3'0"
Nikki Riley
Gem 7h32'19" 20d5'3"
Nikki Robbins
Gem 6h35'14" 27d8'16"
Nikki Rose
Ori 5h30'19" -0d59'39"
Nikki RoseMarie Geiger
And 2h36'22" 47d51'54"
Nikki Rosenhauer's Star
Tau 4h24'50" 23d46'52"
Nikki&Russ
Gem 7h20'19" 26d35'47"
Nikki Sarver
Cap 20h59'12" -17d8'2"
Nikki Schreffler
Uma 8h19'26" 61d20'17"
Nikki Scott
Oph 17h45'1" -0d26'32"
Nikki Sixx
Vir 12h39'27" -9d22'23"
Nikki Somers
And 23h42'10" 45d53'55"
Nikki Stonecipher
Leo 11h30'44" 25d19'27"
Nikki Sweigart
Psc 0h18'15" 17d47'34"
Nikki Taylor
And 1h26'23" 44d25'21"
Nikki Taylor-Compton
Cnc 8h36'30" 8d2'29"
Nikki Tichansky
Aur 5h46'53" 41d27'28"
Nikki Tuttle
Gem 8h6'44" 30d50'32"
Nikki Violetta Lisiecka
Cru 12h37'17" -57d55'39"
Nikki Wolf
Psc 1h42'42" 8d33'36"
Nikki & Xia "The Mother & Daughter"
Aqr 23h52'26" -18d30'7"
Nikki Yours Forever Dylan
Tau 5h56'50" 25d7'15"
Nikkie
Uma 9h24'34" 54d8'24"
Nikkie
Uma 8h38'23" 55d54'29"
Nikkie Nova
Lib 15h12'48" -5d7'58"
Nikkilobe Bauer
Sco 17h51'41" -35d50'48"
Nikki's Booger In The Sky
Psc 1h29'6" 14d28'58"
Nikki's Brilliance
And 23h51'44" 34d54'3"
Nikki's Mom
Ori 5h28'49" 3d11'51"
Nikki's Rose
Cnc 9h19'11" 16d12'40"
Nikki's Shining Star
Cmi 7h25'55" -0d11'23"
Nikki's Smile
Sco 16h9'49" -15d9'46"
Nikki's Song
Lyr 18h28'57" 39d34'53"
Nikki's Star
And 23h51'31" 47d50'48"
Nikki's Star
Mon 6h41'32" -0d16'20"
Nikki's Star
Umi 14h53'13" 73d51'12"
Nikki's Star - Three Toed Cuteopia
Cas 0h21'33" 57d5'43"
Nikki's Twinkle
Uma 11h43'17" 37d28'1"
Nikki's White Rose
Lib 15h9'46" -1d53'32"
Nikko
Aur 4h59'24" 36d6'12"
Nikko
Ori 5h52'52" 8d7'2"
Nikkoli
Ari 2h32'11" 27d44'11"
Nikkos Alexander Kovanes
Per 2h51'50" 38d15'25"
Nikko's Dream Star
Lyn 6h17'3" 55d6'17"
Nikky
Lyn 7h10'34" 59d12'23"
Nikky & Adam
And 1h17'43" 42d57'12"
Nikky - Our Special Dad - 22.12.1954
Ori 6h20'3" 19d9'11"
Nikky Thames
And 0h49'29" 44d46'20"
Nikky's
Aqr 22h9'11" -1d3'47"
Niklas Felix Zwietasch
Uma 9h33'38" 46d48'46"
Niklas Owen
Uma 10h47'18" 64d30'10"
Niko Anthony Miller 12/28/03
Cap 21h54'19" -9d8'14"
Niko David Anderson
Gem 6h16'19" 24d29'26"
Niko James Pickering
Umi 15h2'6" 68d25'2"
Niko Knapp
Cma 7h0'50" -22d9'51"
Niko Topias Rawlins
Cmi 7h11'49" 9d29'52"
Nikola Benesova
Psc 0h32'6" 7d11'28"
Nikola Sikiric
Cyg 20h33'21" 53d57'51"
Nikola the eternal spark of my love
Leo 11h0'56" 2d37'24"
Nikolai
Lib 14h55'22" -17d15'30"
Nikolai
Lib 15h5'4" -28d27'52"
Nikolai and Tatiana
Aqr 22h39'24" 0d35'6"
Nikolai Bieger
Her 16h50'41" 23d30'9"
Nikolai Utochkin
Per 4h35'6" 40d13'33"
Nikolaos Odysseus 23rd July 2005
Ori 5h36'32" -2d44'58"
Nikolaos of Greece
Lib 15h50'27" -18d9'33"
Nikolaos Zarglis
Cru 12h44'53" -57d17'37"
Nikolas
Dra 11h45'2" 71d41'55"
Nikolas
Ori 5h23'53" 5d57'27"
Nikolas A. Infante
Dra 17h11'49" 65d14'56"
Nikolas Alexander DeWeese
Ori 5h30'51" -8d46'54"
Nikolas Beckham Huyser
And 1h25'27" 47d16'46"
Nikolas James Smith
Lib 15h44'13" -9d53'19"
Nikolas Jareb
Psc 0h1'2" 6d2'12"
Nikolas Jenner
Leo 10h59'31" 18d56'14"
Nikolas Kunkle
Uma 14h13'21" 62d7'50"
Nikolas Prentice Dains
Peg 21h47'29" 6d57'28"
Nikolas Richard Doumas
Vir 13h55'24" -2d7'21"
Nikolas Stoulil
Gem 6h48'0" 33d54'35"
Nikolas Thomas Dahlstrom
Lac 22h43'31" 39d48'24"
Nikolas von Wrangell
Uma 8h21'47" 72d24'11"
Nikolaus & Audrey Berceanu
Cru 12h0'56" -62d14'55"
Nikolaus Karl Stoepler, Jr.
Cyg 21h28'59" 36d35'50"
Nikolaus Maximus Pateras
Ori 5h14'27" 15d9'17"
Nikolaus P. Sneen
Umi 14h7'30" 69d42'6"
Nikolay O Grigoryev
Cnc 8h25'22" 22d40'57"
Nikole
Cyg 20h12'33" 33d15'40"
Nikole
And 2h27'20" 45d25'58"
Nikole Lynn Milton
And 0h13'57" 30d10'24"
Nikole Valdamar
And 0h16'36" 46d28'48"
Nikole's Future
Ori 5h21'58" 0d28'24"
Nikolett Lujan
Lyn 7h51'7" 37d17'48"
Nikoletta's Light
Vir 13h19'16" -7d1'43"
Nikolette
Cnc 8h40'19" 19d2'2"
Nikolette Markovna Harris
Uma 9h0'56" 64d16'39"
Nikolya
Crb 15h45'56" 35d17'46"
Nikoo Sadatrafiei
Cas 0h18'17" 53d11'10"
Nikos
Dra 18h33'17" 53d6'15"
Nikos Alexandre Stassinakis
Ori 5h59'15" 17d34'16"
Nikos Nikolopoulos
Uma 10h33'44" 69d47'45"
Nikos Perakis
Uma 9h2'38" 70d44'20"
Nikski
Gem 6h36'50" 13d46'37"
nikstar
Lib 15h9'8" -24d13'42"
Nikul Patel
Her 17h34'40" 41d32'59"
Nikushuri
Uma 13h47'3" 55d33'42"
Niky
Ori 6h12'52" 17d10'50"
Nil Sr
Cas 1h59'24" 60d9'10"
Nila and Alex
Cyg 20h41'22" 45d51'2"
Nila Marie
And 23h8'56" 41d40'37"
Nilani Sriragavan
Dra 17h32'21" 55d51'16"
Nilda Follini
Aqr 21h38'26" 0d48'36"
Nilda Jenkins
Cap 21h42'22" -14d49'17"
Nilda Teresa Toro
Aqr 21h53'52" -0d52'6"
Nile J. Boldt
Leo 9h31'12" 11d25'7"
Niles Donahue
Ori 5h27'18" 13d21'2"
Niles Lee Place
Aur 5h30'55" 42d6'5"
Niles Trumbull Cook
Tau 4h35'26" 3d46'10"
Nilla-N-Dito
Ari 2h10'10" 24d12'55"
Nilla's Star
Gem 7h37'20" 33d44'43"
Nilloc 1226
Uma 8h29'2" 70d13'26"
Nilo Fakaris
Psc 0h32'11" 6d2'54"
Nilofer
Uma 13h39'45" 52d49'11"
Niloofar Nicole Nader Nick
Peg 23h20'25" 32d8'19"
NILOU
Cyg 20h42'55" 46d0'15"
Nils
Uma 9h33'49" 58d26'18"
Nils G Thompson
Umi 14h10'50" 74d33'35"
Nils H. Abrahamsson
Uma 11h27'11" 60d27'25"
Nils H. Nilsson
Ari 2h20'54" 26d13'58"
Nils Hanson
Ori 5h10'28" 8d24'37"
Nils Kristian Haugen
Umi 16h37'49" 80d58'1"
Nils Langguth, 06.09.2005
Cyg 20h7'24" 46d26'8"
Nils Liebich
Cep 22h30'46" 66d59'11"
Nils Marquard
Uma 14h3'37" 55d14'49"
Nils Ole
Uma 10h1'7" 61d28'42"
Nils Volker
Uma 14h3'41" 55d44'42"
Nils Wagner
Leo 10h25'8" 23d40'51"
NilsinaButterflyKissesfromRoseLips
Crb 16h13'10" 25d49'9"
NILSSA FELIX
Cyg 21h40'46" 38d26'25"
Nilsson's Star
Cap 20h25'34" -23d17'58"
Nilufer & Peter Fernandes' Star
Cru 12h40'19" -60d16'6"
Nilvia Judy Franco
Lib 15h4'53" -3d18'28"
Nim Chimsky
Lib 15h4'27" -2d12'43"
Nim Nim
Dra 18h55'57" 68d19'49"
Nima Sarani
Aql 19h42'21" -0d0'46"
Nima-Arpi Sabouri
Umi 19h34'52" 87d58'6"
Nimals
Eri 4h19'26" -31d28'27"
NiMic
Lyn 7h24'59" 57d11'31"
Nimisha Chavda
Uma 8h56'34" 66d20'15"
Nimisha Patel
Sgr 19h41'12" -42d17'22"
Nimisha Patel
And 2h21'34" 41d36'10"
NIMWA
Vir 14h23'41" -0d31'16"
Nina
Umi 15h13'15" 84d25'9"
Nina
Cap 20h23'26" -11d31'55"
Nina
Uma 10h20'9" 63d35'13"
Nina
Cep 0h22'35" 82d8'51"
Nina
Cas 1h53'15" 68d40'51"
Nina
Cas 1h26'39" 69d1'54"
Nina
Uma 9h3'22" 73d0'11"
Nina
Lyn 7h24'30" 56d17'17"
Nina
Dra 17h32'15" 56d42'41"
Nina
Cas 23h27'45" 53d2'9"
Nina
Sgr 18h5'34" -27d5'9"
Nina
Sco 16h42'14" -28d48'26"
Nina
Uma 8h59'59" 51d56'1"
Nina
Lmi 10h7'46" 38d26'37"
Nina
Cyg 20h33'32" 30d10'57"
Nina
And 0h48'27" 40d41'16"
Nina
Tri 2h22'21" 36d26'34"
NINA
Crb 15h44'23" 31d54'35"
Nina
And 23h26'48" 44d51'18"
Nina
And 1h30'58" 47d14'14"
Nina
Cyg 20h34'52" 50d3'4"
Nina
Uma 10h7'32" 49d20'15"
Nina
Uma 10h1'29" 47d24'48"
Nina
Cas 0h24'37" 59d36'58"
Nina
Del 20h30'17" 20d14'20"
Nina
Ori 6h2'35" 19d32'48"
NINA
Aqr 21h34'41" 1d19'51"
Nina
Aqr 22h18'42" 0d13'4"
Nina
Ari 3h14'27" 21d6'23"
NINA
Tau 4h38'31" 18d39'29"
Nina
Psc 0h53'22" 18d24'5"
Nina 21002
Mon 6h49'16" -0d27'46"
Nina A. Davis
Per 3h40'45" 45d25'48"
Nina Alexandra
Mon 6h31'51" 8d10'30"
Nina Alexandra Harbourt
Tau 4h28'45" 30d17'1"
Nina Alleen Edwards
Leo 10h15'14" 13d20'0"
Nina and Jim Huitt
Cyg 21h25'48" 35d55'26"
Nina Angelica Tacut
And 22h59'16" 47d25'3"
Nina Ann
Cas 23h42'36" 56d10'18"
Nina Beana Draf Laf
Vir 12h6'19" -9d22'38"
Nina Bug
Aqr 21h26'34" 0d49'12"
Nina Campbell Madewell
Sgr 18h43'47" -18d37'43"
Nina Campell-Dixon
Cyg 19h37'12" 29d45'59"
Nina Carpenter's Star
Ser 15h52'15" 24d19'54"
Nina Celeste (K.M.B.)
Per 3h14'43" 54d8'51"
Nina Claire Pickell Birthday Star
Vir 14h24'55" 2d34'6"
Nina & Claudio
Uma 11h8'33" 72d37'57"
Nina Danielle Obi
Cas 1h33'28" 64d46'11"
Nina Dayana
Leo 10h28'47" 23d38'44"
Nina Deegan
Cas 0h45'41" 65d4'57"
Nina Dell'Angelica
Lyr 18h46'34" 38d33'18"
Nina Delores Marks
Lib 15h11'1" -16d10'43"
Nina Domech & Maxime Drevet
Uma 10h5'24" 70d7'25"
Nina Eganova
Cas 1h11'9" 58d1'25"
Nina Elizabeth
Lib 14h57'11" -4d40'27"
Nina Elizabeth Beaudway
Ari 2h0'36" 22d14'0"
Nina Elizabeth Hopson
Lyn 8h53'58" 41d34'46"
Nina Ellen Partee Kinworthy
Mon 7h16'56" -0d9'23"
Nina Etenko (Selezneva's Star)
Sgr 18h12'2" -20d1'58"
Nina Evelyn Harris
Cnc 8h32'36" 24d59'46"
Nina Gamber Ich liebe dich
Uma 9h58'23" 44d23'27"
Nina Isabella Hodgson
And 2h16'59" 41d22'56"
Nina Isabelle Smyth
Psc 0h17'11" 19d22'40"
Nina Jade Flores
Uma 10h47'4" 67d55'52"
Nina & Jan
Uma 9h33'58" 63d26'11"
Nina Jane Lewin
And 23h45'52" 34d17'7"
Nina Jessop
Psc 0h9'16" -2d14'39"
Nina Johnson
Crb 15h33'41" 27d52'51"
Nina Judy
Cnc 8h23'32" 20d18'41"
Nina Karolina Chantres
Umi 14h56'9" 70d30'10"
Nina La Linda
And 0h37'9" 27d10'30"
Nina Lilian Watson
Sgr 19h39'12" -15d59'58"
Nina "Little Guava Monkey"
Lyr 19h5'47" 29d48'59"
Nina Lucille Adams
Uma 11h49'42" 57d48'54"
NINA LUMBRERAS
Dra 17h11'10" 54d31'4"
Nina Lynum Øverland
And 2h20'31" 41d26'23"
Nina M. Dalenberg
Cas 0h22'24" 53d34'29"
Nina M. Flores
Leo 9h48'12" 9d9'35"
Nina M. Jennaro
Psc 23h40'21" 2d0'56"
Nina Ma
Sco 16h5'22" -14d3'22"
Nina Marie
Leo 11h39'58" 14d52'5"
Nina Marie Garrett Griggs
And 0h13'11" 44d5'2"
Nina Marie Laudato
Vir 12h3'3" 5d42'22"
Nina Marie Melchiorre
Ari 2h31'40" 11d10'47"
Nina Marie Michaels
And 23h18'10" 46d54'50"
Nina Marie Rasa
And 23h7'25" 38d50'0"
Nina Marie Ruggiero
Uma 8h39'17" 67d35'26"
Nina Marshall
And 0h37'48" 38d39'45"
Nina May
Ori 6h0'58" 9d58'55"
Nina Morean
And 23h26'57" 48d6'57"
Nina Morgan
And 0h28'55" 27d39'35"
Nina "n" Mike
Col 5h43'12" -33d0'15"
Nina Ngo
Ori 5h2'45" 4d1'48"
Nina Nira
Cam 4h0'32" 56d29'36"
Nina Palasdies
Uma 10h29'55" 57d22'34"
Nina Paraloglou
Cnc 8h49'45" 31d29'20"
NINA PEARL
Cap 20h41'33" -22d26'35"
Nina Reyes Garau
Peg 22h7'12" 27d21'34"
Nina Rose Sutherland
Uma 9h19'22" 46d41'33"
Nina Ross Bruno
Cap 20h36'0" -23d21'34"
Nina Roth
Equ 21h3'57" 9d16'13"
Nina Schlosser
Dra 10h2'54" 73d47'40"
Nina Smith
And 1h28'9" 39d36'32"
Nina Sosnicka
Cas 0h9'15" 57d24'39"
Nina Subova
Ari 2h28'26" 21d53'3"
Nina Sue Boling Mom / Susie, Winnie
Sgr 18h9'45" -18d59'32"
Nina Sue Johnson
Ari 2h53'18" 20d30'54"
Nina Sue Steinmetz
Sco 16h12'25" -13d37'7"
Nina Sughrue
Sco 16h6'34" -10d18'16"
Nina Tamminen
Lyr 18h15'56" 39d31'13"
Nina Taylor
Ori 5h53'4" 18d15'4"
Nina & Thomas
Uma 10h45'23" 41d0'51"
Nina+Thomas
Uma 13h2'50" 53d55'59"

Nina Thompson
Lyn 6h47'38" 57d34'48"
Nina, "my baby"
And 2h29'23" 43d50'23"
NINA-90
Uma 9h38'48" 48d27'14"
NinaGlik-
TheBestMotherOfTheUnive
rse
Lyn 8h10'11" 55d12'42"
NINA-J
Vir 12h22'31" -0d3'54"
Nina-Melle-Alix
Ari 3h8'37" 29d5'21"
Nina's
Mon 8h2'42" -0d32'46"
Nina's Star
Uma 12h53'20" 53d13'47"
Nina's Star Of Hope
Vir 13h1'10" 6d27'57"
NinaStar 1
Uma 9h23'15" 46d24'1"
Nine Four
Ari 2h27'46" 22d8'45"
Ninella Mae
Cyg 19h52'40" 38d42'19"
Nineveh Flora Sangari
Vir 12h19'51" -4d45'3"
Ninfa
Tau 4h28'21" 20d47'44"
Ning Chao
Uma 14h10'11" 60d59'52"
Ning Yu
Uma 13h11'0" 52d39'30"
Ningal
Cnc 8h45'26" 21d28'32"
Ningenkougaku Hotetu
Kenkyusho
Sgr 18h19'28" -32d47'23"
Ning-nong
Peg 21h51'49" 8d54'37"
Ninhursag
Psc 0h25'58" 9d4'1"
Nini
Cyg 20h11'55" 35d50'13"
Nini
Uma 12h37'30" 58d4'48"
NINI
Umi 14h30'5" 70d36'7"
Nini
Uma 8h46'50" 72d23'41"
Nini
Lib 15h19'51" -6d25'34"
Nini 54
Umi 15h1'51" 79d22'17"
Nini - Annie Lopes
Sco 17h53'47" -36d9'57"
Nini Johanna
Aqr 20h44'23" 0d54'41"
Nini50-Mom
Ari 3h20'23" 30d27'32"
Ninia
Aqr 22h43'7" -0d56'57"
Niniko
Psc 1h23'48" 33d38'57"
Ninikupenda
Lyr 18h41'22" 40d37'22"
Ninin
Cas 0h29'2" 62d23'7"
Ninin Gagoli - Dragan
Atanasov
Peg 23h57'0" 11d52'35"
NiNi's Glow
Cap 20h56'29" -19d22'56"
Nini's Star
Tau 3h40'36" 27d48'46"
Ninja
Sgr 18h50'28" -33d15'56"
Ninja Princess
Uma 11h31'23" 32d41'21"
Ninja Star
Her 16h35'34" 31d55'40"
Ninlana
Lib 14h54'40" -0d43'17"
Ninna Ricci Creencia
Gem 7h6'58" 25d22'54"
NinnaMama
Psc 0h21'37" -5d20'2"
NINNIE LYONS
Peg 23h20'44" 19d40'4"
Nino
Aqr 22h40'11" 1d6'8"
Nino
Uma 10h34'21" 68d5'38"
Nino and Shayne
Ari 2h12'14" 20d34'57"
Nino Lizé
Uma 10h16'9" 72d5'52"
Nino Maria Antonio de
Noronha
Cnc 8h39'18" 31d16'1"
Nino Moffa
And 0h52'27" 34d22'6"
Nino Sciorra
Aql 19h12'14" -3d22'5"
Nino, 6 septembre 2006
Vir 13h5'23" -12d7'39"
Ninoureyrat
Oph 17h6'14" -28d37'9"
Ninpo
Sco 16h6'33" -10d17'36"
NIOBL 1-5-02
Boo 14h24'48" 11d20'50"
Nioka's Heart
Cnc 8h4'42" 7d6'13"
Niphaporn Artese
Uma 10h29'56" 63d28'32"
Nipper 27
Umi 15h16'7" 68d10'56"
Nippy
Sco 16h12'5" -15d17'11"
Nippy's Star
Tau 4h31'26" 28d45'11"
Niqui
Aql 19h12'4" 5d51'54"
Niraj Chandrakant Patel
1969
Aql 19h49'11" -0d29'7"
Niral Desai
Her 17h47'14" 45d52'7"
Niran
Sco 16h14'49" -32d58'50"
Nirel Leitman
And 0h4'39" 32d7'23"
Nirenberg's Dream Catcher
Uma 12h55'47" 60d8'5"
Nirmal Shewakramani -
50th Birthday
Cnc 9h18'23" 12d52'33"
Niroo Love !!!
Leo 10h44'23" 11d5'50"
Nirupam Gill
Aql 19h45'46" 16d27'21"
~ Nirvana Arcadia ~
Cnc 8h52'31" 14d34'23"
Nirvana, Melanie's star
Sgr 18h12'17" -16d44'55"
Nisa
Vel 9h28'50" -40d50'13"
Nisa LaBella
Gem 7h31'13" 14d28'10"
Nisa Nisa Traboini
Lib 14h37'47" -9d0'56"
Nish - Far
Lyn 6h42'29" 56d24'37"
Nisha
Ori 6h21'9" 13d52'29"
Nisha
Leo 10h53'51" 18d51'25"
Nisha Kiran Patel -12/20/76
Lyn 8h10'15" 41d54'49"
Nisha Sakya
Lmi 9h41'32" 36d42'37"
Nisha Sharma
Cap 21h13'11" -16d17'50"
Nishadi
Cas 0h24'59" 51d54'26"
NISHCHALA HELWING
Lyr 18h53'30" 35d42'16"
Nishelle
Ara 17h34'36" -47d2'35"
Nishi - my vaava kutty
Col 5h51'10" -32d53'47"
Nishu
Vir 13h2'27" 4d56'39"
Nishy-wishy-wishy
Ari 2h42'35" 27d9'52"
Nisi
Lmi 9h45'42" 39d7'13"
Nisrine Choucair 11-29-
1972
Sgr 18h42'37" -18d41'2"
Nissim et Patsy
Ari 2h55'34" 22d4'50"
Nita
And 23h48'13" 44d27'7"
Nita
Sgr 19h35'27" -15d14'33"
Nita
Sgr 19h22'7" -21d41'1"
Nita & AJ Tabaka
Gem 6h46'49" 23d33'25"
Nita Beth
Aqr 21h15'25" -8d49'38"
Nita Casimiro
Lib 15h58'29" -11d41'21"
Nita Gregg
Lyn 7h54'38" 54d26'32"
Nita Liebeskind's
Anniversary Star
Cas 0h6'51" 56d29'4"
Nita Lu
Cyg 20h4'42" 48d15'3"
Nita 'Ma' Sweetalla
Cyg 20h40'34" 36d10'5"
Nita Vera Cucinotta
Cru 12h27'5" -63d38'41"
NitaB
Lib 14h49'58" -3d48'35"
Nitara
Tau 4h36'38" 24d55'11"
Nita's Star
Gem 6h47'57" 26d25'31"
Nitasha and Andres
Sge 19h27'53" 17d32'13"
Nitasha Chopra
Sgr 19h18'43" -17d42'54"
NitaSunilTrivedi
Cap 20h20'57" -23d53'44"
Nite Raiders Brenda Sue
Umi 14h22'52" 69d26'42"
Nithy & Gopi
Gem 7h44'14" 32d45'39"
Niti
Sco 16h4'44" -16d5'51"
Nitica G. Salisbury
Uma 13h17'37" 56d16'52"
Nitin Kulkarni
Mon 6h52'57" -0d22'59"
Nitin Rai Puri
Lyn 6h45'30" 52d34'25"
Nitin Rajan
Cep 0h14'41" 69d42'17"
Nitsa [LemoN] Mouroutsou
And 1h34'12" 49d29'43"
Nitsha
Cas 0h52'41" 61d34'9"
Nittner, Thomas
Ori 6h14'59" 6d14'28"
Nitya Yagnya Nalamothu
Uma 9h51'47" 70d29'3"
Nitza Michelle
Sgr 18h30'25" -27d59'18"
Nitzelle's Eternal Faith
Ari 2h21'47" 24d25'50"
Niunia
And 23h21'21" 47d34'18"
Niuniu
Uma 11h44'8" 50d43'4"
Niurka Moraima Montalvo
Uma 13h37'10" 58d13'32"
Niv
Cap 21h21'33" -17d27'51"
NIVEK
Sco 17h18'22" -43d33'25"
Niveus quod Viridis
Uma 9h23'14" 41d56'46"
NIVIANA
Lyn 8h6'3" 41d6'0"
NIWAGAL
Sco 16h15'29" -23d19'50"
Nix Eutopia
Dra 15h5'52" 60d48'18"
Nixi 87, Edesanya legked-
vesebb lánya
Tau 5h48'57" 23d35'11"
Nixiey
Cnc 9h7'33" 8d21'34"
Nixo
Ori 5h35'50" -0d27'40"
Nixon Chavarria
Cnv 13h22'9" 28d30'9"
Nixon James Karcz
Cap 21h44'39" -20d4'54"
Nixxi
Lib 15h16'18" -10d41'12"
Nixy
Peg 21h42'16" 8d9'30"
Niya Alexandra Haynes
Vir 13h28'15" 11d7'46"
Niyam
Ori 5h4'30" 6d4'8"
Nize Tristao Pietrangelo
Uma 8h17'39" 62d42'23"
NizLai
Uma 9h44'22" 50d11'0"
NJ
Ori 6h5'52" 10d33'12"
nj
Cap 21h37'5" -16d49'54"
NJ State Trooper Philip
Lamonaco
Aqr 21h36'38" 1d26'24"
NJGNLA
Uma 10h37'2" 47d29'2"
NJP
Ari 2h44'57" 23d58'18"
NKK Switches
Dra 19h10'27" 66d7'8"
N'Ma Balanta Mandinka
Leo 9h42'19" 25d56'27"
NMF JEK
Lyn 9h2'33" 34d56'48"
NMV 53
Cnc 8h54'43" 14d55'39"
nnaeel's star
And 1h27'10" 44d31'30"
N.Napoli
Cam 3h47'45" 65d20'34"
NNEO
Ari 3h10'40" 19d27'34"
NNN Kade Major
Her 18h52'59" 24d20'3"
nnstdltbbbsongs&prayers-
dadiloveyou
Ori 5h37'24" -0d16'53"
No Doubt
Cnc 8h39'15" 7d7'37"
No es amor
Per 2h56'38" 47d51'48"
No Longer a Dream
Ori 6h0'27" 6d20'0"
No Matter What
Cyg 20h2'9" 31d44'0"
"No Name"
Ori 6h4'43" 18d10'47"
No one loves u like I do
Gem 6h26'50" 17d9'59"
No Place That Far
Gem 7h23'59" 27d34'36"
No1MomJean
Ari 2h50'43" 22d13'34"
Noa
Cnc 8h42'42" 31d18'55"
noa
Per 4h15'23" 34d37'36"
Noa Caminal Capell
Gem 7h16'16" 19d57'9"
Noa Carolahane Hafia
Ori 6h24'22" 10d7'5"
Noa Kaatjia
Sgr 18h51'16" -25d58'30"
Noa Libby
Psc 1h11'24" 12d50'20"
Noach Ben-Haim
Cep 21h53'9" 65d46'47"
Noah
Uma 8h31'0" 64d35'16"
Noah
Dra 19h41'25" 79d21'52"
Noah
Cap 21h48'35" -11d44'52"
Noah
Cnc 8h51'3" 8d16'44"
Noah
Her 18h21'59" 18d45'4"
Noah
Tau 5h30'23" 23d45'57"
Noah
Her 17h34'17" 26d41'41"
Noah
Per 3h49'47" 41d35'23"
Noah Agpalza
Dra 18h31'46" 63d16'37"
Noah Alexander
Her 18h2'26" 25d24'36"
Noah Alexander Colassaco
Gem 6h51'55" 24d35'56"
Noah Alexander Contois
Lyn 6h49'4" 54d20'35"
Noah Alexander Correa
Uma 11h45'23" 37d45'55"
Noah Alexander
Farnsworth
Tau 4h35'13" 24d56'59"
Noah Alexander Flake
Lmi 10h19'27" 29d51'41"
Noah Alexander Fontaine
Ori 5h33'23" 4d20'11"
Noah Alexander Fontaine,
Eternal
Vir 14h50'10" 4d7'37"
Noah Allan Spargo
Her 18h24'29" 16d45'16"
Noah - Always Papa's Little
Buddy
Tau 4h32'0" 14d57'42"
Noah Andrew Mancilla
Tau 4h24'2" 11d42'36"
Noah Andrew Weyers
Cnc 8h27'37" 28d24'58"
Noah Asher Jacobson
Tau 4h42'52" 26d33'47"
Noah B Curlee
Uma 11h18'22" 36d57'3"
Noah B. Rowland
Lac 21h57'55" 39d5'47"
Noah Bendix
Uma 8h55'2" 49d9'15"
Noah Benjamin Bishop
Ori 4h47'21" -3d40'28"
Noah Benjamin Lumb
Per 4h10'55" 48d23'40"
Noah Benjamin Stinnett
Uma 8h37'35" 52d20'39"
Noah Berry
Uma 9h19'43" 47d21'8"
Noah Birk
Gem 7h19'34" 23d3'1"
Noah Brent Steward
Her 17h26'15" 34d15'8"
Noah Buck's Bright Star
Per 3h24'19" 35d54'52"
Noah Burton Barnes
Umi 15h33'22" 74d3'44"
Noah Byron Vich
Uma 10h4'44" 65d52'27"
Noah Cameron Brooks
Her 16h42'0" 30d9'23"
Noah Cantarinha Tomas
Leo 11h19'43" 15d33'58"
Noah Carter
Leo 9h38'52" 28d37'20"
Noah Charles Brandenburg
Aqr 22h7'3" -1d7'3"
Noah Christian Haber
Tau 3h58'26" 27d35'14"
Noah Christian Ponder
Boo 14h11'17" 17d23'2"
Noah Christopher Martinez-
Lyle
Tau 5h47'13" 21d59'32"
Noah Christopher
Polhemus
Vir 12h30'7" -5d32'51"
Noah Christopher Swenson
Ori 5h31'31" -1d40'12"
Noah Coffey Berg
Aql 19h27'35" 3d54'52"
Noah Cooper
Her 17h51'36" 45d4'21"
Noah **Crikey Mate**
Sco 16h6'54" -18d49'27"
Noah Daniel Gilbert
Ori 5h15'2" 0d52'57"
Noah Daniel Holland
Psc 1h10'30" 31d20'28"
Noah Daniel Lawrence
Lib 15h6'23" -6d22'44"
Noah Daniel Rossler
Ari 2h59'35" 25d5'37"
Noah David Halpern
Her 17h20'20" 38d49'3"
Noah David Ruppel
Vul 19h47'39" 26d7'33"
Noah Deats
Boo 14h52'0" 22d31'31"
Noah Edlund Anderson
Cnc 8h49'59" 26d50'46"
Noah Edward Sands
Lib 15h25'29" -22d19'27"
Noah Edwin Wenzell
Ori 6h11'33" 8d15'10"
Noah Elliot Bates
Cep 23h2'34" 75d43'43"
Noah Elliot Czerwinski
Aqr 21h53'52" 0d17'44"
Noah Elliott
Leo 9h40'53" 30d34'32"
Noah Ezekiel Hall
Ori 5h30'0" 3d46'48"
Noah F.
Aql 19h6'42" 6d17'48"
Noah Fields Farrar
Uma 11h31'14" 63d18'40"
Noah Fox Dawson
Tau 5h46'22" 16d7'5"
Noah Francis Lax
Aqr 23h56'6" -16d43'31"
Noah Francis Raab
Her 18h37'6" 20d52'11"
Noah Gabriel Beere
Lyr 18h42'1" 31d25'51"
Noah Garrett's Baptism
Star
Leo 10h41'2" 16d42'21"
Noah Grace Lorenzo
Equ 21h7'54" 9d5'23"
Noah Grant Kempster
Uma 9h21'36" 47d23'36"
Noah Haines
Cnc 8h35'59" 9d35'12"
Noah Harris Fricks
Cap 20h51'7" -19d36'20"
Noah Henry Blake
Lyn 7h22'40" 46d25'44"
Noah Henry Bucksath
Umi 16h18'49" 77d32'23"
Noah Henry Harvey
Per 4h12'9" 42d32'49"
Noah Henry Jaeger
Cru 12h56'6" -59d52'24"
Noah Howard Meyer
Dra 18h1'21" 79d27'51"
Noah Hutchins Cahill
Aqr 22h54'15" -19d17'25"
Noah Jaden Blondek
Per 3h50'15" 48d7'1"
Noah Jake Horowitz
Ari 2h5'54" 16d28'29"
Noah James
Her 16h49'17" 5d43'12"
Noah James Brown
Uma 11h57'31" 57d17'35"
Noah James Douglas
Uma 11h34'50" 43d4'8"
Noah James Hayes
Cap 21h9'58" -25d7'44"
Noah James Sandoval
Psc 1h19'2" 16d37'35"
Noah James West III
Aur 7h10'56" 42d39'17"
Noah Jonathan Flyboy
Gray
Her 16h29'15" 13d30'25"
Noah Joseph Sworsky
Lyn 7h46'14" 38d9'4"
Noah Kelly Elkaim
Uma 11h48'24" 38d27'21"
Noah Kenyon Gettings
Umi 14h44'21" 73d34'46"
Noah Lacy
Umi 14h21'11" 73d53'41"
Noah & Laura's Amazing
Friendship
Ari 2h52'43" 28d20'28"
Noah Le Huray
Cep 2h25'11" 83d28'36"
Noah Lee Borden 2006
Cap 21h35'20" -16d0'41"
Noah Lennon Duke
Courville
Psc 1h9'44" 11d38'27"
Noah Leydel Loves Dani
DeFilippis
Cyg 21h32'42" 47d34'37"
Noah Lincoln Veres
Her 17h56'12" 22d3'51"
Noah Lo Martire
Uma 9h47'38" 47d35'59"
Noah Loves Kell-Bell 637
Gem 7h15'31" 26d17'11"
Noah Lucas
Cma 6h28'2" -16d24'33"
Noah Luke
Uma 10h59'31" 62d37'50"
Noah M. Concepcion
Lithgow
Umi 14h0'49" 73d44'57"
Noah Matthew
Her 16h25'20" 48d1'5"
Noah Matthew Cook
Aqr 23h52'57" -11d10'20"
Noah McMahon
Aqr 21h45'0" -8d15'50"
Noah Michael Carver
Tau 3h37'12" 23d46'29"
Noah Michael Endsley
Sco 17h4'41" -39d35'33"
Noah Michael Powers
Ori 5h38'52" 0d56'36"
Noah Michael Young
Lib 15h10'44" -12d36'38"
Noah Micheal Ericson
Vir 13h27'8" 9d28'36"
Noah Mueller-Kielwein
Aqr 22h52'40" -12d24'35"
Noah Oley
Cyg 19h57'8" 51d33'28"
Noah P. Bourdon
Umi 14h30'14" 77d40'58"
Noah Parenti-Rusjan
Cnv 13h52'41" 32d5'29"
Noah Patrick Colston
Mon 6h58'12" -0d59'42"
Noah Patrick Craft
Lyn 7h30'2" 58d31'27"
Noah Patrick Gist
Uma 8h34'15" 47d47'4"
Noah Paul
Cra 18h39'38" -42d55'21"
Noah Peter Zimmerman
Sgr 17h54'1" -28d25'46"
Noah Phoenix
Ori 6h20'52" 14d31'19"
Noah & Pop/Pop
Per 3h33'14" 32d49'11"
Noah Quinn Agena Stroud
Ori 5h21'17" 8d13'16"
Noah Ray Santos
Her 17h46'19" 45d1'3"
Noah Raye Milford
Ori 5h19'37" 6d41'19"
Noah Richard Dixon
Leo 11h26'40" 24d52'39"
Noah Riley Jackson
Per 3h13'19" 53d22'28"
Noah Robert Allan-
Holdener
Dra 18h21'30" 59d38'5"
Noah Robert Rothberg
Ori 6h11'16" 21d0'16"
Noah Roberto Mendoza
Sgr 17h49'55" -26d29'45"
Noah Ruffenacht
Tau 5h46'15" 23d4'48"
Noah Scott Middleton
Umi 14h19'35" 72d11'31"
Noah Scott Porter
Sgr 17h51'26" -29d54'26"
Noah Shackleford
Psc 0h11'33" 9d37'16"
Noah Sherod
And 23h15'30" 41d26'18"
Noah Simon Juricic
Lyn 7h44'11" 56d49'20"
Noah Simon Mata
Aqr 23h31'31" -13d8'53"
Noah Smith McMahon
Aqr 23h0'2" -15d23'33"
Noah Stephen Doucette
Uma 11h59'33" 57d0'17"
Noah Stone
Ori 5h25'33" 1d47'50"
Noah Tadhg McCahill
Cep 22h52'40" 71d5'14"
Noah the Great Explorer
Malone!
Ori 5h39'30" -4d44'51"
Noah Thomas
Uma 11h59'29" 38d34'3"
Noah Timothy Hester
Sgr 19h4'11" -29d26'20"
Noah Timothy Thering
Aql 19h26'21" 3d48'17"
Noah V. Magdich
Uma 10h26'41" 60d58'42"
Noah William
Psc 0h18'10" 18d20'4"
Noah William
Psc 0h17'10" 18d47'33"
Noah William Barnett -
Noah's Star
Her 18h49'58" 20d49'13"
Noah William Pratt
Aur 5h56'2" 45d16'38"
Noah Wilson
Uma 10h29'59" 46d35'11"
Noah Wright
Umi 13h36'46" 77d52'3"
Noah Ylang Gabor
Cap 20h38'4" -13d13'44"
Noah Zachary
Ori 5h38'26" 1d19'52"
Noah,
Umi 15h35'51" 77d20'55"
Noah-Kayla
Peg 22h12'49" 16d42'0"
NoahMatthewMichael 1
Umi 15h18'3" 72d59'48"
Noah's Ark
Lyr 18h52'49" 43d51'4"
Noah's Star
Oph 16h40'47" 0d50'56"
Noahstar
Sco 16h4'42" -17d59'8"
Nob Hill Tigers
Lyn 7h31'53" 51d37'11"
Nobby & Betty Golden
Wedding
Cyg 20h15'14" 35d26'40"
Nobili
Peg 23h11'26" 19d53'25"
Noble 10th Anniversary
Star
Lmi 10h15'28" 36d5'52"
Noble and Eleanor Carter
Cyg 19h42'54" 36d12'15"
Noble Earl
Dra 16h24'19" 57d21'12"
"NOBLE NOURISHER" Pat
A Didden
Uma 10h29'59" 65d56'14"
"Noble Star,"In Memory of
Roy Noble
Lib 14h54'33" -1d45'9"
Noble Wind
Uma 8h20'13" 60d30'16"
Noble's Star 7/6/54 -
9/14/03
Cnc 9h8'47" 20d51'37"
noce-lieber
Sco 17h52'42" -38d10'56"
Noces de Diamant - J. & R.
- 2006
Crb 15h57'21" 29d28'17"
Noche de Nolan
Uma 11h56'32" 34d59'35"
Noche Firefly
Umi 15h44'35" 76d5'24"
Nockemann, Paul
Ferdinand
Uma 13h42'17" 60d54'20"
Nocola Williams
Cas 1h37'44" 61d27'46"
Noda
Her 17h9'43" 17d18'38"
Noddy McRothney Rarr
Starr
Ori 6h11'23" 20d39'21"
Nodech
Cnv 13h50'6" 30d45'30"
Nodnarb
Aqr 22h42'5" -8d17'38"
Noe
Ori 6h17'19" 15d4'30"
Noè Audrain
Peg 23h6'3" 28d57'42"
Noé & Isela
Lyr 19h9'3" 27d13'11"
Noe J. Lemus Star
Cyg 21h10'30" 40d12'16"
Noe Marie's Star
Lib 15h42'7" -5d22'30"
Noe & Masaru
Cnc 8h52'34" 19d28'46"
Noé Steck
Sco 17h46'9" -36d42'38"
Noe Vicente
Uma 9h58'50" 44d28'47"
Noée Alena Zgraggen
Dra 17h34'8" 63d9'3"
NoeGailMattCalida
Lmi 10h2'41" 31d59'45"
Noel
And 23h12'49" 40d11'13"
Noel
Lyn 7h56'41" 47d10'33"
Noel
Cnc 7h59'58" 18d44'50"
Noel
Umi 13h38'32" 79d2'55"
Noel
Mon 6h51'49" -0d53'8"
Noel
Cap 21h32'8" -9d15'39"
Noel
Sgr 18h32'3" -22d43'1"
Noel Alexander Hammond
Cru 12h20'16" -57d48'29"
Noel Andrew Peggs 11
Gem 7h38'43" 25d38'59"
Noel B. Monahan
Umi 15h22'10" 71d25'9"
Noel Carson
Lac 22h26'44" 44d12'33"
Noel Condon
Cap 20h22'14" -9d7'29"
Noel Desilets
Cas 0h56'59" 60d15'49"
Noel Evan
Uma 9h0'55" 46d55'44"
Noel Garber
Uma 12h0'59" 52d17'4"
Noel Hanf
Ari 2h14'43" 24d17'35"
Noel Harold Hansen, Jr.
Psc 1h27'28" 15d45'46"
Noel Harris
Vir 11h43'44" 4d24'1"
Noel Hope Millard
Mon 7h18'37" -0d29'46"
Noel Howard Helgoe
Uma 11h42'5" 38d21'8"
Noel J. Cummings
Cas 23h23'49" 53d24'49"
Noel James
Ori 6h18'23" -1d48'3"
Noel James Shaff
Ori 6h11'45" 8d1'6"
Noel Joshua
Uma 8h22'33" 63d38'12"
Noel Lee Reid
Crb 15h45'55" 35d37'4"
Noel Louis Grass
Uma 12h30'19" 53d44'34"
Noel Lynn Bruno
And 2h25'22" 44d57'28"
Noel O'Connor
Ori 6h20'33" -0d59'2"
Noel Pedigo
Umi 14h50'52" 74d35'52"

Noel Perez
Cep 21h35'24" 65d31'30"
Noel Pilli
Cap 20h27'35" -10d42'29"
Noel Quinn's Ladybird
Gem 6h51'15" 18d38'5"
Noel Saavedra
Psc 0h27'37" 2d42'28"
Noel Shannon Maciolek
Sgr 18h0'45" -24d50'57"
Noel Thomas Greig
Cru 12h2'6" -63d11'43"
Noel, Will You Marry Me? Love Neil
Ori 5h37'31" 2d50'54"
Noelani
Sgr 18h4'27" -24d19'32"
Noelani Aaron
Cyg 20h9'11" 40d38'37"
Noelani & Billy
Tau 4h19'15" 7d53'7"
Noelani Bountiful 25th Ward
Cas 1h18'56" 61d47'53"
Noële-Irmawisasajuhe!
Cas 23h42'22" 55d46'48"
Noelene Mary Smith -25 October 1945
Sco 17h26'8" -33d20'45"
Noelia
Sco 17h51'36" -30d52'17"
Noelia
Cam 5h54'58" 69d32'38"
Noelia
Cyg 20h54'28" 48d41'32"
Noelia
Crb 16h20'51" 35d6'13"
Noelia Feliciano
Umi 14h0'28" 70d26'51"
Noelia Lilly Wiggins
Cap 21h27'21" -20d7'38"
Noelia Marie Lerma
And 0h13'10" 27d30'0"
Noelia Marisol Zayas Hayas
Tau 3h47'31" 8d52'5"
Noelia & Marissa
Umi 16h14'1" 82d38'18"
Noelikins
Cru 12h19'8" -56d47'19"
Noeline Louise Claire Glynn
And 0h48'44" 31d20'18"
Noeline Uffer
Lyr 18h35'51" 37d12'34"
Noeline's Birthday Star from Will & Bel
Sge 19h59'34" 19d52'37"
Noell Sisk
Sgr 19h50'14" -15d39'11"
Noelle
Cap 21h5'46" -17d11'20"
Noelle
Sgr 18h31'47" -17d31'4"
Noelle
Lib 15h18'47" -17d23'41"
Noelle
Pho 1h45'54" -47d30'14"
Noelle
Sgr 18h12'5" -34d3'17"
Noëlle
Leo 11h4'54" 7d48'48"
Noelle
Uma 11h55'36" 32d26'51"
Noelle
Lyn 8h16'37" 41d26'35"
Noelle
Cam 4h50'3" 56d8'8"
Noelle
And 23h39'5" 43d53'21"
Noelle Alyn Reimer
Cap 20h33'18" -13d43'6"
Noelle Armstrong Morris
Cyg 21h21'45" 30d15'53"
Noelle Baca
Cap 20h56'36" -21d18'48"
Noelle Belayachi
Dra 19h1'5" 52d58'49"
Noelle C. Durett
Cap 20h22'5" -10d47'2"
Noelle C. Wojnovich
Sgr 19h10'10" -12d53'39"
Noelle Carmella Richard
Lyr 18h51'10" 33d22'18"
Noelle Constance Hauge
Aqr 22h37'37" -4d16'0"
Noelle Denise L'Heureux
Uma 12h2'17" 59d39'51"
Noëlle Dorothy Fairweather
Cap 21h22'18" -23d49'48"
Noelle Elizabeth Braun
Uma 10h49'17" 49d37'39"
Noelle Grace Nguyen
Uma 10h46'14" 42d34'23"
Noelle Jane Waldrom
And 1h19'33" 38d58'41"
Noelle Kainoa
Crb 15h44'20" 36d40'15"
Noelle Love Fleming
And 23h27'54" 46d1'35"
Noelle Luthien Anthony Gallagher
Crb 15h46'22" 34d17'52"
Noelle M. Kasa
And 0h55'44" 39d16'53"
Noelle Modly
Vir 12h50'27" 11d21'33"
Noelle Nicole Wilder
And 0h17'13" 45d8'44"
Noelle & Peter
Lyr 19h17'50" 29d1'51"
Noelle Robin Stillwell
Sgr 18h36'15" -28d20'9"
Noelle Rowley
And 0h14'31" 27d54'20"
Noelle Simone Dryver
Lib 14h45'31" -10d3'36"
Noelle Stephanie Stang
Uma 11h28'30" 59d16'6"
Noelle Vergara
Sco 16h11'19" -17d0'50"
Noelle Wolcin
Mon 6h55'26" -0d54'23"
Noelle,Jess,Tommy,John, & Maggie
Sgr 19h21'42" -15d54'50"
Noelle's Birthday Starr
Sgr 19h53'37" -14d19'20"
Noellie
Cap 20h30'14" -16d8'34"
Noel's Light
And 0h58'34" 40d25'30"
Noel's Star
Dra 15h27'39" 60d32'39"
Noël-Valérie
Umi 15h9'32" 82d8'56"
NOEMI
And 0h58'22" 43d44'30"
Noëmi
Peg 22h4'49" 14d35'33"
Noemi Alvarez Molina
Leo 9h54'24" 7d0'52"
Noemi Antonia Castellano-Garcia
Psc 23h5'7" 1d10'10"
Noemi Casati - Nostra Bellissima Stella
And 2h28'18" 50d32'18"
Noemi & Daniel. Forever loves.
Her 17h16'19" 28d23'23"
Noemi e Massimiliano
Cyg 20h14'45" 35d49'32"
Noemi Elosegui
Psc 1h7'30" 13d58'53"
Noémi és Jankó Szerelemcsillaga
Uma 14h17'28" 57d32'35"
Noemí Faraco
Ori 6h5'47" 4d18'34"
Noemi Kilthau
Cap 21h52'6" -18d52'9"
Noemi & lapo
Lyr 18h46'40" 30d10'35"
Noemi & Nadia Zep
And 0h59'6" 39d49'39"
Noemi Ramona Pfister
Cyg 21h19'0" 33d13'34"
Noemi Samira Hess
Ori 5h58'9" 18d10'12"
Noemiann Adriano
Aqr 22h43'10" -16d54'43"
Noemie
Aqr 22h21'37" -24d28'5"
Noemie
Peg 21h48'8" 22d22'43"
Noémie Crèvecoeur
Cyg 20h14'56" 50d24'7"
Noemie Jenni
Leo 11h19'27" 20d35'42"
Noémie Lameyre
Her 18h21'2" 24d26'22"
Noémie Li Mei
Cyg 20h29'43" 49d0'54"
Noemie Pichette
Aql 19h24'38" 5d3'40"
Noemie Ventavoli
Del 20h49'33" 15d58'14"
NOEMY
Vir 13h16'30" 7d33'12"
Noemy!
And 2h6'11" 40d28'8"
Noetzel, Klaus
Uma 9h4'38" 51d45'59"
Noey
Cap 20h45'48" -22d22'14"
Noga Tarnopolsky
Ari 2h42'15" 15d58'33"
Noggin
Her 18h21'38" 18d50'12"
Nöggu
Equ 21h6'0" 3d46'34"
Noh Hye Jin
And 0h39'9" 25d39'6"
Noha
Eri 4h13'33" -24d49'3"
Nohealani
Psc 1h16'58" 19d38'52"
Nohe's Shining
Leo 10h14'31" 24d52'27"
Noi Amore
Uma 11h45'22" 46d46'24"
Noi Siamo
Gem 6h45'30" 23d17'14"
NoiAi
Cep 0h4'10" 78d19'39"
Noilanie Amira Recustodio
Gem 6h45'57" 24d10'41"
Nojovajo
Nor 16h6'9" -43d31'36"
Nokeitha
Tau 4h15'29" 29d46'54"
nokinaya
Dra 14h58'8" 55d50'9"
Nola
Cas 2h19'56" 71d45'16"
NOLA
Dra 18h12'52" 75d44'5"
Nola Livingston
Lyr 18h39'4" 41d23'9"
Nola Mari
Aqr 20h54'49" -10d28'9"
Nola & Tom's Love
Umi 15h31'30" 74d14'50"
Nolan
Her 17h12'5" 31d11'24"
Nolan Case Adams
Her 16h54'59" 33d57'47"
Nolan Daley Reynolds
Her 17h7'34" 32d37'9"
Nolan David Vetor
Uma 8h43'3" 62d57'47"
Nolan Donald Faust
Her 17h40'46" 32d59'10"
Nolan Drislane
Sgr 18h51'35" -20d31'5"
Nolan Edward Watts
Uma 12h35'5" 62d37'36"
Nolan Edwin
Vir 13h25'12" -22d11'16"
Nolan F. Morales
Uma 11h39'52" 62d17'13"
Nolan Guy Carson
Her 17h43'31" 37d13'11"
Nolan Jackson Millett
Aql 19h18'9" 6d55'17"
Nolan Jacob Miller
Umi 14h16'26" 77d52'17"
Nolan James Donnelly A Star is Born
Cnc 9h9'17" 32d12'38"
Nolan James Timme
Her 17h20'48" 21d44'34"
Nolan Jameson Henwood
Lyn 8h10'50" 54d58'38"
NOLAN JEFFERY
Lyn 8h16'43" 54d25'59"
Nolan John Fisher
Lib 15h40'57" -17d57'42"
Nolan Joseph Harless
Per 3h11'55" 42d2'27"
Nolan Joseph Leon Baker 20/07/1995
Cnc 8h34'50" 6d51'2"
Nolan Martin Schuetz
Dra 18h45'14" 54d50'41"
Nolan Matthew Budig
Leo 11h54'56" 19d32'59"
Nolan Michael Kriegbaum
Uma 8h35'44" 53d52'3"
Nolan Parker Mathis
Sgr 18h24'54" -28d32'40"
Nolan Patrick Finan
Lyn 8h1'22" 55d28'30"
Nolan Peter Curran
Psc 1h41'46" 23d42'59"
Nolan Rainey
Uma 11h28'31" 60d23'34"
Nolan Thomas Shanahan
Ori 5h29'29" 14d57'32"
Nolan Thomas Williams
Per 2h16'28" 54d5'19"
Nolan Thomas Wycherley
Cnc 9h14'8" 9d44'33"
Nolan's Night Light
Vir 12h29'40" 1d2'2"
Nolberto
Vir 12h28'1" 1d32'11"
Nolden, Hermann J.
Uma 9h43'55" 43d37'23"
Noleen Watkins
Cas 1h26'48" 56d9'36"
Nolen
Lib 14h24'34" -13d34'53"
Nolen Calvin Fields
Leo 11h54'57" 18d57'51"
Nolen Hallman
Peg 21h48'44" 11d2'3"
Nomad the Producer
Per 4h18'46" 44d38'34"
Nomar
Vir 13h6'6" 3d30'57"
NOMC FAMILY MEDICINE RESIDENCY
Lyr 18h46'59" 32d25'21"
Nomers
Uma 10h33'58" 44d57'40"
NomiJeanEdwards Best Momma Ever
Uma 10h33'23" 59d56'30"
Nomis
Vir 14h12'40" -20d29'19"
Non Wynne
And 1h40'15" 40d54'17"
Nona
Cyg 21h21'54" 38d32'55"
Nona
Lib 15h51'18" -12d38'52"
Nona
Cas 23h27'40" 53d52'15"
Nona Bellitas
Oph 17h11'11" -0d41'59"
"Nona" In Loving Memory
Cyg 20h22'52" 37d8'7"
Nona Laverne Johnson
Cap 20h32'37" -26d57'10"
Nona Mae
And 0h36'38" 41d38'47"
Nona & Pop's Little Gem in the Sky
Ori 5h25'54" 14d49'24"
Noname Reyes
Gem 7h17'13" 14d12'15"
Noncsi és Balázs 1000 év
Uma 9h7'58" 49d36'8"
Nondus
Ari 2h9'35" 24d13'11"
Nongnard
Sco 17h45'56" -42d53'58"
Noni
Vir 13h7'23" 13d18'21"
Noni Cyd
Uma 11h47'2" 65d27'25"
Noni loves Les
Lib 14h49'50" -20d25'27"
Nonie and Bop
Ori 5h24'17" 2d18'17"
Nonie Bear Wortman
Sco 16h11'58" -20d50'57"
"Nonie" Celina Beston
Leo 10h13'46" 25d0'39"
Nonna
And 0h17'40" 34d28'4"
Nonna Adri
Ori 5h35'22" -0d11'58"
Nonna Giuseppina Gariffo
Uma 13h34'12" 52d56'9"
Nonna' Star
Psc 0h42'40" 7d15'58"
Nonnamirus
Tau 3h32'39" 24d49'2"
Nonna's Light
Lyn 7h44'8" 41d19'34"
Nonni Roli & Mari
Lyr 19h19'41" 41d39'37"
Nonnie
Uma 11h52'52" 28d42'21"
Nonnie
Cas 1h26'49" 63d17'27"
Nonnie (Robyn Catenich)
Lib 15h18'13" -15d5'5"
Nonno Peto's Shining Star
Uma 12h24'44" 56d35'39"
Nonno Sisi
Cyg 21h36'37" 39d39'42"
Nonno Venturi
Her 17h15'41" 28d12'26"
Nonny
Gem 6h35'48" 15d3'29"
Nonny
Crb 16h17'19" 36d35'17"
Nonny's Wishing Star
Ori 5h35'34" -1d18'56"
nono rolli e nonnina mari
Uma 14h4'56" 49d47'28"
Nontiscordardime Bambola
Cma 6h30'49" -31d44'32"
Noodelia
Cnc 8h22'22" 20d22'18"
Noodle
Leo 10h33'19" 24d10'25"
Noodle
Cyg 20h56'7" 47d52'15"
Noodle
Uma 9h37'24" 48d11'18"
Noodle
Mon 7h9'35" -0d44'41"
Noodlebrook
Vir 13h1'10" -20d4'39"
Noodles
Cnc 8h2'30" 24d57'49"
Nook & Cranny
Uma 9h47'10" 65d30'7"
Nookie
Uma 13h37'36" 60d41'3"
Nookie
Cyg 19h49'4" 38d0'40"
Noom'
Uma 9h7'54" 47d19'57"
Noomi
Ori 6h10'5" 15d51'34"
Noo-Noo's Star (Jacqui Ryding)
Peg 21h42'43" 15d55'50"
NOOPIE
Uma 13h34'38" 56d32'47"
Noor
Aqr 22h44'35" -1d41'54"
Noor Bahri Jepsen
Ari 2h11'31" 24d16'0"
Noor Jahan - "Light of the World"
Psc 23h37'30" 2d45'44"
Noor Jahanshahi
Uma 11h31'59" 43d7'21"
Noorjahan ("the forever flower")
Lyn 7h21'19" 44d30'41"
Noorjehan Aziz
Aqr 21h58'1" 0d21'20"
Nooshin
Cnc 8h43'12" 15d26'3"
Nopporn Mandine
Cra 18h23'27" -37d14'27"
Noquisi Christine Mangiantini
Cas 1h34'56" 64d46'36"
Nor Cal
Lyn 7h2'28" 44d50'5"
Nora
Uma 10h2'0" 44d32'46"
Nora
And 0h38'13" 39d27'5"
Nora
And 0h25'30" 37d41'55"
Nora
Cyg 21h13'57" 46d8'12"
Nora
Ari 3h22'10" 28d46'22"
Nora
Dra 15h28'27" 62d14'12"
Nora
Uma 10h13'20" 59d32'28"
NORA
Lib 15h22'36" -11d13'4"
Nora Alice Star
Crb 15h44'26" 36d49'13"
Nora Ann Tarbi
Aqr 23h6'57" -14d2'24"
Nora Anne Goldberg
Sgr 18h50'38" -24d44'30"
Nora Born
Leo 11h47'27" 20d1'20"
Nora Bulnes
Leo 9h30'27" 9d59'14"
Nora Catherine Scheibe
Sco 17h11'4" -33d51'3"
Nora Christina Menth
Lac 22h51'20" 52d46'26"
Nora Cobb
Cam 4h7'17" 57d19'51"
Nora Demnitz
Uma 11h0'6" 41d54'14"
Nora E. Parra
And 1h17'9" 35d16'15"
Nora Eileen
Uma 11h38'25" 62d35'45"
Nora Elizabeth Whetzel
Cnc 8h53'7" 28d43'40"
Nora Ellen
Umi 12h46'1" 85d59'11"
Nora & Esperanza Candas Gillespie
Sge 19h37'36" 17d4'17"
Nora F. Ornelas
And 0h34'27" 28d27'27"
Nora & Franco
And 23h43'16" 47d2'36"
Nora Gail
Cnc 9h10'17" 30d46'10"
Nora Grace
Umi 14h24'45" 74d15'29"
Nora Hilda Vasquez
Sco 16h30'0" -26d15'57"
Nora Jane Wilcox
Cyg 20h29'10" 59d23'47"
Nora Jolie
Vir 12h43'52" -10d34'41"
Nora Knowles, my Mum
Cas 23h43'57" 51d41'28"
Nora Laferney
Vir 13h16'21" 3d57'41"
Nora Lambard
Cyg 20h52'47" 48d23'1"
Nora Lara Medrano
Her 17h29'35" 26d55'36"
Nora Lee McCance
Gem 7h17'32" 32d28'28"
Nora Linda Davis
And 0h32'29" 29d15'56"
Nora Lizet Medina
Vir 14h3'53" -18d11'57"
Nora Louise Campbell
Lib 14h45'56" -12d24'13"
Nora Lynn
Tau 4h36'23" 26d24'11"
Nora Lynn Pierce
Sco 16h20'34" -26d0'41"
Nora M Jones
And 23h1'58" 40d47'45"
Nora Malloy Johnston
Leo 10h38'12" 9d18'40"
Nora Marie Martin Resnick
Aqr 23h3'30" -6d22'38"
Nora May Whitt
Lib 15h39'29" -12d58'13"
Nora McMahon Westol
Crb 15h55'11" 28d16'20"
NORA MEU AMOR
Lyn 7h27'35" 56d38'18"
Nora P Campos
Uma 10h10'51" 51d30'37"
Nora Pace
Cap 20h28'23" -13d35'48"
Nora Pai-Yee Yin
Tau 5h48'6" 27d39'48"
Nora Palou
Aqr 23h18'22" -18d6'10"
Nora Pitaro
Cnc 8h32'48" 16d29'19"
Nora Quinonez
Aql 19h54'24" 15d58'8"
Nora Reardon
Uma 13h39'41" 56d43'42"
Nora Rose Noble
Cap 21h55'54" -18d21'41"
Nora Rose Zaring
Ari 1h52'28" 23d49'8"
Nora Shannon Chan
Cnc 8h35'2" 32d35'36"
Nora Sharp
Cnc 8h18'36" 14d27'49"
Nora Singharath
Psc 1h10'49" 17d11'28"
Nora Susanne
Vir 15h1'2" 5d7'41"
Nora Terese Huber
Uma 8h56'7" 57d41'37"
Nora Walther
Uma 11h48'5" 62d33'11"
Nora Winn
Gem 7h47'21" 32d39'35"
Norah
And 1h40'20" 48d5'8"
Norah
Aqr 21h12'54" -14d6'7"
Norah Elizabeth Light of God
Tau 4h28'20" 20d34'19"
Norah Goldston
Tau 4h33'16" 11d45'29"
Norah Janeann Strong
Aqr 20h43'9" -6d47'9"
Norah & Mehrab forever
Cyg 20h41'22" 42d42'58"
Norah Michelle Wilson
And 0h18'31" 44d6'49"
Norah Olivia Bigler
Lyr 18h48'31" 36d21'52"
Norain Salim
Uma 12h8'21" 44d46'52"
Noralee
Aqr 21h2'15" -6d23'51"
NORALINDA
Leo 11h22'5" 15d7'21"
Norallene
Tau 5h39'50" 24d27'17"
Nora's Beautiful Eyes
Cas 0h50'20" 60d41'27"
Nora's Galactic Love Star
Aqr 20h43'56" 1d11'16"
Nora's Nook
Pho 0h10'11" -47d34'4"
Norbert Brausch
Ori 5h26'15" 8d52'53"
Norbert Fleischmann
Ori 6h23'40" 10d27'25"
Norbert Förster
Uma 14h24'58" 61d35'41"
Norbert Gansen
Ori 5h59'12" 17d10'35"
Norbert Gütschel
Uma 11h52'0" 30d44'53"
Norbert Hoffmann
Uma 9h13'1" 56d25'45"
Norbert Huber
Ori 5h3'15" 5d9'19"
Norbert Keolanui, Jr.
Cap 20h54'28" -18d32'27"
Norbert Nunes Star
Per 3h41'21" 46d44'29"
Norbert Orth
Ori 5h57'41" 17d33'41"
Norbert Peithmann
Uma 9h52'7" 53d35'19"
Norbert Schroedl
Uma 11h29'0" 64d24'43"
Norbert Schulz
Uma 9h29'10" 56d32'18"
Norbert "The Good" Nizze
Ori 6h17'11" 13d50'38"
Norbert Tomaszewski
Ori 6h11'14" 7d55'54"
Norbert Van Der Neut
Sgr 19h52'55" -28d22'30"
Norbert Wagner
Umi 14h38'17" 83d47'26"
Norberto de Los Rios
Aql 19h50'35" 11d36'23"
Norberto Martins
Tau 5h57'28" 23d16'39"
Norbi, Zsuzsi csillaga
Vir 14h9'29" -21d7'20"
Norbona
Vir 14h5'41" -14d56'56"
Norcross Family Star
Cyg 19h41'4" 31d43'9"
Norcsillag
Uma 9h25'28" 58d18'15"
Nordine
Cyg 20h59'24" 46d18'18"
Nordine
Tau 4h41'53" 1d27'44"
Nords Superstar Jasmine Ann McGhee
Cnc 9h5'42" 16d18'3"
Nordvik
Peg 23h59'3" 12d21'12"
Noreah
Cnc 9h20'39" 21d36'2"
Noreen
Tau 4h35'3" 20d53'47"
NOREEN
Cas 0h11'25" 54d1'16"
Noreen Ann Galvin
And 23h29'44" 47d20'48"
Noreen Carberry 2255
Cas 1h26'50" 62d46'3"
Noreen Colette Curtin
And 0h11'54" 46d1'8"
Noreen Dowd
Lib 14h55'30" -2d6'1"
Noreen Garcia
Com 12h28'36" 26d14'58"
Noreen Gravina
Cas 1h25'22" 51d48'27"
Noreen Mahmud
Cas 0h18'46" 61d55'51"
Noreen McCarthy
Vir 13h21'54" -14d1'16"
Noreen Mosher
Gem 6h32'30" 26d53'34"
Noreen Nothstein
Per 4h11'30" 48d0'14"
Noreen O'Rourke
And 23h45'51" 38d43'3"
Noreen Smolinsky
Boo 14h54'53" 36d39'29"
Noreena Lee
Cyg 19h39'10" 28d28'20"
Noreita Claudine Seaman
Uma 13h58'0" 53d4'0"
Noretta McClain
Umi 13h54'51" 74d42'39"
Norhayati <\@> Mrs Shamshul Qamar
Umi 9h38'58" 87d26'20"
NorHenSa 07
Uma 9h48'13" 41d55'13"
Noriane Charbonnier
Umi 13h16'6" 87d1'14"
Norihito and Tae
Ari 1h59'19" 17d48'55"
Norika's Flying Circus (Monty)
Lyn 6h31'38" 57d39'33"
Noriko
Lib 15h4'11" -0d52'34"
Noriko and Anatol forever
Cyg 20h7'37" 30d51'6"
Norina Colatriano
Cam 7h28'10" 77d0'52"
Norine Carlson
And 23h33'51" 44d55'59"
Norine Rochelle Spadaro
Ari 2h43'58" 28d2'19"
Noris
Lib 15h8'22" -3d26'31"
Noris
Aql 20h17'9" -0d8'18"
Norissa
Cyg 20h5'39" 38d56'10"
Noritoyo & Remi
Sco 16h46'43" -41d27'15"
Noriyasu Suzuki
Psc 1h31'9" 12d30'18"
Norlan Delgado
Leo 10h45'1" 19d41'19"
Nor-Lar
Mon 7h21'44" -0d53'45"
Norlen
Her 17h18'7" 34d13'35"
Norm and Eileen's Love
Lyr 19h20'23" 29d13'14"
Norm and Millie's Forever Star
Cam 4h14'13" 68d17'51"
Norm & Bridget Arendas
Cyg 20h30'15" 36d51'28"
Norm & Dorothy Palmer
Cyg 21h15'4" 52d25'19"
Norm Jacobi
Leo 11h52'15" 20d8'38"
Norm John Toulou, Sr.
Her 16h43'48" 6d5'14"
Norm & Lori Douthitt
Cyg 21h20'27" 33d13'1"
Norm Maleng
Uma 13h42'33" 54d30'17"
Norm Sandy
Aur 5h30'32" 56d5'17"
Norm Schultz
Ori 5h44'50" -0d30'46"
Norma
Lib 14h56'38" -5d31'13"
Norma
Cas 23h47'17" 56d40'8"
Norma
Cas 2h33'43" 62d20'42"
Norma
Uma 9h50'53" 63d13'56"
Norma
Nor 16h9'45" -44d32'46"
Norma
Tau 4h26'46" 2d16'21"
Norma #1
Lib 14h37'28" -17d4'31"
Norma "#1 mom" Delgado
Aqr 22h5'34" -2d39'43"
Norma Alicia
And 1h42'7" 38d0'16"
Norma Alicia Cruz
Gem 6h51'29" 19d18'28"
Norma Alicia Mariscal
Gem 6h45'56" 24d43'58"
Norma Allison
Vir 12h28'1" 1d29'49"
Norma Amaya Conneally
Cnc 8h48'45" 31d31'53"
Norma Anne Jones
Cas 0h1'19" 53d38'19"
Norma & Anthony Rizzi
Cyg 20h36'44" 57d45'12"
Norma Ballinger Hurt
Sgr 18h58'58" -34d22'2"
Norma Beatriz Martinez Deuane
Cnc 8h52'35" 28d23'6"
Norma Blasa Torres
Cas 0h30'21" 53d22'27"
Norma Bode Miller
Nor 16h13'15" -47d14'6"

Norma Bolthouse
Tau 3h45'48" 26d26'47"
Norma C. Balais
And 0h58'2" 40d51'17"
Norma C. McGarry
Lib 15h12'13" -12d38'25"
Norma Caldwell Elliott
Cyg 21h39'18" 54d9'39"
Norma Chalmers Mastalir
Cas 0h58'4" 58d52'28"
Norma Chavez
Aqr 20h43'30" -2d6'33"
Norma Conrado
Uma 8h30'4" 64d34'20"
Norma Contryman Stine
Lyr 19h18'24" 29d42'29"
Norma D.
Del 20h34'6" 13d23'23"
Norma Del Valle Tirado
And 0h12'10" 44d43'46"
Norma DeSantis
Uma 10h23'31" 53d57'10"
Norma Eileen
Nor 16h25'10" -51d45'20"
Norma Elena Estrada R.N.
And 23h38'47" 47d38'19"
Norma F. Cossaboon
And 23h1'55" 52d39'50"
Norma Feder
Uma 11h39'3" 31d42'50"
Norma Garcia
And 23h23'5" 51d50'3"
Norma - Gene Nantel Dec. 6th 1986
Cyg 20h25'0" 41d22'43"
Norma H. Yohai
Ori 5h35'34" -8d39'8"
Norma Hale Pearman
Aqr 21h32'11" 2d1'42"
Norma Helding
Lyr 18h43'1" 38d57'46"
Norma Hubert
Uma 11h17'29" 32d28'36"
Norma Ibanez
Cyg 20h9'16" 46d52'9"
Norma Idalia Espinosa Lozano
Nor 16h9'49" -43d30'2"
Norma Iris Castro
Tau 5h43'56" 23d9'41"
Norma Iris Leon
Sgr 18h6'35" -32d4'10"
Norma Isela Morales
Sco 16h21'8" -17d57'7"
Norma J. Miller
Gem 6h13'12" 24d55'16"
Norma "Jane" Welsh
Cyg 19h36'40" 30d58'50"
Norma Jarillo
Cam 4h14'32" 70d50'53"
Norma Jean
And 1h11'0" 36d13'38"
Norma Jean 75
Vir 13h25'53" 5d10'34"
Norma Jean Coughlin
Sco 17h57'7" -41d27'23"
Norma Jean de Roziere
Cas 1h23'11" 62d45'20"
Norma Jean Kelley
Cnc 8h22'11" 13d22'38"
Norma Jeane Brewer
Com 13h11'57" 18d41'4"
Norma Jeane Walton
Cnc 8h15'44" 16d29'18"
Norma Jeanne
And 2h10'14" 41d32'31"
Norma Johanna
And 0h38'6" 26d28'40"
Norma Kloss
Uma 10h56'45" 34d5'3"
Norma L. Pion
Lib 14h51'45" -3d0'29"
Norma L Tooke
Umi 15h4'32" 73d53'34"
Norma Lacey Meade
Vir 14h44'55" 1d58'50"
Norma Lance Pres. Rebekah Assembly
Ari 1h47'14" 18d33'13"
Norma Lee Stretch
Lyn 6h47'26" 57d22'0"
Norma Lehde
Vir 13h0'55" -1d20'48"
Norma Lissette Cuevas
Umi 13h45'43" 76d1'21"
Norma Louise Strauch
Lyr 18h48'1" 41d14'35"
Norma Louise Wigner
Nor 16h27'17" -51d5'14"
Norma Loves Anthony 2iab!
Nor 16h4'13" -43d35'15"
Norma Lucille Hibbert
Per 4h48'41" 45d46'39"
Norma Martinez
And 0h29'47" 43d36'38"
Norma Mikkelsen
Cyg 20h34'21" 37d33'13"
Norma Miner Evans
And 0h44'59" 38d54'12"
Norma Morales
And 1h8'43" 46d6'29"
Norma Munoz Zamarripa
Umi 13h43'2" 76d44'3"
Norma Nicole
And 2h7'16" 46d15'42"
Norma O'Donoghue
Ari 3h25'53" 26d49'13"
Norma O'Sullivan MA MEd
And 2h24'5" 41d16'32"
Norma Pfiester ~ Our Shining Star
Cas 1h56'26" 63d8'54"
Norma Plasencia Ettore
Leo 10h8'41" 11d41'32"
Norma Reyes
Cas 1h2'51" 60d17'10"
Norma Rhoadarmer
Cas 0h41'20" 69d26'34"
Norma Rosalind Elizabeth Gooder
Leo 9h50'56" 22d22'44"
Norma Rosenhain
Cas 0h44'31" 62d5'31"
Norma Ross
Cam 3h29'37" 56d2'9"
Norma Shapiro Reichline
Cas 0h38'14" 63d47'28"
Norma Talmadge Boynton/Wheeler
Tau 4h28'37" 4d53'24"
Norma the Pirate
Nor 16h24'9" -51d20'12"
Norma Underwood Huber
Cas 0h41'37" 65d11'16"
Norma Villarreal
Vir 14h1'36" -14d6'22"
Norma-Daniel
Ari 2h46'57" 26d47'3"
Normador 8355
Leo 11h42'16" 16d35'30"
NormaJean
And 23h24'12" 47d11'9"
Norma-jean Laude
Umi 15h55'19" 76d0'2"
Norman
Aur 6h32'3" 39d5'36"
Norman
Cyg 21h9'52" 31d51'27"
Norman
Her 18h13'27" 26d36'6"
Norman A. Mennes
Cep 5h3'46" 81d53'8"
Norman A Nuessle
Uma 12h54'10" 60d36'27"
Norman Alan 40
Cap 20h30'54" -27d5'56"
Norman Allen
Uma 8h36'19" 72d50'33"
Norman Anthony Lyons
Peg 23h47'4" 26d22'48"
Norman & Beverly Anderton
Per 3h2'57" 54d29'7"
Norman Bieda - July 11, 1966
Her 17h18'25" 39d51'49"
Norman C. Ridley
Lmi 9h29'50" 39d8'38"
Norman Charles Cuthill
Leo 11h21'1" 4d26'32"
Norman & Cleopatra Hasham
Cru 12h38'39" -58d48'42"
Norman D. "Buz" Wilson
Gem 7h15'50" 28d45'56"
Norman Daquioag
Cyg 20h12'8" 35d18'1"
Norman David Lasky
Uma 8h27'39" 68d53'22"
Norman Davidson
Aqr 21h11'47" 0d49'53"
Norman Doo
Tau 4h51'41" 23d29'48"
Norman Douglas Wells
Her 17h22'1" 35d57'47"
Norman E. Jackson
Uma 10h33'21" 44d15'45"
Norman E. McHan Jr.
Ori 6h15'7" 7d43'43"
Norman Edward Herbstreith
Per 4h34'19" 45d40'49"
Norman Edwin Phillips, Jr., #27
Vir 11h49'39" 9d58'43"
Norman Everett Andrew
Crb 16h17'54" 29d45'36"
Norman Francis Talazac
Uma 11h4'51" 53d59'8"
Norman G. Barker
Ori 5h40'22" -0d30'36"
Norman George Dingman
Sco 16h13'45" -10d8'16"
Norman George Lane
Cep 21h55'50" 60d47'3"
Norman Glynn Nowitzky
Her 16h58'33" 33d8'8"
Norman Goldberg
Cap 21h17'15" -25d2'55"
Norman Greely Davis II & III
Aqr 22h18'3" 1d26'19"
Norman Hans Dahl
Ari 3h16'51" 19d13'33"
Norman Helms
Uma 9h57'16" 63d4'53"
Norman Hill, Jr.
Sco 16h46'35" -44d41'0"
Norman Irwin Gordon
Ori 5h18'52" 6d32'24"
Norman J. Couture
Gem 6h44'42" 13d40'32"
Norman J. Miller
Crb 16h22'6" 34d24'40"
Norman J. Wilcke "Our Opa"
Cap 21h41'21" -12d11'16"
Norman James Flemington
Uma 11h24'0" 70d1'0"
Norman Joel Austin "Master Joel"
Her 18h31'49" 17d46'19"
Norman Joseph Hochella
Lyn 7h5'27" 59d39'3"
Norman Kline
Uma 11h17'58" 54d56'49"
Norman L. Bissell
Per 2h50'21" 51d58'2"
Norman Lee Cass Sr.
Aur 5h47'17" 46d1'6"
Norman Lee Groom
Uma 9h29'24" 57d4'51"
Norman Lionel Dufresne
And 0h12'4" 25d18'29"
Norman Loren Gorley, Jr. "Chip"
Boo 14h51'21" 37d36'30"
Norman & Marcella Welsh
Cyg 19h48'21" 32d22'9"
Norman Marcon
Cep 22h11'39" 73d43'0"
Norman Mayer Kaplan, M.D.
Cyg 21h57'32" 49d1'54"
Norman Michie Makay McLeod
Cru 12h45'38" -56d43'4"
Norman & Molly Barton
Lyn 8h39'21" 39d23'27"
Norman Nelson's Eastern Star
Boo 14h52'56" 28d5'13"
Norman Oretskin
Cep 20h40'45" 61d31'25"
Norman "papa" Winch
Lib 15h6'34" -1d37'35"
Norman Paul Creighton & his Heirs
Uma 8h51'58" 50d19'39"
Norman Paul Zolkos
Cnc 8h2'36" 21d23'34"
Norman "Pete" Phelps
Dra 18h52'18" 69d52'48"
Norman Peter Lohstreter
Ori 5h36'22" 9d27'5"
Norman Phillips
Tau 4h31'4" 0d28'34"
Norman Ray Miles
Vir 11h53'26" 3d30'55"
Norman Ray Van Wagoner
Aqr 21h46'48" -1d0'32"
Norman Robert Beebe
Lib 14h58'42" -11d6'52"
Norman Rosenhain
Uma 11h26'44" 59d9'19"
Norman Rossignaud
Cmi 7h30'52" -0d4'28"
Norman Roy Campbell
Uma 10h38'9" 47d25'42"
Norman Roy Marsh
Aql 18h59'11" -0d10'32"
Norman Takata
Her 17h44'27" 29d9'1"
Norman Thuswaldner in the Skies
Lib 15h51'25" -11d32'18"
Norman Wakely
Uma 11h22'55" 30d3'35"
Norman & Wanda Onstad *35th*
Cyg 21h44'23" 39d3'47"
Norman William Healey
Cep 0h4'49" 69d54'12"
Norman Willock
Per 3h43'15" 42d23'43"
Norman, My Guiding Star
Aur 5h43'20" 42d49'21"
Normand Bourdon
Cep 20h39'37" 56d28'17"
Normand Forest Vance
Per 3h19'35" 51d37'44"
Normand Jarest
Uma 12h37'1" 56d13'36"
Normand L. Simoneau
Cnc 7h56'31" 15d47'41"
Norman-Dee
Uma 9h53'17" 52d46'40"
Normandie
Uma 13h38'17" 57d49'27"
Normandy A. Caro
Cyg 20h49'2" 35d0'55"
Normanland
Tau 3h36'51" 15d57'16"
NormanLee
Gem 7h19'42" 26d46'9"
Normanstar
Cen 12h7'37" -51d11'48"
Normaron
Cru 11h59'37" -62d20'43"
Normary
Cyg 20h49'35" 46d59'48"
Norma's Star
Gem 7h27'25" 33d18'38"
Norma's Star
Nor 16h30'0" -50d46'26"
Normie
Lyn 8h22'46" 57d20'54"
Normie
Cyg 19h42'32" 37d11'19"
Normita Star
Ari 3h6'22" 15d48'8"
Normo and Don
And 0h38'57" 25d38'19"
Normund and Tammy Forever
Cyg 20h10'4" 43d33'53"
Normunds & Inara
Uma 8h41'0" 68d10'38"
Normy
Her 17h41'2" 14d35'36"
Nornabelle
Psc 0h18'55" 19d11'19"
Nornie
Cap 20h32'5" -18d38'38"
Norris
Vir 13h16'43" -20d1'20"
Norris - 4851717581
Uma 12h5'54" 45d39'7"
Norris Syverson
Umi 14h25'59" 72d22'8"
Norris W. Davis "Big Sam"
Umi 16h49'37" 77d5'18"
norse nisse
Gem 7h2'5" 17d10'43"
North Anne
And 2h5'35" 38d55'16"
North Beach 2000
Gem 7h41'19" 31d36'22"
North of Davis
Cyg 19h49'40" 32d5'1"
North Star
Umi 16h2'56" 70d45'58"
North Star Academy Class of 2008
Uma 8h33'29" 63d17'41"
North Star R.A.'s Star
Umi 16h37'15" 82d50'13"
North StarGazers Polar Pride 2006
Uma 13h38'39" 57d51'33"
Northern Hero
Uma 9h53'4" 42d4'31"
Northern Nicholas Clemens
Per 2h45'22" 44d58'47"
Northrup-Wendelken
Sgr 19h39'43" -16d40'54"
Northstar Knight
Cnc 8h20'10" 11d2'46"
Northstars #60 Matt Ingra
Uma 10h54'42" 63d2'40"
Northwest Elem. Class of 2006-2007
Umi 15h26'24" 76d45'38"
Norton
Ari 3h11'14" 29d12'21"
Norton and Katie Leaps
Umi 18h7'15" 89d27'36"
Norton "Bestefar" Gaard
Ari 2h42'28" 12d43'36"
Norval Gene Barnes
Sgr 19h15'1" -15d18'10"
Norva-Pam
Lib 15h36'5" -18d20'29"
Norwood And Dawn Patterson
Cyg 19h48'6" 33d11'34"
nos sansfin amour
Crb 16h5'5" 26d1'41"
Nosa
Uma 11h9'28" 36d32'18"
Nosajaivilo
Psc 0h37'25" 17d9'58"
Noshaba & Yusuf
Cru 12h46'8" -56d15'32"
Noslen
Cas 1h22'8" 52d26'57"
Nosso Amor Eterno
Aqr 22h43'47" -3d11'19"
Nostra Amor
Uma 10h6'53" 44d21'46"
Nostro Amore
Cyg 21h25'29" 45d48'52"
Nostro Amore' Infinito
Lib 15h44'57" -8d19'34"
Nostro Striscia
Uma 11h24'46" 32d44'24"
Nostrum Diligo Est Forever
Aqr 22h24'57" -14d2'1"
nostrumdiligoillustro Our Lovelight
Cyg 19h57'56" 41d20'20"
Not as beautiful as Adriel Wong
Peg 22h45'18" 33d29'18"
Not as Pretty as Sara Jane
Sgr 19h17'48" -16d27'46"
Not Fade Away
Vir 11h42'14" 6d35'40"
Not (good enough to be) Gilles
Hya 9h11'57" -0d7'34"
Not The Painter
Sgr 19h7'40" -13d30'56"
Notable Nicole
Uma 12h17'50" 59d29'39"
Nothing Else Matters
Lyr 18h46'36" 38d21'35"
Notna's Christmas
Tau 4h19'37" 3d1'36"
Notorious3432
Aql 19h52'38" 12d1'30"
Notre Amour
Uma 8h43'52" 49d18'12"
Notre Amour
Uma 11h16'53" 49d20'30"
Notre Amour
Cyg 20h40'18" 34d4'59"
Notre Amour
Uma 12h13'34" 61d18'12"
Notre Amour
Sco 16h53'45" -32d51'50"
Notre Amour Brille D'En Haut
Lyr 19h0'52" 31d33'52"
Notre Amour est Eternal
Cyg 21h30'36" 45d29'31"
Notre Anniversaire
Her 18h37'3" 20d4'37"
Notre Creek
Uma 11h7'22" 38d35'59"
Nôtre Erin de Mère
Cyg 20h30'54" 51d29'24"
Notre étoile du pere
Uma 11h2'36" 44d17'57"
NOTRE FUTUR
Cyg 20h43'46" 39d43'58"
Notre Jamais dernier Amour
Cap 20h27'55" -16d2'48"
Notre Papillon
Cas 1h15'44" 67d23'18"
Notre Petit Bébé
Umi 17h5'15" 80d17'13"
Notre un de ces jours l'e-toile
Sgr 19h41'16" -15d44'52"
Notrinfinitude
Lyr 19h12'31" 33d59'4"
Nottingham Football 2005
Ori 6h15'7" 8d7'10"
Nou Nou
Dra 18h47'10" 54d2'2"
N'OUBLIEZ JAMAIS
Ari 2h37'28" 12d37'54"
Nouf AL-Rasheid
Umi 16h13'22" 75d17'0"
Nouf Kanoo
Ori 5h55'32" 3d15'45"
Noula Cochineas and her Agia Elessa
Cru 12h51'21" -58d2'35"
Nouni
Leo 11h13'56" 19d8'24"
Nounou
Uma 8h21'42" 62d3'46"
Nounou
Aqr 22h32'11" -14d12'21"
NounouMyLove
Gem 6h43'42" 25d35'11"
Nounoune
Her 17h33'30" 40d52'26"
Nounours&Nounouche Humery
Uma 12h50'49" 55d24'58"
Nour Fouad Nourallah
And 22h59'34" 47d20'56"
Noura
Cyg 20h10'5" 45d35'47"
Noura
Aqr 20h52'47" 1d24'10"
Noura & Firas
Ori 5h2'37" -0d24'7"
Noura Hussein Abdelaziz
Sco 17h58'7" -30d37'29"
Noura (Jones)
Cas 2h21'49" 74d25'30"
Nouran's Star
Uma 11h37'37" 28d33'33"
Noure Mani
Per 3h27'52" 49d51'42"
Nouri: Star of angelic Grace
And 23h1'3" 52d5'7"
Nour'louni
Cnc 8h42'10" 6d37'57"
Nous
Del 20h36'7" 6d36'14"
Nous rions
Peg 23h10'19" 16d0'45"
Noushka Foo
Vir 13h6'15" 7d22'30"
Nousseiba Kouider
Umi 14h22'3" 66d6'32"
Nova
Uma 9h21'6" 66d34'53"
Nova Danika
And 1h23'2" 47d1'35"
Nova De Eros
Gem 7h10'12" 15d37'10"
Nova Lee Ramsey
Aql 19h5'42" 10d8'1"
NOVA of Virginia Aquatics
Per 2h45'31" 53d27'31"
Nova Provencher
Uma 11h27'19" 59d37'36"
Nova Rose Wilson-Block
And 2h31'35" 38d12'13"
Nova Victoria
Cap 20h59'39" -22d17'6"
Novacrush
Pho 0h45'26" -42d0'44"
Nováky Gergely Tamás
Cap 20h15'8" -11d58'15"
Noval A. Smith, Sr.
Psc 0h33'49" 6d56'59"
Novalee Brin
Lyn 7h0'36" 58d45'40"
Nove
Aqr 22h32'16" -0d2'59"
Novelli ITFE05071958DC
Lyr 19h11'49" 28d47'23"
Novelly Nunez
And 0h11'17" 40d40'29"
November
Cap 21h36'41" -20d2'16"
November
Uma 8h35'49" 65d17'41"
November 22
Lib 15h26'50" -6d30'37"
November Hotel
Lyr 19h11'46" 27d15'4"
Novi
Tau 4h42'50" 14d36'12"
Novia Angela
Sgr 19h45'22" -28d7'34"
Noviah Jale
And 0h28'40" 36d19'22"
Now And Forever
And 2h16'16" 39d16'32"
Now and Forever
Cyg 20h2'50" 55d34'24"
Now and Forever GREYY
Uma 9h16'17" 46d29'17"
"Now and Forever" - Pat and Tonja
And 23h28'21" 43d15'44"
Now I know what love is
Leo 10h43'15" 9d32'33"
Now It Is For Eternity
Ari 2h16'13" 25d44'5"
Now & Later
Lac 22h52'49" 44d24'2"
Now Starring... Zoe Macchiusi
Cep 20h35'20" 65d11'4"
Now Then
Her 18h4'23" 36d58'37"
Now, YouAreAStar, Jon Harlan Bright
Her 17h24'25" 16d20'28"
Nowar Valerie
Uma 9h17'34" 58d30'55"
Nowlan-Haseley
Ori 5h36'50" 9d30'39"
Nox
Lyn 7h17'40" 52d20'46"
Noy
Eri 3h25'12" -4d39'56"
Noydena's Dawn
Vir 13h9'29" 4d24'4"
Nozzie
Leo 9h40'2" 7d14'14"
npcp8677
Lyn 8h57'1" 46d1'20"
Nrc&Krt Grace the skies 4 eternity
Col 6h26'3" -39d45'20"
NSA Martial Arts
Dra 17h49'48" 54d10'14"
NSC JRBD Eternity Star
Lyr 19h12'46" 45d56'33"
N'Shell Dianne Patton
Aqr 22h9'7" -9d7'32"
NSR143F - Panther Prowls
Umi 15h20'6" 68d48'18"
NTID Student Life Team 2005-06
Uma 8h26'43" 62d40'31"
Ntiense
Vir 13h51'30" 3d26'14"
N.T.Janowicz
Leo 9h28'36" 11d9'25"
NTOSH
Cap 21h24'6" -16d28'48"
N.T.P. F.H.L.
Cas 0h44'36" 48d11'9"
Ntronot
Her 16h38'20" 22d35'36"
NTTN
Col 5h31'38" -35d28'51"
Nuala Steele
Cas 23h59'40" 61d43'42"
Nubian King
Her 17h11'11" 46d52'4"
Nubian Queen
Cas 3h13'39" 59d16'25"
Nubis
Aql 19h38'2" 4d59'0"
Nuccio
Cyg 21h11'47" 47d6'54"
Nucharee Perez - My Morning Sun
Cap 20h43'52" -16d9'12"
Nuck
Per 4h46'11" 41d46'49"
Nucky
Umi 13h45'47" 78d43'59"
Nuestra Amor De La Mariposa
Leo 10h5'16" 15d56'59"
nuestra estrella
Vir 12h53'3" 3d31'16"
Nuestra Estrella
Cep 22h13'18" 64d54'17"
Nuestra Estrella
And 1h33'20" 47d44'37"
Nuestra Liaison
Pho 0h7'42" -41d36'24"
Nuestramor amandrew
Uma 10h35'28" 62d47'48"
Nuestro Espera Para Siempre
Per 4h27'4" 46d56'33"
Nuestro Primero Año
And 2h34'14" 49d47'37"
nufasa
And 2h26'47" 49d4'48"
Nug
Boo 14h10'41" 28d45'23"
Nugget
Ari 2h52'51" 25d47'6"
Nugget
Uma 8h39'41" 59d49'30"
Nugget & PeeWee Tauger
Umi 14h34'51" 68d51'55"
NugNhug's Little Twinkle
Cas 0h38'54" 61d44'38"
Nuintara
Uma 9h43'52" 45d59'10"
Nukkastar.
Cha 13h40'8" -79d33'56"
Numa
Tau 5h20'9" 21d22'49"
Numa Numa 12/14/92
Lyn 7h1'5" 47d12'43"
NUMAN AL-NIAIMI
Tau 5h35'27" 27d29'18"
Number 2
Dra 17h29'25" 62d13'31"
Number One Dad
Cyg 19h47'45" 36d45'23"
Number One Gramps 1939 - 2002
Per 3h27'40" 35d1'26"
Number One Grandad Ray
Cep 22h44'28" 65d2'45"
NUMC Advent Festival
Uma 10h19'29" 60d55'24"
numebop
Leo 11h9'55" 3d4'31"
Nummy
And 0h24'52" 37d47'57"
Numori "The Turtle"
Ori 5h59'18" 22d29'6"
Numpty Wood
Cyg 20h31'33" 45d14'30"
Nunc Scio Quit Sit Amor
Uma 9h22'32" 46d3'13"
Nungesser Family Star
Aql 19h18'9" -0d18'43"
Nunnie's Vega
Gem 7h15'42" 32d58'12"
Nunnumol
Ari 2h2'20" 22d16'50"
NUNO MARTINS
Tau 3h47'28" 8d59'33"
Nunquam Sola
Oph 17h11'34" -0d25'24"
NUNU
Cap 21h5'56" -16d3'55"
Nunu b
Cas 0h58'8" 63d43'36"
Nunuchboy
Cas 1h43'5" 69d1'13"
Nunzi & Dee "#1 Grandparents"
Uma 9h37'5" 67d11'51"
Nunziante
Cam 5h32'3" 71d23'48"
Nunziato pour toujours
Aur 6h31'27" 34d40'33"
Nunzio Cerniglia, Sr.
Sco 16h12'31" -12d42'0"
Nuovo IniZio
Cep 4h59'9" 82d23'59"
Nuppy
Boo 14h37'53" 42d54'13"
Nupur Dhamani
Tri 2h28'26" 31d21'38"
Nupur Kanodia
And 0h40'14" 27d32'23"
Nur Al Huda Dorsett
And 23h15'24" 41d3'21"
Nur Hajar Asuad
Vir 14h4'3" 2d52'15"
Nura
Lyn 6h35'4" 61d30'6"
Nuradar - Nursen + Adem + Arife
Uma 10h41'14" 48d38'58"
Nurds
Aur 5h10'57" 42d27'36"
Nuri Vall Rius
Aqr 22h15'35" -16d31'55"
Nuria
Her 18h28'53" 15d25'51"
Nuria Sanchez Colorado. Alias "Nawel"
Cam 6h3'38" 63d34'18"
nuria sara fernandez weid-mann
Umi 13h54'14" 69d27'49"
Nurka
Cas 0h16'14" 55d47'12"
Nurse Jonius Bethious
Aqr 22h3'23" -0d49'13"
Nurse Krista
Aqr 21h19'40" -10d2'31"
Nurse Margaret Mullen
Cas 1h12'30" 57d49'30"

Nurse Mary C Ryan
And 1h40'4" 38d46'3"
Nurse Nicky
Uma 9h5'52" 55d25'22"
Nurse Pamela
And 23h10'5" 42d42'42"
Nurse Sunshine
Leo 9h41'9" 32d1'20"
Nurse Tony Bell
Lyn 8h6'7" 57d24'4"
Nushi
Umi 17h6'11" 85d34'37"
Nusia
Sgr 19h46'57" -15d24'10"
Nußbaum, Dagmar
Sco 16h57'32" -38d36'58"
Nutan
Gem 7h4'57" 17d9'48"
nutmeg
Her 17h58'52" 21d27'47"
Nutmeg
Sco 17h53'51" -35d57'47"
Nutmeg Won Ton Tzu Young
Psc 0h21'3" 16d59'47"
Nutt de Muffinhead
Umi 16h52'39" 81d52'6"
Nutty Alice aka Joan Patterson
Uma 11h30'39" 53d34'45"
Nutty Amor
Ori 5h7'35" 13d54'40"
Nutty Secret Admirer
Lib 15h20'39" -4d15'45"
Nuv U - S & G - 1st November 2003
Cru 12h22'11" -57d53'36"
Nuzzel
And 0h16'15" 29d30'35"
N.Von.Elbing
Aur 6h22'24" 32d30'2"
NW Custom
Per 4h13'16" 51d23'19"
Nwakego Benedicts
Uma 11h4'34" 45d56'16"
Nwal Sara
Umi 14h29'49" 73d51'7"
Nwanneka Osuagwu
And 2h19'53" 49d22'16"
NWR - MMM
Cnc 8h39'12" 22d57'9"
Nya Athena Nichols' Nightstar
Cnc 8h49'47" 30d21'7"
Nya Bakker
Umi 16h41'21" 83d55'4"
Nyadia
And 1h20'13" 50d31'46"
Nyal Matthew Harris
Per 2h53'26" 46d30'10"
Nyasa Lovely
Psc 1h0'30" 17d56'20"
Nycholle Jarvis
Cnc 8h38'19" 18d22'21"
Nycole Bickel My everlasting light
Sgr 19h21'15" -30d4'10"
Nydacen Nilhiri
Dra 18h26'42" 52d3'53"
Nydelig Lise!
Tau 4h37'2" 12d0'21"
Nydia Aigner Tate
Cap 21h18'46" -16d25'21"
NYE2004
Cap 21h45'50" -12d40'13"
Nyella
Cru 12h12'22" -61d54'9"
Nyewilocha
And 1h1'53" 38d8'2"
NYF Corporation
Uma 13h49'19" 52d3'30"
Nyika Chitunda
Lyr 18h55'1" 33d27'34"
Nyist Gábor
Lib 15h20'39" -19d23'54"
Nykita
Ori 5h51'59" -0d0'27"
Nykki Star of Beauty
And 2h15'6" 46d59'37"
NYKKY D
Aqr 22h14'21" 1d1'7"
NYLA
Sco 17h23'17" -43d6'23"
Nyla Dinsmoor
Uma 12h30'52" 53d53'39"
Nyla Elle McNair
Cas 2h51'58" 58d3'25"
Nyla Joy's Chocolate Factory
Uma 13h36'0" 61d37'18"
Nyla Phillipp
Crb 16h6'39" 27d39'48"
Nyla Rese Freeman
Lyr 18h53'54" 37d12'31"
Nyla Samara Swinson
Aql 20h6'17" 15d12'9"
Nylse
And 23h21'34" 52d4'59"
Nyma Schneider
Peg 21h34'54" 21d39'56"
Nyman
Uma 8h47'21" 58d30'14"
Nymphéa
Del 20h46'35" 3d44'5"
nyncuk
Per 3h26'23" 47d16'10"
Nynke Zoutenbier's 2004 X-mas Star
Cas 1h4'50" 63d2'37"
Nyomi Cherrez
Lib 14h53'26" -15d23'4"
Nyouty
Cap 20h34'51" -19d5'10"
Nyrmal Daugherty
Cyg 19h45'22" 33d32'28"
Nyron Persaud
Aqr 21h39'40" 2d16'17"
Nysa Mojica
Cnc 8h55'50" 14d10'20"
Nysa Pallas Amara Erasmus
Vir 12h55'39" -0d36'16"
NYSP-RRKLLA
Cnc 8h31'16" 23d53'50"
Nyssa A. Lansford
Sco 17h27'0" -38d49'1"
Nyssa Corrine Ranney
Vir 13h40'18" 4d11'20"
Nyssa Elisabeth Parmenter
Cyg 19h38'58" 29d58'9"
Nyul csillaga örökké
Gem 7h45'13" 14d31'12"
Nyuszikám
Uma 9h37'25" 44d10'56"
Nyuszkó
Uma 8h43'38" 67d17'3"
Nyuszó
Cas 1h22'16" 53d52'41"
Nyx
Aqr 20h58'54" 0d57'9"
NZ_Pegasus2_Brightly Shining_in_USA
Peg 23h46'28" 14d44'34"
O
Oph 17h36'19" 9d29'57"
'O'
Pup 8h0'13" -34d52'29"
O' Be Joyful
Cas 0h8'0" 58d19'57"
O Bill O
Her 17h22'47" 38d5'51"
O & C Dietrich 28.04.2005
Uma 9h29'59" 65d41'9"
O Capone
Cep 21h31'40" 61d47'53"
O Doris 8221922
Leo 11h14'47" 7d30'53"
O. G. O.
Ori 5h55'36" -0d47'50"
o james musser/ bim
Sgr 18h56'32" -34d41'3"
O. Jane Johnston Smedley
Aqr 23h20'42" -19d10'23"
O L Burns
Dra 16h30'1" 60d51'4"
O. L. Tremoulet, Jr.
Aqr 21h14'55" -8d55'58"
O Leathlobhair
Leo 10h55'9" 19d16'44"
O Presente de Deus
Cnc 8h39'28" 10d55'45"
O Ryan Strangfeld
Tau 3h35'46" 7d34'1"
1 and only
Leo 11h38'50" 20d24'50"
1 Cab 1 Kiss 2getha As 1 Ally - Lenny
Cru 12h29'16" -58d25'59"
#1 Chuckie
Aur 5h34'37" 48d28'41"
1 Corinthians: 13
Ori 5h17'22" 5d47'52"
1 Corinthians 13:13
Uma 12h12'59" 63d9'57"
1 Corinthians 13:4-7
Cyg 20h41'11" 42d17'34"
#1 Dad & Grandpa, Carlton N. Morris
Sco 16h10'4" -11d34'2"
#1 DaDa & Lid To My Pot
Ori 5h53'33" 22d46'38"
#1 Daddy
Gem 7h20'25" 25d55'17"
#1 Daddy LJH IV my Magic Man
Her 16h48'13" 46d56'44"
#1 Gary Barbera #1
Cep 22h29'54" 61d57'50"
#1 Granny In The Whole Milky Way
Cas 0h36'49" 57d8'21"
1 in a Million
Lib 14h51'49" -6d5'54"
1 In A Million - Bubba Lou
Uma 10h0'26" 69d42'49"
1 LT Jason Sanchez
Per 4h11'16" 49d47'19"
1 Million Twix, KPL and CEB Forever
Ari 2h25'33" 25d40'3"
1 Molly 1... You're a Star!!
And 0h26'48" 34d37'29"
#1 MOM in the galaxy
Uma 11h59'7" 44d41'59"
#1 Mom In The World-Carol Kurzweil
Cas 0h40'39" 53d58'30"
#1 Mommy
Cap 20h34'45" -23d49'21"
10/16 Always and Forever
Cap 21h4'15" -20d20'25"
10:24
Sco 16h8'52" -13d18'35"
10 Jahre Astrid + Ernst
Aur 6h32'38" 35d37'20"
10 Kari and Melissa Forever 13
Uma 10h28'37" 65d18'48"
10-07-06
Uma 10h28'50" 66d46'51"
101102
Sgr 18h48'25" -16d11'21"
10-26-04
Uma 11h20'45" 42d3'15"
10282006 A Casey Mathieu
Lyn 6h25'20" 54d30'19"
105PM 32780 Linda Thien's Crossing
Ari 2h19'25" 24d17'55"
10C
Cam 3h59'41" 64d55'33"
10-My princess-07
Cnc 9h3'43" 30d54'10"
10th Anniversary for Dream Techno Co, Ltd.
Ari 2h24'27" 13d32'32"
11
Sco 16h16'39" -11d53'36"
11:11
Lyn 7h49'14" 41d29'3"
11 BLI Class of 2000
Per 2h56'54" 55d41'55"
11 Keep Going 2006
Dra 14h50'40" 55d24'44"
1-1-06
Uma 9h4'51" 54d53'13"
111
Per 2h26'10" 53d56'52"
11-24-1925 Edwards
Sgr 20h23'24" -38d46'47"
1125Catherine1981
Uma 10h44'45" 56d57'37"
1143 Forever
Eri 3h49'15" -11d31'18"
11-4-66, 2-16-90, 3-10-93, 1-1-65
Umi 14h7'40" 66d6'59"
11M25A19T99
Dra 17h16'29" 56d26'10"
12/20
Dra 19h49'13" 63d0'20"
12/5/73 Jason Z. Childers 2/14/05
Per 3h24'28" 48d21'59"
122 Pepper Lane
Uma 9h31'5" 64d31'22"
123!
Uma 8h44'55" 54d42'14"
1.2.3. Autumn Nicole
Lib 15h34'43" -10d8'34"
12968
Cep 23h51'27" 77d29'28"
12th Wedding Anniversary
Psc 0h24'45" 8d18'58"
12tracker30
Cap 20h44'44" -25d5'39"
13 Li-Paul
Uma 11h10'58" 67d33'24"
13051977 - Helen Sarah Gourley
Pho 0h35'22" -49d4'15"
131415 - Walid & Sheena - Our Heaven
Col 6h36'37" -34d44'36"
13th SPAR EXCC S.GIORGIO SU LEGNANO
Per 3h3'23" 40d50'37"
14 Rhonda
Lyn 7h31'18" 38d54'21"
143
Aqr 22h37'4" 0d43'31"
143 Amanda L Boetsch
Cam 5h8'47" 67d33'23"
143 Angie Maznyczenko
And 23h2'14" 48d21'9"
143 Banques "Asparagus"
Peg 21h27'46" 11d50'58"
143 Cindie
Gem 7h43'30" 23d30'34"
143 J.
Cap 20h56'1" -24d40'55"
143 Jeff Johnson
Her 16h21'55" 16d3'17"
1-4-3! Ken Chick
Boo 15h28'36" 44d50'25"
143 Lourdes M.
Lib 15h22'3" -14d42'6"
143 Teresa
Leo 10h38'31" 13d7'34"
143bradley443355233!
And 0h52'19" 39d27'27"
143CaTT
And 23h17'37" 47d23'44"
143-JM Brooks
Cas 0h35'48" 76d36'14"
143-K C Maggio
Ori 4h49'16" 2d56'21"
143RobinMae
Gem 6h54'19" 23d41'47"
143T4ny443v3r
Psc 1h19'19" 24d29'44"
14K
Uma 11h29'22" 59d7'43"
14Rémi Lhomme
Tau 3h44'35" 3d29'50"
14yb1010
And 23h4'2" 48d10'9"
164344 KJT&DSS
Uma 13h13'43" 60d51'19"
16w05k81
Uma 9h49'18" 48d3'29"
"17' Vietnam" Michael R. Mulcahy
Uma 11h29'38" 59d33'24"
176 + 184 = 360 "Together 4 Ever"
Uma 12h33'57" 55d1'22"
17th heaven
Ori 6h16'59" 14d47'26"
1824EVA
Leo 11h15'13" 15d4'16"
* 183 *
Cas 23h54'14" 58d27'9"
18KRAMER43DVS
Ori 5h44'41" 8d3'3"
1941 Rose Queen Sally Stanton
Cas 23h46'55" 54d10'46"
1961 - The Light of Discovery
Leo 10h26'3" 13d54'43"
1996 Wesley Whitman 10
Umi 14h30'2" 75d13'37"
19RBK17
Aqr 22h35'47" -15d20'47"
1-Der
Sgr 18h46'6" -16d44'48"
1-Princess Dena-23
Psc 22h51'27" -0d1'52"
1st Anniversary Star
Cyg 20h27'33" 32d41'57"
1st anniversary, many more to come
Tau 4h20'46" 22d51'13"
1st Lt. Jared Landaker, USMC
Per 3h12'30" 43d30'28"
1st Lt. Joshua M. Palmer
Per 2h14'20" 54d31'9"
1st Lt. Joshua M. Palmer USMC
Cyg 21h41'33" 41d15'46"
1U
Psc 1h17'45" 10d50'36"
1Vroom1
Uma 10h34'19" 46d7'44"
Oak
Uma 9h10'43" 55d7'6"
Oakenwolftalon
Lyn 7h18'59" 44d32'1"
Oaklark Anastasia
Vir 12h45'19" 6d48'33"
Oaklee20
Cnc 8h54'20" 18d55'44"
Oakley Arthur Aylen
Ori 5h22'7" -3d4'7"
Oakley Bear
Lyn 7h37'0" 48d30'45"
Oana Ban- "Banutz"
Her 17h48'11" 45d49'0"
O'Andrew
Aqr 21h37'8" 2d11'51"
OAO
Uma 10h59'23" 52d30'17"
Oath of Yacchin & Micchan
Vir 13h30'38" -6d12'38"
Oatis
Peg 22h3'26" 6d32'13"
Oatis Joel McCrea is Irie
Lyn 7h41'59" 53d57'24"
Oatmeal
Tau 4h17'40" 6d34'22"
Obake
Per 4h4'11" 43d35'8"
Oban Angus Scobbie
Aur 7h28'15" 41d9'10"
Obe
Sgr 19h26'38" -38d33'57"
Obe, Kristle, & Kaden Klopfenstein
Tri 1h51'3" 29d4'35"
Obelix
Ari 3h19'31" 29d56'39"
Oberdorf, Klaus Bernhard
Ori 6h8'28" -1d9'52"
Oberhauser, Martin
Ori 6h18'46" 13d35'45"
Oberheit, Hans
Ori 5h14'34" -8d4'14"
Oberheuser, Hans-Günter
Gem 6h26'35" 20d32'50"
Oberon
Cap 20h18'51" -9d33'5"
Obert
Cep 2h27'6" 84d12'12"
OBGYN BEN
Sgr 18h3'30" -25d34'21"
Obie 2001
Uma 11h1'34" 68d0'6"
Obie William Lawrence
Sgr 18h8'14" -26d5'54"
obie65
Ori 5h30'49" 14d5'2"
OBNZ *Norman*
Ori 6h7'25" 20d59'42"
OBO
Ori 5h27'30" 1d52'26"
Obonheura
Sco 17h47'12" -36d54'34"
Obradovic Alexandra I
Lac 22h50'0" 52d44'24"
O'Brien's Light
Gem 6h54'54" 32d25'10"
Obsidienne
Aqr 21h20'27" -7d42'1"
Obsolete Prime
Ori 5h35'37" 0d0'22"
Occhipinti's dvm racer
Aql 19h25'59" 15d20'49"
Occulus Prime
Per 3h27'45" 50d59'2"
Ocean
Uma 11h27'41" 32d11'29"
ocean
Ari 3h1'10" 22d1'3"
Ocean
Uma 8h28'16" 63d20'4"
Ocean Ahles
Tau 4h29'12" 10d17'16"
Ocean Eyes
Ori 6h1'8" 13d34'13"
Ocean National Bank
Uma 10h42'31" 56d1'30"
Ocean Nicole Clatterbuck
Lib 15h23'54" -7d22'36"
Ocean Prince
Cap 20h36'7" -13d9'10"
Oceana
Cnc 8h49'56" 13d14'3"
Oceana Alicia Lollick
Ari 2h15'24" 23d47'13"
Océane Alain Piché
Peg 22h17'17" 15d9'47"
Oceanna Sunnia
And 0h45'38" 36d27'33"
Océanna-Alizéa
Umi 16h14'19" 70d37'18"
Oceans Forever
Crb 16h23'21" 33d4'32"
Oceans of Love
Sgr 18h17'1" -30d51'5"
Ocho
Pup 8h8'29" -21d32'31"
Ochotzki, Eckhard
Uma 12h50'52" 60d15'50"
OCHSA Star Gazer
Uma 9h35'34" 58d59'24"
Ocie John Gregory 7-9-93
Cnc 8h40'0" 30d30'33"
OCKA
Cyg 20h38'15" 40d35'13"
Ockels' Star
Uma 9h25'48" 54d54'18"
Ocky
Dra 19h21'5" 71d35'46"
O'Connor
Uma 13h36'6" 48d33'33"
O'Connors Folly
Uma 9h29'47" 59d18'35"
O'Connor's Heart
Vir 12h24'10" 3d1'37"
O'connor's Love
Uma 11h23'46" 64d45'15"
O'Connor's Satellite
Leo 10h3'18" 26d43'55"
O'Connor's Shooting Star
Cyg 20h1'22" 54d31'0"
Ocratic (Malaysia) Sdn Bhd
Ori 5h22'43" -2d39'9"
Oct 18, A
Cas 0h52'44" 56d37'5"
(Oct 31, 1952) Ronda (Nov 12, 2004)
Uma 11h16'35" 35d16'42"
Octavia
Ori 4h51'48" 12d19'22"
OCTAVIA
Ari 2h38'7" 20d2'55"
Octavia Ann Jones 50
Mon 6h31'14" -3d1'57"
Octavia Anne Chartier Smith
Uma 10h31'7" 67d19'55"
Octavia Plachkov
And 0h53'46" 36d9'29"
Octavia Smith
Sco 17h27'53" -33d50'39"
Octavia Zayas
Uma 11h42'6" 64d47'6"
Octavio Ballesteros Navarro
Dra 19h47'33" 67d34'4"
Octavius Joseph Belvedere
Per 3h47'47" 33d21'14"
October
Pho 0h39'55" -40d46'47"
October 15th, 2004
Cap 20h28'56" -22d47'16"
October 8, 1971
Lyr 18h44'34" 30d45'50"
October Eyes
Tau 4h26'48" 23d52'20"
October Love - Todd and Erika
Cyg 20h29'10" 52d11'33"
October Nights
Gem 7h14'2" 18d32'8"
October Raye
Sco 17h50'10" -42d0'13"
Octobers Alyssa
Lib 15h26'9" -24d38'13"
Oculus promissio
Leo 11h40'48" 23d57'14"
Oda
Uma 9h33'33" 58d18'48"
Odalis
Ori 5h29'30" -7d37'21"
Odalis
Aql 19h36'56" 11d49'51"
Odalis Milagros Garcia
And 23h16'22" 47d37'52"
Odalys
Sco 16h58'30" -38d11'32"
O'DANE
Lyn 7h36'52" 39d40'45"
Odd Erik "the Skippy"
Gem 6h29'23" 24d39'59"
Odden's Orb
Lyn 7h53'51" 41d8'32"
Oddrey
Ori 6h13'32" 1d58'36"
Oddsocks
Uma 9h48'59" 62d33'28"
Ode to a wonderful mother
Crb 15h39'51" 27d44'26"
Odellanme
Umi 15h15'58" 86d22'0"
Odeny
Lyn 7h16'2" 53d48'57"
Oder, Folker
Uma 10h11'48" 62d11'46"
Odessa
Lyn 7h16'15" 58d8'49"
Odessa & Evelyne Fellows
Cyg 20h24'44" 46d19'26"
Odessa Fowler
Uma 9h43'44" 48d16'36"
Odessa Josephine Barrios
Gem 6h42'40" 23d4'14"
Odette
Cyg 20h11'53" 35d17'33"
Odette
Psc 1h22'29" 33d10'57"
Odette Kirby
Vir 13h23'50" -16d31'12"
Odette Michelle Akers
Cas 23h50'16" 52d52'40"
Odette Ozoa/Bickel
Lib 15h22'28" -15d46'2"
Odette Priestley
Sco 16h57'44" -42d21'52"
Odie
Uma 10h23'23" 60d22'26"
Odie and Shirley
Crb 15h21'35" 32d24'52"
Odile et Denis
Cnc 8h15'40" 9d2'41"
Odin
Ori 5h25'13" 3d55'36"
Odin
Sgr 18h14'20" -19d14'21"
Odin
Sco 16h13'57" -11d47'18"
Odin Arthur Eaton
Per 3h29'42" 43d45'40"
Odin Connor Mevissen
Cen 11h26'40" -64d12'31"
ODIN - NORSE, "GOD OF WAR"
Cnc 8h14'50" 29d1'35"
Odin12282006
Her 17h9'12" 34d31'32"
Odine Fiechter
Umi 14h3'34" 68d30'39"
Odins Valentine 230506
Uma 11h4'54" 33d21'31"
Odo Oliva
Uma 11h44'47" 64d13'55"
O'Donald "Shep" Ashford
Lib 15h23'57" -19d52'3"
"Odranoel" Leonardo Grade School
Ori 5h56'45" 3d39'52"
O'Driscoll's
Cyg 21h29'33" 35d54'38"
Odysseus
Lyr 18h46'59" 43d21'49"
Odyssey Hospice & HealthCare of TN
Uma 11h18'32" 46d22'33"
Oedipus
Gem 7h34'47" 20d12'41"
oelhafen
And 0h9'3" 41d50'4"
Oelschläger, Heinrich
Uma 9h48'56" 64d7'20"
Oelsner, Manfred
Uma 13h58'33" 55d49'0"
O'Erin
Ori 6h18'52" 9d26'57"
Oerni Maag
Uma 11h52'24" 64d3'25"
Of All The Stars In The Sky, MMT
Cyg 21h18'10" 47d5'15"
Of my friends, you shine brightest
Gem 7h46'5" 31d21'26"
O'Farrells' Wishing Star
Cyg 20h2'8" 33d43'30"
OFE
Aqr 21h10'11" -12d39'33"
Ofele, Sheyenne
Uma 11h57'14" 39d20'40"
Ofelia
Cyg 21h44'14" 38d34'51"
Ofelia
Gem 7h40'42" 14d30'31"
Ofelia
Ari 2h25'59" 19d0'55"
Ofelia Aldaz Alvaravo de Saenz
Ori 5h34'7" -0d53'48"
Ofelia Bringas
Uma 10h46'44" 66d20'35"
Ofelia Sanchez Gomora
Cnc 8h44'1" 16d26'43"
Ofelia Solis 9/5/56
Vir 14h18'42" -9d35'9"
Ofelia y Remigio Arencibia
Uma 8h25'29" 65d48'18"
Ofey
Sco 17h24'58" -30d17'26"
Officer Amanda "AJ" Jackson
Uma 10h25'28" 64d30'29"
Officer Gregory B. Matthews
Psc 0h34'58" 12d4'39"
Officer Krauss - Duty.Honor.Country
Uma 11h15'54" 56d47'4"
O'Galvin
Uma 10h19'24" 41d38'14"
Ogarrion
Lyn 8h2'38" 35d30'47"
Ogasawara
Her 17h40'16" 42d13'31"
Ogawa Yoshimi Daicel R21
Cap 20h27'40" -25d49'48"
OGAWBUF - Walter D. Grose 2006
Leo 10h23'39" 21d50'52"
Oggie Boogie / Teddy Bear
Vir 13h49'45" -3d57'24"
Ogg's star
Uma 12h33'31" 54d28'55"
Oghma
Ori 5h58'29" -0d45'16"
Ogni Notte
Del 20h51'19" 15d36'13"
Ogniana
Gem 6h45'22" 26d47'34"
OGNORA
Aql 18h59'39" -5d27'0"
ogre 2676
Lyn 9h16'55" 38d21'5"
Ogunquit Chamber of Commerce
Uma 10h25'41" 64d7'4"
Ogunquit Rotary Club
Aur 5h44'40" 47d39'36"
Oh Boy
Uma 12h41'6" 56d45'40"
Oh Dolly
Sgr 19h53'50" -12d50'26"
Oh For Cute
Cas 1h1'15" 49d27'9"
Oh_Kee_Im
Sgr 18h28'21" -15d58'6"
Oh my
Del 20h39'20" 5d43'18"
Oh My Papa
Ari 2h42'50" 15d57'48"
Oh Tammy How I Love You So!
Her 17h48'39" 42d10'32"
Oh tu...
Peg 22h49'23" 16d29'44"
Oh Vraiment
Uma 13h1'58" 56d53'33"
*OH, Ambur Marie
Vir 15h8'32" 5d21'25"
Oh, my love, my darling
Cyg 20h33'57" 34d7'55"
Oha George
Aql 19h1'52" 14d28'50"
Ohana
Lyr 18h21'14" 44d15'38"
Ohana
And 1h5'5" 45d39'9"
Ohanik
Cep 21h50'21" 61d5'22"
O'Hara Ann Fitzgerald
Vir 14h3'29" -9d53'44"
O'Hara: The Cluster of Angels
Leo 11h28'10" 22d39'39"
O'Hara-G2
Her 16h48'7" 35d11'19"
Ohbadiah [Joey Loves Laura 4-ever]
Cyg 20h59'58" 32d21'39"
Ohenzz
Gem 6h21'23" 21d21'4"
Ohhhhhh, that Dan Henning!!!!!
Cyg 19h43'8" 34d5'27"
Ohiboki Habibati Kawkaw
Ori 5h13'39" -0d31'42"
Ohioganda
Dra 16h36'17" 64d4'0"
Ohm, Jennifer
Gem 7h41'46" 15d56'41"
ohsostormy ilu4e
Psc 1h7'19" 29d3'37"
Ohstonha
Sco 16h13'11" -8d25'47"
Oi
Vir 13h11'40" 5d6'13"

OINKUS
Cam 5h30'31" 75d54'53"
Oisin and Malachi Duggan
Cru 12h33'49" -60d42'29"
Oisin Harris
Umi 17h28'19" 80d21'53"
O.J.
Uma 8h47'50" 69d39'57"
OJ MARSH
Dra 15h47'10" 57d3'45"
Ojanperaj1
Tau 3h38'50" 16d6'4"
Ojazos
Her 17h26'58" 30d53'46"
Ojig Yeretsian
Srp 17h58'47" -0d24'20"
OK Carol
Gem 7h3'14" 33d42'25"
OK Drymond
Cap 20h44'11" -27d13'33"
Okkyn
Cas 0h1'42" 57d36'5"
OKLB
Ari 2h38'34" 29d48'4"
Oksana
Uma 9h41'41" 52d21'27"
Oksana
Aur 5h15'31" 42d5'30"
Oksana
Uma 8h48'6" 54d12'39"
Oksana and Drew's Star
Sgr 19h38'57" -12d17'52"
Oksana Bludchy
Mon 7h15'56" -0d57'26"
Oksana Ferley
Uma 10h33'55" 47d10'14"
Oksana Khilkovich
Leo 11h27'42" 24d29'22"
Oksana Korsakova
Cyg 20h1'45" 39d56'55"
Oksana Kovneva
Gem 6h51'23" 25d33'26"
Oksana Kudelya
And 1h47'2" 50d22'2"
Oksana L. Radko
Cas 1h25'4" 61d53'36"
Oksana Miele
Cyg 21h8'47" 48d21'8"
Oksana Misyevych
And 0h1'42" 43d58'39"
Oksana Sonechko
Cas 23h13'18" 58d28'5"
Oksana Te Quiero
Cam 6h32'22" 67d20'22"
Oksana Wolfson
Sco 17h29'57" -41d47'5"
Oksana, My Love
Cyg 21h47'2" 47d37'51"
Oksana.A.M-Douglas.A.S-Max- <3
Gem 6h31'23" 21d4'6"
Oksanochka Lutsivka
Lyn 7h36'10" 54d18'58"
OkSasha
Leo 11h19'11" 11d29'37"
Oktavian Schäfer
Uma 10h30'0" 48d43'27"
okupski
Cap 20h23'40" -22d56'17"
Ol' Chappy
Cep 21h33'15" 61d44'41"
Ol' Smokey, Malcolm Heathcote
Lyn 6h39'14" 59d27'38"
Ol' Weedge
Uma 10h39'40" 40d11'14"
Ola
Uma 11h46'1" 36d47'23"
Ola Czerniawska
And 1h22'15" 40d38'47"
Ola Elrefai
Sgr 18h29'32" -17d0'20"
Oladapo A. Tomori
Psc 0h12'21" 7d57'57"
Olaf
Com 12h45'35" 27d43'44"
Olaf and Stacey Piesche
Cyg 20h17'13" 37d58'0"
Olaf Bayer
Ori 5h7'30" 12d26'50"
Olaf Friedrich Walter
Tau 5h33'35" 18d20'56"
Olaf Gabriel Bergk
Her 17h15'35" 19d5'57"
Olaf Klein
Uma 8h30'33" 59d54'42"
Olaf Leonard
Per 2h40'54" 53d41'38"
Olaf & Yuma
Ori 6h12'49" 15d40'42"
Olaf-n-Noreen
Vir 12h49'53" 10d56'52"
Olamide Anibaba
Lyn 8h7'19" 55d39'3"
O'Landtha
Her 17h18'42" 32d53'0"
Olanta
Aql 20h1'11" 4d49'9"
Olas
Her 16h45'24" 44d43'28"
Olaum
Vir 13h7'36" -2d24'49"
Olay Phengdara Murphy
Aur 5h31'38" 41d37'43"

Olcote
Cyg 20h56'1" 35d55'25"
Old
Lyn 7h27'34" 53d47'45"
Old Coot
Gem 7h24'19" 19d53'40"
Old Gary Weston
Sgr 17h52'4" -26d12'10"
Old Huck
Eri 3h42'2" -3d2'12"
Old Lady
Uma 11h50'40" 62d20'28"
Old Man
Ori 5h16'40" -5d29'10"
Old Man - Dean
Uma 9h32'56" 58d25'16"
Old Man On An Album Cover
Gem 7h16'58" 24d3'9"
Old Man Tim
Vir 12h32'4" -9d42'29"
Old Poppy
Gem 6h9'41" 22d43'14"
Oldenburger Kinderstern
Uma 11h17'7" 55d41'12"
Oldman5754
Uma 9h21'5" 57d19'36"
Oldrich
Umi 14h33'48" 79d58'6"
Ole Alvar
Uma 9h13'46" 62d46'23"
Ole Bob Pearce
Uma 12h2'19" 48d59'6"
Ole Boyer Hubbard
Leo 10h27'26" 25d36'36"
Ole Kapplegard
Uma 8h17'27" 71d58'6"
Ole Man & The Sky
Leo 9h58'50" 13d34'49"
Ole Man's Mate
Leo 9h59'17" 11d30'24"
Ole Olesen
Vir 11h59'46" -7d12'29"
Ole Yetter
Per 2h19'55" 56d41'1"
Oleander Beach Apts.
Cyg 19h22'45" 49d21'44"
Olechka
And 0h26'17" 42d39'21"
Olechka, My Baby
Sgr 18h11'45" -34d15'26"
Oleg Igorevich Baybakov
Aqr 22h13'7" -14d9'52"
Oleg Sorkin
Uma 11h4'16" 47d12'44"
Oleg - Tanya Solouk
Psc 1h18'20" 19d42'36"
Oleg & Tatiana Boiadji
Uma 11h22'38" 59d18'55"
Olen Ronald Kline
Ori 5h33'15" 9d56'19"
Olenka
Aqr 22h18'27" -12d52'47"
OleoleosUnstuck
Cyg 21h44'32" 36d37'23"
Olesya
Psc 23h57'8" 2d20'30"
Olesya
Vir 12h13'36" 11d39'34"
Olesya And Eric
Uma 10h29'15" 65d54'11"
Olesya Ashley McAllen
Peg 21h40'30" 23d44'44"
Olesya (Kitten) Mysiv
Vir 13h15'32" 13d12'26"
Olfa - mille bisous - ya hobi
Uma 11h31'52" 53d20'41"
Olga
Her 17h56'23" 23d48'45"
Olga
Ari 2h20'23" 26d28'26"
Olga
Psc 1h24'21" 31d19'46"
Olga
Cam 4h20'51" 57d42'13"
Olga
Crb 15h32'36" 37d54'11"
Olga
Cyg 21h25'27" 39d9'26"
Olga
Cas 0h21'34" 50d32'22"
Olga 82
Cnc 8h33'36" 19d19'13"
Olga A. Raschi
Aur 5h14'28" 35d16'15"
Olga Almada
Crb 15h33'57" 37d44'19"
Olga and Jane
Del 20h25'14" 9d54'53"
Olga Annastasha
Sco 17h24'1" -44d21'59"
Olga - B
Lib 15h41'39" -11d31'20"
Olga Boscarello
Aqr 21h23'44" -13d21'55"
Olga Chichkina
Leo 9h50'18" 28d3'12"
Olga Christoffersen
Cra 18h46'59" -39d15'30"
Olga Cunningham
Lyr 19h5'56" 44d17'18"
Olga - DCJ
Lyr 18h43'3" 45d32'58"
Olga Descartes Tosado
Lyn 8h8'58" 46d27'48"

Olga "Dolly" Vannelli
Cap 21h15'50" -15d21'36"
Olga Drozdova
Leo 9h31'56" 28d10'46"
Olga E. Delgado
Leo 11h51'6" 21d0'43"
Olga & Elie
Aur 5h54'28" 29d20'48"
Olga Esther Vanacore
Cas 0h59'10" 57d40'35"
Olga Garbus
And 0h53'17" 39d31'24"
Olga Ippolito
Cas 1h26'6" 52d13'3"
Olga Irizarry
Del 20h42'42" 3d57'34"
Olga Ivanovna Kravtsova
Uma 9h58'10" 66d39'18"
Olga J. Wesolowski
Leo 10h46'55" 9d20'20"
Olga Jakonowska
Vir 13h16'51" 5d40'8"
Olga Jansky
Cas 0h38'3" 64d47'53"
Olga Kazak - 23
Cas 23h56'16" 52d39'2"
Olga Kelskey
Sco 17h52'16" -35d42'7"
Olga Kornienko
Cap 20h26'37" -9d43'11"
Olga Kuptsova
Cas 0h6'23" 54d23'3"
Olga Lucia Zuluaga
Leo 10h3'24" 7d10'38"
Olga Manjarres
Lyn 8h29'59" 56d29'36"
Olga Mantler
Uma 10h17'4" 48d9'7"
Olga Margarita Velez Santana
Cas 1h24'7" 57d24'46"
Olga Maricela Gonzalez
Cap 21h17'45" -18d31'18"
Olga Molnar
Uma 11h24'12" 58d6'50"
Olga N. Crisio
Uma 11h9'5" 55d42'6"
Olga Nova
Tau 4h28'42" 16d58'6"
Olga Ortner
Gem 7h23'24" 16d33'52"
(Olga P. Dimas) Wife, Mom, and Wela
Psc 1h40'30" 5d5'57"
Olga Perez
Uma 9h1'58" 64d53'3"
Olga Selektor
Psc 0h27'25" 8d22'0"
Olga Sevilla Alvarez
Vir 13h18'3" 5d37'59"
Olga Shines 4 Us
Cas 1h28'25" 62d52'41"
Olga Simoni
Lib 15h28'11" -7d5'17"
Olga Slavich
Cas 23h51'42" 53d23'56"
Olga Smirnova
Ari 2h13'57" 26d22'34"
Olga Smuschkina
Uma 14h4'10" 59d50'38"
Olga Sunitsky Star
Leo 10h55'44" 13d2'51"
Olga Tyan a.k.a. Malishka
Tau 3h42'47" 27d34'35"
Olga Vakulyuk
Ori 5h39'55" -3d50'50"
Olga Vasiliev
Tau 4h17'17" 19d40'25"
Olga Venessa Morgan
Psc 1h20'2" 31d38'15"
Olga W. T.
Cas 0h33'18" 53d46'23"
Olga Wojciechowska
Sco 16h8'46" -12d48'16"
Olga Yakhniv
Ari 3h9'20" 28d21'25"
Olga Yasko
Uma 13h5'48" 57d58'24"
Olga Yrievna Konovalova
Psc 1h14'57" 15d48'39"
Olga Zhurbin
Sgr 18h12'33" -19d19'17"
OlgaGrib-81
Cnc 8h35'52" 10d34'4"
Olgalidia Lopez Dearborn
Lib 15h13'38" -11d42'57"
Olgica
Umi 13h51'30" 71d51'54"
Olguita
And 23h17'31" 42d3'0"
Olguita
Lmi 10h17'25" 30d24'34"
Olha Figol
Lib 15h50'53" -14d8'49"
Oli and Danielle
Uma 11h17'29" 36d43'28"
Oli & Denise
Umi 15h23'7" 69d51'38"
Oli Oli Oxen Free Greatest Husky
Cma 6h44'47" -20d53'18"
Olia
Cap 20h22'18" -19d2'33"
Olichka
Ori 6h11'6" 20d45'8"

Olie's Star
Tau 5h43'42" 24d18'59"
Olimpia
Cnc 9h10'14" 8d54'51"
Olimpia
Uma 9h34'10" 42d41'37"
Olin Bennie Davis
Cnc 8h51'53" 12d18'54"
Olina Pandani
Eri 1h50'11" -54d25'6"
O'Linda
Cas 2h22'1" 68d51'27"
O'linde Gutohrlein
Cyg 20h18'11" 43d39'25"
Olinde-Olinda
Uma 9h37'41" 48d46'40"
OLINE
Tau 5h1'3" 23d16'43"
Oli's Prince-Star
Umi 15h38'20" 79d46'35"
Oli's Star
Psc 0h47'36" 4d8'52"
Oli's Wish
Pyx 8h35'46" -26d2'15"
Oliva
Uma 11h9'44" 43d44'6"
Olivaud
Psc 23h9'58" -2d4'56"
Olive and Bob's Diamond Star
Uma 11h42'4" 41d27'2"
Olive Beatrice Pern
And 1h0'40" 42d41'19"
Olive Blanche Lake
Ori 5h53'40" 21d45'38"
Olive Blystone
Uma 11h47'34" 44d22'45"
Olive Edith Gwendoline Painter - 'Nan'
Uma 9h15'51" 59d3'54"
Olive Emil
Tri 2h15'14" 32d55'14"
Olive et Math
Gem 6h52'3" 29d10'34"
"Olive Ewe" (Baaa-aaa)..
Cen 14h15'6" -33d41'7"
Olive Hogan
Cas 3h14'25" 61d17'5"
Olive Jae Lara
Psc 1h11'20" 22d10'27"
Olive Joan Rasmussen Wynkoop
And 0h46'52" 37d55'32"
Olive Juice
Per 3h14'7" 42d36'27"
Olive Juice
Her 16h45'48" 49d2'56"
Olive Juice
Del 20h35'24" 4d14'9"
Olive Juice
Ori 4h50'8" 12d34'34"
Olive juice Star
Cyg 21h57'33" 47d14'47"
Olive Kopaunik
Gem 7h25'1" 29d33'21"
Olive Lee Benson
Vir 12h13'53" -9d58'50"
Olive Marie Ferreira Duffy
Gem 7h38'23" 21d3'39"
Olive Morris Davies, M.D.
Umi 14h34'20" 75d26'9"
Olive Nagle
Psc 23h54'39" -3d11'47"
Olive Oil
Cnc 8h43'40" 24d9'22"
Olive Virginia
Cap 21h32'35" -9d10'33"
OliveJuice
Uma 12h5'45" 61d41'5"
Oliveoyl
Uma 8h28'22" 69d17'30"
Oliver
Uma 8h23'55" 69d3'37"
Oliver
Cas 23h20'56" 58d22'42"
Oliver
Cma 7h7'55" -22d23'26"
Oliver
Cma 6h40'49" -12d55'0"
Oliver
Ori 6h20'21" -1d36'46"
Oliver
Aqr 23h26'55" -17d39'9"
Oliver
Cnc 8h58'22" 16d13'0"
Oliver
Aql 19h53'3" 12d1'29"
Oliver
Her 18h44'29" 14d24'59"
Oliver
Lyr 19h26'48" 37d40'1"
Oliver
Per 3h11'55" 44d1'18"
Oliver
Gem 7h31'3" 34d41'43"
Oliver Alain Corroy
Pup 7h30'23" -24d35'43"
Oliver Alan Wood
Uma 12h18'22" 57d36'52"
Oliver and Rebecca - SEIKAN
Cyg 21h56'1" 52d4'28"
Oliver Andrew Brown
Cep 0h19'18" 69d26'20"

Oliver Andrew Hammond
Umi 14h18'36" 84d3'47"
Oliver Bailey Dicker
Ori 6h14'59" 7d14'5"
Oliver Ben Seaton - Oliver's Star
Lmi 10h13'12" 41d9'52"
Oliver Benjamin Jones Cattley
Cyg 20h3'16" 58d57'17"
Oliver Benjamin Wattley
Umi 15h47'36" 80d54'22"
Oliver Catalina Yorke
Ori 5h41'14" 12d49'47"
Oliver Charles French - Oliver's Star
Her 18h41'10" 16d41'5"
Oliver Christopher Dennis
Dra 14h48'13" 56d17'41"
Oliver Claridge Finks
Per 4h20'18" 48d6'29"
Oliver Colin Long
And 23h52'30" 48d26'59"
Oliver Collyer-Pratt
Uma 10h21'29" 44d12'57"
Oliver D Justice
Vel 9h37'28" -41d58'57"
Oliver David
Umi 17h21'7" 77d40'21"
Oliver David Chapman
Uma 10h39'52" 47d51'2"
Oliver David Howes
Per 4h44'55" 42d51'50"
Oliver David Howes - Ollie's Star
Per 4h9'23" 33d20'29"
Oliver David Richards
Her 16h22'15" 23d37'36"
Oliver Deckwarth 02.09.1970
Uma 9h21'23" 68d21'38"
Oliver Douglas' Wishing Star
Her 15h58'50" 44d13'55"
Oliver Dusterhoff
Ori 5h12'1" 2d52'4"
Oliver Fischer
Her 18h3'16" 16d59'25"
Oliver Flamann
Uma 10h53'5" 44d41'4"
Oliver Gabriel West
Ori 6h15'30" 14d39'43"
Oliver Gessmann
Uma 8h59'2" 63d49'27"
Oliver Hannant's Twinkling Star
Ori 5h59'3" 18d41'0"
Oliver Hauser
Aur 5h40'53" 33d44'33"
Oliver Henry Cureton
Uma 9h28'39" 66d56'8"
Oliver Herbert
Her 16h32'30" 11d27'27"
Oliver Jack Washington
Her 17h41'7" 37d47'56"
Oliver James Booth
Her 18h34'2" 15d37'4"
Oliver James Braid
Ori 6h25'16" 17d24'36"
Oliver James Comery
Her 18h11'48" 23d24'50"
Oliver James Grant
Cep 22h43'16" 86d54'26"
Oliver James Greenaway
Her 17h11'17" 14d51'25"
Oliver James Thomas Leach
Per 2h55'53" 46d2'2"
Oliver John Dodman
Per 4h46'43" 50d17'44"
Oliver John Edward Chalfin
Tau 4h17'7" 23d6'55"
Oliver Jonathan Watts
Ori 5h17'36" -9d11'52"
Oliver Jones
Per 2h59'21" 44d51'51"
Oliver Joseph Peel
Tau 3h45'7" 25d1'54"
Oliver Jumeau Gill
Per 2h32'1" 52d2'45"
Oliver Knoll Clendenning
Umi 15h25'56" 71d40'17"
Oliver Kosack
Ori 5h11'33" -6d54'58"
Oliver Kühne
Dra 16h47'46" 55d9'5"
Oliver Landes
Aql 19h15'21" 1d8'25"
Oliver Lev Brenner
Leo 10h4'8" 23d4'25"
Oliver Liell's Star
Lib 15h41'55" -10d15'34"
Oliver Liley
Peg 21h56'26" 18d15'52"
Oliver Luke Mathews - Oliver's Star
Uma 11h25'28" 48d4'48"
Oliver Ma-at
Cep 3h36'17" 81d3'43"
Oliver Martin Bush
Cyg 21h42'48" 44d14'57"
Oliver Matthew Ballentine
Uma 10h22'39" 52d39'52"
Oliver Max Raskin
Gem 6h39'51" 16d16'20"

Oliver McMahon
Peg 23h49'30" 11d33'44"
Oliver Mendonca
Cma 6h32'58" -20d19'9"
Oliver & Mia
Leo 10h12'11" 23d12'12"
Oliver Michael Keith Irving
Dra 18h43'42" 51d44'38"
Oliver Michael Rush
Per 3h1'31" 47d4'30"
Oliver Nils Fisch
And 2h23'3" 41d30'55"
Oliver O Buck Jr (pappy)
Uma 11h37'2" 64d4'35"
Oliver Owen Ault
Per 2h45'22" 56d57'42"
Oliver "Papa" Bright
Ori 5h44'11" 1d14'14"
Oliver Parish
Uma 11h24'23" 45d1'16"
Oliver Patrick Müller
Com 13h31'24" 22d20'9"
Oliver Paul Hart
Uma 9h29'5" 66d48'38"
Oliver Paul O'Hara
Umi 15h22'11" 70d40'9"
Oliver Pearson
Dra 17h49'22" 57d43'33"
Oliver Pickens
Cma 6h50'8" -16d54'10"
Oliver Reeve Munro Larkin
Leo 11h39'58" 14d55'52"
Oliver Richard Leigh
Cep 21h53'28" 65d8'6"
Oliver & Riza
Cyg 20h50'53" 48d9'34"
Oliver Robert Suckling
Umi 14h44'36" 69d54'45"
Oliver Robert Willsher
Umi 16h56'22" 80d45'17"
Oliver S. Tierney
Lib 15h22'54" -14d55'59"
Oliver Samual Powell - Oliver's Star
Cyg 20h4'56" 39d0'43"
Oliver Samuel Gantz Hartle
Boo 14h46'59" 26d26'53"
Oliver Schnider: Ich Liebe Dich über alles! Deine
And 23h37'13" 48d4'15"
Oliver Schweim
Ori 5h52'47" 19d28'39"
Oliver Simon Ludlow
Per 2h55'24" 40d18'28"
Oliver Stearn
Cnv 13h33'42" 50d38'5"
Oliver Sunderland
Aql 18h50'46" -0d44'6"
Oliver Thomas (Councilman at Large)
Aqr 23h6'54" -8d28'46"
Oliver Visconti
Uma 10h41'37" 58d41'16"
Oliver von Waldenfels, 17.01.1986
Cas 0h5'52" 52d5'6"
Oliver Wall-Panus
Uma 10h35'50" 39d56'41"
Oliver Walter
Uma 9h3'20" 59d41'17"
Oliver William Bryant
Peg 22h18'26" 17d18'18"
Oliver Wohlgethan
Ori 5h28'12" 9d21'55"
Oliver Zimmermann
Ori 6h3'2" 14d7'17"
Olivera Kovacevic
Her 16h48'19" 11d25'55"
Olivera Vujisic
Cyg 19h52'21" 31d36'38"
Oliver's Magic Star
Mon 6h27'37" 8d15'18"
Oliver's Star
Ari 3h16'42" 24d14'17"
Oliver's Star
Uma 8h31'18" 72d14'29"
Oliver's Star
Ori 5h55'38" -0d0'8"
Olive's Love
Vir 13h0'21" -20d48'49"
Olivette E. Desmarais
Lib 15h4'32" -19d14'36"
Olivia
Sgr 19h23'15" -20d35'55"
Olivia
Mon 6h32'50" -4d57'7"
Olivia
Eri 4h51'21" -14d3'13"
Olivia
Aqr 21h38'16" -5d51'10"
Olivia
Umi 15h20'13" 70d25'47"
Olivia
Uma 12h47'14" 52d37'30"
Olivia
Sco 16h59'19" -44d28'35"
Olivia
Sco 17h38'27" -43d22'50"
Olivia
Sgr 18h50'41" -22d41'58"
Olivia
Sco 17h56'29" -31d32'16"
Olivia
Umi 16h19'31" 85d6'43"

Olivia
Ari 2h40'51" 24d40'55"
Olivia
And 0h20'53" 26d15'25"
Olivia
Gem 7h5'23" 20d14'4"
Olivia
Tau 5h42'18" 23d27'44"
Olivia
Mon 6h23'58" 8d13'9"
Olivia
Cnc 8h22'58" 9d32'24"
Olivia
Ori 6h10'4" 6d35'16"
Olivia
Cyg 19h55'19" 31d21'54"
Olivia
Crb 15h36'39" 32d31'24"
Olivia
Uma 9h23'9" 43d1'13"
Olivia
And 23h36'49" 36d38'22"
Olivia
Cyg 20h50'36" 32d46'28"
Olivia
And 23h14'29" 39d32'40"
Olivia
Crb 15h53'0" 37d37'16"
Olivia
Cyg 21h50'9" 50d27'8"
Olivia 3 1 20 8 25
Umi 11h6'1" 88d58'26"
Olivia A. Wasson
Ari 2h6'44" 24d10'20"
Olivia Abigail Buddy
Cap 20h9'37" -17d35'59"
Olivia Adele Carelli
Mon 6h46'29" 7d50'28"
Olivia Alacazar - Rising Star
Lyn 8h30'9" 35d54'12"
Olivia Alexandra Miller
Sgr 17h48'55" -23d48'43"
Olivia Alice Gilbert
And 23h52'3" 40d59'43"
Olivia Alice Wall
Cas 23h28'42" 52d20'1"
Olivia Amy Collett
And 0h44'32" 26d40'0"
Olivia and Lance
Lyn 7h42'49" 51d58'10"
Olivia Anita Elekes aka Oli's Star
Gem 6h39'35" 20d56'40"
Olivia Ann
And 23h36'25" 42d19'2"
Olivia Ann
Cap 21h17'14" -16d31'29"
Olivia Ann
Lib 16h0'13" -8d29'33"
Olivia Ann Golden
Uma 9h55'47" 59d22'47"
Olivia Ann Hamer
Leo 10h23'20" 24d49'10"
Olivia Ann Lenore Sinclair
Peg 22h11'54" 31d4'46"
Olivia Ann Pignatello
Lib 14h34'30" -12d17'59"
Olivia Anne
And 2h21'45" 38d46'11"
Olivia Anne Howie
And 23h58'24" 42d0'37"
Olivia Anne Hughes
Aqr 22h9'36" 1d17'37"
Olivia Anne Kahler
Cyg 19h43'59" 37d0'12"
Olivia Anne Kurman
Sgr 20h23'29" -42d20'24"
Olivia Armstrong
And 23h30'54" 37d39'11"
Olivia Assunta
And 23h47'16" 47d50'11"
Olivia B
Cru 12h14'2" -62d17'32"
Olivia Bailey
And 0h26'56" 42d54'2"
Olivia Barrett Ruffin
Cyg 20h22'11" 52d19'27"
Olivia Benge
And 0h28'2" 46d23'1"
Olivia Beth Tomares
Aqr 22h37'35" -11d12'49"
Olivia Bezmalinovic
Cas 1h30'56" 63d23'6"
Olivia Blair Victoria Ritter
Aqr 22h37'18" -16d54'2"
Olivia Blanco Mullins
Cnc 8h0'8" 17d48'10"
Olivia Bratich
Cyg 19h33'37" 47d15'43"
Olivia Brookes Haller
Boo 14h26'48" 48d21'27"
Olivia Browne
And 2h33'0" 49d10'49"
Olivia Caelestis
Sgr 17h52'4" -29d46'55"
Olivia Caitlin Rowley
Cra 19h9'42" -42d4'8"
Olivia Cate Peterson
Vir 14h14'34" -12d51'14"
Olivia Chloe Chong
Cap 21h33'11" -15d24'29"
Olivia Christine
Cam 4h53'22" 69d39'50"

Olivia Christine Curtis
Sco 16h6'58" -13d32'28"
Olivia Christine Roth Christy
And 23h23'43" 52d58'34"
Olivia Cooper
And 2h35'40" 45d39'24"
Olivia Corrinne Wiebe
Gem 7h5'13" 14d55'17"
Olivia Cross (Wilson)
And 1h11'30" 34d41'5"
Olivia Danielle
And 0h8'59" 29d7'30"
Olivia Dawn
Cas 0h44'0" 63d57'10"
Olivia Dawn Poe
Cnc 7h57'3" 15d19'43"
Olivia De Oliveira
Ori 6h20'33" 14d30'27"
Olivia Delta Rich
Cas 23h30'23" 52d53'35"
Olivia Devan
And 23h28'52" 41d54'29"
Olivia DiNapoli
Leo 9h56'11" 7d53'35"
Olivia E Jenkins, aka Livi
Lyn 8h59'42" 42d2'55"
Olivia Eithne Josephine Gately
Per 2h59'59" 32d46'36"
Olivia Elizabeth
Cam 3h59'33" 69d32'56"
Olivia Elyn Townsend
Dra 15h7'42" 60d1'11"
Olivia Eva Coleman
And 1h44'13" 41d15'34"
Olivia Eve O'Reilly
Lmi 10h28'15" 34d29'49"
Olivia Evelina Dutzer
Aql 19h58'45" 15d30'43"
Olivia Eveline Gosch
Tau 4h26'21" 5d54'11"
Olivia Faye Crowley
And 0h57'6" 40d35'48"
Olivia Florence Calamita
And 2h21'12" 50d11'7"
Olivia Frances Fox
Umi 14h44'13" 72d42'35"
Olivia Frances Lubbock
And 1h9'20" 45d31'33"
Olivia Frances Parry
And 0h21'44" 30d14'56"
Olivia Francesca Chitamun
Cap 21h53'14" -13d18'44"
Olivia Frank
And 2h25'33" 45d48'50"
Olivia & Gabriel's Star
Gem 6h38'35" 17d48'15"
Olivia Gail
And 0h22'56" 41d52'24"
Olivia Garcia ~ Our Loving Mother
Cas 0h34'45" 48d9'5"
Olivia Giancarla Hailè
Lyn 8h54'19" 33d20'25"
Olivia Grace
Lib 15h3'10" -28d15'4"
Olivia Grace
Cra 18h0'0" -37d5'29"
Olivia Grace
Col 6h34'27" -35d18'2"
Olivia Grace
Cas 0h11'54" 51d11'48"
Olivia Grace
And 1h18'10" 33d58'19"
Olivia Grace
Tau 5h57'16" 25d6'46"
Olivia Grace
Peg 22h19'41" 14d25'16"
Olivia Grace - 09 April 2005
Cru 12h47'18" -57d38'19"
Olivia Grace Abbott
Aqr 21h13'23" -14d13'45"
Olivia Grace Beckley
Sco 17h43'46" -33d20'23"
Olivia Grace Blake-Pharr
And 23h5'45" 48d45'9"
Olivia Grace D'Andrea
Cam 4h4'56" 68d36'46"
Olivia Grace Dedert
Peg 21h52'58" 17d10'4"
Olivia Grace Gay
And 0h55'42" 38d59'38"
Olivia Grace Hass
And 0h6'54" 43d51'33"
Olivia Grace Leon's Star
And 1h37'23" 43d45'53"
Olivia Grace Lowry
And 0h26'27" 42d18'14"
Olivia Grace Mitchell
Ori 6h5'47" 18d37'24"
Olivia Grace Mitchell
Tau 5h54'36" 25d6'47"
Olivia Grace Namovicz
And 1h18'2" 49d28'19"
Olivia Grace Obert
Cnc 8h49'20" 28d2'54"
Olivia Grace Phillips 'Twinkle'
And 23h37'0" 34d24'53"
Olivia Grace Queenan
Lib 15h13'18" -22d41'16"
Olivia Grace Rolph-Dickinson
And 2h1'47" 46d15'42"
Olivia Grace Simpson-Braggs
Cam 7h22'35" 71d35'55"
Olivia Grace Taylor
Sgr 18h14'44" -35d6'35"
Olivia Grace Troth - Olivia's Star
And 1h20'47" 37d54'12"
Olivia Grace Valentine
Lib 15h11'33" -27d38'27"
Olivia Grace Whittemore Olson
Col 6h3'49" -29d38'40"
Olivia Grayce Skufakiss
Cas 0h47'34" 64d0'49"
Olivia Green
Ori 5h53'52" -0d55'46"
Olivia Hackbarth
Del 20h30'9" 17d33'40"
Olivia Hamer
And 23h23'11" 43d4'10"
Olivia Heter
Uma 9h32'52" 67d3'41"
Olivia Honor Franzese
Sco 16h6'9" -20d16'33"
OLIVIA HOPE GREER
Cyg 20h48'5" 30d32'6"
Olivia Horn Martin
And 1h2'50" 40d14'9"
OLIVIA HUBER
And 0h15'7" 28d1'58"
Olivia Hutter
Dra 20h28'55" 67d36'56"
Olivia Ione
Lyn 7h45'34" 52d54'40"
Olivia Ione Larson
Uma 11h39'32" 46d38'40"
Olivia Irene Demart
Dra 18h24'34" 53d43'14"
Olivia Irene Elizabeth
Vir 14h14'23" 3d9'17"
Olivia Isabelle
And 0h16'47" 35d23'40"
Olivia J. H. Hartmann
Sco 16h8'21" -12d17'3"
Olivia J. Halikas
And 1h3'8" 34d29'0"
Olivia J. Swerling
Crb 15h34'31" 28d8'48"
Olivia Jaclyn Gioia
Psc 1h20'1" 31d18'58"
Olivia Jane Morton-Smith
And 2h7'7" 42d13'59"
Olivia Jane Terry
Ori 5h6'0" 3d40'54"
Olivia Jane Viale
Leo 9h25'24" 22d49'3"
Olivia Jaye Fuller
And 22h58'26" 50d14'4"
Olivia Jean Hallihan
Lyn 6h59'31" 53d57'28"
Olivia Jean Hopek
And 23h16'22" 41d44'59"
Olivia Jean Whittaker
Lyr 18h15'39" 35d1'11"
Olivia Jeffery Miller
Tau 5h47'32" 22d1'38"
Olivia Jewel
Tau 4h6'2" 23d9'33"
Olivia Joan Brice
Psc 1h43'24" 9d49'8"
Olivia Jolie Kearney
Cru 12h30'19" -60d42'29"
Olivia Josefina Fletcher - 8/22/86
And 0h47'29" 25d5'30"
Olivia Jotautas
Uma 10h43'24" 60d50'35"
Olivia Kate
And 23h15'2" 47d53'54"
Olivia Kate Blakey
Peg 22h55'21" 31d1'20"
Olivia Kate Molko
Vir 14h39'36" -4d41'21"
Olivia Kate Toner
Gem 6h47'14" 19d40'46"
Olivia Katherine Balliet
Gem 7h47'20" 23d27'32"
Olivia Katherine Lewis
Sgr 19h44'23" -22d8'26"
Olivia Katheryn Hoffman
Sgr 18h13'7" -32d11'56"
Olivia Kathleen Umlauf
Pho 0h37'58" -39d52'34"
Olivia Kay Smosarski
Uma 14h6'3" 52d20'40"
Olivia Kay Welch
And 0h39'25" 30d23'57"
Olivia Kennedy Tobin
Gem 6h43'5" 31d14'25"
Olivia Kim Loy
Cnc 8h43'5" 29d13'47"
Olivia Kunkle
Uma 10h38'6" 60d23'24"
Olivia L.
And 0h13'24" 24d51'6"
Olivia L. Ardoin
Sgr 18h19'12" -22d55'0"
Olivia & Lars at Grand Canyon
Uma 8h51'15" 47d28'45"
Olivia Layne
Uma 10h37'48" 68d17'47"
Olivia Lea Crystal
And 0h33'24" 34d14'54"
Olivia Leah Coleman Goedecke
And 23h8'27" 40d16'35"
Olivia Lee Cecka
Cas 1h40'6" 65d45'29"
Olivia Lee Olson
Cnc 8h37'16" 30d38'10"
Olivia Leigh van der Vlugt
Peg 21h37'45" 23d38'40"
Olivia Leighann
Sco 15h56'20" -21d19'4"
Olivia Linda Lattanzi
Uma 12h27'47" 54d45'28"
Olivia Linda Nicole Hernandez
And 1h19'3" 35d59'12"
Olivia Lloyd Johnson - Livvy's Star
And 23h6'12" 38d4'18"
Olivia Louise Evans
And 23h40'4" 47d13'46"
Olivia Lowenstein
And 0h34'59" 39d54'12"
Olivia Luz
Sgr 19h31'14" -38d1'56"
olivia lynn 9/29/04
Lib 15h2'1" -13d58'8"
Olivia Lynn Kluth
Leo 10h15'11" 25d46'40"
Olivia Lynn Mietzner
Ari 2h34'51" 18d58'49"
Olivia Lynn Stantic
Tau 4h2'10" 15d41'29"
Olivia M. Adams
Cyg 20h3'54" 40d3'13"
Olivia M. Bisset
And 0h44'8" 39d33'17"
Olivia Madison DePace
Cyg 20h20'17" 48d34'44"
Olivia Mae
And 23h9'4" 50d40'0"
Olivia Mae
And 23h21'18" 41d15'30"
Olivia Mae Big Sister Tanz-Wood
Peg 21h27'3" 19d2'55"
Olivia Mae Dietrich
Vir 12h7'45" 12d18'30"
Olivia Mae Thompson
And 23h52'5" 39d29'6"
Olivia Mae Viguerie
And 23h46'56" 40d57'57"
Olivia Mai Parise
Aqr 22h44'56" -16d0'57"
Olivia Malin
Lyn 8h2'57" 44d7'11"
Olivia Margaret Shuster
And 0h28'29" 41d52'22"
Olivia Marian Carson, 04022003
Ari 2h18'48" 24d35'17"
Olivia Marie
And 1h12'3" 45d18'15"
Olivia Marie
Cap 21h52'0" -19d41'42"
Olivia Marie Arnold
Vir 11h49'8" 9d28'36"
Olivia Marie Bradburn - Olivia's Star
And 23h53'54" 39d15'29"
Olivia Marie Lock
Leo 11h23'51" 23d37'9"
Olivia Marie Lockbaum
Vir 11h45'50" -0d52'18"
Olivia Marie Thomas
Oph 17h1'59" -0d1'0"
Olivia Marie Walp
Cnc 9h10'58" 9d6'22"
Olivia Marie Zimmerman
Psc 1h28'43" 28d6'11"
Olivia Marie-Claire Calegari
And 1h37'50" 48d33'4"
Olivia Marketa Thomas
Sco 17h29'1" -45d29'35"
Olivia Marlise Gombert
Ori 5h14'48" 8d33'24"
Olivia Mary
And 0h35'41" 21d49'5"
Olivia Mary Robustelli
And 23h6'36" 41d12'33"
Olivia May Harris
Boo 14h22'33" 46d11'14"
Olivia May Rice
And 1h24'41" 35d13'43"
Olivia May Rita Guindi
And 23h14'8" 40d21'47"
Olivia Michele
And 2h37'28" 49d10'16"
Olivia Michelle
And 0h24'1" 46d1'4"
Olivia Migdal (wisieneczka)
Cyg 19h43'59" 33d46'45"
Olivia Morley
And 2h30'50" 49d37'57"
Olivia Morrow // Sir Charles
Uma 10h18'16" 57d53'15"
Olivia Moskunas
Tau 5h47'10" 17d0'2"
Olivia Moussallem
Aqr 21h33'8" 0d20'38"
Olivia Naiya Corral
And 0h14'36" 27d8'53"
Olivia Nancy Hill
Cnc 8h12'47" 26d6'21"
Olivia Nancy Hill
Sgr 19h44'35" -14d2'50"
Olivia Nash
Tau 3h52'45" 13d44'1"
Olivia Nell
And 0h10'32" 36d16'11"
Olivia Nicole Mignone
Ari 1h56'31" 17d35'3"
Olivia Nöelle
And 2h31'25" 49d11'59"
Olivia Oberdorf
Cap 20h36'12" -24d0'38"
Olivia Obiora
And 23h12'12" 47d2'53"
Olivia Orion
Uma 8h15'32" 61d21'58"
Olivia Paige Garber
Uma 11h37'8" 52d32'45"
Olivia Paige Holmes
Sco 16h51'3" -42d7'56"
Olivia Paula Lembo
Aqr 21h6'29" -11d24'22"
Olivia Ponitz
Lib 15h23'24" -7d27'46"
Olivia Preciosa Villarreal
Uma 9h51'4" 56d44'49"
Olivia Printzlau
Tau 5h38'1" 25d48'17"
Olivia R. Barresi (LIBS723)
Leo 9h43'28" 27d32'0"
Olivia Rachelle
Ari 3h21'50" 29d4'0"
Olivia Rae
Tau 5h48'47" 26d13'14"
Olivia Rae
And 1h41'1" 44d18'17"
Olivia Raye Jones
Sco 16h16'19" -15d30'56"
Olivia Rayleene Dandy
And 1h28'0" 43d40'40"
Olivia Rebecca Schwinn
Dra 19h2'5" 77d48'5"
Olivia Reddish
And 0h48'12" 36d12'16"
Olivia Regan
Sco 16h5'21" -12d35'4"
Olivia Renee Ellis 10/11/02
And 1h3'30" 42d27'8"
Olivia Renee Rubio
Aqr 22h6'38" -2d11'0"
Olivia Ritschard
Sgr 18h0'1" -27d49'22"
Olivia Robyn Santos
Uma 8h32'38" 65d6'40"
Olivia Rochelle
And 2h33'55" 46d19'57"
Olivia Rose
Cas 0h44'8" 50d48'48"
Olivia Rose
And 0h33'50" 41d14'2"
Olivia Rose
Tau 4h29'43" 23d25'42"
Olivia Rose
Cam 6h19'9" 63d4'21"
Olivia Rose
Lib 15h35'19" -16d35'22"
Olivia Rose Burke
And 1h12'3" 47d10'3"
Olivia Rose Corpina
And 0h50'36" 35d28'24"
Olivia Rose Graham - 17 Oct 2003
Cru 12h56'56" -59d29'21"
Olivia Rose Hankinson
Crb 15h44'51" 31d14'7"
Olivia Rose Key
Lmi 9h54'35" 37d19'18"
Olivia Rose McCreath - Oli's Star
And 23h57'27" 48d5'34"
Olivia Rose Peachey - Olivia's Star
And 23h2'22" 40d56'44"
Olivia Rose Watson
Cru 12h11'28" -62d15'29"
Olivia Rose Zakas
Vir 13h15'57" -3d39'15"
Olivia Ryann Schwenger
Pyx 8h30'23" -21d30'35"
Olivia Saracino - "Heavenly Star"
Uma 12h10'34" 59d19'5"
Olivia Scarlett Lean
Cas 23h55'39" 56d42'45"
Olivia Scarlett Woolhouse
Cas 23h22'27" 55d27'46"
Olivia Shaw
Uma 11h50'51" 52d39'5"
Olivia Soccio
Del 20h26'0" 16d45'31"
Olivia Solveig Eastman
And 0h20'3" 26d49'33"
Olivia Sophia
And 1h24'20" 47d29'58"
Olivia Stanzani
Her 17h48'49" 39d10'35"
Olivia Stephanie Ner
Mon 6h48'41" -4d43'13"
Olivia Stokes
Ari 3h27'53" 21d58'24"
Olivia Susan Stary
Sco 16h5'7" -10d13'47"
Olivia Susan Williams
Sgr 18h45'59" -32d11'23"
Olivia Taryn Schleeper
And 2h7'9" 38d55'54"
Olivia Taylor Wills
Lib 15h10'22" -13d43'23"
Olivia Theresa Yogus
Uma 9h15'57" 72d32'37"
Olivia Torres - Queen of the World
Cas 0h33'59" 63d42'52"
Olivia VanLenten Crowley
Peg 21h42'8" 23d38'2"
Olivia Victoria Elliott
Peg 21h32'32" 16d46'26"
Olivia Vienne
And 23h51'52" 42d24'42"
Olivia Violet
Lib 15h11'4" -8d23'41"
Olivia Virginia Jonas
Vir 13h24'31" 11d6'12"
Olivia Walden
Cas 0h13'8" 62d28'54"
Olivia Watson
Cyg 19h33'32" 31d31'39"
Olivia Witkop
Lib 15h14'13" -27d58'48"
Olivia Wondrely
Vir 13h19'26" 6d54'30"
Olivia Zusy Cote
And 1h3'57" 44d10'6"
Olivia, My Very Best Friend
And 23h13'28" 51d37'24"
Olivia-Jade Sharpe
And 23h47'57" 36d49'27"
Olivia's Christening Star
And 23h15'40" 39d11'50"
Olivia's Constance Star
Aqr 23h42'9" -18d39'10"
Olivia's First Christmas Star
And 0h25'4" 42d2'9"
Olivia's Grace
Sgr 19h22'10" -30d47'6"
Olivia's Love
And 23h48'19" 46d37'20"
Olivia's Lucky Star
Sgr 19h40'16" -12d11'1"
Olivia's Orion
Ori 5h7'55" 6d23'19"
Olivia's Shining Star
Tau 4h8'8" 20d13'5"
Olivia's Star
Uma 10h14'31" 44d52'43"
Olivia's Starlight
Uma 11h11'28" 39d49'14"
OliviaThompson's Eternal Light Star
Uma 11h45'37" 45d45'32"
Olivier
And 23h33'57" 47d5'9"
Olivier
Her 18h33'22" 25d1'37"
Olivier & Audrey Attia
Uma 10h26'29" 69d39'50"
Olivier Bonhomo
Cas 0h51'58" 56d39'37"
Olivier Briche
Cas 1h22'46" 50d56'22"
Olivier Busslinger
And 1h14'38" 35d31'0"
Olivier Capet
And 2h36'19" 46d17'3"
Olivier Cerinotti
Uma 10h28'31" 42d0'51"
Olivier Choquet
Ori 5h33'2" -5d15'53"
Olivier Devillet Parrain Chéri
Her 18h54'40" 22d15'21"
Olivier Fromont
Cap 20h30'39" -27d9'45"
Olivier Jean
Uma 11h23'9" 60d28'1"
Olivier & Marjorie
Crb 16h22'40" 35d34'17"
Olivier Merlien-Ferol
Uma 11h35'33" 38d55'40"
Olivier Michel Joye
Ori 5h20'44" -4d16'18"
Olivier Mouton
Tau 5h46'40" 17d55'36"
Olivier Perron-Collins
Sco 17h38'56" -34d24'49"
Olivier Renaud Sylvie Bertrand
Uma 9h46'41" 47d58'33"
Olivier Rinaldi
Uma 11h49'27" 43d13'19"
Olivier Schott
Cas 0h35'20" 61d43'44"
Olivier Thépot
Cas 23h10'37" 56d9'43"
Olivier-Massie
Cas 1h34'25" 68d30'17"
Olja Bolja
Cas 23h49'0" 56d43'56"
Olle Wilhelm Svensson
Dra 19h55'1" 79d39'37"
Ollhoff, Wolfgang
Vir 13h35'35" -6d31'33"
Ollie
Boo 14h43'47" 36d8'56"
Ollie
Vir 12h42'43" 6d7'29"
Ollie
Ari 2h38'22" 29d53'52"
Ollie A.
Uma 9h48'50" 48d44'6"
Ollie and Cassie
And 0h5'10" 43d57'48"
Ollie Gale
Per 4h46'29" 42d5'14"
Ollie Hendy Simpson
Dra 17h56'18" 63d17'35"
Ollie James Pearson
Umi 15h47'56" 71d46'49"
Ollie Paul Montellier
Uma 11h20'52" 55d46'46"
Ollie Poole
Umi 22h37'45" 88d42'16"
Ollie T. Snedegar
Ori 6h20'51" 9d23'47"
Ollie-Eileen
Lyn 7h57'57" 36d30'2"
Olly
Cru 12h52'21" -57d54'22"
Olly & Dayna Forever 10/26/02
Cyg 19h36'48" 53d50'16"
Olly Minter
Cep 21h6'44" 59d8'40"
Olmstead_Sonia_38
Her 17h28'56" 46d33'10"
Olof Christer Hjorter
Tau 4h2'21" 19d7'56"
Olof Christer Hjorter
Tau 5h40'12" 23d26'33"
Olof Simonds
Lib 15h3'52" -1d9'37"
Olon Delane Martin
Uma 10h54'45" 60d47'34"
Olubukola Felicia Akinjayeju
Crb 16h7'3" 36d49'7"
Olufemi
Tau 4h14'52" 15d33'27"
Olus 87
Uma 9h44'3" 49d54'5"
Olusina Akande
Her 18h50'38" 23d26'35"
Oluwakemi Adeniyi
Aqr 22h28'15" -3d3'2"
Oluwamakinde, Forever In My Heart
Leo 9h36'32" 32d17'44"
Oly & Fred
Cas 23h20'58" 55d40'11"
OLYA
Ari 3h4'8" 18d29'20"
Olya Drogomiretskiy
Umi 15h3'0" 69d59'10"
Olya Leonova
Lib 15h10'49" -9d30'46"
Olympe Lespagnon
Dra 19h4'26" 53d30'30"
Olympia Gonzales Victoria
Ori 5h37'59" -0d5'1"
Olympia Lagonikos
And 2h13'18" 46d43'50"
"Olympic"-Mark Roitman
Lyn 6h20'29" 54d44'30"
Olyushka
Dra 19h23'23" 70d7'38"
Olyvia Reign
Boo 14h43'40" 30d4'43"
Om
Leo 10h42'59" 18d43'20"
Om
Uma 11h24'39" 53d47'3"
O-ma
Lyr 19h17'2" 29d13'45"
OMA
Vir 11h39'16" 4d58'14"
Oma
Uma 9h24'12" 44d19'16"
Oma and Pap
Cyg 20h16'45" 50d47'57"
Oma Emma Theresa
Uma 11h51'20" 41d47'11"
Oma Ilona & Opa Jim Lintner
And 1h11'25" 34d8'4"
Oma & Opa
Ori 6h17'36" 10d49'34"
Omadean & Nathan Jackson
Lyr 19h19'28" 29d33'41"
Omae
Uma 11h19'48" 32d1'24"
Oma..In ewiger Liebe & Dankbarkeit
Uma 11h32'12" 40d55'2"
Omair
Psc 1h23'48" 28d7'46"
O'Mally
Tau 5h45'58" 25d51'24"
Omar
Ori 5h33'58" 4d32'12"
OMAR
Per 2h40'2" 39d57'41"
Omar
Umi 16h14'47" 75d20'59"
OMAR
Sgr 20h13'16" -43d37'3"
Omar Abou Ezzeddine
Vir 12h9'38" 6d30'30"
Omar Al Bahiti
Leo 11h19'53" 3d33'8"
Omar and Dan's Light of Hope
Pho 1h40'44" -41d19'10"
Omar and Yasmine
Uma 12h13'50" 60d9'27"
Omar Corrales
Ser 15h49'30" 7d13'36"
Omar Donovan
Lmi 10h55'17" 26d6'23"
Omar Elatab
Aql 19h24'42" 8d0'0"
Omar Ernesto Hernandez
Ari 2h47'39" 22d15'24"
Omar Guardia
Dra 16h40'19" 51d51'47"
Omar Howard (Eightball) Jackson
Dra 19h49'13" 76d41'18"
Omar Ibn Ali's Star
Lib 14h37'57" -22d59'40"
Omar Jackson
Her 16h38'33" 33d4'55"
Omar Jahan McBride
Her 17h24'21" 24d3'0"
Omar Kasim Mills
Tau 5h8'10" 24d19'28"
Omar Lazo
Tau 4h41'2" 19d58'14"
Omar Martin
Cep 21h37'4" 64d42'18"
Omar Mhoud Capers
Aqr 22h4'35" -2d31'27"
Omar Miranda * Christine Dao
Uma 9h55'29" 46d1'44"
Omar Soto "04-04-77" "09-23-95"
Uma 11h46'35" 61d45'1"
Omar The Star
Uma 9h13'44" 60d55'53"
Omar-K2
Ari 2h39'57" 20d34'8"
Omar's wish STAR
Lyn 8h6'27" 46d1'18"
Omayra
Gem 7h43'2" 32d56'34"
Omayra Nieves
Cnc 8h13'28" 9d26'1"
Ombretta
Cyg 20h9'53" 36d41'43"
Ombretta
Peg 23h55'16" 27d6'33"
OME - Robert Richard Engelke - 2003
Ori 6h2'52" 18d37'23"
OmedAlona
And 23h11'36" 39d25'32"
Omega
And 0h53'18" 36d50'43"
Omega
Per 3h27'38" 34d54'51"
Omega Amy
Mon 7h16'45" -0d2'54"
Omega James
Cam 7h16'23" 70d20'26"
Omega Mamma
Cas 0h25'17" 61d19'21"
Omega Psi Phi
Uma 13h42'12" 57d50'51"
Omega & Thomas Hewitt "Happy 50th"
Cyg 20h46'29" 31d15'47"
OmegaCesco
Ori 5h23'36" 6d49'55"
Omer Ernest Smiley
Cnc 8h14'24" 12d46'33"
Ömer Güngör
Cas 2h3'20" 59d54'11"
Omer J. Gagnon
Cep 22h35'22" 73d26'42"
Omer Yasar
Tau 4h38'4" 23d34'30"
Omerta
Sco 16h8'33" -16d7'24"
OMGWTFBBQ
Leo 10h38'35" 20d51'17"
Omi
Vir 12h30'59" 7d51'22"
Omi
Uma 10h16'48" 50d45'29"
Omi - 100th Birthday
Psc 0h58'33" 15d23'54"
Omi Ada
Crb 15h50'9" 33d40'29"
Omie's Star
Lyn 7h6'20" 47d24'18"
Omkar
Aql 19h0'49" 16d57'23"
Omnaia Jolie Abdou
Psc 1h3'27" 27d48'25"
Omni Elizabeth Hornedo
Pho 1h23'10" -42d19'42"
Omnia Vincit Amor
Sgr 19h10'38" -24d59'42"
Omnia Vincit Amor
Sgr 19h36'39" -16d7'43"
Omnia Vincit Amor
Her 17h14'9" 29d1'57"
Omnimore
Cap 20h33'0" -14d0'31"
Omnino
Ori 4h53'54" 7d28'12"

O'Monaghan's Emerald Isle
Cyg 20h31'1" 49d10'24"
Omorfee
Leo 10h58'30" 8d26'52"
Omorfos Amy
Cas 1h1'54" 53d45'14"
Omorfoula Adonia
Psc 1h44'34" 14d58'26"
Omriky
Sco 17h57'45" -39d45'11"
On A Night Like This
Sco 16h43'15" -31d28'1"
On butterfly wings,Evan Robert
Cnc 8h48'41" 14d17'35"
On Kim Mee
Leo 11h17'47" 6d25'24"
Ona
Sgr 18h43'41" -17d57'52"
Ona Sonata Weingarten
And 0h54'58" 40d12'14"
Onalee J. Sharp
Aql 19h52'14" -0d0'49"
Onatop - 007
Aql 19h56'56" -0d57'21"
oncapardus
Cnc 8h41'35" 7d49'59"
Once
Lyn 7h40'5" 36d45'55"
Ond Day
Sge 19h36'18" 16d54'32"
One
Leo 9h56'19" 24d37'33"
One and Only
Ori 6h15'4" 2d14'28"
One and Only
Lib 14h47'48" -1d22'49"
One and Only
Leo 11h10'31" -5d36'10"
One Beautiful Ruth
Cas 1h58'44" 65d2'40"
One Day
Leo 11h17'14" -5d25'44"
One Day At A Time! EU Forever...17
Lyr 18h33'57" 36d22'23"
one day closer...to forever
Per 3h51'55" 35d32'3"
One Fine Specimen
Ari 2h14'13" 25d5'5"
One flesh— MC and MP
Cyg 21h10'56" 42d17'53"
One Girl
Sco 17h32'2" -44d18'21"
One Good Thing
Gem 7h18'27" 34d46'24"
One Heart
Ari 3h14'20" 27d35'21"
One Hot Mohamed
Lib 14h52'30" -1d33'20"
One Hull of a Place
Uma 11h4'44" 68d53'23"
One in a Billion
Uma 10h24'6" 44d52'35"
One in a Million
Lyn 7h33'38" 52d37'15"
One in a Million Mama
Umi 14h53'14" 67d43'4"
One in a Zillion
Leo 10h55'34" 2d5'55"
One Life - One Love Mike & RoseAnn
Per 4h6'42" 44d32'2"
One Life, One Love, Forever
Cyg 19h44'58" 34d51'33"
One Love
Cyg 19h50'14" 38d14'30"
One Love
Lyr 18h46'57" 43d11'25"
One Love
Cyg 19h45'53" 54d23'5"
One Love
Cmi 8h1'32" -0d17'18"
One Love- Timothy & Jessica
Sex 10h12'56" 3d17'9"
One - Miracle Dr. Kattner
Her 16h51'23" 35d15'48"
One More Makes Four
Lyn 7h3'20" 53d56'9"
One of a Kind Kid
Umi 14h25'16" 75d7'24"
One of Keith's Dreams
Cnc 8h42'6" 23d15'23"
One Star 4 My One Star Trish Walker
Aur 5h6'4" 46d59'20"
One Thing
Per 3h28'59" 48d14'11"
One True Love
Mon 6h59'32" 9d11'11"
One True Love CBP&JSS
Ori 5h18'14" 7d5'16"
One True North
Gem 7h25'23" 26d23'59"
One Union, One Destiny, One Love
Gem 6h46'25" 29d37'54"
One Whis
Aql 19h3'26" -9d52'2"
One Wish
Her 16h24'8" 47d46'2"
One Year
Tau 5h33'29" 26d14'35"
One Year Star
Leo 9h25'16" 14d25'7"
One Zealous Doctor
Vir 11h41'12" 3d18'6"
Oneda Diaz & Angel L. Reyes Baptism
Ser 15h41'5" 14d8'16"
Oneflip
Cyg 19h53'18" 32d41'54"
Oneida
Ori 5h32'51" 6d16'52"
O'Neil Louis Boulet
Cyg 20h4'26" 46d7'42"
O'neill & Tuddenham Star - December 2004
Pho 1h37'20" -49d11'47"
Oneiro
Psc 0h35'25" 10d15'43"
Oneiroi
Cas 1h28'29" 53d38'56"
Oneisogoustia
Cmi 7h32'16" 3d50'24"
Onelia Martinez (Rob & Joe Sanchez)
Uma 11h22'35" 55d7'52"
Oneofbeauty
Ari 2h16'56" 25d13'50"
Oney Ray Otten
Vir 12h24'57" 7d50'32"
Ong Jun Yu
Lyn 6h59'28" 47d50'6"
Ong Thay Lok
And 0h6'0" 43d56'44"
Ong Tze Guan
Peg 22h45'20" 5d56'7"
Onie
Uma 8h48'0" 71d47'51"
Onin
Vir 14h33'23" 3d43'29"
Onkel Heini
Ori 5h3'33" 15d13'19"
Onkel Willy
Her 17h58'6" 49d43'52"
Only Beaner
And 1h38'24" 43d31'52"
Only Hope
Cyg 20h9'7" 41d8'0"
Only Hope
Uma 12h43'41" 57d21'40"
Only In Heaven 10/2/03
Lib 15h55'14" -5d7'46"
only one
Leo 10h7'25" 25d49'28"
Only One Lifetime is Not Enough
Aql 19h21'55" -0d42'39"
Only One Reason
Cyg 21h11'10" 30d4'39"
Only The Beginning
And 0h52'0" 37d13'59"
Only the Strong Survive
Cyg 19h28'21" 53d46'5"
Only You
Lyr 18h45'10" 39d5'58"
Onnie
Ari 3h7'40" 28d23'21"
Onnie Smith
Psc 1h46'26" 14d34'36"
Onno
Ori 5h55'28" -0d47'13"
Onofrio Santamaria
Cep 3h29'43" 80d39'20"
ONOIR
Gem 6h36'10" 13d36'32"
Onorato Crossroads
Tau 5h47'42" 17d45'42"
Onorina Lynette Bucci
Leo 9h28'12" 23d36'35"
Onri Midgettis
Uma 11h23'24" 52d29'48"
Ons Marianneke
And 0h19'50" 35d50'7"
Onthegreen
Gem 6h33'40" 22d3'40"
Onur Cagigan
Crb 16h19'55" 29d18'11"
Onur Cetin
Per 4h9'6" 41d19'28"
Onyx
Lyn 7h1'14" 48d40'10"
Onyx
Umi 14h38'41" 68d32'34"
Onyx "7-8-2005"
Uma 13h32'35" 56d15'24"
Onyx Alexis
Uma 8h48'13" 58d10'41"
Onze DroomVlucht
Cyg 19h47'19" 29d33'25"
Oo
Sco 16h6'24" -9d27'18"
Ooch
And 1h0'1" 41d39'11"
Oochbay bon Litska
Uma 9h32'1" 59d25'37"
Oodie
Uma 10h31'24" 71d16'45"
oofairytale
Lyn 8h3'18" 55d15'14"
Oogie
Lyn 6h32'42" 57d53'42"
Oogliabooglia
Ori 6h15'27" 10d32'33"
Oogus Bogus Maximilian Schell
Aqr 21h40'1" -1d34'33"
Ooleedoo
Lib 15h27'50" -6d32'29"
OOM
Psc 0h33'1" 4d5'53"
Oompa Ahpoom Stirnaman
Uma 8h36'5" 47d39'20"
Oompa Loompa
Cru 12h35'40" -60d40'7"
Oompah Poohpa
Ori 5h34'44" 3d0'9"
Oona
Tau 5h36'45" 25d59'20"
Oona
Uma 11h26'10" 38d47'32"
Oona Alexandra Bustard
And 0h44'29" 26d38'14"
Oona Mae
Uma 11h38'25" 62d1'50"
Oonagh, Loving Mother
Cas 0h58'28" 58d32'5"
Oopie
Sgr 18h6'29" -26d11'19"
Oopie Richardson
And 23h44'48" 37d55'27"
OOZONO
Gem 7h15'30" 16d40'46"
Opa
And 2h11'12" 47d10'7"
OPA
Uma 9h12'5" 58d0'43"
Opa Fred Rohrer
Per 2h55'36" 54d16'22"
"OPA" Manuel Arthur Raposa
Vir 14h18'37" 5d14'58"
Opa Paul Sonda
Her 17h39'37" 32d28'17"
Opa & Santina
Uma 8h48'57" 47d21'1"
Opal
Cas 23h26'54" 51d28'41"
Opal
Vir 13h20'20" 11d25'58"
Opal
Ari 2h32'30" 26d13'2"
Opal
Cnc 9h13'35" 25d13'2"
Opal
Uma 9h13'16" 55d21'51"
Opal
Cam 5h2'51" 65d25'2"
Opal
Lib 14h53'44" -1d10'34"
Opal Berry
Lyr 18h51'49" 36d56'57"
Opal Charelaine Krout
Sco 17h43'19" -39d47'15"
opal essence
Uma 9h45'0" 60d42'16"
Opal Hoyt
Leo 10h55'48" 18d6'45"
Opal Princess
Sco 17h41'41" -33d32'14"
Opal Priscilla
Uma 11h22'47" 31d49'25"
Opal Schultz
And 0h37'40" 37d17'46"
Opal Trudyann
Uma 8h43'49" 55d2'6"
Opaline
Sco 17h2'5" -41d7'5"
OPE
Sgr 18h45'49" -27d8'19"
Open Correspondence
Boo 14h41'45" 23d48'47"
Opeongo
Uma 9h21'54" 47d59'43"
Opfer, Claudia
Uma 9h9'8" 68d37'41"
Ophalyn Jacob Tran
Vir 12h53'50" 12d45'0"
ophelia
Cap 21h0'41" -17d49'30"
Ophélie Lainé
Tau 3h51'46" 8d5'54"
Ophia Lee Smith
Mon 7h16'0" -0d43'59"
Opie
Leo 9h26'50" 25d9'50"
Opie
Cnv 12h47'37" 39d21'5"
Opper, Udo
Uma 13h50'0" 56d27'39"
Oprah
Cas 2h12'55" 73d45'50"
Oprah Winfrey
Cas 1h24'17" 64d4'43"
Oprah Winfrey
Aqr 22h25'36" 0d42'19"
Opsh
Cnc 8h16'17" 15d40'47"
Opteaman
Aqr 21h59'26" -17d38'1"
OPTIMUS LOVE JONES
Vir 12h1'52" -0d15'4"
Optimus Prime
Uma 8h30'40" 66d53'37"
Optimus Prime
And 1h18'32" 36d45'24"
Opulence
Ori 5h50'58" -0d6'58"
Opus One of Windansea
Uma 12h58'25" 61d19'25"
Ora B. Dublin
Tau 5h19'47" 23d31'25"
ora e per sempre
Mon 6h45'54" 9d25'33"
Ora J. Bonner
Uma 10h44'45" 62d18'0"
Ora Nicole
Psc 0h24'32" -1d55'3"
Oraios Mati
Ori 4h53'24" 11d19'52"
Oralia Tamez Colorado
Mon 6h27'4" 8d46'11"
Oran -n- Vegg
Sco 17h4'2" -38d6'35"
orange
Psc 1h22'1" 23d22'29"
Orange & Blue Victorious
Cma 7h22'2" -11d34'43"
Orange Julius Peck
Lmi 9h49'38" 37d29'56"
Orange Pez
Uma 13h15'38" 56d15'1"
Oranous
And 2h34'15" 45d36'17"
Oras
Umi 15h54'4" 80d1'44"
Orazio Parajon & Tahany Salgado
Ori 5h54'46" -0d0'19"
Orazon
Aur 6h19'14" 53d11'7"
Orb of Coinneach
Gem 6h19'40" 23d14'49"
Orbita
Sco 16h12'44" -9d28'57"
Orbnauticus-Voodooviking
Lyr 19h11'51" 26d19'46"
ORCA the best dog ever
Cma 6h46'29" -15d26'49"
Orce Vesna & Thomas Nikoloski
Cru 12h16'17" -57d45'25"
ORCH Ken Orchard
Uma 11h9'13" 40d44'2"
Orchid & Robert
Sge 19h38'23" 18d47'52"
Orchidea A. Marchezani Corciolli
Leo 11h15'13" -3d23'55"
OrcoDeo
And 23h51'33" 34d1'20"
Ordained 08161997
Aql 19h43'2" -0d6'8"
Ordek
Uma 8h46'43" 57d46'8"
Ordranie Louisa Clarque
Uma 10h28'42" 57d8'40"
O'Really O'Reilly
Sgr 19h34'21" -13d46'53"
ORegCinner
Umi 15h19'25" 72d33'32"
Orel
And 0h38'59" 44d38'6"
Orel W. Divens
Cas 1h35'11" 61d27'35"
Oren DeLacour Pomeroy III
Uma 8h35'25" 59d35'12"
Oren Mi Amor Guzman (O.M.A.G)
Tau 4h27'52" 15d27'57"
Oren O'Neil Clem
Psc 0h17'16" 6d29'59"
Oren Webster
Umi 14h57'14" 73d7'10"
Oren William Clark My True Love
Sgr 17h52'46" -18d9'59"
Orenda-wakan
Uma 13h49'59" 54d49'36"
OREO
Aqr 22h21'40" -7d10'50"
Oreo
Uma 9h49'19" 50d53'50"
Oreo Camacho
Uma 13h58'16" 51d54'57"
Oreo Charisma Stockwell
Cma 6h46'47" -14d33'24"
Oreshley
Uma 12h27'3" 59d41'55"
Oreste Munno
Her 17h49'3" 41d35'27"
Oreste Pizzolitto
Lyr 18h34'41" 32d47'52"
Orestes A. Bihun, Mi Amor Verdad
Mon 7h15'48" -0d46'2"
Orfan-Annie
Pho 1h56'8" -43d9'44"
Orgera
Cnc 8h42'34" 31d28'52"
Ori Baram
Sgr 18h21'9" -32d18'57"
Ori Torbiner
Uma 11h18'24" 46d38'51"
Oria Alexis Romero Ocaranza
Ari 2h15'22" 24d10'3"
Oria Krief
Aqr 21h5'46" 2d7'19"
Oriana
Cnc 8h35'17" 25d20'37"
Oriana Keiko Hogan Taguchi
Umi 14h31'21" 82d39'3"
Oriane Couturier
Psc 0h21'21" 1d51'5"
Orianna Elina
Cas 1h37'24" 61d1'7"
Orias
Ori 6h0'23" 14d5'9"
Orieda Horn Anderson
Sco 17h44'27" -41d13'57"
Orien and Connie Everlasting
Mon 6h50'40" -0d14'4"
Orienta
Cep 20h40'28" 60d37'22"
Orienta
Uma 10h6'17" 64d21'37"
Orietta Magnaterra
Lac 22h56'15" 37d41'50"
Original Spirit
Psc 1h43'17" 18d34'47"
Oriol O.M.
Vir 14h39'15" 7d10'25"
Orion
Cyg 21h9'49" 45d34'2"
Orion
Per 3h27'14" 34d56'41"
Orion Bartholow
Ori 5h26'46" 3d51'37"
Orion Ethaniel Rodriguez
Ori 6h16'38" 16d44'45"
Orion II
Ori 6h3'38" 16d56'7"
ORION III
Per 4h24'22" 35d35'50"
Orion Luis Armendariz
Ori 5h54'4" 18d31'8"
Orion Michael Lawrence McGURRELL
Ori 6h2'0" 12d9'11"
Orion Nickola Swarner
Ori 5h54'9" -0d45'14"
Orion Robinson
Ori 5h28'58" 6d19'9"
Orion Steven Kwong-Yu Crisafulli
Ori 6h12'57" 15d21'47"
Orionona
Uma 12h34'21" 55d43'54"
Orion's Bat
Ori 4h55'22" 10d16'23"
Orion's Belt
Ori 5h32'31" -0d29'22"
Orion's Belt Buckle
Aqr 22h13'34" -8d37'58"
Orion's Eye
Ori 5h20'52" 5d35'13"
Orion's Star
Cap 21h57'45" -23d5'23"
ORIONTAB
Psc 0h15'16" -1d28'26"
Oriseo
Vir 12h6'26" -9d53'39"
Orissa Moulton
Leo 9h28'48" 25d5'2"
Orisunny
And 2h2'54" 46d31'18"
Orit, 15.08.2003
And 2h11'55" 50d7'43"
Orla
And 22h59'47" 39d51'16"
Orla Cabrini
And 22h58'15" 42d11'30"
Orla Elizabeth Sefton
And 1h22'47" 43d18'50"
Orla Rowlands
And 23h23'42" 35d15'47"
Orla, My Guiding Light!
Sgr 18h55'0" -34d59'18"
Orlagh Mae
And 23h4'53" 51d19'43"
Orlaith
And 1h40'53" 36d17'0"
Orlaith Angela Mclaughlin
And 2h28'23" 37d57'42"
Orland Phillips
Aur 6h57'45" 38d13'27"
Orlandito
Vel 9h17'33" -40d34'58"
Orlando
Ari 2h47'13" 13d52'11"
Orlando Aramis
Dra 18h21'15" 56d53'30"
Orlando Charriez
Cnc 8h2'56" 26d26'57"
Orlando Cruz
Per 3h12'5" 55d48'20"
Orlando E. Pizarro
Gem 6h28'49" 22d39'23"
Orlando Leon
Vir 11h57'27" -0d56'54"
Orlando Machin
Aur 5h17'39" 29d37'39"
Orlando Morley
Ori 5h34'24" -0d55'58"
Orlando Rolon
Aur 5h13'29" 32d20'14"
Orlando Salinas
Ari 2h46'9" 15d54'52"
Orlando Tamez
Her 18h5'23" 14d45'24"
Orlando - The Little Prince - Clerc
Col 6h7'54" -42d6'25"
OrlandTuna
Psc 23h19'32" 1d54'51"
Orla's Nanna - Pauline Hammond
And 2h3'17" 45d44'27"
Orlawan One
Her 16h7'50" 24d51'15"
Orlinn
Aqr 20h40'55" -2d44'6"
Orlo
Cir 14h43'57" -63d59'27"
Orly Keren
Sco 16h11'35" -8d49'30"
Orlys and Jamie
Sge 19h26'4" 18d5'32"
Ormaine Szanyi Cecilia
Vir 12h39'48" 6d39'4"
Ormiston
Cra 18h2'21" -37d45'16"
Ormsby / Harloff Families
Uma 14h24'39" 55d44'5"
Orna Goussac
Ori 5h36'12" -2d44'54"
Orna Yakir
Gem 6h45'51" 15d31'27"
ORNAGY
Uma 11h17'22" 44d19'13"
Ornella
Psc 1h21'43" 11d25'36"
Ornella
Umi 16h5'42" 85d9'28"
Ornella
Lib 14h44'0" -14d22'57"
Ornella
Cet 2h12'57" -21d57'40"
Ornella Esposito
Per 3h1'35" 51d40'47"
Ornella Ghelfi
Cas 0h6'32" 59d58'39"
Oro Stella of Shigeru & Masako
Tau 4h28'52" 24d10'44"
Örök barátsággal a mi Ocsinknek
Leo 11h32'11" 10d11'53"
Örök szerelemmel Erikának
Leo 10h34'24" 22d42'39"
Örök Szerelemmel Férjemnek Baracs Jánosnak
Cnc 8h15'59" 20d50'12"
Örök szerelemmel férjem-nek, Gábornak
Gem 7h9'7" 25d34'1"
Örök szerelmünk Csillaga
Ari 2h24'20" 27d3'25"
Örökké és egy, Juhász Zsolt csillaga
Uma 9h39'37" 41d47'27"
Örökké Szeretlek B. Péter csillaga
Cnc 8h9'53" 31d13'25"
OroszlánKirály
Leo 11h25'52" 6d15'41"
Orr Eliyahu
Lib 14h52'0" -9d56'33"
Orr Lauren
Lyn 9h29'2" 40d29'0"
Orren Ed Moore
Lyn 8h45'10" 45d17'25"
Orren Henry Marcotte
Uma 10h2'46" 42d31'23"
Orren Thomas & Isaac McHenry Miller
Umi 14h37'19" 75d14'36"
Orris V. Barden
Ori 5h33'37" -0d53'7"
Orsi
Psc 1h4'31" 21d50'50"
Orsi '75. február 14.
Aqr 21h3'59" -10d39'42"
Orsi csillaga
Cap 20h28'14" -18d36'27"
Orsikámnak Vilitol
Vir 12h29'37" 10d55'20"
Orsola Angelia Bonilla
Peg 21h25'11" 20d4'31"
Orsola (Lena) Zenari
Uma 11h8'50" 30d43'55"
Orthodoxia Stefanou
Uma 11h49'23" 61d33'40"
Ortisi 1
Ori 6h10'5" -0d59'29"
Ortiz
Ori 5h32'3" 2d33'37"
Orton Benson
Uma 11h15'59" 37d7'24"
Ortwin Schäfer
Uma 11h11'32" 42d10'26"
Orus Whittier Dearman Jr
And 1h42'0" 50d9'38"
Orval Ray Riffe
Aql 19h23'49" 2d27'49"
Orvar Kitty-Bell
Cas 0h25'59" 61d21'25"
Orvil Don Farrow
Uma 10h58'35" 72d13'45"
Orville Arthur Paris
Her 18h45'6" 20d32'31"
Orville B. Smith
Her 17h30'14" 39d20'45"
Orville F. Schaudt, Jr.
Vir 13h10'1" 11d34'9"
O-Ryan
Ori 5h55'28" 7d22'29"
O'Ryan
Ari 3h27'6" 27d40'17"
O'Ryan
Lyn 7h38'15" 38d4'11"
O'Ryan Anthony Corwin
Ori 5h25'14" -7d43'31"
ORyan Squeaker
Ori 5h32'1" 6d15'58"
Orza's Rose
Aqr 21h53'52" -1d12'12"
OS
Cas 0h18'59" 53d58'45"
OS
Cyg 21h12'45" 46d52'9"
Osamah El-Rikabi
Uma 14h2'31" 59d18'54"
Osbaldo Hurtado
Uma 13h34'16" 59d43'13"
Osborne's Hope
Aql 19h8'54" 13d1'56"
Osbrinkus
Aur 5h48'8" 37d32'17"
Oscar
Per 3h21'16" 42d27'45"
Oscar
Per 4h9'46" 32d53'47"
Oscar
Per 3h35'27" 45d36'41"
Oscar
Her 18h29'46" 20d24'35"
Oscar
Leo 10h29'57" 17d31'15"
Oscar
Uma 10h34'17" 59d17'57"
Oscar
Uma 8h45'7" 68d15'4"
Oscar
Lep 5h5'51" -11d16'23"
Oscar
Ori 6h14'5" -0d52'38"
Oscar
Sgr 18h38'39" -20d14'22"
Oscar
Cru 12h55'19" -58d47'56"
Oscar Aberkalns #1
Uma 10h46'33" 64d42'55"
Oscar and Annette de la Renta
Cyg 21h11'1" 50d58'10"
Oscar and Judy Lenertz
Lyr 19h20'24" 29d39'46"
Oscar and Melissa Aragon
Lyr 18h41'53" 34d48'35"
Oscar and Rosa Barreto Forever
And 23h15'54" 37d31'45"
Oscar Armando Rodriguez Alvarado
Mon 6h26'53" 1d6'5"
Oscar Blandi
Lib 15h35'40" -15d33'20"
Oscar Bradley Ryan - 14.02.2007
Ori 5h20'29" -1d23'47"
Oscar Colomer
Leo 11h28'33" 27d26'14"
Oscar De La Hoya Winegardner
Cma 7h21'53" -18d20'30"
Oscar Diaz
Ari 2h38'34" 26d11'49"
Oscar E Rodriguez
Ori 5h26'23" 0d0'15"
Oscar el hombre
Her 17h9'8" 32d0'5"
Oscar Fargas
Cap 20h37'51" -16d40'49"
Oscar Frank Spell III
Hya 8h54'10" -1d2'20"
Oscar Gabriel
Lyn 8h20'2" 38d25'35"
Oscar Gabriel Martinez
Her 17h15'0" 35d46'36"
Oscar Gene Hudec
Psc 1h36'32" 20d24'8"
Oscar George Taylor
Cnc 8h31'20" 9d50'56"
Oscar Gillies
Her 17h15'25" 19d16'54"
Oscar & Gina
Umi 13h22'58" 71d4'39"
Oscar Hamilton Field
Ori 5h23'55" 13d17'45"
Oscar Harrison Brook
Leo 11h4'45" -0d29'23"
Oscar & Helen Winchell
Ori 5h51'48" 9d18'16"
Oscar Herrera 1975
Tau 4h32'37" 19d13'15"
Oscar Hottie
Vir 12h31'29" -9d41'22"
Oscar Ian Jon Kelham
Per 3h20'15" 40d49'8"
Oscar James Devine - 13.11.2004
Cru 12h21'35" -61d16'55"
Oscar James Gibson - Oscar's Star
Her 17h47'36" 18d44'8"
Oscar & Jean Messer Happy 60th
Cyg 21h5'12" 47d34'5"
Oscar John English
Cmi 7h13'25" 2d58'18"
Oscar Johnathon
Gem 6h20'43" 21d4'38"
Oscar Lee Cawthorne's Special Star
Uma 11h13'50" 38d33'52"

Oscar Littlefield
Lac 22h28'48" 41d26'11"
Oscar Lopez
Uma 9h57'54" 60d25'37"
Oscar Louis Heitland
Uma 9h23'2" 65d19'14"
Oscar Loves Adriana
Lib 15h59'55" -17d18'46"
Oscar Luis Silveira
Ori 5h28'40" 3d13'46"
Oscar Manuel Cano
Cyg 20h25'56" 46d25'1"
Oscar Marcel Hinze
Cyg 20h23'37" 40d23'12"
Oscar Marino
Lac 22h35'39" 49d56'2"
Oscar Martinez
Psc 1h2'49" 7d35'1"
Oscar Martinez Quintero
Sco 16h46'11" -30d56'14"
Oscar Moreno
Uma 11h48'46" 59d31'27"
Oscar Muldoon Hennessey - Oscar's Star
Umi 15h23'8" 81d33'34"
Oscar "My Star" Katona
Leo 9h48'25" 12d50'9"
Oscar N Kris in the heavens forever
Sco 17h57'18" -35d1'50"
Oscar Ochs
Psc 1h2'45" 27d18'1"
Oscar Ortega Jr.
Leo 10h53'3" 6d56'31"
Oscar Pérez Zapata
Col 6h25'35" -34d24'6"
Oscar Perigault de Maulmin
Uma 8h45'29" 49d55'45"
Oscar Quintana
Vir 13h15'1" 12d43'31"
Oscar Quintero - Tupperware
Her 16h15'45" 18d43'49"
Oscar Robert Edelman
Aqr 20h44'23" -12d5'11"
Oscar Rodriguez
Aqr 21h4'55" -6d3'51"
Oscar Romo
Aql 19h16'38" -8d44'23"
Oscar Roy Beale Flynn - Oscar's Star
Her 16h14'13" 8d48'31"
Oscar Ryan Thacker
Lmi 10h2'50" 29d44'10"
Oscar "Tank" Carter
Ser 15h19'37" -0d57'26"
Oscar & Vi
Ori 6h10'21" 16d0'56"
Oscar Zelaya
Uma 13h46'15" 53d5'34"
Osei Kenyatta Jones
Pho 0h40'37" -41d51'35"
Osenar
Sco 16h46'53" -26d33'43"
Osi
Lib 15h41'7" -16d15'28"
Osidius Emphatic
Cam 5h45'0" 77d12'9"
Osir
Tau 4h38'19" 10d57'25"
Osiris
Del 20h47'52" 15d38'53"
Osiris Zamora
Vir 14h34'16" 4d19'32"
Osito
Cyg 21h28'28" 46d28'4"
Osito
Aqr 22h5'25" -2d24'4"
Osito Flores
Uma 10h40'22" 69d1'52"
Oskar
Cep 20h50'48" 59d50'36"
Oskar
Aqr 22h21'55" 2d33'31"
Oskar Bakke
Ori 5h15'13" -0d11'52"
Oskar Helmer Hauptschule - 2006
Lyr 18h37'54" 39d12'58"
Oski Imperial
Uma 9h32'25" 47d4'48"
Oskorei
And 1h12'0" 38d30'34"
Osman B. Genc
Uma 9h32'26" 51d28'49"
Osmeña
And 0h21'11" 45d35'28"
Osmon Soulmates
Cru 12h24'55" -58d31'43"
Osmooz
Ori 6h20'3" -0d50'0"
OSNAJEN EMT
Lyr 19h7'53" 45d53'34"
Oso
Lyr 19h1'46" 40d16'15"
Oso
Cma 7h21'56" -19d39'1"
Oso's Christmas Star
Uma 11h50'20" 53d52'38"
OSR - Owen Seth Rowley
Her 16h26'42" 44d27'35"
Osros
Ori 6h17'1" 9d35'29"

Ossama Akkary
Her 17h35'35" 41d32'44"
Ossie C. Barnes
Boo 14h35'26" 18d43'28"
Ossie Davis
Cep 22h30'22" 58d6'42"
Ossie Ford
Cyg 21h36'26" 43d36'21"
Ossie Shakir
Cas 1h22'1" 59d51'12"
Osso
Sco 16h43'38" -30d18'14"
Ostap and Natalia Krupa
Cyg 21h15'46" 52d54'35"
Ostara Sundmark
Uma 8h49'12" 46d47'36"
Ostara Willow
Pho 2h12'9" -44d33'58"
Osten Madison Dandan
Uma 10h42'57" 45d35'22"
Ostendorff, Matthias G.
Cap 20h58'15" -25d23'50"
Osterburg, Dieter
Ori 6h0'34" 14d28'35"
Ostwald, Klara Eny
Uma 11h38'45" 29d18'55"
O'Suileabhain Banner
Cyg 21h52'11" 49d49'30"
O'Sullivan's Love
Cru 12h22'41" -60d28'52"
O-SUZI-Q
Aqr 22h21'31" -7d5'39"
Osvaldo
Umi 14h8'28" 67d10'43"
Osvaldo and Silvia Mislavsky
Per 3h39'24" 39d36'51"
Oswaldo Rivera
Gem 7h49'5" 33d11'47"
Otavio
Uma 12h38'31" 61d6'36"
Otch Jack
Psc 1h58'5" 6d27'16"
OTCh Locknor Be A Sparkle Plenty
Cma 6h45'13" -15d17'1"
Otgontugs Turbold King
Uma 11h24'27" 65d41'50"
Otha "Skeet" and Melba Birkner
Cap 20h46'42" -16d18'17"
Othel Maree Henderson
Uma 11h12'13" 49d9'58"
Othella Baldwin
Leo 11h54'16" 22d17'11"
Otho "Art" Smith Jr.
Ari 2h57'20" 24d29'51"
Otho J. (Nick) Nitcholas
Aql 19h50'20" -0d21'44"
Othong Phimmasone
Uma 9h3'50" 51d39'30"
Othur Ray Starr
Her 17h7'13" 35d52'53"
Otillia "Tillie" Stahly
Leo 10h22'31" 17d2'2"
Otis Brison Jr.
Ori 5h23'10" 1d53'4"
otis centuri
Dra 18h50'38" 53d20'23"
Otis Claeys
Aqr 20h53'40" 2d14'25"
Otis Marcus Milne
Dra 18h48'22" 66d25'53"
Otis Reddin
Cyg 21h42'56" 30d14'17"
Otis Thomas Howell
Her 17h16'52" 28d25'43"
Otis Wilson
Uma 10h20'42" 43d35'38"
Otniel Gonzalez
Per 4h31'40" 31d39'18"
Otoniel Figueroa
Uma 8h32'18" 62d34'29"
OTSPA
Ari 3h12'44" 27d29'29"
Ottavia
Psc 0h50'32" 15d2'45"
OTTAVIA 67-KL
Dra 16h8'20" 61d44'34"
Ottavio
Uma 13h40'8" 50d11'15"
Otter
Tau 4h1'35" 30d2'57"
OTTER1006
Lib 15h38'43" -24d41'47"
Otteris
Oph 17h32'32" 6d51'36"
Ottilie Emilie & Alfred Camoin Jr.
Vir 14h34'33" 2d4'47"
Ottmar Schart
Uma 8h12'43" 60d45'28"
Otto
Cma 6h49'0" -15d34'6"
Otto
Cyg 21h39'42" 39d29'42"
Otto and Marion
Cyg 21h14'3" 44d7'57"
Otto Bürger
Uma 11h11'13" 63d47'33"
Otto Duborg
Uma 11h32'0" 36d45'58"
Otto Dunkler
Uma 8h42'13" 56d37'59"

Otto Fischer
Ori 5h25'3" -8d15'54"
Otto Frederick Schick, Jr.
Sco 16h24'7" -28d1'14"
Otto Gonzalez
Aqr 22h12'11" -3d52'7"
Otto H Kopp
Lib 15h14'13" -17d52'45"
Otto Harris
Cep 22h1'7" 68d13'35"
Otto Helsper
Ori 5h57'21" 11d18'37"
Otto Kingsford Jones
Umi 17h2'26" 81d37'24"
Otto Lange
Uma 8h57'55" 57d11'28"
Otto Moritz
Uma 10h13'38" 43d13'5"
Otto Mustang Francis-Jones
Peg 23h57'34" 28d48'31"
Otto P. Morgensen, Our Glowing Star
Uma 11h34'12" 57d30'32"
Otto_Sum
Uma 9h41'46" 62d16'48"
Otto Waltl
Uma 11h7'57" 29d14'3"
Otto Wuhrmann
Uma 10h19'19" 40d25'47"
Otto, Karl
Uma 12h3'5" 50d36'18"
Ottoline.
Leo 10h48'59" 9d52'11"
Ottuki072205
Cyg 19h35'3" 28d2'3"
Otza
Aqr 21h53'31" 1d26'27"
Oui Si Karen
Uma 11h32'54" 56d59'42"
Ouida Darlene
Lmi 9h28'18" 33d51'16"
Ouissem Toumi
Sgr 19h30'0" -32d20'40"
Oulie
Ori 5h58'32" -0d41'33"
Ounjit
Ari 2h47'30" 21d41'31"
Our 20th Wedding Anniversary Star
And 23h11'54" 42d43'21"
Our 3 Angels
Uma 9h59'57" 47d46'22"
Our 4 A Star - Claire Florence Hunt
And 1h11'8" 35d25'49"
Our 5th wedding anniversary
Cyg 19h57'46" 55d26'33"
Our Accidental Beginning
Ori 6h14'44" 18d52'49"
"Our Afghanistan Star"
Lib 15h21'0" -27d25'19"
Our Alexandra
And 0h39'20" 42d59'18"
Our Amazing Grace
Vir 13h23'56" -19d6'39"
Our Andromeda - Rebecca Jayne Curran
And 0h55'31" 43d23'25"
Our Angel
Tri 1h55'50" 31d44'28"
Our Angel
Ari 2h43'54" 26d31'39"
Our Angel
Umi 13h20'11" 73d26'9"
Our angel above Carolyn Marie
And 23h41'57" 47d34'42"
Our Angel Angie
Uma 10h17'30" 48d2'28"
Our Angel Baby - Mason Kade Morgan
Sco 16h26'3" -27d14'32"
Our Angel Carla Modderno
Ori 5h47'41" 8d53'52"
Our Angel Danielle Mucci
And 23h2'32" 45d56'40"
Our Angel Ed
Her 18h0'2" 21d21'23"
Our Angel Ellie
Col 6h31'41" -38d51'13"
Our angel from heaven, Dominic
Vir 14h12'5" 3d13'12"
Our Angel Glynnie
Gem 6h54'28" 23d40'8"
'Our Angel' Hamish James Ingold
Cru 12h0'1" -60d39'3"
Our Angel Jennifer Lynn Thomas
Aqr 22h29'52" -0d28'22"
Our Angel Kristin Marie Rennert
Uma 8h36'10" 60d52'7"
Our Angel Laila Pearl Stott
Cru 12h47'10" -63d10'38"
Our Angel Lisa
And 2h31'21" 43d22'23"
"Our Angel" Sylvia Churchill
Cnc 8h48'39" 22d39'2"
Our Angel, Lewis Christopher Pryor
Per 3h40'34" 40d55'21"

Our Angel, Roland Alexander
Uma 10h14'32" 50d38'36"
Our Anniversary
Cru 12h10'3" -61d19'37"
Our Anniversary Star
Ari 3h14'48" 15d3'6"
Our Anniversary Star (April & Joe)
Ari 1h50'42" 19d15'49"
Our Auntie Sue
Cas 0h26'12" 56d59'7"
Our Baby
Uma 11h35'55" 38d15'51"
Our Baby Boy - Blake Edwin Gray
Col 6h18'44" -34d7'46"
Our Baby Boy, Dylan Dean Bailey
Vir 12h49'4" 4d14'6"
Our Baby Frizzell
Aqr 22h39'24" -1d55'35"
Our Baby West
Peg 21h41'17" 14d15'6"
Our Beagle Bunch
Uma 11h18'10" 35d1'11"
Our Beautiful Angel Nani Janet
Uma 9h16'55" 60d43'14"
Our Beautiful Baby Boy
Umi 16h27'48" 85d39'44"
Our Beautiful Baby Boy - Zachary
Lib 14h24'25" -20d45'37"
Our Beautiful Baby Girl - Hannah
Leo 10h52'8" 23d11'51"
"Our Beautiful Holly Jessica Star"
Vir 12h17'55" -0d54'50"
Our Beautiful Mum - Lyn
Cnc 8h27'14" 14d21'35"
Our Beginning
Sgr 19h18'25" -34d11'24"
Our Beginning
Sco 17h31'44" -40d45'58"
Our Beginning...Michael & Audra Hoy
Aql 19h19'36" 14d17'28"
Our Beloved Darian
Cyg 20h38'17" 36d11'20"
Our Beloved Dog - BEAR
Cma 6h45'49" -13d34'51"
"Our Beloved Flo"
Cas 0h43'16" 64d32'25"
Our Beloved J-Dawg
Cma 7h24'6" -16d0'50"
Our Beloved Layla
Uma 8h25'43" 61d24'28"
Our Beloved Memaw (Elva S. German)
Cyg 20h24'4" 46d53'22"
Our Beloved Mom Gloria A. Hartshorn
Sco 16h5'5" -17d33'34"
Our Beloved Mom & Grandma Esperance
Cas 0h30'40" 52d31'56"
Our Beloved Sooner Girl. Suzanne
Leo 11h51'1" 23d36'10"
Our Beloved, Steve Masterson
Vir 14h16'34" -12d45'4"
Our Beloved, Steve Masterson
Vir 14h14'8" -13d1'59"
Our Big Boy Hal
Aql 19h27'36" 3d52'0"
Our Big Grandad George
Uma 10h51'33" 45d42'51"
Our Blessed Sydney Rain
Cnc 8h25'8" 14d19'41"
"Our Boy Zeke"
Umi 16h33'0" 76d29'45"
Our Bright Friendship
Eri 4h38'5" -0d28'40"
Our Bright Light ,Ideen
Per 3h38'48" 36d21'9"
Our bright star Graham Cannan - Spud
Ori 5h11'50" -9d31'14"
Our Brightest Star - Belle Dale-Wills
Cas 1h56'56" 62d1'27"
Our Brilliant Mother Mandy Swanston
Ari 2h19'9" 13d25'52"
Our Brother Al (Jones)
Gem 6h43'59" 13d53'24"
Our Cath
Cas 1h19'7" 55d44'16"
Our Child, Alison and Paul Griffin
Uma 11h40'43" 53d0'55"
Our Chris
Per 2h11'11" 52d49'28"
Our Dad & Grandpa Bob
Aur 5h49'31" 51d20'43"
Our Dad - John Dolan
Aql 20h11'43" 6d27'58"
Our Dad Stuart Mitchell
Uma 12h1'51" 51d42'39"

Our Dad, Our Star-John W. Conforti
Sgr 20h14'0" -30d1'13"
Our Dad, Ste
Cep 0h7'7" 68d15'52"
Our Daddy, Nicholas Moyer
Sgr 17h58'35" -19d29'15"
Our Daddy, Steven Dixon
Vir 13h20'43" 12d13'14"
Our Daddy's Star
Aql 19h30'31" -0d13'15"
Our Dan
Cru 12h33'20" -58d41'13"
Our Daughter "Jude"
Uma 9h49'49" 54d27'58"
Our Daughter, Allana Barbara Hammel
Cnc 9h6'12" 23d7'25"
Our Dayle
Cru 12h32'14" -55d42'18"
Our Destiny
And 23h39'34" 47d41'9"
Our Dreams: Better Together
Ori 5h46'25" 7d44'19"
Our Dreams...Came True
Uma 11h8'13" 45d33'59"
Our Elaine
And 2h29'4" 50d32'7"
Our Endless Love
Ari 2h54'35" 21d57'42"
Our Eternal Love
Vir 11h38'40" 6d57'14"
Our Eternal Love - Antonia Kozicki -
Cru 12h39'10" -61d58'48"
Our Eternal Love. Kyle and Misty
Ori 6h18'18" 9d53'13"
Our Eternal Love Peter Fisher - Fish -
Cru 12h31'4" -60d27'16"
Our Everlasting...
Cyg 21h40'15" 46d51'38"
Our Everlasting Star of Love
Sco 16h8'32" -10d51'5"
Our Fairy Tale
Uma 9h6'18" 52d57'45"
Our Fairy Tale Love
Ori 6h17'11" -1d1'42"
Our Family Matt, Danielle, & Rita
Uma 11h14'42" 55d29'15"
Our Family Star - Sehorn
Leo 11h20'3" 2d38'21"
Our Family's Shining Light
Gem 7h23'59" 26d36'40"
Our Fire
Lib 15h10'23" -8d31'42"
Our First Anniversary
And 0h36'37" 40d58'18"
Our First Anniversary
Per 4h42'8" 38d45'9"
Our First Christmas
Uma 11h0'16" 45d26'18"
Our First Decade Star
Uma 9h25'29" 71d54'56"
Our First Star
Leo 11h13'31" 24d5'19"
Our Forever Love
Ori 5h39'33" 3d17'47"
Our Forever Love
Sco 16h11'26" -13d49'47"
Our Forever Star
Cas 1h36'33" 65d38'53"
Our Forever Star
Uma 10h32'26" 56d5'16"
Our forever star
Cyg 19h47'8" 38d56'58"
Our friend Elaine
And 0h40'6" 23d2'42"
Our Friend Sweet T
Lib 15h13'6" -26d51'52"
Our Friends
Cyg 20h57'25" 30d2'12"
Our Funny Uncle Punky
Cru 12h18'3" -61d37'52"
Our Future
Uma 12h3'54" 58d46'57"
Our Future Home...
Ori 5h31'45" 9d51'11"
Our Gift from God, Julian
Sgr 18h56'59" -18d56'2"
Our Goddess
Gem 7h38'27" 21d28'52"
Our Godmother Lisa
Ori 5h52'55" 22d40'3"
Our Granny
Col 5h47'37" -35d53'59"
Our Guardian Angel: Mark R Kraynack
Her 17h27'31" 23d56'17"
Our Guardian Angel Robbie
Per 3h20'22" 42d50'31"
Our Guiding Light
Ori 5h24'21" 3d39'58"
Our Guiding Light
Uma 8h25'7" 71d49'57"
Our Guiding Light-Marianne
Leo 11h3'35" 2d34'50"
Our Haiku
Eri 4h19'4" -33d38'39"

Our handsome guy, Stunner
Cnv 12h38'9" 44d19'48"
Our Hans
Cma 7h10'49" -30d18'20"
Our Haven
Uma 11h20'54" 64d8'12"
"Our Heart is Blind"
Gem 6h2'7" 24d35'48"
Our Heart, Little Liam McClintock
Tau 5h48'23" 22d23'32"
Our Heart, Our Love, Our Life
Crb 16h16'52" 29d17'23"
Our Hearts Will Always Shine
Sgr 18h28'41" -18d44'14"
Our Heaven
Umi 16h22'23" 82d19'34"
Our Heaven For Eternity
Tau 4h21'31" 13d6'21"
Our Hero Brian
Per 3h9'31" 51d10'59"
Our Hero Michelle
Umi 16h16'57" 70d55'56"
Our Honey's Light
Cyg 20h12'54" 54d12'46"
Our Hope
Lib 15h26'47" -6d17'24"
Our Hope
Vir 13h50'46" -16d54'32"
Our Infinite Future Found
Cas 2h1'28" 73d42'36"
Our Island Hut
Cam 4h12'31" 67d50'46"
Our "It" Love
Dra 16h38'55" 63d33'53"
Our Jewel
Crb 15h35'29" 26d16'27"
OUR JOURNEY
Cyg 20h21'8" 46d54'8"
Our Journey of Love
Lyn 7h53'18" 38d59'53"
Our Kingdom Amongst the Stars
Ori 5h55'43" 13d15'7"
Our Kiss
Cyg 20h21'44" 45d51'39"
Our Knight Star
Sco 16h58'45" -31d46'58"
Our Lady AEdovi
Ori 5h21'33" 4d35'6"
Our Lady Glenda
Lmi 9h29'6" 33d51'33"
Our Lifestyle
Cyg 20h52'52" 30d8'1"
our light
Pho 23h32'9" -46d46'30"
Our Lil Warrior Jacob Field
Ori 5h21'46" 7d14'34"
Our Little Angel
Gem 6h42'16" 15d17'49"
Our Little Angel
And 0h6'44" 45d20'31"
Our Little Angel
Umi 16h10'24" 80d10'16"
Our Little Angel, Joshua Blackwood
Gem 7h48'50" 19d1'42"
Our Little Giant
Psc 1h19'50" 31d46'46"
Our Little Hero
Per 4h5'56" 42d51'39"
Our little Jamie, Bronwen's brother
Cas 23h22'15" 57d51'20"
Our Little Jerezano Angel
Uma 9h26'15" 57d44'32"
Our Little Luke
Sgr 18h19'32" -23d47'40"
Our little miss Matilda Riddle
Cru 12h26'29" -58d48'26"
Our Little One
Uma 12h4'17" 57d22'9"
Our Little Peanut
Uma 8h41'18" 56d55'17"
Our little Prince Oliver Oxberry
Cep 21h27'51" 85d7'25"
Our Little Secret
Sgr 19h44'56" -18d32'15"
Our Little Star - Macie Ann McCloud
Lib 14h57'5" -11d37'46"
Our Love
Lib 14h31'46" -19d57'7"
Our Love
Cyg 20h18'59" 54d58'7"
Our Love
Uma 13h52'27" 55d27'23"
Our Love
Uma 11h41'0" 64d46'48"
Our Love
Uma 8h30'48" 67d45'20"
Our love
Umi 15h30'44" 73d1'38"
Our Love
Pho 23h45'55" -46d5'29"
Our Love
Lyn 8h12'54" 39d40'43"
Our Love
Lyn 7h39'6" 36d0'45"

Our Love
Cyg 19h47'28" 31d2'19"
Our Love
Cyg 21h15'35" 44d24'46"
Our Love
Cyg 19h59'33" 40d1'13"
Our Love
Cyg 19h57'36" 42d35'19"
Our love
Uma 9h28'14" 45d9'57"
Our Love
And 23h39'13" 47d40'1"
Our Love
Sge 20h7'50" 18d54'1"
~our love~
Tau 3h32'22" 5d32'41"
Our Love
Ori 5h41'31" 7d3'16"
Our Love
Tau 4h31'57" 11d40'1"
Our Love
Aqr 22h11'22" 0d56'51"
Our Love; All thanks to a paper bag
Cyg 20h1'16" 56d19'49"
Our Love Always Brian Kirkman - Jack -
Cru 12h38'55" -62d16'55"
"Our Love" B&M Ray 25 Years 15.04.82
Uma 13h25'48" 62d0'1"
Our Love - Beth & Ray Shepherd
Cyg 20h8'51" 38d16'9"
Our Love Everlasting R & N Mathey
Lyr 18h51'13" 34d52'0"
Our Love FOREVER
Per 4h48'8" 44d34'53"
Our Love Forever
Col 5h44'2" -39d16'10"
Our love is written in the stars
Cyg 19h46'15" 35d34'50"
Our Love Is Written In The Stars MF&AJ
Cyg 19h39'51" 38d24'3"
Our Love - John Hall - A Special Star.
Cyg 21h47'32" 50d12'3"
Our LOVE shines on.....Far Away
Per 3h22'32" 38d32'47"
Our Love Star
Cyg 21h30'30" 51d36'31"
OUR LOVE STAR
Uma 11h46'14" 52d16'28"
Our Love Star - David & Tammy
Aql 19h29'51" 3d1'44"
Our love was written in the Stars.
Psc 1h10'16" 11d2'44"
Our Love Will Always Shine
Sco 17h17'43" -30d52'45"
Our Love Will Light The Heavens
Uma 13h45'44" 52d15'47"
Our Love Will Shine Forever!
Lyn 8h28'11" 47d8'49"
Our Love Will Shine Forever
Cas 0h32'40" 62d53'27"
Our Love, Forever Burning Brightly
Uma 9h4'18" 53d56'4"
Our Love, Life, and New Beginning
Cyg 20h20'12" 33d20'41"
Our Loves Eternal
And 0h42'27" 38d34'6"
Our Love's Illumination
Sex 10h8'46" 4d57'19"
Our Loving Dad Richard Astill
Uma 14h6'10" 54d56'30"
Our Loving Mother Sonia Garner
Crb 15h52'20" 26d42'8"
Our Loyal Friend, Murdock
Cyg 20h10'18" 46d20'35"
Our Lucky Star
Umi 16h35'42" 82d41'26"
Our Lucky Star T.L. & R.H.
Lyn 6h54'16" 50d51'30"
"Our Mama" Lorraine Perry
Cas 0h7'2" 50d23'15"
Our Mamas, Nena
Uma 13h50'43" 60d35'14"
Our Margaret River Night Star
Cru 12h12'37" -60d7'33"
Our Matriach
Cas 0h56'2" 63d30'44"
Our Mema - Iva Strickland Cathcart
Uma 9h33'39" 48d2'56"
Our Mia
Cru 12h27'3" -59d45'51"
Our Michelle
And 1h20'13" 44d49'55"
Our Mike
Umi 15h36'30" 78d3'1"

Our Minute
Cyg 20h52'15" 35d41'24"
Our Mo
Cru 12h17'37" -56d23'39"
Our Mom Clerkin's Star
Leo 9h57'21" 16d2'56"
Our mom Epi is inspiration for life
Aqr 22h7'38" -2d14'16"
Our Mom: Juanita Johnson Rickey
Cap 21h6'57" -18d44'32"
Our Mom Robin Ann
Sco 16h10'6" -11d49'20"
Our Mom: Shirl The Pearl
Lib 14h48'8" -0d59'13"
Our Mom Suzi
Uma 10h28'58" 69d11'47"
Our Mom - The Very Best
Sgr 19h17'30" -13d45'44"
Our Mommy - Christy Jackson
Leo 10h19'37" 26d48'8"
Our Mommy Marie
Ori 5h27'12" 2d0'7"
Our Mommy, Alisan
Aqr 21h59'44" -16d46'1"
our moon
Sgr 18h28'38" -18d39'41"
Our Mother - A Gift To The World
Cas 1h31'24" 66d33'10"
Our Mother Maureen
Lyn 7h52'5" 35d17'32"
Our mother, Joan
Lyn 7h31'15" 35d20'42"
Our Mother's Love - Renee's Star
Sco 17h41'51" -41d47'11"
Our Nan
Leo 11h23'41" 17d2'44"
Our Nan Denise
Cas 23h58'51" 50d15'51"
Our Nana, Kay Evans
Leo 11h36'58" 25d59'11"
Our Nannie
Uma 11h45'55" 56d38'35"
Our New love
Psc 0h50'14" 7d36'38"
Our New Love, Bridget Rain
And 1h29'20" 47d42'38"
Our Newest Angel in the Sky
Umi 15h41'35" 85d57'41"
Our Northern Star
Uma 11h53'9" 34d45'2"
Our Notebook Kristy and Scott
Cyg 20h57'21" 47d51'3"
Our One In A Million Love Star
Sco 16h5'42" -13d58'43"
"Our Only Light in Paradise" SB&SH
Aqr 21h6'38" -9d32'43"
Our ORION
Ori 5h39'54" -0d19'37"
Our Own Shining Star, Linda Kislow
Lyr 18h54'3" 43d45'15"
our own special star "The Paula M"
Sco 16h3'28" -18d25'13"
Our Papá - JT Isely 6/19/15
Lyr 18h32'16" 34d12'6"
Our Papa Star
Aqr 22h50'40" -7d23'35"
Our Perfect Gem
Sgr 19h13'22" -23d30'46"
Our Place
Umi 14h25'22" 78d36'26"
Our Place
Cyg 20h2'28" 37d26'45"
Our Place in Eternity
Cyg 21h38'33" 52d2'36"
Our Place In Heaven
Uma 10h55'2" 36d3'6"
Our Place In The Sky
Uma 9h33'41" 72d11'59"
Our Planet
Per 3h3'29" 51d12'30"
Our Precious Lady
Cap 20h38'45" -18d39'34"
Our precious little angel Tayla May
Cru 12h38'3" -59d30'55"
Our Princess - Nicole Holder
And 2h28'3" 49d44'3"
Our Princess Taffie
Ari 3h7'33" 28d52'2"
Our Promise
Umi 4h33'57" 89d11'4"
Our Rock"Star" Daddy Don
Leo 11h2'22" -6d32'31"
Our Salsa
Umi 16h26'55" 76d7'1"
Our Sara Baby
Uma 9h45'8" 43d2'27"
Our Scott Johnston, a "pi" in the sky
Cru 12h27'41" -55d43'51"

Our Second Anniversary - G C & C W
Cyg 20h7'42" 55d35'53"
Our Shining Star
Cyg 21h45'53" 54d6'54"
Our Shining Star
Cyg 20h22'43" 43d40'11"
Our Shining Star
Her 18h21'31" 23d54'38"
Our Shining Star!
Ari 2h3'31" 14d40'51"
Our Shining Star Cecile
Sgr 19h26'50" -19d16'44"
Our Shining Star - Humphrey O'Donovan
Cru 12h28'40" -58d41'18"
Our Shining Star Johnny
Ari 2h1'42" 19d31'35"
Our Shining Star Samantha Frances
And 0h51'33" 46d11'55"
Our Shining Stars Maggi and Tami
Uma 10h28'17" 52d22'33"
Our Shooting Star
Aqr 21h27'3" 2d6'20"
Our Simkin Angel
Lyr 18h34'19" 27d16'21"
Our Sister Katie Lakiss On Her 30th
Vir 12h1'17" -0d6'28"
Our Smilin' Nanny
Boo 14h15'17" 49d41'28"
Our Soldier Nicholas Brandon Smith
Per 3h26'1" 48d36'31"
Our Son Gabriel
Sgr 18h56'18" -26d39'48"
Our Son Ron Homrich
Her 17h30'1" 44d19'37"
Our Son, Gerald A. Wingerter
Her 16h34'11" 38d24'11"
Our Son, Our Hero-R. Nathan Martens
Aqr 22h0'12" -23d6'6"
Our Sophie
Ori 5h32'4" 5d11'10"
Our Souls Are The Same May 30,1965
Uma 9h30'26" 53d20'17"
Our Special Angel
Lmi 10h31'52" 38d54'32"
Our Special Angel Francis
Ari 1h49'58" 19d26'53"
Our Special Cat - Louis
Lyn 8h2'1" 57d54'34"
Our Special Star
Cyg 20h3'3" 52d42'5"
Our Special Star Joan Clarkson
And 0h39'41" 45d16'59"
Our Spellbound Love 11:11
Lyn 7h14'45" 56d39'33"
Our Star
Uma 8h57'24" 56d52'35"
Our Star
Uma 9h41'28" 56d3'52"
Our Star
Cyg 20h28'41" 54d7'13"
Our Star
Cyg 21h56'51" 54d57'46"
Our Star
Uma 10h22'4" 69d16'24"
Our Star
Leo 11h22'42" -5d43'36"
Our Star
Umi 15h25'18" 78d22'41"
Our Star
Umi 14h48'20" 80d25'21"
Our Star
Cap 21h1'0" -22d24'48"
Our Star
Eri 4h19'37" -32d44'54"
Our Star
Col 5h55'53" -33d20'48"
Our Star
Per 3h12'16" 48d51'30"
Our Star
Cyg 20h30'32" 52d23'52"
Our Star
Cyg 20h32'42" 47d45'2"
Our Star
Cyg 21h18'59" 45d47'47"
Our Star
Cas 0h36'10" 54d45'18"
OUR STAR
Uma 10h32'44" 40d49'20"
Our Star
Cyg 20h27'1" 34d22'39"
Our Star
Crb 15h51'57" 36d49'48"
Our Star
Cyg 19h56'23" 30d16'21"
Our Star
Lyr 18h44'50" 35d3'27"
Our Star
Psc 0h18'13" 15d2'39"
Our Star
Ori 6h9'24" 9d11'26"
Our Star
Ori 6h23'31" 11d13'44"
"Our Star"
Ori 6h16'12" 14d3'46"

Our Star
Cnc 8h18'26" 7d45'34"
Our Star
Vir 12h39'32" 1d22'43"
Our Star
Tau 3h49'45" 24d20'55"
Our Star
Tau 4h7'22" 26d46'46"
Our Star "11:11"
Sco 16h59'42" -33d32'11"
Our Star - Always & Forever
Uma 8h25'23" 62d39'44"
Our Star Angel Danielle
And 0h52'2" 40d51'19"
Our Star Brenda Mantel 2003
Com 12h8'55" 14d47'28"
Our Star D&D
Psc 0h19'1" 15d16'53"
Our Star Derek McDonald
Per 3h47'55" 49d49'57"
Our Star Derek, Julie & Derek Jr.
Uma 9h34'20" 46d31'5"
"Our Star" Diane & Jim
Cyg 21h11'41" 53d3'35"
Our Star - Mam & Nanna, Hilda Dixon
Cas 1h16'39" 68d20'25"
Our Star - Matty and Sem
And 23h38'12" 45d39'13"
Our Star Molly 1491
Cap 20h17'12" -9d18'29"
Our star no matter where we are - C & S
Cru 12h27'30" -60d30'2"
Our Star. No Touching. No Touching.
Tau 4h27'30" 5d0'50"
"OUR STAR" Renée Lorene Russell
Psc 1h17'37" 17d21'6"
Our Star - Roless and Lars
Cnc 8h44'36" 30d59'39"
Our Star Shines - AMS Science Team 6A
Uma 13h18'8" 55d9'58"
Our Star Shines Forever As Our Love
Cyg 20h34'14" 34d39'10"
Our Star Tally Hobbs
Uma 8h14'21" 60d48'34"
Our Star Tamra
Lyn 7h57'44" 53d13'41"
Our star, Allen Simmons
Vir 14h53'32" 4d32'24"
Our Star, Dr Marsha A Marley
Oph 17h24'14" 7d46'22"
Our Star, True Companion
Gem 7h50'18" 18d56'0"
Our Starr
Cyg 20h42'16" 39d18'11"
Our Super Star Dad: Mike
Tau 3h43'24" 16d17'33"
Our Super Star GRAMPS
Leo 10h22'9" 13d4'45"
Our Survivor, Colleen
Gem 7h48'9" 31d42'23"
Our sweet angel Charz 4/4/90-8/4/05
Ari 1h50'26" 19d41'30"
Our Sweet Baby Dominic
Sgr 18h6'12" -27d50'36"
Our Sweet Boy, Adam Henry
Cnc 8h17'6" 16d40'59"
Our Sweet Inspiration~J.E. Bizzell
Cep 21h42'55" 62d32'6"
Our Sweet Judy
Lib 15h16'5" -21d17'20"
Our Sweet Patti "40" 10/14/06
Lib 15h44'43" -12d19'7"
our sweet Willow
Tau 5h23'32" 27d15'5"
Our Tahoe Moon
Cyg 21h53'4" 48d51'26"
Our Three Trees "4-3-01"
Tri 2h9'13" 33d20'36"
Our Toni - April 28, 2004
Umi 9h43'30" 87d1'18"
Our Town
Sgr 18h12'43" -20d2'58"
Our Treasured Find
Lyn 6h42'4" 61d17'8"
Our True Love, Masha & Kirk
Uma 12h4'13" 63d13'1"
Our True North
Uma 11h52'2" 56d43'4"
Our Trust
Cnv 12h40'15" 45d36'13"
Our Twins Baby A & Baby B
Cap 21h17'7" -19d57'0"
Our Unchanging Faith
Uma 9h57'27" 41d46'55"
Our Uncle Kev's Star
Sco 17h31'8" -42d19'11"
Our Uncle Terry
Umi 17h10'37" 78d49'25"

Our Union
Leo 10h45'59" 11d4'38"
Our Valhalla
Uma 11h42'35" 53d3'39"
Our Valiant Hercules
Her 16h57'17" 34d47'55"
OUR Velma Wife Mom Grandma Friend
Uma 8h57'52" 51d33'6"
"Our Wedding" Christine & Bill
Col 5h44'25" -32d54'49"
"Our Wedding" Deborah & Brian
Cyg 20h43'37" 31d19'20"
Our Wee Scalawag
Umi 17h7'26" 77d32'54"
"Our Will" Shining On
Uma 11h33'1" 48d37'29"
Our Window
Cnc 8h36'44" 17d20'51"
Our Wishing Star DTBPRTM Houlihan
Dra 19h25'5" 60d30'57"
Our Wittle Love Star-Danny & Rio
Eri 4h7'56" -35d56'29"
Our Wonderful Grandad - Douglas
Cep 2h30'24" 78d16'32"
Our Wonderful Granddaughter Naseem
And 1h9'35" 42d17'34"
Our Wonderful Mother - Colleen
Psc 23h27'28" 5d30'12"
Ourania & Athanasios 8/10/2006
Cyg 21h52'33" 47d59'20"
Ourania Indira Hephziba
Ari 2h39'23" 27d57'57"
"Ours"
Aql 19h12'11" -0d16'18"
OurShiningStar~NancyT.Harris,PP,PLS
And 1h27'40" 43d13'8"
Ourt
Aql 19h16'31" 5d36'24"
Ourville
And 23h11'36" 43d8'7"
Out of this world Mom Diana McClure
Tau 4h26'7" 17d8'29"
Outer Balewick
Vir 14h40'10" 3d13'54"
Outshining the Bear
Uma 11h20'12" 30d36'10"
OUVYT
Cap 20h12'38" -22d27'47"
OV
Lyr 18h46'48" 37d45'39"
OV Brodskiy
Uma 8h22'1" 66d23'50"
Ovacious
Dra 20h14'32" 64d4'22"
Ovcon
Her 17h36'22" 15d12'7"
Ovella
Aql 19h52'18" 11d13'8"
Over And Over Again
Aqr 22h6'57" -4d0'29"
Overcomer ~ Revelations 21:7 (KJV)
Uma 11h27'10" 45d49'37"
Overdrive (CAROLINE COLE)
Lib 15h29'40" -28d30'16"
Overländer, Fritz
Uma 9h2'22" 68d36'57"
Ovidia
Uma 13h53'1" 54d16'31"
Ovidio
Peg 22h58'32" 17d24'14"
Ovidio Cavalleiro
Cap 20h20'23" -9d20'3"
Ovidiu Dumitru
Dra 17h27'36" 66d20'46"
Owen
Cen 13h55'46" -58d31'38"
Owen
Ori 4h53'30" 2d23'14"
Owen Alexander
Sco 16h46'17" -31d57'41"
Owen and Lynne Anderson
Lyn 8h10'24" 37d18'34"
Owen and Patty Russell
And 0h40'58" 39d52'49"
Owen Anthony - Grandmas Little Man
Psc 1h2'30" 2d56'48"
Owen Bale
Psc 0h55'12" 11d7'38"
Owen Barker
Tau 3h35'24" 4d46'47"
Owen Beck Moriarty
Her 17h24'44" 23d58'2"
Owen Benjamin & Amanda Regina Regal
Lyr 18h28'49" 37d21'13"
Owen Benjamin Cavo
Her 17h21'5" 16d56'49"
Owen Carl Wisniewski
Com 12h49'57" 25d7'8"
Owen Carter Bitz
Aqr 23h25'23" -22d23'5"

Owen Charles Grosenbach
Her 17h21'10" 32d38'4"
Owen Cheung
Leo 9h27'57" 16d16'17"
Owen Christopher Bingham
Gem 6h23'7" 22d48'22"
Owen Christopher O'Toole
Lyn 7h23'28" 58d14'39"
Owen Christopher Ricks
Per 3h5'59" 55d27'28"
Owen Coogan
Uma 9h16'41" 59d18'35"
Owen Craig Beange
Vul 19h37'13" 27d31'53"
Owen Curtis Beesley
Her 17h12'35" 19d24'1"
Owen Daniel Peios
Dra 17h36'4" 67d45'35"
Owen Daniel Vivado
Sco 17h38'50" -32d42'19"
Owen David Buckholtz
Uma 12h17'14" 62d51'54"
Owen Dravenstott, Love Your Angel
Her 18h40'52" 20d32'34"
Owen Emil Graeber
Lyn 9h5'40" 37d46'53"
Owen F. Duffy III
Umi 15h44'6" 78d13'35"
Owen "Father, Husband, Friend"
Aql 19h36'18" 11d38'7"
Owen Francis Burns
Leo 11h43'15" 25d47'20"
Owen Grant Acosta
Ori 5h38'46" 1d59'59"
Owen Gray Wilson
Aqr 21h46'49" -1d5'59"
Owen H. "Sonny" Greer
Uma 9h53'43" 59d17'33"
Owen Harrison Brown
Per 2h56'25" 55d36'1"
Owen Harrison Stephens
Cyg 21h56'21" 43d9'35"
Owen Henry Brown
Per 3h40'54" 46d6'30"
Owen Henry Brown
Leo 11h17'6" 15d8'53"
Owen James Ameo
Lib 15h17'38" -11d51'10"
Owen James Scott
Vir 11h55'49" 4d27'45"
Owen John Dore
Her 17h51'36" 19d59'32"
Owen John Johansen
Tau 5h37'55" 24d16'51"
Owen Joseph
Her 16h23'13" 44d15'55"
Owen Joshua Malcolm (Brave Heart)
Lmi 9h55'0" 41d15'18"
Owen Keddal
Per 4h4'41" 36d41'1"
Owen Kelvin Vellutini
Per 2h58'48" 48d0'18"
Owen L. & Nathaniel B. Hallock
Uma 11h48'44" 58d24'12"
Owen Lester Dunlavy
Uma 9h20'49" 48d23'39"
Owen Lewis Fowler
Aqr 22h29'1" 0d19'25"
Owen & Marian Daly
Uma 9h49'18" 46d38'5"
Owen McIntyre Ray
Lyn 7h26'51" 53d52'35"
Owen Michael
Psc 0h56'43" 32d8'13"
Owen Michael Hall
Sgr 18h28'25" -16d40'7"
Owen Michael Howell
Cep 21h39'47" 58d14'23"
Owen Michael Matatall
Cnc 8h40'27" 31d52'56"
Owen Michael Osterman
Lyn 6h30'43" 57d21'19"
Owen Michael Rhoads
Her 17h14'52" 15d30'44"
Owen Michael Stanfill's Star
Psc 1h5'29" 13d28'38"
Owen Michael Woodson
Cep 21h3'30" 57d26'53"
Owen & Nancy Emery
Cyg 20h36'37" 48d10'1"
Owen Nathaniel Mathias
Per 4h10'43" 44d32'38"
Owen Niesen Anawalt
Peg 22h57'44" 19d13'19"
Owen O'Neill Parlante
Uma 9h21'41" 42d16'49"
Owen Orie Sloan Arnold
Tau 4h25'8" 3d30'47"
Owen Parish
Cyg 20h48'54" 35d8'43"
Owen Parsons
Cep 23h16'3" 81d19'13"
Owen Patrick
Her 16h34'59" 5d13'46"
Owen Patrick Novak
Uma 10h19'19" 43d39'48"
Owen Patrick Thomsen
Per 3h19'38" 54d18'20"
Owen Paul James
Gem 7h6'27" 31d27'54"

Owen Quinn Porter
Ari 2h16'55" 23d7'56"
Owen Raymond
Aqr 23h9'24" -9d27'51"
Owen Reed McClung
Uma 8h59'44" 51d15'17"
Owen Richard Dovey
Cep 21h57'41" 62d38'31"
Owen Richard Phillips
Per 4h24'51" 34d3'3"
Owen Richard Taylor
Cyg 20h46'1" 43d21'17"
Owen Robert Boyce
Lyn 8h10'40" 36d58'1"
Owen Ronald Little
Her 16h34'20" 44d38'28"
Owen S. Gaffey
Ori 6h18'47" 7d20'17"
Owen Sanders Schwab
Her 18h5'31" 35d13'19"
Owen Sean Michael Valentino
Her 16h44'10" 25d15'2"
Owen Sheldon
Ori 5h50'15" 2d13'52"
Owen Stedham's Star
Cam 4h27'10" 66d57'35"
Owen Steven Hussey
Per 3h38'49" 45d32'44"
Owen Stone Bigelow
Lyn 7h22'55" 44d45'48"
Owen Thomas
Lyn 8h0'59" 36d27'14"
Owen Thomas
Tau 5h42'11" 20d0'53"
Owen Thomas
Aql 19h11'53" 11d33'2"
Owen Thomas Kavounas
Lyn 7h43'43" 45d12'11"
Owen Thomas Lee Eastridge
Psc 1h31'21" 14d57'11"
Owen Timothy Kennedy
Gem 6h52'38" 32d39'11"
Owen Vincent Vinciguerra
Uma 10h31'47" 67d23'47"
Owen W Cole
Umi 16h24'4" 80d24'2"
Owena's Super Star
Cru 12h37'39" -60d4'27"
Owendio - Star of Hope
Cyg 21h18'0" 53d21'32"
Owen's Big Dream
Umi 15h24'11" 77d17'20"
Owen's Daddy
Her 17h43'25" 46d35'33"
Owen's dwarf star
Per 3h53'31" 52d39'31"
Owens Family
Uma 9h16'34" 48d30'40"
Owen's Light
Her 16h48'51" 35d45'51"
Owen's Orb
Cru 12h36'15" -60d7'2"
Owie J
Leo 11h21'12" -2d59'5"
Owin Loves Mommy
Umi 14h30'33" 68d39'54"
Owner Robert A. Crown, Crown Dodge
Ari 3h2'20" 19d7'51"
OXANA 19740121
Aqr 21h37'53" 0d57'25"
Oxana Childescu
Gem 6h1'27" 27d6'31"
(oxo) Heather & Jeff (xox)
Ori 5h11'45" 0d17'43"
OxOx Conan & Liz xOxO
Sco 17h43'9" -40d44'14"
Oy Jess
Lyn 7h0'11" 50d58'11"
Oyate Wayan Kapi
Ori 4h54'3" 13d50'15"
Oye Shen Yuen
Psc 23h13'17" -2d24'34"
OZ star
Boo 14h17'53" 51d32'50"
Oza (Frank Mendoza)
Ori 5h41'53" 6d27'47"
Ozan Yildirim + Olessia Niederer
Ori 6h18'11" 15d24'12"
Ozanich Class of 2007
Uma 13h42'2" 55d3'9"
Ozelle Mead Hinton
Psc 23h55'21" 0d36'53"
Ozge Sesenoglu
Aqr 22h32'58" -3d52'45"
Ozie
Cru 11h57'3" -64d0'8"
Özlem
Sco 17h47'1" -42d20'27"
Ozma
Uma 9h33'34" 42d45'3"
OZZ53
Aql 20h1'7" 12d39'39"
Ozzey Yancey
Ari 2h59'44" 27d49'36"
Ozzie
Uma 11h0'46" 35d1'43"
Ozzie
Uma 12h41'47" 60d8'53"
Ozzie Kenneth Murray
Cep 20h48'56" 59d6'23"

Ozzie Martinez
Uma 10h20'48" 51d2'7"
Ozzie & Tyna 9-10-2000 For Ever
Cnc 8h7'32" 25d25'57"
Ozzie & Veni's Star
Cas 0h27'49" 52d8'54"
Ozzie's Twinkle
And 0h24'41" 45d59'25"
P
Tau 4h22'56" 28d3'37"
P
Uma 9h25'32" 66d36'0"
P A ANDRUCZK USA
Cap 21h49'23" -14d54'24"
P. A. M.
Dra 15h24'51" 57d0'9"
P A Miani
Uma 8h38'46" 67d11'48"
P. Ann
Uma 11h27'49" 53d51'15"
P B P Loves Her Nerd
Uma 11h28'12" 60d33'12"
P Bear
Sco 17h54'16" -36d21'45"
P (Big P and the Kitties)
Gem 7h16'28" 28d4'12"
P&C . D&C POLO DE LA RIVA
Uma 8h50'7" 66d46'44"
P. C. S.
Per 4h2'49" 31d36'59"
P. Casseus
Sgr 18h13'29" -29d5'19"
P Daddy
Lmi 10h30'19" 37d44'51"
P Daniel
And 2h11'17" 37d33'14"
P Dazzels K
Cen 13h44'2" -42d51'33"
P Dixon
Aql 19h42'7" -0d1'23"
P & E's Paul Miller Star
Gem 6h29'2" 13d57'32"
P. Evon Culhane
Uma 9h11'26" 69d42'37"
P. G. Mayo
Gem 7h22'58" 32d13'24"
P. J.
Uma 11h35'2" 63d57'20"
P J Butler
Lyn 7h51'24" 42d39'43"
P J K - Best Dad & Friend Ever
Cra 18h48'18" -39d45'6"
P. J. Roup
Cep 20h50'33" 62d1'10"
P. J's Heart
Uma 8h57'17" 64d5'20"
P & J-The Reason
Cnc 8h14'57" 8d4'3"
P L Y - 1956
Ori 5h40'16" -1d49'50"
P. Laureano
Lib 15h48'49" -16d59'19"
P Louie Liska
Ori 5h52'19" 6d54'36"
P Love
Cam 3h47'46" 64d9'6"
P. M. Dobius Maximus
Men 4h13'25" -82d6'31"
P. M. Errico "Always & Forever"
Per 3h2'58" 53d14'14"
P. Mullan Diage
Uma 13h22'7" 61d11'54"
P. O. H.
Umi 14h37'7" 73d51'27"
P Peter Brawn
Aur 5h48'58" 55d36'26"
P. R. Prusiensky
Vir 13h45'4" -13d5'32"
P & S Shmily
Cyg 20h51'4" 49d37'6"
P_SIMON-CANAS1
Psc 0h33'48" 6d52'20"
P und M
Uma 14h4'4" 52d45'12"
P. Weber
Ser 18h20'58" -13d40'32"
P3X-774 - (The Scatena Home World)
Peg 23h10'47" 25d59'57"
Pa
Cep 22h34'28" 75d33'4"
Pa
Cep 22h10'17" 69d32'47"
Pa and Gramma D 20012006
Sgr 19h15'12" -17d46'5"
Pa Haas
Uma 8h24'52" 62d36'24"
Pa Holbourn
Her 16h40'41" 14d13'41"
Pa is a Star
Uma 11h22'37" 58d25'46"
Pa Kinney
Ori 5h11'48" 4d36'32"
Pa Pa
Tau 4h35'31" 22d12'41"
Pa Pa Bear (George)
Uma 13h19'28" 56d9'20"
Pa Pa / Dad Romano
Cnc 8h47'38" 27d51'16"

Pa Pa's " Bushel and a Peck" Star
Cyg 20h9'12" 40d35'48"
'Pa Teakle'
Cru 12h53'54" -60d46'36"
Pa Xiong
Uma 10h16'52" 67d44'49"
Paavo
Cnv 13h35'33" 31d33'43"
P.A.B. 123
Uma 8h29'46" 70d45'59"
Pabete Eldoudou 30 09 87
Cma 7h10'20" -26d12'35"
Pabla Mercedes
Psc 1h11'0" 18d54'59"
Pablo
Per 4h3'47" 32d48'34"
Pablo Data
Ori 5h27'55" 4d51'32"
Pablo Felipe Doria
Cnc 8h24'5" 10d33'49"
Pablo Herrero
Cyg 20h5'36" 30d19'17"
Pablo Javier Agosto
Per 3h30'5" 49d24'52"
Pablo & Kristina
Ori 5h53'17" 7d8'58"
Pablo Trillo
Ori 5h28'52" -0d30'41"
Pablo Wei-Han Liang
Psc 1h21'23" 27d21'18"
Pablo Wellington Weber
Aur 5h29'52" 45d48'26"
Pablo y Miriam
Mon 6h59'55" 2d16'18"
Pablo Zequeira
Cap 20h18'41" -10d8'15"
Pablo-Margot
Mon 6h8'40" -10d59'21"
PAC MAN
Her 16h16'47" 43d38'37"
Pacáiste Réalta
Uma 11h23'46" 33d8'52"
Paccione Star
Cyg 20h4'45" 51d20'58"
Pacha Mama
Uma 9h51'6" 68d16'50"
PACHS Stars 2005 (Room 132B)
Her 17h29'24" 47d39'55"
Pacia Brown's Eastern Star
Cam 5h43'20" 64d48'28"
Pacific International Hotel School
Ori 5h31'3" -2d53'1"
pacificpearl
Pup 7h13'5" -41d2'58"
Pacita Tana
Mon 8h6'58" -5d33'53"
PackRat
Equ 21h16'30" 11d4'38"
Packsum
Uma 11h40'58" 60d1'27"
Packtrick
Uma 13h20'5" 57d35'6"
pacman
Sgr 18h20'2" -32d21'39"
PAC-MAN
Tau 4h22'0" 3d21'28"
Paco Lafuente Bueno
Aql 19h53'59" -0d9'29"
Paco Star
Ori 6h4'50" 19d45'6"
Paco Taco
Aqr 22h36'12" 0d42'42"
Pacscaleve 16 juin 2005
Peg 22h39'8" 19d45'24"
PAD
Uma 11h20'12" 59d48'19"
Padak
Psc 1h16'54" 20d57'20"
padalec
Gem 6h52'12" 14d20'2"
Paddy
Lep 5h13'29" -11d15'37"
Paddy
Cep 0h40'3" 83d2'8"
Paddy "Cha Cha" Dugan
Tau 4h22'27" 14d58'31"
Paddy & Claire Always, Now, Forever
Cyg 21h14'21" 47d19'34"
Paddy John Jennings
Her 17h7'36" 47d50'18"
Paddy K
Vir 13h26'55" 5d28'33"
Paddy King
Cep 0h13'4" 74d31'22"
Paddy m'boy
Gem 7h18'15" 19d20'45"
Paddy Nevin
And 23h24'18" 41d57'56"
Paddy Sherlock
Per 4h6'36" 38d4'51"
Paddy-Dos
Her 16h42'23" 22d44'2"
Paddywack
Uma 10h32'33" 44d56'39"
Padee Khang
Lyr 18h35'40" 36d14'7"
Padma
Cnc 7h59'21" 13d31'18"
Padmini Kamat
Lyn 6h57'16" 44d39'31"

Padraic Felton McNeirney
Lyn 7h5'52" 55d45'9"
Padre
Ari 3h17'50" 28d34'16"
Padre 60
Tau 5h30'51" 19d7'59"
Padre Bernardino
Ori 6h13'3" 15d16'33"
Padre Poppy Frombola
And 0h31'44" 33d21'50"
PADRECITAMOS
Her 17h26'31" 22d27'54"
Pädu & Anja in Love for eternity
Cas 0h59'56" 63d18'41"
Padysmack
Lmi 10h4'5" 36d18'33"
Paetra Pramstaller
Tau 4h34'16" 17d41'37"
Pag
Lib 15h43'29" -28d25'17"
Pagan Scarlet Westbrook
And 1h0'27" 45d32'55"
Page Amanda Gamel
Cas 0h11'1" 50d10'6"
Page Fuller
Sgr 19h44'4" -13d1'34"
Page Jeannette Geach
And 0h35'28" 31d36'7"
Page Patrick
Umi 15h13'13" 78d9'34"
Page Ware
Ori 5h31'29" 9d24'2"
Page Whetsell
Uma 10h32'17" 57d58'11"
Pagel, Erhard
Uma 10h57'3" 42d53'50"
Page's Star
Leo 9h49'39" 9d58'27"
Pagets Love
Lyn 9h35'33" 40d11'50"
Paggie goes to maul
Com 12h47'58" 24d34'44"
Pagius
Cma 6h50'54" -11d37'13"
Pagly Watmough-Scott
Dra 18h49'56" 50d40'19"
Pagoda
Uma 13h35'35" 56d35'49"
Pagoto1939
Uma 10h34'41" 49d21'18"
PAGS II
Gem 6h43'8" 20d11'16"
Pahaliah
Uma 9h18'17" 66d4'58"
Pahla Schoenfeld
Lyr 19h11'32" 26d35'12"
PaHoua Thao Phang
Gem 6h26'5" 25d22'39"
Paige
Gem 7h33'0" 26d47'56"
Paige
Tau 4h21'40" 27d51'32"
Paige
Peg 21h41'51" 23d20'25"
Paige
Cnc 9h20'31" 26d22'39"
Paige
Boo 14h51'52" 14d4'50"
Paige
Cas 0h9'13" 53d47'2"
Paige
Cas 0h13'25" 50d58'10"
Paige
And 2h16'35" 45d0'59"
Paige
Psc 1h2'10" 32d51'49"
Paige
Cas 2h28'41" 62d48'22"
Paige
Uma 10h2'46" 59d21'21"
Paige!
Vir 13h1'56" -16d58'51"
Paige - 1 Today, Love Grampy and Nana
And 23h40'52" 38d5'28"
Paige 2106
And 1h8'37" 37d4'0"
Paige Adele Grinnell
Ari 3h28'51" 22d28'35"
Paige Alese Moomey
And 0h15'52" 36d8'45"
Paige Alexa Mezey
Sco 16h54'34" -38d22'31"
Paige Alexandra Leonard
And 0h46'41" 30d23'2"
Paige Alexandra Pflanz
And 23h26'8" 47d4'24"
Paige Alexandra Spreeman
Cas 3h29'6" 68d30'1"
Paige Alexandra Tebbe
Crb 15h22'39" 25d58'21"
Paige Allison Goren-Levithan Star
Cas 23h27'41" 52d18'19"
Paige Allison W
Gem 6h54'53" 14d31'42"
Paige Alyse Edwards
And 2h12'34" 39d11'16"
Paige Alyson Vitale
Lib 14h52'29" -2d30'23"
Paige Alyssa Reilly
Gem 6h59'22" 34d36'42"
Paige Amber Godard
Psc 22h51'32" 3d45'7"

Paige Anastasia
Tau 3h55'28" 25d42'23"
Paige and Jeff Markun
Mon 7h15'40" -0d19'27"
Paige and Kris
Lib 15h6'48" -7d12'10"
Paige and Nate
Per 4h25'3" 52d32'14"
Paige Anderson
And 2h15'16" 50d19'28"
Paige Ann McCall
And 1h8'49" 43d46'40"
Paige Anne Lapointe
Tau 4h26'48" 13d18'2"
Paige & Annslee Collins BFF
Cyg 19h50'39" 38d11'4"
Paige Ariana Skye Ganoe
Sco 16h19'29" -18d22'12"
Paige Batchelder
Psc 1h9'21" 25d26'51"
Paige Besse
Aqr 23h1'46" -7d25'30"
Paige Christina Jackman
Tau 4h32'38" 20d29'58"
Paige Cole
Uma 12h37'17" 59d18'52"
Paige Collett Mitchell
Tau 4h14'17" 8d35'55"
Paige Covington
Dor 5h19'5" -67d56'41"
Paige Danaé Abney
And 1h9'33" 41d40'3"
Paige & David
Cnc 9h12'47" 7d15'21"
Paige Dawson Kington
Gem 6h44'47" 29d50'15"
Paige Detweiler
Gem 6h34'14" 14d59'38"
Paige Elise
And 23h21'47" 46d31'32"
Paige Elise
Lib 15h19'42" -11d17'11"
Paige Elizabeth
Cas 1h55'53" 61d41'41"
Paige Elizabeth
And 23h17'19" 47d52'41"
Paige Elizabeth
Lyn 7h16'8" 49d0'18"
Paige Elizabeth
And 23h18'49" 35d35'25"
Paige Elizabeth
Uma 11h11'32" 42d30'44"
Paige Elizabeth
Aqr 21h21'30" 0d30'15"
Paige Elizabeth
Leo 9h32'38" 27d20'44"
Paige Elizabeth
Vul 20h27'59" 27d53'41"
Paige Elizabeth Farnam
Cyg 19h36'6" 28d47'17"
Paige Elizabeth Lowe
Psa 22h28'18" -30d16'7"
Paige Elizabeth Pavidis
Ari 2h51'28" 29d51'48"
Paige Elizabeth Reisinger
And 23h7'56" 48d30'44"
Paige Elizabeth Trent
Uma 11h23'43" 46d49'29"
Paige Ellen Crandall
Lyn 7h36'2" 39d12'35"
Paige Elyse Daldy Abraham
Cra 18h45'35" -41d34'51"
Paige Emersyn Begley
Cam 4h25'9" 58d48'37"
Paige Erin Mason
Uma 11h35'23" 57d24'2"
Paige Evelyn Leishman
Psc 1h6'55" 11d35'19"
Paige & Felicita
Del 20h43'25" 16d18'33"
Paige Fisher and Justin Morello
And 1h53'30" 46d24'51"
Paige Godard
Umi 16h17'37" 75d55'32"
Paige Grace
And 0h16'36" 34d28'37"
Paige Hackenberger
Leo 11h46'32" 20d6'18"
Paige Haley
Psc 1h4'43" 13d52'11"
Paige Hall
And 0h40'51" 46d31'49"
Paige Jordan Lauer
Psc 1h28'55" 5d28'2"
Paige Ladawn
And 1h55'44" 44d53'15"
Paige LeeAnn Busdiecker
Ari 2h35'40" 26d50'42"
Paige Leigh Smalling
And 0h32'1" 27d29'48"
Paige Lesley Malter
And 0h31'56" 35d8'6"
Paige Lewis
Sco 16h18'36" -17d35'58"
Paige Lilly Bass
And 22h57'43" 51d14'12"
Paige Lily
And 1h52'57" 36d46'56"
Paige Lily-Ann Jordan
And 0h48'28" 23d39'36"
Paige Mackenzie
Aqr 22h28'57" -0d55'46"

Paige Maclean
Peg 22h38'11" 7d37'2"
Paige Marie
Mon 6h46'8" 7d59'39"
Paige Marie
Sgr 18h14'18" -19d10'59"
Paige Marie Egli
Uma 9h33'45" 43d43'13"
Paige Marie Sandvik
Dra 19h13'18" 73d7'36"
Paige Marie Solomon
Uma 10h47'49" 50d42'21"
Paige Martin
And 1h42'42" 43d55'9"
Paige Matthews
Col 5h57'46" -35d24'14"
Paige Mckenna Dempsey Brown
And 0h41'53" 26d43'11"
Paige McKenzie Goodman
Aqr 23h7'44" -11d45'49"
Paige McKenzie Healy
Uma 12h19'47" 56d59'57"
Paige Meredith Linden
Leo 9h29'11" 28d25'18"
Paige Michelle Kraatz
Cnc 8h13'38" 25d12'28"
Paige Michelle Reischl
And 0h29'23" 37d53'48"
Paige Michelle Winters
Peg 22h41'51" 26d11'24"
Paige Monaghan
Leo 9h23'18" 10d35'24"
Paige Morgan Breneman
Lib 15h21'38" -26d7'33"
Paige Morgan McLeod
Ari 2h50'31" 27d34'16"
Paige Morning Star Corsello
And 1h0'37" 41d53'1"
Paige Nicole
Psc 1h5'13" 7d23'38"
Paige Nicole
Sco 16h10'52" -11d13'46"
Paige Olivia Kessler
Aqr 23h6'55" -8d21'27"
Paige Olivia Wilkins
And 23h6'2" 44d49'44"
Paige Oneto
And 23h37'33" 36d33'37"
Paige P F A Ward
And 2h3'45" 36d36'37"
Paige P Mikkalo
Ari 2h13'46" 25d17'56"
Paige Parker Horan
Sco 17h23'15" -45d10'14"
Paige Pohle
Lib 15h46'41" -29d21'58"
Paige - Pokey
Lyn 6h41'21" 52d18'4"
Paige Possanza
Psc 1h26'26" 13d48'9"
Paige Povlick
Cap 21h42'18" -22d33'15"
Paige Rayann Jerrett
Aqr 23h55'46" -16d51'53"
Paige ReaAnn's My Little Buttercup
Per 3h48'57" 38d45'53"
Paige Rebecca
Uma 8h19'14" 66d55'9"
Paige Roy
Crb 15h48'24" 37d28'58"
Paige Ryan
Cet 1h19'1" -0d48'46"
Paige Sakura Iino
And 0h36'25" 29d26'50"
Paige Sandra Murie Suter
Umi 15h18'45" 74d34'18"
Paige Schmid
And 0h28'6" 42d53'39"
Paige Valerie Thomas
Ori 5h16'52" -4d38'18"
Paige Victoria
Vir 13h35'38" -8d12'23"
Paige Victoria Buen - Heaven's Light
Cnc 8h34'4" 21d0'33"
Paige Victoria Casebere
Vir 12h39'24" 6d55'14"
Paige Whitney Morgan
Aqr 20h55'54" -10d21'48"
Paige Winchester
Ori 6h22'52" 13d56'31"
Paige Wright
Tau 3h33'25" 26d15'19"
Paige-Lee's Sparkling Ray
Gem 6h42'39" 24d9'7"
PaigeNicole830
Ori 6h5'35" 6d47'40"
Paige's Gold Star
Sgr 17h53'39" -29d50'41"
Paige's Shining Star
Aur 5h42'21" 37d1'4"
Paige's Star
And 23h38'40" 41d36'29"
Paige's Star
Aqr 20h53'9" 2d15'13"
"Paige's there she is so big"
Uma 10h11'44" 43d2'44"
Paigian
Lib 15h34'43" -28d2'22"
Paigie - Doodle - Amy
Lyn 7h31'32" 46d46'30"

paihsleah sakers
Lyn 9h11'36" 38d30'20"
Pailey Marrus Vitale
And 2h29'46" 37d51'27"
Paint Eating Soozy
Cyg 20h16'57" 53d37'39"
Paisley Gabrielle Thomas
Sco 17h56'6" -32d58'18"
Paisley Jenkins
Cap 21h1'41" -23d3'51"
Paisley Rain Murray
And 23h11'40" 48d0'10"
Paisley Rose Lutz
Lib 15h3'50" -1d32'44"
Paito
Ori 5h28'0" 3d29'37"
Paiton Elizabeth Newbill
Uma 9h50'35" 69d26'16"
Paityn Elizabeth Gannon
Cas 0h22'56" 54d55'57"
Paityn Rose Novosat
And 0h41'37" 31d14'26"
Paix
Cru 12h29'29" -60d0'14"
paixão
Ari 2h8'45" 14d5'49"
Pak Kam Sit
Lib 14h51'49" -1d28'3"
PakA
Aqr 22h34'42" 0d13'46"
Pakelekia ipo
Uma 13h45'36" 60d42'47"
Paki Akil Mills
Gem 6h50'45" 33d28'30"
Pakis Gustafson
Lmi 10h33'44" 28d54'27"
Pal Daniel Simak
Ori 5h24'48" 3d39'2"
PAL Forever
Aql 19h22'39" -11d14'21"
Palace Resort Shiroishi Zao
Sgr 18h41'40" -19d10'30"
Palagano Gabriele
Cas 0h25'2" 61d40'26"
Palamasi
Dra 16h59'11" 58d50'12"
PALAMOUNTAIN
Cyg 19h36'25" 35d25'4"
Palangga - Vilma F. McGough
Mon 7h32'28" -0d44'26"
Palawa Warrior Craig Shaw
Pho 0h24'24" -54d56'12"
Palco
Ori 5h27'26" 10d33'58"
Palemine's Passion
Crb 15h39'3" 39d20'44"
Palenaki
Mon 7h15'14" -0d51'7"
Palestine
Peg 22h34'51" 7d3'54"
Pálfi Róbert
Umi 16h7'32" 75d28'25"
Palice
Cam 4h36'48" 66d49'34"
Palidan
Pho 0h41'2" -57d22'27"
Palies Diablo
Sco 17h54'20" -35d45'22"
Pallas Galadriel Kennedy
Uma 11h43'53" 33d1'42"
Pallavi
Cas 0h5'35" 56d40'51"
Palm, Matilda
Uma 9h38'40" 57d58'57"
Palma Ciccocelli Selisker
Ari 2h57'4" 20d26'36"
Palmer and Anne Seeley
Lyn 7h53'7" 50d15'47"
Palmer C. McNeal
Cam 4h18'22" 69d32'33"
Palmer Dimitri Quarles
Uma 8h36'22" 71d55'39"
Palmer Milton/Walker
Uma 12h56'28" 54d33'49"
Palmer Nicole Smith
Cnc 9h8'15" 18d5'30"
Palmers Ruby Pentagram
Cyg 19h47'8" 34d4'3"
Palmieri
Psc 0h15'32" 6d1'21"
Palmira Santos
Cap 21h4'50" -20d24'0"
PalmisandNina
Cyg 21h35'9" 38d7'17"
Palmyre Guile
Gem 6h23'14" 21d54'37"
Palo Wing
Del 20h40'18" 14d16'44"
Paloma
Ari 2h15'56" 22d57'11"
Paloma
And 0h31'58" 26d35'40"
Paloma
Her 17h59'28" 50d6'9"
Paloma
Ari 2h58'21" 30d14'23"
Paloma
Sco 16h9'52" -31d16'44"
Paloma Filleris
And 23h16'47" 47d26'24"
Paloma Jasmine Payne - Trigas
Cas 23h2'6" 56d28'48"

Paloma Lozoya
Lib 15h0'4" -22d35'11"
Paloma Medina
Col 5h41'55" -40d11'29"
Palomita
Leo 11h27'18" 26d4'38"
Palvic
Umi 14h29'55" 68d19'14"
Pam
And 23h26'1" 47d54'24"
Pam
Peg 22h22'33" 33d34'54"
Pam A. Medeiros - My Lil' Beauty
And 23h13'17" 43d53'57"
Pam Adiutori
Cas 0h33'53" 62d57'27"
Pam Allen
Vir 13h24'41" 13d4'0"
Pam and Shelby Chermak
Lyn 7h42'15" 59d5'56"
Pam and Vic Star
Cyg 19h37'42" 38d51'21"
Pam Anne Martinez
Tau 5h9'50" 25d18'48"
Pam & Bill Klauser (The Chippitz)
Cyg 21h24'19" 37d18'51"
Pam Brester
Uma 10h32'20" 69d50'24"
Pam Conlin
And 0h36'7" 25d2'41"
Pam cosi dolce
Cyg 19h34'45" 51d30'50"
Pam Danaher
And 23h12'1" 47d55'12"
Pam Fantozzi
Gem 7h43'24" 31d23'16"
Pam Fiske
Lib 15h3'2" -9d53'40"
Pam Gaves
Lyn 9h22'4" 35d35'27"
Pam Gerber
Sgr 19h24'31" -34d10'9"
Pam Graf Nick's Guiding Light
Psa 22h33'59" -25d8'13"
Pam Griffin McMahon
Cyg 21h36'6" 42d46'5"
Pam Gruzynski's Zonar
Dra 18h48'59" 55d11'16"
Pam Handwerk
Cam 3h58'35" 53d50'51"
Pam Harris
Crb 16h10'40" 33d39'5"
Pam Hensley
And 0h22'52" 31d35'11"
Pam Hood
Aqr 22h32'48" 0d48'44"
Pam Hoover
Lib 15h15'48" -12d17'41"
Pam Hopper
Tau 4h4'3" 7d12'32"
Pam Jackley
And 0h41'44" 25d56'53"
Pam & John Devanny - 05 March 1966
Cru 12h31'36" -62d13'20"
Pam & Jules
Uma 11h0'34" 44d15'56"
Pam K. Roberts
Cas 0h23'55" 64d11'8"
Pam Kapolka
Tau 5h22'57" 27d32'35"
Pam Keefer
Cas 0h39'0" 51d34'21"
Pam Kelly
Tau 4h9'8" 16d34'32"
Pam Kelly Best Mom in the Universe
Crb 16h5'51" 33d23'24"
Pam Kent
Tau 4h32'54" 17d29'15"
Pam Kovacevich
Mon 5h57'56" -6d29'8"
Pam "Lamby" Quimby
Uma 11h12'46" 35d31'41"
Pam Leech
Crb 15h34'39" 27d25'51"
Pam Leggat
Aqr 22h35'11" -0d53'50"
Pam Lyn Taveira
Sco 16h4'53" -12d23'8"
Pam MacDonald
Lmi 10h30'59" 33d12'37"
Pam Maliwauki
Cap 20h38'37" -20d21'25"
Pam Melton
Mon 6h44'27" 7d21'24"
Pam Milam
Per 3h45'17" 44d53'18"
Pam "Miss May" Mayfield
Cas 0h28'42" 55d26'8"
Pam Molina
And 23h26'59" 45d42'1"
Pam "MoonStruck"
Cap 20h24'25" -15d39'54"
Pam Moore
Tau 5h28'31" 21d42'31"
Pam Munk
Uma 11h46'55" 31d17'14"
Pam Murphy "My Shining Star"
And 23h15'17" 48d12'7"

Pam Orman
Lmi 10h28'9" 33d12'21"
Pam Perrett
Dra 19h28'52" 62d21'41"
Pam Phillips
Tau 5h0'8" 18d53'41"
Pam Pile
And 0h6'15" 45d34'43"
Pam Rosta
Psc 0h39'12" 9d24'22"
Pam Ryan
Cas 0h18'8" 55d41'41"
Pam Sullivan
Pho 0h27'42" -41d16'47"
Pam Summers
Lib 14h53'36" -4d35'17"
Pam Trautman
Cas 1h43'44" 61d44'40"
Pam Venus
Cru 12h54'8" -64d2'53"
Pam Violet-Nixon
Tau 5h25'19" 22d1'15"
Pam Ward
Uma 10h9'45" 66d10'30"
Pam Wells Shines On
Sco 16h57'28" -42d5'8"
Pam Winters
Uma 9h34'11" 58d48'41"
Pam Yvonne
Cap 21h57'24" -21d1'4"
Pam, Nathan, Callie Daughn-Wood
Sgr 18h59'45" -30d4'19"
Pam, will you marry me?
Ori 5h35'59" 4d13'13"
Pama
Umi 11h47'47" 88d32'4"
Pama Lou R. "Mimi" Pendleton
Uma 12h24'5" 59d45'8"
Pama Luch
And 2h23'17" 49d20'38"
Pamadangdang
Per 3h21'48" 44d58'32"
Pamalamadingdong 04061967
Ari 2h45'23" 12d1'41"
Pamatheena Arlene
Psc 1h30'12" 10d59'47"
Pambearly
Ori 6h12'36" 21d2'9"
Pamela
Gem 7h4'27" 21d22'12"
Pamela
Leo 10h5'41" 17d1'5"
Pamela
And 0h22'17" 25d57'10"
Pamela
Leo 10h16'58" 24d55'26"
Pamela
Tau 4h30'58" 28d47'29"
Pamela
Ari 2h50'5" 13d5'46"
Pamela
Psc 23h32'17" 2d40'1"
Pa-Mel-A
Ori 6h9'35" 3d19'23"
Pamela
Lyn 9h19'24" 34d28'51"
Pamela
And 23h13'13" 48d13'25"
PAMELA
Cyg 20h9'5" 51d56'9"
Pamela
Her 16h41'46" 47d46'37"
Pamela
Uma 10h20'29" 50d48'58"
Pamela
Ori 5h36'34" -1d32'47"
Pamela
Aqr 22h45'35" -21d30'59"
Pamela
Sco 16h16'57" -11d31'9"
Pamela
Lib 15h2'59" -13d21'45"
Pamela
Cap 21h33'15" -8d29'25"
Pamela
Aqr 22h50'24" -8d21'47"
Pamela
Sgr 20h1'39" -44d38'26"
Pamela
Sco 15h55'0" -23d1'5"
Pamela A.
Per 3h19'0" 51d17'34"
Pamela A Dziurgot
Cas 1h46'25" 64d24'9"
Pamela A. Schultz, Super Wife
Oph 16h23'42" -0d38'42"
Pamela Allen
Tau 4h38'11" 28d6'12"
Pamela Amy
Crb 16h14'8" 32d52'58"
Pamela and Arthur Smith
Her 17h17'17" 46d36'2"
Pamela and Daniel
Sco 16h5'27" -18d19'33"
Pamela and Dennis Stuart
Uma 9h6'55" 48d38'1"
Pamela and Lamar
Uma 9h45'53" 67d15'11"
Pamela and Thomas Chess
Cap 21h16'3" -26d32'59"

Pamela Angelus Ori 5h28'59" -4d9'41"
Pamela Ann Lyn 6h54'50" 59d9'5"
Pamela Ann Beesley Cap 20h33'21" -22d9'27"
Pamela Ann Brady Cyg 21h31'36" 37d20'47"
Pamela Ann Doyon & Sam Edward Doyon Tau 5h44'58" 26d27'12"
Pamela Ann Farnsworth Mon 6h50'34" 6d34'9"
Pamela Ann Gray Psc 1h4'34" 27d4'17"
Pamela Ann Happy 50th Birthday Mum Cru 12h38'50" -60d41'18"
Pamela Ann Kelland And 0h39'3" 45d9'17"
Pamela Ann Keller Vir 13h14'56" 5d48'32"
Pamela Ann Meyers Lyn 7h1'53" 56d45'35"
Pamela Ann Phelps And 0h16'38" 28d42'9"
Pamela Ann Wojas Cas 1h36'8" 68d55'47"
Pamela Anne Hutteball Gem 7h23'42" 21d14'52"
Pamela Anne Veronico And 0h32'8" 35d0'27"
Pamela Ashleigh McGinnis Cnc 8h44'30" 10d11'48"
Pamela Ashley Tau 4h22'5" 25d2'23"
Pamela atque Theodoricus Semper Cam 4h22'10" 71d45'27"
Pamela Auspland Wagner And 2h21'40" 50d6'8"
Pamela Battaglia Guillory Leo 9h38'29" 28d36'56"
Pamela Beth Cnc 8h50'39" 10d43'45"
Pamela Beth Dougherty Gem 7h0'19" 34d46'57"
Pamela Beth Krajewski Aqr 22h48'31" -11d11'4"
Pamela Biondio Lib 15h25'14" -26d34'7"
Pamela Blue Dra 16h10'16" 65d10'16"
Pamela Bott Cas 23h31'3" 52d52'55"
Pamela Bowlin Foster And 0h7'46" 35d27'51"
Pamela Byrd And 0h42'1" 32d27'31"
Pamela C. Carney Cyg 21h12'31" 47d3'27"
Pamela C Cremona Lyn 7h3'57" 57d17'54"
Pamela C Hélou, angel Psc 0h6'3" 0d49'16"
Pamela Carlson Psc 1h12'20" 28d28'30"
Pamela Carr's Wild Flower Star Gem 6h56'36" 29d36'38"
Pamela Charlene Elmore Dra 16h11'25" 61d8'30"
Pamela Christmas 70th birthday star And 23h9'27" 42d49'59"
Pamela Clare And 2h11'19" 41d32'46"
Pamela Clark Bradfield Aur 5h9'55" 43d29'7"
Pamela Colette Latimer Holden Peg 23h12'18" 9d56'40"
Pamela Cooke Lyn 7h6'0" 52d28'38"
Pamela Corwin Sgr 18h30'9" -35d5'58"
Pamela Coward Leo 9h31'29" 16d35'43"
Pamela Crystal Kain Cnv 12h41'15" 39d45'19"
Pamela D. Foley Lyn 8h24'37" 37d12'12"
Pamela & Dale Gem 6h46'15" 17d17'40"
Pamela Dawn Birchem Uma 12h21'16" 54d59'0"
Pamela Dawn Bunting Umi 14h24'56" 75d0'10"
Pamela Dawn Hunter Tau 3h45'22" 22d10'10"
Pamela Dea Cap 20h37'39" -16d42'40"
Pamela Dee Camp And 23h7'14" 48d42'28"
Pamela Dee Thweatt Vir 13h33'51" -18d33'41"
Pamela DeMarco Cnc 8h5'11" 14d35'54"
Pamela (Do You Believe??) Langweil Ari 2h55'27" 26d34'7"
Pamela Doris Cas 1h37'57" 62d19'20"
Pamela Drewniak Crb 15h47'28" 36d0'31"
Pamela & Eddie Padgett Lyr 19h15'20" 26d56'8"
Pamela Edwards Peg 22h22'11" 6d45'0"
Pamela Elaine Earnest Cas 1h9'30" 65d59'56"
Pamela Elizabeth Uma 8h38'57" 48d5'58"
Pamela Ellen Peg 22h23'56" 24d9'18"
Pamela "Equus" Kaminska Lyn 8h15'16" 57d18'37"
Pamela Esther Smith Leo 9h29'42" 27d25'23"
Pamela Evelyn Hodge Umi 14h26'35" 66d57'35"
Pamela Faith Fellows Lyn 6h36'34" 57d48'30"
Pamela Fant-Saez Ori 5h34'33" -1d8'23"
Pamela Faye Brimmer Ari 3h19'18" 27d28'5"
Pamela Feldmayer Aqr 21h40'11" -1d6'58"
Pamela Flanagan Mon 7h15'58" -0d48'46"
Pamela Frasure Cyg 21h45'34" 44d1'31"
Pamela G. Castaldi Her 17h0'27" 31d19'58"
Pamela Gayle Cyg 21h42'30" 46d8'44"
Pamela&George Rapp Anniversary Star Cas 0h42'13" 62d11'12"
Pamela Gisele Crawford Cas 0h36'37" 53d41'45"
Pamela Grace Uma 9h13'46" 47d14'27"
Pamela Grace Gem 7h30'12" 16d15'38"
Pamela Grace Allen And 0h21'46" 29d33'55"
Pamela Grace Haworth Uma 10h32'32" 51d2'38"
Pamela Grant Cornell Cyg 20h23'48" 44d47'3"
Pamela Greenyer - 50 & Fabulous Peg 21h38'52" 18d54'50"
Pamela Grigg Cas 1h50'3" 64d54'46"
Pamela Hailey Lyr 18h40'27" 39d38'1"
Pamela Hall And 23h44'27" 38d3'32"
Pamela Hardman-Hennessey Gem 6h43'12" 14d2'9"
Pamela Hathaway Cyg 21h33'12" 49d51'27"
Pamela Hechinger Lyn 8h39'31" 35d51'15"
Pamela Hudson Crb 16h18'1" 35d16'32"
Pamela Irene Kelly Cnc 8h33'3" 10d32'19"
Pamela Irene Marr Ford Vir 14h4'19" -0d51'57"
Pamela J. Bell Cap 21h4'24" -15d41'20"
Pamela J. Clifford Sgr 19h45'43" -20d29'6"
Pamela J. Miller Cas 0h14'25" 53d44'24"
Pamela J. Pearson And 2h34'13" 44d3'47"
Pamela J Perez And 1h52'35" 36d30'3"
Pamela J. Pursh Cas 23h27'36" 53d7'19"
Pamela J. Salassi - My Princess Vir 13h49'34" -5d12'26"
Pamela Jane Sgr 18h12'6" -21d48'55"
Pamela Jane Cnc 7h59'36" 18d5'57"
Pamela Jane & Angela Janet Danner Vir 13h35'41" 10d36'1"
Pamela Jane Carrino "My Love" And 23h56'41" 45d48'51"
Pamela Jane Chadwell Lib 15h25'2" -7d45'18"
Pamela Jane Mabry Crb 15h53'53" 39d34'29"
Pamela Jane Stanton - 20.04.2005 Cru 12h27'38" -59d25'36"
Pamela Jean Sco 16h37'50" -27d21'19"
Pamela Jean Ori 5h22'18" 1d28'31"
Pamela Jean Cyg 19h25'51" 28d36'27"
Pamela Jean Arnold Cyg 21h15'34" 33d32'41"
Pamela Jean Correll Vir 13h20'16" -2d40'44"
Pamela Jean Cowan Ori 6h19'41" 8d37'46"
Pamela Jean Hough Ari 2h21'5" 23d28'55"
Pamela Jean Kidd Crb 15h46'3" 26d58'36"
Pamela Jean Lampis Tau 4h21'33" 26d16'17"
Pamela Jean Limbach Crb 15h45'4" 35d3'52"
Pamela Jean Pulley Cap 21h38'19" -22d30'16"
Pamela Jean Raynak Crb 16h4'58" 26d24'41"
Pamela Jean Reuscher Aqr 23h1'33" -9d25'9"
Pamela Jean Ziegler Cnc 8h21'54" 30d58'55"
Pamela Jeanette Lyn 8h21'58" 55d47'41"
Pamela Jezek Gem 7h10'55" 33d43'24"
Pamela Jo Crb 16h21'11" 37d24'17"
Pamela Jo Uma 10h18'32" 40d34'49"
Pamela Jo Maucher Uma 9h39'57" 48d59'23"
Pamela Jo "P.J." Cap 20h51'17" -21d40'20"
Pamela Jolene Hamilton Cyg 19h35'38" 29d8'30"
Pamela & Joseph Court Cyg 19h45'17" 33d26'19"
Pamela Josephine Serra And 0h18'56" 36d47'13"
Pamela & Joshua ~ December 26, 2005 Per 4h10'29" 49d42'41"
Pamela Joy Cas 2h50'36" 61d0'16"
Pamela Joy Sco 16h8'57" -31d16'19"
Pamela Joy Berg-Meissner Cnc 8h52'1" 15d29'0"
Pamela Joy O'Neal Riddell Dra 18h45'58" 70d16'5"
Pamela June Bagniuk Lib 15h0'43" -0d55'9"
Pamela K Sgr 19h48'22" -14d8'2"
Pamela K. Harper Vel 10h27'48" -42d3'8"
pamela K. maddox Sco 17h50'20" -39d1'21"
Pamela Kay Bailey And 1h42'38" 50d13'51"
Pamela Kay Flohr 1960 Cnc 8h48'45" 17d22'35"
Pamela Kay Miller Vir 14h21'57" -21d17'27"
Pamela Kay Rivers Cam 7h21'20" 62d14'8"
Pamela Kelley Bloodworth Tau 5h34'27" 24d2'30"
Pamela L. Englert Umi 13h21'35" 75d5'32"
Pamela L. Leno Cam 7h28'43" 61d26'11"
Pamela L. Waterston Cas 23h32'37" 55d28'3"
Pamela Leann Clower Bentley Leo 11h34'30" 16d11'12"
Pamela Lee Higgins Pho 1h14'18" -41d59'7"
Pamela Lee Roper-Strouse Cas 1h42'39" 63d3'7"
Pamela Liesl Davidson Uma 9h22'54" 45d34'50"
Pamela Lin Cas 1h28'23" 66d57'5"
Pamela Loomis July 23, 1949 Per 3h21'41" 40d22'18"
Pamela Louise Greenaway Cyg 20h40'44" 51d47'32"
Pamela Louise Mansfield Uma 8h38'5" 59d33'7"
Pamela Lynn Lyr 18h43'38" 35d52'57"
Pamela Lynn Carriker Uma 9h6'47" 65d39'57"
Pamela Lynn Deluca a.k.a. Lina And 0h42'5" 36d48'18"
Pamela Lynn Sears Gem 7h9'4" 21d45'1"
Pamela Lynn Wicks Tau 5h24'22" 18d34'47"
Pamela M Scott Cas 23h34'19" 54d45'8"
Pamela M. Stompoly-Ericson Ori 5h47'34" -2d39'54"
Pamela M Whitehead Uma 11h19'33" 62d59'15"
Pamela M. Wilkening Ori 5h10'54" 15d3'8"
Pamela M. Young Grandchild of IGD Cas 0h25'6" 57d34'53"
Pamela Mae Gem 6h24'20" 23d16'24"
Pamela Marie Cnc 8h35'12" 15d16'8"
Pamela Marie Bowles Lyn 7h57'37" 43d6'17"
Pamela Marie Cassel Cnc 8h45'1" 30d49'44"
Pamela Marie Kiseda Sgr 19h15'34" -21d34'35"
Pamela Marie Rossi And 1h0'24" 34d26'38"
Pamela Marie Tedesco Cas 0h22'7" 61d57'4"
Pamela Mastrosimone Cap 21h42'4" -24d9'12"
Pamela Matteo And 2h2'27" 37d37'21"
Pamela McLaughlin Cas 0h33'25" 58d48'11"
Pamela McTeman And 1h3'27" 44d45'31"
Pamela Michelle Pirkle Vir 13h13'30" -20d34'37"
Pamela Mingolelli Lib 15h10'54" -16d21'55"
Pamela Mongiat Lib 15h7'14" -3d30'37"
Pamela Morasch Lyn 7h53'20" 41d50'26"
Pamela Morroll Cas 23h51'17" 53d25'49"
Pamela Mossey Psc 23h7'25" 1d40'42"
Pamela Nicole Bridges Cnc 9h16'32" 30d27'2"
Pamela Nicole Hughes Sco 17h41'32" -38d28'7"
Pamela Niles-Eternally Yours-Shaun And 0h7'22" 46d29'33"
Pamela Onea Wood Lib 15h39'34" -22d33'26"
Pamela Orbe And 23h4'11" 46d57'50"
Pamela P. Kramlich And 2h27'42" 42d24'49"
Pamela & Patrick Occhino June 4th Cyg 20h15'22" 52d4'0"
Pamela Patton Lyr 18h34'57" 36d36'43"
Pamela Pendleton Cnc 8h55'38" 10d1'18"
Pamela Poirier Com 12h30'33" 28d19'17"
Pamela R. Capel Ari 2h11'26" 23d57'4"
Pamela R. Harris Uma 11h29'47" 59d1'3"
Pamela R. Page Lib 14h52'48" -2d52'30"
Pamela R. Salinas *Hisstellar* Lib 15h41'8" -16d57'6"
Pamela Rea Anderson Xanders Granny Umi 16h9'48" 70d40'22"
Pamela Reynolds Lyr 18h43'12" 34d23'50"
Pamela & Rich Her 18h4'0" 25d16'29"
Pamela Ring Leo 9h35'47" 26d29'13"
Pamela Rose Cru 12h18'27" -56d24'37"
Pamela Roumas Cam 4h58'25" 61d2'53"
Pamela Royer Her 18h7'42" 37d41'11"
Pamela Ruth DeFleron And 0h45'19" 34d9'14"
Pamela Ruth Pitman Cyg 20h48'12" 37d10'29"
Pamela Ruth Vincent Cnc 8h8'1" 26d19'44"
Pamela S. Davis Leo 11h26'1" 26d3'4"
Pamela Sandoval Cas 1h8'59" 66d35'38"
Pamela Sandridge Leo 11h38'49" 14d32'15"
Pamela Sarro Cas 1h48'10" 65d21'25"
Pamela Shields Lib 14h30'2" -23d40'24"
Pamela - Shining Brightly Ori 4h44'19" 4d55'47"
Pamela Smolcznski And 1h30'21" 47d31'30"
Pamela Smolins *Mom* And 0h10'35" 33d37'6"
Pamela Sollitto Lyn 7h38'39" 49d19'23"
Pamela Sommers (Mom) Ori 5h7'12" 15d56'29"
Pamela Southard And 1h29'24" 39d33'19"
Pamela Sowards And 2h6'0" 43d23'16"
Pamela Sowers(Venglar) Oph 17h27'20" -22d40'36"
Pamela Sue Vir 12h36'22" 0d51'16"
Pamela Sue Abbio And 23h4'26" 48d25'49"
Pamela Sue Correa Uma 8h50'9" 65d25'56"
Pamela Sue Dayhuff And 0h42'15" 36d25'51"
Pamela Sue Hart Cyg 19h33'36" 29d50'6"
Pamela Sue Hensley Uma 9h5'33" 56d30'41"
Pamela Sue Higgins Oct. 21, 1944 Lib 15h12'28" -10d33'19"
Pamela Sue Johnston Uma 11h47'37" 35d53'31"
Pamela Sue (Johnston) Butler Vir 11h47'1" -1d27'38"
Pamela Sue Joseph Sgr 19h14'48" -16d32'32"
Pamela Sue Prator Lmi 10h10'52" 32d2'19"
Pamela Sue Trask Lib 15h49'31" -8d26'21"
Pamela Sue Turner Cap 20h23'38" -12d41'34"
Pamela Sue Wade Ari 2h1'25" 14d39'27"
Pamela Susan Alcott And 0h14'27" 39d46'20"
Pamela Suzanne Smith Lib 14h46'18" -8d57'40"
Pamela "Sweetspirit" Denman Peg 21h46'21" 21d49'19"
Pamela Sypniewski Ari 3h23'51" 22d8'58"
Pamela Tanner Hall Ari 2h31'31" 19d25'16"
Pamela Taylor Coddington Lyn 8h31'37" 41d1'14"
Pamela Tharp Cas 1h16'18" 62d0'9"
Pamela "The Mama" Gear Cas 3h37'20" 74d56'18"
Pamela Thornton Yanke Gem 7h18'16" 33d34'42"
Pamela Vali Mullen Leo 10h8'35" 26d26'51"
Pamela Van Gorp Cas 0h56'13" 59d32'39"
Pamela Varkony Gem 6h45'26" 21d25'28"
Pamela Watson Ari 2h37'8" 29d33'53"
Pamela Wells Crb 15h53'43" 28d1'26"
Pamela Wessling Scl 23h29'25" -38d43'18"
Pamela Whitehouse 'Simply The Best' Cas 0h0'37" 59d12'18"
Pamela "Will You Marry Me" Adamson Lyr 19h0'16" 28d53'49"
Pamela Woodrum Cnc 8h13'56" 16d23'15"
Pamela Wornom Cas 0h56'34" 54d59'40"
Pamela Y Simmons Psc 0h15'46" 6d12'54"
Pamela, Lossie's Christmas Light Cma 6h49'40" -21d54'0"
Pamela31 Mon 6h52'30" -0d20'17"
pamelacointin Sco 17h23'25" -45d29'37"
Pamela-Karen Cyg 20h4'5" 39d3'28"
Pamelanyah Crb 15h45'44" 35d58'12"
Pamela's Dream And 1h52'44" 45d47'45"
Pamela's Eternal Light Vir 14h10'31" -5d21'50"
Pamela's Family Star Uma 9h54'33" 48d41'56"
Pamela's Place Tau 4h46'47" 20d17'51"
Pámela's Star Aqr 22h25'46" -22d42'14"
Pamela's Star Stuff Psc 1h22'1" 28d11'37"
Pamella Reiman Crb 15h43'46" 26d50'9"
Pamelove Uma 11h11'27" 31d28'34"
Pami And 1h14'55" 42d42'6"
Pami Kohli Lib 15h53'48" -14d11'15"
Pamie And 0h25'34" 25d22'9"
Pamie and Brent Gem 7h46'41" 20d55'2"
Pamily Leo 9h53'13" 8d34'14"
Pamis N Mausey Lib 15h30'48" -12d35'30"
Pamm & Brack Cox Leo 11h38'31" 14d25'51"
Pammi (Padmini Rao Eladasari) And 23h12'36" 46d44'57"
Pammie Leo 11h31'42" 28d9'13"
Pammy And 1h27'29" 44d46'6"
Pammy Lib 15h19'59" -6d17'57"
Pammy Ori 5h56'31" -0d48'17"
PAMMY STAR ONE Lyn 8h43'22" 35d1'33"
Pammy Yvonne Shanton Elam Lmi 10h24'9" 36d21'2"
Pammy's very own Shining Star Cap 21h34'13" -8d38'4"
Pamojaloma Uma 10h0'38" 48d19'19"
Pamoline Dra 17h15'28" 51d17'6"
Pamollie Cam 4h7'6" 67d10'52"
PAMPI Sgr 19h31'49" -21d43'35"
Pampy Crb 15h46'45" 36d35'39"
Pam's 54th Vir 13h44'26" 2d6'35"
Pam's Galactic Volley Cet 1h12'8" -0d36'59"
Pam's Highway Star Cas 0h18'42" 52d21'35"
Pam's Love Ori 6h18'32" -0d15'54"
Pam's Love Star Aql 18h43'7" -0d13'1"
Pam's One Bright Spot Uma 9h25'41" 48d24'47"
Pam's Smile Tau 4h18'34" 27d24'26"
Pam's "Stairway To Heaven" Uma 11h15'39" 34d38'56"
Pam's Star Psc 1h4'5" 13d54'33"
Pan de la Tierra Ari 2h33'47" 25d10'42"
Pan Haskins Gem 6h8'57" 24d48'22"
Panagiota Julie Sco 17h52'31" -36d34'41"
Panagiota T. Revas-Worrall Psc 0h56'41" 19d28'1"
Panagiotis Boretos Uma 9h33'50" 47d11'33"
Panagiotis Giannakis Uma 9h26'3" 41d35'18"
Panagiotis & Isabel Vel 10h30'50" -53d42'52"
Panagiotis Kioussopoulos Uma 8h48'26" 70d40'14"
Panagiotis Pavlos Labrakis Sgr 17h45'42" -21d29'43"
Panagiotis Siablis Uma 10h17'13" 49d52'39"
Panagos Lemos Uma 11h41'15" 59d58'27"
Panama Al Lib 14h52'42" -24d33'9"
PANAMA - TANZSCHULE Her 16h31'48" 10d11'53"
Panama's Power - The King of Ivy Cep 22h10'7" 60d3'3"
Panayiotaki and Effoula "B & B" Cyg 21h5'4" 36d47'58"
Pancake Vir 12h42'13" 2d12'47"
Pancake Gem 6h44'28" 17d33'20"
Pancakes Uma 9h48'1" 69d28'44"
Panchito Lmi 10h45'40" 34d56'38"
Pancho Lib 14h50'22" -2d56'26"
"Pancho" / Hector F. Puig Ari 2h49'10" 12d52'33"
Pancho Villa Gem 6h6'28" 24d39'19"
Panda Leo 11h37'21" 21d0'40"
Panda And 2h23'24" 45d36'39"
Panda Uma 9h21'58" 63d49'13"
Panda Bear Sco 16h34'6" -34d57'22"
Panda Star Lib 14h45'30" -14d20'24"
Panda's Smile Sco 16h50'30" -39d27'51"
Panda's Star Cnc 9h8'47" 31d6'44"
Pandas-GB Tau 3h43'29" 23d45'17"
Pandey Uma 9h10'24" 66d50'54"
Pandit Ramchandra Kantak Cap 21h31'52" -15d49'47"
Pandora Cap 21h38'58" -14d33'39"
Pandora Cru 12h36'26" -58d6'40"
Pandora Crb 16h4'2" 25d43'47"
Pandora Uma 11h47'0" 34d19'52"
Pandora Lynette Brown Crb 15h41'10" 35d49'6"
PandR74 And 0h9'12" 47d6'25"
Paneena & Baby Haynes Uma 11h13'44" 39d43'2"
Panet Coleman Gem 7h35'24" 29d20'5"
Pang Dee Moua Uma 11h51'14" 39d59'38"
Pang Dee Yang Sgr 18h23'59" -35d16'0"
Pang Houa Vang Cam 4h27'59" 62d50'59"
Pang Khou Yang Vir 14h17'8" 0d7'1"
Pangduner Uma 10h37'45" 55d30'27"
PangHimeNoBakaChanChueBakaSama Ori 6h8'7" 19d5'56"
Pangshua Khang Psc 0h38'26" 10d48'55"
Panier's Pilot Per 3h24'25" 47d23'36"
Panino Ori 5h3'23" 5d22'34"
Panjshir Kitten Vir 12h42'15" 6d47'41"
Pankhuri Chopra Cnc 8h53'6" 23d52'34"
Pankotsch, Rolf Dieter Uma 8h50'19" 49d33'31"
Panny And 1h27'8" 50d32'50"
Pano "Stuff" Leo 9h41'42" 28d22'57"
Panos Georgiou Danos Lyn 7h41'29" 39d2'13"
Panos Kyriakopoulos Uma 11h29'37" 46d35'45"
Panson Te Quiero Mucho~Muchos Besos Cyg 21h14'51" 32d9'27"
Pansy Jo Psc 1h44'11" 9d30'47"
Pansy L. Fengfish Psc 1h17'31" 27d8'19"
Pantea Farnejad Uma 11h36'28" 55d7'39"
Pantelic-Comet Union Cyg 20h45'21" 44d50'53"
Pantelis Panteliadis Uma 13h56'9" 54d47'31"
Pantelis Skulikidis Umi 14h15'45" 75d52'0"
Pantera Noir Uma 13h33'9" 56d12'56"
Panther Angel Zeitner Uma 9h36'7" 44d40'21"
Panthera tigris Gem 6h28'17" 16d5'33"
PantherWoman Cyg 19h41'3" 30d28'55"
Pantotini Agapi Lmi 10h2'8" 28d27'26"
Panza Uma 13h0'11" 58d3'49"
Pao V Vu Cam 4h28'16" 62d56'50"
Paol1 Cam 7h19'28" 78d2'19"
Paola Umi 17h2'20" 77d46'53"
Paola Umi 17h30'51" 80d3'29"
Paola Cam 4h10'34" 73d53'45"
PAOLA Cas 2h10'29" 74d45'9"
Paola Cas 23h10'28" 58d10'49"
Paola Sgr 18h21'41" -32d14'25"
Paola Cas 0h40'17" 54d25'35"
Paola Her 18h55'58" 24d13'48"
Paola Boo 14h56'9" 18d5'14"
Paola Tau 4h3'43" 5d16'32"
Paola Peg 23h12'11" 13d25'52"
Paola Del 20h47'8" 11d47'7"
Paola Boo 14h4'47" 36d6'24"
Paola Crb 16h20'58" 38d59'41"
Paola Cas 0h47'52" 54d23'33"
Paola Alessandro Crb 15h31'17" 36d5'56"
Paola and Tina Peg 22h18'25" 32d55'10"
Paola & Antonio Aur 6h12'21" 53d40'33"
Paola Barbieri Crb 16h23'26" 35d47'7"
Paola Barraza Sgr 19h36'43" -42d25'3"
Paola Bertschy-Duca Lyr 18h56'5" 33d35'48"
Paola Bottone Her 18h39'17" 19d57'27"

Paola Calcagnini
Cep 23h3'13" 75d28'52"
(Paola Capai) - bimba
Uma 10h6'11" 51d21'36"
Paola Chicca
Cas 1h5'58" 69d33'54"
Paola Ciccone
Per 2h52'34" 47d40'38"
Paola D'Adda
Ori 5h0'49" 11d10'52"
Paola De Lumé Mosca
Tri 2h31'14" 30d20'58"
Paola Di Ianni
Umi 14h27'41" 69d59'3"
Paola e Giorgio Manfredotti
Crb 16h21'33" 32d45'5"
Paola Eleonora Politi
Cam 6h40'58" 67d22'22"
Paola Ferri
Crb 16h19'44" 29d54'37"
Paola Floriana
Uma 9h11'10" 55d24'16"
Paola Giovanetti Edstrom
Cnc 8h53'55" 30d50'35"
Paola Guerrasio
Uma 9h10'18" 49d50'54"
Paola & Horacio
Tau 3h54'49" 26d57'49"
Paola Joan Santana Lepe
Sco 16h7'17" -15d39'1"
Paola Larovere
Aur 5h38'13" 33d0'46"
Paola Longo
Umi 16h14'4" 85d5'16"
Paola Lorena Acosta Marin
Mon 6h28'50" -4d45'49"
Paola Maffei
Boo 14h50'57" 23d41'45"
Paola Maria Chiara Maccari
And 0h45'35" 35d56'19"
Paola Marnati
Cyg 20h44'46" 32d40'51"
Paola Moise
Lyr 18h15'17" 35d18'12"
Paola Morales
Aqr 21h37'36" -1d41'48"
PAOLA NAYAR
Aqr 21h59'50" -16d53'15"
Paola Pallotti
Boo 14h40'51" 52d51'4"
Paola Pilocane
Cas 23h38'23" 56d51'53"
Paola Piretti
Ori 6h14'48" 3d20'52"
Paola Poggo
Lyn 6h37'14" 61d18'19"
Paola Simoncini
Ori 5h23'22" -0d36'13"
Paola Solarte Pérez
Lyn 8h30'44" 34d55'51"
Paola Spano
Ori 6h18'43" -1d28'9"
Paola Tosini
Psc 0h29'26" 11d27'8"
Paola Vanessa Hernandez
Gem 6h45'45" 31d56'13"
Paola's Light
And 0h41'41" 36d43'4"
Paoletta
Peg 22h38'1" 27d18'35"
Paoli Paola
Per 3h7'13" 51d41'54"
Paolina Carla Victoria
And 1h7'45" 45d23'27"
Paolina Maso 07/08/1983
Leo 11h15'18" 2d15'54"
Paolita
Lyn 6h46'26" 60d58'12"
Paolo
Cas 0h31'33" 62d25'7"
Paolo
Uma 12h35'20" 56d17'43"
Paolo
Cep 23h15'53" 79d24'0"
Paolo
Umi 13h7'7" 75d41'38"
Paolo
Cyg 19h45'46" 46d37'2"
Paolo
Her 17h48'53" 38d43'48"
Paolo
Aur 4h50'40" 36d48'14"
Paolo Ammendola
Aur 6h36'48" 34d27'44"
Paolo Andrea Malizia 08.06.1940
Cep 23h9'22" 70d26'30"
Paolo Angelo Sabatini & Daniela Giorgini
Peg 22h12'25" 19d29'7"
Paolo Arturo
Sgr 19h16'41" -19d19'1"
Paolo Bandinelli
Cyg 19h47'8" 30d30'53"
Paolo Bertani
Ori 5h54'53" 11d28'13"
Paolo Bonini
Cam 4h0'46" 54d13'54"
Paolo Cassio
Tau 3h24'47" 17d18'2"
Paolo De Ciantis
Cep 22h13'3" 68d20'10"
Paolo Di Valdi
Lib 15h14'5" -20d28'54"
Paolo - Dylan Costa
Peg 21h30'35" 23d30'56"
Paolo e Manuela
Her 18h20'21" 22d40'34"
Paolo Elia Formentini
Aur 5h52'34" 32d32'52"
Paolo Fappani
Umi 17h10'5" 77d12'15"
Paolo&Franci
Cam 3h38'53" 54d24'13"
Paolo Granchi
Sco 16h29'2" -26d20'32"
Paolo Malchiodi
Per 3h38'51" 45d18'12"
Paolo Manzoni
Peg 23h42'41" 29d56'6"
Paolo Nassi
Cas 23h3'33" 54d13'52"
Paolo Pallotto
Cyg 19h50'21" 33d39'32"
Paolo Panero
Crb 16h21'17" 29d39'0"
Paolo Petrolo
Per 3h17'51" 51d17'13"
Paolo Spina
Ari 2h4'26" 12d29'0"
Paolo...Gingerino!
Lyn 8h48'16" 37d45'26"
Pap paw Kenneth & Nana Ruth
Sge 20h1'30" 17d50'4"
Papà
Peg 23h38'25" 26d7'31"
Papa
Peg 22h44'35" 23d10'20"
PAPA
Psc 0h56'46" 9d59'46"
Papa
Aql 19h45'43" 7d49'50"
Papa
Per 4h24'23" 44d2'25"
"PaPa"
Per 3h5'50" 38d57'53"
Papa
Per 3h9'29" 38d25'52"
Papa
Uma 9h46'19" 52d23'16"
Papa
Uma 11h23'40" 58d37'2"
Papa
Uma 9h39'25" 53d30'34"
Papa
Lib 14h50'26" -3d16'54"
Papa
Sgr 18h5'42" -21d21'6"
PAPA
Aqr 22h14'6" -21d55'22"
Papa
Sgr 19h17'41" -13d26'11"
PAPA
Sgr 18h16'23" -27d16'7"
Papa Al
Aql 19h47'4" -0d38'12"
Papa Al G.
Aqr 22h13'36" -1d31'36"
Papa Al Waggener
Ori 5h33'50" 2d24'7"
Papa Alain Simard
Uma 11h26'22" 60d8'9"
Papa Alan
Crb 16h11'29" 39d3'10"
Papa and GG Black
Lyn 7h3'44" 60d39'24"
Papa and Gramma
Cyg 20h8'26" 45d51'16"
Papa and Grandma Gilley
Scl 23h44'36" -28d2'32"
Papa and Jamey's Wisconsin Star
Uma 9h28'4" 59d54'18"
Papa and Maryjane
Sco 16h5'50" -18d2'16"
Papa and Nana
Lyn 8h26'0" 35d27'46"
PaPa and NaNa Grove
Ari 2h22'8" 24d3'59"
Papa and Princey's very own star
Leo 11h35'45" 19d24'38"
Papa and Taylor
Sco 17h22'52" -32d54'19"
Papa Baker
Her 17h32'49" 31d1'44"
Papa Bear
Uma 10h53'28" 41d36'10"
Papa Bear
Uma 10h51'13" 64d53'24"
Papa Bear and Lou Bear
Sco 16h6'11" -12d7'7"
Papa Bear Bill Squibb
Sco 16h6'44" -15d49'16"
Papa Bear Billy Bob
Per 3h23'17" 43d38'20"
Papa Bear "Dad"
Uma 11h30'32" 63d21'53"
Papa Beerling
Cep 22h24'5" 82d8'10"
Papa Bell
Her 16h47'15" 48d42'17"
Papa & Blake's Special Star
Leo 10h15'37" 14d23'30"
Papa Bo
Psc 23h47'18" 5d48'42"
Papa Bob
Cnc 8h40'37" 7d57'56"
Papa Bogey
Sco 16h4'38" -9d8'32"
Papà Brunino
Dra 19h50'22" 79d2'39"
Papa Burkhardt
Cyg 19h48'22" 37d2'17"
Papa Charley
Cep 22h0'35" 65d14'15"
Papa Chaz
Cnc 8h30'20" 28d53'51"
Papa Clyde
Sco 16h6'46" -12d32'39"
Papa Curnett
Leo 11h42'53" 26d6'2"
Papa Dad Prychitko
Cep 20h37'35" 64d5'43"
Papa Dave
Ari 3h13'32" 15d16'0"
Papa Davis
And 1h39'4" 44d26'54"
Papa DeBone
Ari 2h49'1" 28d31'19"
Papa Del Monaco
Ori 6h0'1" 11d8'19"
Papa Dick 6-27-1943
Cep 23h7'54" 75d4'13"
Papa Dick's Star
Dra 19h34'21" 72d42'34"
Papa Don
Per 3h7'58" 53d46'22"
Papa Donald Allen Graves
Lib 15h32'13" -25d4'39"
Papa Dozer
Sgr 19h44'47" -11d55'9"
PaPa Dusty
Uma 8h59'12" 62d57'0"
Papa Ed
Uma 10h11'43" 46d7'46"
Papa Fire aka the big kahuna
Her 16h49'22" 36d31'15"
Papa Frank
Lib 14h53'27" -5d31'2"
Papa Friday
Leo 9h47'58" 16d40'21"
Papa Gene
Ori 5h48'0" 7d11'39"
Papa Gene
Uma 10h33'19" 65d10'22"
Papa Geri
Dra 17h39'10" 69d7'58"
Papà Ghia
Her 15h50'13" 44d45'55"
Papa Haukap
Aql 19h23'16" 13d1'34"
Papa Heath - My Hero
Her 17h50'48" 26d28'20"
Papa Heffernan
Cnc 8h42'50" 23d34'17"
Papa Herb
Cep 22h38'25" 71d25'21"
Papa Jack
Cep 5h5'28" 85d33'2"
Papa Jack Taylor
Tau 5h29'49" 19d58'32"
Papa James Hopkins
Vir 12h48'18" 4d27'26"
PAPA JEC
Ari 2h8'26" 23d30'31"
"Papa" Jerry Beach
Ori 5h11'39" 5d34'38"
Papa J.M Delizee
Col 5h50'33" -29d14'28"
Papa Jo
Cap 20h10'53" -10d46'37"
Papa Joe
Ori 5h7'1" 16d0'20"
Papa Joe Falconetti (The Falcon)
Psc 1h16'5" 32d37'56"
Papa Joe & Nonnie June
Uma 10h10'46" 46d24'26"
Papa Joe Stevens
Ori 5h35'53" -5d5'5"
Papa John
Cep 0h4'45" 69d57'11"
Papa John
Vir 13h25'16" 4d55'46"
Papa John
Aqr 22h8'6" 1d7'37"
Papa John Nutter
Her 18h33'28" 16d3'51"
Papa John Richardson
And 23h20'9" 47d11'12"
Papa John Steele
Ori 5h44'57" 8d57'38"
Papa Jon
Tau 5h19'2" 23d39'7"
Papa Joycie
Leo 9h24'33" 28d23'16"
Papa Ken
Uma 8h58'25" 51d41'56"
Papa Kilpatrick
Dra 15h56'59" 54d12'32"
Papa Lamb
Per 3h25'26" 49d27'23"
Papa Larry Puccinelli
Per 3h43'34" 50d22'10"
Papá Lino
And 23h34'1" 37d7'44"
Papa Lloyd's Eternal Star
Lib 15h20'9" -5d9'43"
Papa Lonzo
Cep 23h5'43" 70d33'16"
Papa MacK
Cep 23h17'50" 75d38'50"
Papa & Mare
Per 4h23'5" 44d55'35"
Papa & Me
Uma 10h35'53" 65d7'45"
Papa Mort
Ori 5h36'5" 5d46'28"
Papa Muckel Uwe 03.03.45-02.09.92
Uma 8h29'1" 64d45'44"
Papa Murphy
Cep 21h16'32" 69d42'46"
Papa Nomo's Retreat
Uma 13h48'26" 55d55'40"
PaPa "O"
Sco 17h45'40" -41d21'34"
Papa O'Day
Cnc 8h17'35" 11d41'26"
Papa of Carson Star
Uma 11h35'8" 52d21'21"
Papa Pat
Tau 5h58'1" 25d32'15"
Papa Paul
Cnc 8h53'9" 28d11'9"
Papa Phelps
Psc 1h19'3" 15d5'59"
Papa Pierre M. Coll!
Cep 0h3'52" 71d41'34"
Papa Purkey
Her 16h21'19" 21d38'26"
Papa (Ray Hermann) Our Shining Star
Gem 7h39'45" 25d21'5"
Papa Renato
Lyn 8h22'9" 54d17'23"
PaPa Richard
Ori 6h5'50" 21d23'36"
Papa Riley
Tau 5h44'26" 18d6'8"
Papa Ron
Ori 5h41'7" 1d45'26"
Papa Ron & Nana Sandra
Ari 2h5'43" 12d38'40"
Papa Ronnie
Cep 3h23'33" 81d25'24"
Papa RoRo
Per 3h11'9" 31d36'17"
Papa Roy
Per 2h43'17" 54d15'25"
Papa Sam
Aql 19h16'46" -0d36'17"
Papa Sam's Stardust
Cep 21h51'31" 61d7'38"
Papa Sly
Vir 13h12'31" 7d51'51"
Papa Spike
Sco 16h15'38" -18d52'21"
Papa - Star
Per 3h14'7" 56d33'58"
Papa Steve
Cnc 8h47'6" 22d30'35"
Papa Styler
Uma 10h11'16" 67d23'39"
Papa Tarzan
Her 16h39'28" 42d8'27"
PaPa Tito
Cep 21h45'57" 63d4'36"
Papa Tom
Uma 8h33'38" 63d58'50"
Papa Tom
Aur 6h24'19" 52d28'13"
Papa TOOCH
Vir 11h49'50" -3d31'4"
PaPa (Victor Tamagna)
Aql 19h7'17" -0d13'11"
Papa Weinstein
Cyg 19h47'50" 35d48'23"
Papa Whiz
Uma 10h26'41" 42d51'22"
PaPa Wollschlager
Per 2h16'1" 56d37'35"
Papa Yanes
Aqr 23h1'14" -6d53'28"
Papa Yoder
Uma 10h17'59" 44d39'10"
Papa Z
Crb 15h29'51" 30d59'48"
Papa Z
Uma 8h45'49" 47d48'57"
Papa Zagerman
Vir 12h27'22" 2d21'5"
Papa Zhenya
Ori 5h37'47" 7d30'24"
Papa-Amma
Lyn 6h43'56" 56d49'39"
PapaBear-MamaBear
Aqr 22h21'58" -12d47'41"
Papadou
Uma 11h55'38" 45d16'22"
Papagano
Her 16h47'7" 33d1'38"
Papageno
Cas 2h41'14" 57d59'47"
Papageno
Cyg 21h19'55" 42d14'19"
Papa-Nooch
Aqr 21h37'8" -3d17'51"
Paparassiliou, Georgios
Uma 13h52'44" 56d27'48"
Papa's Aden
Tau 3h52'5" 24d19'40"
Papa's Erin Nicole Brennan
Vir 13h1'27" 11d3'11"
Papa's Evan James Brennan
Sco 16h50'46" -27d0'50"
Papas Fritas
Tau 3h58'53" 23d5'42"
Papa's Happy 90th Birthday Star
Lmi 11h4'29" 30d48'55"
Papa's Kitten
Uma 10h44'28" 60d32'48"
Papa's Ladies
Dra 18h36'25" 60d36'16"
Papa's Night-light
Her 18h12'31" 20d51'5"
Papa's Oldies
Sco 17h24'18" -44d31'23"
Papa's Star
Psc 1h25'30" 5d30'41"
Papa's Star
Cep 22h57'45" 63d20'39"
Papa's Star
Cap 21h43'6" -11d0'49"
Papa's Star
Her 18h35'35" 20d6'44"
Papa's Star
Tau 3h55'9" 26d39'41"
PaPa's Star
Vir 13h11'47" 12d3'1"
Papa's Star
Cnc 8h24'15" 9d33'41"
Papa's Star
Uma 9h25'44" 44d0'55"
Papa's Star
Per 3h44'54" 48d40'19"
Papa's star
Per 2h59'24" 49d5'42"
Papa's Star
Uma 10h26'16" 47d58'33"
Papa's Star
Cyg 21h24'27" 46d9'44"
" Papa's Star " Bernard G. Earl
Her 17h23'53" 27d17'1"
PaPa's Starlight
Cep 21h33'11" 62d16'24"
Papa-San
Per 2h42'13" 55d34'20"
Papaw
Aqr 21h58'42" -0d24'33"
papaw
Sgr 18h12'40" -27d57'59"
PaPaw Braden
Uma 10h40'31" 50d8'41"
Papaw Edge
Uma 10h45'54" 47d19'54"
Papaw Franklin Delano Spears
Her 16h43'30" 14d5'22"
Papaw Joe
Uma 11h9'58" 48d19'6"
Papaw's star in the sky
Cir 14h47'54" -64d32'45"
Papayo - "Guardian of the Twins"
Gem 6h48'7" 17d20'55"
Papazahariakas
Umi 16h33'21" 77d1'28"
Papazon Vince Yours Forever, Bobbie
Cep 20h45'17" 67d33'54"
Paperclip
Her 17h42'39" 15d8'37"
Paperclip & Magnit
Cyg 20h57'9" 46d25'55"
Paperino
Lac 22h53'47" 38d54'36"
Papi
Dra 19h31'39" 74d19'10"
Papi
And 1h6'47" 46d18'2"
Papi
And 0h56'12" 37d55'21"
Papi
Lyn 9h15'50" 34d47'40"
Papi
Leo 10h40'38" 10d28'20"
Papi Carl
Her 18h47'43" 19d38'19"
Papi Chulo Luis Saez
Leo 9h24'12" 11d27'25"
Papi Francis
Aql 19h37'24" 7d10'45"
Papi In The Mountains-Pete Pequeno
Uma 12h10'1" 57d13'34"
Papi, 11.05.1937
Cam 7h58'59" 72d47'16"
Papichulo JonJon
Sco 16h11'44" -13d8'51"
Papi-Jack
Sco 17h21'51" -33d50'33"
Papillon
Peg 22h9'18" 7d2'44"
"Papillon de Liberte de la Nuit"
Cnc 9h11'7" 24d12'32"
Papineau Porky
Uma 9h36'10" 60d31'33"
Papis
Lac 22h25'8" 47d24'16"
Papi's Princess
And 23h54'53" 43d21'21"
Papito Chulo
Tau 3h52'48" 28d15'23"
Papo
Per 3h31'10" 48d26'39"
Papoo
Aqr 21h50'50" -6d15'33"
Papoon
Lyn 7h50'7" 44d16'51"
Papote
Lib 14h55'0" -18d23'29"
Papotone
Sco 16h2'45" -11d22'24"
Papou Lou
Vir 11h54'35" -3d38'51"
Papou Pappas
Her 18h6'56" 29d44'55"
Papou Theodore Dione Carabelas
Uma 11h20'16" 72d25'46"
Papoun Eternel
Ori 5h18'3" -8d9'42"
Papp, Gertrud *Guggi*
Tau 5h7'14" 25d56'37"
Pappa Bear & Lil Red
Ori 5h32'41" 0d8'25"
Pappa "P"
Per 4h26'14" 42d19'52"
PapPap Gregor the Vigilant Guardian
Uma 13h0'8" 54d47'19"
Pappa's Best Boys - 22081939 - 26112004
Nor 16h23'55" -50d55'19"
Pappas Family Star
Gem 7h18'47" 24d33'48"
Pappy Andy
Tau 4h12'4" 8d35'28"
Pappy Cheech
Ori 5h52'34" 18d16'36"
Pappy Duck
Uma 9h25'45" 72d57'46"
Pappy Jack
Cep 21h34'5" 68d2'57"
Pappy Steve 1/26/46 - 12/14/03
Aqr 22h26'6" 0d9'26"
Paps
Per 3h12'45" 55d41'8"
Paps heavenly eye
Uma 12h51'46" 56d13'44"
Pap's Shining Star ~ Michaela
Cyg 20h11'34" 30d12'32"
Papstar Bro
Sgr 20h17'20" -35d25'47"
Papuzzo e Mimi Bibi
Tau 5h31'44" 28d16'46"
Papy Coco
Uma 8h38'24" 71d13'43"
Papy Doreau
Peg 21h19'33" 16d33'29"
Papy Loulou
Leo 11h30'20" -4d19'42"
Papy Mamy Jean Jacqueline
Aqr 20h57'22" -0d46'9"
Paqua
Leo 10h56'35" 7d13'39"
Paqui Meseguer
Aur 4h50'12" 35d41'38"
Paquita y Bartolomé
Vir 13h12'31" 12d54'28"
PAR CJ Willow
Her 17h16'1" 18d46'25"
Para Mi Bebé
Sco 16h16'29" -9d4'48"
Para mi Rosa, la mas preciosa.
Eri 4h21'47" -22d13'25"
Para sempre o meu anjo de deus
Lib 14h49'4" -4d5'48"
Para sempre seu Rafaela Eugênia
And 0h23'21" 45d51'0"
para siempre soñar
Cas 0h41'17" 51d30'12"
Para siempre tú
Cyg 21h15'12" 45d33'42"
para siempre y siempre
Cyg 20h11'50" 53d26'3"
Paradiddle Chris
Per 3h21'36" 41d23'35"
Paradigm9
Gem 6h44'35" 23d24'5"
Paradis
Mon 7h58'58" -3d47'1"
Paradis Sandra
Aqr 22h35'8" -14d13'30"
Paradise
Uma 11h7'48" 72d18'54"
ParadisE
Tau 4h43'9" 29d20'31"
Paradise
Psc 1h31'3" 13d2'32"
Paradise 1515 - Angela Dawn Orpin
Psc 23h57'4" -4d13'54"
Paramour
Cru 12h25'25" -57d9'11"
Paras Sajjan
Cap 21h39'10" -13d54'6"
Parasempre Meu Amor
Vir 13h19'33" 5d39'35"
Parashakti Sigalit Bat-Haim
Vir 12h36'9" -9d33'15"
Paraskeva Koumanis
Cra 18h52'18" -39d33'44"
Paraskeve F. Kokonezis Gentry
Lyn 6h17'56" 60d58'28"
Paraskevi
Gem 7h25'39" 28d48'56"
Parastoo Jamshiddanaee
Hya 9h25'42" -0d12'41"
Paravox
Lyn 8h58'46" 40d44'0"
Pardeep
Gem 6h14'10" 25d11'3"
Pardis Sahafi
Tau 4h36'1" 9d40'53"
Parham Parto
Tau 5h34'31" 21d4'49"
Pari Passu
Cyg 19h45'10" 31d21'58"
Pari Passu Chels & Cam Price
Ori 5h58'16" 10d2'35"
Pari Sadeghi
Cap 20h29'20" -11d1'47"
Paride
And 23h17'7" 52d0'0"
Paride Pocino
Gem 6h40'57" 24d45'39"
Paridni & Viuek Jindal
Cnc 8h8'14" 20d16'41"
Parilok - " Sarah J's Home "
Umi 13h34'34" 89d21'5"
Paris
Ori 5h23'15" -4d2'23"
Paris
Cma 6h40'19" -14d13'34"
Paris
Sct 18h34'40" -4d27'25"
Paris
Uma 12h9'24" 59d16'23"
Paris
Psc 1h45'23" 13d48'4"
PARIS_1CD
Cyg 21h54'32" 38d54'43"
Paris Apprelle Rose Gaskin
Umi 16h24'49" 73d8'6"
Paris Casey
Uma 10h35'30" 62d53'44"
Paris Dragnis
Cen 11h39'23" -38d26'20"
Paris Ellison
Gem 7h21'57" 22d10'50"
Paris Lyn
And 23h12'39" 39d41'52"
Paris Partee Furcron
Psc 0h53'26" 25d5'19"
Paris Partee Furcron
Psc 1h55'14" 7d8'48"
Paris Renee Crume
And 0h16'46" 25d18'17"
Paris Summer Sondrio 13 03 03
Psa 21h30'31" -36d9'54"
Paris Tsikis
Uma 9h5'42" 50d19'37"
Parisa
Vir 14h6'46" 4d9'1"
Parisa Samimi
Cnc 9h15'6" 15d57'25"
Parisa Syed
Lyr 18h26'13" 31d23'41"
Parish
Psc 1h14'20" 3d37'12"
Parisi J.C. 5-22-93
Gem 6h53'44" 28d42'22"
Parisi Memorial
Lyn 7h40'13" 43d8'4"
Parita Arietis
Ari 2h55'57" 24d39'35"
Park Eungyun
Leo 10h45'23" 21d40'2"
Park Hae Kyung
Ari 3h6'53" 26d4'17"
Park Kyung Ja Kang
Aqr 23h47'43" -18d30'0"
Párkányi Attila
Psc 23h31'7" 5d53'40"
Parker
Vir 15h7'5" 3d31'50"
Parker
Ori 6h19'26" 2d49'10"
Parker
Tau 5h28'25" 27d3'58"
Parker
Uma 11h33'33" 48d18'31"
Parker
Uma 11h25'46" 64d16'43"
Parker
Lib 15h0'52" -5d41'5"
Parker
Umi 15h17'6" 79d33'45"
Parker Allen Koepke
Cap 21h14'3" -24d37'41"
Parker Bentley Stoken
Sgr 19h51'41" -13d22'31"
Parker Brown
Uma 11h2'58" 38d29'4"
Parker Christian
Ari 3h18'58" 29d6'12"
Parker Davidson Lee
Ari 2h48'2" 17d6'4"

Parker Eason
Psc 1h8'17" 28d12'30"
Parker Elizabeth Young
Sgr 18h57'44" -19d25'39"
Parker Everette Knoles
Lib 14h56'40" -18d5'25"
Parker Henry
Lyr 19h27'51" 42d30'32"
Parker Hill
Cnc 8h25'39" 17d54'48"
Parker James Cimaglio
Tau 4h25'48" 21d41'10"
Parker Jane
Cra 18h46'43" -40d24'53"
Parker John
Gem 6h57'8" 19d11'56"
Parker John Borbi
Gem 7h49'13" 31d51'35"
Parker Joseph Hoffman
Ori 6h8'55" 16d38'10"
Parker Josephine Tomlinson
Sgr 18h43'51" -17d46'48"
Parker Joshua Kanan
Umi 15h11'41" 71d22'13"
Parker Jude Brewer
Leo 11h20'51" 14d14'31"
Parker Malone Stroud
Gem 7h19'8" 23d33'9"
Parker Mar... My Guy
Lib 15h33'0" -5d47'17"
Parker Ray McVay
Aqr 23h26'45" -19d32'33"
Parker Reid Quail
Aqr 22h31'19" -9d25'32"
Parker Riley Schroeter
Her 16h48'59" 23d24'57"
Parker Roth Willoughby
Per 3h4'6" 53d35'54"
Parker Scott Hobgood
Tau 3h33'3" 1d58'51"
Parker Sek
Gem 7h11'36" 34d28'43"
Parker Sky McHugh
Tau 5h38'58" 24d57'41"
Parker Star (CRCBRKALMC)
Cyg 19h50'16" 38d10'15"
Parker Thomas
Uma 9h56'51" 46d59'24"
Parker Thomas Ackerman
Sco 16h11'0" -9d34'33"
Parker Thomas Lippstock
Psc 23h0'13" 5d39'15"
Parker Thomas Simpson
Sgr 19h45'31" -12d43'6"
Parker Thomas Smith
Lyn 7h44'51" 41d41'30"
Parker Todd Grovatt
Cyg 20h56'21" 46d31'35"
Parker Troy Olson
Cnc 8h45'40" 23d2'9"
Parker Whitaker Bausman
Aql 19h29'29" -6d19'42"
Parker X-1 Book II
Cyg 19h46'25" 41d10'34"
ParkerAlanSmith
Uma 8h34'28" 50d52'51"
Parker's Daddy
Per 3h34'42" 46d5'17"
Parker's Dream
Sco 17h4'13" -38d31'39"
Parker's Lyon
Lmi 10h33'9" 36d9'56"
Parker's Macey Lea
Umi 14h31'13" 82d6'59"
Parker's Mom
Ori 6h8'58" 9d12'6"
Parkes 369281
Aur 6h19'42" 53d13'7"
Parksters
Sco 16h29'52" -25d6'17"
Parkview II Staff
Uma 8h31'40" 62d59'36"
Parlati Emiliana
Cam 5h37'11" 58d39'36"
Parmida
Leo 10h36'21" 14d7'35"
Parnie's Guiding Star
Cru 12h40'12" -60d25'55"
Parobek929
Vel 9h27'39" -41d52'59"
Parsa jan
Uma 9h0'43" 50d52'42"
Parsème - de maman et fafa
Uma 8h51'2" 61d30'49"
Parte Di Tony's Di Cielo
Sgr 18h22'56" -23d44'37"
Parthier, Astrid
Uma 11h16'1" 63d47'16"
Partners Maureen & David
Cyg 20h28'13" 58d56'46"
Partridge
Aql 19h26'45" 5d2'44"
Paruly
Cnc 9h9'22" 9d18'56"
Parum Agna
Tau 4h25'57" 21d38'43"
"Parva Est Decora"
Umi 15h35'15" 72d1'32"
Parveen's Eye
Lyn 7h37'54" 55d27'34"
Parviz Farahmandi
Tau 4h26'38" 23d3'4"
Parviz Sorouri, M.D.
Uma 9h23'5" 46d16'33"
Parvoneh
Uma 11h43'18" 44d29'50"
Pary
Ari 3h28'30" 27d44'17"
Paryus B. Patel, M.D. and Family
Uma 9h37'33" 54d44'10"
Pa's Star
Per 4h17'55" 41d46'50"
Pascal
Uma 9h28'59" 44d12'1"
Pascal
Uma 12h4'35" 36d11'23"
Pascal
Per 3h30'49" 49d21'35"
Pascal
And 2h33'58" 50d23'15"
Pascal
Dra 14h45'1" 59d22'59"
Pascal
Umi 14h11'7" 69d7'56"
Pascal
Dra 19h1'17" 72d29'20"
Pascal
Umi 16h50'59" 75d14'49"
Pascal
Cap 21h11'46" -22d59'16"
Pascal Arnold
Sct 18h49'21" -12d44'50"
Pascal Bouye
Lyn 7h58'49" 58d1'28"
Pascal Chavaudra
Uma 10h56'43" 59d32'11"
Pascal Daniel Reith
And 1h4'11" 42d15'24"
Pascal Eichmann
Cam 4h3'0" 59d3'7"
Pascal Gabrielli
Leo 10h51'33" 5d28'12"
Pascal Jobert
Cas 0h53'7" 52d12'10"
Pascal Keum-jae
Ari 2h10'11" 25d36'22"
Pascal Lehmann
Lac 22h3'57" 40d47'39"
Pascal Levecque
Vir 13h4'23" -3d41'23"
Pascal Mächler
And 2h38'41" 50d26'5"
Pascal "Menahn" Schnyder
Umi 13h38'40" 76d57'54"
Pascal Niklas
Uma 10h15'6" 44d44'40"
Pascal Paret
Tau 3h26'28" 15d10'4"
Pascal Poussange
Peg 23h43'12" 21d12'0"
Pascal +Sabrina
Lac 22h18'17" 35d40'23"
Pascal & Sarah
Cas 2h21'50" 69d32'4"
Pascal Simonet
Cnv 13h29'35" 31d1'36"
Pascal Stern
Cyg 20h36'7" 42d7'11"
Pascal Stouder
Lac 22h42'57" 41d29'8"
Pascal Winkler
Uma 10h23'18" 42d22'17"
Pascale April
Cyg 19h25'10" 47d3'13"
Pascale Cartolaro
Uma 9h21'41" 43d47'47"
Pascale Courtial
Dra 19h56'56" 59d56'56"
Pascale DeAngello
Aqr 21h9'58" 0d20'30"
Pascale Filiatre
And 0h1'21" 40d16'14"
Pascale Lecosse
Uma 9h21'3" 43d43'50"
Pascale Rummel
Uma 10h2'34" 61d7'21"
Pascale und Christian
Gem 6h48'52" 16d55'34"
Pascale V.
Uma 9h6'26" 65d51'30"
Pascaline Paradis
Sco 17h57'53" -30d50'3"
Pascalou
Her 18h34'28" 14d33'59"
PascalT25091958
Uma 11h50'44" 49d35'14"
Pasch
Crb 15h35'34" 35d42'24"
Pasch, Christiane
Uma 13h59'16" 57d11'43"
Pascha Anzara
Ari 2h48'22" 13d38'57"
PASCHKIN
Ori 6h13'46" 14d10'44"
Paschwitz, Bernd
Uma 9h39'51" 49d30'25"
Pasco Simone
Aql 19h47'18" -0d29'38"
PASCORI
And 1h55'10" 39d43'9"
Päscu
Her 17h52'48" 14d38'47"
Pascual & Carrie Fortanel 10/19/05
Cyg 21h55'43" 47d50'2"
Pascual y Coral
Uma 11h26'14" 40d0'3"
Pase William
Cam 4h42'7" 68d10'4"
Pasfancalyn
Cas 0h53'7" 55d50'42"
Pasha Manley
Psc 0h24'23" 16d59'54"
Pashtoon
Psc 1h24'5" 23d3'10"
Pasimio
Umi 16h12'32" 70d7'13"
PAS-M
Leo 11h30'13" 28d8'17"
Pasqua Domenica
Cap 21h19'38" -25d25'3"
Pasqua Tarantino
Umi 16h5'31" 78d24'21"
Pasquale
Ori 5h35'27" 13d23'36"
Pasquale Buldo
Uma 12h50'26" 61d43'5"
Pasquale Carmine Mattiaccia
Psc 0h56'20" 27d58'23"
Pasquale D'Alessio
Cep 22h52'16" 60d16'27"
Pasquale DeAngelo
Sco 16h42'11" -40d51'28"
Pasquale Delise
Uma 8h26'25" 59d53'44"
Pasquale Esposito
Sgr 19h15'1" -35d20'29"
Pasquale Giugliano
Uma 12h9'35" 45d12'27"
Pasquale Joseph Perfetto
Sco 16h16'29" -8d19'11"
Pasquale Lorizio
Cyg 19h37'54" 38d34'2"
Pasquale M. Riccitelli Jr.
Cnc 8h33'21" 19d41'1"
Pasquale Maffeo
Uma 10h29'24" 62d36'57"
Pasquale Miranda
Gem 6h23'42" 20d43'4"
Pasquale - My Eternal Love
Leo 11h42'23" 15d17'27"
Pasquale Parente
Cyg 19h48'42" 35d52'56"
Pasquale "Pat" Tigani
Uma 9h51'3" 65d6'30"
Pasquale Ricciardi
Per 3h10'33" 56d56'5"
Pasquale Richard Molle
Sco 16h55'36" -39d24'38"
PASQUALETAMARA
Equ 21h2'58" 11d40'38"
Passavant
And 23h58'38" 42d29'52"
Passez la nuit avec moi
Tau 4h21'55" 20d36'2"
Passing Storm
And 2h35'1" 43d37'21"
Passion
And 0h51'15" 35d56'56"
Passion
And 1h33'40" 47d0'33"
Passion
Vir 13h15'21" 7d7'1"
Passion
Cyg 20h23'23" 55d2'50"
Passion
Cap 20h38'18" -20d42'21"
Passion & Poetry
Cap 20h13'42" -16d28'18"
Passionblossom's Pirate
Cam 3h56'39" 63d31'29"
Passionspirit
Uma 9h15'37" 72d50'28"
Passlack, Stephan
Ori 6h16'48" 11d10'41"
Passolino
Umi 14h23'10" 68d40'32"
passus
Her 17h56'49" 26d53'15"
Past,Present,Future:JCW,PIII,PIV,MK
Uma 12h56'53" 62d47'45"
Past,Present,Future-TogetherForever
Vir 14h3'39" -18d40'10"
Pasta Bear
Uma 9h53'46" 71d19'5"
Pastor Charles Willie Collins
Tau 4h21'16" 17d6'35"
Pastor Dennis
Lmi 10h52'21" 25d22'1"
Pastor Douglas "E" Brown
Tau 4h24'49" 26d37'58"
Pastor Gary Nelson
Cep 21h13'24" 67d4'37"
Pastor Jerry Rayburn
Aql 19h42'40" 4d27'27"
Pastor Kevin
Her 17h15'3" 18d18'30"
Pastor Nancy
Vir 13h25'32" 13d35'47"
Pastor Paul Everett/Zion
Tri 2h8'2" 34d1'25"
Pastor Stephen M. Thrash
Cam 3h46'44" 67d4'55"
Pastor Thomas Lee Barrett, Jr
Cap 20h46'47" -17d1'55"
Pastor Val Hood
Ori 5h17'32" -5d33'39"
Pastor Wesley William West
Cep 20h40'14" 59d15'25"
Pasty Stacey
Leo 11h29'27" 26d16'2"
Pasztadorlandia
Dra 18h29'28" 52d47'28"
Pásztor István
Uma 11h37'59" 29d51'22"
Pat
Cyg 19h30'15" 29d42'40"
Pat
Psc 1h9'9" 22d49'49"
Pat
Aur 5h55'44" 36d17'36"
Pat
And 1h48'25" 42d14'53"
Pat
Lyn 7h9'52" 57d13'37"
Pat A. Crowe
Cas 2h8'2" 62d20'45"
Pat and Bill Murphy's Star
Cyg 20h45'12" 31d40'26"
Pat and Julie's Love
Cnc 8h42'0" 16d15'1"
PAT AND JUNE PATTERSON
Ari 2h25'8" 19d21'27"
Pat and Maria
Crb 15h55'17" 30d25'24"
Pat and Peg Ricketts
Her 17h52'35" 21d17'20"
Pat and Sandy Always and Forever
Cyg 19h41'53" 54d39'44"
Pat and Stu 1964
Cnc 8h41'7" 24d20'19"
Pat and Terry Hammell 1976
Cyg 20h0'31" 31d5'1"
Pat and Tom
Ori 5h20'51" -3d57'23"
Pat and Verne Bossie
Vir 12h48'34" 4d8'37"
Pat Armstrong
Uma 9h40'29" 54d54'13"
Pat Auerbach
Dra 16h52'54" 65d49'14"
Pat Austin Adoring Mother & Friend
Mon 8h6'4" -0d47'0"
Pat "Babe" Dixon
Uma 11h42'51" 51d56'47"
Pat Baker
Psc 1h29'4" 26d58'41"
Pat & Barb Wasielewski
Uma 9h24'44" 51d45'23"
Pat Barker Radiant Light For Unity
Cas 0h48'23" 57d6'0"
Pat Barnett
Ori 5h59'23" 3d44'9"
Pat&Ben_Stump
Gem 7h31'11" 29d9'44"
Pat & Bill Bedford Birthday Star
Cyg 20h41'30" 42d48'48"
Pat Bird
Cnc 9h18'21" 16d46'58"
Pat Blackard
Gem 6h48'10" 18d52'37"
Pat Borrelli & Tina Riccardi
Cyg 19h41'15" 35d11'44"
Pat & Brenda
Sge 19h27'16" 18d36'59"
Pat Brown
Cas 1h28'43" 54d55'2"
Pat Bullock
Per 3h40'57" 43d56'45"
Pat Burke
Umi 16h22'6" 70d29'10"
Pat Butcher
Cas 23h59'37" 61d10'37"
Pat Chalmers - "Mama's Star"
Ori 6h11'35" 9d14'22"
Pat (Cheri) Vereb
Cas 0h14'2" 59d20'4"
Pat Christiansen
Tau 5h41'10" 24d49'50"
Pat Cohen
Per 3h9'5" 45d41'40"
Pat Cook
Psc 1h26'8" 32d59'15"
Pat Corey Morrison
Psc 1h5'43" 11d55'45"
Pat Cwiek
Uma 8h24'33" 64d48'51"
Pat DeRessett
Umi 16h3'29" 79d56'30"
Pat Deyer
Cas 0h43'53" 51d26'14"
Pat & Dick Allocca 40th Anniv. Star
Cyg 21h24'26" 31d46'41"
Pat Dougans & Barney
Uma 13h54'41" 62d7'32"
Pat ~ Dream Catcher ~ Hand
Crb 15h53'55" 38d54'43"
Pat Dukat
Aqr 22h55'43" -8d33'28"
Pat & Ed
Cyg 21h34'19" 33d36'51"
Pat Edwards
Sco 16h21'33" -29d42'55"
Pat Faigenbaum-Gevurtz
Uma 9h38'57" 56d13'19"
Pat Fernandes
Cru 12h43'2" -61d23'16"
Pat Fiorito
Cyg 21h14'47" 47d21'18"
Pat Fortenberry
Lib 15h37'17" -29d15'16"
Pat Foster
Tau 5h9'58" 17d53'14"
Pat Gaffney
Tau 4h15'56" 27d56'42"
Pat & Gale
Gem 7h14'0" 34d14'55"
Pat Garrett
Per 3h30'42" 35d34'31"
Pat Gorum
Vir 13h0'45" 10d55'44"
Pat Grimes
Tra 16h14'6" -62d3'31"
Pat Grimsley
Boo 14h32'38" 17d49'25"
Pat Grimsley
Uma 11h40'39" 39d13'18"
Pat H. Simmons II
Sgr 19h35'56" -44d15'45"
Pat Hargrett
Lyn 7h29'37" 52d54'16"
Pat Hartzell/Precious
Ari 2h37'2" 14d33'48"
Pat Henry
Psc 0h41'34" 8d31'47"
Pat - Here's your sign - Drage
Vir 13h9'59" 6d13'47"
Pat Higgins
Lyn 7h36'19" 56d42'16"
Pat Highet
Her 16h21'3" 5d41'52"
Pat Houghton
Psc 1h21'24" 17d51'36"
Pat & Jack Koval
Psc 1h1'58" 27d35'43"
Pat Jean
Sgr 19h39'47" -12d38'19"
Pat Johnson
Cam 4h41'22" 68d38'27"
Pat Johnson
Per 3h48'56" 34d16'52"
Pat Jones
Cap 21h1'28" -20d25'4"
Pat Keller
Cas 1h24'45" 58d2'35"
Pat & Ken Brizel Anniversary 27 Yrs
Cyg 19h50'25" 31d24'55"
Pat Kurz
Gem 7h10'26" 32d35'58"
Pat Kuter
Uma 9h43'19" 64d42'24"
Pat Lanza
Vir 14h4'5" -16d22'31"
Pat Lapin
Psc 23h12'11" 6d2'50"
Pat Lewis
Umi 16h15'39" 85d27'29"
Pat Lisa Ciara & Connor Kelly Stars
Umi 15h27'9" 76d17'15"
Pat Maloney Chiovare
Crb 15h51'36" 34d56'44"
Pat Mann
Uma 12h1'23" 53d26'29"
Pat & Margie's Everlastng Star
Cyg 21h6'47" 36d53'46"
Pat Martin
Aqr 23h12'1" -4d34'26"
Pat Maschari
Uma 9h51'17" 46d27'36"
Pat & Matt
Cnc 8h33'27" 27d43'28"
Pat Mayer
Leo 11h39'12" 17d5'48"
Pat Mazzuca
Uma 13h56'10" 60d43'57"
Pat McCarthy Hauhuth—A Stellar Mom
Psc 23h28'42" 1d45'59"
Pat McGarr
Cas 23h32'43" 51d47'57"
Pat McHenry Webb 12-13-1930
Cep 22h3'58" 70d31'25"
Pat Miller
Cas 1h17'1" 56d39'39"
Pat Mimi McIntire
Mon 6h52'18" -0d20'20"
Pat Mioduszewski
Uma 13h41'55" 54d24'3"
Pat n Lisa
Uma 9h57'34" 61d49'41"
Pat & Nikki
Cyg 20h31'46" 30d1'53"
Pat "Our Angel"
Cep 22h1'11" 72d21'49"
Pat P. Fontana
Uma 11h32'50" 40d0'11"
Pat (Patsy) Bratica
Sco 17h49'13" -42d40'35"
Pat "Patty"
Umi 13h58'18" 75d34'36"
Pat Pendergast
Uma 8h35'40" 67d25'15"
Pat Phillips
Ori 5h35'28" 10d49'1"
Pat Puccio
Uma 12h22'26" 56d8'11"
Pat & Rad Radefeld 60th anniversary
Psc 0h31'15" 17d24'9"
Pat & Richard Foster
Uma 10h42'56" 71d43'56"
Pat Rogers (My Mum)
Cas 0h55'1" 47d39'50"
Pat Romanello
Lyn 7h59'30" 42d37'2"
Pat Ross
Cyg 20h20'36" 59d5'2"
Pat & Sal
Cyg 19h59'24" 47d43'12"
Pat Schaack
Uma 8h38'33" 48d33'50"
Pat Schnell
Uma 10h40'43" 56d56'0"
Pat Spyker
Lyn 6h17'33" 59d54'3"
Pat Spyker
Ari 1h52'20" 17d12'35"
Pat Starr
And 23h39'0" 46d34'12"
Pat Stratford
Tau 4h37'28" 19d32'56"
Pat & Tink
Uma 9h55'16" 51d21'1"
Pat & Tom - Forever Ours
Ori 5h41'36" -1d9'49"
Pat Wharton-Hege
Uma 9h33'32" 66d9'2"
Pat White
Per 3h9'5" 44d48'9"
Pat Williams
Srp 18h8'31" -0d33'4"
Pat & Woody Rawle
Vir 15h1'10" 3d59'45"
Pat, My shining star, My true north
Umi 15h24'47" 74d3'46"
Pata
Lyr 18h31'1" 32d52'11"
Patacchina (Patastar)
Ori 5h56'32" 6d59'1"
Patagy
Ori 5h16'45" 13d3'6"
Pataki Kata
Lib 15h12'5" -23d24'51"
Pataoo
Lib 15h34'43" -19d48'11"
Patapol
Vir 12h17'42" 10d53'15"
Patata
Uma 12h36'59" 57d17'21"
PatBeth
Umi 15h5'40" 70d32'54"
Patcarlot
Lib 15h49'4" -10d38'4"
Patch
Leo 9h32'56" 29d14'13"
Patch
Uma 8h38'6" 51d42'2"
Patcharin 1976
Aql 19h38'55" 10d21'51"
Patches
Tau 4h51'22" 20d54'41"
Patches
Ari 3h3'6" 14d9'18"
Patches
Per 4h36'44" 35d19'0"
Patches
Sco 16h17'44" -12d16'40"
Patches
Cma 6h49'42" -12d48'2"
Patches
Cam 5h39'55" 73d27'20"
Patches
Uma 10h24'46" 62d11'24"
Patches
Uma 12h4'29" 59d57'1"
Patches
Cma 6h57'59" -24d32'46"
Patches and Morris
Vir 12h40'49" 6d12'59"
Patches & Casper
Lyn 6h28'41" 55d53'4"
Patches Newman (Felis catus)
Lmi 9h29'25" 34d51'23"
Patches Randazzo-Youngs
Uma 11h13'33" 50d58'5"
Patch's Passion
Ari 2h27'53" 26d21'1"
Patchy
Lyn 7h44'35" 47d54'32"
Patcorklove
Cyg 20h23'38" 55d18'12"
PatersonPerrigo
Cyg 20h0'17" 46d49'43"
Path
Eri 4h15'41" -22d40'17"
Pathfinder
Psc 0h55'53" 29d24'32"
Pathfinder027 (George Reynolds)
Vir 11h50'21" 6d0'33"
Pathos Kai Dynami
Ori 6h12'47" 18d0'13"
Pathy
Cas 0h48'57" 69d35'42"
Patience
Uma 10h31'36" 68d23'20"
Patience
Ori 5h34'6" 4d0'53"
Patience
Cam 3h54'20" 59d7'7"
Patience
Per 3h34'16" 51d48'31"
Patience
Aur 5h41'10" 47d41'36"
Patience Mae McFall
Crb 15h44'40" 36d24'0"
Patience Nicole Rodgers
Lib 16h0'28" -15d43'3"
Patience Saibre Sauve'
And 0h39'1" 26d45'50"
Patience Salter Stokes 2005
Uma 11h59'16" 61d17'13"
Patient Active TWC-GLV
Crb 15h26'55" 26d16'20"
Patient Eternal Seeker-Sean O'Shea
Ori 6h19'55" 10d17'30"
Patmikdebtinadiantydevcamhailand
Lyn 6h52'50" 53d35'1"
Patna
Uma 13h27'29" 57d56'20"
Pat-n-Eddie's Star
Aqr 21h10'40" -11d48'56"
PatOBrien56
Cnc 8h23'32" 14d42'20"
Patras Patrick Alexandre Louis
Uma 12h26'34" 62d13'42"
Patrese Magliazzo
Crb 16h15'14" 31d53'16"
Patri
Her 18h38'10" 21d39'50"
Patria Altagracia Shenery
Sco 16h57'26" -36d23'18"
Patria Colon
Cam 5h45'41" 58d17'48"
Patria Dulce
Lyn 8h32'21" 34d2'28"
Patria & Greg Forever In The Stars
Sge 19h34'3" 16d54'7"
Patric Cunnane
Lyr 18h30'50" 37d53'44"
Patric Ernest Raftery
Cnc 8h28'23" 10d59'37"
Patric Giess
Dra 16h29'59" 63d56'1"
Patric Lucky Star
Uma 8h43'3" 59d38'11"
Patric Rochat
Tri 1h45'57" 30d55'41"
Patrica
Cap 21h29'22" -9d56'57"
Patrica Ann
Col 5h47'55" -31d10'24"
Patrica Ann Mosley
And 23h18'53" 48d9'43"
Patrica Coykendall
Lib 14h52'28" -4d36'59"
Patrica Holeman - Weyhrich
Cas 0h12'56" 63d13'50"
Patrica L. McLaughlin
Crb 15h54'32" 26d51'54"
Patrice
Leo 10h20'2" 25d27'1"
Patrice
And 0h27'23" 27d6'10"
Patrice
Lyn 7h0'53" 48d13'11"
Patrice
Sco 16h17'50" -21d11'56"
Patrice ab aeternus formsitas
Leo 9h40'19" 15d35'17"
Patrice B. L. Files
Ari 2h47'0" 27d50'58"
Patrice Booth
Lyn 8h52'50" 43d41'7"
Patrice Chipper McAuliffe
Uma 10h34'48" 57d47'20"
Patrice Jan Perez
Sco 17h31'18" -45d7'36"
Patrice L & Virginie N
Cyg 20h49'51" 47d48'11"
Patrice Mayrand
Dra 20h23'35" 72d20'3"
Patrice - Our true star
Crb 15h58'10" 34d20'53"
Patrice Pratt
Lib 15h14'59" -15d20'57"
Patrice Rose
Sco 17h25'56" -36d1'8"
Patrice's Gem
Cyg 20h48'16" 39d45'30"
Patrice's Star Of Hope
Sgr 18h19'59" -22d41'27"
PatriChrista
Leo 11h26'7" 22d30'22"
Patricia
Leo 9h41'37" 26d54'23"
Patricia
Peg 22h28'57" 29d43'33"

Patricia
Crb 15h45'46" 28d5'27"
Patricia
Ari 3h9'3" 29d13'34"
Patricia
Ari 2h41'23" 27d41'44"
Patricia
Sge 19h58'12" 18d57'1"
Patricia
Sge 19h52'54" 17d26'52"
Patricia
Leo 10h5'31" 16d16'45"
Patricia
Tau 4h49'22" 21d38'51"
Patricia
Ari 2h19'6" 18d35'46"
Patricia
Ari 1h49'47" 13d36'49"
Patricia
Lyn 6h59'3" 51d13'58"
Patricia
And 1h39'46" 49d23'14"
Patricia
Cnv 13h1'41" 47d10'57"
Patricia
And 23h18'11" 51d44'0"
Patricia
Per 2h59'13" 55d19'52"
Patricia
Cas 0h10'24" 56d17'37"
Patricia
Uma 11h55'35" 30d43'58"
Patricia
Per 4h6'51" 34d45'29"
Patricia
Per 3h21'6" 44d29'50"
Patricia
And 0h23'8" 37d52'53"
Patricia
And 0h20'30" 38d6'50"
Patricia
And 1h17'57" 42d49'21"
Patricia
And 0h53'57" 39d42'34"
Patricia
And 1h11'43" 37d30'49"
Patricia
Sgr 17h50'46" -29d48'56"
Patricia
Sco 16h38'58" -31d33'11"
Patricia
Sgr 20h22'52" -40d26'9"
Patricia
Cyg 19h36'18" 30d11'16"
Patricia
Vir 14h31'4" -0d55'10"
Patricia
Vir 13h36'21" -9d49'12"
Patricia
Ori 6h1'38" -0d14'1"
Patricia
Umi 14h9'39" 68d13'42"
Patricia
Lyn 7h10'53" 56d29'57"
Patricia
Cas 23h53'29" 57d22'2"
Patricia
Cas 23h15'18" 56d1'32"
Patricia
Boo 14h55'24" 53d18'26"
Patricia 1986
Ari 2h22'45" 14d33'8"
Patricia A.
Cyg 20h43'30" 32d26'3"
Patricia A.
Crb 15h53'58" 38d55'29"
Patricia A. Bedolla
Mon 6h49'55" 3d20'34"
Patricia A. Bibber
Vir 13h22'39" 2d45'57"
Patricia A. Bill
Vir 14h22'10" -5d12'38"
Patricia A. Charles
Cas 0h27'1" 58d1'27"
Patricia A. Christensen
Cas 0h13'23" 62d30'31"
Patricia A. Clements
Gem 7h37'59" 33d34'51"
Patricia A Corbett
Uma 9h23'53" 70d5'25"
Patricia A. Evanchik
Leo 10h58'35" 8d57'19"
Patricia A. Gierloff
Mon 6h36'42" 9d26'48"
Patricia A. Hanson
Sco 17h30'8" -39d29'24"
Patricia A. Henderson
"Love Star"
Ori 6h11'12" 5d40'46"
Patricia A. & Henry P.
Gougelmann
Cyg 21h38'51" 40d50'27"
Patricia A. Lampreda
Leo 11h18'2" -4d37'39"
Patricia A. Pope
And 1h58'23" 43d1'59"
Patricia A. Quintavalle
Tau 4h28'6" 7d58'28"
Patricia A. Smith
Per 2h16'25" 51d16'13"
Patricia A. Trainor
Vir 14h17'2" -16d55'33"
Patricia A. Walker
Cap 20h53'21" -14d52'56"
Patricia A. Wilson56
Cas 0h45'12" 56d30'17"
Patricia A Young
Uma 10h28'37" 69d7'0"
Patricia A. Zarella
Crb 16h19'24" 38d17'17"
Patricia Agresta
Umi 15h31'16" 77d17'33"
Patricia Alberta Manton
Cas 23h3'5" 55d28'52"
Patricia Alberta Wires
Smith
Aqr 22h29'2" -23d51'15"
Patricia Aleene DiFelice
Lib 14h54'15" -3d50'21"
Patricia Allard
Tau 4h14'14" 16d7'57"
Patricia Ambrose "Pat Pat"
And 23h40'30" 42d54'16"
Patricia Amodeo 11-16-
1922
Cas 0h53'48" 58d45'45"
Patricia and Cheryl
Ori 6h3'18" 6d44'4"
Patricia and Dominic
Mercurio
Peg 22h27'37" 30d20'53"
Patricia and Ivan Harding
Umi 15h48'11" 73d44'8"
Patricia and John Reising
Per 2h30'31" 53d52'9"
Patricia and Justin
Cyg 21h10'2" 44d0'47"
Patricia and Richard Gray
Aql 19h45'56" 16d12'48"
Patricia Angel
Peg 22h51'58" 31d18'42"
Patricia Angelelli
Cra 19h5'15" -38d8'49"
Patricia Ann
Umi 13h54'12" 71d16'51"
Patricia Ann
Sco 16h6'17" -11d18'30"
Patricia Ann
Lib 15h31'2" -7d37'39"
Patricia Ann
Cap 21h39'27" -15d3'11"
Patricia Ann
Cap 20h46'40" -18d54'49"
Patricia Ann
Cnc 9h6'0" 31d33'50"
Patricia Ann
And 1h29'23" 49d49'44"
Patricia Ann
Cas 1h16'52" 53d13'22"
Patricia Ann
Cyg 21h4'47" 47d19'48"
Patricia Ann
Ari 2h39'51" 24d29'2"
Patricia Ann
Leo 10h26'27" 21d51'5"
Patricia Ann
Tau 4h11'44" 27d41'29"
Patricia Ann
Ori 5h39'6" 4d36'3"
Patricia Ann
Tau 4h6'52" 17d1'20"
Patricia Ann
Tau 4h26'50" 17d22'52"
Patricia Ann Akizawa
Cyg 21h52'52" 51d35'16"
Patricia Ann Anderson
Ari 2h3'55" 18d51'55"
Patricia Ann Berger
Aqr 23h19'16" -18d42'30"
Patricia Ann Bristow
Cas 23h46'37" 51d26'29"
Patricia Ann Brown Turner
Leo 9h35'54" 13d11'27"
Patricia Ann Chase
Cam 4h25'34" 71d13'21"
Patricia Ann Chlebica
Wallach
Uma 11h45'1" 65d26'42"
Patricia Ann Cockfield
Cnc 8h52'28" 17d53'4"
Patricia Ann Cooke
Crb 15h40'11" 37d3'52"
Patricia Ann Coolahan
Cas 1h24'16" 60d57'17"
Patricia Ann Cox
Leo 9h42'58" 29d19'2"
Patricia Ann Cross
And 0h26'15" 40d5'4"
Patricia Ann De Laurentis
Lmi 10h27'18" 32d17'18"
Patricia Ann Dixon
Aqr 23h22'30" -19d51'35"
Patricia Ann Donahue
Jarrett
Cas 1h34'56" 60d59'24"
Patricia Ann Edgemon
Sco 17h29'31" -45d14'11"
Patricia Ann Elfrink
Psc 1h15'5" 23d1'29"
Patricia Ann Feigenbaum
Vir 11h48'22" -0d30'32"
Patricia Ann Fila Robinson
Psc 1h56'8" 7d16'46"
Patricia Ann Fralix
Ori 6h25'43" 17d8'6"
Patricia Ann Gloede (our
angel)
Cru 12h27'50" -60d0'28"
Patricia Ann Grierson
Cru 12h17'47" -56d25'56"
Patricia Ann Haynes
Cas 1h19'23" 64d17'11"
Patricia Ann Hinton
Uma 12h0'26" 51d58'10"
Patricia Ann Johnson
Ari 2h48'40" 10d46'19"
Patricia Ann Koch
Vir 12h48'8" 5d53'50"
Patricia Ann LoBianco
Ori 5h18'43" 11d19'39"
Patricia Ann Matthews
Lyn 7h43'42" 42d22'23"
Patricia Ann McCarthy
Ari 3h6'5" 27d47'49"
Patricia Ann McLaughlin
Cas 1h28'14" 65d29'13"
Patricia Ann - My Star
Uma 11h58'4" 57d22'26"
Patricia Ann Nielsen
Hardesty
Uma 8h47'5" 50d18'54"
Patricia Ann Nordstrom
Cas 23h44'19" 61d47'51"
Patricia Ann Null
Cyg 20h32'4" 41d31'30"
Patricia Ann *Pat* Clark
Cnc 8h20'19" 10d4'11"
Patricia Ann Perry
Leo 11h2'38" 8d10'56"
Patricia Ann Pittman
Lib 15h7'48" -12d50'20"
Patricia Ann Rahe Weydert
Cap 20h25'18" -12d45'0"
Patricia Ann Robinson
And 1h10'53" 33d49'35"
Patricia Ann Schroeder
Aqr 21h38'5" 2d1'38"
Patricia Ann Simpkins
Pearce
Cas 1h46'45" 60d54'6"
Patricia Ann Slentz Smith
Tau 3h33'39" 6d33'40"
Patricia Ann Snyder
Cnc 8h54'19" 32d16'36"
Patricia Ann Speers
Vir 13h24'36" 8d45'54"
Patricia Ann Spina
And 0h56'16" 39d14'19"
Patricia Ann "SS" Medvetz
Mon 8h0'59" -0d50'13"
Patricia Ann Steinbach
Aqr 22h38'3" -0d15'0"
Patricia Ann Tansey
07/06/58
Cnc 8h13'33" 12d31'11"
Patricia Ann Tiffin
Lmi 9h33'22" 36d53'17"
Patricia Ann Tyler
Leo 11h55'30" 17d58'53"
Patricia Ann Vinal
Aqr 21h38'24" 1d58'15"
Patricia Ann Walsh
Vir 12h51'49" 12d46'43"
Patricia Ann Wells
Ori 5h31'27" -1d12'11"
Patricia Ann Wherry Gruber
Cyg 20h56'33" 30d54'18"
Patricia Ann Wimbish
Tau 3h37'19" 15d52'40"
Patricia Anne
Aqr 21h49'21" 1d9'31"
Patricia Anne
And 0h40'30" 43d13'8"
PATRICIA ANNE
Lyn 8h14'38" 41d2'6"
Patricia Anne
Uma 12h20'58" 58d56'36"
Patricia Anne
Cra 18h0'3" -37d28'29"
Patricia Anne Aurand
And 0h57'54" 40d20'21"
Patricia Anne Bayus
Cap 20h37'2" -21d43'24"
Patricia Anne Cassidy
Tau 5h50'36" 26d35'34"
Patricia Anne Christmas
Star
Cas 23h48'52" 56d50'37"
Patricia Anne Conner
Cam 6h3'17" 63d19'44"
Patricia Anne Donze
Cap 20h26'42" -14d47'35"
Patricia Anne Gallagher
Foell
Sgr 18h12'26" -27d13'29"
Patricia Anne Kelley
Grapes
Cas 2h25'34" 63d15'34"
Patricia Anne Miles
Ari 3h3'42" 13d58'10"
Patricia Anne Owens
Harvey
Ori 5h45'3" 11d18'41"
Patricia Anne Rowe
Cas 23h8'34" 58d44'52"
Patricia Anne Spadafora
Crb 16h13'49" 26d55'58"
Patricia Anne Willdin
Ori 5h23'17" -0d55'13"
Patricia Annette White
Cnc 8h23'24" 23d34'21"
Patricia Antonia Kennedy
Uma 8h38'10" 55d19'35"
Patricia Antoon
Vir 13h20'57" 6d55'4"
Patricia Armenio
Gem 7h9'29" 22d13'12"
Patricia Auger
Psc 1h9'4" 25d30'20"
Patricia Augusta Galli
Sco 17h22'42" -36d47'15"
Patricia Austin- A Shining
Star
Lyr 18h45'51" 30d30'17"
Patricia B. Hunt
Her 17h15'11" 30d0'59"
Patricia B. Stock
And 23h12'14" 51d4'53"
Patricia Bailey
Tau 4h38'1" 18d43'39"
Patricia Baker Latkowsky
Vir 12h24'31" 8d58'53"
Patricia Baldassarre Teti
Cas 0h27'27" 62d54'4"
Patricia Balog Schillaci
And 2h29'37" 44d46'25"
Patricia Barkley
Ari 2h43'19" 22d27'58"
Patricia Barren
Uma 11h47'7" 40d23'40"
Patricia Barthe's Sweet
Pea
Aqr 21h11'16" -3d20'55"
Patricia Barton
Gem 7h2'28" 16d33'22"
Patricia Baskett
Uma 8h49'43" 67d52'54"
Patricia Beattie
Cas 23h24'34" 55d22'54"
Patricia Berg
Tri 2h32'3" 33d50'23"
Patricia - Best Mommy In
The Sky
Sgr 18h51'11" -31d25'22"
Patricia Betro
Vir 12h23'0" 11d0'37"
Patricia Blair
Umi 15h50'30" 71d28'6"
Patricia Blandi
Vir 12h57'56" 9d46'56"
Patricia Bond
Leo 11h33'55" 16d3'21"
Patricia Bradford
Cas 23h41'36" 56d7'18"
Patricia Bradshaw
Uma 11h16'27" 70d22'31"
Patricia Branch April 6,
1941
Ari 3h12'14" 28d44'15"
Patricia Brown
Leo 11h30'44" 5d44'52"
Patricia Budrow's Angel
Umi 15h38'5" 74d12'37"
Patricia C.
Psc 1h22'43" 22d44'16"
Patricia Cahn Chatfield
Tau 4h45'15" 26d26'5"
Patricia Cannon
Cas 23h2'19" 53d57'12"
Patricia Carole MacBride
Lyr 18h33'58" 29d15'29"
Patricia Case
And 23h1'50" 48d7'33"
Patricia Casey Luchsinger
Her 16h26'54" 45d14'41"
Patricia Cassidy
And 0h58'37" 40d57'17"
Patricia Cassidy
Cas 1h10'59" 61d54'12"
Patricia Castillo López
Cas 0h46'37" 64d20'24"
Patricia Chang
Psc 0h55'12" 26d46'5"
Patricia Chaoul
Lyn 7h15'16" 47d21'7"
Patricia Charlene Hobart
Cas 1h28'42" 58d54'20"
Patricia Clancy Varley
Uma 9h43'58" 50d35'35"
Patricia Clini Muscato
Cas 1h41'40" 65d6'48"
Patricia Clory
And 23h35'48" 39d19'41"
Patricia Conibear-Lynch
Uma 12h4'57" 65d24'50"
Patricia Craighead
Lyn 8h17'20" 44d37'17"
Patricia Crowley
And 0h37'50" 40d24'13"
Patricia Cullen
Sgr 19h18'32" -33d50'21"
Patricia D. Clayton
Ari 3h13'30" 28d27'6"
Patricia Dalm-Moreland
Aqr 22h9'23" -12d38'59"
Patricia Darlene
Lyr 18h46'13" 31d25'33"
Patricia Dasch
Cas 1h39'44" 63d56'23"
Patricia Dees
Psc 2h0'37" 8d14'34"
Patricia Delise
Aqr 20h40'27" -3d24'56"
Patricia Dennington
Sgr 18h6'4" -27d35'16"
Patricia Denyce (Kelly)
Walker
Cas 1h29'37" 63d22'8"
Patricia Diane
Crb 16h16'11" 30d23'10"
Patricia Dodge-Torres
Psc 1h20'54" 7d26'2"
Patricia Donohue
Lib 15h1'39" -10d53'30"
Patricia Doris
Cas 23h36'3" 55d51'43"
Patricia Dorothy Kaufman
Block
Crb 15h43'58" 35d36'34"
Patricia Dreame
Aqr 22h29'13" -1d29'54"
Patricia Duffy
Crb 15h38'46" 33d38'17"
Patricia E. Ahto
Cnv 12h56'8" 47d5'17"
Patricia E. Bischoff
Vir 12h43'20" 8d15'35"
Patricia E Dudek
Cnc 8h44'1" 32d31'59"
Patricia E. Kiesler
Sgr 17h59'35" -27d32'21"
Patricia E. Sorensen
Gem 6h52'50" 14d59'54"
Patricia E Stickrod
Vir 13h26'27" 11d37'28"
Patricia E. Sutton
Cas 0h17'45" 59d12'30"
Patricia E. Wert
Cyg 20h15'53" 34d12'6"
Patricia (E.A.G) Gardner
Uma 12h16'1" 58d30'19"
Patricia Eileen
Lyn 8h45'15" 35d32'48"
Patricia Eileen Pruitt
Uma 10h26'13" 55d1'20"
Patricia Eli
Psc 0h3'8" 0d5'20"
Patricia Elise Shoemaker
Her 18h57'39" 13d46'22"
Patricia Elise Silva
Cnc 8h35'16" 7d52'35"
Patricia Elizabeth
Donaldson
Cas 0h43'52" 52d53'23"
Patricia Elizabeth
Schiefelbein
Cam 4h49'15" 54d25'8"
Patricia Elizabeth Winalski
Cap 21h34'33" -8d36'53"
Patricia Ellen Foley
Vir 13h4'53" 9d43'13"
Patricia Ellen Grunigen-
Nagy
Gem 7h39'7" 22d55'31"
Patricia Ellen McLaughlin
Crb 15h26'47" 26d27'26"
Patricia Ellen Taggart
March
Cap 21h37'16" -14d18'55"
Patricia Emily
Sgr 18h17'50" -23d8'37"
Patricia Erler
Ari 3h21'43" 29d41'7"
Patricia Essers
Lyn 9h15'15" 35d8'42"
Patricia F. Echko
Uma 13h40'48" 58d13'6"
Patricia F. Hutson
And 23h19'42" 51d26'16"
Patricia F. Werner
Uma 10h1'54" 58d34'2"
Patricia Falkenberg
Lib 14h53'17" -1d52'41"
Patricia Faye
And 0h51'26" 38d39'4"
Patricia Faye DeVeau
Gem 6h42'36" 23d3'47"
Patricia Faye Nowlin
Vir 12h41'47" 10d41'36"
Patricia Ferguson
Uma 13h29'19" 56d49'35"
Patricia Ferruccio
Psc 0h21'16" 16d51'56"
Patricia Finkelstein
Aqr 22h26'11" -2d25'16"
Patricia Flick
Crb 16h3'0" 27d0'38"
Patricia Foiles
Tau 3h51'7" 21d54'56"
Patricia Foley
Uma 12h0'1" 49d13'6"
Patricia Fontova
Sex 9h44'38" -0d20'34"
Patricia Fopiano
And 0h53'53" 38d51'24"
Patricia Fraley
Lyn 7h32'12" 41d39'53"
Patricia Frazer
Cnc 8h44'45" 12d40'34"
Patricia Frey
Per 3h18'46" 42d24'45"
Patricia Friedman
Psc 0h0'15" 1d12'29"
Patricia Fuentes Nieto
Del 20h47'37" 12d45'9"
Patricia Gaertig
And 1h29'58" 34d22'16"
Patricia Gail
Mon 7h34'28" -0d57'44"
Patricia Gail Hayes
Franco's Star
Ori 5h57'11" 17d47'38"
Patricia Galloway
Cas 0h52'36" 60d24'20"
Patricia Galloway May
Aqr 22h38'53" -15d45'15"
Patricia Garafola
Vir 12h40'37" -1d45'47"
Patricia Garcia Plante
Ori 5h50'41" 12d9'42"
Patricia Gayle Jeanine
Hatfield
Cas 0h56'33" 60d39'2"
Patricia Goethel
Cyg 21h26'15" 31d24'29"
Patricia "Grace" Feingold
Aqr 23h35'37" -24d23'58"
Patricia Grace La Flare
Lib 14h50'54" -1d28'56"
Patricia Grace - Nanny
Del 20h38'27" 15d8'13"
Patricia Graham
Lyn 7h53'37" 42d42'33"
Patricia Greenwood
Psc 0h33'58" 8d15'51"
Patricia Gruber
Vir 13h18'27" 13d8'0"
Patricia (Gurgy) Zwolinski
Leo 11h21'22" -2d40'28"
Patricia Haluna Tayamen
Rhodes
Leo 11h15'36" 10d54'16"
Patricia Harmann
Dra 19h9'33" 62d37'49"
Patricia Harrison's Star
Ser 15h52'3" 24d37'57"
Patricia Helen Vincent
Uma 13h31'35" 53d43'58"
Patricia Henderson
Cam 3h46'41" 55d16'15"
Patricia Hevia
Ori 5h11'4" 6d42'23"
Patricia Higgins Green
Aqr 23h55'50" -3d51'25"
Patricia Hill
Cas 0h23'8" 62d52'33"
Patricia Hinkle (Mama-Mia)
Srp 18h4'3" -0d5'0"
Patricia Hodgins
Mon 7h32'11" -0d45'47"
Patricia Hurst
Cas 0h33'50" 60d36'11"
Patricia Hyatt
Lmi 10h44'59" 37d8'16"
Patricia I. Hope
Cas 0h4'23" 53d26'55"
Patricia Irene Blessie
Sco 17h50'43" -40d5'21"
Patricia Irmscher
Vir 12h38'13" 7d29'27"
Patricia J. Doele
Uma 9h13'18" 59d29'38"
Patricia J. Hamman
Gem 7h16'26" 26d42'33"
Patricia J. Mayo
Vir 12h36'35" 0d49'38"
Patricia J. Riley
Peg 21h29'49" 16d15'25"
Patricia Jane Hannahs
Umi 13h49'19" 79d8'9"
Patricia Jane Pfohl-Smith
And 23h2'54" 48d38'28"
Patricia Janet Panebianco
Cas 1h31'22" 63d19'8"
Patricia JB
And 2h26'12" 46d0'31"
Patricia Jean
Lyn 9h9'50" 45d3'22"
Patricia Jean Cole
And 1h20'20" 40d13'6"
Patricia Jean Haas
Aqr 22h32'54" -0d41'7"
Patricia Jean Strader
Cam 6h34'23" 65d59'35"
Patricia Jean Sullivan Riley
Gem 6h30'24" 17d30'3"
Patricia Jeanne Bitter
Psc 1h10'15" 25d2'48"
Patricia Jenkins
And 0h38'47" 42d11'21"
Patricia Jill D'Amato
Uma 13h55'59" 53d28'18"
Patricia Jo Fairhurst
Cet 1h15'18" -0d4'29"
Patricia Joanne Perlinger
Aqr 22h49'10" -12d51'35"
Patricia & John Stewart
Her 17h51'20" 37d19'4"
Patricia Johnson
And 0h44'50" 43d25'13"
Patricia Johnson Kaltreider
Crb 15h46'30" 27d10'49"
Patricia Joseph
Ori 5h46'25" 8d58'37"
Patricia Joy
And 1h28'20" 43d37'47"
Patricia June
Gem 6h53'40" 28d40'33"
Patricia June
Cnc 8h41'32" 16d38'8"
Patricia June Burdick
Cmi 7h11'26" 11d9'59"
Patricia K Landers
Umi 15h21'46" 68d11'56"
Patricia K Plumstead
Sco 16h12'33" -42d2'38"
Patricia K. Rudolph
Tau 4h53'14" 26d8'44"
Patricia K Valiquette
Ori 5h29'8" 5d55'46"
Patricia Kathryn Henning-
Miller
Gem 7h19'52" 30d40'20"
Patricia Kay
And 1h58'23" 46d22'52"
Patricia Kay Piccolo
Leo 11h35'7" 12d34'9"
Patricia Kelly's Shining Star
Cma 7h20'24" -20d27'11"
Patricia Kelso
And 1h54'40" 39d12'54"
Patricia & Kenneth Lennon
Uma 13h15'4" 58d23'40"
Patricia Kilday
Dra 14h54'14" 59d20'17"
Patricia Kim Ide
Uma 8h10'56" 63d35'13"
Patricia Kossonogow
Col 6h22'36" -34d11'17"
Patricia Kricheff
Uma 9h16'42" 53d28'36"
Patricia Kruse
Ari 2h15'12" 23d57'1"
Patricia L. Allard
Cas 23h9'56" 55d46'32"
Patricia L. Dalton
Uma 8h58'26" 62d52'11"
Patricia L. Hooper
Cas 1h17'21" 57d57'55"
Patricia L. McCarthy
And 0h6'53" 33d5'33"
Patricia L Rose
Uma 8h45'2" 52d33'4"
Patricia Ladd
Cas 0h56'52" 59d41'21"
Patricia Le Galloudec
Cyg 21h14'33" 41d28'33"
Patricia Lea Camerota
Leo 11h4'4" 20d14'3"
Patricia LeClerq
Cas 23h30'26" 59d39'21"
Patricia Lederle
And 2h19'7" 46d17'59"
Patricia Lee
And 2h25'17" 46d31'22"
Patricia Lee
Aqr 21h48'5" -0d3'1"
Patricia Lee Love Eddy and
Amy
Lyr 18h49'27" 39d30'8"
Patricia Lee Tworynsky
Kasij
Cam 6h56'0" 65d29'36"
Patricia Lees
Her 17h0'1" 33d34'43"
Patricia Leigh & Daniel Lee
Masters
Vir 13h10'3" -2d52'42"
Patricia Lentine
Umi 16h13'9" 71d52'42"
Patricia Leonardo
Crb 15h47'59" 36d12'27"
Patricia Levi
Psc 1h43'2" 9d29'1"
Patricia Lewis
And 1h50'15" 46d51'37"
Patricia Lipford
And 0h48'48" 37d53'8"
Patricia Loney-Kaylor
Aur 5h21'38" 33d9'54"
Patricia Lopez - Rising Star
Cas 0h56'20" 56d11'21"
Patricia Lorenz
Umi 14h21'27" 68d30'54"
Patricia Louise Boylan
And 0h59'10" 38d45'46"
Patricia Louise Kennedy
Com 13h10'28" 29d42'42"
Patricia Lucille Akers
Vir 12h27'40" -5d19'7"
Patricia Luiz
Aqr 22h17'50" 0d43'43"
Patricia LuLu Zifferblatt
Her 17h0'58" 23d43'19"
Patricia Lynn
Cas 1h4'54" 49d7'26"
Patricia Lynn 50
Psc 1h19'26" 23d14'55"
Patricia Lynn Alvarez PLA
Peg 22h36'55" 25d6'31"
Patricia Lynn Corbitt
Sgr 19h36'20" -17d53'45"
Patricia Lynn Davis
Uma 8h56'31" 62d5'50"
Patricia Lynn Harsla Smith
Com 13h10'34" 28d1'7"
Patricia Lynn Outram
Ari 3h2'41" 14d43'33"
Patricia Lynn Repose
Sex 10h35'36" -0d51'0"
Patricia Lynn Santora
Crb 15h53'34" 26d29'52"
Patricia Lynn Strickland
Umi 16h1'31" 84d32'58"
Patricia Lynne Thatcher
Ori 5h42'29" -1d53'31"
Patricia M Bennett
Aqr 22h40'0" 2d29'41"
Patricia M. Cooke
Cas 0h27'1" 61d11'11"

Patricia M. Donato
Sgr 17h54'39" -29d5'6"
Patricia M. Lindsey
And 0h7'25" 45d1'48"
Patricia M Long
Tau 3h23'44" -0d20'15"
Patricia M. Lonyay
Ori 6h13'1" 18d58'0"
Patricia M. Mula
Lmi 10h23'11" 34d8'6"
Patricia M Ono
Leo 11h12'12" 1d14'40"
Patricia M. Peña
Crb 15h30'41" 28d47'57"
Patricia M. Schattner
Aqr 22h16'22" 1d28'49"
Patricia M. Smith
Cyg 20h39'30" 45d47'47"
Patricia Mae
Ori 5h54'22" 21d23'55"
Patricia Mae Moody
Sco 16h9'37" -20d40'18"
Patricia Mae Rundell
Vir 12h44'10" 3d28'12"
Patricia Mae Sloan
And 23h13'40" 46d0'33"
Patricia Mae Sloan
Cas 23h31'56" 51d24'37"
Patricia Mae Uhlig
And 0h38'15" 40d44'30"
Patricia & Maggie Mae's
Star
Vir 14h2'28" -17d45'59"
Patricia Maillet Vinet
Cas 23h2'3" 59d35'20"
Patricia Main
And 1h29'29" 35d26'51"
Patricia Maria
Ori 6h2'11" 6d53'30"
Patricia Maria Soberanis
Cap 20h28'17" -13d12'11"
Patricia Marie Beveridge
Curry
Aqr 23h7'42" -7d44'56"
Patricia Marie Cataraso
Tau 5h34'46" 27d18'13"
Patricia Marie Cecelia
And 1h32'32" 41d25'52"
Patricia Marie Dennison
And 2h31'42" 38d54'15"
Patricia Marie Kaufmann
Cnc 8h27'22" 30d43'41"
Patricia Marie Rich
And 0h31'21" 38d53'58"
Patricia Marie Sumlin
Cas 23h54'52" 57d10'3"
Patricia Marlene Suchomel
Ori 5h29'8" 2d52'27"
Patricia Marlow
Gem 7h12'21" 22d52'7"
Patricia Marlow 9/85
And 2h17'54" 47d35'43"
Patricia Marron
Cas 23h8'8" 59d31'42"
Patricia Marshall
Cas 1h26'2" 57d4'56"
Patricia Martin
Gem 7h37'44" 28d2'43"
Patricia Mary Beers
Uma 10h13'24" 70d51'19"
Patricia Mary Finn-Litchfield
Tau 3h43'49" 23d32'23"
Patricia Mary Galloway
May
Aqr 22h27'32" -0d58'24"
Patricia Mary Joyce
And 2h15'27" 50d30'55"
Patricia Mary McCutcheon
Cyg 19h53'50" 35d40'0"
Patricia Mary - Pat's Star
Cnc 7h58'38" 18d23'19"
Patricia Maryclare
Ori 4h54'46" 3d15'38"
Patricia Matagrano
And 23h22'4" 42d33'11"
Patricia Mattingly
Uma 8h47'33" 59d4'16"
Patricia Matuzewski
Savage
Uma 8h57'19" 59d20'38"
Patricia May
Vir 14h9'9" -14d47'4"
Patricia May
Lyn 7h42'9" 37d8'19"
Patricia McGhee - The Star
of Trish
And 1h12'4" 42d1'38"
Patricia McIntosh
Cap 20h50'55" -26d26'53"
Patricia Mckinney
Psc 23h52'14" -0d9'0"
Patricia McNally
Crb 16h4'58" 36d32'45"
Patricia McWhorter
Cas 23h5'39" 59d38'50"
Patricia Mead
Uma 11h37'38" 37d56'51"
Patricia Mercado
Lib 15h12'34" -19d12'27"
Patricia Metzger
Vul 21h14'50" 26d47'31"
Patricia Michelle Allagreen
Sco 16h2'48" -13d54'41"
Patricia Miller
Sgr 19h12'21" -17d33'56"
Patricia Miller
Lmi 10h38'48" 32d40'2"
Patricia Mitchell-Fitzgerald
Cyg 21h16'44" 37d35'26"
Patricia Modjeski 15
Cyg 20h52'33" 35d5'1"
Patricia Molloy Tracy
Cnc 9h6'2" 30d31'5"
Patricia Monica Macalyk
Cru 12h3'52" -61d35'41"
Patricia Montalvo
Cnc 8h4'33" 6d41'25"
Patricia (Moonbeam) Hill
Lyr 19h18'5" 39d28'41"
Patricia Mort Denemy
Leo 9h54'59" 18d25'59"
Patricia Mosser My Little
Bitty
Aqr 23h4'5" -5d17'40"
Patricia Mullins-Thomas
Lyn 7h52'26" 35d25'6"
Patricia Murphy
Cap 20h33'33" -16d29'38"
Patricia My Love
Uma 9h5'14" 47d2'52"
Patricia N. Bottomley
Uma 11h49'28" 53d40'15"
Patricia N. Brackett
Aqr 22h8'55" -9d22'55"
Patricia (Nana Pat)
Uma 13h39'4" 51d25'49"
Patricia "Nana Pat" Berardi
Vir 12h14'29" 0d49'48"
Patricia Nauer
And 23h1'24" 43d46'39"
Patricia Newton Huntoon
Cam 7h45'17" 78d9'12"
Patricia Nicole
Ori 5h35'24" -1d34'9"
Patricia O'Connell
Cnc 8h38'44" 16d35'16"
Patricia Otis
Sco 17h27'32" -41d24'12"
Patricia Packey
Umi 13h41'59" 75d10'18"
Patricia Padovani
Sgr 17h53'27" -29d49'13"
Patricia Page
Cyg 21h17'18" 50d44'19"
Patricia (Pat) Butler
Cap 20h28'44" -22d24'11"
Patricia Pat Mum Trish
Nanny
Cas 1h56'35" 61d28'52"
Patricia & Patrick
Schwarzentruber
Sct 18h26'58" -15d22'54"
Patricia Perez
Lib 15h35'31" -27d52'58"
Patricia Phillips 8/1/86
Leo 10h36'17" 8d5'25"
Patricia Pignoli Colella
And 0h58'8" 37d32'20"
Patricia Pomroy
And 2h19'7" 50d13'37"
Patricia R. Grogan
Cyg 19h59'12" 41d37'40"
Patricia Rasmusson's
Eastern Star
Peg 21h31'16" 12d39'48"
Patricia Reed, Light of my
life.
Lyn 7h7'16" 50d57'57"
Patricia Regnemer
Cas 0h18'59" 64d6'37"
Patricia Renee'
Uma 9h48'18" 52d3'31"
Patricia Reynoso
Lyn 7h30'3" 45d50'38"
Patricia Reynoso
Lyr 19h23'54" 31d3'47"
Patricia Rich
Psc 1h29'56" 27d50'59"
Patricia & Rick Barrie
Pho 23h38'14" -42d11'6"
Patricia Rivers
Uma 11h39'55" 36d10'52"
Patricia Rodriguez
Cnc 9h7'41" 31d45'48"
Patricia Rose Hulin
Psc 0h50'47" 14d5'59"
Patricia Rose Klos Star
Eri 4h37'35" -0d7'36"
Patricia Rose Vanek
Uma 10h51'41" 67d0'19"
Patricia Rosenfeld
And 1h4'34" 43d9'59"
Patricia Rubalcava
And 0h54'30" 37d55'47"
Patricia Rude Van Eck
Cas 1h28'0" 53d18'33"
Patricia Ruengert
Lyn 7h14'58" 47d24'13"
Patricia Rush
Cap 21h56'24" -18d51'0"
Patricia & Russell
Umi 16h57'19" 79d49'6"
Patricia Ryan "Nana"
Uma 11h12'43" 71d23'55"
Patricia S. Mahoney
Cas 0h21'49" 61d21'43"
Patricia S. Torriero
Sgr 19h55'39" -35d30'46"
Patricia Sauder
Lyn 6h31'15" 56d23'16"
Patricia (Schlagie) Ethel
Myrhow
And 0h32'19" 31d59'47"
Patricia Schreiber
Uma 8h52'37" 62d48'10"
Patricia Séguin
Aql 20h17'7" 0d29'6"
Patricia Shambaugh
Lyn 8h26'7" 37d15'13"
Patricia Shannon
Aqr 21h40'40" -0d49'31"
Patricia Sharon Greenfield
Mon 6h46'31" -0d4'6"
Patricia Sheetz Downey
Umi 16h30'41" 76d44'48"
Patricia Smith 7/4/39
Cnc 8h49'30" 15d50'43"
Patricia Soccodato
And 1h32'31" 49d51'53"
Patricia Sonia Allen
And 23h25'46" 52d29'31"
Patricia Spillane
Cas 0h4'43" 56d5'30"
Patricia Spotts
Mon 7h8'33" -0d26'10"
Patricia Starr Bock
Cas 0h51'49" 63d40'38"
Patricia Starz
Psc 1h24'45" 28d51'33"
Patricia Stauffer
And 0h34'55" 21d49'29"
Patricia & Stefan
And 1h7'30" 42d3'7"
Patricia Stoll
Her 17h48'23" 39d17'27"
Patricia Sue
Psc 23h36'37" 3d6'3"
Patricia Sue
Vul 19h4'34" 23d56'30"
Patricia Sullivan
Uma 10h27'40" 42d35'31"
Patricia Sullivan
Aqr 23h14'7" -19d52'1"
Patricia Susan Marsh
Leo 10h22'59" 17d41'16"
Patricia Swee Pee
Psc 0h37'32" 19d15'20"
Patricia T. Fitzgerald
Crb 15h41'10" 37d45'56"
Patricia Tailey McWilliams
Uma 11h38'38" 33d38'44"
Patricia Tallini
Ari 3h6'30" 29d5'44"
Patricia Tananis
Lyr 18h46'56" 36d5'25"
Patricia Taylor
Her 18h32'43" 19d0'43"
Patricia Taylor
Tau 3h47'55" 5d13'2"
Patricia Taylor
Umi 15h58'10" 76d57'24"
Patricia "T.C." Golden
And 23h26'53" 48d22'30"
Patricia Thissen
Ari 1h52'0" 21d53'25"
Patricia Tintari
Uma 12h41'29" 53d9'40"
Patricia Turner
Cyg 20h2'47" 38d11'27"
Patricia Tyrrell
And 1h59'32" 38d28'3"
Patricia und Michael
Dra 19h13'12" 66d40'29"
Patricia Vallandingham
And 23h22'17" 48d9'26"
Patricia Van Gorder
Dra 19h10'21" 65d5'29"
Patricia Van Ness
(Wahdee1)
Psc 1h59'32" 6d36'10"
Patricia Vandivier
Ari 2h1'34" 14d22'25"
Patricia Velazquez
And 2h18'53" 48d14'29"
Patricia Victoria Caputo
Gem 6h47'25" 33d22'50"
Patricia Virginia Rosalie
Schwolow
Sco 16h14'10" -14d15'29"
Patricia Ward Finnell
And 0h30'44" 25d25'7"
Patricia Weiss
Gem 6h32'40" 18d20'15"
Patricia Wesen
Cas 1h29'40" 59d34'39"
Patricia West
Cas 1h40'3" 57d18'47"
Patricia Westerlund
Cas 0h8'31" 53d31'18"
Patricia White
Gem 6h32'15" 12d34'27"
Patricia Whittaker
Ari 2h18'1" 24d46'3"
Patricia Whitten Poole
Sgr 19h29'9" -27d26'59"
Patricia & William Ahearn
Cyg 20h40'48" 46d15'52"
Patricia Williams
Sco 16h13'37" -11d45'29"
Patricia Wilson Cass
Aql 20h11'7" 11d39'47"
Patricia Wood - 16th
January 1945
Cyg 21h21'32" 32d28'10"
Patricia Xiani Chu
Psc 0h28'3" 14d2'45"
Patricia y Gabriela Proa
Uma 9h54'48" 46d59'40"
Patricia Youmans
Uma 11h5'41" 62d58'41"
Patricia Zamora Smith
Aqr 20h56'37" -11d32'9"
Patricia, Daniela, Mauricio,
Gem 7h1'10" 20d55'38"
PatriciaEthel
Cas 1h49'13" 59d0'4"
Patrician Beacon
Uma 9h41'3" 63d46'56"
Patricia-Polish Princess
Lib 15h19'35" -20d9'13"
Patricia's Christmas Star
Lyn 7h55'34" 51d16'26"
Patricia's Heart
And 0h13'8" 39d2'41"
Patricia's Infinite Love Star
Mon 8h2'10" -0d27'23"
Patricia's Kindred Spirit
Lib 15h14'1" -21d32'27"
Patricia's Library in the sky
Uma 11h8'45" 69d14'26"
Patricia's Light.
And 0h20'38" 25d32'35"
Patricia's Little Slice of
Heaven
Psc 1h16'6" 16d48'8"
Patricia's Shining Star
Cas 23h26'34" 55d3'59"
Patricia's Wish
And 0h26'4" 25d45'1"
PatriciaVeronicaMarieAnnB
lake
Uma 11h38'48" 32d27'27"
Patricio Urrutia Sivori
Dra 16h44'42" 51d52'23"
Patrick
Boo 15h6'18" 48d21'32"
Patrick
Uma 9h53'0" 47d3'35"
Patrick
Uma 11h8'54" 45d14'40"
Patrick
Her 17h25'48" 35d12'47"
Patrick
Tri 2h27'14" 31d1'31"
Patrick
Per 3h22'53" 44d58'53"
Patrick
Gem 7h17'16" 17d1'58"
Patrick
Ori 6h2'40" 21d18'30"
Patrick
Ori 4h52'22" 14d10'37"
Patrick
Ori 6h9'10" 7d4'41"
Patrick
Ori 5h48'55" 5d30'18"
Patrick
Uma 14h13'23" 59d18'50"
Patrick
Dra 16h53'11" 55d31'43"
Patrick
Cep 0h12'9" 77d33'59"
Patrick
Uma 9h28'59" 65d46'47"
Patrick
Cap 20h13'13" -14d23'3"
Patrick
Lib 15h14'55" -9d32'21"
Patrick
Aqr 22h29'25" -1d57'20"
Patrick
Lib 14h50'7" -2d30'22"
Patrick
Psc 0h6'26" 7d26'33"
Patrick
Cen 12h10'9" -47d21'44"
Patrick A. Grande
Cep 20h42'23" 65d4'58"
Patrick A J Blair
Uma 9h45'45" 71d18'5"
Patrick Abraham
Dra 10h0'48" 76d18'56"
Patrick Adam
Lac 22h40'49" 54d11'14"
Patrick Adams
Her 17h41'12" 41d52'18"
Patrick Afif
Gem 7h10'2" 21d33'13"
Patrick Aidan Sommer
Uma 10h25'43" 51d2'36"
Patrick "airborn" Chappel
Dra 16h44'6" 51d59'3"
Patrick Alan Casey
Aqr 20h57'2" -12d52'40"
Patrick Alan Newmyer
Uma 8h44'37" 54d17'48"
Patrick Alexander
Dra 18h45'42" 75d22'14"
Patrick Alexander Balson
Uma 8h25'42" 67d54'26"
Patrick Allen Kimbrough
Lib 15h3'42" -22d46'19"
Patrick Allen Murphy
Psc 1h42'56" 23d50'41"
PATRICK always & forever
ASHELY
Lep 5h44'23" -22d40'2"
Patrick and Ally
Uma 9h40'25" 53d23'47"
Patrick and Barbara Marsh
Lyr 18h37'0" 36d3'25"
Patrick and Jennifer Garza
Aqr 23h56'24" -15d23'57"
Patrick and Lydia Peldner
Cma 7h18'56" -14d10'0"
Patrick and Patricia Pease
Cyg 20h24'11" 32d39'18"
Patrick and Rachael
Lyn 7h28'47" 48d0'20"
Patrick and Rachel
Cyg 19h12'22" 53d50'59"
Patrick and Sabrina Ireland
Ori 6h11'40" 21d18'21"
Patrick Andrew Hickey
Sgr 20h0'5" -22d49'57"
Patrick Andrew Rivett
Cep 21h41'38" 64d6'57"
Patrick Anthony Griffiths
Ari 2h28'19" 25d47'23"
Patrick Atcheson
Cyg 21h12'33" 29d40'22"
Patrick Avery Laine
Leo 10h19'49" 10d18'10"
Patrick B. Coyne
Tau 4h16'33" 3d14'21"
Patrick B. Reed * I love you
*
Psc 1h27'21" 16d16'53"
Patrick Baia's Shining Star
Umi 14h14'6" 76d33'32"
Patrick Barnard's Tender
Heart 15
Dra 17h42'7" 65d14'32"
Patrick Barr
Ori 5h27'4" -0d53'3"
Patrick Baumann
And 0h20'42" 29d57'35"
Patrick Bernard Kluesner
Psc 1h9'54" 10d13'55"
Patrick Bochef
Uma 10h45'44" 67d28'20"
Patrick Bonisteel
Cnc 8h12'6" 15d34'19"
Patrick "Boone" M. Britsch
Ori 5h43'33" 1d32'24"
Patrick Bradley Boyle
Her 17h12'55" 34d50'59"
Patrick Brett
Aql 19h38'54" 12d36'52"
Patrick Brian Chavez
Umi 15h34'40" 71d29'50"
Patrick Brian Kenny
Sct 18h40'43" -6d55'34"
Patrick Brian McGahey
Uma 8h28'58" 64d59'40"
Patrick Brian Sullivan
Sgr 19h8'45" -12d56'14"
Patrick Brian Toelken
Aqr 23h25'50" -19d36'15"
Patrick Brugman
Leo 11h40'31" 26d19'26"
Patrick Buob
Umi 13h40'27" 72d16'40"
Patrick Burke
Uma 11h42'55" 36d10'40"
Patrick Butler
Sco 16h8'50" -13d40'29"
Patrick Butler
Cnv 12h37'26" 51d34'3"
Patrick C. Sheridan
Leo 10h17'43" 16d37'29"
Patrick Campbell
Aur 5h41'18" 42d42'5"
Patrick & Charlene Ryan-
Wedding Day
Lyn 7h34'12" 58d28'25"
Patrick Charles Pallisco
Cnc 9h13'56" 22d25'57"
Patrick Christiansen
Lmi 9h48'18" 37d26'6"
Patrick Christopher PACKY
Teahan II
Uma 9h44'52" 59d45'16"
Patrick Christopher Teefy
Uma 8h26'19" 63d42'21"
Patrick Cliquennois
Cnc 8h17'7" 23d34'57"
Patrick Conners
Umi 12h21'37" 86d21'16"
Patrick Conroy
Her 17h36'44" 48d10'57"
Patrick Conway
Aqr 22h59'15" -10d30'44"
Patrick Cruz Gomez Sr.
Boo 14h31'28" 32d30'19"
Patrick D. Boyle
Sgr 19h20'22" -14d19'26"
Patrick D. Carling
Aqr 22h51'23" -16d35'0"
Patrick Daniel Brown
Per 4h35'46" 34d30'7"
Patrick David Bond Jr
Birthday Star
Per 2h59'23" 45d49'10"
Patrick Daws
Sco 16h14'59" -19d36'58"
Patrick Day
Her 18h50'27" 23d16'39"
Patrick Dechello
Ori 5h15'9" 15d34'52"
Patrick DeFilippo
Cap 21h15'41" -16d16'58"
Patrick DeGrosse Jr.
Ori 5h5'7" 11d10'12"
Patrick Demetz McGrath
Per 4h2'32" 32d53'41"
Patrick Dennis Doherty
Sgr 18h4'44" -21d9'30"
Patrick Derosier
Uma 10h10'29" 57d29'45"
Patrick Deschamps mon
superman
Sgr 18h47'21" -35d44'55"
Patrick d'Esperies
Ori 6h16'30" 14d54'37"
Patrick Donahue, Jr.
Her 18h17'31" 14d54'56"
Patrick Donald Borkowski
Umi 15h2'22" 70d56'59"
Patrick Donauer,
12.10.1969
Ori 6h11'45" 20d44'58"
Patrick Dowman's Star
Ari 3h16'59" 29d17'33"
Patrick Drew Denney
Cap 21h42'44" -14d12'54"
Patrick Droogmans
Ori 6h11'57" 5d48'50"
Patrick Düggelin
30.04.2003
And 1h18'9" 44d10'29"
Patrick Dunn-Baker
Aqr 22h26'5" -2d12'49"
Patrick Durkin
Cma 7h17'43" -13d49'57"
Patrick E. Sereno
Psc 1h10'16" 32d43'45"
Patrick E. Thibodeau
Aqr 20h42'45" -7d36'39"
Patrick Edmund Sykes
Cep 22h55'14" 61d53'21"
Patrick Edward Burke
Ori 6h14'32" 15d24'4"
Patrick Edward Carrillo
Her 17h37'8" 27d45'27"
Patrick Edward Madding
Cep 22h47'34" 74d20'49"
Patrick Elasik
Tau 3h47'23" 27d34'13"
Patrick Eldon Good
Uma 10h12'55" 65d34'4"
Patrick Engel
Psc 1h17'2" 18d6'48"
Patrick Eric Riesenberg
Psc 23h48'34" 5d27'22"
Patrick Eugene Widman
Dra 20h38'4" 69d54'25"
Patrick F. Fitzgerald
Cnc 8h20'13" 32d21'26"
Patrick Farrell
Tau 4h9'26" 19d34'59"
Patrick Ferris the
Gentleman
Per 3h48'32" 49d3'16"
Patrick Fink
Cam 7h32'26" 67d47'47"
Patrick Finneran
Uma 8h47'0" 63d8'16"
Patrick Fitzgerald
Rasmussen
Psc 0h6'59" 9d5'13"
Patrick Flanagan
Boo 15h11'11" 42d26'7"
Patrick Flynn
Cep 21h8'15" 58d25'54"
Patrick Fortin
Uma 11h24'46" 60d42'6"
Patrick Fournier
Cyg 20h43'9" 34d8'14"
Patrick Francis Jackson
Leo 10h58'54" 1d24'50"
Patrick Francis Quinn
Vir 12h2'34" -10d17'22"
Patrick Francis Trihey
Lib 15h6'4" -27d47'21"
Patrick Frank Hill
Leo 11h8'21" 20d30'44"
Patrick French
Aur 5h57'22" 36d38'29"
Patrick "Friend, Lover &
Partner"
Uma 8h50'50" 55d6'27"
Patrick Gargano Gomes
Sco 16h11'16" -14d11'23"
Patrick Gatbunton
Ori 6h5'4" 1d25'33"
Patrick Gerard Donohue
Cyg 21h16'13" 44d14'49"
Patrick Glitsch - Rising Star
Boo 14h36'25" 41d25'41"
Patrick "Gus" Cosgrove
Cep 21h45'21" 58d25'26"
Patrick Hadley
Psc 0h31'22" 20d56'47"
Patrick Hammersla
Ori 6h14'4" 9d5'23"
Patrick Harkins
Boo 15h22'21" 32d43'42"
Patrick Healey
Lib 14h50'27" -5d22'34"
Patrick Healy - Thanks
7Up/DNL!
Lyn 9h0'15" 38d5'29"
Patrick Heard
Per 3h41'50" 45d11'59"
Patrick Hemingway
Ori 6h13'57" 15d45'7"
Patrick Hennessy
Tau 4h31'1" 2d44'20"
Patrick Hentzen
Per 3h48'46" 35d14'10"
Patrick Hickey
Uma 11h56'33" 56d29'17"
Patrick Hogan
Sgr 20h2'13" -25d9'16"
Patrick Hughes
Vir 12h59'23" 11d41'33"
Patrick J. Boyle
Aql 19h26'1" 2d17'17"
Patrick J. Harkins
Lib 15h52'19" -6d49'37"
Patrick J. Herdman
Uma 11h40'49" 56d12'20"
Patrick J. Herleman
Ori 4h46'57" 11d31'47"
Patrick J. Joyce
Ser 15h22'5" -0d46'26"
Patrick J. McCullough
Her 17h8'4" 31d17'35"
Patrick J. Murphy
Her 17h25'54" 36d38'18"
Patrick J. Murray
Lib 15h14'0" -22d21'52"
Patrick J. O'Connell
Lyn 7h10'24" 47d44'3"
Patrick J Richgels
Her 16h19'36" 9d10'2"
Patrick J. Schwartz
Ari 3h16'35" 19d27'54"
Patrick Jackson
Sgr 19h23'54" -18d5'2"
Patrick Jacob Moreira
Cru 12h13'25" -58d18'46"
Patrick & Jacqueline
Higgins
Ori 5h59'39" 7d24'24"
Patrick James Edward
Doudican
Dra 18h11'11" 52d35'13"
Patrick James Gannon
Psc 0h58'59" 16d59'44"
Patrick James Kaelin
Cap 20h55'9" -19d57'24"
Patrick James Kirby
Per 3h16'45" 42d58'25"
Patrick James Long
Sgr 19h53'6" -29d54'58"
Patrick James McGuire
Aur 5h16'57" 29d46'58"
Patrick James Morgan
Cep 23h9'27" 70d47'53"
Patrick James Palaszewski
Ori 5h59'26" 17d35'57"
Patrick James Reilly
Ori 5h42'46" -0d26'37"
Patrick Jayson Bernal
Uma 9h8'14" 64d44'48"
Patrick Jero Miguel Castillo
Her 17h21'33" 22d29'14"
Patrick & Jessica Buckles
Forever
Leo 10h5'42" 23d5'12"
Patrick John
Boo 14h29'3" 54d39'14"
Patrick John Budelis
Uma 12h10'7" 59d41'12"
Patrick John Kish
Eri 4h24'58" -36d9'16"
Patrick John Knight
Uma 11h26'36" 71d28'12"
Patrick John Lostaglia
Gem 7h7'33" 21d34'1"
Patrick John Walker
Tau 3h34'7" 5d44'20"
Patrick Joseph
Uma 9h16'42" 63d57'3"
Patrick Joseph Connolly
Per 3h45'31" 48d51'42"
Patrick Joseph Deuschle
Ori 5h59'17" -0d41'50"
Patrick Joseph Grady
Umi 14h22'51" 72d48'6"
Patrick Joseph Hanrahan
Her 17h13'0" 34d2'23"
Patrick Joseph Hoffert-My
Love
Vir 13h7'34" 12d5'10"
Patrick Joseph Hogan
Aqr 22h7'17" -0d1'12"
Patrick Joseph Hoke
Lyn 6h56'37" 54d3'4"
Patrick Joseph Hughes
Sco 17h50'28" -42d41'53"
Patrick Joseph McAteer
Lyr 18h24'25" 32d26'30"
Patrick Joseph Rivero
Isidro
Ari 3h4'4" 12d10'38"
Patrick Joseph Weaver
Aql 19h53'47" -0d14'42"
Patrick Jospeh Mawn
Lyr 18h46'41" 32d18'6"
Patrick J.S. Taylor
Uma 11h28'48" 32d1'25"
Patrick K. Bagley
Psc 0h10'45" 7d54'3"
Patrick Karr
Aqr 20h48'1" 2d22'22"
Patrick Käser
Lac 22h18'59" 46d53'13"
Patrick Kebert
Ori 5h22'49" -0d8'54"

Patrick Keely - Sara's Shining Star
Cyg 19h42'52" 33d2'22"

Patrick 'Kroki' Wyss
Lmi 10h36'28" 36d57'45"

PATRICK KUNKEL
Her 17h20'2" 15d41'0"

Patrick Lafontaine
Per 3h10'40" 52d4'52"

Patrick Laudato
Aur 6h29'14" 39d34'53"

Patrick Laurence Cummings
Her 16h35'39" 18d20'48"

Patrick Lee Gipson
Her 17h24'10" 32d3'8"

Patrick Lee Jenkins
Pho 1h0'23" -43d34'23"

Patrick Leonard Olson "Pat"
Cyg 20h40'33" 40d54'50"

Patrick Leyx
Her 18h10'38" 28d15'17"

Patrick Lloyd Hayes (PatMan!)
Lib 15h45'29" -11d56'38"

Patrick Loren
Her 18h31'13" 12d16'21"

Patrick Loves Amanda
Leo 11h11'14" 15d14'18"

Patrick Loves Rina
Cyg 21h24'30" 43d28'23"

Patrick M. Culligan
Psc 1h19'26" 31d18'38"

Patrick M. Radebaugh
Psc 1h49'1" 7d21'49"

Patrick M. Stewart
Uma 10h39'12" 55d32'28"

Patrick M. Thernton
Cep 22h21'36" 63d51'43"

Patrick MacDonald Horvath
Dra 18h46'59" 61d34'38"

Patrick Mackenzie
Aqr 22h31'1" -1d44'33"

Patrick Mannino
And 2h3'10" 46d33'38"

Patrick Mark Kalisek
Aql 19h47'58" 12d18'13"

patrick marshall
Gem 7h40'39" 32d37'0"

Patrick Martin
Leo 9h36'19" 30d23'33"

Patrick Martin Stump
Tau 5h26'59" 27d24'45"

Patrick Mc Dougall
Cap 20h51'19" -14d44'15"

Patrick McArdle
Uma 12h7'16" 57d17'26"

Patrick McDermaid
Her 16h55'3" 14d39'30"

Patrick McElroy
Cru 12h17'17" -56d52'9"

Patrick McGraw
Uma 13h46'56" 57d19'52"

Patrick McGuire
Cep 20h37'2" 62d0'2"

Patrick McLaughlin
Cap 21h36'4" -15d12'41"

Patrick Meimari
Umi 2h13'5" 89d7'3"

Patrick Michael Bowers II
Umi 13h42'17" 76d7'24"

Patrick Michael Bowers II
Leo 9h48'32" 27d38'8"

Patrick Michael Cooper
Her 17h45'42" 36d40'25"

Patrick Michael Cunningham
Cap 20h39'39" -17d48'45"

Patrick Michael Davis
Crb 16h3'24" 37d54'25"

Patrick Michael Farrell
Her 17h34'2" 36d37'45"

Patrick Michael Francis Tigue
Gem 6h5'59" 22d39'38"

Patrick Michael Grate
Her 17h42'57" 34d19'56"

Patrick Michael Horan
Uma 9h3'30" 54d57'45"

Patrick Michael Meade
Per 3h10'40" 43d41'9"

Patrick Michael O'Hara
Lyn 6h16'55" 54d4'6"

Patrick Michael Pich
Umi 15h49'10" 73d19'46"

Patrick Michael Xavier Byrne
Ori 5h15'30" 6d22'2"

Patrick Miller
Uma 10h22'8" 42d26'9"

Patrick Mills
Her 17h23'43" 36d26'59"

Patrick Monchatre
Uma 9h41'53" 63d49'29"

Patrick Moore
Cap 21h7'24" -25d35'45"

Patrick Murray - Pappa's Star
Uma 11h21'23" 41d57'38"

Patrick "My Angel"
Gem 7h19'17" 33d1'14"

Patrick "My sweet Irishman" Weadick
Leo 11h26'18" 11d52'20"

Patrick - My Treasure
Cru 12h38'58" -60d21'45"

Patrick N. Mackenna, Jr.
Ori 6h9'41" 17d55'56"

Patrick & Nicole
Lyr 19h9'46" 29d6'40"

Patrick Nilsen
Cru 12h52'49" -60d52'14"

Patrick Norris Campbell
Uma 10h26'40" 55d49'28"

Patrick O. Doyle
Gem 7h9'16" 25d42'49"

Patrick O'Donnell
Aqr 22h26'46" -0d18'35"

Patrick Oeschger
Cas 0h15'23" 56d27'20"

Patrick Oliver
Cru 12h46'35" -57d30'45"

Patrick O'Neill
Uma 13h38'49" 56d9'51"

Patrick O'Neill Collins
Tau 4h24'27" 25d5'21"

Patrick O'Reilly Liam Maloy
Psc 0h12'11" 11d8'19"

Patrick & Pamela Saul Sheridan
Per 3h24'31" 46d23'17"

Patrick "Pat" Walsh
Uma 13h35'31" 49d46'15"

Patrick "Patty Cake" Adams
Lac 22h38'29" 43d23'1"

Patrick Peroutka
Aqr 21h10'23" -10d41'57"

Patrick "Pete" Spencer
Uma 11h15'28" 54d21'25"

Patrick Petrolli
Aqr 21h22'31" -0d44'46"

Patrick Pionto
Uma 10h24'3" 39d30'37"

Patrick Prince
Cnc 8h40'51" 32d31'34"

Patrick Quinley
Ori 5h58'33" -0d42'16"

Patrick R Johns
Ori 6h8'42" 15d16'8"

Patrick Rafferty
Crb 16h11'3" 37d53'54"

Patrick Ray Klee, (Mr. P)
Ari 2h42'17" 25d53'43"

Patrick Reed Tomlinson
Cap 21h11'26" -17d14'20"

Patrick Regan
Ori 5h34'47" -2d46'7"

Patrick Richard Metzger
Umi 14h41'17" 76d21'17"

Patrick Riley Rader
Her 18h8'17" 16d17'27"

Patrick Robert Duval
Uma 9h41'3" 60d2'52"

Patrick Robert Layman
Cap 20h27'12" -10d19'55"

Patrick Robert Moon
Umi 16h15'35" 79d5'59"

Patrick Roberts
Psc 1h8'34" 11d16'15"

Patrick Rodney Vigil
Psc 1h24'11" 28d54'20"

Patrick Rosette
Cyg 20h24'41" 30d35'51"

Patrick Ryan
Aur 5h39'39" 44d52'7"

Patrick Ryan Burns
Aur 5h48'2" 48d15'59"

Patrick Ryan Schinzel
Cap 21h37'40" -13d53'54"

Patrick S. Ryan
Boo 14h52'58" 18d18'15"

Patrick Salem Saloom
Lib 15h7'22" -12d52'48"

Patrick Sanders Wieger "TOD"
Lib 14h26'5" -9d18'16"

Patrick Sawyer
Per 4h31'12" 34d11'49"

Patrick Schafer
Per 2h42'3" 56d45'2"

Patrick Seamus Xavier Good
Her 18h41'1" 21d48'35"

Patrick Seamus Xavier Good
Uma 9h7'54" 64d13'39"

Patrick Sean
Vir 12h18'42" 11d48'31"

Patrick Sean Stack
Psc 0h39'22" 19d5'54"

Patrick Seewald
Cyg 22h1'59" 50d42'38"

Patrick Shawn Free
Lac 22h52'57" 51d45'14"

Patrick Sorenson - A Wish For Peace
Lyn 7h38'23" 46d47'57"

Patrick Spengler
Lac 22h50'57" 52d42'7"

Patrick Steinger
Ori 6h5'17" 13d16'6"

Patrick Steven Reed
Tau 5h58'17" 24d55'44"

Patrick Steven Steigauf
Uma 9h47'12" 64d37'12"

Patrick Stillson
Sgr 17h52'55" -29d53'17"

Patrick Thalmann
Lyn 7h35'33" 37d31'4"

Patrick "The Gentle Lion"
Leo 11h44'59" 15d26'28"

Patrick Thomas Brown
Psc 1h35'35" 12d51'42"

Patrick Thomas Luiz
Aur 5h45'27" 41d27'56"

Patrick Thompson
Vir 12h57'58" 12d26'44"

Patrick Timothy Doyle
Cnc 8h39'58" 15d12'29"

Patrick Toland White
Uma 9h48'51" 44d57'48"

Patrick Tommy Cox
Her 17h29'40" 18d50'9"

Patrick Towers
Cen 13h37'23" -41d23'53"

Patrick Towey
Uma 14h22'19" 56d29'59"

Patrick Tracey
Boo 14h28'22" 24d31'32"

Patrick & Traci Rigsby
Cyg 20h34'46" 47d7'5"

Patrick Tran
Sco 17h52'52" -36d12'31"

Patrick T.White
Uma 13h50'58" 57d27'15"

Patrick und Kadiriye
Her 16h35'0" 14d2'51"

Patrick V. McGrath
Per 4h19'42" 52d1'55"

Patrick Valenti
Col 6h3'35" -35d58'28"

Patrick VonBueren,amour à moua
Psc 2h5'37" 5d46'36"

Patrick W. Burke
Psc 1h44'4" 23d3'32"

Patrick W. Callahan
Per 3h40'36" 46d7'1"

Patrick Walter Visgilio
Ari 3h23'2" 29d29'46"

Patrick Wayne Coffman
Vir 14h5'1" 2d58'2"

Patrick White
Uma 11h52'59" 56d9'30"

Patrick Wilbur
Lyn 7h17'46" 51d38'39"

Patrick William Holland
Umi 14h9'33" 65d34'14"

Patrick William McKnight
Sco 16h37'18" -25d54'30"

Patrick William Veresh
Leo 10h58'44" -0d55'20"

Patrick Winchester
Aql 19h4'35" 10d6'34"

PatrickRoryWhelan
Leo 9h31'40" 27d57'4"

Patrick's 40th
Uma 11h24'23" 43d26'42"

Patrick's Chosen One
Lmi 10h31'17" 31d32'42"

Patrick's Doll
Lmi 10h56'47" 27d29'51"

Patrick's Genius
Lmi 11h2'49" 33d11'33"

Patrick's Heart
Lmi 10h54'21" 28d8'53"

Patrick's Joy
Lyn 8h38'15" 35d54'32"

Patrick's Progeny
Per 2h56'14" 53d23'14"

Patrick's Star
Uma 11h18'49" 40d38'37"

Patrick's Star
Cma 6h50'38" -21d59'45"

Patrick's Star
Cap 20h36'50" -18d0'51"

Patricks Star
Lyn 7h14'6" 54d51'1"

Patrik Stricker
Her 18h1'41" 18d1'16"

Patrikon05
Per 3h35'24" 32d7'44"

Patrina
Ari 2h30'42" 25d2'18"

Patrina & Shem
Cyg 20h19'22" 38d41'50"

Patri's Pancho
Cnv 12h41'42" 39d28'43"

Patrizia
Uma 9h44'7" 50d29'35"

Patrizia
And 23h39'0" 33d54'9"

Patrizia
Peg 22h18'34" 17d31'44"

Patrizia
Boo 14h49'4" 14d44'56"

Patrizia
Cam 3h31'49" 61d56'32"

Patrizia
Cam 5h3'52" 67d52'44"

Patrizia
Vir 14h10'20" -15d9'58"

Patrizia
Cam 3h56'11" 77d12'46"

Patrizia
Umi 16h14'6" 85d4'15"

Patrizia
Ori 6h10'39" -2d12'13"

Patrizia Balsamo
Sct 18h36'17" -14d51'11"

Patrizia Bertozzi
Cam 9h13'40" 79d42'54"

Patrizia D'Alessandro
Aur 6h24'59" 40d19'35"

Patrizia Di Nenno
And 0h12'19" 46d32'33"

Patrizia Laginestra
Boo 14h42'1" 15d30'28"

Patrizia & Lorenzo
Cas 1h51'11" 60d54'49"

PATRIZIA + MARKUS
Per 4h33'46" 40d22'38"

Patrizia Martella
Cas 1h35'33" 72d33'32"

Patrizia Mombelli
Cam 3h58'56" 54d25'35"

Patrizia Nievergelt
Uma 10h54'4" 34d4'6"

Patrizia Nurra
Cam 3h24'14" 64d37'19"

Patrizia Pinna
Lyr 18h49'19" 39d39'51"

Patrizia Scali
Vir 13h20'32" 11d42'33"

Patrizia Valnegri
Aur 6h29'43" 34d31'14"

Patrizia Zangiacomi
Crb 15h48'54" 29d50'28"

Patrizia71
And 0h26'7" 36d56'32"

Patrizio
Peg 22h28'30" 19d8'47"

PATRIZIO GIAMPA
Cas 2h41'36" 58d5'55"

Patrocina Garcia
Uma 12h36'11" 57d17'34"

Patrocinio Relucio
Sco 17h51'26" -36d4'26"

Patron Lidia
Umi 14h38'58" 79d51'17"

Patronus
Per 4h10'0" 46d59'45"

Patry&Luca
Ori 5h57'1" 21d40'28"

Patrycja Wojtowicz
Cap 20h35'40" -9d36'30"

Patryk Gawel
Aur 5h12'9" 40d5'39"

Pat's Astellas Star
Ari 2h36'36" 27d45'52"

Pat's Dolphin
Aqr 20h40'8" -12d51'36"

Pat's Guiding Star
Psc 1h30'40" 21d44'59"

Pat's Hope
Uma 12h37'6" 59d58'53"

Pat's Place
Gem 6h50'12" 33d37'34"

Pat's Rose
And 0h18'27" 25d56'17"

Pat's shining star
Cnc 8h20'42" 14d24'41"

Pat's Star
Cap 20h54'56" -23d51'57"

Pat's Star - 'Anam Cara'
Cyg 20h16'22" 58d38'23"

Pat's Star (Grandad's Star)
Uma 11h53'13" 57d13'39"

Pat's Star (Nanny's Star)
Uma 11h40'30" 57d8'16"

Pat's Star of Hillsboro
Per 2h57'5" 46d45'25"

Patsea
Vir 13h21'19" 11d5'1"

Patsie Rose Fannin
And 0h35'11" 41d34'38"

Patsy
And 2h33'37" 39d51'11"

Patsy
Cma 7h15'8" -27d54'36"

Patsy and Marge Tanzillo
Umi 13h38'36" 71d1'19"

Patsy Ann Walls Conley
Dra 17h35'57" 61d52'49"

Patsy Anne
Lup 15h2'12" -53d58'14"

Patsy Bazin
And 2h25'58" 49d44'53"

Patsy Candy Mctigue
Aqr 22h28'52" -2d44'22"

Patsy Crawford
Cas 2h9'50" 64d14'6"

Patsy Deller
Cas 23h32'50" 52d26'4"

Patsy Dunn
Cyg 19h30'32" 28d43'36"

Patsy Garrett Kokinacis
Peg 21h43'59" 24d18'56"

Patsy & Henry Smith
Tau 4h28'43" 14d44'51"

Patsy J. Pingaro
Aqr 23h1'16" -22d49'27"

Patsy Jane
And 0h42'0" 38d45'10"

Patsy June Andersen
Uma 8h28'38" 71d2'29"

Patsy June Massongill
Peg 23h20'48" 33d7'13"

Patsy Lou
Uma 11h11'50" 47d49'40"

Patsy Lynn Newton
Sco 16h8'47" -11d29'59"

Patsy McCabe
Com 12h46'29" 27d3'52"

Patsy Poo
Uma 11h22'53" 51d2'37"

Patsy Ruth Compton
Uma 8h30'46" 64d29'51"

Patsy Ruth Hufford
Cap 20h32'33" -16d20'25"

Patsy Ruth Miller
Tau 5h5'6" 16d22'32"

Patsy Rutland
Cas 23h33'27" 58d18'35"

Patsy Taylor
And 0h23'12" 32d56'3"

Patsy & Thomas Bouziden
Her 17h9'35" 36d14'10"

Patsy W McEnroe
Cyg 20h26'3" 50d23'42"

Patsy Wicker
Uma 10h38'6" 42d58'55"

Patsy William "Chuck" Bertoni
Sgr 18h56'29" -29d58'33"

Patsy & William Probst - 50 years
Uma 11h23'39" 56d53'27"

Patsy Y Joel
Sge 20h2'13" 18d45'39"

Patt Patterson - June 6, 1921
Gem 6h51'51" 25d19'41"

Patterson Dentaire
Per 3h24'15" 49d12'30"

Patterson/SCA3
Cas 1h40'21" 63d45'27"

Patti
Uma 8h53'22" 70d30'38"

Patti
Crb 15h24'59" 30d34'34"

Patti
Cnc 9h3'41" 28d26'53"

Patti 50
Crb 15h41'35" 26d20'11"

Patti A. Benson
Cap 21h41'41" -19d39'0"

Patti Ainsworth
Cyg 21h31'39" 38d19'30"

Patti and Joey Moffa
Aql 19h45'22" -11d36'39"

Patti and John
Cyg 20h42'34" 33d0'57"

Patti Ann & Bob
Vir 13h39'14" -13d5'4"

Patti Ann Jones
Aqr 22h8'41" -0d18'5"

Patti Anne
Psc 23h48'4" 6d14'2"

Patti Ann's Celestial Body
Cyg 20h6'49" 40d10'1"

Patti Baker
Lyn 6h33'57" 61d54'38"

Patti Blackburn
Sgr 19h10'44" -31d22'52"

Patti Brasco
Cas 1h40'40" 62d20'17"

Patti Brittain
Cas 1h22'11" 63d54'33"

Patti Brooks
Leo 11h41'57" 26d16'29"

Patti Burger
Gem 7h34'42" 23d59'21"

Patti C. Bohy
Uma 12h4'30" 58d42'4"

Patti Carter
Cas 1h37'48" 64d38'32"

Patti Dorothy Walker Cheek
Ari 2h4'35" 21d49'26"

Patti Fullen Super Mom
Ari 3h7'43" 20d58'48"

Patti & George Gaves 122905
Cyg 21h41'43" 36d34'32"

Patti Harris "Queen of the Details"
Cas 0h34'20" 47d33'8"

Patti 'Hose' Dallas
Psc 1h10'31" 32d4'54"

Patti J Tietge
Ari 2h58'51" 21d12'32"

Patti Jane Saa
Ori 6h18'30" 15d8'42"

Patti Jo Hopkins
Leo 11h17'17" 18d33'40"

Patti Joan Fry
Tau 5h50'6" 23d10'49"

Patti Kite Geevers
Aql 19h45'11" 9d15'17"

Patti Klein
Psc 0h33'20" 13d0'27"

Patti Korotzer
And 1h7'59" 39d34'40"

Patti Lathrop
Her 18h36'22" 15d26'32"

Patti Lillian Diskin
Aql 18h54'39" -0d5'4"

Patti Lou
Cnc 8h44'50" 18d41'13"

Patti Lou Wright
Lyn 7h41'31" 37d24'59"

Patti Louise Steinberg
Ori 6h6'4" 14d9'30"

Patti Lynn
Crb 15h35'36" 34d7'9"

Patti Lynn
And 0h18'31" 38d7'17"

Patti Martison
Peg 22h1'30" 14d37'20"

Patti "Mom" Benton
Vir 12h26'55" -6d14'38"

Patti Mother Of Three
Cas 23h55'43" 57d47'4"

patti o
Cyg 21h33'20" 39d6'1"

Patti O'Neil
Cas 0h10'40" 54d1'35"

Patti Pascale
Cnc 8h55'14" 23d13'40"

Patti Poore
Cnc 8h9'0" 8d12'52"

Patti Ritter
Aqr 21h2'55" 1d18'22"

Patti Rose
Leo 11h23'46" 24d31'7"

Patti Rose
Crb 15h36'26" 32d41'19"

Patti Rose
Cas 1h39'52" 64d51'47"

Patti Ryan
And 0h48'7" 40d0'10"

Patti S. Cowperthwaite
Sgr 17h55'58" -19d19'7"

Patti Smith
Uma 9h18'14" 47d26'13"

Patti Solomon
Vir 12h53'32" 6d34'35"

Patti Stewart
Psc 1h1'23" 11d47'56"

Patti Sue Culbreth
Tau 4h38'5" 18d22'48"

Patti Warner
Uma 9h19'12" 54d10'50"

Patti Wisman
Cap 20h35'2" -23d31'11"

Patti-Brian
Leo 10h22'39" 21d49'26"

Pattie and David Terry
Cyg 21h53'45" 52d46'35"

Pattie Green Roberts
Cam 3h54'1" 66d2'4"

Pattie Love Forever And A Day
Sgr 19h34'25" -24d34'54"

Pattie Magdik
Vir 13h4'50" -8d11'54"

Pattie Ortiz
Lmi 10h3'11" 33d10'25"

Pattie Zambolla
Psc 0h32'50" 9d25'36"

PattieandDannyReachforth eStars
Cma 6h59'48" -13d19'37"

Pattie's Angel
Psc 23h13'39" -1d11'9"

Pattie's Hopes and Dreams
Lyn 8h40'35" 34d53'7"

Pattigene Long
Crb 16h8'54" 35d51'34"

Pattilou Helen Daum-Henchir
Cyg 20h27'49" 34d44'31"

Pattilyn Jeneen Olsen
Gem 6h55'37" 27d35'50"

Patti's 25th Wedding Anniversary
Cyg 20h22'45" 39d16'59"

Patti's Angel Wings Shine Upon Us
Uma 8h55'26" 63d17'51"

Patti's light of Cassiopeia
Cas 0h58'59" 57d16'9"

Patti's Point
Cas 0h41'22" 56d34'2"

Patti's Shining Star
Uma 10h25'0" 63d46'45"

Patti's Star
Lyn 8h31'12" 45d34'45"

Patti's Star 49ers
Leo 11h26'55" 1d47'50"

Pattricius P
Ari 2h46'33" 14d37'49"

Pattrick "Gadunkadunk" Ryan
Cru 12h7'16" -61d12'58"

Patty
Psc 1h1'32" 2d56'46"

Patty
Cas 23h44'38" 52d44'6"

Patty
Mon 7h45'14" -4d33'31"

Patty
Gem 7h6'18" 24d47'53"

Patty
And 0h41'16" 26d36'59"

Patty
Uma 10h23'48" 45d14'58"

Patty
Uma 11h25'49" 49d1'25"

Patty
And 0h33'9" 41d56'53"

Patty
Lyr 18h31'45" 32d9'17"

Patty A. Rodriguez Barron
Uma 9h51'49" 51d12'5"

Patty Aguilar
Ori 6h15'56" 13d35'37"

Patty Amooor!!
Vul 20h40'51" 21d22'38"

Patty "Amor"
Lyn 7h29'48" 55d2'8"

Patty and Joyce
Cyg 19h56'55" 38d42'9"

Patty and Lee
Cyg 19h43'11" 47d14'35"

Patty and Mal McGowan
Per 3h22'24" 41d28'48"

Patty and Mark forever
Pho 2h20'48" -45d2'46"

Patty And Paul
Leo 9h42'44" 27d35'11"

Patty Anglin and John Taylor 4ever
Uma 10h50'51" 50d20'42"

Patty Ann Kathcart Smith-BRAMM
Sco 16h38'5" -38d26'59"

Patty Armanini
Lyn 8h22'12" 55d10'59"

Patty Beevers
Sgr 19h10'19" -14d45'9"

Patty Berry
Sco 16h7'46" -8d38'1"

Patty Browne
Cas 1h3'4" 53d31'25"

Patty B's Sweet Baboo
Lyn 8h31'11" 34d3'44"

Patty Cake
Uma 10h31'4" 56d51'3"

Patty Cakes
Gem 6h54'34" 33d38'1"

Patty Cardwell
Cyg 21h13'40" 43d50'31"

Patty Cauthen
Cas 0h22'26" 57d13'33"

Patty Collier
Uma 11h24'9" 30d38'50"

Patty Cronin
Psc 1h11'15" 12d38'46"

"Patty Crow" - from the NSICU
And 0h38'20" 27d42'20"

Patty Dougherty
Cas 0h47'42" 60d58'19"

Patty Dunaway Wham
Crb 15h41'55" 36d10'4"

Patty Earls "444"
Uma 9h2'5" 59d50'40"

Patty Feliciani
Psc 0h59'48" 15d55'0"

Patty Fortuna
And 23h39'19" 45d5'2"

Patty & Fred
Uma 9h30'11" 64d36'48"

Patty & Gaby Pimentel
And 2h22'5" 42d25'35"

Patty & George's 50th Anniversary
Cyg 20h17'52" 37d0'12"

Patty Giesken
Uma 8h48'47" 72d7'53"

Patty Gillis, Mom
Psc 0h13'27" 8d17'23"

Patty Gilmore
Lyn 8h49'52" 38d32'52"

Patty Greene
Aqr 22h55'52" -8d17'5"

Patty Harants
And 0h13'59" 27d28'37"

Patty In Heaven
Lib 14h28'35" -13d39'7"

Patty Irene
Cmi 7h29'26" 8d39'11"

Patty J Maples
Cyg 21h29'51" 53d51'34"

Patty Jane
And 23h8'59" 41d22'28"

Patty Jane Goodwine
Peg 22h40'37" 24d19'52"

Patty Jane Spreuer
Uma 11h14'30" 64d38'4"

Patty Jo Sombrio
And 0h14'29" 28d23'37"

Patty & Karl Knutson
Lyn 7h3'58" 57d31'4"

Patty Kitt Gershaw
Psc 0h47'4" 6d9'18"

Patty LaBarbera
Cap 20h24'31" -13d3'47"

Patty Langdon
Gem 6h49'1" 27d26'21"

Patty Laverne Griffin
Ari 3h4'52" 29d25'23"

Patty Lynn 50
Psc 1h24'23" 13d55'22"

Patty Lynn Medeck
Uma 10h47'10" 41d14'28"

Patty Lynn Tindall
Crb 16h3'2" 34d2'15"

Patty M. Saunders
Uma 8h13'43" 67d8'28"

Patty Martens
And 0h30'46" 43d33'32"

PATTY MARTINO
Tau 3h49'25" 30d16'45"

Patty Max
Uma 13h37'38" 50d25'23"

Patty Murphy
Dra 17h34'32" 51d47'6"

Patty Murphy
Psc 0h8'54" 7d14'16"

Patty Murray
Tau 5h45'50" 12d52'25"

Patty Neach
Cyg 20h28'5" 47d37'17"

Patty Omega1
Cas 0h57'32" 60d20'13"

Patty Paramo Cottingham
And 0h48'44" 33d5'12"

Patty Pinto
Ori 5h26'14" 9d57'23"
Patty Pumpkin Phung
Mon 6h52'15" -0d30'48"
Patty Rawlick
Uma 14h27'2" 62d0'53"
Patty Smith
Aql 18h58'49" -0d5'2"
Patty Sofokles
Vir 13h25'50" 12d27'50"
Patty Sue
Cas 23h35'8" 51d52'53"
Patty Tousch
Cnv 13h44'39" 38d23'48"
Patty Trixie 1971
Psc 22h51'47" 2d17'7"
Patty Van Hook
And 0h21'48" 29d5'38"
Patty VanMeter
And 0h18'23" 28d17'22"
Patty Young
Cyg 21h14'27" 46d51'42"
Patty Z
Uma 12h54'2" 62d2'54"
Patty, Will U Marry Me 1 More Year?
Ori 5h33'44" -0d44'46"
Patty-67
Aur 6h12'4" 30d7'57"
PattyCake
Sgr 18h8'57" -28d2'34"
Patty-Cakes
Crb 16h7'8" 31d30'39"
Pattye Staub Johnson
Cnc 8h58'0" 12d33'54"
Patty-Gene Stofer Anniversary 1955
Cyg 20h21'2" 46d9'30"
Patty-LUV
Dra 17h10'55" 65d17'41"
pattymary
Lyr 18h51'0" 43d29'35"
PattynDale
Psc 0h52'26" 13d20'22"
Patty's 50th Birthday Star
Psc 22h53'28" 5d42'24"
Patty's Boy
Boo 14h51'53" 23d10'9"
Patty's Cosmic Butterfly
Sco 17h47'21" -42d54'45"
Patty's Heart
Cas 0h34'18" 57d55'48"
Pattys' Light
Uma 11h5'11" 50d21'2"
Patty's Light On Blake's Life
Umi 15h13'7" 74d21'16"
Patty's Promise
Ari 2h10'42" 22d8'46"
Patty's Shining Beauty
Crb 16h4'33" 37d11'29"
Patty's Star
Gem 6h38'15" 14d48'18"
Patty's Star
And 0h23'6" 32d55'39"
Patty's Star
Umi 13h14'2" 70d14'9"
Patty's Star (Patricia Rae Mullins)
Ori 5h22'55" 9d3'35"
Patty's Unclosing Eye
Ari 3h15'57" 24d13'42"
Patty's Wish
Lyn 8h57'50" 38d0'1"
Patu
Sge 19h43'11" 17d51'41"
Patxi Apellániz
Ari 2h28'40" 11d57'16"
Patxi Garat
Crb 15h55'52" 38d58'34"
Paty
Gem 7h14'44" 18d45'25"
Patzer, Helmut
Uma 11h38'45" 39d54'54"
Paul
Peg 23h23'2" 32d38'27"
Paul
Boo 14h43'31" 30d32'39"
Paul
Per 3h48'54" 33d3'27"
Paul
Tri 2h7'46" 35d17'37"
Paul
Gem 7h13'10" 34d42'44"
Paul
Uma 9h36'5" 47d24'32"
Paul
Ori 5h55'11" 20d57'47"
Paul
Aqr 21h39'19" 1d54'2"
Paul
Vir 12h18'30" 2d11'18"
Paul
Tau 3h37'35" 1d5'22"
Paul
Ori 6h18'0" 15d14'51"
Paul
Tau 5h47'9" 22d10'4"
Paul
Cep 22h52'32" 69d29'57"
Paul
Uma 10h37'0" 68d44'37"
Paul
Cep 22h16'19" 58d55'23"
Paul
Uma 12h30'34" 54d28'6"
Paul
Umi 16h16'49" 80d6'0"
Paul
Sco 15h54'17" -22d6'1"
Paul
Lib 15h26'29" -16d11'9"
Paul A. Campitell
Per 3h9'47" 54d14'10"
Paul A. Findley
Ori 5h34'16" 4d35'47"
Paul A. Lukis
Uma 11h5'2" 46d50'44"
Paul A. Lux
Ori 4h52'45" 12d51'47"
Paul A. "Pat" Tholl
Uma 13h43'39" 50d1'22"
Paul A. Sabo
Per 4h9'53" 51d43'18"
Paul A. Schulz
Uma 9h20'14" 56d52'53"
Paul (a very special star) Lee
Ari 2h41'59" 25d38'27"
Paul A. Walton
Ori 5h32'7" 9d55'15"
Paul Aaron Friedman
Leo 11h9'34" 23d27'35"
Paul Adam Lysczek
Uma 10h18'12" 72d34'46"
Paul Adam Warwick
Dra 18h22'8" 48d50'59"
Paul Aidan Junghahn Shaw
Peg 22h8'36" 33d35'14"
Paul Alexander Cole
Ori 5h33'53" -5d16'33"
Paul Alexander Davis
Sgr 18h59'37" -25d59'54"
Paul & Alice Bourgeois
Per 2h45'27" 54d10'49"
Paul Allen
Uma 9h59'13" 69d18'17"
Paul Allen Bernard
Lyn 6h23'19" 55d4'51"
Paul Allen Howell
Her 16h50'30" 23d19'35"
Paul Alvin Mitchell
Cnc 8h48'54" 31d43'19"
Paul & Amy
Col 6h26'18" -34d49'17"
Paul and Aaron Maye Smith
Cyg 19h38'7" 32d24'47"
Paul and Brandye
Cyg 20h22'33" 36d16'35"
Paul and Brooke
Umi 13h7'34" 72d54'30"
Paul and Charlotte Reid
Cyg 19h57'47" 30d6'26"
Paul and Christine Royce
Cyg 21h13'56" 46d15'26"
Paul and Coleen
Uma 10h29'40" 63d38'30"
Paul and Elaine Anderson
Uma 12h4'45" 65d3'43"
Paul and Emma Anniversary
Cyg 19h41'33" 54d7'28"
Paul and Emmy Forever
Cra 18h20'3" -37d8'44"
Paul and Florence "Flossie" Schorr
Sge 19h39'49" 17d12'38"
Paul and Harley
Lib 15h17'17" -29d1'47"
Paul and Jamise
Gem 6h55'40" 14d7'26"
Paul and Jenifer Hanrahan
Cyg 21h37'54" 38d52'10"
Paul and Julie Gastelum, 9-5-04
Lyr 18h55'45" 26d29'56"
Paul and Julie's Dream
Mon 7h7'35" -0d28'3"
Paul and Karla Neir
Crb 15h40'7" 33d28'46"
Paul and Kerri forever and a day
Uma 11h13'17" 51d40'46"
Paul and Lillian Arendt
Uma 10h22'43" 52d24'37"
Paul and Madeline
Ari 3h9'7" 25d42'10"
Paul and Michelle
Gem 6h57'40" 14d10'21"
Paul and Nancy Gregory
Cyg 21h33'38" 42d0'37"
Paul and Rachel
Lyn 7h50'57" 38d54'48"
Paul and Sandy Anniversary 12 Star
Cyg 20h2'34" 51d35'50"
Paul and Sara Moult
Cyg 21h8'22" 48d0'18"
Paul and Shirley LaChance
Cyg 21h52'35" 44d41'32"
Paul and Siobhan Gallagher
Cas 0h29'3" 51d38'55"
Paul and Tracey Brani
Psc 23h5'35" 7d28'40"
Paul and Tracey Giles 12.08.05
Cyg 19h21'14" 29d39'11"
Paul and Yvette Bodson
Cep 20h12'40" 61d31'11"
Paul & Andie Eternal Love
Uma 8h31'5" 72d0'21"
Paul Andre Prowse
Sgr 17h52'42" -29d20'51"
Paul - Andrew - Katina (PAK)
Cru 12h26'52" -60d12'36"
Paul Andrew Rose
Vir 13h47'31" -8d38'22"
Paul Andrew Shiels
Cas 0h17'48" 52d44'25"
Paul Andrews
Aqr 22h57'46" -4d31'22"
Paul Angelo Gallo, Jr.
Per 3h23'32" 51d39'21"
Paul Angelo Gentile ...... vm
Her 17h3'46" 30d2'17"
Paul Angelo Mangino
Cnc 8h51'1" 31d37'51"
Paul Anthony Clark
And 1h56'23" 41d41'31"
Paul Anthony Laidler
Ori 5h39'0" -0d44'22"
Paul Anthony Lamparillo
And 23h22'5" 35d18'51"
Paul Anthony Lloyd
Cma 6h58'18" -23d9'30"
Paul Anthony Lopez
Vir 13h45'50" 5d13'6"
Paul Anthony Lowell 1963-2003
Lib 15h5'9" -7d26'22"
Paul Anthony Paul
Sgr 18h6'2" -27d38'6"
Paul Anthony Pritchard
Cep 22h45'12" 64d0'21"
Paul Anthony Rannis
Aur 6h35'12" 37d47'26"
Paul Anthony Wallace
Uma 9h52'19" 45d40'35"
Paul Antoine Martin
Lib 15h31'18" -5d8'44"
Paul Aram Kedeshian
Eri 4h0'22" -8d21'14"
Paul Argiriou
Equ 21h23'0" 12d56'19"
Paul Armstrong
Uma 10h33'9" 50d41'41"
Paul Arood and Susan Philbrick
Cyg 21h25'17" 52d16'44"
Paul Arthur Christopher Vandaam
Per 3h30'2" 51d15'57"
Paul Arthur Hadley
Uma 9h13'58" 58d29'41"
Paul Arthur Lemm
Per 3h40'58" 47d57'39"
PAUL ASELIN
Uma 9h19'42" 45d24'28"
Paul Azevedo
Lyn 7h43'36" 43d16'28"
Paul B. Flynn
Aqr 21h5'2" -12d55'29"
Paul B. Killman
Aqr 22h43'48" -3d13'52"
Paul B. MacDonald
Dra 18h22'36" 51d26'42"
Paul B. Matthews "For Heroism"
Per 3h49'35" 33d40'55"
Paul B Mazur
Cap 21h52'40" -10d29'32"
Paul B Young
Uma 10h5'38" 67d55'32"
PAUL BARRY
Uma 9h37'12" 48d9'47"
Paul Barry Baker
Aqr 21h43'53" -8d7'31"
Paul Basile
Uma 13h13'0" 61d34'14"
Paul Bauer Cook
Lyn 7h48'58" 58d46'58"
Paul Beecher
Her 17h38'19" 32d59'35"
Paul Blair Paschall
Her 16h22'1" 22d50'21"
Paul Bocuse
Uma 9h37'35" 42d30'2"
Paul Boesiger
Cas 0h19'22" 55d6'26"
Paul Boger Fink
Sco 15h53'41" -20d39'47"
Paul Boo-Boo Goode
Uma 10h34'21" 40d42'16"
Paul Boulais
Leo 11h32'9" 11d41'33"
Paul Bradley McLaren
Aqr 22h3'18" -8d49'1"
Paul Brian Mattox, II
Aql 19h44'24" 8d3'19"
Paul Brian Musgrove
Ori 5h40'56" -0d18'35"
Paul Briggs
And 23h51'52" 34d32'41"
Paul "Bromers" Bromley
Ori 5h40'31" -0d8'21"
Paul Brower
Uma 9h4'8" 57d52'39"
Paul Bruce Kimmel
Per 3h47'29" 42d54'42"
Paul Bruce Mares
Cam 7h17'27" 60d6'8"
Paul Brückel
Pyx 8h41'19" -27d33'46"
Paul Bruno "Wishing Star Forever"
Per 2h20'26" 55d36'55"
Paul & Bruno's Eternal Star
Cyg 20h0'21" 34d27'3"
Paul "Butch" Brokaw, Jr.
Vir 13h14'43" -4d10'42"
Paul Butcher
Crb 15h52'46" 29d56'50"
Paul Butler
And 1h20'43" 50d9'20"
Paul Byrne
Aur 6h13'1" 40d24'1"
Paul Byron Bonsack
Her 17h45'59" 48d35'48"
Paul C. Oswald, Jr.
Vir 12h46'16" 12d28'38"
Paul Calvin Keener
Per 3h51'11" 43d4'49"
Paul Cameron Watson
Per 3h15'33" 41d30'56"
Paul Canavan
Cyg 21h29'9" 49d35'35"
Paul Carani
Her 16h27'38" 11d10'37"
Paul & Carol's Special Star
Leo 11h18'57" -4d53'6"
Paul Carroll
Boo 14h47'4" 36d38'19"
Paul Carter Lisinski
Cyg 19h31'45" 51d32'50"
Paul Cauchi
Psa 22h8'10" -29d36'34"
Paul "Chalky" Shore
Cru 12h26'14" -60d22'16"
Paul Chandler Matthews
Ori 5h52'17" 20d4'31"
Paul Charles David Wenclawiak
Per 4h18'47" 51d40'43"
Paul Charles Orlando
Tau 4h53'13" 22d4'7"
Paul & Chels Forever
Tau 4h8'17" 12d1'6"
Paul Christian Deniel
Per 3h58'8" 35d51'34"
Paul & Christine Wylie
Cyg 20h58'50" 32d0'45"
Paul & Cindy
Col 5h8'22" -36d9'25"
Paul & Cindy Forever
Per 4h18'9" 34d2'4"
Paul Cipriani
Her 18h21'17" 27d24'32"
Paul Clare
Cyg 20h38'35" 45d17'23"
Paul Clark and Doris Deforge
Cyg 19h47'15" 36d56'13"
Paul Clark's Star
Her 18h44'55" 15d30'47"
Paul & Cleaora
Uma 13h48'43" 51d19'35"
Paul Coffin
Vir 13h19'53" 6d55'58"
Paul Cole
Gem 6h29'42" 16d15'41"
Paul Colucci
Ori 5h28'47" 1d38'29"
Paul Cordero VI
Uma 8h25'15" 66d27'40"
Paul Cox
Sgr 18h11'3" -20d4'2"
Paul Creelman
Cep 22h32'27" 58d5'40"
Paul Cronkhite
Tau 5h49'59" 27d8'34"
Paul Cyr Sr.
Tau 4h15'29" 29d17'6"
Paul D Brooks
Uma 8h58'22" 55d47'36"
Paul D. Coale
Lib 15h36'43" -18d59'1"
Paul D. Leonard Tomah (Big Len)
And 23h10'51" 51d3'24"
Paul D. McLean
Per 2h49'48" 51d32'20"
Paul D. Quay
Cap 20h45'21" -18d35'52"
Paul D. Schmitz
Cyg 19h38'13" 29d50'58"
Paul Daniel Nickerson - 04/30/74
Ori 5h40'30" -0d5'59"
Paul Daniel Simons
Cru 12h38'37" -57d44'5"
Paul Daniels
Aur 5h32'54" 47d54'42"
Paul Darren Blansett
Uma 11h38'40" 37d44'46"
Paul David Barnes Siciliano
Aqr 22h50'6" -7d45'50"
Paul David Harding
Per 4h44'38" 49d32'46"
Paul David Mackenzie
Per 2h12'0" 53d22'45"
Paul David Metcalf 13.02.1969
Cra 18h45'58" -41d31'12"
Paul David Ponath
Her 17h18'32" 18d53'19"
Paul David Richter
Ari 3h2'19" 14d52'18"
Paul David Siederman
Per 3h29'11" 51d0'40"
Paul David Sutton
Gem 7h21'17" 21d27'41"
Paul David Thoreson
Cnc 8h27'15" 27d25'35"
Paul David Werp
Gem 6h45'23" 13d11'35"
Paul David Wilson
Hya 8h55'32" -11d32'39"
Paul David Zaring
Ari 2h1'44" 17d47'52"
Paul Day
Lyn 6h27'16" 60d11'21"
Paul Day, Mi Manchi Ti Amo
Uma 10h6'27" 59d31'56"
Paul de Rozieres
Cep 23h58'16" 86d16'58"
Paul Dee (Gramps)
Per 4h8'19" 50d41'16"
Paul Deedon
Cas 0h19'35" 61d49'58"
Paul Dennis
Uma 9h2'48" 47d41'2"
Paul DeRose
Lib 15h42'52" -20d1'55"
Paul Dever Superstar
Cen 13h32'16" -38d10'5"
Paul & Diane Herrlett
Cyg 20h19'26" 46d40'42"
Paul DiMonte
Uma 9h49'28" 65d22'6"
Paul DiSabatino
Per 4h17'0" 45d25'53"
Paul Dominic Yado "Yankee"
Gem 6h50'56" 33d3'7"
Paul Donald Edmiston 4/28/85-6/7/04
Ori 5h40'58" 1d27'25"
Paul Douglas
Uma 12h8'41" 53d49'12"
Paul Douglas Burris
Sco 16h38'29" -28d22'16"
Paul Douglas Marcian
Her 17h16'49" 20d14'8"
Paul Douglas Vincent
Boo 14h35'53" 27d28'43"
Paul Douglass Connally
Lib 14h52'21" -9d13'50"
Paul Dupuis
Aur 5h39'43" 32d52'25"
Paul Dutch Nordenger
Her 16h37'47" 32d18'46"
Paul Dwight Black
Ori 6h20'52" 9d3'44"
Paul E. Barberi
Cas 0h48'51" 53d40'47"
Paul E. Douglas
And 0h18'34" 33d45'5"
Paul E. Sigler
Tau 3h35'54" 25d33'58"
Paul E. Weiss II
Uma 14h1'53" 48d9'17"
PAUL E WOG
Vir 14h50'13" -0d9'3"
Paul Edmond-Limb
Ori 5h47'35" 12d11'37"
Paul Edouard Bernier
Uma 10h15'37" 57d27'59"
Paul Edward Bartsch
Uma 11h11'58" 51d19'32"
Paul Edward Gover
Vir 12h2'44" 8d34'10"
Paul Edward Gruss
Vir 12h34'40" 0d15'49"
Paul Edward Hilliard
Cru 11h58'26" -61d8'58"
Paul Edwards - The Love of my Life
Per 2h39'13" 52d7'40"
Paul Eli Margulies 01 14 1935
Cep 21h30'42" 61d33'26"
Paul Emanuel Rothschild
Sgr 18h9'33" -19d46'54"
Paul Emerson Blume
Uma 12h2'40" 57d25'11"
Paul Eudaly
Cap 20h48'11" -16d47'55"
Paul Eugene Cunningham
Lyn 8h10'52" 57d26'37"
Paul Eugene Ford
Vir 14h29'14" 1d1'53"
Paul Everett
Per 3h21'42" 42d9'21"
Paul F. Collins
Per 3h9'9" 51d19'45"
Paul F. Covert
Cnc 8h46'50" 31d1'0"
Paul F. Morton
Ari 2h17'55" 26d38'55"
Paul F. Nystrom
Uma 8h22'26" 67d23'24"
Paul F. Putnam
Aql 18h56'7" -0d50'21"
Paul F. Skyles
Psc 0h41'4" 6d23'28"
Paul Field
Ori 5h15'2" -0d47'57"
"Paul Fisher Shining Forever"
Uma 13h36'13" 50d2'38"
Paul & Flora Dunlap
Leo 11h40'21" 15d44'0"
Paul Forrest Midlam
Aqr 23h8'52" -10d44'14"
Paul Fortuna
Cru 12h21'56" -57d5'54"
Paul Francis Allen
Per 4h22'50" 34d8'33"
Paul Francis Hastings Sr.
Uma 9h6'16" 50d54'17"
Paul Frank Levy (Bip-Bip)
Tau 3h40'48" 16d39'12"
Paul Frederick Deering
Uma 11h45'7" 39d33'3"
Paul Frederick Nyberg
Sgr 19h3'34" -25d54'1"
Paul Frollo
Aqr 23h23'51" -21d12'14"
Paul Funfsinn The Shepherd's Keeper
Boo 15h36'17" 46d29'33"
Paul G. Horstmann Jr.
Aqr 21h14'44" 1d30'17"
Paul G. Leitza
Per 4h5'30" 44d1'4"
Paul G. Morse
Her 16h44'12" 40d7'58"
Paul G. Treichler
Gem 7h43'21" 31d36'30"
Paul Gabriel Grogan
Cap 21h24'29" -14d35'4"
Paul Gabriel Ross
Vir 13h40'7" -19d59'23"
Paul Gacioch
Cyg 19h46'3" 32d10'52"
Paul Galeazzi, Jr.
Aur 5h35'25" 40d31'27"
Paul Gallant
Cep 22h34'44" 75d33'26"
Paul Gann
Ori 5h15'30" 2d0'23"
Paul Garnica
Per 4h24'22" 41d53'29"
Paul Garrett Rademacher
Ari 3h8'43" 28d7'9"
Paul Gary Armstrong
Per 3h10'10" 54d59'27"
Paul Gendreau
Cap 20h25'46" -11d10'57"
Paul Georgen
Gem 6h43'48" 31d43'9"
Paul Gerard Spence
Cra 18h0'32" -37d10'17"
Paul Giblen
Leo 9h23'48" 26d36'59"
Paul & Gillian. Forever
Cyg 21h30'49" 55d5'26"
Paul Gioia
Per 2h34'0" 55d28'1"
Paul Giuffreda
Aql 19h27'6" 5d33'37"
Paul Glenn Yonker
Aqr 21h15'2" -14d21'7"
Paul & Gloria Sirianni
Cyg 20h40'12" 46d37'13"
Paul Goddard
Dra 17h57'4" 79d0'36"
Paul Goldberg At Forty
Leo 10h12'25" 14d10'14"
Paul Goldschmiedt Jr.
Boo 14h50'29" 22d19'29"
Paul Goodman
Umi 16h17'35" 73d46'13"
Paul Graniero
Cnc 8h31'39" 20d42'26"
Paul Graves
Gem 6h38'53" 18d37'52"
Paul Gregory McDonald
Dra 18h41'54" 54d31'28"
Paul Guyot
Ori 6h17'7" 15d13'30"
Paul H. -Bud- Tutmarc
Ori 6h1'9" -0d37'58"
Paul H. Fabricius Kjaer
Per 4h46'39" 48d46'3"
Paul H. Handy
Uma 12h57'59" 61d18'0"
Paul H. Jakovac
Cmi 7h28'29" 10d34'29"
Paul H. Poore
Cnc 8h50'3" 14d39'10"
Paul Haakon Barth, Jr.
Her 17h2'19" 28d12'27"
Paul Hake, My Valentine
Her 17h34'8" 26d52'23"
Paul Hamilton
And 23h43'17" 37d22'51"
Paul Harman
Aqr 22h31'58" -15d26'24"
Paul Harry (Antonopulos)
Ori 5h55'41" 18d6'40"
Paul Haworth
Her 18h10'4" 15d40'37"
Paul & Hazel Fouke May 14, 1950
Aur 5h22'21" 32d17'6"
Paul Hebert
Cas 0h4'54" 56d50'51"
Paul Hehir - 11 April 1985
Cru 12h44'53" -59d49'35"
Paul Henry Messens
Vir 13h51'25" -18d2'33"
Paul Hewitt Bennett
Aql 19h56'32" -0d21'56"
Paul Hidlebaugh
Her 16h42'58" 3d50'41"
Paul Hill
Cru 12h14'27" -63d40'4"
Paul Hoffmann
Ori 6h17'27" 7d10'43"
Paul Holdener
Umi 13h45'59" 78d43'46"
Paul Hoyt Kidd
Leo 10h19'16" 17d17'1"
Paul Hugo Bredenbeck
Tau 4h38'4" 14d48'12"
Paul Hulcie Hatfield and Children
Uma 11h28'22" 38d53'44"
Paul Hurton Manning
And 23h16'40" 41d45'40"
Paul Hutton
Boo 14h42'51" 16d4'10"
Paul I Armas
Leo 11h7'29" 25d55'49"
Paul "Iggy" and Connie Reddish
Cyg 20h20'35" 47d3'48"
Paul Ignatius Murphy
Ori 6h23'22" 10d35'31"
Paul Irby's Star
Her 17h48'26" 38d56'19"
Paul Irvine
Uma 9h54'2" 42d57'19"
Paul J. Ahlin
Aqr 22h30'34" -3d28'50"
Paul J. Booher
Uma 13h34'51" 62d12'46"
Paul J. Church
Sco 16h6'33" -16d59'55"
Paul J. Cook
Aql 19h26'34" -0d15'43"
Paul J. Costanzo
Ori 5h40'33" -0d15'27"
Paul J. Dubois
Dra 17h11'2" 68d24'56"
Paul J. Griffo
Her 17h12'24" 22d5'16"
Paul J. Kennedy, Jr.
Sgr 18h24'48" -35d13'13"
Paul J. LeBlanc
Aqr 20h49'50" 0d18'26"
Paul J. Magier
Psc 23h33'49" 5d46'54"
Paul J. Petruccelli
Uma 10h13'54" 62d31'13"
Paul Jacob VanZanten
Leo 9h31'17" 28d34'35"
Paul James
Ori 6h16'27" 9d7'45"
Paul James Martin 122469
Her 18h14'26" 23d37'52"
Paul James Regets
Vir 13h25'11" -8d48'31"
Paul Jan Van de Geer
Ori 6h6'21" 20d48'9"
Paul & Janeice Blankenship
Cyg 20h47'10" 37d36'23"
Paul & Janice Van House
Her 16h50'17" 21d50'12"
Paul Jasmin
Per 3h32'6" 49d50'42"
Paul & Jeanette
Crb 15h54'33" 32d48'22"
Paul & Jennifer Tanguay
Cyg 19h33'46" 52d14'6"
Paul Jeremy (PJ) Watrous
Aql 19h4'35" -0d6'36"
Paul & Joan's Star
Ori 5h37'30" -0d8'33"
Paul John Banks
Cep 21h46'54" 65d4'16"
Paul John Cartlidge
Sco 16h8'33" -12d45'26"
Paul John DeChristopher
Uma 12h3'23" 33d32'52"
Paul John Ferris
Her 18h30'58" 25d9'19"
Paul John Hart
Uma 9h15'1" 68d13'21"
Paul Jonathan Foster
Her 17h29'49" 39d12'58"
Paul Jonathon Lindahl
Cap 21h49'31" -24d32'35"
Paul Jones
Per 4h45'52" 40d2'53"
Paul Jones
Ori 5h23'5" 1d55'13"
Paul Joseph Bastman
Ari 3h6'38" 19d53'56"
Paul Joseph Chomanics
Ari 3h20'53" 23d38'55"
Paul Joseph Dondlinger
Leo 10h7'2" 26d33'41"
Paul Joseph Gaglione
Leo 11h47'13" 21d22'4"
Paul Joseph Gallagher Jr.
Leo 11h54'29" 20d59'12"
Paul Joseph Labelle
Cep 22h12'30" 68d24'2"
Paul Joseph Liguori
Psc 1h12'52" 14d55'51"
Paul Joseph Prudhomme
Cep 22h49'47" 77d5'54"
Paul Joseph Usak
Aqr 20h55'50" -10d36'19"

Paul Joshua Allison
Ori 5h39'39" -0d4'49"
Paul Joubert
Lac 22h26'15" 47d35'28"
Paul Judd Gold
Srp 18h13'23" -0d55'1"
Paul Julier 40
Per 3h47'47" 49d43'35"
Paul Justin Sewell
Cen 13h34'35" -41d4'12"
Paul Kamieniak
Cnc 8h16'32" 24d19'53"
Paul Kang
Psc 1h37'58" 24d53'21"
Paul Kapchan
Sco 16h7'11" -8d53'55"
Paul & Karen Gierer 11/5/03
Cyg 19h33'15" 48d5'38"
Paul & Katie Feeley
Lyn 7h42'59" 36d29'14"
Paul Keenan
Cyg 20h26'8" 40d45'23"
Paul Keller Linford
Per 2h52'51" 38d51'10"
Paul Kelly, father, husband, friend
Her 17h8'16" 31d12'41"
Paul & Kena Butts
Cyg 19h45'24" 55d43'56"
Paul Kern and Leslie Stewart
Ara 17h16'40" -51d4'49"
Paul Kevin Murphy
Ari 3h3'56" 23d30'37"
Paul Kezdi
Aql 19h47'58" -0d49'34"
Paul King
Ori 5h54'16" 18d22'51"
Paul King
Cyg 19h57'32" 39d49'39"
Paul Kohler
Sgr 19h11'17" -16d33'26"
Paul Kohler
Sco 17h4'10" -43d20'26"
Paul Kravagna
Uma 11h9'10" 59d48'27"
Paul & Kristeen
Gem 7h45'19" 34d52'1"
Paul & Kristi Coffman
Lyn 8h26'6" 38d16'38"
Paul & Kristin's Light of Love
Cyg 20h5'40" 30d27'6"
Paul Kupper
Her 18h57'39" 15d32'21"
Paul L. & Estelle (Snookie) Maille
Umi 13h5'51" 74d5'29"
Paul L. Newman
Aqr 21h9'57" -10d34'6"
Paul L. Ryan
Per 3h4'50" 56d50'34"
Paul L. Sartini
Leo 9h40'15" 25d41'34"
Paul L. Vernick
Leo 9h46'58" 27d6'7"
Paul Lamar Chester
Tau 4h12'27" 8d45'6"
Paul Leischuck
Ori 6h6'28" 18d19'15"
Paul Lemieux
Her 16h43'20" 29d46'8"
Paul Leo Lajeunesse
Lyr 18h46'44" 30d24'18"
Paul Leslie Cox - Daddy's Star
Uma 11h31'30" 57d57'59"
Paul Lewis
Lib 15h0'42" -13d20'50"
Paul - Light Of My Life
Cru 12h27'2" -58d30'56"
Paul Linden Owen P.C. 9066.
Uma 11h58'14" 59d52'34"
Paul & Lisa
Uma 13h45'19" 53d41'46"
Paul Loeb
Her 17h2'21" 32d37'55"
Paul Louis
Per 3h37'21" 47d28'48"
Paul Louis Butler
Tau 4h10'9" 4d24'5"
Paul Louis Ricciuti
Sco 16h25'59" -31d48'2"
Paul & Louise Brannon - Our Shinning Star
Per 3h37'24" 44d47'33"
Paul & Louise Webster 50 Years
Cnc 8h22'5" 15d2'41"
Paul Luis Garcia
Gem 6h7'17" 21d44'51"
Paul M. Kahn
Ori 6h14'5" 3d42'2"
Paul M. Latchaw
Vir 14h42'54" 3d40'1"
Paul M. Vayo
Ori 5h45'8" 4d39'30"
Paul M. Violet "Eternal"
Ori 5h24'10" 3d37'39"
Paul Mack
Cap 21h18'51" -17d44'21"
Paul & Mae Plete
Uma 8h12'54" 62d10'8"
Paul Malone - Your Light Shines On
Pyx 8h33'51" -23d53'39"
Paul Maniaci
Per 3h25'17" 51d7'40"
Paul Manitta
Cen 14h18'52" -38d14'15"
Paul Manna
Vir 14h3'9" -20d36'41"
Paul Manzi
Gem 7h35'6" 22d45'54"
Paul Marchant, The Light in My Life
Gem 6h44'56" 25d20'34"
Paul & Marcie - Everlasting Love
Cnv 12h47'5" 38d11'37"
Paul & Marilyn Statham
Cru 12h47'44" -63d48'16"
Paul Mark DiPietro
Ori 5h25'15" 10d30'28"
Paul Mark Lokken
Ori 6h16'25" 13d13'46"
Paul Marrese
Ari 2h25'52" 26d0'49"
Paul Martin Hartigan
Sco 17h58'18" -30d26'52"
Paul & Mary Butler
Uma 11h45'9" 51d29'0"
PAUL MASSEY SR.
Aqr 21h59'41" 0d9'40"
Paul Massey, Jr.
Aqr 23h48'6" -4d27'18"
Paul Matthew Caparros
Her 17h30'58" 47d42'12"
Paul Matthew Lawrence
Vir 13h39'53" -12d54'0"
Paul Matthew Rivas
Psc 23h10'44" -1d58'21"
Paul & Maura
Cyg 21h17'43" 46d11'32"
Paul Maurice Hantke
Uma 9h24'26" 72d1'21"
Paul Mayberry
Her 18h8'29" 22d32'11"
Paul McBroom
Lib 15h5'39" -6d43'17"
Paul McCarton
Per 4h21'55" 45d42'0"
Paul McConchie
Cen 13h12'33" -50d25'18"
Paul McKenzie
Per 2h18'27" 55d56'57"
Paul Meleady Sr.
Sgr 18h0'33" -28d4'53"
Paul & Meredith
Cyg 20h53'1" 34d52'28"
Paul Michael
Uma 9h50'25" 52d15'36"
Paul Michael Abramshe
Uma 11h4'23" 53d13'55"
Paul Michael Jr.
Per 3h52'47" 38d21'30"
Paul Michael Kellner
Ori 5h58'27" -0d45'10"
Paul Michael Kibal My Life's Joy
Cru 12h42'9" -56d43'50"
Paul & Michael McCarthy 3/1/1966
Uma 11h30'36" 57d57'36"
Paul Michael Snoad
Leo 10h29'2" 10d19'42"
Paul Michael Spinella
Ori 5h52'21" 19d59'46"
Paul Miller
Psc 1h12'29" 19d30'11"
Paul Miller
Dra 18h24'36" 57d58'4"
Paul Millili
Her 16h48'35" 47d42'20"
Paul Mirabella
Uma 8h59'31" 68d32'37"
Paul Misan 10051960
Peg 22h3'8" 27d48'7"
Paul Mitchell
Per 4h46'52" 47d20'31"
*~PauL~*Mol~*HunnaY~*
Tau 3h33'51" 19d23'57"
Paul Monfre
Gem 7h30'5" 34d33'7"
Paul Monsch
Ori 5h53'41" 16d49'29"
Paul Moses
Uma 9h0'53" 68d16'3"
Paul "Moz" Morris - Shea's Daddy's Star
Uma 13h39'32" 59d59'44"
Paul Murphy
Her 17h21'36" 16d39'13"
Paul - "My Little Treasure!"
Ari 3h16'48" 24d20'13"
Paul -My Shining Star- D'Arcangelo
Cru 12h36'20" -58d29'30"
Paul Natalini
Per 3h47'42" 32d16'24"
Paul Nathan's Dark Star of Kabaret
Vir 15h0'19" 3d48'49"
Paul Nathenson
Lyn 8h50'3" 33d31'56"
Paul Nevada Frankum
Uma 13h44'36" 48d8'34"
Paul & Nic Always!
Cyg 20h5'17" 32d42'8"
Paul Nicholas Rohach
Ori 5h30'57" 2d28'53"
Paul Nichols "Nick Shining Star"
Vir 12h40'13" -0d42'6"
Paul Nixon
Cep 21h32'47" 65d1'27"
Paul Norbert Lelowicz
Per 3h3'1" 55d10'2"
Paul Norman Macklam
Uma 8h46'13" 62d16'5"
Paul O'Brian Lingerfelt
Lyr 18h50'26" 31d31'40"
Paul Orest Dolocheck
Uma 11h59'23" 37d33'35"
Paul Owens
Per 3h7'35" 54d32'9"
Paul P. Sunnergren
Cnc 9h0'13" 28d33'23"
Paul Pajunen
Uma 10h39'47" 72d0'43"
Paul Parry
Her 17h2'35" 18d28'9"
Paul Pasquale Riggi
Ori 5h55'42" -8d11'1"
Paul Patrick Rush
Oph 17h43'31" -0d33'42"
Paul & Patty Doering
Per 3h32'30" 44d39'54"
Paul Pedevilla
Gem 6h27'30" 25d6'51"
Paul (Pepper) Martin
Sgr 17h55'53" -25d7'2"
Paul Peter Woiciechowski
Cnc 8h48'22" 13d35'21"
Paul Philip Vagnozzi
Cyg 20h49'15" 33d27'58"
Paul Philpott - Thank you honey.
Leo 11h37'41" 28d3'25"
Paul & Phyllis Webb
Crb 15h53'58" 32d24'30"
Paul (PJ) Parker
Uma 9h57'10" 67d20'58"
Paul Por-Wen and Nancy Clark Hung
Lib 15h33'50" -19d2'39"
Paul Prigge
Ori 5h9'31" 15d12'31"
Paul Pruckner
Leo 9h32'3" 10d29'34"
Paul (Pud) Joseph Landry
Per 3h12'2" 53d48'55"
Paul Pullen (Dad)
Her 16h41'32" 6d41'9"
Paul R.
Cep 2h25'36" 81d7'59"
Paul R. Montuori
Lib 15h1'8" -1d3'59"
Paul R. Morris
Aur 5h52'16" 52d1'25"
Paul R. Vandertill
Sgr 19h28'34" -17d55'16"
Paul R. "Wooley" Crosser Jr.
Per 3h19'28" 40d57'39"
Paul Randy Galdames
Ori 5h14'52" 1d57'56"
Paul Raphael Africano
Psc 0h31'40" 18d21'2"
Paul Ravella
Her 17h23'9" 15d34'49"
Paul Raymond Hill
Uma 10h54'42" 72d10'10"
Paul & Rebecca Bishop
And 0h31'48" 24d58'30"
Paul & Rebecca Specht
Cyg 19h58'12" 43d2'3"
Paul - Remo
Umi 14h40'46" 77d54'49"
Paul Rice
Aql 19h59'40" 8d37'47"
Paul Richard Chapman
Her 16h45'44" 8d57'2"
Paul Richard Dagostino
Leo 11h39'44" 26d21'11"
Paul Richard DuFour
Lyn 7h50'41" 35d23'3"
Paul Richard James, Jr.
Cap 21h35'21" -17d34'37"
Paul Richard Kinderman
Aqr 20h48'59" -11d59'0"
Paul Richard Kirkhuff
Vir 13h36'25" -5d21'23"
Paul Richard "Sweet Pea" Schlect
Cep 22h37'47" 67d6'40"
Paul Richard Zaccarine
Tau 5h48'16" 15d37'53"
Paul Richardson Hartman
Lib 15h43'45" -18d2'59"
Paul Rizzo
Dra 9h24'30" 73d19'46"
Paul Robert Driscoll
Per 2h16'35" 57d10'12"
Paul Robert Vickers
Ori 5h46'11" -2d29'29"
Paul & Rochelle - love above all else
Ara 16h39'10" -47d23'2"
Paul Roegner
Cam 3h52'49" 69d6'33"
Paul Ronald Ashby
Uma 8h31'26" 68d58'30"
Paul Rory Walsham
Uma 13h28'27" 53d25'21"
Paul Rosette
Cam 4h33'42" 55d47'44"
Paul Ross - The Brightest Star
Sgr 18h25'2" -22d33'39"
Paul Russell Schwarz M.D.
Leo 11h56'58" 20d22'51"
Paul Ryan McCusker
Tau 4h24'25" 27d3'15"
Paul S. Batchelder
Crb 16h20'44" 33d29'42"
Paul S. Bondy
Sgr 19h43'17" -12d4'36"
Paul S. Doherty
Tau 3h50'31" 26d41'54"
Paul S McCormick
Vir 13h14'56" 10d36'3"
Paul & Sandy Borth
Cyg 21h16'17" 30d20'20"
Paul Sangalli
Per 3h45'9" 38d47'33"
Paul Scaffidi
Lac 22h47'36" 49d56'50"
Paul Schachter
Her 17h24'51" 47d15'25"
Paul Schiffel
Aqr 21h43'1" -6d22'39"
Paul Schüler
Uma 11h7'47" 41d36'23"
Paul - Schulie - Schulenburg
Cru 12h33'8" -60d43'19"
Paul Sealy
Cyg 21h7'3" 40d43'58"
Paul Sean Imber
Her 16h43'36" 19d43'54"
Paul & Sharon
Uma 12h52'46" 58d0'9"
Paul Sheldon Davis
Boo 15h27'49" 44d47'24"
Paul Shepherd
Her 18h29'41" 17d30'4"
Paul Silverio Migliaccio
And 1h2'20" 42d14'35"
Paul Simeone's Star - Love Marisa
Cra 18h29'3" -45d15'40"
Paul Skoda
Per 3h23'15" 52d13'32"
Paul Slovick
Aur 5h53'58" 36d51'56"
Paul Smith "Mite"
Cas 0h56'29" 58d24'15"
Paul Smith - PS: I love you
Vir 11h55'37" 4d11'38"
Paul Snow "Snowman"
Vir 13h20'49" 11d46'9"
Paul Sokolowski
Ori 5h6'54" 4d5'24"
Paul Speck
Dra 14h29'45" 60d12'43"
Paul Stack Jr., Beloved Father
Aqr 22h9'6" 0d26'10"
Paul "Star" Marino
Lib 14h50'52" -1d23'1"
Paul Steger, Sr.
Ori 5h33'52" 1d5'56"
Paul Steinwandtner
Apu 13h56'25" -77d3'33"
Paul Stephan - Rising Star
Umi 17h47'49" 89d6'9"
Paul Stephen Andrews
Per 3h24'27" 51d55'38"
Paul Stephen Dossett
Aur 5h36'49" 45d9'7"
Paul Steven Charlesworth
Sgr 18h3'29" -27d51'3"
Paul Stevenson
Per 3h30'54" 35d19'17"
Paul - "Superman" Donnelly - 15/05/74
Tau 5h11'22" 16d31'49"
Paul & Suzie Forever
Uma 9h7'11" 60d21'56"
Paul T. Martin
Lyn 7h43'33" 39d7'45"
Paul T. Schneider
And 0h44'48" 31d28'39"
Paul Tadrick
Vir 12h36'16" 9d0'44"
Paul Takeo Oshiro
And 0h55'32" 37d52'6"
Paul Tatore
Cnc 8h57'22" 30d59'42"
Paul Tatton
Aql 19h16'13" 7d26'41"
Paul & Teresa's Eternity
Uma 9h57'27" 44d56'2"
Paul Terrell
Ori 6h17'14" 19d39'28"
Paul Terzian * Burning Hot *
Cnc 8h37'45" 23d6'14"
Paul 'The Gov' Nassar
Cru 12h19'31" -62d34'16"
Paul the Legend
Uma 10h28'26" 63d35'56"
Paul "The Man" Hughes, M.D., J.D.
Cnc 8h23'0" 15d26'16"
Paul Theo
Uma 13h57'58" 49d31'8"
Paul Thomas Hubbard
Boo 14h46'22" 53d26'51"
Paul Thomas McDevitt
Uma 12h43'27" 57d23'27"
Paul Thomas Simpson
Her 15h51'51" 42d45'34"
Paul Thomas Wilkinson
Uma 8h17'17" 69d52'28"
Paul Thompson
Ari 2h18'51" 24d38'39"
Paul Thuy Hoang Vu
Uma 10h26'3" 50d27'18"
Paul Tidcombe
Per 2h19'3" 52d17'2"
Paul Trackwell
Aur 6h3'38" 47d10'43"
Paul Tsimalis
Vir 14h41'53" 4d7'51"
Paul V. Pallanich Jr.
Ari 3h4'30" 14d47'21"
Paul V. Profeta
Boo 14h54'19" 21d32'54"
Paul Vecqueray
Dra 17h37'58" 67d24'45"
Paul Venson
Sco 16h17'51" -24d52'58"
Paul & Vera Potash
Uma 11h38'42" 38d27'42"
Paul Verguin
Cas 0h26'57" 61d21'6"
Paul Vetter
Sgr 18h51'25" -19d8'34"
Paul V.H. Halter, III
Her 16h33'35" 38d55'55"
Paul & Vi Mooney
Lyn 8h7'50" 42d51'18"
Paul Vincent Brewer
Uma 11h27'23" 59d54'25"
Paul Vincent Critti
Cap 21h10'1" -19d29'10"
Paul Vincent McCarthy Jr.
Umi 15h16'58" 68d4'36"
Paul Vincent Moossy
Sco 17h56'14" -42d34'25"
Paul W. Boender
Oph 17h48'24" -0d54'31"
Paul W. Cobey
Aqr 21h48'21" -2d26'58"
Paul W. Dennett
Lib 15h39'36" -24d25'38"
Paul W. Nyberg
Cnc 8h43'8" 11d31'0"
Paul W Parkinson - My Shining Star
Psc 0h52'25" 11d10'42"
Paul W. Poff
Leo 9h36'38" 12d20'2"
Paul W. Rowe Sr.
Ari 1h59'59" 14d43'21"
Paul W. Terry
Ari 1h59'43" 22d23'55"
Paul Walter Gil Hernandez
Uma 11h27'1" 32d35'3"
Paul Wardell
Cep 20h49'0" 59d1'13"
Paul Wayne Carter
Per 3h54'30" 38d21'22"
Paul Weck
Sgr 18h55'17" -18d23'9"
Paul Whitley
Lyr 18h27'9" 36d28'49"
Paul Whitney
Lmi 10h57'20" 31d22'14"
Paul Whyte
Per 4h43'28" 39d58'17"
Paul "Widdy" Wydmuch
Col 5h33'0" -41d35'47"
Paul will love Jessica forever
Uma 10h49'31" 58d56'37"
Paul William Goetsch
Aql 19h47'48" -3d41'37"
Paul William Menzies
Cep 23h18'2" 70d3'24"
Paul Wilson
Tau 4h30'22" 26d57'2"
Paul Wilson Taylor
Cyg 19h53'32" 31d22'56"
Paul Woelkers
Lac 22h22'1" 50d25'14"
Paul Zajac
Her 18h56'10" 17d53'33"
Paul Zediker "the star golfer"
Lib 15h32'38" -20d5'40"
Paul Zink
Cas 23h42'40" 58d41'50"
Paul Zink
And 23h2'10" 52d13'9"
Paul, Heather, Skyler, and Hunter
Umi 15h43'13" 77d52'38"
Paul, Nigel & Adrien Rutigliano
Aql 19h21'1" 12d4'31"
Paul, You Complete Me
Ori 6h9'19" -0d56'8"
Paula
Ori 6h1'28" -3d54'3"
P-A-U-L-A
Vir 12h47'1" -10d2'20"
Paula
Sco 17h26'27" -31d44'37"
Paula
Sco 17h51'8" -36d0'28"
Paula
Cnc 8h10'45" 11d44'1"
Paula
Tau 4h11'12" 11d36'45"
!Pàula!
Leo 11h29'55" 5d3'52"
Paula
Leo 10h16'49" 16d27'7"
Paula
Crb 15h27'22" 26d35'34"
Paula
And 23h17'11" 43d24'0"
Paula
Uma 12h5'6" 32d44'11"
Paula
Uma 12h4'20" 35d13'55"
Paula
And 1h19'8" 38d58'7"
Paula 40
Cas 23h9'10" 55d19'34"
Paula 88
And 23h24'56" 42d22'50"
Paula A. Eaton
Sco 16h18'0" -10d13'21"
Paula A. Eisenhart
Tau 4h30'13" 11d10'35"
Paula A. Keane
And 0h34'13" 45d45'44"
Paula Amadio
Com 13h9'38" 30d1'9"
Paula and Jerome Gottesman
Umi 14h37'7" 75d57'59"
Paula Andrea Maya
Ori 6h20'33" 9d29'34"
paula ann
Psc 0h24'10" 7d16'56"
Paula Anne Gray - Only A Star Away
Cru 12h21'29" -59d39'47"
Paula B 40
Mon 7h32'4" -0d44'10"
Paula Baker Allen
Cam 3h53'22" 70d13'35"
Paula Balzer
Cas 1h37'48" 61d13'50"
Paula Barber
Cam 6h46'37" 66d5'37"
Paula Barnett
Lmi 10h10'6" 38d1'10"
Paula Berger
Cas 23h4'7" 59d25'42"
Paula Blaine
Cas 23h28'23" 57d54'57"
Paula Brizuela
Ari 2h46'58" 13d25'23"
Paula C Helm
Cnc 7h55'44" 13d18'37"
Paula Chuck
Gem 6h40'10" 13d34'32"
Paula Cook
Lyn 7h19'6" 50d10'46"
Paula Creech
Cet 1h5'2" -0d54'47"
Paula & Creig thru eternity
Aqr 22h48'31" -16d50'36"
Paula D. White
Uma 11h17'31" 56d48'58"
Paula Dee Paine
Cas 0h47'18" 52d27'13"
Paula Denise Reid
Cas 23h16'40" 59d5'25"
Paula Doherty
And 2h25'53" 42d53'15"
Paula Dusky Eternal
Crb 15h48'28" 28d41'33"
Paula Earnhardt
Cyg 21h59'38" 48d22'17"
Paula & Eduardo
Leo 11h50'42" 20d0'54"
Paula Elaine Robinson
Cas 1h8'16" 63d34'13"
Paula Elizabeth
Aqr 21h0'54" -11d33'4"
Paula Faith
Cen 12h39'14" -34d50'6"
Paula Farkas
And 23h8'17" 35d16'58"
Paula Fawn Sasser
And 0h21'34" 46d27'32"
Paula - Franz Kälin
Tri 2h33'16" 33d45'14"
Paula Freeman
And 2h35'14" 50d13'11"
Paula Galloway: Music of my sphere
Mon 6h28'40" -2d34'3"
Paula García Zarabozo
Sco 17h58'54" -41d35'59"
Paula Gayle Nester Baby
Cyg 21h8'15" 47d26'25"
Paula Gosling
Cap 20h26'25" -13d28'15"
Paula & Graeme
Cyg 21h40'34" 46d28'13"
Paula Green
Cas 0h28'45" 56d9'35"
Paula Grossman Nessenthaler
Mon 6h51'45" 7d2'33"
Paula Horner
Cyg 20h0'57" 46d11'28"
Paula J. Ackerman
And 23h7'0" 37d49'11"
Paula J Norsell
And 0h49'41" 36d22'28"
Paula J Rheault
Mon 6h50'48" -0d5'12"
Paula J. Wright
Leo 9h37'31" 28d48'17"
Paula Jane Harris
Leo 11h4'41" 16d27'15"
Paula Jean
Aqr 22h59'26" -7d53'17"
Paula Jean
Pyx 9h26'39" -35d47'42"
Paula Jean Artemie
Tau 3h43'21" 28d27'37"
Paula Jean Barker
And 23h39'40" 44d43'37"
Paula Jean Bird
Cyg 20h21'45" 43d37'36"
Paula Jean Jergins Jones
Mon 6h38'17" 5d36'42"
Paula Jean Steelman
Cnc 8h57'9" 15d33'33"
Paula Jean Wilshe
Uma 10h18'21" 68d16'42"
Paula Jo Aurich
Cap 21h38'22" -16d43'26"
Paula Joyce Gray
Umi 14h28'32" 73d41'43"
Paula June Jordan
Tau 4h36'5" 7d30'8"
Paula K
Cam 7h48'7" 68d50'3"
Paula K Wilkinson
Leo 9h37'6" 26d29'1"
Paula Kahoury
And 0h11'55" 25d21'41"
Paula Kay
Lyn 7h27'42" 46d21'41"
Paula Kay Lokken
Cnc 8h52'4" 29d17'49"
Paula Kiefer "Forever"
And 1h14'41" 42d21'26"
Paula Kristin
Lib 15h45'42" -6d40'37"
Paula Kristine Mills
And 2h16'5" 46d11'8"
Paula L. Patinella
Pic 5h46'13" -53d29'26"
Paula L. Powers
Cas 23h28'26" 55d49'54"
Paula L. Rodriguez
Gem 6h42'47" 17d10'56"
Paula L. Sechrist
Ari 2h11'48" 24d38'6"
Paula LaRee Ellis
Tau 5h27'5" 28d31'36"
Paula Leona
Cap 20h36'8" -19d12'2"
Paula Leson
Peg 23h30'40" 21d29'20"
Paula Lopez
Sgr 18h30'58" -18d43'35"
Paula Louise McArthur
Sgr 19h19'30" -16d32'7"
Paula Louise Mumford
And 23h4'33" 51d31'48"
Paula Lucille Reuss Schanz
Per 3h46'12" 52d25'57"
Paula M DeFlavis
And 0h24'56" 32d25'4"
Paula M. Pope
Crb 15h28'12" 29d42'16"
Paula Mae Hibberd
Leo 11h49'14" 27d58'8"
Paula Marie
Cnc 9h20'51" 10d40'39"
Paula Marie Evans
Psc 1h43'59" 21d11'9"
Paula Marie Hunt
Cnc 8h5'3" 20d24'22"
Paula Marie Sennes
Ori 6h4'0" 15d31'46"
Paula Marie Sydney
Uma 9h53'38" 52d45'45"
Paula Mary Owen
Sgr 19h42'51" -31d32'38"
Paula & Matt Forever
Gem 7h38'19" 26d24'14"
Paula McClure
Cam 7h17'28" 77d24'12"
Paula Mears Harter
Mon 7h18'37" -0d46'40"
Paula Meyers
Peg 23h24'43" 29d52'37"
Paula Michele
Aql 20h13'12" 13d27'27"
Paula Michelle Jones
Tau 4h39'48" 12d56'25"
Paula Mohns
Uma 13h7'24" 54d27'40"
Paula Momma Momma Livingstone
Leo 11h34'18" 6d15'0"
Paula Moore
Aqr 22h48'45" -16d23'28"
Paula Musco DeAvies
And 23h10'25" 45d29'22"
Paula & Mush
Cyg 21h55'19" 49d50'2"
Paula Packer
Cas 0h17'5" 50d21'36"

Paula & Phil Tenwick Forever
Cyg 21h55'0" 47d1'55"
Paula R. Stucker
Mon 6h44'57" 7d10'54"
Paula Rachel
And 23h7'32" 50d47'53"
Paula Rae
Leo 9h46'14" 8d45'24"
Paula Rae
Srp 18h30'58" -0d21'19"
Paula Raphan
And 0h10'24" 30d55'5"
Paula Reneé
And 0h5'11" 45d32'46"
Paula Ripke
Crb 16h10'41" 37d12'10"
Paula Rose
And 0h12'5" 43d27'15"
Paula Ruby
Gem 6h1'34" 25d56'14"
Paula Ruth Anderson
Psc 1h0'52" 31d10'52"
Paula S. Wallace
Ori 6h15'29" 14d31'1"
Paula Sagerschnig
Leo 9h41'9" 27d48'35"
Paula Sanchez Espino
Lib 15h3'40" -27d17'13"
Paula Satchell
And 23h22'17" 51d55'40"
Paula Shea Gannon
Uma 9h11'58" 51d38'37"
Paula Sinclair
And 0h38'13" 34d55'36"
Paula Sutherland
Ori 6h4'28" 5d17'16"
Paula Suzanne Doody
And 23h56'12" 37d37'56"
Paula - the Special Star in the Sky
And 1h22'42" 49d42'46"
Paula the Tazesgirl
Vir 12h40'8" -7d43'4"
Paula Thomas
Tau 3h42'26" 16d31'18"
Paula Tilton Patteuw
And 2h30'46" 49d30'43"
Paula Trigg
Cas 0h28'37" 58d11'26"
Paula Valverde
And 1h38'12" 39d8'12"
Paula Wagner
Sgr 18h39'5" -29d23'15"
Paula Weingarden
Lyr 18h25'48" 34d2'45"
Paula Werry
Umi 15h38'3" 83d16'12"
Paula & William Marino North Star
Uma 10h57'37" 51d45'49"
Paula Young
And 0h37'46" 22d8'53"
Paula, Mommie Forever
Cyg 20h11'14" 31d43'58"
Paula's Taufstern
Uma 9h55'28" 72d54'22"
Paula3
Umi 14h0'7" 73d16'40"
Paulaann
Gem 6h43'0" 27d50'41"
Paulaboers03221972
Peg 22h48'25" 11d51'18"
Paula-Jane Gilfoyle
Umi 13h41'57" 75d48'8"
PaulaJean Lea
Cas 0h15'14" 57d49'15"
Paula-Kätzle
Ori 5h56'25" 21d33'12"
Paulangelus
Lmi 10h9'56" 36d17'32"
Paulara
Ori 5h46'17" 7d1'22"
Paul-Armand Lechowski
Dra 17h3'19" 68d45'14"
"Paulas Allsehendes Auge"
And 23h38'22" 47d56'51"
Paula's and Carters Star
Umi 15h54'8" 71d1'53"
Paula's Light
And 0h21'15" 38d20'45"
Paula's Light
Tau 5h35'34" 22d12'44"
Paula's Smiling Butterfly
Aql 19h43'50" -0d32'54"
Paula's Sparkle Arkle
Sco 16h52'41" -27d10'18"
Paula's Star
Lib 15h49'49" -17d1'30"
Paula's Tauf- und Glücksstern
Vir 11h42'33" 8d0'28"
PaulaWilliamRobert
Vir 14h26'0" 7d13'35"
Paule Drouin
Cas 1h11'27" 49d11'46"
Pauleen P Lan Tran
Sco 17h37'20" -43d25'28"
Pauleric Hyland
Uma 10h1'13" 52d49'36"
Paulet Wysocki
Crb 15h36'27" 27d43'11"
Pauletta Barth
Gem 6h58'38" 12d53'22"
PAULETTE
Ari 1h57'54" 14d23'11"
Paulette
Leo 11h30'50" 15d11'3"
Paulette
Ari 2h58'54" 25d46'57"
Paulette
And 2h34'40" 44d31'48"
Paulette
Psc 0h16'48" 6d59'15"
Paulette Beatty
Mon 7h16'15" -0d0'8"
Paulette Boyer
Ari 2h5'40" 23d43'43"
Paulette Chicano Tamburro
Leo 10h21'29" 25d40'19"
Paulette Chicano Tamburro
Leo 9h27'0" 10d13'0"
Paulette Jacques
Uma 12h4'43" 65d25'53"
Paulette Milazzo
Cap 20h10'38" -13d49'48"
Paulette Thompson
Psc 1h19'45" 16d15'58"
Paulette Walker
And 1h44'1" 46d0'40"
Paulette Weaver
Ori 5h9'25" 7d26'39"
Paulette,"Pô"
Cyg 20h22'28" 52d58'34"
Pauley Mac
Umi 4h21'42" 88d37'11"
Pauli
Uma 9h51'41" 50d53'24"
Pauli
Lmi 10h15'24" 31d57'26"
Pauli 1934
Uma 10h20'43" 59d3'53"
Pauli Bear
Cnc 8h44'37" 32d10'17"
Pauli Maravilla
And 1h8'14" 42d20'41"
Paulie
Her 17h12'16" 31d1'31"
Paulie
Uma 9h14'30" 61d33'12"
Paulie Demass
Lyn 6h20'34" 59d46'44"
Paulie Donut's Star
Vir 13h24'26" -0d38'1"
Paulie Magua
Cap 20h13'58" -18d9'30"
Paulie S. Hawkins
Vir 13h21'17" -6d50'6"
Paulie-Bear
Ori 5h41'19" 3d36'21"
Paulie's Shining Star
Tau 4h13'53" 27d52'15"
Paulina
Cap 20h35'30" -26d44'6"
Paulina Cooper
Lyr 18h35'43" 37d14'22"
Paulina Elisa Francesca
Umi 14h38'54" 76d46'9"
Paulina Garcia de Leon
Gem 6h48'9" 13d23'24"
Paulina Isabel Walter
Cap 21h48'30" -15d9'13"
Paulina J. Kumala
Psc 23h30'39" 1d12'35"
Paulina Leoni
Ori 6h15'56" 13d16'47"
Paulina Maria Delgrippo
Lyr 19h18'38" 29d58'30"
Paulina my Polish Princess
Psc 0h7'23" -3d49'51"
Paulina Skladzien
Cap 21h4'14" -22d26'51"
PAULINA23
Leo 10h7'27" 9d50'6"
Paulina's Diamond In The Sky
Cas 0h59'0" 62d32'49"
PauLindsay
Her 18h18'16" 23d1'32"
Pauline
Leo 11h50'46" 20d1'7"
Pauline
Peg 21h36'26" 11d36'6"
Pauline
Ari 2h58'43" 30d48'51"
Pauline
And 2h26'40" 50d29'23"
Pauline
Aur 6h5'16" 47d30'27"
Pauline
Uma 13h20'5" 54d0'50"
Pauline
Lib 14h58'38" -16d7'46"
Pauline
Lib 14h37'5" -18d10'24"
Pauline
Umi 12h59'55" 88d24'59"
Pauline
Ori 5h18'27" -4d47'17"
Pauline Abatemarco
Lyn 6h18'25" 56d15'20"
Pauline Agnes
Lyn 7h37'29" 43d21'3"
Pauline Albert
Umi 13h48'29" 77d19'31"
Pauline Ann Caldaralo
And 2h36'39" 44d20'0"
Pauline Ann Ladovich Chmura
Cyg 21h33'3" 55d3'55"
Pauline Ann Murray Bower
Cas 23h50'37" 59d41'6"
Pauline Armijo
Aqr 20h45'21" -9d7'23"
Pauline Bahuaud
Cas 0h23'43" 59d18'23"
Pauline Bennett Hornbeck
Uma 10h59'30" 47d27'41"
Pauline Bourgeois (mem)
Leo 9h41'16" 27d25'0"
Pauline Chalekian
Cas 1h14'24" 53d54'20"
Pauline Cherry
Cam 3h55'54" 52d53'50"
Pauline Cleveland Brumley
Eri 4h22'56" -0d56'32"
Pauline Colson
Leo 10h51'50" 14d11'23"
Pauline Corey Hopper
Cnc 8h22'4" 19d42'17"
Pauline Crinnigan, My Guiding Star
Uma 9h22'37" 67d6'45"
Pauline Danelian
Psc 1h38'56" 4d8'10"
Pauline Deanna
Tau 4h26'10" 16d12'6"
Pauline Ehrminger
Del 20h35'24" 4d6'39"
Pauline Eisinger
Ari 3h4'28" 23d7'19"
Pauline Elizabeth Ann Anderson
Ari 3h13'54" 23d16'56"
Pauline Elizabeth Donnelly
Cnc 9h13'38" 20d17'4"
Pauline Elizabeth Keywood
Her 16h30'25" 45d13'31"
Pauline et Stéphane
Ari 3h22'34" 27d19'45"
Pauline Ethel Ray
Uma 9h55'18" 60d45'39"
Pauline F. Zotos For An Eternity
Cnc 8h39'58" 23d11'32"
Pauline Forbes
Cas 23h37'39" 54d56'55"
Pauline G. Turner
Cyg 19h45'31" 39d57'36"
Pauline Gallien
Aqr 22h17'28" -5d7'12"
Pauline Griffin (babe)
Cap 21h10'47" -23d8'13"
Pauline Holstun
Mon 6h43'14" 7d28'56"
Pauline Jones
Cyg 21h35'34" 38d17'29"
Pauline Jordan
Cas 1h14'7" 57d37'23"
Pauline Joyce Burchell Beasley
Lyr 18h38'39" 35d25'23"
Pauline Judson 17/5/1951-19/1/2005
Col 6h26'15" -37d41'4"
Pauline Krok
Cas 0h33'49" 55d10'56"
Pauline L. (St. Martin) Cornell
Uma 9h35'31" 68d18'37"
Pauline LaDow Tweedy
Leo 11h45'34" 19d28'27"
Pauline Lee
Psc 0h21'24" 4d40'20"
Pauline Lewis
Aql 19h24'27" 0d0'40"
Pauline Lysne
Vir 13h20'24" 3d33'23"
Pauline M. Scott
Uma 9h20'37" 44d1'44"
Pauline Maria
Lyn 8h33'28" 50d3'32"
Pauline Marie Ahing 'PoPo'
Tau 3h40'2" 17d44'54"
Pauline Marino Varone
Lib 15h15'37" -23d21'41"
Pauline Mary Williams
Cas 0h53'25" 50d16'53"
Pauline & Max - Wonderful Parents
Cru 12h6'49" -61d34'47"
Pauline McCarthy
Uma 11h4'16" 48d32'51"
Pauline MCLW
Aqr 21h8'21" -13d15'16"
Pauline Miller
Cyg 19h46'4" 32d37'1"
Pauline "Mimi" Dailey
Cas 23h45'26" 59d44'50"
Pauline MORRIS
Cyg 20h30'58" 59d45'49"
Pauline Morrison
Per 3h4'14" 43d1'42"
Pauline N. Gomez-&-Sandra De Sousa
Sge 19h53'45" 18d46'35"
Pauline Nickerson LaRiviere
Lib 14h27'25" -22d41'19"
Pauline ~ October 18, 1924
Lib 15h31'24" -24d45'12"
Pauline O'hara
And 23h11'15" 50d8'32"
Pauline Pheysey Pichette
Cnc 8h40'32" 16d2'0"
Pauline Plumbe
Umi 14h35'14" 84d58'26"
Pauline R. Lefcoe
Gem 6h56'21" 25d24'37"
Pauline Reeves
And 0h35'33" 36d25'37"
Pauline Regina de Assis
Cap 20h7'40" -10d1'57"
Pauline Rice Neal
Leo 11h17'14" 19d9'50"
Pauline Richards
Lyn 8h32'29" 33d6'15"
Pauline Rose
And 23h5'6" 49d3'23"
Pauline Rose Addis
Cas 2h5'30" 62d31'14"
Pauline Rose Wright
Cas 1h3'59" 67d22'43"
Pauline Selner
Vir 12h23'3" 3d32'58"
Pauline Slater's Light
Tau 5h38'36" 21d20'39"
Pauline Stegmuller
Ori 5h36'23" -1d46'32"
Pauline the Best
Cas 1h1'44" 49d0'34"
Pauline Till
Cas 23h59'42" 66d18'24"
Pauline Tourre Shipman
Cas 1h10'22" 63d9'16"
Pauline Vasiliaskas
Gem 6h7'37" 23d59'10"
Pauline Velazquez
Cyg 21h16'38" 44d11'8"
Pauline Virginia Mossman, "PJ"
Sgr 18h30'4" -36d44'56"
Pauline, "Save the Whales" Frisina
Cet 1h18'44" -19d58'56"
Pauline7537
Gem 7h12'46" 20d52'12"
Pauline-90-2006
Gem 7h12'18" 32d6'17"
Pauline's Flame
Dra 17h52'23" 63d29'4"
Pauline's Rose of the Heart
Cas 0h0'11" 50d6'38"
Pauline's Shining Star
Gem 7h28'16" 29d20'2"
Paulinka Dadej
Sgr 18h51'25" -25d24'17"
Pauli's Star 2 Finish
Crb 15h47'55" 27d44'18"
Paulita
Lac 22h20'41" 45d56'6"
Paulla "Mimi" Stetson
Uma 9h56'43" 55d6'47"
Paullauristarr
Tau 4h40'1" 1d50'25"
paul-lotta
Aqr 21h2'40" -12d31'31"
Paully Denoia
Oph 17h42'30" -0d25'59"
PaulnLaurieDoak
Cyg 19h59'42" 40d26'8"
Paulo
Leo 10h21'13" 7d58'37"
Paulo A. D. G. Araujo
Vir 13h38'0" -19d15'57"
Paulo Arthur Borges
Tau 5h46'45" 26d33'10"
Paulo Calligopoulos
Cen 13h40'5" -63d28'22"
Paulo Cesar Amado
Lac 22h43'59" 53d17'5"
Paulo Costa
Ori 5h12'43" -6d47'32"
Paulo R.J. Puntel
Pho 0h47'36" -42d2'49"
Paulo Trovao
Gem 6h10'2" 26d46'7"
Pauloaxe
Her 16h30'9" 17d52'51"
Paulonia
Aqr 22h28'15" -22d54'9"
Pauls Faith
Lib 15h26'12" -10d4'11"
Paul's Galaxy "CharlieTara"
Cyg 20h4'17" 44d38'34"
Paul's Light
Aur 5h48'32" 51d53'37"
Paul's Light
Sco 16h7'48" -10d18'56"
Paul's NA 15
Aqr 22h37'30" -2d10'46"
Paul's piece of the galaxy
Uma 10h35'54" 58d41'59"
Paul's Planet
Cap 20h43'34" -26d24'51"
Paul's Special Bright Star
Sgr 19h49'52" -13d15'8"
Paul's Star
Ari 2h43'54" 21d28'20"
Paul's Star 56
Her 18h57'43" 13d50'24"
Paul's Valentine
Per 3h16'52" 42d16'20"
Paul's World
Gem 7h47'57" 32d50'2"
Pauly
Per 4h2'24" 39d30'8"
Pauly
Uma 10h42'6" 39d24'36"
Pauly P
Cep 23h50'18" 66d54'26"
paulyne quinsaat funk
Dra 16h46'17" 59d19'45"
Pauly's Star
Peg 22h28'28" 25d37'38"
Pavan Kumar Reddy Nagam
Sgr 19h27'23" -41d4'5"
Pavan Pahal
Uma 11h22'46" 60d12'25"
Pavandeep Bhogal
Uma 13h20'12" 62d57'42"
Pavehawk
Cep 22h49'17" 65d40'38"
Pavel Goncharuk
Cnc 8h55'43" 9d6'44"
Pavel L. Snetkov
Leo 11h32'12" 7d10'13"
Pavel Radutsky
Umi 4h33'48" 88d45'31"
Pavetrany shar
Psc 1h34'35" 13d9'33"
Pavle
Psa 22h2'9" -25d36'59"
Pavle Jaksic
Cnc 8h19'4" 13d5'19"
Pavle Milojevic - Paki
Her 17h51'54" 21d55'50"
Pavlo Levin
Aqr 23h38'18" -16d44'57"
PAW
Umi 14h25'46" 75d33'30"
PAW
Aqr 22h39'10" -7d56'13"
Paw (Carmen, the light of my life)
Cyg 19h43'2" 42d54'5"
Paw Paw
Cyg 20h18'42" 47d3'31"
Paw Paw Bub Taylor
Uma 8h33'0" 60d48'46"
Paw Paw & Nana
Uma 10h2'30" 71d52'35"
Pawel and Magdalena Majewski
Lyr 18h40'25" 40d3'54"
Pawel Iwkin
Ori 5h56'4" 5d55'50"
Pawela, Alfred & Gisela
Uma 9h18'7" 59d29'12"
Pawlack, Wolfgang
Uma 10h24'49" 39d34'20"
Pawlita, Nicole
Lib 15h10'34" -7d27'37"
Pawpaw- Carl E Meitzen
Leo 9h26'45" 18d17'8"
Paws
Leo 9h59'22" 19d36'51"
PAWS
Uma 9h42'16" 67d26'22"
Pax
Dra 18h51'12" 65d27'37"
Paxton
Ari 2h5'44" 26d56'53"
Paxton Cole Graham
Dra 18h35'54" 64d15'35"
Paxton "Pacdaddy" Gilchrist
Gem 6h49'24" 15d53'48"
Pay it Forward
Gem 6h23'0" 21d1'8"
pay pay
Uma 11h9'4" 49d2'29"
Payal
Dra 18h22'46" 59d39'26"
Payal Sonya Razdan
Sco 16h23'32" -29d34'46"
Payam Akhavan-Malayeri
Umi 14h4'53" 69d16'55"
Payaso - My Knight in Shining Armor
Lib 14h58'7" -18d30'57"
Paycheck
Uma 9h9'32" 55d24'41"
Paydakayla
Uma 13h15'31" 55d14'4"
PayDay
Uma 12h32'20" 56d14'21"
Payden Roxy Lynn Hall
Uma 11h31'12" 58d1'6"
Payge Annederson Lane
And 0h15'10" 43d12'31"
Payge Marianna DeMaio
Sgr 19h38'27" -15d51'20"
Payne
Uma 12h18'50" 59d9'40"
Payne Allen Smith
Psc 0h54'37" 26d2'38"
pay-pay the warrior
Sgr 19h34'58" -13d0'50"
Payson Alan Springer
Psc 1h10'52" 27d21'3"
Payton
Psc 0h51'27" 27d31'4"
Payton 12-20-2003
Sgr 19h16'54" -34d47'17"
Payton A. Allen
Ori 5h58'8" 3d12'44"
Payton Alexander Balch
Cap 21h25'2" -17d57'52"
Payton Alexander Croll
Gem 6h36'17" 21d45'19"
Payton Alexander Dryden
Tau 3h45'10" 27d28'31"
Payton Alexandra Kennedy
And 0h18'26" 27d29'48"
Payton Allen
Cyg 19h56'18" 48d12'22"
Payton Anthony Allen
Aqr 21h37'39" -7d45'55"
Payton Areil Blake
And 0h14'14" 45d7'39"
Payton D. Pickard
Her 18h49'21" 12d33'45"
Payton Emma Lussen
Sgr 19h13'54" -15d24'37"
Payton Isabella Melland
Leo 10h40'39" 13d46'43"
Payton Jean Crowley
Uma 9h37'29" 50d24'4"
Payton John Thompson
And 2h25'51" 38d9'33"
Payton Joyce Hansen
Psc 23h26'35" 2d8'50"
Payton Klea Williams
Gem 7h0'32" 17d17'26"
Payton Leigh 10
Psc 0h34'48" 3d46'41"
Payton Louise
Aqr 22h37'36" -9d46'23"
Payton Makenna Wright
Peg 23h31'34" 17d22'10"
Payton Marlene Weisman
Cas 23h14'9" 55d38'16"
Payton McKenzie Begovich
Umi 15h43'38" 77d18'55"
Payton Mckenzie-Grace Burkett
Ori 5h5'39" 10d10'32"
Payton Noelle Sturges
Sgr 19h14'15" -18d18'17"
Payton Reed
Lyn 9h1'23" 33d32'55"
Payton Ryann
Cam 4h5'19" 68d6'39"
Payton Schumacher
Lib 15h12'54" -23d59'59"
Payton Taylor Laney
Lyn 8h32'16" 44d46'39"
Payton Taylor Lithgow
And 1h55'47" 36d6'54"
Payton Victoria Cicerone
Aur 5h39'58" 41d33'52"
Payton's Gift
Vir 13h13'25" 7d0'33"
Pazaaz
Lyn 7h36'12" 48d21'36"
Pazazz
Dra 16h28'44" 66d14'2"
P-B and J
Vir 13h14'48" 5d13'29"
PB & J
Sge 19h18'1" 17d42'48"
PB Wolsey Devitt
Peg 22h16'16" 17d36'1"
PBKTS
Tau 4h34'59" 17d47'43"
PBR Fox
Uma 12h5'51" 59d30'10"
PC
Cyg 21h44'49" 49d40'5"
P.C. Kimbro
Aql 19h56'28" -0d30'41"
P.C. Mr Moo Tommy "Huggly-Bug"
Uma 11h7'5" 45d55'32"
PC Zavas
Aql 19h14'28" -7d48'7"
PCA & HA
Del 20h51'48" 5d47'1"
PCC Royal Rangers Outpost 44
Uma 13h56'48" 52d25'34"
pcgi32406
Del 20h46'16" 15d26'31"
PCH
Sgr 18h35'54" -24d12'48"
pci-bee
Umi 16h24'55" 70d39'4"
PC-JAIMI
Per 1h51'17" 51d15'17"
pcondon4042
Cas 1h45'49" 61d3'56"
PCPA Theaterfest
Uma 11h45'7" 53d49'6"
PCPaulaM
And 1h15'21" 46d8'19"
PCS & JNH...I Love You Babe!!!
Cyg 21h17'34" 42d57'32"
P'Cutenik
Lmi 10h24'47" 35d19'49"
PD7867
Mon 6h37'55" 7d25'45"
PDOC
Uma 11h44'28" 43d48'21"
PDSinc. (Jon Swegarden & John Lund)
Pho 0h13'43" -42d29'18"
Pea
Peg 22h26'49" 20d30'13"
Peace
Uma 11h33'27" 51d33'14"
Peace
Col 5h34'52" -31d43'15"
Peace Be With You
Lyr 18h48'18" 37d42'13"
Peace & Seoul
Gem 6h47'50" 18d21'0"
Peace, Love
Cae 4h38'20" -37d38'53"
Peace, Platt, & Love
Aql 20h9'33" 8d30'51"
PeaceandLoveland
Lyn 9h3'21" 34d21'3"
PeaceBlessingsMercyJesusMaryLoveLiz
Cnv 12h49'19" 38d2'47"
Peaceful
Lib 14h53'14" -3d22'48"
Peaceful Stephanie
Lib 14h54'42" -5d50'14"
Peach
Cyg 20h22'3" 38d54'44"
Peach
Tau 3h51'13" 20d50'19"
PEACH
Tau 4h15'9" 17d39'38"
Peach & Azusa together forever!
Cyg 19h39'23" 39d7'23"
Peach High In The Sky
Cap 20h14'46" -14d45'3"
Peach Warner
Cap 20h18'35" -18d27'2"
Peachels
Lmi 10h21'12" 31d26'48"
Peaches
Lmi 9h25'31" 36d11'33"
Peaches
Leo 11h19'19" 17d57'38"
Peaches
Cap 20h26'35" -24d14'28"
Peaches and Cream
Leo 10h58'39" 18d58'42"
(PEACHES) Linda J. Good
And 0h13'37" 44d6'58"
Peaches Puddles Andregg
Cma 6h36'38" -17d0'13"
"Peachesmatoe"
Vir 13h15'54" -21d35'56"
PeachFuzz "04/05/2003"
Cas 23h36'10" 52d43'8"
Peach's Love
Per 2h54'23" 53d39'12"
Peach's Shimmering Beacon
Ari 3h24'3" 27d3'50"
Peachy
Lib 14h36'43" -19d17'11"
peachy paula
Aqr 21h10'28" -3d26'36"
Peachy & Punk
Lib 14h53'44" -4d15'36"
Pean
Boo 14h42'0" 23d41'52"
Peanie
Crb 15h26'55" 26d55'1"
Peanut
Tau 5h33'12" 26d2'7"
Peanut
Gem 6h10'1" 24d0'37"
Peanut
Ari 2h11'30" 25d5'24"
Peanut
Psc 0h31'9" 18d47'28"
Peanut
Ori 6h9'14" 6d30'7"
peanut
Aqr 21h32'10" 1d15'44"
Peanut
Psc 1h9'14" 10d53'25"
Peanut
And 23h12'6" 51d43'35"
Peanut
Uma 10h30'23" 52d26'24"
Peanut
Cnv 13h5'49" 38d37'6"
Peanut
Per 3h17'5" 46d43'30"
Peanut
Per 4h26'53" 43d11'12"
Peanut
Per 3h13'2" 42d3'29"
Peanut
Aur 5h10'28" 33d2'44"
Peanut
Lep 5h18'31" -11d11'12"
Peanut
Umi 15h8'13" 76d48'19"
Peanut
Umi 16h25'7" 77d8'1"
Peanut
Dra 19h17'24" 59d30'39"
Peanut
Dra 18h29'28" 54d45'49"
Peanut
Uma 8h23'36" 60d36'47"
Peanut Ann
Ari 3h13'37" 28d4'34"
Peanut Brother
Leo 11h42'19" 17d10'7"
Peanut Butter
Leo 11h23'0" 6d56'19"
Peanut Butter
Uma 10h29'34" 65d22'8"
Peanut Butter Cup
Tau 4h15'4" 26d16'48"
Peanut Butter & Jelly
Cyg 20h9'51" 45d25'1"

Peanut Butter N Jelly
Her 17h24'23" 16d2'37"
Peanut&Cupcake
Psc 1h14'40" 20d57'14"
Peanut In The Sky!!
Psc 0h42'37" 16d42'16"
Peanut Smith
Cet 0h50'5" -0d32'35"
Peanut & Stinkybutt
Tau 4h26'46" 22d58'21"
Peanutbutter Cup
Sco 17h54'7" -36d1'41"
Peanut's Lil' Brother
Psc 1h28'4" 18d35'50"
Peanut's Planet
Cap 20h56'51" -18d30'16"
Peanut's Wishing Star
Leo 10h55'6" 5d25'46"
Peapod & Babel
Cyg 20h47'31" 42d59'11"
Peapod's Little Jellybean
Her 16h58'21" 16d17'0"
Pearce 25
Psc 0h40'33" 6d57'29"
Pearce Patrick Bligh
Uma 9h56'28" 59d29'34"
P.E.A.R.L.
Uma 13h7'29" 54d10'55"
Pearl
Leo 9h55'52" 18d25'41"
Pearl
Tau 3h41'0" 23d45'27"
Pearl
Cyg 19h58'6" 39d39'9"
Pearl
Cas 0h23'42" 54d40'25"
PEARL
Cas 1h21'54" 53d29'48"
"Pearl"
Cyg 19h31'34" 30d57'5"
Pearl Braun
Cam 6h15'9" 76d52'11"
Pearl Crossing
Lyr 18h54'33" 43d58'38"
Pearl Doyal
Peg 23h7'53" 17d25'43"
Pearl Elaine Hiscock
Cas 0h26'49" 55d27'24"
Pearl Elizabeth
Sgr 18h1'9" -29d5'38"
Pearl Estelle
Per 3h48'12" 49d49'15"
Pearl Genevieve Ward
Cnc 9h18'14" 30d57'42"
Pearl Gregerson
Uma 11h21'6" 32d40'21"
Pearl Harborine
Sgr 18h8'13" -27d34'51"
Pearl in Heaven
Psc 1h25'44" 25d41'55"
Pearl Keyfetz
Ori 6h11'15" 15d41'18"
Pearl Lannetti
Sco 16h10'21" -9d13'41"
Pearl Mae Dominguez
Lib 14h43'59" -8d52'20"
Pearl Marjorie Steiner
Col 5h40'33" -31d1'44"
Pearl Mavis Somerville
Col 5h45'34" -29d9'25"
Pearl Randall Riggs
Gem 7h40'27" 31d51'24"
Pearl Youngman
Sgr 19h7'41" -16d43'0"
Pearl, a jewel of a friend
And 0h21'29" 39d7'13"
Pearl, The Mamma Bear Star
Umi 15h40'32" 77d44'53"
Pearlantazanopia
Crb 15h46'6" 36d4'28"
Pearlean McPhearson
Uma 11h42'16" 58d29'6"
Pearlene Harris
Vir 12h12'51" 6d0'28"
Pearlie 04/24/2006
Crb 15h26'56" 31d52'49"
Pearlouisa Basilette Shorter
Leo 11h12'56" 13d42'27"
Pearls Before Swine
Cyg 20h5'27" 34d19'30"
Pearls' Star
Uma 8h18'35" 66d4'35"
Pearly Girl
Tau 3h56'46" 28d15'3"
Pearlyn Lim
Tau 3h31'9" 2d34'5"
Pearo "Honey Bee" Tep
Leo 9h46'37" 21d36'58"
Pearson James Carmichael
Ari 3h25'3" 19d47'50"
PEAS
Cyg 21h43'29" 30d5'52"
peas and carrots
Ori 5h17'19" -8d34'56"
PeasCoconut&OurBeans
Ori 5h52'2" 22d1'51"
Peater Joseph Lyon
Her 17h38'1" 37d13'58"
Pebble
Uma 8h46'29" 65d3'52"
Pebble Jean
And 1h17'34" 38d8'48"
Pebbles
Tau 5h49'53" 23d29'13"
Pebbles
Ari 3h24'28" 23d55'47"
Pebbles
Ori 6h17'33" 14d38'36"
Pebbles
Umi 15h31'37" 72d12'39"
Pebbles
Umi 13h54'19" 72d30'39"
Pebbles and Bam-Bam
Cyg 19h51'11" 36d40'7"
Pebbles & BamBam
Uma 11h24'48" 40d21'50"
Pebbles Floyd
Per 3h25'46" 47d13'26"
pebbles2004
Lib 15h6'29" -4d53'42"
Pecher, Alexandra
Uma 11h28'47" 30d46'53"
PECO
Umi 16h17'5" 80d50'58"
Peco & Mimi Logan
Cyg 21h37'18" 46d51'5"
Pecola Harris
Uma 10h39'34" 47d23'15"
Pecos
Ori 5h17'11" 7d3'19"
Pede
Cnv 12h42'48" 39d31'45"
Pedrazzi
Sco 17h15'37" -32d55'40"
"Pedro"
Boo 14h36'54" 51d42'47"
Pedro
Uma 10h9'23" 44d28'19"
Pedro
Psc 1h32'31" 9d38'16"
Pedro A. Muniz-Manzanarez
Sco 17h56'28" -42d30'36"
Pedro Allo Perez
Uma 10h44'49" 51d5'10"
Pedro and Daniella Gonzalez
Her 18h19'34" 17d18'18"
Pedro Antonio Rodriguez
Her 16h18'58" 25d52'50"
Pedro Artur Duarte Pereira
Cen 13h38'45" -37d16'17"
Pedro Cuadros
Per 3h8'21" 51d32'14"
Pedro D. Vazquez
Cmi 7h14'33" 10d3'32"
Pedro Gonzalo Cruz
Ari 2h53'32" 25d36'55"
Pedro Jacob Mompean
Lyn 7h52'58" 48d50'12"
Pedro Jose Cuevas
Cma 7h21'43" -14d2'17"
Pedro Julio Gonzalez
Aqr 21h39'48" 3d6'41"
Pedro Love Matos
Cap 20h32'46" -26d27'3"
Pedro Pick
Aql 19h27'21" 11d20'16"
Pedro Rodrigues
Uma 9h19'4" 52d28'13"
Pedro Sanchez
Cep 21h32'42" 61d8'12"
Pedro Torrico
Ori 5h15'8" -0d29'10"
Pedro Valentine
Aqr 21h37'45" -1d9'28"
Pedro Velasco Vallarta
Dor 4h57'45" -69d5'4"
Pedro y El Gordo
Uma 9h52'7" 65d52'33"
Pedropipas
Lib 15h6'49" -1d15'36"
Pedro's Katie
Cyg 20h6'53" 55d9'28"
PedroX
Vir 13h42'12" -3d21'52"
Pee Pee Louise Kenward
Tau 4h28'31" 21d14'14"
Pee Wee
Cmi 7h30'40" 4d22'17"
Pee Wee
Cnc 8h38'14" 22d25'46"
Pee Wee
Tau 4h36'13" 29d13'41"
Pee Wee & Jenna Leigh
And 0h56'41" 45d36'31"
Pee Wee Toledo
Uma 9h35'58" 69d17'18"
Peebee
Ori 5h59'43" 21d6'25"
Peebie
Tau 5h35'1" 27d9'9"
peebody823
Her 18h39'35" 15d59'39"
Peedie-Roo
Ori 6h20'5" -1d44'49"
Peedle
Lib 15h31'2" -7d5'20"
"Peeeg" 5-4-45 Paul Goldberg
Per 3h18'19" 42d33'35"
Peegee
And 23h33'13" 47d34'15"
Peej and Bets
Cet 1h5'34" -2d48'34"
Peepalina
Dra 9h55'21" 78d2'52"
PeePaw Devereaux
Per 3h23'33" 47d43'27"
Peeper
Cnc 8h44'45" 27d20'17"
Peepers
Sco 16h12'33" -9d29'22"
Peepers
Uma 8h37'31" 62d22'50"
Peepi
Cyg 20h48'43" 36d23'18"
Peep-Po, Alice Patricia
Gem 6h58'48" 14d33'21"
Peep's High Rock
Uma 10h39'3" 71d9'8"
Peeps & Morty Friends Forever
And 0h21'17" 41d53'56"
Peetz, Wolfgang
Ori 6h19'18" 15d47'47"
Peewee
Aql 19h44'52" 14d9'40"
Peewee
And 2h33'59" 44d9'45"
Peewee Maggie
Lyn 7h53'45" 37d47'17"
Peff Bise
Lyn 6h38'12" 61d25'32"
peflrtkf
And 0h44'5" 42d27'28"
Peg
Aql 19h21'29" -0d14'32"
Peg
Sco 16h56'0" -40d18'42"
Peg Bachmann's Wish
Tau 5h18'15" 23d7'20"
Peg & Bob Whittemore
Uma 8h43'57" 70d12'51"
Peg Cadigan
Uma 9h44'31" 42d3'21"
Peg Caruso
Uma 12h30'3" 55d2'57"
Peg Dutcher
Cas 1h14'35" 71d57'43"
Peg French
Crb 16h5'19" 26d57'19"
Peg O'Shea
Sco 16h8'57" -17d5'25"
Peg Potts
Psc 1h19'33" 32d33'5"
PEGA
Aql 19h23'2" 5d43'41"
PEGALICOUS
Leo 10h10'47" 14d28'44"
Pegandron
Lac 22h45'54" 52d20'25"
Pegasus
Vir 13h14'3" 13d11'57"
Pegasus Rose of Frankie
Peg 22h1'18" 16d42'5"
Pegasus Star
Peg 22h44'25" 24d45'39"
Pegeen
Uma 11h45'53" 62d44'27"
Pegeli
Cam 7h54'10" 63d23'26"
PEGGIE
Ari 2h14'18" 23d28'23"
Peggie Ann Simeoli
Cnc 9h11'26" 8d28'14"
Peggie Jane
Lib 15h55'19" -17d39'50"
Peggles
Umi 15h15'21" 71d15'23"
Peggy
Cep 21h22'35" 65d7'44"
Peggy
Cas 1h18'59" 64d18'23"
Peggy
Vir 12h51'57" -0d25'44"
PEGGY
Sco 17h56'11" -30d40'15"
Peggy
Vir 13h6'45" 7d41'44"
PEGGY
Cnc 9h20'23" 10d35'24"
Peggy
Lyr 19h7'53" 28d28'52"
Peggy
Crb 15h50'11" 29d14'36"
Peggy
Gem 6h27'48" 24d24'48"
Peggy
And 0h43'25" 37d23'51"
Peggy
And 1h10'5" 42d44'26"
Peggy 50 Shines Bright
Cas 0h31'14" 65d24'38"
Peggy A Johnson
Ori 5h31'0" 1d22'38"
Peggy A. Rodriguez
Lyn 7h28'22" 44d34'50"
Peggy A.D.
And 1h11'32" 48d26'7"
Peggy and Eric Lieber
Cyg 20h49'56" 33d3'27"
Peggy and Harry
Lyr 19h24'15" 37d38'34"
Peggy and Ray
Lyr 18h30'32" 28d33'40"
Peggy and Shigeo's First Christmas
Cyg 19h21'45" 29d45'30"
Peggy Anderson
Lyn 7h27'27" 52d43'33"
Peggy Ann
And 0h45'33" 42d28'37"
Peggy Ann Harlan
And 23h41'34" 45d59'58"
Peggy Ann Register
Cyg 19h43'0" 40d44'9"
Peggy Anne Murphy Bryant
Sgr 18h9'11" -31d26'42"
Peggy Bertsche
And 1h1'27" 42d48'43"
Peggy Blanz
Lyn 9h9'55" 36d21'31"
Peggy & Buddy Guida
Cap 20h14'1" -13d25'31"
Peggy Burns
Sco 16h6'41" -10d39'4"
Peggy Cassidy
Uma 11h19'50" 42d9'5"
Peggy Chess Thornton
Uma 12h55'24" 56d27'8"
Peggy Chun
Crt 11h23'2" -16d33'38"
Peggy Church - "Peggy's Star"
Umi 14h59'52" 81d13'22"
Peggy Coder
Crb 16h2'34" 36d47'39"
Peggy Dace
Umi 14h54'5" 74d14'51"
Peggy Day
Umi 16h45'52" 80d19'59"
Peggy Dean 7/5/46
Cnc 8h27'26" 27d49'25"
Peggy Deanna Sulkey
Sco 17h37'15" -42d17'13"
Peggy Diane
Sco 17h48'47" -34d35'18"
Peggy Doolittle
Aur 5h31'37" 44d6'38"
Peggy Eames (Ma)
Cas 1h22'23" 61d30'30"
Peggy Ebert
Uma 12h9'22" 62d25'2"
Peggy Ellen Fellman
Cyg 19h58'38" 35d35'52"
Peggy Eva Guerry
Cru 12h57'9" -57d28'46"
Peggy Fallon
Gem 6h49'44" 31d36'30"
Peggy Flowers
Per 3h44'58" 43d50'26"
Peggy Gallagher
Crb 15h29'49" 28d27'14"
Peggy Gates
Lyr 19h15'8" 26d20'32"
Peggy Gene Belnap
Cas 23h10'32" 55d43'19"
Peggy Gentile Van Meter
Sgr 19h26'15" -30d0'58"
Peggy Goranitis
Gem 6h49'57" 33d5'30"
Peggy & Graham
Dra 18h59'14" 52d14'58"
"Peggy" Harriet Fratianne
Umi 13h53'49" 70d7'54"
Peggy Harriman
Ari 2h41'41" 19d23'26"
Peggy Harris
Cas 23h55'21" 53d31'37"
Peggy Hill
And 1h29'33" 47d57'10"
Peggy Huerta
Sco 16h51'7" -38d40'47"
Peggy Jane
Cam 6h19'14" 68d51'17"
Peggy Jean Denson
Cas 23h13'33" 59d25'9"
Peggy Jean Grove
Lib 15h39'21" -7d56'24"
Peggy & Jerry Gedatus 6/15/2006
Her 18h40'39" 21d10'12"
Peggy Jo Smith
Cyg 21h26'38" 44d42'37"
Peggy Joyce
And 23h21'34" 43d37'7"
Peggy Joyce Engebretson
Crb 15h38'57" 38d26'14"
Peggy Joyce Pittman Moore
Lib 15h1'39" -18d32'31"
Peggy Klocke
Leo 11h17'43" 15d6'32"
Peggy Lee
Aqr 21h24'38" 2d17'48"
Peggy Lee Fait
Vir 12h57'33" -14d32'2"
Peggy Leigh Millirons
Vir 12h42'53" 2d32'20"
Peggy Leiter
Dra 20h38'53" 81d2'29"
Peggy Louise Davidson
Uma 13h49'14" 57d7'27"
Peggy Louise Veatch Karns
Leo 10h58'12" 15d36'3"
Peggy Lynn
Leo 11h56'16" 22d11'8"
Peggy Lynn Albair
Leo 11h48'43" 20d16'16"
Peggy Lynn Baker
Vir 14h40'23" 3d0'48"
Peggy Lynn Hart
Ori 5h48'28" 5d46'3"
Peggy Marie Norvelle
Uma 9h7'58" 49d16'48"
Peggy Marie Podboy
Lmi 10h18'9" 29d34'44"
Peggy McGhee -Peg-Leg
Peg 21h39'49" 5d48'50"
Peggy McMillian
Cas 23h0'27" 58d5'34"
Peggy Meents
Vir 14h24'59" 7d11'14"
Peggy Meierhofer
Cas 0h21'21" 59d9'6"
Peggy & Mike
Sge 19h52'7" 18d55'37"
Peggy Moore
Vul 21h0'58" 23d8'6"
Peggy Morin
Vir 13h59'11" -19d30'27"
Peggy Morrow
Cas 0h18'3" 50d10'59"
Peggy Morss light
Aur 5h53'45" 46d29'52"
Peggy O'Connor
Cas 0h49'17" 61d28'22"
Peggy O'Kelley
And 1h11'40" 38d17'43"
Peggy (Old Lady)
Leo 11h36'39" 5d32'52"
Peggy Osland
Cas 0h2'42" 54d6'50"
Peggy Pecqueur de la Motte Foque
Vir 13h8'55" -22d39'4"
Peggy Picaw
Gem 6h28'9" 24d56'1"
Peggy Pierce
Gem 7h29'31" 20d35'17"
Peggy Power
Ari 2h56'45" 25d49'22"
Peggy Rice
And 23h47'41" 34d24'12"
Peggy Ruch
Aqr 21h45'45" 1d40'26"
Peggy S
Sgr 18h20'56" -23d22'36"
Peggy S. Kaketsis
Tau 3h44'1" 22d53'51"
Peggy 's Light.
Ari 3h13'32" 23d9'40"
Peggy Scioscia aka Scios
Cas 1h44'12" 64d21'50"
Peggy Seng O'Donnell
Sco 17h8'17" -38d25'19"
Peggy Sewell
Aql 19h44'13" 16d8'7"
Peggy Shines 4 Us
Cas 1h43'35" 63d11'45"
Peggy Shirer Babcock
Psc 0h16'22" 7d36'32"
Peggy Shurts
Cas 0h43'1" 61d53'12"
Peggy Silvers
Uma 12h28'27" 62d13'4"
Peggy Smith
Cnc 9h11'58" 29d9'36"
Peggy St. Jude Wright
Cnc 8h46'5" 31d39'14"
Peggy Stolz Conti
Aur 7h18'52" 41d33'34"
Peggy Sue
Cnc 8h18'40" 14d49'10"
Peggy Sue
Cma 6h50'40" -21d30'29"
Peggy Sue Black
Cap 20h29'39" -17d11'2"
Peggy Tannenbaum
Cyg 20h25'9" 59d37'32"
"Peggy" The Keeper of My Heart
Equ 21h14'5" 8d40'28"
Peggy & Tom's Star
Uma 11h25'47" 32d7'3"
Peggy Treppa
Psc 23h18'53" 1d14'54"
Peggy VanWinkle Alkinc
Tau 5h45'54" 22d7'54"
Peggy Vesneski
Lmi 10h26'7" 33d46'10"
Peggy Victoria Bearden
Del 20h25'29" 4d36'12"
Peggy Wellman
Lib 15h40'13" -29d17'49"
Peggy Wohlfarth
Boo 15h15'24" 25d56'6"
Peggy Wolfe-Buse
Crb 15h28'54" 29d32'50"
PeggyCharlie
Ori 5h57'15" -0d57'46"
Peggy-Mutz-Mommy, a forever love
Uma 11h13'21" 59d19'57"
Peggys Beauty
Vir 12h34'31" -7d26'14"
Peggy's Blessings
Crb 15h44'6" 33d1'56"
Peggy's Generosity
Lib 15h5'39" -13d13'35"
Peggy's Smile
Crb 15h37'18" 30d48'49"
Peggy's Star
Per 4h46'19" 46d46'37"
Peggy's Star
Cas 0h18'7" 51d57'37"
Peggy's Star
Peg 21h54'25" 8d2'34"
Peggy's Star
Lyn 6h18'59" 56d36'44"
Peggy's Star
Cas 23h5'56" 59d16'54"
Pegie Lynch
Vir 13h44'45" -1d46'48"
Pegi's Star Light, Star Bright
Uma 9h55'39" 45d27'33"
Pegism
Leo 9h45'42" 29d14'7"
PegMoArLes
Uma 11h32'55" 55d50'14"
Pegndon
Lmi 10h34'55" 36d42'41"
Peg's Foot
Ori 6h19'29" 15d55'14"
Pegsouthpres
Gem 6h9'4" 25d59'21"
Pehin Sri Taib Mahmud Sarawak
Ori 6h7'59" 8d6'48"
Pei
Aql 19h18'45" -6d8'13"
Pei Wang
Cyg 20h31'45" 53d19'30"
Peirson Rogers Benson Gallant
Ori 5h51'36" 9d7'4"
Peisang Tsai
Psc 1h10'21" 25d13'29"
Peja
Lyn 8h26'15" 55d8'33"
Pejman Azizi's Shining Star
Tau 4h19'16" 18d47'10"
Pekeño (J.C.Gallego)
Her 18h40'34" 21d59'15"
Pekingese with Angel Wings 3-17-00
Cmi 7h22'28" 9d7'17"
Peko's Star
Lyn 7h40'43" 52d24'1"
Pelican
Tau 4h31'46" 21d8'41"
Pelican
Cam 4h18'35" 69d24'43"
Pelicqiq
Oph 17h5'50" -0d17'1"
Pelin Akgun
Ori 6h16'24" 10d34'26"
Pellegrini
Cyg 19h59'24" 34d6'36"
Pellerin
Cnc 8h50'46" 16d3'51"
Pellett's Big Sky Star
Ari 3h27'23" 28d49'44"
Pellola
Tau 3h48'23" 28d41'17"
Pelosaura
Peg 23h14'15" 31d6'30"
Pelz, Peter
Uma 9h56'3" 46d47'40"
Pema Gamu
Leo 10h55'58" -1d23'48"
pemas<\@>psre.com
Uma 9h4'45" 69d7'23"
Pember's Muse Mahamudra 2867924126
Cru 12h28'14" -59d30'35"
Pembroke Berry
Mon 6h47'2" 8d1'22"
Pembroke Lee King
Sco 17h52'20" -37d8'9"
Pena & Honey Passerello
Cyg 21h52'4" 51d34'39"
Pena Niles
Umi 16h25'35" 72d29'53"
Pendragonia
Aqr 23h52'25" -11d32'0"
Penelope
Uma 11h49'27" 64d13'19"
Pénélope
Cas 23h47'15" 54d20'59"
Penelope
Cas 1h30'38" 65d28'40"
Penelope
Cru 12h24'1" -55d45'12"
Penelope
Sco 17h23'11" -42d14'12"
Penelope
Cas 0h22'51" 52d24'24"
penelope
Lyn 7h46'28" 41d5'28"
Penelope
Aur 5h36'36" 33d25'31"
Penelope
Ori 6h0'37" 10d5'6"
Penelope Cassandra Avila
Cnc 8h41'40" 14d48'6"
Penelope Christine Middleweek
Cru 12h8'13" -63d44'50"
Penelope Davies
Cru 12h40'25" -56d42'36"
Penelope & Elsa Forever
Umi 15h20'19" 74d21'26"
Penelope Freeman
Hya 9h24'33" 5d0'6"
Penelope Irais
Cyg 21h51'20" 54d18'7"
Penelope Jane
Lib 14h49'58" -13d33'41"
Penelope June Larsen
Tau 5h31'49" 18d50'32"
Penelope Lambert
Ori 5h45'18" 7d15'54"
Penelope Nicola Hajiyerou
Mon 6h48'44" 8d6'51"
Penelope Nimwe Niedzwiecki
And 23h3'29" 41d17'26"
Penelope Page
Sco 16h42'11" -31d4'36"
Penelope Rose MacFarland
Uma 9h46'37" 54d16'45"
Penelope Waite
Col 6h7'42" -41d23'49"
Penelope White
Gem 6h31'51" 20d12'0"
Penelope's Dragon
Gem 6h43'11" 16d8'39"
Penelopi Carmenos
And 2h26'57" 43d29'48"
Pengers Love
Dra 12h45'54" 73d54'15"
Pengi-Rican
Cyg 20h11'6" 36d21'45"
Penguin
Uma 8h27'39" 62d52'46"
Penguin
Aqr 22h50'40" -9d24'19"
Penguin
Ori 5h53'35" -0d32'20"
Penguin
Cet 2h46'33" -0d27'49"
Penguin and Pednoi
Aqr 20h45'30" -8d7'47"
Penguin Neff
Vir 14h5'57" -15d17'54"
Penguin Star
Ori 6h4'8" 17d2'43"
Penguins nell'amore
And 0h36'32" 38d55'55"
Penguins & Polar Bears
Ori 5h48'40" 6d14'25"
PenguinStar40307Zuley214 05DJ111304
Ari 1h54'30" 17d39'41"
Penina
Tau 5h28'10" 25d57'47"
Pennewiß, Angilika
Ori 5h22'3" 3d17'43"
Penney Louise Schmidt Elliott
Cam 4h2'6" 57d23'20"
Penni Andrews my shining star
Lib 15h59'48" -18d32'38"
Penni Jane Prates
Ari 2h43'14" 27d2'12"
Pennie 'Mommie' Zonick
Cas 0h44'24" 60d58'52"
Pennigaryius
Umi 13h12'24" 74d8'23"
Penni's Will
Aqr 22h18'43" -2d30'24"
Pennsic Love
Dra 13h34'28" 66d38'38"
Pennstar
Pho 2h23'13" -47d16'45"
Pennsylvania Hospital ICN Staff
Cap 20h25'13" -11d2'5"
Penny
Sco 16h19'30" -11d31'50"
Penny
Umi 15h0'45" 71d41'13"
Penny
Cas 23h51'21" 59d41'54"
Penny
And 0h16'9" 26d6'54"
Penny
Cmi 7h33'41" 4d58'43"
Penny
And 1h2'29" 42d0'6"
Penny A.K.A. Penelope Maxima
Cyg 19h43'42" 28d31'13"
Penny and Baby
And 23h15'13" 40d57'58"
Penny and Bryan Benbow
Eri 4h36'45" -0d59'25"
Penny and Paula
Sgr 18h50'50" -27d44'52"
Penny "Angel" Brewer
Vir 15h6'3" 7d20'36"
Penny Ann
Uma 9h5'15" 46d59'9"
Penny B. King
Crb 15h49'57" 35d25'20"
Penny Baptie's Star
Cap 21h34'30" -13d38'57"
Penny Block
Leo 11h19'50" 7d12'36"
Penny Carducci
Sco 16h51'21" -28d2'57"
Penny Cherry Day 1, 2004 Aug. 2
Gem 6h33'20" 14d10'34"
Penny Cherry Day 2, 2005 Aug. 2
Vir 13h37'39" -15d15'26"
Penny Cherry Day 3, 2006 Aug. 2
Gem 7h39'38" 27d59'54"
Penny Clinton-Jensen
Uma 11h40'23" 57d31'8"

Penny Curtis
And 2h5'2" 36d36'53"
Penny DeWeese
And 1h3'4" 34d22'40"
Penny Elizabeth Judith Deane
Ari 2h41'56" 12d31'9"
Penny Fisher
And 23h23'16" 48d32'3"
Penny Foster
Lmi 9h58'28" 40d21'28"
Penny & Gary - 06 July 1986 -
Cru 12h37'4" -58d1'16"
Penny Gay "My Forever Love"
Lyr 18h29'46" 36d20'10"
Penny & Gerry
Cyg 19h46'59" 54d51'39"
Penny Giannios
Cnc 9h5'42" 23d53'50"
Penny Hanczaruk
Lyn 7h44'10" 38d30'0"
Penny Horan
Leo 10h15'52" 16d32'16"
Penny Hummel
Vir 15h8'8" 4d13'45"
Penny J. Chiminello
Uma 10h52'44" 57d40'40"
Penny Jane McCluskey
Per 1h50'10" 50d44'2"
Penny Jessee
Uma 10h43'28" 57d11'19"
Penny Lautee - My Shining Star
Cas 0h18'56" 53d16'5"
Penny Leigh Willis
And 1h15'13" 39d47'23"
Penny Lenard
Aqr 21h32'17" -7d38'53"
Penny Lorraine Boyer
Crb 15h50'48" 32d21'4"
Penny Lover
Ori 5h31'6" 0d11'0"
Penny Michele Brenner
Umi 10h30'29" 86d44'42"
Penny Mittonette
Cap 20h25'33" -16d57'59"
Penny Moore
Ori 6h11'3" 15d46'15"
Penny Olivia Binn
Vir 12h49'43" 11d41'23"
Penny & Peter Hamerslag
Umi 15h18'45" 75d16'58"
Penny Phillips
Cma 6h33'48" -12d6'3"
Penny "Princess Sparkles" Kemp
Vir 12h43'42" -4d47'39"
Penny Prior Hammerstrom
Cas 0h15'25" 59d28'57"
Penny Purcell
Aqr 21h49'50" -6d11'39"
Penny Robyn Yancey
Sco 17h56'55" -30d23'26"
Penny Sheline
Ari 3h26'33" 23d32'30"
Penny Snow
Ari 3h1'54" 14d0'7"
Penny Sparkles
Lep 5h21'38" -12d14'4"
Penny Spence
And 0h14'47" 31d14'21"
Penny Stabile
Cap 21h46'41" -11d31'46"
Penny Starshine
Cma 7h12'52" -23d41'47"
Penny Sue's Star in Heaven
Uma 11h32'17" 59d10'26"
Penny Thoburn
Com 12h3'16" 20d40'22"
Penny Victoria
Peg 21h49'15" 26d37'12"
Penny Westbury
Crb 15h48'50" 30d36'39"
Penny Zaugg
Psc 2h4'1" 8d20'5"
Pennyann
Aql 19h0'22" 15d58'7"
Pennylane
Uma 11h36'40" 61d41'30"
Penny's [Heart]
Sgr 18h26'52" -16d46'45"
Penny's Simone
Cyg 19h56'7" 41d18'59"
Penny's Star
Ori 5h58'53" 20d52'28"
PennyStar
Lib 15h30'15" -7d59'36"
Penolope Paige Hoover
And 0h29'11" 42d2'9"
PENRO Junior Bowlers
Uma 11h56'46" 57d23'5"
Pensamiento
Lmi 10h45'16" 24d41'22"
Pensando en Ti
Uma 9h32'56" 64d5'35"
Pensonic Sales & Service Sdn Bhd
Ori 5h34'33" 3d58'39"
Pentastar
Uma 9h19'4" 41d51'27"
Pentecost
Uma 11h48'53" 36d8'13"
Pentecost Family
Aql 19h35'23" 8d45'52"
People's Heritage Bank
Uma 11h14'26" 57d33'23"
Peoriastar
Lyn 7h32'22" 46d24'39"
Pepa
Ori 4h51'29" 11d55'8"
Peparo Family Strength
Uma 11h23'33" 48d49'59"
Pepaw Allen
Leo 11h0'58" -0d58'37"
Pepe
Lib 16h1'31" -16d28'54"
Pepe Alessandra
Cas 23h15'44" 55d32'56"
Pepe "Aqui Nadie se Rinde"
Ori 6h5'59" -3d46'35"
Pépé Boington Raymond
Psc 1h19'31" 4d38'5"
Pepe Mato
Aqr 22h44'46" -7d38'40"
Pepe Mayol
Uma 8h14'13" 64d14'32"
"Pepe" Ragovoy
Lyn 6h56'50" 51d29'53"
Pepe Vence
Cnc 8h6'43" 6d53'36"
Pepe y Luis
And 0h54'18" 39d45'10"
Pepe, 03.11.1976
Her 18h46'22" 20d7'4"
Pepe13
Lyr 19h15'45" 28d42'36"
Pepère
Vir 12h17'58" 11d38'13"
Pepère Cauchon
Uma 13h20'47" 58d29'8"
Pépère Guy F. Van de Velde
Aql 19h11'49" -0d31'3"
Pépère Vallier
Ori 5h54'19" 17d27'57"
Pépéron
Uma 10h44'15" 70d21'34"
Pepe's
Tau 4h38'44" 5d22'19"
Peppard
Dra 17h54'17" 51d55'59"
Peppe
Leo 11h5'59" 10d51'38"
Peppe Mou I Love U My 1 & Only Nino
Ari 2h38'53" 21d32'40"
Pepper
And 0h18'31" 37d40'32"
Pepper
Lyr 18h40'52" 34d52'12"
Pepper
Lyn 7h19'10" 56d45'40"
Pepper Ann
Lyr 18h21'12" 39d46'34"
Pepper (Babygirl)
Uma 14h24'34" 60d46'47"
Pepper Bunny Star
Uma 8h56'4" 47d11'31"
Pepper Dey Cote'
Lmi 10h48'20" 30d2'31"
Pepper Pot Signe Hill
Her 18h44'49" 19d9'42"
Pepper Roo Swartz
Cmi 7h36'25" 6d6'48"
Pepperlin
Aur 5h10'50" 46d5'48"
Peppermint Patty
And 23h25'2" 46d50'7"
"Peppermint Patty" Whedbee
Psc 1h24'5" 16d15'31"
Pepperminta Emma
Umi 15h55'23" 81d44'31"
Pep's Dad
Her 18h9'56" 31d39'25"
~Pepsi's and Mija's Shining Star~
Sco 17h18'21" -33d7'34"
Pequeñita
Vir 13h17'52" -14d54'51"
Pequeño Volcan
Aqr 21h14'22" -10d56'18"
Pequita
Uma 9h0'38" 46d50'46"
Per ardua
Lyn 7h48'11" 43d5'46"
Per B. Sorensen
Cap 21h56'32" -14d54'11"
Per E Lindholm
Dra 15h3'40" 64d31'19"
Per Kristian - Papa's Star Forever
Per 2h52'5" 42d26'28"
Per Mare Per Terram
Per 4h49'56" 41d35'29"
Per Oskar Hagard
Tau 4h32'31" 26d38'16"
...per sempre...
Uma 9h23'40" 50d14'8"
Per Sempre
Cyg 20h15'49" 50d21'35"
Per Sempre
Cyg 21h25'44" 50d26'25"
per sempre
Cyg 21h15'57" 38d35'57"
Per Sempre
Uma 9h43'24" 70d42'48"
Per Sempre
Cam 4h15'9" 70d16'48"
Per sempre
Lyn 7h28'50" 53d25'33"
Per Sempre
Cyg 20h8'55" 55d8'7"
per sempre Caterina, 02.01.1979
Dra 20h0'9" 72d35'21"
Per sempre Christian 31.5.1999
And 23h51'10" 43d26'45"
"Per Sempre La Stella" Christina
Uma 12h27'13" 62d46'22"
per sempre Luana
Her 16h44'0" 6d50'37"
Per Sempre Stella
Tri 2h28'35" 30d31'51"
Per Sempre Toni
Ari 3h11'21" 24d4'39"
Per sempre Urs Patrick
Cas 0h25'38" 47d12'45"
Peralta
Gem 7h25'53" 31d51'35"
Perch
Uma 10h10'8" 44d51'33"
Perches
Sco 17h25'0" -42d20'7"
Percival Christopher Van Daam
Per 3h8'53" 51d33'52"
Percy
Ori 5h42'42" 1d18'25"
Percy
Cep 21h50'28" 64d35'59"
Percy Bombin
Sgr 18h34'17" -26d44'2"
Perdita
Uma 10h28'29" 46d47'32"
Pere & Juli
Ari 2h53'55" 22d0'3"
Père Loup
Uma 11h27'21" 59d45'39"
Pere Robert Amadou
Aqr 22h19'33" -14d12'18"
Perecira
Cas 0h47'17" 57d30'33"
Perera
Sgr 17h54'19" -17d55'24"
perfeccione asombrosa-mente
Tau 3h34'17" 27d19'5"
perfect
Uma 9h33'27" 64d35'18"
Perfect Addiction
Aql 18h59'40" -8d7'27"
Perfect Angel, Perfect Mom Linda
Cas 23h4'54" 53d46'10"
Perfect Diane Hunt
Cas 23h20'5" 58d39'52"
Perfect Elizabeth
And 23h52'38" 42d27'12"
"Perfect" H.J.W. & J.L.F.
Psc 0h34'16" 6d26'1"
Perfect Jenni
Cnc 8h38'47" 10d17'38"
Perfect Just Like Cindy
Vir 11h37'44" 2d11'19"
Perfect Lil Pixie - Nikole Michele
Ari 2h49'41" 14d39'0"
Perfect Mary
Cap 20h15'12" -14d41'5"
Perfect Match
Mon 6h50'11" -0d37'36"
Perfect Match
And 23h44'32" 38d34'43"
Perfect P
Uma 13h50'17" 49d53'51"
Perfect Parker
Uma 11h33'9" 42d33'46"
Perfect Pasquale
Uma 12h31'10" 58d30'31"
Perfect Pitch
Her 16h17'47" 48d26'38"
Perfect & Priceless
Cyg 20h13'57" 37d18'1"
Perfect Rainbow
Uma 8h33'12" 63d38'36"
Perfect Sex
Lyn 7h10'7" 54d57'53"
Perfect Spartan
Ari 2h53'24" 18d47'28"
Perfect Unity
Uma 9h28'58" 44d36'0"
Perfect You
Umi 15h51'7" 79d41'28"
Perfection
Cas 1h19'18" 61d40'19"
Perfection
Umi 14h36'42" 73d55'52"
Perfection
Cyg 21h22'2" 31d43'31"
Perfection / Lehfellner - Watson
Cyg 20h39'40" 45d13'54"
Perfection, to shine after the sun
Sgr 19h36'16" -43d13'28"
Perfectly Amazing
Her 18h6'20" 17d28'43"
Perfectly Astronomical Teacher
Sco 16h9'52" -8d46'17"
Perfekt
Gem 6h42'35" 19d42'0"
Pergola
Uma 9h22'45" 47d22'7"
PerHam
Lyn 8h7'0" 56d39'22"
Peri Elizabeth
Cyg 20h36'40" 36d3'26"
Peri Heinnen - 18.09.1992
Vir 12h30'8" 10d38'16"
Peri Kert
Cas 0h34'1" 64d28'45"
Peribu
Uma 14h24'34" 61d4'55"
Periculorum
And 1h9'43" 45d31'55"
Periklis Apostolou
Uma 10h50'0" 40d28'2"
Perita Demarist Frazier
Cnc 8h44'23" 24d30'54"
Periwinkle
Psc 0h9'5" 8d2'55"
Periwinkle 210
Aqr 23h54'35" -12d29'50"
Perkkki
Sco 16h11'24" -10d51'30"
Perl Grinberg 3/1/77
Psc 1h19'20" 17d37'6"
Perl, Elona
Ori 5h7'31" 7d50'48"
Perla del cielo
Cap 20h59'59" -20d7'39"
Perla Elizabeth Romero Martinez
And 0h27'59" 25d16'24"
Perla Montenegro
Cam 3h47'44" 58d40'55"
Perla Somarriba
Gem 7h44'21" 14d34'58"
Perla V. Solis Pulido
Peg 22h29'23" 4d57'14"
Perlin, Jutta
Uma 11h27'20" 31d50'57"
Permanis Sdn Bhd
Ori 5h33'55" 3d52'18"
Pérola do Céu
Psc 0h54'23" 14d38'59"
Perpetua
Leo 10h55'53" 19d21'8"
PERPETUAL LOVE
Gem 7h7'12" 34d23'25"
PERPETUAL PRINCE
Lyn 7h52'19" 47d30'15"
Perpetuus Diligo
Psc 23h19'17" 7d9'28"
Perpetuus Excellentia
Cas 0h38'38" 61d51'3"
Perrenot Beacon
Cam 4h34'58" 64d46'1"
Perret's Lincoln 21
Uma 10h51'13" 45d38'22"
Perrin Owens
Crb 16h8'51" 32d51'49"
Perros Vizcacheros (Sergio Antoine)
Tau 4h30'55" 24d59'18"
Perry
Aur 5h20'21" 44d18'26"
Perry
Cyg 21h39'53" 54d16'5"
Perry Belcher - Our Star
Leo 10h3'31" 23d36'0"
Perry Fethlan
Per 3h13'4" 50d32'42"
Perry Holman
And 1h10'15" 41d46'0"
Perry Jaymes
Sco 16h10'34" -13d46'46"
Perry Jones
Aql 19h45'43" 8d9'44"
Perry Kuipers
Cep 21h7'4" 55d31'20"
Perry L. Roberts
Per 2h56'15" 53d9'7"
Perry Lynn McCool October 29, 1930
Sco 17h0'39" -34d26'4"
Perry Mastro
Uma 9h44'9" 45d26'10"
Perry Miles
Crb 15h37'37" 30d30'40"
Perry Minnich
Aur 5h55'10" 36d55'22"
Perry Parissos
Leo 9h49'0" 32d54'30"
Perry Patrick Anthony
Uma 11h49'11" 43d15'16"
Perry Richard Read
Sco 17h25'55" -33d43'28"
Perry Richard Swanker
Cap 20h41'28" -24d47'34"
Perry T. Denton
Gem 6h54'54" 12d33'37"
Perry "Tarheel" Hovis
Uma 11h31'32" 42d53'54"
Perry - Tax Man
Cyg 21h7'34" 42d55'44"
Perrydise
Sco 16h12'34" -10d35'39"
Perrynoid & The Cosmic Circus
Psc 1h43'20" 23d38'49"
Perry's Star
Peg 21h55'7" 11d35'16"
Perrytwinkle
Cyg 19h41'28" 28d38'37"
Perschon, Alex
Uma 11h26'6" 59d55'25"
Persempre
Lac 22h31'16" 41d57'18"
PerSempreNeiNostriCuori
Uma 11h20'10" 35d50'51"
Persephone
Ari 2h15'38" 22d52'13"
Persephone
Cap 21h32'17" -16d22'24"
Perseverance
Uma 13h48'8" 56d43'30"
Perseverance
Ori 6h0'4" 22d4'11"
Perseverance
Ori 5h23'23" 2d20'12"
Perseverance of Faith and Love
Aqr 23h53'40" -4d27'19"
Perseverance XI "10-12-98"
Umi 13h30'41" 69d51'31"
Perseverance12_31_06
Cap 20h37'59" -15d2'1"
Persie, Ludwig
Ori 6h8'16" 8d53'28"
Perspective
Lyn 6h47'13" 54d5'26"
Pertinentia - la Rosa nel Cielo
Boo 15h18'42" 47d53'28"
Pertty
Sco 16h8'8" -16d36'52"
PERVINCA
And 23h27'27" 35d14'51"
Pervous
Her 17h32'9" 36d41'49"
Peschel, Hannah
Uma 11h56'6" 51d29'58"
Peshie
Umi 16h45'11" 77d41'6"
Pest Family
Sgr 18h26'49" -16d12'44"
Peta Chappell
Cru 12h8'47" -58d7'44"
Petalouda
Peg 22h47'10" 13d17'36"
Petals
Dra 18h57'35" 57d34'33"
Petar - You Light Up My Life - Natasa
Cru 12h40'48" -56d44'41"
Peta-Renea
Cru 12h45'56" -57d48'26"
Pete
Psa 22h26'1" -25d45'0"
PETE
Uma 11h54'25" 58d55'17"
Pete
Her 17h40'50" 40d51'49"
Pete and Amy
Uma 11h50'52" 40d20'26"
Pete and Betty Harlow
And 23h33'53" 44d30'25"
Pete and Ivy
Cyg 20h36'44" 42d21'40"
Pete and Linda Forever
Uma 9h48'54" 57d25'19"
Pete and Michelle Jeanette Beltran
Uma 9h22'26" 51d17'28"
Pete and Stephanie Rutkowski
Cyg 21h43'50" 37d0'50"
Pete and Sue Neville
Cnv 13h58'37" 34d3'39"
Pete and Xena - January 13, 2005
Tau 3h43'22" 28d46'43"
Pete & Arinana
Gem 7h53'16" 14d14'36"
Pete B Floyd
Cnc 8h43'50" 23d21'3"
Pete Bendinger
Cnc 8h55'53" 6d35'20"
Pete Bettina
Her 17h22'33" 32d23'40"
Pete Borders
Sgr 19h4'57" -12d53'6"
Pete Caucci
Psc 23h41'26" 2d41'52"
Pete & Dann Barthelme
Umi 15h55'27" 80d11'4"
Pete Dennis
Uma 8h20'6" 69d9'45"
Pete & Donna Reinsma Forever Love
Cyg 21h59'1" 51d4'21"
Pete Hagenberger
Leo 11h51'19" 24d2'52"
Pete Hardman
Per 2h54'31" 46d52'16"
Pete Hawkes
Cru 12h4'53" -60d44'54"
Pete Holt
Ari 2h13'19" 24d50'37"
Pete In memory of Jerry Vaughn
Uma 10h42'57" 61d31'7"
Pete J. Ramage
Uma 10h56'11" 49d56'19"
Pete & Jane
Sgr 18h11'59" -19d56'23"
Pete Jansons Loves Maureen Doheny
Cyg 21h38'44" 47d37'34"
Pete & Kat Spaghetti
Tri 2h13'18" 35d10'44"
Pete & Katie
Cyg 20h9'17" 38d19'9"
Pete Kearns
Umi 16h0'3" 72d14'10"
Pete Latham
Uma 11h48'26" 41d51'55"
Pete & Lila Otter
Cyg 20h47'51" 40d6'15"
Pete & Linda Minchin Forever Star
Col 6h8'56" -39d57'6"
Pete McPartland, Jr.
Leo 9h57'7" 27d54'3"
Pete Meldrum
Cir 14h56'48" -67d51'45"
"Pete" Peter Michael Kohut
Cnc 9h2'38" 32d25'27"
Pete (PJ) Our Shining Star
Per 3h49'46" 41d44'32"
Pete (PJ) Our Shining Star
Her 17h37'48" 27d1'41"
Pete Pukish / Satori Martial Arts
Dra 15h47'51" 61d27'6"
Pete & Rachel - In Love Forever.
Del 20h32'3" 6d49'25"
Pete & Repeat
Uma 10h44'28" 70d58'3"
Pete & Rhonda Anderson
Aql 19h19'3" -0d29'47"
Pete Rimkus
Per 2h53'19" 54d11'43"
Pete & Ruth moon
Uma 10h51'1" 52d33'0"
Pete Santoro
Uma 11h2'2" 53d16'18"
Pete Sciallo
Vir 13h41'53" -9d19'29"
Pete Smith
Uma 8h10'45" 61d29'40"
Pete Soderman
Ori 6h11'59" 15d16'46"
Pete Sonneville
Per 3h17'48" 53d36'46"
Pete & Stephanie Everhart
Lib 15h34'33" -26d9'41"
Pete the Plumber
Gem 6h53'42" 21d28'24"
Pete Vernier
Her 18h34'9" 15d58'15"
Pete & Virginia Buescher
Uma 9h47'6" 49d38'35"
Pete Viviano
Lyn 6h43'8" 53d44'41"
Pete Whitecross
Cep 21h54'36" 63d54'46"
Pete, Jan, & Chip Moss
Ari 2h20'15" 26d22'56"
Pete, the Dragon, Charles
Dra 18h43'58" 54d36'12"
Peteña
Cas 1h20'58" 56d21'45"
Peter
Sge 19h54'6" 17d52'26"
Peter
Del 20h25'42" 16d30'7"
Peter
Cnc 8h21'46" 12d40'38"
Peter
Uma 9h0'11" 54d30'12"
Peter
Uma 12h30'34" 55d1'40"
Peter
Cep 20h46'45" 61d38'9"
Peter
Cas 2h16'50" 74d11'34"
Peter
Uma 13h0'30" 63d4'5"
Peter
Cep 1h38'15" 86d22'38"
Peter
Sco 16h21'49" -33d23'44"
Peter
Sco 17h58'23" -31d36'58"
Peter
Sgr 19h21'13" -30d42'26"
Peter
Psa 22h14'30" -30d19'47"
Peter 22
Tau 4h25'50" 22d46'25"
PETER 22.03.1957
Uma 13h8'30" 53d17'33"
Peter A. Murray
Uma 10h11'59" 71d22'0"
Peter A. Quigley, Fiberspar Corp.
Umi 14h30'52" 75d46'7"
Peter A. Sokol
Mon 7h34'33" -0d56'53"
Peter A. Spitalieri
Psc 1h19'15" 15d24'21"
Peter A. White
Sco 17h8'11" -42d13'31"
Peter Alan
Boo 14h21'22" 34d17'10"
Peter Alan Stanney
Cyg 21h15'53" 41d25'22"
Peter Alexander Kolodziej
Cap 20h53'23" -20d38'4"
Peter Alfred Civarelli
Umi 14h30'16" 76d21'0"
Peter & Alicia Morales
Lib 15h14'5" -5d5'27"
Peter Allen
Per 3h17'52" 40d42'38"
Peter Allen Knoble's Star
Cap 20h27'53" -10d57'21"
Peter Allen Rush
Per 2h47'41" 46d17'41"
Peter Aluotto
Psc 1h15'16" 11d12'27"
Peter and Danielle Simonian 4-Ever
Cra 19h5'30" -39d0'42"
Peter and Denise Moore
Crb 15h56'37" 31d45'35"
Peter and Jessie
Aqr 22h27'47" -5d45'7"
Peter and Juliana Bunny Love
Tau 4h45'54" 21d0'42"
Peter and Julie's Eternal Flame
Cru 12h34'56" -64d15'4"
Peter and Leah Crouser
Cyg 19h29'34" 52d42'33"
Peter and Lisa Moussa
Cap 21h46'1" -16d3'7"
Peter and Mary Ruebens
Sge 19h55'28" 18d16'55"
Peter and Shirley O'Neill
Cyg 19h55'59" 41d45'28"
Peter and Stephanie Emenecker
Crb 16h11'39" 33d34'22"
Peter Anderson Boyd
Per 4h13'48" 45d36'50"
Peter Andreas Dau
Uma 11h7'27" 57d46'53"
Peter & Anneliese
Umi 15h59'38" 77d35'42"
Peter Anthony Brown
Sgr 18h52'40" -16d26'35"
Peter Anthony De Angelo
Psc 22h55'35" 6d19'26"
Peter Anthony McMullin
Gem 6h30'56" 21d20'4"
Peter Anthony Sarantos, III
Leo 9h22'50" 16d39'56"
Peter Anthony Simon
Ori 5h20'53" 3d32'42"
Peter Anthony Stead
Per 3h35'32" 36d13'1"
Peter Archibald
Ori 5h58'24" 2d8'10"
Peter Ashton Johnson
Aql 19h52'33" -0d58'28"
Peter Baggs
Cep 21h34'53" 63d59'18"
Peter Ball
Sco 16h13'42" -10d8'55"
Peter Barnes - Cosmos Watcher
Nor 16h27'43" -48d13'33"
Peter & Bella Forever Love
Cra 18h48'39" -37d45'35"
Peter Bender
Uma 9h31'11" 46d43'8"
Peter Berger
Per 2h51'56" 53d57'3"
Peter Biczok
Ori 6h17'53" -0d29'45"
Peter Billman
Her 18h43'9" 18d2'50"
Peter Blake
Sco 16h17'55" -16d42'44"
Peter Bluck
Cyg 20h3'14" 37d23'52"
Peter Bohnert
Uma 10h38'31" 40d0'43"
Peter Borzillieri
Gem 7h53'9" 32d55'54"
Peter Bradshaw The Third
Cru 12h25'3" -59d25'51"
Peter Brai
Gem 7h21'20" 35d15'55"
Peter Brandon Lynch
Cen 13h50'31" -37d38'0"
Peter Breeland
Psc 1h32'24" 16d54'41"
Peter Brescia
Uma 11h40'59" 31d6'49"
Peter Brian Gumbrell
Per 4h13'46" 32d9'37"
Peter "Bright Eyes" Pistone
Lib 15h20'34" -28d47'13"
Peter Brunsch
Uma 11h7'4" 44d38'59"
Peter Bryan Creviston
Lyn 7h36'40" 37d58'46"
Peter Bunting Giordano
Aqr 23h10'13" -22d1'42"
Peter Burke
Gem 6h32'5" 24d28'58"
Peter Bury
Per 3h34'15" 47d4'55"
Peter Buschauer
Cap 20h59'23" -21d55'32"
Peter C. Goldmark Jr.
Ori 5h38'27" 2d39'51"
Peter Calangelo
Uma 11h29'53" 36d6'42"

Peter Cameron Newall
Uma 13h42'52" 58d5'25"
Peter Capasso III
Cam 4h8'2" 63d15'45"
Peter Carey
Per 4h49'50" 45d22'30"
Peter & Carlene
Sco 17h2'45" -41d33'15"
Peter & Catherine Dreessen
Cyg 21h18'24" 53d33'2"
Peter Cerny
Uma 8h50'44" 52d45'55"
Peter Charles Prevas
Uma 12h52'0" 57d53'5"
Peter Charles Siska
Cap 21h22'44" -26d1'37"
Peter Chia 143637-Ana & Yeeshine
Per 3h46'51" 41d41'39"
Peter Chiandotto
Cru 12h30'47" -59d52'28"
Peter Cho
Col 5h59'32" -34d37'16"
Peter Chouinard
Aqr 22h7'13" 0d7'25"
Peter Christina Hubeni
Her 18h23'52" 28d56'25"
Peter Christopher
Aqr 22h45'25" -4d12'30"
Peter Clarke
Per 4h46'32" 47d31'6"
Peter Clay
Uma 12h49'33" 62d25'58"
Peter Codella
Her 16h40'7" 28d48'45"
Peter & Conor Quadarella
And 23h35'49" 48d6'26"
Peter Conzatti
Ori 6h15'39" 7d12'35"
Peter Craig Smith
Cyg 21h10'47" 47d29'4"
Peter D Medolan
Cyg 21h59'57" 48d11'6"
Peter Daels
Dra 20h15'13" 71d15'17"
Peter Dale Baker
Cen 11h27'37" -48d30'5"
Peter Dathe
Ori 5h18'31" 15d59'57"
Peter David Bishop
Cep 21h23'28" 57d27'50"
Peter David Henderson
Per 3h22'49" 49d55'19"
Peter David Rees "Sneaky"
Cru 12h27'6" -60d47'11"
Peter Davis
Per 2h40'4" 57d5'39"
Peter Dawkins
Umi 17h21'51" 85d50'55"
Peter Demetri
Dra 17h11'2" 68d24'56"
Peter Desmond Gibbons
Cep 22h47'59" 76d52'11"
Peter Diethelm
Pyx 8h53'29" -31d49'15"
Peter Domenick Libranti, Jr.
Sgr 19h13'55" -22d14'38"
Peter Dominic Koulianos
Sco 16h12'29" -16d15'44"
Peter Dörner
Uma 9h54'39" 57d9'39"
Peter Dunn
Cyg 20h36'23" 55d49'14"
Peter Dutton
Uma 11h8'56" 42d38'41"
Peter Dwain Graham
Uma 8h53'41" 62d59'19"
Peter E. Olsen Jr.
Her 17h20'55" 32d7'49"
Peter Edward Griffiths
Cru 12h45'55" -58d0'40"
Peter Eichinger
Uma 11h4'22" 40d16'57"
Peter Elliott
Tau 4h7'34" 10d10'41"
Peter Elliott
Psc 0h33'23" 4d6'18"
Peter Engel
Cet 3h14'23" 10d20'44"
Peter Enright Jenkins
Uma 9h42'2" 57d41'12"
Peter Eric Daupern
Aql 19h15'56" 5d51'1"
Peter Ernest Lemke
Boo 14h24'0" 40d31'57"
Péter és Judit
Uma 11h6'32" 50d24'45"
Peter Eugene Hockenhull
Ori 6h24'16" 11d0'34"
Peter F. Capuciati
Dra 19h59'52" 62d54'45"
Peter F Morello
Lyn 8h22'43" 34d41'22"
Peter Farkas
Uma 11h37'43" 33d5'4"
Peter Farmer
Uma 10h49'18" 58d59'25"
Peter Fässler
Peg 21h27'13" 15d49'35"
Peter Finnegan
Cep 22h45'21" 65d41'16"
Peter Foldes
Uma 9h3'48" 53d14'52"
Peter & Fonda Philip
Cyg 19h37'5" 51d41'57"
Peter Forster
Lyn 7h27'28" 58d0'49"
Peter Francis Buffer
Sgr 19h41'50" -13d33'4"
Peter Francis Quattro
Vir 12h48'9" 8d54'8"
Peter Francis Roberts
Cep 22h5'19" 60d17'27"
Peter Francis Ross
Lyn 8h31'51" 36d50'52"
Peter Francis Westgate
Gem 7h7'54" 27d21'5"
Peter Franklin Harman
Aql 19h49'26" -0d29'34"
Peter Franz Wolf
Lup 14h33'31" -52d58'36"
Peter Franz Wölfle
Uma 9h7'5" 72d44'46"
Peter Frawley
Vir 12h43'51" -0d53'47"
Peter Fuller
Ori 5h38'15" -3d52'13"
Peter G. McGuinness
Per 4h18'22" 33d36'22"
Peter G. O'Donnell
Uma 8h44'22" 52d24'20"
Peter G Urban - Sensei
Uma 10h28'1" 67d31'22"
Peter Gadzuk" You'll always Shine"
Uma 13h35'5" 55d15'11"
Peter Galletta
Cep 22h30'36" 64d55'28"
Peter Gary McRae
Psc 0h52'24" 26d53'30"
Peter & Gay Giardini's Legacy
Cyg 19h41'9" 36d49'24"
Peter Gayed
Sco 16h48'46" -33d3'58"
Peter Gebauer
Ori 5h42'26" 12d25'41"
Peter Geiger
Psc 0h58'55" 28d57'58"
Peter & Gemma Swallow
Cyg 21h41'3" 33d32'30"
Peter Gentsis
Psc 1h28'38" 24d32'50"
Peter George Karamanos II
Cap 20h12'27" -9d57'40"
Peter Gerard Sell
Lib 14h51'6" -3d43'57"
Peter Gerber
Crb 16h5'4" 32d37'20"
Peter Gilbert
Aql 19h22'49" 7d41'23"
Peter Gili
Tau 4h7'37" 27d19'49"
Peter Glücki
Ori 4h50'38" 10d9'58"
Peter Gornall
Per 3h49'52" 49d48'13"
Peter Gosselin
Her 15h51'8" 49d45'26"
Peter Great Star Thornburgh
Per 3h19'39" 45d46'18"
Peter Gruschovnik
Uma 14h26'44" 55d20'10"
Peter Guilbes
Cep 23h39'32" 74d50'37"
Peter Guthmann
Uma 9h23'12" 48d43'27"
Peter H. Apps
Eri 4h9'54" -33d54'13"
Peter H Kornherr 11.13.1957
Cep 20h47'26" 61d50'51"
Peter H. Schmidt
Cap 21h41'3" -21d35'38"
Peter Haida 22.09.1939
Uma 13h24'15" 55d59'35"
Peter Hambly
Cep 20h40'35" 63d37'8"
Peter Hartmann
Ori 5h6'46" 4d46'21"
Peter Hashim's Palette
Uma 11h48'53" 44d59'16"
Peter Hein
Uma 8h20'9" 65d42'39"
Peter Hein
Uma 10h5'34" 64d41'25"
Peter Helm
Uma 14h23'26" 56d51'26"
Peter Henius Brewster
Psc 1h14'53" 32d46'37"
Peter Henry Storm III
Leo 10h10'38" 11d25'13"
Peter Herman Crider
Lmi 9h30'45" 37d9'43"
Peter Herrmann
Uma 8h37'26" 54d24'27"
Peter Hiatt
Cnc 8h11'42" 17d58'12"
Peter Hinde
Uma 9h48'35" 50d21'3"
Peter Hoffmeister
Uma 10h3'56" 61d40'20"
Peter Hohlbein
Uma 10h23'59" 72d24'35"
Peter Honey
Sco 17h52'29" -41d44'17"
Peter Houseman
Per 4h36'38" 37d16'55"
Peter How - 11.10.1945
Cru 12h55'45" -57d10'51"
Peter Hug 03.08.1946
Umi 14h57'53" 74d31'58"
Peter Hummel
Gem 6h31'2" 27d41'16"
Peter Hunter Robson
Ori 6h6'55" 15d37'55"
Peter Hutton
Uma 8h41'42" 61d52'53"
Peter Hyde Jones
Per 4h11'17" 42d22'13"
Peter Hysen-Burdette
Uma 12h58'31" 58d1'39"
Peter Inderbiethen
Uma 10h2'14" 55d50'31"
Peter & Irma
Lyr 18h46'10" 36d6'32"
Peter Ironman Harding
Ori 5h54'26" 19d21'26"
Peter J.
Cep 21h13'57" 60d32'25"
Peter J. Casale
Tau 5h11'42" 18d24'40"
Peter J. Casella
Uma 11h37'32" 39d33'48"
Peter J. DiCamillo
Vir 14h54'54" 2d44'46"
Peter J. Farrey
Uma 11h19'31" 46d34'15"
Peter J Miller
Per 2h18'10" 55d26'22"
Peter J Mils
Lmi 9h50'28" 37d13'49"
Peter J Ortiz
Cmi 7h27'43" 3d17'20"
Peter J Ratchford 1955
Dra 20h7'18" 71d42'29"
Peter Jackson
Uma 10h32'6" 48d22'29"
Peter James
Uma 9h31'56" 57d6'7"
Peter James Dixon, Jr.
Sco 17h51'51" -35d5'27"
Peter James Fanelli
Uma 11h12'56" 44d23'12"
Peter James Friedman
Ari 3h5'17" 20d4'15"
Peter James (January 19, 2003)
Mon 7h20'35" -0d38'11"
Peter James Lombardo
Her 17h18'18" 36d3'57"
Peter James McFadden
Cep 21h25'9" 65d54'33"
Peter James Rossi
Vir 13h5'32" 8d21'47"
Peter James Soden Forever Free
Aql 19h39'16" 13d32'55"
Peter James Weir
Cra 18h53'34" -39d55'20"
Peter James Wied
Uma 11h24'13" 47d0'31"
Peter James Wied
Uma 10h0'10" 45d26'4"
Peter Jamie Carter
Uma 12h0'34" 51d42'13"
Peter & Janet Puleo
Lyr 19h13'44" 26d22'31"
Peter & Jean Van Hoomissen
And 23h37'31" 37d35'50"
Peter Jeffrey Marcey
Her 17h17'49" 36d51'18"
Peter Jelinek
Cyg 19h55'46" 33d22'58"
Peter Jennison
Uma 10h9'56" 67d46'47"
Peter John
Psc 23h40'21" 1d14'44"
Peter John Buxbaum
Uma 11h16'58" 54d24'11"
Peter John Consalvi
Gem 7h18'45" 32d39'57"
Peter John Elliott - April 5, 1982
Cap 20h31'30" -25d56'46"
Peter John Frontera III
Her 17h11'36" 34d48'50"
Peter John George
Vir 13h24'42" 12d4'40"
Peter John Kremers
Cnc 9h3'0" 28d33'46"
Peter John Lange
Ori 6h15'36" 10d28'16"
Peter John Oeth
Sgr 18h48'44" -16d41'35"
Peter John Sherman
Vir 12h30'44" 12d4'2"
Peter & Joliene
Umi 14h38'55" 73d36'6"
Peter Jonathan Cresci
Ari 2h11'45" 12d10'48"
Peter Jonathan Kelly
Uma 10h12'20" 62d39'24"
Peter Jones
Hya 14h6'35" -26d50'55"
Peter Jordan
Umi 15h43'10" 71d46'59"
Peter & Jordan Anniversary Star
Nor 16h17'17" -55d24'59"
Peter Joseph Allem
Cep 20h32'37" 62d4'45"
Peter Joseph Andolina
Tau 5h49'56" 14d57'39"
Peter Joseph Lumia
Uma 12h47'4" 53d26'50"
Peter Joseph Lumia, Jr
Uma 8h51'12" 59d41'22"
Peter Joseph Marut, Jr.
Ori 6h14'28" 6d35'54"
Peter Joseph McGowan Jr.
Per 3h26'49" 45d34'4"
Peter Joseph Quandt
Umi 15h57'42" 74d11'3"
Peter Joseph Rivera
Aql 19h37'54" 4d56'20"
Peter Joseph White
Ori 6h20'14" -0d25'48"
Peter & June Donelan
Cru 12h14'21" -59d2'55"
Peter "June" Yusi
Leo 10h15'13" 11d9'4"
Peter Jung jun.
Aqr 23h56'1" -15d34'4"
Peter Jurgeneit
Cep 22h18'7" 62d46'43"
Peter K Clanton
Lib 14h51'26" -6d52'27"
Peter K. McCagg
Ori 5h27'52" 3d25'3"
Peter K. Mikaitis Jr.
Aqr 22h38'53" 0d44'3"
Peter Kanner
Dra 18h50'26" 57d57'47"
Peter Kari
Boo 14h48'42" 23d48'40"
Peter Kastanas
Aqr 23h12'23" -21d49'57"
Peter Kay
Cnc 8h54'46" 31d48'4"
Peter Keith
Aql 19h41'42" -0d4'31"
Peter Kellerman
Cma 6h44'57" -21d32'29"
Peter Kenneth McBride
Ari 2h15'55" 27d38'7"
Peter Kessenich
Uma 8h51'51" 61d44'0"
Peter Kielczewski
Per 3h24'35" 50d22'7"
Peter Kim
Aqr 22h12'44" -14d49'34"
Peter Kimon Petro
Lib 15h51'26" -20d9'3"
Peter King
Ori 5h46'3" 3d1'58"
Peter - King of Stars
Cru 12h25'13" -58d43'59"
Peter Klaassen
Cep 23h15'50" 72d53'52"
Peter Knightbridge
Leo 10h30'55" 7d7'54"
Peter Kölbener
Per 3h52'23" 48d51'19"
Peter Krawack
Ori 5h31'5" 10d8'56"
Peter Krogen
Ori 6h7'48" 2d2'16"
Peter & Krystyna - For Eternity
Cru 12h27'57" -60d56'8"
Peter Kulhanek
Cep 22h8'16" 53d29'7"
Peter Kunimünch
Uma 8h20'44" 66d53'48"
Peter Kunz
Boo 13h57'21" 10d18'2"
Peter L. Tsokanis for an Eternity
Tau 4h32'41" 18d2'19"
Peter L. Winter Smiles On His Carol
Gem 7h49'16" 26d52'27"
Peter Lanell's Star
Lyn 7h47'26" 58d22'45"
Peter Lang
Uma 10h6'50" 68d25'11"
Peter & Lauren Margaritis FIL
Cen 12h14'12" -52d43'37"
Peter Laurinc
Uma 10h37'31" 48d46'5"
Peter Lawrence
Her 18h10'26" 18d22'6"
Peter Lawrence DeSciscio - Dad
Per 3h13'48" 52d26'25"
Peter Le Galloudec
Cyg 21h14'35" 38d13'8"
Peter Lechner
Lyn 7h24'45" 51d15'56"
Peter Leditznig
Uma 9h19'26" 49d27'41"
Peter Leefe Jnr
Cru 12h0'25" -62d29'7"
Peter Lewis
Gem 7h9'56" 32d37'15"
Peter Lewis Kingston Wentz III
Gem 7h34'14" 34d38'32"
Peter Lidster
Ori 5h32'44" -4d56'9"
Peter Linke
Uma 8h20'10" 61d58'9"
Peter Lombardo - (The Gentle Rock)
Psc 1h23'4" 20d28'24"
Peter Long
Uma 11h47'43" 53d6'23"
Peter Lotis 1485am Singing Star
Cnc 8h59'33" 32d22'24"
Peter Lüthi
Ori 5h1'2" 13d33'37"
Peter & Lydia
Uma 9h13'54" 59d56'8"
Peter Lynch
Her 16h55'45" 30d57'40"
Peter Lynskey
Per 4h19'48" 46d25'5"
Peter M. Kelly
Cep 23h28'33" 82d19'42"
Peter M. Kociborski & Jamie M. Pyne
Uma 9h29'33" 51d27'51"
Peter M. Rogers
Dra 16h8'19" 58d32'51"
Peter M. Siciliano
Lib 15h20'24" -13d36'51"
Peter Maas
Psc 23h56'16" -0d7'51"
PETER MAERKLIN THE STAR
Cas 0h54'25" 70d55'33"
Peter Marbacher
Umi 13h29'40" 73d43'40"
Peter Marcus Rees
Ori 5h10'28" 15d4'40"
Peter & Maribel Cruz
Uma 11h31'8" 35d59'29"
Peter & Marilynn Sova
Cyg 19h47'52" 32d20'33"
Peter Marshall
Leo 9h40'57" 32d33'35"
Peter Martens
Uma 10h54'19" 49d50'0"
Peter Marti
Boo 13h40'32" 23d48'8"
Peter Maxian
Cep 20h55'19" 69d14'29"
Peter & Maxine Masia's Himrod Star
Cyg 19h17'50" 51d21'25"
Peter Mazzola
Lyn 6h42'23" 50d48'59"
Peter McRobbie
Lyn 6h52'21" 51d48'18"
Peter Melick
Psc 1h21'20" 7d25'24"
Peter Michael Barrett
Tau 5h46'2" 23d54'57"
Peter Michael Bradley
Cyg 21h58'26" 51d4'10"
Peter Michael Close
Cep 21h45'43" 64d34'23"
Peter "Moe" Moran
Cyg 20h39'23" 44d58'36"
Peter Mullin
Gem 7h56'24" 31d25'28"
Peter N. Benz
Uma 11h28'37" 47d0'46"
Peter N. Geilich
Cep 21h38'22" 59d43'39"
Peter Nancy John Joey
Uma 11h5'33" 49d2'40"
Peter Nathaniel
Psc 0h19'59" 7d49'4"
Peter Neil Murphy
Cep 23h57'7" 70d57'38"
Peter Neptune
Del 20h40'14" 16d33'53"
Peter Nicholas Martin
Uma 11h42'6" 32d11'42"
Peter North Hillyer
Lyr 18h27'18" 35d43'27"
Peter of Ladydown
Uma 12h30'18" 55d3'57"
Peter Olejnik
Uma 9h45'46" 72d50'59"
Peter O'Rourke
Per 4h8'34" 52d27'54"
Peter Ortega
Uma 11h13'7" 65d31'4"
Peter O'Toole
Leo 10h0'30" 16d27'40"
Peter Owen Vodonick
Lib 14h27'45" -19d34'35"
Peter Pan Directed by Katie Lebhar
Peg 23h38'27" 23d26'45"
Peter Pang
Dra 18h56'1" 65d45'57"
Peter Pardee
Uma 9h18'31" 43d5'32"
Peter & Patricia Barnes
Lib 15h13'40" -6d50'7"
Peter Paul Bernstein
Cep 20h50'3" 61d46'50"
Peter Paul Maloney
Sco 16h42'19" -37d5'33"
Peter Paul Mastrangelo
Vir 12h53'53" 11d57'54"
Peter Pearce - Dad
Ori 5h28'51" -0d57'58"
Peter Pellegrino
Psc 1h6'57" 31d33'5"
Peter Peschel
Uma 10h27'0" 41d34'7"
Peter Petersen
Ori 5h55'40" 12d42'40"
Peter Plekinski
Uma 8h35'55" 57d56'10"
Peter "PopPop" Murray
Cam 4h29'37" 76d19'2"
Peter & Pumpkin
Vir 14h20'23" 1d35'45"
Peter R Breiner
Cyg 19h57'53" 37d49'24"
Peter R. Schnable
Uma 9h2'38" 62d1'43"
Peter Rafferty
Per 4h4'4" 44d23'42"
Peter Raymond Bass Jr.
Cep 22h18'17" 72d3'39"
Peter Raymond Bass Sr.
Cep 22h43'58" 71d52'21"
Peter Records
Vir 12h41'26" -6d18'21"
Peter Reed
Uma 9h19'27" 58d2'0"
Peter Reich-Rohrwig
Ori 4h53'14" -0d5'58"
Peter Richard Whitfield
Cen 11h39'4" -38d45'58"
Peter Rison
Ori 5h33'28" -0d38'18"
Peter Robert Griffin
Per 2h56'47" 45d56'7"
Peter Robert Johnson
Aqr 22h23'11" -16d21'35"
Peter Robin
Umi 15h0'46" 75d20'58"
Peter Rodney Harris
Uma 13h30'26" 53d4'43"
Peter Rogers (My Dad)
Cep 22h9'8" 67d43'9"
Peter Rojas Pulido
Vir 13h22'33" 7d4'44"
Peter Roman
Boo 15h3'44" 35d2'46"
Peter Rondorf
Cas 23h35'50" 58d31'57"
Peter S Kulka
Cep 23h48'16" 80d56'54"
Peter S. Vita
Psc 1h28'43" 22d46'40"
Peter & Samantha - Our Shining Star
Cru 12h40'9" -62d23'51"
Peter Sanders
Per 3h39'1" 49d50'37"
Peter Saunders Family Star
Leo 9h36'54" 28d50'30"
Peter Schraml - Beda
Uma 9h35'0" 45d28'23"
Peter Schultes
Uma 10h53'41" 58d35'32"
Peter Scott
Gem 7h26'48" 21d26'58"
Peter Scott Lee my Forever Friend
Ori 6h5'6" 19d10'20"
Peter Sebastian Maier
Leo 9h34'54" 10d42'50"
Peter Semetis
Her 17h24'30" 38d9'43"
Peter Serating
Uma 10h47'23" 62d9'42"
Peter & Sheryl - together forever
Cru 12h20'53" -57d19'50"
Peter Shinn
Boo 14h24'20" 36d14'52"
Peter Shirlaw
Per 3h14'39" 46d36'37"
Peter Sidler
Uma 9h42'22" 57d59'50"
Peter Simon McNeal
Psc 0h15'46" 15d14'41"
Peter Simons
And 0h17'56" 39d37'0"
Peter Sloan Binns
Cap 21h3'15" -16d15'38"
Peter Sousa, C.SS.R.
Ori 5h52'42" -8d35'34"
Peter Stafford Wright - 24.11.1950
Cru 12h41'34" -57d57'38"
Peter Steinbach
Ori 5h19'11" -8d36'50"
Peter Stelling (Little Pants)
Cnv 13h41'44" 34d22'26"
Peter Stephen Foreman - Grandad's Star
Per 2h51'34" 41d35'59"
Peter Stock
Lyr 18h53'21" 32d0'50"
Peter Stoddard
Dra 18h53'56" 66d36'36"
Peter Stokes's Anniversary Star
Cru 12h32'32" -56d34'49"
Peter Storch
Aqr 21h48'58" -1d17'12"
Peter Stuart
Uma 11h4'28" 44d1'51"
Peter Stuchlik
Her 16h54'3" 28d29'18"
Peter Suchmann
Ori 5h29'2" 1d54'16"
Peter & Sylvia Hayman - 9.02.1946
Cru 12h33'17" -59d33'45"
Peter Symmonds Undy
Aql 19h23'47" 3d8'10"
Peter T. Maccarrone
Her 16h34'24" 32d34'56"
Peter Terrio
Cet 0h54'58" -0d50'53"
Peter the Cat From Si Ji Dong Daegu
Leo 10h38'11" 16d30'57"
Peter The Golfer
Uma 11h46'12" 37d54'50"
Peter the Great
Her 17h44'57" 39d3'9"
Peter Thorell
Aqr 21h35'58" -0d41'27"
Peter "Tiger" Renick
Uma 9h42'50" 57d46'29"
Peter Tinturin- Song Writer
Lyr 18h48'11" 36d6'29"
Peter Tolzin
Cep 22h6'48" 56d31'2"
Peter Torquil Bloom
Pav 19h38'33" -66d10'42"
Peter Trapani
Per 4h7'4" 35d4'24"
Peter Traverso
Sco 16h14'36" -10d59'17"
Peter Treichl
Uma 10h51'7" 58d44'29"
Peter Trias
Her 18h18'36" 17d42'52"
Peter Turai 2 28 1960
Psc 1h11'55" 28d10'57"
Peter V. Ravioli
Leo 10h19'45" 18d33'43"
Péter Valenta
Uma 11h1'52" 64d57'51"
Peter Van Gerven
Her 17h6'8" 32d41'45"
Peter Van Schaick
Uma 9h53'18" 48d42'5"
Peter Varone, Jr.
Vir 13h29'7" -4d20'45"
Peter Venn
Cyg 21h14'21" 44d59'9"
Peter Vincent Ferrara
Sgr 18h46'1" -16d39'16"
Peter Vincent Hoffman
Leo 11h18'43" -6d0'29"
Peter W. Juras
Per 3h5'2" 49d19'49"
Peter W. May
Per 3h19'23" 52d7'51"
Peter Wallace Orfila
Aql 19h29'31" 4d16'25"
Peter Walter
Lib 15h45'8" -17d48'10"
Peter Ward
Cep 20h44'6" 60d5'26"
Peter Webb
Cep 22h48'36" 65d46'45"
Peter Weiss
Uma 10h37'39" 64d22'8"
Peter West Frank
Gem 7h45'27" 33d35'41"
Peter Wilding
Aur 5h45'53" 53d32'54"
Peter William Barrows
Leo 11h14'31" 16d1'24"
Peter William Beall
Leo 10h53'2" 17d14'1"
Peter William Behrend Eaton
Dra 16h49'40" 54d48'10"
Peter Williams
Umi 15h35'58" 81d32'18"
Peter Wilson Blum IV
Leo 10h16'21" 15d46'58"
Peter Wolff
Uma 11h6'53" 59d54'21"
Peter Worthington
Aur 5h14'1" 43d5'0"
Peter Wyatt Merkel
Umi 14h1'30" 72d7'19"
Peter XIII. - per aspera ad astra
Uma 9h10'21" 64d56'12"
Peter Y.(The Candyman) Goldych
Per 2h54'53" 54d3'34"
Peter Yammine
Ori 6h6'56" 11d10'18"
Peter Yanity
Lyn 7h37'53" 56d47'7"
Peter - You Touched So Many Lives
Per 4h16'47" 46d52'50"
Peter Zimmermann
Vir 12h1'43" 1d23'43"
peter zöch
Her 18h21'56" 18d9'27"
Peter & Zouha
Crb 16h24'5" 32d1'46"
Peter Zubler (Löli)
Boo 14h30'53" 40d42'31"
Peter, Jr.
Cyg 19h52'52" 51d53'35"
Peter, Kylie, Jaffa & Orion
Lib 16h0'2" -5d42'16"
Peter, Prinz von der Mühle
Lac 22h47'14" 53d37'27"
PeterJennyAbbyMorton
Uma 9h36'41" 58d6'59"
Peterman
Ori 5h58'33" 17d42'58"

peTerri Prime Per 3h1'15" 41d39'24"
Peter's Birthday Wishing Star Psc 0h58'2" 10d45'29"
Peter's Comfort Uma 12h34'49" 53d26'56"
Peter's Goldschätzli Peg 23h2'13" 29d13'15"
Peter's Heart Lyn 6h55'0" 55d44'30"
Peters Jazz-and Bluesland Uma 13h6'6" 57d5'17"
Peter's Star Ari 2h49'36" 11d26'33"
Peter's Star Her 17h52'46" 40d58'21"
Peter's Wish Her 16h13'37" 48d27'55"
Peter-Salvatore Canepa Aqr 22h41'42" -16d32'47"
Petersen-Scotto Psc 1h24'42" 6d26'9"
Peter-Terry Burlong Her 17h4'46" 13d39'0"
Peter-TIYOWEH-Kampka Uma 12h31'22" 58d1'47"
PeterWidgie Aldrich Star Per 2h53'55" 55d35'28"
Pete's Big 5-0 And 0h26'34" 38d10'8"
Pete's Dream Star Amy LeeAnn Hughes Psc 1h39'35" 11d21'15"
Pete's Gtfoh Star Tau 4h39'47" 24d7'24"
Pete's Love, Christine Sgr 17h52'45" -17d32'25"
Pete's Pathway Lib 15h4'19" -17d39'23"
Pete's Sake Tau 4h18'42" 4d31'10"
Pete's Smoke Free Zone Uma 8h26'33" 71d6'7"
Pete's Star Per 3h57'8" 52d16'44"
Pete's Twinkle Per 3h12'5" 45d29'57"
Petesjenn Leo 11h7'18" 20d57'55"
Peteulanus Oculas Her 18h40'50" 24d15'46"
Petey Ori 5h23'20" 6d31'15"
Petey Aur 6h33'10" 34d10'6"
Petey Col 6h19'0" -35d4'8"
Petey Boy Hadwin Umi 12h3'33" 86d1'52"
Petheas Aqr 22h25'59" -2d40'39"
Peti & Muci Tau 3h31'26" 12d22'44"
Petie "10-23-42" Uma 10h47'40" 41d51'44"
Petit Lyn 8h34'32" 34d3'3"
"Petit Coeur" Uma 11h47'56" 37d25'49"
Petit Diable Cap 20h28'44" -24d24'49"
Petit Graton Rose Lib 15h58'47" -6d11'54"
Petit Julie Ori 5h52'11" 3d0'52"
Petit loup Uma 11h33'31" 47d29'15"
Petit Lu And 0h23'27" 22d20'40"
Petit Monstre P. L. Sco 17h19'45" -39d43'39"
Petit poisson d'amour Sgr 18h24'29" -23d47'57"
Petit prince Tri 1h47'51" 31d37'29"
Petit Prince du 8 juillet 1984 And 2h11'32" 50d41'19"
petit Prince Gillmore And 1h55'33" 42d55'10"
Petit trèfle et petit caillou Lib 15h46'31" -7d7'14"
Petit Yéti Lep 5h28'18" -13d15'14"
Petite And 1h9'27" 45d44'25"
Petite Bise Ori 5h32'50" -1d57'39"
Petite Chou Mer Ori 5h0'18" -0d20'8"
Petite Clémence Aqr 22h49'1" -17d11'38"
Petite Fanny Ari 3h6'24" 29d10'43"
Petite Fleur Cnc 8h21'7" 17d18'8"
Petite Princesse Tau 5h45'59" 17d8'55"
petite Princesse Cindy And 1h58'8" 37d49'43"
Petite Victoria Cap 20h27'12" -8d39'28"
PetiteChoseQuiBrille-EricDodin Cas 0h13'21" 55d56'39"
petitefleur2moncoeur Tau 5h59'36" 25d17'12"
Petitou Cyg 20h20'28" 51d57'50"
PetKimSarNic1989 Uma 10h28'7" 42d23'16"
Petkovic Marina Lyn 7h49'18" 46d3'24"
Petner Lyn 6h50'23" 53d59'27"
Petr a Sarka Lib 15h16'17" -5d54'32"
Petr Rubacek Pho 1h22'47" -45d16'50"
Petra Sco 17h19'47" -44d35'40"
Petra Aqr 21h43'27" -2d47'13"
Petra Her 16h45'54" 43d1'28"
Petra Uma 9h29'22" 48d48'40"
Petra Uma 12h5'26" 37d14'17"
Petra Aqr 22h11'57" 1d8'47"
Petra Ajdukovic And 1h54'14" 36d53'3"
Petra Annegret Tau 4h47'27" 26d5'17"
Petra Bonita Vir 13h46'56" 4d11'27"
Petra Carwizi Leo 10h51'2" 22d11'52"
Petra Evalina Niiranen And 1h47'44" 48d6'11"
Petra Giselle Sgr 19h39'34" -16d4'53"
Petra Hecker Uma 10h56'9" 58d34'56"
Petra Heiniger 20.2.80-15.30 Uhr Lmi 10h24'59" 28d33'28"
Petra & Jerry Schield Sco 16h10'3" -10d44'16"
Petra Kusche Uma 11h6'51" 42d11'36"
Petra Marie Scott Nielsen And 0h11'55" 25d5'3"
Petra Marinkovic Umi 15h31'4" 84d53'40"
Petra & Michael Droese Ari 1h48'29" 19d59'11"
Petra Michel Her 18h46'24" 20d2'1"
Petra Najib Rayess And 22h59'11" 48d26'3"
Petra Ockain Uma 10h30'23" 49d23'31"
Petra Paula Zammit Cnc 8h20'5" 20d24'58"
Petra Pejchalova Cap 20h30'45" -26d42'9"
Petra + Peter Boo 14h47'9" 16d29'51"
Petra & Roger Cep 20h41'53" 71d50'34"
Petra Rumiskova Halgas Aql 19h11'25" 2d1'43"
Petra Scheibenhofer Dra 18h51'54" 59d34'4"
Petra Schulz Uma 10h4'49" 42d1'18"
Petra Spring Cep 22h19'9" 68d19'33"
Petra und Lukas Stefan Uma 11h19'7" 71d38'27"
Petra Vertes Aql 19h42'44" 3d37'58"
Petrakovitz Silver 25 Her 17h12'51" 19d8'3"
Petraq Vasil Cnc 8h51'47" 30d43'32"
Petras Juodikis Pup 7h42'19" -14d34'30"
Petra`s und Werner`s Stern Uma 12h28'35" 59d53'45"
Petrea Collett Aql 19h30'8" 15d23'48"
Petri Uma 8h58'51" 47d16'37"
Petri Lehtinen Dra 17h8'0" 52d49'31"
Petrie Ori 5h54'14" 20d38'36"
Petrina Supler Lyn 7h11'35" 52d48'55"
Petrisha Sun & Chris Lim Cap 21h32'9" -19d11'29"
Petrita Simona Sgr 17h52'43" -16d38'29"
Petro & Eleni Rebakos Cyg 20h11'12" 43d24'57"
Petronella Uma 9h45'9" 69d54'49"
Petronella Cru 12h19'32" -63d15'44"
Petronella Maria Kersten Ori 5h41'48" -2d10'41"
Petronius Dra 17h39'17" 57d15'27"
Petros Daskalopoulos Uma 10h15'21" 58d48'15"
Petros Hastings Flagg Umi 15h37'10" 71d18'16"
Petros & Rica for another 100 years Lib 15h3'27" -0d45'12"
Petrounka Stoyanova And 0h27'49" 39d11'6"
Petru od Ivane 19.05.2007. Per 3h58'55" 39d47'55"
Petruch Andrea "Habibi" Gem 6h40'33" 16d4'36"
Petruncola Srp 18h33'36" -0d6'58"
Petrus Cru 12h30'41" -59d7'24"
Petrus 27 Sge 19h44'46" 17d50'9"
Petry Uma 10h33'3" 43d0'15"
Pettal Cas 0h56'13" 65d22'21"
Petter Ivo Theodor Løvold Umi 16h38'22" 75d55'25"
Petter Joseph Hohman Gem 6h46'45" 12d55'34"
Pettinger Foundation Crb 16h20'50" 29d11'19"
Petty Officer Barry Tate Ori 6h25'5" 14d38'21"
Petumia Cam 4h22'51" 72d28'14"
Petunia Tau 5h6'54" 27d8'29"
Petunia Lelani Lelani Sibson Her 18h0'47" 24d57'17"
Petus Lib 14h48'35" -0d47'32"
Petya Cyg 20h23'33" 33d17'49"
Petya Dikova Leo 9h34'55" 29d59'34"
Petzbär Uma 9h19'2" 59d41'52"
Petzold, Heiner Ori 6h11'32" 13d27'14"
Peu de fleur de ciel Lilly E. Houle And 0h11'44" 33d6'36"
Peydon Grace Lyn 9h4'1" 42d33'1"
Peyton Ari 2h35'21" 20d21'59"
Peyton Lyn 8h29'30" 54d13'45"
Peyton Cma 7h8'5" -25d16'35"
Peyton A. Rossomando Lyn 7h6'28" 52d17'19"
Peyton Alan Ellis Aur 5h13'54" 46d58'1"
Peyton Alan Nunnelly Dra 17h33'38" 57d13'31"
Peyton Alese Mikowski Umi 16h3'13" 82d41'57"
Peyton Alese Mikowski Crb 16h16'15" 39d6'0"
Peyton Alexander Smith Aqr 22h0'18" -16d43'46"
Peyton Annabelle Gem 6h8'16" 25d11'2"
Peyton Arcieri Gem 7h10'49" 32d38'4"
Peyton Benjamin Mahn Aqr 22h49'42" -8d25'15"
Peyton Blythe Sibert Psc 0h10'49" 7d23'37"
Peyton Breanne Sgr 18h32'43" -16d53'35"
Peyton Brooks Strickland Uma 11h36'36" 50d59'6"
Peyton Buchanan Cnc 8h27'16" 14d35'26"
Peyton Casey Lund Uma 10h36'38" 56d33'22"
Peyton Catherine Scharf And 1h15'7" 45d26'51"
Peyton Chloe Cap 20h14'41" -25d33'18"
Peyton Elisabeth Vrbas (tater-bug) Sco 16h8'59" -13d36'33"
Peyton Elizabeth Priestman Ori 5h34'56" 4d35'2"
Peyton Emery Aydelotte Cyg 20h53'12" 45d32'31"
Peyton Goldston Tau 3h32'0" 1d4'32"
Peyton Grymes, III - GaMPI Star Dra 18h46'13" 53d58'53"
Peyton Hawthorne May Del 20h37'5" 20d34'25"
Peyton Hrnjez Burrell Cap 21h39'43" -11d22'29"
Peyton Imani Thomas Her 17h52'38" 23d54'14"
Peyton Jameson Rice Per 1h53'8" 48d12'26"
Peyton & Jenna Umi 13h29'26" 71d9'20"
Peyton Jo Umi 14h39'23" 75d12'27"
Peyton Jordyn Gem 6h44'29" 32d4'51"
Peyton K. Brown Uma 10h8'12" 44d55'35"
Peyton Lawson Broome Uma 11h5'14" 47d15'50"
Peyton Lynn Burchett Lyr 18h54'20" 36d1'27"
Peyton M. Lee Uma 8h52'14" 47d3'43"
Peyton Mac Crb 15h28'58" 26d11'33"
Peyton Matthew Juhas Uma 9h6'8" 61d45'8"
Peyton Michael Driver Sgr 18h57'2" -31d8'21"
Peyton Miller Bernell Per 2h51'5" 53d3'44"
Peyton N. Daley Ori 5h13'46" 6d47'3"
Peyton Neal Segraves Her 17h36'32" 17d7'2"
Peyton Nicole El-Zoghby Psc 0h39'15" 5d0'17"
Peyton Olivia Lmi 10h34'3" 34d9'36"
Peyton Reed Nir Uma 10h19'59" 39d44'20"
Peyton Rye Cyg 20h43'44" 38d1'40"
Peyton S. Flowers Lep 5h14'41" -12d37'5"
Peyton Scott Stewart Uma 11h23'35" 29d12'37"
Peyton Taylor Lyr 18h49'55" 43d30'8"
Peyton William Chiriatti Per 3h36'38" 51d31'4"
PeytonMichael Eri 3h56'20" -0d6'45"
Peyton's Haven Ari 2h52'3" 11d37'10"
Peytyn Louise Cap 20h31'34" -15d50'48"
Pezzetta Stefania Aql 19h8'54" 7d48'33"
P.F. Uma 12h45'32" 54d39'48"
P.F. Walker Aur 5h35'37" 37d39'0"
PFC Brandon C. Krakowiak Tau 5h32'34" 26d12'44"
PFC Christopher Reed Kilpatrick Boo 14h18'16" 17d9'41"
Pfc. Clint Dowell Her 16h57'13" 30d23'38"
PFC John W. Dearing Cyg 20h52'55" 36d39'40"
PFC Scott Paul Vallely Psc 1h11'8" 12d8'19"
PFC Tyler Ryan MacKenzie "Gigantor" Ori 5h36'6" -1d33'1"
Pfc Tyler Ryan "Monkey" MacKenzie Ori 5h13'22" 15d4'51"
Pfefferkorn, Marianne Ori 6h7'6" 10d12'12"
Pfeifer Tritton Aqr 23h0'53" -8d30'41"
P.F.Kimball Uma 13h22'32" 56d42'57"
Pfleeger Lib 15h21'5" -6d39'35"
Pflugmacher, Helmut Uma 10h47'0" 43d30'47"
Pfüdi And 1h12'33" 47d59'18"
Pfundmeir, Klaus Uma 9h37'33" 49d41'51"
P.G.A. New Jersey Section Her 16h27'52" 10d2'34"
P.G.Dunn Lyn 7h34'18" 52d1'54"
P.G.L.K. Ori 5h25'14" -7d30'50"
Phabiana 092906 Cap 20h24'29" -10d15'54"
Phaedonos Pho 23h52'50" -42d47'19"
Phaedra Sco 16h45'9" -38d5'23"
Phaedra Ann Ori 5h33'10" -0d42'16"
Phaedrus-Major Lyn 6h50'27" 57d5'41"
Phaelan Koock Lib 15h2'29" -4d47'18"
Phaelyn Marie-Helena Kotuby Lyn 7h47'4" 36d15'3"
Phaidra Cates Psc 22h56'24" -0d3'9"
Phailey Cam 4h10'59" 71d6'8"
Phaisuwatana Cap 20h10'18" -13d56'1"
Phalon Lee Russ Aqr 21h44'5" -2d46'17"
Phananza Uma 9h28'1" 42d14'50"
Phang Wah Gem 6h8'47" 25d27'36"
Phanvilay Davis Aqr 22h10'47" -8d5'13"
Pharaol Patrick Meehan Umi 17h39'59" 82d16'14"
Pharbia Clark Umi 14h4'52" 70d59'45"
phaymis Gem 7h42'38" 24d13'34"
PHD Her 17h3'38" 32d5'43"
Phebe Lily Schulz Cnc 8h40'45" 31d38'49"
PheenieFace dadadada Lyn 9h4'25" 33d20'21"
Phelan Family Uma 9h15'58" 54d15'38"
Phelan Robert Perry Birthday Star Sco 16h16'30" -9d26'26"
Phelmon DeSoto Saunders Tau 3h50'32" 9d24'21"
Phenomenal Phillips Crb 15h40'39" 27d33'15"
Pheobe Webb Cyg 19h52'53" 37d43'4"
Pherf Lyn 7h48'19" 44d56'48"
Pherin Aur 4h57'30" 36d1'37"
Pherin Clansalmo Umi 13h24'10" 74d47'46"
Pherty Cnc 8h41'7" 19d26'49"
Pheva Psc 1h8'0" 27d2'7"
PHH Stamm Eri 4h15'48" -33d44'40"
Phi Delta Ori 5h46'21" 5d11'25"
Phi Ho "Special Star" Gem 6h33'45" 21d0'25"
Phi Tieskoetter Per 2h36'29" 52d31'42"
Phia-rin Bellus Aeternus Eternus Gem 7h24'59" 26d27'44"
Phil Ari 2h53'34" 28d33'51"
Phil Peg 22h21'40" 16d35'29"
Phil Her 16h18'56" 6d49'26"
Phil Per 4h4'49" 33d57'25"
Phil Per 3h42'17" 34d25'50"
Phil Lyn 8h30'50" 39d34'42"
Phil Aqr 21h9'40" -12d43'47"
Phil and Bea Tau 5h2'40" 23d54'55"
Phil and Elaine Bullock Uma 8h38'19" 54d16'25"
Phil and Erin Grover Leo 11h9'57" 9d33'34"
Phil and Jan Colla And 1h30'34" 34d41'39"
Phil and Jan's Piece of Heaven Cyg 20h33'44" 54d16'4"
Phil and Tammie Forever Cyg 21h23'40" 45d28'4"
Phil Anderson Her 16h39'31" 6d0'21"
Phil & Arlene Epstein Cyg 21h17'20" 45d3'20"
Phil Arthur Cyg 19h53'33" 39d9'13"
Phil Beck Per 2h46'14" 53d37'57"
Phil Bedel Sco 16h41'12" -30d39'22"
Phil bert's Shiny Star Her 18h32'7" 15d44'20"
Phil C. Patire (The First Knight) Cep 22h43'43" 69d18'30"
Phil Cerrito Sgr 17h58'48" -16d58'33"
Phil & Charle Haverland's Star Gem 7h49'17" 23d59'8"
Phil & Chris Wulliamoz Cep 22h32'0" 61d53'28"
Phil Clothier Aqr 22h17'37" -13d1'3"
Phil Cody Uma 9h46'37" 49d36'44"
Phil Cohen Uma 9h34'54" 57d40'52"
Phil Davis Lyn 9h11'35" 37d0'32"
Phil Doherty (Dad) Ori 6h13'43" 1d51'10"
Phil Downing Leo 10h53'22" 10d37'45"
Phil & Emy 23042005 Ori 6h18'46" 15d8'55"
Phil et Sara Vir 13h15'2" -13d27'43"
Phil Faust Umi 16h31'0" 77d54'24"
Phil Gaha Her 16h18'20" 8d40'30"
Phil Grant Nor 16h7'18" -44d2'44"
Phil Hartman Her 18h16'31" 22d45'46"
Phil & Irene Klosty Cyg 19h59'6" 45d45'9"
Phil & Jenn Perrotte 10/08/2005 Lib 14h41'9" -17d42'38"
Phil & Jill DeVita Peg 21h41'1" 21d6'18"
Phil & Kay Secrist Uma 10h9'45" 47d7'33"
Phil Lantz Uma 9h10'6" 57d18'52"
Phil Lippincott Per 1h43'32" 54d37'47"
Phil Lupo Cep 22h53'45" 74d5'2"
Phil (Madhead) Wisewould Eri 4h8'7" -32d0'27"
Phil mon Papou que j'aime, Ange Psc 22h54'33" 5d9'1"
Phil Morrison Cra 19h4'9" -37d58'24"
Phil Park Lac 22h15'56" 42d45'49"
Phil Parker Cep 22h31'24" 63d42'38"
Phil & Paulene Ori 5h17'38" -4d35'31"
Phil Petro Ori 5h54'50" 18d40'55"
Phil Phillips Tau 5h32'46" 25d32'31"
Phil Randall Jr Umi 16h24'16" 76d11'28"
Phil & Senay - True Love Cyg 21h45'46" 54d51'31"
Phil Shire Per 3h10'27" 54d45'21"
Phil Spano, Sr. 80th Aqr 22h4'28" -11d47'9"
Phil Turner Ori 5h32'36" 0d40'2"
Phil Wells Vir 13h21'3" 1d37'36"
Phil, Maxx, Adriana Cyg 19h41'23" 39d55'0"
Philadelphia Santee Black Cyg 20h32'48" 36d32'11"
Philanda Tuckabob Moon Per 2h43'41" 40d17'48"
Philanise Uma 8h44'41" 72d29'2"
Philanita And 2h15'43" 49d10'31"
Phildog White Sgr 18h51'9" -20d23'40"
Philemon Mwezi Uma 10h28'18" 44d12'54"
Phileo Leo 9h42'56" 21d43'42"
phileo agape Uma 13h51'8" 55d57'8"
Phileris Lyn 6h47'52" 54d47'57"
Phililou Lib 16h0'7" -15d30'2"
Philip Cep 21h21'28" 77d46'33"
Philip Uma 11h56'13" 53d23'47"
Philip Ori 6h16'11" 10d4'31"
Philip A. Farinacci Leo 9h31'32" 19d18'34"
Philip A Hahn Cyg 19h39'50" 38d35'30"
Philip A. Tumminia Car 10h54'5" -59d2'38"
Philip Alexander Terry Aqr 23h50'8" -17d56'47"
Philip Alexandre Sgr 17h51'49" -19d37'4"
Philip Allen Smith Uma 10h19'32" 48d21'13"
Philip and Courtney Ori 5h23'46" 1d39'48"
Philip and Diane La Bonté Cyg 21h22'37" 50d28'52"
Philip and Kim Robinson Leo 11h18'20" -5d51'43"
Philip and Susan Lee Boyle. Cyg 21h55'44" 48d3'3"
Philip and Tessica Umi 15h11'4" 69d35'47"
Philip & Andrew Gordon Psc 22h55'12" 2d39'6"
Philip & Anita Lawler 11/29/1945 Cyg 22h1'26" 50d33'4"
Philip Anthony Davis Cyg 21h46'32" 40d31'14"
Philip Anthony Meaney Per 2h52'41" 50d51'6"
Philip Benedict Joy Sr. (Pop Joy) Uma 11h36'0" 31d9'41"
Philip Bredin Her 16h13'47" 46d55'23"
Philip C. Gilardi Love Forever Star Ori 5h59'25" 1d55'21"
Philip C. Rampy Ori 6h16'3" -0d4'50"
Philip & Cara Basvic Uma 9h32'17" 55d38'43"
Philip Carlton Gem 6h42'58" 19d59'20"
Philip Charles Kostak Gem 7h57'10" 19d56'42"
Philip Charles Stone Cep 22h53'24" 60d30'44"
Philip Colacino Per 2h39'20" 55d58'14"
Philip Corboy Leo 11h26'12" 14d10'59"
Philip Cote Cep 22h34'44" 73d14'10"
Philip DeVliegher Cap 20h39'21" -24d43'13"
Philip DiTullio Per 2h48'46" 52d31'35"
Philip Donald Cotty Dra 19h10'40" 66d52'54"
Philip E. Benton, Jr. And 23h47'16" 45d42'56"
Philip Early Dra 18h30'6" 63d11'30"
Philip Edmond Broussard Aqr 22h10'50" 1d12'34"
Philip Edward Davy Her 16h29'1" 12d50'30"
Philip F. Adams Aur 5h55'6" 42d9'51"
Philip F. Gray, Jr. Per 2h22'48" 55d17'48"
Philip F. Lang Psc 1h19'8" 31d54'18"
Philip F Ragusa Lmi 10h44'38" 28d47'4"
Philip G. Elie, Sr. Cep 22h33'52" 76d49'47"
Philip Galyen's 50th Birthday Boo 15h22'58" 47d45'50"
Philip Gary Goldstein Sgr 18h31'15" -24d3'31"
Philip Gene Aviles Gem 6h50'28" 33d26'13"
Philip George Tomak Lib 15h3'32" -2d29'4"
Philip Geroe Webster, Jr. Leo 10h52'59" 11d0'7"
Philip Glass Ori 6h5'28" 18d37'14"
Philip Goodman Her 16h15'7" 47d45'47"
Philip Grasso Her 17h41'21" 20d54'54"
Philip Hall 05081947 Cru 12h28'30" -60d6'28"
Philip Harrington Per 3h13'54" 46d35'0"
Philip Hausvater Her 16h33'56" 44d57'58"
Philip Heald Anniversary Star Cep 3h52'33" 81d25'38"
Philip Howard Per 4h49'49" 44d23'11"
Philip Howard Her 16h4'49" 17d48'40"
Philip Hungate Ronzone Lyn 7h31'42" 56d29'14"
Philip J. DeLoatch Lyn 6h22'47" 56d54'14"
Philip J. Martin Uma 9h55'3" 47d10'1"
Philip Jackson Cap 20h52'50" -25d17'4"
Philip James And 23h7'19" 40d6'53"
Philip James Ori 6h19'38" 13d54'55"
Philip James Duncan Leo 10h26'32" 21d10'53"
Philip Jang Uma 8h17'15" 68d26'40"
Philip John Fishel Oph 17h55'3" -0d20'35"
Philip John Hawthorn Leo 10h13'58" 24d51'29"
Philip John McConnell Per 2h48'26" 52d41'13"
Philip Jones "Jazzbones" Ori 5h36'26" 10d59'39"
Philip Jordan Vir 15h10'10" 6d59'44"
Philip Josef Dworzanski Aqr 22h18'37" 2d16'33"
Philip Joseph McVey Per 2h25'45" 54d40'7"
Philip & Jude Mueller Sge 20h3'21" 17d30'20"
Philip K Vollmoeller Ori 5h31'17" 1d33'36"
Philip Karr Uma 9h50'19" 51d16'45"
Philip Kelley Gem 7h41'58" 32d40'43"
Philip Klatte Sgr 18h5'6" -23d3'37"
Philip L. Miller Her 18h2'35" 21d40'36"

Philip L. Perkins
And 1h6'7" 45d55'50"
Philip L. Williams
Her 18h46'56" 20d20'23"
Philip Lee Chaleff
Ori 4h43'53" 1d36'22"
Philip Leon Baker
Psc 1h17'46" 25d20'35"
Philip Linerode
Cep 20h51'42" 68d57'5"
Philip Link Hood
Aqr 20h44'27" -10d53'49"
Philip Lowe 30
Cep 20h29'53" 61d48'59"
Philip Luiz Cammarata
Per 2h56'16" 46d58'14"
Philip M J Faulkner
Uma 8h21'7" 63d17'8"
Philip Marc Christensen
Cnc 8h48'56" 31d17'54"
Philip Marcael
Cep 21h17'58" 70d13'53"
Philip Mario DuPont
Aql 19h5'30" -0d44'40"
Philip Martin Unger
Cap 21h6'22" -16d35'30"
Philip Matthew Rebbitt
Ori 6h17'3" 5d47'17"
Philip Medina
Aqr 21h48'47" 0d15'53"
Philip Merlin
Boo 15h26'43" 49d32'42"
Philip Moroney
Cap 21h4'38" -24d41'44"
Philip Nathaniel Bradley - Pip's Star
Ori 5h17'37" 9d6'9"
Philip Norman Croteau
Uma 10h20'50" 56d7'32"
Philip O'Brien
Tau 4h27'3" 10d51'37"
Philip O'Dwyer
Gem 6h53'49" 31d50'18"
Philip Paul Lawrence
Uma 9h56'40" 43d21'15"
Philip Paul Trevor Maxey
Per 3h22'57" 35d27'19"
Philip & Pauline McRory
Cyg 21h16'58" 47d12'46"
Philip & Penelope Feldberg
Uma 10h43'19" 41d1'56"
Philip & Pilar Pees Wedding Star
Uma 10h30'6" 55d0'11"
Philip R Hall III
Lib 14h51'16" -6d26'24"
Philip R. Papa
Gem 7h45'9" 23d49'37"
Philip R. Procida
Lib 14h54'22" -1d59'9"
Philip R. Seaver
Ori 6h22'13" 14d41'14"
Philip Rafferty IV
Cnc 8h59'21" 31d25'31"
Philip Rico Savarino
Dra 17h52'58" 65d19'41"
Philip & Risa Ross
Lyr 18h47'26" 35d42'48"
Philip Robinson Quinn
Her 18h51'26" 24d0'25"
Philip Roxas
Psc 0h48'59" 4d22'47"
Philip Rubis
Gem 6h17'49" 26d9'29"
Philip Ryan
Her 17h49'42" 49d36'57"
Philip Saverio Terio
Cnc 8h11'16" 16d4'21"
Philip Scarfi
Uma 8h11'10" 64d4'49"
Philip Scott
Per 3h11'7" 55d30'30"
Philip *Scot-T* Lucent
Per 4h36'29" 40d39'31"
Philip Scotto
Lib 14h49'58" -15d56'21"
Philip Slater
Uma 11h58'28" 47d48'37"
Philip Smith
Dra 18h37'47" 59d33'46"
Philip Staley
Cen 14h19'2" -58d29'51"
Philip T. Daniels
And 1h37'20" 40d53'14"
Philip T. Silvia
Umi 16h0'55" 80d22'24"
Philip Tamberino
Aur 5h48'46" 55d16'30"
Philip & Tessa Together Forever
Per 4h45'53" 44d34'18"
Philip The Glowing Star of my Life.
Ori 6h13'59" 15d31'5"
Philip Tyler Skop-Bruno
Gem 6h44'59" 34d36'28"
Philip Tyler Zabriskie
Psc 0h32'56" 5d3'14"
Philip Ventura
Umi 15h21'41" 75d9'30"
Philip Vrtkovski's Star 14/11/2001
Cru 12h38'42" -61d22'31"
Philip W Davis
Uma 11h29'10" 51d16'28"
Philip Wells Gianfortoni
Gem 7h41'28" 24d12'45"
Philip & Wendy Sebbens For Eternity
Cru 12h36'55" -59d50'29"
Philip William Prifold Jr.
Lyn 9h16'15" 33d44'55"
Philip Worley Harvey
And 23h41'34" 48d14'42"
Philip Yeo
Sco 17h53'59" -41d52'24"
Philip Yuki
Her 16h43'26" 25d34'40"
Philip,Frances,Bingo&Sash a Usaitis
Uma 9h26'33" 67d39'21"
Philipp
Uma 12h57'40" 62d6'54"
Philipp
Uma 10h40'35" 58d59'10"
Philipp & Anita Boos
Umi 17h45'9" 81d54'34"
Philipp Benjamin Thomas
Uma 9h4'31" 64d54'52"
Philipp Erni, 27.06.1996
Umi 9h23'31" 86d17'14"
Philipp Kempken
Uma 10h39'42" 40d36'14"
Philipp Marco Kern
Ori 5h9'46" 6d29'29"
Philipp - Paul
Ori 5h10'2" 15d37'46"
Philipp Sidler
Crb 16h20'36" 31d16'12"
Philipp Tullius
Uma 11h6'14" 41d16'30"
Philippa
Tau 4h7'28" 6d4'49"
Philippa Grace
And 2h19'28" 38d40'48"
Philippa Jayne Hurst
Cas 0h3'15" 50d26'0"
Philippa Nashe - The Star of P-Nash
Peg 21h16'21" 19d11'15"
Philippe
Peg 22h13'4" 18d18'40"
Philippe
And 2h3'21" 37d43'13"
Philippe
Cep 22h51'22" 71d8'23"
Philippe
Cas 23h7'58" 58d53'43"
Philippe
Tau 3h24'51" 15d22'10"
Philippe
Sgr 18h32'39" -33d6'47"
Philippe Adam
Ori 6h4'48" 13d37'18"
Philippe Carcenac de Torne
Her 17h43'47" 14d24'37"
Philippe Citroën
Ori 5h51'8" 12d28'31"
Philippe Citroën
Uma 11h48'16" 47d12'36"
Philippe et Evelyne Beaume
Uma 10h51'23" 48d13'3"
Philippe et Marie
Aur 6h24'1" 32d14'56"
Philippe" Fillie" Eudaric
Sco 16h9'15" -19d26'27"
Philippe Francois Pettavel
Uma 10h15'26" 58d53'42"
Philippe Gagné
Uma 9h53'17" 45d45'16"
Philippe Gonod
Sgr 19h15'10" -22d14'20"
PHILIPPE GUELAT
Ori 6h14'45" 20d40'42"
Philippe Guerin
Lib 15h22'12" -16d17'55"
Philippe Guidet
Ori 6h11'10" 8d6'48"
Philippe Lechantre
Cas 1h1'31" 63d36'36"
Philippe Lescornez
Cep 22h13'2" 80d49'23"
Philippe Maso 09/10/1954
Lib 14h52'54" -23d58'35"
Philippe Michaud
Ori 6h2'20" 11d1'37"
Philippe Nellissen
Sgr 18h7'50" -32d30'53"
Philippe Neves Laetitia-Cyril
Cas 0h16'29" 54d41'15"
Philippe Rosier
Dra 18h7'31" 76d27'30"
Philippe Sauval
Aqr 21h23'43" -0d17'0"
Philippe Senelonge
And 1h27'7" 46d9'45"
Philippe Starck
Aqr 22h44'34" -9d48'4"
Philippe Thenard
And 0h27'57" 22d10'9"
Philippi, Gabriele
Uma 10h10'24" 42d57'32"
Philipps Stern
Ori 6h13'59" 15d50'42"
Philips
And 2h18'30" 38d20'18"
Philip's Lucky Star
Ori 5h23'46" 1d9'35"
Philip's Shining Light
Per 2h20'58" 57d18'27"
Philipstar
Per 3h37'0" 50d56'26"
Phillip
Cnc 8h44'19" 13d57'34"
Phillip
Crb 16h15'46" 28d4'35"
Phillip
Lib 15h3'6" -2d7'41"
Phillip A
Lib 15h21'1" -27d14'4"
Phillip Alan Bentley
Cnc 8h51'12" 14d22'10"
Phillip Alan Riehl
Ari 2h10'55" 26d19'33"
Phillip Alexander
Cnc 8h46'8" 16d27'41"
Phillip Alexander Lester
Cep 1h43'46" 80d30'58"
Phillip and Anita Shear
Tri 2h22'1" 34d55'10"
Phillip And Anna Camp
Col 5h10'5" -28d25'3"
Phillip and Caroline Kennedy
Dra 18h20'24" 49d1'2"
Phillip and Diane Nash
Crb 16h4'4" 32d47'59"
Phillip and Vanessa
Tau 4h8'30" 18d2'23"
Phillip Andrew Adkins
Vir 13h19'18" -22d2'16"
Phillip Blake Thomas
Umi 16h6'15" 81d35'48"
Phillip Boyes
Cam 11h32'49" 81d53'45"
Phillip Brandon Williams
Uma 10h25'25" 52d50'38"
Phillip Brusca
Her 17h26'24" 34d33'28"
Phillip Butwinski
Aql 19h49'49" 15d43'35"
Phillip C. Warner
Vir 15h7'59" 4d27'36"
Phillip Coghill
Tau 4h25'48" 18d16'42"
Phillip Cohen
Ari 1h53'46" 21d23'43"
Phillip Cross - an eternal love
Cra 18h58'22" -45d11'45"
Phillip Curtis Best
Her 16h32'8" 47d14'52"
Phillip D. Bryant
Boo 14h44'20" 54d10'54"
Phillip D. Quick
Gem 6h27'11" 26d8'11"
Phillip Daniel Bray
Vir 13h11'11" 7d23'36"
Phillip & Deleena
Cyg 20h19'2" 33d53'22"
Phillip Edward Wright
Tau 4h39'42" 19d35'1"
Phillip Glen Bailey
Ori 6h18'35" 13d32'57"
Phillip H. Rezanka
Per 3h13'18" 55d41'5"
Phillip Howden
And 1h26'11" 40d24'57"
Phillip Jack Reynolds
Ori 5h34'42" 2d53'3"
Phillip James Anthony
Her 17h22'42" 34d20'48"
Phillip James Harmon
Tau 5h33'30" 18d28'44"
Phillip James Rue-Hart
Uma 11h22'59" 44d8'32"
Phillip Jerome Ridel
Ori 6h12'52" 19d19'46"
Phillip Jomidad
Cap 21h4'39" -16d2'29"
Phillip Joseph Chapman
Lib 15h6'13" -0d39'37"
Phillip Joseph Harvey
Uma 10h8'19" 64d43'47"
Phillip "Juicy" Burts, Jr.
Cyg 20h57'13" 31d19'39"
Phillip L. Blair
Sco 16h25'28" -33d4'37"
Phillip & Latrice Washington
Cyg 21h38'47" 44d34'8"
Phillip Lockwood
Per 3h5'3" 49d1'15"
Phillip Lorden Mckenna
Gem 7h8'2" 34d50'17"
Phillip Lynn Moyer
Cyg 19h55'8" 43d52'8"
Phillip M. McCracken
Her 18h33'26" 21d57'2"
Phillip Mark Taylor
Pho 0h58'19" -48d45'0"
Phillip Mastropolo
Aqr 22h19'57" -7d47'19"
Phillip Mauro
Leo 10h31'33" 8d9'23"
Phillip & Maxwell Pepin
Tau 5h52'4" 15d17'14"
Phillip McLaughlin
Uma 9h15'45" 62d46'58"
Phillip Micheal Collier
Lyn 7h5'55" 44d56'58"
Phillip N. Gold
Her 16h34'25" 39d26'24"
Phillip our Angel
Umi 16h26'52" 81d47'57"
Phillip P
Cet 0h43'38" -1d28'48"
Phillip & Pam Sebastian
Cyg 21h56'32" 53d5'35"
Phillip Parry
Ori 5h29'12" -5d31'51"
Phillip R Peters
Eri 3h43'37" -0d17'27"
Phillip Rae Beach
Vir 12h0'24" 0d29'27"
Phillip Sarantos #1 Dad
Per 3h48'18" 33d33'9"
Phillip Sontag
Srp 17h58'36" -0d28'32"
Phillip Stavros Ferderigos
Her 16h51'11" 23d14'23"
Phillip Timmer Wolf
Psc 1h32'57" 10d26'35"
Phillip W. Evanzo
Aqr 22h8'10" 2d19'33"
Phillip W. Smith, Jr.
Umi 14h22'49" 89d10'53"
Phillip Wayne Rice "P.W."
Aqr 22h4'19" 0d14'25"
Phillip Willis
Aqr 23h21'9" -17d58'15"
Phillip y Ale por toda eternidad
And 0h43'50" 22d25'47"
Phillip Zenon Therrien
Aqr 21h27'50" 1d12'49"
Phillip, Joachim
Ori 6h7'42" 8d48'24"
Phillipa Kayleigh Simpson
Cas 0h51'35" 63d28'23"
Phillippe Abanil Baburam
Uma 11h34'41" 63d18'34"
Phillip's Aurora
Aql 19h10'13" -0d15'20"
Phillip's Miracle Star
Leo 10h56'11" 13d24'41"
Philly Blue Eyes 12-14
Lib 14h25'35" -17d53'6"
Philly D.
Psc 0h43'19" 7d11'37"
Philly Kelly
And 1h39'46" 40d39'47"
Philly Nxai
Tau 3h43'44" 9d39'23"
Philly Phill
Lyn 8h20'22" 54d27'4"
PhilnPatty
Cyg 20h39'58" 53d27'13"
Philo Kai
Lib 15h26'35" -10d28'20"
Philomena
Cnc 8h39'20" 29d12'42"
Philomena Amodio
Uma 8h50'3" 48d33'45"
Philomena Caputo
Sgr 19h46'36" -13d43'51"
Philomena RCC
Gem 6h26'9" 22d30'55"
Philomena Theresa Fredericks xoxoxo
Gem 6h33'11" 27d33'43"
Philomène
Peg 23h2'7" 16d13'36"
Philopotomus
Ari 2h6'19" 13d6'49"
Phil's Firehouse
Cyg 20h54'46" 32d23'53"
Phil's Passion
Lib 15h9'29" -5d3'40"
Phil's Phantasy 60th Birthday
Gem 7h7'52" 17d38'16"
Philstar
Sco 17h32'36" -38d24'57"
Phim
Cas 1h30'41" 74d3'41"
Phimalion
Cas 0h58'41" 55d2'36"
Phini4us
Aqr 23h23'55" -13d53'17"
phinickx
Uma 9h26'16" 47d59'7"
Phinnaeus Walter Moder
Her 18h14'37" 25d38'40"
Phinneas
Uma 10h17'33" 55d50'53"
Phinneas K. Phry
Uma 11h27'3" 66d52'26"
Phirenova
Lyn 7h49'30" 56d59'22"
Phoebe
Lib 15h33'21" -6d14'5"
Phoebe Alexandra Davies
And 0h58'21" 46d17'38"
Phoebe Alexandra Staab
And 23h39'5" 37d52'39"
Phoebe Allyn Lane
Psc 0h32'22" 3d17'20"
Phoebe Anne Albania Delos Reyes
Vir 12h40'7" 10d43'21"
Phoebe Asam
Psc 1h7'58" 19d9'42"
Phoebe Baylis Luceat Lux Vestra
Vir 11h55'13" -0d9'7"
Phoebe Catherine Andrews
And 23h5'47" 51d27'9"
Phoebe Christison - Beloved Monster
Dor 5h6'47" -68d9'26"
Phoebe Elizabeth
Pho 0h39'58" -50d26'55"
Phoebe Elizabeth Grace Parr
And 2h21'17" 45d36'27"
Phoebe F.P. Chan
Sgr 18h58'34" -34d59'7"
Phoebe Georgia
And 1h56'57" 37d15'49"
Phoebe Georgina
Lmi 10h41'30" 25d3'11"
Phoebe Grace Schneider
And 23h1'57" 42d50'39"
Phoebe Jax
And 2h38'22" 45d54'15"
Phoebe & Jeremy Ho-Price
Cyg 20h46'3" 36d9'47"
Phoebe Katherine Katz
Ari 3h14'19" 27d59'45"
Phoebe Kathryn Costalos
Dra 16h3'4" 68d40'29"
Phoebe Kay Baldwin
Cyg 19h54'7" 33d4'40"
Phoebe Loise Magdalena Ruez
Peg 21h54'9" 12d38'53"
Phoebe Louise
Her 18h52'50" 25d20'26"
Phoebe Lovidge
And 2h25'33" 42d52'11"
Phoebe Mae
And 23h51'4" 47d35'41"
Phoebe Maher Lull
Uma 10h27'54" 52d42'16"
Phoebe Malles
Mon 6h38'9" 5d28'36"
Phoebe Marie Worsnop
And 23h24'2" 41d35'27"
Phoebe Nellie Ostler
Cas 23h32'13" 55d47'40"
Phoebe Parker
Vir 13h36'21" 6d26'9"
Phoebe Rose Holme
And 2h34'52" 50d8'57"
Phoebe Rose Kate Allen
Uma 12h0'43" 59d2'3"
Phoebe Simmons
Crb 15h34'6" 27d12'56"
Phoebe Stewart
Vir 13h15'21" 12d8'47"
Phoebe Susanna May
And 0h56'14" 23d8'35"
Phoebe Taylor
And 0h47'29" 28d42'45"
Phoebe Ting
Gem 7h4'21" 15d6'51"
Phoebe Tucker
Cyg 21h28'41" 34d19'42"
Phoenix
Aur 5h17'38" 41d50'26"
Phoenix
Lyr 18h32'48" 33d5'10"
Phoenix
Gem 7h10'54" 24d33'29"
Phoenix
Ori 5h34'15" -1d1'4"
Phoenix
Pho 0h26'53" -45d10'51"
Phoenix aka Nickolas Mitchell
Cas 23h58'8" 57d33'38"
Phoenix & Dragon - All the Way
Pho 2h15'15" -44d3'40"
Phoenix Fish Rising
Pho 1h34'15" -48d42'45"
Phoenix Gabriel
Pho 0h42'10" -41d21'6"
Phoenix J. M. Ebner
Uma 11h35'12" 64d41'25"
Phoenix Lee Alyea
Aqr 22h4'30" -2d9'6"
Phoenix Marlor
Uma 8h54'2" 67d27'40"
Phoenix Marvel
Lyr 19h7'7" 27d7'32"
Phoenix Raine
Aqr 22h21'9" -19d16'49"
Phoenix Sarah Matarazzo
Pho 0h20'12" -41d58'21"
Phoenix Walter Furness
Pho 0h38'59" -49d53'34"
Phoenixfire
Leo 9h44'43" 9d7'1"
Phohay Boualavong
Uma 11h29'47" 45d59'29"
Phong Tan Luong
Uma 11h27'34" 36d3'39"
Phönix
Ori 6h16'11" 15d26'31"
phorever
Lyr 19h17'57" 29d40'57"
Phoslo
Ori 6h8'11" 15d32'37"
Photios & Vasiliki Stavridis
Cyg 21h21'36" 42d49'30"
Photo Joe
Cnc 8h57'45" 7d59'14"
Phouphet Sayavongsa
Aql 19h23'37" -0d0'54"
Phrankie D Brohm
Sco 16h9'55" -23d59'50"
Phreather V
Lyr 18h51'44" 46d3'58"
Phrosted Phoebe Chenille
Sco 16h7'8" -10d24'9"
PHS Billing Production 2005
Uma 12h42'13" 57d39'41"
Phuangthet & Roland Bolzern
Crb 16h24'59" 27d9'43"
Phung Loves Lynda
Cas 0h40'18" 53d17'41"
Phung Tran
Leo 9h59'2" 12d41'1"
Phuoc 'BuD E' Pham
Lyn 7h45'12" 48d49'33"
Phuong Bich Ho
Gem 6h51'53" 21d49'57"
Phuong Bui
Aqr 21h50'0" -0d14'31"
Phuong Dung Thi Nguyen
Psc 1h44'3" 6d28'32"
Phuong Ho
Leo 10h30'2" 20d58'38"
Phuong & Lawrence Wagner Forever
And 0h21'6" 44d41'22"
Phuong Ngoc Tran-Le
Uma 13h3'36" 53d22'38"
Phydeau Noble Pride
Mon 7h2'26" 7d5'53"
Phylicia I. Hassapis
Ori 5h53'27" 22d7'15"
Phylicia Jessica Bernard
And 0h21'49" 29d32'10"
Phylicia Pena
Umi 14h57'27" 76d45'47"
Phylis and John
Psc 1h32'13" 17d47'16"
Phylis Christmas Barstis
Uma 9h20'34" 43d8'49"
Phylis Francis Simpson
Vel 9h22'25" -42d53'16"
Phylis Josephine Houghton
Lyr 19h17'4" 28d37'47"
Phyliss Douglass
Uma 11h21'10" 34d3'20"
Phyllis
Cnc 8h46'4" 31d29'54"
Phyllis
Lyn 7h28'26" 49d26'22"
Phyllis
Cyg 21h54'20" 38d44'50"
Phyllis
Tau 3h33'33" 25d38'43"
Phyllis
And 0h43'59" 24d59'43"
Phyllis
Leo 11h24'13" 0d2'3"
Phyllis
Sco 16h9'14" -11d52'49"
Phyllis
Cas 1h47'2" 65d44'30"
Phyllis
Dra 15h53'17" 62d19'45"
Phyllis
Cam 4h19'19" 72d27'14"
Phyllis A. Boles
Aqr 20h42'20" -11d21'47"
Phyllis A. Burke
Leo 9h46'34" 32d30'13"
Phyllis A. Hollifield
Lyn 7h42'12" 42d37'49"
Phyllis Akins
Cas 2h14'39" 70d33'59"
Phyllis and Fritz Brown
Cyg 21h57'52" 44d29'47"
Phyllis and Jack
Aql 19h38'39" -11d10'46"
Phyllis and Len Kedson
Uma 13h3'26" 59d59'25"
Phyllis and Paride DeGiulio
Cyg 19h39'16" 51d44'0"
Phyllis Ann
Aqr 22h14'10" 2d14'9"
Phyllis Ann
Ari 3h6'57" 28d32'8"
Phyllis Ann
Sgr 19h24'41" -14d35'9"
Phyllis Ann
Umi 15h58'13" 83d21'19"
Phyllis Ann Sporn-Gerstein
Cas 23h27'18" 52d55'0"
Phyllis Anne
Lyn 8h19'27" 35d43'43"
Phyllis Anne Folz, Our Beloved Mimi
Uma 11h55'3" 54d32'1"
Phyllis Anne Orwasher
Psc 0h31'48" 16d39'51"
Phyllis Arlene Taylor
Cnc 8h9'19" 16d3'53"
Phyllis Audrey Martin
Psc 0h16'34" -2d12'51"
Phyllis Barilla
Lib 15h37'18" -9d37'0"
Phyllis Beverly Beck
Psc 1h10'4" 23d50'13"
Phyllis & Bill Martin
Sge 19h45'5" 17d4'21"
Phyllis Birdie Simms
Uma 13h20'37" 55d51'54"
Phyllis Bliss
And 23h33'24" 38d27'50"
Phyllis Bowen Peterson
Tau 4h29'26" 22d51'41"
Phyllis Carol Kuehl
Gem 6h38'50" 20d52'44"
Phyllis Cohen & Jim Wagner
Uma 10h33'39" 67d30'31"
Phyllis D. Robertson - Mom
Ari 2h59'18" 26d21'39"
Phyllis DiGiovanni
Aqr 21h6'52" -7d6'32"
Phyllis Drescher-Rose
Ari 2h3'56" 24d24'26"
Phyllis Elaine Leece
Vir 13h13'52" -2d31'16"
Phyllis Ercolano
And 0h32'14" 25d57'57"
Phyllis Esther Slicer Morris
Lyn 7h55'55" 39d8'51"
Phyllis Feld
Vir 13h58'21" -14d3'26"
Phyllis Felser
Cyg 19h53'48" 59d35'32"
Phyllis Formisano
Cyg 19h54'0" 51d15'16"
Phyllis Gagnier
Lyn 6h37'51" 56d2'2"
Phyllis Glaser
Lyr 18h52'7" 34d31'38"
Phyllis Gotlieb
Cas 23h45'3" 55d1'46"
Phyllis Graffy
Gem 6h25'0" 27d35'7"
Phyllis Harrsion
Leo 9h57'13" 8d24'53"
Phyllis Heike Bruns geb. 12.11.1999
Uma 11h54'10" 52d1'53"
Phyllis Horne - Mom and Grandma
Cas 1h3'19" 56d47'28"
Phyllis Horne - Pink Grandma
Cep 20h43'55" 64d3'10"
Phyllis Imondi
Uma 10h7'21" 46d33'34"
Phyllis Irene Keegan Richter
Cas 1h44'44" 64d33'35"
Phyllis J. Leach (Ma-Ma)
Dra 19h35'29" 60d25'36"
Phyllis J Powers
Uma 10h38'0" 42d12'24"
Phyllis J. Ryan
Uma 9h23'2" 71d36'34"
Phyllis J. Sukach, P.M.
Psc 0h49'42" 8d58'4"
Phyllis J. Weston
Psc 1h5'33" 24d14'57"
Phyllis Jacobs Anderson
Psc 2h0'28" 6d37'59"
Phyllis Jane Young
And 23h45'2" 44d42'52"
Phyllis Jean
Lyn 7h6'22" 46d24'54"
Phyllis Jean
Uma 9h12'41" 62d30'13"
Phyllis Jean Francis
Lib 15h19'45" -11d34'3"
Phyllis Jean Olson
Srp 18h9'7" -0d59'28"
Phyllis Jean Romsdahl Johnson
Cas 0h3'2" 59d6'18"
PHYLLIS JO CRAWFORD
Cnc 8h56'36" 17d20'28"
Phyllis Joanne Gamble
Lyn 8h6'5" 56d51'50"
Phyllis Julia Adams
Lib 15h49'7" -17d36'40"
Phyllis June Olson
And 23h26'42" 42d37'11"
Phyllis K. Harrison
Lib 15h19'29" -15d33'15"
Phyllis K. Willey 3/6/06-3/14/76
Ari 1h51'17" 17d27'34"
Phyllis Kaminer
Crb 16h8'21" 37d3'35"
Phyllis Key
Cyg 20h22'36" 45d49'14"
Phyllis Killion
And 23h19'0" 43d24'5"
Phyllis Kimberly
Leo 11h12'48" 22d11'23"
Phyllis Kramer
And 2h21'12" 50d28'22"
Phyllis Laurie D&M "5-26-54"
Uma 10h22'11" 46d56'24"
Phyllis Lee Whitney
Leo 11h26'6" 21d57'34"
Phyllis Lilian Nono
Crb 16h0'35" 31d13'17"
Phyllis Lindley
Del 20h40'59" 13d35'38"
Phyllis & Lok Huei
Cap 20h58'35" -17d33'56"
Phyllis Lorraine Hedrick
And 23h23'21" 51d2'52"
Phyllis Lucchesi
Cyg 19h47'1" 33d3'14"

Phyllis Lynn
Aqr 22h8'24" -3d35'26"
Phyllis M. & George H. Zubulake
Cyg 19h22'8" 54d52'45"
Phyllis M. Huseby
Cas 0h45'4" 53d47'19"
Phyllis M. Rizzi
And 0h10'17" 45d12'58"
Phyllis M. Wille
Psc 1h12'29" 4d35'39"
Phyllis Mae aka "Pat"
Cyg 19h38'51" 37d53'36"
Phyllis Marie Antoinette
Gem 6h37'47" 13d48'47"
Phyllis Marie Hatcher
Crb 15h46'53" 27d3'45"
Phyllis Marion Keddie
Cas 23h40'44" 52d41'39"
Phyllis & Mark
Cas 1h11'49" 58d39'43"
Phyllis Massi Zimmerman
Gem 7h49'40" 21d1'30"
Phyllis May
Cas 1h0'22" 49d43'39"
Phyllis McCaskill
Uma 11h34'38" 32d37'13"
Phyllis Michael 25
Crb 15h55'11" 29d1'33"
Phyllis Millikan
And 0h0'22" 34d34'47"
Phyllis Morland
Cas 0h8'31" 58d56'24"
Phyllis Murphy Neal
And 2h5'41" 41d34'22"
Phyllis My Princess
Cnc 8h49'58" 14d46'9"
Phyllis N. Weir
Uma 11h7'44" 69d16'8"
Phyllis "Nana" Maddox
Cyg 19h40'8" 31d42'33"
Phyllis Nolin
Tau 4h25'45" 16d10'14"
Phyllis Norel Halstead
Tau 3h26'54" 14d56'22"
Phyllis & Obie Kinney - 07/07/1956
Cyg 21h30'36" 34d6'43"
Phyllis Parsakian
Cas 0h29'40" 53d43'6"
Phyllis Peters
Lyn 7h7'35" 44d35'46"
Phyllis Petrella
Vir 12h37'45" 9d6'52"
Phyllis Poulson
Cas 1h30'9" 57d33'20"
Phyllis Presta-Lajewski
Cas 1h16'8" 65d57'7"
Phyllis R. Andersen
Uma 10h19'38" 39d49'59"
Phyllis Rowland
Leo 9h23'43" 15d29'57"
Phyllis & Roy Cullinan
Vir 13h32'47" -4d24'50"
Phyllis Schwartz
Ari 2h48'4" 23d45'32"
Phyllis Serafini
Sgr 18h2'52" -19d30'30"
Phyllis Silvers
Cas 0h17'0" 62d9'28"
Phyllis since 1915
Cas 23h34'24" 57d40'56"
Phyllis Sloyer
Per 3h27'13" 32d25'19"
Phyllis Smart
Leo 9h28'1" 18d51'4"
Phyllis Spacarelli-Alexander
Cnc 8h52'11" 30d31'43"
Phyllis & Stan
Ori 5h0'59" 9d51'9"
Phyllis' Star
Vir 13h43'19" -8d1'20"
Phyllis Sue Wong
Cet 1h11'52" -0d57'10"
Phyllis Sutton
Mon 7h30'50" -0d49'36"
Phyllis Szostek Guirlinger
Cyg 19h55'42" 33d34'37"
Phyllis T. Hazard
Cas 0h34'12" 53d57'21"
Phyllis Wallis - Star of My Life
Uma 11h28'54" 51d2'13"
Phyllis Wells Perry
Uma 9h39'49" 56d7'35"
Phyllisa Jayde Star
Crb 15h48'12" 32d35'25"
Phylliscita
Ari 2h48'35" 14d51'22"
Phyllis's Magic Star
Lyn 6h39'55" 55d6'44"
Phyl's deLight
Cap 20h12'19" -26d3'24"
Phyriz
Vul 19h49'55" 28d31'26"
Physmo
Uma 11h38'38" 28d23'49"
Pi Acord
Uma 8h36'36" 64d7'12"
Pi Beta Phi
Uma 10h27'58" 63d49'19"
PI Pearson
Her 18h29'25" 22d20'14"
Pi & Ty
Sgr 18h22'30" -24d3'34"
Pi2K
Umi 15h38'28" 85d21'14"
Pia
Cyg 21h8'52" 46d27'57"
Pia Achermann
Uma 10h58'44" 61d22'39"
Pia Catherine
Vir 13h39'37" 1d21'58"
Pia & Jeff
Crb 16h0'48" 37d17'26"
Pia Josefine
Uma 11h28'44" 36d51'1"
Pia Mezzi
Ori 5h22'31" -0d38'44"
Pia & Pedro 26-8
Umi 15h34'27" 72d34'47"
Pia Slater
Uma 11h53'41" 64d7'0"
Piachta, Inge
Ori 5h8'1" 15d53'56"
Pia-Lorena
Boo 14h32'12" 42d14'15"
Piao-liarng Bipeng
Aqr 21h34'20" -4d17'30"
Piara
Cas 1h15'21" 65d30'58"
Piarjit's Star - "Love Always Wins"
And 0h8'40" 28d51'33"
Piazzerulli
Psc 1h4'24" 5d25'51"
Picasso
Cyg 21h39'11" 49d11'24"
Picasso
Lmi 10h16'58" 36d52'39"
Picci
And 1h44'9" 42d30'37"
Picci
Gem 6h47'32" 23d14'4"
Picciolina
Umi 17h34'14" 80d3'42"
Piccirillo 6
Cyg 21h33'53" 33d30'28"
Picco Serena Maritano
Her 18h3'20" 26d25'38"
Piccola
Lac 22h51'27" 49d49'3"
Piccola Anna
Umi 15h59'46" 76d35'28"
Piccola Bella
Cas 23h4'1" 57d49'52"
Piccola Cerva Bella
Uma 9h51'8" 50d51'58"
Piccola Dany
Aur 5h41'41" 32d36'25"
Piccola Ely
Cyg 21h37'52" 45d57'41"
Piccola Eusty
Umi 13h39'36" 72d7'6"
Piccola Fusco
Umi 16h40'41" 77d0'22"
piccola Guiseppina
Lmi 10h2'16" 31d11'54"
Piccola Principessa
Per 4h47'54" 40d59'36"
Piccola & Pumpkin
Uma 9h41'36" 60d56'33"
Piccola Stella
Tau 5h49'57" 26d19'51"
Piccola Stella - Michela Anzani
Vir 13h51'51" -2d12'31"
Pichia
Col 6h7'25" -28d15'22"
Pichler, Karl
Uma 10h2'8" 54d39'22"
Pichler, Robert
Uma 9h16'30" 71d54'5"
Pichlo, Luca Marie
Tau 3h39'59" 29d52'22"
Pichnette
Her 17h40'51" 39d6'5"
Pichot Noémie ma Chérie
Sco 17h42'24" -40d57'27"
Pickle
Aqr 22h15'26" -23d16'7"
Pickle
Cmi 7h44'26" 0d27'51"
Pickle Nose
Her 18h19'41" 21d12'15"
Picklebiscuit
Cam 6h30'52" 75d58'20"
Pickles
Lyn 7h24'23" 54d16'32"
Pickles
Aqr 22h50'47" -24d5'56"
Pickles
Aqr 22h18'54" 1d43'58"
Picky
And 0h43'10" 41d30'43"
Pico
Uma 10h18'13" 67d31'36"
Picouli
Cnc 8h38'37" 19d30'37"
Picqwicq
Her 18h35'6" 18d19'45"
pics 4ever
Uma 10h24'48" 61d7'24"
Picses 3-19-1982 7:14 AM
Psc 0h58'47" 10d3'7"
Picton Jones
Uma 14h4'50" 60d3'15"
"Picur" Várkonyi Balázs csillaga
Psc 1h42'57" 2d42'57"
Piddály
Uma 10h4'43" 72d25'53"
Pidge
Cnc 8h51'1" 30d35'8"
Pidy
And 0h56'36" 44d16'7"
Pie
Sgr 18h58'15" -31d55'29"
Pie In The Sky
Lyr 19h9'51" 37d28'15"
Pie In The Sky
Ari 2h9'38" 26d5'39"
Piece of Heaven
Cnc 9h12'53" 27d35'59"
Piece of our hearts
Vir 13h28'39" 11d23'14"
Pieces Of A Puzzle
Dra 19h7'33" 73d51'58"
Piechue
Cam 4h11'37" 59d41'58"
Piedra 1
Umi 15h30'25" 74d50'3"
Piekna
Cas 0h52'35" 60d23'11"
Pielnhofer, Eva
Uma 10h7'21" 55d5'4"
Piemme
Aql 20h16'1" -0d8'11"
Pienso en ti
Uma 11h27'9" 40d1'36"
Pientje
Cas 1h52'18" 63d59'15"
Pieper's Silverstar
Uma 9h51'5" 45d36'39"
Pier Paolo
Uma 8h21'14" 60d49'36"
Pier Paolo Graziotti
And 0h28'54" 36d9'52"
Piera Ronca
Pho 0h44'2" -50d37'39"
Pierangelo
Umi 15h41'7" 82d47'35"
Pier-Angelo, 30.06.1960
And 0h48'46" 39d20'58"
Pierce Akin
And 0h18'46" 32d58'11"
Pierce Alexander Lindenberg
Per 3h47'8" 37d11'22"
Pierce Alexander Quinn
Gem 7h42'9" 23d24'34"
Pierce B. Brucker
Vir 12h52'18" 11d51'50"
Pierce Charles Bolduc
Her 18h48'11" 23d28'22"
Pierce Christopher Stotzner
Ori 4h56'4" 10d52'6"
Pierce Elijah LoveGordon
Her 16h47'6" 29d3'2"
Pierce Ferner
Uma 12h2'28" 41d32'11"
Pierce Nevaeh
Lac 22h38'13" 38d21'45"
Pierce Noah Patterson
Cnc 8h24'50" 11d7'47"
Pierce Russell Painter "LVX"
Cnc 8h52'47" 27d54'36"
Pierce William Chastine
Ori 5h45'54" 6d42'24"
Pierelivio
Uma 9h22'11" 44d31'49"
Pierino
Tau 3h40'28" 4d30'24"
Piero
Cyg 19h36'55" 30d9'15"
Piero
Uma 11h17'28" 32d30'7"
Piero
Per 3h7'17" 51d42'12"
Piero e Orietta
Cyg 20h51'14" 53d49'8"
Piero & Isabella
Cyg 20h4'18" 35d2'36"
Piero Masetti
Sex 10h3'13" 0d18'39"
Pierra's Star of Hope
Uma 11h41'55" 34d9'54"
Pierre
Cep 22h11'45" 53d43'49"
Pierre Amalou
Her 17h51'32" 23d43'40"
Pierre Antoine
Sgr 18h36'21" -23d3'22"
Pierre Auge
Dra 12h38'18" 74d29'4"
Pierre Bailleul
Dra 17h8'15" 57d28'50"
Pierre Bélanger
Umi 13h32'46" 86d26'37"
Pierre Cormier
Aur 5h44'57" 47d29'51"
Pierre Couvert
Aqr 21h7'39" 1d40'55"
Pierre d'Abzac
Uma 9h50'30" 45d24'2"
Pierre Duncan
Ori 6h1'13" 10d25'33"
Pierre Dupuis The Light of My Life!
Her 18h17'4" 22d4'21"
Pierre Engelbrecht
Per 2h58'23" 46d8'57"
pierre et colette
Uma 9h13'1" 47d13'13"
Pierre et Emmanuelle Saulnier
Cas 23h30'15" 55d59'52"
Pierre Francesca
Del 20h44'17" 17d35'31"
Pierre Gobeil
Uma 11h29'13" 60d2'16"
Pierre Guy Martin
Cep 23h39'28" 79d36'26"
Pierre & Jessica
Uma 8h54'12" 48d19'40"
Pierre - King of Hearts
Cru 12h25'22" -62d8'20"
Pierre Laporte & Zoé - Maude - Tommy
Uma 9h50'10" 63d39'19"
Pierre Mailloux
Gem 7h15'39" 19d6'15"
Pierre Massé
Cep 0h36'9" 81d48'35"
Pierre Minjollet
Cnc 9h16'42" 15d42'24"
Pierre Morvan
Leo 11h56'42" 22d34'45"
Pierre P. Eno, Jr. - Daddy
Sgr 18h6'6" -27d57'20"
Pierre Paquin
Ori 5h15'34" -7d41'50"
Pierre Pittet
Lib 15h40'6" -26d44'55"
Pierre René Jean Franquet
Sco 17h52'20" -40d30'32"
Pierre Ronchetto
Uma 9h33'3" 42d34'8"
Pierre Soued
Vir 13h33'37" -22d17'54"
Pierre Thibault
Crb 15h34'54" 38d57'25"
Pierre Vanden Driessche
Cas 0h13'18" 59d36'44"
Pierre Vee
Cru 12h21'50" -58d3'49"
Pierre Yakovenko
Sco 16h51'56" -27d14'4"
Pierre-Alain Masserey
Aur 5h21'5" 30d6'29"
Pierre-André Sorbets
Uma 10h20'42" 70d0'12"
Pierre-Emmanuel Niédrée
Psc 0h0'22" -5d3'3"
Pierre-Henri Debray
Her 17h17'4" 17d10'57"
Pierreliane Coudroy 01101955
Umi 13h10'32" 70d33'55"
Pierret
Cas 0h19'36" 55d19'13"
Pierrette
Mon 6h43'58" -0d8'43"
Pierrette Cyr
Cas 23h54'49" 59d56'54"
Pierrette et Marc-André Niquet
Cyg 21h38'42" 37d41'1"
Pierrette Graboski
Uma 11h25'19" 51d34'56"
Pierrette (Noune)
Uma 10h1'49" 51d33'40"
Pierrick Lilliu
Peg 22h38'33" 20d11'9"
Pierrick my best
Umi 16h2'33" 80d4'19"
Pierrot
Aqr 22h10'13" -10d42'3"
Pierrot de la luna
Ori 6h17'25" 15d38'24"
Piers
Her 16h5'21" 40d53'8"
Pierson
Cnc 8h8'48" 9d58'43"
Pierson Fort Melcher
Per 3h27'12" 49d8'56"
Pierson Grey Lipschultz
Aql 19h19'13" 9d50'8"
PIE's Star
Ori 6h14'23" 15d1'47"
Pieser
Ori 5h55'37" 12d29'14"
Pieser
Cas 0h8'46" 53d2'47"
PieStarMum
Uma 11h54'7" 56d28'0"
Pieta
Cnc 8h21'38" 7d10'12"
Pieta Fisher
Cru 12h25'39" -63d18'42"
Pieter George Wybro
Per 4h21'42" 42d48'28"
Pieter Paul
Per 3h5'37" 43d1'14"
Pietro
Cas 0h52'14" 51d54'29"
Pietro
Lyr 19h17'5" 38d57'20"
Pietro Antonio, 21.06.2003
Cyg 19h37'28" 32d26'22"
Pietro Bonasegla
Ori 5h39'30" 1d49'0"
Pietro Carratu
Gem 7h16'16" 34d36'19"
Pietro & Chus 14 marzo 1987
Aur 6h34'36" 41d33'49"
Pietro Giovanni Siragusa
Dra 16h44'10" 63d52'20"
Pietro Rispoli
Peg 23h54'29" 26d40'59"
Pig
Ori 5h53'36" 13d45'37"
Pig Valiant
Leo 11h53'36" 18d59'41"
Pig Wardie
And 0h45'41" 38d6'45"
pigbandit's 25th
Leo 9h56'42" 6d47'7"
Piggers Schwartz
Uma 11h2'29" 45d44'33"
Piggie Love
Leo 9h38'24" 10d13'34"
Piggy
Cnc 9h21'22" 24d49'52"
Piggy
Gem 7h46'46" 33d11'29"
Piggy
Umi 15h20'54" 67d58'44"
Piggy McWiggles
Gem 6h49'31" 27d9'15"
Piggy Princess
And 1h18'20" 49d6'55"
pighead0719
Cnc 8h49'39" 27d57'37"
Piglet
And 1h43'39" 43d49'57"
Piilani
Sco 17h52'20" -40d30'32"
Piiou
Uma 8h25'35" 60d29'29"
PIKA
Tau 4h27'45" 22d11'23"
Pika Blue Swayder Star
Uma 14h19'0" 61d16'26"
Pike
Uma 10h23'37" 57d10'4"
Pilar
Lyn 7h5'29" 53d24'17"
Pilar
Leo 11h12'34" 2d39'50"
Pilar Casado
Dra 19h44'42" 68d3'43"
Pilar Contreras- Shining Star
Mon 6h40'1" 6d28'1"
Pilar Dominguez Sanjuan
Tau 5h31'32" 21d41'15"
Pilar Gamboa
Vir 13h25'45" -11d59'3"
Pilar Garcia Arellano
Vir 13h0'18" -16d45'26"
Pilar Lastra
Cap 21h10'36" -22d14'12"
Pilar M. Herrera
Uma 11h19'45" 52d16'56"
Pilar Mendieta
Uma 14h3'10" 48d17'48"
Pilar Paige
Crb 15h33'45" 38d17'52"
Pilar Pie
Cap 21h34'29" -14d15'36"
Pilar Rodriguez Martinez
Sco 17h52'15" -39d26'42"
Pilawa, Jörg
Uma 11h32'7" 34d2'24"
Pili Collado
Psc 0h56'44" 17d21'24"
Pilialoha
Her 17h19'29" 49d13'50"
Pilialoha
Uma 12h34'0" 55d17'13"
Pilialoha Hoku Claudia Dailey
Uma 11h19'22" 46d30'26"
Piliki Marie
And 23h21'16" 41d18'27"
Pililiaki (BNK)
Vir 13h34'2" 4d29'6"
Pi-Lin
Umi 16h34'17" 82d18'24"
Pillangócska
Lib 15h47'59" -17d51'25"
Pillars of Hercules
Her 17h24'54" 14d32'53"
Pill-poll
Col 5h6'3" -42d5'5"
Pillsbury
Leo 10h10'2" 21d59'18"
Pilly
Dra 18h1'52" 74d21'34"
Pilz
Dra 10h4'24" 78d33'7"
P.I.M (Pia,Isis,Marie)
Del 20h40'18" 5d37'41"
Pimientito Dumesnil
And 23h22'5" 37d32'0"
Pimpa
Ori 5h47'1" 12d31'15"
PINA
Boo 14h47'26" 12d23'8"
Pina
Per 3h25'13" 41d36'22"
Pina Santoro-Petti
Psc 1h22'2" 32d28'26"
Pina Tamburrino
Aur 4h50'30" 36d55'51"
Pinaga
Boo 15h27'41" 42d23'9"
Pinar Siyahi
Gem 7h41'22" 15d44'49"
Pinardo's Star
Uma 10h42'49" 48d17'54"
Pina's Bright Shining Star
Lyn 6h44'2" 58d22'52"
Pinball
Cyg 20h39'6" 44d55'44"
Pincess la mas hemosa
Leo 11h4'15" 16d33'11"
Pinchas Ben Meyer
Her 18h4'4" 14d22'40"
Pinchbeck Star
Cas 0h32'27" 66d51'55"
Pineapple
Umi 15h0'1" 76d23'0"
Ping
Aqr 22h39'39" -5d21'22"
Ping and Pong
Uma 8h47'31" 50d59'49"
Ping Man
Uma 9h29'25" 69d16'47"
Ping Yiu To
Aqr 23h20'17" -18d15'1"
Pingüi-23
Vir 14h14'34" 3d57'24"
Pinguino's Estrella
Pho 0h37'21" -44d16'47"
Pingu's Sparkle
Pho 23h35'21" -48d31'59"
Pinhead & Ms. Piggy
Cyg 19h47'24" 39d20'23"
Pink Daddy
Psc 23h2'40" 6d1'35"
Pink Diamond
Crb 16h6'33" 34d53'46"
Pink Dreams
Uma 13h36'38" 48d15'12"
Pink Flamingo
Cyg 19h28'43" 54d8'16"
Pink Kitty NYC
And 1h33'32" 48d16'0"
Pink Lady 21
Ori 6h6'57" 14d10'47"
Pink Pig Princess
And 23h54'41" 39d17'37"
pink princess
Cap 20h30'55" -15d38'59"
Pinkie
Lyr 18h49'18" 35d12'9"
Pinkie pie
Lyn 6h41'52" 57d54'20"
Pinkney
Her 18h39'8" 22d10'58"
PinkPixie
Cyg 21h30'5" 41d22'17"
Pinkticus Stephen + Lyndsey 4 eva.
And 2h28'4" 45d33'54"
Pinky
Aur 6h28'53" 34d31'58"
Pinky
Psc 1h42'30" 26d56'24"
Pinky
Ari 2h20'5" 26d14'18"
Pinky
Psc 1h12'17" 26d11'6"
Pinky
Tau 5h35'49" 25d43'38"
Pinky
Ari 2h33'43" 14d7'39"
Pinky
Umi 14h33'47" 72d35'27"
Pinky
Cap 21h9'25" -14d57'54"
Pinky
Sgr 19h21'21" -14d21'37"
pinky and teddy
Lib 15h36'0" -11d55'32"
Pinky Lambell
Per 4h48'6" 49d1'53"
Pinky Lee
Lyr 18h52'26" 43d40'43"
Pinky. Patel
Cnc 8h4'38" 25d41'34"
pinky promise
Cyg 21h52'55" 50d8'14"
Pinnacle4
Uma 14h21'52" 59d39'50"
Pino
Boo 14h10'13" 51d41'58"
Pino
Per 4h16'35" 45d6'40"
Pino and Melinda eternally
Cyg 20h25'52" 40d49'38"
Pino e Noemi
Cam 5h2'13" 63d31'24"
Pinocchio
Cnc 8h52'18" 30d36'43"
Pinon's Dream
Lib 15h31'48" -4d56'53"
Pinse
Cnc 8h17'33" 30d31'2"
"Pinselini" Simone Bründler
Lyr 18h16'32" 32d14'47"
Pinson's Point
Aql 19h44'33" 8d13'22"
PINTA
Tau 3h28'14" 14d5'20"
Pintog-dearAMBHfrTRIXIE-11141997
Ori 6h3'44" 1d35'55"
Pinuccia
Ari 2h52'35" 28d27'32"
PinucciaCraig
Vir 14h42'10" -1d42'59"
Pinuzzo
Uma 10h21'42" 57d55'42"
Pio De Nigris
Ori 5h58'0" 22d42'19"
Piotr Pawlowski
Ori 5h17'34" 15d16'46"
Piotr Walter
Uma 13h24'34" 63d6'29"
Piotrowski
Aql 19h35'57" 8d44'55"
PIP
And 1h28'38" 40d39'8"
Pip
Sco 16h52'17" -42d54'28"
Pip 21
Psc 22h59'2" 7d53'24"
Pip Chisholm
Umi 14h40'23" 69d54'51"
Pip Cook
Col 5h59'0" -34d53'54"
Pip - Love Of My Life
Ori 5h56'31" -0d45'49"
Pip N. Smits
Vir 14h46'34" 4d21'52"
PIPE
Her 18h40'28" 25d52'31"
Pipe
Ori 5h58'20" -0d41'51"
Pipe
Aql 18h51'11" -0d46'23"
Pipeandroxyinfinite
Hya 9h25'8" 3d2'21"
Piper
Psc 1h30'10" 26d35'15"
Piper
Per 3h23'41" 48d24'18"
Piper 007 Laidlaw
Uma 9h15'33" 61d11'9"
Piper Bean Karnilaw-Maurice
Cnc 8h39'47" 8d52'26"
Piper & Daddy's Very Own Star
Cam 5h54'22" 65d52'29"
Piper Elise Davis
Sco 16h8'27" -17d0'14"
Piper Fleming
Sgr 18h41'44" -30d43'20"
Piper Grace Taylor
Uma 11h46'44" 59d54'51"
Piper Jean
Tau 3h52'50" 21d34'54"
Piper Kristy Cook
And 23h36'19" 47d0'51"
Piper Lacie Simons
Her 16h11'8" 23d55'20"
Piper Leigh Byrne
And 23h36'37" 47d42'3"
Piper Nicole Hawley 1/9/03
Uma 10h45'22" 49d24'32"
Piper Olivia Willingham
And 0h24'35" 33d53'34"
PiperPerabo
Cas 1h25'2" 65d2'16"
Piper's Star
Lyn 6h59'37" 54d37'33"
pipka
Boo 14h50'30" 32d28'45"
Pipo Casanova
Peg 22h6'8" 5d32'53"
Pipooo
Ori 5h40'37" 12d12'5"
Pippa
And 1h52'16" 45d13'49"
Pippa Bird
Lyn 7h57'30" 58d7'19"
Pippa Frances Catt
And 1h45'28" 48d48'8"
Pippa & Pete 1st Anniversary
Cyg 20h49'56" 35d16'33"
Pippi
Per 4h13'29" 33d7'27"
"Pippi"
Peg 21h16'56" 15d32'44"
Pippi Aurora 98
Tau 4h31'4" 21d38'22"
Pippin
Leo 11h36'39" 17d13'31"
Pipsqueek & Nuclehead's star
Lac 22h20'47" 48d50'11"
Pirate Heaven
Lib 14h49'42" -3d37'26"
Pirate King Grandpa Ken Beekman
Her 16h11'36" 18d42'49"
Pirate Princess Annette
And 1h51'48" 37d39'49"
Pirate Treasure
Sgr 17h50'54" -26d9'32"
Pirate's Inn
Her 16h51'1" 38d12'48"
Piret Mottus
Uma 12h50'27" 61d44'43"
Piri 60
Psc 1h4'49" 14d45'14"
PIRJON TAHTI
Cas 0h55'9" 63d41'5"
Piroska Karakas
Ara 17h28'36" -45d54'23"
Pisamai Mongphet
Tau 5h8'35" 27d51'54"

Piscalko
Sco 16h13'4" -10d58'50"
Pisces Rising
Umi 13h14'27" 70d40'57"
Pish
Sgr 18h5'24" -27d51'55"
Piskatorisz Julianna
Psc 0h33'32" 5d50'40"
Pissant
Pup 8h11'4" -21d38'53"
Pistachio Bread
Uma 13h47'4" 57d58'26"
pistis elpis agape
Ori 6h21'41" 10d41'42"
Pita
Lyn 7h56'2" 53d32'7"
Pitibou
Crb 16h17'13" 29d48'18"
Pitinho y Critinha
Uma 10h27'32" 57d13'3"
Pitons
Gem 6h24'15" 19d14'6"
Pitou d'Amour
Umi 12h25'8" 87d57'32"
Pitt and Linda's Passion
Aqr 22h42'5" -0d35'7"
Pitter Ruth Potter
Lyr 18h39'23" 34d38'59"
Pitufina
Aqr 20h52'28" 1d5'44"
Pityu's Sparkle
Tau 5h34'54" 22d19'42"
Pitz
Cyg 20h12'16" 35d6'52"
Pitzer, Melanie
Uma 11h53'33" 40d44'36"
Pixelworks
Per 4h2'39" 49d40'26"
Pixi Lynn Freeman
And 0h38'44" 39d23'51"
Pixie
Cas 0h24'28" 53d56'14"
Pixie
And 23h25'32" 52d7'25"
Pixie
Vir 12h53'51" 11d6'56"
Pixie
Aqr 23h42'6" -13d8'18"
Pixie
Uma 11h24'37" 62d54'58"
Pixie - Argentum Ursa Glatisant
Uma 9h17'48" 62d23'31"
Pixie B Carter
Aqr 22h37'3" 2d35'32"
Pixie Dust
Lyn 7h49'42" 43d31'29"
Pixie Jean 12/26/2003
And 0h33'40" 41d41'42"
Pixie Lizzy
Lyr 18h47'56" 42d34'6"
PixieDragon
Vir 15h7'59" 0d2'30"
Pixie's Angel
Lyr 18h42'18" 39d57'35"
Pixie's Star
Cap 20h18'56" -11d35'12"
Pixy K
Ori 4h53'19" -0d5'1"
Pizza Hut Area 51 George Torosian
Uma 10h15'20" 66d34'44"
PJ
Uma 13h45'6" 62d21'52"
P.J.
Uma 12h34'10" 56d49'45"
PJ
Uma 13h2'41" 54d44'30"
PJ
Sco 16h5'35" -20d59'21"
P.J.
Uma 11h18'47" 45d1'20"
PJ
Cnc 9h8'27" 30d43'32"
PJ
Leo 11h30'28" 9d45'34"
PJ and Crutch
Crb 16h1'15" 29d56'58"
PJ and Lizzy Lea
Aqr 22h9'13" -12d44'52"
PJ and Sarah
Crb 15h51'2" 27d34'52"
PJ Cannon
Sco 15h51'23" -22d5'24"
P.J. Crowell
Leo 9h28'44" 25d35'17"
PJ Duphily
Uma 9h39'58" 57d3'6"
P.J. Gregg
Mon 7h32'50" -0d51'47"
PJ Johnson
Lyn 6h58'12" 51d8'43"
PJ Lusted
Cra 18h36'33" -42d53'22"
PJ Magik
Cnc 8h48'49" 14d18'30"
P.J. Maynard
Aqr 20h59'42" 1d4'41"
PJ Monson
Uma 10h45'5" 44d7'59"
P.J. Rennie
Aqr 22h25'32" -7d23'4"
P.J. Sanchez
Cnc 8h0'48" 14d46'40"
P.J. Sirl
Crb 15h45'17" 36d42'53"
PJ Smith
Lib 14h30'37" -9d0'8"
PJ & WAL Gittus
Cyg 19h49'3" 31d57'48"
PJG - Papà, Sempre Nei Nostri Cuori
Per 2h15'14" 51d14'8"
PJR
Uma 14h5'1" 51d30'25"
PJ's Everlasting Light
Crb 16h13'59" 38d18'43"
PJ's Light In The Cosmos
Umi 15h26'42" 70d2'11"
PJ's Star
Lib 15h42'50" -29d38'8"
P.J.'s Star
Peg 21h51'35" 12d40'45"
Pjyle
Cam 4h41'22" 53d34'16"
PK
Per 3h10'47" 44d37'42"
PK 60
Leo 9h53'23" 28d38'59"
PK 8.10.01
Cap 21h23'6" -17d27'37"
P.K & Esther, in Love Forever
Cyg 19h35'15" 29d28'7"
/P.K My Brother Patrik Thomas Keim
Aql 19h49'41" -0d54'52"
pk ognora
Ori 6h4'33" 16d58'0"
PK Ranford
Sgr 18h41'49" -32d26'14"
PK the Brightest Star
Dra 18h37'59" 50d57'18"
PK TURBO
Cnc 8h51'6" 30d56'3"
PKD MD Cosmic Cardiac Surgeon
Tau 5h44'11" 17d13'30"
pke*
Lmi 10h7'27" 37d1'38"
PKF20 Coruscant
Lyn 8h15'21" 34d48'4"
PKH
Per 3h13'19" 54d32'9"
PKIAC Osborne
Ori 6h0'15" 16d29'52"
PKLUV 6/7/03
Lyr 19h14'25" 26d34'27"
PKP1952
And 0h45'16" 38d3'7"
PKY-317
Psc 23h48'23" -0d36'14"
Placed By Gideon
Leo 9h45'32" 23d30'28"
Plachetka Glow
Cyg 19h36'23" 53d36'16"
Plagemann, Dirk
Ori 5h58'4" 21d45'32"
Plan 9
Uma 10h48'36" 65d38'41"
Planet Amy "my way"
Leo 9h32'37" 15d35'36"
Planet Bopschki
Tau 3h34'35" 27d40'39"
Planet Clarke, Sue and Derek's Star
Cyg 20h44'54" 44d19'2"
Planet Dawn 240671
Lyn 7h31'38" 47d53'49"
Planet de Audry
Ari 2h55'35" 25d25'45"
Planet Dust
Psc 0h52'9" 6d48'11"
Planet HariPrem
Psc 1h36'20" 27d45'53"
Planet Hayden & Charlotte 2006
Cap 20h23'44" -14d22'5"
Planet Health
Cru 12h38'13" -60d8'29"
Planet Hitler
Gem 6h27'55" 17d42'6"
Planet Holschuh
Cnc 8h49'20" 13d22'55"
Planet Janet
Leo 9h26'26" 26d16'4"
Planet Janet
Sco 16h9'42" -16d30'44"
Planet Jordan Pendexter
Tau 3h42'24" 28d55'44"
Planet Kids
Umi 16h5'45" 76d7'48"
Planet McDonut
Leo 11h10'53" -6d12'29"
Planet Meg
Cra 18h8'3" -37d27'10"
Planet Mike
Uma 11h13'56" 43d26'9"
Planet Monger
Lib 14h26'42" -20d55'0"
Planet Neil
Per 4h0'57" 43d58'29"
Planet Noodle-Bug
Cnc 8h16'1" 30d26'23"
Planet of Midori and Takeshi
Lib 15h29'21" -16d53'20"
Planet of the original "LYNN JANES"
Vul 21h21'3" 27d58'5"
Planet Pacific
Ori 5h23'51" 2d20'45"
Planet Princess Sylvia
Cae 4h51'49" -29d15'27"
Planet Rachael Hooper
And 0h15'55" 39d31'42"
Planet Reece
Per 3h45'17" 42d29'30"
Planet Vernie
Uma 12h33'40" 60d54'23"
Planet William Michael Lawson
Leo 11h7'11" -6d25'53"
PlanetClare
Ari 1h47'50" 20d46'52"
Planète Guillaume 10
And 23h39'17" 47d39'45"
Plan-It Krause-Co
Cru 12h26'58" -60d4'59"
Planner in the Sky
Psc 1h20'4" 27d34'20"
Plant
Leo 11h13'32" 25d54'0"
Plasmodesmata T.F.C. IV
Aql 19h10'57" 8d48'29"
Plassan Mickael
Cap 21h22'13" -27d22'53"
Platinum Circle
Peg 23h7'45" 12d52'48"
Platinum Star-Elena Gochberg
And 1h34'36" 45d55'23"
Platinum Star-Eric Stehley
Aql 19h18'51" 15d0'59"
Platinum Star-Francine Clausen
Cyg 19h55'16" 37d58'39"
Platinum Star-Kerry Huffman
Com 13h33'10" 14d45'18"
Platinum Star-Kristen Hummel
Peg 23h57'0" 28d53'28"
Platinum Star-Melissa Paulson
Lyn 7h11'31" 58d50'23"
Platinum Star-Mike Wicke
Cep 22h43'3" 65d19'38"
Platinum Star-Ralph Kramer
Lac 22h14'28" 49d48'23"
Platinum Star-Steve Eggan
Boo 15h31'29" 44d21'46"
Platinum Star-Susan Butler
And 0h1'25" 48d32'57"
Platinum Star-Suzanne Parkinson
Cam 5h47'43" 58d38'34"
Platinum Star-Tom Mueller
Cma 6h17'11" -26d51'37"
Plato
Cap 21h38'53" -22d46'46"
Platz, Felix Leonard Korbinian
Ori 6h20'49" 11d39'20"
Playa Del Carmen
Uma 8h45'49" 49d29'19"
Playboi's Sweets
Cas 1h14'9" 55d4'7"
Playful Magnificent
Cyg 20h50'8" 33d58'2"
Playfulmouse
Lib 14h22'44" -19d54'5"
PLC-one star
Cyg 21h30'15" 46d57'42"
Pleb
Aur 5h30'16" 40d1'45"
Pleco & Bristlenose
Pho 0h13'46" -42d2'40"
Pleidian
Tau 5h18'43" 17d30'48"
Pleimer, Bianka
Uma 13h12'20" 56d39'17"
Plein, Johann
Uma 11h20'19" 64d45'41"
Plenge, Udo
Cnc 8h18'47" 22d46'15"
Pleuler, Lea
Sco 17h23'31" -32d23'40"
Plewka, Martina
Uma 9h50'30" 67d24'35"
Plewy
Cru 12h24'51" -61d17'33"
Plezhette Alice Gibson - Bennett
Ori 5h13'4" 13d1'25"
PLF
Lib 15h16'30" -10d53'35"
Plifke, Rolf
Uma 10h43'19" 46d20'27"
Plinileidy Abreu
Cam 4h35'11" 66d48'27"
Pliquett, Annegret
Uma 10h10'43" 69d48'24"
Plistschuk, Igor
Ori 5h50'4" 7d16'17"
PLLAMA DEER PRINCESS
Psc 1h5'55" 7d23'11"
PLNDOC
Dra 17h38'39" 68d36'49"
Ploch, Björn
Ori 6h21'41" 16d19'28"
Plopawert
Sco 17h43'12" -38d44'18"
Ployhart
And 1h0'26" 36d48'19"
PLT (Purty Little Thang)
Uma 10h3'51" 50d29'41"
plukovník Ing. Eduard CONGA
Aql 19h39'9" 14d15'58"
plukovník Ing. Radek PECA
Aql 19h13'13" 14d57'9"
Plum
And 23h54'40" 41d28'19"
Plum
Cam 4h6'19" 71d15'11"
PLUMMER FRAN
Aqr 22h36'27" -17d27'41"
plunder bunny
Uma 12h14'15" 54d34'3"
Pluto Helgens
Sgr 19h13'20" -12d55'45"
PLW01191972
Peg 21h39'16" 25d48'50"
PM - Shades of Grey
Gem 6h30'20" 24d50'51"
PM70
Uma 11h30'10" 52d24'1"
PMA Frodo Baggins 42
Leo 11h51'30" 21d44'18"
PMANLJYoung
Aqr 22h20'51" -3d58'46"
PMB Phenomena
Uma 8h45'10" 71d0'50"
PMD50
Ori 5h36'25" 5d33'24"
PMS 2002
Aql 18h52'47" -0d46'28"
PN12698SW
Sco 16h2'59" -29d45'2"
PNH
Gem 6h24'49" 23d48'30"
Pnina.Corestar
Cyg 20h0'32" 35d27'2"
PN-KARLI
Cam 4h3'32" 66d9'41"
P'nut
Cma 6h30'5" -17d22'51"
P-Nut Warner
Uma 11h19'28" 52d41'19"
Po
And 1h39'19" 41d32'51"
Po
Lyn 7h32'51" 36d48'49"
PO
Her 18h37'28" 19d0'6"
Po Chun Cheung
Aql 18h53'15" -0d6'23"
Po On
Lib 14h51'39" -0d38'51"
Po Star 3
Uma 8h18'46" 70d47'32"
Pobanz, Fabian
Uma 11h24'32" 34d57'59"
POB'S
Uma 10h2'9" 49d35'47"
Poca
Sco 17h50'1" -32d11'15"
Pocahontas
Sgr 19h19'30" -20d42'14"
Pocahontas and Nala
Uma 12h14'52" 58d53'12"
Pocha
Leo 10h30'3" 15d25'43"
Poco
Ori 6h4'21" 13d13'11"
Poco
Cma 6h37'8" -15d7'59"
Poco Vigneri
Uma 8h35'11" 66d5'1"
Pocono Emergency Physicians
Per 3h33'6" 46d39'6"
Pócsik János a legfénye-sebb Csillag
Leo 9h22'49" 12d20'12"
Pócsik Sándor - Nagyipa 70
Cnc 8h12'28" 20d40'59"
Podestrella
Ori 5h51'3" 8d59'18"
podplukovník Ing. Dušan PANČÍK
Aql 19h31'22" 12d36'46"
Poet Laura
Cas 1h26'38" 57d34'20"
Poet Lloyd NMN Zeger
Gog Tarzan
And 1h13'52" 41d4'0"
Poets Diamond
Sco 17h53'53" -36d1'45"
pofsc
Dra 17h42'26" 55d34'12"
Pog
Psc 1h9'8" 25d48'24"
Pogi
Umi 17h27'37" 84d27'57"
Pogreba
Umi 15h32'32" 72d14'9"
Poheitai Babe Bessert
Cas 23h1'12" 55d33'27"
Pohl, Brigitte
Uma 10h22'2" 65d24'42"
Pohle, Steffen
Ori 6h7'35" 13d39'45"
Poinsart
Per 2h58'54" 46d4'37"
Point Percy
Eri 3h45'2" -40d2'38"
Poipu Princess Jacqui
Aqr 21h6'42" 2d17'3"
Poiu
Cen 13h22'59" -40d12'29"
Pojita
Cas 0h29'22" 53d14'43"
PoJo
Cyg 21h32'48" 44d2'59"
Poke & Peni
Aqr 22h59'59" -15d54'17"
Pokey
Uma 11h44'32" 52d42'29"
Pokey
Lmi 10h31'54" 31d13'24"
Pokey
Leo 9h22'45" 10d51'28"
Pokey 'T the D' Rumpus
Sco 16h16'13" -15d42'28"
Pokey12141996
Aur 5h41'32" 40d56'38"
P.O.L.
Psc 1h31'55" 21d16'42"
Pol Frippiat
Psc 1h43'59" 6d0'9"
POL&JESS
Cas 23h26'58" 55d42'13"
Pola
Uma 10h47'36" 61d35'21"
Pola Hordynska
Lyn 6h57'4" 52d24'34"
Pola Knight
And 2h11'4" 41d8'24"
Polar Bear
Peg 22h17'49" 33d15'0"
Polar Bear
Leo 11h44'58" 13d48'34"
Polar Brie
Uma 13h30'37" 58d16'52"
Polar Dave
Uma 9h52'15" 42d45'5"
Polar Express
Her 18h44'46" 17d24'27"
Polawyn
Lib 15h27'54" -23d35'33"
Polf, Peter
Cap 20h14'6" -26d18'45"
Polhill
Aql 19h52'1" 6d59'37"
Poli
Psc 23h52'45" 1d27'19"
Police Chief Randy Joe Sonnenberg
Per 3h18'17" 47d44'20"
Polifrone 55
Lyn 8h32'30" 46d21'38"
Polimano
Uma 9h22'43" 59d21'54"
Polin Makertichian
Tau 4h24'29" 17d18'57"
Polina
Tau 3h53'9" 20d42'45"
Polina
Vir 12h13'9" -9d2'1"
Polina Kremen
Aqr 22h7'30" -13d12'8"
Polina Nikolaevna Trishkova Huffman
Ari 2h52'1" 28d38'19"
Polinaland
Leo 11h12'38" 24d23'23"
Polini
Ari 2h16'9" 27d42'43"
Polipo & Angelfish
Crb 16h0'19" 30d39'15"
Polish Pauline
Cas 1h43'57" 64d13'31"
Polish Princess
Crb 15h34'23" 31d21'47"
Pollack
Her 17h36'46" 37d27'21"
Pollenport
Uma 10h35'9" 42d10'53"
Pöller, Brigitte
Uma 11h30'27" 29d39'42"
Polley Green
And 23h14'54" 44d32'35"
Polley's Pride
Uma 12h3'38" 56d3'24"
Polli & Yvi
Uma 9h48'16" 45d39'8"
Pollux
Cap 21h6'59" -18d57'29"
Polly
Ori 5h35'45" -0d53'41"
Polly
Cap 20h50'47" -25d10'18"
Polly
And 1h15'6" 46d9'23"
Polly
And 2h24'52" 38d41'54"
POLLY
Leo 10h39'49" 14d40'29"
Polly and Johnny
Lib 15h15'37" -9d50'9"
Polly Ann
Sco 17h2'35" -39d29'28"
Polly Ann Mitchell
Ori 5h23'32" 3d33'0"
Polly Barbara Hill Tevald
Gem 6h58'11" 13d22'15"
Polly Crowley
Umi 13h39'43" 76d8'9"
Polly Farrington
Equ 21h6'36" 6d53'35"
Polly Jo Gillichbauer
Mon 6h51'36" -6d55'43"
Polly Loves Stan Forever
Cyg 19h47'33" 43d52'17"
Polly May Hedges
Cyg 20h44'3" 34d22'5"
Polly May Walton
And 23h28'47" 35d41'35"
Polly Stern
Psc 1h8'15" 28d40'27"
Polly Webster
Lib 15h47'44" -18d39'48"
Polly Weinstein
And 1h0'45" 45d41'34"
Polly, Angel
And 1h39'34" 41d29'10"
Pollyann Artzer
Vir 12h40'54" 4d8'33"
Polo Man
Cas 23h0'22" 55d27'6"
Polomosha
Aqr 22h40'32" -8d47'44"
PoloNYC Johnny B
Cnc 9h6'52" 25d47'28"
Pom Connection 2006
Psc 1h29'53" 18d25'5"
Pomaika i
Del 20h30'36" 20d9'12"
Pomatamus
Tau 4h50'29" 26d45'45"
Pommarine Hotel
Lyn 8h23'7" 59d8'46"
Pompa Durgan
Per 3h22'2" 39d57'28"
pompi
Leo 10h29'17" 26d5'57"
Pona Pinga
Uma 12h23'15" 54d30'57"
Poncho
Cnv 12h43'52" 46d11'51"
Poncho & Lefty.True Love Never Dies
Cyg 19h45'23" 55d41'2"
Pone Diablo Corleone
Per 3h42'6" 45d50'20"
Ponge
Sgr 19h21'28" -20d38'13"
Pongo
Cma 6h40'24" -11d58'20"
Ponsonbee
Gem 6h43'9" 23d24'54"
Ponvili
Tau 4h7'26" 6d19'35"
Pony
Vir 14h34'26" 4d10'26"
Poo & Poo
Uma 9h5'57" 62d35'10"
"Pooby"
Cyg 20h34'32" 31d35'34"
Pooch 25
Lyr 19h16'38" 28d38'52"
Poochie
Uma 11h15'19" 35d0'3"
Poochie
Uma 10h46'1" 50d58'13"
Poochie Cat 1. Anthony Augustine
Pho 1h9'6" -43d51'51"
Poochie's Star
Uma 12h21'35" 55d50'41"
Pooda Budda - Presley John Rowton
Cnc 8h40'51" 23d0'1"
Pooder Girl, the Star of Texas
Peg 21h40'51" 21d12'57"
Poodi Monzo
Cnv 13h26'47" 48d22'5"
Pooey McPoo
Ari 2h42'45" 27d24'44"
pOoFy YoGa
Tau 4h22'14" 12d41'52"
Poogie and Debbie Vandersnick
Cyg 20h56'6" 39d47'58"
Pooh - 1944 - 2004
Uma 11h19'26" 46d18'48"
Pooh and Piglet
Umi 17h0'19" 79d31'26"
Pooh Bear
Umi 14h9'56" 76d28'56"
Pooh Bear
Cap 20h20'44" -14d1'13"
Pooh Bear
Lib 14h53'33" -10d41'4"
Pooh Bear
Umi 15h9'14" 74d7'31"
Pooh Bear
Cyg 20h48'4" 49d3'13"
Pooh Bear
Vir 14h11'11" 2d43'16"
Pooh Bear's Angel
Lyr 19h3'16" 32d22'28"
Pooh Bear's Pot of Honey
Ori 5h40'30" 11d33'14"
Pooh Bear's Star
Cnc 8h10'55" 13d5'59"
Poohbear & Butterfly
Uma 11h26'25" 63d35'51"
Poohbear- Paul David Hurst
Uma 11h13'2" 71d31'47"
Pooh-Bears/Baby-Cakes
Umi 14h24'11" 74d26'47"
PoohHead
Uma 11h42'48" 63d14'12"
Pooh's Hunny
Cyg 20h23'6" 48d1'49"
Pooh's Sternchen
Uma 8h53'41" 48d22'30"
Pooh's Wee Star
Cep 20h42'15" 61d46'18"
Poohs, Queenie and Jay
Cyg 20h13'33" 42d25'18"
Pooj
Cas 0h57'12" 58d58'59"
Pooja
Sco 17h22'59" -42d57'27"
Pooja & Martin Forever
Peg 23h7'31" 29d25'20"
Pook
Uma 10h19'51" 65d34'36"
Pooka Loves Me
Ori 6h21'16" 13d23'17"
Pookaluke
Sgr 19h41'0" -15d4'1"
PookaMate
And 1h31'41" 47d33'34"
Pooka's Shining Star
Ari 3h8'0" 11d35'54"
Pookers
And 2h33'30" 46d11'10"
Pookie
Lac 22h31'24" 50d45'55"
Pookie
Uma 11h51'42" 37d17'20"
Pookie
Psc 1h3'37" 31d6'23"
POOKIE
Psc 1h17'19" 11d40'13"
Pookie
Cmi 7h24'6" 2d49'53"
Pookie
Ori 5h34'44" 1d21'46"
Pookie
Tau 4h41'49" 17d58'51"
Pookie
Leo 11h47'36" 20d46'36"
Pookie
Sgr 19h40'31" -15d0'56"
Pookie
Sco 16h14'32" -10d43'41"
Pookie
Aqr 22h48'54" -10d27'40"
Pookie
Ori 5h59'20" -0d55'0"
Pookie
Aqr 23h47'22" -3d57'58"
Pookie
Cam 4h26'47" 73d36'12"
Pookie
Uma 14h24'17" 55d3'5"
Pookie
Cap 20h27'51" -27d30'1"
pookie
Cap 21h45'20" -22d48'6"
Pookie
Sgr 18h7'33" -32d56'19"
Pookie
Sco 16h36'54" -29d51'54"
Pookie
Lib 15h15'49" -26d40'18"
Pookie 77
Uma 12h37'6" 58d22'18"
Pookie and Snugglebunny's
Uma 9h33'15" 41d35'36"
Pookie Bear
Del 20h36'52" 15d18'40"
Pookie Boo
Col 5h5'50" -36d20'38"
Pookie BQ Berends
Lyn 7h23'1" 57d58'25"
Pookie loves Pokey /DSF&AIR 4ever
Sgr 18h27'35" -16d28'13"
Pookie My Star
Per 3h13'24" 50d48'17"
Pookie 'N Cupcake
Lib 14h31'51" -18d26'55"
Pookie Ongley
Vir 11h37'53" -1d55'5"
Pookie Toot Illustrious
Sgr 18h25'18" -32d10'14"
Pookie & Tootsie Houghton
Uma 8h12'15" 61d5'45"
Pookie Wookie
Tri 2h15'36" 32d41'41"
PookieBear
Uma 11h38'8" 32d57'41"
Pookies
Del 20h35'21" 6d25'46"
Pookie's Everlasting Love for Brian
Ari 2h18'29" 26d16'44"
Pookie's Fishin Star
Psc 23h30'9" 2d50'10"
Pookie's Spot
Uma 8h35'23" 60d49'12"
Pookie's Star
Lyn 7h43'39" 55d51'39"
Pooky
Dra 16h36'5" 64d9'5"

Pooky
Vir 14h48'46" 5d36'9"
Pooky
Cnc 8h13'54" 10d15'59"
Pooky
Peg 21h15'5" 15d44'22"
Pooky Butt
Cap 20h24'36" -12d23'16"
Pooky Pookers (Alexandria)
And 0h42'14" 42d24'31"
Pookyface Blatherwick
Uma 8h18'51" 71d34'3"
Poolside Dream
Sgr 18h25'22" -34d29'17"
Poonam
Uma 13h59'7" 50d5'39"
Poonam Mandalia
Dra 20h8'57" 70d26'33"
Poongy
Sco 16h40'51" -31d53'32"
Poonita
Lib 15h8'57" -12d45'46"
PooohBear
Vir 12h43'42" -11d32'12"
Poop & Peep
Lib 15h8'1" -1d14'51"
Pooper 1
Ari 2h46'19" 17d11'50"
Poopie
Ari 3h6'59" 14d23'21"
Poopies
Uma 8h40'40" 52d18'38"
Poopiter
Lmi 10h38'9" 34d29'25"
Poopnose
Lib 15h23'1" -19d40'21"
Poops John Oram
Vul 20h57'56" 23d39'4"
POOPSIE Debra Mae Gentry
Vir 12h44'32" 1d11'5"
Poopsie's Star (Ronald A. Gates)
Cam 4h19'55" 69d7'8"
Poopsik
Cyg 21h50'41" 55d13'41"
Poopy
Her 17h12'49" 15d32'52"
Poopy
Gem 7h44'55" 18d37'56"
Poor Old Bill
Umi 14h44'10" 74d17'56"
Poor Papa
Cap 21h23'34" -19d32'39"
Pooshy
Psc 23h8'28" 7d18'29"
Pooter Hearts Pookie
Lyr 18h46'13" 31d0'53"
Pootie
Gem 7h2'50" 23d31'19"
* Pooty *
Per 4h28'12" 33d42'46"
Pooty
Lyn 9h9'15" 41d51'27"
Pooziedoo
Psc 0h20'9" 7d11'15"
Pop
Uma 13h40'48" 60d28'37"
PoP
Per 4h34'20" 48d49'18"
Pop
Crb 15h51'16" 26d11'6"
Pop
Ori 4h57'11" 3d58'22"
Pop Bland
Cap 20h57'53" -21d4'12"
Pop&Dolly
Tau 4h4'18" 25d51'3"
Pop George
Cep 22h21'43" 74d13'35"
Pop Joe Fontana
Her 16h59'17" 33d55'20"
Pop: Nick & Justin's star in heaven
Lib 14h50'30" -3d15'38"
POP (Perfect Offsprings' Parent)
Psc 0h51'23" 18d12'14"
Pop Pop
Leo 11h23'43" 17d10'48"
Pop Pop
Sco 16h13'15" -11d10'35"
Pop Pop Alan Isen
Sco 16h8'17" -11d16'45"
Pop Pop Arno
Leo 11h31'29" -0d42'59"
Pop Pop Bill
Aqr 21h27'3" 1d52'47"
Pop Pop Bob Harry, Sr.
Cyg 20h26'35" 48d48'35"
Pop Pop DiGati
Aur 6h0'48" 48d11'27"
POP POP Frank J. Kolodzinski, Jr.
Gem 7h44'12" 14d21'25"
Pop Pop "Hooch" McGonigal
Sco 17h24'42" -37d28'56"
POP POP KERN
Cep 21h52'57" 85d9'5"
Pop Pop Piccinini
Her 17h11'12" 34d22'46"
Pop Pop Sam
Cep 22h38'18" 71d24'0"
Pop Pops
Cep 23h51'55" 82d55'12"
Pop Pop's Celestial Gazer
Cep 0h0'44" 72d39'20"
Popa
Cra 18h48'25" -41d4'7"
Popazaon Joseph E. DiNardi
Her 16h18'8" 25d41'42"
"Popcorn"
Gem 7h12'15" 22d55'49"
Popcorn
Lyn 9h37'48" 41d24'56"
Pope Benedict XVI
Cep 22h7'14" 63d47'16"
Pope Family
Uma 8h43'29" 61d54'40"
Pope John Paul II
Cep 23h16'14" 79d13'12"
Popescu Catalin della stela Alioth
Uma 9h39'23" 68d14'47"
Pophal, Manuela Sternenmaus
Uma 9h20'4" 68d27'4"
Popi
And 23h49'49" 43d23'24"
Popi Vogiazi
Aql 19h45'31" 16d16'0"
Popita
Uma 11h55'25" 42d4'38"
Poplawski, Diana
Ori 4h58'8" 15d6'4"
Popo
Uma 11h35'2" 61d5'25"
Popo
Umi 15h40'8" 85d51'6"
Poppa
Tau 4h29'53" 20d52'26"
Poppa
Ori 5h58'0" 18d27'57"
Poppa and Gramps
Ori 5h40'36" -6d36'25"
Poppa Billy Glenn May
Lyn 6h55'27" 56d16'31"
Poppa Dennis
Cyg 20h33'3" 41d34'24"
Poppa Gagle
Cep 21h20'14" 58d55'14"
Poppa & GranBee
Gem 6h53'18" 16d44'35"
Poppa Jack
Tau 4h7'54" 4d45'13"
Poppa Joe
Ori 6h18'41" 9d42'21"
Poppa Mack Wright
Cep 0h7'38" 78d12'11"
Poppa McCollum—Poppa's Place
Aql 20h16'11" 5d22'32"
Poppa P
Uma 9h28'18" 43d17'38"
Poppa Paul Saunders
Cep 22h39'21" 74d40'51"
Poppa Roy
Cep 23h58'49" 69d36'3"
Poppa Warren and Nana My
Leo 11h38'9" 16d58'54"
Poppadom Bear
Umi 14h14'15" 66d38'9"
Poppa's Best Boys-22081939-26112004
Nor 16h29'29" -50d58'52"
Poppa's precious
Uma 11h19'52" 54d9'37"
Pöppel
Uma 10h31'53" 46d39'17"
Poppers
Uma 8h46'0" 53d6'50"
Poppet
Cyg 20h8'59" 39d51'48"
Poppi Bill McKelvey
Her 17h23'42" 38d19'39"
Poppi Lorenzano
Aqr 22h7'18" -0d12'26"
Poppie
Lib 15h34'23" -10d10'18"
Poppie
Uma 9h14'39" 49d13'54"
Poppie
Ori 6h18'27" 14d53'4"
Poppie
Boo 14h20'4" 14d30'26"
Poppie's Poppett
Umi 14h30'40" 70d32'43"
Popples Evil Star
Ret 3h50'52" -54d13'2"
POPPOFF with Mary Jane Popp
Leo 10h14'51" 25d32'22"
Poppop
Her 17h15'29" 17d1'32"
Pop-Pop
Cyg 21h48'2" 50d53'21"
Poppop Dell
Cep 22h30'58" 73d15'15"
Pop-Pop Everett
Cap 21h28'23" -23d32'2"
PopPop & Pat's Guiding Light
Lyr 18h47'51" 32d33'31"
Poppop Peter
Ori 5h29'41" 5d54'49"
PopPop Rick
Cep 21h24'4" 60d12'44"
POP-POP Shuman Diamond Star
Lib 15h17'52" -16d11'23"
Poppop-Dan
Cep 21h34'1" 62d15'59"
PopPop's Star (from Grayson)
Cap 20h8'11" -13d43'5"
PopPop's wild golf swing
Aqr 23h8'52" -8d44'21"
Poppy
Aqr 22h36'21" -4d4'51"
Poppy
Ori 5h42'57" -0d7'39"
Poppy
Cep 0h21'8" 72d23'20"
Poppy
Lyn 6h26'40" 54d42'57"
Poppy
Uma 11h27'17" 54d54'12"
Poppy
Vir 12h39'10" 10d57'55"
Poppy
Leo 10h14'45" 8d48'8"
Poppy
Ori 6h8'25" 15d21'42"
Poppy
And 0h15'48" 26d25'57"
Poppy
Ori 6h4'37" 18d21'9"
Poppy
And 2h1'53" 37d2'42"
Poppy
And 2h38'18" 41d0'11"
Poppy
Cyg 21h21'32" 30d8'52"
Poppy
Per 3h9'51" 54d1'19"
Poppy
Lyr 19h2'58" 42d31'52"
Poppy
And 23h53'59" 41d24'33"
Poppy
And 23h43'34" 42d43'21"
Poppy
And 0h53'11" 45d17'0"
Poppy aka Charles and Pops
Uma 12h2'37" 43d2'47"
Poppy Alan
Vol 8h37'59" -72d30'0"
Poppy Allan Space Gardener Hughes
Cma 6h39'56" -17d1'18"
Poppy Bruce - Love Tiarna & Deagan
Cru 12h19'6" -56d49'21"
Poppy Clare Reilly
Apu 15h17'52" -71d1'50"
Poppy Doodie
Leo 11h35'57" 25d29'53"
Poppy Faye Bradley
And 23h45'34" 38d1'26"
Poppy Fusco
Dra 18h21'21" 71d13'5"
Poppy Grace Abbott
Cyg 19h42'32" 31d17'0"
Poppy Jane Johnson
And 2h12'37" 49d22'1"
Poppy John Reckus
Sgr 17h52'38" -28d19'47"
Poppy Larson
Uma 8h15'19" 61d37'14"
Poppy Lonergan
Per 3h41'23" 45d32'0"
Poppy Love
Uma 10h17'29" 47d16'32"
Poppy Mae Haines
And 23h56'13" 47d13'33"
Poppy MegaHertzl
Ari 2h10'15" 21d37'34"
Poppy Milty
Sgr 18h14'16" -35d35'43"
Poppy Parry
Cru 12h52'15" -61d4'17"
Poppy Tia Rutter
And 2h14'43" 42d18'3"
Poppy Trewhella
And 0h41'50" 41d10'57"
PoppyJim
Per 2h46'9" 53d20'25"
Poppy-O
Lmi 11h2'29" 25d10'23"
Poppy's Christening Star
And 1h57'23" 36d26'3"
Poppy's Love Bug
Ari 3h12'12" 12d24'32"
Poppy's Place in the Sky
Cep 20h50'25" 65d27'2"
Poppy's Pride
Ori 5h33'55" 1d13'40"
Poppy's Star
Cru 12h17'36" -57d4'26"
Pops
Cep 21h21'5" 61d37'2"
Pops
Uma 10h32'34" 65d16'55"
Pops
Uma 10h10'27" 56d50'55"
Pops
Boo 14h20'29" 19d8'45"
Pop's Gold Star
Psc 1h15'53" 5d19'33"
Pops n' Mops
Leo 9h38'19" 26d52'22"
Pop's Rock
Ori 6h17'8" 14d37'1"
pops star
Gem 6h50'47" 33d57'16"
Pop's Star
Her 17h23'29" 35d45'39"
Pop's Star
Cru 12h17'55" -57d16'31"
Pop's Star "David Perrie"
Cnc 8h8'45" 32d30'20"
Pops' Star Mem: Harry Stuart Barlow
Pyx 9h24'57" -29d26'33"
Pops Wallick
Cep 22h47'46" 68d39'36"
Popsicle/Lobey
Umi 14h32'43" 74d12'29"
Popsie's Point
Cep 21h53'48" 65d12'22"
Popy
Uma 10h17'5" 44d5'14"
Por Siempre Blanca
Cnc 8h15'50" 12d14'3"
Por Siempre Mi Amor
Lib 15h5'57" -16d48'5"
Por Siempre Senor Maravilloso
Psa 22h54'54" -36d11'2"
Porcelaine RH2007
Cnc 9h0'20" 8d55'28"
Porcelina
Sco 16h55'8" -36d3'56"
Porcelina Laughing Infinite
Uma 9h13'39" 63d35'41"
Porcia Stella Marie Park
And 23h45'44" 40d20'50"
Porcnbeanz
Cyg 19h54'9" 33d4'7"
Porco Bruno
Sco 17h51'44" -44d29'22"
Porfiriadu, Kiriaki
Ori 6h16'31" 15d50'2"
Porgie
Ari 2h31'23" 25d58'45"
Pork Chop
Cam 4h56'27" 70d59'6"
Porkchop
Tel 18h43'59" -45d57'32"
Porky
Pho 23h37'56" -42d20'4"
Porn
Sco 17h54'17" -36d3'9"
Porodica
And 1h51'59" 39d9'15"
Porpise
Vir 11h51'14" 9d41'37"
Porscha DeAnna Bentley
Lyr 18h46'26" 31d16'11"
Porsiel, Andreas-Alexander
Uma 10h49'1" 56d7'45"
Port 30
Per 2h58'41" 54d51'53"
Port of Brisbane Corporation
Cru 12h5'48" -64d31'54"
Port Royal Mariners
Uma 12h8'1" 49d50'48"
Port St. Charles
Uma 8h29'1" 63d14'17"
Porter Allen Rupley Born 3:53 PM
Cap 20h10'37" -12d47'55"
Porter Altemus Hunt
Peg 21h43'59" 10d30'51"
Porter David Chesebro
Cap 21h11'51" -21d44'48"
Porter Hemingway
Lmi 10h53'54" 29d30'24"
Porter J. Heywood
Cyg 20h5'38" 37d2'54"
Porter Jackson Earle
Uma 11h0'11" 50d55'5"
Porter Luv
Sge 19h24'28" 18d29'50"
Porter Schorr
Lmi 10h39'27" 32d43'6"
Porter-Kech
Cas 23h55'54" 56d55'50"
Porterpest
Aur 5h25'4" 29d8'30"
Porter's Point
Uma 11h17'15" 63d41'4"
Porth Y Sêr
Cnv 13h40'48" 30d39'45"
Porthos
Umi 14h12'49" 68d56'7"
Portia Derrig Silva
Ori 4h53'57" 15d26'51"
Portia Jernagin
And 1h16'56" 35d54'15"
Portia Marksbury
Uma 10h6'6" 60d30'8"
Portia Princiotta
Aql 19h33'35" 7d14'34"
Portia's Light
Cyg 20h29'23" 43d7'37"
Portico Plaza
Cru 12h18'17" -57d36'17"
Portinho
Sgr 17h56'10" -28d19'54"
Portobello's Advent Star Hull UK
Ori 5h39'42" -0d41'40"
Poseidon
Uma 10h57'41" 60d13'29"
Poser
Cnc 8h39'39" 7d43'33"
Poser Pie
Cap 20h23'45" -20d17'34"
Posie
Uma 11h31'52" 61d54'42"
Posse Rob
Uma 10h9'9" 56d44'24"
Postal Pam
And 0h33'45" 41d21'28"
Postie Star
Dra 19h42'48" 64d3'3"
Postmaster Mark Dolan
Her 16h17'34" 45d9'45"
POSTULKA
Cyg 20h34'57" 46d29'35"
Potato
Sco 17h58'27" -34d38'32"
Potell J
Dra 16h37'33" 52d21'2"
Potentilla Lilliana
Oph 17h3'3" -4d48'37"
Pothier
Peg 22h38'58" 32d11'59"
Potolina
Lyn 7h51'58" 45d22'1"
Potterstar
Lyn 8h43'56" 34d31'47"
Poughkeepsie Pixie
Cma 7h14'6" -30d30'45"
Pouncho's Furnace
And 0h48'11" 45d50'30"
Poupouille
Gem 7h13'32" 21d33'43"
Pour Grand-Papa Ronald qu'on aime
Uma 14h18'53" 55d24'53"
pour l'amour de Howie et Priscilla
Cyg 21h43'42" 40d33'36"
pour le moment et à jamais
Sgr 18h40'19" -17d34'19"
pour ma soeur
Ari 3h13'50" 29d7'12"
Pour Mon Amour
Aql 20h5'24" 12d3'43"
Pour Mon Ange
Cas 0h56'31" 58d37'21"
Pour tou jours Mackenzie
Leo 10h22'2" 25d9'50"
Pour Toujour Mon Amour
Aqr 21h39'54" 2d17'29"
Pour toujours
Aqr 22h25'37" -21d50'58"
Pour Toujours
Uma 10h27'26" 58d25'21"
Pour Toujours
Uma 9h34'0" 54d58'51"
Pour toujours Amants
Sgr 18h53'20" -27d6'23"
pour toujours amour
Vir 11h54'3" -3d27'28"
pour toujours mien
Cyg 19h14'21" 52d54'37"
pour toujours mon amour
Cyg 21h44'10" 52d5'50"
Pour Toujours: Tara and Lori
Uma 10h31'51" 72d11'37"
Pour vous le décrire...
Umi 2h3'29" 89d8'23"
Poussin...
Cas 23h32'33" 51d59'11"
Poutous
Aqr 23h16'28" -13d6'43"
Pouty
Lib 15h5'26" -0d54'50"
POV MARINA
Uma 8h48'33" 58d55'29"
Povelko
Cyg 20h50'59" 43d55'20"
Povertyneck Hillbillies
Lyn 6h30'45" 60d22'28"
powdero'06
Lyn 9h11'12" 39d59'37"
Power
Per 3h12'33" 46d3'30"
Power Stroke and PITA - 143 Always
And 0h44'8" 30d32'22"
Powley Cowie
Cru 12h28'4" -60d49'52"
POYM
Uma 10h5'4" 47d45'0"
poyo
Her 17h34'18" 19d37'14"
PP Livingstone (my star)
Cyg 20h17'43" 30d10'40"
P.P.A.C. (M) Sdn Bhd
Ori 5h38'42" 3d44'37"
P-Pod Prock
Per 3h14'41" 50d28'13"
PPTLYAJKO
And 2h32'42" 37d24'28"
P.P.Yagambrun.
Sco 17h48'25" -36d22'37"
pgr²
Crb 15h33'3" 33d47'11"
Pr. Areej. M.
Leo 10h56'36" 22d1'45"
Prabha
Psc 1h8'24" 29d24'36"
PrabhakarDhumal
Vir 13h14'34" 0d19'48"
Prabhat Govil
Per 3h5'30" 54d15'27"
Prabhleen Chowdhury
Lib 15h48'5" -20d16'58"
Pracht-Pittelkow, Anita
Uma 12h23'18" 61d54'54"
Pradier Ryke Veronike
Cnc 9h21'1" 30d39'13"
Praecipuus Astrum
Uma 11h38'23" 44d46'25"
Praeclarum Mooshed - * Adam & Kim *
Ori 5h41'31" -2d34'57"
praeclarus
Ori 5h27'58" -7d48'32"
Praeclarus Carisa
Tau 3h38'44" 6d28'48"
praeclarus parca
Sgr 17h52'39" -22d42'11"
Praesidium
Sco 16h6'28" -23d27'17"
Pragna Gautam Mehta
Leo 9h40'24" 14d55'21"
Pragya Kamboj
Sgr 18h16'58" -18d36'37"
Praiseworthy, Leather Worker
Ari 3h5'16" 12d59'47"
Prajakta Bhayade
Sgr 18h27'0" -32d23'29"
Prajwolita
Cam 3h50'43" 55d28'51"
Prakash Hebbar
Lyn 7h26'30" 54d2'57"
Prakash Ram Jassal
Uma 10h32'41" 54d53'42"
Prakriti
Cas 23h41'30" 58d19'37"
Pramilla 3377
Ari 1h57'26" 14d45'6"
Prana-mf
Ori 5h18'18" 5d48'3"
Prandkisa
Col 5h46'49" -31d8'2"
Praneel
Aql 19h44'46" 6d32'16"
Pranis's Miracle
Lib 15h2'23" -21d15'10"
Prashanth
Cyg 20h41'58" 50d48'0"
Pratanas 5
Lib 15h4'48" -13d18'56"
Pratanas 3
Lib 15h24'41" -14d45'0"
Pratham Saxena
Sgr 18h44'6" -24d16'26"
Prather
Per 3h33'6" 44d44'5"
Pratu
Uma 9h18'23" 52d51'20"
Pray For The Soul of Betty
Vir 13h3'17" 4d52'55"
Prayer Wenger
Cyg 20h17'24" 38d34'8"
P.R.C. and R.P.C.
Umi 17h20'56" 86d33'19"
Prealyne Rose Mendiola
Cnc 8h30'41" 29d5'12"
Preben Minor 50
Uma 8h52'58" 67d0'45"
Preciosa
Uma 10h51'17" 59d34'37"
Preciosa
Psc 1h21'22" 16d36'25"
Preciosa Linda
Uma 11h10'40" 48d38'13"
Precious
And 23h4'29" 52d17'21"
Precious
And 0h15'38" 39d41'2"
Precious
Cyg 20h59'2" 32d25'58"
Precious
And 2h8'19" 38d51'4"
Precious
Lyr 18h34'23" 32d43'32"
Precious
Cmi 7h28'19" 2d47'59"
Precious
Ori 5h59'21" 6d45'5"
Precious
Gem 7h4'53" 22d31'8"
Precious
Aql 19h4'3" 15d13'13"
Precious
Peg 22h2'39" 15d17'20"
Precious
Uma 11h30'50" 56d34'16"
Precious
Uma 8h32'9" 67d26'56"
Precious
Lib 15h13'38" -20d26'33"
Precious
Cru 12h4'5" -62d34'37"
Precious Abigail
Cnc 9h6'22" 8d34'2"
Precious Amy
Ari 3h15'20" 23d38'29"
Precious and Beautiful, Diane
Psc 1h22'19" 32d23'36"
Precious Angel
And 1h27'47" 41d6'59"
'Precious Angel'
And 0h59'47" 46d28'28"
Precious Angel
Lib 15h50'9" -17d18'32"
Precious Angel Jack Star
Umi 7h53'54" 88d28'2"
Precious Angel Madison Olivia
Uma 10h20'45" 42d57'58"
Precious Angel, Beautiful Girl -RFE
Uma 10h18'30" 52d0'1"
Precious Anja
And 23h12'8" 46d52'37"
Precious baby girl
Her 17h9'15" 31d58'40"
Precious Baby Girl Sophia
Gem 7h28'26" 18d53'44"
Precious Baby Savannah Callif
Cap 20h17'55" -12d27'18"
Precious Being
Ori 5h31'33" 5d55'34"
Precious Blue Christina
Aqr 22h32'9" -0d2'30"
Precious Butterfly 51299
Vir 15h4'25" 3d46'8"
Precious C J
And 1h20'58" 49d42'28"
Precious CC
Cnc 8h12'54" 12d19'0"
Precious Child Wa Wa
Gem 7h5'6" 21d16'42"
Precious Dalton
Ori 6h17'35" 19d32'8"
Precious Dog Napoleon
Vir 12h40'30" 1d36'3"
Precious Dominion Princess
Sgr 19h9'53" -30d14'25"
Precious Eden
Umi 16h53'51" 77d12'32"
Precious Emma Elizabeth Reinkemeyer
Leo 11h22'46" 10d44'0"
Precious Glyn
Per 3h3'45" 51d2'51"
"Precious Gold"
Lib 15h10'5" -6d48'53"
Precious & His Lordship
Cap 20h42'6" -18d17'34"
Precious Jade - 16
Tau 4h25'40" 20d49'30"
Precious Jade Jacobs
Vir 13h2'1" 0d53'39"
Precious JEM
Pup 8h6'50" -23d9'7"
Precious Jen
And 1h43'25" 43d45'15"
pReCiOuS kAtHy
Lib 14h55'15" -0d42'18"
Precious Kobe
Ori 5h19'27" 9d38'13"
Precious Leanne
Her 16h33'7" 46d38'32"
Precious Letitia I Love
And 2h35'8" 45d24'37"
Precious Little Stagger Lee
Cyg 21h41'24" 51d58'25"
Precious Little Star Isabelle Coit
And 2h2'42" 45d52'52"
Precious Love - Jack & Deb
Uma 9h48'58" 63d52'54"
Precious May - Loved for Eternity
Vir 13h57'40" -19d40'37"
Precious Melina
And 1h42'27" 49d0'16"
Precious Melissa
Ind 20h43'15" -45d57'15"
Precious Mirna Hamza
Uma 11h5'1" 45d29'27"
Precious Moon Squat
Gem 7h13'53" 26d16'32"
Precious Patty Poo
Psc 1h11'40" 16d51'9"
Precious Princess Dawn Ann
And 0h13'1" 39d1'57"
Precious Ruru
Tau 4h26'24" 15d17'50"
Precious Sam
Cas 2h10'6" 62d32'46"
Precious — Sandra L. Wallace
Sgr 19h56'22" -28d20'18"
Precious Serene
Cas 0h56'19" 63d42'53"
Precious Son, Timothy
Leo 9h30'40" 19d37'55"
Precious Star of Lia
Gem 7h3'52" 30d54'5"
Precious Stephanie Hutcheson
And 0h40'26" 28d15'14"
Precious "Sweet-Sweet" Chovanec
Aur 6h29'21" 40d42'32"
Precious Taneysha Michelle
Vir 15h6'46" 3d41'10"
Precious Thing
Umi 15h5'57" 69d36'49"

Precious Trisha
And 23h29'50" 41d17'7"
PreciousLouise1983
Uma 12h2'33" 43d55'34"
Pree T H
Cyg 20h15'3" 37d28'53"
Preeti
Vir 12h47'14" 4d39'38"
Preeti
Gem 6h29'48" 21d37'6"
Preeti
Aqr 22h38'29" -0d22'7"
Preeti Krishan
And 0h1'32" 35d51'32"
Preeti Vijay Chawla
Cnv 12h41'34" 49d14'50"
Preia Rose Hayes
Peg 22h53'54" 21d58'19"
Preisner, Klaus
Ori 6h18'53" 18d50'48"
Prelude
Lib 15h5'3" -17d22'19"
Premier Amour
Uma 12h38'9" 55d59'58"
Premier Gym and Cheer
Uma 13h42'28" 54d18'17"
Premji M. Rabadia
Her 16h49'40" 12d20'21"
Prenseopea Adriana
Del 20h49'40" 17d10'35"
Prentice
Gem 6h6'36" 26d12'46"
prerunner91
And 0h32'31" 33d34'45"
Prescott Andrew Stoops
Cnc 8h10'17" 25d23'27"
Prescott Edwardo Brown
Cnc 8h15'17" 15d5'55"
Prescription Fitness P.T.
Uma 8h43'53" 52d9'9"
President and Mrs. George W. Bush
Cyg 20h18'28" 42d50'11"
President Betty L. Siegel
Aqr 23h24'44" -20d49'15"
President Denise M. Trauth
Sgr 17h53'6" -26d13'6"
President George W. Bush Sucks
Crb 16h3'39" 27d33'14"
President KommR Karl H. Pisec, MBA
Leo 11h38'15" 15d34'12"
President Lindsey Hovland
Lyn 7h6'59" 47d3'13"
President Randy Neverka
Leo 10h1'23" 21d26'57"
Presley Alan Godkin
Cep 0h0'31" 75d32'45"
Presley Eve Carter
Tau 4h20'33" 25d47'34"
Presley Jilleanne Fachko
Cyg 20h49'56" 35d21'4"
Presley Lane Sanders
Sgr 19h16'48" -16d7'25"
Presley Layne Blunt
Sco 15h55'41" -22d52'55"
Presley Lynne
And 0h55'42" 45d30'53"
Presley Paige Aurora
Ari 2h25'6" 24d12'39"
Presley Thomas Christian Brown
Per 3h17'11" 31d18'24"
Presley Yarrow's Sweet Sixteen
Leo 10h51'40" 13d6'29"
Preslie Nicole
And 1h0'7" 37d56'4"
Preslie Wyatt
Cnc 8h45'49" 17d51'14"
Pressina J. Wise
Uma 10h24'38" 64d13'56"
Pressley Mary
Ori 6h2'30" 20d57'37"
Pressley Rankin
Sgr 19h15'3" -27d20'49"
Presti's Star
Uma 13h21'3" 55d20'38"
Preston
Tau 5h49'43" 16d50'42"
Preston
Ori 4h52'25" 13d4'33"
Preston 3
Aur 5h12'5" 28d46'10"
Preston Adam Perrett
And 23h40'15" 42d53'0"
Preston Alexander Bryant
Tau 4h9'11" 5d57'23"
Preston Alexsander 12152004
Sgr 18h50'21" -24d36'10"
Preston and Marthe McDaniel
Uma 13h37'14" 54d2'12"
Preston Andrew Johnson
Psc 1h19'28" 28d13'58"
Preston Brent Fisher
Equ 21h15'43" 11d33'7"
Preston Caroline
Aqr 22h55'31" -9d26'27"
Preston Christopher
Lmi 9h25'41" 38d31'25"
Preston Ellis
Tau 4h33'16" 4d13'21"
Preston Evan Forman
Cam 3h52'23" 61d33'10"
Preston Frank Gosa, Jr.
Per 3h49'52" 42d0'58"
Preston Gage Porreca
Leo 11h2'45" 23d49'0"
Preston George Verness
Leo 9h27'14" 27d50'34"
Preston Geraghty Browne
Her 18h33'44" 19d28'10"
Preston H. Davis
Cep 2h44'7" 83d37'4"
Preston J. Postemski-Lemke
Uma 10h31'25" 68d18'41"
Preston J. Wilkins
Her 18h28'0" 21d43'59"
Preston James Hamrick
Her 18h50'15" 24d21'6"
Preston Joseph Bridger
Her 17h2'25" 15d0'0"
Preston Lee Lance Wagoner
Leo 10h59'22" 14d12'12"
Preston Lehman
Her 17h46'47" 47d24'14"
Preston Noble Lucas
Umi 14h40'49" 77d46'48"
Preston & Parker Hale
Uma 9h40'32" 46d7'17"
Preston "P-nutt" Allen
Sct 18h25'8" -15d13'47"
Preston Schmidt
Her 18h26'40" 26d43'23"
Preston Sunny Rosser
Uma 11h51'52" 40d39'3"
Preston Troy Swaffar
Vir 14h2'44" -5d29'12"
Preston Wade Wimbish
Cyg 20h1'35" 33d49'6"
Preston Werner
Uma 11h53'26" 40d14'36"
Preston Xavier Paoni
Aqr 21h46'10" -1d6'47"
Preston's love
Lib 15h10'18" -16d13'18"
Presul Astrum 7202
Cap 20h13'55" -13d37'31"
Prettiest Star Named Michelle
Ari 3h11'50" 28d50'58"
Pretty Baby Amanda
Sgr 19h18'26" -15d54'34"
Pretty Carla Sue
And 22h59'34" 49d49'34"
Pretty Eyes
Gem 7h31'13" 22d18'7"
Pretty Face
Vir 12h18'54" 12d54'46"
Pretty Fairy Stella of Rie
Cap 20h28'22" -21d53'59"
Pretty Girl
Aqr 22h20'9" -14d45'44"
"Pretty Girl"
Aqr 20h52'27" 1d29'11"
Pretty Girl
And 1h10'47" 36d8'1"
Pretty Girl Danielle
Aqr 20h51'0" -7d1'14"
Pretty Girl Jess
Psc 23h57'45" 10d36'52"
Pretty Girls Star
Ori 5h21'46" 0d35'45"
Pretty Kitty
Cep 21h30'41" 64d17'12"
Pretty Lady
Uma 10h33'59" 54d14'19"
Pretty Lady
Cas 23h26'30" 56d49'59"
Pretty Lady
Vir 13h34'4" -4d36'19"
Pretty Lady
Lmi 9h23'11" 36d34'20"
Pretty Lady
Cyg 19h52'29" 33d10'34"
"Pretty Lady" Angela Gay Sanchez
Psc 0h46'11" 15d48'12"
Pretty Lala
Vir 12h55'29" -7d22'59"
Pretty Mommy
And 0h0'15" 44d44'38"
PRETTY PATSY 12-5-03
Uma 10h53'7" 59d19'32"
pretty pony...tim
Vir 14h35'40" 0d0'8"
Pretty Poohla
Vir 12h41'34" 3d34'12"
Pretty Princess
Tau 5h26'9" 24d25'50"
Pretty Princess
Cam 3h55'38" 55d29'13"
Pretty Princess Wendy Lou
Umi 14h52'10" 79d17'15"
Pretty Star
And 23h19'18" 47d5'30"
Prettylady
Cnc 8h38'58" 22d14'42"
Pretzell, Bernd
Uma 10h54'54" 70d37'35"
Preuß, Jochen
Ori 6h1'40" 10d36'35"
Preziosi Memorial
Uma 8h45'17" 50d1'39"
PRG & BAK
Uma 11h54'38" 61d7'22"
PRHAHC
Sgr 17h52'36" -29d27'37"
PRIANJISHIVAMY 4-10-1-11
Lyr 18h28'26" 35d44'9"
Price 4
Leo 10h22'6" 14d18'42"
Priceless
Cap 20h22'17" -18d15'15"
Priceless Karen
Mon 7h0'5" 7d28'27"
Price's Star
Lyn 7h25'44" 44d41'18"
Pricess Heather R. Roed
Sco 16h24'38" -27d16'36"
Pricey
Cru 12h3'20" -61d53'17"
Pricila
Ari 1h56'57" 18d43'33"
Pride and Joy
Cyg 20h21'45" 52d27'13"
Pride Smith Evans
Lac 22h46'22" 49d56'47"
Priest Family-Thunder's Black Mist
Cnv 12h54'45" 40d55'5"
PRiG
Cyg 20h49'28" 37d7'58"
Priger Tamás (Béb)
Uma 9h44'36" 43d57'52"
Prima Donna Jill
Ori 6h17'33" 16d5'50"
Primativa
Uma 10h25'14" 49d40'53"
Primetime
Umi 16h22'37" 81d3'13"
Primiti Berrios
Sco 17h41'15" -30d56'45"
Primmy
Cnc 8h40'5" 7d10'52"
Primo Amor
Lib 15h52'38" -14d12'49"
Primo Bacio
Leo 10h9'0" 19d19'29"
Primo Maspoli
Cep 21h25'56" 55d29'24"
Primo Parker
Tau 5h37'9" 18d35'48"
Primoris Prognatus
Ori 6h19'9" 16d37'28"
Primoris Verus Diligo
Psc 0h51'38" 6d34'53"
Primps
Mon 7h2'1" 5d54'57"
Prina
Pho 1h7'7" -42d41'19"
Princ - Mirza
Ori 5h40'52" -0d54'25"
Prince
Uma 11h35'51" 50d21'25"
Prince A. Davis
Uma 14h20'16" 55d52'33"
Prince Aldefa
Umi 15h22'7" 68d27'49"
Prince and Cinderella
Gem 7h38'50" 34d21'47"
Prince Andrew Harry
Ari 3h9'56" 11d35'4"
Prince Anthony
Aqr 21h42'6" 1d56'37"
Prince Calin
Cnc 7h59'11" 12d57'34"
Prince Charlie
Lib 16h1'58" -15d48'42"
Prince Charming
Cep 22h20'31" 71d51'7"
Prince Charming
Cnc 8h58'54" 31d59'19"
Prince Charming
Per 4h47'54" 44d17'3"
Prince Charming
And 1h46'34" 39d2'17"
Prince Charming Andrew
Tau 5h34'48" 26d15'7"
Prince Christoph
Cyg 21h38'57" 49d50'55"
Prince Christopher's Virj
Psc 1h10'57" 24d47'17"
Prince Chukas Buddy Roi Nkem
Gem 6h48'43" 33d53'32"
Prince Cody
Uma 10h42'7" 54d13'19"
Prince Daniel James
Ori 5h52'44" 7d14'30"
Prince des Collines
Uma 13h31'7" 58d8'58"
Prince Ephraim III
Sgr 18h27'57" -26d35'36"
Prince Erich M. Mueller
Her 17h52'25" 14d44'56"
Prince Erik James Phair
Tau 5h36'42" 25d48'4"
Prince Fluffyhead
Gem 7h0'52" 13d37'57"
Prince Hans H 'The Noble'
Aqr 22h8'3" -22d55'22"
Prince Jordan
Gem 7h25'24" 28d28'4"
Prince Jordan Alexander Walker
And 0h52'5" 39d19'52"
Prince Matty
Cra 18h4'15" -37d19'53"
Prince Michael Vincent Blume
Sco 16h30'45" -30d37'18"
prince of heart
Cam 6h1'22" 67d2'9"
Prince of Thieves
Her 17h48'16" 41d36'34"
Prince Oren
Cam 3h49'53" 71d1'9"
Prince Parker
Ari 2h56'37" 26d36'41"
Prince Pierre Nicolas Bourlatchka
Per 4h21'29" 40d44'32"
Prince Ramirez of the Highlands
Aqr 22h14'18" -1d29'30"
Prince Rick
Psc 0h58'14" 10d11'37"
Prince Rupert
Ori 5h49'32" 12d4'56"
Prince Ryan
Ori 6h25'34" 16d55'23"
Prince Ryan Charming
Cep 22h23'34" 64d17'5"
Prince Tracy Lee
Ori 5h50'21" 4d4'7"
Prince William
Aqr 22h17'34" -3d33'17"
Prince Zachary Samuel
Ari 3h19'6" 16d2'21"
Prince, Princess, & Paisley Jaques
Ori 5h33'37" -0d44'22"
Princela
Aqr 23h2'20" -5d12'11"
Princes Andrea J Schulz
Psc 1h45'44" 5d18'19"
Prince's Star
Cma 7h6'39" -17d50'34"
Princesa
Sgr 18h42'34" -18d5'18"
Princesa Keyssi
Gem 7h44'36" 26d1'0"
Princesa Lucia A.R.
Col 5h26'24" -29d10'15"
Princesa Margarita
Mon 6h30'57" 7d41'20"
Princesa Marisela
And 0h41'55" 23d1'32"
Princesa Patricia
Sco 16h17'13" -33d14'5"
Princesita
Lib 15h29'47" -24d49'43"
Princesita
And 0h36'16" 27d33'25"
Princesita Lolita
Vir 12h41'1" 11d30'31"
Princess
Vir 13h22'45" 13d39'41"
Princess
Psc 1h29'57" 15d45'7"
Princess
Del 20h40'10" 14d2'9"
Princess
Vir 12h10'49" 4d39'21"
Princess
Ori 5h8'0" 14d8'56"
princess
Ari 2h50'18" 14d20'28"
Princess
Psc 23h8'22" 2d45'2"
Princess
And 0h30'44" 28d27'7"
Princess
Ari 2h32'20" 26d56'26"
Princess
Gem 6h30'19" 18d16'27"
"Princess"
Ori 5h55'56" 17d18'39"
Princess
Gem 6h57'44" 20d47'30"
Princess
Cnc 8h24'47" 28d19'47"
Princess
Cnc 9h9'32" 25d39'56"
Princess
Gem 7h9'24" 26d21'23"
Princess
Tau 5h48'26" 27d38'56"
Princess
Crb 15h37'23" 26d25'39"
Princess
And 2h24'16" 41d53'57"
Princess
And 2h30'7" 41d52'43"
Princess
And 2h35'52" 40d6'21"
Princess
And 2h37'6" 40d43'10"
Princess
And 0h14'12" 44d21'42"
Princess
And 0h40'23" 39d42'40"
Princess
And 0h54'8" 39d41'15"
Princess
And 1h3'33" 42d1'54"
Princess
And 1h39'26" 41d21'53"
Princess
And 23h57'26" 35d24'47"
Princess
And 1h45'18" 36d27'5"
Princess
And 23h59'10" 42d38'58"
Princess
And 0h45'50" 45d53'17"
Princess
And 2h32'21" 47d58'30"
Princess
And 2h17'12" 49d25'5"
Princess
And 23h14'11" 48d37'44"
Princess
And 23h10'14" 50d57'12"
Princess
Uma 13h55'40" 50d23'5"
Princess
Cap 20h42'17" -24d49'38"
Princess
Psc 0h17'41" 4d58'22"
Princess
Pav 17h59'32" -57d11'52"
Princess
Cru 12h17'51" -57d22'51"
Princess
Pic 5h14'55" -50d29'5"
Princess
Cap 20h27'42" -18d17'19"
Princess
Cap 21h2'24" -16d17'37"
Princess
Lib 15h56'24" -18d32'13"
Princess
Sco 16h16'19" -15d25'32"
Princess
Cma 6h49'48" -15d52'3"
Princess
Aqr 22h50'58" -7d56'49"
Princess
Cap 20h14'18" -14d23'11"
Princess
Aqr 20h55'26" -8d25'32"
Princess
Cap 21h36'0" -8d27'41"
Princess
Sco 16h9'22" -11d47'59"
Princess
Cma 6h44'13" -14d46'35"
Princess
Vir 12h55'58" -1d20'53"
Princess
Cas 0h45'31" 74d38'17"
Princess
Uma 14h27'45" 59d23'40"
Princess
Uma 10h56'8" 52d34'39"
Princess
Uma 8h55'57" 52d48'26"
Princess Abbi
And 1h19'47" 49d29'17"
Princess Abbie Gillam
And 23h55'34" 41d41'22"
Princess Abigail
And 0h48'46" 38d6'0"
Princess Aco
Vir 13h16'25" -9d55'53"
Princess Adi
And 1h37'5" 41d12'16"
Princess Aimee
Her 18h56'12" 16d23'47"
Princess Alaina
Ori 5h24'57" -0d1'46"
Princess Alana
And 23h27'58" 47d5'37"
Princess Alana - 21082007
Col 5h46'9" -35d29'20"
Princess Albright
Leo 10h11'56" 14d53'19"
Princess Ale Lafuente
And 1h51'17" 37d56'27"
Princess Alena
Sgr 18h56'13" -34d45'1"
Princess Aletta Felderer
Uma 11h59'21" 33d55'37"
Princess Alexandra Marie Del Sole
Ori 6h5'27" 17d20'39"
Princess Alexandra Michele Silvers
And 0h42'59" 40d59'2"
Princess Alexandria Shea Berry
Vir 13h1'57" -12d56'25"
Princess Alexis
Cnc 8h22'9" 23d36'0"
Princess Ali of Highland
And 1h13'27" 37d11'34"
Princess Alice
And 2h24'8" 39d41'10"
Princess Alice and Pumpkin Tom
And 2h30'53" 50d9'4"
Princess Alison
Tau 4h34'9" 1d42'13"
Princess Alison Klein
Ori 6h19'38" 19d31'41"
Princess Alison's Star
Sco 17h6'20" -42d52'15"
Princess Alita Eghbali
Sgr 19h27'19" -17d54'22"
Princess Allison
And 1h23'26" 38d17'10"
Princess Allison Lenard
Cra 19h3'52" -38d42'39"
Princess Ally
Cas 0h9'6" 58d23'1"
princess ally baby
Cap 21h8'21" -19d51'30"
Princess Ally Willson
And 0h52'34" 38d30'12"
Princess Amalia
Cas 2h0'0" 74d10'56"
Princess Amalia Mladin
Tau 5h53'26" 24d13'43"
Princess Amanda
Psc 1h34'19" 11d3'35"
Princess Amanda
And 0h50'53" 37d35'38"
Princess Amanda
And 0h15'57" 43d41'26"
Princess Amanda
And 0h36'58" 42d1'47"
Princess Amanda
Cyg 19h59'35" 35d1'52"
Princess Amanda
Cam 5h46'24" 61d20'34"
Princess Amanda
Cas 1h7'47" 62d38'6"
Princess Amanda Jean Bradshaw
And 1h28'2" 39d10'1"
Princess Amber
Ori 5h56'10" 7d23'23"
Princess Amy
Leo 9h23'2" 12d26'54"
Princess Amy
Tau 4h10'45" 16d27'43"
Princess Amy
And 2h31'29" 49d57'0"
Princess Amy
And 23h11'59" 40d40'56"
Princess Amy Katherine
And 23h38'7" 43d4'54"
Princess Anastasia
And 0h15'36" 25d8'38"
Princess and Hunnybunny 04-04
And 0h31'13" 42d3'57"
Princess Andrea
And 0h43'16" 25d42'2"
Princess Andrea
Vir 14h13'2" -13d34'13"
Princess Andrea Majdecki
And 1h16'40" 46d9'51"
Princess Angel
And 0h29'54" 28d28'30"
Princess Angel
Ari 2h12'48" 24d52'2"
Princess Angel (Cindy)
And 2h25'27" 39d9'40"
Princess Angel Jennifer
And 2h0'59" 41d40'4"
Princess Angel Kimberly
Cyg 20h47'43" 35d57'55"
Princess Angela Marie
Lib 15h18'51" -4d41'45"
Princess Angela Parrish
And 0h2'48" 40d27'48"
Princess Angelina
And 2h35'36" 39d42'16"
Princess Ani
Lib 14h40'50" -17d44'26"
Princess Ann
Lib 15h2'39" -2d33'47"
Princess Anna
Ori 5h30'35" -7d35'50"
Princess Anna
Sgr 19h24'3" -17d59'37"
Princess Anna
Tau 3h36'5" 12d57'59"
Princess Anna Marie Isaak
And 23h5'58" 47d32'58"
Princess Annette Magnus
Sgr 19h15'13" -15d23'31"
Princess April
And 1h7'18" 41d51'19"
Princess Ariana
Ari 2h29'56" 25d6'9"
Princess Arsenia
And 2h7'33" 42d50'58"
Princess Ashley
And 0h21'13" 25d39'37"
Princess Ashley
Gem 6h51'4" 25d13'20"
"Princess" Ashley Amanda Baquero
And 23h0'12" 37d29'47"
Princess Ashley Auren
Aqr 22h38'46" -20d19'4"
Princess Ashley Paige Niday
And 23h56'18" 38d15'50"
Princess Ashling
And 23h19'24" 52d15'7"
Princess ASHLYN Raine [9th year]
Leo 10h27'13" 13d43'38"
Princess Audrey Lyons
And 0h23'49" 33d40'35"
Princess Autumn
And 0h7'24" 43d42'16"
Princess Autumn Paige
And 23h55'26" 39d48'32"
Princess Avery Leigh
Uma 13h46'54" 56d44'46"
Princess Avgusta
Vir 14h48'16" 3d28'31"
Princess Aya's Shining Light
Psc 23h29'18" 5d31'17"
Princess Azaliyah
And 0h19'11" 26d46'23"
Princess B
And 0h37'5" 36d56'14"
Princess Baby Peanuts 10/19/79
And 2h4'6" 37d30'33"
Princess Baca
Ori 5h56'46" 6d42'4"
Princess Barbara
Lib 15h28'22" -23d29'1"
Princess Bassily
Gem 7h28'14" 20d26'5"
Princess Beanbun
Lac 22h42'53" 36d38'4"
Princess & Bear - Together Forever
Cru 12h38'39" -60d10'6"
Princess Belle
And 0h46'39" 44d20'34"
Princess Betsy
And 0h39'50" 39d10'29"
Princess Bianka - 18.12.1988
Cru 12h30'4" -58d9'12"
Princess Bobbie
And 0h22'30" 36d30'25"
Princess Bobi
And 0h31'2" 29d21'55"
Princess Bonnie Lea B 1941 6262
And 0h16'13" 46d15'58"
Princess Brandy
Aqr 22h36'56" -19d23'41"
Princess Bree's Star
Cap 21h6'3" -20d34'48"
Princess Brenda Diane
And 0h13'4" 28d25'36"
Princess Briana's Estrella
Gem 6h45'46" 17d19'4"
Princess Brianna Arpi Kabalkin
Per 3h37'25" 47d7'38"
Princess Brittani Nicole
Lyr 18h52'19" 34d58'50"
Princess Brittney
Cnc 8h20'54" 13d57'2"
Princess Britty Brannen
Del 20h35'3" 19d20'42"
Princess Brynn
And 0h24'30" 41d32'33"
Princess Bryony Peggy-Jane
Sgr 18h6'52" -27d44'1"
Princess Bubbles
And 23h27'35" 47d34'14"
Princess Butterfly
And 23h39'2" 36d46'35"
Princess Butterfly
Lib 15h18'13" -23d22'54"
Princess Cailee Marie
Tau 5h50'38" 15d5'10"
Princess Caitlyn
And 0h20'12" 28d41'9"
Princess Callie
Sgr 18h54'39" -33d58'53"
Princess Callisto
And 23h27'58" 42d33'58"
Princess Candace Brooke
And 0h44'23" 34d49'33"
Princess&Candy
Ari 2h7'56" 25d16'21"
Princess Candy
Sco 17h56'41" -34d25'36"
Princess Candy - Ying Ying
Cap 21h41'34" -17d6'11"
Princess Cantaloupe
And 0h35'6" 36d5'27"
Princess Cara
And 2h20'52" 39d59'48"
Princess Carol
And 2h18'52" 39d16'14"
Princess Caroline
And 1h4'47" 43d29'34"
Princess Caroline
And 23h2'41" 42d32'45"
Princess Carrie
Per 3h14'14" 55d15'28"
Princess Carrie
And 2h19'9" 39d1'32"
Princess Casey's Pink Star
Ari 3h10'54" 27d30'15"
Princess Cassandra XIII
Cma 6h50'53" -17d37'14"
Princess Cassidy
And 23h24'7" 47d16'36"
Princess Cassie
And 23h52'48" 41d27'48"
Princess Catherine Muir
And 2h16'5" 46d36'46"
Princess Cathleen
And 0h1'17" 44d21'37"
Princess Cathy
And 1h30'33" 47d18'3"
Princess Cera
And 1h37'49" 45d39'17"
Princess Charlotte
And 2h35'55" 49d22'18"
Princess Chelsea
Uma 9h24'22" 51d47'47"
Princess Chelsea Rose
Sgr 19h30'53" -36d8'56"

Princess Chelsea Vassila
And 0h16'23" 40d0'44"
Princess Chelsy Sue
And 23h27'56" 36d14'32"
Princess Chiara Morrison
And 23h46'45" 41d21'43"
Princess Chika 22
Sco 17h24'42" -42d55'10"
Princess Chloe
And 0h17'16" 46d11'58"
Princess Chloe
And 1h35'3" 48d38'44"
Princess Chloe Mae Robinson
And 0h22'15" 44d50'31"
Princess Chloe Pearl
Gem 7h22'53" 33d32'6"
Princess Christal Lee Wilcox III
And 1h10'38" 34d47'15"
Princess Christena
Lyn 7h56'59" 39d53'35"
Princess Christina
Gem 6h42'59" 34d55'28"
Princess Christina Gambale
Ori 5h39'24" -1d11'7"
Princess Christina Garcia
Ari 2h19'34" 22d51'27"
Princess Christina Nicole
And 2h14'19" 45d38'34"
Princess Christina O'keeffe
Vir 13h18'42" 12d15'57"
Princess Christine
Sco 17h54'50" -40d11'41"
Princess Christine of Carlisle
Cap 20h17'44" -13d25'1"
Princess Cindy
Aqr 22h28'16" -2d42'34"
Princess Cody
Aqr 21h9'40" -8d46'19"
Princess ConstancePersistence Grace
And 2h27'24" 45d19'2"
"Princess Cora"
And 1h32'22" 50d37'59"
Princess Corinne
Cam 6h48'10" 65d42'33"
Princess Cory
And 1h28'5" 46d9'3"
Princess Courtney Amber Knipp
Sco 17h50'15" -37d34'16"
Princess Cricket
Vir 13h33'34" -21d39'30"
Princess Cristina Marie Brito
Pho 1h35'55" -42d45'43"
Princess Crystal
And 0h41'58" 41d19'11"
Princess Cyndi Hatcher
And 23h50'54" 43d34'32"
Princess Cynthia Ann Covarrubias
And 2h0'1" 43d40'20"
Princess Dakota
And 1h46'9" 46d38'48"
Princess Dallas
Lib 15h52'29" -10d53'24"
Princess Danielle
Aqr 23h18'58" -18d43'39"
Princess Danielle
Uma 9h34'31" 46d32'0"
Princess Danielle E. Fahey
And 0h59'50" 43d19'52"
Princess Danielle Gorbel
Psc 1h25'26" 27d38'26"
Princess Danielle Marie
Ari 2h44'7" 16d8'32"
Princess Danielle Rosina
And 23h47'58" 41d31'30"
Princess Daphen
Lyn 9h20'35" 37d2'50"
Princess Dara
Sco 16h40'20" -32d40'22"
Princess Dawn
Aqr 22h19'25" -13d39'34"
Princess Deann Michelle
Lib 15h6'59" -14d13'38"
Princess DeAuzhoni
Ori 6h18'21" -0d50'35"
Princess Deborah
And 23h20'17" 52d6'54"
Princess Dee
Sco 17h53'2" -31d34'15"
Princess Dee Dee
Tau 4h25'6" 28d2'17"
Princess Detelina Trendafilova
Leo 11h12'52" 1d15'46"
Princess Diana
Cnc 8h39'50" 15d40'41"
Princess Diana
And 1h39'38" 43d23'37"
Princess Diane of Harper's Harem
And 23h27'23" 36d19'56"
Princess Dianita
And 2h28'20" 50d18'54"
Princess Diljit
And 0h36'1" 26d28'24"
Princess Doi Doi
Ari 1h59'17" 19d16'58"
Princess Dominique
And 23h59'58" 40d36'18"
Princess Donna
And 1h31'2" 49d30'39"
Princess Donna
Tau 5h6'52" 25d12'43"
Princess Dorothy
Lib 15h52'27" -18d52'59"
Princess Dy
Gem 7h53'21" 32d42'13"
Princess Eden06
Aqr 22h1'52" -23d56'47"
Princess Edith R
Sco 17h25'59" -37d26'36"
Princess EJ
And 23h29'17" 47d11'56"
Princess Elena
And 1h41'29" 45d13'19"
Princess Elena
Tau 4h24'27" 3d37'11"
Princess Elena Sambeat
Cyg 20h7'8" 36d18'9"
Princess Elenora
Lib 15h20'0" -6d52'12"
Princess Elisa Schalk
And 0h49'26" 37d25'55"
Princess Elise
Lyn 9h23'26" 40d29'30"
Princess Elise Ann
Cru 11h59'13" -57d24'28"
Princess Elizabeth
Lib 15h23'52" -7d41'50"
Princess Elizabeth
Vir 13h27'52" 13d30'27"
Princess Elizabeth
Crb 15h39'14" 27d33'33"
Princess Ella
And 23h16'22" 50d52'37"
Princess Emilie Jao
And 2h11'20" 37d52'43"
Princess Emily
Sco 16h52'7" -38d42'9"
Princess Emily Ann Finch
Col 5h45'41" -34d39'2"
Princess Emily Sue
Lmi 10h0'23" 39d47'19"
Princess Emma
And 23h56'37" 40d39'34"
Princess Emma
And 23h17'42" 40d29'39"
Princess Emma Rachel Ginsberg
And 1h16'17" 35d7'8"
Princess Erica
Leo 10h12'26" 14d21'41"
Princess Erica
Cnc 8h17'39" 9d10'36"
Princess Erica
Tau 4h8'37" 16d28'19"
Princess Erica Royal
And 2h15'47" 46d45'8"
Princess Erika
Tau 4h14'33" 24d22'29"
Princess Erin
Ari 2h17'31" 13d53'49"
Princess Erin
Cap 20h20'58" -12d51'16"
Princess Erin Ashley Glasgow
Sco 16h11'54" -15d45'4"
Princess Erin Kristine
Lyn 8h38'59" 34d53'13"
Princess Erisa
Aqr 23h13'23" -16d20'28"
Princess Eternal Love
And 2h26'37" 46d51'27"
Princess Eunice Bangit Ong Alonzo
Psc 0h39'23" 13d38'55"
Princess Eva Agnieszka Szalewicz
And 2h24'47" 46d28'48"
Princess Eva Jirsenska
Uma 10h22'17" 71d43'19"
princess fairy bear
Uma 9h4'27" 48d1'58"
Princess Faith
Vir 13h9'6" 11d34'50"
Princess Fiona
And 1h29'58" 44d11'33"
Princess Gabriella
Ori 6h3'6" 20d1'39"
Princess Gabriella
Gem 7h42'44" 15d55'36"
Princess Gabrielle
Lib 15h9'15" -3d52'30"
Princess Gayle
Cas 0h10'22" 53d54'36"
Princess Geen
And 0h34'28" 40d29'2"
Princess Gemma
And 1h20'29" 43d9'6"
Princess Genevieve
Leo 11h35'25" 15d16'27"
Princess Georgia
Uma 9h18'21" 62d12'46"
Princess Georgina
Psc 1h13'40" 10d59'35"
Princess Gillian
Lib 15h3'40" -2d20'10"
Princess Gini
Lmi 10h9'14" 32d37'18"
Princess Giovanna Ariola
Lib 15h3'28" -0d47'7"
Princess Glynis The Supermom
Vir 13h6'57" 11d13'13"
Princess Grace Marie
Cnc 8h19'46" 32d3'36"
Princess Grace Morgan Felix
Sco 16h12'6" -41d22'10"
Princess Guyshia
Psc 0h50'4" 24d47'2"
Princess Gwendolyn
Vir 13h11'13" 8d10'57"
PRINCESS H
Crb 15h38'37" 26d10'58"
"Princess" Haley Angelina
Ari 2h58'47" 26d14'2"
Princess Hannah
Tau 4h29'47" 21d34'25"
Princess Hannah from the Heavens
Uma 9h26'52" 57d37'22"
Princess Haya
And 0h18'39" 37d18'30"
Princess Heather
And 1h12'39" 37d19'9"
Princess Heather
And 2h9'0" 44d23'11"
Princess Heather
Vir 12h41'40" 6d51'29"
Princess Heather
Del 20h41'18" 16d46'47"
Princess Heather
Uma 11h13'0" 28d31'1"
Princess Heather Ariel Barnhart
Vir 14h15'34" -20d39'45"
Princess Heather Lynn
Uma 9h50'59" 42d14'5"
Princess Heather Warren
And 0h36'16" 30d50'41"
Princess Helena
And 1h55'21" 46d11'36"
Princess Helena
Psc 23h43'57" 5d46'58"
Princess H.E.R.
And 2h2'2" 42d54'39"
Princess Hilary
Psc 0h10'0" 6d48'9"
Princess Holly
Sgr 19h45'59" -11d49'27"
Princess Holly Jessica
And 23h58'8" 44d32'0"
Princess Homet
And 1h54'27" 36d36'35"
Princess & Honey Bunny
Cyg 22h0'20" 52d46'6"
Princess Hope Star
Psc 23h4'30" 5d27'33"
Princess HUDA
Ori 5h27'20" -0d16'31"
Princess Hudson
And 23h10'27" 51d3'39"
Princess IDA
Aqr 22h34'14" 2d16'44"
Princess Ingrid
And 2h34'16" 49d7'15"
Princess Irina
Vir 12h43'33" 4d46'54"
Princess Ivett Zesati
Ari 1h54'43" 22d6'51"
Princess Izzy
And 0h54'5" 24d23'56"
Princess Jackie
Leo 10h37'35" 22d18'16"
Princess Jackie
And 0h13'28" 46d32'55"
Princess Jackie
And 2h1'39" 43d58'37"
Princess Jaclyn
Aqr 22h2'31" -13d30'36"
Princess Jaclyn Cianchetti
And 1h5'43" 42d51'57"
Princess Jacqueline
And 2h18'49" 49d28'1"
Princess Jada
And 1h17'11" 47d18'59"
Princess Jagarnauth
And 0h35'2" 29d42'34"
Princess Jai's Star
And 23h4'8" 45d42'28"
Princess Jamie
And 1h4'36" 38d21'21"
Princess Jamie
Cas 23h1'26" 58d50'53"
Princess Jamie
Sgr 18h47'50" -27d9'20"
Princess Jan
Cas 1h29'7" 62d40'32"
Princess Jana
Gem 7h28'7" 26d20'19"
Princess Janatalie
Cnc 8h3'3" 12d48'57"
Princess Janaya Vanel Hernandez
Sco 17h53'21" -38d49'3"
Princess Janet
And 1h58'58" 39d11'15"
Princess Janice Elizebeth
And 23h7'10" 50d57'35"
Princess Janie
Cnc 8h52'27" 22d31'16"
Princess Janie
And 0h16'0" 29d42'31"
Princess Jasmin
And 22h59'54" 48d11'6"
Princess Jasmine Alexis King
Psc 0h9'55" 3d17'29"
Princess Jaya
And 1h17'0" 45d13'56"
Princess Jayani Asha Seth
And 1h40'30" 42d59'17"
Princess Jazz
Vir 12h56'29" -3d2'48"
Princess Jeannie's Laughter
Psc 1h18'16" 11d48'29"
Princess Jen
Cnc 8h54'28" 26d44'52"
Princess Jen
Aqr 22h18'41" -2d21'0"
Princess Jen
Uma 8h30'29" 60d53'21"
Princess Jen
Aqr 23h19'40" -20d26'8"
Princess Jenelle
And 1h23'8" 47d25'35"
Princess Jenn Evans
And 0h52'36" 35d45'12"
Princess Jenna
Aqr 23h15'26" -15d35'0"
Princess Jennifer
Aqr 22h3'51" -16d14'21"
Princess Jennifer
Umi 12h2'16" 86d2'8"
Princess Jennifer
Umi 15h42'56" 70d25'45"
Princess Jennifer
And 2h16'44" 48d15'54"
Princess Jennifer
And 23h31'46" 42d18'40"
Princess Jennsync
And 2h2'2" 40d42'29"
Princess Jenny
Tau 4h48'52" 19d36'48"
Princess Jenny Lee
Lib 15h37'46" -12d45'35"
Princess Jessi
Lyn 7h44'37" 48d36'36"
Princess Jessica
And 0h53'36" 45d39'47"
Princess Jessica
And 0h50'43" 45d10'21"
Princess Jessica
And 23h3'6" 42d38'58"
Princess Jessica
And 2h23'13" 39d25'35"
Princess Jessica
And 0h54'15" 43d23'23"
Princess Jessica
And 0h13'59" 44d41'17"
Princess Jessica
Cnc 9h17'38" 6d30'22"
Princess Jessica
Boo 14h48'2" 29d37'3"
Princess Jessica
Aqr 20h55'57" -9d52'32"
PRINCESS JESSICA
Lib 15h3'13" -3d6'30"
Princess Jessica
Uma 11h34'40" 57d23'50"
Princess Jessica Conner Friday
Uma 8h47'44" 71d31'20"
Princess Jessica Jimenez
And 2h28'32" 44d0'46"
Princess Jessica Lynn Yorker
Vir 12h48'0" 12d28'5"
Princess Jessica Marie Charleston
Lib 15h42'34" -8d15'6"
Princess Jezebel Noel De Salvo
And 0h48'51" 43d41'33"
Princess Jill Rasmussen
Her 17h27'31" 46d26'13"
Princess Jillian
Aqr 21h14'57" -3d14'48"
Princess Joann Whitehurst
Cas 0h29'34" 58d26'16"
Princess Jodi Pie
Cnc 8h53'2" 13d12'45"
Princess Jordy
Gem 6h52'50" 33d5'49"
Princess Josie
Psc 0h11'48" -1d26'16"
Princess Josy
Cap 20h15'13" -19d27'33"
Princess Joy
And 0h37'7" 46d19'41"
Princess Joyti
Uma 9h2'56" 71d38'16"
Princess Juel Joshaw
And 0h36'13" 42d40'27"
Princess Juicy Angel 349164
And 1h37'53" 44d35'6"
Princess Julia Lee Arnold
Aqr 21h44'19" 0d54'57"
Princess Juliana
And 2h26'8" 49d51'30"
Princess Julianna Marie
Lmi 10h45'25" 39d14'11"
Princess Julianne Marton
And 23h27'23" 48d53'1"
Princess Julie
And 1h16'14" 42d59'33"
Princess Julie
Sco 16h29'54" -32d43'24"
Princess JW
And 0h7'51" 41d18'46"
Princess K
And 0h30'5" 30d0'56"
Princess Kailey Mae Ramaker
And 0h28'55" 35d23'54"
Princess Kailyn Rabia 2001
Leo 11h49'30" 26d26'46"
Princess Kaitlin Jade
And 0h14'56" 29d5'6"
Princess Kaitlyn
Gem 7h29'6" 20d32'32"
Princess Kalah
Sgr 19h42'48" -13d22'50"
Princess Kara Guthrie
Cru 12h38'0" -57d2'41"
Princess Karalyn Grace Davis
And 1h40'58" 39d11'3"
Princess Karen
And 23h5'20" 48d28'3"
Princess Karen
Leo 10h35'18" 23d55'15"
Princess Karen
Vir 12h7'40" 5d53'21"
Princess Karen
Cap 20h34'36" -22d38'10"
Princess Karen
Uma 14h4'25" 54d56'36"
Princess Karen Ann Velez
Ari 2h11'12" 23d27'52"
Princess Karey Lynn Haen
Leo 9h47'35" 16d47'52"
Princess Karine
Vir 12h37'2" 4d55'38"
Princess Karriann
Vir 12h49'44" 11d29'22"
Princess Kat
And 0h36'55" 35d45'35"
Princess Katelyn Elizabeth
And 0h15'35" 44d46'21"
Princess Katelyn Marie
Ori 6h7'56" 7d55'28"
Princess Katherine
Sco 16h15'38" -13d25'34"
Princess Kathi
And 0h14'52" 30d2'45"
Princess Kathleen
Tau 4h27'15" 22d56'16"
Princess Kathleen
Peg 22h35'39" 33d23'32"
Princess Kathleen Taelor Of Texas
Leo 11h29'6" 8d59'19"
Princess Kathy
Gem 6h29'19" 24d36'19"
Princess Kathy
And 1h35'38" 48d5'43"
Princess Kati Braier
Del 20h32'8" 15d28'32"
Princess Katie
Vir 13h14'45" 1d14'1"
Princess Katie
And 2h33'29" 42d54'48"
Princess Katie
Cru 12h19'9" -56d44'24"
Princess Katy
And 23h4'5" 41d32'53"
Princess Katy Kat
Lyn 8h50'31" 36d30'28"
Princess Kay
And 1h42'48" 40d53'11"
Princess Kayla
And 0h47'44" 36d2'23"
Princess Kayla Danielle
Sco 16h12'38" -11d42'37"
Princess Kayla Marie Horrigan
And 0h58'22" 35d53'35"
Princess Kayley Lambourne
And 2h37'23" 45d43'1"
Princess Keiko
Vir 14h25'29" 1d3'51"
Princess Kelli
Uma 12h27'7" 56d12'30"
Princess Kelly
Sgr 19h46'35" -37d23'21"
Princess Kelly
Psc 0h40'20" 11d52'31"
Princess Kelly
And 1h18'29" 42d16'45"
Princess Kelly P.
Cap 21h30'13" -9d54'14"
Princess Keri Ann
Lib 15h8'15" -23d33'43"
Princess Kerry
And 0h3'43" 37d37'14"
Princess kidabree
Gem 7h2'12" 22d11'13"
Princess Kiki 82
And 2h8'36" 40d4'41"
Princess Kim
Ori 5h29'32" 0d8'1"
Princess Kimberly
And 23h59'1" 32d35'44"
Princess Kimberly
And 1h15'12" 39d0'7"
Princess Kimberly
Aqr 21h59'10" -13d52'26"
Princess Kimberly
Cas 1h31'5" 67d5'22"
Princess Kiona
Aqr 22h37'42" -21d26'3"
Princess Kistner
Cap 21h52'6" -23d34'15"
Princess Kitten
Leo 11h46'50" 22d13'9"
Princess Kristen
Peg 22h41'35" 17d58'31"
Princess Kristen Angelina
And 1h27'34" 38d30'55"
Princess Kristin
Leo 11h19'12" 16d14'37"
Princess Kristin
Vir 13h24'47" -0d11'36"
Princess Kristin's Special Wish
Leo 10h45'24" 19d10'53"
Princess Kutti
And 0h52'18" 35d26'31"
Princess Kymmie
And 1h54'55" 41d37'35"
Princess Laci
Ari 2h24'26" 25d12'6"
Princess LaLa
Ori 6h12'29" 15d28'24"
Princess Lama
Cnc 8h16'47" 19d11'9"
Princess Landa Mandilawi
And 0h14'55" 28d42'16"
Princess La-Quessa
And 1h15'37" 34d37'7"
Princess Laura
And 23h51'24" 48d28'21"
Princess Laura
Cnc 8h50'29" 10d26'21"
Princess Laura of Sicilia
Cas 1h34'54" 67d50'27"
Princess Lauren
Cyg 21h17'29" 53d9'36"
Princess Lauren
Lib 14h54'45" -4d56'8"
Princess Lauren
Cru 12h11'45" -62d18'57"
Princess Lauren
Gem 6h52'55" 22d16'19"
Princess Lauren Aslynn Myers
And 0h14'3" 28d35'11"
Princess Lauren C. Hoffmann
Sco 16h15'40" -16d18'34"
Princess Lauren J. E. Cafferkey Prempeh
Per 4h46'8" 46d28'36"
Princess Lauryn Dilkes
Lac 22h27'28" 50d14'11"
Princess Lavinia
Ori 6h0'36" 21d29'13"
Princess Lea
And 0h0'8" 41d5'6"
Princess Leah
And 23h44'17" 43d31'55"
Princess Leah
Cap 20h14'44" -15d18'56"
Princess Leah
Aqr 21h32'56" -0d54'46"
Princess Leah Mercedes
Ari 2h59'44" 11d9'44"
Princess LeeAnn
Leo 11h37'56" 26d18'24"
Princess LeeAnne
Gem 7h47'13" 26d17'35"
Princess Lesley Judith
Cap 21h26'24" -15d21'12"
Princess Lexi
Cyg 21h45'47" 53d49'45"
Princess Lia
Cap 21h12'39" -17d53'7"
Princess Lilian Mao
Gem 7h45'56" 24d44'7"
Princess Lilly Grace
And 22h57'56" 42d24'39"
Princess Lindsay
Cyg 19h38'30" 30d21'27"
Princess Lindsay Marie Hardison
And 1h19'27" 36d57'16"
Princess Linz
Uma 10h23'23" 43d4'1"
Princess Lisa
Vir 13h22'41" 11d22'6"
Princess Lisa
Psc 1h41'4" 21d48'36"
Princess Lisa
Vir 14h11'8" -17d2'51"
Princess Lisa Lee
Leo 11h18'45" 26d36'54"
Princess Liz
Ari 2h23'52" 18d34'37"
Princess Liz
And 23h42'40" 47d24'35"
Princess Liza
And 0h16'0" 39d56'38"
Princess Lola
Gem 6h22'0" 18d59'9"
Princess Lola Naden
And 23h51'58" 34d58'14"
Princess Louise
And 23h35'28" 46d55'52"
Princess Lozell
Ari 2h8'59" 27d7'19"
Princess Lucy
And 23h5'8" 45d34'2"
Princess Lucy
Mon 7h29'21" -3d43'42"
Princess Lucy Kate
Col 5h48'39" -32d46'8"
Princess Luisa Vargas
Sgr 18h30'33" -16d20'56"
Princess Lydia Rose Garneau
And 1h8'32" 36d56'45"
Princess Lynn Marie Madera Garcia
And 23h31'6" 38d11'1"
Princess Maddie
And 23h35'1" 41d34'31"
Princess Madison
And 23h21'0" 42d40'29"
Princess Madison
And 0h32'29" 29d1'38"
Princess Madison Nicole
Tau 3h27'28" 8d14'31"
Princess Maggie Elizabeth Schaub
Lyn 9h25'7" 41d19'30"
Princess Maggie Kate
Ari 2h49'44" 25d53'12"
Princess Maha-Naif, with Love
And 0h58'30" 44d4'24"
Princess Mali Bergmann
Cas 0h58'35" 56d40'16"
Princess Mallory
Lib 15h9'7" -11d44'19"
Princess Mandy
Psc 1h10'17" 29d21'25"
Princess Margueritte
And 1h17'39" 37d48'29"
Princess Maria 310506
And 23h30'6" 43d0'46"
Princess Maria Asikis
Aqr 22h14'44" -8d29'27"
Princess Maricela Garcia I
And 0h9'22" 38d41'6"
Princess Mariel
And 0h17'36" 45d2'13"
Princess Mariel
Cnc 8h48'7" 7d29'15"
Princess Marilyn
Cra 18h17'41" -45d17'24"
Princess Marissa Mecir
Sgr 19h6'54" -33d41'1"
Princess Marissa.S.-love of my life
Sgr 19h53'50" -12d19'35"
Princess Marley
And 0h25'55" 29d5'34"
Princess Mary
And 22h58'20" 51d1'30"
Princess Mary Lou
Ari 2h42'50" 29d5'19"
Princess Maryann
Ori 4h50'50" 11d55'24"
Princess Marybeth
Leo 10h54'54" 12d4'4"
Princess Mary's Star
And 0h27'47" 33d11'30"
Princess Marysia
Lib 15h6'3" -7d5'0"
Princess McKenzie Kathleen
Vir 13h10'33" -7d22'32"
Princess McPhee
And 2h36'46" 39d51'11"
Princess Meaghan
Sco 16h8'33" -14d16'46"
Princess Meeta Patel
And 2h15'46" 37d31'49"
Princess Meg
Sgr 19h1'18" -15d45'54"
Princess Megan
Pho 0h50'1" -45d19'42"
Princess Megan
Uma 9h55'34" 51d36'42"
Princess Megan
And 1h34'3" 45d15'55"
Princess Megan
Aqr 21h23'58" 1d30'59"
Princess Megan Benson
Cnc 8h24'0" 32d22'48"
Princess Megan Conchinha
And 1h20'22" 48d34'6"
Princess Megan Tierney
Lib 14h53'32" -1d57'52"
Princess Meghann Aynsley
And 23h41'47" 46d2'39"
Princess Melanie Kaye Miller
And 1h30'48" 49d10'16"
Princess Meli 1436R
Ori 5h22'40" 1d13'57"
Princess Melissa
Tau 4h37'38" 7d52'43"
Princess Melissa
Cnc 8h29'0" 13d11'23"
Princess Melissa
Leo 9h31'56" 27d10'38"
Princess Melissa
Ari 1h57'35" 22d30'15"
Princess Melissa
And 0h17'4" 45d56'10"
Princess Melissa
Uma 9h35'52" 50d23'14"
Princess Melissa Ann
Lmi 9h56'44" 35d6'17"
Princess Melissa Temple
And 2h14'47" 41d48'54"
Princess Mel's Star
Col 5h5'47" -34d40'13"
Princess MeyMey
Aqr 22h49'57" -6d58'51"

Princess Mia
And 1h18'54" 47d56'35"
Princess "Mic" Evans
Cma 7h25'46" -29d45'47"
Princess Michele
Cap 21h39'54" -16d37'46"
Princess Michele K. Quinn
And 1h33'56" 48d30'48"
Princess Michelle
And 23h6'42" 37d21'5"
Princess Michelle
Ari 2h39'30" 21d51'39"
Princess Michelle (Reilly)
And 2h37'36" 49d52'19"
Princess Millie
Ori 5h30'48" -2d37'17"
Princess Millie
Sco 17h52'43" -40d0'27"
Princess Mimi
And 1h11'16" 38d37'34"
PRINCESS Minerva
Vir 12h54'49" 8d54'29"
Princess Miranda
And 1h5'18" 41d33'0"
Princess Mishka - My Moonshine
Ari 2h1'2" 24d31'45"
Princess Mita
Sco 16h14'58" -19d18'11"
Princess Miya
Cap 21h3'23" -19d53'7"
Princess M.J.
Ari 2h19'45" 23d47'8"
Princess Molly
And 23h34'42" 47d30'42"
Princess Monica
Per 3h0'54" 48d36'24"
Princess Monica Lea
And 2h23'45" 46d23'14"
Princess Monkeyface
Uma 9h6'31" 47d56'11"
Princess Mushy - Forever And A Day
Cru 12h56'46" -59d22'14"
Princess Nadia
Gem 6h44'29" 22d8'56"
Princess Nahid Akhtar
Lib 15h29'5" -11d59'35"
Princess Naomi
Cam 4h16'42" 71d3'29"
Princess Nasira
And 1h41'1" 36d37'20"
Princess Natalie
Gem 6h49'54" 14d27'22"
Princess Natalie
Cas 1h6'53" 61d45'34"
Princess Natalie
Cra 19h2'48" -38d43'22"
Princess Natalie Vassilaros
Vir 13h21'55" 2d0'21"
Princess Natasha
Gem 6h45'30" 24d0'15"
Princess Nathalie
Tau 5h44'57" 16d33'4"
Princess Nicki
Cnc 9h5'54" 20d1'27"
Princess Nicola - 10 December 1973
Ori 5h34'3" -4d23'44"
Princess Nicole
Cas 23h21'3" 53d16'35"
Princess Nicole
And 0h27'17" 41d57'54"
Princess Nicole
And 23h27'2" 48d3'22"
Princess Nicole forever
Aqr 20h43'9" -3d15'21"
Princess Nicole Rae
And 0h32'29" 42d46'52"
(Princess) Nicole Renea Glore
Aqr 22h5'26" -13d39'19"
(Princess) Nicole Sagan's Star
And 1h20'35" 49d33'10"
Princess Niece
Cap 21h55'36" -16d42'31"
Princess Niki
And 0h33'47" 35d47'30"
Princess Nikki
And 0h34'22" 38d49'49"
Princess Nikki
Leo 10h14'37" 15d12'49"
Princess Nikki
Leo 10h36'1" 13d46'9"
Princess Nina Légene
Lib 15h1'39" -19d10'35"
Princess - NMS
Lib 14h51'42" -4d55'8"
Princess Noelle
Sgr 18h11'28" -18d6'49"
Princess Nora
Vir 11h43'0" 7d38'7"
Princess Noriko
Tau 3h38'29" 28d2'7"
Princess Nosilla
And 0h12'57" 33d28'11"
Princess Of A Lot
And 23h50'43" 45d1'49"
Princess of Balamb, Nicole Register
Vir 13h46'25" 6d23'8"
Princess of the East Woods
Ori 4h51'58" 11d10'6"
Princess of the Stars Tanya Harris
And 0h28'19" 29d51'7"
Princess of the World AGP
Dra 16h3'23" 56d36'23"
Princess Olivia
And 2h37'34" 49d5'26"
Princess Olivia
Gem 7h14'57" 33d56'35"
Princess Oriana
Sco 16h21'27" -25d44'18"
Princess Oya
Gem 6h55'32" 25d0'35"
Princess Paige
And 2h33'48" 48d15'58"
Princess Pak
Gem 6h42'30" 34d58'32"
Princess Pam
Vir 13h27'38" 5d53'28"
Princess Pamela Foss
Cap 20h59'38" -20d20'1"
Princess Pammy Poo's Star
Lib 15h17'24" -7d43'33"
Princess Paola
And 0h25'36" 42d57'42"
Princess Pat
Vir 11h59'23" -9d13'51"
Princess Patricia 81382 112243 1029
Sgr 17h58'42" -24d23'27"
Princess Patti
Gem 7h36'4" 33d13'6"
Princess Paula
Lyn 8h47'7" 34d37'49"
Princess Paula
Lib 15h18'17" -5d17'28"
Princess Paulette Sue Dennis
Sco 17h37'49" -39d26'20"
Princess Pavelka
Uma 10h48'6" 54d30'46"
Princess Peanut
Sco 16h15'39" -15d7'14"
Princess Peg
Vel 9h56'50" -43d15'24"
Princess Penny
And 23h45'51" 33d47'50"
Princess Pepper Preciosa
Cnv 12h39'49" 42d31'11"
Princess Pergola
Leo 10h49'24" 8d51'0"
Princess Phoebe
And 2h6'30" 47d28'14"
Princess Pickle Starr
And 1h42'11" 46d29'22"
Princess Pig
And 1h54'21" 45d48'6"
Princess Pixie Dust's Star
Gem 6h35'52" 25d16'33"
Princess Prettyweather Mahue
Uma 10h11'17" 59d43'44"
Princess Puppers
Psc 0h25'20" 5d48'58"
Princess Rachael Brady
And 23h22'52" 39d1'39"
Princess Rachel
And 2h15'24" 49d30'3"
Princess Rachel
Lib 15h15'23" -27d47'8"
Princess Rachel - 7.08.1984
Leo 11h6'59" -0d46'13"
Princess Rachel Cecile Bananahammok
Tau 5h37'0" 27d44'20"
Princess Rachel Davis Of Dublin
Dra 14h33'17" 58d17'26"
Princess Rachel Erbaugh
Cnc 7h58'30" 16d18'4"
Princess Rachel Le
Lib 15h14'32" -10d49'59"
Princess Rachelle
Gem 6h51'16" 22d5'58"
Princess Rascallioun
And 0h54'23" 41d4'5"
"Princess" Rashana George Zaklit
Vir 14h24'49" -1d37'20"
Princess Rebecca
And 2h33'12" 45d16'1"
Princess Rebecca Mayton
Cmi 7h37'37" -0d13'34"
Princess Rebecca - Queen of Hearts
Pho 1h10'28" -46d4'27"
Princess Reem
Cas 23h33'17" 57d20'7"
Princess Renuka
Tau 5h48'0" 26d27'6"
Princess Riley
And 23h27'14" 35d45'39"
Princess Rima El-Zein
Sco 16h3'31" -10d50'0"
Princess Risa
Tau 4h36'28" 17d50'9"
Princess Rob
Gem 6h57'55" 27d10'30"
Princess Robin Wong 1985
Ari 2h5'36" 23d8'44"
Princess Rochelle
Psc 1h7'6" 31d7'4"
Princess Rose
And 1h28'8" 40d59'17"
Princess Roweena - 20/08/2004
Ori 5h19'42" 1d35'10"
Princess Roxanne
And 22h58'4" 45d37'50"
Princess Roxy
And 1h7'58" 37d8'10"
Princess Ruby
And 1h19'2" 41d18'12"
Princess Ruth
And 23h15'45" 47d50'7"
Princess Ruth
And 0h17'12" 27d27'10"
Princess Ruthann Shultz
Vir 13h22'37" 12d7'23"
Princess Rya
And 23h55'44" 45d14'0"
Princess Sabra
Ari 3h19'27" 21d11'7"
Princess Salonia
And 23h13'21" 40d22'50"
Princess Samanta
Lac 22h18'0" 51d50'27"
Princess Samiksha
Psc 1h7'47" 11d54'12"
Princess Sandi El Ciaramella
And 1h4'29" 41d44'33"
Princess Sandy
Gem 7h24'42" 28d50'35"
Princess Sara Goel
And 0h31'8" 45d6'21"
Princess Sara Jade
Pho 1h11'55" -45d10'32"
Princess Sara Vassallo
Tau 4h4'4" 6d2'7"
Princess Sarah
Vir 13h3'41" 1d28'18"
Princess Sarah
Ari 2h43'6" 27d23'6"
Princess Sarah
And 23h9'16" 42d46'27"
Princess Sarah
And 2h8'12" 46d27'18"
Princess Sarah
And 1h29'12" 47d9'44"
Princess Sarah
And 1h22'15" 48d32'47"
Princess Sarah
And 23h29'59" 47d19'17"
Princess Sarah
And 1h6'7" 35d49'46"
Princess Sarah
Sgr 18h14'48" -22d45'46"
Princess Sarah
Lib 15h8'56" -7d0'37"
Princess Sarah Elizabeth Broyles
Lib 15h6'53" -2d43'14"
Princess Sarah Frances Pauker
And 2h19'0" 39d8'4"
Princess Sarah Jane
And 1h49'46" 37d24'36"
Princess Sarah Lane Richmond
Aqr 22h23'8" -3d30'3"
Princess Sarah Natae Whipper
Cru 12h47'57" -61d10'53"
Princess Sarah Peters
Psc 1h13'19" 10d4'3"
Princess Sarah Strong
And 1h0'10" 45d26'40"
Princess Sarah's Star
And 23h34'29" 40d34'1"
Princess Sarina
Ori 6h21'38" 15d31'36"
Princess Savannah Jade
Ori 5h54'26" 4d22'33"
Princess Scarlett
Gem 6h55'10" 22d0'36"
Princess Schuyten
Mon 8h2'47" -0d39'58"
Princess Shanee Javaherian
Leo 11h4'16" 3d30'53"
Princess Shannon Horval
Cnc 8h50'17" 27d38'54"
Princess Shantel Rae Friesen
Psc 1h50'12" 6d37'13"
Princess Shari (JAP) Zimmerman
And 0h39'3" 39d16'54"
Princess Shari the Contender
And 23h15'40" 47d56'44"
Princess Sharon
Ori 5h45'39" -8d22'51"
Princess Shawna Garland
Tau 4h51'42" 18d58'9"
Princess Sheena
Cnc 8h35'18" 19d38'10"
Princess Sheila
Lib 15h3'46" -19d53'45"
Princess Shelby Vanessa
Uma 9h0'40" 63d42'19"
Princess Shelly
Cnc 8h48'9" 12d13'57"
Princess Shelly
And 23h19'37" 52d27'54"
Princess Shelsie
Cap 21h30'38" -17d36'44"
Princess Sheree Lynn
Cnc 8h23'11" 32d58'51"
Princess Sherrylynn Jacobsen
And 22h59'48" 50d54'48"
Princess Shooting Star
Uma 10h48'14" 47d56'7"
Princess Shorty
Ari 2h19'50" 13d51'35"
Princess Shugabear
And 1h32'56" 48d0'31"
Princess Sinthu
Ari 2h18'34" 17d10'20"
Princess Sophia
Gem 7h38'50" 16d18'0"
Princess Sophia
Cma 6h40'2" -14d52'55"
Princess Sophia Manjarrez
Cnc 9h5'59" 24d24'48"
Princess Sparkle
Psc 0h59'43" 19d4'55"
Princess Sparkles
And 23h28'14" 47d9'24"
Princess Special K
Ari 2h46'23" 25d47'35"
Princess Sreeya
Sco 16h32'51" -26d10'2"
Princess Stacey
And 23h32'22" 48d40'46"
Princess Stacie of Deerfield
And 1h45'38" 44d32'44"
Princess Stacy
Cnc 9h2'23" 29d0'58"
Princess Stacy
Ari 3h17'57" 20d11'18"
Princess Star
And 23h35'28" 44d18'38"
Princess Star
Her 16h47'40" 38d10'18"
Princess Stefanie
Cap 20h58'51" -18d50'27"
Princess Stephaine
And 2h10'38" 38d59'28"
Princess Stephanie
And 23h37'17" 42d0'55"
Princess Stephanie
And 0h13'39" 46d26'49"
Princess Stephanie
And 23h21'19" 48d19'15"
Princess Stephanie
Cyg 21h42'46" 49d59'32"
Princess Stephanie
Del 20h35'13" 4d43'1"
Princess Stephanie
Cas 23h36'43" 57d55'47"
Princess Stephanie
Dor 4h54'35" -66d38'50"
Princess Stephanie Ann Merrill
And 2h21'50" 46d2'15"
Princess Stephanie Atwood
Uma 9h29'33" 46d6'38"
Princess Stephanie Lynne
Aqr 23h14'36" -7d58'19"
Princess Stephanie of the Galaxy
And 0h57'29" 46d29'38"
Princess Stephanie Raimundo
Ari 1h57'33" 19d40'25"
Princess Stephanie V.
And 1h34'39" 44d51'20"
Princess Summer Lani
Psc 0h53'17" 31d24'8"
Princess Sun
And 1h21'46" 46d50'55"
Princess Sunny
Vir 13h9'7" -10d14'15"
Princess Sunshine
Ari 3h15'1" 22d20'32"
Princess Susan
Sco 16h8'42" -9d47'44"
Princess Susie Ann
Gem 7h12'1" 31d14'52"
Princess Susy
Sco 17h24'36" -42d37'8"
Princess Sweet Pea
Uma 10h9'14" 43d20'6"
Princess Sweet Pea of Cooley
Cnc 8h13'46" 16d2'30"
"Princess Sylvia MacKenzie"
And 1h3'44" 35d34'7"
Princess Taliah's Shining Star
Cru 11h59'41" -64d36'16"
Princess Tanna
Crb 16h15'55" 30d31'23"
Princess Tanya
Sgr 19h17'26" -15d35'2"
Princess Tara
Ori 5h58'2" 11d19'48"
Princess Tara Brickner
Aqr 23h13'28" -7d2'22"
Princess Tasha Nicole Hanna
And 0h54'0" 37d20'48"
Princess Tawny ~ Daddy's Angel
Lyn 8h26'42" 51d19'52"
Princess Taylor
And 23h34'9" 42d0'43"
Princess Taylor
Ori 5h7'4" 15d13'44"
Princess Teal Fantastico
Cnc 8h19'40" 8d3'51"
Princess Teresa M. Powell
And 1h42'11" 44d0'36"
Princess Tiarnne Reneé
Pho 0h39'5" -42d26'43"
Princess Tiffany
Sgr 18h56'52" -27d7'27"
Princess Tiffany
Aqr 22h13'45" 1d32'48"
Princess Tina
And 0h5'28" 36d0'23"
Princess Tinker
Lib 14h52'45" -2d39'53"
PRINCESS TOCCO
Dra 18h22'54" 59d17'37"
Princess Tracy
Sgr 19h16'14" -12d13'19"
Princess Tracy
Lib 15h55'21" -11d6'9"
Princess Tracy
Lyn 8h2'56" 45d34'50"
Princess Tree
And 1h37'6" 36d28'7"
Princess Tyra Stanley
Gem 7h4'25" 10d25'45"
Princess Ula
And 23h36'29" 47d28'49"
Princess Valerie
Lyn 7h53'24" 39d6'23"
Princess Valerie
Gem 6h39'45" 20d41'14"
Princess Valerie Mastro
And 2h36'55" 39d38'30"
Princess Vanessa
Psc 23h42'58" 5d21'44"
Princess Vanessa de la Torre
And 22h59'45" 52d25'42"
Princess Veeka
And 0h40'42" 24d46'40"
Princess Veronica
And 0h30'5" 31d42'43"
Princess Vesna
Leo 10h41'43" 12d32'36"
Princess Vicki
And 23h27'4" 36d2'1"
Princess Vicki Sue
Cnc 9h10'47" 17d31'50"
Princess Vickie Lynn
Psc 0h17'43" 6d32'27"
Princess Victoria
Lib 15h30'20" -20d36'20"
Princess Victoria Grace
Cra 19h1'44" -37d5'19"
Princess Wendy
Psc 1h10'58" 22d30'19"
Princess Winnifred Stephanie Louise
And 0h14'7" 37d29'14"
Princess Xuan
And 0h18'59" 32d9'22"
Princess Yacer
Psc 0h53'12" 6d54'52"
Princess Yuka
Sco 17h49'6" -38d44'27"
Princess Yumiko
And 0h21'11" 37d34'29"
Princess Yung Chen
Lyn 7h29'48" 54d11'4"
Princess Zitácska
Aqr 23h27'18" -10d22'2"
Princess Zoby
Psc 1h17'57" 19d8'6"
Princess Zoe
And 2h0'27" 36d31'15"
Princess Zoe
Lyn 6h26'0" 58d26'35"
Princess ZZ
And 1h21'54" 44d59'22"
princess02121983
Aqr 21h25'29" -11d37'16"
Princess12
Ari 2h45'43" 27d15'42"
PRINCESSA
Lyr 19h18'22" 28d36'58"
Princessa Gina
And 0h24'29" 39d40'5"
Princess-Danielle
And 2h17'58" 47d38'37"
Princess-DeStinè
Uma 10h4'22" 65d9'17"
Princesse
Cep 1h0'16" 80d30'59"
Princesse
Umi 14h38'51" 75d15'8"
Princesse
Her 16h34'38" 21d23'48"
Princesse Agathe
Lib 14h22'58" -19d8'52"
Princesse Alexandra
Vir 11h38'53" 8d19'39"
Princesse Aline boubou
Uma 9h23'1" 58d34'46"
Princesse Aurore
Gem 6h51'41" 18d36'6"
Princesse Capricia, my love
Cma 7h8'22" -27d51'5"
Princesse Cécile
Uma 11h30'8" 36d58'41"
Princesse Chahinez
Gem 6h30'52" 22d0'18"
Princesse d'Amour
Cas 23h38'51" 55d20'4"
Princesse Doudou A.D.M.V
Uma 9h39'10" 55d2'43"
Princesse Jennifer
Peg 21h51'57" 25d35'24"
Princesse Julie
Cas 1h0'3" 61d52'46"
Princesse Laura
Uma 12h3'17" 38d49'11"
Princesse Lilou
Cnc 9h8'3" 15d51'31"
Princesse Lou
Cas 1h27'12" 69d6'1"
Princesse Lou-Anne Metti
Uma 8h20'49" 62d6'45"
Princesse Marie
And 2h8'29" 46d51'56"
Princesse Marina Kustova
Uma 13h59'23" 57d7'49"
Princesse Marine
Ori 6h17'4" 14d6'2"
Princesse Mathilde
Uma 13h40'48" 55d22'51"
Princesse Mil's
Gem 6h19'2" 22d16'48"
Princesse Nathanaëlle
Leo 10h48'2" 12d16'10"
Princesse Sarah
And 23h4'51" 36d55'59"
Princesse Sophie Saladin
Lyn 8h32'42" 56d42'39"
Princesse Sucrée
Crb 16h20'6" 36d38'25"
Princesse Sylvie
Cnc 8h1'25" 22d46'7"
Princesse Tessa
And 23h53'46" 46d49'40"
Princesse Tinette
Sgr 18h3'44" -34d7'53"
PrincessJess07/30/1985
And 1h28'30" 48d39'31"
Princesskary
Sco 16h36'20" -31d31'26"
Princess-Lovey
Cyg 19h42'9" 42d15'7"
PrincessStephanie KeeperoftheStars
Leo 10h16'37" 21d42'29"
Princeton Puppy Boschuk
Aqr 21h50'46" -0d20'5"
Princeza Irena
And 0h42'25" 46d19'49"
Princeza Marina
Peg 23h41'6" 13d58'57"
PrincipAL Silveira
Cap 21h41'39" -14d59'38"
Principal, John Kozusko, GLJH
Her 17h42'59" 17d1'12"
Principecca T One
Sgr 18h8'20" -31d32'38"
Principem pacem
Lib 15h8'42" -29d52'29"
Principessa
Ori 5h42'16" -2d6'47"
Principessa
And 23h51'1" 43d59'38"
Principessa
And 0h5'1" 45d6'43"
Principessa
Crb 15h57'50" 38d37'37"
Principessa
And 23h17'42" 47d15'0"
Principessa
Cas 0h47'19" 58d19'33"
Principessa
Per 3h9'52" 37d0'17"
Principessa
Gem 7h5'18" 34d10'51"
Principessa
And 23h12'52" 37d1'5"
Principessa amabilità Doris 29.07.60
Lac 22h24'3" 47d6'46"
Principessa Ava
And 23h3'19" 40d42'53"
Principessa Eleonora
And 2h16'21" 50d20'35"
Principessa Eleonora
Cnc 8h39'21" 23d38'58"
Principessa Francesca
Vir 12h59'28" 12d15'42"
Principessa Giulia
Aur 6h21'37" 42d10'57"
Principessa Julia My Jaanu
Cap 21h38'11" -15d21'44"
Principessa Kathleen Marie Mushkin
And 1h40'22" 47d58'3"
Principessa Katy
Cnc 9h17'52" 27d14'40"
Principessa KU-01-05-jr
Uma 8h14'58" 59d57'19"
Principessa Ludmilla
Sgr 19h39'38" -20d16'35"
Principessa Nadia
And 1h7'26" 43d59'37"
Principessa Petra
Her 18h11'13" 17d15'51"
Principessa Silvia
Umi 17h8'36" 83d27'9"
Principessa's Stingray
Ari 3h17'0" 27d37'12"
Princy (Joe P's Mom)
Cma 6h58'50" -26d35'55"
Pringle's
Ori 5h48'22" 11d26'59"
Prink
Per 3h18'18" 44d27'44"
Printesa Ramona
Leo 11h44'13" 17d14'24"
Prinzesa Sara
Sco 17h57'54" -31d16'44"
Prinzessin
Uma 13h37'32" 56d4'45"
Prinzessin Gabriela
Uma 11h13'22" 32d11'3"
Prinzessin Johanna V. Hohenadl
Uma 10h12'15" 45d9'23"
Prinzessin Karin Friesenegger
Uma 11h44'14" 34d54'30"
Prinzessin Nina
Uma 14h21'22" 57d4'3"
Prinzessin Sandra Giovanoli
Dra 17h47'20" 69d18'26"
Prinzessli
Crb 15h47'53" 27d47'42"
Prior Sebastian C. Lewis
Umi 15h31'46" 71d33'4"
Prior's Guide Star
Leo 9h43'3" 32d30'20"
Pris+Jon
Sgr 18h4'48" -29d34'8"
Prisca Faessler, from Marco with love
Cas 1h8'16" 48d49'5"
Prisca Sieber
Ori 6h21'4" 10d16'8"
Prisca Steinegger
Vir 14h18'57" -20d54'10"
Priscella Goulette
Cnc 8h48'56" 32d4'33"
Priscila Guerra
Psc 0h55'31" 33d13'58"
Priscila Pereira Schaffert
Gem 6h26'54" 24d52'25"
Priscilla
Peg 22h28'35" 9d19'45"
Priscilla
Leo 9h39'44" 32d22'43"
Priscilla
Uma 11h8'14" 30d24'17"
Priscilla
And 23h4'25" 47d11'19"
Priscilla
Cas 0h29'36" 59d19'30"
Priscilla
Cas 0h34'34" 55d57'24"
Priscilla
Pup 8h13'48" -19d28'45"
Priscilla
Aqr 20h42'37" -8d48'27"
Priscilla
Cas 23h1'54" 57d12'10"
Priscilla A. Salim
Aqr 21h8'41" -13d10'11"
Priscilla and Ed
Dra 10h11'32" 74d49'10"
Priscilla Ann Reynolds
Uma 12h46'57" 58d6'16"
Priscilla Anne Jardine
Aqr 21h9'31" -3d42'8"
Priscilla B. Pelgen 1970 Kitty Esq.
Lyn 7h48'6" 57d43'2"
Priscilla "Baby Girl" Serrato
Psc 1h16'27" 21d58'49"
Priscilla Baird Hart
And 0h2'26" 44d51'36"
Priscilla Bishop
Cas 2h14'8" 62d2'8"
Priscilla Bonner Horton
Leo 10h23'37" 25d18'54"
Priscilla Brooke Eaves
Gem 6h28'59" 24d45'42"
Priscilla Bubeck
Lyr 18h54'34" 33d33'8"
Priscilla Casillas
And 0h9'20" 35d6'16"
Priscilla Dabney
And 23h43'15" 40d38'59"
Priscilla de Jesus
Leo 11h52'0" 22d50'29"
Priscilla Drake Solomon
And 0h23'56" 40d6'0"
Priscilla & Elizabeth Forever 143
Sco 16h15'9" -22d31'57"
Priscilla Favarato Perutti
Gem 6h22'3" 17d36'38"
Priscilla Garnett Shorter Darr
Leo 9h53'6" 12d30'16"
Priscilla Janyll Montanez
Psc 1h8'47" 14d11'22"
Priscilla Jimenez Veneklase
Leo 11h10'15" 4d46'49"
Priscilla J.Miller
And 1h19'2" 37d51'8"
Priscilla Kaster
And 23h38'3" 34d45'57"
Priscilla Kay Murray
Cas 0h25'5" 52d35'48"

Priscilla & Keith As One In The Sky
Cas 1h1'45" 69d55'8"
Priscilla Khoo Ai - Fern
Ari 3h8'48" 23d38'36"
Priscilla Lozano
And 1h38'11" 40d12'11"
Priscilla Marie Rodriguez
Cnc 8h25'13" 18d50'20"
Priscilla Marie Silva
Lib 14h27'31" -19d54'52"
Priscilla Marisol Rodriguez
Psc 1h23'25" 22d48'19"
Priscilla Martelle Walsh
Cas 2h17'15" 72d57'39"
Priscilla Maurice
Leo 9h39'38" 13d47'40"
Priscilla Monique Perez
Cap 20h22'46" -26d25'57"
Priscilla Monreal
Crb 15h45'57" 33d30'50"
Priscilla Polonio
Sco 15h55'34" -20d55'26"
Priscilla Queen of the Sky
Leo 9h46'38" 26d16'46"
Priscilla Rodriguez
Psc 0h26'19" 16d23'30"
Priscilla Rose
Uma 14h10'28" 58d6'7"
Priscilla Stiles - Jersey Farm Girl
Aqr 22h16'45" -0d28'13"
Priscilla the Brightest
Lyn 9h6'17" 34d23'46"
Priscilla the Star of My Universe
Lib 15h23'1" -20d31'58"
Priscilla V. Armour
Uma 11h54'16" 31d28'27"
Priscilla Vivanco
Gem 6h48'50" 27d7'36"
Priscilla Wright Dombek
Cyg 20h27'55" 30d39'57"
Prisilla
Uma 12h31'44" 52d23'53"
Priska
Lac 22h49'5" 49d54'23"
Priska
Sge 19h44'37" 16d36'6"
Priska
Her 16h25'27" 6d42'56"
Priska
Umi 15h31'58" 80d24'27"
Priska
Cep 23h26'5" 86d20'52"
Priska Baumgartner
Dra 18h6'23" 71d40'37"
Priska, 19.01.1953
Cam 6h49'47" 69d12'20"
Prisoner of My Heart
Crb 15h36'45" 26d48'37"
Priss & Guss
Lib 15h5'57" -13d10'25"
Prissy & Bo
Eri 3h58'45" -0d50'1"
Prissy Christ
Uma 10h54'38" 58d27'57"
Prissy Shine
Crb 16h13'3" 39d11'39"
Prissy Sue
Her 18h40'43" 20d29'25"
Prita
Gem 6h44'39" 26d48'28"
Prita Ganatra
Aur 5h35'24" 42d49'48"
Pritchett
Cmi 7h31'16" 7d35'25"
Pritesh Jayesh Patel
Vir 12h56'7" 6d10'48"
Priti
Vir 12h47'26" 12d45'44"
Priti
Lyn 7h0'37" 46d58'26"
Priti R Vemulapalli
Gem 7h19'30" 31d46'12"
Pritty Brittey
Dra 18h42'4" 68d59'16"
Private Sanctuary
Aqr 22h14'34" 1d34'3"
Priya
Lib 15h51'13" -17d57'9"
Priya Badhwar
Uma 10h23'9" 52d53'38"
Priya Dixit-Chitre
Tau 5h28'33" 19d12'52"
Priya Hussain
Leo 10h17'16" 9d47'0"
Priya Margaret Wilson - 8 November 2006
Cru 12h28'34" -60d57'35"
Priya Mathew 1981
Sco 16h10'49" -14d17'9"
Priya Raman
Cas 23h28'43" 51d38'2"
Priya Sebastian
Uma 11h7'44" 65d34'4"
Priya Waney
Uma 12h6'56" 44d50'45"
Priyadarshini
Leo 9h39'5" 28d52'31"
Priyakshi
Pho 2h2'49" -40d58'12"
Priyam
Psc 1h8'19" 7d38'25"
Priyanka
Lyr 18h52'11" 43d54'57"
Priyanka Godbole
Aql 19h26'1" 7d39'57"
Prizznick's Passion
Cep 23h0'26" 76d34'13"
pro angelus <\@>Beate Pietrula
Uma 10h44'45" 47d4'39"
Pro Huny (Pattie Schwatka)
Psc 0h45'30" 16d0'1"
pro meus amor tiffany
Lmi 10h8'7" 34d44'47"
pro nostrum carus abbas
Uma 9h24'25" 67d5'1"
Pro Status
Uma 10h34'32" 68d20'50"
Probitas et Animus
Cru 12h37'3" -58d36'7"
Probst, Ralf
Uma 10h53'48" 42d34'33"
Procter & Gamble (M) Sdn Bhd
Ori 5h30'40" 4d20'20"
ProCurve Networking Alpha
Uma 9h59'36" 43d3'50"
ProCurve Networking Beta
Uma 10h29'27" 43d55'30"
ProCurve Networking Delta
Uma 11h9'11" 49d57'21"
ProCurve Networking Epsilon
Uma 11h29'58" 51d57'7"
ProCurve Networking Eta
Uma 12h44'41" 55d59'34"
ProCurve Networking Gamma
Uma 10h49'45" 46d59'12"
ProCurve Networking Iota
Uma 13h59'28" 59d54'52"
ProCurve Networking Theta
Uma 13h9'44" 58d3'37"
ProCurve Networking Zeta
Uma 12h9'36" 52d53'16"
Prodigiosus
Uma 9h27'13" 64d19'9"
Prodigiosus Lux lucis
Vir 12h47'37" 6d5'45"
Prodyge Crew
Uma 9h40'41" 59d10'17"
Prof
Her 18h41'12" 25d31'47"
Prof. Albert Huch
Uma 9h26'56" 47d57'53"
Prof. Harald Stein
Her 16h53'29" 35d16'26"
Prof. Helaine M. Wasser
Aqr 21h51'50" -5d24'25"
prof. Ing. Antonín KA□DA, PhD.
Aql 19h38'52" 12d10'31"
prof.dr. Koch Sándor, Nagypapi
Cap 20h14'35" -9d30'29"
Profesor Kresha
Peg 22h20'35" 19d41'41"
PROFESSEUR HENRI PUJOL
Cnc 8h4'12" 12d4'45"
Professor Arthur Aldes Teixeira PhD
Ori 5h21'29" -9d6'1"
Professor Chris Ostrowski
Cru 12h33'4" -63d28'10"
Professor E. V. Kadow
Aur 5h15'45" 31d47'57"
Professor Frank Monsour
Gem 7h4'8" 11d37'46"
Professor Gary Wicks
Ori 5h57'9" 12d7'18"
Professor Hawley
Uma 11h30'31" 62d46'1"
Professor Jan Stephen Tecklin
Uma 13h38'38" 55d31'42"
Professor Jay Lance
Ari 2h4'55" 15d30'53"
Professor John Abimbola Kuti
Ori 6h21'40" 10d6'4"
Professor Mervyn Jack
Per 4h13'59" 33d6'42"
Professor Pete
Her 17h50'37" 15d34'47"
Professor Zeigler
Uma 13h57'37" 51d15'32"
Profoundly Magical
Aql 19h44'29" -0d1'36"
Profpoll
Cap 21h44'56" -23d57'43"
Prohaska, Herbert "Schneckerl"
Uma 13h1'37" 60d19'5"
Proinsias
Dra 16h10'27" 51d34'39"
p.ROLF Optimus Maximus
Cap 21h42'34" -8d37'13"
Promessa
Sgr 17h52'30" -28d29'24"
Promessa
Cyg 20h1'17" 37d43'40"
Promessa
Lyn 8h17'37" 33d24'44"
promesse de l'amour rcw2005
Ori 6h12'55" 18d50'22"
Promesse d'une Etreinte
Uma 13h25'34" 53d38'34"
Promethea
Lib 14h54'26" -2d20'13"
Prometto
Lep 6h9'37" -22d59'17"
Promise
Lib 14h51'20" -0d46'49"
Promise
Uma 12h56'55" 61d35'27"
Promise
Leo 11h48'9" 19d26'20"
Promise
Sge 19h29'20" 19d5'18"
Promise
Uma 10h28'50" 41d20'20"
Promise
Per 3h19'7" 46d12'36"
Promise
Cyg 20h39'59" 47d19'20"
Promise Forever
Cyg 21h23'43" 45d43'7"
Promise Lane Smull My Princess
Per 3h7'28" 43d53'55"
Promise Promise
Psc 1h6'17" 7d13'43"
PROMISE PROMISE CA 03
And 23h58'17" 40d11'7"
"Promises." By my side, Roselani.
Ori 6h11'20" 20d37'46"
Promises Keeper
Vir 13h17'50" 13d58'6"
Prommer Katalin
Cas 23h4'14" 54d30'41"
Promod Pratap
Aql 19h13'10" 1d10'29"
Pronay Gyula 1956.02.10.
Aqr 23h35'52" -23d48'20"
Pronga
Tau 4h47'20" 28d36'17"
Properties Alpha LLC
Psc 0h46'4" 6d35'51"
Property of Promise
Dra 18h16'23" 58d32'6"
Prophet- William Grembowicz
Ori 5h50'35" 8d52'18"
Prophetess Gloria Pruitt
Crb 16h23'23" 33d55'8"
Propp, Heinz-Jürgen
Uma 9h10'10" 50d45'10"
Prosha - Prohorova Elena
And 23h36'35" 34d57'30"
Prosper
Gem 6h21'49" 21d6'13"
Prospero
Peg 21h38'31" 18d6'58"
protect Haruto from a nightmare
Ari 3h2'14" 30d28'29"
Protector, Jon Carroll
Ori 5h49'29" 7d23'26"
Proteus Eros
Lyr 19h3'33" 47d12'53"
Proti
Cam 7h40'19" 76d18'56"
Proton7777
Aql 19h7'32" 14d9'11"
Protz
Cnc 8h53'21" 22d45'8"
Prouchista
Tau 5h40'52" 19d16'32"
Proverbs 17:6 Bruce & Mary Kamrath
Uma 11h20'7" 55d11'43"
Providence in Love
Cra 19h2'57" -38d6'52"
Providential Perfect Patti
Psc 1h8'59" 31d17'50"
Prucha
Lyr 18h40'41" 38d29'3"
Prudence
Lyn 7h44'37" 38d42'13"
Prudence Billanti
Cas 22h59'36" 56d4'35"
Prudence Grabovy
Cas 1h16'10" 57d3'1"
Prudence Lasure
Lyn 7h7'35" 48d1'56"
Prudence Rivera
Oph 16h54'26" -0d32'25"
Prudence's 21st
Cnc 8h26'23" 14d58'36"
PRUDHOMMESFP
Ari 3h14'1" 19d43'52"
Prue Lizzy
Cru 12h36'7" -58d42'35"
Prune Saint Léger
Uma 11h4'16" 38d33'23"
Prunella
Uma 9h41'41" 57d25'36"
Pruntos
Ori 5h43'15" 11d18'43"
Prusi
Per 3h44'14" 45d27'54"
Prutske
Psc 0h0'39" 10d21'51"
Pryncess Jess
Cnc 8h29'47" 24d40'43"
przestwor
Agr 22h40'46" -9d1'16"
P.S.
Sco 16h7'28" -12d51'28"
Ps 8:3-5 When I consider..the stars
Cap 20h36'51" -14d29'11"
PS Infinity & Beyond
Crv 12h29'59" -14d37'21"
PS Love
Cyg 19h40'1" 28d27'11"
P.S. We Love You
Lib 14h47'11" -18d2'15"
PSA Lady Stars Coach Andy Salandy
Per 4h25'39" 39d35'12"
Psalm 107:1
Leo 10h28'53" 13d28'39"
Psalm 19:1
Uma 10h21'19" 64d57'41"
Psalm 91: 10-12
Crb 15h42'10" 33d30'9"
Psalms33
Uma 11h39'6" 49d37'54"
[PSB] Pamela S. Barrett
Aqr 20h58'51" 0d25'15"
PSiLUVu17June1995
Lyn 8h33'7" 33d25'29"
PSMcCain
Uma 10h49'50" 45d21'22"
P.S.P. (Princess Stacey Palastrand)
And 0h47'59" 25d26'3"
P.S.P.15.12.72
Uma 9h35'34" 65d17'5"
psst...apples/bananas
Dra 17h30'35" 58d11'45"
Psychosexy
Ari 2h19'27" 22d4'23"
Psycra
Cru 12h41'40" -55d53'19"
Psy-Love
Cas 23h46'53" 54d9'33"
PT phone home
Her 18h27'13" 12d11'31"
Ptah&Sandy070391
Cnc 9h10'46" 28d36'11"
"Ptashka" Natalie Shakhnovsky
Leo 11h16'20" -3d51'10"
Pte's Promise
Ori 5h6'45" 4d52'13"
Ptesan
Per 3h53'18" 33d58'51"
pti a, Jean Flambard
Cap 20h15'47" -13d58'7"
P'tit Bout
Uma 9h55'36" 70d47'12"
P'tit Bout
Gem 6h45'15" 22d59'8"
P'tit cerf volant
Dra 18h36'50" 68d23'42"
Ptite Audrey
Aqr 22h34'50" -11d13'29"
Ptite bourik
Cas 1h59'12" 62d5'19"
Ptolemaeus Alexandre McClusky
Uma 10h34'39" 42d46'20"
Ptolemy J Woodgate
Uma 9h49'56" 57d36'19"
PT-Pam
Leo 10h15'57" 13d14'16"
PTRCSWANEY Always & Forever + 1 Day
Uma 11h2'25" 35d36'21"
Pu
Dra 16h59'22" 58d32'52"
Pu & Lamb
Lib 15h33'36" -19d30'20"
Pualani
Cyg 20h44'34" 34d58'0"
PUALANI SHAUNTEL JACOBS
Umi 14h0'57" 78d41'34"
Puamelia
Col 5h43'5" -34d58'54"
puanani lonetti
Umi 13h26'50" 71d29'46"
Puba
Uma 11h17'30" 62d55'46"
Puba
Gem 6h10'39" 22d46'37"
Publicis-Törmä 1963
Cam 4h38'51" 68d40'10"
Pucca y Polli
Ari 3h6'31" 14d5'36"
Puccetto NTN
Aur 6h36'43" 33d45'13"
Puccinelli Janet
Ori 5h14'55" 0d36'50"
Puch
And 0h6'53" 45d29'57"
Pucha
Sgr 17h54'6" -28d39'10"
Puchax
Ori 5h9'54" 15d17'39"
Puchnick
Umi 15h51'17" 85d37'17"
Puck TreeDancer
Cam 4h30'7" 70d38'55"
Puck TreeDancer
Ari 1h47'3" 24d40'19"
Pud
Lyn 8h15'41" 46d9'47"
PUD
Dra 18h6'40" 70d37'10"
Pud Arnold
Ori 5h28'28" 1d41'8"
Puddin'
Uma 11h52'20" 55d41'43"
Puddin Girl
And 23h58'59" 34d47'47"
Puddin & Noodle
Uma 10h29'37" 51d56'16"
Puddin pop
And 1h15'52" 49d58'25"
Puddin' Pop
Aqr 22h33'10" 2d26'25"
Puddin Victoria Elizabeth Marker
Cam 7h24'12" 66d37'32"
Pudding
Uma 10h41'18" 54d22'12"
Puddle Butt
Aqr 22h11'29" -14d25'3"
Puddles
Ari 2h38'8" 27d6'1"
Puddlington Bear
Ori 5h30'57" -2d41'36"
Puddy
Umi 18h55'51" 86d46'52"
Puddy
Dra 18h38'49" 65d54'50"
Puddy Levenson
Tau 3h51'5" 24d6'41"
Pudge
Cnc 8h13'25" 26d45'29"
Pudge
And 23h8'45" 40d28'42"
Pudge Queen
And 2h33'7" 50d14'57"
Pudgee
Cam 3h58'46" 66d2'22"
Pudgie Hurst
Ori 6h5'17" 19d17'18"
Pudi
Uma 10h13'29" 44d12'51"
Pudums of Ava
Cas 1h20'27" 59d25'51"
Puegot- Kringle's Guide to Heaven
Cma 7h2'44" -26d30'52"
Puerquita
And 23h18'25" 43d17'12"
Puerto Rican Cheese Monkey
Cnc 9h16'38" 19d36'35"
Puerto Rican Papi
Cnc 9h7'22" 24d11'49"
Puf
Cas 0h0'48" 59d8'10"
Puff the Cat
Boo 15h4'44" 35d37'28"
Puffer
Dra 15h25'51" 64d46'10"
Puffin
Dra 12h27'31" 71d24'47"
Puffin
Ori 5h14'42" 0d56'7"
Puffo
Per 2h15'51" 51d33'51"
PUFFY
Oph 17h5'51" -0d26'8"
Puffy Freda
Cnc 8h50'30" 31d38'22"
Pufwheat
Cnc 8h43'58" 32d18'27"
Pug
Cyg 19h51'35" 56d0'36"
Puga
Leo 11h15'36" 5d49'45"
Pugita
And 23h11'31" 41d28'55"
Pugsalina Gizmo Karajgi
Aqr 22h7'47" -0d40'35"
Pugsley
Aur 6h30'50" 34d24'34"
Puh
Uma 10h14'34" 55d33'19"
PUI MAN HO
Lib 15h38'15" -22d48'5"
Puja Goyal
Lib 14h57'49" -18d1'45"
Puja Nayee
Leo 9h51'15" 16d52'2"
Puja, My Shining Star
Sco 17h4'13" -45d18'36"
Puka Star
Leo 10h53'37" 19d58'2"
PULC & PULC
Ori 5h10'59" 7d21'43"
Pulcher Astrum
Uma 11h36'47" 43d1'51"
Pulchrampius Materamo
Leo 9h29'11" 27d4'5"
Pulchrious J.S.
Vir 12h57'55" 12d21'26"
Pulchritudinous Kathryn
Del 20h37'16" 15d16'15"
Pulchritudo Intactilis
And 2h33'37" 44d23'8"
Pulguita
Gem 6h43'25" 23d45'9"
Pullus Saltatricis
Leo 11h35'50" 25d33'2"
Pully
Her 17h49'46" 17d0'44"
Pumba
Sco 16h3'37" -26d30'7"
Pümi
Uma 13h25'33" 55d23'57"
Pumkin Head
Gem 6h58'48" 31d26'29"
Pummira
Cam 4h29'41" 72d14'8"
Pumper
Psc 23h4'42" -0d3'28"
Pumpkin
Umi 13h38'14" 78d34'42"
Pumpkin
Cap 20h17'35" -10d26'36"
Pumpkin
Uma 13h43'29" 56d53'33"
Pumpkin
Sco 16h16'4" -22d37'32"
Pumpkin
Cyg 19h41'27" 37d12'22"
Pumpkin
Lyn 7h58'19" 38d0'49"
PUMPKIN
Aur 6h55'24" 37d48'9"
Pumpkin
Her 16h29'41" 41d46'59"
Pumpkin
Cyg 20h0'56" 52d23'43"
Pumpkin
Leo 9h38'49" 27d7'58"
Pumpkin and Boog's Love Star
Cnc 8h33'36" 29d11'33"
Pumpkin and Pickle
Cyg 21h27'14" 52d33'13"
Pumpkin Bread
Lyn 7h41'9" 37d27'2"
Pumpkin Camilla
Cas 3h23'57" 68d32'51"
Pumpkin Candy
Sco 16h9'43" -17d23'33"
Pumpkin Eagan
Umi 15h23'22" 71d49'8"
Pumpkin Pamela
Vir 12h46'1" -8d13'3"
Pumpkin Tanner
Peg 23h46'8" 22d30'51"
pumpkina
And 1h19'30" 47d42'27"
Pumpking Pie
Cep 21h32'48" 64d26'47"
PumpkinHead
Ori 5h35'48" 3d40'59"
Pumpy Umpy Umkin: LBV
Cnc 8h47'36" 6d44'0"
Puneet Kaur
Gem 7h25'38" 17d17'18"
Puneet Kumar Bhatia (Pumar-ia)
Sco 16h6'26" -8d23'58"
Puniwai Jay
Psc 0h12'17" -2d44'53"
Punk
Col 5h39'10" -27d39'45"
Punk
Lmi 10h40'38" 25d17'51"
Punk Rock Princess
Vir 13h30'12" -21d0'42"
Punk Rock Soldier Medic
Uma 11h8'53" 60d40'40"
Punkin
Her 16h24'28" 46d10'38"
Punkin
Dra 16h36'29" 52d16'28"
Punkin
Gem 6h49'20" 32d42'28"
Punkin
Gem 6h52'29" 34d39'26"
Punkin
Lmi 9h26'41" 33d14'32"
Punkin Butt
Sgr 18h9'52" -20d5'16"
Punkin L.M.D
Tau 3h45'27" 27d34'46"
Punkin' Pie Kinkaid
Tau 5h26'14" 23d24'40"
Punkin Pooh
Cyg 19h31'57" 30d9'57"
Pun'kin & Sunshine
Sco 16h50'43" -33d56'57"
Punkin-Butt
Tau 5h5'0" 23d29'40"
punkin's smile
Cnc 9h1'50" 11d55'46"
Punkin's Star
Dra 18h43'57" 64d56'20"
Punky
Cas 23h21'35" 57d34'43"
Punky
Umi 14h58'18" 84d41'36"
Punky
Ara 17h26'31" -48d15'4"
Punky
Gem 7h27'43" 21d11'59"
Punky Becka Star
Cap 20h59'41" -19d0'19"
punky doodle
And 0h53'47" 36d23'56"
Punky Morey
Uma 9h40'29" 44d48'22"
Punky Resendes
Uma 11h27'28" 61d53'40"
PunkyTheChipperMonkey
Umi 13h38'1" 76d0'17"
Puopolo
Ori 6h2'57" 17d14'19"
Pup Cole
Dra 18h47'3" 52d25'57"
Pupina
Cam 9h25'30" 82d2'23"
Puplet's Yellow Diamond
Uma 9h55'59" 57d53'35"
Puppe
Cep 1h54'50" 77d40'3"
Puppet
Uma 9h49'34" 67d56'13"
Puppy
Uma 12h1'6" 41d58'6"
Puppy
Ari 2h43'57" 16d52'27"
Puppy Bear
Sco 16h57'17" -35d44'59"
Puppy+Kitty
Pup 7h52'23" -28d16'57"
Puppy Tamra
Cmi 8h5'31" -0d0'40"
Puppy's Light
Cmi 7h25'39" 8d9'36"
Pups, unser Stern der ewigen Liebe
Uma 11h19'16" 43d20'24"
Pupstar
Gem 7h2'25" 35d11'22"
pupu
Pup 7h53'16" -25d26'28"
Pupy Dog
Uma 12h52'8" 59d26'2"
Pura Jackie Casas
Tau 3h48'36" 15d3'18"
Pura Vida
Cnc 9h20'39" 30d14'17"
Purcelle
Aqr 21h16'38" -0d0'42"
Purdue
Tau 5h35'14" 26d3'29"
Purdy
Uma 8h33'51" 68d9'41"
Purdy Allison
Cap 20h25'26" -13d44'5"
Purdy Star
Cam 4h8'20" 67d16'3"
Purdy-Zablocki
Cyg 21h30'41" 44d38'34"
Pure Bright Light Of Margie
And 2h15'39" 48d57'59"
Pure Fire
Uma 9h44'32" 50d13'45"
Pure Heart
Per 4h45'28" 43d8'46"
Pure Light
Gem 6h34'17" 22d18'2"
Pure Little One
Lyn 7h16'14" 44d25'33"
Pure Magic "Bill & Melissa's Star"
Cyg 20h49'32" 46d25'32"
Pure Miller Champion
Per 3h7'23" 56d58'49"
PureSoul
Psc 0h59'57" 9d24'0"
Purkey
Peg 22h48'32" 25d43'15"
Purkinite 9.28.40 / 10.16.36
Uma 11h39'35" 45d56'11"
Purnell Gaukstad
Gem 7h0'23" 33d49'22"
Purple
Gru 21h59'18" -39d5'52"
Purple Badger
Gem 7h7'59" 18d6'16"
Purple Dream JP-RR
Ori 5h23'44" -3d8'31"
Purple Flying Elephants
Vir 13h21'0" 8d58'42"
purple mason
Pho 0h34'29" -41d20'50"
Purple Princess
Ori 5h40'37" 6d27'47"
Purple Princess
Cnc 9h11'29" 32d4'59"
Purplerose
And 23h32'26" 37d42'56"
PURPLESTAR-SG-1684
Sgr 19h3'5" -24d3'46"
Purpose
Cas 23h9'39" 58d29'46"
Purrpleshar
Lyr 19h10'4" 38d19'32"
Purshotam Vasani
Cep 5h28'15" 86d4'32"
Purston-Foxfyre
Cyg 21h12'36" 47d1'14"
Purus Amor
Cap 21h42'10" -14d3'11"
Purva & Piyush
Psc 1h38'43" 18d1'2"
Purvi Patel
Uma 11h10'54" 34d21'55"
Pusch, Uwe
Uma 11h21'7" 64d21'12"
Puschel
Uma 11h58'4" 51d6'55"
Puschner, Tim
Leo 11h11'40" 21d21'44"
Pushpa
Sco 16h10'13" -10d17'23"
Pushpa
Col 5h24'20" -41d11'48"
Pushpa Wati
Gem 6h45'26" 14d14'6"

Pushpa Wati Sharma
Aqr 23h35'43" -8d9'41"
PUSHPOP
Dra 19h5'55" 77d57'52"
Push-Pop
Ori 5h48'52" 5d57'39"
Pushu
Cyg 21h19'52" 43d42'58"
Puss
Dra 17h7'18" 67d42'57"
Puss & Pigs
Cyg 21h42'25" 44d32'4"
Pußel, Petra
Uma 9h5'15" 52d15'14"
Pusskin
Peg 21h34'42" 15d55'28"
Pussy Willow
Uma 10h1'5" 48d47'32"
Put Your Name Choice Here!
Cas 3h34'38" 69d53'43"
Putchen
Sco 17h31'13" -43d6'41"
Puthy
Lib 15h1'21" -17d20'2"
Puti
Tau 5h45'2" 24d53'50"
Putney Volkart
Cma 6h58'0" -16d8'19"
Putu Indrawati
Ara 17h30'24" -47d34'18"
Putz
Cam 6h41'26" 76d8'0"
Pütz, Miklas
Ori 5h47'37" -0d3'56"
Puutentwait-817
Leo 10h17'25" 21d18'48"
Puyan Sebastian Frutiger
And 23h28'23" 38d43'55"
Puzzona
Aur 6h37'23" 35d32'44"
Puzzy1990
Cnc 8h12'14" 32d21'16"
PV 09-07-1933
Vir 12h11'22" 2d57'24"
PVT Andrew Cuthbertson
Ori 5h38'27" 1d39'5"
Pvt. Josh Morberg
Aql 19h42'40" 8d13'55"
P.W. Cash
Per 3h33'52" 44d49'4"
Pwednewow13
Peg 0h4'34" 27d37'8"
P.Wick-1
Per 3h26'53" 50d54'32"
Pybas
Umi 17h37'50" 86d31'16"
Pye
Uma 9h20'17" 62d23'54"
Pye Carswell
Pup 6h14'22" -49d19'9"
Pyka
Boo 14h21'31" 16d55'28"
Pyle-Ruybe
Cyg 21h31'1" 54d54'43"
Pynxz's Boy, Tyler Michael
Uma 9h32'56" 47d41'41"
Pyper's Diamond
Gem 6h59'3" 24d32'40"
Pypre 3-10
Psc 1h31'48" 12d14'48"
Pyramis
Ant 10h17'57" -28d54'34"
Pysco Ducky + Hell Bender 4 ever!!!
Psc 0h34'42" 14d46'40"
Pytek
Ori 5h52'12" 18d38'30"
PYTHON Romana
Her 16h49'27" 6d12'55"
PZGZ143
Gem 6h20'22" 19d5'11"
Q
Ari 2h16'40" 23d53'57"
Q
Ari 2h0'7" 21d22'3"
Q
Vir 13h14'21" 12d8'36"
Q
And 2h36'43" 45d17'26"
Q*bert Breasts of Fury
Cohen
Dra 19h47'46" 59d37'39"
Q Dee Dee
Ori 6h5'24" -3d50'23"
Q Rahl
Cnv 12h39'1" 48d23'59"
Q Z
Cas 0h15'10" 59d17'15"
Q-10
Uma 11h0'28" 34d59'8"
Q-101 Super Star
Crt 11h33'51" -20d37'14"
QAB
Gem 7h6'36" 34d23'21"
Qaim Ali
Per 3h9'56" 47d24'35"
Qalb tayyib
Gem 7h19'12" 15d11'59"
Qarr Star
Uma 11h27'1" 71d57'9"
Qaun
Uma 10h15'49" 44d11'17"
Qconnie
Cas 1h40'49" 67d35'25"

Qi&Fei
Lib 15h22'32" -7d2'32"
Qiana Newsome
And 2h23'59" 44d42'46"
QingGuo
Lib 15h50'11" -7d15'0"
Qinyi
Sco 17h56'40" -30d58'10"
qkc
Cyg 21h10'48" 45d20'41"
Qnerfertitichantal
Cap 21h5'29" -21d29'47"
QP
Leo 11h10'31" 10d46'18"
QP BooBoo
Lib 15h6'17" -16d27'8"
QRAT
Dra 12h31'39" 71d26'36"
Qtpie2949
Aqr 22h18'31" -4d17'0"
Qua
Uma 11h42'23" 43d39'59"
Qua Dua
Lyn 8h27'19" 55d9'59"
Quach, Tich Canh
Ori 5h22'11" -4d43'54"
Quack
Crb 16h3'29" 34d59'17"
Quack
Cyg 20h27'35" 42d39'8"
Quack Quack
Pav 20h11'8" -57d13'5"
Quade
Uma 11h57'53" 63d15'1"
Quade Robert Hedges
Per 3h25'28" 37d38'53"
Quail Hill Scout Reservation
Aql 19h49'51" 8d9'32"
Quake
Lyn 6h46'18" 53d52'54"
Quakerman
Sgr 19h20'34" -34d4'33"
Quakertown H.S. Class of 2007
Dra 17h43'0" 55d33'58"
Quane
Lyr 18h48'25" 40d55'34"
Quang Minh
Lmi 10h0'26" 29d18'14"
Quappe Bremner
Aqr 22h7'28" 0d0'25"
Quasar Prince Kyle Matthew
Psc 0h42'51" 7d13'42"
Quate
Lyr 18h51'51" 35d54'50"
Quattro
Umi 13h22'51" 87d35'9"
Quattro Settembre
Vul 19h48'18" 22d0'33"
Quattuor Astrum Matris Deena
Ori 5h36'46" 8d48'28"
Quaymee
Sco 16h40'43" -30d8'7"
Quaynteece
Psc 1h17'31" 21d31'4"
Qubil
Cas 0h38'34" 53d16'44"
Que Sensacion La Soberana
Del 20h41'52" 15d31'39"
Quecha AKA Lucrecia Gatica
Psc 23h26'1" 0d0'50"
Queen Adia
Lyn 8h9'22" 35d35'42"
Queen Aleta
Cas 0h39'41" 57d46'43"
Queen Alexandra
Cas 23h25'30" 56d34'40"
Queen Amanda
Sco 16h17'28" -8d52'4"
Queen Amy NMCB 1223
Cas 1h6'12" 65d18'42"
Queen Angélica Johana Ramirez Cañón
Cap 21h48'10" -17d35'59"
Queen Ann of Heritage Hospice
Cap 21h42'19" -16d47'28"
Queen Ashley
Vir 14h31'50" 6d41'16"
Queen Bea
Cas 1h35'2" 63d39'24"
Queen Bee
Psc 1h15'1" 17d5'24"
Queen BG.
Ori 5h57'7" 21d9'32"
Queen Carole LXXV
Ari 2h22'20" 11d8'36"
Queen Christina L. Appodaca
Aqr 23h35'23" -13d21'54"
Queen Christine's lovely light
Cas 1h35'3" 64d32'15"
Queen Damarliz
Leo 11h7'45" 15d40'8"
Queen Dana
Cyg 19h58'47" 38d27'32"
Queen Daub
Tau 4h37'40" 28d11'9"

Queen Daydaybell
And 23h5'48" 38d48'8"
Queen Dennise, I Love You
Lmi 9h25'46" 36d26'9"
Queen Desi
Cas 1h12'53" 56d22'11"
Queen Di
Lmi 10h33'57" 36d56'6"
Queen Donna
Cas 0h31'38" 60d53'28"
Queen Eleanor Morgan
Psc 0h50'46" 17d47'27"
Queen Esther-Grammykins
Cas 0h36'21" 55d37'25"
Queen Frances Ann Carlen-Jones
Crb 15h38'7" 27d41'43"
Queen Francie
Uma 12h0'1" 65d8'59"
Queen Genevieve
Sco 17h50'50" -32d4'11"
Queen Gillian Jo Steinmetz
Leo 11h24'6" 16d4'41"
Queen Glo
Cas 0h5'27" 58d55'47"
Queen Helene
Ari 3h9'0" 19d8'34"
Queen Jada
Tau 3h49'40" 27d0'0"
Queen Jamie Plocky
Crb 15h46'7" 31d40'27"
Queen Judi
Gem 6h49'42" 16d59'36"
Queen Judy Dominquez
Cas 1h8'1" 60d54'26"
Queen Kristina
Uma 11h14'50" 43d59'49"
Queen Laurie
Ari 2h28'53" 26d48'55"
Queen Laurie Bernard
Cap 20h58'31" -19d53'18"
Queen Lucy the Valiant
Leo 11h23'8" 15d3'13"
Queen Lynnie's Kebrion from Danny
Ori 5h25'3" 2d13'39"
Queen Maddy
Psc 1h51'34" 9d23'56"
Queen Magdela
Leo 10h58'59" 18d56'25"
Queen Maricela
Cas 0h10'22" 62d29'39"
Queen Marija
Cas 0h7'21" 64d37'11"
Queen Master of the Universe
Vir 15h0'57" 3d39'34"
Queen Mom
Vir 12h39'22" 2d40'40"
Queen Nessa
Cnc 8h4'46" 17d59'2"
Queen of California (Claudia)
Cyg 19h51'51" 37d5'14"
Queen of King
Cas 23h41'36" 55d17'47"
Queen Of Moms
Sco 17h47'21" -42d56'23"
Queen Of My Heart Anissa
Leo 10h13'31" 22d5'18"
Queen Of My Heart Esther
Cas 23h43'35" 55d38'33"
Queen of Spades
Cas 1h42'52" 64d11'15"
Queen of the Nile
Lib 14h48'50" -17d13'19"
Queen of the Stars
Cas 0h45'37" 52d7'17"
Queen Patty
Tau 4h29'50" 28d22'38"
Queen Rebecca
Cas 1h7'2" 48d52'16"
Queen Rebecca
Aqr 20h56'13" -11d41'56"
Queen Regina Pealer
Lib 15h4'11" -2d5'9"
Queen Reginia
Cas 0h26'24" 53d3'44"
Queen Reyes
Vir 13h17'1" -12d54'12"
Queen Rita
Cas 1h27'35" 56d6'48"
Queen Robin Ophelia Quivers
Uma 9h43'29" 60d3'5"
Queen Sally Mary
Cas 0h35'26" 55d6'10"
Queen Sasha
Leo 9h35'10" 7d8'41"
Queen Shannon
Cap 20h59'43" -20d48'47"
Queen Shireen
Gem 7h41'42" 25d1'49"
QUEEN SHIRLEY the ONETH
Sco 17h51'45" -42d15'25"
Queen Solinda
Sgr 19h31'20" -30d5'29"
Queen Stacy
Cas 0h10'38" 53d40'28"
Queen Star
Cas 0h8'10" 53d4'46"
Queen Sugar M'lou
Cas 23h15'50" 58d58'6"

Queen Suzanne
Cas 1h25'21" 69d5'44"
Queen Teuta
Lac 22h51'29" 49d2'50"
Queen V
Gem 6h16'45" 21d58'31"
Queen102179
Lib 15h41'34" -17d43'46"
QueenAnneHGlover LifeForce
Tau 3h53'17" 26d34'34"
Queenie
Vir 13h17'10" 5d5'55"
Queenie
Cas 1h15'9" 57d25'10"
QueenKorn
Sgr 18h47'38" -30d15'10"
Queens Knight
Per 3h18'45" 42d21'22"
"QUEENS" Love Me Tender
Cas 1h16'20" 62d34'27"
Queens St. Martin
Leo 10h48'7" 14d24'37"
Queeny
Sco 17h29'13" -41d54'41"
Queezy
Lyn 7h18'15" 45d22'29"
Queißer, Sabine
Uma 10h39'26" 65d51'5"
Quelis Johana
Sco 16h9'54" -26d59'57"
Quello Saggio e Ragionevole
Psc 22h52'2" 3d24'1"
Quenet
Lac 22h44'49" 53d41'57"
Quent DuRant
Tau 3h53'1" 24d6'48"
Quentin
Crb 15h20'10" 26d56'50"
Quentin
Psc 1h52'18" 10d8'30"
Quentin
Lib 15h35'29" -12d27'48"
Quentin Earlen Pickering
Cnc 9h9'41" 32d35'2"
Quentin & Evan Fidance
Ori 5h47'2" -2d22'44"
Quentin G. IV
Cap 20h23'6" -10d13'51"
Quentin Georges 06072005
Cnc 9h2'55" 19d11'58"
Quentin Harrington "Q-Dogg"
Cap 20h30'12" -13d26'59"
Quentin James
Leo 11h54'30" 21d1'14"
Quentin Kaluakalai Caboga
Gem 6h47'44" 33d36'53"
Quentin Leplat
Leo 10h28'39" 16d35'46"
Quentin Levin Braun
Umi 13h22'22" 73d0'0"
Quentin Michael Nichols
Cnc 8h49'49" 27d41'44"
Quentin Spendrup
Ari 3h7'23" 28d25'21"
Quentin Tre' Riley Adan Connors
Dra 17h43'48" 65d57'43"
QuentinD200206
Dra 12h42'15" 73d10'28"
Quentin's 2nd Star
Ori 6h4'48" 17d11'40"
Queri Lorraine
Umi 15h6'4" 84d56'25"
Querida
Tau 5h51'7" 17d40'32"
QuErVo
Ori 5h53'7" 9d26'0"
Queta
Sco 17h40'46" -37d36'37"
Quetzali
Umi 0h31'22" 89d24'24"
Quiana
Lib 15h11'6" -23d43'7"
Quiana Helena Troebs
Aqr 22h30'54" 0d36'52"
Quianna T. Williams
Sgr 18h24'37" -19d43'36"
Quick Carl
Ori 5h35'37" -6d45'12"
Quido
Uma 11h41'43" 47d55'51"
Quiet Man
Cep 21h31'54" 64d30'31"
Quiet Man
Lib 15h26'6" -24d35'20"
Quig
Cyg 20h5'17" 33d11'29"
Quigley
Vir 12h38'3" 4d53'57"
Quigley
Cma 6h47'7" -21d43'17"
Quiktrip
Vir 13h43'56" 1d28'36"
Quila
Uma 10h57'26" 49d23'56"
Quilan
Uma 11h29'10" 63d56'1"
Quilla Bibb Reed
Cnc 8h19'8" 15d35'35"

Quin James O'Grady
Aqr 21h0'3" -6d5'5"
Quin Mill Qiang Mayginnes
Ori 4h52'50" 10d29'47"
Quinceañera J15
Psc 23h43'39" 5d56'10"
Quincey Bartholomew Wahl
Cep 0h34'3" 78d17'27"
Quinci Turner
Ari 2h14'16" 23d57'50"
QuinCin
Cma 6h35'57" -13d19'51"
Quincy
Uma 11h44'28" 62d20'20"
Quincy
And 0h39'27" 28d5'56"
Quincy Celine
Peg 22h20'43" 33d27'36"
Quincy Gabriel Wiley
Sgr 18h24'2" -36d10'45"
Quincy K
Lib 14h52'1" -1d31'21"
Quincy Lee
Ori 5h21'42" 10d30'41"
Quincy Zaria
Lib 14h53'59" -13d35'59"
Quincy's Light Up The Night
Uma 11h32'39" 44d27'44"
Quincy's Twinkler
Gem 6h40'11" 20d6'4"
Quinke, Walter
Cnc 8h16'44" 22d45'59"
Quinlan Murphy Cramer
Leo 11h36'23" 16d0'43"
Quinlan P. B.
Cyg 20h13'54" 54d22'49"
Quinlin Dempsey Stiller
And 1h20'28" 46d28'25"
Quinlin Dempsey Stiller
Her 16h50'40" 37d40'5"
Quinlynne Maddox Nichols
Tau 4h7'8" 19d1'32"
Quinlyn's Heavenly Light
And 0h21'13" 28d34'39"
Quinn
Vir 12h30'1" 8d13'58"
Quinn
Vir 13h21'53" 12d43'55"
Quinn
Ori 5h37'47" 2d59'45"
Quinn
Uma 8h14'20" 66d56'50"
Quinn
Vir 12h14'54" -0d57'15"
Quinn Alexander Avirom
Leo 10h27'26" 12d26'12"
Quinn Alexandra Burger
Her 17h36'2" 16d40'52"
Quinn Asher Buckley
Sco 17h18'36" -31d40'41"
Quinn Avery Griffin
Uma 9h7'25" 52d22'58"
Quinn Bevan Thatcher
Dra 18h3'17" 72d57'27"
Quinn Biever
Aql 19h47'50" 15d7'55"
Quinn Brian Frankel
Aqr 22h44'50" -18d16'52"
Quinn C. Doty
Cap 20h43'56" -18d15'30"
Quinn Carolina Hammond
Psc 1h12'38" 24d52'35"
Quinn Colette Black
Ari 2h49'52" 27d40'21"
Quinn Elizabeth Smith
Ari 3h19'2" 27d42'40"
Quinn Hamm
Cep 22h1'30" 70d23'9"
Quinn Harrist
Umi 19h6'50" 89d0'56"
Quinn Higdon
Ori 5h49'43" 2d43'1"
Quinn J. Sedinger
Vir 13h29'42" -22d36'56"
Quinn Joseph Lasley
Ori 6h8'22" 20d50'0"
Quinn Loran
Umi 14h43'45" 75d22'59"
Quinn Louis
Uma 8h29'47" 71d55'45"
Quinn Lucas Anthony Ricci
Uma 9h59'18" 61d14'24"
Quinn & Luke Ludd
Tau 3h53'41" 30d54'1"
Quinn Lynch
Ori 6h0'41" 17d38'23"
Quinn Marie
Cnc 9h10'5" 8d56'8"
Quinn Michael McNelis
Boo 14h36'31" 19d0'30"
Quinn Michael Oliver
Cet 1h27'44" -12d49'44"
Quinn Nakazawa
Dra 16h8'20" 61d49'11"
Quinn Norman Hannan
Leo 11h10'11" 21d56'39"
Quinn Ordway Angelique
And 23h53'28" 47d52'39"
Quinn Ordway Angelique Waisbrot
Uma 13h32'6" 55d23'23"
Quinn Rayann Thurman
And 0h25'58" 27d20'15"

Quinn Robert Altman
Boo 14h14'56" 25d8'40"
Quinn Thomas Gallagher
Lyn 9h0'23" 36d26'24"
Quinn Trella Medina
Aqr 21h37'42" -8d2'17"
Quinn Valentino
Vir 13h10'15" -14d53'10"
Quinn Wieland
Cep 21h58'22" 70d31'1"
Quinnie Bee
Vir 13h23'26" 7d38'34"
Quinnton Joveill Lawson
Lib 15h44'48" -18d39'30"
Quinntorgan44
Gem 7h20'14" 31d12'13"
QUINO
Ori 5h25'2" 2d33'42"
Quinquaginta Annus Diligo
Lyr 18h45'53" 38d34'3"
Quint and Kerra June 5,2005
Cap 20h7'17" -16d44'9"
Quinta 05-06
Peg 22h52'48" 24d34'0"
Quinta - Sanchez
Sge 20h0'46" 17d51'21"
Quintessa Maria
Cnc 8h21'19" 22d59'45"
Quintin Cacas, Jr.
Boo 14h35'52" 33d24'26"
Quinto Chiarini
And 23h43'43" 43d19'10"
Quinton Hierling
Psc 1h3'23" 9d57'40"
Quinton Lucas Gross
Aqr 21h46'16" -2d51'34"
Quinton Miles Watts
Uma 9h35'9" 53d48'11"
Quiomara Marie Carello
Uma 11h8'23" 44d35'14"
Quiquag
Uma 12h32'25" 58d9'26"
Quique
Peg 21h42'41" 28d2'57"
quiquine<\@>titruc.us
Peg 22h15'52" 23d32'9"
Quirin
Uma 9h4'0" 59d26'1"
Quirkinus
Cir 15h1'56" -60d44'57"
Quiroz
Cyg 20h21'41" 49d19'16"
Quita
Cap 21h39'5" -11d52'47"
Quiteria Maria
Uma 11h18'54" 43d20'52"
Quo Vadis
Vir 12h46'10" 11d39'1"
Quoc Van Nguyen
Sgr 18h4'55" -27d8'13"
Quod fiat, fiat
Cru 12h22'28" -61d47'59"
Quora
Lib 14h54'25" -11d1'40"
Quoven
Ori 5h38'16" 8d41'0"
Quressa Evelyn Thompson
Cabbage
Aqr 23h24'9" -10d16'51"
Quyen Nguyen
Aqr 23h17'45" -10d13'19"
Quyen P Thompson
Cyg 19h56'25" 42d46'59"
Quynh-Mai Nguyen
Ori 5h40'28" 15d19'50"
QVC Sample
Uma 10h28'11" 59d41'53"
"R"
Ori 5h53'57" 21d54'43"
R&A McTamaney - 60
Uma 9h19'33" 67d8'51"
R A Wishing Star THAOF IGH
Lyr 18h29'44" 36d35'35"
R and R Star
Tri 2h17'23" 33d11'22"
R B M
Vir 13h0'1" 4d26'6"
R Beakn
Tau 5h11'40" 16d48'13"
R. Brown Williams
Sco 16h8'50" -17d50'10"
R. Bruce Pelton
Sco 16h43'16" -30d6'49"
R & C
Psc 1h12'40" 28d41'53"
r CodyCat
Crb 16h13'49" 31d35'50"
R & D
Leo 9h29'15" 11d43'36"
R & D Kohler...The Legendary Stars
Dra 19h19'51" 61d18'0"
R. Daniel Ballard
Tau 5h46'20" 14d40'16"
R. Daniel Kelly
Ori 5h9'6" 11d45'49"
R. David English
Aqr 21h30'2" -2d15'54"
R. DeVary
Uma 10h38'51" 62d47'20"
R. Douglas Arquette
Pho 0h22'21" -40d34'58"

R. Drew J.
Ari 3h15'56" 28d24'41"
R. Duane Marshall, D.C.
Cap 20h24'43" -11d22'53"
R. Duffy Holland
And 0h32'39" 28d58'59"
R. E. Fauber
Per 3h8'28" 37d34'28"
R. E. WILLIAMS
Ori 4h46'25" -0d35'16"
R. Edison Hill, Esq.
Her 18h27'26" 20d0'18"
R. F. Willette's Place in the Sky
Uma 12h15'56" 57d53'14"
R. Frank Thornton Jr.
Aur 6h35'6" 34d33'51"
R Garmil 40
Leo 9h29'35" 31d6'42"
R. Garth Ferrell
Vir 14h43'31" 3d31'10"
R. Gregg Lee
Psc 1h4'22" 4d13'37"
R & H McReynolds
Cyg 21h42'32" 45d38'33"
R. HAKAN KIRKOGLU
Ori 6h18'56" 14d21'2"
R. Hawke
Uma 8h24'39" 71d35'38"
R. Hope
Ori 5h32'34" -0d58'30"
R&J Boughton
Lib 15h27'46" -3d51'24"
R & J Complete
Lyn 7h17'3" 47d27'4"
R&J Ferrell
Aql 19h47'47" -0d14'47"
R. J. S.
Umi 16h38'0" 88d11'0"
R. J. Squirrel
Aqr 21h53'0" 0d57'51"
R J T
Dra 17h58'11" 54d55'52"
R & J Wilson's Diamond in the Sky
Ara 17h13'1" -51d42'41"
R Jackie B
Lib 15h12'56" -13d43'18"
R John Kane
Uma 9h59'53" 67d34'29"
R. "Johnny" Wrzesniewski
Cep 22h45'59" 65d14'10"
R. Joseph Heagany V
Dra 16h10'22" 52d29'5"
R K Yeager
Cnc 8h3'7" 21d27'14"
R. Kirk Brown
Cap 21h26'41" -23d53'9"
R & K's Serendipity
Cas 1h12'28" 53d57'5"
R. & L. Jensen Family Wedding Star
Her 17h4'48" 29d56'21"
R L N Toujours 2/27/03
Lyr 18h20'9" 36d20'3"
R L Pare
Cyg 21h21'44" 54d52'21"
R. L. Sharp
Aql 19h50'26" -0d32'6"
R. Leonard Carlton
Cyg 20h46'47" 38d55'38"
R. Linneaus
Sco 17h44'16" -44d37'51"
R little buddy J.J. PIERCE
Tau 4h26'50" 23d5'28"
R - - M - - D O
Col 6h14'26" -33d33'58"
R - - M - - D O
Cap 20h25'48" -19d6'33"
R & M Eternity
Umi 15h31'35" 72d18'14"
R. Mark Rogers
Ser 16h13'21" -0d13'54"
R. Michael Fitzgibbons
Uma 13h36'39" 59d44'45"
R. Myles Allison, June 9, 1924
Gem 6h38'20" 16d16'4"
R&N
Uma 11h9'33" 33d22'4"
R. N. Brearley
Uma 9h30'49" 49d19'25"
R & P
Cyg 21h43'36" 51d34'24"
R. Pierce Reid
Uma 9h16'54" 61d4'50"
r + r
And 2h32'18" 44d39'17"
R & R Forever
Cyg 21h28'3" 46d44'7"
R&R Logan 50
Lib 15h5'14" -2d58'12"
R & R-Always & Forever in the stars
Cyg 21h4'34" 36d27'57"
R S
Cap 21h27'10" -8d48'47"
R S C 2007
Lyn 8h12'16" 39d37'32"
R & S - "Some kind of Wonderful"
Cyg 21h59'1" 51d25'17"
R. Scott Korzen "Soul Mate, Daddy"
Her 17h12'36" 21d15'52"

R. STAR (The Star of Hope)
Ori 5h47'8" 11d26'47"
R. Stephen Wallace
Leo 10h34'28" 15d19'16"
R T J
Leo 11h19'27" 4d28'55"
R T Padget
Lib 15h11'30" -29d6'28"
R W Cornatzer
Uma 8h41'23" 54d18'59"
R. Wesley Dittmer
Eri 4h29'12" -0d55'9"
R. Willow
Peg 22h39'35" 33d4'34"
R Wisch
Uma 10h58'23" 48d13'12"
R, R, M, R Jr 5, K, M, Williams
Uma 12h15'15" 59d52'39"
R,C,M,S,COOKE FAMILY
Ori 5h33'24" 1d35'2"
R•A•P•
Uma 11h23'49" 54d39'34"
R2UOBOY5
Lyn 7h42'45" 40d15'47"
R3
Aqr 22h1'57" -13d59'17"
R69 Kevin and Debbie Forever
Boo 14h17'41" 16d46'46"
Ra Sing
Cnc 8h50'0" 15d28'46"
Ra Stansbury
Sgr 17h58'24" -17d36'34"
RA WIN 2000
Uma 9h30'0" 48d10'31"
Raa Raa
And 23h17'50" 41d59'54"
Raab
Lyn 8h15'49" 34d30'50"
Raab Jürgen
Uma 10h4'20" 46d49'37"
Raab, Norbert
Ori 5h11'40" 0d29'13"
Raajit Lall
Crb 16h4'54" 32d53'18"
Raani Virdee
Cas 0h19'43" 55d41'7"
Raatz, Heinrich
Uma 10h10'58" 51d49'13"
Rabail Sami
Tau 4h23'5" 17d12'45"
Rabbi Balfour Brickner
Her 18h19'32" 16d56'57"
Rabbi David A. Baylinson, D.D.
Cas 23h58'8" 51d6'45"
Rabbi Jonathan Stein
Cep 21h28'7" 63d45'50"
Rabbi Stanley Rabinowitz
Gem 6h44'2" 32d48'36"
Rabbit
Aqr 23h37'24" -16d36'7"
Rabea
Boo 15h30'2" 46d9'10"
Rabia
Uma 11h48'17" 61d18'47"
Rabia Karmali
And 0h0'46" 45d6'59"
Rabia Mashkoor
Dra 17h14'5" 57d40'14"
Rabrab
Leo 9h40'12" 32d23'15"
Rabun Hammock
Ari 2h1'7" 14d16'10"
Racaaf
Uma 10h48'58" 55d33'17"
Race
Aur 5h43'46" 39d11'4"
Race Zander Lance
Ori 5h29'53" 1d21'47"
Raceon
Lib 14h52'50" -1d16'29"
Raceway
Sex 9h50'42" -0d47'51"
rach marie.o146
Ori 5h11'5" 12d18'39"
Rachael
Tri 1h56'32" 31d8'5"
Rachael
Uma 11h45'1" 60d6'13"
Rachael
Uma 10h44'34" 61d52'32"
Rachael 12/22/2000
Cap 20h52'19" -15d46'22"
Rachael Alexandria Lauren Beam
And 0h5'55" 37d17'29"
Rachael Alixandra Barclay
Uma 10h28'10" 49d0'35"
Rachael Always & Forever
And 1h45'33" 42d23'32"
Rachael Amanda Zaring
Uma 9h49'50" 65d5'40"
Rachael and Paul
Cyg 19h42'30" 29d27'2"
Rachael Ann
Cyg 21h11'49" 46d39'6"
Rachael Ann
Uma 8h55'11" 68d49'22"
Rachael Ann Azzinnaro
Lyn 7h36'58" 53d7'17"
Rachael Ann Keeton
Vir 13h23'45" 1d7'25"
Rachael Ann Kennedy
Aqr 22h40'44" 2d12'12"
Rachael Anne Anderson
Cyg 19h46'49" 37d25'40"
Rachael Anne Hagy
Psc 1h15'17" 17d26'59"
Rachael Arianne White
And 1h28'51" 37d57'15"
Rachael Bell
And 0h11'20" 44d40'53"
Rachael Claire Charles
Lib 15h19'10" -4d50'55"
Rachael Courtney Meacham
Mon 6h53'18" -0d16'49"
Rachael Cummins
And 23h12'41" 48d28'53"
Rachael Danielle Peterson
Ori 5h29'45" 2d19'37"
Rachael Dooley
Cap 20h39'1" -25d2'29"
Rachael E. Hazlett
Ari 2h32'15" 19d41'37"
Rachael Edwards
And 1h53'13" 38d16'59"
Rachael Elisabeth Rose Tallman
Ori 5h58'26" 9d59'40"
Rachael Elizabeth
Leo 10h20'17" 10d31'17"
Rachael Elizabeth
Oph 17h31'37" 12d9'6"
Rachael Elizabeth
Uma 11h12'36" 59d57'11"
Rachael Elizabeth Edwards
Vir 12h49'11" -2d51'50"
Rachael Elizabeth Huntley
And 1h22'40" 45d5'28"
Rachael Elizabeth Neeb
And 0h19'4" 43d33'36"
Rachael Elizabeth Rabenberg
Uma 10h6'25" 47d59'12"
Rachael Elizabteth
Leo 9h40'40" 32d10'32"
Rachael Emily Stark
And 23h54'5" 40d14'1"
Rachael Erwin
And 23h44'44" 46d59'23"
Rachael Eubank
Psc 23h53'47" 6d51'27"
Rachael - Forever My Love
Ori 5h59'51" -0d41'11"
Rachael Frasier
Cap 21h31'18" -14d40'20"
Rachael French
Uma 10h8'6" 44d5'43"
Rachael Fuller
Cas 0h35'40" 58d43'2"
Rachael Giambone's Star
Lmi 10h7'10" 33d32'16"
Rachael Grace Deerfield
Lyn 7h48'4" 57d13'6"
Rachael Hamilton
Vir 13h17'26" -3d31'27"
Rachael Hannah Youngworth
Cap 21h11'41" -20d40'32"
Rachael Hildebrand
Lib 15h7'18" -3d33'39"
Rachael Hoffmann
Crb 15h33'29" 38d28'49"
Rachael Irina Stevens
Uma 9h50'12" 42d59'3"
Rachael Isle-Singh
Gem 6h44'53" 30d11'54"
Rachael Jane Gilbert
Lmi 9h43'45" 40d59'7"
Rachael Jewel Bishop
Cas 23h1'15" 56d29'42"
Rachael Jonquil Vigil
Cas 2h8'26" 64d35'6"
Rachael K. Peterson
Uma 9h19'59" 46d31'59"
Rachael Kathleen
Ori 5h25'50" 6d18'31"
Rachael Kathleen Coffey
Gem 6h49'20" 24d25'23"
Rachael Kathryn Bell
Aqr 22h23'46" -12d44'21"
Rachael Katina
Sco 16h57'58" -38d58'32"
Rachael & KC
Umi 15h17'24" 73d3'23"
Rachael Keurajian's excellent star
Psc 1h19'0" 31d15'41"
Rachael -Khara
Sco 16h21'53" -20d31'27"
Rachael Konerama Super Galactica
And 23h4'32" 35d47'2"
Rachael Lambert
And 2h25'32" 49d43'23"
Rachael Lea McCabe
Leo 10h10'44" 22d19'58"
Rachael Leann Miller
Ori 5h20'46" 0d11'3"
Rachael Lee
And 0h23'30" 31d46'59"
Rachael Lee
Sco 16h10'9" -15d31'58"
Rachael Leia Koltun
Ori 5h36'9" 9d53'21"
Rachael Lynn
Tau 4h10'43" 13d12'48"
Rachael Lynn 381
Pho 0h33'33" -40d29'3"
Rachael Lynn Shute
Peg 22h53'43" 8d34'14"
Rachael Lynne Kratz
And 23h18'48" 46d32'54"
Rachael Mannell
Col 5h59'24" -34d47'38"
Rachael Marie
Ari 1h53'37" 22d31'29"
Rachael Marie Betts
Uma 10h45'16" 49d27'37"
Rachael Marie Hanna
Ori 6h22'2" 13d17'50"
Rachael Marie Hottenstein
And 23h11'25" 47d22'34"
Rachael Marie Mcconaghy
Tau 4h15'4" 8d6'11"
Rachael Matthews
Cas 23h40'29" 54d51'58"
Rachael Michelle Morton
Sgr 18h1'3" -29d22'2"
Rachael Mikelson
Aqr 22h10'39" 0d58'32"
Rachael My Love
Cas 23h59'14" 64d29'14"
Rachael - My Starry Eyed Lassie
And 1h24'37" 34d44'28"
Rachael Nichol
And 23h11'39" 45d32'26"
Rachael Nicola Foye
And 1h22'18" 50d7'36"
Rachael Nicole Garacci Robichaux
And 23h5'50" 51d54'12"
Rachael Nicole Livingston
Lyr 18h50'10" 36d36'3"
Rachael Nieto
And 2h16'4" 38d1'0"
Rachael Olivia Carbone
Uma 12h15'32" 59d0'3"
Rachael Olivia Tominaga
Psc 0h49'8" 15d48'11"
Rachael Paige
Cas 1h16'15" 55d55'0"
Rachael Rafferty
Vir 13h9'14" 6d53'22"
Rachael Record & Bryan White 4ever
Uma 10h36'51" 70d52'44"
Rachael Repshas
And 0h37'22" 25d27'48"
Rachael Selena Watson
Cnc 8h33'58" 31d40'13"
Rachael Stoddard
Cap 21h4'31" -16d53'31"
Rachael Stone
Uma 10h13'45" 55d33'32"
Rachael Story Star
And 0h50'15" 40d43'28"
Rachael Stubbs
Sgr 18h27'47" -26d33'47"
Rachael Theresa
Leo 9h26'54" 10d36'41"
Rachael Torricelli
Cnc 8h20'11" 23d11'19"
Rachael Valine Ferrufino
Leo 9h48'25" 8d56'19"
Rachael Victoria Catherine Skellon
Ori 5h43'58" 9d12'11"
Rachael Webster
Lib 15h26'3" -19d0'4"
Rachael Woodfield
And 23h48'51" 34d55'41"
Rachael, F'n'B
Vir 12h46'41" 3d48'24"
Rachael, My special one!
Cnc 8h10'2" 25d36'13"
Rachael, The Love Of My Life
And 23h6'41" 47d17'57"
Rachael's Corona
Sgr 19h42'35" -14d8'41"
Rachael's Kalu
Cas 23h29'27" 59d56'53"
Rachael's Light - Forever Bright
Cru 12h41'1" -57d45'29"
Rachael's Star
Cap 20h16'56" -13d33'19"
Rachael's Wish
Leo 11h37'52" 21d58'27"
Rachal Elizabeth Heil
And 2h25'16" 47d39'45"
Racheal Carr
And 1h12'43" 39d20'44"
Racheal Dawn
Lib 14h27'44" -17d20'9"
Racheal Diane Ragsdale
And 0h35'41" 40d46'8"
Racheal Dotson
Cam 6h54'1" 75d27'29"
Racheal Goodwin
Uma 8h39'23" 58d3'53"
Racheal M. Glock
Gem 7h15'32" 24d56'30"
Racheal M Kincade
Aqr 23h2'22" -9d22'52"
Rachel
Aqr 23h2'34" -13d43'38"
Rachel
Vir 12h56'44" -16d30'33"
Rachel
Vir 13h10'25" -20d0'7"
Rachel
Cam 7h0'52" 78d24'6"
Rachel
Lib 15h1'35" -6d15'52"
Rachel
Lib 15h17'27" -6d14'29"
Rachel
Aqr 22h42'49" -0d57'44"
Rachel
Aqr 22h38'46" -0d33'8"
Rachel
Uma 12h22'42" 59d19'44"
Rachel
Cas 23h30'54" 57d8'30"
Rachel
Cas 23h59'1" 52d42'57"
Rachel
Uma 12h3'31" 62d57'32"
Rachel
Sgr 19h21'31" -30d47'5"
Rachel
Sgr 18h36'27" -23d10'26"
Rachel
Sco 17h40'21" -35d1'13"
Rachel
Gem 6h42'1" 29d13'56"
Rachel
Tau 4h3'57" 26d19'41"
Rachel
Tau 4h19'1" 29d41'18"
Rachel
Leo 11h28'54" 27d30'45"
Rachel
And 0h14'3" 29d37'29"
Rachel
Psc 1h0'11" 23d2'54"
Rachel
Vir 14h54'23" 3d31'45"
Rachel
Vir 14h58'18" 7d16'33"
Rachel
Aqr 21h30'42" 1d3'19"
Rachel
Aqr 22h39'20" 0d37'17"
Rachel
Peg 21h10'27" 12d54'56"
Rachel
Ari 2h7'50" 20d24'38"
Rachel
And 0h54'29" 39d6'7"
Rachel
And 0h40'2" 41d3'51"
Rachel
And 1h15'9" 42d53'18"
Rachel
And 2h15'58" 37d52'6"
Rachel
Gem 7h19'36" 31d16'58"
Rachel
And 0h51'10" 36d58'42"
Rachel
And 1h5'59" 34d51'33"
Rachel
Aur 6h33'28" 34d44'40"
Rachel
Cyg 19h45'5" 35d43'1"
Rachel
Cyg 19h40'52" 32d42'3"
Rachel
And 2h38'29" 50d6'50"
Rachel
And 1h22'3" 49d54'4"
RACHEL
And 2h14'55" 45d30'3"
Rachel
Lyn 7h39'10" 48d27'44"
Rachel
Lyn 7h39'29" 49d11'41"
Rachel
Cas 0h18'54" 51d33'0"
Rachel
And 23h8'22" 50d43'31"
Rachel
Cam 4h59'48" 54d12'50"
Rachel
Dra 17h6'46" 51d21'59"
Rachel 10/24
Sco 17h52'38" -36d28'14"
Rachel A. Boseke "Queenie"
Cas 2h10'15" 65d10'0"
Rachel A Guymer
Ori 4h57'40" 11d8'39"
Rachel A Monger
Lyr 18h48'38" 31d3'59"
Rachel A. Rutherford
And 0h54'45" 22d6'35"
Rachel A. Sia
Cas 0h17'1" 63d39'32"
Rachel A. Taylor
Uma 11h3'38" 45d10'48"
Rachel- A wonderful sister
Aqr 22h21'32" -0d34'26"
Rachel Ada Rodway
Uma 11h11'35" 55d48'48"
Rachel Alexandra Snyder 10312004
Sco 17h53'26" -35d53'18"
Rachel Alexis Kolton
Leo 10h28'10" 18d50'5"
Rachel Alyssa Core
Psc 0h51'27" 15d27'27"
Rachel Amanda Hanson
Vir 11h59'18" 6d26'21"
Rachel Amanda Mochan
Cap 20h27'39" -14d46'56"
Rachel Amber Clark
And 23h51'44" 41d28'54"
Rachel Amey Quinajon
Aqr 22h57'32" -4d32'3"
Rachel - Anam Cara (F. xxx)
Cas 23h53'32" 58d38'3"
Rachel Anastasia Osborne
Cyg 21h32'53" 32d0'41"
Rachel and Alex 10-19-04
Cyg 20h7'8" 55d45'40"
Rachel and Bubba 1016 Always
Gem 6h51'34" 25d52'43"
Rachel and Derek's Star
And 0h52'23" 37d37'55"
Rachel and Isaac Anniversary Star
Cyg 21h30'21" 46d20'40"
Rachel and Ismael
Pho 0h7'49" -42d57'44"
Rachel and Lillian Rose
Uma 11h57'5" 38d18'48"
Rachel and Nick's Star
And 1h9'58" 36d27'11"
Rachel and Rocky
Lyn 7h50'26" 38d15'49"
Rachel and Shawn Collier
Cyg 21h42'54" 45d29'8"
Rachel and Zach
Vul 20h31'11" 26d30'0"
Rachel Angel
Aqr 22h27'43" -1d54'42"
Rachel "Angel" Kappel
Vir 14h34'21" -0d21'14"
Rachel Ann
Gem 7h24'13" 26d31'34"
Rachel Ann
Tau 5h30'23" 27d4'2"
Rachel Ann
And 0h28'4" 41d57'8"
Rachel Ann Allen
And 0h21'27" 32d0'38"
Rachel Ann Barnes
Leo 11h45'42" 25d43'15"
Rachel Ann Brennan
Sco 17h10'37" -31d24'53"
Rachel Ann Butor
Vir 13h7'30" -6d1'52"
Rachel Ann Chism
Ari 2h21'50" 21d49'22"
Rachel Ann Glines
Cap 21h5'56" -19d8'53"
Rachel Ann Hanrahan
Dra 17h55'42" 69d34'4"
Rachel Ann Miller
Vir 12h24'21" -0d41'58"
Rachel Ann Parker
Cap 20h30'9" -15d2'39"
Rachel Ann Reed 10/06/1983
Lib 15h4'57" -19d1'43"
Rachel Ann Schipano
Lyn 8h17'19" 44d37'17"
Rachel Ann Schmalenberger
Leo 11h16'24" 15d39'56"
Rachel Ann Stephens
Gem 8h7'0" 31d20'41"
Rachel Ann Thomson
Uma 8h18'55" 66d32'56"
Rachel Ann Young
Lib 15h19'12" -13d16'35"
Rachel Anne
Lib 15h32'21" -7d41'13"
Rachel Anne
And 0h43'3" 41d47'4"
Rachel Anne
Cyg 20h27'23" 48d16'58"
Rachel Anne
Cnv 12h46'25" 37d30'32"
Rachel Anne Bowman Cole
Psc 1h6'22" 25d25'5"
Rachel Anne Breque
Com 12h51'38" 22d37'26"
Rachel Anne Kent
Lep 5h51'7" -24d29'11"
Rachel Anne Linden
Lyn 7h38'42" 56d15'54"
Rachel Anne Willoughby
And 0h53'3" 34d15'33"
Rachel Annette
Tri 2h11'45" 35d3'47"
Rachel Annette Brannon
Dra 17h17'2" 53d15'30"
Rachel Anyce Narlock
Srp 18h12'34" -0d11'56"
Rachel Atlanta Frye
Cnc 9h4'24" 15d1'41"
Rachel Ayala-Conkley
Leo 10h48'47" 12d26'18"
Rachel B. Ettlin
Sgr 17h57'30" -16d53'5"
Rachel Baker
Gem 7h33'52" 24d11'5"
Rachel Barta
Cas 1h13'57" 59d43'26"
Rachel Bauer
Vir 12h46'23" 10d46'53"
Rachel Bautista Menor
Gem 6h45'21" 24d37'39"
Rachel Beth "Miss Rachel"
Her 17h43'15" 46d0'52"
Rachel Beth Sakofs
And 23h46'4" 36d29'24"
Rachel Beverly Nicholson
Uma 8h16'5" 64d34'3"
Rachel Blair Baker
Mon 6h48'49" -0d8'21"
Rachel Boccio
Cyg 21h43'28" 45d47'26"
Rachel Braddock
Gem 6h47'4" 17d5'59"
Rachel Bradford
And 1h4'36" 42d35'10"
Rachel Brooke Hill
Aqr 22h37'29" -3d1'49"
Rachel Broxterman
Uma 11h33'4" 45d41'44"
Rachel Bubb
Cru 12h39'56" -59d34'46"
Rachel Buchanan
Per 2h56'12" 44d36'20"
Rachel Buck Dondero
Mon 6h50'12" 7d48'2"
Rachel Buckbee
Gem 6h49'57" 33d13'9"
Rachel Byrne
Cyg 21h55'16" 37d8'31"
Rachel Carr
And 23h56'34" 48d0'52"
Rachel Casey
Lyn 7h17'47" 50d37'33"
Rachel Castaneda Saiza
And 1h42'16" 38d55'21"
Rachel Catherine Vuchinich
And 22h59'28" 52d53'53"
Rachel Catherine Whitesitt
Uma 12h2'56" 57d19'38"
Rachel - Cecilia
Cnc 8h45'17" 23d15'20"
Rachel Chambers
Psc 23h14'8" 6d39'11"
Rachel Chaney
Leo 10h14'30" 20d33'43"
Rachel Chaya Rachel Cary Baker
And 22h58'48" 50d41'18"
Rachel Christine Bishop
Psc 1h14'2" 4d52'34"
Rachel Christine Wang
Uma 10h51'30" 71d45'28"
Rachel Christy
Cas 1h27'5" 63d12'29"
Rachel Clare Olivey
Uma 11h13'11" 33d17'0"
Rachel Clark
Lib 15h10'44" -15d30'28"
Rachel Connell - Pisces Queen
Cas 23h49'22" 53d27'34"
Rachel Cook
Peg 21h42'2" 27d44'56"
Rachel Coppick
And 1h57'54" 45d31'50"
Rachel Costagliola Moulin
Uma 8h23'49" 65d14'32"
Rachel Counsell
And 1h30'17" 48d52'49"
Rachel Cristine
Uma 11h51'43" 31d11'31"
Rachel Crockett
And 1h11'50" 42d8'37"
Rachel Czech Star Teacher
Cas 1h45'34" 65d21'1"
Rachel D. Emerine
Leo 10h17'49" 16d15'21"
Rachel D. Larkin
Crb 15h39'59" 37d39'32"
Rachel D. Young
Cnc 8h0'51" 14d13'14"
Rachel Dale
Cyg 21h15'42" 45d55'57"
Rachel Daone Kaufman
Leo 10h18'44" 14d11'18"
Rachel Darling
And 23h6'10" 51d14'15"
Rachel Dawn
Ori 6h17'27" 16d30'39"
Rachel Dawn (Foreverything)
Del 20h39'21" 15d23'32"
Rachel Dawn Greenley
Cas 2h18'23" 73d15'53"
Rachel Dawn Schreiber
Her 17h54'17" 14d19'3"
Rachel DeMarall
And 23h42'12" 45d57'56"
Rachel Dene Dieterich
Sgr 20h6'4" -37d6'28"
Rachel Derby
Cap 20h31'46" -15d48'59"
Rachel DiBuono
Lib 15h26'46" -19d16'22"
Rachel Dick
Cas 0h32'21" 60d6'2"
Rachel Dodge
Cap 20h45'45" -18d12'42"
Rachel Dolores Wilson
Lib 14h45'1" -11d3'30"
Rachel Dorothy
And 0h23'2" 33d21'14"
Rachel Downing
Vir 13h1'27" 0d40'23"
Rachel & Drew
Cyg 21h32'19" 51d20'8"
Rachel Dye
Lib 14h43'34" -15d18'36"
Rachel E.
Lac 22h22'25" 46d32'33"
Rachel E. A. Preziosi Adkisson
Cas 0h7'22" 55d41'11"
Rachel E. Bielinski
Cas 1h32'22" 63d26'44"
Rachel E. Payne 12/3/85 - 9/12/04
Cep 23h42'4" 74d41'35"
Rachel E. Stedina
Cnc 8h12'33" 32d13'52"
Rachel E. Williams - Ab imo pectore
Cnc 8h19'10" 23d38'51"
Rachel Eileen
And 23h15'17" 44d19'20"
Rachel Eilidh Lockhhead
And 2h11'35" 42d24'56"
Rachel Elaine Hadzik
Umi 16h19'50" 73d38'29"
Rachel Elaine Meeks
Ori 6h9'7" 9d8'26"
Rachel Eleri Griffiths
And 0h26'5" 38d40'33"
Rachel Eliana
Ori 5h59'8" 21d48'8"
Rachel Elisabeth Mackenney
Lib 15h20'29" -23d44'58"
Rachel Elizabeth
Cam 5h47'23" 65d2'0"
Rachel Elizabeth
Lyn 6h47'49" 53d28'12"
Rachel Elizabeth
And 0h16'11" 25d26'36"
Rachel Elizabeth
Lyn 7h36'17" 48d37'22"
Rachel Elizabeth
Uma 9h3'36" 49d40'45"
Rachel Elizabeth Ann Farrelly
Uma 12h2'20" 36d12'44"
Rachel Elizabeth Bishop
Peg 23h55'25" 10d17'7"
Rachel Elizabeth Carella
And 2h32'17" 49d58'40"
Rachel Elizabeth Crothers
Psc 1h21'50" 17d7'39"
Rachel Elizabeth Fair My 1 and Only
Leo 10h51'14" 14d0'39"
Rachel Elizabeth Fesmire
Cep 0h10'24" 71d26'49"
Rachel Elizabeth Howard
And 23h12'15" 52d30'50"
Rachel Elizabeth King
Cas 0h27'21" 54d27'58"
Rachel Elizabeth Laura
And 0h52'22" 44d5'55"
Rachel Elizabeth Margaret
And 2h37'27" 46d47'48"
Rachel Elizabeth Morgan
Aqr 23h53'48" -12d10'49"
Rachel Elizabeth Nau
Gem 6h23'37" 20d35'58"
Rachel Elizabeth Seidel
And 2h0'1" 46d29'27"
Rachel Elizabeth Shoemaker
Dra 17h34'6" 66d52'51"
Rachel Elizabeth Steele
And 23h44'26" 42d15'46"
Rachel Elizabeth Swart
And 0h54'32" 35d35'53"
Rachel Elizabeth Van Tassel
Cas 0h24'47" 51d46'24"
Rachel Elizabeth Vitello
Lib 14h48'9" -2d40'26"
Rachel Elizabeth Winter
Com 12h50'39" 22d14'23"
Rachel Elizibeth Hutchison
Cap 20h12'12" -21d0'48"
Rachel Ella Turpy - Ella's Star
And 2h16'19" 39d13'17"
Rachel Ellen
And 1h7'16" 36d40'15"
Rachel Elliot
Cnc 8h13'0" 15d56'53"
Rachel Emma
Cnc 8h21'40" 11d50'50"
Rachel Emslie
Cra 18h5'37" -43d19'56"
Rachel Erin Schaefer 10-16-94
Lib 15h6'16" -0d56'50"
Rachel Estelle Jerman
Uma 10h55'33" 64d21'1"
Rachel Ester Groom
And 1h29'0" 41d9'20"
Rachel Estrella Rothman
Leo 11h17'12" 14d50'24"
Rachel Evans
And 22h58'56" 35d56'10"
Rachel F. Gary
Lyn 7h51'31" 39d15'5"
Rachel Faith Meyers
Vir 13h55'11" -9d16'25"

Rachel Falbo
Sgr 18h24'24" -35d56'47"
Rachel Frances Dunmoyer
Lyr 18h48'19" 34d7'55"
Rachel Frances Shaw
Ori 5h37'19" 2d56'47"
Rachel Frei, 21.10.1993
Cas 0h2'39" 52d25'4"
Rachel Gabrielle
Vir 12h48'46" 10d55'0"
Rachel Garcia Christofferson
Sco 16h17'11" -11d45'52"
Rachel Gena Bennett
Aqr 21h52'1" -6d35'58"
Rachel Glassheim
Ari 3h5'34" 20d7'59"
Rachel Grace Berkshire
Sco 16h8'50" -14d15'32"
Rachel Grace Clutterbuck
And 2h21'22" 38d51'2"
Rachel Grace Grettum
Uma 10h14'32" 44d5'56"
Rachel Grace Seeterlin
And 23h21'20" 48d45'25"
Rachel Gwynn
Leo 9h23'26" 22d4'38"
Rachel H. Ostrow
Mon 6h40'19" -6d5'35"
Rachel Haddad
Psc 0h8'2" 4d51'45"
Rachel Haferkamp
Com 13h11'44" 18d44'37"
Rachel Haley Gontarski
Lib 14h54'10" -3d48'29"
Rachel Hannah Adler
Cyg 20h35'54" 42d28'54"
Rachel Hardy
Cyg 21h55'16" 48d21'44"
Rachel Hargrove
Lyn 8h45'31" 36d17'22"
Rachel Hartz
And 0h12'14" 36d35'0"
Rachel Hayes
And 0h23'27" 45d27'17"
Rachel Helen Parker
Sco 17h52'35" -36d19'18"
Rachel Heller
And 0h57'30" 44d11'35"
Rachel Hickman-Baker
Peg 21h57'47" 12d51'8"
Rachel Hill
And 2h28'23" 41d10'2"
Rachel Hinson
Vir 12h45'20" 7d56'46"
Rachel Hobbs
And 0h50'4" 46d34'1"
Rachel Hope
Uma 10h3'55" 55d53'47"
Rachel Hughes
And 2h37'34" 44d12'59"
Rachel Husong
Uma 10h56'12" 47d0'14"
Rachel Isabel Burton
And 2h35'13" 49d25'7"
Rachel J. Dudas
Ori 5h47'31" 1d51'17"
Rachel J. Ferrari
Gem 6h47'45" 30d38'24"
Rachel Jackman
And 23h5'7" 51d22'23"
Rachel Jade Lizzy Shepherd
Psc 1h57'47" 2d58'4"
Rachel James
Cas 0h40'55" 61d45'1"
Rachel Jamie Halasz
Cas 2h28'59" 65d38'17"
Rachel Jane Collier
Vir 13h31'32" -21d9'34"
Rachel Jane Kelleher Deveau
Mon 6h48'10" -7d28'47"
Rachel Jane Miller
Psc 1h23'58" 19d29'36"
Rachel Jane Winn 051187
And 2h2'42" 46d5'44"
Rachel & Jared
Ori 6h9'32" 7d29'25"
Rachel Jessup & Todd Gerarden
For 1h48'28" -30d27'46"
Rachel Jo Fifer
Uma 11h54'16" 63d26'5"
Rachel Jo Ottaway
Her 16h33'2" 36d46'59"
Rachel Joanna Field
Cnc 8h27'42" 14d17'49"
Rachel Jones
Psc 1h30'0" 15d48'53"
Rachel Joy Okura
Tau 4h2'52" 24d10'38"
Rachel Julia Brooks
Vir 12h6'9" -2d56'53"
Rachel Karmen
Vir 13h0'47" -1d20'55"
Rachel Kate Alesso
And 2h28'37" 49d20'8"
Rachel Kathryn
Aqr 20h45'44" 1d22'38"
Rachel Kathryn Tierney
And 23h42'12" 42d11'13"
Rachel Kavazanjian
And 23h20'4" 45d50'58"
Rachel Kay Stein
Aqr 22h17'42" -3d28'43"
Rachel Kaye Grabel's Star
And 2h6'7" 39d30'51"
Rachel Keating Rott
Cap 21h41'41" -14d8'24"
Rachel Kehrberg
Aqr 22h38'34" 2d15'31"
Rachel Kelly Hansen
Uma 10h54'13" 60d16'0"
Rachel Kimberly Larreta
Sgr 18h6'13" -27d51'15"
Rachel King
Cyg 19h37'42" 31d6'44"
Rachel Kleinman
Psc 1h34'31" 24d44'54"
Rachel Korpanty's Star
Oph 17h21'27" -22d56'53"
Rachel Kurinij
Gem 6h23'20" 25d19'40"
Rachel L. Kuebler
And 23h17'44" 46d5'5"
Rachel L Stackle
Tau 4h9'46" 16d15'27"
Rachel La Motto
And 23h19'3" 47d16'16"
Rachel Lagarde
And 1h47'10" 41d25'34"
Rachel Lambros My Beloved Wife
Pho 0h22'59" -40d57'31"
Rachel Lane
Vir 13h23'12" 12d37'43"
Rachel LaPaglia
Cnc 9h6'39" 32d16'56"
Rachel Lauren
And 0h56'56" 44d6'10"
Rachel Lauren
Cyg 20h35'56" 51d47'5"
Rachel Lauren
Sco 17h25'11" -39d31'26"
Rachel Lauren Kelly-Kellner
Dra 17h40'17" 52d5'33"
Rachel Lauren Lavalley
Lyr 18h45'11" 39d57'13"
Rachel Lea
Tau 5h35'33" 22d5'54"
Rachel Lea Bergen
Uma 10h53'13" 49d26'13"
Rachel Leach
Sgr 19h20'8" -17d33'17"
Rachel Leah Grandis
And 2h34'24" 45d7'20"
Rachel Leah MagShamhráin
And 2h29'24" 45d35'48"
Rachel Lee
And 0h21'56" 38d14'38"
Rachel Lee
Uma 12h0'29" 58d53'36"
Rachel Lee Berryhill
And 1h24'44" 42d58'17"
Rachel Lee Scott
Vul 19h47'44" 24d34'39"
Rachel LeeAnn Wall
Sgr 18h48'6" -30d51'47"
Rachel Leigh Brailov
Uma 12h43'53" 55d1'31"
Rachel Leigh Friedman
And 0h57'27" 39d39'49"
Rachel Leigh Rosenberger
Gem 7h9'58" 26d50'25"
Rachel Leigh's Super Star
Vir 13h13'39" -0d22'28"
Rachel Lewis
Ari 2h8'56" 21d21'5"
Rachel Liming
Aql 19h51'57" 16d17'25"
Rachel Lindstrom
Gem 6h42'16" 27d36'19"
Rachel Linnea Larson
And 0h40'11" 41d9'29"
Rachel Loraine Chatham
And 1h29'34" 39d43'51"
Rachel Louise
And 0h7'35" 30d12'12"
Rachel Louise
Uma 9h45'2" 56d57'9"
Rachel Louise Baltera
Mon 7h22'4" -0d55'22"
Rachel Louise Best
And 23h13'2" 35d31'10"
Rachel Louise Hawkes
And 23h33'13" 42d15'32"
Rachel Louise Pierce
Ari 2h16'45" 13d44'3"
Rachel Louise Wagner Ronan
Uma 9h57'24" 51d13'1"
Rachel Lucas
Uma 10h33'59" 63d0'37"
Rachel Lyche Goodwin
Peg 23h41'24" 15d16'24"
Rachel Lyga
Cnv 12h43'14" 41d34'38"
Rachel Lynn
Lyn 8h4'29" 36d10'33"
Rachel Lynn
Ari 3h20'50" 27d45'13"
Rachel Lynn
Mon 7h18'40" -0d22'44"
Rachel Lynn
Sgr 17h58'47" -16d55'50"
Rachel Lynn Borden
And 1h35'33" 45d40'56"
Rachel Lynn Hayward
Sgr 20h1'11" -31d7'36"
Rachel Lynn Henry
And 2h13'39" 42d22'43"
Rachel Lynn Pennywitt
And 0h16'27" 29d22'36"
Rachel Lynn Porter
And 22h58'44" 45d42'9"
Rachel Lynn Sickler
Sgr 19h19'33" -16d42'30"
Rachel Lynn Stackle
Tau 4h16'31" 23d44'28"
Rachel M.
And 23h59'44" 45d18'59"
Rachel M. Burr
Ari 3h20'7" 19d39'34"
Rachel M. Hunnicutt's 153 Star
Cyg 20h19'51" 39d28'13"
Rachel M. Lawlor
Cas 1h22'26" 69d23'19"
Rachel M. Rodriguez
Cas 1h21'45" 67d19'4"
Rachel M. Salomonson
Tau 3h53'30" 24d50'36"
Rachel Mac Isaac
And 0h31'54" 46d37'47"
Rachel Mackenzie
And 2h24'21" 38d51'42"
Rachel Madison Calhoun
Her 17h53'17" 21d59'10"
Rachel Mae Luther
Tau 5h54'12" 25d27'44"
Rachel Mae Rose
Leo 9h33'58" 24d31'46"
Rachel Maree Dawson
Cru 12h34'51" -58d27'53"
Rachel Marie
Cas 1h13'39" 69d41'40"
Rachel Marie
Ori 5h35'52" 1d13'1"
Rachel Marie
And 0h23'53" 42d0'15"
Rachel Marie
Uma 11h45'52" 41d22'54"
Rachel Marie
And 23h18'58" 45d42'36"
Rachel Marie Berendzen
Sco 17h25'47" -31d54'53"
Rachel Marie Bergman
Cnc 8h15'53" 21d14'14"
Rachel Marie Brickwedel
Umi 15h7'2" 71d2'17"
Rachel Marie Conner
Lyn 7h37'8" 52d44'30"
Rachel Marie Fahim
And 0h41'34" 32d5'35"
Rachel Marie Green
Crb 16h5'11" 37d21'18"
Rachel Marie McCloud
Uma 8h57'1" 68d51'55"
Rachel Marie Miller's Star
Uma 12h5'30" 45d2'49"
Rachel Marie Oxenfeld
Sgr 18h49'37" -21d19'0"
Rachel Marie Peterson
And 1h37'21" 48d8'47"
Rachel Marie Price
Crb 15h41'35" 26d42'59"
Rachel Marie Smith
And 0h50'22" 35d40'25"
Rachel Marie Thomas
Crb 15h42'33" 35d32'20"
Rachel Marie Woolley
Pho 1h19'58" -42d57'16"
Rachel Marie Wuzzle Watson
Lyn 8h36'46" 43d19'16"
Rachel & Mark Hughes
Uma 12h7'9" 49d47'49"
Rachel Marsico
Tri 2h11'59" 34d24'9"
Rachel Martin
And 1h11'37" 40d6'31"
Rachel Maurer
And 2h16'38" 43d47'8"
Rachel Maxwell
Lyn 8h15'36" 45d3'12"
Rachel Maya Sennitt
Cas 1h25'15" 53d57'22"
Rachel McConnell Nichols
Vul 19h4'19" 23d49'44"
Rachel McLeod's Midnight Wish
Vir 13h17'1" 5d2'53"
Rachel McMarr
Lib 16h0'16" -17d6'31"
Rachel Mentzel
Lyn 8h12'24" 49d21'46"
Rachel Mezak
Vir 12h42'46" 7d6'46"
Rachel Michele
Lyn 6h45'55" 51d37'52"
Rachel Michelle
Tau 4h7'16" 13d26'13"
Rachel Michelle Glover
Vir 14h40'12" 0d33'20"
Rachel Michelle Marcinkus
Cnc 8h23'34" 15d50'10"
Rachel Miller (Buttercup)
Sco 16h8'38" -17d37'39"
Rachel & Missy ~ Best Friends
Lib 15h43'12" -15d14'42"
Rachel Mitchell
And 0h16'17" 32d11'23"
Rachel Mitchell
Tau 4h2'43" 8d5'38"
Rachel Moen
And 0h44'59" 34d33'1"
Rachel Mon Namour 05052005
Uma 10h12'9" 51d48'33"
Rachel "Munchkin" McKee
Per 3h29'51" 36d51'33"
Rachel Murphy
Leo 10h1'11" 21d39'50"
Rachel "My Pumpkin"
Vir 14h11'5" -13d30'59"
Rachel N. Labadie
Lib 15h9'3" -6d26'21"
Rachel Navarro/Loving Mother
Uma 10h13'58" 52d58'18"
Rachel Nesmith
And 0h39'38" 38d51'23"
Rachel Newling
Cyg 21h47'28" 46d25'21"
Rachel Nicole 3-23-1984
Crb 15h46'24" 32d6'54"
Rachel Nicole Fulford
Sco 17h52'43" -35d54'48"
Rachel Nicole Lankenau
Cap 20h31'16" -22d35'48"
Rachel Nicole McGavic
Tau 4h3'15" 5d34'33"
Rachel Nicole Palmer
Lib 15h25'52" -22d37'56"
Rachel Nicole Ruddle. Thank You. CM
Uma 11h35'26" 45d15'18"
Rachel Nicole Ryan
Aur 5h24'11" 31d2'36"
Rachel Nicole Tama
And 2h9'40" 38d59'49"
Rachel Noel Schriner
And 1h46'58" 38d57'3"
Rachel Noelle
Com 13h5'48" 27d44'2"
Rachel Noelle
Uma 13h37'15" 55d40'12"
Rachel & Nurney
Lib 15h9'8" -23d52'51"
Rachel O. Esperdy
Sco 17h22'0" -42d57'5"
Rachel O'Donnell Rodenbaugh
Cyg 20h36'0" 37d8'6"
Rachel Olivia
Cap 20h16'14" -14d37'15"
Rachel O'Neil
And 0h46'47" 41d37'8"
Rachel O'Shea
And 1h57'39" 45d27'42"
Rachel O'Toole
Cas 23h56'19" 52d52'15"
Rachel Overholt
And 0h46'55" 43d4'30"
Rachel P. Mendelsohn
Uma 13h55'56" 55d33'22"
Rachel Pacylowsky
Gem 7h18'7" 19d42'30"
Rachel Paige Thav
And 0h31'36" 29d12'2"
Rachel Palmer Dallas Star
Aqr 21h40'56" 1d39'4"
Rachel Pascale
Cnc 8h11'55" 9d45'48"
Rachel Paula Gerber
And 1h41'30" 44d14'16"
Rachel Payeur-Narine
Ari 2h39'49" 22d6'19"
Rachel Pearson Gordon
Psc 0h59'33" 6d22'54"
Rachel Peckham
And 23h12'27" 43d53'25"
Rachel Peek "Rae's Star"
And 1h21'24" 36d17'14"
Rachel Pendry
Cyg 21h46'10" 44d5'10"
Rachel Perfect Girl Uptergrove
Tau 4h16'45" 26d22'9"
Rachel Petruzzi
Mon 7h3'16" -0d50'48"
Rachel Piltser
Ori 5h8'49" 9d12'4"
Rachel Piotrzkowski
Mon 6h50'49" -0d25'6"
Rachel Prince
Uma 9h33'15" 50d53'38"
rachel.... princessinthesky
And 1h28'23" 40d8'55"
Rachel Prinz
Psc 0h53'27" 16d38'34"
Rachel Purdy
Per 2h24'30" 51d6'14"
Rachel "Racheee" Sugar
Uma 13h18'3" 61d2'53"
Rachel Rebecca Guilford
Leo 10h28'13" 18d21'58"
Rachel Rebecca Jones
And 0h20'19" 29d13'59"
Rachel Reddy
Sco 16h16'27" -11d25'8"
Rachel Reeves
Ari 2h29'35" 18d42'28"
Rachel Rene
Mon 6h27'54" -4d26'47"
Rachel Renee
Sco 16h8'4" -16d23'3"
Rachel Renee
And 0h35'19" 29d5'36"
Rachel Rix Berk
Cnc 8h47'55" 31d6'42"
Rachel Roller
Vir 13h7'42" 7d58'5"
Rachel Rose
Psc 22h54'40" 5d35'50"
Rachel Rose Jackson
Cas 1h47'46" 61d19'27"
Rachel Rothe's Christmas Star
Her 16h50'49" 15d41'2"
Rachel Rountree Freeman
Com 12h46'42" 16d16'44"
Rachel Ruth
Uma 9h24'47" 47d8'39"
Rachel S. Dennis
Lyn 7h1'12" 49d39'19"
Rachel Samantha Smith
Uma 11h37'19" 60d27'44"
Rachel Samantha Smith Birthday Star
Sgr 18h45'2" -25d2'6"
Rachel Sariah Barnhart
Vir 13h48'48" -7d29'36"
Rachel Scafe
And 2h24'10" 46d25'1"
Rachel Schaeffer
Sgr 18h8'58" -19d44'53"
Rachel Schaeffer
Vir 13h2'44" -20d26'20"
Rachel Sciortino
Cas 0h30'53" 63d1'48"
Rachel Scott Kendrick
Cas 0h10'15" 59d47'20"
Rachel Sellars
Aqr 21h45'33" -2d18'58"
Rachel "Sephi" Gerichter
Aql 19h14'47" 11d38'52"
Rachel Shanaman
Psc 1h49'25" 7d12'31"
Rachel Sian Topper
And 23h38'55" 47d28'4"
Rachel Simon
Psc 0h37'1" 7d21'22"
Rachel Smith
Leo 9h31'5" 15d11'40"
Rachel Song 19820723
Leo 10h21'48" 8d12'56"
Rachel Spitzley
Ari 2h47'43" 17d16'48"
Rachel Sprock
And 2h14'19" 45d59'31"
Rachel Stephanie
Vir 12h36'22" 1d24'40"
Rachel Stevenson
Cas 1h5'1" 59d24'24"
Rachel Stilwell and Louise DiAbundo
Leo 9h28'58" 16d37'45"
Rachel Stirling
Tau 3h41'0" 23d51'50"
Rachel Stockstill
Dra 16h21'19" 57d50'32"
Rachel Strubin
Leo 9h37'4" 28d32'40"
Rachel Sullivan
Psc 2h5'42" 8d49'5"
Rachel Suneetha Fenn
Vir 14h6'48" 6d44'33"
Rachel "Sunshine" Clark
Dra 9h35'25" 77d13'28"
Rachel Suzanne Byers
Cnc 8h38'38" 31d43'11"
Rachel Swit
Uma 12h59'21" 57d43'37"
Rachel Taylor Jacob
Psc 23h49'10" 6d29'40"
Rachel Terhune's Rock Star
Uma 11h57'4" 49d58'52"
Rachel the Rocket - 16
And 0h18'27" 27d39'19"
Rachel Thomson
Cyg 19h37'54" 37d42'2"
Rachel Thongtavee
Her 17h37'42" 39d9'0"
Rachel Tighe
Lib 15h39'16" -25d39'56"
Rachel Treble Hunt Forever
Aqr 23h29'24" -8d30'35"
Rachel Tresa
And 1h52'27" 38d18'28"
Rachel Troxtel
And 1h8'42" 45d23'40"
Rachel V
Ori 5h19'0" 0d13'36"
Rachel Valencia
And 23h33'17" 50d16'24"
Rachel Varkoly
Lyn 7h24'12" 52d49'26"
Rachel Velasquez
Cyg 20h49'1" 36d39'36"
Rachel Vene Barragan
Gem 6h36'1" 27d31'18"
Rachel Vivian Womack
Lib 14h37'59" -17d33'43"
Rachel Ward
Gem 6h52'56" 22d22'7"
Rachel Ward 8/3/1988
And 23h23'41" 52d6'34"
Rachel Wartman
And 23h3'33" 49d14'5"
Rachel Weiss
Cas 0h34'59" 61d29'49"
Rachel Zimmerman's Special Star
Uma 13h10'45" 55d10'32"
Rachel Zoe Major
And 1h34'19" 41d6'19"
Rachel, my love
Ori 4h50'5" 6d32'11"
Rachel, Ray & Sadie
Ori 5h56'11" -1d54'14"
Rachel13
Boo 14h16'28" 20d13'39"
Rachel32795
Cas 23h33'43" 52d49'55"
Rachelann Mae Horak
Uma 13h5'22" 63d4'35"
RachelBeth
Cas 1h13'34" 50d16'18"
RACHEL-BRE-KIM-AL-MARIE
Sgr 18h37'15" -24d26'28"
Rachelbubby
Cnc 8h20'37" 9d43'32"
Rachele Catherine Brown
And 2h21'1" 50d8'14"
Rachele Kopp
Uma 9h55'1" 46d49'29"
Rachele Smith
Cap 20h20'15" -9d24'35"
Rachell Allysa Andrews
Psc 0h34'4" 7d28'25"
Rachell Besina Hadassah Smith
Cas 1h4'29" 49d41'30"
Rachelle
And 0h53'57" 39d54'26"
Rachelle
Ori 4h50'58" 14d31'8"
Rachelle
Cyg 19h21'23" 28d48'39"
Rachelle 7
Lyr 18h46'53" 31d45'9"
Rachelle and Joe
Per 3h38'53" 46d54'25"
Rachelle & Andrea: Best Friends
Cas 1h25'39" 59d16'1"
Rachelle Buschbacher
Dra 10h52'1" 75d51'28"
Rachelle Christine Swartley
Gem 6h48'58" 20d44'2"
Rachelle Dawn Legoo
Uma 9h58'43" 50d2'58"
Rachelle Douglas
Cas 1h39'34" 64d35'56"
Rachelle Eva Holloway
Cas 1h8'52" 56d17'36"
Rachelle Flora Danielson
Cnc 9h9'38" 32d39'24"
Rachelle Gage
Ari 2h2'13" 19d57'49"
Rachelle Haley Voitic
Umi 17h9'5" 84d56'13"
Rachelle Hazel
Cas 1h24'3" 56d36'53"
Rachelle Janelle Hanna
Vir 14h8'10" -15d47'19"
Rachelle & Jennifer Eternity Star
Ori 6h0'2" 14d4'29"
Rachelle Jobe
Uma 10h48'37" 54d28'44"
Rachelle Lea Wright
Uma 11h50'4" 38d28'54"
Rachelle Lynn Imlay
Lib 14h49'45" -7d57'46"
Rachelle Lynn O'Neil Tolleson
Aqr 22h1'51" -4d18'37"
Rachelle Marie Brookhart
Aqr 20h43'13" -0d40'44"
Rachelle Marie Villarreal
And 0h18'0" 27d56'15"
Rachelle Nicole
And 2h21'46" 50d4'58"
Rachelle R. Rogers
Gem 6h47'43" 22d2'34"
Rachelle Rose Tatro
Psc 1h22'8" 24d36'55"
Rachelle Sayed
Gem 7h11'46" 23d4'14"
Rachelle Teruko Hart
And 23h0'22" 41d18'5"
Rachelle Yvonne Davis
Uma 9h11'5" 52d5'32"
Rachelle-Marie Nguyen
Tau 4h35'55" 20d36'12"
Rachelle's Night Light
Lib 15h34'38" -7d1'55"
Rachel's
Cyg 20h1'36" 39d55'43"
Rachel's Angel
Aqr 22h20'56" 1d29'44"
Rachel's Baby Owen
Cep 22h12'59" 65d6'39"
Rachels Bisou
Tau 4h50'15" 15d46'11"
Rachel's Dreamcatcher
Crb 15h37'14" 29d43'57"
Rachel's Dreams
And 23h42'15" 39d58'32"
Rachel's Hope
Lyn 8h1'9" 58d56'58"
Rachel's light for a dreary night
Cap 20h8'14" -25d29'1"
Rachel's Paradise
And 2h30'51" 37d27'53"
Rachel's Pounce
Cyg 21h44'40" 45d46'48"
Rachel's Shining Star
And 0h3'53" 35d32'22"
Rachel's Shining Star
Aqr 22h15'48" -3d32'12"
Rachel's smile
Aqr 21h1'43" 2d26'21"
Rachel's Star
Tau 3h57'39" 10d18'40"
Rachel's Star
Tau 5h45'43" 26d41'2"
Rachel's Star
And 23h37'11" 49d58'37"
Rachel's Star
And 1h31'8" 49d5'33"
Rachel's Wish
Cyg 20h57'16" 31d1'39"
Rachel's Wish
Tau 3h43'31" 24d8'26"
Rachel's Wish Well
Lyn 6h58'43" 61d16'37"
Rachel's Wonder
Ori 5h31'29" 10d35'2"
Rachels-Mark
Sge 20h6'33" 20d43'31"
Rachel-Theresa
Cnc 9h11'39" 20d20'41"
Rachelthu
Psc 0h27'24" 10d16'29"
Rachelus Smithula
Cyg 20h42'53" 53d31'34"
Rachie Kozuszek
Peg 22h21'26" 17d17'5"
Rachkova Sneja Roumenova
Umi 14h51'6" 85d16'51"
Rachna
Umi 15h15'46" 67d46'4"
Rachna Dar
Cap 21h55'25" -18d22'1"
Rachny Chhun_SH 20050505
And 1h16'8" 36d42'11"
Rachua Benttley Levison
Cas 0h37'42" 57d13'28"
Rachy Rache's Sparkly Sparkly Star
And 23h12'54" 38d12'8"
Racine Victoria Perez
Tau 4h17'58" 2d45'11"
Rackel
Uma 10h41'55" 51d54'8"
Rackel & Yael
Umi 15h18'48" 76d2'50"
Racquel
Pho 0h15'15" -39d24'48"
Racquel Alexandra Doherty
Sgr 18h22'45" -26d49'53"
Racquel Balsamo Sommer
Sco 16h53'29" -44d10'29"
Racquel G. Ranger
Lyn 7h43'55" 39d23'32"
Racquel Harris Mason
Lib 14h51'17" -5d31'50"
Racquel M. Brandt
Crb 15h45'38" 33d9'36"
Racquel S. Young
And 23h17'38" 43d40'50"
Racquel-Cutest Girl in NJ-Buglioli
Sco 16h13'20" -11d16'36"
Rácz József szerelmemnek
Leo 10h34'13" 20d56'46"
RAD
Uma 12h5'47" 45d5'37"
Radachka & Tigrenok
Cyg 20h18'58" 33d13'45"
Radames Martinez, Sr.
Cmi 7h41'15" -0d8'8"
Radar 888
Cam 4h15'12" 69d28'6"
Raddatz C
Uma 10h32'45" 44d0'55"
RaDeanna
Cnc 8h46'2" 32d25'0"
Rademacher, Herbert Paul
Uma 10h39'18" 72d13'39"
RADE-Sijaj i osvetli nas-EUROTEER
Per 4h16'50" 42d52'14"
Radford B Anderlot
Sgr 18h4'16" -27d57'18"
Radha Nadkarni
Umi 20h40'7" 88d39'18"
Radha Vijai
Uma 11h27'36" 29d41'20"
Radhames
Sge 20h5'53" 20d36'4"
Radhika
Leo 9h25'9" 26d38'39"
Radhika Mandalam
Gem 6h29'12" 26d7'26"

Radhika Surapaneni
Psc 1h14'56" 17d40'36"
Radia Bensarsa
And 1h6'35" 45d57'3"
Radia Driouach
Uma 9h11'8" 59d58'51"
Radiance dei Dawna
Psc 1h26'1" 24d37'16"
Radiance of Charlotte
And 0h52'32" 44d52'50"
Radiant and Steadfast
Jeanne
Lyr 18h35'7" 34d5'42"
Radiant Diane
Sgr 18h43'29" -22d11'54"
Radiant Jodie
Lib 15h9'53" -29d13'33"
Radiant Julie
Cap 21h21'43" -17d15'54"
Radiant Love
Cyg 19h47'39" 34d6'51"
Radiant Marie Elena Nane
Rogers
Cnc 9h9'49" 32d9'30"
Radiant Mother Traci
Mathers
Crt 11h37'59" -18d22'10"
Radiant Rachel
Uma 11h41'22" 60d59'13"
Radiant Richelle
Vir 14h25'56" 4d44'12"
Radiant Ronny
Gem 7h49'16" 18d51'10"
Radiant Roseann, Majestic
Woman
Sco 16h13'7" -12d55'44"
Radiant Star Janelle
Frances Casper
Cas 23h8'7" 56d35'14"
Radiant Star of Faith D
Forbes
Cap 20h28'16" -18d21'20"
Radical Black Cat Ralph
Sekerak
Aur 5h48'9" 29d55'31"
Radio Boy
Vir 14h38'8" -6d59'6"
Radke, Gerhard
Ori 6h18'50" 14d14'27"
Radko
Dra 17h40'20" 65d22'31"
Radostar
Sco 17h1'34" -36d49'54"
RadTen
Umi 15h27'29" 67d33'51"
Radule - Pile
Per 4h24'27" 45d33'14"
Rae
Leo 10h22'34" 25d28'22"
Rae
Aqr 22h22'7" -24d4'51"
Rae Alter
Sgr 19h23'57" -35d36'8"
Rae Amelia Osborne
Cma 6h21'9" -30d43'38"
Rae & Bear
Umi 13h57'36" 76d35'29"
Rae Delman
Uma 13h4'39" 58d48'24"
Rae Gee 114
Eri 4h42'30" -0d59'29"
Rae Gina Perrin
Uma 9h13'37" 58d36'24"
Rae Jacobson
Rosengarten
Uma 11h35'38" 33d31'29"
Rae Livingston
Apu 14h37'29" -73d30'59"
Rae Lynn
And 23h14'27" 47d36'19"
Rae Lynn McFarlin
Aqr 21h51'20" -2d50'45"
Rae Lynn Oleavia
"Pumkyn"
Psc 0h45'44" 14d28'23"
Rae Marie Quadra Estrella
Lyn 7h40'22" 36d9'26"
Rae Meus Pulchrae
Per 2h21'18" 57d5'7"
Rae of Tranquillity
Uma 10h4'13" 47d37'46"
Rae Rae
Uma 10h50'40" 65d43'38"
Rae Rae
Psc 1h0'1" 7d24'1"
Rae Ramos
Lyr 18h39'34" 38d43'43"
Rae Tauber
Aqr 21h20'9" -14d25'2"
Rae'ah Sanchez
Ari 2h23'30" 26d20'24"
Raeann Bowman
Vir 13h22'5" 9d6'57"
Raeanna Marie
Cap 20h40'30" -17d2'25"
RaeAnne Lorenzen
Cnc 8h13'1" 15d31'0"
Raeannie
Cyg 19h34'15" 29d27'35"
Raebear
Lyn 7h13'55" 58d30'38"
Raechelle Anne Magpantay
Ori 5h7'36" 11d22'34"
Raechel-Rae Boston
Psc 0h52'55" 17d12'50"
Raecode
Cnc 8h20'5" 14d20'1"
Raed Dallow
Sco 16h13'35" -12d58'57"
Raeder Love
Aqr 21h58'38" 0d5'56"
Raeeseh Azzeh
Psc 1h27'30" 28d1'43"
Raef & Riley's Starfire
Umi 15h52'5" 79d30'2"
Raegan Conroy
Lib 15h40'0" -12d23'13"
Raegan L. deOliveira
Ari 2h54'15" 29d16'49"
Raegan Leigh deOliveira
Ari 2h54'52" 29d35'6"
Raegan Madison Cook
Cap 20h21'22" -19d56'8"
Raegem's Hope 222
Gem 6h45'33" 15d0'45"
Raegen Ross Cool
And 23h16'1" 36d27'54"
Raejean
Cas 0h30'47" 53d55'5"
Raejung
Ari 2h40'2" 29d33'36"
Raelaine
Lyn 9h1'4" 33d26'39"
Raelan Jiade
Tau 3h34'41" 11d46'10"
Raelene Loves Ray -
Always
Cru 12h2'14" -62d1'5"
Raelene lynn
Tau 4h58'45" 24d35'7"
Raeley May Branson
And 1h56'54" 45d44'25"
RaeLin
Lyr 18h49'14" 39d30'45"
Raella, Wabbit Love
Forever
Lyn 7h48'3" 57d37'55"
Rae-Lyn Limerick
Peg 22h23'22" 20d52'27"
Raelynn
And 23h6'13" 37d12'26"
Rae-Lynn
Aqr 22h6'8" -3d14'12"
RaeMarieTyson78
Uma 12h21'42" 57d1'29"
Raemond G Taylor
13.05.1967
Tau 3h50'44" 0d44'18"
Raena Alonzo
Cnc 8h1'22" 17d9'40"
Raena - Forest - Rain
Uma 11h8'10" 62d31'3"
Raena K
And 0h13'49" 40d59'4"
Raena & Michael
Gem 7h23'27" 24d49'8"
Raequawn Smitherman
Uma 8h22'22" 64d22'47"
Raeven
Ari 2h56'44" 14d39'12"
Raevyn Maer
Cap 20h51'55" -18d42'0"
Raey Fountain
Sco 17h24'3" -37d38'28"
Raf
Uma 11h6'28" 69d27'14"
Raf en Meinoutje
Ori 6h12'0" 15d39'17"
Rafael
Ari 2h18'1" 17d11'25"
Rafael
Ori 6h10'22" 7d10'19"
Rafael
Her 17h31'17" 36d31'24"
Rafael Acosta
Dra 17h22'20" 55d51'56"
Rafael Alejandro Hopf
Her 17h16'34" 48d21'53"
Rafael Alejandro Solano
Cnc 8h38'46" 24d31'22"
Rafael Alonso Bazan
Lib 15h9'28" -24d16'29"
Rafael Antonio
Tau 5h24'13" 28d31'53"
Rafael Bagoyan
Ari 2h49'39" 26d0'12"
Rafael Casillas
Leo 9h49'59" 7d52'25"
Rafael Cepeda
Gem 7h8'29" 15d41'35"
Rafael Chavez, Jr.
Cnc 8h54'35" 10d57'45"
Rafael Gabriel
Uma 8h59'0" 48d47'32"
Rafael Künzle
Umi 17h26'51" 87d52'52"
Rafael Luis Mendoza
Leo 11h36'38" 9d54'15"
Rafael Michael Escoto
Aqr 22h33'10" 0d50'40"
Rafael Pazan - fortitudo
immortalis
Ori 6h15'10" 12d48'6"
Rafael Rivas
Gem 7h28'13" 25d52'0"
Rafael T. Tejada
Cru 12h6'0" -60d46'30"
Rafael, Haisee,Raingel
Cas 1h22'10" 62d1'19"
Rafael, Judy, and Tommy
Sco 17h57'40" -37d33'1"
Rafaela Martell
Psc 1h11'22" 27d13'48"
Rafaela Perez
And 23h1'46" 52d18'25"
Rafael"Cano"Rodriguez
Cyg 20h9'30" 54d34'50"
Rafaella Calista Merland-
Roa
Dra 18h33'19" 53d0'3"
Rafal
Vir 13h13'7" 13d34'9"
Rafal Rozycki
Lyr 18h41'0" 35d19'38"
Rafal W. Zbrzezny
Lyn 6h52'40" 57d36'7"
Rafeal Martinez
Lib 15h32'20" -12d43'14"
Rafer Brel Gluyas
Sco 17h45'10" -37d26'36"
Raff
Uma 11h6'29" 35d10'1"
Raffael & Liliane
And 0h8'49" 45d20'43"
Raffaela
Uma 11h45'22" 38d11'22"
Raffaela
Her 18h2'5" 17d40'1"
Raffaela Anna Zoe
Antoniou
Uma 14h23'7" 57d52'25"
Raffaela DeZuzio
Uma 10h14'47" 71d40'24"
Raffaela Maria Niedermann
Cep 1h24'38" 78d28'35"
Raffaela Villella
Sgr 19h7'31" -22d22'14"
Raffaele Boldi
Ari 3h22'3" 20d34'51"
Raffaele Campaniello
Her 16h38'12" 29d38'36"
Raffaele Canepa
Dra 19h26'46" 65d31'10"
Raffaele Tizzano
Peg 22h30'17" 17d6'12"
Raffaella
Her 16h39'55" 23d51'35"
Raffaella
Per 3h3'9" 44d40'49"
Raffaella Cancian
Cma 7h16'25" -31d6'32"
Raffaella Conti
Lyr 18h25'59" 41d58'56"
Raffaella Mink
Tau 3h55'38" 30d49'54"
Raffaella & Thomas forever
Boo 13h57'11" 24d29'2"
Raffaella Zamagni
Lac 22h37'24" 41d56'51"
Rafferty
Pup 8h5'12" -34d49'53"
Rafferty Lambe
Boo 15h31'55" 45d11'5"
Raffie
Uma 9h8'13" 64d25'20"
Raffindia
Lyr 19h18'35" 39d17'53"
RaffitaBink
Tau 4h24'9" 20d6'0"
Raffy
Aql 20h12'24" -0d0'23"
Raffy Haddadin
Aqr 23h26'37" -7d25'47"
Rafiki
Ori 5h40'54" 8d28'19"
Rafo, Salam Aelis
Ori 6h11'52" 15d58'2"
RAG212
Psc 0h33'0" 12d33'43"
Raga Malaty
Vir 15h6'57" 5d27'41"
RAGAN
Gem 6h49'51" 27d55'41"
Ragan Marina Riley
Vir 12h12'29" -0d59'1"
Ragan P. DeMaso
Uma 9h37'36" 53d27'28"
Ragan Roncka
Lac 22h4'59" 45d38'47"
Ragazza Bella
Tau 4h29'56" 20d18'47"
Ragdoll 417
Lyn 7h49'58" 57d43'7"
Rage and Rhi Forever
Cap 21h46'24" -9d26'12"
Ragen Hamane
And 0h46'1" 42d46'6"
Raggy
Vir 14h3'47" -14d55'55"
Raghavachary
Gem 7h7'52" 31d41'14"
RaghavCaitanya
Per 3h6'1" 39d11'38"
Ragheb Alame
Boo 13h49'0" 10d15'5"
RAGHNOR
Ori 5h55'28" 6d30'10"
Ragin' Cajun Love
Cyg 20h58'6" 48d28'40"
Ragini
Leo 10h15'33" 27d0'19"
Ragnhild Venbakken
Dra 14h56'4" 57d32'20"
Ragnhild Viola Nelson
1902-1987
Aqr 22h25'26" -2d51'12"
Ragnhilds Stjerne
Cyg 21h55'23" 49d38'11"
Rags' Butterfly
Ari 3h8'14" 28d4'23"
Ragsbie P. Rabbit
Per 3h45'28" 43d48'34"
Ragusa
Uma 11h28'47" 38d49'41"
Ragyogás
Uma 9h9'48" 49d34'51"
RAH
Sco 17h46'20" -41d23'36"
Rahat
Psc 1h25'19" 10d42'23"
Rahel
Uma 10h39'25" 48d11'13"
Rahel Ramseier
Umi 14h14'10" 71d39'23"
Rahel und Flo
Cas 1h48'36" 68d52'15"
Rahel, 21.06.2002
Ori 4h54'56" 9d48'4"
Raheleh K. Fazollah
Umi 15h38'11" 72d12'39"
Rahmah
Lyn 7h45'2" 48d40'18"
Rahman
Lyr 18h41'4" 26d48'2"
Rahman Nafeeza
Uma 11h27'44" 61d28'28"
Rahmati, Christian
Uma 13h45'54" 55d45'40"
Rahnert, Erhard
Uma 12h12'14" 61d21'31"
Rahul Baweja - Star of the
Universe
Per 3h17'6" 53d22'52"
Rahul Ghosh
Umi 17h23'11" 78d45'9"
Rahul Mehra
And 2h37'27" 45d42'9"
Rahysa Catalina Gomez
Vargas
Vir 14h25'35" 3d0'13"
Rai
Sco 17h56'0" -30d42'46"
RAI of Light
Ori 4h53'17" 12d35'44"
Raidder King Miller
Cep 21h30'53" 61d8'56"
Raidon Orion
Uma 11h4'44" 49d21'46"
Raif Anthony Pino
Cap 21h32'47" -11d39'46"
Raige Trinidy
Peg 22h10'27" 33d29'55"
Raija Anita Lahti-Alhussaini
Sco 17h8'12" -33d5'3"
Railay
Cnv 12h36'31" 41d40'57"
Railen Blaize Rozzell
Psc 1h9'9" 9d55'30"
Raimee Lynn Cooper-Delta
Gamma LSU
Psc 23h5'54" 0d9'18"
Raimee Takaki
Cnc 8h9'28" 7d31'11"
Raimi Anita Austin
Psc 0h58'57" 28d32'21"
Raimo Johannes Hakala
Uma 10h3'57" 55d58'58"
Raimo R1 Haakana
Umi 14h54'59" 70d35'37"
Raimond Alfonse
Strautman
Lmi 10h25'40" 28d29'38"
Raimundo e Alberto
Vir 12h23'6" 11d44'58"
Raimundo Sevillano
Navarro
Gem 6h12'50" 23d28'30"
Rain
Leo 9h30'26" 11d1'3"
Rain
Uma 9h19'32" 62d20'13"
Rain
Lyn 7h10'33" 59d2'54"
Rain On Me
Lyn 7h18'5" 59d24'12"
Rain Storm Akasha
Sco 16h4'56" -28d53'38"
RAINA DAWN BOAL
Vir 12h50'54" 1d59'35"
Raina Diane
And 23h37'40" 47d39'24"
Raina Dione Presley
Summers
And 1h18'57" 48d37'30"
Raina Nicole
Ari 3h27'33" 21d45'46"
Raina Russell
Leo 9h34'6" 27d42'6"
Raina Skye Hewlett
And 1h31'29" 50d23'22"
Raina Tzina
Uma 10h6'6" 69d0'1"
Raina Vasilchik
Uma 11h22'58" 53d36'30"
Rainasia
Mon 6h42'6" -1d10'5"
Rainbird
Cyg 20h51'48" 37d13'51"
Rainbow
Lyn 9h4'39" 45d15'1"
Rainbow
Ori 5h20'5" -0d48'4"
Rainbow Angel
Vir 13h54'48" -18d44'11"
Rainbow Kids
Uma 9h3'9" 51d7'6"
Rainbow & Peaches 1
And 23h48'24" 45d21'51"
Rainbow Reef Beach Hotel
Tri 2h9'1" 35d50'0"
Rainbow Trail Lutheran
Camp
Umi 8h47'17" 88d9'30"
Rainbow7
Psc 1h9'51" 11d48'9"
Rainbow's End
Cap 20h41'35" -25d57'31"
Raindrop Gummy Bears
Umi 14h17'41" 76d20'7"
Raine
Tau 5h44'39" 21d27'57"
Raine
Lyn 8h22'11" 35d38'10"
Raine Marie
Gem 6h51'47" 25d56'5"
Raine Melody
Cyg 21h56'56" 52d30'8"
Rainee C. Marshall
Lib 14h52'57" -7d10'55"
Rainee & Todd Facey
Lyr 18h52'51" 32d26'50"
Rainee's Starr
Gem 6h52'48" 17d45'34"
Rainer
Ori 6h19'4" 19d46'44"
Rainer (Bubs)
Her 16h46'1" 9d5'44"
Rainer Dick
Uma 9h10'18" 68d33'4"
Rainer (Duke) Krauss
Aur 6h34'55" 35d17'33"
Rainer Eberhard
Uma 9h24'1" 64d28'44"
Rainer Gehring
Ori 5h49'21" 8d22'23"
Rainer Hünich
Ori 5h56'30" 18d17'51"
Rainer K.Schaaf - a
Fantabulous Man
Cyg 20h7'26" 46d33'22"
Rainer Müller
Uma 10h12'33" 44d0'3"
Rainer Otto
Uma 8h55'0" 47d35'55"
Rainer Schubert
Uma 8h17'26" 62d27'10"
Rainer Seidel
Ori 4h45'53" 4d17'55"
Rainer Sigleur
Ori 5h56'12" 3d27'42"
Rainer-Thomas Böckelen
Uma 13h36'48" 61d40'30"
Rainforth Major
Per 2h24'38" 56d48'46"
Raini Lei
Leo 9h37'28" 26d52'48"
Rainie Grace Summers
Psc 0h50'20" 12d10'43"
Rainier's Gem
Sgr 19h46'22" -13d44'36"
Rainier's North Columbia
Academy
Crb 16h11'42" 36d46'55"
Rainna Kennon Stapelfeldt
Uma 11h13'38" 34d12'3"
Raintazz
Vir 13h10'17" 3d38'9"
Rainwater
Gem 7h29'29" 20d25'36"
Rainy night
Tau 4h5'14" 16d4'37"
Rairigh
Uma 12h7'31" 52d28'44"
RAISA
Lib 15h11'58" -14d5'15"
Raisa Fareen Chowdhury
Vir 12h14'37" 11d53'8"
Raisa Novikova
Lyn 7h0'35" 60d48'11"
Raisa Pearl Bruner
Com 12h45'2" 31d9'37"
Raisah Silverwind
Pho 2h15'2" -45d44'39"
Raissa
Aqr 20h43'48" -2d26'3"
Raissa Jäger
Uma 12h11'21" 58d10'13"
Raistlin Patrick Terry
Dra 12h22'49" 71d2'19"
Raj
Cas 0h36'23" 50d38'14"
Raj _ Ee Ling
Lyn 9h22'25" 41d10'49"
Raj Nair
Umi 14h9'18" 73d42'45"
Raja ki Rani
Per 3h33'3" 33d19'20"
RAJA LEDA
Cas 23h47'42" 57d55'42"
RAJA SAEED
Sco 16h17'40" -10d26'45"
Rajakumari Solly Pravinata
21/01/86
And 1h9'38" 47d53'29"
Rajalakshmi
Leo 10h17'10" 22d3'17"
Rajasi Punit Patel
Ari 2h44'35" 26d50'25"
Rajeane Little=Kane
Psc 0h41'43" 7d40'0"
Rajesh L. Motwani
Cyg 19h32'40" 27d50'2"
Rajesh Mohan
Ari 3h16'5" 27d58'16"
Rajeswary Harrington
Cep 23h42'14" 74d15'37"
Raji Randhawa
Cnc 8h40'20" 28d55'20"
Rajiv Goswami
Per 4h27'3" 44d20'33"
RAJJL 081305
Lyr 18h27'36" 33d36'55"
Rajmund
Aqr 21h1'12" -10d45'56"
Rajnai Attila csillaga,
RaMaAtAk60
Ari 2h30'47" 25d20'56"
Rajneel Rajiv Chand
Lyn 7h30'59" 40d29'34"
Rajneesh K Vijh
Cap 20h30'12" -11d10'21"
Rajni Lucienne Jacques
Mon 7h18'20" -7d25'9"
Rajnita Sharma Schell
Crb 16h7'57" 38d9'47"
Rajpaul's shining moment
Uma 10h26'6" 49d46'13"
Raju. I will Luv U to
Eternity. Priyu
Ori 5h58'32" 22d45'6"
Raka Leanne
Ari 2h53'25" 25d58'3"
Rakas Moona-Aiti
Crb 15h31'27" 25d37'22"
Rakesh C. Tailor
Dra 19h32'37" 63d29'17"
Rakesh Duggal
Cyg 20h55'34" 41d13'28"
Rakhe Alexander
And 0h17'8" 35d6'40"
Rakhee
And 0h20'28" 27d12'0"
Rakhee Patel
Sco 16h13'7" -11d15'44"
Rakhi
Sco 17h56'24" -30d7'52"
rakk187th /Sadin
Psc 1h28'0" 7d23'4"
Rakóczki Laci csillaga
Psc 23h15'25" 3d26'25"
Ralee
And 1h18'51" 49d21'56"
Raleigh
Uma 14h17'33" 56d28'33"
Raleigh & Kathy
Cyg 20h51'22" 43d10'25"
Raleigh Keagan's Star
Leo 9h53'11" 6d49'41"
Raleigh-Elizabeth Smith
Aql 20h11'17" 13d10'38"
Ralene
Vir 13h44'26" -11d28'1"
Ralene Johnson
Lyr 18h34'18" 33d7'24"
Raley
Lib 14h50'22" -1d30'50"
Ralf
Tri 2h16'23" 36d46'31"
Ralf Adolf Danowski
Uma 8h41'1" 69d29'58"
Ralf Antony
Ori 4h47'20" -2d52'52"
Ralf Bahr
Ori 6h19'15" 7d20'35"
Ralf Büttner
Ori 5h58'44" -2d16'20"
Ralf Gerd Lichtenberger
Ori 6h12'16" 8d34'29"
Ralf Heidenreich
Uma 9h7'45" 72d33'2"
Ralf Peter Medardus
Schiejok 15.04.1969
Uma 10h14'16" 41d51'29"
Ralf Richter
Ori 4h44'10" 1d46'55"
Ralf Scheiding
Uma 12h2'35" 48d11'47"
Ralf Zschäckel
Uma 8h39'52" 51d42'27"
Ralf, mein Glücksstern
Uma 9h6'35" 69d57'42"
Ralfina Marinelli
Vir 14h38'36" -6d24'17"
Rallie
Lyn 7h33'35" 48d12'2"
Rally Azar
Cyg 21h35'41" 42d44'2"
Ralph
Per 3h21'55" 45d58'28"
Ralph
Cyg 20h44'23" 46d15'3"
Ralph
Tri 2h10'1" 36d21'37"
Ralph
Com 12h31'44" 26d24'31"
Ralph
Aql 19h32'13" -11d5'27"
Ralph
Dra 19h12'23" 61d34'30"
Ralph "60"
Uma 8h52'30" 64d16'2"
Ralph A. Hartley
Aur 5h42'0" 45d4'31"
Ralph A. Lindblom "2-4-32"
Her 17h22'19" 38d33'49"
Ralph A. Wize
Cnc 9h6'32" 23d35'3"
Ralph A. Zumwalde
Leo 9h29'29" 29d42'1"
Ralph and Elise's Star
Del 20h35'20" 18d13'47"
Ralph and Margaret Craton
9/15/1950
Cas 23h6'23" 54d33'44"
Ralph and Mary Heinauer
Cas 23h32'22" 51d49'20"
Ralph and Mary Stanford
Uma 11h22'27" 38d58'48"
Ralph and Rose Zetterberg
Cyg 21h22'30" 35d5'47"
Ralph And Shirley Rose
Cyg 19h38'34" 34d6'3"
Ralph & Andrea our love to
no end
Cyg 20h43'32" 53d19'24"
Ralph Angelo Teneriello
Per 3h26'20" 52d2'39"
Ralph Anthony Falco
Her 17h30'23" 31d22'11"
ralph auguste
Uma 10h13'23" 46d29'32"
Ralph Bremer
Uma 9h1'0" 49d5'20"
Ralph Brown Gemini 621
Cyg 19h56'53" 44d48'45"
Ralph "Bud" Berrett, Jr.
Uma 8h22'26" 64d34'35"
Ralph C. Tatoian, Trooper
Cnc 8h50'8" 12d43'25"
Ralph & Camille Misiti
Cyg 19h38'28" 53d36'49"
Ralph & Carol Venuto
Cyg 20h32'17" 57d27'32"
Ralph Castillo
Dra 18h48'5" 54d49'48"
Ralph Charles Toerper
Cyg 21h41'46" 43d49'28"
Ralph Christopher &
Thomas John
Uma 11h6'9" 48d14'42"
Ralph Chumbley
Lib 15h18'32" -20d3'12"
Ralph Clark
Uma 13h4'11" 60d30'39"
Ralph Daniel Sarenac
Her 18h1'5" 21d17'36"
Ralph & Doris Adams
Cas 23h38'8" 57d23'2"
Ralph E. Cognion
Cep 22h1'59" 56d39'48"
Ralph E. Murray, Sr
Cep 3h37'26" 82d35'2"
Ralph E. Paluska
Uma 10h2'51" 68d18'16"
Ralph E. Rand Jr.
Ori 5h36'21" -1d16'1"
Ralph E. Schneider
Cam 5h0'47" 57d50'19"
Ralph E. Swihart
Cyg 20h13'30" 39d48'11"
Ralph & Edna Clouser
Sge 19h26'59" 17d56'47"
Ralph Embler
Uma 8h55'57" 62d56'7"
Ralph Emil Decker
Aql 20h7'13" 2d17'33"
Ralph Firsick
Cep 21h23'17" 55d57'51"
Ralph & Fran Brown
Cyg 21h22'57" 45d49'47"
Ralph & Francis Powell
Uma 9h1'32" 59d46'52"
Ralph G. Richard
Sco 16h6'33" -9d13'43"
Ralph George Patino
Vir 14h41'33" 6d27'52"
Ralph Grippo, Jr.
Her 16h52'30" 27d35'52"
ralph gysin
Cyg 21h13'9" 38d39'41"
Ralph H. & Jeanne A.
LaFond Viner
Ori 5h31'14" -0d7'24"
Ralph H. Ward
Sco 16h12'55" -34d38'0"
Ralph Haltinner
Per 4h44'16" 45d41'17"
Ralph Harold Owens Jr.
Ori 6h20'28" 6d10'29"
Ralph Hartmann
Uma 14h6'0" 56d47'10"
Ralph Henry Culwell
Leo 10h21'14" 8d35'25"
Ralph Henry Pflugh
Lib 14h54'0" -4d10'40"
Ralph I Bean B & E
Forever & Ever
Peg 21h48'48" 15d10'45"
Ralph Isaiah Rooker
Ori 5h51'59" 6d19'32"

Ralph J Hiester (Pop)
Tau 3h46'33" 29d5'45"
Ralph J Jones
Cyg 20h37'24" 51d41'0"
Ralph J Santucci
Uma 9h52'20" 60d29'8"
Ralph J. Wikner
Aql 19h52'25" -0d58'28"
Ralph James DeMarco
Cap 21h47'10" -14d19'33"
Ralph James Hosier (Dragon)
Dra 15h48'44" 63d2'3"
Ralph James Moore Senior 1/29/1940
Uma 11h31'19" 60d8'2"
Ralph Jeffery Marro
Per 2h25'2" 55d13'36"
Ralph & Juleann
Uma 9h36'13" 41d28'22"
Ralph Kemper
Uma 8h11'7" 62d23'35"
Ralph Krauss
Cap 21h45'10" -9d38'42"
Ralph L. Cote
Aql 19h58'54" 10d41'18"
Ralph Lauren
Sco 16h6'18" -11d18'54"
Ralph Lee Whitney
Lyn 8h21'29" 35d31'35"
RALPH LELAND WEB-STER
Ori 6h19'50" 5d56'31"
Ralph Liscio
Aqr 20h41'58" 1d9'17"
Ralph Ludwig
Lyn 9h19'31" 40d10'38"
Ralph Mannheimer
And 22h59'15" 46d59'40"
Ralph Maresco Jr
Cap 21h47'52" -13d49'53"
Ralph Martini
Vir 12h59'2" 10d47'15"
Ralph Meincken
Uma 8h54'42" 48d3'14"
Ralph & Michelle Fegely
Lyr 18h26'42" 35d1'23"
Ralph mit grosser Dankbarkeit
Dra 16h26'52" 58d28'13"
Ralph & Nalani
Crb 15h32'32" 36d33'52"
Ralph Neely
Aql 19h48'35" 14d0'36"
Ralph Novak Family Star
Uma 11h10'4" 34d44'16"
Ralph O. True
Her 17h12'56" 33d31'22"
Ralph Obenauf
Ori 6h7'30" 21d5'27"
Ralph O'Bier
Uma 8h11'40" 62d46'42"
Ralph & Odie Fontanez
Lib 15h4'18" -3d15'48"
Ralph Oliver
Uma 12h4'4" 42d54'41"
Ralph Otto Anderson
Cyg 20h47'31" 30d44'2"
Ralph P. Juliano, Jr.
Aql 19h48'31" -0d44'4"
Ralph P. Kass
Cnc 7h59'26" 18d18'9"
Ralph "Papa Bear" Rucktashel
Uma 9h35'56" 42d30'27"
Ralph & Pat Dicenzo
Ori 6h0'6" 1d17'8"
Ralph & Paula McFarland
Cyg 20h11'13" 45d17'3"
Ralph Pierre Chalifoux
Cmi 7h24'31" 2d35'57"
Ralph R Laderoot
Per 4h18'37" 41d48'6"
Ralph R. LePage
Aur 5h39'33" 33d34'47"
Ralph S. Soehngen
Uma 9h5'25" 51d31'46"
Ralph Sidney Llewellyn Pritchard
Ori 6h18'5" 8d39'28"
Ralph (Sonny) Imondi
Uma 10h4'21" 44d4'48"
Ralph "star quality" Floyd
Sco 16h5'51" -25d33'13"
Ralph Stephen Pica
Per 3h33'6" 51d12'41"
Ralph & Tammy Zampini
Uma 9h58'14" 52d48'25"
Ralph & Terri Humphlett III
Uma 11h20'48" 45d13'23"
Ralph W. Broom III
Uma 9h39'37" 52d54'53"
Ralph W. Farley
Her 16h56'22" 32d25'6"
Ralph W. Schubert
Lib 14h49'34" -2d18'21"
Ralph Wade
Cep 22h11'13" 67d22'26"
Ralph Waldo Emerson IV
Aql 19h42'35" 12d20'5"
Ralph Walter Zippel
Uma 11h39'23" 37d32'13"
Ralph Wenger
Her 17h51'55" 48d4'47"
Ralph Wilcke
Uma 10h4'45" 51d9'6"
Ralph Zuniga
Umi 13h8'46" 70d11'35"
Ralpha Susari
Sgr 18h55'55" -16d20'52"
Ralphbone The Magic Kiddog
Ari 1h59'12" 23d34'26"
Ralphi
Cyg 19h59'15" 38d23'54"
Ralphie
Uma 9h32'46" 46d5'10"
Ralphie
Tau 4h20'56" 30d53'2"
Ralphie
Gem 6h49'0" 25d33'50"
Ralphie
Her 16h43'32" 11d40'27"
Ralphie
Aqr 23h12'57" -7d12'46"
Ralphie Ferrara, Jr.
Ori 5h33'39" -0d25'6"
Ralphie Man
Uma 8h19'23" 66d9'3"
Ralph's and Eunice's Heavenly Light
Cyg 20h3'18" 32d1'6"
ralvic
Lib 15h46'59" -17d37'36"
RaM
Tau 4h10'59" 5d27'57"
RAM 111
Aql 19h42'2" -0d51'39"
RAM Builders, Inc.
Sco 17h41'15" -34d56'45"
RaMaKosBlinkiBill
Uma 12h32'31" 59d40'55"
Ramakrishna
Tau 3h59'27" 11d3'53"
Ramakrishnan, Doris
Uma 8h19'31" 67d41'57"
Ramal Khaled Maad
Vir 13h9'55" -10d43'27"
ramalama
Cnc 8h21'5" 19d46'28"
RAMAMURTHI
Aql 19h11'16" 14d55'1"
Raman mamu
Psc 0h32'35" 5d25'42"
Ramanathan Sarathkumar
Uma 11h34'39" 35d15'5"
Ramanbhai Sanabhai Patel
Uma 11h0'11" 70d30'44"
Ramar & Jenny
Cnc 8h41'33" 11d5'33"
Rambam
Oph 16h57'50" -6d35'2"
Ramble
Cap 20h22'25" -27d30'22"
Ramble on
Ori 5h7'30" 7d2'13"
Rambling Rose Violet Bowman Manley
Ori 5h39'9" 4d27'12"
Rambo P. Davis
Uma 11h28'51" 53d49'38"
Ramdeo Hurhangee
Uma 9h13'42" 56d43'55"
Rameek Swinton
Uma 8h44'38" 62d48'34"
Ramel K. Werner
Sgr 19h33'38" -15d12'19"
Ramelow, Wilhelm
Ari 2h22'56" 18d40'47"
Ramesh
Eri 4h23'6" -0d50'6"
Ramesh Ramakrishnan
Lyn 7h44'0" 36d59'22"
Ramesh Ujam
Vir 13h10'23" -2d23'19"
Rameshgopal
Psc 1h16'5" 32d30'39"
Ramey D. Buchholz
Uma 9h8'20" 63d4'2"
Ramgelnubop "10-18-45"
Lyn 6h36'31" 54d13'28"
Rami
Uma 8h51'22" 64d43'31"
Rami Attia
Uma 9h43'16" 57d55'38"
Rami H. M. Schwartzer
Psc 1h21'10" 10d48'20"
Rami Isam Ayyoub Nusar Hilal Haddad
Cnc 9h1'14" 27d35'29"
Rami Levy's Lucky Star
Psc 0h11'10" 2d53'30"
Rami & Lucie
Eri 4h14'10" -27d17'14"
Rami Maghrbi
Psc 1h15'10" 22d43'59"
Rami my baby
Ori 6h0'35" 10d50'51"
Rami Richani
Uma 8h19'26" 60d51'48"
Ramie Gutierez
Uma 11h36'25" 43d24'42"
Ramina's Shooting Stars
Vir 12h48'10" 3d17'25"
Ramiro Alexander Riley
Lib 14h59'19" -4d14'39"
Ramiro Contreras
Uma 10h49'49" 39d49'28"
Ramiro & Michelle Flores Eternity
Uma 11h28'58" 42d48'54"
Ramiro N. Hinojosa, Sr.
Psc 0h21'33" -2d14'32"
Ramiro R. Guerra
Ori 5h38'57" -2d4'55"
Ramkase & Tymoon Siew 25th Jan 1964
Uma 12h12'22" 62d36'11"
RAMlovesALBeternally330 4
Cyg 20h17'40" 52d14'18"
Ramma Bans
Uma 13h51'5" 54d6'49"
Ramnath Krishaswamy
Uma 13h34'31" 54d25'56"
Ramnik Maneeth Lal
Pho 0h39'14" -49d47'42"
Ramo and Aida Imperioso
Gem 7h24'23" 32d18'13"
Ramon
Psc 1h36'13" 26d18'2"
Ramon
Lmi 10h17'9" 29d0'35"
Ramon
Dra 18h14'57" 79d5'40"
Ramon Alfredo Murillo Ibarra
Cap 20h18'16" -9d47'51"
Ramon Alvarez Abreu
Boo 14h43'10" 12d50'6"
Ramon Chavez Jr. (Honey)
Psc 1h41'15" 8d33'34"
Ramon Dean Moats
Tau 4h28'15" 24d1'30"
Ramon Erkamp
Psc 0h25'51" 13d20'15"
Ramon Harris Enright "White Knight"
Aqr 22h37'14" -1d10'42"
Ramon Jose Vega
Uma 9h39'22" 47d1'21"
Ramon Louis Ortega
Aql 18h50'53" -0d54'59"
Ramon Luis & Anaima
Col 5h49'42" -33d9'32"
Ramon M. Sabers
Lib 15h57'59" -18d42'25"
Ramon Vincent McGlown
Ori 5h31'42" 11d48'28"
Ramon Vontobel
Vul 19h52'6" 23d1'46"
Ramona
Crb 15h48'55" 28d23'26"
Ramona
Tau 5h57'19" 24d57'14"
Ramona
Ori 6h18'19" 16d38'14"
Ramona
Her 17h31'13" 47d37'12"
Ramona
Cyg 20h0'1" 43d9'20"
Ramona
Cas 0h47'41" 50d48'54"
Ramona
Lyn 7h45'22" 36d28'50"
Ramona
Uma 11h31'4" 38d43'24"
Ramona
And 2h22'50" 38d30'27"
Ramona
Vir 14h6'49" -16d24'51"
Ramona
Cyg 20h23'22" 58d36'32"
Ramona Anne Laffler
Vir 12h10'42" 12d6'50"
Ramona Benz 10.10.1988
Lmi 10h40'4" 32d41'26"
Ramona Bowers
Crb 15h50'14" 34d52'32"
Ramona C. Browne
Ori 6h1'30" 16d58'8"
Ramona Dahl
Uma 12h9'53" 60d37'35"
Ramona Gabbert
Cas 23h6'53" 56d33'28"
Ramona Gay Loynd
Sco 17h16'3" -32d27'36"
Ramona Jarmin
Gem 7h31'53" 24d26'16"
Ramona little star
Cap 21h12'2" -16d30'34"
Ramona Lopez Garcia
Cap 20h34'35" -19d34'46"
Ramona & Marc
Peg 23h21'34" 32d4'9"
Ramona Marcia Claire Beckerman
Lib 15h15'37" -6d50'15"
Ramona Mullins
Sgr 19h0'38" -17d5'53"
Ramona Ochoa
Tau 4h39'12" 7d32'12"
Ramona Patricia
Dra 18h53'53" 55d36'58"
Ramona Resto
Dra 15h57'0" 61d11'10"
Ramona "SpiderWoman", "DouceDemon"
Tau 4h27'7" 9d8'34"
Ramona Tacoronte
Gem 6h47'28" 21d14'3"
RamonAlexis
Uma 11h32'24" 57d53'3"
RamonB1
Uma 11h9'31" 56d27'40"
Ramoncita Martinez
Aqr 21h43'15" -2d13'20"
Ramone's and Leslie's Love
Tau 4h19'56" 3d16'22"
Ramon-Luca Nobs
Lyn 7h56'1" 34d15'25"
Ram's Heart
Uma 11h26'52" 52d4'2"
Ramsden
Cnc 8h39'7" 10d17'30"
Ramsey
Umi 16h47'5" 82d38'36"
Ramsey
Psc 0h53'17" 7d0'38"
Ramsey Family - Dorothy, JH, Toni
Tri 1h52'18" 28d12'30"
Ramsey Harpell
Uma 10h31'26" 62d31'20"
Ramsin Pauls
Ori 5h36'47" -1d13'48"
Ramy Khoriaty
Her 17h44'17" 29d58'47"
Ramya
Psc 23h34'55" -2d12'33"
Ramzi Issa
Tau 4h35'4" 9d17'5"
Ramzi, my darling husband
Cyg 20h56'57" 35d50'32"
Ran
Aqr 21h43'23" 1d27'27"
Ran~d~Erin
And 2h28'43" 44d18'51"
Rana Erkan
Cnc 8h59'32" 15d45'19"
Rana Franich Aguirre
Lyn 8h12'38" 56d28'59"
Rana Lynn
Uma 12h6'0" 54d19'52"
Rana M. Zeidan
Cyg 20h24'18" 52d57'6"
Rana Mio Amore
Gem 7h46'20" 33d2'44"
Rana My Tweetheart
Cyg 20h2'24" 33d30'13"
Rana Sameer Abd Al-Kareem Al-Lamee
Sco 17h1'20" -43d52'55"
Ranada
Leo 9h33'48" 7d0'1"
Ranae Ailene Gall
Leo 11h29'15" 7d20'2"
Ranbir
Aql 20h35'13" -3d25'53"
Rancho Mongo
Uma 9h8'32" 58d55'50"
Rand Kmiec
Uma 8h59'12" 64d18'6"
Randa
Cap 21h15'16" -16d11'24"
Randa
Aql 20h11'32" 12d53'17"
Randa Kachef
Cap 20h53'21" -20d47'13"
Randal and Marilyn
Gem 7h42'1" 31d51'18"
Randal Barber
Gem 7h23'17" 25d50'44"
Randal Bennett Kline
Ori 6h12'45" 21d21'0"
Randal Faith
Cnc 8h35'52" 27d57'55"
Randal John Pease
Sgr 18h39'15" -22d38'53"
Randal Joseph Rabon
Per 4h24'45" 43d59'27"
Randal "Randy" Pommerenk
Aqr 21h39'9" -2d11'39"
Randal Steve Davis
Lib 15h5'24" -6d38'52"
Randal Thomas Zell
And 0h34'22" 39d8'40"
Randall A Dockham
Gem 6h52'39" 26d20'21"
Randall A Mulkins
Leo 9h30'4" 28d30'29"
Randall & Abigail Patterson
Aql 19h49'37" 12d6'55"
Randall Alan Gray
Uma 9h20'17" 53d39'17"
Randall Allen Morrow, Jr.
Uma 8h38'9" 70d51'49"
Randall and Alissa Rhodes
Cyg 20h7'4" 33d11'26"
Randall and Christina Zimmerman
Pyx 8h38'44" -33d20'28"
Randall and Phyllis White
Cyg 21h46'8" 44d5'28"
Randall & Anne Coleman~Music is Joy
Lyr 18h47'29" 34d41'28"
Randall Anthony Zuccalmaglio
Aql 19h5'23" 15d43'20"
Randall Bernard Shields "Grandpa"
Dra 18h25'47" 57d56'0"
Randall Breann Carter
Cmi 7h50'2" 1d32'46"
Randall Brown
Her 16h40'37" 26d41'30"
Randall Bryce Smith
Psc 23h49'6" 7d28'50"
Randall C. Krieg (the second)
Her 17h2'26" 19d49'25"
Randall Caldwell Groves
Her 16h21'50" 3d44'29"
Randall Carroll Walker
Sex 10h49'30" -0d9'6"
Randall Carter Wyatt
Ari 2h18'27" 23d8'49"
Randall Curtis
Leo 10h21'19" 9d12'58"
Randall D. "Mac" McPherson
Aqr 21h51'53" -1d33'15"
Randall David Blackledge
Aqr 22h35'22" -8d41'14"
Randall David Smith
Lib 15h12'29" -7d47'1"
Randall Degony Priest III & Family
Uma 11h44'28" 38d14'51"
Randall E. Mitchem
Dra 16h27'48" 56d16'57"
Randall E. "Randy" Blackburn, Jr.
Per 4h10'18" 51d29'2"
Randall F. Stewart
Cyg 20h8'19" 43d54'36"
Randall Frank Hed
Her 16h57'33" 24d53'13"
Randall Frank Koenig
Cep 22h39'16" 67d37'34"
Randall G. Baldwin
Boo 14h35'48" 42d34'27"
Randall G. Bishop
Leo 10h35'31" 13d7'20"
Randall G. Park
Vir 13h45'35" -13d13'9"
Randall Gene "Bud" Wallace
Aql 19h50'34" -0d44'40"
Randall Gregg Mason
Uma 10h38'5" 62d45'40"
Randall John Fedon
Sco 17h22'20" -37d18'22"
Randall Keith Puckett
Hya 9h8'29" -0d33'9"
Randall Kent Kramer
Sco 17h57'12" -30d47'9"
Randall L. Mills
Cap 21h37'26" -11d12'39"
Randall Lee Cartner
Sgr 18h0'14" -27d47'32"
Randall Lee Connolly Jr.
Cnc 9h17'2" 8d8'18"
Randall Lee Rawson
Per 4h47'54" 49d34'22"
Randall Lee Wiseman III
Sgr 19h22'1" -15d4'31"
Randall Leroy Carter
Her 17h41'14" 38d28'8"
Randall Lewis Merrifield
Gem 7h47'23" 32d37'48"
Randall Lloyd Armstrong
Cnc 8h44'27" 11d37'25"
Randall Louis Zuccalmaglio
Cap 21h45'34" -10d19'18"
Randall Miller Gomoll
Psc 1h16'15" 10d9'16"
Randall N Brentley 4-ever
Cnc 8h30'56" 20d15'47"
Randall Paul Thompson
Her 17h42'56" 27d39'10"
Randall R Mulkins
Leo 9h35'33" 27d56'38"
Randall R. Romig
Her 18h2'4" 21d38'32"
Randall Ramsey Fedon
Sco 17h26'16" -35d50'42"
Randall Scott
Lyn 7h10'38" 59d10'18"
Randall Talton
Her 17h5'27" 23d12'30"
Randall Thomas Jagoe
Cyg 21h44'1" 39d56'14"
Randall W. Scofield II
Her 17h22'47" 27d52'20"
Randall Whetzel
Per 4h14'29" 50d48'19"
Randall Wright Patterson
Per 2h54'59" 54d50'32"
Randall's Star
Her 17h55'34" 23d54'19"
Rande Lewis ednaR love Matt Yedlin
Cyg 20h7'51" 34d35'30"
Rande, Blake, Drew and Margo Turner
Uma 11h58'19" 59d2'59"
Randee Sue Roberts
Uma 9h20'53" 56d15'15"
Randee Ulsh
Ari 3h14'44" 26d30'19"
Randee, With All My Love, Cody
And 0h36'31" 42d35'52"
Randel Downing
Lib 15h19'8" -14d32'8"
Randel Santangelo
Uma 8h52'55" 55d44'15"
Randell & Breanne Together Forever
Sco 16h6'5" -9d15'57"
Randell Mark Seaver
Gem 7h23'9" 28d59'9"
Randi
Her 18h29'34" 12d15'18"
Randi
And 1h0'10" 39d59'34"
Randi
Cap 20h46'42" -21d37'50"
Randi
Vir 13h25'18" -0d53'35"
Randi
Uma 11h41'38" 53d24'0"
Randi
Cap 20h25'54" -24d31'5"
Randi 103099 F.I.L.
Psc 1h26'33" 25d40'36"
Randi and Sam Facchani
Cyg 19h58'27" 40d18'49"
Randi Ann Lokelani Buza
Vir 13h14'16" -2d25'40"
Randi Auerbach
Cnc 8h6'37" 9d16'24"
Randi Bolick
Leo 10h11'10" 21d29'12"
Randi BooBoo Rowins
Vir 13h5'19" -11d23'44"
Randi Danielle
Leo 11h45'56" 21d22'30"
Randi Danielle Hudgens
Leo 11h9'9" 9d16'58"
Randi Denise Castle
Cas 0h49'54" 48d56'37"
Randi Ellen Haas
Cyg 20h22'50" 33d44'43"
Randi Gaiser
Umi 14h26'9" 75d16'20"
Randi Greig
Vir 11h48'17" 7d46'10"
Randi Howell 8-14-1996
Umi 14h17'48" 69d17'30"
Randi Israel
Psc 0h7'13" -2d59'29"
Randi Kay
Ari 3h3'3" 14d7'16"
Randi L Tripp
Sco 17h1'58" -36d32'30"
Randi L. West
Ari 2h3'0" 20d50'37"
Randi Leigh Intili
Psc 1h4'0" 28d15'4"
Randi Lyn
Cnc 9h2'44" 28d54'1"
Randi M. and Justin V. Forever
Uma 10h30'0" 62d44'36"
Randi Mason McNab 5/23/63-6/21/05
Uma 10h36'34" 60d55'53"
Randi Perry
Com 12h44'9" 14d14'42"
Randi Rae
Tau 4h53'26" 17d45'44"
Randi Rothe Raimondo
Cas 1h20'36" 62d49'37"
Randi S Risely
Tau 4h29'5" 17d43'40"
Randi Sue Den Besten
Tau 5h26'47" 25d56'25"
Randi Taylor
Uma 10h9'16" 52d43'58"
Randi Vetvik and Andre Gagnon
Cyg 19h46'17" 34d27'27"
Randi Zacour
Uma 10h36'18" 49d53'20"
Randie Faith Langdon
Psc 0h18'4" -0d44'52"
Randie Laine 2002
Aql 20h0'36" -0d31'24"
Randilee
Tau 3h51'33" 25d21'17"
Randilee Serra
Aql 19h14'29" 7d3'48"
RandiLynn
And 2h22'57" 49d46'19"
Randi's 50th Birthday Star
Cas 0h26'53" 54d5'32"
Randi`s Star in the Heavens
Peg 22h39'16" 13d41'54"
Randle Carlyle Ferguson
Pho 0h45'53" -41d6'7"
Randola Virginia Beatrixia VDBasa
Ori 5h14'42" 8d37'11"
Randolph
Cap 20h27'3" -9d32'28"
Randolph Charrington
Per 2h46'48" 54d18'53"
Randolph Edison Shelton, Jr.
Dra 18h40'50" 50d54'20"
Randolph John Smith
Aqr 22h29'26" -23d19'20"
Randolph Joseph Hill
Uma 8h53'35" 46d55'19"
Randolph P Tate
Lib 15h14'13" -22d4'44"
Randolph Ray Hesselberg
Ori 6h3'36" 19d51'28"
Randolph Ryan Hilman
Ori 5h39'23" 8d23'24"
Randon
Per 2h18'36" 54d0'55"
Randulie Sparrow Cableasley
Cep 21h13'9" 67d11'53"
Randy
Cep 22h54'40" 70d48'49"
Randy
Cyg 19h50'18" 52d34'55"
Randy
Sco 16h55'5" -42d27'12"
Randy
Cra 18h9'8" -39d49'31"
Randy
Uma 8h49'26" 47d28'58"
Randy
Per 3h25'14" 52d1'11"
Randy
Her 16h36'59" 32d58'31"
Randy
Gem 7h42'10" 33d15'28"
Randy
Ori 6h8'22" 7d17'10"
Randy
Equ 21h14'12" 11d34'18"
Randy
Leo 11h40'2" 15d53'52"
Randy
Tau 5h38'19" 27d41'51"
Randy "A Gentle Giant" Krieg
Uma 13h37'56" 48d34'6"
Randy A Ratajczyk's Shining Star
Her 18h44'50" 20d5'56"
Randy Alan Faber
Gem 7h13'34" 15d54'20"
Randy and Bernadette Brumley
Ara 17h45'58" -50d8'51"
Randy and Jeanne Pohlman
Uma 8h42'25" 60d13'43"
Randy and Katy Dancer
Crb 15h42'38" 32d32'27"
Randy and Tammy Johnson
Leo 11h31'25" 17d14'44"
Randy Armstrong
Cap 21h35'34" -15d58'55"
Randy Baker
Vir 14h1'58" -16d4'6"
Randy Bergum
Vir 13h37'44" -5d51'5"
Randy & Beverly
Uma 10h25'45" 51d45'37"
Randy Bret Spitler
Sgr 19h4'17" -31d42'13"
Randy Broadway
Sgr 20h28'3" -42d18'24"
Randy C.
Sco 16h11'44" -8d48'9"
Randy C. Holman
Uma 11h34'45" 35d33'23"
Randy C. Smith
Psc 0h55'20" 10d20'50"
Randy Carlos Cavanagh
Lib 15h46'42" -9d52'35"
Randy Carlson
Vir 14h4'30" -15d25'18"
Randy Chew 1V
Aql 18h53'35" -0d3'37"
Randy & Chris
Aqr 22h59'39" -11d54'43"
Randy Combs
Her 17h43'53" 15d58'50"
Randy Congdon
Aql 19h29'31" 7d25'12"
Randy Crowder
Her 18h6'34" 31d56'14"
Randy D. Ayers
Peg 21h13'10" 16d24'38"
Randy DeWayne Wesley
Psc 1h26'58" 17d41'51"
Randy Donalds
And 23h5'55" 48d21'54"
Randy Dowden
Her 16h37'10" 42d18'53"
Randy Eaton
Sgr 19h33'37" -12d43'45"
Randy Edward & Jessica Lee Gagné
Ori 5h3'41" -0d13'20"
Randy Faith
Aqr 21h41'4" 2d3'50"
Randy Floyd Bradeen
Psc 0h0'47" 7d52'24"
Randy Gagan
Sgr 19h25'2" -17d35'47"
Randy Geffon
Aql 20h4'0" 4d3'11"
Randy Gonyeau and Cathy Gidney
Lyr 18h54'32" 35d53'56"
Randy Guill
Lib 15h17'34" -27d43'14"
Randy Harmsma
Uma 11h23'56" 56d3'16"
Randy Hidlebaugh
Cnc 8h58'26" 15d12'26"
Randy Holloway
Her 18h47'51" 20d20'44"
Randy J Ackerman, my wonderful love
Cyg 20h28'20" 40d48'51"

Randy J. McDermott
Cap 21h45'21" -11d49'56"
Randy James
Uma 10h56'25" 39d6'17"
Randy Joel Rugh
Cnc 8h29'19" 29d1'58"
Randy Jr.
Umi 14h54'43" 73d46'31"
Randy & Judy Johnson
Eri 4h20'59" -0d23'51"
Randy & Julie (Jordan) Char
Boo 14h21'22" 42d59'21"
Randy & Katie's Shining Love
Uma 9h26'14" 49d1'2"
Randy Kennett - Superstar!
Uma 9h59'35" 68d34'49"
Randy Kinkade
Lib 15h8'56" -10d33'30"
Randy Kuhn (Ereinion Elensar)
Uma 11h7'29" 66d27'45"
Randy L. Hinkelman
Aqr 22h35'38" 2d21'15"
Randy Lee
Cmi 7h8'15" 9d35'4"
Randy Lee Gillott
Her 17h15'59" 16d13'40"
Randy Lee Simmons
Her 16h35'11" 5d50'47"
Randy Lehl
Lac 22h13'49" 49d45'52"
Randy Liddell
Her 16h42'15" 36d27'20"
Randy & Lisa's Crystal In The Sky
Cyg 20h9'17" 30d50'37"
Randy Loper
Boo 14h39'51" 22d8'5"
Randy Loves Amy
Cyg 19h51'34" 39d52'17"
Randy Loves Amy
Sgr 18h4'20" -27d3'17"
Randy L.P.
Per 3h23'11" 35d47'9"
Randy Lyn Pounders
Aur 7h15'49" 42d30'2"
Randy Lynn Nelson
Uma 12h0'1" 59d41'46"
Randy & Maggie
Cyg 20h22'29" 43d31'57"
Randy Martin DeWolfe
Leo 11h9'15" -0d27'18"
Randy Mason
Boo 14h50'25" 25d18'42"
Randy Massey
Aur 6h28'14" 35d13'26"
Randy McClure
Lac 22h22'25" 43d14'54"
Randy McDaniel
Aql 19h27'55" 15d14'52"
Randy Means
Leo 10h3'53" 22d25'6"
Randy & Melanie
Lyr 18h39'36" 41d39'15"
Randy Miller
Gem 7h50'41" 31d0'26"
Randy Mock
Her 17h45'51" 46d12'2"
Randy my love
Cnc 8h57'46" 19d4'44"
Randy n Angel
Leo 11h0'25" 17d5'48"
Randy Noblin
Ari 2h13'56" 12d51'4"
Randy Nuuhiwa
Aqr 21h7'36" -8d56'12"
Randy Orton
Her 18h44'8" 19d1'33"
Randy P. Peppers
Lyn 7h55'20" 38d44'52"
Randy Pandy
Aqr 22h36'26" 1d16'0"
Randy Paul
Per 4h15'42" 51d17'38"
Randy Pièrre Guerrette
Mon 7h53'29" -6d3'57"
Randy Place
Lib 15h49'38" -12d42'6"
Randy R. Larson
Aqr 22h8'16" 0d16'55"
Randy R Seel
Vir 11h57'1" -11d24'13"
Randy "Ranger" Kippen
Pyx 8h55'22" -34d53'55"
Randy Ray Collins
Aql 19h17'55" -7d32'33"
Randy Ray Russell
Per 4h49'30" 47d23'50"
Randy Rex Houston
Uma 8h43'6" 71d26'11"
Randy Robbins
Lyr 19h18'43" 28d41'48"
Randy "Rock - a - Billy" Blumke
Cnc 8h42'44" 16d4'45"
Randy Rosier
Cyg 21h10'45" 48d22'45"
Randy S. Coolbaugh
Tau 5h9'8" 21d29'43"
Randy S. Molnar
Uma 11h23'40" 31d3'43"
Randy Seale
Uma 11h50'24" 65d9'20"
Randy Shannon
Per 3h18'57" 51d2'9"
Randy Shelton
Aql 19h34'41" 13d16'13"
Randy Sommerfeld
Ori 6h4'43" 20d50'8"
Randy Springer
Her 16h30'13" 8d0'11"
Randy & Staci Rohde
Lyr 18h51'43" 29d53'15"
Randy V. Williams
Uma 10h47'4" 54d22'12"
Randy W. Long
Uma 14h6'6" 50d34'9"
Randy Wayne Griffin
Ari 2h20'58" 11d19'10"
Randy White-The Bandit
Aql 19h44'2" -0d7'24"
Randy William Knapp
Ari 3h18'50" 29d38'50"
Randy "Wolvie" Gross
Per 3h4'35" 48d47'51"
Randy Worley Shines Brightly
Her 17h20'58" 38d27'18"
Randy's Love
Aqr 20h47'2" 0d2'20"
Randy's Star
Ori 5h46'12" 7d21'7"
Randy's Star
Leo 11h41'45" 25d18'51"
Randy's Star
Dra 19h18'3" 67d51'1"
Randy's Star
Sco 17h20'23" -41d40'35"
Randy's Wish
Cap 20h15'49" -9d24'43"
Randyvance
Uma 10h33'57" 48d43'32"
Rane Alexandria Castillo
Lmi 10h41'42" 31d32'52"
Ranee Vejmaneesri
Sgr 18h46'20" -18d25'37"
Ranessa Marie Fortes
Lib 15h9'35" -1d48'18"
Raney Alexandre
Cyg 20h58'31" 46d30'54"
Raney Ellen Ausborn
Uma 13h15'18" 57d8'38"
Ranger
Umi 15h24'36" 77d26'39"
Ranger
Sgr 17h54'5" -17d56'21"
Ranger Lou Rules
Per 4h24'15" 40d16'58"
Rangie (AKA Angela Williams)
Sco 17h24'49" -30d44'56"
Rangika Lowe
Ari 3h19'34" 30d16'8"
Rani
Umi 14h56'56" 78d54'17"
Rani
Umi 16h27'45" 76d7'58"
Rani Penny
Sgr 18h14'38" -27d36'55"
Rani Van Gogh
Uma 9h30'48" 65d25'54"
Rani Young
Sco 17h34'9" -37d28'11"
Rania Avat and Joe Howayek
Cyg 21h56'1" 50d57'22"
Rania Dounou
Uma 11h32'53" 47d17'7"
Rania, Roroti... Say Yes?
And 2h18'21" 45d14'44"
Ranipulgui YS
Cas 0h28'42" 61d33'8"
Rani's Heart
Sgr 18h29'7" -27d21'38"
Ranita Pipi
Psc 1h51'11" 9d39'32"
Ranjita Muna Mishra
Peg 22h13'3" 10d20'40"
Ranlin
Uma 10h38'19" 66d50'54"
Rannel Sashnee Naidoo
Cru 12h56'36" -64d29'37"
Ranny and Pappy
Uma 10h47'9" 49d44'45"
Ranny Williams
Uma 11h49'45" 64d47'43"
Rano Mamedova
Uma 11h54'19" 58d50'13"
Ranya Gemayel
Boo 13h51'59" 11d20'40"
Raoul Briffa
Psc 23h29'50" 6d22'11"
Raoul Buin
Uma 9h52'33" 59d40'26"
Rapa 0107
Aqr 21h11'32" -11d9'17"
Rapelli
Per 2h17'46" 51d28'24"
Raphael
And 2h32'26" 46d36'19"
Raphael
Boo 14h41'38" 37d58'58"
Raphael
Psc 0h39'40" 16d11'46"
Raphaël
Ari 3h5'48" 24d16'30"
Raphaël
Sco 17h35'27" -39d24'30"
Raphael Adams Fox
Cra 19h3'47" -38d21'58"
Raphael Augustinus Girardoni
Dra 19h12'38" 79d33'7"
Raphael Augustus Wright
Uma 10h0'30" 58d38'28"
Raphael Botticelli Hubert
Umi 13h39'46" 74d39'48"
Raphael Cushnir
Boo 14h38'52" 19d44'46"
Raphael de Padua
Aqr 21h45'51" 1d48'30"
Raphael Gilbert
Leo 10h24'46" 27d23'11"
Raphael Henri Ford
Leo 10h15'3" 22d53'25"
Raphael Jacob
Cnc 9h2'24" 19d58'9"
Raphael K. Osei, M.D.
Sco 16h8'31" -12d1'27"
Raphael Kohlenberg
Cep 4h52'45" 83d41'12"
Raphael Kotanjyan
Vir 12h35'32" 10d30'42"
Raphaël Lameyre
Mon 7h3'30" -7d29'5"
Raphaël Lamothe
Uma 11h29'3" 59d41'7"
Raphael Lopes
Lac 22h36'12" 49d35'30"
Raphael S. Kendall
Psc 1h9'39" 20d39'26"
Raphael & Sandra Frydman Love
Cyg 20h4'16" 57d51'53"
raphaela BOOTZ
Uma 11h28'18" 53d14'18"
Raphaela, 15.02.2003
Umi 17h13'39" 82d40'41"
Raphaella
Uma 11h18'26" 36d56'26"
raphaella menna
Ori 5h56'50" 11d18'11"
Raphaëlle Groulx
Uma 9h41'25" 67d20'26"
Raphel "Mose" Nichols
Umi 14h52'19" 78d43'9"
Raphi und Simone
Umi 16h43'58" 76d39'7"
Raphlili Goureman
Gem 6h2'31" 23d29'4"
Raphoun
Peg 21h36'1" 26d16'13"
Rapp, Anne & Hermann
Uma 11h18'4" 38d46'7"
Raps Star
Crb 16h8'37" 33d51'39"
Rapthi
Ori 5h56'12" 17d11'29"
Raptor08
Lib 14h56'59" -7d31'45"
Raquel
Cap 21h22'9" -16d12'50"
Raquel
Uma 10h25'4" 61d25'16"
Raquel
Crb 15h43'42" 36d52'4"
Raquel
And 0h26'10" 34d19'2"
Raquel
Gem 7h34'4" 32d54'17"
Raquel
And 23h56'22" 32d35'16"
Raquel
And 23h32'15" 43d0'16"
Raquel B. Alvarez
Sco 16h9'26" -14d10'31"
Raquel B. Aponte
Lib 15h6'41" -1d45'16"
Raquel Barata
Lep 5h28'30" -14d4'44"
Raquel C. Aguilera
Her 17h42'48" 45d6'39"
Raquel+Cecilia
Uma 10h51'1" 71d39'48"
Raquel Evian Bates
Lyn 7h29'48" 53d31'12"
Raquel Fresco
Del 20h28'4" 16d7'56"
Raquel Garcia
Gem 7h32'47" 16d20'11"
Raquel Guzman
Ari 2h54'56" 17d55'38"
Raquel Jane Ramero
Aqr 22h31'14" -22d53'43"
Raquel King
Dra 15h14'10" 57d44'58"
Raquel Lynn Wilhelm
And 23h4'15" 50d46'17"
Raquel Maria Santiago
And 1h10'40" 37d40'3"
Raquel Merino
Psc 1h2'47" 5d22'6"
Raquel mi amor eterno
And 2h20'28" 42d41'50"
Raquel Montalvo
Tau 5h52'12" 15d39'40"
Raquel Moreton (08/Enero/2000)
Cnc 8h50'22" 27d11'53"
Raquel Napolitano
Tau 4h38'47" 27d38'36"
Raquel Napolitano
Tau 5h23'12" 18d52'23"
Raquel Rosado
And 23h8'54" 42d43'25"
Raquel "Sassy" Reynolds
And 0h13'50" 29d49'4"
raquel the love of my life
Lyn 8h15'0" 51d2'45"
Raquel will you marry me?
And 1h36'59" 47d54'29"
Raquelita
Ori 6h4'7" 19d48'36"
Raquelle Warren
Vir 13h46'44" 3d24'43"
Raquel..........mommy
Ari 3h6'41" 23d6'53"
Raquel's Star
Crb 16h7'22" 34d38'15"
Rara
Cyg 19h44'15" 37d34'51"
RaRa
Lep 5h37'4" -12d44'46"
RaRa831
Cnc 8h20'28" 23d17'24"
Rare Gem
Tau 4h41'8" 23d50'0"
Rarra Avis
Cnc 8h55'44" 29d14'9"
Rary
Umi 14h2'2" 69d25'18"
RA'S BUCKWHEAT STAR 44
Lib 15h15'15" -20d34'37"
Ra's Guiding Light
And 2h9'15" 43d55'45"
Ras Obasi
Tau 5h44'31" 26d12'0"
Rasa S. Kay
Crb 15h43'4" 34d47'59"
Rasa Vanilla Sky
Gem 6h48'54" 20d39'6"
Rasa Vitkute, Neabyknavenaja
Ari 3h25'17" 23d41'9"
Rasaja Taika
Cnc 8h55'2" 29d26'54"
Rascal
Gem 7h16'8" 21d46'3"
Rascal
Uma 11h24'55" 42d38'8"
Rascal
Per 3h39'15" 46d7'32"
Rascal
Aqr 20h48'46" -12d18'41"
Rascal
Lyn 7h12'54" 54d0'15"
Rascal and Briar's Star
Umi 13h24'16" 69d40'58"
Rascal Jones
Cmi 7h33'27" 4d49'11"
RascalPup
Cma 7h27'9" -15d32'56"
Rasche, Julia
Uma 8h13'43" 64d20'11"
Raschi
Boo 13h52'53" 22d36'41"
Rasco
Tau 5h13'53" 18d1'45"
Rasha
Lib 15h48'29" -9d5'44"
Rasha Nakouzi
Uma 11h30'27" 63d32'58"
Rashaad Scott
And 0h51'4" 38d46'59"
Rashael von Mosch
And 23h23'3" 41d32'40"
Rashaud Malik Ford
Leo 9h53'38" 26d16'37"
Rasheed
Lib 14h52'38" -7d35'45"
Rasheed "RARA" Jackson
Gem 7h21'18" 33d20'17"
Rasheeda Mitchell
Cam 4h31'22" 56d13'0"
Rashel
Tau 3h51'13" 17d33'1"
Rashel Manoukiyans
Aqr 21h7'41" -3d19'20"
Rashele Marie Wright
Cyg 21h47'17" 44d31'20"
Rashell Mae Hughes
Sgr 17h56'25" -29d18'47"
Rashelle
Uma 8h33'9" 60d14'10"
Rashelle Ann Curry
Sco 17h40'50" -43d5'21"
Rashelle Lee Rook
Uma 10h17'41" 42d18'8"
Rashelle Renee Vogus
Gem 7h16'47" 33d32'32"
Rashere
Lib 14h49'49" -11d16'2"
Rashida
Sgr 19h0'2" -30d44'47"
Rashida Best
And 0h10'46" 28d37'10"
RaShida K. Williams
Cap 20h23'5" -8d59'37"
Rashida Lana' Carr
Umi 14h19'22" 71d16'35"
Rashmi Dalvi
Gem 7h40'57" 22d33'48"
Rashmi Nanji Kakad
Aur 7h25'45" 40d54'46"
Rasl Duke
Dra 18h57'52" 64d48'40"
Rasmia
Sco 16h12'54" -14d0'9"
Rasmia Hourani
Cru 12h31'25" -60d44'30"
Rasonya
Lyn 7h11'24" 58d31'4"
Rass & Christina
Cyg 19h46'0" 47d20'17"
Rassfeld, Kerstin
Uma 11h0'55" 64d51'56"
RastaMikeCoppin333
Cnc 8h19'30" 17d33'49"
Rastas
Dra 16h6'56" 59d9'26"
Rastatt's Favorite
Umi 15h8'13" 76d18'14"
Rastus & Toots
Cas 0h21'24" 51d57'20"
Raszipovits Heni
Peg 22h57'41" 20d52'29"
ratana
Dra 17h28'15" 56d30'23"
Ratasha Martinez
Lyn 8h40'10" 35d15'30"
Rathbun
Per 3h16'26" 43d6'33"
Rathje, Ulrich
Uma 13h9'44" 60d9'1"
Rath-McIntosh
Aql 19h50'41" -0d14'10"
Rati
Sco 17h53'43" -36d7'20"
Rational Mastermind
And 2h21'49" 48d23'57"
Ratius Libuserius
Lib 15h31'44" -26d52'51"
Ratna
Cnc 9h11'9" 10d44'22"
Ratnakar Mahadeo Phatak 'Nana'
Psc 1h43'8" 5d37'25"
ratsekim
Uma 14h2'27" 60d57'26"
Ratty Natty
Tau 5h18'50" 21d15'14"
Raul A. Perez, Sr
Uma 8h34'30" 47d37'42"
Raul and Roy
Aqr 22h7'58" -3d23'44"
Raul Ayala
Her 17h5'50" 34d26'43"
Raul Eduardo
Leo 11h5'32" 0d30'10"
Raul Feria
Her 18h56'55" 24d12'49"
Raul J. Perez
Per 3h5'21" 39d35'24"
Raul J. Villarreal, Sr. (1948-2003)
Gem 6h10'51" 27d6'47"
Raul Rodriguez
Her 18h29'32" 16d5'17"
Raul Vasquez 9-22-73
Vir 12h48'39" -11d7'37"
Raúl Zuara Grondona
Sco 17h48'1" -39d48'1"
Raul's Dimple
Uma 11h35'39" 62d38'15"
Raupach, Gerhard
Uma 8h37'8" 59d28'3"
Rausch, Holger
Sgr 17h54'57" -28d11'6"
RAV- 1
Crb 16h14'6" 37d36'14"
Ravan Star
Cyg 20h8'38" 50d53'41"
RaVani
Tau 3h37'50" 2d51'49"
Ravely Arias
Sgr 18h9'15" -31d39'57"
Raven
Uma 9h17'52" 57d34'46"
Raven
Mon 6h46'15" -0d19'5"
Raven
Lib 15h26'6" -14d14'56"
Raven
Ori 5h15'19" 6d14'30"
Raven
Ori 6h9'45" 15d47'38"
Raven
Ori 6h8'15" 17d29'48"
Raven and Janisse Forever
Lyn 7h11'45" 47d36'28"
Raven Branham
Aur 5h53'59" 41d38'23"
Raven Campbell
Ari 2h11'17" 24d8'17"
Raven Guliford
Cap 20h11'19" -25d46'49"
Raven Jean McGeen
Lyr 18h55'56" 33d39'48"
Raven Jordan
Cnc 8h38'17" 7d32'31"
Raven La'Dale
Per 4h38'37" 40d37'0"
Raven Lee
Tau 5h29'4" 20d7'31"
Raven M. VanWinkle
And 0h17'42" 45d2'52"
Raven MacRae Campbell
Psc 0h45'23" 8d17'35"
Raven Michelle Crowl
Cnc 8h50'24" 12d19'59"
Raven Nevar
Sgr 18h9'5" -28d34'54"
Raven Noel Herron
Crb 15h52'7" 38d30'39"
Raven Persephone
And 2h35'47" 49d4'52"
Raven Sky Rorher-Barr
Ari 2h11'43" 26d10'13"
Raven VanLoan Cooley
Cru 12h28'58" -56d47'31"
Raven Yvonne
Sco 16h36'21" -28d16'15"
Raven66
Tau 5h11'59" 24d48'38"
Ravenblood
Aqr 23h13'21" -22d16'39"
Raven's Dreams
Cam 4h0'42" 67d15'44"
Ravens Eye
Cap 20h11'1" -10d13'37"
RavenStarBellamy
Uma 9h43'26" 46d39'40"
Ravi
Uma 11h34'39" 43d52'46"
Ravi
Umi 15h7'33" 69d44'48"
Ravi G. Malik
Her 17h19'12" 47d32'31"
Ravi - Sheena
Pyx 8h56'14" -34d28'9"
Ravi Subramaniam
Ori 5h59'36" 10d31'14"
Ravi, my lovestar for eternity
Cyg 20h56'47" 39d26'5"
Ra-vid
Cyg 20h13'51" 34d9'59"
Ravin
Aql 19h10'58" -0d43'51"
Ravindra Rishiraj Sharma
Tau 4h20'56" 18d34'8"
Ravissante Sonia
And 23h32'59" 37d56'0"
Ravneet Bhogal
And 2h17'2" 42d4'1"
Ravonne
Cas 0h44'31" 54d18'3"
Ravonne R. Lukonen
Lyn 6h34'37" 57d9'4"
Ravyn Ella
Del 20h39'20" 13d26'20"
R.A.W
Uma 11h9'47" 32d9'20"
Raw Dogs Queen of Dreams 381
Cas 0h54'21" 65d16'30"
Rawan
Vir 13h27'29" 2d10'2"
Rawan Sadaka "7ayaty"
And 23h34'35" 41d16'14"
Rawana
Uma 9h28'19" 71d39'54"
Rawiah S. Al Nanih
Cap 20h43'16" -23d12'40"
Rawlio
Cep 22h56'33" 71d7'45"
Rawn and Roshini
Tau 6h0'14" 27d53'12"
Rawr
Her 18h24'29" 14d17'8"
rawr
Lyn 7h19'44" 57d32'5"
Raxicoricophalipitorious
Sgr 18h56'35" -36d29'28"
Ray
Cep 22h57'48" 70d28'15"
Ray
Cep 0h42'32" 81d22'56"
Ray
Cyg 19h56'57" 33d17'21"
Ray
Per 4h3'57" 38d22'53"
Ray & Alisha Gremli
Uma 10h27'8" 62d22'12"
Ray & Amina
Cyg 21h9'17" 42d22'4"
Ray and Alex Cotillo
Uma 11h37'11" 40d57'18"
Ray and Betty Dever
Cam 4h5'37" 70d20'9"
Ray and Delois Liming Heavenly Star
Gem 7h11'59" 17d21'0"
Ray and Janet Dillard
Uma 8h59'51" 71d2'27"
Ray and Janice Orlando
Uma 10h2'16" 57d6'46"
Ray and Joyce Spindler
Cyg 21h37'56" 47d25'15"
Ray and Melissa
Cyg 20h46'20" 39d3'5"
Ray and Michael Beebe Dad and Son
Her 16h38'36" 24d35'58"
Ray and Shirl Gluski,my parents
Cyg 19h42'3" 34d42'0"
Ray Anthony Davis 2514
Ari 2h50'29" 12d14'24"
Ray Arnold Boos
Psc 0h54'55" 8d3'50"
Ray Bear
Lib 15h44'20" -6d4'33"
Ray Brown
Uma 8h9'46" 63d1'5"
Ray C. Meyers
Uma 9h38'18" 48d26'48"
Ray C. Meyers
Uma 11h20'40" 51d46'17"
Ray & Carole
Lyn 8h56'21" 36d16'32"
Ray & Carole Edwards
Uma 10h44'49" 41d47'16"
Ray & Carrie Sedano
Cyg 21h25'51" 51d53'58"
Ray Charles Brimhall
Psc 1h44'44" 19d53'33"
Ray Clarke
Per 3h21'19" 39d12'7"
Ray Crosby
Tau 3h43'14" 23d18'41"
Ray Dean Hunter
Ari 2h18'12" 25d4'1"
Ray DeBernardis
Cep 22h30'29" 64d22'54"
Ray & Dolores Stiles *60th*
Cyg 20h14'7" 46d34'24"
Ray E. Horton
Uma 12h10'22" 57d5'41"
Ray E Smith
Dra 16h43'37" 59d24'43"
Ray Earl Johnston, Jr.
Aql 19h49'27" 8d55'16"
Ray & Earlene Cargill
Uma 9h20'2" 45d11'11"
Ray Edward Clark, III
And 1h2'49" 38d40'35"
Ray Egan
Aqr 21h39'33" -3d46'19"
Ray Faulkenberry
Cnc 8h44'7" 18d30'40"
Ray Franklin Gerardo Bautista
Aqr 21h57'15" 0d27'46"
Ray Garval
Per 3h50'8" 32d40'32"
Ray Giles
Per 3h28'52" 52d28'39"
Ray Ginocchio
Uma 11h6'45" 60d26'14"
Ray Glass
And 0h32'37" 28d8'37"
Ray Godkin - Cycling
Cru 12h50'48" -61d25'19"
Ray Goley
Uma 9h23'36" 62d6'20"
Ray "Grand Daddy" Bowen
Sgr 19h3'5" -18d32'19"
Ray Guy Smart
Sco 16h9'15" -18d10'21"
Ray IV and Jasara
Per 2h47'40" 35d46'35"
Ray J. Bryan
Uma 8h55'27" 59d44'21"
Ray James 23-09-04
Cep 22h12'22" 67d33'46"
Ray & Jann Sutton
Del 20h46'51" 3d23'22"
Ray & Jeanneane Adamini
Psc 0h18'16" -5d22'34"
Ray Johnson, Jr.
Umi 14h38'47" 71d17'43"
Ray K. Davidson
Cep 2h5'52" 81d44'56"
Ray Karnes
Lyn 7h58'7" 50d10'20"
Ray Karosas
Ari 2h13'58" 14d1'55"
Ray & Kay Hustad
Per 3h32'21" 48d41'5"
Ray & Kim
Cap 20h46'34" -21d28'49"
Ray Knight
Ari 2h31'44" 11d5'26"
Ray Koziel
Per 3h11'30" 51d5'38"
Ray & Laura
Ori 5h38'45" -2d33'58"
Ray Lesch
Uma 12h56'2" 59d45'46"
Ray Light
Ori 5h31'52" -0d58'38"
Ray Lopez
Her 16h29'36" 23d13'37"
Ray Luebs
Boo 14h42'11" 18d57'18"
Ray M. Nix, Sr.
Ari 2h36'55" 14d57'51"
Ray & Marissa
Cyg 19h49'15" 29d57'22"
Ray McAlinden
Cep 21h23'17" 69d25'39"
Ray Medred
Lib 15h4'13" -27d58'57"
Ray Meyer
Leo 11h11'36" 11d51'38"
Ray Moscardina
Cep 21h34'11" 67d22'50"
Ray "Muscle" Latulipe
Cas 22h57'33" 57d42'25"
Ray Nunn
Pho 0h42'23" -46d35'7"
Ray of Hope
Aqr 23h11'13" -22d25'46"
Ray of light
Lep 5h55'11" -18d55'34"
Ray Penzel & Randy Casper
Cma 7h14'5" -23d50'1"

Ray Piatt, Jr.
Umi 15h25'39" 68d24'28"
Ray Proffitt - 7/12/1976
Ori 5h39'31" -2d52'53"
Ray Ray
Sgr 18h17'57" -19d26'35"
Ray Ray
Vir 13h26'49" 1d59'4"
Ray Ray
And 0h52'0" 45d17'40"
Ray Ray Hans
Cra 18h5'41" -38d17'21"
Ray Ray & Marky for ever & ever
Cru 12h21'29" -60d42'3"
Ray Ray's Star
And 0h48'31" 45d20'16"
Ray Rhoades
Uma 10h33'22" 51d21'49"
Ray Riza
Ari 3h2'51" 15d26'16"
Ray & Ruth
Her 16h20'30" 45d27'34"
Ray S. Castle
Psc 0h20'16" 6d18'44"
Ray Scott's Dock
Psc 0h54'6" 4d55'47"
Ray & Sharon Hustad
Per 3h15'37" 44d20'6"
Ray Spivey & Ed Mayhew
And 0h34'35" 39d57'38"
Ray Spunky Combest
Cnc 8h39'28" 31d37'56"
Ray Stanchfield
Per 4h34'54" 40d57'23"
Ray Stuart, Husband,Papa,Dad,Hero
Cep 22h15'4" 60d53'26"
Ray~The sky sparkles with your love
Per 3h34'30" 48d21'19"
Ray Thompson
Gem 6h54'44" 25d55'7"
Ray V. Fierro
Cnc 8h36'13" 28d46'48"
Ray & Vera Vanderpool
Cyg 21h35'31" 40d5'40"
Ray Vet
Per 2h51'36" 46d37'35"
Ray W. Dick
Leo 9h55'9" 31d30'31"
Ray W. Palmer
Lib 15h58'30" -17d3'57"
Ray & Wanda Withers 4ever
Dra 19h33'58" 66d20'4"
Ray Waskoviak
Per 3h6'36" 45d5'47"
Ray Webb, Jr.
Her 18h44'0" 21d31'0"
Ray, 1-15-05, Love You!
Ori 5h34'35" -7d31'6"
Raya
Cep 23h18'2" 79d16'35"
Raya Christine Bernauer
Vir 11h59'51" -0d27'17"
Raya Kiperman
Leo 11h3'30" 1d57'43"
Raya Lyon Capel
Uma 8h16'46" 64d42'1"
Raya Maren Cox
Uma 9h19'47" 44d6'25"
Rayah Al-Sabah
Tau 5h18'16" 16d14'13"
Rayan Elawar
Uma 13h6'58" 59d26'24"
Rayan Jibril
Psc 1h24'4" 32d45'49"
rayana05
Dra 19h6'3" 60d29'40"
Rayangel
Cnc 8h41'31" 19d0'39"
Rayann Waddey
Umi 15h34'52" 71d4'1"
Raybear
Cas 0h53'20" 58d21'57"
Ray-Bob 50
Lyn 7h33'54" 37d40'17"
Rayburn Charlton
Tau 5h9'34" 23d44'47"
Rayce Christopher Olson
Ari 2h59'37" 14d55'57"
RAYCELIA
Per 2h23'0" 52d4'18"
Raychael
Cnc 8h45'51" 17d59'45"
Raychel Ann
Cap 21h40'3" -12d45'43"
Raychelle-Ann
Sgr 17h58'21" -28d58'45"
Ray-Dor817
Gem 6h55'53" 15d28'25"
Raye and Ed
Cyg 20h27'14" 31d18'30"
Raye Raye
Lyn 8h7'33" 39d48'35"
Rayed Yasin
Lyn 9h8'53" 36d49'8"
Rayell Eileen Brokaw
Psc 1h9'52" 4d5'55"
Rayelynn Majoewsky
Lyn 6h51'17" 52d4'10"
Rayenetta Aireanne
Ari 2h49'38" 29d9'39"

Rayette & John Insolia
Cyg 19h55'53" 59d38'11"
Raygen & Thomas, Amor Vincit Omnia
Cyg 20h16'49" 51d48'47"
RAYGIN
Aql 19h3'48" 14d51'37"
Rayhaneh
Cnc 8h17'59" 28d38'41"
Rayj Rivas
Umi 16h6'47" 81d19'56"
Rayla Malynn Haag
Ori 5h47'40" 7d27'22"
Rayleen
Lyn 7h53'17" 55d28'47"
Rayleen I. Williams
Cas 0h46'5" 56d22'56"
Rayleena Leslie Angelica Cruz
And 1h43'0" 46d6'37"
Raylene Ann Rosas
Ori 5h34'23" -1d5'23"
Raylene Buehler
Cnc 9h15'28" 9d13'25"
Raylene H. Brewer
Lyn 7h39'54" 44d58'48"
Rayma R.Castiglione
Cam 5h12'39" 72d26'13"
Raymo Sanchez
Cnc 8h20'20" 8d24'12"
Raymon Allen Kocol
Uma 9h17'38" 55d8'7"
Raymon Erroffeff
Ori 5h0'10" 5d30'10"
Raymond
Aql 19h11'23" 13d14'1"
Raymond
Ari 3h20'46" 16d26'17"
Raymond
Boo 13h48'19" 16d20'40"
Raymond
Uma 8h32'27" 69d55'57"
Raymond
Cep 22h52'0" 77d28'32"
Raymond A. Charles
Her 17h32'41" 36d35'25"
Raymond A. Crisio
Uma 11h41'20" 53d42'11"
Raymond A. Eaton
Aql 19h53'0" -0d15'56"
Raymond A. Hodgson
Vir 12h5'7" 3d23'7"
Raymond A. Johnson Jr.
Aql 18h50'31" -1d24'11"
Raymond A. Maffei
Cep 21h40'54" 67d58'26"
Raymond A. Ramirez
Aqr 22h39'26" -1d43'53"
Raymond A Swanson
Her 16h37'39" 34d26'4"
Raymond Adrian
Leo 10h35'19" 14d19'25"
Raymond Albert Noger
Per 2h55'4" 55d4'55"
Raymond Alexander Dodge
Sco 16h9'55" -16d24'30"
Raymond & Amber
Cyg 20h18'27" 40d29'29"
Raymond and Bob Smith
Uma 11h47'37" 33d53'18"
Raymond and Clara Louise SR
Per 3h25'11" 47d20'6"
Raymond and Deborah Hartman
Ari 2h41'43" 31d0'57"
Raymond and Diana 50th
Cyg 20h38'23" 58d12'11"
Raymond and Joan Chapdelaine
Uma 9h55'27" 66d17'46"
Raymond and Peggy Poole
Car 6h46'24" -51d49'35"
Raymond Andrew Ferraro
Aqr 22h27'48" -23d51'34"
Raymond Andrew Ricci
Leo 11h34'0" 25d37'57"
Raymond Anthony Wiesen
Per 3h52'10" 32d53'35"
Raymond Arthur Leudesdorff
Cap 21h29'59" -8d42'58"
Raymond Arthur Yerg, Jr. "Ray Dog"
Cma 7h21'42" -18d25'30"
Raymond Baisarry
Ori 5h55'9" 7d31'32"
Raymond Bennett Ridgeway
Ori 6h16'42" 15d5'41"
Raymond Bergen, Jr.
Tau 4h11'28" 27d57'34"
Raymond Bertuglia
Her 16h38'40" 48d11'36"
Raymond Bongiovi
Cyg 19h44'21" 29d4'10"
Raymond Bouchard
Boo 14h47'28" 46d6'12"
Raymond Brian Fredericks
Sco 17h56'2" -40d24'50"
Raymond C Coccia
Sgr 18h16'11" -23d28'15"
Raymond C. Hawker
Uma 10h34'3" 53d55'58"

Raymond C. Jackson
Aql 19h17'59" -0d2'6"
Raymond C Leffler Jr family Star
Lyn 9h11'49" 33d55'9"
Raymond Carmadello IV
Gem 6h50'30" 31d52'37"
Raymond Carter Davis
Gem 6h30'24" 15d18'7"
Raymond Charles Rowe 19th Jan 1954
Cru 12h17'56" -58d8'48"
Raymond Christopher Angie
Cap 21h32'31" -14d33'17"
Raymond Clarence Johnson
Uma 11h28'13" 63d30'49"
Raymond Clifton Grove
Per 3h50'21" 48d52'36"
Raymond & Colleen
Cyg 20h28'15" 36d24'50"
Raymond Cooper, Sr.
Uma 11h26'20" 35d7'10"
Raymond Curtis Luger
Uma 11h17'46" 60d10'15"
Raymond D.
Uma 10h54'15" 68d22'15"
Raymond D Henderson
Uma 10h28'19" 61d6'34"
Raymond D. Slavin
Cap 20h28'33" -9d45'17"
Raymond Danial(eternal light)Nunez
Cep 20h50'19" 56d30'37"
Raymond Davenport
Uma 9h50'11" 67d39'37"
Raymond Dean
Aql 19h25'21" -11d30'1"
Raymond Dean Manke
Ori 5h53'52" 22d20'49"
Raymond Dean Manke
Ari 1h58'9" 13d36'54"
Raymond Dean Rose
Aur 5h28'59" 43d13'49"
Raymond & Debbie Bluff Forever
Lyn 7h12'14" 58d46'0"
Raymond deFiesta
Leo 9h24'32" 14d53'26"
Raymond Dennis Rodriquez
Uma 9h4'40" 61d29'6"
Raymond Déry
Psc 23h23'11" 6d3'18"
Raymond Donald
Per 3h18'54" 52d44'22"
Raymond Drew Lewallen
Umi 15h14'53" 68d39'35"
Raymond E. "Ram" Wheeler
Her 17h26'16" 32d31'20"
Raymond E. Roswell
Ori 4h52'23" 14d42'57"
Raymond E. Worsdale
Cep 22h31'46" 62d49'37"
Raymond E. Withelder
Vir 12h20'45" 12d34'16"
Raymond Eaves
Umi 13h29'14" 75d13'40"
Raymond Ebacher
Uma 9h42'26" 70d35'20"
Raymond Edd
Aur 6h23'48" 48d27'35"
Raymond Edward Streib
Uma 8h32'14" 66d44'14"
Raymond Edwin Kempf
Aql 19h49'19" -0d43'39"
Raymond Emery Weaver II
Her 17h19'7" 32d19'14"
Raymond Ernest Fyler II
Gem 6h35'4" 22d32'19"
Raymond Everlet
Leo 11h11'12" 26d36'12"
Raymond F Strahley Jr
Aqr 22h22'31" -0d25'3"
Raymond F. Swift
Uma 13h41'11" 50d13'20"
Raymond Ferrari
Tau 4h28'54" 11d54'32"
Raymond Fox
Psc 23h3'37" -0d36'50"
Raymond Francis Bower
Lib 14h52'19" -4d57'58"
Raymond Frank Velasquez
Sco 17h58'19" -30d9'6"
Raymond G. Canfield
Uma 10h57'52" 49d10'35"
Raymond G. Dom III
Cep 21h34'59" 76d1'2"
Raymond Geiger
Uma 10h38'57" 61d9'0"
Raymond George Michna
Ari 2h40'9" 29d0'47"
Raymond George Steele
Cyg 21h55'31" 51d20'36"
Raymond Gimmi
Ori 5h19'16" 0d4'54"
Raymond Golden
Psc 1h33'42" 9d41'27"
Raymond Grove Noel
Aur 5h10'18" 46d58'41"
Raymond Gruman 7/1/1965-12/24/2000
Ori 5h49'14" 6d33'58"

Raymond Guy
Cap 21h40'21" -15d48'22"
Raymond Guzman III
Her 18h57'6" 23d32'26"
Raymond H. Bayer III
Lib 14h50'27" -4d34'54"
Raymond Hanrahan
Peg 21h44'45" 24d26'58"
Raymond Hassell
Cep 21h17'7" 66d12'54"
Raymond Hay
Tau 3h25'47" -0d53'47"
Raymond Henry Hendrix
Per 2h12'0" 57d24'52"
Raymond J. Mazzella
Per 3h2'11" 41d11'8"
Raymond J. McNulty
Per 3h15'20" 47d5'2"
Raymond J. Moore
Uma 10h16'34" 61d8'34"
Raymond J. Vreatt
Aqr 22h3'19" -1d56'24"
Raymond Jay Olson
Uma 10h27'57" 61d52'21"
Raymond & Jennifer Renna
Uma 8h37'44" 62d38'48"
Raymond John Fanelli
Umi 15h53'14" 78d22'12"
Raymond Johnson
Uma 8h37'12" 63d37'56"
Raymond Joseph Friend
Psc 1h3'28" 3d41'38"
Raymond Joseph Harrington
Cep 22h49'27" 72d4'20"
Raymond Joseph Miller, Jr.
Sgr 17h45'14" -17d35'2"
Raymond Joseph Panteney
Umi 16h2'47" 73d30'28"
Raymond Joseph St. George
Oph 17h33'58" -22d33'46"
Raymond J.T. Boyles - Lone Star
Lyn 8h11'28" 35d13'51"
Raymond & Juanita Green
Sge 19h28'55" 18d17'17"
Raymond Kaczmarek
Ori 5h50'58" -0d5'32"
Raymond Keeton
Cep 20h54'27" 60d45'44"
Raymond Kenneth Swanson
Per 4h1'2" 45d27'9"
Raymond Kenneth Swanson
And 1h29'42" 47d47'48"
Raymond L. Eades
Her 17h39'16" 17d14'26"
Raymond L. Kaitfors
Sco 16h17'20" -12d19'16"
Raymond L. Perez
Sco 16h10'49" -17d10'37"
Raymond Lawson
Gem 6h27'5" 22d3'17"
Raymond Lea Cook, Big 50
Uma 11h39'8" 60d51'19"
Raymond Lee
Aqr 20h53'38" 2d3'25"
Raymond Lee Gordon
Cep 20h51'13" 62d2'37"
Raymond Levi Starkey
Aur 5h46'27" 49d47'5"
Raymond Lewis Ezell
Cmi 7h31'0" 4d0'10"
Raymond Lindeen
Cyg 21h24'30" 31d11'44"
Raymond & Louise Belanger
Cep 22h13'59" 73d41'6"
Raymond M Farrow Jr.
Aqr 22h28'7" -17d49'43"
Raymond M. Peslar III
Ari 2h11'25" 23d56'23"
Raymond Madigan Sr.
Cap 20h26'1" -24d37'48"
Raymond Madison Wilkins
Ori 5h0'19" -1d19'17"
Raymond & Marie Erickson
Cyg 19h28'40" 48d26'19"
Raymond Marrero
Lyn 8h8'32" 49d0'15"
Raymond Martin Chemnick
Sco 17h1'40" -36d57'23"
Raymond Matta
Vir 11h42'21" 9d55'28"
Raymond Matthew Dickscheid
Gem 7h44'13" 32d30'36"
Raymond Mazza
Equ 21h8'9" 12d2'35"
Raymond Michael Trindade Jr.
Per 3h19'17" 48d21'17"
Raymond Mikesell
Leo 10h21'2" 25d5'57"
Raymond Miles Baker
Gem 6h27'29" 20d36'12"
Raymond Miller
Uma 10h44'35" 56d57'22"
Raymond Mills
Boo 14h38'0" 16d53'39"
Raymond Mullins, Best Brother Ever!
Aqr 22h8'4" -21d50'28"

Raymond - My Shining Star
Cnc 8h30'45" 28d24'2"
Raymond N. McEckron, Sr.
Ari 3h18'39" 29d5'6"
Raymond Noll Keith
Ori 6h16'42" 18d20'7"
Raymond Oneill Wallace
Leo 9h54'55" 24d2'16"
Raymond Pagán Hernández
Pyx 8h45'23" -28d35'34"
Raymond & Pamela Forslund
Uma 10h58'25" 46d6'8"
Raymond "Papa" Clements
Cep 21h16'0" 59d44'8"
Raymond Perez
Cap 20h37'25" -26d51'48"
Raymond Peterson
Umi 14h31'14" 75d26'43"
Raymond Phillip Langevin
Ari 2h20'58" 17d41'38"
Raymond Reed
Her 17h23'36" 35d56'47"
Raymond Richard Osa Memorial Star
Aql 19h48'26" -0d53'15"
Raymond Robertson
Leo 11h47'56" 16d58'58"
Raymond Rodney Baer (Papa Baer)
Sgr 18h56'53" -23d58'34"
Raymond & Rosetta 50th Anniversary
Eri 4h21'54" -31d54'38"
Raymond S. Korzen "Dad"
Her 17h13'19" 22d7'56"
Raymond Santiago
Boo 14h36'11" 30d49'25"
Raymond Scott Gribben 16111966
Ori 5h42'14" -0d10'52"
Raymond Sean Finley
Cep 22h20'22" 73d7'11"
Raymond T. Sporar, Jr.
Her 16h55'10" 19d53'59"
Raymond Tanguay
Leo 10h5'41" 21d49'20"
Raymond Tashjian
Cyg 19h50'51" 31d54'15"
Raymond (The Bear) Swartz
Uma 10h17'15" 64d2'49"
Raymond V. Trotta
Psc 1h41'39" 10d32'57"
Raymond Vigil
Uma 11h52'37" 61d32'42"
Raymond Vincent Moser
Umi 15h35'43" 77d41'30"
Raymond W. Downs
Psc 1h34'15" 15d55'17"
Raymond W. Guck
Sgr 19h25'1" -14d15'58"
Raymond W. Hutchinson Sr.
Gem 6h53'38" 22d19'10"
Raymond W. Kabuss
Sgr 18h4'8" -25d41'56"
Raymond Willett
Cep 23h2'52" 70d11'0"
Raymond William Eich
Uma 9h49'2" 54d36'24"
Raymond Wolfe "My Eternal Love"
Cap 21h23'58" -25d30'56"
Raymond Zane, Sr.
Cep 20h48'16" 60d39'5"
Raymond Zarro
Vir 12h35'12" 9d8'22"
Raymond "Zeke" LeClair
Uma 10h45'27" 65d4'10"
Raymond Zeltner
Vir 13h19'5" 6d6'6"
Raymond, Michael, Douglas
Tri 2h3'15" 33d53'16"
RaymondLRosell
Aqr 22h31'10" -1d48'53"
Raymond's Star
Uma 9h1'43" 58d7'59"
Raymund
Tau 5h51'32" 13d39'5"
Rayn
Leo 10h13'22" 14d4'36"
Rayna
Aqr 22h14'46" 2d24'13"
Rayna
And 0h16'18" 27d21'6"
Rayna
Lyn 8h13'5" 57d17'48"
Rayna
Uma 13h11'45" 55d32'10"
Rayna and Paul Love Endures Star
Sgr 19h58'43" -16d8'39"
Rayna Beth Legg
Aql 19h1'12" -0d22'26"
Rayna Crespo-Butler
Gem 7h22'18" 28d4'41"
Rayna Jeanne
And 23h21'4" 44d49'33"
Rayna & Jen's Shining Star
Cyg 19h47'50" 34d57'14"
Rayna Marie Kendall
Peg 21h41'3" 7d14'27"

Rayna Marie Sabella "our sunshine"
Peg 22h23'8" 33d23'5"
Rayna Solomon
Cam 5h10'7" 64d16'34"
Raynald Paquette
Aql 19h19'37" 0d17'40"
Raynard Delatori
Eri 3h28'36" -21d12'49"
Rayne Zachary Sealy
Cap 21h30'56" -14d59'6"
Raynelle Marie Gipson
Tau 5h10'26" 21d59'7"
Raynelle Slaten Star
And 1h28'25" 44d42'51"
Raynette Murrell-Knight
And 23h1'9" 52d1'2"
Raynor
Uma 10h9'13" 45d34'50"
Raynox
Leo 10h17'31" 26d20'29"
Rayo Loves Flecha
Cyg 19h48'33" 43d27'58"
Rayon de Roussel
Ori 6h18'7" 14d47'28"
Rayong Sangsod
Aur 6h3'27" 36d30'54"
Rayoush
Uma 8h52'47" 51d0'58"
Rayray
Leo 10h47'38" 18d45'58"
RaY-RaY
Lib 15h6'45" -5d51'16"
rayray
Aqr 22h14'13" -2d16'34"
Ray-Ray
Ind 21h16'57" -51d45'41"
Ray's Ardent Eyes
Her 16h36'27" 38d15'58"
Rays of Love
Ori 5h37'11" -0d31'52"
Ray's Star
Cep 22h19'15" 72d46'26"
Ray's Star
Lyn 6h49'41" 50d50'56"
Rayveion Zeaphyr-Cambraile Mann
Aqr 22h30'40" -3d11'42"
Raywood's Starlite Star Brite
Lyn 9h40'35" 41d4'15"
Rayya Jawhar
Cyg 20h26'0" 53d23'46"
Rayya, World's Best Organizer
Uma 10h41'4" 68d45'55"
Rayza Martinez
Sgr 18h42'37" -21d59'20"
Raz
Vir 13h48'24" -4d49'47"
RAZ
Uma 10h38'8" 65d54'17"
Razan Moghraby
Boo 14h39'27" 40d34'25"
Razia Shaikh
And 0h40'34" 43d16'49"
Râziel
Gem 6h5'41" 24d5'49"
Razimky
And 0h9'43" 39d25'41"
Raziya
Lib 15h8'6" -26d44'27"
razmaroo
Cas 23h42'48" 58d20'53"
Razmi
Mon 7h33'7" -0d21'56"
Razor
Vir 14h13'10" -10d16'56"
Razor
Uma 11h43'26" 45d36'25"
Razor D. Rockefeller
Lmi 10h31'4" 37d19'48"
Razz
Peg 22h40'4" 25d21'23"
RB and Teen Hon
Eri 4h20'33" -8d14'23"
RBG-Chromatographer1
Uma 9h8'42" 55d36'6"
RBH+KHH.12.12.1981
Cyg 21h11'48" 32d9'21"
RBK Love4Ever
Ori 5h21'23" -6d52'56"
RBS - 1
Ori 5h51'5" 6d22'33"
RBS - 2
Peg 22h38'0" 21d25'48"
RBS - 3
Dra 14h34'29" 56d32'25"
RBS - 4
Eri 3h51'9" -37d9'16"
RBS - 5
Aql 19h12'20" -7d16'29"
RBS (Robert, Barbara, Samantha)
Uma 12h41'1" 57d50'45"
RC Leah
Uma 10h46'56" 69d52'57"
R.C. Willis
Uma 9h29'37" 44d57'39"
R.C. Woodward
Uma 12h22'14" 60d6'54"
RC, Son of Mary Heavenly Star
Psc 2h1'41" 9d52'6"

RCJCB Taira
Lyn 8h4'55" 42d10'3"
RCL-570
Oph 17h3'22" -0d8'57"
RCM & JLM
Cyg 19h42'42" 47d34'34"
RC's Shooting Star
Cas 0h40'41" 69d46'30"
RCS13
Lib 14h52'52" -1d28'39"
RCV 1
Lyn 6h48'20" 54d4'1"
R.D. Ivey
Uma 9h30'58" 51d3'34"
R.D. Nordstrom
Lyn 6h43'9" 60d7'19"
R.D. Qualls
Ori 5h38'58" 5d7'2"
RD-90
Ori 5h55'43" 6d6'7"
RDBII-080246
Cyg 19h44'49" 33d23'7"
R.deC.
Aur 6h4'46" 52d33'29"
rdemo53
Leo 10h23'56" 24d31'13"
RDF the one that shines out !
Lib 15h18'39" -27d25'46"
RDG 082005
Per 2h21'36" 56d57'8"
RDJSH Astonishing Father & Husband
Her 16h33'14" 6d5'34"
Rdy
Cap 20h29'16" -14d36'19"
RE 18
Cyg 20h26'51" 32d21'8"
RE/MAX Realty One
Tri 2h9'34" 35d13'55"
Re Re
Uma 11h44'44" 51d55'59"
Rea Simpson Van Fosson
Uma 11h26'8" 46d45'5"
Rea Walshe
Cas 23h11'10" 58d41'3"
Readd-McSpadden's Camp
Uma 8h45'27" 64d11'17"
Reade Elinor Fenner
Lyr 18h50'1" 34d27'10"
Read's Star
Ori 5h46'39" 2d36'5"
Readyeddy
Cnc 8h52'42" 32d29'32"
Reagan
Cnc 9h2'15" 10d14'29"
Reagan
Mon 6h49'27" -0d51'0"
Reagan Alysse Norwood
Lib 14h53'3" -13d58'8"
Reagan Corinne 091799
Ori 6h20'49" 5d52'23"
Reagan Elise Scissell
And 0h48'40" 44d6'50"
Reagan Elizabeth Basham
Psc 23h9'7" 5d53'40"
Reagan Elizabeth Beardsley
Vir 13h15'44" -21d5'25"
Reagan Elizabeth Kennedy
Psc 1h12'51" 14d23'50"
Reagan Elizabeth Lyman
And 1h51'12" 41d35'2"
Reagan Emily
Vir 12h9'56" -8d22'12"
Reagan Faith James
Cas 23h21'1" 55d8'14"
Reagan Helena
Cap 20h10'17" -10d34'15"
Reagan I. McIntosh
Dra 17h57'5" 64d58'25"
Reagan Irene Martzloff
Uma 12h15'20" 61d18'3"
Reagan J. Gervais
Ori 5h6'41" 4d30'26"
Reagan Katz
Lyn 8h29'25" 57d51'54"
Reagan Kiley
And 0h2'57" 32d6'43"
Reagan Louise Wille
And 2h34'33" 45d8'18"
Reagan Lynn Nutter
Del 20h28'25" 18d34'28"
Reagan Makenna Davis
Ari 3h14'25" 19d47'11"
Reagan Marie
Cyg 21h6'53" 30d19'54"
Reagan Michelle
And 23h40'50" 43d40'20"
Reagan Sasser
Uma 14h11'43" 56d30'0"
Reagan True Davis
Gem 6h49'0" 33d7'14"
Reagan Walker Crann
Her 16h29'21" 49d8'10"
Reagan Wille
Cep 20h48'55" 60d36'51"
Reagan's Light
Uma 11h32'8" 48d5'25"
Reagan's Twilight
Tau 5h59'35" 23d15'46"
Reaghan Vivian Ridge
And 23h54'25" 42d22'20"
Reah Kapoor
Cyg 21h34'41" 36d11'3"

Reah's Ray of Light !
Lib 15h31'25" -26d41'36"
R.E.A.L
Uma 11h31'4" 48d3'21"
Réal Circé
And 1h3'34" 41d55'2"
Real friends are forever
Umi 15h7'47" 68d38'8"
Realt Aedan
Pho 1h5'17" -42d54'53"
Réalt Rebecca Ryan
And 1h18'25" 34d44'19"
Réalta Bhreandáin
Lyr 18h37'23" 27d2'29"
Realta Isobel Solan
And 23h0'29" 45d52'26"
réaltín thaibhseach Aisling
And 0h7'57" 40d55'45"
Realtor
Cas 1h50'58" 65d22'47"
Reanna Cappelle
Cnc 9h3'39" 29d3'7"
Reanna Leigh Rusling-Powell
And 2h33'48" 41d3'12"
Reanna Lynn Barbey
Leo 11h23'55" 21d46'11"
Reanna Rowe
Psc 1h18'26" 15d53'41"
Reannah Nicole Villafana
Gem 6h44'12" 15d56'34"
Reannyn
Leo 10h9'33" 17d17'58"
Rear Admiral Erich Topp
Umi 20h20'56" 89d29'5"
Reasie Ballinger
Uma 11h56'51" 50d44'5"
Reason
Uma 8h17'0" 61d21'9"
Reatha Simon
Lyn 6h59'36" 50d33'39"
Reavis
Cma 6h35'30" -13d24'35"
R.E.B
Peg 23h10'52" 19d10'32"
Reba Carolyn Noski
Ari 2h9'43" 16d1'44"
Reba Faye Garnes
Cas 0h38'13" 57d57'4"
Reba H. Larson
Lyr 18h38'1" 31d2'17"
Reba Hunter
Uma 14h27'36" 55d33'25"
Reba Lee Huntly
Cnv 13h44'10" 34d24'49"
Reba Wilkinsky-90
Cas 1h5'58" 61d26'9"
Reba's Glow
Crb 15h32'20" 27d5'59"
Rebbecca Evans
Uma 13h35'2" 56d17'28"
Rebbeclathor.
Tel 19h16'20" -50d41'26"
Rebeca Alba Pareja
Ori 6h11'18" 20d59'52"
Rebeca Cruz-Alvarez
And 2h20'20" 45d34'2"
Rebeca de Almeida Pezzolo Clarke
And 2h5'29" 46d20'28"
Rebeca Diez Alarcia
Peg 22h22'47" 24d21'4"
Rebeca Eugenia Ponce Wallace
Uma 9h5'0" 65d53'44"
REBECA GIL SANCHEZ
And 23h16'42" 47d55'25"
Rebeca Gomez Galindo
Gem 6h31'5" 25d38'48"
Rebeca Pinet
Peg 21h41'54" 23d24'27"
Rebeca Valenzuela de Martinez
Cyg 19h52'36" 37d46'35"
Rebecah Kirk
Uma 9h21'14" 56d10'15"
Rebecca
Lyn 7h55'29" 58d7'12"
Rebecca
Cas 0h12'14" 63d13'0"
Rebecca
Cas 1h47'13" 65d58'56"
Rebecca
Lib 15h12'24" -12d44'58"
Rebecca
Cap 20h31'40" -10d23'46"
Rebecca
Sgr 19h16'5" -16d52'56"
Rebecca
Aqr 22h36'24" -15d54'44"
Rebecca
Cru 12h56'26" -64d39'47"
Rebecca
Lib 15h9'56" -27d39'38"
Rebecca
Sgr 18h1'6" -29d13'9"
Rebecca
Crb 16h11'58" 38d57'47"
Rebecca
And 23h45'28" 44d29'27"
Rebecca
And 23h26'9" 38d53'39"
Rebecca
And 2h22'55" 49d4'57"
Rebecca
Uma 10h15'55" 45d22'21"
Rebecca
And 1h3'53" 33d52'30"
Rebecca
And 1h28'14" 37d15'38"
rebecca
Per 3h16'40" 41d31'52"
Rebecca
Lmi 10h36'57" 38d5'55"
Rebecca
And 0h31'3" 38d0'44"
Rebecca
And 1h24'48" 42d45'40"
Rebecca
Gem 6h2'42" 23d57'57"
ReBecca
Gem 6h33'44" 17d58'52"
Rebecca
Ori 6h0'42" 18d9'26"
Rebecca
Gem 7h15'55" 16d12'55"
Rebecca
Del 20h30'58" 16d33'1"
Rebecca
And 0h57'38" 23d47'3"
Rebecca
Psc 1h30'42" 18d19'37"
Rebecca
Tau 4h35'3" 19d43'20"
Rebecca
Cnc 8h5'17" 14d32'54"
Rebecca
Mon 6h40'5" 7d47'17"
Rebecca
Ori 6h7'30" 13d17'31"
Rebecca
Her 16h30'39" 13d44'38"
Rebecca
Tau 4h41'11" 5d39'57"
Rebecca
Psc 1h45'20" 13d10'49"
Rebecca
Ari 3h21'12" 14d47'26"
Rebecca A. Aguilar
And 0h39'16" 42d21'21"
Rebecca A. (Craven) Marino
Cas 0h28'24" 61d24'18"
Rebecca Adam
Aql 19h34'13" 10d41'24"
Rebecca Aijian Hughes
Lyn 8h28'27" 57d53'3"
Rebecca Aimee Davies
And 1h15'34" 49d43'7"
Rebecca Alandt
Tau 4h17'37" 7d59'35"
Rebecca Alice
Gem 6h32'55" 24d56'23"
Rebecca Alice Cross
And 23h23'56" 48d32'38"
Rebecca Alison Smith
Ari 2h49'49" 12d49'5"
Rebecca Alma Kalish
Lib 15h20'42" -11d26'57"
Rebecca - Always & Everything
Cas 0h40'16" 50d56'12"
Rebecca Amato
Sgr 19h37'9" -37d35'10"
Rebecca and Daniel Randall
Ori 6h4'12" 10d13'7"
Rebecca and Gladys
Lmi 10h45'38" 38d36'56"
Rebecca and Keith Forever
Cyg 20h23'6" 55d30'54"
Rebecca and Kieran
Cyg 20h12'4" 38d55'0"
Rebecca and Robert Forever
Cyg 21h43'56" 47d48'9"
Rebecca and Shane
Cyg 19h46'22" 35d9'56"
Rebecca and Steven Wahl
Cnv 12h55'4" 40d38'48"
Rebecca Anderson
And 23h22'49" 52d1'19"
Rebecca & Andrew Lake
Uma 9h22'11" 53d7'28"
Rebecca Ann
And 23h37'46" 48d13'23"
Rebecca Ann
Tau 5h46'43" 27d31'1"
Rebecca Ann Bishop
Gem 7h18'43" 19d41'18"
Rebecca Ann Carter
Cap 20h13'25" -10d38'44"
Rebecca Ann Cramer
Psc 23h58'29" 0d49'17"
Rebecca Ann Daoud
Sgr 18h25'3" -16d1'37"
Rebecca Ann Dincans
Lyr 19h23'2" 37d53'5"
Rebecca Ann Floyd
Cnc 8h47'16" 8d59'48"
Rebecca Ann Hankinson
Cap 21h18'33" -24d6'39"
Rebecca Ann Leonard
Cnv 12h30'13" 38d9'19"
Rebecca Ann Meyer
Ori 6h16'1" 13d54'22"
Rebecca Ann Pinedo
Cas 23h53'54" 53d59'15"
Rebecca Ann Pittuck
And 1h12'14" 42d22'36"
Rebecca Ann Ridgeway
Mon 7h17'49" -9d18'19"
Rebecca Ann Roark
Cas 0h58'56" 62d16'13"
Rebecca Ann Schwartz
Uma 10h21'46" 51d44'45"
Rebecca Ann Staten
Sgr 19h1'5" -34d4'59"
Rebecca Ann Tukel
And 0h9'54" 30d28'10"
Rebecca Anne
Aqr 22h29'45" -22d34'3"
Rebecca Anne Christensen
Lyr 18h43'23" 35d49'3"
Rebecca Anne Colbert
And 1h37'12" 45d44'59"
Rebecca Anne James
Aqr 22h14'34" 2d15'14"
Rebecca Anne May
And 0h40'45" 43d10'26"
Rebecca Arnett
Cnc 8h12'46" 7d18'24"
Rebecca Aron
Vir 12h21'20" 13d9'29"
Rebecca Ashley Rosenblatt
Tau 4h48'0" 29d19'16"
Rebecca "Babe" Rosenthal
Lyn 8h31'23" 33d34'2"
Rebecca babygirl Bernas
Del 20h31'7" 18d21'5"
Rebecca (Beccy) Schoof
And 23h21'44" 47d30'26"
Rebecca "Becky" Garcia 4ever2gether
Umi 15h16'8" 71d32'43"
Rebecca "Becky" O'Dell
Psc 23h53'21" 0d12'15"
Rebecca "Becky" Tyndall
And 2h20'18" 48d45'24"
Rebecca Behar
Tau 4h17'31" 14d39'13"
Rebecca Benny Verette
Psc 0h50'35" 14d20'3"
Rebecca Bighill
Tau 4h3'55" 23d12'24"
Rebecca Blackwell
Crb 15h58'51" 34d49'22"
Rebecca Bondar
Lyn 6h37'50" 56d19'53"
REBECCA BOO PAYNE
Cyg 20h49'35" 48d43'10"
rebecca BOO pearcy
Psc 0h37'41" 8d20'44"
Rebecca/Bookies
Cyg 20h27'1" 56d43'54"
Rebecca "Bridge-Builder" Main
Eri 2h39'17" -40d10'31"
Rebecca Briles
Sgr 18h34'37" -24d41'36"
Rebecca Büchele, 06.05.2003
Uma 10h55'1" 33d59'37"
Rebecca C. Dumont. (Shooting Star)
Dra 19h13'2" 60d54'59"
Rebecca C. Schultz
And 0h36'11" 32d25'11"
Rebecca C. Scott
And 1h36'22" 44d51'4"
Rebecca C. Smith
Vir 14h8'42" -12d3'26"
Rebecca C. Williams
And 0h47'16" 31d24'44"
Rebecca Camille O'Mara
And 1h45'17" 45d48'36"
Rebecca Caroline Chancey
Aqr 22h57'14" -6d47'38"
Rebecca Caroline Tomlinson
And 0h27'14" 46d30'54"
Rebecca Catherine Kistner
And 0h46'16" 45d46'41"
Rebecca Catherine Logue
Tau 4h26'59" 19d17'55"
Rebecca Catherine Sorensen
Ori 5h37'43" 7d18'16"
Rebecca Caulboy
And 0h15'37" 43d33'38"
Rebecca Cecelia Major
Peg 21h29'11" 15d27'1"
Rebecca Cecelia Schwartz
And 1h27'54" 43d37'52"
Rebecca Chadsey
Cap 20h33'29" -16d14'17"
Rebecca & Charles M. Vest, MIT
Umi 19h38'12" 88d30'45"
Rebecca Cliff's Little Red Wagon
Lyn 7h13'48" 55d22'55"
Rebecca Colleen Tucker-Dusseau
And 2h3'49" 41d43'22"
Rebecca Conners
Sco 16h57'23" -38d38'48"
Rebecca Cornett
And 1h12'11" 42d28'13"
Rebecca D. Satterly
And 0h35'33" 44d58'21"
Rebecca Danielle
Lyn 9h10'53" 38d18'45"
Rebecca Dart
Uma 9h1'46" 58d27'49"
Rebecca Davis
And 1h47'48" 43d36'53"
Rebecca Davis Freeh
Cnc 8h38'3" 21d29'2"
Rebecca Dawn
And 23h22'34" 48d17'3"
Rebecca Dawn Clay "01/03/1997"
Cap 21h26'18" -19d35'8"
Rebecca Dawn - Cleeka
Ori 5h11'34" -10d34'24"
Rebecca Dawn Frankosky
And 23h17'40" 44d5'31"
Rebecca Dawn Gulley
Her 17h54'21" 14d46'32"
Rebecca Dawn Nemeth
And 1h33'31" 41d53'29"
Rebecca Deal
And 23h46'59" 46d3'0"
Rebecca Denae Schall
Psc 23h14'26" 1d21'53"
Rebecca Denise Queen
And 0h43'1" 36d27'51"
Rebecca Dettman
Col 5h59'56" -34d49'28"
Rebecca & Dewayne Meeks
Cyg 19h42'50" 31d21'18"
Rebecca DiCamillo
Cyg 21h51'13" 39d27'45"
Rebecca Dunne
Ori 5h36'15" -2d12'59"
Rebecca Duvall
Ari 2h34'35" 21d33'32"
Rebecca E. Meyers
Uma 12h8'17" 59d56'19"
Rebecca East
Del 20h35'2" 7d10'42"
Rebecca Edith Ham King
Cap 21h40'55" -23d58'43"
Rebecca Eli
Cnc 8h8'46" 15d59'17"
Rebecca Elizabeth Eckstein
Per 1h43'48" 50d43'34"
Rebecca Elizabeth Harvath
Tau 4h39'14" 6d4'22"
Rebecca Elizabeth Hopkinson
Cas 23h16'4" 58d42'53"
Rebecca Elizabeth Rodway
Uma 11h37'53" 53d7'39"
Rebecca Ellen
Uma 11h50'1" 33d46'44"
Rebecca Ellen Sims
Psc 1h4'25" 32d58'30"
Rebecca Eller
Mon 8h2'7" -4d16'28"
Rebecca Elliot Skipper
Aqr 22h8'48" -1d14'20"
Rebecca Emily Mikes
And 22h57'54" 35d13'10"
Rebecca Estelle Macy
And 1h42'22" 45d56'2"
Rebecca Esty
And 1h30'21" 50d17'35"
Rebecca Evans
Lyr 19h18'45" 28d27'52"
Rebecca Fahringer
Per 3h12'28" 43d21'55"
Rebecca Fairall
Uma 13h38'51" 48d28'13"
Rebecca Faye Collyer
Gem 7h16'30" 33d54'50"
Rebecca Flanigan Matt's Angel
Cyg 20h49'9" 45d3'34"
Rebecca Ford
Sco 15h56'27" -24d13'50"
Rebecca Franklin-Porter
Ori 6h24'25" 10d25'12"
Rebecca G. Stvan
Cyg 21h39'8" 35d45'19"
Rebecca Gan Holman
And 23h45'25" 35d5'1"
Rebecca Ganell Riley
Cap 20h18'41" -12d16'20"
Rebecca Garner Star
Cyg 20h4'30" 33d58'41"
Rebecca Gavidia
Crb 16h13'0" 34d47'27"
Rebecca Gayle
Psc 0h47'47" 14d43'50"
Rebecca & Gerry Mckay
Lyn 7h36'36" 38d32'43"
Rebecca Goldfarb
Hor 4h16'16" -47d39'24"
Rebecca Gontko
Cap 20h18'40" -18d52'41"
Rebecca Goodman
Cap 21h16'50" -17d31'42"
Rebecca Grace Fones
Lyn 8h6'28" 40d52'7"
Rebecca Grace Hadsell
Uma 11h52'13" 45d3'57"
Rebecca Grace Horn
Ari 3h8'53" 27d28'16"
Rebecca Grace Sanchez
Psc 1h0'26" 16d50'8"
Rebecca Gracelynn
Psc 1h28'16" 13d1'4"
Rebecca Greene
And 0h16'33" 33d37'17"
Rebecca Hampton
Cap 21h23'58" -20d43'22"
Rebecca Harlow
And 2h6'16" 46d14'41"
Rebecca Harmon
Sco 16h57'14" -41d53'36"
Rebecca Harmony Connolly
Lib 14h46'52" -12d47'35"
Rebecca Hart
And 23h5'44" 49d57'0"
Rebecca Hermiz
Cam 6h45'27" 63d43'10"
Rebecca Hicks
And 0h22'30" 25d38'8"
Rebecca Himes
Aqr 22h18'24" 1d0'14"
Rebecca Holub Parham
Leo 11h19'7" 15d22'29"
Rebecca Hood
And 1h9'17" 45d20'21"
Rebecca Hunsinger
Psc 1h7'18" 29d13'32"
Rebecca Hussey
Leo 9h55'48" 27d16'34"
Rebecca Irene Broker
Tau 5h38'25" 27d8'5"
Rebecca "Isabel"
Crb 15h54'23" 28d43'1"
Rebecca J. Denault
Lib 14h42'22" -10d31'1"
Rebecca J. Hozubin
And 2h35'31" 47d40'32"
Rebecca J. Schauer Petrovich
Mon 6h51'5" -0d51'21"
Rebecca Jacobs
Ari 3h12'2" 26d39'53"
Rebecca Jane
Ori 6h13'38" 18d9'56"
Rebecca Jane Geary
And 23h10'58" 50d44'18"
Rebecca Jane O'Brien
And 0h0'29" 46d27'2"
Rebecca Jane Paulos
Crb 15h26'27" 30d16'23"
Rebecca Jane Picard
Cnc 8h8'43" 12d35'32"
Rebecca Jane Quenby
And 22h59'7" 51d43'46"
Rebecca & Jarek
Cyg 21h37'51" 39d46'1"
Rebecca & Jason Kitchens
Tau 5h41'55" 20d58'40"
Rebecca Jayne Dickson - Rebecca's Star
Uma 11h23'19" 58d0'11"
Rebecca Jayne Mangan
Eri 4h34'22" -36d27'54"
Rebecca Jean Adair
Umi 14h29'21" 74d42'14"
Rebecca Jean Anderson
Tau 5h39'6" 27d43'3"
Rebecca Jean Bernstein
Lib 15h43'57" -11d45'35"
Rebecca Jean Carlson
Psc 1h31'48" 10d2'54"
Rebecca Jean Grant
Vir 13h18'7" -11d20'50"
Rebecca Jean Hannibal
And 1h47'5" 41d39'8"
Rebecca Jean Stimson
And 1h0'10" 44d1'31"
Rebecca Jean Wharton
Uma 12h48'44" 60d20'25"
Rebecca Jeanne
Ori 5h42'53" 2d8'29"
Rebecca Jo
Lyr 18h29'38" 37d19'12"
Rebecca Jo Campbell
And 23h29'52" 46d53'29"
Rebecca Joann Henry
Crb 15h37'50" 26d54'45"
Rebecca Joe Larson Daniels
Cas 1h36'44" 64d54'0"
Rebecca & Joel
Sgr 19h9'18" -22d23'39"
Rebecca JoHannah Moyes
Ori 5h41'39" 7d12'39"
Rebecca Johnson
Gem 7h2'50" 26d32'41"
Rebecca & Jonathan: Eternal Love
Ori 6h6'7" 5d24'59"
Rebecca Jordan
Cas 0h28'46" 55d5'34"
Rebecca Jordan Kane
Ari 2h58'39" 28d42'38"
Rebecca Joy
Ori 5h2'50" -0d3'56"
Rebecca Joy Hatef
And 1h41'9" 43d15'10"
Rebecca Joy Ortwine
Uma 9h27'19" 41d36'5"
Rebecca Joy Quilala
Crb 15h34'20" 30d36'59"
Rebecca June Pennington
Gem 6h49'5" 23d4'38"
Rebecca K Frurip
And 1h36'18" 39d42'58"
Rebecca K. Nordberg
Tri 2h30'30" 28d37'15"
Rebecca Katheryn Madsen
Uma 11h22'54" 64d24'9"
Rebecca Kathleen Juliana Stone
Leo 10h58'34" 7d14'17"
Rebecca Kathryn Jones
Aqr 22h47'11" -8d3'16"
Rebecca Kay Lambert
Tau 5h55'34" 27d3'47"
Rebecca Keene
Del 20h33'31" 13d24'34"
Rebecca Kelly
And 23h16'32" 49d22'30"
Rebecca Kestenbaum Klafter
Gem 7h33'18" 32d52'4"
Rebecca Kim Phillips
Leo 11h1'8" 16d11'44"
Rebecca Kincade
And 1h7'51" 42d17'12"
Rebecca King Maclean
Sco 16h28'50" -33d0'16"
Rebecca Koshnick
Cet 1h20'31" -0d51'58"
Rebecca Kristin Gary
Aqr 22h30'34" -2d17'9"
Rebecca Kurz
Cap 20h22'14" -26d21'43"
Rebecca L. Bennett
Cas 0h58'16" 59d24'56"
Rebecca L. Crane
Sgr 18h45'45" -16d46'1"
Rebecca L. Cullars
Sgr 18h6'58" -27d32'39"
Rebecca L Lincoln
For 3h32'14" -26d28'13"
Rebecca L. Miller
Cam 3h46'42" 66d10'54"
Rebecca L. Taylor
Cas 0h20'49" 54d35'57"
Rebecca L'Amour de Ma Vie
Leo 10h32'10" 18d56'45"
Rebecca Lane Johnson
Cnc 8h47'42" 30d32'38"
Rebecca Lauren Cole
Vir 14h57'46" 4d41'30"
Rebecca Leann Sturgis
Sgr 18h42'21" -23d32'34"
Rebecca Lee
Vir 13h22'56" 8d15'9"
Rebecca Lee
Leo 9h25'2" 12d48'25"
Rebecca Lee
Crb 15h45'45" 33d4'15"
Rebecca Lee
And 2h17'53" 47d39'23"
Rebecca Lee Parsons
And 0h27'19" 32d51'6"
Rebecca Lee Sanchez
Tau 5h24'22" 18d50'32"
Rebecca Leigh
Cnc 8h12'15" 28d48'51"
Rebecca Leigh Furlan
Crb 15h50'10" 27d49'20"
Rebecca Leigh Hendle
Uma 9h23'55" 46d9'6"
Rebecca LeMay
Lib 14h28'12" -18d49'46"
Rebecca Lin Swanson
Leo 10h56'45" 14d4'30"
Rebecca Lindsay Nelson
Psc 1h13'57" 27d29'8"
Rebecca Linnea Sjölund
Lib 15h44'33" -16d11'57"
Rebecca Linville
Psc 1h17'41" 10d25'39"
Rebecca "little tommy" Melanson
Sco 17h21'54" -41d24'33"
Rebecca Lorraine
And 23h45'48" 45d8'43"
Rebecca Lorraine Garraton
And 22h59'45" 44d56'50"
Rebecca Louise
And 2h37'15" 50d13'20"
Rebecca Louise
And 22h58'49" 51d37'15"
Rebecca Louise
And 1h4'24" 41d57'16"
Rebecca Louise
Peg 22h17'30" 17d13'24"
Rebecca Louise Jayne Hogben
Cas 0h18'1" 56d13'21"
Rebecca Louise Susan Wilkinson
And 0h4'33" 39d55'6"
Rebecca Louise, Lee Thomas, & Jordon
Cyg 21h11'10" 41d24'18"
Rebecca Lowe & Fermin Martin
Cyg 20h46'33" 54d20'55"
Rebecca Lucile Schaeffer
Cam 3h30'34" 60d21'37"
Rebecca Lynn
Cas 0h36'27" 52d9'42"
Rebecca Lynn
And 1h16'1" 45d17'10"
Rebecca Lynn
And 0h4'36" 39d33'30"
Rebecca Lynn
And 0h44'51" 40d23'55"
Rebecca Lynn
And 1h51'20" 37d56'8"
Rebecca Lynn
And 2h13'5" 44d4'14"
Rebecca Lynn
Leo 10h9'14" 24d49'52"
Rebecca Lynn
Tau 5h5'10" 25d40'40"
Rebecca Lynn
Tau 4h34'55" 18d15'49"
Rebecca Lynn Armistead
Tau 4h29'36" 20d53'9"
Rebecca Lynn Baker
Sgr 18h25'30" -26d55'55"
Rebecca Lynn Cossey
Cap 21h8'51" -24d22'56"
Rebecca Lynn Cutler
Uma 9h10'35" 63d7'27"
Rebecca Lynn Davidson
Sco 17h25'3" -32d37'11"
Rebecca Lynn Dempsey
And 1h46'59" 37d52'55"
Rebecca Lynn Donaho
Ari 2h42'22" 15d9'10"
Rebecca Lynn Garthright "9-28-00"
Lyn 8h0'8" 40d37'5"
Rebecca Lynn Graziano
And 2h27'0" 47d43'33"
Rebecca Lynn Groman
Del 20h51'11" 10d8'13"
Rebecca Lynn Headrick
Ori 5h6'6" -0d20'41"
Rebecca Lynn Hyman
And 0h45'27" 33d46'17"
Rebecca Lynn Meszaros
Her 17h49'17" 42d0'1"
Rebecca Lynn Pope
Crb 15h39'30" 25d41'4"
Rebecca Lynn Rust
Lyn 7h41'39" 49d56'28"
Rebecca Lynn S.
Sco 16h43'27" -35d38'34"
Rebecca Lynn Sangwin
Tau 4h18'39" 17d59'23"
Rebecca Lynn Schwam
Sgr 19h7'59" -15d54'46"
Rebecca Lynn Steger
Leo 9h29'59" 15d18'40"
Rebecca Lynn Stone
Tau 4h43'5" 27d14'10"
Rebecca Lynn Symes
Ori 6h14'51" 6d15'30"
Rebecca Lynn Thomas
Psc 0h49'21" 14d13'7"
Rebecca Lynn Vaughn
Cnc 8h43'19" 16d45'30"
Rebecca Lynn Wakenight
Leo 9h27'50" 17d16'25"
Rebecca Lynn Fleming
Cam 4h3'15" 53d47'18"
Rebecca Lynne
Lyn 7h59'59" 49d19'18"
Rebecca Lynne
Cap 20h18'40" -15d49'27"
Rebecca Lynne Otten
Cap 20h59'32" -17d37'43"
Rebecca Lynne Patterson
And 0h55'5" 41d21'2"
Rebecca Lynnette Windle
Cam 3h51'23" 66d48'19"
Rebecca Lynsey Battle
Uma 10h5'12" 64d51'47"
Rebecca M. Carter
Ari 3h23'26" 26d58'48"
Rebecca M Kuzel
Gem 6h40'31" 15d23'35"
Rebecca M Nicholson
Cam 4h15'13" 63d27'52"
Rebecca M. Walsh
And 0h19'33" 46d13'21"
Rebecca Machorro Westra
Aqr 22h28'52" -12d20'21"
Rebecca Maddux Lamb
Aqr 23h5'53" -8d20'43"
Rebecca Margaret Olson
Tau 5h7'18" 18d6'49"
Rebecca Margaret Vlisides
Uma 9h50'8" 56d11'39"
Rebecca Marie
Gem 6h31'20" 24d58'31"
Rebecca Marie
And 23h11'15" 42d40'29"
Rebecca Marie
And 1h44'29" 50d36'9"
Rebecca Marie
And 0h32'21" 41d30'0"
Rebecca Marie Catanzaro
Leo 11h39'9" 17d0'44"
Rebecca Marie Cook
Tau 3h36'5" 24d28'9"
Rebecca Marie Davi
Vir 13h19'0" 6d20'27"
Rebecca Marie Francis McTeague
Uma 11h3'28" 48d40'1"
Rebecca Marie Gale
Lyn 7h9'46" 59d4'58"
Rebecca Marie Gale
Cas 1h45'59" 61d12'12"
Rebecca Marie Lucas
And 0h43'30" 36d18'18"
Rebecca Marie Nevin
And 0h6'46" 40d18'59"
Rebecca Marie Schmidt (Wife)
Cyg 20h50'2" 46d33'55"

Rebecca Marie Walczak
And 1h41'58" 41d50'14"
Rebecca & Mark
Cyg 20h21'3" 54d59'46"
Rebecca Martinez: My love forever
Lmi 10h25'5" 31d24'31"
Rebecca Mary Hazelden
And 23h42'45" 37d11'49"
Rebecca Mary Mayo
Vir 13h36'1" 1d18'23"
Rebecca Mata
Vir 13h20'14" -14d17'45"
Rebecca Maureen Du Bets
Cnc 8h40'29" 7d56'49"
Rebecca Maureen Gordon
Ari 3h9'14" 26d45'35"
Rebecca May
Cas 23h47'19" 57d55'51"
Rebecca McKnire
Cnc 8h34'1" 9d23'42"
Rebecca McSweeney
And 0h15'4" 48d35'10"
Rebecca Mezyk
Vir 13h23'15" -6d2'2"
Rebecca Michelle Wilkins
And 2h2'49" 41d49'1"
Rebecca Minozzi 7.12.2006
Crb 16h23'32" 38d17'17"
Rebecca Morgan Fisk
Sco 16h17'40" -16d33'33"
Rebecca Morris
Cnc 8h30'5" 21d25'28"
Rebecca Muireann II
Ari 2h58'39" 21d36'26"
Rebecca Munoz
Lyn 7h26'10" 56d38'18"
Rebecca 'My Chicken' Serratore
Cru 12h38'6" -58d22'25"
Rebecca My Love
And 2h16'42" 50d10'6"
Rebecca Myriah
Vir 12h42'25" -0d59'36"
Rebecca Nadine Andelin
Leo 10h49'47" 12d33'1"
Rebecca Nadine Geer
Leo 11h36'4" 9d36'19"
Rebecca Napoletano
Crb 15h51'49" 26d30'55"
Rebecca Nicole Kutait
Cas 0h3'12" 53d46'47"
Rebecca Nicoli
And 0h39'35" 38d12'9"
Rebecca Noelle Worthington
And 0h1'38" 45d7'5"
Rebecca Novotny
Dra 16h11'27" 60d38'17"
Rebecca Nussear
Ori 5h52'8" 8d34'59"
Rebecca Nuttall 18
Cyg 21h0'3" 53d19'18"
Rebecca O'Brien
Cas 0h30'15" 60d8'15"
Rebecca Paige
Sgr 18h36'31" -36d12'5"
Rebecca Pamela's Star
And 0h36'41" 45d52'11"
Rebecca Panek
Peg 21h55'28" 33d59'17"
Rebecca Patricia
Lyr 18h57'17" 33d16'0"
Rebecca Pauline
Ari 2h38'36" 25d54'3"
Rebecca Pauls
Cyg 21h37'27" 43d1'30"
Rebecca Peguignot
And 0h31'52" 31d14'51"
Rebecca Perkins
Com 13h9'6" 28d25'5"
Rebecca Perry
Cnc 8h43'46" 31d37'7"
REBECCA PIGLET ROTH
Mon 7h9'18" -0d17'47"
Rebecca Probyn
And 23h35'34" 38d5'9"
Rebecca R. Rhodes
Tau 4h29'26" 22d28'52"
Rebecca Rae
Tau 4h38'26" 20d16'7"
Rebecca Raffo
Lyn 8h59'36" 45d31'51"
Rebecca Raine Law
Per 3h53'17" 36d18'27"
Rebecca Raman
Sco 16h59'38" -41d59'10"
Rebecca Ramirez
Psc 0h40'35" 17d32'53"
Rebecca Randolph
Dra 18h37'5" 51d10'31"
Rebecca Ray
Vir 14h13'3" -12d4'51"
Rebecca Reed
Cas 2h6'6" 59d58'18"
Rebecca & Reese
Umi 16h18'50" 73d25'9"
Rebecca Renee Brown
Lyr 19h9'50" 37d26'59"
Rebecca Renee Godri
Gem 7h8'8" 24d25'20"
Rebecca Revoire
Tau 4h36'55" 23d12'34"
Rebecca Rhodes
Ori 6h7'0" 17d44'2"
Rebecca Rookard aka Tootsiepop
And 1h26'39" 48d19'25"
Rebecca Rose
And 0h12'32" 40d56'59"
Rebecca Rose
Cnc 8h41'23" 15d22'49"
Rebecca Rose
Vul 20h4'55" 29d0'33"
Rebecca Rose
Vir 13h23'59" 11d51'8"
Rebecca Rose
Aqr 22h32'17" 0d12'1"
Rebecca Rose Brisson
Lyn 8h31'58" 57d34'16"
Rebecca Rose Davis
And 1h9'59" 44d16'37"
Rebecca Rose Starr
Vir 11h59'56" -0d32'53"
Rebecca Rose Thornhill
And 23h8'8" 48d20'44"
Rebecca Rose Ward
Uma 11h28'30" 35d37'53"
Rebecca Ross
And 23h35'54" 43d19'27"
Rebecca Rothenberg
Cap 20h29'45" -11d5'56"
Rebecca Ruck
Ari 3h18'24" 15d16'22"
Rebecca Ruiz
Leo 10h18'45" 6d43'8"
Rebecca S. Malinas
Peg 22h27'48" 30d44'52"
Rebecca 'sady' Bostain
Sgr 18h16'45" -22d49'55"
Rebecca Salzhauer's Shining Star
Tau 5h30'36" 27d44'30"
Rebecca Sarwi
And 1h14'23" 41d42'5"
Rebecca Scarlett Costanzo
Tau 3h50'53" 27d6'29"
Rebecca Schwab
Sco 17h49'48" -42d46'36"
Rebecca Scout Cannon
Gem 6h33'9" 25d54'41"
Rebecca Sevigny
And 0h27'4" 33d53'58"
Rebecca Seymann
Lyr 18h36'54" 27d14'30"
Rebecca Shepherd
Lyn 9h11'2" 34d46'4"
Rebecca Sheppard
Leo 10h46'47" 16d43'11"
Rebecca Simpson
Cap 20h26'18" -11d28'4"
Rebecca Sparks
And 0h49'16" 45d12'25"
Rebecca Speckman
Gem 7h32'23" 30d11'53"
Rebecca Spencer
And 2h12'56" 47d31'32"
Rebecca Stewart
Aqr 23h19'44" -18d21'27"
Rebecca Sue
Gem 7h28'9" 18d57'1"
Rebecca Sue Courtney
Ari 2h22'47" 26d20'31"
Rebecca Sue Duncan
Uma 11h12'32" 47d46'43"
Rebecca Sue Rein
And 0h41'52" 44d22'56"
Rebecca Susan Del Greco
Cap 20h53'22" -15d43'57"
Rebecca Susan Moan
Per 4h31'54" 40d39'17"
Rebecca Susanne Yost
Cap 20h23'43" -24d19'17"
Rebecca Suzanne Guarino
Sco 17h21'18" -39d24'30"
Rebecca Suzanne Scott
Ari 2h39'46" 30d24'45"
Rebecca Sweeney
Cra 18h35'47" -45d18'23"
Rebecca Sykes
Uma 10h11'37" 48d53'7"
Rebecca Taos
Cam 7h47'36" 64d18'2"
Rebecca & Taylor
Sgr 18h5'14" -27d12'30"
Rebecca Termine "Becca's Blaze"
Cas 1h42'42" 64d36'56"
Rebecca & Thomas
Gem 7h37'17" 21d58'20"
Rebecca Thorne
And 23h30'6" 42d32'17"
Rebecca & Tiffany Baker
Uma 11h29'33" 46d29'43"
Rebecca Toh Lui Key
Leo 11h10'1" 1d21'31"
Rebecca Tonjes
And 23h50'14" 35d21'32"
Rebecca Townsend
Cap 20h29'17" -9d34'59"
Rebecca Tron
Uma 9h51'30" 59d45'0"
Rebecca Tutewohl
Ari 3h25'39" 22d13'16"
Rebecca Vance - Big Sweetie
Cas 23h20'1" 57d46'29"
Rebecca Victoria 21
And 1h25'11" 34d13'8"
Rebecca Viens
And 0h22'43" 38d12'53"
Rebecca Waters
Lyr 18h53'42" 28d5'35"
Rebecca Wellborn Maury
Ari 2h2'54" 24d19'2"
Rebecca Whaley
Uma 11h26'46" 60d16'17"
Rebecca Wight
Vir 11h39'3" 3d59'42"
Rebecca Wilder
And 2h19'33" 50d1'19"
Rebecca Wilma Egger
Ari 2h41'49" 16d46'15"
Rebecca Wong - Reb's Star
Cas 1h0'57" 49d11'14"
Rebecca Wythe Simmerman
Cas 0h34'54" 53d31'46"
Rebecca Yousef
Cap 20h56'23" -21d10'3"
Rebecca Zacher
Ari 2h43'5" 15d56'40"
Rebecca Zahn
Uma 11h4'41" 46d17'49"
Rebecca Zanero
Aqr 22h26'20" -0d2'59"
Rebecca Zoe Admave-Eberhart
And 0h21'9" 45d18'45"
Rebecca, Mathew and Mikhayla
Gem 6h16'38" 26d43'51"
rebeccaolbrantz
Ari 3h5'47" 28d12'32"
Rebecca-Olivia
Uma 9h43'19" 52d24'3"
Rebecca's Birthday Star
Cas 1h22'8" 56d13'19"
Rebecca's Diamond
Ori 6h14'1" 15d46'51"
Rebecca's Eye
Sgr 18h42'49" -16d33'31"
Rebecca's Flight
Crb 15h41'15" 35d36'31"
Rebecca's Glow
Lib 15h53'20" -9d52'44"
Rebecca's Grandad
Uma 11h27'42" 31d20'57"
Rebecca's Heart
Vir 13h19'30" 10d0'41"
Rebecca's Radient Rays
Lib 14h51'26" -2d52'50"
Rebecca's Rays of Treasured Light
Uma 9h53'26" 45d22'49"
Rebecca's Shining Star
Aqr 23h9'24" -17d2'41"
Rebecca's Smile 143
Cyg 20h47'45" 39d46'41"
Rebecca's Spirit
Vir 14h9'39" -14d51'6"
Rebecca's Star
Umi 16h52'27" 75d9'24"
Rebecca's Star
Per 2h45'46" 54d15'15"
Rebecca's Star
Aqr 22h45'16" 2d17'41"
Rebecca's Star
Cnc 8h29'57" 15d40'44"
Rebecca's Wish Star
Cas 1h36'51" 66d27'49"
Rebecka Early
Cnc 9h1'58" 8d35'23"
Rebecky
And 23h39'31" 40d41'56"
Rebeka Jane
Sgr 19h19'45" -22d2'40"
Rebekah
Lib 14h50'21" -7d10'30"
Rebekah
Tau 4h28'6" 16d13'15"
Rebekah
Psc 0h52'47" 25d16'47"
Rebekah
And 0h21'17" 30d56'17"
Rebekah Ann Stephens
Cyg 19h40'31" 29d16'13"
Rebekah Ashley
And 2h24'53" 49d8'43"
Rebekah Bowe Milsted
Cas 0h49'37" 62d33'49"
Rebekah Brown
And 2h6'42" 39d52'17"
Rebekah Brown 917
Vir 11h44'38" -0d34'21"
Rebekah Buck
Uma 11h34'42" 52d3'17"
Rebekah Capers
Gem 6h26'4" 16d32'42"
Rebekah Colleen Hill
Lmi 10h36'9" 38d56'15"
Rebekah Dawn
And 1h14'48" 49d51'57"
Rebekah Elanie Strickland
Umi 17h2'43" 83d40'14"
Rebekah Elisabeth Schwarz
And 23h1'44" 42d14'33"
Rebekah Elizabeth Abercrombie
Cap 21h31'54" -8d53'53"
Rebekah George
And 0h23'58" 25d5'37"
Rebekah Grace Hernandez
Lyn 7h33'45" 38d44'28"
Rebekah Hagar
Cas 0h12'2" 50d49'10"
Rebekah Henson
Uma 8h55'15" 54d51'58"
Rebekah James
Uma 10h31'33" 68d12'11"
Rebekah Jane Smith
Cyg 20h7'57" 51d56'30"
Rebekah Jasmine-May Kennedy
Leo 9h46'56" 25d12'33"
Rebekah Jasmine-May Kennedy
Leo 10h13'42" 22d47'8"
Rebekah Jean
Uma 11h18'50" 40d30'57"
Rebekah Jean Byerly
And 0h50'57" 42d6'35"
Rebekah JoAnn Lust
Psc 1h25'10" 16d13'17"
Rebekah Joy Garrett
And 23h13'54" 47d54'4"
Rebekah Joyce Wharton
Peg 22h24'32" 7d19'14"
Rebekah Lee
And 0h25'10" 38d24'59"
Rebekah Leigh
Cnc 8h26'11" 13d28'55"
Rebekah Lipsky
Crb 15h52'2" 29d11'12"
Rebekah Lynn Langley
Cnc 8h15'5" 30d4'15"
Rebekah Lynn Smith
Peg 22h49'18" 26d54'36"
Rebekah Mae (my pretty kitty)
Cyg 19h59'44" 36d51'3"
Rebekah & Marc Beneteau
Cyg 19h20'58" 50d53'10"
Rebekah Maroun's Harmony Star
Leo 10h7'40" 8d51'56"
Rebekah McCorvey
And 0h38'13" 33d7'3"
Rebekah Michelle
Tau 4h38'20" 7d53'57"
Rebekah Michelle
Ori 5h54'38" 18d22'26"
Rebekah My Love
And 23h23'15" 42d1'9"
Rebekah Nelsen
Cas 0h42'40" 48d48'15"
Rebekah Whitaker (DUCK)
And 0h18'37" 25d42'15"
Rebekah4
Lib 14h53'53" -11d7'23"
Rebekah's Dawning Light
Cnc 8h0'29" 17d2'21"
Rebekah's Radiance
Cnv 13h55'37" 33d59'23"
Rebekka Betters
Aqr 21h18'9" -7d23'56"
Rebekka - Christin
Ori 6h12'29" 13d50'34"
Rebekka und Gianni "Rund um die Welt"
And 2h1'16" 42d50'24"
Rebekkah Lowe
Uma 11h13'8" 28d21'2"
Rebekkah Sue Chenier
Per 3h47'49" 42d33'9"
Rebel
Cnv 13h42'58" 35d56'56"
Rebel
Tri 2h37'32" 35d9'55"
Rebel
Cnc 8h38'8" 18d45'26"
Rebegah Kay Wind
Vir 14h2'7" -13d56'55"
Rebfeb-1945
Cyg 19h41'55" 33d38'31"
Rebi & Ali
Vir 14h36'40" 5d45'58"
Rebin Sunny
Leo 10h26'57" 12d2'44"
RebnDanFee
Pho 2h15'2" -46d29'52"
Rebrunettie
Leo 11h5'18" 20d34'34"
Rebstar
Cet 2h27'17" -3d5'23"
Rebuca
Uma 10h59'33" 37d42'21"
Recanto Aborior Cor Cordis Pexum
Cyg 21h28'25" 32d24'50"
Rechilda and Jay's Wedding Day Star
Leo 11h36'13" 22d59'37"
Reci
Sco 17h17'6" -44d36'28"
Reckitt Benckiser (M) Sdn Bhd
Ori 5h37'45" 4d3'45"
Recuerde el amor Verdadero
Aqr 22h44'41" -11d27'24"
Réczi Sándor (Emo)
Uma 9h1'48" 62d17'54"
Red Ashley
And 2h32'18" 44d32'2"
Red Baron
Ari 2h45'2" 28d8'56"
Red Beard
Per 4h49'24" 46d27'59"
Red & Dee Colangelo ... Always
Cyg 19h55'16" 31d8'23"
Red Diamond of Sylviane
Uma 10h59'10" 33d24'48"
Red Dog
Cnv 12h39'41" 43d29'55"
Red Dog
Per 3h8'56" 47d7'53"
"Red Dog"
Leo 10h18'51" 17d57'0"
Red Dog
Ori 6h4'40" 21d6'15"
Red Dog
Cma 6h59'57" -30d7'24"
Red Flower
Lib 15h37'58" -15d3'29"
RED from the purple planet
Crb 15h43'38" 26d24'12"
Red Gilson
Her 18h56'44" 24d7'3"
Red Heather
Psc 1h13'18" 17d29'27"
Red Hot Chilli Mama - Pamela Brown
Cru 11h56'29" -56d23'55"
Red Jantzen
Uma 10h19'42" 69d0'44"
Red Masque
Psc 1h30'59" 23d26'49"
Red McManus *Uncle Red*
Lib 14h48'58" -12d24'39"
Red Mountain - 07-13-05
Cyg 21h32'51" 55d14'32"
Red Mountain Spa Healing Star
Uma 10h23'23" 63d50'22"
Red Pony
Gem 6h23'21" 19d56'27"
Red Raider Flint
Lib 15h30'1" -25d45'30"
Red Raven
Cam 4h52'33" 69d3'26"
Red Reese
Ori 6h6'7" 6d15'38"
"Red" Richards
And 2h21'42" 46d8'51"
Red Rose Major
Uma 13h31'52" 57d27'43"
Red Sonja
Sco 16h12'43" -15d7'17"
Red star of hope for humanity
Ori 6h1'35" 13d13'17"
Red Sunshine- Denise S Leech
Gem 7h18'40" 25d16'6"
Red Velvet Jennifer
Ari 3h21'41" 29d30'23"
Reda petit Prince
Cyg 20h50'38" 46d57'3"
Reddy Cabrera
Aqr 20h53'8" 0d36'5"
Rede
Vir 12h43'58" 6d15'16"
Redgy Lee Garver
Leo 9h58'48" 9d32'50"
Red-headed Ranger
Cnv 12h42'29" 41d47'1"
Redican
Ori 6h20'29" -1d52'30"
REdiego
Leo 11h0'40" 17d54'56"
Rediscovered Love
Leo 9h29'36" 32d25'28"
Red-Kap Sales
Cnv 12h47'18" 43d7'4"
Redmond
Uma 11h38'55" 64d0'6"
Redmond R22 IP
Her 18h45'6" 18d49'3"
Redmond Voss
Cma 6h43'39" -18d26'2"
Redneck Rock Star
Gem 6h52'4" 32d15'43"
Rednek Rokstar
Uma 9h3'17" 48d21'29"
Red-n-Jam
Cnc 8h14'45" 16d16'24"
Redpipe
Uma 13h47'54" 52d25'30"
Red's Star
Uma 9h24'39" 56d7'46"
Red's Wish...My Dream...Inevitable.
Gem 7h28'6" 16d17'16"
Redsonian Skye
Ari 2h45'27" 14d33'20"
Redtail
Gru 23h9'42" -55d41'13"
Redtail Ravine
Uma 10h35'54" 49d25'18"
Red-Winged Blackbird - Fly!
Cap 21h47'16" -12d7'35"
Redwood Memorial Hospital
Uma 9h37'31" 54d41'49"
Ree Ree
Vir 14h39'37" -6d55'30"
Ree The Beautiful
Cam 7h31'55" 69d24'50"
Reebok
Aql 19h13'49" -7d39'41"
Reece
Umi 12h27'8" 87d52'56"
Reece Alan O'Brien
Psc 23h49'37" 5d51'2"
Reece and Coleman Robinson
Ari 3h20'23" 24d2'45"
Reece Anthony Ragusa
Crb 15h46'41" 39d13'3"
Reece Bogart Murray
Uma 11h17'17" 31d58'29"
Reece Bradford Gale
Leo 11h20'10" 23d0'2"
Reece Christopher Logan
Uma 12h9'22" 54d13'44"
Reece Danielle
Leo 9h43'44" 30d9'16"
Reece Danielle Staggs
And 0h21'26" 37d58'47"
Reece David Avern
Ori 6h19'20" 7d50'38"
Reece Duckett
Cru 12h22'4" -57d10'53"
Reece James Parker
Ori 5h35'51" -9d12'35"
Reece James Trimmer - Reece's Star
Ori 6h8'26" 6d57'2"
Reece Joey Patrick
Umi 14h54'28" 70d51'1"
Reece Kieran Ellis
Per 3h9'5" 32d46'7"
Reece Lewis
Cap 21h23'22" -14d49'59"
Reece Mason Adams
Peg 21h47'25" 21d44'51"
Reece Patrick James - Roo's Star
Uma 11h24'56" 67d1'27"
Reece Robert FitzPatrick
Gem 7h29'45" 33d13'20"
Reece Robert Stengle
Psc 1h25'35" 27d31'48"
Reece Steve Parr
Umi 14h32'44" 77d42'46"
Reece W. Winton
Ori 6h18'14" 7d17'58"
Reed
Lyn 8h44'59" 34d34'8"
Reed
Lyr 18h40'2" 36d8'42"
Reed Asher Egli
Uma 9h23'33" 41d53'37"
Reed C. Bruegman
Lmi 10h28'36" 36d6'46"
Reed Daniel Christianson
Peg 22h24'26" 11d32'34"
Reed David Teitler "Reedie Pie"
Gem 6h50'38" 32d53'46"
Reed Dinan
Her 16h13'1" 20d17'39"
Reed Electric
Uma 12h27'26" 59d19'0"
Reed Justin Laviolette
Cep 22h40'10" 67d48'7"
Reed Lagerfeld Willig
Ori 5h9'32" 13d1'11"
Reed Leslie
Cam 4h33'37" 73d35'18"
Reed Louis Esposito
Peg 23h29'11" 27d9'26"
Reed Nalu North
Psc 23h4'56" 1d25'17"
Reed & Ryan Weber
Lyr 18h44'28" 40d6'40"
Reed Tucker Grogan
Ari 2h23'43" 11d46'0"
Reed William Lawhorne
Vir 13h37'38" -10d46'47"
REEDA-53
Psc 1h30'44" 21d32'37"
Reeder o6 Addison Chase Reeder
Aqr 22h45'46" -12d0'42"
Reed's Radical Realm
Cnc 8h36'56" 31d9'22"
Reef Charles Billingham
Umi 16h7'14" 81d34'29"
Reega Beega Conlin
Vir 14h29'42" 1d58'54"
Reelit
Cyg 21h21'13" 36d52'9"
Reem
Lyr 18h19'28" 41d27'5"
Reem Lahham
Oph 17h40'13" -0d9'51"
Reem The Mother of Love
Tau 3h40'40" 27d13'40"
Reema
Sco 16h16'44" -23d38'10"
Reema
Sgr 18h50'55" -26d11'48"
Reema Al Sudairy
Cas 2h13'12" 73d13'58"
Reema Samaha
Cnc 8h32'7" 23d16'52"
Reemi, my love
Uma 11h59'8" 37d56'31"
Reena
Cyg 20h17'55" 52d48'25"
Reena Mehta
Ari 3h16'6" 22d23'39"
ReenaBawa
Umi 16h39'18" 82d27'30"
Reeni
Uma 10h3'20" 51d8'46"
Reenie Butt
Umi 16h6'21" 70d55'8"
Reeny
Cra 19h2'48" -37d44'36"
Ree's Light
Cnc 8h39'48" 22d26'21"
Rees Ormond
Umi 16h24'13" 70d29'33"
Reese
Aqr 20h44'8" -10d35'41"
Reese
Sco 16h14'29" -15d59'35"
Reese
Gem 6h53'8" 22d13'22"
Reese
Del 20h45'34" 15d50'45"
Reese
Lmi 10h48'48" 22d59'30"
Reese Adams Moore
Tau 4h22'7" 26d20'23"
Reese Christopher Logan
Uma 10h1'59" 57d56'0"
Reese Elizabeth
And 23h25'23" 46d52'42"
Reese Elizabeth Buchert
And 0h20'22" 28d48'55"
Reese Elleyna DeShaw
Cas 23h39'17" 55d24'13"
Reese Emma
Vir 13h23'3" -12d37'28"
Reese J. Hardman
Leo 9h36'8" 28d25'54"
Reese Jolie Miller
Ori 5h54'23" 18d12'29"
Reese Lily
And 2h34'32" 44d13'47"
Reese Madison
Vir 13h5'2" 8d5'29"
Reese Marie Jefferson
Cap 21h45'56" -23d45'40"
Reese Martin
Uma 9h18'15" 49d14'10"
Reese Masita
Ari 3h9'19" 18d10'2"
Reese McKenna Kintz
And 0h25'30" 32d36'7"
Reese Michelle Hitchings
Sco 16h25'51" -40d37'1"
Reese Morgan
Del 20h38'50" 8d14'27"
Reese Murphy
Cet 2h27'17" 8d57'28"
Reese - Sweet Cheeks - Benadon
Ari 2h33'37" 24d55'49"
Reese Tyler
Psc 1h14'40" 11d49'36"
Reeselynn 72055
Mon 6h47'36" 7d7'49"
Reeser's Star
Aqr 22h45'18" -16d32'58"
Reet
Uma 8h58'38" 58d28'22"
Reet104
Gem 6h46'11" 23d39'7"
Reeta Ali
Sgr 18h15'16" -22d26'9"
Reflection 12/25/03
Uma 10h55'11" 48d13'16"
Refra
Vul 20h24'18" 26d40'57"
Reg & Chell
Cyg 20h41'43" 45d58'22"
Reg+Joan Anniversary Star 1964-05
Uma 10h20'52" 41d9'13"
Reg & Margaret Richardson
Cyg 19h50'11" 36d46'3"
Reg of Sunshine, Ray of Hope
Cyg 20h47'48" 34d37'41"
Reg & Sheila Moisey
Umi 16h5'12" 79d21'2"
Reg Walker - 'Eternity'
Cep 21h22'0" 64d51'27"
Rega
Cnv 13h30'46" 32d7'55"
Regal Hepatus Anders
Aql 19h44'40" -0d5'6"
Regan
Gem 7h36'16" 20d37'13"
Regan
Ori 5h26'51" 14d56'48"
Regan
Ori 6h17'24" 15d18'46"
Regan Addison Barker
Ari 2h25'28" 26d22'32"
Regan Ann Carter
Umi 16h51'49" 75d15'29"
Regan Ann McDevitt
And 1h37'36" 46d0'38"
Regan Christine
Sco 16h18'28" -10d7'58"
Regan Daniel
Her 16h41'48" 5d9'42"
Regan Davis
Leo 9h53'32" 24d26'9"
Regan deVictoria
Leo 11h3'12" 2d16'55"

Regan Elizabeth Thayer
Uma 11h6'50" 52d21'40"
Regan Faye Blackwell
Peg 22h50'45" 5d38'48"
Regan Lee Morrissey
Tau 4h48'12" 27d16'5"
Regan Lynn James
Lyr 18h45'25" 39d44'58"
Regan Marie Hunt
Sgr 18h35'28" -17d42'42"
Regan Mary Notter
And 23h25'0" 38d30'43"
Regan Nicole
And 0h49'35" 44d42'19"
Regan Noel Brown
Uma 11h18'25" 46d8'54"
Regan Rae Taylor
Lyr 18h45'36" 34d46'32"
Regan Reed Gabriel
Vir 14h42'23" 3d41'37"
Regan Reynolds
Umi 13h31'24" 74d51'47"
Regan's Night Rays
Uma 8h40'7" 64d7'4"
Regan's Star
Tau 5h34'18" 26d44'16"
Regenia Cornell McCall
Tau 5h49'15" 15d1'48"
Reggi & Josephine
Cyg 21h42'2" 36d1'47"
Reggie
Vir 13h21'21" 2d48'1"
Reggie Bledsoe
Cap 20h25'22" -23d13'37"
Reggie Dean Market
Per 2h32'26" 53d12'20"
Reggie in the Sky with Diamonds
Leo 10h18'24" 23d26'17"
Reggie Lee deNicola
Uma 8h41'57" 66d38'55"
Reggie Monygomery
Cnc 8h23'20" 9d6'24"
Reggie "My Mystery Man"
Dra 18h56'53" 59d59'31"
Reggie Ramirez
Aql 19h35'53" 7d13'46"
Reggie Rohe, BSOBL
Ori 6h6'56" 12d40'36"
Regibear
Cnc 9h3'8" 20d54'16"
Regina
Cnc 8h55'27" 19d32'0"
Regina
Leo 11h20'59" 24d32'2"
Regina
Peg 22h9'25" 12d6'42"
Regina
Psc 0h23'0" 18d21'16"
Regina
Cas 0h18'59" 54d57'41"
Regina
And 23h0'39" 49d32'9"
Regina
Uma 10h9'39" 49d1'35"
Regina
Uma 10h44'44" 40d28'48"
Regina
Crb 15h26'30" 31d43'50"
Regina
And 0h18'54" 36d56'49"
Regina
Uma 9h35'25" 67d11'55"
Regina 0429
Com 12h18'26" 26d14'12"
Regina and Joe Forever
Cyg 22h1'14" 52d41'52"
Regina And Willard Johnson
Ori 5h22'45" -7d51'45"
Regina Ann Davis Saenz
Ori 5h37'17" -1d29'33"
Regina Ann Hukill
Aql 19h36'57" 13d14'59"
Regina Ann Krol
Cnc 8h0'40" 22d5'53"
Regina Ann Land
Cap 21h28'21" -10d20'57"
Regina Beate
Ori 6h1'43" 10d40'52"
Regina Calderone
Cyg 20h43'38" 52d56'8"
Regina Celia Miguel
Sco 16h17'7" -9d28'53"
Regina Cerine
Uma 11h24'3" 40d57'41"
Regina Clark
Gem 6h53'14" 23d45'42"
Regina Cook-Pfeiffer
Psc 1h29'10" 13d35'32"
Regina Crenden
Crb 15h41'50" 36d17'24"
Regina Crouch Chester
Sgr 18h28'55" -17d4'7"
Regina Deak
Sgr 19h40'54" -15d50'48"
Regina Dreyer Thomas
Ori 6h6'13" 19d26'8"
Regina Duffy
Sgr 17h52'35" -29d56'19"
REGINA E N 23
Vir 13h20'2" 11d28'40"
Regina E. Yanok
Dra 17h25'52" 55d10'32"
Regina Erin Burns Edwards
Aqr 21h29'38" -1d54'54"
Regina és Bendegúz
Uma 10h49'35" 46d54'2"
Regina Felicia Enright
And 2h4'15" 42d5'21"
Regina Gabrielle Varesio
Cap 20h42'1" -22d4'42"
Regina (Gigi) Clinton
And 23h6'56" 47d36'19"
Regina & Hacki
Uma 13h29'50" 56d13'45"
Regina Haddad Nero
Uma 12h3'55" 42d56'16"
Regina Henderson
Aqr 21h34'35" -8d0'3"
Regina (Jean) Welch
Mon 6h52'52" -0d15'28"
Regina L.
Ori 5h35'0" 8d10'15"
Regina L. McCallum
And 2h20'58" 50d11'17"
Regina Lachemann
Gem 7h19'21" 19d54'5"
Regina Laurice
Cnc 8h50'36" 27d51'42"
Regina Leah Cantu
Ari 3h18'0" 20d55'29"
Regina Leoryl (Starkey) Suthers
Cap 21h41'34" -14d28'23"
Regina Lucinda
And 23h45'38" 37d27'57"
Regina M. Sinopoli
Cyg 20h53'35" 35d45'30"
Regina Mae Shaffer
Cas 1h43'15" 69d4'29"
Regina Magna DDavis 54
And 1h23'32" 37d39'8"
Regina Marie
And 2h14'23" 39d12'17"
Regina Marie Stern
And 23h16'14" 46d46'30"
Regina Martinez
Aqr 22h13'1" -1d43'40"
Regina Mary Mattson
Uma 8h45'55" 67d29'43"
Regina Micheline Hiester
Sgr 18h55'50" -35d15'54"
Regina Michelle Kauffman
And 1h33'26" 48d9'52"
Regina Murphy Gates
And 0h40'24" 25d21'28"
Regina Nevin Caulfield
Cas 1h21'59" 56d51'28"
Regina Norton
Ara 17h20'1" -52d8'55"
Regina Ona - Star
Sco 17h41'33" -42d47'52"
Regina Papilio
Uma 9h34'45" 54d50'36"
Regina Pascoe - Regi
Vir 12h55'53" -3d12'0"
Regina "Puddin'Pop" M. Gilliam
Uma 9h24'8" 61d4'48"
Regina Puryear
Lib 15h4'49" -1d45'20"
Regina Raffaele
Cas 0h12'46" 63d13'42"
Regina Renee Brooks
Crb 15h39'13" 27d20'53"
Regina Roberts
Cas 0h32'16" 56d27'50"
Regina Rocha Guzmán
Cnc 8h49'38" 15d28'50"
Regina Rose
Com 12h42'1" 13d55'19"
Regina Rose
Lyn 8h57'42" 42d25'50"
Regina Rose Jasiewicz
Vir 13h57'21" 3d58'37"
Regina Rushing Humberson
Cyg 20h54'33" 55d17'38"
Regina Russell
Uma 9h2'44" 53d50'45"
Regina Ruth Pinto
Psc 0h29'47" 8d22'14"
Regina Ryder "Midnight star"
Ari 2h58'20" 25d1'22"
Regina Salzman
Lyn 8h12'7" 39d29'34"
Regina Scott
Ari 2h12'20" 22d38'33"
Regina Soulnier
Cap 21h44'50" -10d22'12"
Regina Theresa Maul
Ori 5h45'40" 9d21'30"
Regina Torelli
Cap 20h25'52" -10d10'56"
Regina Trautmann
Ori 5h57'34" 17d13'47"
Regina Turner
And 0h35'26" 46d4'10"
Regina Virginia Kershaw
Cas 0h36'14" 58d2'16"
Regina Webber
And 0h33'36" 44d49'3"
Regina Willis Star
Cas 0h54'4" 60d29'47"
Regina Yamrick Hoffenstein
Aqr 22h9'2" 2d19'16"
Reginald A. Vitek "RAV65"
Tau 5h59'30" 25d54'50"
Reginald Allen Lee
Cma 6h24'55" -28d26'30"
Reginald Benedict Lewis
Pho 0h17'27" -41d48'28"
Reginald Gillespie
Sgr 18h12'27" -29d35'42"
Reginald Harrison Berry
Uma 8h35'15" 70d43'1"
Reginald Jordan Williams
Uma 13h24'3" 53d56'31"
Reginald Joseph Allen
Her 16h16'10" 6d27'37"
Reginald L. Sandifer
Cep 20h39'22" 55d57'13"
Reginald Leonard Tomes - Gramps
Psc 23h58'1" 1d13'23"
Reginald & Leticia
Vel 10h50'41" -56d25'58"
Reginald Mason
Dra 17h38'28" 58d53'53"
Reginald Mitchell
Boo 14h38'41" 31d34'56"
Reginald Owen Dalton
Cma 7h20'0" -30d45'49"
Reginald R. Lane
Tau 4h19'44" 30d19'49"
Reginald Victor Pope 9-10-1920
Cru 12h46'14" -64d36'14"
Reginald's Compass
Pyx 8h35'31" -31d4'42"
Reginall Michelle Garcia
Gem 6h19'29" 25d39'3"
Régine
Cap 20h32'51" -18d41'36"
Régine et Xavier
Uma 9h3'42" 50d7'47"
Régine Fonn
Ari 2h55'54" 28d43'18"
Régine Lecours
Cas 0h7'23" 53d34'18"
Regis Bertram O'Neil
Her 17h11'56" 24d44'58"
Régis Cholley
Cyg 21h40'10" 43d57'51"
Regis & Marnetta Larrimer-Eternally
Uma 10h33'8" 43d1'47"
Regis & Mary Ellen Gallagher
Cas 0h32'1" 63d2'15"
Regis May Leven
Tau 4h5'21" 27d37'10"
Regi's Michael
Ori 5h15'31" -4d30'30"
Régisseur de Comètes
Ori 6h16'29" 10d40'9"
Réglisse
Cmi 7h32'41" 2d59'24"
Reguana
Cnc 8h26'17" 14d9'40"
Regula
Her 17h15'33" 16d8'41"
Regula Barbara Bobby
Lac 22h4'30" 53d1'37"
Regula Jeger von Ah
Uma 10h12'58" 47d32'11"
Regula Oberli
Peg 21h32'56" 21d2'19"
Regula Schwab
Aur 5h24'30" 39d49'29"
Regulus Ron Anjard
Leo 9h42'8" 21d59'43"
Reha & Kathleen
Cyg 21h33'52" 48d50'31"
Rehema
Lyr 19h14'37" 35d16'33"
Rehmanessance
Sgr 18h2'7" -29d59'54"
Rehmanessance
Sco 17h53'38" -42d3'32"
Rehn, Erik Christoph
Uma 9h19'2" 49d30'58"
rehsuleman
Her 18h45'9" 23d28'6"
Rei'Anne Patricia Vancil
Psc 1h36'6" 18d7'21"
Reichel, Ulrike
Uma 9h12'40" 52d28'53"
Reichelt's Love
Lyn 6h36'33" 58d5'1"
Reichert, Dirk
Uma 9h47'59" 42d32'29"
Reid - 4
Uma 10h57'5" 49d37'31"
Reid Alexander Irwin
Her 17h50'8" 15d46'35"
Reid and Victoria
Cnc 8h29'11" 33d3'14"
Reid Arthur
Leo 11h44'28" 24d19'35"
Reid B. (Buddy) Hughes, Jr.
Peg 23h9'20" 30d35'13"
Reid B. Hughes
Cma 6h38'49" -17d56'3"
Reid Davis
Ori 5h56'52" 17d30'15"
Reid Garret Johnson
Aqr 21h8'18" -2d20'7"
Reid Hollister
Her 18h12'33" 27d43'57"
Reid Louis Berean
Cap 21h58'13" -21d15'23"
Reid Louis Rumack
Cnc 7h59'19" 16d15'22"
Reid Matthew Alicanti
Cnc 8h19'0" 11d7'33"
Reid Michael Fisher
Cnc 8h48'5" 20d48'14"
Reid Owen Willett
Her 18h8'55" 32d16'59"
Reid Randon Miller
Her 18h29'46" 13d37'28"
Reid Riecken
Aql 19h8'33" 12d50'11"
Reid Robert Sacco
Leo 9h32'26" 27d16'39"
Reid Sarah Lyons Dounias
Vir 13h44'56" -7d26'25"
Reider
Her 16h17'46" 7d15'24"
Reidl, Susanne
Ori 6h11'24" 16d2'48"
Reif Jordin Andreason
Sgr 18h28'58" -26d28'44"
Reighan Connor Rose Demers
Cyg 19h58'25" 35d48'58"
Reign Beau
Gem 6h41'54" 16d13'56"
Reign Josiah
Lmi 10h29'30" 34d50'11"
Reigna B. Hopkins
Uma 13h3'29" 53d39'51"
Reiiel
Leo 9h39'17" 27d39'35"
Reika Oshima
Leo 10h55'58" 4d9'13"
Reiko
Cas 1h10'41" 66d42'30"
Reiko Asaoka Ruiz
Lmi 10h5'2" 34d20'54"
Reikowsky, Günter
Uma 10h21'1" 65d0'19"
Reileigh Nicholle Lyst
Lib 15h24'18" -10d11'25"
REILEY ANN BRAGG
Psc 1h24'45" 21d30'47"
Reiley Madison Zeiders
Umi 14h22'16" 78d38'44"
Reilly
Leo 10h29'14" 14d37'8"
Reilly
Psc 23h48'49" 5d41'22"
Reilly
Aur 5h48'59" 54d36'39"
Reilly Ann
Umi 15h11'19" 78d31'54"
Reilly Ann Jamison
Lmi 10h9'25" 37d55'25"
Reilly Dog
Cnv 13h56'56" 33d49'58"
Reilly Eleanor O'Canna
Uma 9h36'36" 58d15'49"
Reilly Hamilton
Lib 15h41'39" -19d6'34"
Reilly & John's Wedding Star
Tau 5h33'25" 26d27'39"
Reilly Maria
Psc 23h51'26" 2d25'30"
Reilly Marie Jones
Uma 11h41'5" 47d33'9"
Reilly Maurer
Boo 14h44'6" 35d44'47"
Reilly Nicole
Gem 6h50'3" 18d34'2"
Reilly Peters
Sgr 18h15'50" -29d29'25"
Reilly Rose Galten
Cas 1h8'34" 54d43'36"
Reilly Thomas Boyne
Her 16h18'9" 9d7'16"
Reilly's Star
Aqr 20h42'15" -5d53'57"
Reiltín McCarthy Doyle
Lyr 18h34'33" 39d49'39"
Reily Kane
Uma 11h39'41" 51d31'7"
Reimann, Annett
Ori 6h0'38" 10d4'31"
Reimann, Steffen
Ori 5h58'13" 21d32'54"
Reimann, Werner
Ori 6h3'15" 10d15'25"
Reina
Tau 4h18'30" 7d44'33"
Reina and Arik
And 1h27'18" 50d20'9"
Reina Demoss
Dra 16h14'57" 60d36'59"
Reina Jasmir
Gem 6h56'24" 34d38'26"
Reina M. Rivera
Uma 8h41'21" 69d52'23"
Reinald+Cordelia+L+F+L+F Fischer
Ori 6h18'30" 9d33'8"
Reinald Müller
Ori 5h36'23" 14d44'59"
Reinaldo Ferrer
Uma 10h58'37" 61d53'5"
Reinardt's Rhapsody
Sgr 19h8'39" -18d4'6"
Reinauer, Tim
Uma 14h11'37" 56d15'2"
Reine Abdulahad
Cas 1h23'49" 52d47'42"
Reine Karine
Peg 23h31'1" 17d35'56"
Reine Stevenson
Cas 1h29'0" 68d26'51"
Reine Wojcik
Dra 18h21'42" 54d21'55"
Reiner
Uma 11h30'54" 31d36'15"
Reiner Bihlmaier
Ori 6h23'7" 16d13'42"
Reiner Kaesbach
Ori 6h16'53" 6d17'22"
Reiner Schott
Uma 9h45'24" 62d1'18"
Reiner Schulte
Uma 8h40'40" 48d23'44"
Reiner Wirth
Ori 6h18'5" 20d1'32"
Reinette
Psc 23h47'11" 5d47'30"
Reinhard
Uma 9h48'34" 48d14'41"
Reinhard Baur
Uma 11h35'1" 33d53'42"
Reinhard Gebke
Uma 8h46'40" 54d49'0"
Reinhard Inführ
Umi 15h11'21" 75d40'47"
Reinhard Kegel
Uma 8h36'21" 50d20'26"
Reinhard Paslack
Uma 9h11'14" 63d12'3"
Reinhard Zethner
Psc 0h55'41" 6d47'21"
Reinhardt, geb. Feltes, Eva Maria
Ori 6h13'29" -0d38'40"
Reinhold Becker
Ori 6h25'8" 10d1'1"
Reinhold Konrad Fritsch
Uma 8h10'55" 65d43'12"
Reinhold Reker
Ori 5h23'0" -8d44'37"
Reinhold Wild
Ori 6h21'13" 16d44'29"
Reinhold Zintgraf
Uma 9h19'53" 49d36'54"
Reinhold, Delia & Volker
Uma 9h41'54" 69d31'26"
Reinig-Leinen
Cam 4h26'29" 67d0'1"
ReinMill
Lyn 7h24'30" 46d26'17"
Reinmillers 50th WeddingAnniversary
Cyg 20h47'59" 46d54'57"
Reino
Uma 9h3'50" 52d39'8"
Reisa Aletha
Lyn 8h57'23" 38d12'55"
Reisa Lynn Weiner
Uma 10h18'28" 39d28'55"
Reisa Urman
Cnc 8h55'52" 12d33'45"
Reischl, Doris
Uma 14h6'15" 52d27'45"
Reiss Allan Sleeper
Per 3h28'57" 48d30'37"
Reiss Niedecken
Sco 17h11'32" -35d41'53"
Reiter, Claudia
Ori 6h20'53" 9d43'54"
Reitzenstein, Karl-Ulrich
Ori 6h19'36" 16d39'52"
Réjane
Cyg 19h26'33" 36d2'32"
Réjane & Mathieu, the Second
And 0h25'40" 39d51'15"
Rejean Arsenault
Cep 23h5'6" 76d28'28"
Rejina Lebesco
And 2h17'23" 49d36'50"
Rejoice Jimenez Charles Van Corpus
Lib 15h9'39" -6d16'41"
ReKai Dawn Dennis
Tau 3h58'14" 6d38'16"
Rekha Vara
Psc 0h15'53" 3d21'38"
Relais Bonne Eau
Lyn 6h44'59" 60d59'13"
Relda Ann Mason
Psc 0h11'9" 8d54'51"
Rella
Uma 9h2'19" 59d18'39"
Rella
Vir 13h9'54" -20d5'44"
"RElle"
Gem 6h21'11" 27d39'0"
relyt
Sgr 18h17'43" -16d32'6"
R.E.M.
Cas 1h43'33" 61d0'6"
Rema
Uma 9h21'23" 48d1'20"
ReMae T. George
Umi 15h33'12" 71d6'14"
Reme Butterfly
Eri 3h49'37" -9d10'40"
Remember Emmanuel&Karine
Psc 1h43'54" 6d2'10"
Remember I Love You
Lyr 19h10'16" 46d15'47"
Remember I love you! Duane Marshall
Ori 6h0'49" 17d46'4"
Remember Joy
And 0h19'54" 41d20'49"
"Remember When" Adam & Lyla 11/8/03
Cyg 19h38'50" 31d36'8"
Remembered Promise
Sco 17h26'36" -35d11'32"
Remembrance
Aqr 22h12'26" -8d24'50"
Remembrance
Psc 1h21'37" 31d38'40"
Remembrance
Uma 10h32'29" 45d21'51"
Rémi
Uma 9h48'37" 50d19'8"
Remi Antoinette
Vir 13h13'59" -18d16'54"
Rémi et Gaelle
Leo 9h40'48" 29d56'46"
Remi Evans
Del 20h34'3" 19d26'2"
Remi Marie
Crb 15h51'49" 27d39'58"
Remi Taylor Cole
And 2h33'57" 49d3'26"
Rémi Trépanier
Lyr 18h48'52" 31d22'42"
Remick Merry
Tau 5h22'42" 27d34'3"
Remie Ann Russell
Lyr 18h46'33" 32d33'37"
Reminder for Tony
Per 2h16'27" 51d30'5"
Remington
Umi 15h0'29" 71d16'7"
Remington
Uma 10h41'50" 64d26'12"
"Remington" - A Wish Come True
Psc 0h50'40" 21d1'21"
Remington M. Greger
Cap 21h45'59" -12d58'40"
Reminiscing On Forever
Peg 22h39'17" 15d57'3"
Remle Diggs Sherman
Vir 13h2'55" 9d32'29"
Remo
Her 18h55'4" 23d59'20"
Remo DiGirolamo
Umi 15h59'55" 76d3'51"
remo gasser
Cam 9h15'33" 77d50'15"
Remo Roland Mazenauer "Gehrer"
And 23h37'25" 33d34'2"
Remo & Valeria "side by each"
Lyn 8h16'41" 38d15'29"
Remon "Sweetheart" Estafanos
Lib 15h31'22" -26d46'30"
Remora Peterson
Uma 13h31'57" 56d14'39"
Remosa 25 aniversari
Vir 11h46'58" -6d9'11"
Remy
Uma 11h15'11" 36d28'11"
Rémy
Lmi 9h39'45" 37d11'49"
Remy
And 1h44'1" 47d28'35"
Remy <3#
Cnc 9h3'52" 31d58'28"
Remy Bamieh
Aqr 21h53'43" 1d30'4"
Remy Carriero
Psc 1h12'11" 24d0'25"
Remy Cowie Wilkinson
Her 18h50'19" 20d57'48"
Rémy & Emilie forever
Cyg 19h37'6" 32d3'4"
Remy Kane
Vir 13h26'0" 12d27'12"
Remy Saunders
Leo 11h49'34" 19d34'59"
Rémy Stoll
Umi 13h57'5" 73d16'9"
Ren
Sco 16h15'56" -15d25'41"
Ren
Peg 21h30'36" 15d38'29"
Ren Angel
Crb 16h7'44" 36d55'28"
Ren Minutola
Ori 6h18'53" 7d8'42"
Ren Nikita
Gem 6h40'12" 14d59'19"
Ren Sean Ryan
Cen 14h6'58" -40d23'31"
Ren, Our Beloved Chihuahua
Crb 15h36'31" 29d28'51"
Rena
Leo 10h15'19" 25d51'39"
RENA
Tau 4h34'50" 17d0'28"
Rena
Gem 7h1'56" 34d49'1"
Rena
And 23h0'46" 48d54'15"
Rena
Lyn 7h12'7" 54d6'1"
Rena Ash
And 23h5'54" 41d58'40"
Rena Bell
Uma 10h10'30" 68d40'46"
Rena DelliCarpini
Gem 6h9'13" 27d23'28"
Rena Elizabeth Bailey Broughton
And 0h53'2" 42d9'36"
Rena Jacobs
And 0h37'42" 38d46'35"
Rena M Graham
Sgr 18h31'29" -22d26'25"
Rena Mae 10
Cyg 19h49'28" 32d41'49"
Rena Marie Barsocchini
Apu 17h30'13" -69d22'50"
Rena Noray Brand Siegel
Lib 14h44'28" -10d45'3"
Rena Phillips -The "MA" Star
Cas 0h4'21" 59d45'16"
Rena Rachel
Dra 16h10'47" 56d12'35"
Rena Sheree
Uma 10h59'23" 65d13'41"
Rena & Ted Star of Everlasting Love
Cyg 20h48'31" 41d47'30"
RENA75
Tau 4h15'13" 27d1'19"
Renae
Uma 14h15'20" 61d6'54"
Renae Allison Hesse
And 0h49'28" 36d26'58"
Renae Ann Sinn
Vir 14h13'49" -1d3'4"
Renae Anne Denton
Boo 15h7'41" 34d36'11"
Renae Bryant
Mon 6h50'51" -0d27'55"
Renae Degnan
Gem 7h48'23" 26d33'42"
Renae Elise Sneddon
Cyg 19h38'34" 38d16'11"
Renae Johnson
Sgr 17h54'20" -17d13'32"
Renae Mitchell
Cru 12h26'33" -59d13'39"
Renae "My Shining Light" xox Wade
Cru 12h33'24" -61d3'14"
Renae Natalia Juniper Beam
Cas 0h26'50" 64d9'8"
ReNae Patterson
And 23h36'12" 48d11'14"
Renae Silva
Vir 11h52'52" 6d15'55"
Renae Sue Bergh
Her 16h46'7" 38d4'35"
Renae Warren
Vir 12h40'28" -4d57'58"
Renae's Love
And 1h32'22" 49d35'36"
Renaissance Man Crescenzo
Cep 22h25'40" 72d8'39"
Renaissance Woman
Cas 2h20'35" 62d30'7"
Renald & Jacqueline St-Pierre
Uma 10h1'33" 68d52'29"
Renaldo Schwerin
Uma 10h18'27" 60d19'11"
RenaLisaVoisine
Ari 2h5'23" 10d42'51"
rename
Ori 5h4'0" 8d33'30"
rename
Per 4h20'48" 46d17'49"
Rename
Ori 5h23'55" -8d36'57"
Rename
Ori 4h59'0" -3d43'40"
Rename
Ori 4h58'7" -3d17'30"
Renarda
Sco 16h13'20" -20d41'40"
Rena-ReRe-Kid
Tau 3h40'45" 21d51'30"
Renat
And 0h1'30" 44d53'15"
Renata
And 0h42'2" 38d4'47"
Renata
And 23h32'21" 42d21'52"
Renata
Cnc 8h34'3" 7d56'15"
RENATA
Ari 2h54'55" 28d32'49"
Renata
Leo 9h54'46" 28d24'44"
Renata
Umi 13h58'36" 88d15'59"
Renata Citarella
Crb 16h17'30" 27d48'45"
Renata deDonato deFlorio
Gem 7h3'43" 17d1'30"
Renata E Dyczewski
Cnc 8h33'34" 16d10'14"
Renata e Roberto
Aur 5h19'37" 30d40'46"

Renata Giserman
Psc 23h48'50" 0d7'6"
Renata Lucena Queiroz
Sco 17h56'2" -31d6'43"
Renata Maniaci
Crb 16h22'57" 26d49'5"
Renata Maria Bognar-Baptista, 30.08.1954
Umi 14h1'46" 78d56'18"
Renata Petraglia
Ari 2h47'44" 18d25'12"
Renata Šolcová
Cas 23h35'33" 55d50'21"
Renata Solorzano de Souza
Vir 14h53'8" 4d44'31"
Renata Star
Sco 17h52'53" -38d13'51"
Renate
Uma 12h39'47" 57d32'8"
Renate
Uma 9h10'51" 65d15'46"
Renate
Cam 5h46'54" 69d55'24"
Renate
Boo 14h19'28" 12d40'11"
Renate
Cyg 21h13'4" 47d16'13"
Renate And Matt
Lib 14h55'54" -16d8'1"
Renate Demuth
Uma 13h55'4" 53d45'30"
Renate Francke
Uma 8h25'23" 63d3'9"
Renate Gray
Psc 0h54'1" 32d52'28"
Renate Horscht-Ofenloch
Uma 11h18'11" 33d46'27"
Renate Karin Blakeley
Uma 9h28'29" 47d8'8"
Renate Mileka Kaaa (Didi/BBIG)
Cnc 9h11'27" 25d59'24"
Renate Schneider
Uma 8h20'44" 61d54'21"
Renate und Wolfgang Scheibe
Uma 9h40'50" 69d38'11"
Renate-Margarete Kopp
Uma 9h17'23" 66d6'14"
Renate's Star
Gem 7h16'58" 34d3'12"
Renatka
Cap 21h35'25" -12d20'52"
Renato
And 2h28'38" 44d15'42"
Renato Barizzi
And 23h36'38" 40d32'42"
Renato Burg 3/25/1989
Ari 2h12'9" 18d37'20"
Re.na.to Cara.va
Uma 13h43'9" 56d16'50"
Renato Di Tommaso
Uma 11h41'36" 44d20'35"
Renato Ochoa
Ori 5h57'34" 5d46'47"
Renato Rafael Pacheco Mogrovejo
Umi 14h21'29" 72d51'14"
Renatta Rose
Gem 7h46'56" 34d16'31"
Renatta Scales
And 0h2'50" 34d14'15"
Renaud Charpenay
Cyg 20h43'46" 36d8'26"
Renaud Larocque
Cep 21h29'9" 56d4'19"
Renault 6
Uma 11h33'14" 63d5'6"
Renay
Cra 18h8'19" -39d10'59"
Rency
Cyg 19h43'26" 34d21'7"
Renda Kay Trahan Daughdrill
Aqr 23h4'10" -7d41'53"
Rendez-vous de notre éternité
Lyn 7h56'59" 33d58'36"
René
Tri 1h46'22" 30d46'37"
Rene
Cas 0h25'0" 48d25'42"
René
Uma 9h7'40" 50d38'49"
Rene'
Vir 13h34'58" 3d33'47"
Rene
Ori 5h32'13" 11d23'3"
Renë
Cnc 8h51'50" 14d40'47"
René
Sct 18h29'52" -15d22'36"
Rene
Cas 2h41'33" 65d20'36"
Rene 7
Gem 7h21'34" 28d6'46"
Rene A. Zakhour
Her 17h15'48" 27d24'46"
René Alaina
Sgr 18h2'31" -23d9'5"
Rene Alberto Ramos Duran
Ari 2h52'20" 29d40'30"
René Allemann
Uma 10h3'39" 44d33'46"
RENE' ALMAGER
Gem 6h16'9" 26d57'3"
Rene Andre VanderGroen
Her 17h25'22" 33d58'48"
Rene B. Cavazos
Uma 9h2'28" 68d22'0"
Rene Beth Lynch Castro
Aqr 22h4'48" -0d51'40"
Rene Broit
Cru 12h39'38" -57d40'33"
Rene Cejas
Her 16h44'36" 9d6'50"
Rene Chavez
Ori 6h18'18" -2d25'53"
Rene Chavez
Uma 11h9'2" 69d17'39"
Rene Dante Ponce
Leo 11h53'49" 27d39'33"
René Dault
Peg 21h28'38" 16d20'48"
René Dupont
Crb 16h8'5" 38d58'37"
Rene Flores
Uma 11h57'7" 48d24'26"
Rene Garcia
Aql 19h47'52" 15d3'44"
Rene Garcia
Aql 18h55'56" 7d51'6"
René Gilgen
Cyg 19h36'24" 30d3'46"
Rene Gonzales
Lyr 19h14'31" 27d5'24"
Rene Greene
Aql 19h43'52" -0d2'23"
Rene H. Leclerc
Uma 10h35'5" 68d31'57"
Rene Hawkey
Gem 6h46'17" 33d21'48"
Rene Haygood Turner
Vir 12h9'34" 8d11'42"
René Krüger
Uma 12h24'11" 58d21'39"
Rene L. P. Van Dessel
Lmi 10h42'36" 23d21'38"
René Lerf
Boo 14h41'2" 12d8'11"
Rene' M. Hooper
Uma 11h55'24" 38d37'2"
Rene McMaster
Cas 0h12'2" 52d58'34"
Rene Michelle
And 23h36'36" 42d8'41"
Rene Michelle
And 1h0'15" 42d23'36"
Rene Nana Banana Burgess
Cas 1h40'10" 61d51'54"
René + Nicky
Cap 20h14'4" -10d24'18"
Rene Raatjes
And 1h53'44" 44d52'39"
René Rauser "Der Froschkönig"
Lyr 19h8'49" 46d18'17"
Rene' Refi
Cap 20h30'58" -11d14'14"
Rene' Robledo
Lib 15h10'11" -9d27'59"
René Rosado
Vir 13h1'48" 9d43'21"
Rene Rubalcaba
Cap 20h27'11" -14d34'52"
Rene Ruiz
Cmi 7h28'55" -0d2'35"
Rene Russo
Aqr 22h39'8" -1d58'17"
René Schramm
Uma 10h5'23" 64d57'41"
René Signer
Ori 6h8'57" -0d3'3"
Rene "Star" Williams
Sgr 19h28'23" -15d24'39"
René Stern
Boo 14h10'2" 27d58'48"
Rene Vallee
Tau 4h40'51" 21d46'45"
René Villemure
Sgr 18h1'8" -32d25'20"
René W. 31.08.1956
Per 3h41'41" 45d17'38"
René Westbrook
Cas 23h28'46" 53d3'11"
Rene Zinsmeister
Uma 11h13'53" 43d49'23"
RENE50
Vir 13h23'56" 13d50'16"
Renea
And 0h47'25" 40d5'16"
Renea Ann Smith Hill
Lmi 10h42'53" 30d32'5"
Renea & Mike
Cap 20h30'23" -11d27'45"
Renee
Lib 15h52'21" -13d3'49"
Renee
Aqr 22h41'59" -21d44'8"
Renee
Sgr 18h4'21" -19d47'31"
Renee
Sgr 17h46'58" -19d47'2"
Renee
Lib 15h14'39" -20d16'23"
Reneé
Aql 19h45'57" -0d1'9"
Renee
Lyn 7h12'58" 53d11'55"
Renee
Sco 17h37'31" -35d45'15"
Renee
Lib 15h19'57" -23d40'23"
Renée
Pho 0h53'15" -54d20'32"
Renee
Psc 0h15'54" 2d17'39"
Renee
Gem 7h41'59" 32d40'53"
Renee
And 0h30'33" 34d31'37"
Renee'
And 1h56'1" 37d29'27"
Renee
And 1h41'34" 41d22'10"
Renée
Cyg 20h59'23" 36d58'42"
Renee
Cas 0h8'43" 53d1'0"
Renee
Cnc 8h54'51" 13d36'9"
Renee
Del 20h42'27" 9d48'8"
Renee
Aqr 22h14'48" 2d15'15"
Renee
Gem 6h43'38" 26d23'38"
Renee
Psc 1h39'57" 27d45'32"
Renee
Leo 11h41'13" 19d3'4"
Renee Abernathy
Dra 16h22'34" 56d0'26"
Renee AG06
Uma 9h34'21" 44d42'16"
Renee Aileen
And 0h49'31" 40d6'25"
Renee Alyece Hanzlik
Dra 16h55'58" 51d31'51"
Renee Alyse Bernardo
Vir 12h16'3" -2d54'38"
Renee and Drew
And 1h9'20" 38d46'47"
Renee and Kylie
Lib 14h59'6" -16d33'26"
Renee and Miles Lane
Vir 13h14'3" -0d41'26"
Renee and Ronnie
Uma 8h15'50" 66d27'31"
Renee and Scott
Cyg 20h51'3" 48d14'46"
Renee and Sean Hambacher
Aql 20h28'51" -7d14'11"
Renee and Tim
Cyg 19h19'10" 51d59'38"
Renee Angela Middagh
Cnc 9h21'18" 16d37'50"
Renee Ann
Col 5h24'3" -39d19'28"
Renee Ann Galgano
Psc 23h10'33" 1d17'11"
Renee *Astraea* Vaupel
Aqr 20h57'14" 0d26'38"
Renee Atkinson
Tau 4h13'57" 27d20'58"
Renee Babygirl Nguyen
Mon 6h51'12" -0d59'10"
Renee Baldanado
Leo 10h11'41" 10d59'48"
Renee Barnard Hubbard
Cen 13h49'59" -31d30'51"
Renee Benyon
And 23h19'40" 51d3'8"
Renee' Berger
Ari 3h21'20" 29d31'36"
Renee Bigham
Cnc 8h18'22" 17d30'44"
Renée Boehnke
Aur 6h39'5" 38d55'38"
Renee Bradley
Lib 15h31'15" -12d48'0"
Renee C. Coviello
Gem 7h19'24" 19d48'26"
Renee' C. Paulauskas
Lmi 10h25'4" 37d45'0"
Renee C. Watson
Cap 20h53'35" -26d22'25"
Renée Carole Yancy 4004
Cas 1h20'4" 62d9'21"
Renee Cathryn Matzner
Uma 9h11'47" 46d51'31"
Renee Clark
Cyg 19h55'51" 41d42'17"
Renee Cruz
Tau 5h56'15" 25d43'56"
Renee Daniels
And 0h10'2" 45d30'28"
Renee Degutis
Tau 4h26'14" 7d47'5"
Renee Denise Cook
Crb 15h19'33" 29d19'2"
Renee DeRosa (Fitchette)
Uma 8h18'34" 68d3'47"
Renee DeSegur
Lyr 18h40'17" 37d53'45"
Renee Dorothy Ader
And 1h15'12" 46d42'30"
Renee Dumais
Psc 1h16'39" 16d26'24"
Renee E. Jones
Sco 16h42'50" -31d14'44"
Reneé Elizabeth Porter
Sgr 19h22'36" -25d43'6"
Renée Eva Doucette
Vul 20h39'9" 23d11'11"
Renee Famiglietta
Ari 2h10'23" 16d20'1"
Renee Faye Rose Toedtemeier
Lib 15h31'53" -20d34'18"
Renee Foti
Gem 6h39'42" 27d39'53"
Renee G.
Cyg 20h46'48" 31d46'0"
Renee G. Mandy
Leo 10h36'11" 13d12'1"
Renee Gabriella Cruz
Vir 13h35'53" 3d48'16"
Renee Gamble
Uma 8h40'49" 60d1'16"
Renee Gilmore - The Ideal Mother
Lib 14h36'34" -18d3'6"
Renee Guerrieri
Lyn 8h31'33" 38d58'3"
Renee Helen
Vir 12h49'37" 12d47'12"
Renee Helen Woods
Leo 11h39'40" 10d19'51"
Renee Herman
Uma 11h17'45" 48d5'46"
Renee' Hertzog
Sgr 17h59'55" -17d21'54"
Renee HOH!
Sgr 18h52'52" -16d59'31"
Renee Hulse-Vervest 050164
Cas 0h11'41" 53d41'1"
Renee Hutson
Cap 21h30'44" -15d13'13"
Renee Imrek's Star of My Heart
Ori 5h35'22" -0d30'35"
Renee Iris Irene
Sco 16h11'46" -11d37'28"
Renee Ives
Lyn 9h8'10" 34d22'23"
Reneé J. Gardner 4/9/83
Ari 2h7'48" 25d5'45"
RENEE JKN
Dra 18h22'14" 53d4'6"
Renee Joffrion
Cam 4h4'39" 66d40'2"
Renee' Jones
Boo 14h32'48" 39d6'55"
Renee Joy Garfinkle
Lib 14h53'44" -6d19'37"
Renee Kathleen Brady
Cas 1h39'51" 63d34'39"
Renee Klar
Sco 16h8'10" -16d42'55"
Renee L. Escobedo
Uma 13h35'21" 56d9'2"
Renee L Hardin
And 1h40'32" 41d22'51"
Renee Labate
Sgr 18h30'17" -15d55'17"
Renee Lavoie
Uma 8h14'40" 71d38'11"
Reneé Lena Shelden
Uma 10h8'33" 52d17'41"
Renee & Lenny Rudner
Cyg 20h1'24" 59d16'38"
Renee Louise
Leo 10h36'4" 9d49'10"
Renee Loveless
Ori 5h49'50" -4d15'23"
Renee loves Sara forever
Psc 1h9'1" 4d25'42"
Renee Lyn Castro
Ari 3h21'24" 28d28'50"
Renee Lyn Milville
And 0h14'47" 27d50'0"
Renee Lynn
Uma 9h31'48" 48d56'25"
Renee Lynn Gennarelli
Sgr 19h39'47" -42d33'10"
Reneé Lynn Stein
Cas 0h37'37" 53d19'47"
Renee M. Amirault
Vir 12h41'22" 9d22'45"
Renee M. Anderson
Cas 0h42'44" 54d20'44"
Renee M. Pry
Cap 20h23'12" -13d34'46"
Renee M. Wanyo
Crb 15h25'0" 26d39'53"
Renee Madeline
Vir 13h22'19" -13d20'21"
Renee Magdangal "Raven Star"
Uma 9h7'42" 53d58'41"
Renee Marchione
Cam 3h30'18" 63d27'29"
Reneé Marie 1952
Per 3h20'56" 41d32'51"
Renee Marie Foiada
And 23h23'47" 48d3'40"
Renee Marie Nicole Giroux
Gem 7h17'15" 14d35'14"
Renée Marie Rouleau
And 2h32'45" 37d28'31"
Renée Marie Smith
Gem 7h50'37" 15d14'52"
Renee Marie Tesar
Cam 3h43'14" 57d58'50"
Renée Marie Waters
Lib 14h51'53" -7d29'42"
Renee Martin
Ari 3h14'35" 28d40'8"
Renee Maryse Vella
Vir 11h46'42" -6d31'26"
Renée Massey
Lib 14h41'26" -9d11'46"
Renee Matthews
Tau 4h27'46" 22d9'41"
Renee Mazzarella
Aqr 23h1'28" -12d28'26"
RENEE MICHELE ARSENAULT
Psc 1h0'34" 26d47'43"
Renee Michelle
Aqr 22h2'37" 0d8'27"
Renee Michelle Brown 1/1/76
Cap 20h28'55" -13d0'33"
Renee Michelle DePhillips(McGinnis)
Cyg 19h40'40" 39d56'43"
Renee Michelle Frank
Psc 1h28'45" 26d44'43"
Renee Michelle Hupp
Sgr 18h56'4" -34d35'33"
Renee Michelle Royer
Sgr 19h10'21" -19d56'38"
Renee Millo
Cas 23h45'23" 55d22'22"
Renee Moira Martin
And 23h19'53" 41d50'45"
Renee & Mr. Jones
Uma 11h23'19" 63d19'14"
Renee Murakami
Mon 8h9'59" -0d48'3"
Renee Murray
Psc 1h2'12" 4d7'29"
Renée N Peter = Grandparents' Star LUM x
Cyg 20h24'38" 51d35'4"
Renee Nanette Iris Irene
And 23h33'6" 48d51'54"
Renee Nicole
And 23h34'59" 42d13'8"
Renee & Norman Super
Cru 12h19'3" -57d25'19"
Renée Panagos
Uma 12h58'28" 58d6'36"
Renee Parrack
Lyn 7h6'51" 59d39'34"
Renee Perry
Aqr 22h35'22" -2d23'23"
Renee Peterson
Cyg 20h37'14" 37d33'5"
Renee Pizzarusso
Sgr 19h37'18" -17d35'24"
Renée Poletti
Lib 15h9'6" -28d48'21"
Renee Prisco
Lyn 8h25'17" 38d1'54"
Renee Reenny
Gem 6h38'41" 16d22'1"
Renee Revello Bibbens
Leo 10h11'29" 25d22'28"
Renee Rose
Lib 14h52'24" -2d24'5"
Renee Rucker Machen
Gem 7h35'35" 25d10'57"
Renee S. Ritten
And 0h44'11" 39d49'1"
Renee Sachs
Cas 1h35'0" 62d34'29"
Renee Scarpino
Uma 11h30'38" 55d42'19"
Reneé Schiefer
And 0h39'6" 36d29'53"
Renee Shawe
Sge 19h18'38" 17d15'17"
Renee Smoltz
Vul 20h34'25" 28d5'6"
Renee Staton's Star
Del 20h33'19" 17d13'45"
Renée Terrie Rosen
Cap 20h43'48" -14d52'48"
Renee the One and Only for Jer
Aur 5h36'43" 38d41'2"
Renee Vai
Sgr 18h17'46" -21d20'40"
Renee Veres
Aqr 21h13'54" -11d10'0"
Renee Vessels
Lib 15h12'29" -21d37'39"
Renee Wieczorek
Lyr 18h28'52" 39d57'29"
Renee Yeung Ka Ki
And 0h31'26" 33d9'45"
Renee Yu Hang Sun
Vir 14h33'23" 2d21'20"
Renee's Guiding Light
Ari 2h41'44" 14d26'49"
Renee's Hjerte Hjem
And 23h39'30" 42d12'10"
Renee's Shining Star
And 1h28'50" 37d42'5"
Renegade
Aur 5h55'41" 36d28'4"
René's Lovely Light
And 1h49'54" 38d49'36"
Rene's Star
Sco 16h7'51" -10d12'24"
RENIA
Cyg 21h23'14" 37d24'42"
Renico
Per 1h45'41" 54d23'33"
Renie
And 0h56'8" 35d27'22"
Renie
Gem 7h20'6" 13d40'8"
Renie
Ari 2h21'31" 23d48'22"
Renita 11th February 2005
Cru 12h49'0" -60d5'36"
renjam
And 23h12'41" 35d36'38"
Renne Hague/Always Right/I Loveu
Umi 14h26'59" 89d37'55"
Renne Ratcliffe
Uma 11h19'10" 58d32'29"
Renne Suzanne Pink
Psc 1h11'50" 8d5'40"
Renni
Cap 20h40'46" -26d26'20"
Rennie F. Halsband
Uma 11h30'21" 34d33'58"
Rennie McKie
Umi 17h5'52" 77d10'38"
Rennie's Heart
Sgr 19h17'21" -23d25'57"
Renny Renkins
And 23h35'56" 46d39'2"
Reno Hess
Cnv 13h46'59" 30d28'48"
Reno Robels
Uma 10h10'9" 65d52'57"
Reno-kun
Uma 9h22'30" 56d49'32"
Rensso & Alejandra
Cyg 20h17'38" 36d42'34"
Renu-Kala
Ori 6h10'46" 20d5'13"
Renwick
Uma 11h6'47" 63d17'27"
Renz
Tau 3h30'49" 2d12'30"
Renzo
Cam 5h3'4" 64d26'26"
Renzo et France
Aqr 20h40'29" -10d3'54"
Renzo Raggi
And 1h54'46" 45d37'56"
Renzo Ruiz
Ori 5h36'38" 0d46'52"
Renzy Lisbeth Acevedo
Vir 12h45'24" 0d18'22"
REO
Cnv 12h46'22" 38d57'19"
Reoh Darwell
Uma 11h26'46" 47d59'0"
Repletion Tutelary
Cap 21h22'56" -14d53'52"
REPP1
Tau 4h28'33" 26d1'34"
Republic of Belarus
Cep 21h27'49" 60d53'48"
Republican Convention 2004
Vir 13h54'52" -0d43'43"
Requel
Psa 22h34'53" -26d34'42"
Requetecontrailodatro
Umi 16h59'42" 80d41'48"
(RES) DA - middle of W
Cas 0h55'33" 62d23'33"
Re's Star
Psc 0h32'43" 3d17'10"
Resa Jane Spencer
Leo 10h58'58" 2d3'46"
Resa My Love
Ori 5h13'22" -0d41'51"
Reschke, Arthur
Uma 9h19'1" 44d7'38"
Resciniti's Southern Cross
Aqr 22h16'51" -7d55'22"
rescue gal 80
And 23h14'48" 41d9'37"
Réséda
Psc 0h50'42" 15d26'38"
"Resergam"
Ori 6h3'18" 6d34'30"
Reser's Fine Foods
Per 3h47'9" 33d28'27"
Reserved for George Matthews
Uma 10h3'14" 48d48'15"
Reshard A. Beverly
Aqr 22h32'17" -1d35'47"
Reshma
Sgr 18h11'41" -19d17'30"
Reshma
Peg 22h41'32" 13d29'18"
Reshma
Ori 6h10'43" 19d26'38"
RESHMA BALANI
Uma 11h43'11" 43d32'14"
Reshma Matadeen
Cap 21h43'39" -9d24'4"
Residence Life Staff 2005-2006
Uma 12h42'32" 57d19'36"
Resiliance
Ori 5h44'37" -0d1'33"
Resilience MKRSIY 50
Per 2h24'14" 55d21'52"
Resistenza
Lyn 7h21'10" 45d53'19"
Respect for People
Mic 21h10'10" -32d19'12"
Resqnine204
Psc 1h20'44" 17d5'18"
Resse Cup
Uma 12h33'39" 60d55'59"
Restless Spirit
Ori 5h33'21" 1d2'45"
Reszetar-Dragun
Cyg 21h43'3" 47d6'55"
Ret Maj USAF D. Wayne Sauls
Her 16h51'36" 33d43'11"
Reta Armstrong
Cas 1h22'21" 64d56'32"
Reta Baran "Reta's Reign Star"
Cas 1h49'5" 60d19'47"
Reta Jones Loreman
Vir 15h6'40" 3d53'59"
Retagene Alinen Hanslik
Vir 13h47'3" -5d55'14"
RETARD
Cam 4h19'28" 73d17'18"
Retha
Uma 11h2'14" 47d44'33"
Retha Felkner WGM 2005
Cap 21h35'32" -9d8'12"
RethaBree1138Rick
Uma 13h59'0" 58d6'12"
Retineo
Vir 13h53'28" 6d12'53"
Retired Fire Chief "Gramps" Moran
Lib 15h13'8" -15d42'54"
Retired Vulcan - L.J. Prehoda
Crb 15h49'46" 29d0'56"
Retirement Star Of James W. Butler
Ori 6h7'5" 18d18'45"
Reto
Dra 17h12'24" 66d54'9"
Reto
Uma 10h43'9" 60d20'39"
Reto Camenisch
Cas 1h53'55" 77d25'27"
Reto Caseri
Cnv 13h56'49" 36d0'57"
Reto Jaggi
Uma 8h20'1" 59d57'3"
Reto & Karin
Umi 15h32'33" 77d44'11"
Reto Lütolf
Dra 20h40'46" 69d4'17"
Reto Portmann
Sgr 18h46'38" -26d35'13"
Reto Roman Leser
Ori 6h12'35" 14d11'2"
Reto und Silvana
Her 16h45'33" 9d5'34"
Reto Weber
Ori 6h8'15" 20d41'4"
RETSOF
Uma 9h20'34" 61d18'52"
Rett
Aql 19h42'38" -4d30'37"
retta
Tau 3h41'55" 16d28'29"
Retta Clements Hughes
Lyn 8h30'16" 45d19'50"
Retty Ong
Uma 11h57'5" 49d20'30"
Returning2U122104
Uma 13h42'43" 51d42'51"
Retus Craft
Aql 19h41'38" 12d13'12"
Reuben
Ari 2h14'43" 24d48'22"
Reuben
Ori 5h54'23" 22d43'24"
Reuben
Umi 17h26'3" 79d33'29"
Reuben E. Cannon
Aqr 22h41'31" -1d36'25"
Reuben Jacob
Umi 17h0'4" 76d15'2"
Reuben James Harris
Cyg 21h20'53" 54d38'42"
Reuben Jeevan Bissessar
Her 17h17'54" 18d14'35"
Reuben Joseph Hayes
Her 16h15'16" 46d41'59"
Reuben Mwenda Gitonga
Cnc 8h51'31" 31d40'6"
Reuben Philip Baby Walter
Per 2h51'3" 50d38'59"
Reuben Ramkissoon
Sgr 18h29'12" -31d29'59"
Reuben Schantz
Cap 20h25'10" -16d58'36"
Reuben Schlossberg
Umi 18h1'22" 88d47'35"
Reuben Sid
Umi 17h4'48" 84d59'56"
Reuel Samuel Berg
Per 4h1'36" 33d41'37"
Reunion
Leo 9h49'18" 20d30'27"
"Reuniting Star" = Our Eternal Love
Lyr 18h46'42" 39d8'37"
Reuschel, Konrad
Uma 13h2'46" 62d3'41"

Reuter, Andreas
Ori 5h7'48" 8d46'27"
Reuter's Stars "Elevate Elkdom"
Cas 0h40'3" 53d57'53"
Rev. Adam Ogorzaly
Cyg 21h56'28" 46d4'29"
Rev. Albert Van Dyke
Ari 2h1'51" 21d37'16"
Rev. Colonel James J. Puchy
Per 3h15'3" 53d39'47"
Rev. Dieter Punt
Gem 7h5'58" 30d34'51"
Rev. Dr. Abram Jones Cox, III
Lib 14h59'58" -9d55'7"
Rev. Dr. Cecelia Williams Bryant
Crb 16h21'10" 33d26'44"
Rev. Dr. Emily Chandler
Uma 11h53'38" 56d33'57"
Rev. Dr. Robert L. Veon
Ori 5h39'46" 2d10'59"
Rev DreamWitch
Ari 2h19'53" 24d57'32"
Rev. Efford & Lucille Haynes
Cyg 20h40'53" 47d12'0"
Rev. Father Andrzej Skrzpiec
Cep 23h6'45" 76d20'37"
Rev. Francis P. Groarke
Aqr 22h28'20" -5d27'49"
Rev. Franklin J. Pollard
Vir 12h40'5" 12d8'36"
Rev. Harold Niedzwiecki, OFM
Aql 20h6'35" -0d26'38"
Rev. James Aquino
Cep 21h23'59" 60d27'55"
Rev. James Melville Clark
Leo 9h55'33" 21d40'34"
Rev. Jeremiah A. Wright Jr.
Cep 21h57'48" 62d20'4"
Rev. Joseph M. McLafferty
Cyg 19h50'32" 41d38'30"
Rev. Linda L. Boes
Pyx 8h43'33" -31d7'20"
Rev. Lyn Dean-Dennie
Cep 22h44'28" 68d18'46"
Rev. Merlin
Per 4h35'2" 31d43'13"
Rev. Michael J. Stumpf
Vir 13h8'40" 7d11'5"
Rev. Msgr. Austin P. Bennett
Aur 5h36'22" 53d9'10"
Rev. Msgr. Herbert A. Bevard
Cep 21h41'48" 62d57'10"
Rev. Randy Kanipe & Family's Star
Aql 19h49'9" 8d55'19"
Rev. Raymond Goodwin Jr.
Cyg 19h57'2" 45d44'8"
Rev. Robert A. White
Cyg 19h39'52" 29d22'22"
Rev. Romaine Norma Williams
Gem 7h6'29" 24d39'9"
Rev. Stanley J. Aksamit
Gem 6h10'22" 23d47'39"
Rev. Susan Gunther
Cyg 20h59'0" 49d36'27"
Rev. Terence O. McAlinden
Per 3h22'11" 47d7'57"
Rev. Velma Dela Pena
Leo 9h31'42" 11d44'4"
Rev. William C. Rousseau
Cep 22h17'10" 61d39'42"
Rev. William J. Anderson
Ori 4h49'33" 11d57'4"
Rev. William Shields
Lib 15h24'52" -15d23'50"
Reva Dorn
Vir 13h27'12" 11d35'15"
Reva Leanna Rodriguez
Sco 17h31'18" -38d43'54"
Reva M. Trembly
Crb 15h42'6" 26d6'52"
Reva Mae Wolfe
Vir 14h34'55" 6d28'59"
Reva Shackelford
Lyn 8h5'22" 44d16'46"
Reva Wanda Shirley McCord
Cnc 8h31'40" 18d6'13"
Revan McKinnon
Tau 3h39'57" 29d29'33"
Rêve éternel
Psc 1h49'21" 6d28'10"
Revee's Star
Lyn 6h29'37" 60d17'8"
Revelle Monds Young
Leo 10h41'8" 17d42'56"
Revera (Vee) Tolochko Kahn Wayburn
Cap 20h24'56" -20d49'54"
Reverand Thomas Philipp
Uma 8h50'21" 51d20'52"
Reverend Edwin and Mrs Beverly Fitz
Gem 7h32'8" 29d9'14"
Reverend Jason Michael Henderson
Uma 12h39'15" 59d32'30"
Reverend Keith Mozingo
Per 3h46'57" 41d31'44"
Reverend LeRoy Walker Jr.
Gem 7h10'50" 24d16'14"
Reverend Lureen Carryl-Slater
Vir 13h43'28" 1d53'42"
Reverend Nathan VanderWerf
Ari 2h7'2" 17d32'44"
Reverend Paul Armstrong
Cap 21h0'39" -14d56'9"
Reverend Richard Lavoie M.S.
Cep 22h22'44" 64d7'15"
Reverend Sisyphus
Cep 0h4'4" 84d57'16"
Reverend Thomas Gresham
Her 16h45'22" 34d13'52"
Reverie
Leo 10h23'3" 19d7'15"
Reverie
Uma 9h16'11" 54d6'13"
Rêveur
Dra 18h43'56" 71d4'35"
Rêveur Plien d'espoir
Sge 20h5'5" 17d50'50"
Revital Star-Vivian Chow
Sco 17h26'46" -43d6'2"
Revonda
Lyr 18h49'53" 31d46'57"
Revvy Barnes
Dra 18h23'52" 76d55'17"
Rex
Umi 14h10'31" 70d13'31"
Rex
Uma 13h49'33" 54d27'2"
Rex
Leo 10h46'14" 18d28'19"
Rex Adrian Jarrett, Jr.
Vir 12h20'25" -5d2'41"
Rex Albertson-A Good Kind Wise Man
Tau 3h52'46" 24d31'0"
Rex Alexander Kelly
Sgr 18h20'10" -35d36'16"
Rex Bierko
Ori 5h7'28" 4d58'1"
R.E.X. Boswinkel
Umi 17h39'5" 80d38'21"
Rex Brownie
Ori 5h34'48" 2d9'59"
Rex Davis
Ari 2h16'54" 26d15'19"
Rex Dearmont Rust
Vir 12h42'12" 1d36'5"
Rex E. Schindler
Cnc 8h16'10" 20d23'7"
Rex Gadd
Ori 6h14'18" 3d20'52"
Rex James
Cep 21h36'35" 61d36'49"
Rex Jerome Hughes GSD
Uma 11h19'41" 35d42'0"
Rex Kerschner
Her 16h45'51" 38d26'44"
REX (KING) RAY
Ari 2h7'35" 24d0'56"
Rex Lavern
Sco 16h22'0" -19d22'9"
Rex M. and Loni G. Christensen
Lyn 7h24'55" 44d21'45"
Rex Michael Staudacher
Umi 14h24'8" 76d41'55"
Rex Narvaez
Cnc 8h19'1" 15d29'55"
Rex Nathan Marling
Cnc 8h59'19" 8d16'14"
Rex Phibbs
Cma 7h13'45" -30d22'14"
Rex Thwaits
Uma 11h11'46" 38d8'1"
Rex Waldron
Aur 5h5'7" 34d7'29"
Rex William Hickman
Umi 15h10'32" 84d7'43"
Rexie
Gem 7h12'39" 23d30'48"
Rex-O
Lyn 9h19'39" 38d54'50"
Rextar 1
Umi 16h16'1" 80d21'4"
Rexval Valerio
Aqr 22h33'14" 1d13'42"
Rey Garcia - My love for Eternity
Psc 23h30'28" 2d57'18"
Rey - Ney
And 0h6'6" 45d50'16"
Reya Marie Sytsma
Leo 9h55'54" 25d56'10"
Reymundo
Cam 6h17'44" 75d30'38"
Reyna
Sco 16h47'29" -44d3'4"
Reyna Angulo
Sco 17h37'34" -40d35'52"
Reyna Arevalo
Uma 10h25'0" 44d15'44"
Reyna "Delilah" Mendoza
Lib 15h14'19" -21d3'41"
Reyna Hafliger
Her 18h14'8" 25d22'44"
Reyna Hernandez
Lib 14h33'48" -24d56'27"
Reyna Laura
And 0h2'27" 38d6'57"
Reyna Mendoza
Aqr 20h42'46" -8d34'36"
Reynaldo F. Marine
Umi 14h4'41" 71d49'45"
Reynaldo Tamez, Jr.
Leo 10h35'36" 25d38'41"
ReynaLinda
Dra 19h40'17" 71d35'30"
Reynolds Alexander Cochrane
Uma 11h53'18" 63d8'18"
Rey's JIB Star
Tau 3h51'58" 26d29'34"
Reysa M. Lamourtte
Gem 7h0'13" 14d37'42"
Reyus Big Bear Hawkins
Per 3h18'5" 47d30'16"
Reza Arthur Parsey
Vir 13h11'41" 3d46'28"
Reza Masudur Rahman
Psc 23h25'39" 2d45'36"
Reza & Yalda Madani
Cyg 21h22'38" 45d10'39"
Rezeda Mullayanova
Sgr 19h11'20" -22d14'4"
Reznor
Uma 9h25'50" 49d19'47"
RF 1999
Uma 10h55'11" 54d58'20"
RFHRHS Class of 2005
Cas 1h21'57" 62d50'4"
Rfyre
Leo 9h58'48" 14d27'36"
RG 77
Uma 10h43'28" 71d56'31"
R.G. Heaven's Hot Rodder
Uma 8h43'33" 54d47'46"
RG&REUnderwood's 50th Jehovah Jire
Uma 11h42'37" 60d5'48"
RGS012751
Aqr 22h21'9" -3d47'49"
RGS1 12-16-03
Lyn 7h38'48" 53d41'46"
RGW9302006
Sge 20h5'39" 18d15'24"
Rhaam Rozenberg
Umi 15h29'27" 81d21'18"
Rhachel
Lyn 7h15'47" 58d19'39"
Rhamiel-Angel of Empathy
Gem 7h24'55" 33d25'30"
Rhande Osborn 50
Uma 8h24'49" 71d6'1"
Rhandee Gortney
Aqr 23h44'58" -15d23'49"
Rhapsody Flores
Psc 1h4'33" 3d39'26"
Rhapsody's Star
Vir 14h41'46" -0d6'41"
Rhawnie Morris
Sgr 19h18'5" -14d12'10"
RHB 143
Ori 5h29'17" 10d32'10"
Rhea
Cas 0h36'35" 62d31'3"
Rhea Guilbert
Tau 5h11'49" 18d26'33"
Rhea Lorraine Dominguez
Cnc 8h50'33" 12d19'40"
Rhea Lovus
Lyr 18h34'13" 28d7'38"
Rhea Mae Clyburn
Lyn 8h54'13" 46d15'0"
Rhea Michele
Tau 5h25'22" 19d19'59"
Rhea Murtagh Stafford
And 0h49'4" 46d6'58"
Rhea Nicole Schoustra
Vir 14h40'50" 4d23'27"
Rhea Quichocho
Cyg 20h49'12" 34d43'52"
Rhea Sauve
Lyr 18h20'53" 45d12'30"
Rhea & Shazam
Cyg 20h22'57" 47d37'16"
Rheanna
Vir 13h18'51" 8d19'20"
Rheanna Bellomo
Lac 22h25'37" 51d1'57"
Rheanna Turner
Lmi 10h30'38" 29d29'34"
Rheanne
And 23h25'35" 50d42'45"
Rheannon
And 0h47'28" 36d43'47"
Rheannon Lee Hill
Cnc 9h7'30" 32d57'29"
RHEBA
Tau 4h31'49" 25d28'54"
Rheco HC Blehr
Boo 14h41'12" 36d44'52"
Rheeanna Grady
And 1h37'20" 49d42'33"
RhetOracle
Ari 3h28'11" 28d54'59"
Rhett Addison Sikes
Uma 9h37'29" 42d12'35"
Rhett and Tink
Tau 5h58'35" 25d6'21"
Rhett Davis
Uma 8h51'8" 57d16'7"
Rhett Evans Duvall
Sco 17h35'23" -38d12'27"
Rhia und Danny
Ori 5h8'10" 16d0'57"
Rhian Hastings
Dra 17h48'41" 52d13'27"
Rhian K. Littlechild
Cam 7h7'30" 74d37'31"
Rhiana - My Pum'kin Little
Peg 22h36'28" 3d8'31"
Rhiana's Light
Cas 23h46'34" 56d50'8"
Rhianna
Psc 1h29'3" 32d59'58"
Rhianna Alexis Ramshaw
Sgr 18h10'43" -30d8'25"
Rhianna Duri
Cyg 20h4'25" 35d12'27"
Rhianna Isabella
And 0h45'7" 41d3'46"
Rhianna Lorraine
Lyr 18h51'1" 38d16'19"
Rhianna Marie Wright
And 1h47'43" 40d49'24"
Rhiannon
Cyg 20h52'22" 36d7'8"
Rhiannon
Her 17h18'37" 36d1'21"
Rhiannon
Peg 22h21'34" 20d9'12"
Rhiannon
Uma 10h23'48" 62d52'1"
RhiAnnon
Sgr 19h20'22" -15d47'34"
Rhiannon Alexia Gray
Cru 12h25'5" -57d13'45"
Rhiannon and Bronwyn's 1st Christmas
And 23h57'24" 40d23'3"
Rhiannon and Sean - 05/24/2006
Lyn 7h0'58" 52d6'29"
Rhiannon Blue Eyed Baby
Lmi 10h39'1" 28d55'56"
Rhiannon Chiacchiaro
And 23h57'28" 41d34'55"
Rhiannon "Daughter" Carman
Cru 12h18'24" -56d36'10"
Rhiannon Elaine May
Crb 16h8'12" 28d39'44"
Rhiannon Elizabeth
Psc 1h27'0" 24d41'20"
Rhiannon Kelsey Rae Karr
Leo 11h10'50" 24d22'9"
Rhiannon Kinsley Davis
Psc 1h18'46" 14d59'36"
Rhiannon Layelle Morell
Peg 23h11'22" 30d58'59"
Rhiannon Lee Eccleston
Lib 15h27'21" -24d24'31"
Rhiannon Lujan
Dra 19h14'47" 64d30'49"
Rhiannon Marie-Louise Talbot
And 23h29'44" 35d49'5"
Rhiannon Packer
Cas 2h29'30" 68d35'46"
Rhiannon Price Jackson Jones
Lyr 19h23'43" 38d2'38"
Rhiannon Shmee
Psc 1h23'0" 24d57'32"
Rhiannon Sunshine
Vir 12h22'31" 10d21'55"
Rhiannon Townsend
Del 20h45'51" 3d40'13"
Rhiannon's Star
Peg 23h24'0" 20d12'32"
Rhianon Felicity
And 2h18'27" 38d1'25"
Rhianwen Davies
Cru 12h56'5" -59d10'54"
Rhina
Lib 15h13'14" -12d50'45"
Rhince "2-17-06"
Uma 13h47'9" 55d16'30"
Rhinestone Meg
Peg 22h24'54" 33d12'23"
Rhizome Cowboy
Cnc 8h45'52" 17d55'13"
Rho Vezzoso
Ori 5h12'20" 2d18'49"
Rhoads Less-Traveled
Leo 10h9'40" 9d39'37"
Rhoads's Family Star
Uma 13h40'54" 56d59'1"
Rhoadster
And 2h16'31" 50d4'4"
Rhoda
Crb 16h10'57" 38d24'6"
Rhoda
Gem 6h54'56" 17d0'46"
RHODA
Lyn 6h19'7" 58d5'29"
Rhoda Doris Hamilton-Smith
Cnc 8h14'21" 13d2'53"
Rhoda Joy
Cnc 9h7'46" 31d22'31"
Rhoda Rubin
Cam 6h1'43" 60d56'4"
Rhoda Seldes
Cas 2h20'0" 65d56'16"
Rhoda Sutter Rarick
And 23h45'58" 44d53'54"
Rhoda Tam
Uma 14h2'41" 52d50'28"
Rhoda Williams
Ari 3h26'50" 26d5'31"
Rhodalynn Marages
Lyn 8h10'44" 42d26'54"
Rhodanthe
Gem 6h28'34" 13d7'32"
Rhoda's Light
Col 5h49'28" -38d19'23"
Rhoda's Star
Ori 6h15'42" 14d31'19"
Rhodes's Journey
Cyg 21h27'58" 32d0'9"
Rhodie
Uma 8h41'3" 62d53'43"
RHOLDERKSWJKN
Dra 17h51'5" 67d8'7"
Rhona
Cyg 21h52'25" 49d16'32"
Rhona Jane Phoenix
Cas 23h1'49" 55d12'51"
Rhonalie
Aql 19h30'42" 11d39'22"
Rhonda
Aqr 21h30'59" 1d28'31"
Rhonda
Cnc 8h20'10" 25d19'55"
Rhonda
Gem 6h43'13" 15d39'45"
Rhonda
Psc 1h8'29" 26d7'29"
Rhonda
Lyn 7h54'14" 38d32'9"
Rhonda
Uma 11h24'23" 58d48'18"
Rhonda
Aqr 22h21'17" -4d55'35"
Rhonda A. Pafford
Lyn 7h32'47" 47d42'21"
Rhonda Allen - 21/08/1964
Cas 23h58'28" 56d10'42"
Rhonda always in my heart love, Dad
Uma 9h37'28" 53d11'40"
Rhonda & Angela
Lib 15h36'17" -6d1'32"
Rhonda Ann Chamberlain
Lib 15h47'39" -18d23'18"
Rhonda B. DeZuzio
And 1h9'28" 36d2'4"
Rhonda Blacharski
Lib 15h13'43" -19d32'37"
Rhonda Borealis
Gem 7h26'20" 26d39'14"
Rhonda Chalmers Lynch
Gem 8h2'14" 32d14'27"
Rhonda Clark
And 1h38'31" 49d51'6"
Rhonda Crown
Del 20h35'42" 13d21'25"
Rhonda DeClue
And 0h54'22" 40d21'25"
Rhonda Denise Grice
Tau 4h19'7" 3d11'20"
Rhonda Dorsey ~ My Angel
And 0h36'20" 39d34'36"
Rhonda E Walker
Uma 13h13'44" 54d44'53"
Rhonda Faye Waguespack Gaughan
Peg 22h36'26" 7d38'43"
Rhonda Fuechsel Phillips
Leo 11h51'15" 21d35'49"
Rhonda Gail
Lib 15h19'35" -23d16'2"
Rhonda Gail Smart Breit
Cap 20h13'31" -17d13'20"
Rhonda Gayl
Leo 10h3'49" 26d43'19"
Rhonda Gayle
Cap 21h23'50" -16d9'51"
Rhonda Gettig's Star
Ori 5h33'17" 0d16'26"
Rhonda Giordano
Cam 3h58'39" 58d26'2"
Rhonda H. Chambers
Lyr 19h17'6" 34d33'54"
Rhonda Hartwell Honeycutt
And 1h32'36" 45d4'28"
Rhonda Iacona
Cap 21h9'29" -21d6'7"
Rhonda Jane
Crb 15h40'48" 30d53'13"
Rhonda Jean
Cyg 19h51'21" 47d0'13"
Rhonda Jean
Uma 10h4'22" 71d50'46"
Rhonda Jean Howard
Gem 7h26'28" 29d54'54"
Rhonda Jo
And 23h6'16" 43d23'6"
Rhonda Jo
Lyn 6h38'25" 54d19'6"
Rhonda Jo Hart
Tau 5h10'47" 16d52'1"
Rhonda Jones 08271975
Vir 13h17'49" -5d46'2"
Rhonda Joy Martin
Sco 16h36'42" -29d56'43"
Rhonda K. Huston
Aqr 22h34'5" -17d43'59"
Rhonda Kay
Cas 1h42'7" 66d23'1"
Rhonda Kay Malott
Vir 13h58'37" -3d6'37"
Rhonda Kay Miller-Pabo
Leo 9h25'7" 11d59'1"
Rhonda L. Coulston
Uma 9h55'27" 72d8'24"
Rhonda Lease
And 1h26'4" 49d26'24"
Rhonda Lee
And 2h33'55" 49d26'9"
Rhonda Lee Free
Cap 21h19'44" -14d37'44"
Rhonda Lee Glover
Psc 1h5'49" 31d59'27"
Rhonda Louise
Cas 23h48'12" 50d7'59"
Rhonda Louise Dow
Cyg 20h47'10" 48d3'41"
Rhonda LuRae Davidson
And 0h28'41" 38d52'4"
Rhonda Lynn
Lib 14h52'57" -3d35'34"
Rhonda Lynn Gilmore
Vir 13h15'55" 8d10'42"
Rhonda Lynn Stum
Aql 20h37'8" -7d36'28"
Rhonda M. Brewer
Uma 8h20'12" 72d34'44"
Rhonda Mae Bencosme Moir
Ori 5h36'43" -1d18'54"
Rhonda Mae Grenier
Aqr 22h54'22" -10d10'27"
Rhonda Marie
Vir 12h54'46" 12d41'16"
Rhonda Marie
Peg 21h42'39" 13d42'11"
Rhonda Marie Hudson
Psc 1h7'24" 33d4'42"
Rhonda Mary MacNeil
Cas 23h59'59" 59d5'11"
Rhonda McKaig
Psc 1h8'7" 15d6'0"
Rhonda McMullen
Cam 4h22'32" 77d57'34"
Rhonda Meyers
Ori 5h13'50" 9d20'7"
Rhonda My Love
Ari 3h12'59" 16d38'4"
Rhonda Price
Lmi 10h34'29" 38d8'11"
Rhonda Randell
And 1h4'10" 39d30'10"
Rhonda Reneigh Niles
Crb 15h16'24" 29d26'40"
Rhonda Rohelia
Crb 16h6'31" 35d9'8"
Rhonda Sears (Hyper 1)
Sco 17h51'19" -35d47'56"
Rhonda Stewart - Rising Star
Lyn 9h20'54" 34d29'54"
Rhonda Struble - Sousley
And 23h51'34" 40d41'21"
Rhonda "Thumper" Marts
Uma 8h47'48" 69d41'37"
Rhonda & Tom Jonet
Ori 5h0'22" 5d8'20"
Rhonda Towns
Sco 17h17'56" -33d0'17"
Rhonda- Vern, Liz, Kate
Uma 11h18'45" 65d40'47"
Rhonda Wilson
Cnc 8h42'45" 14d31'0"
Rhonda Winfield
Equ 21h12'10" 9d8'34"
Rhonda Z. Locke
Ari 3h8'22" 28d13'6"
Rhonda, Ann Ramsay
Psc 1h27'27" 33d30'1"
Rhonda98
Cam 5h46'54" 59d54'26"
Rhonda's Heart
Cas 1h5'52" 60d28'15"
Rhonda's Hope
And 23h24'59" 42d54'3"
Rhonda's Little Piece Of Heaven
Uma 9h30'41" 48d39'58"
Rhonda's Magic Star
Dra 19h37'12" 63d14'6"
Rhondastar
Cyg 21h56'56" 52d43'51"
Rhondelle's Star
Cas 1h10'48" 50d56'33"
Rhu
Cma 7h24'15" -23d33'23"
Rhyan Close - Rising Star
Lyr 18h47'18" 39d53'42"
Rhyan Forman and Bo Kime
Uma 11h32'22" 43d54'39"
Rhyan Marie Stoecker
Aqr 22h38'3" -22d53'3"
Rhydz
Per 4h6'24" 32d48'1"
Rhylee Starr
Uma 10h8'33" 57d11'36"
Rhyli Hannah Brown
Sgr 18h35'17" -32d58'2"
RHYMA
Sgr 18h21'29" -32d49'10"
Rhymster22
Her 16h16'3" 44d32'21"
Rhys Algie - Eternal Ray of Sunshine
Dor 5h48'20" -62d41'37"
Rhys Anthony Edwards
Dra 16h54'20" 52d29'3"
Rhys Charlie Glover
Her 18h12'47" 19d24'3"
Rhys Christopher Hodges
Umi 17h23'22" 75d48'23"
Rhys Edward Byers
Vir 13h4'12" -2d21'53"
Rhys Elliot Thomas Gregory
Ori 5h47'38" 12d12'13"
Rhys Jackson Birkett 2nd Aug '04
Leo 11h13'49" -0d22'16"
Rhys James
Cra 18h35'18" -41d23'49"
Rhys James Allen
Lmi 9h43'36" 40d57'6"
Rhys James Gasson
Per 2h52'59" 47d38'55"
Rhys James Harris
Ori 5h59'32" 11d22'48"
Rhys James Smith
Gem 6h53'58" 21d35'43"
Rhys Peter Bretschneider
Cnc 9h16'19" 8d18'26"
Rhys Phillips
Leo 11h29'48" -3d25'57"
Rhys Pope - You're My Star
Gem 7h41'8" 21d11'53"
Rhys The Man
Cep 23h9'30" 71d2'20"
Rhys William John Gray
Ori 6h19'1" 16d47'55"
Rhyse
And 2h20'22" 49d56'44"
Rhysie Fearnhead Football Genius!
Cep 22h44'48" 65d59'34"
RIA
Uma 10h12'0" 53d21'29"
Ria
Aqr 23h5'56" -10d55'49"
Ria
Sco 17h36'33" -41d26'47"
Ria
Lup 15h17'42" -41d38'48"
Ria
Cap 21h12'48" -26d33'56"
Ria
Lmi 10h14'22" 38d30'9"
Ria
And 2h13'10" 44d46'54"
Ria Beadham
Cru 11h59'18" -62d41'37"
Ria E. Intal
Uma 11h7'57" 66d21'32"
Ria Louise Barney
Ori 5h15'24" 15d20'7"
Ria *My Shining Star*
Tau 5h19'19" 16d49'16"
Ria Player
And 1h36'54" 49d10'4"
Ria Sood
And 23h18'10" 51d56'9"
Ria717
Cnc 9h6'6" 31d11'13"
Riabear
Sgr 19h22'26" -30d12'39"
Riadh I Abbas
Vir 13h17'21" -7d4'23"
Riain O' Halloran
Lyn 7h56'23" 53d25'15"
Rian
Uma 8h12'8" 69d29'57"
Rian Elizabeth
Cap 20h23'5" -9d42'20"
Rian Elizabeth McKay Nagel
Aqr 22h19'43" -23d24'0"
Rian Kier Morgan-Smith
Aqr 21h9'37" -8d54'30"
Rian McMahon
Umi 11h22'33" 88d28'9"
Rian Patrick
Ori 6h5'48" 4d36'9"
Rian Teise Sherman
Ori 6h13'47" 5d51'25"
Riana
Uma 9h12'39" 52d40'13"
Riana
Dra 16h52'16" 55d13'35"
Riana Josephine Dominguez
Tau 4h22'41" 23d26'0"
Riana MacKenzie
Sco 17h48'12" -42d10'17"
Rianna J. Sellars
Lmi 10h5'36" 33d46'9"
Rianna June DeGraaf
Umi 14h25'27" 75d52'18"
Rianne30
Ori 6h18'20" 19d27'58"

Riannon Lynn
Sco 16h5'5" -24d47'54"
Riarv
Psc 1h16'3" 6d40'53"
Ria's Star
Leo 11h37'15" 9d56'53"
Riata Indigo Nothling Baker
Uma 11h19'57" 60d39'16"
Riawn
And 0h48'16" 25d46'44"
Riaz
Gem 6h47'33" 12d27'38"
Ribel
Lib 15h34'4" -25d30'14"
Ribo Chebeir - Lion Heart 28
Leo 10h9'52" 26d4'42"
Ribsy29
Ori 6h16'19" 7d29'53"
Ric and Lin
Cyg 20h25'48" 52d15'18"
Ric "Flying Fools"
Uma 13h50'18" 52d3'22"
Ric J. "ricmir" Volenec
Uma 12h1'55" 59d12'5"
Ric (Katie's Guiding Light)
Cyg 21h26'19" 45d46'35"
Ric Laue
Aql 20h27'29" -0d54'27"
Ric & Linda
Cyg 19h54'24" 58d6'20"
Ric the Troubleshooter 14.08.1998
Col 5h49'35" -39d6'20"
Ric Vitagliano
Ari 3h17'45" 20d6'34"
Ric Wiskotoni
Ori 5h26'52" 1d52'40"
Rica
Uma 10h7'52" 42d45'49"
ricanlunap
Lib 15h38'14" -3d55'49"
Ricarda
Leo 9h33'46" 7d19'18"
Ricarda Gschaider
Gem 7h24'13" 21d5'27"
Ricardo
Her 18h32'17" 19d49'17"
RICARDO
Cnc 8h2'33" 25d12'22"
Ricardo
Ori 5h28'23" 0d7'21"
Ricardo
Her 16h47'32" 9d8'27"
Ricardo
Per 4h13'36" 52d2'26"
Ricardo A. Trevino
Cnc 8h56'30" 27d56'40"
Ricardo Alfonso Hurtado Valdez
Peg 21h48'14" 16d40'14"
Ricardo Balanzategui Jr.
Uma 9h47'18" 70d51'46"
Ricardo Cardoso dos Santos
And 22h58'12" 37d20'55"
Ricardo & Daluza
Cnc 8h31'36" 31d35'43"
Ricardo Frei Torres
Cep 21h11'19" 57d14'31"
Ricardo J. Hernandez
Gem 6h46'11" 19d52'9"
Ricardo Jason De La Mora Estrada
Her 16h16'42" 44d24'4"
Ricardo & Jeanneth
Vir 14h9'5" -5d43'56"
Ricardo Jose Martinez
Tau 4h20'52" 8d42'55"
Ricardo Jurado
Ori 5h18'13" -6d55'4"
Ricardo L Bailey
Lyn 7h6'46" 58d46'30"
Ricardo Logun De La Mora Autran
Per 3h7'19" 36d53'37"
Ricardo Loya
Aur 5h11'18" 31d29'3"
Ricardo Loya
And 0h50'44" 36d7'8"
Ricardo & Maggie (Bebe/Nene)
Sge 19h39'54" 17d9'23"
Ricardo Manuel Valenzuela
Her 17h26'10" 16d56'0"
Ricardo Morales
Cap 20h52'30" -25d6'30"
Ricardo Quintero
Aql 19h47'48" 13d10'32"
Ricardo Rene Suarez
Lib 14h49'9" -18d33'42"
Ricardo Salgado
Lyr 18h54'30" 43d13'26"
Ricardo Webb
Uma 13h59'5" 60d1'47"
Ricardo y Linda, Un Corazón
Gem 7h29'20" 16d5'26"
Ricardo's Joy
Uma 12h41'27" 54d40'41"
Ricca
Lyn 7h46'5" 37d25'21"
Riccah
Leo 10h10'40" 24d16'24"
Riccardo.
Cnc 8h18'10" 8d43'20"
Riccardo
Lyn 8h55'54" 39d17'51"
Riccardo
Cas 0h14'56" 53d34'21"
Riccardo
Uma 8h12'54" 72d11'48"
Riccardo e Silvia
And 1h56'14" 46d39'36"
Riccardo Frusteri
Per 4h20'21" 39d17'22"
Riccardo Girardelli
Umi 17h31'45" 80d12'21"
Riccardo Giudici
Umi 9h49'57" 86d14'35"
Riccardo (il mio Bambolotto)
Aur 6h29'50" 40d38'16"
Riccardo Morello
Tau 4h34'30" 28d4'4"
Riccardo Panzeri
Cam 4h10'34" 73d53'45"
Riccardo Pierno
Uma 12h10'9" 46d38'27"
Riccardo Rinaldo Grossen
And 1h12'29" 47d0'42"
Riccardo Rotini
Crb 16h15'53" 35d28'28"
Riccardo Vitanza
Aur 6h4'31" 35d50'3"
Ricchi Nachbur-Greco
Lmi 9h35'56" 37d14'1"
Ricci Anthony LaRiccia
Her 16h57'15" 30d47'29"
Ricci Neal Woody
And 2h16'14" 47d30'30"
Riccia
Umi 14h52'57" 72d8'38"
Ricco Donley
Aqr 23h10'47" -19d15'57"
Ricco, don't you go changing!
Col 6h26'47" -35d30'44"
Rich
Cap 20h47'2" -25d54'53"
Rich
Lib 15h44'26" -3d40'44"
Rich
Her 18h37'26" 13d3'46"
Rich
Psc 1h24'56" 18d17'52"
Rich 64
Her 16h36'34" 48d24'18"
Rich and Helen's Piece of Heaven
Lmi 10h44'2" 24d33'51"
Rich and Jess
Cyg 19h35'47" 53d36'19"
Rich and Jess's star
Lyr 18h48'39" 43d57'25"
Rich and Julie Montenegro
Sco 16h18'20" -29d45'15"
Rich and Kathleen's Star
Cyg 21h5'47" 54d34'54"
Rich and Maeva's Future
Lib 14h55'2" -3d31'48"
Rich and Sam The Star of Ammerdown
Cyg 21h42'2" 46d21'12"
Rich Attard 50
Her 17h17'39" 19d26'39"
Rich Becerra
Boo 14h21'45" 13d30'53"
Rich Belau
Her 16h44'9" 46d9'3"
Rich Bertucci
Ari 2h32'26" 26d44'52"
Rich Bette Willie
Lyn 8h29'58" 33d45'12"
Rich & Cathy Eiden
Col 5h35'34" -30d0'25"
Rich & Chris Nowak
Uma 10h36'25" 61d21'21"
Rich & Christy
Cyg 19h41'14" 29d8'23"
Rich Collins
Lmi 10h4'17" 28d12'53"
Rich Collins and Cathy Broderick
Her 16h42'44" 6d12'27"
Rich Connine Always & Forever Star
Sgr 18h12'47" -25d2'5"
Rich Cornhill
Uma 10h35'44" 63d59'39"
Rich & Dari 4-ever
Lmi 9h49'5" 38d59'58"
Rich & Dawn Tinsey and Family
Aql 20h11'21" 11d52'51"
Rich Franz
Sgr 19h1'33" -22d17'52"
Rich Garbee
Psc 1h22'58" 17d7'42"
Rich George
Uma 14h15'3" 59d15'5"
Rich Hegerich
Cep 22h4'44" 65d13'0"
Rich Heil
Leo 11h22'47" 17d26'3"
Rich & Jen
Uma 8h14'49" 65d29'26"
Rich & Jen Hollien-20 Happy Years
Cyg 20h2'31" 36d54'28"
Rich Johnson
Her 17h13'24" 36d26'1"
Rich & Kim Buszek
Cyg 20h39'54" 53d7'0"
Rich Kleitman
Ori 5h28'34" -1d13'5"
Rich Mack
Uma 13h51'0" 54d41'25"
Rich Mastriano
Tau 4h24'31" 19d57'33"
Rich Mayo
Ari 2h7'5" 24d50'46"
Rich Montgomery
Cep 22h29'10" 63d41'26"
Rich - My Gorgeous Boy
Cep 0h1'33" 72d32'23"
Rich - My True Shining Star
Gem 7h32'36" 27d26'11"
Rich n' Rach
Uma 11h6'15" 52d57'9"
Rich - Nagielski
Ori 6h9'22" -0d4'59"
Rich Nussbaum
Sgr 17h57'12" -19d9'39"
Rich Rydarowski
Aqr 22h32'58" -3d27'38"
Rich Smart
Leo 11h19'20" 23d54'38"
Rich Stremme
Aqr 21h22'19" -13d13'58"
Rich Temple
Ori 5h27'6" -10d5'23"
Rich & Trudy Shontz
Cyg 21h37'25" 49d49'43"
Rich Whiteford
Her 16h54'16" 33d24'55"
Rich Zanzoth - The Love of My Life
Ori 6h11'22" 15d53'30"
Rich, Elena loves you
Vir 13h19'17" 11d31'16"
Richa Derashri
Boo 14h29'30" 14d33'6"
RICHALDEN
Lib 14h49'40" -18d23'12"
Richard
Aqr 22h50'58" -5d9'38"
Richard
Uma 11h30'28" 59d9'8"
Richard
Cep 22h4'2" 68d24'40"
Richard
Sco 17h24'2" -37d33'33"
Richard
Psc 23h57'46" 9d0'21"
Richard
Boo 14h36'35" 27d26'14"
Richard
Uma 11h6'13" 43d46'6"
Richard #1 Son
Aur 6h3'33" 47d51'16"
Richard A. and Joanne L. Seaman
Per 3h16'55" 52d34'56"
Richard A. Beam
Her 17h52'41" 28d51'47"
Richard A. Caurso
Her 18h32'13" 19d48'8"
Richard A. Cooper
Gem 7h23'3" 15d15'34"
Richard A. Holland
Per 3h20'20" 44d51'10"
Richard A. Mather, CSB
Uma 9h21'40" 46d40'50"
Richard A. Mikkelson
Cnc 8h35'51" 7d23'23"
Richard A. Miller
Her 18h50'31" 22d54'50"
Richard A. Nemmers, Sr.
Cep 22h49'24" 67d9'4"
Richard A. Patriacca
Psc 1h10'52" 32d22'11"
Richard A. (Ricky) McGregor
Sco 16h35'31" -38d19'33"
Richard A. Sather
Dra 19h35'29" 68d58'45"
Richard A. Stewart
Uma 8h28'2" 68d45'15"
Richard A. Tortora, Sr.
Cep 21h35'19" 70d46'48"
Richard - A Very Precious Pearl.
Per 4h47'44" 38d51'54"
Richard A. Waybright
Aqr 22h12'24" -24d13'47"
Richard A. Wilson
Tau 4h56'3" 26d7'22"
Richard A. Woehler
Sco 16h4'59" -8d46'3"
Richard Addeo
Her 17h58'52" 22d25'42"
Richard a.k.a. Megan Black
Psc 1h29'59" 6d59'40"
Richard Al Spradling
Umi 17h7'29" 80d20'12"
Richard Alan Brock
Per 4h44'32" 46d43'21"
Richard Alan Conaway
Vir 12h7'58" 7d54'47"
Richard Alan Hart
Lyn 7h55'53" 58d28'2"
Richard Alan Lebow
Cnv 12h42'22" 40d2'25"
Richard Alan Martin, Jr.
Leo 10h13'16" 24d58'12"
Richard Alan Watkins - Laughing Bear
Ara 16h35'54" -52d48'6"
Richard Alan Winkler II
Uma 9h18'57" 55d19'2"
Richard Albert Last
Gem 6h58'26" 23d18'50"
Richard Alexander Whiteley
Cep 22h33'30" 62d0'41"
Richard Allan
Col 5h32'49" -33d31'0"
Richard Allan Creutz
Cep 21h32'28" 64d31'53"
Richard Allan Swanson
Lib 14h40'40" -11d47'31"
Richard Allen
Uma 10h26'41" 50d35'34"
Richard Allen Belzer
Per 3h40'41" 49d41'50"
Richard Allen Berg
Aql 19h54'42" -0d26'53"
Richard Allen Bowyer
Uma 9h15'32" 64d56'41"
Richard Allen Daniels Jr.
Tau 5h9'51" 17d29'0"
Richard Allen Smith, Sr.
Aur 5h29'47" 40d43'26"
Richard Allen Williams
Ori 4h51'26" 9d39'22"
Richard & Amanda
Pyx 8h41'11" -20d23'54"
Richard & Amanda Owens
Gem 6h59'25" 22d16'46"
Richard and Bernadette Marshall
Uma 10h36'57" 55d33'7"
Richard and Carol Davis
Cyg 20h48'6" 35d37'25"
Richard and Christine Hunsinger
Vir 13h58'37" -13d14'30"
Richard and Elaine Owen
Per 3h49'23" 42d51'2"
Richard and Julia Lyn
Cyg 21h18'0" 31d18'39"
Richard and Kerry
Cru 12h36'11" -56d37'54"
Richard And Linda Marty
Ori 4h47'19" 13d19'13"
Richard and Marianne Yeider
Crb 15h30'46" 30d58'0"
Richard and Marie Procida
Aqr 20h44'22" -2d55'54"
Richard and Paula Oram
Cyg 19h58'46" 46d29'23"
Richard and Samantha
Ari 3h7'20" 27d28'33"
Richard and Tramy Luong
Cnc 8h46'12" 27d19'58"
Richard and Vicki Prather 5/1/1982
Cyg 19h45'32" 32d34'4"
Richard Anderson
Crb 15h48'19" 26d23'20"
Richard Anderson
Psc 1h21'2" 24d4'43"
Richard Anderson Roberts
Pho 1h16'22" -46d29'59"
Richard Andrew Biby
Gem 7h12'39" 16d15'51"
Richard Andrew Sweatt
Cep 0h48'36" 83d15'45"
Richard Angelini Ortiz
Oph 17h44'10" -0d32'28"
Richard Angell
Cap 21h12'44" -14d35'54"
Richard & Angie Buzas
Cru 12h39'32" -59d41'23"
Richard & Ann Wynne
Cyg 21h52'5" 42d15'34"
Richard & Anne - Eternal Love
Cru 12h4'19" -61d18'40"
Richard Anthony
Uma 11h11'48" 33d21'52"
Richard Anthony Fasanelli (Poppy)
Gem 6h46'22" 14d29'4"
Richard Anthony Petrak
Her 18h35'0" 24d33'51"
Richard Anthony Wakefield
Psc 22h53'4" 3d51'52"
Richard Armstrong
Cap 20h17'43" -9d17'27"
Richard Aronson
Tau 5h34'20" 27d43'41"
Richard Arthur - The Best Dad
Ori 6h2'16" 9d59'12"
Richard Asaro
Boo 14h56'38" 27d13'6"
Richard at 60
Uma 11h40'26" 31d40'8"
Richard Attias
Uma 10h22'47" 45d52'18"
Richard Axell
Per 2h41'11" 54d13'36"
Richard B Bradley's Big Shining 40
Psc 1h3'17" 3d31'50"
Richard B. Holmes Jr.
Ari 2h51'54" 27d31'6"
Richard B. Kirkland
Lyn 7h21'20" 50d0'49"
Richard B. Wasilewski Memorial Star
Aqr 22h40'18" -20d32'44"
Richard B. Zug
Her 18h3'18" 31d54'42"
Richard Bachstetter
Psc 0h46'5" 16d6'40"
Richard Bailey Gordon Dayhuff
And 0h23'18" 32d50'34"
Richard Baker
Uma 8h52'46" 50d31'5"
Richard Barone
Uma 10h52'17" 58d33'34"
Richard Barrett
Psc 0h8'43" 8d55'16"
Richard Barton Maxwell
Aqr 20h55'37" -8d55'53"
Richard Bate
Lyn 9h19'24" 33d37'26"
Richard Becker
Lyr 18h37'39" 31d36'19"
Richard Belliveau
Sco 17h57'44" -30d56'15"
Richard Bennett
Ori 5h36'55" 13d27'27"
Richard & Betty Dusablon
Aql 19h0'44" 16d5'22"
Richard Bianco's Christmas Star
Umi 16h9'5" 75d49'30"
Richard Bielski
Ori 6h16'54" 13d35'17"
Richard Billings Kocak
Umi 15h39'9" 72d34'12"
Richard Blakeley
Uma 9h26'14" 45d38'23"
Richard Bolenbaugh
Aur 5h16'39" 30d2'40"
Richard Bonnage 'Tiger'
Boo 15h21'32" 47d51'32"
Richard & Bonny
Tau 4h44'58" 26d45'15"
Richard Borbon
Gem 7h28'44" 25d24'9"
Richard Bouchard, Sr.
Cep 20h50'50" 59d38'48"
Richard Boyajian A Stellar Nurse
Per 3h22'15" 42d45'37"
Richard Boyd
Dra 16h52'27" 58d34'27"
Richard & Brenda DeSantis
Cyg 19h56'18" 44d53'6"
Richard Brenneman
Psc 0h23'38" -3d45'48"
Richard Brenton Holliday Sr.
Her 17h31'6" 46d14'28"
Richard Brian "R.B." Smith
Ori 6h15'12" 15d25'30"
Richard Burd
Her 17h16'17" 13d34'23"
Richard Byron Ellinor
Ori 5h38'59" -1d32'58"
Richard C. & Cecelia A. Schoppaul
Cyg 20h37'19" 59d31'0"
Richard C. Collins
Uma 8h51'29" 65d27'37"
Richard C. Garbarino
Per 2h25'3" 53d36'4"
Richard C. Neal
Ori 5h29'35" 2d28'15"
Richard C. Pieper
Lib 15h5'11" -29d9'5"
Richard C Pitts, US Navy
Aql 19h5'18" 14d59'6"
Richard C. Rummelhart, Clea & PWM
Per 3h11'11" 47d51'23"
Richard C. Sapp
Ori 5h52'39" 5d40'54"
Richard Call
Sgr 18h14'50" -31d54'21"
Richard Campbell
Uma 11h59'38" 43d59'16"
Richard Carl Hall
Her 17h38'59" 33d48'34"
Richard Carl Howell
Leo 11h7'57" 3d0'37"
Richard Carl Segal
Cnv 13h51'52" 38d5'37"
Richard Carlo Salituro "Rick"
Per 4h6'48" 34d25'14"
Richard Carmen Giorgi
Uma 11h37'53" 61d23'17"
Richard Cartagena
Uma 9h8'58" 60d33'32"
Richard Casimier Skarup II
Her 18h3'58" 18d9'0"
Richard Charles Burns
Cep 22h50'40" 66d43'45"
Richard Charles Harbaugh JR
Uma 12h3'49" 52d43'44"
Richard Charles Maertz
Aqr 20h38'37" -2d10'46"
Richard Charles O'Connor
Her 17h12'44" 27d7'22"
Richard Cheeseman
Cep 22h33'25" 61d58'7"
Richard Christian Appell
Cep 22h41'51" 64d53'50"
Richard & CJ Byrne
Uma 12h14'24" 61d9'55"
Richard Clapper
Lyn 7h48'5" 36d17'14"
Richard Cline
Gem 6h35'56" 21d19'7"
Richard Cole & Gina Paige Yancey
Ori 4h53'23" 12d46'11"
Richard Conaway
Aur 5h35'39" 44d19'46"
Richard Cook
Uma 10h14'43" 43d41'21"
Richard Coppol
Cep 22h13'41" 73d44'35"
Richard Coueffin
Vir 13h8'3" -3d17'1"
Richard Coyne
Tau 3h50'43" 22d51'12"
Richard "Crab-Man" Bennett
Psc 0h52'17" 15d49'31"
Richard Craig Embry
Cyg 20h47'18" 31d22'50"
Richard Craig Steinfeld
Tau 4h27'34" 18d24'29"
Richard Curley
Aqr 22h40'47" -1d17'8"
Richard Curran
Uma 10h21'25" 62d4'43"
Richard D. Chewning, MD
Tau 4h19'40" 8d39'59"
Richard D. Chimblo
Sco 16h13'12" -8d41'45"
Richard D. Cummings
Her 17h29'41" 39d6'15"
Richard D. Denehy
Her 16h21'42" 19d6'43"
Richard D. Gray
Lib 14h52'8" -4d21'2"
Richard D. Moss
Psc 0h18'6" -5d26'16"
Richard D. Rasmussen
Aql 19h53'3" 3d44'36"
Richard "Dad" Honnas
Sgr 19h4'11" -24d34'55"
Richard "daddy" Bunker
Sgr 18h35'55" -17d48'13"
Richard "Daddy & Papa" Mucelli
Per 4h32'21" 39d22'4"
Richard Dale Land
Tau 5h58'15" 25d26'5"
Richard Dalessandro
Per 2h27'40" 52d29'58"
Richard Dan Jennette
Lmi 10h34'31" 35d46'36"
Richard Daniel
Tri 2h27'43" 37d4'30"
Richard Daniel Clark
Pho 0h23'22" -42d26'48"
Richard Darroll "Striker" Davison
Her 17h53'3" 23d26'15"
Richard David MacRaild
Her 17h12'35" 33d41'48"
Richard David Tiffany
Per 2h55'37" 56d9'3"
Richard David Waritz
Vir 12h29'18" -6d29'11"
Richard David Wilkins
Psc 1h30'38" 15d31'31"
Richard Davis Ahrens
Tau 5h32'18" 26d54'1"
Richard & Dayna's Wedding Star
Ori 5h39'50" -2d53'1"
Richard Dean Boes 1977 - 2003
Her 16h32'40" 34d40'16"
Richard Dean Brown
Ori 5h56'4" 22d51'8"
Richard Dean Bukoski
Cep 22h19'45" 67d21'40"
Richard Dean Parsons
Ori 6h11'56" 3d28'39"
Richard Dean Schlepp
And 23h50'15" 39d42'28"
Richard Dean Whitcomb
Ori 6h7'25" 0d11'10"
Richard & Deanna Gilbert
Del 20h46'20" 11d30'50"
Richard DeBlis
Cas 1h20'19" 65d23'34"
Richard DeCarlo
Her 17h19'20" 15d23'50"
Richard Deen-Francis Bruce
Ori 5h41'32" -0d21'1"
Richard DeJong
Her 17h33'14" 34d2'6"
Richard & Denise's Anniversary Star
Her 17h42'10" 40d15'28"
Richard Dennis Auten (Rick)(Ricky)
Ari 2h38'35" 25d4'12"
Richard Dewey "My Valentine"
Leo 11h57'36" 22d6'21"
Richard Diamond
Cep 22h9'45" 56d30'5"
Richard Dicharry Jr. NOLA
Psc 23h51'7" 6d30'11"
Richard "Dick" F. Ficke, Jr.
Ari 2h17'39" 10d35'4"
Richard "Dick" Fox
Cep 22h40'56" 74d50'16"
Richard "Dick" Hayes
Uma 10h31'15" 48d47'26"
Richard "Dick" Orkin
Boo 14h33'28" 26d57'41"
Richard "Dick" Welton
Per 3h5'39" 53d8'16"
Richard Diecker
And 1h24'14" 47d28'15"
Richard DiSalle
Cep 20h57'53" 57d47'9"
Richard & Dolores Polancyak
Cyg 20h40'33" 41d43'16"
Richard Dorfman
Cap 21h36'21" -13d11'45"
Richard & Dottie's Sunny
Lyn 8h38'6" 38d42'13"
Richard Douglas
Sco 16h12'55" -15d59'41"
Richard Douglas Haefner
Ori 5h22'53" 4d19'44"
Richard Douglas Lacey Sr.
Leo 10h47'28" 14d23'58"
Richard Drake Geller
Ari 3h19'13" 30d47'36"
Richard Drew
Uma 11h57'30" 37d59'30"
Richard Drury
Per 2h38'6" 55d50'18"
Richard Duane Sibley
Her 16h42'17" 36d20'2"
Richard "Dude" Avery June 22, 1961
Per 2h52'52" 52d52'1"
Richard Duke
Gem 7h16'0" 23d18'8"
Richard Dunbar
Ari 2h34'41" 14d21'29"
Richard E. Beard
Lib 15h17'13" -6d30'8"
Richard E. Desio
Her 17h13'45" 39d1'31"
Richard E. Gray Sr.
Cep 20h54'59" 60d8'37"
Richard E. Hartley, The Sax Man
Tau 4h29'28" 19d25'46"
Richard E. Lumsden
Psc 1h49'29" 6d30'28"
Richard E. MacNeil
Ori 6h3'18" 21d25'57"
Richard E & Mary E. Holden
Uma 9h8'1" 57d30'36"
Richard E. McLoughlin
Ari 3h18'19" 22d5'56"
Richard E. Murphy
Sco 17h6'45" -38d41'22"
Richard E. Newman
Uma 8h22'33" 67d35'14"
Richard Earl Durr
Lib 15h17'11" -5d28'29"
Richard Earl Griffin
Ari 2h4'25" 23d36'22"
Richard Earl Mitchell
Vir 15h7'0" 4d36'25"
Richard Earle Kirby "Papa's Star"
Aql 20h16'35" 1d33'42"
Richard Edmond Hawley
Cep 22h48'53" 70d17'29"
Richard Edward Dayhuff
And 1h31'48" 43d36'23"
Richard Edward Nulman
Her 18h23'38" 18d2'14"
Richard Edwin Saunders
Lib 15h47'9" -9d20'33"
Richard Ehrhart
Psc 1h29'19" 27d48'23"
Richard Eisenberg
Her 17h58'52" 44d7'8"
Richard & Eleanor Bloomer
Ori 6h18'10" 7d5'11"
Richard Ellis Stirk
Uma 12h28'32" 58d17'24"
Richard Emil Madsen
Cep 23h20'59" 77d13'25"
Richard Emil Wolfram
Leo 9h50'56" 24d0'1"
Richard Emlin Reed
Cnc 9h15'15" 8d25'10"
Richard Ennen
Cyg 21h31'38" 44d21'14"
Richard Eric Bradley
Brenda's Love
Ari 2h8'19" 23d59'31"
Richard & Esmeralda
Gem 7h6'27" 32d47'59"
Richard Eugene
Per 4h9'7" 33d52'23"
Richard Eugene Barnes
Vir 13h53'54" -18d12'59"
Richard Eugene Roberts
Her 16h55'45" 34d41'45"

Richard & Eva
Lyr 19h2'29" 28d55'53"
Richard F. Bertrand Jr.
Sco 16h10'48" -10d50'16"
Richard F. Brauneisen
Sco 16h10'13" -12d58'19"
Richard F. Coles " Hail Redskins"
Her 17h9'6" 25d19'36"
Richard F. Fleming
Sgr 18h53'52" -24d34'17"
Richard F. Moore
Her 16h39'19" 45d42'43"
Richard F. Peone: special to us all
Aqr 22h35'28" -16d24'54"
Richard Fairgrieve
Lyn 8h13'22" 51d28'23"
Richard Falcone
Vir 13h3'43" 2d50'23"
Richard Felix Addor
Cas 23h2'32" 56d34'6"
Richard Ferris
Ori 6h7'8" 1d24'32"
Richard Files
Sco 17h45'55" -35d18'46"
Richard & Frances Bell
Aql 19h46'46" 11d55'1"
Richard & Francine
Cas 0h56'43" 57d45'49"
Richard Francis & Alicia Marie
Cyg 20h23'12" 54d34'36"
Richard Francis Barrett
Ori 5h25'55" 10d44'56"
Richard Francis Christian Kent
Aqr 21h47'35" -5d37'53"
Richard Francis Kienle
Ari 1h59'58" 14d38'39"
Richard Frankovic
Cep 21h32'1" 64d46'8"
Richard Frederick Nettell
Lib 15h7'31" -11d41'40"
Richard French
Ari 2h45'26" 16d17'25"
Richard Fry
Cra 18h39'48" -42d45'1"
Richard G Casper
Cep 22h41'34" 65d25'33"
Richard G. Haylock
Ori 5h36'21" 3d12'27"
Richard G. Parker Jr.
Lib 15h4'19" -23d48'57"
Richard Garay
Aur 5h19'49" 44d25'54"
Richard Gary Glasco 3/1/50-5/16/05
Per 3h34'48" 47d41'16"
Richard Gary Zell
Per 4h38'42" 43d29'45"
Richard Georg Rudolf
Uma 14h15'6" 60d39'2"
Richard George Skowronski
Sco 16h16'39" -13d27'7"
Richard George Trevena
Uma 9h13'22" 63d1'0"
Richard Gerald Sanborn
Psc 23h19'43" 2d11'5"
Richard Gerald Vento
Ori 5h22'24" -7d5'39"
Richard Giglio
Per 4h48'7" 45d44'44"
Richard Giichi Suzuki
Aqr 20h39'23" -11d16'19"
Richard Gilbert Quick
Per 2h16'37" 55d40'7"
Richard Giles Wells
Cap 20h13'22" -17d19'30"
Richard Glenn Fister
Uma 10h42'17" 48d26'36"
Richard Glenn Pittwood
Aqr 23h23'7" -19d25'11"
Richard Goodman Alexander, Sr.
Vir 13h4'1" -11d29'27"
Richard Goodwin
Cet 1h14'49" -15d0'6"
Richard Gordon Bastow
Ori 5h35'54" -1d44'30"
Richard Gordon Miller
Leo 9h24'29" 19d4'2"
Richard Gordon Partridge
Leo 10h56'1" 10d44'50"
Richard Gordon Poage
Cap 21h49'32" -19d52'0"
Richard Gould
Sco 16h40'25" -29d39'55"
Richard Grandpa Farrell
Per 3h8'26" 52d43'28"
Richard Grinstead
Cnc 8h35'55" 22d46'11"
Richard Grizzell
Gem 6h54'30" 23d2'47"
Richard Groetzinger
Hya 9h37'52" -0d51'21"
Richard Gromoll
Dra 18h39'11" 76d28'25"
Richard Grzybowski
Lyn 8h19'34" 35d8'48"
Richard Gutierrez, Jr.
Cap 21h47'16" -23d5'19"
Richard H. Grant
Leo 10h32'18" 20d8'3"
Richard H. Heil
Uma 9h40'24" 68d31'52"
Richard H. Hoffman
Psc 23h44'34" 6d54'11"
Richard H. Koch
Per 2h34'16" 55d41'51"
Richard H Miller
Gem 6h43'11" 18d31'22"
Richard H Nye
Cnc 8h22'52" 7d9'8"
Richard H Sargent Jr.
Lib 15h21'18" -25d52'15"
Richard H. Stevens
Lib 14h42'48" -11d39'58"
Richard Hale Good
Her 17h8'32" 32d8'39"
Richard Hall
Uma 9h29'28" 63d18'45"
Richard Hamilton
Aqr 21h32'20" 1d17'23"
Richard Hangley
Uma 11h30'46" 53d12'45"
Richard Harless
Cep 21h32'21" 63d50'54"
Richard Harold Whyte
Boo 15h46'53" 51d38'25"
Richard Harries
Ori 5h53'43" 15d15'4"
Richard Heath Woods
Uma 10h3'43" 68d40'56"
Richard & Helen Zeff
Lyr 19h15'35" 29d7'43"
Richard Helms
Aql 18h53'18" -0d6'28"
Richard Hendershot
Aur 5h50'42" 52d19'6"
Richard Henry McKinnon
Cyg 19h34'5" 29d6'34"
Richard Herbert Sansweet
Sco 17h49'20" -30d36'1"
Richard Holmes
Cam 7h18'36" 63d4'48"
Richard Huesman
Per 4h0'41" 42d3'50"
Richard Hugh Kimball
Her 17h17'31" 15d25'19"
Richard Hunt
Leo 9h40'12" 29d12'11"
Richard "Hunter" Campbell
Uma 12h24'53" 54d13'23"
Richard Hunter Reinhart
Aqr 21h45'49" -2d50'3"
Richard I
Her 17h9'50" 32d48'56"
Richard I. Lynch
Cap 21h31'20" -16d31'56"
Richard J. Childs
Psc 23h46'23" 5d0'48"
Richard J. DiPietro
Cet 2h46'57" 8d22'46"
Richard J. Dougherty
Cep 22h33'25" 78d1'27"
Richard J. Farrell
Her 18h44'45" 20d9'10"
Richard J. Fisher
Per 1h47'37" 50d59'5"
Richard J. Franz
Uma 9h56'52" 45d45'57"
Richard J. Gallego
Gem 7h44'30" 26d56'35"
Richard J. Green, Jr.
Uma 11h36'46" 52d47'37"
Richard J. Harjung
Cam 4h17'13" 68d30'20"
Richard J. Kopp
Cep 20h38'44" 61d11'23"
Richard J. Krautbauer
Her 16h59'3" 30d39'28"
Richard J Kuehn
Cyg 21h3'51" 50d39'25"
Richard J Kyle
Cyg 21h53'18" 42d47'16"
Richard J. Latta
Her 16h22'55" 10d11'29"
Richard J McKinnon
Ari 2h31'18" 12d23'1"
Richard J. Mooney
Boo 15h17'26" 33d44'53"
Richard J. Osmon Sr.
Psc 0h54'59" 9d30'30"
Richard J. Passabet
Uma 11h43'36" 42d24'29"
Richard J Schlumpf
Cnc 8h35'45" 21d52'29"
Richard J. Schmidt
Uma 12h48'33" 53d13'48"
Richard J. Smith
Uma 9h20'2" 64d9'13"
Richard J. Smith
Her 17h48'21" 27d55'48"
Richard J. Stephan
Aur 5h41'26" 40d42'47"
Richard J. Tucker
Gem 7h16'36" 14d5'26"
Richard J. Varadi
Sgr 18h8'51" -25d22'27"
Richard J Zammer "My Rick"
Cep 22h23'0" 86d18'34"
Richard Jackson
Leo 10h24'14" 27d16'53"
Richard Jackson Horlocker
Sgr 18h36'40" -22d54'57"
Richard Jacob Harvey
Ori 5h23'0" 0d49'31"
Richard James Claxton
Umi 14h11'47" 68d37'7"
Richard James Francis Kusak II
Uma 9h23'38" 60d19'2"
Richard James Kaercher
Lib 15h2'2" -0d57'23"
Richard James Montero Sr
Gem 6h23'48" 21d56'26"
Richard James Peebles
And 23h46'45" 46d52'40"
Richard James St. Clair
Cnc 8h15'15" 10d41'35"
Richard James Westbrook Allen
Uma 9h42'18" 60d16'41"
Richard James Williamson
Leo 9h29'42" 11d49'47"
RICHARD & JANNETTE
Lib 15h14'1" -5d14'18"
Richard Jared Salome
Cyg 21h47'30" 55d11'20"
Richard Jason Halliday
Ori 6h18'45" 19d6'6"
Richard Jeannin, Sr.
Sgr 18h19'6" -18d5'19"
Richard Jeffery Jones
Lmi 10h25'30" 30d10'44"
Richard Jeffrey Welsch
Ori 5h29'29" 2d41'13"
Richard Johann Boerst
Cyg 20h14'24" 43d28'1"
Richard John Haus Sr.
Ori 6h19'55" 18d14'39"
Richard John Hunter
Lib 14h57'41" -17d48'21"
Richard John Kelly
Cep 21h49'14" 55d36'13"
Richard John Kilts
Psc 1h16'48" 31d33'36"
Richard John Kinnard - Dickie's Star
Ori 5h42'13" -4d14'31"
Richard John Liotta
Cas 0h48'43" 52d53'55"
Richard John Nasta
Gem 7h16'55" 33d51'46"
Richard John Robert Cembalisty
Aqr 22h13'17" -0d21'50"
Richard John Roberts 100275
Cru 12h17'31" -62d36'45"
Richard John Wright
Cep 22h48'6" 76d22'10"
Richard Johnson
Cep 20h41'42" 64d51'49"
Richard Joseph
Lib 15h6'8" -3d0'56"
Richard Joseph Beaton
And 2h35'49" 45d38'40"
Richard Joseph Couture
Gem 7h24'1" 26d44'20"
Richard & Joseph Craig
Per 4h16'35" 51d27'18"
Richard Joseph Filosi
Cap 21h48'55" -12d27'5"
Richard Joseph Gilbert
Uma 9h54'15" 51d56'43"
Richard Joseph Herzog
Leo 9h22'23" 27d59'4"
Richard Joseph Kauffman
Cap 21h34'6" -15d7'8"
Richard Joseph McLaughlin
Per 3h40'44" 43d47'45"
Richard Joseph Patrick McKenna
Lyn 8h28'41" 38d26'7"
Richard Joseph Sgrignoli
Per 3h18'46" 54d31'12"
Richard Joseph Spear
Uma 8h55'8" 48d58'13"
Richard Joshua Hernandez
Lyn 7h57'3" 38d31'15"
Richard Julius
Cep 23h12'42" 70d46'28"
Richard K. Collard
Sgr 19h47'14" -11d44'34"
Richard K Gumbert
Gem 6h40'44" 20d49'52"
Richard K. Retherford Jr.
Ari 2h19'44" 22d52'11"
Richard K. Vanhorn, Jr.
Ori 5h55'8" 20d53'21"
Richard Kaltenbach II
Her 17h49'47" 49d38'56"
Richard & Karen soul to soul
Cyg 21h57'11" 40d18'52"
Richard Kavity
Aur 5h47'31" 37d40'26"
Richard Kendall Barron
Her 18h11'28" 32d6'19"
Richard Kenneth Fraser
Per 4h4'13" 47d50'20"
Richard Kenneth Lewis
Lyn 8h48'0" 33d8'56"
Richard Kepler VanEvera
Cep 21h9'7" 65d57'52"
Richard Kettlewell
Per 3h39'47" 44d45'14"
Richard Kevin Hughes
Her 18h19'23" 15d13'13"
Richard King Fisher of Bass Cottone
Vir 12h57'31" -3d36'26"
Richard Kissane
Cyg 21h42'19" 38d25'15"
Richard Klinger
Leo 10h9'41" 12d15'34"
Richard Knott Woolley
Leo 10h7'54" 11d0'1"
Richard Kocur
Ori 5h46'4" -0d26'3"
Richard Koebel
Ori 5h41'6" -2d16'40"
Richard Krommenhoek
Lib 15h13'33" -12d18'52"
Richard L. Alexander
Vir 14h18'25" 1d37'14"
Richard L. Chace, Jr.
Vir 12h29'17" -1d16'46"
Richard L. Clark
Ori 5h5'29" 13d19'28"
Richard L. Duppler
Cep 22h32'32" 57d35'34"
Richard L. Fitzpatrick
Cap 21h11'35" -20d55'30"
Richard L. Gibbons
Sgr 18h54'9" -16d0'12"
Richard L. - "My Way" - Ansley
Aql 19h8'38" -0d24'33"
Richard L. Nugent
Psc 1h16'29" 18d37'26"
Richard L. Penn " Rick"
Uma 9h46'34" 56d46'4"
Richard L. Plasch
Per 3h14'58" 44d21'57"
Richard L. Wallace, M.D.
Lyn 6h19'0" 54d13'35"
Richard L Winkelbach
Umi 15h33'52" 69d44'24"
Richard L. Yale
Eri 4h30'35" -0d44'7"
Richard L. Young
Per 2h54'36" 51d12'53"
Richard Lack
Vir 13h23'57" 11d47'16"
Richard Lance Listiak
Cyg 21h49'16" 38d32'50"
Richard "LANCE-BOB" Warren
Uma 9h25'19" 43d23'59"
Richard Landau
Ori 5h40'36" 15d31'36"
Richard Lang
And 23h36'27" 47d47'37"
Richard Lawlor, Jr.
Per 2h5'53" 53d31'50"
Richard Lawrence Baker
Lmi 9h54'15" 37d41'21"
Richard Lawrence Hershey, III
Her 16h54'56" 17d7'41"
Richard Lawrence Kleiman
Cap 21h0'55" -24d9'24"
Richard Lawrence Listman, Jr.
Ari 3h21'35" 27d0'5"
Richard Lee Barrett
Cap 20h15'36" -16d12'2"
Richard Lee Brown "Happy 58th"
Uma 11h1'37" 44d52'50"
Richard Lee Froio BD 10-12-55
Lib 15h52'15" -12d4'46"
Richard Lee Garrett A.K.A. Snow
Uma 11h30'1" 39d3'10"
Richard Lee Lock
Her 16h22'43" 47d42'40"
Richard Lee Reynolds
Lib 15h16'2" -23d15'27"
Richard Lee Ritz
And 1h23'8" 44d7'14"
Richard Lee Robinson
Ori 6h10'1" 11d54'14"
Richard Lee Rowsey
Lib 15h48'3" -17d47'58"
Richard Lee Schenck II
Cnc 8h17'30" 20d41'39"
Richard Lee Stiles
Leo 10h36'53" 10d53'14"
Richard Lee Uhlman
Cas 23h27'4" 56d5'11"
Richard Lee Valerio
Her 17h40'59" 39d4'42"
Richard Lee Wiliams
Aur 6h23'26" 41d40'37"
Richard Lee Wilson
Uma 10h54'18" 54d51'37"
Richard Legoza
Cen 11h43'13" -59d3'5"
Richard Leitner
Lyn 7h1'31" 53d14'53"
Richard Lemar Straney
Uma 11h18'14" 42d37'10"
Richard Lenz
Tau 4h14'40" 20d9'19"
Richard Leon Rouisse
Leo 11h1'16" 16d54'52"
Richard Leon Stein
Per 2h46'12" 53d48'37"
Richard Leonard Harvey
Uma 13h16'57" 57d32'9"
Richard Leroy Wilson
And 2h21'5" 45d51'40"
Richard Li
Dra 18h48'24" 72d50'37"
Richard & Lisa Hollins
Cas 0h12'20" 53d14'1"
Richard & Lisa (Vogel) Ohlmeyer
Cen 14h21'27" -32d39'27"
Richard Louis Knopman
Uma 11h56'18" 32d24'12"
Richard Louis Peddicord
Boo 15h39'40" 39d38'9"
Richard Louis Schepacarter
Sco 17h53'48" -30d17'46"
Richard Louis Thomas
Cnc 8h20'44" 15d0'59"
Richard Loves Jade
And 23h44'20" 42d39'19"
Richard Love's Wit, Wisdom and Love
Aql 19h49'52" -0d5'42"
Richard Lubner
Cru 12h33'58" -60d28'57"
Richard Lukas Jr.
Her 16h9'6" 46d33'36"
Richard Luszcz
Cep 21h29'22" 64d58'40"
Richard Lutgen
Uma 9h54'8" 47d32'44"
Richard Lutz
Cnc 9h6'52" 32d21'2"
Richard Lyle Lomax
Uma 10h4'44" 70d32'13"
Richard Lyles
Aql 19h21'47" 16d11'29"
Richard Lynch
Uma 8h50'3" 50d44'45"
Richard M Cesta
Aql 20h22'23" 7d54'22"
Richard M. DeMell
Per 3h51'10" 48d54'32"
Richard M. Furniss
Psc 0h50'26" 3d43'7"
Richard M. Neiter
Her 16h39'44" 26d58'52"
Richard M. O'Connor
Her 17h26'36" 44d53'20"
Richard M. Patterson
Lyn 8h6'35" 54d57'37"
Richard M. Pavich
Boo 14h52'15" 52d49'7"
Richard M. Surface
Per 2h20'49" 52d43'5"
Richard MacMillan
Ari 3h16'19" 21d43'3"
Richard Malcom Gordon Scott
Cnc 8h13'17" 32d17'36"
Richard Manning
Vir 13h12'41" -14d59'59"
Richard Marc Garzon
Lyn 7h58'14" 40d27'33"
Richard Marc Kershow
Cam 4h25'20" 66d36'1"
Richard Marion Jaillett, Jr.
Gem 6h21'58" 18d9'1"
Richard Mark Aston
Leo 11h11'47" -1d17'53"
Richard Martin Yahnian
Leo 9h44'41" 29d2'9"
RICHARD MARTINEZ
Uma 11h19'7" 49d2'51"
Richard Martland Blaney
Psc 0h50'46" 21d4'29"
Richard & Mary Louise Adam - 50th
Uma 9h26'3" 67d45'54"
Richard Matthew Grant
Her 17h45'10" 42d44'44"
Richard Matthews
Gem 6h51'16" 32d24'2"
Richard Maurice O'Brien
Cyg 21h40'53" 37d10'59"
Richard Max J. C. C.C. King
Per 3h21'3" 47d17'58"
Richard Medlock
Uma 11h18'9" 40d36'23"
Richard & Melanie Golding
Cyg 21h53'39" 43d21'0"
Richard Mel's Mo Quinonez
Crb 15h20'28" 26d8'52"
Richard Merckx
Ari 3h25'31" 28d33'32"
Richard Merrill Bull
Her 17h38'27" 32d24'55"
Richard Messruther
Aqr 21h4'51" -10d50'51"
Richard Meurer
Uma 11h37'55" 44d14'15"
Richard Michael
Cap 20h47'36" -26d0'47"
Richard Michael Gladman
Aqr 22h6'36" -16d27'42"
Richard Michael Heath
Cep 21h39'25" 56d37'50"
Richard Michael Heeth
Aqr 22h4'7" -13d35'38"
Richard Michael Lambert
Uma 10h10'36" 45d54'6"
Richard Michael Lanza
Dra 16h8'38" 54d35'17"
Richard Michael Press
Uma 11h31'25" 63d49'36"
Richard Michael Walka
Cap 20h11'1" -20d41'32"
Richard Miller
Tau 3h49'51" 27d4'9"
Richard Miner
Aql 19h8'42" 7d0'54"
Richard Mitchell
Cyg 20h30'40" 50d32'20"
Richard Moberg
Cep 2h25'56" 81d2'32"
Richard Monastersky
Psc 0h20'11" 15d27'10"
Richard Moore, Sr.
Leo 9h29'15" 27d6'10"
Richard Morrissey
Ori 5h36'24" 3d4'39"
Richard (My Boy) Keith
Uma 8h59'46" 56d16'11"
Richard Myers
Vir 12h13'32" 1d31'1"
Richard N. Flora (Uncle Rich)
Uma 8h42'39" 63d48'43"
Richard N. Miller March 22, 1950
Ari 2h33'45" 30d36'23"
Richard N Peery
Dra 16h40'47" 53d34'4"
Richard Nejman
Psc 1h7'3" 32d17'5"
Richard Nesper, Light of my Life
Cnc 8h1'21" 12d55'34"
Richard Nichols Boland
Aql 19h33'35" 12d28'25"
Richard Noble Watson
Per 3h26'8" 52d25'9"
Richard Nolan Pittman
Aur 6h24'53" 41d18'48"
Richard Normen Hugon
Her 17h38'19" 28d30'33"
Richard Norris - "Skinny Rich"
Uma 13h1'51" 60d13'33"
Richard Northmore Little Jr.
Ari 2h48'15" 29d0'53"
Richard Nyle
Cep 20h45'6" 59d56'24"
Richard Ochs
Uma 11h34'13" 39d9'23"
Richard O'Connor
Uma 11h37'19" 43d33'30"
Richard Olin Loyd
Per 3h41'14" 43d40'29"
Richard Oliver Becker
Cap 21h36'53" -14d41'2"
Richard O'Mahoney Lupton 9/10/1915
Vir 12h31'10" -1d39'24"
Richard Omer Bredahl 9/10/1936 NJ
Crb 16h12'38" 36d23'52"
Richard Ordner
Aqr 21h11'51" -3d8'31"
Richard Oxley
Ari 2h44'30" 16d21'40"
Richard P. Bryant
Her 17h20'54" 16d4'50"
Richard P. Krause
Cnc 8h59'36" 18d34'57"
Richard P Love III
Cnc 8h45'7" 20d14'39"
Richard P. Savitt - "Star Runner"
Sgr 19h0'14" -35d28'45"
Richard P. Tessi - My Shining Star
Lyn 7h44'59" 44d54'3"
Richard P. Wellman
Her 17h16'27" 35d52'41"
Richard "Pa" Ellerby
Ori 6h8'58" 15d31'23"
Richard "pa" Wishart
Ari 3h22'52" 21d34'44"
Richard Palmer
Vir 13h8'37" 11d53'13"
Richard "Pappy" Barton
Psc 0h17'47" 8d19'35"
Richard Pappy Napier
And 0h55'46" 39d12'46"
Richard Paterson Whitelaw
Uma 14h0'49" 47d57'17"
Richard Patrick Tracey
Cep 22h23'15" 62d51'50"
Richard Patterson
Lib 15h34'31" -7d59'27"
Richard Paul Carroll
Leo 9h44'37" 27d19'45"
Richard Paul LaRue
Lib 15h26'53" -20d5'42"
Richard Paul Opsuth
Tau 3h33'42" 25d27'10"
Richard Paul Pellegrino
And 23h4'55" 48d58'38"
Richard Paul Penrod
Cnc 8h57'12" 7d10'21"
Richard Paul Whan
Cyg 21h56'48" 46d24'24"
Richard & Paulette Beck Star 1
Umi 15h50'50" 75d12'9"
Richard Perry White
Her 16h49'3" 32d59'0"
Richard & Peter
Cas 23h30'52" 57d42'16"
Richard Peter Lee
Aql 19h41'29" -0d5'52"
Richard Petrone, Jr.
Her 17h29'30" 43d0'51"
Richard Pfennig Family
Uma 8h18'54" 70d13'1"
Richard Phillips
Lyr 18h42'54" 34d43'39"
Richard Picha
Her 17h25'17" 34d32'40"
Richard Pickering
Vir 11h59'51" -1d22'54"
Richard Pokorski
Uma 11h48'52" 36d59'53"
Richard Pote
Cam 4h5'16" 59d56'27"
Richard Pratt
Uma 10h11'1" 45d21'57"
Richard Prince
Aql 19h42'14" 6d47'4"
Richard 'Prince Ka' Stierl
Uma 11h44'54" 31d16'45"
Richard Propper
Cep 20h52'57" 55d57'30"
Richard Putney
Ori 5h43'54" 6d53'25"
Richard R. Barnes
Lyn 8h48'22" 39d45'59"
Richard R. Busto 7-27-1950
Tri 2h35'15" 33d55'39"
Richard R Demitros
Aql 19h45'11" 3d46'9"
Richard R Harden
Per 3h59'39" 35d8'40"
Richard R. Neumann
Lyn 7h51'46" 57d40'1"
Richard R. Olds
Uma 12h5'52" 57d15'56"
Richard R. Towle
Cas 0h22'37" 57d56'47"
Richard Ray See Sr.
Aql 19h19'54" 8d46'34"
Richard Reece Stokes
And 23h22'56" 51d50'42"
Richard Rees
Cyg 20h4'3" 33d38'22"
Richard Reilly
Psc 1h33'31" 12d59'16"
Richard Rhoades
Her 18h46'39" 19d35'10"
Richard "Rich" Garlick
Sco 17h34'25" -42d44'11"
Richard "Richy" Dysart
Ari 1h55'53" 18d2'40"
Richard "Rick" Burwell
Cap 21h41'24" -17d15'23"
Richard "Rick" Houghton
Uma 10h55'29" 66d54'59"
Richard (Rick) J. Dale
Lyn 6h50'9" 51d48'3"
Richard Rivas
Ari 3h5'42" 18d48'26"
Richard R.Jennings
And 0h6'6" 32d19'26"
Richard Roati
Sco 16h7'33" -16d39'31"
Richard Robert Stackwell
Cyg 21h5'36" 37d54'22"
Richard Robert Williams II
Tau 4h14'35" 13d11'41"
Richard Robinson
Her 18h10'32" 22d18'27"
Richard Rohr
Tau 3h47'45" 9d46'44"
Richard Rowe
Cnc 9h20'11" 14d32'2"
Richard Rowe - A Blessed Road
Ant 9h58'13" -38d9'26"
Richard R.R.R.3
Ari 2h24'23" 22d4'35"
Richard Rudolph Garcia, Sr.
Cnc 7h58'7" 16d57'33"
Richard "Rudy" LaFlesh
Uma 11h56'6" 42d54'13"
Richard Russell Martin's Galaxy #13
Cep 22h46'39" 67d38'22"
Richard Russell Smith
Aql 19h7'12" -0d35'43"
Richard S
Cep 20h47'59" 61d28'39"
Richard S. Bizon
Her 17h24'49" 31d58'5"
Richard S. Doyle
Leo 10h3'8" 21d30'2"
Richard S. Dvorin
Leo 11h22'22" 25d55'7"
Richard S. Ferro
Tau 5h50'45" 15d56'34"
Richard S. Hiltz
Cep 20h35'9" 61d36'16"
Richard S. Hoffmann
Cap 20h22'35" -16d42'52"
Richard S. Levine
Psc 1h10'25" 28d19'18"
Richard S. Rogerson
Ori 5h17'31" 5d4'38"
Richard S. Watt
Boo 15h29'32" 33d24'49"
Richard S Wilshin
Cru 12h23'3" -58d5'6"

Richard S. Wing
Aql 19h20'27" 14d27'49"
Richard Samuel Wilcock
Per 3h6'32" 46d46'10"
Richard Sana
Uma 10h5'4" 53d36'52"
Richard Sanfelippo
Aur 5h4'31" 36d5'16"
Richard & Sara Kate
Cyg 20h45'43" 54d4'20"
Richard Schmidt
Uma 13h5'19" 55d33'32"
Richard Schmidt
Aql 18h51'29" -0d1'4"
Richard Schwall
Aur 6h26'9" 41d18'39"
Richard Scott
Leo 10h15'49" 20d26'0"
Richard Scott Arthur
Ari 2h36'49" 25d52'55"
Richard Scott Crouse
Dra 19h39'37" 64d47'49"
Richard Scott Keirsbilck
Sgr 19h11'31" -14d31'1"
Richard Scott Murray
Per 3h34'0" 45d49'2"
Richard Scott Thoma
Her 18h46'34" 16d9'40"
Richard Scott Vaughan
Ari 2h23'37" 23d50'56"
Richard Shane Meder
Ari 2h6'24" 13d46'53"
Richard & Sharon Monrean
True Love
Cyg 21h47'45" 38d33'41"
Richard Sheppard
Vir 13h30'53" 2d47'15"
Richard Sher ~ My Eternal
Love
Cnc 9h2'30" 15d5'21"
Richard "Shiva" Leone
Uma 8h42'11" 57d21'33"
Richard Snow Pettegrow
Uma 8h40'47" 52d3'4"
Richard "SONNY" Sutton
Cyg 21h10'4" -15d56'40"
Richard Spencer Wojtech
Per 2h40'12" 54d42'47"
Richard "Spike" Braese
Her 16h45'17" 45d4'25"
Richard Stahl
Vir 14h38'23" 3d49'3"
Richard Stanley Tindale
Uma 8h59'1" 46d48'24"
Richard Steimer
Uma 11h53'6" 41d51'15"
Richard Stephen Youshock
Her 16h22'29" 26d17'14"
Richard Steven DiDonato
Ari 2h8'55" 10d32'1"
Richard Stockton
Cep 22h43'57" 79d19'9"
Richard Strong
Sgr 18h39'30" -17d4'2"
Richard Sullivan's "Paper
Moon"
Ori 5h58'41" 17d11'11"
Richard Surdyk
Aql 19h10'12" -0d5'21"
Richard Sweeney
Peg 22h27'59" 31d17'31"
Richard T. Brown
Uma 11h58'42" 31d19'1"
Richard T. Byrnes
Cet 0h52'15" -0d47'52"
Richard T. Colarusso
Cap 20h11'26" -9d42'47"
Richard T. Doherty
Per 2h19'15" 57d14'51"
Richard T. Huber
Sco 17h38'27" -41d47'22"
Richard Taft Webb
Per 3h26'4" 43d32'49"
Richard Tanis Coes III
Sgr 19h3'20" -19d34'10"
Richard Taylor
Her 17h24'20" 38d40'55"
Richard Taylor
Aql 19h35'49" 15d59'51"
Richard Taylor
Cunningham
Uma 11h52'17" 64d35'48"
Richard Taylor Dad & Dad
2
Her 18h5'24" 38d45'48"
Richard Taylor "Dick"
Horton
Cap 21h18'30" -20d18'47"
Richard Tebbetts
Cmi 7h28'9" 7d23'4"
Richard - "The Fisherman"
Sgr 20h25'12" -44d23'35"
Richard The Great
Vir 12h44'10" 4d13'33"
Richard The Great
Vir 12h53'9" 6d8'45"
Richard Theodore "RT"
Anderson
Ori 5h39'39" 1d39'44"
Richard Theodore
Schoerner
Cnc 9h21'30" 28d47'50"
Richard Thiede
Aql 19h58'37" -0d31'25"
Richard Thomas Allor
Lib 15h7'20" -2d57'31"
Richard Thomas Gray
Her 18h4'37" 17d38'19"
Richard Thomas Halloran
III
Cnc 9h18'33" 7d53'48"
Richard Thomas Nuzum
Per 3h15'34" 43d40'47"
Richard Thomas Petty
Boo 15h23'41" 49d28'39"
Richard Thompson
Leo 9h35'9" 22d35'12"
Richard Thompson
Umi 15h9'13" 75d49'35"
Richard Thornton Senegor
Ori 5h13'40" 8d51'17"
Richard Thum Star of
Honor & Joy
Cap 21h44'35" -23d35'42"
Richard & Tina Sliwoski
Leo 10h8'6" 26d15'34"
Richard & Tori Rowland
Cyg 21h28'11" 49d10'57"
Richard Torres
Per 4h5'36" 44d22'3"
Richard Tracy
Per 3h28'27" 35d3'27"
Richard & Tressia Rieder
Uma 9h43'51" 44d2'1"
Richard Tritley
Tau 4h28'13" 21d48'22"
Richard Troy Bangs
Tau 5h21'0" 27d33'42"
Richard Troy Harper
Aql 19h19'28" -0d6'44"
Richard Truche
Aqr 22h43'2" -10d13'32"
Richard Turcotte
Per 4h8'22" 31d27'42"
Richard Turner
Aqr 23h0'59" -8d32'47"
Richard V. Heffran
Aur 5h17'46" 33d23'59"
Richard V. Kennedy, Jr.
Umi 16h11'41" 75d48'35"
Richard V. McCoy
Aur 5h30'52" 42d43'38"
Richard V. Stewart
Cep 22h19'54" 61d3'23"
Richard Valerio
Per 3h19'54" 53d14'34"
Richard Vandevander
Semper Fi
Cyg 21h48'44" 46d35'37"
Richard Vernon Baptiste
Cyg 21h31'42" 31d59'14"
Richard Vincent Cecchi
Per 3h56'59" 36d45'3"
Richard Voglhofer
Umi 13h49'34" 73d3'26"
Richard W. Baldwin
Tau 5h52'13" 28d20'2"
Richard W. Breunich
Ori 5h33'13" 3d16'15"
Richard W. Donovan
Uma 10h23'25" 55d19'28"
Richard W. Faretra
Aur 5h48'42" 50d45'12"
Richard W. Kositzke
Ori 5h58'28" 6d49'7"
Richard W. Olson
Lib 15h46'20" -28d28'0"
Richard W. Rader
Cap 21h41'34" -9d22'36"
Richard W. Wagner
Psc 22h56'47" 2d38'0"
Richard Walter Pretty
Lib 14h51'3" -2d16'50"
Richard Wander
Her 18h34'0" 13d28'43"
Richard Ward Gray
Ori 6h18'49" 15d37'14"
Richard Warfield Vath, Sr.
Psc 0h23'7" 7d52'38"
Richard Warren
Blankenship
Uma 9h12'30" 59d30'45"
Richard Waylon Lemmon
Her 17h44'48" 47d21'18"
Richard Wayne Staley
Uma 10h29'33" 59d43'31"
Richard Wendell Bell
Lib 14h53'2" -0d33'34"
Richard Wenderoth
Uma 9h54'54" 69d36'55"
Richard Wenk
Psc 0h43'20" 17d21'9"
Richard Wesley Bossie
Leo 10h38'27" 17d31'24"
Richard White
Aql 19h37'59" 14d5'4"
Richard White
Sgr 19h46'47" -15d10'57"
Richard White/Best
Husband & Dad
Cep 2h43'53" 78d31'49"
Richard Whitney
Sgr 19h42'18" -39d46'58"
Richard & Whynonia Lang
Uma 10h28'0" 70d55'26"
Richard William Noble
Uma 10h58'2" 39d48'14"
Richard William Setchell
Cep 22h51'15" 68d4'50"
Richard Wilson
Vir 13h43'53" -18d32'11"
Richard "Woody" Woodin
Aqr 22h28'10" -5d39'35"
Richard Wright
Uma 11h20'57" 41d32'27"
Richard Wynne Hughes
Her 17h19'53" 19d4'39"
Richard & Yasemin
Cyg 19h49'27" 33d26'53"
Richard & Yvonne
Lib 15h2'44" -25d20'22"
Richard ZIG Champion
Her 16h46'42" 37d30'37"
Richard, Starr K.. Ashley
Belowsky
Umi 16h2'51" 80d26'3"
Richard-Kerr Oliver
Gem 7h18'40" 35d10'51"
RICHARDS*ACTY*2004
Uma 9h57'30" 57d0'20"
Richard's Guiding Light
Cep 21h52'58" 61d39'13"
Richard's Little Bit
Lyn 8h4'26" 41d11'15"
Richard's Loco Moco
Leo 10h14'41" 16d45'57"
Richard's Star
Cnc 8h41'40" 32d33'54"
Richard's Star
Ori 5h22'0" -4d58'1"
Richardson's Paradias
Cap 20h8'42" -8d49'45"
Richea LaFawn
Cap 20h31'52" -18d33'1"
Richele & Gav's Star
Sge 20h3'0" 17d28'20"
Richele Panther
Sgr 19h1'40" -30d58'9"
Richelle Elizabeth
Sco 17h54'55" -39d14'32"
Richelle Elizabeth Kelly
And 23h42'45" 47d6'53"
Richelle Jamison
Lyr 18h29'40" 36d41'39"
Richelle Rose
Aqr 22h27'37" -3d35'58"
RichelleAndrus
Cas 1h2'14" 49d36'22"
Richelle's Jewel & Sam's
Vertex
Cyg 20h23'45" 57d54'5"
Richemont-Gemini
Uma 11h23'54" 59d0'50"
Richfield & Emily Chang
Cap 20h27'1" -12d17'46"
Richia Castro-Larsen
Mon 6h40'8" 8d10'28"
Richie
Tau 3h58'41" 4d56'46"
Richie
Psc 1h22'49" 24d53'23"
Richie
Leo 9h46'49" 28d16'30"
Richie
Vir 14h13'52" -19d53'42"
Richie
Vir 13h25'38" -6d22'16"
Richie
Uma 12h47'13" 55d6'2"
Richie Anderson
Gem 6h32'56" 19d57'15"
Richie Bethie
Gem 7h30'31" 19d59'18"
Richie Carter
Cap 20h31'5" -11d15'34"
Richie Christie's Valentine
Star
Cap 20h29'26" -11d27'23"
Richie & Debbie
Cap 21h36'40" -19d44'16"
Richie Gomez
Cap 20h58'33" -15d17'12"
Richie Hillyard
Psc 0h22'27" 8d1'59"
Richie Michael
Aql 19h12'9" 3d38'3"
Richie Neal
Crb 16h2'3" 39d20'56"
Richie Pariseau
Uma 12h8'10" 60d30'16"
Richie Peck
Leo 10h12'16" 15d57'13"
Richie Perronne 1-21-65
Cap 20h59'6" -20d25'27"
Richie Richard Earl
Harrison
Ori 5h26'6" -0d29'0"
Richie Richie
Tau 4h7'56" 27d28'30"
Richie Rodriguez
Per 3h33'25" 50d1'27"
Richie Scott's Rock Star
Cep 21h11'11" 68d46'43"
Richie Shannon
Per 3h10'51" 54d9'5"
Richie Shannon
Ari 3h10'55" 19d7'0"
Richie & Sue - 40th
Anniversary
And 23h12'8" 48d9'21"
Richie the Great
Cap 20h24'6" -20d31'16"
Richie, Alex, Hannah
Uma 10h43'21" 59d54'20"
Richie's Star
Sgr 19h25'57" -14d2'14"
RICHIP
Leo 11h49'43" 10d41'17"
RichMax
Cru 12h44'18" -56d28'36"
Richmcbride1062
Aql 19h45'33" 5d17'28"
RichnDot1947
Vir 13h12'8" 3d4'22"
RichnJul
Cyg 20h34'43" 34d36'37"
Richoo
Cas 2h8'11" 62d34'7"
Rich's Carol Agnes
Ori 5h55'18" -0d5'6"
Rich's "You Are So
Beautiful" Star
Leo 11h41'26" 25d44'21"
Richter, André
Ori 6h15'56" 16d22'57"
Richter, Hartmut
Lib 14h36'5" -11d13'57"
Richter, Helga
Lib 15h17'15" -8d50'44"
Richter, Knut
Ori 5h14'41" -7d51'10"
Richter, Manfred
Uma 9h37'4" 61d2'23"
Richter, Thomas
Uma 13h7'32" 54d40'18"
Richy "Pop" Behren
Uma 11h35'1" 38d49'25"
Richy V
Lib 14h51'6" -8d32'36"
Richy VanLuven (9/24/55-
8/11/02)
Uma 10h39'15" 47d18'50"
Rick
Aur 5h50'21" 41d41'32"
Rick
Her 16h31'57" 14d15'7"
Rick
Psc 0h48'43" 14d35'20"
Rick
Uma 11h16'43" 54d11'28"
Rick
Sco 16h59'7" -41d51'52"
Rick
Sgr 18h36'22" -26d19'49"
Rick Ahee
Psc 1h25'10" 24d33'19"
Rick Alan
Tau 4h35'31" 20d15'5"
Rick and Ashlee
Eri 3h51'49" -0d48'47"
Rick and Cathy Mauro
Cyg 20h14'43" 55d4'31"
Rick and Cher
Uma 12h26'58" 54d19'6"
Rick and Debra Carpenter
12/28/94
Sco 17h52'47" -34d40'59"
Rick and Dee's
Cyg 20h50'42" 46d40'18"
Rick And Heather Jeffery
Peg 22h41'19" 23d58'48"
Rick and Jenn Forever
Sgr 19h6'44" -15d33'2"
Rick and Linda Beam
Cyg 20h43'57" 33d17'59"
Rick and Lisa Haux
And 0h12'32" 44d52'45"
Rick and Marissa
Cyg 20h30'47" 52d17'58"
Rick and Rayell Forever
Uma 8h43'15" 66d26'24"
Rick and Reney Hargrove's
Star
Mon 7h15'40" -0d46'54"
Rick and Tricia's Lucky
Star
Ori 6h17'41" 8d56'6"
Rick Augsburg aka My
Dad!
Cap 21h32'3" -10d42'43"
Rick Baker
Tau 4h19'2" 26d49'38"
Rick Banchero
Crb 16h0'7" 35d0'18"
Rick Basden
Vir 12h13'45" 4d26'38"
Rick Beaumont Beloved
Dad & Husband
Cyg 21h3'43" 49d8'1"
Rick Black
Gem 7h20'15" 26d54'41"
Rick Bohte
Cyg 20h8'15" 34d45'23"
Rick Bowers
Ori 5h27'14" 13d29'43"
Rick Braun
Her 17h42'1" 17d10'3"
Rick Buesch
Uma 9h51'52" 70d55'7"
Rick Butterfield, Jr.
Cap 21h51'57" -8d42'28"
Rick Buttle
Boo 15h30'10" 40d58'30"
Rick Cardona
Her 17h48'3" 45d10'28"
Rick&Carolyn 9/19/1991
Gem 6h36'15" 27d11'20"
Rick Castillo
Aqr 22h40'19" -1d57'7"
Rick Cavalli
Cyg 21h53'49" 41d44'40"
Rick & Colleen Bullen
Aqr 23h4'15" -7d4'41"
Rick Combs
Ori 5h41'24" -3d24'46"
Rick Constable
And 0h1'3" 45d54'30"
Rick Crownover
Per 3h47'28" 46d34'42"
Rick Culley
Per 4h12'24" 51d22'19"
Rick Danial Sargent
Per 3h25'54" 51d6'52"
Rick Davis
Sco 17h58'49" -39d9'59"
Rick DeArman
Aql 19h13'36" -0d55'54"
Rick & Deese Love and
Memories
Uma 11h11'28" 58d17'43"
Rick Dougherty
Ori 5h2'43" 15d1'45"
Rick E Rose
Cap 20h7'32" -20d14'2"
Rick & Elaina Briemle
Cyg 21h5'4" 44d2'49"
Rick F. Maly
Uma 13h42'3" 55d42'19"
Rick F. Yarbrough
Vir 12h54'49" -6d5'11"
Rick Fanning
Cyg 20h30'57" 55d14'45"
Rick Farina, Mr.Baseball
Her 16h20'26" 43d57'44"
Rick Fords Wishing Star
Ori 6h14'24" 15d49'30"
Rick Fowler, Sr.
Uma 11h10'12" 50d45'5"
Rick G. Brown
Lmi 10h21'56" 34d57'46"
Rick G Whaner
Uma 10h41'29" 66d32'12"
Rick & Glenda Macrae
Lyr 18h54'20" 32d8'30"
Rick Heimgartner * Karissa
Blucher
Eri 3h50'8" -0d26'25"
Rick Holmes
Per 2h15'21" 55d38'39"
Rick Inman
Ori 5h32'39" 3d52'31"
Rick J. Hans
Aqr 22h46'16" -18d16'57"
Rick Jackson
Per 2h50'34" 52d4'30"
Rick & Janet Tobin - Super
Parents
Cyg 21h19'35" 46d58'22"
Rick & Janna Stars togeth-
er
Cyg 21h16'20" 43d32'5"
Rick & Jennifer Strong
Lyr 18h34'25" 37d39'12"
Rick K. Knierim
Her 16h33'33" 48d36'39"
Rick & Karen Ulbright
And 0h49'2" 37d14'14"
Rick&KathySquires-
YearOne
Cyg 21h23'33" 36d2'9"
Rick King Jr.
Leo 11h27'23" 13d12'0"
Rick Kmet
Uma 14h12'49" 56d33'57"
Rick Kunkel
Uma 11h50'55" 30d51'2"
Rick L. Healy
Lmi 10h20'3" 36d14'50"
Rick La Bar, your own star
!!
Dra 19h10'40" 61d4'24"
Rick Lallish
Dra 19h30'44" 62d22'35"
Rick & Lindy Brownell's
Brilliance
Uma 11h33'20" 31d28'30"
Rick&Lorraine Best Friends
Forever
Lib 14h51'35" -1d42'52"
Rick loves Bee
Uma 8h37'44" 64d16'10"
" Rick Mangogna "
Ori 6h1'5" -1d40'51"
Rick & Mary Corbet-
Gonzales
Uma 8h20'59" 66d4'46"
Rick Mattis
Vir 13h18'47" 12d27'16"
Rick Mazzella
Leo 10h49'11" 12d25'58"
Rick McNichols
Gem 7h37'38" 14d31'29"
Rick Meade
Vir 14h33'33" 5d58'14"
Rick Moroski
Cir 15h22'16" -57d16'50"
Rick My Sweetheart
Gem 7h38'34" 26d17'49"
Rick N Tina
Crb 15h53'12" 35d4'7"
Rick Needham
Per 2h23'11" 51d39'28"
Rick Nelson; Beloved
Husband & Dad
Her 18h47'24" 20d27'34"
Rick Nichols
Leo 10h14'59" 14d8'3"
Rick Osterman & Ron
Diggs
Uma 11h4'55" 48d18'5"
RICK PETRICCA
Ori 5h15'55" 1d30'41"
Rick Phillip Gutierrez
Cap 21h56'49" -19d17'27"
Rick "Pyle" Dibiase
Ori 6h12'8" 18d27'33"
Rick & Randi
And 2h27'51" 45d6'19"
Rick Raymaker
Aqr 23h21'18" -11d42'41"
Rick ~ Ricky ~ Lewis
Her 18h47'41" 23d55'37"
Rick & Rita Bennett
Aql 19h54'44" -0d17'26"
Rick Rocheleau
Psc 1h10'4" 28d54'34"
Rick & Rose Fox
Cyg 19h42'10" 39d53'57"
Rick Salapong
Aql 19h36'36" 12d28'59"
Rick & Scott
Uma 13h36'25" 55d21'8"
Rick & Sheryl Johnson
Cyg 21h53'11" 53d24'48"
Rick Simpson
Her 17h57'59" 19d38'24"
Rick Snyder
Sgr 17h56'17" -16d55'56"
Rick Stanger
Lyn 8h30'3" 52d51'47"
Rick Steves
Her 17h51'51" 46d34'40"
Rick Thomas Ackerman
Cnc 8h42'49" 29d5'21"
Rick & Todd
Dra 15h53'59" 54d36'26"
Rick Underwood
Tau 4h27'29" 29d36'51"
Rick V. Garcia
Umi 15h56'45" 81d9'29"
Rick Vogt's Defying Gravity
Tau 5h54'8" 25d6'16"
Rick W. Patterson
Uma 10h55'9" 65d47'32"
Rick Woods
Cep 0h49'37" 79d2'33"
Rick Wright
Ari 2h20'49" 18d2'23"
Rick Yelverton
Lyn 7h24'17" 46d17'4"
RickDani
Her 17h31'49" 49d14'25"
Ricke
Uma 9h7'14" 58d57'41"
RicKelli
Cyg 21h35'13" 36d37'20"
Rickert, Thomas
Uma 13h41'37" 62d2'4"
Rickey Burdine
Crb 15h56'12" 29d41'14"
Rickey Davis
Psc 1h14'40" 18d56'58"
Rickey Duarte
Per 3h37'57" 48d45'21"
Rickfire - 7 May 1964
Pho 23h34'41" -54d2'26"
Ricki
Vel 9h16'30" -39d59'52"
Ricki Carroll
Sco 17h57'48" -30d4'14"
Ricki Lynn
Aqr 23h7'7" -22d24'16"
Ricki McCuiston
Lyn 7h27'36" 51d53'35"
Ricki Nicole Borek
Ari 2h33'40" 24d46'37"
Ricki V Silveria
Aqr 23h3'18" -10d27'41"
Ricki White
And 1h37'27" 38d10'34"
Rickie
Hya 9h22'25" -0d53'2"
Rickie Allen Agnew
Leo 10h19'6" 6d53'38"
Rickie James Pope
Umi 16h58'57" 75d28'39"
Rickie Michael Morrow
Lmi 9h52'23" 34d6'59"
Ricki-Lee
Cru 12h38'8" -58d40'36"
Rick-My Daddy and Hero
Ori 6h13'38" -0d59'46"
Rick'n'Daryl
Ori 6h5'56" 15d51'1"
Rick's Eternal Night Light
Lib 15h13'26" -7d21'41"
Rick's lil piece of Heaven
Tau 5h45'47" 22d19'51"
Rick's Lucida
Vir 13h22'43" 5d1'57"
Rick's Lucky Star
Aql 19h52'10" -0d21'56"
Rick's Revenge
Cap 21h57'46" -10d36'31"
Rick's Shining Star
Umi 14h34'34" 72d48'57"
Rick's Star
Boo 14h59'16" 28d7'45"
Rick's Star
Gem 7h54'55" 31d2'11"
Rick's White Dragon
Ori 5h51'51" 9d7'54"
rickshaw
Dra 18h33'15" 72d51'46"
Rickstar
Uma 8h15'50" 71d49'54"
Ricky
Cap 21h45'34" -13d14'40"
Ricky
Vir 12h31'3" 10d57'2"
Ricky
Cnc 8h21'19" 24d8'51"
Ricky
Ari 2h3'55" 23d29'58"
Ricky
Leo 9h22'31" 21d46'1"
Ricky
Aur 5h41'15" 40d59'50"
Ricky #22
Boo 14h32'43" 53d16'4"
Ricky Allen Hughes
Her 17h30'8" 26d23'44"
Ricky Allen McGowen
Aql 19h52'2" -0d40'47"
Ricky Alonzo
Per 4h37'40" 49d47'26"
Ricky Austin
Cru 12h36'43" -57d49'26"
Ricky & Briana
Gem 7h41'14" 16d34'59"
Ricky Carter
Psc 0h47'2" 17d51'45"
Ricky Chris Ramoutar
Dra 18h39'0" 74d54'10"
Ricky Colburn
Cnc 8h45'53" 7d47'40"
Ricky Crossley
Cnc 8h26'52" 28d5'51"
Ricky D. Hernandez
Ori 5h9'20" 2d15'47"
Ricky D. Redcay
Her 18h56'51" 15d52'13"
Ricky & Dora Renderos for
Eternity
Ori 5h58'20" 6d29'15"
Ricky Dunne Neegard
Aur 5h52'26" 32d26'1"
Ricky Earl McLaughlin
Uma 9h9'14" 68d25'10"
Ricky Esposito
Ori 5h58'35" 18d17'27"
Ricky Farmer
Uma 11h41'54" 28d38'44"
Ricky Fout
Her 16h58'47" 26d19'44"
Ricky Giden
Her 16h37'39" 36d55'44"
Ricky Giguere
Cas 23h22'9" 53d31'45"
Ricky Gillies
Ori 6h20'47" 14d1'21"
Ricky Hiton
Leo 11h49'31" 16d21'57"
Ricky Hulvey
Cam 4h11'9" 67d51'49"
Ricky... I have this feeling
XXXOOO
Tau 4h18'52" 0d46'9"
Ricky & Ila
Uma 9h30'10" 57d11'50"
Ricky James Jackson
Aur 6h31'45" 33d55'25"
Ricky Joe Ridley
Cep 22h5'5" 61d20'53"
Ricky Lee Burchett
Cap 20h59'20" -27d20'3"
Ricky Lee Wighton
12021971
Cru 12h24'43" -61d47'17"
Ricky Lee Wiginton
Umi 15h48'6" 72d49'34"
Ricky Lombardo
Leo 11h15'37" 19d15'26"
Ricky LP 3
Tau 4h14'19" 16d40'45"
Ricky Lynn Jackson
Psc 0h52'29" 9d11'11"
Ricky Michael Labine
Her 18h50'49" 25d32'16"
Ricky Michael Scotto
Sgr 18h24'3" -18d10'35"
Ricky Mudford
Umi 15h30'33" 81d49'39"
Ricky Murdock IV
Tri 1h57'22" 30d43'8"
Ricky Ricardo
Leo 11h34'30" 19d53'50"
Ricky & Risa
Her 17h3'14" 31d34'38"
Ricky Robert Barr
Uma 9h39'6" 56d8'34"
Ricky RockStar Etter
Dra 17h45'27" 50d52'54"
Ricky Sheridan
Her 16h25'32" 15d26'25"
Ricky Shults
Aql 19h41'37" 6d17'6"
Ricky 'Soul' Cheng
Uma 9h6'49" 57d32'21"
Ricky Staley
Ari 2h3'50" 20d19'38"

Ricky Sumera
Leo 10h26'55" 22d51'9"

Ricky -The Light Of My Life- Tamera
Cru 12h43'31" -60d33'14"

Ricky Theisen
Per 2h43'47" 54d16'21"

Ricky Thomas McGrail
Cen 13h30'43" -41d44'19"

Ricky "Thunder" Prendergast
Uma 10h5'8" 55d43'44"

Ricky Tucker's Magical Wish Star
Dra 18h57'41" 60d49'29"

Ricky Young
Uma 9h28'31" 43d49'5"

Rickylove
Aqr 21h59'11" -13d35'20"

Ricky's Allstar
Peg 22h39'36" 26d55'53"

Ricmic
Lib 14h49'4" -15d21'38"

Rico
Cap 20h34'36" -21d11'7"

Rico
Her 17h10'49" 25d11'27"

Rico
Ori 5h40'12" 4d27'51"

Rico and Catherine
Cyg 19h34'34" 54d6'30"

Rico Macias
Uma 11h23'17" 55d0'36"

Rico Suave
Sgr 18h13'27" -19d29'26"

Ricochet's Heroine
Sgr 18h12'53" -20d1'18"

Ricoloveskim
Uma 10h42'2" 56d45'33"

Ricordi
Ori 5h38'0" -2d23'52"

Rico's place
Cma 7h19'46" -32d14'27"

Ridalls
Cma 6h56'17" -20d39'17"

Rider Valentine Umphenour
Vir 13h46'57" -18d26'25"

Riding on a Goldwing and a Prayer
Aur 6h37'41" 42d54'24"

Ridley Oliver Berger-Sacks
Psc 1h33'40" 23d6'27"

Ridley Oliver Sacks
Uma 10h48'49" 64d36'34"

Rie Shima
Uma 12h41'38" 56d12'12"

Riechel, Karl-Heinz
Ori 5h54'21" 22d49'20"

Rief William Kramer
Tau 4h26'44" 20d48'44"

Rieger, Richard
Uma 10h53'36" 57d19'55"

Riegger, Christiane
Uma 9h59'4" 65d9'49"

Riegl Lászlóné Sauter Katalin
Leo 11h57'4" 23d16'59"

Riek, Fritz
Ori 6h19'33" 3d1'57"

Rieko M.
Lyn 8h16'0" 42d10'34"

Riel, Sven Carsten
Uma 9h38'48" 68d4'33"

Riese Carol Flood
And 0h21'54" 31d45'43"

Riese Declan Gurr
Cru 12h57'25" -59d14'16"

Riftwing
Uma 9h41'40" 57d10'32"

Rigatti Luca
Lyr 19h25'5" 43d20'22"

Rigby
Cyg 21h44'13" 50d7'40"

Rigel
Vir 13h3'59" 9d16'41"

Rigel
Ori 5h19'58" -7d6'58"

Rigel & Me
Leo 10h21'52" 14d14'1"

RigelBobba
Ori 6h0'56" 17d48'3"

Riggolliott
Sgr 18h11'6" -21d4'22"

Riggs-Hansen
Aqr 23h30'7" -9d2'5"

Rightbringer
Peg 23h19'20" 27d14'10"

Rightmyer Family Star
Cas 0h34'52" 63d53'13"

Rigler Ibolya Anita
Cas 1h6'38" 69d39'16"

Rignall's Rising Star
Uma 8h21'14" 60d30'48"

Rigo
Cnc 8h24'32" 11d7'34"

RIGO
Per 3h42'1" 45d30'51"

Rigo Franco "Batman"
Lyr 19h7'29" 27d56'17"

Riikka Peltola
Lyn 7h56'0" 35d10'38"

Riina Kiisu
Mon 6h45'6" 8d46'59"

Riise's Piece Of The Cosmos
Lyn 8h14'4" 38d47'57"

Rijon's Love
Ori 6h19'39" 16d24'48"

Rijucala
Cam 5h27'57" 65d21'8"

Rik Anthony
Leo 9h25'38" 26d6'42"

RiKa
Sge 19h41'19" 18d51'13"

RiKa2
Sge 19h50'46" 17d14'31"

RikandFran
Sge 20h5'46" 16d47'5"

RIKE-072404
Her 17h15'20" 28d56'54"

Riker
Crb 16h9'29" 36d12'37"

Riki
Crb 16h14'38" 38d30'17"

Riki & Rino
Sgr 18h24'43" -32d51'21"

Rikke
Per 3h1'33" 47d47'52"

Rikke Hoffmeyer Boas
Uma 8h59'4" 61d8'52"

Rikke Monjor "My Dream Day"
Umi 13h46'44" 75d50'44"

Rikke & Tommas
Cyg 20h16'18" 52d9'58"

Rikki
Leo 10h50'38" 13d49'55"

Rikki Anne
And 0h35'22" 36d3'57"

Rikki Chernoff
Cep 22h45'53" 66d58'10"

Rikki Jo Pease
Lib 15h43'29" -29d25'34"

Rikki Kirschner
Uma 8h18'49" 67d50'44"

Rikki Lewin
Umi 16h53'32" 86d47'20"

Rikki Lichty
Sgr 18h34'10" -17d41'43"

Rikki Marie Schlappich
Ori 6h0'19" 10d36'36"

Rikki Nicole
Sgr 17h59'33" -27d43'46"

rikki rockett
Leo 10h30'33" 18d53'52"

Rikki S. Celis
Lib 14h58'40" -14d46'39"

Rikki Stellar
Ori 6h14'30" 16d21'13"

Rikki's Star
Vir 13h25'26" 11d30'42"

Rilee Danielle Jones
Cap 21h32'24" -18d27'37"

Rileigh
Sco 17h55'0" -32d5'2"

Rileigh Christine Messner
And 23h18'44" 45d11'37"

Rileigh Elizabeth
And 1h28'51" 37d6'34"

Rileigh Elizabeth Boyle
Sgr 19h39'46" -14d7'35"

Riley
Cma 6h40'42" -12d59'36"

Riley
Uma 13h44'35" 54d11'45"

Riley
Sco 16h48'36" -33d4'39"

Riley
Aur 6h27'29" 52d29'56"

Riley
And 1h32'15" 45d6'26"

Riley
Cyg 20h43'7" 43d22'23"

Riley
Gem 6h51'6" 27d20'38"

Riley Adamson
Pup 8h15'45" -23d10'54"

Riley Alan Johns
Uma 10h50'26" 46d47'33"

Riley Alexis
And 1h18'32" 46d19'30"

Riley Alexis
Gem 7h46'29" 23d22'22"

Riley and her Daddy
Uma 11h24'40" 59d27'24"

Riley Anne Nowling
Cma 6h18'37" -17d17'12"

Riley Anthony Stainback
Cep 22h47'11" 71d40'34"

Riley Austin Brady
Leo 10h18'12" 26d31'49"

Riley Bender
And 0h40'15" 46d34'34"

Riley Bird
Umi 14h30'22" 76d2'52"

Riley Brian Chin
Cap 20h28'1" -10d41'34"

Riley Brook
Mon 6h33'58" 10d51'3"

Riley Brooke Spohn
And 2h6'35" 42d38'10"

Riley Cecile
Umi 15h43'26" 82d16'46"

Riley Christian
Ori 4h50'6" 14d23'51"

Riley Church
Uma 9h1'19" 59d10'15"

Riley Ciandella
Uma 10h19'20" 59d7'27"

Riley Cole Newport
Sgr 19h14'1" -34d12'17"

Riley Cooper Smith
Cap 20h42'40" -25d56'35"

Riley Corbin Forsyth
Umi 16h21'40" 78d26'22"

Riley Delany
Lyn 8h34'17" 57d28'36"

Riley Dias & Melinda Gauthier
Cyg 20h1'47" 30d42'18"

Riley Elise
Vir 13h14'29" -2d52'5"

Riley Elizabeth
Sgr 19h6'19" -17d51'24"

Riley Elizabeth Simms
And 0h15'22" 31d53'17"

Riley Elizabeth Smith
Cnc 8h35'6" 10d38'11"

Riley Erinn Foye
Tau 4h29'51" 22d10'19"

Riley Fegley
Psc 23h48'47" 0d48'58"

Riley Finn Welsh White
Umi 13h50'21" 71d40'19"

Riley Gale Horgan
Lyn 8h4'46" 38d5'10"

Riley Galu
Ari 2h8'45" 16d47'36"

Riley Grace Griebe
Gem 7h11'37" 34d28'15"

Riley Grace - Our Miracle
Col 6h11'35" -35d29'31"

Riley Grace-Gods Shining Miracle
Gem 7h29'25" 19d56'21"

Riley Henry Linker
Umi 14h45'19" 75d52'3"

Riley Holstine
Cyg 20h42'22" 37d22'38"

Riley Isabel
Vul 20h16'5" 22d53'49"

Riley J & Hailey M Fitzsimmons
Cyg 21h53'51" 43d38'27"

Riley J. Hardman
Leo 9h28'2" 25d52'15"

Riley Jack Nathan March
Dra 15h28'16" 64d30'30"

Riley James Cunico
Aqr 23h29'1" -8d18'20"

Riley James Fitzpatrick
Oph 17h35'25" -0d29'52"

Riley James Gill
Uma 11h23'22" 70d12'58"

Riley James McCallum
Sco 16h15'1" -13d45'53"

Riley James Mead
Tau 4h17'43" 9d10'13"

Riley James Robinson
Lyn 7h26'20" 58d19'43"

Riley James Timmerman
Tau 5h52'40" 24d45'48"

Riley Jane
Vir 13h13'17" -16d0'3"

Riley Jane Rachel
Uma 13h47'3" 61d1'1"

Riley Jane Vanderpool
Hya 12h12'13" -27d6'46"

Riley Jane-Marie Perkins
Lib 15h25'28" -7d18'30"

Riley & Janice Johns Family Star
Umi 9h43'8" 86d42'43"

Riley Jennifer Hand
Lyn 7h57'40" 59d22'27"

Riley Joan Walsh
And 1h21'59" 47d14'14"

Riley Joseph
Her 16h35'2" 17d41'54"

Riley Judelson
Sgr 19h46'22" -12d38'5"

Riley Judith Olsen
Leo 11h27'0" -5d43'5"

Riley Kain Nelson
Ori 5h20'38" 12d9'34"

Riley Kate
Leo 10h50'7" 19d21'56"

Riley Kathryn Pegher
Sco 17h24'17" -42d20'35"

Riley Kay Flynn
And 1h21'32" 44d37'43"

Riley Kekona Samaniego
Lib 15h51'19" -17d29'18"

Riley Kopke
Cam 3h41'37" 61d56'37"

Riley Lara Heffron
Umi 13h5'14" 75d52'21"

Riley Laud
Leo 10h19'36" 20d1'42"

Riley Lee Taflinger
Cyg 21h52'28" 45d47'7"

Riley London Clark
Umi 13h19'4" 70d57'48"

Riley M. Kelly ~ Love Yvonne
Per 4h27'1" 43d12'4"

Riley Madison Storey
Lyr 18h48'5" 37d12'55"

Riley Marcus Patterson
Umi 15h52'45" 70d19'16"

Riley Marshall
Cap 21h44'56" -19d19'13"

Riley Mathew Dennis
Cyg 19h49'11" 31d56'38"

Riley McDowell
Ori 6h10'27" 12d42'55"

Riley Michael Phillips - 5.08.2005
Cru 12h17'10" -63d31'6"

Riley Nathan Gilbert
Umi 14h30'58" 71d27'2"

Riley Nichole
Lmi 10h31'42" 33d42'16"

Riley Nicole Mucke
And 1h16'6" 46d16'33"

Riley Olivia
Lyn 8h24'55" 37d6'53"

Riley Ormerod
Cma 6h45'37" -13d41'2"

Riley Osborne
Gem 7h3'35" 15d49'32"

Riley Paige Diamond
Gem 7h43'39" 34d10'38"

Riley Paige Kintz
And 0h31'13" 31d42'6"

Riley Patrick Boyle
Uma 9h27'44" 49d42'53"

Riley Patrick Boyle
Umi 14h22'44" 85d8'34"

Riley Payton Maloy
Sco 16h30'51" -38d19'19"

Riley Preston
Cma 6h38'3" -15d57'36"

Riley Richard Sneddon - Riley's Star
Her 17h30'30" 17d21'54"

Riley Rose Buck
And 1h29'50" 48d42'12"

Riley Rosentein
Uma 10h15'40" 65d3'2"

Riley Routledge
Uma 11h42'21" 59d44'32"

Riley Scholl
Vir 14h11'4" -7d53'7"

Riley Shay Keeble
Dra 19h8'31" 74d30'37"

Riley Shea
Peg 21h21'25" 20d13'19"

Riley Star
And 2h15'16" 37d50'44"

Riley Stewart
Cru 12h3'40" -62d51'56"

Riley Stone Carpenter
Lyn 7h52'11" 37d48'28"

Riley Sullivan
Ori 5h54'8" 17d36'0"

Riley T. Shaw
And 2h14'16" 39d52'53"

Riley Tanner
Cnc 8h2'1" 25d32'38"

Riley Tate
Ori 5h53'10" 5d51'30"

Riley The Courageous
Uma 11h52'1" 31d14'41"

Riley Thomson's star
Aqr 23h13'15" -21d52'48"

Riley Voss
Uma 8h47'20" 67d30'59"

Riley Wayne Huff
Ori 5h43'53" 7d46'40"

Riley Witter Bond
And 2h13'17" 46d10'7"

Riley Yin
Ori 4h52'15" 13d24'17"

Riley-Ann
Cap 20h36'3" -23d9'25"

Riley'nin Gözyasi
Lep 5h18'46" -12d33'35"

Riley-Rae's Starr
Psc 1h2'57" 12d5'28"

Riley's guiding star
Col 5h42'43" -33d59'47"

Riley's HoPE
Cyg 20h43'36" 37d24'43"

Riley's Star
Uma 11h10'0" 46d43'18"

Riley's Star
Tau 5h9'34" 18d24'32"

Riley's Star
Eri 4h30'6" -2d34'3"

Riley's Star
Uma 10h17'44" 67d35'40"

Riley's Star on Her First Christmas
Lyn 7h59'27" 39d45'3"

RileySophia Winston
And 23h41'47" 47d22'17"

Riley-Toby-Raven
Tau 4h32'42" 17d13'57"

Rilke, Herbert
Aqr 20h44'0" -3d35'49"

RILYA
Lyn 6h52'42" 58d36'21"

Rilyn Kate
And 0h9'9" 45d0'53"

Rima
Cas 23h59'20" 56d6'41"

Rima Ghamrawi
Uma 11h58'36" 29d22'17"

Rima Rajput's Star
Cas 23h18'31" 54d20'18"

Rimini
Cyg 21h31'56" 55d14'24"

Rimmi Gulati
Lib 15h0'23" -12d9'50"

Rimo
Pho 23h47'26" -45d35'43"

Rimondi Maria Luisa
Umi 13h19'9" 70d33'2"

Rina
Uma 9h58'39" 58d55'0"

Rina
Lib 15h5'47" -15d24'17"

Rina
Sco 17h32'4" -45d29'9"

Rina
Sco 16h37'31" -33d51'15"

Rina
Peg 0h13'51" 16d21'2"

Rina
And 0h50'59" 44d45'16"

Rina & Augie Longo
Cyg 20h41'25" 37d34'47"

Rina Rosazza
Cru 12h33'39" -58d21'46"

Rina Rosazza
Uma 10h32'25" 64d23'38"

Rina, my only daughter
Cnc 8h24'30" 23d48'57"

Rinaki
Sco 17h58'18" -30d50'50"

Rinaldo & Elisa Maria (Isa)
Umi 17h6'59" 79d21'28"

Rina's Radiance
Umi 14h29'2" 69d6'45"

Rinchen Sangmo 1965
And 0h11'31" 45d49'52"

Rinda Lynn Avon
Uma 11h22'30" 58d18'50"

Rindy Carryn
Lib 15h46'19" -6d48'18"

Ring Of Fire
Crb 16h9'54" 39d29'36"

Ring & Serpent
Aqr 22h26'59" -2d19'15"

RingierTV
Mon 7h36'3" -0d37'4"

Ringle ~ Schwegman Wedding Star
Cyg 21h28'19" 51d22'46"

Ringo 1973
Leo 10h12'16" 8d59'39"

Ringo Armstrong
Cmi 7h34'20" 4d13'14"

Ringold Agape
Tri 2h5'49" 34d25'6"

Rinion
Tau 5h11'20" 18d41'38"

rinkusschonesmadchen
Del 20h42'20" 14d0'1"

Rinna E. Hernandez
Ori 5h49'30" 12d31'34"

Rinna18
Cap 20h12'49" -9d16'49"

Rino
Lac 22h22'20" 47d1'13"

Rino & Minja
Cap 20h14'10" -17d38'13"

Rio
Gem 7h3'53" 21d11'18"

Rio Elementary
Vir 12h16'28" 0d54'41"

Rio Ellis Smith
Ari 3h29'1" 27d40'46"

Rio Vermelho
Lyn 9h33'36" 40d49'55"

Rion
Sgr 18h27'44" -16d20'23"

Rion
Dra 16h17'15" 59d25'47"

Riordan Robert Carey
Umi 16h18'21" 72d39'37"

Rioux Style
Uma 9h4'44" 52d13'40"

RIP <3 CAM - Live.Love.Laugh.
Cnc 8h45'15" 31d2'16"

RIP <3 Corporal JRP [4648]
Vir 12h33'29" 12d2'56"

RIP Helen Dundas 31Oct 47-11Sep 05
Cru 12h36'17" -59d26'10"

Rip Oneill
Uma 11h3'42" 51d59'10"

Ripamonti Marta Virginia
Cam 5h48'37" 61d8'32"

Ripley Ellen Berg
Peg 22h4'12" 36d3'58"

Ripley Paige Tippett
Cru 11h58'55" -59d29'25"

Ripper
Ori 5h58'53" -0d13'38"

Rippy, Our Baseball Star
Aqr 20h59'44" -9d54'14"

Rique Rivera
Cnc 8h53'46" 10d38'4"

RiRi
Tau 4h10'0" 6d51'28"

RiRi
Lib 15h15'35" -16d14'48"

RIS
Sco 16h5'22" -14d32'43"

Risa
Sgr 18h24'59" -16d42'35"

Risa
Cnc 8h46'48" 11d39'55"

Risa
Leo 11h46'57" 10d21'15"

Risa
Leo 10h10'19" 24d39'11"

Risa
Gem 7h23'28" 21d12'42"

Risa
And 0h42'17" 26d27'20"

Risa & Anders
Lyn 7h55'13" 46d31'20"

Risa Haas
Gem 7h24'40" 15d33'26"

Risa Pamela Goodman
Del 20h39'30" 15d9'13"

risa randa
Ori 5h59'44" 20d58'23"

Risa Sasha-Anne Dunahoo
Gem 6h31'10" 18d41'17"

Risa Sue Breckman
Psc 23h16'25" 5d9'54"

Rise
Lyn 9h9'32" 38d55'53"

Risele
Uma 9h25'24" 67d30'14"

Risgae's Irish Connection 3/31/01
Uma 9h19'31" 57d17'30"

Rishanda Lawrence
Tau 3h32'32" 27d9'51"

Rishanne
And 23h35'56" 40d19'56"

Rishi
Umi 12h32'14" 89d32'38"

Rishi Dhiri (Monkey)
Lep 5h39'59" -17d26'40"

Rishidhara
Cam 4h16'28" 79d24'4"

Rishika Arun
Aqr 22h31'14" -24d10'28"

Rishka
Cnc 8h51'20" 11d47'0"

Rishona Abegail Alberts
Cas 23h41'41" 55d10'22"

rishuajoshay
Cnc 9h11'55" 26d55'22"

Rising Melodies of Alexis
Tau 4h31'8" 29d56'36"

Rising Star 2006 - Chris Elliott
Dra 18h20'13" 58d53'30"

Rising Star 2006 - Cory Smith
Dra 19h46'0" 66d31'13"

Rising Star Canine Rehab Center
Uma 11h24'2" 47d12'19"

Rising Stars: Marshall & Isabella
Pho 1h22'15" -49d34'49"

Rising Sunflower Power
And 0h26'50" 41d52'53"

Riss' Passion
Aqr 22h6'49" -8d52'30"

Rissa
Ari 2h43'56" 28d52'11"

RISSA
Tau 4h31'39" 20d1'12"

Rissmann, Darina
Uma 12h33'36" 53d21'36"

Rissoles Galactica
Cru 11h59'25" -61d24'1"

Rissy
Ori 5h38'59" -2d4'30"

Ristness
Vir 13h17'33" 2d18'31"

Rita
Her 17h10'53" 20d14'13"

Rita
Ori 5h55'17" 18d21'40"

Rita
And 1h6'41" 40d52'40"

Rita
And 2h25'48" 41d50'18"

Rita
Crb 16h7'54" 35d54'51"

Rita
Crb 16h7'47" 31d33'13"

Rita
Cyg 20h41'4" 46d28'50"

Rita
Lac 22h41'57" 49d12'20"

Rita
And 1h6'16" 45d1'30"

Rita
And 2h32'9" 46d20'33"

Rita
Aql 20h8'43" -0d0'49"

Rita
Cap 21h55'42" -22d24'5"

Rita
Uma 9h35'23" 55d31'10"

Rita
Uma 9h36'39" 58d2'21"

Rita
Cas 23h4'22" 58d44'1"

Rita
Cas 22h59'47" 56d43'10"

Rita
Uma 8h44'38" 70d42'27"

Rita
Cep 1h22'42" 78d39'17"

Rita
Cas 1h33'15" 66d52'15"

Rita A. Alderete: BELLA
Her 18h30'2" 13d7'53"

Rita A. Benton
Aqr 22h42'9" 1d10'17"

Rita A. Carroll
Psc 0h22'17" -3d1'10"

Rita A. Corrigan
Cap 21h21'12" -26d40'32"

Rita A. Kolasinski
Leo 9h39'25" 31d45'21"

Rita Adelman
And 23h20'24" 43d49'16"

Rita Agnes Norgren
Umi 15h47'39" 70d23'47"

Rita Alice
Tau 5h27'19" 27d15'45"

Rita and John Gosling
Cyg 19h39'51" 52d35'7"

Rita and Morty Bishop
Cnc 8h10'33" 9d0'10"

Rita and Stanley Kaplan
Uma 13h37'18" 57d6'32"

Rita Angel
Ori 5h45'39" 3d26'56"

Rita Angelica Paolino
Lib 14h47'38" -8d5'19"

Rita Ann Brown
Per 3h49'32" 43d57'1"

Rita Ann Karlosky
And 0h48'30" 40d30'2"

Rita Ann Labate
Cyg 19h40'29" 39d49'26"

Rita Arianna
Cap 20h30'50" -14d41'34"

Rita Arlene Montell
Ori 5h6'26" 12d4'48"

Rita Armitage
Uma 11h13'45" 65d14'36"

Rita B. Contreras
Peg 23h38'9" 13d29'6"

Rita B. Kusinitz
Sgr 19h16'47" -20d59'37"

Rita (Babe) Stowell
Cas 0h59'55" 69d24'53"

Rita Banicki
Tau 5h41'26" 26d8'16"

Rita Baxter
Uma 11h45'28" 53d7'6"

Rita Bernier
Tau 3h36'24" 3d38'35"

Rita Bhagchandani
And 23h27'58" 35d40'56"

Rita Bhan Zutshi
Uma 9h32'31" 65d31'52"

Rita (Bobbie) Coulombe
Leo 9h27'49" 25d47'41"

Rita Botzenhardt
Cnc 8h17'46" 28d24'13"

Rita Brenner - Our Special Blessing
Cyg 21h39'27" 51d54'40"

Rita Brown
Sco 17h41'14" -37d45'14"

Rita Burchall
Lib 15h20'45" -20d6'50"

Rita Burns Boutin
Vir 13h54'49" -8d39'35"

Rita C. Paulus Love Shines Forever
Cnc 8h32'29" 29d3'52"

Rita C. Russo-Fednick
Vir 15h0'36" 4d10'15"

Rita C. Tuttle
Cyg 19h48'38" 38d57'7"

Rita Carmichael
Umi 14h44'17" 71d53'24"

Rita "Chickie" Yager
Cas 0h16'57" 51d22'1"

Rita Chiodo
Mon 6h48'28" -5d3'53"

Rita Chovan
Sgr 19h22'59" -27d26'13"

Rita Couchie
Psc 1h10'31" 12d27'38"

Rita D. Wible
Gem 6h45'2" 32d42'40"

Rita & David
Uma 10h54'8" 57d5'5"

Rita DeBenedetto
Leo 9h39'56" 31d7'39"

Rita DeGonza
Cas 1h3'29" 53d56'5"

Rita Donahoe Gorman
Sgr 19h10'40" -15d18'29"

Rita Dorothy Driver
Lib 15h10'48" -9d0'58"

Rita Doyle
Cyg 20h30'31" 58d47'55"

Rita e Lucio
Ori 6h4'54" 6d46'25"

Rita Eileen Swanson
Cyg 20h44'26" 44d14'30"

Rita Elizabeth Patterson
Uma 11h25'38" 51d49'42"

Rita Ellen
Cyg 20h27'1" 34d36'39"

Rita és Bálint csillaga
Cas 23h1'50" 58d55'0"

Rita Estephan
Cyg 19h36'9" 29d54'25"

Rita et Gilles
Umi 12h18'17" 88d6'15"

Rita Evans
And 0h42'50" 40d27'8"

Rita Faye
Leo 9h30'22" 29d50'17"

Rita Fiona
Cas 23h12'52" 59d24'23"

Rita Forgione
Uma 8h37'15" 54d49'13"

Rita Fort
Tau 4h19'16" 29d6'27"
Rita Francesconi
Gem 6h19'54" 21d30'5"
Rita Frankel
Sco 16h26'23" -41d52'55"
Rita Garlinski
Her 17h26'35" 24d2'15"
Rita Gee
Cas 1h55'26" 63d56'36"
Rita Gentile
Uma 9h29'5" 60d18'12"
Rita - Good night and God bless
Eri 2h14'22" -51d27'18"
Rita Harvey
Tau 4h42'43" 28d48'56"
Rita Hawa
And 23h28'4" 35d47'50"
Rita Haynes
Aql 18h50'23" -0d56'22"
Rita Hunter
Cas 0h47'39" 61d18'55"
Rita J. Noyes "1935-2006"
Gem 7h29'28" 27d17'5"
Rita Jane
Lyn 7h49'30" 55d7'52"
Rita Jean
Leo 9h32'41" 29d14'25"
Rita Jean Fazekas
Gem 7h19'13" 16d37'8"
Rita Jean Porter
And 0h13'17" 44d46'30"
Rita Jean Spell
Hya 8h31'29" -1d39'12"
Rita Julia Delgado
Cnc 8h34'25" 24d39'6"
Rita June Gregory
And 1h43'41" 49d27'58"
Rita Kathleen's Key
Ori 5h36'14" -3d10'14"
Rita Kay
Cap 20h35'0" -20d24'20"
Rita Kerns
Cas 0h24'26" 61d35'40"
Rita Krut
Vir 14h6'59" -16d56'48"
Rita Kunk, Nurse Practitioner
Lyn 7h45'28" 48d40'21"
Rita Lepuri
Ori 5h3'21" 5d25'28"
Rita LoPresti
Vir 12h47'22" 8d8'2"
Rita Lucille McCormick Riley
Leo 11h1'45" 10d55'46"
Rita Lucretia Herzfeld
Cas 0h33'39" 64d4'15"
Rita M. Aversa
Uma 11h46'5" 57d22'52"
Rita M. H. Itschert geb. Zachmann
Cas 23h27'51" 59d39'15"
Rita M. LaRocca
Uma 9h40'34" 44d27'50"
Rita M. Rennell
And 1h52'4" 46d45'45"
Rita Mae Gantt
Aqr 20h42'26" 0d7'43"
Rita Mae Schulz
Lmi 9h46'56" 36d5'6"
Rita Mairead Mc Cabe
Psc 0h46'58" 2d58'4"
Rita Maria Swain
Lib 14h34'47" -9d52'33"
Rita Marie
Gem 7h31'1" 28d36'56"
Rita Marie Banfield
Lib 14h49'50" -1d20'58"
Rita Marie Daoust Murphy
Lib 15h38'10" -28d8'9"
Rita Marie Oglesby
Uma 11h1'24" 64d16'46"
Rita Mary 327
Crb 15h44'47" 26d53'13"
Rita May
Psc 0h25'59" 0d30'36"
Rita May Briddon - Loveliest Star
And 1h13'51" 41d23'49"
Rita Mazzella
Vul 19h36'45" 24d54'39"
Rita Micciche
Gem 6h45'20" 33d59'42"
Rita Michelle Lewellen
Crb 15h48'39" 35d51'15"
Rita Montague
And 0h25'58" 25d44'2"
Rita Moon Moen
Sco 17h4'46" -33d12'8"
Rita Moore
And 23h9'51" 42d35'5"
Rita & Morris Young
Cap 20h34'8" -10d17'43"
Rita Murphy Kyle
Uma 14h28'27" 57d18'24"
Rita My Angel, My Everything
Cas 0h26'20" 61d32'39"
Rita Nanfara
Cmi 7h27'15" 10d50'55"
Rita O'Neill
Lib 15h6'45" -3d6'39"

Rita Passio
Cyg 21h47'14" 50d0'56"
Rita Pearl McNamar
Aqr 22h24'5" -4d52'12"
Rita "Pinky" St. Clair
Cyg 19h58'2" 35d31'46"
Rita Price
Cas 3h14'33" 63d51'27"
Rita Rae Howe
Vir 11h42'4" 3d44'27"
Rita Redina
Cas 0h5'37" 53d17'59"
Rita Reichmuth
Cyg 19h45'20" 35d24'53"
Rita Renae
Uma 9h19'52" 56d21'4"
Rita Rexford Zinni
Cam 5h2'9" 62d51'28"
Rita Rey
Aur 6h32'46" 34d54'21"
Rita Rigenerato Pollaro
Cas 0h42'42" 65d5'17"
Rita Roberts
Cas 1h12'9" 58d4'30"
Rita Rose
Del 20h38'14" 15d57'19"
Rita Rose Carluccio
Lyr 18h31'0" 31d5'1"
Rita Rossetti 12/1/1975
Ori 6h18'15" 13d53'2"
Rita Russell
And 0h29'40" 33d58'54"
Rita Russo
Sco 16h9'55" -10d30'23"
Rita Santi
Cyg 21h40'50" 55d23'37"
Rita Sarandria
Cap 20h41'42" -16d27'15"
Rita Schoen
Psc 1h10'32" 3d12'35"
Rita Segal
Crb 16h12'26" 36d10'23"
Rita Sisti
And 0h52'47" 35d54'41"
Rita Skovhus
Per 3h17'33" 42d8'53"
Rita Soldano
And 1h0'32" 45d36'38"
Rita Stahl
Vir 12h59'34" -15d50'34"
Rita "Sucha" Klopner 1949-2007
Gem 6h47'8" 18d12'31"
Rita Sunshine Pollack
Aqr 23h4'26" -7d46'52"
Rita Trimeri
Peg 22h10'10" 12d25'38"
Rita Trujillo Mejia
Gem 7h13'57" 24d49'59"
Rita Ursula Kelly
Uma 11h26'36" 33d56'10"
Rita Van
Lyn 8h46'51" 36d3'43"
Rita Veronica
And 2h6'36" 42d5'36"
Rita Weissmann Happy At 75
Vir 12h8'43" -8d12'14"
Rita Y Chang
Aqr 21h6'24" -14d25'23"
Rita Yvonne Hayes Parham
Gem 6h6'52" 24d48'13"
Rita Z
Uma 12h45'59" 55d31'39"
Rita Z.
Sco 16h17'3" -32d23'38"
Rita Zachariadou
Uma 8h59'22" 72d4'0"
Rita Zaengle
Lyr 18h42'27" 30d21'19"
Rita, 10.12.1954
Uma 9h42'4" 44d31'46"
Rita89
And 23h5'59" 49d16'8"
Ritacco's Eyes
Cap 20h19'25" -18d53'34"
Ritachka
Cnc 8h5'35" 19d51'18"
RITAllion beauty
Leo 9h46'12" 31d13'28"
Ritamae Seth Lane
Uma 10h37'21" 64d55'46"
Rita's husband, Tommy
Uma 12h16'12" 60d38'9"
Rita's Light - Our Mother Star
Sco 17h47'32" -39d35'29"
Rita's Memory
Uma 10h47'4" 50d46'47"
Rita's Spirit
Lac 22h4'50" 48d33'20"
Rita's Star
Crb 15h47'9" 35d36'37"
Rita's Tigger Bean
Uma 12h42'37" 56d30'2"
Richard Murphy
Peg 21h50'46" 8d48'34"
Ritchie Alan Erickson
Gem 6h55'42" 14d33'50"
Ritchie Alan Young
Leo 11h22'17" -4d13'17"
Ritchie Mangas
Ari 3h16'18" 28d55'35"

Ritchie Raymonds
Uma 9h30'38" 62d26'52"
Ritchie West
Her 18h55'12" 24d21'19"
Ritha & Emmanuel
Gem 7h35'47" 23d57'33"
Ritika's Butterfly Light
Pho 0h51'45" -48d29'36"
Ritish Gullapally
Umi 16h2'50" 75d21'23"
Rito e Renata
Crb 15h58'27" 29d5'41"
RITO1107
Uma 11h48'43" 36d51'34"
Riton
Uma 9h50'58" 47d32'53"
Ritsa N. Prasinou
Ari 2h50'19" 27d39'23"
Rittenbruch
Uma 9h52'0" 60d9'7"
Ritter
Ori 5h39'48" 0d48'56"
Ritter, Frank & Kirstin
Uma 8h46'5" 48d16'51"
Ritter-Fern
Uma 9h49'2" 62d17'50"
Rittler Family Star
Sgr 18h19'45" -24d24'3"
Riulà
Peg 22h51'19" 27d44'0"
Riva Si
Leo 9h39'37" 6d54'13"
Riva, Anika and Sahil Pallan
Tri 2h29'21" 30d39'54"
RivaAmorynBillie together eternally
Eri 4h29'30" -0d47'58"
Rivaleah
Lyn 9h12'47" 37d42'41"
Rivard-Pringle
Cyg 20h18'8" 54d27'37"
RIVENDELL
Mon 7h18'25" -0d25'22"
River
Uma 9h1'22" 66d12'46"
River Albion Elijah Cohen
Vir 14h24'36" -0d28'15"
River Archer McDole
Ari 3h8'12" 25d59'28"
River Blue
Gem 6h55'5" 16d53'13"
River Jack
Pav 19h44'52" -57d12'5"
River Jordan Most
Uma 10h40'21" 50d6'35"
River Lee Beck
Eri 3h34'4" -42d21'40"
River of life
Dra 10h8'41" 77d41'5"
River Rat
And 0h41'52" 36d36'16"
River Sky Starlight
Lyr 18h52'38" 36d48'41"
River Varland
Sgr 18h4'55" -21d38'1"
Riverview Class of 2011
Leo 9h48'42" 28d35'55"
riverwide984
Lib 15h18'42" -8d41'20"
Riveyra Tay
Cru 12h37'27" -58d14'23"
Rivicah
Uma 11h30'16" 39d8'46"
Riviera Daisy Brown
Crb 15h58'34" 36d23'43"
Rivky & Moshe's Light
Uma 9h52'36" 61d4'42"
Rivmo
Leo 11h14'57" 16d12'55"
RIWAJEKI
Aql 19h46'36" -0d34'19"
Riya
Tau 3h26'48" 3d38'29"
RIYA DEVI DUA
Vir 12h53'1" 12d17'39"
Riya Nikore
Leo 9h22'32" 11d17'57"
Riyal
Sgr 17h43'37" -18d23'51"
Riyan Sullivan
Ori 5h40'46" 3d11'58"
RiyaNuY
Sgr 20h5'36" -34d13'47"
Riz
Psc 0h1'6" 3d53'6"
RizAdo
Cra 18h8'26" -37d8'30"
Rizi & Tom's Everlasting Star
Cyg 21h10'57" 42d17'11"
Rizia Moreira
Del 20h27'31" 16d39'52"
Rizielina Roncal-Castro
Sco 17h37'6" -42d19'2"
Rizk
Sex 9h42'50" -0d20'24"
RizRich
Cyg 20h11'17" 51d38'8"
Rizzo
Boo 14h36'25" 33d2'1"
Rizzo
Cas 2h4'26" 62d13'21"
RJ
Uma 11h6'56" 34d12'23"

RJ
Per 3h1'41" 54d50'29"
RJ
Vir 13h38'16" 4d32'33"
R.J.
Psc 0h30'45" 10d47'2"
R.J.
Peg 23h8'55" 13d11'7"
RJ and Kara Forever
And 22h59'38" 40d37'32"
R.J. Bobin
Per 3h26'1" 45d13'37"
RJ Butterfield
Aur 5h51'16" 48d2'3"
RJ Inocencio
Del 20h34'14" 15d32'48"
R.J. & Jewels
Psc 0h45'54" 14d11'10"
R.J. & Kay <\@>5
And 0h32'39" 36d43'26"
RJ Manfre
Uma 8h39'12" 56d8'45"
R.J. McClelland "11-23-1925"
Uma 12h33'42" 56d39'28"
R.J. Navas
Umi 16h14'24" 72d14'9"
RJ Perrine
Col 5h53'26" -36d15'24"
R.J. Pettit
Tri 2h37'28" 33d57'13"
"RJ" Ronald John Palmer Mannix
Umi 16h49'43" 78d3'3"
R.J. Selbie
Umi 16h29'13" 81d19'22"
RJ & Steph
Uma 11h41'15" 34d16'35"
RJ Veenstra
Lyn 6h40'53" 60d48'18"
RJ122106MoonStarsMoreThanLifeItself
Uma 11h13'33" 32d13'11"
RJB
Sgr 19h50'8" -13d22'49"
R.J.B.O. 21879 "Star Of Compassion"
Uma 11h26'52" 52d49'29"
RJJ - 50
Vir 13h29'47" 11d12'14"
RJK Ray of Hope
Per 3h36'26" 51d27'58"
R.J.Kunkle
Per 3h34'8" 43d50'8"
RJMAX33
Lyr 19h8'56" 26d15'29"
RJMW & MMK - Immortalis Amor
Gem 7h17'1" 19d59'26"
RJS 170904
Peg 21h54'45" 19d42'57"
RJ's Birthday Star
Psc 1h16'0" 10d1'8"
rjtsce
Pyx 9h25'38" -28d54'9"
RJZ3
Cnc 8h43'39" 24d26'44"
RK
Aql 19h21'38" -10d44'39"
R.K. Hagebusch
Uma 9h2'8" 69d33'38"
RK Star
Uma 8h57'4" 69d27'55"
RKA's Diane A
And 2h37'38" 44d54'40"
R.K.E.
Leo 11h48'10" 25d56'0"
RKE # 48
Uma 11h6'20" 44d32'5"
"<\@>rkle"
Sgr 18h4'22" -17d5'50"
RKMT-Roberta's Star
And 23h10'15" 43d11'40"
R.Krieger58
Cnc 8h38'46" 10d19'47"
RKVAN27
Leo 10h1'10" 24d48'47"
R.L. Fuerst
Uma 12h27'26" 57d52'7"
R.L. Giovinazzo
Her 17h17'52" 34d7'33"
RLD21 EAW BFF
Psc 0h41'12" 8d22'11"
RLH77
Crb 15h30'49" 29d18'8"
R-Lincoln 05244
Umi 14h49'20" 75d0'19"
R.L.I.P.4EVER
Her 17h44'39" 22d35'17"
R.L.N. 1980
Aqr 23h12'32" -11d36'10"
RLW III
Aql 19h48'35" 5d9'3"
R.M.+D.P.
Crb 15h48'50" 25d52'54"
R.M. McNeel
Her 18h31'45" 22d32'14"
RM&MJ
Pho 23h38'52" -41d45'45"
R.M.A
Ari 3h14'43" 29d53'29"
RMB Watching Over His Daughter
Her 17h0'34" 27d23'8"

RMB051792
And 0h44'7" 36d7'58"
R.M.C.G. Binky Mikula
Gem 7h37'17" 16d2'24"
RMD Shines
Crb 16h14'10" 38d5'47"
R.M.L. 5-20-1957
Uma 12h15'3" 60d9'18"
RMM "70"
Uma 8h52'44" 54d52'46"
RMMR277
Aqr 22h10'40" -1d24'45"
RMS RMS: Gift from God
Per 2h29'47" 52d59'1"
RMSentz 04-10-96
Ari 3h23'5" 24d35'42"
RN Soulmates
Cyg 20h13'47" 49d23'5"
RnAFoReVeR
And 0h29'44" 35d38'56"
RNJ Norfolk - Light my way - Tintoela
Ori 5h41'14" -2d29'36"
RNSchneider 1546-60
Lyn 8h16'13" 33d20'25"
Ro
And 23h14'40" 48d27'45"
Ro
Aqr 21h17'54" -8d10'16"
Ro Barisciano
Uma 11h22'52" 59d24'30"
Ro & Mel Mel Together Forever
Peg 21h40'15" 10d53'7"
Roald
Aqr 22h8'32" -0d59'37"
Roald Edrick Riva
Uma 9h41'24" 72d21'58"
Roamin' Love
Dra 18h47'11" 57d40'24"
Roan Christopher Holsten
Per 3h35'57" 43d57'27"
Roan Niels McGowan
Vir 12h44'32" 12d43'18"
Roanna
And 0h54'50" 22d47'13"
Roanne Lee Whanger
Lib 15h7'13" -20d40'31"
Roarie Daisy Anderson
Tau 4h23'0" 26d22'43"
Rob
Cnc 8h45'6" 27d4'25"
Rob
Lyr 18h51'57" 26d51'27"
Rob
Gem 7h26'50" 20d11'28"
ROB
Leo 11h40'6" 21d39'47"
Rob
Per 3h6'4" 51d46'10"
Rob
Ori 4h57'39" -0d53'56"
Rob
Uma 10h33'22" 65d45'46"
Rob 06-18-06
Lyr 18h50'30" 42d49'6"
Rob Alexander
Uma 11h24'50" 60d17'0"
Rob and Adele Wylie
Cyg 21h53'20" 40d20'2"
Rob and Amy Skwirut
Ori 5h34'42" 8d42'15"
Rob and Brianna Ware
Cyg 19h58'13" 41d28'53"
Rob and Caitlin's Star
Cyg 19h53'12" 37d24'6"
Rob and Gina's Grasshoper Star
Sge 19h42'54" 17d8'25"
Rob and Heather Steward
Cyg 21h15'49" 41d22'55"
Rob and Jean - Love Always
Cyg 19h42'14" 29d33'0"
Rob and Jen
Cyg 20h25'19" 55d4'44"
Rob and Loren Eternally
Ara 17h33'56" -47d35'0"
Rob and Margeaux Rumbel
Sge 20h2'58" 17d11'7"
Rob and Mel Florio
Lyr 18h45'49" 39d33'6"
Rob and Mere's Star
Ori 6h4'38" 19d36'18"
Rob and Sarah
And 0h56'26" 23d17'3"
Rob Aney
Leo 10h36'5" 20d16'41"
Rob & Angela Dixon's Anniversary Star
Cyg 19h35'38" 28d52'21"
Rob - A.W.T.W.B.
Cap 20h32'41" -14d54'19"
Rob Berger
Dra 18h50'25" 65d43'1"
Rob Burns
Per 3h46'2" 48d48'19"
Rob & Catie's Polish Love Star
Boo 14h20'44" 47d25'21"
Rob & Charity 7/9/98
Vir 12h34'15" -1d13'55"
Rob & Chris
Cyg 21h29'27" 51d38'6"

Rob & Christina
Aqr 23h5'51" -15d25'28"
Rob Cline
Psc 0h33'13" 7d27'0"
Rob Corbin
Ari 3h28'40" 25d47'44"
Rob Cunsolo *831-2002*
MMM L
Her 16h51'27" 23d55'36"
Rob Dannegger
Lac 22h10'57" 49d30'31"
Rob Dare Traveling Minstrel
Sco 16h37'6" -29d53'43"
Rob Davis
Sco 17h29'13" -45d24'3"
Rob Denton
Cnc 8h39'49" 23d11'52"
Rob Dillard
Per 3h28'58" 45d32'19"
Rob Doig
Dra 20h26'36" 62d30'9"
Rob Drysielski
Sgr 18h34'45" -23d37'2"
Rob Duane Walker
Leo 9h40'13" 12d35'2"
rob dunham
Lyn 8h5'54" 43d21'55"
Rob Dyer's Star of Hope
Aqr 21h20'52" -10d1'56"
Rob Elliott and Kayla Wilson Forever
Aqr 20h48'34" 0d9'28"
Rob Elliott and Sarah Jones Forever
Gem 6h27'40" 22d40'10"
Rob & Erin Forever
Lac 22h47'38" 52d32'59"
Rob Flintom
Boo 14h27'35" 37d15'39"
Rob Goldstein
Her 16h43'22" 18d13'28"
Rob Gower
Cyg 20h51'7" 47d21'12"
Rob Harrison
Cas 23h36'51" 55d14'28"
Rob Helean - rayon de soleil
Del 20h37'5" 15d28'59"
Rob & Jen forever with all my heart
Cyg 19h39'8" 29d49'12"
Rob & Jennifer Edmonds
Sge 20h4'13" 19d5'37"
Rob & Jess's wedding star
Cra 18h19'52" -39d24'2"
Rob Jøran Hansen
Cep 21h49'58" 64d59'9"
Rob Kegley
Lib 14h59'29" -11d18'58"
Rob Kidney
Psc 1h10'51" 9d33'23"
Rob & Laura
Umi 16h6'3" 74d57'32"
Rob & Laura 'True Love'
Cyg 20h11'11" 47d36'16"
Rob Laury
Sgr 18h36'37" -18d40'13"
Rob Lawson
Cep 22h8'45" 68d25'20"
Rob Leichter-DelGreco
Per 2h8'58" 56d14'8"
Rob & Lorry Hager
Uma 9h37'32" 61d51'55"
Rob Loves Patti TLA
Cyg 21h37'32" 52d2'46"
Rob & Mandie 07/09/05
Cyg 19h52'38" 32d24'21"
Rob & Mandy Forever 10/11/96
Cyg 21h6'38" 36d51'18"
Rob & Mandy Vernons' Wedding Star
Cyg 19h52'25" 32d21'18"
Rob Maple
Vir 13h9'25" 6d47'51"
Rob & Marcy's Endless Love
Aqr 21h43'23" 0d38'7"
Rob Matera
Her 16h30'4" 47d50'5"
Rob Mckimm
Lyn 7h42'29" 49d36'45"
Rob McNamara - Celestial Angel 1987
Cru 12h29'59" -60d39'33"
Rob & Michelle - 10 Sep 2005
Pho 0h21'53" -53d11'47"
Rob - my Forever Love
Cyg 20h50'22" 31d24'19"
Rob 'n' Anne-Marie - Soulmates Forever
Cyg 21h42'15" 31d59'27"
Rob Nej
Cyg 20h56'47" 45d30'31"
Rob & Nicole's Piece Of Heaven
Psc 0h52'15" 28d29'8"
Rob Notting
Gem 7h15'54" 20d28'3"
Rob Noxon
Per 4h38'52" 37d53'23"
Rob O'Reilly
Uma 12h28'56" 52d23'46"

Rob P Grotts
Uma 10h24'42" 63d30'13"
Rob Patterson
Dra 19h11'44" 74d36'32"
Rob Pierce
Umi 15h28'5" 74d34'23"
Rob Price
Leo 10h11'23" 14d4'29"
Rob Roy's My Fair Lady
Uma 8h22'4" 68d25'10"
Rob Sandberg, M.D.
Aqr 22h42'9" -1d5'47"
Rob Schasel
Crb 16h5'51" 26d6'4"
Rob Sis- My love for all eternity
Psc 0h14'9" 9d30'46"
Rob & Tara - Forever & A Day
Col 6h23'14" -35d20'52"
Rob Tillman
Pic 5h32'0" -43d54'35"
Rob & Ty Thompkins
Tau 4h25'2" 17d7'28"
Rob Varenica - My Shining Star
Aqr 23h52'26" -19d14'48"
Rob Webster
Boo 14h57'28" 34d28'11"
Rob & Wendy Conte
Cyg 20h12'18" 36d43'36"
Rob Yurkanin & Christine Le
Sgr 19h6'55" -14d24'12"
Rob Zino
Cap 20h12'12" -20d49'13"
Rob Zuuring
Psc 23h14'7" -0d31'41"
Rob-Ann-04-09-40
Cyg 21h59'15" 45d28'25"
Robb Hammond
Her 16h37'59" 12d34'6"
Robb & Jen Zimmerman 2-15-98
Ori 6h19'3" 13d33'57"
Robbert O'Neal Davis
Umi 15h18'45" 68d26'0"
Robbi Alyne
Ari 2h24'13" 24d17'3"
Robbi Loves Nate Always and Forever
Cyg 21h47'19" 41d25'46"
Robbi Shanker Shrestha
Psc 1h14'23" 27d7'11"
Robbie
Ori 5h16'36" 9d7'8"
Robbie
Cam 3h39'38" 59d20'10"
Robbie
Per 3h22'17" 43d30'51"
Robbie
And 0h33'56" 31d52'48"
Robbie Allen Mudrak
Her 18h10'16" 18d41'22"
Robbie and Ashley
Vir 13h26'0" -4d16'30"
Robbie and Dani Snelson
Cnc 8h38'17" 24d39'7"
Robbie and Kara's Wedding Star
Leo 9h38'24" 27d55'52"
Robbie Bertram Schutz
Lib 15h7'47" -4d46'56"
Robbie Breen
Uma 8h32'29" 72d10'9"
Robbie "Bubba" Ross
Her 17h11'34" 39d26'14"
Robbie Dominic Nelson
Per 3h42'5" 41d42'55"
Robbie & Fallon Forever and Always
Cyg 21h32'34" 52d55'44"
Robbie Frank
Uma 9h12'48" 70d10'35"
Robbie Franklin
Dra 17h19'34" 64d23'11"
Robbie Groth
Uma 13h51'7" 50d19'40"
Robbie Heier 9/18/2001
Uma 10h8'15" 45d56'12"
Robbie J. Perilloux Jr.
Tau 3h31'3" 9d36'17"
Robbie Jo Price
Lib 15h14'20" -9d6'30"
Robbie Johns
Cep 21h20'50" 65d36'37"
Robbie light of my life
Lib 15h56'14" -11d1'42"
Robbie & Lori Gibson
Peg 22h50'59" 27d10'43"
Robbie MacIntyre
Uma 11h25'12" 54d38'2"
Robbie Mcl
Per 2h22'33" 55d28'44"
Robbie Punter
Ori 6h19'29" 18d56'12"
Robbie Ruibal
Uma 9h20'22" 53d3'31"
Robbie Schaare
Ori 6h9'27" 20d52'39"
Robbie & Shannon
Cyg 20h56'25" 34d25'36"
Robbie Simon Wiley
Her 17h40'41" 24d37'44"

Robbie Stevens
Her 16h52'41" 37d39'46"
Robbie Sue
Gem 6h45'36" 26d43'49"
Robbie & Teresa Stahl
Sgr 18h54'59" -26d26'18"
Robbie White
Cyg 20h31'26" 31d2'30"
Robbie's Girl
Ori 5h59'59" 16d53'32"
Robbie's Light
Uma 11h7'28" 32d56'53"
Robbie's Place
Uma 8h59'24" 47d57'22"
Robbie's Star
Cep 22h47'25" 63d25'7"
Robbie's Star
Umi 16h9'55" 71d6'21"
Robbin Ann Leighty
Cyg 20h56'59" 35d5'20"
Robbin Manning
Psc 0h22'23" 6d23'7"
Robbin Reneé 459
Lyn 6h20'48" 56d22'39"
Robbins McClintock
And 1h13'1" 38d46'26"
Robbins to Renick
Dra 16h16'21" 62d5'23"
Robbins02/02/02
Sgr 18h0'10" -27d43'51"
Robblee
Umi 14h37'22" 73d28'57"
Robby
Lac 22h26'8" 53d2'10"
Robby
Cnc 9h3'52" 31d46'9"
Robby
Lmi 10h48'34" 29d10'7"
Robby 16
Lyn 7h46'34" 44d16'21"
Robby Allan
Uma 13h32'51" 54d28'28"
Robby Holbert
Aql 18h52'42" -0d14'12"
Robby "My Superman"
Per 4h4'21" 32d27'55"
Robbyn
Lmi 10h43'22" 27d53'29"
Robemma
Cyg 21h38'50" 30d19'46"
Roberson's Razorback
Tau 4h35'52" 5d5'31"
Robert
Hya 8h53'0" 5d43'26"
Robert
Tau 3h27'55" 9d4'26"
Robert
Ori 6h20'48" 13d27'23"
Robert
Ari 2h42'27" 20d21'11"
Robert
Cnc 9h7'39" 29d18'5"
Robert
Tau 4h42'9" 22d35'57"
Robert
Boo 14h52'43" 32d6'23"
Robert
Per 2h48'30" 56d52'56"
Robert
Vir 12h20'6" -3d3'47"
Robert
Sgr 18h43'40" -23d6'40"
Robert A. Abate
Cnc 8h10'0" 28d46'8"
Robert A. Choinard
Uma 9h12'38" 65d10'38"
Robert A Durant Jr
Ari 2h7'35" 23d33'20"
Robert A. Filson
Umi 16h26'14" 75d34'50"
Robert A Fossett
Gem 6h55'12" 16d13'4"
Robert A. Gardner
Cap 21h39'21" -14d34'47"
Robert A. Golden
Ori 6h20'56" 13d44'29"
Robert A. Irwin
Per 3h13'25" 53d58'5"
Robert A. Kaminski II
Aql 19h46'45" -0d54'26"
Robert A. Laughlin
Per 2h44'17" 54d12'25"
Robert A. Meszaros
Her 16h30'57" 32d2'12"
Robert A. Miller
Cep 22h19'32" 63d31'33"
Robert A. Milton
Aql 19h51'13" 11d55'51"
Robert A. Mosher, Jr.
Gem 7h26'25" 34d51'56"
Robert A Nader
Sco 17h49'27" -37d39'51"
Robert A. Olkowitz
Her 16h39'36" 32d46'31"
Robert A. Pacenti
Uma 9h56'59" 54d34'39"
Robert A Petersen, Sr.-Our Captain
Tau 5h10'33" 21d36'36"
Robert A Reznick
Cep 20h36'29" 63d59'1"
Robert A. Smutek
Gru 23h15'53" -40d36'49"
Robert A. Wakeley 3
Cnv 13h49'46" 34d25'13"

Robert A. Warner
Aqr 22h42'48" -0d58'58"
Robert A. Wcislo, Jr.
Dra 17h1'22" 63d5'18"
Robert A Wicklund
Cep 21h41'57" 64d14'56"
Robert Aaron Arvid Jansson
Peg 21h29'47" 16d16'37"
Robert Aaron Mitchell
Crt 11h33'48" -23d18'17"
Robert Abbate
Her 17h35'25" 36d46'47"
Robert Acheson Goldstein
Per 4h23'26" 47d39'32"
Robert Adam and Sara Louise Moore
Psc 1h0'19" 26d43'43"
Robert Adams
Lyn 9h40'50" 39d46'56"
Robert Aeolus
Ori 4h53'9" -0d17'45"
Robert Alan Anderson
Ori 6h13'32" 16d21'32"
Robert Alan Coomer: My Special Star
Cep 20h54'51" 66d27'51"
Robert Alan Fletcher
Per 3h41'25" 46d26'58"
Robert Alan Goldman
Gem 7h21'48" 25d54'33"
Robert Alan Kreitzer Jr.
Aur 6h1'59" 47d57'14"
Robert Alan Levine
Ari 2h45'9" 25d12'51"
Robert Alan Pipik Jr.
Sco 16h50'24" -40d50'57"
Robert Alan Sugarman
Gem 6h41'47" 17d29'29"
Robert Albert Frazier Jr
Cap 21h26'40" -16d22'54"
Robert Albert Petrak
Her 16h39'25" 37d51'7"
Robert Albertson
Boo 14h47'48" 20d55'36"
Robert Alex Baker
Uma 13h33'19" 57d46'59"
Robert Alexander Brooks
Tau 4h27'53" 14d33'9"
Robert Alexander LeMire
Cyg 20h22'58" 44d54'48"
Robert Alexander Rodriguez
Cnc 8h14'35" 16d33'6"
Robert Alexander Sawicki
Aur 5h47'36" 51d13'16"
Robert Alfred Cappa
Gem 6h51'13" 26d48'32"
Robert Alfred Johnson
Her 16h27'33" 48d45'15"
Robert & Alita Piorun
Cyg 21h53'31" 44d46'31"
Robert Allan Erb
And 0h59'49" 37d25'55"
Robert Allan Lewis
Cep 21h28'49" 64d45'4"
Robert Allen Cover
Vir 11h40'6" -0d54'7"
Robert Allen Evans
Ori 4h50'12" 4d43'34"
Robert Allen Moskovitz
Sco 17h46'54" -42d44'5"
Robert Allen Rhodes
Uma 10h27'28" 39d26'39"
Robert Allen Schmittthenner
Cmi 7h37'40" -0d5'2"
Robert Allen Schneider
Aql 19h2'19" 16d2'15"
Robert Allen Schreiber III
Vir 11h48'18" -3d3'9"
Robert Ambroziak
Lmi 9h24'12" 38d59'9"
Robert & Amy Carson
Cyg 21h29'15" 32d44'15"
Robert Anconetani
Psc 1h6'26" 12d53'48"
Robert and Andrea 2005
Cyg 19h35'27" 52d51'19"
Robert and Bethany Gallegos
Cnc 9h17'25" 10d41'20"
Robert and Beverly's Star
Psc 1h19'4" 19d9'31"
Robert and Brienne Litchfield
Cyg 20h20'24" 46d12'11"
Robert and Caroline
Her 18h54'52" 23d44'42"
Robert and Celeste Johnston
Cyg 19h33'34" 28d33'22"
Robert and Dorothy Cully
And 2h11'13" 38d12'46"
Robert and Eleanor Wolven
Cyg 19h32'6" 54d12'10"
Robert and Eleni
Umi 15h30'16" 74d55'37"
Robert and Elizabeth
Lib 15h9'18" -6d6'59"
Robert and Jacqueline Hamilton
Sco 17h26'56" -40d42'36"
Robert and Janette Frock
Uma 11h52'13" 49d37'29"

Robert And Joy's Everlasting Star
Umi 14h3'46" 66d21'22"
Robert and Kate Kite
Lmi 10h49'46" 35d33'6"
Robert and Kathleen Dworak
Lmi 10h17'58" 37d10'53"
Robert and Kathleen Pease
Cnc 8h37'29" 15d16'33"
Robert and Kerri's love is eternal.
Ori 4h50'33" 5d10'20"
Robert and Kristin
Cyg 20h28'10" 52d22'2"
Robert and kTs Star
Gem 7h9'24" 15d11'19"
Robert and Leona Wollenberg
Lib 14h46'15" -10d51'18"
Robert and Lorene's 10 year star
Sge 20h10'23" 18d55'37"
Robert and Lydia Crenshaw
Cyg 21h28'16" 46d57'21"
Robert and Margaret McGregor
Lyn 8h2'31" 44d33'49"
Robert and Marian Pappenfus
Uma 10h16'55" 60d7'15"
Robert and Maureen LaBonte
And 0h37'31" 35d16'12"
Robert and Paula Wade
Sge 19h25'49" 17d1'46"
robert and rachel
Aqr 20h39'51" -2d26'18"
Robert and Regina Wathier
Cap 20h51'43" -21d0'15"
Robert and Renee of Sarasota
Cas 1h12'51" 53d1'9"
Robert and Sandra Johnson
Vir 12h59'4" 13d0'31"
Robert And Sharon's BrAnDi Bear
Ari 2h10'44" 24d0'45"
Robert and Susan McCormick
Gem 6h57'15" 31d46'50"
Robert and Susan Wolfe
Per 2h49'26" 40d42'27"
Robert and Toni True Love
And 23h14'27" 45d14'52"
Robert Anderson
Cen 13h39'34" -41d43'0"
Robert & Andrea McDonald
Cap 21h7'48" -17d39'25"
Robert Andrew
Lyn 9h14'8" 38d13'29"
Robert Andrew Blessing
Her 18h9'56" 29d2'33"
Robert Andrew Caldwell
And 0h22'26" 28d28'28"
Robert Andrew Kane
Aur 5h29'6" 47d38'51"
Robert Andrew Klein
Lyn 6h24'57" 56d36'0"
Robert Andrew Orr
Umi 16h26'17" 71d16'46"
Robert Anthony
Ori 6h16'38" -1d36'15"
Robert Anthony
Per 3h17'45" 52d2'0"
Robert Anthony Aguilera
Sgr 18h11'21" -18d27'8"
Robert Anthony Detrick
And 23h13'34" 52d45'25"
Robert Anthony Drabik a.k.a."RAD 1"
Boo 14h32'46" 40d23'19"
Robert Anthony Ely Stanley
Boo 13h48'22" 17d18'7"
Robert Anthony Fineo
Her 18h30'47" 21d48'25"
Robert Anthony Imbrogno
Per 3h46'14" 51d44'22"
Robert Anthony Malyn - Tony's Star
Gem 6h4'45" 26d25'22"
Robert Anthony Marotta
Sco 16h23'42" -24d57'29"
Robert Anthony Petty
Aur 5h48'34" 54d59'53"
Robert Anthony Sabatino
Umi 14h31'40" 73d54'13"
Robert Arak
Gem 6h49'43" 33d32'57"
Robert Archuleta
Aur 5h23'59" 36d18'5"
Robert Argyle McAllen
Her 16h21'33" 5d41'33"
Robert Armando DeLaRosa
Gem 6h53'41" 27d39'14"
Robert Armet
Aql 20h0'32" 6d54'46"
Robert Armstrong
Psc 23h46'8" 7d19'22"
Robert Armstrong
Ori 5h27'50" 5d2'39"
Robert Arnold Paton
Pho 0h40'38" -47d3'46"

Robert Arthur Barrett
Her 18h42'55" 14d57'23"
Robert Aubrey Aus
Per 2h41'23" 53d29'33"
Robert & Audrey Ewell
Cnc 8h41'28" 7d7'23"
Robert Austin Hattan
Tau 4h34'54" 3d50'29"
Robert Austin Tierney
Lib 14h30'29" -19d4'39"
Robert B. Dillon
Ari 2h57'59" 28d16'31"
Robert B. Itri
Cep 23h5'12" 80d12'32"
Robert B McGeachie - His Star
Cru 12h36'51" -58d39'37"
Robert "Babe" Laas
Gem 6h9'20" 27d26'24"
Robert Baker
Cyg 21h41'16" 46d3'26"
Robert & Barbara McPeek
Cyg 19h52'48" 56d11'48"
Robert Barrett Crawford II
Boo 14h42'30" 22d43'53"
Robert Barrette
Lyn 9h1'13" 34d59'27"
Robert Bartner
Ari 2h2'57" 19d20'59"
Robert Bathgate Denholm
Uma 8h58'27" 55d0'10"
Robert Beldegreen
Uma 11h49'10" 57d11'25"
robert belluscio
Leo 10h13'29" 7d24'28"
Robert Beloved Father/Grandfather
Uma 10h1'4" 49d52'15"
Robert Benjamin Bongiardino
Sgr 18h25'18" -32d5'51"
Robert Bennett Lackey
And 0h16'22" 28d17'53"
Robert & Benoit Grondin
Cep 22h35'29" 67d7'32"
Robert Berman
Cma 6h43'28" -15d18'16"
Robert Bernard Yelinek
And 1h29'29" 50d21'42"
Robert Berry Jr.
Gem 7h25'11" 31d25'44"
Robert Betts
Scl 0h50'10" -33d31'2"
Robert Big Daddy
Her 17h18'12" 24d43'45"
Robert (Bijan) Kayvon
Lyn 7h20'0" 56d57'3"
Robert "Bill" Thompson
Ori 6h14'25" 2d22'32"
Robert (BJ) Anthony Gomez, Jr.
Lyn 8h0'3" 36d5'30"
Robert Blake, Jr.
Crb 16h11'29" 39d33'14"
Robert Bluhm
Uma 9h20'11" 47d31'28"
Robert "Bo" Rawson
Aql 19h17'7" 6d4'28"
Robert "Bob" Daniel Shaver
Leo 10h26'0" 23d27'8"
Robert "Bob E. Hale"
Her 16h38'39" 48d45'54"
Robert "Bob" Heilman
Lib 15h39'33" -29d49'29"
Robert "Bob" Hoff
Uma 11h20'32" 60d30'47"
Robert "Bob" Kahn
Cnc 9h12'59" 17d45'25"
Robert "Bob" & Nancy Brown
Ori 5h53'6" 20d9'14"
Robert "Bob" Siff
Leo 10h19'8" 11d20'47"
Robert (Bob) Skelton
Uma 10h55'43" 46d44'14"
Robert (Bob) W. Jarnagin
Gem 7h44'42" 24d46'59"
Robert "Bobby" Allen - 12/8/1958
Cep 1h54'37" 82d20'47"
Robert Bobby Boo Goji Boy Dunlop
Col 6h33'50" -34d22'22"
Robert (Bobby) Morrow
Cep 21h29'26" 59d7'4"
Robert "Bobby" P. Dunn
Her 18h50'1" 21d37'26"
Robert "Bobby" Rush
Uma 11h14'19" 52d42'6"
Robert (Bobby) W. Kizer
Per 2h25'52" 52d8'55"
Robert Bodmer
Per 3h10'8" 46d46'8"
Robert Bordeau
Cnc 8h40'48" 15d7'58"
Robert Boucher
Cap 21h11'48" -19d45'2"
Robert Boushley
Aql 18h57'10" -0d36'19"
Robert Bower
Lac 22h21'43" 52d4'25"
Robert Boyer
Uma 11h7'19" 60d34'55"
Robert Boyle
Per 2h52'55" 53d30'21"

Robert Bradley Starling
Her 18h11'24" 15d22'53"
Robert Bradley Whitlock
Uma 13h31'10" 54d46'6"
Robert Braithwaite ~ 90th Birthday
Lib 15h0'58" -13d52'36"
Robert Bräu
Uma 9h3'35" 69d15'15"
Robert Bravo's Star
Uma 10h16'37" 72d45'49"
Robert Brazeal
Ori 5h17'24" 6d46'7"
Robert Breth
Ari 2h9'50" 15d0'24"
Robert Brown
Per 3h12'33" 55d5'30"
Robert Bruce Ascue, III
Lmi 9h38'7" 37d39'0"
Robert Bruce Crowe
Sgr 18h38'57" -26d38'44"
Robert Bruce Derrick
Ari 3h0'45" 28d6'11"
Robert Bruce Graham
Ori 6h5'46" 15d46'53"
Robert Bruce Kenney
Uma 14h5'13" 49d0'51"
Robert Bryan Bollinger
Ori 5h24'13" 2d35'50"
Robert Bryan Dabney
Aql 19h7'56" -8d50'29"
ROBERT BRYAN RICE
Lib 15h42'10" -17d43'45"
Robert Bryan Somewhere in Space
And 0h47'36" 33d56'43"
Robert Burba Benham
Lib 15h27'51" -27d52'53"
Robert Burgess
Uma 10h16'33" 53d0'18"
Robert Buttach
Uma 10h53'6" 40d15'6"
Robert C. Anderson
Per 4h10'11" 34d29'33"
Robert C. Bergner Jr " My Soulmate"
Her 17h36'2" 37d29'55"
Robert C. Bogle
Aqr 22h40'46" 0d2'29"
Robert C. Chalmers Sr. "Bob"
Uma 8h11'42" 70d13'36"
Robert C Cione
Leo 10h23'35" 13d1'57"
ROBERT C. DEBIASE
Sgr 19h27'57" -13d19'36"
Robert C. DiChiara
Lyn 7h50'21" 47d1'27"
Robert C. Dirnbeck
Psc 1h4'11" 4d43'50"
Robert C. Downs, Jr 3/17/60
Psc 0h50'27" 29d26'1"
Robert C. Edwards
Tri 2h7'56" 31d39'17"
Robert C. Everich
Per 2h57'45" 43d27'18"
Robert C. F. Wolfe
Psc 0h40'37" 12d16'16"
Robert C. Fischer Jr.
Ori 6h17'6" 10d53'35"
Robert C. Gelwicks
Cep 3h21'13" 78d1'3"
Robert C. Goldrick
Cnc 8h50'7" 32d25'30"
Robert C. Gombar
Sco 16h28'49" -27d11'53"
Robert C. Heed
Her 16h50'16" 23d34'47"
Robert C. Keever
Aur 7h13'6" 42d9'31"
Robert C Lakin
Tau 5h35'20" 22d34'12"
Robert C. Lane
Cet 1h44'20" -6d50'6"
Robert C Makin
Uma 11h23'32" 58d19'38"
Robert C. Maurer
Sgr 19h43'50" -13d52'59"
Robert C. Neason, Jr
Tau 5h48'38" 21d59'38"
Robert C. Penny "Bob"
Her 17h52'0" 29d58'28"
Robert C. Sanderson Jr.
Cap 21h43'23" -15d33'48"
Robert C. Schamel D.D.S.
Her 17h12'52" 28d57'31"
Robert C Skaggs Jr.
Pho 0h17'55" -47d55'41"
Robert C. Thompson
Uma 10h59'47" 40d57'35"
Robert C. Tooley
Uma 10h10'9" 45d28'8"
Robert Caban
Sco 17h56'25" -44d59'22"
Robert Cadzow Jr.
Cnc 8h35'38" 13d8'37"
Robert Cagney Hartmann
Uma 8h41'40" 58d55'25"
Robert Caigan Wade
Tau 5h8'31" 17d21'1"
Robert Callaway
Uma 11h37'56" 35d48'32"
Robert Callery
Ori 5h46'18" 2d21'28"

Robert Carl
Uma 9h48'46" 54d24'14"
Robert Carl Wilkins
Cep 22h11'59" 59d19'31"
Robert Carroll Boyd
Ori 6h12'52" 19d21'49"
Robert Cash
Uma 8h57'26" 62d8'19"
Robert & Catherine Summers
Gem 6h57'50" 22d12'36"
Robert Celentano, Jr.
Sco 17h54'34" -35d38'3"
Robert Céline
Psc 0h58'43" 28d24'50"
Robert Cerny
Lib 14h51'44" -22d32'39"
Robert Chambers — Forever My Star
Gem 7h18'8" 24d5'12"
Robert "Champ" Dugan May 3-1940
Her 17h51'14" 22d26'53"
Robert Charles
Uma 9h3'41" 55d12'48"
Robert Charles Aronson
Her 16h11'20" 47d24'26"
Robert Charles Bradley
Per 3h23'53" 49d34'40"
Robert Charles Dee III
Ori 6h10'15" 16d9'41"
Robert Charles Dixon
Ari 2h19'7" 17d40'20"
Robert Charles Doty
Ori 6h19'4" 10d44'47"
Robert Charles Loucks
Aql 19h31'56" 10d0'28"
Robert Charles Maples
Uma 10h41'46" 55d18'56"
Robert Charles Parsley
Gem 7h34'34" 33d6'18"
Robert Charles Wandel III
Uma 10h47'33" 61d7'49"
Robert Charles Worth
Cnc 8h50'57" 27d53'48"
Robert Chavez
Lyn 7h45'21" 41d37'53"
Robert Christian Acker
Aqr 21h54'38" 0d38'20"
Robert Christian Hazell - 'Bob'
Cap 21h35'14" -20d35'41"
Robert & Christine Always & Forever
Cyg 19h48'54" 33d31'19"
Robert Christopher Ferraro, Jr.
Her 18h7'0" 28d33'28"
Robert Christopher Fincati
Ari 2h56'24" 22d50'33"
Robert Christopher Pierce
Tau 5h11'45" 19d56'2"
Robert Christopher Pope II
Per 2h31'37" 54d5'50"
Robert Christopher Upton
Per 2h43'13" 52d39'4"
Robert Clappison
Per 3h10'42" 52d20'8"
Robert Clark
Her 16h52'32" 34d42'55"
Robert Clark Mabry
Lyn 6h55'49" 51d51'17"
Robert Clarke
Uma 11h41'49" 33d33'19"
Robert Clarkson Lunsford
Uma 10h42'53" 50d52'6"
Robert Clifton
Her 16h37'49" 37d4'1"
Robert Clinton Brown
Sgr 18h35'20" -22d55'14"
Robert Clinton Ziegler
Ori 5h24'1" 3d41'0"
Robert Clyde McCone
Her 17h9'13" 32d16'8"
Robert Cody Hansen
Hya 9h23'33" 5d17'56"
Robert Cole "Bobby" Nelson
Ori 5h30'59" 0d23'40"
Robert Cole Rodden
Aqr 22h34'1" -0d18'28"
Robert Cole Sawyer
Per 3h49'45" 42d22'52"
Robert Conlin Murphy
Her 17h25'25" 15d34'1"
Robert Constantine
Boo 14h48'23" 24d54'29"
Robert Cooperman
Srp 18h30'47" -0d55'1"
Robert Côté
Tau 4h26'58" 21d28'58"
Robert Cousin
Uma 8h29'52" 67d31'4"
Robert Coussan
Sco 17h16'36" -33d4'20"
Robert Craig Merrigan
Her 17h26'42" 35d32'34"
Robert Cree
Per 4h26'9" 49d31'20"
Robert Crismond
Aqr 22h43'44" 1d18'44"
Robert Cross
Uma 8h23'51" 63d53'46"
Robert Curran aka Pookie
Sco 16h52'9" -42d16'37"

Robert C.V.
Gem 6h49'1" 23d54'14"
Robert D. Adams
Gem 7h4'13" 14d56'17"
Robert D. Bloing MD
Psc 1h23'26" 17d58'21"
Robert D. (Bob) Rukavina
Uma 11h31'3" 34d21'15"
Robert D. Carl, III
Cap 21h46'55" -19d31'13"
Robert D Corsaro
Sco 17h58'20" -30d30'41"
Robert D. Gess III
Aur 5h52'35" 50d47'41"
Robert D. Gleason
Boo 13h53'9" 24d7'35"
Robert D. Hamilton
Per 4h21'51" 41d48'20"
Robert D. Hopping
Cyg 20h14'20" 50d9'1"
Robert D. Morales
Tau 5h27'41" 27d37'21"
Robert D. O' Connell
Aqr 21h17'11" -7d30'4"
Robert D. Perkins
Leo 11h56'1" 22d6'3"
Robert D. Puett
Cnc 8h56'44" 17d3'7"
Robert D. Ruble Happy Anniversary
Cep 22h19'24" 58d52'21"
Robert D Shoumaker, M.D.
Per 3h26'0" 36d24'1"
Robert D. Sleeper, Jr.
Vir 13h25'0" 11d48'5"
Robert Daddezio
Her 16h50'39" 27d33'55"
Robert Dahmer
Umi 14h9'51" 74d9'43"
Robert Dakota Ray
Uma 9h23'14" 72d11'7"
Robert Dale Bonnell
Crb 16h5'21" 37d46'47"
Robert Dale Henton
Sgr 19h10'31" -14d2'38"
Robert Dallas
Aqr 23h29'31" -13d53'53"
Robert Damico
Her 17h59'5" 29d52'50"
Robert & Dana Boyd
Cyg 20h52'59" 35d2'57"
Robert Daniel Conrad
Tau 5h27'47" 24d44'21"
Robert Daniel Cowart
Uma 9h1'3" 64d28'21"
Robert Daniel Nichols
Her 16h16'31" 6d57'22"
Robert David Ace Marlin
Uma 10h5'4" 63d18'41"
Robert David Adams II
Her 17h12'30" 32d28'20"
Robert David Gates
Per 3h36'44" 31d47'0"
Robert David Godley 26052004
Her 18h27'28" 12d33'13"
Robert David Goodbar, Jr.
Her 16h46'42" 30d57'51"
Robert David Goodbar, Sr.
Her 16h39'7" 37d40'51"
Robert David Irwin
Her 17h8'53" 30d17'29"
Robert David Lyman
Leo 9h39'43" 6d44'48"
Robert David Morgan - Rob's Star
Dra 18h11'48" 72d47'3"
Robert David Morse
Sgr 19h22'15" -26d40'44"
Robert David Pickett
Uma 10h17'25" 64d4'13"
Robert David Siller
Ori 5h55'2" 6d33'29"
Robert David Span
Cnc 8h47'3" 21d45'2"
Robert David Stubbs
Per 4h33'58" 40d57'32"
Robert David Webber Jr. (Bobby)
Uma 10h55'21" 40d59'55"
Robert Davis
Aqr 22h38'34" -5d20'17"
Robert Daze
Cap 20h47'53" -27d9'53"
Robert Dean Edwards
Aql 19h10'38" 12d15'5"
Robert Dean & Patricia Ann Schmidt
Cyg 20h14'9" 50d24'23"
Robert Dean Staples
Dra 20h21'58" 74d44'0"
Robert Dean Stroud
Psc 1h27'56" 17d6'53"
Robert Dean Stroud
Her 16h21'35" 5d52'49"
Robert DeBarbra
Psc 1h24'12" 32d56'3"
Robert & Deborah Jordan
Lyr 19h16'56" 35d15'0"
Robert Decrescenzo
Ori 5h53'20" -0d45'47"
Robert DeFelice
Aur 5h49'20" 48d27'16"
Robert DeFoe
Lib 14h59'28" -14d14'48"

Robert Deighan
Vul 19h56'17" 23d2'25"
Robert DeLoach Holtville's Star
Uma 9h28'49" 45d50'32"
Robert DelTorto
Aqr 22h57'12" -9d26'59"
ROBERT DENIS ROOKS
Lyn 6h35'33" 54d31'44"
Robert Dennis Bellamy III 10-31-03
Ori 6h9'10" 15d11'55"
Robert Denorris Cableton
Aqr 22h38'10" -22d36'19"
Robert & Dianne Nocco Eternal Love
Cas 2h50'7" 60d23'19"
Robert DiPonti
Sge 19h42'51" 18d8'16"
Robert Dittman
Uma 8h53'44" 61d34'56"
Robert Dominguez
Her 18h16'35" 28d27'8"
Robert Dominguez
Boo 14h46'48" 35d9'27"
Robert Donald Hale
Ori 5h38'4" 1d11'5"
Robert Donald Mackay
Aql 19h5'40" 15d13'3"
Robert Donelan *The Donelan*
Ori 5h6'55" 10d58'54"
Robert & Donelle Kellinger
Uma 11h38'43" 56d44'36"
Robert Donn Shryock
Aqr 22h14'53" -1d32'30"
Robert & Donna Harry
Cyg 19h43'30" 54d6'4"
Robert & Doris Krulls 65th.
Cyg 19h57'4" 31d22'1"
Robert Douglas
Uma 8h10'46" 62d16'41"
Robert Douglas Macom
Ori 6h16'56" 13d38'27"
Robert Douglas Patrick
Per 3h16'40" 55d6'45"
Robert Douglas Ramsey
Her 18h9'35" 47d46'25"
Robert Drummond
Cnc 8h45'4" 21d44'56"
Robert Duane Larson
And 23h10'8" 49d2'37"
Robert "Duck" Hutchins
Tau 4h31'32" 21d46'46"
Robert Duffy & Skie Ocasio
Cyg 20h53'45" 39d22'45"
Robert E. Anderson
Cas 0h55'34" 58d28'14"
Robert E. Barber Jr.
Uma 9h30'8" 44d32'48"
Robert E. Barry
Per 4h12'44" 45d44'22"
Robert E. "Bob" Drenttel
Cep 20h35'10" 62d59'18"
Robert E Cabral
Sco 16h15'37" -11d36'18"
Robert E Carey
Ari 3h15'42" 28d23'3"
Robert E Cuber Jr.
Cyg 20h42'57" 39d6'14"
Robert E. Dahmen
Her 17h33'1" 41d30'36"
ROBERT E. FENTON
Sco 17h34'50" -45d11'22"
Robert E. Figueiredo
Tau 5h48'46" 23d51'29"
Robert E. Gillies
Psc 0h22'51" 15d0'25"
Robert E. Golden Jr.
Tau 5h53'59" 25d8'24"
Robert E. Higgins
Uma 12h9'48" 58d21'57"
Robert E. Hogg : Fisherman Daddy
Aql 18h47'28" -0d6'49"
Robert E. L. Talley
Aqr 23h4'51" -7d38'42"
Robert E. Lamb
Tau 3h45'38" 18d16'51"
Robert E. Langer
Ori 5h57'58" 10d9'57"
Robert E. LeBlanc
Cap 20h26'48" -8d50'39"
Robert E. Lewis
Cap 20h9'29" -14d59'0"
Robert E. Malzacher, Sr.
Leo 10h25'33" 11d34'24"
Robert E. Martin
Uma 9h40'17" 44d18'26"
Robert E. Mayer
Sco 16h16'44" -23d55'45"
Robert E. McIntyre, Sr.
Ori 5h54'26" 8d53'1"
Robert E. Mikonowicz
Per 3h18'42" 52d14'47"
Robert E. Oakman II
Sco 16h40'55" -28d56'14"
Robert E. Philbrick
Aur 5h34'38" 43d34'49"
Robert E. Phillips
Cap 21h35'6" -23d10'31"
Robert E. Platt
Her 18h52'51" 24d38'22"
Robert E. Reiser
Tau 5h44'6" 22d17'1"
Robert E. Rodriguez
Ori 5h52'11" 20d56'18"
Robert E. Rosenberg
Uma 8h51'37" 66d12'3"
Robert E. Routch
Cyg 21h34'7" 30d29'24"
Robert E. Runyon (Bobby)
Cep 22h39'43" 65d55'45"
Robert E. Saunders III (Bobs)
Per 2h37'53" 55d38'36"
Robert E. Schleier : Forever Loved
Dra 18h26'52" 56d49'23"
Robert E. Toledo
Dra 17h42'53" 66d28'34"
Robert E. Wainer, Jr.
Cap 21h16'18" -16d7'15"
Robert E. Westendorf
Cep 20h33'54" 60d50'15"
Robert Eagles' 50th Birthday Star
Cru 12h50'28" -64d35'39"
Robert Earl Henderson
Leo 11h7'0" 20d16'55"
Robert Earl Hunt
Ori 5h27'15" 3d19'57"
Robert Earl Johnson
Tau 4h42'58" 17d50'40"
Robert Earl Mackenzie
Cyg 19h40'28" 36d29'6"
Robert Earl Patz
Tau 5h23'26" 19d15'57"
Robert Earl Price
Uma 11h37'46" 48d17'4"
Robert Eddie Dickerson Jr.
Tau 5h40'36" 21d13'46"
Robert Edgar Allen (Bubby)
Per 3h21'15" 51d43'57"
Robert Edward Caprarella
Umi 14h33'37" 75d45'17"
Robert Edward Conrad-Gendloff
Sgr 19h17'10" -12d41'9"
Robert Edward Donohue
Sco 16h21'7" -10d1'34"
Robert Edward Hryszko, Jr.
Vir 14h17'15" -0d12'3"
Robert Edward Morgan
Aur 4h50'27" 39d35'2"
Robert Edward Murray
Her 17h18'6" 46d0'47"
Robert Edward Piele
Vir 14h0'36" -20d38'32"
Robert Edward Sakowski
Per 4h43'32" 36d46'52"
Robert Edward Sievers
Per 2h35'23" 53d57'25"
Robert Edward Watts "I Love U Star"
Per 4h14'31" 45d14'54"
Robert Edward Wilms
Sgr 19h23'47" -30d47'41"
Robert & Eileen's 20th Anniversary
Ari 2h48'20" 28d27'41"
Robert & Elizabeth Entersz Forever
Cyg 20h27'56" 46d4'39"
Robert & Elizabeth Havener
Cyg 21h51'29" 50d36'5"
Robert Elizah Jones
Sco 16h8'19" -13d9'28"
Robert Ellis Jones Jr.
Lib 15h50'39" -7d0'27"
Robert Emmett Nolan Seraydarian
Cep 21h48'42" 65d57'24"
Robert Emmett Towey
Sco 16h16'26" -41d30'7"
Robert Emory Smitley
Lyn 8h52'24" 38d32'49"
Robert Episcopo
Per 3h12'29" 54d26'42"
Robert Erdkamp
Ori 6h2'59" 9d56'34"
Robert Eric Schmitt
Cyg 19h32'30" 29d8'29"
Robert Eric Scott
Cap 21h44'25" -13d0'18"
Robert Ernest Savela
Ori 5h55'32" -0d21'36"
Robert et Colette Longpré
Cyg 19h44'13" 29d29'26"
Robert Ethan Roberts
Cam 3h55'16" 55d30'19"
Robert Eudell Growden
Her 17h40'39" 17d14'0"
Robert Eugene "Bob" Coles
Uma 9h56'15" 43d53'0"
Robert Eugene Chastain
Umi 15h8'23" 80d42'16"
Robert Eugene Hamilton
Dra 18h49'23" 70d47'46"
Robert Eugene Hensel 771930
Uma 11h54'10" 59d29'50"
Robert Eugene Paul 101443
Lib 14h55'9" -22d18'45"
Robert Eugene Sawicki
Aur 5h51'48" 52d22'8"
Robert Eugene Sifuentes II
Sco 16h45'3" -32d8'49"
Robert Eugene Spiotti
Aqr 20h49'46" -9d1'56"
Robert Eugene Stoneberger
Ori 5h55'7" 7d12'41"
Robert Euphoria Weiss
Cas 0h14'2" 57d4'2"
Robert Evan Kennett McGregor
Cep 0h13'17" 76d22'21"
Robert Evan Schlissberg
Ori 5h52'19" 0d1'21"
Robert Everett Underwood
Lib 15h16'21" -6d41'2"
Robert F. Beckwith
Uma 13h8'7" 56d44'59"
Robert F. Brown
Lib 14h49'54" -2d27'8"
Robert F. Dansby III
Per 3h20'54" 32d54'25"
Robert F. deOliveira
Tau 4h15'2" 19d7'18"
Robert F. Dowling, Jr.
Uma 11h7'28" 51d47'37"
Robert F. Ennis
Her 18h41'27" 19d23'4"
Robert F. Jensen
Vir 13h10'6" 12d16'42"
Robert F. Marler
Uma 12h22'21" 62d58'55"
Robert F. Petitto
Ari 2h4'37" 23d29'55"
Robert F. Place
And 0h14'54" 26d36'48"
Robert F. Rossa
Ori 4h48'45" 11d1'4"
Robert F. Spetzler Star
Sco 16h54'27" -38d41'10"
Robert F. Stark
Per 4h20'19" 41d49'4"
Robert F. Tiernan
Per 3h10'15" 57d4'55"
Robert F. Torino
Crb 15h44'34" 29d9'3"
Robert Fano
Vir 13h20'40" -17d38'55"
Robert Feldman
Cma 6h32'12" -17d25'53"
Robert Filson
Uma 12h36'11" 56d19'48"
Robert Fischer
Ori 6h16'11" 15d37'52"
Robert Fitzgerald
Her 16h37'51" 39d14'17"
Robert Flores Martinez
Cnc 8h23'35" 10d11'15"
Robert Flötzl
Uma 8h53'58" 62d37'35"
Robert Foltz
Uma 9h41'46" 57d48'31"
Robert Francis Bordeau
Dra 15h6'7" 63d19'3"
Robert Francis Costa
Uma 12h35'33" 62d11'22"
Robert Francis Dowden 3-11-1956
Dra 18h57'4" 55d1'39"
Robert Francis Gerber
Lmi 10h0'58" 39d24'42"
Robert Francis Katz Jr.
Cyg 20h1'17" 52d21'28"
Robert Francis Samuel Gardner
Ori 6h1'23" 10d32'31"
Robert Frankenthal
Vir 14h24'30" 6d2'43"
Robert Franklin Burke
Gem 6h42'36" 25d44'59"
Robert Franklin Chatham
Leo 9h24'10" 25d28'36"
Robert Frey, Jr.
Cnc 8h51'22" 31d59'19"
Robert Friedl, JR.
Her 18h46'58" 19d25'41"
Robert Fultz
Uma 11h16'8" 62d12'22"
Robert Furtwengler
Cap 21h9'20" -17d37'40"
Robert G. Betts, CMSgt. Ret. US AF
Aql 19h10'26" 13d4'37"
Robert G Crockett
Ori 4h48'37" 0d31'49"
Robert G & Elizabeth Wilcox Forever
Cyg 20h7'1" 52d45'12"
Robert G Fine
Gem 7h28'12" 16d11'44"
Robert G. Goulet
Vir 11h55'53" 8d58'39"
Robert G. Hall
Sco 16h15'37" -14d55'7"
Robert G. Hallas
Her 16h41'9" 18d17'28"
Robert G. Hayes
Per 3h22'32" 52d6'6"
Robert G. Knupp, Jr
Uma 12h28'58" 52d29'14"
Robert G. Marsh Jr.
Tau 4h26'5" 15d20'38"
Robert G. McCormick
Uma 10h13'18" 70d49'10"
Robert G. Rice IV
Her 16h57'48" 34d12'22"
Robert G. Sturgeon
Ari 2h35'56" 27d21'12"
Robert G. Trombino, Jr.
Sgr 18h20'38" -22d47'3"
Robert G. Turnage
Psc 23h59'38" -0d56'9"
Robert Gaines
Cep 23h29'18" 79d27'40"
ROBERT GALE AND NINA RUTH SIGMAN
Aqr 22h50'45" -9d23'3"
Robert Gallagher
Her 17h39'38" 46d58'15"
Robert Galliher Lebarron
Uma 11h6'20" 36d51'30"
Robert Garcia
Sco 16h6'34" -22d53'2"
Robert Garrett Jr.
Umi 13h6'34" 74d44'22"
Robert Garth Bailie
Cep 1h4'59" 85d31'50"
Robert Gary Schumann Jr.
Aqr 20h44'52" -13d51'48"
Robert Gaylord Ravenal DeSipio
Psc 1h4'14" 31d9'54"
Robert Gedaliah
Cap 21h34'36" -8d39'13"
Robert Gene Grice
Crb 15h52'23" 34d32'0"
Robert Genest
Cap 21h35'19" -10d46'0"
Robert Geoffrey Gegwich
Lyn 9h15'53" 41d2'1"
Robert Geoffrey Packer my star
Cru 12h39'49" -59d31'16"
Robert George Kaufmann
Tau 5h52'56" 28d9'31"
Robert George Schmidt
Aqr 23h48'51" -19d23'28"
Robert George Shino
Per 2h42'48" 56d21'35"
Robert George Supplee
Psc 1h10'47" 30d56'28"
Robert George Winer
Aql 19h39'26" 8d11'38"
Robert Gerald Benson
Uma 10h25'57" 47d29'52"
Robert Gerald Medd
Uma 9h51'3" 67d31'51"
Robert Gerson
Boo 14h51'51" 27d59'15"
Robert Gianella
Cap 20h41'45" -21d0'30"
Robert Girard
Boo 15h36'8" 43d26'23"
Robert Giroux
Ari 3h17'2" 28d12'26"
Robert Glenn Magee
Sgr 18h26'45" -32d36'3"
Robert Gonzales
Ori 4h46'23" 3d45'29"
Robert Gonzalez
Sco 16h13'4" -13d50'25"
Robert Gonzalez & Norma Aranda
Her 16h10'43" 23d56'46"
Robert Goodwill - 21st April 1935
Tau 5h11'28" 18d13'55"
ROBERT GORDEN MILLER
Per 3h14'20" 47d3'23"
Robert Gordon
Vir 13h59'28" -7d5'1"
Robert Gordon Brown
Cep 22h52'45" 60d43'8"
Robert Gordon Burdick
Aqr 22h35'9" 0d3'4"
Robert Graham
Per 3h4'21" 51d50'37"
Robert Graziano Jr.
Per 3h8'58" 52d49'45"
Robert Gregory Courtenay
Gem 6h40'34" 30d52'32"
Robert Griswold Christina Roco Star
Cep 0h13'41" 72d52'20"
Robert Gruder
Cet 2h18'27" 6d53'28"
Robert Gruen
Hya 8h48'24" -2d32'40"
Robert Guarnieri
Cyg 21h49'25" 46d5'3"
Robert Gustafson
Sco 17h35'25" -32d19'40"
Robert Guy
Sco 17h21'42" -41d6'18"
Robert H. Furino
Uma 10h37'21" 56d21'30"
Robert H. Hernandez
Cnc 7h58'37" 15d4'32"
Robert H. Jarvis
Aql 19h14'32" -7d6'58"
Robert H. Johnson
Leo 9h23'19" 26d46'13"
Robert H. Lavender
Cep 22h7'48" 56d25'8"
Robert H. Lavoie
Aql 19h16'12" -0d5'52"
Robert H. Leland
Ori 5h57'34" 14d58'7"
Robert H. Lewis-Upper Pavillion
Lib 15h6'33" -13d24'14"
Robert H. & Margaret Ann B. Hageman
Uma 11h47'2" 33d38'58"
Robert H. Meyers
Per 2h39'47" 52d59'58"
Robert H. Rawlings
Aqr 22h28'25" -1d1'43"
Robert H. Rudolph
Uma 11h43'42" 47d35'0"
Robert H. Thompson
Ori 6h3'32" -3d55'32"
Robert Hague
Cep 21h29'45" 63d55'8"
Robert Hahn II
Cma 6h35'44" -29d52'26"
Robert Hall
Uma 9h21'31" 68d37'34"
Robert Halligan
Vir 12h48'23" 12d30'27"
Robert Hamill
Cnc 8h33'6" 12d42'17"
Robert Harold Mohr
Her 16h35'31" 37d14'4"
Robert Harold Noe
Cnv 12h47'32" 39d8'18"
Robert Harold Spicher
Her 17h24'17" 16d7'7"
Robert Harrison Donovan
Lib 15h39'53" -15d11'53"
Robert Harry Swafford
Tau 4h37'18" 28d0'21"
Robert Hayes Walker
Tau 5h22'30" 22d40'43"
Robert Henning
Pup 8h15'31" -43d0'33"
Robert Henry Czapinski
Her 16h59'50" 30d21'7"
Robert Henry Heygster
Lyn 7h49'22" 43d0'5"
Robert Henry Ramirez
Sco 17h8'53" -38d44'57"
Robert Henry Vonderohe 2004
Uma 10h32'30" 62d23'16"
Robert Henry Wagner II
Gem 7h1'47" 16d45'12"
Robert Hoffmeyer Sr.
Cap 21h35'20" -18d43'23"
Robert Holas
Dra 19h2'56" 66d58'52"
Robert Hollis
Cma 6h59'11" -16d34'42"
Robert Horrigan
Ori 6h1'23" 21d19'53"
Robert Houghton
Lmi 10h6'11" 30d32'10"
Robert House
Her 17h14'5" 32d37'12"
Robert Hoy
Aqr 22h40'54" -9d50'19"
Robert Hugh Cunningham, Jr.
Aql 19h50'52" -0d34'20"
Robert Hulse
Ari 1h47'56" 23d3'39"
Robert Hunter Downie (Snaz)
Tau 5h3'38" 24d54'7"
Robert Hutchinson
Umi 16h49'22" 82d48'17"
Robert I. Rondello
Ori 5h38'48" -0d27'51"
Robert Ian Dale
Aur 6h28'7" 42d48'41"
Robert Ingram
Uma 10h33'49" 58d52'31"
Robert Istvan Incze 3-1-1981
Uma 12h9'15" 60d48'43"
Robert J. Antonelli
And 0h9'4" 35d28'41"
Robert J Astrab, Jr
Psc 23h42'26" 2d0'51"
Robert J. Barclay
Gem 6h45'12" 26d39'54"
Robert J. Bland
Per 3h15'53" 45d52'13"
Robert J. Brennie
Uma 10h59'50" 68d2'43"
Robert J. Bublak
Uma 11h15'56" 44d53'0"
Robert J & Christopher J Stewart
Aql 19h41'10" 6d10'2"
Robert J Clough
Leo 10h18'26" 25d40'11"
Robert J. Combs, Jr
Umi 13h10'44" 71d33'8"
Robert J. Dalrymple
Sco 16h15'5" -10d22'19"
Robert J. Damiani
Per 4h25'45" 39d5'58"
Robert J. Forgey
Cep 21h43'18" 76d58'55"
Robert J. Franz, Sr.
Gem 7h4'33" 34d16'14"
Robert J. Galli
Uma 10h56'31" 70d3'3"
Robert J. Gibson
Lib 14h54'41" -4d53'37"
Robert J. Gilliam
Cep 22h45'43" 73d21'59"
Robert J. Graper / Daddy
Uma 11h14'27" 29d55'27"
Robert J. Grycowski
Dra 20h12'54" 75d12'23"
Robert J. Hancharick
Cep 21h46'1" 57d20'38"
Robert J. Henrichs
Ari 2h46'41" 25d36'57"
Robert J. Holiday, Sr.
Gem 7h6'26" 33d16'43"
Robert J. Hudock
Aql 20h33'40" -1d58'14"
Robert J. King, Sr.
Her 18h54'41" 24d20'28"
Robert J. LaCosta
Eri 4h27'10" -1d30'57"
Robert J. Matteo
Sco 17h53'12" -43d35'2"
Robert J. Medina
Aur 6h32'56" 35d59'44"
Robert J. Miller
Cyg 20h18'1" 42d3'59"
Robert J. Moser Sr.
Ori 5h31'50" 1d35'59"
Robert J Murphy
Her 17h5'44" 30d12'45"
Robert J. Norrby
Cep 20h37'55" 64d25'3"
Robert J. O'Connell
Per 3h11'45" 55d31'20"
Robert J. Palmieri
Gem 7h19'21" 35d3'48"
Robert J. Post
Uma 13h43'15" 53d35'41"
Robert J. Postler
Uma 10h5'33" 51d17'33"
Robert J Powell
Uma 13h49'39" 54d48'57"
Robert J. Ritter Jr.
Tau 4h32'23" 1d17'42"
Robert J. Saltarelli
Sct 18h50'2" -9d53'14"
Robert J. Sanator
Leo 11h12'16" 2d1'9"
Robert J. Schuitema
Uma 11h10'51" 32d2'37"
Robert J. Sears, Jr.
Lyn 7h35'17" 50d27'29"
Robert J. Shank
Tau 4h35'28" 30d33'32"
Robert J. Shannon
Sgr 18h27'25" -16d36'24"
Robert J. Shay
Pho 23h52'34" -39d39'34"
Robert J. Stapleton
Lib 14h53'44" -16d29'32"
Robert J Starkie
Cen 13h26'25" -60d9'15"
Robert J. Stephen
Ori 5h30'50" 2d38'18"
Robert J. Tirrell 10-17-1936
Per 3h5'50" 53d46'46"
Robert J. Van Trieste
Uma 8h36'58" 47d51'8"
Robert J. Wise
Tau 3h25'27" 0d19'27"
Robert J. Zambon
Ari 2h25'29" 22d3'40"
Robert Jacob
Ari 2h31'36" 24d53'0"
Robert & Jacqueline McKinney
Men 5h29'49" -71d17'54"
Robert Jaffee
Lib 15h13'9" -5d40'32"
Robert Jakway
Tau 4h20'20" 26d51'58"
Robert James Allor
Cap 20h36'57" -18d8'44"
Robert James Best
Ori 6h1'50" 18d4'0"
Robert James Briggs
Ari 3h10'44" 29d17'30"
Robert James Cant
Uma 8h40'28" 47d53'38"
Robert James Dahmer 12:25 A.M.
Aqr 21h36'34" 2d7'50"
Robert James Fisher
Uma 10h32'47" 55d5'36"
Robert James Griffin
Her 17h24'27" 15d28'46"
Robert James Hart
Cap 20h35'11" -11d3'55"
Robert James Hastings
Per 4h20'29" 45d48'16"
Robert James Ihnat
Tau 4h52'29" 20d37'37"
Robert James Johnson
Sco 16h5'50" -17d38'57"
Robert James Kolibas
Sgr 18h39'46" -22d10'55"
Robert James Lanphear
Uma 11h20'24" 44d7'17"
Robert James Leppert
Aur 5h43'35" 47d33'4"
Robert James MacKenzie, Sr.
Her 17h44'46" 46d29'13"
Robert James Mitchell
Aql 20h12'55" 8d24'45"
Robert James Pilla
Cep 22h55'44" 79d38'34"
Robert James Rusnell
Leo 11h41'26" 11d46'43"
Robert James Spella
Cap 21h36'21" -18d37'3"
Robert James Summers
Peg 22h46'37" 24d49'32"
Robert James Webb
Cep 21h24'24" 59d46'54"
Robert James Yasick Sr.
Cyg 21h14'46" 45d58'28"
Robert & Jane Kauffman
Uma 10h56'23" 47d53'47"
Robert Jaron Kubala
Aql 19h46'28" -0d27'11"
Robert Jay Hayes
Her 16h27'29" 6d49'17"
Robert Jay Meany
Aur 5h18'25" 31d0'1"
Robert Jay Morgan
Her 17h53'51" 31d25'0"
Robert Jay Roth
Per 3h38'28" 33d16'19"
Robert Jay Zimmermaker
Her 18h53'27" 24d11'58"
Robert & Jaylene VanDerTuin
Eri 3h25'2" -22d0'7"
Robert & Jazzy
Cyg 20h37'3" 57d47'21"
Robert Jean Pierce
Aqr 22h15'9" -18d33'4"
Robert/Jeanne 101092
Lib 15h41'51" -26d7'16"
Robert Jefferson Emrey
Tau 5h51'14" 16d47'1"
Robert Jefferson Holl
Vir 14h37'19" 3d41'21"
Robert Jeffery Edwards
Cep 21h36'43" 56d29'34"
Robert Jeffrey Byers
Sgr 19h48'8" -12d26'17"
Robert Jelcz
Sgr 19h45'52" -20d42'51"
Robert Jennifer and Jillian Elias
Aqr 22h32'22" -1d52'32"
Robert & Jennifer Rhode
Cyg 19h45'29" 41d32'59"
Robert Jerry Wyatt
Cnc 8h10'32" 25d3'41"
Robert & Joan Altmaier
And 0h56'54" 39d28'59"
Robert Joesph Cardamone
Cep 20h47'15" 68d8'52"
Robert John Anderson
Per 3h7'2" 56d55'20"
Robert John Birkhaug
And 0h49'36" 36d55'1"
Robert John Boening
Per 3h3'56" 45d50'2"
Robert John Boyd Petersen
Ori 5h11'1" -0d39'43"
Robert John Butterworth
Lup 15h13'50" -44d58'29"
Robert John Dery
Uma 10h13'49" 44d54'11"
Robert John "Duke" Kiesel
Sgr 18h10'13" -26d26'51"
Robert John Flood
Ori 5h51'15" -0d35'2"
Robert John Gleichert
Uma 11h55'55" 39d19'20"
Robert John Kerle
Aqr 23h12'37" -21d15'0"
Robert John Kirk
Cyg 20h40'0" 50d55'10"
Robert John Klinger, Jr.
Her 16h31'22" 15d38'56"
Robert John Martell
Cyg 19h47'55" 31d23'26"
Robert John Metz III
Sgr 19h43'56" -11d55'23"
Robert John Montero
Col 5h43'30" -31d8'39"
Robert John Nelson
Uma 10h35'19" 40d56'2"
Robert John Palmaccio
Per 3h31'44" 51d17'17"
Robert John Paul Karol Deibert
Her 16h9'42" 45d28'40"
Robert John Shaw
Aqr 22h36'16" 0d25'14"
Robert John Sutton
Per 4h3'23" 35d32'18"
Robert John Weiss
Dra 14h58'51" 56d58'53"
Robert John Williams Born 1/29/1946
Aqr 21h39'12" -8d3'8"
Robert Johnson Jr
Uma 9h44'31" 65d11'2"
Robert Jon
Lib 15h6'7" -1d50'7"
Robert Jon Corradi
Ori 4h49'50" 13d19'49"
Robert Jones, father & role model
Uma 8h42'58" 66d52'1"
Robert Joseph
Dra 9h34'44" 74d21'52"
Robert Joseph Baerga
Cnc 8h50'16" 31d29'56"
Robert Joseph Bencic
Vir 13h7'1" 6d11'46"

Robert Joseph Booterbaugh
Per 3h14'16" 53d36'55"
Robert Joseph Brown
Uma 11h52'11" 36d25'28"
Robert Joseph Burkholder
Vir 13h16'59" -2d36'35"
Robert Joseph Canterino
Ari 3h21'53" 19d37'53"
Robert Joseph Cartelli
Lyn 8h42'31" 41d42'16"
Robert Joseph De Kleine
Cyg 20h32'5" 44d27'39"
Robert Joseph Doheny Jr.
Ori 6h2'14" 19d44'1"
Robert Joseph Flannery, Jr
Cap 20h26'9" -12d9'26"
Robert Joseph French
Uma 11h5'37" 53d9'5"
Robert Joseph Gaissert
Ori 5h57'1" 19d19'51"
Robert Joseph Gauthier
Sco 17h58'14" -37d51'59"
Robert Joseph Gordon
Umi 15h36'5" 75d13'19"
Robert Joseph Guerrina
Aqr 22h30'18" -2d29'8"
Robert Joseph Harvey
Cep 21h39'10" 61d50'57"
Robert Joseph Hayes
Uma 9h43'17" 42d13'36"
Robert Joseph Heier/STINKYPANTS
Vir 12h43'36" -3d6'4"
Robert Joseph Johnson
Uma 9h4'6" 68d10'39"
Robert Joseph Leone
Cnc 9h20'20" 27d44'18"
Robert Joseph Page, Sr.
Leo 9h25'27" 26d16'15"
Robert Joseph Rohde
Tau 4h10'57" 17d30'30"
Robert Joseph Sutherby
Ari 2h57'52" 30d0'1"
Robert Joseph Valinoti
Gem 6h46'2" 34d40'22"
Robert Joseph Wagner III
Lib 15h30'51" -7d52'50"
Robert Joshua Reed
Uma 12h38'28" 62d12'4"
Robert & Judy Krahulec
Cnc 9h8'40" 30d48'19"
Robert & Julia
Her 17h44'30" 48d30'7"
Robert Julius Lamz
Leo 11h28'14" 10d44'0"
Robert K Aasen
Tau 3h46'17" 20d18'0"
Robert K Barr
Lib 14h51'51" -1d8'23"
Robert K. Bartholomew
Uma 11h15'52" 44d17'48"
Robert K. Bohn
Per 3h29'41" 52d49'57"
Robert K. Bowman
Uma 9h14'9" 60d33'2"
Robert K. Jordan
Boo 14h22'43" 46d12'25"
Robert K. Orsino
Per 4h6'40" 40d36'27"
Robert & Katie
Cnc 8h22'31" 26d24'18"
Robert & Katie Love Never Fails
Cyg 19h48'29" 43d38'0"
Robert Keith Juranek
Tau 5h41'11" 24d51'31"
Robert Keller
Per 3h12'45" 54d30'50"
Robert Keller
Cep 22h32'20" 57d52'7"
Robert+Kendra
Cyg 21h31'2" 41d0'5"
Robert Kenneth Donald Stewart
Vir 13h6'30" 2d3'51"
Robert Kenny
Uma 12h35'49" 53d32'14"
Robert Kevin Hannigan
Psc 0h50'9" 9d59'1"
Robert Kevin Lewis
Cap 21h41'15" -14d56'35"
Robert King
Cep 20h47'52" 64d59'27"
Robert Kipnis
Aqr 22h26'6" -0d37'50"
Robert Kirk Maldonado
Gem 6h57'31" 13d10'8"
Robert Klein
Per 3h46'48" 44d44'2"
Robert Knight
Uma 9h59'28" 51d12'1"
Robert Kohl
Aql 19h38'8" 2d36'2"
Robert Konopka
Ari 3h23'27" 23d29'40"
Robert Kososki
Umi 15h22'5" 71d44'19"
Robert Kupchak
Cyg 20h29'30" 34d33'9"
Robert Kyle Scott
Per 3h33'50" 44d2'46"
Robert Kyran Morgan
Per 2h21'8" 56d13'20"
Robert L. Archie
Ari 1h50'27" 18d25'59"
Robert L. Atchinson III
Lib 14h40'53" -10d59'7"
Robert L. Bahara
Sgr 18h11'52" -19d11'32"
Robert L. Benton
Psc 1h10'5" 23d1'20"
Robert L. Berkebile
Vir 12h49'37" 8d54'12"
Robert L. Bull, Jr., MD
Gem 7h49'30" 29d19'3"
Robert L Burford
Cap 20h20'46" -14d22'34"
Robert L. Chandler Jr.
Ori 5h37'49" -0d3'41"
Robert L. Cline 3/13/29-2/1/84
Ari 1h50'0" 21d10'25"
Robert L. Cody
Uma 10h48'45" 67d57'25"
Robert L Elmendorf
Her 16h16'34" 5d37'1"
Robert L. Ford Family
Her 18h36'56" 19d40'45"
Robert L. Franson
Umi 14h23'53" 74d22'29"
Robert L. Hargrove
Lib 15h6'37" -7d23'5"
Robert L Harlan
Cap 20h7'57" -24d22'39"
Robert L. Hart, Jr.
And 0h45'19" 39d18'44"
Robert L. Howard, Jr.
Cyg 21h15'38" 42d28'5"
Robert L. & Juliane W. DeZarn
Per 3h10'5" 54d25'33"
Robert L. Knight
And 23h24'11" 52d3'52"
Robert L. Lindsey
Per 4h24'11" 43d26'26"
Robert L. Lorden
Aql 19h12'27" -0d6'3"
Robert L. Love
Aql 19h16'47" 9d15'46"
Robert L. Marien
Lyn 7h3'1" 50d5'23"
Robert L. Pettigrew
Cas 0h32'27" 58d5'56"
Robert L. Phillips
Per 3h51'42" 37d47'14"
Robert L. Rice
Per 3h47'19" 42d47'44"
Robert L. Root
Per 4h13'45" 51d27'51"
Robert L. Roy
Psc 23h56'35" 7d31'22"
Robert L Schlott
Uma 10h10'30" 63d45'10"
Robert L. Seiler
Her 17h4'28" 32d24'37"
Robert L. Skiladz
Uma 9h35'23" 69d45'55"
Robert L. Slusher
Aql 18h55'20" -0d7'39"
Robert L Smith IV
Gem 6h43'22" 21d55'54"
Robert L. Smith, Jr.
Ori 6h19'50" 19d58'11"
Robert L. Turner III
Per 3h42'33" 42d44'44"
Robert L. Warford
Lyr 18h37'55" 31d42'44"
Robert L. Weinreis
Uma 11h15'46" 72d6'6"
Robert L. Weisholtz
Cep 20h59'53" 64d23'4"
Robert Labbe
Boo 15h8'13" 48d41'16"
Robert Lafromboise, Sr.
Uma 11h39'23" 62d46'42"
Robert Lamb Jr. Human Rights Champ
Per 2h30'13" 52d35'40"
Robert Lampley, Jr.
Leo 10h55'53" 7d15'40"
Robert Langenberger
Uma 10h54'30" 63d44'42"
Robert Langridge
Cru 12h19'56" -56d45'29"
Robert Larkins
Uma 13h51'26" 61d31'13"
Robert Larocca
Uma 11h2'12" 48d25'46"
Robert Laszewski
Uma 8h56'11" 71d55'48"
Robert Laughlin
Ori 5h59'26" -0d59'25"
Robert Lawes
Uma 11h58'57" 33d19'51"
Robert Lawlor, Jr.
Cep 22h36'52" 64d25'12"
Robert Lawrence Kepner
Psc 23h3'40" 5d47'30"
Robert Lawrence Ryan
Sco 16h50'3" -38d57'14"
Robert Lawrence Wehrmann Sr.
Aql 19h58'28" 15d13'51"
Robert Lazarewicz
Ari 2h4'33" 22d30'14"
Robert Lazarus
Uma 10h13'34" 56d25'48"
Robert Lazo
Ari 2h20'28" 17d33'14"
Robert LeBlanc
Per 4h11'24" 39d31'5"
Robert LeClerc
Uma 8h21'31" 67d47'53"
Robert Lee
Per 4h8'30" 51d30'17"
Robert Lee Allen
Uma 9h2'46" 55d9'37"
Robert Lee Anderson, Jr.
Ari 3h14'11" 15d23'25"
Robert & Lee Ann Patterson
Cyg 21h28'21" 53d20'8"
Robert Lee (Bob) Knight
Sco 17h54'51" -42d43'28"
Robert Lee Carrico
Her 18h13'1" 21d34'45"
Robert Lee Charley
Her 17h36'29" 28d40'39"
Robert Lee Cohen
Cep 22h35'27" 72d42'50"
Robert Lee Hanes
Ori 5h9'39" 15d14'5"
Robert Lee Hawkins III
Ori 6h11'16" 15d7'35"
Robert Lee January
Gem 7h27'56" 26d34'12"
Robert Lee Larson
Leo 10h31'43" 17d58'11"
Robert Lee Lyon's Bobby's World
Dra 18h23'49" 58d3'22"
Robert Lee Michaelis
Cyg 20h25'20" 44d54'31"
Robert Lee Nicholson
Vir 13h12'17" 5d44'44"
Robert Lee Ramirez, Jr.
Sgr 19h25'22" -39d6'44"
Robert Lee Shaw November 23, 1990
Tau 5h38'13" 20d13'27"
Robert Lee Shindle
Umi 15h30'16" 75d25'15"
Robert Lee Sims
Her 16h54'33" 32d29'47"
Robert Lee Smith
Pho 1h0'27" -41d23'48"
Robert LeGrand Simpson, Jr.
Ori 4h51'45" 11d4'14"
Robert Leigh Dillard
And 0h36'46" 29d26'45"
Robert Leo Beaudry
Uma 11h32'29" 55d4'58"
Robert Leo Evans, Jr.
Cep 22h3'2" 71d33'26"
Robert Leonard Hoey
Ori 5h5'16" 15d38'48"
Robert Leroy Chiodo- Good Friend
Uma 9h41'40" 68d23'2"
Robert Leroy Reed
Her 17h5'46" 25d27'17"
Robert Leslie Alexander
Cru 12h33'50" -58d9'4"
Robert & Leslie Bocchino
Cyg 20h4'0" 46d40'3"
Robert & Leslie Normand
Ori 5h13'51" 15d7'47"
Robert Lewis
Per 3h47'3" 48d6'4"
Robert Lewis McAllister
Per 3h11'14" 41d21'49"
Robert Lewis Stutzman Senior
Per 2h41'51" 52d29'19"
Robert Lewis White
Aur 6h35'14" 34d4'10"
Robert Limou
Psc 23h44'33" -0d58'15"
Robert & Linda O'Gorden 03/08/1975
Cyg 20h34'49" 41d36'36"
Robert & Lisa Burns
Sgr 19h18'0" -27d40'5"
Robert Logan Nance
Her 18h19'0" 16d57'15"
Robert Lohr
Cnc 8h56'55" 6d35'17"
Robert Longster
Per 4h48'4" 40d26'11"
Robert Louis Byrn
Ori 6h18'19" 14d27'17"
Robert Louis DePaulis
Aur 6h10'32" 48d26'17"
Robert Louis Gunn, Jr.
Sgr 18h11'10" -20d7'14"
Robert Louis Hodge
Lyn 8h5'49" 37d59'45"
Robert Louis Jungels
Aql 19h53'21" 2d4'20"
Robert Louis Nielsen
And 1h5'15" 39d49'34"
Robert Louis Schaefer Jr.
Vir 13h54'42" 5d24'43"
Robert Louis Southerland, Jr.
Lyn 6h22'10" 55d45'7"
Robert Love
Per 3h42'44" 37d49'56"
Robert Loves Musiati Forever
Ari 3h10'48" 27d11'21"
Robert Lucien Pratscher Happy 60th!
Gem 7h17'11" 13d27'14"
Robert & Lucille Prim
Cyg 21h30'53" 45d2'58"
Robert Luther Glyn
Uma 11h9'4" 49d39'45"
Robert Lyle Hiechel
Dra 17h5'17" 66d32'47"
Robert Lynn Lenker
And 0h51'41" 37d41'8"
Robert Lynn Megas
Her 17h58'6" 23d51'23"
Robert Lynn Williams
Sgr 18h37'50" -28d39'58"
Robert & Lysa Mackenzie
Vir 13h4'33" -11d44'58"
Robert M. Charvoz
Cyg 20h30'55" 52d22'32"
Robert M. Chretien
Cas 1h0'40" 69d33'39"
Robert M. Coghill
Uma 10h27'53" 44d27'3"
Robert M. Cronin
Cep 21h31'30" 63d17'57"
Robert M. Henshaw
Psc 0h9'34" 6d2'28"
Robert M. Hough, Jr.
Leo 10h59'56" -2d18'39"
Robert M. Kalmanson
Sco 17h46'28" -38d25'18"
Robert M Munoz
Per 2h47'54" 55d20'20"
Robert M. Nagle
Ari 2h56'34" 25d46'22"
Robert M. Rasulo - Rising Star
Boo 15h23'32" 42d13'14"
Robert M. Santosuosso
Ori 5h19'20" 15d40'6"
Robert M. Shugert
Cnc 8h48'15" 8d25'15"
Robert M. Stearns II
Sco 17h12'10" -42d7'56"
Robert M. Van Fleet
Leo 10h12'21" 12d1'41"
Robert M Williams
Cas 0h38'8" 51d25'4"
Robert Magnus Johnson
Uma 8h15'55" 72d25'20"
Robert Malenchek
Tau 4h36'41" 25d37'11"
Robert Mantilla
Tau 5h59'11" 24d55'56"
Robert Marchandt
Uma 9h35'39" 46d23'42"
Robert & Margarita Evertsen
Ori 6h7'31" 18d6'23"
Robert Mark Anthony Diaz
Sgr 17h52'15" -17d20'8"
Robert Mark Davies - Always
Ori 6h18'1" -0d4'50"
Robert Mark Fiser
Aql 19h30'14" -0d2'43"
Robert Mark Stanton Jr.
Aur 5h40'10" 49d54'13"
Robert Mark Thomas
Dra 18h44'1" 62d33'9"
Robert Mark Thompson
Gem 7h2'55" 34d20'13"
Robert & Marlene Parker
Uma 10h54'22" 40d27'20"
Robert & Marlene Rienzo 25 Years
Cyg 20h1'38" 41d47'9"
Robert Martin "Dr. Bob" Schwartz
Psc 22h58'43" 7d47'48"
Robert & MaryLou Harrison
Sgr 18h33'39" -23d42'19"
Robert Matthew Whitman (Lovey)
Per 4h20'18" 48d37'20"
Robert Mattie Buckman
And 23h34'29" 45d23'4"
Robert Maurice Jones
Cet 0h55'14" -0d7'54"
Robert Maxmilian Peck
Uma 10h30'39" 65d33'28"
Robert Maxwell
Her 17h21'10" 34d8'51"
Robert Maxwell Siegel
Ori 5h32'27" 2d51'32"
Robert Maxwell Stuart Bidstrup
Cra 18h59'12" -40d6'17"
Robert McCaffery-Lent
Leo 11h1'23" 0d19'5"
Robert McCarrell, Jr.
Leo 11h23'24" 2d8'58"
Robert McCarthy
Sgr 18h57'57" -27d47'25"
Robert McCool
Lib 14h51'10" -2d15'54"
Robert McFarland
Ari 2h33'48" 28d6'48"
Robert McFarlin, Jr.
Vir 14h35'19" 4d0'27"
Robert McGregor Shaw
Leo 11h2'49" -3d1'26"
Robert McKenzie
Cyg 20h49'3" 35d46'19"
Robert McKenzie Mark
Vir 13h30'45" -18d6'14"
Robert McKinley Welsh
Per 3h37'59" 37d30'45"
Robert McLaurin Boote
Lib 15h31'10" -19d3'21"
Robert McNair
And 0h55'46" 39d42'42"
Robert McPhail Cole
Leo 11h19'4" -0d26'53"
Robert Meese Raimondo
Cep 22h38'1" 74d45'31"
Robert & Melissa Harriss 10/03/2007
Cru 12h31'53" -61d51'37"
Robert Mello
Crb 15h41'59" 26d20'35"
Robert & Meredith Lubking
Cyg 19h44'31" 31d30'3"
Robert Merson
Ari 2h17'30" 12d20'14"
Robert Michael Baker
Per 3h28'25" 51d1'55"
Robert Michael Barrett
Per 3h36'51" 51d51'48"
Robert Michael Barrow Jr
Aqr 22h24'5" 0d21'28"
Robert Michael Cerullo
Cnc 9h14'19" 9d7'36"
Robert Michael Collins
Sgr 18h50'55" -19d6'6"
Robert Michael Del Gatto
Gem 7h7'2" 17d55'25"
Robert Michael Flink
Sco 16h23'32" -25d23'48"
Robert Michael Johnston, Jr. "Mike"
Aqr 23h13'16" -7d19'25"
Robert Michael Lowe- Specialist ARNG
Per 3h34'30" 48d0'49"
Robert Michael MacKercher
Lmi 10h43'21" 26d58'11"
Robert Michael Malouf
Ari 2h53'38" 29d2'39"
Robert Michael Markham
Ori 6h17'55" 16d18'3"
Robert Michael McCluskey
Cyg 19h51'41" 37d54'39"
Robert Michael McMillan Jr.
Uma 11h18'50" 37d44'19"
Robert Michael Medcalf
Per 3h16'51" 43d5'37"
Robert Michael Osman
Tau 4h39'36" 7d20'34"
Robert Michael Rabena
Cep 23h15'8" 75d56'42"
Robert Michael Slaby, Jr.
Cap 21h16'8" -14d29'3"
Robert Michael Thomas McNamara
Dra 18h58'8" 73d5'16"
Robert Michelsen
Lyn 7h9'36" 49d2'27"
Robert Miles Barrett
Uma 12h51'57" 53d3'19"
Robert Miller
Sco 17h8'8" -38d15'0"
Robert Miller
Her 16h38'46" 48d11'27"
Robert Miner 51
Per 3h44'10" 43d0'44"
Robert Mitchell III
Cnc 8h53'6" 31d55'58"
Robert Molvaer
Cap 21h29'51" -24d33'1"
Robert Morris Lee
Leo 9h45'52" 27d29'26"
Robert Morris Shawler
Her 17h12'40" 31d8'32"
Robert Morris Watson
Sgr 18h40'57" -25d22'25"
Robert Morrow
Ori 5h10'39" 3d29'32"
Robert Mosca
Tau 4h24'7" 3d2'9"
Robert Moss Rowe - Uncle Bob
Aqr 21h4'33" -13d9'56"
Robert Moya
Ari 2h15'40" 24d34'26"
Robert Moyer
Uma 10h51'11" 63d13'53"
Robert Muir
Uma 13h42'23" 51d11'45"
Robert Murray
Aqr 22h35'39" -2d13'26"
Robert my shining star
Aqr 23h53'29" -10d27'4"
Robert N. Massi
Psc 1h31'5" 16d36'44"
Robert Nanni
Uma 9h9'0" 53d30'41"
Robert Nels Kyllo
Tau 5h57'25" 24d17'33"
Robert Nelson
Dra 18h42'10" 66d30'35"
Robert Newman
Gem 8h0'19" 25d48'50"
Robert Newman
And 1h5'39" 45d58'40"
Robert Nicholas Cotoia
Vir 13h3'49" 12d2'44"
Robert Nicholas Logan
Cep 22h12'30" 73d38'59"
Robert & Nicole Garcia
Cyg 20h33'4" 59d43'2"
Robert Nimmo
Gem 7h6'8" 22d0'49"
Robert Noel Balcombe
Cyg 19h35'47" 34d42'34"
Robert Noel Byg
Aqr 22h2'24" -14d48'24"
Robert & Norma Cox 10/11/57
Lyr 18h47'3" 38d29'19"
Robert & Norma Featherston
Cyg 20h24'49" 51d6'0"
Robert Norman
Per 4h0'11" 48d29'53"
Robert Norman Shaw xxx
Cep 21h24'34" 69d5'14"
Robert Norton "Angel's Light"
Aqr 22h10'15" -21d4'48"
Robert O. Feeney
Lib 15h20'14" -10d40'23"
Robert O. Loftis, Jr.
Leo 9h49'37" 6d32'8"
Robert O. Whitesell
Uma 10h52'37" 44d51'36"
Robert Ohlmeier
Ori 6h20'50" 8d48'33"
Robert Owen Hudson
Ori 5h37'26" -0d57'48"
Robert P C Wu
Psc 0h53'19" 32d16'8"
Robert P Cartwright
Ori 5h39'31" 12d0'28"
Robert P. Hennessey
Cap 21h9'30" -25d33'21"
Robert P Klipper
Dra 18h32'48" 54d33'51"
Robert P. Maurer
Uma 14h26'28" 59d17'48"
Robert P McKnight
Ori 4h54'59" -0d4'13"
Robert P. Monahan
Per 3h49'40" 36d51'25"
Robert P. Roeding
Per 2h30'6" 54d55'45"
Robert P. Tunney, Sr.
Her 16h52'26" 17d17'34"
Robert P. Valentine
Her 18h43'38" 14d11'35"
Robert Padgug
Ari 2h41'18" 29d27'30"
Robert Palmieri
Gem 6h47'28" 33d20'0"
Robert Palmisano "My Babe"
Cap 20h27'24" -19d20'48"
Robert Parker, 'Bertie'
Ari 1h56'49" 15d4'56"
Robert Parker, Our Eternal Eagle
Aql 19h5'7" 14d46'19"
Robert Paterson
Uma 11h18'41" 33d49'53"
Robert Patrick Finn
Psc 0h15'10" 16d59'4"
Robert Patrick Ranalla III
Gem 6h50'12" 28d55'59"
Robert Paul
And 0h9'58" 45d0'58"
Robert Paul Burkey
Aur 6h5'38" 47d35'25"
Robert Paul Clemente Johnson
Per 4h48'3" 47d28'5"
Robert Paul Evans
Ori 5h44'28" 7d34'33"
Robert Paul Fay
Uma 8h27'1" 66d58'38"
Robert Paul Francois Billaud
Psa 22h29'56" -31d1'22"
Robert Paul Geiss
Gem 7h18'44" 14d39'19"
Robert Paul Guidetti Jr.
Cep 22h55'37" 58d44'1"
Robert Paul Henry
Lib 15h48'39" -17d19'10"
Robert Paul Hudson
Gem 7h48'21" 33d10'26"
Robert Paul Lavin
Tau 4h32'10" 19d17'27"
Robert Paul McMahon
Sco 16h14'11" -19d13'2"
Robert Paul Sjogren
Her 16h46'27" 29d52'28"
Robert Paul Stedeford, Sr.
Aqr 21h33'59" 1d58'39"
Robert & Peggy Rowland
And 1h33'10" 40d55'47"
Robert & Penelope Scott
Cnv 12h32'50" 43d37'57"
Robert Penn
Cep 0h6'53" 72d22'48"
Robert Perkins
Dra 19h30'26" 60d1'43"
Robert Perrin
Psc 23h24'55" -2d31'1"
Robert Peter and Ruth Ann Wade
Cas 23h35'54" 58d48'47"
Robert Phaneuf
Cap 20h22'42" -10d29'32"
Robert & Phyllis Whiting
Cyg 19h54'52" 39d41'50"
Robert Pike Jr.
Per 4h40'0" 48d49'12"
Robert Pineda
Her 18h49'47" 13d29'24"
Robert Pineda
Cep 21h26'19" 61d50'14"
Robert Platek
Uma 9h31'38" 45d54'16"
Robert "Pop" Barry
Aqr 22h24'38" -1d12'44"
Robert (Popi) Nyland
Uma 10h50'45" 61d52'30"
Robert Pryce
Cru 12h31'41" -59d53'44"
Robert Psareas
Aql 19h19'11" 2d44'48"
Robert Quenneville- Clairmont
Leo 11h28'57" -6d3'37"
Robert Quinn
Boo 15h7'3" 44d55'37"
Robert R. "Bob" Thomas
Her 18h45'40" 20d21'13"
Robert R. Catino Sr.
Aqr 21h55'42" 0d47'39"
Robert R. Coffman
Aql 19h52'56" -0d14'50"
Robert R. Cornwell, M.D.
Ori 5h26'21" 2d11'35"
Robert R. Fuchs
Psc 1h4'50" 3d54'24"
Robert R. Greene
Gem 7h17'59" 17d46'59"
Robert R. Kenny
Per 3h30'48" 43d52'50"
Robert R. Krenos
Cep 22h17'40" 63d1'46"
Robert R Leonard
Aqr 21h50'56" -7d46'13"
Robert R. Miller
Cam 5h33'20" 71d40'26"
Robert R. Rickert, MD
Dra 15h57'1" 60d5'43"
Robert R. Rudolph
Uma 11h51'18" 51d16'36"
Robert R. Swartz
Ari 2h7'22" 25d56'18"
Robert / Rachael Goings... 50 Years
Sgr 19h28'11" -14d3'29"
Robert Rahlfs
Uma 11h8'6" 37d25'40"
Robert Raleigh Davis
Cnc 9h9'20" 25d9'23"
Robert Ralph Zoppelt
Uma 11h5'40" 65d13'12"
Robert Rathjens
Uma 11h14'6" 72d47'8"
Robert Ratten - Shining Star
Cma 6h43'18" -19d17'22"
Robert Rau Jr.
Ori 5h7'31" 15d3'43"
Robert Rawl
Cep 22h5'54" 69d47'49"
Robert Raymond Demers
Ori 5h16'1" 5d2'6"
Robert Raymond Scott, Jr.
Cnv 13h43'6" 39d46'4"
Robert Raymond Silverthorne IV
Ori 5h27'42" 2d38'35"
Robert & Rebecca Carrillo
Cyg 21h32'10" 53d44'28"
Robert & Rebecca Horvath
Cyg 21h15'5" 46d48'30"
Robert Reed Marquis
Ori 5h43'52" 2d5'14"
Robert Reed Slorance
Tau 5h50'2" 23d56'37"
Robert Reeves
Ori 5h6'23" 3d36'49"
Robert Reid Jantz
Vir 14h17'28" 0d28'57"
Robert Reid Kent
Aqr 22h10'3" 0d14'12"
Robert Reid Redmond
Uma 8h43'12" 57d54'7"
Robert Reiprich
Uma 11h57'12" 32d30'53"
Robert & Renee Epstein
Cyg 21h20'42" 37d22'57"
Robert & Renee Rusnell Forever
Tau 3h59'57" 22d56'13"
Robert Reynolds
Uma 10h14'4" 53d9'11"
Robert Reynolds Clifton
Cap 21h39'53" -16d10'7"
Robert Rhee and Jessica Ham
Cap 20h30'41" -13d3'50"
Robert Rhoades Quimby
Cep 22h4'51" 57d2'0"
Robert Rhodes
Ori 5h51'53" -0d2'3"
Robert Ridder
Per 4h37'18" 42d27'9"
Robert Ridley
Uma 10h27'33" 59d31'2"

Robert Riley
Ori 5h42'50" 7d34'15"
Robert Rising
Dra 15h52'10" 53d36'36"
Robert & Rita Heim 60th Anniversary
Cyg 19h41'11" 52d29'41"
Robert (Robbie) Sowers
Psc 0h25'56" 15d8'35"
Robert Rocco Compono
Her 16h10'0" 41d34'42"
Robert Roemer
Tau 4h21'39" 23d17'58"
Robert Romboy
Dra 17h19'58" 52d56'43"
Robert Ronald Daugherty
Vir 12h48'36" -11d24'42"
Robert Ronald Kane
Ori 6h17'38" 14d37'48"
Robert Rooster Kelley
Cen 13h43'26" -32d31'41"
Robert Ropke
Uma 11h50'8" 55d33'41"
Robert & Rose Curtis
Cyg 21h25'40" 39d39'19"
Robert & RoseAnn Rogacki
Leo 9h24'19" 25d30'10"
Robert & Roumiana Popp
Tau 3h48'40" 24d27'20"
Robert Rousselle
Per 3h20'36" 45d21'26"
Robert & Roxanna Miele Forever
Aqr 21h45'24" -2d32'2"
Robert Roy Taylor, III
Sgr 18h0'8" -23d56'58"
Robert Russell Sause
Per 3h1'48" 53d9'36"
Robert Rutkowski
Dra 17h32'25" 60d38'19"
Robert Ryan Brandt
Leo 9h31'58" 23d16'48"
Robert Ryan Brooks
Ari 2h4'38" 22d5'28"
Robert S. Bellomy
Aur 7h17'59" 38d6'20"
Robert S. Bos
Tau 4h32'31" 16d21'42"
Robert S. Diaz
Tau 5h46'29" 15d20'36"
Robert S Downes
Uma 10h46'51" 59d3'8"
Robert S. Ehrlich
Cep 20h46'0" 61d48'19"
Robert S. Frick
Lib 15h46'56" -13d9'42"
Robert S. Kelly
Cap 20h27'34" -15d43'5"
Robert S. Louden
Uma 12h9'4" 58d55'48"
Robert S. Martin
Psc 23h58'35" 1d31'0"
Robert S. Miles
Uma 12h6'10" 65d43'24"
Robert S. Schur
Aql 19h28'23" -11d12'14"
Robert S. Walder
Per 3h4'57" 51d22'39"
Robert S. Wolcott
Uma 9h4'3" 56d52'44"
Robert Sachs
Aur 5h58'47" 53d55'18"
Robert Salmon
Leo 9h33'5" 29d43'4"
Robert & Sandra Seeley
Cyg 21h40'26" 53d6'25"
Robert Saraceno
Her 17h44'7" 22d21'26"
Robert Schleeweiss
Sco 17h27'58" -41d41'23"
Robert Schultz, Jr.
Cap 20h17'24" -10d14'46"
Robert Schunneman A Durned Fine Dad
Cep 22h37'53" 72d35'15"
Robert Schuster
Gem 6h44'16" 15d0'34"
Robert Schuyler
Ori 6h13'41" 15d11'29"
Robert Sckalor
Per 2h39'37" 53d21'56"
Robert Scott Barnes, Gemini Star
Gem 6h58'3" 15d26'7"
Robert Scott Davis
Cap 21h45'15" -13d45'17"
Robert Scott Parker
Her 17h30'54" 36d8'30"
Robert Sean
Her 16h37'32" 35d45'48"
Robert Sean Little 'Robbie'
Umi 13h5'26" 74d3'1"
Robert Sebastian Dudzik
Gem 6h48'45" 23d45'23"
Robert Seraydarian
Uma 12h2'28" 38d21'55"
Robert Seymour
Uma 10h42'18" 50d55'27"
Robert Shannon O'Brien
Psc 0h24'44" 7d36'6"
Robert & Shannon Thompson
Cyg 20h58'39" 40d51'33"
Robert Sharkey
Uma 11h8'40" 48d43'54"
Robert & Sharon McLaughlin
Ari 3h11'21" 26d43'35"
Robert & Shasta
Lyr 19h3'12" 42d40'15"
Robert & Shauna O'Connell
Aql 19h50'2" 14d19'32"
Robert Shawn Cooper
Per 4h22'6" 44d36'9"
Robert Shawn Harland
Lyn 6h36'47" 61d50'27"
Robert & Shelba Henke
Cyg 21h30'25" 43d57'48"
Robert Shield
Uma 11h25'45" 51d8'52"
Robert Shining Star Of Schonfeld
Aqr 21h15'26" -8d49'28"
Robert & Shirley Janota
Cyg 21h15'34" 47d7'30"
Robert & Shirley Palmer - 50 years!
Cyg 20h59'58" 44d14'38"
Robert Sieggen
Cam 7h50'58" 65d16'20"
Robert Silas
Uma 10h24'27" 44d27'27"
Robert Simpson
Per 3h37'53" 45d55'31"
Robert Simpson
Cep 20h40'40" 63d11'57"
Robert Singletary
Uma 11h27'12" 38d34'12"
Robert (SKINNER) Coffin
Vir 11h57'36" 3d18'58"
Robert "Skip" Gilbert
Per 2h55'4" 54d28'52"
Robert Slewett
Her 16h20'56" 19d14'31"
Robert Smee
Aqr 22h38'10" -0d19'14"
Robert Smith
Cep 22h1'11" 68d47'44"
Robert Snyder
Uma 8h18'17" 62d46'57"
Robert Sommerville Wilkerson Given
Ari 2h34'59" 22d23'18"
Robert Soukup
Aql 19h54'46" -0d26'25"
Robert Speaks
Psc 23h54'45" -0d7'1"
Robert Spies, geb. 21.05.1994
Ori 6h15'43" 10d7'45"
Robert Springfield
Ori 5h58'51" 11d29'37"
Robert Stagnitto
Lib 15h29'33" -24d31'8"
Robert Stanley BBB Czachorski III
Psc 0h37'14" 8d0'8"
Robert Stanley Girschick
Uma 11h34'53" 36d14'52"
Robert Starks Higgins
Uma 11h22'27" 38d21'11"
Robert Starr Waite
Her 18h6'41" 22d41'11"
Robert Stephen
Psc 1h24'27" 11d39'6"
Robert Sterling Moore II
Tau 5h28'2" 27d34'24"
Robert Steven Moore
Uma 10h1'57" 42d46'15"
Robert Stevenson
Umi 13h42'49" 74d18'28"
Robert Steyaert
Gem 6h47'31" 34d47'29"
Robert Stikeman
Boo 14h32'35" 19d38'8"
Robert Street
Ori 5h41'5" 8d2'18"
Robert Stuart Darby III
Cap 20h11'2" -25d43'35"
Robert Sven Swenson
And 2h34'23" 40d37'33"
Robert & Svetlana
Uma 9h16'18" 65d33'42"
Robert Swearingen
Her 16h33'57" 36d21'35"
Robert Swensgard
Pho 1h39'21" -50d59'40"
Robert T. Davis, Sr.
Per 3h9'49" 52d52'46"
Robert T. Eisenman - Mr. MiffMiff
Gem 6h44'4" 21d0'35"
Robert T. Heise
Lib 14h48'12" -5d13'17"
Robert T. Schlachter
Ari 3h13'27" 27d42'5"
Robert T. Volenec
Aur 6h38'30" 50d39'52"
Robert Tate
Uma 9h2'29" 56d37'22"
Robert Tayron Brooks
Boo 14h47'4" 46d15'31"
Robert Terence Vickery
Cep 20h38'40" 65d16'1"
Robert & Theresa Keehfuss & Family
Cap 20h13'38" -10d58'55"
Robert Thomas Berryman
Umi 15h16'30" 69d12'38"
Robert Thomas Bowers II
Per 3h53'44" 40d24'17"
Robert Thomas Donovan
Gem 6h27'46" 25d36'46"
Robert Thomas Kampfman
Uma 13h47'55" 61d7'30"
Robert Thomas King
Uma 10h46'22" 48d13'35"
Robert Thomas Miller II
Sgr 19h13'42" -13d28'19"
Robert Thomas Pastena
Aqr 22h26'16" -0d1'13"
Robert Thomas Sellers
Cnc 8h51'41" 27d4'24"
Robert Thomas Whittington
Gem 6h42'37" 29d54'21"
Robert Thompson
Her 18h38'26" 17d14'52"
Robert Thorndike Poor
Aql 20h20'18" 3d14'14"
Robert Timothy McInerney
Per 3h14'48" 39d23'39"
Robert Tomas Lucero
Lmi 9h47'31" 35d18'17"
Robert Toogles Addie
Cyg 19h46'31" 56d12'3"
Robert Totte
Cnv 13h23'11" 29d19'50"
Robert Trail
Cma 6h41'46" -19d43'1"
Robert Treadway Jr.
Lib 15h6'7" -18d55'24"
Robert Tripi
Uma 10h29'27" 46d26'48"
Robert Trippany
Eri 3h51'16" -0d5'46"
Robert Troy Henderson
Psc 0h59'49" 28d33'16"
Robert Troy Teague
Sco 17h52'36" -36d25'4"
Robert V. Armagost, Jr. - 4/14/1958
Uma 11h9'13" 43d15'41"
Robert V Bazzano
Crb 15h38'16" 37d52'2"
Robert V. Pack
Cap 21h37'24" -16d56'23"
Robert & Valerie Germain
Cyg 20h43'40" 53d21'56"
Robert Vance Coburn
Cap 20h43'55" -18d33'55"
Robert & Vanessa
Pyx 8h40'39" -33d38'18"
Robert Vea Clifton
Ori 5h59'14" 17d11'30"
Robert Vencil Young
Uma 13h28'39" 53d29'48"
Robert Vern Mechling Dec 12, 1946
Her 17h42'55" 37d58'50"
Robert Vernon Steptoe
Uma 11h54'50" 33d26'22"
Robert Vigneau
Cnv 13h54'51" 31d13'29"
Robert Vincent Mellon
Ori 5h58'8" 6d47'36"
Robert Vincent Wagley
Cet 3h13'45" -0d14'30"
Robert Virelli
Ori 5h39'6" 9d19'45"
Robert Virgil Burke
Ori 5h43'58" 11d48'43"
Robert & Virginia Jarolin
Sco 16h13'44" -18d46'14"
Robert Visinand
Lmi 11h0'19" 27d27'49"
Robert Vogel
Uma 9h41'19" 61d53'42"
Robert Volpe
Sgr 19h13'28" -34d12'50"
Robert Volynsky
Dra 17h44'26" 53d35'36"
Robert Vorreiter
Uma 8h38'16" 69d38'31"
Robert W. Barrett
Cnc 8h30'19" 22d14'10"
Robert W Bolten
Tri 2h40'57" 33d52'27"
Robert W Boos II
Uma 13h47'8" 56d39'4"
Robert W. Courtney II
Aql 19h40'24" 15d31'25"
Robert W. Cubala
Per 3h9'20" 47d42'3"
Robert W Darrow III
Leo 9h35'22" 28d42'16"
Robert W Devonshire Jr
Vir 13h28'53" -16d49'36"
Robert W. Flynn-The Window God
Her 16h31'46" 34d32'19"
Robert W. Grammick
Aqr 22h22'22" -2d24'0"
Robert W. Holzhauer
Leo 9h26'5" 28d36'20"
Robert W Kaiser
Her 16h42'0" 34d0'13"
Robert W Keeling
Uma 13h38'36" 57d27'10"
Robert W. Lee Jr.
Leo 11h31'19" 23d2'30"
Robert W Manka
Uma 8h28'36" 66d59'26"
Robert W. McClurg
Her 16h42'2" 36d31'50"
Robert W Mullenax
Uma 10h11'45" 70d40'32"
Robert W Netherway
Ori 6h23'51" 16d52'17"
Robert W. Ochs
Lib 14h40'36" -18d21'30"
Robert W. Payne
Gem 6h53'41" 26d28'18"
Robert W. Peltzer
Her 18h4'36" 36d50'11"
Robert W. Raleigh, Jr.
Gem 7h47'50" 25d22'5"
Robert W. Ruban (1918 - 1999)
Umi 15h30'12" 73d21'55"
Robert W. S. Anderson
Uma 10h37'32" 71d37'57"
Robert W. Schoch
Tau 4h35'17" 12d49'54"
Robert W Shannon
Lyn 8h38'4" 38d17'45"
Robert W. Siegmund
Her 17h44'43" 16d47'1"
Robert W. Smith
Aur 5h55'45" 37d26'48"
Robert W. Stanek
Vir 11h59'42" 6d49'41"
Robert W. Terraglio
Per 3h23'46" 45d16'5"
Robert W. Ulmer
Uma 11h28'27" 64d9'13"
Robert Wade Britten
Cap 21h16'51" -19d3'58"
Robert Wagner
Ori 6h19'53" 9d57'54"
Robert Walker
Dra 18h49'23" 50d22'52"
Robert Wallace Palmer
Cnc 8h36'44" 24d28'33"
Robert Walter Jackson
Her 18h6'3" 33d11'53"
Robert Ward Kidston
Uma 9h10'15" 53d36'15"
Robert Warner Corey
Ari 2h42'4" 16d56'4"
Robert Warren Dodd
Aqr 23h5'16" -8d45'43"
Robert Warren Holder, Sr.
Leo 10h34'32" 18d50'58"
Robert Watkin Sr
Sgr 19h1'44" -29d44'44"
Robert Watt
Ori 4h55'29" 3d11'38"
Robert Wayne Burris
Ari 2h52'43" 18d45'35"
Robert Wayne Burt
Uma 9h40'30" 48d8'19"
Robert Wayne Cornell
Per 3h22'5" 42d9'10"
Robert Wayne Davis
Vir 12h30'11" 3d3'49"
Robert Wayne Edwards
Uma 11h0'39" 69d48'6"
Robert Wayne Elen
Tau 5h45'44" 21d1'20"
Robert Wayne Willey
Sco 17h40'12" -40d29'45"
Robert Welch's Eastern Star
Cep 22h11'17" 67d36'21"
Robert Wesley Daniels
Tau 3h59'51" 18d41'8"
Robert Wesley Kallmerten
Gem 6h45'1" 26d59'48"
Robert Wesley Klavins
Leo 10h12'55" 10d21'28"
Robert Wesley Stecher Junior
And 23h22'56" 42d18'8"
Robert Westfield Nelson
Uma 11h6'33" 67d22'0"
Robert "Whitey" Miller
Cep 22h4'58" 57d16'24"
Robert Whitfield
Lyr 18h45'29" 40d48'47"
Robert Wilkes Witty
Tau 4h32'19" 5d32'54"
Robert Willard Robinson
Aqr 22h25'15" -2d56'55"
Robert William Adams III
Per 3h36'30" 43d36'47"
Robert William Austin Halligan
Cep 23h9'5" 77d42'15"
Robert William Campbell
Sco 16h43'35" -30d9'33"
Robert William Christman
Uma 11h26'30" 63d57'55"
Robert William Dengerd II
Aqr 22h29'34" -4d22'7"
Robert William Kilcoyne
Per 3h24'35" 50d42'15"
Robert William Lavelle
Cnc 8h14'11" 13d0'37"
Robert William Lesley
Uma 13h25'2" 54d54'14"
Robert William Letkeman-Prouse
Uma 11h25'15" 59d26'50"
Robert William Martin
Psc 0h31'45" 16d52'49"
Robert William Noyes
Ori 5h58'7" 21d39'54"
Robert William Planchard
Cep 22h58'31" 79d33'32"
Robert William Sambursky, Jr.
Lib 14h24'31" -15d13'30"
Robert William Schneider
Cap 21h39'54" -15d35'32"
Robert William Viggiani
Lib 15h5'47" -0d42'55"
Robert William Young
Per 3h3'26" 46d49'41"
Robert Wilson
Per 3h25'49" 34d57'12"
Robert Wilson Perks
Per 2h22'19" 51d44'3"
Robert Winfrey
Uma 12h2'36" 52d3'9"
Robert Woodrow Hamilton, Sr.
Cnc 8h42'15" 7d25'16"
Robert Wyrick
Boo 14h43'21" 26d35'45"
Robert Yengibaryan
Per 3h12'43" 41d8'7"
Robert Yodka, Jr.
Gem 7h1'29" 18d34'23"
Robert Yott
Lyn 6h20'59" 54d24'3"
Robert Yule-Young
Cru 12h25'17" -56d44'16"
Robert Zarcone
Sco 16h22'12" -26d59'37"
Robert Zimmerman
Tau 3h50'38" 10d16'45"
Robert Zitto
Cep 23h18'33" 78d19'1"
Robert Zwiers
Umi 16h20'29" 75d40'31"
Robert, Gian and Giovanni
Crb 15h45'12" 32d54'39"
Robert, Helen, Sara, & Amanda Mcgee
Dra 17h57'57" 59d36'9"
Robert, mio eternal amore
Cyg 20h20'35" 54d23'44"
Robert, Virginia & Ziti Rosasco
Ori 6h3'3" 18d49'6"
Roberta
Leo 11h22'12" 22d40'36"
Roberta
Aur 4h52'39" 39d30'6"
Roberta
Cyg 20h42'39" 52d5'37"
Roberta
Cyg 21h39'40" 51d1'38"
Roberta
Boo 14h30'15" 39d5'16"
Roberta
Aqr 23h10'8" -16d30'57"
Roberta
Sgr 17h55'39" -17d14'26"
Roberta A. Steele
Crb 16h16'41" 31d6'37"
Roberta and Sameer
Ori 5h52'47" 21d44'28"
Roberta Anderson
And 0h17'40" 45d9'21"
Roberta Ann Phillips
Gem 7h13'53" 31d32'26"
Roberta Ann Tabolsky
Tau 5h51'29" 26d49'5"
Roberta Annunziata
Cap 20h56'34" -21d29'12"
Roberta Bailey
And 0h45'44" 37d17'22"
Roberta Biagini
Umi 15h0'54" 67d51'41"
Roberta "Bobbi" Hamilton Eliskovich
Apu 16h32'24" -70d43'36"
Roberta "Bobbie" Foley
Uma 8h30'19" 64d27'11"
Roberta "Bobbie" Kay Boles Askins
Cap 21h30'28" -16d2'34"
Roberta (Bobbity) Nina Murfitt
Gem 6h53'21" 21d24'3"
Roberta Bubina Mia
Peg 23h6'59" 31d25'34"
Roberta Buble
And 23h11'14" 51d49'28"
Roberta Buckwalter
Cnc 8h10'56" 20d51'33"
Roberta Cairone
Peg 22h42'28" 23d22'2"
Roberta Cloud
Lmi 10h44'1" 35d14'28"
Roberta D'Amore
Lyn 7h52'12" 49d42'51"
Roberta D'Arrigo
Cyg 19h53'43" 35d11'2"
Roberta & Donald Parks
Cyg 20h22'32" 58d7'14"
Roberta Dye
Cas 0h1'54" 53d40'31"
Roberta Eileen
And 1h47'30" 37d14'48"
Roberta Ezzell
Her 16h14'45" 18d25'36"
Roberta Greenberger
Cas 0h7'38" 59d20'9"
Roberta Hankins
Cam 6h1'36" 65d9'57"
Roberta Hilson
Tau 3h34'59" 21d3'27"
Roberta J. Babb
Peg 22h28'24" 9d2'55"
Roberta Jean Ait (Bert)
Cyg 20h9'44" 42d49'5"
Roberta Jean Gates
Cas 0h45'31" 51d48'26"
Roberta June
Lyn 6h45'42" 53d44'45"
Roberta King The Best Mom Ever !
Cas 23h23'13" 59d45'12"
Roberta Kowalke
Gem 7h26'35" 13d7'27"
Roberta L Resecker
Ari 2h12'19" 27d7'53"
Roberta Langella
Cas 0h32'52" 62d21'17"
Roberta Lewis-Solar
Gem 6h41'14" 18d35'29"
Roberta Louise Ready Allen
Lib 15h6'1" -29d26'17"
Roberta Lynn
Vir 12h14'30" 11d48'47"
Roberta Lynn Cline
Mon 6h29'37" 8d20'24"
Roberta Marian Bronson
Vir 12h35'40" -9d7'25"
Roberta Mariquita Gonzalez
Uma 8h55'54" 51d52'1"
Roberta McDaniel
Leo 10h13'25" 14d2'26"
Roberta Mia Elizabeth
Cap 21h15'31" -17d24'34"
Roberta Montesi
Tau 5h53'15" 25d14'27"
Roberta Motta Grandi
Per 3h3'16" 51d41'51"
Roberta Nancy
Leo 11h33'25" 6d12'27"
Roberta Nixon
Lib 15h5'10" -23d59'22"
Roberta Ragan
Cnv 12h48'47" 37d49'35"
Roberta Rice
Sco 17h26'4" -32d5'55"
Roberta "Ro" Doran
Cyg 21h14'30" 30d2'43"
Roberta Roberts
Ari 2h59'35" 24d40'0"
Roberta Rose
Lyn 6h33'59" 54d26'20"
Roberta - Satin
Aur 6h12'53" 48d49'23"
Roberta Schartung
Tau 4h34'37" 16d59'34"
Roberta Schönenberger
And 23h1'58" 40d23'25"
Roberta Tavolotti
Lyn 8h23'3" 44d10'16"
Roberta W. Sposato
Uma 8h53'52" 65d24'56"
Roberta Williams Pratt
Lyn 7h22'54" 58d1'22"
Roberta Woolf
Crb 16h13'55" 38d44'42"
Roberta Yates
Cas 0h53'22" 60d29'50"
Roberta, My Love
Cas 0h19'32" 50d21'33"
Roberta7
Aur 5h36'29" 32d53'37"
Roberta-Christian-Dennis-Brad
Cas 23h23'10" 57d20'22"
Robertainez
Uma 13h1'16" 56d56'47"
ROBERTANDMARIE
Psc 1h2'11" 15d8'59"
Roberta's FAB Star
Uma 13h50'1" 53d57'58"
Roberta's Rainbows
And 0h53'6" 44d14'46"
RobertHanh2004
Lmi 10h40'56" 34d40'8"
RobertIreneMaryMadalene Spezialetti
Crb 15h44'6" 26d33'25"
Robert-Jan Boelen
Sco 17h23'50" -30d31'26"
RobertLPappagalloAdAstra PerAspera
Uma 11h37'34" 47d35'48"
Roberto
Cas 2h10'16" 59d43'9"
Roberto
And 1h58'4" 39d4'3"
Roberto
Ori 5h0'21" 5d19'42"
Roberto
Uma 9h20'50" 66d55'14"
Roberto 30 maggio 2003
Per 3h22'36" 50d21'7"
Roberto and Bruttissima Nov. 2005
Cap 21h46'47" -10d34'13"
Roberto Antonio Carmona "Bee"
Cnc 8h43'33" 20d14'49"
Roberto Araba Fenice II
Vir 11h50'33" 6d9'2"
Roberto Cabellos
Uma 11h10'50" 64d57'5"
Roberto Carlos Braga
Ari 2h26'59" 25d41'40"
Roberto Cavagnero
Aur 5h7'51" 40d33'30"
Roberto Cohen
Leo 9h40'28" 27d45'10"
Roberto Consoli
Her 17h28'48" 36d37'4"
Roberto De Cervo
Umi 17h2'56" 80d26'35"
Roberto de Jesus
Her 16h35'46" 13d14'41"
Roberto e Nietta Franchi
Lyr 19h26'53" 42d51'10"
Roberto e Vanessa 22.11.2003
And 23h10'30" 35d30'9"
Roberto et Anne-Laure Tojeiro
Her 16h15'5" 44d34'45"
Roberto Feroldi
Cas 0h10'11" 52d13'0"
Roberto Francisco Soto Chavez
Uma 10h6'18" 48d51'14"
Roberto & Jessica
Cnc 7h55'57" 18d31'39"
ROBERTO LOVE
And 23h56'10" 36d51'56"
Roberto M. Fieg
Cyg 21h0'9" 48d15'57"
Roberto Malinconico
Cam 4h8'39" 56d52'0"
Roberto Martone
Crb 16h11'33" 39d31'44"
Roberto Morgia
Uma 9h30'25" 53d25'27"
ROBERTO - NICOLA (RABBIT)
Per 4h35'7" 39d23'39"
Roberto Olla
Umi 14h35'20" 80d36'4"
Roberto Patruno
Her 17h24'44" 32d0'26"
Roberto Pozzi
Uma 9h23'53" 44d43'4"
Roberto Rossi 30 luglio 1958
Per 4h4'21" 31d23'26"
Roberto - skorpion king - Fantasia
Peg 22h0'23" 13d7'52"
Roberto Torres
Leo 11h45'30" 26d6'48"
Roberto with Miry
Ori 5h0'58" -0d37'15"
Robertomary
Uma 13h41'53" 55d12'59"
Roberts
Uma 11h26'57" 63d56'26"
Robert's Family Get Away
Lyn 6h22'40" 57d12'19"
Robert's Joy
Mon 7h7'28" -6d32'5"
Robert's Melody
Sgr 18h25'4" -32d10'50"
Robert's Reach
Leo 10h21'2" 26d15'53"
Robert's Reliance
Gem 7h11'50" 33d35'4"
Robert's Star
Per 3h10'22" 45d30'28"
Robert's Star
Uma 9h3'15" 53d57'30"
Robert's Star in Heaven
Aql 19h49'8" -0d18'23"
Robert's Venice
Vir 13h47'20" 1d27'
Robert's Vision
Uma 11h40'47" 48d59'19"
Robert-Sabina
Vir 13h19'36" 5d48'58"
Robert-Simon Edilber
Sco 16h46'16" -26d44'21"
Robertson and Perry
Lmi 11h4'38" 28d34'25"
Robertson-Legg
Uma 9h52'42" 50d17'3"
RobertTillemaOne
Uma 10h47'37" 64d58'52"
Robi Landolt
Per 4h11'57" 35d21'16"
Robi Rob Rob
Scl 23h27'52" -32d20'28"
Robi the Lion
Ori 5h2'4" 4d27'30"
Robia
Lib 14h54'25" -7d22'44"
Robianna Jones
Lyr 18h54'8" 36d52'16"
Robim
Uma 8h33'16" 68d46'2"
Robin
Aqr 23h38'36" -7d12'14"
Robin
Cap 20h16'31" -9d18'34"
Robin
Cma 7h1'20" -26d34'18"
Robin
Uma 11h14'17" 34d15'48"
Robin
Cnv 12h23'12" 33d29'3"
Robin
Crb 15h43'7" 31d39'49"

Robin
Per 3h49'53" 32d5'10"
Robin
And 23h42'52" 47d34'57"
Robin
And 2h21'31" 47d50'56"
Robin
Lyr 18h40'6" 39d28'28"
Robin
Lyr 18h46'1" 43d23'14"
Robin
Cyg 20h33'24" 42d4'11"
Robin
And 0h39'49" 45d37'45"
Robin
Ori 5h52'58" 5d45'10"
Robin
Aql 19h4'18" 3d2'36"
Robin
Tau 3h35'4" 14d11'9"
Robin
Mon 6h52'20" 7d51'34"
Robin A. Gould Allen
Sco 17h23'13" -37d32'45"
Robin A. Olschewske
Per 4h29'24" 39d2'36"
Robin A. Walker
And 1h14'22" 46d15'59"
Robin Abigail Waugh
Cap 21h32'53" -13d41'49"
Robin & Adam Posnack
Lyn 7h24'15" 53d8'27"
Robin Adele Peterson
Cam 4h28'11" 54d39'31"
Robin aka "Bunkie"
Ari 2h43'23" 29d43'9"
Robin and Alan
Vir 13h33'22" 2d18'23"
Robin And Jenn's Forever Star
Cyg 21h15'55" 45d14'56"
Robin and Katie Forever
Cyg 19h17'52" 52d37'20"
Robin and Sherry Hood
Uma 11h52'52" 42d56'53"
Robin & Andy
Cyg 20h41'38" 45d49'3"
Robin Ann
Tau 4h6'19" 8d20'6"
Robin Ann
Leo 9h40'54" 27d57'1"
Robin Ann 1-14
Cas 0h40'46" 56d59'14"
Robin Ann Lewis (Baby Girl)
Cap 20h41'28" -25d4'12"
Robin Ann Poure
And 23h9'5" 36d31'32"
Robin Armstrong Mitchell
Uma 9h20'54" 45d41'33"
Robin August Teacher
Uma 13h18'49" 56d19'15"
Robin Aurora
And 0h0'38" 44d5'25"
Robin Baxley
Lyn 7h35'17" 54d15'40"
Robin(beautiful)Freeman
Leo 11h10'17" 7d17'24"
Robin Beth O'Hagan
And 0h18'24" 36d29'20"
Robin Bobb
Aqr 21h5'16" -8d38'16"
Robin Bobbin Bob
Uma 9h18'1" 51d25'1"
Robin Brindle
Lmi 10h42'14" 37d26'46"
Robin C Davis
Aqr 23h29'22" -13d48'29"
Robin Carla Porter
And 1h41'44" 43d13'47"
Robin Carol Cartwright
Uma 10h25'58" 48d20'47"
Robin Catherine Toye
Aqr 22h18'31" 1d36'54"
Robin Chandler Cox
Psc 0h13'36" 3d57'40"
Robin Chasity Ray Lambe
Cnc 8h20'21" 13d47'39"
Robin Cheaney
Psc 0h38'52" 7d56'57"
Robin & Chris
Sge 20h5'42" 18d36'47"
Robin & Chris Kramer
Cyg 20h5'10" 35d7'28"
Robin Christine
Cyg 20h29'41" 39d58'33"
Robin Christine Philbrook
Gem 7h24'43" 24d4'50"
Robin Christine Rohrer
Ari 3h19'38" 15d30'9"
Robin Colleen W S
Leo 10h17'21" 26d53'39"
Robin Corigliano
Vir 13h0'25" -11d5'31"
Robin Costenbader-Jacobson
Gem 7h3'38" 25d40'32"
Robin D.
Psc 23h28'45" 7d35'14"
Robin D. Hardwick
Ser 15h17'8" -0d31'36"
Robin Della-Vedova
Lyn 7h11'43" 58d50'51"
Robin Demartino
Cas 23h48'58" 55d46'43"
Robin Dilio
Uma 9h9'29" 47d49'20"
Robin & Dorothy Curtis
Sco 17h29'10" -39d17'4"
Robin E. Martin
Aql 19h46'57" -0d30'55"
Robin E. Wright
Uma 8h41'7" 58d2'25"
Robin E Yanchitis
Aqr 21h31'8" 1d32'51"
Robin Elaine King
Tau 5h45'40" 17d41'36"
Robin Elizabeth
Tau 3h29'35" 19d4'13"
Robin Elizabeth LaManque
Umi 16h20'50" 86d15'5"
Robin Emma Sneyd - Robin's Star
And 1h55'29" 38d34'13"
Robin Everly
Ari 2h57'21" 25d29'21"
Robin Eves Star
Lyn 8h10'39" 41d24'9"
Robin Faller
Crb 15h32'59" 27d40'23"
Robin Fixmer
Oph 17h25'30" -0d17'42"
Robin Forman
Uma 8h55'59" 66d1'17"
Robin Franks
Her 17h56'44" 35d19'33"
Robin Frederick
Lib 14h56'30" -18d36'40"
Robin Hall
Uma 8h46'35" 60d50'32"
Robin Hayes Gurganus
Ari 2h30'26" 18d49'59"
Robin Heather Rainwater
Cyg 19h36'20" 31d2'38"
Robin Hoarau-Legeay
Gem 6h18'11" 25d38'58"
Robin Holmes Poole Wife & Mother
Ari 2h33'1" 27d21'32"
Robin Hood Bruce
Aql 18h52'55" -0d28'15"
Robin Hunziker Smith
Cam 6h19'56" 70d44'11"
Robin I. Craig
Car 10h56'45" -58d26'48"
Robin Immerman
Umi 15h27'41" 79d7'0"
Robin J Flaig
Tau 3h47'40" 27d41'18"
Robin Jane Sigel
Crb 15h59'26" 26d34'5"
Robin Jean Folken
Umi 16h11'6" 74d50'57"
Robin John Webb
Cru 12h42'7" -61d24'49"
Robin Jones Gunn
Cam 6h34'8" 75d17'46"
Robin K. Dove
Leo 9h26'27" 28d39'42"
Robin Kathryn Winkler
Peg 21h42'13" 16d40'54"
Robin Katker
Uma 12h17'26" 60d24'39"
Robin Kemmerer
Uma 11h55'13" 60d59'42"
Robin Kimberly Taves
Uma 11h30'47" 55d59'3"
Robin L. Camara
Cas 1h22'30" 61d7'48"
ROBIN L. FERGUSON
Leo 11h53'25" 20d57'14"
Robin L. Miller
Leo 9h24'42" 17d55'2"
Robin L. & Nick Farrantini
Lib 15h17'10" -9d24'58"
Robin L. Romano
Her 16h20'47" 48d11'29"
Robin L Whitmore
Ari 2h20'30" 24d10'54"
Robin Lanning
And 1h24'13" 43d47'22"
Robin Law
And 23h17'46" 44d17'44"
Robin Lee
Cyg 21h54'56" 54d19'12"
Robin Lee Hochtritt
Sgr 18h49'30" -16d46'5"
Robin Lee Rudd
Uma 13h20'39" 52d29'10"
Robin Leigh Liles
Leo 10h9'45" 13d48'43"
Robin Lelevier
Dra 20h38'15" 83d58'1"
Robin Leslie
Ari 3h19'37" 28d10'2"
Robin Leslie Pierce
Lyn 6h46'59" 51d30'26"
Robin Levin
Tau 4h38'57" 19d9'22"
Robin LGJ
Peg 22h56'24" 12d8'31"
Robin Long
Per 3h52'33" 34d7'29"
Robin Lou Branz
Uma 10h12'37" 53d44'19"
Robin Lucky
Psc 0h37'36" 10d6'41"
Robin Lukas
Ori 5h15'42" -1d37'11"
Robin Luu
Vir 13h21'34" 9d26'57"
Robin Lynn
Psc 1h26'28" 27d44'44"
Robin Lynn
Cas 0h58'55" 60d28'20"
Robin Lynn
Aqr 22h20'23" -24d30'5"
Robin Lynn Cotherman
Lib 15h7'4" -10d45'21"
Robin Lynn Hanousek
Aqr 23h0'24" -13d52'28"
Robin Lynn MacGruder Wiley
And 1h7'13" 33d51'15"
Robin Lynn Mason
Mon 6h35'44" 8d20'52"
Robin Lyons Sands
Cnc 8h47'44" 17d48'32"
Robin Margaret
Lib 15h27'39" -23d28'20"
Robin Marie
Sgr 18h36'22" -19d43'21"
Robin Marie
Tau 5h15'41" 22d55'36"
Robin Marie 09/15/1966
Vir 12h36'14" 10d56'47"
Robin Marie Angst
Crb 15h39'14" 38d32'8"
Robin Markland
Dra 17h41'35" 64d36'53"
Robin Marla Lipsky Cohen
Gem 6h44'52" 27d41'33"
Robin Matuseski
Vir 13h49'2" 3d22'40"
Robin McCafferty
Her 18h33'19" 15d23'54"
Robin McCall's 50th
Uma 11h27'35" 48d0'8"
Robin McGinnis
Umi 14h39'46" 70d38'50"
Robin McWherter
Cam 4h4'35" 57d32'46"
Robin Michelle Walker
Aqr 23h54'41" -14d57'6"
Robin Mock
And 2h6'49" 42d52'18"
Robin "Mom" Hernandez
Dra 18h26'47" 74d0'48"
Robin Müller
Uma 11h0'3" 66d15'57"
Robin "My Love"
Cas 0h19'16" 57d10'17"
ROBIN MY PRINCIPESSA BELLA
Aur 6h24'12" 31d26'26"
Robin N. SanGiacomo
And 0h19'14" 36d19'49"
Robin Newlove
Cep 23h14'55" 70d11'31"
Robin Nicole Moore
Ari 3h6'27" 12d55'51"
Robin O. Crewes
Uma 13h56'37" 50d48'7"
Robin Orr
Tau 4h39'22" 23d35'20"
Robin Oster 6-26-1953 - 9-19-2003
Cnc 8h9'39" 15d3'10"
Robin Ostlund
Aqr 22h11'19" -13d25'15"
Robin Paul Zentner
Lib 15h28'57" -16d0'24"
robin payne
Lac 22h27'16" 44d36'31"
Robin Peggy Hale
Leo 9h35'41" 26d36'6"
Robin "POP" Francis
Her 18h52'6" 25d0'26"
Robin Rajaniemi
Uma 12h42'0" 62d9'52"
Robin Reilly
Uma 10h8'15" 44d50'11"
Robin René Kiff
Tau 5h58'41" 25d11'36"
Robin Renee
And 1h8'40" 37d21'36"
Robin (ROB) Kimble
Ari 2h34'8" 27d4'19"
Robin "Rob" Proctor
Per 3h29'12" 47d22'5"
Robin Rochelle Hill's Shining Star
Aqr 22h58'35" -7d42'1"
Robin Roman Stein
Aqr 23h14'55" -7d54'57"
Robin Ruppenthal
And 0h42'22" 43d9'28"
Robin Rutchik
Psc 1h14'1" 29d16'21"
Robin S. Mason
Leo 11h32'41" 7d48'37"
Robin Sandra
Cyg 19h18'39" 53d27'15"
Robin Schafer
Ari 2h23'16" 11d16'23"
Robin Scott Kyle
Leo 10h18'27" 11d45'43"
Robin Selyf Edwardes
Her 17h49'12" 15d18'27"
Robin Shull
And 2h36'44" 44d23'36"
Robin Snaden
Lmi 10h19'3" 30d40'21"
Robin Spanjersberg Mitchell
Umi 14h22'47" 76d12'45"
Robin Stahel
Ori 5h42'18" -3d14'22"
Robin Sue H B
Lyn 7h24'14" 53d27'39"
Robin Suzanne Colvin
Gem 7h17'11" 18d29'17"
Robin Suzanne Harper
Ari 2h10'44" 24d8'13"
Robin Sweeney
Gem 7h0'37" 17d26'54"
Robin Swenson
Oph 17h33'5" 7d21'31"
Robin - T. O. A. O.
Ari 2h15'55" 26d7'40"
Robin Tellier 05-09-81
Nena Choudri
Cyg 19h44'29" 56d23'58"
Robin Vasicek
Cru 12h35'11" -55d59'8"
Robin Veikko Metsaots
Cnc 8h19'34" 19d51'3"
Robin Victoria Jackson's Star
Leo 9h57'36" 12d35'52"
Robin 'Vov' Huxley
Ori 5h51'57" 3d20'27"
Robin Waddington - Our Special Star
Umi 15h25'28" 81d27'54"
Robin Wade
Aqr 21h14'5" 1d44'32"
Robin Webel
Psc 0h43'32" 18d4'48"
Robin Wendy Mahon
Cas 1h17'43" 68d59'33"
Robin Wheeler Jackson
Cap 20h32'30" -13d57'52"
Robin Williams
Cep 21h25'41" 64d35'6"
Robin Wilson
Lyr 18h39'20" 31d0'38"
Robin Wolfe
Lyn 8h11'42" 50d45'24"
Robin, 45888
Aqr 22h46'5" -4d8'37"
Robina D'Arcy-Fox
Vul 21h18'33" 26d26'49"
Robina Maxima Noblelini
Aqr 22h31'24" -5d22'18"
Robinandlainey
Umi 14h29'9" 74d18'15"
Robin-Hope
Lib 14h50'54" -1d38'52"
Robinia
Ori 6h10'9" 6d31'58"
Robin's Angel
Lmi 10h24'0" 34d24'15"
Robin's Baby Boy
Ari 2h4'0" 22d48'36"
robin's christmas star
Lib 15h26'26" -20d57'53"
Robin's courage
Per 3h23'16" 49d48'14"
Robin's Essence
Uma 11h22'20" 53d53'46"
Robin's Gift
Sgr 18h20'6" -18d23'8"
Robin's Guiding Light
Per 3h6'17" 51d13'9"
Robin's heart
Lyn 7h57'36" 40d12'15"
Robins Life
Lyn 7h45'49" 54d15'15"
Robin's Light
Cam 5h38'47" 66d51'37"
Robin's Light
Gem 6h38'27" 25d45'27"
Robin's Lyte
Vir 13h26'10" -0d50'33"
Robin's Mark
Aur 5h49'33" 49d37'23"
Robin's Star
Per 2h59'16" 57d4'58"
Robin's Star
Peg 23h26'36" 29d10'4"
Robin's Star
Tau 4h19'54" 6d41'13"
Robin's Star
Cap 21h2'31" -19d30'11"
Robin's Star
Dra 16h42'16" 59d34'2"
Robin's Twinkle
Sco 16h14'23" -17d18'40"
Robinson 5
Umi 19h20'3" 87d15'57"
Robinson Elementary School
Ori 5h33'41" 1d11'0"
Robinson's Love
Ori 6h0'18" 18d42'0"
Robinson's Ruby
Cyg 19h55'19" 40d30'28"
Robjob
Lyn 9h16'46" 40d31'18"
Robleigh
Peg 23h24'44" 10d12'20"
Roblena Daphne Tate
Cap 21h25'53" -16d5'51"
ROBO
Tri 2h21'59" 36d36'23"
roboticwillyroberts
Sco 16h8'17" -11d48'53"
Robroy Centennial Star
Ari 2h40'39" 23d21'33"
Robroy Richard Orr "My Soulmate"
Ori 5h50'47" 6d14'0"
Rob's Astellas Star
Leo 9h46'35" 27d38'21"
Rob's & Brandi's Lovelight
Crb 15h40'34" 35d54'32"
Rob's Drop of Jupiter
Lyn 8h11'57" 44d40'6"
Rob's Girl
Lyn 7h30'15" 57d12'11"
Rob's Laurie
Ori 5h29'26" 1d16'39"
Rob's Place in the Stars
Ori 6h14'41" -0d57'56"
Rob's Star
Ori 5h32'11" 5d49'57"
Rob's Star
Cyg 20h45'57" 51d55'41"
Rob's *Star* Home
Cyg 19h35'39" 36d19'37"
Rob's Star - One in a Million
Per 3h40'20" 42d5'28"
Robson
Tau 3h35'20" 1d8'43"
Robstar
Cru 12h27'54" -58d43'53"
Roby
Uma 14h23'50" 55d26'15"
Roby
Cyg 19h53'32" 35d7'21"
Roby
Boo 14h31'13" 38d51'13"
Roby Oliver David DeGroat
Her 17h0'14" 28d43'32"
Roby, Du Sourire Léger
Sgr 19h50'23" -17d19'43"
Robyn
Vir 13h59'13" -0d53'12"
Robyn
Gem 6h37'3" 27d2'37"
Robyn
Tau 5h17'2" 26d9'20"
Robyn
Psc 1h20'11" 24d0'5"
Robyn
Gem 7h35'18" 20d39'23"
Robyn
Vir 12h31'54" 3d53'25"
Robyn
Ari 2h34'46" 18d24'19"
Robyn
And 2h30'6" 43d32'20"
Robyn
Cnc 8h33'31" 30d54'12"
Robyn 1569
Psc 1h18'29" 16d26'56"
Robyn Alice Bronner
Ari 2h32'34" 26d21'12"
Robyn and Madison (my two stars)
Tau 4h27'11" 17d4'2"
Robyn Ann
And 23h18'55" 44d45'38"
Robyn Anne Turner - A True Angel.
Aqr 22h47'18" -14d57'27"
Robyn Ashley Hesse
And 0h42'59" 38d5'8"
Robyn August
And 0h15'4" 29d28'40"
Robyn Aurelia Gabrielle 4.6.03
Ari 2h34'37" 27d31'53"
Robyn Badolato
Tau 5h49'34" 17d7'35"
Robyn Batson
Cra 18h39'36" -40d4'36"
Robyn Blair Fisher
Ori 6h12'48" 8d15'27"
Robyn Blyth
Mon 7h31'53" -0d31'16"
Robyn Buhr
Cas 0h45'0" 67d15'29"
Robyn Careen Clark
Aql 19h24'17" 10d49'47"
Robyn Charlton-Vickers
And 2h21'41" 46d11'51"
Robyn Clare
And 1h24'39" 36d30'20"
Robyn & Connie Paterson
Cyg 19h29'2" 31d14'34"
Robyn Craig
Gem 7h46'24" 32d2'7"
Robyn Cunningham
Lyn 7h59'27" 58d33'14"
Robyn Davis
Cas 23h44'50" 53d59'10"
Robyn Day
Ari 2h38'41" 24d55'53"
Robyn DeMonto
Uma 11h32'18" 52d15'40"
Robyn Denicola
Lyr 18h47'26" 31d35'26"
Robyn Diaz
Leo 11h18'10" 25d40'24"
Robyn E. Souza
Ori 5h38'49" -0d49'38"
Robyn Elaine Wilson
Cnc 8h17'4" 21d16'58"
Robyn Elizabeth Carson
Sco 17h22'40" -32d8'3"
Robyn Elizabeth Clark
And 2h22'4" 42d23'0"
Robyn Elizabeth Welch Mother Sister
And 1h3'6" 36d25'39"
Robyn Emma Bennett
And 23h11'57" 45d0'16"
Robyn English
And 0h4'54" 39d34'20"
Robyn Fisher Andrews
Uma 11h6'22" 57d4'20"
Robyn Gansmer
Gem 6h53'15" 26d33'6"
Robyn Glover Star Light Star Bright
Cra 18h47'17" -41d37'52"
Robyn Gorecki
Uma 10h56'11" 55d20'51"
Robyn Haye
Cas 23h50'7" 53d56'40"
Robyn Helena
And 1h52'9" 45d8'2"
Robyn Hiler
Lib 15h18'2" -8d2'5"
Robyn Hope
Aur 5h11'31" 38d12'11"
Robyn Hughes
Cru 12h18'4" -60d56'50"
Robyn in the Heavens
Cap 21h51'29" -19d38'19"
Robyn Inch
Boo 14h17'29" 47d8'4"
Robyn Jane Willie
Ori 6h25'10" 16d56'55"
Robyn Joan - "Grandma"
Ari 2h44'42" 12d11'31"
Robyn Jones
And 0h20'8" 26d39'3"
Robyn K
Uma 9h53'45" 70d42'50"
Robyn Kaye's Star
Uma 11h2'14" 53d35'8"
Robyn Lea Humara
Cas 0h26'1" 46d53'15"
Robyn Lee Erro
Cap 20h20'9" -15d32'53"
Robyn Leigh Woodrow
Ori 6h0'5" 7d9'18"
Robyn Leslie Lund
Lyr 18h34'10" 30d59'18"
Robyn Lynn
Cap 20h38'12" -26d21'6"
Robyn Lynn Villegas
Psc 1h25'56" 16d36'53"
Robyn M. Lahey
Lib 14h57'58" -16d27'6"
Robyn Mae
And 2h7'14" 45d56'41"
Robyn Margaret Hay
Cru 12h54'30" -56d44'7"
Robyn Marie
And 2h18'55" 38d23'23"
Robyn Marie Campbell
Sco 17h9'30" -44d57'57"
Robyn Marie Flake
Del 20h41'16" 13d50'3"
Robyn Marie Williams - Robyn's Star
And 23h38'50" 47d34'39"
Robyn McCoy Fields
Dra 18h26'54" 77d13'29"
Robyn McCrimmon
And 0h25'2" 33d32'0"
Robyn Michele Schroeder
Cas 23h13'25" 54d36'8"
Robyn Michelle
Uma 9h33'9" 43d14'59"
Robyn Michelle Carrick
Uma 11h32'5" 32d34'36"
Robyn Michelle Morrow
Cyg 20h2'12" 30d1'12"
Robyn Michelle Ward
Sco 16h18'30" -11d37'18"
Robyn Norman
Lib 15h36'40" -26d47'51"
Robyn Parrott
Cnc 8h19'51" 8d38'48"
Robyn Renee
Lyr 18h48'16" 43d33'48"
Robyn Renee 34
Gem 6h42'21" 15d15'29"
Robyn Renee Crosier
Cyg 20h21'38" 51d22'33"
Robyn Renee Verga
Gem 7h3'5" 16d25'53"
Robyn Rivera
Leo 10h27'18" 23d1'58"
Robyn Roback
Cyg 21h21'45" 39d30'44"
Robyn S. Race
Cas 1h56'4" 60d19'49"
Robyn Schwarzwaelder-Cook
Ori 6h2'29" -3d53'8"
Robyn Sciacca...fifty and fabulous!
Cas 3h12'9" 58d47'39"
Robyn Slattery McAreavy 50
Cnc 8h54'6" 24d58'44"
Robyn Smith - Guardian Love Star
Vir 12h48'48" 3d46'48"
Robyn Sprock
Crb 16h11'18" 38d19'2"
Robyn Starbright
And 0h33'13" 28d53'16"
Robyn Telles
Cyg 21h36'9" 39d41'44"
Robyn Turner
And 2h30'25" 48d57'5"
Robyn Watt
Sgr 20h1'4" -26d2'55"
Robyne Kathleen
Hya 9h20'57" -0d44'44"
Robynlira
Lyn 7h30'31" 52d41'28"
Robynn
Ori 6h17'59" 10d29'55"
Robyns Angel
Cas 0h48'24" 53d11'40"
Robyn's Glisten
Cas 0h28'42" 61d45'56"
Robyn's Radiance
And 0h15'31" 32d11'59"
Robyn's Star
And 2h17'4" 49d50'11"
Robyn's Star
And 0h39'36" 42d37'13"
Robyn's Wishing Star
And 23h16'2" 47d49'43"
Robynthia 1997
Ori 5h53'34" 18d38'12"
ROC 1
Cap 21h13'4" -25d24'34"
Roc & Lissa Hill "Star for Life"
Cyg 20h32'53" 42d42'33"
Roc Star
Aur 5h33'49" 42d54'16"
Roca Pony
Peg 22h16'45" 17d12'19"
Rocco
Ari 2h48'32" 16d33'22"
Rocco
Psc 0h42'38" 9d50'11"
Rocco
Cam 3h37'7" 55d43'30"
Rocco
Uma 9h46'10" 65d13'15"
Rocco
Sgr 18h49'12" -16d59'14"
Rocco Benne
Tau 4h24'3" 8d54'30"
Rocco Campanella
Psc 23h23'59" -2d54'27"
Rocco Cimino
Her 16h59'12" 33d44'20"
Rocco Del Mastro
Psc 0h21'48" 9d13'56"
Rocco Grey Taylor
Aql 19h46'51" 13d51'41"
Rocco Jack Tumberello
Aql 19h19'24" 11d26'32"
Rocco Joseph Favata, Jr.
Sgr 19h39'3" -16d47'13"
ROCCO - JULIE
Uma 11h50'39" 55d21'13"
Rocco Mazzeo
Her 16h10'19" 48d36'24"
Rocco Memole
Lib 15h4'34" -2d40'36"
Rocco Schiavello
Sco 16h7'53" -8d38'32"
Rocco Vitacca
Sco 16h50'53" -34d59'54"
Roch
Psc 0h51'42" 25d22'48"
Rocha
Lib 15h2'34" -1d1'27"
Roche
Lac 22h56'39" 50d7'22"
Rochel My Aussie Princess Nerenberg
Cru 12h5'46" -58d33'53"
Rochele Neer
Leo 10h52'40" 0d29'2"
Rochelle
Vir 12h8'55" 7d17'11"
Rochelle
Vir 13h36'36" 1d42'54"
Rochelle
Psc 2h1'48" 3d18'21"
Rochelle
Her 18h46'58" 19d38'10"
Rochelle A Mindala
Vir 12h49'21" 5d46'28"
Rochelle Adele 869
Sgr 18h11'48" -18d16'34"
Rochelle Ann Dutchover
Leo 10h52'36" 19d40'31"
Rochelle Ann Falkenberg
Leo 11h13'44" 9d44'17"
Rochelle Cleal MVSFB
Ori 5h59'9" -1d29'15"
Rochelle E. Sweet
And 0h47'45" 27d6'37"
Rochelle Giovinco
Lyn 7h52'26" 50d59'6"
Rochelle Graveline
Crb 15h29'51" 32d2'7"
Rochelle Hoffman
Cyg 19h33'37" 52d59'6"
Rochelle Irene
Lmi 10h46'3" 30d33'25"
Rochelle Katelyn Traina
Cyg 21h49'41" 41d58'22"
Rochelle LeeAnn Reno
Mon 6h43'45" 9d26'45"

Rochelle Leslie Litke
Leo 11h24'48" 17d27'23"
Rochelle Marie Perucca
Crb 15h31'59" 31d10'59"
Rochelle McElhiney
Sgr 18h28'6" -16d22'48"
Rochelle Nakanishi
Cet 1h23'50" -0d10'10"
Rochelle Phillips
Cnc 9h12'10" 21d37'14"
Rochelle Ramet
And 0h53'25" 42d18'40"
Rochelle (Rocky) Walter Hill
Her 17h58'54" 24d4'24"
Rochelle Smith
Cam 3h26'12" 59d29'50"
Rochelle Tuttle
Lyn 8h10'16" 36d12'12"
Rochelle's Star
Cyg 19h58'17" 37d53'12"
Rochellie Aime Letellier
Sgr 18h47'38" -21d52'58"
Rochette Sharie
Lib 15h3'20" -12d8'33"
Rochy
Vir 14h10'54" -12d53'50"
Rocio
Uma 12h37'11" 56d2'50"
Rocio
Cas 3h15'41" 58d58'18"
Rocio
Gem 6h46'43" 34d1'19"
Rocio
Aur 6h12'10" 31d19'22"
Rocio
Aqr 22h8'8" 0d23'22"
Rocio Gamez Rodriguez
Vir 12h16'6" 4d39'12"
Rocío González Segura
Lmi 9h39'48" 36d23'54"
Rocio Jauregui
Cas 2h8'55" 62d53'30"
Rocío la amada de Paulino
Ari 3h6'35" 28d37'47"
Rocio Lopez
Cap 20h57'32" -25d50'36"
Rocio Martinez Jansen
Uma 9h8'15" 60d1'58"
Rocio Mendoza
Leo 10h56'25" 7d7'2"
Rocio Pareja
Ori 5h49'32" 0d39'54"
Rocio Salvadores Osuna
Vir 13h11'3" 5d17'44"
Rocio Tobias
Dra 18h28'23" 50d39'30"
Rock
Her 18h40'30" 17d16'44"
Rock
Uma 11h36'51" 62d49'28"
Rock
Cap 20h29'42" -10d39'44"
Rock and Nadia O'Neal
Aql 20h8'10" 6d4'24"
Rock Angel 23
Umi 16h27'44" 81d56'46"
Rock Boss
Dra 19h32'2" 61d51'39"
Rock Dog
Uma 11h55'5" 28d47'38"
Rock L. Zahm
Cap 20h46'37" -17d52'32"
Rock Star
Sgr 19h54'49" -31d28'17"
Rock Star by Benjamin Invelito
Ori 5h42'21" -2d15'18"
Rock Star Melanie McElroy
Lib 15h49'43" -12d34'25"
Rock $tar
Uma 11h52'20" 55d38'29"
ROCKANDQUEENMARY
Uma 9h34'32" 60d48'40"
Rockee
Lib 14h51'54" -0d45'34"
Rockefeller Oreo Fitch Byrd 8/6/02
Aql 19h31'32" 12d47'36"
Rocket
Peg 22h39'17" 5d24'13"
Rocket
Uma 11h3'17" 42d38'50"
Rocket
Psc 1h50'15" 2d56'25"
Rocket Man Don
Dra 16h53'52" 63d23'54"
"Rocketman" Victor Crawford, Jr.
Per 2h44'33" 53d35'48"
Rockets Red Blair
Per 3h52'18" 42d51'30"
Rockford's Promise
Uma 10h45'26" 60d41'30"
Rockie Heggie
Uma 11h9'6" 64d2'52"
Rockin 9
Tau 5h6'53" 17d29'24"
Rockin' Rob
Her 17h17'2" 33d58'42"
Rockin' Steve
Lyr 18h45'27" 39d16'54"
RoCkiNdUcKie84
Lib 15h4'19" -12d27'43"

Rocking C.A.W
Aql 18h44'27" -0d7'10"
Rockley Plum Tree
Peg 23h20'19" 34d9'29"
Rockn-Roll "10-11-1957"
Lyn 7h29'8" 44d30'41"
Rock's Star
Cru 12h44'45" -56d32'53"
ROCKSOFFT
Aqr 23h0'25" -8d44'3"
Rockstar
Uma 10h20'48" 65d8'31"
Rockstar
Uma 12h37'54" 58d18'23"
RockStar 60
Psc 1h15'56" 13d22'44"
Rockstar KARA Kuszmar
Ari 1h49'12" 23d13'22"
Rockwell
Her 16h56'9" 22d31'24"
Rocky
Leo 10h53'42" 18d51'5"
Rocky
Vir 11h54'17" 2d21'13"
Rocky!
Psc 0h25'5" 21d20'2"
rocky
Aur 5h36'28" 41d54'12"
Rocky
Per 4h20'47" 40d10'43"
Rocky
Uma 12h2'47" 61d19'20"
"Rocky"
Uma 10h22'43" 63d33'0"
Rocky
Dra 18h19'43" 76d42'7"
Rocky
Cma 7h12'8" -22d30'3"
Rocky & Amy Our love shines forever
Cyg 20h30'27" 38d17'16"
Rocky and Mary Diilio
Uma 12h39'23" 53d23'29"
Rocky and Rebekah
Psc 0h17'51" 21d30'29"
Rocky and Twink Brown
Ori 5h29'55" 7d10'46"
Rocky Berman
Cep 23h15'15" 73d4'42"
Rocky Dude Boggs
Lyn 7h25'17" 47d48'14"
Rocky Flanders
Lyn 8h36'4" 44d25'18"
Rocky & Joyce
Cyg 19h39'17" 39d1'51"
Rocky Landisi
Leo 11h35'47" 19d55'42"
Rocky Lane Parr
Cap 20h23'8" -12d6'53"
Rocky Martin
Ori 5h29'44" 2d0'38"
Rocky Robilotto
Per 2h39'1" 57d8'46"
Rocky Rutenberg
Cnc 8h39'59" 31d15'40"
Rocky-Lukey-Gracey
Cru 12h21'14" -57d51'41"
Rockyn Robyn
And 1h26'23" 44d38'48"
Rocky's Star
Aqr 22h49'51" -5d24'58"
Rocky-Sam
Lyn 7h37'40" 36d29'37"
Rocudens
Mon 6h48'9" 7d54'41"
Rod
Aql 19h18'36" 6d49'39"
Rod
Per 2h22'33" 55d13'44"
Rod
Per 3h9'46" 55d20'27"
Rod Akers
Ori 6h17'26" 7d6'22"
Rod And Alyssa Underwood
Cyg 21h27'30" 51d42'6"
Rod and Jenny Gordon
Gem 7h24'24" 26d15'32"
Rod Coaltrain Swedburg
Uma 8h23'9" 60d46'31"
Rod G. Meadows
Lyn 7h11'52" 58d22'3"
Rod & Ginny Tisdale
Cyg 21h12'30" 38d34'25"
Rod McCarthy
Tau 4h35'55" 16d9'40"
Rod Nimmons
Per 2h56'10" 55d57'24"
Rod O'Steel aka Rob Looney
Per 3h56'12" 52d17'37"
Rod & Peg Zorn
Cyg 21h13'12" 43d23'22"
Rod Stephenson
Aur 6h55'44" 37d32'55"
Rod Welling Our Shining Star
Tau 3h30'11" 26d7'3"
Roda Johnson
Lyn 7h57'9" 57d49'24"
Rodan Ferguson
Umi 16h24'5" 74d8'39"
Rodare
Ari 3h22'37" 29d5'49"

Roddar
Aqr 22h17'25" 1d23'6"
Roddy Capps
Leo 10h16'56" 23d25'44"
"Roddy" Rodney R Baca
Boo 14h35'2" 36d24'14"
Roddy Sammon
Per 3h49'9" 32d48'46"
Rode to Stardom- John E. Rode
Psc 0h4'57" 8d35'3"
Rodella Dove Campbell
And 23h49'41" 48d28'52"
rodéo.tour.
Uma 9h44'8" 55d40'58"
Roderic O. Ballance
Boo 15h8'44" 34d45'41"
Roderica Mones Banana
Cap 21h41'51" -14d53'23"
Roderick Carl Diamond
Leo 11h26'27" 13d30'43"
Roderick D. Luckie
Her 16h33'6" 13d38'34"
Roderick Hennessy
Aqr 21h27'58" -2d15'4"
Roderick J. Kotylo
Cep 21h25'24" 66d25'6"
Roderick K. Gaines
Cap 21h6'41" -25d17'32"
Roderick Macrae III
Per 3h9'29" 56d4'45"
Roderick McGeoch - Share the Spirit
Cru 12h31'39" -60d22'25"
Rodge n Nikki Townsend
Cyg 21h18'58" 53d5'24"
Rodger D Toliver
Tau 3h38'48" 22d39'22"
Rodger Haun
Sco 16h12'46" -24d25'39"
Rodger M. Mitchell
Ari 3h27'2" 26d29'8"
Rodger -N- Billie Jo Forever
Tau 4h29'4" 20d45'52"
Rodger R. Kneip & Robert H. Kinley
Uma 12h30'18" 56d45'59"
Rodger S.Hoyt
Uma 9h27'10" 63d20'18"
Rodger Stephen Thomas
Cru 12h38'24" -64d4'50"
Rodger's Star
Aqr 21h26'32" -0d13'55"
Rodica
Leo 11h1'45" 0d39'20"
Rodica Irimia
Sgr 18h13'22" -28d5'58"
Rodjer Blue
Her 17h33'1" 34d59'53"
Rodman Loch Dewitt-Rogers
Leo 11h21'32" 16d13'0"
RodNBarbWilliams
Cyg 20h18'44" 55d57'21"
Rodney
Aqr 21h15'27" -13d12'45"
Rodney Alexander Graden
Per 3h5'57" 54d0'7"
Rodney and Caroline's Dream
Uma 10h56'47" 40d29'55"
Rodney and Heather Borders 8-23-03
Cyg 21h5'56" 46d59'26"
Rodney B. Stoops, Jr.
Gem 7h4'0" 32d49'9"
Rodney "Babe" Mooney
Cep 22h13'10" 56d35'10"
Rodney Baber
Per 3h11'24" 45d54'35"
Rodney Batchelor
Col 5h43'51" -42d9'30"
Rodney Blair Mack
Sco 17h49'24" -41d38'44"
Rodney C. Jung, M.D.
Lib 15h45'3" -18d25'48"
Rodney Carol Coster
Psc 23h13'11" 0d47'37"
Rodney Carr Sutphin
Cyg 21h41'41" 42d10'7"
Rodney & Charlotte Saunders' Star
Cyg 20h13'27" 41d54'0"
Rodney Daniel Scott
Her 17h11'23" 32d26'20"
Rodney Downs, Sr.
Tau 4h9'53" 19d56'42"
Rodney Eugene Harris
Gem 7h41'25" 31d4'55"
Rodney Glen Hiner
Aur 5h44'49" 38d22'30"
Rodney Gordon Campbell Humphrey
Vir 14h7'26" -15d52'49"
Rodney Hackney
Ari 2h37'5" 13d16'56"
Rodney James
Her 17h31'47" 44d43'13"
Rodney Jason William Buhler
Cyg 21h13'14" 51d48'1"
Rodney Kees
Cnc 8h52'40" 24d34'3"

Rodney Lee Berriker
Uma 10h53'25" 41d35'39"
Rodney Lee Johnson PD HotRod 121453
Her 17h22'32" 38d17'29"
Rodney Lewis Roadifer
Cep 21h44'46" 57d22'26"
Rodney Lynn
Her 18h48'26" 19d17'30"
Rodney Marvin Kubler
Per 3h47'58" 44d2'8"
Rodney Mayfield
Lac 22h18'36" 54d11'53"
Rodney Paul Luckey
Ori 6h1'14" 20d46'41"
Rodney Robert Ray
Uma 9h17'4" 62d45'19"
Rodney Shane Fillingame
Her 17h27'47" 17d36'57"
Rodney Snyder
Uma 8h21'22" 60d36'23"
Rodney Steven Krawczynski
Crb 15h42'39" 28d24'35"
Rodney Ted Evans II
Boo 15h24'41" 49d34'10"
Rodney Wade Butler
Per 3h15'33" 53d14'58"
Rodney's Wish
Per 2h49'8" 53d37'9"
Rododendren (Michelle Lynn Britton)
Leo 10h38'15" 10d34'15"
Rodoleau
Per 4h32'53" 31d47'36"
Rodolfo Alexandro Gomez IV
Lyn 8h3'11" 37d49'29"
Rodolfo & Aracely M Loqui-Canzanese
Vel 9h12'5" -44d23'44"
Rodolfo C. Soriano
Sgr 19h1'27" -32d34'14"
Rodolfo Garcia Triviso
Cyg 20h49'24" 30d20'17"
Rodolfo Gentile
Cas 0h21'2" 55d26'9"
RODOLFO GONZALEZ, SR.
Aql 19h13'43" -1d39'0"
Rodolfo Javier Llamas
Tau 5h13'1" 24d21'18"
Rodolfo "Lefty" Sarabia
Psc 0h41'15" 13d31'10"
Rodolfo Rodriguez
Her 17h14'38" 14d41'50"
Rodolfo Sacchi
Ori 6h19'45" 8d6'18"
Rodolphe et Annabelle
Del 20h46'22" 13d50'42"
Rodoula Pitrou
Leo 9h31'2" 23d21'29"
Rodric J. DiPietro
Ori 5h45'40" 4d10'3"
Rodrick Gregory
Sgr 19h50'3" -28d19'32"
Rodrick R Kaiser
Her 16h46'7" 37d33'47"
Rodrigo Andres Valenzuela Dastres
Psc 23h7'45" 5d20'30"
Rodrigo Damasceno
Psc 1h29'59" 6d10'17"
Rodrigo Fernandez Esquivias
Leo 10h13'20" 15d22'3"
Rodrigo Liberato
Lib 15h45'36" -7d0'47"
Rodrigo Rios
And 2h37'40" 46d12'47"
RodrigoeJúlia Amor Infinito 270202
Psc 23h43'24" 1d21'54"
Rodrigues
Uma 11h55'4" 28d35'41"
Rod's Star
Cru 12h56'22" -57d9'46"
Rody N Mushy
Aql 19h28'54" 7d25'40"
ROE
Ari 3h24'22" 26d21'50"
Roe
Ari 1h53'15" 22d52'30"
ROE 17
Per 3h10'15" 54d59'46"
Roe East
Uma 12h30'3" 61d40'36"
Roe Hartmann
Leo 9h48'43" 28d22'49"
Roeetha
Psc 0h34'31" 8d26'46"
Roei Kashi
Leo 11h20'47" 21d52'31"
Roel
Leo 10h47'25" 13d3'37"
Roelle
Uma 10h25'28" 49d36'51"
Roena
Leo 9h59'5" 7d53'5"
Roe's Star
Tau 5h17'40" 27d40'42"
Roesch Xaver
Uma 13h36'30" 57d46'11"
ROG-10
Uma 9h54'13" 58d45'35"

Rogelio
Cet 2h49'1" 6d37'13"
Rogelio Consuegra
Cap 21h25'40" -23d51'12"
Rogelio Roberto Maese
Ari 2h42'47" 13d4'24"
Roger
Sgr 18h41'55" -29d52'51"
Roger
Psc 0h3'11" 3d54'6"
Roger
Dra 15h2'15" 58d14'23"
Roger
Lep 5h33'47" -14d0'33"
Roger
Ori 5h30'23" -1d39'56"
Roger A. and Celia P. Bills
Uma 11h34'54" 52d53'31"
Roger A. Craig
Lib 14h48'52" -2d46'4"
Roger A Meloun
Aur 5h26'1" 45d19'0"
Roger A. Strunk
Aql 20h15'6" 4d42'12"
Roger Alexander Meier
Cep 22h9'50" 67d43'13"
Roger Allan White
Uma 9h3'45" 70d51'42"
Roger Allen Lucas
Her 17h9'43" 34d0'39"
Roger and Cindi Silverstein
Aqr 22h26'53" -2d5'6"
Roger and Drue Kendell
Cyg 19h51'46" 38d19'32"
Roger and Elaine's Forever Star
Sge 20h4'42" 18d42'3"
Roger and Julie Clapp
Ori 6h4'46" 14d3'24"
Roger and Linda Witten
Gem 6h41'9" 20d53'31"
Roger and Unmi
Her 16h50'54" 31d7'53"
Roger Arthur Erekson
Ori 4h49'51" 2d37'50"
Roger Avery
Per 3h8'41" 55d17'29"
Roger B. Pierson Jr.
Aql 19h32'35" 11d29'41"
Roger Barnhill
Per 4h19'42" 37d20'26"
Roger Beak Stern
Cap 20h50'45" -20d13'6"
Roger Beaulieu
Per 3h4'49" 50d24'48"
Roger Besu
Lyn 7h50'5" 59d31'12"
Roger Blaise Bonnett
Per 3h3'16" 45d53'58"
Roger Borowski - K9RB
Oph 16h44'47" -0d53'17"
Roger Bryan Kerr
Cnc 9h17'37" 32d21'56"
Roger C. Frankenfield
Cnc 8h35'39" 21d28'51"
Roger Candanoza, Sr.
Leo 11h37'5" 25d46'26"
Roger Cassandra
Gem 6h51'25" 23d21'58"
Roger Chambers Star of Leadership
Per 3h7'23" 51d53'7"
Roger & Colleen Gorton
Sco 17h36'46" -45d21'39"
Roger "Corky" Mundy
Lib 15h40'43" -22d0'38"
Roger Dale Howard
Leo 10h15'6" 26d57'50"
Roger David
Psc 0h7'45" 7d8'15"
Roger David Tucker
Uma 10h47'28" 63d44'8"
Roger Dean Atkins
Vir 13h34'4" 3d16'40"
Roger Dechene
Cyg 21h35'3" 36d56'35"
Roger & Dee Carter
Sge 19h56'36" 18d59'43"
Roger Delgado "Koki"
Lyn 7h36'29" 52d38'37"
Roger Donald Hancock
Cep 22h53'42" 70d36'33"
Roger & Dorothy Laghezza
Cyg 21h18'1" 50d47'31"
Roger Dowiat
Cas 0h29'57" 52d26'51"
Roger Driscoll
Sgr 18h36'16" -23d58'13"
Roger Duong
Per 3h36'27" 46d33'20"
Roger E Gray
Cas 23h38'32" 55d26'36"
Roger E. Ruvalcaba and Baby Girl
Her 16h56'27" 33d57'55"
Roger Earl Smith, Grandpa of Nicole
Her 17h30'38" 45d59'23"
Roger & Elizabeth - Ruby Wedding
Eri 3h56'7" -38d6'16"
Roger Erixon
Uma 9h42'46" 60d52'58"
Roger F. Greenslet
Cnc 8h43'4" 19d27'9"

Roger Francis Dwyer
Cep 22h28'2" 73d30'57"
Roger G. C. Meloun
Vir 12h33'2" -2d39'7"
Roger G. Halleran
Ori 4h47'33" -0d36'7"
Roger Gadient
Cyg 20h55'40" 46d54'32"
Roger Gaudreault
Uma 12h1'57" 45d34'55"
Roger Gibbs
Cep 21h30'24" 68d14'0"
Roger Goldbeck
Cnc 8h36'58" 18d25'42"
Roger Goodwin
Ori 5h29'20" 7d54'45"
Roger Goose
Lib 14h57'11" -5d44'26"
Roger Grant Nalley
Aur 6h29'33" 34d13'5"
Roger H. Peterson Sr.
Sgr 18h15'45" -20d44'36"
Roger Haakinson
Cap 21h43'5" -10d11'31"
Roger Hagan
Her 18h30'55" 20d59'53"
Roger Hall From Frederick Hall
Per 3h39'55" 47d10'52"
Roger Hansen
Ori 5h30'44" 1d22'51"
Roger Harrigan
Cep 22h11'8" 54d8'34"
Roger Henderson
Uma 11h28'11" 30d13'26"
Roger Hill
Cam 3h53'19" 67d21'43"
Roger Houtman
Dra 17h59'42" 63d47'33"
Roger I. Hosford
Tau 5h45'11" 22d9'7"
Roger Ian Relfe
Psc 0h54'16" 10d55'41"
Roger J. Cummings
Lib 15h8'11" -15d55'15"
Roger James Chingitchcook Kirkman
Sco 17h24'6" -42d12'0"
Roger James Dwyer, Jr.
Aql 19h5'38" 3d20'3"
Roger James Kirkman
Aql 19h38'36" -10d40'8"
Roger James Van Wyngaarden
Per 3h48'44" 49d39'39"
Roger & Jane Burgett
Sco 15h53'21" -25d58'33"
Roger & Joan Pinkham
Uma 12h17'53" 62d39'15"
Roger Joseph Pierre
Cep 0h6'55" 71d12'37"
Roger Kachel
Ori 5h18'54" 6d19'58"
Roger Keith Tobin
Tau 5h36'54" 25d21'0"
Roger Kirts
Lyn 7h26'6" 44d22'9"
Roger Knox Aletras Jr.
Cap 20h43'35" -14d47'44"
Roger L. Heintz
Psc 23h54'6" 5d14'6"
Roger L. Kringen
Psc 1h14'22" 23d4'3"
Roger L Leblanc Sylvain
Uma 11h28'4" 58d24'42"
Roger Lamar
Her 17h20'26" 13d39'40"
Roger Landry
Cyg 19h54'37" 31d16'54"
Roger Lee
Col 6h16'44" -36d21'35"
Roger Lee Ash
Cnc 8h58'59" 31d18'52"
Roger Lee Kile, Jr.
Cep 21h24'7" 61d0'17"
Roger Lee Knight
Per 2h48'47" 53d30'21"
Roger Lee Mitchell
Cep 22h43'53" 74d34'29"
Roger Lee Rainey
Her 16h36'43" 36d16'4"
Roger Leo Vallee
Aur 5h38'53" 40d48'45"
Roger Leon McClure
Tau 5h3'51" 25d47'34"
Roger & Loretta LeDuc Sr.
Equ 21h13'40" 7d35'23"
Roger Madoff
Per 3h7'52" 53d26'20"
Roger Maestas
Uma 11h44'34" 41d7'18"
Roger Marcelin
Gem 6h50'18" 25d39'20"
Roger Marsh Wagner MD
Aqr 22h1'47" -22d34'7"
Roger & Mary Hare
Gem 7h42'13" 33d17'9"
Roger McAdams
Aur 5h24'13" 32d15'17"
Roger McIntyre
Sgr 18h58'16" -36d26'32"
Roger Michael Sisler
Her 18h37'43" 21d16'4"
Roger Michael Striegel
Vir 13h40'44" 7d7'42"

Roger Millikan
Ori 5h23'9" 2d11'33"
Roger Moore
Her 16h45'12" 43d0'33"
Roger Morello
Ari 2h34'46" 29d42'20"
Roger Neil Ennis
Her 18h8'18" 27d44'38"
Roger Neil Lafeen
Per 3h33'15" 49d48'31"
Roger Neuberg
Uma 8h53'5" 62d53'53"
Roger Nowak
Tau 5h46'51" 14d15'13"
Roger Nye Lockwood
Cep 21h50'21" 69d47'34"
Roger O. Craig
Aql 19h6'9" 4d18'34"
Roger Passaro
Lac 22h48'15" 52d3'16"
Roger Patrick Schmuck
Ari 2h43'6" 26d40'7"
Roger Pellerin
Uma 11h26'16" 60d26'34"
Roger Pessidous
Peg 22h21'49" 21d29'14"
Roger Pritt
Boo 14h30'4" 40d59'11"
Roger Reynolds Mee
Uma 11h31'53" 39d19'43"
Roger Riggle III
Aqr 22h18'41" -17d52'34"
Roger Rohlfing
Leo 10h52'37" -3d20'4"
Roger S. Lerner ~ 4 Star 5 Star
Per 2h44'7" 56d26'39"
Roger S. Wilder
Gem 7h50'26" 16d16'35"
Roger & Sandy Wilkin
Uma 13h41'30" 59d22'27"
Roger Savoy
Cep 1h33'31" 81d45'12"
Roger Schürch min Schatz
Lyn 8h10'14" 51d26'32"
Roger Selby
Lyn 7h46'58" 56d46'14"
Roger Sierra El REY DE LOS HUITRE
Lib 14h48'53" -10d43'0"
Roger Sierra Luna
Gem 6h44'6" 12d18'37"
Roger Simard dit Bonsoir
Umi 15h35'16" 79d17'25"
Roger Smith
Dra 18h31'32" 72d37'38"
Roger Solem
Ori 6h6'2" -0d16'46"
Roger Sorenson Pyne
Tau 3h28'38" 2d31'34"
Roger St. Martin, Sr.
Cnc 8h13'9" 25d16'13"
Roger Stephens
Per 2h39'51" 55d40'19"
Roger Stonebank
Aur 5h12'44" 42d8'18"
Roger "Synchro" Sparks
Sco 17h49'2" -40d25'12"
Roger T Kelley With Love Son Steve
Cap 21h24'1" -22d0'52"
Roger T. Rashid
Uma 8h32'38" 66d28'45"
Roger & Terry Gorman-Heaven Eternal
Cyg 19h48'24" 35d58'47"
Roger Thomas Watrous Jr
Aql 20h5'20" 6d31'12"
Roger Thöny
Tri 2h31'59" 28d22'56"
Roger " Tiger "
Tau 5h18'42" 28d35'4"
Roger Tinkham
Dra 17h33'45" 65d59'29"
Roger W. Beuerman
Lib 15h18'47" -19d16'27"
Roger Wallimann
Uma 10h42'33" 69d9'45"
Roger Walter, 07.05.1976
Uma 12h0'49" 47d11'27"
Roger Wayne Shadwell
Sco 17h17'47" -43d45'46"
Roger Wells
Ori 6h11'49" 18d6'50"
Roger Williams
Equ 21h9'0" 11d31'30"
Roger Wm Esquerra II
Uma 8h24'4" 71d10'10"
Roger, my eternal love 07-14-05
Her 17h10'34" 34d13'9"
Rogerdog
Leo 10h33'42" 22d7'26"
Roger—God's Warrior Psalm 18:32
Ori 6h9'18" 16d26'33"
Roger-Morena-Noël
Uma 9h23'11" 47d37'50"
Rogers Ambres
Ori 5h35'18" 1d51'25"
Roger's Dreams-Forever and Always
Tau 4h12'36" 7d11'31"
Roger's Eternally
Ori 5h52'21" 7d17'49"

Rogers Lee Mooney
Gem 6h53'53" 31d0'12"
Roger's Star
Tau 4h31'47" 8d58'2"
Roget
Sgr 19h18'38" -18d23'8"
Roget and Doug
Uma 11h31'30" 29d41'35"
Roghieh&Mostafa
Asadzadeh AzariMajd
Pho 1h5'32" -48d37'11"
ROGJAQNATMUELLER-HEAVENSCENT
And 1h54'56" 45d12'40"
Rogozin Pavel Pavlovich
Cep 0h18'36" 71d17'53"
Rogue's Guardian
Lib 14h56'43" -2d32'0"
Rohan
Psc 0h32'29" 6d18'15"
Rohan
Ari 2h32'39" 21d20'22"
Rohan and Maura
Del 20h19'23" 9d23'56"
Rohan David Marion
Sgr 18h18'11" -34d13'50"
Rohan Grace
And 23h38'5" 49d0'20"
Rohan Hajisattri
Uma 11h55'55" 55d40'51"
Rohan Joel Burman
Leo 11h16'5" 27d8'50"
Rohan Joshua Charles
Gem 6h42'27" 28d50'21"
Rohan William James
Talbot
Dra 10h35'35" 78d22'32"
Rohina
Gem 6h49'2" 15d44'47"
R.O.I.
Umi 14h30'46" 77d25'51"
Roia Atchekzai
And 1h23'26" 45d29'23"
ROIANNE M. McDANIEL
Lib 15h12'10" -6d11'44"
Roii Gurevitz
Uma 10h47'55" 62d45'16"
Róise Sadhbh Maxwell
And 2h37'48" 48d51'46"
Roisin
Umi 16h42'33" 76d56'48"
Roisin Ann Deak
And 23h24'59" 47d12'26"
Roisin Martina Scully
Lib 15h6'4" -1d59'24"
Roisin Murphy
Sgr 17h53'41" -29d57'48"
Roisin's Star
And 1h25'50" 49d34'18"
Roitzsch, Barthel
Uma 8h56'53" 47d41'41"
Rojan
Cam 4h6'7" 67d8'46"
Rojas Robert
Psc 0h55'48" 28d17'56"
Rojejoas
Peg 22h42'3" 23d6'7"
Ro-Joe Santamaria
Lmi 9h58'25" 36d13'43"
Rokaia Jasmine Mohamed
Sco 16h51'4" -45d16'49"
Roko Nosh Camaj 9-11-01
Umi 14h28'14" 73d41'1"
ROKSANA
Uma 11h55'55" 57d52'56"
Rolaboodiforever
Uma 11h48'37" 45d39'11"
Rolana
Cyg 20h6'19" 39d4'33"
roland
Cas 1h13'0" 50d35'50"
Roland
And 0h38'47" 26d7'43"
Roland
Ori 5h25'57" -3d45'58"
Roland Alfred Rudi
Dornbusch
Uma 11h3'49" 58d7'35"
Roland Allin
Uma 10h18'38" 47d41'49"
Roland and Susan Maher
Leo 9h42'2" 25d28'44"
Roland Arthur White (1940)
Cep 22h18'26" 62d42'57"
Roland Babayev
Lyn 7h34'54" 42d3'50"
Roland Becker
Ori 5h55'42" 11d30'59"
Roland & Bina
Peg 21h46'26" 25d28'30"
Roland C Scott IV
Her 17h56'39" 16d37'15"
Roland "Chräbeli"
Umi 15h43'53" 72d33'29"
Roland Deschain of Gilead
Her 17h10'37" 28d57'41"
Roland Dion
Aur 5h49'53" 44d44'11"
Roland Dwight Tanner
And 2h19'8" 38d21'43"
Roland / E
Uma 10h18'59" 42d29'10"
Roland E. Laprey
Cep 21h31'53" 64d41'51"

Roland Eugene Martin, Sr.
Uma 9h57'11" 69d59'58"
Roland Galvan
Vir 13h53'53" -5d38'21"
Roland Horend
Uma 12h44'30" 52d39'53"
ROLAND J. POZZI,
CSWOY
Cyg 21h19'26" 41d14'40"
Roland Jendros
Uma 9h25'13" 56d21'5"
Roland John Minogue
Ori 6h20'6" 13d57'26"
Roland Joseph Mombaur
Uma 11h59'6" 31d35'47"
Roland Kent Rutherford
Ari 3h6'49" 13d39'5"
Roland Kern
Cas 23h45'16" 55d20'4"
Roland Kirsch
Ori 5h56'39" 18d4'56"
Roland Kormann
And 1h54'27" 39d30'39"
Roland Lorenz
Uma 9h10'25" 56d51'37"
Roland Martin
Uma 12h5'58" 65d14'51"
Roland Mesidor
Cnv 13h55'48" 36d18'27"
Roland Minda
Uma 13h40'50" 58d12'42"
Roland Munoz
Aql 19h43'7" 7d14'43"
Roland Murmann
Uma 9h2'30" 47d37'55"
Roland - My Daddy -
Gillard
Cnc 8h40'16" 8d16'22"
Roland Neal Hanson
Lmi 10h25'50" 33d23'6"
Roland Olivier
Leo 11h33'38" 28d0'40"
Roland "Our Dad"
Per 4h21'40" 33d47'43"
Roland Pfeuffer
Ori 6h15'34" 5d48'6"
Roland "Poppy" Bennett
Umi 14h55'48" 71d29'17"
Roland R. Cloutier Jr.
Cnc 8h51'43" 30d27'25"
Roland Reist 28.07.1959
"Reistini"
Ori 5h31'52" -8d29'47"
Roland "Ro-Ro" Eckstein III
Per 3h37'37" 43d54'54"
Roland Speer
Uma 9h5'32" 58d26'37"
Roland Trainor
Cma 7h12'56" -14d30'22"
Roland V. Leroux
Aqr 21h41'46" -6d6'19"
Roland Weidisch
Uma 10h11'23" 64d34'18"
Roland Wesley Burnett
Larsen
Psc 1h9'42" 27d39'12"
Roland Woessner
Uma 12h8'37" 58d26'29"
Roland Zöller
Uma 11h34'37" 60d48'15"
Rolanda Dominguez
Cyg 21h41'44" 48d19'8"
Rolande Boucher
Ori 6h19'52" 14d29'46"
Rolando (RP) Pecos
Vir 12h34'42" -2d24'40"
Rolandrea "2-7-83"
Del 20h48'56" 5d49'50"
Rolf B612
Boo 13h47'19" 23d54'55"
Rolf Blanke
Uma 9h21'1" 63d40'40"
Rolf Brandenberger
Umi 16h23'17" 71d58'10"
Rolf Büteführ
Ori 6h24'15" 14d13'14"
Rolf F. Hänni
Her 18h16'19" 27d6'48"
Rolf Graf von Hardenberg
Uma 9h56'4" 69d20'22"
Rolf Grob
Dra 17h36'8" 59d11'54"
Rolf Gruss
Tri 2h21'17" 32d50'51"
Rolf Hartung
Uma 11h52'37" 41d14'58"
Rolf Heinemann
Ori 6h19'55" 15d39'50"
Rolf Inge Bertil Bernklev
Lup 15h1'47" -43d42'54"
Rolf Kaiser
Uma 10h41'1" 59d21'30"
Rolf Meier
Cep 21h11'33" 66d54'53"
ROLF Michl
Tri 2h39'13" 35d35'31"
Rolf Nußbeck
Uma 10h14'44" 44d15'58"
Rolf & Pauline 72492
Sco 16h30'51" -27d37'29"
Rolf Peter Schätz
Ori 6h18'59" 8d31'22"
Rolf Ranik Halle
Cep 21h11'25" 58d23'31"

Rolf Rüedi
Cas 1h33'39" 59d54'39"
Rolf Schenk
Uma 11h50'5" 28d57'6"
Rolf Steiger
Cep 22h17'21" 66d24'15"
Rolf Sulger
Cyg 21h33'16" 49d39'2"
Rolf Zaugg
Ori 6h11'45" 10d6'49"
Rolland Rachel
Cnc 8h7'45" 10d43'0"
Rollande
Cas 23h47'36" 56d44'16"
Rollande & Gerald Brouard
Uma 10h22'49" 54d40'2"
Roller
Uma 8h33'9" 65d9'21"
Rollie
Uma 11h1'6" 60d58'54"
Rollie
Lyn 7h13'17" 52d56'55"
Rollie Douglass Reeves - #1
Vir 13h19'36" 11d17'25"
Rollin L. Culp
Her 17h14'59" 19d18'9"
Rollin R. Archibald
Cyg 19h27'48" 45d13'43"
Rollins
Boo 14h40'16" 36d42'33"
Rolly DeVore
Ori 5h55'2" 20d56'30"
ROLO
Cyg 21h57'25" 40d8'34"
Roloff, Götz
Uma 9h59'15" 53d0'32"
Rolonda Everly
Tau 5h14'55" 21d15'17"
Rolouan
Ori 6h7'31" 5d51'50"
Rolph Storto
Cep 23h30'27" 64d44'1"
Roly Berger Maximus
Leo 9h56'27" 13d55'12"
Roly & Natty
Tau 4h5'49" 6d34'0"
Roma
Ori 5h22'18" 13d47'35"
Roma
Cyg 21h52'56" 43d7'47"
Roma
Cas 0h55'9" 54d28'44"
Roma
Dra 19h11'17" 74d37'20"
Roma McArthur
And 23h13'53" 35d27'43"
Roma Renee Jadick
Cas 2h43'56" 58d2'29"
Romain
Aqr 23h30'47" -14d12'1"
Romain 1982
Uma 10h38'40" 41d15'42"
Romain Ducrocq
Peg 23h46'59" 15d42'12"
Romain et Chloé
Sgr 18h51'35" -15d52'40"
Romain et Eleonore =Amour
Cas 23h34'48" 51d9'18"
Romain Krummenacher
Her 18h23'1" 14d59'12"
Romain Pétillon
And 2h19'43" 47d24'7"
Roma-Mindy
Cap 21h53'3" -19d58'48"
Roman
Umi 17h1'56" 78d47'37"
Roman
Cnc 9h10'45" 17d55'9"
Roman
Cas 0h8'42" 53d44'34"
Roman
Boo 15h10'39" 49d41'25"
Roman
Uma 12h1'9" 31d4'58"
Roman
Tri 2h13'38" 31d53'27"
ROMAN - 30.09.2005
Ari 3h15'35" 18d51'44"
Roman and Julia Galloza
Umi 15h50'8" 77d24'13"
Roman and Ryley Baines
Aql 19h55'16" 0d0'39"
Roman and Tiffany
Per 3h48'49" 33d35'56"
Roman Augustus
Tau 5h42'38" 22d14'48"
Roman Bachmann
Eri 3h50'37" -36d8'40"
Roman Blaha
Per 4h28'42" 33d12'25"
Roman Casabona Toledo
Uma 10h57'16" 39d59'26"
Roman Christopher
Drumgoole
Her 17h45'40" 42d23'18"
Roman Dante' Castagna
Sgr 18h48'16" -23d43'57"
Roman Douglas Ware
Vir 12h25'18" -4d9'33"
Roman Ellis Michael Hill
Gem 7h14'11" 22d33'17"
Roman Gennadievich Liufa
Uma 11h27'39" 55d21'15"

Roman Inks
Per 4h8'34" 42d15'34"
Roman James Pennella
Her 18h44'4" 12d58'1"
Roman Julian Kalba
Greggs
Tau 5h54'24" 27d16'11"
Roman Karapogosyan
Ari 2h42'28" 27d47'12"
Roman Kedgley
Peg 23h13'21" 18d23'35"
Roman Kling
Dra 15h23'54" 56d9'33"
Roman M. Bell
Psc 0h40'28" 16d45'47"
Roman Marasco Kane
Gem 7h35'52" 32d55'53"
Roman Panczyszyn
Lib 15h25'58" -8d36'29"
Roman Patrick Salvati
Lyn 8h34'13" 46d25'21"
Roman Roy Horvath
Lib 15h13'26" -27d40'34"
Roman Saglio
Mic 20h30'55" -32d59'8"
Roman Sky
Aqr 23h7'34" -10d7'59"
Roman Treystman
Per 3h49'22" 38d14'2"
Roman & Valentina
Katsnelson
Ori 6h20'16" -0d42'54"
Roman Vicary
Uma 9h59'14" 67d7'44"
Roman Vincent Rocco
Uma 8h34'40" 71d3'22"
Roman Weiler
Uma 10h26'29" 49d54'26"
Roman Zurek
Uma 9h37'58" 53d38'4"
Romana Fischer
Uma 9h43'2" 50d28'7"
Romana Parisi
Cas 23h24'44" 58d3'58"
Romance
Lib 15h5'24" -7d14'31"
Romance
Lyr 19h3'59" 42d7'3"
Romane
Cas 0h58'40" 54d21'0"
Romane Juncas
And 0h28'29" 21d46'0"
Romane Latrasse
Cas 23h50'12" 58d56'47"
Romane Lefebvre
04.09.2003
Umi 16h59'34" 79d24'12"
Romania Hill
Sgr 20h13'55" -38d32'32"
Romanian Princess Maria
Panica
And 0h18'41" 24d35'34"
Romano
Peg 22h39'14" 26d54'25"
Romano Frediani
Aur 5h48'56" 33d50'27"
Romano Pasqual
Cnc 9h11'18" 17d1'54"
Romans 13:10 Coons (
Allie's Path)
Crb 15h41'33" 28d44'2"
Romanski, Martin
Uma 10h43'26" 46d21'44"
Romantical & Untamed
Peg 23h44'49" 12d4'24"
Romantically His
Cas 0h38'50" 55d51'36"
Romany Rose Sage
Lib 15h19'19" -9d40'25"
ROMA'S
Leo 11h27'38" 18d14'41"
Romayne J. Shepherd
Uma 11h48'24" 59d19'15"
Rome Alexander Robbins
Gem 6h33'2" 15d0'10"
Romel C. Castejon-Gomez
Ori 6h15'1" 15d5'17"
Romel Racoma Sanchez
Her 18h4'54" 33d0'13"
Romelle Jones
And 0h41'2" 39d16'28"
Romeo
Ori 6h16'17" 12d39'15"
Romeo
Cnc 8h17'7" 16d51'33"
Romeo
Leo 10h28'5" 17d32'44"
Romeo
Leo 9h33'39" 29d25'43"
Romeo
Umi 13h28'16" 73d46'19"
Romeo
Cma 6h44'33" -16d35'33"
Romeo and Juliet
Cyg 21h47'35" 44d6'13"
Romeo Cadieux
Tau 4h52'54" 22d3'31"
Romeo Devirgiliis
Ori 5h7'38" 4d38'54"
Romeo Martello
Lyn 7h59'24" 42d38'5"
Romeo Oscar Bravo Bravo
India Echo
Ori 5h54'21" 12d51'15"

Romeo Smith
Per 3h8'37" 39d25'36"
Romeo Tan
Sgr 19h38'55" -11d45'28"
ROMEO23
Uma 11h51'4" 61d35'30"
RomeoGumbi Island
Uma 11h18'34" 51d51'41"
Romeopapa
Psc 0h57'17" 7d11'22"
Romeo's Jammy Malbec
Cnc 8h38'55" 24d36'0"
Romeo's Light
Psc 1h28'52" 33d7'50"
Romeromar
Gem 6h49'44" 16d53'44"
romero-michel
Ori 5h22'27" -7d47'53"
Romie
Tra 16h13'43" -62d11'59"
Romie Elizabeth Wesson
Oph 17h18'5" -0d2'26"
Romilda Barbara
Lac 22h54'12" 38d35'50"
Romilda M. Coniglio
Cas 1h18'28" 60d54'58"
Romina Antonia
Coppolecchia
Uma 8h45'26" 55d4'53"
Romina Di Noto
Umi 13h11'10" 73d35'28"
Romina Ferrauti
And 1h19'56" 48d20'10"
Romina Julienne
Cyg 21h17'44" 45d54'4"
Romina kai Nikolas
Cnc 8h59'16" 13d31'56"
Romina La Fosso Amer
Mon 6h49'8" -2d35'35"
Romina Locatelli
Cas 2h2'44" 58d18'29"
Romina Mazzotti
Cam 5h52'15" 59d5'54"
Romina Paula Callerio
Psa 22h10'0" -32d49'47"
Rominka a Kubko- the Oxs
Cnc 8h25'57" 22d51'47"
Rommel Creasman
Umi 12h16'26" 88d43'27"
Rommel, my shining star
Love Renata
Gem 6h46'30" 23d50'25"
Rommell
Uma 10h0'43" 44d54'52"
Rommey
And 2h23'57" 48d53'38"
Romnia Suzan Bakkar
Cnc 8h32'42" 12d20'0"
Romolo Civile
Ori 5h23'54" -0d3'47"
Romolo Venturi
Cam 3h24'56" 64d50'40"
Romona Baldwin
Lib 14h34'21" -12d36'0"
Romona Stow
And 2h16'59" 46d47'50"
Romonique Edwards,
Tinker Bell
And 1h18'23" 42d19'29"
Romuald Gayet
Tau 4h18'0" 5d20'28"
Romuald Moulin
Peg 22h23'24" 21d6'47"
Romualdou et Karinou
Cnc 8h35'41" 10d0'16"
Romulus Stellare
Cap 21h34'28" -12d41'31"
Romy
Sgr 18h16'31" -21d34'20"
Romy
Dra 17h35'54" 67d28'43"
ROMY
Sco 17h16'37" -37d57'59"
Romy
Sco 16h42'11" -31d13'9"
ROMY 4-1-1984
Uma 10h33'51" 53d56'57"
Romy Eigenmann,
26.05.1976
Peg 22h38'32" 26d31'29"
Romy Zoe Steinlein
Uma 11h28'16" 30d15'39"
Ron
Per 3h55'28" 46d18'8"
Ron
Ori 5h47'26" 1d30'9"
Ron
Cep 22h8'35" 57d9'25"
Ron
Cma 7h22'1" -12d52'17"
Ron
Aqr 22h28'31" -1d13'34"
Ron 05
Uma 12h28'6" 54d8'56"
Ron A. Arnold
Aur 5h50'58" 54d14'21"
Ron & Al
Cyg 20h41'23" 46d17'2"
Ron Alan Smith
Leo 9h39'5" 27d4'43"
Ron Alexander Andring
Lib 15h56'12" -11d11'30"
Ron and Bob
Leo 9h49'29" 10d38'13"

Ron and Gwen for eternity
Cyg 21h51'46" 50d4'17"
Ron and Jackie Skwirut
Ori 5h29'39" 5d8'2"
Ron and Judy the best
grandparents
Peg 22h45'8" 21d39'56"
Ron and Lainie Stowell
Dra 16h1'2" 62d11'58"
Ron and Lindsay Theobald
Cyg 19h19'21" 51d32'32"
Ron and Lindy's
Cyg 20h59'42" 35d7'52"
Ron and Lori Singleton
Cyg 21h57'48" 50d5'37"
Ron and Lynn
Cas 1h44'7" 61d53'23"
Ron and Penny Tiehen
Psc 1h41'31" 7d57'25"
Ron and Sharon Hines
Cyg 20h48'33" 35d36'6"
Ron and Susan Maré
Cyg 21h37'6" 36d28'36"
Ron and Tim's Forever Star
Her 17h17'47" 36d51'5"
Ron Angelo Jr.
Cnc 7h56'28" 15d54'12"
Ron & Ann Lensmire 7-28-2004
Cyg 21h18'48" 33d38'11"
Ron & Barbara Hudson
Cyg 20h36'33" 49d19'38"
Ron Beard
Mon 6h52'20" -1d59'20"
Ron Bell
Aql 19h2'14" -0d9'55"
Ron Benson
Cyg 20h50'26" 46d21'50"
Ron & Betty Howard
Lyr 18h48'21" 36d34'54"
Ron & Bev
Gem 6h5'15" 21d46'38"
Ron Blumberg
Cnc 9h19'53" 15d36'28"
Ron Bottcher
Sco 17h2'49" -42d2'48"
Ron Briggs
And 23h57'27" 41d37'20"
Ron Buddha Ramirez
Cam 4h14'28" 71d31'59"
Ron "Butch" Cook Forever
Remembered
Leo 10h11'19" 14d39'18"
Ron & Carlene
Uma 8h55'18" 68d2'18"
Ron Carley
Uma 8h51'33" 67d41'48"
Ron & Carol Barwick's 65th
Birthday
Umi 13h10'57" 75d4'14"
Ron Carr - 2004
Boo 13h54'40" 24d28'16"
Ron & Cathy Boucher
Boo 14h37'22" 18d50'56"
Ron Chafy
Uma 11h56'35" 47d32'3"
Ron Christopher Steadman
II
Leo 11h22'54" 16d50'11"
Ron Currington
Uma 10h58'39" 60d42'8"
Ron Dalton
Ori 6h8'10" 15d37'26"
Ron Davis
Cap 21h57'29" -19d26'26"
Ron Debock
Psc 23h41'5" 6d8'23"
Ron & Debra Bowen
Cyg 20h5'12" 30d47'0"
Ron Dosh
Tau 3h49'19" 24d15'12"
Ron " Dragon-6" Newman
Tau 5h52'58" 17d41'52"
Ron Earle's Legacy - 27
Sept 1926
Ori 5h39'59" -1d53'54"
Ron Edward Charles
Lowley's Birthday Star
Cap 21h44'22" -19d55'1"
Ron Ellis
Aur 5h49'22" 40d41'52"
Ron & Eunice - Golden
Oldies
Cru 12h28'53" -61d0'25"
Ron Freedman
Cas 0h38'24" 64d14'37"
Ron G. Vestal
Per 2h37'38" 55d53'44"
Ron Goralski
Her 18h53'2" 23d13'30"
Ron Hedstrom
Her 17h29'36" 35d15'3"
Ron & Hollynn, Endless
Love
Ori 5h37'17" 13d1'23"
Ron Horne-Reno
Vir 12h41'56" 8d53'28"
Ron John Buchan
Cyg 20h4'16" 40d34'53"
Ron John & Diane Koczon
10-27-01
Sco 16h8'40" -9d10'35"
Ron & Kathy Jensen
Lyn 7h53'14" 56d56'34"

Ron & Katrina Lee
Cyg 21h32'34" 47d12'6"
Ron Keller
Her 18h13'38" 15d16'9"
Ron Kizer
Uma 11h20'21" 49d6'7"
Ron Kurtz, Jr.
Aqr 23h52'7" -12d22'40"
Ron Legendre
Uma 9h37'0" 43d25'50"
Ron Lehner
Lib 15h0'1" -18d22'21"
Ron Leo Tomasso
Psc 0h20'35" 16d23'37"
Ron & Lisa Jennifer Kirsty
Kaitlin
Lyr 19h16'12" 29d41'1"
Ron Lord
Vir 14h2'2" -3d32'9"
Ron Loves Melissa
Lyn 7h30'34" 35d47'5"
Ron & Lynn's star of thanks
Cru 12h18'20" -64d13'9"
Ron MacLeod
Per 4h25'44" 39d41'6"
Ron & Margaret Rush
Her 16h22'1" 12d35'48"
Ron & Marge Sittner
Uma 9h34'53" 55d52'44"
Ron/Marsha/Grandpa/Gran
dma Moeller
Eri 3h46'16" -0d20'44"
Ron Mathews
Her 17h35'7" 27d35'41"
Ron Matteson
Sgr 18h46'2" -28d38'59"
Ron McAfee
Aql 19h13'44" 7d25'13"
Ron McInerney
Per 3h47'9" 32d58'1"
Ron Mecklenburg
Lib 15h9'37" -5d9'12"
Ron Milrot
Her 16h45'25" 41d34'16"
Ron Monsen
Her 17h21'14" 17d19'54"
Ron & Nan Hampton
And 23h47'38" 34d9'50"
Ron & Nancy * Teamo
Supreamo
Cma 7h26'12" -22d45'50"
Ron Oates
Per 2h25'30" 55d14'25"
Ron Opey Patrick
Cep 21h58'45" 55d28'24"
Ron Orchard
Umi 13h47'35" 72d56'14"
Ron & Pat's Star of Eternity
Umi 16h18'36" 76d53'31"
Ron Pendergrass
Lib 14h45'40" -11d55'40"
Ron Peterson
Aur 5h54'19" 53d54'10"
Ron Poppy Jones
Leo 10h33'47" 14d20'10"
Ron Richard Smith
Lib 15h3'53" -16d39'23"
Ron Roberts for Always in
All Ways!
Leo 10h31'53" 9d20'31"
Ron Rocket OBrien
Vir 13h28'17" 12d9'9"
Ron Ron
Psc 1h15'49" 26d15'59"
Ron Ronald Ervin
Scholastica Seibel
Uma 11h31'15" 62d10'43"
Ron Rospierski
Uma 11h33'30" 53d2'54"
Ron S. Busby
Her 17h13'46" 29d12'37"
Ron Sanders-Mountaintop
Lighthouse
Psc 0h32'30" 17d52'24"
Ron Simpson
Her 18h25'37" 18d19'32"
Ron Snyder "Papa's Star"
Cep 21h52'7" 61d3'54"
Ron Solon
Sgr 19h29'41" -16d5'21"
Ron~Sonia~Sarah~Jacob
Ungerman
Uma 9h23'25" 59d25'45"
Ron & Stacey
Lib 14h57'21" -2d4'24"
Ron Stauffer
Aur 5h39'16" 37d21'15"
Ron Steiner
Vir 12h22'2" -7d5'14"
RON STELLA - Forever My
King
Sgr 18h28'49" -28d12'46"
Ron Stewart
Uma 14h15'13" 59d8'0"
Ron Sturgis
Her 17h39'38" 19d10'33"
Ron Suarez
Ori 5h40'29" 12d8'17"
Ron & Susan Church
Aqr 22h7'18" -0d49'24"
Ron Taylor - Ron's Star
Uma 8h37'8" 59d57'54"
Ron & Vicki Hartinger
Uma 11h53'58" 29d3'19"

Ron W. Faulk
Leo 10h36'3" 13d32'42"
Ron Warmingham
Ari 2h30'55" 12d11'41"
Ron Wheat
Aur 5h15'9" 43d7'21"
Ron Winters
Leo 10h40'28" 14d50'46"
Ron Yogi
Psc 0h11'26" -2d2'51"
Ron, Karla & Deke Jacob
Uma 9h49'56" 60d7'4"
Rona
Sgr 18h32'17" -27d58'30"
Rona
And 0h9'16" 44d33'19"
Rona and Arthur Gelfand
Del 20h48'1" 17d49'31"
Rona Dadyan
Psc 1h7'11" 17d21'48"
Rona McKnight Cassel
Crb 15h54'55" 28d3'14"
Rona Rochell
Ari 2h57'44" 26d9'55"
Rona Therese
Tau 4h16'25" 24d48'27"
Rónai Levente
Uma 9h54'50" 50d16'53"
Ronak Christian
Dra 15h36'49" 57d11'13"
Ronald
Cas 23h52'21" 59d6'1"
Ronald
Tau 5h52'23" 24d19'54"
Ronald A. Bethke II
Aur 6h8'52" 38d12'47"
Ronald A. McClellan
Her 16h43'34" 27d51'10"
Ronald Aaron Free
Ari 3h21'17" 28d4'34"
Ronald Adonay Romero
And 23h30'52" 41d22'48"
Ronald Aladeniyi
Psc 1h22'14" 27d3'17"
Ronald Alan Kemper
Leo 11h25'12" 19d45'17"
Ronald Alan Young 38-Special
Sco 17h56'51" -33d8'8"
Ronald Allen Perez-October 8, 1995
Lib 15h14'16" -11d49'3"
Ronald Allen Thompson
Cnc 8h39'28" 18d25'27"
Ronald and Donna Sendak
Cas 1h19'31" 56d32'35"
Ronald and Elizabeth Davis
Cyg 20h41'38" 42d50'10"
Ronald and Joanne Panicucci
Lib 14h50'29" -0d44'19"
Ronald and Kevin Huffman
Her 17h27'42" 36d16'33"
Ronald and Sharon Brown
Cyg 19h54'20" 38d35'34"
Ronald Andrew Machingo
Lac 22h41'55" 51d59'10"
Ronald Anthony Hudek
Her 16h40'0" 9d42'23"
Ronald Arroyo
Per 4h20'7" 43d37'22"
Ronald Arthur Sparks
Her 17h16'59" 35d15'38"
Ronald Arthur Viney III
Her 17h42'52" 20d6'0"
Ronald Astarita
Tau 3h36'49" 6d23'33"
Ronald B Blakley
Lib 14h56'19" -11d44'17"
Ronald B. Siegel
Psc 1h5'21" 10d0'13"
Ronald Bailey
Tau 4h26'58" 10d45'23"
Ronald Barnett, Jewish Prince
Ari 3h19'38" 24d2'8"
Ronald & Beverly Catanese
Lyr 18h16'29" 30d44'19"
Ronald "Big Pop" Gratale
Her 16h48'58" 39d17'22"
Ronald " Big Star" Hayes
Lyn 7h35'45" 43d5'36"
Ronald Bonassar
Uma 8h56'50" 66d6'55"
Ronald Boumans
Her 17h37'27" 37d10'42"
Ronald Brown
Uma 11h26'22" 51d42'56"
Ronald Bruinsma
Umi 16h26'48" 77d35'10"
Ronald & Bunnie Zeske
Peg 23h7'50" 27d54'36"
Ronald C Lieb
Ari 3h28'27" 26d27'6"
Ronald Cairns
Cep 22h52'5" 65d53'42"
Ronald Carlson
Vir 13h41'58" -6d39'12"
Ronald Carlson's 60th
Her 16h47'14" 42d12'52"
Ronald Carter Harris
Sgr 18h3'39" -28d0'44"
Ronald Cary Geiger III
Leo 11h27'20" 26d37'49"
Ronald Chaghoury
Uma 10h47'5" 68d50'7"
Ronald Charles Bissell
Cnc 8h39'36" 8d27'38"
Ronald Charles Bubany
Uma 9h55'32" 45d39'34"
Ronald Clyde Jetton
Aql 18h44'26" -0d31'37"
Ronald Colin Curnutt
Aql 19h13'0" -0d38'37"
Ronald Connally
Ori 4h48'43" 4d49'22"
Ronald Cooke
Cep 23h56'59" 74d7'50"
Ronald Craig Balser
Ori 6h16'53" 1d13'37"
Ronald Craig Nowinski
Vir 14h28'30" 1d7'36"
Ronald Creevy
Tau 4h51'58" 30d1'6"
Ronald Crescenzo
Cnv 12h44'21" 41d56'15"
Ronald Curtiss Berry "Handsome"
Per 3h40'42" 46d11'30"
Ronald D Anderson Andy
Her 17h5'36" 36d36'2"
Ronald D. Hart, M.D.
And 0h2'26" 45d15'39"
Ronald D. Palmer
Aqr 22h30'2" -2d6'47"
Ronald Dale Cearlock
Her 17h17'20" 46d46'9"
Ronald Dale Lokey
Ori 5h35'4" -2d12'53"
Ronald & Darlene Wells
Cyg 21h12'16" 44d20'57"
Ronald David Cloutier
Cyg 20h26'32" 37d28'15"
Ronald Davies
Uma 13h34'2" 52d40'26"
Ronald Dean Andrew
Uma 10h13'25" 56d30'8"
Ronald Dean Gregory
Uma 9h4'36" 64d17'7"
Ronald Douglas Widdoes
Gem 7h7'59" 33d12'8"
Ronald Duane Dickinson
Cyg 19h47'9" 29d37'3"
Ronald Duke Forbes
Cma 7h24'7" -16d38'40"
Ronald Dwain Morris
Sgr 17h55'55" -20d45'57"
Ronald Dwight Whiting
Tau 4h43'41" 22d8'55"
Ronald E. Adams
Sco 16h15'57" -11d29'40"
Ronald E. Adams
Sco 16h13'50" -12d2'34"
Ronald E. and Suzanne Bujold
Her 18h35'4" 18d57'12"
Ronald E. Hall
Ori 5h8'52" 10d49'56"
Ronald E. Miller Forever Shining
Her 16h36'38" 15d19'11"
Ronald E Porep
Peg 23h26'32" 32d31'45"
Ronald E. Raines Jr.
Vir 13h14'58" 7d56'9"
Ronald E. van Ingen
Cnc 8h40'5" 16d9'8"
Ronald Earl Block, Jr.
Sco 17h34'39" -32d59'36"
Ronald Eason
Leo 11h41'40" 13d21'55"
Ronald & Edith Welsby
Cyg 19h41'57" 29d29'28"
Ronald Edward Robert Scholten
Aql 19h34'48" 2d30'47"
Ronald Ellard, Sr.
Ari 3h13'11" 19d16'1"
Ronald Emlyn Jones
Uma 10h28'24" 66d36'41"
Ronald Esquard Curtis
Ari 2h22'15" 13d39'57"
Ronald Eugene King Junior
Tau 4h28'38" 17d51'34"
Ronald Eugene Roy Senior
Cap 20h8'50" -10d40'11"
Ronald Eugene Wiseman
Ori 5h38'9" -0d40'59"
Ronald F. and Linda S. Campbell
Uma 11h22'26" 42d48'43"
Ronald F. Gordon
Tau 4h28'56" 15d37'12"
Ronald Flam ~ Master Wood Carver
Aql 19h30'0" 12d1'45"
Ronald Francis Staley, Sr.
Cap 21h47'42" -17d52'38"
Ronald G Ercolani
Uma 10h29'50" 65d9'58"
Ronald G. Frey
Uma 10h1'47" 65d42'8"
Ronald G. Joyce Jr.
Per 4h47'32" 46d35'11"
Ronald G. Paul
Uma 9h35'10" 62d28'20"
Ronald G Snatchko
Uma 8h9'6" 62d20'17"
Ronald Gary Hawker
Uma 9h39'14" 62d10'51"
Ronald Geistfeld
Aql 18h53'25" 10d54'47"
Ronald Gene Phipps
Ori 6h18'11" 15d3'49"
Ronald George
Cnc 8h27'30" 13d26'27"
Ronald Gustave Benkert
And 2h22'27" 49d9'49"
Ronald H. Puent
Her 16h35'2" 20d38'43"
Ronald H. Stern
Ari 2h0'36" 17d49'36"
Ronald H Walters, Jr.
Uma 9h33'33" 44d44'59"
Ronald Harrison Lee
Gem 6h42'9" 19d37'47"
Ronald Hipolito
Lib 15h11'20" -23d29'3"
Ronald Hochman
Uma 13h22'37" 56d59'19"
Ronald Howard McConaughy
Cep 22h38'51" 64d0'54"
Ronald Huerta
Cmi 7h11'47" 10d54'46"
Ronald & Irene Kmiotek
Tau 3h34'48" 19d58'17"
Ronald J. Feitzinger
Gem 7h13'28" 27d34'54"
Ronald J. Kroll
Aql 20h8'27" 10d26'12"
Ronald J. Kroon
Aur 6h12'57" 45d8'21"
Ronald J. Michaud
Lib 15h21'56" -11d36'52"
Ronald J. Peltier forever...Holly
Aqr 22h38'36" -4d31'16"
Ronald J. Pflum
Uma 11h7'43" 56d8'19"
Ronald J. Radosevich
Uma 11h57'33" 58d48'1"
Ronald J. Ruszczyk PhD
Lib 15h25'4" -13d54'30"
Ronald J. Sucharzewski
Her 17h16'29" 35d54'38"
Ronald J. Tamashkin
Uma 11h0'48" 67d27'17"
Ronald J Wise
Uma 8h57'50" 53d11'1"
Ronald James Bean
Per 3h39'55" 42d11'6"
Ronald Jeffrey Matthews
Psc 1h24'35" 21d43'33"
Ronald Jesse Furlow
Her 16h36'54" 36d25'24"
Ronald Jesse Medeiros 60th Birthday
Aqr 22h4'10" 0d42'57"
Ronald & Jessica United 4Everafter
Cap 21h36'36" -10d41'30"
Ronald John Heromin, M.D., P.A.
Leo 11h12'5" -2d33'8"
Ronald Jose Flores
Aqr 22h30'26" -2d17'26"
Ronald Joseph Catalano
Cnc 8h26'10" 30d45'13"
Ronald Joseph Cuevas
Per 4h39'40" 41d12'2"
Ronald Joseph Micheli
Psc 0h2'40" -5d3'50"
Ronald Joseph Rossetti
Psc 1h41'17" 18d47'53"
Ronald Joseph Tippit
Psc 1h5'41" 3d57'0"
Ronald J.Sorenson
Uma 10h21'46" 67d14'19"
Ronald K. Beaton
Leo 11h26'35" 18d12'23"
Ronald K. Thompson
Crb 15h47'12" 39d5'10"
Ronald Katz
Cnc 8h14'13" 11d31'29"
Ronald L. Cardinale
Uma 10h20'53" 49d32'51"
Ronald L Diebold
Leo 11h46'29" 26d50'26"
Ronald L Myers
Pho 0h1'19" -44d2'59"
Ronald Lassin
Per 3h11'18" 55d12'11"
Ronald Lee Hammer
Vir 14h26'21" -1d44'57"
Ronald Lee Keesling
Her 17h24'8" 34d16'19"
Ronald Lee Marquess
Ari 2h17'32" 17d7'23"
Ronald Lee Rhode
Psc 0h18'9" 15d24'16"
Ronald Lee Stafford
Umi 17h3'47" 81d20'55"
Ronald Lorenz
Cep 22h19'25" 63d16'2"
Ronald Louis SoRelle
Gem 6h37'29" 27d9'54"
Ronald M. Borer III
Vir 13h29'44" 12d29'11"
Ronald M. Mederski, Sr.
Sco 16h14'56" -10d11'8"
Ronald M Phillips
Pho 0h50'54" -47d26'58"
Ronald Marcus Davenport
Ari 3h9'16" 17d31'12"
Ronald Martin Bondy
Lyr 18h32'50" 35d53'42"
Ronald Michael Pflug
Gem 6h48'23" 21d44'40"
Ronald Milton Berg Jr. 08221954
Leo 9h31'21" 13d32'12"
Ronald Modesto
Cnc 8h42'28" 14d1'2"
Ronald Moran and Jean Mackin
Vir 13h41'56" -5d25'38"
Ronald Moreno Gallegos
Gem 7h39'34" 32d12'4"
Ronald Morse
Lyn 7h42'46" 48d21'3"
Ronald Mosley
Lyr 18h46'6" 31d51'9"
Ronald Musumeci
Ari 3h26'52" 27d13'57"
Ronald Neal Benvie
Ori 4h48'47" -3d35'1"
Ronald Norman Stein
Aqr 20h41'54" -8d32'46"
Ronald P. Langlois
Dra 15h46'27" 63d26'4"
Ronald P. Stanton
Crb 15h59'46" 39d21'27"
Ronald "PaPa" Miller
Tau 5h26'38" 24d52'33"
Ronald & Patricia Forever
Cyg 20h35'51" 30d4'11"
Ronald Paul Serratore
Per 3h32'51" 32d33'36"
Ronald Paul Ziel
Her 17h18'12" 20d14'8"
Ronald R. Rensing
Per 3h28'39" 50d45'12"
Ronald R. Richter
Ser 18h15'34" -13d40'14"
Ronald R. Smith
Uma 9h44'11" 63d20'57"
Ronald Ray Kool III
Leo 11h18'23" 7d3'11"
Ronald Raymond Powell, Sr.
Ari 2h9'52" 22d2'33"
Ronald Redpath Young
Cep 21h46'2" 59d35'59"
Ronald Rizo
Cnc 8h19'30" 9d2'58"
Ronald Roberts
Sco 16h30'55" -35d18'1"
Ronald S. Senykoff a.k.a. Dad
Cep 20h54'29" 64d18'35"
Ronald Schlesch
Cyg 21h1'17" 46d4'24"
Ronald Sher
Vir 15h0'47" 2d9'17"
Ronald Simone
Psc 1h6'34" 28d44'52"
Ronald Slosek
Uma 11h37'31" 62d55'42"
Ronald Steele
Leo 11h13'44" 24d43'30"
Ronald & Stephanie
Cyg 20h55'40" 44d55'27"
Ronald Steven Kaysen
Uma 10h48'13" 60d3'8"
Ronald Terrill
Cap 21h16'35" -25d42'24"
Ronald Thomas Bettencourt
Sco 16h55'14" -30d21'45"
Ronald Thompson My little soldier
Per 3h20'16" 42d48'37"
Ronald Timm
Psc 23h30'15" -0d14'58"
Ronald Todd Shafer
Lib 14h54'26" -1d51'18"
Ronald Turney
Aql 19h6'1" 9d19'49"
Ronald Tyler Farmer
Per 3h13'48" 47d29'2"
Ronald V. Williams
Ori 5h54'9" -0d40'51"
Ronald Vandel Thoreson
Lyn 8h34'23" 45d19'35"
Ronald Vaughn Blankenship, Jr.
Lib 15h11'25" -24d20'7"
Ronald Victor Piacentini
Leo 11h11'22" -5d53'32"
Ronald W Beal
Psc 2h1'27" 9d43'56"
Ronald W. Gebur
Lib 15h22'24" -6d38'38"
Ronald W. Milne
Per 2h42'20" 40d1'48"
Ronald W. Sherwood
Her 17h17'52" 47d0'0"
Ronald W. Whitley, Jr.
Tau 4h48'45" 24d35'49"
Ronald Wagner
Uma 10h32'52" 51d47'44"
Ronald Walker
Lyr 18h26'15" 31d57'43"
Ronald Weaver
Lyr 18h42'39" 41d5'37"
Ronald Welty
Per 4h29'57" 35d10'59"
Ronald Yap
Aur 5h38'52" 33d13'38"
Ronald Yarbrough
Aql 19h31'18" 11d41'1"
Ronaldo Murphy-Saunders
Cep 21h49'6" 65d22'55"
Ronald's Home
Cep 22h38'20" 70d46'55"
Ronalee Matthews & Family
Uma 12h39'32" 58d19'29"
Ronan Andrew Daglish
Gem 7h7'9" 22d46'32"
Ronan Bear
Vir 15h7'20" 4d31'12"
Ronan Bluett
Aur 5h54'13" 33d4'26"
Ronan McCarthy - Great Nana's Star!
Cas 1h12'15" 54d45'54"
Ronan O'Lenskie
Per 4h8'48" 45d2'43"
Ronan Quinn
Cnc 9h4'25" 16d1'43"
Ronan Thomas Rigg June 16 2006
Per 4h29'47" 51d28'11"
RONandSUE
Sge 19h32'28" 17d30'59"
RonAnn
Uma 9h43'33" 52d11'56"
Ronan's Light
Aqr 22h57'13" -10d23'26"
Ronda Bonney
Com 12h0'59" 21d49'8"
Ronda Downs
And 0h53'32" 35d19'43"
Ronda Dureault
Cnc 8h40'1" 18d38'44"
Ronda Gay
Gem 6h17'48" 21d47'7"
Ronda "GypsyRose" Brown
And 0h10'43" 35d46'11"
Ronda Hope Roderiques
Aql 19h36'13" 11d55'30"
Ronda Lea
Uma 13h49'5" 58d0'51"
Ronda Lee
Aqr 23h36'25" -8d36'20"
Ronda Lin
Gem 6h41'57" 31d48'46"
Ronda Louise Hecker
Cyg 21h43'41" 40d55'26"
Ronda M. Tilander
Lyn 8h5'57" 46d47'59"
Ronda Michelle Lamont
Uma 12h37'37" 56d44'54"
Ronda Montagne
Cnc 8h16'34" 10d11'33"
Ronda Mwangi
And 1h21'47" 39d24'46"
ronda prien
Uma 11h52'22" 40d10'23"
Ronda Rachel Minor "1958-2006"
Gem 6h51'7" 30d42'33"
Ronda Sue Lang
Ari 2h48'54" 17d22'53"
Rondal
Aql 19h37'0" 11d32'3"
Rondalyn Sue
Leo 9h28'38" 25d21'51"
Ronda's Wish
Cyg 20h7'1" 40d44'19"
Rondeau
Ori 5h1'40" 6d19'43"
Rondee
Cas 0h35'56" 48d58'51"
Rondene Horizon 10071938
Lib 15h31'35" -27d13'59"
Rondeve
Sge 19h45'12" 17d24'48"
Rondi
Crb 15h34'0" 30d30'56"
Rondi Marie Crawford 03/16/39
Cmi 8h10'12" 3d25'0"
Rondine
And 1h26'52" 46d49'28"
Rondo
Cnc 8h30'18" 28d1'32"
Rondo Curtis
Uma 11h23'39" 43d46'48"
Ronelyn
Aqr 22h5'28" -3d37'44"
Ronen Kowalski
Sgr 19h13'50" -18d33'23"
Ronette
Sco 16h54'44" -37d20'47"
Roney's Riddle
Ori 6h20'6" 8d7'49"
Ronfot Papou
Leo 10h14'48" 10d48'58"
Roni
Ori 5h43'14" 9d11'28"
Roni
Cru 12h39'21" -60d42'33"
Roni Kendall
Tau 3h45'5" 27d33'49"
Roni Leigh Brown
Sco 16h7'2" -17d25'4"
Roni Nahum
Psc 0h54'23" 31d43'9"
Ronie-Marie
Tau 5h41'0" 26d20'0"
RONILAS3
Lyn 8h42'43" 40d50'23"
Ronilla 1271
Sgr 18h59'54" -32d33'16"
Ronin Kai
Cnc 8h20'13" 20d57'47"
Ronin Lee Ross
Gem 7h8'34" 33d46'33"
Ronis Orellana - Shining Star
Dra 16h28'57" 55d58'56"
Ronja
Uma 8h45'41" 50d55'7"
Ronja Lisa
Umi 14h14'16" 77d56'54"
Ron-Jan
Lib 15h21'9" -9d10'31"
Ronk
Cnc 8h43'12" 32d25'25"
Ronmar
Lyn 7h57'48" 42d45'35"
Ronmichelle
Lyn 7h41'32" 59d10'29"
Ronn
Psc 1h1'23" 10d1'0"
Ronna Robinson
Umi 15h11'47" 70d5'13"
Ronnette
Cyg 20h45'16" 51d57'55"
Ronni
Ari 2h7'54" 17d59'10"
Ronni
Uma 10h43'51" 65d10'2"
Ronni Moore I Love You
Ori 6h10'4" 2d23'25"
Ronnica Lea
Lyn 7h20'39" 58d42'44"
Ronnie
Uma 8h19'28" 60d19'1"
Ronnie
Cep 21h31'7" 67d56'52"
Ronnie
Aqr 22h7'0" -14d32'23"
Ronnie
Sco 17h28'25" -44d44'7"
Ronnie
Sco 16h39'58" -31d5'50"
Ronnie
Cma 7h5'8" -29d20'20"
Ronnie
Gem 7h25'12" 16d38'12"
Ronnie
Ari 3h10'11" 27d19'15"
Ronnie
Her 16h6'31" 46d14'42"
Ronnie and Leanne
Cyg 21h16'54" 34d26'11"
Ronnie Anne Eddy
Tau 5h51'47" 26d42'6"
Ronnie Argo
Lib 15h41'45" -6d37'44"
Ronnie B 80
Sco 17h57'7" -40d36'3"
Ronnie "Baby" Conroy
Vul 19h35'55" 25d11'14"
Ronnie Brian Concoby
Her 17h39'0" 31d59'53"
Ronnie Coletti
Per 4h37'39" 32d37'27"
Ronnie Conroy
Psc 0h20'31" 16d44'57"
Ronnie Deer
Vir 11h37'37" 5d0'24"
Ronnie DiGregorio
Cas 23h32'39" 56d23'38"
Ronnie E. Lehnert 5/13/57-12/18/96
Tau 3h44'35" 22d36'18"
Ronnie Esmail
Uma 11h55'5" 39d5'18"
Ronnie & Gerard Keane
Cas 23h29'38" 53d32'52"
Ronnie Hazelwood
Her 17h10'24" 31d4'24"
Ronnie Jackson
Per 3h54'15" 38d31'16"
Ronnie & Judy Ray
Cyg 19h58'26" 47d17'24"
Ronnie Lopez Jr.
Vir 13h4'59" -3d35'34"
Ronnie Mayer
Psc 0h25'11" 15d37'43"
Ronnie Munze
Uma 8h50'52" 61d5'56"
Ronnie Nye
Uma 10h0'53" 58d49'2"
Ronnie O. Enfinger
Uma 11h13'15" 42d25'2"
Ronnie P Suiter
Uma 13h44'44" 55d37'54"
Ronnie Petersen
Her 16h13'9" 25d6'32"
Ronnie Porter
Cep 21h45'3" 65d45'40"
Ronnie Pullen
Umi 14h7'9" 68d15'47"
Ronnie Richard Stewart
Cnc 8h14'2" 22d56'8"
Ronnie & Rochelle Phyffer Forever
Cap 20h58'2" -20d27'19"
Ronnie Scott Jr.
Uma 13h48'53" 51d21'3"
Ronnie Smith
Sgr 19h44'15" -15d53'4"
Ronnie Stroup
And 23h21'58" 44d33'6"
Ronnie "Super Star" Vaughan
Lib 15h37'4" -7d20'52"
Ronnie Talbert's 40th Birthday star
Ari 2h9'25" 16d49'12"
Ronnie Teall
Aqr 23h10'42" -11d20'15"
Ronnie the Mimesabe
Gem 6h48'51" 27d0'59"
Ronnie W. Tippens
Psc 0h57'56" 14d16'33"
Ronnie Wayne Goodwin
Aqr 22h39'19" -1d43'39"
Ronnie Weinberger
Cep 23h10'44" 70d45'19"
Ronnie Zientek
Vir 11h44'47" -0d53'55"
Ronnie's Dreams
Uma 12h0'0" 44d4'44"
Ronnie's Little Star
Sco 16h41'39" -33d4'9"
Ronnie's Star
Sgr 18h10'1" -30d55'36"
Ronnie's Twinkle
Peg 21h35'4" 16d26'31"
Ronniqueka "Nikki" Covington
And 0h30'2" 40d16'57"
Ronny
Cyg 19h39'44" 44d55'5"
Ronny Eberhard Bergk
Uma 8h35'32" 54d25'40"
Ronny George
Uma 11h5'43" 46d45'49"
Ronny Kolb
Ori 6h17'7" 8d58'45"
Ronny's Star
Her 16h33'14" 33d34'39"
RonPosnack
Her 17h57'22" 14d53'22"
Ron's Destiny
Lib 14h56'15" -2d13'46"
Rons L. Pence
Her 16h45'5" 13d34'26"
Ron's Legacy
Uma 10h52'13" 54d58'0"
Ron's Raider
Sco 16h9'40" -9d1'47"
Ron's Star
Her 18h5'39" 48d33'39"
Ronsstar 060526
Cru 12h37'22" -57d18'34"
Ronstar
Uma 10h28'11" 65d16'59"
Ronstar
And 0h11'42" 39d49'8"
RONTAN 5689
Aql 19h35'22" 12d39'25"
Röntsch, Karola
Uma 10h21'10" 39d34'37"
&-Roo
And 1h13'3" 41d41'44"
Roo
Lmi 10h27'13" 32d14'11"
Roo
Cma 6h18'11" -29d32'56"
Roo Tson
Cnc 9h20'5" 12d19'43"
Rood, Günter
Uma 14h1'11" 55d20'10"
ROOFER
Uma 10h43'2" 52d15'52"
Rooflock
Cas 0h10'7" 54d4'59"
Roohan aka Mrs. Farhan Mustafa
Cap 20h14'59" -9d45'39"
Rooney
Uma 12h36'3" 57d59'22"
Rooney Buck
Uma 12h7'59" 61d55'58"
Roonie
Uma 10h43'14" 68d46'28"
Roopika Nayer
Uma 12h50'7" 52d51'0"
Roos
Lmi 10h33'51" 32d25'13"
Roos, Dennis
Uma 13h39'12" 61d30'21"
Roosevelt Benton
Aur 6h2'33" 47d44'4"
Roosimba
And 0h40'59" 23d13'11"
"ROOSLI" FRIENDS FOREVER
Lyn 7h54'35" 35d17'16"
Rooster
Cap 21h40'36" -14d12'53"
Rooster, Grammadog
Cyg 21h43'31" 46d10'20"
Roper's Star
Ori 5h7'57" 6d21'32"
Rori
Lyr 18h48'21" 34d10'38"
Rori Kemp Destefano
Vir 12h49'48" 10d59'43"
Rori Marie
Leo 11h31'26" 14d57'32"
Rori Renee Rangel
And 2h28'59" 44d47'44"

Rorienne
Uma 11h58'28" 35d0'23"
Rori-Lou
Pic 5h33'55" -44d36'20"
Roro
Cap 20h17'7" -14d17'28"
RoRo Dear
Leo 10h25'58" 26d46'58"
Rorry..
Eri 3h43'40" -14d41'30"
Rory
Sco 17h4'1" -33d42'19"
Rory
Cnc 8h20'10" 23d51'27"
Rory
Lyn 8h58'44" 41d47'45"
Rory
Per 2h51'21" 50d9'34"
Rory Alexander Laird - Rory's Star
Her 18h43'10" 13d32'47"
Rory and Tricia Always and Forever
Vir 14h7'18" -1d36'28"
Rory B '50'
Uma 11h5'43" 69d42'41"
Rory Connor McMillan
Her 17h31'0" 30d21'45"
Rory Dami
Aur 5h54'12" 32d24'42"
Rory Farrington Treuhaft
Cnc 8h52'26" 13d23'45"
Rory Hernandez
Sco 17h44'54" -41d58'41"
Rory I Lampert
Lyn 6h45'50" 54d17'14"
Rory Jared
Umi 17h23'15" 81d59'3"
Rory JJ King
Cep 21h22'34" 63d24'46"
Rory Joseph O'Dwyer
Uma 8h45'21" 54d35'59"
Rory Joshua Lee
Aur 6h31'24" 51d45'9"
Rory & Kathy MacLennan - Potentate 2005
Uma 11h30'58" 56d20'12"
Rory Kenneth Kearns
Aqr 22h58'23" -10d17'25"
Rory Lipede
Ari 2h46'43" 25d4'20"
Rory Lochlan Bushong
Umi 15h9'36" 71d33'48"
Rory O'Connor
Ori 5h52'22" 8d34'35"
Rory O'Dowd Learey
Lib 15h53'53" -19d47'48"
Rory Parsons
Uma 12h2'53" 34d51'14"
Rory Patrick
Uma 8h26'19" 68d43'15"
Rory Ratcliffe
Cep 1h49'32" 80d25'50"
Rory 'Rawr' Taylor
Uma 10h14'53" 47d49'13"
Rory Schenck
Vir 13h28'36" 5d47'16"
Rory Timothy McLaughlin
Peg 22h55'39" 16d27'19"
Rory Tolunay
Psc 1h50'12" 5d43'28"
Rory West Johnson
Uma 13h0'37" 57d38'30"
Rory William Gemmell - Rory's Star
Her 16h17'9" 19d53'40"
Roryborealis
Ori 5h53'15" -0d12'50"
Rory's Boy
Uma 9h13'46" 59d27'21"
Rory's Star
Umi 14h34'27" 79d51'25"
Rory's Star- Forever Special
Col 5h55'5" -36d41'16"
Ros' Babe
Peg 22h10'28" 27d18'7"
Ros Kay Plunkett
Leo 10h21'14" 17d45'51"
Ros Puckridge - SGKS
Sco 17h57'6" -31d47'7"
Ro's Star
Boo 14h48'38" 15d42'0"
Rosa
Lyr 19h19'2" 29d36'1"
Rosa
Peg 23h39'46" 14d29'48"
Rosa
Uma 8h43'45" 47d8'26"
Rosa
Cyg 19h38'12" 31d36'9"
Rosa A. Cordova
Uma 11h55'47" 45d13'58"
Rosa Alba Barrientes
Tau 5h33'20" 17d13'8"
Rosa Alicia Velasquez-Romero
Ari 3h8'32" 29d30'7"
Rosa and Giovanni Rechichi
Cyg 20h41'3" 55d13'1"
Rosa Arguelles
Uma 9h32'27" 46d49'14"
Rosa B. Greenberg
Cas 0h27'32" 61d39'33"
Rosa Balistreri
Peg 23h20'58" 32d44'52"
Rosa Bisson Savoie
Uma 11h58'6" 45d14'3"
Rosa Carbonell
Ori 4h44'44" 4d11'50"
Rosa da Silva
Her 17h45'28" 47d29'13"
Rosa De Gennaro
Her 17h38'48" 39d10'55"
Rosa E. Rodriguez
Sco 17h25'17" -42d11'23"
Rosa Elena Munoz de O'Keeffe
Her 16h17'49" 26d5'34"
Rosa Elena Rocha
Cas 0h21'27" 61d2'27"
Rosa Eliyah
Psc 0h53'35" 10d32'35"
Rosa Elizabeth Maygar
Uma 9h21'51" 62d9'29"
Rosa Eneyda
Cyg 21h16'12" 38d46'32"
Rosa Fernandez aka bebo face
Peg 21h23'36" 18d0'25"
Rosa Galvis
Ari 3h22'24" 27d16'54"
Rosa & Giuseppe Regina
Cyg 20h18'58" 31d2'47"
Rosa Gogo
Crb 15h37'39" 32d47'47"
Rosa Ho
Tau 4h24'5" 17d48'46"
Rosa Irene Lopez
Lyn 6h57'3" 55d10'59"
Rosa Jane strawberry
Sco 16h11'3" -14d11'33"
Rosa Jimenez Sanchez
Umi 13h39'36" 75d58'48"
Rosa Jordán Pascual
Ari 3h10'7" 28d38'53"
Rosa & Jordi
Peg 21h57'7" 14d39'6"
Rosa Julio
Uma 9h11'24" 54d50'41"
Rosa L. Holguin
Cyg 20h11'56" 46d51'22"
Rosa Lattuca
Gem 7h6'57" 34d21'29"
Rosa Lee Allen
And 1h14'43" 37d4'20"
Rosa Lee Beall
Uma 10h25'16" 47d42'25"
Rosa Levy Aharoni Mardirossian
Uma 13h16'52" 58d51'26"
Rosa Lia
Sco 17h16'22" -43d24'44"
Rosa Lucia Puliafico/Brown/Stewart
Crb 15h47'35" 35d31'9"
Rosa Luise Kampheuer
Cnc 8h42'13" 32d58'21"
Rosa Luxemburg
Cas 1h44'8" 61d22'4"
Rosa Mae Dale
And 23h51'58" 41d37'14"
Rosa Maria Baquero
Tau 4h49'51" 18d29'27"
Rosa Maria Juanola Canals
Cas 0h41'50" 69d33'35"
Rosa Maria (Pelusaaa)
Lmi 9h56'20" 35d34'34"
Rosa Mercedes
Lib 15h20'23" -15d41'7"
ROSA MI CORAZON
Uma 8h45'41" 70d10'25"
Rosa Moran-Loqui & Violeta E. Loqui
Cas 0h20'18" 64d33'53"
Rosa Musgrave
Cas 1h7'22" 49d34'35"
Rosa Ochs-Brechbühler
Sge 19h49'55" 16d43'16"
Rosa Paula
Cas 23h3'55" 58d17'11"
Rosa Pin
Uma 12h24'44" 58d26'11"
Rosa Pin
Lyn 8h0'1" 38d18'10"
Rosa Rita Pirovano
Ser 18h18'40" -14d11'47"
Rosa Stella
And 0h5'36" 47d50'46"
Rosa Taglialatela Vinci
Ari 2h40'43" 30d42'14"
Rosa Terry
Cam 7h20'13" 75d36'1"
Rosa Todd
Aqr 21h42'54" -0d52'45"
Rosa Tuuli Marika
And 23h28'16" 41d19'48"
Rosa Valencia
Lib 15h25'47" -7d32'52"
Rosa Villalobos
Sgr 17h53'35" -28d30'10"
Rosa, Tito, Guido
Her 17h36'49" 32d1'36"
RosaAna
Sgr 18h29'23" -16d50'32"
Rosabel Leilani Brose
Tau 4h18'41" 26d36'7"
Rosabella
Ari 2h53'12" 18d0'43"
Rosabella Ann Juni
Crb 15h46'16" 31d19'14"
Rosae-Karina Fuellemann
Cas 1h1'26" 54d9'12"
Rosaidaliz
And 1h25'47" 35d26'3"
RosaLaura
Vir 12h40'44" 0d40'33"
Rosalba
Peg 22h10'18" 10d22'30"
Rosalba
Lib 15h27'3" -26d17'4"
Rosalba Bravo - Shining Star
Crb 16h7'19" 33d32'23"
Rosalba Mangini
Peg 0h11'57" 15d24'21"
Rosalba Ybarra Graham
Cap 21h36'43" -17d33'7"
Rosalee Frantz Furr
Lib 15h16'30" -43d27'27"
Rosalee Lynn DeMaro
Lyn 7h42'10" 59d17'0"
Rosaleen Delaney
Lyr 19h18'41" 28d22'23"
RosaLeeWhitehead
Sco 17h53'53" -42d33'27"
Rosales
Tau 4h35'40" 6d42'4"
Rosalia and David
Cyg 21h4'7" 54d18'7"
Rosalia De Torre Luna
Psc 0h41'5" 11d7'4"
Rosalia Lynn Lumia
Uma 14h9'46" 56d29'50"
Rosalia Siu
Cas 0h40'49" 62d6'22"
Rosalie
Cas 0h25'45" 61d44'37"
Rosalie
Cas 1h39'55" 64d14'51"
Rosalie
Vir 13h11'7" 1d14'18"
Rosalie
And 1h11'6" 43d4'52"
Rosalie
Cas 0h10'53" 53d9'39"
Rosalie Alice Watson
Sgr 19h18'52" -14d59'30"
Rosalie and Richard
Cyg 21h43'57" 31d41'50"
Rosalie Ann Gonzalez
Uma 10h21'21" 55d46'29"
Rosalie Branchick
And 1h11'44" 33d58'57"
Rosalie deForest Cargill
And 0h48'52" 36d51'51"
Rosalie Detch
Lib 15h51'40" -16d59'0"
Rosalie G. L. 143
Ari 2h4'27" 22d47'47"
Rosalie Hernandez
Psc 0h57'21" 8d30'5"
Rosalie Howe
Uma 13h27'46" 62d15'10"
Rosalie Hughes Steinsiek
And 1h23'41" 47d10'27"
Rosalie J. Kane (Nanny)
Vir 13h11'15" 12d39'6"
Rosalie Joyce Paton
Cyg 21h32'38" 32d22'3"
Rosalie Kesnow
Lyr 18h48'32" 31d39'27"
Rosalie Macaluso
Ari 2h47'36" 27d24'18"
Rosalie Maiorana
Cyg 20h3'13" 33d32'3"
Rosalie Montalbano
Cas 23h26'11" 53d12'7"
Rosalie Rodriguez Verdugo
Vir 13h58'17" 0d0'9"
Rosalie S. Donovan
Sgr 19h42'14" -14d49'26"
Rosalie Seugnette Knight
And 2h5'47" 43d19'2"
Rosalie Tarquinio
And 23h40'41" 47d17'32"
Rosalie-Headoverheelsrnm
Cnc 8h18'22" 18d56'32"
Rosa-Lina
Uma 11h22'22" 71d38'23"
Rosalina Marie Battagliola
Cas 1h15'14" 64d54'39"
Rosalind
Umi 9h0'5" 86d35'25"
Rosalind
Vul 20h39'7" 24d36'37"
Rosalind Alster
Cap 21h44'30" -11d34'44"
Rosalind Dunn
Cas 23h43'47" 55d27'44"
Rosalind Mary Armstrong
Tau 5h11'33" 18d35'7"
Rosalind & Milton Sabel
Cyg 19h44'4" 55d3'12"
Rosalind Musarra
Uma 13h53'26" 55d28'30"
Rosalind My Love
Lib 15h23'11" -7d7'8"
Rosalind Sheree Bell
Leo 10h55'7" 9d32'1"
Rosalinda & Gilberto Armendariz
Lyn 7h55'26" 56d38'43"
Rosalinda Maldonado
And 1h45'33" 48d32'19"
Rosalinda S.
Leo 11h40'59" 16d15'43"
Rosalinda S. Gallanosa
Lib 15h23'14" -19d53'44"
Rosalind's 50th Birthday Star
Tau 3h37'32" 25d14'44"
Rosaline
Aql 19h58'22" 0d22'30"
Rosaline Smithers
Uma 12h8'33" 56d28'23"
Rosalva Araico
Dra 19h43'21" 66d13'22"
Rosaly
Ori 6h19'45" 10d3'58"
Rosalye Thibault
Cas 23h9'35" 59d13'7"
Rosalyn Braverman
Lyn 8h11'44" 56d33'57"
Rosalyn Janae
Aqr 23h5'17" -6d18'23"
Rosalyn Katz
Cnc 8h39'12" 18d43'54"
Rosalyn M. Bristow
Uma 9h41'5" 43d36'17"
Rosalyn Martin
Ari 2h59'56" 30d45'24"
Rosalyn Thorpe
Aql 19h59'55" 11d36'50"
Rosamel
Lib 15h29'0" -5d12'14"
Rosamond Mitchell
Uma 8h20'39" 64d58'55"
Rosamunde Vincius Thro
Umi 15h44'41" 74d45'49"
Rosana
Cas 0h34'29" 47d56'0"
Rosana Mendez Cabrera, M.D.
Mon 7h31'20" -0d30'53"
Rosanas Glücksstern
And 0h41'49" 41d42'26"
RosAndrew
Cyg 21h24'54" 37d49'10"
Rosann Coogan
Uma 8h43'11" 58d19'49"
Rosann Marie Lozano
Com 12h33'21" 17d5'43"
Rosann Santi
Lyn 8h8'49" 45d49'33"
Rosanna
Per 4h21'12" 38d44'28"
Rosanna
Lyr 19h0'43" 27d11'35"
Rosanna
Lyr 19h18'41" 29d29'10"
Rosanna
Her 17h53'28" 28d57'30"
Rosanna
Uma 9h26'8" 60d44'48"
Rosanna
Sco 17h29'34" -35d8'59"
Rosanna Evangelou
Cnc 8h26'21" 29d51'10"
Rosanna Gordon
And 23h42'19" 47d36'0"
Rosanna Handy
Gem 7h39'9" 24d48'58"
Rosanna in the sky with diamonds
Cap 21h46'40" -19d39'44"
Rosanna Jeanette Schroeder
Ori 6h3'19" 17d21'1"
Rosanna Jeanette Schroeder
Ori 6h6'33" 13d17'41"
Rosanna L Witt
Lyn 7h2'7" 51d22'18"
Rosanna Lynn Hall "Ron's Rose"
And 0h19'20" 43d31'41"
Rosanna Simeone
Cas 23h16'20" 58d26'32"
Rosanna-Star Queen
Crb 15h44'58" 28d58'51"
Rosanne Santos
Cyg 21h37'42" 50d36'9"
Rosanne T. Eastman
And 1h1'45" 46d37'33"
Rosaria
Cas 23h27'31" 56d39'30"
Rosaria Caruso 20 Years HVB
Uma 13h48'28" 51d57'34"
Rosaria LaRosa
Cas 1h37'33" 62d3'47"
Rosaria My Love
Ari 3h17'43" 27d39'28"
Rosaria "Sara" Faith
And 23h56'30" 48d31'29"
Rosarin
Tau 4h31'39" 8d32'15"
Rosario
Leo 9h38'25" 27d46'52"
Rosario
Cas 2h40'58" 57d34'13"
Rosario
Uma 9h40'6" 54d41'16"
Rosario
Uma 12h31'46" 54d30'46"
Rosario Casado Borregon
Del 20h37'22" 14d27'21"
Rosario Cazares - My Best Friend
Gem 6h40'20" 21d2'45"
Rosario Conelli
Oph 17h38'32" -0d30'26"
Rosario Deleon - Moreno
Aql 19h51'48" -0d40'33"
ROSARIO EROS
Uma 11h25'10" 33d41'54"
Rosario Gonzalez
Gem 6h40'9" 20d13'54"
Rosario "Princess" Rosalez
Leo 9h30'34" 16d57'8"
Rosario Principe Delle Stelle I
Psc 1h23'6" 25d24'50"
Rosary Rodriguez-Herrera
And 0h47'0" 37d37'40"
Rosa's James Holmes
Cnc 8h24'23" 26d17'35"
Rosaura Bonilla
Lib 14h53'22" -2d41'48"
Rosco & Lucille Harris
Peg 22h26'9" 10d31'42"
Roscoe
Cma 6h49'11" -21d52'21"
Roscoe and Ajoke Simpson
Cyg 19h41'7" 31d3'55"
Roscoe Scarborough
Uma 11h34'5" 52d41'9"
Roscoe TC Youngblood
Lmi 10h15'11" 35d23'59"
Roscoe's Gate to Heaven
Cma 6h30'56" -16d10'13"
Rose
Cas 1h35'26" 62d53'2"
Rose
Cyg 19h58'50" 34d17'52"
Rose
Cam 3h57'34" 53d48'38"
Rose
Cas 0h35'42" 55d32'15"
Rose
Cyg 20h31'5" 47d34'29"
Rose
And 2h6'59" 46d15'1"
Rose
Lyn 7h12'54" 48d46'56"
Rose
Cnc 7h56'39" 13d47'13"
Rose
Aql 19h23'8" 6d17'34"
Rose
Aql 19h30'40" 3d38'3"
Rose
Vir 13h21'18" 6d56'48"
Rose
Aql 19h22'13" 0d40'17"
Rose
Vir 14h38'22" 3d31'57"
Rose
Ori 5h13'42" 6d13'3"
Rose (3/18/34-5/19/05)
Psc 1h19'17" 21d51'50"
Rose A. Demola
Crb 15h24'35" 27d54'30"
Rose A. Rowell
Cyg 20h47'21" 33d28'22"
Rose A. Stowell
And 0h51'51" 38d58'24"
Rose Adrian Star
Aql 18h55'23" -0d8'30"
Rose and Daniyel's Star
Cnc 8h29'2" 27d41'20"
Rose and Karl
Cyg 20h12'27" 34d45'31"
Rose and Michael
And 23h31'53" 35d56'14"
Rose Angel
Cru 11h57'26" -64d2'46"
Rose Ann
Lib 15h9'3" -17d40'34"
Rose Ann
Aqr 20h58'18" 0d7'13"
Rose Ann Boatwright
Ori 5h43'0" 3d32'1"
Rose Antonacci
Cas 0h28'49" 64d11'16"
Rose B.
Uma 8h38'13" 68d59'15"
Rose B R M D S B T 16 Grandchildren
Sco 15h59'31" -20d24'53"
Rose Bilicki
Cas 0h15'1" 54d32'41"
Rose Borella
Lyn 7h59'36" 58d58'8"
Rose Callen
Cyg 19h44'2" 37d42'1"
Rose Cantillon
And 23h59'54" 43d0'44"
Rose Caputo at 100
Sco 16h10'12" -12d36'48"
Rose Carleo
Uma 11h30'25" 58d51'4"
Rose Carroll
Tau 3h46'14" 12d50'1"
Rose Castellitto
Uma 10h47'2" 49d43'9"
Rose Cathleen Fitzgerald
And 2h37'29" 38d49'43"
Rose Cecelia St Thomas
Cyg 20h52'54" 36d52'52"
Rose Cody
Uma 9h59'2" 66d25'35"
Rose Cuni
Cas 1h38'11" 64d1'10"
Rose de Sable Roswitha
Lmi 10h55'12" 29d24'11"
Rose Dearing
Cam 3h47'50" 57d9'51"
Rose E. Croshier
Cam 7h37'0" 64d58'0"
Rose Elizabeth Hughes - Rosie's Star
And 0h48'29" 28d20'35"
Rose et Richard
Uma 11h28'9" 58d32'18"
Rose Fishman
Peg 22h17'31" 4d2'0"
Rose Forever After
Uma 11h37'55" 41d35'49"
Rose Frasca
Vir 12h13'12" 12d54'29"
Rose Gelormino
Leo 10h9'28" 24d48'53"
Rose Guadalupe Rodriguez
Leo 11h38'4" 19d27'16"
Rose Guarni
Cas 1h17'12" 63d58'10"
Rose Hannah
Ari 2h24'42" 22d42'28"
Rose Hannah Wiley
Srp 18h19'53" -0d21'50"
Rose Hester
Lib 15h10'38" -23d41'46"
Rose Hunter
Lyr 18h26'1" 35d0'25"
Rose Hurley LaBreck
Ari 3h11'44" 29d26'28"
Rose Immordino Chuva
Crb 16h9'34" 32d34'48"
Rose Jeannine Choquette
Mon 7h32'18" -0d48'2"
Rose Johansen
And 1h4'27" 43d46'31"
Rose Jung
And 23h8'10" 35d14'16"
Rose Kathleen
And 22h59'26" 51d20'55"
Rose Kenney G.G.
Vir 12h56'51" 5d42'55"
Rose Kunny
Cam 5h48'24" 56d33'50"
Rose L. Alvarado
Uma 13h27'1" 55d59'42"
Rose LeSchack
Cas 23h45'57" 57d20'5"
Rose Lights Up The World
Psc 1h4'38" 11d8'15"
Rose Louise
Ori 6h4'15" 6d28'32"
Rose Lucia Johnson
And 23h47'11" 39d17'53"
Rose M. Ventorino
Uma 10h31'13" 62d37'39"
Rose M. White
Crb 16h17'21" 27d58'9"
Rose Magnana
Lmi 10h40'41" 36d15'49"
Rose Margulies
Per 4h12'55" 31d13'47"
Rose Marie
Cyg 19h48'41" 37d11'28"
Rose Marie
Cyg 19h57'48" 38d43'34"
Rose Marie
Lib 15h8'11" -14d49'36"
Rose Marie
Sgr 18h37'41" -25d36'30"
Rose Marie Day
Cyg 19h49'8" 31d38'37"
Rose Marie "Dolly" Cavalli
Umi 9h34'23" 86d6'14"
Rose Marie Fox
And 0h37'27" 42d16'48"
Rose Marie Gentile
Ori 5h36'22" 9d30'31"
Rose "Marie" Hampson - 23/08/1947
Cas 0h56'53" 48d6'13"
Rose Marie Horsley
Uma 10h14'46" 51d56'10"
Rose Marie Iski
Uma 11h13'24" 62d13'3"
Rose Marie L. Owens
Leo 9h48'30" 20d40'38"
Rose Marie Lackey
Cnc 8h18'2" 14d23'37"
Rose Marie Lawson
Lyr 18h51'9" 32d10'47"
Rose Marie Liddy Supple a.k.a Rita
Leo 10h48'44" 22d45'42"
Rose Marie Martin
Uma 13h56'0" 62d9'18"
Rose Marie Matley
Cas 1h18'41" 57d21'19"
Rose Marie Mayfield
Sco 17h28'7" -37d39'46"
Rose Marie Redfern
Cas 0h54'51" 57d29'15"
Rose Marie Saggione
Cas 0h40'54" 60d54'12"
Rose Marie Serio
Psc 23h59'51" 0d48'47"
Rose Marie Tropf
And 0h41'17" 28d57'17"
Rose Marie Vecchiolla Tursi
Psc 0h58'48" 28d29'41"
Rose Marie Zambori
Lyr 18h54'0" 38d58'33"
Rose Mary Khoury
And 23h6'50" 40d20'49"
Rose Mary Parr
Cru 12h55'58" -57d54'14"
Rose Mary Perry
And 1h17'13" 35d5'56"
Rose Mary Smith
Lyr 18h34'56" 39d56'33"
Rose McKenzie
Uma 10h10'1" 67d16'3"
Rose Mickie "January 11, 1928"
Cyg 19h46'40" 52d56'59"
Rose Montanari
Gem 6h21'2" 26d8'12"
Rose Morgan
Leo 9h38'5" 11d53'11"
Rose "Nana" Jackson
And 0h19'41" 32d0'23"
Rose Neuringer
Cas 1h10'13" 58d52'50"
Rose Oden
Com 12h43'21" 18d31'26"
Rose of Erin
And 2h13'56" 41d24'22"
ROSE OF SHARON
Umi 14h34'35" 80d58'39"
Rose of Steel ~ Rosie
Cma 7h1'22" -31d33'2"
Rose of the Night
Aqr 23h39'20" -9d37'53"
Rose O'Keefe
And 2h18'27" 39d38'54"
Rose Olexy
Tau 4h23'42" 29d14'21"
Rose Ono
Gem 6h42'46" 20d12'24"
Rose Orenstein
Cas 0h11'53" 57d13'33"
Rose Palmer-Torres
Per 3h29'23" 47d41'38"
Rose panach'ee
Lib 15h29'37" -7d28'55"
Rose Perkins
And 2h9'50" 44d12'38"
Rose Quaglino
Uma 9h45'38" 67d17'40"
Rose Rita Ferrario
Cnc 8h54'54" 8d31'49"
Rose Rita - Tempus Fugit Amor Manet
Uma 8h22'34" 67d5'17"
Rose (Rosh) Whelan
Cas 0h36'56" 54d27'22"
Rose S. Juhnke
Cap 21h37'6" -22d53'33"
Rose Scher Bey
Crb 15h39'5" 27d39'17"
Rose Sharon
And 0h15'40" 45d14'45"
Rose Siegel
And 0h37'59" 37d42'41"
Rose Slack
Cas 23h59'17" 54d15'26"
Rose Sottile
Uma 9h32'19" 47d36'52"
Rose "spdstress" Muraoka
Cap 20h49'2" -25d2'23"
Rose Speedy's 21st - Love Nigel
Cru 12h19'54" -57d35'49"
Rose St. Clair
Lac 22h24'51" 47d42'48"
Rose Stroble
And 0h1'56" 38d3'21"
Rose Strong
Psc 23h5'58" 1d10'28"
Rose Stumpo
Uma 11h6'10" 35d39'37"
Rose Turrin
Crb 15h26'47" 29d42'22"
Rose Van
Uma 9h13'22" 47d30'35"
Rose Villasenor-Hammerson
Uma 9h48'43" 47d14'23"
Rose Walck
And 23h57'2" 45d18'4"
Rose White Haag
Cnc 9h4'20" 8d5'28"
Rose "You're The Star Of My Life"
Uma 13h57'6" 54d45'26"
Rose, I will always love you.
Uma 11h27'21" 38d48'40"
Rosealeen
Lib 15h16'17" -4d8'51"
Roseann
Cas 1h12'27" 66d43'11"
Roseann
Lyn 9h6'39" 43d42'25"
Roseann
Ari 2h33'46" 12d54'40"
Roseann
Leo 9h51'40" 27d43'1"
RoseAnn and Charles Tyree,Sr.
Cas 0h25'23" 64d14'18"

Rose-Ann El-Ghaly's Star B.M.I.T.U.
Cyg 20h9'53" 50d42'14"
Roseann Loves Tim
Cyg 21h28'42" 46d33'59"
Roseann Shields Strelkow
And 0h23'56" 35d46'3"
Roseanna and Matt's Star
Cyg 20h45'42" 37d55'33"
Roseanna Marina Giovinazzo
Cas 0h36'23" 64d36'45"
Roseanna My Love
And 0h25'50" 46d5'50"
Roseanne E. Costa
Sgr 19h46'14" -34d7'18"
Roseanne Stewart
Uma 12h22'1" 62d35'42"
roseanneisthebestmomon-earth
Uma 9h29'20" 52d43'37"
Rosebud
Uma 10h28'8" 61d14'43"
ROSEBUD
Uma 10h28'56" 68d26'4"
Rosebud
Aqr 23h11'56" -19d16'23"
Rosebuds
And 23h2'57" 35d56'45"
Roselia Ruiz Arevalo
Aqr 22h34'26" -3d46'11"
Roseline Joseph
Lyn 6h52'31" 53d44'47"
Rosella
Lyn 6h31'10" 58d6'12"
Rosella Buel-Bingham
Gem 6h0'58" 23d32'18"
Rosella Maxton
Lyr 18h44'57" 39d11'25"
Roselle McChesney
Crt 11h4'51" -18d2'34"
Rosellina
Umi 17h29'27" 80d27'26"
Rosellina Di Bona
Per 2h17'38" 52d44'17"
Rosellini
Psc 1h39'21" 18d26'46"
RoseLyn Flores
And 23h45'44" 47d27'42"
Roselyn (Lynn) Singer
Cas 2h42'4" 64d59'13"
Roselyn Williams
Dra 19h34'17" 66d40'32"
Roselyne
Cnc 8h38'43" 8d5'42"
Roselynn Helen Deleo Ruso
And 23h38'18" 34d19'5"
Rosemar 8/30/1967
Vir 12h34'15" -1d23'42"
Rosemarie
Dra 18h27'56" 78d37'25"
Rosemarie
Uma 11h50'42" 63d18'24"
Rosemarie
Cyg 20h26'49" 59d52'54"
Rosemarie
And 0h45'25" 39d9'28"
Rosemarie
Cas 1h17'57" 53d55'6"
Rosemarie
Uma 11h4'33" 48d34'58"
Rosemarie - 15th May 1933.
Cas 0h44'3" 63d14'14"
RoseMarie Agnello
Dra 16h12'9" 58d7'57"
Rosemarie Aquilone
Lac 22h33'24" 52d11'39"
Rosemarie B.
Crb 16h5'10" 37d9'17"
Rosemarie Bertauski
Sgr 18h51'15" -18d32'43"
RoseMarie Callaghan
Aqr 22h2'49" 0d19'15"
Rosemarie Cugini
Cas 1h3'46" 62d10'16"
Rosemarie DiSanza
Lyn 8h16'30" 37d16'56"
Rosemarie "Dootsie"
Crb 15h44'16" 32d49'54"
Rosemarie E. Doran
Tau 4h27'2" 15d33'18"
Rosemarie Granger
Ari 3h3'43" 27d44'47"
Rose-Marie Hernandez-Nuninga
Cam 4h28'18" 77d8'26"
Rosemarie Horowitz
Vir 13h18'13" 5d55'34"
Rosemarie Houlihan
Lib 14h40'18" -11d49'6"
Rosemarie Hummel Bonbrest
Uma 14h5'56" 54d9'4"
Rosemarie Inzinga-Patnoe
Aur 4h59'6" 36d4'22"
Rosemarie J Noronha Guterres
Cru 12h14'11" -62d6'21"
Rosemarie Kreuder
And 0h29'58" 41d7'42"
Rosemarie & Matthew
Umi 17h38'30" 88d59'33"
Rose-Marie Montler
And 0h22'37" 44d37'26"
Rosemarie Nickel
Gem 7h47'31" 16d53'29"
Rosemarie Puccinelli
Uma 11h57'34" 64d59'26"
Rosemarie Rhue Klinker
And 1h29'2" 39d24'11"
Rosemarie Scaccia
Cas 2h24'59" 66d13'30"
Rosemarie Schrader
Uma 12h2'23" 36d32'30"
Rosemarie Sienkiewicz
Leo 11h17'8" 5d59'50"
Rosemarie (Sullo) Kazokus
Cas 0h50'16" 74d8'35"
Rosemarie Thorhauer
Uma 11h23'31" 40d52'43"
Rosemarie Trainacher
Eri 4h34'16" -17d5'39"
Rosemarie Virginia Bass
Cep 22h41'2" 72d29'20"
Rosemary
Dra 18h9'17" 70d29'17"
Rosemary
Cas 1h32'6" 60d56'18"
Rosemary
Mon 7h31'24" -0d55'7"
Rosemary
Psc 0h15'52" 1d48'38"
Rosemary
Ori 5h57'20" 17d34'32"
Rosemary - A Perennial Star
Pho 1h23'27" -52d21'46"
Rosemary and Burgie's Star
Lyn 9h18'18" 35d39'36"
Rosemary And Edwin's Star
Lib 15h46'10" -17d34'47"
Rosemary April Poulson
Tau 5h27'23" 25d24'11"
Rosemary Arnold
Tau 5h51'39" 25d11'50"
Rosemary Barra
Aqr 21h11'32" 1d31'50"
Rosemary Blue-Eyes
And 0h48'7" 26d41'53"
Rosemary Bower
Cas 23h2'35" 56d59'9"
Rosemary Bowers Shea
And 23h28'55" 48d31'17"
Rosemary Branson, DVM
Sco 17h45'38" -43d51'44"
Rosemary Brescher
Crb 15h34'9" 36d36'52"
Rosemary Cathryn Williams Seminoff
Her 16h35'34" 45d36'14"
Rosemary Cavallieri
Ori 6h16'29" 10d56'47"
Rosemary Cole from Kuna, Idaho
Cas 1h17'2" 59d52'11"
Rosemary Cox MBE
Cmi 7h55'6" 11d43'19"
Rosemary Crocoll
Her 16h13'30" 44d28'55"
Rosemary Darder
Ori 4h54'46" 12d36'1"
Rosemary Dunn - Our Super Star
Cas 23h43'56" 59d55'42"
Rosemary Elaine Carroll
And 22h57'59" 48d6'4"
Rosemary Elisabeth
Lyn 6h23'56" 57d0'48"
Rosemary Elizabeth Willardson
Vir 13h15'25" 11d17'28"
Rosemary Esteves
Cyg 20h53'17" 38d51'35"
Rosemary Figueroa
Per 2h41'13" 53d22'44"
Rosemary Fogell
Tau 4h36'19" 18d19'28"
Rosemary Gallagher
Uma 10h13'11" 52d17'5"
Rosemary Gaylord
And 2h22'49" 42d14'27"
Rosemary Grow
Cnc 8h38'58" 27d43'21"
Rosemary Haas
Uma 10h28'37" 40d5'42"
Rosemary Harris
Umi 15h12'52" 76d10'1"
Rosemary Hurley
Cnc 9h6'30" 31d53'52"
Rosemary Jackson
Sge 19h34'33" 18d58'4"
Rosemary Jane Kind
And 23h10'38" 42d2'7"
Rosemary Jo Walters
Uma 9h36'10" 61d20'41"
Rosemary Joan Beecroft - Rosemary's Star
Cas 23h41'20" 55d37'59"
Rosemary June Fedon
Sco 17h27'39" -37d6'49"
Rosemary Kane
Cyg 21h51'32" 49d8'31"
Rosemary Kevil
Pho 0h48'48" -49d18'20"
Rosemary Krstevski
Cru 12h26'9" -60d18'58"
Rosemary LaBarbera
Cas 1h17'53" 64d42'13"
Rosemary Lyric's
Lyr 19h3'30" 47d1'23"
Rosemary Marguerite Sugrue
Cas 1h13'28" 53d12'59"
Rosemary Matthews 7/85
Crb 16h10'16" 37d35'5"
Rosemary Maxine Whitten (Mommy)
Tau 4h28'14" 18d3'45"
Rosemary McKinnon
And 2h19'12" 45d20'54"
Rosemary Messer
Cas 23h28'26" 53d9'9"
Rosemary Nielsen Hendler
Lib 15h19'15" -22d49'36"
Rosemary O'Connell Doyle
Cas 0h32'56" 63d59'8"
Rosemary Patricia Catena Pepe
Boo 14h24'5" 14d35'35"
Rosemary Ponist
Uma 13h46'32" 59d4'24"
Rosemary Raimondi
Umi 14h5'26" 77d59'43"
Rosemary Ray
Cas 1h24'56" 63d31'57"
Rosemary Romaine Arnold "6-26-1919"
And 1h11'33" 46d15'53"
Rosemary Shovlin
Cas 0h23'20" 61d24'35"
Rosemary Sieders
And 2h21'20" 49d25'33"
Rosemary Slagan
Gem 7h24'29" 32d27'8"
Rosemary Spano
Cas 0h49'8" 54d58'37"
RoseMary St. Elizabeth Spinelli
Del 20h41'41" 16d35'13"
Rosemary T. Rath
Uma 10h3'17" 47d0'4"
Rosemary Tisch
Com 12h48'52" 28d25'17"
Rosemary W. Damon
Vir 14h36'6" 5d29'5"
Rosemary Welsh
Aqr 22h41'9" -1d46'9"
Rosemary Wieszt
Sco 17h56'49" -31d25'29"
Rosemary-Jacob
Cyg 21h47'53" 38d56'52"
Rosemary's Angel
Lyr 18h50'52" 43d48'23"
Rosemary's Crystal Rainbow Star
Ara 17h12'57" -54d12'41"
Rosemary's Light
Aqr 23h11'13" -10d13'3"
Rosemary's Second Chance
Cam 3h59'28" 56d18'25"
Rosemary's Star
Cyg 19h41'55" 35d43'59"
Rosemary's Strength Star
Ari 3h6'47" 26d44'58"
Rosemary's Window
Her 18h54'38" 24d18'18"
RoseMichael
Uma 11h18'8" 64d34'13"
Rosen
Cnc 8h23'50" 26d9'12"
Rosen
Her 17h17'44" 15d25'3"
Rosenberg
Crb 15h45'24" 33d37'5"
Rosenberg 3
Ari 2h58'53" 26d18'36"
Rosenburg, Claudia
Uma 10h38'6" 66d23'1"
Rose-n-Dale
Lyn 6h58'9" 50d3'39"
Rosendo Ruvalcaba
Sgr 19h20'10" -19d30'46"
Rosenkranz, Tino
Uma 12h27'37" 53d25'17"
Rosenmüller, Nina
Uma 9h8'43" 51d59'20"
Rosenny
And 23h8'32" 47d54'4"
RoseNRob4EverAlwaysTogetherThankGod
Lyn 8h10'33" 35d12'40"
Rosepetal Gordon
And 0h19'29" 41d36'22"
Rose's Amor
Cas 0h37'40" 53d3'29"
Rose's Deli and Bakery
Per 2h25'17" 53d16'19"
Rose's Star
Umi 15h31'24" 75d29'37"
Rose's Wish
Cnc 8h35'48" 17d47'16"
Rosetta Baynard
Uma 14h19'39" 60d43'56"
Rosetta Jackson Lewis
Ari 3h9'24" 11d52'23"
Rosetta Marie
Lib 14h47'30" -17d49'27"
Rosetta Nadine
Psc 0h14'12" 8d50'21"
Rosette & Dominick Bertellotti
Cyg 19h32'12" 30d12'6"
Rosette Stalder
Lib 15h19'42" -19d24'58"
Rosey
Cyg 21h33'2" 49d52'17"
Rosey June Joy Yates
And 23h27'38" 47d48'2"
Rosey McDonald
Cyg 21h44'30" 53d5'10"
Roshan
And 0h39'28" 41d3'28"
Roshan
Ori 6h12'25" 15d37'27"
Roshan Kaushal
Leo 11h30'29" 19d41'57"
Roshan Nandkumar
Ind 21h54'44" -70d27'54"
Roshaun Dominic Ghanie
Her 18h5'0" 21d16'52"
Roshelle
Psc 0h51'36" 27d40'14"
Roshiah L. Jefferson
Uma 8h55'2" 64d8'48"
Roshini Neduvelil
Cyg 19h29'35" 31d2'38"
Roshko
Cnc 8h41'25" 13d43'13"
Roshni
Cyg 19h45'7" 35d8'49"
Roshni Devi Singh
Lib 14h40'19" -10d57'4"
Rosh-N-Losh
Tau 3h34'39" 11d54'28"
Roshontia, I love you
And 23h13'31" 38d28'44"
Rosi
And 23h21'56" 48d43'16"
Rosi
And 23h56'9" 35d57'21"
Rosi Accardi
Lyn 6h34'35" 54d25'17"
Rosia Ann Granger
And 23h37'19" 41d24'22"
Rosie
And 23h9'11" 43d20'57"
Rosie
And 23h24'36" 47d38'20"
Rosie
Uma 10h55'26" 49d43'59"
Rosie
Uma 10h56'45" 52d6'6"
Rosie
And 1h37'57" 41d21'27"
Rosie
Uma 11h49'57" 41d56'54"
Rosie
Crb 15h24'41" 30d47'16"
Rosie
Ori 6h17'5" 7d23'35"
Rosie
Vir 12h32'19" 12d57'1"
Rosie
Ori 6h17'10" 16d17'9"
Rosie
Tau 4h27'30" 15d29'10"
Rosie
Ari 3h3'54" 29d26'14"
Rosie
Tau 3h47'30" 29d1'31"
Rosie
Tau 4h39'3" 27d57'56"
"Rosie"
Leo 9h35'1" 26d35'46"
Rosie
Uma 9h19'21" 54d31'37"
Rosie
Sgr 19h6'37" -11d58'46"
Rosie
Psc 0h13'41" -4d11'43"
Rosie
Vir 13h27'48" -3d10'23"
Rosie
Cma 6h19'30" -30d40'21"
Rosie 5/11/03
Umi 15h33'57" 71d29'4"
Rosie Alicia Morales
Lib 15h39'23" -17d46'54"
Rosie Alvarado
And 1h19'51" 36d13'53"
Rosie Bear
Psc 0h21'17" -3d20'41"
Rosie Begley
Gem 7h39'57" 24d4'45"
Rosie Cheeks
Leo 11h16'0" 14d9'59"
Rosie Desmond Miller
Cas 23h20'16" 59d3'25"
Rosie Elizabeth Burke
Cep 3h22'51" 82d30'3"
Rosie Espino
Del 20h43'3" 8d49'1"
Rosie Estrella Rubio
Cyg 19h55'29" 39d2'21"
Rosie Foldvari
Cas 0h6'21" 53d9'23"
Rosie "Good Dog" Radebaugh
Equ 21h12'34" 9d1'54"
Rosie Green
And 23h4'28" 49d43'41"
Rosie Harvey Stovall
Cap 21h22'21" -15d56'8"
Rosie Hefter
Gem 7h44'52" 23d23'35"
Rosie Hostetler
Uma 10h48'31" 50d34'33"
Rosie In The Sky With Diamonds
Ari 2h33'14" 27d31'1"
Rosie K.
Leo 11h41'23" 26d31'54"
Rosie Ling
And 0h11'27" 47d31'17"
Rosie Loy
Mon 6h37'42" 5d34'32"
Rosie Lynn
Sgr 19h25'32" -21d8'59"
Rosie Lynne
And 22h58'17" 39d28'21"
Rosie Macaluso at 100
Ari 3h19'42" 20d5'23"
Rosie Mae Rice
And 1h51'6" 45d55'18"
Rosie Marriott
Lyr 18h49'51" 34d46'5"
Rosie May Peacock
Cyg 21h40'30" 36d18'55"
Rosie Mccormick
Ori 6h24'22" 13d55'24"
Rosie Mo Tuesday
And 0h29'25" 42d7'16"
Rosie Oldham
Sco 16h8'31" -16d59'52"
Rosie Ortiz
Peg 21h23'48" 18d15'24"
Rosie Pennington Smith
Aqr 22h15'5" -0d39'59"
Rosie Posie
Leo 11h39'29" 28d7'9"
Rosie R. Martinez
Cyg 20h35'31" 30d44'15"
Rosie Roe
Cas 23h40'9" 55d25'36"
Rosie - Springer Spaniel - with love
Cnv 13h45'45" 38d55'54"
Rosie T.
Peg 23h18'57" 26d43'4"
Rosie T.
Peg 22h46'19" 4d26'25"
Rosie Valentina Rivas
Lmi 10h27'8" 33d15'5"
Rosie Volpe
Vir 13h16'38" -12d58'8"
Rosie Ward
And 23h37'17" 37d15'46"
Rosie Warland
Umi 13h39'29" 76d2'11"
Rosie Whitesell
Lib 14h55'40" -3d55'8"
Rosie Wrigley
Uma 11h54'45" 63d3'43"
Rosie "Xiao Momma" Gonzales
Sgr 18h40'37" -16d1'57"
Rosie, Our Shining Star
Lib 14h53'21" -24d15'3"
RosieandMatt
Lib 15h15'17" -5d0'44"
Rosie's Rhinestone
Cyg 20h34'29" 55d30'48"
Rosiland Perlmutter
Aql 19h17'50" -8d28'48"
Rosilyn Mitchell
Lyr 18h49'22" 29d41'26"
Rosilyn - The Light Of My Life
Uma 9h33'25" 60d21'13"
rosimon . in ewiger Liebi
Lac 22h57'48" 38d18'5"
Rosina
Uma 9h27'8" 66d51'32"
Rosina Bella
Lyn 8h5'11" 42d58'39"
Rosina May Tao
Tau 4h14'11" 27d8'12"
Rosina (Rose) Scarantino
Mon 7h17'58" -0d1'43"
Rosinha La luz de Johnny
Vir 13h53'13" 1d7'27"
Rosio Mendoza
Cnc 8h32'4" 24d41'51"
Rosita
Vir 13h18'1" 5d59'35"
Rosita & Mario (Rose & Murray)
Lyn 7h20'13" 44d41'19"
Rosita Montano
Lyr 19h20'5" 29d9'40"
ROSITA-GT
Crb 15h31'5" 30d19'29"
Rösli Suter
Dra 18h52'40" 59d45'44"
Roslyn 49
Uma 11h16'4" 47d16'58"
Roslyn Denning O'Toole
Lib 15h33'21" -19d40'59"
Roslyn M. Council
And 0h51'42" 39d7'52"
Roslyn Mary Fontenot
Cas 0h43'11" 53d31'36"
Rosmahwati
Cap 20h28'54" -27d23'44"
Rosmaliz Valdes Bous
And 0h36'19" 42d8'57"
Rosmari Galli
Col 6h26'0" -35d53'2"
Rosmarie Andres-Käch
Umi 10h5'15" 86d49'30"
Rosmarie Graf Schranz
Oph 17h40'6" 12d45'3"
Rosmarie Müller
Umi 16h13'6" 79d57'29"
Rosmarie Pel
And 1h44'23" 43d1'32"
Rosmarie Spörri
Ori 5h0'55" 4d36'52"
ROSMARTIN
Uma 10h15'36" 49d51'31"
ROSMIKEH
Uma 11h23'39" 59d48'27"
Rosó
Vir 13h23'21" -19d15'21"
Ross
Sco 16h38'29" -37d8'0"
Ross
Sgr 18h42'38" -24d10'48"
Ross
Psc 0h40'41" 5d2'17"
Ross
Vir 13h47'15" 5d54'39"
Ross
Peg 22h17'32" 16d59'15"
Ross Alan Cherry
Her 17h52'58" 46d55'25"
Ross Alexander Layden
Sco 16h10'11" -25d56'51"
Ross Alexander Mckenzie Mcgovern
Her 16h19'12" 21d1'15"
Ross and Angela Terry
Ori 4h59'52" 2d54'53"
Ross and Ashley
Aql 19h1'41" 14d58'23"
Ross and Cherie
Lyr 18h20'35" 31d59'54"
Ross and Jennifer Dalgleish
Cyg 20h51'15" 32d4'2"
Ross Andrew Miller
Her 16h38'4" 35d58'35"
Ross Bernard
Uma 9h2'43" 63d21'26"
Ross 'Big Boy' Brown
Umi 15h5'41" 76d33'39"
Ross Brake
Lib 14h51'31" -3d28'46"
Ross Chartier
Uma 11h7'46" 41d59'56"
Ross Ciavarella's Magic Monkey Star
Lib 14h49'0" -24d44'24"
Ross' City of Lights
Tau 4h7'35" 27d1'11"
Ross & Courtney
Sgr 17h52'48" -28d47'4"
Ross Ducept
Uma 11h5'23" 61d58'57"
Ross Edward Geren
Aql 19h0'56" -0d29'39"
Ross Edward Humes
Lib 15h24'59" -27d40'2"
Ross Elliot Hunt
Lyr 18h42'1" 38d55'35"
Ross Erickson
Ari 2h17'21" 22d48'57"
Ross Eugene Miller
Sco 17h54'17" -36d43'31"
Ross Evan
Ori 6h2'47" -0d44'26"
Ross&Frederique Marks 4 Life
Cyg 20h43'48" 34d3'12"
Ross James Sneddon
Uma 10h47'44" 69d36'56"
Ross & Jen
Uma 11h15'25" 46d57'42"
Ross Jonathan Mlnarik
Leo 11h1'24" 11d34'56"
Ross Kaya Jess
Tau 4h24'6" 19d8'22"
Ross & Kris's Eternal Love
Uma 13h38'57" 56d37'28"
Ross Leach
Sco 17h34'53" -31d58'44"
Ross Lee Morgan
Ari 3h19'9" 24d37'8"
Ross Lloyd Fogarty - 22.12.1945
Cap 20h28'10" -15d38'25"
Ross Logan Collord
Dra 18h39'33" 51d59'3"
Ross loves Jo forever!
Cas 0h6'21" 58d21'15"
Ross Malet
Uma 9h59'57" 59d24'10"
Ross Martin Patrick Nevin
Uma 11h13'42" 30d11'52"
Ross McCleary Wheeler
Her 17h52'57" 37d21'41"
Ross McMartin
Uma 9h5'45" 48d39'8"
Ross Michael Olson
Tau 4h23'0" 27d0'37"
Ross Morgan Hallo
Uma 9h19'54" 49d21'33"
Ross Morgan Ruben
Her 18h41'41" 24d23'42"
Ross Myron To Infinity and Beyond
Umi 15h19'54" 68d8'25"
Ross Noble
Cep 20h38'58" 61d16'7"
Ross Nolen Harms
Cap 21h42'39" -15d14'22"
Ross Pawley-Kean
Her 16h11'50" 10d44'24"
Ross Peterson
Umi 5h9'38" 88d57'54"
Ross' rendezvous
Uma 11h57'8" 56d46'26"
Ross "Rusty" Toyo Snyder
Tau 4h42'44" 6d23'54"
Ross Terry
Ori 5h31'6" 3d38'47"
Ross Wilson Beattie
Per 4h32'33" 32d42'49"
Ross Word
Ori 5h27'19" -0d50'31"
Ross012006
Aqr 23h52'28" -12d54'6"
Rossana
Sco 16h17'56" -11d55'7"
Rossana
Ori 5h23'46" -3d53'8"
Rossana
Cyg 19h58'27" 33d9'0"
Rossana Rocchi
Peg 22h43'2" 4d57'32"
Rossanna
And 0h11'59" 38d59'17"
Rossella
Her 18h11'31" 14d32'25"
Rossella
Her 18h57'16" 15d2'48"
Rossella
Ori 6h10'17" -2d11'34"
Rossella
Cma 7h14'34" -31d28'22"
Rossen Marinov
Uma 9h20'42" 56d13'37"
Rossi
Gem 7h4'27" 21d49'50"
Roßkothen, Kurt
Uma 11h29'45" 29d53'10"
Rößler, Hans Dieter
Gem 6h28'54" 19d21'23"
Rosslyn.
Uma 10h1'54" 65d11'45"
Rossoll, Hans-Jürgen
Uma 12h59'31" 60d15'10"
ROSSPAR
Per 2h10'29" 52d44'0"
Rosstopher
Her 16h46'10" 7d16'29"
Rossy
Uma 11h1'44" 36d11'54"
Rossy
Cap 20h33'8" -26d33'49"
Rostrevor Apts.
Lac 22h17'12" 52d59'47"
Roswell Weil
Aqr 22h41'40" -2d27'28"
Roswitha
And 1h27'40" 41d0'18"
ROSWITHA
Psc 0h47'17" 9d45'7"
Roswitha Janson
Uma 9h49'41" 68d41'55"
Rosy
Cyg 19h44'46" 46d35'29"
Rosy
Ori 4h57'17" 15d6'3"
Rosy Betty Boop
And 2h33'16" 39d37'38"
Rosy Leyva
Her 16h39'42" 7d16'10"
Rosy Salgado
And 1h56'11" 41d17'13"
Rosy Salgado
Cas 0h43'59" 48d10'22"
Rosy Salgado
Cas 0h44'30" 48d1'41"
Rosy-sixty
Ori 5h45'59" -3d12'57"
Rota Y. Tan
Cam 4h33'16" 66d19'23"
Rota-Anderson
Sge 19h48'40" 17d3'10"
Rotary Club of York
Lyr 18h50'32" 34d48'47"
Roth Family Star
Cyg 20h30'21" 55d33'32"
Roth-Bottoni (Wishes Do Come True)
Psc 1h18'20" 25d16'33"
Rothe
Vir 13h6'19" -6d39'57"
Rothe, René
Uma 11h3'55" 55d50'5"
Röttger, Fabienne
Uma 11h32'44" 31d16'55"
Rotzler, Roland
Ori 5h11'46" 0d31'55"
Rouba
Ari 3h10'26" 11d44'12"
Roubal
Tau 5h35'31" 18d50'21"
Roudoudou
Cas 0h4'8" 53d36'41"
Roula E. Salim
Aur 6h9'13" 53d49'7"

Roula & Marwan
Aqr 20h53'54" 0d46'4"
Roula &Tony Abi Nasr
Vir 11h51'44" 2d57'56"
ROULAKI
Leo 10h35'23" 7d31'21"
Round House
Cep 0h36'47" 86d56'9"
roundhead..wishes do come true
Crb 16h21'44" 30d12'1"
Roundrock Apartments on Sea
Cnv 12h22'39" 48d32'33"
Rouquiah Al-Rifae
Uma 11h49'40" 59d37'53"
Rousana Razey
Psc 1h10'57" 26d25'4"
Rousse
Gem 7h39'23" 24d1'59"
Rovalon
Umi 18h29'50" 88d13'3"
ROVATAZATOLOU
Aql 19h43'20" 3d58'10"
Row & Chels
Cru 12h29'25" -64d32'45"
Rowan
Cnc 8h28'46" 23d27'4"
Rowan
Gem 7h52'51" 32d6'0"
Rowan Alexander Gannon
Per 3h39'15" 48d6'0"
Rowan Alexandra
And 2h25'1" 48d6'38"
Rowan and Sally Dore
Uma 9h5'45" 72d2'15"
Rowan Constant
Uma 10h32'19" 45d39'51"
Rowan Edward
Peg 22h34'42" 29d15'54"
Rowan G. Naicker
Cru 12h49'22" -63d52'12"
Rowan Groff
Leo 11h32'10" 14d8'26"
Rowan Hewawissa - Always A Star
Pho 23h40'44" -43d19'7"
Rowan Martin Gurr
Per 2h47'57" 50d2'59"
Rowan Rachel Binkley Jones
Leo 10h22'9" 8d35'57"
Rowan Sabine Dykstra
Umi 14h38'3" 73d21'2"
Rowan Sabine Dykstra
Umi 14h35'38" 72d49'30"
Rowan Saunders
Gem 7h19'13" 26d24'7"
Rowan Silverwolf
Ori 5h27'19" -0d37'11"
Rowan Thomas Morgan
Gem 6h48'5" 35d7'49"
Rowan Travis
Ori 5h53'5" 20d25'9"
Rowdy Ty Soileau
Psc 1h32'14" 9d13'31"
Rowdy Walker
Leo 10h3'17" 25d6'46"
Roweina Morgan McNeil
Crb 15h51'26" 36d11'58"
Rowen Matthews
And 2h30'53" 46d15'23"
Rowena
Cas 0h13'49" 52d23'17"
Rowena
Uma 12h13'50" 62d39'58"
Rowena
Uma 10h19'56" 59d56'31"
Rowena
Cas 23h14'0" 59d4'27"
Rowena Bradnam's 40th Birthday Star
Cas 2h39'38" 64d30'46"
Rowena Horne
Vul 20h40'42" 25d51'5"
ROWENA LACANIENTA
Aqr 20h40'37" -2d11'39"
Rowena Paras Jocson
Leo 10h32'1" 26d31'41"
RowenaD
And 1h4'3" 38d29'25"
Roweyda C. "Sandy" Pearse
Cyg 21h14'31" 42d26'50"
RowMer 110204
Uma 10h24'24" 60d23'29"
ROWNAN LEVI MOLONEY 14072007
Psa 22h59'49" -29d57'57"
Rox
Uma 14h18'51" 59d15'36"
Rox and Jess
Sco 17h1'36" -39d49'18"
Rox & John Buonauro
Cam 3h54'41" 70d26'47"
Rox Star
Ari 3h7'2" 26d55'53"
Roxan Le Metayer
Cam 3h33'4" 68d18'7"
Roxana
Sgr 18h28'8" -34d48'27"
Roxana
Ari 2h42'41" 27d23'11"
Roxana Consuelo Barajas
Leo 11h31'22" 18d22'46"
Roxana Enid Cintron
Lib 15h3'42" -9d2'2"
Roxana Hodges
Aql 19h28'13" 12d20'52"
Roxana Maria Horsch
Crb 15h40'16" 26d39'45"
Roxana Quintero
Sgr 19h16'2" -16d36'20"
Roxana Santamaria
Ori 5h32'23" 3d41'29"
Roxana Sills
Lyn 8h27'33" 51d13'24"
Roxana Toma
Vir 13h17'50" -21d44'7"
Roxana y Michael 03-22
Ari 2h48'48" 24d48'41"
Roxane
Leo 9h31'18" 29d18'27"
Roxane
Gem 6h18'0" 25d20'19"
Roxane
Psc 0h21'41" 3d38'31"
Roxane & Andrew (The Dreamers)
Sco 16h52'31" -44d40'38"
Roxane et René
Uma 11h55'7" 40d25'3"
RoxAnn Delmonto
Per 3h29'2" 45d42'5"
Roxann K Rode Miller
Aqr 21h40'28" -0d55'55"
ROXANN "Light Of Lanny's Life"
Uma 8h35'4" 61d6'5"
Roxann & Ray Unis
Cyg 19h40'50" 54d55'7"
Roxanna Bonita
Vir 14h15'16" 6d26'48"
Roxanna Grace Cooper
Aqr 21h28'47" -2d42'25"
Roxanna Jeanne
Gem 7h39'19" 27d27'59"
Roxanna Khidan
And 0h54'56" 38d19'17"
Roxanna Lee
Uma 11h53'12" 55d4'28"
Roxanna Rose
Ari 3h22'18" 22d40'30"
Roxanna Rose The Star
Cas 0h58'29" 57d0'44"
Roxanna Short
Uma 10h34'19" 72d31'37"
Roxanne
Cas 2h7'26" 65d30'28"
Roxanne
Umi 16h24'33" 76d26'14"
ROXANNE
Cru 12h44'18" -62d50'7"
Roxanne
Cas 0h8'35" 59d34'34"
Roxanne
And 23h36'24" 50d38'41"
Roxanne
And 2h16'20" 49d5'20"
Roxanne
Cyg 19h41'39" 38d31'19"
Roxanne
And 1h19'48" 37d40'36"
Roxanne
And 0h31'42" 39d26'59"
Roxanne
And 1h9'54" 36d48'16"
Roxanne
Gem 7h18'9" 15d52'52"
Roxanne
Vir 13h1'46" 6d23'47"
Roxanne
Leo 9h55'13" 7d1'6"
Roxanne
Ori 5h20'5" 4d22'25"
Roxanne A. Economous
Ori 5h50'10" 11d22'4"
Roxanne Alexia Mourinho
Leo 11h21'50" -3d49'11"
Roxanne Alexis
Tau 5h36'2" 26d7'33"
Roxanne Allen
Vir 12h46'16" 8d20'53"
Roxanne and Mailey
Cyg 19h59'11" 36d11'23"
Roxanne Camacho
Crb 16h14'29" 26d17'0"
Roxanne DeBolt Hernandez
Cas 3h12'22" 65d21'43"
Roxanne Dewalt Butz
Cas 2h25'54" 64d42'34"
Roxanne Elaine
Vir 13h18'26" 5d48'16"
Roxanne Geller
Sco 16h15'31" -12d29'20"
Roxanne Hunter
Cam 3h20'13" 67d58'8"
Roxanne Kiel
Tau 4h6'55" 6d16'43"
Roxanne L. Robbins
Crb 16h18'56" 34d39'44"
Roxanne Malbrough
Aqr 22h36'49" -16d2'23"
Roxanne Marie
And 23h13'19" 45d14'44"
Roxanne Marie Fernandez
Com 12h43'41" 27d20'55"
Roxanne Marie Scheuer
Lib 15h42'0" -17d59'29"
Roxanne Martinez
Cas 23h58'53" 59d53'45"
Roxanne McCahan
And 23h44'54" 42d25'50"
Roxanne Nicole Ryken
Leo 11h39'26" 16d28'39"
Roxanne Othy
Psc 1h20'25" 6d30'10"
Roxanne Painter
Aqr 23h29'7" -19d45'43"
Roxanne Painter
Aqr 21h13'15" -14d16'18"
Roxanne Penny
Ori 5h40'37" 3d59'58"
Roxanne Rau's "Radiance"
Cam 3h32'56" 64d56'51"
Roxanne Ruth Newman
Sgr 19h38'47" -44d1'3"
Roxanne Sandra
Ori 5h17'11" -8d51'46"
Roxanne Snell
And 23h41'49" 42d53'28"
Roxanne & Tiffany
Lyr 18h49'49" 31d52'15"
Roxanne Torrez Ablog
Aqr 20h43'22" -9d12'21"
Roxanne Umphrey
Hya 9h0'26" -0d19'34"
Roxanne Veronica Hernandez
Uma 9h12'57" 63d28'39"
Roxanne Wright
And 23h22'13" 51d12'16"
Roxanne101004
Leo 11h2'15" 16d25'8"
RoxannePizzolo
Cas 1h47'22" 64d12'40"
Roxanne's Flutterby
Leo 9h56'32" 8d45'5"
Roxi
And 23h53'22" 48d10'31"
Roxi M.
Crb 15h49'28" 39d32'28"
RoxiBob
Ari 3h19'43" 27d32'18"
Roxie
And 23h35'27" 49d42'46"
Roxie
Cnv 13h54'12" 31d41'15"
Roxie
And 1h29'58" 35d25'45"
Roxie Baby
Aqr 22h29'8" -3d10'24"
Roxie Carol
And 0h8'22" 41d35'25"
Roxie Jo
Lyn 7h46'43" 57d49'30"
Roxie la Doxie
Oph 16h53'24" -0d54'52"
Roxie Lopez "My Spanish Rose"
Uma 9h1'32" 52d8'36"
Roxie M "#1 Mom and Grandma"
Per 2h58'3" 41d56'53"
Roxie May Lewis Delyea
Cnc 8h42'14" 7d31'37"
Roxie Parham
Uma 12h10'39" 48d29'57"
Roxie Robinson
Cas 1h25'36" 55d27'57"
Roxie the Cricket Queen
Crb 15h36'48" 28d24'44"
Roxie's Ray
Uma 11h37'31" 34d2'22"
Roxie's Star Light
Leo 9h57'36" 24d56'24"
Roxi's Star
Tau 5h19'11" 23d42'37"
Roxsan Albanese
Uma 9h1'0" 64d46'59"
Roxstar
Lyr 18h49'7" 37d32'59"
Roxxy Daniela Beltran
And 23h12'9" 37d34'19"
Roxy
And 23h11'42" 43d17'47"
Roxy
Gem 7h21'20" 31d53'42"
Roxy
Gem 6h58'35" 25d56'11"
Roxy
Cmi 7h32'59" 11d8'54"
Roxy
Vir 13h27'46" 12d18'21"
Roxy
Uma 13h59'39" 57d40'48"
Roxy
Cir 14h51'39" -65d41'26"
Roxy Bat Dog
Cma 6h36'36" -23d21'27"
Roxy Claire
Cru 12h28'46" -62d5'34"
Roxy Ridgeway
And 2h33'38" 37d37'31"
Roxy (Rocky) Blair
Lyn 8h16'42" 45d55'3"
Roxy, always our Pretty Girl
Cap 21h42'38" -9d52'8"
Roxy082369
Vir 13h11'58" 10d52'1"
Roxy331
Ari 2h47'58" 26d17'3"
Roxzy Mclean
Cnc 8h30'36" 23d5'50"
Roy
Her 16h51'14" 25d56'49"
Roy
Uma 9h38'50" 62d13'29"
Roy
Dra 17h28'59" 68d11'24"
Roy Alexander Di Vittorio
Tau 4h36'56" 15d59'18"
Roy Allen Davis
Pic 5h56'28" -55d59'37"
Roy Allen Sawyer, 111
Aql 18h54'7" -0d47'4"
Roy Allen Thieme, Jr.
Her 16h32'12" 31d18'0"
Roy and Edith Dean
Uma 11h58'3" 63d38'21"
Roy and Hazel Carver
Cyg 20h19'55" 47d31'27"
Roy and Kimberly Skeins
Tau 4h23'58" 27d26'54"
Roy and Phyllis Lewis
Eri 3h54'51" -0d25'13"
Roy and Sibyl Lewis
Sge 19h54'50" 18d15'37"
Roy Andrew Barela
Cyg 19h54'20" 36d3'1"
Roy Atkins & Jack Seabright
Cyg 21h46'17" 49d13'40"
Roy B. Woolsey
Gem 6h46'43" 13d27'5"
Roy Balduf
Boo 14h45'11" 52d57'3"
Roy Benin
Cep 21h16'3" 64d16'3"
Roy Boardman
Cep 3h58'30" 81d35'16"
Roy Boy Fairy
Lyn 6h54'41" 55d26'30"
Roy "Moody Blue" Finney
Aur 5h25'22" 43d31'15"
Roy Brown
Cep 20h20'39" 60d34'16"
Roy Butler
Psa 22h52'0" -28d47'40"
Roy C. Monsour
Ori 6h11'45" -0d21'9"
Roy Call
Uma 11h41'41" 44d3'14"
Roy Carpenter
Aur 5h31'39" 48d19'44"
Roy Carter
Uma 11h3'45" 42d5'40"
Roy Charles Simpson
Aqr 22h24'57" -0d31'6"
Roy Christian
Dra 15h14'42" 56d52'12"
Roy Cocovski - 31 January 1956
Aqr 21h20'7" -2d39'3"
Roy Corson
Leo 9h28'52" 19d9'33"
Roy D. Albert
Per 3h10'2" 51d44'4"
Roy D. Jacobson-God's Shining Star
And 2h26'18" 49d21'25"
Roy & David
Ori 6h19'46" 14d33'28"
Roy Dean
Per 3h27'55" 41d46'11"
Roy Derek Trugler
Uma 10h27'16" 63d6'24"
Roy Dormer & Victoria Baumann
Uma 10h53'49" 49d36'35"
Roy Dupuis
Tau 3h42'18" 17d19'33"
Roy E. Rogers
Uma 10h16'18" 62d25'19"
Roy Edwin King
Tau 3h56'24" 8d24'24"
Roy Eugene Burrows III
Peg 23h38'30" 12d15'23"
Roy Eugene Roberts
Aql 18h55'53" -0d22'40"
Roy F. Meyer's Obsession
Vir 13h22'14" 5d30'20"
Roy Fisher
Ori 6h11'14" 15d13'40"
Roy Fong's Lucky Star
Her 17h43'3" 38d22'29"
Roy Fong's star
Psc 1h43'45" 18d4'42"
Roy Forrest Gorton
Boo 14h56'31" 19d56'37"
Roy Generoux; Dear Husband, Dad, Poppa
Uma 11h32'35" 56d16'26"
Roy Gordon Cole
Leo 10h7'4" 20d18'37"
Roy Gustafson, Astronomer
Aql 19h41'33" 7d53'8"
Roy H. & Dorothea N. Cooper
Aqr 22h9'22" 2d13'48"
Roy H. Watson
Ori 4h51'24" -0d38'30"
Roy Hall
Ori 5h0'0" -0d15'32"
Roy Irvin Pate
Per 3h55'4" 31d31'58"
Roy Jaerae Butler
Gem 6h56'11" 14d19'3"
Roy James Bell
Leo 9h28'48" 25d1'34"
Roy Joseph Flery
Uma 11h2'9" 64d30'16"
Roy Joseph Llewellyn
Lib 14h51'25" -0d57'54"
Roy Kahn
Cyg 21h16'24" 46d0'32"
Roy Kellon Smith
Psc 0h46'21" 10d19'0"
Roy & Kelly
Uma 10h11'22" 41d55'46"
Roy Kresse
Uma 11h6'46" 43d10'22"
Roy L Austin #50
Aqr 23h7'58" -16d34'41"
Roy L. Rucker, Sr.
Uma 8h37'24" 51d33'36"
Roy Lantz
Aur 5h29'6" 41d19'16"
Roy Lee
Cnc 8h12'14" 31d36'47"
Roy Lee Hammond
And 0h42'11" 25d22'15"
Roy Little 1947
Cru 12h29'42" -61d16'49"
Roy Lynn Mayfield Family
Uma 9h44'26" 47d59'52"
Roy Main
Aur 5h57'44" 36d57'43"
Roy Marshall
Her 16h31'35" 29d59'27"
Roy Mason
Cyg 20h10'16" 51d3'8"
Roy Michael Elliott (Scorpinock)
Sco 16h10'48" -15d47'56"
Roy Michael Wallis
Uma 8h51'44" 65d52'37"
Roy Moehrke
Aqr 22h28'46" -1d7'29"
Roy Moss
Uma 10h38'37" 70d58'56"
Roy Neideigh
Sco 16h16'16" -13d31'53"
Roy Nesler - Rainman
Boo 14h20'34" 48d53'29"
Roy Nordin
Lib 15h16'46" -20d55'40"
Roy Oneto
Psc 1h16'29" 17d36'50"
Roy P. Watmough JR
Vir 14h42'40" -1d48'31"
Roy Paradis
Boo 13h39'13" 22d28'5"
Roy Paul Marshall
Her 18h42'26" 15d40'13"
Roy Phillips
Cap 20h33'23" -22d48'48"
Roy Ricketts
Sco 17h22'58" -42d53'0"
Roy Robert Everingham Sr. * Laura
Leo 10h19'27" 22d35'38"
Roy Sanders
Aqr 22h58'29" -9d12'6"
Roy Smales
Aqr 22h14'36" -19d37'10"
Roy Steven Gillum
Boo 14h28'27" 26d2'3"
Roy T. Collins
Uma 10h26'41" 66d35'31"
Roy Thaddeus Janowiak
Aur 5h37'57" 50d17'28"
Roy Thompson (1955-2006)
Vir 14h32'33" 1d43'46"
Roy Tobuk III
Col 5h51'13" -33d41'42"
Roy Truman Chapman
Ori 5h27'29" 8d20'38"
Roy Urban, my one and only love
Uma 10h46'50" 52d33'38"
Roy V. Lohndorf, Jr.
Ori 5h3'33" 3d17'8"
Roy Victor
Cas 1h0'51" 63d40'46"
Roy Victor Bowditch
Per 3h13'39" 47d25'39"
Roy Wilson
Uma 9h51'49" 66d51'55"
Roy Wilson Altman Jr.
Uma 9h46'30" 51d47'28"
Roy Wilson, Jr.
Uma 10h59'8" 55d39'57"
Roya Dinbali
Dra 16h24'17" 51d38'11"
Roya Venencia Zaucha
Ari 2h18'25" 14d26'26"
Royaa
Ori 5h6'21" 9d58'0"
Royal Dad
Ori 6h19'6" 8d22'39"
Royal Martia Shelton
Lyn 7h0'35" 48d39'47"
Royal Pride
Ori 6h6'26" 16d54'28"
Royal Princess Madeline
Peg 22h30'37" 7d58'51"
Royal Star Lux
Ori 4h53'15" 4d36'36"
Royal Thomas- ONeill's Star of Hope
Leo 11h51'59" 22d6'0"
Royal Westmoreland
Per 4h36'26" 52d2'18"
Roya-le
Cyg 19h45'54" 35d54'52"
Roy-Alexander Schmid, 29.10.1981
And 1h29'50" 50d21'32"
Royalty
Dra 18h43'16" 56d4'13"
Royanna Marie Robejsek
Ori 5h40'9" -0d32'27"
Roya's Star
Sgr 19h33'6" -36d58'34"
Royce G Woolever
Ori 5h30'57" -3d47'20"
Royce Milton Colby
Ori 5h8'54" 4d11'31"
Royce & Nicole
Uma 10h12'25" 70d42'18"
Royce R. Howell
Uma 11h36'50" 59d30'50"
Royce & Teresa
Cyg 19h51'25" 30d1'8"
Royce Thunder - MM/RS
Ori 5h55'20" 16d40'10"
Royce Whiddon
Uma 9h57'35" 59d35'3"
Royce Wood
Aql 19h47'5" -0d32'53"
Royee Gak
Psc 0h33'32" 18d58'1"
Roy-G-Bov
Lib 15h17'50" -26d23'17"
Roylex
Uma 9h56'47" 42d32'19"
Roy's 40th Birthday Star
Uma 9h18'57" 58d32'26"
Roy's Hole in One
Her 17h39'56" 16d47'34"
Roy's Horizon
Leo 10h34'22" 21d53'19"
Roy's Princess
Dra 17h36'48" 51d50'21"
Roz
Tau 4h14'38" 3d46'28"
Roz Horwitz
Uma 11h39'54" 62d5'11"
Roz(Roc) Brown
Cnc 9h18'2" 8d38'25"
Rozafa Noelani
Apu 16h32'54" -70d26'56"
Rozalia Kocjan
And 2h16'1" 45d30'31"
Rozalia Michél
Cas 0h29'13" 61d21'43"
Rozalin & Justin
Cyg 20h24'49" 37d15'19"
Rozanne Shoshi Solomon
Cas 1h16'20" 56d50'44"
Rozanne Snoberger
Cnc 8h5'10" 23d51'21"
Rozi
Gem 7h48'1" 19d34'30"
Rozi & Rudi
Cyg 21h32'21" 51d16'36"
Roziczky Pál 50
Cas 1h21'52" 68d3'10"
Rozier Ford Lincoln Mercury, Inc.
Uma 10h14'56" 48d39'48"
Rozina
Ori 6h18'13" 19d7'46"
Roz's Dad
Ori 5h39'20" 8d51'35"
ROZTOMILKA
Cam 4h45'26" 66d30'5"
Rozy Gevorkian
Sgr 18h13'23" -24d47'45"
Rozy I Love You
Leo 11h49'4" 20d18'12"
Rozzina
Psc 1h46'52" 22d3'26"
R.P. Mangogna
Per 4h14'38" 36d4'54"
RP2006
Tau 4h52'36" 16d44'54"
rpaige -n- ming
Lib 15h56'32" -12d40'21"
RPE - Rhonda's Light
Cru 12h6'48" -61d41'47"
RPGILES
Ari 3h15'5" 27d25'43"
R.P.M. Grisoni
Peg 21h21'39" 15d38'24"
RPM06
Cam 4h31'9" 66d0'39"
RPS, Insurance Star
Aur 5h9'52" 47d0'24"
RPSJR21 -Ronald P. Smith, Jr
Psc 23h18'12" 0d31'37"
RQ Cordell 2
Ari 2h41'13" 13d44'34"
R.R. "Putzi"
Ori 6h17'41" 11d9'40"
RRCIII
Sgr 19h54'40" -34d33'3"
Rrooxdanneey
Uma 8h49'38" 70d59'43"
RSC2005
Ori 5h33'9" -0d34'32"
RSC2006
Uma 11h33'44" 47d50'40"
rsjohn01
Lib 14h52'20" -5d27'0"
R-souls
Uma 9h25'2" 45d44'3"
RSR
Psc 0h18'28" 5d1'46"
RsRc030504
Gem 6h52'52" 21d51'14"
R.T. Beard, III
Her 17h9'40" 15d32'15"
Rt. Rev. Lambert Reilly,O.S.B.
Boo 14h23'26" 14d51'44"
RT3
Umi 16h17'12" 76d53'42"
RTFlach
Per 3h11'35" 52d41'14"
R-Tinks
Cyg 21h41'8" 34d14'52"
RTJ Mahoney
Aqr 22h22'36" -14d48'34"
RTT mon éternel fiancé
Psc 23h39'21" 3d3'44"
Ru
Gem 6h41'48" 19d33'25"
Ru Ru
Aqr 21h5'34" -6d38'49"
Ru Shen
Lib 15h19'39" -4d37'15"
RU.5*
And 1h19'6" 41d44'41"
Ruairi Collier McNeill - "Ruairi's Star"
Her 16h43'46" 8d30'11"
Ruairi Oisin McCracken
Cep 22h9'42" 61d13'35"
Ruaraidh Potts (Super Nova)
Cas 1h14'4" 67d31'51"
Rubano Super-Star
Ori 5h33'3" 1d4'42"
Rubber Ducky
Ari 2h32'18" 27d41'19"
Rubber Ducky
Gem 6h47'14" 26d55'57"
Rubber Starlet
Lyn 7h42'35" 36d39'43"
Rubber Toe
Cam 4h22'51" 57d27'59"
Rubel, Monika
Uma 9h30'31" 48d54'12"
Ruben
Per 4h5'34" 45d26'54"
Ruben
Ori 6h18'19" 11d25'19"
Rubén
Uma 12h4'1" 52d35'18"
Ruben Alfonso
And 0h31'22" 29d47'53"
Ruben and Farrah Mora
Umi 15h31'27" 73d16'7"
Ruben and Jenn
Sge 19h6'59" 16d52'3"
Ruben Anthony Salinas
Her 18h30'50" 13d11'24"
Ruben Ayala, Jr.
Cyg 20h56'46" 44d38'21"
Ruben Benitez
Cnc 9h2'1" 17d41'14"
Ruben Brañas Ferrandiz
Cas 0h16'46" 51d9'3"
Ruben Cipriano Mejia
Lyn 7h25'44" 46d7'27"
Ruben Davis Croft
Per 2h15'58" 55d33'59"
Ruben Ferreras
Her 17h10'42" 26d12'2"
Ruben Manriquez
Leo 11h17'45" 22d49'9"
Ruben Richard Villarreal
Leo 10h11'22" 20d33'25"
Ruben Sharma
Peg 22h2'33" 18d37'38"
Ruben Sigron
Cas 1h1'56" 63d45'54"
Ruben William
Sco 17h53'24" -36d34'10"
Rubens Gomes Filho
Leo 10h36'35" 26d45'4"
Ruben's Star
Uma 8h33'14" 68d2'14"
RuBerry
Cnc 8h37'52" 28d56'31"
RUBES
Cnc 9h5'46" 31d47'42"
Rubes
Cas 23h1'6" 55d58'29"
Rubi Loves You
Cyg 20h15'41" 45d30'54"
Rubi Magaly Fernandez Arias
And 2h10'19" 44d59'29"
Rubiel Mercado
Aqr 21h2'4" -12d16'6"
Rubina Khan
Uma 8h44'32" 51d0'30"
Rubio Ray
Uma 8h48'5" 64d8'10"
Rubunia Kinga
Sgr 18h55'5" -34d5'13"
Ruby
Cas 23h15'57" 55d41'18"

Ruby
Lyn 7h8'33" 58d42'31"
Ruby
Cap 20h31'2" -14d43'51"
Ruby
Cma 6h46'3" -12d59'14"
Ruby
Vir 13h33'17" -7d52'8"
Ruby
And 23h1'0" 43d20'53"
Ruby
Crb 16h15'27" 37d30'43"
Ruby
And 2h31'25" 44d50'31"
Ruby
And 1h54'53" 43d6'41"
Ruby
Tau 3h35'54" 25d4'3"
Ruby
Peg 21h48'16" 25d45'51"
Ruby
Tau 4h29'8" 19d56'4"
Ruby A. Miller
Aql 20h8'47" 4d53'25"
Ruby A Perez
Ari 2h13'35" 14d45'32"
Ruby Adella Shannon
Uma 9h3'9" 49d56'26"
Ruby Agnes Lilly Mueller
And 0h9'0" 43d53'5"
Ruby Alice Kelly - Ruby's Star
And 2h8'16" 41d25'33"
Ruby Amelia
Uma 11h38'46" 62d45'55"
Ruby and G.B.
Eri 3h41'46" -0d1'23"
Ruby Andrade
Cam 5h16'2" 62d29'3"
Ruby Ann Esposito
Peg 23h54'25" 27d9'12"
Ruby Anne Reichel
Cas 1h28'7" 72d18'43"
Ruby Arctic Rose
Vir 14h42'29" -4d42'25"
Ruby Ava Wordsworth
Cas 23h54'38" 56d22'11"
Ruby Baker
Uma 9h2'35" 59d21'7"
Ruby Belen Dickson
Sco 16h10'50" -13d34'40"
Ruby Blue Williams
And 0h29'5" 26d41'14"
Ruby Brown Scott Van Rooyen
Gem 7h20'8" 18d38'15"
Ruby Burchell
Aql 19h21'37" 7d23'12"
Ruby Catherine Sheehan
And 2h31'33" 50d15'53"
Ruby Christina Johnston
And 1h54'40" 37d14'49"
Ruby Christine James
And 0h6'13" 45d41'17"
Ruby Davis
Cmi 7h49'53" 4d30'24"
Ruby Elizabeth
Psc 1h6'31" 3d55'36"
Ruby Elizabeth Balmer
Cyg 20h56'24" 37d47'8"
Ruby Ellen
Cnc 8h4'46" 12d14'56"
Ruby Florence Rose
Agr 21h20'44" -2d4'37"
Ruby Frances Ernestine Hill Redford
Lyn 6h54'49" 50d53'22"
Ruby Gene - Too beautiful for Earth
Cru 12h22'57" -61d14'34"
Ruby Geocadin
Gem 6h47'4" 23d8'37"
Ruby Grace
Dra 16h51'33" 56d31'20"
Ruby Grace Boylan
And 23h27'46" 49d25'18"
Ruby Halsworth
Ori 6h3'28" 20d41'45"
Ruby Hamilton
Cas 23h36'38" 53d50'54"
Ruby Hansen
Lyn 7h31'22" 36d46'34"
Ruby Hasse
Vir 12h32'6" -1d55'59"
Ruby I
Cas 1h18'50" 61d58'2"
Ruby II
Cas 0h34'22" 57d45'39"
Ruby Irene Morgan
Psc 0h7'11" -2d50'5"
Ruby Isabelle Earls
Uma 12h21'54" 58d18'52"
Ruby Jane
And 23h23'11" 51d10'0"
Ruby Jane Brown
Umi 16h14'31" 74d50'58"
Ruby Jane Kennedy
Gem 7h9'42" 22d41'48"
Ruby Jane Westerfield
Tau 3h39'16" 14d44'41"
Ruby Jane Willemijn Mehldau
Gem 6h40'33" 16d29'10"
Ruby Jayne Davies
And 0h49'23" 22d18'26"
Ruby Jean Dabrowski
Crb 15h40'50" 31d55'3"
Ruby Jean Robinson
Tau 4h25'24" 25d59'32"
Ruby Jo Latham
And 23h43'3" 38d8'50"
Ruby Johnson
Gem 6h19'39" 22d24'59"
Ruby Jordan
Eri 3h21'40" -5d11'45"
Ruby Jordan Katherine Brown
Cas 0h39'46" 62d2'28"
Ruby Josephine Eleanor Arthur
Cas 2h26'42" 65d16'48"
Ruby Jowers
Uma 9h37'32" 46d52'34"
Ruby Joy Figueroa-Gino
Leo 11h37'4" 24d42'45"
Ruby Kwan Yu Yeung
Lib 14h53'15" -4d32'56"
Ruby L Young
Vir 13h18'54" 5d49'54"
Ruby Lee Edman
Sco 16h17'55" -16d15'7"
Ruby & Lexi
Gem 6h55'9" 24d13'33"
Ruby Lily
And 23h55'37" 42d58'31"
Ruby Lorraine Smith (Petie)
Cyg 20h5'39" 32d43'58"
Ruby Lucille Gallant
Sco 17h52'35" -37d31'39"
Ruby M.
Sco 17h34'35" -39d31'41"
Ruby M. Clark
And 1h54'20" 39d47'59"
Ruby Mae
And 2h24'31" 37d43'37"
Ruby Mae Keenan
Ori 6h17'53" 9d45'44"
Ruby Mae Shaw - Ruby's Star
And 23h11'56" 38d57'15"
Ruby Mae Vaughan
And 23h30'6" 37d15'42"
Ruby Mansell
And 23h52'35" 42d5'33"
Ruby Marie
Psc 1h46'28" 14d4'49"
Ruby Martha Ann Bretos
Cyg 21h26'36" 54d7'58"
Ruby May Hannigan
And 2h28'23" 37d30'27"
Ruby May Sadler
And 2h17'0" 39d9'39"
Ruby May's Christening Star
And 23h17'2" 42d1'5"
Ruby Michelle Blythe
And 0h3'46" 34d32'55"
Ruby Orchid
Lyn 8h31'2" 50d21'19"
Ruby Payne
Uma 9h27'41" 47d37'17"
Ruby Pearl
Boo 14h14'45" 34d24'4"
Ruby Rae
And 1h45'8" 49d4'55"
Ruby Reid
Peg 21h58'53" 34d1'3"
Ruby Reis
Cru 12h27'53" -56d5'5"
Ruby Rianna Allen
Sco 16h16'31" -13d37'11"
Ruby Rimmer
And 23h45'3" 34d7'29"
Ruby Rose Richardson
And 2h4'32" 42d4'4"
Ruby Rye
Lyn 8h43'1" 40d59'52"
Ruby Sadie Elizabeth
Sco 17h23'7" -42d44'17"
Ruby Sherman
Lib 15h13'10" -16d8'20"
Ruby Shuman - Sunshine
And 2h24'24" 41d19'4"
Ruby "Snowflake" N. Amerson
Peg 22h3'29" 6d51'57"
Ruby Star Fisher
Aqr 23h53'19" -3d58'33"
"Ruby & T" - Friends Forever
Lyn 7h56'38" 35d0'5"
Ruby "Teenie" Zeuch
Mon 6h50'11" 6d59'43"
Ruby Terner Rios
Uma 8h40'6" 64d35'57"
Ruby - The Brightest Star
Cru 12h33'30" -59d17'31"
Ruby Uptain 12/07/1931
Hya 10h16'4" -25d41'51"
Ruby Virginia McDaniel Erwin
Cnc 8h16'9" 12d29'41"
Ruby Wright Brown
Cas 1h35'36" 62d53'15"
Rubylyn Moore Hoganas
Tau 4h28'53" 15d24'51"
Rubyphylena Elizabeth Fleck
And 0h52'32" 41d30'31"
rubyrose
Uma 9h59'52" 47d39'25"
Ruby's Star
Uma 11h28'45" 47d32'17"
Ruby's Star
Aqr 21h2'1" 0d7'54"
Ruby's Star
Uma 11h52'27" 28d28'13"
Ruby's Star
Lib 15h29'51" -11d3'58"
Ruby's Wishing Star
Gem 6h31'12" 25d44'29"
RubyV
Srp 18h16'19" -0d14'14"
Ruch, Brigitte
Uma 9h33'52" 64d46'10"
Ruchi Rawlley
Vir 12h2'8" -9d7'17"
Rücker, Robert
Ari 3h8'43" 19d24'31"
Rucksana Hussain
Cas 23h10'32" 54d23'52"
Rudder in the Heavens
Lyn 7h34'55" 39d21'29"
RUDE DOG
Tau 5h11'8" 21d51'42"
Rudel DeCastro
Umi 15h37'55" 70d58'7"
Rudella for Joan, 10/05
Umi 14h15'52" 74d25'17"
Rudey
And 23h28'23" 44d54'9"
Rudi
Aur 7h16'42" 42d31'59"
Rudi
Cmi 7h48'19" 4d49'38"
Rudi loves Justine
Cyg 20h42'10" 54d3'11"
Rudi Roth
Vir 11h58'49" -0d7'24"
Rudi, Two Shoes
Cyg 20h54'7" 45d55'10"
Rüdiger Gilles
Uma 10h25'4" 48d42'17"
Rüdiger Groß
Ori 6h8'44" -1d45'24"
Rüdiger Großkopf
Psc 1h2'5" 12d38'59"
Rüdiger Keudel
Uma 10h48'54" 64d24'23"
Rüdiger Kopplin
Uma 11h58'43" 30d20'22"
Rüdiger Noculak
Uma 11h57'12" 33d40'57"
Rüdiger und Alexandra
Uma 13h37'43" 61d48'40"
Rudolf
Lyn 7h32'16" 41d33'29"
Rudolf Apfel
Sco 16h14'45" -16d32'20"
Rudolf Bräu
Uma 9h47'56" 43d1'24"
Rudolf Copoolse
Uma 9h52'30" 44d35'9"
Rudolf & Elfriede Szabo
Cas 0h5'16" 56d39'35"
Rudolf Gwerder
Vul 19h53'19" 23d17'36"
Rudolf Lenhart
Uma 9h34'35" 58d0'14"
Rudolf Lodewick
Aqr 21h59'27" -13d57'3"
Rudolf Merten
Uma 11h14'41" 72d36'45"
Rudolf Meyer
Uma 10h45'14" 48d1'17"
Rudolf P. Kohlmeister
And 23h37'57" 39d27'34"
"Rudolf Taschner"
Dra 20h12'31" 72d22'25"
Rudolf William Thomas Kelly
Vir 12h18'28" 10d59'23"
Rudolfine Tichy
Cas 23h17'9" 62d16'59"
Rudolph A Sandoval
Per 3h45'49" 51d3'48"
Rudolph and Annie Caruso
And 0h57'33" 44d22'27"
Rudolph Frank Kwiatkowski
Lib 15h27'42" -6d45'32"
Rudolph Gassner
Uma 10h1'8" 62d59'19"
Rudolph J. Okrzesik
Per 3h29'39" 47d51'47"
Rudolph J Richman
Gem 7h18'9" 33d12'23"
Rudolph Revenue Hefner
Riley III
Her 18h3'54" 24d53'37"
Rudolph Richard Izzie, Jr.
Cnc 8h47'6" 17d43'59"
Rudolph, Jan
Ori 6h17'42" 16d41'7"
Rudy
Aql 19h56'4" 4d11'0"
Rudy
Uma 11h31'4" 34d33'9"
Rudy
Cnv 12h40'17" 47d34'7"
Rudy
Cas 0h44'39" 54d39'52"
Rudy
Uma 9h21'0" 67d49'22"
Rudy
Umi 13h36'4" 70d43'4"
"Rudy"
Uma 12h2'27" 57d3'3"
Rudy
Uma 10h26'43" 57d13'35"
Rudy
Cru 12h2'2" -63d16'21"
Rudy Alexis Ortiz
Leo 11h2'48" 23d4'39"
Rudy and Ru Ru Princi
Uma 9h19'43" 65d41'53"
Rudy & Brenda's
Umi 19h20'43" 87d4'28"
Rudy Brown Eternal
Uma 8h29'5" 72d3'31"
Rudy Castillo
Her 17h28'21" 30d15'45"
Rudy Castillo
Aur 6h6'5" 39d15'57"
Rudy Crew
Dra 18h55'38" 69d54'51"
Rudy Edghill
Per 3h28'26" 38d19'29"
Rudy Flubacher
Tau 6h0'23" 25d22'58"
Rudy & Javi
Per 3h40'41" 43d46'34"
Rudy Kundig
Uma 8h37'28" 72d9'28"
Rudy L. Lorenson
Leo 11h45'21" 26d39'10"
Rudy Loesel
Boo 14h45'16" 54d28'37"
Rudy Martinez
Eri 4h9'24" -0d11'23"
Rudy Padilla
Dra 18h42'14" 58d33'7"
Rudy Petit Prince
Dra 17h31'45" 50d52'15"
Rudy Pramstaller
Lib 15h12'15" -15d19'32"
Rudy Prieto
Her 17h0'6" 46d4'46"
Rudy Rivas
Uma 10h27'17" 46d44'14"
Rudy Rock Haywood
And 23h25'34" 48d31'50"
Rudy Theale, Jr.
Umi 14h37'47" 85d33'15"
Rudy Torres
Tau 3h31'0" 12d56'11"
Rudy West
Per 4h39'44" 38d59'6"
Rudy's Love
Peg 22h26'18" 24d7'52"
Rudy's Star
Uma 12h48'26" 58d2'4"
RudyV-2006-JennyC
Cyg 19h52'28" 36d20'5"
rudziaczek
Lib 15h21'30" -13d21'29"
Rue Moyer
Ori 5h15'41" 12d49'45"
Rueben James Brown
Peg 23h42'19" 15d48'16"
Ruedi Bättig
Cyg 19h40'52" 35d12'16"
Ruedi Lienhard
Tau 5h43'2" 19d45'28"
Ruedi + Vreni
Cep 22h24'27" 67d17'34"
Ruege
Crb 16h9'59" 29d8'18"
Ruel M. Aspacio MD.
Uma 10h46'42" 56d45'44"
Rueske Maneuver
Cam 5h28'22" 75d31'55"
Ruey
Lyr 18h25'31" 33d15'40"
Ruf, Peter
Uma 8h21'55" 69d56'26"
Ruff Family Wishing Star
Cyg 20h0'45" 55d4'57"
Ruffins
Lyn 7h35'57" 56d32'21"
Ruffy
Uma 12h40'41" 54d17'36"
Rufina
And 2h15'14" 42d55'59"
Rufus
Uma 10h26'8" 58d58'17"
Rufus Love
Cep 22h57'55" 84d8'17"
Rufus Parkinson
Cep 0h23'12" 84d27'11"
Rufus Parkinson
Aql 19h5'9" 1d39'50"
Rugby Michael Scruggs
Cap 20h22'36" -12d21'34"
Ruger
Cma 7h23'4" -16d32'45"
Ruger
Hya 11h34'24" -28d10'53"
Ruggeberg, Hannah
Ori 6h2'15" 13d28'25"
Ruggero e Sofia
Cam 6h51'50" 69d35'33"
Ruggiero Giuseppina
And 23h10'29" 44d12'4"
Rugs
Per 3h5'3" 42d21'53"
Ruhl, Sabine
Uma 9h39'24" 62d35'38"
Rühmann, Jette Marla
Ori 6h10'37" 9d0'36"
RUHT
Ori 5h36'0" -8d47'15"
Rui Austin
Leo 10h10'44" 22d46'2"
RUI & JIADONG
Lib 15h28'38" -5d47'31"
Ruiz Jessica
Uma 9h7'53" 64d6'7"
Rujmol
Umi 14h55'46" 67d32'27"
RUKHSAAR
Uma 9h15'3" 49d6'5"
Rukmini Chowdhury
Ari 2h58'3" 28d58'47"
Ruksina Neyplub
Crb 15h55'46" 33d59'32"
Rula & B.
Ser 18h21'15" -11d18'34"
Rules Roost
Her 16h20'50" 44d51'15"
Ruling Angel
Ori 6h6'11" 9d29'28"
Rulon Hatch
Her 18h2'13" 21d25'42"
Rumchnaroo
Sco 16h18'41" -11d5'8"
Rumea Agnieray
Vir 12h33'20" 10d37'18"
Rumi
Cap 20h29'55" -11d15'20"
Rumor (Wynwood's Talk of the Town)
Cma 6h59'28" -24d12'30"
Rumore Deeter
Uma 8h44'6" 67d8'27"
Rumplef*$!
Ori 6h4'55" 21d8'41"
RUN
Umi 15h25'51" 83d31'0"
Run Over By Ernie's Red Truck star
Ori 6h20'9" 21d13'32"
Rundle/Mattleman Star
Leo 11h0'51" 18d39'14"
Rune
Uma 8h42'38" 51d32'47"
running bear
Uma 11h5'16" 40d35'50"
Running Fawn
Sco 16h11'56" -13d53'22"
Runzheimer, Joost
Ori 6h19'15" 14d29'39"
Ruokanen
Lyn 8h25'44" 36d49'51"
ruoy ttub sllems
Vir 13h9'16" 3d18'23"
ruoy ttub sllems
Vir 12h9'46" -2d18'55"
Rupal
Vir 13h12'23" -4d37'48"
Rupal
And 0h6'49" 47d0'27"
Rupal Patel
Gem 6h18'18" 22d50'55"
Rupert Bear
Uma 10h31'45" 56d42'28"
Rupert Bear 1979
Umi 15h35'8" 67d57'19"
Rupert Gruber
Ori 6h18'3" 15d47'15"
Rupert H. Stephens
Uma 10h43'10" 62d56'28"
Rupert Jack Alexander Tillet
Ori 5h50'6" 18d57'16"
Rupert John Hewitt, 1943-2003
Leo 11h9'48" -0d5'5"
Rupert Lattner
Aur 5h26'1" 39d32'22"
Rupert Lebmeier
Uma 9h43'34" 57d59'25"
Rupert Murray
Sge 19h2'19" 18d47'39"
Rupio Barnes
Uma 10h6'22" 47d30'0"
Rupp, Elke
Ori 5h22'5" 3d14'15"
Rupp, Viktor
Uma 9h54'33" 46d26'52"
Ruppelt, Ingo
Ori 6h9'53" 16d30'56"
Ruqiyya-Wadi
Cyg 20h28'32" 51d54'10"
Ruscetti's Legacy
Aqr 23h24'49" -18d25'2"
Rüschen, Gertrud
Ari 2h45'45" 11d43'22"
Ruselo Christian Comer
Cap 20h56'35" -19d58'21"
Rush
Ori 5h57'12" 6d27'12"
Rush Digravio
Uma 13h19'25" 56d40'1"
Rush Hunter Baldwin
Ori 6h19'24" 10d51'7"
Rush Limbaugh
Cap 20h22'10" -10d44'16"
Rush7-02112
Cnc 8h38'20" 9d7'37"
Rushanna Mukhamedianova
Psc 1h28'5" 18d32'44"
Rushi
Vel 9h31'13" -41d21'7"
Rushika A. Kumararatne
And 22h59'0" 45d48'57"
Rusnak
Uma 11h0'45" 66d27'38"
Russ
Gem 6h42'56" 12d30'16"
Russ
Aql 20h7'6" 4d1'50"
Russ Arbuckle
Cep 22h45'47" 66d46'47"
Russ Burr My Star
Peg 22h15'4" 27d36'1"
Russ Davoren
Ori 5h3'56" 1d56'43"
Russ & Elise Baslow
Cyg 19h16'44" 54d9'51"
Russ Gilardi
Cap 20h14'58" -11d24'28"
Russ Haas
Aql 19h58'0" -0d25'40"
Russ Howard
Per 2h56'38" 43d36'5"
Russ & Janet Ogburn
Leo 11h11'21" 9d8'21"
Russ Klar
Uma 9h39'38" 43d46'13"
Russ Klein
Her 17h7'36" 14d30'10"
Russ Marchner
Psc 1h28'32" 6d6'2"
Russ Meneve
Cep 21h41'51" 61d37'55"
Russ Miller
Gem 7h46'7" 16d8'40"
Russ Muth
Aur 5h45'37" 49d39'12"
Russ Opferkuch
Her 17h18'32" 45d15'16"
Russ & Penny Miele
Cyg 19h55'50" 37d37'50"
Russ Sanchez
Aur 6h9'49" 28d49'33"
Russ & Susan Temple
Per 3h50'16" 33d53'44"
Russ Turk
Her 18h51'57" 23d11'49"
Russ, My Brother, My Hero, My Heart
Gem 7h13'38" 20d50'38"
Russbell
Cmi 7h31'23" 5d5'3"
Russel Beitzel
Uma 8h10'19" 64d16'31"
Russel C. Norrie With Love
Uma 9h30'57" 67d37'18"
Russel Lisonbee
Cap 20h53'46" -15d48'15"
Russel Mackinlay Porter
Leo 10h0'20" 20d38'0"
Russel Newell Elliott
Boo 15h33'44" 48d28'52"
Russell
Leo 11h9'34" 13d35'22"
Russell
Ari 2h23'45" 26d50'1"
Russell
Tau 4h23'11" 23d33'40"
Russell A. Foulk
Per 3h33'40" 49d28'31"
Russell and Jennifer's Star
Cyg 20h47'39" 35d10'5"
Russell and Susan's Wishing Star
Lyn 9h19'10" 39d1'55"
Russell and Vicki Roeder
Psc 23h14'27" 6d43'23"
Russell Armstrong
Aur 5h13'16" 46d59'4"
Russell Aron Medrano
Her 17h22'1" 25d45'41"
Russell Ashley Brazier
Lib 15h2'14" -9d40'48"
Russell Ashley Spurrell
Cru 12h41'58" -60d27'25"
Russell Bartholemew Kelly
Uma 13h9'29" 52d46'41"
Russell Bartlett
Gem 6h47'47" 15d4'44"
Russell Berger
Psc 22h52'12" -0d3'42"
Russell Bland
Uma 8h43'12" 64d45'51"
Russell Bramley
Psc 23h21'36" 4d53'26"
Russell Bray Butcher
Cru 12h40'7" -57d33'7"
Russell Bucci
Cep 22h38'15" 72d25'2"
Russell Cangialosi
Cep 21h29'35" 64d2'1"
Russell Charles Taylor
Uma 11h31'36" 53d14'12"
Russell Chrupalyk
Sco 16h48'33" -31d15'23"
Russell Clark
Per 3h28'53" 46d42'40"
Russell Collin
Ari 3h22'29" 29d33'9"
Russell Cottrell Family
Aql 20h12'57" 14d38'40"
Russell Crawford
Cap 20h17'33" -14d43'25"
Russell D. Miller (Rascal)
Aqr 22h16'38" 1d17'42"
Russell Dale Gilman
Umi 15h4'27" 75d35'59"
Russell Dean Callaghan
Lib 15h3'55" -21d4'36"
Russell Driver
Sgr 19h9'28" -27d20'24"
Russell E Mohring
Vir 12h7'59" 12d1'11"
Russell E. O'Connell
Psc 23h2'31" -1d1'10"
Russell Edward Beckley Jr. IV
Aqr 20h42'34" -9d7'14"
Russell Edward Forristall
Uma 11h45'59" 32d25'51"
Russell F. Myers, Jr.
Gem 6h49'12" 15d45'17"
Russell Fitzgerald Karli
Boo 14h14'42" 16d23'15"
Russell Galichowski
Uma 11h29'5" 59d3'9"
Russell Gallant
Lmi 10h38'37" 39d7'29"
Russell Gene Frazier Jr
Her 17h40'51" 36d19'18"
Russell H. Schrank
Her 17h0'12" 17d25'19"
Russell Howard Samson
Psc 0h59'13" 10d4'44"
Russell Hunter
Cyg 20h7'32" 41d16'51"
Russell Irvin Spurlock Jr.
Her 17h54'3" 16d46'31"
Russell James Kelson
Cep 22h19'40" 62d40'50"
Russell John Grindon
Uma 10h6'17" 58d29'49"
Russell John Truber
Sco 17h38'9" -40d1'42"
Russell Judd
Aql 19h12'14" 5d41'0"
Russell K Brown
Ori 5h25'4" 2d43'21"
Russell K. Stewart
Aql 20h3'25" 1d47'11"
Russell L. Anderson
Tau 3h54'40" 1d35'47"
Russell L. Berry/Jessica D. Warner
And 0h51'17" 43d20'25"
Russell L. Renaud
Uma 10h30'32" 40d46'10"
Russell Laing
Cep 0h12'50" 77d2'58"
Russell Lane Crowley
Boo 14h35'24" 50d6'35"
Russell & LeAnn Carmichael 6-1-05
Cyg 19h45'17" 41d40'29"
Russell Lee - 16.02.1956
Cru 12h23'36" -63d34'47"
Russell Lee Elliott, III
Leo 11h9'42" -1d36'31"
Russell Lee Hughes
Lib 15h6'37" -11d30'51"
Russell Lee Pantleo, Sr.
Men 5h13'55" -76d41'44"
Russell Lee Rigsby
Aqr 21h20'23" -11d11'40"
Russell Lee Stafford, Sr
Cyg 19h56'8" 47d41'38"
Russell Long
Cep 23h4'29" 78d24'32"
Russell Lutz
Boo 15h30'50" 43d35'59"
Russell Lyn Qualls
Cas 0h51'57" 52d21'49"
Russell Maginnis
Per 3h34'5" 46d52'15"
Russell Malone
Aql 19h14'21" -11d25'12"
Russell Martin Petrak
Per 4h49'34" 42d42'32"
Russell McGregor
Boo 14h53'52" 23d43'43"
Russell Meade Durgin
Dra 18h31'19" 70d8'29"
Russell Norman Reiling
Gem 7h33'44" 34d42'18"
Russell O. Scarborough
Uma 10h55'40" 53d34'30"
Russell -"OV R.E.B."
Cap 21h7'33" -15d8'10"
Russell Powers
Aqr 21h4'20" -13d56'28"
Russell R. Conley
Psc 1h19'52" 32d26'37"
Russell & Rachael
Cyg 19h54'51" 38d3'19"
Russell Reno
Aqr 23h42'28" -16d27'32"
Russell Rivera Jr
Crb 15h35'5" 31d9'52"
Russell Robin Ware III
Cap 20h51'46" -18d7'47"
Russell "Rusty" Reed
Ori 4h53'58" 2d41'14"
Russell Seay Davis
Aqr 21h42'42" -1d1'38"
RUSSELL SHANNON HALL
Uma 10h44'13" 58d24'17"

Russell Simmons
Her 17h21'36" 40d2'39"
Russell "Smartassticus" Bauer
Ari 2h29'31" 22d12'0"
Russell Sorenson
Aql 19h35'32" -7d0'37"
Russell Stanford Warner
Boo 14h49'1" 36d26'26"
Russell Stephen Frisch Jr.
Per 4h32'29" 39d44'5"
Russell Sweeney
Per 3h18'12" 41d42'57"
Russell T. Amerson
Per 3h17'56" 47d43'42"
Russell T. Molling,Sr.
Cep 21h22'42" 59d52'4"
Russell Tanoue
Cnc 9h5'23" 19d53'9"
Russell "the one I love" Dill
Umi 13h52'53" 78d23'39"
Russell Thomas Langham
Cru 12h37'46" -57d56'54"
Russell Thomas Phillips
Her 16h43'55" 6d52'37"
Russell Thomas Sohlberg
Uma 9h35'2" 59d48'10"
Russell Van Conroy
Tau 3h32'53" 23d29'9"
Russell Vie Eschbacher
Ori 5h26'50" 0d5'8"
Russell W. Beard
Uma 11h30'54" 47d10'55"
Russell Wagner, Always!
Gem 7h6'27" 16d39'55"
Russell William Fausher
Lib 14h26'36" -17d27'32"
Russell William Mitchell
Uma 10h2'46" 65d30'59"
Russell Witt
Uma 10h11'30" 69d9'45"
Russellech, mine forever
Lac 22h11'5" 38d7'23"
Russell's Light of the Galaxy
Leo 10h44'24" 17d33'29"
Russell's Little Flower
Cyg 21h38'51" 48d27'40"
Russell's Star
Cep 21h43'34" 61d52'10"
Russellus Barbarae
Cas 1h30'14" 63d55'48"
Russhelle Sands
Lib 15h8'27" -8d38'42"
Russo
Vir 13h28'30" 13d19'1"
Russo-Schassburger
Uma 13h44'32" 55d13'7"
Rustee Reeves "1912-2006"
Cas 1h1'53" 69d25'29"
Rustem and Sophie Demiraj
Cyg 19h38'42" 54d5'38"
Rusti Beall
Cap 21h43'42" -10d58'17"
Rusty
Umi 17h33'10" 82d49'7"
Rusty
Vir 13h29'44" -3d7'36"
Rusty
Uma 8h30'18" 69d2'55"
Rusty
Umi 15h36'15" 72d2'4"
Rusty
Sgr 19h19'13" -33d43'15"
Rusty
Sco 16h7'48" -23d16'4"
Rusty
Ori 5h35'54" 10d22'22"
Rusty
Tau 5h57'58" 24d0'55"
Rusty
Leo 9h30'34" 25d40'30"
Rusty
Aur 5h43'37" 41d39'20"
Rusty Alter-Rey
Sco 17h9'50" -37d13'10"
Rusty Dovell
Uma 10h43'6" 42d29'35"
Rusty Goforth
Ari 2h4'51" 20d23'16"
Rusty Holien
Aur 5h38'54" 41d33'53"
Rusty Joel Awalt
Vir 13h46'54" 3d10'42"
Rusty (Ruby Tuesday) Sheppard
Dra 18h40'35" 59d30'45"
Rusty Wallace
Uma 11h19'42" 68d26'59"
Rustys Star
Sco 16h13'57" -10d3'45"
Rusty's Twinkle
Ari 2h42'22" 16d14'33"
Ruta
Mon 7h1'58" 8d5'0"
Rutger Kuan Lu Broek
Ari 3h22'52" 28d52'57"
RutgerTillemaOne
Uma 9h36'33" 65d8'41"
Ruth
Cam 4h7'23" 65d42'27"
Ruth
Uma 9h20'9" 52d42'17"
Ruth
Sco 16h39'5" -29d0'52"
Ruth
Crb 15h30'55" 28d4'17"
Ruth
Ari 2h58'48" 26d7'35"
Ruth
And 2h14'7" 41d26'46"
Ruth
Boo 15h43'26" 45d10'51"
Ruth
And 23h37'13" 42d54'17"
Ruth
And 23h21'27" 39d13'14"
Ruth A. Collins
Ari 3h22'45" 27d7'48"
Ruth A Heaner
Uma 11h49'13" 29d29'33"
Ruth A. Urquhart
Crb 15h44'38" 34d16'7"
Ruth Abernathy
Cas 1h29'30" 60d14'46"
Ruth Abigail
Crb 15h43'9" 34d37'32"
Ruth Adams
Lmi 10h0'32" 37d48'45"
Ruth Alice Dubinsky
Cas 1h11'13" 54d18'56"
Ruth Allison Murray
Lib 15h7'1" -27d35'7"
Ruth Alyce Jersey
Cas 1h40'43" 65d18'15"
Ruth Amneris
And 2h23'50" 47d47'37"
Ruth and Chuck Plett
Cyg 21h15'11" 47d7'19"
Ruth and Lloyd Brown
Cyg 21h59'29" 47d46'21"
Ruth Angeline Girt
Leo 10h9'28" 20d5'19"
Ruth Ann
Her 16h34'0" 20d51'33"
Ruth Ann
Cas 2h4'23" 63d50'20"
Ruth Ann
Cas 1h21'9" 60d44'42"
Ruth Ann C.
Uma 10h59'1" 64d45'54"
Ruth Ann & Eugene Moriarty
Vir 14h7'15" -8d9'39"
Ruth Ann Hall
Gem 6h38'31" 23d39'33"
Ruth Ann Huff
Uma 10h2'24" 41d37'36"
Ruth Ann Huszar
Cas 1h31'2" 66d38'13"
Ruth Ann Inman
Aqr 23h12'27" -7d23'17"
Ruth Ann Koroch
Sco 17h16'35" -30d14'40"
Ruth Ann Medeiros
Sgr 19h14'35" -32d14'12"
Ruth Ann Miller
Cam 3h32'24" 54d27'47"
Ruth Ann Moore
Lib 15h53'9" -18d44'24"
Ruth Ann Motz Bradley
And 23h7'41" 41d21'49"
Ruth Ann My Eternal Shining Star
Cas 23h28'19" 52d18'49"
Ruth Ann Nagucki-Pollack
Gem 6h47'35" 29d29'18"
Ruth Ann Negrete
Cyg 20h5'46" 44d43'42"
Ruth Ann Nunn
And 23h51'15" 47d0'48"
Ruth Ann Posluszny
Per 3h36'27" 42d54'47"
Ruth Ann Powles
Peg 22h35'54" 12d32'54"
Ruth Ann & Robert Naun
Cyg 21h52'35" 50d40'57"
Ruth Ann Southworth From Whitney
Cas 0h58'27" 57d44'24"
Ruth Anne Jager
Leo 10h22'41" 8d58'24"
Ruth Anne Shoben
Cyg 20h35'54" 52d29'56"
Ruth Anne Thorning
Sco 17h2'21" -39d17'14"
Ruth Audrenne Ward Redmond
Cyg 20h14'11" 49d31'4"
Ruth Audrey Rundquist
Gem 7h26'2" 21d38'14"
Ruth Bailey
Cap 20h47'8" -21d13'8"
Ruth Barbagallo
Leo 9h39'46" 27d8'32"
Ruth Benowitz
Psc 1h26'27" 23d23'33"
Ruth Birnbacher
Uma 9h42'7" 48d9'18"
Ruth Blanch Dearing Irwin
Peg 21h33'33" 25d0'54"
Ruth Brofsky
Cas 1h28'55" 66d51'30"
Ruth Brown
Ari 2h55'57" 18d56'55"
Ruth Brownlie 24:01:75, A True Star
Cyg 21h20'10" 32d59'9"
Ruth C. Milgram shine bright 4ever
Ori 5h51'14" 7d0'3"
Ruth C Patti-Greatest Generation
Lib 14h51'54" -13d43'16"
Ruth Campbell Presley
Leo 10h26'15" 13d24'17"
Ruth Carter
Cas 1h17'6" 54d58'5"
Ruth Cristy Wilson
Cnc 9h16'5" 13d5'41"
Ruth Crowley
Aur 7h15'48" 37d54'11"
Ruth Curtis
Uma 9h46'53" 70d41'27"
Ruth D. Riddle
Cas 0h25'10" 63d46'5"
Ruth Dabbs
Vir 14h7'17" -14d45'51"
Ruth Dalia Lott
Ari 2h43'17" 23d10'9"
Ruth Dooley
Boo 14h50'32" 27d31'27"
Ruth E. Brehant
Cas 2h51'20" 60d14'44"
Ruth E Murphy
Tau 4h15'53" 18d46'21"
Ruth E. Pitts
Gem 7h17'19" 33d26'22"
Ruth E. Schell
Uma 13h15'13" 52d41'14"
Ruth E. Speechly
Mon 6h50'47" 8d30'41"
Ruth & Edward Breliant 80th B'day
Cyg 20h1'0" 56d50'42"
Ruth Elaine Brewer
Cnc 8h43'38" 6d45'24"
Ruth Elaine Warner
Cap 20h15'4" -23d44'49"
Ruth Elisabeth
Lmi 10h6'13" 29d13'56"
Ruth Elizabeth
And 2h4'18" 41d49'16"
Ruth Elizabeth Brooks
And 23h12'58" 46d47'2"
Ruth Elizabeth Easton
Lmi 10h14'6" 31d22'32"
Ruth Ellen
Equ 21h8'22" 4d43'36"
Ruth Ellen
Cam 6h17'21" 67d32'20"
Ruth Ellen Johnson
Uma 10h20'51" 48d53'46"
Ruth Ellen Kaney Beckman-Great Mom
Cas 23h38'24" 53d17'26"
Ruth Elmore Ratcliffe
Cas 0h15'19" 63d47'10"
Ruth & Elwood Capelle
Lyr 18h15'45" 35d0'10"
Ruth F. Newmaker
Cap 21h1'16" -17d33'17"
Ruth Fannon Moore
Cas 23h6'36" 58d27'5"
Ruth Ferry
Sgr 18h58'17" -25d33'16"
Ruth Foertmeyer Shank
Uma 11h42'51" 31d47'10"
Ruth Frances Andrews
And 0h31'54" 42d12'50"
Ruth & Freddy Boeschen
Cnc 8h21'18" 10d59'38"
Ruth Fridlund
Uma 8h39'14" 58d7'9"
Ruth Friend
Lyn 7h39'6" 48d16'32"
Ruth G. Blumrosen
Uma 13h22'25" 58d49'48"
Ruth Gonsalves
Cas 1h8'15" 63d49'45"
Ruth Gonzalez
Lyn 6h41'2" 52d11'53"
Ruth Grace Lennon Singletary
Cas 0h28'11" 54d32'15"
Ruth Gratz
Crb 15h29'4" 30d55'43"
Ruth Gregory
Sgr 19h12'44" -13d24'57"
Ruth Guest
Psc 0h14'21" 6d27'51"
Ruth H. Moody
Crb 15h53'20" 28d15'58"
Ruth Hardin
Psc 1h10'51" 29d17'28"
Ruth & Harry
And 2h7'2" 42d25'41"
Ruth Heeg
And 1h29'46" 41d12'15"
Ruth Heery
Cas 23h37'40" 53d22'33"
Ruth Hemeon
Cas 23h16'21" 55d35'24"
Ruth Hollars
Uma 12h53'34" 56d21'53"
Ruth Hultman-DeSmidt
Vir 13h39'25" -7d50'8"
Ruth Idsia
Leo 11h21'40" -1d13'15"
Ruth Ileen Robinson
Cas 1h25'2" 57d43'53"
Ruth J. Gozemba
Cas 0h33'59" 58d6'0"
Ruth J. Pfeiffer
Uma 10h40'59" 57d4'1"
Ruth J. Randle
Ari 2h11'24" 15d59'25"
Ruth Jackson
Uma 10h39'6" 53d25'41"
Ruth Jane Riley Dickens
Gem 7h5'16" 24d56'51"
Ruth Jean Angove
Lyr 18h54'14" 31d28'0"
Ruth Joan Colonero
Cap 20h37'39" -19d43'48"
Ruth Joan Sullivan
Leo 11h46'31" 23d5'35"
Ruth Joann Frye Thom
Cas 1h46'26" 61d14'40"
Ruth & John Overholser
Lyn 7h38'33" 53d19'57"
Ruth Katherine Stanton
Leo 10h21'17" 26d6'30"
Ruth Klimpl
And 23h29'22" 47d38'20"
Ruth Klock
And 0h32'31" 31d40'29"
Ruth Kyoung Stepp
Lyn 7h10'57" 53d54'54"
Ruth L. Stadler
Cnc 8h27'50" 26d2'17"
Ruth Lara
Lyn 8h40'3" 35d24'4"
Ruth Lavery
Crb 15h37'16" 27d35'7"
Ruth & Lenny
Gem 7h44'57" 32d24'59"
Ruth Lester
Cyg 20h30'33" 45d9'26"
Ruth Limb
Gem 7h32'18" 24d8'0"
Ruth Little Sparkle
Crb 15h36'7" 26d9'6"
Ruth Louise Thompsen
Psc 23h9'36" 6d1'34"
Ruth Louise Wood
And 0h48'10" 22d27'37"
Ruth "Loveliest Person" Eden
Umi 16h46'16" 79d48'17"
Ruth Lowenstern Weinstock
Aqr 22h23'8" -23d7'53"
Ruth M. Judson
Crb 15h50'42" 37d22'48"
Ruth M. Pearl
Uma 8h39'55" 53d12'57"
Ruth M. Petrucci
Sgr 18h32'23" -17d31'5"
Ruth M Winstanley*
Cap 20h22'24" -22d34'36"
Ruth Malone - L'ange à mon coeur
Lib 14h38'42" -9d39'41"
Ruth Marie
Cas 23h51'11" 50d32'2"
Ruth Marie Eltrich
And 1h19'23" 38d35'38"
Ruth Marie Kraus Maicke
Umi 13h7'1" 70d48'54"
Ruth Marie Reid Cheek
Psc 1h7'38" 20d39'55"
Ruth Marie Tamlyn Bailey
Lyn 8h2'29" 34d59'49"
Ruth Marie's Wish Angel
Lib 14h51'9" -7d36'43"
Ruth Mary
Cas 1h56'4" 62d35'9"
Ruth Mays
Cas 0h58'15" 58d19'20"
Ruth McCann
Tau 4h13'29" 27d36'48"
Ruth McLendon
Ari 1h52'10" 22d57'40"
Ruth Mefford
And 1h36'52" 43d59'5"
Ruth Mitchner's Star
Eri 4h12'54" -0d6'25"
Ruth "Mom-mom"
And 0h55'3" 36d49'10"
Ruth Mueller - My Divine Light!
Cru 12h31'9" -61d56'58"
Ruth Myers "Ruthie"
And 23h18'57" 47d6'4"
Ruth N. Neely
Lib 15h30'2" -12d7'26"
Ruth Nana Akaba
Leo 11h38'20" 23d15'21"
Ruth Naomi Banks Hackett
Lmi 10h32'22" 38d8'2"
Ruth Nellie Kampras
Cap 21h47'4" -12d47'39"
Ruth Nur Karima
Cas 23h31'8" 52d6'53"
Ruth Ojalvo-Bodmer
Ori 6h5'12" 11d1'9"
Ruth Olivia Gosnell-Robinson
Psc 1h12'3" 9d2'9"
Ruth Ormond Krebs Mattox
Mon 7h46'58" -0d32'59"
Ruth Palmer
Cas 23h13'17" 55d36'56"
Ruth Pearl Harrison
Tau 4h59'13" 19d52'23"
Ruth Pennock Indrelunas
Lyn 7h11'20" 57d30'56"
Ruth Presselmayer
Apu 16h1'50" -76d40'20"
Ruth Price
Per 2h39'41" 52d43'4"
Ruth R. Reed
Cnc 9h5'10" 25d37'18"
Ruth Rabin "Our Shining Star"
Cas 23h9'1" 58d32'28"
Ruth Rabinowitz
Cas 23h22'56" 55d30'14"
Ruth Raiguel Schadel
Psc 1h28'15" 10d16'51"
Ruth Reblitz's Star
Lib 14h51'3" -7d32'49"
Ruth Regine Egger-Büschi
Umi 16h21'35" 72d13'48"
Ruth Reinhard
Tri 1h35'14" 35d37'56"
Ruth Rigsby Ardman
Crb 15h50'42" 27d44'1"
Ruth Rivera
And 1h7'5" 38d48'11"
Ruth Rogers
Vir 12h41'20" 11d31'13"
Ruth Rose
Cam 5h55'24" 56d28'52"
Ruth Roy
Cas 1h18'22" 53d23'7"
Ruth Rubin
Ori 5h36'41" 5d25'5"
Ruth S. Matthews
Cnv 12h46'56" 38d5'30"
Ruth S. Pinder
Cas 0h4'56" 57d0'5"
Ruth S. Revell
Uma 10h45'47" 65d31'19"
Ruth S Sarmiento
Leo 11h46'53" 26d17'12"
Ruth Saenz
Psc 2h0'50" 9d26'57"
Ruth Saul (Rue)
Cas 2h18'28" 72d23'37"
Ruth Schleiden
Uma 9h50'45" 68d44'57"
Ruth Shirley White
Uma 10h38'14" 49d40'36"
Ruth Shwam
Leo 11h49'45" 20d5'59"
Ruth Slivken
Cas 1h34'11" 66d36'1"
Ruth Spano 80th
Gem 7h38'25" 32d39'7"
Ruth Spencer Smith
Uma 14h23'36" 58d5'3"
Ruth Star Chappa
Vir 13h40'55" -7d23'6"
Ruth Staub
Lyr 18h52'57" 43d36'36"
Ruth Stefanie
Uma 8h48'57" 64d37'7"
Ruth Stephenson Falleroni
Sco 17h42'0" -35d56'50"
Ruth Stevens
And 0h14'8" 27d10'32"
Ruth Strader
Cap 20h7'3" -14d16'35"
Ruth Sunley
Gem 6h44'42" 13d18'0"
Ruth Townsend Huff
Umi 15h33'48" 67d47'19"
Ruth Ultmann
Cnc 8h4'19" 7d35'19"
Ruth Veenema
Cam 4h33'20" 58d5'8"
Ruth Violet Bartlett
Cas 1h33'29" 57d55'34"
Ruth Wayne
Lyr 18h36'52" 34d47'39"
Ruth Whitt
Lyr 18h41'46" 35d1'18"
Ruth & Will Forever
Umi 16h47'37" 81d50'0"
Ruth Willey
Sgr 17h48'52" -17d24'46"
Ruth Wills
Ari 1h54'24" 25d26'50"
Ruth Wilson
Sco 17h17'29" -32d58'51"
Ruth Wolff & Martin Bloom
Lyn 7h42'58" 36d9'24"
Ruth Wood
Cap 20h44'52" -21d58'33"
Ruth Workman
Sgr 19h14'14" -34d43'14"
Ruth Wyers
Cas 2h51'1" 61d24'49"
Ruth Yvette Caple-Slice
And 0h31'49" 36d41'1"
Ruth Zangerl 25.04.1965
Uma 11h43'28" 59d9'8"
Ruth Zillefrow Dudley
Cas 0h2'28" 55d35'5"
Ruth, 17.06.2003
Umi 14h53'16" 67d52'13"
Ruth, Karyn, Mark, James, Ethan, Samuel
And 23h42'41" 36d42'8"
Ruth, Star of Dominick
Uma 13h42'7" 58d9'1"
Ruth-Ann
Tau 3h52'30" 24d34'29"
Ruthann
Ori 6h16'58" 14d24'6"
Ruthann Gieseke
Psc 1h11'14" 16d37'0"
Ruthann McCarthy
Ori 6h18'17" 14d13'42"
Ruthann Palmer
Ari 2h6'29" 21d40'9"
Ruthanna
Umi 13h50'44" 75d11'45"
Ruthellen
Aqr 21h33'53" -0d51'39"
Ruther-Glover's Regals: K,S,M &K
Ori 5h40'32" 4d35'16"
Ruthi HM-1
Ari 2h35'26" 31d1'38"
"Ruthie"
Tri 2h23'47" 34d28'25"
Ruthie
And 0h44'29" 38d54'21"
Ruthie
Lyn 7h33'8" 38d37'4"
Ruthie
Vir 12h53'22" 2d17'30"
Ruthie
Tau 3h48'17" 22d9'12"
Ruthie
Vir 13h20'20" 7d56'28"
Ruthie
Leo 9h33'1" 13d53'39"
Ruthie
Uma 11h56'58" 28d34'56"
Ruthie
Peg 22h41'25" 26d27'42"
Ruthie
Gem 6h50'32" 17d2'17"
Ruthie
Aqr 21h5'11" -8d45'37"
Ruthie A. Goodman
Boo 14h50'12" 35d13'37"
Ruthie and Roger Pernick
Uma 13h3'27" 58d51'48"
Ruthie Ann Westbrook
Lyn 6h17'59" 61d0'11"
Ruthie Bachman
Sgr 18h33'57" -26d34'1"
Ruthie Byrnes
Lyn 9h36'8" 40d5'35"
Ruthie Deviney
And 2h20'38" 50d0'46"
Ruthie Freed
And 0h48'15" 36d23'30"
Ruthie "Girl Of My Dreams"
Cas 1h19'55" 59d51'0"
Ruthie Jewel Lindquist
Psc 0h57'42" 26d10'46"
Ruthie & Louie
Tau 4h21'45" 28d0'47"
Ruthie Mae 05/01/1923
Tau 4h45'42" 17d58'36"
Ruthie Norman - Our Angel in Heaven
Cas 23h38'13" 58d9'32"
Ruthie Ogden-Stevens
Cas 1h36'56" 62d8'50"
Ruthie Phillips
And 0h04'32" 25d52'8"
Ruthie & Rita
Sgr 19h22'9" -20d50'15"
Ruthie Schoentrup
Cas 1h45'31" 63d53'35"
Ruthie's Jewel
Cyg 21h24'45" 35d49'46"
Ruthie's Star
Cru 12h30'23" -59d2'30"
Ruth's Light
And 0h40'40" 42d17'18"
Ruth's Star
Sgr 19h21'16" -22d36'18"
Ruth's Stern
Uma 9h15'38" 64d21'10"
Ruth's Twinkle - Susan's Star
And 1h41'26" 38d56'53"
Ruthvk80
Cas 1h41'12" 67d42'11"
Ruthy Cordelia McCarty,My True Love
Cnc 8h2'57" 14d42'16"
Ruthy Darling Gorgeous
And 1h20'34" 34d44'51"
Ruthy L. Jacobs
And 1h15'28" 42d47'54"
Ruti
Lyn 9h40'3" 39d29'55"
Ruti
Uma 11h13'42" 40d59'26"
Rutilia
Lyn 7h51'58" 45d22'1"
Rutilus Lauren
Ori 5h26'2" 2d2'47"
ruuscheli
Uma 10h31'19" 69d48'15"
Ruve
Gem 6h39'41" 17d22'9"
Ruz
Per 4h44'11" 49d12'3"
Ruza - In Aeternum
Lyn 8h53'45" 38d28'25"
Ruza & René
And 0h39'11" 30d15'49"
Ruzena
And 1h33'21" 48d3'8"
Rúzsa Magdi
Uma 11h23'58" 29d53'21"
RV - 23062007
Cyg 19h52'29" 38d45'46"
R.V. Lucas
Umi 15h9'44" 69d48'40"
Rview
Del 20h40'29" 12d44'7"
RVL
Leo 11h31'4" 27d25'58"
RVMA
And 23h25'11" 42d24'2"
RV's Light
Uma 9h57'53" 58d24'3"
R.Wayne Rogers a.k.a Best Dad Ever
Her 16h40'3" 21d45'22"
RWD-50
Vir 13h55'8" 4d4'7"
RWKAEwald
Cas 0h58'45" 63d42'18"
RWS STAR BEYOND EXCELLENCE 4EVER
Aql 19h10'4" -7d48'36"
RWT 43
Crb 15h57'42" 34d5'1"
'Rwy'n dy garu di
Sco 16h43'26" -31d53'0"
RX18
Aqr 21h41'56" 1d58'46"
Ry Ry O Ryann
Cnc 8h25'26" 14d47'45"
Rya
Crb 15h53'6" 34d44'57"
Rya
Lib 14h47'44" -7d36'28"
Rya
Cep 21h44'34" 58d39'23"
Rya Sky
Leo 10h23'4" 22d9'52"
Ryah Skye Thompson
Umi 16h45'29" 82d16'39"
Ryall
Umi 14h17'10" 75d17'5"
Ryan
Eri 4h21'27" -2d17'41"
Ryan
Ori 5h55'48" -0d8'57"
Ryan
Aql 19h49'3" -0d28'33"
Ryan
Aql 19h2'8" -8d37'38"
Ryan
Cap 21h38'19" -14d0'26"
Ryan
Uma 11h44'21" 53d17'46"
Ryan
Dra 20h21'57" 67d27'59"
Ryan
Sco 17h26'46" -35d48'0"
Ryan
Sco 17h52'52" -35d16'36"
Ryan
Sco 17h24'14" -40d2'35"
Ryan
Sco 17h28'37" -38d13'20"
Ryan
Sco 17h28'9" -43d3'49"
Ryan
Ori 6h0'12" 19d3'56"
Ryan
Ori 6h2'36" 19d22'12"
Ryan
Ori 5h54'23" 19d41'34"
Ryan
Ori 6h1'35" 19d58'33"
Ryan
Ori 6h0'59" 19d3'21"
Ryan
Her 18h14'54" 20d5'4"
Ryan
Ari 2h19'31" 24d48'51"
Ryan
Her 16h49'36" 25d23'37"
Ryan
Ori 6h20'1" 10d15'1"
Ryan
Ori 6h2'39" 14d21'11"
Ryan
Psc 0h19'11" 7d59'57"
Ryan
Ori 5h52'15" 9d13'14"
Ryan
Ori 4h52'52" 11d4'49"
Ryan
Tau 4h10'39" 2d58'37"
Ryan
Ori 5h55'36" 7d8'13"
Ryan
Cnc 8h44'48" 30d17'1"
Ryan
Uma 10h48'46" 48d28'5"
Ryan
Uma 10h11'44" 49d55'32"
Ryan 2004
Sgr 19h14'40" -21d47'4"
Ryan 21
Leo 9h22'29" 12d25'8"
Ryan A. Boring "Super Star"
Cnc 8h33'53" 31d10'9"
Ryan A. Law
Lyn 6h17'6" 56d52'44"
Ryan Abbott
Ori 6h4'34" 19d16'6"
Ryan Adelman
Leo 10h33'57" 18d36'43"

Ryan & Adrian Coffey's Wedding Star
Vir 11h47'20" 5d29'46"
Ryan Adrien Bukvich
Lyn 7h45'35" 56d53'1"
Ryan Alan
Sco 17h49'47" -31d39'23"
Ryan Alan Glover
Vir 12h54'3" 3d2'39"
Ryan Alex Bass 5/1/2006 - 21/2/2007
Cap 20h58'4" -20d28'23"
Ryan Alexander Jameson
Ori 5h53'45" 8d39'11"
Ryan Alexis Fitzgerald
Leo 10h16'17" 14d8'34"
Ryan Allen Mantey
Her 16h18'33" 10d45'55"
Ryan Allen Wolfenbarger
Per 3h10'26" 55d45'7"
Ryan & Alyson Flemming 10/01/2005
Cyg 19h56'10" 44d52'28"
Ryan and Alyssa
Cap 21h28'34" -14d41'19"
Ryan and Angie
Crb 15h37'5" 28d44'46"
Ryan and Brandi Sullivan
Gem 7h29'13" 35d7'10"
Ryan and Brittany, forever in love
Cyg 20h49'44" 53d37'50"
Ryan and Crystal Ann
Sgr 18h6'26" -29d41'38"
Ryan and Danielle Kriesch
Leo 10h10'40" 12d1'16"
Ryan and Emily
Lib 14h35'37" -23d27'43"
Ryan and Hannah
Sgr 18h20'18" -33d6'29"
Ryan and Heather Smithson
Lyr 18h49'10" 44d1'4"
Ryan and Jennifer's Star
Oph 17h46'8" 13d39'8"
Ryan and Jessica's Lasting Light
Cyg 20h16'15" 47d2'52"
Ryan and Julia forever in love
Gem 7h12'4" 30d38'50"
Ryan and Katy's Star
Cyg 21h26'35" 53d3'22"
Ryan and Kelly Powal
Uma 9h23'24" 69d44'51"
Ryan and Kimberly's Wedding Star
Cyg 19h35'14" 54d5'20"
Ryan and Libs Star
Cyg 19h52'21" 39d55'29"
Ryan and Lynne Wimpenny
Cyg 20h13'7" 56d46'32"
Ryan and Mary Forever
Mon 6h54'26" -0d19'10"
Ryan and Meredith
Cyg 19h35'51" 53d34'26"
Ryan and Michelle
Sco 17h44'39" -37d38'57"
Ryan and Nicole Henry
Vir 13h7'41" 3d10'14"
Ryan and Rachel Lode
Gem 6h43'32" 29d47'26"
Ryan and Rosey
Leo 10h15'43" 17d52'36"
Ryan and Samantha Forever
Cas 0h5'37" 54d47'59"
Ryan and Shannon's Love
Crb 15h33'29" 27d56'16"
Ryan and Shawnett Gillig
Sco 17h29'22" -41d45'4"
Ryan and Stephanie Engolio
Per 3h12'0" 32d10'20"
Ryan and Tera
Cyg 19h35'1" 55d0'3"
Ryan and Tina's Star
Cyg 21h24'41" 36d24'57"
Ryan Andersen Buerman
Her 18h28'28" 17d4'0"
Ryan Andrew Brookhart
Uma 12h2'31" 55d15'58"
Ryan Andrew Brown
Uma 9h25'23" 59d31'50"
Ryan Andrew Curry
Sco 17h56'32" -41d53'49"
Ryan Andrew Klumph
Her 17h53'54" 27d47'53"
Ryan Andrew Madera
Per 3h11'0" 53d1'10"
Ryan Andrew Rogers
Umi 14h43'52" 73d25'27"
Ryan Andrew Salerno
Umi 13h23'28" 75d29'54"
Ryan Andrew Vander Schaaf
Vir 13h0'52" 11d2'6"
Ryan "Andy" Anderson
Leo 11h21'18" 12d51'4"
Ryan & Andy Gilstanley
Uma 9h3'43" 48d13'35"
Ryan Angel Unger
Dra 17h48'47" 55d25'19"
Ryan Angell
Ori 6h14'53" 15d7'9"
Ryan & Anje's Dream
Cyg 21h38'32" 54d1'57"
Ryan Anne Yandell
Lyn 8h37'37" 33d33'38"
Ryan Anthony
Vir 12h22'43" 11d53'29"
Ryan Anthony
Vir 12h12'26" 4d12'55"
Ryan Anthony Cooper
Sgr 18h56'49" -21d25'8"
Ryan Anthony Futo
Dra 9h52'27" 77d9'20"
Ryan Anthony Jackson
Psc 23h55'25" 0d49'31"
Ryan Anthony Kalman
Vir 13h56'18" 6d25'44"
Ryan Anthony Kelemen
Aqr 20h39'34" -13d3'22"
Ryan Anthony Newman
Psc 1h17'48" 32d32'54"
Ryan Anthony Stein
Ori 6h1'22" 19d51'37"
Ryan Anthony Wells
Dra 18h26'14" 52d40'48"
Ryan Anthony-Paige Stewart
Cas 23h50'7" 61d49'26"
Ryan Antonio Moats
Aqr 22h7'18" -1d29'20"
Ryan Archembault
Uma 11h41'0" 33d41'27"
Ryan Arnold Kellner
Equ 21h7'41" 11d50'1"
Ryan Arnow Raimondo
Per 2h18'5" 56d38'48"
Ryan Arthur Mamrot
Aql 19h47'41" 16d4'2"
Ryan Arthur Scott
Psc 1h42'35" 7d19'32"
Ryan Arvy Cunanan
Ori 5h28'9" 5d50'20"
Ryan Ashley Silverthorn
Cnc 9h17'13" 26d2'48"
Ryan Ashton
Uma 9h53'41" 72d26'11"
Ryan Atherton
Ori 5h59'51" 11d58'51"
Ryan August Lange
Cep 23h24'30" 80d21'54"
Ryan August Smith
Boo 14h47'36" 21d41'41"
Ryan Aydan Ziegler
Gem 7h43'32" 34d20'15"
Ryan B Inge
Ari 2h36'18" 25d35'11"
Ryan Baiocco
Tau 5h48'34" 20d13'15"
Ryan Baker
Sgr 18h27'7" -27d29'21"
Ryan Bauer
Lyn 7h13'44" 58d37'47"
Ryan Beck
Per 3h47'55" 44d53'32"
Ryan Beihl
Ori 4h56'31" 4d37'7"
Ryan Bethell Jr.
Cyg 20h53'36" 47d49'22"
~Ryan & Bobbie Lynn Marks~
Ori 4h51'13" 6d24'35"
Ryan Bonner Smith
Ari 2h48'11" 30d4'49"
Ryan + Bonnie
Sge 19h51'20" 18d41'24"
Ryan Boy
Ori 5h22'47" -7d6'1"
Ryan Brackman Wahler
Vir 11h38'17" 3d27'12"
Ryan Brady
Cep 22h27'53" 74d57'42"
Ryan Brittanie
Aqr 22h17'44" -24d50'40"
Ryan & Brooklynn
Cyg 20h28'14" 37d24'35"
Ryan Brooks
Ari 2h22'10" 17d40'42"
Ryan Brothers
Lib 15h22'59" -24d30'53"
Ryan Brown
Peg 23h46'9" 25d24'9"
Ryan Brown
Per 2h42'5" 50d58'51"
Ryan Bubba Horner, Jr.
Per 3h32'10" 49d26'21"
Ryan (Bubba) Shield
Ori 4h50'44" 5d17'35"
Ryan Burgess
Ari 2h56'21" 11d29'35"
Ryan Burkle
Aql 19h6'27" 3d45'32"
Ryan C.
Tau 3h28'45" 9d18'33"
Ryan C. Lewis
Aqr 23h9'52" -17d53'15"
Ryan C. Moesel
Cap 21h3'52" -19d7'59"
Ryan C. North
Cyg 21h31'41" 36d0'15"
Ryan C. Sakowski
Sco 16h13'51" -13d7'4"
Ryan Cady
Ori 5h1'56" 15d23'11"
Ryan Calabrese
Sco 17h13'27" -43d26'43"
Ryan Calhan
Her 17h23'17" 19d27'31"
Ryan & Carolyn
Per 3h48'5" 51d39'9"
Ryan Casey
Leo 11h37'9" 21d53'55"
Ryan Chad Colgan
Uma 10h50'19" 55d25'8"
Ryan Charles Hart
Gem 7h40'42" 27d41'8"
Ryan Charles Plouffe
Ori 6h16'17" 10d55'3"
Ryan Charles Schad
Peg 22h38'49" 31d37'36"
Ryan Chase Clow
Tau 4h1'9" 8d27'27"
Ryan Chase Vanderwork
Ori 5h48'50" 1d16'21"
Ryan Chighizola
Uma 10h42'36" 68d7'23"
Ryan Christian Sakhleh
Aqr 22h4'54" -23d45'52"
Ryan Christopher
Uma 9h2'6" 56d18'21"
Ryan Christopher
Aqr 21h37'0" 0d37'1"
Ryan Christopher 21
Her 18h28'22" 25d58'1"
Ryan Christopher Cahill
Her 17h11'27" 47d53'50"
Ryan Christopher Harper
Vir 12h58'7" -2d43'40"
Ryan Christopher Hughes
Umi 14h43'33" 73d44'15"
Ryan Christopher Lydon
Gem 6h55'35" 22d8'31"
Ryan Christopher Mallard
Uma 8h55'2" 52d34'55"
Ryan Christopher Neal
Dra 20h25'59" 68d23'10"
Ryan Christopher Pontius
Ori 6h3'8" 19d14'21"
Ryan Christopher Friday
Sgr 19h28'36" -42d16'45"
Ryan Christopher Sullivan
Per 3h35'36" 47d28'15"
Ryan Christopher Sweeney
Ori 5h50'42" 21d2'26"
Ryan Christopher Utter
Vir 14h38'22" -4d18'51"
Ryan Churchward
Cen 13h42'16" -42d39'49"
Ryan Clancy
Umi 14h34'6" 74d4'24"
Ryan Clark
Her 17h16'37" 34d56'51"
Ryan Clarke
Her 17h58'7" 29d14'58"
Ryan Clay Campbell
Vir 13h22'19" 3d43'8"
Ryan Clint Russell
Uma 11h43'28" 49d3'22"
Ryan Cochran - Star Dad
Her 16h22'26" 22d55'5"
Ryan Cole Curtis
Uma 8h39'21" 62d36'59"
Ryan Conner Wight
Her 16h40'37" 44d7'38"
Ryan Cook
Per 3h15'51" 47d16'24"
Ryan Cook
Per 3h32'50" 46d40'35"
Ryan "Cookie Monster" Gray
Tau 5h22'35" 20d34'43"
Ryan Cory Ford
Aur 6h23'7" 41d18'10"
Ryan Cutshall
Uma 9h16'51" 47d30'54"
Ryan D Campbell
Uma 10h54'47" 60d27'7"
Ryan D. King
Aql 19h29'19" 10d10'22"
Ryan D. McGuire
Aql 19h47'14" 15d25'27"
Ryan D. Winslow
Her 17h29'50" 18d17'31"
Ryan Daniel Dance
Cep 21h32'8" 65d32'42"
Ryan Daniel French
Her 16h13'21" 46d13'10"
Ryan Daniel Murphy
Lib 15h38'36" -17d44'6"
Ryan "Daniel" Starnes
Cep 23h15'40" 71d39'9"
Ryan Dannenhauer
Lib 14h30'17" -22d12'28"
RYAN DAVID
Ori 5h54'22" 20d44'47"
Ryan David
Tau 4h40'36" 5d21'34"
Ryan David Adkins
Cnc 9h2'30" 27d25'13"
Ryan David Corey
Ari 3h13'29" 27d48'21"
Ryan David Hill
Ori 5h25'12" -4d40'25"
Ryan David McCurdy
Uma 10h34'7" 63d20'18"
Ryan David Smith
Ori 5h9'7" 0d32'49"
Ryan David Wooten, Jr.
Vir 13h26'29" 12d8'25"
Ryan Davis
Ari 2h49'14" 27d52'21"
Ryan Dean
Her 16h38'31" 45d40'46"
Ryan Dean
Dra 18h44'38" 79d14'14"
Ryan Dean Cogle
Ori 5h58'48" 21d44'2"
RYAN DEAN ZAPHERSON
Her 18h14'57" 15d33'17"
Ryan Dennis DeHaan
Lmi 10h29'51" 37d45'13"
Ryan Depledge
Umi 16h56'10" 85d45'27"
Ryan Dewar
Per 3h46'51" 48d57'53"
Ryan Dex Ferro
Tau 4h37'42" 15d9'38"
Ryan Dinesh Roodal
Vir 11h42'42" -6d6'54"
Ryan Dixon
Umi 16h55'11" 85d6'3"
Ryan Donald Guba
Uma 11h18'9" 34d1'36"
Ryan Donnelly
Uma 8h42'56" 50d26'28"
Ryan Doran
Cep 20h44'14" 60d42'22"
Ryan Douglas
Ori 5h33'46" -0d58'2"
Ryan Douglas Christensen
Her 16h31'56" 46d11'10"
Ryan Douglas Greer
Uma 11h31'29" 58d20'38"
Ryan Durham Brolliar
Per 3h38'55" 44d48'47"
Ryan E. Carter
Vir 11h38'25" 3d42'29"
Ryan E. Cooke 18th Star
Aqr 22h28'48" 1d7'9"
Ryan E. Robinette
Ori 5h19'49" -8d21'0"
Ryan Edward Haas
Psc 23h55'48" 5d45'30"
Ryan Edward Kaine
Crb 15h57'4" 26d34'18"
Ryan Edward Stokes
Ari 2h24'58" 26d27'44"
Ryan Edward Vance
Per 4h32'40" 40d33'8"
Ryan Elliot Bell
Vir 14h51'21" 2d35'13"
Ryan & Emily
Cas 1h11'30" 51d9'9"
Ryan Engelbrecht
Sco 16h17'41" -13d0'46"
Ryan & Erica
Cyg 20h51'4" 53d31'35"
Ryan Evans
Lib 14h43'32" -15d39'16"
Ryan Everett Walker
Psc 0h36'13" 4d39'35"
Ryan Fekete
Peg 22h29'8" 30d3'4"
Ryan Fenway Malvey
Gem 7h31'45" 19d43'54"
Ryan Fitzgerald - Rising Star
Aur 5h27'48" 46d12'6"
Ryan Flynn Dillman
Tau 5h6'31" 18d29'9"
Ryan Foley
Cep 22h0'15" 56d38'39"
Ryan Francis
Ori 6h18'31" 14d24'4"
Ryan Francis Basilio
Ori 5h32'33" -1d12'10"
Ryan Francis Vreeland
Aur 5h26'7" 39d38'25"
Ryan Frederick Letz
Boo 15h35'45" 39d55'2"
Ryan Fuchs
Ari 2h1'31" 20d39'45"
Ryan G. Rudge
Tau 3h26'53" 0d11'1"
Ryan G. Scheel
Ori 5h18'54" -0d18'24"
Ryan Geho
Uma 11h21'19" 29d0'54"
Ryan Geoffrey Fenner
Dra 20h3'0" 72d42'1"
Ryan George Cassar - Ryan's Star
Gem 7h33'38" 15d2'5"
Ryan George Riley
Tau 4h40'31" 23d35'19"
Ryan George Watts
Uma 10h53'43" 69d5'3"
Ryan Gerber
Ori 6h5'56" 7d2'18"
Ryan Gibbons
Uma 9h5'7" 46d34'49"
Ryan Glaser
Ori 5h47'55" 6d9'47"
Ryan Glen Christianson
Mon 7h32'6" -0d49'3"
Ryan Goodall
Uma 8h51'16" 49d25'20"
Ryan Gossack-Keenan
Uma 11h28'15" 58d52'57"
Ryan Grace
Psc 1h26'41" 26d35'42"
Ryan Gregory
Ori 5h54'46" 19d19'1"
Ryan Gruber
Ori 5h52'30" 20d57'45"
Ryan Gspandl
Ori 5h12'7" 12d47'33"
Ryan Guthrie
Dra 18h32'29" 51d44'50"
Ryan Gyllenhammer
Ori 5h53'54" 20d53'13"
Ryan Hanrahan
Cyg 20h45'16" 38d46'56"
Ryan Harris
Uma 11h36'22" 60d43'58"
Ryan Hawkins, aka WHITE LION
Vir 14h0'34" -14d48'32"
Ryan Hayes Elliott
Aur 5h21'9" 31d35'17"
Ryan & Heather Our Luv 4Ever Shines
Cyg 20h35'23" 54d11'49"
Ryan Henry Harrower
Ori 6h2'5" 17d44'21"
Ryan Hewitt
Per 3h33'40" 40d12'4"
Ryan Hoffman Woodrow
Cap 21h9'50" -15d12'43"
Ryan Hotler
Psc 1h13'0" 24d25'13"
Ryan Houston Crosley
Cap 20h33'8" -21d23'9"
Ryan Hunter Korson
Ori 6h5'42" 9d56'27"
Ryan Hunter's Prime 5
Ori 5h38'14" -0d33'46"
Ryan is the Bestest
Cnc 8h23'8" 14d22'38"
Ryan Isaac Shimony
Her 18h56'28" 17d54'13"
Ryan J. Chrostowski "9-28-05"
Her 17h39'9" 48d31'48"
Ryan J Feustel
Lyn 8h11'38" 56d18'38"
Ryan J. Hinkson
Uma 11h16'48" 40d57'30"
Ryan J. Jablonski
Ari 2h36'46" 26d33'38"
Ryan J. Michaelsen
Ori 5h55'37" 8d13'56"
Ryan J Rees
And 23h21'22" 44d57'10"
Ryan "Jack" Ellis
Cru 12h8'59" -60d29'54"
Ryan Jackson
Peg 21h44'51" 7d14'22"
Ryan Jacob
Cyg 19h49'15" 30d20'41"
Ryan Jacob Miller
Ori 6h17'21" 15d24'32"
Ryan Jacob Ridley
Psc 0h17'14" 9d57'56"
Ryan James
Sco 16h12'27" -10d57'22"
Ryan James Bayer
Cap 21h50'30" -9d33'30"
Ryan James Billingsley
Cyg 20h48'1" 31d12'22"
Ryan James Brooks My Eternal Love
Aur 5h46'7" 48d53'24"
Ryan James Burr
Ori 6h8'23" 12d48'35"
Ryan James Covillo
Lib 15h37'16" -11d54'58"
Ryan James Doolan
Per 2h27'20" 53d23'3"
Ryan James Fulton
Gem 7h13'29" 21d57'12"
Ryan James Harman
Cep 22h32'28" 75d37'2"
Ryan James Harrington
Ori 6h7'15" 10d28'9"
Ryan James Harris
Uma 11h38'2" 28d32'58"
Ryan James King
Ori 5h37'47" 0d49'12"
Ryan James Leavay
Uma 10h7'25" 71d10'10"
Ryan James Leiendecker
Ori 5h51'12" 9d22'10"
Ryan James Mendenhall
Cep 22h52'10" 68d38'5"
Ryan James Nicholas
Ori 6h17'24" 14d25'49"
Ryan James Schweitzer
Cap 21h18'59" -22d24'16"
Ryan James Sheedy
Ori 6h9'16" 15d28'19"
Ryan James Sweet
Ori 5h21'6" 4d52'37"
Ryan James Thomson
Her 18h16'2" 19d46'41"
Ryan James Wilson
Her 17h10'50" 22d21'18"
Ryan & Jamie (12-31-04)
Uma 11h59'23" 39d14'56"
Ryan Jarrod Kennelly
Dra 15h49'42" 59d40'53"
Ryan Jeffrey Hornacek
Leo 10h16'13" 14d45'29"
Ryan Jeffrey Rockwell
Uma 10h54'52" 39d56'8"
Ryan Jensen
Ori 6h4'27" 10d37'43"
Ryan Jinx Walton
Lyn 8h37'29" 43d47'46"
Ryan John Baum
Cnc 8h7'48" 7d17'37"
Ryan John Bjornson
Gem 6h23'33" 26d46'6"
Ryan John Langdon
Mon 7h20'26" -0d15'42"
Ryan John McDonough
Cep 22h32'23" 73d57'28"
Ryan John McDonough
Leo 11h34'2" 27d29'26"
Ryan John Vincent Quinn
Leo 10h19'16" 10d51'20"
Ryan Jones
Aqr 20h44'5" -8d43'23"
Ryan Joseph Furlow
Her 17h26'9" 36d22'11"
Ryan Joseph Gallagher
Umi 15h26'55" 71d44'5"
Ryan Joseph Gaynor
Peg 23h37'52" 15d58'8"
Ryan Joseph Greenlee
Sgr 19h16'43" -26d39'25"
Ryan Joseph Holland
Tau 5h45'10" 27d7'54"
Ryan Joseph Jaffry
Aqr 22h59'7" -12d31'39"
Ryan Joseph Leavitt
Ori 5h40'8" 2d7'14"
Ryan Joseph Leveille
Tau 5h54'50" 23d54'19"
Ryan Joseph Nadig
Ori 5h31'34" 0d24'30"
Ryan Joseph Ouellette
Aql 19h6'15" -9d37'2"
Ryan Joseph Sherrieb
Umi 14h26'14" 76d30'13"
Ryan Joshua Bloom
Vir 12h59'27" -13d43'33"
Ryan K Bailey & Dragonfly
Eri 3h50'41" -0d28'9"
Ryan Karrasch
Cep 20h55'27" 68d54'25"
Ryan & Katherine
Ori 6h1'12" 20d12'42"
Ryan & Kayla
Lyr 18h48'46" 32d29'24"
Ryan Kearney
Aql 19h15'29" 8d41'35"
Ryan Keith Thompson
Cap 20h28'28" -27d34'51"
Ryan Kellan Wood
Cru 12h37'44" -59d5'30"
Ryan Kelly
Cap 21h1'22" -16d2'40"
Ryan Kenna
Ari 2h44'53" 15d11'28"
Ryan Kennelly
Lyn 7h22'12" 52d45'24"
Ryan Kimball
Uma 10h50'50" 67d53'10"
Ryan King
Ori 5h8'50" 9d6'19"
Ryan Klym
Her 17h4'34" 32d25'18"
Ryan Kyle Lewis
Cnc 8h23'21" 16d4'9"
Ryan L. Boober
Sco 16h10'29" -11d34'53"
Ryan L. Browne
Uma 11h3'48" 65d4'32"
Ryan L. Conover
Lyn 6h52'47" 51d28'19"
Ryan L. Sills
Umi 15h14'35" 72d33'51"
Ryan Larkins
And 1h22'37" 48d46'10"
Ryan & Laura Forever
Cyg 20h9'13" 33d34'3"
Ryan Lax
Tau 5h20'21" 28d23'45"
Ryan Lee
Uma 8h48'34" 47d19'3"
Ryan Lee Crenshaw
Ori 4h52'0" 8d7'22"
Ryan Lee Crist
Sco 17h46'30" -41d25'36"
Ryan Lee (RL)
Per 2h59'13" 55d7'34"
Ryan Lee Roberson
Uma 9h57'57" 52d40'51"
Ryan Lee Robinson
Cep 20h53'56" 60d15'52"
Ryan Lee Stadler
Dra 19h27'25" 64d21'2"
Ryan Lee Wikete
Her 18h37'33" 16d54'25"
Ryan Levi Rains
Lib 15h2'50" -24d32'42"
Ryan & Lisa
Cyg 20h21'29" 45d37'3"
Ryan Loves Amanda
Leo 11h6'33" 18d36'8"
Ryan Lynch - The Legend
Per 4h41'46" 43d0'36"
Ryan Lynn Easterwood
Umi 15h46'38" 88d25'59"
Ryan M. Arnold
Per 3h41'9" 43d54'36"
Ryan M. Kuenzner
Ori 5h37'39" -0d15'1"
Ryan M. Maganzini
Leo 11h54'52" 22d23'38"
Ryan M. Meyer's Star
Aqr 22h9'37" 1d3'7"
Ryan M. Sutton
Gem 6h30'34" 20d32'58"
Ryan M. Wilme "Our Special Star"
Dra 18h28'51" 76d34'26"
Ryan Magana and April Opp
Lyr 18h31'18" 36d46'51"
Ryan Mallory
Aur 5h39'41" 37d28'49"
Ryan Manning
Ori 5h21'38" 8d2'14"
Ryan Marc Nadel
Sco 17h50'56" -43d59'50"
Ryan Marie Szymanski
Uma 12h35'53" 57d32'37"
Ryan & Marion
Ori 5h19'15" 12d43'42"
Ryan Mark Lega
Uma 9h40'4" 62d44'12"
Ryan Marks
Cyg 20h30'18" 45d55'1"
Ryan Martin
Cen 14h11'23" -60d44'28"
Ryan Marynowski
Her 18h46'32" 21d46'37"
Ryan Mathew
Cnc 7h59'12" 16d13'28"
Ryan Matthew
Ari 3h24'19" 28d1'24"
Ryan Matthew
Cep 22h7'41" 67d21'59"
Ryan Matthew Bode
Ori 6h16'42" 10d32'58"
Ryan Matthew Corzine
Ori 5h28'51" 14d50'39"
Ryan Matthew Ehrie
Aql 19h55'45" 6d2'28"
Ryan Matthew Herren
Ori 5h37'0" 9d15'22"
Ryan Matthew Levy
Sco 16h36'9" -29d23'9"
Ryan Matthew McGlothlin
Sgr 17h50'33" -19d56'18"
Ryan Matthew O'Dea
Lib 15h0'17" -21d45'54"
Ryan Matthew Scott
Uma 10h55'21" 41d29'11"
Ryan Matthew Shelbourne
Cyg 19h51'20" 53d0'42"
Ryan Matthew Shirar
Lib 15h17'18" -25d5'29"
Ryan Matthew Zimmerman
Sgr 17h45'21" -16d47'40"
Ryan Maurice Sawdy, Jr.
Ori 5h46'35" 7d41'25"
Ryan McCauley
Per 4h12'14" 42d58'51"
Ryan Mcgee is forever with me
Her 17h21'3" 37d40'6"
Ryan McKee
Per 2h45'51" 52d21'9"
Ryan McKee
Boo 14h35'57" 19d3'35"
Ryan&Melanie Stone Mud On The Tires
Psc 1h21'1" 24d18'56"
Ryan & Melissa - 1 yr of perfection
Sgr 17h52'29" -28d46'12"
Ryan Michael
Sco 16h51'53" -38d17'47"
Ryan Michael Anderson
Cnv 12h43'44" 47d12'49"
Ryan Michael "Bubo" Newton
Boo 14h35'19" 37d0'16"
Ryan Michael Duclos
Lib 15h6'17" -3d23'24"
Ryan Michael Fatta
Vir 13h26'3" 12d21'37"
Ryan Michael Faubert
Gem 7h41'10" 34d4'6"
Ryan Michael Harrison
Lib 15h19'23" -5d52'7"
Ryan Michael McCauley
Per 3h23'27" 33d8'58"
Ryan Michael Ritchie Koster
Uma 9h49'52" 60d23'48"
Ryan Michael Romel
Leo 11h7'51" 2d53'12"
Ryan Michael Sargent
Uma 9h39'47" 57d22'54"
Ryan Michael Sullivan
Uma 8h21'15" 65d2'30"
Ryan Michael Tedore
Ari 2h3'58" 24d23'25"
Ryan Michael Tobey
And 0h30'39" 33d42'27"
Ryan Michael Wall
Sco 16h11'28" -10d2'0"
Ryan Michael Woods
Cnc 8h23'39" 22d58'27"
Ryan Micheal Lillard
Ori 5h38'46" -7d47'42"
Ryan Michelle Bahr
Leo 9h49'51" 21d47'2"
Ryan Mitchell Jones
Ori 5h7'52" 11d36'54"
Ryan Mitchell Kemp
Cru 12h36'30" -61d6'50"
Ryan Mitchell Warden
Ori 5h50'23" 11d53'17"

Ryan Mon Monde
Cep 21h51'55" 65d47'30"
Ryan Moore
Aqr 20h49'32" -12d23'57"
Ryan Mowry
Aql 19h11'43" -0d43'4"
Ryan Mulhern, Our Lost Hero
Her 17h9'7" 26d38'57"
Ryan Murphy
Aur 6h23'36" 43d3'45"
Ryan Murray
Ori 5h44'33" -3d38'40"
Ryan Musumeci
Gem 6h51'17" 32d46'15"
Ryan - My "Knight" In Shining Armor
Her 17h8'17" 36d32'38"
Ryan N. Valett
Cyg 20h0'50" 32d28'50"
Ryan Napoli
Cap 20h35'50" -17d37'58"
Ryan Nicholas
Cep 23h57'51" 69d30'27"
Ryan Nicholas
Umi 13h9'11" 70d49'6"
Ryan Nicholas Brown
Gem 6h46'38" 34d16'9"
Ryan & Nickie Trott
Cyg 21h31'42" 34d26'9"
Ryan Nicolas Taylor
Lib 14h53'37" -13d17'41"
Ryan of Warrenton
Ori 5h38'30" 8d57'7"
Ryan Oksenhorn
Leo 11h56'9" 16d59'26"
Ryan O'Neal Farrell
Leo 11h24'20" 15d20'6"
Ryan ooo Taryn
Cru 11h56'40" -63d46'17"
Ryan O'Rourke
Cnc 9h9'44" 16d15'43"
Ryan*Our Love Can Do Miracles :)
Vir 13h50'44" 6d38'21"
Ryan P. Wildman
Her 16h9'24" 49d16'12"
Ryan P Winkelman
Ori 5h38'41" 1d39'40"
Ryan Padraic McCue
Leo 9h29'46" 16d1'48"
Ryan Page
Lyn 7h39'9" 49d8'21"
Ryan Palmer
Her 16h31'58" 12d51'59"
Ryan & Pamela Pollard
Uma 11h7'57" 43d33'48"
Ryan Paratore
Tau 4h2'44" 18d23'30"
Ryan Parker
Uma 10h58'23" 51d24'48"
Ryan Parks Medlock
Ori 6h4'42" 9d24'16"
Ryan Patrick
Lmi 11h6'8" 32d51'11"
Ryan Patrick
Vir 12h28'35" -0d16'47"
Ryan Patrick & Alethea Christine
Ari 2h36'31" 27d19'10"
Ryan Patrick Allen's Twinkle
Tau 4h11'56" 23d2'23"
Ryan Patrick Cyr
Ori 6h0'0" -0d49'52"
Ryan Patrick Finnie
Sco 16h58'14" -35d2'26"
Ryan Patrick Gaishin
Her 17h18'41" 26d56'48"
Ryan Patrick Gockel
Ori 6h15'37" 9d36'1"
Ryan Patrick Gula
Ori 6h3'14" 18d12'10"
Ryan Patrick Hawkins
Her 16h9'4" 47d54'39"
Ryan Patrick Kavanagh
Uma 11h44'26" 60d39'10"
Ryan Patrick Kilroy
Her 17h41'45" 45d27'0"
Ryan Patrick Mahan
Cap 21h7'51" -23d41'32"
Ryan Patrick McMillion
Leo 11h30'52" 11d9'31"
Ryan Patrick Meehan
Ori 5h38'13" 11d22'49"
Ryan Patrick Murnane
Cnc 9h4'33" 11d42'41"
Ryan Patrick Regan
Ori 6h1'40" 18d3'22"
Ryan Patrick Shidler
Vir 12h42'10" 11d53'35"
Ryan & Patti
Ori 4h51'13" 11d34'38"
Ryan Paul Peck
Her 17h58'11" 25d43'42"
Ryan Paul Redmond
Sgr 18h5'21" -26d19'33"
Ryan Paul Turner
Aqr 21h20'52" -5d31'36"
Ryan Peter McMullin
Vir 12h41'56" -0d22'50"
Ryan Peter Schlossmacher
Ori 5h56'4" 21d4'45"
Ryan Philip Gerace
Uma 9h4'4" 62d18'0"
Ryan Philip Skora
Ori 5h41'49" 1d2'37"
Ryan Phillips
Her 17h7'43" 36d45'50"
Ryan Pip Mullany
Cnc 9h4'26" 14d27'20"
Ryan Pooh
Uma 9h23'58" 59d36'59"
Ryan Quirion Guthrie
Cep 22h43'30" 64d2'45"
Ryan R. Clark
Uma 9h49'2" 71d38'36"
Ryan R. Roettger
Ori 5h31'17" -0d29'29"
Ryan Raines
Her 17h10'19" 47d54'44"
Ryan Ralph Benjamin Brown
Per 4h16'56" 45d53'17"
Ryan Ramirez
Sgr 18h44'12" -27d22'59"
Ryan Rapko
Dra 15h20'11" 62d23'39"
Ryan Renee Stout
And 23h13'26" 47d45'50"
Ryan Richard Jaros
Per 2h14'31" 54d33'38"
Ryan Richard Rucoba
Uma 11h16'21" 54d0'6"
Ryan Ripple
Vir 13h57'32" -15d58'40"
Ryan Rippy # 29
Aql 19h20'49" 13d25'38"
Ryan Robert Hay
Tau 5h8'7" 25d29'26"
Ryan Robert MacCluen
Vir 13h8'29" -1d22'57"
Ryan Rocheleau "1983-2006"
Gem 7h28'14" 34d11'34"
Ryan Roger Zaccour
Cnc 8h49'1" 30d50'18"
Ryan Ronald Couse
Aur 5h42'2" 49d55'59"
Ryan Ronald Gendreau
Ori 5h42'55" -0d24'48"
Ryan Sabara
Ori 6h3'4" 19d52'48"
Ryan Salvador Marquez
Uma 8h48'31" 53d15'1"
Ryan Samuel Telford
Per 1h31'2" 51d51'41"
Ryan Sands
Cap 21h32'29" -8d31'7"
Ryan & Sarah Together Forever
Gem 7h27'54" 21d39'59"
Ryan Saturday
Men 5h14'41" -70d18'44"
Ryan Savoy
Lib 14h53'38" -2d9'26"
Ryan Schackne
Per 3h38'40" 50d38'35"
Ryan Schatzel
Her 17h8'12" 39d21'51"
Ryan Scolaro
Dra 19h48'21" 64d28'33"
Ryan Scott Bayliss
Cap 21h28'29" -18d32'9"
Ryan Scott Emerson
Cas 23h22'33" 57d47'49"
Ryan Scott Gardner
Sco 17h28'37" -41d36'15"
Ryan Scott Harris
Ori 5h2'5" 7d6'47"
Ryan Scott Johnson
Aur 5h30'16" 46d27'4"
Ryan Scott Lee
Cnc 8h32'4" 14d43'23"
Ryan Scott Milsom
Per 4h8'22" 45d15'16"
Ryan Scott Nally
Vir 11h59'2" -0d9'32"
Ryan Scott Palombo
Gem 7h53'56" 31d23'13"
Ryan Scott Rudolph
Her 18h45'35" 16d53'33"
Ryan Scott Smith, My True Love xxxx
Cyg 21h54'36" 51d54'35"
Ryan Scott Woolley
Lmi 10h51'31" 33d41'12"
Ryan Scott Worcester
Cep 21h27'38" 64d13'28"
Ryan Sebastian Coons
Leo 11h32'22" 25d46'0"
Ryan Servetnick OLS Superstar
Sct 18h31'37" -3d59'55"
Ryan & Sharni - A Dream Come True
Cru 12h26'17" -58d40'22"
Ryan & Sheri Hollien-10 Happy Years
Cyg 20h5'48" 37d9'5"
Ryan & Shira - Forever & for Always
Ari 3h12'18" 15d9'23"
Ryan Shuster
Tri 2h17'14" 32d46'35"
Ryan Sjaarda
Aur 6h39'19" 39d13'32"
Ryan Skellington
Cap 21h51'54" -9d42'49"
Ryan Small Family
Uma 9h0'24" 49d45'39"
Ryan Smith Jackson
Her 17h41'5" 44d32'19"
Ryan Smith Jackson
Umi 17h24'53" 78d54'10"
Ryan Sorensen
Per 3h50'50" 33d29'58"
Ryan ~ Sparkle ~ Lil' Sandusky
Umi 14h39'28" 67d30'59"
Ryan Speier
Aql 20h11'14" 13d55'29"
Ryan & Staci 4 EVER
Leo 11h49'37" 20d29'34"
Ryan Stana
Ori 6h2'27" 18d8'34"
Ryan Stephen
Her 17h16'20" 28d37'43"
Ryan Steven Elsea
Her 17h33'55" 33d6'49"
Ryan Steven Hamblin
Ori 5h52'59" 21d10'9"
Ryan Steven Marcisz
Vul 20h30'32" 23d26'40"
Ryan Stevens Harris
Sco 17h31'32" -39d55'54"
Ryan Stoffer
Vir 12h48'11" 7d53'43"
Ryan Stone
Leo 9h29'25" 11d20'18"
Ryan Storm
Uma 11h28'7" 42d44'36"
Ryan Sturt Osmond
Leo 11h31'56" 10d38'34"
Ryan Taylor-Garrett Duran
Gem 6h42'40" 34d26'23"
Ryan Tejaratchi A.K.A. Ronin
Gem 7h9'55" 18d23'31"
Ryan Thomas
Gem 7h2'41" 34d4'45"
Ryan Thomas Bonnell
Peg 22h38'33" 23d7'28"
Ryan Thomas Gosling
Her 17h27'21" 31d30'53"
Ryan Thomas Jones
Ori 4h55'54" 10d3'26"
Ryan Thomas Pestrichella
Tau 3h37'13" 21d10'58"
Ryan Thompsett
Gem 6h1'12" 27d11'51"
Ryan & Tiffany- Our Love Is Forever
Cnc 8h52'48" 24d48'12"
Ryan Todd Moore
Sgr 19h23'7" -27d35'25"
Ryan Tooma's little bit of Heaven
Cru 12h56'25" -57d43'52"
Ryan Torigiani
Ori 5h52'19" 8d57'33"
Ryan Trewin
Ori 4h59'8" 15d1'21"
Ryan Tuinila
Tau 5h50'3" 25d55'37"
Ryan Vondra
Cep 1h55'27" 77d58'48"
Ryan Vuillemot
Cap 20h53'5" -21d0'23"
Ryan W. Ferguson
Lib 14h52'39" -2d8'10"
Ryan W. Scott
Uma 11h26'49" 46d55'44"
Ryan Wade
Ori 6h3'12" 19d16'26"
Ryan Walker
Sco 17h29'0" -37d33'3"
Ryan Wallace Shirlaw
Umi 15h4'20" 76d23'33"
Ryan Warren Towery
Vir 12h38'9" 3d25'12"
Ryan Wayne Bruce
Cnc 8h47'13" 14d56'4"
Ryan Wayne Bruce
Her 17h55'53" 29d41'43"
Ryan Wayne Shaw
Ori 6h2'25" 10d17'20"
Ryan Wendelin Pass
Tau 4h32'23" 7d21'41"
Ryan Wheeler
Psc 1h21'1" 16d27'35"
Ryan Whitfield
Sco 16h13'42" -13d18'35"
Ryan William
Dra 9h58'58" 78d16'23"
Ryan William
Boo 14h38'42" 37d25'33"
Ryan William Considine
Aur 5h38'20" 41d7'44"
Ryan William Daly
Lib 14h53'42" -3d26'53"
Ryan William Hughes
Lib 15h25'7" -10d48'35"
Ryan William James Tyas
Psc 1h25'50" 32d33'20"
Ryan William Labus
Umi 16h45'45" 81d8'4"
Ryan William Marrone
Per 2h20'7" 51d6'12"
Ryan William McLaughlin
Uma 12h4'47" 49d0'24"
Ryan William Powell
Uma 9h48'26" 41d40'50"
Ryan William Thomas
Psc 1h7'25" 10d22'32"
Ryan William Verhey
Cnc 9h4'49" 23d35'20"
Ryan Williams
Lac 22h31'21" 38d46'18"
Ryan Woodruff
Ori 6h0'58" 17d1'52"
Ryan Worth
Cra 18h5'30" -43d23'1"
Ryan Young
Psc 1h13'51" 17d55'44"
Ryan you're my star-crossed lover.
Her 18h50'23" 20d22'5"
Ryan Z
Ori 6h23'24" 17d10'54"
Ryan Zachary Kelly
Lib 15h32'37" -19d37'3"
Ryan Zachary Slaven
Umi 14h19'32" 76d43'20"
Ryan Zarn
Uma 9h39'24" 59d28'4"
Ryan, Kathe's Angel June 22, 1980
Gem 7h18'38" 32d31'48"
Ryan, Terri, and Jeromy Carroll
Cyg 21h51'33" 43d28'56"
Ryan51205
Ori 5h53'40" 18d24'57"
RyanandMegan
Cyg 20h26'26" 56d14'40"
Ryanathan
Cnc 8h39'17" 18d35'55"
Ryane Chrisler
And 2h8'59" 42d52'1"
Ryane Elizabeth
And 0h31'35" 25d20'37"
Ryane Zniber
Her 17h57'8" 25d41'1"
Ryangela
Uma 11h21'49" 40d59'39"
RyanLauraMetcalf
Cyg 19h24'22" 48d6'46"
RyanM.Duffy
Ori 6h4'10" 10d12'12"
Ryann
Sgr 19h40'2" -13d23'58"
Ryann and James
And 0h47'57" 39d58'38"
Ryann Dorian Florence Nolan
Cnc 8h50'8" 32d26'13"
Ryann Leigh *7:51pm*
Gem 7h28'59" 32d31'24"
Ryann Marie Schlaline
Ori 5h38'12" -1d35'2"
Ryann Moon Savior
And 23h12'16" 51d15'5"
Ryann Piper
And 23h22'32" 41d19'2"
Ryanna G. Romano
Psc 23h3'2" 7d17'10"
Ryanna Violet Law
Cas 3h21'29" 72d4'56"
Ryanne Clingersmith 1/23/1985
Aqr 22h42'14" -8d17'58"
Ryanne Murcko
And 23h12'0" 49d57'37"
Ryan-Rateb Abu Nasra
Ori 6h7'40" 11d1'26"
Ryan's 4
Uma 10h25'26" 62d38'47"
Ryan's Babygirl
Cyg 21h27'12" 45d36'18"
Ryan's Beauty
Cyg 19h57'42" 30d15'32"
Ryan's Dale
Uma 10h23'38" 52d28'12"
Ryan's Hope
Cap 21h0'44" -27d5'44"
Ryans Kikis Shining Star Forever!
Ori 5h12'0" 11d21'49"
Ryan's Light
Sco 16h7'34" -12d3'52"
Ryan's LuViEz Star
Umi 15h39'39" 72d12'35"
Ryan's Seoid
Aql 19h41'56" -0d1'57"
Ryan's Star
Ori 4h54'1" -2d39'48"
Ryan's Star
Lyn 8h29'0" 56d16'39"
Ryan's Star
Sco 16h47'47" -44d29'11"
Ryan's Star
Ori 5h34'16" 9d23'43"
Ryan's Star
Aql 19h25'0" 8d34'47"
Ryan's Star
Uma 8h46'9" 47d34'48"
Ryan's Star of Love
Cyg 19h45'10" 35d14'45"
Ryan's Sweet Dream
Ori 5h28'42" -0d19'2"
Ryan's Zitro Star
Sco 16h18'25" -42d3'20"
Ryanstar
Her 18h3'23" 21d56'23"
Rybaby Patrick
Uma 10h22'4" 58d42'14"
Rybinder31
Cam 4h41'30" 61d44'58"
RyBugs
Vir 13h44'13" -22d30'58"
Rychlik Wojciech csillaga
Sco 17h15'37" -43d48'59"
Rycott
Cyg 21h34'44" 40d58'40"
Ryde in the Sky
Dra 18h56'5" 50d37'37"
Rydell Family
Uma 9h53'50" 45d29'3"
Ryder
Sgr 19h21'17" -15d3'40"
Ryder Anthony Panico
Hya 9h14'39" -2d34'6"
Ryder Anthony Panico
Cnc 8h53'14" 13d22'40"
Ryder Blaze Chaney
Gem 6h45'23" 30d50'43"
Ryder Hamilton Hay
Cru 12h24'18" -56d4'6"
Ryder James Griffin
Uma 9h20'0" 56d57'6"
Ryder James Hayes
Boo 14h33'56" 15d31'55"
Ryder Lee Senko
Aqr 22h54'6" -6d38'54"
Ryder Louis Kawohio'kalani Briley
Ori 5h41'4" 4d11'12"
Ryder Owen
Lyn 6h59'50" 55d51'59"
Ryder Thomas Robinson
Sco 17h22'7" -31d45'32"
Ryder Wesley Cullen
Uma 8h32'25" 70d18'1"
Ryder's Little Gem
Cnc 8h52'22" 13d42'44"
Ryder's Shining Ray
Gem 6h35'20" 24d8'37"
Rydr Michael Lasky-Blattmachr
Gem 7h15'55" 17d29'29"
"Rye" Jazmyne Reed and John Skvarek
Vir 12h20'43" 3d9'16"
Rye & Joni
Sge 20h4'47" 16d48'36"
Rye Matthew Henley
Lyn 7h53'1" 57d0'40"
Ryedan Lee Ashworth
Cam 7h37'42" 78d42'33"
Ryeley Morgan Gravelding
Vir 12h23'19" -4d42'8"
Ryen Amir
Vir 14h41'32" 5d22'46"
Ryen M Cameron "Pumpkin"
And 0h10'20" 33d10'50"
Ryken Doc Oldfield
Uma 8h50'41" 47d51'28"
Ryker and Cole Snarr
Lmi 10h30'56" 37d4'23"
Ryker Lee Andersen
Umi 14h39'54" 76d37'10"
Ryker Luke Autajay
Uma 10h27'12" 42d56'18"
Ryknows
Uma 11h47'44" 52d28'4"
Ryla Mae Heins
Ari 2h28'14" 10d54'43"
Rylan Chase Baldwin
Lib 14h40'2" -10d15'47"
Rylan Darney Swingle
Sgr 18h54'35" -28d53'35"
Rylan Jackson Lambert
Ori 6h3'4" 19d40'46"
Rylan- Kristin
Uma 12h40'35" 58d27'7"
Rylan Raine
Gem 7h20'34" 32d41'19"
Rylan Reid Jones
Tau 3h27'36" 7d22'57"
Rylan Trulson
Ori 5h58'29" 17d55'16"
Ryland McKinney Stewart
Umi 14h23'19" 75d12'53"
Ryle Katelyn Bent
Cyg 20h54'26" 46d24'39"
Ryleah Julia
Lib 15h46'9" -29d45'45"
Rylee
Ori 5h55'21" 18d36'37"
Rylee Addison Wemhaner
Vir 13h22'27" 11d28'45"
Rylee Annora
Sco 16h5'18" -16d31'21"
Rylee Blanchard
Gem 7h30'7" 20d34'41"
Rylee Catherine
And 0h46'16" 39d25'3"
Rylee Emma Warman
And 2h16'4" 45d45'54"
Rylee Gallego Haag
Ori 5h28'42" 8d41'36"
Rylee Irene King Murphy
Lib 14h49'34" -11d50'41"
Rylee K. Brokaw
Gem 7h15'40" 29d6'42"
Rylee Kay Marcus
Umi 14h11'51" 77d45'42"
Rylee Kealohilani Egbert
Eri 3h48'1" -13d49'49"
Rylee Lynn Scott
Lyn 8h27'57" 55d18'26"
Rylee Mae Tighe
Ari 2h5'46" 25d2'25"
Rylee Mayer
Aqr 22h8'1" 0d55'9"
Rylee Murtaugh Butler
Dra 19h8'46" 65d49'39"
Rylee Nicole Robertson
And 2h36'13" 48d33'43"
Rylee Patton Buchert
Ori 4h53'48" 3d40'56"
Rylee Rain Roberts
Uma 13h7'58" 60d34'22"
Rylee Rebecca Hill
Crb 16h21'12" 38d15'37"
Rylee Simone Rotsma
Umi 16h26'3" 77d46'11"
Rylee Suzanne
And 23h26'22" 47d22'37"
Rylee Yu Star
Vul 19h30'42" 23d4'26"
Rylee's Little Light
And 1h45'3" 42d16'48"
Rylee's Star
Sgr 17h53'51" -29d44'53"
Ryleigh
Tau 5h51'20" 17d52'46"
Ryleigh Aaron Shelton
Cyg 19h59'15" 36d15'37"
Ryleigh Jane Travers
Tau 5h9'6" 24d27'17"
Ryleigh Kresse
Cam 5h40'53" 62d25'48"
Ryleigh Madison Aragi
Oph 16h55'43" -0d5'10"
Ryleigh Marie Gagnon
Psc 1h4'39" 33d1'32"
Ryleigh Nicole Maxwell
Psc 0h35'3" 14d9'11"
Ryleigh Shay Johnson
Cap 20h20'13" -9d32'30"
Ryleigh T. Crawley
Gem 6h6'9" 23d47'4"
Ryleigh's Rainbow Room
Ari 2h52'31" 13d15'9"
Ryley Daniel Robinson
Umi 14h52'54" 77d6'48"
Ryley Isabel Navarro Cote
Ari 2h32'32" 11d29'58"
Ryley James Bucknall
Her 16h55'25" 34d46'50"
Ryley Nichele Olson
Aqr 22h26'55" -19d18'17"
Ryli
Tau 4h13'56" 5d49'27"
Ryli Mariellen Bland
Umi 15h14'14" 69d0'34"
Rylie Brooke Ebersole
Cas 2h14'50" 72d41'1"
Rylie Elizabeth Burke
Cnc 8h39'5" 19d2'45"
Rylie Heusner
Sco 17h48'58" -36d17'6"
Rylie Jayden
Gem 7h5'43" 27d36'9"
Rylie Jo-Lynn Sherrill
Uma 11h56'17" 39d14'37"
Rylie Kate
And 1h28'38" 37d51'58"
Rylie Kate Watts
Cru 12h16'36" -60d12'31"
Rylie Kay Madison
And 0h59'14" 45d9'34"
Rylie Lin
Cap 21h52'1" -9d8'46"
Rylie Madison
Cyg 21h31'55" 33d37'49"
Rylie T. Rodriguez
Leo 10h35'53" 12d44'23"
Rylina
Lyr 19h17'40" 28d51'31"
Rylyn
Crb 15h51'31" 29d39'37"
Ryne Adam Brock
Uma 14h14'18" 57d56'56"
Ryne Keith Goldsworthy
Lyn 7h25'48" 53d34'17"
Rynne Brandt
Dra 18h30'27" 71d5'40"
Rynok
Ori 5h33'9" 3d59'36"
Ryo
Ori 6h13'44" 20d51'49"
Ryoichi & Miho
Lib 15h44'49" -15d50'41"
Ryon
Sco 17h56'27" -41d30'33"
Ryon Antony Barr
Umi 15h9'41" 69d8'43"
Ryon Provencher
Ori 5h32'50" 1d48'52"
Ryonda
Uma 11h14'28" 29d22'26"
RYonne
Ori 4h5'52" 9d31'10"
RyRoJa
Vir 13h27'51" 8d23'9"
RyRy
Leo 10h9'50" 20d24'4"
Ry-Ry 34876404
Lmi 10h23'28" 36d41'13"
Ry-Ry's Baby-Girl
Uma 11h3'7" 60d12'21"
Ry's Dippy
Vir 13h16'35" -2d59'36"
Ryson Jeffrey Haag
Ori 5h28'3" 9d5'48"
Ryszard Widelski
Aur 6h29'18" 51d32'18"
Ry-Tan
Cas 1h20'13" 55d23'36"
Ryu Dorei
Leo 10h51'32" -0d4'5"
Ryuhs
Sgr 18h56'26" -29d29'2"
Ryun Borchers
Leo 11h16'57" 16d49'32"
Ryunson Terheggen
Boo 14h33'11" 53d59'1"
RYV
Uma 8h33'3" 60d11'0"
Rzepucha, Werner
Uma 11h14'8" 64d20'17"
S & S
Cyg 21h43'57" 43d55'33"
S A Brazier
Cyg 20h12'26" 30d19'58"
S A C H A
Vir 11h47'10" 3d41'27"
S a d i a
Cnc 8h16'39" 15d45'5"
S. Allan Dubow
Her 17h52'36" 23d41'52"
S and J
Uma 11h3'26" 47d50'57"
S. B. Monteith
Cyg 21h14'34" 43d19'45"
S Bug
Per 3h15'18" 51d51'42"
S. C.
Cas 3h12'35" 60d41'48"
S & C Casagrande
Peg 21h35'3" 25d50'44"
S&C CULLEN
Aqr 20h40'43" -9d0'8"
S C Tartan Memory
Uma 9h52'3" 58d12'4"
S. Castlen
Lyr 18h39'8" 36d33'35"
S. Chase Collett
Tau 4h40'25" 12d14'14"
S. D. Fairbairn
Cap 20h34'23" -11d4'41"
S&E Beginning of Eternity
Her 18h39'22" 15d48'53"
S. Earl Sansing
Per 3h22'33" 51d17'17"
S F Pawlyszyn
Cnc 7h56'46" 12d32'40"
S G C 75
And 2h28'37" 49d31'44"
S.:.G.:.C.:. Vasileios PATKAS, 33o
Ori 6h6'59" 17d54'53"
S. Gene Hayes
Aur 5h55'17" 37d12'42"
S. George Crossland, Jr.
Ori 6h17'50" 14d44'1"
S. H. M. I. L. U.
Col 5h28'38" -35d53'28"
S I S K A
Cru 12h24'13" -63d14'39"
S&J Golladay
Tau 5h6'58" 28d15'18"
S. J. Taylor Price
Cma 6h35'50" -16d40'3"
S. Jack Payne
Ari 3h16'47" 22d24'19"
S & K Cometa
Lmi 10h9'52" 30d13'36"
S K Perera
Lib 14h58'39" -9d39'28"
S. Lupe Silva
Tri 2h13'10" 33d50'34"
S. Lynda Rudolph
Cyg 21h31'10" 46d59'58"
S&M
Lyn 7h33'43" 35d38'12"
S. M. Lasiter
Lac 22h22'37" 41d29'29"
S. Maryann Goch
Tri 2h6'29" 33d31'21"
S. McCrae
And 23h30'3" 37d52'3"
S. Milton Piuma
Ori 4h49'11" 12d47'23"
S. O. K.
Psc 1h2'52" 13d53'57"
S O L Y Basta
Uma 8h37'44" 64d56'3"
S. O. S.
Umi 16h32'37" 77d44'11"
S. P. G. C.
Dra 18h30'2" 61d51'30"
S & P Real Estate
Uma 10h50'21" 60d23'51"
S. Pallus Warren (Omi)
Leo 9h36'55" 12d27'12"
S. R. E. C. Clarrage
Uma 11h16'19" 65d39'24"
S. Rand Werrin
Cnc 8h39'43" 30d48'50"
S&S
Uma 11h22'48" 61d53'5"
S S Sambo
Her 17h21'54" 29d22'43"
S & S Wrapped Forever
Lib 15h17'46" -15d48'21"

S&SMc
Cyg 20h23'4" 39d32'40"
S & T Pacey's Team Buh Wedding Star
Cyg 20h35'43" 61d19'9"
S & T Sylvester
Umi 15h8'30" 73d11'56"
S. Terri McKenzie
Cyg 20h26'57" 35d2'48"
S&Ts Tyler
Dra 14h45'56" 59d9'8"
S u a n a
Aqr 22h29'43" 1d18'52"
S. W. Phillips Rock & Roll Galaxy
Uma 8h21'15" 64d46'47"
S. Whitehead
Ari 3h18'27" 18d50'22"
s2d2
Gem 6h3'16" 27d59'6"
S4KTX
And 1h16'23" 49d17'19"
6 of us in arthurs chariot
Uma 10h19'1" 65d21'0"
612
Uma 13h47'26" 53d15'24"
6-18 Crystal Rodgers
And 0h23'34" 45d6'12"
6-3-00 Blackbliss Forever
Umi 13h30'46" 74d3'41"
65 in 2005 RJ & Margaret Henderson
Umi 15h21'53" 76d1'10"
65 Jahre Hans-Jürgen Knapheide
Ori 6h22'56" 14d49'50"
65BeaMeyer12140
Cyg 19h36'22" 31d59'8"
68Feb05Marie37
And 1h0'25" 40d13'20"
7 Iron
Lyn 8h25'37" 38d30'46"
7 Virtudes
Vir 12h50'30" 6d9'25"
#70 Michael S. Sumney
Her 16h41'42" 34d41'59"
7002 Chevy
Aql 19h24'22" 7d59'29"
710
Uma 11h18'42" 66d41'27"
749
And 23h24'31" 52d2'15"
75JWW1930
Ori 5h32'10" 7d13'32"
77 Muffin 77
Vir 13h46'19" -5d54'6"
78 Bumpersticker
Mon 6h41'1" 7d16'55"
7amada's Angel
Psc 0h44'32" 6d14'31"
7B47B Cast & Crew 2006
Uma 11h39'11" 57d34'56"
7HF
BestFriendsUntilTheStarsAreGone
Tau 5h33'30" 18d58'27"
Sa Posada Don
Per 4h49'26" 45d40'3"
Sa&Sa
Boo 14h32'22" 19d30'21"
SA682005
Peg 21h55'7" 11d36'48"
Saadat Foroushani
Cyg 21h55'16" 39d14'24"
Saadeh and Darla Al-Jureidini
Ari 2h11'27" 25d15'6"
Saal, Roswitha
Uma 10h14'32" 59d43'50"
Saani Syed
Lyn 6h58'12" 48d16'59"
Saara Bowen - My heaven on earth
Cru 12h26'51" -62d39'47"
Saara Leventhal
Leo 10h36'2" 17d36'25"
Saarah's Dark Blue Star in the Nite
Cnc 8h5'56" 7d16'27"
sab&diam
Cas 1h13'47" 51d28'51"
Sab011990
Cap 21h43'34" -8d54'36"
Saba
Cas 0h14'34" 56d17'28"
Saba Alzaydi
Lmi 10h8'51" 37d54'30"
Saba gorgeous midget Banie
Cyg 21h32'25" 36d23'18"
Saba Rauf
Cas 0h12'15" 55d15'25"
Sabah
Per 2h56'7" 41d36'46"
Sabah
Vir 13h54'59" -17d41'22"
Sabastian Michael Rommel Alcobendas
Uma 11h10'11" 50d20'23"
Sabbath Virginia Marlene Janet
Umi 14h10'8" 74d23'46"
Sabeen
Ari 2h55'2" 25d21'21"
Sabeena Setia
Cnc 8h50'51" 32d40'26"
Sabi Hellson
Leo 11h7'42" 16d41'3"
Sabia
Cnc 8h22'19" 20d55'8"
Sabiha Nikhat Khan
Uma 9h53'52" 58d22'31"
Sabin
Ori 5h34'53" -9d47'32"
Sabina
Uma 13h3'9" 54d5'45"
Sabina
Sco 17h58'44" -39d6'6"
Sabina
Aur 6h24'46" 31d15'4"
Sabina
And 23h3'5" 50d31'2"
Sabina Ahmad
Psc 1h9'46" 11d18'21"
Sabina & Carlos
Cyg 19h42'28" 29d32'25"
Sabina Jablonski
Uma 9h35'10" 54d34'49"
Sabina Marie Gerstenberger
Ari 2h39'25" 28d5'37"
Sabina Matin
Peg 22h7'45" 7d48'33"
Sabina's star
And 1h48'29" 43d58'11"
Sabine
And 1h18'17" 42d0'44"
Sabine
Uma 11h24'5" 34d20'52"
Sabine
Cnc 8h16'7" 9d15'45"
Sabine
Boo 14h40'30" 16d1'23"
Sabine
Umi 13h20'39" 72d39'9"
Sabine
Hya 10h40'58" -22d24'28"
Sabine 10.12.1985
Her 18h25'24" 12d50'56"
Sabine Arzt
Ori 6h2'20" 3d49'38"
Sabine Baumgartner
Cam 3h37'41" 60d13'38"
Sabine Bennett
Leo 10h27'45" 26d18'53"
Sabine Bergen
Uma 8h37'36" 57d25'2"
Sabine Camille Crossen
Sgr 17h57'51" -24d40'23"
Sabine & Hans-Peter
Ori 5h58'16" 13d54'2"
Sabine & Jan
Ori 6h20'17" 6d24'10"
Sabine Jugert (Schätzelein)
Uma 10h0'16" 42d35'19"
Sabine Künzler
Umi 13h51'33" 78d35'2"
Sabine Liebetrau
Psc 2h2'12" 9d1'29"
Sabine Marie Munoz
Umi 14h21'0" 74d56'59"
Sabine Mathieu 23-06-65
Uma 9h41'36" 57d53'10"
Sabine Mende
Her 17h25'2" 35d41'28"
Sabine Rösler
Ori 5h33'17" 14d4'49"
Sabine Ruth
Dra 18h59'4" 64d49'22"
Sabine Schwarz
Dra 12h41'29" 70d52'24"
Sabine & Tobias
Her 18h38'15" 19d54'53"
Sabine Uhrmacher
Uma 8h57'13" 48d58'42"
Sabine Wicha
Psc 1h40'42" 21d5'4"
Sabine, * 11.07.1965
Uma 10h2'17" 53d35'39"
Sabine, sousouyté
Uma 13h6'37" 59d50'52"
SabineAlbanFlorianAlexianeBbDD
Uma 8h47'44" 59d49'15"
Sabino Petrignani
Lyn 8h31'43" 39d48'54"
Sabino Ranaudo III
Lib 15h3'17" -17d35'43"
Sabita Ryder
Vir 11h38'41" -3d50'39"
Sabita Viana
Lac 22h35'43" 42d38'53"
Sable
Cnc 8h1'16" 22d36'23"
Sable Ann
Uma 9h29'24" 45d9'43"
Sable Bajana
Gem 7h46'29" 29d46'2"
Sable Kathryn Ayre
Crb 15h38'23" 26d26'55"
Sabra
Vir 13h4'18" 8d58'7"
Sabra
And 23h21'23" 52d55'53"
Sabra Ariel Gaskill
Uma 9h44'14" 52d31'50"
Sabra Eve DeMarco
And 1h3'13" 39d50'46"
Sabrin
And 0h32'31" 28d6'55"
Sabrina
Ori 6h20'14" 19d41'3"
Sabrina
Lmi 10h50'19" 29d21'25"
Sabrina
Cyg 19h34'51" 29d25'10"
Sabrina
Crb 15h44'2" 28d38'53"
Sabrina
Leo 11h27'59" 13d21'51"
Sabrina
Tau 4h40'25" 19d7'50"
Sabrina
Ari 3h11'18" 11d1'25"
Sabrina
Psc 0h50'59" 9d51'28"
Sabrina
Vir 15h8'59" 5d42'24"
Sabrina
Cyg 20h57'8" 33d21'22"
Sabrina
Lyn 8h14'48" 40d17'36"
Sabrina
Crb 15h51'45" 34d49'16"
Sabrina
And 1h47'36" 36d54'34"
Sabrina
And 1h3'52" 36d3'0"
Sabrina
Lyn 7h40'39" 36d35'20"
Sabrina
Uma 9h58'2" 46d45'35"
Sabrina
Cas 23h47'32" 51d23'51"
Sabrina
Cas 0h23'9" 54d41'41"
Sabrina
Cas 0h6'52" 55d26'44"
Sabrina
Cas 3h15'43" 58d4'24"
Sabrina
Cam 4h31'53" 58d12'29"
Sabrina
Cyg 19h42'29" 39d20'20"
Sabrina
Lyr 18h49'34" 39d15'5"
Sabrina
Cas 1h33'56" 62d39'16"
Sabrina
Uma 14h10'23" 55d2'16"
Sabrina
Dra 20h21'7" 73d49'30"
Sabrina
Cam 3h17'37" 67d57'44"
Sabrina
Uma 8h30'31" 71d13'33"
Sabrina
Cam 5h8'53" 62d42'24"
SABRINA
Ori 5h42'42" -2d22'43"
Sabrina
Cma 6h49'39" -11d32'23"
Sabrina
Cma 6h35'6" -13d54'38"
Sabrina
Cap 20h9'17" -14d28'7"
Sabrina
Sgr 19h48'0" -12d28'50"
Sabrina
Peg 21h29'35" 23d32'51"
Sabrina
Sct 18h54'15" -12d39'52"
Sabrina
Peg 21h30'12" 23d28'44"
Sabrina
Sgr 20h19'44" -40d19'46"
Sabrina
Sco 16h19'52" -34d31'36"
Sabrina
Sgr 19h21'5" -22d30'53"
Sabrina 21
Tau 3h52'46" 24d16'46"
Sabrina Aepli
Uma 13h35'44" 51d25'50"
Sabrina Akhtar
Crb 15h36'49" 37d48'56"
Sabrina Alaina Menough
Aqr 22h7'24" 1d33'7"
Sabrina Alexis Werley-"My Hugger"
Psc 0h52'44" 25d47'37"
Sabrina Allison DiNapoli
Crb 16h5'59" 38d39'58"
Sabrina Anais Tereo
And 1h51'11" 37d20'22"
Sabrina And Thinh
Uma 11h59'6" 58d30'48"
Sabrina Ann Aldridge
Sco 15h55'16" -21d18'5"
Sabrina Ann Dokas
Aqr 21h0'53" 1d5'52"
Sabrina Ann Lockwood
Cyg 21h19'12" 45d5'11"
Sabrina Ann Mary DeJesus-Stoddart
Lyn 6h49'16" 50d39'18"
Sabrina Ann Poszich Sutton
And 23h39'55" 47d33'7"
Sabrina Arauz Quinonez
Leo 11h27'43" -5d6'32"
Sabrina Ashjian
And 1h12'57" 39d17'9"
Sabrina Brabetz
Lyn 7h55'33" 57d48'31"
Sabrina Branch Hunt
And 0h46'32" 28d10'33"
Sabrina (Bree) Snyder
Lib 15h24'8" -6d25'24"
Sabrina Cardoza Boyd
And 23h10'39" 37d2'52"
Sabrina Carroll
Lyn 9h6'13" 35d41'35"
Sabrina Celeste
Cas 23h10'21" 58d39'29"
Sabrina Censoni Barbieri
Uma 8h38'17" 64d43'31"
Sabrina Charlotte Pineda
Umi 16h7'5" 77d42'21"
Sabrina & Christian
Ori 5h16'22" 3d26'30"
Sabrina Christina Ambriz
Tau 5h47'48" 14d13'6"
Sabrina Christine Morgan
And 0h21'38" 37d1'17"
Sabrina Ciaramello
Peg 21h11'50" 14d32'27"
Sabrina Danielle
Ori 5h17'26" -0d8'31"
Sabrina Davis Sorabella
Cas 1h33'52" 63d58'45"
Sabrina Donato
Aur 5h29'22" 36d6'32"
Sabrina Emilia Noto
Lib 15h56'29" -7d33'32"
Sabrina G.
Cap 20h37'7" -22d53'21"
Sabrina G.
Cnc 8h4'0" 22d40'28"
Sabrina G. Betancourt
Crb 15h30'2" 32d32'0"
Sabrina Giangravè
Umi 14h1'4" 74d22'19"
Sabrina Gusmano
Lyr 18h29'54" 37d0'40"
Sabrina Hannon
Com 12h35'47" 24d27'49"
Sabrina Harbin
Vir 13h49'32" -6d19'22"
Sabrina Heimsch
Dra 17h3'36" 57d39'51"
Sabrina Hillstrand
And 1h38'1" 40d49'20"
Sabrina Hirsch
Sgr 19h50'57" -13d7'55"
Sabrina Horn
Leo 9h36'5" 28d52'36"
Sabrina Ireland
Ori 5h56'18" 17d5'8"
Sabrina Isabel
Uma 10h31'13" 47d14'47"
Sabrina & Jennifer
Lyr 19h4'20" 42d12'47"
Sabrina Josephine Varga
Sco 17h42'58" -35d23'31"
Sabrina Josse
Cap 21h33'12" -24d27'0"
Sabrina Kate Schindelheim
Lac 22h52'8" 51d21'45"
Sabrina Katherine King
Ori 5h31'44" 0d47'29"
Sabrina Kathleen
Gem 7h39'41" 24d11'52"
Sabrina Kei Chaffee
Aql 19h28'46" 6d0'48"
Sabrina & Kirstens
And 0h45'6" 37d31'30"
Sabrina Lea Weaver
And 0h24'12" 28d17'58"
Sabrina Lee Howard
Psc 1h3'25" 24d23'54"
Sabrina Leigh Dano
Cnc 8h34'6" 27d46'58"
Sabrina Librandi
Lib 15h53'52" -6d2'47"
Sabrina Lorenna Bitelli
Ari 2h53'11" 29d22'28"
Sabrina Lorraine Perrotta
And 23h17'49" 47d39'10"
Sabrina Malia
Cas 0h36'31" 58d56'54"
Sabrina Marie
And 2h20'58" 50d30'41"
Sabrina Marie
Ari 3h3'46" 27d26'59"
Sabrina Marie
Psc 0h22'38" 10d56'31"
Sabrina Marie Baarda
Lyn 7h38'42" 38d25'26"
Sabrina Marie Daufel DofRN Bardsley
Aqr 22h32'49" 1d24'19"
Sabrina Marie Dunton
Cmi 7h21'59" 9d50'45"
Sabrina Marie Schupp
And 0h7'9" 43d49'27"
Sabrina Marie Strumpher
Ari 2h46'21" 26d8'15"
Sabrina + Michi Grond
Cas 1h18'5" 50d52'15"
Sabrina Milarta
Uma 12h51'44" 60d49'43"
Sabrina Musella
Her 17h53'8" 20d40'25"
Sabrina Noelle
Uma 11h28'7" 32d12'31"
Sabrina Perfect Princess
And 0h35'10" 36d15'8"
Sabrina Puccianti
Per 3h35'7" 45d18'9"
Sabrina Renee Ippolito
Aqr 23h0'49" -6d5'40"
Sabrina Renee Welch
Cyg 21h39'33" 43d1'36"
Sabrina Rhea Law
Uma 9h14'3" 56d37'59"
Sabrina Sandberg
Ori 6h11'54" 19d12'15"
Sabrina Sankar "Boobar"
Cyg 20h10'0" 38d13'40"
Sabrina Seelig
Crb 16h5'57" 36d17'58"
Sabrina Sheehy-David
Vul 19h53'43" 22d56'50"
Sabrina Sierra
Aqr 22h46'59" -9d14'15"
Sabrina & Sofia Sanford
Umi 13h16'35" 72d33'37"
Sabrina Soledad Adducci
Ori 6h13'15" 3d29'4"
Sabrina Starlit
Mon 6h48'47" -0d31'22"
Sabrina Stephan
Lyn 8h53'38" 39d19'53"
Sabrina Straub
Ori 6h4'27" 9d36'7"
Sabrina Strazzanti
Cas 0h45'58" 53d58'34"
Sabrina Summer Brooks
Cnc 8h25'1" 18d15'20"
Sabrina und Stefan
Uma 12h44'13" 59d6'35"
Sabrina und Tom
Uma 9h13'2" 49d38'22"
Sabrina Wojnarowicz
Sgr 18h44'18" -21d27'57"
Sabrina Zorro Fumagalli
Cam 7h18'9" 78d4'24"
Sabrina,Will you marry me?Love Eric
Tau 5h23'36" 26d50'27"
SabrinaAli-28.05.2005
And 0h26'51" 22d5'56"
Sabryna Marie
Psc 1h7'49" 21d37'1"
Sabs
Uma 8h40'20" 53d46'2"
Sabs
Lyn 6h42'45" 56d15'4"
sabsystren
Uma 9h38'35" 67d9'58"
Sabz_T-022505
Aql 19h30'2" 8d2'37"
sac0n
Tau 3h45'51" 8d39'43"
Sacaka
Tri 2h28'16" 30d27'11"
Sacchetta
Crb 16h18'59" 32d12'46"
Sacchi Patel
Tau 5h49'49" 16d22'14"
Sacha
Sco 17h51'57" -37d8'16"
Sacha
Sco 16h5'3" -29d2'56"
Sacha Angelique
And 0h36'43" 26d8'22"
Sacha - Jérémy
Umi 4h20'33" 89d10'42"
Sacha Kai Dillon
Psc 0h57'28" 9d39'33"
Sacha Lisa King
And 2h29'9" 39d14'57"
Sacha und Patricia
Cyg 21h35'27" 34d54'40"
Sachdeva
Sco 16h21'25" -42d8'13"
Sachie
Ori 5h34'31" 3d36'59"
Sachiko
Aqr 23h4'12" -6d54'11"
Sachiko
Dra 16h35'28" 54d57'37"
Sachiko & Asuka
Gem 7h27'8" 26d58'3"
Sachiko & Chris Entwistle
And 23h17'52" 45d43'37"
Sachin and Taneshia
Aql 19h48'55" 2d37'52"
Sachin's Smile
Cru 12h6'0" -57d36'15"
Sachi's Shine
Leo 11h49'33" 21d51'40"
Sachu
Cnc 8h44'21" 7d44'4"
Sack Forever
Ori 6h9'7" 20d39'1"
Sacnite
Ari 2h14'24" 23d49'27"
Sacred Union of Warrior and Purity
Ori 5h40'20" -1d6'59"
Sadaf Hanif
And 1h50'59" 37d55'18"
Sadaj
Cru 12h33'32" -62d1'47"
Sadal Cores
Uma 10h37'31" 48d2'1"
Saddi
Ari 2h8'6" 25d29'48"
Sad'e
Leo 11h47'30" 19d42'6"
Sade Lee Nixon
And 1h53'30" 45d58'55"
Sade O. Knox
Psc 1h27'7" 26d32'41"
Sade Tina Marie Moralez
And 1h19'29" 36d48'43"
Sadhna Gummi Bear Patel
Leo 11h21'10" 2d54'35"
Sadhvi Mahendru
Per 4h21'59" 33d42'45"
Sadi Lee Piepkow
Cap 20h14'21" -15d42'42"
Sadia
Sgr 18h19'33" -23d15'29"
Sadia
Crb 15h34'50" 37d28'9"
Sadie
Ari 2h47'0" 22d30'0"
Sadie
Sgr 18h9'33" -35d9'53"
Sadie
Cyg 20h37'47" 52d30'34"
Sadie
Cas 1h38'8" 61d37'33"
Sadie Addison Smith
And 2h15'42" 45d25'37"
Sadie Allison Copas Hope
Psc 0h8'45" 6d20'43"
Sadie Ann Haley
Tau 4h39'50" 21d8'32"
Sadie Austin Maierhofer
Peg 22h29'32" 6d12'26"
Sadie Belle Porchowsky
Umi 14h12'17" 68d8'42"
Sadie Catherine Gaudenti
Uma 10h22'54" 56d13'24"
Sadie Hobbs
Boo 14h43'44" 30d26'47"
Sadie Hoffman-Deming
Uma 11h59'16" 41d14'0"
Sadie Hope
Lyn 7h40'57" 49d49'47"
Sadie Jayne Probst
Uma 11h57'14" 33d52'20"
Sadie Kate
And 0h16'16" 40d56'45"
Sadie Langston
Cnc 9h9'19" 17d22'33"
Sadie Lorraine
Vir 14h39'16" 3d32'7"
Sadie Louise Penix
Cnc 8h1'14" 21d19'31"
Sadie Lynch
Cma 7h12'43" -31d40'6"
Sadie M. Diana
Sco 16h18'44" -29d53'0"
Sadie Marie Champion
Uma 8h13'21" 65d52'10"
Sadie Marie Our Shining Star
And 0h31'23" 29d23'52"
Sadie Maye
Lyn 7h46'18" 40d55'52"
Sadie Mo Dunsker Leiman
And 2h23'33" 43d24'20"
Sadie Odessa Taylor
Peg 21h46'38" 9d25'41"
Sadie Ondich
Uma 10h27'34" 52d39'20"
Sadie Renee Bowser
Cas 1h37'30" 60d55'31"
Sadie Rose Dvorachek
And 0h7'37" 45d37'15"
Sadie Rose Harlow
Gem 7h12'11" 22d49'42"
Sadie Ruth
Cas 0h25'8" 66d7'57"
Sadie Sadie Moonlady
Lup 15h45'56" -32d34'19"
Sadie Salaets
Lib 15h8'12" -4d43'57"
Sadie Sheridan
Cas 1h9'29" 69d40'27"
Sadie Sicuro
Uma 12h41'24" 53d23'40"
Sadie Virginia Wood
Cas 3h28'44" 70d16'37"
Sadie Yesbeck Ramey
Gem 7h9'44" 26d47'57"
SadieDon
Lib 15h43'33" -25d56'13"
Sadie's Guardian Angel
Sgr 18h10'31" -16d31'40"
Sadie's Spitfire
Lyn 7h28'33" 51d21'19"
Sadie's star
Crb 16h12'51" 33d31'28"
Sadie's Star
Cas 23h9'48" 58d42'44"
Sadie's Star
Cru 12h30'25" -56d16'55"
Sadie's Sweet Dream
Uma 10h1'38" 62d23'59"
Sadilek, Zdenek
Ori 6h16'47" 10d22'33"
Sadye "Tulip" LaBarbera
Cap 20h29'44" -15d10'17"
Saeger, Kai
Uma 10h58'44" 62d54'26"
Saeid Badie
Vir 13h37'41" 5d41'27"
Saeko
Uma 9h19'37" 70d55'36"
Saema Rahmany
Leo 9h59'5" 31d22'34"
SaEr
Cnc 8h49'49" 11d5'2"
Saeric
Cyg 20h44'40" 30d59'50"
Safari, Ali
Sgr 18h1'20" -27d44'47"
Safea
Uma 11h40'42" 62d27'14"
Saffaya Cecilia Minot Anne Battle
And 23h2'29" 46d5'35"
Saffron Peel
Cap 20h15'48" -13d4'29"
Safina
Lyn 7h33'25" 37d49'13"
Safira
Ari 3h17'41" 28d52'7"
Safiya Lee Blount
Lyn 8h41'2" 33d40'45"
Safiya Nur
Lyr 18h52'7" 32d6'56"
Safiyyah Khan
And 23h30'40" 47d50'24"
Safraz Gajraj
Vir 12h9'23" -7d38'58"
Safron
Lyn 7h32'2" 39d46'19"
Sagan Alan Traugh
Aur 5h40'16" 39d35'23"
Sagan Richelle
Lyr 19h11'5" 27d16'32"
"S'agapo"
Ari 3h22'46" 26d17'58"
s'agapo
Gem 6h47'0" 26d13'4"
Sagapo
Eri 4h57'29" -4d53'12"
S'agapo
Uma 10h28'58" 58d31'43"
SAGARENA
Cam 7h43'25" 63d39'29"
Sagarika
Vir 12h0'20" 6d42'19"
Sagarika
Uma 10h5'58" 47d24'25"
Sagas
Ori 6h7'17" 4d47'38"
Sage
Ori 6h18'37" 14d58'57"
Sage
Leo 11h39'22" 12d38'39"
Sage
Tau 4h44'59" 23d11'16"
Sage
Sco 17h37'11" -41d35'31"
Sage A. Ceja
Dra 18h25'29" 73d42'28"
Sage Alexandrea McClain
And 0h5'48" 32d9'49"
Sage and Aspen Barrett
Aqr 22h31'48" -0d58'16"
Sage and Jenna's Connecting Star
Gem 6h53'28" 23d33'27"
Sage & Angie Forever
Cyg 20h59'2" 46d15'28"
Sage Faye Peregrin
Ari 2h12'57" 25d58'57"
Sage Florentino
Her 16h32'19" 14d16'30"
Sage Garriss
Ari 3h29'19" 28d2'29"
Sage Haley
Cnc 8h41'17" 31d42'25"
Sage Hilel
Aqr 22h5'24" -15d5'14"
Sage Marie Blankenfeld
And 0h24'27" 42d32'27"
Sage McGuire
Leo 11h27'30" 21d30'40"
Sage Morris-Greene
Ori 4h51'28" 3d36'17"
Sage Pond Haley
Uma 8h48'41" 49d38'32"
Sage Roe
Sgr 18h40'57" -28d44'38"
Sage Tedeschi
Leo 11h45'22" 23d8'24"
Sage Vincent-Amadeo Haney
Uma 8h17'7" 70d6'51"
Sagen
Tau 4h18'51" 5d51'14"
Sagilibras
Ara 17h19'4" -51d12'37"
Sagio
Tri 2h34'25" 33d59'37"
Sagittal Arch
Ori 6h2'24" 19d30'43"
saguaro
Her 16h44'43" 48d14'10"
Sahar
Gem 6h57'5" 22d39'36"
Sahar
Lib 15h2'39" -25d36'48"
Sahar
Aqr 22h20'34" -3d28'51"
Sahar Ghazale
Dra 17h17'21" 53d47'53"
Sahar & Jad
Uma 10h6'52" 51d36'57"
sahar osman
Cam 4h37'44" 62d16'28"
Sahara El Rhibear
And 23h33'6" 42d13'8"

Sahara Leveille
Cnc 8h42'49" 22d10'33"
Sahara's Shining Star
Cap 20h32'14" -14d52'0"
Sahel Siminoo & Rayan Songhorian
Cyg 20h56'57" 32d22'39"
Sahily
Uma 9h52'8" 55d41'1"
Sahira
Vir 14h50'1" 3d52'15"
Sahiti
Aur 6h24'34" 31d0'40"
Sahtenmoonf 's
Aqr 22h35'13" 2d35'37"
Sai Lup Lup
Cap 20h49'41" -26d34'11"
Sai Santosh Sreenivasan
Boo 15h3'55" 41d36'27"
Saiai Tokoya
Cnc 8h47'6" 16d56'59"
Said Aly
Aql 19h50'26" -0d35'57"
Said I. Jacob
Vir 11h37'48" 4d15'45"
Said I Loved You...
Per 2h25'29" 54d26'59"
Said Shamekh
Cam 4h40'7" 64d45'27"
Saida Carlson
Cnc 8h58'1" 17d15'52"
Saideh
Ari 2h50'7" 22d4'3"
saidoufoo - Carly Ng
Vir 12h56'57" -17d26'44"
Saige Brooklyn Venable
And 0h24'43" 43d58'38"
Saige Elizabeth
Mon 7h0'14" 6d48'59"
Saige Marie Ciarcia
Umi 21h22'54" 89d52'50"
Saige Rovero
Sco 16h13'21" -11d15'10"
Saige Ryan - Little Light of Mine
Sco 17h31'21" -45d5'39"
Saijal Sabine
Aqr 21h4'34" -11d12'49"
Saik, Renate
Uma 13h45'34" 55d14'55"
Saika Shafi
Uma 11h22'9" 61d24'40"
Sailaweigh
Cnc 8h37'12" 13d50'52"
Sailboat
Vir 14h38'41" 3d9'4"
Sailor David A. Grodt
Her 16h21'51" 19d10'41"
Sailor Moon
Ari 2h34'41" 18d22'10"
Sailormark
Psc 23h49'16" 6d47'40"
Sailor's Admiration
Cas 0h51'27" 47d39'29"
Saima
Lib 14h54'5" -0d33'41"
Saima Maher
Leo 11h27'35" 9d2'53"
Saint Adam Kristopher Doellman
Cnc 8h37'38" 24d47'35"
Saint Anthony
Vir 12h56'40" 11d19'9"
Saint Brevin
Del 20h26'40" 15d12'10"
Saint Catherine 'Mammy' Union
Per 3h34'18" 39d21'24"
Saint Christopher James Schneider
Leo 9h27'59" 26d34'1"
Saint Doris
Psc 1h17'8" 28d6'27"
Saint James
Tau 5h41'23" 24d53'26"
Saint Joseph
Her 16h54'10" 32d24'48"
Saint Lucia
And 0h46'19" 35d55'32"
Saint Marie Vilgrain
Lib 14h22'45" -10d11'17"
Saint Mary Louise
Sgr 18h35'40" -24d14'33"
Saint Nick
Vir 14h16'20" -12d30'43"
Saint Nick
Dra 15h20'7" 63d55'41"
Saint Sally/Our Shining Light
Lyr 18h46'23" 38d23'34"
Saint Susan Aurora Ferro 1111
Tau 5h29'16" 22d23'15"
Saira the Butterfly
Leo 11h23'38" 12d46'11"
Sairahiniel
Psc 1h20'24" 23d27'22"
Sairina
Umi 14h59'25" 69d46'41"
Saisha & Angel
Vir 11h41'47" 3d11'15"
Saiyan & Sayla Baccam
Sgr 18h33'40" -17d9'31"
SAJ125
Umi 14h6'31" 75d5'40"
Sajado
Uma 12h5'12" 36d27'26"
SaJen Kootaka
Uma 9h21'13" 57d32'34"
Saji
Peg 0h7'14" 17d2'2"
Sajith Raj
Ori 5h12'41" -3d20'26"
Sakdu Bindi
Cnc 8h41'54" 30d58'34"
Saki & Susie
Aqr 21h53'52" 1d43'0"
Sakis Kalogirou
Psc 1h13'30" 4d6'10"
Sako Nazarian
Aqr 20h48'23" 2d6'1"
Sakoni
Sgr 19h21'41" -39d45'33"
Sal
Psc 1h22'39" 5d54'33"
SAL
Sgr 18h36'42" -22d57'48"
SAL
Lyn 6h24'37" 54d38'26"
Sal
Uma 9h55'21" 55d22'20"
Sal and Deborah Croce
Cyg 20h0'20" 33d24'30"
Sal and Gert Carrubba
Cyg 19h37'19" 53d48'48"
Sal Bo
Cyg 21h22'7" 54d33'59"
Sal Coppolino
Tau 3h35'48" 19d50'40"
SAL G.
Cnc 9h9'9" 22d50'19"
Sal & Karin
Uma 10h49'14" 60d7'0"
Sal & Kel's Night's Kiss
Cru 12h0'47" -60d27'53"
Sal & Lo's 1 Year - I Love You Babe
Tri 2h18'54" 33d46'25"
Sal & Maggie Now & Forever
Cyg 20h48'46" 42d12'41"
Sal & Naycie Hodgson
Cyg 20h34'52" 58d24'13"
Sal Olivo
Uma 9h43'59" 42d25'59"
Sal Palmeri
Leo 10h55'40" 11d9'44"
Sal Porcino, Jr.
Psc 0h15'3" -0d50'17"
Sal Pulice
Uma 12h22'54" 55d30'48"
Sal Quezada
Leo 11h4'28" 23d30'19"
Sal - Shining Always - Tonu Kanapu
Cru 12h40'54" -64d25'27"
Sal Tedesco
Lac 22h52'13" 51d58'44"
Sal Velardi
Her 17h53'53" 47d56'23"
Salad Tosser
Uma 9h36'32" 61d9'5"
Salah Abrar Siraj
Dra 18h13'26" 59d29'34"
Salah Hosny
Cap 20h41'50" -26d45'20"
Salakesh Ananda
Ori 5h19'38" -7d26'25"
Salam Nakouzi
Her 17h2'43" 30d44'18"
Salameh Fareed Dugum
Vul 19h33'38" 24d13'54"
Saldare
Aqr 22h17'52" 2d0'51"
Sale Ellen Mahady Buchanan
And 0h16'9" 44d2'20"
Saleen M. Faul
Gem 7h40'30" 15d45'45"
Saleena Cruz
And 0h47'18" 40d7'38"
Salem
Cyg 19h59'53" 36d16'40"
Salem A. Al.-Jaber
Gem 6h49'43" 21d57'2"
Salem Alnoaimi
Tri 2h40'24" 33d49'28"
Salem M&F Diamond Stars 2005
Sco 16h32'55" -35d11'54"
Salena Jade Ortiz
Uma 11h23'15" 54d56'37"
Salena Paez
Ori 5h26'21" 3d33'51"
Saletmaier, Uwe
Ori 5h15'6" -8d54'20"
Salh & Olivia's Star
Cyg 21h34'43" 51d42'49"
Salida del Sol
Ori 5h41'16" -7d20'47"
Salima
Aql 18h46'11" 8d46'30"
Salina Marie 08-15-79
Leo 10h39'22" 22d43'21"
Salina Monte Castillo
Crb 16h6'32" 35d4'55"
Salina Villanueva
Aql 20h2'5" 2d33'37"
Salina Yung Kim
And 0h19'58" 46d11'34"
Salisha Khan
Lib 15h37'48" -6d11'55"
Saljo
Cam 14h18'37" 83d31'3"
Salla Heinaro
Sco 17h54'8" -35d51'53"
Sallee Anne
And 1h18'31" 36d59'32"
Sallee Berlin
Uma 8h50'51" 60d40'19"
Sallerina
Umi 4h54'5" 89d7'27"
Sallie
Cas 23h3'43" 57d10'31"
Sallie Anne Lewis
Sco 17h17'18" -43d54'39"
Sallie Billinghurst
Cas 0h16'4" 56d9'51"
Sallie Louise Nicholls
Ari 2h48'17" 27d40'19"
Sallie Mae Loan Star
Lyn 7h47'56" 36d40'0"
Sallie Marie McCormick
Lmi 10h23'59" 33d16'31"
Sallie Marie Strueby
Sgr 19h9'18" -31d5'36"
Sallie May Warzecha
Aqr 21h59'12" 1d48'6"
Sallie Star
Pho 1h7'43" -41d4'11"
Sallows-O'Neill
Cyg 19h28'29" 36d6'1"
Sally
And 1h14'10" 34d9'5"
Sally
Psc 1h10'1" 32d34'48"
Sally
Cas 2h42'2" 57d45'32"
Sally
Cyg 20h49'41" 46d21'18"
Sally
Tau 5h50'53" 17d17'53"
Sally
Gem 6h45'23" 15d9'17"
Sally
Com 12h35'53" 28d6'9"
Sally
Cas 0h36'48" 62d21'46"
Sally
Lib 14h32'4" -14d54'13"
Sally 1936 Jerry 1931 Desmond
Ari 2h12'52" 24d9'43"
Sally "4Ever & A Day"
Aqr 23h54'14" -10d25'8"
Sally A. Herd
And 2h31'51" 49d54'6"
Sally and Daniel
Sge 20h4'34" 18d17'15"
Sally and Elaine
Gem 6h17'34" 22d2'51"
Sally and Larry Brash
Uma 10h43'25" 52d40'49"
Sally and Nicks star
Tau 4h28'25" 18d22'14"
Sally Ann
Aqr 20h45'43" 2d19'4"
Sally Ann
Aqr 22h41'45" 1d12'37"
Sally Ann
Leo 11h44'23" 19d4'39"
Sally Ann
Leo 11h17'5" 16d37'1"
Sally Ann Block
Sco 16h57'29" -32d35'9"
Sally Ann Bowron
Cas 0h29'57" 51d15'11"
Sally Ann Dearth
Dra 17h44'11" 53d58'4"
Sally Ann & Felix's Love...
And 2h36'47" 45d33'30"
Sally Ann Grim
Ari 2h34'53" 17d53'44"
Sally Ann Harkins
Her 17h39'56" 19d14'52"
Sally Ann Jankowski
Cas 1h22'55" 58d47'23"
Sally Ann Kasper
Lyn 7h55'12" 57d17'8"
Sally Ann Mullinix
Uma 10h54'36" 50d0'22"
Sally Ann Santora
And 0h45'16" 40d38'52"
Sally Ann Straw
And 23h8'36" 43d32'51"
Sally Anne
Dra 18h20'35" 58d19'17"
Sally Anne Leming - "Lemingland"
Her 16h32'14" 13d38'44"
Sally Araujo
Aqr 21h49'56" 1d4'39"
Sally B. Wyner
Tau 3h45'27" 28d3'18"
Sally Baker
Cas 23h31'10" 55d52'16"
Sally Beaman
And 2h27'51" 50d1'32"
Sally Beatrice Long
Psc 0h25'31" 7d35'35"
Sally Behr Pettit Star
And 23h12'14" 49d19'2"
Sally Blevins
Gem 6h52'48" 13d4'49"
Sally Brent Davies
Cas 1h31'58" 65d27'34"
Sally Bridge
Uma 10h37'8" 63d31'48"
Sally Bryan
Cas 1h23'38" 52d25'19"
Sally & Buddy Beres
Lyr 18h45'40" 37d42'14"
Sally C Newton
Cam 3h58'32" 53d21'7"
Sally Cato
Vir 12h39'3" -8d17'30"
Sally Cooper
Ari 2h11'44" 22d59'48"
Sally De Los Santos
Peg 22h9'4" 8d31'31"
Sally Descisciolo
Uma 10h30'56" 72d3'20"
Sally Dietz's Star
Sgr 19h16'56" -31d55'7"
Sally "Doodles" Field
Uma 12h2'16" 59d15'50"
Sally Duquette
Lyn 8h24'46" 42d26'26"
Sally Elisabeth White 21.02.1978
Cru 12h26'45" -59d39'41"
Sally Elizabeth
Cyg 19h36'22" 35d46'0"
Sally Elizabeth Daley
Cyg 21h38'53" 41d33'7"
Sally Elizabeth Russell
Cas 2h14'56" 73d38'2"
Sally Elizabeth Sewell
Uma 14h9'12" 56d23'57"
Sally Ella Zhang Cavellier 4-14-03
Cam 4h21'22" 57d45'14"
Sally Eunice Louise Wood
And 0h9'2" 29d19'29"
Sally Evers
Vir 14h14'21" 3d31'48"
Sally Favaloro
Uma 10h57'14" 40d25'8"
Sally Fernandez
Lyn 7h34'42" 38d59'11"
Sally Fowler's Star
Oph 16h33'9" -0d0'18"
Sally Garrard
Cap 21h0'35" -21d52'58"
Sally Gebler
Gem 7h26'55" 16d28'13"
Sally "Grandy" Hoffman
Leo 11h16'5" 20d24'8"
Sally H. Martin
Uma 10h9'3" 49d22'49"
Sally Hanson
Dra 15h53'23" 63d46'42"
Sally Harris
Leo 10h33'7" 13d49'0"
Sally Hatch
Her 18h43'39" 17d35'38"
Sally Hoyt
Cas 23h20'16" 56d55'37"
Sally Irene
Gem 6h34'58" 12d27'22"
Sally J. Hitchcock UK 16th Birthday Star
Psc 1h21'10" 22d14'9"
Sally Jane Terry
Sgr 18h40'59" -28d53'7"
Sally Jayn Dick
Aqr 22h5'17" 1d48'45"
Sally Jean George
Aqr 22h40'48" 2d10'15"
Sally Jo
Uma 10h35'12" 45d20'20"
Sally Jo Porter
Del 20h38'22" 17d38'52"
Sally Jones-Sawyer
Cyg 20h8'24" 32d15'39"
Sally June
Uma 8h44'11" 53d58'48"
Sally "Kailluna" Strom
Cnc 8h44'50" 17d5'24"
Sally Kate Henderson
Vir 13h40'14" -11d54'41"
Sally Kay Brahman
Uma 12h47'34" 57d29'24"
Sally Kelly Etebari
Uma 11h51'30" 38d27'6"
Sally & Kenny Ryan - Together Always
Lib 15h42'23" -17d53'27"
Sally Kirkby
Cas 1h28'15" 53d32'53"
Sally Lawson
Sgr 19h0'19" -34d50'46"
Sally Lee
Cas 0h8'46" 56d13'52"
Sally Lee
Per 4h19'10" 50d29'23"
Sally Linn Moceyunas
And 1h26'21" 34d48'8"
Sally Lippert
Leo 10h9'11" 13d42'58"
Sally Lorraine
Tau 4h37'45" 27d41'43"
Sally M. Smith
Uma 9h28'36" 50d51'1"
Sally Macbeth
Cap 21h37'46" -9d50'4"
Sally Mae
Tau 5h56'3" 25d21'35"
Sally Marie
Psc 23h56'51" 2d9'51"
Sally Marie
Sgr 17h58'59" -17d51'58"
Sally Marie Armiger
Lyn 8h46'54" 42d28'37"
Sally Marie Ruelas
Ari 2h5'54" 12d13'35"
Sally Marrows
Lib 15h13'9" -28d56'2"
Sally Mary Schuler Nolan
Gem 7h34'40" 23d14'30"
Sally Maxwell, the Princess
Peg 22h11'16" 16d41'6"
Sally May Drake
Tau 5h33'42" 17d13'43"
Sally Melissa Day
Pav 18h43'13" -58d1'47"
Sally Meyer
Cra 18h9'17" -39d26'21"
Sally Michele
Ori 4h51'39" 13d22'52"
Sally Monaghan
Leo 11h47'55" 26d16'37"
Sally Moreno
And 0h21'47" 43d16'0"
Sally My Sweet Angel
And 2h1'23" 46d14'30"
Sally Nelson
Gem 7h26'0" 33d47'43"
Sally Nemeti
And 1h13'39" 37d55'19"
Sally Newall-Sanderson
And 1h49'53" 40d24'1"
Sally O'Conner
Cas 0h34'40" 62d16'43"
Sally Oelbaum
Cas 0h58'16" 60d57'49"
Sally Oola Last
And 1h20'23" 35d12'8"
Sally Patricia Finnis
Ari 3h18'41" 23d6'25"
Sally Peterson
Vir 13h18'55" 11d9'40"
Sally Price
Mon 7h39'59" -1d8'48"
Sally Psycho Thomas
Psc 23h6'7" -1d4'19"
Sally Randall
And 2h12'7" 43d25'27"
Sally Risley
Sco 16h17'48" -16d38'50"
Sally Rosen
Lyn 7h20'49" 52d12'27"
Sally Ryan
Uma 9h24'12" 67d10'49"
SALLY SADEK
Lyn 7h18'39" 56d55'0"
Sally Shealey (Schmoldt)
Uma 13h47'29" 51d47'56"
Sally Siciliano-Kachner
Lib 15h38'39" -19d57'58"
Sally Six O
Cam 4h44'53" 55d55'57"
Sally Sofia
Tau 5h31'27" 20d39'56"
Sally Sookhai
Uma 14h27'12" 61d3'18"
Sally Star
Cyg 19h45'2" 37d24'17"
Sally & Steve Harfoot
Cyg 20h15'44" 33d16'6"
Sally Sue
Cas 2h26'12" 64d33'2"
Sally T
Mon 7h31'5" -0d37'40"
Sally Tapp's right to twinkle!
And 23h59'28" 32d38'37"
Sally Tina Joe
Gem 6h37'56" 26d26'52"
Sally Victoria Fraser Maloney
And 23h21'52" 42d1'57"
Sally von Muralt
Lyr 19h5'23" 42d41'14"
Sally Walker
Tau 3h56'46" 22d0'40"
Sally Whitlock
Cas 0h40'1" 62d39'7"
Sally Wilcox
Cyg 21h27'36" 36d47'6"
Sally Wolf
Psc 0h33'7" 16d12'55"
Sally Zyra Cleveland
Lib 15h7'38" -13d5'58"
Sally88
Sgr 19h14'24" -17d41'1"
SallyAnne McCartin
Crb 15h41'54" 37d21'37"
Sally's Lucky Star
Vir 14h19'39" -21d44'42"
Sally's Mars
Lyr 18h40'10" 39d55'52"
Sally's Star
And 0h41'58" 40d44'32"
Sally's Star
Cyg 19h35'52" 29d47'15"
Sally's Star-Mustang Sally
Lyn 6h58'20" 44d36'24"
Sally's Wish Upon A Star
Tau 4h42'22" 23d10'5"
Sallywacker8.SallyBasl.NV.USA.AUS-1
Cra 18h46'52" -39d50'33"
salma
Leo 11h17'26" 6d40'20"
Salma F. Fares
Cas 0h0'45" 64d56'48"
Salman Ahmed Al Noaimi
Tau 3h46'18" 12d24'47"
Salman Baba
Per 2h59'4" 55d25'6"
Salman Rushdie
Pho 23h40'13" -44d40'17"
SalNicolAntonia
Lyn 6h23'40" 59d41'24"
Sal-Ogami7
Tau 4h27'18" 18d17'9"
Salohcin
Cnc 8h20'23" 7d41'23"
Salomé
Cnc 8h9'36" 25d13'26"
SALOME
Her 18h25'5" 25d9'7"
Salomé
Cas 0h51'0" 53d3'50"
Salome & Andy, 14.08.1987
Her 17h45'0" 15d48'23"
Salome Jens
Cas 0h26'16" 56d27'35"
Salomée-Héline
Leo 10h46'44" 14d40'40"
Salomeh
Oph 17h7'6" -14d53'4"
Salomon Felix- Ec.4:10-12 Pro.27:17
Ori 5h31'57" 6d25'56"
Salomon & Jennie Johanson-Maya
Cyg 21h11'55" 44d12'39"
Saloni the Moon Princess
Lib 15h21'0" -9d38'49"
SaLori
Lyn 7h38'41" 38d19'45"
Salote's Star
Uma 9h21'16" 44d35'3"
Saloua
Sgr 18h49'16" -16d27'25"
Sal's Reflection
Cas 22h59'58" 57d0'12"
Sal's Star of Kathryn
Uma 11h24'1" 47d13'39"
Sal's Starlight
Ori 4h56'22" -0d31'10"
SalsAlina
Uma 12h16'22" 54d27'49"
Salsalita
Ari 2h32'12" 27d1'31"
Saltanat Kusherbaeva
Cyg 19h42'47" 29d19'31"
Saltans Astrum
Cyg 20h37'43" 36d43'45"
Saltire
Gru 21h33'40" -36d56'25"
Salty
Lyn 7h20'24" 50d6'36"
Salty Debb 55
Lib 14h52'31" -9d39'7"
Saluki
Vel 9h22'29" -41d41'28"
SALUKI
Vir 13h22'12" 11d53'57"
Saluki of Arabia
Eri 3h43'13" -13d21'44"
salus
Cir 15h9'14" -57d42'54"
Salute to C.D. Moody Construction
Cas 23h52'15" 56d16'8"
Salute to Clipper Corporation
Cas 0h28'8" 50d19'44"
Salute to Costa Nursery Farms
Cas 23h3'23" 58d34'48"
Salute to D-Unique Tools
Cas 23h57'38" 56d54'34"
Salute to H.J. Russell & Company
Cas 22h59'1" 58d48'36"
Salute to HR Now!
Cas 1h8'31" 66d55'25"
Salute to Metasys Technologies
Cas 0h4'24" 54d52'49"
Salute to SupplierGATEWAY
Cas 23h30'34" 51d1'39"
Salute to The Vidal Partnership
Cas 1h27'50" 52d16'31"
Salute to Williams Capital Group
Cas 23h48'37" 58d28'12"
Salva
Psc 0h50'10" 16d14'56"
Salvador
Cyg 19h40'48" 33d8'28"
Salvador Alier
Sco 16h19'46" -41d15'45"
Salvador O. Chavarria Delgado
Uma 11h41'13" 38d33'7"
Salvador Soler Casacuberta
Ser 15h18'24" 13d33'31"
Salvador y Perla
Cru 12h27'26" -56d13'18"
Salvador Yunez M.D.
Cep 21h55'25" 58d43'56"
Salvatore
Uma 8h48'28" 70d43'51"
Salvatore
Gem 7h33'13" 14d45'16"
salvatore
Ori 6h10'6" 7d36'6"
Salvatore
Her 18h35'48" 15d41'15"
Salvatore
Uma 11h57'34" 43d31'13"
Salvatore Alvaro 3905 M/RSCS
Cnc 8h48'8" 28d45'23"
Salvatore Aronica "6"
Cas 0h52'55" 52d16'0"
Salvatore B. Marinucci
Cyg 19h48'37" 45d58'28"
Salvatore Battaglia
Cap 21h50'48" -17d37'16"
Salvatore Blair
Cep 2h50'1" 82d3'25"
Salvatore Brancato Sr.
Cep 21h27'34" 58d58'56"
Salvatore Caiola
Cyg 21h59'56" 45d56'24"
Salvatore Cannizzaro
Peg 21h30'47" 23d31'32"
Salvatore Cassar Sr.
Vir 12h48'18" 2d23'46"
Salvatore Cazzolla
Uma 10h25'19" 53d7'46"
Salvatore Chianello *13.März 1965
Uma 9h52'9" 44d30'8"
Salvatore Civiletti
Umi 16h53'33" 75d37'49"
Salvatore De Angelis
Cyg 20h43'46" 33d32'21"
Salvatore Del Bove
Ori 6h18'30" 8d55'11"
Salvatore F. Bello, Jr.
Per 3h5'51" 35d29'23"
Salvatore F. Culotta
Vir 13h36'29" -4d59'13"
Salvatore Frank Santaniello
Her 16h29'24" 19d16'15"
Salvatore Galle
Ari 3h17'12" 25d43'22"
Salvatore Gambino
Her 17h55'23" 14d40'18"
Salvatore Grassia
Tri 2h41'4" 33d52'59"
Salvatore & Heather
Leo 11h10'17" -2d15'0"
Salvatore "June" Corso Jr.
Her 16h47'40" 6d8'35"
Salvatore Lograsso
Her 18h52'59" 25d2'42"
Salvatore & Manuela forever
Uma 8h55'40" 67d2'14"
Salvatore Massaro
Ari 2h15'26" 24d6'44"
Salvatore Mauro Jr.
Ari 2h48'54" 29d39'6"
Salvatore & Melissa Trazzera
Vir 13h33'59" -4d54'48"
Salvatore Michael Lonobile
Tau 4h36'31" 25d1'30"
Salvatore "Nonno" Scali
Leo 11h47'40" 17d54'29"
Salvatore Parisi-Shine On The World
Cru 12h23'56" -57d45'32"
Salvatore Patalano "Nov.8,1915"
Sco 17h46'14" -42d37'20"
Salvatore Peter Nicoletta
Uma 11h59'56" 55d44'8"
Salvatore Pirrello
Sgr 18h23'50" -25d12'23"
SALVATORE PUPO
Dra 15h8'49" 61d22'5"
Salvatore "Sam" Pereca
Per 2h50'43" 32d59'6"
Salvatore (Tory) Cimino
Umi 15h4'8" 69d30'37"
Salvatore & Valerie Grillo
Per 2h20'59" 56d4'18"
"Salvi"
Uma 11h14'35" 29d43'55"
Salvi Costanzo
Uma 13h33'25" 58d37'13"
Salvi & Martin T.A.P.S. 10. Juli 2001
Umi 13h12'39" 72d25'49"
Salvo&Cri
Crb 15h57'55" 32d31'18"
Salvo Pulvirenti
Aql 20h17'12" -0d34'23"
Salvoldelli Eugenio
Ori 5h1'15" -0d55'19"
Salwa
Aur 7h14'38" 41d56'28"
Salwa El-Ache
Pho 23h57'39" -54d45'55"
Salwan Kassab
Umi 14h13'34" 69d58'17"
Salwolke, Karl
Uma 8h54'18" 48d17'24"

Salyndia L Johnson "My Boo's Star"
Umi 17h33'40" 80d7'44"
Salzarulo
Vir 13h19'10" 6d12'43"
Sam
Vir 15h5'26" 3d1'15"
Sam
Ori 5h45'6" 2d48'25"
Sam
Ari 3h11'37" 21d23'21"
Sam
Vul 20h16'17" 23d2'18"
" Sam"
Leo 10h18'8" 15d21'58"
Sam
Her 17h45'46" 17d1'55"
Sam
Peg 22h20'38" 20d53'32"
Sam
Lyn 7h23'53" 49d55'3"
Sam
Cyg 20h49'4" 48d40'9"
Sam
And 2h11'31" 41d16'34"
Sam
Per 3h8'16" 44d43'3"
Sam
Cyg 20h58'24" 32d11'49"
Sam
Umi 16h5'8" 78d40'54"
Sam
Lib 14h58'43" -0d38'3"
Sam
Cma 6h13'33" -21d0'38"
Sam
Uma 9h35'3" 60d58'56"
Sam
Uma 13h35'28" 56d36'54"
Sam
Uma 12h16'3" 54d53'50"
Sam
Uma 10h9'24" 55d6'9"
S.A.M. 2007
Ori 5h41'5" 2d46'34"
Sam A. Lewis-Wright
Sco 17h52'24" -35d52'15"
Sam a.k.a. Wendie Kaye
Cam 4h27'27" 70d19'54"
Sam & Alexandra Sankari - Eternity
Col 5h47'49" -36d5'24"
Sam Allen
Cnc 8h39'34" 16d29'37"
Sam and Adam
Uma 12h0'10" 54d0'30"
Sam and Bert
Cma 6h54'39" -22d9'2"
Sam and Charlotte
Umi 14h7'30" 76d33'37"
Sam and Crystal Thompson
Uma 11h8'43" 47d7'34"
Sam and Cyp's Lonestar
Cyg 19h30'37" 29d54'58"
Sam and Dale
Her 17h15'0" 45d25'2"
Sam and Lynzie's Star
Cyg 20h6'34" 50d41'3"
Sam and Millers Dimple
Aqr 22h55'11" -10d36'5"
Sam and Nelda Blanton
Uma 10h25'37" 47d39'50"
Sam and Paula Elias
Lyn 8h18'14" 34d12'16"
Sam and Sue Fair
Lib 15h8'25" -26d59'54"
Sam and Tori Forever
Umi 17h2'42" 84d25'39"
Sam & Andrea
Cra 18h20'2" -45d22'28"
Sam Andrew Catanese
Boo 14h10'39" 27d47'42"
Sam & Anna Norris / Glasgow KY USA
Cas 1h45'25" 67d14'43"
Sam Anthony Miller
Per 3h37'12" 41d26'40"
Sam Arnsby's Star
Lyn 9h3'7" 38d31'36"
Sam B.
Aqr 21h59'30" -13d21'12"
Sam B 21
And 2h18'44" 39d16'17"
Sam Bam and Nae Nae
Sco 16h53'34" -42d33'50"
Sam Baxas
Eri 3h58'25" -0d59'7"
Sam & Ben's Wedding Star - 26.08.06
Cra 18h4'51" -39d34'48"
Sam (Big Wilf)
Uma 9h39'15" 63d49'42"
Sam Blackburn
Ori 4h49'43" 7d19'46"
Sam Brady
Uma 10h9'21" 45d16'32"
Sam Brannen
And 2h33'49" 48d57'14"
Sam & Bruno
Oph 17h4'26" -0d54'7"
Sam Builta
Cyg 21h42'12" 41d27'44"
Sam Cairns 40th Birthday
Per 4h30'8" 47d7'50"
Sam Cassaniti - my star forever
Psc 1h26'10" 11d50'18"
Sam Cast 7/1/97
Uma 11h10'9" 48d20'50"
Sam Corbo Loves Kristine Wellins <3
Cnc 9h11'54" 32d38'47"
Sam Darla Lukens
Cam 3h51'1" 69d25'4"
Sam David Gerstein
Per 3h29'50" 48d47'15"
Sam Dean
Her 17h41'33" 22d19'18"
Sam Dell Douglas
Aur 5h3'51" 46d58'37"
Sam Depledge
Per 2h44'42" 50d20'9"
Sam & Dorothy's Wishing Star
Cyg 21h16'44" 40d18'41"
Sam Dravo Pettit Star
Her 17h43'22" 46d29'58"
Sam Driessen
Aur 5h53'25" 53d21'47"
Sam Dylan Hein
Leo 9h27'43" 18d46'59"
Sam Easter
Leo 9h38'6" 29d46'36"
Sam & Ella Migliaccio
Umi 16h44'4" 76d12'53"
Sam Ferlazzo
Uma 11h2'56" 50d39'33"
Sam Frohman
Ori 5h28'59" 6d18'24"
Sam Gaffney
Cas 2h58'56" 60d34'47"
Sam Garbo, KOVGB
Ori 5h33'3" 4d32'44"
Sam Garrett Olson
Aur 5h10'16" 43d5'47"
Sam Gibner
Per 3h4'58" 54d15'34"
Sam Goosey Liska
Her 17h26'17" 36d31'24"
Sam Grace Jr.
Leo 11h48'2" 11d29'4"
Sam Habbi Yono II
Aql 19h9'58" 3d7'42"
Sam Hamilton Scott
And 2h16'12" 45d59'45"
Sam Harrison Pozner
Vir 14h28'41" -3d28'59"
Sam Hawksley
Cas 0h39'37" 56d12'12"
Sam Hemphill
Cep 21h47'24" 64d2'6"
Sam Herbert Bird
Umi 16h12'5" 72d50'19"
Sam Hervey
Aql 19h9'21" -0d4'7"
Sam Hill
Cnc 8h45'59" 19d47'41"
Sam Houston Abell
Aqr 22h40'12" -3d39'0"
Sam Hunter
Lyr 19h18'11" 29d23'12"
Sam I Am
Vir 14h41'30" 3d44'16"
Sam I Am
Ori 5h42'36" 9d14'51"
Sam I Am
Dra 17h47'49" 67d52'15"
Sam I am
Uma 10h35'49" 63d27'25"
Sam I Am
Sco 16h48'33" -31d30'8"
Sam "I Am" 5-3-7
Uma 10h0'48" 72d53'53"
Sam I Am 717
Lyn 8h5'48" 40d54'50"
Sam Ian Blumenfeld
Her 17h26'4" 20d15'7"
Sam & Ida Mandell
Cyg 21h42'43" 44d23'36"
Sam is Very Cool
Del 20h40'54" 15d57'32"
Sam & Jade Bradley
Umi 14h52'4" 70d45'31"
Sam James McCarthy
Uma 11h46'40" 60d22'30"
Sam Janis
Tau 3h34'18" 17d0'15"
Sam & Jayne Klein
Lyr 18h51'32" 37d0'50"
Sam & Jeremy - Twinkling Together
Cyg 19h10'42" 50d42'24"
Sam Johnson
Her 17h9'14" 32d47'32"
Sam Kinison, Samsbyterian's Star
Srp 17h43'48" -15d35'43"
Sam Lealofi
Per 3h16'20" 42d20'38"
Sam Lee Lemons
Leo 11h28'36" 15d49'11"
Sam Lentine
Cnc 7h57'46" 11d33'32"
Sam Lewis "A special gift"
Dra 19h7'30" 70d44'8"
Sam Liu
Cep 22h48'15" 77d48'29"
Sam loves Karen
Lyr 19h11'22" 45d36'0"
Sam Loves Tina
Cyg 21h11'40" 36d57'24"
Sam MacDonald
Umi 14h46'40" 73d24'40"
Sam & Mary Ann
Cnc 8h1'0" 19d20'1"
Sam Max Martinez
Aqr 21h14'56" -7d52'24"
Sam:MayUAlwaysHowlToU rHeartsContent
Cnv 12h47'4" 39d20'16"
Sam McKenzie Cartwright
Lmi 10h6'59" 32d39'54"
Sam Mehrizi
Aqr 22h23'56" -9d11'29"
Sam & Mel
Ori 5h34'45" -5d25'4"
Sam Meyer
Aqr 22h1'49" -22d51'24"
Sam Michael Iafrate
Sco 16h14'6" -28d33'1"
Sam & Mikala Henning
Cyg 19h38'49" 34d27'20"
Sam & Monk - Always & Forever
Leo 11h35'38" 17d56'19"
Sam N. Esses
Sco 16h12'48" -8d32'12"
Sam n' Kel
Lib 14h51'20" -1d46'13"
Sam 'n Kerry
Cyg 20h53'4" 31d14'26"
Sam O'Connor
Ori 5h12'49" 6d33'43"
Sam Ott (Beloved Basset)
Uma 13h48'47" 61d6'50"
Sam Palmer DeBerry
Cnc 7h56'18" 17d48'19"
Sam & Pat Dovidio
Her 18h21'17" 16d56'6"
Sam Patrick Hopkins
Cap 20h36'45" -11d9'0"
Sam (Pop Pop) Hill
Her 16h17'8" 18d40'12"
Sam Popovski
Umi 13h53'25" 69d40'11"
Sam Richard Scott
Cep 20h35'19" 64d20'56"
Sam Rodman
Sco 16h8'7" -14d51'59"
Sam Rust
Sgr 19h52'11" -13d36'6"
Sam & Ruth Young's 60th Anniversary
Cyg 20h13'51" 41d46'6"
Sam Sam
Gem 7h44'15" 24d27'20"
Sam & Sam
Cma 7h12'13" -17d48'14"
Sam & Sarah Kerzner
Uma 11h26'37" 58d39'21"
Sam Sargent
Leo 11h13'54" -5d29'24"
Sam Scoles
Vir 11h50'59" 7d7'29"
Sam Segond
Lib 14h52'13" -0d33'14"
Sam Shakeri
Uma 11h46'2" 52d35'7"
Sam Star
Uma 12h26'26" 53d14'8"
Sam Steven Poland
Lmi 9h57'42" 36d32'33"
Sam & Susan Wilder
Cyg 19h45'28" 30d51'35"
Sam Swan
Umi 13h41'25" 78d5'22"
Sam Tate
Lyn 8h19'49" 46d35'59"
Sam Thomas Cumisky
Per 2h33'43" 55d51'18"
Sam Thomas Watson
Umi 15h25'38" 78d19'34"
Sam & Tommie Ashcroft
And 1h5'49" 46d14'57"
Sam Truzzolino
Aur 5h14'58" 41d56'27"
Sam Tso
Aqr 22h33'50" 2d6'59"
Sam V. & Linda M.
Cyg 20h23'25" 54d24'55"
Sam Valentine Bissell
Psc 1h27'39" 26d44'5"
Sam von Will
And 0h41'29" 33d59'51"
Sam Ward
Umi 16h9'45" 77d14'10"
Sam Waterston
Her 17h23'50" 36d41'9"
Sam Weyrauch
Cap 20h23'36" -10d7'53"
Sam White
Lib 15h8'20" -7d54'36"
Sam Zausner
Gem 6h49'5" 26d52'2"
Sam2599
And 1h30'54" 49d54'33"
Sama
Dra 15h48'50" 64d9'5"
Samah Zbibi
Eri 3h39'50" -10d52'44"
Samaira Sunjay Kapur
Psc 1h24'23" 16d15'4"
Samalix Noemi Carvajal (Bebe)
Sgr 18h43'54" -16d18'14"
Saman
Uma 13h21'58" 52d39'23"
Saman Mustafa Barzani
Lmi 9h46'26" 40d8'59"
Samanda
Uma 9h29'54" 42d5'46"
SamandKristin4ever
Cyg 19h45'58" 54d48'10"
SamAndTom
Cap 20h14'38" -23d18'39"
Samangelo
Cap 20h34'8" -24d25'32"
Samanntha's Hideaway
Psc 1h19'39" 24d26'28"
Samantha
Ari 2h8'33" 22d44'59"
Samantha
Leo 11h44'20" 19d58'51"
Samantha
Gem 7h21'17" 24d34'38"
Samantha
Tau 5h16'42" 27d53'8"
Samantha
Tau 3h53'21" 23d2'56"
Samantha
Vul 21h15'8" 26d47'59"
Samantha
Her 18h51'16" 24d6'30"
Samantha
Psc 1h27'55" 15d37'36"
Samantha
Ari 3h13'29" 15d2'20"
Samantha
Cyg 20h57'53" 31d13'55"
Samantha
And 0h11'14" 41d39'10"
Samantha
Lyr 19h13'35" 34d44'37"
Samantha
Cyg 19h41'27" 36d29'10"
Samantha
Cnv 13h45'41" 31d42'5"
Samantha
Crb 16h7'0" 32d23'30"
Samantha
And 1h7'24" 37d3'37"
Samantha
And 1h45'51" 37d2'31"
Samantha
Lyn 7h53'35" 36d9'27"
Samantha
Cnc 8h8'45" 32d30'20"
Samantha
And 1h37'10" 47d24'23"
Samantha
Lyn 8h13'39" 48d4'6"
Samantha
Cas 0h34'27" 47d11'56"
Samantha
Cas 1h28'43" 58d26'44"
Samantha
Cas 0h59'21" 56d22'51"
Samantha
Cyg 21h15'27" 46d57'58"
Samantha
Uma 10h13'1" 49d44'34"
Samantha
Uma 9h40'37" 46d38'39"
Samantha
Sco 17h48'9" -36d1'54"
Samantha
Lib 15h21'22" -25d31'51"
Samantha
Cma 7h5'11" -24d6'34"
Samantha
Cru 12h0'58" -63d10'59"
Samantha
Dra 17h3'16" 57d47'44"
Samantha
Cas 0h40'46" 61d18'5"
Samantha
Cas 1h11'31" 62d30'52"
Samantha
Lyn 6h58'8" 57d8'3"
Samantha
Umi 15h19'53" 71d8'1"
Samantha
Cep 0h28'59" 78d47'56"
Samantha
Sgr 18h50'23" -19d32'44"
Samantha
Aqr 22h1'13" -21d40'46"
Samantha
Cap 21h39'1" -15d19'37"
Samantha
Aqr 23h27'19" -13d48'17"
Samantha
Aqr 23h12'45" -10d9'25"
Samantha
Cap 20h26'13" -13d48'48"
Samantha
Lib 15h30'30" -6d46'50"
Samantha
Mon 6h54'57" -0d24'24"
Samantha 4/2/2001
Ari 2h6'17" 22d44'18"
Samantha '04
Uma 9h56'54" 46d50'53"
Samantha A. Frost
Cnc 8h51'10" 25d4'11"
Samantha A. Griswold
Vir 11h48'33" 8d34'7"
Samantha A. Malone
Crb 16h13'20" 34d48'19"
Samantha A Montoney
Ori 5h5'9" 10d57'5"
Samantha A. Moravec
Cyg 19h54'40" 37d6'37"
Samantha Absmeier
And 0h20'43" 22d8'43"
Samantha Aeriel's Inspiration
Gem 7h41'35" 33d46'43"
Samantha Aimee Rodriguez
Tau 4h3'8" 16d55'26"
Samantha Alfirin Elbereth of Menel
Col 6h0'53" -42d41'18"
Samantha Algaze
And 23h54'28" 38d52'0"
Samantha and Dale's Undying Love
Ari 1h51'52" 17d52'34"
Samantha and Kevin Cotter
Cas 0h48'2" 57d2'25"
Samantha and Kevin Mahoney
Vir 14h17'1" -0d59'12"
Samantha and Matt
And 0h41'9" 35d23'38"
Samantha and Rachel Barkowski
Umi 15h2'54" 80d1'34"
Samantha Anderson
And 1h47'48" 41d56'0"
Samantha & Andrew Olsen
Pho 1h4'57" -40d28'55"
Samantha Angela Doheny
Ori 6h2'6" 18d39'40"
Samantha Ann
And 23h24'30" 52d12'51"
Samantha Ann
Cas 0h47'51" 60d58'26"
Samantha. Ann . Burton
Aqr 22h35'53" 0d25'22"
Samantha Ann Collins
Leo 11h0'1" 14d8'16"
Samantha Ann Fick
Leo 9h37'21" 13d53'46"
Samantha Ann Hicks
Tau 5h59'30" 26d45'13"
Samantha Ann Jonas
Lyn 7h45'31" 36d59'48"
Samantha Ann Leach
Cyg 21h48'59" 50d12'28"
Samantha Ann McCoy
Uma 9h13'5" 66d39'12"
Samantha Ann Muller
And 2h16'59" 48d11'27"
Samantha Ann Squires Tenicki
Psc 22h51'56" 5d17'59"
Samantha Ann Varela-Traficante
Cyg 20h31'42" 59d7'36"
Samantha Anne Connell
Cnc 8h44'33" 32d47'3"
Samantha Anne McKeithan
And 23h56'44" 45d5'28"
Samantha Anne Ramsey
Ari 2h48'20" 29d38'25"
Samantha Anne Straughton
Ari 1h56'9" 13d37'35"
Samantha Ariel Massmann
Dra 14h45'6" 56d59'4"
Samantha Ashlei
Vir 12h24'11" 11d3'36"
Samantha Ashley
Gem 6h27'12" 23d6'41"
Samantha Ashley
Cas 0h16'30" 62d50'19"
Samantha Ashley
Cas 2h20'50" 63d54'58"
Samantha Ashley Crespo
Lyn 8h40'48" 33d44'29"
Samantha Ashley Martin
Lyn 9h17'37" 46d0'0"
Samantha Ashley Schrack
Cnc 9h11'54" 10d52'39"
Samantha Ashlynn Wiseman
Ari 1h48'56" 22d15'30"
Samantha B. - Always and Forever
Lib 14h57'53" -1d27'16"
Samantha B. Fischer
Lyn 9h8'17" 38d9'36"
Samantha Babe'
Equ 21h13'32" 10d22'15"
Samantha Baby
Sco 17h10'3" -38d9'20"
Samantha Baby Girl Banks
Cap 20h44'14" -17d15'13"
Samantha "Bantoo" Wendy
Sgr 19h8'23" -27d16'12"
Samantha Barnes
Uma 9h11'8" 46d52'40"
Samantha Bates
Vir 14h6'57" -15d17'4"
Samantha Bauguess
And 1h27'34" 44d46'19"
Samantha Baxter
Cam 4h32'34" 67d21'7"
Samantha Behl
And 23h31'4" 49d29'53"
Samantha Bertrand
Sgr 18h3'18" -28d56'33"
Samantha Blackhurst
And 0h30'35" 42d57'25"
Samantha Blaha
Uma 10h19'12" 40d14'35"
Samantha Blake Paston
Sco 16h13'1" -8d35'26"
Samantha Bradley
Aqr 22h24'28" -12d45'18"
Samantha Bravo (I Love You)
Sgr 18h14'11" -20d5'9"
Samantha Brooke Alden
Uma 13h5'23" 54d17'56"
Samantha Brown
Leo 11h6'38" 9d0'12"
Samantha Broyles-A Dream Come True
Cap 21h24'44" -15d31'42"
Samantha & Bryan
Vir 14h4'22" -20d11'16"
Samantha & Bryan
Vir 13h47'5" 6d25'10"
Samantha Bryant
Cas 0h48'41" 63d48'10"
Samantha Bryce
And 1h7'22" 40d7'48"
Samantha Burger
Cam 4h40'46" 59d36'17"
Samantha C. Berek
Uma 10h11'37" 45d54'54"
Samantha Campbell
Mon 7h32'55" -0d32'31"
Samantha Carita
Lyr 18h29'18" 37d57'15"
Samantha Carrola Landry
Cet 1h20'45" -0d41'5"
Samantha Caudill
And 1h43'52" 50d10'9"
Samantha Cavenaugh Taylor
Tau 4h26'43" 24d7'25"
Samantha Chan
Vir 12h49'16" 6d7'57"
Samantha Chan
Lib 15h42'3" -29d10'23"
Samantha Char Seger
Ari 3h3'34" 10d28'14"
Samantha Charpentier
Cnc 8h44'43" 17d41'27"
Samantha Cheeks Ralph
Lib 15h13'23" -16d30'7"
Samantha Christie
Ari 2h23'13" 17d46'20"
Samantha Christine
And 1h2'11" 42d44'44"
Samantha Christine Lund
Cnc 8h13'0" 16d33'44"
Samantha Church
Lyn 7h31'59" 51d26'41"
Samantha Ciavarella
Cap 20h57'27" -25d28'48"
Samantha Claire Henley
Her 16h58'27" 31d4'23"
Samantha Clare Butterwick
And 23h19'48" 47d44'59"
Samantha Clare Weinraub
Lyn 6h46'19" 56d46'54"
Samantha Cole
And 23h24'34" 48d2'33"
Samantha Collins
Umi 14h34'6" 68d6'5"
Samantha Condon
Psc 2h5'24" 6d44'8"
Samantha Corey
Lib 14h52'52" -5d24'9"
Samantha Cox
Sco 17h51'6" -35d43'7"
Samantha & Craig Everlasting Love
Crb 16h0'54" 30d49'22"
Samantha Crandall
And 23h0'27" 40d31'7"
Samantha "Daisy" Hebling
Uma 12h15'32" 53d59'55"
Samantha Daszkiewicz
Lyn 8h19'58" 37d23'43"
Samantha Daun
Cyg 21h56'8" 51d8'21"
Samantha Dawn
Uma 9h2'42" 47d2'32"
Samantha Dawn Halpin
And 2h20'57" 50d13'58"
Samantha Dawn Murphy
Cas 1h12'47" 49d0'35"
Samantha Dawn Rochford
Tau 3h58'41" 14d58'52"
Samantha Dawn Starling
Aqr 22h59'6" -10d50'59"
Samantha De Jersey
And 22h59'33" 51d23'18"
Samantha Dee Meadows
Cnc 8h43'43" 18d27'22"
Samantha Defeo
Cam 4h2'17" 70d24'23"
Samantha del Canto
And 1h13'56" 36d44'31"
Samantha Delores Simatos
Cap 21h28'12" -9d18'36"
Samantha & Derric
Uma 11h41'21" 33d44'3"
Samantha Devin
Tau 4h42'21" 17d35'8"
Samantha Diaz
Lyn 7h49'18" 38d9'4"
Samantha Donaghey
And 2h17'38" 39d20'23"
Samantha Elizabeth
Cas 23h40'20" 57d31'38"
Samantha Elizabeth Fast
Sgr 19h42'58" -14d1'39"
Samantha Elizabeth Helmer
Ari 2h43'12" 29d33'9"
Samantha Elizabeth Madasz
Gem 7h15'1" 27d11'13"
Samantha Elizabeth Stoneham
And 23h11'9" 48d38'15"
Samantha Elle Froehle
And 1h3'53" 44d17'26"
Samantha Ellen Montgomery
And 0h37'21" 33d31'10"
Samantha Ellen Weisman
And 23h11'29" 48d59'27"
Samantha Elliott
Ori 6h4'54" 20d52'3"
Samantha Elyse
Lib 15h9'35" -14d5'6"
Samantha Emily Grace Pope
Lyn 7h24'22" 59d8'41"
Samantha Erdt
Cnc 8h4'21" 14d13'20"
Samantha & Eric's Everlasting Love
Uma 11h47'37" 53d25'38"
Samantha Escalie
Uma 9h42'52" 64d17'17"
Samantha Everett
Lyn 7h11'47" 53d10'54"
Samantha Faye
Cap 20h27'29" -14d7'9"
Samantha Fink
Uma 9h30'28" 60d40'18"
Samantha Flannigan
And 0h50'43" 44d41'41"
Samantha Florence Dixon Phelan
Sco 16h11'25" -10d50'12"
Samantha Freeman
Leo 10h0'7" 9d29'32"
Samantha Gaiera
Cnc 8h6'24" 14d35'44"
Samantha Gail
Aqr 21h25'24" 0d14'18"
Samantha Garber
Leo 11h6'6" 21d52'53"
Samantha Gayle Eldridge
Her 16h6'8" 49d36'57"
Samantha Gayle Evans
Sco 16h38'31" -30d54'42"
Samantha Gean Kulick
And 23h13'45" 42d12'46"
Samantha Geile Lymburn
Vir 13h9'26" -16d59'38"
Samantha Godwin
Dra 16h6'35" 66d56'18"
Samantha Goldswain
And 23h34'54" 50d20'29"
Samantha Goodwin
Psc 0h37'56" 7d8'57"
Samantha Grace
Aqr 21h15'14" -8d53'3"
Samantha Grace
Lib 14h51'15" -1d20'9"
Samantha Grace
And 0h31'47" 28d53'12"
Samantha Grace Ethier
Uma 12h53'21" 61d15'24"
Samantha Grace Maguire
Lib 16h1'3" -15d50'45"
Samantha Grace Miner
Gem 7h50'16" 33d1'12"
Samantha Grace Sharp
Uma 9h44'40" 52d29'10"
Samantha Grace Verrico
Cap 21h1'22" -16d52'41"
Samantha Gregory
Lib 15h34'26" -11d1'28"
Samantha Groene
Uma 12h36'40" 56d14'53"
Samantha Guay
Aql 19h23'26" 8d31'53"
Samantha H Mallinson
And 1h9'57" 42d42'48"
Samantha H. Silva
Vir 13h24'18" -8d8'29"
Samantha Haddock
Aql 19h55'36" 8d50'19"
Samantha Haffley
And 2h18'32" 47d44'20"
Samantha Haley Potter
And 2h12'58" 51d1'5"
Samantha Hantman
Vir 12h12'54" -10d43'50"
Samantha Harding
Uma 9h2'18" 47d7'57"
Samantha Harry
Cnc 8h15'6" 32d1'3"
Samantha Heaning Capinpin
Cnc 8h34'48" 7d39'36"
Samantha Heckathorn
And 0h46'15" 26d48'25"

Samantha Hemler
Lib 14h54'33" -2d50'32"
Samantha Hertzog
Cas 1h44'10" 60d30'51"
Samantha Hmelovsky's Star STELLA
And 0h6'47" 36d25'34"
Samantha Hofmann *Meine Traumfrau*
Her 18h42'4" 19d46'31"
Samantha Holt
Uma 13h6'30" 54d31'27"
Samantha Hooker
Uma 12h48'47" 58d21'41"
Samantha Hoover
Cas 1h42'23" 61d54'41"
Samantha Houlihan
Sco 16h59'24" -36d44'55"
Samantha Hrubiec
Cyg 20h29'5" 52d47'41"
Samantha Inez Ucciferri
Lib 15h18'0" -18d59'4"
Samantha J. Hudson
Sgr 18h50'56" -16d10'17"
Samantha J. Runyan
And 0h54'10" 39d50'4"
Samantha J. Snyder
Uma 12h5'9" 58d35'12"
Samantha Jacalin Pichichero
Crb 15h40'50" 27d20'44"
Samantha Jackman
And 2h38'33" 49d7'4"
Samantha Jade Reese
Cnc 8h4'22" 13d32'1"
Samantha Jae Rogers
And 1h4'54" 47d19'35"
Samantha James Ellis
Vir 13h30'6" 3d40'5"
Samantha Jane
And 1h31'6" 37d43'0"
Samantha Jane Allen
Sgr 18h6'37" -35d13'28"
Samantha Jane Fiala
Aqr 21h39'30" -0d53'18"
Samantha Jane Major With Love
Cap 21h29'39" -16d40'7"
Samantha Jane Stuart
And 23h41'7" 46d4'35"
Samantha & Jason Campbell
Cyg 20h38'56" 45d46'5"
Samantha Jean
Ori 5h56'43" 18d33'50"
Samantha Jean
Vir 14h7'3" -15d58'29"
Samantha Jean Gmuer
Uma 10h9'42" 67d54'30"
Samantha Jean Lumia
Uma 11h37'24" 48d48'53"
Samantha Jean Turner
Leo 10h19'36" 10d52'37"
Samantha Jeanne
Psc 0h45'24" 4d0'15"
Samantha Jennings
Lyn 7h4'16" 54d13'22"
Samantha Jewel
Leo 11h18'43" 18d23'32"
Samantha Jillian
Crb 16h16'0" 38d28'23"
Samantha Jo Attkisson
Sgr 19h45'12" -13d7'54"
Samantha Jo Bokor
Psc 23h6'2" -0d40'37"
Samantha Jo McCombs
Lib 15h24'56" -3d47'2"
Samantha Jo Morales
Cas 1h37'46" 63d55'25"
Samantha Jo Susan
Cnc 8h54'26" 32d9'22"
Samantha Joan Branson
And 0h14'19" 40d8'6"
Samantha Joanne
Tau 3h25'52" 9d56'24"
Samantha Joe
And 0h37'39" 32d29'40"
Samantha & John <3
Cyg 21h24'0" 44d47'58"
Samantha Jordan
And 2h19'40" 47d35'35"
Samantha Josefine
Sco 16h52'25" -32d40'47"
Samantha Josephine Breese: My Baby
And 0h15'20" 33d22'15"
Samantha Josephine Coe
And 0h0'12" 37d37'28"
Samantha & Joshua Forever
Uma 9h12'45" 48d30'3"
Samantha Joy
Lib 15h44'4" -10d22'23"
Samantha Joy Bordoff
Uma 12h26'32" 57d42'46"
Samantha Joy Griffin
Lyn 7h30'6" 46d24'28"
Samantha Joy Rieks
Ari 2h39'34" 22d7'4"
Samantha Judkins
Cnc 8h43'38" 11d9'40"
Samantha K
Gem 7h39'38" 33d48'14"
Samantha K. Barnett
Vir 12h41'58" 7d26'31"
Samantha K. Kirklin
Gem 7h44'33" 23d50'14"
Samantha Kammerer
Sco 16h41'6" -37d5'7"
Samantha Karine Wise
Cap 21h37'31" -15d44'42"
Samantha Kate
Cnc 8h18'6" 7d47'7"
Samantha Katheryn
Cnc 8h44'44" 25d31'21"
Samantha Kathryn
Psc 0h41'11" 9d25'25"
Samantha Kay Leeming - Always My Love
Cru 12h28'32" -59d13'25"
Samantha Kay Williams
Lyn 8h12'59" 38d15'54"
Samantha Kendal-Williams
And 1h22'28" 47d41'31"
Samantha & Khalaf
Uma 13h3'54" 54d6'52"
Samantha Kimball Crocker
And 0h16'1" 28d6'51"
Samantha Kirstie Powell
Cnc 9h11'14" 29d44'41"
Samantha Klemann
And 2h35'2" 47d27'11"
Samantha Klipp
Vir 13h53'28" -12d5'21"
Samantha Knoell
Sgr 18h41'6" -18d7'50"
Samantha Knowwell
Lyn 7h57'58" 37d49'30"
Samantha L. Archer
Sgr 18h24'23" -22d35'33"
Samantha L. Gross
And 2h21'43" 48d17'6"
Samantha L Meyer
Ori 6h2'35" 17d28'55"
Samantha Lacy
Uma 12h3'32" 53d9'1"
Samantha Lakshmi Nagin
Lmi 10h8'26" 31d17'5"
Samantha Lane Marsh
Per 4h31'40" 40d38'39"
Samantha Lanza
And 23h26'11" 47d58'58"
Samantha Laura Hull
Cyg 20h24'44" 42d18'58"
Samantha Lauren Davies
And 1h44'35" 38d59'9"
Samantha Lauren Grenadier
And 2h38'24" 43d40'15"
Samantha Lauren Larsen
Aqr 22h17'44" 1d17'31"
Samantha Lea
Lyn 6h53'35" 50d43'5"
Samantha Lea Elkan
And 2h22'39" 46d54'48"
Samantha Lee
Lib 15h9'58" -23d15'18"
Samantha Lee Dodick
Vir 12h8'26" 11d0'28"
Samantha Lee Sherman
Aqr 23h15'27" -13d23'57"
Samantha Lee. Slorance
And 23h46'28" 45d5'12"
Samantha Leigh
Uma 10h34'26" 42d11'26"
Samantha Leigh
Cnc 8h8'12" 22d12'38"
Samantha Leigh
Aqr 22h6'34" -13d58'0"
Samantha Leigh Belz
Cnc 8h38'30" 23d53'22"
Samantha Leigh Novak
Tau 5h9'16" 24d58'46"
Samantha Leigh Thomas
And 0h29'4" 31d38'13"
Samantha Leigh Waggoner
Umi 14h34'10" 74d32'52"
Samantha Leonie Robson
Leo 10h28'16" 14d58'54"
Samantha Leslie
Cam 11h46'37" 77d8'1"
Samantha Lesteberg
Sco 16h5'48" -10d19'52"
Samantha Lin Hammis
Umi 13h11'38" 73d48'34"
Samantha Lindsey Kemp
Ori 5h39'49" 8d55'31"
Samantha Lininger
Lib 14h54'45" -0d45'54"
Samantha Long
Cas 1h28'18" 54d7'18"
Samantha Lopez
Lib 15h5'17" -10d18'13"
Samantha Lorraine
Tau 4h9'3" 7d38'36"
Samantha Louis Giordano
Del 20h41'30" 17d45'14"
Samantha Louise Bailey-Loomis
Uma 9h41'5" 51d35'51"
Samantha Louise Duffy
Aqr 22h0'30" -17d29'30"
Samantha Louise Knights
Cas 23h24'7" 55d58'44"
Samantha Louise Newbold - Max's Mommy
And 23h9'42" 43d9'34"
Samantha Louise Reid
Uma 10h57'40" 64d51'46"
Samantha Louise Wiley
And 2h30'22" 39d46'34"
Samantha Lowe - Graduation Day
Cru 12h48'6" -63d32'52"
Samantha & Luis
Aql 19h12'22" 2d43'37"
Samantha Lynn
Vir 12h35'9" 12d47'18"
Samantha Lynn
Tau 5h25'34" 26d52'31"
Samantha Lynn
Tau 3h50'31" 27d55'38"
Samantha Lynn
Aqr 23h11'51" -20d2'48"
Samantha Lynn
Umi 14h22'7" 67d49'33"
Samantha Lynn
Lyn 8h6'19" 57d59'31"
Samantha Lynn Brooks
Psc 1h21'1" 25d23'20"
Samantha Lynn Chiavetta
And 1h13'46" 43d45'16"
Samantha Lynn Ciociola
And 23h14'30" 47d22'20"
Samantha Lynn Clay
Psc 1h27'33" 15d7'53"
Samantha Lynn Dent
And 0h41'22" 39d28'55"
Samantha Lynn Kirby
And 0h41'20" 37d17'44"
Samantha Lynn Koerperich
Psc 1h37'36" 13d57'22"
Samantha Lynn Pierce
Vir 14h49'28" 0d20'34"
Samantha Lynn Policki
Sgr 19h18'28" -22d34'5"
Samantha Lynn "Queenie's Star"
Cas 23h54'0" 53d45'59"
Samantha Lynn Risickella
And 2h21'38" 47d58'36"
Samantha Lynn Schlaud
Cap 20h15'11" -18d3'33"
Samantha Lynn Swigart BFFFNMW
Sco 17h36'38" -40d25'52"
Samantha Lynn Velez
Lib 14h51'7" -1d0'28"
Samantha Lynn Wheeler
Crb 16h6'57" 35d56'5"
Samantha Lynne
And 23h50'1" 44d49'9"
Samantha Lynne Briggs
Sgr 20h5'42" -38d23'39"
Samantha M. Corsi
Aqr 21h41'10" 1d57'18"
Samantha M. Ray
Vir 14h35'43" -1d34'38"
Samantha Maconochie &Paul Galligan
Ari 3h5'19" 28d11'9"
Samantha Madeline Mascera
Gem 6h53'22" 15d13'31"
Samantha Mae
Psc 1h23'40" 24d8'59"
Samantha Mae McCabe
Sco 16h9'1" -15d18'21"
Samantha Malia Solis
Uma 13h55'9" 52d23'27"
Samantha - Mamita
Tau 5h35'35" 27d12'57"
Samantha Manahkala
And 23h29'23" 48d40'56"
Samantha Manning
And 2h36'47" 44d37'26"
Samantha Marchner
Lib 15h30'49" -26d51'1"
Samantha Maria Stuart
And 0h24'47" 46d28'59"
Samantha Marie
And 1h22'43" 42d7'18"
Samantha Marie
Tau 5h33'25" 27d6'0"
Samantha Marie
Peg 21h37'36" 17d46'32"
Samantha Marie
Lib 15h15'48" -19d13'30"
Samantha Marie
Uma 12h26'11" 53d16'31"
Samantha Marie 13
Gem 7h16'18" 23d16'6"
Samantha Marie Anderson
Ari 2h53'42" 22d25'23"
Samantha Marie Beaudette
Cyg 21h25'46" 34d5'5"
Samantha Marie Britton
Lib 14h49'18" -7d33'56"
Samantha Marie Carder
And 2h16'34" 45d21'15"
Samantha Marie Dodge (Sweet Star)
Lyr 18h52'46" 32d56'10"
Samantha Marie Gonzalez
Sco 17h4'1" -36d48'4"
Samantha Marie Green
Tau 4h11'47" 28d4'0"
Samantha Marie Grimaldi
Aqr 23h53'17" -19d16'9"
Samantha Marie Hagan
And 1h59'21" 37d44'48"
Samantha Marie Klenczar
Gem 7h20'36" 32d49'7"
Samantha Marie Paonessa
Tri 2h32'28" 36d17'59"
Samantha Marie Perez
And 0h30'35" 26d10'0"
Samantha Marie Rix
Vir 13h20'28" 12d43'22"
Samantha Marie Saeger
Cas 0h37'58" 52d53'43"
Samantha Marie Smith
Cas 2h29'39" 67d25'22"
Samantha Marie Ventresca
Gem 6h54'27" 35d5'16"
Samantha Marie Wells
And 0h11'17" 43d35'45"
Samantha Marie Yeoward
Lib 15h3'46" -29d48'38"
Samantha Marshall
Vir 13h35'17" -12d53'43"
Samantha & Martin Henry Family
Cyg 19h28'35" 36d31'30"
Samantha & Matt Forever
Col 5h52'55" -38d31'53"
Samantha May Scharff
Gem 7h47'0" 33d57'11"
Samantha May Sumstine
Lyn 9h33'54" 40d40'0"
Samantha Mc Grath
And 1h56'53" 37d40'33"
Samantha McAleer
Psc 1h43'48" 21d43'58"
Samantha McCain
Dra 18h57'38" 67d54'51"
Samantha McCastle
And 23h12'51" 45d56'27"
Samantha McCloud
Cas 1h22'10" 55d15'19"
Samantha Menard
Ari 3h17'45" 27d26'37"
Samantha & Michael
Peg 21h22'12" 16d55'42"
Samantha Michelle Thompson
Psc 23h45'37" 1d6'29"
Samantha Michelle Ward
And 1h29'57" 40d14'10"
Samantha (Monce) Jarvis
Sco 16h39'22" -38d16'52"
Samantha Montanez
Uma 11h51'41" 33d26'10"
Samantha Morgan Snyder
Cam 12h47'28" 77d25'0"
Samantha "Munchkin" Scism
And 0h15'45" 48d13'43"
Samantha Munson
Sco 16h24'4" -33d9'27"
Samantha Murphy Soprych
Umi 11h42'17" 89d6'33"
Samantha My Epiphany
Mon 6h48'44" -0d7'28"
Samantha N. Bowling
Tau 3h53'56" 27d14'8"
Samantha N Buccellato (My Love)
Sgr 18h54'28" -21d10'33"
Samantha Nash
Cnc 8h49'19" 13d56'57"
Samantha Neubauer
Lyn 7h45'11" 41d40'44"
Samantha Nichole
Com 12h8'2" 15d52'39"
Samantha Nichole Schaffter
Psc 0h39'17" 4d23'10"
Samantha Nicole Fineo
And 2h11'34" 49d0'30"
Samantha Nicole Kempisty
Cas 23h47'49" 56d24'19"
Samantha Nicole Kirby
Crb 15h42'51" 32d27'54"
Samantha Nicole Micciche
Gem 6h45'34" 33d2'40"
Samantha Nicole Michlitsch
Leo 11h26'36" 5d9'54"
Samantha Nicole Pualani Miller
And 1h6'29" 45d51'29"
Samantha Nicole White
Cas 0h32'26" 64d35'51"
Samantha Niles
Crb 15h59'30" 38d17'27"
Samantha Orona & Riley Kay Daken
Ari 2h8'53" 21d51'55"
Samantha O'Shea's Spot!
Lyn 8h29'34" 39d53'17"
Samantha - Our Twinkling Angel
Cru 12h31'48" -60d8'39"
Samantha Page
And 0h5'11" 47d52'22"
Samantha Paige Gulline
Gem 6h13'13" 26d39'14"
Samantha Paige Kilpatrick
Gem 6h30'21" 25d46'36"
Samantha Paige Lutz
Cam 6h12'40" 61d48'28"
Samantha Paige Scott
Cnc 9h16'7" 10d20'19"
Samantha Paolini
Psc 1h2'44" 29d22'8"
Samantha Patterson
Leo 10h36'45" 15d21'40"
Samantha Pauline Evelyn Powell
Vir 13h36'45" 4d1'25"
Samantha *Peanut* - Sweet 16
And 2h15'22" 45d15'50"
Samantha "Peanut" Kriews - Sweet 16
Mon 6h43'42" 9d19'0"
Samantha Pedersen
Uma 11h16'1" 48d9'46"
Samantha Phyllis Henson
Cyg 21h40'37" 52d45'24"
Samantha Plastino
And 2h24'4" 47d26'23"
Samantha Pontillo
Cnc 8h11'31" 31d55'40"
Samantha Proegler
Cas 1h5'54" 63d51'5"
Samantha Quinlan Stone
Gem 7h26'59" 21d26'7"
Samantha Rachel
Sco 17h48'24" -31d38'19"
Samantha Rae
Com 13h6'26" 17d49'3"
Samantha Rae
And 0h37'24" 26d34'47"
Samantha Rae Aguilera
Cam 6h1'15" 68d28'24"
Samantha Rae Christopher
And 0h28'48" 32d22'18"
Samantha Rae Figueroa
Lyn 7h34'21" 35d49'21"
Samantha Rae Garrett
Vir 13h55'40" 5d20'38"
Samantha Rae Hersh
Vir 11h43'20" 9d57'34"
Samantha Rae Kimler
Lyn 9h13'13" 41d49'23"
Samantha Rae Tubbs
Sgr 18h4'6" -27d49'33"
Samantha Rae Wright
Cnc 8h51'52" 11d7'16"
Samantha Raven
And 23h27'16" 48d39'31"
Samantha Raye Dellobuono
And 0h40'27" 38d58'7"
Samantha Rebeckah Faires
Lib 15h29'55" -6d55'22"
Samantha Redden
Tau 3h51'4" 2d40'10"
Samantha Redmond
Gem 6h35'39" 18d4'29"
Samantha Rendon (Sammycakes)
Cnc 8h23'44" 22d6'30"
Samantha Rene Strassburg
Aqr 22h33'36" -0d10'40"
Samantha Renee Clifford
Tau 4h0'17" 13d44'10"
Samantha Rey Ross
Cnc 8h24'39" 10d5'4"
Samantha Rhea Hamilton
Cyg 21h53'2" 47d38'0"
Samantha Ricketson
Lib 14h31'55" -15d35'16"
Samantha Riley Beschell-Babel
Ari 2h31'55" 25d53'46"
Samantha Ripkin
Psc 1h25'7" 20d45'37"
Samantha Ritchey
Lib 15h6'16" -7d20'9"
Samantha Roberts
Umi 13h58'2" 71d41'59"
Samantha Robinson
Peg 21h53'21" 15d41'54"
Samantha Rodriguez
Lib 14h51'50" -3d21'43"
Samantha Rose
Dra 16h44'5" 53d10'1"
Samantha Rose
Psc 1h38'55" 12d1'12"
Samantha Rose
And 23h0'6" 42d42'58"
Samantha Rose 2004
Umi 15h26'56" 68d54'40"
Samantha Rose Bowyer
Uma 12h20'25" 53d57'29"
Samantha Rose Cox Stone
And 0h17'28" 43d45'41"
Samantha Rose Folken
Crb 16h16'37" 37d40'22"
Samantha Rose Irwin
Sgr 18h46'13" -24d31'20"
Samantha Rose Katz May 25, 2004
And 1h42'17" 39d1'58"
Samantha Rose Pierce
Psc 1h16'36" 16d43'17"
Samantha Rose Pope
Cas 23h59'14" 56d49'21"
Samantha Rose Seals
And 23h59'31" 45d12'25"
Samantha Rosenbaum
Gem 6h53'1" 32d30'19"
Samantha Rousseau "Jade"
Uma 11h29'18" 52d50'14"
Samantha Roxanne Beautiful
Psc 0h50'36" 10d37'37"
Samantha Ruan
Cam 4h37'2" 59d18'45"
Samantha Rubin
Peg 21h26'35" 17d1'50"
Samantha Ruggiero
Ant 10h54'48" -36d33'52"
Samantha Ruth Cerney
Sgr 18h8'11" -22d10'9"
Samantha "Sami" Patterson
Cap 21h57'44" -11d59'38"
Samantha "Sammy" Jean
Crb 16h15'26" 38d57'26"
Samantha "Samsonite" Kuhl
Leo 11h13'13" 15d19'21"
Samantha Sauter
Crb 15h49'2" 28d14'43"
Samantha Schneider
Psc 0h51'7" 16d3'6"
Samantha Schupp
And 0h47'25" 42d45'44"
Samantha Sealey
Leo 11h52'50" 18d28'10"
Samantha Sean
Cap 21h37'7" -8d57'46"
Samantha Segal
Cap 21h38'4" -11d0'12"
Samantha Sevigny
And 0h18'51" 32d39'4"
Samantha Skye Smith
And 1h30'3" 49d45'17"
Samantha Smith Johnson
Uma 10h46'16" 56d20'13"
Samantha Spalding
Leo 10h42'59" 14d8'51"
Samantha "Squish" Brook
And 1h55'45" 37d49'0"
Samantha Staley
And 2h12'0" 42d59'43"
Samantha Strickland
And 23h5'36" 42d25'40"
Samantha Strutt
And 22h58'15" 49d58'34"
Samantha Sturgeon
Cas 23h27'13" 58d27'11"
Samantha Sue Dimpsey
Cas 0h40'47" 64d50'13"
Samantha Sue Marler
Sgr 19h28'41" -18d7'38"
Samantha Sue Schubert
Oph 17h30'51" -0d53'48"
Samantha Sue Suchy
Leo 9h39'7" 28d9'13"
Samantha Suzanne
Aqr 22h42'11" -23d32'34"
Samantha Suzanne Billson
Ori 5h0'36" 15d28'52"
Samantha "Sweetpea" Tallarico
Lyn 7h51'25" 58d29'32"
Samantha Taylor
Sco 16h54'15" -41d55'30"
Samantha Taylor
Tau 4h13'36" 8d44'22"
Samantha Termine "Sam's Star"
Cas 1h31'24" 62d20'46"
Samantha Thompson
And 23h34'26" 38d1'2"
Samantha Thomson
And 2h32'31" 50d7'33"
Samantha Tilley
And 2h11'20" 41d28'18"
Samantha Tina Marie Towner
Cas 2h34'33" 69d12'17"
Samantha Toni
And 0h58'24" 43d11'16"
Samantha Toole
Cnc 8h15'53" 10d33'40"
Samantha Trebus
And 23h48'16" 46d56'36"
Samantha Virginia Backle
And 23h27'13" 39d0'5"
Samantha Vlado
Mon 6h32'2" 3d24'2"
Samantha Vogeltanz
Lib 15h22'7" -12d51'37"
Samantha Walker
Sco 16h6'37" -12d40'57"
Samantha Ward
Gem 6h59'15" 20d57'4"
Samantha Weaston
Cnc 9h15'3" 14d53'6"
Samantha Webb
Lmi 10h31'39" 28d6'3"
Samantha Willow Crosbie
Tau 4h18'35" 20d3'50"
Samantha Wilson
And 2h16'57" 38d10'16"
Samantha Woerner
Gem 7h39'57" 33d8'23"
Samantha - XP Pony with Heart
Peg 21h41'54" 27d38'8"
Samantha Yoshiko Ikari
Uma 10h33'59" 56d18'8"
Samantha Zazycki A.K.A Rising Star
Leo 10h4'26" 14d33'40"
Samantha Zombie
Ari 2h8'15" 21d10'30"
Samantha4ever
Cas 23h10'22" 55d16'8"
SamanthaKay
Mon 6h52'14" 7d45'0"
Samantha-Louise
Uma 9h5'49" 68d17'34"
SAMANTHAMAE
Cnc 8h46'46" 12d23'2"
Samantha's 1st Birthday
Aqr 22h39'25" -0d14'28"
Samantha's Diamond Rocks
Gem 7h49'46" 29d30'13"
Samantha's Fire
Cru 12h16'23" -64d21'48"
Samantha's Key
And 1h37'52" 42d53'15"
Samantha's Shining Star
Crb 15h53'35" 27d9'48"
Samantha's Shining Star
Cas 0h29'53" 61d46'56"
Samantha's Smile
And 1h21'14" 47d59'44"
Samantha's "Something Star"
Aqr 22h48'9" 1d32'58"
Samantha's Star
And 23h24'12" 45d54'43"
Samar
Cam 5h18'51" 58d11'59"
Samar
Gem 7h4'29" 22d0'55"
Samar Zegar
Lyn 8h7'57" 56d3'0"
Samara
Per 3h11'38" 31d44'17"
Samara Danielle Wellman-Bratton
Tau 3h55'26" 27d9'10"
Samara Fox Brown
And 0h29'10" 38d1'14"
Samara Lillioja
Uma 11h43'17" 58d16'42"
Samara M Walden
Uma 9h54'24" 49d47'21"
Samara Renee Sims
Umi 16h57'17" 81d36'19"
Samara "The Special Lady"
Vir 15h1'48" 3d36'4"
Samara und Markus
Peg 22h54'15" 18d28'44"
Samara Zoe Murray
And 0h15'51" 46d23'1"
Samarra Jane
Ari 3h0'44" 25d40'49"
Sam-a-saurus Rex, a wonderful soul.
Sco 17h16'9" -32d42'54"
Samatha Jean
Aql 19h7'12" 8d56'1"
Samatha Marie Waters
Lyn 6h48'46" 52d17'15"
Samatha Renee Sforza
And 0h16'6" 29d16'10"
Samatha Webster
Ari 2h23'29" 17d42'36"
Samathos
Uma 14h26'45" 56d0'58"
SaMaV
Gem 7h53'18" 27d34'59"
Samaya's Angel
Aql 18h53'40" 8d28'30"
Sambey
Cyg 20h11'12" 34d58'49"
Sambo
Tau 5h26'7" 25d12'42"
Sambo
Uma 12h33'6" 61d34'17"
Sambo Labeda
Uma 11h29'56" 52d2'14"
Sambucca Chief Wahoo
Ori 6h19'40" 13d7'50"
SamDel
Cyg 19h38'25" 52d48'32"
Sam-e
Leo 10h15'47" 19d3'30"
Samedi D'amour
Aqr 21h38'13" -8d2'49"
Samee Jayne
Ari 2h2'2" 18d27'55"
Sameena Ahmed
Cnc 9h12'13" 32d35'10"
Sameena Corinne Khan
Vir 13h18'44" 3d16'12"
Sameena Tabassum
And 2h10'12" 39d16'20"
Sameer Khan
And 0h18'20" 42d13'58"
Sameera
Uma 9h19'48" 68d45'51"
Samela
Leo 11h41'32" 26d36'17"
Samer
Aur 5h45'19" 47d52'53"
Samer Aaron Salim
Leo 11h56'56" 14d56'17"
Samer Masoud
Vir 12h18'5" -9d47'51"
Samer Salameh
Ori 5h17'27" -5d37'7"
Samer Salameh
Ori 5h22'16" -5d20'44"
Samerica
Umi 13h30'18" 72d11'51"
Samhain Weldon
Uma 8h37'24" 65d21'8"
Sam-Hubbard-291086
Sco 16h35'56" -37d13'45"

sami
Ant 9h33'18" -25d0'1"
Sami
Uma 11h48'26" 62d8'31"
Sami
Lyn 6h26'22" 55d8'42"
Sami
Cap 20h25'51" -9d33'51"
Sami
Vir 13h2'3" 5d27'40"
Sami
Ari 2h19'6" 14d20'21"
Sami
Crb 15h37'40" 28d37'43"
SAMI
Cnc 9h8'45" 31d6'19"
Sami
Aur 5h36'17" 32d29'48"
Sami and Gezime
Sgr 17h44'6" -20d5'30"
Sami Bear
And 2h2'44" 40d41'50"
Sami Douglass
Aqr 22h0'30" -15d3'52"
Sami Elkouhen
Per 3h3'16" 46d58'59"
Sami Ernest Hélou
Cap 20h11'16" -21d54'6"
Sami Johnson
Per 4h50'47" 50d31'30"
sämi jucker, 07.05.1995
Ori 5h57'29" 3d30'34"
Sami Marie
Uma 11h10'4" 48d5'48"
Sami Rouvray
Cru 12h13'18" -59d40'43"
Sami & Stella Forever
Aqr 21h55'21" 2d2'0"
Samia
Cnc 9h7'55" 15d50'46"
Samia Boultame
Cas 0h1'21" 56d58'38"
Sâmia Rassi Camara
Uma 14h20'25" 62d7'30"
Samia's Dawn Star
Leo 10h12'14" 26d22'24"
SamieJeb
Vir 12h45'47" 6d48'25"
Samigullin, Rustem
Uma 11h38'35" 30d25'53"
samin
Ari 3h22'40" 29d0'57"
Samir Madany Nebaly
Bouadi
Ori 5h57'5" -0d45'10"
Samir Tiongson Saxena
Aql 19h4'21" -10d26'21"
Samira
Cam 5h32'0" 60d6'17"
Samira
Aqr 21h26'7" 0d42'12"
Samira
And 23h8'29" 47d6'55"
Samira
Uma 9h46'1" 48d16'26"
Samira Joy Nukho
Crb 16h4'11" 38d45'52"
Samira Kauthar
Cnv 12h47'32" 40d0'36"
Samira Shahabuddin
Col 6h7'56" -31d29'28"
Samira. The Princess
Leo 9h43'7" 27d16'10"
Samire Idrizi
Ari 2h42'43" 13d58'45"
SamiRose Viviano
Uma 10h22'37" 62d30'25"
Samka Amaru
Ori 5h10'1" -0d28'31"
SAMKRN
And 0h45'50" 36d2'7"
Samlet
Lib 15h35'59" -10d16'35"
Samm Marie Davidson
Uma 11h23'4" 60d49'10"
Samma
Uma 11h6'5" 67d1'11"
Sammac
Uma 11h12'37" 33d12'29"
Sammango
Umi 16h34'32" 77d14'27"
Sammantha
Lyn 8h22'40" 51d24'50"
Sammantha Gwinn
Uma 11h5'54" 46d14'39"
Sammantha Marie Starkey
And 0h38'58" 33d20'36"
Sammantha Mills "Sammie"
Cnc 8h31'14" 23d59'34"
Sammantha Nichole
And 1h32'0" 38d36'7"
Samma's fiery star of wish-
ing
Leo 9h38'23" 31d37'43"
Samme - Williams
Aql 19h18'43" 1d12'4"
Sammer
Per 3h39'38" 44d8'59"
Sammi
Lyn 8h57'51" 46d0'22"
Sammi
Cnc 8h38'15" 15d35'41"
Sammi Baby
Tau 5h43'32" 18d8'53"
Sammi Boswell
Uma 10h42'12" 51d45'43"
Sammi Chiu
Ori 5h30'37" 1d5'47"
Sammi Girl
Vir 13h0'32" -4d59'23"
Sammi Gorney
Cnc 8h36'47" 16d21'58"
Sammi Jo
Uma 10h17'1" 67d6'0"
Sammi Jo Burke
Cap 21h45'31" -11d31'31"
Sammi Leigh
Sco 16h25'37" -32d15'22"
Sammi Lou
Cyg 20h7'23" 39d53'25"
Sammi & Vincent
Her 17h4'13" 28d29'21"
Sammie
Psc 1h5'50" 3d50'43"
Sammie
Gem 6h32'55" 20d36'35"
Sammie and Chloe
Cas 0h30'43" 62d7'11"
Sammie Jo
Cnv 12h40'43" 47d42'25"
Sammie JoAnne Dietz
Cam 3h26'18" 65d1'14"
Sammie pooh
Lib 15h21'18" -14d19'1"
Sammie's
Ori 5h48'33" 8d16'10"
Sammie's Shining Star
Lyn 7h34'59" 39d15'51"
Sammi-Jo Allen
Leo 11h20'0" 28d15'56"
Sammi's Cosmo Home
Uma 12h35'55" 55d39'5"
Sammi's Vision
Cas 23h47'13" 56d22'58"
Sammita
Lib 15h0'1" -19d15'31"
Sammuel & Dorothy
Jefferson
Tau 4h33'52" 30d49'11"
Sammy
Lyr 18h41'19" 34d41'49"
Sammy
Uma 11h31'34" 45d15'9"
Sammy
Uma 9h29'13" 48d10'38"
Sammy
And 23h37'47" 38d34'1"
Sammy
Her 18h52'42" 23d3'28"
Sammy
Ari 2h57'10" 18d27'36"
Sammy
Lib 15h20'36" -19d34'13"
Sammy
Lib 15h3'14" -0d52'2"
Sammy
Cma 6h46'49" -13d11'36"
Sammy
Uma 11h9'19" 55d4'58"
Sammy
Uma 9h20'14" 53d15'8"
Sammy
Uma 11h25'37" 64d57'43"
Sammy
Umi 13h38'56" 74d25'2"
Sammy
Cru 12h51'47" -62d25'44"
Sammy Allen
Uma 11h36'27" 62d52'57"
Sammy Barckhausen
Vir 13h20'9" 11d4'7"
Sammy Bennett
Leo 10h29'11" 25d58'53"
Sammy Benson
Cma 6h34'49" -30d43'32"
Sammy Cammarota
Dra 17h45'34" 53d59'43"
Sammy & Crystal's Special
Star
Ori 6h8'58" 2d35'2"
Sammy Franklin Doyle
Gem 7h9'42" 33d4'10"
Sammy G. Stover
Boo 14h34'2" 40d40'18"
Sammy Gallardo
Cnc 9h5'46" 11d15'23"
Sammy Gonzalez
Her 18h32'26" 18d47'41"
Sammy Jam
Leo 10h3'58" 14d54'0"
Sammy Jim
Ori 5h5'14" 10d22'52"
Sammy Jo
Tau 5h11'20" 18d18'40"
Sammy Jo
Ari 2h15'42" 22d18'56"
Sammy Jo Peanut
Lyn 7h24'24" 51d10'18"
Sammy Jo Peanut
Uma 11h10'47" 47d46'3"
Sammy Kaufman
Uma 10h44'4" 43d54'12"
Sammy King
Lib 15h22'57" -26d34'38"
Sammy Marie
And 1h14'11" 46d29'31"
Sammy McCaughey
Uma 11h27'44" 58d42'43"
Sammy Mezo
Her 17h13'23" 16d22'38"
Sammy Padgett
Uma 10h35'46" 48d58'30"
Sammy Rae
Leo 9h40'58" 12d12'48"
Sammy Rodriguez
Crb 15h55'42" 34d44'32"
Sammy Salfino
Aqr 22h59'56" -15d2'16"
Sammy "Sam Sam"
Bonnete
Peg 22h12'7" 7d48'13"
Sammy Scheunemann
Uma 12h23'39" 54d19'59"
Sammy Settos Hughes
Sgr 19h20'31" -23d24'14"
Sammy & Sohaib
Her 18h20'42" 17d20'31"
Sammy T.
Umi 14h24'53" 77d57'8"
Sammy T. Blair
Ari 3h23'15" 19d6'24"
Sammy Vasta
Per 2h25'14" 55d16'13"
Sammy Young
Mon 7h34'53" -0d27'8"
sammyanthaly
Sge 20h9'2" 18d5'28"
SammyLouise
Leo 10h28'8" 8d53'0"
Sammy's Lucky Star
Uma 9h9'30" 58d29'30"
Sammy's Star
Psc 1h28'42" 25d35'11"
Sammy's Star
Tau 5h59'57" 24d57'22"
Samone Lorraine Duster
Psc 0h53'53" 24d35'0"
Samone Nelson
And 0h22'54" 45d11'1"
Sample
Uma 10h32'42" 52d0'28"
Sample
Cyg 20h4'59" 46d32'30"
Sample
Lyn 8h8'46" 39d35'47"
"Sample"
Lyn 7h20'47" 44d46'36"
Sample
Ori 4h51'26" 3d53'22"
Sample
Uma 9h37'23" 56d45'59"
Sample
Dra 15h6'52" 57d56'29"
Sample
Dra 19h38'52" 64d41'35"
Sample's Star
Gem 7h13'3" 34d27'21"
Sampson the Great
Per 3h14'33" 44d54'22"
Sampson the Great
Sco 16h38'31" -28d54'55"
Sampson William Hundelt
Her 17h32'30" 41d31'7"
Sampson's Keeper
Cas 1h14'47" 53d40'58"
Samra
Psc 1h20'22" 10d45'10"
Samrah
Sco 17h45'57" -39d16'21"
Samrock06
Lyn 9h1'20" 36d40'39"
Sam's Light
Uma 9h1'31" 65d10'41"
Sam's Little Emo Star
Ari 3h9'42" 29d19'27"
Sam's Love
Ari 2h51'45" 14d35'6"
Sam's Love For Janell
Del 20h32'12" 12d52'48"
Sam's Mom
Aql 19h46'28" -0d6'3"
Sam's Star
Ari 2h55'6" 31d10'31"
Sam's Star
Uma 11h54'54" 38d27'12"
Sam's Star
And 1h32'40" 50d14'59"
Sam's Star 11-18-84
Uma 12h35'59" 52d28'50"
SamSam
Aqr 22h31'51" 0d4'39"
Samsarah
Cnc 8h27'8" 31d37'31"
Samsel1966
Leo 11h39'14" 25d16'38"
SamSirius
Cma 7h24'12" -13d48'0"
Samsissy
Umi 14h40'28" 77d44'50"
Samson
Cep 21h49'34" 55d43'38"
Samson
Cmi 7h33'5" 4d25'25"
Samson
Uma 11h34'28" 44d32'6"
Samson
Uma 10h42'36" 46d41'44"
Samson Emanuel Cornejo
Lib 15h4'28" -22d25'59"
Samson Inskeep
Uma 11h31'13" 45d12'14"
Samson Lynch
Cma 7h20'37" -19d14'13"
Samson Neosirius
Ori 5h51'53" 6d54'27"
Samson, forever in the
sky......
Cyg 21h10'55" 45d20'49"
Samstar
Aur 6h4'24" 30d29'9"
Samstar
Uma 10h33'34" 66d56'2"
Samual A. Mackie
Lib 15h39'55" -17d11'36"
Samual David Morrow
Her 18h12'59" 27d35'48"
Samual Joseph Vincent
1989-2005
Uma 9h57'35" 51d37'32"
Samual L Wheaton
Uma 10h0'4" 43d29'31"
SamuAm
Sgr 18h20'3" -21d3'15"
Samuel
Ori 6h5'7" 21d8'52"
SAMUEL
Ori 6h4'21" 13d24'54"
Samuel A. Joachim Wilson
Umi 15h28'14" 72d6'45"
Samuel A. Martello
Uma 10h34'25" 49d8'15"
Samuel A. Swift
Lyn 6h43'20" 59d1'32"
Samuel Aaron Hecht
Lib 15h36'11" -11d25'42"
Samuel & Aimee
Tau 4h7'18" 7d20'50"
Samuel Alan Joachim
Wilson
Her 17h15'26" 36d14'34"
Samuel Aleksandr
Her 17h27'30" 18d10'20"
Samuel & Alessandra
Uma 10h1'25" 70d5'20"
Samuel Alexander Johns
Per 4h25'17" 45d55'54"
Samuel Alexander Tarker
Gem 7h39'45" 27d6'22"
Samuel Alvin Gross
Umi 16h15'55" 75d44'45"
Samuel - An Angel on a
Star
Col 6h4'34" -42d28'17"
Samuel Andrew
Aeschliman
Tau 4h30'32" 27d27'43"
Samuel Armstead Waitt
Sco 16h46'25" -41d44'52"
Samuel Austin Eastman
Ori 5h1'51" 7d15'21"
Samuel B. Love, Sr.
Aql 18h45'48" 7d47'58"
Samuel B. Tavelman
Lib 15h21'34" -26d50'43"
Samuel Bac
Uma 9h27'20" 43d47'5"
Samuel Baillargeon
Leblanc
Per 4h23'10" 34d6'16"
Samuel Beckett "Goose"
Leonard
Tau 4h34'46" 4d40'32"
Samuel Benjamin
Armstrong
Umi 16h15'38" 71d54'55"
Samuel Black
Boo 14h22'41" 16d5'3"
Samuel Bond
Ori 6h14'16" 6d46'54"
Samuel Boyd
Uma 9h13'44" 53d17'53"
Samuel Brodhagen
Uma 11h35'53" 60d40'37"
Samuel Bryce Pollock
Lib 15h47'47" -18d41'45"
Samuel C. & Lillian G.
Sperlazzo
Ori 6h10'51" 15d44'35"
Samuel C. Riolo
Cep 21h50'45" 62d20'54"
Samuel Caleb Myers
Vir 11h51'38" -3d4'22"
Samuel Caleb Owens
Uma 11h17'51" 36d59'8"
Samuel Carlisle
Cep 22h8'29" 64d16'29"
Samuel Carroll
Uma 9h36'20" 45d9'48"
Samuel Charles Cowlard
Uma 11h13'20" 31d59'42"
Samuel Charles Hill
Sco 17h20'31" -37d46'18"
Samuel Christopher Allard
Sco 16h13'35" -18d9'57"
Samuel Christopher
Sobeck
Gem 7h12'29" 34d48'0"
Samuel Colin Usdan
Tau 6h0'19" 23d52'4"
Samuel Cowen
Vir 14h13'13" -8d43'31"
Samuel Craig Millard
Leo 11h10'30" 20d9'13"
Samuel Curtis Ordean
Beam
Her 17h47'29" 32d21'28"
Samuel Dale Imperial
Chappell
Per 3h44'42" 45d39'57"
Samuel David Parry
Per 2h50'6" 52d5'33"
Samuel David Zuckernik
Her 18h8'34" 22d47'36"
Samuel Dean-McCuaig
Umi 15h51'10" 76d17'22"
Samuel Descoteaux né 6
mars 2004
Umi 13h58'5" 72d41'42"
Samuel Devon (Samuri)
Becker
Uma 9h39'35" 67d40'33"
Samuel Diaz
Cap 21h46'50" -13d24'50"
Samuel Douglas
DiGiovanni
Ori 5h14'6" 7d47'43"
Samuel Douglas Price
Lmi 10h24'42" 36d33'50"
Samuel Duncan
Her 17h46'2" 32d2'34"
Samuel Dylan Bayley
Cru 11h59'4" -61d7'57"
Samuel E Deibler
Leo 9h34'29" 28d14'45"
Samuel E. Garcia
Vir 12h54'20" 5d12'34"
Samuel E Klessel
Cnv 13h55'15" 38d44'4"
Samuel E. Norwood
Per 4h47'40" 46d43'33"
Samuel Edward Elliston
Her 16h33'55" 17d3'52"
Samuel Edward Navin
Ari 3h7'29" 29d13'45"
Samuel Elias
Aqr 22h37'44" -1d9'25"
Samuel Erik Tellepsen
Uma 11h14'29" 58d55'34"
Samuel Ethan Trissel
Cep 22h51'17" 80d26'6"
Samuel Eugene Scott
Cap 20h41'12" -24d30'38"
Samuel Everett Miller
Cnv 13h45'6" 31d14'50"
Samuel F. Buxton
Lyn 8h50'25" 34d55'20"
Samuel F. Winsper
Tau 4h21'28" 25d20'52"
Samuel Firman Biederman
Sgr 19h12'49" -17d13'53"
Samuel Garcia - Shining
Star
Dra 19h31'48" 62d55'37"
Samuel Garrick True
Umi 6h23'2" 88d24'35"
Samuel George Ferguson
Uma 13h41'37" 55d23'46"
Samuel George Gay
Aur 5h23'57" 47d19'26"
Samuel Gibran
Aqr 22h45'58" -12d21'23"
Samuel Glenn Johnston
Dra 19h31'24" 64d18'30"
Samuel Glenn Studer, Jr.
Eri 4h26'32" -0d7'45"
Samuel Godbolt
Uma 11h46'21" 50d12'39"
Samuel Griffin Bart
Uma 9h24'37" 66d48'47"
Samuel Guillermo Perez
Ori 5h24'10" 2d33'13"
Samuel Henry McGee
Lib 14h50'18" -3d44'28"
Samuel Hodge
Ari 3h6'45" 28d55'19"
Samuel Hughes Bruketta
Her 16h53'19" 33d25'55"
Samuel Hunter Cole
Ari 2h45'6" 29d4'18"
Samuel Isaac Elior Malkiel
Ori 6h22'0" 11d12'36"
Samuel J. Abruzzo
Per 3h31'9" 48d33'44"
Samuel J. Taylor
Uma 8h26'37" 70d10'36"
Samuel Jack Saunders
Umi 15h47'31" 74d34'59"
Samuel Jacobson
Aql 19h51'27" 13d11'31"
Samuel Jäggi
Per 2h15'2" 51d16'54"
Samuel James
Ori 5h54'18" 21d34'20"
Samuel James Easton
Lyn 8h9'34" 43d38'5"
Samuel James Frisby
Her 16h58'29" 30d39'45"
Samuel James Harvanek
Her 17h59'38" 14d50'49"
Samuel James Judy
Uma 9h21'0" 70d38'56"
Samuel James Lea
Umi 16h49'33" 82d49'27"
Samuel James Norval
Cep 23h2'6" 71d8'48"
Samuel Jan 16.12.2006
Sgr 18h2'1" -27d0'45"
Samuel Jason Obi
Her 16h34'52" 49d1'24"
Samuel & Jennifer Devins
Cyg 20h59'52" 45d5'48"
Samuel Jerome Maffei
Ari 3h22'53" 21d48'43"
Samuel John Ackroyd
Ori 5h52'18" 9d17'28"
Samuel John Allan McCart
19/04/97
Umi 13h15'48" 87d45'40"
Samuel John Barry Larter
Uma 8h12'41" 61d24'21"
Samuel John May
Uma 11h17'12" 33d55'18"
Samuel John Willis
Uma 9h18'42" 42d36'59"
Samuel Jonathan White
Aur 6h30'0" 34d57'21"
Samuel Jones Strickland
Uma 9h24'42" 57d42'57"
Samuel Joseph
Her 16h54'31" 16d5'57"
Samuel Joseph Baylis
Gem 7h10'37" 28d35'50"
Samuel Joseph Brook
Her 18h40'7" 22d46'28"
Samuel Joseph Grocott
Per 4h34'55" 32d2'57"
Samuel Joseph LeBlanc,
Jr.
Her 17h36'45" 20d34'38"
Samuel Joseph Thomas
Her 16h16'57" 19d6'1"
Samuel Joshua Riggs(08-
02-1989)
Per 4h17'45" 40d30'37"
Samuel Julian Anderson
Lac 22h24'43" 48d4'40"
Samuel Kolloff
Umi 14h37'57" 73d30'52"
Samuel L Little
Boo 14h49'35" 20d14'19"
Samuel L. Mattison, Sr.
Sgr 19h35'53" -30d53'4"
Samuel L Willcocks
Vir 13h56'49" -8d31'12"
Samuel Lawrence Acosta-
Holder
Lyn 7h29'59" 50d35'26"
Samuel Lawrence Wheaton
Tau 5h9'0" 26d45'27"
Samuel Leatherland
Cep 22h1'20" 57d5'33"
Samuel Lee Britain
Tau 4h1'34" 22d25'55"
Samuel Lee William
Heitman
Gem 6h50'13" 22d7'9"
Samuel Leibovitz
Cep 22h15'30" 72d19'53"
Samuel LeMar
Vir 14h7'42" -17d51'44"
Samuel Lennard
Per 3h4'57" 46d59'43"
Samuel Lesher
Per 4h49'31" 46d45'52"
Samuel Lewis Hemsley
Gem 6h46'8" 33d49'18"
Samuel Lewis Mordecai
Cnc 8h5'26" 10d52'15"
Samuel Lindholm
Cnc 8h31'5" 23d42'11"
Samuel Lister
Cru 12h21'23" -56d21'22"
Samuel Louis Luketich
Tau 4h11'36" 16d8'10"
Samuel Louis Van Marter
Cap 21h12'32" -22d25'21"
Samuel Lucas Mitchell
Cyg 20h14'9" 48d40'20"
Samuel Luis Santos
Boo 14h58'41" 22d39'13"
Samuel M. Moss
Mon 6h48'18" -0d20'52"
Samuel M. Walden
Ari 2h50'47" 25d27'52"
Samuel Maddox
Gem 6h13'44" 24d49'15"
Samuel Manning
Cep 22h26'30" 83d27'31"
Samuel Marshall Keltz
Her 17h20'30" 36d22'48"
Samuel Martin Redvers
Mason
Her 18h46'26" 13d10'7"
Samuel Matthew Busby
Lmi 9h37'6" 38d6'24"
Samuel Matthew Fleeman
Uma 8h36'53" 49d24'45"
Samuel Matthew Hudson
Her 16h10'47" 17d21'55"
Samuel Meyer Pizer
And 23h13'5" 42d23'12"
Samuel Michael Anderson
Her 17h17'56" 34d46'50"
Samuel Michael Piscitelli
Sco 16h11'39" -41d21'35"
Samuel "Mikey" Melendez
Gem 6h43'46" 12d47'6"
Samuel Mills
Per 3h28'36" 50d49'12"
Samuel "Molly" Botros
Pho 23h48'5" -41d22'36"
Samuel Montana D'Olimpio
Aqr 23h22'20" -19d19'55"
Samuel Montana D'Olimpio
Aqr 22h3'23" -19d20'10"
Samuel N. Jones -
Cikatricis
Aqr 22h56'49" -7d23'10"
Samuel Nathan
Schwalbach
Uma 8h54'9" 57d19'16"
Samuel Nicolas Rago
Her 16h20'27" 19d7'50"
Samuel Noble Hunt
Her 16h32'24" 45d48'38"
Samuel Oliver Biddle
Ari 2h36'19" 27d10'11"
Samuel Parker Dobbel
Sco 17h54'40" -38d5'24"
Samuel Pastore
Ari 3h10'30" 29d39'12"
Samuel Patrick Coombs
Ser 15h56'27" -0d49'33"
Samuel Patrick Fischer
Lyn 7h56'39" 35d19'3"
Samuel Patrick Porch
And 23h25'25" 42d21'24"
Samuel Paul
Her 18h8'57" 49d18'44"
Samuel Paul Solper
Sco 16h52'2" -29d26'35"
Samuel Peder Munch
Aqr 22h29'50" -8d14'59"
Samuel Peter
Her 17h53'1" 22d32'9"
Samuel Peter Alec Fingeret
Her 17h3'3" 18d24'48"
Samuel Peter Johnson
Cnc 8h5'54" 17d31'23"
Samuel Phillip Keys
Uma 13h5'56" 57d47'14"
Samuel Pierre Kingscote
Uma 10h41'15" 62d59'31"
Samuel Randolph Fox
Cnc 9h10'49" 31d16'2"
Samuel Richard
Leo 11h0'52" -4d39'41"
Samuel Richard Evans
Per 4h9'51" 48d9'52"
Samuel Richard Hill-
Morriss
Per 3h1'21" 42d36'7"
Samuel Rizal
Ori 6h16'13" 19d52'48"
Samuel Robert Gillette
Tau 4h35'21" 24d49'17"
Samuel Robert Khavinson
Cap 21h37'53" -9d11'47"
Samuel Robert McCloskey
Aql 18h47'28" 12d6'32"
Samuel Robert Moreira
Cru 12h16'27" -58d0'24"
Samuel Robert O'Brien
Uma 10h53'49" 58d6'30"
Samuel Robert Richardson
Ori 5h33'55" 3d5'7"
Samuel Rogers
Cnc 8h42'1" 32d24'23"
Samuel Ron James
Per 2h41'3" 52d42'34"
Samuel Rosenau
Aql 19h41'12" -0d0'44"
Samuel Ryan Kaska
Cae 4h34'13" -37d18'47"
Samuel Ryan O'Sullivan
Lib 15h9'0" -5d30'44"
Samuel Ryder
Cru 11h58'22" -62d47'25"
Samuel (Sam) Compton
Uma 11h24'4" 64d4'52"
Samuel Shackelford
Lib 14h47'35" -12d17'35"
Samuel Skelhorn
Uma 9h28'45" 57d51'13"
Samuel Smith
Cep 22h55'33" 70d42'7"
Samuel Stephen Morykwas
Per 3h3'38" 43d48'17"
Samuel Stewart Jones
Crb 15h53'30" 30d2'34"
Samuel T. Campos
Her 16h25'7" 11d8'35"
Samuel T. Otis
Vir 13h12'59" -6d30'7"
Samuel T. W. Davidson, III
Boo 15h36'48" 49d50'43"
Samuel Tait
Lib 15h10'37" -6d22'47"
Samuel Tate
Lyn 7h31'17" 41d6'50"
Samuel Taylor Farley
Ori 5h38'52" 1d15'54"
Samuel Taylor Johnson
Vir 14h15'30" 4d40'11"
Samuel Todd Southerland
Lyn 7h33'43" 58d52'47"
Samuel Troy Amidy
Cru 12h42'41" -58d21'45"
Samuel Victor Cream
WRIGHT
Leo 11h55'37" 21d42'42"
Samuel Vincent Evans
Leo 11h1'24" 8d30'57"
Samuel Vincent Hemenway
Lib 14h54'42" -1d1'52"
Samuel Walker Plociniak
Uma 9h25'7" 65d27'23"
Samuel William Castray
Sgr 18h4'3" -27d47'21"

Samuel Winner
Aqr 23h1'17" -5d3'56"
Samuel Wolf Kentridge
Cru 12h31'21" -61d51'0"
Samuel Wright
Lmi 9h48'17" 36d54'8"
Samuel Wyatt Rains
Per 3h55'10" 39d7'50"
Samuel You Will Always Shine
Ari 1h49'23" 18d40'8"
SamuelCarter
Lib 14h48'41" -1d42'42"
Samuele Lennon Artz
Ari 2h39'58" 29d45'58"
Samuel's Star
Cyg 20h58'6" 45d52'54"
Samuel's Star
Umi 15h15'22" 78d16'0"
Samuel's Wish
Uma 13h27'14" 57d45'46"
Samuili Katya
Gem 7h22'45" 14d2'56"
Samuli Santeri
Uma 9h49'0" 49d26'32"
Samurai's Serendipity
Uma 11h24'59" 43d47'45"
SamWoodka
Tau 4h22'37" 17d26'43"
Samy
Sco 17h55'44" -44d35'2"
Samy J
Uma 12h32'49" 57d19'57"
San
Cap 21h19'1" -15d50'54"
San - Alva - Char
Lib 15h56'39" -9d37'42"
San Antonio Lone Stars
Her 16h47'56" 32d24'24"
San Antonio Sandi
Lyn 8h50'7" 40d5'9"
San Felipe
Lib 15h47'58" -7d16'28"
San Juanita Diana Garza
And 0h43'36" 41d43'57"
San Juanita Loza
Ori 5h41'15" 0d46'44"
San Juanita Mata
Lib 14h23'7" -17d23'14"
Sana
Uma 10h55'53" 66d19'7"
Sana and Rabia's Eternal Star
Cyg 21h15'25" 47d42'43"
Sana Gilani Makhzoum
Cyg 19h52'59" 32d23'43"
Sana & Misha's Eternal Love
Cnc 8h25'56" 10d39'22"
Sana Mothana Saeed
Aqr 22h24'40" -21d38'56"
Sana Prime
Gem 7h14'2" 15d29'15"
Sanaa
Cap 21h26'2" -19d10'54"
Sanaa
Cap 20h28'13" -12d47'52"
Sanaa
Sgr 19h10'48" -25d19'7"
Sana'a Antoine Khalil
Cyg 20h52'35" 47d19'57"
Sanai Jade Kaufman
Umi 16h12'54" 73d42'54"
SanAndres
Aqr 22h38'59" -15d51'37"
Sana's Shining Star
Crb 15h46'53" 30d56'57"
Sana's Star
And 0h39'21" 46d9'1"
Sana's Star
Hya 14h15'26" -26d14'12"
Sanber
Ori 5h32'34" 3d22'55"
SanBorn Darling
Crb 15h24'20" 25d43'31"
Sancassani Emanuela
Umi 13h12'40" 72d14'22"
Sancha & Paul
Ari 3h28'47" 21d6'1"
Sanchez
Lyr 18h33'56" 36d0'43"
Sanchez-Marlotica, Francisco
Uma 11h13'16" 49d26'49"
Sanchez's Serenity
Sgr 18h20'56" -25d46'47"
Sanchi S.
Cma 6h43'51" -17d3'19"
Sancho de Vilaverde Correia
Umi 13h33'59" 73d41'52"
Sanci
Lib 15h50'25" -14d23'54"
Sanctimonia Allie
Col 6h21'36" -34d46'40"
SanD
Gem 6h29'12" 25d19'45"
Sand Acres Hotel
Per 2h46'44" 53d31'22"
Sanda Jo
Lib 15h6'56" -7d28'25"
Sanda Star Brite
Cas 0h41'19" 63d59'32"
Sandar's Star
Sgr 18h43'26" -34d19'41"
san.d.b.
Sco 16h9'24" -12d59'1"
Sandchörnli
Boo 14h43'8" 12d11'28"
Sandee Mallow
Leo 9h51'47" 31d7'53"
Sandeep Kaur Purewal
Cnc 8h14'43" 32d9'52"
Sandeep Singh
Ori 5h25'27" 2d21'18"
Sandeep Tungare
Vir 14h41'53" 0d39'24"
Sander Romero
Lib 15h26'41" -10d0'51"
Sanderellie
Lib 15h5'17" -5d52'9"
Sanders 50
Cap 21h18'9" -19d0'49"
Sanderson Stellaris
Cra 18h33'47" -42d23'55"
Sandi
Aqr 22h15'15" -17d49'53"
SANDI
Lib 15h54'38" -10d29'42"
Sandi
Uma 10h38'9" 71d11'59"
Sandi
And 1h17'8" 35d3'32"
Sandi
Peg 23h21'56" 34d5'15"
Sandi Austin
Mon 7h33'23" -0d28'16"
Sandi Cooper
Cyg 21h49'21" 48d20'8"
Sandi Dekker
Sco 17h51'32" -36d1'27"
Sandi Fruchey
Tau 4h0'57" 30d54'43"
Sandi Impiazzi's Diamond In The Sky
Pyx 9h1'1" -34d27'43"
Sandi K.
Crb 16h21'43" 39d30'55"
Sandi Kaiser
Uma 11h45'52" 52d46'52"
Sandi Kupperman
Cyg 20h59'35" 38d5'13"
Sandi Lee
Cyg 20h3'57" 35d32'14"
Sandi Lee
Cap 21h32'0" -15d53'35"
Sandi Locklear
And 0h35'0" 32d40'49"
Sandi Lyn
Ari 2h8'43" 13d12'51"
Sandi Lynn Hemingway
Com 12h28'43" 28d14'19"
Sandi M. Lariviere
Cap 20h44'35" -25d20'27"
Sandi Nelson
Vir 14h56'24" 4d8'9"
Sandi Real
Cas 0h0'36" 59d55'31"
Sandi Smith
And 1h11'53" 38d12'23"
Sandi, Whiskers' Shining Star
Cap 20h11'29" -8d41'48"
Sandi1950
Lyn 8h15'47" 54d12'49"
Sandie
Lib 15h33'42" -3d55'30"
Sandie brat Kristensen
Lyn 7h19'59" 50d14'17"
Sandie Caine Whitman
Cam 3h42'32" 54d31'6"
Sandie & John Hall's Ruby Love
Cyg 21h30'41" 41d36'0"
Sandie Rice
Cyg 19h41'47" 36d5'9"
Sandie Smith
Cas 23h6'15" 56d48'31"
Sandie Stoddard
Aqr 22h40'58" -1d15'9"
Sandie1
And 23h38'18" 42d51'36"
Sandiek
Ari 3h0'40" 10d59'32"
SandieRabbit1969
And 23h18'53" 47d36'27"
SanDino's Grace
Cyg 21h22'2" 34d23'30"
Sandira Calviac
And 23h50'54" 35d18'43"
Sandi's Nova
Ori 6h10'45" 3d34'9"
Sandi's Star
Lup 15h57'22" -37d11'12"
Sandman's Pleasure
Oph 18h19'44" 11d17'28"
Sandone Family
Umi 16h3'16" 75d10'42"
Sandor
Umi 13h32'43" 74d10'46"
Sandor Judit
Uma 13h43'32" 53d32'48"
Sandor Loevy
Uma 11h5'14" 64d25'52"
Sandor & Patti Forever In Love
Cyg 20h16'37" 39d29'51"
Sandors Of The South
Aqr 22h20'8" -0d56'1"
Sandown North Elementary School
Uma 11h20'0" 69d40'38"
Sandoz Chanta
Her 17h16'31" 13d57'17"
Sandra
Her 17h17'9" 14d39'58"
Sandra
Tau 3h46'35" 6d48'19"
SANDRA
Ori 4h54'50" 10d16'36"
Sandra
Gem 6h26'27" 25d53'39"
Sandra
Gem 6h32'2" 20d57'35"
"sandra"
Ori 5h52'47" 22d23'16"
Sandra
Her 18h51'43" 21d51'2"
Sandra
Cyg 20h37'18" 38d21'56"
Sandra
Lyr 19h4'56" 41d44'41"
Sandra
Lyr 18h19'40" 38d59'58"
Sandra
Lyr 18h28'41" 39d39'29"
Sandra
And 23h14'7" 42d5'39"
Sandra
Cas 1h2'34" 54d9'56"
Sandra
Uma 10h21'47" 47d5'10"
Sandra
And 0h29'24" 38d57'30"
Sandra
And 0h30'32" 42d28'14"
Sandra
And 1h26'19" 43d12'8"
Sandra
Uma 9h55'51" 43d11'40"
Sandra
Uma 10h45'10" 44d3'35"
Sandra
Per 3h17'2" 42d40'30"
Sandra
Lmi 10h37'30" 34d45'50"
Sandra
Uma 11h14'51" 30d46'0"
sandra
Cep 20h45'42" 69d20'35"
Sandra
Uma 10h26'32" 68d53'47"
Sandra
Uma 10h19'1" 65d20'45"
Sandra
Uma 13h11'54" 60d21'14"
SANDRA
Uma 9h40'14" 62d59'50"
Sandra
Uma 13h28'55" 58d40'28"
Sandra
Lac 22h52'26" 53d26'48"
Sandra
Ori 5h15'50" -4d2'24"
Sandra
Ori 6h6'51" -2d6'6"
Sandra
Cap 20h34'22" -19d39'6"
Sandra
Sgr 18h50'25" -27d10'30"
Sandra
Cas 0h19'30" 62d17'10"
Sandra 5562
Lmi 10h48'25" 26d53'2"
Sandra - A
And 2h21'19" 48d48'58"
Sandra A. Marino
And 2h37'34" 37d18'46"
Sandra A. McClelland
Psc 23h17'3" 6d38'34"
Sandra ~ A Star in Her Own Right
Cam 5h45'28" 56d4'37"
Sandra A.K.A. Babe
And 2h5'37" 42d42'2"
Sandra Aldarequia
Cas 3h12'47" 66d0'6"
Sandra Alter
Uma 10h43'33" 59d7'41"
Sandra and Dave's Star
Cnc 9h21'39" 32d20'4"
Sandra and Harold Kitchen
Cyg 20h33'18" 58d15'29"
Sandra and Jessie
Lyn 7h2'15" 44d48'10"
Sandra and Kevin
Cyg 20h10'13" 56d29'32"
Sandra and Mark Steele
Psc 1h33'36" 22d28'8"
Sandra and Matt
Del 20h30'44" 9d55'43"
Sandra and Shane's Eternal Luminary
Vir 13h47'16" -6d12'3"
Sandra Angel
Ari 2h58'24" 30d56'51"
Sandra & Angel
Uma 10h48'19" 45d48'6"
Sandra Ann
And 1h7'30" 46d37'3"
Sandra Ann
Leo 9h56'23" 23d12'47"
Sandra Ann
Cap 20h46'40" -16d48'18"
Sandra Ann
Lib 15h11'52" -15d12'8"
Sandra Ann Buchholtz
Tau 4h21'49" 5d35'34"
Sandra Ann Howell
Leo 10h20'27" 11d21'52"
Sandra Ann Miller
Vir 11h58'0" -0d6'23"
Sandra Ann Poznecki
Cyg 21h55'44" 44d48'9"
Sandra Ann Shatzel
Uma 10h57'59" 33d24'4"
Sandra Ann Wright
Ori 5h27'19" -0d38'46"
Sandra Asay Huston
And 2h23'10" 41d27'37"
Sandra Atkins
Cas 0h27'43" 48d14'16"
Sandra B
And 2h0'21" 46d28'8"
Sandra B O'Leary
Psc 1h7'52" 6d45'1"
SANDRA BACHMANN
Umi 16h17'57" 73d36'16"
Sandra Ball
Uma 11h31'45" 47d49'24"
Sandra Barraclough
Cas 0h5'58" 54d49'44"
Sandra Barreiro Xuxu&3G Toujours
Ari 2h37'39" 13d56'33"
Sandra Beagan
Vir 14h18'26" 2d19'17"
Sandra Beebe 56
Gem 6h48'55" 20d43'20"
Sandra Bejjani
Cru 12h35'38" -58d48'13"
Sandra Belina
Vir 14h19'29" 0d43'58"
Sandra Bella
Aqr 22h48'50" -6d39'20"
Sandra Biange Baker
Lib 15h49'13" -11d30'37"
Sandra Birdie
Aqr 22h37'1" -10d44'55"
Sandra Blackwell
Cas 23h59'52" 60d36'56"
Sandra Boehmer
Per 3h38'40" 45d38'31"
Sandra Böhme
Uma 10h3'19" 56d35'22"
Sandra - Bright And Shining Forever
Cyg 21h44'24" 41d23'5"
sandra britschgi
And 1h40'37" 38d42'7"
Sandra Brogl
Gem 7h8'58" 16d12'29"
Sandra Brumbaugh's Sparkling Star
Lyn 6h45'33" 56d25'0"
Sandra Butcher
Cap 21h20'41" -27d16'51"
Sandra By Moonlight
And 1h28'49" 49d28'4"
Sandra Calatayud
Her 18h20'40" 15d32'29"
Sandra Camerire (Mom)
Cnc 8h21'41" 10d33'17"
Sandra Carina
And 0h46'21" 31d34'28"
Sandra Catherine Nix
Cap 21h32'40" -17d11'38"
Sandra Cathrine Jordan Dishman
Ori 5h17'12" 6d44'51"
Sandra Cecchetto
Aqr 22h58'7" -24d24'4"
Sandra Ciallella
Sgr 18h40'47" -28d39'48"
Sandra Clickner
Uma 8h42'17" 56d39'27"
Sandra Cowan
Uma 10h39'29" 46d7'47"
Sandra Czajka
Tri 2h11'22" 34d51'38"
Sandra D. Linthicum
Aql 20h10'36" 13d14'30"
Sandra D. Zimmerman "Little Bitty"
Umi 14h10'30" 68d22'55"
Sandra Dafoe
Mon 6h52'42" -0d45'51"
Sandra Darling
Cas 1h12'45" 64d6'8"
Sandra DeAnne Cruse
Lib 15h10'7" -12d9'7"
Sandra Dee
Ari 3h6'59" 25d20'53"
Sandra Dee Chess
Vir 13h7'41" 2d1'2"
Sandra Dee Kuulei Matarrese
Leo 9h24'27" 26d24'43"
Sandra Dee Miller
Cyg 20h4'46" 39d23'40"
Sandra Dee Sorrell
Lyn 7h37'30" 37d26'3"
Sandra Delaney
Uma 8h22'41" 64d53'54"
Sandra Deloach
Cnc 8h39'59" 19d20'7"
Sandra Denise Clements
Del 20h37'45" 13d6'25"
Sandra Dent
Gem 7h32'28" 15d3'26"
Sandra Dewing
Uma 8h38'8" 61d14'36"
Sandra Di
Cnc 8h26'28" 7d2'7"
Sandra Diane Smith
And 0h21'19" 28d2'8"
Sandra Diane Ziniewicz-Cureatz
Cas 23h27'38" 57d35'18"
Sandra Diblanc
Umi 16h42'48" 75d24'22"
Sandra Dickinson
And 0h53'39" 36d55'9"
Sandra -Die Miez- Böttcher
Ori 4h54'16" 11d11'23"
Sandra Dillard
Lyn 7h19'55" 44d18'12"
Sandra Dörig
Her 17h7'45" 18d21'52"
SANDRA DREAMS
Sco 16h19'34" -19d43'1"
Sandra Dudley
Tau 5h47'21" 22d29'8"
Sandra e Alessandro
And 23h4'22" 42d18'32"
Sandra & Edi Ott
Lac 22h48'38" 52d52'9"
Sandra Elaine Harrington 1960
Gem 7h47'20" 19d29'48"
Sandra Elaine Lenzen
Lyr 19h10'27" 26d56'38"
Sandra Elizabeth
Cnc 8h21'47" 15d36'42"
Sandra Elizabeth Duff
Gem 6h55'22" 25d41'14"
Sandra Elizarraraz
Leo 11h2'25" 17d7'27"
Sandra Escaler
Gem 6h30'49" 17d33'6"
Sandra Evelyn
Vir 13h25'0" -13d35'34"
Sandra F. KinCannon
Leo 10h5'42" 17d21'12"
Sandra F. Wietzel
Leo 11h15'1" 3d27'23"
Sandra&Fabian
Ori 5h23'22" 11d1'14"
Sandra Fae
Uma 11h8'41" 33d44'13"
Sandra Faith, Hope, Love
Sgr 19h10'25" -13d25'13"
Sandra Finlay
Cra 18h19'37" -37d1'55"
Sandra Fleischli
Cnv 13h55'58" 35d5'55"
Sandra + Florian
Uma 8h46'16" 59d32'28"
Sandra Forever
Cep 22h11'22" 68d1'25"
Sandra Frank
And 2h24'32" 42d35'2"
Sandra G. Nino de Perez Negron
Ori 5h44'42" 7d12'0"
Sandra Gail
Sco 16h48'35" -32d1'33"
Sandra Gail Spaulding
Cnv 12h44'20" 37d56'46"
Sandra & Gary's Silver Star
Cyg 21h17'42" 47d21'36"
Sandra Gaudet
Cas 0h21'54" 51d47'19"
Sandra & Geoffrey Hazell
Cru 12h35'34" -58d36'21"
Sandra Gilmore
Sex 9h49'24" -0d35'3"
Sandra Giron
Lyr 19h7'15" 27d4'45"
Sandra Glick
Cas 0h38'32" 57d41'20"
Sandra Godwin
Cnc 8h45'26" 14d53'47"
Sandra Gottschalk
And 2h30'44" 43d37'30"
Sandra Grace Salter Gilmore
Sco 16h9'53" -19d40'1"
Sandra Granitto
Sgr 19h32'25" -18d12'36"
Sandra Greenberg
Uma 12h6'40" 60d20'45"
Sandra Griffith
Cyg 21h23'51" 50d28'29"
Sandra Guarino
Vul 19h47'22" 28d5'29"
Sandra *HABI* Habegger
And 1h9'6" 47d29'40"
Sandra Hagen
Cas 0h52'37" 57d40'54"
Sandra Harter
Leo 11h41'23" 26d5'44"
Sandra Hazel Frank
And 0h30'29" 39d38'40"
Sandra High Farrow
Cyg 19h49'16" 31d2'6"
Sandra Hiraoka
Gem 6h42'8" 25d48'47"
Sandra Holland
Lyr 18h40'35" 31d34'44"
Sandra Hoover
And 2h22'51" 41d24'51"
Sandra Huaman
And 0h33'40" 41d57'31"
Sandra Huber
Aur 6h29'36" 34d28'4"
Sandra Huber
Boo 15h42'48" 46d14'19"
Sandra Huey
Cyg 20h45'8" 36d6'23"
Sandra Hutchison
Lmi 10h36'18" 34d37'54"
Sandra I Love You Forever
Cam 6h56'2" 69d30'45"
Sandra I Marko
Lib 15h49'33" -18d46'56"
Sandra Ilze Harsh
And 23h15'35" 48d24'44"
Sandra Isabel Morais Fonseca
Umi 14h46'32" 69d46'32"
Sandra J
Leo 10h36'16" 9d17'18"
Sandra J. Barreto
Lmi 10h30'22" 37d56'33"
Sandra J. Emerson
And 0h18'23" 26d12'7"
Sandra J. Haldeman "Queen Of Sales"
Cas 0h43'28" 58d3'36"
Sandra J. Kiessling
Crb 15h32'10" 26d32'38"
Sandra J. Ruark
Psc 0h12'58" 8d44'18"
Sandra Jane Pompei "1951-2005"
Tau 3h49'0" 26d41'20"
Sandra Jane Powell
Cas 0h55'45" 60d38'59"
Sandra Jean
Lyn 7h14'53" 53d13'14"
Sandra Jean
Psc 1h21'18" 16d13'49"
Sandra Jean
Psc 0h53'47" 18d26'43"
Sandra Jean Andree-Barbier
Dra 18h34'6" 54d26'29"
Sandra Jean Duvall
Mon 8h6'37" -0d40'43"
Sandra Jean (H.O.M.K.)
Leo 10h56'3" 7d58'29"
Sandra Jean Marshall
And 2h22'58" 48d58'43"
Sandra Jean Pecorino
Cas 1h19'49" 52d47'47"
Sandra Jean Wright Linter
Cap 21h51'9" -14d26'52"
Sandra Jeanne Warszawski
Aqr 21h0'36" -10d20'45"
Sandra Jo
Ori 5h34'57" 8d23'37"
Sandra Joann Wolters
And 0h12'24" 25d3'55"
Sandra Johnecheck
And 23h50'40" 39d51'44"
Sandra Johnson
And 0h32'46" 35d4'6"
Sandra Johnston
Cnc 8h12'48" 27d52'7"
Sandra Jones
Cam 3h31'5" 62d47'50"
Sandra Joseph Robicheau
Sgr 18h38'32" -27d21'51"
Sandra Joy Bren
Psc 0h52'48" 26d40'26"
Sandra K. Dee
Psc 23h55'38" 6d47'34"
Sandra K Evans
Cnc 8h42'33" 13d49'5"
Sandra K. Nelson
Ari 2h37'41" 20d10'42"
Sandra Karin
And 1h27'9" 33d51'33"
Sandra Kay
Sco 17h7'5" -45d4'54"
Sandra Kay Cowsert
Sgr 19h38'30" -13d8'53"
Sandra Kay Crowe
Gem 6h55'54" 28d49'8"
Sandra Kay Deetz
Cap 21h39'46" -16d6'45"
Sandra Kay Godwin Coley
Vir 12h53'47" 11d35'43"
Sandra Kay Goodweiler-Ripp
Sco 17h21'16" -31d4'32"
Sandra Kay Long
Del 20h42'59" 4d6'40"
Sandra Kay McCroskey
Ari 2h22'49" 25d59'44"
Sandra Kay Nelson
Leo 10h57'38" 17d7'48"
Sandra Kay Olson
Com 12h46'27" 26d24'27"
Sandra Kay Schubert
Cas 23h29'48" 57d29'57"
Sandra Kay Viera
Gem 6h34'11" 18d38'46"
Sandra Kaye
Psc 1h5'32" 14d54'12"
Sandra Kaye Fanrak
Cnc 9h20'32" 7d58'3"
Sandra Kaye Fritz
Psc 1h10'1" 10d21'8"
Sandra & Ken Taylor
Sgr 17h44'8" -26d52'1"
Sandra Kingsley
Psc 0h46'47" 6d38'12"
Sandra "Kizzy" Sandcroft
And 23h22'2" 51d2'34"
Sandra Kobs
Leo 11h4'32" 4d17'45"
Sandra Koch
Uma 11h1'21" 61d0'21"
Sandra Kotsopoulos
Uma 11h30'27" 56d13'22"
Sandra L Bishop My Shining Star GJB
Crb 15h38'57" 28d29'14"
Sandra L. Brunetti
And 0h34'44" 41d23'27"
Sandra L. Buttery
Cas 1h25'24" 62d41'40"
Sandra L. De La Vergne
Ori 5h32'54" 15d22'57"
Sandra L. Lonardo
Tau 4h29'53" 26d4'11"
Sandra L. Lynch
Sco 17h28'24" -37d39'7"
Sandra L. Mekita "Happy 21st Bday"
Cnc 8h34'9" 21d19'12"
Sandra L. Simmons
Sco 16h11'8" -19d17'21"
Sandra L. Vieira
Ori 5h58'8" 22d30'2"
Sandra Lea
And 0h5'11" 36d27'10"
Sandra Lea
Cas 0h12'37" 54d13'15"
Sandra Lee
And 23h5'37" 37d47'52"
Sandra Lee
And 23h34'14" 36d30'9"
Sandra Lee
Tau 3h42'48" 20d8'9"
Sandra Lee Buono
Cap 21h58'4" -19d33'25"
Sandra Lee Clark
And 0h39'33" 36d13'8"
Sandra Lee Iverson
Lyn 7h46'14" 36d43'41"
Sandra Lee Kesckes
Cnc 8h13'59" 7d27'45"
Sandra Lee Levy
Crb 15h32'0" 29d30'25"
Sandra Lee Martin
Mon 6h46'59" 6d25'41"
Sandra Lee Miller
Cyg 19h42'50" 33d14'39"
Sandra Lee Morris
Psc 0h18'45" 10d6'38"
Sandra Lee Ortiz
Lib 14h29'16" -12d35'58"
Sandra Lee Pumm
Uma 13h21'41" 55d27'35"
Sandra Lee Tannenbaum
Psc 1h39'22" 11d34'43"
Sandra Lee Thomas Lepore
Lyn 8h25'58" 42d19'34"
Sandra Leendert Mandy Elizabeth
Cas 1h38'36" 62d52'45"
Sandra Leigh Austin
Crb 15h38'10" 35d1'11"
Sandra Leigh Manzi
And 2h21'1" 48d45'49"
Sandra Leiko Bounds
Sco 16h18'47" -17d10'35"
Sandra Leonarda Fiech
Uma 10h0'25" 57d10'37"
Sandra Leonhardt.. Meine Liebe
Cap 21h32'51" -13d47'56"
Sandra LoBue
And 1h50'20" 45d10'31"
Sandra Longo
Umi 17h33'53" 81d51'50"
Sandra Lonsfoote
Leo 11h8'15" 10d37'20"
Sandra Lopez
Umi 13h6'18" 75d33'41"
Sandra Louise Wilkins
Cas 1h37'24" 62d14'39"
Sandra Lovkin
Psc 23h40'4" 5d42'21"
Sandra Lütolf
Uma 13h29'8" 54d38'5"
Sandra Lynn
Ori 5h53'37" 17d46'48"
Sandra Lynn
And 2h18'59" 45d8'37"
Sandra Lynn Cottle
Lyn 7h7'45" 59d30'58"
Sandra Lynn Cox
Dra 17h15'26" 65d26'3"
Sandra Lynn Ternyik
Psc 1h18'38" 6d48'49"
Sandra Lynne's Shining Star
Cas 23h37'43" 55d18'51"
Sandra M
Sgr 19h32'37" -16d15'37"
Sandra M. Cucinotta
Sco 16h43'58" -28d18'48"
Sandra M. Glazier
Cnc 8h47'8" 32d14'11"

Sandra M. Martinez
Cas 0h13'57" 60d45'30"
Sandra M. Morgan
Cas 0h46'49" 53d34'48"
Sandra M. Russell 1985 - 2005
Cas 1h8'32" 66d34'39"
Sandra MacQuillan
Cas 0h6'33" 59d41'44"
Sandra Mae
Vir 12h5'42" -9d14'49"
Sandra Maggiore
Cyg 19h44'1" 28d20'18"
Sandra Maginn
Cas 0h41'44" 64d45'5"
Sandra - Major
Uma 12h45'3" 59d47'17"
Sandra Mannino
Gem 6h56'1" 12d41'28"
Sandra Manolescu
And 1h53'1" 38d0'51"
Sandra Margarita Mayorquin
Cas 1h0'36" 60d32'44"
Sandra Marie
Ori 5h22'31" 4d24'28"
Sandra Marie Allred
Pho 1h8'26" -43d17'23"
Sandra Marie Garcia
Cas 0h17'0" 47d24'57"
Sandra Marie Leaney West
Sgr 18h10'47" -30d4'8"
Sandra Marie Mizer-Burk
Cas 1h28'42" 56d46'38"
Sandra Marie Perucci 9/18/1961
Vir 14h7'43" -15d8'40"
Sandra Marie Sygo
Cas 1h28'12" 64d51'12"
Sandra & Marshall Brown
Ori 6h14'2" 9d7'56"
Sandra Mei
Cas 1h23'39" 60d58'39"
Sandra Mi Amor
Cam 6h46'48" 62d49'49"
Sandra "Mi Gordis..." Guevara
Gem 7h19'36" 18d18'12"
Sandra Michaud
Crb 15h37'57" 28d30'1"
Sandra Milena Flowers
Cap 21h41'58" -15d56'13"
Sandra Minor
Cas 1h48'53" 63d43'29"
Sandra Monika Gust
Aqr 22h10'0" 1d24'35"
Sandra Morales González
Aqr 22h44'1" -9d50'42"
Sandra - Mother of Raya
Vir 13h27'10" 12d5'8"
Sandra Müller
Lib 14h42'56" -8d36'10"
Sandra My Blue Star
Crb 15h47'50" 37d15'13"
Sandra MY Star
Cen 13h5'21" -47d10'35"
Sandra (Nana) McCarthy
Aqr 21h43'23" 1d13'19"
Sandra Nicole McGourty
Sco 16h37'28" -38d9'6"
Sandra Nunes
Gem 7h15'20" 32d8'45"
Sandra O'Toole Halibrand
Lyn 6h22'36" 60d43'39"
Sandra P. Hill
Gem 7h32'8" 31d13'46"
Sandra & Pascal
Per 3h19'51" 42d54'3"
Sandra Pender
And 2h11'9" 43d5'3"
Sandra Perez
Vir 14h19'12" -20d44'27"
Sandra Peters
Ori 6h14'16" 7d0'22"
Sandra Playford
Cas 0h32'41" 51d26'35"
Sandra Prado Dundon
Tau 5h39'6" 22d40'30"
Sandra Racine Orenstein
Uma 11h43'58" 43d46'4"
Sandra Ramirez
Cyg 20h59'34" 31d59'44"
Sandra Ranberg LeBarron Florio
Lmi 10h24'37" 29d0'24"
Sandra Raughton
Uma 11h26'4" 43d44'56"
Sandra Raymond
Cas 1h49'50" 60d26'48"
Sandra Reaidy
Leo 9h32'46" 13d36'57"
Sandra Reilly Hammond Kerby
Uma 13h0'15" 62d50'27"
Sandra Renae
Aqr 22h41'39" -17d25'11"
Sandra Renee Barnes
Aqr 22h53'23" -20d40'5"
Sandra Rigatuso
Ari 2h26'7" 25d0'58"
Sandra Rivera
Lyr 18h44'56" 33d45'45"
Sandra Roddy
Sgr 18h54'33" -34d19'55"
Sandra Rose
Sco 16h4'7" -12d45'3"
Sandra Rose
Cas 0h17'41" 58d59'14"
Sandra Rose Michael
Vir 12h35'23" 2d5'27"
Sandra Rose Walper
Cam 5h29'57" 69d12'0"
Sandra Roy
And 23h47'46" 34d21'40"
Sandra Russo
Lmi 10h10'55" 31d36'12"
Sandra Ruth Larsen
And 0h32'6" 31d4'18"
Sandra Sacs
Sco 15h52'6" -20d47'22"
Sandra Salomone
Dra 16h41'16" 66d46'34"
Sandra Sanders
Uma 13h28'12" 59d56'20"
Sandra " Sandy " Leinen
Tau 3h33'13" 25d5'10"
Sandra "Sansanz" Silvino
Cnc 9h11'48" 26d46'16"
Sandra Schultz
And 23h35'20" 48d39'48"
Sandra Skrzypiec
Gem 6h57'9" 16d59'31"
Sandra Slightom
Lmi 10h5'30" 31d49'14"
Sandra Soto
Sco 17h53'42" -35d50'47"
Sandra Spillane
Vir 12h45'7" 6d20'51"
Sandra Stanisic
Cas 23h34'0" 52d23'10"
Sandra Sualocin
And 1h37'57" 41d51'41"
Sandra Sue
Lmi 10h42'2" 26d43'47"
Sandra Sue Bain
Com 12h52'14" 27d57'0"
Sandra Sue Stanich
Aqr 21h31'40" 2d28'35"
Sandra Sue Weidner
Gem 7h27'27" 35d9'48"
Sandra Sutherland
Cas 1h19'20" 54d52'37"
Sandra Swan: THE Irish-Celtic Witch
Cyg 21h31'33" 50d13'2"
Sandra Tacke
Cap 21h32'2" -17d8'43"
Sandra Tanner Parker
Lyn 7h2'8" 48d6'47"
Sandra Tarnarider
And 1h18'16" 46d12'3"
Sandra Taylor
Lyn 7h52'12" 59d11'0"
Sandra Theresa Ryan
Pyx 8h39'37" -23d22'33"
Sandra Thomann
Cep 22h24'48" 67d8'49"
Sandra Thomason
Crb 15h58'45" 34d11'52"
Sandra Thompson Haviland
Cas 1h36'5" 64d25'59"
Sandra Tomlinson, My Mother
Tau 3h38'20" 27d55'39"
Sandra Townsend
Cnc 8h8'48" 13d47'50"
Sandra Trixie Mazur
And 0h42'22" 40d12'58"
Sandra Tucker
Cam 7h11'3" 72d12'54"
Sandra und Detlef
Uma 9h5'56" 60d50'21"
Sandra und Robert
Uma 10h53'10" 70d6'24"
Sandra und Stefan Stern
Uma 8h42'23" 60d51'51"
Sandra und Urs Borer
Cas 0h10'22" 59d33'42"
Sandra & Valters 25th Anniversary
Cyg 20h32'12" 45d47'3"
Sandra Van Fossen
Lib 14h52'19" -12d26'45"
Sandra Vernace
Tau 4h38'46" 9d27'6"
Sandra Verne Olson
Cyg 20h50'5" 35d24'43"
Sandra Vicioso
Cas 1h25'47" 66d30'17"
Sandra Vilson
Del 20h49'38" 6d11'34"
Sandra Vizcarra
Cnc 9h6'22" 28d30'57"
Sandra von Allmen
Cas 1h21'32" 69d2'33"
Sandra Von Doran
Cas 0h27'48" 60d25'7"
Sandra Wasserman
Leo 10h50'15" 19d1'20"
Sandra Wilson
Uma 9h5'32" 50d47'19"
Sandra Y Elias Rubio Amor Eterno
Leo 11h24'49" 5d47'29"
Sandra Youngman
Lib 15h13'21" -14d12'46"
Sandra Yuvonne Alcock Walters
Vir 13h21'40" 13d15'36"
Sandra's Glücksstern
Ori 6h15'50" 10d55'37"
SandraAngelica
Cnc 8h49'58" 32d2'21"
SandraHet
Ori 6h19'12" 10d11'18"
Sandra.K3
Cep 21h42'44" 55d38'43"
SandraKayTallerico
Uma 11h11'0" 41d11'45"
Sandrala
Vir 14h13'33" -13d4'23"
SandraLee Mother of 3-Loved by Me
Aqr 22h21'7" -0d10'27"
SANDRAMEDA
Ori 5h6'51" 7d48'5"
Sandras
And 0h58'50" 40d13'30"
Sandras & Bernds Hochzeitsstern
Uma 10h51'10" 44d53'30"
Sandras Glücksstern, 24.09.1982
And 2h7'47" 42d12'6"
Sandra's Glückssternli
Cam 7h35'7" 82d1'5"
Sandra's Special Star x x x x x
And 1h4'28" 44d58'27"
Sandra's Star
Psc 23h11'17" -2d23'19"
Sandratra Ravelojaona
Psc 0h23'58" 0d23'25"
Sandrea
Sgr 18h56'19" -17d44'27"
Sandridge Beach Hotel
Vul 20h47'44" 28d19'36"
Sandrina
Dra 17h54'53" 64d45'43"
Sandrina Aschmann
Boo 15h35'30" 46d15'20"
Sandrine
Lyr 18h16'45" 37d16'24"
SANDRINE
Vir 12h23'16" 12d23'14"
Sandrine
Ari 3h1'37" 18d5'21"
Sandrine
Lac 22h52'49" 54d2'49"
Sandrine
Cas 23h41'54" 56d6'38"
Sandrine
Cen 13h41'15" -37d16'37"
Sandrine Baudinot
Cas 0h40'19" 61d47'37"
Sandrine "Bigou d'Amour" Brisson
Umi 16h51'53" 78d55'59"
Sandrine Céleste
Her 18h50'41" 22d20'0"
Sandrine Courtois
Leo 11h36'20" 1d19'0"
Sandrine Meyer
Sco 17h14'7" -31d54'55"
Sandrine Moppé
Ari 2h0'6" 18d38'31"
Sandrine Pages "Mon petit ice love"
Cas 0h58'24" 56d42'51"
Sandrine Seger
Vir 13h22'41" -20d32'9"
Sandrine Souris
Uma 10h19'19" 43d27'26"
Sandrine Vanden Eynden
Aqr 22h49'59" -15d53'50"
Sandrinha60
Aqr 22h5'21" -2d13'2"
Sandrini, Riccardo
Vir 12h53'8" 2d9'20"
Sandro
Crb 15h57'15" 29d29'50"
Sandro
Lac 22h19'28" 47d41'9"
Sandro Alexandre
Her 18h36'15" 21d30'18"
Sandro & Alice
Crb 15h58'17" 27d16'47"
Sandro Caludio Koch
Umi 17h6'25" 75d10'16"
Sandro et Genevieve
Umi 15h35'49" 79d13'42"
SANDRO FORSTER
Umi 15h49'1" 79d41'12"
Sandro Giraudo
Lyn 8h23'11" 38d33'51"
Sandro Hirschi
And 1h16'22" 42d18'26"
Sandro M.
Her 18h33'24" 19d18'3"
Sandro Manuel Schiess
Her 16h13'22" 6d22'14"
Sandro Personeni
Dra 19h38'31" 65d9'46"
Sandro Toma
And 1h19'30" 48d51'11"
Sandro Wolf
Boo 13h56'46" 22d45'10"
Sandwich
Lyn 8h14'48" 54d7'1"
Sandy
Cas 1h17'18" 62d39'15"
Sandy
Cas 1h38'5" 60d54'32"
Sandy
Uma 9h26'14" 68d13'53"
Sandy
Umi 14h31'29" 72d45'22"
SANDY
Uma 12h47'33" 60d22'4"
Sandy
Uma 10h9'22" 62d54'32"
Sandy
Psc 1h7'27" 2d46'15"
Sandy
Crb 15h51'29" 26d3'23"
Sandy
Crb 15h42'54" 27d41'27"
Sandy
Gem 7h5'49" 22d23'55"
Sandy
Her 16h17'12" 7d28'25"
Sandy
Ori 5h43'26" 2d18'22"
Sandy
Del 20h34'29" 14d23'26"
SANDY
Tau 4h58'38" 20d37'18"
Sandy
And 23h24'29" 52d18'18"
Sandy
Uma 9h13'13" 49d25'11"
Sandy
Lyn 8h22'46" 38d4'3"
Sandy
Crb 15h46'46" 32d33'21"
Sandy
Gem 7h5'47" 33d35'48"
Sandy 2 22 42
Cas 1h40'57" 65d33'59"
Sandy 6-11-63
Cas 1h50'20" 67d45'24"
Sandy 72
Uma 12h49'38" 55d53'32"
Sandy A Sander
Cnc 8h45'23" 7d23'50"
Sandy Adams My True Love
Cam 4h44'25" 67d19'40"
Sandy Alexandra
Lib 15h21'23" -9d41'53"
Sandy Alexandria Head
Lmi 10h22'58" 32d50'46"
Sandy Allen: L Y L N O A
Tau 4h34'40" 15d12'33"
Sandy and Brian Smith
Col 5h44'16" -32d10'40"
Sandy and Colin's Silver Star
Cyg 21h54'38" 50d22'54"
Sandy and her 1/2 Carl 1107
Sco 16h56'51" -45d3'6"
Sandy and Kim Forever
Cyg 19h43'37" 38d21'10"
Sandy and Scott Katzman
Cyg 20h7'12" 54d57'37"
Sandy Anne Vancil
And 0h47'45" 40d57'31"
Sandy Antunez
Lyn 8h46'6" 33d48'11"
Sandy B
And 1h17'44" 47d8'39"
Sandy Banner
Leo 11h36'34" 13d20'42"
Sandy Beach Island Resort
Lmi 9h43'56" 38d36'19"
Sandy Bennett
Cyg 20h52'46" 35d14'52"
Sandy Boules
Leo 9h42'52" 29d43'49"
Sandy Brandmeier's Excel Together
Her 17h43'34" 30d58'50"
Sandy Browdie
And 2h33'2" 50d4'50"
Sandy Brown
Vir 14h16'56" 3d35'38"
Sandy (Bunny) Jacke
Ari 2h0'7" 13d3'20"
Sandy Burns
Uma 11h27'17" 70d50'40"
Sandy Campagnolo
Lyn 7h58'34" 48d59'14"
Sandy Cantu
Vir 12h37'8" 0d42'59"
Sandy Chappell
Sgr 18h13'3" -19d25'50"
Sandy Chen
Cyg 20h45'34" 44d59'20"
Sandy & Chung
Ori 6h10'58" 15d13'27"
Sandy Collier
Dra 18h36'56" 56d59'5"
Sandy Craig
Gem 7h45'3" 33d2'36"
Sandy Czerwinski Daughter Friend
And 1h45'0" 50d24'34"
Sandy Czerwinski Mom Mimi Friend
Cas 1h36'52" 61d51'45"
Sandy Darling
Tau 4h19'16" 22d31'40"
Sandy Dawn
Aqr 22h11'27" -13d11'4"
Sandy DeAndrea's Wishing Place
Tau 5h30'22" 25d2'9"
Sandy DeCarlo
Lib 14h54'24" -18d39'18"
Sandy Dimos
Cas 1h40'28" 61d1'48"
Sandy Element
Aql 20h19'40" 0d33'15"
Sandy & Ervil & Gram & Gramps
Uma 9h54'58" 70d2'31"
Sandy Faltaous
Ari 3h8'44" 29d23'6"
Sandy Ferguson
And 0h35'0" 38d25'51"
Sandy Ferrara
And 0h25'3" 25d40'58"
Sandy Five-O
Cas 1h17'29" 65d45'48"
sandy forever
Umi 17h2'19" 78d58'24"
Sandy Fountain
Aqr 23h9'21" -9d56'54"
Sandy Gainforth Barr
Umi 15h32'4" 68d43'39"
Sandy Garcia
Crb 15h37'15" 36d2'4"
Sandy Goodstein
Ari 2h53'18" 29d0'59"
Sandy Grewal
Uma 9h35'27" 71d34'16"
Sandy Griffitts
Dra 18h49'25" 56d4'4"
Sandy Guenther
Cyg 20h19'37" 55d12'2"
Sandy Hambel
And 23h20'52" 44d49'1"
Sandy Hapoienu
Cyg 19h46'3" 31d33'52"
Sandy Hoover McHone
Sgr 19h0'16" -18d45'51"
Sandy Jae Cook
Cnc 8h16'58" 20d59'4"
Sandy Jean Williams
Crb 15h37'0" 31d0'18"
Sandy Johnson
Vir 14h59'36" 4d10'31"
Sandy Joy Ohebshalom
Cnc 8h34'21" 7d54'8"
Sandy K Princess of Darkness
Leo 10h12'56" 21d45'27"
Sandy Kalanquin
Aqr 22h38'13" 1d50'45"
Sandy Katzman
Sgr 19h55'29" -12d28'45"
Sandy Kleinstein
Srp 18h5'32" -0d9'11"
Sandy Klinski
Vir 13h10'30" -5d0'7"
Sandy Kobaly
Ori 6h17'46" 13d17'25"
Sandy Komiliades
Mon 6h46'58" -0d2'32"
Sandy Krystyna Long
Cam 7h46'47" 69d49'4"
Sandy L Martini
Cap 20h36'32" -18d0'43"
Sandy Lam
Lyn 7h39'36" 44d41'52"
Sandy Lane
Uma 14h26'34" 56d9'28"
Sandy Lane Hotel
Lyn 7h34'14" 48d2'27"
Sandy Leckie
Cas 23h50'1" 57d35'0"
Sandy Lee
Ori 5h15'6" -8d54'20"
Sandy Lee
Gem 7h41'38" 18d55'35"
Sandy Lee Froeming
Lib 15h41'14" -9d27'21"
Sandy Leigh Renfrow
And 1h31'30" 42d52'29"
Sandy Lerner
Cnc 8h49'10" 32d29'38"
Sandy Liberg
Cas 23h39'2" 57d55'53"
Sandy Lou
Aqr 21h6'28" -9d19'22"
Sandy Lucille Rodeghero
Sco 17h52'13" -36d6'51"
Sandy Marchione
Leo 10h28'20" 17d53'36"
Sandy Mariani
Boo 13h49'54" 24d54'49"
Sandy Marie
Cas 23h45'16" 61d48'48"
Sandy McCoy
Leo 10h7'0" 24d7'58"
Sandy McKinnon
Eri 4h23'43" -29d37'16"
" Sandy Mello Our Shining Star"
Umi 15h34'48" 74d53'51"
Sandy Melvin
Cnc 9h17'23" 9d56'31"
Sandy(Mom, Memaw, Babe) Aaron
Crb 15h48'55" 26d36'5"
Sandy Moon
Ari 3h10'4" 19d47'48"
Sandy Nassau
And 2h30'2" 37d24'57"
Sandy Ottey
Gem 7h27'30" 28d35'40"
Sandy Palmer
Cam 4h41'17" 58d59'47"
Sandy Pants
Tau 5h10'29" 25d24'31"
Sandy Peng
Gem 7h32'44" 25d51'59"
Sandy Place
Sco 16h12'17" -16d52'12"
Sandy Pooh
And 23h23'14" 50d6'43"
Sandy Prime
Aqr 22h24'3" -3d57'16"
Sandy R. farah
Uma 9h13'33" 55d21'11"
Sandy' Red
Ari 2h35'27" 13d0'34"
Sandy Renae
Cas 0h3'8" 55d47'12"
Sandy Reynolds
Psc 22h52'59" 6d32'44"
Sandy Robison
And 0h49'32" 42d50'57"
Sandy Rogers
Vir 13h8'28" 7d13'42"
Sandy & Rosie Sorce
Cyg 19h36'36" 32d16'22"
Sandy Ross
And 22h57'34" 51d43'26"
Sandy Rottari
And 0h21'53" 38d16'35"
Sandy Royle
Cyg 19h57'47" 43d24'13"
Sandy S Cheung
Cnc 8h39'44" 31d26'59"
Sandy Sabha
Uma 14h23'49" 55d43'34"
Sandy "Sammy" Byers
Ori 6h11'29" 15d9'58"
Sandy Sanders 1192nd
Cam 3h58'55" 70d38'18"
Sandy Sandhar
Leo 11h13'2" 13d1'38"
Sandy Seaton & John Berklich
Cyg 20h51'7" 34d47'27"
Sandy Shiner
Psc 0h19'41" 15d6'23"
Sandy Sidorsky
Umi 15h2'35" 71d42'4"
Sandy Silver
Tri 2h9'20" 34d53'10"
Sandy Solomon
Cyg 19h58'2" 41d59'22"
Sandy Sorce
Cyg 19h53'42" 31d11'33"
Sandy Sorter
Uma 12h28'35" 55d11'17"
Sandy Spano
Tau 5h26'0" 26d55'45"
Sandy & Steven
Cyg 21h47'16" 42d21'55"
Sandy & Stu's Retarded Star
Cyg 21h39'13" 39d16'19"
Sandy Sundling, beloved
Uma 13h18'13" 57d22'43"
Sandy Tate
Tau 4h5'42" 9d25'46"
Sandy - The Most Beautiful Star
Ari 2h7'49" 23d24'42"
Sandy The Tucson Star
Cnc 9h21'27" 10d24'55"
Sandy - Thy Light Bestows Love
Vir 13h6'32" -20d17'36"
Sandy Trieu Gacioch
Cyg 19h42'10" 33d44'41"
Sandy Twinkle Brite
Aqr 22h52'49" -6d47'40"
SANDY WARRIOR STAR
Sgr 18h33'23" -28d2'39"
Sandy Webbercollins
Cap 21h43'12" -10d29'26"
Sandy Willey
Aqr 22h44'51" -16d7'49"
Sandy Wistey
Lib 15h9'17" -23d15'37"
Sandy Woosnam
Sco 16h5'1" -11d25'4"
Sandy Wright
And 0h15'14" 33d23'17"
Sandy Xu
And 22h59'28" 51d36'26"
sandy yahmanib
Psc 23h11'32" 1d26'24"
Sandy Yost
Peg 23h18'3" 22d15'0"
Sandy Youngfert
Uma 10h15'45" 55d46'54"
Sandy011754
Dra 16h38'40" 57d15'6"
Sandydean
Lyr 19h10'14" 38d17'51"
Sandye "Dermie" Richards
And 2h13'46" 42d57'59"
SandyluvsJas
Ari 2h36'8" 22d13'2"
Sandy's 60 Carat Diamond
Lib 15h45'10" -24d13'32"
Sandy's Beta
Uma 10h19'35" 44d37'56"
Sandy's Guiding Light
Ari 2h56'13" 25d32'34"
Sandy's Light
Uma 11h32'25" 48d33'28"
Sandys Lucky
Cam 6h21'47" 65d40'47"
Sandy's Shining Star
Uma 12h4'38" 61d54'48"
Sandy's Smile
Crb 15h46'24" 36d12'43"
Sandy's Star
Crb 15h39'32" 33d18'5"
Sandy's Star
Uma 10h21'32" 40d2'2"
Sandy's star
Sgr 18h14'24" -18d4'32"
Sandy's Star
Car 9h44'30" -60d21'33"
Sandy's Valentine Louise
Cyg 19h57'41" 51d53'12"
sandystar
Vir 13h25'19" 3d20'7"
Sané
Tau 4h21'31" 6d25'22"
Sanée Scavo
Gem 7h26'18" 27d43'12"
Sanek
Crb 16h19'6" 37d57'19"
Sanela M.
Sco 16h10'16" -29d38'35"
Sanele Sobantwana
Aqr 21h57'29" -7d56'25"
SANEL-FRKI
Cnc 8h14'5" 15d32'14"
Sanford Rising Penn Jr.
Uma 11h43'37" 58d24'16"
Sanford Steven Hermann
Her 18h22'10" 17d20'30"
Sang
Gem 6h26'30" 24d40'0"
Sang Sengkeomysay Yeh
Tau 4h5'35" 16d57'21"
sangeeta
Sco 16h18'42" -13d18'7"
Sänger, Noah
Uma 10h58'19" 54d30'57"
Sanguine Vista
Cyg 19h57'13" 36d36'9"
Sanhita and Rohit
Del 20h34'30" 13d9'6"
Sanhita Sheth
Del 20h35'59" 6d46'9"
Sani & Benny
Uma 14h23'57" 58d57'0"
Sanita Amor
Uma 11h13'57" 41d18'46"
Sanja
Leo 10h26'7" 24d10'36"
Sanja Dragicevic
Peg 23h51'16" 29d40'30"
Sanja Klebansky
Cam 5h54'48" 60d22'46"
Sanja Marinkovic
Cyg 20h43'1" 46d3'2"
Sanja Maros
Cap 20h26'49" -26d49'44"
Sanja Papic
Cas 1h16'48" 53d21'41"
Sanja Petrovich
And 1h0'46" 41d20'54"
Sanja, angel of Lyon
Cas 1h26'38" 60d37'43"
Sanjana
Lib 15h29'27" -7d27'7"
Sanjay Jain
Sco 17h42'45" -32d10'51"
Sanji and Nami
And 23h41'33" 47d40'33"
Sanjiv K. Patankar
Aur 6h1'26" 31d34'39"
Sanjiweni Shetal - Shining Star
Lyn 6h46'40" 51d10'42"
SanJuana the Queen
Gem 6h12'19" 22d18'9"
Sanna Gaffney
Mon 6h59'33" -0d54'8"
SANNA - Susanna Faye
Ori 6h0'58" 9d59'19"
Sanne Jasmin
And 23h9'51" 37d7'20"
Sanne - Stern meines Herzens
Uma 9h15'5" 61d32'12"
SanochkaArens
Aqr 22h31'29" 1d34'35"
Sano's star
Cma 6h55'49" -13d24'31"
Sans Cire Tania
Her 17h33'12" 17d15'17"
sansanee
Ori 5h54'31" -0d41'16"
Sanse
Uma 10h25'40" 63d42'55"
Sansonetti the Juggler
Sco 15h54'15" -20d23'26"
Santa
Ari 2h20'40" 25d14'3"
Santa
Cyg 21h36'26" 49d12'26"
Santa Badami
Lyn 7h51'12" 50d36'30"
Santa Barbara Limole
Tau 4h21'58" 1d50'21"

Santa Christiano
Uma 9h27'35" 67d44'29"
Santa Dave Simon My Friend My Love
Cap 20h31'45" -14d34'28"
Santa Delicata
Lyr 19h17'54" 29d33'52"
Santa Irene Aseveldo
Cap 20h27'19" -18d9'57"
Santa M. Villarreal
Ori 5h48'16" 2d0'38"
Santa Monica
Cyg 21h23'47" 33d20'29"
Santa Terribile
Umi 8h50'34" 86d33'10"
Santa - "The light of our life."
Aqr 22h59'29" -11d6'36"
Santa Valenzano
Aur 4h52'2" 33d39'47"
Santal Doreen Engli
Sco 17h52'23" -40d27'44"
Santana
Uma 11h27'58" 44d7'2"
Santana
Per 2h53'33" 53d17'48"
Santana
Psc 0h37'21" 12d17'52"
Santana Herda-Camacho
Sco 17h44'46" -32d38'18"
Santangelo
Cnc 8h47'1" 32d20'33"
Santaure
Sgr 17h51'49" -29d52'44"
Sante Romaldini
Lib 15h24'51" -5d36'26"
Santford Craig
Per 3h54'25" 37d12'9"
Santhony Johnson Sr.
Ori 5h29'42" -5d23'31"
Santiago Acosta
Gem 7h13'1" 17d53'0"
Santiago Francisco Ortiz
Vir 14h19'40" 3d6'55"
Santiago Lopez Cisneros
Aqr 21h51'32" 1d17'7"
Santiago Ramos Cooper
Gem 6h21'30" 21d9'18"
Santiago Sandoval
Cyg 20h1'18" 43d52'38"
Santimyer Star Forever
Cap 21h11'25" -22d17'44"
Santina Caruso
Leo 11h28'17" 9d35'44"
Santina Maria
Tau 4h30'40" 8d30'2"
Santina "Mein Liebchen"
Gem 6h33'30" 20d11'44"
Santina Teresa
Leo 9h39'40" 14d14'38"
Santine D.L.C
Peg 21h59'56" 22d40'8"
Santino
Cnc 9h5'28" 23d8'2"
Santino
Ari 3h11'1" 26d30'32"
Santino and Sarah
Sge 19h11'23" 19d47'43"
Santino Gemellaro
Cam 4h6'44" 68d4'29"
Santino Raffaele
Uma 13h47'27" 49d53'10"
Santino Star
Boo 15h32'28" 45d22'21"
Santle L Perrotto
Umi 16h20'17" 71d25'27"
Santo D.L.C
Peg 23h59'47" 21d17'40"
Santo Mercuri's Guiding Star - smsd8
Uma 11h27'50" 60d46'2"
Santo "Sandy" Galasso
Uma 11h49'15" 62d10'14"
Santoro Bianca
Ori 6h3'56" -2d9'16"
Santoro Kirkell
Dra 18h50'54" 66d44'21"
Santos Medrano
Crb 15h20'52" 27d17'50"
Santos Silva Gomez
Uma 12h33'17" 55d8'51"
Santosh K. Boddikuri
Ari 1h47'33" 23d40'20"
Santosha
Uma 9h41'57" 67d8'29"
SanVed
And 0h41'7" 31d16'33"
Sany del Carmen Nunes-Ramirez
Uma 9h41'53" 53d56'57"
Sanya Hooper
Vir 13h40'42" -4d5'6"
Sanya's Hope
Uma 10h57'24" 47d55'16"
Saofai Vaili
Uma 11h18'46" 59d36'46"
Saoirse and David
Cas 23h54'42" 56d35'46"
Saoirse Hannah Hogan
And 22h58'6" 41d40'41"
Saoirse Kearney
And 2h17'35" 46d53'48"
Saoirse - Star of Freedom
Aql 19h51'40" -0d30'51"
Saosin
Ori 5h54'30" 21d20'44"
Sapajoy
Gem 6h46'11" 32d43'21"
Sapana Patel
Cyg 19h47'8" 54d20'41"
Sapiens Aquila Dr. Stephen J Harvey
Aql 20h0'20" 14d45'51"
Sapienza 052147
Tri 2h30'36" 30d32'3"
Sapna
Srp 18h39'9" -0d46'8"
Sapna Janveja Mann
Cam 4h42'38" 76d19'33"
Sapora Fega My Lovely
Tau 3h55'47" 26d51'9"
Sapphira
Uma 11h29'36" 39d26'28"
Sapphire
And 1h1'4" 44d25'33"
Sapphire
Del 20h24'0" 9d43'35"
Sapphire
Cas 23h37'26" 57d31'8"
Sapphire Alivia Blue Bailey
Lib 14h52'20" -7d17'34"
Sapphire Dragon's Eye
Uma 13h0'6" 59d38'44"
Sapphire Moon
Crb 15h39'55" 36d21'7"
Sapphire Ocean Amanda
Vir 12h32'4" 0d55'37"
Sapphire86
Ari 3h11'15" 19d36'0"
Sappho
Uma 10h26'2" 65d3'4"
Sar Bear
Vir 12h49'56" 0d15'2"
SAR Berns
Ori 5h27'16" 6d7'54"
Sar Max
Ori 5h52'57" 18d44'28"
Sar Sar
Lyn 7h8'31" 58d25'29"
Sara
Lyn 7h30'1" 53d31'48"
Sara
Cas 0h33'47" 60d1'40"
Sara
Cas 23h3'20" 54d51'15"
Sara
Cas 23h32'33" 55d36'33"
Sara
Cas 0h38'3" 60d16'27"
Sara
Cas 0h31'14" 62d26'6"
Sara
Cam 7h59'6" 69d17'52"
Sara
Cam 5h34'37" 71d25'37"
Sara
Aqr 22h27'31" -1d25'21"
Sara
Aqr 22h9'16" -1d21'41"
Sara
Umi 15h59'40" 79d36'47"
Sara
Ori 5h28'39" -1d23'14"
Sara
Sgr 19h37'22" -17d53'24"
Sara
Cap 20h8'33" -13d21'9"
SARA
Psc 0h44'31" 2d46'58"
Sara
Sco 16h8'12" -23d38'4"
Sara
Ori 6h4'12" -2d11'32"
Sara
Cas 0h52'42" 56d39'59"
Sara
Cas 0h31'33" 62d25'7"
Sara
Cnc 8h37'28" 19d37'3"
Sara
Her 17h36'54" 21d16'10"
Sara
Tau 3h49'37" 29d32'35"
Sara
Tau 3h48'19" 24d3'42"
Sara
Ari 3h16'9" 28d39'46"
Sara
Cyg 19h33'25" 27d50'22"
Sara
Cyg 19h36'47" 27d58'23"
Sara
And 0h24'56" 32d56'23"
Sara
Ori 4h49'20" 5d31'31"
Sara
Vir 12h24'38" 6d0'41"
Sara
Mon 6h32'44" 7d4'34"
Sara
Aqr 21h27'28" 0d41'9"
Sara
Psc 0h27'45" 16d53'22"
Sara
Vir 12h10'12" 10d5'5"
Sara
Her 16h44'16" 9d8'26"
Sara
Cyg 21h21'44" 34d42'16"
Sara
And 2h6'49" 41d54'33"
Sara
Per 3h24'4" 41d31'32"
Sara
Per 3h28'9" 44d11'51"
Sara
Lyn 7h31'23" 37d46'47"
Sara
And 1h17'10" 46d34'16"
Sara
And 1h57'56" 45d11'10"
Sara
Lyr 18h44'28" 39d32'29"
Sara
Lac 22h56'46" 46d42'12"
Sara
And 23h34'24" 47d34'39"
Sara 23091970
And 0h6'38" 35d41'41"
Sara A. Dietz
Ari 3h24'20" 22d28'9"
Sara Agner's Star
Ari 2h35'36" 17d44'55"
Sara Ahmed
Cas 0h49'36" 50d2'2"
Sara Aldaco
Cap 20h10'58" -13d45'20"
Sara Alexandra Pisak
Cas 0h26'19" 62d35'1"
Sara Alexandria
Cnc 9h2'14" 24d41'50"
Sara & Ali
Cyg 19h51'21" 37d58'13"
Sara Allison
Gem 7h17'48" 33d22'14"
Sara Al-Mutawa
Leo 9h54'2" 15d43'54"
Sara Alta
Sgr 18h32'24" -27d8'8"
Sara Amanda Godwin
Peg 0h13'8" 17d12'35"
Sara Amber
Lyn 7h33'43" 57d10'15"
Sara An Walter
And 23h47'30" 37d34'2"
Sara and Chris's Bright Future
Cnc 8h2'25" 17d10'24"
Sara and Dilly forever
Vir 12h11'39" 12d43'35"
Sara and Jason Eischeid
Cyg 20h49'31" 45d1'21"
Sara and Jason's Love Star
Tau 5h12'58" 22d52'8"
Sara and Jessica
Sco 16h16'2" -14d59'49"
Sara and Lincoln's star
Gem 7h6'29" 22d15'59"
Sara and Ryan forever
Aqr 22h36'9" 0d46'54"
Sara and Stephen
Uma 8h54'27" 49d39'9"
Sara Andon
Uma 9h43'44" 46d42'44"
Sara Angel
And 1h3'24" 41d50'26"
Sara Ann
Cnc 8h29'25" 30d41'40"
Sara Ann
Lyn 7h9'44" 49d20'29"
Sara Ann
Uma 9h23'26" 56d15'1"
Sara Ann Coltoniak
Tau 3h55'21" 29d8'50"
Sara Ann DePietri
Umi 15h24'18" 69d28'53"
Sara Ann Francis Mouland
Lyr 18h29'17" 39d42'46"
Sara Ann Gramling
Tau 3h34'29" 13d45'56"
Sara Ann Hermanson
Lmi 10h7'36" 29d28'4"
Sara Ann Mills
Cap 20h22'23" -11d29'21"
Sara Ann Murray
Tau 5h42'13" 22d46'34"
Sara Ann Sacson
Lib 15h48'33" -11d42'18"
Sara Ann Schettler
Ari 3h8'6" 14d4'59"
Sara Ann Zalvis
Sco 17h13'13" -42d59'3"
Sara Anna Marie
Tau 5h49'14" 27d44'26"
Sara Anne
Ari 1h51'47" 23d43'12"
Sara Anne Mulrooney
Aqr 21h30'51" 1d47'16"
Sara Anne Zuccalmaglio
And 2h18'27" 48d51'20"
Sara Annina
Umi 17h0'5" 75d29'10"
Sara & Anthony's Forever Star
Cyg 19h58'58" 41d2'26"
Sara Asaf Ohringer
Uma 13h31'58" 54d43'43"
Sara Ashley Miller
Peg 23h1'20" 17d17'57"
Sara Austin
Sgr 18h57'2" -22d19'10"
Sara Azzaline
Cnc 9h21'7" 16d56'20"
Sara B. Teel
Uma 8h39'55" 48d8'23"
Sara Beara
Sgr 19h13'2" -33d44'42"
Sara Belle
Sco 16h4'56" -19d1'57"
Sara Berentsen
Cam 5h13'35" 63d8'58"
Sara Beth Coluccio
Aqr 22h58'34" -8d43'24"
Sara Beth Friede
Uma 11h33'21" 44d46'0"
Sara Beth Lanier
Ori 5h59'32" 20d45'16"
Sara Beth Reeves
Cnc 8h54'43" 14d27'23"
Sara Beth Williams
Mon 6h31'43" -0d6'14"
Sara Bilbe
Cru 12h45'6" -56d36'4"
Sara Bouras
Del 20h40'27" 15d10'23"
Sara Brandon Gresham
Cam 4h23'33" 76d34'28"
Sara Brunkhorst
Leo 10h33'25" 14d26'47"
Sara Brunscheon
Cnv 12h41'50" 43d18'49"
Sara Bun Marie
Uma 11h44'27" 41d37'31"
Sara Bunny Wolff
Lep 5h37'58" -25d55'29"
Sara Burrage
Aqr 23h6'50" -5d34'49"
Sara C. Supel "Be Bop"
Lyr 18h38'28" 35d17'46"
Sara C. Tapp & D. Rodney Tapp
Uma 13h4'13" 58d14'41"
Sara Cage
And 2h10'11" 41d57'44"
Sara Catherine Ball
Lmi 10h45'45" 26d40'16"
Sara Catherine Carlson
Lyr 18h53'34" 34d56'17"
Sara Catherine Tabor "Angel Beauty"
Tau 4h26'45" 17d45'47"
Sara Chaya & Avi Forever!
Psa 22h50'15" -35d23'30"
Sara & Chris
Cma 6h47'4" -13d37'58"
Sara Christine Willcox
And 1h38'41" 45d40'29"
Sara "Coachy" Domeier-Wills
And 0h13'26" 41d58'30"
Sara Connoll's Star
And 2h17'53" 39d6'17"
Sara Corbett
Cam 6h46'32" 67d43'53"
Sara Cottrel
Gem 6h55'54" 27d53'49"
Sara Cowan
And 2h35'32" 44d28'48"
Sara D. Calvert
Gem 6h40'56" 12d18'40"
Sara D. (Cutecycle)
Vir 13h45'34" -18d18'15"
Sara Del Carmen
Ari 2h42'55" 28d1'47"
Sara Denise Tickell "Carebear"
Mon 6h28'0" 8d0'37"
Sara Desirae Ware
Leo 10h18'35" 26d4'30"
Sara Destiny Madelynn
Tau 5h57'41" 24d52'49"
Sara D'Haeseleer
Vir 12h14'23" 3d4'21"
Sara Di Pasquale
Umi 14h41'57" 85d3'4"
Sara Dylan - Shining Star
And 23h37'57" 38d57'46"
Sara E. Mann
And 1h11'14" 35d17'47"
Sara E. Morrow (angel eyes)
Vir 13h6'26" 10d44'37"
Sara E Powell
Ari 2h13'25" 21d0'29"
Sara E. Snyder
Sgr 19h12'5" -30d8'21"
Sara E Steinmetz
Leo 11h8'58" 15d43'25"
Sara Edblom
Lib 15h1'3" -13d26'9"
Sara Eldien
Gem 7h12'56" 32d42'54"
Sara Elin
Cas 23h2'43" 53d43'43"
Sara Eliza
Tau 4h32'11" 12d20'56"
Sara Elizabeth
Cnc 8h8'52" 7d1'31"
Sara Elizabeth
Uma 11h1'55" 35d1'16"
Sara Elizabeth
Uma 11h11'15" 44d56'37"
Sara Elizabeth
And 1h38'28" 40d22'56"
Sara Elizabeth
Cas 1h13'32" 49d27'29"
Sara Elizabeth
Cyg 21h31'0" 51d8'21"
Sara Elizabeth
Aqr 22h41'30" -1d22'54"
Sara Elizabeth Bordeleau
Gem 7h53'33" 15d23'45"
Sara Elizabeth Brown
Leo 10h11'40" 23d5'55"
Sara Elizabeth Brown
Uma 11h12'22" 46d13'48"
Sara Elizabeth Eaton
And 0h42'11" 27d13'16"
Sara Elizabeth Halstead
Crb 15h44'48" 29d7'42"
Sara Elizabeth Hardyman
Cyg 20h2'5" 31d50'3"
Sara Elizabeth Henrichs
Lmi 10h22'8" 34d35'31"
Sara Elizabeth & Katie Leigh
And 0h47'50" 39d5'59"
Sara Elizabeth Leisure
Uma 9h29'20" 47d21'48"
Sara Elizabeth Martin
And 1h7'50" 45d36'59"
Sara Elizabeth O'Brien
Cas 1h20'50" 64d29'33"
Sara Elizabeth Ouellette
Sgr 18h10'36" -18d1'25"
Sara Elizabeth Siciliano
Cas 2h28'55" 67d44'29"
Sara Elizabeth Thrower
Sgr 18h51'48" -16d56'28"
Sara Elizabeth Whyte
Sco 16h54'6" -32d46'41"
Sara Ellen Daugherty
And 0h30'57" 29d7'13"
Sara & Eric Forever
Cyg 20h4'22" 36d14'5"
Sara Falotico
Cyg 19h38'9" 37d54'17"
Sara Fernandes Cunha
Per 4h22'50" 50d20'59"
Sara Fernandez
Cap 20h18'8" -10d51'4"
Sara Flowers
Psc 0h48'7" 9d41'47"
Sara Francescangeli
Ori 5h11'5" 2d17'36"
Sara Francoise Poguet
Vir 12h56'54" -16d20'25"
Sara Friedman
Ori 5h34'48" 2d31'38"
Sara Garcia Crespo
And 23h4'10" 51d59'12"
Sara Grace
Cyg 21h39'38" 39d52'42"
Sara Grace Star
Cyg 19h46'26" 29d19'48"
Sara Grace Vincitore
Uma 9h26'9" 50d54'38"
Sara Grajeda
Lyn 7h41'40" 42d56'4"
Sara "Grandma" Narvaez
Cas 1h9'44" 67d28'6"
Sara Grasty
And 2h15'5" 46d44'20"
Sara Gryphon
Leo 10h12'12" 19d15'19"
Sara Gürtler
Uma 10h56'41" 56d11'0"
Sara Halterman
And 2h27'56" 45d8'55"
Sara Hayet Fares
Per 4h16'34" 40d0'13"
Sara Hayward
Sgr 19h19'9" -35d26'5"
Sara Heart
And 1h18'14" 37d33'40"
SARA HERNANDEZ
And 0h43'34" 40d5'2"
Sara Hosseina
Uma 8h57'45" 66d12'25"
Sara J. Hill
And 0h37'45" 37d49'41"
Sara J Smith - GaMPI Shining Star
Cam 7h50'6" 78d51'37"
Sara Jane
Cas 0h38'24" 61d21'51"
Sara Jane
Psc 0h13'14" 6d51'21"
Sara Jane
And 0h30'1" 36d25'11"
Sara Jane Feather
Cas 1h33'24" 60d40'33"
Sara Jane Needham
Cap 20h21'2" -12d22'31"
Sara Jayne Monahan
Lyr 19h14'40" 27d4'34"
Sara Jazmin Zarate
Tau 3h43'38" 11d8'8"
Sara Jean
Peg 22h25'38" 22d50'53"
Sara Jean Adair
Gem 7h13'22" 21d20'44"
Sara Jean Bolden
Cap 20h25'0" -10d32'32"
Sara Jean Bucci
Aqr 22h17'40" 0d19'23"
Sara Jean Knepp
Sgr 19h4'10" -13d13'1"
Sara Jean Lawrence
Tau 4h19'40" 10d57'44"
Sara Jean Manglos
Cas 1h39'57" 64d21'56"
Sara Jean Monahan
Ari 2h51'53" 27d12'31"
Sara Jean Reese
Cap 21h7'37" -26d14'11"
Sara Jeanne Linaberger King
Ari 2h12'5" 21d56'46"
Sara J.Elton
Lib 14h50'29" -5d4'27"
Sara Jenkins
Uma 10h29'17" 66d39'36"
Sara Jessica Holloway
Ori 4h50'44" 5d15'58"
Sara Jill
Uma 11h29'37" 58d51'29"
Sara Jill
Mon 7h18'18" -0d3'53"
Sara Jo
Cas 1h32'36" 64d19'10"
Sara Jo
And 2h17'34" 44d11'46"
Sara Jo Truman
Psc 1h31'15" 4d2'23"
Sara JoAnne
Psc 0h39'26" 14d47'48"
Sara Johnson
Tau 4h28'12" 21d38'0"
Sara & Jon Normandin
Cyg 20h27'29" 45d12'32"
Sara Jordan LeWinter
And 1h28'58" 39d48'37"
Sara Josephine's Star
Dra 16h32'51" 51d50'24"
Sara Joy
Ari 3h5'57" 11d58'41"
Sara Joyce Rohrback
Vir 13h24'5" -21d42'32"
Sara Judiff
And 23h9'11" 48d7'45"
Sara K. Wells
And 23h28'36" 49d42'50"
Sara Kastin
Gem 7h32'8" 32d40'7"
Sara Kate
Com 13h22'31" 27d47'46"
Sara & Kate Celik
Cyg 19h50'12" 30d21'6"
Sara Kate Welch
And 2h33'7" 47d43'54"
Sara Katherine Lucero
And 1h30'16" 42d57'46"
Sara Kathleen
Peg 22h59'25" 13d37'46"
Sara Kathryn
Lib 14h32'40" -9d58'21"
Sara Kayleigh Lord
Sgr 18h35'34" -27d40'57"
Sara Keefer
Cnc 8h39'18" 8d18'56"
Sara Kelly Rainer
Vir 14h23'15" 0d58'42"
Sara Kennedy
Sgr 18h50'37" -28d23'15"
Sara Kilchoer
Cas 3h22'42" 72d13'5"
Sara Kristina Adams
And 1h55'40" 38d52'13"
Sara Kristine Gallienne
Lyn 8h33'52" 51d26'10"
Sara Kwong
Sgr 18h57'33" -19d37'32"
Sara L. Brewer
And 0h47'31" 42d45'59"
Sara L. Leake
And 1h43'3" 42d4'3"
Sara L. Smith
Lib 15h21'56" -6d14'18"
Sara Lai-Ming Rose
Gem 7h26'55" 23d49'17"
Sara Lane
Uma 11h59'0" 49d40'38"
Sara Langhenry
Cyg 21h7'46" 55d15'19"
SARA LAYLA
Ori 5h3'44" 13d39'9"
Sara Leba
Cas 1h33'35" 61d25'34"
Sara Lee
Lib 15h9'50" -1d30'12"
Sara Lee
Ari 1h59'47" 12d54'59"
Sara Lee Brodhun
Tau 5h50'11" 14d35'26"
Sara Lee (Snoojer) Hendrix 11/19/90
Sco 17h14'0" -34d1'20"
Sara Lee, Nicholas Jackson Lockhart
Cyg 19h39'23" 51d20'57"
Sara Leonard
Aqr 22h13'34" -0d51'50"
Sara Lewis
Uma 11h16'15" 66d12'13"
Sara Lilie Familie
Psc 0h8'1" 8d38'10"
Sara Lindgren
Gem 7h13'49" 24d19'44"
Sara Lindsay
Uma 12h20'58" 62d11'38"
Sara & Linnette
Her 17h18'33" 48d7'11"
Sara Lisabeth Davis
Tau 4h29'2" 19d23'41"
Sara Lisiewski-aris
Psc 0h54'44" 31d44'38"
Sara Longhenry
Tau 4h20'15" 27d0'47"
Sara Lorraine Williams
Cnc 8h46'59" 13d49'5"
Sara Louise Kneeland
Cap 20h20'52" -10d49'39"
Sara Louise Porterfield
Cyg 20h35'53" 59d47'59"
Sara Louise Riebli
Sco 17h53'22" -39d1'48"
Sara Lovecchio
Her 17h12'48" 49d27'14"
Sara Lovely Schultz
Cap 20h31'35" -26d19'5"
Sara Loves Berto
Umi 16h25'33" 71d7'17"
Sara Lucia
Ori 6h19'0" 19d35'53"
Sara Lullaby
Cep 20h52'9" 69d53'46"
Sara Lydia Bottoni
Lyn 9h20'2" 38d34'4"
Sara Lynn
And 23h1'36" 47d46'14"
Sara Lynn
Cap 21h10'29" -16d1'31"
Sara Lynn Harris
Uma 10h1'57" 71d28'29"
Sara Lynn Hoege
Com 12h21'45" 28d6'39"
Sara Lynn Kendall
Psc 1h7'27" 32d19'12"
Sara Lynn Rondeau
Vir 11h47'0" -5d15'42"
Sara Lynn Sparks
Lib 15h47'2" -7d45'44"
Sara Lynn Ward
Cnc 8h23'46" 18d26'25"
Sara Lynn Weitz
Sgr 19h22'0" -14d27'29"
Sara Lynne Cole "Our Angel"
And 2h3'18" 37d55'22"
Sara M. Nickell
Cnc 9h4'16" 17d10'50"
Sara M. Pollock
Tau 4h24'48" 25d42'50"
Sara (Ma)
Aqr 21h45'10" -1d37'8"
Sara Mae
Vir 14h2'49" -16d14'17"
Sara Maggs
And 2h16'58" 46d17'11"
Sara Maitland
Srp 18h34'37" -0d33'32"
Sara Manzula
Leo 9h39'47" 26d46'31"
Sara Margery Elbaum
Psc 2h2'26" 9d35'34"
Sara Maria Lauto
Sgr 17h53'33" -29d22'37"
Sara Marie
Cma 7h4'16" -30d33'19"
Sara Marie
Aqr 21h12'52" -10d53'13"
Sara Marie
Uma 11h42'40" 32d2'48"
Sara Marie Addessi
Lyn 7h3'35" 57d58'31"
Sara Marie Anzallo
And 2h14'9" 50d27'56"
Sara Marie Blanton
Gem 6h57'14" 25d6'10"
Sara Marie Muro/Princess
Uma 10h35'35" 50d14'31"
Sara Marie Smith
Lib 14h58'27" -21d25'14"
Sara Martello
Aqr 22h43'41" -22d4'44"
Sara May Chagnon
Lib 15h34'42" -6d27'52"
Sara McClure
Leo 10h20'48" 13d22'19"
Sara Mccoma's Kingdom
Psc 0h41'8" 17d51'8"
Sara McKnight
Psc 0h48'32" 18d26'54"
Sara Mefford
Uma 10h46'15" 56d52'33"
Sara Megginson
Umi 13h47'23" 78d35'46"
Sara Meles
And 2h23'15" 41d38'20"
Sara Melissa Fusco
Cnc 8h49'53" 23d48'51"
Sara Melissa Triplett
Tau 4h25'35" 15d24'52"
Sara Melynne
Leo 10h48'48" 15d58'49"
Sara Merrill Turner
Per 2h24'5" 55d13'5"
Sara & Michael Iskrzycki
Cyg 20h36'50" 38d48'10"
Sara Michele Linton
Lib 15h11'11" -13d2'7"
Sara Michelle
Her 18h40'58" 18d30'15"
Sara Michelle Austein
And 0h18'4" 27d19'53"
Sara Michelle Welch
And 2h11'7" 43d55'48"
Sara Minhea Choi
Gem 6h59'2" 21d14'23"
Sara Morgan Gaffney
Aqr 21h37'46" 2d38'49"

Sara Morning Star The Touti Fangare Sco 17h53'44" -35d43'46"
Sara Muni Tau 4h37'2" 20d47'23"
Sara Neeck Ari 2h10'23" 24d11'30"
Sara Nesmith Cas 1h1'24" 58d24'42"
Sara Neubert 1 Ari 1h53'56" 21d45'48"
Sara Nicola Pickering Cas 23h13'1" 54d34'3"
Sara Nicole Andrews Psc 0h52'11" 25d30'40"
Sara Nicole Hulbert Cas 1h49'5" 64d56'18"
Sara Nicole Wilson Ari 2h50'4" 27d22'28"
Sara Niedzwiecki And 1h58'26" 44d14'30"
Sara Ontiveros Aql 19h10'22" 3d38'30"
Sara Osowski Gem 7h15'39" 20d24'27"
Sara P. Worley Cas 1h15'51" 73d1'11"
Sara Pabst Cyg 21h40'26" 54d29'20"
Sara Palleschi Cas 23h37'17" 54d37'40"
Sara Panizza Crb 16h22'13" 36d50'52"
Sara Panizzo And 23h23'18" 41d17'45"
Sara Parlagreco Gem 7h18'27" 31d12'14"
Sara Pasini Cas 0h27'21" 62d23'20"
Sara (Perfect mom, wife & friend) Gem 7h17'44" 16d52'45"
Sara Polite And 0h27'20" 43d13'56"
Sara Pralle Uma 9h28'12" 61d11'41"
Sara Pulliam And 1h51'52" 38d45'21"
Sara Q Tau 4h38'44" 27d33'55"
Sara Quinn Burgess Ori 5h54'39" 17d24'46"
Sara Rachael Brushaber Uma 9h3'10" 57d22'35"
Sara Rachel Yi Tao Sigler Chapman And 0h9'58" 32d44'55"
Sara Rae Lyn 7h53'2" 53d38'16"
Sara Rae Fogliasso Lib 15h11'47" -6d13'50"
Sara Rae Haines Gem 7h40'41" 33d32'23"
Sara Rae Lynn Purdon Lyn 7h7'38" 58d18'8"
Sara Raechel Yong-Ja Johnson Sgr 18h39'50" -26d4'44"
Sara Raimondi Cyg 19h40'32" 34d3'45"
Sara Rebecca Bentz Lib 14h56'31" -14d48'3"
Sara Rehring Leo 9h30'1" 26d46'14"
Sara Rene Hackman Crb 16h11'30" 30d34'26"
Sara Renee Psc 2h4'28" 8d17'42"
Sara Renee Heckman Uma 13h52'48" 52d20'4"
Sara Renee Tolbirt And 0h16'45" 25d20'17"
Sara Renee Williams Cnc 8h13'18" 30d37'49"
Sara Reuland Lyn 8h45'12" 36d41'12"
Sara Robison Gem 6h36'40" 17d56'19"
Sara Rocklin Crb 16h1'18" 34d20'23"
Sara Romelli Cas 23h17'6" 59d54'46"
Sara Rose And 0h2'49" 45d14'30"
Sara Rose & Bradly James Poore Cyg 20h52'35" 49d4'41"
Sara Rose Chamboullides Uma 11h12'29" 42d35'15"
Sara Rose Christenson Lib 14h23'48" -14d43'33"
Sara Rose Klevatt Sgr 18h34'6" -23d8'2"
Sara Rose Mirshamsi Sgr 19h12'14" -24d9'25"
Sara Rubira Srp 18h18'46" -0d33'16"
Sara Ruggiero And 1h33'20" 47d44'37"
Sara Ryn Cap 20h35'2" -14d48'47"
Sara Salena Salcido And 0h10'29" 35d0'29"
Sara Santucci Vir 13h41'59" -6d42'3"
Sara "Sarabear" Tierce-Hazard Tau 4h31'59" 18d18'41"
Sara Schenck Cas 0h47'24" 50d34'21"
Sara Seitz Cyg 21h22'36" 35d57'27"
Sara - Sera Tau 4h32'36" 21d16'9"
Sara Sheehan's Star - 23/03/1986 Cru 12h26'20" -55d54'42"
Sara Shundari Banerjee Cap 21h56'34" -10d5'13"
Sara Sidney Leone Ari 2h53'16" 28d45'17"
Sara Sky And 0h44'6" 37d5'20"
Sara Spaugy And 23h7'11" 48d18'8"
Sara Star Gem 7h13'37" 34d9'1"
Sara Star Vir 12h54'3" 1d17'26"
Sara Stemler Leo 11h13'9" 15d50'26"
Sara Sundermeyer Crb 16h12'19" 35d18'53"
Sara Suzanne Keith And 1h17'9" 42d14'10"
Sara Swan Cyg 20h54'10" 31d57'12"
Sara T. Anderson Leo 11h13'47" 25d2'52"
Sara & T.J. Aqr 22h19'18" 1d4'32"
Sara Torrisi Lac 22h55'15" 49d31'53"
Sara und Abder Ramdane Uma 9h21'25" 44d36'53"
SARA UZUN Del 20h41'3" 10d52'26"
Sara Valentine Sgr 18h48'27" -26d15'12"
Sara Ventura Cma 7h14'52" -30d4'34"
Sara Victoria Pohner Cnc 9h9'42" 32d38'53"
Sara Virginia Parrish Cnc 8h46'43" 17d38'9"
Sara Walzer And 23h47'28" 34d32'30"
Sara Weasle Uma 10h56'24" 61d45'40"
Sara Weiss And 0h42'11" 43d45'43"
Sara White And 23h15'46" 43d27'11"
Sara Zimmerman Lyr 18h26'49" 39d31'21"
Sara, Caylin, Brianna Her 17h3'11" 33d9'10"
Sara1995 Tau 5h33'48" 26d8'56"
Sara545 Com 13h9'0" 26d5'23"
Sarabeth Cam 4h23'45" 56d54'26"
Sarabeth Caldwell Waller Peg 21h21'46" 23d35'11"
Sarabi Lmi 10h3'57" 37d51'52"
SaraBoo Cnv 13h13'22" 48d21'38"
Saraboo Tau 4h43'10" 21d3'1"
Sara-Don Forever Uma 10h32'6" 63d2'27"
Sarados Umi 4h58'8" 88d30'33"
Sarah Umi 15h1'20" 82d6'55"
Sarah Mon 6h45'12" -0d16'37"
Sarah Oph 16h54'48" -0d18'21"
Sarah Lib 15h30'33" -3d52'20"
Sarah Aqr 21h19'38" -5d20'40"
Sarah Aqr 20h41'55" -3d7'28"
Sarah Cap 20h24'7" -12d21'51"
Sarah Sco 16h6'9" -14d28'6"
Sarah Cap 21h43'2" -21d20'58"
Sarah Lib 14h29'26" -17d39'3"
Sarah Sco 16h11'16" -17d8'12"
Sarah Sco 16h12'4" -15d38'6"
Sarah Lib 15h51'54" -18d11'35"
Sarah Cam 4h4'52" 67d10'43"
Sarah Cam 7h12'48" 73d26'10"
Sarah Umi 14h19'13" 70d3'46"
Sarah Dra 16h27'8" 53d42'15"
Sarah Uma 9h40'59" 52d57'14"
Sarah Uma 10h19'21" 58d20'4"
Sarah Col 6h25'11" -34d21'6"
Sarah Sco 17h33'40" -37d6'14"
Sarah Cru 12h22'26" -59d49'36"
Sarah Tau 3h56'12" 15d20'25"
Sarah Uma 13h57'1" 57d41'14"
Sarah Tau 5h4'14" 17d4'51"
Sarah Tau 4h26'5" 15d2'29"
Sarah Tau 3h46'8" 19d42'42"
Sarah Ari 2h45'27" 21d18'47"
Sarah Ari 2h54'4" 15d50'45"
Sarah Psc 0h50'2" 15d27'12"
Sarah Psc 1h36'1" 21d35'5"
Sarah Leo 9h30'57" 12d41'37"
Sarah Cnc 8h51'40" 12d1'50"
Sarah Gem 6h42'4" 14d17'0"
Sarah Psc 1h46'16" 13d31'26"
Sarah Ori 4h52'13" 10d59'28"
Sarah Vir 12h15'7" 4d11'41"
Sarah Vir 14h17'34" 4d2'33"
Sarah Her 16h17'42" 25d47'54"
Sarah Crb 15h54'36" 27d16'58"
Sarah Crb 15h43'26" 26d19'33"
Sarah Cnc 9h5'55" 28d14'31"
Sarah Ori 5h58'11" 22d44'2"
Sarah Gem 7h6'53" 21d36'4"
Sarah Ari 2h54'32" 27d56'12"
Sarah Ari 2h48'50" 27d26'29"
Sarah Ari 2h44'11" 27d29'43"
Sarah Ari 2h9'7" 23d29'43"
Sarah Psc 1h25'8" 27d38'18"
Sarah Psc 1h21'8" 24d42'0"
Sarah Dra 18h38'29" 51d45'3"
Sarah Uma 11h26'3" 46d26'10"
Sarah Uma 10h52'31" 46d20'12"
Sarah Uma 10h28'32" 46d21'58"
Sarah Cam 5h38'32" 59d28'0"
Sarah Cas 0h15'47" 53d25'40"
Sarah And 23h56'10" 38d0'53"
Sarah And 22h58'27" 42d23'0"
Sarah And 23h7'5" 42d39'19"
Sarah And 23h45'42" 44d50'30"
Sarah And 1h11'22" 47d49'41"
Sarah And 1h2'0" 46d37'54"
Sarah And 1h21'46" 45d25'14"
Sarah And 1h28'31" 46d30'29"
Sarah And 1h47'17" 46d1'42"
Sarah And 2h18'38" 45d42'2"
Sarah And 2h35'28" 45d13'31"
Sarah And 2h17'2" 50d29'44"
Sarah Per 3h15'32" 45d47'56"
Sarah Lmi 10h35'55" 39d4'27"
Sarah Uma 11h19'9" 40d8'55"
sarah And 2h10'12" 40d2'4"
Sarah And 0h47'37" 44d23'18"
Sarah And 2h10'7" 38d24'4"
Sarah And 0h57'33" 41d34'16"
Sarah Her 17h3'47" 33d11'0"
Sarah 143 Gem 7h41'14" 15d24'56"
Sarah 7.24.02 Cereceres Lib 15h44'39" -9d8'6"
Sarah A. 10 Beursken Psc 0h4'30" 2d13'31"
Sarah A. Amason Ori 5h10'24" 2d55'58"
Sarah A. Moyer Uma 13h46'40" 58d18'39"
Sarah A. Rahman Leo 10h46'50" 18d51'48"
Sarah A. Sheffield Lyn 8h36'44" 45d3'31"
Sarah A. Trobaugh Lyn 7h58'33" 38d57'35"
Sarah Abigail Pohl And 23h27'55" 48d7'31"
Sarah Abigail Thompson Aqr 22h9'28" -1d7'19"
Sarah Abt Cnv 12h27'5" 31d22'29"
Sarah Addison - Sparkle of my Life Col 5h28'45" -35d48'25"
Sarah Addobati Ari 2h39'10" 25d52'39"
Sarah adel al Kharafi Uma 9h43'29" 70d39'46"
Sarah Adragna And 0h51'4" 43d36'8"
Sarah Agnes Cassidy Cap 21h43'41" -13d25'34"
Sarah Ahmed And 0h47'13" 39d51'33"
Sarah Aileen Lochridge And 0h20'47" 28d39'5"
Sarah Aileen Mann And 0h22'44" 40d1'54"
Sarah Alberstein Tau 4h23'49" 22d13'9"
Sarah Albert Cep 21h30'27" 61d37'40"
Sarah Albrighton Cas 23h38'58" 56d40'41"
Sarah Ali Umi 14h24'1" 68d35'13"
Sarah Alia Hosny Dra 18h27'59" 50d19'32"
Sarah A'lice Aqr 21h47'31" -1d45'40"
Sarah Alice Allan Vir 14h9'43" -0d4'8"
Sarah Alice Gilkey Uma 10h33'1" 57d36'42"
Sarah Allison Slaven Cam 7h48'8" 66d33'7"
Sarah Allsup Sco 16h10'21" -30d21'0"
Sarah Alyse And 2h5'2" 41d55'47"
Sarah Amanda Margaret Sinclair Lyn 9h2'57" 39d40'33"
Sarah "Amazing" Wergin Cnc 8h40'50" 21d34'39"
Sarah Ananda Wolchock Sco 16h19'31" -13d52'26"
Sarah Anathea Findlay Cyg 20h22'28" 40d40'24"
Sarah and Alex And 1h30'30" 41d8'14"
Sarah and Brendan Ori 5h41'6" 13d4'2"
Sarah and Brian Cyg 20h3'16" 47d18'1"
Sarah and Daniel's Love Lib 15h23'51" -21d45'13"
Sarah and Doug(las)'s Star Cyg 21h37'49" 37d58'13"
Sarah and her dad Ari 2h7'1" 24d20'1"
Sarah and James Ellis Cyg 20h29'34" 38d12'47"
Sarah and Jason Ori 4h48'58" 11d46'14"
Sarah and Jon And 1h34'17" 46d8'46"
Sarah and Joseph Reavley Vir 12h11'50" 3d53'27"
Sarah and Josh Lyn 6h30'34" 56d57'30"
Sarah and Keith's Wish Cyg 19h36'22" 52d33'47"
Sarah and Kevin Deal143 Sge 20h3'55" 16d34'19"
Sarah and Malia - My 2 ladies Per 3h50'6" 38d10'9"
Sarah and Michael Ori 6h17'38" 9d47'13"
Sarah and Nathaniel's Star Sgr 18h32'54" -16d57'34"
Sarah and Stuart Barnett And 0h50'47" 37d59'45"
Sarah and Tess And 23h11'47" 48d3'19"
Sarah And Tommy Forever Ori 5h31'3" 1d41'9"
Sarah & Andrew Rudkosky Cyg 20h0'7" 43d4'19"
Sarah & Andru Uma 12h37'42" 60d19'46"
Sarah Angela Krump Uma 9h9'58" 60d51'17"
Sarah Angrisani Cyg 21h41'31" 45d38'48"
Sarah Ann And 1h30'12" 46d4'56"
Sarah Ann And 2h25'38" 46d23'40"
Sarah Ann And 23h3'38" 35d56'45"
Sarah Ann Lib 15h35'17" -20d33'49"
Sarah Ann Lib 15h44'56" -18d49'43"
Sarah Ann Aqr 22h31'11" -2d6'27"
Sarah Ann Coble Aqr 22h5'35" -13d31'26"
Sarah Ann Dalgetty Cas 0h16'20" 59d6'3"
Sarah Ann Ford Aqr 21h47'49" -0d2'41"
Sarah Ann Fraze Ari 3h19'31" 27d30'45"
Sarah Ann Gnatek Fournier Ari 2h52'28" 28d50'16"
Sarah Ann Graybill Cas 1h29'4" 66d2'51"
Sarah Ann Grunst Sgr 19h28'58" -17d59'9"
Sarah Ann Hansen Sgr 18h5'21" -27d9'36"
Sarah Ann Haverstock Umi 14h13'2" 68d7'57"
Sarah Ann Johnson And 2h12'5" 50d3'54"
Sarah Ann Lake Gem 6h43'6" 33d44'17"
Sarah Ann Law Cap 21h27'7" -17d7'15"
Sarah Ann Mendoza Vir 12h26'44" 9d52'42"
Sarah Ann Mullenbach Uma 10h39'19" 54d21'8"
Sarah Ann Quine Cap 20h42'39" -22d28'33"
Sarah Ann Raeburn And 0h56'17" 23d16'45"
Sarah Ann Richards Lyr 18h46'8" 40d52'48"
Sarah Ann Sanders Cas 2h20'47" 62d26'37"
Sarah Ann Sommer Uma 8h40'24" 56d16'53"
Sarah Ann Wilson Lyr 19h16'36" 39d37'0"
Sarah Annalena Julia Leo 10h36'30" 9d24'5"
Sarah Anne And 2h15'43" 46d43'17"
Sarah Anne Lib 15h56'50" -19d4'35"
Sarah Anne Adams Ori 4h51'8" 12d43'8"
Sarah Anne Brock Sco 17h19'10" -40d1'0"
Sarah Anne Byrd Aql 19h46'48" 12d13'46"
Sarah Anne Frazelle And 0h7'11" 41d52'57"
Sarah Anne Koerbel Cap 20h42'38" -21d38'11"
Sarah Anne Kooken And 0h33'48" 38d51'20"
Sarah Anne Martin And 23h46'59" 46d21'30"
Sarah Anne McNeil Ari 2h6'38" 11d33'40"
Sarah Anne Nitsche Sgr 18h46'11" -32d21'1"
Sarah Anne Radzavicz And 2h27'9" 45d58'43"
Sarah Anne Rivers Cyg 20h29'34" 45d59'3"
Sarah Anne Ross Mon 7h38'13" -0d44'22"
Sarah Anne Schwartz Umi 15h35'20" 69d30'0"
Sarah Annemarie Karkowski Ari 2h21'59" 19d3'56"
Sarah Antonia Zomaya Vir 13h25'43" 5d40'9"
Sarah Antosik Lib 14h48'43" -2d46'16"
Sarah Arbore Cam 7h52'23" 67d28'56"
Sarah Arlene DeWitt Cam 4h28'57" 56d28'48"
Sarah Ashlee And 2h26'24" 46d51'57"
Sarah Ashley Cap 21h17'33" -17d31'20"
Sarah Ashley loves Aaron Brent Cyg 21h32'3" 52d35'40"
Sarah Ashley Sager And 23h30'44" 41d32'56"
Sarah Ashlynn Fealey Dra 16h52'12" 55d36'38"
Sarah Asilde Uma 10h25'13" 41d5'29"
Sarah Astrea Kruger-Poires Psc 23h6'47" 8d7'47"
Sarah B. Gem 7h14'31" 26d3'22"
Sarah B Cam 4h31'6" 59d14'19"
Sarah B. Clark Lyr 18h58'21" 26d24'3"
Sarah Baberra And 2h18'35" 44d50'55"
Sarah Baethge Psc 1h1'53" 27d36'55"
Sarah Baker Tau 3h49'32" 5d9'39"
Sarah Barah Quintenz Psc 23h53'45" -0d34'21"
Sarah Barbuscak And 0h31'45" 45d34'19"
Sarah Bartlett Cas 0h36'14" 48d21'11"
Sarah bat Rivka Cyg 20h12'17" 37d59'53"
Sarah Baum And 22h58'28" 47d10'48"
Sarah Bauman And 23h4'36" 43d2'47"
Sarah Bear Cas 0h27'12" 52d6'51"
Sarah Bear And 23h3'17" 52d11'56"
Sarah Bear Ori 6h17'28" 13d34'38"
Sarah Bear is Loved Lib 15h48'25" -5d42'45"
Sarah "Bear" Kershaw Umi 14h58'36" 70d16'53"
Sarah Beara Cam 6h37'9" 81d11'17"
Sarah Beara's Special Star Aqr 22h21'24" -5d56'3"
Sarah Bear's Shining Star Umi 15h11'13" 69d49'6"
Sarah Beattie And 0h28'52" 26d8'7"
Sarah Bebe Leo 10h10'57" 26d41'5"
Sarah Belfield Nielsen And 23h47'46" 41d33'58"
Sarah Belisle Lyn 6h25'28" 60d32'30"
Sarah Bell Uma 10h5'4" 47d22'58"
Sarah Bella Leo 11h7'7" 21d13'48"
Sarah Bella Vir 14h9'51" -13d15'48"
Sarah Bentley And 1h34'5" 45d20'19"
Sarah Beth And 2h32'56" 50d5'12"
Sarah Beth Tau 4h21'45" 19d1'3"
Sarah Beth Cap 21h52'43" -13d10'10"
Sarah Beth Uma 11h15'50" 57d17'15"
Sarah Beth Aqr 23h41'37" -15d57'58"
Sarah Beth Bissell: Star of Life Aqr 22h19'29" -17d34'10"
Sarah Beth Chavez Psc 1h17'49" 24d54'41"
Sarah Beth Dillon Aqr 23h37'36" -6d49'37"
Sarah Beth Dorman, our princess And 0h16'56" 38d17'54"
Sarah Beth Giusto Cas 1h37'14" 62d56'0"
Sarah Beth Jerasi Per 2h23'39" 55d13'1"
Sarah Beth Landon And 0h38'46" 38d4'14"
Sarah Beth Lantry Ari 3h22'2" 29d34'24"
Sarah Beth McClintock Lyn 6h29'35" 57d13'55"
Sarah Beth Pitchford Aqr 23h44'28" -10d52'35"
Sarah Bethany Kolbinsky Pyx 8h49'54" -25d52'24"
Sarah Bettens Cas 0h18'5" 62d2'15"
Sarah Betzabé And 1h56'31" 45d29'48"
Sarah Billingham 20-02-1982 Psc 1h35'33" 13d6'27"
Sarah Binder Uma 11h46'36" 42d3'42"
Sarah Birnbaum Tau 4h46'11" 22d19'1"
Sarah Bladen Leo 11h52'35" 22d8'49"
Sarah Blake Ori 6h15'58" 5d48'6"
Sarah Blasko Sgr 19h16'42" -15d39'0"
Sarah Blowers Ori 4h52'19" 14d47'54"
Sarah Boeck Ari 2h20'12" 25d28'41"
Sarah Bones Tau 5h23'49" 18d55'54"
Sarah Bredin Cas 0h32'23" 62d57'32"
Sarah Breen And 0h57'23" 43d58'5"
Sarah Bremgartner Lac 22h20'6" 41d29'42"
Sarah Briney Ari 2h41'6" 25d30'19"
Sarah Brittiny And 1h20'2" 40d18'2"
Sarah Brodeur Cas 2h23'14" 68d51'13"
Sarah Brood Cas 23h35'18" 52d37'13"
Sarah Brooke Denning Uma 8h59'58" 58d10'0"
Sarah Brooks Col 5h59'59" -34d50'34"
Sarah Brook's (Super-Star) And 23h1'23" 48d15'58"
Sarah Brooks Toaldo Ari 2h30'46" 18d27'54"
Sarah Brown And 23h0'18" 50d43'34"
Sarah Brown Sgr 19h4'37" -21d46'9"
Sarah Bruce - My Shining Star And 2h25'1" 46d49'34"
Sarah Bryden-Brown Col 5h59'14" -34d50'55"
Sarah Buckley Aqr 21h54'9" 1d2'34"
Sarah Bühler Boo 13h48'43" 10d58'3"
Sarah Burns (Sarah Folan) 1919-2005 Sgr 17h55'50" -17d48'1"
Sarah Butler Psc 1h23'41" 22d8'37"
Sarah C A Marwick Ori 5h20'33" -4d17'54"
Sarah C. Farias Crb 15h40'40" 26d29'55"
Sarah C. Marron Sco 17h24'33" -43d25'29"
Sarah C. White Cap 21h37'42" -13d27'32"
Sarah Cadence Cas 1h12'11" 59d45'15"
Sarah Caitlin Tau 4h32'17" 9d47'13"
Sarah Calabro Sco 17h43'9" -41d39'20"
Sarah - Carla & Rebecca's Mummy x And 0h54'57" 44d3'7"
Sarah Carley Sco 16h14'16" -11d16'28"
Sarah Carlson Aql 19h27'1" 3d45'4"
Sarah "Carmen" Johnson Cyg 20h4'5" 47d49'20"
Sarah Caroline King And 23h40'32" 47d30'7"
Sarah Caroline Sowers Crb 15h46'37" 25d43'12"
Sarah Carpenter Tambling Car 10h48'7" -62d41'38"
Sarah Cassidy Price Cap 21h31'22" -16d25'57"
Sarah Catherine Catalano Tau 3h25'57" 7d1'25"
Sarah Catherine Cluff And 0h30'10" 28d8'30"
Sarah Catherine Gemma Stevens Gem 6h23'21" 19d59'31"
Sarah Catherine Mccollum Cap 20h8'24" -25d45'11"
Sarah Catherine Nuzzi Aql 19h13'20" 15d16'23"
Sarah Catherine Torkelson Aqr 22h33'43" -0d52'18"
Sarah Celest Emerick Gem 7h28'51" 15d31'36"
Sarah Cencius Tau 4h9'58" 9d55'26"
Sarah Chandler Newsome's Star Psc 1h25'34" 3d44'57"
Sarah Christina Allenbach Dra 14h45'54" 56d29'31"
Sarah Christina Whitaker Vir 13h20'23" -4d18'4"
Sarah Christine Gem 6h51'45" 27d2'17"
Sarah Christine Psc 1h21'10" 30d51'11"
Sarah Christine Her 17h11'8" 31d58'37"
Sarah Christine Carroll Tau 5h59'6" 24d55'12"
Sarah Christine Rahe Psc 1h46'11" 18d43'51"
Sarah Churchill Dysart Ari 2h4'46" 24d34'7"
Sarah Claire Moore Aqr 21h44'26" -0d10'41"
Sarah Clarisse Heaster Mon 7h19'25" -0d30'46"
Sarah & Clark Lalliss - Tupperware Cyg 21h23'50" 31d36'49"

Sarah Clarke
Leo 11h7'47" 12d2'33"

Sarah Cohen
Psc 1h22'32" 32d51'52"

Sarah Collier
And 0h16'51" 45d52'53"

Sarah Connelly
Ari 2h7'41" 23d20'0"

Sarah Connolly
Leo 10h35'56" 14d17'28"

Sarah Corona
Mon 6h45'15" -0d9'27"

Sarah Coryell
Lyr 18h44'9" 38d59'3"

Sarah Creighton
Ori 5h59'58" 11d23'56"

Sarah Crowley
Peg 21h39'28" 15d11'0"

Sarah Curry
Ari 2h4'32" 24d48'30"

Sarah D. Bliss
Lyn 6h57'21" 60d20'57"

Sarah Dache
Lib 15h3'25" -18d14'43"

Sarah & Dale
And 23h10'23" 52d59'17"

Sarah Daly
Aqr 23h23'5" -19d55'26"

Sarah "Dalz" Alyas
Mon 8h6'34" -0d35'43"

Sarah Damphousse
Vir 13h31'29" -17d47'2"

Sarah & Dan
Cyg 21h57'24" 43d22'14"

Sarah Danielle Blasko
Peg 23h13'10" 27d59'36"

Sarah Danielle Coty
Cma 6h50'19" -17d5'53"

Sarah Danielle Hudgins
Uma 10h15'2" 60d37'55"

Sarah Danielle Peebles
Cnc 8h14'43" 23d43'17"

Sarah Darlene Merz
Lib 14h54'13" -4d6'17"

Sarah Davis
Vir 14h23'42" 5d54'54"

Sarah Davis
Lmi 10h33'46" 37d56'54"

Sarah Davis Smith
Psc 1h3'35" 13d49'4"

Sarah Dawley
Tau 4h43'37" 26d43'0"

Sarah Dawn Rowland
And 23h2'2" 48d12'59"

Sarah Dear
Ori 5h36'1" -0d4'26"

Sarah Deguzman
Lyn 7h46'53" 35d52'23"

Sarah Del Greco-"Beyond the sunset"
Vir 11h39'41" 2d57'56"

Sarah Delia
And 23h13'46" 47d51'52"

Sarah Delieu
And 2h29'6" 43d26'18"

Sarah Denise Espinosa
Leo 11h13'10" 10d4'17"

Sarah Denise Fortune
Vir 13h12'36" -17d49'48"

Sarah Detar
Sgr 19h20'11" -41d19'21"

Sarah Devin Britt
Peg 21h43'37" 14d28'58"

Sarah Diane Greenaway
Dra 17h17'3" 56d29'26"

Sarah Diane Smith Birthday Star
Vir 13h52'52" 3d54'26"

Sarah Dillenback
Lyn 7h46'50" 56d53'17"

Sarah Doherty 2007
Uma 11h23'7" 58d50'18"

Sarah Douag
Ari 3h29'5" 21d45'12"

Sarah & Douglas Love-Star
And 1h1'6" 44d21'26"

Sarah Drew
Ari 3h5'44" 29d1'27"

Sarah Dudé
Ori 5h54'16" 16d41'16"

Sarah Dumaual
Lib 15h21'24" -26d19'37"

Sarah Dupuy
Aqr 22h25'46" -1d47'19"

Sarah & Dylan's Star
Leo 9h41'46" 10d40'57"

Sarah E. A. Allman
Uma 9h12'34" 69d36'27"

Sarah E. Anderson
Leo 10h20'25" 15d3'59"

Sarah E. Britt
Sgr 19h17'24" -14d7'49"

Sarah E. Busby 071078
Per 3h21'43" 42d51'10"

Sarah E. Daly
Lib 14h52'10" -8d57'44"

Sarah E. F. Miller
And 1h29'50" 46d30'12"

Sarah E. Kilduff
Crb 15h53'33" 38d33'1"

Sarah E. Knouse
Leo 9h43'30" 25d20'23"

Sarah E Lentin
Cas 2h33'22" 65d28'36"

Sarah E. Martin
Aqr 22h11'53" -1d27'12"

Sarah E. Starsoneck
Crb 16h23'51" 34d4'55"

Sarah Eble
Sgr 19h12'57" -18d7'52"

Sarah Eden
Cyg 20h14'5" 49d10'50"

Sarah Eichler
And 0h17'52" 44d32'18"

Sarah Eileen Watson
Gem 6h3'54" 22d33'3"

Sarah Elaine 71602
Per 3h34'14" 45d22'49"

Sarah Elaine Friend
Cnc 8h3'34" 20d30'8"

Sarah Elaine O'dea
Lyr 18h25'7" 32d14'44"

Sarah Elaine Rodriguez
Cap 20h57'7" -20d49'26"

Sarah Elise Fink
Lib 15h48'31" -18d13'54"

Sarah Elise Ticer
Aqr 20h54'18" -2d48'17"

Sarah Eliza Veerman
Ari 3h5'21" 12d16'47"

Sarah Elizabeth
Vir 12h32'13" 3d13'12"

Sarah Elizabeth
Tau 4h36'17" 22d25'16"

Sarah Elizabeth
Gem 6h50'51" 25d57'54"

Sarah Elizabeth
And 1h13'4" 36d20'21"

Sarah Elizabeth
Cyg 20h14'38" 41d45'0"

Sarah Elizabeth
And 23h31'55" 38d48'38"

Sarah Elizabeth
Cap 20h36'25" -20d36'28"

Sarah Elizabeth
Cas 1h13'43" 69d30'21"

Sarah Elizabeth
Uma 9h47'56" 56d11'32"

Sarah Elizabeth
Uma 11h39'46" 54d7'53"

Sarah Elizabeth
Lyn 6h32'3" 56d50'0"

Sarah Elizabeth
Lib 15h9'52" -22d36'42"

Sarah Elizabeth Allen
Cnc 8h58'45" 13d53'26"

Sarah Elizabeth Alpha One
And 22h58'38" 50d58'24"

Sarah Elizabeth Banducci
Cas 0h38'56" 64d44'42"

Sarah Elizabeth Barrett
Leo 11h34'38" 15d6'31"

Sarah Elizabeth Besse
Ori 5h28'53" 6d34'4"

Sarah Elizabeth Borchers
Peg 21h54'58" 14d27'38"

Sarah Elizabeth Braam
Tau 5h52'14" 27d5'54"

Sarah Elizabeth Brattolli
And 1h9'55" 42d1'3"

Sarah Elizabeth Bunday
And 0h36'17" 28d44'22"

Sarah Elizabeth Catenac
Peg 21h43'22" 14d42'29"

Sarah Elizabeth Cooper
Leo 10h14'18" 8d0'30"

Sarah Elizabeth Crowder
Per 3h8'15" 38d40'35"

Sarah Elizabeth Cupkin Crowell
Uma 9h3'28" 46d35'40"

Sarah Elizabeth Cupkin Crowell
Uma 9h0'39" 46d58'8"

Sarah Elizabeth Daniel
Gem 6h38'26" 12d31'31"

Sarah Elizabeth Dunlap
Leo 11h48'25" 21d51'26"

Sarah Elizabeth Ebersole
Cas 1h31'50" 65d36'36"

Sarah Elizabeth Eddy
Cas 23h0'39" 58d32'56"

Sarah Elizabeth Fleming
Cyg 20h56'17" 39d29'24"

Sarah Elizabeth Foley
And 0h39'43" 41d9'10"

Sarah Elizabeth Gear
Psc 1h18'44" 30d39'11"

Sarah Elizabeth Ginsberg
Tau 4h37'8" 10d36'28"

Sarah Elizabeth Hall
Uma 10h37'25" 43d4'28"

Sarah Elizabeth Hargis
Cam 4h3'59" 67d43'45"

Sarah Elizabeth Harris
And 2h36'55" 45d32'12"

Sarah Elizabeth Holton
And 23h42'11" 34d25'58"

Sarah Elizabeth Kendrick Kidder
Leo 9h53'14" 6d43'42"

Sarah Elizabeth Kuchar
Leo 10h48'54" 15d35'43"

Sarah Elizabeth Kulscar
Cnc 8h2'43" 26d7'35"

Sarah Elizabeth Kurrasch
And 23h48'27" 39d15'54"

Sarah Elizabeth Latham
Sco 17h27'25" -41d42'45"

Sarah Elizabeth Liddell
And 0h4'54" 35d24'2"

Sarah Elizabeth Little
Cap 21h20'13" -24d44'51"

Sarah Elizabeth Mardis
Crb 15h33'2" 35d14'44"

Sarah Elizabeth Mattix
Cyg 20h19'5" 39d10'43"

Sarah Elizabeth Meisinger
Tau 5h49'53" 23d8'15"

Sarah Elizabeth Misslin
Crb 15h54'18" 34d16'59"

Sarah Elizabeth Mitchell
Psc 0h56'26" 28d50'9"

Sarah Elizabeth Mullan
Cas 2h50'17" 60d31'6"

Sarah Elizabeth Norton
Ori 5h54'24" 12d24'13"

Sarah Elizabeth Nunn
Cas 0h18'45" 54d13'44"

Sarah Elizabeth Parfet
Vir 12h23'38" 11d44'20"

Sarah Elizabeth Rehm
Tau 4h36'59" 1d13'30"

Sarah Elizabeth Rhodes
Uma 13h58'20" 60d43'49"

Sarah Elizabeth Riblon
Sco 16h40'58" -38d34'19"

Sarah Elizabeth Schaefer
Uma 10h6'33" 49d52'49"

Sarah Elizabeth Schroeder
Lyn 6h38'33" 56d3'13"

Sarah Elizabeth Shairs
Gem 7h41'15" 32d59'54"

Sarah Elizabeth Smith
And 2h21'51" 46d4'19"

Sarah Elizabeth Socia
Lib 15h43'46" -6d43'16"

Sarah Elizabeth St. John Hostrander
Peg 22h19'52" 11d25'45"

Sarah Elizabeth Steinfeld
Cas 2h18'20" 69d29'59"

Sarah Elizabeth Strawn
Cnc 9h18'51" 11d49'23"

Sarah Elizabeth Strickland
Gem 6h47'20" 26d29'23"

Sarah Elizabeth Sundberg
Cas 1h40'13" 61d5'52"

Sarah Elizabeth Thomas
Crb 15h53'38" 35d48'34"

Sarah Elizabeth Tirres
Cap 21h57'1" -19d31'12"

Sarah Elizabeth Twinkles
Vir 14h4'45" -4d29'32"

Sarah Elizabeth Vital
Aqr 22h17'46" 2d2'46"

Sarah Elizabeth Welch
Uma 11h14'19" 56d30'12"

Sarah Elizabeth Werner
Cap 20h27'50" -10d1'34"

Sarah Elizabeth Whitehead
Uma 10h31'59" 47d22'53"

Sarah Elizabeth Wood
And 23h26'56" 48d57'21"

Sarah Elizabeth Wood
Sgr 19h12'53" -27d30'54"

Sarah Elizabeth Woods
Umi 17h3'28" 83d7'29"

Sarah Elizabeth Yarwood
Ari 3h17'48" 27d24'22"

Sarah Ella
Cas 0h32'8" 64d3'47"

Sarah Ellen Catherine Murphy
And 23h23'28" 35d24'12"

Sarah Ellen Gzesh
Cap 21h37'26" -23d26'36"

Sarah Ellen Marsh
Tri 2h22'52" 33d2'55"

Sarah Emily Jayne Iannone
Vir 13h38'13" -18d26'44"

Sarah Emily Mrowka
Lyn 9h39'18" 41d18'44"

Sarah Engelhard
And 23h27'39" 49d3'52"

Sarah & Erica (Our Love Forever)
Sge 19h53'13" 18d52'10"

Sarah Eriksson
Leo 10h29'47" 24d18'58"

Sarah Estelle Edwards
Tau 4h26'13" 26d43'31"

Sarah Estelle Kabatt
Lib 15h45'25" -8d48'33"

Sarah Esterline Rhueark Hall
Aql 20h11'26" 14d46'22"

Sarah Esty
Crb 15h39'37" 38d28'24"

Sarah Eternal
And 2h19'30" 46d34'28"

Sarah Faye
And 2h24'40" 46d14'6"

Sarah Ferraz
Sgr 19h20'22" -22d20'28"

Sarah Fielitz
Cnc 8h25'38" 21d40'46"

Sarah Fiona Purser
Lib 14h40'55" -12d48'11"

Sarah Flanagan 3:16am
Ori 5h38'29" -0d5'15"

Sarah Flieg
Cas 2h16'22" 74d7'6"

Sarah Frahm
And 23h27'56" 42d1'2"

Sarah Frances Lentine
Cas 0h28'15" 53d54'34"

Sarah Frazer
Ari 3h9'0" 23d25'7"

Sarah Frieda Beller
Cyg 20h33'16" 37d53'35"

Sarah Fritchman Larkin
Leo 11h29'6" 14d52'17"

Sarah G.
Uma 9h9'35" 65d23'45"

Sarah Gaither
Uma 9h17'13" 55d59'17"

Sarah Garner Avery
Lmi 10h45'23" 27d46'20"

Sarah Glatfelder
Leo 10h15'21" 14d22'33"

Sarah Glick
Gem 6h32'39" 22d21'25"

Sarah Gloria Robison (SGR)
Aqr 21h5'54" -13d31'10"

Sarah Goldstein - Shining Star
And 23h41'59" 42d33'32"

Sarah Grace
And 0h19'16" 44d45'23"

Sarah Grace
Vir 14h2'9" -6d51'24"

Sarah Grace
Sco 17h25'37" -33d22'52"

Sarah Grace Anderson 2006
And 23h27'3" 40d3'25"

Sarah Grace Austin
Umi 20h1'18" 88d56'52"

Sarah Grace Edwards
Cnc 8h49'29" 30d22'30"

Sarah Grace Gaissert
And 0h42'49" 40d45'27"

Sarah Grace Hingley
Ari 3h9'37" 29d0'5"

Sarah Grace Hoffman
Leo 9h47'21" 19d24'31"

Sarah Grace Satterfield
Peg 22h31'58" 7d13'36"

Sarah Grace Wainscott
And 23h24'21" 38d18'53"

Sarah Grace Wilson
Peg 22h9'42" 34d22'8"

Sarah Graham
Peg 22h6'55" 13d7'24"

Sarah Graley
Aqr 20h53'41" -11d8'54"

Sarah Grant
Tau 4h5'56" 5d12'48"

Sarah & Grant's Island
Uma 13h5'9" 57d43'10"

Sarah Grisman, Forever Loved
And 1h16'18" 46d6'12"

Sarah "Guapacita 5000" lhmoud
Gem 7h24'43" 28d34'34"

Sarah Guidetti
Uma 8h48'34" 60d9'36"

Sarah Guillemin
And 2h13'19" 42d54'20"

Sarah Gwisdalla
And 1h44'42" 42d45'58"

Sarah Haddock
And 1h42'50" 41d18'43"

Sarah Haeberlin
Lyn 6h39'7" 58d4'23"

Sarah Hall
Uma 11h49'8" 50d44'40"

Sarah Hamby
Lyr 18h49'52" 34d58'33"

Sarah Hamilton Guion
Aqr 23h7'1" -6d33'55"

Sarah & Harley
Cyg 19h31'9" 29d3'10"

Sarah Hasegawa
Del 20h34'50" 15d54'5"

Sarah Hayes Lauric
Dra 18h40'23" 75d16'32"

Sarah & Heather Fodor Love Mom
Cyg 20h46'23" 32d24'24"

Sarah Hecksteden
Sgr 18h11'4" -19d25'26"

Sarah Helen, will you marry me?
And 2h5'45" 43d34'9"

Sarah Hendrix
Cnc 8h38'27" 29d6'44"

Sarah Herbert
Lyn 8h13'7" 34d53'32"

Sarah Hermanski
Tau 4h36'13" 18d27'52"

Sarah Hill
Uma 11h46'27" 35d56'34"

Sarah Hing
Lyn 7h39'45" 56d31'18"

Sarah Hood
Cas 0h16'50" 51d11'27"

Sarah Hope
Cru 12h5'6" -62d24'30"

Sarah Hope Siudut
Leo 11h54'16" 26d49'48"

Sarah Howard
And 23h13'50" 38d24'5"

Sarah Howe
Leo 11h37'53" 21d37'35"

Sarah Howenstein
Lyn 7h26'55" 45d33'50"

Sarah Hubbard
And 0h39'32" 39d49'16"

Sarah Huffman
Dra 19h28'37" 79d48'17"

Sarah Huntley
And 2h12'2" 45d18'34"

Sarah Hyland Oneill
Sco 17h23'57" -42d34'45"

Sarah & Igor
Cru 12h34'41" -60d33'53"

Sarah. Il mio pezzo di cielo.
Aqr 21h33'49" -5d23'50"

Sarah in the sky
Uma 8h12'59" 68d55'27"

Sarah Inman
Cam 4h7'25" 57d51'20"

Sarah Irene
Cam 4h53'46" 61d16'48"

Sarah Isabelle
And 1h53'14" 36d59'23"

Sarah Israel
And 23h4'45" 45d24'48"

Sarah Ita Yellen
And 23h46'53" 35d35'15"

Sarah J. Berrio
Cas 23h38'24" 54d8'44"

Sarah J. Moulden
Cap 21h11'46" -19d31'46"

Sarah J Patton
Vir 12h25'2" -1d14'6"

Sarah J Williams
Aqr 22h31'59" -2d29'35"

Sarah Jade
Ari 2h55'17" 27d48'15"

Sarah James Hampe
And 1h11'14" 46d41'14"

Sarah Jane
And 23h29'17" 47d35'21"

Sarah Jane
And 0h23'46" 36d29'50"

Sarah Jane
Cnc 9h3'0" 26d17'3"

Sarah Jane
Ori 6h10'25" 15d34'39"

Sarah Jane
Ori 6h17'6" 14d40'33"

Sarah Jane
Lib 15h1'36" -20d45'5"

Sarah Jane 21061968
And 0h4'24" 38d36'1"

Sarah Jane Barnes
Leo 11h39'4" 26d21'43"

Sarah Jane Booth
Cas 1h23'8" 54d49'41"

Sarah Jane Collins
Psc 1h5'14" 28d35'46"

Sarah Jane Cooper
Ari 3h10'49" 19d45'12"

Sarah Jane De Paul
Uma 10h49'16" 46d2'26"

Sarah Jane Frankel
And 23h27'50" 47d52'22"

Sarah Jane Franklin
And 1h33'41" 41d25'20"

Sarah Jane Hamby
Lmi 10h3'4" 38d48'33"

Sarah Jane Hills
Ari 2h22'0" 23d59'16"

Sarah Jane Kimble
And 1h1'58" 37d54'28"

Sarah Jane Kish
Lyn 7h10'50" 58d20'3"

Sarah Jane Markulics
And 0h42'54" 43d22'52"

Sarah Jane Munn
Psc 1h49'58" 7d29'38"

Sarah Jane Prak
And 0h16'55" 44d59'35"

Sarah Jane Schechter
Per 4h41'42" 46d49'46"

Sarah Jane Smith Hobart
Cas 0h8'41" 54d5'10"

Sarah Jane Steindl
And 23h44'42" 47d26'36"

Sarah Jane Stubbings née Mayes
And 1h26'8" 43d55'21"

Sarah Jane Wardell
Uma 8h18'4" 70d34'42"

Sarah Janet Gronefeld
Ori 6h12'7" 16d50'54"

Sarah Jayne Butler
Cas 23h52'10" 52d31'45"

Sarah Jayne O'Sullivan
And 1h26'22" 40d17'17"

Sarah Jean
And 0h36'11" 34d49'0"

Sarah Jean
Gem 7h44'32" 34d3'54"

Sarah Jean
Cap 21h38'21" -11d4'37"

Sarah Jean Albright
Lmi 10h13'4" 28d0'58"

Sarah Jean (Angel)
And 0h39'15" 36d5'25"

Sarah Jean Bales
Cnc 8h20'3" 32d14'46"

Sarah Jean Elizabeth James
Cyg 19h31'56" 29d13'30"

Sarah Jean Foggon
And 23h43'32" 43d5'44"

Sarah Jean Hicok
Tau 5h32'52" 20d7'32"

Sarah Jean Hoffman
Sgr 19h37'51" -13d29'48"

Sarah Jean Holland
Her 18h8'37" 15d25'59"

SARAH JEAN MAHONEY
Cyg 21h10'28" 46d49'32"

Sarah Jean Sandt-Hill
And 1h47'56" 46d46'52"

Sarah Jean Vazquez
Uma 10h33'16" 59d40'20"

Sarah Jean Veneklasen
Gem 7h40'45" 32d34'9"

Sarah Jean Veneklasen
And 0h31'17" 38d13'47"

Sarah Jean'ne Turner
And 0h45'11" 45d12'9"

Sarah Jeanne Wielgos
Cap 20h26'23" -19d2'21"

Sarah Jillian Kuhn
Cas 23h35'53" 56d30'23"

Sarah Jo
Leo 10h23'50" 18d39'54"

Sarah Jo Bramblett King
Psc 0h28'16" 15d36'39"

Sarah Jo Gets Her Wings
Leo 10h55'45" 19d19'51"

Sarah Joan Viney O'Neill
And 1h56'7" 45d58'14"

Sarah Joelle
Gem 7h44'0" 20d19'7"

Sarah Joette Moore
Uma 11h27'6" 55d27'31"

Sarah Johnson
And 0h14'51" 43d34'10"

Sarah Johnson
Lmi 10h34'57" 33d21'30"

Sarah Josephine
Sco 17h47'21" -36d4'55"

Sarah Josephine Alohalani Cook
Leo 10h56'5" 5d9'46"

Sarah Josephine Higgins
And 2h35'53" 50d26'48"

Sarah & Josh
Ori 5h12'58" 9d2'50"

Sarah Joy
Ari 3h15'48" 28d11'26"

Sarah Joy Blemings
Psc 0h44'5" 8d35'47"

Sarah Joy Dicks
And 2h15'13" 50d36'24"

Sarah Joy Marier
And 23h7'15" 52d58'24"

Sarah Joyce
Aql 19h22'6" 6d51'44"

Sarah "Jugglozombie" Martino
And 0h33'16" 38d3'44"

Sarah Julia Brown
Lib 14h44'9" -10d45'13"

Sarah Julia Mandy
Lyn 9h15'50" 38d16'56"

Sarah Julian Daniel
Psc 23h41'11" 5d35'25"

Sarah Jung
Dra 19h32'35" 69d4'57"

Sarah K Georgeau
Lyn 7h32'54" 49d35'16"

Sarah K Pringle
And 0h18'17" 36d49'25"

Sarah K. Rosenson
And 0h30'23" 40d2'4"

Sarah Kaitlyn Ball
Cap 21h30'11" -14d35'25"

Sarah Karrigan
Uma 8h16'54" 69d13'35"

Sarah Kate Hodges
Pho 0h37'12" -50d57'45"

Sarah Katherine
Ser 16h0'21" -0d13'7"

Sarah Katherine
Cas 0h18'17" 57d42'0"

Sarah Katherine Burns Johnson
Ari 2h7'58" 25d7'32"

Sarah Katherine Heller
Lyn 8h20'25" 33d52'28"

Sarah Katherine Menzies
And 23h12'34" 42d27'25"

Sarah Katherine Smith
Cep 21h9'52" 61d14'31"

Sarah Kathleen
And 1h48'20" 38d1'23"

Sarah Kathleen Blair - My Lilly
Ori 4h53'6" 5d24'52"

Sarah Kathleen Guttierrez
Tau 4h27'32" 12d30'57"

Sarah Kathrine Macrander
Cnc 9h4'17" 29d50'20"

Sarah Kathryn Bolan
And 1h4'3" 37d53'47"

Sarah Kathryn Kolb
Cnc 8h40'47" 14d35'32"

Sarah Kay
Uma 13h41'27" 48d54'49"

Sarah Kay
Cap 20h40'44" -27d3'21"

Sarah Kay Bean, The Butterfly Star
Cap 21h24'43" -18d35'36"

Sarah Kay Drew
Ori 6h7'15" 21d19'7"

Sarah Kay Higgins
And 0h37'40" 44d57'16"

Sarah Kay Kershner
Dra 17h23'27" 52d4'28"

Sarah Kensley Kellahan
Gem 7h8'32" 32d45'4"

Sarah Keys
Vir 15h4'1" 4d53'42"

Sarah Kiley
Cyg 20h27'28" 49d58'46"

Sarah Kip Miller
Psc 0h16'42" 5d53'35"

Sarah Kissinger
Cam 5h21'16" 68d8'53"

Sarah Kolb
Leo 9h44'46" 27d25'1"

Sarah Kristi Sonier
Lib 15h22'38" -19d41'11"

Sarah Krüsi - Stern der Liebe
Uma 10h39'48" 71d25'52"

Sarah L Adcock
Leo 11h19'56" 14d6'30"

Sarah L. Arnold
Dra 20h26'7" 69d58'31"

Sarah L. Phillips Princess Star
Psc 23h39'23" 4d30'39"

Sarah L. Redding
Ari 2h50'5" 14d0'32"

Sarah L. Shogren
Aqr 22h25'7" -1d31'43"

Sarah L Wilson
Crb 16h17'39" 31d44'11"

Sarah la Princesse de mon Coeur
Sco 16h16'38" -12d21'19"

Sarah Lalueza
And 2h33'0" 43d20'30"

Sarah Lane
Uma 14h24'29" 55d30'16"

Sarah Lane
Ara 17h57'29" -51d36'28"

Sarah Laura Zweber
And 1h53'28" 45d59'18"

Sarah Laverne
Peg 21h12'51" 16d2'10"

Sarah le merveilleux
Aqr 22h17'24" -3d38'24"

Sarah Leanne Curtis Edgar
Sgr 17h54'3" -29d6'41"

Sarah Leclair
And 23h34'35" 48d38'39"

Sarah Lee
Lyn 7h44'37" 42d19'51"

Sarah Lee Cron
Crb 16h10'7" 29d2'16"

"Sarah Lee" Sarah Lee Morehead
Gem 7h3'52" 20d43'34"

Sarah Lee Woodhouse
Cyg 20h19'0" 53d40'27"

Sarah LeeAnn Endsley
Ari 2h37'18" 27d35'44"

Sarah Lees
Mon 6h59'51" 0d34'12"

Sarah Leigh
And 22h59'22" 51d11'41"

Sarah Leingang
Mon 6h24'32" 0d15'27"

Sarah Lenore
And 2h16'21" 47d44'6"

Sarah Lesley Bowlus
And 0h41'5" 25d23'55"

Sarah Leticia Hamann
Cam 3h32'15" 61d39'29"

Sarah Lewe Terenzi
And 2h14'25" 45d57'20"

Sarah Linda Smith
Umi 17h8'3" 85d31'5"

Sarah Litchfield
Tau 5h51'30" 14d55'4"

Sarah Litz
Cyg 19h38'19" 28d24'54"

Sarah Lloyd
Aql 20h4'26" 4d9'6"

Sarah Lois Morrison
Aql 19h23'41" 6d27'29"

Sarah Lopeman
Cas 2h12'40" 62d2'43"

Sarah Lorne Lindquist
Cam 5h40'14" 56d29'14"

Sarah Lorraine
Crb 15h33'36" 26d14'44"

Sarah Lorraine - My Jacqueline
Cra 19h5'37" -39d13'58"

Sarah Lorraine Powell
Lib 15h8'49" -4d31'23"

Sarah Lou Campbell
Com 12h29'43" 27d59'18"

Sarah Loughman
And 1h50'11" 38d4'21"

Sarah Louise
Lib 15h14'41" -9d46'1"

Sarah Louise Anderson
And 1h42'46" 50d12'25"

Sarah Louise Bakos
Cas 0h15'4" 51d59'25"

Sarah Louise Corkett
And 23h18'11" 52d28'3"

Sarah Louise Cronin
Cas 23h28'52" 52d18'23"

Sarah Louise Doss
And 1h10'25" 40d55'42"

Sarah Louise Fox
Cas 23h50'5" 53d45'9"
Sarah Louise Hunt
Cas 0h32'34" 66d27'26"
Sarah Louise James
And 0h21'42" 39d51'15"
Sarah Louise Knight
Cyg 21h36'19" 34d23'29"
Sarah Louise Marshall 043083
Tau 4h0'49" 26d44'20"
Sarah Louise Patton
And 23h42'22" 39d13'0"
Sarah Louise Quinnell
Cas 23h8'20" 55d33'58"
Sarah Louise Rebold "Cuddles"
Vir 14h12'42" -19d9'13"
Sarah Louise Rose
Ori 5h19'6" 11d59'26"
Sarah Louise Walsh
Aqr 23h18'41" -23d7'47"
Sarah Louise Warren
Gem 6h36'9" 20d22'25"
Sarah -love u2 the stars &back- Ian
Cru 12h33'43" -59d31'3"
Sarah Loves Jeff
And 23h15'28" 35d29'38"
Sarah loves Nathan
Cyg 20h4'34" 33d27'26"
Sarah Lowe
And 1h30'42" 50d25'23"
Sarah Loy
And 0h42'22" 29d38'31"
Sarah Lucille Rodriguez
And 1h29'44" 34d35'20"
Sarah Luisa Torres Ramos Mullins
Psc 1h20'5" 15d17'20"
Sarah Lydia Ruth Arrington
Mon 6h28'7" -3d26'27"
Sarah Lynn
Peg 23h26'49" 12d29'53"
Sarah Lynn
Cnc 8h9'42" 16d39'9"
Sarah Lynn
Ari 3h13'25" 26d52'26"
Sarah Lynn
And 1h34'22" 48d33'50"
Sarah Lynn Alverman
Uma 12h9'31" 47d12'29"
Sarah Lynn Bauer
Aur 5h14'39" 33d8'29"
Sarah Lynn Fink
Uma 10h24'33" 55d23'10"
Sarah Lynn Givens
Cap 20h9'40" -10d6'16"
Sarah Lynn Grantham
Cas 0h15'6" 55d11'42"
Sarah Lynn Gray
Cnc 8h50'53" 32d7'26"
Sarah Lynn Gray
Lib 15h45'8" -28d59'34"
Sarah Lynn Grunewald
Cyg 20h5'43" 39d4'5"
Sarah Lynn Haque
Cyg 21h53'12" 50d38'48"
Sarah Lynn Rose
Cyg 19h43'32" 34d31'5"
Sarah Lynn Simpson
Lib 15h12'23" -20d19'2"
Sarah Lynn Smith
Lyn 6h26'13" 56d38'20"
Sarah Lynn Spreitzer
Cnc 8h37'33" 25d40'59"
Sarah Lynn Weber
Ari 1h52'8" 17d43'31"
Sarah Lynne
And 0h15'40" 45d19'33"
Sarah Lytle
Cnc 8h50'33" 17d54'58"
Sarah M. Balthasar
Cnc 8h53'58" 24d41'11"
Sarah M Bosken
Ori 6h1'48" 18d9'56"
Sarah M. Carpenter
Cas 1h26'8" 61d11'57"
Sarah M. Chaffin
Aqr 22h58'21" -21d51'47"
Sarah M Davis
Sgr 19h20'37" -17d45'52"
Sarah M. Green
Ori 6h14'34" 15d2'31"
Sarah Mack
Psc 1h17'9" 15d59'25"
Sarah Maclean
And 1h14'32" 34d14'50"
Sarah Madeline Pound
And 0h40'23" 31d1'36"
Sarah Mae
Dra 18h46'50" 62d17'52"
Sarah Mae
Cru 12h19'53" -63d13'40"
Sarah Mae Ayers
And 2h11'17" 44d2'32"
Sarah Mae Burgess
Cnc 8h24'17" 18d40'14"
Sarah Mae Fox
Sgr 19h47'42" -12d48'19"
Sarah Mair
And 1h27'55" 47d13'57"
Sarah Manion
Ori 5h17'41" 7d40'54"
Sarah Margaret Barnes
Tau 5h9'59" 25d49'15"
Sarah Margaret Caroline Clive
And 23h14'46" 47d7'34"
Sarah Margaret Cave
And 23h24'20" 48d36'45"
Sarah Margaret Cave
Cas 0h32'8" 61d20'34"
Sarah Margaret Michaels
Cas 0h35'38" 57d32'57"
Sarah Margaret - Taylor
Lyn 6h28'31" 57d9'27"
Sarah Maria
Aql 19h58'35" -0d32'16"
Sarah Marie
Lib 15h6'33" -7d13'50"
Sarah Marie
And 0h34'34" 42d49'39"
Sarah Marie
Gem 7h4'16" 30d25'7"
Sarah Marie
Her 17h36'2" 35d18'2"
Sarah Marie
Leo 10h15'1" 26d42'39"
Sarah Marie
And 0h14'28" 32d16'58"
Sarah Marie
Ari 2h40'2" 24d48'5"
Sarah Marie
Ari 2h18'4" 10d42'19"
Sarah Marie
Vir 11h38'4" 2d40'58"
Sarah Marie
Cnc 7h59'33" 13d2'49"
Sarah Marie Boland
Gem 7h39'57" 27d38'21"
Sarah Marie Brown
Umi 13h29'31" 73d11'8"
Sarah Marie Craft
Sco 17h28'1" -39d34'6"
Sarah Marie Dodge
And 23h51'7" 41d45'50"
Sarah Marie Elizabeth Villari
Lib 15h54'37" -15d29'23"
Sarah Marie Ferguson
Gem 7h19'11" 19d7'41"
Sarah Marie Hoff
Lyn 7h29'49" 47d5'8"
Sarah Marie King
Lmi 10h29'57" 38d59'31"
Sarah Marie Kohn
Gem 7h59'0" 32d9'0"
Sarah Marie Kozlowski
Ari 3h28'52" 20d1'55"
Sarah Marie Krumm
Tau 3h31'8" 18d36'56"
Sarah Marie Malone
Cap 21h39'33" -15d47'9"
Sarah Marie Mazzoni
Ind 21h55'3" -70d8'27"
Sarah Marie Olson
Cam 3h44'35" 71d57'41"
Sarah Marie Pattison
And 1h40'56" 50d36'19"
Sarah Marie Pistininzi
Psc 1h32'15" 3d49'25"
Sarah Marie Pochy
Lyr 19h15'54" 29d25'21"
Sarah Marie Price
Ari 1h58'36" 23d43'29"
Sarah Marie Springsteen
And 0h33'53" 41d29'24"
Sarah Marie Stolp
Cnc 9h2'23" 20d56'48"
Sarah Marie Vanderlip
Aqr 22h16'28" 1d49'4"
Sarah Marie Warren
And 1h9'48" 37d12'53"
Sarah Marie Wiese "Wiggles"
Uma 10h52'45" 64d27'58"
Sarah Marie Witt
Cas 0h50'23" 50d45'31"
Sarah Marie York
Sgr 17h55'24" -17d46'23"
Sarah Markey
And 23h30'28" 42d44'53"
Sarah Marples
Cas 23h36'48" 55d55'54"
Sarah Marrero
Vir 12h32'14" -10d42'27"
Sarah Marriott
And 23h13'40" 51d20'35"
Sarah Marshman
Cru 12h53'49" -60d20'57"
Sarah Martin
Lib 15h2'40" -22d4'17"
Sarah Mary Martinson
Cnc 8h36'0" 24d2'59"
Sarah Maslov
Uma 10h50'56" 70d30'42"
Sarah Matthews
Gem 6h54'20" 33d10'8"
SARAH MAXINE HINEBAUGH
Leo 11h21'4" 15d27'38"
Sarah May
Boo 14h54'31" 13d36'53"
Sarah May
And 0h5'23" 44d18'14"
Sarah May and Stanley's Diamond
Cyg 19h44'42" 27d50'10"
Sarah McDowell
Cam 4h23'42" 56d1'7"
Sarah McEntire
Leo 11h2'4" 20d41'19"
Sarah McGilvray
Psc 0h21'53" 16d55'27"
Sarah McQueen Eitzen
Aqr 22h33'12" 0d50'10"
Sarah Mehaffey
Psc 1h29'44" 21d14'36"
Sarah Melissa Rosario
Cnc 9h17'28" 14d22'48"
Sarah Melissa Thomasee
Aqr 21h44'33" -2d35'43"
Sarah Mellendorf
Cyg 20h48'42" 47d18'0"
Sarah (Meoh Meneh) Morning Star
Lyn 7h18'20" 48d3'17"
Sarah Mercier
Uma 11h7'30" 61d15'56"
Sarah Merovitch Dubman
Crb 15h35'37" 28d17'18"
Sarah- Mi Cielo
Psc 0h31'9" 17d2'58"
Sarah & Michael—Together in Love
Cnc 8h22'20" 11d30'37"
Sarah Michele Kiniery
Ari 3h25'45" 22d30'14"
Sarah Michele Kravette
Crb 15h48'46" 38d12'50"
Sarah Michele Page
And 0h10'9" 32d42'20"
Sarah Michele Steudle
Leo 10h38'21" 21d52'44"
Sarah Michelle
Ari 2h36'24" 25d43'38"
Sarah Michelle Almodovar-Kosek
Ori 6h24'53" 16d57'51"
Sarah Michelle Barrett
Lib 15h47'34" -4d44'29"
Sarah Michelle Danielson
Ari 3h13'19" 29d19'1"
Sarah Michelle Henry
Sco 16h16'17" -16d13'6"
Sarah Michelle Hughes
Dra 17h19'6" 63d46'11"
Sarah Michelle Luban
And 23h27'38" 47d7'24"
Sarah Michelle MacNeil
Aqr 22h42'22" 1d55'33"
Sarah Michelle McNamara
Tau 3h36'28" 22d31'33"
Sarah Michelle Rudstrom
Uma 8h21'29" 63d3'32"
Sarah Mihoa Brunick
Cas 0h14'43" 63d3'13"
Sarah Mildenberger
Cnv 13h34'10" 30d44'59"
Sarah Miranda Hildebrand
Pho 23h33'45" -47d24'31"
Sarah Monk
Tau 5h46'6" 16d34'45"
Sarah Moore and Zack Brooks' Star
Gem 7h33'15" 14d51'25"
Sarah Morgan CC
Psc 1h11'38" 26d28'29"
Sarah Morris
Com 12h23'30" 27d6'15"
Sarah Moussavi
Uma 9h11'40" 46d38'23"
Sarah My Angel
Cnc 8h53'15" 12d22'34"
Sarah my anjyl
Cnc 8h29'32" 27d46'43"
Sarah: My Beautiful
Uma 9h30'58" 72d34'31"
Sarah My Beautiful Fuzz
Lib 14h52'13" -2d57'3"
Sarah - My Love
Gem 6h56'44" 19d25'10"
sarah my texas angel
And 2h19'12" 50d10'56"
Sarah*MySSF*Wesley
And 1h37'38" 43d58'14"
Sarah N Lane Briggs Star Baby #1
Uma 11h26'4" 58d58'13"
Sarah Nakata
Mon 7h10'57" -1d5'17"
Sarah Nancy Jenkins
And 1h42'19" 45d21'38"
Sarah Neo No-E Hays
Ori 6h9'20" 7d19'17"
Sarah Nevett, Brightest Star
Cas 0h24'32" 56d54'21"
Sarah Nichole Cullen
Vir 12h18'43" -2d22'16"
Sarah Nickerson
Tau 4h21'16" 27d49'38"
Sarah Nicole
Lyn 8h16'29" 55d52'19"
Sarah Nicole
Sgr 18h3'0" -28d8'42"
Sarah Nicole Akers
Aqr 22h48'55" 2d37'40"
Sarah Nicole Hays
Dra 16h28'9" 58d45'53"
Sarah Nicole McNerney
Cas 1h34'8" 61d59'13"
Sarah Nicole Oates
Cas 23h29'43" 53d2'25"
Sarah Nicole Portela
And 0h26'27" 42d35'46"
Sarah Noelle
Vir 13h17'41" -12d7'47"
Sarah Norton's Star
Leo 10h10'19" 10d16'40"
Sarah Novack Williams
Lib 14h51'7" -3d47'28"
Sarah Nunziata Wrobel
Sco 16h20'41" -17d38'50"
Sarah Nyer
Lyn 8h0'31" 43d51'55"
Sarah Oates
And 0h38'59" 26d18'12"
Sarah Olivia
And 1h30'12" 50d1'9"
Sarah O'Reilly
Psc 1h2'30" 19d21'54"
Sarah O'Rourke LeCates
Leo 11h49'54" 10d30'47"
Sarah Ostroski
Aqr 23h56'7" -7d27'33"
Sarah Ostrowski Royster
And 1h46'28" 46d46'53"
Sarah Owen
And 23h6'41" 43d1'40"
Sarah Owen McGarity
Sco 17h18'58" -42d38'22"
Sarah & Padra Swimming in the Sky
Cyg 21h25'58" 30d59'7"
Sarah Paff
And 0h21'50" 24d55'56"
Sarah Paige Morgan
And 2h31'43" 49d1'26"
Sarah Panayiotou
Lyn 9h18'17" 33d14'53"
Sarah Papay
And 22h58'29" 52d24'21"
Sarah Patrica Lornson
Cas 0h32'34" 47d31'44"
Sarah Patricia Addi
And 23h12'18" 52d48'15"
Sarah & Patrick Myers
Cyg 21h21'4" 36d23'55"
Sarah & Paul
Cyg 20h7'31" 35d29'15"
Sarah Paul
And 23h37'4" 44d17'16"
Sarah Pearse
Peg 22h43'5" 4d51'33"
Sarah Peaslee
And 23h23'7" 48d21'33"
Sarah Perry
Peg 23h9'25" 13d49'42"
Sarah Pershing
Aqr 22h53'27" -16d31'42"
Sarah Pickford
Vir 13h50'13" -17d56'1"
Sarah Pie In The Sky
And 1h48'15" 42d16'7"
Sarah Plotkin
Leo 11h37'58" 12d48'56"
Sarah Polous
Cap 20h29'51" -22d39'55"
Sarah Price
And 1h3'24" 41d30'47"
Sarah "Princess" Arnold
And 0h36'11" 40d5'58"
Sarah (princess) Dworsky
Sgr 18h36'7" -26d45'39"
Sarah "Princess" Morrison
Leo 10h25'52" 21d9'52"
Sarah Propeck
Cas 0h44'21" 65d28'27"
Sarah "Punk" LaFleche
Lyr 18h29'18" 33d3'4"
Sarah Rachael Crowther
And 23h22'18" 50d19'46"
Sarah Rachel Van Horn Lerner
Gem 6h52'16" 24d36'50"
Sarah Rachelle
Cap 20h28'43" -14d29'54"
Sarah Rachelle McNulty
Psc 1h14'2" 30d39'21"
Sarah & Ralph
Ori 6h0'24" 14d36'50"
Sarah Raquel Sieburg
Gem 6h1'2" 26d26'15"
Sarah Rasi
Cam 7h55'30" 76d26'17"
Sarah Raynolds
Lmi 9h32'34" 34d37'22"
SARAH REBECCA
Cas 1h25'5" 59d42'34"
Sarah Rebecca
Tau 4h21'47" 22d21'2"
Sarah Rector
Sgr 18h41'42" -33d38'53"
Sarah Rene' DeArman
Psc 0h13'25" 7d37'43"
Sarah Rene Hall
Leo 10h44'42" 14d34'0"
Sarah Renee
Psc 0h32'45" 11d35'21"
Sarah Renee Coleman
And 0h55'39" 38d4'1"
Sarah Renee Kelly
Ari 3h4'50" 25d50'23"
Sarah Renee Melton - Granddaughter
Lib 15h8'23" -24d13'54"
Sarah Reynolds
And 0h48'50" 30d43'42"
Sarah Rhiannon
Lyr 19h14'49" 26d57'15"
Sarah Rib's Star
Cnc 8h11'4" 26d20'37"
Sarah Rocks
Her 17h45'33" 38d47'16"
Sarah Rogers
And 2h22'54" 48d50'0"
Sarah Romero
Lyn 6h58'47" 61d39'35"
Sarah Rondinelli
Cas 23h47'51" 51d17'58"
Sarah Rooks
Uma 9h23'15" 52d28'40"
Sarah Rosalia May
Psc 0h7'17" 8d18'6"
Sarah Rose
Tau 4h11'9" 15d57'53"
Sarah Rose
Lyn 8h38'42" 34d2'51"
Sarah Rose
Gem 7h9'56" 32d13'8"
Sarah Rose
Lib 14h51'55" -4d22'44"
SARAH ROSE
Lib 15h6'1" -11d45'5"
Sarah Rose Capizzo
Cnc 8h7'4" 17d11'35"
Sarah Rose Grace
And 0h3'9" 43d43'53"
Sarah Rose Gunstream
Cap 21h55'43" -19d14'45"
Sarah Rose Irene Maule
Cnc 8h44'26" 32d7'8"
Sarah Rose Jiminez
Cas 1h23'45" 57d14'5"
Sarah Rose Kelly
And 23h23'43" 39d16'19"
Sarah Rose Morehouse
Sgr 19h8'17" -19d29'4"
Sarah Rose Zarba
Lib 14h22'10" -17d34'18"
Sarah Rosemary
Psc 1h4'36" 3d57'8"
Sarah Rowe
And 2h23'12" 39d21'31"
Sarah Ruth
And 0h47'36" 46d35'9"
Sarah Ruth Yau - Sarah's Star
And 1h5'15" 45d42'59"
Sarah Ruth Zuba
Sgr 19h4'14" -12d46'49"
Sarah Sammon
Per 3h53'1" 39d0'36"
Sarah Sanchis
Uma 13h27'35" 53d12'58"
Sarah 'Saren' Allardice
Ori 5h42'9" -1d26'57"
Sarah Scheurer God's Angel #6
Lib 15h44'21" -6d21'40"
Sarah Schinkel
And 2h1'11" 39d34'54"
Sarah Schinkel
And 1h16'35" 39d51'46"
Sarah Scott
Aqr 23h44'9" -9d19'42"
Sarah Scroggie
And 0h5'18" 37d27'26"
Sarah (SEA)
Aqr 21h8'41" -6d10'15"
Sarah See's Little Nuggett
Sco 16h14'16" -16d31'6"
Sarah Self
Lyr 18h52'47" 36d28'33"
Sarah Setz
Lac 22h55'10" 46d33'16"
Sarah Sha Woo
Cap 21h21'22" -19d9'38"
Sarah Sheikha Khalil
Gem 6h58'23" 16d44'35"
Sarah Sheyanne
And 0h16'34" 27d0'45"
Sarah Shrum
Cap 21h32'45" -17d37'6"
Sarah & Silvan
Cas 3h13'26" 65d2'34"
Sarah Silverman 83051XOXO
Vir 12h34'31" -1d39'55"
Sarah Sims Bryant
Mon 6h32'29" 7d57'9"
Sarah Skin Babe
Cas 23h38'58" 53d34'43"
Sarah Smile
Leo 9h58'18" 20d3'45"
Sarah Smiles
Umi 15h5'42" 68d49'36"
Sarah SnookieDollard James Anderson
Gem 7h7'38" 26d25'18"
Sarah Sorbello
Uma 9h4'34" 48d51'44"
Sarah Sparkle
Uma 13h43'12" 51d42'9"
Sarah Speckman
Oph 16h43'24" -0d54'55"
Sarah Spencer
And 2h20'58" 42d42'3"
Sarah Staples
Dra 19h14'40" 61d34'35"
Sarah Star
Uma 8h45'7" 59d51'59"
Sarah star
Peg 23h55'4" 29d6'48"
Sarah Star Higgins
Gem 6h53'20" 21d51'55"
Sarah Stephinson
Leo 11h21'8" -2d29'42"
Sarah Stockton
Mon 6h46'54" 6d19'15"
Sarah Stone
And 0h45'22" 26d31'5"
Sarah Susana Wilson
Psc 1h21'9" 15d45'1"
Sarah Sutter
Lac 22h56'50" 43d55'55"
Sarah Suzanne Neighbors
Cet 2h21'50" -2d10'46"
Sarah Szabo
Ori 6h3'57" -3d49'34"
Sarah Tafler Koplik
Ori 5h31'31" -0d12'3"
Sarah Tam Oi Pui
Cnc 8h32'16" 19d42'35"
Sarah Tansy Blake
Crb 16h3'28" 35d11'2"
Sarah Taylor
Gem 7h33'25" 31d8'56"
Sarah Terinoni
Psc 0h28'57" 8d28'1"
Sarah Terranova
Lib 15h1'38" -17d4'49"
Sarah Tewksbury
Lyr 19h20'4" 28d36'24"
Sarah Thaemert
And 0h25'28" 42d35'52"
Sarah The BEAutiful
Vir 13h21'53" 2d43'10"
Sarah "The Queen"
Sco 16h16'3" -28d2'47"
Sarah Theresa and Melinda Caroline
And 0h12'20" 44d51'47"
Sarah Theresa Estes
Tau 4h30'30" 21d2'34"
Sarah Theresa Kawalek
Lib 15h43'48" -25d46'42"
Sarah Thompson
And 23h57'53" 42d31'40"
Sarah Thow the bestest friend
And 23h18'5" 52d58'30"
Sarah Time
Cra 18h8'13" -37d2'36"
Sarah Tindle's 21st Birthday Star
Tau 5h49'24" 24d10'11"
Sarah Tobin Smith
Cnc 8h36'54" 9d21'53"
Sarah Tojo
Gem 7h19'4" 28d47'47"
Sarah Torsell
Cam 3h50'40" 63d9'41"
Sarah Trish Gooss
Ori 6h4'38" 18d22'22"
Sarah Trump
Lyn 9h12'28" 41d48'29"
Sarah Turner
Aqr 22h41'53" -0d43'55"
Sarah Tutty
Cyg 20h12'20" 41d30'5"
Sarah Urbas
Cap 20h29'23" -19d20'43"
Sarah Utley Stone
Sgr 19h0'42" -29d25'13"
Sarah Vandenberg
Lib 14h59'47" -12d15'55"
Sarah Victoria Tsacoyannis
Sco 16h8'40" -8d33'19"
Sarah Virginia Firestar 2935
Gem 6h38'59" 21d51'26"
Sarah W. Nathan
Cas 0h45'57" 62d12'17"
Sarah Wale
Lmi 10h10'6" 33d20'29"
Sarah Walsh
Cyg 19h45'46" 33d14'18"
Sarah Watson
Gem 7h38'27" 21d38'25"
Sarah Watterson
Cnc 8h5'52" 9d20'6"
Sarah Watts
Cas 1h26'31" 59d59'37"
Sarah Weatherby Stephens
And 1h26'23" 34d45'39"
Sarah Webster
Aqr 22h30'49" -20d2'41"
Sarah Welter - Traum und Realität
Uma 13h59'13" 56d9'39"
Sarah & Wesley Mills' Wedding Star
Cyg 20h6'47" 51d18'32"
Sarah Westley
Cru 11h58'17" -62d45'33"
Sarah White
Umi 17h19'15" 82d54'51"
Sarah Whitney Hooker
Lyn 8h24'5" 55d47'31"
Sarah Williams
Cas 23h4'16" 54d48'17"
Sarah Wilt
Cnc 8h50'29" 25d13'13"
Sarah Winn
And 1h33'31" 48d28'34"
Sarah Wood
Tau 5h32'0" 18d31'19"
Sarah Wright's Star
Ser 15h51'58" 23d32'52"
Sarah Young
Lib 14h28'40" -10d5'57"
Sarah Younglove
And 0h26'28" 44d10'17"
Sarah Zales
And 2h11'10" 45d11'6"
Sarah Zouari - M
Peg 22h9'27" 21d35'4"
Sarah, Grace, Dana & Tom Theiss
Uma 13h14'50" 59d52'21"
Sarah, Madison, Seany's Star
And 0h40'1" 26d40'30"
Sarah, minha amorinha, Medeiros
Leo 11h29'58" 8d46'28"
Sarah, my one and only
Vir 13h6'14" 9d28'54"
SARAH, NIZHONI SHINING STAR
Ari 3h28'9" 20d59'31"
Sarah28031980
Ari 2h29'15" 27d32'13"
Sarah61894
Gem 7h24'38" 25d22'45"
Saraha Dennis
Mon 6h58'19" -5d11'49"
Sarahbabes
And 0h8'55" 28d59'13"
SarahBaby
Psc 0h45'53" 15d33'49"
SarahBaby
Cap 20h22'7" -9d17'39"
SarahBear
Aqr 20h51'34" -8d56'13"
Sarahbear
Ari 2h2'19" 12d57'26"
Sarahbel
Cyg 19h53'23" 37d40'39"
Sarahbell
Sco 16h8'30" -33d17'41"
Sarahbelle Cullison
Sgr 18h34'21" -22d46'34"
SarahBeth
And 0h44'55" 26d30'26"
Sarahbeth
Tau 3h51'24" 24d40'0"
SarahBob
Gem 6h42'18" 35d4'25"
Sarahdipity
Lyn 7h55'51" 48d20'35"
Sarahdipity
Ari 2h42'21" 28d6'42"
SarahDougWillSophieAlex Carrie
Vir 13h12'2" 5d51'26"
SarahFaithStar
Cas 0h29'14" 59d25'14"
Sarah-Jane Beel
And 2h4'13" 43d35'26"
Sarah-Jane Marie
Lyn 7h11'3" 44d49'56"
Sarah-Jane & Ryan Alcock
Ara 17h10'53" -51d52'26"
SarahJanePowers070590C elestial
Cnc 8h50'56" 28d19'25"
Sarah-Kate Helen Maria Murphy
Cas 0h51'43" 54d5'36"
Sarah-Kelly
Psc 1h3'15" 2d58'54"
SarahL
Gem 7h26'6" 21d26'12"
Sarah-Marie Angela Steer
Cyg 21h35'28" 48d57'7"
Sarah-Michelle Palladino
Uma 9h41'38" 55d10'36"
Sarah-Nicole Young 12:42 PM
And 1h11'34" 37d12'50"
SarahRB
Aqr 21h16'29" -10d32'12"
Sarah-Rose Milner
Uma 11h44'8" 49d25'26"
Sarah's
Uma 9h51'58" 44d24'15"
Sarah's
Com 12h31'38" 23d50'48"
Sarah's Angel
And 1h12'11" 36d52'24"
Sarah's Beauty
Sco 16h7'40" -10d19'54"
Sarah's Christmas Star
Peg 22h21'35" 20d54'29"
Sarah's Dream
Cyg 20h49'48" 31d11'39"
** Sarah's Etoile **
Ari 1h56'57" 24d2'16"
Sarah's & Gerry's Star
Umi 16h34'44" 84d4'27"
Sarah's Guardian Angel
Ari 2h13'6" 13d26'43"
Sarah's Guardian Angel
Cyg 21h9'15" 30d16'15"
Sarah's Guiding Star
Psc 1h16'37" 28d30'36"

Sarah's Heart
Ori 5h27'47" 1d42'45"
Sarah's Kako
Crb 15h47'7" 29d30'47"
Sarah's Light
Lib 15h21'49" -11d22'23"
Sarah's Light
Sgr 18h13'36" -28d5'27"
Sarah's Little Angel
Lyr 18h50'27" 36d58'30"
Sarah's & Marco's Stern der Liebe
Uma 8h53'49" 65d6'32"
Sarah's Neverland
Uma 9h21'8" 56d40'24"
"Sarah's piece of heaven" 27-10-06
Cen 13h57'28" -60d52'20"
Sarah's Place
Ori 5h34'22" -1d10'46"
Sarah's Reach
Tau 5h37'23" 25d19'7"
Sarah's Shining Star
Cas 0h50'42" 61d41'39"
Sarah's Smile
And 0h17'35" 27d23'31"
Sarah's Special Sparkle
Sgr 19h58'23" -25d24'31"
Sarah's Star
Sgr 18h5'18" -26d57'11"
Sarah's Star
Aqr 22h43'5" -22d39'47"
Sarah's Star
Cas 23h0'51" 54d23'52"
Sarah's Star
Cas 23h26'32" 53d6'48"
Sarah's Star
Cas 23h26'38" 57d23'28"
Sarah's Star
Uma 13h39'11" 61d39'18"
Sarah's Star
Aqr 22h52'30" -7d19'32"
Sarah's Star
Lib 15h19'12" -7d28'59"
Sarah's Star
Lib 14h53'28" -1d11'45"
Sarah's Star
Vir 13h23'41" -4d32'40"
Sarah's Star
Sco 16h15'46" -14d23'36"
Sarah's Star
Del 20h34'56" 15d23'39"
Sarah's Star
And 0h37'49" 27d28'26"
Sarah's Star
Ari 3h12'10" 28d22'29"
Sarah's Star
Ori 6h18'2" 2d11'12"
Sarah's Star
Tau 3h43'21" 7d35'22"
Sarah's Star
And 0h56'39" 22d10'45"
Sarah's Star
Cyg 21h42'19" 31d15'54"
Sarah's Star
And 23h50'24" 35d24'5"
Sarah's Star
And 23h26'6" 36d52'30"
Sarah's Star
And 0h53'31" 37d42'46"
Sarah's Star
And 2h19'25" 37d37'39"
Sarah's Star
Uma 11h14'58" 52d15'32"
Sarah's Star
And 23h59'37" 45d5'56"
Sarah's Star
And 2h30'35" 50d9'8"
Sarah's Star
Cyg 20h30'43" 43d0'4"
Sarah's Star
And 23h19'44" 44d59'25"
Sarah's Star (",)
Aqr 23h43'32" -4d6'33"
Sarah's Star of Eternal Light
Cap 21h20'5" -19d50'5"
Sarah's Sweet 16 Star
Cnc 8h4'35" 8d19'46"
Sarah's Three
Cas 1h17'13" 57d25'40"
Sarah's Wish
And 23h26'25" 48d7'41"
Sarah's Wishing Star
Lib 15h4'40" -26d30'47"
Sarahstar
Vir 12h49'19" 2d8'25"
SarahStar 10-29-2004
And 1h43'46" 43d49'26"
SarahSweetheart
Leo 10h42'52" 14d18'37"
Sarah.......Tigereyetw
Lyn 9h13'32" 35d16'27"
Sarai
Ori 5h31'2" -5d37'28"
Sarai Elina
Cas 0h52'38" 63d29'12"
Sarai Elizabeth
Leo 11h2'25" 21d39'58"
Sarai La Hermosa
Lib 14h54'7" -2d2'15"
Sarai Nicole
And 0h14'5" 29d16'51"
Sarai Samathanam Henry-John
Sgr 19h6'58" -16d22'42"
Sarajane
Lib 15h18'49" -9d42'36"
Sarajean
Leo 9h30'50" 29d22'53"
Saraka
Aql 19h16'45" -0d16'24"
Saralee Ahlswede
Aql 19h24'13" 2d19'45"
Saralee June
Gem 6h35'32" 20d14'34"
Saralina
Cap 20h11'23" -9d42'33"
Saralyn Landan
Cap 20h51'12" -22d24'5"
SaraMarie
Gem 7h53'48" 27d48'58"
SARAMIA
Cyg 20h19'8" 52d19'15"
Sarandkarjeff
Aur 5h35'6" 46d3'48"
Sarang
Ori 5h47'7" 7d21'19"
Saranna Superlative
Gem 6h36'40" 21d32'18"
S.A.R.A.P.
Lep 5h9'55" -11d8'25"
Sara's Dancing Star
Aqr 22h2'49" -3d59'58"
Sara's Dreams
Lib 15h11'32" -6d52'8"
Sara's Eclipse
Psc 23h24'46" 4d37'44"
Sara's Hart
And 1h42'1" 42d10'18"
Sara's Heart
Vir 13h29'43" 2d4'10"
Sara's Heart
Ari 2h23'5" 18d24'20"
Sara's Heart
Gem 6h8'25" 26d48'50"
Sara's Love
Tau 4h5'8" 14d56'19"
Sara's Moon
Sco 16h7'11" -20d32'13"
Sara's Oasis
Gem 6h52'19" 22d1'2"
Sara's Radiance
Psc 0h58'19" 9d42'5"
Sara's Smile
Ari 3h21'3" 28d15'37"
Sara's Smile
Uma 12h32'59" 57d27'6"
Sara's Special Star
Gem 7h5'51" 25d0'54"
Sara's Star
Gem 7h56'38" 29d59'37"
Sara's Star
And 0h15'17" 23d57'13"
Sara's Star
Mon 6h29'28" 5d1'58"
Sara's Star
Ori 6h19'22" 15d8'15"
Sara's Star
And 0h8'18" 44d49'0"
Sara's Star
And 0h23'56" 41d33'58"
Sara's Star
Cyg 19h47'32" 31d40'10"
Sara's Star
Cas 1h42'7" 61d12'41"
Sara's Star
Uma 10h34'9" 60d38'42"
Sara's Star
Cap 20h31'27" -11d45'2"
Sara's Star
Lib 14h53'0" -6d6'20"
Sara's Star
Aqr 23h17'36" -18d17'9"
Sara's Star for Carley
Umi 14h36'22" 75d10'0"
Sara's Sunshine
Umi 16h6'24" 71d54'17"
Sara's Super Star
Sco 16h17'57" -13d3'16"
Sara's Twitterpation
Cnc 8h17'48" 21d50'56"
Sara's Very Own
And 2h37'49" 39d8'34"
Sarath
Dra 17h30'47" 64d30'30"
SaRaya
Gem 7h27'35" 33d8'38"
Saraya Skye
Sgr 18h43'7" -28d24'13"
Sarb
Aql 18h51'14" -0d55'36"
Sarbear
Sco 17h33'6" -37d20'7"
Sardar Bhatti
Mon 7h35'10" -0d19'11"
Sardar Jarnail Singh Nijjar
Uma 11h36'43" 31d47'16"
Sardonix
Sgr 19h13'32" -14d6'21"
Sarebear
Aqr 22h42'20" -0d23'17"
Sare-Bear
Peg 22h50'41" 27d13'5"
Saree B
And 23h19'35" 47d11'4"
Saree Erin Anderson
Ari 3h15'7" 28d43'21"
Sareena Soraya
And 2h9'14" 37d32'8"
Sarena
Com 13h3'7" 23d31'34"
Sarey and Sam Bernstein
Cyg 21h33'31" 52d36'34"
Sarge
Psc 1h26'1" 2d47'15"
Sarge & Jenny Billings & Family
Uma 9h21'32" 64d7'7"
Sari
Lyn 8h47'39" 34d11'32"
Sari Ann
Cyg 21h23'22" 32d42'39"
Sari C. Braun's Star of Splendor
Tau 3h41'1" 22d34'49"
Sari Jacoba
Uma 9h4'33" 51d34'3"
Sariah
And 23h57'50" 43d10'45"
Sariah RaNay Echols
And 1h1'27" 41d52'52"
Saribel Torre
Cas 0h32'12" 51d40'4"
Saric
Lyn 7h32'21" 47d34'34"
Saric
Uma 8h12'13" 70d14'5"
Sarieee
Her 18h3'3" 37d5'53"
S.Ariel.S
Ari 2h50'41" 26d59'42"
Sarika Bhatia
Umi 14h21'20" 66d38'58"
Sarill
Cyg 21h33'13" 44d55'15"
Sarim
Tau 4h39'25" 18d44'31"
Sarin
Per 4h47'7" 49d40'2"
Sarina
Cnc 8h55'35" 22d37'2"
Sarina Fee
Vul 20h20'9" 23d33'38"
Sarina Isabella
Gem 6h49'14" 23d13'1"
Sarina la piculeta gionti
Cyg 20h58'56" 40d14'20"
Sarina Patel
Leo 11h13'59" 25d9'28"
Sarina Rose Edwards
Vir 13h8'9" 7d2'41"
sarinabean1
Leo 10h30'20" 26d55'6"
Sarina-Domenica, 13.09.1981
Dra 17h13'6" 66d15'26"
SARINE
And 23h32'57" 41d14'57"
Saris
Lib 15h10'1" -17d9'56"
Sarit
Vir 13h19'36" -2d39'59"
Sarit
Cra 18h1'16" -37d5'50"
Sarit Zadeki
Gem 6h51'15" 23d3'32"
Sarita
Lib 15h15'0" -18d40'30"
Sarita
Sco 15h54'52" -20d37'34"
Sarita and David
Uma 10h42'0" 51d13'21"
Sarita Margaret Bahnsen
Cyg 19h58'26" 31d33'58"
Sarita, Jean and James
Ori 5h53'15" 10d28'0"
Sarita's star
Cap 20h35'26" -15d15'9"
Sarka
Uma 10h29'30" 58d30'20"
Sarkis
Leo 11h16'8" 14d7'48"
Sarkis SAKO Nazarian
Aqr 22h53'39" -3d46'43"
Sarobotica
Ari 3h28'15" 25d25'26"
sarol121466
Sgr 18h36'27" -23d53'37"
Sarolt Hoven
Aqr 21h43'43" -4d37'14"
Saron
Cap 21h34'12" -8d28'37"
Sarona
Peg 22h37'56" 15d36'39"
Sarongeli
Umi 15h17'39" 80d11'48"
Sarp Adiyaman
Uma 13h28'35" 55d13'43"
Sarra Fay
And 23h46'43" 39d15'54"
Sarrah
And 2h15'6" 48d48'28"
Sarrah Elizabeth Fulcher
Cas 1h29'6" 61d22'0"
Sarresa
Ori 5h45'3" 12d29'17"
Saruul's Light
Sco 16h12'48" -11d44'9"
SARVA03
Ari 1h56'10" 20d22'25"
SAS
Uma 11h13'38" 28d22'26"
SAS
Sgr 19h36'49" -11d55'13"
Sas
Umi 13h16'59" 72d3'29"
SAS 2004-2005
Her 17h19'49" 45d32'25"
SAS TURD
And 1h48'29" 43d15'28"
Sasa
Uma 11h39'30" 61d11'15"
Sasà Battaglia
Lac 22h11'47" 36d24'16"
Sasa Mitrovic
Lyn 6h34'22" 59d21'16"
Sasa Vann
And 0h38'35" 36d37'7"
Sascha
Lac 22h42'52" 36d52'49"
Sascha
Ori 5h59'45" 3d23'15"
Sascha
Dra 14h51'28" 57d10'13"
Sascha BOGDANOV
Uma 9h40'39" 50d21'55"
Sascha Florian Hlubek
Leo 10h1'6" 21d20'37"
Sascha Frank
Dra 20h27'6" 71d22'45"
Sascha Kalan
Gem 7h28'42" 24d38'38"
Sascha Kupferschmid
Uma 9h23'43" 55d5'28"
Sascha Macias
Mon 6h52'57" -0d24'29"
SASCHA *MON BON-HEUR*
Umi 15h59'52" 70d25'29"
Sascha + Paula
And 23h14'29" 42d50'30"
Sascha Reimann
Uma 11h6'8" 66d53'33"
Sascha-Rene
Uma 8h38'32" 52d21'14"
SASED'S LIGHT
Her 16h49'30" 37d43'31"
Sasha
Per 4h6'50" 45d2'4"
SASHA
Uma 10h28'37" 46d55'9"
Sasha
And 1h28'59" 42d55'7"
Sasha
Lyn 7h45'36" 41d18'15"
Sasha
Lmi 9h46'50" 39d23'49"
Sasha
Cnc 8h56'50" 9d46'33"
Sasha
Cyg 20h5'59" 57d44'3"
Sasha
Leo 11h4'26" -4d15'14"
Sasha
Vir 12h0'17" -0d8'52"
Sasha Aleksander
Dra 16h9'17" 62d15'35"
Sasha Amaris Marie Ponce
Mon 6h42'42" -0d21'38"
Sasha and Chris
Uma 10h17'14" 62d32'9"
Sasha Baron Monks
Vir 12h46'6" 7d54'15"
Sasha Belle
And 0h20'12" 33d27'12"
Sasha Biggins
Lib 15h34'56" -19d42'29"
Sasha Dell'Elce
Gem 6h45'19" 22d15'56"
Sasha Elizabeth Johnson
And 1h46'48" 41d36'15"
Sasha Graddy
Pho 0h56'28" -48d47'4"
Sasha Jensen
Ori 6h7'59" -1d32'0"
Sasha Katzen Warburg
Cas 0h29'3" 58d56'3"
Sasha Kayley Deans
And 1h3'31" 41d55'18"
Sasha Kleyner
And 23h43'23" 42d41'41"
Sasha Knox
Cra 18h49'17" -40d11'32"
Sasha Korsunsky
Uma 8h38'17" 68d28'0"
Sasha Lyn Henderson
Aqr 22h15'25" -23d12'0"
Sasha Lynne
Uma 11h27'49" 62d43'49"
Sasha Marie
Cmi 7h21'18" 9d6'58"
Sasha Marie Hatch
And 0h19'6" 40d45'5"
Sasha & Mary Bennett
Vir 13h24'5" 11d40'25"
Sasha Monnard
Ori 5h48'47" -3d51'59"
Sasha Mossberger
Sco 16h9'26" -9d22'11"
Sasha Myagkov
Sco 16h54'4" -41d45'49"
Sasha Nicole Reed
And 1h4'54" 46d37'55"
Sasha & Nina Mironov
Tau 3h41'37" 7d42'14"
Sasha Owen
Uma 11h20'42" 53d11'26"
Sasha "Pretty Baby" Serago
Umi 9h55'43" 88d22'26"
Sasha Quinn Hartman
Vir 13h5'17" 12d0'18"
Sasha Renee Lehocky
Gem 7h43'38" 32d13'25"
Sasha Rusetskaya
Cnc 8h45'11" 32d49'28"
Sasha Shawn
Cnc 8h27'17" 25d52'44"
Sasha Sulski
Cas 0h1'45" 57d3'42"
Sasha Synoah Orosco
Leo 9h55'52" 6d57'5"
Sasha Szafir Meneghel
Del 20h20'11" 15d41'33"
Sasha Velarde
And 2h22'25" 50d34'36"
Sasha Yvonne Gunther
Cas 1h43'3" 61d34'9"
sashalena
And 1h5'7" 39d5'51"
Sashamooncat
Lyn 6h40'34" 57d44'26"
Sasha's Blessing
Tau 4h31'44" 15d2'29"
Sasha's Heart
Sgr 19h29'19" -16d23'20"
Sasha's Kari - Eternally
Pyx 8h40'27" -27d45'32"
Sasha's Star
And 0h42'58" 39d26'33"
SashaSarahbellum
Peg 23h15'13" 15d54'42"
Sashechka
Vir 14h11'50" -15d44'42"
Sasheen Aisha
Umi 14h21'22" 78d19'21"
Sashenka Zayarskaya
Vir 12h49'46" 7d18'14"
Sashka
Gem 7h35'5" 23d5'42"
Sashy K
Lyn 9h9'40" 41d16'57"
Saskia
Ori 6h9'29" 8d7'15"
Saskia der schönste Stern
Vir 12h50'48" 11d22'47"
Saskia Isabelle
Uma 13h3'19" 56d25'42"
Saskia Melissa
Sgr 19h6'20" -18d8'20"
Saskia Natascha Mani
Lyn 7h19'52" 46d43'8"
Saskia Scarlet Denbury Preston
Cyg 20h43'9" 33d4'12"
Saskia Sikking
Cas 0h51'37" 60d41'19"
Saskia Sury
Vir 12h12'36" 3d27'17"
Saskia Verschoor
Cyg 20h10'57" 38d34'49"
Saskia vom Halligalli
Lyr 19h13'40" 34d33'2"
Saskuyn 'evesur
Uma 10h25'43" 63d57'57"
Sasooneh
Gem 7h23'47" 26d40'34"
Sassafras
Sgr 18h4'19" -22d5'15"
Sassan Rostamipour
Uma 8h42'0" 68d22'13"
Sassano Maria Teresa
Cam 3h52'12" 54d29'19"
* Sasse *
Cas 23h26'3" 54d35'47"
Sassiest Star Ever - Florentine
Ori 5h40'58" 0d26'53"
Sass-Pfaff
Lyr 19h17'15" 29d14'23"
Sassstar
Lyn 6h20'12" 56d22'13"
Sasssy
Uma 10h30'26" 44d15'3"
Sassy
Cmi 7h37'55" 5d53'28"
Sassy
Gem 6h30'51" 12d15'36"
Sassy
Cap 21h2'12" -18d5'53"
Sassy 85
And 23h20'11" 46d58'30"
Sassy and Kent's baby
Her 18h36'14" 16d19'54"
Sassy DeLellis
Aqr 22h40'50" -4d49'21"
Sassy Galaticia
And 1h30'53" 37d6'15"
Sassy Shiraz
Tri 1h50'53" 30d16'12"
Sassy Stephanie
Gem 7h0'33" 27d36'17"
Sassy Turner
Crb 16h3'10" 26d2'37"
Sassy, Juicy, Chick
Cas 0h36'23" 62d40'46"
Satanasso
Cyg 19h58'56" 39d15'2"
Satara
Cma 7h9'36" -24d14'6"
"Satchman"
Ori 4h58'10" 11d12'5"
Satellite
Her 18h9'37" 29d30'24"
Satfire Sunflower Ann Nelson
Cyg 20h59'58" 46d57'2"
Satgin Seraj
Leo 10h56'50" 16d25'53"
Sathea
Ori 5h45'14" 8d37'54"
Satine
Vir 11h42'49" 4d52'24"
Satine
Apu 15h43'4" -79d15'35"
Satine 8
Dra 20h7'19" 71d43'44"
Satine & Erik Philipp
Uma 12h51'16" 53d41'47"
SATISH
Cnc 9h0'54" 10d2'58"
Satokie and Remington Remembered
Cyg 19h46'21" 35d29'40"
Satomi
Uma 8h14'33" 70d52'14"
Satomi Elsie Hofmann
Mon 6h42'39" -1d3'19"
Satomi Marks
Cyg 20h25'49" 46d0'41"
Satoru & Yayoi's Happy Life
Psc 1h0'38" 21d43'11"
Satuki-NC
Sgr 18h40'17" -20d13'29"
SatusAmmonsus
Cyg 19h36'59" 54d8'40"
Satya Clermont
Sco 15h58'9" -23d0'41"
Satya Sitara
Cnc 8h43'29" 10d58'55"
Satya Wati-Prasad
Lyn 7h34'41" 43d43'7"
Satyajit Deb
Lyr 18h52'19" 36d22'28"
Satyam
Cyg 21h20'45" 39d26'19"
Sauce
Cap 20h13'5" -10d37'50"
Saucier Family Star
Lyn 7h37'16" 47d42'35"
Saucy Minx
Lyn 8h14'29" 44d33'17"
Saud A. Sadiq 06/29/1955
Cnc 8h20'53" 14d37'6"
Saud S. Hashmi
Lyn 9h27'31" 40d45'34"
Saudade
Peg 23h31'14" 17d39'37"
Sauer, Frank H.
Vir 13h26'40" 7d19'55"
Sauer, Hans Peter
Uma 11h34'28" 32d0'59"
Sauer, Udo
Uma 10h20'38" 39d32'38"
SAUL
Leo 10h26'11" 22d31'47"
Saul Aaron Jones
Gem 6h24'12" 23d6'49"
Saul and Rosalyn Infeld
Lyn 8h9'27" 34d30'51"
Saul Vaca
Cma 7h15'1" -26d38'6"
Saul Z
Aql 19h9'25" 14d24'5"
Saul, Ange, and Melina Espinoza
Tri 2h9'9" 33d25'3"
Saule
Cyg 20h30'15" 33d30'25"
SaulGraz
Aur 6h13'19" 45d26'38"
Sauman
Psc 1h4'41" 32d31'44"
Saunders Lawson Wilson IV
Cnc 8h52'33" 26d10'15"
Saundersons 18 - Ivana & Alistair
Peg 23h48'11" 14d23'39"
Saundra
Peg 22h17'28" 16d24'56"
Saundra Elaine
And 1h27'16" 44d0'20"
Saundra J. Gaitten-Brennan
Cap 21h29'11" -14d37'33"
Saundra Kay
Sgr 17h58'46" -16d58'34"
Saundra Kay Braswell Spry
Umi 21h6'39" 89d1'51"
Saundra Vandeven "9-6-1944"
Lyn 8h56'53" 39d4'54"
Saundra Yvonne Lewis
Psc 0h41'51" 12d12'22"
Saundra's Star
Lib 14h22'8" -13d42'41"
Saundra.W
Cam 4h12'0" 55d34'42"
Sauro Nora
Umi 15h59'13" 82d31'2"
Sauveur Mon Coeur
Cma 7h16'42" -14d34'35"
Savage Farms North Roy Frank Ralph
Sgr 18h3'26" -35d46'5"
Savahanna Dawn Chivers
And 23h24'52" 51d26'16"
Savana Davies - Savana's Star
And 23h31'39" 35d26'2"
Savana Devan
And 1h31'23" 40d13'35"
Savana Flores
Uma 9h7'17" 62d51'36"
Savana Jade 28082000
And 0h10'27" 37d41'13"
Savana Renay
Leo 9h40'46" 31d3'42"
Savana Rose Weaver
Lyn 6h49'52" 54d33'17"
Savana Summer Slater
And 2h37'19" 40d14'39"
Savanah Ranilla
Aqr 22h33'54" 1d14'18"
Savanna
Cas 1h0'53" 62d17'38"
Savanna
Cap 21h10'36" -23d22'15"
Savanna Belle Sirmon
And 0h14'22" 28d38'39"
Savanna Cumbie, Teen Angel
And 23h36'43" 42d8'23"
Savanna Faith Gayle Cole
Uma 9h52'53" 61d17'33"
Savanna Foote
And 0h42'2" 41d13'52"
Savanna Gail Terrell
Aql 19h4'2" 0d47'20"
Savanna Jaydon
Lmi 10h27'58" 36d51'52"
Savanna K Moody
Mon 7h15'23" -0d13'30"
Savanna Victoria Fuentes
And 0h39'10" 34d37'7"
Savannah
Psc 1h23'35" 31d4'26"
Savannah
Cyg 20h58'54" 35d12'10"
Savannah
And 2h37'35" 39d14'36"
Savannah
Crb 15h35'57" 37d31'0"
Savannah
Lyn 7h39'52" 49d28'6"
Savannah
Uma 11h18'3" 51d15'47"
Savannah
And 0h36'58" 22d19'40"
Savannah
Ari 2h43'29" 15d53'28"
Savannah
Ori 5h58'35" 21d4'59"
Savannah
Tau 4h46'17" 23d38'28"
Savannah
Tau 4h36'31" 27d1'56"
Savannah
Cap 21h9'38" -15d50'4"
Savannah
Cap 20h57'20" -20d47'16"
Savannah
Cas 1h16'12" 66d0'22"
Savannah
Cam 4h33'49" 66d12'5"
Savannah
Umi 13h39'11" 71d52'5"
Savannah
Umi 14h4'50" 72d41'54"
Savannah
Uma 10h25'58" 56d40'26"
Savannah
Dor 4h12'46" -56d1'5"
Savannah 822
Ori 6h12'31" 21d15'56"
Savannah Anne-Marie
And 0h51'14" 35d54'2"
Savannah Ariel Morris
And 1h15'28" 40d51'30"
Savannah Aspen
Ori 5h54'58" 12d32'26"
Savannah Banana
Uma 12h5'2" 53d29'0"
Savannah Brooke
And 23h12'35" 47d12'53"
Savannah & Caroline Christmas 2006
Peg 22h44'37" 8d25'29"
Savannah Chase Harlan
And 1h30'8" 50d26'56"
Savannah Cheyanne Feinberg
Leo 10h34'28" 9d14'14"
Savannah Claire Jones
Gem 6h28'58" 27d3'15"
Savannah D. Reeves
And 23h5'58" 39d9'37"
Savannah Danielle Rankin
Vir 13h22'27" 13d1'55"
Savannah Dennise Roberts
And 23h3'28" 50d57'54"
Savannah Elizabeth
Sco 16h6'41" -14d19'10"
Savannah Elizabeth Johnston
Cnc 8h21'12" 10d6'11"
Savannah Fae Joy Paul
And 2h13'5" 45d23'53"
Savannah Faith Sides
Tau 4h34'49" 11d46'12"

Savannah Goodman
Ari 2h24'21" 27d39'41"
Savannah Grace
Her 17h50'59" 46d52'46"
Savannah Grace Bond
Birthday Star
Cyg 21h30'20" 35d20'22"
Savannah Grace Engert
Cnc 8h46'26" 16d40'17"
Savannah Haley Christoffel
Lib 15h46'20" -4d50'8"
Savannah Heape
And 0h19'44" 29d23'28"
Savannah Hope Williams
Lib 15h1'34" -12d18'20"
Savannah Hotel
Lac 22h23'54" 48d53'46"
Savannah House
Leo 11h23'23" 6d7'13"
Savannah Isabel Gonzalez
And 23h8'50" 37d11'46"
Savannah Jane
Sco 17h46'15" -36d41'11"
Savannah Jean
And 23h34'43" 37d37'1"
Savannah Jeffries
Sco 17h16'15" -34d14'9"
Savannah Jewel Bonner
Uma 11h32'7" 41d38'40"
Savannah Jo Johnson
And 0h39'12" 35d43'42"
Savannah Kate Malarney
Ori 5h52'41" 21d8'42"
Savannah Kayle Booker
Peg 22h40'48" 16d41'27"
Savannah Lane "5-16-02"
Uma 13h25'43" 58d43'51"
Savannah Lee Lacy
And 0h0'59" 35d33'6"
Savannah Leigh Carman
And 2h22'44" 48d1'37"
Savannah Leigh Elliott
Cas 1h2'38" 63d36'5"
Savannah Leigh Poole
Uma 11h6'25" 64d41'17"
Savannah Lily
Cyg 21h17'3" 34d44'3"
Savannah Lily
Ari 3h5'27" 11d14'43"
Savannah Louise Hill
Peg 22h58'49" 28d41'49"
Savannah Lynee Willis
Crb 15h39'24" 26d12'48"
Savannah Lynn
And 0h10'5" 44d7'57"
Savannah Mabie
Crb 15h51'4" 34d23'31"
Savannah Mae
And 2h29'3" 41d20'38"
Savannah Mae Baldwin
And 1h12'4" 35d58'11"
Savannah Marie
Ari 2h49'46" 14d20'47"
Savannah Marie
Aqr 21h20'14" -1d20'39"
Savannah Marie Duarte 12-28-2003
Lyn 8h21'49" 52d31'39"
Savannah Marie Klug
Cyg 21h55'10" 48d58'40"
Savannah Marie Woodard
And 23h3'11" 52d26'22"
Savannah Mary
And 0h14'46" 42d58'40"
Savannah Maureen Hill
Vir 14h14'38" -17d17'3"
Savannah May Boyd
Tau 5h47'10" 12d49'20"
Savannah Michelle Reese
Cap 20h32'3" -17d46'33"
Savannah Morgan
Cnc 7h55'53" 18d59'16"
Savannah Morgan Jones
Cap 20h37'54" -14d15'15"
Savannah Murray Kruse
Cap 21h15'19" -21d8'2"
Savannah Nicole Veras
And 23h27'32" 47d34'45"
Savannah Noel
Ori 5h33'14" 1d19'31"
Savannah Rachelle Byers
Uma 14h25'23" 56d40'34"
Savannah Rae Taylor
And 0h59'39" 45d33'24"
Savannah Renee Green
Peg 21h47'9" 14d6'28"
Savannah Renee Smith
Uma 10h48'59" 60d32'42"
Savannah Riley O'Quin
Psc 1h11'6" 3d45'17"
Savannah Rose
Ori 6h3'14" -0d44'5"
Savannah Rose
Del 20h41'20" 16d43'46"
Savannah Rose Fedon
Sco 17h19'30" -37d1'14"
Savannah Rose
Genenbacher
And 0h31'14" 27d27'30"
Savannah Rose Lobel
Ari 2h38'22" 25d47'23"
Savannah Rose - Munchkin
Vir 14h5'3" -8d36'44"
Savannah Ruth Marsceill
Uma 11h24'41" 35d16'53"
Savannah S Chandler
Cap 20h12'26" -10d32'27"
Savannah Sage Mintz
Gem 6h28'43" 21d0'45"
Savannah Scout Barnes
Tau 4h34'31" 24d34'24"
Savannah Smiles
Uma 9h41'23" 71d16'2"
"Savannah Starr"
Cam 5h58'20" 60d25'16"
Savannah Taylor Hill's
Shining Star
Uma 11h31'49" 48d47'11"
Savannah Udeen
Leo 9h36'50" 7d24'34"
Savannah Vecchiarelli
Cas 1h34'21" 68d47'51"
Savannah Waller
Ori 5h52'18" 6d11'39"
Savannah Wells
Lib 15h10'10" -5d54'8"
Savannah Whitehead
Sgr 19h38'15" -22d12'53"
Savannah Wright
Dra 16h18'25" 59d21'41"
Savannah, my own bright
star.
Ari 2h42'18" 25d49'7"
Savannahrose
Sgr 19h55'6" -19d20'3"
Savannah's GodFather
Psc 0h10'46" 7d52'0"
Savannah's Moon
Cas 0h55'18" 52d8'7"
Savannah's Star
Lyr 18h46'17" 34d15'25"
Savannah's Star
Vir 13h29'1" -3d46'42"
Savanna's Star
Psc 1h6'3" 17d46'4"
Savelly N. Perez
Aqr 22h33'45" -1d35'37"
Saverio Bitonti
Lyr 19h18'38" 28d15'42"
Saverio DeFino
Cmi 7h32'48" 8d2'54"
Saverio DeLuca
Boo 15h31'55" 46d36'38"
SAVI
Cru 12h6'1" -62d29'48"
Savi Dale Mooney
Uma 11h24'49" 47d22'39"
Savina and Dayal Rakha
hyphen Patel
Cyg 20h21'51" 33d48'41"
Saving Grace
Leo 10h30'48" 15d25'43"
Savino Terranova
Dra 17h54'52" 53d4'30"
Savitri Ganesan
Sgr 19h45'56" -16d48'31"
Savka & Torat
Uma 9h12'39" 47d28'18"
SAVM
Gem 6h32'52" 27d32'49"
SAVO
Ari 3h21'38" 28d54'7"
Savoeun Pen
Gem 6h53'3" 26d27'31"
Savonne
Cnc 8h59'37" 24d10'25"
Savoroski
Ori 4h49'40" 13d7'21"
Savva
Per 3h32'30" 48d27'49"
Savvina Porfiriadou
Dra 17h9'25" 63d45'12"
SAVVY
Lyn 6h44'18" 58d0'18"
Savvy Snowstar
Cam 3h56'37" 65d4'39"
Savvy.Artistic.Resourceful.
Advocate
Uma 11h51'41" 39d57'13"
SAW
And 0h14'8" 28d17'14"
SAW1966
Ari 2h12'36" 22d26'8"
Sawa Michelle Hashimoto
Ori 5h59'0" -0d44'12"
Sawah
Uma 9h52'13" 46d9'41"
Sawara
Cru 12h20'44" -57d48'18"
Sawdust
Cyg 21h3'9" 51d30'49"
Sawyer Beasley
Lyn 9h22'2" 33d1'48"
Sawyer Daniel Gall
Her 17h44'43" 46d31'59"
Sawyer Hayes Sills
Per 4h11'59" 49d26'54"
Sawyer Sylvester Ruggiero
Umi 14h41'33" 74d57'48"
Sawyor May Plath
Tau 4h21'58" 29d51'4"
Saxby
Vir 14h19'50" 2d18'14"
Saxena, Shyama
Uma 10h19'9" 65d14'21"
Saxon Kiesewetter
Uma 11h27'13" 64d41'57"
Saxophone
Ari 3h12'41" 28d33'13"
"SAY ANYTHING..Jen"
Uma 12h15'9" 59d33'19"
Say Goodnight Gracie
Leo 9h23'52" 10d41'30"
Say Sophia Xiong
Gem 7h19'31" 26d6'57"
Saya Tanaka
Cam 7h50'41" 70d42'18"
Sayanh Vongkhamchanh
Vir 12h58'5" -5d41'3"
Sayann Alecia Yun
Aqr 23h6'39" -10d28'22"
Saydie's Daddy, David
Uma 11h24'54" 58d41'37"
Sayed
Uma 11h26'37" 35d49'14"
Sayed Mahmoud Fathi
Saber
Uma 10h23'56" 49d37'18"
Sayeed
Aqr 22h44'23" -17d5'50"
Sayer Steeley
Lyn 7h37'9" 57d26'37"
Sayla
Uma 10h17'19" 48d24'2"
Saylor Francesca Episcopo
Cnc 8h26'54" 25d55'18"
Saylor Paige Smith
Psc 0h13'29" 11d39'19"
Sayra P. Villegas
Cyg 20h50'34" 48d10'4"
Sayra Starer-Blumberg
Sgr 18h24'22" -32d6'39"
Sayra Sue Mansolino
And 0h8'1" 42d24'6"
Sayre
Lyn 6h21'19" 60d2'9"
Sayu o Miteiru
Psc 1h55'46" 5d17'53"
Sayuring
Cap 20h17'35" -16d12'5"
Sayward Alexis Staggs
Tau 4h22'34" 9d15'15"
SB + CM
Sco 16h6'53" -16d57'31"
SB143HB
Lyn 7h47'33" 48d33'25"
Sbarbaro The Great
Defender
Dra 18h58'44" 55d9'58"
SBC
Gem 6h39'58" 26d9'16"
S.B.C & D.S.D.
Cyg 20h47'8" 51d54'15"
SBLS
Vir 13h29'52" -13d48'55"
SBShrankkkkNB
Her 16h39'22" 6d47'2"
SBT
Uma 12h58'32" 57d55'52"
sbt2668
Aqr 22h46'21" -8d11'38"
SBubbie Star
Vir 13h14'3" -0d44'18"
Sbutternut
Lib 15h47'10" -29d28'4"
SC
Uma 9h40'35" 61d48'37"
Scalarman
Vir 14h21'57" -0d6'10"
Scales Elementary
Umi 20h26'49" 89d43'57"
Scaling
Pyx 8h33'6" -21d17'37"
Scally
Cma 7h18'22" -13d35'57"
Scalumchuous
Aql 19h10'11" 14d54'23"
Scalz
Vir 13h18'25" 4d9'13"
Scamper
Leo 11h9'6" 4d15'48"
Scandal
Per 2h58'18" 48d13'19"
ScanDarly
Ori 5h38'11" 7d40'50"
Scanzi Nadine
Psc 22h52'54" 4d17'53"
SCAR
Uma 13h56'21" 59d5'29"
Scaramouche
Lyn 7h57'57" 34d2'19"
Scarlet
And 23h14'9" 43d5'37"
Scarlet
Umi 15h10'46" 68d53'14"
Scarlet Angel
Cas 0h49'43" 65d30'29"
Scarlet Angela Fuentes
And 0h11'34" 36d0'55"
Scarlet Diaz
Sco 16h5'44" -11d50'15"
Scarlet & Hollis Anderson
Cap 21h31'15" -13d51'35"
Scarlet Louise Oliva
Tau 5h50'32" 14d38'32"
Scarlet Musette Teeter-
Prasanchum
Aqr 23h18'59" -22d17'37"
Scarlet Rae
Ori 5h8'48" 9d21'9"
Scarlet Rose
Cyg 21h56'6" 48d5'13"
Scarlet Rose
Cas 2h23'19" 70d41'54"
Scarlet Rose Faulkhard
Cyg 20h7'31" 30d25'14"
Scarlet Rose Heap
Cam 5h24'29" 57d28'56"
Scarlet Ruane
Cas 23h18'41" 54d17'32"
Scarlet Sophia Grace
Strasser
Cnc 8h3'41" 17d12'29"
Scarlet Yeung
And 1h25'6" 43d24'21"
Scarlett
And 0h23'15" 46d8'10"
Scarlett
Uma 13h18'56" 60d0'11"
Scarlett Allen
And 1h11'30" 46d21'3"
Scarlett Bayard
Crb 16h8'21" 28d13'2"
Scarlett D'Agostino
Leo 11h38'48" 24d7'9"
Scarlett & Eddie
Cyg 21h51'42" 44d1'29"
Scarlett Elizabeth
Sgr 18h12'22" -28d10'48"
Scarlett Ellen Wight
Cas 23h47'6" 50d50'10"
Scarlett Eloise Welch
Lemieux
Lyn 8h27'42" 50d54'56"
Scarlett Eris Maguire
And 23h1'18" 49d18'31"
Scarlett Grace Harrison
And 1h29'36" 34d10'14"
Scarlett Jewel Hughes
Aqr 21h4'25" -10d47'1"
Scarlett Key
Pho 1h10'23" -41d1'15"
Scarlett M. Cerreta
Lyn 7h57'5" 36d39'46"
Scarlett Marie-Jacqueline
Patricia
Pho 0h34'20" -47d37'23"
Scarlett May Burkitt
Psc 0h2'46" 1d38'6"
Scarlett Mia Bayley
Cru 12h56'46" -57d44'41"
Scarlett Noël Pomers
Crb 15h58'5" 27d19'3"
Scarlett Olivia
And 22h59'38" 39d56'29"
Scarlett Rose
Cma 6h37'12" -18d12'6"
Scarlett Rose Sparks
Cra 18h4'40" -40d32'32"
Scarlett Stone
Psc 1h1'9" 28d8'28"
scarlett, 16.05.67
Dra 19h29'56" 70d3'27"
Scarlette Sherice Tapia
Lyn 8h12'31" 36d29'29"
Scarpetta
Uma 8h30'5" 66d8'32"
Scarpini Laura
Per 3h1'10" 51d38'30"
Scar's Lady Vivamus
Aql 20h4'36" 12d4'19"
Scarsella Domenica Emilio
Fred Tony
Uma 12h32'10" 62d14'3"
Scary
Leo 11h27'22" 24d6'44"
Scattolini Loredana
Cyg 21h42'1" 53d4'40"
Scatty Natty
Psc 1h10'35" 28d17'21"
SCB
Umi 16h57'31" 82d39'37"
S.C.C.
Lmi 10h37'19" 36d20'27"
SCDI- Friends and Lovers
Forever
Uma 8h31'59" 66d10'8"
SCEB
Uma 10h28'3" 45d8'54"
Scelena
Uma 11h51'3" 36d24'8"
Scelfo Shining Lite
Lyn 8h7'50" 55d40'46"
SCENIC
Lac 22h33'21" 38d9'23"
SCER
Cnc 9h10'29" 24d52'43"
Schaad-Soder
Lac 22h39'57" 38d9'44"
Schaaf, Sylvia
Uma 10h38'13" 66d14'47"
Schaar Star
Cap 20h11'17" -17d6'1"
Schächer, Michael
Uma 13h59'29" 61d38'5"
Schacht, Willi
Uma 10h19'23" 54d57'21"
Schade - 4
Uma 14h0'53" 59d37'16"
Schaefer, Wolfgang
Ori 5h14'51" -8d12'16"
Schaeffer
Ori 5h30'21" -6d5'29"
Schaeffer
Cnc 8h20'28" 22d49'20"
Schäfer, Helmut
Psc 1h14'55" 2d53'59"
Schäfer, Jens-Peter
Uma 9h46'50" 54d35'22"
Schaffran, Helmut
Uma 13h51'27" 55d54'50"
Schafstall, Franz
Uma 11h1'4" 37d31'49"
Schägner, Gerd
Ori 5h20'25" 3d39'40"
Schakarian, Ruben
Ori 6h16'22" 19d17'17"
Schallenberg, Hubert
Lib 15h42'46" -17d50'44"
Schander
Uma 9h18'47" 60d48'18"
Schantel Annette Shelby
Vir 13h55'56" -5d1'27"
Schappele, Hubert
Ori 5h23'36" 10d10'0"
schäpperli
Lac 22h49'8" 52d58'5"
Scharizer
Dra 20h19'58" 62d30'19"
Scharlene A. Burton
Lyn 8h13'13" 43d38'25"
SCHATJE
Cnc 8h52'7" 32d6'9"
Schatzi
Leo 10h44'0" 12d48'7"
Schatzi
Leo 9h42'1" 17d43'39"
Schatzi
Uma 9h5'46" 54d9'41"
Schatzi
Lyn 6h48'0" 56d42'52"
Schatzi
Umi 16h4'46" 75d34'48"
Schatzi Silvano
Her 17h34'44" 32d23'18"
Schatzie Pauline
Uma 9h51'5" 52d28'38"
SCHATZINGER, Doris
Tau 5h1'42" 25d18'36"
Schätzu
Ori 6h8'30" 7d18'22"
Schatzy
Psc 1h4'13" 30d12'1"
Schelle, Roland
Uma 13h4'26" 62d19'56"
Schellenberger, Herbert
Uma 9h2'2" 60d56'15"
Schelwat, Christian
Ori 6h11'40" 15d49'6"
Schemmer, Bernard
Uma 8h37'58" 47d36'35"
Schenk Patric
Leo 10h27'59" 16d36'34"
Schenky's Star - David
John Schenk
Ori 5h27'27" -1d42'42"
Scherie Haley and Samuel
Uma 10h3'2" 55d55'56"
Scherie Kay
Cnc 8h41'24" 18d49'16"
Scherkenbach
Uma 11h10'0" 58d15'13"
Scheubert, Günther
Ori 6h10'23" 16d50'39"
Scheuer, Wolfgang
Ori 6h18'26" 14d31'45"
Scheuermann, Siegbert
Uma 8h43'37" 48d13'55"
Schich, Margot
Uma 11h3'29" 63d42'41"
Schiefelbein, Frank
Ori 5h34'56" 10d17'39"
Schimpf, Markus
Uma 12h45'18" 60d6'51"
Schink, Peter
Uma 11h36'20" 28d55'40"
Schippers
Cap 20h42'56" -21d3'18"
Schirmer 96
Uma 11h34'31" 60d41'21"
Schirmer, Roland
Ori 5h26'2" 13d9'18"
Schirow, Harry
Uma 13h16'31" 60d28'25"
Schirp, Manfred
Uma 10h30'34" 41d44'27"
Schlanda Taylor
And 0h34'56" 31d42'33"
Schlebusch, Jan
Uma 10h34'3" 67d30'20"
Schletti
Her 17h54'20" 14d39'5"
Schließner, Hans-Jürgen
Ori 6h13'15" 8d58'49"
Schlitzohr
Uma 9h25'3" 47d41'6"
Schloopy
Uma 12h41'19" 62d30'6"
Schloss Velden
Aur 5h15'3" 44d14'19"
Schlosser: Master of the
UNITverse
Gem 6h37'48" 18d9'50"
Schlosser, Marleen
Uma 10h9'35" 68d57'42"
Schlösser, Marlen Luisa
Helga
Ori 6h8'32" 6d6'28"
schlumilu
Per 3h43'55" 45d13'12"
Schmedlina
And 1h30'52" 49d54'21"
Schmeißer, Luca Walter
Ori 5h13'38" -8d18'55"
Schmelzer, Detlev
Uma 9h33'51" 47d43'8"
Schmetterling
Uma 13h11'46" 53d14'58"
Schmeusel
Per 2h24'57" 51d6'12"
Schmid
Crb 15h36'0" 37d3'57"
Schmidbauer, Werner
Ori 6h10'20" -0d43'57"
Schmidhammer, Ludwig
Uma 8h55'58" 47d52'14"
Schmidi
Cnv 12h51'9" 41d37'58"
Schmidmayer, Anna
Uma 9h7'46" 55d55'41"
Schmidt, Dirk
Ori 6h20'28" 9d43'10"
Schmidt, Eckard
Ori 4h57'7" 11d9'3"
Schmidt, Ernst Hermann
Ori 6h17'25" 9d42'57"
Schmidt, Frank-Peter
Ori 6h11'50" 2d6'59"
Schmidt, Friedrich-Wilhelm
Karl
Ori 6h18'21" 11d12'5"
Schmidt, geb. am
13.06.1965, Birgit
Uma 9h13'41" 50d29'56"
Schmidt, Gunther
Uma 11h34'32" 53d28'40"
Schmidt, Hans-Joachim
Uma 11h5'12" 57d10'28"
Schmidt, Helmut
Uma 8h27'40" 64d13'50"
Schmidt, Herbert
Ori 6h16'5" 16d51'43"
Schmidt, Luisa
Uma 9h4'12" 68d36'30"
Schmidt, Michael
Uma 13h58'59" 55d53'17"
Schmidt, Morten Erik
Ori 6h10'10" 2d33'9"
Schmidt, Sonja
Ori 6h16'7" 10d18'13"
Schmidt, Thomas
Uma 12h14'34" 61d29'59"
Schmidt, Werner
Uma 8h13'36" 72d52'55"
Schmidt, Willi
Uma 9h14'0" 52d8'37"
Schmidtmeister, Hans-
Walter
Vir 12h51'7" 2d23'30"
Schmiedel, Walter
Ori 6h18'25" 15d43'24"
Schmitt Family A Gift of
LOVE
Cas 0h53'57" 59d20'13"
Schmitt, Dieter
Uma 8h38'26" 63d42'30"
Schmitthausen, Dieter
Uma 8h21'57" 68d35'20"
Schmitz Family
Uma 10h17'44" 52d0'23"
SCHMLANT
Per 3h37'7" 47d21'49"
Schmoellermann
Uma 8h46'2" 60d59'27"
Schmölz, Helmut
Uma 11h43'9" 37d56'28"
Schmoo & Smootchie For
Eternity
Cyg 20h9'55" 35d58'23"
Schmoopdromeda 18
Uma 11h16'35" 64d15'56"
Schmoopee
Cyg 21h21'7" 51d47'10"
Schmoopie
Leo 11h19'24" 5d18'59"
Schmoopie
Dra 19h31'15" 61d10'25"
Schmoopie Haas "9-24-
2000"
Uma 10h41'33" 44d52'16"
Schmoopsie Poo
Gem 6h54'41" 33d53'55"
Schmoopy
Vir 12h6'47" 8d16'50"
Schmoopy
Ori 5h54'40" 18d37'13"
Schmoopy
Cnc 8h17'58" 22d32'56"
Schmoopy Love
Sgr 19h20'3" -15d35'36"
Schmötzer, Karl-Heinz
Uma 9h49'9" 42d31'17"
Schmusebär
Uma 13h33'3" 55d26'54"
Schmusebär &
Härzkäferchen
Lac 22h54'2" 38d6'30"
Schmusemaus
Uma 8h21'25" 63d38'47"
Schmusi (Nicole Blazy)
Uma 11h21'56" 35d51'36"
Schmuts
Lyn 6h21'14" 61d19'26"
Schnäggeler
Crb 16h20'35" 31d44'31"
Schnätzubär
Umi 15h34'32" 68d42'58"
Schneby
Lib 14h48'22" -17d18'8"
Schneck's Odyssey
Uma 12h30'51" 56d56'16"
Schneemann, Eckhart
Uma 9h16'38" 64d10'7"
Schneggli (Bigi Stucki)
Uma 11h20'46" 31d41'7"
Schneider, Heiko
Ori 6h10'25" 15d32'25"
Schneider, Manfred
Uma 9h33'28" 66d6'25"
Schneider, Margarete
Ori 6h14'54" 1d56'1"
Schneider, Uwe
Uma 13h4'48" 57d19'48"
Schneiders Flowers
Uma 11h28'9" 63d30'45"
SCHNITZ
And 2h26'48" 37d29'5"
Schnookie
Ori 5h40'49" 1d39'8"
Schnooks
Aqr 21h38'4" -0d23'37"
Schnookumbs
Ari 2h50'41" 26d45'55"
Schnoutelle
And 23h17'26" 41d54'57"
"Schnucki" Susanne
Giesecke
Ori 6h24'29" 10d48'46"
(Schnuckums) Nikki D
Daffron
Uma 8h44'16" 54d48'35"
Schnüfel
Cam 10h52'39" 79d44'21"
Schnuff
Mon 7h1'40" 4d59'44"
Schnuff
Sge 20h16'39" 19d4'2"
Schnuffelbaer
Uma 9h45'25" 62d54'42"
Schnuffelbärchen
Uma 8h48'59" 47d33'17"
Schnüffi
Dra 18h19'52" 72d18'48"
Schnuffi Kristin
Uma 8h51'10" 69d54'5"
Schnuge Bolle
Mon 7h0'52" 5d32'23"
Schnügel
Lac 22h46'54" 48d51'37"
Schnüger
Cam 6h20'11" 66d1'30"
Schnuggel, 27.07.2002
Ori 5h2'8" 13d16'5"
Schnuggibuts
Lyr 19h21'17" 37d32'59"
Schnugi
Uma 12h57'31" 54d58'26"
Schnukulupulus
Leo 9h58'35" 21d20'36"
Schnuppe
Lac 22h57'5" 46d8'26"
Schnuppel
Uma 12h51'19" 62d11'36"
Schnuppi
Uma 10h53'55" 58d51'39"
SCHNURLIOPA
Umi 16h55'48" 84d30'46"
SCHNURZL2111
Uma 9h23'8" 69d58'2"
Schober, Gisela
Uma 10h40'26" 47d9'51"
Schöber, Leon Paul
Tau 4h29'56" 29d21'27"
Schoch Star
Pho 0h14'27" -39d51'54"
Schoeny
Uma 13h32'35" 62d58'25"
Schoetzau, Jens
Ori 5h6'21" 5d47'59"
Schöler, Rolf
Uma 12h50'34" 52d58'2"
Scholte, Klaus
Ori 6h18'42" 19d54'9"
Scholtes, Erwin
Ori 4h56'15" 15d15'26"
Scholtes, Leonhard
Uma 11h34'37" 38d54'53"
Schommer, Harry
Uma 9h37'33" 54d10'6"
Schön Kristin
Crb 15h41'37" 31d41'54"
Schön Leben
Cap 21h15'54" -15d0'2"
Schönborn, Sabrina
Ori 6h16'26" 15d56'5"
Schöne, Mathias
Ori 5h24'31" 10d47'24"
Schöne, Sebastian Oliver
Christian
Uma 12h17'20" 60d43'19"
Schoner Traumer
Ari 2h7'20" 24d50'7"
Schönfelder, Eva
Uma 11h39'6" 40d55'34"
Schoolgirl
Leo 10h35'9" 7d59'32"
Schooner M. Krown
Uma 13h50'47" 55d2'24"
SCHOONY'S LANDING
Vir 12h8'2" 11d28'51"
Schöpf Fam. Forever, Ursi,
Kü, Zita, Beat
Crb 15h50'27" 27d51'15"

Schöps, Andreas Psc 1h53'10" 6d11'34"
Schoss Cnc 8h9'2" 24d57'5"
Schossböck, Richard Uma 11h8'28" 63d54'53"
Schotz Guy Aqr 22h39'32" 0d54'17"
Schotzcess Cnc 8h20'11" 14d41'34"
Schpazel Ori 6h17'8" 15d14'12"
Schpenke Uma 11h14'19" 36d26'2"
Schrader-Nahoom Lps And 23h45'33" 46d54'21"
Schrage, Wolfgang Uma 9h44'53" 49d14'12"
Schraivogel, Siegfried Uma 8h53'51" 50d59'36"
Schramme, Guido Uma 8h47'10" 48d1'47"
Schree DeLoreto Uma 10h36'19" 49d32'54"
Schreen Raboza Mon 6h28'23" 6d31'3"
Schreiber, Günter Ori 5h13'22" -8d18'25"
Schreiber, Olaf Uma 12h49'31" 62d14'14"
Schreiber, Rolf Ori 6h14'0" 9d1'9"
Schreiner, Antje Uma 8h10'30" 62d59'12"
Schreiner, Ina Ori 6h2'33" 10d52'17"
Schreiner, Marita Psc 1h9'30" 2d50'19"
Schreiner, Zoe Marie Uma 8h39'43" 59d13'49"
Schröder, Nadine Uma 11h28'25" 33d54'11"
Schrovany Ori 6h16'17" 14d37'0"
Schubert, Bernd Uma 9h42'3" 64d3'29"
Schubert, Dieter Ori 5h51'42" 7d21'52"
Schubgalsond Cam 3h48'57" 53d15'7"
Schubring, Ernst Jürgen Uma 11h15'7" 34d20'40"
Schukies, Helga Uma 12h58'20" 52d52'25"
Schuler & Sparkles Love Star Cap 21h23'48" -17d34'45"
Schuller, Anne Uma 11h16'4" 64d1'43"
Schulte, Martin Uma 12h8'59" 50d19'52"
Schultz, Carsten Uma 8h16'9" 64d35'9"
Schultz, Gabriele Ori 5h1'22" 0d14'22"
Schulz, Erhardt Uma 10h31'16" 69d31'43"
Schulz, Horst Uma 13h37'41" 56d48'50"
Schulz, Kathleen Uma 12h35'35" 58d10'12"
Schulz, Steffen Uma 14h3'18" 56d13'51"
Schulze Föcking, Luc Ori 5h18'10" -4d23'31"
Schulze, Barbara Uma 10h5'36" 51d54'15"
Schulz-Menningmann, Uwe Gem 6h27'52" 20d7'26"
Schumann, Wolfgang Uma 8h53'23" 61d28'44"
Schupp, Jürgen Uma 8h58'54" 68d52'44"
Schuster & Johnson-Whatley Umi 15h36'52" 73d9'47"
Schuster, Josef Ori 6h19'15" 16d42'50"
Schuster, Julia und Florian Ori 6h10'18" 16d43'52"
Schuster, Klaus Uma 11h14'34" 63d50'34"
Schütte, Malte Ori 6h4'6" 10d12'49"
Schutzengel Uma 10h40'23" 58d44'10"
Schutzstern Cyg 20h42'29" 45d44'26"
Schuyler Chubet Crb 16h14'3" 26d19'22"
Schuyler Haynes Per 2h37'8" 55d22'17"
Schuyler Peterson Pho 0h57'36" -49d25'0"
Schuyler Rose Ori 5h28'52" 13d31'53"
Schuyler's Gold Her 17h17'16" 16d31'24"
Schwab Family Ori 5h11'10" 5d57'31"
Schwarcz Erzsébet Cas 0h6'19" 54d55'53"
Schwarcz Miklós csillaga Uma 10h57'19" 68d45'37"
Schwarczné Misicza Julianna csillaga Uma 11h0'42" 68d48'48"
Schwartz Family Star Uma 11h7'34" 50d7'0"
Schwartz, Friedrich Ori 6h19'39" 16d10'53"
Schwartz, Gerold Uma 8h49'12" 57d54'2"
Schwartz-stock Lmi 9h50'52" 36d40'49"
Schwarz, Peter Uma 8h50'27" 48d55'15"
Schwarz, Rudolf Uma 12h55'27" 61d7'52"
Schwarze, Torsten Uma 8h40'59" 59d9'51"
Schwarzin, Werner Uma 10h59'40" 46d36'15"
Schwarz's Star Cap 21h30'40" -24d28'8"
Schwazche Ori 5h31'5" 2d29'8"
Schwegler, Jörg-Oliver Uma 13h51'53" 56d39'0"
Schweiger, Georg Sco 17h19'4" -32d33'13"
SchweigerWillUBeMine Tau 4h52'24" 23d7'32"
Schweigreiter, Leonhard Ori 5h14'25" 5d57'3"
Schweitzer, Eugen Ori 5h38'6" -2d28'50"
Schweizer Kolpingwerk And 1h16'58" 49d19'32"
Schweizer, Thomas Uma 11h16'35" 34d37'48"
Schwens, Marco Uma 10h38'47" 65d50'27"
Schweppes And 0h41'54" 41d24'32"
Schwer & Keiko 9years memory Sgr 18h21'15" -32d48'57"
Schwetge, Hansjörg Ori 6h18'48" 18d51'21"
Schwinger, Maximiliane Uma 9h59'10" 55d41'7"
Schwinny Lyn 7h49'3" 42d15'36"
SCI & SCII Lib 15h57'8" -5d22'20"
Sciacca Antonina and Girolamo Cyg 19h14'59" 51d6'30"
SciBishop Tau 5h52'56" 26d47'29"
Sciolino Cam 6h49'34" 69d37'41"
Scioni Nicola Giovanni Uma 9h24'38" 61d42'16"
Scip Steorra's Victory Cas 23h54'52" 53d11'33"
Scola Psc 0h41'4" 8d44'52"
Sconia Volva Boo 15h24'17" 34d7'24"
Scoobi Uma 9h22'26" 48d55'52"
Scoobie Doo Cyg 21h16'19" 52d40'17"
Scooby Hogge Pup 7h36'0" -28d56'56"
Scooby-Doo Per 3h43'25" 44d3'50"
SCOOMOO Lyr 18h45'14" 35d18'32"
ScoopyMoni Cyg 21h51'22" 37d46'43"
Scoot Boo 14h36'26" 33d1'20"
Scootch Sgr 19h10'26" -15d19'19"
Scooter Vir 13h59'25" -1d6'47"
Scooter Umi 15h33'20" 84d47'24"
Scooter Uma 13h36'44" 55d44'42"
Scooter Uma 12h44'59" 57d4'46"
Scooter Uma 11h36'51" 59d48'30"
Scooter Lyn 7h39'20" 52d48'30"
Scooter Her 17h8'55" 33d44'55"
Scooter And 1h57'52" 42d10'3"
Scooter Uma 13h51'6" 49d12'51"
Scooter Cmi 7h26'14" 8d44'1"
Scooter Ori 5h56'50" 18d20'51"
Scooter Ori 6h13'21" 20d54'42"
Scooter 1977 Leo 9h38'47" 27d6'31"
Scooter Bug Per 3h21'27" 42d9'13"
Scooter Byars...GOD's CREATION... Boo 14h19'38" 16d47'3"
Scooter Fais Cnc 8h48'29" 25d24'15"
Scooter Pie Cmi 7h36'22" 3d31'57"
Scooter Rentner Ori 6h10'42" 15d14'9"
Scooters Enraptured Serendipity Mon 6h51'35" 7d4'25"
Scooter's Love Sgr 18h1'18" -20d43'34"
"Scooters" Star Uma 10h30'41" 56d27'45"
Scooter's Star Cmi 7h30'7" 5d6'23"
Scooter's Star - Scott M. Dykstra Her 17h30'18" 28d3'50"
Scorpion Queen Sco 17h47'23" -32d45'51"
Scorpion09 Boo 14h20'10" 22d33'32"
Scot and Tiffany McLean Ori 5h44'26" 11d47'12"
Scot Crews Vir 15h4'14" 5d39'32"
Scot n Brenda McCreary Cyg 19h56'14" 31d26'11"
Scot R. Lee Cnc 8h18'14" 19d45'44"
Scot Shively Lib 15h47'55" -8d37'10"
Scot W. Taylor Cnv 12h50'12" 43d33'45"
Scot, my love Her 17h26'42" 24d49'39"
"Scotch"- Sugar and Spice V Uma 11h43'12" 58d4'39"
Scotlynn Vaughn-Gjonaj Crb 15h19'29" 30d1'41"
Scott Cyg 19h56'37" 36d51'36"
Scott Her 16h39'38" 48d41'35"
Scott Boo 14h39'3" 18d18'15"
Scott Gem 6h45'41" 22d27'42"
Scott Ari 3h29'16" 22d3'42"
Scott Uma 12h42'9" 61d27'49"
Scott Umi 18h16'40" 88d30'34"
Scott Vir 11h42'27" -5d55'2"
Scott Aqr 20h40'26" -2d12'28"
Scott "#32" Stein Her 17h24'6" 45d6'31"
Scott A. Fournier Lib 15h3'46" -2d5'20"
Scott A. Gormley Uma 11h6'52" 60d12'59"
Scott A. Kitkowski Psc 0h51'50" 20d43'14"
Scott A. Neugroschl Sco 16h15'36" -14d49'44"
Scott A. Polleveys Gem 8h1'19" 32d42'7"
Scott A. Walters Success Uma 9h17'30" 55d39'38"
Scott Aaron Pina Uma 13h32'37" 57d58'0"
Scott Adam Poll - Scott's Star Per 3h46'0" 38d55'56"
Scott Addison Hunter Ori 5h39'44" 0d37'37"
Scott Agnew Tau 4h55'12" 23d32'54"
Scott Alan DeVine Per 3h24'59" 49d13'33"
Scott Alexander Shindle Tri 2h9'26" 34d29'43"
Scott & Alfie Eagling Aql 19h45'18" 16d14'30"
Scott Allan McLane, II Her 17h6'49" 31d25'26"
Scott Allen Boo 14h32'26" 35d29'3"
Scott Allen Ackerman Tau 5h51'20" 17d16'45"
Scott Allen Campbell Per 4h10'48" 51d18'40"
Scott Allen Couch Star Her 17h36'10" 28d18'33"
Scott Allen Emerson Her 17h28'37" 45d14'12"
Scott Allen Gwizd, Jr. Umi 16h51'48" 80d23'42"
Scott Allen Howell Ori 5h11'26" 3d40'19"
Scott Allen Veith Gem 6h45'30" 27d2'33"
Scott Allen White Cnc 8h59'45" 15d1'31"
Scott & Amy Crummy Lib 15h18'7" -5d46'3"
Scott & Amy Diamond Cyg 20h48'13" 47d10'17"
Scott and Adam Cnc 8h31'47" 23d38'28"
Scott and Andrea McCain Ori 5h33'1" 2d1'48"
Scott and Anne Widman-Bray Cyg 20h9'43" 32d9'35"
Scott and Ashley Escarze Vir 14h25'54" -0d7'28"
Scott and Chris Burkitt and the Zoo Aql 19h13'31" 1d5'21"
Scott and Emily Sweeney Ori 5h8'15" -0d22'29"
Scott and Erin Eri 3h44'2" -13d6'16"
Scott and Jennifer Beecher 1/8/2005 Her 18h45'22" 19d3'45"
Scott and Jessica Always Her 17h58'48" 14d53'33"
Scott and Jessica Orth Uma 11h41'33" 60d23'52"
Scott and Kasi Uma 10h26'36" 46d9'12"
Scott and Kate Forever x Cyg 19h36'31" 30d8'15"
Scott and Katie Lyn 7h52'27" 42d19'25"
Scott and Kimberly Thompson Peg 21h35'59" 22d4'57"
Scott and Lina McGeoch Cru 12h3'29" -62d29'33"
Scott and Lise And 0h46'41" 43d59'19"
Scott and Maleia Cyg 21h44'11" 42d6'34"
Scott and Marie Dixon And 23h38'25" 44d37'8"
Scott and Rachel Stell Sco 16h40'2" -35d53'25"
Scott and Sally Umi 14h59'24" 72d50'25"
Scott and Sandra Thatcher Aqr 20h40'27" -7d47'26"
Scott and Sarah Sge 19h52'18" 17d10'29"
Scott and Sharon Campbell Forever Cyg 21h49'4" 44d14'30"
Scott and Sherri Nutt Cyg 20h18'11" 53d48'41"
Scott and Suzie Cyg 21h28'56" 47d33'6"
Scott and Tammy Burrows Lyr 18h31'44" 28d52'39"
Scott and Vicki's Twilight Place Cyg 19h47'13" 43d53'36"
Scott Anderson Her 16h36'6" 36d41'8"
Scott Anderson Tulloh Mills Ori 5h35'38" -10d7'4"
Scott Andrew De Freitas Aqr 23h56'18" -14d4'0"
Scott Andrew Franklin Uma 11h37'3" 44d31'59"
Scott Andrew Geller Cap 20h24'52" -18d8'7"
Scott Andrew Korobkin Her 17h29'43" 46d1'26"
Scott Andrew Ougheltree Vir 13h26'59" -18d39'25"
Scott Andrew Owens Cap 21h40'46" -14d34'25"
Scott & Ann Michele Langlinais Star Ori 5h55'12" 12d30'14"
Scott Anthony Cnc 9h17'7" 25d36'7"
Scott Anthony Cocca (Scotter) Tau 5h19'12" 18d26'11"
Scott Anthony Milton Ori 5h39'51" 12d52'5"
Scott Anthony Oden May 6, 1963 Uma 10h6'1" 60d31'5"
Scott Anthony Tye Aqr 22h10'20" 0d19'10"
Scott Anthony Walker Aqr 22h39'32" -0d40'18"
Scott & Ashley Pho 23h52'27" -52d49'33"
Scott Avedisian-Remember Bahamas! Per 2h56'48" 55d3'6"
Scott B. "800" Baker Cap 20h31'13" -24d12'24"
Scott B. Robins Aqr 21h33'44" -4d53'59"
Scott B. Skeat Cap 20h7'25" -11d14'14"
Scott Baron Vir 12h50'31" 7d22'55"
Scott Bates Boo 14h36'13" 53d17'4"
Scott & Becky's Shootin Star Equ 21h8'14" 4d50'0"
Scott Belknap-Numerus Unus Pater Ori 5h55'57" -0d43'2"
Scott Bennet Ori 5h36'14" -3d51'54"
Scott & Beth Crb 15h27'13" 30d35'22"
scott&bill Uma 9h46'25" 47d37'31"
Scott Blackburn Hammon Cyg 19h55'51" 39d20'28"
Scott Bleeks Ori 6h1'54" 13d53'50"
Scott Bodenstedt Ori 5h57'17" 6d44'37"
Scott Borenstein Uma 9h21'48" 65d15'0"
Scott Bouchie Doan Psc 1h8'45" 17d13'17"
Scott Bowers Tau 4h46'6" 18d58'4"
Scott & Britts Isle Of Devine Bliss Lyn 6h18'38" 54d49'10"
Scott Brownlee Leo 11h11'7" 8d56'28"
Scott Brown's Star for Madison Uma 14h13'18" 62d15'23"
Scott Buchholz Hya 9h46'18" -20d31'17"
Scott Buddy Kidd Sco 16h6'20" -20d55'27"
Scott C. Frank Cep 21h8'51" 67d11'47"
Scott C. Palmer Per 3h18'17" 44d35'11"
Scott Caldwell Boo 14h37'57" 28d20'42"
Scott Caplan Aur 5h31'8" 46d57'21"
Scott&Carey Lyr 18h45'17" 40d8'20"
Scott Charles Tonks Uma 11h35'37" 56d11'32"
Scott Charles Underwood Her 18h9'35" 18d1'57"
Scott & Chelle Hall's Shining Star Sge 19h27'56" 17d8'19"
Scott & Christina Melkonian Cyg 20h16'59" 55d20'0"
Scott & Christine St. Onge Cyg 20h19'2" 55d3'21"
Scott & Christine Wilkinson Forever Sge 20h4'33" 17d11'37"
Scott Christopher Dyer Cnc 8h12'27" 15d25'12"
Scott Christopher Engel Leo 9h56'54" 19d46'59"
Scott Christopher O'connor Dra 19h44'22" 62d56'54"
Scott Christopher Williams Cap 21h15'56" -15d37'25"
Scott Clark Sco 16h52'55" -42d7'53"
Scott Clarke Per 3h1'50" 51d28'34"
Scott Clinton Wyne Cep 21h11'53" 56d6'4"
Scott Corobotiuc Boo 14h32'56" 19d39'15"
Scott "Cradl" Craven Tau 5h52'36" 26d36'36"
Scott D. Wells Del 20h44'24" 9d3'44"
Scott Daniel Dra 19h16'16" 64d7'42"
Scott & Danielle Umi 15h18'19" 75d36'18"
Scott David Ciociola Per 4h42'4" 49d31'20"
Scott David Eugene Cartwright Uma 11h32'49" 56d25'23"
Scott David Harvey Ari 2h50'41" 11d1'45"
Scott David Scranton Aql 19h4'29" 13d22'51"
Scott DeAlvarez Cnc 8h22'26" 14d38'29"
Scott Dennis Vella Per 2h52'9" 54d6'55"
Scott Derek Palmer Ori 5h29'21" 9d58'18"
Scott Donaldson Her 16h20'58" 3d51'41"
Scott Donato Per 2h39'47" 55d7'30"
Scott Douglas Pitzer Her 17h50'48" 22d31'41"
Scott " Duck" Burks Leo 9h59'42" 31d26'5"
Scott Dyck: Forever Loved Leo 10h17'27" 21d46'0"
Scott E. Bernstein Lmi 9h55'51" 37d34'41"
Scott E. Birch Uma 9h9'47" 55d35'19"
Scott E. Eells Ori 6h9'54" 9d21'56"
Scott E. Folmer Aur 5h51'48" 53d49'58"
Scott E. Friedman Leo 11h43'56" 19d53'41"
Scott E. Gallagher Per 2h26'14" 56d12'12"
Scott E. Huber Her 17h30'28" 45d56'11"
Scott E. Kargman Ori 6h11'51" 7d8'12"
Scott Easton Umi 17h9'7" 76d16'55"
Scott Edward Her 17h6'34" 26d56'35"
Scott Edward Foster Tau 4h27'55" 12d34'0"
Scott Edward Hartup Leo 11h50'29" 19d4'52"
Scott & Elaine's Love Eternal Ara 17h23'5" -52d6'24"
Scott Ellis Rozzell Vir 14h50'15" 3d59'19"
Scott Ellis Spellerberg-Olsson Boo 14h41'50" 34d31'57"
Scott Emory Dennison Cap 21h23'56" -14d37'55"
Scott Ericson Dra 18h29'3" 50d21'7"
Scott Escue And 0h22'46" 40d22'28"
Scott Ethridge Lib 15h7'33" -26d57'0"
Scott Evan Levy Ari 1h53'50" 22d7'7"
Scott Evans Magill Uma 8h27'16" 62d18'42"
Scott Everett Agans Ser 15h22'55" -0d53'44"
Scott Fagan - My True Love Leo 11h18'59" 14d27'28"
Scott Francis Her 16h25'15" 45d1'21"
Scott Fuller Cap 21h40'52" -9d26'27"
Scott Furtado Sco 16h3'0" -25d46'11"
Scott G. Anderson Ari 3h25'15" 26d42'5"
Scott G. Hodge Psc 0h37'52" 9d18'49"
Scott G. Price Lyr 18h35'24" 33d16'12"
Scott George Baum Lib 15h4'44" -0d30'50"
Scott Gerber Her 16h45'38" 28d56'50"
Scott Giles Uma 8h15'35" 64d6'55"
Scott (Giroud) Horbury Jr Uma 11h30'8" 50d11'5"
Scott Glenn Mattson Cap 20h26'25" -10d25'54"
Scott Goldbaum Her 16h35'43" 29d7'21"
Scott Grady Squires (#1 Daddy) Cep 22h55'1" 70d9'25"
Scott H. Woodruff Per 3h14'44" 51d43'21"
Scott Hagen's Shining Star Lyn 8h0'43" 55d17'14"
Scott Hale Psc 1h21'55" 16d7'8"
Scott Hall Ori 5h55'43" 21d9'48"
Scott Harris Lac 22h16'32" 41d24'48"
Scott Harris Dra 15h3'4" 62d36'58"
Scott Harrison Whitmarsh Cnc 8h2'49" 20d53'22"
Scott Hawley Per 3h9'19" 54d26'51"
Scott & Heather Bohannon Sge 19h35'41" 18d50'43"
Scott Henderson Cnc 9h15'33" 16d15'43"
Scott Henry Lyn 9h5'7" 37d49'55"
Scott Henry Aql 19h17'53" -0d0'27"
Scott Hinman Ori 5h21'54" 1d50'8"
Scott Hoffman Uma 11h24'36" 51d19'29"
Scott Holloman Per 3h7'25" 50d30'54"
Scott Holmberg Gem 6h59'48" 24d6'14"
Scott Howlett Cru 12h39'45" -58d38'51"
Scott (Husband) Vaccarella Cnc 8h1'47" 11d27'23"
Scott Hutchins Uma 9h28'45" 62d14'22"
Scott J. Garmon Aql 19h44'28" 5d37'4"
Scott J. Hagloch Cap 20h27'36" -12d34'47"
Scott J. Pigaga Aur 6h32'25" 34d19'21"
Scott Jacobs Cam 4h26'36" 73d6'30"
Scott Jacobson Lyn 7h38'6" 59d1'8"
Scott James Smith Leo 10h32'5" 13d36'27"
Scott & Jenna Cnc 8h39'3" 31d44'54"
Scott & Jennifer Dyer Family Star Eri 4h21'39" -7d58'45"
Scott Jensen Tau 4h37'58" 24d43'6"
Scott Jerome 5-1-94 Uma 11h35'18" 53d56'29"
Scott John Boocock Cru 12h1'2" -63d26'3"
Scott John Frank Mediz Cnc 8h55'13" 13d21'33"
Scott John Reisser Uma 9h50'11" 55d17'55"
SCOTT JOHNSON Lib 14h24'56" -9d24'31"
Scott Johnson Her 16h45'46" 28d18'11"
Scott Johnson Cas 1h25'51" 53d32'43"
Scott Joseph Bugai Her 16h35'5" 43d49'48"
Scott Joseph Kennebeck Cnc 8h51'22" 13d4'4"
Scott K. Sechler Cep 22h12'29" 69d48'45"
Scott Kaplan Per 3h18'40" 52d39'43"
Scott & Karen Wood Lyr 18h27'10" 39d30'31"
Scott & Kathy "Forever" Uma 10h11'51" 47d46'10"
Scott & Kati McGuire Sge 19h43'57" 18d56'51"
Scott & Kelly White Uma 10h58'23" 38d42'46"
Scott Kevin Schroeder Her 18h34'45" 14d11'54"
Scott & Kim Anderson Pho 1h10'55" -48d31'41"
Scott Kim & Zack Rogers Her 18h30'53" 14d7'51"
Scott Kingsley "The Chief" Her 17h34'12" 45d3'51"
Scott Kiyoshi Ashlock Leo 9h57'17" 6d30'42"
Scott Klenke Aur 5h43'24" 55d5'53"
Scott Konetski Her 16h34'13" 37d41'25"
Scott Kugel Vir 13h16'0" 6d45'19"
Scott L. Dostal Aql 19h58'14" -0d27'45"
Scott L. Webb 09/23/58-01/06/98 Lib 14h40'46" -11d19'12"
Scott Lackie Aql 19h8'38" -4d51'29"
Scott Larned And 0h38'29" 40d3'32"
Scott Lawrence Abramowitz Her 16h32'51" 31d41'35"
Scott Lawrence Parker Cnc 8h49'18" 31d44'51"
Scott Lee Her 18h48'57" 26d4'45"
Scott Lee Cobbs Tau 3h36'29" 12d53'6"
Scott Lee Dupree Vir 13h22'58" -17d39'47"
Scott Lee Erickson Cnv 13h38'8" 34d20'43"
Scott Lee Garland Ari 2h49'44" 19d32'25"
Scott Lee Sadger Leo 11h3'28" 20d10'35"
Scott Lee Styles Per 3h47'16" 49d1'59"
Scott Lew Lib 15h30'56" -6d49'39"
Scott Lewis Hicks Ari 2h44'51" 21d11'9"
Scott & Liz Spence Eri 3h52'28" -0d56'47"
Scott Loedeman Cen 13h45'54" -43d26'22"
Scott M Benson Per 3h10'53" 52d45'18"
Scott M. Freeman Tau 5h43'18" 23d7'45"
Scott M. Johannesmeyer Uma 9h0'28" 56d9'23"
Scott M. Johnson Cep 0h2'37" 67d4'5"
Scott M. Knabusch Uma 10h34'6" 56d8'38"
Scott M. Reekie Per 4h5'15" 48d45'5"
Scott M. Williquette Her 16h37'10" 38d3'14"
Scott Mackenzie Cole Uma 10h38'51" 61d28'8"
Scott Maines Cap 21h43'27" -17d10'33"
Scott Mandarino Gem 7h14'35" 16d34'22"
Scott & Mandy Cyg 20h34'7" 41d29'4"
Scott Marshall Aql 18h58'6" -0d7'59"
Scott Martin Nadel Lyn 7h43'58" 56d46'5"
Scott Massey Vir 12h33'37" -9d43'28"

Scott Matthew
Aql 18h56'6" -0d19'3"
Scott McLaughlin Ferguson
Peg 22h59'56" 8d30'3"
Scott & Megan
Her 16h50'55" 33d34'36"
Scott Mehalko
Leo 10h49'11" 19d35'42"
Scott Mendlinger
Tau 3h27'19" 15d28'43"
Scott Michael
Ori 6h1'0" 17d55'37"
Scott Michael Brent
Uma 9h6'21" 68d10'15"
Scott Michael DeGraffenreid
Cnc 8h52'11" 28d15'1"
Scott Michael Pierce (Magician)
Ser 16h16'40" -0d1'26"
Scott Michael Ramp
Cap 20h20'3" -11d12'50"
Scott Michael Spino
Per 3h10'40" 42d46'55"
Scott Michael Weyhrich
Leo 10h10'27" 25d13'35"
Scott Michael Wright
Dra 16h32'4" 62d1'55"
Scott & Michele
Cyg 19h29'50" 53d54'39"
Scott Millan
Aqr 22h56'43" -9d36'35"
Scott Miller a Memory of Love
Uma 10h0'5" 62d43'37"
Scott Mitchell
Psc 1h22'40" 28d8'20"
Scott Morgan - My Heaven My Destiny
Cra 18h9'48" -43d24'25"
Scott Mors - My favorite Godson
And 2h36'13" 44d43'0"
Scott Muffin Cupcake Chelsey Pepper
Cnv 13h54'49" 39d45'4"
Scott Murphy
Per 2h45'12" 53d46'24"
Scott Neill
Lyn 7h26'12" 53d20'19"
Scott Niedecken
Aqr 20h47'7" -13d17'21"
Scott O. Lowe
Uma 14h18'34" 59d38'21"
Scott Oltman
Cnc 8h28'47" 19d39'20"
Scott Orlosky Star
And 1h40'7" 44d12'56"
Scott Ousdahl
Per 3h15'50" 31d33'57"
Scott Ov
Her 18h17'24" 21d39'36"
Scott P. Larned
Cnc 9h9'35" 22d0'56"
Scott P. Tarantino
Uma 11h49'24" 62d17'32"
Scott Palmer
Cyg 19h46'57" 32d54'44"
Scott & Pam
Cyg 20h57'3" 52d7'30"
Scott Patrick MacDonald
Dra 17h7'25" 54d32'14"
Scott Patrick McNichols
Aur 5h37'48" 48d52'58"
Scott Paynton
Her 17h23'50" 38d4'16"
Scott Pearson
Cru 11h58'57" -56d7'9"
Scott Philip Elsky
Lib 15h38'42" -29d18'14"
Scott Philip Martin
Boo 15h38'13" 46d49'1"
Scott Philip Steers - Scott's Star
Uma 10h36'37" 72d42'4"
Scott Porter
Lyn 8h3'23" 36d28'42"
Scott Proctor
Uma 10h30'43" 39d50'22"
Scott Prosuch
Sgr 18h31'51" -29d21'56"
Scott Purcell
Per 2h43'30" 55d30'6"
Scott Q Dukes
Gem 7h47'27" 33d39'55"
Scott R. Bowman
Dra 16h31'22" 63d29'0"
Scott R. Marcantonio
Boo 14h30'34" 19d24'39"
Scott R. O'Donnell-Knik03
Per 3h3'49" 45d48'3"
Scott R. Reynolds
Cep 20h51'21" 59d13'47"
Scott & Rachel
Lyr 18h30'31" 36d22'54"
Scott Randolph Snader "T-Bone"
Lib 15h3'47" -25d51'15"
Scott Randy Defebaugh
Cnc 8h45'38" 15d16'32"
Scott Ray Fleming
Uma 8h39'43" 60d36'27"
Scott Reavis
Tau 5h9'27" 18d1'20"
Scott Redler
Per 2h54'32" 53d23'36"
Scott "Rhyno" Eisenbrandt
Leo 10h21'46" 14d42'0"
Scott Rhys Woollaston
Cru 12h40'21" -62d17'50"
Scott Richard Garcia
Sgr 17h53'24" -28d40'2"
Scott Richard Joa
Aql 19h52'6" -0d13'53"
Scott Richard Moore, My Best Friend
Per 2h53'46" 51d30'36"
Scott Richardson Lemmon
Aql 19h45'8" 14d2'44"
Scott Robert Folken
Ori 6h12'50" 15d37'48"
Scott & Robin - Afterall
Cyg 20h18'47" 45d34'28"
Scott Robins
Per 3h21'23" 39d59'16"
Scott Romano
Sgr 18h32'35" -21d19'52"
Scott Russell Hughes
Uma 13h38'9" 55d24'29"
Scott Ryan McGauvran
Her 16h55'23" 33d58'10"
Scott Ryen Worner
Her 17h24'36" 25d39'26"
Scott S. Pride
Gem 7h40'56" 32d7'31"
Scott 's Star
Dra 19h48'14" 73d47'28"
Scott Sabo
Ari 2h31'22" 27d8'3"
Scott Samuel Oakes
Cmi 7h37'19" 5d11'23"
Scott & Sarajane Baumgardner
Crb 15h59'54" 26d12'44"
Scott Sargeant+ Jenna Tress Forever
Leo 10h0'52" 16d32'18"
Scott "Scoot" Frazier
Pho 1h23'5" -47d20'5"
Scott Sea Yuen Chan
Uma 10h36'44" 57d0'29"
Scott Sedam
Cnc 8h11'36" 21d12'3"
Scott Sedlaczek
Ori 4h50'21" 4d32'57"
Scott Seiwert
Eri 3h47'47" -0d23'26"
Scott Sellers
Uma 10h57'47" 52d44'45"
Scott Shaw
Uma 12h12'24" 55d4'1"
Scott Shining Star
Psc 0h52'3" 21d16'4"
Scott Sievwright
Uma 8h53'0" 47d5'3"
Scott Simpson - Piranha
Cru 12h19'5" -57d20'4"
Scott "Skinny" Jeffery Larson
Uma 8h56'23" 68d51'11"
Scott & Stacy Hall
Gem 7h16'15" 32d52'40"
Scott Stephen Spencook
Her 16h35'49" 46d48'44"
Scott Stephens
Ari 3h9'54" 25d14'52"
Scott Steven Andersen
Boo 14h21'6" 14d28'4"
Scott Stowe Jr the 2nd Man in Black
Uma 10h9'19" 43d28'42"
Scott Strathearn 1
Uma 12h0'54" 56d31'20"
Scott Strickland
Aql 19h50'9" -0d46'50"
Scott Sullivan
And 1h48'25" 37d26'32"
Scott&Suzy12252006
Cyg 19h49'54" 32d40'36"
Scott Swarner
Sco 17h57'47" -30d22'29"
Scott T B Papa
Vir 13h9'51" 4d22'8"
Scott T. Lewis
Per 3h25'42" 51d19'13"
Scott T. Pierce
Uma 11h38'2" 48d34'23"
Scott Tanoory
Sco 16h33'35" -25d6'31"
Scott & Tara's Wedding Star
Cyg 20h37'14" 41d40'1"
Scott Terence Hodgson
Cas 1h58'0" 60d44'26"
Scott & Teresa
Aqr 22h56'34" -9d25'54"
Scott & Teresa's Anniversary Star
Pup 8h15'38" -36d59'21"
Scott Thibodeau
Per 2h17'44" 55d29'30"
Scott Thissen
Ari 2h30'16" 26d27'35"
Scott Thomas McConnell
Sco 16h16'53" -17d56'55"
Scott Thomas Para
Gem 6h48'35" 30d28'3"
Scott Thompson My PunkinHead
Vir 12h54'1" -0d20'55"
Scott & Toby Durchslag May 10, 2004
Sge 19h28'7" 17d27'39"
Scott Truesdale
Her 16h43'15" 11d4'30"
Scott Tyler Cook
Gem 7h10'5" 34d10'1"
Scott Tyler Malson
Psc 0h47'33" 16d47'55"
Scott Vaughn
Ori 5h6'44" 9d4'59"
Scott Vernon Buckley
Sco 17h45'4" -37d11'45"
Scott W. Gooden
Her 18h43'47" 25d16'6"
Scott W. Maurer
Del 20h37'0" 15d33'12"
Scott W Quigg
Sgr 17h53'24" -25d14'58"
Scott W. Salyers
Uma 10h7'0" 48d30'50"
Scott W. Schallhorn
Per 3h42'36" 41d13'32"
Scott W Sorenson
Dra 20h17'43" 67d4'7"
Scott W Wilson "Bomber"
Leo 10h30'9" 16d44'3"
Scott Wade Harootunian
And 0h46'6" 44d3'22"
Scott Wadsworth
Vir 13h18'28" -13d58'37"
Scott Walker Sample
Aqr 21h31'31" -0d46'24"
Scott Walsh
Cmi 7h25'24" -0d1'17"
Scott Warnock Rafferty
Ori 5h58'9" 22d6'15"
Scott Watson
Ori 6h25'37" 17d6'15"
Scott Wayne Williams Jr.
Lib 15h19'31" -21d32'54"
Scott Webster Nolley
Lyr 18h37'38" 30d9'57"
Scott & Wendy Payne
Eri 3h52'3" -0d25'36"
Scott Wermerskirchen
Ori 6h19'31" 15d48'1"
Scott Wert and Emery Nicoletti
Umi 14h42'31" 69d15'2"
Scott Wesley Dunham
Ari 3h6'56" 15d50'41"
Scott Wiard
Her 16h52'48" 35d29'37"
Scott Wicklund
Gem 6h59'1" 12d52'41"
Scott Wielgus
Tau 4h19'29" 0d52'49"
Scott Wiesner
Lmi 9h25'17" 38d35'42"
Scott William Bartlett
Ori 5h34'34" -2d56'27"
Scott William Gibson
Uma 11h9'49" 35d9'0"
Scott William Hershey
Leo 9h30'3" 10d7'22"
Scott William Page
Leo 9h47'22" 31d30'27"
Scott William Patterson
Her 18h48'30" 17d20'24"
Scott Williams
Cep 0h18'37" 67d42'30"
Scott Wilson
Tau 5h28'4" 25d30'15"
Scott Wood Davison
Ari 2h49'4" 27d24'53"
Scott "Wulls" Edwin Wullbrandt
Lib 15h43'29" -26d59'10"
Scott Yarwood
Ari 3h16'33" 28d28'7"
Scott Zifferer
Per 4h33'11" 32d33'6"
Scott, Elvira et Arnaud
Cyg 19h35'36" 33d56'2"
Scott, hic et ubique
Uma 10h40'0" 60d1'25"
Scott, you light up my life.
Ori 5h8'24" 4d8'54"
scottandbrenda
Cyg 20h39'4" 38d29'57"
scottangelatirzahgracealex-grubaugh
Uma 11h16'8" 36d48'4"
ScottBakay
Aqr 23h48'53" -18d37'19"
Scott-Cook 30th
Psc 1h20'59" 30d41'45"
ScottFredrickLehman
Leo 10h48'16" 16d18'49"
Scottie
Lmi 10h40'45" 26d24'43"
Scottie
Ori 4h54'36" 14d21'4"
Scottie
Lyn 7h53'57" 46d56'46"
Scottie&Coley4Ever
Lac 22h20'25" 46d44'16"
Scottie R. Laird
Lyn 7h45'20" 42d58'26"
Scottish Thistle
Lyn 7h27'41" 53d6'3"
Scott-n-Dee 2004
Cyg 21h9'30" 47d57'59"
Scotto Family
Her 16h46'56" 37d41'33"
Scott's *Lucky* Star
Per 3h19'12" 46d35'8"
Scott's Star
Psc 1h24'13" 33d9'29"
Scott's Star
Psc 1h30'13" 14d11'2"
Scott's Star
Tau 3h40'27" 28d4'34"
Scott's Star
Uma 10h10'42" 63d41'39"
Scott's Sunshine
Her 18h2'34" 24d43'58"
Scottsburg
Lib 15h3'28" -13d12'22"
Scottt Sward
Her 16h39'38" 36d41'21"
Scotty
Ari 2h45'39" 25d31'14"
Scotty
Her 17h56'28" 15d36'47"
Scotty
Psc 2h5'44" 4d1'24"
Scotty
Leo 9h46'10" 12d54'24"
Scotty
Cnc 8h20'32" 8d23'20"
Scotty
Ori 5h25'45" -4d6'11"
Scotty
Dra 19h16'12" 68d59'32"
Scotty
Pho 0h17'39" -39d45'8"
Scotty Alfie Stanley Payton
Ori 5h36'7" -2d16'50"
Scotty Boy Hayes
Cru 12h44'24" -57d17'18"
Scotty D
Leo 9h35'0" 29d13'21"
Scotty F. Jonna
Tau 5h39'0" 26d12'6"
Scotty Joe
Uma 12h8'29" 51d45'7"
Scotty Loves Kathy
Tau 5h22'12" 19d34'7"
Scotty Magee
Cru 12h38'11" -59d43'18"
Scotty Nicole Ortego Townsley
Sco 17h6'22" -36d31'58"
Scotty Panici
Sco 16h42'14" -30d6'21"
Scotty Reaves
Her 17h20'41" 31d50'45"
Scotty Stoi
Her 18h51'47" 25d15'56"
Scotty Waters
Cnc 9h10'53" 31d29'0"
Scotty32779
Uma 13h51'47" 59d3'17"
Scottys Jade
Uma 11h28'15" 49d17'11"
Scotty's Shining Light
Umi 15h32'27" 77d7'27"
Scout
Uma 8h40'45" 61d12'40"
Scout Boys D5P708C10
Aql 19h41'24" 7d36'43"
Scout Hawksley
Cas 0h22'45" 57d16'39"
Scout Lauren Wilkins
Lib 15h16'49" -23d8'0"
Scout Nicole Herndon
Uma 9h20'20" 44d21'47"
Scout R.
Ori 5h43'9" 8d46'40"
Scrapalot
Psc 1h42'31" 7d9'51"
Scrapper Smith-Davis
Sgr 17h58'52" -30d6'42"
Scratch
Umi 15h40'34" 77d14'55"
Screaming Eagle Weaves The Sky
Aql 19h44'23" 9d46'45"
Scrigno magico
Umi 16h31'7" 75d1'29"
scriptschool
Uma 9h32'44" 54d4'43"
Scrubby
Aqr 22h29'28" -10d18'17"
Scruff
Uma 11h20'54" 65d29'52"
Scruffie Wuffiekins
Aqr 22h4'53" -0d4'20"
Scruffy
Uma 13h40'48" 56d27'16"
Scruffy
Ori 5h56'34" 6d38'19"
Scruffy Dootle
Uma 9h40'50" 42d20'50"
Scruffy Dootle
Uma 11h42'11" 37d13'46"
Scruffy's Spark (Daryl Reardon '81)
Psc 0h22'38" 12d26'4"
Scrumptious
Ori 6h5'33" 10d44'9"
Scrumptious
Cnc 9h16'14" 18d21'43"
Scrumpy's Star
Uma 11h33'46" 45d16'56"
SCRUMS
Ori 5h43'43" -0d46'14"
Scuba Steve
Per 3h17'33" 47d49'2"
Scubahood
Peg 22h26'7" 6d50'17"
Scuber Lives
Leo 9h54'15" 13d29'49"
Scucci-Kellie
Umi 16h37'39" 82d45'45"
Scuddle Star
Uma 10h32'57" 40d46'27"
Scuhllie
Lib 15h32'38" -8d2'54"
Scungio's Legacy
Dra 19h43'26" 65d22'57"
Scutari
Dra 19h3'45" 51d57'52"
Scutch
Psc 1h7'18" 32d20'16"
SDC3532120604
Ori 6h21'19" 16d43'22"
SDP'n ME
Cyg 19h17'8" 50d45'52"
Sdrawde Mot
Tau 5h56'3" 23d30'41"
SE Scott
Aql 19h21'25" -0d42'57"
SEA
Uma 11h50'38" 50d28'1"
Sea Babe
Del 20h28'10" 20d5'2"
Sea Barb Forever
Cyg 21h25'13" 33d8'17"
Sea Biscuit
Boo 14h16'8" 52d38'39"
Sea Breeze Bch Hotel
Crb 15h33'16" 36d8'36"
Sea Cow
Ori 5h34'27" -5d24'22"
Sea Squirt Gill
Boo 14h15'34" 16d57'59"
SEA18
Cra 18h46'38" -37d50'38"
Seabass
Umi 10h2'35" 86d3'22"
SEABERT101
Umi 16h22'13" 77d34'48"
Seabiscuit
Uma 11h10'38" 48d52'46"
SeabrightSeeker
Umi 15h23'48" 84d44'10"
Seabrook Philip
And 0h23'8" 33d25'40"
Seafoam Haciendas
Tri 2h23'50" 34d36'15"
Seagate
Cyg 20h56'48" 31d0'5"
Seagoville Wee Dragons
Dra 14h32'32" 59d40'15"
Seahawk
Aql 19h34'59" 11d15'58"
Seaira Angelle Moore
Cnc 8h44'26" 27d43'51"
Seamas William Comer
Aql 19h17'38" 15d42'44"
Seamus Galvin
Umi 14h30'21" 78d19'4"
Seamus Henry
Psc 1h6'27" 7d6'31"
Seamus J. P. Nyland
Tau 5h9'45" 20d46'28"
Seamus James Micheal Ericson
Uma 11h57'45" 56d9'20"
Seamus "Jim" Brennan
Her 17h25'5" 36d34'26"
SEAMUS KILCREASE
And 1h42'5" 37d44'13"
Seamus M.W. Turner
Ari 2h14'33" 24d33'7"
Seamus O Cléirigh
Per 4h39'36" 41d6'8"
Seamus The Superior
Lib 15h11'42" -21d7'46"
Sean
Cap 20h8'44" -9d44'59"
Sean
Cam 4h17'26" 65d43'49"
Sean
Lyn 8h2'13" 40d7'21"
Sean
And 23h35'30" 45d26'5"
Sean
Per 4h6'22" 52d1'37"
Sean
Her 17h49'51" 20d46'29"
Sean
Leo 9h30'5" 18d38'32"
Sean
Cnc 8h55'54" 17d36'47"
Sean
Ori 6h17'9" 7d29'32"
Sean 72
Vir 13h5'28" 12d26'26"
Sean A Williams
Leo 10h21'22" 21d55'20"
Sean Aleki McConnell
Cam 7h5'54" 78d3'29"
Sean Alexander Victorino
Aql 19h28'47" 7d16'57"
Sean Allan Niewoehner
Her 17h11'13" 34d8'45"
Sean Alto
Per 2h20'4" 55d35'34"
Sean & Alyssa
Sge 19h42'57" 18d54'57"
Sean and Angie
Uma 11h49'36" 49d12'36"
Sean and Beth's Wedding Anniversary
Ori 5h33'52" 9d53'3"
Sean and Caitlin Gallagher
Per 3h39'16" 44d56'37"
Sean and Christy
Tau 5h25'16" 23d13'2"
Sean and Cynthia's
Cyg 19h26'50" 36d26'37"
Sean and Debbie Wood
Sge 19h12'30" 19d3'59"
Sean and Erika Forever
Uma 11h42'0" 28d19'44"
Sean and Gina Fessler
Cnc 8h13'12" 9d25'3"
Sean and Hunter
Ori 6h4'58" 16d9'30"
Sean and Jen Corbet
Vir 13h15'22" -21d36'40"
Sean and Jillian Hovenkamp
Her 16h14'57" 13d24'26"
Sean and Julia
Lyr 18h26'43" 39d54'44"
Sean and Kamron Glassford
Uma 11h19'38" 40d54'2"
Sean and Kristen
Uma 8h57'40" 46d44'51"
Sean and Reyna
Cyg 19h44'4" 39d18'32"
Sean and Serena
Lib 15h11'59" -22d22'47"
Sean and Stephanie 2005
Umi 15h30'4" 78d59'28"
Sean and Theresa
Uma 8h39'20" 56d6'24"
Sean Andrew Lampl
Aql 20h5'4" 10d20'53"
Sean Andrew Miller
Lib 15h44'10" -26d6'2"
Sean Anthony Navas
Ari 3h19'10" 21d36'32"
Sean Anthony Tucci
Aql 19h52'6" 10d48'24"
Sean Anthony Vergara
Boo 14h42'4" 19d55'36"
Sean Anton Johnson
Tau 4h16'48" 18d38'39"
Sean Arterburn Then, Now, & Forever
Lmi 10h3'5" 37d17'56"
Sean Asquith
And 23h22'14" 52d5'35"
Sean B. Rose
Aqr 22h27'8" -7d1'17"
Sean Barnes
Uma 9h29'38" 71d41'41"
Sean Barry
Ori 5h28'29" 2d28'10"
Sean Bartolo
Aur 5h2'12" 35d48'23"
Sean Bell 604
Her 16h10'4" 46d3'29"
Sean - Big Time - Farrell
Ari 2h8'19" 23d22'23"
Sean Blumenthal
Lyn 7h11'43" 52d47'5"
Sean Body
Per 4h5'36" 43d48'48"
Sean Brady
Lyn 8h27'32" 36d18'51"
Sean Brian Beach
Cyg 19h44'54" 40d32'28"
Sean Bugler
Lib 15h15'18" -5d56'30"
Sean Calpini
Ori 6h19'59" 13d27'41"
Sean Calvin
Cyg 21h56'3" 50d15'55"
Sean & Caroline Kavanagh
Cyg 21h18'41" 42d42'33"
Sean Cattanach
Sco 17h50'15" -30d36'32"
Sean Chavious
Per 2h33'2" 55d14'55"
Sean Chili Pepper McConville
Her 16h21'42" 11d13'3"
Sean Christopher Suskevich
Dra 17h47'40" 65d51'40"
Sean Clancy Doyle
Tau 5h17'28" 24d56'28"
Sean Clay
Cep 20h18'56" 60d31'17"
Sean Clinton
Cap 21h9'57" -22d25'19"
Sean Conrad Loves Tiffany Marie
Cyg 20h57'0" 47d16'6"
Sean D. Essen
Uma 11h19'20" 31d8'18"
Sean D. Kammerlohr
Sco 16h18'57" -18d48'10"
Sean (Daddy) May
Her 17h30'15" 36d55'15"
Sean & Danielle
Cyg 20h51'2" 35d0'2"
Sean David Northwood
Leo 9h35'11" 29d49'3"
Sean David Tuson
Ari 2h5'23" 14d58'43"
Sean Dean
Cap 21h32'59" -13d53'2"
Sean Devlin
Cnc 8h53'16" 12d44'54"
Sean Doepp
Ori 6h5'59" 13d55'41"
Sean Dogg
Sco 16h7'29" -16d50'18"
Sean Dolman
Her 18h20'43" 27d19'20"
Sean Douglas Rucano
Psc 1h45'58" 13d23'7"
Sean Douglas Sparks
Aql 19h50'6" 6d31'32"
Sean d'Rohan Foley (Bah)
Uma 11h55'19" 54d20'21"
Sean Dunsmore
Umi 13h46'19" 69d38'3"
Sean Edward
Cep 22h30'38" 58d5'0"
Sean Edward Doughty
Psc 0h26'58" 18d59'51"
Sean Edward Kearney
Uma 8h59'48" 60d5'22"
Sean Eugene McCarthy
Uma 10h42'18" 53d56'58"
Sean Faherty
Ori 6h6'4" 21d1'35"
Sean Farrell
Boo 14h55'2" 37d1'52"
Sean Francis Ryan
Uma 10h6'13" 70d44'5"
Sean Garrett Joseph Mills
Her 17h39'6" 35d28'26"
Sean Geery (Kermie)
Lib 14h52'21" -2d36'8"
Sean Gilleece
Umi 14h13'11" 73d13'17"
Sean Gregory Augustine
Ori 5h52'28" 6d20'53"
Sean H. Cook
Sgr 18h26'3" -16d21'49"
Sean Harold Hilton
Her 16h45'46" 28d58'55"
Sean Henry Griffin
Umi 17h10'45" 83d16'3"
Sean Humphrey
Ori 5h33'56" -2d10'2"
Sean Jackson - "U and Me"
Gem 6h58'11" 28d21'1"
Sean Jacob Milner
Uma 11h53'18" 41d53'21"
Sean James Benjamin
Vir 13h23'19" 11d4'42"
Sean James Blakeslee
Lib 14h59'0" -21d51'54"
Sean James Callaghan
Gem 6h40'39" 16d8'8"
Sean James Michael Edward O'Keefe
Aqr 22h41'44" 1d20'41"
"Sean" John Christopher Halpin
Lac 22h29'43" 44d36'27"
Sean Johnson Kwatra
Lyn 6h32'34" 61d28'30"
Sean Jones
Uma 9h37'16" 48d41'40"
Sean Jordan
Peg 22h16'50" 19d41'39"
Sean Joseph Bennett
Tau 3h34'44" 23d24'4"
Sean Joseph Doyle
Tau 4h17'14" 18d3'4"
Sean Joseph Krause
Sco 16h10'41" -17d53'48"
Sean Joseph Kreuz
Per 2h19'4" 57d5'22"
Sean Joseph Morrison
Cnc 8h45'2" 23d28'57"
Sean Joseph Ryland
Her 18h35'16" 19d55'54"
Sean Josias Navedo
Vir 13h36'22" -8d29'8"
Sean & Julie & Grace Byron
Aql 19h38'4" 5d0'19"
Sean & Katie
Cyg 19h50'30" 31d55'22"
Sean Kavanagh Super Star
Dra 18h31'35" 50d8'36"
Sean Kelly Richardson
Cnc 9h5'12" 28d0'16"
Sean & Kenny
Her 17h13'6" 28d8'56"
Sean Kevin Rankin
Vir 14h15'12" -7d42'0"
Sean Khan's Guiding Light
Cap 20h20'18" -25d34'11"
Sean King
Per 4h17'20" 50d43'6"
Sean Lewis
Her 16h33'24" 31d59'43"
Sean & Lisa M F E O
Ari 2h34'5" 20d46'41"
Sean Lowry Pellegrino
Cap 20h25'23" -11d32'10"
Sean & Lucinda
Uma 11h14'45" 52d55'10"
Sean & Lucy Handley
Cyg 19h36'48" 30d8'51"
Sean Luke Stahursky
Sgr 19h53'26" -12d41'4"

Sean M. Sheehan
Per 3h38'3" 35d21'16"
Sean M. Williams My only true love
Sco 17h41'35" -41d27'18"
Sean Mackenzie Clark
Aur 5h47'36" 38d59'23"
Sean MacMillan
Cyg 20h26'15" 51d9'48"
Sean Maher
Per 3h26'19" 49d13'33"
Sean Marc Heaton
Ori 5h56'1" -0d43'21"
Sean Marshall Powell
Ori 4h50'5" 1d4'29"
Sean Matthew Swiatek
Lib 15h5'4" -27d3'9"
Sean McCall
Gem 7h10'26" 23d9'6"
Sean McGuire Healey
Leo 11h11'37" 11d11'44"
Sean McNeill Travis
Ori 5h19'0" -3d10'15"
Sean Michael
Lib 14h41'5" -9d23'10"
Sean Michael 20
Sge 20h19'4" 17d48'17"
Sean Michael Carton
Ari 3h4'37" 22d56'39"
Sean Michael Caselli
Ari 2h11'55" 26d19'18"
Sean Michael Clark
Cap 20h20'20" -20d19'34"
Sean Michael Comerford
Psc 2h3'21" 5d14'39"
Sean Michael Crispin, My Love
Lyn 6h32'55" 59d7'10"
Sean Michael Duffy
Umi 16h16'30" 80d31'54"
Sean Michael Engler
Leo 11h30'17" 12d3'49"
Sean Michael Farrell
Ori 6h17'28" 16d9'12"
Sean Michael Flynn Jr.
Uma 12h29'22" 53d38'30"
Sean Michael Frances Jackson
Ori 5h41'55" 2d33'17"
Sean Michael Heffernon
Her 17h36'31" 37d54'34"
Sean Michael Herold
Dra 17h49'38" 67d54'22"
Sean Michael Maloy
Tau 4h24'36" 8d29'19"
Sean Michael McCarthy
Gem 6h37'20" 20d54'13"
Sean Michael McMullen
Lib 15h44'49" -5d56'5"
Sean Michael Roach
Cnc 8h19'57" 28d50'53"
Sean Michael Rosney
Uma 10h53'25" 53d38'28"
Sean Michael Steward
Her 17h29'40" 32d24'33"
Sean Michael Thompson
Cru 12h30'28" -59d7'55"
Sean Michael Vincent Kenny
Her 17h28'27" 37d58'46"
Sean Michael Williamson
Per 3h41'40" 40d58'7"
Sean Michael Wood
Boo 15h19'6" 41d29'49"
Sean Monney
Tau 5h47'17" 16d59'9"
Sean Moran Horvath
Dra 18h23'22" 62d17'35"
Sean Natalo
Vir 13h26'41" 6d23'38"
SEAN &ND SAMS LOVE
Sgr 18h25'16" -31d58'21"
Sean & Nilou
Cyg 19h45'36" 55d42'10"
Sean Norris
Uma 11h3'55" 68d33'43"
Sean O' Connor
Her 18h33'29" 17d41'1"
Sean O'Boyle
Sco 16h27'4" -33d39'17"
Sean O'Brien
Ori 5h52'5" 20d46'48"
Sean O'Cain
Cep 22h14'7" 68d15'3"
Sean (Oke) Teahan
Gem 7h12'39" 32d11'14"
Sean Owens Schola
Aur 5h23'58" 30d53'27"
Sean P. Carver
Per 3h5'32" 56d45'32"
Sean P McElroy
Her 18h37'36" 21d18'44"
Sean P. Moore - Shining Star
Cep 22h45'1" 68d55'25"
Sean Patrick
Leo 9h33'5" 27d29'19"
Sean Patrick Buckley
Her 17h49'32" 46d1'20"
Sean Patrick Connelly
Ori 6h17'0" 15d37'58"
Sean Patrick "Crittermon" Likens
Ari 1h48'22" 17d52'21"
Sean Patrick Donovan
Leo 11h45'54" 22d15'54"
Sean Patrick Dwyer
Leo 10h53'53" 16d48'29"
Sean Patrick Early
Sgr 18h36'21" -24d2'4"
Sean Patrick Fabian Lowney
Her 17h10'1" 33d34'13"
Sean Patrick Fisher
Aur 7h23'44" 40d21'31"
Sean Patrick Gwin, March 1, 1959
Uma 12h29'15" 56d47'47"
Sean Patrick Hartje
Gem 6h33'59" 23d53'40"
Sean Patrick Hines
Her 16h44'44" 29d31'29"
Sean Patrick Huey
Vir 13h10'15" 9d9'28"
Sean Patrick Lucas 08-20-99
Her 18h7'34" 17d34'33"
Sean Patrick Maher
Psc 1h26'3" 16d2'20"
Sean Patrick Maher
Psc 1h8'26" 9d54'33"
Sean Patrick Nelan
Cnc 8h12'19" 32d24'7"
Sean Patrick Quilty
Per 3h53'59" 46d54'38"
Sean Patrick Toner
Tau 4h3'49" 28d27'19"
Sean Patrick Wallace
Lyn 7h6'50" 59d51'11"
Sean Patrick Warnecke
Leo 9h42'2" 28d32'34"
Sean Patrick's Star of Light & Love
Aqr 21h3'20" -12d54'34"
Sean Patton
Lib 15h0'59" -14d17'1"
Sean Paul Costello
Boo 15h34'30" 45d12'40"
Sean Peter Connett
Lmi 9h47'33" 40d13'44"
Sean Phillip Callahan-Mitsch
Gem 7h37'57" 32d46'25"
Sean Phillip Dexheimer
Lib 14h39'52" -8d51'56"
Sean Pickering
Sgr 18h21'20" -32d49'56"
Sean Preston Warner
Ori 5h37'56" -0d48'39"
Sean Raphael Arcens
Umi 17h29'20" 80d35'56"
Sean Reid
Her 16h44'46" 47d17'44"
Sean Richard Buchtler
Crb 15h53'57" 32d46'32"
Sean Richard Maddix
Leo 11h44'19" 20d5'20"
Sean Richardson
Pho 23h58'4" -45d22'43"
Sean Robert Aubert
Aql 18h57'56" -0d5'6"
Sean Robert Cook Poole
Sco 16h13'51" -10d55'18"
Sean Robert Joyce
Lac 22h24'17" 47d54'0"
Sean Robert Kent
Ori 6h25'24" 16d56'45"
Sean Robert Marsh
Uma 8h46'52" 56d45'47"
Sean Robert Tavernier
Sco 17h43'49" -37d48'23"
Sean & Robin's Light of Adoration
Cyg 21h8'47" 38d16'39"
Sean Roger Hollingdale
Cru 12h26'48" -62d23'50"
Sean Rowan Hill
Her 17h42'51" 24d54'52"
Sean & Roxanne Mathis
Cnc 8h49'11" 15d3'13"
Sean S.
Uma 10h48'53" 62d50'39"
Sean S. McNulty, Jr. "Gentle Giant"
Her 17h21'25" 23d57'48"
Sean Scannell
Psc 0h52'38" 27d39'28"
Sean Silk
Cma 7h17'12" -14d36'44"
Sean Simpson - Rising Star
Uma 10h7'17" 57d5'4"
Sean Sirkin
Aqr 22h29'45" 0d34'42"
Sean Star
Cap 20h22'12" -14d43'11"
Sean Stearns
And 1h8'17" 42d12'49"
Sean Stephen Thompson
Leo 10h11'17" 17d8'26"
Sean Stephen Wood Cornell
Lib 14h49'26" -3d4'18"
Sean Steven Webb
Vir 14h6'13" -13d14'53"
Sean Swearns
Ari 2h17'8" 23d55'19"
Sean T Barrow
Psc 0h49'30" 20d45'10"
Sean T. Pritchard
Ori 4h51'22" -0d8'59"
Sean Taylor
Uma 10h25'55" 68d50'33"
Sean Taylor Harrison
Uma 12h5'51" 48d33'12"
Sean Tepfer
Tau 4h6'44" 13d1'0"
Sean The Great
Lyn 9h33'52" 40d37'8"
Sean Thomas
Leo 10h34'35" 19d28'45"
Sean Thomas Connolly
Boo 15h10'2" 47d35'15"
Sean Thomas Fetter
Leo 9h39'15" 27d14'26"
Sean Thomas Hill
Her 18h51'32" 23d39'54"
Sean Thomas Hill
Umi 14h18'54" 66d44'23"
Sean Thomas Karcher
Gem 7h15'58" 32d29'44"
Sean Thomas Lauren
Ori 6h16'9" 14d54'47"
Sean Thomas Neill
Sco 16h3'41" -15d32'32"
Sean Thomas Reynolds Jr.
Uma 11h10'2" 63d38'31"
Sean Thomas Saydah
Uma 10h43'42" 49d32'54"
Sean Thomas Smith
Ori 5h42'1" 8d7'4"
Sean & T.J.
Cyg 19h57'17" 58d3'33"
Sean Vincent Richie
Ari 3h5'59" 14d22'40"
Sean Wagner
Del 20h43'3" 5d2'10"
Sean Wall BoBo Monkey
Uma 8h51'18" 69d23'1"
Sean Weems, the Neptunite
Lib 14h50'10" -1d12'51"
Sean Wesley Vick
Her 17h21'50" 24d10'10"
Sean White
And 2h34'17" 44d31'8"
Sean William
Leo 11h6'1" 7d12'56"
Sean William Hughes
Aql 19h35'41" 6d52'1"
Sean William Hull
Lib 15h21'14" -13d45'38"
Sean William Irving
Vir 13h25'30" 11d12'25"
Sean William Rudisill
Vir 13h15'1" -20d3'58"
Sean Wilson
Gem 6h48'26" 34d1'25"
Sean Winslow Turner
Her 16h14'22" 47d14'27"
Sean Wiseman
Her 16h31'10" 48d58'47"
Sean Woods
And 1h12'4" 34d25'59"
Sean Youngblood
Oph 17h10'44" -0d18'51"
Sean, Love Always M
Uma 8h34'36" 54d1'42"
Sean, Until Forever, Christy
Per 3h26'44" 35d3'26"
Seana
Lyr 18h49'39" 29d39'8"
Seana
Vir 14h6'55" -8d5'49"
Seana Ann Paige Dunn
And 1h59'12" 36d56'55"
Seana Marie Osowski
And 23h19'33" 43d27'46"
Seanan Amelia Erin
Cru 12h2'13" -62d12'18"
Seanchrisjo
Cnc 8h31'53" 12d56'11"
SeanChristopher10/28
Sco 17h16'52" -32d54'16"
Seandra
Leo 10h31'21" 22d55'17"
Seanean Of My Heart
Vir 11h45'48" -3d37'35"
Seaneen
Ari 2h7'47" 23d52'20"
SEANETTLES
Lyn 6h33'13" 54d33'14"
Seanfish
Eri 3h24'23" -21d13'55"
Seaniemaster Gleason
Her 16h48'49" 27d40'44"
SeanKate
And 1h9'11" 34d23'59"
Seanna Lee Arana
Tau 4h20'57" 11d58'43"
Seanna Marie Lopez
And 0h32'40" 29d57'23"
Seanna Patrice Riley
Cap 20h30'36" -12d54'5"
Seanne Neal Carney
Per 3h20'54" 42d10'57"
Seannery Jade Tennimon
Psc 1h46'5" 23d29'8"
Seanny Ryan
Cnc 8h50'6" 31d59'13"
Sean-Patrick M. Hillman
Cnc 9h9'30" 18d36'20"
SeanRyerson
Sgr 18h15'7" -20d8'58"
Sean's Hopeful Heart
Cas 1h25'2" 64d1'22"
Sean's Star
Boo 14h38'13" 19d45'19"
Sean's Star
Uma 11h49'39" 38d14'57"
Sean's White Shadow
Cyg 20h52'25" 42d57'14"
Seanshine
Lyn 8h50'39" 34d57'2"
Seanski
Mon 6h33'47" 1d51'56"
SeanTara
Aql 19h5'12" -0d1'8"
Seany, Our Special Star
Lyn 7h41'9" 46d27'22"
Seareen
Uma 8h35'22" 51d45'13"
Searles for Eternity
Cyg 21h58'48" 52d57'38"
SEARRIASREID
Uma 10h43'55" 46d31'35"
Seaside's Elite Jellybean (Murphy)
Sco 16h15'43" -11d46'6"
Season
Tau 5h34'47" 18d55'36"
Seasons
Leo 9h54'40" 24d7'2"
Seasons
Cap 21h27'34" -21d4'47"
Sea-Starr Cardinal
Uma 12h31'40" 57d52'37"
Seaton
Peg 22h19'18" 17d36'58"
Seattle Poet Marion Kimes
Crv 12h31'54" -16d14'2"
Sea-U Private Guest Hse
Uma 8h37'55" 72d31'17"
Seaver Bros. - TDG
Per 2h45'9" 56d24'53"
Seay Star - Austin M Seay 29Jan1940
Aqr 23h48'50" -4d18'58"
Seb
Aqr 21h48'31" 2d10'32"
Seb & Angel
Uma 8h27'41" 66d47'34"
Seb Myles Lear
Ori 5h41'16" 12d10'0"
Seba
Peg 21h43'28" 27d32'5"
Sebaba
Uma 9h45'11" 55d47'3"
sebamed 5,5
Ori 5h0'54" 7d42'8"
Sebastiaan Hendrikus Dyckerhoff
Ari 3h16'5" 16d44'59"
Sebastian
Ori 5h44'3" 1d58'44"
Sebastian
Cnc 9h6'7" 28d37'6"
Sebastian
Gem 6h43'49" 16d45'53"
Sebastian
Per 2h18'54" 57d2'22"
Sebastian
Per 2h16'47" 57d0'22"
Sebastian
Vir 13h59'22" -8d20'57"
Sebastian
Ori 6h20'5" -0d37'15"
Sebastian
Aqr 23h43'58" -13d37'38"
Sebastian - "7"
Umi 14h24'21" 73d57'38"
Sebastian A Melgoza
Psc 1h37'7" 14d39'54"
Sebastian Alistair Scott
Lyn 8h32'26" 48d7'51"
Sebastian Arbogast
Cmi 7h33'36" 4d54'46"
Sebastian Charles Donald Fellingham
Peg 22h51'46" 19d16'46"
Sebastian Charlie Emsley Anderson
Her 18h54'15" 20d44'33"
Sebastian & Eleonora - 1 year Anniversary
Cru 12h16'9" -58d45'33"
Sebastian Finck Wir lieben Dich
Uma 14h4'7" 58d15'31"
Sebastián Galilea Sola
Cru 12h28'28" -60d7'28"
Sebastian George
Gem 6h53'16" 30d35'5"
Sebastian George Taylor
Lib 16h0'23" -16d42'30"
Sebastian Grover
Her 16h38'25" 7d5'26"
Sebastian Herrmann
Ori 6h15'13" 8d0'26"
Sebastian J Guterres Brix-Nielsen
Cru 12h14'5" -62d8'53"
Sebastian Jackson Stone
Leo 10h25'59" 8d36'57"
Sebastian James Vickers
Lib 14h57'27" -19d52'47"
Sebastian Jordan
Dra 19h6'25" 65d41'7"
Sebastian Kirkby
Umi 16h9'13" 80d45'47"
Sebastian Klaus Depenbrock
Lmi 10h32'4" 28d3'51"
Sebastian Kyle
Sgr 19h22'32" -17d22'13"
Sebastian Leal Perez
Aqr 22h36'51" -1d43'33"
Sebastian Lewis
Lib 15h24'23" -14d28'21"
Sebastian Lewis
Cep 0h1'41" 71d51'42"
Sebastian Lombardo's Supernova
Sgr 18h16'7" -18d15'3"
Sebastian Los
Uma 12h14'53" 61d12'15"
Sebastian Milano Colley
Per 3h24'47" 39d46'58"
Sebastian Mohandas
Umi 14h58'58" 74d38'29"
Sebastian P. Janoski
Cnc 8h56'4" 31d2'56"
Sebastian Padilla
Vir 11h53'57" 6d26'34"
Sebastian Preston Simon Estradas
Cmi 7h35'29" 3d58'41"
SEBASTIAN R.
Cap 20h43'48" -15d43'46"
Sebastian Ross Brown
Tau 4h31'9" 16d32'56"
Sebastian Rzepa
Vir 12h16'46" 11d8'40"
Sebastian Salazar
Ari 2h33'54" 25d59'5"
Sebastian Sozzi and Elan Rivera
Vir 13h23'31" 10d11'23"
Sebastian Steven Betters
Aqr 22h41'53" 2d21'25"
Sebastian Tomas
Uma 9h27'28" 46d3'40"
Sebastian Tyler Carrillo
Uma 8h34'25" 51d17'40"
Sebastian Vasquez Griffeth
Ori 5h8'7" 12d26'23"
Sebastian Xavier Esoof
Cep 22h33'17" 61d56'45"
Sebastiana and Aniello Ferrara
Ori 6h12'6" 17d41'6"
Sebastiana "Rio De Janeiro"
Cas 0h36'34" 48d21'30"
Sebastiano
And 2h13'19" 44d53'2"
Sebastiano Manno
Cyg 21h27'11" 35d49'17"
Sebastiano Marcozzi
Ori 5h48'24" 5d0'56"
Sebastiano Micheli
Umi 16h0'17" 73d41'52"
Sébastien
And 23h57'31" 48d27'46"
Sébastien Brière
Per 2h43'59" 41d8'52"
Sébastien Bruchon
Lyn 7h40'45" 36d29'36"
Sébastien Cras
Sgr 18h11'21" -19d50'30"
Sébastien David Arsequel
Uma 9h34'19" 45d6'45"
Sébastien Lajeunesse
Cyg 21h46'2" 43d59'56"
Sébastien Mennet 08.05.2004
Umi 14h42'33" 68d20'53"
Sebastien Miron Milot
Cap 21h1'28" -26d27'1"
Sebastien Nantel
Aur 5h47'21" 47d47'22"
Sébastien-Lina
Her 18h18'40" 15d3'50"
Sebastion Cabot
Dra 17h31'6" 67d48'17"
SEBASTJAN
Sgr 17h52'16" -29d52'22"
Sebastri-X
Sgr 18h24'52" -21d7'51"
Sebert St. Oliver Goodall
Aqr 22h0'13" -8d3'58"
Sebiha Arms
Aql 19h9'37" 7d7'41"
Seble Gameda
And 23h43'30" 35d30'17"
Sebu
Ari 2h29'30" 19d1'0"
Sechriest 20
Her 17h23'34" 34d2'41"
Second Chance
Tau 4h30'57" 7d47'1"
Second Chance
Vul 20h19'32" 23d37'11"
Second Chance
Cyg 21h52'24" 53d58'2"
Second Chance at Love SD JD
Uma 11h23'36" 63d0'52"
Second Chances
Vir 12h43'8" -3d17'30"
Second Star on the Right
Peg 21h12'26" 13d9'9"
Second Star to the Right
Psc 0h37'52" 8d13'23"
Secondhome
Vir 13h40'30" 5d36'37"
Secret
Peg 21h35'28" 19d49'16"
Secret Hynes
Aur 7h2'49" 42d26'53"
Secret Love
Per 4h13'40" 41d13'4"
Secret Teresa
Cap 21h7'19" -17d24'54"
Sectra
Uma 9h58'22" 61d36'25"
S.E.D.
Umi 13h35'55" 71d15'19"
Seda
Uma 13h5'29" 52d40'1"
Seda & Sude Esiyok-Brubaker
Lyr 18h22'31" 32d34'25"
SEDAT TORUNLAR in Liebe Sonne
Uma 9h31'9" 48d47'51"
Sedgstar 17
Cyg 19h47'15" 39d9'20"
Sedi
Tau 4h51'57" 21d25'53"
Sedona Estrella Robinson
Lib 15h42'30" -7d5'20"
Sedona Faith Brickley
Lyr 19h20'25" 28d56'6"
Sedona Strong (AKA Sharon Cheri)
Lyn 8h0'14" 34d20'20"
Sedquadyt
Ori 6h9'34" 10d14'43"
Sedric Altman-Employee of the Year
Ori 6h6'11" -0d32'34"
Sedric's Journey to his Babii <3
Per 3h14'40" 43d32'33"
Sedway Shines
Per 4h18'0" 40d35'56"
See you later alligator
Her 17h7'34" 30d13'8"
See Yuh
Peg 22h21'18" 16d41'59"
Seebandt, Andreas
Uma 10h37'38" 45d23'34"
Seeber
Uma 8h57'41" 51d27'17"
Seed of Pearl
Cap 21h12'37" -18d6'16"
Seefy
Aql 19h48'53" -0d25'5"
Seeing Heidi
Umi 14h57'13" 74d15'13"
Seek Not Afar For Beauty, Karissa
Lib 15h31'6" -27d30'22"
Seeling, Jochen
Uma 10h52'14" 49d16'56"
Seema
Uma 11h18'12" 68d57'32"
Seemi Sitara Siddique
Cnc 9h17'26" 27d41'54"
Seemie Xavier
Gem 6h47'36" 19d37'12"
SEENA
And 2h25'29" 45d4'48"
Seena Spindel
Sgr 20h12'24" -37d36'7"
Seenetha Gonzalez
Lyn 8h24'32" 50d49'22"
Seeta
Cas 1h17'34" 55d57'53"
Seeta Seeram
Leo 11h32'21" 10d48'21"
Seffie Bakker
Lib 14h46'58" -9d13'23"
SeFierAMonAmourJeAuxN eJamaisFeindre
Uma 9h21'15" 48d40'7"
Sefys Forever!
Cyg 20h51'6" 47d24'55"
Segilola Atine Jolaosho
Umi 14h40'23" 77d56'41"
Ségolène
Peg 22h57'7" 19d32'50"
Sehaam and Sameer
Per 3h21'45" 44d53'1"
Seham
Ara 17h5'10" -48d58'31"
Sei J Yee Jr.
Eri 4h30'46" -23d52'59"
Sei la luce nella mia vita
Umi 15h57'57" 79d20'3"
Sei Martina
Crb 15h22'27" 29d53'28"
Seifert, Fabienne
Uma 13h4'39" 60d15'39"
Seifert, Lutz
Uma 12h5'54" 39d3'47"
Seiffert, Ludger
Ori 5h1'9" 6d23'14"
Seifried, Bettina & Lutz
Uma 10h16'22" 57d45'20"
Seija Helena Hakala
Uma 11h32'4" 37d26'45"
Seika and Takuya
Sco 17h23'57" -42d9'47"
SEIKO
Ari 1h58'19" 13d44'16"
Seiler, Wolf Dietrich
Psc 1h50'28" 5d27'59"
Seils
Cnc 9h0'56" 6d43'47"
Seimoon
Uma 10h50'46" 39d30'59"
Seinsch, Walther
Uma 13h24'27" 59d12'52"
Seirra
And 0h50'16" 36d2'48"
Seishu Sono of Ashiya, Hyogo, Japan
Lib 15h40'3" -24d45'2"
Seja
Sco 16h10'8" -16d21'38"
Sejal
Ori 6h17'33" 7d13'53"
Se'Jara
Psc 23h28'36" 4d29'34"
Sek Gek Cher
Sco 17h14'58" -37d53'55"
Seka 24
Lib 15h20'49" -5d36'45"
Sekai Rey Lyles Parks
Uma 10h41'47" 45d28'13"
Sekler, Alex
Uma 9h4'11" 50d24'47"
Sel
Del 20h33'11" 10d11'59"
Selam T. Haile
Uma 10h38'59" 57d42'28"
Selamawit Maria Zibra
Cyg 19h49'11" 31d56'42"
Selanetti
Cas 0h9'16" 55d16'10"
Selbach, Oliver
Uma 8h40'9" 59d25'0"
Selda
Lyn 8h2'35" 41d57'47"
Selda Cavegn
Ori 6h19'1" 10d42'34"
Seldon S. Hill
Aqr 22h33'32" 1d32'16"
SELEAN
Uma 9h37'56" 49d43'16"
Selen & Hakan
Lyn 7h17'21" 57d42'50"
Selena
Cap 21h49'18" -8d51'14"
Séléna
Lib 15h32'12" -24d52'25"
Selena
Cap 21h50'28" -22d54'16"
Selena
And 0h20'1" 45d45'49"
Selena
And 2h5'40" 41d43'16"
Selena
And 1h39'3" 37d34'38"
Selena Alexis Morley
Gem 7h41'45" 17d30'12"
Selena "Angel" Marie
Ori 6h12'32" 17d35'11"
Selena Beth Devault
Ori 5h30'4" -0d33'43"
Selena Bryant Sharpton
And 0h20'0" 35d13'21"
Selena Charity Crayon
Gem 7h28'52" 26d37'15"
Selena Gjolaj
Tri 2h9'21" 34d12'51"
Selena I. Beaverson
Vir 11h42'3" 2d32'51"
Selena Kaczmarek
Aqr 21h25'20" 2d9'46"
Selena L. Fraiser
Cap 21h40'41" -13d41'58"
Selena Loren McQuarrie
Tau 5h9'1" 20d5'49"
Selena Marie Dawn Stillions
Tau 4h27'8" 14d40'38"
Selena Marie Lundy
And 23h11'55" 47d32'36"
Selena Martin
And 2h38'0" 37d44'0"
Selena Williams
And 2h23'49" 42d20'37"
Selena Zepeda's Heavenly Risen
Aqr 23h21'11" -9d39'2"
Selena's Freedom
And 0h48'30" 39d30'55"
Selendrae
Mon 6h52'28" -0d13'44"
Selene Aguilera
And 0h6'28" 47d53'44"
Selene Hernandez
And 2h13'0" 43d19'16"
Selene Sanchez
And 0h20'54" 33d24'17"
Seleste Salamanca
Cam 6h20'34" 69d53'46"
Selin
Psc 1h13'51" 31d25'21"
Selina
Lac 22h57'22" 44d0'30"
Selina
And 2h32'44" 45d4'46"
Selina
Uma 9h23'17" 46d1'25"
Selina
Vir 13h19'54" 12d55'21"
Selina
Dra 18h35'5" 72d31'42"

Selina Ashland
Psc 22h51'58" 5d3'4"
Selina Black (MacDonald)
Cas 1h27'14" 69d57'15"
Selina Duffy
Cas 1h27'45" 54d1'52"
Selina Gotch
Lyn 6h51'49" 58d29'30"
Selina Herlihy
Cru 12h31'45" -59d59'46"
Selina Jackson
Cas 1h24'13" 64d23'55"
Selina K. Raymond
Cas 0h45'58" 60d46'45"
Selina Lydia
Dra 17h32'31" 58d4'45"
Selina Michelle
Tri 1h45'20" 31d28'3"
Selina Morris
Lib 15h6'11" -5d39'34"
Selina Nee
Sge 19h38'42" 18d29'2"
Selina Pirinccioglu
Vir 12h53'55" 11d7'14"
Selina + Stefan
Her 17h32'54" 36d0'1"
Selina + Stefan
Uma 11h23'59" 31d38'58"
Selina V Iniguez
Sgr 18h13'20" -19d31'15"
Selina Yingqi Xing
Vir 12h53'36" 11d31'12"
Selina-Angelina Hägler
And 0h13'47" 46d2'30"
Selina-Jane Spencer
And 23h47'23" 39d8'3"
Selinda Feiertag
Uma 10h15'32" 50d11'11"
Seline
Per 1h45'26" 54d33'33"
Selita Daniele Brown
Garrison
Uma 10h58'32" 42d28'6"
Sellars-SW_ENG04
Uma 12h0'28" 55d33'24"
Sellotapeface's Star
And 0h3'34" 38d37'46"
Selma
Cyg 20h29'43" 55d10'28"
Selma Marion
Cnc 8h30'22" 31d24'55"
Selma Schwartz
Crb 15h41'17" 36d25'43"
Selma Vaca
Lib 15h45'1" -12d30'14"
Selma Victoria Evango
And 0h27'34" 32d27'17"
Selna Dawn Miller
Sgr 17h58'16" -18d45'2"
Selvey
Lib 15h6'45" -2d47'6"
Selvi
Lyr 18h47'38" 40d35'15"
Selwood-Starr
Cyg 20h55'23" 34d5'4"
Selwyn
Aql 19h55'39" 0d35'59"
Selwyn Epstein
Cam 4h41'0" 75d3'59"
Sem Kepers
Leo 11h51'7" 15d31'58"
SEMA
Cap 20h41'43" -21d17'25"
Sema
Uma 12h23'55" 58d50'3"
Sema Boodram
Peg 23h27'54" 18d58'12"
Semeraro
Cyg 19h36'58" 55d4'21"
Semi
Uma 11h37'38" 46d18'51"
Sempai Glenn H. Fischer
Sgr 19h33'41" -14d8'31"
Sempcthebrighteststargave
earth2suns
Uma 10h46'27" 65d23'57"
Semper
Cas 0h15'59" 57d26'26"
Semper
Lyn 8h1'37" 40d6'13"
Semper Amabo Te Jennifer
Gem 7h4'44" 18d2'29"
Semper Amemus
Crb 16h22'39" 33d51'0"
Semper Amemus
Cyg 19h48'54" 42d30'26"
Semper Amemus
Uma 10h27'20" 58d29'37"
Semper Amemus ~ Forever
Love
Crb 15h39'35" 37d3'50"
Semper Atqui Infinitas-Mi
Amor-HSM
Uma 10h45'38" 61d29'44"
Semper et Forem: rch &
anb
Crb 15h48'16" 27d56'30"
Semper Fi
Cam 4h0'44" 67d23'27"
Semper Fidelis
Uma 8h33'51" 70d13'23"
Semper Fidelis
Uma 10h32'53" 47d53'1"
SEMPER FIDELIS
Patrick&Géraldine
And 2h27'49" 49d45'38"
Semper Mom
Cap 20h34'40" -17d58'54"
semper nena
Cyg 20h26'0" 34d31'5"
Semper Pariter
Per 2h20'31" 51d11'29"
Sempiternus Amor
Cyg 20h0'17" 46d6'56"
Sempiternus Amor
Ara 17h37'39" -47d29'10"
Sempiternus Emma
060207BS
Uma 10h11'7" 70d23'2"
Sempiternus Nela 190506S
Uma 9h41'22" 47d14'13"
Semplicemente Bella
Peg 23h19'27" 17d31'39"
Semplicemente Stefania
Ori 6h4'23" 16d15'16"
Sempre
Cyg 20h24'0" 46d38'13"
sempre alla sinistra <3
Ori 6h9'22" 20d51'11"
Sempre e Dovunque
Sco 17h0'46" -44d37'21"
sempre il mio mito
Uma 9h23'2" 62d26'2"
sempre insieme
Cyg 20h32'0" 36d24'14"
Sempre nel mio cuore
Psc 1h8'34" 27d45'15"
sen ve ben
Gem 7h34'48" 20d43'25"
Séna
And 0h16'2" 26d12'9"
Sena
Ori 5h15'42" 0d26'26"
Senaa Al Jalahema
Leo 9h41'54" 21d53'38"
Senan & Joan Collins -
Ruby Star
Cyg 21h19'59" 41d1'57"
Senan S Gormley
Leo 11h0'41" 14d19'56"
Senara Mendez
Lyn 7h58'10" 58d24'45"
Senator Bob
Per 2h55'12" 53d51'7"
Senator Bunky Huggins
Sco 16h54'48" -38d44'36"
Senator J.J. Exon
Crb 16h0'55" 36d24'34"
Senay
And 1h19'9" 43d24'36"
Seneca Cheyenne Allen
Uma 9h5'0" 56d55'59"
Senet Eric
Leo 11h39'56" 17d41'43"
Sengül, Ömer
Uma 12h2'59" 61d35'4"
Senia: The Light Of My Life
And 2h28'35" 45d11'52"
Señior Hillhouse
Cnc 8h38'16" 23d4'18"
SENKAN
Sgr 19h7'54" -12d39'10"
Senny Ly
Psc 0h43'36" 16d22'16"
Senor Puppaloni
Uma 11h53'59" 55d14'35"
Señor Verde The "G" Star
of M.T.W.
Cru 12h25'32" -63d22'12"
Senora Iris
Cas 1h24'41" 57d20'24"
Senora Tolbert
And 2h18'1" 50d15'22"
Senorita Lynn
Uma 11h53'33" 36d39'41"
Sen-ptah
Aur 6h38'16" 35d20'36"
Senrab - MJ
Cmi 7h8'28" 10d9'56"
Sensation's Fancy Free
Peg 22h10'35" 34d57'40"
Sensei Babyok
Sco 17h18'42" -41d46'24"
Sensei Yvonne Montalvo
Plus & One
Umi 16h32'18" 77d18'41"
Sensi Alvaro Byrne
And 23h8'57" 43d12'34"
sent
Her 16h20'51" 20d33'21"
Sent From Above For Us
To Love!
Leo 10h12'30" 17d34'13"
Senta Stumer Stich
Leo 9h46'39" 25d30'22"
SenTan-40
Leo 11h22'57" -5d38'30"
Sentinel-Maya & Karesz
Cas 23h33'42" 54d50'56"
Seo Yeon Hee
Sgr 18h31'8" -36d27'36"
Seok Chin (Alicia), Hong
Pho 0h24'44" -51d5'52"
Seona
Uma 13h41'56" 55d15'52"
Seon-ae & Andrew - 1105
Forever
Cru 12h29'2" -61d34'34"
Seonag Richardson
And 23h27'11" 50d38'52"
Sepehr Fatulahzadeh Rabti
Uma 11h32'43" 57d11'43"
Seperate Chaser
Per 3h4'27" 45d32'53"
Seperoth
Psc 23h49'12" 5d8'41"
Sepha
And 0h46'20" 33d57'38"
Sepheryn
Aql 19h16'52" 3d0'5"
Sephrenia
Uma 10h59'14" 36d45'3"
Sepideh Norouzi
Cnc 8h52'14" 11d2'46"
Sepin, Paul
Ori 6h16'46" 10d34'16"
Sepp Altmann
Uma 9h8'38" 54d46'18"
SEPPI
Uma 13h12'3" 57d17'29"
SepraStar
Lyn 7h18'49" 53d22'46"
September
Tau 4h14'50" 27d7'56"
September
Gem 6h51'32" 16d56'35"
September 11, 2001 Lest
We Forget
Uma 13h12'55" 55d52'7"
September Connley
And 0h39'36" 27d5'56"
September Jade Johnson-
Goldberg
Leo 10h43'42" 16d2'43"
September Rene
Tomlinson
Crb 16h16'59" 30d9'10"
September Zamora
Her 18h41'34" 22d16'52"
Septième Ciel
Uma 8h49'42" 55d4'24"
Septimus Charles Rivers
20/2/1966
Cru 12h17'46" -62d28'27"
Sepyon
Umi 13h35'31" 76d15'18"
Sequana Blondell
Umi 13h49'48" 69d37'19"
Sequestered Utopia
Umi 15h25'5" 77d19'29"
Sequin Nicole Clark
Psc 1h10'53" 22d49'28"
Sequoia
Cru 12h25'14" -62d31'9"
* Serae Danielle Edwards
5292 *
Dra 16h0'28" 56d38'25"
Serafin Melendez II
Gem 7h29'30" 18d35'27"
Serafina
Vir 14h28'55" 5d13'17"
Serafina
Uma 9h41'20" 57d12'17"
Serafina
Uma 9h37'42" 70d22'56"
Serafina Joy Codignotto
And 2h23'7" 48d6'7"
Serafina Rosa
Cam 9h19'50" 73d38'30"
Serafina Sofia
Ori 5h42'7" -1d36'34"
Serafino Cereste
Peg 21h44'29" 21d47'9"
Serah Lyn Verrier
Ari 3h16'13" 29d30'5"
Seraina
Lac 22h19'54" 48d25'39"
Seraina
Cas 23h33'17" 54d32'19"
Serana Elizabeth
Lib 15h45'18" -5d42'33"
Serana Pellegrino
Leo 11h51'37" 20d30'6"
Seraph
Cyg 19h41'20" 31d31'49"
Seraphim
Leo 10h27'53" 23d47'35"
Serata Claudia
Aur 5h18'35" 41d8'57"
Serdar Altan
And 23h22'19" 42d26'40"
Serdar Sunay
Psc 0h49'56" 29d18'27"
Serega
Peg 22h54'33" 15d22'29"
Sereina Tanner
Tri 2h19'26" 32d5'33"
Seren Abbey Hughes
Jones
And 1h7'10" 46d36'31"
Seren Brammer
Umi 15h4'26" 72d2'42"
Seren Emory
Cam 4h23'52" 67d24'26"
Seren Jolene Lewis
Leo 10h54'45" 14d38'22"
Seren Leah Millan
And 1h45'45" 39d49'1"
Seren Lee
Uma 10h42'3" 42d2'50"
Seren Palfrey-Adams
And 0h21'34" 46d36'49"
Seren Rose Harries
Cru 12h19'2" -57d33'51"
Seren Yogi
Uma 10h50'23" 67d19'3"
Serena
Uma 11h20'29" 65d7'52"
Serena
Cam 5h10'1" 65d39'5"
Serena
Uma 9h30'23" 60d11'6"
SERENA
Aqr 22h51'31" -4d49'36"
Serena
Vir 13h57'50" -15d53'32"
Serena
Sco 17h56'11" -41d59'19"
Serena
Cyg 19h47'27" 46d53'13"
Serena
Peg 21h43'27" 21d1'52"
Serena
Crb 15h59'6" 25d40'54"
Serena Ashlynn "Punchy
Girl"
Mon 6h53'54" -6d38'45"
Serena Bell Coffey
Psc 0h22'42" 7d8'25"
Serena Beth Kerr
Cap 21h55'44" -19d55'50"
Serena Cenciarelli
Per 4h1'7" 46d56'40"
Serena Doree Arrabito
Pho 0h34'48" -50d29'26"
Serena Elizabeth Pierce
Aqr 23h12'28" -11d28'36"
Serena F.W. Soffer
Aur 5h32'10" 46d40'37"
Serena Garofalo
Lyr 18h49'31" 30d4'26"
Serena Giles
Lyr 18h40'47" 35d35'9"
Serena Grace
Crb 15h30'54" 29d10'16"
Serena Grey
Uma 10h6'58" 56d51'40"
Serena Indira Cutugno
Cnc 9h0'44" 32d55'24"
Serena & Jadon
Lyr 19h6'7" 43d1'7"
Serena Jinn
Sco 16h50'58" -27d14'51"
Serena Joy Gardell
Cas 23h16'12" 55d13'41"
Serena Keeler
Uma 11h12'16" 34d26'45"
Serena Khorsandian
Vir 12h53'51" -7d54'2"
Serena L. Swiatek
Tau 4h18'26" 26d33'14"
Serena Leopardi
Cam 4h17'11" 56d15'35"
Serena Lisa Holzberg
Tau 4h10'25" 27d26'36"
Serena Louise Gosling
Cra 17h59'31" -37d8'21"
Serena Louise Hanor
Cap 20h17'42" -27d10'48"
Serena & Michael Koss
Cep 21h8'43" 58d34'45"
Serena ~ My Princess
Lyn 8h20'29" 44d31'12"
Serena Nordio
Umi 13h44'17" 77d28'8"
Serena Paige
And 1h20'47" 49d3'28"
Serena Puccio
Uma 10h8'42" 44d11'38"
Serena R. Abouhalkah
Lyn 7h24'47" 46d9'11"
Serena Rem
Ari 2h12'18" 25d39'13"
Serena Rose Pope
Uma 12h13'50" 61d2'51"
Serena Simone
Vir 13h16'15" -21d54'34"
Serena Yang
Psc 0h7'24" 11d35'5"
SERENA143
Aqr 22h31'55" 1d22'29"
serenagirl
Uma 10h12'36" 66d17'46"
Serena's Beauty
Cyg 20h14'52" 43d19'38"
Serena's Eternal Light
Sco 16h32'40" -34d30'47"
Serena's Star
Aqr 21h45'7" -2d51'46"
Serendipitous Soulmate
Uma 13h52'35" 51d9'54"
Serendipitous Susan
Leo 11h52'26" 24d39'21"
Serendipity
Leo 11h56'11" 25d54'24"
Serendipity
Crb 15h32'32" 28d58'28"
Serendipity
Her 17h23'59" 24d29'34"
Serendipity
Cnc 9h7'30" 27d1'56"
Serendipity
Gem 6h21'53" 22d58'55"
Serendipity
Ari 2h34'0" 25d21'56"
Serendipity
Peg 23h15'33" 17d27'13"
Serendipity
Aql 19h25'49" 2d28'39"
Serendipity
Vir 14h35'52" 3d51'32"
Serendipity
Sex 10h22'5" 1d53'40"
Serendipity
Ori 5h21'20" 4d46'39"
Serendipity
Ori 5h56'27" 14d13'44"
Serendipity
Peg 22h43'23" 12d17'28"
Serendipity
Ori 6h3'47" 15d47'14"
Serendipity
Ori 6h16'56" 16d26'15"
Serendipity
Her 17h52'9" 48d5'9"
Serendipity
Cyg 21h15'29" 46d59'43"
Serendipity
Her 16h54'35" 39d18'20"
Serendipity
Cyg 21h13'12" 44d41'42"
Serendipity
Lyn 7h11'3" 49d29'45"
Serendipity
Lyn 8h1'44" 44d12'36"
Serendipity
Per 2h54'55" 42d49'30"
Serendipity
Cyg 21h28'49" 37d29'46"
Serendipity
Cyg 20h6'43" 33d3'18"
Serendipity
Lyr 18h47'30" 33d15'48"
Serendipity
Lyr 18h52'35" 36d21'0"
Serendipity
Crb 16h15'31" 33d42'53"
Serendipity
Lyr 18h23'51" 35d49'27"
Serendipity
Gem 7h4'52" 33d20'58"
Serendipity
Gem 7h9'1" 34d27'6"
Serendipity
Aur 4h59'13" 32d43'50"
Serendipity
Mon 6h49'44" -0d7'18"
Serendipity
Mon 6h46'28" -0d30'24"
Serendipity
Ori 5h7'7" -0d20'1"
Serendipity
Cap 21h41'4" -13d41'53"
Serendipity
Uma 10h55'56" 63d17'5"
Serendipity
Uma 11h55'21" 65d43'12"
Serendipity
Cep 22h13'53" 69d33'52"
Serendipity
Umi 15h59'12" 73d22'36"
Serendipity
Cam 4h14'37" 73d31'48"
Serendipity
Boo 14h40'41" 53d25'32"
Serendipity
Uma 13h42'19" 54d1'16"
Serendipity
Cas 23h20'52" 57d24'13"
Serendipity
Uma 9h44'29" 53d53'42"
Serendipity
Sco 16h42'2" -30d41'30"
Serendipity
Sgr 17h51'53" -29d54'16"
Serendipity
Sco 17h50'14" -41d29'11"
Serendipity
Sco 17h47'9" -42d38'6"
Serendipity Angel 150979
Cru 12h51'36" -62d3'30"
Serendipity Ditto
Uma 8h49'37" 68d36'5"
Serendipity SNS 11-22-03
A&F
Sgr 20h5'5" -26d10'12"
Serendipity-Elizabeth and
Marc
And 1h13'27" 44d12'29"
Serendipity's Paradise
Tau 4h44'48" 18d21'16"
SerenDips01
Ori 5h54'14" 8d34'7"
Serendipty
Cas 1h37'32" 66d6'11"
Serendipty::9/11/05::I <3
You
Cas 23h45'29" 51d19'1"
Serene Lucian
Ori 6h19'18" 10d6'54"
Serenella
And 1h16'29" 38d52'15"
Serenella Buzi
Uma 10h33'37" 64d44'21"
Sereneyes
Uma 11h26'26" 62d38'16"
Seren-Haf Menna Hill
Cas 0h57'59" 63d39'57"
Serenity
Uma 11h27'15" 59d8'20"
Serenity
Umi 16h34'6" 82d6'8"
Serenity
Aqr 20h40'49" -2d12'13"
Serenity
Lib 15h55'58" -6d2'41"
Serenity
Sgr 19h15'4" -22d32'0"
Serenity
And 1h38'53" 43d10'18"
Serenity
Uma 11h38'45" 41d55'30"
Serenity
Lyn 7h38'53" 37d12'0"
Serenity
Cnc 8h21'52" 31d0'46"
Serenity
Lyn 8h28'8" 35d59'31"
Serenity
Lmi 10h9'25" 35d40'1"
Serenity
Crb 15h48'37" 35d21'51"
Serenity
Crb 15h37'56" 30d5'56"
Serenity
Per 3h9'41" 53d8'43"
Serenity
Lyn 8h7'48" 45d2'41"
Serenity
And 1h59'7" 46d36'41"
Serenity
Tau 4h28'18" 10d41'49"
serenity
Her 17h12'35" 20d0'48"
Serenity
Lyr 19h15'54" 29d44'40"
SERENITY
Leo 10h19'59" 24d13'56"
Serenity Granted
Cyg 21h41'16" 45d8'8"
Serenity Lynn
Psc 1h50'52" 5d12'28"
Serenity & Matthew
Gem 7h18'17" 20d32'40"
Serenity Rae Moberg
Crb 15h32'14" 27d6'51"
Serenity Rae Siefer
Lyn 8h1'22" 40d43'55"
Serenity Star of
JTBMcBride
Dra 19h44'54" 61d24'56"
Serenity Starr "Bug" Perry
Lyn 8h24'20" 41d55'43"
Serenity Vanderpool
Peg 23h57'44" 23d44'16"
Serenity's Miracle
Mon 6h50'2" 6d21'19"
Seren's Star
And 0h4'43" 38d54'58"
Seren's Surprise
Dra 14h54'26" 55d5'55"
Serephim
Uma 10h11'30" 45d53'55"
Seresa
Lib 14h48'2" -1d46'10"
Serg, Monica, & Tootles'
Star
Tri 1h50'56" 29d10'11"
Serge Binkert
Crb 16h5'2" 36d28'33"
Serge de Groote
Ari 3h14'37" 12d6'41"
Serge Duclos
Cep 21h42'33" 64d33'33"
Serge et Daisy
Umi 15h25'56" 80d30'7"
Serge et Michelle Vallée
Aql 19h8'51" 13d1'50"
Serge Gbedevi
Uma 13h37'40" 58d28'53"
Serge Kampf
Gem 6h28'25" 27d27'17"
Serge Nicolas Evanow
Vir 12h26'54" 12d24'30"
Serge - Nicole
Uma 10h17'45" 43d29'43"
Serge Yannick Gbedevi
Uma 9h8'41" 54d16'50"
Serge, mon accord majeur
étoile
Lyr 18h38'21" 28d11'56"
Sergeant Jeremy
Wolfsteller
Per 4h9'35" 31d13'28"
Sergeant Wayne Horton
Gem 6h20'58" 21d13'47"
Sergey Belashov
Cep 22h32'42" 68d32'2"
Sergey Ermolenko
Dra 17h27'19" 52d15'44"
Sergey Gary Podwalny
Ori 5h40'22" 2d32'6"
Sergey Sorkin
Uma 9h1'31" 61d24'58"
Sergey Toni
Vir 13h44'23" -16d52'26"
Sergey Vodvud
Aur 4h52'18" 32d20'2"
Sergio
Cyg 19h54'24" 45d33'41"
Sergio
Vul 19h30'46" 24d36'51"
Sergio
Lib 14h52'17" -6d16'50"
Sergio & Adrian
Gem 6h45'50" 27d51'16"
Sergio Alberto Vanegas
Arbelaes
Lyn 6h28'3" 57d31'23"
Sergio and Denise
Ari 2h35'6" 25d15'38"
Sergio Ayala
Cnc 8h41'53" 22d39'27"
Sergio Brandao
Uma 13h39'21" 58d8'14"
Sergio Cattaneo
And 23h52'0" 34d2'1"
Sergio Daricello
Lyn 9h20'6" 38d50'22"
Sergio & Elise 1999
Sge 19h24'34" 18d31'40"
Sergio Fonseca
Cru 12h10'38" -63d19'37"
Sergio González Perea
Psc 1h16'1" 28d0'20"
Sergio Ignacio
Lyn 7h34'46" 57d52'57"
Sergio J. Vega, Sr.
Ori 5h21'36" 5d11'34"
Sergio Jimenez Marin
Lyn 6h59'55" 58d43'7"
Sergio Lebrija
Tau 4h26'33" 13d37'44"
Sergio Melendez
Uma 11h46'23" 45d30'27"
Sergio Mena
Uma 8h44'59" 64d55'23"
Sergio Ricardo Montalvan
Lyn 8h13'16" 54d36'53"
Sergio Roche & Family
Per 4h11'19" 51d36'43"
Sergio V. Proserpi, M.D.
Aql 19h41'22" 7d53'11"
SergioGalicia & Elizabeth
Gettelman
Sge 20h4'31" 16d33'38"
Serguei A. Lapin, Me
Amate
Hya 9h36'2" -0d15'58"
Sergul Oktaymen Pekmezci
Leo 11h7'26" 17d13'11"
Serhen
Umi 14h14'25" 69d5'45"
SeriAble
Lep 5h13'55" -11d39'37"
Seridion
Sge 19h58'0" 17d2'21"
Serilious
Uma 8h54'6" 49d30'11"
Serina
Ori 5h59'45" 21d26'23"
Serina Lynn
Cas 23h49'10" 54d12'34"
Serina Pererya Saldivar
Cas 23h27'57" 58d29'3"
Serissa Rose
Cyg 20h32'41" 38d39'28"
Seriva
Uma 10h33'52" 44d12'39"
Serivia
Psc 1h23'53" 33d17'54"
Serkan Altin
Cam 7h49'58" 63d25'6"
Seronia
Cap 21h40'55" -9d19'25"
Serop Aghyarian ~ "Baboo"
Boo 15h39'38" 41d54'28"
Serpa
Cap 21h44'19" -10d40'40"
Serra Tansel "Happy 18th
Birthday"
Psc 0h45'17" 14d28'24"
Serramor
Peg 22h28'0" 15d5'53"
Serrina Marie Cedarleaf
And 0h41'12" 45d33'32"
Serrrgio-Legion
Cep 22h4'36" 61d1'42"
Servanne
Uma 9h32'6" 45d53'36"
Seryn Makaela Schmitt
Desrosiers
Lyr 19h13'41" 45d56'33"
SES Dolphin Star
Del 20h36'41" 13d30'18"
seshem_nefer
Tau 4h25'53" 13d16'0"
Sesilu
Uma 9h25'18" 61d22'56"
Seslee Elise Skrabanek
Sco 16h17'36" -18d10'7"
SESO Anniversary Star
Leo 10h16'34" 22d33'24"
Setara
Sco 15h56'10" -22d13'17"
Setel Billing Department -
2005
Uma 12h9'22" 62d25'58"
Seth
Psc 1h54'42" 6d8'50"
Seth
Tau 3h38'52" 28d35'45"
Seth
Leo 10h30'58" 18d55'10"
Seth
Psc 1h8'1" 12d49'41"
Seth A. Huffman
Cap 20h13'9" -14d4'14"
Seth A. "Spike's Star"
Pierce
Uma 11h12'13" 46d51'25"

Seth Aidan
Aqr 21h55'50" 0d33'19"
Seth and Carolyn's Virgini star
Cyg 19h47'32" 47d33'6"
Seth And Chrissy
Dra 15h40'44" 57d27'22"
Seth and De
Aqr 22h33'48" -24d4'27"
Seth and Jennifer
Vir 14h2'11" -22d12'38"
Seth and June Roy
Gem 7h22'53" 14d43'46"
Seth and Nora
Psc 0h12'5" 5d12'14"
Seth and Sally Carpenter
Ori 5h24'58" -3d45'30"
Seth Andrew Carlson
Her 17h7'7" 34d21'37"
Seth Andrew Young
Per 3h8'29" 53d34'15"
Seth Bailey
Aur 5h42'41" 49d59'17"
Seth Benjamin Daines
Lib 15h1'12" -2d5'12"
Seth Benjamin Macneal
Uma 10h52'27" 55d23'30"
Seth Bird
Per 2h46'14" 53d15'45"
Seth Boswell
Cru 12h47'48" -64d34'32"
Seth Charles Witter-Merithew
Leo 9h46'41" 23d6'17"
Seth Cogar & Jeanie Stump
Aqr 21h12'2" -8d47'4"
Seth Daniel
Uma 10h7'33" 57d29'6"
Seth Daniel Hall
Psc 1h38'13" 4d11'6"
Seth David
Lmi 10h35'46" 39d4'6"
Seth Davidson Stuart - 'Seth's Star'
Psc 0h25'48" -0d11'59"
Seth Edson
Umi 16h19'55" 72d35'5"
Seth Elias Pipakis
Cru 12h12'12" -62d58'4"
Seth Elliot Been
Cap 21h39'10" -10d44'8"
Seth Feit
Tau 4h39'39" 23d1'0"
Seth Frances Golay
Per 2h50'28" 37d18'28"
Seth Garrett
Her 18h45'0" 19d8'57"
Seth Grey (Bananas)
Her 17h18'53" 27d42'45"
Seth Gwin
Per 3h16'18" 54d48'0"
Seth Harrington Howard
Ori 4h50'18" 14d19'34"
Seth Howard
Aql 19h8'34" 4d53'44"
Seth Hunter Weatherfield
Ori 5h42'42" 11d55'18"
Seth Jackson Cook
Her 16h44'28" 36d15'59"
Seth James Honchar
Peg 21h29'46" 15d52'22"
Seth Jerome Baldwin
Her 17h37'7" 26d52'51"
Seth John Misale
Per 3h46'23" 33d27'58"
Seth Jordan Mackay
Dor 5h29'50" -63d59'32"
Seth Joseph Dalton
Sco 16h13'19" -20d22'2"
Seth & Kelly Pfeiffer
Uma 10h25'52" 61d24'48"
Seth Kenneth
Aql 19h55'34" -0d29'24"
Seth Liss
Vir 13h17'13" 3d58'2"
Seth Logan Randolph
Psc 1h14'36" 23d21'32"
Seth love "The Love Doctor"
Per 3h31'51" 41d38'24"
Seth Lunsford
Uma 9h53'47" 67d50'56"
Seth Martin Hufstedler
Ori 4h51'30" 9d42'37"
Seth Martin Souza
Equ 21h9'44" 11d27'4"
Seth Michael
Boo 14h32'10" 33d39'52"
Seth Michael Massari
Umi 14h32'57" 78d53'53"
Seth Michael Myers "anam cara"-KMK
Her 16h27'8" 23d26'10"
Seth Michael Weingarten
Cnc 8h52'49" 31d59'44"
Seth Neil Nicodemus
Uma 11h33'59" 49d3'44"
Seth & Nigel
Col 5h52'33" -37d45'30"
Seth R. Okin
Ori 5h7'2" 3d11'6"
Seth Sruplock
Lib 15h39'24" -17d8'3"
Seth Thomas McFadden
Dra 19h49'20" 66d23'41"
Seth Thomas Sadowski
Psc 0h26'20" 6d39'8"
Seth Tomlin Pound
Uma 11h19'15" 64d12'35"
Seth W. Hampton
Uma 13h45'40" 58d25'57"
Seth W. Jones
Ori 6h2'15" 14d44'36"
Seth Woollet
Lyn 6h57'16" 51d47'22"
Seth, Ethan, & Ellie Sandstrom
Tri 2h17'33" 33d0'0"
Seth1992
Ari 3h20'0" 27d25'38"
setharix
Uma 11h39'3" 59d18'49"
Setoyama of Go
Leo 18h20'46" -32d37'50"
SeTrouver
Ori 5h8'20" 3d46'21"
Setsuko
Ori 5h55'43" 21d9'8"
Sett
Umi 16h37'7" 82d17'49"
Sette Sorrisi
Ari 2h51'42" 11d11'45"
Setting Sail
Cep 23h24'59" 78d21'45"
Settlers Beach Villa Hotel
Cas 0h39'44" 58d2'56"
Setzer Family Star
Her 16h42'35" 36d46'15"
Seufert, Edwin
Uma 11h17'25" 64d57'4"
Seung Me Kwon RamsDell
Ori 5h47'3" 5d44'54"
Seung Tak Il
Uma 9h35'23" 64d33'58"
Seung-Hee Woo, a person whom I love
Sco 17h26'28" -42d23'52"
Seute, Walter
Lib 15h41'35" -17d26'25"
Sevali
Oph 17h2'21" -0d23'31"
Sevan Karinca
Lyn 6h42'41" 56d58'10"
Sevanna Angelina Marina Coelho
Tau 5h45'33" 26d24'38"
Sevda
Tau 3h36'20" 8d18'12"
Sevda Hamidova
Sco 17h36'24" -30d55'46"
Sevda Hashimova
Uma 11h30'56" 58d3'28"
Seven
Cam 4h45'39" 62d13'6"
SEVEN
Cam 4h9'32" 66d5'18"
Seven
Dra 17h2'7" 60d36'48"
Seven
Lib 15h5'3" -6d49'24"
Seven
Lib 14h55'47" -9d48'28"
Seven
Tau 4h8'42" 6d6'45"
Seven
Uma 12h4'55" 42d10'0"
Seven of 9
And 2h28'37" 44d46'44"
seventeen
And 1h9'20" 34d7'46"
Seventh Heaven
Her 18h36'7" 15d19'32"
Seventh Son
Sgr 19h20'23" -15d53'11"
Severin
Cas 23h53'47" 59d30'52"
Severin & Joel
Uma 13h19'10" 60d50'59"
Severin Oertig
Cnv 12h46'46" 46d14'20"
Severine
Crb 15h16'34" 29d42'29"
Séverine Cavan
Uma 8h56'25" 56d7'38"
Severine C.L.G. Cailleteau
Uma 10h9'18" 64d32'19"
Séverine Delorme
Aqr 22h50'41" -22d41'4"
Severine Fulliquet
Umi 14h8'34" 71d6'43"
Séverine Mathis
Del 20h35'7" 7d17'30"
Séverine Perales
Cnc 8h45'17" 6d48'6"
Severine Puce Mia Eternelle
Aql 19h4'24" -7d57'35"
Séverine Quoniam
Cap 20h35'6" -11d7'31"
Séverine Roy
Com 12h40'54" 16d47'7"
Séverine, Stärn i mim Läbe
Ori 4h59'32" 14d10'59"
Seversons
Crb 15h51'34" 29d1'39"
Severus
Oph 17h34'14" 7d7'15"
Sevgi
And 23h6'37" 47d16'44"
Sevie
Lyn 8h15'47" 33d17'21"
Sevil Rahimova
Cnc 8h39'12" 18d43'55"
Sevim Avcil
Her 17h44'13" 37d38'41"
Sevina Cécile
And 23h14'2" 42d27'45"
Sew Buttons
Uma 12h20'46" 62d42'13"
Sexton
Sex 10h50'19" 3d22'10"
Sexy and Sugarlips Wishing Star
Ori 5h41'35" 11d55'53"
Sexy B
Uma 11h9'52" 34d55'33"
Sexy Billy's Star
Aql 19h3'59" 14d58'43"
Sexy Blue Eyes
Cyg 21h43'12" 34d50'18"
Sexy Chele
And 2h24'30" 39d16'29"
Sexy Foot Doctor
Uma 11h24'8" 55d58'53"
Sexy Heather Michels
Sco 17h3'40" -33d49'25"
Sexy Leo
Uma 8h24'51" 70d4'34"
Sexy luv
Cap 21h3'23" -25d50'52"
Sexy Lux
Dra 17h7'17" 59d55'38"
Sexy Mama
Cas 1h24'32" 55d51'37"
Sexy Persian Stallion Ehsan Afaghi
Vir 11h40'21" 3d34'50"
Sexy Rachel
Vir 12h57'44" 3d45'42"
Sexy Rexy Rayda
Sco 17h20'42" -42d6'50"
Sexy Sally
Lyr 18h35'16" 38d52'26"
Sexy Susie
Cap 21h47'16" -20d11'48"
Sexy Thing
Uma 11h18'2" 53d56'59"
Sexy Tyler
Her 17h22'32" 38d15'38"
Sexy Zoe
And 2h14'39" 41d24'8"
sexykness
Sco 17h18'21" -33d8'10"
Sexy's Star
Sco 16h8'3" -13d57'39"
Seychelles - MLK
Uma 11h32'53" 43d30'53"
Seyf
Aqr 21h28'42" -8d17'25"
Seyhan Musaoglu
Cnc 8h56'42" 28d5'46"
Seymour
Lyn 6h24'0" 59d54'38"
Seymour Ballos
Ori 6h6'28" -0d2'12"
Seymour Berger
Lib 14h53'27" -9d32'43"
Seymour & Paulina Katz
Cap 21h16'56" -15d47'32"
Seymour's Star
Cep 22h16'44" 61d30'6"
Sezen
Psc 1h15'7" 27d13'29"
Sezen
Gem 7h36'43" 34d39'33"
Sezz - Star Princess
And 1h33'32" 49d52'52"
SFC. Leisa D. Williams
Uma 11h40'30" 51d5'22"
Sfc. Patrick D. Moon
Ori 5h41'22" 3d33'38"
SFC Psycho
Ori 5h58'37" 21d2'27"
SFerguson2006
Dra 15h7'49" 63d58'48"
SFLG
Vir 12h14'25" 3d26'26"
'Sforzo Perfezi'one
Pho 0h50'6" -47d22'18"
SFPA The Jenny
Cap 21h46'55" -18d18'16"
S.G. Junior
Sco 16h10'57" -13d39'45"
S.G. Rita Bergin
Lmi 10h27'30" 39d14'18"
SG19W
Aql 20h34'27" -7d26'30"
sg91605
Uma 11h57'25" 38d31'27"
SGAM
Cyg 21h30'7" 46d11'52"
S.G.C. Bernard Star
Tau 4h26'39" 22d1'4"
SGH Katrin
Uma 8h12'49" 63d40'0"
SGM Golomboski
Sgr 19h13'8" -27d2'26"
SGM Scott S. Dickmann
Uma 11h20'52" 58d31'39"
Sgsp/101460 (Susan's Star)
Ori 5h25'40" 4d20'42"
Sgt. Andy Stevens
Cyg 21h25'42" 55d4'59"
Sgt. Angela Belle Kuczynski-Weaver
Cnc 8h16'12" 17d29'5"
Sgt. Brian J. Walkowiak
Per 3h0'14" 55d36'9"
Sgt E W Lehrmann (TDm)
Ori 5h31'55" -3d21'50"
Sgt. Eric W. Mueller
Aql 19h24'41" 5d9'3"
Sgt & His Princess
Sco 16h8'37" -11d29'2"
Sgt. Hunter Crocker
Cma 7h5'3" -24d3'57"
SGT James R Kettner 82nd ABN SAPPER
Sco 16h55'25" -42d54'44"
Sgt. Jamie Michalsky
Aql 19h40'35" 12d21'55"
Sgt. Javier J. Garcia
Her 17h5'28" 44d47'52"
Sgt. Johnny Stewart II
Per 2h23'56" 55d16'58"
Sgt. Joseph W. Perry
Per 2h12'10" 54d5'28"
Sgt. J.R. Hatch
Tau 4h18'31" 7d21'59"
Sgt Kelly S Morris
Leo 11h16'26" 10d4'59"
Sgt. Kirk J. Bray U.S. Marine Corps
Per 2h20'53" 53d54'54"
Sgt. Major Loreto "Al" Almazol
Ori 5h31'20" 0d38'53"
Sgt. Major V.R. Robinson
Aql 18h53'6" -0d43'53"
Sgt Michael Dune Milne, NYPD, Ret.
Ori 5h42'27" 1d40'49"
Sgt. Michael W. Wooten
Aql 19h44'3" 7d27'51"
Sgt. Poopah Bear
Uma 11h59'43" 44d41'24"
Sgt. Reichert
Uma 11h20'53" 58d35'9"
Sgt. Richard Roberts U.S. Army
Per 3h4'59" 52d36'30"
Sgt. Rock
Lib 15h23'11" -7d37'47"
Sgt Roger Eugene Lewis son of Miles
Lib 15h20'49" -15d58'13"
Sgt. Squirrelmonkey
Her 18h2'31" 36d48'52"
Sgt. T. Ned
Aql 19h17'59" -0d40'44"
Sgt. Tyler Dee Prewitt
Aql 19h41'20" 6d22'1"
SgtMaj Ronald "Devil Dog" Wise
Per 4h13'22" 34d25'57"
sh4d0w_r3b3l_c0wb0y
Aqr 21h8'11" -9d19'48"
Sha Deniece
Sco 16h8'54" -9d25'17"
Sha Linn Wade
Lyn 7h57'44" 48d40'30"
Sha Lohm
Per 3h22'5" 42d12'33"
ShaaBay
Lib 15h41'49" -18d44'43"
Shaan Deanne Foster
Vir 13h31'55" -15d9'15"
Shaax
Uma 13h16'44" 60d58'18"
Shabaz - 30 October 2004
Cru 12h22'59" -56d58'58"
SHABAZZ
Crb 15h59'33" 34d40'18"
Shabjoon
Aqr 22h38'9" -2d48'55"
Shabneez
Cas 23h3'43" 54d55'51"
Shaboonce
Leo 9h30'19" 27d0'23"
Shabre - Special 2 United 4 Eternity
Cru 12h17'23" -56d15'30"
Shabreen
Ori 6h17'37" 13d52'41"
Shabrel Cheri Johnson
Vir 12h26'0" -3d27'55"
Shabs
Cnc 9h17'52" 7d10'57"
Shacara
And 23h27'49" 41d33'23"
Shachar
Ori 5h54'32" 11d45'28"
Shackleford - Duke of Dinglesden
Gem 6h18'16" 25d58'7"
Shad Azimi
Per 4h17'54" 45d16'9"
Shad Dahlquist "Golden 11"
Uma 10h14'29" 45d54'44"
Shad Edward Higgins
Gem 6h44'6" 17d46'33"
Shadaè Peta Marinos
Vir 14h3'26" -13d51'21"
SHADANA
Tau 5h33'52" 27d28'5"
Shaddai Yuki Hildegard
Cma 6h44'49" -14d6'3"
Shade
Leo 11h12'30" -5d58'32"
Shade Louisiana
Leo 10h36'50" 12d34'48"
Shade of my Heart
Cas 1h39'54" 64d14'35"
Shaded Dragonfly
Cam 3h37'57" 62d0'35"
Shadera's Dream
Tau 4h17'16" 30d7'2"
Shadetree
Crb 16h1'22" 36d24'59"
Shadi Khodadoost
Uma 10h39'30" 45d14'37"
Shadia Salah Tekko
Leo 10h4'19" 26d6'16"
Shadow
Psc 1h3'22" 29d10'26"
Shadow
Equ 21h19'6" 9d3'29"
Shadow
Uma 9h38'50" 46d37'11"
Shadow
And 23h18'31" 47d36'24"
Shadow
Per 3h28'1" 52d15'25"
Shadow
Cyg 20h54'45" 37d29'43"
Shadow
Uma 9h28'44" 61d38'21"
Shadow
Uma 13h23'53" 54d6'31"
Shadow
Uma 9h5'40" 56d42'25"
Shadow
Cap 20h52'53" -17d45'11"
Shadow Logue
Uma 11h18'20" 62d40'47"
Shadow Walter
Uma 9h4'12" 54d28'18"
Shadowbrook
Her 17h17'32" 32d44'5"
Shadowcreek
Ari 2h12'27" 22d33'52"
Shadrah
Uma 8h39'42" 53d59'19"
Shady
Uma 11h30'2" 63d37'22"
Shae
Sco 16h6'47" -18d31'24"
S.H.A.E.
And 0h30'48" 43d5'26"
Shae Alexanderia Pilcher 10
Lib 14h32'47" -18d38'17"
Shae' and Dustin's Star
Sgr 18h16'33" -20d31'54"
Shae Daniel Hovey
Per 3h9'24" 52d20'11"
Shae il mio amore luminoso
And 1h8'10" 44d46'42"
Shae Law
Lyr 19h6'3" 36d29'43"
Shae Star
Cyg 21h12'32" 38d40'47"
Shaelan Mae Clayton
Cyg 19h56'23" 47d8'34"
Shaelee Louise
Cyg 20h8'4" 40d43'41"
Shaeleigh Tiernon Payne
Leo 10h7'23" 27d48'40"
Shaelyn Francis Crawford
Vir 13h47'43" -10d35'19"
Shaelyn Villareal
Vir 14h43'0" 3d20'18"
Shaelynn Michelle Morrison
Vir 11h58'55" -0d9'7"
Shaemara
And 1h5'19" 41d28'40"
Shaena & Tim
Cnc 8h7'12" 19d45'41"
SHAERLAEHN
Eri 4h10'1" -0d12'54"
Shaffin Lalani
Uma 9h18'48" 47d10'13"
Shaga
Boo 14h16'56" 7d30'5"
Shaggy Carey
Lib 15h28'2" -15d40'6"
Shaggy the dog
Cma 6h58'17" -13d58'12"
Shaghayegh Marashi
Uma 9h42'46" 71d47'35"
"Shags" Kirk McCarthy
Apu 15h33'4" -71d14'34"
Shahab Issa Abu Shahab
Sco 16h29'29" -26d9'51"
Shaheda Tanay Khan
Psc 1h31'47" 13d48'52"
Shaheen Ali
Leo 11h52'20" 27d35'19"
Shaheena
Lib 15h28'27" -15d16'57"
Shahela Perveen
Lib 14h39'7" -9d10'29"
Shaherezaad
Srp 18h30'44" -0d18'34"
Shahin I. Assarpour
Cap 21h42'8" -14d28'35"
Shahine Racoulman Ayan
Ori 5h1'45" 9d38'50"
Shahira
And 0h50'50" 38d19'59"
Shahla
Gem 7h11'40" 14d47'17"
Shahlo
Lyr 18h44'36" 39d46'11"
ShahMathur
Cam 4h30'41" 66d3'34"
Shahnaz
And 2h34'59" 50d6'21"
Shahnaz & Paul ~ July 3, 2005
Cyg 21h40'27" 39d17'38"
Shahrouz Dehghahi
Lyn 6h58'33" 55d48'19"
Shai and Timeri Anderson
Cyg 20h42'9" 52d21'3"
Shai - Baby
Ori 6h7'5" 15d56'58"
Shai Laurel 9
Aqr 23h53'36" -7d27'15"
Shai Mon Petit Chou
And 0h36'55" 32d38'19"
Shaie
Sco 16h16'27" -11d1'8"
Shaikah Al Mandeel
Cyg 20h18'52" 52d49'23"
Shaila
Ori 6h6'12" 10d23'46"
Shailaja and Y.S. Ram Rao
Tau 5h5'50" 17d51'34"
Shailee Amanda Smith
And 2h8'48" 44d24'44"
Shailini & Shishma - "Sisters Forever"
Pho 2h20'10" -40d32'48"
Shailyn Brooke Francis
And 0h16'3" 28d19'6"
SHAIN 17.04.2004
Uma 11h40'28" 55d24'45"
Shain D Eighmey
Cyg 19h56'19" 37d8'10"
Shaina
Cnc 8h0'43" 26d23'48"
Shaina Leigh Tobin
Cap 20h22'51" -25d31'25"
Shaina Lynn
And 2h27'43" 50d8'20"
Shaina Rae Mowery
Uma 11h26'23" 36d27'17"
Shaina Yael My Shining Star of Life
Psc 0h49'36" 20d19'58"
Shaina's Love
Cma 7h20'56" -16d21'12"
Shaina's Star
Cyg 20h3'25" 38d32'7"
Shairaz Yasin
Dra 19h1'52" 73d21'32"
Shairoby Rossana Sarmiento Trillos
Aqr 22h1'26" -15d36'5"
Shaitra My Belle
And 2h26'57" 47d35'32"
Shaka Zulu
Sgr 18h58'42" -28d13'11"
Shakara a.k.a. Lil Momma
Umi 14h27'39" 73d4'42"
Shake Hovespian
Gem 7h34'49" 25d30'13"
SHAKEL24
And 0h49'33" 23d4'59"
Shakera Monique Smith
Cas 1h30'45" 64d43'12"
Shakes
Gem 6h45'8" 19d1'19"
Shakir Teal Prime
Cnc 8h22'22" 32d26'58"
Shakira
Aqr 20h58'10" -12d48'22"
Shakira (Nichte)
Ori 5h45'7" 6d40'15"
Shakira Rae Adams
Equ 21h16'9" 8d58'56"
Shakti
Gem 7h8'50" 15d39'47"
Shaktiputohm
Aql 20h11'54" 12d59'40"
Shakuntla Chugh
Uma 11h35'28" 45d8'40"
Shala
Lyn 8h41'18" 44d17'17"
Shala & Mark Harper always & 4-ever
Uma 11h20'40" 49d53'30"
Shalae
And 23h48'26" 46d24'40"
Shalae
Cyg 21h36'34" 48d13'9"
Shalah
And 0h29'10" 40d5'45"
ShaLaLa
Del 20h27'39" 16d36'46"
Shalan
Dra 17h12'2" 64d7'18"
Shalan Rae Koehn
Uma 8h56'4" 53d21'4"
Shalandra Leia
Leo 10h50'35" 18d21'13"
Shalane L. Bulis
Peg 21h45'2" 10d28'39"
Shalane's Lasting Love
Aqr 22h29'54" 2d33'32"
Shalece Daniele Melelana Thompson
Uma 9h37'56" 45d47'52"
Shalee Nicole Vance
Her 17h46'17" 23d51'34"
SHALEE NORAGER
Cyg 20h5'29" 41d15'24"
Shaleen
Uma 8h24'8" 61d55'32"
Shaleen Fresh
Vir 13h51'5" 1d13'36"
Shaleena's Light
Cam 7h3'31" 77d44'57"
ShaLeesa Jade Stevens
Lib 14h51'32" -2d44'46"
Shaleigh
Crb 16h12'30" 37d30'47"
Shalena
Leo 11h24'50" 23d56'43"
Shalena
Peg 22h15'51" 19d53'31"
Shalena-McKaye
Lib 15h3'11" -10d12'51"
Shaleska R.W. Julian
Cnc 8h20'16" 15d11'30"
Shalia Kris Dean's Angel Star
Psc 23h49'28" 7d24'42"
Shalin
Peg 22h36'24" 27d20'45"
Shalin Connelly Chada
Dra 19h22'9" 60d11'56"
Shalini
Peg 22h24'46" 29d49'57"
Shalini
Gem 6h30'43" 24d57'35"
Shalini Persaud
Cam 3h59'26" 67d51'33"
Shalisha Hammond - Rising Star
Uma 13h20'5" 56d46'34"
Shallon M. Silvestrone
Aqr 22h20'56" 0d20'28"
Shalon Tenille
Cas 23h18'42" 59d18'51"
Shalona
Uma 12h0'58" 40d57'34"
Shalosh
Cam 6h19'49" 62d13'27"
Shalva Raya Snyder
Aqr 21h6'57" 0d42'11"
Shalyn Kay Boone Smith
Mon 7h16'4" -0d59'49"
Shalynn Stacey Smith
Aqr 22h23'34" -1d23'41"
Shalyns Shining Star
Aqr 22h23'32" -1d58'15"
Sham (Sowbhagwattie) Maraj
Uma 13h20'30" 54d26'43"
Shamana Ansari
Sco 16h6'58" -12d49'33"
ShamanHKW
Aql 19h49'53" -0d12'20"
Sham'ath
Sco 17h9'58" -33d13'1"
Shambo-aeiou
Ari 3h13'42" 27d41'33"
Shambra Lee Rodriguez
Gem 6h41'1" 31d26'56"
Shamee & Our Little Star
Umi 14h23'24" 76d4'58"
Shameka Stanley
Lib 15h38'45" -27d40'18"
Shameme Adam's Star
Ser 15h51'51" 21d49'54"
Shamiah Tierra Pittman
Cnc 8h36'51" 23d43'18"
Shamika Hamilton
And 2h21'28" 50d25'42"
Shamim Raja
Per 4h46'3" 45d4'39"
Shamin Lona Elizabeth Marquis
Oph 16h48'37" -0d24'34"
Shamiqua
Uma 11h35'40" 51d39'51"
Shamiqua Walker
Umi 14h11'31" 66d59'57"
Shamiree and Daniel
Per 4h20'12" 33d36'4"
Shamiyana
Lib 15h21'17" -7d4'40"
Shammi
Peg 21h43'25" 23d36'30"
Shammy
Vir 11h49'46" 2d38'0"
Shamrell
Psc 0h18'56" 8d34'29"
Shamrez Shaikh
Cnc 8h41'6" 32d32'29"
Shamrock
Lyn 6h54'26" 51d9'15"
Shamrock Getzler
And 23h13'33" 52d47'7"
Shamus
Cep 20h40'20" 63d50'31"
Shamus
Cmi 7h32'27" 6d1'52"
Shamus
Tau 5h54'53" 24d2'22"
Shamus the Dreamer
Lyn 9h13'18" 41d51'13"
Shan
Sco 16h17'54" -19d40'11"

Shan Allen Roper
Per 2h19'51" 55d17'34"
Shan & Devyani
Pic 5h36'56" -44d57'58"
Shan Kelley
Her 18h3'56" 26d7'57"
Shan Shan in starry sky
Tau 4h53'54" 26d43'4"
Shana
Tau 3h58'20" 27d50'7"
Shana
Psc 1h10'51" 23d31'48"
Shana
Ari 2h58'8" 21d24'53"
Shana
Leo 9h36'48" 10d37'57"
SHANA
And 2h21'6" 50d15'16"
Shana
Lyn 7h54'19" 37d58'5"
Shana
Vir 13h28'33" -1d10'23"
Shana
Lyn 6h55'3" 54d55'57"
Shana Burkhart
And 0h14'32" 39d23'37"
Shana Christensen
Psc 23h46'51" 6d23'40"
Shana Cissé
Leo 10h3'23" 25d9'16"
Shana Cornfield
Lmi 10h7'50" 38d45'38"
Shana Dobson
Cnc 8h50'34" 26d54'49"
Shana Freedman
Lib 15h4'8" -7d7'41"
Shana Gabrielle Kassel
And 1h47'4" 46d40'5"
Shana GT12 Momin
Vir 14h37'57" 4d21'52"
Shana Hawkins
Ari 2h9'47" 23d46'56"
Shana Jenkins, from Planet Earth
Tau 5h51'17" 17d1'50"
Shana Joy Bartlett
Mon 6h46'14" -0d27'3"
Shana L. Hackworth
Sco 16h6'27" -14d4'6"
Shana Leigh
Vir 12h39'14" 1d26'34"
Shana Loven
Ari 3h18'58" 29d3'19"
Shana Lynn Sanders
Lac 22h46'42" 51d14'26"
Shana Marie
Mon 8h3'42" -4d9'25"
Shana Punim
Gem 7h55'57" 30d44'16"
SHANA ROWLETT
Uma 11h41'33" 53d34'51"
Shana Rutschmann-Niederhauser
Cas 1h8'36" 50d48'3"
Shana Shackleford
Leo 11h18'34" -5d3'36"
Shana Sieve
Ori 5h34'36" 6d45'23"
Shana The Star of My Life
Umi 10h39'11" 87d36'39"
Shanae L. Brewer
Aql 19h14'6" 5d28'12"
Shanahan Gayle Elmore the Tator Bug
Her 17h34'54" 43d26'36"
Shanaia Celise Galarza
Crb 15h48'22" 26d8'23"
Shanaiki
Uma 9h35'1" 48d3'17"
Shanamandor
Tau 3h40'39" 28d39'44"
Shanan D Pickett
Uma 11h27'41" 53d12'41"
Shanan Okalani Atienza Sabagala
Leo 9h59'29" 23d5'31"
Shananigans
Tau 5h4'15" 21d22'18"
Shana's Mykaila Grace
Lyn 9h15'16" 36d55'2"
Shanay Rose
And 0h1'20" 43d45'31"
Shanaylee
Lyn 7h25'19" 47d42'50"
Shanchita
Sco 17h23'14" -33d12'40"
Shanda
Tau 3h48'24" 10d5'35"
Shanda Kreuzer
Ori 5h28'55" -0d43'49"
shanda sue simpson
Cam 5h15'25" 64d53'22"
Shanda "Sweetheart" Schrader
Sgr 18h17'1" -32d46'9"
Shandanken Jesse
Uma 10h9'16" 53d34'53"
Shandcox
And 2h27'40" 44d39'43"
Shandel Pastuch's Shooting Star
Psc 0h30'33" 17d3'11"
Shandia
Psc 0h40'5" 19d6'40"
Shandie
And 0h43'45" 37d14'3"
Shandra
Ori 5h41'41" 4d59'7"
Shandra
Gem 6h37'51" 26d0'4"
Shandra Renee Hopkins
Leo 10h58'2" 7d25'31"
Shandra Rose Silvia
Sco 17h4'55" -38d39'12"
Shandy Moore
Cma 6h24'28" -17d55'13"
Shandyn Leesley
And 0h54'7" 38d56'1"
Shandys Shooting Star!
Umi 14h2'6" 71d18'3"
Shane
Sco 17h24'33" -42d22'43"
Shane
Sgr 19h6'42" -24d38'10"
Shane
Cyg 20h9'14" 45d17'51"
Shane
Vir 11h44'33" 2d39'2"
Shane
Psc 23h30'46" 2d20'47"
Shane
Aqr 22h32'8" 0d21'30"
Shane
Aql 19h55'33" 14d35'59"
Shane
Tau 4h47'17" 28d57'21"
shane a. & tami jo miller
Cyg 21h38'11" 50d50'25"
Shane Abergel
And 1h27'0" 46d16'49"
Shane Alexander Chazin
Lyn 6h47'23" 52d6'59"
Shane and Amber Pilarczyk
And 0h41'16" 45d13'58"
Shane and Ashlee
Tau 5h52'36" 13d5'27"
Shane and Jessica Stephens
And 1h41'24" 39d4'31"
Shane and Liz
Cyg 21h37'32" 41d5'59"
Shane Andrew Gibson
Col 5h59'23" -39d25'21"
Shane Anthony Ciccozzi
Cru 12h15'51" -61d12'40"
Shane Beck
Cnv 12h27'23" 33d9'57"
Shane Beloved Son
Ari 2h8'9" 25d0'26"
Shane Bock
Per 3h39'59" 45d30'33"
Shane Bonner - Aim High
Psc 1h10'50" 23d31'55"
Shane C. Swanberg
Aql 19h2'9" 15d31'22"
Shane Calaway Jamison's Lucky Star
Dra 19h33'15" 69d56'38"
Shane Casey
Tau 5h52'16" 25d17'11"
Shane Clanin Swanberg
Uma 13h54'13" 54d12'52"
Shane Colin Wagner
Umi 17h25'40" 82d34'6"
Shane Corrie Simmons
Her 17h28'8" 27d17'37"
Shane Crawford
Lib 14h59'39" -6d47'47"
Shane Dale Warrick
Tau 4h34'30" 28d16'35"
Shane Daniel
Umi 16h15'25" 76d59'55"
Shane Daugherty
Dra 18h47'39" 50d28'24"
Shane David Eggenberger
Umi 17h41'43" 80d13'44"
Shane David- The Beloved
Lib 14h52'38" -4d21'15"
Shane & Diane Sawyer
Crb 15h35'15" 30d7'39"
Shane Dodd
And 23h24'56" 52d29'11"
SHANE DORAN
Eri 3h58'36" -0d43'27"
Shane Doyen
Per 3h15'8" 34d37'35"
Shane E. Morrill
Vir 12h15'8" 11d3'15"
Shane Ellis
Tau 4h44'44" 21d54'2"
Shane Emily
And 23h28'42" 44d45'54"
Shane Feltham B.Sc.
Uma 11h44'9" 41d41'9"
Shane Fenner
Uma 11h30'25" 55d54'50"
Shane Foreman
Uma 8h55'16" 54d0'59"
Shane Graham
Sgr 18h47'46" -34d49'38"
Shane Hamilton Hackett
Tau 3h47'25" 27d50'21"
Shane Hardy
And 0h15'22" 26d32'17"
Shane Henry Gebhard
Uma 8h58'59" 71d1'45"
Shane Hopfensperger
Sco 16h50'46" -27d34'45"
Shane John Stevenson
Dra 17h27'38" 67d59'58"
Shane Joseph Creeley
Cnc 8h48'24" 32d8'25"
Shane Joseph Halligan
Tri 1h50'37" 31d33'21"
Shane Joseph Halligan
Per 3h6'15" 54d10'46"
Shane K. Haldorsen
Sco 16h7'9" -15d3'50"
Shane & Kelly
Cas 0h18'41" 55d14'35"
Shane & Kerri Rowley
Uma 10h27'38" 66d43'42"
Shane Kobialka
Umi 16h18'33" 76d54'11"
Shane Lance Brethorst
Ori 5h48'56" 12d31'6"
Shane Leonard Smith
Dra 17h22'21" 60d36'44"
Shane M. Pfaffly
Ari 3h10'21" 26d54'20"
Shane*Maegan-A Match Made In Heaven
Vir 14h1'55" -19d24'49"
Shane Marshall Donner
Ori 5h17'10" 1d6'8"
Shane & Martina
Cyg 20h53'10" 31d25'50"
Shane Matthew Frasier
Tau 5h5'14" 21d45'49"
Shane Mcdonough
Lib 15h7'2" -12d40'26"
Shane Medrano
Ari 2h1'12" 21d17'41"
Shane Michael Aiello
Gem 7h47'28" 33d18'48"
Shane Michael Alston
Lyn 6h45'45" 53d55'51"
Shane Michael Holm
Gem 7h30'48" 16d2'49"
Shane Michael Nash
Cru 11h56'52" -59d38'45"
Shane Michael Sande
Cep 23h49'30" 67d42'46"
Shane Michael Wier
Her 18h53'18" 23d13'26"
Shane Miller
Lyn 9h37'33" 40d55'20"
Shane Moore
Cyg 19h54'56" 32d10'10"
Shane Moritz
Aqr 22h29'57" -2d13'21"
Shane Murphy
Cru 12h36'55" -59d19'32"
Shane -N- Brandi
Aqr 21h51'23" -1d0'14"
Shane Nelson
Sco 16h40'43" -26d32'47"
Shane Nicholas
Uma 9h15'37" 59d50'46"
Shane Noel Morita Ockander
Uma 13h30'50" 52d41'49"
Shane Patrick Littlefield
And 0h35'46" 25d56'39"
Shane Patrick Moschberger
Sgr 18h12'36" -21d22'49"
Shane Patrick Raftery
Gem 7h4'14" 16d56'35"
Shane Paul Castle
Per 4h16'24" 43d43'27"
Shane Paul William Finkel
Umi 16h47'3" 78d17'25"
Shane Peros
Sco 16h15'35" -9d33'18"
Shane Robert Sietsema
Per 4h38'0" 40d36'30"
Shane Ryan Clawson
Uma 10h28'13" 56d40'9"
Shane & Sarah
Cyg 19h56'15" 36d9'4"
Shane Scott
Aqr 21h45'37" 1d53'25"
Shane Sea Forever Be Mine
Cyg 19h35'56" 34d54'55"
Shane (Shamus) Wike
Cet 1h27'55" -0d18'19"
Shane Sigston - Love endures all
Psc 1h37'32" 24d42'44"
Shane & Terrilyn
Sex 9h52'1" -0d37'49"
Shane the STAR of my life Jamieson
Cru 12h38'7" -59d2'3"
Shane Thomas
Tau 5h40'44" 18d58'48"
Shane Thomas Fisher
Uma 11h53'9" 41d23'13"
Shane Thomas Schuster
Tau 5h51'42" 23d34'53"
Shane & Tiffanie The Star of Love
Cyg 21h2'0" 46d14'1"
Shane Torren Ceri Naveky
Ori 5h23'24" 1d42'45"
Shane Triumphant
Uma 11h36'57" 33d7'7"
Shane Tyler Sexton
Leo 10h27'34" 17d2'11"
Shane Vincent Stesner
Per 2h34'22" 51d37'37"
Shane William Davis
Tel 18h46'59" -46d18'32"
Shane Williamson
Lmi 10h54'4" 29d4'12"
Shane2007
Aqr 21h31'12" -4d57'6"
Shanee
Cap 21h14'16" -22d56'2"
Shanee Marie Ascarrunz
Sgr 18h18'32" -24d42'35"
Shaneeka Perkins
Uma 14h15'35" 59d38'35"
Shaneez Wazir
And 1h49'16" 45d3'21"
Shanekun Narumi McConnell
And 23h6'4" 51d37'23"
Shanel Anne Neaum
And 1h4'51" 45d16'21"
Shanel Nicole Libby 143
Vir 12h13'8" 10d47'44"
Shaneman
Ori 6h9'0" 15d35'32"
Shaneman
Dra 17h33'9" 51d32'27"
Shane....My Love, My North Star
Leo 9h27'57" 6d43'7"
Shane"O"
Ori 5h32'5" 3d11'15"
Shane's and Ashley's First Love
Uma 11h6'18" 57d53'18"
Shane's Irish Magic
Aur 6h7'46" 37d10'14"
Shane's Star
Aqr 21h53'22" 1d49'42"
Shane's Star
Aqr 22h20'13" -3d57'36"
Shangri-La
Tau 4h9'42" 14d12'58"
Shani & Gary Tyson
Cyg 19h42'41" 52d11'33"
Shani Marshall
And 0h46'13" 24d42'41"
Shania
Her 16h49'26" 9d7'3"
Shania and Immi
Ori 5h56'21" 13d11'20"
Shania Shields
Lmi 10h47'13" 28d33'3"
Shaniah-Raé
And 2h2'13" 42d16'6"
Shanian Chen
Uma 14h16'52" 56d29'29"
Shanice 11.12.2001
Sgr 19h16'58" -22d47'35"
Shanice Marie Martinez
Tau 3h34'34" 18d10'48"
Shanie
Sgr 18h31'26" -16d6'54"
Shanie Fox Pai
Tau 5h55'49" 24d45'51"
Shanika Layton
Cam 4h27'18" 56d34'22"
Shanikkokoa of Guammy Paradise
Gem 6h32'10" 24d4'38"
Shanita McPherson
Lyn 9h12'56" 36d58'31"
Shanita Oliver
Com 12h46'59" 28d11'47"
Shanleigh's Star
Psc 1h13'23" 2d46'2"
Shanlyn 40
Dra 16h43'36" 52d37'29"
Shanlynell
Ori 6h1'18" 19d39'59"
Shann
Cnc 8h54'51" 22d46'28"
Shanna
Gem 6h36'19" 20d49'12"
Shanna
Mon 6h43'31" 8d10'54"
Shanna
Cnc 8h7'31" 8d39'11"
Shanna
And 0h54'46" 38d15'10"
Shanna
Uma 11h15'50" 45d22'53"
Shanna
Lyn 6h36'42" 60d2'53"
Shanna
Sco 16h9'13" -20d21'54"
Shanna
Aqr 21h43'57" -2d57'2"
Shanna
Aqr 22h50'9" -4d44'53"
Shanna
Cma 7h2'15" -24d0'2"
Shanna Alexander
Tri 2h36'33" 31d30'47"
Shanna and Linda Fick
Sco 17h16'14" -30d41'57"
Shanna Andee 12.06.99
Lib 15h51'51" -14d33'11"
Shanna D. Langgle "My Precious"
And 23h38'57" 48d16'27"
Shanna Kay Upton
Ari 2h12'55" 23d24'51"
Shanna King
And 2h0'26" 45d20'5"
Shanna L. Sosa
Cas 0h12'56" 51d27'22"
Shanna Leigh
Tau 4h36'30" 21d11'30"
Shanna Litterst
Cnc 8h50'14" 26d58'1"
Shanna Longenecker
Gem 7h18'16" 19d55'7"
Shanna Lyn Ginder
Ori 6h16'55" 2d56'4"
Shanna Lynn
Cap 20h27'49" -12d44'30"
Shanna Lynn Dreiling
Sco 16h7'32" -15d40'5"
Shanna Marie Belton
Ari 2h23'58" 22d52'20"
Shanna Marie Jones
Vir 12h42'23" 7d54'9"
Shanna Michelle Lusk
Psc 23h49'31" 0d37'22"
Shanna R Conroy
Lyn 6h34'25" 57d35'25"
SHANNA xoxo ADAM 213
Cam 5h57'51" 58d49'43"
Shanna Yeubanks
Uma 11h42'10" 52d23'22"
Shannah Anne
And 1h42'10" 43d56'27"
Shannah Marie Ylinen
Cam 3h17'7" 60d3'28"
Shannalanigon
Psc 0h46'4" 16d52'16"
Shannan Lee Anderson
Psc 1h19'31" 31d25'8"
Shannan Lee Lloyd
Cas 1h14'49" 67d6'20"
Shannan Marie
Cap 20h35'43" -26d54'22"
Shannan Yano
Ari 3h21'34" 24d10'3"
Shannan Yorton
Gem 7h37'29" 25d59'38"
Shannan's Hope
Lyr 18h46'31" 38d42'46"
Shannantia Atoya Hopkins
And 2h23'52" 45d10'34"
Shanna's Star
Aqr 22h9'27" 0d39'14"
Shannel Emily Louise Gooch
Peg 23h57'56" 22d25'42"
Shannel McCrory
Cap 21h28'57" -24d36'14"
Shannen Doherty
And 1h44'12" 45d6'41"
Shannen Lei
Cam 7h2'10" 78d13'22"
Shannen Marie Erenberger
Gem 7h45'8" 34d26'55"
Shannen McKean
Lmi 9h59'47" 41d23'51"
Shanne's Little Piece of Heaven
Ori 5h33'58" 1d39'42"
Shannie E Waltenbury
Umi 15h52'16" 73d35'55"
Shanni's Star
Leo 10h56'1" 18d6'0"
Shannon
Gem 6h52'11" 19d15'58"
Shannon
Ori 6h11'39" 20d11'5"
Shannon
And 0h20'24" 25d33'33"
Shannon
Ari 2h20'51" 24d25'56"
Shannon
Leo 11h36'32" 24d23'35"
Shannon
Crb 15h47'38" 26d23'13"
Shannon
Psc 0h25'27" 7d57'12"
Shannon
Tau 5h4'56" 20d51'59"
Shannon
Tau 5h45'51" 17d8'24"
Shannon
Ori 6h15'45" 13d46'24"
Shannon
Uma 9h40'40" 42d42'32"
Shannon
Aur 5h57'38" 42d40'10"
Shannon
Lyn 7h47'46" 38d38'9"
Shannon
And 1h44'51" 42d26'38"
Shannon
And 1h47'5" 40d49'45"
Shannon
And 1h20'9" 41d10'19"
"Shannon"
Cyg 19h50'19" 37d6'47"
Shannon
Lyn 7h55'8" 34d3'51"
Shannon
And 1h25'11" 35d16'51"
Shannon
Crb 16h11'40" 34d10'15"
Shannon
Lyn 7h26'27" 49d56'13"
Shannon
Lyn 7h35'25" 50d11'34"
Shannon
Crb 15h53'56" 38d55'1"
Shannon
Cas 0h22'50" 48d44'23"
Shannon
Uma 11h9'7" 47d42'32"
Shannon
And 23h5'42" 51d41'43"
Shannon
Lyn 7h46'43" 53d57'56"
Shannon
Uma 9h18'42" 54d19'38"
Shannon
Uma 11h50'33" 57d31'19"
Shannon
Vir 11h39'53" -1d6'14"
Shannon
Sgr 18h14'58" -20d35'7"
Shannon
Aqr 22h58'15" -7d40'13"
Shannon
Sco 16h7'6" -10d36'24"
Shannon 7597
Vir 11h52'49" 3d2'48"
Shannon Adams
Cap 21h5'18" -19d38'18"
Shannon - Aidan
Uma 11h54'36" 37d48'54"
Shannon Albright
Psc 1h9'26" 11d30'50"
Shannon Aldrich
Leo 9h49'57" 23d38'27"
Shannon Aleene
And 1h44'12" 48d58'5"
Shannon Alene Keenan
Psc 0h23'41" 15d32'15"
Shannon Allan Meek
Aql 19h5'48" 10d51'44"
Shannon Alyssa Redmond
Crb 15h53'59" 38d49'53"
Shannon Amanda Jones
Her 18h56'57" 23d39'3"
Shannon Amber Steffler
Sco 16h15'4" -23d34'2"
Shannon & Amy Ruis
Lyr 19h23'3" 37d38'1"
Shannon and Bobby's Star
Leo 10h4'57" 23d50'14"
Shannon and Grant's star
Tau 4h19'34" 7d14'39"
Shannon and Lauren Cookle
Cyg 20h34'46" 52d19'47"
Shannon Andrea
And 0h15'47" 27d59'15"
Shannon Ann Keenan
Sgr 20h4'45" -30d24'2"
Shannon Atcheson
Cyg 20h44'53" 33d27'24"
Shannon "Babygirl" Harkins
Ori 5h20'24" -0d36'57"
Shannon Bard
Lac 22h21'47" 47d18'32"
Shannon Belflower
Cnc 7h59'28" 13d50'36"
Shannon Blakesley
Crb 15h45'48" 27d41'18"
Shannon Bloom
Psc 0h22'1" -1d25'14"
Shannon Blote
Leo 10h15'11" 24d32'18"
Shannon Brissing
Sgr 18h7'15" -26d14'33"
Shannon Brooke
Aqr 22h37'25" -0d12'38"
Shannon Cainstellation
Aqr 23h16'7" -15d6'7"
Shannon Canty's Star
Vir 11h49'1" -3d18'32"
Shannon+Cappy 10/9/05
Cyg 20h21'2" 55d23'29"
Shannon Christine Downing
Vir 12h49'14" 9d42'8"
Shannon Christopher Leto
Pho 1h17'8" -43d11'15"
Shannon Coates
Sgr 19h12'6" -24d49'32"
Shannon Cobb
Lyn 6h32'0" 56d38'14"
Shannon Cobb
Lyr 18h39'53" 34d11'49"
Shannon Cook
Sgr 18h51'44" -16d4'40"
Shannon Cooper
Ori 4h45'56" 11d47'36"
Shannon Corey
Leo 11h24'1" 14d54'35"
Shannon Courtney
Leo 10h7'5" 24d37'0"
Shannon Cristi
Sgr 18h54'29" -17d22'25"
Shannon Cristina Woo
Umi 10h56'26" 87d26'16"
Shannon Crystal Pigeon
Sco 16h9'43" -16d7'4"
Shannon Cupp
Leo 10h46'33" 16d43'59"
Shannon D. Calvert
Ori 4h58'55" -3d28'34"
Shannon Dale and Nicole Jean
Cra 18h37'31" -38d35'9"
Shannon Dale Hickman
Lib 15h26'41" -8d8'32"
Shannon & David Gillette 3/21/04
Gem 7h3'12" 16d5'28"
Shannon Deann
Cam 4h17'2" 70d44'46"
Shannon Debra Stanley
Peg 23h29'33" 24d5'52"
Shannon Delia Mulhern
Ari 2h49'52" 19d9'26"
Shannon Denise Marlin 143264
Cnc 8h50'51" 28d5'44"
Shannon DeSantis
Cnc 8h43'43" 29d2'3"
Shannon Devone Hass
Vir 14h7'36" -0d10'14"
Shannon Diane McMahon
Tri 2h8'45" 34d48'30"
Shannon Diane Nordvik
And 0h50'20" 22d29'24"
Shannon Dipeppe Whalen
And 2h32'46" 39d0'32"
Shannon Donehoo
Leo 11h30'37" 25d43'18"
Shannon Dorothy Clark
Sco 15h54'25" -20d38'7"
Shannon "Edge" Skinner
Sgr 19h32'31" -15d8'59"
Shannon Elizabeth
Lyn 6h24'45" 61d11'44"
Shannon Elizabeth
Cas 1h19'52" 62d7'53"
Shannon Elizabeth
Sco 17h51'24" -35d45'8"
Shannon Elizabeth Beach
Umi 14h49'41" 68d15'32"
Shannon Elizabeth Butler
Cas 23h47'56" 52d59'4"
Shannon Elizabeth Fitzpatrick
Vul 19h53'32" 29d4'57"
Shannon Elizabeth Gazecki
And 2h20'2" 42d0'30"
Shannon Elizabeth Lyle Croom
Tau 4h20'57" 13d16'39"
Shannon Elizabeth MacDonald
Aqr 23h5'51" -10d2'21"
Shannon Elizabeth Mason
Lyn 8h18'43" 50d59'19"
Shannon Elizabeth Stigall
Cnc 8h42'4" 23d12'40"
Shannon Fahey
Lyn 8h19'12" 50d31'41"
Shannon Faith
Mon 6h55'26" -0d48'49"
Shannon Falvey
Uma 10h23'10" 61d34'36"
Shannon Fay Delaney
Tau 3h43'48" 25d26'6"
Shannon Faye
And 0h13'39" 48d37'43"
Shannon Finley
Psc 0h59'50" 14d52'8"
Shannon Frances
Cma 6h32'23" -31d5'36"
Shannon Frank
Lyn 7h45'49" 55d51'56"
Shannon Frost Benecke
Umi 16h6'28" 73d49'54"
Shannon Gabrielle Holderness
Cnc 8h25'46" 20d52'19"
Shannon Game
And 0h7'28" 44d37'11"
Shannon Gayle Tolley
Leo 9h30'52" 29d17'30"
Shannon Good
Mon 6h46'52" 9d39'42"
Shannon Greene
Gem 7h42'30" 32d38'30"
Shannon Gritton
Lyn 6h48'49" 50d44'50"
Shannon H. Frank
Ari 2h23'50" 25d26'30"
Shannon Hackett (My Perfect Angel)
Vir 13h3'32" -18d19'54"
Shannon Haida Walsh
Gem 7h23'37" 13d41'35"
Shannon Hamby
Uma 12h37'7" 61d59'22"
Shannon Havelka Sherman
And 0h50'41" 33d46'50"
Shannon Hoban
Uma 13h41'44" 57d32'13"
Shannon & Hope
Ari 2h21'44" 11d53'21"
Shannon Hunsinger
Uma 11h46'58" 30d57'51"
Shannon Irene Mathieu
And 22h59'6" 48d21'33"
Shannon Jaie
Cap 21h6'19" -19d11'50"
Shannon & Jason—Forever One
Cyg 20h21'9" 55d3'22"
Shannon Jean
Cas 0h22'55" 48d24'40"
Shannon Jean Iles - Shannon's Star
And 23h41'18" 37d6'58"
Shannon Jeffers
Ori 5h38'29" -0d59'14"
Shannon Jo
Sgr 18h12'22" -19d40'43"

Shannon Jo Guzman
Aql 19h7'56" 6d27'10"
Shannon Jordal
And 23h15'1" 44d46'47"
Shannon Juenger
Lib 15h20'35" -8d3'48"
Shannon K. Shinn
Ari 2h52'29" 29d10'58"
Shannon Kalvig
Her 17h49'34" 39d14'36"
Shannon Kangas-Swedberg
Col 5h9'33" -36d0'14"
Shannon Katherine
Leo 11h41'57" 21d43'23"
Shannon Kathleen
Tau 4h42'0" 3d14'5"
Shannon Kathleen
Leo 11h30'57" 11d13'30"
Shannon Kay-common name Shannykins
Pho 23h36'47" -55d28'5"
Shannon Kaye & Nichlaus Hank
Cyg 19h49'6" 43d40'6"
Shannon Kelly Falkner's Baby Blue
Cyg 20h51'53" 43d58'59"
Shannon Keri Rice
Gem 6h53'27" 32d30'38"
Shannon Kiernan
Per 4h30'42" 37d30'16"
Shannon King & Enrico Contosta
Ori 6h0'51" 17d32'48"
Shannon Kristin Leary
Sco 17h15'6" -42d8'26"
Shannon L. Bresnahan
Cas 0h19'39" 63d0'32"
Shannon L. Kuklak Whiteside
And 1h8'1" 38d7'56"
Shannon L. McGrandy
Ari 2h51'5" 20d2'48"
Shannon L. Rae
Psc 1h7'13" 10d13'44"
Shannon L Schultz
Lib 14h52'34" -3d20'10"
Shannon Lamanna
Lyn 7h38'23" 38d4'17"
Shannon Lamond Smith Birthday Star
Sco 16h15'24" -12d6'40"
Shannon LaRay
Uma 10h37'29" 57d5'3"
Shannon Lauf
Ori 6h6'31" 21d0'36"
Shannon Le Simmons
Aql 20h14'22" 8d30'19"
Shannon Leah Fee
Leo 9h34'53" 29d22'37"
SHANNON LEE
Crb 15h34'30" 27d40'41"
Shannon Lee
Gem 6h50'26" 25d23'18"
Shannon Lee
Lib 15h48'54" -6d36'30"
Shannon Lee Bitzer
Per 3h1'11" 47d30'57"
Shannon Lee Dale
Cas 0h21'45" 47d32'16"
Shannon Lee Dillon
Gem 6h28'19" 15d33'49"
Shannon Lee Faris
Cap 20h19'21" -12d18'38"
Shannon Lee Leonowicz
Aqr 22h54'27" -9d4'47"
Shannon Lee McAneney
Sgr 19h3'15" -31d37'44"
Shannon Lee Michaud
Umi 16h18'32" 75d41'48"
Shannon Lee Parker
Lib 15h25'3" -26d50'24"
Shannon Lee Polk
Gem 7h22'39" 24d41'33"
Shannon Lee Priddle
Gem 6h21'19" 22d0'49"
Shannon Lee (Sweet Pea)
And 1h19'23" 34d48'3"
Shannon Leigh Costa
Leo 9h55'55" 23d26'48"
Shannon Leigh Edwards
Cnc 8h28'27" 23d24'53"
Shannon Leigh Ressler
Tau 5h57'41" 26d57'38"
Shannon Leigh Sigafoos
Cas 2h43'54" 58d4'31"
Shannon Lindmeyer
Lib 14h44'55" -11d18'54"
Shannon Louise Boorstein
Lyn 8h17'28" 45d56'5"
Shannon Louise Sturgeon
Cnc 8h53'2" 31d32'13"
Shannon loves Eli mostest!
Ari 2h43'30" 28d25'14"
Shannon Lucia
Gem 6h48'47" 16d55'18"
shannon lydick
Uma 11h49'44" 54d10'8"
Shannon Lyn Cagle
And 23h21'59" 42d21'44"
Shannon Lyn Connor
Lib 15h37'52" -11d26'49"
Shannon Lynn Anastos
Vir 14h5'17" -14d48'24"
Shannon Lynn Sandridge
Sco 16h33'56" -27d43'39"
Shannon M.
Com 13h1'53" 17d37'12"
Shannon M. Martinez
Cyg 20h5'8" 40d2'56"
Shannon M. Medders
Oph 17h54'5" -0d42'6"
Shannon M. Richardson
Cam 5h18'11" 72d37'37"
Shannon M. Rose
And 22h58'44" 52d4'35"
Shannon M. Stoll
Gem 7h3'40" 29d21'53"
Shannon M Thomas
Crb 15h36'7" 30d51'3"
Shannon Mackenzie Daly
Gem 7h5'44" 34d10'51"
Shannon&Malika(star of love)Hurley
Cyg 19h36'24" 39d4'10"
Shannon Marie
Gem 7h52'25" 32d26'46"
Shannon Marie
And 1h48'18" 41d16'6"
Shannon Marie
Psc 0h49'38" 9d34'6"
Shannon Marie
Uma 8h46'51" 58d23'44"
Shannon Marie
Cap 20h51'46" -27d22'49"
Shannon Marie Artale
Vul 21h19'11" 20d56'33"
Shannon Marie Barrus
Her 16h41'12" 45d8'57"
Shannon Marie Broesder
Aqr 22h12'34" -1d53'19"
Shannon Marie Cone Bonnell
Crb 16h7'46" 38d41'25"
Shannon Marie Duffy
Tau 4h49'4" 29d39'6"
Shannon Marie Fredricks
Lib 16h0'54" -7d23'20"
Shannon Marie Gindt
Del 20h30'22" 5d17'14"
Shannon Marie Higgins
Cap 20h43'2" -18d44'25"
Shannon Marie Lamoureux
Cap 21h11'57" -19d43'58"
Shannon Marie Lantz
Uma 11h41'32" 37d51'3"
Shannon Marie McKennna
Leo 11h7'2" -1d31'0"
Shannon Marie Moss
And 1h48'32" 46d35'25"
Shannon Marie Nutt
Uma 10h11'12" 53d15'30"
Shannon Marie Owen
Cnc 8h40'20" 23d4'17"
Shannon Marie Sweeney
And 2h12'46" 44d22'5"
Shannon Marjoribanks
Dra 20h3'46" 73d15'51"
Shannon Marlene
Lyn 7h1'48" 59d43'0"
Shannon Maureen
Cas 2h14'42" 64d19'15"
Shannon Maureen
And 0h46'46" 36d27'56"
Shannon Maureen Taylor
Lyn 6h39'32" 57d3'49"
Shannon McCormick
Lyn 7h21'4" 51d26'20"
Shannon McDonald
Cyg 20h8'8" 35d48'6"
Shannon McKeon
Uma 13h56'14" 58d48'58"
Shannon Melissa Goodrich
Lib 14h22'23" -17d40'26"
Shannon Michael Amaro
Ori 5h44'29" 8d27'30"
Shannon Michelle Belanger
Gem 6h46'23" 32d17'39"
Shannon Michelle Luckett
Psc 1h24'38" 31d28'1"
Shannon Michelle Smith
Cap 21h52'33" -26d38'25"
Shannon Michelle Williams
Tau 5h44'26" 26d30'28"
Shannon Miller 06/24/06
And 0h19'34" 32d52'7"
Shannon & Minky Steinfadt
Uma 8h52'31" 65d29'32"
Shannon Mitchell
And 1h42'26" 40d48'37"
Shannon Mohen
Peg 23h53'25" 28d58'6"
Shannon Molani Pabo
Gem 6h56'18" 27d20'4"
Shannon Monroe
Mon 7h19'6" -0d48'41"
Shannon Moody
Lib 14h41'26" -17d11'22"
Shannon Moran
Cnc 9h0'3" 22d46'4"
Shannon Murphy
Lmi 10h36'3" 31d46'36"
Shannon Murphy
Psc 23h59'13" -3d23'19"
Shannon Murphy
Cas 23h38'40" 55d28'37"
Shannon my Beautiful Love
And 0h30'49" 33d13'34"
Shannon My Love
Sgr 19h8'19" -14d0'30"
Shannon N. Head
Aur 6h54'37" 38d30'43"
Shannon N. Kasprzyk
Ori 5h58'24" 19d22'5"
Shannon Nicole
And 23h32'15" 42d40'30"
Shannon Nicole Kelly
Cnc 9h4'37" 17d11'58"
Shannon Nikkole Haley
Aur 5h11'25" 29d10'34"
Shannon O'Leary
Tri 1h52'46" 31d1'48"
Shannon Patricia Peake
Lib 15h38'28" -14d6'50"
Shannon Patricia Wagner
And 1h25'58" 38d55'15"
Shannon Patrick Reilly
Boo 14h47'28" 51d33'4"
Shannon Petty
Cap 21h44'3" -14d38'59"
Shannon Phipps
Uma 11h32'43" 43d25'45"
Shannon Pickard
Psc 0h53'23" 10d39'39"
Shannon Pressley
Psc 1h38'30" 24d39'6"
Shannon Pritchard
Del 20h32'2" 18d59'21"
Shannon R Guthrie
Ari 2h48'59" 18d22'32"
Shannon R. McCarthy
And 1h48'3" 36d29'1"
Shannon Rae
Psc 0h57'40" 31d45'22"
Shannon Rae
And 0h21'14" 46d36'32"
Shannon Ray Wood
Cas 0h19'8" 62d35'35"
Shannon Raye Williams
Lyn 8h37'16" 41d8'37"
Shannon Regan
Cas 23h55'3" 59d8'29"
Shannon Reinagle
Cas 23h29'36" 52d6'30"
Shannon Rene Knipf
Uma 11h50'0" 31d38'20"
Shannon Renee
Cap 20h34'31" -16d57'32"
Shannon Renee Miller
Aqr 23h24'0" -19d6'24"
Shannon Rich Yow
Lyn 8h2'17" 34d20'28"
Shannon Robards
Pho 23h58'28" -41d7'12"
Shannon Robertson
Uma 10h43'25" 41d56'20"
Shannon & Ron Ducharme
Mon 8h0'23" -0d55'33"
Shannon Rose Neary
Cas 1h36'55" 60d52'37"
Shannon Rose Nemeth
Gem 7h39'21" 24d15'4"
Shannon Sadler
And 1h28'20" 49d23'27"
Shannon & Sarah
Cyg 20h7'4" 58d54'54"
Shannon Saylor
Sgr 18h46'58" -20d40'12"
Shannon Schwarz
Lib 14h58'7" -6d27'24"
Shannon & Sean O'Malley
Cyg 21h29'8" 46d4'42"
Shannon & Sean Swayne
Cyg 19h59'11" 44d13'25"
Shannon "Sexy Face" Mintz
And 0h24'13" 30d5'17"
Shannon Shay
Cnc 8h16'45" 20d30'9"
Shannon Simmons
Gem 7h27'13" 26d14'0"
Shannon Smith
Cas 1h55'29" 67d52'57"
Shannon Stafford
Lep 5h55'6" -12d7'18"
Shannon Stefanie Magid
Tau 4h25'43" 22d57'15"
Shannon Stellar Wansley
Vir 14h15'24" -12d59'43"
Shannon Steppello
Cnc 8h43'16" 12d11'6"
Shannon Stolorski
Gem 7h24'59" 21d13'52"
Shannon Stone
Sgr 19h29'2" -13d21'11"
Shannon Stovall
Uma 9h12'14" 55d34'57"
Shannon Thornberry
Lyn 7h45'45" 37d57'35"
Shannon & Tiana Jackson
And 1h29'19" 47d45'36"
Shannon Toner
Vir 13h23'13" -8d4'24"
Shannon Toni
Vir 12h57'26" -2d41'44"
Shannon Troy Krueger
Lib 14h42'39" -9d15'12"
Shannon V.
Ari 2h10'32" 19d1'0"
Shannon Vail
Peg 22h37'16" 12d52'48"
Shannon W. Wright
Sgr 18h2'41" -27d25'27"
Shannon Walk
Vir 14h14'3" -18d26'27"
Shannon Weisler
Aqr 22h39'57" 0d55'37"
Shannon Wilson
Aqr 21h57'13" 0d17'20"
Shannon Wood
Cas 1h13'20" 63d9'12"
Shannon Yehl
Psc 0h44'33" 8d22'24"
Shannon Yvonne Jaunsen
Uma 8h47'57" 68d34'46"
Shannon, Dorian's Heaven on Earth
Aql 19h1'20" 16d54'17"
Shannon, Lee & Ralph Daily
Uma 9h48'32" 52d36'59"
Shannon, My Love
Gem 6h30'0" 21d18'29"
Shannon, The Best Mom Ever!
Cas 2h41'4" 63d50'44"
Shannon43075
Dra 16h26'33" 56d58'21"
ShannonBrad8112006
Lyn 8h9'20" 33d10'44"
ShannonGail
Sgr 18h12'15" -35d25'24"
ShannonLeigh80
Dra 20h6'32" 63d24'16"
Shannon's Eternal Guiding Light
Gem 6h53'41" 27d1'11"
Shannon's Eternal Love
Umi 14h49'51" 79d31'52"
Shannon's Everlasting Star
Lmi 10h32'10" 32d41'30"
Shannon's Eyes
Tau 5h45'56" 26d21'14"
Shannon's Field of Dreams
Cap 21h21'20" -22d16'42"
Shannon's funk
Cyg 21h47'18" 50d54'40"
Shannons Home World
Cam 5h42'48" 58d31'0"
Shannon's Kiss
And 1h42'8" 38d37'33"
Shannon's Light of Inspiration
Cnc 8h55'12" 10d41'28"
Shannon's Star
Mon 6h28'32" 8d8'49"
Shannon's Star
Gem 6h45'18" 18d46'20"
Shannon's Star
Lyr 18h38'4" 31d46'57"
Shannon's Star
Sco 16h10'44" -15d44'50"
Shannon's Star
Umi 13h54'11" 89d18'41"
Shannon's Star
Umi 14h6'7" 67d59'19"
Shannon's Star
Cam 4h29'45" 65d49'39"
Shannon's Star
Cas 0h14'28" 61d16'16"
Shannon's Star
Cas 23h34'21" 57d36'23"
Shannon's Star
Cru 12h26'27" -60d45'47"
Shannon's Star
Pho 23h51'32" -48d21'43"
Shannon's Star of Love
Leo 11h23'37" -1d11'53"
Shannon's Wish
Sgr 19h0'56" -17d57'20"
Shannon's Wishing Star
Vir 11h49'39" -3d2'31"
ShannStar
Sgr 19h5'49" -31d43'19"
Shanny Bananny BF3E
And 0h19'42" 30d49'49"
shanny2006
Tau 4h37'15" 27d53'31"
Shannyn's Star
Lib 15h2'51" -17d56'45"
Shano & Chez
And 0h25'28" 25d25'18"
Shanon
Lib 14h54'59" -1d21'28"
Shanon Chaiken-Huyter
Psc 0h0'0" 0d0'0"
Shanon Dawn.Star Bednar
And 1h43'45" 36d20'58"
Shanon Dione Hardman Badders
Ari 2h34'2" 14d51'21"
Shanon M. Rush
Ari 3h1'52" 28d24'29"
Shanon R. Hungerford
Lib 14h49'42" -10d35'56"
Shanon R. Stroud
Sgr 20h22'15" -34d8'34"
Shanooks
Her 17h59'8" 29d54'35"
Shans 397
Cyg 19h48'8" 33d10'0"
ShanStar
Cyg 20h24'55" 59d26'57"
SHANT ARAKELIAN
Uma 9h12'3" 61d57'10"
shantala
Psc 0h35'4" 12d26'8"
Shantavia
Tau 5h17'20" 28d13'17"
Shantay Sherray & Aaron Gerard
Cyg 19h41'43" 29d4'50"
Shantel My Love
Lib 15h6'12" -26d42'53"
Shantel Tinsley
Tau 5h13'8" 22d5'38"
Shantell
Cyg 21h35'42" 30d38'33"
Shantelle
Uma 13h20'36" 62d11'20"
Shantelle
Lib 14h52'27" -6d25'35"
Shanté's Love
Psc 1h5'47" 16d42'40"
shanti
Cnc 8h40'52" 14d58'42"
Shanti
Lyn 8h57'14" 41d47'37"
Shanus
Her 17h17'37" 32d42'0"
Shanyn Shanley
Cam 4h38'36" 68d35'50"
Shaolin Sun Wu-Kong Hind
Ori 5h35'40" 1d55'27"
Shaona Land
Crb 15h48'23" 35d58'37"
Shar Cummins
Lyn 8h11'23" 52d47'44"
Shar Star
Cap 21h32'53" -23d29'59"
Shara
Cas 0h53'32" 50d6'58"
Shara Kumari
Lep 5h30'39" -21d48'14"
Shara L.C.
And 23h42'23" 46d2'11"
Shara Lyn Johnson
Lyn 7h54'52" 50d27'3"
Shara R. Richter
Tau 5h48'35" 26d5'36"
Shara, John, Jacob, Richard Northup
Ori 5h38'30" -2d56'57"
SharaB
Tau 3h47'39" 29d22'42"
Shara-Belle
Mon 6h43'43" -0d19'21"
Sharaf Homa
Uma 11h0'6" 34d43'57"
SHARAH
Sge 19h25'23" 18d35'30"
Sharaj
Umi 14h24'48" 67d48'16"
Sharalee A Miyama
Vir 14h41'44" 2d14'32"
Sharalie Nichole
Tau 4h38'46" 22d25'11"
Sharan May McCarthy
Lmi 10h50'22" 27d30'16"
Sharan Shetty,the Eternal Superstar
Tau 5h48'26" 23d0'4"
Sharareh
Ari 2h7'56" 13d22'19"
Sharaul's Star
Umi 16h17'48" 72d55'8"
Sharayah Marie Christmas
And 2h14'26" 44d22'6"
Sharayah Noel
And 0h40'10" 37d12'24"
Sharayas Heavenly Angel
Lyr 18h53'16" 42d36'12"
Sharbear
Lib 14h49'29" -6d25'35"
Shardae and Darrell
Cyg 20h38'38" 41d48'27"
Shardane
And 0h20'4" 27d20'31"
ShareBear
Uma 10h12'34" 54d48'42"
Shared Moments
Cyg 20h20'54" 48d2'50"
Sharee Nicole Rinker
Gem 7h29'43" 20d2'29"
Sharee Reynolds
Tau 3h57'5" 25d32'30"
Shareka M. Galloway
Psc 1h17'1" 7d1'14"
Shareka "Pebbles" Cuthbert
Sco 16h21'43" -32d58'17"
Sharen Parker
Lib 15h23'9" -6d34'34"
Sharen Ruth McColley
Vir 13h48'20" -12d5'20"
Sharese Marie Pace
Lyn 8h28'21" 38d33'3"
Shari
And 23h20'16" 43d1'55"
Shari
And 0h20'24" 45d8'13"
Shari
Vir 13h14'31" 5d58'3"
Shari
Vir 11h57'30" -4d33'28"
Shari
Aql 19h22'39" -11d7'4"
Shari A. Sellars
Tau 4h33'44" 16d20'18"
Shari Adamczyk-Barton
Tau 4h38'56" 21d41'16"
Shari Ananian
And 0h26'54" 34d48'16"
Shari Bernstein
Uma 8h24'45" 68d30'8"
Shari Beth
And 2h26'2" 41d26'0"
Shari Boone Lockwood
Gem 7h10'17" 27d29'39"
Shari Broomberg
Vir 13h2'9" 3d13'43"
Shari Cannon
Ari 1h58'37" 17d47'25"
Shari CJ Arguello
Psc 0h45'29" 14d3'52"
Shari E. Linnane
Sco 17h51'27" -37d14'4"
Shari E. Loges
Leo 9h28'53" 22d55'34"
Shari Eva Kay
And 23h27'54" 41d32'2"
Shari G SS&D
Ari 2h22'44" 27d34'57"
Shari Gitlin
And 2h20'30" 46d21'18"
Shari Gower
And 1h44'26" 45d0'23"
Shari L. Piccione
Sco 17h36'35" -42d54'54"
Shari L. Schulz
Cnc 8h21'13" 12d8'40"
Shari Lynn
Lup 15h30'55" -41d27'53"
Shari Lynn
Sco 16h21'6" -26d28'34"
Shari Lynn Beaty
Lyn 8h10'38" 40d11'40"
Shari Lynne Flanagan
Gem 6h31'25" 15d7'0"
Shari Raynor
Tau 3h43'43" 6d34'24"
Shari Robin Levy
Sco 16h49'47" -34d29'42"
Shari Rosen - The Love of My Life
And 0h16'58" 25d50'11"
Shari & Shawn Beckman
Cyg 20h45'38" 33d26'4"
Shari Thieman Greene
Sco 17h50'41" -42d8'4"
Shari Tushi Hartman
Gem 6h44'50" 24d46'10"
Sharic
Cnc 8h37'46" 31d36'7"
Sharice
Cap 21h6'27" -16d6'21"
Sharié Perry
Cas 1h27'59" 62d17'5"
Sharieffa Naeemha Anderson-Watson
Crb 15h40'40" 27d20'25"
ShariHecate
Ari 3h4'4" 14d3'15"
ShariLyn Ruth Cox
Cep 23h29'41" 80d27'37"
Sharing A Soul
Sge 19h52'4" 17d24'24"
"Sharing The Night Together"
Sge 19h41'38" 17d18'9"
ShaRinn Minor
Cyg 20h34'56" 40d38'18"
Sharis Boettcher
Leo 10h17'53" 20d22'43"
Shari's Hope
Lyn 7h26'12" 50d22'36"
Sharissa
Vir 12h38'27" 0d54'22"
Sharissa Leanne Evans
Lyn 7h33'51" 36d19'40"
Sharita Nicole Swift-Williams
Ari 2h21'18" 23d21'25"
Sharjo
And 23h20'58" 42d6'3"
Sharkbait
Ori 5h31'38" 4d29'31"
sharkey
Aur 7h15'57" 42d26'21"
SharkeyRomero
Per 3h1'27" 55d39'32"
Sharky
Cap 21h2'19" -20d24'7"
Sharla
Sco 16h15'33" -10d20'9"
Sharla Morgan Ford
Cas 1h51'29" 60d23'48"
Sharlan Irene Skrupa
Psc 1h15'17" 12d2'1"
Sharla's Star of Love,Hope & Belief
Per 4h43'52" 40d43'3"
Sharlea Jane Anderson
Cra 18h2'46" -45d20'9"
Sharleen
Umi 15h39'26" 68d16'35"
Sharleen
And 0h35'31" 39d22'57"
Sharleen Cooper Cohen
Lyn 7h40'1" 41d23'24"
Sharleen Marie Ward
Cnc 8h54'57" 15d0'16"
Sharleine Lemon
Vul 20h16'21" 22d45'21"
Sharlene E. Manduriao "Shar Star"
Cas 2h30'58" 67d15'54"
Sharlene Gail Strott
And 0h33'14" 27d45'24"
Sharlene Morrison - Austin
And 0h47'0" 34d38'9"
Sharlene My Angel
Ori 5h8'59" 15d45'3"
Sharlene Puckett
Umi 14h26'45" 69d17'15"
Sharlene Sandra Howell
Ari 2h50'17" 15d34'33"
Sharlene W. Klimo
Gem 7h26'38" 15d21'13"
Sharlene's 60th Shines for Eternity
Cas 1h28'57" 57d42'17"
Sharlene's Star
And 2h14'1" 42d54'29"
Sharlet Anna
Per 3h49'36" 41d40'16"
Sharline Lambrecht Unsere Süsse
Uma 11h7'50" 28d40'35"
Sharline Rodriguez
And 1h18'50" 49d8'12"
Sharlotte
Peg 21h59'33" 13d5'10"
Sharls David
Pho 0h38'18" -49d16'14"
Sharlyn
Per 3h6'46" 43d2'33"
Sharlyn Ann Jarosek McMurray
Ori 5h30'44" 13d35'26"
Sharlyn J Troilo
Com 12h36'2" 23d38'25"
sharmane meredith
Sgr 18h22'30" -22d35'7"
Sharmayne Kelley
And 23h42'39" 34d0'14"
Sharmila R Rambissoon
Vir 14h15'7" -6d26'0"
Sharnelle
Ari 2h10'49" 25d1'25"
Sharnelle Louise Johnson
And 1h6'14" 45d12'21"
ShaRob
Vel 9h59'53" -51d18'46"
Sharolyn Yvonne
Com 12h51'27" 22d0'59"
Sharon
Leo 10h17'35" 21d52'6"
Sharon
Cnc 8h36'50" 16d31'22"
Sharon
Cnc 8h38'51" 15d45'38"
Sharon
Aql 19h0'44" 16d11'7"
Sharon
Leo 10h33'52" 25d54'5"
Sharon
Vul 20h18'23" 24d15'27"
SHA-ron
Cyg 19h39'44" 27d49'32"
Sharon
Tau 3h37'37" 26d16'45"
Sharon
Tau 3h37'40" 14d31'10"
Sharon
Leo 10h46'42" 13d51'6"
Sharon
Lyn 7h44'4" 48d49'19"
Sharon
Crb 15h37'4" 37d48'27"
Sharon
And 2h37'39" 40d23'21"
Sharon
Lyn 8h0'31" 42d17'48"
Sharon
Lmi 9h32'21" 36d47'53"
Sharon
Her 18h1'8" 35d49'54"
Sharon
Pho 1h7'11" -55d31'43"
Sharon
Sgr 18h19'3" -33d2'18"
Sharon
Cap 20h26'47" -15d57'22"
Sharon
Vir 13h28'23" -19d0'55"
Sharon 11
Leo 10h57'53" 1d37'58"
Sharon 5
Lmi 10h21'26" 32d25'5"
Sharon A. Beliveau
Uma 8h33'4" 63d9'2"
Sharon A. Fraum
Sgr 18h33'41" -34d11'29"
Sharon A. Kleva
Sco 16h9'34" -9d18'31"
Sharon A. Palermo
Cnc 8h43'31" 24d25'24"
Sharon A. Rendueles
Peg 21h43'56" 8d43'11"
Sharon Abbott
Mon 7h49'48" -0d43'31"
Sharon All
Vir 13h54'45" -1d58'4"
Sharon All
Vir 13h51'58" 2d36'53"
Sharon Allen Baker
Leo 10h11'48" 20d10'5"

Sharon Alvarado Cnc 8h25'13" 13d32'5"
Sharon Anastasia Cnc 8h39'40" 18d8'34"
Sharon and Annie Kelly And 2h8'59" 45d6'18"
Sharon and Bob Cyg 19h56'4" 38d13'0"
Sharon and Dennis Forever Tau 4h31'37" 18d31'10"
Sharon and Jacqui's Wishing Star Leo 11h30'59" -1d25'55"
Sharon and John DePalma Cyg 20h49'17" 39d48'7"
Sharon and John star Tau 4h41'46" 27d18'10"
Sharon and Kovi Cyg 19h40'14" 38d56'47"
Sharon and Robert Miller Per 3h1'17" 47d31'23"
Sharon Anderson Cha 9h8'42" -78d22'18"
Sharon Anita Zahara And 0h41'31" 25d43'26"
Sharon Anmuth Ari 2h36'54" 13d51'10"
Sharon Ann Lyn 7h9'42" 54d15'45"
Sharon Ann Barone Cap 20h27'12" -9d32'18"
Sharon Ann Chickering Cyg 21h45'19" 49d43'13"
Sharon Ann Harrington 1954-2006 Vir 13h31'24" 3d56'22"
Sharon Ann Jhoslien Wheeler Cnc 8h43'8" 17d59'45"
Sharon Ann Jostes Lib 15h14'16" -8d11'59"
Sharon Ann Kincaid Uma 11h30'43" 44d48'18"
Sharon Ann McDade Uma 10h23'35" 57d18'38"
Sharon Ann Rivera Aur 5h13'3" 32d49'7"
Sharon Ann- The Gracious Vir 12h10'42" 5d6'54"
Sharon Ann Valerio Cas 0h15'49" 62d49'39"
Sharon Ann Webb Uma 11h39'26" 33d58'13"
Sharon Anne Bostick Leo 11h9'8" 10d11'26"
Sharon Anne Fralick Bower Umi 15h2'27" 71d4'36"
Sharon Anne Henderson Psc 0h33'4" 7d29'51"
Sharon Anne Henrikson Uma 14h0'16" 53d31'7"
Sharon Asciutto Lyn 8h26'56" 33d14'34"
Sharon B. Gennett Cyg 20h41'24" 46d23'13"
Sharon Baker Peg 21h36'24" 21d9'27"
Sharon Bannen Lyr 19h12'37" 38d52'15"
Sharon Beall Lmi 10h22'28" 32d35'27"
Sharon Bell Cas 23h28'34" 56d7'31"
Sharon Benz Uma 10h20'26" 67d47'11"
Sharon Bibby Crb 15h49'30" 30d5'34"
Sharon Blazejewski Born: Sept 13/1948 Uma 11h32'12" 57d40'27"
Sharon Bogie And 1h32'29" 47d56'11"
Sharon Borland Wilson Cas 23h45'23" 56d54'45"
Sharon Bradley And 23h17'41" 48d25'36"
Sharon Broschk Umi 15h20'59" 69d5'59"
Sharon Bullard Cyg 19h39'21" 34d10'26"
Sharon Caponegro Lyn 7h34'46" 58d28'42"
Sharon Carchidi Lyn 7h41'50" 57d37'22"
Sharon Carmany Bowman Eri 4h25'59" -0d32'19"
Sharon Carole Her 18h43'14" 25d22'50"
Sharon Catedrilla And 0h42'12" 41d43'5"
Sharon Cerelle Konits Sgr 19h25'48" -41d17'8"
Sharon Charolette Ann Evenson Lyn 7h32'31" 47d33'24"
Sharon Cheryl Ng Cheong Ton Ori 6h5'1" 7d6'31"
Sharon "Chiron" Klug Cas 1h47'49" 61d27'1"
Sharon Christian Rando Ari 2h34'55" 24d46'35"
Sharon Chung-Yan Hui Psc 1h1'21" 11d59'47"
Sharon Conomos - Happy 40th Cru 12h49'8" -62d29'8"
Sharon Cusack Uma 10h23'53" 52d29'45"
Sharon Darden Cyg 20h5'54" 58d51'36"
Sharon Darlene Gish Cam 8h52'54" 76d12'41"
Sharon Darlene Welch Sgr 19h17'19" -21d25'52"
Sharon Delgado Cnc 8h18'28" 19d29'1"
Sharon Denise Corbett And 1h38'7" 39d30'41"
Sharon Denise Vance Crb 15h36'10" 30d56'49"
Sharon Desiree Bland Stoll Lmi 10h51'12" 26d41'18"
Sharon DesJardins Uma 8h58'9" 53d20'34"
Sharon Diane Dal Porto Vir 12h29'16" 4d53'9"
Sharon Diane King Gem 7h37'29" 24d46'10"
Sharon Dick Ant 9h58'46" -38d8'43"
Sharon Dickerson Crb 15h44'0" 31d3'28"
Sharon Dixon Cas 3h38'18" 76d55'46"
Sharon E Sperry Cnc 8h15'6" 28d31'9"
Sharon Elaine Cas 1h9'52" 69d44'39"
Sharon Elaine Irby Uma 10h4'30" 63d15'52"
Sharon Elizabeth Aqr 22h19'44" -23d30'17"
Sharon Elizabeth Davis Cas 23h49'31" 53d16'24"
Sharon Elizabeth Erlund Vir 13h45'8" -8d26'11"
Sharon Elizabeth Good Uma 9h45'10" 53d5'46"
Sharon Elizabeth & Saffron Leia Ralph Cas 0h25'22" 61d27'49"
Sharon Emily Lombard Vir 11h42'10" -5d3'5"
Sharon Emirali Cru 12h29'6" -59d49'33"
Sharon Erickson Cas 2h16'6" 71d25'47"
Sharon Esther Chan Cyg 20h14'54" 51d37'30"
Sharon Etta Gerber Aqr 23h4'54" -14d35'53"
Sharon Etzman Aqr 21h24'42" -8d9'41"
Sharon F. Davis And 0h43'53" 45d35'39"
Sharon F. Jacobus Leo 10h33'2" 15d48'25"
Sharon Fitzpatrick And 0h7'52" 33d8'9"
Sharon Foy Cas 23h6'38" 54d27'21"
Sharon Frances McLoughlan Tau 5h57'33" 26d7'53"
Sharon Frith Per 3h46'22" 47d14'24"
Sharon Fulks - Star Educator Vir 13h17'40" 12d26'3"
Sharon G. Bray Cas 1h25'33" 51d46'3"
Sharon & Garhett 9/26/2000 Crb 16h21'0" 38d25'12"
Sharon Gehman Lib 14h51'11" -5d25'56"
Sharon Gillen And 2h27'32" 43d47'18"
Sharon Gray Lyn 7h36'15" 56d22'21"
Sharon Grenham Lyn 8h26'0" 46d52'3"
Sharon Hagen Vul 19h47'34" 27d57'6"
Sharon Haig Lyr 18h46'25" 36d4'39"
Sharon Hammann Sgr 18h24'1" -33d0'4"
Sharon Hancock Tau 5h9'37" 20d46'59"
Sharon Hargrove Leo 11h10'28" 16d38'18"
Sharon Hartley Lyr 18h58'21" 26d50'39"
Sharon Hines Psc 1h22'42" 17d29'55"
Sharon Holt Eri 4h33'1" -0d45'2"
Sharon Holtzner *Baby* Tau 4h32'22" 24d57'28"
Sharon Hossler Cyg 19h57'29" 37d25'51"
Sharon Houghton Bellows Cap 20h16'6" -13d39'0"
Sharon Huesca Peg 23h11'17" 31d39'51"
Sharon J. LaDuke Uma 8h21'57" 66d31'54"
Sharon J. Lovett Ori 5h16'49" 0d1'35"
Sharon J. & Mark W. Wright Cyg 20h11'4" 54d7'29"
Sharon J Stanczak Vir 12h35'3" 12d3'53"
Sharon Jean Ellis Vir 12h43'33" 7d43'34"
Sharon Jean Moyer Cyg 19h52'30" 44d22'50"
Sharon Jennifer Samara Aqr 22h13'9" 2d25'32"
Sharon & Jeremy Forever Cyg 20h1'25" 58d58'54"
Sharon Jessrock1 "10-2-48" Cas 0h46'17" 51d44'30"
Sharon Johnson Leo 11h26'7" 12d26'53"
Sharon Jokinen Snyder Ari 2h20'10" 13d4'13"
Sharon & Juan Restrepo Ori 5h53'22" 8d10'24"
Sharon K. Dohn Leo 10h29'36" 13d12'53"
Sharon K. McLaughlin Sco 16h22'31" -31d59'14"
Sharon K. Melear Lyr 19h20'49" 37d37'56"
Sharon K. Sanders Dra 20h19'12" 67d7'57"
Sharon K. Sanders Vir 13h28'6" -4d11'28"
Sharon Kane Aqr 21h36'24" -8d10'46"
Sharon Kantor 4/4/43 Psc 1h23'52" 12d22'43"
Sharon Katherine Hays-Jenkins Vir 12h46'24" -4d33'31"
Sharon Kay Sco 16h7'19" -11d24'40"
Sharon Kay Uma 11h47'8" 60d15'4"
Sharon Kay Aql 19h50'13" 3d39'4"
Sharon Kay Algar And 0h39'0" 42d30'15"
SHARON KAY CHAMBERS QUESENBERRY Gem 6h43'42" 34d21'19"
Sharon Kay Judge-Bennett Ori 5h2'47" 3d39'17"
Sharon Kaye Stringfield Gem 7h32'41" 27d25'16"
Sharon Klopfenstein And 1h13'31" 45d8'39"
Sharon L Ansley Sco 16h7'0" -17d42'5"
Sharon L. Crispin 031875 Psc 0h44'4" 7d19'19"
Sharon L. Gaudenti Vir 13h18'51" 3d35'11"
Sharon L. Gonzales Cam 5h56'3" 58d25'25"
Sharon L. Micinski And 23h27'10" 45d48'3"
Sharon L. Ostler Crb 16h0'51" 32d38'13"
Sharon Lavens Tau 3h47'2" 4d0'22"
Sharon Lea Hanson-Olson Uma 9h46'17" 54d38'25"
Sharon Leah Zalusky Cap 21h39'24" -19d40'58"
Sharon Lee Cap 20h20'20" -11d51'37"
Sharon Lee Leo 10h42'45" 11d42'18"
Sharon Lee Ancona Cyg 21h59'26" 51d45'32"
Sharon Lee Anderberg Lmi 10h3'54" 36d18'13"
Sharon Lee Brockman And 2h20'9" 50d12'5"
Sharon Lee Fadelici Cas 0h17'18" 61d39'28"
Sharon Lee Ginsburg Crb 15h28'9" 27d42'26"
Sharon Lee Holloway Cyg 21h36'27" 39d22'22"
Sharon Lee Oldfield Psc 1h8'5" 27d30'45"
Sharon Lee Shields Tau 5h26'56" 28d11'22"
Sharon Leigh Crb 16h17'36" 28d8'51"
Sharon Leslie Schuttloffel Ori 5h27'20" 4d8'35"
Sharon Louise Lib 15h48'49" -6d25'41"
Sharon Louise Hunter Cyg 20h20'20" 46d50'13"
Sharon Louise Russell Cyg 20h39'52" 46d23'39"
Sharon Louise Severin Aqr 23h22'39" -20d8'49"
Sharon Lynn Ori 5h34'47" 2d29'28"
Sharon Lynn Ari 2h24'57" 12d57'32"
Sharon Lynn Barcal Ari 2h33'50" 27d30'4"
Sharon Lynn Butcher Crb 16h6'32" 34d34'5"
Sharon Lynn Nixon Sco 16h13'4" -13d43'57"
Sharon Lynn Park Lib 15h17'29" -15d57'50"
Sharon Lynn, Always Uma 8h26'53" 61d24'43"
Sharon Lynne And 23h33'50" 47d14'15"
Sharon M. Cap 20h17'12" -9d42'41"
Sharon M. Farag Uma 9h38'41" 67d21'53"
Sharon Magers Ramirez Leo 11h44'44" 24d48'8"
SHARON MAKABENTA Dra 16h16'32" 61d15'54"
Sharon Marie Uma 13h31'9" 56d35'31"
ShaRon Marie Sgr 19h21'45" -16d11'3"
Sharon Marie Sco 17h11'18" -43d32'50"
Sharon Marie Per 3h17'16" 34d14'27"
Sharon Marie Alcalde Sgr 19h19'44" -21d25'58"
Sharon Marie Best Giusti Uma 10h57'24" 56d41'52"
Sharon Marie Christoph Cas 1h8'32" 54d46'32"
Sharon Marie Knop Tau 3h31'40" 14d55'13"
Sharon Marie Mitchell Meis And 0h44'41" 39d49'49"
Sharon Marie Morison Hanks Cma 6h38'48" -22d37'49"
Sharon Marie Murphy Cam 8h33'56" 83d58'56"
Sharon Marlisa Mickey Mon 6h52'28" 8d26'31"
Sharon Mary Connors Ari 2h17'38" 27d10'7"
Sharon Mary Sullivan Tau 4h0'33" 1d39'42"
Sharon Matz Leo 10h32'23" 18d6'10"
Sharon Mayes Pho 23h27'18" -40d46'55"
Sharon Mayfield Lyn 7h29'31" 52d38'20"
Sharon McCain Tau 4h14'27" 18d18'56"
Sharon McCarten Gem 7h39'0" 34d27'6"
Sharon Mckenna And 2h29'19" 50d43'19"
Sharon McPherson Lyr 18h33'32" 32d16'37"
Sharon Miller Uma 12h46'43" 56d27'11"
Sharon Morgan Cas 0h49'54" 51d52'57"
Sharon Mummpy Fahrion Tau 5h27'45" 27d6'41"
Sharon "My Love" Lmi 10h32'17" 31d45'51"
Sharon N Cam 3h37'41" 58d58'53"
Sharon "Nana" Simms Ari 3h6'23" 26d43'36"
Sharon Nicole Glover Dra 19h37'0" 60d3'10"
Sharon Norman Kiker Aqr 22h9'55" -8d29'9"
Sharon Norton Sgr 18h5'47" -16d34'16"
Sharon "Number 1 Mum" Elliot Cra 19h4'0" -41d5'35"
Sharon Oliva And 1h12'57" 34d56'58"
Sharon Osbourne And 2h20'55" 45d26'59"
Sharon P. Bode Ari 2h9'18" 21d52'32"
Sharon P. Carpitella Vir 12h56'18" 10d44'27"
Sharon P. Griffith Cyg 19h8'20" 52d57'55"
Sharon Parker And 1h39'10" 50d36'58"
Sharon Perry Crb 15h49'40" 28d13'13"
Sharon & Pop Dyson's Star Cra 18h58'2" -41d14'59"
Sharon Rasiul Vir 13h12'2" -4d29'51"
Sharon Reubell Lair Srp 18h33'41" -0d12'44"
Sharon Robinson Lib 15h30'45" -8d39'43"
Sharon Rogers Cas 0h1'48" 56d48'44"
Sharon Rose And 0h9'37" 34d52'8"
Sharon Rose And 1h43'31" 43d16'12"
Sharon Rose Bright Star Of Beauty And 1h46'37" 49d34'6"
Sharon Rose Henke Pho 23h53'29" -51d9'58"
Sharon Rose Tice Lib 14h51'37" -9d32'7"
Sharon Rose Wood Ari 2h16'24" 16d19'50"
Sharon Roslund Lyr 18h44'12" 31d8'27"
Sharon Ruhland My Love Gem 6h49'43" 31d21'21"
Sharon Ruth Shining Star Sgr 18h11'7" -34d6'4"
Sharon S. Richardson Cyg 21h16'6" 41d59'11"
Sharon Sager Weir Hya 9h19'2" -0d44'30"
Sharon Scro Cas 23h23'30" 58d0'47"
Sharon Shaffer Gem 7h37'1" 23d34'25"
Sharon Sherris Sco 17h28'49" -38d22'26"
Sharon Shi Gem 6h50'41" 27d35'27"
Sharon Shines Eternally Psc 1h45'6" 13d14'26"
Sharon Silverman: Smiling Forever Tau 4h32'34" 28d53'5"
Sharon Simpson Uma 11h43'31" 30d45'43"
Sharon Sisson Umi 15h59'59" 80d44'57"
Sharon Smith Cas 1h27'20" 64d35'56"
Sharon Sue Speegle Cam 5h39'39" 66d49'40"
Sharon The Rose of Peaceful Valley Gem 7h5'20" 15d5'42"
Sharon Theresa Winfield Major And 2h38'28" 42d44'52"
Sharon Theriac Peg 23h57'3" 29d2'33"
Sharon Thomas Lyr 19h13'11" 27d6'21"
Sharon & Tonya Jones (As One) Cap 21h39'45" -16d29'59"
Sharon Treaster Uma 8h18'53" 60d10'27"
Sharon V. Ori 6h20'55" 7d30'47"
Sharon Van Grant Cap 20h28'21" -12d17'6"
Sharon Vincent Uma 9h28'36" 44d41'44"
Sharon W Foremsky Vir 11h38'58" 4d11'38"
Sharon Walker Cas 0h40'51" 69d31'1"
Sharon Wallace Crb 16h15'10" 34d33'53"
Sharon Woodmansee Cap 20h17'46" -14d41'45"
Sharon Wylie McIntosh Cru 12h4'29" -62d59'58"
Sharon Y Hansen Leo 9h22'37" 16d11'14"
Sharon Yvonne Grove Per 3h34'28" 33d15'7"
Sharon Yvonne Smith Cnc 9h8'45" 32d38'1"
Sharon2004October22Debbie Cas 23h10'39" 54d28'49"
Sharon3257 Cap 20h18'4" -11d1'39"
SharonAnn Srp 18h18'24" -0d6'51"
SharonDanCaitlinDJ Lyn 7h0'46" 44d27'1"
Sharon's Uma 11h46'6" 43d33'35"
Sharon's Birthday Star And 2h35'59" 49d36'25"
Sharon's Birthday Star Peg 0h4'38" 11d15'58"
Sharon's celestial body Cnc 8h49'55" 28d13'18"
Sharon's Glow Uma 10h28'31" 62d3'20"
Sharon's Heart Leo 11h37'23" 13d12'14"
Sharon's Hummer Haven Cyg 20h37'33" 33d29'28"
Sharon's Light Cas 0h37'20" 55d14'12"
Sharon's Lucky Star Uma 13h44'28" 60d12'15"
Sharon's Star Psc 23h59'35" -2d29'59"
Sharon's Star Peg 22h6'15" 30d44'2"
Sharon's Star Per 3h39'5" 44d16'23"
SHARON'SHONALEE64 Psc 1h28'40" 25d21'3"
Sharonski Shahnau Cnc 8h19'1" 14d0'53"
Sharples Umi 14h30'40" 75d50'41"
Sharpy Umi 17h32'19" 80d15'30"
Sharpy's Eye And 0h54'26" 23d50'41"
Sharr Crb 16h6'49" 36d32'18"
Sharra Krise Dra 18h31'20" 55d49'46"
Sharren Rathburn Sgr 17h51'57" -20d51'23"
Sharri Lynn Lmi 10h48'56" 31d11'50"
Sharrie Dra 17h41'45" 60d31'26"
Sharron A. Higgs 12/03/2004 Per 2h51'4" 38d51'11"
Sharron and Terry Forever Uma 9h38'26" 54d3'43"
Sharron Annette King Kirby Ori 5h50'10" 6d22'1"
Sharron Catello Aqr 22h14'52" 1d31'47"
Sharron Lynne Distefano Lyr 19h26'27" 42d24'48"
Sharron & Perry Forever Dreaming Cyg 20h29'43" 30d15'53"
Sharron & Peter Uma 11h3'22" 48d29'44"
Sharron Taylor Price Vel 8h51'43" -39d54'49"
Sharry & Jacques Option Per 2h59'22" 45d40'1"
Sharry Jane Gem 6h38'14" 21d31'43"
Sharryse Aql 19h41'32" -11d23'12"
Shar-Tyna Uma 10h57'37" 65d31'37"
Shary Kasper Lyr 19h3'21" 42d40'24"
Shary Poushin Tau 5h38'47" 26d39'34"
Sharyn Lyr 18h47'13" 37d33'52"
Sharyn Uma 10h36'0" 49d19'39"
Sharyn Christine Phelan Cnc 8h45'58" 21d22'2"
Sharyn Elizabeth Keith-Monsour Cru 12h22'31" -58d14'53"
Sharzad Gem 6h33'16" 14d58'34"
Shasdy A. Velasco Mon 6h24'3" 9d12'46"
Sha-Sha Leo 10h19'26" 26d29'53"
Shasha Thorley Uma 11h34'48" 47d32'29"
Shashi Del 20h20'33" 9d50'48"
Shashush Cyg 21h46'53" 51d34'42"
Shaskan Family Star Sgr 18h48'36" -23d36'43"
shasme Cap 21h9'14" -17d7'5"
Shasta Sgr 18h47'53" -17d53'40"
shasta Sco 16h58'1" -41d22'33"
Shasta Vir 13h40'9" 2d0'23"
Shasta Leo 10h0'50" 20d56'48"
Shasta Daisy Sgr 19h6'14" -15d34'0"
Shasta Douglas-Daviault Umi 15h42'27" 77d57'9"
Shasta Emma Storm Umi 14h59'26" 82d27'54"
Shasta Leigh Ari 2h3'30" 18d3'25"
Shasta Savannah Spencer Peg 22h42'35" 11d16'12"
Shasten Snellgroves Crb 16h8'26" 38d39'30"
Shatana Lib 15h47'5" -27d27'16"
Shatara, Demetres Jr, & JaVoskia Jr Ori 6h14'26" 15d5'50"
Shatha Lib 15h22'11" -4d36'11"
Shathi Sge 19h53'1" 18d46'30"
Shauency Uma 10h29'59" 60d19'48"
Shaughn Taylor Angelo Russell Leo 11h43'44" 26d19'48"
Shaughn's Star Of Love Lib 15h38'12" -18d50'46"
Shauma Mama Ari 3h20'45" 29d34'47"
Shaun Ari 2h36'59" 14d3'43"
Shaun Alan Fyfe Psc 1h2'53" 11d21'18"
Shaun Alexander Robinson Ori 5h35'34" 5d20'34"
Shaun Allen Thielman Her 16h35'20" 33d39'20"
Shaun and Christy Uma 14h16'15" 57d3'33"
Shaun Anthony Carroll Psc 1h41'3" 18d58'3"
Shaun & Brittni Aur 5h53'6" 53d5'11"
Shaun Brown Sgr 18h36'30" -33d1'0"
Shaun Christopher Harris Cep 22h11'32" 58d47'14"
Shaun Christopher Murphy Uma 10h26'24" 49d33'45"
Shaun Coleman Cep 5h9'5" 81d55'9"
Shaun Dolan Uma 11h50'7" 53d33'31"
Shaun Hayes Her 16h46'45" 47d47'23"
Shaun Ingram Peg 22h59'3" 21d27'3"
Shaun Jon Christensen Sgr 18h23'10" -35d16'58"
Shaun Kaiman Uma 10h45'25" 63d48'27"
Shaun & Kell - 25 August 2006 Pho 1h12'13" -45d6'14"
Shaun Kenneth Seastrand's Star Per 4h38'49" 36d31'8"
Shaun Knowles Cru 12h4'3" -62d17'46"
Shaun Landon Leo 10h56'5" 8d43'36"
Shaun Leyson Leo 11h41'51" 10d57'27"
Shaun M. Gray Ori 6h11'18" 15d30'20"
Shaun McGarry Per 3h4'38" 45d56'19"
Shaun Michael Davis Per 3h13'7" 52d30'30"
Shaun Michael Milan Cnc 8h47'1" 16d51'1"
Shaun Myles Austin Vir 12h33'37" -0d11'25"
Shaun O'Reilly Uma 12h56'39" 61d54'48"
Shaun Paddock Equ 21h23'53" 8d29'46"
Shaun & Pamela Bailey Cyg 21h18'34" 52d23'43"
Shaun Patrick Rodriguez Aur 5h31'24" 45d53'26"
Shaun Paul Matlock Lyn 6h44'28" 57d24'27"
Shaun Paul Rowan Per 4h28'0" 34d46'5"
Shaun Peter Burgmann Cap 21h5'25" -20d52'57"
Shaun Pilling Cep 0h3'8" 71d7'37"
Shaun Preston Mitchell Ori 6h11'22" 15d51'28"
Shaun R. Kennedy And 0h19'4" 26d8'33"
Shaun R. Laughran Cam 3h59'36" 67d54'52"
Shaun & Randi Lib 14h48'38" -3d48'10"
Shaun River Cox Cyg 19h31'32" 44d4'9"
Shaun Ross Creely Psc 0h9'11" -1d54'56"
Shaun Sanders Umi 14h47'4" 74d29'48"
Shaun Seymour Psc 23h51'0" 7d21'3"
Shaun Turcotte Umi 13h53'11" 76d9'34"
Shaun Whited Hya 8h38'17" -12d27'21"
Shaun William Morey Cru 12h17'11" -56d15'36"
Shauna Ori 5h8'42" 1d3'16"
Shauna Cnc 9h19'28" 11d18'52"
Shauna Cyg 19h53'25" 37d42'25"
Shauna And 0h34'7" 45d51'15"
shauna And 0h23'46" 38d3'7"
Shauna And 1h16'29" 41d22'39"
Shauna Alice Bogues And 2h25'22" 43d41'29"
Shauna and Dustin Wyatt Cyg 20h13'2" 39d20'17"
Shauna Anita Haynes Sawyer Tau 4h7'37" 15d43'48"
Shauna Ann Bump Uma 9h8'46" 48d48'48"
Shauna Bee Leo 10h16'22" 22d9'43"
Shauna Fisher Morse Aqr 22h9'3" -0d52'27"
Shauna Hartrich Leo 9h26'57" 25d45'22"
Shauna Jackson Her 16h46'48" 48d7'46"
SHAUNA JANINE Cam 3h48'17" 55d27'14"
Shauna Kaye Shaefer Gem 7h27'37" 28d1'42"
Shauna L. O'Mahoney Aur 5h38'43" 44d31'15"

Shauna Lee Fox
Cas 1h4'37" 48d50'10"
Shauna Louise
Sco 16h4'46" -12d52'5"
Shauna Love
Lyn 8h24'24" 45d22'26"
Shauna Macdonald 1975
Cnc 8h42'30" 32d14'3"
Shauna Maguire
And 23h0'22" 51d13'24"
Shauna Marie Matlen
Ari 2h58'14" 21d50'6"
Shauna Meyers
Gem 7h48'56" 26d33'19"
Shauna Na Na
Sco 16h57'44" -41d54'1"
Shauna Potts
Cap 20h22'58" -12d28'13"
Shauna Richardson
Vir 12h15'19" -5d8'31"
Shauna Sisu Vennum
Leo 10h27'0" 23d54'27"
Shauna Sue Nicholson
Aqr 22h37'7" 0d23'24"
Shauna, star of beauty and kindness
And 1h35'36" 47d12'26"
ShaunaKathleen
Tau 3h58'15" 27d20'5"
Shaunananana
Uma 14h8'48" 58d0'59"
Shauna's Star
Aqr 21h30'0" -0d14'29"
Shaunasmithee
Dra 14h53'52" 56d34'54"
Shaundra Kim Garcia
Cas 1h15'40" 62d4'41"
Shaundricka Hope, Teacher of Stars
Uma 11h19'42" 35d49'35"
Shaundrika Quentet Stevenson
Ari 3h14'10" 27d3'2"
Shaun-hanan
Her 16h7'23" 46d28'29"
Shauni
And 1h27'14" 50d33'34"
Shaunie
Uma 10h35'47" 44d24'33"
Shaunita E M Ortiz
Sgr 20h14'21" -37d34'18"
Shaun-Marie
Sgr 19h39'22" -14d46'36"
ShaunMayo
Aqr 22h32'25" -4d3'15"
Shaun-Michael Edward Woody
Tau 4h3'0" 22d57'36"
Shaunna I Love You
Cyg 19h32'49" 32d24'33"
Shaunna Lynn
Cas 0h18'1" 51d24'9"
ShaunnaMarie143
Sgr 19h46'42" -14d6'10"
Shaunny Ryan
Srp 18h28'53" -0d23'18"
Shaun's Angel
Cap 20h45'2" -26d28'51"
Shauntae Fowler
Psc 1h42'26" 8d41'57"
Shauntay
Gem 6h53'17" 30d8'39"
Shauny & Kristi Together Forever
Psc 1h24'12" 10d56'40"
Shausta Marie Johnson
And 0h23'37" 31d31'22"
Shauwn Marie
Peg 23h51'57" 20d1'42"
Shaver-Taylor
Gem 6h37'24" 15d10'5"
Shavon A. Lee
Vir 13h56'58" -17d59'59"
Shavon Danielle Rosales
Aqr 21h52'12" -8d8'24"
Shavon Skelton
Cnc 8h40'39" 31d25'36"
Shavonne Marie
Cap 20h38'15" -26d38'13"
Shavoynne
Tri 2h41'17" 33d51'29"
Shaw Cooke
Aql 18h53'54" -0d46'58"
Shawan Brantley
Her 17h39'48" 48d36'51"
ShaWanda Magnificent Star
And 0h10'58" 29d46'4"
Shawinca
Lyn 7h53'45" 52d45'5"
Shawn
Umi 14h31'13" 73d41'11"
Shawn
Sco 17h37'19" -31d53'30"
Shawn
Uma 10h31'32" 50d14'59"
Shawn
Per 2h47'55" 54d19'17"
Shawn Alan Thomas
Ori 5h32'54" 1d40'12"
Shawn & Amber Hilley 12242006
Sgr 19h16'23" -18d15'46"
Shawn and Diana's Eternal Flame
Aql 20h10'41" 4d39'7"
Shawn and Emily Forever.....7/6/200
Vir 12h49'36" 4d25'32"
Shawn and Jordann Forever
Tau 4h4'6" 5d57'50"
Shawn and Karen
Cyg 21h32'23" 41d14'3"
Shawn and Katie Yanke
Cyg 21h58'20" 49d10'24"
Shawn and Tiffany
Umi 13h41'33" 70d33'54"
Shawn and Yvette Taylor
Uma 10h16'50" 46d3'21"
Shawn Anthony Orfe
Uma 13h32'2" 56d5'58"
Shawn & Becky Perkins
Crb 15h51'14" 29d42'39"
Shawn Borowski
Vir 14h15'59" 0d55'49"
Shawn Bowman
Ori 4h51'0" 0d43'54"
Shawn "Bud" Brown
Per 2h43'52" 54d34'2"
Shawn C Fischtziur
Ari 3h5'46" 12d9'16"
Shawn Cacioppo
Cas 2h3'55" 62d23'29"
Shawn & Cammie's Star
Cyg 20h12'53" 33d32'1"
Shawn Charles Day
Ari 2h46'14" 28d49'1"
Shawn "chase" Maxwell
Her 17h44'38" 43d53'3"
Shawn Coelho
Cep 22h37'3" 57d57'31"
Shawn Connelly Gray
Aqr 21h41'40" -2d26'57"
Shawn Conners
Her 17h45'28" 36d8'17"
Shawn D. Naugle
Cep 20h41'5" 59d23'58"
Shawn D. Smallwood
Tau 4h13'54" 15d25'31"
Shawn Dalton Quiles
Uma 8h46'28" 50d50'2"
Shawn Daniel Kiernan
Tau 3h50'11" 20d56'21"
Shawn & Danielle
Her 16h42'46" 3d56'26"
Shawn Daryl
Cep 22h20'3" 68d18'27"
Shawn David
Uma 11h28'44" 60d44'30"
Shawn David Fabian
Boo 14h48'30" 24d15'39"
Shawn Deborah Hunsberger
Peg 21h38'37" 16d51'16"
Shawn Donald Williscroft
Ori 6h11'29" -0d33'30"
Shawn Douglas Clover
Sco 16h12'30" -16d11'51"
Shawn Dwayne Nicholson
Uma 11h37'32" 37d52'59"
Shawn Earl Hopson
Per 3h29'1" 44d40'39"
Shawn Feltus
Tau 4h41'4" 7d19'23"
Shawn Foley
Per 3h13'55" 53d6'31"
Shawn Fondal
Cap 20h14'12" -10d1'25"
Shawn Franklin Burnette
Lyr 18h33'12" 33d43'58"
Shawn Gaver
Lib 14h54'29" -2d28'2"
Shawn Gavin
Dor 4h19'5" -51d41'39"
Shawn Godshall
Aqr 21h54'21" 0d17'4"
Shawn Greenberg
Cyg 21h17'23" 43d32'24"
Shawn Groleau
Ori 6h13'2" 15d8'12"
Shawn Harry McDonald
Umi 16h17'40" 71d31'56"
Shawn Hatem
Aur 5h44'57" 50d20'56"
Shawn Hilton
Lyn 6h23'18" 61d25'49"
Shawn Holland & Elena Pray
Psc 1h29'54" 19d44'47"
Shawn Hotnisky
Gem 7h19'17" 33d27'58"
Shawn J. Waked
Per 4h24'25" 43d36'53"
Shawn Jeffrey Miller
Ori 5h22'39" 1d25'18"
Shawn & Jill Trealout
Sge 20h10'25" 18d42'26"
Shawn Johnson
Uma 10h10'21" 59d18'10"
Shawn K 1
Cnc 8h49'55" 26d47'48"
Shawn & Karen Pa
Her 17h52'27" 48d23'57"
Shawn Killingsworth
Sco 16h42'50" -29d32'26"
Shawn Knadler
Tau 3h42'58" 18d30'58"
Shawn L Ramsey
Her 17h37'7" 26d24'31"
Shawn Langley
Uma 9h57'12" 59d40'14"
Shawn Lazarian The Adorable
Cnc 8h24'59" 9d18'36"
Shawn Lee Wilburn
Cnc 9h7'33" 31d8'28"
Shawn & Lezley
Ori 4h48'38" 10d55'34"
Shawn & Libra: An Infinite Love
Apu 17h23'29" -69d42'12"
Shawn Lynch
Per 3h0'2" 55d30'25"
Shawn Marie
Sgr 18h13'16" -16d25'7"
Shawn Marie Midnight
Ari 2h36'34" 26d46'44"
Shawn Marie Rippee
Aqr 22h16'53" 0d33'9"
Shawn Marie & Wandering Wolff
Lyr 18h37'12" 26d9'20"
Shawn & Marnel
Umi 15h20'56" 74d39'56"
Shawn Mathieu Ayotte
Uma 8h44'2" 48d25'7"
Shawn Matthew Walsh
Crb 15h34'48" 30d2'54"
Shawn McVittie - Husband & Father
Uma 11h23'39" 58d39'23"
Shawn Michael
Ori 5h45'11" 11d38'3"
Shawn Michael 4/16/1971
Uma 11h19'49" 34d21'22"
Shawn Michael Lovett
Uma 8h56'19" 56d59'52"
Shawn Michael O'Neil
Crt 11h32'34" -7d45'26"
Shawn Michael Patrie
Psc 1h10'25" 28d18'22"
Shawn Michael Sisson
Cyg 19h28'56" 31d0'58"
Shawn Michael Stoneham
Per 3h17'7" 51d57'25"
Shawn Michael Wyne
Eri 4h34'9" -0d1'44"
Shawn Michelle Danesi
Uma 9h23'20" 50d42'8"
Shawn Niethammer
Her 17h53'34" 28d22'51"
Shawn P. Liegl
Sgr 18h44'19" -30d15'28"
Shawn P. Seidel
And 0h18'57" 32d0'5"
Shawn Patrick Galligan
Dra 19h36'50" 68d19'44"
Shawn Patrick Murphy
Her 16h33'39" 26d36'0"
Shawn Patrick Sheppard
Gem 6h31'6" 13d11'57"
Shawn Patrick Sullivan
Tau 5h30'4" 21d27'17"
Shawn Patrick Sutton
Lmi 10h37'12" 36d11'2"
Shawn Patrick Vásquez
Aqr 22h37'33" -1d16'59"
Shawn Paul Scott
Lib 14h54'33" -16d14'0"
Shawn Phillip Odom
Uma 10h59'29" 45d16'49"
Shawn Piper
Her 18h49'4" 15d28'45"
Shawn Portmann
Aur 6h49'54" 43d9'56"
Shawn Powell God's Angel #2
Sgr 19h22'21" -26d47'26"
Shawn Prescher
Psc 23h58'34" -0d59'33"
Shawn Prutsman
Per 2h38'12" 51d22'58"
Shawn R. Bahuaud
Tau 5h50'51" 27d54'49"
Shawn R. Hartnett 7/3/96
Sco 17h22'28" -32d9'25"
Shawn Rapach
Sco 16h51'45" -27d9'6"
Shawn Ray Kilgore
Ori 6h16'51" 13d18'16"
Shawn Reed
Gem 6h32'8" 19d47'21"
Shawn Renee Zorita Maley
And 23h21'43" 47d44'13"
Shawn Ropson
Boo 15h22'22" 40d8'43"
Shawn Schmersey
Aql 19h47'58" 11d36'3"
Shawn Skillings
Cap 20h25'19" -12d16'6"
Shawn Spivey "Rockstar"
Cep 23h20'43" 71d39'27"
Shawn Stoness
Uma 10h33'19" 59d32'27"
Shawn Sykora (YogaShawn/Bcam)
Aqr 21h51'3" -2d4'29"
Shawn Thomas Cooley
Per 4h6'49" 43d53'28"
Shawn Thompson -683-
Uma 11h0'57" 63d30'45"
Shawn Trahan
Aql 20h3'47" 13d2'7"
Shawn Turcotte "Shamus"
Uma 12h21'59" 62d32'5"
Shawn Tyler Baltazor
Sco 16h13'0" -12d22'6"
Shawn William
Cep 22h12'24" 72d2'40"
Shawn William Hooper
Cap 21h53'30" -15d23'1"
Shawn Yonley
Sco 17h20'1" -41d2'39"
Shawn Zavadil
Cam 4h24'19" 63d10'41"
Shawna
Lib 15h12'10" -21d3'6"
Shawna
Sco 16h10'50" -11d9'3"
Shawna
Cap 20h22'31" -11d22'5"
Shawna
Per 3h27'6" 49d55'56"
Shawna
Uma 9h4'9" 48d36'59"
Shawna Alex Crawford
Cam 6h54'57" 65d18'24"
Shawna And Lance 7/05
Cyg 20h51'12" 46d16'40"
Shawna and Reid Beuchler
Her 17h59'49" 24d4'22"
Shawna Ann April
Cnc 8h36'40" 20d36'36"
Shawna Baumgardner
Psc 1h20'26" 16d48'30"
Shawna Bussell
Cap 20h46'9" -16d9'33"
Shawna Cyr
Cnc 8h46'7" 7d43'50"
Shawna Deering
Mon 6h24'6" 9d5'25"
Shawna Elizabeth Walker
Leo 10h26'21" 24d18'55"
Shawna Joy
Pho 0h45'11" -39d44'7"
Shawna K
Uma 10h26'28" 62d9'54"
Shawna Katherine Nixon
Uma 10h57'1" 70d2'27"
Shawna Kaye Courts
Lyn 7h8'9" 45d2'3"
Shawna Lee Huffman
Vir 12h33'19" -0d38'34"
Shawna Leeann Cleveland
Vir 14h46'36" 6d26'18"
Shawna Leigh
Uma 9h59'32" 52d20'41"
Shawna Liane
Leo 11h14'59" 16d41'52"
Shawna Linn
Lib 14h53'42" -1d44'36"
Shawna Lynn Johnston
Cas 1h36'46" 73d4'9"
Shawna M Trahan
Sco 17h17'10" -32d51'20"
Shawna Mae Little
Cas 1h12'9" 67d58'40"
Shawna Marie Crews
Cas 0h28'30" 51d16'1"
Shawna Marie Gardner
Cam 3h37'13" 65d25'41"
Shawna Marie Goss
Cnc 9h2'13" 18d12'20"
Shawna Marie McCowan
Cnc 8h50'38" 27d18'42"
Shawna Marie Shifflett
Lib 15h13'53" -15d55'57"
Shawna N Jones
Aql 19h20'19" -8d3'32"
Shawna Nichole Beeman
Cas 1h48'36" 61d20'49"
Shawna P G Medusa
Vir 13h13'43" 12d40'28"
Shawna Rae
And 23h54'52" 40d49'22"
Shawna Ray
Uma 11h27'9" 65d17'58"
Shawna Salazar
Col 5h58'28" -33d15'19"
Shawna Shiel
Gem 7h47'18" 14d25'51"
Shawna Sue's & Little Paxton's Star
Ori 4h48'19" -3d34'1"
Shawna Sundshine Star
Lmi 10h12'0" 35d56'9"
Shawna Wallace
Crb 15h53'25" 26d52'4"
Shawna Zeilstra
Vir 12h13'59" 3d14'26"
ShawnaDee
Psc 1h24'20" 17d43'40"
shawnalee
And 23h4'58" 48d21'30"
Shawnallyn
Uma 13h43'42" 54d34'11"
Shawnara
Uma 13h16'18" 61d27'59"
Shawna's Star
Psc 23h2'11" -0d59'42"
Shawna's Wishing Star
Uma 10h19'54" 53d58'49"
Shawnathon St. Jude
Ari 2h19'41" 25d5'12"
Shawnda
Lyn 7h44'47" 46d41'41"
Shawnda Lee
And 2h36'53" 49d16'26"
Shawndell Seabrook
Sco 16h50'11" -42d4'22"
Shawnee
Tau 4h20'21" 20d47'33"
Shawnee
Cnc 8h19'25" 9d55'45"
Shawnee
Vir 13h49'8" 2d44'1"
Shawnee Rodriquez
Aqr 20h57'12" 1d43'31"
Shawnee Rose Sloop
Leo 11h10'55" -2d10'35"
Shawnee Smith
Crb 16h4'19" 36d23'56"
Shawnell Mitchell
Lmi 10h30'59" 37d42'26"
Shawnette (Lioness) Johnson
Leo 11h28'54" 25d14'12"
Shawnette Moore
And 23h24'8" 42d41'10"
Shawney Nicolette DeLana
Tau 5h13'39" 18d1'50"
Shawn-FBS
Per 3h17'18" 31d18'17"
Shawnie & Mattie Forever
Gem 6h45'8" 26d22'31"
Shawnie's Star
Cam 5h2'6" 57d53'19"
Shawnna Renee Smith
Cnc 8h30'50" 9d51'58"
Shawnna Stroke
Psc 0h53'8" 33d29'1"
Shawn's Eternus Lux Lucis
Boo 14h29'1" 41d42'46"
Shawn's Smile
Her 17h10'2" 31d30'17"
Shawn's Star
Psc 1h22'16" 31d40'0"
Shawn's star
Vir 13h37'6" -1d27'41"
Shawn's Undercover Angel
Lyn 7h53'42" 37d30'3"
Shawntae Irene Wilkins
Vir 12h42'51" 7d16'32"
Shawntel L. Key
Cap 21h48'7" -22d59'23"
ShawntelleAdler011469
Cap 20h25'31" -19d38'47"
Shawntray Dwayne Kelly
Uma 10h42'8" 45d15'38"
Shawny's Star
Per 3h44'34" 45d44'13"
Shay
Tau 5h40'10" 19d8'50"
Shay
Uma 11h41'29" 52d42'12"
Shay and Deanna
Crb 15h45'7" 28d14'42"
Shay "belle e'toile" Schofield
Cnc 9h4'19" 30d50'24"
Shay Bo
Leo 11h22'45" 4d22'20"
Shay G Morgan
Her 16h45'37" 4d34'34"
Shay Grice
Cyg 21h33'56" 36d2'19"
Shay Groenewold
And 0h4'15" 45d23'25"
Shay L. Hulburt
And 0h25'54" 38d36'30"
Shay "Lenn"
Sco 17h16'43" -33d55'36"
Shay Patrick Keelan (Wibbly Pig)
Uma 9h53'41" 58d51'4"
Shay Ray I Love You, Happy 15th
Leo 11h10'28" -3d18'40"
Shay Shay
Gem 6h59'11" 15d37'4"
Shay W. Nelson
Gem 7h9'8" 21d26'46"
Shayan Merchant
Crb 15h18'57" 27d19'0"
Shayana Lyon
Uma 11h23'51" 33d10'54"
Shayda Zaerpoor
Gem 7h44'10" 20d53'23"
Shaye
Uma 10h18'43" 43d22'52"
Shaye Ellen Callanan
Sgr 18h37'33" -27d17'3"
Shaye Emily Robinson
Vir 12h48'31" 0d25'1"
Shayla
Vir 13h1'0" 10d5'28"
Shayla
Leo 11h6'10" 20d38'5"
Shayla
And 23h42'4" 44d46'56"
Shayla Ambrose
Ari 3h4'13" 24d22'2"
Shayla and Jimmy
Cyg 20h19'33" 48d34'50"
Shayla Caitlyn King
Pho 0h4'24" -57d19'23"
Shayla Diane Maher
Uma 10h36'31" 45d47'37"
Shayla Evelyn
Sco 17h6'26" -36d42'15"
Shayla Falon 7
Sco 17h22'43" -42d21'57"
Shayla LaDawn Moseley
Cnc 8h4'27" 26d19'54"
Shayla Marie
Cnc 8h46'2" 30d26'34"
Shayla Marie
Lib 14h52'41" -0d45'27"
Shayla Marie
Uma 8h24'9" 64d39'39"
Shayla Marie Christensen
Cnc 8h57'57" 29d37'1"
Shayla Rae
Aqr 23h33'37" -22d47'19"
Shayla "Red" Howard
Uma 9h15'43" 62d29'57"
Shaylan&Kevin-The Forever Love Star
Cap 21h43'47" -15d42'41"
Shayla's Eden
And 23h4'27" 49d18'37"
Shaylee Dalyn
Lyn 7h41'21" 36d17'39"
Shaylee Danielle
Cru 12h0'43" -62d31'31"
Shaylee Starr
Tau 4h10'5" 6d30'8"
Shayleigh Campbell Carlson
Aqr 22h18'14" -24d24'30"
Shaylene
Tau 5h11'54" 18d6'46"
Shaylie May Brown
And 23h5'32" 51d42'18"
Shaylin Allyssa Baum
Cas 1h40'9" 68d3'40"
Shaylin Rose
Uma 10h49'0" 51d12'25"
Shaylyn Love
Uma 11h58'34" 59d40'33"
Shayma Jaan and Noorie Baby
Lib 15h7'2" -25d0'56"
Shaymen
Peg 23h35'24" 31d38'11"
Shayna
Cnc 8h39'57" 23d57'59"
Shayna
Cma 6h40'54" -21d55'28"
Shayna Alexis Denn
Boo 14h17'55" 48d14'17"
Shayna Alise Casilla
Peg 22h15'35" 34d43'40"
Shayna Desjardins Born: Aug 24/1987
Uma 11h31'39" 56d48'15"
Shayna Kaplan
Vir 12h36'33" 2d9'57"
Shayna Kovacs
Lyn 7h25'1" 54d5'52"
Shayna Lee
Uma 9h7'27" 49d9'4"
Shayna Long
Cnc 9h14'35" 16d48'59"
Shayna Lynn
Tau 4h11'58" 26d58'30"
Shayna Lynn Becker
Aqr 22h17'52" 0d27'8"
Shayna Marie Baer
Tau 4h10'11" 13d34'55"
Shayna Nichole Davidson
Lmi 10h24'45" 32d11'17"
Shayna Nicole Brown
Ari 3h1'48" 13d24'30"
Shayna Ranae Linneen
Cnc 8h11'38" 12d3'35"
Shayna Susan Daugherty
Sgr 19h17'47" -14d20'13"
Shayna Zeitlin
Tau 3h42'1" 27d46'41"
Shayna's Star of Guidance & Wisdom
Sgr 18h29'45" -32d54'36"
Shayne
Lyn 7h31'12" 38d20'47"
Shayne Alpha
Cyg 21h54'34" 39d27'47"
Shayne and Jennifer Boswell
And 1h14'20" 41d46'45"
Shayne and Susan's Forever Star
Cyg 21h10'10" 48d8'36"
Shayne Andrew Isgrigg, Jr.
Aql 19h12'30" -1d35'27"
Shayne Burgess *Odd* ~ I love you!
Lib 14h52'33" -1d24'51"
Shayne Daniel Norman
Her 16h1'57" 18d25'14"
Shayne Del
Leo 10h55'28" 9d49'37"
Shayne Denise Perkins
Cnc 8h34'31" 7d23'4"
Shayne Elizabeth Sebold
Mon 7h3'40" -0d10'15"
Shayne MaKenna Hylleberg
Vir 13h19'43" 7d9'20"
Shayne Taylor Benz
Peg 22h46'13" 33d8'36"
Shayneh Paunim
Cas 1h2'28" 61d29'9"
Shayne's Star
Leo 9h33'48" 23d32'21"
Shaynika
Cnc 9h17'24" 15d29'32"
Shayonna Roberts
Cam 4h40'30" 55d47'43"
shayra matos
Vir 12h14'9" 13d2'16"
Shayrelle
Col 5h11'52" -39d37'11"
Shayzalyn Moani
Sco 16h53'32" -37d7'10"
SHAZ
Cnc 8h27'1" 10d9'35"
Shaz Hollitt - The Light of My Life
Sco 17h28'46" -35d30'15"
Shazaam - The Stokes Star
Cru 12h55'35" -56d43'43"
Shazi and Mare's Star
Cra 19h9'28" -42d31'37"
Shazia Shabnam Afgan
Cas 23h0'36" 54d55'14"
Shazia Tariq
Vir 13h17'43" 4d41'0"
Shazza Star
Cap 21h26'21" -20d47'19"
Shazzo
Psc 1h9'39" 25d58'26"
SHC
Lmi 10h41'55" 30d57'33"
She 5-0
Mon 7h18'52" -0d49'8"
She Caught the Katy Murdoch
Her 17h21'51" 36d23'59"
She: Sarah H. Eichner
Uma 13h32'49" 62d25'8"
She She
Vir 13h25'40" 12d10'8"
She' She' Yancy
Gem 6h52'20" 16d57'24"
shea
Gem 7h39'37" 21d33'49"
Shea
Uma 11h24'55" 36d47'23"
Shea
Uma 10h18'18" 44d40'10"
Shea
Uma 9h20'4" 45d36'40"
Shea Alexander Richards
Cnc 8h59'56" 7d34'45"
Shea and Coco
Cyg 21h16'50" 31d15'50"
Shea E. McDonough (Shea Babie)
Cyg 19h37'58" 35d48'55"
Shea Gilchrist
Sco 17h51'24" -37d50'35"
Shea Hunter
Ori 5h12'8" -8d47'39"
Shea Kelly Welden
Ori 5h34'10" 4d52'14"
Shea L. Gruenberg
Cam 4h1'10" 65d50'51"
Shea Lynn Permoda
Pho 0h6'42" -47d6'45"
Shea Martin
Ori 5h23'29" 0d51'31"
Shea Michael Norris
Ori 6h19'7" 14d37'57"
Shea Needham
Boo 14h12'36" 17d24'49"
Shea & Nik: Best Friends Forever
Aqr 22h55'59" -10d3'11"
Shea O'Shea
Leo 11h24'20" 15d45'47"
Shea Pearson
Crb 16h23'42" 35d1'9"
Shea Turima Coleman
Ari 2h9'44" 25d43'58"
Shea W Cleghorn
Lyn 6h18'48" 60d6'55"
Shea Weir a.k.a Suga
Cap 20h26'45" -14d10'30"
Shea Whitney
Cma 7h21'42" -15d13'44"
Shealin Theresa Doyle
And 2h23'30" 39d48'43"
Shealiopia
Sco 16h13'55" -15d56'15"
Sheal's Enchantment
Peg 23h16'43" 17d26'57"
Shealynn Rose Doody
And 2h1'6" 43d51'57"
Shealynnn
Ari 3h20'24" 27d30'1"
Sheard Star - 50 Golden Light Years
Leo 11h0'49" -0d11'45"
Sheareen
Ari 2h12'53" 25d46'36"
Shearer
Uma 13h22'46" 59d17'41"
Sheba
Uma 12h35'59" 58d16'28"
Sheba
Uma 8h28'53" 64d16'1"
Sheba
Cma 6h39'24" -25d6'38"
Sheba
Uma 11h30'18" 42d17'49"
Sheba Palmer
Tau 4h44'15" 24d21'10"
Shebly (I love you!)
Lyr 18h51'51" 32d1'57"

Sheeba
Crb 15h30'23" 31d28'56"
Sheebo
Psc 1h21'11" 4d5'3"
Sheela
Cap 20h34'17" -10d3'21"
Sheela
And 2h22'27" 46d30'59"
Sheela Casper
Ari 2h2'7" 23d33'0"
SHEELA VENKAT
Uma 11h36'55" 39d46'44"
Sheelagh's Star
And 0h12'57" 25d47'14"
Sheena
Lyr 19h8'55" 26d54'59"
Sheena
And 0h57'4" 35d43'18"
Sheena
Vir 13h18'37" -6d57'57"
Sheena
Cma 6h18'29" -30d43'54"
Sheena aka flippy
Sco 16h31'10" -42d6'46"
Sheena and Atif's Star
Ori 6h15'57" 10d39'32"
Sheena Angelique Presbaugh
Leo 10h33'12" 21d39'8"
Sheena Ann Smith
Sco 16h50'1" -32d25'40"
Sheena Anne Walker
Lib 14h47'55" -7d53'2"
Sheena Beauty and Fire
And 0h36'43" 35d18'29"
Sheena Bernard
Uma 13h53'59" 51d27'17"
Sheena Bush
Umi 15h33'39" 71d35'22"
Sheena Chan
Tau 3h38'10" 8d18'59"
Sheena Chmielewski
Aqr 21h32'38" 2d31'40"
Sheena Clowater
Cas 0h51'40" 48d4'3"
Sheena Dane Makinster's Europa
Sgr 18h27'14" -27d2'36"
Sheena Douglas
And 2h9'0" 42d8'57"
Sheena Hourtovenko
Ori 5h18'24" 9d13'8"
Sheena Lor
And 2h26'27" 50d33'49"
Sheena Marie
And 2h26'38" 48d45'21"
Sheena Marie Arsenault
Lib 15h25'58" -20d14'32"
Sheena Marie Gilbert
Lyn 7h42'52" 49d52'37"
Sheena Meredith
Ari 2h13'22" 22d53'11"
Sheena "My Love" Fellie
Vir 13h9'15" 2d20'51"
Sheena Parikh
Sco 16h16'36" -11d1'26"
Sheena Reneé Turner
Lyr 18h47'49" 39d54'46"
Sheena & Shaun Together Forever
Uma 9h32'13" 46d4'8"
Sheena Wanke
Uma 11h32'6" 43d11'31"
Sheeny
Uma 11h42'39" 62d7'29"
Sheeny Meeny
Lyn 9h14'30" 45d14'24"
Sheeolai
Per 4h4'35" 38d45'56"
Sheep<3Panda
Ari 2h51'49" 28d11'27"
Sheepuuu
Psc 1h19'18" 16d40'17"
Sheerce Mullins
And 0h4'38" 40d50'50"
Sheets
Cap 21h4'7" -16d51'32"
Sheeza
Gem 7h27'38" 26d32'19"
Shehrezad
Leo 10h57'18" 14d56'55"
Sheik Neerooa
Lyn 8h2'39" 38d35'23"
Sheikh Khalifa
Aql 19h19'15" -7d30'34"
Sheila
Umi 16h16'18" 70d58'50"
Sheila
Cam 5h56'24" 69d31'26"
Sheila
Cas 1h32'2" 60d24'13"
Sheila
Her 15h59'26" 49d30'58"
Sheila
And 0h9'33" 46d1'5"
Sheila
Vir 14h58'34" 4d31'59"
Sheila
Mon 6h49'1" 7d8'32"
Sheila
Gem 6h5'28" 22d32'23"
Sheila A. Maude
Ari 2h41'22" 29d4'25"
Sheila A. Oliver
Leo 11h15'48" 2d6'18"
Sheila Abshire
Lyn 7h58'26" 59d21'57"
Sheila and Dylan's Star
Mon 6h52'4" -0d36'37"
Sheila and Joe
Cnc 8h7'2" 11d49'47"
Sheila Ann Brash
Cas 0h47'1" 48d5'28"
Sheila Ann Hayne
Tau 4h20'30" 7d20'35"
Sheila Anne Chitwood
Tau 5h18'10" 25d36'42"
Sheila Aveyard
And 23h59'41" 40d8'16"
SHEILA B. - LEUOMGY SRIBHJAV
Sco 17h23'34" -42d22'58"
Sheila Barcello
Ori 6h1'47" -3d48'49"
Sheila Bell
Ori 5h44'49" 3d18'56"
Sheila Bienemann
Cas 1h47'33" 64d43'45"
Sheila Bouck
And 1h59'16" 38d58'32"
Sheila Brackett
Psc 1h8'28" 24d49'37"
Sheila Bridget
Cas 2h20'25" 65d58'11"
Sheila "Bug" Musgrave
Aqr 23h4'51" -8d28'56"
Sheila Burt
Cas 2h39'51" 72d48'3"
Sheila C. Clevenger
Lyn 8h3'54" 33d33'22"
Sheila Camponi Brown
Cas 1h28'0" 61d15'58"
Sheila Casady
Cas 2h24'44" 74d35'21"
Sheila Childs
Lyn 6h43'4" 53d49'38"
Sheila Connelly's Shining Star
Lyr 18h26'24" 32d52'43"
Sheila Cordoza
And 1h5'33" 43d4'22"
Sheila Cornerstone To Our Family
Uma 10h58'27" 53d38'26"
Sheila Croal
Cas 0h7'44" 58d37'38"
Sheila Debra Mallinger
Cam 4h59'11" 58d36'21"
Sheila Denise Inman 76
Psc 1h30'59" 20d35'3"
Sheila "Diamond" Cormier
Lib 14h52'41" -3d41'12"
Sheila Diane Berry
Uma 10h12'25" 45d21'22"
Sheila Dianne Rawls
Cnc 8h27'29" 18d0'56"
Sheila E. Haydock
Vir 14h4'34" -13d56'35"
Sheila E. Kiviat
Lib 15h14'55" -6d52'39"
Sheila Echols
Ori 5h31'12" 6d5'24"
Sheila Garcia
Com 12h30'9" 27d37'50"
Sheila Goode
Cap 20h18'21" -15d18'39"
Sheila Graf
Cas 1h7'47" 65d17'20"
Sheila & Graham Noble
Uma 10h5'22" 45d37'14"
Sheila Griffin Stewart
Uma 9h16'57" 66d5'25"
Sheila Hatch
Cas 23h30'22" 53d34'43"
Sheila Healey
Cas 22h57'28" 56d43'42"
Sheila Isabelle
Vir 14h22'27" 6d4'6"
Sheila J. Stevens
Aqr 22h12'4" -2d17'44"
Sheila James Birth Alpha "1-4-1956"
Crb 15h28'55" 28d2'40"
Sheila Jane 1947
Cru 12h28'8" -60d47'9"
Sheila/Jose (light of love) Nunez
Cas 0h58'16" 55d17'0"
Sheila Karen Caldwell
Psc 1h30'40" 14d55'15"
Sheila Kay
Cnc 8h46'47" 24d19'51"
Sheila Kay
Ari 2h32'14" 25d24'23"
Sheila Kay
Sco 16h35'9" -34d25'12"
Sheila Krauss
Cas 23h29'20" 59d13'10"
Sheila L. Cassidy
Psc 0h1'51" 5d44'17"
Sheila London
Ari 3h25'54" 27d17'26"
Sheila Lynn
Lyn 8h51'3" 41d23'42"
Sheila Lynn Baxter
Cnc 8h27'45" 22d52'29"
Sheila M. Blackwell
Leo 9h45'16" 23d14'57"
Sheila M Janiszewski
Cyg 21h41'13" 44d35'36"
Sheila Mae
Cnc 8h59'31" 14d49'37"
Sheila Marca
Tau 4h30'6" 26d39'57"
Sheila Margaret Margiotta
Cas 1h33'58" 64d13'6"
Sheila Marie Gleason
Leo 9h35'2" 23d32'14"
Sheila Marie Wiegman
And 0h51'47" 40d32'21"
Sheila & Markus
Cep 22h7'47" 68d44'46"
Sheila Mary Murray
Uma 12h40'28" 55d57'7"
Sheila Maureen Weissman
Tau 5h33'51" 20d14'41"
Sheila Max Wise
And 22h58'36" 47d35'30"
Sheila Maxine
Psc 1h9'32" 28d0'23"
Sheila Maxwell
Uma 11h48'32" 37d32'36"
Sheila McBroom
Psc 1h29'37" 17d25'17"
Sheila McCleery Kelly
Cas 1h35'50" 63d56'52"
Sheila Milano
Ari 2h13'28" 22d59'21"
Sheila Monson
Gem 7h13'20" 19d58'12"
Sheila Moriarty
Cas 1h33'46" 62d5'9"
Sheila Nanette Tate Willis
Leo 10h32'26" 20d51'33"
Sheila Naomi Revels
Lyn 7h58'25" 38d9'25"
Sheila Nick Brian Vis For Life
Uma 13h19'43" 55d43'11"
Sheila Nicodemo-McKee
Lyr 18h39'9" 34d45'1"
Sheila Nicole Jackson
And 0h46'23" 35d39'25"
Sheila Oestreich
Cnc 8h48'41" 32d16'18"
Sheila One
Uma 8h44'35" 52d23'39"
Sheila Renee
Cyg 21h31'25" 33d17'41"
Sheila Reynolds
Aql 19h11'6" 6d24'6"
Sheila Ritter
And 2h37'19" 41d3'45"
Sheila Rose Balthrop-Flynn
Cas 2h26'11" 67d18'33"
Sheila S
Tau 3h38'44" 16d4'2"
Sheila S. Garrett
Crb 15h55'41" 27d20'32"
Sheila S J Scott
Cas 0h24'38" 56d55'56"
Sheila Sellers
Cas 0h28'39" 58d36'14"
Sheila Shining Bright Star
Vir 14h28'33" 7d1'40"
Sheila Smoochie Boushee
Sgr 18h48'5" -30d48'5"
Sheila Soini Normansell
Ari 3h6'8" 29d35'55"
Sheila Sophia Martin
Uma 12h7'22" 59d5'53"
Sheila - Star of Motherly Love
Gem 6h35'30" 22d29'0"
Sheila Starr Cutting
Aqr 22h38'57" -6d57'11"
Sheila Sustrin
Lib 15h2'56" -1d26'1"
Sheila The Pretty Redhead
Leo 9h44'6" 27d32'41"
Sheila "The Saint"
Uma 8h57'15" 62d13'50"
Sheila Thomas Moehlenkamp
And 2h27'40" 39d19'15"
Sheila Tom
Sco 16h35'27" -37d24'27"
Sheila Tomas deCardenas
Psc 23h21'29" 3d45'2"
Sheila Tomkinson - Del's Mum
Per 4h31'48" 42d9'34"
Sheila Touchet
Lyn 9h16'59" 37d42'37"
Sheila Vanderwende
Sco 17h53'23" -35d58'36"
Sheila Vixen
And 23h35'17" 41d41'7"
Sheila Webb
Lmi 10h18'45" 28d51'35"
Sheila Wright
Cas 23h3'23" 54d56'36"
Sheila Zimmerman
Sco 16h9'30" -24d57'53"
Sheila, The Light of My Life
Aur 5h12'54" 42d34'10"
SheilaAndMaury27081955
Vir 13h30'56" -19d0'20"
Sheilagh Ruth
Psc 1h11'52" 11d48'2"
Sheilah Matilda Sarpong
Sgr 19h8'41" -25d32'10"
Sheilajim urtloml
And 22h59'48" 49d35'22"
Sheila's Bowtye
Cyg 20h24'12" 43d57'59"
Sheila's Heavenly Body
Cas 0h38'44" 62d33'24"
Sheila's Perfect Star
Ari 2h6'36" 23d4'15"
Sheila's Star
Per 3h13'43" 41d26'4"
Sheila's Star
Lyn 7h44'59" 36d10'18"
Sheila's Star
Lyn 7h31'59" 35d41'54"
SheilaShining
Cap 20h39'14" -20d50'2"
Sheilastar
Peg 21h53'12" 16d51'58"
Sheilda Lee
Tau 5h51'9" 24d13'12"
Shekinah Colson
Sgr 18h22'8" -32d15'18"
Shelagh Angharad Colledge
Gem 7h44'31" 15d8'9"
Shelana
Uma 10h2'17" 43d1'57"
Shelbert 25
Uma 11h21'36" 65d10'15"
Shelbi Rose
And 1h39'39" 44d5'58"
Shelbie Nicole Petersen
Vir 14h35'44" -0d0'49"
Shelby
Aql 19h47'37" -0d4'7"
Shelby
Cam 4h15'5" 66d40'31"
Shelby
And 23h25'47" 52d58'50"
Shelby
Psa 22h16'50" -32d24'15"
Shelby
Cyg 20h5'28" 36d3'19"
Shelby
Cyg 21h47'57" 37d20'12"
Shelby
Cyg 20h52'46" 46d1'40"
Shelby
Cet 2h29'42" 7d50'15"
Shelby
Tau 5h44'21" 18d17'48"
Shelby Ann
Lyn 7h22'20" 46d11'33"
Shelby Anne Alexander
Uma 11h24'4" 68d20'12"
Shelby Archer
Umi 15h43'11" 71d17'47"
Shelby Bailey
Aqr 21h57'57" 0d7'3"
Shelby Baird Shines Forever
Per 4h25'19" 43d29'1"
Shelby Boo
Cmi 7h27'45" 7d24'37"
Shelby Brianne Faller
And 23h4'59" 46d15'1"
Shelby C Nadolsky, My Snookumberry
Sgr 19h14'43" -16d45'41"
Shelby Cahill
Cap 20h14'15" -18d40'24"
Shelby & Chris
Vir 12h31'44" 8d45'37"
Shelby Cohen
Uma 9h48'54" 62d34'19"
Shelby Cooper Tritz
Ori 6h13'2" 11d42'6"
Shelby Daeffler's Wishing Star
Cas 0h23'40" 54d43'47"
Shelby Deann Wetter
Cam 3h59'48" 54d52'26"
Shelby Denson
Lyn 7h11'7" 58d53'53"
Shelby Elizabeth Blackwood
And 2h20'27" 50d3'16"
Shelby Elizabeth Ratliff
Psc 23h35'41" 4d57'13"
Shelby Erin Watts
Uma 10h32'55" 47d22'0"
Shelby Faye McDonald
Lyr 18h46'56" 38d26'10"
Shelby Fraiser
Lyn 7h46'43" 38d16'22"
Shelby Hand
Sgr 19h3'44" -18d53'46"
Shelby Harlan
And 2h38'31" 44d42'24"
Shelby Hojio
Cap 21h46'49" -11d22'28"
Shelby Hope Kamminga
Cnc 8h16'25" 19d20'53"
Shelby J Vigil
And 23h17'40" 41d45'30"
Shelby Jean Etheridge
Uma 8h24'0" 64d5'39"
Shelby Jeanne Dunlap Whitman
Sco 16h43'4" -33d53'52"
Shelby & John's Love
Tau 4h0'46" 24d21'31"
Shelby Johnston
Cam 4h5'20" 66d12'25"
Shelby Jones Pollock
Aql 19h22'2" 3d7'14"
Shelby Kathryn Cooper-Wallace
Mon 6h40'25" 7d2'31"
Shelby Kay Powell
Vir 13h25'49" 13d5'34"
Shelby Kelin
Vul 19h22'58" 23d16'14"
Shelby L. Harwell
Gem 8h0'54" 29d58'12"
Shelby LaFon-Hales
Sgr 17h50'32" -29d17'48"
Shelby Leigh
Uma 11h30'16" 57d1'58"
Shelby Leigh
Vir 13h31'25" -15d40'56"
Shelby Leigh Smith
Cas 1h19'9" 60d59'41"
Shelby Lou
Cap 21h58'28" -9d39'10"
Shelby Luckey
Ori 5h22'11" 6d44'0"
Shelby Lynn
And 0h51'30" 39d19'18"
Shelby Lynn
Mon 6h54'0" -0d19'31"
Shelby Lynn Anderson
Gem 6h52'30" 34d19'14"
Shelby Lynn Belk
And 0h19'41" 41d59'46"
Shelby Lynn Hawkins
Vir 13h25'7" 11d3'8"
Shelby Lynn Jones
Uma 9h50'26" 47d0'28"
Shelby Lynn Steward
Ori 5h32'56" -0d28'37"
Shelby Lynn Thorne
Cyg 20h2'17" 40d54'43"
Shelby Lynne Rose Acosta
Cnv 13h4'18" 36d55'21"
Shelby Marie
And 2h14'40" 45d41'3"
Shelby Marie
Psc 1h28'46" 26d32'42"
Shelby Marie 7-15-89
Cas 1h0'32" 60d32'27"
Shelby Marie LaSalle
Crb 15h53'32" 27d49'43"
Shelby *Mom's Everlasting Star*
And 1h56'27" 38d49'45"
Shelby Nakamura
Eri 4h13'57" -22d34'35"
Shelby Nichole Magee
Cam 5h47'34" 60d7'36"
Shelby Nicole Carr
Cnc 8h24'18" 9d29'38"
Shelby Nicole Horrighs
Lib 15h38'1" -7d10'2"
Shelby Nicole Howard
Tau 5h8'38" 17d31'47"
Shelby Noreen Dyson
Cnc 8h52'3" 20d17'16"
Shelby Outlaw & Walter Bauer
Cyg 20h58'43" 33d36'53"
Shelby Paige Mastrogiovanni
Gem 7h25'47" 26d20'59"
Shelby Perkins
Sco 16h50'55" -38d24'47"
Shelby Pierson
Umi 13h12'36" 72d59'46"
Shelby Rae Rattray
Aqr 21h47'56" -2d36'55"
Shelby Ray Schaefer
Sgr 19h42'11" -12d51'43"
Shelby Reed Carver
Mon 6h51'31" -0d7'11"
Shelby Renee Cameron
Sgr 18h4'30" -20d42'48"
Shelby Renee Keehan
Psc 2h2'19" 7d10'19"
Shelby Richter
Aqr 22h54'34" -15d58'47"
Shelby & Scott Robinson
Cyg 20h21'10" 46d52'44"
Shelby Stark
Lyn 8h8'36" 34d35'28"
Shelby Starr
Cmi 7h30'49" 8d35'3"
Shelby Suzanne
Uma 11h9'18" 42d47'14"
Shelby Taylor
Crb 15h37'44" 30d48'35"
Shelby Taylor Miller
Cyg 19h48'41" 33d18'22"
Shelby Vien Hutchins
Tau 4h46'45" 21d47'43"
Shelby Wilson
Tau 5h49'24" 21d18'56"
Shelby Wynne Boatwright
Uma 12h5'9" 57d36'34"
Shelby Yan Duell
And 23h5'47" 50d42'53"
ShelbyBaby
Lyn 6h53'41" 51d39'14"
Shelbylee Ann Tolley (Neemoo)
Aql 19h52'31" 10d38'31"
Shelby's Beautiful Star
Gem 7h3'12" 14d50'1"
Shelby's Star
Leo 10h19'31" 8d13'34"
Shelby's Star
Sco 17h44'17" -35d52'30"
Shelbys Star Leo
Leo 11h36'9" -1d46'51"
Shelda L. Orioles
Uma 11h47'23" 34d39'8"
Sheldon Bobrow
Cep 20h35'36" 65d9'4"
Sheldon Cvont
Tau 5h26'10" 27d1'52"
Sheldon Hine
Aqr 21h27'50" -0d11'35"
Sheldon Isaac
Uma 10h19'29" 72d4'50"
Sheldon Krupp
Psc 1h39'27" 17d55'32"
Sheldon Neil Stronach
Her 17h10'1" 28d49'34"
Sheldon Riley Parris
Ari 2h15'21" 11d23'56"
Sheldon Terry "The Shan"
Sgr 19h5'1" -30d32'38"
Sheldon's Wishing Star
Lmi 10h47'21" 28d8'32"
Shelene Adana Nichol Smith
Sco 16h8'25" -15d33'47"
Shelfy
Uma 9h31'14" 62d52'47"
Sheli
Vir 12h30'45" -0d12'12"
Shelia
Uma 10h27'5" 56d23'0"
Shelia A. Page
Sgr 19h46'13" -12d28'45"
Shelia and T.J Fitzgerald
Cyg 20h43'52" 50d19'52"
Shelia(Angelface)Burt
And 0h13'0" 29d2'26"
Shelia B. Soloe
Uma 10h47'1" 45d29'30"
Shelia Beck
Sgr 19h12'33" -13d7'36"
Shelia Diana Favorite
Cnc 9h7'31" 21d1'28"
Shelia Hotard
Mon 6h32'27" 8d36'34"
Shelia Rei's GMS Star
And 2h16'11" 46d46'58"
Shelia Ross
Ari 2h50'5" 27d27'23"
Shelia Wood
Sco 17h22'34" -32d57'5"
Shelina
Col 6h19'36" -40d54'38"
Shelit777
Per 3h47'2" 32d15'40"
Sheliz Ismail
Cas 23h26'47" 55d29'19"
ShelJac
Cyg 20h20'9" 55d40'18"
ShelJacJoe
Tri 2h6'49" 33d13'14"
Shell
And 23h38'27" 48d2'50"
SHELL
Vir 12h42'56" 2d22'21"
Shell Bell
Lyn 7h50'2" 58d23'28"
Shell Bell
Lib 15h2'0" -14d3'12"
Shell Campbell's Star
And 23h27'10" 42d24'0"
Shell The Boss
Umi 14h34'38" 86d39'23"
Shell, your my one and only love 4T
Aql 19h50'4" -0d8'58"
Shella Rae Embree
And 23h29'45" 42d24'14"
Shellbabes
Dra 18h49'2" 65d19'11"
Shellbee Lynn Saenz
Per 3h25'30" 54d33'8"
Shellbell
Sgr 19h14'50" -21d48'41"
Shellean
And 23h20'6" 48d19'14"
Shellee Lyn McKeon
Aqr 20h50'37" -13d32'55"
Shellee Wileman
Lyr 19h13'12" 27d22'46"
Shelley
And 0h19'37" 33d11'27"
Shelley
Gem 6h20'46" 21d59'58"
Shelley
Vir 13h18'30" 12d37'29"
Shelley
Cas 0h48'39" 57d48'14"
Shelley
Lyr 18h49'22" 43d39'23"
Shelley
And 2h36'41" 45d46'37"
Shelley
Lyn 9h11'45" 34d32'15"
Shelley
Cap 20h33'42" -20d18'17"
Shelley
Cas 1h28'2" 67d48'48"
Shelley
Cam 7h1'41" 74d24'28"
Shelley
Cra 18h54'26" -38d54'45"
Shelley A. Mastro
Uma 11h33'32" 31d12'40"
Shelley Adrienne Sink
Leo 9h48'22" 8d50'48"
Shelley and Halle Carr
Gem 7h11'11" 16d30'9"
Shelley and James' star of devotion
Her 17h26'11" 23d53'36"
Shelley & Andrew
Cyg 21h42'17" 54d26'44"
Shelley Ann Hoskings Zambardi
Cas 0h18'43" 54d0'16"
Shelley Ann Williams
Leo 11h32'20" 6d44'56"
Shelley Anne Burns
Uma 10h9'48" 48d20'52"
Shelley Bonelli
Cmi 7h25'38" -0d2'12"
Shelley Bursik
Dra 15h56'46" 52d35'14"
Shelley Carner
Tau 4h34'26" 24d58'13"
Shelley Cohen
Cyg 19h45'39" 33d39'11"
Shelley Cousins Superstar
Leo 10h52'39" 7d3'23"
Shelley Danise
Uma 9h29'57" 47d37'9"
Shelley Dowd
And 0h16'46" 26d57'8"
Shelley Elaine Browning
Uma 9h26'5" 44d40'15"
Shelley Esposito
Dra 18h57'38" 57d0'49"
Shelley Faye O'Neill
Psc 1h18'49" 23d29'0"
Shelley Fenton-Fox
Vul 20h24'30" 23d18'25"
Shelley Flores
Ari 2h47'24" 24d43'27"
Shelley Grace
Ari 2h32'43" 20d32'29"
Shelley & Greg Mepyans
Cyg 20h26'32" 52d3'54"
Shelley H Price
Uma 10h54'0" 41d7'11"
Shelley Jean Clay
Uma 10h21'18" 66d50'30"
Shelley Jefferson
Cas 22h59'4" 55d38'9"
Shelley Jennifer Hobbs
Ari 1h51'49" 20d2'19"
Shelley Jo
Cas 2h10'40" 59d44'2"
Shelley Joyce Kensler
Peg 22h10'2" 33d43'24"
Shelley Kathleen Anderson Powell
Lib 15h27'46" -23d55'40"
Shelley Kohn
Cru 12h25'25" -63d15'21"
Shelley Kugi
And 1h12'51" 41d34'38"
Shelley L Meacham
Aqr 21h45'22" -7d29'22"
Shelley L. Schruff
Lmi 10h13'58" 28d4'31"
Shelley Lee
Sco 17h51'48" -35d38'22"
Shelley Lynn
Lyn 6h38'22" 57d43'27"
Shelley Lynn Hamel
Lib 14h54'11" -1d20'35"
Shelley Lynn Szczepanski 11-14-1972
Sco 17h21'2" -42d56'57"
Shelley Lytle
Crb 15h26'38" 26d34'20"
Shelley M Grubbs
Cyg 19h51'1" 58d26'35"
Shelley Marie Ford
Peg 23h7'16" 17d1'24"
Shelley Marie Lauer
Sco 16h59'54" -38d51'28"
Shelley Marielle Guerin
Umi 14h57'48" 74d30'9"
Shelley Mefford
Uma 10h48'4" 58d57'2"
Shelley My Kansas Star
Lmi 10h27'55" 33d12'0"
Shelley & Nathan Hancock
Uma 11h4'8" 34d7'8"
Shelley Osborn
And 1h11'55" 43d35'21"
Shelley Oteachus
Leo 9h39'36" 25d11'8"
Shelley & Peter Nussbaum
Lyn 7h46'27" 36d46'26"
Shelley Pleace
Cyg 20h42'8" 45d36'53"
Shelley R. Abraham
Vir 13h27'2" -16d31'50"
Shelley Rae
Cap 21h17'8" -15d59'59"
Shelley Ray
Tau 4h29'39" 19d19'28"
Shelley Renee
Psc 23h56'10" 1d28'21"
Shelley Rinkle
Cyg 19h54'48" 30d59'41"
Shelley Robinson
Uma 12h32'18" 61d24'37"
Shelley Schmersahl
Ari 2h16'16" 26d31'27"

SHELLEY SCHORSCH
Cnv 12h47'6" 39d2'49"
Shelley (Sheesh)
Cas 0h23'13" 53d54'41"
Shelley Smith
Vir 12h46'41" 5d3'56"
Shelley Stapleton
Ari 3h1'17" 28d11'42"
Shelley Stockwell-Nicholas
Psc 0h48'26" 14d9'1"
Shelley Tillman
Uma 10h54'59" 46d47'27"
Shelley Tokiko Fujiwara
Tau 4h44'8" 28d47'18"
Shelley Tye Dyed
Ari 2h59'2" 25d33'36"
Shelley W. Patterson
Mon 7h30'51" -6d30'19"
Shelley, Deb, Carolyn & Laura
Uma 13h43'32" 57d52'8"
Shelley's piece of heaven
Pho 1h14'41" -46d32'8"
Shelley's Sparkle
Cap 21h32'29" -24d41'38"
Shelley's Star
Col 5h39'21" -35d16'37"
Shelli
Cyg 19h53'11" 56d44'46"
Shelli and Frank's place
Cyg 21h26'10" 31d14'58"
Shelli Coman
Eri 3h58'53" -0d17'39"
Shelli Marie
And 1h38'45" 46d34'42"
Shelli Ulrich
Cnc 9h8'15" 30d50'45"
Shellie
And 23h22'19" 41d41'24"
Shellie Amanda Limsky
Lib 15h17'44" -5d51'56"
Shellie D.
Crb 15h34'2" 31d41'32"
Shellie Danielle Cornette
And 2h6'22" 42d12'13"
Shellie Rene Klocker
Ori 5h18'57" 15d3'6"
Shellie Small
And 0h46'26" 36d57'33"
Shellie Winkler
Cap 21h31'9" -16d37'59"
Shellissa Rae
And 23h16'17" 39d22'24"
shell-n-be
Leo 11h37'40" 27d0'58"
Shells Family
Uma 11h20'20" 48d48'36"
Shell's Tiara
Sgr 19h51'7" -11d57'18"
Shells' Wishing Star
Crb 15h22'25" 29d17'37"
Shellshe
Col 5h45'34" -34d15'34"
Shellstar
Lib 15h12'38" -19d6'9"
Shelly
Cap 21h42'25" -14d6'25"
Shelly
Vir 14h0'39" -14d1'12"
Shelly
Cam 3h22'45" 61d22'56"
Shelly
Uma 11h42'43" 50d37'31"
Shelly
And 23h3'3" 51d39'33"
Shelly
Cas 0h22'11" 50d44'57"
Shelly
And 1h34'51" 45d35'43"
Shelly
And 1h0'16" 45d5'9"
Shelly
Gem 6h49'44" 31d4'6"
Shelly
And 2h38'29" 42d50'15"
Shelly 3184
Cap 20h35'44" -17d58'53"
Shelly A. Badger
Uma 12h29'54" 62d23'30"
Shelly Alexander
And 23h12'6" 43d52'35"
Shelly Allen
Cap 20h27'27" -12d2'51"
Shelly Alligood
Lyn 8h43'42" 45d52'58"
Shelly and Nucy
Cyg 19h48'54" 33d36'23"
Shelly and Royee Forever
Aqr 23h29'8" -13d14'4"
Shelly and Scot McCulloch
Col 5h38'43" -36d14'9"
Shelly Ann
Vir 12h17'17" 12d0'27"
Shelly Ann Simpson
Tau 5h41'54" 19d2'29"
Shelly Baker
Ari 2h34'54" 17d46'30"
Shelly Barber
Per 3h50'54" 32d54'55"
Shelly Bell
Lib 15h6'40" -6d26'0"
Shelly Bold
Uma 10h46'29" 56d39'42"
Shelly Bordas
Vir 12h54'56" 9d45'59"
Shelly Dawn
And 1h40'37" 42d32'0"
Shelly Dawn Burnside Maskus
Dra 20h5'15" 71d32'38"
Shelly Duelley
Cam 7h29'58" 61d36'6"
Shelly Eddy
Leo 11h13'0" 23d57'31"
Shelly Hughes
Uma 11h11'6" 53d46'31"
Shelly Irene Potter
Lib 15h7'7" -9d57'52"
Shelly Jean
And 1h9'12" 38d11'7"
Shelly Jennifer Emanuel
Ori 6h1'34" 21d16'36"
Shelly & John Nelson
Sge 19h49'15" 18d27'27"
Shelly Kaplan
Lib 15h47'32" -12d51'10"
Shelly Kay Tschirhart
Cas 23h22'53" 54d56'21"
Shelly Komer Zechory
Leo 11h3'12" 21d48'43"
Shelly Krim
Gem 7h29'17" 32d2'50"
Shelly Lorraine Nelson
Crb 16h9'49" 37d52'3"
Shelly Lynn #12
Vir 13h14'17" 5d30'48"
Shelly Lynn Ray 555
Uma 11h20'34" 44d35'31"
Shelly M. Tidwell
Vir 13h16'1" 4d54'38"
Shelly Malcolm
Uma 10h12'37" 45d5'44"
Shelly "Mama" Williams
Psc 23h46'21" 1d10'42"
Shelly Marie
Sco 16h27'40" -32d55'7"
Shelly Marie Avila
Ari 2h37'48" 17d51'42"
Shelly Marie Garretson
Uma 10h23'27" 54d27'59"
Shelly Mattson
Cnc 9h9'37" 31d26'5"
Shelly Moulton
Crb 15h45'42" 26d11'4"
Shelly Myers
Tau 4h27'17" 22d4'53"
Shelly O'Reilly
Ori 4h50'51" 1d55'28"
Shelly "Our Angel" Smits
Pho 0h7'14" -42d48'11"
Shelly Perkins
Ari 2h25'46" 27d13'15"
Shelly R. Reed
Cyg 21h37'42" 42d22'5"
Shelly R. Robertson
Cap 21h17'54" -14d30'42"
Shelly Rae Cousineau
Lyr 18h47'36" 31d14'41"
Shelly Renee
Gem 6h58'27" 25d36'35"
Shelly Renee Chernoff
Tau 4h46'34" 29d13'56"
Shelly Stetson
Lib 14h49'24" -6d30'19"
Shelly Stevens Williams
Cas 0h27'23" 54d24'19"
Shelly Turnbull
Ari 2h39'12" 12d31'52"
Shelly Vogelhut
Aqr 22h14'50" -0d55'28"
Shelly Wagner
Cnc 9h8'59" 15d38'12"
ShellyAnn
Gem 6h47'52" 20d13'3"
ShellyJeff2005
Cyg 20h23'9" 58d31'50"
Shellylee
Ori 5h41'17" -2d31'39"
Shellys Forever Star
Cam 4h49'8" 63d7'43"
Shelly's Magic
And 1h0'45" 45d41'31"
Shelly's Nose
Lib 15h3'43" -17d13'22"
Shelly's Star
Leo 9h39'36" 27d19'45"
Shelly's Tao
Aql 18h51'29" 9d51'19"
Shelly's Voice In Heaven
Cyg 20h38'43" 56d13'37"
Shellz
Cnc 8h14'38" 16d38'49"
Shelton D. Bailey
Leo 11h20'26" 11d39'17"
Shelton Doerksen
Uma 11h56'55" 43d6'26"
Shelton Edwards Green
Cyg 20h31'1" 39d55'19"
Shelton Howard Baker
Psc 0h32'28" 5d51'31"
Shelton & Marie Smith
Lyr 18h56'52" 33d23'11"
Shelva Jane Fleming Bieber
Lib 15h23'5" -6d14'31"
Shelvye W Foutz
Tau 4h18'30" 11d48'58"
Shely Bean
Lmi 10h31'1" 32d41'19"
Shemariah Ann Zacher
Aur 5h56'55" 43d5'47"
shemesh mâchôwl
Aqr 22h1'50" -14d54'9"
Shen
Cnc 8h14'54" 25d56'11"
Shena
Gem 7h26'55" 25d47'35"
Shenandoah Lorraine Lone Eagle
Sco 16h12'27" -19d30'24"
Shenanigans
Per 2h51'4" 43d40'46"
Shen-Chia Chu
Vir 12h3'30" 6d55'46"
Shendil Pyarilal
Tau 4h44'8" 18d33'2"
Sheneate
Aqr 21h50'52" -0d30'33"
Sheneice Hinton
And 2h20'2" 41d25'23"
Shenika Swint
Com 13h5'24" 26d7'36"
Shenita & Jose
Mon 6h55'36" 1d48'24"
Shenita Mitchell
Peg 22h32'24" 9d4'3"
Sheniyah Denise Hackworth-Shavers
Lib 15h9'4" -6d42'20"
Shenkabbleem
Uma 11h40'39" 44d15'46"
Shenña
Crb 16h9'56" 34d58'14"
Shennon Bell
And 1h43'56" 46d39'41"
Shenoy Godwin
Leo 11h21'52" 2d56'11"
Shen-Shen
Uma 9h17'59" 56d6'2"
Sheona
Lib 15h59'33" -17d7'47"
Shep
Cep 22h58'18" 71d43'28"
Shep
Lyr 18h41'35" 37d34'33"
Shepard
Uma 11h14'12" 45d57'15"
Shepard's Heaven for Hope
Uma 10h20'21" 66d10'12"
Shepherd's Star
Uma 13h19'34" 59d54'59"
Sheppard B. Fowler
Lyn 8h45'12" 40d43'36"
Sheppard-Miller Ognora
Crb 16h10'35" 38d49'1"
Sher Bear
Cas 23h31'45" 59d45'2"
Sher L. Cupolo
Lyn 8h32'32" 47d10'20"
Sher & Mike Z
Ori 6h0'21" -0d37'36"
Shera Smith
Cap 21h26'59" -25d31'52"
Sherali Valji
Per 3h7'25" 40d29'58"
Sheran McCants
Lyr 19h8'51" 27d9'56"
Sheray K. Morrison~Sunshine's Star
Cas 1h40'35" 61d39'27"
Sheray Schultz
Aqr 23h9'23" -8d27'9"
sherbert
Sgr 18h11'16" -22d30'56"
Sherdeane Kinney, LBWB
Dra 17h15'2" 54d0'42"
Shere
Aql 20h1'0" -9d24'50"
Sheree
Sgr 18h5'6" -25d27'52"
Sheree
Sgr 19h46'53" -22d53'34"
Sheree
And 0h40'54" 27d26'37"
Sheree
And 0h19'6" 36d41'1"
Sheree
Uma 10h50'27" 51d1'53"
Sheree Belinda Mueller
Psa 23h0'25" -33d13'54"
Sheree Estrella
Uma 11h7'21" 38d21'0"
Sheree Fetkin
Com 12h35'46" 28d3'2"
Sheree G Burton & Makena G Burton
And 0h41'38" 39d29'44"
Sheree Joann Sherman
Cyg 20h23'1" 51d34'49"
Sheree Jolley
Sgr 19h1'17" -21d12'7"
Sheree Martarano
Cnc 9h6'12" 31d47'57"
Sheree Matarano
Cnc 8h59'55" 29d26'48"
Sheree the Beautiful
Cas 1h7'15" 60d33'5"
Shereen
Cep 22h51'53" 65d3'59"
Shereen
Lib 15h20'42" -16d17'48"
Shereen
Pyx 8h44'23" -29d41'25"
Shereen & John Barouky
Cru 12h25'24" -57d44'50"
Shereen Khan
Lib 14h50'16" -0d56'40"
Sheree's Beacon
Peg 21h52'0" 12d57'18"
Sheree's Star
And 2h9'47" 46d26'38"
Sherell
Cap 20h18'33" -13d40'54"
Sherene John
Lib 15h4'43" -19d46'3"
Sheri
And 0h33'32" 43d3'1"
Sheri
Ori 6h10'11" 6d10'24"
Sheri Amor
Gem 7h30'1" 20d25'48"
Sheri Ann
Leo 11h29'31" 17d5'54"
Sheri Ann
Cnc 8h2'16" 25d59'0"
Sheri Ann Griffith
Lyr 19h16'22" 29d6'54"
Sheri Annette Wood Craft
And 1h11'27" 41d35'39"
Sheri C. Nilsen
Dra 16h1'24" 53d6'37"
Sheri Chapman
Sco 17h53'10" -36d0'33"
Sheri Creekmore
Crb 15h47'44" 29d58'34"
Sheri D
Gem 6h29'30" 21d33'19"
Sheri & Drew Trimmer Forever Love
Ori 5h55'59" 3d5'44"
Sheri E. Capes
And 0h11'11" 32d58'28"
Sheri Ellen Jobe
Lyn 7h38'47" 56d23'24"
Sheri Grey Price
Vir 12h48'48" 7d3'2"
Sheri Hamilton - Our Love Star!
Sco 16h40'11" -41d27'54"
Sheri Henthorn
Psc 0h51'47" 16d40'58"
Sheri Hogan
Cas 1h23'47" 52d21'9"
Sheri' Holmes
And 2h16'22" 44d17'10"
Sheri K. Herman
Per 2h43'53" 51d19'27"
Sheri L. Coleman
Vir 13h10'59" 6d52'55"
Sheri L. McGough
Tau 3h59'20" 11d33'18"
Sheri Lee
Cap 20h16'13" -25d18'39"
Sheri Lee Burgess
Cnc 9h14'27" 32d57'14"
Sheri Leigh Logue
Lmi 10h24'32" 39d12'5"
Sheri Liska
And 0h46'57" 36d17'38"
Sheri Louise Traina
Uma 10h10'30" 49d48'57"
Sheri Lynn
Vir 14h38'23" -2d47'21"
Sheri Lynn Wood
Tau 5h19'6" 27d18'38"
Sheri Mackay
Cas 1h31'59" 67d6'17"
Sheri Manda-Lynn
Sco 16h35'53" -28d31'34"
Sheri McCormack
And 0h46'42" 45d5'27"
Sheri McCoy
Vir 13h38'41" -4d49'16"
Sheri Morshedi
And 0h21'1" 44d52'52"
Sheri Peplau
Cap 21h7'52" -17d18'42"
Sheri René Collins
Uma 11h59'52" 33d54'32"
Sheri Ridgeways Star
Aqr 23h1'26" -7d12'24"
Sheri, with love from Michael
Tau 4h19'26" 10d20'0"
Sheridan
Aql 19h49'54" 0d15'53"
Sheridan
Aqr 22h23'16" -0d5'14"
Sheridan Brooke
Psc 1h1'48" 25d13'21"
Sheridan Elizabeth Koch
Gem 7h39'49" 33d57'29"
Sheridan Lynn
And 2h6'28" 40d52'37"
Sheridan Reese Crisafulli
Scl 0h41'57" -32d58'14"
Sheridan's Light
Sgr 18h20'24" -22d31'5"
Sheridan's Star
Cru 12h44'40" -58d55'0"
Sheridon Robertson-Sands
Leo 9h40'37" 12d0'43"
Sherie and Phillip Catlett
Gem 7h51'10" 16d8'34"
Sherie Ann Fabrizio
Tau 3h44'1" 27d50'20"
Sherie Ann Miller
And 2h10'20" 41d19'49"
Sherie Bear Friedlander
Aqr 21h32'50" 1d57'13"
Sherie Graham
Sgr 18h44'15" -16d29'3"
Sherie Louise Johnstone
Cru 11h57'20" -62d43'53"
Sherie's Butterfly Star
Lib 15h43'2" -19d58'52"
"Sheriff" Charlie Wells
Sco 16h16'50" -16d39'24"
Sheriff Chris Daniels
Per 3h30'45" 33d24'31"
Sheriff Koth's Star of St. Thomas
Per 3h29'6" 51d47'22"
Sheriff Max
Ari 2h32'57" 26d11'29"
Sherilicious
Leo 9h38'58" 27d21'9"
Sherilyn
Gem 6h51'48" 13d17'33"
Sherilyn Adams
And 23h20'46" 48d24'23"
"Sherilyn Janese Derstine"
Ori 4h51'5" -0d29'35"
Sherilyn Sullivan
And 1h59'33" 41d36'43"
Sherin
Tau 3h51'59" 23d24'13"
Sherine Sadeghi
Ori 5h13'17" 12d54'33"
Sherinne
Crb 15h38'8" 30d54'35"
Sherious
Cyg 21h15'44" 47d28'20"
Sheris AspenStar
Psc 0h45'58" 9d50'22"
Sherisa Marie Bolt
Lib 15h40'38" -4d26'27"
Sherken-40
Umi 13h38'42" 72d49'20"
Sherleen
Lib 15h24'40" -9d40'31"
Sherlock Gaydos
Ori 5h37'17" 12d1'35"
Sherlock Herbert Griffith
Per 3h28'18" 48d44'59"
Sherly
Uma 9h27'40" 67d45'51"
Sherly Halina Huallpa
Ori 5h43'54" -3d24'47"
Sherlyann
Cas 1h33'27" 67d5'29"
Sherm, my love forever. Doggie
Cyg 19h26'44" 52d19'13"
Sherma Kaye Johnson Andrus
Cam 6h14'28" 78d57'3"
Sherman
Uma 13h52'36" 53d47'49"
Sherman and Kiyo Morii Elwood
Uma 10h55'34" 63d25'52"
Sherman & Jane Shay
Lyr 18h49'53" 35d47'59"
Sherman Lee & Lottie Mildred Greer
Vir 13h43'32" -18d39'47"
Sherman R. McGrew
Tau 4h18'56" 27d33'0"
Sherman "The Truth" Suter
Her 17h52'38" 28d58'9"
Sherman Wilson
Cep 21h46'31" 56d27'8"
Sherman-Shostak
Lyr 18h30'26" 34d1'45"
Shermia McHenry
Umi 15h7'50" 70d14'44"
Sher-Nik
Mon 7h34'58" -0d42'7"
Sheron
Her 18h19'32" 15d33'35"
Sheron & Curtiss Gardner
Cas 0h51'30" 60d18'45"
Sheroy - 15.5.1951
Cyg 19h40'9" 32d36'1"
Sherradley
Lyr 18h19'8" 32d46'14"
Sherrae Maillet
Uma 11h12'48" 69d55'23"
Sherray's Twinkle
Uma 9h27'14" 68d22'12"
Sherrey Lynne
Psc 1h26'2" 28d35'34"
Sherri
And 0h19'10" 46d21'53"
Sherri
Uma 9h18'29" 70d51'53"
Sherri
Uma 12h6'43" 53d48'37"
Sherri 03/12/52
Umi 15h45'56" 77d57'55"
Sherri A
Lib 14h51'52" -2d22'40"
Sherri A. Macpherson
Uma 13h45'44" 56d7'31"
Sherri and Kevin Forever
Uma 9h33'58" 56d13'47"
Sherri Ann
Aqr 23h3'51" -9d7'42"
Sherri Ann
Lmi 10h27'6" 35d33'35"
Sherri Ann
Cnc 7h57'41" 12d57'33"
Sherri Ann Sutton
Sgr 18h54'39" -20d9'27"
Sherri Argo
Lyn 7h21'39" 48d52'32"
Sherri Aubin's Star
Aqr 23h13'7" -8d50'31"
Sherri Carlene
Leo 10h6'20" 14d36'53"
Sherri Carpenter, my Mother
Ari 3h7'50" 27d39'47"
Sherri Clark
And 2h31'47" 44d58'36"
Sherri E. Gainey
Tau 3h57'39" 18d23'7"
Sherri E Parris
Aqr 20h41'25" -12d18'18"
Sherri Foster
Lyn 7h49'30" 45d29'57"
Sherri G.
Sgr 17h53'46" -29d37'29"
Sherri Hall
Com 13h13'22" 26d9'13"
Sherri & Joel 09/04/05
Cyg 21h32'26" 51d14'35"
Sherri Karabelski
Leo 11h0'27" 11d30'31"
Sherri L. Teafatiller Evans
Cnc 8h48'38" 9d11'36"
Sherri Lee Felix
Vir 12h39'46" -1d50'59"
Sherri Lee Warnkin
Lib 14h50'46" -1d6'38"
Sherri Leigh Nix "My Amazing Mom"
Leo 9h29'27" 9d33'55"
Sherri loves Perry
Cyg 20h37'23" 46d27'59"
Sherri Lyn Cassario
Cyg 19h48'10" 35d58'44"
Sherri Lynn
And 1h7'50" 37d36'27"
Sherri Lynn Anderson
Her 18h44'31" 20d18'13"
Sherri Lynn Holcomb
Uma 9h57'52" 47d53'19"
Sherri Lynn Mack
Cyg 20h16'7" 40d48'19"
Sherri Lynn Standard
Cas 0h27'48" 57d7'36"
Sherri Lynne
Aqr 20h47'30" -11d33'16"
Sherri Maksimovich
Lmi 10h33'13" 32d10'50"
Sherri Michele
Lib 14h31'12" -10d21'3"
Sherri Renee
Dra 16h13'59" 65d12'47"
Sherri Schweigert
Mon 6h49'17" -0d30'37"
Sherri Star
Tau 3h41'18" 28d31'57"
Sherri Steinford-Dove of Peace
Col 6h11'3" -34d18'5"
Sherri Tribout is loved by Kevin
And 1h14'8" 41d55'47"
Sherri Trollinger
Cap 21h20'59" -25d50'25"
Sherri Wilcox
Uma 11h4'26" 61d16'51"
Sherrice Colter
And 1h11'2" 41d51'33"
Sherrice Kirby
Cam 3h49'17" 56d42'23"
Sherrie
Cam 6h17'30" 62d11'39"
Sherrie A. Taylor - A Fairies' Guide
And 2h37'28" 49d48'46"
Sherrie Ann Roth Cronin
Crb 15h18'50" 31d1'44"
Sherrie Colby
Crb 16h7'43" 26d33'44"
Sherrie Dove, always in my heart.
Umi 15h3'12" 77d9'59"
Sherrie Frick
And 23h44'31" 34d12'19"
Sherrie Gunn
Uma 11h8'52" 34d58'42"
Sherrie Helen
Com 12h1'14" 21d21'36"
Sherrie Keese
Tau 3h25'42" 9d38'42"
Sherrie L. Brar
Ari 2h47'3" 21d17'58"
Sherrie L. Miles
Cru 12h26'39" -61d1'31"
Sherrie LaRosa
Cas 22h59'14" 57d51'6"
Sherrie Lashaun Johnson
Aqr 21h0'41" -10d38'15"
Sherrie Lynn Graham
Tau 3h40'51" 27d32'50"
Sherrie Lynn Robles
And 0h41'7" 22d17'53"
Sherrie Lynn Seaman Brar
Cas 1h30'23" 63d19'3"
Sherrie Myers Mary Huff loves you!
Tau 5h54'34" 25d36'32"
Sherrie Shellito
Cam 9h11'6" 76d4'43"
Sherrie Soleymani
Cnc 9h15'45" 28d46'55"
Sherrie Taylor
Cnc 8h27'5" 14d23'27"
Sherrie's Star
Vir 13h5'33" 11d13'10"
Sherrie's Star
Cyg 21h14'41" 50d47'59"
Sherril Rivera's
Uma 11h12'21" 60d37'47"
Sherrill
Uma 12h39'3" 58d36'12"
Sherrill
Lyr 18h22'34" 37d54'19"
Sherrill
Cyg 19h35'49" 35d27'21"
Sherrill
Uma 9h29'3" 43d32'40"
Sherrill Griffin Turner
Cyg 20h41'49" 52d17'23"
Sherri's Golden Arches Star
Uma 9h37'18" 43d2'7"
Sherri's guiding star
Cnc 7h56'1" 13d45'43"
Sherri's Serenity
And 23h40'47" 46d6'13"
Sherri's Star
Gem 6h49'26" 21d26'23"
Sherri-S3
Uma 11h30'30" 53d56'40"
Sherrny
And 1h30'42" 37d57'39"
Sherron
Peg 21h34'25" 16d22'58"
Sherry
And 0h18'44" 26d52'19"
Sherry
Gem 7h39'57" 31d39'21"
sherry
And 23h20'46" 47d14'6"
Sherry
And 2h2'23" 46d4'53"
Sherry
Cas 23h38'11" 55d44'34"
Sherry
Lyn 6h40'42" 60d15'23"
Sherry
Vir 13h9'58" -1d52'8"
Sherry
Aql 19h4'14" -0d58'13"
Sherry
Psc 0h55'4" 7d24'36"
Sherry
Cap 20h38'48" -26d2'21"
Sherry 1955
Uma 13h41'50" 56d17'56"
Sherry A. Palmer
Mon 6h52'28" -0d32'21"
Sherry Adler's Shining Star
Ori 5h7'45" 9d4'1"
Sherry and Emalee
And 1h40'42" 47d26'44"
Sherry Anderson
Cnc 9h20'16" 23d23'29"
Sherry Ann Cini-Putnam
Ori 5h56'8" -0d26'0"
Sherry Ann Groves(Grammie Moonbeam)
Lib 14h57'21" -13d56'46"
Sherry Ann Lamm
Lib 15h6'24" -2d17'26"
Sherry Ann Miranda
Cma 6h47'43" -13d40'0"
Sherry Anne Gallucci
Ori 4h50'24" 7d53'59"
Sherry Archer Happy 50th. Birthday
Uma 9h0'49" 66d31'45"
Sherry B Coventry
Sgr 18h3'31" -20d41'6"
Sherry Bajwa & Raymond Inda
Uma 8h42'28" 52d21'6"
Sherry Belinda Harper
Dra 17h40'52" 60d5'23"
Sherry Beth
Gem 7h27'10" 28d20'52"
Sherry Blaine Casper
Ori 5h28'20" 4d32'15"
Sherry Bowen's Shining Star
Gem 7h6'40" 22d21'10"
Sherry Brady
Ori 5h52'49" 20d37'45"
Sherry Bright
Cyg 19h39'37" 31d34'40"
Sherry Cameron II
Cas 1h13'37" 69d29'38"
Sherry Cameron III
Cas 0h28'12" 53d54'14"
Sherry Catherine
Srp 18h4'56" -0d2'28"
Sherry Chen
Cas 1h48'7" 62d37'4"
Sherry Coats
Cam 5h39'8" 75d7'36"
Sherry Cochran Eiles
Cas 1h12'12" 58d23'10"
Sherry Cossette World's Best Mom
Leo 11h6'3" 22d18'58"
Sherry D.
Psc 1h10'24" 10d54'27"

Sherry & David Rothenberg
Cyg 19h35'45" 37d48'1"
Sherry Davis Wheale
Ari 1h59'1" 17d40'12"
Sherry Dawn Huffman
And 2h34'21" 45d8'29"
Sherry Dawn Underhill
Cam 5h26'10" 67d8'52"
Sherry Duarte
Crb 15h30'17" 29d22'10"
Sherry E. Showalter
Sco 16h19'17" -18d48'29"
Sherry Einhorn Weiss
And 2h25'29" 46d2'28"
Sherry Evans
And 1h11'0" 38d17'28"
Sherry Ferrero
And 1h23'24" 40d29'48"
Sherry Fitton
Cas 0h29'42" 61d39'24"
Sherry Fitzpatrick
And 1h45'1" 50d35'18"
Sherry Fortune
Lyr 18h43'53" 39d49'31"
Sherry Francis Wolf
Aqr 21h49'11" -7d21'44"
Sherry George
Umi 16h28'41" 75d18'8"
Sherry Hicks
Lib 15h3'38" -25d55'3"
Sherry Howell
Dra 19h30'35" 75d21'58"
Sherry Idell
Sco 16h50'33" -33d26'56"
Sherry & John Raggio
Lib 15h57'19" -11d47'3"
Sherry Johnson
Sgr 18h57'6" -33d59'44"
Sherry K. Patrick
And 2h25'29" 49d12'42"
Sherry Kay
And 0h41'44" 45d48'27"
Sherry Kaye Ford
Cas 1h8'31" 55d53'20"
Sherry Kim Buchans
Lib 15h46'45" -28d20'16"
Sherry Kimiko Stanley
Cma 6h42'38" -15d6'17"
Sherry L.
Uma 8h45'59" 53d2'10"
Sherry L Hills
Sgr 18h25'44" -27d9'42"
Sherry L. Lightfoot
Tau 5h49'23" 16d58'17"
Sherry L. Roysdon - Best Friend-PFL
Leo 9h34'33" 24d40'33"
Sherry L. Sampson
Uma 13h44'19" 62d16'52"
Sherry Lange
And 0h11'33" 34d47'36"
Sherry Leanne Stephens
Cyg 21h17'52" 40d29'29"
Sherry Lee
Cnc 8h58'25" 7d31'54"
Sherry Lee Caillouette R.I.P. Mom
Uma 13h5'2" 62d42'25"
Sherry Lee Lloyd
Lyn 8h20'14" 52d11'6"
Sherry Lemoine
Tau 3h32'9" 8d30'57"
Sherry Lindsley
Sgr 18h4'18" -27d59'19"
Sherry Lojewski
Lyn 6h55'50" 55d42'55"
Sherry Loveless
Cyg 19h50'20" 38d12'22"
Sherry Lyn Boland
Uma 10h6'44" 49d31'27"
Sherry Lynett Dillehay
Lyn 7h35'22" 54d21'49"
Sherry Lynn
Ori 5h19'2" 1d49'10"
Sherry Lynn Golubic
Aqr 22h19'10" -12d45'26"
Sherry Lynn Kumer
Leo 11h11'35" 0d47'35"
Sherry Lynn Miller My Lady
And 2h32'25" 41d1'14"
Sherry Lynn Schneider
Vir 14h22'26" 1d23'42"
Sherry Lynn's Angel of Light
And 0h36'2" 30d32'4"
Sherry M. Metcalf
Aqr 21h56'21" -2d34'33"
Sherry M. Vines
Tau 4h18'2" 23d13'32"
Sherry Melissa McNeal
Tau 4h41'14" 5d1'13"
Sherry Milad
Vir 13h25'34" -10d52'34"
Sherry Minh Tran
Sgr 18h26'50" -17d46'32"
Sherry "My Princess" McMillion
And 23h0'29" 40d34'58"
Sherry - Remy
Cyg 21h38'45" 51d44'5"
Sherry Steele
Uma 9h10'22" 62d33'6"
Sherry Stemple
Cas 0h43'7" 65d30'54"
Sherry Stull
Cas 1h18'46" 57d46'7"
Sherry Thompson
Cyg 21h33'17" 48d26'32"
Sherry Von Spies
Aqr 22h39'57" -3d24'55"
Sherry Voss
Lyn 6h29'24" 60d16'32"
Sherry Workman
Cyg 20h24'15" 31d7'52"
Sherry Wozniak
And 1h27'57" 39d26'40"
Sherry Yvonne Townsend
Leo 11h34'14" 16d4'32"
Sherry, Queen of My Heart
Vir 12h40'58" 2d47'44"
Sherry82104
Cnc 8h15'51" 17d49'40"
Sherryayn Perpetua Lingenfelter
Cnc 8h40'16" 16d23'38"
SherryDawn
Aql 19h26'0" 7d59'48"
Sherrydwen
Lib 14h22'46" -9d15'4"
Sherryl Ann Davis
Lyr 19h8'19" 26d19'0"
Sherryl Ann Rimer
Sco 16h6'23" -11d26'16"
Sherryl Hegge
Psc 0h37'23" 11d8'51"
Sherryl Lynn Smith, We Love You!
Uma 10h58'40" 33d46'55"
Sherryn's Star
Psc 2h4'1" 6d48'28"
Sherrys Jewel
And 1h23'8" 46d29'17"
Sherry's Morning Glory
Lib 15h4'46" -2d38'59"
Sherry's Star in the Sky
And 1h18'11" 36d5'1"
SHERRYSSTAR
Lib 15h14'24" -15d20'25"
Shersohn 6/29/35
Cnc 8h52'15" 12d50'1"
Sherwin
Umi 14h22'39" 68d18'14"
Sherwood 26.8.1945
Vir 13h56'1" -7d41'7"
Sherwood F. Maki
Uma 13h15'50" 61d15'48"
Sherwood Morris Krogan
Cyg 20h44'33" 38d25'2"
Sheryl
And 1h45'38" 48d57'40"
Sheryl
Uma 11h56'34" 45d49'34"
Sheryl
Com 12h49'12" 21d58'0"
Sheryl
Aqr 20h51'5" -7d0'33"
Sheryl
Lib 15h8'34" -13d51'24"
Sheryl A. Bauer
Lyn 6h57'54" 54d19'30"
Sheryl A. Heiner
Cas 1h34'44" 62d1'38"
Sheryl Ann Embrey Warner
Cas 1h29'12" 59d53'28"
Sheryl Ann Sullivan
Per 3h30'41" 43d31'8"
Sheryl Ann Wiens
Com 12h0'39" 21d38'11"
Sheryl Anne Schmidt
And 23h18'26" 40d30'8"
Sheryl Anne Suplee
Cyg 21h53'12" 38d0'55"
Sheryl Bridal
And 0h25'31" 27d1'41"
Sheryl Canon-Hill
Cap 21h57'43" -17d40'10"
Sheryl Depropriis
And 0h39'55" 42d26'24"
Sheryl DeVoe
Sgr 19h6'22" -27d27'31"
Sheryl Elise Buckley
Her 17h40'58" 48d6'4"
Sheryl Flores
Uma 13h1'36" 63d10'48"
Sheryl G. Garcia
And 0h9'21" 32d58'52"
Sheryl Gabrielle Linden
Uma 8h32'34" 66d15'11"
Sheryl Gleim
Tau 5h55'8" 24d59'40"
Sheryl Goff
Lib 15h9'33" -7d28'24"
Sheryl Gonsar
And 0h43'41" 32d31'17"
Sheryl Graham
Cas 1h8'13" 63d27'4"
Sheryl Heiner
Cyg 20h30'43" 38d17'20"
Sheryl Irene Embree
Gem 6h51'50" 33d48'9"
Sheryl & Joe Dettling 7121975
Cyg 20h54'31" 46d29'1"
Sheryl Kaye Berg
And 0h36'37" 42d38'42"
Sheryl Kimiko
Aqr 22h0'44" -13d32'56"
Sheryl Lavery
Cas 1h36'43" 68d36'51"
Sheryl Lee Green Fowler
Crb 15h36'3" 27d50'43"
SHERYL LIN
Lyn 8h23'15" 41d41'19"
Sheryl Lynn
Cas 23h38'32" 55d23'8"
Sheryl Lynn Sofia
Gem 7h7'38" 34d18'15"
Sheryl M. Rhoades
Cyg 21h54'24" 51d45'16"
Sheryl Mae Ovalles
And 0h45'50" 24d21'56"
Sheryl Marie
And 1h58'37" 46d3'40"
Sheryl Meccariello
Sgr 19h19'44" -20d59'47"
Sheryl Mendoza
And 23h58'30" 34d57'8"
Sheryl Rae Bryan
Tau 4h35'30" 0d47'48"
Sheryl Rego Morley
Ori 5h32'18" 9d26'44"
Sheryl Renee Carson
Vir 11h43'5" 9d27'12"
Sheryl S. Crum
Ori 6h3'13" 12d37'26"
Sheryl S Meyers
Cyg 21h50'45" 47d7'17"
Sheryl Sanderson
Sgr 18h29'38" -34d46'12"
Sheryl Seagren
Com 12h16'41" 27d53'52"
Sheryl Sims Mallard
Cnc 7h58'52" 10d27'4"
Sheryl Witkowski
And 2h24'3" 46d50'4"
Sherylannshiloh
Gem 7h21'25" 31d6'13"
Sheryle Fonzi
Uma 9h25'11" 61d15'59"
Sheryle Irene Nelson
Leo 11h13'33" 21d12'8"
Sheryl's Own
Cyg 20h9'22" 40d49'48"
Sheryl's Star
And 23h9'18" 39d32'18"
Sheryl's Star
Lyr 19h12'38" 45d57'1"
Sheryl's Star
Lib 14h55'28" -4d22'15"
She's Mum's KKK Magestic DJ Passing By
Cru 12h24'58" -58d39'18"
She's Star
And 1h2'1" 40d1'20"
Shesh and Shap
Uma 10h8'36" 46d4'36"
Shesha
Lib 14h57'11" -11d1'9"
Shesly Clara Levasseur
Sco 16h5'20" -18d2'52"
Shevonne Marie Nuttal
Cas 0h56'27" 65d36'13"
Shewana Skinner
Sco 17h25'31" -45d13'58"
Sheyen's Star
Crb 16h13'55" 38d52'41"
Sheyn Malekh Shtern or JRM's Star
Aqr 22h46'16" 1d32'56"
Shi Bruggette Barber Serna
Gem 6h40'34" 31d15'11"
Shia Lanier Castillo
Sco 17h50'16" -38d17'58"
Shiama Groff
Leo 10h52'4" 23d0'23"
SHIANE
Her 18h2'22" 17d41'19"
Shianna Marie Edwards
Crb 16h11'57" 32d8'19"
Shianna The Angel From Heaven
Uma 9h57'16" 63d29'51"
"Shian-Read"
Ori 6h11'18" 20d45'10"
Shibby
Ori 5h20'30" -5d23'8"
Shichiomi
Uma 9h29'41" 51d35'10"
Shiela
Sco 16h3'11" -10d12'35"
Shiela and Chuck's 50th Anniversary
Cap 21h25'2" -15d47'19"
Shiela Hancock
Cam 5h46'42" 56d13'27"
Shiela's Piece of Heaven
Col 5h44'19" -41d27'26"
Shield of Victory-Randy/Nichol
Crb 15h50'40" 38d8'24"
Shields Family
Umi 15h2'48" 76d39'9"
Shi-Fawn Raechel
Uma 12h1'14" 44d54'41"
Shigeo with Chiharu
Psc 0h52'37" 18d17'25"
Shiggle Biggle
Lyn 7h59'28" 34d19'27"
Shiho Takedomi
Lib 15h42'54" -9d6'50"
Shika
Uma 11h48'47" 56d43'59"
Shikha-Cindy
Uma 13h27'45" 52d33'52"
Shikha-Cindy
And 2h36'58" 46d23'52"
Shilah & Frank - All My Love Always
Psc 1h22'38" 18d13'42"
Shilarna
Leo 10h59'59" 9d20'51"
Shileyn and Erik
Ari 2h50'58" 26d10'21"
Shiljac - Guardian Angel
Peg 21h37'35" 16d11'50"
shilo 25
Uma 12h26'16" 55d14'4"
Shilo Dawn
And 0h9'56" 31d36'20"
Shiloah Matheny
Lyr 18h33'28" 37d11'34"
Shiloe Marissa
Dra 16h28'36" 66d31'11"
Shiloh
Cma 6h44'33" -16d29'51"
Shiloh
Lib 14h49'12" -2d26'20"
Shiloh
Gem 7h16'45" 26d53'25"
Shiloh C. Divine
Cnc 8h20'55" 19d47'13"
Shiloh Ebb Wolf
Gem 7h33'35" 20d6'31"
Shiloh Kime
Ori 6h19'45" 13d23'30"
Shiloh Marie Lona
And 0h35'4" 27d28'2"
Shiloh Shannon
Equ 21h10'25" 10d16'51"
SHILOH-DORK
Uma 10h38'18" 52d2'14"
Shilohs (Punkin pie) Peace
Lib 14h55'33" -22d57'28"
Shilpa
Cap 21h53'34" -9d53'50"
Shilpa
Uma 11h45'9" 43d48'48"
Shilpi
Ind 20h39'24" -47d18'40"
Shimeka Green
Mon 6h33'59" 10d58'59"
Shimen
Cas 0h25'0" 62d22'42"
Shimgua
Lmi 10h10'35" 31d26'17"
Shimmer
Lib 14h31'30" -9d23'0"
Shimmer
Aql 19h43'49" -0d1'6"
Shimmer of Hitomi & Keita
Sgr 18h59'51" -33d53'3"
Shimmering Lisa
Cap 20h46'47" -14d38'36"
Shimmering Shel
Cas 23h59'46" 53d5'10"
Shimmering Success Star
Uma 13h51'5" 61d16'22"
Shimmer's Eternal Light
Lib 15h23'19" -7d30'41"
Shimmins
Vir 12h29'31" 10d41'39"
Shimmy
Lmi 10h25'37" 37d53'43"
Shimon
Lib 15h59'41" -10d14'40"
Shimon Elbaz
Tau 5h49'19" 21d47'42"
Shimon Foreman
Uma 11h42'12" 59d10'54"
ShimStar 1
Peg 21h38'25" 10d52'11"
Shin Chang
And 0h37'21" 39d5'24"
Shin Sung Woo
Leo 10h45'29" 21d43'40"
Shin Tsudome
Lib 15h9'10" -6d22'42"
Shina Lee
Uma 9h35'28" 41d32'37"
Shinali Patel aka 'Little Gangsta'
And 0h35'27" 25d43'57"
SHiNE BRiGHT AMOR
Tau 5h28'19" 27d10'38"
Shine Brightly for Billie
Tau 4h11'37" 14d26'21"
Shine for Kai
Leo 11h38'25" 26d32'51"
Shine For You - Irene Cornish 19.4.56
Cru 12h40'31" -57d14'29"
shine forever
Ari 2h49'44" 22d23'0"
shine n mike
Ori 5h28'32" 14d10'48"
Shine on Dear Sonia! Love You, C&L
Crb 16h21'0" 37d29'28"
Shine on Linda
Cnc 8h13'22" 31d1'31"
Shine On William 'BunnyLove' Morris
Cen 12h47'16" -39d41'21"
Shine On You
Uma 9h29'55" 64d21'42"
Shine On You Crazy Diamond, Toni
Ari 3h26'33" 26d19'38"
Shine On, Den
Ori 5h27'46" -5d17'53"
Shine with a smile Susi
Ori 5h54'0" 11d16'3"
Shinee
Uma 10h4'42" 43d36'59"
Shinera-Paul
Sgr 18h32'10" -27d19'19"
ShinesLikeTheSun & WalksByTheMoon
Lib 15h27'56" -23d55'51"
Shiney Roo
Pho 0h20'55" -44d38'18"
Shinin Hero Star
Per 3h49'53" 43d49'6"
Shining at Fifty
Per 3h14'52" 53d5'16"
Shining Beautiful Irina
And 0h50'1" 35d45'48"
Shining Beauty
Agr 22h27'46" 1d51'6"
Shining Beauty
Uma 11h8'47" 63d31'35"
Shining Bright Alexandria Gallagher
Col 6h30'54" -33d56'5"
Shining Carrie
Gem 7h49'6" 28d13'44"
Shining Cutie Jeanette Star
Gem 6h31'13" 16d33'58"
Shining D'Arne 21
Cru 12h38'2" -59d53'12"
Shining Emma J
Cap 20h56'9" -19d58'2"
Shining For You
Gem 7h44'17" 21d55'56"
Shining for you Hunny
Cnc 8h9'39" 24d53'39"
Shining Forever NPG 17th June 1986
Gem 6h49'5" 23d44'44"
Shining Goddess
Leo 10h10'27" 11d19'39"
Shining Godsons Ryan & Tyler
Cyg 21h24'7" 47d1'49"
Shining Harriet
Gem 6h35'58" 21d44'49"
Shining House in the Heavens
Umi 15h13'53" 68d25'34"
Shining Jill
Sgr 18h9'46" -29d46'7"
Shining Jules
Leo 9h26'23" 24d24'40"
Shining Knight
Her 17h30'12" 16d35'17"
Shining Leanne
Peg 21h41'44" 22d38'29"
Shining Leif
Uma 10h10'1" 50d20'36"
Shining Life Vest
Uma 11h57'17" 39d27'41"
Shining Light of Leonard Rubio
Gem 6h46'52" 17d32'6"
Shining Lisa Yu of the Vast Skies
Cnc 8h43'42" 21d27'19"
Shining Love Gma n Gpa Bliven-Megan
Per 4h31'9" 39d40'16"
Shining Love,Sam & Margaret Bennett
Sge 20h12'29" 18d8'53"
Shining Marquis
Uma 11h9'28" 53d0'6"
Shining McCulloch
Umi 13h8'8" 75d25'19"
Shining Michelle
Ari 1h57'11" 23d56'31"
Shining Nelly
Dra 19h30'33" 64d31'58"
Shining Noble Alex Elly Jake & John
And 2h20'30" 46d47'10"
Shining Nora
Leo 11h17'40" 21d59'47"
Shining Nour
Uma 11h11'10" 30d44'50"
Shining Penny
Tau 3h37'18" 16d16'22"
Shining Reminder of Love
Cyg 20h17'33" 40d30'50"
Shining Sam
Ori 5h53'20" -0d44'21"
Shining Schumann
Cnc 8h28'8" 23d17'21"
Shining Seto V
Cnc 8h37'55" 7d46'19"
Shining Shelley
Cnc 9h10'19" 26d22'23"
Shining Shelly
Gem 6h48'43" 29d40'30"
Shining Sherwood Dreamweaver
Ari 2h47'22" 16d42'17"
Shining Shugs Wade
Uma 11h18'28" 40d1'14"
Shining Silver
Sco 16h12'41" -12d20'7"
Shining Siobhan
Vir 14h3'27" -14d57'40"
Shining Star
Sgr 19h22'59" -27d40'2"
Shining Star Chigusa Nagayo
Lib 15h24'49" -28d19'15"
Shining Star- Dixie Duke-Turnquist
Psc 1h14'21" 14d28'31"
Shining Star June
Uma 10h16'32" 42d49'2"
Shining Star Linie and her son Ryan
Dra 19h48'12" 59d12'15"
Shining Star of Blue Water
Aqr 22h8'50" -16d10'36"
Shining Star of "Nanase Hoshii"
Sco 16h54'0" -37d36'7"
Shining Star of "Nanase Hoshii"
Sco 16h54'0" -37d36'7"
Shining Star of Sandra S Lindenberg
Tau 4h25'57" 26d53'15"
Shining Star of Texas, R. Munsey
Ori 6h12'13" 5d42'12"
Shining Stars
Ari 2h39'33" 17d33'10"
Shining Stars of 2017
Her 17h17'35" 28d26'22"
Shining Stevens
Her 17h27'51" 31d36'23"
Shining Sweetly Candice Brooke C
Ari 3h12'36" 29d11'57"
Shining Thomas
Ori 5h39'17" -0d43'3"
Shining Tomik
Cru 12h36'22" -60d7'51"
Shining Zhoogie (For Katie Pines)
Per 2h40'56" 41d9'49"
Shino WIlson
Uma 11h40'2" 45d17'5"
Shinobi Wolf
Vir 12h17'53" 9d15'57"
Shiny Prakash
Sgr 19h19'27" -42d55'40"
Shiny Purple Star
Dra 15h46'10" 61d52'28"
Shinyxlilxstar
Uma 10h23'40" 68d3'1"
Shioban - A star for lovers
Ari 2h9'39" 23d10'2"
Shiobhan Kelly
Vir 13h16'29" 7d13'49"
Shion
Gem 7h31'41" 33d1'25"
SHIRA
Ari 2h5'16" 24d14'17"
Shira Cherns
Crb 16h9'7" 35d56'38"
Shirald Lori
Psc 1h14'45" 17d36'45"
shiranndromonous
Aqr 22h34'20" 1d37'24"
Shireen
Cyg 19h47'0" 31d14'4"
Shireen Mann
Psc 23h14'37" 6d44'26"
Shireen Practically Perfect Star
Eri 3h13'56" -3d21'26"
Shireen Solomon
Sgr 18h35'38" -24d2'56"
Shirell and Janyce Willis
Eri 1h49'2" -52d0'44"
Shirell Lynn Shaw
Lyr 19h12'58" 27d20'54"
Shirene
Crb 16h4'33" 34d16'2"
Shiri
Ori 5h11'17" -9d51'47"
Shiri SandraSue Howell
Ari 2h37'2" 14d59'0"
Shirin
Uma 10h31'59" 55d26'49"
Shirl - Les' Guiding Light
Cru 12h19'32" -62d36'45"
ShirLee
Lyn 7h4'42" 48d59'42"
Shirlee Henning Brewer Caldwell
Lyn 8h0'0" 41d37'24"
Shirlee L. Bliss
Sco 16h15'58" -15d19'32"
Shirlee M. Sheckells
Cam 6h29'1" 62d6'7"
Shirleen
And 0h35'36" 32d54'54"
Shirlene Everhart
Com 12h23'26" 26d46'52"
Shirlene Jahzahrah
Psc 1h8'31" 5d4'50"
Shirlene Rosman
Lyn 8h4'32" 53d12'21"
Shirlene's Love and Light
Her 18h14'53" 18d35'25"
Shirley
Cnc 8h38'55" 18d24'28"
Shirley
Ari 2h50'35" 18d36'31"
Shirley
Vir 11h42'47" 9d0'11"
Shirley
Cas 1h26'20" 63d35'30"
Shirley
Cas 1h46'32" 64d13'4"
Shirley
Vir 12h52'37" -9d7'33"
Shirley
Cru 12h18'23" -57d56'30"
Shirley
Vel 10h10'7" -40d25'41"
Shirley 100
Uma 10h20'33" 46d35'15"
Shirley A. Bridger
Leo 9h22'53" 28d59'40"
Shirley A. Brock
Cam 4h3'18" 57d9'50"
Shirley A. Bruno O'Brien
Cas 0h14'36" 57d33'25"
Shirley A Conroy (05-18-1960)
Tau 5h9'8" 25d26'29"
Shirley A Gabaree
Leo 11h44'35" 25d25'17"
Shirley A. Kersey
Uma 11h28'53" 44d1'14"
Shirley A Torres
Psc 0h55'15" 30d11'27"
Shirley A. Wilson
Leo 10h26'46" 13d43'53"
Shirley Ames Clayton
Uma 8h53'31" 62d44'52"
Shirley and Larry Biggs
Cyg 20h27'8" 43d59'6"
Shirley and Matt Hughes
Uma 10h46'14" 42d26'35"
Shirley and Nate
And 1h2'37" 40d55'38"
Shirley and Richard's Golden Star
Cyg 20h22'1" 56d4'18"
Shirley and Steve Dye
Cyg 19h47'56" 30d42'39"
Shirley Ann
Lmi 10h14'39" 38d39'0"
Shirley Ann
Ori 6h13'17" 15d43'46"
Shirley Ann
Leo 9h51'28" 24d36'16"
Shirley Ann
Cnc 8h15'15" 22d42'32"
Shirley Ann
Sco 16h4'7" -24d27'6"
Shirley Ann Bailey
Ari 3h14'1" 30d11'11"
Shirley Ann Bocian
Vir 14h0'3" 3d32'9"
Shirley Ann Bolotta
Tau 3h42'34" 10d15'38"
Shirley Ann Cox
And 0h37'39" 34d3'12"
Shirley Ann Dahlstrom
Uma 11h10'15" 41d52'37"
Shirley Ann Hill
Crb 15h35'32" 30d27'43"
Shirley Ann Larson
Sco 16h33'53" -38d37'52"
Shirley Ann Madrid
Per 2h26'51" 52d43'47"
SHIRLEY ANN MILLER
Crb 16h8'20" 36d49'32"
Shirley Ann Myers
Ori 5h50'15" -3d30'21"
Shirley Ann Rouselle Clark
Uma 9h43'28" 65d42'33"
Shirley Ann Schaale
Aqr 21h37'51" 0d51'26"
SHIRLEY ANN SHOT-TROFF
Psc 0h20'11" -3d24'2"
Shirley Ann Simonson
Cap 20h21'19" -10d48'8"
Shirley Ann Sneed
Lyr 19h3'7" 33d33'14"
Shirley Ann Stapp
Aqr 21h41'10" -7d48'21"
Shirley Ann Stein
Lib 15h2'30" -10d12'58"
Shirley Ann Swick
Sco 16h7'56" -14d47'18"
Shirley Ann Thompson
Crb 16h11'58" 26d50'42"
Shirley Annette Moragne ("SAM")
Cap 21h40'27" -14d19'27"
Shirley Annmarie Gilsback 1932-2005
Tau 4h42'20" 27d8'11"
Shirley Arbour
And 23h18'49" 41d23'9"
Shirley & Arthur Levin
Uma 10h36'29" 53d32'48"
Shirley Ash
Mon 6h53'29" -0d0'24"
Shirley Back
Psc 23h14'48" 1d57'12"
Shirley Baker
Tau 3h38'44" 27d29'0"
Shirley Ballantyne
Ori 5h59'1" 17d36'41"
Shirley Bayla Neidorf
Cnc 8h22'39" 11d4'29"
Shirley & Bob Brodsky
Cyg 19h53'2" 33d11'26"
Shirley Bolton Peacock
Gem 7h23'0" 22d52'7"
Shirley Boyle 60
Ari 2h37'17" 12d34'49"

Shirley Brenner
Sgr 18h32'54" -20d54'5"
Shirley Brown
Cas 23h59'57" 53d49'56"
Shirley Brown
And 0h34'1" 36d54'7"
Shirley Brown Foster
And 1h50'42" 43d50'36"
Shirley Callihan
Cap 21h35'55" -8d42'14"
Shirley Clyman
Ari 2h38'51" 23d9'20"
Shirley Cooper
Leo 9h30'45" 9d25'53"
Shirley Cronshaw
Lyn 8h24'47" 39d50'45"
Shirley Curtis
Gem 7h15'56" 19d3'11"
Shirley Czaplewski
Cas 22h58'51" 59d38'53"
Shirley D. Bonaccorso
Vir 13h50'51" -3d26'26"
Shirley D. Davis
Gem 6h53'26" 18d55'52"
Shirley D. Harris
Uma 10h35'22" 45d15'47"
Shirley Darlene Trodden
Cas 0h51'40" 52d40'23"
Shirley Dawson
Cas 23h34'41" 54d59'52"
Shirley Ding
Com 12h42'5" 14d11'38"
Shirley Don
Cap 21h1'0" -17d32'58"
Shirley E. Watson
Aqr 22h25'45" -0d31'51"
Shirley Elizabeth Dolezsar
Cyg 19h39'13" 29d17'29"
Shirley Eveline Detterline
Cas 0h43'23" 52d20'54"
Shirley F. Combs
Cyg 20h5'48" 35d38'11"
Shirley FB
Gem 7h21'41" 22d45'36"
Shirley Fleischhauer's Shining Star
Lyr 18h27'15" 39d50'43"
Shirley Foraker
Ori 6h16'33" 10d55'22"
Shirley G. Dickinson
Lib 15h8'43" -22d31'29"
Shirley G Fala
Cas 0h42'53" 62d38'0"
Shirley Gaw
Leo 11h6'59" 28d11'4"
Shirley Gay Goetz
Cap 21h1'49" -26d6'18"
Shirley Girl
Aur 5h15'13" 30d49'59"
Shirley Girl Banks
Cas 0h47'26" 61d15'32"
Shirley & Glenn
Umi 15h25'22" 76d8'50"
Shirley Gulick
Uma 11h52'30" 38d21'35"
Shirley Handsaker-shining star
Lib 15h35'45" -19d22'45"
Shirley Hanson
Ari 2h36'42" 27d24'20"
Shirley Heusel
Uma 11h58'14" 43d53'5"
Shirley Hotwagner
Vir 12h54'16" -5d47'42"
Shirley Huett
Cnc 8h14'11" 8d40'46"
Shirley Hunt
Cas 2h13'15" 71d43'52"
Shirley - in Memory of Pat & Les
Cyg 19h30'41" 28d54'40"
Shirley J. Doyne
Umi 13h49'44" 74d25'37"
Shirley J. Parker
Ori 5h41'12" 0d11'20"
Shirley J. Teeters
Uma 8h54'58" 65d16'39"
Shirley James
Lyn 9h18'16" 39d27'49"
Shirley Jean
And 1h14'30" 36d48'30"
Shirley Jean
Cas 1h36'2" 60d34'22"
Shirley Jean Brawley
Psc 1h10'19" 20d4'23"
Shirley Jean Brooke
And 0h59'11" 37d36'42"
Shirley Jean Phillips
Lib 15h59'13" -12d16'20"
Shirley Jean Rodgers
Cas 1h39'55" 62d54'56"
Shirley Jean Swanson
Cyg 19h59'13" 31d16'56"
Shirley Johnson
And 1h38'42" 48d13'21"
Shirley Jones
Aur 5h0'1" 40d4'1"
Shirley Joyce Sanchez
Cas 0h34'53" 53d19'39"
Shirley Juanita Bridges
Gem 7h26'24" 27d58'6"
Shirley June Hedge
Cas 22h59'0" 53d30'5"
Shirley June Lee DeAlessio
Lep 5h20'12" -12d43'15"
Shirley K. Bullard
Cam 4h28'0" 58d19'7"
Shirley Kiacz DiPaolo
Uma 8h54'44" 46d50'1"
Shirley Kline Bennett
Gem 7h26'23" 25d54'56"
Shirley L. Bishop
Uma 8h14'27" 66d21'48"
Shirley L. Diamantoni
Dra 18h51'51" 69d4'24"
Shirley L. Noble
Cas 0h47'48" 66d49'54"
Shirley L. Rux
Cas 2h1'25" 59d8'24"
Shirley L. Whitmer
Tau 4h5'26" 19d33'23"
Shirley Lee Croddy
Cnc 8h5'13" 13d9'33"
Shirley Lee Saddler
Cas 1h24'35" 68d38'59"
Shirley Long
Leo 10h13'6" 11d54'9"
Shirley Long Culbreth
Lyr 18h45'22" 40d4'44"
Shirley M. Bartee
Cnc 8h43'4" 15d10'33"
Shirley M. Bass
Cyg 21h16'40" 55d8'3"
Shirley M Carter
Peg 21h47'1" 6d37'11"
Shirley M. Comer
Cap 21h41'12" -13d0'3"
Shirley M. Fisher
Lyn 7h40'2" 39d14'41"
Shirley Mae
And 1h28'18" 49d54'49"
Shirley Mae
Vir 13h42'28" -8d19'46"
Shirley Mae Bessinger
Cnc 8h36'25" 24d41'16"
Shirley Mae Brown
Uma 11h4'45" 52d23'30"
Shirley Mae Burr
Cap 20h25'29" -11d33'56"
Shirley Marcus Feinberg
Tau 5h48'7" 26d46'52"
Shirley Marie
Cnc 8h30'32" 28d56'39"
Shirley Marie Ashcraft
Psc 0h16'0" 16d55'0"
Shirley Marie Corzo
Psc 1h33'49" 9d53'17"
Shirley Marie Schetgen
Gem 7h3'58" 32d20'24"
Shirley Marie Stephenson
Ari 1h56'42" 24d7'19"
Shirley Marie Stevenson
Ari 2h38'50" 12d32'56"
Shirley Marie Therese
Ori 5h20'0" -7d47'48"
Shirley Mary Pearson Hatfield
Uma 11h25'51" 52d0'48"
Shirley Mathew-Rogers
And 1h39'14" 41d44'34"
Shirley May DeWree
Tau 4h48'7" 19d22'27"
Shirley Miller
Cas 2h31'25" 67d50'30"
Shirley "Mom" Temple
Uma 13h44'37" 55d13'18"
Shirley "Momma Bear" Roberts
Cyg 20h4'56" 30d45'53"
SHIRLEY "MONGIE" JOYCE
Cnc 8h36'12" 28d39'5"
Shirley Mount Hufstedler
Ori 4h47'30" 7d11'10"
Shirley Muriel Rowe
Mon 6h38'8" -1d28'7"
Shirley Mutzfeld
Cyg 21h9'18" 53d13'58"
Shirley - My Shining Star
And 23h48'1" 44d9'0"
Shirley Neimark
Uma 10h38'55" 50d44'41"
Shirley Patricia Wright
Ori 5h57'25" 22d40'15"
Shirley Pepin's star
Sgr 18h26'19" -16d23'40"
Shirley Pomeroy Herndon
Vir 12h29'7" -6d21'37"
Shirley R Shaskas
Sco 16h8'25" -11d48'26"
Shirley Race
Lyn 7h41'33" 38d23'46"
Shirley Rae
Cas 0h25'43" 51d18'31"
Shirley Rhae
Uma 11h44'53" 51d44'42"
Shirley & Ronald Roddy Everlasting
Cyg 20h7'16" 34d32'11"
Shirley Rose
Aqr 21h23'6" 0d50'17"
Shirley Rose Schiestle Villalpando
Leo 10h9'23" 20d9'51"
Shirley Ruby Newton
Cyg 19h42'54" 28d30'53"
Shirley Ruth
And 2h30'5" 40d3'22"
Shirley Ruth Thiele
Sgr 19h28'27" -23d16'9"
Shirley S Roberts
And 0h32'31" 27d29'51"
Shirley & Sheldon Lasher
Vir 13h47'25" 6d54'32"
Shirley Shepherd
Cas 1h7'45" 49d18'18"
Shirley Siu-Ki Chen
Tau 5h35'19" 19d59'18"
Shirley Smith
Cas 0h9'43" 59d37'2"
Shirley Smith
Cas 1h3'7" 63d24'42"
Shirley Smith
Uma 9h5'43" 69d34'51"
Shirley Snyder "Nanah's Star"
Mon 6h47'24" -5d4'28"
Shirley Stern
Cru 12h39'21" -57d46'50"
Shirley Stetson
Lyr 19h13'55" 38d28'52"
Shirley Stevens-McEckron
Cnc 8h44'53" 32d5'44"
Shirley Tiso
Lib 14h51'38" -23d15'14"
Shirley Tracy
Vir 13h19'6" 12d17'38"
Shirley Turner
And 0h56'47" 40d46'20"
Shirley Underwood
Cap 21h57'9" -9d13'57"
Shirley Unger & Nutmeg
Umi 15h19'14" 72d0'54"
Shirley V. Warren
Sgr 19h13'22" -16d30'58"
Shirley W.
And 0h44'29" 39d48'40"
Shirley W. Kriner
Cas 0h41'53" 66d34'2"
Shirley Waddill
Cnc 9h6'45" 30d52'34"
Shirley Wagner
Uma 10h30'7" 45d59'15"
Shirley Wahl
Cas 23h2'54" 59d17'9"
Shirley Wallen
Cas 1h49'36" 61d41'12"
Shirley Wencis
And 23h14'0" 52d24'47"
Shirley Whitney
Sco 17h4'22" -38d34'54"
Shirley=Wife=Mom=Mom-Mom=Love
Psc 0h35'51" 19d31'28"
Shirley Winters
Cas 1h44'26" 65d29'59"
Shirley Y. Mentzer-Myers
Leo 11h37'17" 25d35'27"
Shirleyann Gloria Ouren
Cyg 20h10'27" 39d12'5"
Shirleygold
Cap 21h5'35" -16d15'6"
ShirleyMae
Uma 9h4'43" 47d27'31"
Shirley-Penny
Psc 0h27'52" 15d27'14"
Shirley's Adam
Sco 17h52'14" -36d25'47"
Shirley's Bright Laughter
Ari 3h19'20" 10d55'57"
Shirleys Diamond Star
Cap 20h12'14" -9d48'53"
Shirley's Eternal Light of Love
Sco 16h3'17" -21d30'45"
Shirley's Goodness
Cas 1h10'11" 53d37'41"
Shirley's Grace
Lib 15h36'36" -7d39'52"
Shirley's Phil Paz
Uma 10h35'26" 70d31'54"
Shirley's Star
Cam 3h25'49" 65d12'7"
Shirley's Star of Fame
Crb 15h52'3" 27d53'52"
Shirley's Starr
Ari 2h3'40" 12d45'37"
SHIRLIE JUNE
Psc 0h42'58" 14d4'58"
Shirlnise
Ori 6h14'53" 15d35'57"
Shirl's Star
Vir 13h22'15" 6d27'44"
shiro+katsue+daisuke+yumiko YAMADA
Sco 17h57'41" -38d0'9"
Shisher
And 1h57'59" 37d41'27"
Shi-Shi
Lmi 10h23'20" 33d30'27"
Shisley
Ari 2h31'33" 10d40'45"
Shiu Cheung & Karen Li
Psc 0h14'2" -5d26'27"
Shiu Kumari Sarup
Lyr 18h39'47" 37d57'38"
Shiu-Chuan Chang
Lac 22h23'34" 53d11'26"
Shiuli Rehman - Happy Anniversary xx
Cyg 19h44'59" 29d1'24"
Shiv
Ari 2h6'55" 26d31'18"
Shivani
Cnc 9h19'49" 10d43'13"
Shivani Parmar
Sgr 18h11'26" -23d18'13"
SHIVAR89
Cap 20h49'8" -15d1'23"
Shively
Lyn 7h53'41" 56d33'47"
Shivinder Singh Grewal - Shiv's Star
Peg 22h36'44" 21d13'29"
Shivon A. Sukhu
Cyg 21h9'13" 54d53'9"
Shi-Voo-Ahh
Sco 16h7'59" -19d13'38"
Shiya Linn
Leo 9h44'51" 24d58'5"
Shiyanne Dawn Thornell
Uma 9h20'14" 60d1'59"
Shk Mohamed bin Rashid Al Maktoum
Her 17h56'0" 42d7'43"
Shkemanemanin
Umi 14h3'42" 65d49'36"
Shlanka-Belle
Vir 15h4'4" 7d10'13"
shlee my love
Tau 4h32'52" 28d38'23"
Shlomi & Nornor
Cas 0h8'55" 51d0'42"
Shmaggie
And 2h26'6" 50d12'35"
Shmatee
Ari 2h51'1" 27d52'21"
SHMICAY
Per 4h10'33" 50d55'7"
Shmily
Per 3h14'52" 45d58'20"
SHMILY
Cas 0h24'57" 52d24'6"
SHMILY
Cyg 21h23'43" 47d40'1"
SHMILY
Cnc 8h43'16" 21d10'45"
Shmily
Gem 6h46'1" 27d38'57"
Shmily
Her 17h21'51" 29d48'56"
Shmily
Cyg 21h19'39" 54d3'10"
Shmily
Sco 16h6'55" -21d58'52"
SHMILY
Cap 21h49'40" -19d29'56"
SHMILY
Dor 6h24'1" -66d9'35"
Shmily 06
Uma 12h39'7" 54d53'37"
S.H.M.I.L.Y. Carlson
Cas 1h41'1" 69d11'12"
SHMILY Jenn
And 0h16'45" 32d54'43"
SHMILY ONION
Psc 0h2'22" 6d3'37"
SHMOOKIE
Cyg 20h9'6" 34d25'59"
Shmooobledie Bumpkins
Leo 10h28'23" 6d23'18"
shmoop-a-loop
Per 2h40'7" 40d14'47"
Shmoopie and Babyhead Forever
Sco 16h8'26" -11d33'47"
Shmoopy - Aaron and Kelly
Cyg 21h6'19" 47d56'33"
Shmoopy Anne
Lac 22h46'57" 50d1'47"
Shmoopy Poops
And 23h21'53" 42d21'0"
Shmooshy Beautifus
Uma 10h31'41" 66d7'39"
Shmu B. Grant
Her 17h27'18" 32d28'24"
Shmumpkie
Tra 15h46'49" -60d44'33"
Shmylie & Shmoralie
Cas 0h31'22" 54d28'26"
Shnookums
Lyn 6h27'14" 60d41'12"
Shnoopy
Tau 3h47'53" 29d12'0"
Shnootal Bear Star
Ori 6h23'43" 14d17'6"
Shnuggempuff
Umi 15h7'53" 71d59'16"
Shnuggie - "Snoopy"
Uma 10h5'12" 50d9'2"
Sho & Ty
Cnc 8h53'24" 26d40'37"
Shobana Venkat
Sco 16h24'11" -26d42'25"
Shock
Cnc 8h6'23" 26d15'3"
SHOCKER
Cma 6h23'0" -15d8'16"
Shoe
Uma 9h42'29" 66d47'7"
Shoe~Pizza~Silverbum
Uma 10h24'26" 66d45'52"
Shoe's Shiner
Sgr 19h26'36" -17d5'57"
SHOGGIES23
Gem 7h18'3" 15d36'31"
Shohana K. Pathan
Sgr 18h24'43" -16d11'0"
ShohilandSam4ever
Ari 2h1'28" 13d8'45"
Shoira Saidkarimova
Lib 15h19'6" -9d46'40"
Shoji & Miho
Vir 14h30'32" -0d34'25"
Shoko Katsuragawa
Ari 3h17'40" 26d13'34"
Shoko Ozawa
Srp 18h13'19" -0d20'33"
Shomi Lee
Tri 2h25'9" 34d23'0"
Shomik
Lib 15h47'4" -23d31'24"
Shon and Haley
Ori 5h30'20" -3d8'1"
Shona
Aur 5h18'17" 42d33'25"
Shona
Uma 10h43'30" 46d2'46"
Shona
Psc 0h34'6" 17d32'42"
Shona Angus
Dra 19h46'20" 61d34'3"
Shona Ann
Lib 15h42'47" -7d58'17"
Shona/Eric
Cyg 19h42'1" 53d26'16"
Shona Lee
And 1h14'45" 50d34'10"
Shona Lewis
Cas 0h37'9" 54d3'21"
Shona Mary Wear
Cap 20h12'16" -22d11'38"
Shona Renee Lubken
Uma 11h17'52" 44d34'23"
Shonagh Walker
Col 5h59'2" -34d48'58"
Shonathon
Psc 1h31'59" 16d56'23"
Shonda Always and Forever
Lib 15h57'53" -19d41'47"
Shonda J. Muschany
Uma 8h47'13" 47d16'13"
Shonda Paige
Lyn 9h10'50" 37d31'14"
Shondrella
Cam 5h41'30" 75d45'28"
ShonetteEthan
Peg 23h29'30" 28d41'20"
Shonna Lee Bosco
Cnc 8h18'25" 14d35'12"
Shonne and Shotas' Heavenly Light
Psc 1h8'2" 11d5'22"
Shon-n-Jeff
Cyg 20h32'48" 39d40'17"
Shonsearea Dawn Halstead
Vir 14h26'27" 2d38'20"
shooshy
Leo 10h21'25" 14d11'36"
SHOOTER
Umi 15h26'41" 77d19'53"
Shooting Star
Crb 16h6'17" 32d57'39"
Shooting Star International Agency
Crb 16h16'42" 28d32'42"
Shooting Star Kelly Michelle Watson
Aqr 22h7'56" -0d28'3"
Shopna -Our amazing Mother and Wife
Umi 13h39'47" 77d29'23"
Shopping Holly
Aqr 21h32'10" 1d44'49"
Shora Codie Ingram
Vir 12h23'44" 10d52'23"
Shoriful A Mahi
Psc 1h36'22" 22d33'45"
SHORTCAKE
Cyg 21h10'30" 42d32'52"
Shorteduwop
And 0h48'2" 36d8'18"
Shortie
Ari 2h45'18" 24d38'0"
Shortridge49njHigbie
Ori 5h33'10" 13d58'44"
Shortstop
Lyr 19h11'5" 39d14'58"
shortstuff-mern
Lib 14h59'14" -12d8'50"
Shorty
Sgr 19h33'36" -15d38'6"
Shorty
Umi 14h39'11" 75d2'8"
Shorty
Umi 14h18'49" 74d55'14"
Shorty
Vir 14h23'38" 1d46'8"
Shorty C
Aql 19h8'26" 5d38'0"
Shorty Terry
Uma 11h28'48" 36d11'42"
SHORTY16
Psc 1h4'26" 9d44'0"
Shorty46
Pho 0h37'9" -46d42'54"
shorty's little star
Uma 11h25'3" 33d51'27"
Shoshana Emily Oppenheim
And 23h24'55" 43d24'40"
Shoshana, Lily
And 23h55'16" 34d50'4"
Shoshanna
Lmi 10h19'12" 36d48'36"
Shoshanna
Cnc 9h4'56" 26d18'29"
Shoshanna lily
Cas 0h15'59" 54d0'37"
Shoshia
Tau 4h38'24" 10d54'38"
Shotaro Takiguchi
Gem 7h59'30" 22d45'8"
Shotgun
Ori 6h1'44" 19d40'4"
Shotlips
Uma 11h40'24" 52d35'43"
Shotze
Lyn 7h17'4" 58d20'47"
Shoua Vue
Cap 20h24'56" -19d23'15"
Shouhanbou
Cnc 8h32'51" 10d16'26"
Shouq
Crb 16h24'12" 28d59'12"
Shouruq
Cyg 21h32'29" 46d9'1"
Show Me Heaven
Per 3h16'54" 44d33'29"
Show Me The Money
Uma 8h12'39" 71d57'7"
Show*Stopper
Per 3h2'27" 43d52'25"
Showow
Per 4h41'35" 49d55'51"
Shows the Way
Leo 11h43'53" 13d43'4"
Shrena
Del 20h42'49" 15d17'28"
Shri Sadguru Babaji Ramnarine
Cep 20h44'0" 57d52'17"
Shriya
Umi 14h17'50" 77d15'38"
SHROGAL
Sco 16h34'58" -28d25'37"
Shrump
Lib 15h39'26" -21d0'7"
Shruti Ahluwalia
Tau 3h46'55" 17d8'34"
Shruti Bhardwaj
And 1h58'3" 45d34'44"
Shruti Chaganti
Dra 18h12'44" 61d2'41"
Shterlina 35
Lib 15h8'9" -13d39'35"
Shuang Light Sunset
Sgr 18h54'45" -34d14'52"
Shubaby
Cnc 8h37'18" 16d26'21"
Shufei Chiu
Dra 16h18'30" 64d8'38"
Shug
Dra 17h32'39" 63d39'24"
Shug and Hondo
Tau 3h54'16" 8d33'32"
Shug Roma
Uma 10h38'29" 65d17'14"
Shuga
Leo 9h35'42" 17d37'32"
Shukri
Ori 6h17'6" 10d50'29"
Shumaila Pirani
Lyn 6h50'20" 58d25'38"
Shumba Famba
Cas 0h0'27" 55d5'1"
Shumin
Psc 1h1'35" 8d32'52"
Shumpei Yoshizawa & Michele Worthy
Tau 5h13'53" 25d22'10"
Shun Ling
Dra 15h48'7" 56d7'37"
Shune Min
Tau 3h44'9" 30d23'6"
Shupp's Shining Star
Her 18h35'52" 19d4'38"
Shuri Salazar-Gem
Srp 18h40'36" -0d11'22"
Shurooq
Cnc 8h24'25" 6d54'50"
"Shut Up"
Dra 17h14'56" 51d42'12"
ShuTonne Stewart
Uma 12h32'51" 55d5'48"
Shutterbug
Cyg 21h15'38" 35d25'3"
Sh'Vaughn Elise Heath
And 23h23'10" 52d6'55"
Shwan M Saieed
Uma 10h5'45" 49d51'41"
SHWEE DZ
Tau 3h41'58" 10d3'36"
ShweePee
Sgr 18h50'26" -16d1'44"
Shweetie
Uma 9h16'55" 49d38'28"
Shy ann from Cheyenne
Peg 22h2'31" 12d4'52"
Shy Cowboy
Cyg 19h29'55" 30d49'23"
Shyam Patel
Gem 7h17'44" 16d33'21"
Shyama
Psc 1h20'49" 32d57'18"
Shyamali N. Joseph
Lib 14h53'37" -2d43'43"
Shyan
Sco 16h3'50" -13d39'34"
SHYAN
Ori 5h56'26" 18d26'30"
Shyane Nichole Hatcher
Cyg 20h14'31" 34d41'29"
Shyann Rivers Rutherford
Vir 13h19'57" 14d11'40"
Shyann Starr McDowall
And 23h12'41" 47d6'29"
Shyanne
And 2h27'20" 47d33'51"
Shyanne Kaye Marie Haag
Psc 0h54'53" 17d30'45"
Shygirl Cyndi
Gem 6h50'22" 31d41'41"
SHYLA
Cma 6h55'28" -11d30'7"
Shyla Loves Pryde
Tau 4h13'54" 25d3'15"
Shyla Reno
Uma 10h30'35" 44d58'36"
Shyla Torres
Lib 15h22'2" -11d41'2"
Shylah
Psa 22h56'26" -36d0'50"
Shylah Joy
Psc 0h48'31" 5d48'32"
Shylo Marie Harlan
And 23h28'37" 44d23'49"
Shyne
Ari 2h38'32" 14d3'41"
Shyra Louise Early
Per 3h16'52" 42d26'17"
Shyrle Lee Applegate
Lib 15h20'0" -24d4'38"
Shysie
Aqr 22h8'20" -1d2'35"
Si & Jen Bunning - Soulmates
Cyg 21h45'3" 54d27'21"
Sia Howe
Gem 7h31'6" 30d25'17"
Siadbh Stack
Lyr 18h48'50" 32d33'45"
Siam
Psc 1h28'1" 29d37'41"
Siamese Dream
And 1h5'4" 43d3'54"
Sian Anne Clougherty
And 0h0'22" 34d52'31"
Sian Cerys Connolly 090904
And 1h46'44" 45d6'59"
Siân Ellen
Lyn 6h31'51" 56d25'25"
Siân Jean Tanner
And 1h32'0" 50d6'5"
Sian Josephine Moore
Pav 19h42'25" -65d41'50"
Sian Shereen Reindl
Sco 17h10'42" -31d53'25"
Sian V. Warren
Uma 9h52'44" 71d2'15"
Sian Williams
Cas 0h31'21" 66d41'41"
Sian-Ady Long
Dra 16h25'24" 63d1'14"
Siara Lyden
Uma 10h31'25" 65d41'5"
Sibby "Lala" Bennett
Uma 10h39'29" 47d0'29"
Sibel
Com 13h25'49" 17d52'38"
Sibley Sol Thomson
And 23h59'3" 45d35'30"
Sibyl Juanita
Vir 13h30'17" 4d29'39"
Sibylla Schulz
Ari 3h27'26" 21d2'21"
Sibylle
And 22h59'54" 38d15'50"
sibylle brönnimann
Cyg 21h10'19" 31d15'43"
Sibylle Bruggmann
Uma 11h17'13" 54d14'46"
Sibylle Dötsch
Uma 13h50'23" 61d56'46"
Sibylle Kessler
And 23h4'32" 49d9'23"
Sibylle Knöpfel
Crb 16h24'22" 33d53'8"
Sibylle & Patrick
Uma 9h22'41" 47d12'7"
Sibylle & Tom
Uma 11h24'46" 53d20'30"
Sic Note
Lyr 18h49'32" 34d54'56"
SiCaBo
Sgr 18h16'26" -28d5'54"
Sichwart, Juri
Uma 9h42'32" 63d33'7"
Sicilian Princess
Leo 9h28'43" 11d28'55"
Sicilian Rose
Tau 4h55'18" 21d54'18"
Sickly Puppy
Cma 7h21'6" -15d19'1"
SICU Staff Indiana University Hosp.
Per 3h21'54" 46d46'45"
"SID"
Cyg 20h26'3" 44d2'0"
Sid Addison
Her 16h17'20" 6d23'57"

Sid and Eddie
Cyg 21h31'29" 49d5'9"
Sid and Joan 50th Anniversary
Uma 11h39'29" 58d9'54"
Sid and Nancy
Her 18h6'1" 47d13'49"
Sid Glover
Ori 6h7'36" 17d19'10"
Sid Michael Book
Sgr 20h18'34" -33d35'11"
Sid Rutty's Valentine's Star
Cru 12h2'0" -63d41'56"
Sid Sprague
Cap 20h15'20" -17d19'15"
Siddhi and Chintan Forever
Psc 0h57'26" 19d18'0"
Sidekick
Lyn 8h32'40" 46d25'55"
Sidelle
Uma 11h43'2" 39d38'37"
Sidney
Lyn 7h26'37" 44d41'4"
Sidney
Aur 5h57'10" 37d8'16"
Sidney
And 23h20'47" 51d32'5"
Sidney
Her 17h0'46" 13d35'2"
Sidney & Alicia
Uma 10h23'59" 43d51'21"
Sidney Allen Woodard
Lib 14h54'30" -6d18'21"
Sidney Anaya Brown
Leo 11h16'32" 24d13'48"
Sidney and Ellen Levine
Cyg 19h36'18" 39d29'43"
Sidney Ann
Tau 5h49'18" 27d5'6"
Sidney Antoine Ordoyne IV
Uma 9h29'12" 71d33'37"
Sidney Bolter, M.D.
Peg 22h22'8" 9d27'58"
Sidney Byrd "The Byrdman"
Sgr 19h18'22" -22d32'18"
Sidney C. Riester
Lib 16h1'28" -5d54'53"
Sidney E. and Frances D. Reason
Sco 16h16'22" -12d53'26"
Sidney Elizabeth
Aqr 22h32'47" 0d26'29"
Sidney George Whittle
Her 16h38'4" 6d56'46"
Sidney & Ivy Churchill
Ari 3h10'0" 14d9'55"
Sidney Jo Lewis
Lib 15h2'16" -14d51'19"
Sidney Joan Kiner
Ari 3h3'24" 28d27'24"
Sidney Kay
Cam 4h29'45" 59d42'27"
Sidney Kohl
Cyg 21h21'7" 38d50'39"
Sidney Mallard
Cyg 20h42'40" 33d4'9"
Sidney Margaret Michelle Stanton
Ari 2h23'47" 24d7'39"
Sidney & Marilyn Weisman
Cyg 20h57'41" 47d55'1"
Sidney Mason Pinner
Lib 15h51'46" -6d24'14"
Sidney Nicole Carbajal
Cyg 19h58'44" 34d40'17"
Sidney P3
Cra 18h54'16" -37d58'45"
Sidney Poitier
Uma 11h30'11" 57d10'58"
Sidney R. Levine
Sco 16h11'52" -12d14'22"
Sidney Robert & Oma Jean Wandvik
Umi 15h31'46" 73d54'41"
Sidney Rock
Ori 6h16'17" 15d9'8"
Sidney Rofey
Uma 11h29'29" 65d31'24"
Sidney Sockman
Cnc 8h26'36" 19d1'38"
Sidney Stalnaker
Uma 11h15'4" 37d18'26"
Sidney Walter Adamson
Uma 14h4'39" 61d22'37"
Sidney Waymire
Per 3h4'4" 42d37'36"
Sidney Wilson - Sid's Star
Cnv 12h44'14" 39d37'30"
Sidney Zdenek Vacek
Per 4h27'32" 34d6'20"
Sidney's Jello
Vir 13h24'15" 13d22'46"
Sidonia
Lib 15h34'47" -11d11'14"
Sidonia & Stefan forever
Cam 6h33'38" 65d44'39"
Sidonie Gent
Dra 19h12'28" 74d5'40"
Sidonie Ruth Rhodes
Leo 10h56'50" 19d32'7"
SidraMikhiel
Crb 16h8'5" 37d58'31"
Sid's Dream
Her 17h29'20" 34d8'17"
Sidus Adamo
And 0h58'30" 40d22'22"
Sidus Adria-E293
Ori 5h35'17" -3d40'54"
SidusMariae XLVI
Cas 0h59'20" 61d22'22"
Sieanna Christina Thissen
Sgr 19h8'38" -13d5'21"
SIEBEN
Sco 17h52'41" -38d11'47"
Sieben, Jana
Uma 10h1'20" 44d23'7"
Sieber Cédric
Sct 18h26'56" -15d31'48"
Siebersma
Ori 5h41'13" -2d2'17"
Sieburg, Julia Tamara
Uma 8h38'55" 46d53'47"
Siegberg, Erica
Leo 10h27'7" 18d56'49"
Siegfried
Ori 5h23'29" 4d14'10"
Siegfried
Psc 1h7'55" 2d58'24"
Siegfried and Shirley
Aql 20h0'2" 15d29'15"
Siegfried Brandenburger
Ori 5h44'7" -5d26'53"
Siegfried Dämmig
Uma 11h1'29" 58d26'19"
Siegfried Karl Scheffzik
Uma 9h15'13" 54d40'53"
Siegfried Sundmark
Uma 8h33'4" 70d16'47"
Siegfried Thiel
Uma 9h50'16" 43d24'7"
Siegi - der hellste Stern am Firmament
Uma 12h48'35" 53d6'20"
Sieglinde Quaatz
Uma 10h32'7" 39d58'5"
Siegmund Frey
Ori 4h50'11" -2d21'25"
Siegrid Levin
Uma 8h43'2" 63d5'0"
Siemny Chhoun
Tau 4h25'27" 15d13'1"
siempre brille a tres
Tri 1h40'37" 29d5'35"
Siempre David
Gem 7h12'59" 22d26'13"
SIEMPRE de Alexis Michel
Lib 15h13'51" -9d6'47"
Siempre estare contigo...
Aql 19h41'53" 3d29'4"
Siempre Isela
Tau 4h58'20" 22d32'0"
Siempre Te Querre
Uma 9h32'51" 57d43'55"
Siempre y por eternidad
Eri 4h31'1" -3d14'43"
siempre y por siempre
Cap 21h42'47" -13d15'36"
Siena Angelina Sadovnik
Cnc 8h10'18" 7d41'49"
Siena Camille Sparacin
Ari 2h29'32" 21d50'43"
Siena Dawn
Leo 11h34'0" -5d16'58"
Siena Jade Livingston
Cam 6h3'38" 62d50'54"
Siena Marie Schönborn
And 23h17'5" 49d40'48"
Siena Reece
Leo 9h24'51" 18d3'52"
Siena Ria Hagens
Cru 12h31'17" -60d1'51"
Sienna
Sgr 17h58'19" -23d4'5"
Sienna
Lyn 6h35'39" 55d54'27"
Sienna
Ari 2h45'8" 20d14'11"
Sienna
Uma 8h54'32" 50d19'38"
Sienna
And 2h23'43" 44d47'44"
Sienna 06
Col 5h47'56" -34d29'4"
Sienna Amy Mason
Crb 15h43'58" 35d22'1"
Sienna Ava
Car 9h48'44" -60d23'44"
Sienna Bailey Rowden
And 2h32'54" 44d27'14"
Sienna Clark
Vir 12h19'53" -2d56'30"
Sienna Fucci
And 1h14'59" 45d27'53"
Sienna Grace Copeswinnerton
Sgr 19h19'34" -16d48'3"
Sienna Joy Fraser
Cru 12h18'9" -63d1'13"
Sienna Lappin
And 0h16'47" 41d3'18"
Sienna Lee Nordstrom Wanagas
Aqr 23h16'10" -15d48'46"
Sienna Lily
And 23h29'19" 47d29'48"
Sienna M. Baird
And 0h53'38" 38d35'51"
Sienna Rio Emily Bailey
Vul 19h39'2" 27d37'32"
Sienna Rose
Tau 3h51'58" 3d22'36"
Sienna Rose
Gem 7h38'6" 33d16'22"
Sienna Vidya Sindwani
Tau 4h35'26" 23d28'21"
Sienna Willow
Uma 12h20'1" 54d58'16"
Sienna & Zianna
And 23h24'10" 48d11'43"
Siera
Tau 4h37'53" 17d10'28"
Siera Coyne
Uma 11h22'39" 71d34'45"
Siera Marie
Leo 9h30'27" 32d52'55"
Siere Eve
Col 5h48'24" -35d11'27"
Sierra
Sgr 18h4'29" -35d54'36"
Sierra
Psc 1h57'31" 7d25'33"
Sierra
Uma 9h59'0" 66d56'33"
Sierra
Lyn 8h4'8" 53d29'38"
Sierra
Lib 14h29'30" -19d0'19"
Sierra
Gem 7h36'52" 34d8'30"
Sierra
Uma 11h20'5" 36d46'44"
Sierra
And 0h49'48" 39d18'46"
Sierra
Uma 11h27'45" 42d38'1"
Sierra
Uma 10h39'4" 48d50'3"
sierra
Cyg 19h40'29" 39d12'27"
Sierra
Peg 22h5'13" 10d17'4"
Sierra
Ori 5h15'29" 1d12'32"
Sierra
Psc 23h15'19" 7d2'30"
Sierra
Ari 3h4'37" 28d45'19"
Sierra 95
Uma 9h1'52" 52d51'25"
Sierra A. Feaster
Lyr 18h49'16" 31d26'10"
Sierra Alyssa Lyons
Sco 16h11'15" -13d39'51"
Sierra Andrea Schmitt
And 23h39'33" 42d17'19"
Sierra Anne Listug-Lunde
Sco 17h38'59" -32d56'51"
Sierra Ashley Schwarz
Umi 14h44'25" 85d50'54"
Sierra Beach Hotel
Per 3h8'14" 51d29'56"
Sierra Boyack
And 1h17'36" 36d25'33"
Sierra Dawn
Cap 20h16'47" -9d43'2"
Sierra Dawn Clemons
Cnc 8h53'52" 27d1'52"
Sierra Delfinia
And 0h30'32" 32d13'18"
Sierra Elizabeth Jendrzey
Cap 21h9'54" -16d8'54"
Sierra Elizabeth Pedri
And 0h37'43" 41d16'30"
Sierra Francis Stewart
Crb 16h2'46" 37d39'31"
Sierra Grace Willems
Uma 11h24'28" 64d20'23"
Sierra Hanley
Uma 12h59'45" 59d40'30"
Sierra Kay Roberts
And 0h18'36" 29d16'30"
Sierra Kelley Razor
Del 20h48'2" 17d2'14"
Sierra L D Ortiz
Cap 21h22'53" -14d37'37"
Sierra Lumber & Fence Company
Lyn 9h7'22" 38d3'24"
Sierra Lynn Bard
Vir 13h17'26" -2d41'14"
Sierra Lynn Cerrone
Lib 15h6'40" -18d37'42"
Sierra Magick
Psc 1h18'49" 17d55'34"
Sierra Marie
Uma 11h49'32" 33d50'41"
Sierra Marie Daigle
Cam 5h51'42" 66d20'30"
Sierra Marie Mercado
Sco 16h17'39" -10d49'53"
Sierra McPeak
Psc 0h56'31" 31d45'32"
Sierra Megan Papp
Leo 9h50'20" 24d44'39"
Sierra Michelle Williams
Uma 13h54'7" 51d6'33"
Sierra Monique Brown
Vir 13h21'9" 12d24'10"
SIERRA NICOLE
Tri 1h56'14" 34d0'5"
Sierra Nicole Bond
Cyg 20h18'21" 44d42'25"
Sierra Nicole Stewart
Lib 15h12'44" -12d31'54"
Sierra Ona Messier
Cap 20h8'26" -22d12'59"
Sierra Rae Mikula
Sgr 18h54'36" -16d12'15"
Sierra Rain 22605
Psc 1h59'43" 7d22'51"
Sierra Renea McCallum
Leo 11h1'50" 11d54'11"
Sierra Rouyese Bell "Mommy's Angel"
Sco 16h14'26" -20d25'39"
Sierra Sharron
Ari 3h1'9" 18d32'44"
Sierra Skye Crowe
Crb 16h16'25" 31d54'40"
Sierra Tandyk
Gem 6h48'21" 14d53'30"
Sierra T.R. Cass
Lyn 8h39'47" 37d1'0"
Sierra Whitlee Muse
Sco 16h38'35" -29d0'9"
Sierra, FJ
Cas 23h32'26" 52d16'55"
Sierrah Michelle Fernandez
Lyn 8h6'7" 59d27'49"
Sierra's Shining Star
Tau 3h55'13" 29d43'24"
Sierra's Star
Lyn 7h10'29" 48d15'1"
Sierra's Wish
Ari 2h43'36" 21d18'39"
Sievert, Melanie
Sgr 18h3'51" -27d55'10"
Sievo
Tau 4h20'40" 25d36'51"
Sifl
Lyn 8h36'47" 46d30'25"
Sifu Raymond Wong
Her 17h45'59" 27d20'45"
S.I.G. Greenebaum
Sco 17h43'2" -44d29'14"
Sig. Mario Falanga
Per 1h43'56" 50d45'27"
Siggy & Yola
Lyr 19h18'53" 29d47'22"
Siglas
Uma 11h24'58" 39d45'53"
Sigma Delta Tau - Beta Phi Chapter
Uma 13h54'44" 51d15'59"
Sigma Omega Nu. Love, Iota Class 06
Cas 0h39'32" 56d56'3"
Sigmund Stanis
Ori 4h50'41" 3d0'6"
Sign of Sheri (S.O.S.)
Uma 11h47'38" 64d59'12"
Signe G-det meste vakker stjernen
Uma 11h57'33" 59d6'36"
Signe Lauren Johannes
Cas 1h36'22" 68d42'14"
Signe M. King
Leo 9h53'34" 15d49'31"
Signe Marie Cady
Uma 11h33'50" 62d48'50"
Signo Frederico
Cap 20h21'43" -11d30'29"
Signs of Eternity M to K
Vir 14h32'23" -1d2'46"
Sigolène F.
Ari 2h7'45" 25d40'7"
Sigorney's Dream
And 1h58'11" 39d21'0"
SIGOS
Uma 11h47'52" 37d15'56"
Sigourney Gray
Col 5h59'58" -34d50'24"
sigrid
Lyn 7h57'49" 49d35'34"
Sigrid
Leo 10h21'25" 26d5'55"
Sigrid
Ori 5h2'44" 13d8'14"
Sigrid Allgaier
Uma 9h10'43" 54d37'16"
Sigrid Borchardt
Ori 5h43'11" -4d43'45"
Sigrid & Horst Pelzer
And 1h46'59" 42d3'38"
Sigrid Lou Battjes May Baby Galaxy
Tau 5h9'7" 27d48'30"
Sigrid, La Mas Brillante De Todas
Psc 1h14'5" 15d29'28"
Sigrid's Courtyard
Sgr 18h15'52" -32d41'25"
Sigrud Lebray
Tau 5h12'58" 17d6'23"
Sigrun
And 1h7'55" 45d51'19"
Sigune für unser geliebtes Kind u. Mama
Uma 13h18'57" 62d37'52"
SiHads
Cyg 21h38'3" 33d15'27"
Sihame Laoulida
Aqr 23h6'53" -20d9'48"
Sihavong-Pengsavath
Cyg 19h41'18" 51d21'45"
Sihaya
Ori 6h0'6" 21d55'12"
Siilverain the Wind Caller
Cap 20h55'14" -25d49'14"
Sikandar (mera baby)
Aqr 22h35'50" 0d52'35"
Sikes Super Stars
Uma 11h4'31" 65d22'9"
Sikina Shawntivia Hunter
Tau 3h53'53" 22d25'8"
SIL
Her 18h17'24" 21d7'10"
Silas Anthony Riley
Sgr 19h9'35" -20d57'48"
Silas Hersh
Cyg 21h51'34" 44d21'30"
Silas Holloway
Uma 11h51'6" 52d57'44"
Silas Marshall
Vir 12h10'16" 11d44'19"
Silas Morgan Charles Lancashire
Cep 21h20'39" 78d2'50"
Silas Nithaniel Luciano "Big"
Cnc 8h33'28" 24d27'1"
Silas Robert
Her 17h22'10" 35d25'30"
Silber Star
Psc 1h6'16" 24d4'23"
Silchester
Dra 17h31'7" 52d22'18"
Sileese
Cep 22h57'47" 71d2'8"
Silent Night
Uma 11h25'49" 41d49'9"
Silent Strong Z-Man
Ori 4h53'14" 4d10'39"
Silent Warrior
Cap 20h31'9" -12d7'10"
Silentghost
Cru 12h11'13" -55d44'10"
silently she...
Leo 10h30'59" 8d2'4"
SILGIU
Umi 15h21'3" 83d46'7"
siliconklaus
Peg 23h23'44" 32d59'20"
SiLiz
Uma 13h23'15" 54d5'56"
Silje Brinck Nicolaisen
Uma 11h26'52" 52d7'45"
Silje Odland
And 2h24'37" 39d8'38"
Silje Therese Vaktdal
Cas 1h37'30" 71d30'42"
Silk
Cap 21h44'31" -13d3'54"
Silke
Uma 9h22'36" 70d8'47"
Silke
Uma 8h12'12" 64d16'57"
Silke
Vir 12h10'44" 11d7'23"
Silke - 23.11.1975
Tau 4h42'15" 24d12'45"
Silke Dettmann
Uma 9h59'21" 41d41'32"
Silke Ingebrand
Uma 8h42'53" 50d35'59"
Silke Jäger
Uma 8h19'5" 66d29'12"
Silke Jansen
Uma 10h3'27" 67d56'25"
Silke Töpfer
Ori 5h51'31" -4d1'18"
Silke und Gernot VIDONJA
Ori 6h11'38" -3d52'24"
Silke, 17.9.1973
Tri 2h29'59" 34d12'39"
Silkes Glücksstern
Uma 11h57'21" 38d7'1"
Silkey
Dra 18h42'44" 63d2'13"
SilkynTall
Mon 7h3'13" -0d4'27"
Sille
Sgr 19h36'1" -15d15'45"
sillo celestia lumino
Lyr 18h30'48" 28d54'16"
Silly
Sgr 18h53'43" -19d14'11"
Silly Berry
And 0h54'2" 40d56'12"
Silly Girl Kristy Lynn
Lyn 9h9'8" 36d44'12"
Silly Goose
Umi 14h34'39" 70d16'21"
Silly Jilly
Her 18h46'33" 24d36'24"
Silly Meice McBride
Cep 22h14'59" 61d34'34"
Silly Old Bear
Cap 20h25'17" -13d9'24"
Silly Papa
Per 3h14'17" 43d46'41"
Sillyons
Ori 5h3'2" 6d41'49"
Silopi
Ari 2h18'13" 25d24'43"
SiLou
Uma 8h11'18" 68d57'24"
Silpa Lettica
Uma 10h26'9" 43d43'12"
SILU
Psc 1h30'45" 9d5'13"
Silva
And 1h12'9" 38d32'56"
SILVA KA
Tau 5h18'42" 25d38'21"
Silva Maia
Uma 9h6'16" 51d26'8"
Silvan
Umi 10h42'28" 88d31'19"
Silvan my love forever
Crb 16h2'28" 35d9'1"
Silvan Nicola
Uma 11h26'32" 71d9'18"
Silvana
Uma 11h16'32" 61d21'0"
Silvana
Uma 8h12'8" 62d19'26"
Silvana
Lib 14h53'22" -4d20'0"
Silvana
Per 3h24'21" 41d36'38"
Silvana
Tau 4h45'44" 28d51'9"
Silvana
Tau 4h39'46" 11d3'18"
Silvana
Ori 6h10'26" 3d34'18"
Silvana
Her 16h33'53" 5d12'33"
Silvana Alejandra Aguayo Lundin
Ori 4h53'13" -1d53'2"
Silvana Avila
Uma 11h48'24" 46d2'55"
Silvana Berteramo
Cas 0h25'26" 61d19'52"
Silvana Bravi
Per 3h5'10" 51d43'15"
Silvana Frenette
And 2h23'9" 41d51'26"
Silvana Pearn-Rowe
Ori 5h7'57" -0d17'8"
Silvana Villacreses
Aql 19h3'48" -8d19'41"
Silvanna
And 0h43'53" 40d52'46"
Silvano
Cep 3h47'39" 82d29'10"
Silvano Turco
Uma 11h53'0" 40d9'14"
Silver
And 0h47'22" 44d4'53"
Silver
Cnv 13h46'51" 31d20'37"
Silver Bear
Aqr 22h41'35" -15d0'44"
Silver Belle-143
And 23h41'39" 42d14'29"
Silver Caine
Cyg 20h2'5" 41d30'4"
Silver Fox
Lmi 10h32'12" 31d43'9"
Silver Fox
Dra 18h36'26" 62d44'21"
Silver Fox
Sco 17h1'37" -38d48'9"
Silver Hoop
Aql 20h1'46" -0d51'5"
Silver Lake (06/30/06)
Tau 4h11'54" 24d19'42"
Silver Montgomery
Gem 7h36'54" 15d54'17"
Silver Rock Resort
Vul 19h37'4" 26d27'47"
Silver Sands Resort
Aur 5h49'59" 54d13'36"
Silver Serenity
Cas 0h22'37" 50d52'25"
Silver Sirois
Ori 6h15'44" 15d13'39"
Silver Sister
Uma 9h39'28" 57d51'51"
Silver Smith
Cyg 20h50'18" 34d14'22"
Silver Soulmates
And 23h48'45" 38d19'13"
Silver Star
Ari 3h3'29" 30d8'25"
Silver Star for Elise
Mon 6h39'32" 5d51'21"
silver wedding
Uma 11h12'8" 70d16'39"
Silver Wings
Aql 19h13'35" -0d30'41"
Silverbells Longs Legger
Tau 5h28'46" 25d2'43"
silverdaddy065
Lyn 6h42'9" 57d58'43"
Silver's Star
Aur 5h59'21" 54d13'23"
Silverstein Star
Per 4h2'39" 33d35'45"
Silvet
Umi 15h4'26" 73d21'46"
Silvia
Cas 23h28'1" 57d26'48"
Silvia
Uma 9h11'24" 54d48'58"
Silvia
Peg 23h17'35" 31d5'57"
Silvia
And 0h37'54" 33d34'18"
Silvia
And 1h38'45" 42d18'29"
Silvia
And 1h39'26" 42d33'21"
Silvia
Lyn 7h50'32" 42d21'21"
Silvia
Aur 6h41'9" 44d8'48"
Silvia
Cam 5h59'45" 56d41'49"
Silvia
Cas 2h46'34" 58d4'14"
silvia
Cas 2h42'0" 58d5'51"
Silvia
Cas 0h52'42" 56d0'35"
silvia
Lac 22h56'25" 44d51'26"
Silvia
And 23h30'46" 41d38'20"
Silvia
Lyr 19h9'3" 44d24'18"
Silvia
Aur 5h55'50" 46d5'19"
Silvia
And 2h14'39" 49d58'33"
Silvia
Mon 6h33'32" 6d52'14"
Silvia
Ori 5h52'25" 6d45'42"
Silvia 10/12/99
Ori 4h43'43" 1d21'51"
Silvia Adriana Estay Cabrerra
Dra 16h56'43" 57d6'58"
Silvia Albertin
Lyr 18h32'23" 27d19'32"
Silvia Alexandra Stroescu
Her 17h0'3" 29d39'26"
Silvia and Rob Perretta
Cap 20h48'21" -21d56'19"
Silvia Bernardinetti
Aur 6h27'32" 40d17'13"
Silvia Biancucci
Boo 14h6'24" 9d26'12"
Silvia Buchs
Lyn 6h45'23" 54d59'25"
Silvia Chenevert
Uma 12h2'2" 53d13'15"
Silvia Coimbra
Gem 7h5'21" 33d0'25"
Silvia Corona
Lyn 7h32'56" 37d56'45"
Silvia De Zen
Aur 5h36'45" 33d7'4"
Silvia & Devin 5-9-02
And 1h32'36" 42d24'56"
Silvia Dimitrov
Cap 21h5'43" -19d4'53"
Silvia Eliana Valenzuela Feijoo
Vel 8h57'22" -46d46'2"
Silvia et Benjamin
Ari 3h11'23" 11d13'56"
Silvia Fogliani
Per 3h24'10" 41d39'1"
Silvia Fredianelli
Cas 0h32'52" 62d21'17"
Silvia Gabriela Martinez
Uma 10h35'29" 55d1'27"
Silvia Gardenghi
And 1h45'53" 48d57'59"
Silvia & Gianmarco
Cas 1h25'40" 59d25'37"
Silvia Gillespie
Lib 15h5'32" -25d31'40"
Silvia Grace Galindo-Donovan
Uma 11h32'20" 56d13'42"
Silvia & Hans ILY
Uma 9h29'51" 57d42'39"
Silvia Helena
Dra 16h5'16" 65d15'31"
Silvia Hsu Cheng
Leo 10h14'54" 14d33'25"
Silvia & Josef
Uma 9h13'41" 51d58'1"
Silvia KarinaTorres
Lib 15h43'6" -17d55'51"
Silvia Knutti
Cas 23h22'52" 55d28'12"
Silvia Krummenacher
Uma 12h11'23" 55d55'18"
Silvia Kuehne
Dra 20h28'13" 67d38'2"
Silvia Künzle
Aur 6h37'2" 39d3'48"
Silvia Lady of the Woods
Aqr 20h39'34" 0d3'0"
Silvia Ludueña Lipkofker
Cas 1h20'43" 67d21'50"
Silvia Maria Ocaña
Per 3h43'25" 43d1'54"
Silvia & Massimiliano
Sex 10h3'0" 2d10'22"
Silvia Montano y Everardo Bucio
Umi 17h3'8" 83d11'12"
Silvia N. Encinas
Lyn 8h56'44" 34d5'57"
Silvia Neroni
Cyg 19h59'15" 40d17'3"
Silvia Prandelli
Lyr 18h29'35" 41d38'27"
Silvia Reichelt
Cnv 13h57'2" 35d53'19"
Silvia Ruano Vigueras
Ori 6h16'50" 2d3'31"
Silvia Santi
Dra 10h23'22" 74d54'18"

Silvia & Simon
Uma 11h37'17" 39d1'26"
Silvia & Simone
And 23h14'16" 36d20'32"
Silvia Spescha
Her 18h41'31" 19d3'30"
Silvia Stutzky
And 2h19'24" 48d28'50"
Silvia+Thereza+Ricardo 3corações
Cru 12h45'7" -55d57'17"
Silvia und Bänz
Lac 22h46'1" 54d10'23"
Silvia Weber Bobbl
Uma 10h13'48" 42d20'16"
Silvia y Israel
Ori 5h23'46" -5d26'37"
Silvia's Gabrielle
Peg 23h34'16" 12d49'31"
Silvia's Star
And 23h7'48" 49d13'35"
Silvie
Lyr 18h58'35" 33d2'8"
Silvie Herdinova
Lib 14h56'9" -17d43'48"
Silvietta
Lyr 18h26'42" 41d32'59"
Silvina "JaxSil" Orosco
Lib 15h37'15" -28d47'30"
Silvina y Miguel Angel con JuanmiB
Cyg 21h21'13" 38d55'20"
Silvio Ascheri
Cam 5h17'49" 56d31'0"
Silvio Berlusconi
And 1h54'37" 45d36'19"
Silvio Bonaccio
Leo 11h16'19" 15d15'20"
Silvio "DD" Mazzella
Dra 18h20'18" 59d17'36"
Silvio Marcigliano
Cru 12h4'51" -63d15'4"
Silviu
Cnc 8h46'42" 31d16'3"
Silvius Dornier
Cma 6h44'10" -16d8'15"
Sim 1
Psc 23h50'52" -3d4'8"
Sim Brooks- Rising Star
Dra 19h15'16" 61d41'23"
Sim Pleaz Bowling
Sco 16h4'57" -14d44'22"
Sim & Russ Always 19/03/2006
Cru 12h29'25" -58d30'52"
Sima and Russell Ort
Cap 20h24'30" -10d39'10"
Sima Auerbach
Umi 17h8'37" 79d9'41"
Sima Tova Kerman 9/12/90
Vir 11h47'3" -0d57'6"
Simaan Kaleem Barday
Cap 21h58'33" -8d37'56"
SimAndJim
Lib 14h51'15" -3d24'23"
Simba
Sgr 19h12'33" -16d7'6"
Simba
Per 2h24'25" 55d13'39"
Simba
Ori 6h19'52" 19d30'25"
Simba
Lmi 10h29'7" 29d39'51"
Simba
Leo 9h42'43" 26d44'10"
Simba
Ori 6h7'55" 15d11'28"
Simba Higgins
Uma 9h18'43" 51d45'30"
Simba Sue
Leo 10h36'15" 14d6'23"
Simbalisa
Ori 5h23'5" -0d10'10"
Simba's Kitty
Lyn 9h12'54" 36d5'11"
SIMEI
Lyr 18h38'46" 37d36'38"
Simen Arnesen Lunden
Lmi 10h49'16" 37d12'56"
Simeon
Lib 14h58'46" -4d13'14"
Simeon Earnest
Uma 8h34'53" 64d6'28"
Simeon Rainbow
Uma 8h16'46" 60d29'21"
Simeon Robert Schmitt
Gem 6h19'8" 21d42'54"
Simera King - Mom, Wife & Friend
Cas 0h17'51" 57d46'55"
SIMEVA
Ori 6h20'34" 9d26'22"
Simfronia Cabusas
Mon 6h55'9" 3d33'12"
SiMi
Vir 13h14'25" -14d31'2"
Similove
Uma 11h28'6" 33d1'37"
Simi's Sweet 16th
Cru 12h43'46" -61d57'53"
SimkiAnish
Cyg 21h28'31" 36d35'54"
Simmone Pannunzio
And 1h15'29" 45d10'17"

SimmonsU.P.S.
Uma 10h31'39" 68d16'56"
Simmsylvania
Lyn 8h22'37" 44d4'54"
Simmy
And 0h0'30" 44d54'25"
Simmy & Betta
Her 18h14'50" 18d37'16"
Simmy Richardson
Col 5h7'38" -37d47'48"
Simo.....
Lyr 19h27'19" 41d9'9"
Simoes Herminio
Psc 0h36'14" 18d37'36"
Simon
Cyg 20h39'17" 39d10'24"
Simon
Her 15h51'17" 40d2'30"
Simon
Cma 6h58'34" -22d42'42"
Simon
Lac 22h45'19" 54d2'52"
Simon
Uma 12h25'18" 58d1'20"
Simon Allso
Cep 23h26'0" 83d19'4"
Simon and Annette Nolen
Psc 1h21'14" 17d48'55"
Simon and Annie
Cru 12h23'14" -58d26'2"
Simon and Melissa Alvarez
Cyg 20h3'40" 36d2'4"
Simon Andrew Snyder
Her 16h56'57" 34d30'52"
Simon Anthony Moore
Uma 13h43'18" 61d28'2"
Simon Arthur
Cep 3h38'11" 83d34'27"
Simon & Asa's Forever Star
Vel 8h22'25" -45d46'52"
Simon Atkinson
Aqr 22h54'7" -12d4'44"
Simon Bernstein
Cyg 19h55'34" 37d38'35"
Simon Bert
Aqr 22h39'26" 2d11'43"
SIMON BIECHLER
Umi 15h51'37" 82d9'29"
Simon 'Big Si' Hayes
Per 3h3'50" 42d35'1"
Simon Blewett - Orla's Daddy
Ori 6h11'49" 8d33'29"
Simon Bridges
Lyn 6h35'19" 57d2'42"
Simon Brunner
Umi 16h1'29" 79d37'34"
Simon Calabrese
Cyg 20h11'12" 35d49'2"
Simon Challenger
Per 2h13'40" 51d15'0"
Simon Christopher
Cap 20h30'29" -10d24'51"
Simon Clarke
Uma 14h2'19" 56d15'51"
Simon Cooper
Per 3h25'39" 38d47'48"
Simon D. Dobbins
Tau 3h42'38" 28d39'23"
Simon Dale Bradley
Uma 11h37'7" 50d35'8"
Simon & Danielle's Star
Cru 12h9'26" -62d20'8"
Simon deMolitor
Uma 11h20'9" 30d32'5"
SIMON&DIANA
Uma 8h20'13" 69d35'9"
Simon Dominic and Tracey Watts
Cyg 21h58'57" 48d14'40"
Simon Ehrlich
Lib 15h29'44" -17d18'10"
Simon Elliott
Cep 20h46'30" 59d37'47"
Simon Escott
Leo 10h57'25" 1d13'37"
Simon F. Carson
Tau 3h42'42" 7d7'44"
Simon "Finny" Maurer
Aqr 21h22'48" -0d29'45"
Simon Franz Isenring
Boo 14h12'42" 29d13'10"
Simon Frederick Walters
Cep 22h52'37" 63d41'56"
Simon G. Mortlock
Cep 0h12'46" 85d7'56"
Simon George Lakosky
Cep 22h42'9" 67d27'31"
Simon Haldemann unser Freund und Helfer
Her 16h32'28" 18d46'40"
Simon Harrison Siena
Tau 4h12'26" 27d24'11"
Simon He
Aur 7h7'28" 38d1'29"
Simon Heath
Per 4h13'38" 47d29'36"
Simon Hess Loves Scott Stafford
Psc 1h7'7" 28d24'48"
Simon Hobson Marchese
Per 4h20'42" 52d39'36"
Simon James King
Lib 15h0'34" -2d49'0"

Simon James Martin
Cep 22h8'40" 73d6'16"
Simon James Moore - "Our Hero"
Per 4h13'44" 43d14'28"
Simon James Rutledge
Tau 4h20'9" 11d18'54"
Simon John Booth
Cep 21h36'58" 55d29'52"
Simon John West
Cyg 20h51'38" 46d6'59"
Simon Jonas Winterhalter
Uma 11h42'45" 32d3'10"
Simon Jones
Vir 12h55'13" -3d18'47"
Simon Joseph
Umi 15h19'58" 72d19'56"
Simon & Kath
Cyg 21h47'32" 44d41'44"
Simon Koza
Leo 11h48'6" 15d21'11"
Simon Land - 21 Today
Per 3h29'48" 40d16'39"
Simon Leigh-Smith
Ori 5h35'21" -1d31'34"
Simon Lejeune
Ori 6h11'48" -0d45'31"
Simon Lewis Callahan
Cep 22h49'31" 70d34'47"
Simon Major
Uma 11h23'48" 42d13'54"
Simon Man-Kit Choi
Psc 23h26'21" 1d49'2"
Simon Mark Druce
Uma 13h33'30" 62d49'34"
Simon Michael Schmitt
Her 16h24'9" 17d9'0"
Simon Michel
Her 18h36'54" 19d10'5"
Simon Mouat
Cru 12h37'36" -61d2'35"
Simon "Nobby" Brown
Sco 17h43'11" -39d15'9"
Simon & Peeya Butler
And 2h35'29" 50d2'19"
Simon Peter Burns
Per 4h26'6" 33d11'14"
Simon R. Ross
Per 2h43'11" 56d43'35"
Simon Rankin Natalis Quinquagesimus
Per 3h18'38" 42d55'27"
Simon & Regina
Cyg 19h59'28" 39d20'21"
Simon Remillard
Gem 6h40'18" 21d4'43"
Simon Richardson
Ari 2h31'28" 13d10'31"
Simon Scicluna
Cnc 9h7'23" 32d18'38"
Simon Shirley
Aql 19h12'52" -11d28'48"
Simon Smith's Star
Uma 11h10'18" 34d21'11"
Simon & Stephanie Linnell
Cyg 21h24'47" 33d46'53"
Simon&SuzaneHay
Sco 17h57'23" -37d29'25"
Simon & Tamara - Children of Light
Ari 1h58'41" 20d34'50"
Simon Tobias Meier
Uma 9h22'59" 67d11'50"
Simon Varley
Her 16h57'44" 44d10'52"
Simon Verkaik
Psc 1h27'38" 24d2'39"
Simon Vos
Cru 12h42'49" -57d18'52"
Simon Walters
Ori 5h29'41" 13d29'37"
Simon Weinberger
Uma 9h25'36" 54d29'23"
Simon Weinstein
Dra 19h42'58" 72d55'30"
Simon Wendall
Ori 6h19'53" 10d25'43"
Simon Wheeler
Vir 13h12'53" 12d6'37"
Simon William Bakker
Cyg 21h51'23" 47d48'1"
Simon William Ginders
Cru 12h53'40" -56d43'6"
Simon Wooster Ames
Umi 16h27'1" 77d22'27"
Simon Wüst
Ori 6h2'41" 21d5'26"
Simona
Crb 15h24'20" 25d37'56"
Simona
Crb 15h56'37" 25d44'27"
Simona
Her 17h41'15" 44d45'41"
Simona
Cyg 20h8'35" 36d40'49"
Simona
Lyn 9h20'22" 39d27'25"
Simona
Cyg 19h39'0" 30d1'43"
Simona
Aur 5h37'2" 32d30'21"
Simona
Aur 6h23'12" 33d16'16"
Simona
Umi 15h40'27" 83d16'50"

Simona
Cma 6h52'58" -11d22'57"
Simona Bongiorno
Her 18h10'48" 18d16'33"
Simona Ceglie
Aur 6h50'14" 43d59'30"
Simona Chiales
Umi 15h7'55" 79d13'58"
Simona e Dante
Peg 23h14'15" 31d6'30"
Simona e Maurizio
And 2h37'20" 39d45'36"
Simona Jakubcova
Cyg 20h35'6" 45d23'28"
Simona Jana
Ori 6h2'1" 21d4'24"
Simona Lanzolla
Aur 7h23'49" 40d15'47"
Simona Maccari
Ori 5h45'26" 1d55'18"
Simona Marie Procell
Lib 16h1'12" -16d40'37"
Simona Moraru
Cyg 21h37'4" 42d3'43"
Simona Patitucci
Ari 3h28'9" 20d7'51"
Simona per sempre
Per 3h25'16" 50d9'25"
Simona Sammaciccia
Umi 14h1'18" 75d0'38"
Simona Sarati
Lyr 18h45'41" 31d3'45"
Simona Scibilia
Cyg 19h33'50" 30d4'50"
Simona Sozzè Bricco franco
Uma 10h19'10" 50d37'8"
Simona Tasco
Cyg 19h35'6" 30d0'16"
Simona & Vincenzo
Cyg 21h34'13" 33d56'43"
Simoncini
Cam 4h17'27" 68d51'42"
SimonDi
Cyg 21h19'53" 32d34'48"
Simone
Uma 9h47'19" 42d28'33"
Simone
Cnc 9h10'49" 32d24'19"
Simone
Uma 9h52'44" 49d49'2"
Simone
Cas 0h5'42" 57d6'7"
Simone
Per 2h50'16" 48d10'20"
Simone
Cyg 19h53'34" 37d59'20"
Simone
Tau 5h39'16" 21d18'17"
Simone
Uma 11h22'38" 67d51'24"
Simone
Cas 23h30'43" 57d57'54"
Simone
Psc 1h17'25" 2d56'34"
Simone Alane
Vir 14h26'3" 3d26'15"
Simone Alexandra
Leo 10h40'16" 15d35'24"
Simone Alice, 05.06.1986
Her 17h5'34" 13d36'37"
Simone Alpha Andromeda
And 23h15'49" 44d13'1"
Simone Andrea
Cyg 20h52'44" 36d16'34"
Simone Brambilla
Cas 23h27'10" 56d37'9"
Simone Cook (Sim1)
Pho 2h21'43" -46d56'46"
Simone Dimtsas-Ritchie
Vir 13h43'14" 4d17'5"
Simone Dolly
Cyg 21h38'52" 46d4'49"
Simone e Paolo
Cas 0h31'56" 54d59'44"
Simone e Tania
Uma 13h3'9" 57d49'12"
Simone Eva Payette
Crb 16h5'59" 37d23'6"
Simone Gariépy
Uma 9h52'30" 64d38'6"
Simone Harmina Prince My Bella
Cru 12h7'11" -61d23'2"
Simone Haubensak
Sge 19h52'36" 18d44'52"
Simone Ivis Brennan
Psc 0h17'10" -3d42'50"
Simone Jan, 6, 1999
Cmi 7h28'46" 8d38'42"
Simone Kiah Sanders-May
Gem 7h35'25" 24d15'44"
Simone Knoff
Ari 2h45'48" 28d6'35"
Simone Larson
Cnc 8h17'8" 10d13'33"
Simone & Laurence Plaskow
Cyg 20h17'24" 50d21'9"
Simone Lee
Ari 3h17'46" 19d31'39"
Simone Loosli
Cas 23h32'47" 53d47'0"
Simone Lüscher
Aqr 22h41'22" -15d34'56"

Simone Luu
Lib 14h47'3" -13d44'24"
Simone Marchese
Cas 23h54'2" 56d49'22"
Simone Mignacca
Gem 6h46'19" 12d52'45"
Simone "My Chini"
Vir 15h1'30" 4d24'30"
Simone - My Wife and Lover
Tau 5h49'16" 14d45'9"
Simone Nicole
Vir 13h35'19" 3d53'15"
Simone Nicole Giddens
Cap 20h19'32" -27d20'45"
Simone Oneto
Gem 7h44'56" 25d1'33"
Simone R Bélanger
Psc 1h41'35" 24d11'18"
Simone Recasner
And 2h27'44" 37d19'58"
Simone Rose Courtright
Cap 20h21'32" -15d28'47"
Simone S. Gearhart
Tau 4h27'6" 13d10'59"
Simone Schad-Smith
Uma 8h47'24" 69d12'29"
Simone Simons
Cas 23h21'7" 58d40'12"
Simone Smoochie is Loved
Tau 4h50'52" 28d57'49"
Simone Spitzenberger
Uma 10h4'34" 63d27'21"
Simone & Stanley
Ari 1h47'46" 11d51'41"
Simone und Holger
Cap 21h20'15" -17d7'44"
Simone Van der Zandt
Lib 15h43'44" -12d30'42"
Simone Walliser
Per 3h41'18" 45d21'35"
Simones Majestic Heavenly Radiance
Pav 20h18'45" -69d21'39"
Simone's Star
And 2h17'12" 41d53'27"
SiMONE-Schatz Optima Maxima
Cap 20h18'45" -16d33'25"
Simonetta
And 2h22'6" 42d50'23"
Simonidesz Csaba csillaga
Sco 17h4'4" -30d54'25"
Simonik Forever Kiddo'
Lib 14h52'57" -3d1'2"
Simonne Bryan
Lyr 19h19'15" 28d7'8"
Simon's 6
Uma 11h20'51" 59d46'59"
simons glücksstern
Cep 3h24'18" 81d6'56"
Simon's Kensington
Psc 0h42'40" 17d15'43"
Simon's Star
Per 3h39'19" 31d48'1"
Simon's Star
Aql 19h0'2" -7d57'38"
Simonson
Cyg 19h57'17" 35d56'58"
SimonViktorMarvin
Her 17h9'48" 43d44'54"
simor ambrosius
Umi 14h25'2" 71d47'43"
Simp
Uma 11h25'45" 43d36'36"
Simplicity
Lyr 18h30'55" 28d41'49"
Simply Amazing
Cas 0h56'50" 61d42'15"
Simply Amazing K.S.Mason 1
Ori 5h31'47" 1d32'35"
Simply Carol
Lyn 6h45'37" 53d49'42"
Simply Divine
Gem 6h38'44" 26d32'44"
Simply Our Star
Cyg 20h24'52" 36d35'43"
Simply Sensational
Cep 3h15'47" 82d30'8"
Simply The Best
Lyn 8h54'37" 34d13'27"
Simply Wonderful
Lyn 8h17'21" 59d26'6"
Simpson Middle School 8th Graders
Per 4h48'48" 38d49'48"
Simpson Supercells
Her 17h14'48" 27d6'28"
simpsy
Cam 4h21'38" 63d31'15"
Simran Heer
Tau 5h45'57" 15d24'48"
Simran Sanjiv Amin
Uma 8h33'57" 67d11'23"
Simrun Lila Bista
Cap 20h49'9" -16d25'37"
Simryn Covarrubias
And 1h11'59" 45d19'47"
Sims Love
Uma 8h53'30" 51d23'20"
simtropolis
Ari 2h46'47" 19d47'51"
Sin Fung
Lib 14h53'59" -1d9'7"

Sina
Uma 10h20'28" 65d46'35"
Sina
Uma 9h8'50" 56d33'19"
Sina
Tau 4h21'9" 13d1'51"
Sina
Lmi 11h6'0" 24d56'11"
Sina and Jessica
Sge 19h10'1" 19d43'42"
Sina Ceye
Umi 14h24'43" 69d15'2"
Sina Kim
Psc 1h25'36" 10d46'59"
Sina Marie Graham
Tau 4h23'31" 13d8'53"
Sina Noemi Sahli
Her 17h0'32" 18d23'39"
Sina Rahel
Mon 7h33'53" -0d50'59"
Sina Vacca
Cam 5h7'57" 68d46'13"
Sina Wickenheisser
Psc 23h44'13" 2d16'37"
Sinai Tzabari
Umi 14h19'54" 76d16'24"
Sinara Shines
Gem 7h4'56" 28d16'54"
sincerity
Cap 21h43'37" -8d30'27"
Sinclair
Tau 5h48'21" 27d54'3"
Sindi Lee Dawn
Cyg 20h16'42" 32d53'26"
Sindu Shukla
Lmi 9h41'49" 34d4'9"
Sinduja
Cap 20h58'12" -16d40'26"
Sindur's Mystique
Cnc 8h29'17" 27d50'55"
Sindy
Crb 16h17'53" 30d31'29"
Sindy Ann Sawyer
Tau 5h58'31" 23d52'12"
Sindy Hartman
Cas 1h27'18" 52d3'54"
Sindy Lee Strickland
Mon 7h8'37" -0d30'40"
Sindy Marie Lee
Lib 15h14'47" -21d43'32"
Sine Suksdorf Bjarnason
Ori 5h31'47" 10d20'59"
Sinead
Uma 10h14'29" 51d13'50"
Sinead
Uma 11h13'2" 41d21'52"
Sinead Bryant
Uma 11h36'13" 63d42'50"
Sinead Carol McVey
Crb 15h58'30" 32d37'18"
Sinead Cross
And 2h18'57" 37d43'1"
Sinead Epiphany Curran
And 23h15'8" 52d16'40"
Sinead Grace Hennelly-Bridges
Cas 23h11'47" 59d5'36"
Sinead Hegarty
Umi 14h19'17" 68d57'35"
Sinead Murphy
Psc 0h48'59" 17d39'29"
Sinead O'Neill
Cas 0h46'36" 48d11'34"
Sinead Poppy
Sco 17h18'40" -33d5'55"
Sinèad & Scott Bewley
Mon 7h17'14" -0d3'42"
Sinead Stewart
Lib 15h28'45" -18d28'10"
Sinead Uden
Cnc 9h5'46" 9d33'32"
Sinead-"My Gem"
Uma 10h25'17" 45d4'31"
Sinee
Tau 5h49'19" 22d16'46"
Sinemma
Del 20h33'42" 14d41'13"
Sing
Sco 16h4'19" -19d44'53"
sing ting
Aqr 22h29'13" -2d26'0"
Sing4God
Aur 6h30'8" 51d36'41"
Singer Star
Cap 21h45'59" -9d26'24"
SiNic
Cyg 21h28'47" 32d5'33"
Sinisa
Uma 8h48'32" 51d45'15"
SINISA
Boo 13h49'15" 18d21'6"
Sinisa Ljustina
Peg 21h47'6" 8d13'17"
Sinka József
Cnc 9h3'5" 23d52'28"
Sinn, Bernhard
Ori 6h4'44" 10d44'55"
Sinn, Stefan
Uma 11h29'24" 28d32'37"
sinnlicher Büffel
Equ 21h23'25" 6d51'22"
Sinobu 13 Kurono
Gem 6h19'38" 24d26'50"
Sinolickij, Sergej
Ori 6h17'36" 16d36'57"

Sinrock31
Cnc 9h0'15" 29d41'42"
Sinthia
Del 20h33'38" 14d8'43"
Siobhan
Ari 2h19'35" 11d56'31"
Siobhan
Leo 10h22'41" 27d9'40"
Siobhan
Cas 23h6'1" 58d30'14"
Siobhan
Cyg 20h20'1" 55d59'32"
Siobhan
Uma 13h37'33" 54d6'53"
Siobhan
Aqr 22h27'47" -24d22'19"
Siobhan 2/25/88
Ori 5h50'51" -0d13'11"
Siobhan and Justin Clark
Sge 19h50'34" 18d58'29"
Siobhan Ann Jackson
And 23h26'50" 36d33'52"
Siobhan Bourke
Vir 13h11'38" -20d44'1"
Siobhan Carroll
Vir 13h28'35" -20d8'30"
Siobhan Catherine Loughran
Uma 11h33'32" 63d6'3"
Siobhan Delilah Doherty
Sgr 18h24'21" -34d24'13"
Siobhan Devlin
Aqr 22h9'51" 1d50'10"
Siobhan Dineen Carey
Psc 0h38'18" 17d43'48"
Siobhan Eileen Hull
Cas 23h52'0" 59d12'14"
Siobhan Gerdin
Tau 5h54'33" 25d0'30"
Siobhan Grace
Cyg 20h42'33" 36d51'31"
Siobhan Hickey
And 2h37'17" 50d36'7"
Siobhan Mary Knightly
Lyr 19h11'20" 27d18'26"
Siobhan Maureen
Lyn 6h18'38" 60d5'18"
Siobhan Nvwela
Cap 20h41'19" -21d50'12"
Siobhan Patricia Cormican
And 1h40'37" 42d33'38"
Siobhan & Patrick - March 1995
Cru 12h37'50" -59d44'38"
Siobhan "Robin" Hennessey
Uma 14h2'2" 54d20'44"
Siobhan's Star
Boo 14h23'11" 45d28'51"
Siobian Jones
Cap 21h51'59" -9d53'2"
Siodhan Brigid Hebda
Sco 16h57'56" -44d12'1"
Síofra Beth Murphy
And 1h19'44" 49d16'1"
Sion Arwyn
And 23h29'29" 42d31'12"
Sion Sebastian Vick
Boo 14h26'54" 24d37'28"
SionMoore
Dra 16h24'44" 59d0'13"
sioraiocht
Uma 10h57'7" 63d57'31"
Sippé
Cyg 21h31'45" 41d24'32"
Sippy Laster
And 1h4'23" 41d3'41"
Sir
Lyn 7h34'42" 36d10'27"
Sir Alexander Star
Aur 5h40'54" 42d14'52"
Sir Ando
Ari 3h8'30" 20d0'36"
Sir Anthony
Psc 23h19'24" 5d27'31"
Sir Anthony of Labato
Cnc 8h37'14" 11d30'12"
Sir Artur Iasso
Uma 11h0'45" 47d23'16"
Sir Bob, A Knight in Shining Armour
Ori 5h15'16" 12d35'47"
Sir Brian & Princess Joan
Umi 15h25'49" 74d5'6"
Sir Brody Gusar
Lyr 18h52'20" 33d38'53"
Sir Casey Kamir
Tau 4h30'4" 19d10'50"
Sir Charles Barthalamule
Cep 22h10'26" 61d36'44"
Sir Charles McLaughlin
Her 17h38'14" 27d16'43"
Sir Charles Paul Haggarty
Cnc 8h41'31" 17d29'26"
Sir Charles Poppie Rothstein
Per 4h44'13" 46d51'53"
Sir Christopher "Kiki" St. Jones
Sco 16h11'54" -13d35'10"
Sir Dan of 1977
Tau 4h25'22" 21d43'19"
Sir David L. Scott
Aqr 22h3'59" -14d59'20"

Sir David R. Campbell
Durie KCMG
Cep 22h11'23" 57d16'46"
Sir Daylon of Loomis (my
love)
Uma 10h27'30" 46d9'11"
Sir Doug
Lib 15h58'25" -13d0'50"
Sir Dusty of Ferry Brook
17.09.1993
Ori 5h2'57" 15d19'15"
Sir Fool
Uma 10h33'57" 46d21'8"
Sir Francis Duffy
Cap 20h25'27" -12d8'37"
Sir Francis Stephen
Ari 2h11'28" 25d7'13"
Sir Frank W.
Uma 10h56'31" 59d18'59"
Sir Fredrick N Ballantyne
Cep 22h38'46" 73d29'4"
Sir Gavin
Psc 23h17'30" -2d57'15"
Sir Harrison September 29,
1991
Ori 5h1'38" 5d37'9"
Sir Ivan's Star
Cyg 21h52'16" 54d35'25"
Sir Jacob Baccetti
Lib 15h2'14" -1d48'17"
Sir James II
Lib 14h32'13" -19d42'2"
Sir Jeff
Cep 21h25'11" 62d31'34"
Sir Joe
Cnc 8h57'10" 23d42'54"
Sir Johnly Colbeck
Cep 21h44'8" 56d45'14"
Sir Lance Luetkemeyer of
Capricorn
Aur 7h15'23" 42d39'42"
Sir Lancelot
Dra 10h0'53" 77d45'18"
Sir Lancelot
Sco 17h23'48" -31d18'29"
Sir Lickensnout
Aql 19h7'42" 9d58'28"
Sir Mark William Lucherini
Sco 16h9'52" -12d19'35"
Sir Matthew Woodruff
Lib 15h17'1" -6d41'35"
Sir Mel Mann
Aql 20h2'12" 10d48'36"
Sir Mousie of Rockwood 13
Dra 15h39'41" 59d23'0"
Sir Nibor Yelnats Von
Ekard
Aql 19h32'25" 16d12'37"
Sir Otis of Southpaw
Umi 15h52'31" 78d40'10"
Sir Page - My Knight In
The Stars
Vir 14h34'45" 0d54'45"
Sir Patrick
Gem 6h18'37" 22d59'39"
Sir Philip
Gem 7h41'50" 34d24'9"
Sir Poppie
Her 18h45'57" 19d13'8"
Sir Richard
Gem 6h25'0" 18d54'35"
Sir Richard
Cep 20h38'8" 67d0'27"
Sir Richard Chan
Umi 14h22'49" 66d42'21"
Sir Richard Oliva
Aqr 22h33'46" -0d34'1"
Sir Richard Phelps
Leo 11h43'59" 18d22'37"
Sir Ripley
Cep 22h1'36" 56d15'53"
Sir Robert of Lebron
Sco 16h12'45" -14d57'10"
Sir Robert the Brave
Sco 16h5'36" -12d15'23"
Sir Rockefeller Orevello Hill
Cmi 7h30'29" 8d43'25"
Sir Ronald D. Wolf
Lib 15h7'38" -15d18'9"
Sir Rudolph Rudy Valentino
Coffield
Ori 5h37'51" -1d42'6"
Sir Slick Battle
Per 4h24'2" 44d13'47"
Sir Stelios Haji Iannou
Aqr 21h4'2" 1d43'7"
Sir Steve
Ori 5h41'46" 1d44'44"
Sir Talboy Candyman
Lib 15h4'28" -19d24'46"
Sir Terry Snookie Bear
Psc 1h2'27" 10d7'18"
Sir Thomas and His Little
Girl
Lib 15h42'36" -7d25'57"
Sir Thomas Fuller
Her 18h41'26" 20d35'53"
Sir Timothy Michael Costa
Sco 16h13'50" -11d10'3"
Sir Todd
Per 3h35'31" 43d29'10"
Sir Wilheim "Willy"
Cep 22h43'2" 74d41'37"
Sir Willard Lyman Elmsley
Ori 5h24'10" -8d35'20"
Sir William Arthur Greaves
Leo 11h29'43" 7d19'6"
Sir William Mueller
Leo 9h23'5" 19d9'29"
Sira Lani
Peg 23h37'23" 9d1'10"
Sirahn Toufayan
Lyn 7h31'30" 53d12'48"
Siranush Hovhannisyan
Uma 10h18'14" 68d9'14"
Siree Lynn
Ori 5h19'54" -3d57'44"
Siren
Sco 16h15'23" -9d48'16"
Sirena
Cyg 21h17'27" 52d56'41"
Sirena
And 23h36'3" 48d11'53"
Sirena Jennifer
And 1h57'46" 41d24'15"
Sirenita Fanny
Uma 9h54'56" 49d4'52"
Siren's Sandbox
Ori 6h7'27" 14d12'43"
Siressa Lea Ivers
Aqr 21h47'17" -6d35'40"
Siri
Cma 7h1'14" -24d26'9"
Siri
Lac 22h55'56" 44d52'3"
Siri Evolin Busby
Hya 14h23'45" -28d44'46"
Siri Lizbeth Bright
Ori 5h41'40" 4d17'1"
Sirikun 7
Crb 16h0'7" 33d8'42"
Sirimel
Cas 0h8'34" 52d15'5"
Sirio Cicchetti
Her 17h33'5" 48d26'54"
Siriporn Thummikanon
Crb 16h23'54" 34d34'54"
Sirisha
Leo 10h42'5" 14d46'25"
Siriunas JM
Cma 6h59'51" -16d18'37"
Sirius
Cma 6h34'34" -16d14'34"
Sirius 25
And 1h18'10" 50d2'42"
SiriusDreams ~ Denise &
Bob Miller
Cma 6h35'1" -16d37'16"
Sirka's Star
Gem 7h45'30" 26d0'54"
Sirléi Tetzner
Aqr 22h41'33" -20d0'37"
Siro Leonardo Gambini
Boo 13h53'14" 24d2'16"
SIROD
Lac 22h52'35" 46d10'42"
SIRRONDRAHCIR
Ari 3h19'20" 28d59'36"
Sir's Lady
Cmi 7h32'0" 2d51'27"
SIRUS
Sco 17h22'29" -32d58'32"
Sis
Uma 12h10'49" 58d48'12"
Sis Carr
Lyn 8h23'40" 43d10'31"
Sis.:. Dina
MICHALOPOULOU, 32o
Dra 18h33'12" 59d34'31"
Sis & Lorraine: Stars of our
hearts
Aqr 22h9'14" 0d43'8"
Sis & Rob "After All These
Years"
Lyr 18h46'19" 32d30'47"
Sisarukset Arto,Asko,Asta
Salminen
Cas 0h48'37" 57d15'48"
Siska
Dra 15h0'29" 58d27'13"
Sissel
Psc 1h2'33" 5d26'3"
Sissou
Uma 11h30'12" 60d37'55"
Sis-star
And 23h31'16" 42d15'19"
Sis-Star
And 0h45'2" 37d10'2"
Sissy
Aur 5h57'11" 37d4'45"
Sissy
Uma 12h2'2" 43d50'32"
Sissy
And 2h13'42" 41d28'52"
Sissy
Uma 10h5'40" 60d41'19"
Sissy and Ted Geballe
Uma 9h12'6" 53d5'57"
Sissy Carissa
Uma 10h12'31" 65d1'16"
Sissy Kathy
And 0h20'14" 37d50'10"
Sissy's Spirit
Ori 5h39'32" -0d25'27"
Sissy's Star: Tammy 1
Ori 5h53'59" 12d49'24"
Sista Pat
Cap 20h26'46" -10d9'31"
Sista Sue
Gem 7h26'40" 33d4'0"
Sister
And 23h51'0" 45d25'41"
Sister Carol Therese
Johnson
Cas 0h39'23" 47d37'4"
Sister Cathy Doherty
Gem 7h13'9" 26d56'17"
Sister Celestin Maurer
Cyg 19h49'45" 31d19'18"
Sister Cin
Sco 17h34'38" -33d28'14"
Sister Heidi Marie Krack
2004-2005
And 0h43'7" 40d21'25"
Sister Joan Therese
Per 4h33'22" 41d7'27"
Sister Joann Boneski
Sgr 19h11'11" -23d45'57"
Sister Julia Costello D M J
Uma 9h3'1" 48d41'27"
Sister Karen Hawver
Cas 1h44'32" 63d35'3"
Sister K-K 49
Cam 4h58'57" 52d56'54"
Sister M. Celeste
Sgr 18h28'43" -21d32'13"
Sister Margarita Rico
Ari 2h49'7" 24d52'53"
Sister Maria
Cap 20h31'59" -18d14'39"
Sister Marie Maurice
Hickey, IHM
Cas 1h22'9" 62d5'10"
Sister Martha Lockwood
Kaufman Star
Cas 0h33'26" 55d33'24"
Sister Mary Beata
Uma 12h35'25" 61d17'28"
Sister Mary Jean Ryan
F.S.M.
Tau 5h29'44" 21d10'4"
Sister Mary Jules Berger
Uma 11h51'9" 56d29'5"
Sister Mary Louise
Jurewicz
Psc 0h58'52" 9d25'9"
Sister Mary McNulty, I.H.M.
Uma 10h47'21" 52d45'20"
Sister Mary Paul
Psc 0h59'54" 24d39'54"
Sister Mary Ross PhD
Per 4h9'23" 47d46'26"
Sister Mechtilde
Psc 1h5'45" 22d16'34"
Sister René Drolet
Cas 2h15'39" 62d15'50"
Sister Rie Crowley
Dra 18h11'58" 74d3'53"
Sister Rita Ann Podhola,
RSM
Uma 13h13'6" 62d56'38"
Sister Rose Michael Hillery
Uma 10h37'41" 67d38'20"
Sister Sharon
Crb 15h51'45" 36d25'34"
Sister Shine *Shirley*
Vir 13h51'42" -3d15'2"
Sister Star
Cap 20h18'8" -11d16'48"
sister ~ star04
Crb 15h20'19" 31d51'33"
Sisterli-Andrea
Nussbaumer
Uma 8h20'18" 61d52'36"
sisters
Lib 14h55'46" -9d42'48"
Sisters
Cas 0h3'13" 53d30'34"
Sisters At Heart
And 0h5'7" 45d31'44"
Sisters Forever
Uma 9h26'3" 65d0'42"
Sisters Forever, Sharla &
Tori
Vir 13h14'45" 10d13'20"
Sister's Poppy
Vir 12h52'6" -0d52'17"
Sisty Doll
Psc 1h11'43" 21d42'4"
Siszer Balázs
Psc 0h34'54" 5d11'2"
Sita
Sgr 19h25'6" -25d35'54"
Sita
Uma 8h20'12" 66d39'27"
Sita Legac
Sgr 18h30'27" -16d20'14"
Sital Bhambra
Cas 0h3'40" 53d35'19"
Sitara
Leo 10h9'15" 20d24'34"
S.I.T.E.S.
Crb 15h38'42" 32d11'2"
Sith Lord Sean
Peg 23h14'7" 15d56'13"
Siti Sarah
Eri 4h37'11" -20d21'58"
Sitiacokta Ekim & Kcihple
Annod
Gem 6h45'17" 27d36'3"
Sitora Azimova
Cap 21h36'42" -12d37'51"
Sittmann, Gerlinde
Uma 10h16'26" 70d49'38"
Sitto-Minnie Abdun-Nur
Uma 10h49'53" 68d56'48"
siu wing wing
Cnc 8h46'29" 15d21'54"
Siu Yin Wong
Uma 9h30'8" 64d55'35"
Siu-Ling Jeng
Ori 6h23'22" 10d51'32"
SiuMai Ng
Uma 8h31'26" 60d53'29"
Siv TG1
Per 2h35'16" 54d29'52"
Sivan
Ari 2h11'42" 25d41'40"
Sivan Gamliel
Sco 16h9'37" -20d26'47"
Sivan Kaymak
Aql 19h46'4" 16d12'26"
Sivan Shemesh
Sgr 18h2'59" -16d54'29"
Sivan Wilson
Mon 6h54'12" -0d26'41"
Sivan Zlovitch Suzi
30/12/82
Cap 20h55'37" -19d54'29"
Sivanne
And 0h15'15" 25d1'52"
Siwatu
Sgr 17h58'41" -25d45'48"
Six
Uma 11h30'23" 65d45'37"
SIX - SISSI
Sco 17h55'34" -41d32'54"
Six Two And Even
Peg 22h17'58" 24d16'1"
Sixela
Mon 6h51'12" -0d25'24"
Sixth Star
Sex 10h6'7" 4d15'56"
Sixto
Uma 11h15'43" 38d30'55"
Sixto and Leonor
Cyg 20h35'43" 41d2'28"
Sixto Ortiz
Cyg 19h52'41" 32d52'57"
"Size Does Matter" Social
Sci Soph
Uma 9h3'4" 52d2'45"
Sizlayin
Vir 11h48'11" -1d5'51"
Sizzle & Jizzle
Umi 14h38'53" 68d10'59"
Sizzlean 74
Tau 4h22'4" 4d19'17"
Sizzlin' Kitty
Lyn 7h41'25" 48d49'21"
S.J.
Peg 23h47'23" 29d10'50"
SJ Freshwater 60
Cep 0h9'35" 72d34'37"
S.J & L.J H.
Cap 20h54'38" -17d2'23"
SJ Martin
And 1h11'40" 39d7'25"
SJA Buschini
Uma 9h56'25" 53d38'45"
SJObrst
Aql 19h44'32" 6d17'7"
SJRay88
Vir 13h13'37" -20d47'54"
SJS * Sciarabba
Lmi 10h4'42" 28d56'47"
SK Luvs Her SWB 8-20-05
Sge 20h1'21" 18d37'10"
S.K. Paul Barrera, Jr.
Cam 4h14'40" 61d15'19"
SK8ERGRRL
And 0h16'6" 29d42'51"
Ska
Aql 19h24'24" 2d29'19"
Skadi76
Leo 10h36'8" 12d0'24"
Skander Goucha
Sco 17h57'6" -30d47'58"
Skanky-Bear
Uma 11h28'49" 32d47'16"
Skansing-Goldgabber
Her 18h48'41" 20d15'39"
Skar Star
Cas 2h3'25" 74d18'53"
Skatertaylor
And 0h35'58" 27d49'20"
Skatie
And 2h24'22" 45d50'46"
Skeeball 1995
Umi 13h59'51" 74d53'27"
Skeeter
Cap 20h30'13" -19d45'16"
Skeeter
Cyg 19h55'53" 37d11'45"
Skeeter Dean
Psc 0h21'37" 17d50'25"
Skelwith
Cyg 20h29'34" 33d39'13"
Skephi
Dra 18h40'9" 51d28'6"
Sketch Sarah Dittyum
Ori 6h9'52" 19d53'27"
SKG55
Crv 12h17'55" -19d26'36"
"Ski" Jean Gail Strayer
Cap 20h12'23" -22d41'10"
SKIEEZZAA
Cas 1h27'1" 61d57'34"
Skigan 3/10/02
Uma 12h3'13" 55d40'57"
Skinner Star
Her 16h28'41" 14d6'42"
Skinny George
Peg 22h25'57" 3d44'31"
Skinny Kitty
Cnc 8h44'48" 26d57'39"
Skinny loves Shae
Ori 5h50'4" 6d9'26"
Skip
Her 16h42'59" 45d7'20"
Skip
Cyg 20h34'39" 50d13'4"
Skip
Aur 5h55'33" 53d39'58"
Skip
Psc 1h10'17" 32d33'37"
Skip
Sco 16h16'37" -9d27'48"
Skip
Aqr 22h41'14" -0d40'48"
Skip 143 Sior
Cnv 13h49'51" 33d37'19"
Skip And Pip Walker
Cyg 19h58'25" 39d55'19"
Skip & Betsi
Eri 3h36'11" -11d33'17"
Skip Curran
Sgr 19h25'15" -12d49'6"
Skip Gilbert, Jr.
Umi 14h21'30" 75d45'36"
Skip Gise
Cep 22h42'54" 66d27'37"
Skip & Judy Butler
Cyg 19h41'28" 32d0'20"
Skip Leighton
Vir 12h30'39" 12d38'30"
Skip & Michael
Lyn 6h35'18" 56d41'30"
Skip Oliver
Ari 3h15'23" 23d10'17"
Skip Seaman
Ori 6h24'24" 17d0'16"
Skip & Sonny
Cyg 20h25'54" 52d58'7"
Skip Winfield Combs
Leo 11h55'54" 22d20'50"
Skip, Always A Star To
Me!!!!!!
Her 17h47'17" 43d6'29"
Skip, Love of my Life and
#1 Dad
Cep 22h15'38" 69d59'56"
skip-and-judith
Uma 10h56'6" 51d41'35"
Skiporzac - Grandma's
Truth Seeker
Sco 16h8'50" -14d1'54"
Skipper
Uma 10h25'34" 62d30'1"
Skipper
Cru 12h46'14" -57d10'5"
Skipper Brady Sanders
Aql 19h17'57" 13d17'17"
Skipper Scott
Del 20h38'2" 15d36'9"
Skipper Wyatt McCallum
Sco 16h22'11" -28d30'31"
skipper14
Leo 9h58'2" 30d14'35"
Skippy
Tau 4h22'13" 4d20'17"
Skippy
Sgr 18h48'14" -20d23'26"
Skippy and Shana's
Shining Star BFF
Ari 3h2'51" 15d33'47"
"Skippy" Jin Haw Tan
Her 17h5'32" 15d3'49"
Skippy Skiba
Umi 16h21'54" 76d30'5"
Skip's Cinnamon Star
Per 3h16'29" 47d55'9"
skipy
Aql 19h47'54" 16d7'34"
Skitterpup
Tau 5h33'47" 24d36'45"
Skittle Beedeeboo
Leo 10h20'59" 27d15'20"
Skittles
Uma 11h17'13" 47d8'35"
Skittles
Lyn 6h19'48" 54d24'6"
Skivilla
Mon 7h16'55" -0d46'11"
SkizzleLizzle Juggalo
Sco 17h45'55" -35d7'19"
Sklyer Maureen Testa
And 1h37'15" 47d13'55"
Skoczek, Peter
Ori 4h59'31" 2d36'7"
Skong
Lyr 18h28'34" 31d54'3"
SkorpivoB
Sco 17h51'31" -36d19'5"
Skot Harris
Oph 16h26'13" -0d5'52"
Skot Tamburri My Loving
Husband
Vir 12h18'42" 12d26'9"
SKRIEN
Lmi 10h16'56" 37d2'11"
SKSparky
Crb 15h35'16" 32d30'8"
SKT072904
Per 2h53'43" 35d37'31"
Skulan
Aql 19h51'34" -0d58'13"
Skully Mad Dog Barks
Tau 4h55'35" 25d7'26"
Skunky
Cas 0h35'37" 47d46'2"
Skutz
Lyr 19h20'16" 37d36'24"
Sky
Ori 6h15'49" 10d9'52"
Sky
Lib 15h19'16" -14d29'9"
Sky
Uma 8h23'26" 66d9'16"
Sky Angel
Crb 15h50'47" 36d20'1"
Sky Bird
Aqr 22h26'1" -1d33'39"
Sky Dancer
Uma 10h57'1" 62d21'24"
Sky Financial Group
Umi 15h33'21" 74d26'8"
SKY for Skyler Nicole
Murphy
Dra 17h34'2" 69d50'35"
Sky Green Crowell Wilkins
Wyman
Lib 15h10'44" -24d4'16"
Sky High Ridings
And 2h10'35" 46d46'51"
Sky Kalani Tucker
Dra 17h5'26" 56d28'31"
Sky Kay Vernon
Sco 15h57'45" -20d53'1"
S.K.Y. & K.M.H.
Tau 5h6'55" 24d29'37"
"SKY LARK" Larry R.
Kitchen
Per 2h40'43" 54d1'3"
Sky Michael Michelic
Cru 12h15'16" -61d43'15"
Sky Shantell Lowe
And 1h14'50" 35d7'34"
Sky12
Tau 3h57'34" 15d2'36"
Skybaby
Cam 3h56'16" 65d6'40"
Skybar
Ori 6h3'7" 21d29'49"
Skye
Cyg 21h30'54" 36d37'19"
Skye
Cas 0h9'55" 53d56'57"
Skye
And 23h14'44" 48d43'34"
SKYE
Umi 15h49'42" 70d49'17"
Skye "11-22-01"
Umi 15h26'57" 71d2'31"
Skye Alexis Wommack
Psc 1h54'33" 7d44'16"
Skye and Richie
Uma 9h5'3" 59d52'1"
Skye Ann
Tau 5h7'16" 27d9'35"
Skye Bunny 1
Cnc 8h13'20" 25d4'1"
Skye Catherine Radulic
Cap 20h46'35" -19d17'15"
Skye Danielle
Vir 14h7'4" -5d53'55"
Skye Darche Fowler
Lib 14h37'11" -19d29'47"
Skye Daxton Samuels
Vir 12h49'52" 0d9'14"
Skye Haas
Mon 6h56'45" -0d26'16"
Skye Hannah Drunkin
Munkey
Lib 14h52'43" -3d14'29"
Skye Homer - Skye's Star
And 23h44'37" 42d50'12"
Skye Losson
Cyg 20h7'35" 56d46'54"
Skye Mae Cannedy
Aqr 20h41'36" -10d29'9"
Skye Marie McAllister
Addison
And 2h17'16" 38d6'32"
Skye Moreau
Vir 14h1'53" -6d50'30"
Skye N. Summers
Umi 13h56'1" 71d35'55"
Skye Nicole Mallinson
And 0h50'1" 44d56'1"
Skye Patricia McCaw
Psc 1h7'2" 9d27'28"
Skye Pepper
And 23h3'51" 41d50'40"
Skye Reider Barle
Cmi 7h35'23" 7d5'35"
Skye Suzanne Livingston
And 23h35'59" 37d53'9"
Skye Taylor Stansbury
Crb 15h45'48" 37d8'26"
Skye Thomas Dahl
Psc 0h56'13" 28d19'53"
Skye Victoria Rosas
Ari 2h21'26" 22d6'10"
SkyeAnna Grace
Tau 4h46'6" 18d39'49"
Skyelar Marquise Star
Vir 13h4'0" -1d15'14"
Skyemax Eve
Uma 12h32'32" 55d25'23"
Skye's Star
Cra 18h55'26" -41d4'19"
Skyezy
Vir 13h46'46" 3d50'59"
SkyFox Jackson
Cas 1h15'53" 57d33'27"
Skyla
Cas 0h2'50" 59d2'59"
Skyla Avalos Tackett
Leo 9h36'4" 32d18'53"
Skyla Rose DeGratto
Sgr 19h5'45" -21d19'27"
Skyla Rose Frank
Aqr 22h59'49" -9d5'51"
Skyla Starr
Ari 2h22'40" 27d46'41"
SKYLAH ROSE
Lac 22h26'43" 48d49'17"
Skylar
And 23h20'19" 41d24'5"
Skylar
Cap 21h9'23" -15d46'6"
Skylar
Vir 12h25'15" -4d26'58"
Skylar Adison Middleton
Lyn 9h8'34" 43d19'35"
Skylar Alexander Rhine
Ronda
Uma 11h47'59" 59d42'36"
Skylar Annaliese Iris
Falconer
And 2h5'10" 39d7'4"
Skylar B. Ellis
Sco 16h15'17" -12d28'33"
Skylar "Bee Bop" Jackson
And 0h50'56" 43d39'0"
Skylar Blake
Gem 7h25'30" 26d32'12"
Skylar Brooks "Mommy's
Shining Star
Umi 16h36'51" 77d48'58"
Skylar Brynn Hicks
Sco 16h9'19" -16d2'49"
Skylar "Bubba" Marie
Leo 10h15'5" 20d27'26"
Skylar Canavan White
Cnc 8h18'15" 19d50'15"
Skylar Caprice Lund
Vir 14h12'22" 7d0'45"
Skylar Carol Ann
And 23h23'55" 48d13'31"
Skylar Christopher
Del 20h39'57" 15d3'12"
Skylar Clementine Borof
Psc 1h25'54" 25d54'38"
Skylar Conover-Linnane
Cyg 19h54'32" 37d2'37"
Skylar D. Ross
Mon 5h57'31" -8d14'9"
Skylar Davis
Psc 1h11'10" 23d46'18"
Skylar Doege
Lmi 10h25'59" 32d22'23"
Skylar Elizabeth Fife
Psc 1h27'30" 29d36'14"
Skylar Elizabeth Little
Sco 16h51'22" -29d57'15"
Skylar Elizabeth Sexton
And 2h10'29" 43d10'15"
Skylar Elle Louisa Fierens
Vir 13h14'13" -17d46'5"
Skylar Faith Poteet
Lib 15h24'28" -17d32'12"
Skylar Harlan
And 2h15'3" 38d23'27"
Skylar James Malleck
Ari 3h21'40" 30d27'44"
Skylar Jay Prehodka
Sco 17h58'5" -43d5'2"
Skylar Jayne Ellington
And 2h33'31" 38d59'1"
Skylar Kristen Dann
Per 3h54'23" 38d54'22"
Skylar Lanier
Lyr 18h48'33" 38d58'23"
Skylar Little Lion King
Leo 11h8'32" -2d44'11"
Skylar Mackenzie Marie
Klitzing
Cam 3h24'21" 56d12'34"
Skylar Maddox Downey
Ari 3h19'39" 28d46'10"
Skylar Marie Stephenson
Umi 15h58'41" 83d28'58"
Skylar Mia Moore
Gem 7h29'34" 28d56'14"
Skylar Moon Berry
Uma 10h37'21" 44d0'6"
Skylar Moon Berry
Sco 16h51'29" -43d53'27"
Skylar Rae Stock
Psc 1h21'13" 16d28'22"
Skylar Ray Hoehn
Umi 16h27'33" 77d12'16"
Skylar Rayne Benson
Leo 11h31'34" 2d13'52"
Skylar Rose Hahn
And 0h49'2" 40d22'16"
Skylar Sonn Tancredi
Psc 0h47'40" 10d49'38"
Skylar Tancredi
Uma 12h32'56" 58d8'52"

Skylar Victorelli
Sgr 18h17'7" -17d58'46"
Skylar Wick Verhamme
Sgr 18h14'30" -22d49'9"
Skylar Wingate
Umi 15h8'35" 81d55'2"
SKYLAR9
Tau 4h6'49" 7d35'33"
Skylar's Sky Light
Psc 23h21'1" 4d58'21"
skylaura-618
Leo 10h42'47" 17d57'37"
Sky'Leigh Ann Jones
Psc 1h17'6" 31d10'51"
Skyler
Lyn 8h21'50" 37d36'40"
Skyler
Vir 14h40'18" 2d54'46"
Skyler
Cap 20h50'23" -15d18'30"
Skyler
Uma 12h20'49" 58d46'32"
Skyler
Dra 15h59'21" 57d5'44"
Skyler
Umi 14h18'34" 69d53'39"
Skyler Adora a Diana
Peg 21h43'2" 26d13'9"
Skyler Alex
Aqr 22h14'34" -13d34'26"
Skyler Anderson
Vir 13h27'30" 12d26'38"
Skyler Anne Maki
Uma 12h52'10" 53d18'23"
Skyler Bear
Her 16h49'39" 48d33'15"
Skyler Charles Evans
Cap 21h52'9" -23d52'9"
Skyler Faye LeMaster
Cam 7h23'34" 61d8'57"
Skyler Hayes Herring
Ori 5h37'52" 0d6'25"
Skyler Jane
Leo 9h40'5" 27d38'51"
Skyler Katz
Lib 15h3'5" -21d3'52"
Skyler LaDawna DeJarnatt
Vir 12h24'2" 12d12'16"
Skyler Lark Morgan
Aqr 23h39'54" -9d10'0"
Skyler Lee Kim
Ori 6h3'3" -0d12'22"
Skyler Logan Szot
Uma 11h14'52" 64d23'56"
Skyler Marie DeGarmo
And 1h36'5" 46d27'16"
Skyler Marka Raycene Goff
Lyn 9h11'57" 37d36'48"
Skyler McLarney
Gem 7h11'24" 26d32'11"
Skyler Murphy
Her 17h46'36" 32d47'24"
Skyler Noelle Schaible
And 0h17'51" 46d27'11"
Skyler Rae Schaffnit
Ari 2h45'42" 25d0'38"
Skyler Rogers
Vir 12h48'25" -3d59'29"
Skyler Rose
And 0h29'18" 42d18'50"
Skyler Rosenhain
Vul 19h25'48" 24d57'1"
Skyler Rowan
Uma 9h17'46" 48d51'13"
Skyler Sorensen
And 2h19'50" 48d56'40"
Skyler Thomas Bender
Sgr 17h59'25" -27d42'42"
Skyler Urban Holubowicz
Aqr 22h37'1" 1d42'42"
Skyler William Beaubien
Uma 11h34'40" 47d18'33"
Skylerlily
Cra 18h21'12" -37d11'12"
Skyler's Star
Lyn 7h57'13" 36d40'42"
Skymax
Uma 11h22'35" 36d16'21"
Sky's Starr
And 22h59'58" 47d17'0"
Sky's The Limit
Lib 14h48'8" -3d9'27"
Sky's The Limit L
Aqr 23h32'41" -21d37'2"
Sky's The Limit- LA Team 2004
Aql 19h40'5" 2d46'36"
Skyway
Leo 11h42'0" 25d21'31"
Skyy Dorian Reese
Mon 7h9'48" -3d28'50"
SL Love
Cas 1h7'38" 61d42'5"
Slabon, Stefanie
Uma 12h31'51" 57d59'45"
Slada forever
Uma 13h39'52" 57d21'55"
Sladana Jakovljevic
Cru 12h37'31" -59d50'36"
Slade
Aur 6h35'33" 52d24'55"
Slade Alan Deister
Cap 20h45'29" -19d46'43"
Slade and Lindsay
Gem 6h52'22" 22d21'52"
Slade Anthony Eide-Ettaro
Her 17h34'44" 27d29'29"
Slade Jaxson
Psc 0h47'6" 6d11'54"
SLADJANA
Uma 13h51'56" 55d30'45"
SLADJANA 1
Lyn 7h1'0" 46d55'32"
Slaki
Cas 23h32'17" 54d32'11"
Slant
Sgr 19h46'23" -29d22'21"
S.L.A.P.
Sgr 18h57'25" -22d10'10"
Slappy Slap Slap
Uma 14h9'7" 60d4'32"
SLAS1016
Uma 12h48'13" 61d52'47"
SLASWTNSS
Uma 9h54'54" 47d59'48"
Slate
Ori 4h54'36" 3d47'50"
Slattengren Station
Tau 3h45'14" 12d40'26"
Slaughter Soulmates
Aql 19h2'56" 8d50'6"
Slava Keselman
Sco 17h45'37" -41d30'24"
Slave Shari Murray
Lyn 7h40'44" 57d30'57"
Slavica
Uma 14h5'46" 48d34'26"
Slavik Anush
Crb 16h6'49" 39d32'35"
Slavka
Aur 6h8'39" 39d36'7"
Slawomir Paskiewicz
Cma 7h22'19" -14d57'38"
Slayden's Wish
Per 2h21'57" 54d31'42"
Slayer
Ori 4h53'54" 1d19'34"
S.L.C.
Cas 23h25'30" 57d36'18"
Sleep Sweet Jaffer
Tau 4h17'5" 14d17'11"
"sleep with the angels"
Cas 1h52'2" 63d27'34"
Sleeping Beauty
Leo 11h20'2" -0d17'15"
Sleeping Beauty
Gem 6h48'39" 14d31'14"
Sleeping Beauty
Tau 5h34'23" 27d38'40"
Sleeping Beauty
And 0h57'19" 24d4'3"
Sleeping Beauty And The Beast
Cyg 19h35'2" 53d8'5"
Sleeping with the Angels*Jill Hirst
And 0h17'18" 44d20'18"
Sleepy Head
Cen 14h10'14" -41d16'7"
Sleepy Head Sarah
And 1h36'33" 41d4'16"
Sleepy Kitty
Psc 1h44'18" 6d14'11"
Sleepy Little Levi
Uma 10h10'18" 54d2'12"
Sleepyhead Tang
Vir 13h16'24" -22d1'2"
sleff
Cnc 8h42'2" 28d23'42"
Sleshi
Uma 10h57'17" 55d3'31"
SLF
Tau 4h15'11" 15d48'57"
SLG 02-03-1980
Aqr 21h9'18" -13d10'6"
S.L.H 7-02-1975
Cnc 8h44'4" 7d21'40"
SLH Mommie Dearest
Ori 5h54'38" 21d15'38"
Slice of Heaven
Lyr 18h53'26" 32d50'47"
Slick
Her 17h19'0" 27d25'13"
Slick
Umi 15h57'17" 74d35'23"
Slick Nick Scalisi
Per 2h38'42" 52d43'53"
Slick Rick Coronado
Cha 12h14'0" -80d46'19"
Slick Rikki
Uma 8h27'40" 70d46'32"
slick-duck
Lac 22h14'18" 49d38'25"
Slider
Dra 18h42'20" 53d14'23"
Slim and Shawtty
Cas 1h18'11" 54d59'21"
Slim Longerbeam
Uma 12h19'7" 62d54'19"
Slim's Light Shines Forever
Cma 6h41'9" -16d40'48"
Slinky Fishin' 007 1/8/2006
Uma 10h23'11" 66d10'24"
Slipaway Star
Cas 0h56'38" 57d12'42"
SLIPSTREAM
Cap 20h14'53" -26d45'39"
Slique
Ari 3h7'6" 21d14'11"
"sliver" in memory of Paul A Garcia
Ori 5h42'49" 2d26'28"
SLJ7979
Cnc 8h45'45" 7d28'53"
SLK-Sharon Lee Kridle
Cnc 8h28'29" 24d13'11"
S.L.L. 12-25-04
Ori 6h10'35" 15d57'22"
Sloan
And 0h25'37" 45d33'43"
Sloan Brielle Dunlap
Uma 10h38'14" 72d6'17"
Sloan Grace Garabedian
Lib 14h53'23" -4d24'25"
Sloan Stroud
Cnc 8h57'25" 10d12'37"
Sloan Washburn
Leo 11h28'32" 5d41'55"
Sloane Caporale
Uma 12h6'21" 49d11'41"
Sloane Elizabeth Kutach Richter
Uma 11h47'28" 42d18'24"
Sloane Kathleen Bergner
And 0h32'49" 45d10'26"
Sloane Marie Droll
Sgr 18h25'18" -31d47'45"
Slobodan Ajdukovic
Ari 2h6'27" 17d51'55"
Slobodan Kucevic
Umi 13h53'52" 74d52'54"
Sloppy Joe Beans 8/3/03
Lyn 7h57'5" 41d31'23"
Slosser
Vir 14h57'55" 6d18'1"
Slot Machine Super Stay
Sgr 18h12'50" -26d13'21"
Slow Rollers
Lyr 18h49'13" 30d36'33"
Slowly...
Ori 5h57'10" 5d54'21"
SLRHS Class of 2006
Umi 17h0'17" 80d5'53"
SLRHS Class of 2007
Together As One
Lyn 8h11'27" 38d6'55"
SLS
Cnc 8h49'40" 26d16'32"
SLSB143B
Uma 11h26'33" 71d30'51"
SlSDcK
Leo 10h6'29" 20d5'44"
S.L.S.NO.4
Sgr 19h0'23" -29d52'9"
Slugger Gatsik
Cap 20h26'29" -10d28'15"
Sluggo
Umi 15h26'29" 79d16'53"
Sly Tyger
Peg 23h41'49" 20d28'5"
Sly, Wookie & Techno Boy's Wonder
Her 16h47'18" 29d36'4"
Slynn Quinlan
Ori 5h32'50" 14d47'34"
Smac
Dra 17h14'50" 62d45'11"
SMAK
Umi 14h15'32" 71d43'12"
Small Family Star 2006
Uma 10h30'33" 44d31'28"
Smalley
Ori 5h8'39" 5d58'12"
Smallish Bear
Cas 1h32'48" 58d4'41"
Smalls
Leo 9h54'0" 32d47'46"
Smalls' Sanctuary
Lmi 10h39'31" 25d12'59"
Smallz
Per 3h24'51" 43d56'18"
Smama's Star
Uma 11h40'51" 63d26'22"
Smart Star
Tau 5h13'18" 23d37'14"
"Smarty Arty" (Arthur J. Cutillo)
Her 16h56'33" 18d37'6"
smarty44
Her 18h21'10" 26d29'57"
Smash
Sco 17h52'12" -36d50'7"
Smash 2
Uma 12h14'33" 61d50'59"
Smashley
And 0h38'48" 38d13'18"
Smauk!
Cyg 19h39'13" 43d16'53"
"SMD Harter"
Ori 5h6'32" 5d0'4"
SME Dukey Hunny Bun Viking Squirrel
Cyg 21h12'46" 45d51'13"
SMECKLES
Uma 12h34'19" 56d27'29"
SMEE
Uma 11h16'58" 29d33'17"
Smejat
And 23h15'49" 35d19'46"
Smidge
Lib 15h51'58" -10d31'45"
Smidovich, Kirill
Uma 12h45'2" 53d48'50"
SMIKO7
Lib 15h11'20" -16d47'6"
smile
Her 18h42'22" 18d13'6"
SMILE
Ori 5h55'2" 2d21'2"
Smile in the Sky
Leo 9h30'16" 26d42'55"
Smile Jim
Vir 14h43'16" 3d19'56"
Smile of Naoko Sugita
Sgr 19h20'17" -34d26'46"
Smile Sunshine
Psc 23h27'37" 7d48'53"
"Smiley"
Ari 2h5'16" 22d19'53"
Smiley
Cyg 20h52'9" 47d58'25"
Smiley
Sco 16h10'2" -16d31'4"
Smiley Face
Peg 22h49'23" 22d3'36"
Smiley Mc Kay
Cyg 19h30'7" 32d0'30"
Smiley Miley
Umi 14h35'18" 74d40'42"
Smiley Sarior's 18th
Uma 9h29'37" 56d20'7"
smiliestar21
Lyn 7h17'21" 52d13'13"
Smilin' Bob Woodin
Uma 8h51'22" 56d31'19"
Smilin Jack
Her 17h22'23" 32d22'28"
Smilin R
Cas 0h55'12" 61d23'47"
Smiling Goat Moon
Ari 2h3'53" 10d56'33"
Smiling Holly Rees
Cnc 8h7'20" 10d11'46"
Smiling Jim Weaver
Per 4h21'52" 43d38'19"
Smiling Jom
Gem 6h45'1" 16d37'34"
Smiling Kostelnik
Psc 0h52'13" 9d41'50"
Smiling Miles
Umi 15h16'36" 78d51'38"
Smiling Sam
Cep 21h46'49" 63d13'28"
Smilja Vucic
Sgr 18h6'43" -18d43'42"
Smita - as you wish forever
Lyn 8h23'17" 36d56'2"
Smitch-Cheeps
Pho 0h3'51" -39d59'27"
Smith Edward Calhoun, Jr.
Gem 7h42'16" 26d37'52"
Smith Pass
Uma 8h54'12" 72d28'28"
Smith, Alexander
Ori 5h54'6" 21d42'6"
Smith50
Lac 22h19'34" 41d50'55"
Smith's Present
Psc 1h29'31" 12d51'18"
Smith-Smith
Cyg 21h16'42" 38d1'10"
Smithy
Uma 8h42'8" 54d3'26"
Smitten
Per 4h33'8" 31d49'18"
Smitty
Gem 7h9'27" 34d8'13"
smitty
Her 17h20'7" 34d5'4"
Smitty
Uma 11h28'37" 41d56'16"
Smitty
Boo 14h23'59" 18d37'44"
Smitty
Gem 6h49'29" 26d47'18"
Smitty
Uma 8h38'9" 72d20'52"
Smitty
Uma 10h43'5" 69d0'48"
Smitty's Star
Tau 4h25'42" 25d9'32"
Smitty's Star
Her 18h35'37" 14d51'43"
S.M.L. 222
Umi 15h35'24" 71d18'44"
Smmcisme
Per 4h21'41" 52d31'21"
SMOC SEE
Cap 21h55'6" -10d28'7"
Smoke
Tau 5h51'0" 23d47'28"
Smokey
Uma 13h15'46" 58d50'38"
Smokey Bear
Cyg 19h42'30" 30d56'3"
Smokey Joann Parker
Cap 20h23'49" -12d38'36"
Smokey Joe
Uma 11h33'12" 61d39'0"
Smokey Merkich
Cma 6h39'57" -16d56'36"
Smokey Scahill Wolanski
Uma 8h59'56" 67d58'56"
Smokey Sornberger
Uma 11h37'54" 61d57'41"
Smokie
Per 3h10'1" 43d21'6"
Smokie
Lmi 10h37'5" 26d27'41"
Smokin' Sam
Per 4h23'1" 41d39'28"
Smok'n Smolen: The Perfect Package
Sco 17h20'23" -33d10'53"
Smoky Reinert
Aur 6h30'41" 34d53'51"
SMOO 1
Gem 7h29'42" 32d36'44"
"Smooch"
Cap 20h30'7" -11d33'22"
SMOOCHIEPOO
Lyn 6h58'48" 51d14'22"
Smoody
Lyr 19h10'12" 27d53'52"
Smoopee (Richard and Kristi)
Pav 19h8'57" -58d33'43"
Smooshie
Lib 15h40'22" -16d19'18"
smootchie pie
Cap 21h30'25" -8d37'35"
Smoove
Ori 5h32'27" 4d34'36"
Smores
Uma 13h19'18" 54d0'48"
SMR "Ra"
Cam 4h37'1" 69d26'47"
S.M.R.'s Star
Lyr 19h11'51" 27d21'7"
Smruthi R. Garlapati
Uma 9h15'26" 56d23'43"
SMS
Gem 6h29'0" 15d17'1"
S.M.S. Lady Warriors
Her 16h5'41" 46d45'21"
S.M.S. Warriors
Her 16h31'52" 46d48'38"
SMSgt. Linas K. Venclauskas
Aql 19h43'52" 9d26'52"
Smu Star
Lib 15h10'55" -5d54'35"
SMUDGE
Sgr 18h6'17" -27d52'0"
Smudge Curtis Wheeler
Lyn 8h23'10" 56d26'55"
Smudge & Doc
Cyg 20h5'12" 42d30'38"
Smudger
Umi 13h44'56" 74d19'2"
Smugtooth3000
Dra 16h45'1" 58d44'27"
Smukke Camilla
Ari 2h19'45" 25d41'47"
Smur-d-mur
Lyr 18h51'30" 36d17'7"
Smuschlypuh
Ari 2h53'38" 29d18'50"
Smythwood & Sons
Dra 13h41'51" 68d40'29"
S.N. Pringle
Sco 17h33'6" -39d57'48"
SNAGA
Peg 22h22'17" 27d30'6"
Snage-Star
Crb 16h12'58" 37d22'23"
Snake Griffin
Sgr 18h6'59" -17d49'31"
SnakeFish ROL
Psc 23h29'4" 5d2'40"
Snake's Kitten
Sgr 19h16'14" -16d24'19"
Snapper
Uma 8h46'51" 72d19'12"
Snapper06
Tau 4h24'18" 22d48'49"
Snappy and Snapalicious
And 1h1'1" 38d4'0"
Snappy White
Pup 8h19'39" -11d40'18"
Snap's Star
Uma 11h16'43" 65d44'50"
Snatole
Aqr 21h48'29" -2d20'42"
Snax Forever
Lyn 9h11'48" 36d3'27"
SnBFugue8
Cyg 20h52'52" 38d17'21"
Sneakers
Lmi 10h25'35" 34d59'44"
sneaky peaky
Ori 6h18'54" 21d18'13"
Snef1216
Her 16h43'34" 6d25'55"
SNEGUROCHKA
Lmi 10h33'27" 34d42'35"
SnehPraful
Sgr 17h52'30" -28d41'45"
Snejina
Uma 14h9'50" 59d27'16"
Snelten
Sgr 18h11'24" -29d34'40"
Snezana Markovic
Gem 7h38'16" 15d57'31"
Snezana Sicovic
And 0h6'32" 29d11'21"
Snickerdoodle
And 23h2'5" 47d36'20"
"SnickerDoodle" light of our lives
Gem 7h5'6" 10d33'34"
Snickers
Uma 9h20'52" 57d11'35"
Snickers
Aqr 21h14'3" -13d31'47"
Snickers
Aqr 20h55'12" -8d50'42"
Snickers and Jennifer
Cas 1h40'26" 64d16'9"
Snickers on the Run
Ari 3h20'24" 29d2'12"
Sniffy
Cap 21h42'55" -15d5'31"
Sniggy's Star
Aqr 21h30'42" -0d23'38"
SNINE100704
Vir 13h19'23" 11d24'25"
SnL eternal
Gem 6h44'16" 21d10'39"
SNL50905
Tau 3h57'19" 18d37'58"
S-n-M
Cyg 20h14'57" 50d36'55"
SnM4L
Lib 15h3'55" -17d11'30"
Snooka
Vir 12h19'48" -0d52'33"
Snookems
Psc 0h53'36" 7d14'52"
Snookerdoodle W. L. D
Sco 16h10'37" -9d20'3"
Snookles
Lib 15h13'2" -16d42'1"
Snooklies
Cra 18h25'16" -37d10'34"
Snooks & Snooks Forever & Eternity
Cyg 19h59'58" 31d41'8"
Snook's Sparkling Light
Cru 12h29'13" -58d25'49"
Snookums
Cet 2h41'24" -0d28'43"
Snookums
Lib 15h6'17" -1d40'36"
Snookums
Uma 11h37'49" 56d11'7"
Snookums (aka Lynn Cooke)
Vir 12h19'9" 4d0'15"
Snoopy
Boo 14h46'47" 10d55'7"
Snoopy
Cnc 8h38'39" 7d41'11"
Snoopy
Uma 8h55'0" 50d13'14"
Snoopy
Cma 6h49'17" -14d40'51"
Snoopy
Sco 16h2'50" -18d51'5"
Snoopy 1925
Crb 15h34'47" 26d52'11"
Snoopy 71
Uma 10h37'57" 40d7'13"
SNOOPY (Annick Peeters)
Uma 9h24'11" 52d35'40"
Snoopy Mitchell
Cmi 7h27'58" 9d31'4"
Snoopy poopy
Her 16h57'9" 24d7'50"
Snoopy143
Cma 7h0'45" -22d39'50"
Snoose
Gem 6h42'10" 25d21'31"
Snoozeanne
Uma 11h25'32" 67d12'7"
Snoozecruz
Leo 9h26'3" 25d6'32"
Snorch Star
Ari 2h1'5" 12d50'39"
Snouty Fish
Cyg 19h30'52" 31d51'37"
Snow
Uma 9h12'17" 55d6'15"
Snow
Sco 16h15'39" -10d42'49"
Snow Angel Dydee
Cas 0h51'42" 56d32'25"
Snow Bunny
And 23h31'36" 42d5'4"
Snow Star
Umi 15h23'34" 70d14'19"
Snow Twins
Sgr 18h13'18" -27d3'8"
Snow Weiss
And 0h10'56" 43d57'56"
Snow White
And 0h56'31" 43d18'14"
Snow White
Ari 3h15'56" 29d22'14"
Snow White
Dra 9h58'55" 75d3'22"
Snow White & Prince Charming
Cap 21h11'15" -22d19'50"
Snow Wind's May Day 20.07
Cma 7h10'43" -26d32'5"
Snowball
Cma 7h20'20" -14d5'13"
SnowBall
Ori 5h28'33" 8d0'38"
Snowcaps
Cas 23h48'45" 50d0'26"
Snowcaps Legend
Peg 22h53'37" 21d2'39"
Snowcone's Light in the Sky
Cap 20h7'23" -21d15'27"
Snowdrop
Aql 19h1'11" 15d45'31"
Snowe Melinda Saxman
Lyn 7h53'52" 41d42'23"
Snowflak3
Lyn 9h0'35" 38d46'7"
Snowflake
Cyg 20h26'23" 43d17'11"
Snowflake
Tau 4h25'54" 19d40'9"
SNOWFLAKE
Umi 14h32'49" 74d57'14"
Snowflake
Uma 10h44'21" 59d48'48"
snowflake
Lib 15h34'1" -25d34'36"
Snowflake 11696
Cnc 9h11'43" 31d3'3"
Snowflake & AJ
Gem 6h31'29" 19d31'39"
Snowflakes Love
Ari 2h29'47" 10d29'40"
Snowflower
Umi 14h42'45" 68d22'41"
Snowy
Tau 4h11'29" 5d47'46"
Snowy White
And 1h21'38" 39d58'39"
SNR143
Sco 17h29'19" -42d28'16"
SNS Property Finance
Uma 9h25'0" 49d52'13"
SNS, no matter what.
Ari 1h55'33" 20d50'34"
Snuff & Elmo's Place
Sco 17h57'54" -39d42'53"
Snuffi & Engel
Uma 10h0'4" 52d43'50"
Snuffleupagus
Psc 1h3'9" 27d44'26"
Snuggle Bear
Umi 15h38'40" 76d22'40"
Snuggies
Her 18h47'19" 19d21'20"
Snuggle Bear Finnegan
Sgr 19h41'53" -12d39'3"
Snuggle Buddy
Cyg 21h26'14" 47d48'40"
snuggle bum
Ori 6h15'21" 10d57'31"
Snuggle Bunny
Ari 2h52'37" 17d53'15"
Snuggle Bunny
Ari 3h14'22" 27d42'44"
Snuggle Bunny
Uma 8h29'12" 67d45'45"
Snuggle Bunny
Uma 13h50'16" 61d37'4"
Snuggle Puppy
Leo 11h16'10" 16d59'8"
Snuggle Time
Ori 5h11'52" 1d38'13"
Snuggle Wuggles
And 1h45'47" 49d45'31"
Snugglebear
Leo 10h33'29" 26d48'25"
SnuggleBug
Per 3h19'42" 42d12'16"
Snugglebum
Ori 5h33'43" 7d46'4"
Snugglebum (Angela)
Cas 1h56'39" 60d46'11"
Snugglebunny
Cnc 9h7'15" 25d40'23"
Snugglenose
Uma 9h28'35" 64d51'19"
Snuggles
Umi 13h52'49" 75d32'2"
Snuggles
Aqr 22h51'24" -7d21'24"
Snuggles
Sco 17h18'47" -32d33'30"
Snuggles
Cnc 8h43'30" 24d13'25"
Snuggles
Cnc 8h35'9" 18d56'46"
Snuggles
Cnc 9h3'8" 16d10'51"
Snuggles
Psc 23h29'11" 3d11'43"
Snuggles
Tau 4h35'35" 1d38'10"
Snuggles
Cyg 19h44'20" 33d44'43"
Snuggles - King of Cats
Cnc 8h39'59" 28d10'7"
Snuggly
Tau 4h18'23" 3d15'19"
Snugs
And 2h34'58" 39d34'47"
SNUGZO
Umi 15h20'30" 73d44'3"
SnV0301
Ari 2h23'43" 24d19'8"
SNY
Cru 12h44'12" -58d34'8"
Snyder's Star
Ari 3h11'33" 20d25'35"
SO
Her 17h18'57" 46d3'15"
So Much More Than Love
Her 17h54'8" 24d18'24"

So Special and Beautiful - Mairi MacD.
Cyg 21h16'40" 46d33'53"
So Young Youn
Vir 15h9'42" 6d0'59"
Soapy's Star
Aqr 23h6'50" -6d29'57"
SOAR
Aql 19h44'48" 13d19'40"
Soaring Eagle
Aur 4h58'47" 31d55'25"
Sobes
Leo 11h16'42" 24d38'44"
Sobiety
Cnc 8h37'7" 21d59'41"
Sochurek, Hans Heiner
Uma 8h56'52" 48d29'3"
Societa' Modenese Di M.S. 1906-2006
Cep 21h23'30" 59d51'21"
Socky
Cnc 8h54'25" 10d7'53"
Socorro
Ori 6h11'45" 15d15'48"
Socorro Rampas
Crb 16h1'34" 26d29'57"
Socorro Rubalcaba
Uma 9h46'17" 61d12'19"
Socorro's Star
Cap 20h59'11" -26d22'3"
SOCRATES. A Scholar and Gentleman.
Lyn 7h50'15" 39d29'20"
Soeren Krause
Uma 8h36'20" 53d11'48"
Sofee
Aqr 23h6'27" -22d26'29"
Sofi Lambert
Uma 9h56'7" 57d57'4"
Sofia
Cas 23h36'37" 53d24'43"
Sofia
Aqr 22h32'29" -10d30'1"
Sofia
Sgr 18h41'50" -22d16'39"
Sofia
Aqr 22h0'5" -16d0'56"
Sofia
Per 4h21'10" 41d26'14"
Sofia
And 0h31'18" 39d59'16"
Sofia
And 0h2'15" 44d35'0"
Sofia
Crb 16h19'25" 36d30'47"
Sofia
Cyg 20h30'27" 37d50'52"
Sofia
Lyr 18h51'48" 42d23'31"
Sofia
Cas 1h23'11" 56d34'48"
Sofia
Leo 9h50'24" 27d21'58"
Sofia
Her 18h12'29" 15d9'13"
Sofia
Her 17h55'3" 14d28'19"
Sofia
Ari 2h20'5" 13d31'0"
Sofia 4,5,6
Ari 2h43'33" 26d3'25"
Sofia Almonte
Her 17h27'13" 19d39'50"
Sofia Carina Velasco
And 23h28'8" 44d56'21"
Sofia Carus Sidus
Ari 2h17'58" 14d44'26"
Sofia Catherine Sherzai
Sco 17h47'58" -38d29'32"
Sofia Cecilia Alvarez
Uma 8h34'52" 60d3'39"
Sofia D' Alessio
Ori 6h23'53" 14d4'16"
Sofia Elena Babin
Sco 17h10'15" -34d32'34"
Sofia Elise Pitkin
Uma 9h16'41" 66d1'7"
Sofia Elizabeth Rachael Kezmarsky
Lyn 7h35'44" 48d36'45"
Sofia Escalante
Col 6h10'52" -29d1'36"
Sofia Fils-aime
Mon 7h19'53" -10d52'55"
Sofia Fowsar
Lyn 7h42'13" 36d54'23"
Sofia Gabriela Guerrero
Sgr 18h39'49" -34d8'33"
Sofia Grace Cozonac
Cas 1h22'17" 68d17'41"
Sofia Gray Falsetto
Sgr 18h52'57" -15d57'22"
Sofia Guimaraes Evans
And 22h59'26" 51d2'26"
Sofia Harper
Lyn 7h12'37" 58d52'4"
Sofia Heart Pistolesi
Sco 16h5'23" -12d13'22"
Sofia Iren Fields 4/30/2006
And 0h12'38" 43d51'11"
Sofia Isabela Velasco Jenkins
Vir 14h4'54" -3d16'58"
Sofia Isabelle Miller
And 0h10'22" 34d0'26"
Sofia Jane Best
Crb 15h54'36" 27d6'20"
Sofia Joy Buompane
Sgr 18h54'57" -36d38'56"
Sofia Julianna de Melo
Cas 0h13'49" 58d6'6"
Sofia Leone
Uma 8h13'39" 60d29'51"
Sofia Leonor Stockwood
Lyr 18h47'44" 36d32'27"
Sofia Mallaci-Bocchio
And 23h29'7" 38d39'49"
Sofia Marianneve C Dirusso
Cyg 21h44'39" 48d13'45"
Sofia Marie
Lyr 18h37'55" 27d47'0"
Sofia Marie
Sgr 19h3'47" -15d54'56"
Sofia Marie Hutchins King
Lib 15h15'0" -27d35'30"
Sofia Marie Rinaldi
Lib 14h53'4" -2d10'9"
Sofia Mercedes Hoven
Dra 18h50'31" 66d53'7"
Sofia Merree
Cnc 8h39'18" 7d2'20"
Sofia mi hermosa enamorada
Ori 5h35'21" -6d55'6"
Sofia Milanese
Cas 0h27'21" 62d23'13"
Sofia Murphy Gazzola
Aqr 22h1'13" -19d16'19"
Sofia Olivia
Cas 1h38'7" 62d44'0"
Sofia p
Uma 10h8'24" 63d1'18"
Sofia Papakonstantinou
Uma 8h33'53" 61d6'9"
Sofia Petrovna Polyakova
Uma 8h52'11" 67d46'49"
Sofia Pilar Wing
Psc 1h10'43" 30d0'48"
Sofia Rhiannon
Cas 0h30'33" 63d7'45"
Sofia Robles Tapia
Aqr 22h7'32" -15d15'51"
Sofia Rocha
Cas 23h28'12" 57d51'9"
Sofia Rosa Calvanese
Umi 14h32'57" 76d40'8"
Sofia Rose
Cap 21h0'42" -20d18'0"
Sofia Rose Benedetti
Sgr 19h44'43" -13d14'11"
Sofia Rose Corpina
Cas 0h57'19" 51d37'4"
Sofia Rose DeMasi
Sco 16h8'58" -17d47'33"
Sofia Rose Garreffa 8-10-2004
Cru 12h37'11" -57d50'21"
Sofia Rose Grosso
Lyn 7h34'39" 36d40'14"
Sofia Rose Mukhtar
Cnc 8h3'27" 10d45'34"
Sofia Rose Santorelli
Sco 16h50'26" -26d30'38"
Sofia Ryan Masino
Uma 11h21'54" 36d24'46"
Sofia Socorro
Cyg 21h11'3" 45d40'5"
Sofia Sosa
Ari 2h38'3" 27d8'49"
Sofia Stemberga
Cyg 21h43'28" 46d6'1"
Sofia Tseliou
Gem 7h16'0" 19d1'23"
Sofia Volpe
And 0h14'47" 31d35'0"
Sofia Williams
Lyn 9h3'32" 34d13'36"
Sofia Yasmine Karageorgos
Vir 13h0'34" -19d38'55"
Sofia0919
Vir 12h9'0" 7d55'41"
Sofiane
Sgr 18h18'24" -23d24'21"
Sofiane et Laetitia
Sgr 19h15'49" -20d56'59"
Sofiane Hassaine
Tau 5h35'14" 25d18'50"
Sofie Bille-Steenberg
Cyg 19h41'16" 40d11'24"
Sofie Boutsen
Psc 1h26'3" 17d16'31"
Sofija Marianna Ivansons
And 23h16'36" 45d22'6"
Sofija Rose
Uma 11h2'20" 44d17'48"
Sofilya
Cas 1h19'49" 59d44'27"
Sofiya Isabella Escareno
Dra 18h55'23" 72d28'39"
Sofos Raymond
Sco 16h14'12" -29d40'19"
Sofra Bray
Cnc 9h6'58" 7d26'2"
SofrankoHead
And 2h25'4" 46d37'14"
Sofronio Lagahit Cosculla
Lyn 8h20'1" 52d27'46"
SoftHearted SARA
And 0h33'24" 27d22'23"
Sofu loves Ben
Cyg 20h34'31" 31d4'48"
Sofya Korotyanskaya
Sgr 18h12'57" -20d12'51"
SO.GE.VA snc di Caliciotti F.
Umi 16h56'20" 79d24'6"
Sogni
Cnc 8h0'10" 16d56'35"
SoGo
Uma 10h34'37" 46d13'58"
Sogyal Rinpoche
Lyn 8h15'41" 51d1'16"
Soha
Psc 0h27'17" 17d59'53"
Soha Frangié 49
And 22h57'35" 51d47'47"
Soha Salah Mansour
Cap 20h29'0" -12d9'22"
Soham's Taraa
And 0h30'10" 36d21'12"
Soheyl Bakhtavar
Tau 3h57'45" 8d32'2"
Sohia Eileen Olson
Sco 17h10'3" -43d7'35"
Soho
Her 17h56'21" 49d22'41"
Soi & Gary
Cyg 19h41'13" 37d49'14"
Soili S.O. Sutherland
Ori 5h43'12" -0d20'20"
Sojourner
Cap 20h41'29" -21d15'44"
Sojurn
Psc 1h10'49" 26d36'38"
Sokol-Weisberg
Cyg 20h47'41" 43d13'18"
Sokunthea
Cap 20h29'50" -14d9'0"
SOL
Aqr 22h40'19" -12d18'13"
Sol
Uma 11h13'59" 65d49'21"
Sol And Candace
Cyg 20h31'12" 42d14'0"
Sol de Tiempo de Noche
Sco 16h16'21" -12d26'45"
Sol DeLee
Per 3h19'16" 51d59'28"
Sol Farrell
Cep 22h46'12" 66d8'43"
Sol G Plante
Leo 11h17'18" 10d15'25"
Sol Leon Wyatt
Peg 22h24'41" 23d41'9"
Sol Mate
Tau 5h30'22" 17d44'6"
Sol Mate
Ori 5h59'33" -2d24'35"
Sol Pumpkin
Uma 10h16'22" 44d50'48"
Sol Riou
Cas 1h45'42" 54d42'29"
Sol Yellenberg
Dra 19h58'54" 71d4'41"
Sola virtus invicta
Sgr 18h52'56" -32d55'34"
SOLACH
Del 20h33'18" 13d15'49"
Solana
Leo 9h50'59" 14d3'46"
Solange
Leo 9h42'22" 17d49'43"
Solange
Psc 23h8'42" -1d55'20"
Solange Mámá
Psc 1h1'14" 23d22'2"
Solange Mikaelian
Aql 19h1'34" 9d4'50"
Solange Olivia
Ori 5h37'5" 12d23'47"
Solange Valentine
Leo 10h28'1" 21d24'37"
Solange50
Leo 9h47'40" 28d31'47"
Solanges Lévesque
Aqr 23h51'28" -8d19'6"
Solarancio
Umi 17h36'37" 84d40'31"
SolarKenRG44
Sgr 19h13'43" -16d5'5"
Solarski Family Star
Cyg 21h50'36" 54d34'42"
Solben Juin Vignt Cinq
Tri 2h28'53" 30d39'38"
Solberg, Erna
Uma 12h44'2" 62d27'47"
Soldier Nicholas Gambale
Ori 5h32'57" -0d28'24"
Sole
Vir 12h8'43" -4d19'25"
Sole e Completamento
Leo 9h37'30" 32d13'2"
Sole Mio Balaam Nicola
Umi 17h10'28" 81d9'11"
Soledad
Lyn 7h12'42" 53d38'52"
Soledad
Sco 17h52'27" -41d20'31"
Soledad Martinez Belljar
Gem 6h50'0" 29d25'26"
Soledad Pena
Del 20h37'35" 8d53'3"
Soledad Somewhere Over The Rainbow
Ari 2h0'57" 19d24'10"
Soleil
Ari 2h11'30" 16d48'20"
Soleil
Uma 11h42'19" 43d53'12"
Soleil de mes Nuits
Peg 22h21'25" 21d15'41"
Soleil Lana Richards
Psc 22h52'59" 3d21'23"
Soleil Preciosa
Sgr 18h24'14" -28d22'24"
Solena Fabios
Aqr 22h22'1" -13d45'38"
Soléna - Star of Scorpius
Sco 16h35'39" -42d6'51"
Solène
Ori 6h12'43" 7d9'30"
Solène
Uma 9h43'49" 41d57'52"
Solenne & Julien pour la vie
Uma 10h51'32" 72d39'9"
Soler
Lyr 18h36'35" 39d24'31"
Solia Maxine
Per 4h48'33" 36d15'42"
Solida
Cas 0h15'55" 50d55'29"
Soline
Peg 23h29'12" 19d20'27"
Solis Susana Ogdieum
Aql 19h49'32" -0d24'57"
Solisch, Peter
Ori 5h52'20" 7d8'6"
Solitude
Ari 2h12'37" 13d13'36"
Solly Star
Uma 10h50'44" 46d35'22"
Solmich
Tau 5h47'34" 22d59'35"
Solnyshko Morris Friedman
Cap 21h26'0" -19d12'30"
Solo Tu
Peg 22h40'18" 12d26'37"
Solo tu Bon
Sgr 18h7'13" -27d27'37"
Solo Un Para Mi
Eri 3h46'56" -12d39'12"
Soloman
Uma 13h16'0" 61d16'40"
Solomiya Stefanyshyn
Vir 12h44'41" 3d26'35"
Solomon
Uma 11h28'17" 30d34'4"
Solomon Advertising
Her 16h40'48" 24d27'3"
Solomon and Milo Leaman
Dra 10h48'54" 76d39'53"
Solomon Faust Schick
Sco 16h59'39" -38d52'2"
Solomon Joshua Dixon
Tau 5h39'6" 20d6'40"
Solomon & Papa's Star
Cap 20h35'47" -17d32'7"
Solomon "Sid" Botcher
Cep 22h3'36" 57d23'12"
Solomon the Great
Sco 17h13'55" -38d54'23"
Solomonic
Ori 5h51'50" 19d47'1"
Solomon's Sparkling Star
Umi 13h41'38" 78d33'2"
Solon Dixon Dyas
Cep 22h48'0" 65d50'26"
Solouid
Sco 16h3'57" -26d13'5"
Solucki
Uma 9h37'40" 57d33'53"
Solumar
Ori 4h50'39" 3d30'19"
Solus Spes
Lib 15h25'27" -26d23'53"
Solveig och Lennart's Stjerna
Uma 11h34'10" 46d41'57"
Sølvi Løken-Vickers
Psc 0h40'9" 15d49'50"
Solway-Gluck
Cyg 20h25'20" 33d25'34"
Sólya Ádám
Uma 9h12'25" 60d11'52"
Solyamante
Sco 16h46'55" -44d48'50"
Som Paul Paudel
Ori 5h20'37" 5d40'46"
Soma
Lyr 18h51'32" 36d8'38"
Somar Ramos
Sgr 19h45'34" -35d22'9"
Sombrenal
Uma 9h58'35" 43d8'32"
Some Kind of Wonderful
Cyg 21h24'49" 46d18'36"
Some Kind of Wonderful...
Per 2h48'31" 55d6'23"
Some Kinda Playgirl
Per 3h48'0" 33d56'58"
somebody
Sco 17h20'38" -33d13'14"
Somebunny Star
Leo 9h45'19" 21d48'30"
Someday
Ori 6h13'44" 2d36'38"
Someday
Sgr 18h45'49" -30d50'56"
"Someday"
Pic 5h39'2" -44d24'29"
Someday
Uma 11h21'28" 57d51'35"
Someday
Cap 21h49'20" -10d3'23"
Someday - Forever
Uma 11h43'13" 56d44'2"
Someday, my Love
Cap 20h40'3" -20d36'48"
Someichu-gong
Cru 12h50'33" -63d50'59"
Somer
And 23h5'41" 46d53'44"
Somer Breeze
Aqr 21h55'55" 0d32'18"
Somer Marie Stahl
Vul 19h26'22" 25d8'7"
Somer Smith
Cnc 8h33'9" 15d47'4"
Somer Star
Psc 1h34'6" 21d40'28"
Something Special For My Love, Erin
Uma 11h34'3" 53d28'38"
Sometimes You Just Know...
Cnc 9h9'48" 15d33'8"
Sometimes You Just Know...
Cyg 20h20'16" 46d15'59"
Somewhere Out There...
Lyr 18h51'3" 35d32'25"
Somewhere Over the Rainbow
Lyn 7h33'12" 58d5'9"
Somewhere We'll Always be Together
Ari 3h18'19" 16d0'49"
Somi & Doug
Cap 20h37'55" -13d51'9"
Sommer
Aqr 22h47'18" -13d18'51"
Sommer Ann Good
Com 12h21'7" 27d47'22"
Sommer Dawn
Aqr 20h48'45" -8d52'44"
Sommer Leigh Williams...I Love You.
Sgr 18h36'21" -34d48'11"
Sommer Lovin'
Cap 21h27'17" -18d8'35"
Sommer Swanzy Dance Academy
Lyr 18h53'7" 33d10'38"
Sommerhäuser, Lothar
Uma 8h16'1" 64d24'33"
SommerLynn Nicole Patterson
Leo 10h42'6" 17d20'55"
Sommerstern
Uma 9h35'7" 64d55'0"
Sommi Ambra
Cam 3h29'38" 68d13'34"
Sommi Giorgia
Uma 13h54'29" 53d1'1"
Somnium Peregrinus Ingo
Vir 13h4'38" 7d44'58"
Somnus Laetificum Quam Mater
Leo 11h7'52" 11d56'35"
Son of Man
Pyx 8h50'49" -24d49'30"
Son of Mark, The Star Kyle Kriese
Cma 7h11'27" -15d59'45"
Sona Polakowski
Uma 11h48'54" 64d42'44"
Sonai Chatterjee
Uma 14h3'34" 57d30'33"
Sonal A. Doshi
Vir 14h18'49" -21d4'31"
Sonali
Cap 21h55'26" -16d20'32"
Sonali
Vir 13h7'5" 12d8'33"
Sonam & Charu - Soulmates Forever
And 0h35'1" 26d8'27"
Sonam Kapoor
Aqr 22h15'42" -1d31'59"
Sondia Bell
And 23h36'34" 47d0'36"
Sondra
Aqr 22h49'48" -24d42'14"
Sondra 9.22.35
Vir 12h50'43" -8d33'45"
Sondra Brackman
Psc 1h41'37" 5d51'54"
Sondra Cher Insua
And 0h51'39" 37d19'12"
Sondra Goldfarb
Cam 3h59'8" 56d2'26"
Sondra Lauren
And 1h3'28" 43d50'3"
Sondra Lynn Collins
Gem 6h53'29" 26d24'35"
Sondra Lynnette Williams
Leo 9h25'58" 8d42'21"
Sondra "Our Star" Schweitzer
Cas 1h32'6" 60d59'33"
Sondra Segal-Sklar
Uma 9h7'36" 62d49'26"
Sondra Thiederman Ph.D.
Cyg 21h10'17" 46d35'7"
Sondra & Thomas Seastrand
Cyg 20h20'58" 58d15'3"
Sondra Warren
Ari 3h18'2" 21d32'59"
Sonechko
Vir 12h28'54" -6d5'56"
Soner
Sge 19h48'11" 16d47'2"
SONEVA
Cyg 21h10'7" 51d4'7"
Song and Sun Park
And 23h22'58" 42d5'26"
Song Jian Feng
Sgr 18h29'28" -26d53'52"
Song Meng yuan: Monse Marie Song
Uma 8h41'36" 69d51'0"
Songbird
Uma 10h30'37" 66d55'9"
Songbird Mr & Mrs J P Kerr
Uma 10h42'40" 71d22'27"
songbird716
Cnc 8h41'24" 12d37'39"
Sonia
Lac 22h31'33" 41d29'32"
Sonia
Per 3h26'21" 50d4'5"
Sonia
And 1h16'57" 49d22'18"
Sonia
And 23h24'20" 47d15'25"
Sonia
Cyg 20h10'50" 35d48'11"
Sonia
And 2h36'30" 40d5'28"
Sonia
Cnc 9h3'2" 32d15'46"
Sonia
Dra 9h28'49" 73d37'48"
Sonia
Cas 23h19'53" 60d47'41"
Sonia
Lyn 7h25'19" 53d30'3"
Sonia
Cam 6h2'21" 84d36'57"
Sonia
Lyn 8h23'3" 44d10'16"
Sonia<3
Psc 1h49'4" 6d24'37"
Sonia and Gerardino
Crb 16h11'28" 27d3'50"
Sonia Azaouzi
Cyg 21h5'1" 49d7'18"
Sonia Azaouzi
Cyg 21h5'1" 49d7'18"
Sonia "Baby" López
And 0h23'41" 23d47'50"
Sonia Bartoli
Cyg 20h11'52" 36d5'39"
Sonia Ben Romdhane, 08.08.1981
And 23h17'1" 40d18'7"
Sonia Bessa
Cap 20h16'52" -24d10'39"
Sonia Caltvedt
Ori 6h14'43" 2d26'58"
Sonia Chabra
Uma 10h35'55" 42d11'42"
Sonia Colette Kramm
Ari 3h23'9" 20d0'7"
Sonia Combs
Uma 11h59'30" 58d31'56"
Sonia Denise
And 0h53'32" 36d37'59"
Sonia E. Cruz Ramos
Tau 5h48'14" 17d55'13"
Sonia e Geraldo
Cru 12h48'9" -56d23'6"
Sonia Elizabeth Lee Scapito
Cam 3h19'24" 67d59'52"
Sonia Enloe McCauley
Lyn 7h57'47" 57d55'22"
Sonia Escobar
Sco 16h6'36" -18d13'51"
Sònia Esteve
Vir 14h15'19" -2d38'51"
Sonia Faris
Leo 9h33'56" 6d27'40"
Sonia & Gabriele
Ori 5h59'13" 9d5'36"
Sonia Gill
Lib 15h57'18" -11d16'11"
Sonia Grover
And 23h37'25" 42d17'32"
Sonia H Mattoon
Lib 15h12'16" -15d44'51"
Sonia I. Pena
Psc 1h26'51" 14d20'46"
Sonia Irene De La Mora Autran
Lyr 18h54'2" 43d13'48"
Sonia J. McTaggart
And 0h2'35" 34d56'10"
Sonia Jaen
Psc 2h1'11" 5d0'41"
Sonia Jonathan
Peg 21h20'38" 18d10'5"
Sonia Labadie
Cnc 9h17'40" 11d21'19"
Sonia Lady Ge
Aqr 23h4'7" -17d8'13"
Sonia M. Anaya's Unforgetable Star
Uma 11h47'42" 57d41'18"
Sonia Magnifico
Cyg 21h38'50" 33d24'19"
Sonia Maia
Umi 15h4'12" 73d40'24"
Sonia Marie Didiano
Ari 2h4'37" 27d46'30"
Sonia Marina Gonçalves Correia 79
Vir 13h6'34" -20d33'33"
Sonia Nicholas
Sco 17h52'42" -41d42'28"
Sonia & Noel Montero
Cyg 21h38'11" 40d4'19"
Sonia Rani Parekh
Aqr 22h49'26" 0d26'9"
Sonia Rodriguez
Lib 14h44'15" -12d37'43"
Sonia Rosemary Elizabeth Blake
Cas 0h7'49" 53d13'37"
Sonia Schenkel (Mamour)
Cnc 8h30'38" 30d36'42"
Sonia Senosiain
Cyg 19h46'1" 36d4'45"
Sonia Smith
Gem 7h45'35" 33d46'50"
Sonia Smukler
Lyr 18h34'41" 37d15'16"
Sonia Spagnol
Psc 0h44'3" 15d25'42"
Sonia 'Starr' Howard
Uma 10h31'18" 71d14'18"
Sonia Tabriz
Leo 10h55'21" -2d18'50"
Sonia Torres
Psc 1h45'58" 6d48'15"
Sonia Ursenbacher
Uma 14h22'17" 58d54'21"
Sonia Vanessa Thompson-Serna
Gem 6h58'19" 13d55'42"
Sonia Vosbikian
Psc 0h59'27" 8d23'4"
Sonia Yvonne
Cap 21h56'24" -13d59'11"
Sonia Yvonne Garcia
Lyn 7h7'51" 53d5'13"
Sonia - Zendegim
Vir 13h23'27" -16d58'0"
Sonia, my Gift from the Angels
Cyg 20h25'34" 50d42'38"
Soniarl
Umi 11h34'28" 86d29'15"
Sonia's Glow
Cam 7h39'32" 77d19'50"
Sonita Loyd
Cap 20h12'18" -9d51'33"
Sonja
Sco 16h9'55" -11d58'23"
Sonja
Aqr 20h59'27" -0d9'34"
Sonja
Cas 0h28'19" 75d1'4"
Sonja
Uma 13h17'57" 62d5'24"
Sonja
Uma 9h53'44" 48d38'13"
Sonja
And 23h55'45" 43d10'34"
Sonja
Crb 16h16'59" 38d8'1"
Sonja
And 2h27'13" 46d39'38"
Sonja
And 2h35'32" 50d28'23"
Sonja
Tri 2h21'51" 36d8'48"
Sonja
And 0h16'0" 41d55'28"
Sonja
Uma 11h14'20" 42d51'35"
Sonja
Aql 19h35'10" 7d9'11"
Sonja
Ari 2h40'31" 13d32'17"
Sonja
Peg 21h33'49" 21d56'2"
Sonja & Ali
Umi 13h14'11" 71d30'51"
Sonja Amstuz
Aql 20h4'50" 11d54'37"
Sonja Arnold
Lmi 10h26'28" 30d13'55"
Sonja Boulware
Ori 5h14'25" 1d44'36"
Sonja Bryan-Rorabaugh
Uma 12h40'24" 60d28'34"
Sonja+Carsten
Uma 9h13'31" 51d19'11"
Sonja & Chelsea Hohnhorst
Cyg 20h15'24" 51d52'5"
Sonja Curnew
Cas 0h58'3" 58d52'39"
Sonja Gerstenberger
Cam 4h4'57" 69d26'0"
Sonja "Guinevere" Kurt
Lyr 18h39'56" 38d42'0"

Sonja H. - The Mule Star
Cyg 21h18'55" 42d38'24"
Sonja & Holger
Ori 6h15'11" 16d30'35"
Sonja Huang
Cnc 8h43'12" 18d34'24"
Sonja & Jost Wenzel
Cyg 21h13'43" 42d35'6"
Sonja Khristian Ortiz Wood
Ori 4h46'9" 6d27'24"
Sonja Kleih
Sgr 19h27'16" -31d29'57"
Sonja & Martin
Cas 0h6'11" 53d42'58"
Sonja Mazour
Cas 1h13'44" 53d46'26"
Sonja McCreath Costello
Cyg 19h17'29" 51d12'59"
Sonja Meijer
Crb 16h22'9" 37d45'1"
Sonja & Michèl
Uma 9h28'55" 57d44'42"
Sonja Müller
And 0h24'59" 31d29'54"
Sonja Murphy
Psc 22h52'56" 6d50'9"
Sonja Namtvedt
Aqr 21h41'9" -0d7'34"
Sonja Nicole Kallam
Lib 15h13'28" -16d44'51"
Sonja Niederberger-Zeiter
Her 17h10'8" 20d29'48"
Sonja Pie
Aqr 21h22'19" 2d13'32"
Sonja Pützer
Lib 15h12'45" -7d23'10"
Sonja Rodenbrügger
Cas 23h0'47" 59d22'1"
Sonja Romain
Aqr 22h10'11" -21d37'0"
Sonja Schalbetter
Vul 20h26'0" 27d26'18"
Sonja Shark
Uma 13h49'47" 53d20'24"
Sonja (SonSon) King
Cmi 7h25'4" 8d44'26"
Sonja the Honey Bunny
Cap 21h10'39" -21d0'11"
Sonja Theresia Yvonne Van Mastrigt
Com 12h29'18" 24d57'49"
Sonja und Claude Siepmann
Cyg 20h42'23" 45d48'50"
Sonja und Markus
Cyg 20h11'55" 48d53'7"
Sonja Winkler-Hess, 23.08.1962
Equ 21h2'1" 4d58'2"
Sonja Zaugg
Aur 5h25'56" 40d16'29"
Sonja, Steve, Sophie & Callum Black
Uma 10h53'39" 72d39'6"
Sonja's eigener Stern
Uma 14h4'12" 59d15'14"
Sonja's Inspiration
Leo 9h26'13" 10d11'18"
Sonja's & Markus Hochzitsstern
Uma 9h27'30" 43d27'40"
Sonja's & Roman's Hochzeitsstern
Cas 23h5'2" 56d57'44"
Sonjia Monique Jaime Williams
Per 3h13'18" 51d34'56"
Sonk
Uma 11h25'32" 54d49'31"
Sonka, Kathrin
Ori 4h56'0" 2d40'37"
SonMoonStar
Cnc 8h3'27" 8d18'35"
Sonnenschein
Uma 13h56'28" 49d40'15"
Sonnenschein Britta Kreye
Uma 10h4'14" 64d48'25"
Sonnenschein Carola
Ori 6h11'27" 7d18'10"
Sonnenschein - Katharina
Uma 10h9'34" 57d41'48"
Sonnenschein Tanya
Ori 5h43'24" 8d37'25"
Sonneschein Adu
Cep 22h18'38" 67d28'31"
Sonnet 116
And 1h32'19" 43d47'44"
Sonnia Pepe
Lyn 7h10'0" 58d18'56"
Sonnie
Sgr 18h5'14" -21d47'23"
SonnieMarie (Guglielmini) Lebbing
Gem 6h18'35" 22d7'57"
Sonny
Uma 10h12'59" 51d45'44"
Sonny
Lib 15h33'59" -15d29'16"
Sonny
Cap 20h42'52" -18d13'30"
Sonny
Umi 15h48'53" 75d54'7"
Sonny
Lib 15h5'42" -22d44'31"
Sonny 77
Uma 9h2'36" 47d14'43"
Sonny & Allison
Her 16h15'26" 6d14'47"
Sonny and Laurie Wilson
Cyg 20h21'41" 54d6'58"
Sonny and Mary Louise Jones
Uma 13h36'0" 54d15'52"
Sonny Armstrong
Per 3h35'54" 45d55'5"
Sonny Bommeliene Huwada
Uma 9h2'54" 54d1'43"
Sonny Bunny - Mr Poo
Cep 20h43'10" 59d44'42"
Sonny Carroll
Tau 3h43'40" 23d47'25"
Sonny Dixie Thomas
Her 16h49'0" 4d22'11"
Sonny & Grace McQuirt
Leo 10h11'48" 18d25'43"
Sonny James Cowen
Cep 20h55'0" 62d8'19"
Sonny & JessaLee
Cyg 20h31'45" 30d14'56"
Sonny Joe Fama
Aqr 22h29'27" -4d16'32"
Sonny Kruse
Sco 17h2'9" -44d21'24"
Sonny M. Armstrong
Psc 1h7'29" 31d45'30"
Sonny Napolitano
Her 16h28'35" 48d19'2"
Sonny Pawloski's Memorial Star
Aql 19h37'4" 2d50'55"
Sonny (Poppy) Geyser
Cep 22h20'8" 63d14'53"
Sonny Rider
Uma 10h37'25" 68d41'21"
Sonny Roberts
Cep 5h9'16" 81d37'50"
Sonny Trainum
Cep 22h50'8" 60d23'52"
Sonny, Poppy, Dad
Per 2h49'5" 54d11'21"
SonnyJoan
Aqr 23h50'26" -10d11'23"
SONODA
Vir 14h20'59" 3d40'36"
Sonoma
Uma 11h37'1" 54d1'41"
Sonoma Rose Parlagreco
Umi 14h12'22" 68d40'32"
SonomaMellon2007
Uma 13h46'6" 60d51'22"
Sonona Al-Lan Williams
Uma 9h52'22" 52d57'5"
Sonora Maxim
Cru 12h48'14" -61d7'40"
Sonray
Ari 1h57'33" 24d18'6"
Sons of Mary Health of the Sick
Cyg 20h20'23" 43d42'8"
Sons of Ronnann
Cru 12h3'30" -62d29'17"
Sonsuz ask
Uma 9h59'11" 62d23'32"
SONU
Aqr 22h28'55" -1d40'31"
Sonu and Yesha
Umi 15h18'23" 74d50'44"
Sonu 'oney
Lib 15h10'53" -4d3'8"
SONULIK
Cap 20h37'21" -9d54'8"
Sonu's Shining Star
Gem 6h23'9" 20d12'17"
Sony Bear
Uma 14h11'4" 57d0'58"
Sonya
Lyn 8h21'5" 52d41'23"
Sonya
And 23h38'46" 44d39'47"
Sonya
Per 2h54'27" 33d19'58"
Sonya 40
Cnc 8h41'1" 30d51'28"
Sonya Alison Gyffard
And 2h34'29" 43d56'6"
Sonya Ayayay
Uma 9h21'55" 55d47'44"
"Sonya" Barg
Sgr 19h3'31" -21d16'29"
Sonya Blue
Vir 12h13'12" 5d48'34"
Sonya Brady
Sgr 19h11'52" -16d41'11"
Sonya Burnside
Com 12h20'7" 17d24'8"
Sonya C Adams
Lyn 7h26'27" 44d17'48"
Sonya Contino
Ari 2h37'47" 27d55'9"
Sonya Degani
Aql 19h32'21" 10d21'34"
Sonya Denise Hampton
Aqr 21h45'33" -8d1'58"
Sonya Dhanani
Cnc 8h8'32" 28d42'58"
Sonya Dumouchelle
Per 4h9'28" 44d14'39"
Sonya Durkee
Psc 0h5'47" 7d36'35"
Sonya Genua mi amor
Cam 7h6'8" 72d36'5"
Sonya Gilbert Watts
Lib 15h36'56" -5d58'9"
Sonya Grisham
Uma 10h46'45" 42d19'34"
Sonya Gugliara
Sco 17h22'43" -37d53'45"
Sonya Jean Gilman
Ari 2h8'26" 23d27'1"
Sonya Jean Peterson
Cnc 8h41'34" 22d9'36"
Sonya Johnson
Com 12h27'30" 28d28'36"
Sonya Kylie Jones
Leo 9h26'32" 15d11'10"
Sonya L. B&B
And 23h17'44" 47d23'56"
Sonya L. Mcoy's Sunshine
Lyn 6h58'10" 51d43'45"
Sonya Lehman
Gem 7h21'6" 25d18'49"
Sonya Lynn
Sgr 17h54'36" -29d39'3"
Sonya Marie
Tau 5h42'52" 26d28'8"
Sonya Maya
And 0h42'26" 26d42'5"
Sonya Morrow Holland
Cam 7h17'28" 73d27'59"
Sonya (My Peanut)
Ori 5h58'20" 12d33'51"
Sonya Neely
Lyr 19h16'40" 28d25'32"
Sonya Renee Tincher
Gem 7h46'48" 24d21'49"
Sonya Suezzett Kamer
Uma 9h40'3" 57d46'42"
Sonya Syth
Aqr 23h37'2" -24d46'10"
Sonya313
Psc 1h12'20" 30d15'17"
Sonya's Secret
Sgr 18h29'52" -20d41'35"
Sonya's Special Star
Ari 3h28'38" 29d22'34"
SonyBalony
Vir 13h41'4" -0d33'50"
Sonyétoile
Del 20h34'51" 7d50'51"
Sonyuk
Leo 9h31'46" 10d50'41"
Soo Choi
Tau 4h30'7" 12d31'30"
Soo Ho & Christene
Uma 9h25'23" 65d9'13"
Soo San Choi
Tau 4h35'3" 21d49'55"
Soo Sevilen
Vir 14h27'28" -0d0'40"
Sooki
Uma 11h40'2" 55d17'58"
Sooki
Lib 15h29'27" -28d49'12"
Soóky József
Uma 10h27'41" 62d52'7"
Soókyné Farkas Mimi
Uma 10h19'1" 63d16'42"
SooMee & Steve's Fourth Decade Star
Cyg 19h39'11" 34d39'59"
Soon Dam and Ian Bae
Sco 16h36'50" -34d56'43"
Sooner E Plageman
Lib 15h59'20" -5d51'22"
Sooner's Hell Yeah
Cnc 8h43'15" 18d1'4"
Soonie S. Mulford
Uma 9h7'59" 56d54'25"
Soós Noémi Noccsy
Psc 1h49'30" 3d23'20"
SooSee
Crb 15h42'6" 30d45'54"
Soosel
Dra 10h49'45" 77d55'20"
Soosipantalones
Umi 15h7'8" 72d15'19"
Soot Thee Rak
And 0h35'1" 46d27'54"
Sooty
Cmi 7h33'31" -0d11'28"
Sooty Hetrick
Cyg 20h42'55" 35d58'10"
SoozeyQ
Sgr 18h34'37" -18d11'31"
Sooz's Piece of Heaven
Ari 3h10'51" 27d30'44"
Sop Sofofo Wandji Jean-Marie
Ori 5h35'10" -5d38'24"
SOPES
Dra 18h42'33" 62d57'35"
Sophak So
Uma 9h40'19" 57d32'29"
Sophal Star
Sco 16h15'43" -14d40'31"
SophanRiensey Susan Chin
Uma 9h55'48" 46d41'54"
Sophary Khun Ung
Sco 17h14'35" -42d38'11"
Sophia
Sco 17h47'36" -37d30'58"
Sophia
Ara 17h23'21" -55d39'55"
Sophia
Cru 12h28'50" -61d13'21"
Sophia
Sgr 18h2'18" -27d48'45"
Sophia
Cap 20h8'41" -8d43'46"
Sophia
Sgr 19h14'42" -17d27'45"
Sophia
Ori 5h23'1" -0d5'11"
Sophia
Aqr 22h39'27" -2d55'48"
Sophia
Uma 14h17'44" 59d47'25"
Sophia
And 23h32'51" 40d58'42"
Sophia
Crb 15h38'32" 39d23'10"
Sophia
And 1h32'13" 48d34'40"
Sophia
And 0h32'32" 37d38'3"
Sophia
Psc 1h9'4" 31d39'26"
Sophia
Tau 4h31'15" 22d48'20"
Sophia
Gem 7h47'46" 26d35'16"
Sophia
Gem 6h21'52" 21d17'19"
Sophia
And 0h44'29" 27d15'37"
Sophia
Ori 5h28'20" 5d40'28"
Sophia Abigail
Tau 5h46'23" 27d55'3"
sophia alexandra eisel
Cyg 19h50'45" 39d52'17"
Sophia Alexandra Landers
Cam 6h0'23" 60d28'17"
Sophia Alexandra Walter
Vir 14h3'13" -20d24'39"
Sophia Ally Garavan
And 0h27'24" 46d19'19"
Sophia Amelia
Tau 3h46'29" 29d24'27"
Sophia Anees Miller
And 1h16'11" 46d12'38"
Sophia Angel Whited
And 1h45'29" 44d23'40"
Sophia Ann McCabe
Vir 15h6'51" 6d46'35"
Sophia Anne Battaglia
And 23h21'58" 48d35'24"
Sophia Anne Juneski
Cas 1h21'40" 51d51'51"
Sophia Apostolatos
Cas 23h36'14" 57d17'4"
Sophia Arabani
Lyn 6h23'57" 57d43'40"
Sophia Ashley
Cyg 21h21'40" 39d28'22"
Sophia Ava O'Hara
And 0h30'3" 33d55'48"
Sophia Avia
Cnc 8h47'38" 17d58'0"
Sophia Avia Rosone
And 1h36'2" 48d29'54"
Sophia Bartlett Rogers
Sgr 19h46'34" -24d28'45"
Sophia Bell
Cas 1h36'6" 61d56'30"
Sophia Bella
Umi 14h5'59" 75d15'7"
Sophia Bluebell McKeeve
And 23h49'15" 45d57'15"
Sophia Bonas
Sco 16h56'53" -40d19'43"
Sophia Boo
Leo 11h25'41" 8d44'37"
Sophia Boyiu Van
Lyn 7h19'33" 52d12'6"
Sophia Brielle Yeakel
Cas 0h19'41" 62d13'36"
Sophia Brynn Jimenez
Psc 1h20'58" 7d50'59"
Sophia Brynne
And 2h7'45" 38d24'26"
Sophia Campari-Hand
Uma 8h40'52" 55d13'53"
Sophia Catherine Schmitz
And 0h47'13" 36d43'1"
Sophia Charlotte
Ori 5h27'18" -3d49'32"
Sophia Cho
Ari 2h11'14" 22d22'43"
Sophia & Christopher
Lib 14h54'44" -1d57'18"
Sophia Claire
And 23h24'4" 46d59'34"
Sophia Claire de Hennin
Umi 14h27'21" 71d15'57"
Sophia Colette
And 2h20'40" 47d29'44"
Sophia Danielle Schmid
And 0h27'58" 43d31'22"
Sophia Danielle Willens
Peg 22h6'24" 35d9'30"
Sophia Danielle Yormark
Lib 15h20'39" -21d29'26"
Sophia Dawn Cunningham
Lib 14h30'57" -13d22'11"
Sophia DeGiosa
Uma 11h58'26" 65d39'18"
Sophia Donna Nehr
And 2h27'31" 44d11'24"
Sophia Dugan
And 0h44'20" 40d40'8"
Sophia Dutterer
Cap 20h38'53" -10d59'47"
Sophia E. Ross
Lyn 8h4'35" 41d40'16"
Sophia Eileen
Sco 16h9'36" -20d3'25"
Sophia Eliza Michaelena
Ari 2h51'46" 25d44'21"
Sophia Elizabeth Baer
Uma 10h50'31" 57d23'23"
Sophia Elizabeth Colley
Cas 2h33'4" 65d33'24"
Sophia Elizabeth Dussault
And 0h24'42" 45d6'43"
Sophia Evelyn Nagel
Aql 19h27'45" 4d49'40"
Sophia Feeley
Psc 1h41'15" 5d25'53"
Sophia Feldberg Adelman
Lyn 7h50'47" 36d40'58"
Sophia Florence
And 0h36'54" 46d25'47"
Sophia Florence
And 23h22'30" 44d34'17"
Sophia Florence Fanello
And 0h7'11" 40d29'55"
Sophia Florence Fanello
And 0h15'11" 30d7'50"
Sophia Francesca Adams
And 2h31'38" 50d0'29"
Sophia Gabriella Sturgeon
Sco 17h52'38" -35d56'16"
Sophia Grace
Uma 10h3'27" 52d15'22"
Sophia Grace
And 1h23'16" 44d14'51"
Sophia Grace
Leo 11h44'40" 19d42'37"
Sophia Grace Heer
Peg 21h47'53" 11d29'51"
Sophia Grace Mahon
Ari 2h55'11" 27d41'55"
Sophia Grace Maritato
Cap 20h10'45" -18d37'55"
Sophia Grace Milazzo
Cas 2h16'12" 63d57'0"
Sophia Grace Robinson
Sco 17h26'28" -42d48'22"
Sophia Grace Thompson
And 1h11'55" 48d29'53"
Sophia Graham
Cas 0h32'39" 61d7'45"
Sophia Grey
Cas 23h24'58" 58d42'19"
Sophia Hannah Starr
Cnc 9h15'8" 10d52'9"
Sophia Hope Boggan
Cnc 8h3'16" 15d10'37"
Sophia Isabella Keyser
And 1h15'46" 45d55'27"
Sophia Isabella Mezzatesta
Sco 17h16'26" -39d55'50"
Sophia Isabelle Budd
And 23h6'22" 50d58'43"
Sophia Isadora Lindley-Kessler
And 0h45'0" 37d3'24"
Sophia Jean Turtzo
And 0h38'16" 46d31'33"
Sophia Judge
Uma 10h11'44" 53d33'22"
Sophia Juliet
Cas 1h36'46" 60d21'4"
Sophia June-Raquel Salomao Schmidt
Tau 4h32'17" 29d3'51"
Sophia Kay
Vir 11h50'36" -0d34'6"
Sophia Kearns Fitzpatrick
Uma 8h57'5" 53d19'5"
Sophia Kelsey Burke
Vir 12h43'48" 11d27'9"
Sophia Kennedy Davidson
And 23h23'22" 43d52'2"
Sophia Kouros
And 1h2'50" 38d17'6"
Sophia L. Hopkins
Uma 12h36'6" 55d27'14"
Sophia Lael Naughton
Eri 3h15'51" -8d22'15"
Sophia Lal
Uma 11h56'51" 59d14'57"
Sophia Laven
Sgr 18h33'53" -35d0'20"
Sophia Lee Hollcraft Johnson
And 1h38'1" 46d24'47"
Sophia Leonora Mendelsohn
Psc 0h18'24" 15d28'17"
Sophia - Little Angel
Uma 8h47'2" 47d34'8"
Sophia Long
Uma 11h42'4" 28d48'41"
Sophia Lotkowski
Ori 5h18'50" -4d34'48"
Sophia Louise Legg
Cas 23h35'57" 55d6'17"
Sophia Louka
Aqr 22h53'34" -16d52'15"
Sophia Macbeth O'Steen
And 1h29'34" 47d49'51"
Sophia Mae
Umi 16h54'46" 83d52'39"
Sophia Mae Vest
And 23h6'17" 51d0'5"
Sophia Maria de la Paz George
Uma 11h45'27" 58d24'26"
Sophia Maria Hamel
Sco 16h10'28" -14d56'18"
Sophia Mariam Marogi
Ari 2h22'20" 14d38'15"
Sophia Marie
Psc 0h49'25" 12d22'44"
Sophia Marie
Gem 6h32'36" 14d57'43"
Sophia Marie
Lyr 18h55'9" 33d40'55"
Sophia Marie Bohrer
Tau 4h2'15" 27d18'49"
Sophia Marie Covert
And 2h38'33" 37d43'27"
Sophia Marie DiCicco
Crb 16h8'38" 34d23'53"
Sophia Marie Donato
Sco 16h16'20" -13d48'7"
Sophia Marie Mason
Uma 9h26'33" 59d12'54"
Sophia Marina
Leo 9h36'27" 26d36'2"
Sophia Mary Ann Trey
Vir 11h49'43" 8d46'14"
Sophia Maryella Murphy
And 2h8'35" 45d7'1"
Sophia Mavros Danson
Tau 4h27'14" 17d55'24"
Sophia May Kangas
Dra 18h50'32" 65d46'51"
Sophia & Michael
Cyg 21h31'45" 38d28'36"
Sophia Michelle
Cas 0h16'57" 58d18'10"
Sophia Michelle Grise
And 2h6'0" 46d31'49"
Sophia Michelle Vodov
Sco 16h8'57" -15d58'13"
Sophia Molinaro
Lyn 7h41'13" 59d8'47"
Sophia Murray Boucher
Aqr 22h12'25" -18d14'58"
Sophia Nicola Jones - My Precious Sophia
And 1h33'4" 40d43'48"
Sophia Noel Wilcox
Cap 20h12'30" -9d33'41"
Sophia Noelle Blakely
Uma 13h11'35" 62d56'17"
Sophia Oliver
Uma 9h1'40" 63d37'34"
Sophia Pesa
And 0h9'31" 45d35'3"
Sophia Rebeca Best
Vir 13h6'26" 11d18'44"
Sophia Reece Goeke
Ari 2h17'17" 12d37'30"
Sophia Renae
Sco 16h13'16" -9d29'16"
Sophia Ria Hoetmer
Cas 0h56'13" 53d35'36"
Sophia Rose
Cnc 8h42'49" 17d57'46"
Sophia Rose
Aqr 22h59'12" -10d45'37"
Sophia Rose
Umi 13h50'54" 71d32'6"
Sophia Rose DeSimone
Leo 9h40'42" 29d33'20"
Sophia Rose Finneran
And 1h8'27" 47d36'3"
Sophia Rose Fromm
And 23h1'18" 35d41'41"
Sophia Rose Hamblion
Cap 21h22'40" -19d55'44"
Sophia Rose Mazziotta
And 1h7'45" 36d47'51"
Sophia Rose Sylvestre
And 1h36'26" 45d19'33"
Sophia Rose Tranquch
Psc 1h27'7" 17d0'37"
Sophia Rueckerl
Dra 17h56'54" 60d38'23"
Sophia Ruffini
And 23h54'25" 38d18'59"
Sophia Ruth
Cnc 9h11'46" 10d50'6"
Sophia & Ryan
Uma 8h43'10" 54d15'29"
Sophia Schupp
Her 18h14'16" 25d41'44"
Sophia Shaw
Uma 9h10'33" 65d7'52"
Sophia Shinn
Boo 14h37'3" 37d15'9"
Sophia Sinclair Krinsky
Psc 1h7'49" 10d36'14"
Sophia Stoyanka Manalov
Sco 16h59'44" -38d18'25"
Sophia T. Bauer
Crb 15h54'11" 25d55'37"
Sophia T. Sgroi
Cnc 8h47'33" 18d33'7"
~Sophia Tasnadi 10/30/51-08/12/90~
Sco 16h10'44" -11d18'2"
Sophia Vahr Hyldborg
And 23h1'47" 45d5'31"
Sophia Veronica DeMaio
Lib 15h4'29" -2d11'48"
Sophia Villar
Crb 15h46'34" 36d20'3"
Sophia Wolf Quadracci
Sgr 18h5'14" -19d35'31"
Sophia Wynn Shartzer
Cyg 19h55'25" 40d48'32"
Sophia Zuzunitsa
And 0h19'28" 46d36'51"
Sophian Baron
Umi 15h33'37" 81d56'15"
Sophia's Mystic Enchantment
Cnc 9h3'58" 28d9'44"
Sophia's Star
Tau 4h19'10" 4d51'19"
Sophie
Cnc 8h28'57" 26d32'57"
SOPHIE
Her 17h20'47" 44d54'17"
Sophie
Lmi 10h47'36" 34d13'3"
Sophie
Lmi 10h41'3" 35d35'3"
Sophie
And 1h39'54" 37d27'50"
Sophie
And 0h22'22" 42d54'45"
Sophie
And 1h55'41" 37d52'2"
Sophie
And 0h51'47" 39d34'18"
Sophie
Aur 6h57'22" 43d50'57"
Sophie
Lmi 9h45'39" 40d24'5"
Sophie
Lmi 10h34'6" 37d36'27"
Sophie
Umi 14h55'25" 78d57'41"
SOPHIE
Uma 9h27'19" 64d37'3"
Sophie
Cas 1h33'20" 67d47'59"
Sophie
Uma 9h58'58" 53d52'16"
Sophie
Uma 9h4'49" 55d6'36"
Sophie
Cas 1h22'9" 64d17'48"
Sophie
Sgr 19h32'2" -36d23'28"
Sophie
Sco 17h28'22" -39d17'0"
Sophie
Sgr 18h0'3" -33d53'32"
Sophie
Lup 15h49'50" -32d50'13"
Sophie
Sgr 17h58'53" -24d34'22"
Sophie 19
And 0h24'31" 25d10'44"
Sophie Alexandra Cook
Aur 5h56'20" 42d23'47"
Sophie Alexandra Naylor
Cas 1h0'51" 49d46'36"
Sophie Alexandra Norman
Pav 17h56'46" -57d36'48"
Sophie Alfiero
And 23h12'7" 44d22'41"
Sophie Alice Mary
Cmi 7h55'48" 11d7'58"
Sophie Amelia Bennett
And 23h56'26" 44d1'2"
Sophie and Luc's Valerie
Uma 12h36'37" 59d43'37"
Sophie Anna Lamin
Sco 16h3'43" -9d2'15"
Sophie Anna Lamin
Sco 17h50'53" -43d52'8"
Sophie Anne Arber
Cas 23h42'54" 55d48'13"
SOPHIE AURELIA
And 0h51'33" 40d44'42"
Sophie Bambridge
And 2h30'11" 43d28'26"
Sophie Bella
Sco 16h10'24" -10d17'59"
Sophie Benette
Vir 12h49'53" 5d1'43"
Sophie Benker - Forever Carl's Star
Cru 12h6'43" -62d31'45"
Sophie Benton and Chloe Benton
And 23h14'27" 43d11'25"
Sophie Beresiner
Cam 7h43'24" 69d24'42"
Sophie Bergeron
Uma 9h22'54" 59d47'27"
Sophie Beth
And 23h13'24" 50d43'59"
Sophie Blauvelt
Uma 12h1'44" 53d31'15"
Sophie Bradburn
Psc 23h51'36" -3d18'2"
Sophie Briffa
Cnv 12h45'53" 33d19'37"

Sophie Camphausen
Sgr 18h23'9" -32d14'25"
Sophie Carole Adkins
Uma 10h33'33" 42d35'36"
Sophie Castilho Conway
Cas 23h31'13" 51d33'13"
Sophie Catherine Jamil
And 23h0'42" 51d12'27"
Sophie Catto
And 1h41'18" 45d23'2"
Sophie Cecchetto
Aqr 22h10'6" -21d25'6"
Sophie Christine Mashiter - Sophie's Star
And 1h30'52" 34d34'17"
Sophie Claire
And 1h57'22" 37d36'23"
Sophie Claire'
Gem 7h12'23" 25d20'19"
Sophie Cole Stavropoulos
Tau 4h4'10" 29d20'14"
Sophie Corbett
Cas 0h35'19" 58d30'13"
Sophie Daniele Millar
And 23h16'10" 35d34'35"
Sophie Dawn
Uma 11h51'42" 52d36'57"
Sophie Diana
Lib 15h26'29" -12d21'53"
Sophie Diane Milan
And 23h57'57" 36d25'15"
Sophie Dianne Ortyl
Ori 5h55'54" 18d31'43"
Sophie Doolaghe
Del 20h25'35" 16d28'16"
Sophie/Dorothy Wasus/Hess/Hanley
Uma 11h35'17" 47d50'54"
Sophie DV
Tau 3h37'53" 18d26'43"
Sophie Elena Casuso Wheeler
Cyg 19h46'55" 41d3'35"
Sophie Elise Straniere
Uma 8h37'43" 60d46'57"
Sophie Elizabeth
Leo 11h21'19" 17d37'33"
Sophie Elizabeth Allen
Cas 23h51'58" 52d54'41"
Sophie Elizabeth Davis
Cas 0h51'48" 52d28'43"
Sophie Elizabeth Gillian Seed
Cyg 20h19'23" 45d32'26"
Sophie Elizabeth Gorvett
And 0h12'21" 25d11'28"
Sophie Ellen D'Arcy - Sophie's Lucky Star
And 2h26'51" 42d42'18"
Sophie Eloise
Sgr 18h39'46" -23d52'32"
Sophie English
And 23h46'8" 36d29'6"
Sophie F -R 172604051975
Aqr 23h42'3" -10d34'27"
Sophie Fatale
Sgr 19h0'17" -13d16'21"
Sophie Forte
Mon 6h47'55" -4d42'34"
Sophie Gibson
Leo 10h16'49" 24d27'19"
Sophie Girl
Sco 16h26'51" -41d21'21"
Sophie Goodwin
Vul 19h37'49" 27d46'55"
Sophie Grace
And 1h9'6" 34d21'32"
Sophie Grace Bell
Umi 13h47'8" 71d41'29"
Sophie Haigney
Cas 0h26'22" 56d4'36"
Sophie Hannah
And 0h22'29" 46d35'45"
Sophie Hannah
And 2h12'47" 41d55'54"
Sophie Harvey Rivera
Cyg 20h52'8" 36d56'17"
Sophie Hayley Hussey
And 1h55'3" 38d16'38"
sophie hofmeester
Ori 5h24'26" 14d30'49"
Sophie Hunter
Ori 5h53'8" 6d8'12"
Sophie Isabella
And 1h31'15" 42d7'53"
Sophie Jean Howgate
And 0h24'28" 37d20'36"
Sophie Jean Mallett born 27-2-2004
Pho 0h53'29" -48d33'43"
Sophie Johnson
Lmi 9h46'49" 39d52'48"
Sophie Julia Ryan-Thorup
Sgr 18h0'34" -20d23'30"
Sophie K.
Ori 5h12'33" -1d11'24"
Sophie Kamil Vardi
Cas 0h40'34" 57d11'36"
Sophie Kanati
Cru 12h39'10" -59d11'39"
Sophie Kathleen Alliston
Cru 12h35'46" -61d9'10"
Sophie Keeley
Cyg 19h28'29" 57d27'11"
Sophie Kirk
Cma 6h49'51" -27d58'32"
Sophie la brillante
Ari 2h32'56" 11d24'7"
Sophie Landry
Cyg 21h54'12" 45d22'45"
Sophie Laryn Headrick
Psc 1h28'25" 9d6'31"
Sophie Lauren
Cas 23h41'17" 57d25'17"
Sophie Laws
Cyg 19h41'27" 34d33'34"
Sophie Lee Swenson
Aqr 21h59'39" -0d0'58"
Sophie Leigh Barlow
And 23h16'50" 48d22'43"
Sophie Levitt Halpern
Cam 5h16'3" 71d46'45"
Sophie Lightbody Lindsay
Umi 16h6'26" 74d9'20"
Sophie Lillian
And 23h13'17" 51d20'55"
Sophie Lin Smale
Uma 11h17'45" 65d4'54"
Sophie Liptak
Aqr 22h53'29" -22d6'55"
Sophie Louis - Quetel
Crb 15h22'23" 26d15'9"
Sophie Louisa Jessica Minster
And 1h40'23" 42d8'50"
Sophie Louise
And 23h33'55" 38d34'47"
Sophie Louise Cavanagh
And 0h1'7" 33d3'47"
Sophie Louise Knight
And 0h6'53" 38d28'19"
Sophie Louise Leonard
And 23h13'11" 35d23'14"
Sophie Lovos
Sgr 18h36'16" -27d13'5"
Sophie Lynn
Uma 11h14'4" 45d9'57"
Sophie Mackenzie Fitzpatrick
And 1h47'39" 40d32'1"
Sophie Mae
Umi 16h19'49" 70d52'35"
Sophie Marie
Cam 3h36'5" 62d2'11"
Sophie Marie Garrity
And 1h48'19" 46d36'18"
Sophie Marie Nevin
Uma 11h21'11" 39d25'33"
Sophie Marie O'Sullivan - Sophie's Star
Ori 6h23'17" 16d0'7"
Sophie Marie Reichenbach
And 1h8'51" 36d25'19"
Sophie Marin
Uma 8h43'11" 69d10'19"
Sophie Marlies Cuthbert
And 1h33'35" 36d14'6"
Sophie Mary Sterbenz-Braun 1-22-07
Cas 1h13'41" 64d17'12"
Sophie Maureen Doak
And 23h39'56" 41d56'2"
Sophie Miller
And 2h23'52" 48d37'30"
Sophie Molly Paye
And 1h3'32" 45d15'1"
Sophie mon étoile
Sco 16h50'22" -27d16'1"
Sophie Mophie
Tau 4h14'6" 16d30'27"
Sophie Nel
Ari 3h28'55" 26d30'10"
Sophie Noble
Lyn 7h1'11" 47d16'20"
Sophie O'Donnell Duffy
Tri 1h45'49" 29d48'21"
Sophie Ouellet Morasse
Aql 18h57'47" 8d11'40"
Sophie Overall
Sgr 19h14'7" -22d12'29"
Sophie Palmer
And 1h1'37" 46d17'7"
Sophie Patricia
Aqr 20h46'31" -7d4'20"
Sophie Rachel James
And 1h31'46" 49d18'29"
Sophie Reynolds
And 1h58'13" 39d20'21"
Sophie Rhys Jones
And 23h20'38" 48d57'19"
Sophie Ritter
And 1h39'43" 40d55'30"
Sophie Rose
Leo 10h25'24" 18d21'9"
Sophie & Sam Davies
Cyg 20h29'35" 41d32'48"
Sophie Schmidt
And 0h43'13" 35d32'44"
Sophie Short
And 0h12'34" 45d4'54"
Sophie Silberstern
Ori 6h17'43" 14d5'29"
Sophie Skaggs
Uma 9h28'31" 64d57'0"
Sophie Superstar
And 2h38'41" 45d44'23"
Sophie Sutherland
Boo 14h18'54" 47d7'41"
Sophie, Daddy's Girl Forever
And 23h45'39" 41d2'30"
Sophie-Amélie
Vir 14h15'44" -6d38'32"
Sophie-box Jackson
Cmi 7h47'49" 5d23'58"
Sophieluvjeff
Tel 18h30'59" -46d22'17"
Sophie's Mommy
Aqr 21h48'18" 1d21'29"
Sophie's Splender
Sco 16h7'43" -14d43'20"
Sophie's Star
Peg 22h37'47" 15d50'25"
Sophie's Star
Peg 21h44'17" 24d28'5"
Sophie's Star
And 2h30'44" 48d25'55"
Sophie's Wishing Star
Gem 7h47'33" 14d54'39"
Sophis Eskel Jensen
Aur 6h7'1" 50d25'47"
Sophy Abbot Larkin
Leo 11h15'41" 14d57'36"
Sophy Holland
Vir 13h13'22" -22d16'39"
Sophy's Star
And 0h42'1" 22d6'22"
Sopi1
Cep 21h44'31" 61d45'52"
SOPS
Umi 18h9'32" 88d18'21"
SORA
Sgr 17h54'7" -28d52'24"
Sora Myojo
Lib 15h1'42" -14d21'4"
Sora ~ Truly Loved Heart & Soul
Crb 15h37'25" 38d44'57"
Soraya
Vir 13h19'2" 10d23'55"
Soraya
Cnc 9h19'1" 10d17'34"
Soraya
Psc 0h14'44" -2d20'2"
Soraya and David's Petals
Cyg 19h45'47" 32d48'53"
Soraya Jackson
Uma 9h41'56" 59d56'20"
Soraya Nauffal
Uma 13h56'43" 51d21'26"
Soraya Ortega Alvarez
Leo 9h51'20" 29d55'30"
Soraya Rose
Uma 11h8'44" 51d31'7"
Soraya Samai Silva
Leo 11h52'9" 24d8'41"
Soraya Sanchidrian
Sco 17h47'48" -41d9'23"
Sorcha Harris
Cyg 21h49'33" 46d13'0"
Sorcirer "sunshine" Trejo
Cyg 21h12'12" 45d41'23"
Sorella
Gem 7h40'11" 33d8'49"
Sorella
Sco 17h9'53" -34d27'34"
Sören
Ari 2h9'2" 25d19'19"
Soren J. Fitje
Per 2h17'45" 57d4'47"
Søren Jacobsen - January 26, 1996
Per 4h10'49" 32d3'6"
Soren James
Cnc 8h39'11" 18d55'49"
Sorgen's Star
Del 20h42'58" 7d24'22"
Sorger-Kemm, Ingeborg
Uma 11h17'45" 64d14'6"
Sori and Brian
Ari 3h9'4" 28d10'28"
Sorina Damita Gervasi
And 2h18'11" 45d7'12"
Soroosh Varahramyan
Leo 10h30'58" 18d25'47"
Soroptimist Star Doreen Sparks
Vir 12h45'58" 2d49'4"
Soror Meus
Her 18h23'55" 12d28'27"
Sorphea Pho
Lyn 7h31'2" 35d53'9"
Sorrell
Lyn 8h55'48" 35d34'45"
Sorrell Anna
Dra 10h51'58" 78d21'45"
Sorrell Napolean Brown
Cap 21h47'45" -18d19'3"
Sorresse
Leo 9h46'45" 32d39'33"
Sorriso brillante di Brynn
Tau 3h31'15" 24d46'24"
Sorriso-Bello
Ari 3h14'40" 26d33'10"
Sorta Fairytale
Leo 9h25'32" 27d2'32"
Sosaie
Aqr 20h55'27" -1d14'45"
SOSI
Cep 22h4'35" 72d31'39"
Sosinsky Star
Uma 10h33'59" 48d4'52"
Soso Toloraia
Per 3h17'51" 55d5'46"
Søssen
And 2h27'20" 44d57'6"
Sosso
Lib 15h5'27" -20d35'36"
Sosso Tennistar
Uma 14h20'26" 57d18'1"
Sosso Veneziana
Cnc 9h17'50" 25d28'41"
Sossy Shirikian
And 0h0'13" 33d8'24"
Sot Staples
Uma 9h31'45" 63d42'21"
Sotiris Belias
Her 16h22'33" 8d51'6"
Sotka and Polis Families
Uma 9h37'41" 49d4'16"
Soto
Lyn 8h31'23" 51d22'44"
SotoSoft
Gem 7h36'26" 14d43'32"
Soufiane
Sgr 18h3'51" -16d15'19"
Souheil Abouhamad
Ori 5h53'14" 11d36'19"
Souhelen
Lib 15h1'41" -8d21'58"
Soul
Lyn 7h58'55" 34d53'44"
Soul compass
And 23h21'34" 52d4'59"
Soul Def
Her 16h43'25" 39d7'53"
Soul Mate
Sge 19h8'23" 19d8'54"
soul mate
Cap 21h10'30" -17d28'50"
Soul Mate Star
And 1h5'20" 35d51'59"
Soul Mate Star
And 0h3'46" 44d3'37"
Soul Mate, The Eternal Voyage
Cra 19h1'16" -38d29'59"
Soul Mate, Voyage Through Eternity
Tau 5h11'16" 24d34'8"
Soul Mates
Vir 13h17'24" 6d27'20"
Soul Mates
Cyg 20h32'45" 50d50'6"
Soul Mates
Sgr 18h39'2" -22d46'35"
Soul Mates Found - Scott & Nicole
Col 5h46'12" -27d47'38"
Soul Mates - Kyle & Treasure Radice
Cyg 19h45'40" 48d10'7"
Soul Mates - Nathan & Trisha Robbe
Lyn 7h2'52" 49d53'45"
Soul Matey
Gem 7h28'4" 20d31'39"
Soul Mateys
Crb 16h4'2" 34d28'16"
Soul Meets Soul
Gem 6h27'8" 27d7'7"
Soul One's Bryan and Lindene
Dra 9h50'15" 75d52'56"
Soul Sister
Ori 5h21'30" -4d15'22"
Soul Sister
And 1h39'28" 46d7'59"
Soul2Soul
Uma 10h21'2" 45d19'6"
Soulbaby
Tau 4h32'25" 13d37'13"
Soulchild - Tyrese Satha
Aur 6h21'10" 32d18'9"
Soulm8
Cyg 21h25'56" 44d47'1"
Soulmate
Lyr 18h40'1" 37d57'39"
Soulmate
Cas 0h33'33" 53d39'47"
Soulmate
Psc 2h4'36" 10d30'45"
Soulmate
Her 17h43'31" 15d41'3"
Soulmate
Ori 5h21'18" -4d49'20"
Soulmate
Uma 13h34'26" 60d57'21"
Soulmate
Cyg 20h20'36" 54d40'23"
Soulmate
Uma 9h35'17" 54d50'25"
Soulmate 1
Pho 1h13'2" -43d57'52"
Soulmate - Przeznaczeni Sobie
Cep 23h4'27" 59d50'33"
Soulmate - Suzi Gulan
Cru 12h27'27" -60d51'32"
Soulmate143
Tau 4h20'13" 14d15'1"
Soulmates
Gem 6h30'40" 24d50'19"
Soulmates
Cyg 20h22'2" 56d5'34"
Soulmates
Uma 11h58'29" 62d19'33"
Soulmates
Uma 10h20'43" 61d1'44"
Soulmates
Umi 15h37'15" 77d58'53"
Soulmates Mary & Matt Forever
Aqr 22h12'13" 0d25'16"
Soulmates ~ Sean and Julie
Psc 0h23'35" 2d10'25"
SoulmatesCnKn10.16.99
Peg 21h38'45" 16d50'46"
Soulmates-KT
Cyg 20h14'23" 49d45'43"
Soulneska
Psc 1h8'8" 5d59'39"
Soul-Sol "Forever"
Uma 11h32'14" 40d38'28"
Soumaya
Psc 23h21'56" 4d47'21"
Soumyaa Santa Ray
Sgr 18h7'43" -29d47'22"
Soupanie
Psc 0h19'46" 4d13'42"
Soups, Angel Love
Cyg 19h48'40" 31d53'11"
"Source Properties"
Del 20h36'43" 15d27'37"
Souren and Sara
Gem 6h50'25" 13d40'54"
Sourire Lumineux
Sgr 18h58'8" -23d35'44"
Sour-Toe
Crb 15h51'4" 29d38'11"
South Beach Resort
Cam 6h37'19" 66d21'34"
South Gap Ocean Hotel
Dra 15h36'43" 57d48'43"
Southern Cross April 2007 Millhowse
Lib 15h1'31" -12d9'27"
Southern Grace
Lmi 10h27'15" 35d51'12"
Southern Nights
Per 3h16'47" 41d27'3"
Southern Palms Beach Club
Her 17h52'25" 37d34'22"
SouthPaw
Umi 14h17'12" 73d51'39"
Southrons
Ori 5h37'15" 8d17'26"
Sovann
Uma 11h3'49" 69d13'37"
Sovereign's Ace
Gem 6h45'49" 28d12'23"
Sowers-Thompson
Lyr 18h31'1" 31d23'11"
Sowinski, Bartosz Zygmund
Uma 9h30'44" 49d33'44"
Sox
Cas 0h25'37" 64d8'0"
S.P.
And 2h33'7" 37d38'52"
sp1- my love my life
Ori 6h6'29" 5d56'30"
Space Nugget 10/14/87
Lyn 8h3'33" 34d24'7"
SpaceMonkey Star
Uma 12h33'55" 56d40'19"
SpaceTraveler MRB
Sgr 19h27'0" -34d21'25"
Spädde
Dra 9h25'9" 74d19'59"
Spadge
Ari 2h44'55" 27d10'28"
Spagackle
Uma 14h15'54" 61d2'56"
Spagnola
Her 16h43'59" 6d55'59"
Spahzo Riley
Mon 6h42'35" 8d40'51"
SPALL JS 05
Vir 11h46'26" 9d57'9"
Spamgeezer
Lmi 10h18'34" 36d20'52"
Spamgeezer
Lmi 10h31'27" 36d37'59"
Spamgeezer
Lmi 10h4'57" 40d18'30"
Spami
Uma 10h23'7" 49d11'15"
Spammons
Umi 14h35'31" 75d20'54"
Spanbaum 11/23/01
Lmi 10h10'29" 30d57'5"
Spandau, Michael
Ori 5h8'4" 15d59'21"
Spangle
Crb 15h48'54" 35d23'52"
Spang's Star
Leo 10h13'54" 7d48'58"
Spankeetang 2005
Cnv 13h46'34" 35d26'47"
spanky
Vir 14h23'20" 0d32'4"
Spanky
Gem 6h50'52" 20d52'46"
Spanky
Sco 17h56'53" -30d25'56"
Spark in my heart
Cyg 20h22'46" 43d39'17"
Sparkey
Uma 11h20'8" 61d52'44"
Sparkie - our little Lion
Lmi 10h13'23" 29d49'57"
Sparkie's Star
Cnv 12h47'44" 39d14'53"
Sparkle
Lmi 10h13'37" 31d33'45"
Sparkle
Cnc 9h11'48" 32d42'45"
Sparkle
Ari 2h10'3" 23d45'55"
Sparkle
Uma 14h9'10" 61d54'4"
Sparkle
Cra 19h2'29" -38d39'40"
Sparkle Astralis
Tau 3h59'35" 20d25'36"
Sparkle by the Sunshine
Lyn 7h53'38" 49d52'11"
Sparkle in My Love's Eye
Crb 15h51'47" 28d6'22"
Sparkle L. Gulley
Sgr 18h38'42" -33d51'44"
Sparkle of Lauren's Eye
And 2h21'37" 45d9'52"
Sparkle of Our Fantasy
Leo 11h41'14" 20d47'29"
Sparkle Plenty (Mom)
Sco 16h9'15" -12d24'18"
Sparkle!Sparkle!
Ari 2h42'55" 26d48'6"
Sparkle Williams
Peg 23h12'21" 17d35'31"
Sparklepins
And 2h2'28" 46d31'11"
Sparklepup
Uma 10h33'1" 43d43'22"
Sparkler
Aur 5h15'8" 31d45'29"
SPARKLES
Uma 11h23'37" 47d24'2"
Sparkles
Crb 15h28'46" 28d36'53"
Sparkles
Cma 6h51'1" -17d44'57"
Sparkles
Cap 21h27'26" -17d16'54"
Sparkles Greco
Equ 21h11'9" 10d10'13"
Sparkles JMS
Ari 2h10'22" 27d49'0"
Sparklesparkle
Lyn 8h0'51" 40d41'52"
Sparklicious XXIX
Ari 3h14'34" 28d54'4"
Sparkliee Rose
Tau 5h43'19" 25d31'6"
Sparkling Alice
Cas 0h2'56" 53d26'2"
Sparkling Ami Star
Lyn 7h26'22" 59d42'25"
Sparkling Doodle Helen Marie Lyons
Sgr 18h56'32" -17d20'23"
Sparkling Emily
Cra 18h28'27" -38d33'23"
Sparkling 'Iona Molly' Star
And 0h19'17" 46d21'17"
Sparkling Jeannie
Leo 9h29'44" 14d16'54"
Sparkling Joan at 60!
Sco 16h3'23" -18d47'32"
Sparkling Julie Anne Dayeh 17Sept69
Cru 12h38'45" -60d44'26"
Sparkling Lorain
Cam 5h57'51" 61d26'35"
Sparkling Mr. C
Boo 14h54'25" 21d19'40"
Sparkling N&G Steiner
Cyg 21h52'17" 38d41'48"
Sparkling Nan
Uma 10h55'19" 40d15'4"
Sparkling Riya
Gem 6h57'57" 15d39'10"
Sparkling Sarah Leech
And 1h21'21" 41d9'51"
Sparkling Suzanne
Tau 4h6'28" 23d16'37"
SparklingBeautyAmber
And 22h57'39" 51d6'46"
Sparkly Kitten
Ari 3h29'25" 26d52'46"
Sparkly Renae
Ori 6h2'28" 15d31'5"
Sparks
Aqr 23h8'10" -7d15'46"
Sparks Tressa
Uma 9h31'3" 64d52'9"
Sparky
Uma 8h22'47" 63d52'55"
Sparky
Uma 8h29'28" 71d49'8"
Sparky
Uma 11h44'0" 52d37'51"
Sparky
Dra 16h52'24" 54d10'5"
Sparky
Umi 17h36'50" 80d18'9"
Sparky
Lib 14h23'33" -17d9'54"
Sparky
Ori 5h33'15" 0d2'36"
Sparky
Ori 5h4'29" 12d28'20"
Sparky
Aql 20h1'21" 3d44'30"
Sparky
Cnc 8h47'10" 27d12'41"
Sparky
Lac 22h35'1" 37d55'49"
Sparky
Per 3h19'8" 43d31'24"
SPARKY
And 0h49'17" 37d3'54"
Sparky and Lucy
Ari 3h20'21" 21d31'42"
Sparky & Bobby, our loved Dad & Mom
Uma 11h30'52" 55d8'10"
Sparky (Carol D. Volpe)
And 0h56'0" 24d16'43"
Sparky - Eternal Love Foreva
Col 5h17'54" -39d7'53"
Sparky II
Cap 20h24'8" -23d27'48"
Sparky Kort
Aqr 21h13'24" -7d23'13"
"Sparky" Robin
Uma 11h2'50" 45d27'4"
Sparky's Bright Light
Cmi 7h20'47" 1d14'17"
Sparky's Heidout
Ari 2h7'38" 13d52'21"
Sparrow of Knowledge
Cyg 19h38'9" 30d27'36"
Spartacus
Lyn 8h50'51" 36d41'30"
Späth, Rainer
Uma 10h14'6" 53d5'55"
Spats
Uma 11h20'30" 61d5'39"
Spatz
Lyn 7h19'0" 50d29'8"
Spatzi - Angelika
Uma 11h56'45" 64d35'53"
Spatzter111048
Uma 11h13'15" 49d3'56"
Spaulding's Star
Uma 10h47'9" 49d9'27"
Spauly's Star
Uma 14h24'48" 61d29'46"
SPC Curtis Wayne Jackson
Dra 20h26'33" 66d47'34"
Spc. Jeremy F. Regnier
Uma 13h39'9" 53d20'51"
Spc. Joshua Justice Henry
Per 3h7'39" 53d46'52"
Spc. Robert (Bob) E. Hall, Jr.
Per 3h29'23" 46d16'30"
Specht, Anja & Sebastian
Ori 5h15'51" 3d27'47"
SPECIAL
Vir 13h36'35" -3d38'40"
Special A
Leo 11h32'20" 19d31'5"
Special Agent Dale Cooper Lager
Cyg 20h56'12" 36d10'42"
Special Agent Star 07
Umi 15h27'39" 77d10'16"
Special Angel Danielle
Cyg 20h9'22" 36d11'40"
Special Angel Grace
Sco 16h54'7" -42d0'3"
Special Angel Peter Dom McLaughlin
Umi 14h59'34" 67d48'52"
Special & Beautiful, Doreen MacLean
And 2h2'25" 38d31'46"
Special Daddy Brian
Leo 9h53'4" 18d57'2"
Special Daddy Steven R. Gower
Lyn 8h10'39" 33d57'14"
Special Double D
Uma 11h9'39" 50d53'17"
Special ED
Lyn 8h16'18" 41d30'25"
Special First Anniversary
Tau 4h14'54" 1d43'55"
Special Friend Victoria Rachel Lee
Sgr 20h21'14" -42d18'0"
Special (Joann Davis)
Lib 15h29'13" -19d37'50"
Special K
Sgr 18h30'5" -16d24'6"
Special K
Cap 20h24'17" -21d7'28"
Special "K"
Umi 14h11'24" 69d34'37"
Special K
Uma 9h5'0" 56d47'58"
Special. K .
Lib 15h40'45" -28d32'41"
Special K
Cap 21h38'35" -23d16'19"
Special K
Tau 4h16'22" 21d10'5"
Special K
Leo 11h51'37" 21d35'43"
Special K
And 1h55'37" 41d24'38"
Special K
Crb 15h48'18" 31d35'22"

Special "K" #05
Cyg 21h57'9" 42d16'54"
Special K's
Leo 10h56'33" 12d12'24"
Special Lady
Peg 21h24'16" 13d5'1"
Special Little Magic Star
Cep 21h31'29" 63d45'0"
Special Nan & Grandad Duggan
Cyg 21h47'2" 38d33'34"
Special Number One Bucket
Psc 1h17'27" 16d19'43"
Special Smile
Ori 5h11'53" 15d43'45"
Special Star Of Tranea
And 1h53'41" 38d52'44"
Special Supus Dupa Kel
Lmi 10h48'26" 27d23'57"
Specialist Jacob S Fletcher US Army
Per 3h21'55" 48d9'34"
Speck
Uma 11h18'40" 65d9'46"
Speck
Mon 6h45'44" -0d13'38"
speckhawley
Sco 17h44'28" -39d52'34"
Spectacular Miracle
Uma 12h46'47" 57d48'19"
SpectroJack
Psc 23h22'19" 5d27'33"
specula
And 23h18'4" 43d52'37"
Sped
Gem 6h46'50" 32d24'29"
Speed Grandest Grandad
Cep 21h55'35" 63d54'46"
Speedo
Tau 3h29'46" 22d34'28"
SPEEDY
Sco 17h13'38" -37d26'8"
Speedy Gonzalez
Aql 19h10'30" -0d31'41"
Speedy's Star
Aqr 20h55'12" 1d39'43"
Spelkaster's Rock-A-Dile Red
Gem 7h24'33" 29d49'16"
Spenance
Lmi 10h31'20" 36d58'10"
Spenc
Aur 5h45'42" 29d20'58"
Spence
Aql 20h1'47" 10d55'51"
Spence and his Lady Gwen
Crb 16h23'5" 33d22'20"
Spence -N- Syleste
Sgr 20h26'7" -33d59'18"
Spence Willis
Lyn 7h8'6" 58d52'0"
Spencer
Uma 8h42'21" 54d44'20"
Spencer
Aql 19h56'53" -10d0'21"
Spencer
Lyn 7h57'10" 37d15'7"
Spencer
Cyg 20h23'38" 44d54'22"
Spencer
Boo 14h12'2" 29d36'36"
Spencer A. Soglin's Special Star
Umi 14h34'14" 74d15'43"
Spencer Aaron Locke
Ori 6h4'22" -0d22'9"
Spencer Alec Friebel
Ari 3h15'0" 11d10'39"
Spencer Alvin Wood
Cep 22h4'6" 64d15'45"
Spencer Bazzano
Umi 16h36'20" 77d49'9"
Spencer Caspian Ilanlou
Cep 22h7'40" 56d38'22"
Spencer Charles Howard Striegler
Ari 3h22'49" 25d31'20"
Spencer Clark
Her 17h51'29" 40d37'20"
Spencer Cole
Del 20h49'42" 13d25'1"
Spencer Copeland
Cep 0h16'9" 73d4'8"
Spencer Dani Davimos
Ari 2h39'57" 13d25'44"
Spencer David
Psc 1h31'13" 4d58'11"
Spencer Donovan Cagle
Leo 9h25'4" 11d6'22"
Spencer Flagg
Aur 6h35'1" 35d20'46"
Spencer Hamblin
Cnc 8h41'12" 32d4'40"
Spencer James Spurrell
Per 4h9'52" 41d46'10"
Spencer John Daley
Dra 14h42'42" 55d25'33"
Spencer John Gittings
Lmi 10h56'3" 28d6'35"
Spencer John Mayorga
Sco 16h8'44" -13d53'17"
Spencer Kaufman
Cnc 8h24'30" 14d48'8"
Spencer Keith Brown
Sco 16h13'35" -13d31'33"
Spencer Layne Ronvel
Umi 4h19'13" 89d2'39"
Spencer Lee Phillips
Dra 17h42'27" 54d15'58"
Spencer Marshall
Cep 23h47'50" 73d41'12"
Spencer Marshall Ebersbach
Per 3h55'17" 38d52'56"
Spencer Matthew Cote
Cyg 19h45'33" 36d29'39"
Spencer Maynard
Her 18h48'32" 21d34'21"
Spencer Michael Jones
Tri 2h40'33" 34d45'27"
Spencer Mitchell Schock
Cnc 8h49'23" 32d28'0"
Spencer Nicole Goldberg
Tau 4h12'51" 19d36'58"
Spencer Pullen
Psc 1h1'55" 4d13'27"
Spencer Ryan
Leo 11h43'13" 17d11'56"
Spencer Ryan Bounds
Uma 8h27'11" 69d10'53"
Spencer Ryan Doyle
Aqr 22h16'26" -20d0'24"
Spencer Scott Zwick
Ori 5h44'59" 5d8'23"
Spencer Sean McCauley
Her 17h15'36" 20d6'38"
Spencer T. Thompson
Cap 21h39'51" -18d55'52"
Spencer Thomas
Ori 5h46'15" 3d39'12"
Spencer Viviano
Uma 9h57'15" 63d3'43"
Spencer Wayne Krenke
Cas 0h38'1" 51d49'39"
Spencer William
Sco 16h6'16" -11d40'33"
Spencer Witty
Per 3h12'4" 57d16'55"
Spencer Zubrow
Psc 1h21'17" 16d51'20"
Spencer, The Knight of Stars for 2
Leo 9h58'54" 12d43'7"
SpencerDale Highway
Uma 9h28'36" 54d26'5"
Spencer's Comet Blast
Uma 8h35'31" 72d35'55"
Spencer's Glimmer
Sgr 17h52'10" -28d42'11"
Spencer's Princess Katie
Cyg 20h23'15" 54d50'42"
Spencer's Wishing Star
Vir 12h43'44" 12d25'31"
Spence's Prophesea
Cas 23h48'1" 57d35'57"
Spenser Bohlender 1-18-1997
Cap 20h20'22" -9d58'9"
Spenser Jordan Fossell
Tau 3h45'5" 30d25'41"
Spenser Scharfman
Umi 16h33'11" 75d53'49"
Spenser-Davies
Cyg 20h1'2" 46d39'14"
Speranza
Ori 6h3'50" 19d38'53"
speranza
Crb 16h18'6" 27d6'56"
Speranza
Umi 14h32'3" 68d50'39"
Sperata My
Dra 17h37'5" 58d18'45"
Spero Antonio Strates
Leo 9h48'22" 27d1'13"
Speros Astrum
Cyg 20h58'16" 33d11'50"
Sperry's Shriend
Cnc 8h35'7" 23d6'9"
Spes ac Fides
Lib 14h26'32" -17d24'57"
Spesh
Cap 20h49'18" -24d56'2"
Spesial Olychka
Uma 9h27'5" 47d12'57"
Spew
Cyg 19h43'9" 41d10'49"
Speyday & Sweetpea-A love eternal.
Per 3h7'4" 54d30'21"
Spez - The stars are shining!
Col 6h20'17" -35d29'58"
SPHB
And 0h54'16" 38d30'41"
SPICK
Lyn 7h9'56" 45d57'5"
Spicy Peanut
Cap 21h21'5" -22d51'43"
Spider
Uma 14h7'40" 60d58'7"
Spider
Cyg 19h39'47" 33d49'13"
Spielberger, Hans-Peter
Uma 9h3'54" 53d23'43"
Spiggymonmaneater
Cap 20h14'55" -16d39'4"
Spika
Tau 3h54'58" 21d52'37"
Spike
Aur 6h59'10" 37d37'33"
Spike
Uma 12h36'32" 57d25'41"
Spike Angel
Lyr 19h3'26" 36d52'38"
Spike Blalock
Cnc 7h59'29" 12d3'43"
Spiker 051903
Ori 5h30'54" 1d55'51"
Spiker 19
Gem 6h4'17" 27d22'22"
Spikes Sparkle
Cha 10h43'26" -76d9'0"
Spike's Tang Li
Sgr 19h23'25" -19d14'1"
Spiky
Lyr 8h6'23" 44d6'26"
Spilly Mac
Uma 10h42'28" 54d21'15"
Spilson
Umi 16h4'51" 74d36'4"
Spinachios
Tau 4h11'46" 24d19'23"
Spindlefire
Boo 14h52'26" 53d52'38"
"Spinelli's" Star
Ori 6h3'47" 19d58'19"
Spinner
Ori 6h10'11" 8d53'15"
Spinney
Cas 1h43'6" 61d23'18"
Spinney
Lib 15h54'49" -17d7'27"
Spiridonova Natalya Evgenievna
Vir 13h30'9" 11d26'17"
Spirit
Per 3h39'25" 46d17'1"
Spirit
Uma 11h45'2" 52d21'43"
"Spirit"
Umi 15h51'16" 74d31'1"
Spirit Fracchia; Vacaville, CA
Psc 1h23'24" 16d16'49"
Spirit of Britten
Dra 18h29'50" 57d40'58"
Spirit of Karen
Leo 11h4'12" 6d39'10"
Spirit of Meghan Rose
Tau 3h48'26" 22d39'5"
Spirit of Peace
Aql 20h32'7" -8d7'45"
Spirit of Peace Baptist Church 2005
Cas 0h53'44" 59d33'13"
Spirit of Shawna Hooke
And 23h29'21" 39d3'31"
Spirit of Tate
Leo 9h38'20" 28d17'11"
Spiritual Mother Migdalia
And 23h12'18" 51d40'37"
Spiritus Angeli
Psc 0h6'20" 10d11'51"
Spiritwind
Uma 12h32'10" 61d42'31"
Spiro-Kostos
Dra 20h32'45" 74d34'1"
Spiros Pantazis
Leo 11h13'27" 14d6'18"
Spishak Star
Uma 13h35'22" 48d31'57"
Spit Fire
Cnc 9h3'11" 9d12'6"
Spitfire
Ari 2h6'28" 14d1'23"
Spitty Kitty
Lyn 8h29'51" 34d5'38"
Spitzbarth, Anja
Cnc 8h22'46" 18d56'5"
Spitzer 3
And 23h21'44" 51d47'26"
SPIVA
Cyg 19h34'6" 27d54'19"
Spladecki
Cap 21h46'18" -8d46'44"
Splendore di Serena
Cnv 12h18'48" 33d22'21"
Splendour
Apu 16h25'58" -73d57'22"
splinto
Cet 1h22'27" -12d13'19"
Splisaence
Ori 5h56'0" 17d14'10"
SPO
Ori 6h14'50" 8d29'26"
Spock
Mon 7h0'45" -6d0'16"
Spodymek, Georg Josef
Ori 5h41'51" -2d29'52"
Spokane Gary Gourneau
Ori 6h15'31" 13d47'11"
Spokane Produce, Inc.
Uma 12h26'44" 61d32'18"
Spokane Pump
Uma 8h51'19" 47d5'26"
Spolick
Leo 9h40'27" 28d40'18"
Sponderelli
Peg 21h44'27" 11d24'4"
Spontaneous
Umi 15h32'32" 69d24'21"
Spontaneous
Umi 19h2'24" 88d33'23"
Spookie
Cnc 7h57'0" 17d59'35"
Spooky
Cma 6h43'38" -28d54'13"
Spoon!
Del 20h37'31" 15d38'19"
Spoonarius Glomminus
Uma 9h49'32" 65d33'50"
Spoonie Luv
Umi 15h14'12" 81d11'50"
Spoonin Forever
Cyg 19h57'6" 32d0'17"
Spot
Boo 14h53'21" 31d32'14"
Spot
Psc 1h26'33" 18d10'19"
Spot's Star
Psa 22h28'34" -31d12'2"
Spottie
Psc 1h21'33" 33d16'29"
Spotts-Marino
Dra 18h41'19" 73d45'12"
Spottswood's Faith Journey
Ori 5h32'22" 5d39'45"
Spotty
Cap 20h49'50" -24d25'44"
Spreckelmeyer Spellmann
Psc 1h31'38" 3d3'11"
Sprenger, Josef
Uma 10h32'10" 65d40'42"
Spring Star
Ori 5h49'42" -4d7'24"
Spring, Summer, Winter and Fall
Uma 10h6'42" 67d22'43"
Springer, Sabine
Uma 12h55'50" 53d11'2"
Springfield - Celestial Celebration
Uma 11h20'58" 58d58'22"
Springstone
Peg 22h12'18" 13d4'58"
Sprinkles
Tau 5h32'23" 18d10'45"
Sprinkles
Cnc 8h20'28" 12d32'5"
Sprinkles
Cap 21h33'47" -16d29'0"
Sprinkles Hogge
Pup 7h57'7" -24d25'21"
Sprocket
Boo 14h47'29" 32d19'6"
Sproule's Star
Per 2h10'0" 53d48'7"
Spry
Crb 16h16'27" 28d59'0"
Spud
Cap 20h29'39" -20d21'2"
Spud
Eri 4h41'12" -0d13'13"
Spud Grant
Per 4h11'50" 39d30'7"
Spuggie
Dra 15h9'33" 60d13'25"
Spunk-52
Dra 19h14'31" 65d9'39"
Spunky
Tau 4h1'8" 11d51'49"
"Spunky" Loucile Arlene Anderson100
Psc 0h34'1" 14d9'6"
Spunky Luv
Umi 16h54'4" 75d35'51"
Spunky Monkey's
Cru 12h40'39" -57d23'4"
Spurney-Michaelides:Tomjanekrystom
Ori 5h21'37" -4d36'10"
Spur's Gold N' Bold
Peg 23h40'50" 23d12'36"
Sputnick Pam
Del 20h40'2" 6d57'33"
Sputnik
Psc 1h6'56" 29d11'39"
"Sputty"
Ari 2h36'42" 12d30'47"
Spy
Ori 6h25'5" 10d50'45"
Spyros Dessylas
Uma 11h47'33" 59d2'35"
Spyros Pantenas
Uma 9h37'52" 58d45'27"
Squabble the Star
Dra 17h16'48" 54d46'15"
Squadron Leader Neil Anderson
Peg 22h28'31" 30d55'5"
Squannacook Elementary School
Uma 11h34'33" 59d2'58"
Square-D
Uma 13h57'2" 57d40'53"
Squash Josh Nadvornik
Her 18h45'0" 19d2'57"
Squawking Duck
Ori 5h6'48" -1d5'13"
Squeak
Leo 10h20'8" 23d30'19"
Squeak
Uma 10h29'46" 43d22'22"
Squeak Bear
Vir 13h22'26" 5d44'8"
Squeaker
Ari 3h8'26" 22d35'42"
Squeaker's Star
Aql 18h42'41" -0d13'2"
Squeaks
Uma 11h20'22" 46d19'1"
Squeaky
Peg 23h56'49" 23d10'38"
SQUEAKY
Tau 5h10'58" 27d19'1"
Squeaky
Umi 18h35'51" 86d50'30"
Squeek
Ori 6h0'13" 11d2'33"
Squeeker and Squishy's
Tau 4h23'2" 21d39'22"
Squeeky
Umi 16h30'29" 77d18'19"
Squeeky Wheels Wild Bill Cornelius
Uma 9h23'38" 43d7'32"
Squeeze
Ori 5h57'38" 3d28'59"
Squeezel
Gem 6h43'33" 34d7'29"
Squidge
Leo 11h22'38" 15d13'16"
Squie
Tau 3h50'3" 20d43'32"
Squink
Lib 14h39'32" -16d58'30"
Squirrel
Ari 3h6'46" 29d57'13"
Squirrel and Twister
Eri 3h44'15" -0d8'1"
Squirtly
Sgr 18h23'10" -29d29'14"
Squirt's Dad Stanley
Cnc 9h15'27" 14d13'16"
Squirttle
Lyr 18h53'43" 43d38'15"
Squishy
Lyn 7h41'52" 47d44'32"
Squishy
Crb 15h58'9" 34d49'53"
Squishy
Cnc 8h46'5" 8d35'51"
Squishy
Ori 5h21'3" 4d21'30"
squishy
Cen 13h34'57" -38d13'29"
Squishy and Shmuffin
Psc 1h21'41" 32d1'54"
Squishy Loves Punky Face
Cnc 8h36'29" 24d8'33"
Squishy Luvs
Cyg 21h32'21" 45d39'17"
Squishy Sunshine
Tau 4h57'31" 20d18'25"
Squishy Wiggles
Psc 1h26'11" 18d35'50"
Squishy's Heart
Lib 15h6'59" -12d54'19"
Squonk
Peg 22h27'47" 33d19'19"
Sr. Antoinette Traeger, O.S.B.
Uma 11h20'54" 56d56'37"
Sr. Barbara Cardinal, IHM
Lyn 8h3'33" 41d33'55"
S.R. Coontz
Aql 19h58'44" 10d18'34"
Sr. Dorothy Stang, SNDdeN
Uma 9h51'38" 68d24'50"
Sr. M. Johnice Rzadkiewicz
Uma 11h32'31" 36d15'18"
Sr. Regina Rausch, O.S.B.
Uma 11h53'16" 57d57'4"
Sra. Montserrat Sierra Martin
Uma 13h33'34" 57d19'18"
Srahoen
Gem 7h28'21" 20d35'36"
SREDJR1996
Lmi 10h48'19" 27d50'34"
Sree Vidya Bhaktavatsalam
Psc 0h54'59" 13d9'5"
Sri Dewi Tiurmala - T/M #1
Uma 11h18'58" 64d23'10"
Srijan&Evie
Cyg 19h48'46" 35d13'35"
Srini
Cap 21h49'46" -23d4'20"
Srividya Aur Vyom
Uma 11h5'57" 36d11'1"
Srividya Gogineni
Aqr 21h12'18" -8d10'30"
Srpski Story
Cas 0h4'0" 54d9'43"
SRWILLI5762
Uma 11h42'4" 51d32'54"
SRZ1955
Uma 11h15'55" 53d0'56"
SS Annmarie Brock
Cap 20h39'50" -15d32'10"
S.S. Coach
Psc 1h40'33" 11d57'23"
SS&J on Valentine's Day
Sco 16h15'56" -15d55'44"
SS Jaksch
Cap 21h9'23" -16d57'20"
S.S. Karstenson
Eri 3h46'5" -39d29'8"
S.S. Panhead
Cyg 21h32'23" 34d48'21"
ss Pedretti
Uma 11h4'39" 52d46'0"
S.S. Yokobosky
Aqr 21h51'38" 2d2'49"
SS, All my life, forever your wife
Cru 12h26'1" -58d48'10"
SS501
Gem 6h32'10" 16d3'47"
SSA Chachel 2199
Tau 4h39'1" 9d53'30"
SSAMMAYER081200
Lyn 7h58'22" 41d18'13"
S.S.C. 97
Uma 9h25'2" 64d41'0"
SSCartagena
Ori 6h11'24" 19d21'21"
SSCherry
Her 17h26'12" 33d50'35"
S.Seychelles & A.Sion for ever
Uma 9h34'55" 60d36'52"
SSFriedstrom IMTL
Lyr 18h49'4" 33d53'51"
SSG Glen Burkhart
Aql 19h43'36" 5d18'40"
SSG Tony Bruce Olaes
Her 16h54'15" 34d21'28"
SSG.Christopher Scott Potts
Lib 15h57'1" -18d23'13"
SSGT Duane A. Parisek, Sr.
Aqr 22h24'8" -19d10'24"
SSGT Kerry Read
Sco 16h10'21" -11d29'46"
S.S.L.S. Cant
Her 16h56'48" 19d58'40"
SSMoeller
Sco 16h6'40" -18d32'3"
Ssnoopy
Uma 10h48'45" 53d44'7"
SSOIII
Lyn 7h16'47" 44d31'40"
sssshhh!
Ori 6h2'36" 11d58'14"
SSSvalina4ever
Ari 2h15'2" 25d51'40"
St. Andrew
Ari 2h36'2" 25d27'23"
St. Anthony's Hospice
Uma 10h26'31" 53d24'30"
St. Benedict , The Happy Wanderer
Ori 5h58'14" 17d0'34"
St. Cloud
Tau 4h27'50" 4d50'56"
St. Cloud - Celestial Celebration
Uma 11h50'19" 56d46'50"
St. Cricki's Comet
Mon 7h18'9" -1d27'31"
St Cyrus' Son
Ori 6h21'38" 16d49'39"
St. Estelle
Sgr 19h28'5" -27d46'22"
St. Jennifer Agnes
Cas 1h40'30" 64d17'48"
St. John Catholic Church, (Westmst)
Psc 1h17'13" 27d25'35"
St. Joseph's 3rd Grade All-Stars
Uma 11h23'53" 51d27'37"
St. Lawrence Beach Apt Hotel
Boo 15h18'54" 44d20'49"
St. Linus LIFE TEEN Epiphany 2007
Lyn 7h54'36" 41d47'3"
St. Louis Beauty
Gem 7h33'36" 27d29'2"
St. Maria Cecilia Mallorca
Cap 21h49'9" -15d28'31"
St. Maria Roddy
Per 3h14'55" 42d16'18"
St. Mark's Lions Pride
Cyg 21h47'3" 36d13'31"
St Martins
Dra 16h59'16" 64d57'21"
St. Mary School-Gilroy CA
Uma 9h2'30" 57d57'2"
St. MiMi
Cnc 9h9'10" 31d39'36"
St Naz
Umi 14h30'19" 77d43'52"
St. Paul's
Cep 21h16'42" 66d30'42"
St*r St*r Belle
Ori 5h28'6" 6d31'23"
St. Rose Super Stars Class of 2008
Uma 9h3'58" 56d34'34"
St. Valentine
Ori 6h9'4" 3d19'25"
ST x
Uma 8h43'58" 50d52'4"
Sta & Mike
Cyg 19h30'53" 29d30'43"
Stäbe, Herwig
Ori 6h14'49" 13d18'58"
Staca-Peta-Pixie-Sagie-Put-Put52472
Gem 7h47'45" 21d45'37"
Stace and Libra Reading
Ori 5h11'54" 3d44'41"
Stace King
Lyn 7h45'23" 44d9'48"
Stacedi & Moxy
Umi 14h8'26" 75d41'21"
Stacee Lee
Dra 16h43'5" 59d33'40"
Stacer
Cap 21h57'43" -19d27'34"
Stacey
Umi 17h29'58" 80d21'54"
Stacey
Vir 13h4'49" -3d48'40"
STACEY
Lyn 6h32'51" 55d50'21"
Stacey
Uma 10h35'19" 60d52'59"
Stacey
Lyn 8h46'53" 40d37'22"
Stacey
Cas 0h15'22" 51d27'57"
Stacey
Cas 0h48'47" 56d31'8"
Stacey
Psc 1h30'16" 11d29'0"
Stacey
Tau 4h46'42" 19d12'56"
Stacey
Cnc 8h6'53" 24d49'36"
Stacey A. Young Grandchild of IGD
Cas 0h29'41" 57d54'21"
Stacey and Josh
Vir 12h23'22" -5d41'1"
Stacey & Andrew
Cyg 20h3'13" 55d3'39"
Stacey Ann
And 0h29'6" 42d47'28"
Stacey Ann
And 0h52'16" 37d20'43"
Stacey Ann Chaplin
And 23h12'7" 52d16'26"
Stacey Ann Gagliano *MOM*
And 0h55'59" 43d17'48"
Stacey Ann Gray
Peg 22h28'5" 33d47'50"
Stacey Ann Hartmann
Vir 13h52'54" -1d10'3"
Stacey Anne
Sgr 19h14'55" -15d50'22"
Stacey Anne Bent
Cas 23h7'0" 58d49'2"
Stacey Bellinger
Uma 8h36'55" 49d53'3"
Stacey Biggar
Tau 3h37'59" 27d47'54"
Stacey Brickner & Eric Sutters Star
Gem 7h17'34" 19d49'58"
Stacey Bullock, will you marry me?
Tau 4h25'0" 22d42'7"
Stacey Cappozzo
Cyg 20h28'18" 44d25'38"
Stacey Cooper
Psc 0h10'18" -2d52'58"
Stacey Corveleyn
Uma 10h23'14" 71d21'35"
Stacey Cummings
Lib 15h16'55" -20d9'52"
Stacey Cunningham "Stac's Star"
Lyn 7h51'53" 50d19'50"
Stacey Dallas
Vir 13h41'19" -5d59'15"
Stacey & Danielle Rucker
Leo 10h1'55" 18d45'7"
Stacey Dazley x
Tau 4h38'1" 28d35'31"
Stacey & Donald
Cyg 21h53'16" 53d42'2"
Stacey Elisabeth Schroeder
And 0h14'19" 43d48'25"
Stacey Femino
Cru 12h2'51" -62d13'27"
Stacey Freize
Vir 13h12'32" 8d46'21"
Stacey "Frog" Allen
Lib 14h51'31" -16d2'6"
Stacey Furler
And 23h28'38" 38d9'29"
Stacey Gilchrist
Cap 20h39'11" -20d35'25"
Stacey Gretchen Morrow
And 0h57'21" 22d56'44"
Stacey Grinalli
And 0h24'38" 42d10'18"
Stacey Grossman
Cap 20h22'14" -9d41'30"
Stacey Hale Middlebrook
Vir 11h41'25" 2d54'17"
Stacey Helen Gaslard
Cas 2h12'2" 72d25'59"
Stacey Hernandez
Umi 15h48'56" 83d49'11"
Stacey Hynes
Ari 2h41'17" 22d20'12"
Stacey Ilyse Dudley
Cru 12h6'52" -61d16'39"
Stacey Jackson
Her 17h58'58" 46d37'42"
Stacey James Monson
Boo 14h39'39" 44d10'19"

Stacey Jane
Sco 16h42'37" -31d19'53"
Stacey Jane Murial Parker
Sco 16h12'14" -28d54'3"
Stacey Joanne Allen
Mon 6h50'40" -0d32'25"
Stacey & Joe
Vir 12h34'8" -9d47'52"
Stacey K
Vir 14h44'13" 3d9'28"
Stacey Kalynn Moreno
Cas 0h20'30" 64d0'5"
Stacey Kasting & Carlos Valdez
Cyg 19h54'33" 36d6'34"
Stacey Kemp & Adam Yates' Star
Cyg 20h17'8" 33d22'45"
Stacey Kilps
And 1h29'21" 48d54'26"
Stacey Kinman
Cam 3h32'27" 59d10'1"
Stacey Knight
Her 16h49'21" 7d22'58"
Stacey Kristine Gawrys
And 1h23'15" 49d55'42"
Stacey Krizan - GaMPI Shining Star
Lyn 8h34'19" 41d30'13"
Stacey L. Bailin
Gem 7h42'59" 33d44'8"
Stacey L DePasquale
Cas 0h19'12" 64d16'22"
Stacey L. Griffin
Lyr 18h26'48" 32d11'35"
Stacey L. Slagan
And 0h21'5" 33d12'23"
Stacey L. Smalls
Psc 1h8'51" 7d0'42"
Stacey L Smith
Psc 1h35'38" 15d53'6"
Stacey L. Welch
Uma 11h19'30" 48d33'57"
Stacey Laryn Sims
Cyg 19h46'59" 36d7'20"
Stacey Lea "TE-TE-B"
Uma 9h43'27" 55d52'47"
Stacey LeBlanc
Lyn 9h9'29" 44d49'52"
Stacey Lebron
And 0h42'50" 42d36'48"
Stacey Lee Cooper
Sco 17h26'14" -41d2'22"
Stacey Lee Henke
And 1h25'59" 43d15'29"
Stacey Lee Our Shining Star
Aqr 23h17'49" -14d37'18"
Stacey Lee Sanderson (Hefe)
Per 3h15'16" 52d3'23"
Stacey Lewis
Aqr 21h13'40" -9d41'39"
Stacey Liesen
Lmi 10h7'46" 36d26'47"
Stacey Loree Moore
Uma 10h20'34" 44d46'27"
Stacey Louise
And 2h25'44" 44d29'48"
Stacey Louise Fisher
And 22h59'29" 48d28'21"
Stacey Louise Smith
And 0h14'3" 26d21'35"
Stacey Louise Wade Petrowski, R.N.
Cas 0h42'31" 64d19'28"
Stacey Lynn Bell
Cnc 8h12'41" 21d32'0"
Stacey Lynn Catanzaro
Tau 4h42'50" 23d2'52"
Stacey Lynn DeBuck
And 0h38'34" 28d52'28"
Stacey Lynn Doherty
Cas 1h12'15" 59d12'15"
Stacey Lynn Gnatowski
Cyg 19h45'19" 31d46'34"
Stacey Lynn Griffel
Leo 10h41'0" 14d44'8"
Stacey Lynn Hall
Gem 7h15'58" 31d28'19"
Stacey Lynn LaFarge Ackerman
Lib 15h50'29" -4d34'0"
Stacey Lynn Lake
Aqr 22h37'42" 0d38'4"
Stacey Lynn Marino
Aqr 20h39'54" -11d31'45"
Stacey Lynn Palmieri-18
Tau 5h19'49" 27d21'19"
Stacey Lynn Reimer
Crb 15h44'2" 32d50'16"
Stacey Lyons
And 2h22'22" 46d42'58"
Stacey M. Pryce
Tau 4h8'49" 28d30'53"
Stacey MacDiarmid's Brilliance
Lmi 10h19'59" 36d18'5"
Stacey Marie
Cyg 21h21'56" 33d1'1"
Stacey Marie
Gem 6h23'48" 22d57'33"
Stacey Marie
Ori 5h29'28" 7d12'15"
Stacey Marie
Aqr 22h26'7" -6d39'19"
Stacey Marie Gabriel
And 1h27'6" 44d47'35"
Stacey Marie Holdren
And 0h9'7" 45d7'0"
Stacey Marie Holl
Cnc 8h47'19" 15d49'6"
Stacey Marie Schrom
Psc 0h16'34" 17d14'57"
Stacey Marie Tutunjian
Psc 0h19'14" 15d39'17"
Stacey Mason
Leo 11h24'50" -5d49'45"
Stacey McCormick
Gem 6h46'42" 34d49'41"
Stacey MF Lea
Psc 0h40'59" 5d46'4"
Stacey Mi Amor
Vir 14h35'33" 0d43'10"
Stacey Michelle Alleborn
And 0h50'51" 36d3'2"
Stacey Michelle Catterall
Ori 5h45'0" -2d35'25"
Stacey Michelle Winston
Cyg 19h47'18" 51d28'14"
Stacey Mirabile
And 1h21'29" 40d32'4"
Stacey (Mom) Kubalek ~ Magnificent
Cas 1h57'25" 66d10'27"
Stacey Nadine Cameron
And 1h54'35" 42d17'44"
Stacey: Nothing shines as bright.
And 2h22'34" 37d32'55"
Stacey Padiernos
Cnv 13h3'9" 45d3'33"
Stacey Phillips
Ori 5h35'20" 5d29'12"
Stacey Pope
Ari 2h40'35" 13d0'0"
Stacey Rae Lynn
Lib 14h50'16" -0d48'38"
Stacey Ranelli
Cru 11h56'58" -61d40'15"
Stacey Reames
Lmi 10h46'34" 32d40'47"
Stacey Reed
Cyg 20h49'57" 30d55'43"
Stacey Register
Sex 9h51'31" -0d42'40"
Stacey Rene
And 23h15'46" 48d38'7"
Stacey Renee Griffin
Gem 7h29'2" 26d25'16"
Stacey Renee Jones
Aqr 22h5'10" 0d54'43"
Stacey Ret
Tau 5h18'47" 25d57'11"
Stacey Rivas
Ari 2h16'28" 26d5'3"
Stacey Roxanne
Sgr 18h52'33" -29d24'52"
Stacey Sargent
And 23h12'23" 51d3'39"
Stacey & Scott Aulinskis
Lyr 19h3'55" 34d56'18"
Stacey Seeds
Sco 17h28'13" -37d56'50"
Stacey Shym
Cnc 8h52'25" 32d13'8"
Stacey Smith
Tau 4h27'14" 18d11'48"
Stacey Sue Ellis
Cam 5h51'27" 60d18'12"
Stacey Sue, I Love You
Uma 9h24'43" 54d53'49"
Stacey Swieter
And 1h46'46" 50d19'27"
Stacey Takis
Leo 10h28'18" 16d27'52"
Stacey The 18th Angel
Lib 15h53'40" -20d4'31"
Stacey Troger Eustace
Aqr 21h12'37" 0d55'39"
Stacey Trohanis
And 0h59'46" 35d37'42"
Stacey Yvette Hall
Leo 9h25'59" 26d19'1"
Stacey-Ann
Gem 6h48'41" 32d56'58"
Staceybrooke
Tau 5h50'19" 17d55'26"
Stacey-Jay Carmichael
Ori 5h37'46" 3d15'8"
Stacey-N-Paul
Ari 2h26'34" 15d47'35"
Stacey's Dreams
Sgr 18h13'8" -18d6'22"
Stacey's Hope
Umi 15h37'46" 79d58'13"
Stacey's Rose Petal
Cas 1h19'55" 64d49'55"
Stacey's Star
Pho 0h31'4" -46d22'52"
Stacey's Star
And 1h0'47" 46d10'41"
Stacey's Star
And 1h21'40" 46d48'6"
Stacey's Wish
Gem 6h3'16" 23d40'19"
Staci
Cnc 8h49'12" 25d32'5"
Staci
And 0h20'48" 36d46'21"
Staci
Lyn 8h25'18" 42d18'29"
Staci Ann
Ari 1h49'20" 19d3'33"
Staci Ann Chen
Mon 6h34'44" 8d15'21"
Staci Bowermeister
Leo 11h29'52" 9d32'8"
Staci Erica Laurette
Her 17h51'26" 45d37'43"
Staci Falzon
Sco 17h37'12" -31d39'9"
Staci Jean Townsend
Mon 6h52'53" -0d38'16"
Staci Jo
Leo 11h19'21" 2d4'37"
Staci Jo Banks Wertman
Lmi 10h5'27" 33d52'15"
Staci Jo Wertman
Leo 10h54'27" -1d53'12"
Staci Leigh Chase
Ori 5h30'50" 11d30'17"
Staci Marie
Sco 16h15'26" -14d39'32"
Staci McAllister
And 0h43'5" 39d19'8"
Staci & Mike's
Umi 15h45'14" 75d32'47"
Staci & Noah Columbo
Cyg 19h51'44" 32d17'55"
Staci Nolin
Ori 5h37'48" -3d34'32"
Staci Rene Plickebaum
Sgr 18h14'55" -17d22'50"
Staci Renee
Cnc 8h20'2" 8d53'10"
Staci Switzer
And 0h14'7" 27d1'44"
Staci Winecoff
Lib 15h5'48" -27d58'3"
Staci, The Love Of My Life
Cas 2h6'37" 62d17'0"
Stacia
Lib 15h18'46" -5d11'48"
Stacia
And 1h10'13" 38d4'29"
Stacia
Uma 10h40'2" 40d59'26"
Stacia Ann Kimler
And 23h17'46" 47d17'56"
Stacia Jean
Lib 14h53'36" -18d27'45"
Stacia Katrin
And 2h21'56" 46d44'50"
Stacia L
Ari 2h5'38" 19d35'7"
Stacia Leilani Carrier
Psc 23h59'21" 2d29'19"
Stacia Papas
And 0h8'44" 34d56'44"
Stacia Star
Ari 2h52'29" 17d50'47"
Stacia's Birthday Star
Ari 2h11'46" 26d51'41"
Stacie
Crb 15h52'38" 36d9'39"
Stacie
Nor 16h5'21" -44d11'42"
Stacie Ann~Kristen's Heart
Cas 2h27'57" 67d0'37"
Stacie Athene
And 0h33'28" 33d32'45"
Stacie Bond
Crb 15h27'14" 32d12'45"
Stacie & Gregg Schwartz 2006
Cap 21h13'59" -14d57'10"
Stacie Guan
Psc 0h35'44" 16d30'24"
Stacie Hummel
And 0h16'32" 28d49'58"
Stacie Jo heart in the sky
Sgr 20h0'34" -38d19'59"
Stacie L. Deagan
Uma 9h33'18" 56d45'36"
Stacie Lee
Dra 19h39'24" 73d57'19"
Stacie Lyn Cone
Crb 16h11'44" 37d34'57"
Stacie Lynn Hermes
Psc 1h1'27" 3d23'8"
Stacie Lynn SRT10
And 2h16'45" 48d13'14"
Stacie Lynne Rosier
Psc 1h11'34" 10d35'38"
Stacie Matthews
Aqr 22h37'37" 2d0'37"
Stacie Michelle Kaminski
Cas 0h21'0" 61d56'57"
Stacie Michelle Miles
Gem 7h35'40" 21d36'48"
Stacie Nicole Burris
And 0h20'26" 28d23'46"
Stacie Nowell
Psc 1h37'51" 20d17'55"
Stacie Rae
Cas 1h24'6" 68d37'28"
Stacie Rene Green
And 23h17'9" 48d36'0"
Stacie Roth
Ari 2h10'3" 25d10'28"
Stacie's Smile
Sgr 18h31'25" -16d43'44"
Stacie's Star
And 2h30'38" 45d29'39"
Stacitootle
Gem 6h51'38" 23d21'43"
Stacy
Cnc 8h54'49" 27d0'4"
Stacy
Tau 5h51'27" 23d17'18"
Stacy
Ori 6h1'22" 21d28'35"
Stacy
Cnc 8h42'56" 18d42'36"
Stacy
Ori 5h23'22" 6d21'59"
Stacy
And 0h52'33" 45d37'45"
Stacy
And 23h20'29" 46d26'18"
Stacy
Uma 14h2'0" 48d50'1"
Stacy
Lyn 8h13'36" 41d7'11"
Stacy
And 0h33'33" 42d20'36"
Stacy
Uma 12h22'20" 53d34'26"
Stacy
Cru 12h41'2" -55d50'48"
Stacy
Sco 17h18'7" -32d36'0"
Stacy Alford
Cnc 8h46'43" 14d11'57"
Stacy and Albert's Star
Cyg 21h28'6" 47d30'39"
Stacy and Devaroo 1
Cyg 20h38'48" 46d35'13"
Stacy and German Forever
Ori 5h37'22" 8d0'52"
Stacy and Mark
Ori 5h46'51" 12d23'44"
Stacy and Rob 1-30-06
Col 6h29'55" -40d48'43"
Stacy Ann
And 23h16'53" 52d11'1"
Stacy Ann Casto
Lib 14h51'30" -7d54'44"
Stacy Ann Grashaw 12/20/66-09/18/75
And 2h17'46" 46d30'19"
Stacy Ann Shuman
Leo 9h48'24" 22d41'22"
Stacy Ann Smith
Cyg 19h42'45" 31d12'41"
Stacy Anne Anderson Cardinale
Gem 6h50'7" 23d9'5"
Stacy Anne Brown & Kevin Moore
And 23h18'41" 48d18'29"
Stacy Anne Chase
Cap 21h41'3" -20d38'7"
Stacy Anne Conrad
And 0h41'58" 32d22'11"
Stacy Anne Mikol
Psc 0h58'46" 26d30'43"
Stacy B. Veitenhens
Her 16h27'46" 48d37'31"
Stacy Baar
Uma 11h31'4" 61d25'23"
Stacy Balloun
Aqr 20h50'53" -10d36'38"
Stacy Barlow
Cas 2h47'52" 65d27'29"
Stacy Beckett's Star
Sge 20h9'11" 19d17'28"
Stacy Blas
Psc 1h45'6" 13d12'58"
Stacy Brea
And 1h26'1" 34d59'26"
Stacy & Brent Livingwell
Cma 6h42'56" -15d16'13"
Stacy Buster
Uma 10h56'38" 68d58'50"
Stacy Butler
Vir 12h35'40" -9d23'19"
Stacy Campbell
Sco 16h21'23" -30d14'47"
Stacy Cappiello
Cas 1h40'53" 64d4'7"
Stacy Caroline
And 23h19'14" 43d59'40"
Stacy Caroline Rutz
Cnc 8h38'46" 9d25'44"
Stacy Cates
Cas 0h52'22" 53d52'32"
Stacy Cavagnaro
Mon 8h10'3" -0d35'45"
Stacy Cummins
Cnc 8h44'58" 32d18'59"
Stacy D. 1229
Ori 5h38'11" 0d4'25"
Stacy & David Unconditional Forever
Cas 23h56'49" 53d24'17"
STACY DAWN ILU
And 2h22'29" 47d10'26"
Stacy Deem
Gem 6h57'5" 14d53'33"
Stacy Denise G Johnson
Cas 0h2'53" 54d17'24"
Stacy & Derek Hafner
Cyg 20h38'51" 47d30'16"
Stacy Di
Cas 23h7'9" 59d17'6"
Stacy Dorr
Vir 13h6'25" 12d24'1"
Stacy Drapikowski
Cas 0h57'29" 60d26'1"
Stacy Dye beautiful & unselfish
Cyg 20h40'59" 36d10'36"
Stacy Eileen Schroeder
Aqr 21h35'25" -0d40'18"
Stacy Elizabeth
Com 12h44'5" 17d20'32"
Stacy Fay Sankar
Uma 11h35'47" 59d17'55"
Stacy Gayhart
And 0h24'52" 44d14'28"
Stacy Hatfield AST
Lib 15h32'40" -7d32'35"
Stacy Homer
Uma 11h31'54" 61d14'6"
Stacy J. Bovee
And 23h23'29" 44d28'13"
Stacy & James
Cyg 19h42'38" 31d14'49"
Stacy Janette
Tau 5h57'13" 24d56'6"
Stacy Jean
Lib 14h49'36" -2d36'14"
Stacy Jessica Porter
Lmi 9h47'56" 35d59'6"
Stacy Jo Butenhoff
Cam 3h42'50" 53d26'58"
Stacy Jo Krout
Cyg 20h29'14" 45d38'7"
Stacy Jo Wicker
Ari 2h38'12" 13d9'56"
Stacy Jo's Star
And 0h33'15" 34d17'37"
Stacy June LUJS
Psc 1h7'48" 28d13'37"
Stacy Kay
Gem 8h6'44" 30d43'25"
Stacy & Kelly- 2 Become 1
Gem 6h8'5" 24d25'39"
Stacy King Marshall
Tau 5h18'1" 21d45'17"
Stacy L Anderson
Tau 5h26'46" 22d10'42"
Stacy L. Harrell-Mathers
And 2h34'49" 45d35'12"
Stacy & Lane
Sge 20h5'35" 20d28'48"
Stacy Lee McGann
Gem 7h6'52" 27d41'44"
Stacy Lowery
Oph 16h34'57" -0d27'0"
Stacy Lucadamo
Cap 20h11'0" -16d48'3"
Stacy Lynn
Sgr 19h17'6" -20d57'25"
Stacy Lynn
Vir 13h14'12" -22d16'59"
Stacy Lynn
Tau 3h47'54" 16d45'21"
Stacy Lynn and Aryn Gabriel
Aql 19h20'2" -0d40'18"
Stacy Lynn Anderson
Tri 1h48'17" 31d30'39"
Stacy Lynn Ayers
And 1h50'53" 38d43'49"
Stacy Lynn Bunch
Cas 1h36'21" 68d52'29"
Stacy Lynn Freshour
And 0h46'25" 37d21'14"
Stacy Lynn Hall
Cam 3h27'17" 67d49'40"
Stacy Lynn Hasenpat
Cnc 8h22'37" 14d9'59"
Stacy Lynn Shultz
Gem 6h48'41" 23d10'17"
Stacy Lynn Willis
And 0h46'32" 45d39'19"
Stacy Lynne
Cas 1h14'56" 54d50'20"
Stacy M. Fabrizio
Aqr 21h57'58" 1d35'36"
Stacy Mae
Aqr 22h32'48" -17d37'10"
Stacy Mahadeo
Cap 21h3'54" -19d40'41"
Stacy Marie
Dra 18h44'36" 74d9'45"
Stacy Marie
Lyn 6h33'5" 57d2'36"
Stacy Marie
Ari 2h39'1" 14d41'25"
Stacy Marie
Lyn 6h58'3" 48d1'8"
Stacy Marie Cox
Lib 15h12'2" -7d3'46"
Stacy Marie Howe
Cnc 9h0'58" 32d39'3"
Stacy Marie Pudder
Lyr 18h53'47" 39d9'49"
Stacy McCants, The Angelic
Ari 3h7'29" 12d25'28"
Stacy Mensch
Cas 23h26'9" 53d0'27"
Stacy Meyer
Gem 6h5'19" 25d30'0"
Stacy Michelle Miron
Leo 9h29'21" 7d27'9"
STACY MING
Lib 15h46'22" -18d45'9"
Stacy My Beautiful
Crb 15h42'37" 28d12'27"
Stacy Neltnor
Cap 20h12'23" -11d11'18"
Stacy Nichole Graber
Uma 9h45'10" 42d28'44"
Stacy Nicole Andrews
And 0h28'17" 45d34'26"
Stacy Padgett
And 0h3'44" 42d34'16"
Stacy Patricia Best
And 23h53'29" 35d8'36"
Stacy/Pooh
Oph 17h51'44" -0d44'7"
Stacy R. Hoppes B.A.
And 1h11'16" 45d4'45"
Stacy Rena Brubaker
Vir 13h16'46" 2d22'22"
Stacy Rene Rafferty
Vir 14h3'44" -18d59'48"
Stacy Renee Gapper
Dra 16h7'40" 54d24'27"
Stacy Rima
Uma 10h6'19" 45d3'19"
Stacy R.M. Mattson
Crb 16h15'32" 38d31'30"
Stacy Rogers
And 2h20'11" 49d56'23"
Stacy Rose Baumgart
And 23h1'50" 41d11'54"
Stacy Russell
Ori 6h25'39" 17d5'28"
Stacy & Ryan
And 1h41'21" 43d53'9"
Stacy Ryan Locklear
Aql 20h14'15" 9d58'53"
Stacy S. Fagan
And 23h10'44" 44d13'33"
Stacy Salazar
Gem 6h50'46" 18d21'42"
Stacy Sarata
Cnc 9h3'41" 21d13'8"
Stacy Scharf
Cas 1h10'34" 66d13'49"
Stacy Souza
Ari 2h38'19" 28d21'37"
Stacy Stelmat
Aqr 22h25'53" -3d21'10"
Stacy Tanzberger
And 23h6'22" 44d27'40"
Stacy Taylor
Uma 8h22'1" 62d11'22"
Stacy - The Most Beautiful Princess
And 1h19'38" 47d9'26"
Stacy & Thorin
Tau 4h23'7" 12d5'15"
Stacy Victor
Cap 21h28'7" -17d59'33"
Stacy & William's Anniversary Star
Sge 19h12'15" 17d54'39"
Stacy Wood
Cas 1h32'35" 63d25'27"
Stacy Y Richan
Cnc 8h38'10" 7d37'29"
Stacy Yvonne McAuliffe
And 0h52'15" 36d21'12"
Stacy Z. Loves Mark P. Christensen
Dra 16h21'0" 68d21'42"
Stacy3873
And 1h11'30" 40d5'26"
Stacye Jones
Pho 0h21'35" -55d25'34"
StacyLady
And 1h16'19" 37d39'27"
StacyLeigh
Ari 3h26'7" 23d49'42"
Stacy-N-Allen
Lyn 7h40'3" 58d21'4"
Stacys 30 yrs
Leo 9h50'54" 28d20'56"
Stacy's "baby" Star
And 23h43'20" 37d16'33"
Stacys Bellus Nitidus Astrum Janice
Psc 23h1'19" 1d6'8"
Stacy's Dance
Her 17h51'7" 15d15'18"
Stacy's Light
Ari 2h44'34" 25d37'0"
Stacy's Little Twinkle
Vir 13h16'53" -13d26'39"
Stacy's Smile
Sco 16h26'47" -26d23'58"
Stacy's Star
Lib 15h52'41" -18d50'52"
Stacy's Star
And 23h10'20" 43d58'23"
Staddikry
Uma 9h24'28" 66d15'13"
Staff <\@> Park Place Nursing Facility
Lyr 18h49'54" 35d16'55"
Staff Sgt. Joseph Michael Weiglein
Crb 15h48'3" 29d12'11"
Staff Sgt. Robert C. Hobbs
Aql 19h48'54" 12d22'43"
Staff Sgt Thomas Douglas Robbins
Ori 5h32'29" -4d30'6"
Staffan 50
And 0h38'56" 39d13'42"
Stafford Crossman
Uma 10h59'4" 63d49'28"
Stafford Wade Primeaux
Dra 17h25'29" 57d55'17"
Stafford's Lady Abigail Elizabeth
Uma 9h46'56" 68d21'5"
Stage Tyler Atkins
Vir 13h17'43" 13d49'16"
Stahr's Smile
Her 17h32'45" 34d14'43"
Stainless
Her 17h38'46" 26d49'38"
Stairway To Heaven
Vir 11h49'8" 7d2'17"
Stale Nordgard optimus pater
Gem 6h54'29" 18d33'21"
Stalen
Lyn 8h25'57" 33d19'8"
Stalia Gagliarda
Lyn 6h46'19" 56d2'58"
Stalin
Sco 16h59'4" -42d38'21"
Stalker
Cyg 21h58'7" 48d34'5"
Staller's Star
Uma 12h36'34" 54d48'48"
Stallings
Ori 5h19'33" 15d41'52"
Stallings Family
Uma 9h13'24" 49d42'27"
Stallion
Dra 19h40'26" 71d2'46"
stamatia
Sgr 18h17'12" -16d46'21"
Stamer, Erich
Aqr 21h25'30" -0d8'18"
Stampe
Uma 11h29'44" 54d28'21"
Stan
Sco 16h19'33" -12d20'54"
Stan
Lyn 7h43'4" 49d49'32"
Stan
Lmi 9h39'25" 35d35'54"
Stan <3 Meghan
Uma 11h47'57" 62d40'30"
Stan and Carol Tompkins
Cyg 21h21'0" 52d55'58"
Stan and Debie Seifert
Cyg 20h45'25" 42d20'2"
Stan and Linda Trott
Her 17h14'23" 19d16'58"
Stan Ball
Sco 17h18'54" -31d22'49"
Stan Bernstein
Per 2h30'47" 54d9'19"
Stan Brehaut Jr
Lyn 8h25'5" 56d20'23"
Stan Clutton
Boo 14h44'14" 34d20'54"
Stan Heart Of My Heart
Uma 11h30'7" 53d36'19"
Stan & Jelica Horvat
Cru 12h21'44" -58d53'43"
Stan & Lois Nielsen Family Star
Cyg 20h4'4" 38d56'52"
Stan Lorys
Aur 6h8'47" 52d12'30"
Stan Loves Polly Forever
Cyg 19h57'9" 42d31'5"
Stan Motl 003
Umi 16h36'57" 83d36'14"
Stan & Nina Walls
Cyg 20h17'45" 40d28'49"
Stan Oram
Vir 14h3'56" -13d10'19"
Stan Osment
Cet 1h6'14" -0d51'30"
Stan Roderick - our Grandpa's a Star
Ori 5h38'54" -2d49'30"
Stan Rote
Cep 3h45'5" 81d3'25"
Stan Schlosser
Aur 6h1'45" 47d58'48"
Stan Swanstrom
Aur 6h23'28" 41d16'52"
Stan Tyrrell
Per 3h29'14" 36d10'23"
Stan Watson
Sgr 18h25'5" -32d18'0"
Stan Williams
Per 2h50'24" 45d22'8"
Stan Zogas Jr.
Aur 5h29'46" 46d23'51"
Stana Koutroumanis (Rados)
Gem 6h55'53" 19d56'9"
Stanace Ray Carter
Uma 11h29'23" 48d26'30"
StanBoyd
Her 17h23'27" 38d43'57"
Stanci
Ori 6h4'26" -0d37'24"
Stand By Me
Cyg 20h45'15" 43d15'42"
Stander, Ludwig
Uma 13h1'45" 60d22'33"
Standing Elk
Gem 6h32'23" 26d33'36"
Standing Still
Tau 3h45'56" 26d57'24"

Standish Bridget Millennas Oph 17h22'15" 7d54'15"
Stanford Antonio Orfila Cam 4h14'25" 70d37'12"
Stanford John Harris Per 3h29'7" 54d6'13"
Stanford L. Sirak 1923-2006 Uma 11h31'31" 55d16'14"
Stanford Perlman Aql 19h35'24" -0d0'6"
Stangelyne Her 17h16'11" 26d47'6"
Stania Gomolakova And 23h42'12" 42d44'36"
Stanique Ari 2h34'24" 30d7'57"
Stanis Pho 1h27'34" -52d2'43"
Stanislas Her 17h16'1" 19d24'30"
Stanislav a Jaroslava Lyr 19h24'10" 31d32'56"
Stanislav09021984 Vir 12h11'59" 5d50'18"
Stanislava Draganova Psc 0h10'22" 6d9'42"
Stanislaw J. Strycharz Mon 6h50'21" 3d9'25"
Stanislaw Jopek Lib 15h46'17" -18d21'2"
Stanislawa Lipski Tau 5h27'55" 21d35'9"
STANIZZI Lyr 18h26'3" 34d48'38"
Stanley Gem 7h9'42" 33d49'53"
Stanley Psc 1h2'14" 17d35'40"
Stanley Dra 19h34'56" 64d50'57"
Stanley and LeeAnne Singing Star Cyg 21h35'37" 46d52'6"
Stanley & Ashley: Forever Soulmates Per 3h9'24" 43d27'42"
Stanley B. Barnett IV Vir 12h16'43" 7d54'41"
Stanley B. Gelman Umi 4h40'28" 89d17'8"
Stanley & Becky Young Per 3h27'56" 46d19'45"
Stanley Bowen Vir 12h17'7" -0d18'2"
Stanley Bryan DeVore Her 18h51'58" 22d39'58"
Stanley Cecil Gross Vir 13h27'13" 13d47'54"
Stanley Cholminsky Cap 21h30'3" -9d46'18"
Stanley Clarke And 2h20'58" 38d36'0"
Stanley Cometz Cap 20h30'4" -21d8'2"
Stanley Cowan Per 4h24'8" 51d54'39"
Stanley Czaplinski Vir 13h4'32" 11d11'2"
Stanley Douglas White III Cnc 8h33'33" 24d32'23"
Stanley Emerson Her 18h4'1" 36d27'54"
Stanley F DeForest Gem 7h1'19" 24d18'10"
Stanley F. Mercer Tau 5h17'47" 22d4'1"
Stanley Friedman Cap 20h49'26" -16d11'58"
Stanley Gabriel Tau 4h44'4" 18d2'40"
Stanley George Heap Cam 4h46'38" 52d48'42"
Stanley & Gladys Davis Uma 13h27'40" 59d7'25"
Stanley Hanford Scofield Cas 22h59'24" 57d46'16"
Stanley III Umi 14h45'52" 75d13'54"
Stanley J Oquist Uma 10h17'12" 66d47'16"
Stanley Jay Whitlock Uma 9h39'0" 51d21'5"
Stanley John Pryzby Psc 1h24'10" 15d8'10"
Stanley K. Kwok Lib 15h56'13" -12d56'49"
Stanley Kent Hart Cas 0h37'19" 66d23'47"
Stanley Kowalski Sct 18h31'21" -12d35'13"
Stanley Krogol Uma 11h36'31" 41d1'0"
Stanley L Nice Sgr 18h27'4" -20d48'33"
Stanley Lewandowski Uma 13h59'43" 50d7'14"
Stanley Lewis Hagenston Uma 10h40'15" 62d59'26"
Stanley M. Bell Her 18h7'23" 32d30'55"
Stanley Merves-Mervo1 Uma 8h38'21" 58d37'16"
Stanley Messina Uma 10h18'14" 39d32'15"
Stanley Michael Karandanis Ari 3h9'48" 30d52'41"
Stanley Oh & Angel Au Lib 14h53'28" -3d32'39"
Stanley Ostrum Cep 20h40'55" 59d59'42"
Stanley P Mokrycki Per 3h7'14" 53d56'56"
STANLEY(PAPA)OLSZEWSKI Dra 19h4'3" 64d39'39"
Stanley & Pat Ng Pho 0h50'12" -42d0'18"
Stanley Paul Kott, Sr. Tau 4h2'37" 16d56'30"
Stanley & Peggy Ayres Crb 16h24'23" 34d7'28"
Stanley R. Besaw Uma 10h6'54" 45d47'32"
Stanley R. Tylecki Per 3h12'0" 55d41'52"
Stanley Raub Aqr 22h28'8" 0d56'21"
Stanley Raymond Crow Sco 17h49'24" -36d40'34"
Stanley Raymond Mealy Uma 9h1'57" 46d36'19"
Stanley Rodriguez Aqr 23h21'27" -18d12'28"
Stanley Rose Per 2h59'36" 44d31'27"
Stanley Ross Shelton Jr. Aqr 23h52'12" -14d21'57"
Stanley S. Stone Aur 5h47'56" 49d56'26"
Stanley & Sandra Kent Cyg 21h15'21" 42d8'47"
Stanley "Stan the Man" Santos Per 4h14'39" 49d37'12"
Stanley Strompf * Happy 100th! Aqr 23h49'11" -16d36'3"
Stanley Thomas Hiser Ori 6h15'40" 11d7'26"
Stanley Triplets Uma 13h38'40" 48d46'23"
Stanley V. Karling's Star Aur 5h24'10" 41d8'18"
Stanley Vodicka Per 3h15'38" 47d50'15"
Stanley W. Winslow, Sr. Her 17h17'25" 29d44'42"
Stanley Washburn, Jr. Uma 13h53'5" 58d43'28"
Stanley Williams Milne Cep 3h26'57" 81d57'27"
Stanley Wolper Her 16h43'3" 32d25'6"
Stanley Zawadzki Lib 15h48'43" -9d38'5"
Stanley Zychowka Ori 5h46'6" 5d54'41"
Stanleyklecknerius Cep 22h56'3" 71d42'44"
Stanley's Light in the Sky Cnv 12h46'2" 40d52'42"
Stanley's Little Doll Ari 3h25'20" 26d36'30"
Stannye Jo Nelson Leo 10h58'20" 14d10'46"
Stan's Chopin Aql 18h55'40" 12d3'44"
Stan's Home In Heaven Peg 23h36'26" 21d13'7"
Stan's Krypton Her 16h43'36" 8d40'44"
Stan's Shooting Star Cru 12h46'45" -63d56'36"
"Stan's Star" Uma 11h12'46" 70d5'2"
Stan's Star Cap 21h48'50" -14d9'25"
Stanton Prince Of South Philly Cap 21h40'56" -14d46'21"
Stanton T. Friedman Cam 5h44'52" 59d26'30"
Sta-Paul Love Star Cas 0h27'20" 59d7'9"
Star Cnv 13h40'13" 40d53'39"
Star Umi 15h37'19" 81d16'34"
Star 2010 Dra 19h22'9" 72d50'33"
Star 21 Lib 14h22'13" -24d24'19"
Star 29 Lyn 8h9'2" 42d50'30"
Star 30 Ori 5h1'59" 15d4'53"
Star 5 Section 3 Tri 2h7'42" 30d44'37"
Star 513 Cnv 12h48'9" 38d55'41"
Star above the mountains Ori 5h3'43" 6d48'4"
Star Alabama-Allyson Mary Wright Del 20h38'53" 16d55'54"
Star Alexis Lib 15h31'50" -19d39'55"
Star Alix Monique Morales Vir 11h48'30" 0d2'56"
Star Allen And 23h25'23" 47d12'44"
Star Amelia - Love Gran & Granddad Cru 12h28'50" -60d1'26"
Star Americano Psa 21h30'44" -35d59'57"
Star Amy Uma 9h5'14" 52d51'30"
Star Amy Faye Toedtemeier Lib 16h0'45" -17d49'37"
Star Andrea And 0h19'44" 43d23'17"
Star Anne - My Mom Aqr 21h48'6" -1d27'40"
Star Anny Ellen Toedtemeier Cap 20h17'24" -15d14'16"
Star Arcelia And 1h14'42" 50d10'0"
Star Athletics Allstars Her 18h53'46" 20d56'33"
Star Bace 9 Aql 19h26'57" -0d7'28"
Star Base Corie Hughes Williamson Tau 4h49'15" 22d13'20"
Star Bear Uma 10h19'43" 44d42'36"
Star Belonging To Sherrance Jane And 2h18'12" 46d22'41"
Star "Betty Austin" Uma 14h1'53" 61d58'14"
Star Bezanson And 0h42'41" 41d22'16"
Star Blessed Pho 2h20'0" -41d1'37"
Star Boros Tau 5h38'9" 18d50'52"
Star Brandon Kaplan Cep 20h46'56" 59d13'46"
Star Bray Uma 9h43'58" 61d57'1"
Star Brian Uma 10h12'44" 59d46'36"
Star Bright Daisy White Cyg 21h18'51" 51d10'4"
Star Bux And 1h25'25" 45d46'30"
Star Caitlin Lyn 6h23'51" 61d24'41"
Star Cambria Sco 17h52'40" -30d19'54"
Star Camille Vir 13h4'15" 8d40'49"
Star Cananbershise 50 Cyg 20h15'17" 49d35'44"
Star Carol Spindle Hueter Ori 6h2'42" 10d30'13"
Star Cary Dianne LouiseToedtemeier Lib 14h55'42" -18d23'18"
Star Catchers Tau 4h37'13" 30d45'2"
Star Chan Vir 13h20'52" 8d11'57"
Star Charles Surfy Chuck Kadlec Vir 14h6'7" 4d38'52"
Star Charley Zimmerman Ari 2h12'50" 20d22'18"
Star Chuck Grandad Keef Cas 0h34'18" 56d48'32"
Star Cole Johnson Mattingly Her 18h15'38" 18d43'52"
Star Cross (Dax, Kylie, Zeke, Wes) Cru 12h25'33" -59d18'48"
Star Dad Her 18h16'25" 15d0'45"
Star Daddy-O Cep 22h13'59" 71d7'33"
Star Daigo Tau 5h40'8" 21d25'26"
Star Daniella Dascoli Marques And 0h19'47" 36d27'26"
Star Dazzle Cas 0h13'4" 56d48'25"
Star de la Bastide-Prior Uma 9h29'58" 65d22'19"
Star Dena Drotar Goddess of Love And 0h1'23" 33d46'41"
Star Denise Cas 0h28'41" 56d22'16"
Star Dog 2 Simba & Rocky Cma 6h30'17" -14d52'35"
Star Dolly 'Lenore D. Hutton' Lyr 18h44'40" 39d42'36"
Star Doud Ari 2h11'12" 11d2'13"
Star Drifter Gem 7h14'14" 24d50'8"
Star Eckel Gem 6h51'21" 22d19'44"
Star Eileen And 0h47'33" 25d11'9"
Star Eileen Uma 9h59'48" 56d35'59"
Star Elaine And 1h30'17" 40d53'47"
Star Elijah C Aql 19h12'0" 10d52'2"
Star Elizabeth Lyr 18h26'16" 32d52'46"
Star Erica Pho 0h35'25" -39d54'26"
Star Ernest Boyd Hueter Ori 5h57'22" 8d19'33"
Star Ernest Ross Hueter Ori 5h52'25" 6d21'30"
Star Ernest Tyler Hueter Ori 5h49'47" 4d49'50"
Star Erwic Lyn 8h0'39" 34d41'16"
Star for Daddy Shmunes Aql 19h47'34" -0d34'1"
Star Fournier Eric & Stacy Forever Cas 0h29'3" 60d20'38"
Star "Fowler" place of Dreams Cap 21h14'38" -26d23'15"
Star Frank J.N. 75 Her 18h44'7" 19d55'40"
Star Gazer E Cnc 8h33'44" 31d12'53"
Star Gentry Boo 14h41'44" 26d51'37"
Star George Sgr 20h18'51" -30d30'59"
Star Grace Vir 14h3'36" -14d55'57"
Star Grier Uma 10h58'2" 34d5'26"
Star Großmama Uma 11h15'19" 35d27'50"
Star Guardian - Eastern Dist. BSA Aqr 23h46'20" -14d27'19"
Star Harvey Silverman Cap 20h18'51" -24d14'45"
Star Helen Tau 3h43'57" 20d38'28"
Star Heslewood Ori 5h38'51" -8d11'32"
Star Inge Lep 4h55'20" -26d24'5"
STAR JAMIE Tau 4h59'11" 21d59'4"
Star Janine Lib 15h39'11" -23d6'42"
Star Je And 2h37'40" 50d6'53"
Star Jennkitty Aqr 21h46'45" -3d4'3"
Star Jessamyn Tau 5h46'8" 27d2'11"
Star Joan LeBrun Hueter Ori 6h0'40" 9d23'18"
Star John Stephen Goldfus Cep 22h27'58" 64d54'46"
Star Judith Ellen Sco 15h52'4" -22d17'14"
Star Judith Jardon Lib 14h57'46" -17d54'30"
Star Julian Uma 9h57'21" 56d15'9"
Star K Sco 16h16'48" -10d12'33"
Star Katherine Cas 0h6'14" 55d35'26"
Star Katherine Anderson Hueter Ori 6h8'3" 12d57'7"
Star Kristen Cas 0h37'36" 56d40'56"
Star Kristin Joan Hueter Ori 6h3'41" 11d36'56"
Star La Ari 2h15'3" 24d5'22"
Star Laberta Grandmother Keef Cas 0h29'35" 57d22'37"
Star Laikin's Ori 5h22'4" 5d28'36"
Star Langston And 1h47'38" 49d10'47"
Star Lawrence 4Ever Mine 04/19/2005 Gem 7h23'6" 32d15'42"
Star Leavers 2007 Peg 23h2'10" 17d21'37"
Star Lesley Cas 1h23'3" 67d43'14"
Star Lewis Leo 10h51'11" 9d11'53"
Star Light Trey Bright Gem 6h59'43" 28d47'53"
Star Ling Vir 13h21'12" 5d55'57"
Star Lisa Cas 23h50'15" 50d35'29"
Star Liya Lyr 18h53'4" 41d5'49"
Star Looie -Lindsey Carey Wright Cyg 20h24'29" 52d10'49"
Star Loon Psc 0h36'26" 8d10'7"
Star Luf Dara & John Cyg 20h33'26" 51d42'18"
Star Luka Forty Four Her 16h47'38" 38d48'45"
Star Luna Nichole Flame) Sgr 19h13'28" -22d14'1"
Star Lynn And 23h13'31" 48d59'25"
Star Maddie Lib 14h47'53" -10d25'27"
Star Major-Russell Kirk Layne-Cupid Tau 4h38'7" 22d32'55"
Star Mallory Wedlock Her 16h41'43" 32d35'59"
Star Mandy Crb 15h45'18" 32d8'51"
Star Marie and Mike And 1h32'11" 43d30'47"
"Star Mart" Stewart-Smith Ori 6h4'47" 21d9'53"
Star Matilda - Shine in our Hearts Tau 5h24'47" 24d16'30"
Star Megan Elizabeth Tau 3h32'41" 9d22'43"
Star Memphis Crb 15h51'14" 25d58'55"
Star Meredith That Can't Be Dimmed Ori 6h8'52" 19d44'31"
Star Michelle Housen Lyn 7h41'47" 56d29'41"
Star Midnite (Bear's Eternal Flame) Cru 12h35'42" -59d6'37"
Star Mimi named for Milly Spencer Crb 15h51'4" 28d34'2"
Star Mindy Merit Toedtemeier Lib 16h1'4" -17d40'59"
Star Minori Iwao Cap 20h32'34" -21d41'25"
Star Morgan Brown Mattingly Crb 15h19'50" 29d18'30"
Star Mouska 'Kelly K. Steinmann' Cas 0h58'18" 63d40'29"
Star Nana Josephine Cru 12h37'39" -58d0'8"
Star Nanny Dra 12h44'16" 70d42'15"
Star Nastja Kim Cep 23h34'10" 86d24'6"
Star Nestor Her 17h49'50" 46d48'18"
Star Newton And 23h40'54" 37d25'40"
Star Nikki Cap 21h47'12" -19d38'31"
star nunya And 23h22'52" 38d57'27"
Star Of A. T. Clemmons Per 3h18'30" 44d32'14"
Star of Aaron Per 3h52'52" 48d46'10"
Star of Abigail And 0h22'18" 33d35'52"
Star of Adam And 1h2'59" 44d14'53"
Star of Adam Her 17h24'23" 34d53'18"
Star of Adnan Aqr 22h0'51" -1d50'22"
Star of Aiko's family Cap 20h8'16" -8d33'36"
Star of Aki Ogawa Sgr 18h43'12" -20d31'22"
Star of Akihiro Kamei & 29th Ari 3h14'57" 27d7'58"
Star of Akiko Murokawa Sco 17h25'46" -41d36'31"
Star of Akira & Asami Gem 7h25'7" 23d39'7"
Star of Akira & Ayumi Ogihara Cnc 8h5'45" 25d7'0"
Star of Akira Namba Sgr 18h27'6" -27d48'17"
Star of Alayna Lee DeBattista Cru 12h25'32" -57d45'38"
Star of Alex And 1h26'55" 37d49'17"
Star of Alex Riley Cru 12h3'8" -62d52'35"
Star of Alexander 13 September 1994 Cru 12h11'18" -63d25'17"
Star of Alexis Tau 4h12'4" 16d53'33"
Star of Alice Vir 12h8'27" 1d57'4"
Star of Aliki Foods Per 3h48'8" 36d55'45"
Star of Aly Cas 1h54'44" 63d55'22"
Star of Alyssa, Happy and Bright Ari 2h54'50" 24d27'46"
Star of Amelie Rose Treweek Cru 12h39'14" -60d34'26"
Star of Ami 7.29 Leo 10h21'28" 14d4'56"
Star of Amy Aqr 21h55'15" 1d10'8"
Star of Amy Aql 19h13'38" -9d51'58"
Star Of Amy Elizabeth Bessemer Cru 12h35'52" -64d28'1"
Star of an Angel Uma 11h21'37" 61d40'24"
Star of Anders Umi 14h40'9" 72d16'25"
Star of Andrew Uma 13h53'35" 55d49'28"
Star of Andrew Uma 9h45'38" 46d36'22"
Star of Andrew Uma 11h15'50" 48d50'23"
Star of Andrew William David Smith Lib 14h57'51" -1d3'57"
Star Of Angel Aqr 20h56'2" -13d18'58"
Star of Angel Sarah And 1h8'28" 46d8'36"
Star of AnnaMarie and Jason's Love Aqr 22h9'0" -9d37'19"
Star Of Anthonett Leo 10h23'46" 12d38'41"
Star of Anthony Col 6h13'55" -42d30'21"
Star Of Antonia Caterine Cigna Ori 6h15'19" 10d19'46"
Star of April And 0h39'20" 34d2'45"
Star Of Ardath Mildred Grotto Cas 0h0'57" 53d39'56"
Star of Art Aur 6h36'24" 28d3'23"
-=<< Star of Ashley >=- And 23h6'34" 51d5'21"
Star of Ashley And 22h59'8" 47d17'0"
Star of Ashley Uma 11h23'42" 58d11'54"
Star of Atsushi & Sanae Cnc 8h50'43" 13d35'13"
Star of Augie aka Larry Cermak Leo 11h2'53" 0d36'46"
Star of Aurora Ari 2h14'8" 19d15'59"
Star of Ayumi Aqr 22h11'18" -10d50'31"
Star of Barbara Meyer Sgr 18h33'54" -23d48'17"
Star of Beatrice - 28.09.1946 Cas 23h43'22" 53d37'57"
Star of Beau '82 Cru 12h27'4" -58d50'42"
Star Of Beauty Uma 9h3'10" 47d21'23"
Star of Becca Uma 12h3'7" 60d21'37"
Star of Benjamin Cru 12h25'36" -60d51'47"
Star of Berglind Aqr 23h0'0" -8d11'45"
Star of Bernadette C. Allen Tau 6h0'17" 27d42'43"
Star of Bernard Uma 10h19'1" 42d0'48"
Star of Berway Ori 6h10'57" 18d59'52"
Star of Bethany Mullen Sge 20h9'13" 17d49'14"
Star of Betty & Damon Sge 19h53'50" 18d39'36"
Star of Biddie Vir 11h50'33" 6d34'6"
Star of Blacky Cru 12h53'3" -61d11'38"
Star Of Bob & Regina Johnston Ori 5h34'42" 13d29'11"
Star of Bonnie & Abe Cyg 19h43'50" 56d1'52"
Star Of Braidon Ori 6h12'29" 1d57'42"
Star of Brea Cas 0h30'53" 52d13'5"
Star Of Brendan; Heart of a Hero Cyg 21h15'40" 46d59'51"
Star of Brian and Dawn Scheele Her 16h42'20" 46d33'21"
Star of Brian Cagley Per 3h15'46" 41d23'20"
Star of Brian & Jane Cra 18h58'40" -39d43'0"
STAR OF Brighton Leo 11h5'43" 12d2'45"
Star of Brittney and Keith Vir 13h49'9" 5d15'6"
Star of Broz Nor 16h14'27" -59d52'42"
Star of Bruce Sco 17h23'46" -45d33'52"
Star Of Bryce Cru 12h17'39" -56d29'37"
Star of C.A. Kenichiro Gem 7h5'44" 20d9'27"
Star of Camelot Uma 11h30'41" 64d46'34"
Star of Cardwell Cru 12h17'33" -57d10'17"
Star of Carmel Ari 2h34'28" 25d57'35"
Star Of Carol Psc 1h29'2" 14d36'14"
Star of Catharine Cyg 19h40'2" 31d54'56"
Star of Chanel Ewen Vir 13h34'20" 8d20'59"
Star Of Charlene Crb 15h48'26" 25d45'56"
Star of Charlotte Psc 1h10'12" 26d28'23"
Star of Charlotte & Wes Uma 8h18'42" 63d46'24"
Star of Chelsea Tau 3h29'42" 17d56'24"
Star of Chiharu Cap 20h16'57" -17d32'30"
Star of Chiharu Soejima Gem 6h24'50" 24d26'0"
Star of Chikashi Psc 1h20'44" 8d5'58"
Star of Christian Jaiden de la Pena Tau 4h10'8" 5d40'9"
Star of Christine Vir 14h21'56" 0d40'50"
Star of Christine Gem 7h41'47" 33d22'42"
Star of Christopher Lyn 7h58'23" 43d1'30"
Star of Christopher Sco 17h48'42" -39d43'22"
Star of Christopher Col 5h55'19" -42d13'47"
Star of Christopher Cep 22h35'36" 76d44'45"
Star of Christopher Lowell Ori 5h49'37" 6d12'1"
Star of Christopher Noel Power Cru 12h39'55" -58d7'9"
Star of Cindi Lib 15h2'36" -25d48'54"
Star of Cindy And 0h29'53" 42d9'57"
Star of Cindy 30 And 1h38'58" 45d6'50"
Star Of Cindy Battreal Aqr 23h7'20" -8d51'4"
Star of Claudia Leo 10h48'50" 8d31'4"
Star of Claudia Leo 10h28'9" 20d25'44"
Star of Claudio Uma 13h50'39" 54d44'21"
Star of Colleen Cnc 8h51'14" 22d10'19"
Star of Colleen Cas 0h28'5" 50d53'18"
Star of Colter Cru 12h37'2" -58d27'26"
Star of Constance and Don Babbitt Cas 1h25'31" 53d54'27"
Star of Constantina Psa 22h7'54" -27d53'5"
Star of Cori Uma 10h27'48" 65d59'3"
Star Of Corrinne & Chris Uma 9h47'52" 48d1'11"
Star of Courtney Louise Davies Ari 2h39'49" 13d8'9"
Star of Courtney Phillips Aur 5h28'8" 46d14'41"
Star of Curt Brenden Aur 5h52'28" 37d29'0"
Star of Cynthia Gwen Uma 11h19'41" 63d0'21"
Star of D E N Per 3h26'17" 50d44'29"
Star Of D. S. G. - F. M. C. Ori 6h16'27" 16d42'31"
Star of Daijiro Cap 20h13'47" -9d7'15"
Star of Dainsy And 23h48'8" 34d1'12"
Star of Daisuke & Rikaha Psc 0h52'5" 32d39'34"
Star of Daizaburo love from Miho Leo 10h34'50" 10d20'4"
Star of Damian Per 3h14'46" 52d57'26"
Star of Daniel Lyn 7h12'17" 58d57'36"
Star of Daniel Uma 9h18'8" 54d5'23"
Star Of Danielle "LuLu" Kistner Umi 23h21'58" 88d55'49"
Star of Danny Hutcherson Aqr 21h45'34" -0d10'11"
Star Of Danny & Kimmie Umi 16h8'38" 70d45'26"

Star of Darcy Ella Pearson
Col 6h22'41" -33d10'37"
Star of Dave Richards 28-10-66
Psa 21h52'26" -29d44'6"
Star Of David
Cma 7h24'11" -24d45'35"
Star of David
Cru 12h41'12" -56d24'29"
Star of David
Uma 9h32'1" 58d47'23"
Star of David
Lyn 7h50'40" 57d7'47"
Star of David
Boo 14h52'24" 52d30'51"
Star of David
Umi 14h51'37" 75d44'27"
Star of David
Cap 21h14'57" -15d2'58"
Star of David
Her 17h55'5" 49d50'45"
Star of David
Cyg 19h58'53" 51d12'55"
Star of David
Per 3h37'47" 49d42'46"
Star of David
Per 3h37'0" 45d17'39"
Star of David
Her 15h54'27" 42d33'32"
Star of David
Aql 19h56'52" 8d58'0"
Star of David
Ori 4h53'36" 2d17'3"
Star of David
Ori 5h18'4" 0d30'26"
Star of David
Ori 5h56'53" 17d3'3"
Star of David
Ari 2h50'4" 27d53'49"
Star Of David
Her 17h19'2" 29d38'14"
Star of David #2 7/12/93
Per 3h21'37" 46d44'36"
Star of David (Fried)
Aur 5h43'14" 47d49'57"
Star of David Goldenberg
Cnc 8h50'10" 31d10'2"
Star of David II
Aql 19h20'29" 4d32'9"
Star of David M. Zimmer, MD
Cnc 9h6'9" 26d24'55"
Star of Dawn
Cyg 19h22'6" 46d56'28"
STAR OF DEANGELIS****
Cyg 20h47'42" 47d17'32"
Star of Demitra Saravanos
Peg 22h43'51" 12d21'6"
Star Of Denise Lynn Dunco
And 1h39'13" 39d39'55"
Star of Destiny
Uma 9h14'20" 63d41'5"
Star of Dietra Hachinsky
And 0h3'9" 34d31'34"
Star of Direction ~ James B. Harris
Leo 10h40'24" 15d31'59"
Star of Doherty
Sge 20h9'43" 16d38'32"
Star Of Domenick & Marlon
Her 17h19'37" 45d33'56"
Star Of Dominick Paul
Psc 23h46'10" 7d22'22"
Star of Don Kopfman
Per 3h5'49" 45d19'30"
Star Of Donald Edward McMullen
And 0h38'8" 26d27'45"
Star of Donald & Judith Leppala
Pho 1h25'36" -41d27'56"
Star Of Donald Scharninghausen
Uma 14h3'48" 60d1'42"
Star of Donna "jean" Rossi
Sgr 18h1'59" -35d30'52"
Star of Donnell
Sgr 19h28'13" -14d34'1"
Star of Doodie
Tau 5h53'22" 24d13'17"
Star of Douglas
Aur 5h47'6" 45d34'56"
Star of Douglas Wayne Perkins
Sco 16h15'13" -16d14'52"
Star of Dr. Paul Ira Schneiderman
Cep 21h32'33" 64d43'10"
Star of Duke
Uma 11h49'50" 42d56'32"
Star Of Dylan Paul King
Aqr 22h54'24" -8d35'29"
Star Of E. Lynn Champion
Sgr 18h15'8" -23d12'51"
Star of Eden
And 1h14'58" 39d23'46"
Star of Eiko
Lib 15h45'7" -16d36'24"
Star of Eimear & Morgan
Cas 23h58'3" 57d13'35"
Star of Eishi Moriyama
Cnc 8h26'8" 8d35'39"
Star of Eivy
Cap 21h39'50" -13d11'9"
Star of Elaine Stallman
Cnc 8h41'29" 21d13'41"
Star of Eli
Per 4h39'53" 40d40'43"
Star of Elijah
Cyg 19h52'55" 33d9'42"
Star of Elijah
Tau 5h34'39" 26d19'34"
Star of Elijah Phillips
Aur 5h52'13" 46d23'27"
Star of Elzie Monique Velasco
Peg 21h55'54" 34d15'51"
Star of Emily
Cap 20h50'5" -18d10'25"
Star of Emily Ann
Sco 17h19'21" -39d53'9"
Star of Emotions
Per 4h48'34" 49d2'58"
Star of Eri
Gem 6h45'59" 23d33'13"
Star Of Eric Dragun
Ori 6h0'18" 10d38'46"
Star of Erica
Cam 3h32'34" 61d39'40"
Star of Erik
Per 3h26'43" 51d48'23"
Star of Erika & Nyachi-Nyachi
Aqr 22h47'10" -5d20'1"
Star of Erin
Tau 5h47'29" 25d43'5"
Star of Ernestine Hinei Taria
Cru 12h45'8" -57d55'44"
Star of Eternal Del and Carly
Her 16h39'9" 35d18'13"
Star of Ethan Cace Bolton
Cnc 8h52'34" 23d5'13"
Star of Ethan D. Boone: 8.24.2003
Uma 12h10'1" 48d40'11"
Star of Etsuko & Shigeko
Gem 7h24'31" 23d30'25"
Star of Evelyn
Peg 22h4'16" 8d10'18"
Star Of Evette
Sco 15h56'13" -23d47'11"
Star of Evie Grace
Psc 0h56'8" 10d50'10"
Star of Faith
Ori 5h33'56" 2d9'15"
Star of Faith
Cru 12h24'35" -58d40'44"
Star of Faith Eve
Aqr 22h28'14" -5d47'14"
Star of Farder
And 23h50'5" 34d33'34"
Star of Francis
Uma 8h34'24" 71d10'21"
Star of Frank A. Stenson
Sco 17h19'8" -32d59'9"
Star of Fumie Tachiki
Sco 16h45'35" -41d35'0"
Star of Fusako Udaka
Cnc 8h9'41" 22d45'8"
Star of Gabrielle Leigh Capasso
And 23h3'42" 45d25'5"
Star of Geari Viney
Crb 15h35'52" 35d53'0"
Star of Genji & Miho
Psc 1h35'33" 27d14'20"
Star of George
Gem 7h27'51" 27d47'52"
Star of George
Boo 14h54'34" 22d51'9"
Star Of George
Ori 5h9'0" 7d13'41"
Star of George Pancheri
Gem 7h39'23" 22d12'12"
Star of Georgia
Uma 10h35'30" 57d9'28"
Star of Georgia and Kobe Tea
Cru 12h42'27" -61d43'11"
Star of Gibson
Cyg 20h17'42" 47d16'49"
Star of Gibson
And 23h14'27" 42d47'35"
Star Of Glenn E. Leichliter
Per 3h44'33" 39d5'34"
Star of Grace
And 0h55'7" 22d2'0"
Star of Grace
Cas 1h25'27" 63d23'28"
Star of Graeme
Sco 17h15'46" -45d19'40"
Star of Grandma Bear
Hor 4h9'47" -41d41'15"
Star of Greg Sexton
Cep 22h14'12" 61d11'33"
Star of Gwen Henzi
Vir 13h22'52" -4d34'27"
Star of Hamilton
Psc 0h43'34" 16d47'29"
Star of Hana
Gem 6h45'43" 22d32'48"
Star of Hannah
And 1h20'37" 44d36'48"
Star of happiness of eternity of Jun
Lib 14h29'39" -13d32'50"
Star of Harrison Bailey Veitch
Cru 12h43'20" -57d54'29"
Star of Haruaki Nakai
Gem 7h33'44" 19d34'7"
Star of Haruhisa & Aki
Tau 4h56'1" 26d47'30"
Star of Haruka Shiozawa
Aqr 22h7'55" -5d14'34"
Star of Haruko & Hiroyuki Namba
Gem 7h10'18" 24d49'37"
Star of Harumi
Ari 2h21'29" 17d13'40"
Star of HB
Cap 21h48'12" -14d56'21"
Star of HBW
Her 16h25'59" 17d20'35"
Star of Heaton
Sge 20h9'8" 18d9'26"
Star of Hetal
Uma 11h41'9" 44d10'47"
Star of Hideaki & Yoko
Psc 1h16'27" 7d44'15"
Star of Hideki
Gem 6h46'31" 24d10'12"
Star of Hideki & Yukiko
Psc 0h15'14" 13d44'15"
Star of Hilary
Sco 17h18'32" -32d45'8"
Star of Hina November 28th
Sgr 18h42'48" -19d9'24"
Star of Hiro & Hitomi
Leo 10h37'22" 8d34'7"
Star of Hiro & Iku
Lib 14h44'33" -13d15'39"
Star of Hirokazu & Satoko
Psc 23h58'49" 2d49'50"
Star of Hiroki & Mayumi
Vir 12h28'54" 6d24'40"
Star of Hiromitsu & Yuko
Ari 3h5'51" 26d54'46"
Star of Hirotaka & Kaori
Vir 12h21'10" 5d50'51"
Star of Hiroyuki & Makiko Abo
Sgr 17h51'52" -43d56'26"
Star of Hisataka
Psc 0h51'12" 19d4'17"
star of hitoshi & youko
Psc 1h1'21" 22d51'51"
Star of Hope
Cnc 8h8'23" 32d30'23"
Star of Hope
Gem 6h53'34" 33d52'14"
STAR OF HOPE AND GOOD FORTUNE
Per 1h46'12" 50d42'53"
Star of Hope & Faith
Per 4h40'11" 36d20'4"
Star of Hosoya
Tau 3h37'30" 28d52'27"
Star of Hoxie
Sco 16h14'48" -18d18'20"
Star of Humperdink and Hank
Cas 1h3'25" 62d46'54"
Star of Ichiro & Keiko
Sco 16h46'46" -26d33'42"
Star of Ida Gale ~ 4.21.2007
Uma 11h0'52" 48d29'55"
Star of Iqbal
Per 3h10'5" 50d17'19"
Star of Irene-Bruno
Cep 20h43'8" 71d55'0"
Star of Isao
Sgr 18h20'56" -32d5'51"
Star of Ivona
Vir 13h1'20" -15d16'34"
Star of J. C.
Lyn 8h3'0" 40d15'7"
Star Of Jaana
Gem 7h5'39" 19d33'55"
Star of Jack
Aqr 22h59'34" -10d30'43"
Star of Jackie
Crb 15h42'22" 32d45'2"
Star of Jacob
Per 3h14'38" 46d15'10"
Star of Jacqueline
Crb 15h34'1" 30d35'59"
Star of Jacqueline B. Lage
Uma 8h41'59" 63d34'42"
Star of Jake
Col 5h26'10" -29d57'22"
Star Of James Michael Ragsdale
Gem 6h7'19" 27d21'24"
Star of James Skinner
Cnc 8h48'4" 10d48'37"
Star of Jane
Cap 20h38'13" -11d18'24"
Star of Janet
Cas 1h24'19" 55d20'14"
Star of Janus-Our Mommy and My Wife
Cas 1h26'56" 57d23'56"
Star of Januz Dervishi
Cap 20h35'27" -25d29'23"
Star of Jasmin
Cru 12h37'28" -59d45'10"
Star of Jayden
Sgr 19h13'25" -13d56'50"
Star of Jazmine D Sosa
Sgr 19h15'3" -22d9'32"
Star of JD, Rarni & Blue - Nov 2005
Cru 12h0'5" -61d3'50"
Star of Jeneva
Cnc 9h11'28" 21d35'51"
Star Of Jenna Christine Lerch
And 23h24'2" 42d26'36"
Star of Jennifer
And 0h17'54" 36d53'57"
Star of Jennifer & Michaelangelo
Cyg 20h20'11" 52d14'29"
Star Of Jennifer S. Rodman
Psc 1h20'6" 17d16'10"
Star of Jennifer Yee
Uma 10h5'12" 67d46'3"
Star of Jenny
Sco 17h22'24" -44d13'27"
Star of Jenny's 50th
Cru 12h35'44" -60d41'22"
Star of Jeremiah
Cnc 9h0'44" 15d30'16"
Star of Jeron
Uma 9h17'10" 49d9'50"
Star of Jesse
Aqr 22h35'46" -1d12'43"
Star of Jesse David Hurley
Her 17h26'15" 23d39'16"
Star Of Jesse May Victor
And 0h14'21" 44d16'21"
Star Of Jessi and Tom
Uma 13h41'32" 49d33'57"
Star of Jessica Erin Muellner
Sgr 19h55'13" -43d38'29"
Star of Jessie
Aqr 22h12'38" 1d18'9"
Star of Jill Davis
And 2h2'10" 46d45'3"
Star of Joanne Kossar
Uma 10h29'18" 68d42'58"
Star of Jodi
Lyr 18h44'48" 34d35'8"
Star of Joe, the light in my heart.
Psc 1h23'21" 32d15'10"
Star of John Khoury - The Bomb
Pho 23h28'58" -41d42'48"
Star Of John M. Powers
Ori 4h52'23" 1d34'35"
Star of Jon
Her 16h33'46" 48d17'59"
Star of Jon
Her 16h37'15" 47d2'21"
Star of Jonah
Cet 2h17'55" 0d45'22"
Star of Jonah
Uma 8h54'24" 56d10'27"
Star of Joni
And 0h31'2" 30d27'38"
Star of Jordan (aka Snookies)
Boo 14h44'28" 35d34'4"
Star Of Josef & Lisa Kleine
Ori 5h36'36" 3d2'17"
Star of Joseph Anthony Vitale
Cap 21h39'31" -13d30'19"
Star Of Joshua & Lara Mayoral
Cyg 19h35'10" 47d12'49"
Star of Joy
Psc 1h4'46" 32d53'8"
Star Of Joy
Vir 14h51'3" -0d12'52"
Star of Joya
Lyn 8h59'27" 42d16'48"
Star of Jozsef & Satoko
Vir 14h8'55" 1d11'52"
Star of Julia
Cyg 20h31'53" 38d18'32"
Star of Julian 12.15.2003
Uma 8h57'17" 62d15'4"
Star of Julie
And 1h3'21" 35d26'5"
Star of Julie Goldklang
Cas 1h28'40" 57d55'0"
Star of Julie Robert
Gem 6h29'11" 15d56'27"
Star of Jun Imai
Tau 4h18'11" 25d19'58"
Star of Jun Nagasawa
Gem 7h9'55" 20d12'5"
Star of Junko Kuma-San
Vir 12h27'36" 5d58'43"
Star of Junya
Cap 20h28'21" -24d27'0"
Star of Justice for Derek Matthew
Cma 6h55'39" -12d13'1"
Star of Justine
Cru 12h19'16" -56d34'23"
Star of Kako & Mako
Cap 20h18'55" -16d11'50"
Star of Kamata
Leo 10h31'15" 17d0'23"
Star of Kanon
Cap 20h12'31" -17d6'36"
Star Of Kaori & Tamiki
Leo 11h16'25" 17d3'23"
Star of Karam
And 23h31'24" 36d38'12"
Star of KareBear
Uma 12h38'31" 59d2'18"
Star of Karen
Col 5h49'11" -29d39'36"
Star of Kathleen Atley 26.9.39
Cru 12h55'20" -63d2'17"
Star of Kathryn
And 1h27'35" 38d55'49"
Star of Kathryn and Benton
Mon 6h47'48" -0d46'21"
star of KATIE
Cru 12h4'51" -60d38'26"
Star of Katrina Anne
Lib 15h3'37" -23d33'24"
Star of Katrina & Christopher
Her 18h21'19" 22d17'15"
Star of Katsumi & Kyouko
Sco 16h51'58" -39d15'35"
Star of Katsuys & Naomi
Cnc 8h16'10" 14d25'0"
Star of Kazuhiko & Mao
Sgr 18h58'27" -33d55'46"
Star of Kazuhiro & Kumiko
Vir 11h47'44" 0d48'35"
Star of Kazuya & Rino
Tau 4h32'51" 23d48'24"
Star of Kazuyo
Sco 16h52'59" -41d29'0"
Star of Kellis Day
Her 17h25'10" 39d21'40"
Star of Kelly Ann
And 23h18'39" 48d23'18"
Star of Ken
Ari 3h7'51" 11d13'50"
Star of Kenjiro & Ayako
Aqr 21h39'15" -0d29'46"
Star of Kent the best dad-star ever
Per 3h37'36" 44d6'0"
Star of Kenta
Cnc 8h5'34" 16d28'40"
Star of Kezrar
Uma 9h24'24" 49d4'39"
Star of Khrystal
Crb 15h38'20" 27d36'34"
Star of Kieran
Sgr 19h16'56" -20d55'34"
Star Of Kiley
Tau 4h5'15" 16d9'13"
Star of Kim; Angel on Earth
Aqr 23h44'33" -23d29'59"
Star of King
Tau 5h48'33" 25d8'4"
Star of Kokubun & Haruna
Gem 7h31'58" 27d31'5"
Star of Kota & Hiromi
Cnc 8h48'56" 11d8'33"
Star of Kouki
Leo 10h39'58" 17d29'43"
Star of Koyuki
Cap 20h27'43" -25d50'1"
Star of Kozo & Kanako
Lib 14h59'59" -24d39'33"
Star of Kristen Marie Baker
Gem 6h57'35" 12d21'40"
Star of Kristin, Beautiful and Wise
Lib 15h3'43" -24d34'2"
Star of Kristyn
Cen 12h18'28" -37d47'17"
Star of K.T.S.
Cnc 8h1'52" 12d8'47"
Star of Kumi & Nori
Cnc 8h49'32" 10d47'6"
Star of Kuniaki & Natsumi
Leo 11h12'53" 13d20'4"
Star of Kunigoro Yuguchi 1910-1994
Vir 13h28'12" 4d9'46"
Star of Kunihiko & Emiko
Sco 17h52'36" -43d10'19"
Star of Kunihiko & Shizuko
Ari 2h38'23" 23d47'49"
Star of Kyoko and Takayuki
Cas 1h26'40" 52d43'34"
Star of Kyoto Augustine
Leo 9h32'50" 21d15'30"
Star of Lachlan - 2 June 2005
Gem 7h47'18" 30d54'4"
star of Lailanie
Aqr 20h43'18" -10d26'15"
- Star of Lakshani -
Col 6h3'59" -33d4'19"
Star of Larisa Anaya
And 0h36'6" 42d4'19"
Star Of Larry & Anita Johnston
Ori 5h23'28" 13d57'16"
Star of Larry, Super Troop
And 0h38'59" 30d19'50"
Star of Laura
Tau 3h33'51" 25d14'52"
Star of Laura (La La) (Harris)
Tau 3h48'9" 8d4'34"
Star of Laura V.
Lyn 7h28'54" 44d48'9"
Star of Lauren
Gem 6h3'13" 26d30'59"
Star of Lauren (i.l.y.t.i.a.b.)
Uma 12h48'20" 61d9'32"
Star of Lauren Alexis Garner
Vul 20h22'30" 27d47'13"
Star of Lauren Jae Vitale
Ari 3h14'57" 28d43'35"
Star of Leah
And 1h41'26" 37d4'56"
Star of Leanne
Tau 4h28'32" 28d37'10"
Star of Leo C. Amilkavich
Leo 10h10'7" 22d13'1"
Star Of Lia
And 0h22'24" 29d30'41"
Star of Liam Isaacs
Ori 5h15'6" 6d20'13"
Star of LINCOLN
Uma 11h53'7" 60d39'30"
Star of Linden Pothier
Lib 15h38'16" -28d48'45"
Star Of Lisa
Psc 0h16'22" 8d44'22"
Star of Lori
And 23h32'6" 37d58'58"
STAR OF LORI MICHELE'S SPIRIT
Lib 15h33'17" -14d34'13"
Star of LouVonne
Crb 16h24'13" 36d25'1"
Star of love "Bruno"
And 1h15'23" 47d48'4"
Star of Love (Jennifer's Star)
Tau 5h6'44" 28d16'16"
Star of Lucci
Vir 13h37'44" -15d13'48"
Star of Luna Vista
Crb 16h15'9" 33d46'7"
Star of Lynn Forever and Always JJJ
Lyn 7h56'4" 39d38'48"
Star Of M. J.
Cas 23h20'55" 53d14'1"
Star of Mai
Gem 7h32'22" 19d2'13"
Star of Maiko 1984
Gem 7h26'53" 26d49'7"
Star of Makayla and Ashley
Gem 6h6'36" 23d31'10"
Star of Malcarne
Uma 11h47'1" 54d40'26"
Star of Mallory
Uma 11h19'21" 68d43'6"
Star of Mamma Mia Kanai
Aqr 20h49'28" -9d45'18"
Star of Mamoru Miyanaga
Cap 20h54'13" -17d9'12"
Star of Mandy Dunlop
Tau 4h28'50" 10d10'20"
Star of Manuel
Boo 14h17'48" 47d6'36"
Star of Maon Fujioka
Gem 6h44'19" 24d16'6"
Star of Marilyn Jean Webb 15.9.1955
Cra 19h2'41" -42d56'11"
Star of Marissa Raynne
And 0h3'41" 45d30'19"
Star of Mark Stanton
Lyn 8h13'39" 41d30'29"
Star of Marric
Ori 6h11'4" -0d8'10"
Star of Marshall
Aqr 21h4'39" -13d51'20"
Star of Martha Scott
Cap 20h43'3" -14d48'45"
Star of Mary
Cas 1h2'2" 54d9'6"
Star of Mary Arden and Dave Harris
Uma 11h45'11" 52d27'40"
Star of Mary & David
Lib 14h59'58" -2d37'5"
Star Of Mary Elizabeth Harvey
And 0h37'11" 38d11'31"
Star of Mary L Thomas
And 0h48'21" 28d2'51"
Star of Masachika & Saki
Sgr 18h58'47" -34d13'42"
Star of Masayo & Tatsuyuki with Babe
Gem 6h45'38" 24d9'55"
Star of Masayuki & Miyako
Leo 9h50'2" 25d59'43"
Star of Matthew
Leo 10h27'5" 20d16'30"
Star of Matthew Niegel
Her 17h10'24" 34d20'0"
Star of Matthew, The Prince
Ari 2h46'32" 30d46'3"
Star of McCarthy - 22.02.1963
Psc 1h54'34" 2d58'29"
Star of McGill
Uma 9h59'11" 48d56'36"
Star Of McGrew/Vallejo
Cnv 12h37'43" 42d53'17"
Star of Melissa - Angel In Heaven
Ari 1h48'25" 18d31'28"
Star of Mgazma
Uma 10h50'25" 56d2'53"
Star of Michael
Ori 5h42'22" -1d40'20"
Star of Michael Ann & Bozworth
Umi 15h1'36" 69d45'13"
Star Of Michael Of The Wolf Clan
Ori 5h34'15" 0d21'10"
Star Of Michael Patrick Mooney
Uma 8h26'55" 70d39'8"
Star Of Michael's JOY<3 jjongheehee
Sco 17h30'38" -43d18'40"
Star of Miki Oya
Sco 16h50'4" -38d8'55"
Star of Mikiko Minami
Leo 10h18'38" 17d58'12"
Star of Mimi
Tau 5h43'1" 23d21'15"
Star of Minton
Aqr 22h19'31" -1d39'51"
Star of Mirth
Vir 13h4'38" 9d37'51"
Star of Miss Jessica
Psc 1h10'8" 11d9'53"
Star of Misuzu Shinmura
Lib 15h33'27" -21d0'3"
Star of M.J. Martha Jimenez
Vir 13h8'58" 6d38'47"
Star of Montgomery
Leo 10h24'17" 22d55'36"
Star of Morg
Sco 16h49'8" -36d19'5"
Star of Motherhood and Love - Janet
Ari 2h14'37" 26d22'35"
Star of Motoo & Haruna
Ari 2h25'45" 15d15'5"
Star of MSN
Cap 20h37'10" -23d23'52"
Star of my angel Carrie Musick
Mon 7h17'55" -0d31'5"
Star of My Heart
Tau 5h39'1" 24d52'47"
Star of my life, Kelsey
And 0h16'40" 43d13'36"
Star Of My Loving Wife Veronica
Ari 3h27'35" 27d10'43"
Star of Nacchi
Leo 9h47'31" 10d59'40"
Star Of Nadia Petruccelli
Ari 2h35'56" 13d1'41"
Star of Nadine
Cas 2h14'5" 62d31'45"
Star Of Nadine Calantropio
Uma 14h27'59" 56d25'12"
Star of Nadya
Gem 7h10'48" 20d43'31"
Star of Nagahara Family
Vir 11h57'49" 10d13'20"
Star of Nagisa
Aqr 22h6'34" -5d9'20"
Star of Nana
Cap 20h37'9" -8d48'38"
Star Of Nanny & Linda
Ori 6h12'53" 9d13'5"
Star of Naoko & Fuyuki
Gem 6h52'46" 21d21'27"
Star of Naoshi and Sayaka
Cyg 21h26'58" 34d31'8"
Star of Naotaka 2006
Leo 10h43'8" 10d1'14"
Star of Naoya & Mariko
Ari 2h24'37" 13d13'52"
Star of Natalie Richards
Pho 0h37'21" -49d53'43"
Star of Nathan
Aur 5h43'42" 41d51'26"
Star of Nathan Lee Ford
Cru 12h28'22" -58d59'48"
Star of Nathan McNeilage
Cra 18h1'51" -37d5'48"
Star of Natumi's Dream
Cap 20h12'26" -9d51'19"
Star of Neal
Her 18h21'28" 13d53'1"
Star of Neva
Uma 12h38'46" 59d22'21"
Star of Nicholas
Umi 14h42'8" 69d37'34"
Star of Nicholas V.
Psc 1h32'1" 24d3'7"
Star of Nicole L Sosa
Cnc 9h8'47" 25d46'6"
Star of No Brand Dragon
Lib 15h26'48" -23d40'42"
Star of Noah T
Sgr 19h28'28" -30d43'13"
Star of Nono 79
Cap 20h20'20" -9d10'33"
Star Of Noreen
And 23h9'50" 42d3'28"
Star of Nozomi
Sgr 18h57'50" -35d16'19"
Star of only one Midori
Vir 12h26'17" 6d38'35"
Star of Opal Sharmini - 5 May 1982
Cen 13h23'3" -54d35'46"
STAR OF OSCAR
Sgr 18h30'59" -28d8'21"

Star of OZ
Aql 19h28'23" 0d7'2"
Star of Pagueli
Psc 1h6'21" 21d45'53"
Star of Pasquale and Marie
Cyg 21h33'18" 50d43'35"
Star of Patricia L. Stenson
Lib 15h5'51" -9d52'33"
Star of Patrilia
Aql 19h0'24" -8d11'32"
Star Of Payne
Vir 14h0'57" -9d18'50"
Star of Peace
Lyr 18h46'26" 37d47'15"
star of peach & Family
Sgr 18h21'34" -33d2'39"
Star Of Peter
Vir 13h22'41" -0d3'27"
Star of Peter and Angela Holmstrom
Cyg 19h51'38" 39d27'26"
Star of Phezman 2005
Umi 16h34'25" 76d0'18"
Star of Pietro
Cnc 8h34'26" 15d56'54"
Star of Pinkney and Carol Morrow
Per 4h32'40" 38d42'23"
Star of Precious
Cru 12h24'31" -61d50'10"
Star of Princess Aiko
Vir 11h54'28" -0d57'35"
Star of Princess Sarah B. II
Uma 11h29'12" 31d40'43"
Star Of R. E. K.
Uma 11h29'47" 42d57'10"
Star Of Rachael Lauren Bohn
Umi 14h9'54" 71d9'0"
Star of Rachel
And 1h8'1" 39d14'53"
Star of RAM
Uma 10h0'44" 41d28'13"
Star of Raycraft
Cra 18h55'42" -38d19'25"
Star of Rebecca Ennis
Lib 14h37'14" -17d8'29"
Star of Reggie - 12.04.1942-26.03.2006
Cru 12h37'3" -60d12'23"
Star of Rheta
Psc 1h54'7" 6d43'24"
Star of Rhiannon
Uma 13h11'20" 58d48'59"
Star of Richard Elliot Tohl
Tau 4h45'22" 22d54'28"
Star Of Richie & Patricia Johnston
Ori 5h30'18" 14d21'20"
Star Of Ricky & Kathy Nastri
Ori 5h22'7" 14d10'54"
Star of Rie 1962-0216 No 1
Aqr 21h49'5" -5d2'8"
Star of Rob and Amanda - 31/10/2004
Cru 12h23'26" -57d34'21"
Star of Robbie Daniel
Ari 2h34'3" 27d59'45"
Star of Robert Logan
Col 6h24'56" -34d34'11"
Star of Robert S. H. Presuhn
Lyr 18h48'19" 35d15'37"
Star of Ron and Mindy
Cap 20h10'13" -13d8'31"
Star of Ronald Hachinsky
Cep 21h56'17" 61d1'54"
Star of Ronald Lindsay Foster
Pyx 9h1'14" -23d4'19"
Star of Rose
And 22h58'51" 52d12'2"
Star of Rose Mary
Lib 14h35'13" -18d31'34"
Star of Rosemary
Lyn 7h58'56" 59d21'42"
Star of Rum
Her 17h39'39" 28d1'20"
Star of Rumi 1017
Lib 15h49'30" -16d22'39"
Star of Ryo
Cap 20h33'52" -23d19'0"
Star of Ryo & Yuki
Vir 12h19'39" -5d2'55"
Star of Ryogo & Asuka
Gem 6h19'47" 24d30'29"
Star of Ryoichi & Kimiko
Psc 0h49'19" 19d23'13"
Star of Ryoko
Cnc 8h11'24" 20d14'16"
star of Ryousuke
Cnc 8h7'26" 23d50'16"
Star of Ryuichi &Minako
Cnc 8h9'52" 20d44'54"
Star of Sachi Yuguchi 1916-2005
Ari 2h37'21" 18d21'9"
Star of Sally Whennell-Warner
Cru 12h27'8" -60d23'59"
Star of Sam Dragovitch
Cru 12h32'13" -62d48'27"
Star of Samaria
Psc 1h10'41" 19d35'3"
Star of Samson
Aqr 22h4'28" -3d10'41"
Star Of SaraBoo
Vir 13h26'6" -6d15'43"
Star Of Sarah E. Thomas
Peg 22h42'54" 27d6'44"
Star of Sarah O3231
Lyr 18h26'24" 46d2'33"
Star of Sasaki Family
Sgr 18h23'10" -32d33'48"
Star of Satoru & Kimie
Leo 11h13'47" 12d51'19"
Star of Satoshi & Hitomi
Vir 13h10'6" 7d50'35"
Star of Satoya
Aqr 21h40'40" 1d42'5"
Star of Saunders
Her 17h30'44" 16d9'10"
Star Of Savanna
Vir 13h14'36" 5d10'7"
Star of Sayaka
Cnc 8h12'22" 22d14'21"
Star of Schalan
Dra 18h9'18" 77d49'58"
Star of Scoffone - Big Al
Lib 15h10'42" -7d38'49"
Star of Scott 18 yrs-joie de vivre
Vir 13h22'40" -9d18'6"
Star of Sean Daniel
Gem 7h6'7" 27d15'53"
Star of Setsuko & Yasuji
Ari 2h27'34" 24d52'18"
Star of Shannon
Cyg 21h48'6" 48d13'30"
Star of Shara
Leo 11h10'33" -0d27'7"
Star of Sharon
Cap 20h21'54" -12d50'55"
STAR OF SHARON
Cap 20h28'17" -23d10'45"
Star of Sharon
Cas 0h27'35" 53d35'49"
Star of Sharon
Boo 14h40'56" 19d43'14"
Star of Sheila
Psc 0h32'24" 8d13'55"
Star of Sheila Kay
Sgr 18h7'37" -27d56'41"
Star of Sheila O'Malley
Cas 23h50'55" 50d36'5"
Star of Shelby
Tau 4h43'36" 9d5'21"
Star of Shelley Jenks 05.04.84
Ari 3h18'30" 15d19'34"
Star of Sherertz
Lib 15h41'13" -5d2'28"
Star Of Sherri Trevino
Uma 11h57'36" 42d14'39"
Star of Sherry E. Dye
Mon 7h55'29" -4d51'8"
Star of Shey
Ori 5h27'0" 2d9'6"
Star of Shigeki 7
Cap 20h30'21" -26d42'49"
Star of Shigeko & Masaru
Vir 13h8'4" 7d42'13"
Star of Shigeru & Emiko
Cnc 8h13'34" 14d34'49"
Star of Shihoko
Sco 16h6'0" -28d40'11"
Star of Shine
Her 16h53'19" 31d42'11"
Star of Shinichi Takeshita Family
Cap 20h29'45" -26d12'54"
Star of Shokotan
Cap 20h38'14" -20d5'38"
Star of Shuji & Mika & Rino
Psc 1h38'31" 25d19'3"
Star of Shuri
Cap 20h35'50" -17d37'58"
Star of Shuta Aoki
Sgr 19h36'42" -38d2'47"
Star of Siwon
Lib 14h57'45" -17d7'52"
Star of Skye
And 0h19'33" 26d36'42"
Star of Sonia
Vir 13h24'0" -13d8'9"
Star of Sophia
Vir 13h32'52" -4d47'56"
Star of Sophia Lee
Cru 12h21'45" -58d38'38"
Star of Sophie Harrison
Tri 1h57'34" 33d53'32"
Star of Sotiris
Cru 12h13'32" -63d41'27"
Star of Souju
Gem 7h7'10" 24d25'32"
Star of Soul
SkoopOnSomebody
Psc 1h17'17" 19d19'30"
Star of Stash and Moe
Aql 19h22'1" 16d20'41"
Star of Stephanie
Cra 18h8'10" -37d4'16"
Star of Steven
Pho 1h16'16" -44d43'43"
Star of Steven
Psc 1h23'42" 23d59'4"
Star Of Steven
Uma 11h34'40" 47d29'9"
Star of Susan
And 23h2'30" 51d45'52"
Star Of Susan Elizabeth Kavanaugh
Sco 16h14'55" -9d59'28"
Star of SWABB
Gem 7h0'58" 24d4'23"
Star of Sylvia Edna Valle 1919-2003
Cru 12h1'47" -62d20'58"
Star of T & K
Gem 7h8'31" 19d12'25"
Star of Tadaharu & Hanako
Leo 3h34'34" 21d40'29"
Star of Takahiro & Naho
Leo 9h39'18" 8d11'48"
Star of Takashi & Chikako
Aqr 21h41'48" 1d44'53"
Star of Takayuki Ishikawa with Dream
Aqr 21h38'26" 1d33'41"
Star of Takeo & Itsuyo
Tau 3h57'46" 24d18'43"
Star of Tammy & Jay
Cyg 20h29'39" 56d50'43"
Star of Taressa
Lib 15h51'4" -17d6'3"
Star of Tarymir
Uma 10h42'18" 55d13'56"
Star of Teachers, Audrey Gilbert
Uma 10h53'5" 52d11'4"
Star of Ted
Tau 4h10'17" 27d52'56"
Star of Terrence Patrick O'Neil
Lyn 8h22'41" 45d43'48"
Star of Terunosuke
Ari 2h55'14" 25d35'53"
Star of Tessie
Ind 21h50'6" -67d14'18"
Star of Tetsuhito & Machiko
Ari 2h39'43" 17d30'48"
Star of Thanks & Love
Cnc 8h17'14" 27d20'10"
star of the Enchanted Cove
Crb 16h4'28" 28d8'16"
Star of the Mermaid
Lyn 7h0'10" 52d27'9"
Star of the Princess Johana
And 2h36'50" 50d9'58"
Star of the Sea
Aql 19h10'7" 3d13'16"
Star of the Sob Divas
Ori 5h38'1" 1d6'28"
Star of Theresa
And 23h58'15" 47d55'51"
Star of Thousand Hearts
Lib 14h54'24" -1d41'17"
Star of Tiffany Kane
Psc 1h36'20" 27d32'26"
Star of Tolerance
Cyg 21h15'47" 43d39'33"
Star of Tomo, Nori & Yui
Aqr 21h39'12" -0d13'54"
Star of Tomoaki-Ogura
Gem 7h8'10" 20d28'38"
Star of Tomoko
Leo 11h11'43" 12d42'1"
Star of Tony and Amy Undernehr
Pho 0h57'59" -43d46'54"
Star of Toru & Kiwa
Cap 20h38'17" -20d1'6"
Star of Toshio & Mariko forever
Sgr 18h54'9" -18d51'49"
Star of Toshitaka & Mana
Leo 9h42'7" 26d13'32"
Star of Travis
Pho 0h34'26" -42d31'6"
Star of tribute to Ai
Psc 0h25'45" 21d24'47"
Star of Tricia
Uma 12h50'14" 53d33'16"
Star of Tristan
Cyg 20h13'9" 57d45'4"
Star of Trixy
Leo 10h7'41" 22d45'58"
Star of Tsuchimaki
Aqr 20h57'57" -9d3'59"
Star of Tsunashi & Yukiyo
Lib 15h33'36" -21d7'41"
Star of Tugiko Saji Hashimoto
Aqr 23h20'53" -16d17'30"
Star of Ukyou
Cnc 8h10'13" 22d36'16"
Star Of Ulana And Mlke Nosal
Uma 14h11'38" 58d27'35"
Star of Umehara prof
Lib 15h20'24" -14d26'41"
Star of unyuchan
Cap 20h12'53" -8d39'26"
Star Of Valerie
And 23h47'52" 47d8'53"
Star of Valerie
And 0h49'14" 43d41'18"
Star of Vannary
Uma 9h33'35" 47d51'53"
Star of Vernon
Lib 15h10'45" -29d26'56"
Star of Vicki Behm
Sgr 18h7'8" -25d41'8"
Star of Vivia Victoria
Uma 11h42'17" 35d57'47"
Star of William A Rogers
Aqr 21h13'31" -10d5'57"
Star of Woofie
Uma 10h36'40" 65d48'58"
Star of Y Namba great CCD architect
Aqr 21h15'56" 1d49'47"
Star of Yang Xiaoyang
Tau 5h38'45" 23d12'20"
Star of Yasuhiro
Gem 7h8'21" 24d56'16"
Star of Yasutaka & Misaki
Cnc 8h54'26" 13d21'43"
Star of Yba
Aqr 20h43'53" -7d57'25"
Star of Yoko & Takayuki
Cnc 8h50'47" 16d1'30"
Star of Yoshihiro & Wakako
Cnc 8h32'47" 11d4'33"
Star of Yoshihito & Ikuko
Leo 11h11'40" 13d11'52"
Star of Yoshimasa & Takami
Sgr 18h40'35" -18d45'37"
Star of Yoshinori & Miyuki
Cnc 8h3'2" 15d4'45"
Star of Yosuke Seki
Sco 17h29'31" -42d37'54"
Star of Yuji & Machiko
Leo 9h46'17" 26d20'44"
Star of Yuka
Leo 9h46'59" 25d58'14"
Star of Yuka
Sco 16h48'8" -26d50'54"
Star of Yuka & Yoshi
Sgr 18h41'52" -20d16'30"
Star of Yukiya Ibaraki
Tau 3h37'32" 28d58'22"
Star of Yuko & Ryo
Ari 3h0'16" 30d31'18"
Star of Yuko Usewax
Psc 1h21'17" 7d57'35"
Star of Yulia
Leo 11h46'38" 25d30'53"
Star of Yumiko & Hideki
Gem 7h8'8" 24d37'10"
Star of Yunagi Shinmura
Cnc 8h43'35" 9d12'56"
Star of Yusuke Mizuta
Leo 10h56'32" 2d58'16"
Star of Yuuya and Hitomi
Gem 7h27'30" 22d42'46"
Star of Yvonne Carmen Salotti
Tau 5h34'21" 22d17'17"
Star Of Zachary Robert Stack
Cep 21h43'53" 63d25'3"
Star of Zack
Per 3h3'0" 47d27'5"
Star of Zander
Uma 10h20'25" 54d3'27"
Star Olga
Tau 4h23'17" 25d15'51"
Star Over The Bay
Peg 22h43'23" 31d46'7"
Star Pappy
Aur 6h27'43" 41d17'53"
Star Path
Uma 12h0'58" 40d57'34"
Star Paula
Cap 21h8'33" -25d16'11"
Star Peachcake 'Lenore A. Graff'
Cyg 19h55'17" 33d33'55"
Star Pop Pop
Uma 12h55'44" 54d3'39"
Star Rachyl
And 1h20'36" 44d17'41"
Star Ray Star Ray Night
Gem 6h54'57" 18d55'42"
Star Reilly Nuckolls
Uma 11h14'58" 36d6'22"
Star Robinson
Uma 12h2'40" 34d15'42"
Star Ronaldus Ferrier
Ari 1h55'17" 18d41'46"
Star Rubio
Lib 15h2'46" -23d47'29"
Star Ryan
Ori 6h14'32" 6d57'16"
Star Ryan Forgue
Per 3h30'53" 40d57'34"
StaR Saleeby #1
Uma 12h22'45" 60d29'51"
STAR SARA-MIRJAM
Cep 1h42'20" 84d18'33"
Star Scercy
Uma 12h9'37" 54d12'32"
Star Schepis
And 1h7'13" 45d54'53"
Star Scout Tyler Collins
Uma 9h28'39" 44d6'34"
Star Searcher CM
Uma 13h15'33" 54d54'5"
Star Shaddock
Vir 13h56'34" -20d24'2"
Star Shadow
Uma 11h59'39" 58d20'29"
Star Shannon
Psc 0h50'23" 12d2'3"
Star Shannon
Vir 12h6'53" 11d51'3"
Star Sherry Lynn Toedtemeier
Aqr 23h42'44" -13d34'55"
Star Shirey
Cyg 20h31'7" 36d52'31"
Star Siddhesh
Uma 11h16'41" 47d38'24"
Star Sillins
Uma 11h39'6" 60d16'0"
Star Simonic
Cyg 20h53'28" 46d58'6"
"star sisters"
Vir 13h40'46" -5d54'56"
Star Skooter 'Corinne L. Graff'
Umi 14h14'24" 72d21'47"
Star Sophia
And 1h34'26" 48d11'0"
Star Sophia
Ori 5h54'46" 21d50'15"
star sophie
Cyg 19h33'59" 28d55'41"
Star Soprano N K R
Aqr 21h50'59" -0d24'15"
Star Spero
Lib 15h47'38" -16d13'39"
Star Spixee (Ruth Murphy Hay)
Lyn 7h25'0" 46d30'26"
Star Splendiferous Carol-Ann
Leo 10h52'30" 21d35'20"
Star Sumer Rae
Sco 17h20'1" -41d57'24"
Star Talon Skye
Ori 6h11'46" 3d21'1"
Star Tess
Sgr 18h21'32" -20d42'53"
Star Tony
Lib 14h37'34" -10d38'4"
Star Tracee Corene Dillard
Aqr 21h14'11" -10d10'2"
Star Trek Christopher K. 12/26/1964
Aqr 22h43'4" -7d54'16"
Star Tyler Christian Raring <3
Psc 0h28'25" 12d0'28"
Star VanBora Larry Teasdale I Lov M
Cap 20h32'34" -20d43'35"
Star Warrior Nicholaus A. Baldone
Sgr 19h16'51" -35d19'46"
Star Watkins
Sgr 19h49'23" -14d1'29"
- Star White -
Pyx 8h31'55" -36d12'45"
Star Wright Star Bright
Psc 23h6'43" 5d34'1"
Star Yvonne Mahoney
Gem 6h25'22" 25d27'6"
Star Zachary
Dra 18h7'54" 59d38'6"
Star Zee, A Heavenly Work Of Heart
Vir 14h11'39" -15d34'34"
Star Zoe
And 0h49'0" 25d29'1"
Star219—Alli's Star
Uma 11h44'40" 64d27'57"
Star4Lisa
Ari 2h4'21" 24d18'18"
Star4Sam
Cap 20h36'21" -18d38'45"
Stara
Tau 4h11'30" 6d2'51"
Starah
And 1h50'1" 46d47'22"
StarAiden
Vir 14h23'16" -1d31'28"
!Staranaid!
Uma 11h47'25" 42d12'3"
Star-Bernard 12août 1946
Ori 5h35'0" -8d32'55"
Starbird Lee
Psc 1h2'58" 27d13'10"
Starboy
Leo 10h11'40" 8d9'7"
StarBridge 052621-011305, R.I.B.
Per 4h13'38" 33d6'30"
Starbright
Tri 2h32'21" 33d58'52"
Starbright Gigi
Ari 3h7'53" 28d28'27"
Starbright Mel
Ari 3h22'13" 28d58'8"
Starbuck
Cma 6h38'20" -14d50'38"
Starburst
Psc 23h49'11" 7d31'0"
Starburst Sonia
Vir 12h48'34" 5d41'9"
Starchild Sara
Lyn 7h37'25" 54d13'20"
Star-Crossed Lovers!
Cyg 21h36'38" 39d20'37"
stardate 5943.7 All Our Yesterdays
Uma 11h34'58" 54d16'23"
Stardog
Dra 16h59'56" 64d35'38"
Stardust
Uma 9h25'5" 62d29'31"
Stardust
Psc 1h2'53" 10d32'52"
Stardust 918
Uma 12h32'57" 57d17'30"
Stardust Sprinkled Moon Pie
Peg 23h9'51" 16d52'8"
Stare at a Memory
And 22h59'20" 52d28'41"
Starfire Sports
Lyn 7h8'57" 48d37'21"
Starfish
And 23h31'26" 47d47'9"
Starfish
Her 17h41'20" 24d6'47"
StarFish FastLane
Dra 19h30'16" 60d5'4"
Starflores
Sco 16h51'3" -41d22'9"
Stargate
Eri 2h4'0" -56d31'26"
Stargate Lisa Lorraine#01-20-1969
Cas 1h43'50" 65d3'29"
Stargazer
Lib 15h28'50" -10d52'25"
Stargazer Lily
Car 8h26'41" -57d23'52"
Stargazer New Year MLP
Lyn 7h34'56" 38d19'38"
Stargazer Rick
Uma 13h21'3" 58d20'24"
STARGAZRJIM
And 1h40'36" 36d32'43"
STARGIRL 32961
Ari 2h33'48" 24d32'8"
Stargirl Larisa
Ann...Forever Loved
Sgr 17h58'56" -18d9'19"
Starguar De-Montre "Tré" Cochran
Lyr 18h54'43" 33d36'17"
StarHole Brendan James Haynes
Vir 12h24'35" 10d36'47"
stARI
Gem 7h45'24" 32d20'5"
Stario
Ori 5h35'28" 1d22'16"
Stario
Gem 7h22'46" 26d32'21"
Starissa
Vir 13h26'18" -0d46'11"
Starjiwoo
Gem 7h11'42" 26d7'57"
StarKart
Lyn 6h52'55" 52d9'12"
Starkey
Psc 0h35'15" 8d28'46"
Starkey Lee Tharrington, Jr.
Cyg 19h30'36" 29d54'25"
Starkey's Love
Ari 2h14'14" 11d17'38"
Starkman Family
Uma 13h50'40" 52d42'39"
Starkweather
Cyg 19h46'29" 29d33'19"
Starla
Ari 3h21'21" 28d56'11"
Starla
Ori 6h5'47" 17d34'20"
Starla
Vir 12h28'42" 8d26'3"
Starla
Lac 22h21'14" 46d32'46"
Starla
Dra 17h22'21" 51d21'5"
Starla
Crb 15h51'15" 37d25'45"
Starla Alice Houghton
And 23h16'56" 41d32'39"
Starla Angelique Stolk
Cnc 9h1'20" 24d0'4"
Starla Anne Calvert
Uma 11h9'24" 59d14'54"
Starla Cocio
And 0h40'45" 28d1'36"
Starla Jeanne Anderson
Psc 0h25'35" 10d30'31"
Starla Jones Jacobs
Cap 20h10'15" -13d28'53"
Starla Joy
Sgr 18h24'18" -23d4'51"
Starla Leeann
Crb 15h28'6" 29d18'10"
Starla Marie Brown
Peg 0h10'52" 17d48'51"
Starla McCauley
Tau 4h21'52" 27d16'11"
Starla Zemelis
Uma 10h49'51" 46d33'47"
starla1316
Gem 6h29'33" 25d18'33"
Starlantia
Cnc 8h16'52" 17d31'21"
Starla's Eyes
And 0h45'45" 40d8'5"
Starlee Jean Smith
Cap 21h54'47" -22d53'26"
Starlena Meeks
Aqr 23h1'26" -9d54'47"
Starlene
And 0h58'23" 41d23'15"
Starlene
Aqr 22h40'10" 0d53'13"
Starlet Setareh
And 23h9'56" 51d58'37"
Starlett Dawn De Graffenreid
Leo 9h25'14" 16d16'43"
Starlight
Her 18h5'25" 17d39'57"
Starlight
Tri 2h34'11" 35d43'17"
Starlight
Lyn 7h41'1" 54d10'28"
Starlight for Shanna
And 23h9'56" 35d26'15"
Starlight Lindsey
And 1h53'56" 44d15'33"
Starlight Madelynn Claudia
Psc 1h20'19" 10d32'7"
Starlight Sonata
Aqr 22h38'52" -0d42'1"
Starlight Starbright
Umi 17h27'3" 82d16'30"
Starlight Terri
Cas 0h50'15" 64d4'4"
Starlight Vineyard's Linda
And 2h30'15" 45d13'2"
Starling
Cyg 20h40'58" 45d45'59"
Starlite
Psc 0h33'10" 15d0'30"
Starlitious "2-20-87"
Uma 11h44'48" 43d36'37"
StarLiz531
Lyn 8h1'39" 41d57'41"
Starlone Thomas II
Leo 11h40'25" 24d54'2"
STARLOS
Her 16h36'59" 47d49'55"
Starlotte Charlotte
And 0h10'9" 44d28'33"
StarMala
And 0h55'24" 39d44'18"
Starman
Aur 5h30'8" 36d7'32"
Starman Andrew Michael Miller
Psc 1h22'48" 26d4'53"
Star-MoMo
Vir 13h7'23" 7d37'4"
Stärn
Cas 23h47'39" 51d53'31"
Starness
Sgr 18h48'31" -24d51'16"
Stärnli Céline
Cep 22h50'22" 66d50'55"
stARo
Uma 13h17'10" 58d3'14"
StarOfSueBrubaker
Uma 8h42'19" 55d41'21"
Starphanie
Sco 17h47'59" -42d31'32"
Starr
Aqr 21h50'34" -3d50'3"
Starr
Lyr 18h39'25" 34d23'26"
Starr Ann
Tau 5h5'42" 24d5'36"
Starr Bravo
Lib 14h57'3" -0d48'16"
Starr Collister
And 0h22'16" 38d32'19"
Starr Crabb
Uma 13h58'11" 61d47'53"
Starr & Jimmy
Uma 10h36'27" 61d35'2"
Starr Lee
Aur 5h38'4" 45d0'31"
Starr Luvly
Psc 1h24'16" 18d45'14"
Starr Lynn
Psc 1h6'11" 29d33'11"
Starr Lynn Shannon
Sgr 19h36'53" -15d19'9"
Starr M. Bradley
And 22h59'1" 50d53'43"
Starr Miller
Tau 4h25'48" 23d33'3"
Starr Mom
Cas 1h40'11" 61d12'52"
Starr Nolan
Cnc 8h30'23" 14d16'45"
Starr Of Karen
And 0h11'6" 47d55'58"
Starr of Minna
Uma 9h8'56" 67d57'15"
Starr Petree
Uma 10h44'37" 57d18'11"
Starr Powers
Uma 11h32'22" 63d11'7"
Starr R Bruce
Per 2h19'18" 57d7'29"
Starr Spangler
Leo 10h23'11" 18d37'52"
Starr Torres
Sgr 19h24'13" -19d8'14"
Starr Trapanese
Crb 15h17'32" 26d41'52"
StarRazor
Uma 9h23'5" 52d26'11"
Starring Alan Knieter!
Her 17h21'36" 23d17'3"
Starring Rich & Luanne
Tri 2h29'57" 31d23'25"

Starrissa
Aqr 20h46'6" -9d28'9"
Starrla Manista
Cap 21h30'55" -18d48'20"
Starrlene
Leo 10h2'40" 15d40'59"
StarrLynn - Forever & Always
Ori 6h11'27" 13d19'16"
StarrMom
Gem 7h12'51" 26d16'49"
Starr's Birthday Star
Uma 11h38'59" 42d14'26"
Starr's Star
Sgr 19h0'55" -18d54'35"
Starry
Uma 9h59'56" 58d29'50"
Starry
Lyn 7h53'19" 47d31'58"
Starry
Tau 3h33'58" 27d57'24"
Starry Night
Cyg 20h1'18" 58d58'23"
Starry Rodriguez
Ari 2h3'2" 19d20'19"
Starry269 (Amy)
And 23h37'47" 45d44'30"
Starry-Eyed Ranger 2/24/02
Cam 4h31'59" 66d41'13"
Starry-Eyed Surprise
Lib 15h33'41" -22d16'49"
Starrz2u
Psc 0h50'31" 9d43'12"
Stars and Love are Forever
Psc 1h19'49" 28d20'11"
Stars Are Falling, All For Us
Ori 5h4'26" 3d46'48"
Stars In Love
Cyg 21h40'39" 36d19'21"
Stars Never Fall
Cnc 8h51'35" 10d44'41"
Star's777Star
Tau 5h18'36" 18d38'56"
starscappy
Cnc 7h55'54" 19d21'8"
Starsha Ann Ten Elshof
Cnc 8h3'6" 13d30'5"
Starsha Ora
Umi 15h49'2" 73d29'4"
StarShawn
And 0h25'51" 37d48'12"
Starshine
Psc 1h18'50" 24d45'5"
StarshinE*
Aqr 21h16'14" -1d17'7"
StarShine:Celebrating Jim & MaryAnn
Cyg 20h29'29" 40d25'2"
"Starshine" Darnell Wedding Star
Cru 12h33'53" -57d36'20"
Starship McCarthy
Uma 11h51'54" 41d22'33"
Starship Sarah O'Brien
Cyg 20h49'56" 52d18'31"
Star-Skip Anubis
Ori 4h51'1" -0d20'41"
Starsky...Kevin Michael Wilmoth
Per 3h36'42" 51d13'10"
StarSteppers
Lyn 7h55'37" 51d15'26"
Starstruck
Mon 6h39'16" 5d22'7"
StarTEK 50
Tau 5h24'30" 18d11'25"
Starter 73
Umi 15h48'41" 84d53'31"
StarTex Title Company
Uma 9h39'53" 59d14'6"
StarTico
Uma 10h10'29" 51d33'46"
Starveya
Crb 15h35'11" 30d37'50"
Starward
Uma 11h57'33" 29d40'54"
Starwars
Dra 16h37'49" 60d28'18"
Stary-Stary Star 5/6/98
Crb 16h22'52" 32d44'6"
Stasch, Erich
Uma 10h50'44" 57d3'58"
Stash
Ori 4h49'5" -0d29'49"
Stash Bolc
Dra 18h44'33" 60d4'4"
Stasha
Cas 0h52'38" 64d13'3"
Stasha Price
Aql 19h52'48" 12d19'54"
Stasi
Lyn 7h41'14" 41d36'52"
Stasi Prandalos - A Shining Light
Cru 12h39'28" -59d18'12"
Stasia Lynn Dussault Shimkus
Uma 8h37'18" 66d12'30"
Stasia Marie
And 0h35'25" 45d10'27"
Stasia Skowron
Cyg 19h47'41" 54d20'31"
Stassia0706
Cas 23h21'12" 58d20'32"
Stathis
Cep 21h45'15" 63d56'28"
Static Electra
And 0h45'59" 25d3'53"
Statira
Uma 13h49'22" 52d17'58"
Stato Jo Zebro
Lmi 10h27'25" 38d16'56"
Staton Langston Awtrey
Ori 5h7'24" 3d19'12"
Statsy The Star
Cyg 19h51'21" 33d8'18"
Statua, spes, quod occupo dies
Cas 0h18'21" 56d40'25"
Stauber
Mon 6h49'27" -0d4'23"
Stäuble
Dra 18h17'53" 75d7'50"
Staudenmier & Cutler
Cyg 21h15'10" 53d30'27"
Stauske TM
Uma 9h55'14" 43d40'34"
STAV RON "8-30-93"
Aql 19h55'22" 16d11'9"
Stavros
Per 4h16'43" 51d15'7"
Stavros
Cap 20h32'50" -21d48'27"
Stavros
Sco 17h56'41" -31d20'44"
Stavros Chatziantoniou
Cam 4h13'9" 70d35'36"
Stavros Dapias
Aur 5h4'22" 46d57'29"
Stavros Demetrios Naltsas
Boo 14h34'26" 43d20'31"
Stavros Evripidou
Cep 22h12'56" 66d0'9"
Stavros George Christias
Uma 9h9'29" 70d44'29"
Stavros Georgia 4/24/1955
Vir 14h41'59" 4d27'16"
Stavros Ioannou
Peg 21h42'59" 15d53'48"
Stavros Tsibiridis
Lyr 18h50'2" 43d15'1"
Stavroula
Gem 6h26'43" 27d39'5"
Stavroula Ioannis Bouris
Gem 7h2'27" 35d1'54"
"Staying The Course" Blue Ocean II
Uma 11h45'54" 61d14'57"
Ste&Mik
Peg 21h27'53" 21d40'25"
Stea Elena
Sco 17h17'25" -40d35'16"
Steadies K
Tri 1h50'34" 29d19'7"
"Steady" Jason Rehmert
Per 3h14'23" 52d53'8"
Steafanno Anthony "Starfanno"
Cru 12h44'15" -58d29'11"
Steaua lui Ilie Retezatu
Lmi 9h57'19" 41d0'10"
stee
Peg 22h53'47" 25d55'23"
Steedman McKay Jenkins
Aqr 22h14'7" 1d41'46"
Steele
Aur 5h38'59" 33d43'15"
Steele
Cma 6h40'23" -13d10'11"
Steele Family
Dra 20h54'49" 80d57'21"
Steele Thornton Shapiro
Psc 0h9'9" -4d51'20"
Steeli Morgan Sellers
Mon 6h24'39" -3d54'48"
Steely
Uma 8h33'10" 71d56'42"
Steena & Jonner
Ori 6h9'23" 14d5'9"
Steenie
Ser 18h1'46" -10d30'13"
Steen's Star Alpha
Lyn 6h45'27" 53d44'5"
Steer, Alfred
Uma 13h12'54" 61d50'33"
Steeve Dahan (mon dada)
And 23h35'1" 41d29'9"
Steeve Dubois
Ori 5h21'31" 9d11'53"
Steeve Estatof
Sco 17h3'25" -38d22'9"
Stef
Ari 2h18'50" 19d50'37"
Stef & Bryan Friends Forever
Uma 10h40'40" 59d5'31"
Stef Forever
Lyr 18h52'37" 26d55'52"
Stefan
Ari 2h33'39" 19d51'6"
Stefan
Per 2h14'36" 55d59'24"
Stefan
Per 3h22'32" 41d0'3"
Stefan
Uma 10h11'26" 58d11'55"
Stefan
Uma 13h2'47" 53d6'30"
Stefan
Ori 5h48'24" -0d20'17"
Stefan Affeltranger
Cam 5h55'9" 71d14'52"
Stefan Alexandru Panin
Sgr 17h57'42" -16d58'48"
Stefan Anthony Miller
Cap 20h41'40" -24d23'52"
Stefan Aronfeld
Her 16h51'29" 37d44'31"
Stefan "auf Ewig"
And 0h15'31" 22d12'40"
Stefan Daniel Mohrmann
Psc 0h55'51" 29d25'10"
Stefan Dudas
Cam 6h53'26" 68d29'41"
Stefan Dzwonnik
Pav 19h28'21" -67d53'59"
Stefan Eberhard
Ori 5h59'34" 7d7'26"
Stefan Friedrich
Uma 8h23'50" 64d43'43"
Stefan Gabelt
Ori 6h16'57" -0d5'52"
Stefan Gieryn
Lyr 18h27'17" 29d29'57"
Stefan Giokas
Gem 7h43'21" 33d59'28"
Stefan Heinert - Familystar
Uma 9h2'28" 53d31'9"
Stefan Hermes
Uma 9h14'55" 57d42'3"
Stefan Hoj Jensen
Her 18h47'51" 19d1'26"
Stefan Homer Zachar Jr.
Vir 12h54'39" -8d57'23"
Stefan - in Liebe Deine Sandra
Uma 11h8'6" 54d38'36"
Stefan Jackson
Boo 15h15'47" 40d46'18"
Stefan Koch
Ori 5h19'38" 1d6'46"
Stefan Lænner
Lib 15h49'2" -13d26'33"
Stefan Lange
Uma 9h41'42" 72d40'31"
Stefan Laszlo-Lehni
Uma 8h24'1" 64d33'25"
Stefan Leuenberger
Aql 19h34'17" 6d8'21"
Stefan Lohse
Ori 5h56'18" 21d57'22"
Stefan & Milena
Ori 6h4'55" 7d8'12"
Stefan Petrov
Lyn 7h28'29" 45d43'5"
Stefan & Sabine
Com 13h6'19" 18d8'58"
Stefan Schulte
Uma 11h2'28" 58d43'39"
Stefan & Simone
Lib 14h44'26" -18d39'34"
Stefan Spiro Dedes
Cnc 8h47'57" 24d23'36"
Stefan Tatransky
Aur 5h22'49" 47d8'49"
Stefan Triemer
Uma 8h46'16" 60d27'45"
Stefan und Christiana
Uma 8h50'17" 72d38'47"
Stefan und Fränzi
Vul 19h29'30" 24d34'40"
Stefan und Michaela
Uma 13h35'29" 49d51'36"
Stefan Weckler
Ori 5h57'12" 2d16'43"
Stefan Weir
Aur 6h1'50" 38d23'16"
Stefan Wiese
Uma 9h15'42" 55d21'19"
Stefan Wiesmann
Lac 22h42'0" 53d54'16"
Stefan Wintermeyer
Uma 9h16'35" 54d31'48"
Stefan Wolf
Uma 9h24'31" 56d59'51"
Stefan Yurica
Aqr 22h41'1" -10d28'18"
Stefan Zimmermann
Uma 10h1'47" 59d43'13"
Stefani
Ori 5h5'41" 15d33'1"
Stefani Andersen
Umi 15h9'7" 79d47'43"
Stefani Gabriella Vail
Uma 10h28'51" 62d30'14"
Stefani L. Infante
Mon 7h21'36" -1d53'37"
Stefani Micheal Wilcox
Cap 21h45'52" -12d0'29"
Stefani Nicolette Griffith
Aqr 23h38'38" -14d48'52"
Stefani Renee
Com 13h6'43" 16d51'41"
Stefani Taylor Wren
Cas 0h51'35" 63d43'18"
Stefania
Lac 22h49'53" 53d44'2"
Stefania
Umi 14h4'23" 69d22'52"
Stefania
Lib 14h53'55" -1d3'42"
Stefania
And 1h36'14" 47d44'16"
Stefania
Lmi 10h42'20" 31d46'1"
Stefania 13/8/1965
Umi 17h42'51" 80d2'2"
Stefania Aionios
Lyn 8h47'17" 42d29'6"
Stefania Aversano
Tau 3h48'45" 29d23'36"
Stefania Cimino
Per 3h15'30" 33d56'16"
Stefania & Clayton Eve of Dec 03
Umi 15h23'57" 74d41'19"
Stefania Crucitti 1/5/1984
Uma 8h38'52" 58d1'46"
Stefania Milani
Cam 4h44'22" 59d44'5"
Stefania Mongelli
Cyg 19h40'24" 46d40'29"
Stefania Neculce
Ori 5h10'31" -0d26'3"
Stefania P. Kuehner
Vir 13h26'44" 11d10'1"
Stefania Palazzo
Uma 12h6'55" 44d40'4"
Stefania Pitondo
Her 18h37'55" 18d22'46"
Stefania & Roland
Uma 12h51'56" 56d13'13"
Stefania Scarfo
Lyn 8h41'41" 36d15'25"
Stefania Vallesi
Cas 2h42'58" 57d41'28"
Stefania Venneri
Tau 4h31'55" 4d59'38"
Stefania Zappulla
Tri 2h32'12" 31d24'39"
Stefanie
And 1h27'39" 39d59'43"
Stefanie
Tau 5h43'34" 15d56'48"
Stefanie
Ari 3h19'49" 18d51'17"
Stefanie
And 0h19'50" 25d38'56"
Stefanie
Lib 15h14'39" -7d14'5"
Stefanie
Lib 15h38'21" -9d42'0"
Stefanie
Lib 15h29'43" -23d19'38"
Stefanie A. Donegan
Leo 11h7'27" 2d32'47"
Stefanie Alissa Young
Lib 15h8'28" -3d37'49"
Stefanie Alynn
And 0h56'4" 39d17'57"
Stefanie Alynn
And 0h56'56" 46d5'5"
Stefanie Anawald
Tau 5h28'34" 27d33'26"
Stefanie Anne
Sco 16h21'9" -18d51'40"
Stefanie Anne Vinsel
Psc 1h25'30" 16d43'6"
Stefanie Arts
And 23h24'35" 51d11'16"
Stefanie Baker
And 23h28'50" 46d28'40"
Stefanie Casebier
And 1h40'14" 47d31'57"
Stefanie Danielle Schneider
And 0h42'2" 30d19'54"
Stefanie Dawn Strauss
Cyg 20h27'47" 34d15'22"
Stefanie DeAngelis
Gem 6h53'54" 33d36'30"
Stefanie Diane Schuff
And 0h29'22" 31d43'41"
Stefanie Elena Wright
And 23h24'37" 46d55'43"
Stefanie Erika Williams
Ari 1h51'56" 22d50'21"
Stefanie Eyer
Umi 15h42'40" 74d42'8"
Stefanie Fay
Uma 10h1'42" 49d20'47"
Stefanie Faye Layton
Gem 6h2'50" 24d26'44"
Stefanie Genza Joan Adolfsen
Cas 0h59'59" 76d54'7"
Stefanie Hallgren Ankerdal
Peg 23h22'34" 27d9'22"
Stefanie Hamid
Cnc 8h18'35" 22d14'11"
Stefanie in the Sky
And 23h8'32" 40d51'38"
Stefanie Jean Allen
And 0h41'42" 39d6'4"
Stefanie Jo Abercrombie
Vir 14h6'40" -16d46'13"
Stefanie Jolene Sherman
Gem 7h17'18" 31d39'6"
Stefanie Jon Kirkwood
Psc 23h44'26" 7d28'57"
Stefanie Kay
Ori 4h51'46" 8d10'16"
Stefanie Kovacic
Uma 9h15'1" 65d17'7"
Stefanie Landon
Gem 7h21'3" 29d28'4"
Stefanie Lehmann, 15.06.1977
Umi 13h50'59" 72d23'59"
Stefanie Little
And 1h12'1" 38d12'33"
Stefanie Loren Dussault
Uma 11h23'21" 43d34'18"
Stefanie Louise Kump
Cyg 19h42'30" 28d54'58"
Stefanie Lovallo
Cam 3h51'5" 59d35'30"
Stefanie Lynne Bogsch
Lib 14h56'22" -6d40'10"
Stefanie Marie Cecchi
Cnc 9h1'2" 14d35'23"
Stefanie Marie Rulis
Uma 10h54'49" 44d12'23"
Stefanie Mary Tammaro
Uma 9h15'45" 50d19'41"
Stefanie McCallum
Cas 0h59'51" 56d32'50"
Stefanie & Michael
Umi 13h38'13" 73d54'48"
Stefanie + Michael
Uma 8h54'15" 68d2'46"
Stefanie Miyuki Sugawara
Cap 21h9'46" -17d9'26"
Stefanie Müller
Sge 19h50'35" 16d49'56"
Stefanie Nicole Crawford
Cnc 8h47'18" 32d35'27"
Stefanie Rea Policki
Lyn 8h3'3" 40d28'38"
Stefanie Regina
Ari 3h20'20" 22d58'17"
Stefanie Rene Pickart
Ori 5h22'25" 0d45'13"
Stefanie Renee Builtron
Cnc 7h58'49" 16d45'28"
Stefanie Renee Mickels
And 2h21'4" 47d56'3"
Stefanie Richter
Uma 11h53'5" 30d59'20"
Stefanie Robinson
Vir 12h47'22" -1d0'23"
Stefanie Rose
And 1h22'15" 43d12'58"
Stefanie Sacha
Cnc 8h11'9" 10d29'3"
Stefanie Schneider
Cam 3h52'11" 70d21'2"
Stefanie Star
Ari 3h8'32" 27d16'8"
STEFANIE STAR
And 1h40'51" 38d50'27"
Stefanie Toffoletto
Uma 11h32'51" 58d8'30"
Stefanie Urciuoli
Leo 9h32'57" 27d54'3"
Stefanie Werner 31/07/1984
Del 20h54'45" 6d35'25"
Stefanie, 15.12.1978
Dra 17h35'37" 51d15'59"
Stefanie's Star
And 23h11'54" 48d25'7"
Stefanie's Star
And 2h7'4" 42d41'43"
Stefanie's Star
Aqr 22h24'6" -6d17'14"
Stefanie's Star
Sgr 19h29'30" -21d23'58"
Stefanie's Twinkling Star
Cas 1h33'14" 62d10'13"
StefanieV2005
And 1h14'45" 50d19'35"
Stefani's Christmas Star
Vir 13h19'5" 11d35'52"
Stefani's Passion
Cnc 8h52'15" 24d41'49"
Stefano
Lyr 18h44'17" 31d43'37"
Stefano
Lac 22h47'56" 52d58'36"
Stefano
Cas 22h58'56" 54d15'3"
Stefano
Ori 5h1'21" -0d13'18"
Stefano
Umi 15h49'40" 85d31'36"
Stefano
Umi 15h14'26" 84d1'12"
Stefano
Umi 15h49'20" 80d55'37"
Stefano
Per 3h0'14" 51d39'29"
Stefano Anthony
Per 2h46'32" 48d23'47"
Stefano Berra 9.02.1979 - 9.11.2005
Cru 12h13'14" -62d51'58"
Stefano Civati
Aur 4h53'22" 33d12'18"
Stefano Damian
And 1h44'20" 49d49'15"
Stefano Di Cesare
And 0h27'18" 36d8'6"
Stefano e Bianca
Cam 7h34'38" 63d59'41"
Stefano e Sara
Cas 23h11'5" 54d35'56"
Stefano Favot
Her 17h51'40" 23d38'9"
Stefano Fiore
Uma 10h58'40" 60d46'36"
Stefano Forti
Lac 22h13'10" 43d1'26"
Stefano Galastri
And 0h15'0" 47d1'12"
Stefano Moran-Guiati
Ari 2h20'13" 13d15'26"
Stefano Panicucci
Cam 11h54'53" 77d8'54"
Stefano Panno
Per 3h27'21" 34d9'35"
Stefano Zoia
Umi 15h47'40" 75d28'2"
Stefanos Panagiotis Stathis
Uma 10h36'58" 46d48'33"
Stefans & Martinas Hochzeitsstern
Ori 5h55'24" 21d42'47"
Stefans Radiant Hope
Lib 14h42'51" -20d29'45"
Stefan-Vassil Star
Ori 5h14'41" -0d37'45"
Stefany Bushey
Psc 22h59'32" 2d41'14"
Stefany & Friendy-I Love U Forever
Vir 11h44'35" 10d9'27"
Stefany L. Ortiz
And 0h51'45" 36d10'31"
SteFany Leui
Leo 10h5'53" 23d25'45"
SteFany Leui
Lib 14h55'14" -24d56'48"
Stefany Nicole Sargent
And 22h57'43" 52d2'13"
Stef-a-rama Lawlor
Cas 1h50'32" 61d23'58"
Stefeny Velasquez
Psc 0h46'29" 19d37'31"
Steff
Cep 3h39'24" 81d13'47"
Steff Wilson
Eri 4h8'9" -33d50'45"
Steffaery
Umi 18h39'56" 88d27'49"
Steffan Tucker's Shining Star
Uma 11h43'50" 45d33'49"
Steffan Wynn Thomas
Dra 18h27'50" 53d32'52"
Steffanie Ann Majors
Umi 15h12'12" 70d50'20"
Steffanie Chau
Crb 16h3'22" 37d14'2"
Steffanie Janelle Kotik
Psc 1h18'0" 24d35'23"
Steffanie Solitro
Cru 12h2'30" -60d53'41"
Steffany Elise Millage
Gem 6h2'47" 27d59'53"
Steffen and Muffin's Star
Umi 14h18'19" 67d35'30"
Steffen Eichhorn
Ori 6h14'13" 8d34'46"
Steffen Freund
Ori 5h11'51" -7d17'2"
Steffen Henssler
Lup 15h10'46" -40d17'17"
Steffen Karl Peabody
Ori 5h34'20" 5d18'13"
Steffen "La Hombre Grande" Krieg
Pho 0h36'11" -50d15'35"
Steffen Peter Josef Fröhling
Uma 8h45'21" 63d34'14"
Steffi
Ari 3h3'0" 18d28'20"
Steffi
Her 18h10'31" 28d44'22"
Steffi & Peter
Uma 8h39'37" 57d51'9"
STEFFIE 20.02
Uma 12h45'34" 55d52'14"
Steffie von Böer
Ori 4h59'46" 14d30'55"
Steffu
Lib 14h51'33" -0d39'48"
Steffy My Star
Per 3h5'18" 56d19'48"
Steffy Scott
Lib 14h30'47" -16d42'40"
Stefi
Her 18h18'22" 22d52'4"
STEFI I LIABA DI
Tri 1h47'38" 30d57'34"
Stefi's Star of Love
Sgr 18h3'48" -17d38'8"
Stefolas
Ari 2h1'31" 18d28'18"
Stef's brite lite
Aqr 22h38'42" -24d1'59"
Stef's Star
Uma 10h20'48" 45d11'46"
Stefunny
Cnc 9h6'5" 21d30'31"
Stefy
And 23h47'19" 43d23'50"
Stefy
Umi 17h43'33" 81d0'7"
Stefy e Ale
Uma 11h7'43" 42d47'1"
Stefy & Max: la nostra stella! :)
Cam 5h1'26" 69d26'35"
Stefy70
Ori 5h40'28" 1d56'16"
Steidl, Helmut
Uma 10h56'21" 43d39'25"
Steinbach Viktória (Cuni)
Sco 17h18'31" -43d25'26"
Steinel, Heinrich W.
Uma 13h13'57" 52d52'57"
Steinemann, Anton
Gem 7h32'7" 20d3'7"
Steiner
Aql 19h43'9" 6d0'26"
Steiner, Peter
Uma 12h54'27" 53d50'31"
Steinhart, Wilhelm
Ori 6h19'54" 16d47'4"
Steini's 2005er
Uma 9h32'32" 58d33'14"
Steinkamp, Jörn
Gem 7h30'55" 20d2'49"
Steka Ventie
Tau 5h57'32" 27d4'47"
stelachrishan
Sco 16h8'15" -19d8'32"
Stelasa
Tau 4h36'10" 16d30'36"
Stelios Sklavenitis
Uma 10h16'10" 49d58'10"
Stell of Nobuyuki Tanaka
Vir 13h14'17" -9d46'19"
Stella
Umi 14h22'28" 76d35'26"
Stella
Ori 6h14'26" -1d6'5"
Stella
Ori 5h39'28" -2d7'13"
Stella
Mon 6h59'37" -0d46'59"
Stella
Vir 11h44'5" -2d31'6"
Stella
Sco 16h13'24" -17d9'18"
Stella
Cap 21h51'38" -18d7'6"
Stella
Lib 15h4'56" -12d38'26"
Stella
Cas 23h18'0" 58d26'58"
Stella
Cas 0h54'28" 74d58'34"
Stella
Uma 14h2'20" 60d46'58"
Stella
Cas 0h19'52" 55d30'43"
Stella
Lac 22h45'57" 49d44'36"
Stella
Per 4h46'22" 43d52'35"
Stella
And 0h24'40" 42d46'10"
Stella
And 0h29'13" 42d0'39"
Stella
And 23h48'17" 35d24'36"
Stella
Lyn 7h37'54" 36d2'31"
Stella
Crb 16h7'16" 34d57'6"
Stella
Lyr 19h14'36" 33d48'30"
Stella
Lyr 18h38'17" 35d53'7"
stella
Tau 4h25'14" 18d58'25"
Stella
Ari 2h47'29" 16d41'29"
Stella
Tau 5h27'37" 20d27'46"
Stella
Gem 6h37'25" 12d28'55"
Stella
Crb 16h5'39" 26d45'40"
Stella
Leo 11h42'52" 26d10'10"
Stella
Gem 6h25'38" 19d38'15"
Stella
Leo 10h15'14" 21d16'6"
Stella
Psc 0h54'8" 24d51'44"
Stella
Ari 2h47'35" 25d18'14"
Stella 5
Cas 0h56'29" 56d42'44"
Stella Agnesis
Sco 16h17'38" -29d58'28"
Stella Alma Ea
And 1h50'21" 46d18'14"
Stella Amodeo
Ori 5h56'0" 19d59'53"
Stella Amor
Sco 16h19'29" -17d43'6"
Stella Amore
Cyg 20h22'50" 58d9'9"
Stella amoris
Boo 14h41'55" 37d24'2"
Stella Amy Pelin
Lib 15h40'55" -28d32'25"
Stella and Meg, Friends forever!
Cas 0h42'54" 50d53'36"
Stella Ann Mary Kennedy
Tau 4h54'16" 20d52'59"
Stella Ann Pries
Sco 17h45'20" -33d42'39"

Stella Ann Spindler
Tau 4h15'14" 27d43'39"
Stella (Anne Estelle) Rafalo
Leo 10h44'55" 18d15'40"
Stella Anne Mavis Patterson
Vir 13h58'18" -16d52'15"
Stella Armoris Dietmari A. Muelleri
Uma 9h58'11" 44d23'7"
Stella Arnot
Lyr 19h11'9" 41d30'55"
Stella Balzano
Sgr 18h11'5" -19d52'25"
Stella Bella
Lyn 8h13'11" 34d47'56"
Stella Bella
Ori 5h24'20" 2d31'12"
Stella Bella - Our Beautiful Star
Sco 16h33'24" -29d15'54"
Stella Bell'amore
And 0h33'51" 36d2'17"
Stella Bellissimo Rossi
Scl 0h13'20" -37d30'53"
Stella Bertha Theodosiou
Ari 2h54'19" 21d52'37"
Stella Blume Major
Tau 4h24'49" 29d0'33"
Stella Bommeris
Peg 21h46'14" 11d34'11"
Stella Brady
Cas 0h28'47" 54d50'31"
Stella Busch's 100th Birthday Star
Ori 5h18'20" 7d59'24"
Stella Bustos
Vir 13h27'25" -0d44'2"
Stella By Starlight
Gem 7h13'4" 34d21'51"
Stella By Starlight/A Mother's Love
Psc 0h54'46" 31d33'30"
Stella C. Walcker
Cnc 8h11'11" 21d18'29"
Stella Cacciatore
Cnc 8h14'16" 15d11'27"
Stella Cadente
Del 20h49'38" 9d42'56"
Stella Callum
Tau 5h3'46" 23d47'53"
Stella Candidi Signi
Sgr 17h54'51" -17d24'35"
Stella Cardile-Dimondo
Cru 12h17'41" -57d16'28"
Stella Christensen
Gem 7h54'52" 28d11'38"
Stella Cirigliano
Uma 11h44'7" 46d15'38"
Stella Cruz-Foy
Lyr 18h48'2" 31d10'20"
stella d'amore
Uma 10h55'10" 63d10'15"
Stella d'amore Nicole & Chris
Sct 18h25'31" -15d21'26"
Stella D'Amore Steffi & Markus
Uma 10h43'36" 59d59'36"
Stella danzante
Cas 0h26'11" 62d27'57"
Stella Dawn Rigby
Ori 5h24'40" 1d56'36"
Stella del Amy
Peg 22h9'14" 16d15'52"
stella di amore
Leo 9h36'17" 31d42'20"
Stella di Ashley's
Ori 6h17'40" 14d38'59"
Stella di Ella
And 1h31'10" 34d42'13"
Stella di Escott
Cma 6h51'32" -12d25'32"
Stella Di Jacque
Ari 2h48'11" 15d59'3"
Stella di Kobe e Anna per sempre
And 2h3'13" 38d36'49"
Stella di mare
Aur 6h23'21" 48d17'26"
Stella di Marsilio
Aqr 22h35'2" -23d18'53"
Stella di Megu
Sco 16h44'27" -27d42'8"
Stella Di Natalee
Cap 21h48'36" -14d35'42"
Stella di Papá
Cru 12h23'23" -60d53'29"
Stella di Romeo
Leo 9h42'7" 30d38'58"
Stella di Ryohei e Maki
Leo 9h44'8" 11d37'33"
Stella di Sale
Per 3h8'20" 40d57'21"
Stella Di Umberto Vittorio Comana
Per 3h45'13" 40d59'52"
Stella di Yuugo Morimoto
Gem 6h18'52" 24d51'27"
Stella Dio-Orso Pietra e Rotolare
Uma 9h0'45" 60d38'25"
Stella Dombrowski
Psc 1h31'6" 21d14'0"
Stella Duanea
Vir 12h14'16" 1d4'55"
Stella Ellina
Cas 23h39'34" 52d59'39"
Stella Emeritus SA
And 1h32'7" 49d17'14"
Stella Erin Kline
Lib 14h46'56" -17d1'4"
Stella Eternalis
Eri 4h33'24" -0d33'21"
Stella Ethel Phyllis
Cru 12h1'28" -61d11'28"
Stella Fitzgerald
Crb 15h46'13" 34d20'34"
Stella Floriana - Forever Shining
Col 6h33'46" -42d54'34"
Stella Fortunata Felix
Cap 20h41'14" -18d34'8"
Stella Georgina Keegan
Sgr 19h24'15" -15d49'44"
Stella Grace Reutimann
Umi 14h27'17" 67d40'12"
stella grace santos
Cnc 9h18'29" 15d23'30"
Stella Guida - Louise Hamilton
Pho 23h28'47" -49d11'38"
Stella Harris's Star
Ari 3h9'59" 17d52'50"
Stella I. Stoutenburgh
Aqr 22h39'5" 1d59'3"
Stella IV
Lib 15h34'4" -15d43'32"
Stella Jane
Sco 17h40'20" -41d57'59"
Stella Jemima Rose
Peg 21h39'54" 22d13'0"
Stella John Charles Hicks
Uma 10h14'17" 70d49'52"
Stella Judith Miner
Cas 0h7'51" 60d32'53"
Stella & Justin
Cyg 19h43'0" 53d31'2"
Stella Kana
Aqr 22h54'31" -22d59'49"
Stella Katerina Yost
Uma 12h42'59" 61d27'43"
Stella Kay Main
Gem 6h55'5" 19d14'58"
Stella Kayoko
Cap 20h25'51" -22d4'24"
Stella Kyoko
Psc 1h2'55" 23d9'25"
STELLA LAJOIE
Sgr 18h45'0" -28d4'42"
Stella Legare
Vir 13h36'45" -0d52'32"
Stella Leone
Cyg 21h19'8" 41d50'56"
STELLA LEONIS
Cep 21h44'5" 72d25'30"
STELLA LOUISE
Cyg 21h28'40" 34d45'10"
Stella Lucente
Cap 21h36'22" -17d9'58"
Stella Lucille
Uma 11h13'33" 57d40'9"
Stella Lucille Fava
Cas 0h37'18" 56d47'35"
Stella Luminosa
Uma 14h18'3" 56d24'28"
Stella Luminosa del TK
Ori 5h6'45" 7d55'24"
Stella Luna
Crb 15h50'7" 35d49'0"
Stella Lymberopoulou (Sterlisia)
Cap 20h12'23" -12d49'50"
Stella M.
Aqr 21h22'31" -0d44'46"
Stella M. Conrad's Shining Star
And 2h4'18" 42d2'0"
Stella M0928
Lib 14h48'3" -10d30'59"
Stella Mae Jastrebski
Psc 22h53'46" 4d15'41"
Stella Magnolia
And 23h21'6" 44d23'46"
Stella Mahoney
Cas 0h34'46" 47d32'16"
Stella Manos Greven
Cnc 8h30'2" 10d57'40"
Stella Marcelle Munoz
Umi 14h46'17" 73d43'20"
Stella Marcia Ostwald
Aqr 22h12'46" 0d21'9"
Stella Mares
Leo 10h3'20" 9d54'13"
Stella Maria
Cyg 19h55'23" 33d20'58"
Stella Maria Thurn
Uma 10h43'54" 43d46'17"
Stella Marie
Per 3h9'37" 54d21'25"
Stella Marie
Crb 15h54'25" 27d38'5"
Stella Marie
Cap 21h42'28" -12d24'20"
Stella Marie Kinsburg
Cnc 8h30'15" 12d36'11"
Stella Marie Moore Spliethof
And 23h19'0" 38d38'12"
Stella Marie Naleway
Ari 2h57'39" 26d10'8"
Stella Marie Stearman
Tau 4h19'0" 17d5'32"
Stella Maris 081598
Crb 15h50'24" 38d43'2"
Stella Maris Hardware
Uma 11h13'6" 58d10'46"
Stella Maris - Mary Bolling
Cru 12h36'12" -59d55'7"
Stella Mary Anne
Leo 11h31'26" 27d2'47"
Stella Maximianus
Sco 16h8'23" -28d50'28"
Stella Mayfield Granny Mae
Cas 1h10'18" 54d56'9"
Stella McCabe Cavara
Cnc 8h29'58" 30d36'15"
Stella Meena Lee
Vir 13h27'23" 6d41'40"
Stella Merjave - Beloved Mother
Tau 4h19'26" 28d26'15"
Stella Merschel
Gem 7h37'29" 23d0'44"
Stella Michel
Crb 15h48'37" 31d42'28"
Stella Michèle
Peg 23h0'50" 26d10'37"
Stella Michnik
Lyn 8h23'43" 52d1'18"
Stella Mimi
Cas 0h13'37" 57d38'54"
Stella 'MoB' The Stars' Fairy
Cyg 20h4'58" 37d30'37"
Stella Morag Hinshelwood - Stella's Star
And 2h14'19" 49d21'29"
Stella Mulas
Psc 22h58'31" 4d49'27"
Stella My Star Lite
Mon 8h3'53" -0d45'38"
Stella Namour
Gem 6h38'29" 27d17'48"
Stella Niagara
Lyn 7h56'27" 39d31'37"
Stella Notte Pollack
Gem 6h49'56" 26d43'7"
Stella of 1-17
Cap 20h37'25" -23d52'5"
Stella of AI Akira & Kaori
Cnc 8h50'57" 13d10'13"
Stella of Ai*ce
Ari 3h7'34" 30d27'43"
Stella of Ai Sagami
Cnc 8h26'53" 8d57'39"
Stella of Akane. Y
Sgr 18h20'4" -32d35'42"
Stella of Akari
Cap 20h11'21" -9d14'13"
Stella of AKI
Cnc 8h59'54" 14d31'48"
Stella of Aki
Sgr 19h28'10" 30d58'36"
Stella of Akifumi-Ishikawa
Lib 15h28'20" -24d0'36"
Stella of Akihiro & Tomomi
Psc 1h3'41" 23d12'58"
Stella of akinori & aihua
Lib 14h41'5" -13d14'8"
Stella of Akito Bandoh
Aqr 21h22'14" -12d25'0"
Stella of Ami
Sco 16h48'27" -26d55'26"
Stella of Annie
Sgr 19h20'5" -37d32'11"
Stella of Ayako Suwazono 1982 0310
Psc 0h23'0" 20d56'36"
Stella of Chiaki
Aqr 22h2'17" -5d54'18"
Stella of CHie Suzuki
Tau 3h48'23" 0d40'21"
Stella of Chieko Okamoto
Aqr 20h41'50" 1d55'46"
Stella of Chiharu
Psc 0h18'48" 6d38'53"
Stella of Chika Ura
Aqr 22h41'38" -6d14'20"
Stella of Chikao Yamashita
Cnc 8h10'20" 31d15'41"
Stella of cosmo venus Hiroki
Sgr 18h24'43" -32d53'46"
Stella of Daisuke & Hisako
Lib 15h40'19" -19d56'56"
Stella of daisuke & natsuko
Sco 17h28'50" -41d30'23"
Stella of Dora Katsuma
Vir 13h11'7" -8d56'22"
Stella of Eiko
Ari 2h37'28" 29d16'22"
Stella of Emi
Sco 16h8'5" -28d47'22"
Stella of Emiko Nakamura
Vir 14h9'35" -21d43'51"
Stella of Eriko
Vir 13h53'19" -14d45'15"
Stella of Eriko Higashi
Aqr 22h35'24" -12d35'25"
Stella of Go
Cap 20h29'46" -23d41'47"
Stella of Hanako
Ari 2h43'28" 16d58'9"
Stella of Happy Wedding Hiro & Meg
Lib 15h31'52" -21d38'19"
Stella of Haruta's
Vir 14h6'32" 0d44'54"
Stella of hatabou with Merit team
Cap 20h26'47" -8d32'44"
Stella of Hatsukaze
Sco 17h51'23" -43d32'30"
Stella of Hideaki & Ayako
Psc 23h55'3" -1d13'23"
Stella of Hidekito Yumi
Aqr 21h37'56" 1d39'58"
Stella of Hideo
Lib 15h27'22" -23d40'3"
Stella of Hideo Kawamura
Gem 7h39'27" 16d6'2"
Stella of Hiroe Mochida
Lib 14h46'2" -13d18'10"
Stella of Hiroki
Sgr 18h22'34" -32d56'31"
Stella of Hiroshi & Sumie
Lib 15h38'16" -15d13'39"
Stella of Hiroyuki & Rika
Cap 20h14'4" -9d14'59"
Stella of Hisako Miki
Sco 16h51'22" -38d10'31"
Stella of Hisashi Shinozuka
Psc 0h42'55" 4d8'5"
Stella of Hisen Amanokawa
Lib 15h29'28" -21d28'0"
Stella of Hitomi
Cap 20h20'47" -9d4'24"
Stella of Inspi-Atsushi Sugita
Vir 13h24'6" 2d54'45"
Stella of Iwane Family
Tau 5h31'20" 19d43'32"
Stella of JJ Asuka
Cap 20h31'30" -22d6'0"
Stella of JJ Tsukasa
Sgr 18h47'46" -20d37'21"
Stella of Jun & Natsumi
Lib 15h43'54" -16d13'46"
Stella of Junko
Aqr 21h26'35" -1d42'30"
Stella of Junpei Ohshio
Sco 16h14'8" -42d8'30"
Stella of Junzo Makino
Tau 3h37'11" 29d10'22"
Stella of Kana Nagase
Cap 20h31'25" -23d23'52"
Stella of Kanae
Vir 11h43'44" 1d7'45"
Stella of Kaori
Ari 3h7'2" 30d32'14"
Stella of Kaori
Vir 13h32'23" -5d40'30"
Stella of Kaori Morita
Cnc 8h4'1" 15d2'35"
Stella of Katsuki Takahashi
Leo 11h36'27" -3d10'13"
Stella of Katsuyuki & Kyoko
Psc 0h16'31" 12d34'20"
Stella of Kawori Yamamoto
Sgr 18h25'27" -26d47'56"
Stella of Kayoko Yamashita
Tau 4h51'3" 28d45'0"
Stella of Kazuhiko & Yuko
Gem 7h55'50" 21d3'24"
Stella of Kazuki
Cnc 9h7'20" 30d2'44"
Stella of Kazuki
Aqr 22h43'18" -5d50'35"
Stella of Kazuko Kobayashi
Cap 20h31'28" -27d22'11"
Stella of Kazuko Yamamoto
Sgr 20h25'46" -42d38'33"
Stella of Kazuna
Gem 6h42'26" 24d16'27"
Stella of Kazuo & Tomiko
Sco 16h53'29" -38d43'13"
Stella of Kazuyo Uenishi
Sco 17h54'56" -44d55'25"
Stella of Keiichi
Leo 11h31'22" -3d14'17"
Stella of Keiju
Sco 16h49'57" -27d35'12"
Stella of Keiko Tago
Lib 15h13'58" -13d58'52"
Stella of Keiko.S
Leo 10h32'17" 27d40'7"
Stella of Keita & Yuka
Tau 4h17'18" 26d14'3"
Stella of Kenichi & Rie
Leo 9h41'4" 9d25'16"
Stella of Kenji Kato & M
Cap 20h16'43" -16d12'0"
Stella of Kiyo
Cnc 8h46'2" 10d36'57"
Stella of Kiyonobu & Tomoko
Lib 15h27'24" -14d4'2"
Stella of Kiyotaka and Satoko
Gem 7h7'17" 24d40'23"
Stella of Ko
Sco 17h54'9" -44d40'4"
Stella of Koichiro & Kumiko
Tau 5h16'43" 17d7'28"
Stella of komazawa Family
Aqr 21h17'57" 1d32'48"
Stella of Kumiko
Aqr 23h25'3" -16d24'31"
Stella of Kunito & Yukiko
Lib 14h47'25" -19d12'27"
Stella of Kurumi. Aoi. Hanako
Leo 10h36'1" 16d55'55"
Stella of Kyoko
Leo 10h34'45" 17d28'47"
Stella of Licht
Cnc 8h10'7" 20d8'58"
Stella of Lupine
Ari 2h32'58" 14d9'45"
Stella of Makoto & Ai
Aqr 22h8'38" -10d46'28"
Stella of Manabu & Ikuko
Aqr 19h0'33" -34d21'1"
Stella of Manon
Vir 13h26'55" 3d20'59"
Stella of Masa & Rena
Sgr 19h18'39" -34d2'21"
Stella of Masanari Kojima
Tau 4h1'24" 3d16'2"
Stella of Masanori Kurosawa
Lib 14h38'10" -19d44'6"
Stella of Masao
Sgr 19h14'34" -35d17'38"
Stella of Masaru Arakawa
Aqr 23h48'55" -21d21'18"
Stella of Masaru & Yuki
Cap 20h13'33" -16d9'43"
Stella of Masashi Sato
Sgr 19h51'12" -12d6'30"
Stella of Masatoshi
Aqr 23h11'14" -24d30'9"
Stella of Masayasu & Kanako
Gem 6h48'33" 24d0'2"
Stella of Matsuda Family
Gem 6h49'44" 22d34'20"
Stella of Mayumi
Cap 20h25'41" -9d39'36"
Stella of Mayumi
Aqr 22h45'52" -5d53'33"
Stella of Michie Sato
Sco 17h58'5" -44d59'28"
Stella of Michio Sato
Sgr 20h8'13" -35d25'22"
Stella of Miho & Masahiro
Gem 7h54'24" 21d37'27"
Stella of Mika Nozaki
Ari 3h0'3" 30d23'54"
Stella of Mina
Vir 14h0'29" 1d11'29"
Stella of Misa
Gem 7h56'22" 21d20'53"
Stella of Misaki
Tau 5h36'3" 23d18'23"
Stella of Misao
Vir 13h34'52" -5d45'19"
Stella of Mitsuru & Akiko
Vir 13h31'1" -6d42'34"
Stella of Momoka
Sgr 18h43'15" -20d6'56"
Stella of Motoo Nakanishi
Ari 2h43'28" 15d21'45"
Stella of Motowo & Taeko
Aqr 21h22'37" -12d26'18"
Stella of Muneshige & Kazuko Yamamoto
Sgr 19h28'56" -30d58'48"
Stella of Mutsumi
Psc 1h18'46" 19d38'9"
Stella of Nahoe
Cnc 8h55'51" 25d47'24"
Stella of Naoki & Kaoru
Lib 15h42'37" -15d32'28"
Stella of Naomi Ono (Smile Star)
Ari 2h24'23" 17d10'50"
Stella of Naoyuki & Shinobu
Psc 0h48'5" 4d36'47"
Stella of Nasu Yuuji
Ari 2h40'18" 15d48'19"
Stella of Natsuki
Vir 13h32'17" 7d5'20"
Stella of Naturally Hidenori & Taeko
Vir 11h58'39" 10d43'4"
Stella of Niko Mitsumura
Tau 4h37'3" 14d30'17"
Stella of Nobuhiro & Eiko
Sco 16h44'40" -41d39'45"
Stella of Norihiko & Miki
Lib 15h4'46" -29d21'18"
Stella of Noriko & Eiji
Gem 7h13'54" 24d49'35"
Stella of Noriko Tanaka
Vir 13h11'13" -7d51'58"
Stella of Ogushi Family
Gem 7h10'12" 25d2'7"
Stella of Reiko & Masahide
Sco 16h52'15" -38d42'12"
Stella of Rena
Cap 20h27'31" -8d46'5"
Stella of Rika
Leo 9h50'8" 11d15'25"
Stella of Ruriko
Aqr 23h17'31" -16d8'50"
Stella of Ryo-chan
Leo 10h41'49" 21d57'36"
Stella of Ryouhei
Cap 20h7'30" -9d8'25"
Stella of Ryousuke Saurus
Sgr 18h21'13" -32d17'53"
Stella of Ryusei K
Tau 4h2'50" 4d0'9"
Stella of Sachi
Tau 5h15'7" 16d59'17"
Stella of Sae Shirasaki
Gem 6h49'7" 22d45'39"
Stella of Sakiko Arai
Psc 1h6'34" 5d10'2"
Stella of Saori & Hirokazu
Cnc 8h10'31" 21d57'33"
Stella of Satoru & Mayumi
Sgr 18h55'3" -34d44'22"
Stella of Seiji.S
Cap 20h26'51" -26d51'44"
Stella of Seishi Tsukioka
Cnc 8h8'0" 30d25'25"
Stella of Sena
Ari 2h36'24" 29d2'3"
Stella of Shigeo & Sachiko
Lib 15h15'49" -14d2'28"
Stella of Shiho Atsuta
Sgr 19h25'7" -30d57'58"
Stella of Shingoro Sato
Psc 1h44'26" 26d39'4"
Stella of Shinich Chiaki
Aqr 22h46'50" -5d9'59"
Stella of Shion Maeda
Vir 13h29'3" -1d0'35"
Stella of Shoji
Leo 9h46'46" 26d20'34"
Stella of Shouchan
Gem 6h17'0" 25d30'5"
Stella of Shu & Hiroko
Gem 7h3'43" 33d14'57"
Stella of Shuji & Mayumi
Tau 4h29'14" 24d15'16"
Stella of Sog Yoona
Gem 7h9'53" 25d14'5"
Stella of T & M
Ari 3h11'38" 27d22'17"
Stella of Taichi Utani
Leo 9h35'49" 21d15'11"
Stella of Taichirou
Cnc 9h21'2" 23d59'44"
Stella of Taiji Kobayashi
Sgr 18h24'16" -32d38'4"
Stella of Taiki Hase
Sco 16h56'12" -37d55'38"
Stella of Takayoshi Tosaka
Cnc 8h1'7" 15d1'28"
Stella of Take & Miwa
Leo 9h44'3" 11d14'24"
Stella of Takehiro Fukui
Aqr 20h51'21" -4d28'23"
Stella of Takeshi Doi
Ari 3h1'53" 30d6'45"
Stella of Tamae Yamasaki
Tau 3h43'58" 0d50'53"
Stella of Taro & Saya
Ari 2h33'10" 13d17'5"
Stella of Tashiro Family
Sco 16h47'34" -41d26'29"
Stella of Teruaki Ueno
Sgr 19h18'6" -34d8'33"
Stella of Teruo Kuroda
Aqr 21h26'34" -12d33'30"
Stella of Tetsuo
Cap 20h28'31" -9d59'41"
Stella of Tetsuo Nakagawa
Lib 15h31'38" -20d46'58"
Stella of Tetsuya Nakamura
Cnc 8h31'2" 8d36'12"
Stella of The Yellow Monkey
Cap 20h31'4" -26d51'32"
Stella of Tomoaki & Chikako
Ari 2h27'15" 13d18'43"
Stella of Tomoko
Leo 10h30'11" 16d55'14"
Stella of Toshie & Yuji
Vir 11h49'17" 0d32'52"
Stella of Toshihiko & Emiko Yoshimi
Sco 17h29'56" -42d54'57"
Stella of Toshiki
Leo 9h33'50" 21d40'55"
Stella of Toshinori
Gem 7h58'29" 21d7'44"
Stella of Toshio & Sachiyo
Cnc 8h3'40" 18d39'20"
Stella of Toshiro
Gem 7h59'20" 21d33'56"
Stella of Yasuhiro Doi
Cnc 8h17'18" 12d0'32"
Stella of Yasuo &Kimiko
Gem 7h27'38" 27d54'10"
Stella of Yoshie Okuhara
Psc 1h37'53" 17d7'12"
Stella of Yoshiko Fujie
Leo 11h24'31" 28d17'18"
Stella of Yoshinori Torii
Ari 2h44'43" 21d2'16"
Stella of Yoshio & Yuka
Leo 11h8'47" 12d57'5"
Stella of Yui
Leo 11h11'48" 27d47'23"
Stella of Yuichiro Ryuzaki
Cnc 8h17'43" 14d8'20"
Stella of Yuka
Leo 10h57'59" 3d28'9"
Stella of Yuka Takahashi
Gem 6h23'34" 24d40'10"
Stella of Yuki
Gem 6h45'50" 22d36'11"
Stella of Yuki
Vir 13h26'20" -1d57'14"
Stella of Yukino
Sgr 18h14'19" -19d46'21"
Stella of Yukito
Psc 23h59'59" 2d50'34"
Stella of Yuli Love Angel
Cap 20h27'32" -8d36'33"
Stella of Yumiko
Ari 3h5'22" 30d28'6"
Stella of Yuto
Lib 15h19'23" -13d38'50"
Stella of Yuuma
Aqr 22h3'38" -7d12'56"
Stella of Yuzo Oshima
Psc 1h33'37" 25d10'34"
Stella Oglapius
Ori 6h5'26" 18d9'54"
Stella - our shining star
Car 10h7'15" -62d55'20"
Stella Pacaj
And 0h35'17" 27d8'59"
Stella Paige Melucci
Aqr 23h4'15" -10d12'55"
Stella Pannazzo
Gem 7h33'46" 34d9'5"
Stella Parker Reynaga
Com 12h38'24" 28d22'30"
Stella Pauline Kogan
Uma 9h53'59" 60d22'24"
Stella Pearl
And 0h4'44" 40d35'44"
Stella Pelissari
Psc 1h30'31" 18d29'21"
Stella Perfetta
Ori 5h52'49" 6d40'38"
Stella Petruzzi
Gem 6h43'13" 15d35'47"
Stella Piccioli
Crb 16h1'26" 27d25'43"
Stella Pie
Uma 10h34'52" 39d42'22"
Stella Rivera
Com 12h49'7" 28d16'43"
Stella Romina
Sge 19h52'27" 17d9'36"
Stella Rosa-Maria Rasera
Uma 9h47'46" 54d8'13"
Stella Rose
Uma 11h33'23" 57d29'35"
Stella Rose
Cnc 8h14'15" 6d43'15"
Stella Ruri
Sco 16h57'49" -37d40'12"
Stella Russell
And 0h44'21" 36d24'33"
Stella Sajiun
And 1h25'4" 48d29'56"
Stella Sardo Faicco
Peg 22h8'16" 33d22'4"
Stella Sewell
Cas 23h55'57" 54d17'10"
Stella Shu
Sco 16h59'6" -37d54'15"
Stella Simonetti
Uma 8h14'38" 59d51'52"
Stella Sira
And 0h10'50" 45d17'41"
Stella Sofonio
Tau 4h14'17" 27d21'57"
Stella "Starr" Hochbaum
Cap 20h13'22" -18d10'29"
stella STRUNZILLOOOO STRUNZILLAAAAA
Gem 6h43'52" 15d17'40"
Stella T
Crb 15h41'12" 27d29'41"
Stella Terese Tschach
Cas 0h2'12" 59d56'56"
Stella the Wonderful Dog
Cma 7h17'24" -30d30'22"
Stella Theresia
Cas 1h14'41" 53d1'51"
Stella Trüby
Lyn 6h45'50" 58d9'23"
Stella Tutera
Uma 12h0'31" 31d56'51"
Stella Una
Cas 23h19'51" 55d6'32"
Stella Vanegas
Com 13h10'18" 28d46'55"
Stella Victoria
Lib 15h14'31" -9d44'4"
Stella Virginia Wrubel
Cru 12h48'53" -56d6'32"
Stella Weaber
And 0h19'29" 37d57'56"
Stella Werchowicki Luff
Gem 7h13'22" 33d38'27"
Stella & William Donovan
Cyg 19h39'29" 29d15'26"
Stella Zayas
Cam 8h35'44" 74d59'31"
Stella Zio Rolando
Crb 15h28'50" 29d58'1"
Stella Zoe
Cnc 8h55'45" 16d0'32"
Stella Zoe
Cet 1h23'30" -0d11'23"
Stella, Our Star
Ari 2h18'23" 24d17'22"

Stellae Alexandria Angelus
Lyn 7h42'27" 48d49'31"
StellaElaineCullity
Tau 3h49'16" 26d2'11"
StellaLou
And 1h5'29" 37d29'29"
Stellaluna
Vir 14h42'57" -5d29'14"
Stellamadre
Umi 15h47'48" 84d15'49"
Stellamber
Cyg 21h13'9" 41d21'43"
stellamona
Aqr 23h12'6" -9d29'35"
Stellar
Uma 9h29'57" 68d38'50"
Stellar
Sco 16h34'18" -27d31'15"
Stellar
Uma 10h31'6" 47d59'56"
Stellar
Cyg 21h33'6" 30d33'33"
Stellar
Gem 7h11'8" 17d14'21"
Stellar
Aqr 22h10'44" 0d58'40"
Stellar
Gem 7h53'26" 13d37'20"
Stellar Abrahan
And 1h3'24" 38d26'10"
" Stellar Enigma " 4 Dustin Findley
Per 2h53'36" 41d41'56"
Stellar Helena
Sgr 18h7'48" -30d18'53"
Stellar Leader Karen
Gem 7h44'15" 34d8'21"
Stellar Light
Crb 15h34'43" 32d23'26"
Stellar Michele
Cas 0h1'14" 54d9'29"
Stellar Mom - Julie Kay Boren
Vir 13h12'55" -2d9'47"
Stellar Oso
Lib 15h50'15" -18d56'13"
Stellar Parents
Cyg 20h16'49" 44d21'18"
SteLLaR RosE
Uma 12h5'58" 45d55'50"
Stellar Silvie Majova
Cas 23h32'0" 55d32'52"
Stellar Sprintz
Uma 10h21'54" 43d46'2"
Stellar Stella
Lyn 8h47'41" 40d28'20"
Stellar Stella
Psc 1h9'55" 24d4'56"
Stellar Summer Larsen
Gem 6h21'15" 27d50'57"
Stellar VanKirk
Cam 4h1'8" 71d32'37"
Stellar Wife Amanda Troeger
Uma 11h11'8" 46d0'9"
Stellar Xeller
Cam 4h23'3" 69d37'15"
StellaRae11032006
Sco 17h40'40" -41d26'1"
Stellaris di Bruno La Terra
Ori 5h40'47" -1d15'51"
Stellaris Diane Vera
Psa 23h0'13" -34d33'24"
Stella's Christmas
Sco 16h11'24" -12d16'47"
Stella's Sparkling Star of Dreams
And 2h17'36" 41d27'22"
Stella's Star
Cas 1h23'54" 56d54'38"
Stella's Star
Cyg 21h43'52" 43d55'43"
Stella's Star
Leo 9h22'46" 24d45'30"
Stella's Star
Cmi 7h36'58" 9d48'23"
Stella's Star
Cap 21h50'29" -23d34'0"
Stellatus Regina Stevie Jean
Cap 21h34'7" -20d18'7"
Stellina
Oph 17h13'11" 4d17'29"
Stellina
Per 2h23'30" 53d41'44"
Stellina Katiuscia
Cep 22h8'9" 56d35'48"
Stellini Sabrina
Umi 15h27'9" 67d42'49"
Stellystar
Vir 12h10'48" 12d4'33"
Stelna
Uma 10h33'31" 44d37'20"
Steluta
Gem 7h3'7" 16d40'4"
Stempel, Elli
Uma 11h44'51" 28d18'56"
Sten & Savi's Forever Star
Umi 14h27'46" 85d48'10"
Sten Wilhelmus Johannes Theodorus
Lib 14h57'57" -19d56'50"
Stender, Jacob
Uma 10h54'22" 33d39'9"
Stenners Star
Psc 0h40'20" 19d27'5"
Steno Babe
Cap 20h37'26" -10d34'47"
Stenzel, Uwe
Ori 6h9'16" 6d54'0"
Step
Umi 13h48'15" 70d36'23"
Step1
Lmi 10h24'54" 36d29'13"
Stepam
Umi 14h19'25" 70d27'41"
Stepan Voevudski
Cap 21h45'56" -12d18'20"
Stepan Zimin 10/6/03
Uma 11h50'16" 31d21'26"
Stepanida Bleicher
Uma 8h23'46" 62d36'14"
Steph
Dra 16h56'13" 57d15'54"
Steph
Cap 21h13'59" -25d13'18"
Steph
Sco 17h56'49" -31d14'55"
Steph
Gem 7h4'41" 33d50'42"
Steph
Cyg 21h27'34" 39d52'18"
Steph 4ever
Uma 10h57'43" 62d19'6"
Steph and Alan Morrison
Vir 14h9'54" 4d36'10"
Steph and Ian
Cyg 21h41'29" 31d39'29"
Steph and Jenny
Lyr 19h6'45" 32d8'59"
Steph and Matt
Uma 9h30'38" 67d42'6"
Steph and Steve
Sgr 19h25'14" -12d50'36"
Steph Anderson's Star!
Equ 21h8'12" 6d13'1"
Steph Bannon
And 1h19'1" 38d58'51"
Steph & Craig
Cyg 20h17'50" 43d42'26"
Steph Domske NMB
Sgr 18h32'16" -17d25'40"
Steph Fitzpatrick's Star
Sge 20h9'10" 16d41'32"
STEPH LEGAULT
Lib 15h5'44" -27d10'50"
Steph Leonard
Lyr 19h13'45" 26d33'5"
Steph & Rach
Cyg 21h11'4" 38d41'47"
Steph Schwartz
And 2h6'10" 46d47'48"
Stephaine Gail
And 2h37'4" 49d16'33"
Stephaine Recart
Ari 2h36'52" 26d15'33"
Stephan Abrami
Cas 3h2'1" 72d23'21"
Stephan Adam Dyson
Gem 6h47'21" 15d21'29"
Stephan Adamsky
Per 2h19'11" 54d25'28"
Stephan Alexander Friedrich
Uma 8h51'48" 49d2'47"
Stephan David Barnes
Per 3h47'40" 41d42'34"
Stephan DeVita "Mr. D"
Psc 0h57'20" 14d41'4"
Stephan Dietze
Uma 11h24'29" 39d46'21"
Stephan Eicher
Lyr 18h27'40" 32d31'22"
Stephan Eugster
Aqr 22h29'8" -0d7'44"
Stephan Haag
Tau 4h3'36" 15d33'25"
Stephan J. and Patricia M. Voller
Lyn 6h40'31" 57d21'54"
Stephan Kinzl
Cnv 12h29'36" 31d57'6"
Stephan Krampitz
Uma 8h48'49" 68d41'13"
Stephan Küppers
Uma 9h17'24" 55d14'56"
Stephan Martelly
Sgr 20h10'37" -39d6'15"
Stephan my love
Cas 23h35'52" 51d2'0"
Stephan & Nichole Milik
Lyr 19h12'12" 27d11'20"
Stephan Pahlitzsch
Ori 5h55'58" 17d16'13"
Stephan Ray Jaskula
Umi 14h45'8" 72d55'55"
Stephan Schad
Uma 10h22'41" 41d24'33"
Stephan Schneider
Per 4h12'20" 45d2'38"
Stephan von "Sunny"
Uma 9h12'24" 70d15'23"
Stephan, Ulrich
Uma 8h58'52" 50d50'50"
Stéphane
Tau 4h0'47" 26d1'52"
Stéphane
Cap 21h42'56" -13d58'27"
Stéphane
Aqr 23h49'2" -11d33'30"
Stéphane «b' d' Amour» Picard
Sgr 18h36'33" -23d56'26"
Stéphane Bilodeau
Ori 5h52'0" 9d8'43"
Stéphane Canals
Cap 21h32'41" -22d34'53"
Stéphane Cherki
Aqr 23h5'11" -9d30'17"
Stéphane Comeau
Crb 15h51'13" 31d10'15"
Stéphane et Héléana
Dra 19h54'39" 62d11'47"
Stephane Filmyer
Tri 2h40'30" 34d47'32"
Stéphane Lambiel
Her 18h34'50" 20d14'3"
Stéphane Martin
Uma 11h30'0" 60d9'42"
Stéphane Mercuri
Uma 9h58'10" 54d16'19"
Stéphane Miron
Uma 8h34'5" 64d24'52"
Stephane Pilon
Umi 16h39'2" 83d17'9"
Stéphane Yamina 1041992
Uma 8h53'1" 59d13'47"
Stephanee (My love forever burns)
Tau 4h30'47" 27d30'4"
Stephanee Sheree
Aqr 23h14'15" -14d0'27"
Stephanese
Cap 20h34'16" -11d14'5"
Stephani
Tau 5h34'26" 26d15'2"
Stephani Jewel Tuck
Leo 9h42'46" 22d11'58"
Stephani Marie
Lib 15h16'7" -22d18'56"
Stephani "Sunshine" Springer
Uma 9h27'4" 53d7'0"
Stephania
Sco 17h44'42" -42d10'37"
Stephanie
Sgr 19h50'2" -36d31'56"
Stephanie
Vel 9h0'50" -39d17'24"
Stephanie
Sco 16h55'20" -40d45'24"
Stephanie
Aqr 22h33'11" -22d49'49"
Stephanie
Sgr 18h19'28" -23d20'17"
Stephanie
Uma 9h36'5" 59d0'50"
Stephanie
Lyn 7h33'36" 58d53'2"
Stephanie
Dra 17h17'35" 58d8'46"
Stephanie
Cas 0h31'25" 61d16'37"
Stephanie
Cas 23h41'21" 55d44'4"
Stephanie
Cas 0h48'12" 64d20'42"
Stephanie
Cas 1h56'32" 63d48'36"
Stephanie
Cas 3h15'14" 64d21'29"
Stephanie
Uma 8h13'45" 70d3'0"
Stephanie
Lib 15h30'15" -15d16'42"
Stephanie
Lib 15h1'13" -17d34'55"
Stephanie
Lib 15h34'47" -22d7'49"
Stephanie
Sco 16h15'7" -17d52'57"
Stephanie
Cap 21h29'28" -17d1'0"
Stephanie
Sco 16h16'52" -9d42'55"
Stephanie
Sco 16h4'4" -14d42'14"
Stephanie
Lib 14h59'41" -4d43'3"
Stephanie
Leo 9h25'16" 21d54'23"
Stephanie
Gem 7h22'6" 16d4'5"
Stephanie
Gem 6h58'27" 17d54'40"
Stephanie
Del 20h59'36" 15d12'50"
Stephanie
Ari 2h18'12" 23d9'25"
Stephanie
Psc 1h15'29" 24d0'44"
Stephanie
Psc 1h4'14" 24d0'25"
Stephanie
Gem 7h30'16" 26d18'25"
Stephanie
Gem 7h49'39" 29d58'26"
Stephanie
Gem 6h56'52" 27d41'9"
Stephanie
Gem 7h10'32" 23d34'3"
Stephanie
And 0h29'46" 32d51'42"
Stephanie
Leo 11h34'5" 24d22'22"
Stephanie
Aqr 21h22'30" 2d7'3"
Stephanie
Vir 14h22'7" 4d40'35"
Stephanie
Vir 14h31'17" 2d13'34"
Stephanie
Her 16h42'30" 7d25'55"
Stephanie
Vir 12h42'39" 3d3'48"
Stephanie
Vir 12h50'47" 1d48'37"
Stephanie
Ori 5h56'52" 6d58'12"
Stephanie
Ori 5h25'26" 3d54'59"
Stephanie
Ari 2h22'35" 18d52'12"
Stephanie
Gem 6h28'37" 13d31'18"
Stephanie
Ori 6h17'57" 8d55'59"
Stephanie
And 1h4'12" 35d3'22"
Stephanie
Psc 0h53'37" 31d17'54"
Stephanie
Lyn 7h55'35" 34d4'48"
Stephanie
Gem 7h43'20" 32d2'25"
Stephanie
Cnc 8h43'11" 31d21'49"
Stephanie
Crb 15h46'14" 34d0'17"
Stephanie
Uma 11h16'14" 34d30'40"
Stephanie
And 0h16'15" 44d40'8"
Stephanie
Cyg 21h11'25" 31d14'37"
Stephanie
Uma 10h47'33" 47d36'5"
Stéphanie
Boo 15h20'5" 46d49'3"
Stephanie
And 23h33'44" 47d58'5"
Stephanie
And 23h14'44" 45d38'50"
Stephanie
Cas 1h27'10" 59d23'32"
Stephanie
Cas 0h50'27" 56d26'11"
Stephanie
Crb 15h52'7" 39d23'19"
Stephanie
And 23h37'2" 39d0'42"
Stephanie A. Kimball
Psc 0h17'11" 19d29'48"
Stephanie A. LaMarine
And 23h20'43" 41d27'6"
Stephanie A. Straface
Cap 20h40'56" -24d39'17"
Stephanie A. Young
And 23h53'39" 32d57'22"
Stephanie Abbett
Ari 2h25'14" 26d47'10"
Stephanie Acevedo
Tau 4h31'19" 17d37'18"
Stephanie Actipis
Com 12h41'48" 17d7'47"
Stephanie Adams
Her 18h23'6" 18d14'46"
Stephanie Adams
Sgr 18h17'19" -25d45'17"
Stephanie Aguilar
And 0h21'14" 43d30'50"
Stephanie Aho
Uma 11h28'35" 60d31'57"
Stephanie Aie-Ting Wong
Del 20h36'0" 18d1'5"
Stephanie Alechman
Uma 9h14'2" 54d4'29"
Stephanie Alexandria Walker
And 23h12'15" 41d30'22"
Stephanie Almond
Tau 4h26'53" 11d27'37"
Stephanie Altieri Hancewicz
Cas 0h27'45" 65d7'15"
Stephanie Amanda
Ori 6h21'16" 10d18'4"
Stephanie Amanda Therese
Aqr 23h39'27" -24d33'8"
Stephanie and Alex
Leo 11h6'22" 17d11'38"
Stephanie and Daniel's
Cyg 19h37'52" 53d40'24"
Stephanie and Darren
Psc 1h7'28" 27d31'43"
Stephanie and Devlin Scofield
Cyg 21h30'26" 53d51'34"
Stephanie and James Star
Cyg 20h11'9" 38d36'9"
Stephanie and Jeff's Forever Star
Aqr 22h36'6" -16d28'27"
Stephanie and Jeremy
Vir 13h59'17" -6d16'9"
Stephanie and Kristy
Lyn 8h28'44" 40d28'11"
Stephanie and Ray, Together Forever
Cap 21h34'18" -9d45'21"
Stephanie and Scott Buczek
Lib 15h5'29" -3d3'10"
Stephanie and T.J.
Sge 19h14'22" 19d31'4"
Stephanie Anderson
Sco 16h10'51" -11d12'32"
Stephanie "Angel" Denkins
Lib 15h45'59" -28d13'52"
Stephanie - Angel On Earth
Cap 20h37'33" -9d14'9"
Stephanie Ann
Leo 11h13'16" -3d40'41"
Stephanie Ann
Lyn 8h16'11" 57d41'19"
Stephanie Ann
Cnc 9h13'13" 15d55'3"
Stephanie Ann
Gem 7h2'53" 21d11'10"
Stephanie Ann
Ari 3h4'40" 12d21'36"
Stephanie & Ann
Gem 7h13'50" 34d29'45"
Stephanie Ann
And 0h8'55" 45d23'0"
Stephanie Ann
Uma 10h1'8" 51d50'45"
Stephanie Ann 20
Gem 7h49'25" 15d36'2"
Stephanie Ann Balogh
Ori 5h21'37" 1d44'11"
Stephanie Ann Dearest
Cap 21h33'1" -15d41'47"
Stephanie Ann Doran
Peg 22h24'0" 33d49'58"
Stephanie Ann Eppley
Ari 2h27'57" 19d46'9"
Stephanie Ann Freed
Cas 1h40'28" 63d5'54"
Stephanie Ann Louie
Ori 5h59'6" 21d14'57"
Stephanie Ann Manemann
Uma 13h50'24" 60d30'42"
Stephanie Ann Manriquez
Sco 15h53'33" -20d55'31"
Stephanie Ann McMillan
Sgr 18h53'19" -33d5'3"
Stephanie Ann Muller
Boo 15h36'42" 46d46'50"
Stephanie Ann Reynolds
Aqr 21h57'7" 0d6'27"
Stephanie Ann Roman
Sco 17h37'1" -41d56'50"
Stephanie Ann Smith
Sgr 17h58'58" -16d59'18"
Stephanie Ann Toscano
Lib 14h49'18" -1d44'35"
Stephanie Ann Vowles
Lib 15h51'39" -12d32'17"
Stephanie Anne
Sco 16h8'6" -10d31'7"
Stephanie Anne
Vir 13h20'23" -20d3'44"
Stephanie Anne
Ori 6h5'13" 13d53'33"
Stephanie Anne
Ori 6h9'15" 14d48'21"
Stephanie Anne < <3
Uma 11h31'54" 49d42'6"
Stephanie Anne Belanger
Sgr 19h42'19" -16d54'8"
Stephanie Anne Bireley
Leo 9h32'51" 28d22'31"
Stephanie Anne Butori
Cyg 21h41'21" 49d40'27"
Stephanie Anne (Fairyprincess)
Gem 7h35'24" 34d6'44"
Stephanie Anne Houlden
And 1h23'44" 41d18'21"
Stephanie Anne Ingram
Lyr 18h29'51" 39d1'51"
Stephanie Anne Marie Lowther
Uma 11h47'28" 40d58'8"
Stephanie Anne Thurber
Leo 10h52'40" 7d10'37"
Stephanie Anne Wacik
Ori 6h7'14" 19d10'32"
Stephanie Anne Yoxen
Vir 12h2'59" -1d16'43"
Stephanie Annette
Tau 4h37'3" 5d13'4"
Stephanie Antoinette
Aqr 22h31'45" -2d7'7"
Stephanie Antonuccio
Cyg 21h9'23" 53d29'26"
Stephanie Apple Chozick
And 1h32'43" 49d34'34"
Stephanie Apter
Ari 2h6'57" 22d53'34"
Stephanie Arthur
And 0h13'0" 32d54'46"
Stephanie Autumn Wohrle
Lib 14h45'25" -15d1'31"
Stephanie B 18
And 1h43'53" 45d34'10"
Stephanie 'B' Boito
And 23h3'31" 43d16'53"
Stephanie B. Frank
And 2h20'12" 45d20'41"
Stephanie Baker
Ari 3h20'0" 27d48'38"
Stephanie Bakula
Cnv 12h44'34" 39d24'49"
Stephanie Balsamo
Cnc 8h10'48" 32d1'35"
Stephanie Beck & Derek Chui Always
Ori 4h54'10" -0d40'31"
Stephanie Begg
And 1h3'5" 42d44'54"
Stephanie Benavidez
Mon 7h20'49" -0d28'24"
Stephanie Beth Hanson
Sgr 18h31'41" -27d39'23"
Stephanie Blaker
Cam 4h8'17" 54d48'35"
Stephanie Blyth
Lyn 6h41'33" 51d31'14"
Stephanie Bobersky Fedyshin
Cas 1h26'4" 62d30'49"
Stephanie Bohler
Ori 4h49'56" 15d41'7"
Stephanie Bongailas
Cru 12h44'34" -57d15'25"
Stéphanie Bontemps
Tau 4h29'16" 16d36'42"
Stephanie Bowen
Lyn 6h27'19" 59d32'57"
Stephanie Bowman
Leo 11h14'10" 23d26'12"
Stephanie Brandmeyer
Vir 12h29'22" 1d6'52"
Stephanie Brigitte Knafelman
Vir 13h58'13" -14d13'1"
Stephanie Brigman
Peg 23h25'56" 29d15'22"
Stephanie Brodie
Cnc 8h57'5" 10d46'22"
Stephanie Brooke
Cnc 8h19'28" 14d36'25"
Stephanie Brooke Knue
Gem 7h22'57" 24d17'40"
Stephanie Brown
Ori 5h30'54" 11d59'40"
Stephanie Brown
Aqr 22h18'2" -2d16'17"
Stephanie Bryant
Cas 0h19'14" 55d32'31"
Stephanie Burton
Uma 10h41'47" 64d39'45"
Stephanie Bush
Vir 12h37'46" 10d44'58"
Stephanie C.
Cap 21h31'54" -13d55'33"
Stephanie Calos Hicks
Gem 7h43'29" 31d37'17"
Stephanie Camille
Cap 20h28'6" -9d41'59"
Stephanie Cara Carlin
Cas 0h15'15" 61d24'49"
Stephanie Carol
Ari 3h18'34" 29d42'45"
Stephanie Carola
Sco 17h49'24" -32d53'23"
Stephanie Carrasco
Cas 1h7'55" 55d50'47"
Stephanie Casey
Umi 15h40'51" 84d49'47"
Stephanie Cash
Leo 10h31'54" 13d34'17"
Stephanie Cassandra Nardelli
And 23h15'8" 40d40'15"
Stephanie Castro
Ori 6h4'59" 10d38'16"
Stephanie Celeste Clescere
Ori 5h32'14" -4d8'27"
Stephanie Cerise Saathoff
Cyg 20h3'26" 34d19'29"
Stephanie Chang
Gem 7h5'11" 14d43'22"
Stephanie & Cheyenne
Cyg 20h51'0" 47d30'10"
Stephanie Chicklet Magid
Cyg 19h58'35" 30d26'55"
Stephanie Ching / Strphcturtle
Cyg 19h34'43" 28d24'5"
Stephanie Christine
Ari 3h7'42" 26d26'33"
Stephanie Christine
And 2h36'44" 42d53'52"
Stephanie Christine
Lib 15h46'16" -5d34'58"
Stephanie Christine Blagaich
Tau 5h23'41" 27d51'39"
Stephanie Christine Peecher
Cnc 8h42'39" 31d11'43"
Stephanie Christine Tasker
Per 3h18'53" 50d45'22"
Stephanie Cianchetta's Star
Cas 1h24'15" 57d33'16"
Stephanie Cisneros Burton
Aql 18h42'4" -0d12'23"
Stephanie "Clementine" Leal's Star
Gem 6h42'11" 18d14'26"
Stephanie Cone
Cas 23h35'35" 55d52'29"
Stephanie Corinne Habura
And 23h27'32" 47d14'8"
Stephanie Costa
Psc 22h54'7" 2d40'37"
Stephanie D. Blick
And 23h25'33" 51d39'12"
Stephanie D. Russow
Mon 7h23'0" -5d34'0"
Stephanie Da siliva
Lib 14h50'21" -12d30'49"
Stephanie Dang & Phat Vo, In Love
Ori 5h36'31" -3d1'21"
Stephanie Darling
Col 5h59'26" -34d50'33"
Stephanie Davenport
Cap 20h25'34" -13d17'11"
Stephanie Davis
Cap 20h52'7" -16d20'21"
Stephanie Davis
Lyn 8h20'3" 35d15'9"
Stephanie Dawn
Aql 20h9'42" 8d32'57"
Stephanie Dawn Wood
And 0h33'24" 30d49'51"
Stephanie DeAnne
Tau 5h56'55" 26d51'45"
Stephanie DeBoer
Uma 8h52'13" 61d31'59"
Stephanie Delgado
Cas 23h36'32" 57d13'18"
Stephanie DeMarco
Gem 7h8'6" 15d2'8"
Stephanie Denise Allen
Lyn 8h33'17" 47d17'11"
Stephanie Denise Feeney
Sgr 19h45'11" -41d24'49"
Stephanie Denise Jones 10.08.1983
Lib 15h25'54" -4d32'48"
Stephanie Denise Woods
Ari 2h33'23" 24d33'1"
Stephanie & Dennis Delgado
Leo 9h54'43" 16d6'38"
Stephanie Dewey's Star
Mon 6h46'29" -0d31'17"
Stephanie Diane Clift
Uma 10h44'48" 50d19'49"
Stephanie Diane Lipari
Ori 5h58'48" 5d50'55"
Stephanie Diane Ponnett
Sgr 19h40'4" -14d28'9"
Stephanie Dianne Martin
Vir 14h38'57" 4d24'38"
Stephanie DiSilvestro
Aql 19h48'21" -0d49'22"
Stephanie Divonzo
Cas 23h27'56" 57d45'28"
Stéphanie Docoche
Mon 6h59'5" -4d6'22"
Stephanie Doherty
Psc 1h10'28" 33d6'41"
Stephanie Dyane
Sgr 18h7'39" -26d18'16"
Stephanie E. Duvall
And 2h8'30" 41d35'23"
Stephanie E. Hancock
Cyg 21h41'26" 51d45'8"
Stephanie E. Howard
Cap 20h49'23" -16d32'56"
Stephanie E. Keno
Peg 23h4'9" 9d59'59"
Stephanie E. Perez
Leo 9h34'41" 26d57'26"
Stephanie E. Wilcox's Shining Star
Uma 11h21'0" 40d55'7"
Stephanie EHS 06
Sgr 18h35'42" -17d16'38"
Stephanie Elise Martin
Cnc 8h48'36" 14d7'23"
Stephanie Elizabeth
And 23h5'48" 48d19'6"
Stephanie Elizabeth Glass
Leo 11h40'56" 25d46'51"
Stephanie Elizabeth Gray
Cas 0h11'12" 65d7'55"
Stephanie Elizabeth Ramirez
Lib 14h29'36" -22d15'27"
Stephanie Elizabeth Taylor
Lib 14h34'43" -14d42'46"
Stephanie Elizabeth Vallejo
Vir 13h17'34" 11d31'15"
Stephanie Elizabeth Welch
And 1h58'41" 47d24'28"
Stephanie Elyse
Leo 10h26'15" 20d26'1"
Stephanie Epps
And 23h21'48" 42d6'8"
Stephanie & Eric 1 Yr. Anniversary
Umi 15h24'52" 74d7'34"
Stephanie Erin Lindley
Cyg 20h13'2" 31d42'26"
Stephanie Esquibel
Cap 20h58'8" -19d58'1"
Stephanie Eve
Crb 15h33'48" 27d31'8"
Stephanie Eve Clark
Peg 21h36'48" 26d16'58"

Stephanie Faith
Ori 6h0'25" 19d53'40"
Stephanie Faith Beatty
Cas 0h22'43" 54d35'26"
Stephanie Farnum
Leo 9h29'1" 25d36'36"
Stephanie Fawaz
Crb 16h11'18" 36d47'47"
Stephanie Feehley
Lyn 7h35'40" 56d53'19"
Stephanie Figueroa
Ari 2h43'4" 14d37'30"
Stephanie Fish
And 0h41'13" 40d43'8"
Stephanie Fisher
Ori 5h56'43" 6d37'34"
Stephanie Flores
Crb 16h21'40" 37d52'25"
Stephanie Fowler
Tau 4h5'14" 7d28'32"
Stephanie Gable
Uma 8h23'43" 60d4'45"
Stephanie Gail Ryan
Leo 10h29'34" 6d23'55"
Stephanie Galluzzo
Del 20h32'35" 13d3'8"
Stephanie Garcia
Tau 4h31'4" 29d55'41"
Stephanie & Gary's Star
Uma 9h12'38" 46d38'25"
Stephanie Garza
Cyg 20h58'27" 35d3'5"
Stephanie Garza
Leo 10h14'34" 6d25'42"
Stephanie Gaston
Leo 11h28'48" 9d33'19"
Stephanie Gaye Ellis
Sgr 19h11'16" -15d52'6"
Stephanie Geisheimer
Uma 9h5'1" 56d54'53"
Stephanie Genevieve
Cyg 21h30'47" 30d56'40"
Stephanie Gentry
And 2h18'46" 49d48'52"
Stephanie George
Sgr 18h54'54" -33d47'45"
Stephanie & George's Wedding Star
Uma 11h35'27" 58d33'12"
Stephanie Gifford
Uma 11h13'17" 63d22'59"
Stephanie Glazer
Peg 22h51'2" 31d28'5"
Stephanie G-Love
And 0h27'43" 27d40'47"
Stephanie Grace
Lmi 10h17'44" 36d45'20"
Stephanie & Gracie Schaefer
And 23h42'12" 41d25'33"
Stéphanie Graisier
Cyg 21h9'55" 47d19'9"
Stéphanie Grauer
Uma 9h12'17" 59d17'5"
Stephanie Gray
Aqr 23h0'19" -9d55'58"
Stephanie Guzman
Leo 9h30'1" 11d15'11"
Stephanie Hall
Tau 3h35'58" 14d29'34"
Stephanie Hartung
Uma 10h59'56" 72d10'30"
Stephanie Herkenhoff
And 0h38'19" 37d58'56"
Stephanie Hernandez
Lib 14h54'40" -0d50'57"
Stephanie Herr
Cap 20h12'19" -15d45'1"
Stephanie Hicks
And 1h27'16" 38d11'27"
Stephanie Higueras Dollente Niebla
Umi 15h35'11" 71d27'54"
Stephanie Hope Martin Kobs
Cap 21h55'57" -13d32'32"
Stephanie Hope Morabito
Cyg 21h14'58" 43d26'44"
Stephanie Horn
Cap 20h15'33" -13d9'35"
Stephanie Huerta
Lib 14h36'52" -22d6'59"
Stephanie Hui May Chow
Vir 13h23'5" 12d17'18"
Stephanie Hunneke
Cap 20h33'9" -18d47'26"
Stephanie Huszar
Cam 5h4'33" 63d55'10"
Stephanie Hylit Tan
Sco 17h41'21" -35d57'7"
Stephanie "I Love You Forever"
And 0h39'59" 42d36'33"
Stephanie Iacono
Cap 21h51'34" -19d55'15"
Stephanie Ibarra
And 0h7'59" 43d59'59"
Stephanie Imm
Lyn 9h16'59" 38d59'27"
Stephanie J
Peg 22h22'2" 4d59'22"
Stephanie J. Galeazzi-Barnes
And 2h19'58" 40d4'23"
Stephanie J. Guaraldi
Sco 16h12'22" -12d35'4"
Stephanie J. Rinear
Cas 23h29'14" 51d23'34"
Stephanie J Segovia
Psc 1h22'22" 24d8'50"
Stephanie J. Smith
Aqr 22h18'40" 1d47'44"
Stephanie Jacob
Cam 5h4'44" 68d32'35"
Stephanie Jade Kew
And 23h21'37" 52d53'27"
Stephanie Jaggers "Babygirl"
And 2h11'27" 45d53'23"
Stephanie Jan Wilson
Psc 0h2'51" 1d52'30"
Stephanie Janae
Crb 15h29'29" 28d7'25"
Stephanie Jane
Leo 10h58'56" 17d53'15"
Stephanie Jane Hall
Vir 13h50'50" -19d44'58"
Stephanie Jane Oliver
Cas 23h41'17" 56d56'58"
Stephanie Jane Smith
Cas 23h8'21" 58d10'9"
Stephanie Jarrell
Cnc 9h16'9" 32d53'14"
Stephanie & Javier
Cap 21h52'57" -19d38'24"
Stephanie Jayne Mullin
And 23h40'55" 38d27'19"
Stephanie Jean McLaughlin
Gem 6h55'17" 15d5'51"
Stephanie Jean Steller-Sharp
Uma 8h40'7" 69d34'25"
Stephanie Jean Watts
Cru 12h38'21" -61d26'58"
Stephanie Jill
And 0h51'58" 40d47'28"
Stephanie Jo Keena
Cyg 21h44'8" 46d43'19"
Stephanie Jo Phillips Wishing Star
Cap 20h51'24" -15d51'57"
Stephanie Jo Schindel
Sgr 19h12'42" -14d20'30"
Stephanie Joanne Cartwright
Cas 23h33'29" 52d0'42"
Stephanie Joanne Daly
Cam 5h36'25" 66d12'43"
Stephanie Joelle
Tau 5h8'52" 18d1'33"
Stephanie & John Hakun
Cyg 20h30'43" 58d23'10"
Stephanie Johnson-Rolle
Ari 2h3'14" 19d2'41"
Stephanie Jossellinne Soriano
Psc 1h39'57" 24d5'26"
Stephanie Joy Barone
Ari 3h3'44" 30d6'22"
Stephanie Joy Hutcherson
Cyg 20h19'47" 39d57'28"
Stephanie Juliana Cox
Cyg 20h54'41" 32d48'20"
Stephanie K. Chrzaszcz
Cap 21h22'7" -14d35'4"
Stephanie K. Polangco
And 2h29'34" 42d29'46"
Stephanie Kanavos
Cas 0h14'16" 54d57'42"
Stephanie Kasmiroski
Oph 17h3'29" -0d48'7"
Stephanie Kate Waldeback
Cra 19h1'48" -39d58'20"
Stephanie Kay Hernandez
Sco 17h52'31" -40d50'26"
Stephanie Kay Sheetz
Psc 0h0'21" -2d36'44"
Stephanie Kay Smith
Ari 2h59'21" 21d51'25"
Stephanie Keelin
And 1h41'25" 42d38'8"
Stephanie Kemp
Cyg 20h8'9" 55d35'51"
Stephanie Kerr
Cas 0h13'59" 50d43'49"
Stephanie Kim
Peg 21h11'26" 18d3'8"
Stephanie "KITTY" Kinman
Cnc 9h6'19" 30d26'44"
Stéphanie Kochalski
Lib 14h32'27" -11d46'28"
Stephanie Kotlarz 2/23/87
Psc 1h29'58" 32d3'51"
Stephanie L. Gibson
Gem 6h45'52" 30d3'17"
Stephanie L Hackenburg
Cas 23h35'22" 51d15'56"
Stephanie L. Lollis
Ori 6h17'10" 19d7'42"
Stephanie L. Opdahl
Uma 10h3'45" 58d1'25"
Stephanie Lannette Kirkland Shaw
Ori 5h42'10" 1d24'54"
Stephanie Lapham
Cas 0h42'36" 64d15'23"
Stephanie Laure Parma
Cnc 8h55'22" 15d52'36"
Stephanie Lauren Lazaro
Cnc 9h17'31" 29d43'46"
Stephanie Lauren Solove
Lib 15h22'20" -24d47'41"
Stephanie Lea
Lib 14h54'21" -14d47'22"
Stephanie Leah
Aqr 21h25'38" -8d24'56"
Stephanie LeAnne
And 0h37'4" 42d24'38"
Stephanie Lee
Sgr 19h29'7" -15d39'47"
Stephanie Lee Dewhirst
Cas 0h25'49" 56d32'38"
Stephanie Lee Karnes "6-6-67"
Crb 16h9'7" 35d43'3"
Stephanie Leigh Koch
And 1h14'44" 37d15'18"
Stephanie Leigh McClain
Cnc 8h13'18" 16d54'33"
Stephanie Letourneau
Lyn 7h37'15" 53d14'13"
Stéphanie Lienhart
Cap 21h35'45" -22d43'59"
Stephanie Lim
Aqr 22h36'25" -0d13'1"
Stephanie Lisa Farkas
Lyr 19h12'53" 45d40'36"
Stephanie Long
And 0h42'15" 45d12'19"
Stephanie Loraine
Cnc 8h21'40" 12d5'2"
Stephanie Louise
Pyx 8h39'48" -32d36'53"
Stephanie Louise Nohe
Vir 13h23'39" 6d54'55"
Stephanie Louise Smith
Cap 20h8'48" -22d18'1"
Stephanie Love Of My Life
Sco 16h14'32" -15d59'51"
Stephanie Lozada
Cap 21h23'56" -23d5'42"
Stephanie Lucas
Cas 2h24'32" 67d36'22"
Stephanie Lucinda Elwes
Sco 17h53'18" -35d41'17"
Stephanie Lynn
Sgr 19h19'4" -17d3'31"
Stephanie Lynn
Cap 20h27'37" -9d9'49"
Stephanie Lynn
Lib 14h53'48" -1d27'21"
Stephanie Lynn
Dra 19h52'7" 78d52'29"
Stephanie Lynn
Vir 12h39'12" 6d0'44"
Stephanie Lynn
Leo 11h46'43" 21d7'28"
Stephanie Lynn
Ari 2h40'59" 24d30'57"
Stephanie Lynn
Cnc 8h50'54" 22d38'29"
Stephanie Lynn
Lyr 18h47'25" 31d55'54"
Stephanie - Lynn "Angel"
Lyr 19h6'52" 42d52'52"
Stephanie Lynn Cartwright
Lyn 8h50'15" 43d29'51"
Stephanie Lynn Gruber
And 23h40'17" 45d16'23"
Stephanie Lynn Kubus
And 1h1'36" 39d0'23"
Stephanie Lynn Lima
Tau 4h20'42" 13d13'18"
Stephanie Lynn Macary
And 1h11'10" 35d49'25"
Stephanie Lynn Moore
Uma 10h22'37" 63d48'28"
Stephanie Lynn Perez
And 23h31'11" 36d30'29"
Stephanie Lynn Raasch
Gem 6h34'2" 19d34'49"
Stephanie Lynn Shoemake
Peg 22h31'17" 3d32'15"
Stephanie Lynn Snead
Cyg 21h13'23" 44d5'12"
Stephanie Lynn Sneed
And 1h47'47" 42d37'4"
Stephanie Lynn Suzanne Salinas
Sco 16h38'26" -25d2'57"
Stephanie Lynn Wicker
Cyg 21h34'4" 42d51'33"
Stephanie Lynn Williams
And 23h9'51" 52d15'40"
Stephanie Lynne Angioletti
Gem 6h7'35" 25d29'35"
Stephanie M.
And 23h21'0" 49d10'54"
Stephanie M Ayotte
Ari 3h25'25" 29d39'42"
Stephanie M. Hagelin
And 0h40'49" 36d3'36"
Stephanie M. Jones
Cas 1h12'56" 69d25'5"
Stephanie M. Koepke
Dra 17h56'56" 65d48'28"
Stephanie M. Woodard
Aqr 22h16'19" 1d29'2"
Stephanie Madelyn Lauder
Gem 7h44'14" 33d31'50"
Stephanie Mae Strahan
Vir 12h38'30" -0d44'12"
Stephanie Mae Winkler
Lyn 9h1'24" 34d17'12"
Stephanie Manoff
Lib 14h52'7" -2d56'44"
Stephanie Maoli
Psc 1h4'24" 31d20'48"
Stephanie Mara Pilnacek 92305
Cas 1h16'43" 64d47'43"
Stephanie & Marc Pianfetti
Cyg 19h38'25" 30d14'25"
Stephanie Marchese
Aql 19h31'19" 3d25'7"
Stephanie Marchetti I love You
Ori 5h30'37" 0d51'55"
Stephanie Mareen Evans
Sgr 19h53'44" -26d30'1"
Stephanie Marie
Cap 20h43'26" -20d45'4"
Stephanie Marie
Ari 3h23'21" 28d38'41"
Stephanie Marie
Lyn 8h17'6" 38d4'4"
Stephanie Marie
And 2h15'9" 42d3'51"
Stephanie Marie
And 23h31'5" 46d51'28"
Stephanie Marie Bachman
Ari 1h56'39" 18d17'34"
Stephanie Marie Begley
And 23h18'22" 52d24'41"
Stephanie Marie Bertuzzi
Psc 1h25'31" 32d8'11"
Stephanie Marie Burruss
Vir 12h51'38" -0d23'28"
Stephanie Marie Couch Star
And 0h31'36" 34d39'18"
Stephanie Marie Donato
Leo 10h19'0" 24d27'12"
Stephanie Marie Fay
Ari 2h59'22" 21d13'2"
Stephanie Marie Ferrer
Vir 14h15'56" 3d14'56"
Stephanie Marie Figgemeier
Uma 10h19'36" 39d40'11"
Stephanie Marie Goncalves
Aql 20h10'16" 9d17'6"
Stephanie Marie Keller
And 2h7'53" 45d44'25"
Stephanie Marie Knoop
Cyg 20h7'34" 35d54'18"
Stephanie Marie Lewin Marsh
Leo 10h46'0" 9d6'9"
Stephanie Marie Mercado
Vul 19h52'59" 24d18'32"
Stephanie Marie Moore
Psc 0h45'42" 14d58'19"
Stephanie Marie Nieves
Cnc 8h50'49" 31d32'28"
Stephanie Marie Phan
Lib 14h36'56" -18d24'28"
Stephanie Marie Roosa
And 0h26'9" 40d52'54"
Stephanie Marie Simon
Cas 1h24'6" 57d24'13"
Stephanie Marie Simon
Ori 6h16'28" 3d19'42"
Stephanie Marie Singrin
Sco 16h54'40" -41d8'31"
Stephanie Marie Thompson
Aql 19h21'5" -8d4'24"
Stephanie Marie Tickner
And 0h49'14" 42d4'45"
Stephanie Marie Williams
Lyr 19h14'15" 27d21'31"
Stephanie Martin
Cap 20h24'49" -12d55'53"
Stephanie Martinez
Sco 16h54'41" -45d31'47"
Stephanie Marx & Clay Peacock
Col 5h42'5" -32d30'27"
Stephanie & Matthew
Uma 13h30'33" 54d21'24"
Stephanie Maupin
Del 20h39'37" 16d18'44"
Stephanie May Oats
Cru 12h15'46" -63d34'44"
Stephanie McBrayer
Uma 9h48'2" 64d3'32"
Stephanie McCole
And 0h19'6" 36d24'19"
Stephanie McGahey
Cap 21h5'52" -18d48'48"
Stephanie McKendrick
Vir 12h17'19" 1d45'4"
Stephanie McMahon
Lib 14h50'50" -0d45'27"
Stephanie McNeal
Cyg 20h42'13" 43d15'20"
Stephanie Mehall
And 23h16'46" 51d35'4"
Stephanie Meleady
Mon 6h50'54" -0d10'40"
Stephanie & Michael
Uma 9h24'15" 49d10'23"
Stephanie Michele Quinajon
Cap 21h41'5" -18d44'21"
Stephanie Michelle
Vir 12h56'42" -13d59'0"
Stephanie Michelle
Leo 10h37'52" 13d28'37"
Stephanie Michelle
Ori 5h54'42" 18d20'7"
Stephanie Michelle Chalef
Aqr 21h15'13" -10d5'43"
Stephanie Michelle Gomez
Ori 6h19'33" 18d51'26"
Stephanie Michelle Kozatek
Aql 18h43'8" -0d53'4"
Stephanie Michelle Pokorney
Cyg 21h57'21" 49d0'0"
Stephanie Michelle Smith
Gem 6h42'27" 21d36'14"
Stephanie Michelle Tiliakos
Tau 4h15'58" 18d33'54"
Stephanie Miller
Ari 2h57'55" 30d1'52"
Stéphanie Minassian
Uma 8h48'40" 56d33'22"
Stephanie Minton
Lib 15h17'9" -8d48'47"
Stephanie Mio Ciccina
Tau 5h43'54" 25d52'27"
Stephanie Mireles
Lac 22h21'37" 45d53'5"
Stephanie Moondancer
And 23h32'17" 47d10'54"
Stephanie Morin
And 23h30'27" 41d49'55"
Stephanie Mouawad & Dan Merino
And 0h37'1" 29d10'12"
Stephanie Munro
Tau 4h47'14" 21d4'10"
Stephanie Murphy
And 0h43'47" 25d28'36"
Stephanie my lovely blue eyed bride
Cyg 21h0'0" 32d25'28"
Stephanie My Princess
Tau 5h42'24" 22d16'30"
Stephanie n Baby Nicholas Foley
Her 17h11'53" 23d13'43"
Stephanie N. Cochran
Ori 5h53'22" 17d16'36"
Stephanie N. Landes
Cas 1h36'7" 61d11'6"
Stephanie Natalie Gutierrez
And 2h36'32" 45d48'21"
Stephanie Navarro
Gem 7h40'47" 18d21'25"
Stephanie Nicholas
And 23h15'21" 48d50'38"
Stephanie Nicole
Boo 14h34'0" 44d33'27"
Stephanie Nicole
Lmi 10h27'35" 32d39'55"
Stephanie Nicole Asmar
Gem 6h4'54" 27d42'39"
Stephanie Nicole Campbell
Uma 11h53'20" 52d14'37"
Stephanie Nicole Gallo
Cas 1h37'54" 66d5'37"
Stephanie Nicole Kienberger
Cas 1h25'44" 56d1'53"
Stephanie Nicole Knapp
Sco 16h5'22" -14d30'8"
Stephanie Nicole Santillan
Lyn 9h19'6" 34d38'0"
Stephanie Nicole Stiglich
And 0h10'31" 45d18'16"
Stephanie Nicole Talbot
Lib 15h0'11" -20d46'24"
Stephanie Nix
Vir 13h4'29" 4d10'30"
Stephanie Noel Chase
Sgr 18h26'5" -26d55'44"
Stephanie Noel Tanner
Per 3h22'8" 45d37'19"
Stephanie Noreen
Lyn 7h43'12" 54d8'56"
Stephanie Norris
Gem 7h14'42" 16d23'11"
Stephanie Nucci
And 2h18'46" 40d0'54"
Stephanie O'Donohoe Armstrong
Cas 0h30'51" 54d37'58"
Stephanie Oliveras
Lib 15h5'0" -2d47'28"
Stephanie Our "Rock Star" Forever
And 1h6'53" 39d19'12"
Stéphanie Palazzo
Vul 21h3'25" 22d51'41"
Stephanie Palermo
Cas 1h56'40" 60d17'27"
Stephanie Parica
Cyg 21h11'17" 40d41'28"
Stephanie Parra Ortega
Tau 5h27'54" 27d56'18"
Stephanie Patricia
Uma 9h29'51" 62d33'28"
Stephanie Payne
Sco 16h4'53" -11d7'15"
Stephanie Pearson
Vir 13h40'9" -6d3'53"
Stephanie Pernice
Vul 19h55'54" 25d37'10"
Stephanie Persing
And 2h31'7" 44d46'4"
Stephanie Phillippy
Lyr 18h34'46" 46d52'57"
Stephanie Pierce
Ari 2h38'11" 14d1'21"
Stephanie Poree
Crb 16h20'53" 27d2'54"
Stephanie "Precious" Teboe
Uma 10h58'13" 62d28'9"
Stephanie "Princess" Davis
Gem 7h32'59" 20d4'45"
Stephanie Prohaska
Cyg 21h35'24" 36d58'49"
Stephanie Pruchnick Pickle
Lyr 19h20'29" 38d10'53"
Stephanie Pusic & Andrew Houston
Cyg 20h8'5" 54d20'17"
Stephanie R Grizzell
Gem 7h18'18" 24d22'55"
Stephanie R. Jensen
Cnc 9h3'8" 28d23'54"
Stephanie R. Monk
Ori 6h2'58" 10d26'17"
Stephanie R. Rodriguez
Cap 20h53'9" -25d54'2"
Stephanie R. Troutman
Uma 10h32'40" 47d3'7"
Stephanie Rae
Lyn 7h37'20" 41d22'36"
Stephanie Rae
Crb 15h41'35" 33d38'32"
Stephanie Rae
Cnc 8h21'8" 18d17'38"
Stephanie Rae Bowman~RN
Vir 12h43'44" 3d50'34"
Stephanie Rae Cline
And 0h40'49" 39d28'6"
Stephanie Rae Vasquez
Cnc 8h17'26" 19d7'43"
Stephanie Ray Marie Cameron
Ari 3h21'44" 16d17'14"
Stephanie Raye Fritz
Gem 7h40'58" 33d46'56"
Stephanie Reisnour
Leo 11h43'34" 25d46'13"
Stephanie Rena Smith
And 1h16'37" 37d20'47"
Stephanie Renae Eddlemon
And 0h17'19" 26d37'36"
Stephanie Rene
And 2h26'1" 42d10'58"
Stephanie Rene
Ori 5h37'32" -0d34'32"
Stephanie Rene Puder
Sco 17h41'7" -35d0'50"
Stephanie Renee Berry
Mon 6h52'53" -0d18'1"
Stephanie Renee Lawson
Aqr 22h5'33" -13d43'52"
Stephanie Renee Moran
Gem 6h21'19" 18d34'17"
Stéphanie Renier
Cas 3h14'35" 67d7'52"
Stephanie Rhodes
Cas 1h27'28" 60d34'13"
Stephanie Riddle
Com 13h4'14" 19d30'35"
Stephanie Rinaldi
Cap 20h23'14" -13d22'10"
Stephanie Rittenberg
Ori 5h40'49" -2d28'57"
Stephanie Rockwell
Ari 1h52'52" 23d42'42"
Stephanie Rodriguez
Lib 14h49'31" -12d37'48"
Stephanie Rogl
Uma 9h0'38" 67d13'20"
Stephanie Roper
Cas 1h19'47" 57d41'51"
Stephanie Rose Anderson
Cam 5h43'40" 65d21'32"
Stephanie Rose Harris
Cap 20h54'22" -16d2'53"
Stephanie Royeton
Cam 5h14'3" 61d40'49"
Stephanie Ruiz
And 0h50'18" 45d23'33"
Stephanie & Rythm
And 1h53'0" 45d26'26"
Stephanie Sankey
And 0h18'7" 39d45'28"
Stephanie Sarah
And 0h34'14" 27d16'33"
Stephanie Sarah Covey
Cas 0h29'33" 62d39'41"
Stephanie Schloesser
Vir 13h11'5" 8d48'29"
Stephanie Schroeder
Cas 0h18'11" 51d42'46"
Stephanie Serrano
Vir 13h4'32" 7d11'36"
Stephanie Shari Saffi
Cnc 8h25'32" 11d39'58"
Stephanie Sharp
Aqr 21h26'45" -0d44'47"
Stephanie Shubert Randis
Aql 19h13'43" 5d26'37"
Stephanie - Sibling & Marcus
Cru 12h41'22" -58d58'40"
Stephanie Sick
Sgr 18h50'43" -16d13'34"
Stephanie Sills
Cas 2h15'3" 66d36'35"
Stephanie Simmons
Tau 4h42'45" 7d21'4"
Stephanie Sims
Vir 12h8'9" 3d54'0"
Stephanie Skertich
Gem 7h9'12" 33d4'45"
Stephanie Smith
Ori 5h30'55" 3d35'11"
Stephanie Smith
Psc 1h5'48" 28d35'4"
Stephanie Soderstrom
Lyn 7h31'49" 36d17'10"
Stephanie "STAR" Arce
Cas 23h25'47" 59d41'55"
Stephanie Star del Pino
Peg 22h40'23" 12d17'13"
Stephanie Stark
Cap 20h23'41" -22d49'6"
Stephanie Stephens
Lib 14h40'53" -18d19'17"
Stephanie 'Stephie' Scanlan
Cas 23h42'49" 54d51'0"
Stephanie "Stephy" Ann Ried
Vir 13h16'48" 2d13'27"
Stephanie Steve
Lib 15h9'35" -28d3'33"
Stephanie & Steve Makowski
Lib 15h48'0" -5d43'30"
Stephanie Stokes-Buzzelli
Uma 8h54'4" 56d45'23"
Stephanie Streeter - Shining Star
Crb 16h7'17" 26d30'43"
Stephanie Sue Bronsted
Vir 12h34'0" -6d47'41"
Stephanie Sue Sherman
Sco 16h12'49" -21d19'31"
Stephanie Sue Stratton
Gem 6h26'21" 21d41'53"
Stephanie "SunShine" Pentz
Gem 6h29'35" 20d36'26"
Stephanie Suzi Richter
Lyr 19h0'41" 37d35'14"
Stephanie Sylvia
Uma 13h46'43" 52d31'27"
Stephanie T. Medina
Tau 4h24'4" 21d57'13"
Stephanie Tabor
And 23h37'53" 38d0'55"
Stephanie Thomas
Ara 17h45'21" -47d53'24"
Stephanie Thompson
Cas 23h50'3" 53d47'11"
Stephanie Thompson Parks
Uma 11h36'57" 33d25'58"
Stephanie Tikki Brown
Leo 10h43'36" 22d9'58"
Stephanie Tipler
And 1h44'20" 47d33'8"
Stephanie & Todd's Star
Ori 5h56'45" -0d6'43"
Stephanie & Tristan, Mother & Son
Sco 16h13'31" -13d34'8"
Stephanie Turner
Psc 0h18'32" 6d27'3"
Stephanie V. Snow
Crb 15h54'34" 26d32'24"
Stephanie Vanderham
Cap 21h2'10" -18d35'7"
Stephanie Vest
Aqr 23h51'35" -11d40'58"
Stephanie Villanueva
Cnc 8h0'30" 20d30'26"
Stephanie Vitale
Crb 15h36'50" 38d54'10"
Stephanie Volkart
And 2h19'47" 40d21'45"
Stephanie Waldschmidt
Umi 15h30'37" 75d42'50"
Stephanie Walker
Lyn 7h15'51" 54d0'29"
Stephanie Walters
And 1h17'7" 50d12'12"
Stephanie Weissmann
Cam 7h47'16" 73d20'43"
Stephanie Whalen
Vir 11h48'9" 5d57'46"
Stephanie Wiebe
Gem 7h2'51" 19d43'3"
Stephanie Williams
And 23h19'58" 47d0'32"
Stephanie Willman - Graduation Day
Cru 12h47'56" -63d25'29"
Stephanie Wright
Cas 1h21'23" 51d21'13"
Stephanie Wright
And 0h36'0" 35d32'18"
Stephanie Xiomara Tinsley (Davila)
Aqr 22h2'48" -3d13'24"
Stephanie you are my shining star
Cas 0h16'27" 59d33'53"
Stephanie Young
And 2h29'42" 45d47'53"

Stephanie Yvonne
And 0h56'47" 39d4'49"
Stephanie Zauggarius Coopernia
Sgr 19h30'49" -18d4'9"
Stephanie Zefferino
Cap 21h43'8" -13d48'34"
Stephanie, Friendship of a Star
And 1h22'51" 43d40'29"
Stephanie, Isabel & Julia
Lyr 18h28'12" 28d46'44"
Stephanie, Jasmine and Justice
Cas 23h12'37" 58d49'33"
Stephanie, Sophie & Paul
Uma 12h5'12" 63d38'24"
Stephanie-Angel
And 23h49'22" 45d33'42"
Stephanie-Ann Ramos. My True Angel
Mon 6h50'10" 7d20'24"
Stephanie-Hunter
Lib 15h36'42" -25d26'18"
stephanielynncox23
Vir 13h40'16" 1d20'54"
Stephanie-Marie
Lyn 6h32'19" 57d7'50"
Stephanie's Blue Eyes
Lyn 8h16'36" 57d10'6"
Stephanie's Bunch Walnuts
Lib 14h57'29" -2d12'24"
Stephanie's Diamond
Uma 11h9'53" 62d31'7"
Stephanie's Eye
Sco 16h13'40" -9d44'47"
Stephanie's Heart
Leo 10h44'29" 14d8'34"
Stephanie's Inspiration
Crb 15h50'25" 35d49'15"
Stephanie's Lego
Ari 2h13'46" 25d21'14"
Stephanie's Light
Cnc 8h53'32" 13d38'15"
Stephanie's Light
Cyg 21h51'33" 37d33'38"
Stephanie's Light
Lib 15h47'51" -10d42'2"
Stephanie's Love
Tau 4h48'32" 18d40'52"
Stephanie's Massive Ball of Gas
Uma 11h13'7" 29d32'54"
Stephanie's Sapphire
Lib 15h43'25" -17d19'14"
Stephanie's shooting star
Cap 20h39'47" -19d44'39"
Stephanie's Solar Snowflake
Sco 16h5'41" -17d38'50"
Stephanie's Sparkling Irish Eye
And 0h24'14" 32d52'42"
Stephanies Star
Leo 9h24'39" 16d38'44"
Stephanie's STAR
And 23h42'5" 46d43'47"
Stephanie's Star
And 1h57'8" 42d57'48"
Stephanie's Star
And 0h38'51" 42d34'14"
Stephanie's Star
Aqr 23h34'56" -13d1'7"
Stephanie's Star
Aqr 21h37'58" -8d7'24"
Stephanie's Star
Vol 8h12'18" -68d4'30"
Stephanie's Wish
Dra 16h33'58" 51d33'41"
Stephanie..Shining Brightly Forever
Cnc 8h42'27" 10d28'34"
Stephanie-Vanessa (LT-Butterfly)
Gem 6h45'57" 25d48'31"
Stephanita
Cap 21h0'24" -26d8'16"
Stephanja's Shining Star
And 0h1'10" 37d16'22"
stephanjel
Uma 14h12'52" 59d36'16"
Stephan's Star
Per 3h0'6" 40d58'24"
Stephany Cala
Ori 5h35'16" -1d59'36"
Stephany Darlene Corr Constellation
And 2h5'31" 41d58'8"
Stephany & J.D.
Cyg 20h21'31" 31d35'5"
Stephany Joy
Cnc 8h29'5" 29d12'22"
Stephany (luminarium decorus)
Ari 2h35'13" 25d45'49"
Stephany Marie Perales
Uma 10h3'3" 54d41'28"
Stephany Nicole Papadakis
Mon 6h44'43" -0d17'32"
Stephany R. Schultz
Ari 2h44'47" 30d33'0"
Stephany Rachel Anne Sexton
Uma 9h19'5" 68d34'49"
Stephany Vo
Aqr 21h44'20" -1d20'35"
StephAri Manevitch
Aqr 22h38'9" -3d33'3"
Stephbunny
Tau 5h31'17" 17d20'6"
Stephen
Psc 1h17'19" 9d28'52"
Stephen
Uma 11h54'41" 36d31'11"
Stephen
Her 16h12'6" 47d50'14"
Stephen
Uma 9h55'55" 46d23'23"
Stephen
Uma 9h47'49" 71d10'45"
Stephen
Cep 0h6'25" 76d59'0"
Stephen
Cru 12h30'50" -58d27'1"
Stephen
Cen 13h44'43" -42d4'0"
Stephen A. Hicks 05061982
Tau 4h19'36" 26d24'3"
Stephen A. Miele, Jr.
Tau 5h50'56" 26d24'49"
Stephen A. Ott, SR
Aql 20h0'15" 15d55'10"
Stephen Adrian Sides
Uma 12h5'49" 51d23'9"
Stephen Alan Sweat
Sgr 18h25'1" -30d50'25"
Stephen Albanese
Boo 14h42'59" 48d41'42"
Stephen Albert Weste
Per 3h19'16" 44d9'27"
Stephen Alexander Bory
Tau 3h28'53" 8d1'55"
Stephen Alexander Doyle, 1965
Lac 22h16'6" 37d28'55"
Stephen Allen Eddington
Dra 16h58'47" 62d7'12"
Stephen & Amanda
Cyg 21h23'0" 51d29'6"
Stephen and Carrie-Anne Elsley
Cyg 20h5'51" 40d15'8"
Stephen and David Kiss
Per 3h22'12" 48d2'37"
Stephen and Deborah Forever
Cru 12h0'19" -62d0'54"
Stephen and Ghada Beal
Cyg 21h37'17" 38d47'22"
Stephen and Jenny
Lyn 7h14'45" 58d48'55"
Stephen and Joshua Lavery's star
Uma 9h14'34" 57d9'52"
Stephen and Kathryn's
Cyg 20h8'23" 33d34'1"
Stephen and Kirsty Reynolds
Cyg 20h11'36" 31d12'31"
Stephen and Kristine McKean
Apu 16h17'12" -74d47'34"
Stephen and Maura
Cyg 21h48'2" 43d52'24"
Stephen and Stephanie Shea
Her 16h44'30" 41d10'16"
Stephen Andrew
Her 17h25'56" 38d20'22"
Stephen Andrew Eagleton
Aqr 22h53'47" -8d31'39"
Stephen Andrew Kocerha
Cnc 8h43'55" 24d25'56"
Stephen Andrew Lee
Uma 12h48'16" 53d37'43"
Stephen Arlee Schiller
Her 17h24'46" 16d55'42"
Stephen Armstrong
Aur 5h26'29" 44d24'44"
Stephen Arthur Soggee
Ori 5h42'25" -1d12'41"
Stephen B. Cisterna
Aur 6h23'53" 40d11'16"
Stephen B. Irvine
Ori 5h24'59" 0d54'37"
Stephen B. Liggett
Sco 16h13'4" -13d40'30"
Stephen B Pollari
Cnc 9h4'2" 30d18'10"
Stephen Barker, Byron Hunter
Cnc 8h33'55" 8d36'41"
Stephen Barry Edwards - 1955
Aql 19h9'27" 0d6'23"
Stephen B.Brengard
Uma 12h15'44" 52d31'3"
Stephen & Becky
Cap 20h38'58" -14d38'51"
Stephen & Bethann's Star
Uma 11h0'57" 35d42'24"
Stephen Biancuzzo
Boo 15h31'44" 42d9'27"
Stephen Bikofsky
Sco 16h45'35" -29d20'24"
Stephen Birch
Per 4h44'50" 45d42'56"
Stephen Birt Slater
Sgr 17h55'50" -29d33'27"
Stephen Blake Barton
Aqr 22h57'3" -8d42'11"
Stephen Bratichak
Per 3h11'46" 38d29'15"
Stephen Brodhead Flagler
Psc 1h16'16" 4d19'28"
Stephen Byrne
Uma 11h45'21" 41d33'28"
Stephen C C Jacko
Per 3h17'38" 42d50'31"
Stephen C. Daily
Per 3h11'37" 31d11'17"
Stephen C. Katz
Her 17h1'26" 35d58'47"
Stephen Carrie LaFlamme
Uma 10h27'25" 53d3'35"
Stephen+Casey
Her 17h44'25" 37d55'59"
Stephen Charles Busschaert
Uma 11h34'13" 34d18'45"
Stephen Charles Crouch
Sco 16h56'34" -33d18'8"
Stephen Charles Fajen
Aur 6h55'10" 37d43'44"
Stephen Charles Galen, Jr.
Lyn 8h27'10" 39d12'22"
Stephen Charles Heckman
Uma 12h48'31" 59d34'29"
Stephen Charles Holser
Vir 13h18'43" -3d27'43"
Stephen&Christine Kusznir 2-7-2005
Cru 12h43'4" -57d14'56"
Stephen Christopher Cannon
Her 16h41'56" 28d18'54"
Stephen Clark
Cyg 20h30'12" 48d40'31"
Stephen Clifford Jackson
Cyg 21h21'42" 32d33'49"
Stephen Colavito
Cep 20h52'57" 64d22'14"
Stephen "Cole" Corkern
Ori 5h33'29" 2d24'7"
Stephen Condon
Cru 11h57'57" -63d29'24"
Stephen Conrad Waaks
Ori 5h53'14" 20d58'21"
Stephen Cost
Peg 23h15'37" 33d25'45"
Stephen Craig and Karen Fosse Kline
Cru 12h50'44" -60d52'54"
Stephen & Cynthia Foulger
Her 17h27'10" 14d48'54"
Stephen D. "Dad" Albright
Cep 22h16'33" 55d43'28"
Stephen D. Germond, MD
Cep 21h0'39" 58d27'38"
Stephen D. Griffin
Uma 10h1'3" 56d20'5"
Stephen Dana Lewis
Cas 0h6'27" 56d57'0"
Stephen Dandridge
Aql 19h12'17" 12d57'9"
Stephen Daniel Colby
Cyg 21h22'1" 38d30'28"
Stephen Daniel Smith
Cyg 19h48'2" 33d24'36"
Stephen Daniel Wilson
Sco 16h15'48" -23d55'37"
Stephen David Ayala
Sco 17h49'38" -39d18'34"
Stephen David Hulkower
Her 17h10'53" 45d53'54"
Stephen David Santalesa
Aur 5h48'42" 54d13'27"
Stephen David Varma
Cyg 20h55'34" 46d5'53"
Stephen Dean Young
Cnc 8h7'17" 7d41'44"
Stephen DelGrande
Uma 11h34'6" 28d21'37"
Stephen Dempsey 1926 to 2006
Uma 11h46'47" 50d56'45"
Stephen Denaro
Her 17h23'14" 28d52'52"
Stephen Donald Tilson
Vir 12h23'1" 7d8'32"
Stephen Douglas Lord
And 0h10'52" 45d2'19"
Stephen Douglas McAdams
Ari 2h37'16" 25d3'33"
Stephen Douglas Rabow
Vir 13h44'36" -10d42'21"
Stephen Dunlap
Tau 4h34'14" 27d47'41"
Stephen Dupuy
Per 3h9'53" 40d21'41"
Stephen E. McLain
Her 17h43'44" 36d13'10"
Stephen Edward Garnett
Cep 5h28'42" 83d23'43"
Stephen Edward Hilbert
Lyn 6h40'47" 60d11'16"
Stephen Edward Kocsis
Ori 5h33'14" 4d2'33"
Stephen Edward Mertes
Psc 1h2'34" 23d36'23"
Stephen Eliot Dobrynski
Ari 2h13'11" 14d32'29"
Stephen Ellison
Sco 16h52'4" -34d18'19"
Stephen Emerson Phillips
Dra 18h41'8" 82d58'27"
Stephen Eric Cohen
Lib 15h10'29" -5d22'35"
Stephen & Erica
Cnc 8h5'1" 14d27'16"
Stephen "Esteban" Snedden
Cnc 7h58'37" 12d57'46"
Stephen F. Baker
Aqr 22h48'57" -3d38'43"
Stephen F. Shaw, Sr.
Psc 1h41'18" 27d36'27"
Stephen Fehr Loves Barb Keith
Eri 3h57'12" -0d42'47"
Stephen Fischer
Psc 0h37'59" 4d48'34"
Stephen "Fish" Herring
Vir 12h42'15" 2d39'49"
Stephen Floyd Morris-Podzamsky
Aqr 22h21'8" -0d17'23"
Stephen Folk
Uma 12h41'34" 56d57'18"
Stephen Foreman
Uma 10h46'29" 70d48'11"
Stephen Foster
Cyg 20h26'53" 48d51'59"
Stephen Fowler
Psc 1h22'53" 25d23'7"
Stephen Francis Davidson
Cru 12h21'0" -56d42'42"
Stephen Frank
Sco 17h31'14" -44d39'51"
Stephen Frank Shaw
Gem 6h55'58" 29d7'49"
Stephen Franklin Hales
Uma 11h58'15" 43d54'26"
Stephen Frey (The Best Dad)
Sco 17h14'20" -45d10'1"
Stephen Gabriel Miller
Ari 2h42'58" 14d39'49"
Stephen Gallo
Per 2h12'53" 57d19'9"
Stephen Gallo
Cap 21h57'9" -23d16'37"
Stephen George Fallon
Vir 13h35'14" 1d31'16"
Stephen George Nowakowski
Sco 16h7'18" -9d32'37"
Stephen George Skomorucha
Her 18h17'47" 28d16'28"
Stephen Glatfelter
Sco 16h4'55" -9d11'48"
Stephen Govan
Lib 15h29'2" -8d32'52"
Stephen Goytil
Uma 11h17'49" 28d36'54"
Stephen & Grace
Sco 17h6'58" -33d59'34"
Stephen Gregory Ludwig
Col 6h11'55" -34d21'26"
Stephen Halliday
Uma 8h57'12" 57d56'43"
Stephen {Handsome Prince}
Aql 19h44'16" -0d2'17"
Stephen Hanrahan
Uma 11h56'44" 44d1'6"
Stephen Harold Stein
Ari 2h7'2" 23d8'38"
Stephen Harris
Boo 15h23'4" 41d21'2"
Stephen "harry"
Ori 6h20'59" 9d29'31"
Stephen Harvey Murgatroyd
Uma 11h24'59" 34d37'35"
Stephen Heyes - HMS Ardent 21 May 1982
Lib 15h55'36" -17d15'0"
Stephen Hogg
Ari 2h48'58" 27d17'47"
Stephen Holland Superstar
Boo 14h40'40" 33d45'51"
Stephen Holmes
Dra 16h23'51" 58d9'15"
Stephen Howson
Uma 9h22'35" 51d58'50"
StePHen Hyer
Cnc 9h4'28" 16d16'52"
Stephen I. Donnelly
Lmi 10h35'52" 36d11'16"
Stephen J. Bach
Crb 15h49'17" 38d56'0"
Stephen J Boucher
Per 3h10'48" 55d44'26"
Stephen J. Bundra
Ori 5h9'36" -3d21'50"
Stephen J. Cole
Sco 16h29'26" -27d33'20"
Stephen J Connelly
Cru 12h27'24" -60d42'29"
Stephen J. Consales
Aur 5h53'27" 52d26'16"
Stephen J Dix
Leo 11h18'58" 26d8'47"
Stephen J Earley
Tau 4h23'37" 26d52'29"
Stephen J. Hutchings
Vir 14h45'14" 2d5'1"
Stephen J. Kaluzny
Cyg 20h14'6" 44d13'53"
Stephen J Leckband
Uma 9h14'13" 53d22'15"
Stephen J. Romeo, Sr.
Cep 22h27'32" 63d45'10"
Stephen Jackson Smith III
Lmi 9h37'51" 38d11'20"
Stephen James
Hya 9h55'57" -25d46'5"
Stephen James Crimes
Cep 3h44'42" 78d9'34"
Stephen James Crosby
Sco 17h15'48" -37d46'51"
Stephen James Dewar
Cyg 20h29'31" 48d17'0"
Stephen James Diggins
Cen 13h42'57" -41d53'40"
Stephen James Fenn
Cep 22h11'22" 56d19'35"
Stephen James Hole BSc (Hons)
Uma 9h24'25" 52d18'32"
Stephen James Kent
Uma 8h55'11" 55d17'24"
Stephen James Michie
Umi 14h43'40" 80d24'14"
Stephen James Pharez
Her 16h25'45" 15d1'51"
Stephen James Saganich
Aql 19h58'59" 12d44'49"
Stephen James Young
Cyg 21h46'0" 42d58'2"
Stephen Jandecka
Aql 19h8'37" -0d0'36"
Stephen & Jeana Bahl
Lmi 10h51'8" 29d23'26"
Stephen & Jeannette Zupko
Cyg 19h38'56" 52d37'15"
Stephen Jeffery Peacock
Tau 4h3'37" 21d57'18"
Stephen & Jennifer Hunt
Ori 6h5'23" 0d35'4"
Stephen & Jennifer Nicholas
Cyg 20h26'20" 45d47'51"
Stephen Jerrold Hayden
Lib 15h8'14" -5d55'23"
Stephen John Babjak
Uma 10h25'54" 41d44'4"
Stephen John Bashaw
Tau 4h7'34" 2d12'37"
Stephen John Brooker
Cas 3h29'13" 77d4'39"
Stephen John Kuduk
Her 17h21'31" 15d30'5"
Stephen John Kuhn
Her 17h16'20" 35d4'54"
Stephen John Logue Jr.
Lib 15h15'26" -20d19'21"
Stephen John Manville - Dad's Star
Ori 5h56'15" 7d3'8"
Stephen John Oliva
Cap 20h25'44" -26d45'16"
Stephen John Oram
Uma 10h6'56" 49d25'56"
Stephen John Proud
Uma 10h32'12" 49d20'34"
Stephen John Smith
Umi 14h53'40" 72d11'41"
Stephen John Warwick Family Star
Per 4h13'51" 52d0'24"
Stephen John Wolstenholme
Psc 0h46'49" 6d12'51"
Stephen Johnson
Tau 3h25'20" 8d59'59"
Stephen Joseph Crawford
Cra 18h51'54" -40d57'9"
Stephen Joseph Hagan
Gem 6h34'51" 19d0'37"
Stephen Joseph Jeffrey
Cep 23h18'2" 72d19'50"
Stephen Joseph Knox
Lmi 10h26'26" 34d55'40"
Ste"pH"en Joseph Kovacs
Ari 1h59'21" 14d54'49"
Stephen Joseph Malec
Cep 22h41'50" 73d46'51"
Stephen Joseph Minichiello
Uma 13h52'10" 56d51'10"
Stephen Joshua Thompson
Boo 14h39'4" 54d44'43"
Stephen Jr.
Her 16h23'52" 45d45'14"
Stephen K. Freeman
Her 17h48'58" 38d49'49"
Stephen K. Potapa
Vir 13h56'23" 4d40'41"
Stephen K. Smith
Aql 20h4'0" 7d23'25"
Stephen Karl Owens
Cap 21h25'56" -23d17'49"
Stephen Karl Wilson
Uma 8h36'16" 59d2'52"
Stephen Kasirajan
Cyg 20h48'49" 46d31'21"
Stephen & Katie Beth Hall
Aqr 22h14'39" -3d37'47"
Stephen Kelley
Ori 5h6'47" -1d12'28"
Stephen Kennedy
Aur 5h22'48" 48d51'40"
Stephen Kenneth Wood
Ori 6h10'58" 15d46'26"
Stephen Kerr Allison
Per 3h42'22" 41d23'22"
Stephen Kopcha
Per 3h15'15" 52d46'0"
Stephen Korenek
Aqr 22h16'17" -4d47'55"
Stephen Kropp
Cma 7h23'9" -30d6'5"
Stephen Kueffner
Her 17h55'47" 18d26'22"
Stephen Kyle McDermott
Aql 19h43'55" -0d6'26"
Stephen Kyle Teafatiller
Her 16h28'25" 19d48'23"
Stephen Kyle Veater
Uma 12h47'34" 62d57'33"
Stephen L. Bowen DVM Eagle Scout
Aql 19h7'41" -0d23'50"
Stephen L. Cleaver
Uma 11h6'0" 59d57'22"
Stephen L. George
Per 3h43'20" 46d49'54"
Stephen L. Higgs
Cep 22h23'37" 64d51'17"
Stephen L. Jacobs
Her 17h22'6" 38d46'36"
Stephen L. Seamon
Her 18h3'58" 37d26'37"
Stephen L. Williams
Uma 11h27'47" 48d11'42"
Stephen L Zackoski
Lib 15h48'38" -18d27'15"
Stephen & LaCeta Walters 2006
Lib 15h11'39" -28d25'31"
Stephen Lane
Cyg 19h59'31" 37d25'20"
Stephen Lawrence Hoelscher
Boo 14h51'45" 54d27'37"
Stephen Lee Jackson
Tau 5h20'46" 22d38'13"
Stephen Lee McDaniel
Uma 12h56'38" 54d36'36"
Stephen Leonard Griffith
Cas 2h27'56" 67d58'4"
Stephen Leroy Gronseth
Leo 11h57'13" 27d2'54"
Stephen & Lisa
Sge 20h3'20" 17d9'2"
Stephen & Lisa Sands
Uma 13h40'35" 58d34'10"
Stephen Little
And 23h26'24" 48d39'30"
Stephen Love
Uma 9h1'19" 56d18'0"
Stephen & Lucy Calder
Cyg 20h18'14" 53d6'44"
Stephen M. Rems
Per 2h58'43" 54d52'56"
Stephen M. Toth
Uma 11h53'3" 46d8'27"
Stephen M. Wilson (my forever love)
Vir 12h13'20" 4d2'11"
Stephen Mark Brown
Per 4h27'16" 43d15'35"
Stephen Mark Jokela, Jr.
Her 18h4'37" 36d10'26"
Stephen Martel
Sgr 19h4'48" -33d8'38"
Stephen Matthew Schwartz
Per 4h1'9" 41d21'19"
Stephen Maxwell Tkach
Crb 15h47'14" 38d8'5"
Stephen McArthur
Sgr 17h48'48" -21d58'28"
Stephen McDonough
Cyg 19h59'55" 30d9'35"
Stephen McShane
Per 2h24'40" 54d1'45"
Stephen Meteyer
Per 2h18'53" 54d9'23"
Stephen Michael Brown
Tau 5h30'23" 23d29'51"
Stephen Michael Crane
Crb 16h8'58" 31d20'6"
Stephen Michael Gill
Tau 3h53'30" 27d48'19"
Stephen Michael Haigh Jr.
Psc 0h54'15" 10d32'54"
Stephen Michael Keogh
Leo 11h9'53" 24d41'2"
Stephen Michael McClain 9/06/56
Ori 5h47'51" 11d15'14"
Stephen Michael Osborn
Uma 9h26'10" 55d59'1"
Stephen Michael Stahl
Tau 5h45'56" 21d18'39"
Stephen Michael White
Aql 20h12'49" 6d13'13"
Stephen Micheal Brown
Her 17h5'27" 32d43'31"
Stephen Millington
Per 2h11'4" 54d20'49"
Stephen Miron 40 Years
Tau 5h11'33" 18d32'28"
Stephen Moltenbrey
Lib 14h58'2" -0d52'27"
Stephen Montgomery Black
Uma 13h40'28" 49d13'52"
Stephen Montgomery, (my boy)
Her 18h26'54" 18d1'51"
Stephen Murphy
Her 16h29'21" 48d44'0"
Stephen Murrin
Sco 16h5'55" -11d36'36"
Stephen Neil Skinner
Cep 21h33'50" 83d1'56"
Stephen Noble's Super Nova
Cam 4h29'55" 76d15'28"
Stephen Noel Ashbarry
Cap 20h53'37" -20d24'43"
Stephen O. Tate
Ori 6h15'28" 13d48'19"
Stephen O'Brien
Ori 5h51'36" -3d13'15"
Stephen Olges
Cyg 21h39'26" 50d41'2"
Stephen Olinsky
Ari 2h7'38" 26d13'11"
Stephen P. Kledaras
Boo 14h54'31" 22d31'6"
Stephen P Leone Our Shining Star
Uma 9h51'7" 49d45'9"
Stephen Palyocsik
Per 3h48'22" 40d41'48"
Stephen Papaloizou
Ori 5h16'42" -8d53'48"
Stephen & Patricia Aldred
Cyg 20h42'1" 32d43'5"
Stephen Patrick Daubel
Per 4h14'0" 51d20'46"
Stephen Patrick Durtschi
Car 9h27'46" -63d1'11"
Stephen Patrick Nicholas Boyle
Uma 9h18'48" 71d21'22"
Stephen Paul Lombardi
Per 3h41'55" 38d8'28"
Stephen Paul Nix
Cep 21h29'29" 60d52'15"
Stephen Peter Wilson
Hya 9h36'40" -11d40'17"
Stephen Pettit
Aur 5h39'5" 48d53'59"
Stephen Phelps & Allen Reed United
Tri 2h30'43" 31d8'15"
Stephen Phillip Kelly
Gem 7h11'17" 33d20'28"
Stephen Pierre Lepine
Umi 14h42'34" 75d8'35"
Stephen Piotrowski
Lib 15h4'30" -11d23'9"
Stephen Povelko
Ori 5h19'11" 15d28'2"
Stephen Powers
Ori 6h4'18" -0d21'9"
Stephen Powilatis
Boo 15h7'40" 40d54'16"
Stephen Prescott Donohue
Psc 0h36'38" 7d6'39"
Stephen Pypnioskie
Aur 5h32'39" 38d53'9"
Stephen R. Barker
Cam 4h19'45" 68d19'11"
Stephen R. Bobbett
Tau 3h46'2" 9d0'5"
Stephen R. Fava
Sgr 18h57'21" -21d10'24"
Stephen R. Grater
Gem 7h0'38" 28d53'22"
Stephen R. Hesse
Ari 2h56'24" 21d13'31"
Stephen R. Tittle, Sr.
Aur 5h10'9" 41d59'30"
stephen ray finnell
Uma 11h5'1" 56d8'29"
Stephen Raymond Franke
Uma 10h55'59" 60d6'9"
Stephen Retenski
Psc 22h58'42" -0d1'37"
Stephen Rhoads
Aql 19h56'40" 4d29'23"
Stephen Richard O'Daniels
Cnc 9h6'34" 8d0'40"
Stephen Robert
Vir 12h50'4" 0d58'41"
Stephen Robert McCulloch
Gem 7h11'19" 21d27'51"
Stephen Robert Minns
Psc 1h59'23" 2d50'59"
Stephen Robert Williams
Aqr 22h15'53" 2d16'19"
Stephen Rocco Scotti
Per 2h6'39" 54d2'33"
Stephen Roy Mischley (1948)
Vir 14h40'26" 4d0'21"
Stephen Russ
Sco 17h56'17" -30d19'50"
Stephen Ryan Foskey
Uma 9h12'11" 63d40'41"
Stephen Ryan Hall
Per 4h20'2" 34d1'41"

Stephen Ryan Kelly Her 18h22'59" 27d16'45"
Stephen & Sandra Murphy Cru 12h23'39" -58d36'2"
Stephen & Sarah Pratt "Forever" Cyg 21h43'37" 41d55'22"
Stephen & SarBear Vir 13h3'56" 3d3'54"
Stephen Scomis Per 2h48'37" 54d11'45"
Stephen Scott Abernathy Her 17h36'11" 36d10'54"
Stephen Seftar II Umi 15h30'40" 71d43'23"
Stephen Singh - Forever In My Heart Her 16h30'9" 18d26'51"
Stephen Smith Uma 13h8'6" 63d19'20"
Stephen Smith 40 Per 3h21'24" 37d50'50"
Stephen Stanko Vir 11h49'35" -3d29'18"
Stephen (Star of Sweet!) Tau 5h25'55" 27d16'22"
Stephen "Steve" DeGregorio Lib 15h43'8" -13d12'30"
Stephen (STP) Andrews Sco 17h37'16" -39d37'53"
Stephen Stumpf Sco 16h15'23" -12d39'43"
Stephen Superman Waterson Cru 12h33'1" -56d57'6"
Stephen & Susan Ranous Dra 18h48'56" 67d28'41"
Stephen & Suzanne Dickinson Lyn 6h37'38" 55d31'37"
Stephen T. Leone, Jr. Aur 5h33'46" 41d25'35"
Stephen Thayer Martin Aur 5h40'14" 37d5'33"
Stephen the Great Pho 0h38'32" -50d6'2"
Stephen Thomas Groves Ari 2h10'49" 27d26'42"
Stephen Thomas Milford 1949-2004 Gem 7h55'53" 31d4'47"
Stephen Thomson Cap 21h43'35" -11d38'17"
Stephen Tolman Glass Per 2h27'42" 56d12'19"
Stephen "Tucker" Belangia, Jr. Ori 5h58'2" -0d43'4"
Stephen Valentini Aqr 22h11'32" -0d16'23"
Stephen Verl Brown Ori 6h6'47" 17d11'39"
Stephen Viramontes Per 2h46'55" 43d46'22"
Stephen W. Chase Per 3h22'32" 32d12'36"
Stephen W. Hassard Her 17h0'29" 34d33'17"
Stephen W. Hensley Cep 0h10'2" 69d2'23"
Stephen W. Hoots Ori 4h50'20" 12d10'27"
Stephen W. Knipf Uma 11h38'8" 35d17'4"
Stephen W. Morrill Uma 10h50'5" 39d38'27"
Stephen Walsh Per 3h10'48" 45d37'9"
Stephen Watson Boo 14h40'55" 23d5'40"
Stephen Wayne Holmbeck Leo 10h58'11" 16d18'57"
Stephen Wayne Kent Aqr 22h20'7" -24d12'16"
Stephen Weaver And 0h18'14" 37d36'28"
Stephen Westwood Uma 10h9'19" 60d18'25"
Stephen William Cable Per 2h52'31" 49d16'11"
Stephen William Kilian "01/15/2004" Cap 20h26'35" -10d58'27"
Stephen William Phillips Umi 15h12'40" 81d27'39"
Stephen William Skillman Her 17h47'49" 38d7'26"
Stephen Wolsh Her 17h7'24" 31d34'34"
Stephen Woods Aql 18h58'51" 15d38'20"
Stephen (Woody) Woodward Cyg 21h2'35" 31d41'53"
Stephen, the Dad Star Sgr 18h3'49" -28d36'3"
Stephen1701 Ori 6h6'35" -3d50'43"
Stephen-831 Per 3h46'17" 48d24'2"
StephenAnthonyChristopher McKearnin Cnc 8h35'40" 32d53'14"
Stephendeborah Pho 2h15'46" -46d24'51"
Stephenie Sgr 19h35'27" -15d24'14"
Stephenie Carla Cason Psc 1h11'24" 16d30'38"
Stephenie Lei Menzies Ari 2h50'34" 14d12'47"
Stephenie Su And 0h49'3" 39d50'17"
STEPHENS Cnc 8h49'27" 27d15'50"
Stephens Lib 15h22'39" -7d42'50"
Stephen's 351XC Star In The Sky Cru 12h23'45" -63d19'42"
Stephen's Astellas Star Ari 3h8'37" 28d28'0"
Stephen's Spirit Ori 5h5'40" 4d12'45"
Stephen's Star Gem 7h11'4" 20d46'24"
Stephentina Cyg 19h37'17" 33d53'54"
Stephery Adam Cethoute Peg 22h45'48" 13d30'12"
Stephi Cnc 8h29'24" 15d7'8"
Stephi3481 Psc 1h26'7" 32d2'52"
Stephie Sco 17h26'2" -40d19'27"
Stephie and Kyle's Star Ari 2h41'42" 27d55'36"
Stephie Angel Lyn 8h0'49" 40d47'5"
Stephie Bear And 0h39'7" 31d19'40"
Stephie Brown Lyn 7h32'21" 49d59'30"
Stephie Wong 84 Sco 17h53'34" -38d43'18"
Stephielle Lib 15h44'7" -9d19'52"
StephieLovesChris Uma 8h43'52" 55d42'27"
Stephie's Star And 0h39'46" 22d56'40"
Stephi's Fish Psc 2h0'29" 7d10'19"
StephLyons And 2h33'58" 49d29'8"
Stephnie Uma 9h53'26" 61d47'37"
Stephnie Vir 13h49'47" -6d29'55"
StephNR Cnc 8h44'4" 18d41'54"
Stephs One And Only Star Sco 16h55'46" -38d38'49"
Steph's place in the heavens. Cap 21h35'23" -15d42'59"
Steph's Second Star Lyn 7h22'56" 53d41'21"
Stephs Star Tau 4h12'5" 3d48'19"
Steph's Star Ori 5h45'1" 8d47'37"
Steph's Star And 1h14'7" 37d30'56"
Stephsar Haffbrow Lyn 7h33'50" 36d17'45"
StephStevJesJoMoore Lyn 9h42'25" 40d54'11"
Stephy And 2h11'53" 38d47'8"
STEPHY Ari 2h9'22" 16d16'33"
Stephy's Star Vir 12h5'44" -6d16'28"
Stepien Her 16h50'5" 47d44'26"
Steppy Cru 12h32'52" -61d39'29"
Sterffy Crb 16h8'7" 39d2'44"
Stergios Kalogiros Uma 14h5'56" 54d27'23"
Sterki George Lac 22h22'43" 47d31'36"
Sterki Werner Her 18h43'26" 20d9'51"
Sterl And 23h6'22" 51d2'17"
Sterling Cas 23h37'13" 52d12'49"
Sterling Lib 15h21'6" -13d48'38"
Sterling Sgr 18h19'24" -32d24'38"
Sterling Dale Fuentes Ori 4h47'44" 12d5'54"
Sterling Elizabeth Crb 15h40'42" 26d43'36"
Sterling Grace Norris Cnc 8h15'54" 9d27'56"
Sterling K. Schaefer Uma 10h57'45" 59d12'56"
Sterling Michael Robert Malone Tri 2h15'3" 33d3'7"
Sterling & Michelle's 1st Year Uma 9h19'45" 45d36'45"
Sterling Rose Ari 2h19'43" 20d7'48"
Sterling Silver Lib 15h13'28" -22d23'48"
Sterling Silver Lady Cas 0h51'38" 60d39'19"
Sterling Travis Golden Cap 21h30'12" -14d14'58"
SterlingMephisto Leo 10h6'15" 24d50'51"
Sterling's Star Her 16h56'7" 17d43'34"
Stern Christel Tau 4h23'26" 13d34'32"
Stern der 2 Knödelchen Ari 3h2'43" 18d53'39"
Stern der Adams Uma 13h3'17" 57d24'53"
Stern der Liebe von Nadja & Georges Uma 9h25'11" 63d10'38"
Stern der Schmusekatze Uma 10h42'31" 51d7'17"
Stern der Weisen Uma 13h56'13" 49d27'26"
Stern des Herzens Wolle und Maja Ori 6h3'4" 13d46'14"
Stern des Lebens Crb 15h36'20" 36d49'46"
Stern Stefanie J.G. Ori 6h4'15" 14d7'1"
Stern Toni München Uma 8h35'9" 52d38'12"
Stern von Markus J. Gräßler Uma 10h31'31" 58d54'51"
Stern7 Sonja und Dominik Ori 6h18'3" 16d24'3"
Sternberg, Jörg & Jacqueline Ori 6h21'56" 15d58'48"
Sterndli forever Umi 16h23'24" 73d10'37"
Sterne Ochsi Umi 15h9'23" 70d26'23"
Sternenlicht Simone Uma 12h39'59" 58d33'56"
Sternenprinzessin Elissavet Sgr 17h52'7" -28d45'42"
Sternenzauber von Stephan und Franziska Crb 15h51'10" 26d32'19"
Sternli & Engeli And 23h35'59" 40d2'40"
Sternschnuppe #19 Cas 23h13'11" 59d38'11"
Sternzauber Cnc 8h15'43" 16d51'35"
Steron Col 5h35'46" -34d6'54"
Sterrantino, Giorgio Ori 6h20'19" 15d59'50"
Steso Umi 15h11'2" 76d21'29"
Stevan Watkin Southgate Dra 11h16'2" 77d41'8"
StevCathBD Tau 4h16'22" 10d36'35"
Steve Vir 12h41'56" 0d37'13"
Steve Tau 5h8'3" 21d22'48"
Steve Aql 19h4'55" 15d47'39"
Steve Leo 10h29'5" 26d47'38"
Steve Her 17h6'50" 28d48'28"
Steve Cyg 21h26'55" 39d23'15"
Steve Aur 5h42'37" 46d56'35"
Steve Cas 23h32'15" 51d20'43"
Steve Umi 17h30'12" 80d12'14"
Steve Lib 15h5'1" -0d57'37"
Steve Lib 15h6'20" -1d13'17"
Steve Sgr 19h5'47" -18d23'33"
Steve Cep 0h14'40" 71d35'42"
Steve A. Cruz Cmi 7h32'47" 8d17'5"
Steve Adams Psc 1h10'10" 29d31'5"
Steve Albershardt Nor 16h7'38" -57d57'0"
Steve Albright and Family Uma 10h21'49" 63d58'15"
Steve Allen Cyg 21h22'16" 35d16'13"
Steve Allen Rios Vir 12h43'34" -6d52'43"
Steve Aloi Tau 4h11'53" 15d43'36"
Steve and Anne Van Lue Lib 15h13'51" -6d18'57"
Steve and Beka Aql 19h54'47" 8d54'58"
Steve and Bern's Star Cru 12h36'40" -56d23'3"
Steve and Bette Peck Cyg 20h33'38" 43d14'49"
Steve and Denise Ori 5h11'55" -0d32'40"
Steve and Diana Fusco-Tompkins Cnc 8h49'32" 8d13'18"
Steve and Hollie Gorczany Crb 15h29'27" 28d53'39"
Steve and Irene McLaughlin Peg 22h54'40" 19d18'46"
Steve and Jacque Her 16h39'22" 36d40'7"
Steve and Jerri Sge 20h5'45" 16d43'57"
Steve and Joan Tumin Sge 19h45'5" 17d4'21"
Steve and Katie Cap 20h39'9" -21d19'8"
Steve and Kim Cyg 21h19'15" 38d3'31"
Steve and Kristie's STAR of LOGAN Tau 3h26'24" 4d44'0"
Steve_and_Lauren_52507 Per 2h17'52" 55d53'14"
Steve and Linda Her 18h57'11" 16d47'16"
Steve and Lindsays Star Crb 16h20'44" 33d21'30"
Steve and Megan Forever Cyg 21h54'58" 37d34'26"
Steve and Melissa's twinkling star Cyg 20h13'39" 49d55'45"
Steve and Niki Sgr 18h21'52" -22d31'20"
Steve and Raquel's Star Cyg 20h49'48" 44d33'59"
Steve and Sally Kelly Eri 4h36'0" -0d9'7"
Steve and Susie Forever! Ori 5h20'16" 0d35'18"
Steve and Tami Seimers Cyg 20h27'3" 44d39'38"
Steve And Tarnya Welsch Pho 0h11'46" -43d40'10"
Steve and Val Yang Cyg 19h41'17" 54d15'39"
Steve & Andrea Sipes Per 2h41'19" 53d25'51"
Steve Andrew Culwell Ori 5h54'3" 17d57'12"
Steve Avery Connally, Sr. Ori 5h39'29" 4d50'25"
Steve Azar Ari 2h19'43" 25d28'58"
Steve Baxter Boo 14h55'8" 26d54'14"
Steve Berkowitz Psc 0h14'10" 8d4'52"
Steve Berman 63 Sco 16h48'39" -33d13'14"
Steve Berzansky Cep 21h14'44" 58d20'41"
Steve & Betty Per 3h16'50" 37d22'27"
Steve Black - Shining Star Boo 14h48'27" 23d51'33"
Steve Bradley Her 17h14'15" 19d45'19"
Steve Brady Per 3h8'22" 52d39'29"
Steve Brashear Ori 6h21'28" 13d20'26"
Steve Britt Her 16h35'46" 35d23'4"
Steve Brock Her 17h21'24" 31d59'13"
Steve Burylo 12/1/02 Per 2h22'55" 54d38'7"
Steve C. Cook Per 3h10'5" 53d19'43"
Steve C. Solis Leo 10h7'22" 25d17'29"
Steve & Candice's Shooting Star Uma 9h7'15" 54d42'49"
Steve Carlos Barber Vir 14h6'23" 5d8'41"
Steve & Caroline Fry's Wedding Star Cyg 21h31'40" 55d19'24"
Steve & Cassie's Star Uma 10h43'53" 61d2'51"
Steve & Cath Her 16h53'28" 35d15'49"
Steve Caylor Tau 4h54'14" 22d31'25"
Steve & Cheryl Cyg 20h50'2" 35d59'49"
Steve & Cheryl's Forever Happiness Cyg 19h47'26" 35d34'36"
Steve Childs - Happy 40th Cru 12h57'10" -59d10'54"
Steve Chizmadia Aqr 22h50'29" -1d38'37"
Steve Cholodnuik Cep 20h37'40" 61d36'28"
Steve Collings Aqr 21h47'17" -0d56'4"
Steve Copeland Her 17h0'19" 35d2'44"
Steve Cothran And 2h2'22" 42d8'22"
Steve Crowe Cep 21h51'19" 56d14'20"
Steve & Cynthia 12 Ori 5h16'54" 0d22'49"
Steve Czyznik Uma 11h56'30" 35d21'43"
Steve Danchak And 0h41'15" 23d28'39"
Steve Dartnal Peg 23h28'3" 15d7'47"
Steve Davaris III Uma 8h33'51" 71d27'44"
Steve & Deb Parelman Cyg 20h19'4" 54d34'51"
Steve & Debbie Van Natta Cyg 20h7'29" 30d34'12"
Steve DeBoer Uma 11h57'4" 58d15'53"
Steve Del Rosso Vir 14h30'49" 5d6'15"
Steve & Demi Tau 3h53'13" 13d12'16"
Steve Deveney Dra 17h10'45" 65d36'29"
Steve Dickey Lib 14h48'21" -1d1'45"
Steve Dobbertin Her 17h32'22" 28d36'26"
Steve Doll Cap 21h41'55" -15d39'5"
Steve & Donna Marriott Cyg 20h57'43" 37d34'41"
Steve Douglas-World Greatest Father Ori 4h52'20" 13d19'20"
Steve Doyle Vir 12h42'9" 6d41'4"
Steve Eckert Cap 20h18'36" -11d41'49"
Steve Edmundson Uma 9h53'22" 61d1'31"
Steve & Elena And 1h43'45" 42d44'32"
Steve & Emily Leo 9h37'49" 29d25'50"
Steve Erny Cep 21h50'3" 64d18'3"
Steve Esterbrooke Cyg 19h34'0" 29d41'31"
Steve & Evelyn Forgach Ori 5h57'35" 11d29'21"
Steve & Evelyn Redding Ori 5h7'22" 15d1'3"
Steve F. Fish Uma 11h24'47" 37d32'31"
Steve F. Polgar Uma 9h11'8" 68d41'56"
Steve Fausett Jr. Per 3h37'20" 41d46'33"
Steve & Freda Galloway Ori 5h58'29" 18d14'35"
Steve & Gabby Forever Cru 12h18'48" -58d45'13"
Steve Gallo Sco 17h23'26" -30d23'58"
Steve Garrett Johnson Ari 2h23'41" 24d34'37"
Steve Gates Her 17h49'26" 38d41'23"
Steve G.Carazo Aqr 21h5'23" -11d18'45"
Steve Gillespie Leo 11h56'4" 23d6'44"
Steve Ginn Her 18h16'49" 15d3'4"
Steve Glassgow Gem 6h39'37" 22d53'52"
Steve Golden Lyn 9h18'47" 36d47'3"
Steve Grayson Sex 10h28'40" -2d58'14"
Steve Greathouse Ori 5h35'30" 10d32'10"
Steve Haberman 2006 Uma 13h51'10" 53d21'14"
Steve Hammond Umi 14h10'33" 73d11'41"
Steve Harris Lyn 8h9'12" 39d26'13"
Steve Haverly Cnc 8h30'48" 21d21'31"
Steve Heinz Cnc 9h11'32" 16d16'2"
Steve Hemmings Cnv 13h41'43" 40d34'50"
Steve Hoenniger Ori 5h41'34" 2d47'22"
Steve Hopkinson Uma 11h28'4" 37d12'32"
Steve Howard Cep 1h56'31" 84d41'49"
Steve Hrycak Gem 6h51'46" 32d40'12"
Steve Irwin Cru 12h30'56" -61d23'28"
Steve Irwin - Dad Husband Croc Hunter Ori 5h41'29" -1d38'23"
Steve Irwin - Wildlife Warrior Cru 12h32'30" -57d44'48"
Steve J. Condrack Aur 5h37'53" 33d17'38"
Steve J. Vandevelde Lib 15h29'23" -6d13'47"
Steve James MacMillan Lyr 18h34'30" 39d44'35"
Steve Jameson Cru 12h5'50" -63d30'22"
Steve & Jennie Wickens in luv 4ever Cyg 21h42'2" 36d55'6"
Steve Johnson Her 17h5'49" 13d27'40"
Steve Joseph Boo 14h24'8" 47d22'9"
Steve & Kathy Clair Kruse Lyr 19h23'32" 37d31'45"
Steve Keane - Orion Consulting, Inc. Ori 5h10'45" -0d36'50"
Steve & Kelly Armitage Leo 11h2'54" 21d38'34"
Steve & Kelly Bryant Cyg 21h16'53" 43d35'17"
Steve Kennedy's Star Cnc 8h40'21" 30d40'18"
Steve Kern Uma 10h49'31" 71d26'1"
Steve Kim Pat Whitney Connor Castro Umi 13h40'16" 76d46'40"
Steve Kitch Ori 5h49'23" 5d23'57"
Steve & Kristen - For All Eternity Ori 5h39'39" -1d51'58"
Steve Kurt Shoaf Aur 7h12'36" 42d49'5"
Steve L Freeman Her 17h51'26" 25d57'35"
Steve Labossier Her 17h51'53" 46d10'8"
Steve Lappin Ari 3h3'35" 16d11'59"
Steve & Laura Cyg 20h48'41" 35d43'59"
Steve LeBrun Aqr 22h11'25" 0d10'39"
Steve & Lisa Smith Cyg 19h38'5" 34d53'35"
Steve & Liza Cyg 20h49'13" 35d22'32"
Steve Lockyer Eri 4h27'36" -35d17'7"
Steve & Lori Brockett Uma 11h57'1" 61d0'1"
Steve Love Aur 5h42'26" 40d26'12"
Steve Loves Camila (V-day 2006) Cyg 20h35'57" 44d7'38"
Steve Loves Helen Star Tau 4h11'42" 23d54'23"
Steve M. Sladek Ori 5h31'24" 10d19'38"
Steve Madison Her 17h12'23" 32d6'52"
Steve Maggart Dra 19h7'26" 70d53'20"
Steve & Marcia Walsh Cyg 21h21'24" 35d40'29"
Steve Marks Her 16h17'39" 6d49'36"
Steve Marouf Psc 0h4'50" 10d3'42"
Steve Marston And 0h26'32" 32d57'38"
Steve Mavrides Per 4h27'18" 49d20'26"
Steve Mavrides Lib 15h49'12" -9d23'59"
Steve Mc Nally, alias DD Cyg 19h55'21" 32d23'17"
Steve McElligott Sco 16h29'51" -25d7'49"
Steve McGill Happy 1st Anniversary Per 4h8'2" 45d19'14"
Steve McGowan Cap 20h24'34" -11d11'20"
Steve McKay Uma 10h59'33" 46d40'13"
Steve Michael Taylor Her 17h38'44" 39d27'49"
Steve Miller Free at Last Uma 11h8'32" 54d12'15"
Steve "Moe" Karol Admonis Per 4h40'45" 47d23'21"
Steve Monaco Cas 0h15'4" 58d43'24"
Steve Moses Aql 19h27'36" -7d50'17"
Steve My wish upon a star 021597 Her 17h35'48" 38d10'29"
Steve n Sophia Cnc 8h41'25" 23d17'27"
Steve & Nancy Lorenzo Uma 12h24'26" 54d44'39"
Steve Neff Aqr 22h41'2" -23d28'25"
Steve Nyman Gem 7h39'51" 33d28'15"
Steve Osborne Her 17h15'24" 15d52'32"
Steve Oshrin's Bed Bug Ori 6h17'32" 10d31'30"
Steve "Pappy" Orenyak Uma 11h53'47" 42d26'28"
Steve Park Gem 6h52'8" 22d19'28"
Steve Parker Aql 18h58'20" -0d11'42"
Steve & Patricia Berkley Cam 3h41'56" 57d33'32"
Steve Pawlewicz Cas 0h11'11" 66d40'52"
Steve "PawPaw" James Gem 7h35'31" 25d52'33"
Steve Peters' Star Uma 13h41'17" 56d15'21"
Steve Post Boo 14h41'36" 34d1'25"
Steve Primeau Crb 16h14'23" 30d13'18"
Steve Prokes Aqr 22h3'41" 0d45'33"
Steve & Rachael Clancy Ori 5h59'56" 18d27'40"
Steve+Rachel Fanning Endless Lovers Her 18h34'31" 20d5'47"
Steve Rice Cyg 21h5'50" 41d2'34"
Steve & Rita Browy Cyg 19h34'5" 27d54'5"
Steve Roberts Ori 5h27'20" 1d30'24"
Steve Robinson Lib 15h17'38" -15d35'53"
Steve & Rochelle Montoya Lib 16h1'39" -6d20'13"
Steve Rochna Dra 17h47'9" 62d33'22"
Steve Rogers Aql 19h45'56" 16d25'8"
Steve Rooklidge Lac 22h18'41" 51d59'21"
Steve & Ruth Ann Faszholz Umi 15h25'32" 73d24'42"
Steve S. Lenox Per 3h30'49" 51d41'30"
Steve & Sammy Col 5h45'39" -32d36'14"
Steve Sanchez III Boo 14h19'3" 18d3'51"
Steve Sayang Cisca Ari 2h19'23" 11d32'38"
Steve Schaaf Peg 22h42'21" 26d27'11"
Steve Sedelko Lmi 10h46'4" 28d35'8"
Steve Settle Ori 5h59'19" 6d18'48"
Steve Shafer Cep 3h21'43" 78d39'36"
Steve Shearer Uma 12h2'24" 55d18'48"
Steve Shearer Uma 13h32'46" 53d9'24"
Steve Shields Her 16h32'38" 38d3'28"
Steve & Shirley Forever & Always Cyg 20h50'41" 46d43'38"
Steve Siddals Cep 20h54'45" 69d4'12"
Steve Siegmann Ari 2h38'45" 18d18'13"
Steve Simmons Gem 7h22'24" 14d39'30"
Steve Smith Her 16h46'18" 37d29'5"
Steve Smith Dra 20h20'28" 71d36'47"
Steve Smith "My Mr. Destiny" Gem 7h54'41" 13d37'3"
Steve Stevenson Aql 20h13'29" 14d56'17"
Steve Stout Dra 20h35'50" 68d59'18"
Steve Sullivan 50 Year Star Sgr 17h51'43" -29d55'57"
Steve Sumpter Cnc 8h42'28" 16d10'35"
Steve & Suzi LFD Soulmates Forever Cyg 20h2'44" 44d38'28"
Steve Szybowski, Musician & Artist Uma 9h26'34" 67d13'57"
Steve TenBroeck Vir 14h54'50" 1d23'19"
Steve "Tevie" Hawes Leo 11h14'29" 14d10'38"
Steve The Dinosaur Cet 3h7'25" 5d49'10"
Steve "The Meanie" Holmes Her 18h7'55" 36d44'25"
Steve Thorson: Husband and Father Per 3h25'25" 43d25'47"
Steve & Tiffany Cramer Gem 6h44'7" 15d43'28"

Steve & Tracy Moorcroft
Cyg 19h48'39" 59d2'1"
Steve Travis AKA Elvis
Vir 14h16'37" 5d48'38"
Steve V. Johnson
Leo 9h52'20" 14d16'52"
Steve Valdez
Gem 6h41'28" 25d19'49"
Steve Vanoy
Her 17h48'10" 34d17'57"
Steve Vilinsky
Cnc 8h44'36" 7d3'29"
Steve VonHolle
Gem 6h54'34" 17d43'21"
Steve W. Crittenden
Ori 5h30'6" 8d27'5"
Steve Wallis
Ori 5h20'48" -1d26'4"
Steve Webber
Uma 8h39'28" 67d41'50"
Steve Welch
Cma 7h18'47" -12d31'14"
Steve White
Cmi 7h23'52" 3d8'50"
Steve Whiteman
Lib 14h56'19" -10d14'21"
Steve Williams
Aql 19h48'18" 13d28'26"
Steve Williams' Solarius
Cet 1h10'32" -0d0'19"
Steve Wilson
Cnc 8h38'59" 16d2'38"
Steve Worthington
Sgr 19h8'0" -17d44'28"
Steve Wyatt
Lib 15h8'34" -6d45'54"
Steve- You are my shining star
Gem 7h4'52" 33d48'26"
Steve Zigler
Ori 6h4'40" 5d0'51"
Steve Zumsteg
Boo 14h48'10" 37d30'59"
Steve, Karen, Colin & Colton Duell
Lyr 19h4'49" 42d2'5"
Steve, Kristin, and Jillian
Leo 11h4'12" 8d36'36"
Steve, my favorite democrat!
Boo 14h10'54" 28d47'12"
Steve, My Shining Star
Uma 9h27'54" 47d1'33"
Steve,Michele,Alissa&Nicole Juliana
Umi 16h12'18" 74d23'59"
Steve80
Per 4h33'35" 41d38'20"
SteveChar
Uma 10h8'20" 50d57'46"
SteveCrystalScrappy
Lyn 8h0'55" 36d26'51"
SteveG-SteveG
Lyn 7h48'51" 57d44'7"
Stevekelly Star
Uma 10h14'9" 46d44'46"
StevElena Gillis
Umi 15h10'50" 78d43'35"
Steven
Umi 14h54'26" 86d43'38"
Steven
Per 4h15'57" 44d7'1"
Steven
Leo 10h26'7" 24d38'32"
Steven
Cnc 8h55'56" 18d48'7"
Steven
Psc 0h59'22" 25d31'12"
STEVEN
Peg 22h51'4" 13d52'52"
Steven 07-05-71
Cnc 8h43'46" 13d8'36"
Steven 2/14/00 Alois
Ori 6h5'46" -0d8'56"
Steven <3 Sarah
Aur 5h38'33" 32d1'23"
Steven A. Barrett
Ori 5h35'52" -4d15'13"
Steven A DeBonis 7/18/1984-9/4/2003
Umi 15h45'10" 74d48'27"
Steven - A Fathers Pride
Leo 11h46'25" 20d14'16"
Steven A. Jake Buffington
Uma 9h10'23" 72d40'30"
Steven A. Marshall
Aql 19h13'58" 9d15'13"
Steven A. Mittleman
Dra 18h46'49" 56d41'56"
Steven A Simmonds
Uma 10h15'6" 55d43'44"
Steven A. Stuart
Uma 13h45'5" 54d49'4"
Steven A. Wild
Her 18h38'39" 12d39'2"
Steven Adam Ellison
Psc 1h23'51" 17d22'23"
Steven Aja
Tau 4h8'19" 8d2'24"
Steven Aja
Tau 3h38'52" 8d46'45"
Steven Alan Conrad
Lib 15h33'45" -15d30'9"
Steven Alan Geddes
Uma 11h18'8" 45d40'50"
Steven Alan Hunter
Aur 5h50'3" 46d21'46"
Steven Alan Weber
Lyn 7h7'39" 59d30'26"
Steven Albert Peterson
Cyg 19h46'29" 38d45'6"
Steven Alvarez
And 22h59'10" 38d57'7"
Steven & Amanda
Sco 17h36'34" -41d20'23"
Steven and Ashley 12/21/04
Sge 20h6'43" 17d41'21"
Steven and Brittany Wike
Cyg 20h52'4" 35d11'58"
Steven and Emma Munro Forever
Cyg 21h54'58" 49d28'36"
Steven And Heidi Forever
Cyg 20h26'3" 54d34'7"
Steven and Henrietta Marchant
Cyg 21h36'35" 45d20'27"
Steven and Jaclyn
Cyg 19h48'38" 38d0'54"
Steven and Jeannie Holt
Ori 5h42'21" 7d47'52"
Steven and Jessica
Cyg 21h23'15" 37d9'55"
Steven and Nathalie Scott
Cyg 21h43'7" 48d53'26"
Steven and Sheila Starlight
Vir 12h45'15" 6d49'24"
Steven and Susan Wilson
Cnc 8h16'34" 18d39'22"
Steven Andrew Biddinger
Uma 11h54'25" 58d7'43"
Steven Andrew Kowal
Uma 10h36'24" 43d21'35"
Steven Andrew Sanderson
Tau 4h47'0" 27d20'12"
Steven Anthony Gabriel Reese
Cam 3h19'51" 59d41'1"
Steven Anthony Mitts
Her 18h47'15" 20d10'40"
Steven Anthony Ockander
Gem 7h19'33" 24d52'31"
Steven Anthony Piepiora
Leo 9h48'22" 28d41'39"
Steven Anthony Spidalieri
Sgr 18h33'4" -27d24'58"
Steven Asparagus
Per 3h5'46" 38d57'13"
Steven B. Belkin
Per 2h20'7" 55d17'47"
Steven B. Chapman
Cep 3h10'30" 85d37'48"
Steven Ballard
Per 4h3'54" 31d58'31"
Steven Barkus
Per 3h50'21" 33d32'54"
Steven Baum
Ori 6h6'25" 21d3'47"
Steven Berget
Leo 10h42'32" 17d41'13"
Steven Bernard Getsie
Aqr 21h38'23" 1d59'38"
Steven Biasatti
Ori 6h18'13" 14d4'30"
Steven Bieganousky
Leo 11h23'0" 5d17'58"
Steven Bjerke
Uma 11h9'32" 49d29'21"
Steven Blair Thompson
Uma 11h33'5" 52d25'32"
Steven "Bob" Robertson's Star
Ori 5h27'1" 5d1'48"
Steven Bonanno
Per 3h8'2" 55d4'49"
Steven Boymel - #60
Her 17h27'27" 38d40'47"
Steven Bradley Smith
Psc 0h26'49" 13d40'49"
Steven Branchaud
Sgr 19h12'56" -15d46'36"
Steven Brett Conway
Tau 4h35'50" 6d16'37"
Steven Bricio Balderas
Lib 15h15'18" -21d57'55"
Steven Bronsink
Uma 14h26'5" 58d5'12"
Steven Bruce Brown
Per 3h6'31" 50d58'18"
Steven Burdette
Psc 0h58'55" 32d26'27"
Steven Burnette
Uma 11h57'45" 47d12'14"
Steven Bye - Steve's Star
Uma 13h49'11" 61d51'20"
Steven C. Cable
Uma 9h4'46" 62d53'42"
Steven C. D'Ooge
Per 3h34'19" 48d5'29"
Steven C. Greenman DDS
Ari 2h10'26" 11d35'5"
Steven & Caitlin
Lyr 18h23'56" 38d20'50"
Steven Charles - Love Forever
Cru 12h4'53" -61d16'9"
Steven Christopher Marshall
Umi 15h24'21" 70d46'30"
Steven Christopher Sissenstein
Sco 16h10'26" -16d21'14"
Steven Churney
Her 17h36'40" 31d33'14"
Steven Clayton Fulghum
Sgr 18h32'37" -17d44'59"
Steven Clinton Smith
Cap 20h14'49" -14d47'46"
Steven Cole
Psc 0h53'30" 13d14'13"
Steven Constantiner
Her 16h53'53" 14d58'31"
Steven Craig Emmons
Tau 5h44'3" 27d16'59"
Steven Cross
Sgr 18h53'58" -27d37'14"
Steven & Cynthia Strum
Vir 14h16'38" 2d48'42"
Steven D. LeClair
Boo 14h19'44" 29d35'58"
Steven D. Lee
Ori 4h46'35" -2d24'49"
Steven D. McKee
Aql 19h10'30" -0d6'5"
Steven D. Rhoades
Cyg 19h46'40" 35d32'52"
Steven D. Sandven
Cnc 8h45'35" 31d8'7"
Steven D'Aguanno
Her 17h20'33" 32d53'11"
Steven D'Alessio & Nina Ruiz
Lib 14h50'13" -1d37'19"
Steven Daniel Kerrigan
Her 17h13'31" 14d34'46"
Steven Daniel Luedtke
Cap 20h48'14" -15d24'0"
Steven & Daniella
And 2h35'14" 44d42'59"
Steven&Danny
Lyn 7h37'56" 48d38'13"
Steven Darrel Jernigan
Her 17h7'0" 31d33'6"
Steven David Smith
Cyg 21h23'24" 50d42'28"
Steven DeBoer
Uma 11h35'28" 60d43'12"
Steven Dominguez
Per 3h21'48" 51d39'44"
Steven Donald Brune
Cyg 21h57'20" 39d15'31"
Steven Douglas Hazel
Her 17h57'11" 25d5'5"
Steven Douglas Kelsall
Ori 6h15'57" 14d43'58"
Steven Douglas Newton
Tau 4h33'28" 24d55'23"
Steven Douglas "Number One Dad"
Per 3h32'52" 48d45'48"
Steven E. Hanley
Lyn 6h32'46" 61d22'11"
Steven E. Hellman
Gem 6h4'40" 26d7'20"
Steven E. Taylor
Vir 12h31'21" 4d12'50"
Steven Eamerson Semple
Ser 16h15'18" -0d46'22"
Steven Eames - forever shining bright
Pav 20h1'55" -61d41'37"
Steven Edward Balga
Ori 6h12'25" 6d7'6"
Steven Edward Flake
Aql 19h52'46" -0d53'17"
Steven Edward Legg
Umi 16h43'5" 89d8'19"
Steven Eric Rosenberg
Ori 5h56'25" 18d5'9"
Steven & Erica
Uma 12h10'36" 59d56'27"
Steven Eugene Bea
Sgr 20h23'10" -34d0'59"
Steven F. Meagher
Vir 13h30'50" 9d1'11"
Steven F Touloumis
Her 16h42'8" 42d20'5"
Steven F. Walsh
Ori 6h18'21" 14d55'51"
Steven Falk
Cep 22h42'25" 76d8'20"
Steven Fekete
Leo 9h23'15" 26d23'42"
Steven Frank Masi
Umi 14h45'33" 74d25'8"
Steven Frank Roger Wolski
Lyn 8h1'54" 50d8'17"
Steven Frank Ross
Sgr 18h32'24" -27d51'36"
Steven Fred Malzone
Sco 17h55'2" -45d16'54"
Steven Frederick Ruleau, Jr.
Uma 14h20'59" 56d16'34"
Steven Fredrich
Uma 9h8'0" 69d31'6"
Steven G. Akers
Uma 8h47'48" 69d17'11"
Steven G. Dombrowski
Gem 7h36'37" 27d30'9"
Steven G. Jones- Lil Steve Reno
Dra 18h53'43" 74d55'56"
Steven G. Makrakis
Psc 1h5'58" 24d57'12"
Steven G. Santos
Uma 10h15'0" 44d55'33"
Steven G Smarsh
Psc 1h7'55" 32d46'42"
Steven Garrett LeDuc
Sco 16h5'51" -16d16'1"
Steven Gary Miller
Aur 5h40'48" 38d23'10"
Steven Gaus
Dra 16h50'28" 59d17'51"
Steven George Munro
Ori 5h49'35" 12d5'1"
Steven Gooch... Timeless
Cap 20h29'1" -25d57'7"
Steven Goodstein
Per 3h1'1" 56d0'43"
Steven Graham Winter Stingley
Leo 11h44'3" 23d44'39"
Steven Green
Uma 11h52'20" 56d59'41"
Steven Gregory Burns
Lac 22h48'52" 53d54'6"
Steven Gregory Voisin
Ori 5h3'43" 15d24'39"
Steven Grosskopf
Ari 2h55'16" 27d27'36"
Steven H. Chang, R.Ph., C.C.N.
Lib 15h23'16" -14d23'56"
Steven H. Goldstein
Psc 0h21'13" 8d46'12"
Steven H. Henke
Cyg 19h51'29" 37d22'1"
Steven H. Horstman
Ori 5h55'45" 6d41'46"
Steven Halkett
Gem 6h54'58" 21d54'51"
Steven Hanna
Psc 1h14'46" 24d45'11"
Steven Hazell
Lyn 7h28'23" 53d37'52"
Steven Hedrington
Tau 4h13'27" 9d7'10"
Steven Henry Cooper
Cyg 19h59'13" 51d53'8"
Steven Hoffman
Tau 3h48'17" 29d0'54"
Steven Hoffman
Uma 8h34'29" 64d33'9"
Steven Hollingsworth
Her 18h0'19" 17d28'24"
Steven Howard White
Del 20h24'57" 20d5'29"
Steven Hurst
Ori 5h12'27" 5d45'18"
Steven I King
Sco 16h7'4" -11d27'7"
Steven is my shining star
Her 16h52'53" 31d59'15"
Steven Istvan Foldes, M.D.
Leo 11h45'59" 24d42'5"
Steven J. Bolich
Uma 9h49'41" 50d27'33"
Steven J. Coveleski Sr.
Aur 5h45'38" 33d1'19"
Steven J. Giamundo 20 Years HVB BDB
Umi 15h32'21" 81d58'36"
Steven J. Gladwin
Tau 4h26'38" 12d48'6"
Steven J. Juliano
Per 3h10'53" 56d4'52"
Steven J. Kumble
Cnc 8h34'16" 30d40'9"
Steven J Malovich III
Mon 6h48'31" 6d55'45"
Steven J Nanni
Cnc 8h19'34" 9d10'46"
Steven J. Rixmann
Her 16h49'28" 34d10'19"
Steven J. Shannon
Tau 4h41'1" 19d39'26"
Steven James Chandler
Per 3h8'6" 54d30'31"
Steven James Fightmaster
Her 17h35'41" 33d21'15"
Steven James Hauser
Her 16h12'57" 49d15'32"
Steven James Nugent
Umi 15h49'10" 73d29'20"
Steven James Oxenrider
Vir 12h48'33" 11d30'18"
Steven James Silva
Cep 22h58'28" 70d50'40"
Steven James Thielman
Tau 5h21'47" 23d33'59"
Steven Jay Chiavola
Uma 10h52'23" 67d13'48"
Steven Jay Gibson
Uma 9h47'59" 43d51'24"
Steven Jay Levine
Lib 14h44'33" -11d24'14"
Steven Joel (Stevie) Hansen
Uma 12h1'43" 51d19'57"
Steven John Breuker
Cep 22h31'30" 84d4'4"
Steven John McNiece
Aqr 21h37'41" -5d45'1"
Steven Joivstan Lengkoan
Vir 13h1'31" 10d12'19"
Steven Jon Davis
Uma 9h35'49" 46d8'46"
Steven K Garrod - 21 Years
Cep 3h47'55" 81d25'25"
Steven & Kari
Cyg 21h56'18" 46d0'19"
Steven Keith Collins
Per 3h8'28" 55d39'8"
Steven Kelleher
Her 17h58'8" 27d33'3"
Steven Kent Miller (Sierra Steve)
Lib 14h40'41" -17d24'37"
Steven Kent Okamoto
Aql 19h33'18" 9d18'20"
Steven Kip Taggart
Her 17h45'44" 17d6'44"
Steven Klug
Boo 15h16'12" 49d33'46"
Steven Kumin "Happy 63rd. Birthday"
Sco 16h10'12" -9d36'25"
Steven Kwan
Her 18h6'24" 35d6'34"
Steven L. Corte, Jr.
Lac 22h13'22" 41d30'58"
Steven L. Dettwiler
Cep 22h9'5" 61d10'22"
Steven L. Kady
Lyn 9h22'2" 33d26'11"
Steven L. Satcher
Per 2h44'54" 56d21'38"
Steven L. Toth
Ori 6h8'17" 17d4'1"
Steven LaFayette Myatt
Cnc 8h41'13" 23d1'34"
Steven Landon Best
Boo 14h31'49" 18d28'44"
Steven Lapekas 1972-2006
Her 18h41'41" 22d20'49"
Steven & Lauren
Uma 10h34'0" 53d40'59"
Steven Lawrence Scott
Cyg 19h47'11" 33d40'45"
Steven Lee Bosch
Sco 16h8'43" -23d16'56"
Steven Lee Butts - 'The Surfing Shrew'
Uma 10h18'33" 56d27'53"
Steven Lee Carney
Aql 19h24'8" -11d5'12"
Steven Lee Fether
Ori 5h53'10" 22d47'50"
Steven Leodegario Espinosa
Boo 15h20'19" 47d58'23"
Steven Levinson
Per 2h38'50" 53d55'41"
Steven Lewis
Per 3h11'52" 45d34'53"
Steven Lightnin' Smithson
Gem 6h46'28" 25d15'21"
Steven Likidis
Vel 9h23'43" -42d8'23"
Steven & Lisa Hinsley
Uma 9h21'30" 42d23'38"
Steven & Liz Forever
Cyg 21h35'50" 46d36'51"
Steven Louis Sinatra
Gem 7h33'0" 17d39'48"
Steven - Lover of Long Shadows "5"
Dra 19h48'23" 70d14'22"
Steven Lucas
Her 18h0'8" 22d13'7"
Steven Lynn Snyder
Tri 1h58'38" 34d7'9"
Steven M. Bortz
Leo 10h12'51" 26d13'16"
Steven M. Carbaugh
Cep 21h43'38" 62d14'8"
Steven M. Engel
Lib 15h18'36" -27d27'45"
Steven M. Gerber "Coach"
Cep 20h48'28" 60d32'55"
Steven M. Mudd
Ori 4h47'2" -2d4'33"
Steven M. Pangilinin
Vir 12h40'55" 5d42'48"
Steven M. Ross
Dra 19h10'7" 72d9'11"
Steven Marcus Melchior
Uma 12h10'33" 53d7'15"
Steven Mark Davis - Shining Bright 11
Aql 19h50'30" -0d30'55"
Steven Mark Potter
Lyn 7h33'7" 39d14'0"
Steven Marousek
Her 17h52'54" 23d12'43"
Steven Matthew Caufield
Sco 17h35'0" -37d2'11"
Steven Matthew Farinella
Uma 11h39'12" 55d0'8"
Steven Maurice Jones
Sco 16h4'24" -11d46'31"
Steven Maximus Szabo
Per 1h43'58" 51d24'17"
Steven McGibbon Optimus
Uma 11h5'31" 60d34'16"
Steven Medeo
Uma 8h46'46" 61d36'22"
Steven & Megan
Leo 11h55'45" 25d45'50"
Steven Mendes "Srg"
Cep 0h16'50" 70d1'39"
Steven Michael
Per 3h36'0" 51d21'35"
Steven Michael Beck aka "Dad"
Aql 19h45'59" 7d26'16"
Steven Michael Croce
Lib 14h48'6" -0d43'6"
Steven Michael Gyorki
Her 16h50'53" 44d20'20"
Steven Michael McEvoy
Lmi 9h46'56" 40d50'40"
Steven Michael Posluszny
Per 3h22'16" 43d24'35"
Steven Michael Rivkin
Vir 12h39'53" -1d46'0"
Steven Michael Schooley
Cap 20h8'58" -18d33'46"
Steven Michael Zaitz
Sco 16h12'32" -14d40'24"
Steven Michale Welsh
Per 3h35'36" 46d21'3"
Steven Morabito
Cnc 9h14'17" 27d31'53"
Steven Morris Pool
Boo 14h46'41" 37d6'6"
Steven Morton
Oph 16h34'15" -6d17'6"
Steven Murison
Cap 20h55'21" -24d16'41"
Steven Murray Jopson - 26.01.1965
Aqr 21h34'4" 0d29'14"
Steven Nels Javert
Uma 13h1'10" 55d16'7"
Steven Nicholas
Uma 11h56'53" 34d23'24"
Steven Nicholas Stehle
And 2h17'6" 45d1'35"
Steven Novak
Ori 6h6'35" 17d27'21"
Steven O'Rourke
Peg 22h48'27" 4d12'10"
Steven Oswald Mortimer
Cru 12h29'56" -60d29'20"
Steven Our Angel
Lib 15h9'30" -4d22'6"
Steven Owen Swanson
Tau 5h35'59" 26d2'56"
Steven P. Ellison
Per 4h15'37" 48d30'8"
Steven Patrick Gierek
Her 17h44'11" 44d37'19"
Steven Patrick Ryan
Ari 2h33'0" 25d30'22"
Steven Paul Mayo
Cyg 19h46'54" 33d5'24"
Steven Paul Walasek
Ori 5h1'33" 9d39'32"
Steven Paul Wallace
Vir 12h35'13" 4d14'55"
Steven Peter DiBiasi
Aql 20h23'21" 4d59'3"
Steven Peter Swift
Umi 16h46'23" 75d7'8"
Steven Pianelli
Crb 16h21'24" 31d54'51"
Steven Pittman
Ori 5h22'11" 1d30'50"
Steven Poe
Her 16h33'58" 36d53'26"
Steven Pollard
Per 3h18'27" 52d7'53"
Steven Poscente
Per 3h3'45" 33d7'11"
Steven Proulx Foberg - Tiger
Lmi 10h8'41" 38d22'20"
Steven Przybylowicz
Ari 3h17'16" 16d42'47"
Steven R. Abell
Gem 7h22'28" 15d24'53"
Steven R. Baco
Leo 11h39'48" 12d9'7"
Steven R Ipp Canyon Star
Ori 5h6'58" 4d55'36"
Steven R. Isko
Hya 9h22'7" 4d22'52"
Steven R. Russo
Leo 11h43'57" 22d44'5"
Steven Ramos
Cep 22h44'36" 77d58'50"
Steven Ray
Uma 10h34'58" 67d37'51"
Steven Ray
Per 3h40'54" 37d31'5"
Steven Ray Hansen
Uma 13h36'14" 54d53'38"
Steven Ray Shell
Vir 11h57'9" 3d50'54"
Steven Raymond
Vir 12h20'34" -5d18'22"
Steven Raymond Arnold
Tau 4h32'7" 17d34'24"
Steven Raymond Kroeger
Cnc 8h50'41" 32d32'52"
Steven Richard Hammers
Her 18h6'30" 49d20'0"
Steven Richard Leach
Ori 5h24'48" 0d31'20"
Steven Ripley Young
Tau 3h46'28" 5d58'17"
Steven Ritchie
Cnc 8h31'38" 25d38'23"
Steven Rivellino
Per 2h10'20" 56d35'49"
Steven Robert Mullins
Her 17h38'46" 27d32'19"
Steven Robert Spencer
Gem 7h16'57" 24d56'50"
Steven Robert Valente, Jr.
Psc 1h15'52" 5d35'36"
Steven & Robin
Cyg 20h6'41" 35d46'45"
Steven Rodriguez
Lib 14h41'57" -11d4'46"
Steven Rodriguez Fallano
Ari 3h23'5" 27d16'23"
Steven Ronald Giammona
Sco 16h2'32" -22d54'40"
Steven Roseman
Her 16h32'34" 24d14'35"
Steven & Rose's Little Star
Ori 5h24'49" -8d25'35"
Steven Russell Shea
Her 16h30'21" 45d53'57"
Steven Russell Wodock Jr.
Peg 22h40'53" 34d5'6"
Steven Ryan Black
Uma 11h21'8" 64d28'11"
Steven Sanchez
Aql 18h51'53" 8d1'20"
Steven Sanders
Lyn 8h30'10" 35d25'10"
Steven Sattem
Sgr 18h6'18" -23d36'9"
Steven Sauerhoff
Ori 4h55'48" -0d54'1"
Steven Schermetzler
Cap 20h50'13" -14d32'35"
Steven Schiller
Leo 10h41'58" 15d51'46"
Steven Scott Graves
Ori 4h53'5" 14d31'8"
Steven Scott Mickens
Leo 9h55'7" 7d28'49"
Steven Scott Riley - 7.09.1975
Vir 14h8'16" 6d1'36"
Steven Shears
Lac 22h14'42" 45d17'29"
Steven Sibley
Sco 17h7'6" -34d22'9"
Steven Spade
Her 17h18'12" 26d8'41"
Steven Spitzer & Daralyn Calderón
Gem 6h54'40" 14d0'35"
Steven Stewart Delmar
Tau 5h31'15" 21d45'15"
Steven Stojanovic
Her 16h39'2" 48d31'59"
Steven Symington
Per 2h14'29" 55d45'42"
Steven T Filipowicz
Per 3h46'17" 52d4'5"
Steven & Tammy, Eternally Entwined
Sge 20h5'14" 16d40'2"
Steven Tankle
Aql 19h36'25" 11d35'25"
Steven Tarozzi
Leo 11h3'9" 15d17'54"
Steven "Tato" Lambkin, Jr.
Gem 7h14'46" 24d30'48"
Steven Taylor McSweeney
Ari 3h9'38" 24d56'13"
Steven Thomas Helge
Sgr 19h40'38" -12d26'8"
Steven Thomas Marbach
Cap 20h23'30" -11d57'44"
Steven Thomas Pace (SteveO)
Dra 20h20'3" 62d17'19"
Steven Thomas Phillips
Gem 7h8'41" 15d48'25"
Steven Thomas Susbauer
Aur 5h53'20" 52d41'48"
Steven & Tiffany Forever
Crb 15h42'42" 36d13'51"
Steven Tiffen
Per 3h45'22" 48d50'45"
Steven Todd Hunter
Lib 14h57'7" -4d23'19"
Steven Todd Nill
Aqr 23h1'57" -9d48'30"
Steven Todd Sullivan
Uma 13h9'51" 58d14'40"
Steven Tomaselli
Gem 7h47'21" 32d18'9"
Steven Tomlinson
Sco 16h6'48" -18d1'10"
Steven & Trish Together
Cru 12h55'35" -58d40'44"
Steven Turner - Always & Forever
Cyg 21h20'14" 51d54'16"
Steven Tyler
Gem 7h19'47" 33d47'51"
Steven Tyler
Vir 12h16'42" 12d26'31"
Steven Tyler Jensen
Uma 13h41'24" 53d19'50"
Steven V and Charlotte C Staar
Sge 19h27'19" 18d7'13"
Steven Verssen
Per 4h15'23" 52d2'51"

Steven Victor
Her 18h46'41" 19d32'12"
Steven Victor Sisson
Aql 19h30'4" -11d10'27"
Steven Victor Tallarico
And 23h6'24" 43d4'31"
Steven W. Chandler
Cep 21h43'41" 62d46'55"
Steven W. Gulley
Aur 7h25'21" 40d52'27"
Steven W Palmer
Gem 6h20'39" 17d50'4"
Steven W. Sloan
Gem 6h46'46" 28d1'1"
Steven W. Whitehead
Cas 0h39'51" 53d27'54"
Steven Wachtel
Sgr 18h1'7" -25d19'43"
Steven Waite Newell
And 2h19'5" 45d51'17"
Steven Ward White
Leo 10h35'25" 17d40'55"
Steven Wayne Bordok Junior
Sgr 18h9'32" -22d42'20"
Steven Wayne Jacks
Ari 2h47'35" 16d41'35"
Steven Wayne Loftis
Boo 15h29'29" 44d48'8"
Steven Wilensky
Cep 21h41'53" 69d12'42"
Steven William Sepeda
Sco 16h16'44" -21d43'0"
Steven William Toner
Ari 1h55'29" 23d45'53"
Steven Woolf
Gem 7h45'57" 32d41'10"
Steven Youngblood
Cep 0h3'24" 78d16'34"
Steven Yuki Bunch dob 11/13/03
Umi 14h40'1" 72d53'3"
Steven, Forever Our Guardian Angel
Her 18h39'37" 23d40'37"
StevenLove1
Lib 15h4'2" -0d55'57"
StevenRitzTheWizard
Per 3h2'28" 55d41'48"
Steven's Angel Star
Ori 4h49'39" -0d33'39"
Steven's Colby
Umi 16h20'43" 77d15'3"
Steven's Ebullience
Ori 6h10'24" 7d3'13"
Steven's Generosity
Uma 9h32'23" 52d12'17"
Stevens Schaller
Cnv 13h59'55" 35d47'59"
Steven's Silver Star
Cap 21h0'21" -18d16'3"
Steven's Skateboard
Per 3h34'50" 33d41'55"
Steven's Star
Per 4h31'33" 34d15'15"
Steven's Star
Cyg 20h8'22" 32d23'19"
Steven's Star
Uma 9h34'13" 50d8'25"
Steven's Star
Uma 13h59'17" 51d12'42"
Steven's Star
Ori 5h58'58" 4d16'48"
Steven's Star
Equ 21h16'58" 8d30'36"
Steven's Star
Sco 17h20'18" -44d49'27"
Steveo
Oph 17h9'12" -4d10'48"
Steve-O
Cep 1h51'23" 78d6'33"
Steve-O
Her 17h8'53" 29d0'49"
Steveo
Gem 6h18'33" 24d11'16"
StevErin Like A Star
Aqr 22h14'56" -2d8'30"
Steve's Heaven
Vir 12h53'26" -4d37'43"
Steve's Home
Cru 12h37'30" -57d32'31"
Steve's Odyssey
Car 10h11'44" -62d49'48"
Steve's Place in Heaven
Aur 5h18'42" 44d16'46"
Steve's Sanctuary
Uma 9h4'33" 66d34'41"
Steve's Shining Guardian Angel
Boo 15h25'29" 49d35'9"
Steve's Shining Light
Ori 6h6'29" 13d6'51"
Steve's Star
Equ 21h5'57" 12d10'12"
Steve's Star
Uma 9h55'27" 53d52'55"
Steve's Star
Lyn 7h3'17" 53d50'57"
Steve's Star
Cap 21h45'27" -15d30'54"
Steve's Star of Peace
Uma 11h51'14" 41d32'12"
Steve's Way
Per 4h36'23" 39d39'43"
Steveshelleus
Ari 2h12'3" 27d50'3"
Stevey Lynn
And 23h23'56" 48d12'38"
Stevi Lynn
And 0h19'36" 41d18'6"
Stevi N Guaraldi
Tau 4h22'24" 18d24'21"
Stevie
Cnc 8h43'53" 7d15'16"
Stevie
Her 18h8'53" 37d21'37"
Stevie
Cyg 22h0'19" 45d38'20"
Stevie
Cap 20h17'52" -16d34'11"
Stevie
Aqr 23h49'28" -11d8'28"
Stevie
Aql 19h42'37" -0d6'59"
Stevie
Mon 6h50'20" -0d46'56"
Stevie
Uma 8h35'16" 66d15'9"
Stevie
Uma 10h10'45" 71d3'36"
Stevie and Andrew's Star
Tau 3h40'29" 16d5'2"
Stevie and Jessica
Lyr 18h41'10" 39d41'40"
Stevie - B
Uma 8h33'16" 62d17'11"
Stevie B
Sco 16h49'48" -31d7'16"
Stevie Bene't Reilly
Psc 1h59'30" 5d53'30"
Stevie Christina's Star
Crb 16h4'1" 32d15'31"
Stevie D
Ori 5h48'5" 2d36'49"
Stevie Elizabeth
And 1h27'18" 47d4'16"
Stevie Haberfield
Cyg 19h57'26" 54d10'16"
Stevie Huber
Vir 14h54'29" 5d5'56"
Stevie Jade 06111992
And 0h12'53" 39d11'14"
Stevie Joe
Tau 5h47'28" 25d37'55"
Stevie Jones Tyneside No. 1 CSC.
Uma 14h17'27" 56d30'27"
Stevie L Schlenker
Per 3h22'3" 46d1'44"
Stevie Leigh Hicks
Aqr 23h54'58" -10d59'28"
Stevie lights up the universe!
Vir 12h37'25" 12d28'56"
Stevie Lynn Fisher
Gem 7h9'47" 17d41'45"
Stevie Marie
Cnc 8h19'54" 28d41'31"
Stevie Marie Woodard
Aql 20h14'12" 14d48'11"
Stevie N. Majewski
Ori 5h17'3" 16d5'40"
"Stevie" Oneal Uetrecht Jr.
Psc 1h27'30" 17d57'55"
Stevie Radwan
Uma 10h14'22" 42d49'44"
Stevie Rita Levesque
Umi 15h55'51" 85d55'54"
Stevie Rosanna
Ori 5h27'44" 7d2'38"
Stevie Spring
Cep 21h15'35" 55d55'32"
Stevie "The Duke Of Earl"
Cru 12h3'21" -60d41'53"
Stevie Weevie Preston Grayson
Uma 12h24'24" 54d59'20"
Steviegay and Honey Peaches
Sgr 19h25'31" -17d7'7"
Stevie-Jay
And 0h52'30" 24d21'1"
Stevie-Lee Hansen
Cap 20h56'19" -20d14'11"
Stevielee Langan
Crb 15h43'32" 31d58'39"
Stevie's Amazing Brilliance
Cas 23h6'59" 54d59'36"
Stevie's Star
Aqr 21h10'29" -11d18'57"
Stevie's Star
Per 3h1'16" 37d21'21"
Stevie's Star
Crb 15h42'42" 29d46'50"
Stevikat
Aqr 22h7'36" 1d53'35"
Stevious Ignoramusious
Aqr 22h26'58" -22d56'50"
Stewart A. Daniels
Cep 3h14'16" 83d29'31"
Stewart and Ann
Cyg 20h35'17" 47d27'28"
Stewart B. Dittmeier
Sco 16h13'10" -15d39'5"
Stewart Bailey * Director of Bands
Lyr 18h42'38" 39d44'13"
Stewart C. Broberg
Gem 7h10'15" 14d15'21"
Stewart Epstein
Cep 21h55'23" 60d30'34"
Stewart Francis Jewel
Pyx 8h36'4" -31d7'55"
Stewart Gray Wyne
Boo 15h14'58" 41d26'21"
Stewart Guymon
Ori 6h8'55" 17d21'20"
Stewart Hayes 03/03/45
Cep 22h9'18" 72d53'25"
Stewart Ian Rafalo
Sco 16h18'55" -15d54'46"
Stewart James Richardson Lawrie
Uma 11h30'3" 28d49'34"
Stewart Little's Star
Ori 6h12'31" 21d4'58"
Stewart Marcus
Cep 21h46'49" 60d35'54"
Stewart McKee 'Dad'
Her 16h41'46" 43d13'30"
Stewart Stars Saxton
Lyn 7h42'27" 50d46'59"
Stewart's Doxology
Lyn 7h42'11" 37d57'59"
Stewart's Star
Aur 5h35'5" 43d34'28"
Stewart's Star
Her 18h51'5" 23d25'53"
Stewart's Star
Ori 5h34'2" 14d52'51"
Stewart's Star
Uma 9h25'58" 66d40'14"
Stewli
Uma 9h26'12" 41d49'19"
Stewy
Lib 15h38'43" -6d35'15"
Stian Aksel Tvedten
Tau 5h19'13" 24d20'48"
Stice's Beacon
Gem 7h6'40" 34d16'58"
Stich Schafer
Leo 11h1'47" -5d28'33"
Stiches
Lmi 10h36'8" 38d49'3"
Stichler, Bernd
Psc 1h53'36" 7d29'31"
Stichting Kinderopvang Irene
Umi 15h45'0" 76d54'24"
Stief, Katharina
Cnc 9h2'13" 16d39'59"
Stiffel
Boo 14h18'21" 40d31'9"
Stifu
Peg 22h25'51" 23d31'30"
Stig (Stephen Phillips)
Cep 22h8'28" 53d35'44"
Stig William Franzén
Aqr 23h6'48" -15d22'2"
Still Leto BTC
Peg 23h12'43" 17d35'27"
stillwater
Aql 19h28'34" 7d25'56"
Stillwater's Fire Up North
Sco 16h34'45" -28d57'10"
Stilts, My Beloved Pet
Lyn 6h49'53" 54d13'52"
Stimo's eye
Cap 21h55'26" -10d4'17"
Stimpert
Lyn 6h43'16" 61d38'2"
Stina
Lib 14h50'5" -19d12'50"
Stina
Vir 14h35'17" -0d59'33"
Stina
Uma 11h11'31" 29d48'36"
Stina
Uma 11h57'27" 44d2'20"
"Stina Bina and Tika"
Leo 9h39'57" 26d4'51"
Stina Chouinard
Hya 9h16'41" -0d0'40"
Stina's Star
Tri 1h55'37" 27d46'22"
Stine
Ari 3h27'9" 27d56'12"
Stine
Uma 8h58'0" 53d48'4"
Stine Tangen
Dra 11h48'50" 73d24'10"
Stingray's Track
Ori 5h17'1" 9d3'55"
Stink
Vir 12h33'50" -0d18'28"
Stink loves Brat.
Tau 4h25'3" 16d19'49"
Stinker Bug
Umi 15h26'57" 77d46'5"
STINKERBOY DEAN
Her 16h42'27" 28d50'8"
Stinker's Star
Aqr 23h3'48" -4d49'52"
Stinky
Sco 16h15'11" -12d46'25"
Stinky
Sco 16h44'38" -32d20'28"
Stinky
Cra 19h1'58" -37d54'54"
Stinky
Her 18h57'0" 23d50'36"
Stinky
Cnc 8h49'31" 23d44'16"
Stinky & Brownie Forever
Sgr 19h37'58" -24d45'1"
Stinky Face
Leo 10h27'17" 18d28'55"
stinnee
Cyg 21h58'51" 52d58'17"
Stiofan Roibeard Crump 50 2006
Tuc 0h26'17" -63d12'28"
Stipulation Star
Uma 10h28'24" 42d20'29"
Stirfry
Crb 16h9'53" 32d31'17"
Stirling
Ori 5h39'0" 0d9'47"
Stitch
Uma 10h37'6" 45d17'38"
Stitch's Papa
Cap 20h33'20" -19d25'51"
Stjepan and Mary Grgecic
Uma 8h49'43" 55d53'59"
Stjepan Gajcevic
Per 2h47'39" 41d3'31"
Stjerne Jepsen
Cyg 21h59'17" 51d4'46"
St.Matthew
Cap 21h58'6" -20d29'8"
Stobie Pyle
And 23h32'14" 47d2'34"
Stochniol, Roland
Ori 6h18'38" 10d50'10"
Stöcker, Josef
Uma 8h54'36" 50d9'16"
Stockhausen, Christiane
Uma 9h19'43" 42d38'59"
Stocking
Mon 7h19'10" -0d25'31"
Stoeana Juhlin my loving wife
Cas 0h36'17" 58d3'7"
Stoffelini
Aqr 23h11'34" -8d16'7"
Stoical Star Karley Gale
Lib 15h27'52" -6d16'3"
Stoker
Cnc 8h4'33" 20d26'28"
Stokes Family Star
Uma 11h50'48" 56d34'36"
Stoke-Verhey
Uma 8h28'10" 63d14'25"
Stoli Wolf
Cma 7h0'36" -31d32'58"
Stoll, Nadine
Uma 10h44'6" 66d38'2"
Stoll, Wilhelm Karl
Uma 13h3'43" 54d39'53"
Stone
Umi 13h58'19" 71d32'0"
Stone Adastral Pictor
Tau 5h9'25" 18d34'45"
Stone Henderson Filipovich
Per 3h10'54" 41d46'50"
Stonebridge
Eri 4h21'29" -33d35'52"
Stonehouse Wish
Uma 9h19'40" 52d8'29"
Stoneman Falcon Martin
Ari 2h20'14" 23d51'58"
Stoney Hering
Cyg 21h42'22" 48d54'47"
Storbeck, Marita
Uma 9h0'34" 61d23'39"
Stori Layne - Princess of my Heart
And 1h37'50" 36d28'51"
Stork
Lyn 8h14'18" 35d49'9"
Storm
Boo 15h10'30" 40d27'48"
Storm
Cmi 7h44'5" 1d1'26"
Storm
Lib 15h44'38" -28d52'14"
Storm Lee
Aur 7h28'48" 39d29'41"
Storm Sailor
Tri 2h6'17" 33d48'44"
Storm Sundown Mitchell
Cnc 8h43'23" 32d3'28"
Storm Wallace Our Heavenly Angel
Cap 21h30'33" -16d22'1"
Stormbrigga
Sco 17h26'51" -41d22'6"
Storme Jeremy Charette
Lyn 7h17'55" 59d20'36"
Storme Kennedee Camie
And 23h47'16" 47d29'1"
Stormi Bell
Sgr 18h28'51" -16d21'25"
Stormi Marie
Leo 9h45'7" 24d40'18"
Stormie Taylor Stevens
Lib 15h3'36" -17d46'30"
Stormin Norman
Tau 5h35'23" 23d23'54"
Stormin' Norman
Uma 11h52'11" 40d16'26"
Stormi's Dream Sweet 16
Mon 8h3'46" -0d49'38"
Storms As She Walks
Roxanne Loget
Lyn 7h20'16" 54d13'29"
Storm's Little Madeleine
And 22h59'41" 39d58'10"
Storms Locket
Peg 22h5'48" 7d36'28"
Stormy
Uma 11h19'22" 34d27'40"
Stormy
Lib 15h14'14" -17d37'37"
STORMY and CLOUDY
Uma 9h59'42" 70d24'58"
Stormy Silkey
Gem 7h14'54" 26d55'44"
Story Oreo
Del 20h29'19" 14d3'32"
Störzel, Uwe
Uma 11h11'24" 70d50'41"
Stowedaddy
Ari 3h6'19" 28d14'54"
Stoycho Danev
Sco 17h22'40" -32d22'30"
STP 09
Uma 9h25'41" 65d7'55"
STP Glofire
Cap 21h32'36" -13d57'27"
Stradivari-Koo Ramos Sibulboro
Srp 17h56'59" -0d12'52"
Strahlendes Funkeln der Melanie
Uma 10h13'12" 71d45'41"
Straighty
Uma 8h15'47" 66d1'23"
Strait From Texas
Aql 19h29'53" 4d32'49"
Strangers in the Night
Cyg 19h35'29" 53d55'13"
Strategy Three
Uma 9h9'26" 55d35'48"
Strathallen
Cru 11h57'40" -58d40'25"
Stratos and Blondie Star
And 23h30'10" 37d58'58"
Stratton Family
Uma 8h35'56" 55d9'19"
Stratton Stella
Aqr 22h29'58" -0d38'4"
Stratton Whitaker
Lyn 8h38'49" 38d1'32"
Straub, Walter
Ori 5h39'59" -2d22'55"
Straubie
Uma 11h34'58" 63d7'9"
Strause-Marrero Forever One
Lib 15h0'1" -22d25'54"
Strawberry
Cap 21h4'34" -16d35'43"
Strawberry
And 0h24'59" 36d27'13"
Strawberry Short Cake
Lmi 10h26'32" 35d56'54"
Strawberry Shortcake
Lyn 7h59'25" 34d30'56"
Strawberry Wine
Uma 13h39'59" 57d27'48"
StrawberryMi
Gem 7h23'49" 23d19'20"
Strazzamgloriax
Dra 18h15'51" 57d28'18"
Streich, Tino
Ori 5h39'15" -1d53'54"
STREIT
Crb 16h13'8" 39d34'42"
Strenger, Philip
Ari 2h21'26" 18d56'29"
Strength
Uma 11h40'38" 52d37'8"
Strength & Courage
Psc 0h18'8" 20d2'20"
Strength of Bonded Souls
Umi 14h21'51" 74d57'59"
Strength of Heart
Cyg 20h10'54" 39d26'46"
Stretch
Cyg 19h39'36" 28d15'49"
Stretch
Lyn 7h31'16" 54d21'48"
Stretch
Psc 0h48'33" 6d56'49"
Stretch (Derek Rowe)
Cru 12h30'33" -58d57'50"
STREZA
Vir 13h39'11" -16d40'1"
Strezmilowski
And 2h25'59" 46d19'53"
Strib and Scotty
Lyn 8h53'24" 45d14'34"
Strick9 (Victor Slaven)
Uma 10h33'18" 45d6'8"
Strictly Dance
Lyr 18h41'30" 39d12'5"
Strike
Ori 6h5'49" 6d3'24"
Strino Guglielmo
And 2h32'4" 41d10'23"
StRobacey
Cam 4h6'28" 68d12'59"
Strong as Steel ~ Smooth as Velvet
Her 16h24'49" 16d4'5"
STRONG LIKE BULL
Tau 5h47'51" 17d43'51"
STRONGLY
Cas 0h38'38" 69d44'38"
Stroop the Rock Star.
Lyr 18h50'7" 32d8'31"
Strozewski, Günter
Ori 6h17'20" 10d5'42"
Struan magnuficus sexagensum astrum
Cen 13h40'10" -42d13'25"
Strung
Uma 12h0'54" 32d35'19"
Strutzl
Uma 12h57'13" 58d20'19"
Struzina, Werner
Uma 12h58'36" 52d21'10"
Stryganek
Cyg 20h51'20" 30d14'13"
Stfomit
Cnv 12h45'23" 34d16'6"
Stu
Cap 20h26'59" -9d36'59"
Stu & Gem - Forever
Ara 17h54'5" -51d16'32"
Stu Son
Her 16h45'14" 10d22'15"
Stuart
Hya 8h47'56" 5d7'46"
Stuart
Psc 23h40'21" -0d28'4"
Stuart Allan Kirk
Psc 1h40'25" 20d20'13"
Stuart and Candace
Cas 0h54'8" 53d56'19"
Stuart and Kat in love at La Croix
Lyn 7h26'10" 54d7'33"
Stuart and Katrina Hignell
Cyg 19h57'8" 46d53'56"
Stuart Anorina
Cru 12h40'49" -60d29'41"
Stuart Athelston McHardy
Aur 6h35'47" 52d9'7"
Stuart Black: My Little Star
Ori 6h19'15" 5d40'48"
Stuart Buchanan
Per 3h3'31" 49d54'17"
Stuart Charles 50
Cnc 8h22'12" 9d19'50"
Stuart Charles Ames
Cam 4h2'1" 68d29'32"
Stuart Coltrane
Cnc 8h47'1" 32d34'41"
Stuart Crayk
Per 4h28'0" 32d4'37"
Stuart Curtis Browning
Gem 7h4'17" 25d27'16"
Stuart Daniel
Sgr 18h30'50" -16d33'17"
Stuart David Swanson, Loved by all
Her 18h22'42" 23d9'39"
Stuart Dinn
Cru 12h34'6" -61d36'47"
Stuart & Donna Levers 4C's Forever
Aql 20h8'4" 10d39'16"
Stuart Douglass
Uma 9h58'56" 58d40'27"
Stuart Edward Warren
Sgr 17h55'43" -29d30'47"
Stuart Feather
Per 3h42'35" 40d17'18"
Stuart Findlay
Her 16h36'17" 17d59'50"
Stuart Frost's 60th Birthday Star!
Uma 11h52'48" 42d2'36"
Stuart G. Bart's Everlasting Star
Gem 6h44'35" 33d13'10"
Stuart Glenn Swanson
Tau 4h22'11" 18d43'49"
Stuart H. Baker
Ori 5h17'53" -5d16'36"
Stuart Hardesty
Lyn 8h12'0" 46d29'28"
Stuart Heslop
Cep 22h48'20" 69d56'54"
Stuart I Spiegel
Leo 11h13'30" 17d12'4"
Stuart J. McNab
Uma 10h39'55" 57d14'2"
Stuart James
Tau 5h9'56" 18d31'7"
Stuart James Fleming
Umi 15h39'24" 70d7'4"
Stuart John Cooper -My Love My Life
Cnc 8h11'39" 32d54'23"
Stuart Kershaw's amazing star!!!!!!
Sco 17h28'1" -42d0'37"
Stuart L. Holmes
Psc 1h41'17" 20d36'55"
Stuart M Harris - Stuart's Star
Uma 8h33'28" 62d30'10"
Stuart M. Hollanshead
Psc 1h30'35" 15d53'31"
Stuart Mackie Robertson
Ori 4h46'31" 1d12'2"
Stuart & Margie Chase
Psc 1h8'27" 18d52'19"
Stuart & Marlene Hodes
Cyg 21h35'32" 39d31'30"
Stuart McCallum
Vir 12h55'9" 5d32'40"
Stuart McKay
Umi 14h46'7" 73d1'17"
Stuart Michael Quan
Per 3h52'15" 38d48'15"
Stuart Michael Vaughan
Cap 20h30'7" -11d28'57"
Stuart & Patricia Edwards
Sco 17h52'31" -36d32'35"
Stuart R. Deans
Uma 10h56'3" 68d42'12"
Stuart R. Septoff
Cam 4h38'34" 74d18'34"
Stuart R. Sincleair
Leo 11h47'57" 11d0'4"
Stuart & Rachel's Star
Cyg 19h44'38" 27d51'26"
Stuart Richard Paul
Lib 14h25'21" -9d51'11"
Stuart Rogers
Leo 9h25'33" 16d57'23"
Stuart S. Asch
Per 3h30'4" 41d33'3"
Stuart Struhl
Per 3h40'42" 51d57'1"
Stuart the Gentle Giant
Her 15h57'3" 42d31'19"
Stuart Wayne Andrews
Ari 2h32'58" 18d45'57"
Stuart Wheelans
Aur 6h33'6" 47d15'19"
Stuart Woodward Davis Johnson
Lib 14h48'4" -10d32'7"
Stuart's Bright Light
Cyg 20h5'23" 31d8'8"
Stuart's Super Star
Dra 18h56'42" 51d59'36"
Stuart's Vibrant Gem
Umi 14h6'21" 76d37'34"
Stuart's Wish
Col 5h59'49" -35d37'4"
Stuch, Walter
Uma 9h35'39" 49d36'43"
Stuckmann, Gerhard
Uma 10h38'24" 66d4'25"
Studio Niccolò Lucchini
Uma 11h48'57" 48d4'52"
Studley
Uma 9h33'24" 65d36'56"
Studly Scott
Her 17h39'48" 47d2'29"
StudMuffin Forever
Umi 14h39'50" 75d6'4"
Studor
Uma 11h2'50" 59d58'16"
Stueart Allen Pennington
And 0h42'26" 28d55'5"
Stuffy
Peg 22h28'5" 6d37'26"
Stuie B. Carns
Her 18h33'25" 15d50'14"
Stuka Genghis
Uma 10h20'50" 62d13'34"
Stukel Star
Vir 12h43'47" 5d11'55"
Stullboy & Kindgirl
Sge 20h8'1" 20d36'26"
Stumby
Lyn 7h12'12" 56d51'24"
stumie
Psc 1h43'53" 12d14'34"
Stumps and Gram
Lib 14h50'37" -4d42'37"
Stunning
And 1h36'38" 49d1'40"
Stunning
Her 16h55'59" 35d1'19"
Stunty
Ori 5h30'29" -2d44'9"
Stupenda Giovanna
Peg 22h43'56" 27d1'19"
Stupore
Eri 4h21'33" -0d41'0"
Sture Lindstrand
Her 16h47'49" 35d8'14"
Sturgislover
Dra 18h3'11" 59d52'47"
STURM Claudia
Peg 22h56'38" 9d15'20"
Stu's Light - Stu McGregor 1963-2007
Cru 12h28'56" -60d53'27"
Stu's Shining Star
Cnc 8h15'42" 16d43'58"
Stute
Her 18h54'27" 24d19'10"
STVENJO
Lyn 7h56'58" 53d56'46"
Stylz 01201983
Aqr 22h31'50" -10d14'46"
Su
Umi 13h41'1" 73d58'5"
Su Ann Wong
Cra 18h48'54" -39d21'28"
Su Graf
Lyr 18h33'11" 36d19'8"
SU Qi, Angel
Sgr 18h21'43" -16d16'2"
Su Schatz
Lyr 19h16'52" 35d19'17"
Su Skjersaa Lukinbeal
Uma 9h31'18" 44d56'10"
SUAN THIP RIVERSIDE
Crb 15h36'27" 25d45'35"
SuAnn Oberholtzer
Uma 8h15'38" 63d46'35"

Suanne My Princess
Lib 14h32'59" -14d47'13"
Suave's Only Star
Tau 5h32'55" 24d0'21"
Subb
Uma 12h16'55" 58d37'13"
Subba Gollamudi
Lib 15h32'39" -7d42'45"
Subbalakshmi
Cnc 8h43'56" 10d55'57"
Sub-Commander T'Pol
Ori 5h37'17" 8d57'2"
Subeen Nawaz
Psc 0h33'44" 18d50'30"
Subi Kooner
Cap 21h12'42" -23d21'12"
Sublime
Cyg 21h0'56" 47d31'11"
Sublime 4 Joel
Tau 4h45'34" 18d55'15"
Subramanian Anantha Parameswaran
Aur 6h29'59" 34d21'1"
Subrina Herrera
Cam 5h46'6" 61d2'3"
Substrate
Lyn 7h58'29" 46d34'34"
Such Great Heights
Sgr 18h48'45" -30d10'50"
SuCha
Tau 5h29'26" 27d0'4"
Sucka
Cap 21h19'42" -23d4'54"
sucker
Dra 15h34'8" 59d59'55"
Sucky
Cep 21h48'8" 62d58'22"
Sucric
Ori 5h59'53" 22d44'35"
SUD (Sara Oleskiewicz) Chambers
Gem 7h37'45" 32d7'18"
Sudaba Rabinovich
Gem 7h39'14" 14d47'5"
Sud-A-Belle
Peg 22h9'49" 34d0'28"
Sudal
Sco 17h25'46" -45d6'32"
Sudarius Virtuous
Uma 9h25'25" 42d31'42"
Sudeera
Uma 9h27'36" 57d59'32"
Sudha "Baby"
Cyg 19h47'7" 35d19'49"
Sudha Baxter
Lup 15h30'57" -39d48'34"
Sudie Marie
And 0h21'7" 26d22'7"
Sudina Silverwolf
Sgr 18h54'59" -16d7'59"
Sue
Cap 21h47'20" -10d59'24"
Sue
Cas 23h47'3" 52d50'54"
Sue
Uma 13h38'24" 61d12'44"
Sue
Peg 21h38'39" 21d41'33"
Sue
Leo 9h48'45" 18d32'6"
Sue
Cnc 8h48'24" 26d37'13"
Sue A. Nodine
Gem 6h51'5" 31d35'36"
Sue A. Werb, CSEP
Cas 1h28'4" 62d17'50"
Sue and Christy
Peg 22h41'21" 26d1'27"
Sue and Chuck Seton
Cyg 20h5'52" 35d56'12"
Sue and Lloyd Neal
Cyg 21h28'8" 52d29'25"
Sue and Mike "3-21-03"
Cyg 20h56'55" 39d3'29"
SUE AND PAUL LOVE STAR
Uma 13h47'19" 48d46'36"
Sue and Stan Friedman
Aqr 22h18'45" -5d18'26"
Sue Anderson
Cyg 19h41'39" 36d59'15"
Sue Ann
Uma 9h24'9" 48d18'19"
Sue Ann
Vir 11h55'49" -4d28'12"
Sue Ann Bruse
Ori 6h2'26" 6d4'16"
Sue Ann Dale
Psc 23h44'53" 4d48'15"
Sue Ann Potts
Uma 10h3'38" 54d25'41"
Sue Ann Staake-Wayne
Gem 7h36'19" 24d8'14"
Sue Ann Swenberg
Uma 10h31'21" 48d31'4"
Sue Anna
And 0h19'21" 33d2'4"
Sue Anne King
Uma 9h33'45" 48d24'37"
Sue Anne Putt
Cas 0h57'30" 64d10'5"
Sue B
Sge 20h14'13" 18d21'48"
Sue Baby
Uma 8h23'44" 64d47'34"
Sue Bertalmio
Psa 22h6'59" -29d32'3"
Sue Bloom
Psc 1h32'42" 4d21'33"
Sue Bottoms
Gem 6h27'43" 13d53'43"
Sue Breecker
And 2h20'6" 46d48'48"
Sue Bruesewitz
Aqr 20h57'33" 0d31'29"
Sue Byster
Cas 1h23'58" 63d7'7"
Sue Calverley
Lyr 18h43'51" 28d27'26"
Sue Carol Bowman
Uma 11h55'39" 59d18'42"
Sue Carol Lee
Uma 11h17'31" 28d48'56"
Sue Carper
Uma 10h24'56" 49d42'59"
Sue Casford
Uma 11h33'29" 37d52'45"
Sue Chaar
Lib 14h52'59" -4d40'50"
Sue Chassen
Dra 17h25'17" 57d33'47"
Sue Cobb
Hya 10h27'30" -14d11'30"
Sue Collins
Cyg 21h42'0" 54d55'41"
Sue Connell Collins
Lib 15h9'34" -4d23'15"
Sue Conti
Mon 6h53'53" -0d30'9"
Sue C.'s Q
Eri 4h41'51" -0d59'32"
Sue Culpepper
Sco 16h39'41" -26d40'3"
Sue "Custard Bunny" Chambers
Lep 5h26'37" -24d6'34"
Sue Damron
Gem 6h59'0" 25d18'24"
Sue D'Arcy
And 0h7'8" 44d34'2"
Sue & Dennis Chong
Cyg 21h31'34" 39d27'47"
Sue Devereaux
Aqr 23h38'17" -24d19'5"
Sue Dibnah
Cnc 8h26'59" 30d21'37"
Sue Eileen Jacques
And 1h42'46" 42d1'58"
Sue Ellen Fraser
Psc 1h32'20" 5d52'24"
Sue Ellen Glorianne Monroig Cruz
And 0h51'39" 39d45'45"
Sue Elliott
Aqr 21h44'18" 2d33'18"
Sue Epstein
Gem 6h51'34" 26d37'3"
Sue Eye Fry
Tau 4h31'3" 21d58'54"
Sue G. Webster
Cap 21h3'13" -22d18'53"
Sue Gambles, a Star to us forever!
Cas 22h58'16" 56d28'23"
Sue Gargano
And 1h35'26" 48d21'53"
Sue Geary Star - Twinkling Memere
Leo 9h50'53" 28d17'10"
Sue Gerard
Cru 12h28'44" -60d0'37"
Sue Gilkes - (My Heart is Yours)
And 1h59'26" 46d52'42"
Sue Goddard
Vir 12h37'24" 2d19'41"
Sue Gordon
Uma 9h30'21" 44d13'14"
Sue Haas
Mon 6h39'46" -0d34'45"
Sue Hall
Cas 1h30'14" 65d45'36"
Sue Halladay-Kichula
Cas 0h15'34" 52d55'28"
Sue Hancock
Cyg 20h57'14" 47d32'7"
Sue Heath Wilemon
Uma 11h15'46" 35d33'34"
Sue Hudson
Cas 1h40'57" 64d47'57"
Sue - Indee, 07.05.1983
Tri 2h13'47" 32d52'25"
Sue Ivey
Peg 21h57'26" 10d12'42"
Sue Jacobs
Crb 16h18'46" 32d49'18"
Sue Jarrett
Uma 11h36'33" 40d9'54"
Sue Jarrett
Uma 9h58'39" 45d12'3"
Sue Jennings
Aqr 23h8'22" -4d56'3"
Sue & Joann forever a star
Lyr 19h6'22" 31d53'49"
Sue & John
Ori 5h36'26" -2d42'16"
Sue Jupp
And 0h2'8" 43d46'27"
Sue K Howe MY Love
Uma 9h49'51" 66d7'39"
Sue K. "Poncho" Rumble
Ori 5h53'50" 20d51'0"
Sue Karen McMullin
Ari 3h3'43" 26d48'24"
Sue Kinoshita
Peg 0h3'20" 25d19'16"
Sue Kruse
Cas 23h32'24" 59d10'22"
Sue Kudrick
Uma 8h36'47" 51d7'58"
Sue L. Richards
Her 17h14'7" 18d48'55"
Sue Lancaster
Uma 10h17'4" 49d55'26"
Sue Leugers
Sco 17h39'54" -39d3'6"
Sue Libby's Star
Sco 16h14'12" -12d52'31"
Sue Lowe
Crb 16h9'56" 37d16'24"
Sue Macy
And 0h27'5" 32d43'39"
Sue "Mama" Reddell
Sgr 17h45'0" -29d36'32"
Sue & Mark Jones
Cyg 20h45'56" 53d41'50"
Sue Merry Lightworker
Cyg 19h53'22" 39d33'17"
Sue (Mom) Harrington.
Leo 9h58'47" 18d44'53"
Sue "Mom" Wells
Cap 20h48'43" -20d2'13"
Sue Moses
Crb 15h53'42" 26d26'25"
Sue Murphy
Aqr 21h47'34" -6d51'22"
Sue Nanu Nanu Fudge Freedom French
Tau 5h47'57" 14d42'9"
Sue Nawrocki
Cas 23h40'59" 52d34'58"
Sue Neal Carpenter
Uma 10h17'19" 50d0'43"
Sue & Nick Bell
Cru 12h20'59" -59d0'59"
Sue Nuttall
Lyn 7h7'6" 53d27'49"
Sue O'Connor
Psc 1h7'12" 10d8'18"
Sue & Phil Milazzo
Cyg 19h36'8" 38d17'41"
Sue Phillips Viper Allstars
Lyn 8h30'31" 36d3'25"
Sue Ramsden
And 2h14'32" 49d17'41"
Sue Roberts - Austin & Dylan's Mom
Uma 11h34'9" 55d53'18"
Sue Robinson
Cyg 20h48'12" 54d16'58"
Sue Rogers
Lyn 9h6'25" 40d23'6"
Sue Rosan Pellowski
Mon 6h49'15" 7d46'55"
Sue Roy "Shines Forever"
Tau 3h37'26" 16d6'52"
Sue S. Cannon
Cas 0h48'27" 61d12'46"
Sue Scott
Cyg 20h5'49" 33d27'32"
Sue Sherwin
Cnc 8h8'25" 22d28'14"
Sue Shields Evans
Lib 15h55'46" -18d32'47"
Sue Shimer
Lib 15h5'26" -23d34'11"
Sue (Shmo) Peterson
And 1h47'18" 38d37'19"
Sue Simpson - A Beautiful Moment
And 2h14'47" 41d41'7"
Sue (Soheir El Banna)
Ari 2h49'0" 20d24'39"
Sue Surotchak
Sgr 19h18'12" -14d53'40"
Sue Thompson
Cyg 19h37'56" 32d21'31"
Sue Turner
Ari 2h55'47" 10d57'51"
Sue Wald
Cas 1h36'49" 66d30'17"
Sue & Walter Wessel
Cyg 20h30'50" 55d6'21"
Sue Walters
Cas 1h16'44" 56d19'39"
Sue Ward
And 2h12'4" 45d55'45"
Sue Warren
Cas 23h59'58" 54d5'59"
Sue Weinberger
Leo 10h24'37" 15d9'55"
Sue Wilder
Tau 4h35'44" 21d57'43"
Sue Willows
Umi 14h6'28" 76d39'45"
Sue Zycherman
Uma 11h16'46" 67d17'51"
Sue, forever yours..
Sco 16h8'38" -10d25'56"
sueandneil
Sge 19h38'59" 17d29'9"
Sueann March Fa'amausili
Peg 21h44'34" 23d49'35"
Suebee
Cap 20h25'11" -12d44'59"
SueBob2004
And 23h29'54" 39d19'8"
SueEllen Monfra Penouilh
Tau 5h52'9" 25d8'15"
SUEHANS
Peg 22h29'38" 18d48'10"
Suehay
Cas 3h15'46" 59d26'58"
sue-jim-rajarojo
Vir 12h23'2" -3d37'34"
Suelily
Crb 16h21'47" 34d0'24"
Sue-Linh
Aur 6h24'19" 41d42'54"
Suellen Griffin
Sgr 18h0'1" -31d41'48"
Suellen Hampton
Aqr 20h48'41" -11d25'19"
Suellen Inwood
Vir 12h15'4" -0d56'58"
Suellen Marie
And 23h19'50" 48d25'29"
Suellen Martin Hall
Uma 11h31'32" 34d56'1"
Suellen Surman
Sco 17h37'8" -41d36'56"
SueLou
Cyg 21h10'26" 36d37'35"
Suema
Leo 11h3'21" 23d31'16"
Suemiss
Ari 2h22'10" 24d15'42"
Suena
Uma 10h3'49" 41d58'16"
Suenan
Uma 10h53'10" 63d32'31"
Suenas Dulces JS
Vir 13h43'52" -17d0'32"
Sueño
Pho 0h46'23" -48d40'12"
sueño
Peg 22h43'35" 2d50'6"
Sueño de Sarah
Cnc 9h14'37" 14d37'56"
Sueño Grande
Dra 15h5'16" 60d42'25"
Sueno hermoso (Beautiful Dream)
Cru 12h2'40" -61d0'3"
suepons
Ari 2h42'1" 14d9'10"
Suerte
Uma 11h25'33" 55d21'8"
Sue's Fabulous 50 Star
And 1h33'34" 44d47'25"
Sue's Little Star
And 0h39'32" 37d6'34"
Sue's Shining Star
Peg 21h38'36" 12d45'45"
Sue's Shining Star
Lyn 7h10'58" 57d37'3"
Sue's Smile
Lib 15h14'56" -28d26'28"
Sue's Star
Sco 16h35'26" -32d3'5"
Sue's Star
Umi 15h30'23" 68d41'36"
Sue's Star
Lib 15h2'1" -19d42'22"
Sue's Star
Del 20h38'38" 14d43'54"
Sue's star
Ori 5h19'20" 5d59'36"
Sue's Star
Ari 2h26'36" 24d51'47"
Sue's Star
Gem 6h32'34" 21d56'55"
Sue's Star
Gem 6h25'6" 20d16'31"
Suesann Richins
Aqr 22h43'36" -18d24'55"
Suesse, Joachim
Ori 5h7'12" 5d39'15"
Suesser Stern
Uma 9h38'5" 58d32'37"
" Süessi Regi "
Tri 1h46'10" 31d36'26"
Sue-Sue
Aqr 22h50'23" -7d55'13"
Sueweirdiemama
Cas 23h34'19" 55d28'59"
Sueybell Dake
Cas 0h28'13" 58d43'2"
Suezanne M. Carabellese
Uma 9h48'35" 48d10'44"
sue-z-ku
Cnc 9h19'34" 27d27'53"
SUEZQT1771
Cap 20h29'42" -15d54'36"
Sufian Barakat
Sgr 19h24'29" -32d7'46"
Sufyan
Per 3h21'57" 52d12'1"
sug
Cnc 8h35'51" 32d8'40"
Suga
Uma 9h36'48" 69d9'52"
Suga
Uma 8h26'32" 62d21'40"
Suga Bum
Vir 14h12'29" 3d39'47"
Suga Momma
Vir 13h44'11" -0d23'33"
Sugandha
Aur 5h47'28" 41d0'42"
Sugar
Cap 20h55'8" -20d25'8"
Sugar 27.02.1976 (Katrin Pranschke)
Ori 5h49'23" -0d10'40"
Sugar Bear
Umi 16h50'55" 80d37'48"
Sugar Bear
Lib 15h35'15" -11d40'47"
Sugar Bear
Uma 10h5'10" 60d21'8"
Sugar Bear
Uma 9h35'30" 60d34'54"
Sugar Bear
Sco 17h36'27" -40d18'44"
Sugar Bear
Uma 11h53'18" 30d14'38"
Sugar Bear
Leo 11h29'31" 4d2'22"
Sugar Bear Taylor
Aqr 21h34'8" -5d14'5"
Sugar Booger Linda
Vir 13h22'28" 11d34'4"
Sugar Booger, Emily Crow
Sgr 19h20'11" -16d57'5"
Sugar Bug
And 0h53'56" 40d57'0"
Sugar Bugar
Her 16h50'29" 33d0'22"
Sugar Go Lightly
Lib 15h37'52" -23d26'57"
Sugar Love
Lib 15h44'9" -12d59'30"
Sugar Magnolia
Crb 16h8'51" 27d17'20"
Sugar Mamma
And 23h19'8" 51d46'36"
Sugar Muffin Emily
Dra 16h48'23" 58d26'24"
Sugar Pie Azarian The Greatest dog.
Uma 9h12'14" 70d3'3"
Sugar Pig
Gem 7h1'18" 34d41'47"
Sugar Plum
Uma 11h13'35" 50d29'41"
Sugar Plum
Aqr 21h3'35" -11d23'12"
Sugar Plum Pamela
Cnc 8h14'59" 32d1'44"
sugar poofy bear catfish
Aqr 20h40'13" -9d14'20"
Sugar Poop
Uma 11h18'17" 37d50'30"
Sugar Pop
Ari 3h10'39" 29d5'47"
Sugar Puddin
Uma 11h32'19" 63d33'27"
Sugar Ray
Boo 14h35'31" 37d28'50"
Sugar & Spice
Uma 11h4'9" 67d9'54"
Sugarbabe Lambright
Cma 6h24'4" -28d23'3"
SUGARBABY
Dra 18h57'57" 57d12'24"
Sugarbooger
Uma 10h36'36" 52d41'54"
Sugarbunns Gaze
Cep 23h20'39" 76d59'16"
Sugarfoot
Vir 13h22'35" 7d28'15"
Sugarlips
Sgr 19h22'46" -21d4'41"
Sugarmarie
Her 16h42'58" 35d52'5"
SugarPeach
Dra 17h4'28" 52d30'6"
Sugarpop
Sco 16h15'0" -14d8'4"
Sugar's Star
Lyn 7h21'24" 53d42'52"
Sugarus Infinitum
Cas 0h8'55" 59d5'30"
Sugarwho
Lib 15h24'58" -11d3'46"
Sugeli Flores
Gem 7h31'47" 25d9'36"
suger
Cnc 9h4'51" 31d44'2"
Suger Star
Cap 21h46'45" -11d23'4"
Suha
Crb 15h52'21" 34d15'10"
Suhail Alhreish
Sgr 18h16'8" -33d6'9"
Suhee Lim
Cap 20h30'41" -16d3'52"
Suheyla
Aql 19h32'8" 11d29'23"
Suhsen Family - Stella di Sogno
Uma 8h33'59" 64d4'36"
Sui Generis
Cnv 13h33'43" 39d12'20"
Suits' Star
Peg 23h8'19" 28d48'35"
Sujatri
Tau 3h56'19" 18d30'16"
Sujeire
Uma 11h12'50" 54d54'37"
Sukayna Alugaili
Uma 8h36'59" 60d35'0"
SukhaRam
Lyn 7h54'21" 42d9'4"
Sukhdeep Kaur Dharna
Vir 12h52'51" -0d32'38"
Sukhninder
Sco 16h19'46" -13d28'38"
Suki Man
Aqr 23h41'16" -11d15'30"
Sukina Mucchina
Sgr 19h29'11" -42d40'16"
Sukvinder Bhamra
And 23h5'15" 51d16'44"
Sukyoung
Psc 1h15'44" 3d7'44"
Sula Elizabeth
Cap 21h21'43" -25d41'36"
Sulabh H. Shroff
Per 4h26'45" 44d55'42"
Sule inalli-in Love Burcu,Noyan,31.01.1953
Cep 20h47'50" 67d43'16"
Sulean Ilie
Ori 6h8'13" 18d27'20"
Suley de Alexis
Lyr 18h35'47" 36d1'32"
Suleymi Najor
Cam 3h27'55" 62d14'46"
Suley's Star
Lyn 8h0'42" 38d54'11"
"Suli"
Vir 14h39'2" 0d38'31"
Sulian
Umi 15h49'26" 80d47'50"
Sulieta Ang
Cyg 20h53'22" 31d45'36"
Su-Lin Kim Trepanitis
Psc 0h37'25" 19d33'6"
Sullina Louise Becker
Cma 7h25'42" -19d53'49"
Sullivan
Cma 7h5'1" -14d30'9"
Sullivan 70
Ara 17h13'21" -53d0'38"
Sullivan C. Palermo
Uma 9h51'54" 49d30'24"
Sullivan Cade Amick (Stinky)
Leo 10h40'26" 17d0'37"
Sullivan James Diercks
Uma 14h18'37" 57d25'7"
Sullivan Trigg
Tau 4h13'43" 29d7'21"
Sullivan's Celestial Spirit
Sgr 19h1'0" -28d20'51"
Sullivans Star
Sgr 18h55'10" -21d23'42"
Sullman Major and Minor
Uma 11h45'5" 29d46'45"
Sullo
Cyg 20h19'45" 38d29'58"
Sully
Umi 13h25'49" 74d43'39"
Sully Jereidini
And 1h12'28" 37d35'13"
SULOR
Dra 18h25'39" 52d1'5"
Sultan Ahmed "Danu" Khan
Uma 9h51'31" 49d10'34"
Sultan and Sweets
Cyg 19h48'50" 37d54'1"
Sultan Nitin
Lib 15h2'54" -4d2'11"
SULTAN SAID BITAR
Umi 13h10'19" 71d21'5"
Sultana Krakana
Uma 14h13'17" 60d17'45"
Sultana Nicodemus
Lib 15h20'8" -11d27'12"
Sum Moe Twinkle
And 23h14'23" 42d21'33"
Suma
Aql 19h46'30" -0d2'9"
Sumair & Kiran
Ori 5h53'11" 6d42'4"
Sumalee Irene Turner
Vir 13h40'55" 3d49'36"
Suman Pallan
Uma 10h34'44" 43d7'7"
Suman Shunmel
Gem 7h33'26" 27d55'51"
Sumanna Dhalla
And 1h29'21" 34d31'54"
Sumanth Varma
Uma 10h2'47" 48d30'31"
Sumawuscha
Boo 14h12'15" 7d45'21"
Sumedha A Bahri
Psc 1h23'42" 31d59'13"
Sumeet THE HAWK Sehgal
Aql 19h35'53" 5d15'37"
Sumera M. Ahmed
And 23h49'30" 34d43'57"
Sumer's Twinkling Angel Dust
And 0h30'51" 34d47'45"
Sumi
Aur 6h26'28" 41d20'7"
Sumi
Cas 1h2'20" 48d44'5"
Sumi
Del 20h54'41" 10d59'40"
Sumire
And 1h52'28" 46d50'2"
Sumit Banerjee
Her 18h2'49" 21d29'44"
Sumit Kumar
Leo 11h42'49" 27d2'47"
Sumiyya
Dra 20h26'14" 74d3'41"
Summa Keepin
And 2h30'21" 37d57'43"
Summa Louise
And 23h57'53" 41d3'25"
Summer
Lyn 8h19'53" 45d56'49"
Summer
And 23h20'35" 47d36'57"
Summer
Tau 4h10'40" 14d54'51"
Summer
Umi 14h2'20" 79d3'55"
Summer Ann Horan
Aqr 21h6'21" 2d15'59"
Summer Anne Wilson
Aqr 22h29'6" 1d39'51"
Summer Bessie Meadowcroft Porter
Dra 16h40'24" 62d12'23"
Summer & Brandon
Uma 11h26'28" 35d15'8"
Summer Brooke
And 0h40'46" 35d12'16"
Summer Daniell
And 1h44'42" 43d37'35"
Summer Dawn
Psc 0h50'49" 14d2'45"
Summer Dawn
Tau 4h24'36" 23d58'20"
Summer Dawn Layng
Crb 15h50'0" 38d27'59"
Summer Elizabeth Maginley
And 2h11'46" 45d8'31"
Summer Elizabeth Ronner
Cas 0h16'26" 54d41'57"
Summer Elsie Hutson
And 1h14'50" 44d16'58"
Summer Eva
Tau 4h32'28" 5d11'25"
Summer Faye Davis
Ori 5h27'2" -7d57'31"
Summer Garsko
Psc 23h1'14" 7d20'52"
Summer Grace Garcia
And 1h9'29" 38d43'23"
Summer Grogan
Tau 5h32'44" 21d40'20"
Summer Hagdahl
Aql 19h43'48" 4d38'13"
Summer Haltom
Sgr 19h5'1" -24d2'46"
Summer Hernandez
Psc 23h56'46" -4d32'37"
Summer Hope Barnes
Vir 12h47'44" -10d9'23"
Summer Hope Foundation
Uma 13h37'0" 58d6'39"
Summer Hubbard
Dra 15h55'1" 62d51'15"
Summer Jade Murphy
And 23h33'41" 39d14'19"
Summer & Jason Ross
Cyg 20h2'49" 49d39'45"
Summer Joy
Uma 11h41'35" 51d38'8"
Summer Joy
And 23h22'21" 51d58'43"
Summer Kay Lundregan
And 0h2'14" 40d23'48"
Summer Kempton
Uma 8h37'50" 48d16'20"
Summer Lauren Cooke
Cnc 9h7'26" 12d12'16"
Summer Lee
Cam 3h59'46" 64d13'2"
Summer Leigh Fields
Gem 7h24'20" 15d11'38"
Summer Love
Sco 16h11'59" -14d58'21"
Summer Lue Lather
And 2h0'49" 42d59'49"
Summer Lynn
Cnc 8h12'0" 24d35'10"
Summer Mae
And 1h54'38" 45d30'54"
Summer Millhouse
Cas 1h21'5" 50d56'19"
Summer (Muzzy) Goforth 6-21-1986
And 0h56'12" 39d37'19"
Summer & Nicholas
Aql 20h8'24" -0d48'36"
Summer Noelle
Cnc 8h40'21" 32d34'52"
Summer Paris Klinman
Lib 15h2'46" -6d40'48"
Summer Performing Arts Company~SPA
Uma 11h35'17" 63d6'10"
Summer Rae Beasley
Cas 23h44'44" 63d48'33"
Summer Rain
Cnc 8h16'2" 29d51'39"
Summer Renee Tate
Cap 20h22'55" -9d44'12"
Summer Rose
Uma 10h55'50" 60d24'25"
Summer Rose
Lmi 10h25'14" 31d8'25"

Summer Smith
And 23h6'48" 51d37'25"
Summer Solstice
Cyg 20h34'25" 55d50'37"
Summer Star
Leo 11h35'32" -6d15'44"
Summer Star
And 1h34'50" 50d37'36"
Summer star
Tau 5h44'58" 22d7'49"
Summer Victoria Sesty
And 0h45'57" 30d1'53"
Summer - You are my Shining Star
And 0h16'50" 27d51'13"
Summer, Light of My Life
Vir 12h4'48" 7d10'4"
Summer2005
Cas 23h40'51" 53d38'30"
Summerainsleydale
And 0h31'10" 30d54'14"
Summermoon
Gem 6h37'50" 13d22'33"
SummerNicoleSmith
Vir 14h20'26" -20d55'48"
Summer's Guardian Angel
Gem 7h18'26" 24d16'29"
Summer's Light
Lib 15h23'7" -23d18'59"
Summer's Own
Psc 1h9'21" 21d27'20"
Summer's Star
Leo 10h34'17" 12d51'14"
Summers Star
Lmi 10h37'31" 27d4'24"
Summerset Apartments
Uma 14h13'47" 57d7'25"
Summerwind
Cam 3h54'55" 67d39'36"
Summit
Cap 20h30'5" -10d36'16"
Summit
Tau 4h51'53" 20d50'32"
Summo
Per 3h26'5" 47d7'35"
Sun Delia
Cru 12h9'50" -61d59'53"
Sun Hee Grinnell
Uma 11h58'31" 30d58'35"
Sunan Ammann
Uma 8h56'26" 71d6'59"
SunChaFlynt
Cyg 19h42'33" 42d52'13"
Suncica Radin - moj san i java
And 2h5'16" 41d22'29"
Sundance
Peg 23h31'1" 25d8'33"
Sundance
Sco 17h49'5" -41d58'8"
Sundance Poppin Jack
Cma 7h17'34" -30d29'56"
Sundari
Uma 9h47'14" 42d8'39"
Sunday Morning
Aqr 20h53'4" -8d29'28"
Sunday Twilight
Dra 18h36'50" 51d6'55"
Sundays Hossanna Hunt
Per 3h25'8" 54d34'12"
Sundee
Uma 11h47'32" 41d45'53"
Sundy Geiser
Umi 14h38'39" 82d35'11"
Sune A. Carlson
Sgr 17h53'39" -29d12'52"
Sune's Star
Cnc 8h57'8" 12d59'17"
Sung Hi Lee
Ari 2h10'23" 22d47'49"
Sung Hoon Sonny Pang
Psc 0h41'28" 7d22'41"
Sung Hyun
Ori 5h58'12" 2d59'37"
Sung Won
Aqr 22h26'53" -7d28'19"
Sung Yong Lee
Tau 4h26'27" 9d57'0"
Sunglasses
Cas 0h32'16" 66d19'15"
Sungoo's Star
Psa 21h32'19" -27d42'55"
Suni
And 23h49'32" 48d33'3"
Suni Alexander
Psc 1h19'31" 15d19'36"
Suni Ewell
Crb 15h26'35" 32d5'57"
Sunidia
Ori 5h22'53" 9d16'42"
Sunie's Star
Ari 2h52'3" 20d1'43"
Sunil Chacko
Umi 13h42'30" 77d43'38"
Sunil Prasad
Leo 9h44'42" 28d17'22"
Sunil Rajadhyksha
Cap 21h36'21" -15d43'59"
Sunilda Colmenares
Mon 7h2'19" 8d10'44"
Sunita
Dra 20h13'29" 72d37'8"
Sunita Menon
Aqr 21h42'7" 0d47'24"
Sunita Romani
Uma 9h18'7" 42d8'9"
Sunjay
Psc 22h52'29" 4d49'41"
Sunnäschy
Dra 18h58'48" 62d4'36"
Sünneli
Cyg 20h43'26" 45d45'26"
Sunneschi
Cep 23h10'55" 74d44'5"
Sunneschi 12.01.04 22:58
Dra 18h18'31" 71d42'57"
Sunneva
Umi 10h33'18" 88d47'7"
Sunni
And 2h31'16" 40d32'49"
Sunni
Aql 20h8'30" 5d58'4"
Sunni Lee Rafanan
And 1h11'24" 37d47'31"
Sunni Mary Gapac
Uma 9h28'46" 61d18'0"
Sunni Rae
Vir 12h41'45" 5d11'34"
Sunnie and Jeff
Sgr 18h22'36" -32d25'25"
Sunnie Bear Day
Sgr 18h10'18" -30d53'0"
Sunnie Page's Star
Vir 12h58'53" 3d43'18"
Sunny
Ari 2h19'49" 10d45'15"
Sunny
Ori 6h22'25" 10d55'47"
Sunny
Leo 11h11'25" 13d30'58"
Sunny
Vul 20h36'32" 24d15'11"
Sunny
Her 17h14'31" 16d9'19"
Sunny
Gem 6h58'59" 31d31'59"
Sunny
And 23h21'9" 48d38'57"
sunny
Uma 11h15'25" 71d33'13"
Sunny
Umi 15h34'28" 76d37'28"
Sunny
Vir 14h41'37" -0d34'55"
Sunny&Angeli
Tri 2h19'56" 33d4'23"
Sunny "Angelic" Barnett
Crb 16h2'27" 38d56'29"
Sunny Brooke Snyder
Lyn 8h20'22" 36d27'56"
Sunny Chenoa Alder
Cnc 8h3'32" 13d9'20"
Sunny Cole
Lyn 6h51'26" 52d47'23"
Sunny Day Stratton
Uma 10h52'35" 54d58'9"
Sunny Ina Cohen
Ari 2h10'11" 19d30'50"
Sunny Jules
Dra 19h7'32" 75d3'17"
Sunny Kay
Cas 2h28'2" 68d6'30"
Sunny Longhitano
Vir 14h4'17" -17d4'26"
Sunny Lynn Klinker Hampsey
Uma 14h25'36" 56d39'34"
Sunny Ming Siong Chong
Gem 6h51'49" 15d59'14"
Sunny O'Brien
Uma 10h29'16" 55d50'20"
Sunny Smile Lisa
And 1h39'18" 43d5'29"
Sunny Starr
Dra 16h19'19" 59d21'6"
Sunny25
Aqr 22h13'16" -1d36'12"
Sunnycho
Psc 0h44'53" 18d23'2"
Sunnydell Weimaraners
Ari 2h11'18" 27d10'53"
Sunny's Pooky
Psc 0h8'6" 3d38'51"
Sunny's Summer Place
Umi 15h40'39" 85d27'28"
Sunrise 13
Lyr 19h8'38" 42d20'46"
Sunset
Uma 10h22'53" 42d22'34"
Sunset Aloha 35
And 0h43'34" 44d45'6"
Sunset at Four
Vir 13h16'40" -22d37'21"
Sunset Sondra Gockel
Ari 2h29'15" 21d0'18"
Sunshine
Peg 21h19'29" 13d7'21"
Sunshine
Tau 4h42'23" 18d8'24"
Sunshine
Ori 6h13'18" 15d30'6"
Sunshine
Cnc 9h18'14" 8d28'27"
Sunshine
Ori 5h37'24" 11d42'40"
Sunshine
Ori 5h7'38" 6d1'3"
Sunshine
Her 17h12'23" 18d10'24"
Sunshine
Gem 6h35'11" 19d12'10"
Sunshine
Her 16h16'36" 26d6'45"
Sunshine
Her 17h17'35" 24d30'33"
Sunshine
Leo 9h29'58" 27d52'39"
Sunshine
Boo 14h46'10" 27d37'17"
Sunshine
Cnc 9h15'15" 25d34'29"
Sunshine
Ari 3h10'19" 28d34'51"
Sunshine
Tau 5h7'10" 27d42'22"
Sunshine
And 0h40'17" 42d29'9"
Sunshine
And 1h15'28" 41d38'43"
Sunshine
Cnc 8h45'57" 32d2'7"
Sunshine
Gem 7h29'31" 30d58'25"
Sunshine
Gem 7h39'25" 30d15'29"
Sunshine
Gem 7h56'44" 31d32'20"
Sunshine
Gem 7h43'59" 33d51'55"
Sunshine
Tri 2h10'18" 33d16'59"
Sunshine
Crb 15h31'15" 36d49'39"
Sunshine
Crb 15h50'34" 33d52'16"
Sunshine
Uma 11h0'17" 37d18'22"
Sunshine
Crb 15h41'54" 39d23'30"
Sunshine
And 23h24'55" 41d48'59"
Sunshine
Aur 5h38'43" 49d43'49"
Sunshine
Cyg 21h32'53" 48d58'17"
Sunshine
Lyr 19h4'49" 47d2'42"
SunShine
Sco 17h44'26" -36d47'52"
Sunshine
Cap 21h17'51" -24d38'36"
Sunshine
Cru 12h39'4" -57d33'36"
Sunshine
Col 6h20'39" -41d7'6"
sunshine
Umi 16h39'51" 81d26'55"
Sunshine
Dra 18h44'33" 80d45'2"
Sunshine
Ori 5h36'59" -3d22'3"
Sunshine
Eri 4h28'51" -2d36'17"
Sunshine
Ori 5h32'35" -0d59'2"
Sunshine
Aqr 22h25'58" -0d6'1"
Sunshine
Vir 14h46'3" -1d39'46"
Sunshine
Lib 15h33'1" -6d26'34"
Sunshine
Sgr 18h46'55" -17d19'13"
Sunshine
Sco 16h8'11" -11d41'37"
Sunshine
Cyg 21h31'10" 54d19'3"
Sunshine
Cyg 19h56'26" 54d12'57"
Sunshine
Cas 0h48'47" 62d26'50"
Sunshine
Cas 1h25'56" 62d39'33"
Sunshine
Uma 9h54'31" 55d13'12"
Sunshine
Uma 12h13'55" 56d18'34"
Sunshine
Uma 12h22'23" 54d27'28"
Sunshine
Uma 13h18'3" 57d11'47"
Sunshine
Cam 4h14'15" 63d15'17"
Sunshine
Uma 11h53'53" 62d36'31"
Sunshine 4 Nana
Ari 2h38'49" 26d28'3"
Sunshine Acres Volunteers
Uma 8h36'55" 51d27'46"
Sunshine Angel Rancho Sequoia
Sco 16h46'32" -35d20'53"
Sunshine Baby!
Peg 21h30'36" 18d3'56"
Sunshine Belle "11-11-03"
Lyn 8h23'33" 38d53'20"
Sunshine - Brandy B. Collins
Lyn 7h14'49" 51d57'24"
Sunshine Cremer
Aqr 21h7'43" -5d59'33"
"Sunshine" Dawn Robinson
Aqr 23h9'42" -12d47'47"
Sunshine DeCamp
Cnc 8h37'2" 14d5'54"
"Sunshine" - Erika Ehrhardt
Sgr 19h39'41" -15d45'29"
Sunshine Erinn Mikeska
Tau 5h2'21" 24d31'58"
Sunshine Flower
Lib 15h4'25" -9d47'8"
Sunshine II
Leo 9h57'20" 27d57'12"
Sunshine Jenni Lee "Nikko"
Uma 11h9'52" 28d40'56"
Sunshine Leah
Lmi 10h32'45" 38d33'3"
Sunshine Mari
Sgr 19h14'15" -22d8'19"
Sunshine Marie Davies
And 1h11'25" 41d32'44"
Sunshine McGinley
Lib 15h10'28" -16d1'35"
Sunshine Melina
Uma 10h13'46" 56d53'17"
Sunshine on Texas
Tau 4h14'26" 16d19'21"
Sunshine & Poncho's Sparkle
Cyg 19h40'46" 38d45'44"
Sunshine Queen
Sco 16h16'57" -21d52'35"
Sunshine Raden Carnonie
Uma 8h41'3" 68d57'30"
Sunshine Ray Loveface
Leo 11h30'40" 0d22'43"
Sunshine Sugar Bear
Cap 21h6'1" -21d48'14"
Sunshine256
Lib 14h50'40" -2d18'56"
SunshineCL-14
Psc 0h34'30" 18d43'24"
Sunshine-Love
Tau 3h41'59" 26d6'53"
SunshineMom
Lib 15h3'55" -6d58'38"
Sunshine-Musketeer
Cas 0h52'22" 62d54'46"
Sunshines and Starlights Milestone
Boo 14h10'40" 29d15'26"
SunshnNikki
Vir 13h18'33" 6d45'41"
Sunspot Mercury
And 23h45'26" 46d59'25"
SUNTAN
Tau 4h4'18" 4d43'12"
Sunter's Star
Cyg 20h57'36" 54d48'14"
Suntronic - The most beautiful
Uma 11h8'44" 65d22'13"
Sunuttha "Nat" Suphat
Uma 13h2'42" 61d13'4"
SuPa
Cnc 8h29'30" 15d0'22"
Supa Dupa Poopa Pie
Cyg 21h19'27" 42d35'42"
Supa Zed
Ori 5h34'44" -4d27'20"
SupaDave
Leo 9h31'24" 27d13'45"
SupaKupaStar
Cam 4h4'9" 68d31'31"
Supalak
Cam 4h25'54" 70d40'49"
Suparman, The Hang
Ori 5h35'27" 9d59'46"
Super Boo
Her 18h8'25" 48d17'57"
Super D
Dra 16h11'6" 55d49'34"
Super Band
Cep 23h12'38" 79d14'46"
Super Daddy
Her 17h22'0" 39d14'13"
Super Daddy NCH
Uma 11h22'38" 63d7'31"
Super Daddy Star
Her 17h37'22" 36d46'33"
"Super" David Duane
And 0h23'50" 27d29'39"
Super Huchet
Del 20h25'2" 6d43'32"
Super Jordan Wronzberg
Cep 20h35'49" 65d12'20"
Super Laura
Lyn 7h48'39" 41d50'25"
Super Monica Romero
Lyn 8h24'45" 33d54'42"
Super Monkey Star
Lyn 8h19'32" 34d11'43"
Super Mutti
Gem 6h46'22" 23d10'8"
Super Nova Bob
Ori 5h49'41" 20d25'44"
Super Nova Kaori
Ori 4h52'51" 10d33'8"
Super Russ Berndt
Her 18h23'12" 24d8'45"
Super Sandy
Lyr 18h36'0" 30d2'28"
Super Shafter
Ari 3h11'37" 28d36'4"
Super Sig
Uma 9h55'59" 71d59'42"
super special
Sgr 18h10'46" -31d4'6"
Super Special Mommy's Star
Cas 1h11'15" 61d56'31"
'Super Star BK' - Brian Elliott 1945
Psc 1h54'8" 2d58'39"
Super Star Dad
And 0h17'8" 26d30'54"
Super Star Estrella
Uma 9h28'34" 67d47'55"
Super Star Eva Jane Clark
Ari 2h30'44" 22d17'43"
Super Star Hugo
Lib 15h16'37" -8d49'33"
Super Star - Jay Sparks
Cyg 20h6'12" 32d27'51"
Super Star Jimmy Ferraro
Leo 11h34'49" 26d35'33"
Super Star Karen
Sgr 18h17'8" -20d49'47"
Super Star Marci
Ari 3h9'59" 27d48'51"
Super Star Sharayah Emilie
Lyn 6h49'48" 54d16'39"
Super Star Susie
Sgr 18h20'27" -23d50'49"
Super Steve
Aqr 21h21'59" 2d14'4"
Super Tara Tara Fritsch
Lyn 8h14'11" 41d21'54"
Super Yutty
Lib 15h37'51" -17d58'33"
Superb Grand/Parents Stan & Chris Minch
Uma 8h36'38" 59d26'56"
Super-Engi
Ori 6h11'56" 16d4'3"
SuperFrank
Vir 14h42'8" 3d57'41"
SUPERGARY
Sgr 17h57'25" -28d20'39"
"Supergirl"
Ari 3h11'12" 28d7'15"
Supergirl
Cnc 8h18'2" 20d25'45"
-Supergirl- (Rachel & Neil Forever)
Cru 12h3'35" -61d59'24"
Supergirl Rebecca
And 0h43'27" 30d54'48"
Superhelden
Uma 14h17'32" 59d20'18"
Superlips
Uma 10h2'44" 62d17'43"
SuperM Road Star
Cnc 9h16'40" 32d41'56"
Superman
Per 4h12'9" 42d45'27"
Superman
Her 16h46'42" 38d39'2"
Superman aka Scott Rifkin, MBA
Aql 18h42'21" -0d48'12"
Superman George 9th March 2005
Per 2h50'9" 40d42'53"
Superman Kerry Murphy
Sco 16h54'47" -41d9'23"
Superman my PC- My Honor is My Life
Lup 15h46'59" -35d35'51"
Superman's Angel
Uma 11h2'19" 65d25'38"
Superman's Sunshine 5-19-06
Her 17h21'42" 34d2'5"
SuperMegaLin 5000
And 2h18'20" 46d44'44"
Supernaut
Psc 0h36'20" 7d5'27"
Superne Astrum
Boo 14h53'2" 24d6'49"
Supernova de Masanobu Tokunaga
Lib 15h2'15" -14d2'39"
Supernova3001
And 22h59'34" 51d22'13"
supersawa
Lib 14h56'8" -2d51'55"
Superstan
Uma 10h3'37" 49d44'39"
Superstar
Leo 9h34'21" 30d9'1"
Superstar
Cyg 20h51'41" 31d38'53"
Superstar
Cnc 8h38'16" 7d11'20"
SUPERstar
Sco 16h17'16" -10d42'28"
SuperStar Eileen Burgess
Mon 6h49'28" -0d2'59"
Superstar John Warburton I Love You
Cru 12h18'46" -56d49'9"
Superstar Joseph Otto Jr.
Her 17h50'48" 47d23'18"
SuperStar Kristy
Tau 4h25'18" 20d15'20"
Superstar Leo - 21.05.2004
Cru 12h50'14" -57d54'52"
SuperStar Leylim
Psc 0h11'35" -0d24'56"
Superstar McKenna Ann 080805
Cas 23h38'34" 55d5'54"
Superstar Mom
Cas 0h32'53" 53d3'27"
Superstar Morgan Nicole 112104
And 2h21'33" 39d9'4"
Superstar Mum - Joanne
Cas 1h9'20" 51d0'0"
Superstar Nick Mallick
Vir 13h37'30" 1d49'15"
SuperStar OMAR SIDER
Ari 3h6'31" 28d45'35"
Superstar Shula Bright & Beautiful
Uma 12h24'53" 53d22'31"
SuperStar Sis Mem
Sco 17h39'11" -37d1'29"
SuperStar Stacey Kids
Lyn 7h44'2" 36d51'28"
superstar007
Aqr 22h14'18" 2d6'59"
Supervisor
Cap 21h4'39" -17d53'3"
Suphan Driscoll
Tau 4h15'28" 29d12'31"
Supicastar
Uma 12h1'4" 52d8'53"
Supola Melitta
Sgr 17h58'57" -22d14'29"
Sura Chansrichawla
Car 7h51'23" -61d58'50"
SURANJAN'S STAR
Lib 15h18'8" -23d6'42"
Suraya
Lib 15h24'15" -6d22'29"
Sureau Cécile
Del 20h30'38" 6d41'54"
Suren
Aql 19h13'25" 1d10'17"
Suren Hegde
Aqr 22h51'5" -7d36'3"
Surena, My Beautiful Princess
Vir 12h30'13" 7d2'20"
Suresh Hemakumara Seneviratne
Vir 13h3'15" 12d24'38"
Sureyya
Cas 1h5'8" 63d55'43"
Surf star 12-26 Kelly Harp
Cap 21h48'25" -21d30'17"
Surfdom06
Uma 13h41'50" 49d9'57"
Surfergirl
Uma 10h23'44" 70d40'54"
Surfin' Urban
Gem 6h36'17" 22d17'32"
Surge
Uma 11h36'33" 61d47'8"
Suri
And 23h54'16" 36d13'3"
Suri Imre
Uma 13h57'55" 51d57'6"
Suri Sirivongsack
Dra 15h55'55" 52d31'21"
Surilda Lynn Clark-Sturm
Cyg 20h13'59" 31d50'18"
Surinder Kaur Dhillon
Ori 5h22'32" -0d24'33"
Surinya Reynolds
Cra 18h45'31" -40d11'52"
Surma, Bodo
Ori 5h11'38" 15d54'18"
Surprise
Lyr 18h52'46" 38d40'10"
Surprise
Uma 12h40'2" 57d30'36"
Surprise Band
Cas 0h50'23" 75d7'46"
Surridge Shiner
Com 12h57'5" 20d22'45"
Surverne Watkins- Miller
Col 5h11'29" -28d23'38"
Survival
Aql 19h53'40" -0d32'38"
Surviving Angel Tere
Lyr 18h48'36" 39d6'15"
Susan
Cas 0h16'51" 47d28'21"
Susan
Cas 23h28'38" 52d7'2"
Susan
Lyr 19h12'20" 30d23'21"
Susan
Lyn 9h3'24" 34d50'5"
Susan
Tau 4h32'19" 30d53'40"
Susan
And 2h10'52" 39d21'29"
Susan
Per 3h17'35" 43d16'38"
Susan
Gem 6h55'29" 15d24'30"
Susan
Tau 5h51'44" 25d19'44"
Susan
Gem 7h49'13" 26d7'54"
Susan
Ari 2h24'4" 17d53'41"
Susan
Leo 11h13'39" 12d1'6"
Susan
Vir 13h32'36" -12d36'6"
Susan
Ori 5h13'5" -6d30'57"
Susan
Umi 18h49'36" 86d56'41"
Susan
Sgr 18h33'37" -24d15'7"
Susan
Sgr 19h15'24" -30d50'31"
Susan
Sgr 18h58'49" -35d25'54"
Susan
Psc 0h40'9" 6d15'38"
Susan (4-13-45)
Ari 2h4'4" 18d24'50"
Susan A. Alfano
And 0h37'9" 42d51'37"
Susan A. Christine
Sgr 19h29'31" -13d52'0"
Susan A Vatalaro
Tau 3h50'11" 9d17'19"
Susan A. Weber
Aql 20h6'13" 15d58'51"
Susan A. Young
Com 13h6'48" 25d15'5"
Susan & Alan Borislow
Vir 12h48'8" 11d25'18"
Susan Albertson
Peg 23h18'26" 13d38'26"
Susan & Allen ~ Happy 25th!
Cma 7h11'33" -26d11'11"
Susan Aller McNiell
Cas 23h22'7" 53d21'19"
Susan Allers
Uma 11h57'10" 39d45'23"
Susan Allison Cianfrogna
Psc 0h55'4" 8d15'9"
Susan Allyson George
Cyg 20h31'57" 38d43'35"
Susan Amanda
Cnc 8h34'6" 16d17'40"
Susan Amy Benoit Roy
Tau 4h13'3" 29d2'35"
Susan Amy Kittleson
Uma 10h17'33" 52d41'19"
Susan and Dana Peterson
Crb 16h9'37" 32d7'47"
Susan & Andon
Sge 19h52'18" 17d15'5"
Susan Angela
Cru 12h39'10" -64d41'15"
Susan Ann
Ari 3h22'47" 29d3'8"
Susan Ann
Cas 0h8'14" 53d56'47"
Susan Ann Clark
Sgr 18h50'18" -20d31'11"
Susan Ann Janet Helin
Uma 11h28'10" 30d22'59"
Susan Ann Klinger
Sco 17h53'39" -42d31'8"
Susan Ann Lyon Spadaro
Vir 12h36'21" -8d12'16"
Susan Anne
Ari 2h57'6" 26d10'29"
Susan Anstey
And 23h45'53" 45d44'29"
Susan Ashley Ranson
Uma 9h22'2" 56d48'42"
Susan Atkin
Psc 0h39'2" 2d46'57"
Susan Atkins
Cas 1h25'32" 71d48'38"
Susan Atkinson
Ori 5h0'19" 5d57'34"
Susan Au-Yeung
Tau 6h0'42" 25d31'40"
Susan Avis Dickerson
Per 2h57'26" 32d14'36"
Susan Avis Harris
Gem 6h26'17" 20d58'42"
Susan Aylwin MacKinnon
Uma 9h24'2" 45d30'47"
Susan B. Friend
Ari 2h41'15" 26d20'51"
Susan B. Rubin "Sweet Sue's Star"
Cas 1h27'54" 56d11'2"
Susan "Baba" Spivack
Ori 6h3'6" 18d47'20"
Susan Baker
Oph 16h25'3" -0d5'51"
Susan Baker Wright
Vir 13h59'0" -14d52'0"
Susan Baltz Field of Dreams
And 1h20'13" 44d33'59"
Susan Barbara Lopez
And 1h42'10" 38d46'4"
Susan Barbara Stanitski
Umi 17h6'53" 76d14'21"
Susan Barish
Lyn 6h47'44" 51d4'29"
Susan Barnes
Aqr 21h6'34" -6d57'20"
Susan Barry Black
Cnc 8h21'35" 8d26'30"
Susan Beech
Lib 15h26'2" -5d3'50"
Susan Begley
Cas 1h34'24" 60d13'39"
Susan Bellamy
Cas 23h30'24" 54d27'36"
Susan Beringer
Cnc 9h8'32" 25d35'36"

Susan Beth
Cap 20h24'12" -22d39'2"

Susan Beth Hiner
Cyg 19h54'51" 31d36'58"

Susan Bialoblocki
And 1h50'49" 41d59'17"

Susan Billington
Sco 16h5'49" -11d28'32"

Susan Blankenburg
Lib 15h3'46" -12d57'19"

Susan Bloom
Psc 1h56'4" 5d38'0"

Susan Boser
Cas 0h36'34" 64d35'27"

Susan Boyes
And 0h5'5" 45d11'14"

Susan Bradford
Sco 17h22'59" -40d34'42"

Susan Branch
Gem 7h50'8" 15d30'57"

Susan Brennan
Cas 0h53'37" 49d15'42"

Susan Brianne
Cap 20h31'9" -10d39'19"

Susan Brooks
Col 6h18'18" -36d34'35"

Susan Burns
Cap 20h22'54" -19d57'40"

Susan Butcher
Cma 7h1'54" -22d3'54"

Susan Butsicaris
Tau 5h10'36" 18d44'39"

Susan Butterfield
Psc 0h38'22" 17d1'39"

Susan Byers
Pho 2h15'19" -40d9'32"

Susan C. Bacigalupo
Lyn 7h38'31" 59d21'0"

Susan C. Beers
And 0h17'52" 25d58'15"

Susan C. De Bella
Cyg 20h21'38" 33d35'29"

Susan C. May
Cas 23h55'39" 53d50'52"

Susan C. Ruble
Dra 18h41'44" 57d27'3"

Susan Caldwell
Uma 9h23'16" 46d8'26"

Susan Canada
Vir 13h17'42" 1d45'15"

Susan Carol
Vir 12h47'3" 3d1'46"

Susan Carol Fullum
Lyr 18h53'56" 43d46'58"

Susan Carotenuto
Tau 3h50'20" 23d52'54"

Susan Caryl
And 23h13'43" 45d38'9"

Susan Catherine Mary Cox - Sue's Star
Uma 13h36'4" 53d52'13"

Susan Charles
Ari 2h50'6" 14d30'33"

Susan Chiang-Furtuna
Uma 11h31'6" 62d45'42"

Susan Chien
Sgr 18h33'47" -30d44'28"

Susan Chiverton
And 0h42'15" 31d43'11"

Susan & Chris Patterson Eternally
Lyr 18h49'9" 31d48'36"

Susan Chrisman
And 2h35'54" 40d36'33"

Susan Ciarra Rodriguez
And 1h4'7" 39d25'55"

Susan Ciurczak
Psc 1h18'51" 10d56'52"

Susan Claire Johnson Bormolini
Cyg 20h18'27" 51d39'12"

Susan Clardy
Cas 1h5'59" 63d3'47"

Susan Clark
Cyg 21h6'27" 54d36'49"

Susan Clive
Sco 16h57'28" -37d43'38"

Susan Coddington
Vir 13h27'7" 10d34'47"

Susan Colleen Anderson
Cyg 19h41'48" 35d5'8"

Susan College
Vir 13h7'59" 11d39'49"

Susan Conley
Cas 23h35'36" 55d44'50"

Susan Conthan Morgan
And 1h20'16" 48d32'20"

Susan Cooper
Sgr 18h34'37" -23d39'10"

Susan Copley Jarvis
And 0h33'59" 27d3'13"

Susan Cortopassi
Sgr 18h16'23" -17d27'44"

Susan Cullen Leiphart
Lib 15h8'28" -6d21'56"

Susan Cunningham
And 0h8'53" 31d47'24"

Susan D. Ehrenberg
Lyn 7h11'12" 45d35'47"

Susan D. Foster's Everlasting Star
And 1h15'24" 36d56'23"

Susan D Hertz - Worlds Best Grandma
Vir 13h7'15" 8d57'27"

Susan Dale (Harlich) Oberstar
Ari 3h21'51" 29d11'38"

Susan & Damon
Aqr 23h55'6" -10d18'50"

Susan & Dan Rosa
Cyg 20h26'13" 56d50'11"

Susan Date
Uma 11h18'36" 38d31'16"

Susan Davies
And 0h28'1" 35d42'24"

Susan Davis
Lib 15h13'57" -4d44'26"

Susan Dawn Ellington
Cyg 21h21'56" 54d53'42"

Susan Dec
Sgr 18h5'16" -27d56'1"

Susan Decker
Per 3h21'35" 46d55'17"

Susan Dee Petersilie
Cap 21h25'16" -25d43'55"

Susan DeLap
Lib 15h1'57" -18d23'27"

Susan DeMarzo
Crb 15h33'38" 38d0'41"

Susan Denise Chachere
Peg 22h37'29" 21d17'48"

Susan Denise Moon
Boo 14h33'22" 26d19'25"

Susan DeRoose
Cyg 20h14'41" 48d49'28"

Susan Deveau
And 0h47'15" 45d21'31"

susan DI ane
Ori 6h10'34" 16d11'18"

Susan Diane
Crb 15h39'59" 27d25'24"

Susan Diane
Lyn 9h1'15" 33d10'17"

Susan Diane Honey - 'Susie Wong'
Col 5h48'31" -35d48'11"

Susan Diane Weinstein
Lib 15h5'1" -6d22'50"

Susan Diann Greene
Sco 16h51'40" -44d41'0"

Susan Dias Domingues
Cas 0h24'40" 53d22'6"

Susan Dineen
Tau 4h35'9" 6d35'55"

Susan DiPalma
Aqr 22h24'15" 2d7'27"

Susan Doucette
Aqr 23h45'5" -10d56'25"

Susan E Bergeron
Lyn 8h32'42" 35d23'12"

Susan E. Bricher
Vir 11h51'58" -0d26'57"

Susan E. Goral
Lyn 7h41'8" 37d40'4"

Susan E. Gran Brugger
Aqr 22h51'38" -16d42'45"

Susan E. Harris
Cas 0h32'15" 54d43'18"

Susan E. Hatfield
Cnc 8h42'47" 23d27'0"

Susan E. Howard
And 1h46'42" 49d56'9"

Susan E. Maness
And 0h32'13" 37d31'37"

Susan E. Presland Stein
Cas 0h55'31" 54d25'2"

Susan E. Rivers
And 2h23'46" 41d8'45"

Susan E. Saccardi
Cnc 9h4'7" 26d22'49"

Susan E. Smith
Cap 21h36'55" -11d41'43"

Susan E. Wilson
Lyr 18h35'56" 32d45'19"

Susan E. Zeluff
Leo 11h46'11" 20d47'46"

Susan Edwards
And 23h59'24" 39d53'54"

Susan Elizabeth
And 0h29'50" 42d31'0"

Susan Elizabeth
Gem 6h8'56" 26d15'52"

Susan Elizabeth
And 0h19'58" 32d38'42"

Susan Elizabeth Berg
And 23h40'4" 44d20'41"

Susan Elizabeth Brown
Uma 12h42'34" 59d11'26"

Susan Elizabeth Darr
Lib 14h34'22" -9d8'13"

Susan Elizabeth Deck
Cap 20h47'7" -24d58'35"

Susan Elizabeth Hamilton
Cas 23h5'6" 58d37'5"

Susan Elizabeth Jacobs
Uma 13h15'51" 61d6'46"

Susan Elizabeth Judd
Per 3h23'31" 33d5'49"

Susan Elizabeth Reis Freedman
Uma 11h22'2" 51d17'4"

Susan Elizabeth Skindzier
And 2h20'17" 46d17'57"

Susan Elizabeth Suarez Brown
Aqr 21h2'0" -6d29'46"

Susan Ellen Koontz
Cas 1h38'11" 63d32'35"

Susan Elliston Fuller
Vir 12h9'53" 11d51'41"

Susan Elste
Uma 10h27'11" 46d34'3"

Susan Emily Plemons
Sgr 18h23'46" -32d26'45"

Susan Emma Louise Neilly
Uma 12h1'8" 35d28'44"

Susan Englisis
Cnc 8h41'58" 32d27'40"

Susan Eppard
Peg 22h54'24" 30d35'52"

Susan Euker's Ophelia
Tau 3h57'41" 21d13'15"

Susan Eva Hardy-Rieger
Sgr 20h21'4" -42d3'28"

Susan Evans
Pho 0h59'1" -50d42'38"

Susan Eve Parsons
Per 3h21'3" 32d16'34"

Susan F. Sagrera
Lib 15h14'30" -27d8'25"

Susan Fassett
Gem 6h54'31" 18d57'52"

Susan Feder
Dra 16h22'37" 54d9'52"

Susan Fellner
And 22h57'54" 47d3'42"

Susan Finlay
Sgr 18h22'59" -23d9'49"

Susan Fiorito
Cyg 20h52'20" 42d8'21"

Susan Foley
And 2h34'48" 39d26'29"

Susan Foote
Psc 0h23'10" 8d39'19"

Susan "For My Angel"
And 2h20'0" 50d29'6"

Susan Forbes
Cas 3h15'25" 64d6'44"

Susan Frances Whymark
And 22h59'31" 48d23'50"

Susan Fussell
Aqr 22h52'59" -7d45'47"

Susan G
And 0h31'25" 41d5'42"

Susan G
And 1h30'38" 34d42'20"

Susan G. Brown
Lmi 9h28'40" 34d17'28"

Susan G. Edwards
Cap 20h36'15" -27d30'30"

Susan G. Neville
Ori 6h5'48" 5d16'51"

Susan G. Tucker
Sco 16h13'35" -16d39'28"

Susan Gail
And 1h30'0" 34d28'2"

Susan Gail Jester
Aqr 22h1'13" 0d44'6"

Susan Gale Jones
Cam 7h42'21" 69d25'29"

Susan Gale McNeal
Uma 12h1'24" 32d45'17"

Susan Garcia
Tau 5h30'7" 21d59'47"

Susan & Gary
Cyg 20h36'14" 34d11'28"

Susan Gearheart Marshall
Leo 11h33'7" 25d30'15"

Susan Geisler
Cas 23h10'19" 56d12'56"

Susan George
Cas 23h46'38" 57d41'43"

Susan Geraci
Gem 6h28'27" 19d29'36"

Susan Gigliotti 01-04-1920
Cap 21h46'46" -18d39'43"

Susan Gillis Bailey
Uma 11h39'12" 47d18'48"

Susan Gin Duell
And 23h19'23" 51d39'22"

Susan Goochey
Sco 17h18'24" -33d10'29"

Susan Goode
And 23h37'51" 48d13'14"

Susan Grosse
Tau 5h45'0" 16d59'21"

Susan Gutridge "Suzie-Q"
Crb 15h29'25" 29d27'34"

Susan Gwyn Lessa
And 2h22'2" 42d19'28"

Susan H.
Lyn 7h7'52" 55d19'38"

Susan H. Chan
Cap 21h39'42" -14d19'48"

Susan H. Hendricks
Cap 20h9'52" -27d31'45"

Susan H. Majors
Ari 2h57'11" 12d5'55"

Susan Hale Munson Reikes (Sudie)
Lyn 7h33'26" 54d18'5"

Susan Hansbury
Ori 5h15'37" 8d35'20"

Susan Hardesty
Uma 10h49'59" 57d13'32"

Susan Harris
Lyr 18h55'48" 37d19'10"

Susan Hayward-Blanchard
Lyn 8h33'47" 42d6'16"

Susan Heather Williams
Sco 16h9'39" -9d34'36"

Susan Hein
And 0h23'55" 42d14'22"

Susan Helen Nairn
Del 20h36'42" 15d27'7"

Susan Helmstetter
Ari 2h54'26" 28d21'5"

Susan Herman
Vir 12h32'41" -4d2'3"

Susan Hester
And 1h40'16" 49d45'22"

Susan Hill Goforth
Leo 11h14'11" 6d20'50"

Susan Hinkel Snow - "Mom"
Ari 2h12'5" 12d21'48"

Susan Hobbs
Cnc 7h59'50" 15d51'24"

Susan Hodson
Cas 0h11'49" 53d31'19"

Susan Holden
Cas 0h35'30" 53d56'13"

Susan Hope Pine
And 23h23'39" 41d32'20"

Susan Howlett Butcher
Uma 12h11'28" 52d34'2"

Susan Hudock
Dra 16h53'58" 62d32'39"

Susan I. Katz
Uma 10h2'12" 68d22'50"

Susan Iatesta
And 1h41'5" 48d2'7"

Susan Irene Eriksen
Lib 15h56'21" -6d44'58"

Susan Irene Higby
Umi 16h54'9" 78d8'3"

Susan J. Duncan
Ari 2h56'18" 11d48'2"

Susan J. Ghisson
And 23h22'30" 42d18'37"

Susan J. Horner
Lyr 18h32'51" 36d24'22"

Susan J. McCarthy
Leo 9h40'33" 28d25'29"

Susan J. Sanders
Vir 12h48'26" 11d28'42"

Susan Jacqueline La Bella 12/8/1940
Umi 14h45'56" 74d14'19"

Susan Jane
Sgr 17h58'29" -16d58'49"

Susan Jane
Ori 4h44'40" 4d23'3"

Susan Jane
Cas 0h57'7" 54d58'35"

Susan Jane Baker
Psc 0h58'35" 14d1'34"

Susan Jane Harmon
Lyn 7h13'47" 58d28'10"

Susan Jane Parker
Gem 7h4'44" 10d56'24"

Susan J.E. Stephens
Lib 15h48'46" -18d35'30"

Susan Jean Boone
Uma 10h28'48" 48d42'16"

Susan Jean Collier
Cep 22h47'49" 74d34'35"

Susan Jean Gerencser
Her 18h56'23" 16d23'35"

Susan Jean Petruzzi
Ori 5h45'15" 9d16'21"

Susan Jean Shupe
Crb 15h39'57" 28d37'57"

Susan Jean Wagner
Aql 18h55'51" 9d29'16"

Susan Jo Gettler Conley
Tau 3h36'45" 16d13'49"

Susan Jo Rozok
Crb 16h14'28" 30d56'13"

Susan Jones
Leo 10h17'52" 24d59'49"

Susan Jones
Sgr 18h25'17" -18d58'6"

Susan Jones
Lib 15h8'20" -15d23'28"

Susan Joy
Gem 7h3'29" 33d44'45"

Susan Joy Corona Birthday Star
Sco 16h9'55" -10d53'36"

Susan Julia Berenbaum
Vir 12h25'42" 7d13'43"

Susan Julia Pierce
Cas 0h17'35" 58d10'13"

Susan K. Cordano
Aql 20h7'40" 4d8'1"

Susan K. Knoll
Uma 10h1'57" 48d37'3"

Susan K. Mabie
And 2h14'3" 42d5'14"

Susan K. Maertens
Lyn 7h52'58" 49d6'25"

Susan K. Mayer
Sco 16h5'9" -23d49'18"

Susan K Retz
Aqr 22h14'40" -23d23'5"

Susan Kalua
Peg 22h25'48" 5d4'50"

Susan Kathleen
Lib 15h7'18" -5d24'5"

Susan Kathleen Cellini Link
Uma 11h16'49" 52d24'7"

Susan Kay
Cas 23h37'37" 55d13'32"

Susan Kay Ipsen
And 1h30'22" 46d26'29"

Susan Kay Keith
Cyg 21h13'55" 51d19'3"

Susan Kay Schmidt
Psc 0h12'15" 6d7'24"

Susan Kaye Andrews
And 2h22'34" 37d49'48"

SUSAN KEOHANE
Cas 0h49'28" 63d46'47"

Susan Kessen
Vir 13h18'26" 7d3'39"

Susan Kieferdorf
Cnc 9h8'28" 32d0'53"

Susan King
Cnc 9h15'15" 14d26'55"

Susan King
Dra 15h28'41" 56d31'24"

Susan Kirkland
Cas 0h54'37" 62d57'29"

Susan Kirklin
Vir 14h25'44" 5d19'57"

Susan Knapp Schulman Star
And 23h58'18" 40d18'53"

Susan Kok
Ari 2h35'40" 26d43'54"

Susan Koller
Cnc 8h39'20" 24d59'25"

Susan Korzenko
Sco 16h13'31" -11d57'20"

Susan Kvalnes
Uma 11h51'24" 62d55'48"

Susan L. Cohen
Vir 11h56'29" -5d7'26"

Susan L. DeSimone
Leo 9h55'57" 29d23'56"

Susan L Foong
Ari 2h12'19" 11d19'21"

Susan L. Kirshenbaum
Gem 7h14'57" 32d44'13"

Susan L. Nordengren
Cas 1h13'24" 54d38'28"

Susan L Platts
Sgr 18h8'21" -20d4'12"

Susan L. Rushton
Ari 2h45'26" 20d12'1"

Susan L Sullinger
And 0h28'41" 37d29'24"

Susan Lanette Johnson
Vir 13h21'24" -14d57'42"

Susan Lanzetti
Cas 1h1'0" 63d37'44"

Susan Larsen
Lyn 8h35'2" 38d32'45"

Susan Laura Appley
Cnv 12h43'1" 38d12'26"

Susan Lea Lucie
Uma 10h11'42" 50d57'34"

Susan LeAnn Palmer
Cyg 20h36'11" 35d12'54"

Susan Lee Patterson
Com 12h36'0" 23d40'34"

Susan Lee Ware Gumbiner
Per 4h15'41" 52d21'57"

Susan Leeper Gerrard
Ori 5h48'26" -3d16'21"

Susan Leigh Balch
Aqr 22h7'30" -13d28'3"

Susan Leigh Stevenson-Hill
Gem 6h35'22" 16d46'25"

Susan LeSage
Uma 14h1'33" 52d4'43"

Susan Leslie MacDonald
Crb 15h45'35" 31d0'3"

Susan Lister's Star of Happiness
Cas 0h12'44" 56d56'14"

Susan Lorraine
And 2h26'23" 42d13'36"

Susan Lucci
Cap 21h41'12" -16d45'31"

Susan Lujan
Cam 5h51'40" 61d3'1"

Susan Lyn
Psc 1h2'18" 13d0'30"

Susan Lynch
Cam 7h29'14" 61d49'54"

Susan Lynn Akers
Per 4h15'12" 50d57'18"

Susan Lynn Barsness
Aqr 22h44'39" -7d19'20"

Susan Lynn Behr-Sutherland
Psc 0h49'17" 17d0'12"

Susan Lynn Christian
Sgr 19h49'7" -16d44'44"

Susan Lynn Dotson
Uma 11h27'37" 46d56'55"

Susan Lynn English
Ori 5h36'26" 14d8'23"

Susan Lynn Pomerenk
Lib 15h8'36" -27d15'1"

Susan Lynn Ritter
Vir 13h16'17" -0d30'40"

Susan Lynn Strevens
Sgr 20h21'45" -33d52'26"

Susan Lynne Callaghan
And 0h13'25" 37d48'32"

Susan Lynne Lallement
And 1h54'7" 38d55'12"

Susan Lyons Chesterman
Cam 13h8'56" 77d54'12"

Susan M. Barbetti
Crb 15h34'20" 34d30'8"

Susan M Benway
Aqr 21h39'29" 0d49'22"

Susan M Bisco-Anderson
Psc 1h20'38" 31d55'43"

Susan M. Bowden
Leo 11h15'34" 0d37'17"

SUSAN M. BURNETT
Sco 17h54'21" -38d34'17"

Susan M. Cutro
And 23h34'39" 47d15'47"

Susan M. Dymowski
Ari 3h8'27" 29d19'25"

Susan M. Fenelon Kerr
Psc 23h7'20" 7d46'28"

Susan M. Golden
Lib 15h19'39" -9d54'52"

Susan M. Hughes 2006
Sgr 19h33'59" -16d36'42"

Susan M. O'Shea
Ori 5h37'50" 6d30'31"

Susan M. Pawlick
And 2h20'2" 45d20'7"

Susan M. Rary
Lib 14h52'5" -17d58'37"

Susan M. Vandervoort
Psc 1h4'12" 25d4'57"

Susan M. Walker
Psc 1h26'47" 15d53'0"

Susan M. Wege
Cas 1h23'11" 63d39'41"

Susan M. Woolsey
Cas 1h39'11" 66d59'24"

Susan M. Yameen
Aqr 22h21'37" -14d37'8"

Susan MacKain
Cas 23h45'36" 58d49'5"

Susan Maguire
And 0h21'4" 45d29'42"

Susan Mai Chen
And 23h15'32" 51d37'39"

Susan Maiden
And 2h19'24" 48d34'37"

Susan Malone
And 23h51'34" 39d24'41"

Susan Malone
Leo 11h18'25" 10d49'10"

Susan Margaret Biddle
Lib 15h25'57" -20d37'35"

Susan Marie
Uma 9h33'18" 62d27'16"

Susan Marie
Umi 15h15'4" 68d26'50"

Susan Marie
Leo 9h29'29" 24d5'27"

Susan Marie
And 0h59'5" 43d18'51"

Susan Marie
And 0h37'38" 42d24'15"

Susan Marie Best Mom There Could Be
Cyg 21h50'5" 49d8'21"

Susan Marie Bruecken
Tau 4h27'48" 24d30'12"

Susan Marie Burns
Her 16h32'57" 32d57'30"

Susan Marie Cook
And 2h13'16" 44d17'11"

Susan Marie Flores
Umi 15h52'40" 70d53'58"

Susan Marie Hoelter
Aqr 22h2'25" -14d35'16"

Susan Marie Pierscionek
Tau 3h48'31" 3d9'26"

Susan Marie Reiley
And 1h14'46" 40d13'7"

Susan Marie Rotter
Cas 23h36'29" 55d40'21"

Susan Marie Rufo
And 0h18'31" 45d51'15"

Susan Marie Scharnhorst
Lyn 7h14'13" 49d37'12"

Susan Marie Shreve
Leo 9h37'52" 25d53'19"

Susan Marie Walters
Cas 0h50'38" 60d36'3"

Susan Marie Warner
Cap 21h2'54" -16d25'15"

Susan Marie Wright
Uma 8h45'55" 49d26'26"

Susan Marie Wright Muegge
Aqr 20h58'13" -0d14'43"

Susan Marshall Greene
Ari 2h7'33" 23d19'0"

Susan Martha Brooks
Cas 1h7'20" 52d32'45"

Susan Mary
Lyr 19h24'39" 41d24'25"

Susan Mary Romano
Cas 1h34'12" 65d8'9"

Susan Mary Towler
Aur 5h58'36" 36d25'27"

Susan Mauntel
Cas 23h26'22" 58d11'13"

Susan Maxfield Bergeron
Crb 15h42'38" 36d30'0"

Susan May DePaul
Vir 12h59'51" 4d4'56"

Susan May Pudney
Lmi 10h45'38" 29d31'44"

Susan McCarthy
Cyg 21h53'42" 51d4'49"

Susan McConnell
Aqr 22h44'26" 0d29'37"

Susan McLeer
Lib 14h42'24" -11d17'12"

Susan McLeod
And 1h6'14" 46d4'17"

Susan McMahon Julian
Aql 19h50'53" 15d2'14"

Susan Mele
Tau 3h34'46" 11d17'14"

Susan Melia
Cas 1h20'29" 72d2'48"

Susan & Michael
Lib 15h5'58" -27d27'48"

Susan Michele Cornish
Dra 18h55'23" 58d39'43"

Susan Miller
Lib 15h44'31" -16d55'15"

Susan Mills Walsom
And 23h57'8" 40d36'3"

Susan Mitchell
Ori 5h7'13" 11d41'44"

Susan Modesto
Cnc 8h10'4" 15d30'51"

Susan (Mom) Meyer
Per 4h11'19" 45d20'11"

Susan Moore - 1st November 2006
Sco 16h39'20" -39d41'2"

Susan Mufson
Gem 6h28'31" 22d56'45"

Susan Murphy's Celebration of Life
Her 16h59'49" 29d39'31"

Susan Musa 1952
Gem 7h4'34" 16d27'23"

Susan My Love
Aqr 23h39'39" -10d38'10"

Susan Myra Andrews
Lib 15h42'27" -4d30'40"

Susan N. Byington
And 1h1'19" 46d25'49"

Susan N. Comegys
Cas 0h14'48" 63d38'57"

Susan N Parrott
Cas 0h31'37" 50d13'14"

Susan Nakao & Ray Ward
Cyg 20h30'42" 34d1'0"

Susan Nance Carlisle
Cas 1h2'57" 64d26'45"

Susan Ngo
Ari 2h15'3" 24d49'58"

Susan Nichole
Sco 16h13'39" -22d27'40"

Susan Nichole Wikete
Lyr 18h31'19" 34d25'51"

Susan Nila Saxe
Cas 1h30'30" 57d39'27"

Susan O. Allen
Mon 6h27'25" 5d13'23"

SUSAN O ROSAS
Lib 14h24'27" -18d1'44"

Susan Ostrander
Cas 0h23'13" 61d55'57"

Susan Overdorf
And 23h59'46" 39d32'13"

Susan Overfield
Dra 18h59'35" 66d33'16"

Susan Paisner
Cas 23h35'15" 56d20'43"

Susan Palasota
Uma 11h39'39" 55d54'46"

Susan Palmer
Com 13h24'46" 24d26'48"

Susan Park
Tau 4h35'50" 18d50'28"

Susan Parrott Ward
Ori 6h15'22" 10d2'16"

Susan Pearce
Cnc 8h19'35" 18d28'40"

Susan Pfisterer Barnett
Aqr 22h4'4" -0d39'39"

Susan Phelan's Eternal Brilliance
Umi 14h34'51" 86d31'58"

Susan & P.J. Eisma
Aql 18h58'54" 7d2'58"

Susan Potter
Cyg 22h1'26" 45d24'8"

Susan Price
Psc 1h11'4" 17d12'50"

Susan "Princess Suzykins" Banner
And 1h33'15" 42d30'23"

Susan R.
Vir 14h32'59" 3d41'17"

Susan R
Uma 8h46'23" 52d48'58"

Susan R. Granstrom
Uma 10h17'58" 48d54'45"

Susan R. Litersky
Aqr 23h27'39" -19d53'2"

Susan Rae
Ari 3h17'1" 29d8'46"

Susan Rae Halpern Sheehan
Vir 13h21'31" 13d12'32"

Susan Rae James
Ari 2h33'38" 14d49'19"

Susan Rae Kauffman
Uma 11h22'13" 53d13'39"

Susan Rae Nelson
Vir 12h42'35" 4d0'11"

Susan Reasoner
Lib 15h33'9" -10d21'24"

Susan Regina Sushko
Cas 1h43'26" 64d12'46"

Susan Reibsome
Leo 11h6'13" 8d32'16"

Susan Remiszewski
Aqr 22h7'34" -14d46'39"

Susan Renee
Eri 4h35'57" -2d2'1"

Susan Renee Lide Lloyd
And 2h32'16" 44d56'47"

Susan & Richard's Silver Star
Cyg 20h40'54" 45d52'21"

Susan Rigby
Ari 2h30'53" 26d34'46"

Susan Roberts
Tau 4h19'38" 24d52'24"

Susan Rogers
Lyn 8h24'7" 55d38'11"

Susan Rosas
Lib 14h58'19" -12d5'58"

Susan Rose and Bryce Fauble
Cyg 19h29'50" 52d56'53"

Susan Rouse
Lyn 7h12'44" 51d59'33"

Susan Rush
And 0h35'26" 41d37'26"

Susan Ruth Meyers
Lep 5h45'18" -21d53'31"

Susan Rynd Benjamin
Sco 17h13'41" -44d1'6"

Susan S. Lichterman
Uma 10h55'43" 59d55'47"

Susan S. Lundy
Lib 15h30'7" -23d43'47"

Susan S of AZ
Pho 23h35'24" -42d59'34"

Susan Schreder Petrak
Cas 1h57'22" 60d9'22"

Susan Sentance
Ori 6h16'33" 10d44'48"

Susan Sherlock Field
Ari 3h6'3" 23d14'6"

Susan Silverman-Roati
Cap 21h9'46" -15d40'15"

Susan Skutchy Love
Sgr 19h40'6" -13d42'24"

Susan Slali
Cap 20h24'56" -12d32'24"

Susan Soper
Tau 5h11'55" 18d6'42"

Susan "Soul Mates for Life"
Leo 9h25'16" 32d40'10"

Susan Star of the Family
And 0h16'56" 29d23'20"

Susan & Stephen Averill
Cyg 19h34'25" 29d16'39"

Susan Stout
And 2h21'35" 50d26'18"

Susan Stromquist
And 23h43'18" 46d51'13"

Susan "Sue" B. Riley
Uma 10h41'35" 56d41'22"

Susan Sullenberger
Uma 12h37'31" 62d50'29"

Susan Sutherland
And 0h59'53" 43d26'59"

Susan Swig Watkins
Psc 23h14'50" 1d46'54"

Susan Sydney Jayda Kreller
And 0h22'1" 25d7'24"

Susan Szymanski
Ari 2h9'17" 26d42'37"

Susan T. Sprawls
And 2h38'0" 44d17'35"

Susan Takata
And 0h20'51" 32d17'20"

Susan Terzi
Uma 8h29'9" 64d3'26"

Susan Thayer : A Reminder of God
Uma 10h24'52" 70d26'7"

Susan the Hitchhiker
Aql 19h56'4" -0d32'13"

Susan Thompson
Lib 14h54'41" -5d40'52"

Susan Tilstone
Uma 10h44'31" 68d36'59"

Susan Trinder
Sgr 17h55'1" -24d53'34"

Susan Trussell
Cet 0h43'6" -4d15'56"

Susan Turner
Cas 0h47'46" 60d25'0"

Susan Uberti
Leo 11h53'23" 25d59'48"

Susan Violet Cousineau
Cnv 12h37'47" 36d59'17"

Susan Vivian-Villalpando
Aqr 22h40'26" 0d7'1"

Susan W. Tillis
Cas 0h40'13" 63d52'30"

Susan Walls Heard
Sco 16h36'15" -35d11'14"

Susan & Wayne Kelso "Forever"
Cyg 20h18'50" 53d26'27"

Susan Wiener
And 1h23'1" 48d3'10"

Susan Wilhelm Wibbels
And 0h24'55" 28d17'16"

Susan Williams
Cas 23h39'46" 58d46'53"

Susan Wilson
Umi 15h31'53" 67d31'11"

Susan Wilson- Special Daughter
Gem 7h42'11" 25d27'26"

Susan Witman
Aur 5h43'28" 52d20'11"

Susan Wolfson
Cas 1h26'27" 63d36'53"

Susan Wood
Ari 2h42'48" 30d5'28"

Susan Wright You're Twenty One!
And 23h24'19" 41d37'25"

Susan Zabka Rohrbaugh
Lyn 8h48'0" 34d43'22"

Susan Zamer
Uma 9h28'41" 42d18'22"

Susan Zammikiel (reach for the sky)
Cnc 8h11'16" 21d59'10"

Susan Zaring Baker
Cnc 9h7'7" 15d28'31"

Susan Zuber
Cas 23h4'21" 53d45'47"

Susan, Gaye, Ninness
Vir 13h24'2" -8d53'49"

Susan, the brightest star in my sky
Lyn 7h16'1" 53d24'26"

Susan0826
Psc 23h25'55" 7d36'58"

Susana
Cep 21h0'29" 63d11'36"

Susana
Ori 5h13'13" -10d35'2"

Susana
Lib 15h7'41" -23d30'22"

Susana Bradley
Cas 0h20'27" 56d59'27"

Susana Brito
Her 16h49'21" 41d40'6"

Susana Cancino
Cas 1h3'24" 62d49'45"

Susana Carrero Jordán
Del 20h49'32" 6d43'49"

Susana Chavez
Cnc 9h5'53" 15d11'13"

Susana Farolito Guillen
Cnc 8h19'25" 18d38'18"

Susana Freire
Psc 0h46'34" 16d30'58"

Susana González Jiménez
Tau 4h35'56" 14d8'28"

Susana Herran-Young
Vir 13h58'59" -13d56'59"

Susana Huerta Villanueva
Aqr 23h38'4" -13d47'56"

Susana + Isaac /50/2006
Del 20h35'32" 14d37'23"

Susana JC101001
Vir 13h9'43" -6d20'52"

Susana Lopez
Cap 20h35'16" -21d57'33"

Susana Maria Galvez
Vir 12h58'7" 12d28'21"

Susana Marquez
Sgr 18h2'58" -35d28'11"

Susana Monica Kacanas 1.8.57
Pho 0h37'6" -48d46'25"

Susana Montes Lopez
Uma 11h30'5" 35d41'31"

Susana Ortiz-Arriaga
Cas 23h58'56" 56d42'30"

Susana Rodriguez-43531359H
Per 4h7'55" 44d31'46"

Susana Taide Palmero Boldt
Crb 15h49'7" 34d20'26"

Susanaestephane
Uma 13h3'49" 55d5'53"

Susan-Cody
Ori 5h11'2" 15d42'43"

SusanElizabethGreenleaf
Ari 2h36'32" 21d7'11"

Susanica
Ari 3h16'33" 28d47'4"

SusanJames
Lib 15h10'16" -4d57'32"

SusanJean
And 1h33'30" 39d33'26"

susanjohn
Ori 6h10'49" 3d26'27"

susankirkegaard
Vir 13h16'53" 5d15'38"

Susann
Uma 10h59'32" 33d49'0"

Susann Asotta
Gem 7h36'27" 26d31'4"

Susann Belitzer - the teaching star
Uma 11h14'11" 35d4'53"

Susanna
Psc 0h14'10" 6d10'9"

Susanna Ashley Gripton
And 1h54'33" 37d50'26"

Susanna Di Sessa
Lyn 6h43'33" 50d25'4"

Susanna Frieda Strahm-Morgenthaler
Lac 22h45'32" 36d14'32"

Susanna Kempin
Uma 12h52'14" 56d54'29"

Susanna L. E. Mikoula
Uma 11h23'45" 58d20'57"

Susanna Marilyn Nicol
Psc 1h18'3" 24d41'40"

Susanna Narinesingh
Cas 0h38'48" 61d9'8"

Susanna "Sunny" Joy Hickin
Leo 10h8'27" 18d25'31"

SusannaElizabeth Vermeulen 19710201
Ori 5h35'36" -1d49'40"

Susannah
And 22h59'43" 47d47'0"

Susannah Bayshore McIntyre
Cyg 21h4'52" 55d2'22"

Susannah Catherine Metzger
And 2h27'40" 47d28'48"

Susannah Hyde
Sgr 19h14'36" -21d31'23"

Susannah Mary Caston
And 23h33'19" 41d25'1"

Susannah May
And 2h38'25" 37d43'20"

Susannah Taylor
Cas 1h36'13" 60d4'36"

Susanne
Uma 12h47'5" 53d54'12"

Susanne
Umi 13h15'14" 72d58'54"

Susanne
Uma 13h34'53" 61d47'58"

Susanne
Mon 6h45'47" -0d52'20"

Susanne
Uma 10h28'40" 41d40'37"

Susanne
Crb 16h19'50" 38d30'5"

Susanne
Uma 10h45'30" 47d58'20"

Susanne
Com 13h11'22" 19d32'30"

Susanne
Leo 11h7'43" 21d8'27"

Susanne
Crb 15h35'44" 27d27'33"

Susanne and Alie
Ori 6h10'28" 16d12'11"

Susanne Arnusch
Uma 11h11'34" 42d4'32"

Susanne Berner
Cyg 19h44'35" 35d8'59"

Susanne & Daniel in ewiger Liebe
Uma 11h56'51" 61d13'28"

Susanne De Faveri
Cyg 21h25'24" 32d34'16"

Susanne Dolan
Uma 9h35'33" 62d41'2"

Susanne Faith Manheimer
Cmi 7h27'50" -0d10'5"

Susanne Frances Papa
Gem 6h37'15" 18d19'22"

Susanne Herzog
And 1h25'14" 43d27'29"

Susanne Hodgdon's Memorial Star
Cas 0h24'37" 52d56'46"

Susanne J
Ari 2h49'45" 28d43'10"

Susanne Lee
Aqr 22h25'42" 0d24'18"

Susanne Leuenberger
Umi 14h39'37" 81d46'14"

Susanne Louise
Aqr 23h5'23" -9d31'6"

Susanne Lucretia Paulina Bakker
Cas 0h38'28" 58d26'3"

Susanne Lytle
Crb 16h3'38" 33d59'1"

Susanne Maria Barker
Aqr 21h0'24" 0d43'28"

Susanne Melanie Snyder
Vir 13h18'32" -13d20'29"

Susanne "Mia amata anima gemella"
Lmi 10h26'20" 34d12'43"

Susanne Müller
Cep 20h39'43" 55d44'18"

Susanne Paduck
Uma 10h15'14" 46d11'8"

Susanne Rogers
Dra 16h31'50" 53d38'49"

Susanne Rüsgen "Engelchen"
Uma 9h53'54" 41d44'27"

Susanne Schneidereit
Ori 5h18'42" 3d16'59"

Susanne Suzy
Cap 21h27'52" -24d30'12"

Susanne Szalai
Cas 0h44'13" 53d43'23"

Susanne Tholle
Psc 0h15'51" 16d5'24"

Susanne Thuma
Cas 0h44'45" 52d59'48"

Susanne Thuma-Aigner
Uma 11h51'43" 31d18'42"

Susanne & Tony
And 1h24'54" 35d22'23"

Susanne und Christian
Boo 13h36'24" 20d12'3"

Susanne und Thomas's Liebesstern
Sge 20h2'15" 17d8'20"

Susanne und Vinzenz
Cas 1h23'51" 52d53'16"

Susanne Uzun
Boo 14h9'13" 27d59'10"

Susanne Werner
Uma 8h42'19" 49d36'25"

SusanPatriciaALPHA
Psc 0h58'36" 26d47'10"

Susan's Beacon 2004
Per 2h26'3" 54d52'20"

Susan's Guardian Angel
Sgr 18h58'20" -34d31'36"

Susan's Healing Radiance
Cnc 9h21'16" 6d39'25"

Susan's Heart
Aqr 22h25'25" -14d10'37"

Susan's Hope
Leo 11h42'3" 20d16'31"

Susan's Inspiration
Aur 6h18'32" 45d4'45"

Susan's Jubilee
Vir 12h59'56" 12d18'15"

Susan's Light
Cas 1h19'5" 69d0'41"

Susan's Little Star
And 0h34'5" 33d40'38"

Susan's Love
Cam 3h54'17" 58d3'59"

Susan's Miracle
Cyg 21h22'33" 35d18'48"

Susan's Place
Uma 11h3'32" 63d17'16"

Susan's Private Escape
Gem 7h1'35" 29d14'23"

Susan's Star
Psc 1h7'52" 28d35'48"

Susan's Star
Ari 2h11'35" 22d31'7"

Susan's Star
Peg 21h29'26" 16d16'6"

Susan's Star
Vir 14h41'29" 3d46'29"

Susan's Star
Uma 12h39'5" 62d0'38"

Susan's Star
Sgr 19h26'39" -14d52'6"

Susan's Star
Aqr 21h52'30" -6d5'19"

Susan's Star (12)
Cyg 21h51'38" 44d25'29"

Susan's Star of Hope
Uma 10h51'7" 60d58'53"

Susan's Stella Mater
Ori 5h56'0" 18d18'54"

Susan's Thank You Star
Vir 12h41'16" 11d23'14"

Susan's Valentine Star from Bob
Ari 2h12'47" 26d4'0"

Susanstar
Lmi 9h28'58" 34d57'25"

Suse
Ori 6h6'19" 15d39'22"

Suse und Robi
Cnc 8h24'55" 23d10'37"

Sushi Arevalo
And 1h31'54" 41d18'31"

Sushi Bopp
Cma 6h53'8" -16d11'21"

Sushil & Nikki
Cyg 19h43'45" 47d49'8"

Sushmita Kolhapurkar
Sco 16h4'39" -13d3'5"

Susi
Aqr 22h10'12" -0d52'8"

Susi
Cas 0h29'47" 61d26'53"

Susi
Uma 8h53'24" 57d30'47"

Susi
Uma 10h20'13" 47d31'38"

Susi
Cnc 8h11'16" 24d36'28"

Susi Allten
Cap 21h15'5" -26d7'58"

Susi and Sallie
Dra 20h30'54" 69d41'59"

Susi B. Lawicki
Lep 5h16'54" -11d55'26"

Susi & Fabio
Lyr 18h46'1" 31d25'27"

SUSI PLATTER
Ori 6h4'54" 9d47'3"

susi sonnenschein
Umi 15h51'30" 79d5'5"

Susi: The delicate little flower
Ori 5h58'59" 17d31'28"

Susie
Vir 13h6'52" 11d10'14"

Susie
Leo 10h22'52" 14d48'10"

Susie
Cnc 8h46'35" 32d24'7"

Susie
And 0h39'41" 37d20'34"

Susie
Tri 2h24'14" 31d49'10"

Susie
And 0h13'17" 43d43'58"

Susie
Cyg 21h7'2" 43d6'10"

Susie
Aqr 21h43'41" -5d26'45"

Susie
Lib 15h20'57" -9d34'41"

Susie and Justin
Tau 5h49'35" 24d13'47"

Susie and Nick's Evenstar
And 0h44'39" 39d47'35"

Susie Bethell-Larimer
Sco 17h51'36" -36d14'4"

Susie Bond
Leo 11h23'18" 12d54'11"

Susie Cano
Vir 13h30'29" -0d50'20"

Susie Clark
And 1h10'54" 42d21'59"

Susie D
Crb 15h48'49" 33d20'32"

Susie & Darrell (BF 143)
Dra 16h25'58" 54d35'0"

Susie & Dave's Star
Cyg 20h22'18" 39d7'30"

Susie & Eric
Crb 15h43'25" 29d40'28"

Susie G
Tau 4h54'32" 16d40'6"

Susie Gumberts
Tau 5h50'45" 17d19'51"

Susie Harriet Whitfield
Cam 6h33'52" 68d55'35"

Susie Hope
Crb 15h45'14" 27d13'28"

Susie Karas - Rising Star
Cas 0h21'58" 52d32'48"

Susie Krampf
Aur 5h28'21" 42d22'26"

Susie Krueger
Uma 11h27'22" 46d5'56"

Susie Mannese Taddeo
Cnc 8h37'30" 16d10'51"

Susie Mills Clark
Uma 10h31'15" 49d2'18"

Susie... my angel
Vir 13h49'51" -17d20'7"

Susie "Nana" Stafford
Uma 11h3'10" 57d35'22"

Susie & Nico
And 0h36'47" 36d55'36"

Susie "Penguin" Johnson
Aqr 22h35'45" 0d36'49"

Susie Q
Leo 11h7'11" 23d9'23"

Susie Q
Tau 4h51'39" 26d51'24"

Susie Q
Psc 1h6'40" 30d34'40"

Susie Q.
Uma 11h9'36" 34d41'2"

Susie (Q)
And 0h28'2" 42d2'16"

Susie Q
Lib 15h3'10" -8d20'46"

Susie Q
Cap 21h33'23" -9d29'25"

"Susie Q" ~ born on Aug. 06, 2001
Cas 0h30'40" 52d40'16"

Susie Rae "Taxman" Eisenberg
And 0h12'43" 28d5'26"

Susie Renee Madsen
Crb 16h16'26" 32d29'22"

Susie Rosenstein
And 0h34'52" 36d40'29"

Susie Rushton
Peg 22h25'42" 29d29'2"

Susie & Scott Verdonck
Cyg 19h28'9" 53d46'3"

Susie Shimon
Mon 6h45'0" -0d3'34"

Susie Snoflake
Cas 1h6'11" 49d34'12"

Susie Spence
And 0h57'40" 36d32'18"

Susie Stokes (Bolop 1)
Cas 1h5'57" 58d10'55"

Susie Suarez-Making Memories of Us
Peg 23h59'40" 21d39'15"

Susie T. Rios
Aqr 22h34'22" 0d39'36"

Susie The Brightest Star In My Life
Ari 2h36'36" 27d20'4"

Susie Wallace
Uma 11h57'59" 30d53'14"

Susie Woodrum
Aqr 22h15'22" 0d52'46"

Susiebeauty
Crb 15h45'1" 32d35'12"

SusieGator
Ori 5h36'30" -0d19'47"

SUSIELB1
Vir 12h34'31" -1d14'21"

SusieQ
Leo 10h23'41" 26d3'24"

SusieRuston1977
Uma 8h36'44" 46d47'57"

Susie's Angel
And 1h3'46" 35d8'1"

Susie's Angel
And 23h11'22" 35d15'50"

Susie's Inspiration Shines On
Cas 1h24'17" 57d23'58"

Susie's Star
And 1h50'37" 40d39'12"

Susie's Star
Cnc 8h18'10" 31d15'2"

Susie's Star
Gem 7h30'14" 15d57'43"

Susie's Star From Her Loving Family
And 2h34'45" 48d47'34"

SusiesLatramper
Uma 9h30'50" 72d37'15"

susmarie
Sgr 19h33'1" -13d57'3"

Suso Rebaque
Crb 15h44'46" 36d31'4"

Suspiri
Vir 11h40'52" -1d54'15"

Sussebass
Lyn 6h49'20" 53d38'24"

Sussy
Lyn 6h37'2" 54d5'27"

Sussy A. Santana
Cam 7h24'5" 60d22'2"

Susu
Umi 14h52'57" 81d26'27"

Su-Su
Sgr 17h59'6" -17d24'50"

SuSu
Sco 17h25'24" -45d10'19"

SuSu
Lyr 18h55'44" 33d44'48"

Susu1014
Peg 23h55'0" 18d43'29"

Susy
Cyg 20h9'50" 36d41'27"

Susy
And 1h14'40" 49d46'54"

Susy D Russell
Lib 15h46'7" -28d13'54"

Susy Johanna Labhardt-Gütlin
Cas 1h46'6" 59d14'22"

Susy Lynn
Ari 3h8'33" 27d22'27"

Susy Mesa
Leo 11h15'5" 23d46'30"

Susy Osman
Cas 22h58'54" 55d50'2"

Susy Simc
Psc 0h55'49" 5d4'48"

Susy Werre
Aqr 22h15'24" -16d20'13"

Susyna
Aur 5h12'9" 41d0'11"

Suszan Warner Berkich
Sgr 19h27'44" -36d46'51"

Sutaa of Christine Yamashita
And 23h31'37" 37d50'53"

Sutchie
Uma 10h44'22" 69d38'0"

Suthasinee Buddhinan
Cam 4h48'28" 71d38'53"

Suthathip Anantakul Runyan
Dra 19h45'4" 74d40'33"

Sutni Sethi aur Futti Pui hamesha!
Uma 11h33'28" 58d23'29"

Sutton Blake
Dra 16h22'56" 62d49'49"

Sutton Troy
Cas 22h59'38" 57d44'10"

Sutton's Star
Uma 11h38'24" 60d45'22"

Suvi
Leo 9h55'50" 29d39'36"

Suvi
Tau 3h26'7" 6d24'4"

Suvi Elaine
And 0h40'51" 36d17'17"

suwi19
Sge 19h56'20" 17d24'47"

Suwimol Ju Toungvutigul
Sgr 19h41'30" -15d18'51"

Suyu
Vir 13h40'21" -0d35'31"

SUZ' HEART
Peg 21h48'51" 15d14'21"

Suzabella
Ari 3h8'18" 19d45'37"

Suzan A. Wallace
Ori 6h5'3" 18d16'58"

Suzan B.
And 0h1'29" 46d2'9"

Suzan Carlisle
Cnc 8h50'2" 30d43'0"

Suzan Champagne
Cyg 21h46'5" 44d22'57"

Suzan Eleanor Bigalke Steele
Cap 20h20'51" -11d33'52"

Suzan Ferguson
Mon 6h53'34" -0d50'29"

Suzan Hinds
Crb 15h34'59" 34d52'4"

Suzan Hope Vigilante
Ori 6h7'47" 11d45'15"

Suzan Ready
Equ 21h14'20" 8d56'56"

SUZANA SMILYANICH
Gem 7h40'24" 20d42'17"

Suzana -Thomas ' Love Star
Uma 14h25'43" 57d24'0"

Suzana Tran Holland
Cnc 8h41'23" 18d30'17"

Suzann - Center of my Universe
Leo 10h11'46" 22d46'7"

Suzann Lane-Kinsey
Peg 21h33'53" 19d48'29"

Suzann M. Riester
Cap 20h17'30" -23d3'0"

Suzann Maybin
And 1h12'46" 35d37'1"

Suzanna
Cas 0h43'33" 59d43'45"

Suzanna
Tau 3h33'56" 3d1'50"

Suzanna
Uma 10h23'40" 69d10'17"

Suzanna Carolyn Phillips
Cnc 8h36'43" 8d29'27"

Suzanna Joy Ramsey
Cas 0h0'48" 57d27'16"

Suzanna Kathryn Duckett
Mon 8h3'40" -0d40'3"

Suzanna Rachelle
Cyg 21h36'44" 36d18'45"

Suzanna Rose Harman
Sco 16h11'2" -13d12'39"

Suzanna Talitha Pequeno
Leo 11h33'41" 6d0'37"

Suzanna Vik Superstar
Psc 1h8'38" 28d1'23"

Suzanna Villanueva Busquets
Cap 21h56'32" -14d23'29"

Suzannah Kate King
Cnc 8h13'30" 17d15'12"

Suzannah Lee Laulusa
Psc 0h11'46" 6d14'4"

Suzannah Rogers
Ori 6h13'32" 8d30'44"

Suzanna's Star of Youme
Uma 8h37'57" 71d50'41"

Suzanne
Aqr 22h35'27" -1d36'37"

Suzanne
Cnc 8h59'53" 11d42'45"

Suzanne
Gem 7h28'32" 22d23'46"

Suzanne
Ori 5h57'5" 16d57'41"

Suzanne
Leo 11h29'11" 27d39'6"

Suzanne
Cnc 9h7'19" 22d42'57"

Suzanne
Gem 6h15'2" 27d10'32"

Suzanne
And 1h11'57" 41d49'4"

Suzanne
Lmi 10h14'27" 37d48'31"

Suzanne
Cas 23h46'23" 51d35'5"

Suzanne 01161956
Cap 21h45'43" -12d44'41"

Suzanne and Brendan Keatley
Eri 4h37'2" -0d56'20"

Suzanne and Dan
Cyg 21h9'47" 46d40'36"

Suzanne and Jeremiah Brown
Gem 6h38'19" 26d45'23"

Suzanne and Michael Forever
And 0h55'23" 37d16'6"

Suzanne Aroon
Cma 7h10'44" -24d46'50"

Suzanne Baker
Sgr 18h31'57" -16d12'9"

Suzanne Bannon-Kohn
Del 20h38'41" 17d33'42"

Suzanne Biven MacDonald
Cam 7h43'3" 67d14'9"

Suzanne Boyett
Cap 20h27'57" -12d26'19"

Suzanne Brooks
And 1h6'38" 44d11'58"

Suzanne & Bryan Peterson
Sgr 18h31'51" -26d34'8"

Suzanne Buchanan
Crb 16h15'51" 34d0'49"

Suzanne Byrne
Aqr 22h0'27" -9d23'20"

Suzanne C. Earle
And 23h9'4" 48d34'50"

Suzanne Cannini
Ori 5h39'0" 3d18'59"

Suzanne Carroll
Cap 21h24'20" -18d10'31"

Suzanne Charlotte Materko-Ellison
Her 16h44'19" 44d29'37"

Suzanne Christine Allen
And 0h39'35" 32d35'39"

Suzanne Coddington
Cap 21h42'40" -23d49'37"

Suzanne D Konior
Sco 17h51'2" -36d7'8"

Suzanne D. Stone-Vine
Cas 23h52'28" 54d5'6"

Suzanne Dancause
Leo 9h42'50" 26d31'35"

Suzanne & David's Wedding Star
Cnc 8h52'24" 28d28'34"

Suzanne Dawn Clayton
Umi 14h22'53" 75d16'34"

Suzanne DeDominicis
Ari 2h17'50" 20d25'4"
Suzanne Dee Wingo
Lib 15h26'3" -19d52'50"
Suzanne Denice
Psc 1h32'31" 11d25'2"
Suzanne Detzner
Cas 0h37'43" 58d35'14"
Suzanne Divine
Sgr 18h50'18" -22d1'4"
Suzanne Dixon
Cru 12h32'28" -60d33'51"
Suzanne Dodge
And 23h28'32" 47d5'43"
Suzanne Doucette
Aql 19h49'52" -0d21'8"
Suzanne E. Burke "Granny"
Cas 0h19'38" 61d50'21"
Suzanne E. Nelson
Sgr 18h43'20" -29d20'0"
Suzanne Eichenlaub Baby Girl
Aql 18h57'7" 9d24'29"
Suzanne Elaine Mole Pegrum
Cru 12h36'1" -59d56'56"
Suzanne Elizabeth Regis Cook
Aur 5h3'25" 33d45'9"
Suzanne Eng
And 23h40'34" 48d7'5"
Suzanne et André Génin
Uma 8h24'45" 62d4'12"
Suzanne Fallon
Mon 7h36'3" -0d37'4"
Suzanne Farrelly
Ser 15h51'17" 21d20'35"
Suzanne Fraley Paddock
Cas 0h34'58" 63d56'12"
Suzanne Frost
Vir 14h15'4" -0d38'51"
Suzanne Furlow
Leo 11h26'57" 2d6'6"
Suzanne Gocke
Cas 23h38'34" 57d12'42"
Suzanne Goodwin
Psc 23h48'53" 5d47'28"
Suzanne Grether
Lyr 18h32'22" 36d54'54"
Suzanne Gruber
And 23h24'54" 48d17'16"
Suzanne H. Brockhaus
Sco 16h12'7" -29d50'33"
Suzanne Hance's Star
And 0h20'38" 28d40'42"
Suzanne Harper (03/18/1942)
Lyr 19h18'12" 29d42'44"
Suzanne Higgs Brown
And 0h29'29" 32d11'17"
Suzanne Hussein Abdelaziz
Lib 15h10'0" -20d33'45"
Suzanne Jaffe Bloom
And 0h39'9" 38d20'30"
Suzanne Janet Wizs
Cam 3h36'33" 66d48'53"
Suzanne Janssens
Lib 16h0'30" -10d28'51"
Suzanne & Jason Locklair
Boo 14h41'39" 37d19'43"
Suzanne Jenise Reinsch
Peg 21h41'21" 25d50'49"
Suzanne Jokai Lo
Ori 5h53'7" 12d31'28"
Suzanne K. Thomas In Loving Memory
Cyg 19h40'55" 30d37'19"
Suzanne K. Walthour
Uma 12h2'14" 52d15'54"
Suzanne Kaiser
Tau 4h35'28" 27d48'23"
Suzanne Karen Smith
Uma 11h2'39" 38d40'58"
Suzanne Kiser
Lib 15h23'22" -4d42'50"
Suzanne Koch
Umi 13h18'13" 74d36'2"
Suzanne Koehl
Cnc 8h5'16" 27d36'52"
Suzanne Kurmas Lewandowski
Cas 23h58'32" 61d32'56"
Suzanne Lackey
Leo 10h21'41" 12d51'50"
Suzanne Laurie Letton
Crb 15h39'52" 35d39'33"
Suzanne Lawrence
Cap 20h31'36" -9d35'46"
Suzanne LeCraw Baker
Psc 0h40'34" 15d28'7"
Suzanne Leitner "Our Star Teacher"
Her 17h34'38" 47d56'48"
Suzanne Leslie
Leo 9h42'22" 6d42'57"
Suzanne Lester Drasutis
Cas 23h37'33" 54d44'8"
Suzanne Lombardi
Aqr 20h51'21" -12d51'38"
Suzanne Lynn
Her 17h11'51" 18d38'28"
Suzanne M. English
And 1h33'1" 48d32'2"
Suzanne M. Verlicco
Sgr 18h12'10" -30d2'38"
SUZANNE MADDIX HURTADO
Mon 6h59'41" -0d50'20"
Suzanne & Malcolm
Cyg 20h50'17" 35d8'21"
Suzanne Marie
Uma 10h31'36" 62d38'35"
Suzanne Marie Cruse
Cas 0h21'2" 58d17'37"
Suzanne Marie & David Scott
Mon 6h47'57" -0d10'30"
Suzanne Marie Eighty One
Lib 15h26'26" -26d18'6"
Suzanne Marie Justice
And 23h48'18" 47d9'44"
Suzanne Marie Kemp
And 2h18'50" 41d51'42"
Suzanne Marie McGee
Cas 0h18'2" 55d52'8"
Suzanne Marie Thompson
And 0h46'47" 35d58'1"
Suzanne Martinetti
Sgr 18h17'21" -33d45'6"
Suzanne Martinez
Cas 1h17'27" 62d24'20"
Suzanne Mascorro
And 2h18'4" 48d50'12"
Suzanne Michelle Farndon
And 2h21'21" 44d47'18"
Suzanne Moore
And 23h52'50" 43d33'38"
Suzanne Murphy
Dra 16h11'54" 68d7'32"
Suzanne. My Eternal Love
Cyg 19h59'30" 38d55'50"
Suzanne Narain
Uma 11h27'34" 58d39'31"
Suzanne Noble
Ari 2h15'34" 14d49'1"
Suzanne Offit
Cas 0h25'55" 62d53'28"
Suzanne Patricia O'Donnell
Aqr 21h36'1" 2d3'51"
Suzanne Percival Pacini
Uma 13h33'30" 52d42'49"
Suzanne Petra Williams
Cyg 19h41'17" 27d58'16"
Suzanne Plotkin-Young
Sgr 19h49'33" -36d16'26"
Suzanne Price
Cap 21h57'22" -13d31'26"
Suzanne Rae DeMarcus
Sco 17h45'15" -41d44'26"
Suzanne Rhodes
And 1h12'48" 43d5'10"
Suzanne Roberts
Uma 10h59'57" 64d38'37"
Suzanne Robyn
Gem 6h25'42" 27d33'6"
Suzanne Roff
Cyg 19h35'36" 31d16'19"
Suzanne & Ronan
Cyg 21h18'11" 38d10'6"
Suzanne Ross Guest
Peg 22h55'14" 13d19'6"
Suzanne Roy
And 2h36'11" 49d54'0"
Suzanne Ruth Ann
Sco 17h45'48" -37d3'27"
Suzanne Ryan
Uma 11h46'5" 35d55'25"
Suzanne Seitz
Cyg 21h46'42" 50d11'11"
Suzanne & Simmy Forever
Cyg 20h23'43" 55d23'15"
Suzanne Simpson
Cyg 20h37'39" 40d37'15"
Suzanne Smith
And 0h34'49" 41d31'3"
Suzanne So & Perry Ip
Tau 3h44'12" 29d46'44"
Suzanne Squires
Cyg 21h3'50" 48d43'27"
Suzanne Steele Talley
Uma 10h27'40" 53d36'53"
Suzanne Swanson
Cas 2h50'4" 60d58'20"
Suzanne "Sweetness" Thatcher
Cnc 8h47'30" 21d8'57"
Suzanne Taverna
Cas 1h50'13" 61d42'20"
Suzanne Taylor Sabia
And 0h12'57" 44d23'17"
Suzanne Taylor-Olon
Uma 9h30'54" 49d59'33"
Suzanne the best wife in the world
Aqr 23h48'55" -22d53'6"
Suzanne Toma Breecker
Gem 6h49'15" 23d22'3"
Suzanne Tripp
Crb 15h32'59" 26d58'23"
Suzanne W. Fisher
Ori 5h28'35" 7d15'17"
Suzanne Waffle
Mon 7h30'41" -0d53'53"
Suzanne Wangmann
Col 5h59'9" -34d48'12"
Suzanne Ward Harriman
Aqr 23h25'54" -12d49'3"
Suzanne & Wayne 9/20/80
Per 3h21'44" 42d59'51"
Suzanne Weil 50
Tau 5h52'12" 16d39'41"
Suzanne Weitzel
Vir 14h3'54" -0d54'8"
Suzanne Wood
And 0h33'38" 39d6'46"
Suzanne Zuloaga
Sco 16h58'50" -31d37'3"
Suzanne31806
And 0h34'16" 27d49'9"
SuzanneBain
Leo 9h53'2" 21d21'48"
SuzanneP
Mon 6h33'43" 7d47'8"
Suzanne's Hope
And 0h48'48" 44d33'33"
Suzanne's Immortal Hope
Pho 0h53'3" -48d24'27"
Suzanne's night light
Sgr 18h25'42" -32d34'55"
Suzanne's Smile
Ori 5h20'47" 11d42'41"
Suzanne's Wish
Dra 19h11'21" 63d25'3"
Suzanness Mastories
Cas 0h27'44" 57d17'7"
Suzannia
Uma 10h26'41" 55d9'14"
Suze & Jerry 50
Ori 6h13'43" 9d14'41"
Suzemarie1983
Sgr 17h53'28" -29d42'12"
Suzette
Aqr 22h34'28" -1d20'13"
Suzette
Peg 22h55'30" 30d6'25"
Suzette Amber " Sunshine "
Vir 13h58'9" -5d45'41"
*Suzette *Ann *Martinez
Leo 10h58'14" -2d29'38"
Suzette Davis O'Neal
Aqr 22h5'19" -15d30'38"
Suzette Hunter
Cma 7h26'58" -25d16'14"
Suzette Kinkle
Vir 12h59'51" -3d36'50"
Suzette Rae Dykes
Lib 14h45'5" -10d30'27"
Suzette Ruth Zema
And 1h45'52" 38d43'25"
Suzettes Eyes
Psc 1h10'44" 33d21'48"
Suzette's Star
Srp 18h8'56" -0d49'14"
Suzhou Tourism School China
Ori 6h20'6" 18d40'9"
Suzi Cash
Lmi 10h23'36" 37d58'5"
Suzi Conwell
And 2h26'15" 46d52'19"
Suzi * Dafnis
Cra 18h46'9" -40d20'11"
Suzi McQuinney-Frank
Cas 0h10'37" 58d13'50"
Suzi Q
Aqr 21h41'59" -2d17'50"
Suzi R Decker
And 0h40'37" 33d6'53"
Suzi Tommey
Cyg 20h38'18" 31d31'8"
Suzie
Cnv 13h50'51" 34d14'48"
Suzie
Leo 11h46'42" 19d59'26"
Suzie
Cmi 7h29'19" 9d12'33"
Suzie
Psc 0h26'51" 12d2'27"
Suzie
Cmi 7h29'59" 7d1'51"
Suzie
Vir 11h56'6" -6d2'41"
Suzie <3 Dean
Cam 3h44'34" 54d38'16"
Suzie Apodaca
Sgr 19h39'17" -15d42'52"
Suzie Bernard
Gem 7h39'56" 32d7'51"
Suzie Coops
Cra 18h46'6" -40d32'8"
Suzie (don't #?*! with me) Kelly 27567
Cru 12h28'58" -60d30'6"
Suzie in the Sky
Cnc 9h11'30" 15d52'16"
Suzie J.
Tau 4h19'24" 29d10'47"
Suzie Mollichelli
Cma 6h54'23" -22d2'47"
Suzie Padilla
And 0h26'40" 38d28'6"
Suzie Q
Cas 0h1'13" 57d12'57"
Suzie Q
Ari 2h27'18" 26d32'34"
Suzie Shining Forever
Mon 6h27'3" -4d10'31"
Suzie Sunshine my Starlight Angel
Pho 23h59'9" -45d33'47"
SUZIE WEAVER
Tau 3h40'52" 28d38'59"
Suzie Wisz-Chippy
And 0h31'2" 45d30'3"
SuzieQ
And 0h31'10" 26d17'12"
Suzies Charm
Uma 8h20'9" 68d21'12"
Suzie's Everlasting Star
Cyg 20h28'39" 55d58'51"
Suzie's Star
And 0h11'43" 45d39'47"
Suzie's Star
And 0h18'46" 35d47'26"
Suzimmy Love
Uma 12h43'56" 58d29'3"
Suzi-Q
Peg 22h1'29" 15d22'58"
Suzuko Koshi
Sgr 18h12'3" -19d24'49"
Suzy
Lib 15h27'44" -6d32'6"
Suzy
Uma 13h35'45" 59d9'15"
Suzy
Lib 15h10'12" -28d17'4"
Suzy
Ari 3h27'16" 28d13'13"
Suzy
And 1h0'37" 36d27'38"
Suzy
Crb 15h39'22" 30d31'49"
Suzy and Scott Silver
Cyg 20h47'10" 35d3'58"
Suzy Ann
And 2h22'19" 37d51'47"
Suzy Barron's Star
Psc 1h42'8" 23d33'54"
Suzy Barry
Psc 1h45'12" 17d54'22"
Suzy Buzan
Lib 15h39'34" -19d50'8"
Suzy Georges - Extraordinary Woman
Col 5h8'20" -38d43'18"
Suzy Jackson
And 0h3'52" 39d46'47"
Suzy Juice
And 23h23'2" 52d11'1"
Suzy Koshier
And 23h6'34" 48d23'25"
Suzy Ponchukyan
Com 12h21'4" 30d0'36"
Suzy Q
Lyn 8h23'5" 34d13'1"
Suzy Q
Tau 5h43'22" 20d31'59"
Suzy Q
Cap 21h42'34" -18d27'25"
Suzy Q
Aqr 21h7'52" -2d0'40"
Suzy Q
Uma 9h9'27" 68d12'9"
Suzy Q Fedeli
Cyg 21h59'59" 50d43'57"
Suzy Q-C Piatt
Ari 2h3'49" 13d4'38"
Suzy Schulman
Cap 20h14'55" -25d13'43"
Suzy Star
And 23h49'23" 46d20'49"
Suzy & Vari
Cyg 19h52'5" 39d14'50"
SuzyQ
Lib 15h34'48" -24d56'36"
SuzyQ
Cap 21h55'21" -13d55'29"
SuzyQ 82
Uma 12h45'42" 58d59'41"
Suzy's Star
Peg 23h13'54" 27d4'3"
Suzzy
Cap 21h31'45" -14d29'55"
svdcrs
Uma 9h29'44" 44d38'15"
Svein Tore Aurud
Tau 5h48'55" 18d20'54"
Svelte Machine
Tau 5h32'31" 21d33'49"
Svemir Milos
Leo 11h10'2" -0d41'33"
Sven 11.07.1980
Peg 22h51'9" 17d29'27"
Sven & Anja
Lyr 18h47'6" 36d31'8"
Sven Frutiger
Dra 19h54'17" 70d34'59"
Sven Gohla
Ori 5h24'2" -8d2'55"
Sven Hagemeier
Cep 22h13'45" 67d12'28"
Sven Harke
Uma 8h37'59" 53d19'56"
Sven Hauser
Uma 10h57'44" 63d47'32"
Sven Hoffmann
Uma 10h10'58" 42d39'26"
Sven Homan
Vir 13h7'38" 3d41'23"
Sven Liermann
Uma 10h7'44" 47d26'20"
Sven Of My Soul
Per 2h23'1" 57d7'16"
Sven Rapp
Peg 22h49'6" 13d10'37"
Sven Raul, 17.09.2004
Uma 10h52'46" 43d33'2"
Sven Schuhmacher(Svenyboy)
Ori 6h16'51" 20d0'34"
Sven Sverdrup
Cep 22h21'29" 62d44'9"
Sven Tim
Uma 11h24'14" 39d8'43"
Sven Torsten
And 1h48'40" 37d49'42"
Sven Wiederholt
Pho 0h32'21" -47d6'10"
Sven Willi Wenk
Uma 8h18'8" 66d46'36"
Svend Akselsen
Per 3h33'29" 51d43'22"
Svenja
Tau 3h36'11" 21d54'15"
Svenja
Uma 9h9'29" 53d41'21"
Svensworld Est. 11/20/01
Cmi 7h19'48" 8d32'54"
SvenTigro
Sco 17h34'13" -38d14'10"
Svenya
Ari 3h1'40" 13d46'57"
Svetik
Sgr 18h36'53" -24d5'15"
SvetlaDang
Cap 20h19'55" -11d52'30"
Svetlana
Uma 9h31'52" 67d40'1"
Svetlana
Cnc 9h10'11" 20d51'33"
Svetlana
Her 17h8'25" 17d3'42"
Svetlana
And 23h19'56" 48d2'3"
Svetlana Demidova
Cnc 8h21'30" 32d21'20"
Svetlana Djordjevic
Cyg 19h49'12" 52d27'21"
Svetlana Krakhmalnik
Cas 1h23'44" 54d43'54"
Svetlana M. Lipsey
And 0h21'15" 46d17'27"
Svetlana Manning
Cnc 8h41'41" 14d20'33"
Svetlana Maslova
Sgr 20h5'16" -30d13'4"
Svetlana Petrova
And 23h2'8" 40d55'35"
Svetlana Royter-Vaiman
Leo 11h5'55" 16d54'1"
Svetlana Sakhanenko
Sgr 18h23'37" -33d10'55"
Svetlana Strelnikova
Sco 16h15'8" -21d0'12"
Svetlana Terecik
Uma 10h1'11" 47d44'30"
Svetlana, Little Kitten
And 0h33'42" 36d53'55"
Svetlanochka
Lmi 10h52'6" 35d32'13"
Svetsellen
Uma 10h37'33" 61d1'0"
Svetty
Sgr 18h0'25" -35d42'3"
Sveva Bellocchio
Umi 15h46'40" 84d9'35"
SVM2505
Per 4h34'33" 40d39'7"
Swaddle Bug
Cap 20h48'4" -21d17'50"
Swallow
Cyg 21h21'12" 31d24'4"
Swamp Siren
Lib 14h42'25" -16d25'16"
Swan
Uma 14h2'54" 58d15'29"
Swan
Lyn 7h43'39" 41d45'50"
Swan
Cyg 21h22'15" 39d15'24"
Swan 50
Psc 1h31'44" 13d33'44"
Swan Swan H
Cyg 19h58'10" 43d6'15"
Swanee
Cyg 20h48'38" 36d42'1"
Swanic
Gem 6h52'37" 21d45'28"
SWANK 206
Cyg 20h39'10" 34d59'17"
Swann Raines
Cyg 21h24'15" 37d12'20"
Swann Samuel Cukier
Vir 12h37'19" -5d21'4"
Swannie
Aqr 22h29'33" -7d22'33"
Swanson 062406
Cyg 20h15'35" 51d22'31"
Swanson Farms
Aur 5h58'19" 28d59'59"
Swapynette
Her 18h14'27" 16d35'8"
Swaran
Cyg 21h57'28" 55d1'16"
Swartz - 4
Uma 9h58'18" 59d44'31"
Swati
Eri 4h29'41" -21d30'30"
Swati Murdia
Sco 16h23'39" -27d15'12"
Swayze
Lyn 8h26'38" 42d12'23"
Swazzertot Hoppertop
Peg 23h24'39" 32d45'19"
SWB Morningstar
Ori 6h18'26" 16d31'1"
SWebster 050967
Uma 9h50'47" 46d51'46"
Swede-Heart Star
Uma 11h31'12" 50d32'17"
Swee' Pea
Umi 12h12'55" 86d12'59"
Sweep 929
Gem 6h40'30" 20d43'55"
Swee'pee
Lmi 10h4'19" 40d55'22"
Sweet
Cnc 8h0'10" 12d49'15"
Sweet 107
Mon 7h18'14" -0d18'39"
Sweet 16 Amy Neill
Sco 16h56'2" -41d55'25"
Sweet 16 Hanging on a Moonshadow
Aqr 23h3'3" -8d11'42"
Sweet Aftyn
Leo 9h46'32" 9d13'35"
Sweet Alanna Parisi
Vir 13h29'31" 11d44'45"
Sweet Alecia Beth
Gem 6h57'22" 23d10'26"
Sweet Alisha
Vir 12h5'21" -9d47'57"
Sweet Amanda Thompson
Lyr 19h10'15" 45d51'46"
Sweet and Sour Lyna
Cyg 20h35'49" 39d50'42"
Sweet Angel
And 23h36'15" 48d5'34"
Sweet Angel
Vir 13h19'20" 5d46'7"
Sweet Angel
Uma 11h53'30" 63d49'8"
Sweet Angel Boy Evan
Umi 15h48'52" 85d47'16"
Sweet Angel Deanna Lee
Lib 14h28'9" -9d29'38"
Sweet Angel Stefanie
Uma 8h34'12" 59d55'0"
Sweet Angel Trenton
Vir 13h46'1" -1d18'44"
Sweet Angela
Sgr 19h49'24" -38d15'56"
Sweet Angela
And 0h35'35" 35d27'20"
Sweet Annamarie
Psc 1h38'8" 24d47'13"
Sweet Annette
Gem 7h9'39" 23d31'28"
Sweet Audrey's Super
Boo 14h35'18" 35d7'8"
Sweet Baby
Mon 8h0'32" -0d25'19"
Sweet Baby 13
Aqr 22h47'1" -11d45'27"
Sweet Baby Angel
Umi 14h25'11" 76d41'48"
Sweet Baby Angel Doll Kitten Karla
Ori 6h0'14" 10d48'19"
Sweet Baby Chuckie
Peg 21h54'12" 13d31'21"
Sweet Baby Girl
Lib 15h25'15" -24d56'45"
Sweet Baby James
Cyg 21h11'6" 47d16'20"
Sweet Baby Josie
Umi 14h24'0" 76d11'24"
Sweet Baby Love
And 0h39'56" 37d3'54"
Sweet Baby Matthew
Umi 16h27'39" 82d52'59"
Sweet Baby Maxey Bear
Leo 9h44'25" 24d56'33"
Sweet Baby William
Lib 14h54'46" -11d22'59"
Sweet Beat Rice
Uma 11h34'36" 53d6'28"
Sweet - Beata
Cnc 8h53'50" 32d29'19"
Sweet Beautiful Tracey
Aqr 20h55'53" 0d27'35"
Sweet - Bella
Uma 10h13'28" 47d36'2"
Sweet Brave Sara
Aur 6h11'45" 30d20'16"
Sweet B's Sparkle
Aql 19h51'29" -0d12'53"
Sweet Cali
Sgr 18h16'52" -23d13'55"
Sweet Carleigh
And 0h51'25" 41d4'51"
Sweet Caroline
Uma 10h40'13" 40d23'34"
Sweet Caroline
Lyn 7h44'52" 42d58'36"
Sweet Caroline
Psc 0h21'24" 16d52'38"
Sweet Caroline
And 0h30'3" 29d3'19"
Sweet Caroline
Lyn 7h28'38" 54d46'2"
Sweet Caroline #15
Lyn 7h40'27" 58d7'30"
Sweet Carolyn
Crb 16h4'11" 32d36'50"
Sweet Charlotte
Sco 17h29'7" -39d52'1"
Sweet Charlotte My Love
Cnc 7h57'34" 11d46'55"
Sweet Charolett
Uma 10h35'54" 43d59'42"
Sweet Cheeks
Lmi 10h11'36" 38d36'57"
Sweet Cheeks
Her 16h49'35" 17d36'36"
Sweet Cheeks
Lyn 8h13'11" 54d16'0"
"Sweet Cheeks" Kristen
And 0h18'47" 28d1'54"
Sweet Cheryl
Cam 7h49'16" 67d0'47"
Sweet Christine
Sgr 19h21'20" -15d5'22"
Sweet Christy
Cap 20h42'10" -21d12'41"
Sweet Cielito
Cap 20h27'57" -11d42'22"
Sweet Cindy Sue
Lib 14h53'7" -7d1'15"
Sweet Clara
Cap 21h28'12" -20d36'32"
Sweet Colleen
Uma 11h20'50" 47d40'43"
Sweet Colleen
And 23h59'1" 40d34'9"
Sweet Daddy
Per 2h13'46" 56d5'4"
Sweet David
Sex 10h15'49" 3d13'31"
Sweet Deanna
Crb 15h32'5" 36d59'42"
Sweet December
Cnv 12h47'20" 39d0'19"
Sweet Dog Wood
Mon 7h19'57" -0d15'25"
Sweet Dreams
Vir 12h36'35" -4d28'12"
Sweet Dreams
Uma 10h26'31" 54d45'35"
Sweet Dreams
Cyg 21h4'40" 36d55'55"
Sweet Dreams
Ori 5h34'42" 3d21'22"
Sweet Dreams Made For Jaclyn Reed
Lmi 10h25'20" 28d33'21"
Sweet Dreams, Swish
Aqr 22h29'8" 1d20'45"
Sweet Dreamweaver
Cas 1h0'16" 64d45'44"
Sweet Elizabeth
Psc 0h23'7" -4d8'23"
Sweet Emely Perez
Gem 7h30'32" 18d49'8"
Sweet Emily
Peg 21h49'10" 15d18'48"
Sweet EMS (Emily Ernest)
Cnc 8h38'50" 18d15'57"
Sweet Gayla, My Love
Peg 22h43'43" 16d39'4"
Sweet Gene
Lib 14h26'52" -19d46'45"
Sweet George's Smile
Sco 17h50'45" -41d23'30"
Sweet Gilda
Crb 16h19'8" 31d15'19"
Sweet Grandma Jean
Cyg 19h53'14" 37d31'3"
Sweet Heart
Vir 13h18'52" 12d10'55"
Sweet Heart Stacie
And 1h17'51" 48d52'38"
Sweet Honey Bee
Lmi 10h30'16" 35d24'43"
Sweet Ida Mae
Tau 3h45'57" 14d45'28"
{sweet_intoxication}Midnight~
Aqr 22h15'8" 2d14'27"
Sweet Irina Gambarina
Cap 21h51'54" -9d59'5"
Sweet Jayne
Uma 10h39'44" 59d38'14"
Sweet Jenni
Uma 9h49'44" 60d59'5"
Sweet Jenny Anne Keator
Psc 0h43'22" 15d19'39"
Sweet Jenny Lynne
Vir 14h7'30" -7d35'13"
Sweet Jesus Al
Uma 11h23'2" 47d36'2"
Sweet Joe
Uma 8h37'23" 60d30'57"
Sweet Joy
Uma 12h45'43" 57d24'34"
Sweet Judy Blue Eyes
Cas 2h15'8" 63d49'38"
Sweet Kacia
Leo 10h21'47" 17d45'13"
Sweet Kassandra
Sco 17h30'46" -39d42'56"
Sweet Katie
And 23h21'50" 46d21'44"
Sweet Katie 40
Aqr 23h56'20" -11d4'46"
Sweet Katie G's Eyes
Lib 14h52'26" -1d37'32"
Sweet Katie IWBW
Dra 18h42'54" 65d35'46"

Sweet Kirstine
Cas 23h55'48" 53d33'6"
Sweet Kristina
Uma 12h0'24" 33d4'20"
Sweet Lara Amelie
Uma 9h56'18" 60d48'24"
Sweet LeRoy
Ori 5h50'20" 2d58'18"
Sweet Lisa's Lucky Star
Lib 14h59'45" -2d29'17"
Sweet Lori Ann
Lyr 18h38'48" 37d1'51"
Sweet Lorraine
Gem 7h42'16" 23d39'31"
Sweet Lorretta
And 2h15'12" 42d18'58"
Sweet Love
Umi 14h7'49" 78d27'29"
Sweet Luann
Psc 1h17'32" 32d7'18"
Sweet Madison
Vir 14h27'48" 5d32'52"
Sweet Malak
Lyn 7h16'9" 54d40'23"
Sweet Mamma
Uma 8h52'10" 49d20'36"
Sweet Man
Aql 19h30'27" 8d30'45"
Sweet Man
Sgr 18h19'40" -32d25'14"
Sweet Mandy Grace
Umi 16h34'28" 84d19'12"
Sweet Maria
Cnc 8h49'41" 22d49'5"
Sweet Marie Lynn
And 1h43'1" 39d20'26"
Sweet Marina
Sco 17h21'13" -31d53'2"
Sweet Marissa's Sparkling Light
And 0h12'13" 29d0'55"
Sweet Marjie Marie
Lyn 8h19'4" 47d7'0"
Sweet Mary
Cyg 20h33'27" 30d10'10"
Sweet Mary Ann
And 0h32'9" 41d48'30"
Sweet Matched Pair Geatly In Love
Aql 19h49'4" 14d21'50"
Sweet Meagan
And 1h19'13" 38d51'59"
Sweet Melissa
And 1h14'19" 44d16'26"
Sweet Melissa
And 2h12'33" 38d46'38"
Sweet Melissa
Cyg 21h26'25" 36d27'9"
Sweet Melissa
Crb 16h3'53" 33d3'44"
Sweet Melissa
And 2h33'41" 50d21'42"
Sweet Melissa
And 23h12'20" 40d32'50"
Sweet Melissa
And 0h2'15" 45d31'59"
Sweet Melissa
Vir 12h46'4" 3d19'8"
Sweet Melissa
Her 18h28'42" 21d13'18"
Sweet Melissa
Crb 15h34'12" 25d52'49"
Sweet Melissa
And 0h24'51" 33d6'53"
Sweet Melissa
Sco 17h51'11" -35d57'38"
Sweet Melissa
Cas 23h50'3" 56d29'57"
Sweet Melissa Star
Lyn 7h14'0" 50d54'26"
Sweet Melissa, My Girl
Lyr 18h29'44" 28d47'14"
Sweet Memories
Sgr 18h56'47" -21d29'31"
Sweet Mimi Helen
Crb 15h48'0" 26d2'20"
Sweet Mother Ethel Kramer
Cyg 20h58'47" 48d39'56"
Sweet Mr Jay
Cnc 7h58'27" 10d4'22"
Sweet Natalie
Ari 3h29'25" 27d21'25"
Sweet Nickle
Dra 17h53'18" 56d36'34"
Sweet November
Uma 9h14'35" 62d46'27"
Sweet Olya
Uma 9h37'54" 56d55'22"
Sweet P
Lyn 8h2'20" 48d53'46"
Sweet P
Crb 16h3'21" 35d39'24"
Sweet P Crigler
Mon 7h21'57" -0d41'27"
Sweet Patricia
Lib 15h41'54" -19d42'25"
Sweet Pea
Vir 13h35'4" -19d23'42"
Sweet Pea
Aqr 23h54'3" -10d14'22"
Sweet Pea
Umi 16h58'47" 79d6'28"
Sweet Pea
Umi 13h58'8" 75d33'30"
Sweet Pea
Lib 15h27'34" -5d42'37"
sweet pea
Uma 10h14'18" 54d46'41"
Sweet Pea
Lyn 6h52'17" 55d23'27"
Sweet Pea
Uma 11h44'3" 55d51'5"
Sweet Pea
Uma 12h25'46" 56d23'1"
Sweet Pea
Uma 10h37'38" 56d23'6"
Sweet Pea
Cas 23h17'23" 58d44'35"
Sweet Pea
Uma 14h3'35" 61d31'44"
Sweet Pea
Uma 14h2'37" 61d49'31"
Sweet Pea
Uma 9h30'27" 70d43'36"
Sweet Pea
Cas 2h27'35" 68d18'11"
Sweet Pea
Sco 16h8'0" -32d36'17"
Sweet Pea
Sco 17h53'11" -40d27'43"
Sweet Pea
Psc 1h21'48" 3d38'21"
Sweet Pea
Crb 15h18'5" 31d46'45"
Sweet Pea
And 1h3'45" 36d6'2"
Sweet Pea
Psc 1h18'34" 31d52'38"
Sweet Pea
Cnc 8h45'27" 32d24'20"
Sweet Pea
Gem 7h22'17" 32d6'28"
Sweet Pea
And 0h8'8" 44d39'2"
Sweet Pea
Per 3h20'33" 42d59'3"
Sweet Pea
Uma 10h34'13" 41d31'59"
Sweet Pea
Uma 11h23'58" 38d36'44"
Sweet Pea
And 2h19'43" 46d20'29"
Sweet Pea
And 2h34'48" 46d30'15"
Sweet Pea
And 23h15'6" 41d46'50"
Sweet Pea
Cyg 21h38'15" 38d2'26"
Sweet Pea
Cyg 20h0'42" 47d25'25"
Sweet Pea
Lyr 18h56'56" 45d36'25"
Sweet Pea
Cam 3h59'36" 58d0'29"
Sweet Pea
Cnc 8h34'51" 29d11'0"
Sweet Pea
Cyg 19h40'47" 28d26'38"
Sweet Pea
Ari 2h37'53" 28d1'8"
Sweet Pea
Psc 0h50'58" 15d18'55"
Sweet Pea
Tau 4h48'25" 18d54'48"
Sweet Pea "21"
Cap 21h34'49" -19d48'21"
Sweet Pea Christine
And 23h59'37" 45d37'9"
Sweet Pea Hale
Uma 11h54'42" 45d12'56"
Sweet Pea Lori
Cas 1h24'27" 68d23'12"
Sweet Pea Loves Brown Sugar Forever
Boo 14h39'7" 14d31'55"
Sweet Pea Newman
Cma 7h4'35" -31d14'18"
Sweet Pea Utton
Dra 16h16'10" 56d41'38"
Sweet Pea Walker
Psc 1h30'43" 12d46'15"
Sweet Pea's Star
Cnc 8h53'27" 25d40'54"
Sweet Petra
Cnc 8h54'8" 15d57'52"
sweet petunia
Leo 11h23'3" 3d6'45"
Sweet Potato
Cam 4h38'26" 68d22'18"
Sweet Precious Smiling Nancy
Lib 15h47'50" -17d41'9"
Sweet Princess 0416
Aqr 21h47'38" -2d41'55"
Sweet Putzie Krahenbuhl
Lib 15h3'54" -6d5'50"
Sweet Rebellion
Vir 15h5'8" 7d1'55"
Sweet Reita
Lib 15h5'28" -0d40'48"
Sweet Rellish & S.E.B. Love Star
Gem 7h25'29" 32d2'31"
Sweet Rhonda G
Cap 21h28'33" -10d46'26"
Sweet Rob
Pho 1h31'43" -43d2'4"
Sweet Ron
Lyn 7h7'2" 55d39'1"
Sweet Samantha
Cam 6h2'41" 61d39'46"
Sweet Sammy
Cnc 9h8'12" 15d42'59"
Sweet Sarah
Uma 9h9'58" 48d12'12"
Sweet Sharon Kent
Ari 2h59'32" 10d57'44"
Sweet Shelly
And 1h27'30" 46d43'37"
Sweet Shirley
Psc 0h2'7" 3d39'1"
Sweet Simone
Cyg 19h50'58" 36d56'42"
"Sweet Sixteen" - Lindsay Spears
Tau 5h50'18" 26d13'2"
sweet smile
Uma 10h11'13" 49d8'59"
Sweet Sophia
Lyr 18h44'28" 31d40'49"
Sweet Spirit
Ori 5h55'57" 17d52'57"
Sweet Star
Leo 10h34'1" 19d4'8"
sweet sue
Ari 3h10'13" 29d20'7"
Sweet Sue's Bus 7
Aqr 22h56'38" -4d51'0"
Sweet Sugi
Uma 12h32'54" 61d48'11"
Sweet Susan
Cap 20h10'3" -14d0'51"
Sweet Susy B.
Uma 8h12'15" 69d45'47"
Sweet Svetlana
Gem 6h39'51" 18d41'57"
SWEET SWEET BUCY
Uma 12h5'38" 48d28'42"
Sweet Sweet Jeanette
Cam 7h30'11" 68d36'16"
Sweet Sweet Julie Luchini
Lib 14h57'23" -15d30'5"
Sweet T
Cam 4h36'45" 67d25'40"
Sweet T Zorr
Vir 13h17'54" 9d18'29"
Sweet Tiff
Uma 8h47'13" 64d56'38"
Sweet Tifffany
Cma 7h6'55" -22d2'22"
Sweet Toni
Sgr 19h14'9" -34d43'31"
Sweet Trouble - J. Cholewinska. KTF
And 1h8'17" 45d38'15"
Sweet Virginia
And 0h22'3" 23d20'18"
Sweet Waif ~ 4-09-2003
Cyg 20h58'10" 46d19'14"
Sweet William
Lyn 8h21'30" 35d23'56"
Sweet William and The General
Uma 14h4'24" 48d46'18"
Sweet William Caulk
Uma 9h56'35" 70d45'59"
Sweet Wonderful Micalle
Lib 15h33'32" -28d51'4"
Sweet Wonderful You Patricia Barron
Cyg 21h10'5" 46d53'56"
Sweet Yala
Tau 3h45'44" 28d6'49"
Sweetbaum
Aql 19h0'24" 9d24'19"
sweetcheeks
Lyn 6h47'58" 50d41'20"
Sweetcheeks' April Eyes
Leo 11h9'12" 3d41'43"
Sweetdream Star
Lmi 10h23'3" 33d32'4"
Sweetest Christina Magnifico
Lyn 7h33'22" 39d36'35"
Sweetest Dreams
And 2h27'28" 44d46'1"
Sweetest Luck
Vir 13h9'43" -15d26'24"
Sweetest Mark
Cmi 8h5'42" -0d13'59"
Sweetest of Peas
Gem 7h17'2" 33d50'58"
Sweetest Sparkle Michele
Leo 11h40'51" 15d44'49"
Sweetheart
Her 17h36'43" 19d17'56"
Sweetheart
Psc 1h10'58" 26d44'11"
Sweetheart
Cmi 7h24'54" 9d34'12"
Sweetheart
Psc 0h43'14" 20d27'56"
Sweetheart
Lyn 8h44'15" 34d55'25"
Sweetheart
Cyg 21h56'50" 54d15'29"
Sweetheart
Sco 16h21'52" -39d14'11"
Sweetheart 143
Uma 11h49'47" 60d6'29"
Sweetheart Dre
Ari 3h6'47" 26d33'33"
Sweetheart Erin K.
Aqr 22h24'38" -0d8'16"
Sweetheart "Jack"
Aqr 21h35'12" -0d1'58"
Sweetheart Livvie
Lyn 6h45'53" 50d54'28"
Sweetheart Natoleon
Uma 11h41'52" 31d57'17"
Sweetheart Susann
Aqr 23h55'38" -4d22'10"
Sweethearts sweet sixteen
Aqr 22h58'28" -9d5'30"
Sweetie
Umi 15h52'55" 71d36'56"
Sweetie
Sco 16h15'2" -22d41'0"
Sweetie
Cyg 21h32'40" 41d1'33"
Sweetie
Her 17h24'49" 45d27'22"
Sweetie
Cnc 8h34'31" 21d33'51"
Sweetie
Peg 22h20'48" 8d4'44"
Sweetie
Ori 5h12'40" 15d9'21"
Sweetie Elizabeth
And 23h7'2" 45d34'14"
Sweetie Laura
Leo 11h30'29" 15d34'42"
Sweetie Mark
Vir 12h48'55" 11d34'25"
Sweetie McSquishy
Vir 12h49'25" 3d11'40"
Sweetie Pie
Psc 1h4'54" 7d22'15"
Sweetie Pie/ Becky
Uma 11h22'19" 68d0'10"
Sweetie Pie Pumpkin Face
Psc 23h2'57" 2d14'19"
Sweetie Veraltyd
Leo 9h35'32" 28d42'47"
Sweetie Wonderland 16.06.2005
Uma 11h55'6" 34d11'46"
Sweetiebride
Psc 0h0'44" 9d17'22"
sweetiekins
Umi 14h57'59" 80d25'36"
SweetiePie
Crb 15h50'11" 34d18'33"
Sweetiepuss82
Tau 5h19'3" 18d1'40"
SweetLadi Kim
Lib 15h6'44" -1d14'20"
Sweetnes
Uma 10h28'36" 64d42'3"
Sweetness
Vir 12h33'35" -1d9'16"
Sweetness
Aqr 22h38'56" -0d22'10"
sweetness
Cap 21h41'15" -8d44'44"
Sweetness
Cru 12h3'40" -62d17'43"
Sweetness
Leo 10h17'55" 14d23'4"
Sweetness
Aqr 20h46'35" 0d39'47"
Sweetness
Vir 12h54'17" 2d8'44"
Sweetness
Lyr 19h6'39" 32d0'13"
Sweetness and Lucky's Star
Ori 5h59'59" 21d1'45"
Sweetness Karen L.T.A.G.U.
Mon 7h3'26" -7d19'7"
Sweetness Unparalleled
Gem 7h8'32" 34d50'15"
Sweetpea
Cnc 8h18'59" 32d24'19"
Sweetpea
Per 3h51'57" 41d11'26"
Sweetpea
Lyn 9h23'50" 41d2'56"
Sweetpea
And 23h32'34" 36d26'14"
Sweetpea
And 0h59'59" 38d52'34"
Sweetpea
Cyg 21h26'40" 46d27'17"
Sweetpea
And 2h24'51" 49d18'22"
Sweetpea
Leo 9h57'19" 15d6'5"
SWEETPEA
Gem 6h55'24" 27d31'28"
Sweetpea
Cnc 8h31'35" 7d19'36"
Sweetpea
Ori 6h18'19" 6d37'3"
Sweetpea
Ori 5h6'39" 4d59'26"
SweetPea
Psc 1h24'31" 12d9'39"
Sweetpea
Peg 21h43'7" 8d18'4"
Sweetpea
Lib 15h40'37" -20d42'41"
SweetPea
Cap 20h25'40" -18d56'50"
Sweetpea
Uma 10h10'6" 66d51'57"
Sweetpea
Cru 12h31'39" -56d53'14"
Sweetpea
Sco 17h44'39" -41d45'32"
Sweetpea 1
Psc 0h56'35" 15d47'31"
Sweetpea, Danielle
Cam 5h7'10" 64d43'50"
Sweetpea1990
And 0h35'11" 32d2'39"
SweetPea5262001
Uma 11h21'45" 50d10'46"
SweetPeasAreJewellsForKeeps
Cyg 20h36'27" 38d40'0"
Sweets
And 0h48'56" 41d14'36"
Sweets
Tau 3h59'11" 26d49'14"
Sweets
Her 17h36'2" 19d18'18"
Sweets
Psc 1h18'56" 21d34'53"
Sweets
Aql 19h43'50" -0d3'3"
Sweets
Sco 16h52'50" -39d24'11"
Sweets Star
Cnc 8h2'58" 23d27'51"
Sweets TB
Leo 10h6'56" 24d41'19"
SweetspiritBoo
Cnc 8h45'53" 30d46'56"
Sweettee Gale
Sge 20h0'55" 17d25'20"
SweetThing
Aqr 22h36'19" -0d48'46"
Sweetu's Starlet
Lyn 9h9'14" 34d33'31"
Sweety Carlie
Uma 11h48'56" 58d15'5"
Sweety Dawn
Leo 11h9'16" 2d23'17"
Sweety Honey Major
Aql 19h7'6" -0d1'21"
Sweety Star
Sgr 19h4'40" -16d44'50"
Sweety Stevan
Umi 15h43'45" 81d24'48"
Sweety's Star
Vir 12h17'18" 11d57'52"
Swen Voggenberger
Cas 0h31'54" 52d30'30"
Sweta
Cnc 8h51'35" 13d46'10"
Swetlana
Ori 5h36'54" -2d5'50"
Swette
And 2h31'24" 45d40'31"
SWG
Uma 10h0'0" 55d56'55"
swillie
And 23h7'33" 52d12'5"
Swimsquirt
Psc 1h15'22" 15d24'23"
Swin of the Stars
Cap 20h55'26" -16d43'50"
Swinda Koch
Ori 6h20'2" 5d52'7"
SwingDrouhne18
Sgr 18h1'50" -17d45'13"
Swingman - TTT - 03
Eri 3h45'11" -0d11'35"
Swiss Babe LAWB1958
Sgr 18h58'44" -34d48'53"
Swiss Gisele Schall
Leo 11h18'32" 17d2'52"
Swiss Pearl
Uma 8h55'25" 51d43'21"
Swisscom Fixnet
And 0h43'11" 41d36'20"
Switalla, Hans-Jochen+Heide
Psc 1h0'42" 3d4'48"
Switzler-Burke
Lyr 18h54'37" 43d46'4"
S.W.M.B.O.
Cnc 8h42'33" 6d35'44"
Sworn Silence 2006
Com 13h20'25" 30d34'7"
SWR All Stars 2004
Ori 5h58'46" 10d18'11"
Sy Cohen
Her 18h4'38" 32d26'17"
Sy Simon
Ori 6h17'16" 16d24'18"
Sy Trabulsy
Lyr 18h36'6" 28d52'4"
Syamka
Uma 8h48'6" 57d51'19"
Syan Ngo
Umi 13h19'29" 85d58'56"
Syana
Aql 19h15'41" 3d4'34"
Sybil Berneking
Uma 11h41'16" 48d38'47"
Sybil Gene Silcox
Aqr 23h25'14" -11d14'25"
Sybil L. Roberts
Com 12h27'54" 22d44'44"
Sybil Mansfield Bateman Russell
Lib 14h42'2" -14d6'57"
Sybil "Sam" Lawson
And 0h47'32" 45d50'53"
Sybil Smith
Vir 15h7'7" 3d57'47"
Sybil V. Greene
Leo 11h17'10" 14d54'24"
Sybil VanDyke
Aqr 22h50'59" -4d45'16"
Sybille Aschwanden-Wallimann
Peg 23h19'38" 32d50'18"
Sybille H Sebreny
Uma 9h26'35" 45d47'25"
Sybille Kathe
Uma 8h38'45" 48d32'23"
Sybrina's Angel Star
Vir 13h17'17" 12d28'27"
Sybur
Crb 16h22'40" 35d48'21"
Sycamore Canyon School - Dr. Chasse
Psc 1h6'7" 6d38'14"
Syd Leach
Per 2h58'15" 47d24'34"
Sydell Lemar Wilson
Aqr 22h11'0" -14d27'16"
Syd-lynn
Lyn 7h16'39" 48d35'59"
Sydne Jenae
Sgr 20h14'7" -38d31'50"
Sydne Mandigo
Uma 11h22'14" 62d55'49"
Sydne Marie Hudson
Tau 5h9'28" 18d23'29"
Sydnee's Star 07-10-2004
Lib 15h26'47" -17d56'38"
Sydney
Sco 15h55'50" -21d18'47"
Sydney
Cap 21h43'5" -13d0'58"
Sydney
Cmi 8h4'8" -0d8'44"
Sydney
Uma 13h46'58" 55d9'37"
Sydney
Pho 1h0'19" -44d42'39"
Sydney
Ari 3h2'53" 19d59'53"
Sydney
Ori 6h12'15" 12d53'27"
Sydney
Uma 13h0'48" 52d20'34"
Sydney
Lyn 8h28'5" 33d14'44"
Sydney
Ari 2h52'29" 30d6'25"
Sydney Aker
Tau 3h27'41" 8d57'31"
Sydney Alexandria Jackson
Lyn 9h1'44" 42d19'32"
Sydney Alexis Yuille
And 23h49'13" 42d39'35"
Sydney Alison Harmon
Lyn 7h37'2" 38d24'56"
Sydney Alison Wood
Sco 16h9'29" -14d47'9"
Sydney Allison Sigler
Vir 13h45'40" -5d30'33"
Sydney and Aryannah
Cyg 19h33'43" 53d11'21"
Sydney Anela Gabrielski
Ori 6h14'30" 7d12'40"
Sydney Ann
Tau 4h9'41" 6d44'25"
Sydney Ann Gregg
Her 18h7'13" 26d34'39"
Sydney Ann Schrader
Cyg 20h56'5" 51d41'16"
Sydney Anne
And 1h35'55" 48d12'55"
Sydney Anne Baker
Ari 2h10'25" 24d13'36"
Sydney Anne Cirincione
And 0h58'13" 39d33'6"
Sydney Ashman
Uma 10h15'59" 43d22'26"
Sydney Ashton
And 1h3'16" 48d38'47"
Sydney Autumn Perez-March 24, 2000
Ari 2h13'40" 25d18'50"
Sydney B. Gibbs
Aql 19h0'42" -0d24'1"
Sydney Bernard Shines Forever
Umi 15h42'46" 76d57'16"
Sydney Bristow-Smith
Cap 21h5'52" -17d33'7"
Sydney Brook Weisberg 4/19/03
Vul 20h24'17" 28d4'43"
Sydney Brooke Wortham
Ori 5h28'21" 7d9'20"
Sydney C. Saponaro "God Bless You"
And 0h46'53" 30d26'21"
Sydney Carita
Lyr 18h33'38" 37d29'7"
Sydney Catherine Watson
Vir 13h42'47" 3d34'30"
Sydney Chelsea Badway
Gem 6h55'33" 28d5'39"
Sydney Christine Ohly
Leo 9h41'48" 28d13'58"
Sydney "Cookie" Belitz
Per 2h38'8" 55d39'55"
Sydney Curry
Aqr 21h52'38" -1d46'59"
Sydney Dale Corletta
Dra 15h27'17" 56d17'0"
Sydney & Darby's Star in The Sky
And 0h54'20" 38d28'17"
Sydney E. Pennell
And 1h25'39" 35d15'41"
Sydney Easton Wright
Cnc 8h49'0" 27d5'5"
Sydney Elena Gomezcoello
Cap 21h43'5" -12d37'50"
Sydney Elise Beebe
Gem 6h29'48" 24d15'45"
Sydney Elizabeth Carnow
Ari 2h37'31" 29d54'47"
Sydney Elizabeth Floyd
Mon 8h0'32" -0d37'6"
Sydney Elizabeth Olson
And 0h14'4" 43d6'0"
Sydney Elizabeth Richardson
Gem 7h25'0" 32d0'27"
Sydney Estelle Tuplin
Psc 22h57'12" 7d37'31"
Sydney Geil
And 23h26'6" 36d54'4"
Sydney Genevieve Bilpush
Tau 4h10'40" 5d23'17"
Sydney Grace Donley
Umi 14h42'53" 74d23'24"
Sydney Grace Hildebrandt
Vir 12h56'26" -9d59'3"
Sydney Grace Marino
And 0h28'7" 41d52'22"
Sydney Grace Reynolds
Psc 1h51'31" 7d31'31"
Sydney Helena
Dra 19h32'17" 63d36'8"
Sydney Huntermark
Umi 14h43'10" 73d36'23"
Sydney Jo Davidson
Vir 13h42'9" 4d47'46"
Sydney K. 1995
Tau 4h31'45" 18d48'30"
Sydney K. Carter
Lib 14h51'18" -1d38'22"
Sydney Kamea Griffiths
Peg 21h40'22" 26d41'16"
Sydney Kardell, Pye in the Sky
Uma 9h29'30" 54d24'5"
Sydney K'arrot Kang [SKK]
Gem 6h37'47" 12d50'2"
Sydney Katharine Townshend
Psc 0h22'50" 16d50'31"
Sydney Kay La Rue
Cma 6h57'16" -31d52'17"
Sydney Kennedy Lynn
Lyn 8h59'25" 34d2'4"
Sydney Lauren
Lib 14h51'21" -1d44'14"
Sydney Lauren Kuhn
And 0h57'35" 44d25'10"
Sydney Lee
Ori 5h25'55" 3d6'31"
Sydney Leppert
Per 3h4'27" 52d13'24"
Sydney Lof Gager
Ari 3h22'9" 29d46'18"
Sydney Louise Wright
Sco 16h10'33" -9d21'54"
Sydney Lynn Ian Robert
Uma 9h37'39" 55d53'30"
Sydney Lynn Reik
Cnc 8h40'3" 27d38'36"
Sydney M E 3
Psc 1h1'46" 3d39'10"
Sydney M. Johnson
And 2h0'3" 46d14'50"
Sydney Major
Leo 10h52'56" 15d28'24"
Sydney Marie Aven
Ori 5h32'23" 7d28'25"
Sydney Marie Casserly
Ari 2h5'4" 23d15'30"
Sydney Marie Johnson
And 1h17'51" 45d27'4"
Sydney Marie Johnson
Lyr 18h47'56" 30d56'4"
Sydney Marie Manning
Psc 1h8'29" 32d50'52"
Sydney Marie Moreland
Sgr 18h26'13" -25d51'5"
Sydney Marie Urps
And 1h41'32" 42d28'53"
Sydney May Rossiter
Tau 5h44'9" 21d18'11"
Sydney McCarthy
Ori 6h2'42" 7d28'8"
Sydney McCluney
Tau 5h34'37" 25d24'43"
Sydney Mckinzie Duprass
Cam 3h37'58" 64d58'30"
Sydney Milivoje Patrick Maitland
Aur 5h18'36" 42d17'35"
Sydney Montgomery
And 1h4'47" 41d32'19"
Sydney Morgan Williams
Vir 14h5'23" -16d21'31"
Sydney Neal
Uma 9h42'35" 56d52'12"
Sydney Nicole
And 23h30'6" 45d3'59"

Sydney Nicole Bosnic-Nicholson
Sgr 18h50'12" -28d44'33"

Sydney Nicole Boyce
Vir 11h50'48" 6d37'3"

Sydney Nicole Freedline
Tau 4h32'0" 19d41'18"

Sydney Nicole Lipply
Cap 20h29'2" -26d35'32"

Sydney Paige
And 0h45'8" 44d44'43"

Sydney Paige Kasulinous
Vir 13h11'49" -4d14'48"

Sydney Paige Monaco
Lmi 10h29'32" 36d21'53"

Sydney Paige Saville
And 0h12'42" 29d26'18"

Sydney Paige Simonds
Cam 4h51'42" 59d45'56"

Sydney Pie Cato
Peg 22h38'39" 33d42'7"

Sydney Ramunno
Sco 17h33'46" -45d13'37"

Sydney Reece
Ari 3h18'51" 22d48'9"

Sydney Regan Hess
Tau 4h20'11" 17d14'6"

Sydney Renee Ellison
And 0h1'30" 48d39'59"

Sydney Renee Rosenberger
Sco 17h18'57" -38d15'6"

Sydney Renee- The Reborn
Tau 4h21'13" 26d33'50"

Sydney Ringer
Cnc 8h20'30" 14d11'12"

Sydney Robin Starr
Cam 5h17'56" 64d31'50"

Sydney Rose
And 23h45'40" 35d9'54"

Sydney Rose
And 0h24'30" 38d35'6"

Sydney Rose Hatton
And 0h38'20" 34d0'31"

Sydney Rowley
And 1h21'31" 37d4'16"

Sydney Salazar
Tau 3h40'39" 29d10'13"

Sydney Schabacker
And 1h19'2" 38d38'44"

Sydney Shotkoski
Uma 10h51'36" 51d57'31"

Sydney Sinclair Noteboom
Aqr 22h0'41" -16d15'41"

Sydney Sky Brock-Mugsy
Ori 5h17'27" -4d46'26"

Sydney Spence
Dra 19h39'2" 68d1'57"

Sydney Tala Feldman
Lyn 7h55'32" 38d56'51"

Sydney Taye Callaway
Leo 10h12'33" 26d4'37"

Sydney Taylor Huddleston
Tau 5h4'13" 22d0'17"

Sydney Turner
Psc 0h6'19" 8d41'40"

Sydney Veronica Pedrosi
And 1h45'44" 45d24'29"

Sydney Victora
Lyn 8h2'0" 42d44'52"

Sydney Wiencek
Gem 7h15'20" 16d29'53"

Sydney9/7/2005
Vir 14h44'16" 4d47'59"

Sydney's Birthday Star
Cap 21h50'58" -11d28'48"

Sydney's Dancing Starshine 1/22/96
Aqr 21h46'27" -0d33'37"

Sydney's Star
Leo 11h50'27" 15d35'8"

Sydney's Star
Gem 7h21'41" 33d12'23"

Sydni
Uma 10h42'13" 44d56'15"

Sydni Alyson
Umi 15h24'1" 68d52'0"

Sydni Harper Pistolesi
Sco 16h17'16" -15d21'15"

Sydni Jo
And 0h31'50" 29d10'31"

Sydni Lee
Cam 3h33'54" 58d39'0"

Sydni Moler
Crb 16h11'47" 36d43'31"

Sydni Renee Harper
Cnc 9h18'53" 31d35'6"

Sydni Rose
Uma 11h39'22" 58d57'42"

Sydnie Grace
And 2h2'6" 36d49'1"

Sydnie "Sunshine" Lorenz
Leo 9h56'43" 23d49'50"

Sydra
Umi 15h52'13" 77d40'43"

Syed B. Zaidi
Gem 6h28'57" 21d17'34"

Syed Hashem Reza
Aql 19h38'44" 4d47'48"

Syed Hashmi
Ori 5h30'58" -4d41'20"

Syed Imran Ahmed
And 23h20'28" 37d47'55"

Syed Mustafeez Hashmi (Uncle Ji)
Tau 4h46'45" 23d0'30"

Syeda Atifa Safia
Gem 7h3'29" 20d42'33"

Syeda "MeMe" Bostwick
Aqr 22h7'4" -1d14'46"

Syeta Justin
Tau 5h3'42" 24d40'30"

Syleste Cecilia Reid
Uma 13h2'19" 55d18'58"

Sylke
Uma 9h54'45" 53d14'18"

Sylke Ursula Reichardt
Gem 7h57'13" 21d53'12"

SylliFürlmmer
Per 4h9'29" 45d22'7"

Sylph
Gem 7h38'3" 30d7'55"

Sylva
Aql 20h20'18" -0d50'23"

Sylvain Agrati
Tau 4h26'41" 22d5'16"

Sylvain Coulombel
Ori 4h59'59" -2d53'8"

Sylvain Lorier - 24.04.80
Dra 17h20'24" 52d25'59"

Sylvain Ramsay
And 0h58'3" 37d29'33"

Sylvain Simard 24.06.1974
Aql 19h6'42" 9d2'55"

Sylvan Lefcoe
Gem 7h11'29" 28d14'39"

Sylvana
Peg 22h10'35" 11d58'57"

Sylvane
Uma 10h39'33" 51d32'46"

Sylvania
Tau 3h56'32" 9d16'5"

Sylver Anderson
Uma 12h14'29" 56d24'50"

Sylver Dawn
Uma 9h50'6" 62d42'14"

Sylvester
Cas 1h41'16" 64d11'1"

Sylvester
Leo 10h14'45" 14d8'0"

Sylvester A. Vogia
Gem 7h41'29" 32d0'39"

Sylvester and Tweety
Vir 12h43'52" 5d38'44"

Sylvester J. Jaramillo
Uma 14h5'22" 61d18'10"

Sylvester Morris
Per 4h17'10" 39d55'30"

Sylvester Paul Coen
Oph 17h30'3" -0d47'22"

Sylvester Theopolous Kerly
Ori 5h23'56" 4d22'4"

Sylvi
Uma 11h4'55" 71d41'48"

Sylvi Linn Tryggestad
Cap 21h3'49" -20d11'7"

Sylvia
Lib 15h12'28" -4d22'13"

Sylvia
Dra 17h20'32" 66d9'46"

Sylvia
Cas 22h58'56" 56d29'36"

Sylvia
Leo 10h17'3" 11d55'51"

Sylvia
Ari 2h57'46" 21d43'0"

Sylvia
Lyr 18h50'30" 26d58'30"

Sylvia
Leo 10h5'43" 22d1'58"

Sylvia
Cnc 8h23'15" 19d23'17"

Sylvia
Ari 3h17'44" 24d23'34"

Sylvia
And 2h31'51" 44d14'38"

Sylvia
Cam 4h55'42" 54d55'36"

Sylvia A. Couch
Lyn 7h53'8" 56d32'29"

Sylvia A. Mulder Miller
Crb 15h42'33" 26d12'19"

Sylvia and Ken - Everlasting Love
Cyg 19h56'3" 54d5'57"

Sylvia Ann
Per 2h23'35" 51d9'38"

Sylvia Ann Powis Collier Brown
Leo 10h12'15" 13d42'5"

Sylvia Blitzer
Psc 0h40'57" 13d23'29"

Sylvia & Bob Stern
Psc 0h48'57" 9d42'23"

Sylvia Boddeker
Aqr 22h22'33" -24d40'47"

Sylvia Button
And 2h32'4" 40d33'13"

Sylvia C.
Tau 4h10'1" 22d17'55"

Sylvia Catherine
Psc 1h25'19" 11d40'46"

Sylvia Chambers
Ari 2h10'39" 24d43'37"

Sylvia Christine Musial
And 1h59'37" 38d17'23"

Sylvia Clare
And 0h12'41" 43d39'0"

Sylvia Cooley Winslett
And 23h47'1" 47d21'17"

Sylvia Dainty
Lyr 18h35'30" 37d14'15"

Sylvia de Almeida Braga
Tau 3h57'32" 27d25'45"

Sylvia DeRider Star Mom!
And 0h16'53" 43d11'57"

Sylvia Dinkova Banova
And 1h39'11" 42d36'2"

Sylvia Dowling
And 0h7'30" 46d13'53"

Sylvia E. McCabe
Cas 1h37'51" 60d50'16"

Sylvia Edwards
Cap 21h13'31" -20d26'14"

Sylvia Faye Holt
Uma 11h0'23" 38d15'24"

Sylvia Gianacopolos' Shining Star
Cyg 19h38'53" 46d3'57"

Sylvia Grace Cockerham
Cas 1h10'41" 55d54'3"

Sylvia Hardman
Uma 11h49'31" 33d37'13"

Sylvia Hoffer
Lyn 6h17'46" 59d59'46"

Sylvia J. Morrow
Lyn 8h10'25" 38d6'48"

Sylvia Jacoby
Cas 1h21'39" 64d18'43"

Sylvia Jacquemin De La Latte
Cas 0h15'16" 53d56'33"

Sylvia Jonice Arban
Per 3h46'10" 32d46'40"

Sylvia kai Giorgos
And 0h16'5" 45d8'50"

Sylvia Karkus Furash
Gem 6h38'5" 24d13'3"

Sylvia Kay
Per 3h10'32" 40d39'10"

SYLVIA KAY LARGE
Vir 13h39'6" -3d58'28"

Sylvia L. Cates
Crb 15h53'40" 35d13'49"

Sylvia L. Johnson
Lyn 8h22'13" 45d47'43"

Sylvia L. Mansfield
Leo 9h33'4" 31d45'46"

Sylvia Leah Rose Snow
Cas 1h41'46" 62d52'25"

Sylvia Lee McPherson
Cyg 19h57'47" 49d50'27"

Sylvia & Lefteris
Leo 11h29'33" 11d37'51"

Sylvia Lexow
Psc 1h29'30" 10d32'17"

Sylvia Lucas
Her 16h20'46" 19d8'28"

Sylvia M. Bebinger
Crb 15h28'51" 26d39'32"

Sylvia M. Bernat
Cnc 8h23'52" 6d46'17"

Sylvia M Bigelow
Lyr 18h46'26" 31d4'33"

Sylvia M. Sharp
Cyg 21h29'25" 39d15'6"

Sylvia Marie Arencibia
Sco 16h12'19" -21d10'4"

Sylvia Marie Ebbert
Lyn 9h14'28" 34d5'27"

Sylvia Marie Genevieve Lobdell
Cyg 21h53'14" 51d43'33"

Sylvia Marie Olson
Cas 22h58'15" 54d26'43"

Sylvia May Brown
Cas 23h0'3" 59d10'59"

Sylvia Meredith
Vir 14h15'39" -20d18'9"

Sylvia Nester King
Leo 9h34'28" 24d12'33"

Sylvia Olynska
And 1h39'42" 42d2'47"

Sylvia Osewalt
Uma 10h34'30" 40d5'20"

Sylvia & Peter Saggin - Our Beloved
Ara 17h55'11" -56d53'22"

Sylvia Rhodes
And 0h33'24" 42d4'9"

Sylvia Rose
And 23h46'26" 44d42'30"

Sylvia Ruiz
Her 17h48'39" 39d17'22"

Sylvia S. Finch
And 1h16'41" 35d2'3"

Sylvia Sager
Cas 0h41'47" 58d6'12"

Sylvia Sanderson PhD., ARNP
Com 12h39'45" 17d5'29"

Sylvia Schäfler
Boo 15h17'10" 46d37'34"

Sylvia Sheridon
Cas 23h11'20" 54d10'20"

Sylvia & Sidney Herberman 80-80-60
Uma 11h37'13" 64d46'17"

Sylvia Siebel
Uma 8h58'8" 69d50'21"

Sylvia Stern
Psc 23h58'49" -3d21'17"

Sylvia Striano
Sco 16h10'46" -26d5'50"

Sylvia Teel Eternal
Sgr 18h16'32" -29d32'8"

Sylvia Thomas
Lac 22h29'29" 51d7'29"

Sylvia Torres
Crb 15h46'38" 32d50'46"

Sylvia & Tyke
Cyg 20h29'49" 43d35'17"

Sylvia Wachs Around My Heart
Cas 1h59'53" 62d10'50"

Sylvia Whorton
Srp 18h15'44" -0d7'2"

Sylvia Xhudo & KMS
Cnc 8h23'15" 16d57'58"

Sylvia Young
Uma 10h55'11" 33d35'29"

Sylvia's Chelsea True Blue
Uma 10h11'28" 42d14'30"

Sylvia's Giggle
And 23h23'19" 47d29'21"

Sylvia's Light
Ari 2h25'14" 11d44'1"

Sylvia's Star
And 23h49'51" 43d1'23"

Sylvia's Star
Cas 23h30'16" 53d26'48"

Sylvia's Star
Uma 8h13'59" 66d16'44"

Sylvie
Uma 9h38'4" 67d39'11"

Sylvie
Aqr 23h46'51" -14d2'56"

Sylvie
Cas 23h39'39" 51d33'42"

Sylvie
Lyn 8h49'23" 34d38'20"

Sylvie A. Laflamme
Vir 13h12'56" -1d11'0"

Sylvie B (Amour de ma vie)
Sco 16h56'45" -40d36'35"

Sylvie Barman
Cyg 21h12'20" 46d22'21"

Sylvie Bernard
Crb 16h19'18" 26d53'59"

Sylvie Gingras
And 23h13'21" 52d5'50"

Sylvie Kerkow
Ori 5h59'26" -2d57'30"

Sylvie Mocellin
Cap 21h32'35" -22d30'39"

Sylvie Poitrenaud
Sco 17h2'16" -41d11'48"

Sylvie Prat
Cas 0h29'40" 61d37'21"

Sylvie Rodrigues Lisboa
Gem 6h54'21" 14d38'12"

Sylvie SCM
Umi 16h1'6" 73d32'25"

Sylvie Simon
Crv 12h44'4" -13d20'12"

Sylvie T. Tremblay
Dra 20h22'4" 63d18'41"

Sylvie, 16.08.1971
And 23h37'7" 47d27'59"

Sylviehebral
Ari 3h4'43" 19d9'34"

Syl-vietta
Apu 16h12'4" -76d18'13"

Sylvou (03 juillet 1985)
Aql 19h4'16" -9d12'1"

Sylwia Antczak
Ori 5h57'3" 5d58'49"

Sylwia Moje Sloneczko
Ari 2h5'42" 21d28'59"

Sylwia Romaniewicz
Cap 20h20'58" -11d1'3"

Sylwia Roszkowska
Leo 9h44'13" 26d8'58"

Sylwia's Shining Star
And 23h8'1" 47d23'26"

Symba
Ori 5h36'37" -1d46'1"

Symon Zachary York
Leo 9h47'9" 14d20'33"

Symone
Crb 16h7'4" 37d5'22"

Symphany Segura
Lyr 18h53'53" 39d30'19"

Symphoney
Uma 11h14'14" 36d24'15"

Symposium
And 1h31'54" 41d17'46"

Syna
Uma 10h21'8" 55d37'18"

Syna Astrid
Dra 18h31'27" 74d40'42"

Syndia Wong
Sgr 19h10'0" -17d0'52"

Syndie
Sgr 18h0'20" -29d44'34"

Synergy
Cen 11h20'46" -54d6'42"

Synthesis
Uma 10h49'20" 44d4'2"

Synthia
Sgr 18h41'39" -26d11'15"

Syreeta Monique Elmore
Gem 6h51'3" 13d4'5"

Syreeta's Love
Uma 9h42'12" 56d32'48"

SyrekA.D.AJ
Equ 21h4'28" 9d9'3"

Syren
Cam 4h42'54" 65d27'38"

Syrena
Ori 5h53'46" 12d35'41"

Syrena's Star
Cyg 21h21'10" 39d41'20"

Syriah Denton
Vir 14h41'45" 3d48'6"

Syriana
Sgr 19h16'34" -15d54'32"

Syrup Sirup
Uma 9h26'40" 68d29'52"

Syvila Love
Leo 9h33'47" 32d53'44"

Szabó Dávid
Leo 9h51'21" 22d1'19"

Szabó József csillaga
Uma 10h16'52" 70d17'45"

Szabo Julia
Cnv 13h57'5" 37d26'25"

Szabó Mónika
Uma 10h55'53" 47d12'33"

Szakács Feri 50
Lib 15h44'22" -17d3'4"

"Szalacinski Fire"
Uma 9h34'22" 42d14'31"

Szalai Dávid Imre
Uma 13h59'4" 50d49'37"

Szalóki Anett, "Nettike"
Cnc 8h32'50" 9d0'3"

Szalony Al 33
Lyr 19h14'8" 26d20'42"

Szamóca
Sco 16h11'29" -35d15'3"

Szandi
Sgr 18h19'53" -18d10'26"

Szandus
Cas 23h58'9" 59d50'38"

Szapu
Uma 9h21'26" 56d19'52"

Szatmári Andrea
Cas 23h42'32" 51d32'3"

Szczepan Goral
Mon 6h39'15" 5d51'24"

Sze Ku on 60th Birthday
Cam 4h3'15" 72d14'28"

Sze Wing Pang
Cas 2h21'17" 73d5'8"

Székács Luca
Cas 1h12'45" 69d1'58"

Szekely
Cnc 8h40'0" 24d20'18"

Szépség és a Szörnyeteg
Cas 23h34'12" 54d49'9"

Szerdi Szabina
Tau 3h36'45" 5d16'23"

Szerelmem, Hegyes Ildi
Lib 14h46'53" -24d2'7"

Szerelmemnek Bélusnak M és O
Tau 4h25'5" 10d14'21"

Szerelmemnek, Egerszegi Tamásnak
Gem 7h3'10" 28d18'52"

Szerelmemnek, Sziklay Péternek
Uma 9h3'58" 72d22'28"

Szerelmünk gyümölcse, Marci
Ari 2h18'3" 17d22'11"

Szerelmünk Örök Csillaga
Uma 10h23'31" 61d37'30"

Szerelmünk Örök Jelképe
Leo 9h30'46" 12d13'51"

Szeretett Édesanyánk, Margit
Uma 11h57'44" 56d10'3"

Szilvi áés Tomi orangyalcsillaga
Uma 14h3'12" 54d20'55"

Szilvi és Laci csillaga
Uma 13h20'11" 61d47'32"

Szilvi és Zoli házasságcsillaga
Cas 23h40'4" 51d32'2"

Szilvia Fazekas, Tibor's Sweetheart
Uma 14h8'21" 57d41'46"

Sziszka Hercegno
Uma 8h59'17" 62d25'29"

SziTuGáMa házasságcsillaga
Cas 2h45'28" 65d26'46"

SZIVI - 560604 - 50
Gem 6h27'6" 25d39'38"

Szklarewicz-Perea
Cyg 20h57'39" 34d35'51"

Szkopanska, Dorota
Uma 10h0'27" 66d25'51"

Szombat Esti Láz
Cas 0h45'53" 60d52'2"

Szonyi Tibor és Miksa csillaga
Uma 11h33'42" 63d14'56"

SZORIAN
Cas 1h1'57" 63d12'50"

Sztace [Stacy Marie]
Cas 0h5'49" 55d17'40"

Szu Keang Trading Sdn Bhd
Ori 5h38'37" 4d8'26"

Szucs Gábor, a Bramac csillaga
Cnc 8h10'45" 31d24'56"

Szucs Laci
Uma 10h47'13" 72d19'24"

Szüleink, Erika és Miklós
Uma 9h48'42" 50d5'13"

Szunyogh Dániel
Uma 11h56'54" 55d37'59"

Szuszi & Kobold nyugodt világa
Tau 3h29'50" 18d19'36"

"T"
Crb 15h47'23" 27d34'17"

T
Cap 20h50'14" -20d45'10"

T & T
Per 3h40'29" 43d55'31"

T. 4. E.
Ori 5h42'12" -1d38'29"

T & A Forever
Umi 16h2'45" 80d40'45"

T & A "Miracle"
Psc 0h54'13" 25d55'49"

T. A. Moe
Uma 8h14'41" 67d5'8"

T A N T I V Y
Uma 12h42'32" 60d18'51"

T A Y
Cnc 8h39'28" 7d32'10"

T amo mi nena hermosa/Karen Carrera
Vir 13h53'59" -10d53'40"

T. and A. Douglas
Uma 12h48'47" 57d27'45"

T and J's first Christmas
Sco 16h10'34" -16d19'48"

T B - 3 9
Cnc 8h40'27" 7d51'18"

T B Star
And 0h28'35" 37d38'10"

T Baranya
Cnc 8h53'18" 8d31'22"

T. Bobo48
Ori 5h34'1" -0d23'15"

T & C '05
And 0h36'22" 37d13'38"

T. C. Bluhm 10.01.05
Cyg 19h39'35" 37d58'20"

T E Jake
Lib 15h26'41" -20d20'11"

T E O D O R A M A L T C H E V A
Cas 23h41'14" 57d5'59"

T. Edward Fling
Her 17h4'12" 14d35'12"

T. Eloise Austin
Vir 13h18'52" -3d20'9"

T. F. N.
Umi 13h12'12" 70d22'25"

T. Fukuma - Supernova.999
Tau 5h33'26" 19d44'2"

T H E W S U S - Gift of Courage
Cru 12h28'21" -58d29'51"

T. Harrison, Always +7 & 3/8ths
Ori 5h32'52" 2d19'6"

T IV
Psc 1h44'11" 8d41'38"

T J
Leo 9h49'39" 14d27'10"

T. J. Jensen - God's Gift To Jack
Aql 19h43'19" -0d2'18"

T. & J. J's Star of Love
Lyr 19h18'17" 29d20'2"

T & J's Irish Rovers
Cyg 20h22'5" 56d33'21"

T & K
Cyg 20h42'48" 37d4'44"

T. K Lucky
Cyg 20h45'28" 36d24'42"

T Kay Pierce
Cyg 20h0'24" 35d52'12"

T. Kei. & BoYoon Forever
Aur 5h24'12" 32d34'50"

T Kellcy
Sgr 18h51'28" -20d38'58"

T L C 2229
Peg 22h29'32" 34d32'38"

T. L. O. M. L.
Her 17h49'35" 46d15'36"

T- Lady Forever
Leo 11h17'52" 14d18'14"

T. Larry Edmondson
Uma 12h42'32" 62d17'18"

T. Lee Senn
Leo 9h28'47" 7d16'12"

T Love
Her 16h59'52" 33d44'0"

T & M Rowling - éternel amour -
Pho 0h38'44" -47d53'32"

T Mai
Ari 2h34'35" 25d51'29"

T & M's Mountain Tops and Valleys
Cru 12h23'30" -57d31'47"

T+N
Equ 21h4'8" 4d2'15"

T & N
Cyg 20h55'29" 35d5'53"

T N A
Vir 12h59'16" -1d54'3"

T & N - Bult
Eri 4h0'37" -38d32'24"

T * N * T
Vir 12h45'21" -5d58'26"

T 'n' T
Umi 15h29'39" 70d30'59"

T N T S A L
Cas 1h26'17" 51d54'41"

T R Jones, 3-Star Father
Cyg 20h1'29" 50d31'7"

T. Rauch - Aufbruch zu neuen Ufern
Uma 11h40'29" 45d35'20"

T Rex
Psc 1h11'8" 10d46'35"

T Rocks Tumpties
Dra 18h59'48" 61d18'28"

T Ryan
Leo 11h18'26" 17d31'3"

T & S Broadbent Golden Supernova 94
Cru 12h34'44" -59d28'2"

T Star
Cyg 19h53'25" 51d38'10"

T & T
Ari 1h52'2" 19d32'9"

T & T
Umi 14h6'24" 70d39'55"

T. T. Dill
Per 2h40'11" 50d33'31"

T & T (Tommy & Tony)
Aqr 21h47'4" -8d9'3"

T the Big Kitty
Cnc 8h10'36" 10d4'30"

T. Tyler Bostic
Ori 5h56'28" 5d44'41"

T. Vyse
Per 3h9'57" 42d50'51"

T X
Cyg 20h32'18" 44d32'1"

T-1433
Umi 15h17'23" 73d42'47"

#2
Cam 3h35'17" 67d27'15"

2/24 Marines USMCR-Chicago Illinois
Per 3h22'8" 47d36'13"

2 Be Loved Is Only 2 Be Loved By U: EG
Ori 6h6'37" -0d13'57"

2 B's
Crb 15h44'22" 27d10'56"

2 Com-ee-go
Vol 8h40'2" -66d36'42"

2 Dream Linda Sutherland
Aqr 23h14'36" -16d24'28"

2 Gems
Lyn 6h31'53" 60d32'19"

2 Jesse's Dance
Tau 5h25'51" 22d27'23"

2 Lobsters 4 Ever
Vir 13h39'56" -11d52'34"

2/m
Uma 11h53'14" 34d37'42"

2 My Star Jordan Cuthbert - Love DT
Uma 11h28'1" 59d52'44"

2 Souls Become 1, Lindsay and Andy
Uma 12h15'17" 58d59'23"

2 Special Princesses
Lib 15h45'42" -16d38'45"

20+1
Cru 12h30'58" -58d23'12"

2005 Dr Pepper "Top Performers"
Umi 11h59'53" 89d0'46"

2005-2006 Hickory Ridge Hound Dogs
Cmi 7h46'24" 4d51'50"

2005USConference: Unity & Diversity
Cas 0h36'33" 57d32'2"

2006 Forest Hills Competitive Cheer
Crb 16h4'53" 36d45'50"

2006 Star Volunteer Dorothy Dennis
Lyn 6h46'30" 50d59'13"

2006Hickory Swim&Dive Region Champs
Her 17h11'6" 23d41'42"

2007 Paglialunga
Ori 6h0'1" 10d31'20"

2010: a solutions odyssey
Cru 12h24'21" -59d7'23"

2011 Tiffany H
Cap 20h18'9" -19d41'27"

21
Per 3h48'3" 45d25'16"

21 Phoenix Rose 2004
Cam 4h27'49" 65d53'5"

#21 Rocket Boy (F. Joshua Marshall)
Per 3h29'11" 33d48'23"

21 to the 33 forever star
Aqr 22h48'26" -6d17'44"

21.12.2003
Dra 19h3'12" 53d58'16"

21BeautifulJCPerry121884
Sgr 18h0'51" -28d38'19"

21st Upchurch Dreamer
And 0h21'3" 35d38'36"

22
Ari 2h52'28" 22d44'48"
222-222-2222
Tau 5h30'20" 27d16'42"
2264 Meredith and Shawn
Gem 7h4'15" 31d52'47"
22Love
Umi 14h15'49" 74d17'51"
2437 Andrew
Ari 2h16'1" 22d28'26"
24JAUGA97 - TB6/1
Uma 11h42'5" 36d52'36"
24Skiddles
Uma 11h31'51" 58d2'27"
25 coffees and a bowl of soup
Uma 11h23'25" 33d49'45"
25 Silver Linings of Love
Tra 15h53'29" -64d48'27"
25 Years of My Boopie Poo
Cyg 20h31'3" 35d42'16"
25 years till forever
Cyg 20h58'16" 31d23'45"
254
Lyn 8h10'8" 46d58'42"
"25AndreA07"
Cam 3h31'7" 63d27'23"
260671 Anja Kuhles
Uma 9h20'44" 59d53'13"
26tara
Tau 5h45'45" 13d31'43"
2711
Ori 5h59'38" 17d13'39"
2F8A2T5E3
Leo 9h46'58" 6d45'49"
2Gether 4Ever
Cru 12h38'31" -60d39'8"
"2gether 4ever" John & Linda
Ori 5h42'24" -1d45'22"
2nd Lieutenant Morris
Her 16h24'15" 48d16'40"
2nd LT JT Wroblewski
Aur 5h11'32" 42d1'47"
2pretty and 2hansome . . . 2perfect
Lyn 7h20'13" 50d22'20"
2shea Avina
Leo 11h39'5" 14d14'21"
2XU=US
Umi 14h26'1" 67d37'54"
<3<3<3<3<3Jean and Justin<3<3
Uma 11h27'41" 31d58'42"
<3 <3 Caleb Thomas Story <3 <3
Her 16h38'40" 7d40'2"
<3 Bonnie Jean <3
Ori 5h14'34" 15d39'43"
3 CP 34/04
Uma 11h54'50" 49d24'26"
<3 Edd!e & JeNn <3
Cyg 20h58'6" 45d18'3"
<3 Kevin and Brittany Forever <3
Per 3h40'26" 34d2'27"
3 Königsstern
Cas 23h35'28" 57d8'28"
3 K's Shooting Star
Vir 13h21'3" 2d0'54"
<3 Leigh P <3
Crb 15h54'53" 27d23'25"
<3 Natalie
Cas 23h3'4" 54d34'40"
<3 Rachel Hoban <3
Sgr 19h40'58" -13d9'5"
3. Sek. 04 - 05
Uma 8h15'6" 66d35'12"
3 Times is Forever
Lmi 10h19'54" 37d2'10"
<3 Trisha Jean and Ryan Douglas <3
Ori 6h3'35" 17d13'26"
3 Years of Heaven on Earth
Cyg 19h44'41" 38d10'25"
306 Conewango
Uma 10h13'52" 44d12'0"
31 Brittany 32
Uma 13h21'4" 52d54'14"
311 - 10/08/76
Lyn 8h39'51" 34d15'31"
35 A Glimpse of Heaven
Dra 17h1'44" 55d23'45"
35 Years of Love,Terry & Nancy Moon
Cyg 20h52'25" 49d25'1"
35th Anniversary Star
Crb 16h22'12" 39d17'0"
361945 Dan DiPardo 60
Psc 1h12'40" 11d36'13"
381JED1970
Eri 3h50'10" -0d11'4"
3bsi & 3bdoosh
Leo 9h45'54" 8d17'43"
3ddd
And 2h21'23" 50d36'9"
3rd Pond
Ari 2h48'59" 13d25'27"
3-Schu Class of 2004-2005
Pho 0h36'31" -41d9'45"
3taps
Ari 3h17'4" 19d57'38"
T619 Shoreview
Cam 6h40'15" 65d52'19"
T.A. Hockett
Uma 10h44'1" 46d43'13"
Ta & Jana Faith
Sge 20h7'28" 20d29'26"
Ta Joon Jun
Cap 21h36'20" -10d44'37"
Tá mé chomh mór sin i ngrá leat
Umi 16h25'56" 70d15'1"
Tá mo chroi istigh ionat
Leo 9h29'55" 13d31'58"
T.A. Vernon - Our Dad & Grandpa
Cru 12h29'38" -60d32'16"
TAA Futuris
Cru 12h50'3" -60d38'56"
Taana (Tahna) Running
And 0h36'0" 45d3'6"
T.A.A.T.
Cas 0h33'8" 52d45'49"
Tab Lucas Sommer
Cap 20h47'38" -14d57'51"
Tabard
Sct 18h55'10" -4d41'56"
Tabatha
And 2h9'8" 46d46'51"
Tabatha
And 0h43'44" 43d0'39"
Tabatha A. Toppins
And 0h38'4" 41d47'57"
Tabatha Aurora May
Gem 6h48'19" 23d37'39"
Tabatha Elizabeth Pepin
And 23h38'31" 43d11'12"
Tabatha Harris
Uma 9h27'52" 70d3'54"
Tabatha & LaVerne Carroll
Uma 12h51'9" 58d2'1"
Tabatha Marie Shelnutt
Sgr 18h3'35" -16d55'57"
Tabatha Monks
Lyn 8h6'28" 56d5'9"
Tabatha Moscone
Vir 12h54'6" 0d32'36"
Tabatha Nicole Kennedy
Ari 2h9'56" 20d53'43"
Tabatha Taylor
And 23h23'32" 49d20'33"
TabbeyRoad I Am TheEggman
And 23h11'37" 51d56'45"
Tabbie Riley
Gem 7h10'36" 24d9'41"
Tabbitha Ann Rogers
Vir 12h53'31" 0d43'24"
Tabby
Tau 5h54'2" 24d29'8"
Tabby Ann
Uma 11h37'30" 32d15'7"
Tabby Baby Doll
Lyn 7h47'0" 50d7'29"
Tabby Kat
Crb 15h36'11" 26d50'34"
tabbymolina
Tau 5h21'20" 26d24'10"
Tabea
Umi 17h0'34" 80d9'23"
Taber H.E. 13
Vir 13h55'5" 1d12'37"
Taber Jeston William Graham "T.J"
Uma 9h33'34" 57d12'39"
Tabi
Lyn 6h33'55" 54d17'50"
Tabitha
Sgr 17h53'48" -29d31'36"
tabitha
Sgr 20h7'35" -28d32'34"
Tabitha
Sco 16h52'10" -36d38'49"
Tabitha
Tau 4h11'53" 8d56'3"
Tabitha "7-26-97"
Umi 16h20'17" 76d23'56"
Tabitha Ann
Lyr 18h48'44" 35d40'52"
Tabitha Baby Girl Serviss
Sgr 17h52'59" -17d42'18"
Tabitha Christie Mahoney
Leo 10h26'51" 9d37'40"
Tabitha Easton
And 0h35'50" 42d45'52"
Tabitha Elizabeth Shearer
And 1h22'56" 50d16'25"
Tabitha Ellen Cole
And 0h14'30" 40d35'50"
Tabitha G. Marie Martinez
Leo 11h14'9" -5d5'17"
Tabitha Giunta
Lib 15h4'47" -10d48'29"
Tabitha Haley
And 0h50'49" 45d30'41"
Tabitha Heidi Neal
And 23h21'1" 48d16'16"
Tabitha Hunter
And 0h33'0" 43d13'3"
Tabitha J. McGrath
And 1h15'13" 50d1'2"
Tabitha Jean
Sgr 20h17'6" -40d31'56"
Tabitha Jewel of the Galaxy
Aqr 22h34'2" 0d20'54"
Tabitha L. Cleveland
Lyr 19h7'42" 31d25'53"
Tabitha L. Morgigno
Leo 11h56'42" 21d26'41"
Tabitha Lee Bond
Psc 0h27'17" 16d47'58"
Tabitha Lynn Johnston
Ari 1h55'54" 24d57'29"
Tabitha May Whittle
And 0h22'24" 27d22'54"
Tabitha; Mother of Angels
Sco 17h46'46" -43d32'27"
Tabitha Nichole Stalvey
Cap 21h28'18" -18d56'2"
Tabitha Nicole
And 0h28'12" 42d44'35"
Tabitha Riley
And 2h29'19" 44d32'38"
Tabitha Rose Sacco
Ari 2h55'25" 25d2'42"
Tabitha Ruth
Uma 10h25'39" 54d34'44"
Tabitha Scapanda
Aqr 22h24'56" -4d24'38"
Tabitha Scott Whitlow
Lyn 7h22'39" 52d4'38"
Tabitha Sears
Cas 1h38'24" 67d5'12"
Tabitha Sellers
Uma 13h30'57" 58d12'53"
Tabitha Sibel
Tau 5h47'0" 20d5'37"
Tabitha Smith
And 23h23'40" 49d33'49"
Tabitha Szalewski
Lmi 10h48'43" 34d52'44"
Tabitha "Tubby Tater Tot" Koda
And 23h24'37" 52d28'21"
Tabitha Turner
Ari 2h4'51" 23d48'12"
Tabitha05
Uma 11h8'58" 39d26'14"
Tabitha's Birthday Star
Peg 21h51'31" 26d43'48"
Tabitha's Mommy
Cas 1h7'16" 64d50'58"
Tabitha's star
Sco 16h5'21" -17d20'54"
Taboo
Ori 5h53'6" 22d48'30"
Tabor "Fat Dog" Edwards
Sco 16h50'31" -39d11'30"
TabTex
Eri 4h12'42" -24d30'0"
TAC 2006
Dra 18h44'53" 52d33'3"
Tacha Pilamaya
Sgr 17h51'39" -23d54'3"
Tachamani
Uma 11h44'22" 32d57'26"
Tachi
And 0h4'8" 45d37'52"
Tacy Vanzwol
Uma 10h24'51" 58d4'35"
Tad Alexander Peterson
Aur 5h18'54" 31d12'30"
Tad Buhman
Cap 21h31'6" -14d4'45"
Tad "Cappy" Walkley
Per 3h25'43" 51d11'13"
Tad Jones
Per 3h44'37" 45d12'36"
Tad Mattia
Aur 5h52'12" 47d55'28"
Tad Pedersen
Uma 12h42'43" 61d17'27"
Tad Robert Anderson
Per 3h4'28" 50d21'28"
Tad Skotnicki
Uma 10h31'14" 67d0'1"
Tad Stearns
Cnc 9h2'29" 18d27'58"
Tadahisa-Angelique
Cap 20h16'6" -19d14'50"
tadao
Sge 20h1'52" 17d25'34"
Tadashi Andrew Dodge
Cnc 9h6'21" 29d24'7"
Tad-cu
Cep 21h29'38" 64d40'12"
Tadeu Toshio Kimura
Cas 0h37'38" 51d9'19"
TADEUSZ
Ari 2h52'38" 26d52'10"
Tadeusz A. Wyka
Uma 10h7'15" 46d0'35"
Tadpole
Cnc 8h42'32" 8d4'45"
Tadpole
Dra 19h0'2" 74d55'58"
Tae Alexander Tran
Uma 13h48'29" 56d39'44"
Taebi-Nevaeh Gray
And 22h59'28" 38d25'54"
Taedan Riley
Lmi 10h34'3" 38d36'38"
Taefy
Lyr 18h47'2" 40d4'30"
Taeliac
Lac 22h33'29" 49d24'38"
Taelor Vogler
Cyg 20h32'39" 30d56'30"
Taelson Kekahi Aloha Lani
Lib 14h49'32" -3d45'47"
Taelyn
Vir 14h15'0" -14d1'7"
Taelynn
Lyn 7h4'29" 54d6'9"
Taeo
Gem 7h32'59" 26d46'1"
Taera Batovsky
Aqr 23h38'36" -7d3'11"
Taeray
Ari 3h13'20" 27d3'2"
Taevia
And 0h47'29" 37d14'7"
Taf G. Paulson
Gem 7h28'7" 28d9'14"
Tafari Malcolm
Cep 21h26'53" 60d8'49"
Taffy
Lib 14h47'21" -8d53'33"
Taffy Ann Bennett
Umi 16h19'13" 82d46'31"
Taf's Muse
Cyg 20h30'24" 41d53'54"
Tag
Ori 5h19'22" 6d11'26"
Tagada Tsoin Tsoin
Lib 15h58'8" -17d28'29"
Taghi Seyyedi
Ori 5h14'41" -5d52'11"
Taghi Taherzazeh Samani
Uma 10h8'21" 59d49'47"
Taghrid Deza
Lyn 8h40'5" 43d39'53"
Tagman
Her 18h53'21" 24d19'27"
Tah
Cnc 8h50'20" 31d54'39"
Tahimy
Lib 14h51'49" -3d15'16"
Tahir
Sco 16h22'35" -36d13'6"
Tahir i Marina
Cyg 21h33'12" 34d45'48"
Tahirah Niomi 12/30/1998
Cyg 20h39'43" 53d53'7"
Tahitia's Shining Star
And 2h17'12" 48d50'1"
Tahlena
Sco 16h49'1" -26d57'1"
Tahli Jasmin Harper
Cra 19h2'24" -39d18'51"
Tahlia
Aqr 23h31'41" -17d0'59"
Tahlia Angel Nimoh Akyereko
Vir 13h20'35" -16d29'11"
Tahlia Catherine Martin
Cru 12h4'45" -58d23'55"
Tahlia J
Umi 15h53'29" 79d28'38"
Tahlia Jennifer Lowe
Psa 21h28'35" -35d58'25"
Tahlia - My little star
Gem 6h52'46" 13d40'41"
Tahlia Rose
Sco 16h49'45" -27d6'0"
Tahlia's Great Grand Papa Tony
Cru 12h45'4" -60d44'22"
Tahmahkera
Vir 13h45'2" 2d5'31"
Tahmeena Alam
And 0h20'55" 26d0'58"
Tahmina
Vir 14h17'0" 3d4'52"
Tahnee Jo Moody
Psc 0h22'45" 7d33'28"
Tahnee M Land
Lib 15h40'46" -5d53'35"
Tahni
Col 5h46'24" -33d53'41"
Tahoe at Logan Shoals
Cma 6h43'48" -15d17'24"
Tahrisha Lexie Allen
Lyn 6h42'45" 53d20'54"
Tahvia Danise Angela Gopaul
Psc 23h18'2" 5d42'27"
TAI
Lyn 6h38'23" 58d36'15"
Tai Eason
Ari 2h35'0" 20d5'54"
Tai Mee
Ori 5h37'24" -3d57'58"
Tai Pang
Cnc 8h19'29" 24d29'16"
Tai & Ry Big Leap
Ori 5h34'28" 3d41'52"
Tai Shung (KI) Enterprise Sdn Bhd
Ori 5h37'41" 3d48'34"
Tai Tau B FaFa
Lyn 7h40'57" 47d53'4"
Tai Washington
Ari 2h9'6" 23d12'6"
Tai, R.N.
Cas 1h28'5" 59d46'12"
TaiAnna Thielke
And 1h27'23" 48d25'47"
Taiba and Abdallah
Ori 6h16'33" 10d23'49"
Taiga Jack Colton (7:26 p.m.)
Eri 4h14'34" -1d20'31"
Taigen Christine 2004
Gem 6h27'13" 26d6'23"
Taija Analicia
Peg 21h45'34" 13d57'44"
Taildragger Bill O'Neill
Aur 5h37'35" 33d12'14"
Tails
Uma 8h29'41" 62d47'46"
Tai-Lu "Best Mom " Yeh
Psc 1h21'34" 32d35'10"
Taimiah
Leo 11h9'31" 16d38'26"
Taina
And 0h50'31" 36d11'4"
Taina Jennifer Taylor Soltero
Ari 2h33'25" 31d11'34"
Taina Yvette Cruz
And 1h11'29" 41d37'58"
Tainamarie
Aql 19h24'38" 6d13'1"
Tair & Michaela
Cyg 20h34'33" 39d7'36"
Tairra
Uma 11h19'12" 48d8'30"
T.A.I.S Dawn Marie Jennings
Aqr 22h53'1" -8d23'44"
Taisa Gold's Smiling Star
Pho 2h17'43" -43d18'38"
Taisha
And 0h19'43" 44d40'34"
Taishar Makhir
Aqr 21h48'11" 1d28'32"
Tait
Her 16h44'40" 19d45'32"
Tait Kmentt
Per 3h25'5" 45d17'26"
Tait Maria Brennan - Tait's Star
And 1h25'55" 33d55'37"
Taite David Lehov
Aqr 20h50'36" -11d8'20"
Tait-Russell
Uma 11h20'45" 43d19'48"
taixiu
Leo 9h46'49" 26d51'55"
Taiylor Evans
And 0h52'30" 43d3'24"
Taj Monroe Tallarico
And 1h5'50" 45d4'42"
Taja
Cyg 20h51'8" 32d0'43"
Taji
Her 18h6'6" 33d4'4"
TAK+KNR2007
Tau 3h30'4" 22d9'4"
Tak Tsukuda
Tau 5h41'9" 22d40'54"
Taka Maka Maka Ma
Mon 6h52'9" 9d23'7"
Taka & Tomo
Cnc 8h37'8" 29d45'21"
Takako
Tau 4h45'38" 29d15'45"
Takao & Ayumi
Lib 15h14'11" -13d59'54"
Takao Mizokawa
Leo 11h27'9" -6d8'22"
Takara & Adam
Cyg 20h40'34" 43d52'5"
takashi kaisaka
Sgr 18h41'50" -20d33'42"
Takashi Love Miyuki
Vir 14h39'29" 4d51'58"
Takashi Nishizawa
Cyg 21h10'58" 36d42'24"
Takayuki Fujinaga
Uma 11h40'46" 61d36'10"
Take A Breath and Make A Wish.
Tau 4h40'8" 20d56'47"
Take a chance on Alicia
Vir 13h17'13" 5d12'57"
Take me to the clouds above NK&AZ06
Cru 12h17'4" -56d36'45"
Take My Breath Away
Ori 6h1'1" 18d2'52"
Takeshi
Lib 15h54'27" -19d18'45"
Takis Antonakopoulos
Uma 9h11'39" 69d59'4"
Tal
Ari 2h24'47" 18d37'48"
Tal Marrone
Cnc 8h52'27" 14d20'7"
Tal Rozenberg
Uma 13h5'32" 53d41'7"
Tal Tal Hamud
Uma 13h45'5" 56d3'56"
Tal Tal Hamud
Umi 14h18'57" 75d18'56"
Tala
Gem 7h4'5" 11d3'41"
Tala and Faysal
Cyg 19h52'55" 36d43'29"
Tala Anne Bernhardt
Cap 20h21'37" -22d0'29"
Tala Pascua
Sco 17h54'9" -38d30'5"
Talal Peret
Aqr 20h45'9" -8d58'6"
Talamajacara
Umi 16h11'29" 73d53'42"
Talamante-Pohlmann
Cnc 8h51'45" 14d36'15"
Talamn
Ari 2h28'12" 21d16'11"
Talan James Walstrom
Lyn 8h7'33" 38d13'41"
Talan Ray
Umi 15h21'54" 68d56'54"
Talan Richard Mullins
Ori 5h19'0" 1d42'40"
Talbot
Cma 6h40'35" -20d24'2"
Talea Nicole Jones
Peg 22h18'50" 12d0'25"
Taleen Aghabekian
Aqr 22h54'34" -8d47'46"
Talena
Gem 6h52'1" 14d59'19"
Talena Ann Mobley
Tau 3h31'51" 9d15'39"
Taletha
And 0h38'1" 26d10'6"
Talia
Ari 2h20'9" 23d40'13"
Talia
Leo 11h34'46" 22d4'25"
Talia
Psc 0h43'29" 7d40'55"
Talia
Tau 4h31'29" 17d43'8"
Talia
Lyn 8h40'37" 41d0'12"
talia
Lib 15h59'21" -11d50'51"
Talia
Uma 13h45'9" 59d7'47"
Talia And Pookie Bear 373 . . .
Cnc 9h0'11" 14d57'52"
Talia Bellah Manchester
Lyr 19h13'19" 36d0'39"
Talia Elizabeth Barrow
Uma 9h14'19" 57d35'19"
Talia Elizabeth Kruzhkov
Sco 16h13'23" -13d46'8"
Talia Erin Oberfeld
And 0h47'23" 42d20'20"
Talia Joelyn
Ari 2h13'7" 16d26'48"
Talia Josey Schneder
Crb 15h19'46" 27d2'0"
Talia Kate Anderson
Cas 2h25'52" 67d37'3"
Talia Keri Coughlin
Pic 5h16'25" -49d25'57"
Talia Kristine Cuffari
Cnc 9h20'18" 17d29'41"
Talia Maria Sandoval Montoya
Tau 5h56'32" 26d46'42"
Talia Marie Danastorg
Lib 15h6'40" -1d30'50"
Talia Ojeni Avedian
Lyn 6h34'27" 56d39'26"
Talia Rose
And 0h47'14" 40d35'8"
Talia Rose Bukhman
Lyn 7h41'2" 45d44'9"
Talia Sroka
Lyn 8h15'0" 37d15'55"
Taliana Roberta
Per 3h0'10" 51d36'58"
Taliasha
Tau 4h14'45" 28d37'53"
Talicia Teal
Sco 16h24'47" -41d50'0"
Taliesin
Lmi 10h42'18" 26d56'27"
Talin and Emin
Cyg 21h21'36" 45d49'48"
Talin Theodore Harry Schlachet
Psc 1h11'22" 6d50'44"
Talina Al- Dabbous
Lyn 7h21'32" 55d45'26"
Talisano
Mon 7h5'12" -0d14'39"
Talisa's Talisman
Lib 15h20'56" -8d3'12"
Talisha
Gem 7h28'41" 26d4'34"
Talisha Joyner
Cam 5h17'22" 62d51'50"
Talissa Shines
Gem 7h53'26" 31d27'30"
Talita Emmanuelly de Souza
Tau 5h10'23" 16d27'18"
Talitha
Ari 2h19'14" 13d37'7"
Talitha
And 0h15'40" 41d50'53"
Talitha Banks
Leo 11h50'35" 22d14'38"
Talitha Marikit
Aqr 22h56'43" -20d39'12"
Talitha May
And 23h43'50" 36d15'26"
Tall Americano
Lac 22h37'10" 43d25'44"
Tall Joan
Dra 16h15'37" 61d43'53"
TALLA
Sgr 17h45'2" -26d32'18"
Talley Hopson Gregg
Ori 6h6'22" 10d45'10"
Talley-Verde (TV) Star
Ori 6h24'47" 14d57'16"
Tallhead Daddy
Lib 15h47'5" -28d42'50"
Tallie Bingley
Uma 9h57'55" 48d7'59"
Tallie Denison Gailey
Cyg 19h42'34" 38d55'19"
Tallino
Cam 4h27'29" 57d20'45"
Tallis
Vir 13h58'12" -16d0'10"
Tallon
Boo 14h48'30" 50d52'30"
Tallulah
Lyn 8h5'47" 46d41'23"
Tallulah Angelina Benedetto
Sgr 18h0'9" -27d57'1"
Tallyn Scott-Rhodes
Cru 12h39'40" -58d54'2"
Talmadge B. Ault II
Uma 11h27'28" 44d43'29"
Talmadge John "Tom" Nash
Lib 14h53'41" -2d59'26"
Talochka
Ari 1h48'26" 17d40'1"
Talo'fa ia te oe
Lyn 8h30'6" 43d39'22"
Talon
Vir 11h52'40" 9d0'48"
Talon Holmes
Dra 18h44'46" 56d7'11"
Talon Michael Romero-Eide
Her 16h31'50" 29d20'9"
Talon Ryan James
Vir 14h40'49" -3d18'49"
Talon Tope
Cap 21h15'17" -21d53'55"
Talonda D Thomas
Uma 12h29'53" 56d52'38"
Talor Grace
Cas 1h35'4" 62d7'24"
TalSpence
Mon 6h51'52" -0d32'36"
Taltinov Maxim
Aqr 20h41'16" -12d28'49"
Talton
Aql 18h52'29" 8d17'29"
Talton Lea
Crb 16h1'30" 33d27'56"
Talya Elizabeth Trias
And 0h11'33" 28d18'18"
Talya Grace
Cas 0h59'18" 58d43'2"
Talya Lattanzio
Uma 11h42'8" 33d4'21"
Talya Toledo
Cas 1h12'48" 57d6'5"
Talyn Tina Moon
Ari 2h43'3" 25d19'14"
Tam Bevon
Uma 10h40'5" 48d35'4"
Tam Duong
Cap 20h13'1" -14d55'21"
Tam Duy Tieu XXVI
Aqr 23h9'16" -10d48'56"
Tam Le
Sco 17h50'25" -41d7'2"
Tam M Vu
Gem 7h12'20" 20d11'17"
Tam Tam
Uma 12h28'57" 55d52'2"
Tam The Journey
Leo 11h27'38" 10d32'31"
Tama Gonen
Mon 7h15'10" -0d42'51"
Tamaki & Seiji & my friend forever
Sgr 19h15'37" -35d20'30"
Tamalpais
Ari 3h10'15" 29d53'40"
Taman Powell
Gem 6h38'7" 25d56'54"
Tamandrew
Cyg 19h35'51" 36d56'22"
Tamar & Jason's Wedding Star
Uma 9h56'43" 42d8'2"
Tamara
Lyn 7h54'57" 41d17'58"
Tamara
And 1h13'47" 38d15'38"
Tamara
Cyg 19h54'22" 33d7'43"
Tamara
Gem 7h47'25" 34d27'10"
Tamara
Tri 2h8'35" 32d52'51"
TAMARA
Cyg 21h55'8" 50d50'31"
Tamara
Gem 7h30'35" 26d4'27"
Tamara
Gem 6h47'21" 25d35'18"
Tamara
Gem 6h41'20" 17d39'31"
Tamara
Vir 14h39'11" 2d40'22"
Tamara
Vir 11h59'23" -7d34'12"
Tamara
Sgr 19h41'34" -12d19'14"
Tamara
Lib 14h47'26" -17d13'33"

* Tamara*
Cas 23h21'27" 54d24'28"
Tamara
Dra 16h27'33" 58d7'59"
Tamara
Uma 11h21'34" 65d58'26"
Tamara
Cam 5h42'32" 66d56'44"
Tamara
Cam 7h42'53" 63d44'44"
Tamara 16.06.1981
Peg 22h25'15" 23d6'1"
Tamara Aliyeva
Gem 6h34'31" 20d58'34"
Tamara Amestoy
Ori 5h37'39" -0d45'46"
Tamara Ann
Ori 5h49'45" 11d21'15"
Tamara Ann Brown
Cyg 20h49'37" 44d40'4"
Tamara Ann Hays
Cam 7h21'8" 73d5'15"
Tamara B.
Cep 2h21'19" 81d8'18"
Tamara Bennett
Lyn 8h26'12" 52d56'4"
Tamara Beth
Sgr 19h20'47" -41d55'34"
Tamara Bolzani e
Salvatore Fioretto 10 aprile
2001
Crb 15h57'22" 25d46'41"
Tamara Bongulielmi
Lac 22h23'2" 48d22'57"
Tamara Brillante
Sco 16h9'10" -20d19'19"
Tamara (Chäferli),
01.08.2002
Ori 5h33'22" 10d8'43"
Tamara Dawn Bannister
Cnc 9h14'48" 14d40'56"
Tamara Demidenko
Cet 1h20'45" -0d56'31"
Tamara Deyton Boskofsky
Crb 15h39'15" 27d17'25"
Tamara Figlia
Her 16h50'2" 33d59'3"
Tamara Fontaine
Ori 6h11'43" 21d5'47"
Tamara Henshaw Wise
Sco 17h51'27" -35d46'55"
Tamara Ibinarriaga
Ari 2h10'3" 26d10'59"
Tamara Jade
Sco 17h55'51" -37d29'29"
Tamara Joy Clarke
Uma 10h47'23" 56d16'30"
Tamara Kathleen Hartigan
Sgr 19h42'14" -17d12'10"
Tamara Kay Sandoval
Ari 2h6'46" 11d56'16"
Tamara Kaye Kusleika
Vir 12h37'2" 3d2'32"
Tamara Keller
Cas 23h34'32" 51d40'34"
Tamara Kochman
Cam 7h54'40" 67d18'4"
Tamara Krol
Ari 1h53'55" 17d32'59"
Tamara L Gallegos
Lib 15h9'6" -16d10'7"
Tamara L Israel
Psc 1h7'20" 31d22'33"
Tamara L. Werner
Uma 8h30'25" 70d53'28"
Tamara Lanktree
Psc 1h20'46" 10d30'30"
Tamara Leah Nolte
And 2h25'15" 48d57'22"
Tamara Lee Monahan
Cnc 8h15'11" 15d58'28"
Tamara Leigh Cronk
Lib 14h41'22" -9d4'43"
Tamara Lenore Starr
Aur 6h12'55" 31d18'56"
Tamara Lyn Holder
Srp 18h21'28" -0d0'38"
Tamara Lynn
Sgr 18h41'16" -27d26'36"
Tamara Lynn Fisher
"Tammy"
Sco 17h21'38" -30d59'42"
Tamara Lynn Kavall
Ori 6h13'21" 7d16'35"
Tamara Lynn Pearson
Crb 15h32'38" 31d37'30"
Tamara Lynn Scarborough
Tau 5h49'11" 26d52'0"
Tamara Lynnette
Crb 15h38'35" 33d35'43"
Tamara M. Tsilimingras
Ari 3h24'45" 19d47'46"
Tamara Marie
Cap 21h0'30" -27d14'19"
Tamara May Lowe
And 23h15'45" 42d44'54"
Tamara Michael
And 1h44'23" 45d26'17"
Tamara & Michael - 19th
June 2005
Ori 6h16'49" 20d34'31"
Tamara Michelle Rodriguez
Sco 16h16'17" -10d40'27"
Tamara Navarino
Ari 2h27'55" 24d41'1"

Tamara Nielsen
Ser 15h42'50" 18d20'3"
Tamara Noelle
Cyg 21h26'27" 49d6'9"
Tamara Ojai
Vir 12h41'33" 6d57'9"
Tamara: Olympus in Terra
Cnc 8h8'51" 24d59'9"
Tamara S.
And 0h45'21" 34d57'47"
Tamara Sara Lynn Skarda
Leo 10h17'23" 21d46'24"
Tamara Scavino
And 1h54'37" 45d36'19"
Tamara Schumacher
Crb 15h58'35" 27d37'6"
Tamara & Scott 021382
Cep 21h28'56" 65d0'4"
Tamara Scott Pitts
Sgr 20h4'36" -44d44'45"
Tamara Shada
And 23h11'9" 40d3'3"
Tamara Sonja Victoria
Cooke
Cnc 9h6'53" 17d9'56"
Tamara Stratton
Uma 10h27'26" 46d34'22"
Tamara Sue
Uma 8h20'35" 67d43'35"
Tamara "Tamy" Schreiber
Aur 5h21'22" 30d1'45"
Tamara Terell Teryniak
Aqr 22h3'13" 0d47'11"
Tamara Vevrik
Tau 4h48'27" 20d5'23"
Tamara Wayand
Sco 17h52'59" -36d16'48"
Tamara Webb
And 1h42'17" 42d18'49"
Tamara Weller
Lyr 18h50'37" 43d36'16"
Tamara78
Sco 16h57'34" -38d20'56"
Tamaragi Ellen Fruits
Sco 16h51'15" -33d48'21"
Tamara's Guidin' Light
Sgr 19h11'52" -26d2'37"
Tamara's Guiding Light
Cap 20h28'35" -13d53'52"
Tamara's Vision
Cas 0h37'0" 60d29'45"
TamaraScott
Cyg 21h20'10" 52d4'14"
Tamare Esperes
Crb 15h47'14" 27d26'49"
Tamargo
Tau 5h59'45" 26d6'42"
Tamarind Cove Hotel
Peg 22h30'26" 31d4'29"
Tamaris Lynn Davis
Cap 21h38'12" -10d21'54"
Tamarisk Rose Catanzaro
Lyn 6h53'2" 54d39'20"
Tamarka - Kocham Che
Mama!!
Uma 8h47'17" 46d51'0"
Tamás Eva
Cap 20h14'45" -12d12'9"
Tamás Kiscsillaga, Evi
Hópihe
Uma 9h53'12" 60d5'8"
TamasT
Ori 5h36'19" 14d43'57"
Tamatha
And 1h28'48" 44d27'43"
Tamatha Jo Mosley
Cyg 20h54'11" 34d56'50"
Tamayar, Mehmet Seref
Vir 13h28'11" 7d20'50"
Tambara
Lyr 19h12'45" 26d32'17"
TamBeau
Lib 14h26'9" -12d16'52"
Tambra
Cas 0h26'0" 61d15'7"
Tambra Dee Beardsley
Mon 7h19'14" -0d31'38"
Tambra Holroyd
Vir 12h13'20" 12d28'52"
Tameka
Lyn 8h49'43" 38d32'15"
Tameka Bates
Cnc 9h4'28" 16d47'42"
Tameka " Black Orchid "
Tau 4h19'12" 7d40'7"
Tameka Faith Hickey
Col 5h14'35" -31d16'32"
Tamekia Robinson
Aql 19h56'45" -0d59'1"
Tameko Star
Leo 10h16'51" 15d5'13"
Tamela Pressley Carr
Ari 2h11'44" 26d15'28"
Tamela Rose
Aqr 21h31'24" 0d25'8"
Tamer Hussein Abdelaziz
Vir 12h46'25" 11d5'47"
Tamera Lynn
Sgr 19h57'3" -30d20'25"
Tameris Diane Rought
Sgr 18h24'57" -32d26'25"
Tamgaris
Cru 12h40'49" -57d13'28"
Tami
Her 18h47'14" 21d12'45"

Tami
And 0h41'15" 44d0'56"
Tami Ann Rich
Cnc 8h34'42" 10d12'44"
Tami D. Cunha
Cam 5h9'40" 63d27'45"
Tami Deron Munafo
Crb 15h46'43" 27d7'36"
Tami Estrella
And 0h57'4" 40d21'22"
Tami Fickle-Sparks
Sco 16h46'16" -31d33'56"
Tami Iida - Rising Star
And 2h10'22" 45d15'9"
Tami Jean Mayers
Aqr 20h41'34" -8d44'46"
Tami K
And 0h9'46" 45d1'23"
Tami Keiser
Cyg 20h31'9" 58d57'20"
Tami & Kelven's Beautiful
Baby Girl
Aqr 20h57'53" -10d44'9"
Tami Kilger
Aqr 23h23'44" -5d40'40"
Tami L. Bell
Tau 4h5'57" 14d41'42"
Tami LaFrinere Lawrence
Cyg 19h50'28" 33d19'8"
Tami Lea Star
Aqr 22h45'23" -8d7'1"
Tami Louise Hahn
Cam 5h46'28" 58d54'33"
Tami Lynn Brown
Sgr 17h55'13" -29d29'23"
Tami Lynn Himes
Umi 13h41'37" 75d7'15"
Tami Marie Schiavo a.k.a
my poo
Gem 7h36'29" 14d39'25"
Tami McHugh Star
Daughter
And 2h23'25" 45d20'46"
Tami Michelle Kenney
Vir 11h47'51" 3d53'22"
Tami Munson
And 0h39'15" 45d44'4"
Tami My Love Allways
Vir 13h5'50" -20d53'35"
Tami Olin
Lyn 7h30'55" 35d41'3"
Tami Redes
Lib 15h32'44" -5d25'55"
Tami Renae
Uma 12h6'6" 58d31'12"
Tami Renee Eshelman
Leo 10h18'28" 24d48'1"
Tami Schimdt
Cnc 8h40'57" 23d13'26"
Tami Sharee
And 0h9'27" 39d58'20"
Tami Statham
Uma 11h31'48" 52d5'32"
Tami The Great
Sgr 18h18'0" -23d55'20"
Tami Wentz
And 2h10'55" 42d11'4"
Tami Wilhelm
Lyn 8h59'28" 38d32'0"
Tamia Delgadillo
Uma 10h43'49" 58d30'21"
Tamiamies Freddy
Uma 11h24'48" 38d32'34"
Tamie
Cas 1h30'45" 57d7'24"
Tamie
Leo 11h57'8" 20d55'44"
Tamie
Lib 15h58'48" -13d45'16"
Tamie Hu
Aur 6h25'49" 42d7'32"
Tamie Maxfield
Uma 11h41'13" 60d26'48"
Tamie Wold
Lib 14h56'41" -0d43'4"
Tamie's Christmas Star
And 0h43'6" 32d33'53"
Tamika Whitmore
Lyn 9h12'31" 42d10'27"
Tamilina Sue
Leo 9h22'43" 16d32'2"
Tamilla
Vir 13h25'52" 3d18'3"
Tamina
Leo 9h41'9" 31d50'54"
Tamina maza
And 1h53'36" 39d47'23"
Tämi's Guardian
Uma 13h34'41" 54d6'32"
Tami's Star
Lmi 10h8'49" 34d33'2"
Tamitha Denise Napier
And 2h32'14" 39d38'23"
Tamitha Grindell (a beauti-
ful star)
And 1h2'10" 39d47'0"
Tammaris Marie Rivera
Martinez
Leo 10h1'40" 21d7'43"
Tammer Helena
Umi 15h38'57" 78d15'11"
Tammera
Cas 0h30'20" 50d47'38"

Tammery Lynn Frugé
Myers
And 0h19'55" 40d9'59"
Tammey L. HB Gohl
Ori 5h9'10" 15d27'44"
Tammi
Cnc 8h40'16" 8d33'10"
Tammi Jo
Eri 4h13'7" -23d42'5"
Tammi Johnson
Gem 6h35'18" 21d10'16"
Tammi Lea
Tau 3h42'53" 27d58'12"
Tammi Leigh
Psc 1h21'23" 28d28'20"
Tammi Leigh Erwin
And 0h20'21" 45d59'59"
Tammi Lynn Lohr
Umi 17h12'41" 81d34'30"
Tammi R. Blanchard
Lyn 7h22'19" 56d20'1"
Tammi Rombach and
Benita Petruck
Uma 8h36'48" 47d6'8"
Tammie
Uma 11h53'25" 32d13'11"
Tammie D. Troutmans
Sgr 20h14'0" -38d50'45"
Tammie Doran
Uma 11h29'6" 58d54'6"
Tammie Frazier
Lib 15h4'14" -2d36'50"
Tammie Renee Dowling
Moore
Cap 21h58'20" -12d17'0"
Tammie Simpson
Ori 6h15'23" 9d45'49"
Tammie, Brian's love
Uma 13h38'10" 56d3'26"
Tammie's Star.
Cap 20h15'4" -24d12'6"
Tammra Sterling
And 23h28'44" 48d36'2"
Tammy
Per 3h26'52" 45d14'46"
Tammy
Uma 12h1'41" 30d59'34"
Tammy
Lyn 8h41'7" 33d50'43"
Tammy
And 0h29'1" 36d32'14"
Tammy
And 1h44'6" 43d16'24"
Tammy
Ari 2h15'55" 23d42'55"
Tammy
Cnc 8h25'31" 17d28'54"
Tammy
Lyr 19h13'21" 27d5'11"
Tammy
Sgr 19h55'30" -29d1'31"
Tammy
Sgr 17h51'39" -29d17'38"
Tammy
Aqr 23h22'46" -18d45'16"
Tammy
Sco 17h31'30" -35d42'59"
Tammy
Cas 23h34'56" 56d40'12"
Tammy
Uma 12h41'5" 58d54'28"
Tammy
Cas 2h23'59" 65d21'24"
Tammy (2727) Jackman
Lyn 7h48'0" 42d34'50"
Tammy Abbey
Cas 23h46'7" 59d14'14"
Tammy Adrian Rivenbark
Sco 17h24'53" -38d32'50"
Tammy and Gavin
Randazzo
Uma 8h13'7" 71d28'40"
Tammy and John
And 1h17'27" 46d29'34"
Tammy and Mickey's Lucky
Star
Cyg 19h36'12" 44d58'34"
Tammy and Romuald
Olejnik
Cyg 19h57'37" 53d59'40"
Tammy and Romuald
Olejnik
Cyg 20h25'5" 58d29'59"
Tammy Ann
Uma 10h15'54" 52d39'6"
Tammy Ann Zielinski
Tau 3h28'34" -0d46'46"
Tammy Beckett
And 0h38'6" 30d42'22"
Tammy Beverly Southard
And 0h31'56" 31d10'33"
Tammy Brindley
Aql 19h39'14" 15d13'55"
Tammy Brzezniak
Cas 1h26'30" 60d36'11"
Tammy C. Hanson
And 2h6'16" 44d8'32"
Tammy Carnes
Crb 15h47'3" 26d39'45"
Tammy Causer
Uma 8h36'21" 58d7'12"
Tammy Celestial Body
Sco 16h12'8" -16d43'5"
TAMMY COOK/HOLLI
Leo 10h25'32" 15d54'17"

Tammy D. Powers
And 0h44'53" 37d19'8"
Tammy Daniele
Ari 2h32'22" 24d48'21"
Tammy Davis
Cnc 8h11'41" 9d41'35"
Tammy Dawn Smith
Uma 10h55'33" 52d37'13"
Tammy Denise Barfoot
And 0h39'10" 33d4'52"
Tammy Dias Forever Our
Shining Star
Aqr 22h29'53" 0d1'54"
Tammy Fullerton
Uma 11h6'43" 60d59'18"
Tammy Gaffney
Ari 2h52'37" 28d10'55"
Tammy Giannobile
Aqr 21h15'37" 2d29'23"
Tammy Giannobile
Aqr 23h23'37" -11d54'1"
Tammy Griffin Roller
Peg 22h27'17" 29d45'53"
Tammy Grubb Lee
Cap 20h25'49" -24d15'45"
Tammy Harvey
And 0h30'3" 45d8'52"
Tammy Hevner
Vir 13h4'13" -18d29'39"
Tammy Holtmann
Sco 16h7'28" -12d4'4"
Tammy Hopkins
Psc 1h17'29" 4d1'40"
Tammy Huynh
Sgr 18h1'5" -18d20'43"
Tammy J Grotts
Uma 10h50'25" 63d11'53"
Tammy James
Ori 6h11'38" 8d25'35"
Tammy Jarecki
Uma 11h6'4" 45d56'16"
Tammy Jean
Vir 13h14'55" 10d27'4"
Tammy Jean Bryant
Lyn 6h42'18" 60d17'21"
Tammy Jean Voorhies
Ari 2h32'21" 25d11'14"
Tammy & Jeremy
Sge 20h3'20" 20d33'36"
Tammy Jo
Crb 15h34'57" 26d2'48"
Tammy Jo
Sco 17h28'2" -40d57'6"
Tammy Jo Flohrs
Cnc 8h39'43" 22d52'32"
Tammy Jo Jacobs
And 0h48'57" 45d11'58"
Tammy Jo Skeens
Sco 16h8'51" -15d58'23"
Tammy Jo Willis
Aqr 21h13'13" -7d55'58"
Tammy Joe Vierheller
Ori 5h48'0" 2d8'33"
Tammy Johnston
Vir 13h45'33" -3d19'19"
Tammy Joy Eliza Kohan
Cap 21h8'19" -24d41'26"
Tammy K (Mi Brujita)
Gem 7h21'57" 31d24'16"
Tammy & Karl, The
Gabrielson Star
Cyg 19h50'15" 33d11'16"
Tammy Kate Mandy
Miranda
And 0h53'6" 45d30'29"
Tammy Kelly
Tau 5h37'16" 25d42'46"
Tammy L. Summers "LOVE
4 LIFE"
Vir 12h47'25" 5d14'10"
Tammy Lea Baker
Sge 19h51'22" 18d32'53"
Tammy Lea Kerr
Tau 5h9'45" 27d50'8"
Tammy Lee
Lmi 10h19'39" 28d53'58"
Tammy Lee
Leo 10h2'33" 13d2'5"
Tammy Lee Jones
12.08.1976 - Our Angel
Cru 12h29'13" -60d14'12"
Tammy Lee Ryneer
Aqr 22h2'28" -2d6'52"
Tammy Littlefield
Uma 11h3'29" 53d31'50"
Tammy Lou
And 1h44'1" 49d36'25"
Tammy Louise White
Massey
Aqr 20h49'40" -11d49'32"
Tammy Lynn
Sgr 17h52'42" -17d20'5"
Tammy Lynn
Cru 12h34'17" -58d9'21"
Tammy Lynn
Tau 5h0'19" 28d28'25"
Tammy Lynn Heinle
Leo 11h1'55" 4d37'34"
Tammy Lynn Lovings
Per 3h39'28" 32d32'25"
Tammy Lynn Morelock
Sco 16h8'16" -18d7'51"
Tammy Lynn Mosher
Ori 6h2'19" 17d25'59"

Tammy Lynn "My Angel"
SEU 06
Ori 5h59'38" 21d30'44"
Tammy Lynn "My
Christmas Angel"
Cnc 8h57'24" 12d56'24"
Tammy Lynn Shaw
Aqr 23h52'42" -4d44'58"
Tammy Lynn Texiera
And 0h31'6" 42d21'3"
Tammy Lynne Place
Cas 1h50'48" 61d36'32"
Tammy Lynnette Hartman
Cnc 8h43'27" 10d18'4"
Tammy M
Leo 10h18'14" 26d35'10"
Tammy MacIntosh - A Star
In Our Eyes
Ori 5h10'36" 7d9'43"
Tammy Mainez & Aaron
Gillen
Cyg 21h29'45" 43d57'52"
Tammy Marketti
Sco 16h11'58" -11d59'0"
Tammy Mason
Ori 5h50'27" -0d6'34"
Tammy & Mat
And 23h16'17" 47d53'9"
Tammy Maude
Aqr 22h48'6" -18d39'1"
Tammy Michelle Porter
Del 20h28'1" 6d30'41"
Tammy Milam
And 23h6'20" 42d40'13"
Tammy Milton
Crb 15h27'55" 29d0'53"
Tammy Minky Rosseter
Tau 3h46'19" 21d40'18"
Tammy Montgomery
Uma 11h2'43" 55d44'47"
Tammy Montgomery
Heitman
Cam 7h19'46" 65d31'40"
Tammy Moreno
Uma 9h9'19" 56d21'7"
Tammy Morgan
Oph 17h43'6" -0d29'24"
Tammy Norton
Vir 11h53'26" -6d3'34"
Tammy Peace Carvin
Leo 10h53'24" 15d41'48"
Tammy Pearce
Aqr 22h13'53" -7d48'22"
Tammy (Queen) & Alison
Cas 1h37'1" 61d15'58"
Tammy Ray
Aqr 22h35'11" -1d40'11"
Tammy Ray Ballew
Lyr 19h20'1" 29d42'2"
Tammy Rena Bacigalupi
Sco 16h44'38" -34d11'43"
Tammy Rodriguez Price
Cnc 8h39'17" 23d45'37"
Tammy Rose
Cnc 8h32'20" 22d31'5"
Tammy Ruth
Ari 3h27'39" 27d41'38"
Tammy Ruth Deer
Crb 16h17'15" 31d42'35"
Tammy Ryan
Vir 12h20'36" -6d38'57"
Tammy S. Belkin
Leo 9h28'32" 19d13'36"
Tammy Shoemaker
Tau 4h34'9" 0d20'33"
Tammy Starr
Mon 8h2'4" -0d29'15"
tammy staser
Lyn 8h26'44" 34d13'2"
Tammy & Steven
Cyg 20h25'27" 50d26'52"
Tammy Sue
Lyn 8h15'12" 41d37'24"
Tammy Sue Andrews/Hall
(TAZ)
Cnc 8h57'57" 16d30'13"
Tammy Sue Midgett
Psc 0h16'21" 6d52'19"
Tammy Synder Murphy
Uma 14h22'29" 61d47'18"
Tammy "Tameramers"
Fitzhugh
Cyg 19h52'44" 38d2'23"
Tammy Taylor
Psc 0h28'4" 15d24'55"
Tammy Tegler
And 1h20'8" 45d11'52"
Tammy Thompson
And 2h12'33" 46d41'58"
Tammy Trost "The Twinkle
of my Eye"
Psc 1h45'35" 9d53'2"
Tammy Trzcinski's Wishing
Star
Tau 5h44'43" 13d20'34"
Tammy Twardus
Ori 5h37'3" -1d5'8"
Tammy Vietta
Ari 2h14'26" 14d33'51"
Tammy Voclain
Cas 0h51'10" 52d47'41"
Tammy Wise
Tau 3h39'1" 29d17'29"

Tammy, Amber & Shannon
Willett
Umi 15h59'47" 70d19'19"
Tammy, The Light of My
Life
And 1h4'49" 36d27'42"
Tammy7866
Cnc 8h47'9" 32d19'59"
Tammy-Baby Butterflies
Mama
And 23h20'28" 46d49'40"
TammyGaye
Mon 7h56'25" -0d53'54"
Tammy-Leigh Martin
And 0h29'4" 46d39'55"
Tammy's Light
Ari 2h22'30" 24d59'8"
Tammy's Shimmering Star
Cnc 8h51'43" 11d46'37"
Tammy's Shining Star
And 2h11'41" 38d2'23"
Tammy's Special Star
Ari 2h5'53" 24d51'37"
Tammy's Star
Cnc 8h28'51" 32d5'2"
Tammys Star
Cyg 20h53'37" 41d37'4"
Tammy's Star of Hope
Gem 7h16'20" 17d0'4"
Tammy's Wish
And 1h32'40" 47d2'47"
TAMO Nico
Uma 11h27'19" 30d23'43"
Tamoosh & Farkoosh's
Private Star
Sco 17h5'54" -42d45'45"
Tamora Sue Bouwman
Lyn 7h32'19" 51d43'48"
Tamra
Uma 10h20'3" 47d8'36"
Tamra
Cap 20h25'47" -19d59'20"
Tamra Bolander
Peg 21h51'12" 14d17'22"
Tamra Byerly
Uma 12h4'32" 33d6'52"
Tamra Holt
Uma 13h53'39" 51d44'24"
Tamra Jean McCoy Your
Heavenly Star
Gem 6h20'55" 17d59'47"
TAMRA JO 12/6/1958
Sgr 19h7'1" -30d7'56"
Tamra Joan Poteet
Tau 4h18'2" 14d5'49"
Tamra Joy Peters
Cnc 7h57'46" 16d31'38"
Tamra K
Vir 12h15'52" 9d23'43"
Tamra L. Ritchie
Aqr 22h27'27" -1d0'16"
Tamra Lea Farah
And 2h23'16" 46d10'36"
Tamra Lynn Rector
Leo 9h39'55" 10d57'8"
Tamra Pierson
Lyn 7h56'0" 39d38'19"
Tamra Shelnutt
And 0h45'51" 45d10'19"
Tamra Trevino
Mon 6h49'10" -0d6'20"
Tamsen Renee
Aqr 23h3'19" -6d41'42"
Tamsin Baldwin
And 2h38'10" 48d45'58"
Tamstar
Lyn 8h5'36" 56d10'59"
Tamyra Diane
Sgr 18h58'52" -24d21'47"
Tamzin Slade
Aqr 23h37'46" -12d31'32"
tamzntrev2007
Cnc 8h24'49" 17d27'41"
Tan
Psa 22h8'4" -27d55'44"
Tan Duc Nguyen
Ori 6h7'4" 16d7'10"
Tana
Cnc 8h38'15" 8d24'20"
TANA
Sco 16h54'45" -34d30'46"
Tana 38
Vir 11h57'39" -0d10'45"
Tana Dawn
Cas 23h2'45" 56d24'5"
Tana Justine
And 0h19'43" 31d18'44"
Tana K. Starkey
Lyr 18h48'35" 39d14'30"
Tana Lee
Leo 11h21'40" 7d25'22"
Tana Lee's Diamond In
The Sky
Cas 0h57'44" 63d4'51"
Tana Lynn Smith
And 0h32'52" 41d32'10"
Tana My Love Forever &
Always
Cas 23h19'43" 55d39'25"
Tana Norodom
Dra 16h7'26" 52d33'53"
Tana Sirois
Sco 16h8'29" -25d14'9"
Tanaka Mahachi
Ori 5h29'27" -5d17'14"

Tanaki
Ari 3h8'39" 22d56'6"
Tana-Nashe
Uma 13h6'1" 55d57'3"
Tananda Darling
Ari 2h49'57" 11d42'42"
TANaPP5r1.3.4Uf...sOs420 WingsMtl
Uma 11h25'31" 44d27'55"
Tanaquils Hyson CD, CGC, AAA/AAT
Cnv 12h38'6" 46d40'22"
Tanaya
Cas 1h16'43" 54d1'20"
Tance Wingate
Tau 4h48'15" 18d56'8"
Tanchick1112
Cnc 8h31'29" 21d20'38"
Tanchum
Gem 6h5'3" 24d9'28"
Tandi
Psc 1h48'34" 6d7'44"
Tandus
Uma 12h3'44" 57d12'18"
Tandy
Uma 10h58'42" 50d51'56"
Tane
Cyg 21h11'26" 48d41'24"
Tanechka
Lmi 9h50'26" 40d39'57"
Tanechka Schohot
Cnc 8h49'59" 30d25'52"
Tanesha L. Chever
Cnc 8h17'20" 11d46'31"
Taneshia
And 0h18'1" 40d15'29"
Tang cho nguoi dep cua anh, Ngan
Psc 1h14'54" 20d39'15"
Tang Lung
Cap 20h39'17" -20d40'21"
Tanga Radford
Lib 14h54'35" -5d14'38"
Tange Dion
Cas 23h18'53" 55d54'2"
Tangela
Uma 11h29'46" 39d1'30"
Tangela Renee Atkins
Sgr 18h24'27" -16d11'52"
Tangerine
Vir 12h57'31" 12d10'37"
Tangi
Cnv 12h35'38" 46d52'45"
Tangible Sweetness
Lyn 7h58'38" 49d32'51"
TANGIDA KHANAM RANU
Cap 21h9'34" -19d33'8"
Tango 10
Uma 10h39'10" 41d11'24"
Tango Mike
Aqr 21h50'49" -2d23'26"
Tanha Nalini Ramoutar
Aqr 22h53'30" -22d34'6"
Tani Kennedy
Lib 15h55'43" -18d19'21"
Tani Mihaela Arau
And 2h34'15" 45d19'2"
Tania
And 0h9'33" 45d2'24"
Tania
And 0h20'38" 46d3'22"
Tania
And 23h26'2" 46d20'26"
Tania
And 0h0'1" 44d59'57"
Tania
Leo 10h23'7" 11d15'0"
Tania
Peg 22h42'25" 23d17'28"
Tania
Umi 16h9'24" 76d39'3"
Tania
Umi 14h16'4" 78d7'9"
Tania Cherry
Gem 6h47'25" 23d53'14"
Tania Corene Hanthorn
And 1h49'50" 37d19'32"
Tania De Voght
Lyn 8h9'41" 52d19'18"
Tania Dean
And 23h25'2" 52d1'49"
Tania Elizabeth Galindo
And 2h18'21" 45d36'27"
Tania Espada
Aql 20h5'8" 9d34'10"
Tania et Jean-François Trudeau
Cas 1h19'6" 54d35'7"
Tania Glynn's Angel Light
Car 9h53'42" -60d3'54"
Tania Gowandan
Aqr 22h24'46" -7d28'44"
Tania Joseph
Lib 14h56'36" -6d28'43"
Tania Maria Bakar
Tau 3h48'6" 7d53'3"
Tania Marie Willoz
Uma 11h10'39" 33d39'3"
Tania Michelle Lyerly
Peg 22h44'15" 29d17'26"
Tania Montalvo & Brian Kohles
Lmi 10h54'31" 24d54'9"
Tania Nicole Jerome
Lyn 7h48'56" 50d47'21"
Tania Ojos Finos Yanes
Ari 3h11'28" 30d55'18"
Tania & Roger
Cep 3h56'25" 81d0'35"
Tania Rose Morales
Vir 13h18'21" 4d30'13"
Tania Saint Amand
Umi 13h18'49" 73d31'57"
Tania Santos
Vir 12h14'58" 6d3'41"
Tania Saunders
Vel 9h58'35" -52d22'21"
Tania Vancini
Tau 5h51'56" 16d56'55"
Tania's lil' piece of Heaven
Sco 17h15'8" -34d29'24"
Tania's Shining Star of Virgo
Vir 13h4'10" 10d9'17"
Taniesha Kierstyn Harvey
Vir 15h0'57" 4d12'5"
Tanis S. Martin
Vir 11h38'14" 4d42'19"
Tanisa K. Smith
Gem 7h20'17" 19d53'44"
Tanisha Colette Moore
Cam 3h30'59" 59d32'9"
Tanisha Hanley
Aqr 21h5'42" -11d45'11"
Tanisha Jones
Cam 4h43'16" 55d13'40"
Tanisha Marie Dixon
Cas 23h51'1" 50d2'34"
Tanisha Pallerla
Aqr 22h58'27" -7d37'31"
Tanisha Palvia
Ori 6h7'44" 3d38'13"
Tanisha T. Rogers
Lib 15h5'3" -7d46'0"
Tanita Maxwell
Aqr 22h13'59" -9d15'40"
Tanith Harris
Equ 21h5'40" 5d7'13"
Taniya Nayani
Ori 5h35'26" 9d31'34"
Tanja
Ori 6h15'46" 16d35'13"
Tanja
Her 17h41'7" 15d43'24"
Tanja
Uma 9h35'2" 51d7'11"
Tanja
And 23h52'40" 43d8'18"
Tanja
Her 17h56'6" 40d7'58"
Tanja
Uma 9h20'39" 71d41'46"
Tanja
Uma 14h4'40" 56d52'38"
Tanja
Cas 23h3'30" 53d32'21"
Tanja 20
Uma 11h37'56" 29d6'13"
Tanja Alexandra
Lac 22h36'16" 38d10'37"
Tanja Antonijevic - 05.09.1971.
Cyg 19h46'23" 35d3'19"
Tanja Aufdermaur
Peg 22h56'12" 31d28'18"
Tanja Buchholtz
Vir 14h16'18" 2d16'18"
Tanja Bunjac
Tau 5h44'2" 24d30'35"
Tanja -Die Frau meines Herzens!
Uma 11h7'58" 37d43'27"
Tanja Dina
Dra 20h0'29" 72d56'56"
Tanja Driesen
Uma 11h45'43" 54d24'2"
Tanja Etzenberger
And 23h30'12" 50d30'4"
Tanja Flühmann
Umi 13h11'13" 70d37'8"
Tanja Hristovski Dekaric
Cnc 9h5'42" 31d40'14"
Tanja & Jens Rosen
Uma 9h25'41" 44d34'2"
Tanja Junge
Ori 5h55'12" 12d39'41"
Tanja Kovacevic
Lmi 10h37'40" 37d8'10"
Tanja Lambert
Uma 12h49'16" 56d55'21"
Tanja Nicole Malone
Sgr 19h8'50" -12d56'46"
TANJA & ROLAND
Cas 0h17'20" 53d39'58"
Tanja Schibig
Cas 2h50'48" 60d27'5"
Tanja Sophie Hansen
Cas 1h36'10" 61d30'22"
Tanja Stojkovic
Leo 11h32'44" 2d20'5"
Tanja und Klaus
Ori 5h11'11" 0d33'23"
Tanja Veer
Uma 10h36'39" 40d25'9"
Tanja - Vesna
Ori 5h6'9" -0d31'26"
Tanja Volmer
Cru 12h10'19" -63d43'55"
TANJA,
Uma 13h58'46" 61d22'41"
Tanja-Lea Wullschleger
Aur 6h35'6" 34d8'15"
Tan-Jam
And 23h15'44" 47d55'29"
Tanja's Engelsstern
Uma 9h5'22" 53d55'1"
Tanjena
And 23h28'10" 47d33'7"
Tanjica
Tau 4h26'25" 15d35'9"
Tank Hinkley
Uma 10h3'17" 66d59'55"
Tankersley
Per 3h11'44" 51d20'57"
Tanler Wade Schnarr 12-21-02
Her 17h51'57" 35d12'28"
Tanmaye
Leo 11h45'22" 24d27'35"
Tanna 24
Uma 9h29'34" 69d56'57"
Tanna M. Collins (PB & Jellybean)
Cap 20h53'29" -16d21'22"
Tanna Rae
And 0h42'31" 43d50'11"
Tannaz Machhi
Aql 20h2'56" -0d22'30"
Tanne
Psc 1h16'40" 28d47'10"
Tanner
Dra 20h17'5" 62d18'23"
Tanner Andrew Kalinowski
Her 17h33'59" 16d51'21"
Tanner Anthony Marzullo
Sgr 18h7'29" -25d38'17"
Tanner Atwood
Umi 16h24'29" 77d39'5"
Tanner Blaize
Psc 1h12'45" 29d14'4"
Tanner Charles
Ori 5h35'54" 3d39'42"
Tanner Colby Dueker
Ori 5h44'1" 1d57'14"
Tanner Cox
Her 16h47'49" 33d33'5"
Tanner Daniel Blaydon
Gem 6h45'56" 17d23'4"
Tanner David Dunivent
Her 18h28'28" 19d2'20"
Tanner Fulmer
Uma 13h13'30" 58d34'19"
Tanner Gregory Moore "Tan Man"
Umi 14h36'58" 75d39'3"
Tanner Heckel
Uma 11h19'4" 52d8'14"
Tanner James Odle
Lyn 8h15'21" 35d8'1"
Tanner James Wincott
Uma 10h27'34" 68d58'20"
Tanner Lee Gwinn
Cnc 8h58'21" 13d16'43"
Tanner Lucas Smith The Great
Ori 5h9'12" 2d15'57"
Tanner Nathan Tolbert
Tau 5h12'6" 26d33'1"
Tanner Patrick Ryan Werner
Sco 17h56'50" -31d14'38"
Tanner Richard Crowley
Aql 19h23'16" 4d28'41"
Tanner Ryan Reichenberg
Per 3h31'59" 47d8'20"
Tanner Sean Toohig
Ari 2h40'27" 23d57'24"
Tanner Weaver
Her 17h54'36" 23d24'40"
Tanner Weston Holladay
Psc 0h44'13" 6d20'28"
Tanner Wildon McClelland
Leo 9h23'17" 24d47'30"
Tanner Winston Davis
Lmi 10h40'30" 27d6'10"
Tanner's Great Nana-Valerie Law
Umi 16h19'36" 77d57'10"
Tanner's Smile
Lmi 10h17'49" 35d15'33"
Tanner's star
Tau 4h36'44" 24d26'5"
Tanner's Star of Wonder
Umi 15h35'27" 84d20'41"
Tannous Family Star
Leo 11h13'39" 15d25'24"
Tanny
Umi 13h6'47" 70d39'58"
Tanor and Lisa
Umi 14h46'35" 77d14'26"
Tanor Dracon Hearthfire
Sco 17h26'4" -41d22'4"
Tanqueray
Tri 2h21'44" 32d10'14"
Tansel & Juliet
Tau 5h23'49" 19d23'20"
Tansy
Cas 22h57'24" 56d12'51"
Tansyn Mychel Cook
Ori 6h20'22" 9d27'59"
Tantbury
Tri 2h16'9" 33d14'5"
Tante Christa, Klopstein
Umi 17h38'24" 80d6'53"
Tante Irène
Umi 5h17'45" 89d22'18"
Tante Vera
Gem 7h0'16" 25d58'45"
Tanto Amore
Tau 4h44'47" 27d9'17"
Tantor
Ori 5h30'47" 1d33'52"
Tantrum
Umi 16h49'8" 77d35'36"
Tantum in Somnium
Psc 0h6'53" 3d2'54"
Tanushka Orlova
Leo 10h42'14" 10d51'36"
Tanvi
Eri 3h41'2" -13d22'49"
Tanvi Vattikuti
Uma 10h27'0" 45d16'13"
Tanya
Lyr 18h29'26" 46d18'3"
Tanya
And 1h47'25" 49d27'57"
Tanya
Lyn 8h4'24" 38d44'30"
Tanya
And 2h17'45" 44d0'58"
Tanya
Ari 2h46'42" 20d45'34"
Tanya
And 0h15'33" 22d16'45"
Tanya
Psc 0h43'38" 8d42'47"
Tanya
Peg 22h44'59" 17d9'23"
Tanya
Mon 6h46'5" -0d19'31"
Tanya A. Miller
Sge 19h42'6" 17d21'53"
Tanya and Fiona Moore
Uma 10h8'32" 49d34'36"
Tanya and Lou Russo
Per 3h14'41" 51d34'55"
Tanya and Rob Forever
Uma 11h30'30" 61d4'3"
Tanya Ann Black
Leo 11h35'52" 19d49'4"
Tanya Annarilli
Cap 21h2'10" -20d53'40"
Tanya Arlene Gore
Vir 12h40'32" -9d4'25"
Tanya & Brian Rickard
Lyn 8h33'29" 33d48'35"
Tanya Bug Superstar
Ari 3h22'53" 27d3'40"
Tanya Catmull
Cas 0h6'22" 52d30'14"
Tanya Dalusio
Ari 1h52'56" 18d58'4"
Tanya Danielle Davis
Uma 9h22'47" 50d13'35"
Tanya Darlene Wolf
Ari 1h52'58" 18d55'16"
Tanya Deveau 7/29/2006
Leo 9h51'12" 28d17'50"
Tanya Edith Moore
Leo 11h1'32" 17d13'58"
Tanya Firnstahl
Cnc 8h19'7" 14d31'1"
Tanya & George Madina
Uma 10h47'48" 43d52'22"
Tanya Goldie - One In A Million
Cru 12h22'42" -62d42'17"
Tanya Hope Leach
And 0h18'23" 31d7'16"
Tanya Jennifer Blackburn
Sgr 19h22'9" -30d46'25"
Tanya Jones
Vir 12h41'57" 12d38'34"
Tanya L. Battle
Cas 23h45'1" 57d33'9"
Tanya Lee
Tau 5h7'34" 18d24'24"
Tanya Lee
And 0h12'49" 39d46'24"
Tanya Lee Maddaloni
Psc 23h50'42" -3d8'24"
Tanya Louise
Cru 12h45'43" -56d44'20"
Tanya Louise
And 23h54'16" 35d30'56"
Tanya Louise
Leo 9h45'48" 28d0'0"
Tanya Loyola
Lep 5h21'56" -22d14'24"
Tanya Lynn Dryja
Dra 19h21'28" 67d37'35"
Tanya Lynn Holtzman
Lib 15h47'2" -19d55'16"
Tanya M. Himmelburger
Cap 20h18'52" -9d45'33"
Tanya Maier
Ori 6h19'45" 17d56'12"
Tanya Marie
Vir 13h41'11" -10d49'29"
Tanya Marie
Cas 1h25'16" 64d40'51"
Tanya Marie Dwyer
Uma 10h41'55" 69d3'36"
Tanya Marie Eaton
Leo 10h26'2" 14d9'25"
Tanya Marie Joan Savage
And 1h12'18" 45d54'38"
Tanya Marie Shanks
Uma 14h2'7" 61d2'28"
Tanya Marie Windle
Aqr 22h56'32" -10d25'6"
Tanya Melvin-Holiday
Tau 5h54'43" 24d11'3"
Tanya Meyer
Tau 5h58'23" 24d47'26"
Tanya & Michael Schweitzer
Cyg 20h29'41" 53d3'12"
Tanya Michelle Hutchinson
And 1h9'49" 46d47'17"
Tanya Morgan
Uma 11h26'33" 35d35'52"
Tanya Morgenshteyn
Ari 3h9'47" 29d10'58"
Tanya Moric
And 0h39'24" 26d12'50"
Tanya Moskowsky Pettus
Cas 23h22'43" 55d1'21"
Tanya Muñoz
Cap 21h41'17" -15d9'12"
TANYA MUNSON
Tau 5h52'51" 28d22'30"
Tanya Nicole
Ari 1h50'16" 18d29'20"
Tanya Pincivero
Uma 11h21'42" 29d4'42"
Tanya Portalla-You Are Our Star
And 23h31'9" 47d55'8"
Tanya Priscilla Smith
Gem 7h25'30" 33d53'35"
Tanya R. Hobbs
Uma 9h5'13" 57d56'33"
Tanya R Jones
Sco 17h27'37" -36d46'23"
Tanya R. Whitlock
Gem 6h30'49" 17d28'8"
Tanya Rae Hemington
Ori 5h57'13" 10d24'58"
Tanya Renee
Lib 15h38'18" -18d46'43"
Tanya Renee King
Tau 3h37'37" 19d12'46"
Tanya Rochelle Brown
Cap 20h29'35" -18d5'13"
Tanya Roque
Ari 3h26'26" 20d12'31"
Tanya Sabio
Crb 16h6'55" 35d4'43"
Tanya & Sash Forever
Uma 11h37'52" 62d7'7"
Tanya Shariff
And 23h19'7" 44d14'18"
Tanya Smets
Gem 7h25'35" 29d25'58"
Tanya Superstar Five-0
Cas 0h13'31" 55d37'35"
Tanya Tadeo
Lep 5h4'35" -23d29'32"
Tanya & Terry
Lib 15h15'22" -27d18'59"
Tanya Torrellas
Cam 3h21'44" 57d11'22"
Tanya Tricia Magney
Vir 13h42'36" 0d28'6"
Tanya Walters
Cas 0h6'11" 56d54'37"
Tanya Washnik
Vir 13h32'24" -16d26'59"
Tanya Wells
And 0h9'1" 39d57'2"
Tanya Williams
Cyg 20h17'36" 36d38'2"
Tanya Wulff
Leo 9h44'5" 7d36'57"
Tanya, my Guiding Star
Sco 16h50'14" -27d30'51"
Tanya-Dawn
Psc 23h59'46" 9d39'55"
Tanya's Mooray Moon Star
And 23h24'1" 39d18'43"
Tanya's Smile
Ori 5h38'22" 6d22'10"
Tanya's Star
Lib 14h31'15" -8d56'28"
Tanya's Sunshine
Lyn 8h12'19" 41d18'57"
TANYELI
Aur 5h14'49" 29d19'40"
Tanzania
Uma 8h43'41" 49d5'51"
Tänzer, Brigitte
Uma 9h10'30" 55d20'50"
Tao Xu
Per 3h48'57" 42d12'5"
taoja
Uma 10h18'23" 48d17'41"
Tapolcsányi Krisztina
Vir 13h1'19" -22d12'24"
Tar Szilvia
Uma 10h19'25" 63d34'50"
Tara
Uma 10h16'50" 63d12'49"
Tara
Cas 0h44'58" 61d30'57"
Tara
Cyg 20h9'34" 53d44'43"
Tara
Sco 16h15'33" -17d8'38"
TAR-A
Lib 14h42'14" -11d35'47"
Tara
Aqr 23h53'25" -3d46'54"
Tara
Aqr 22h28'36" -7d6'11"
Tara
Lib 14h50'25" -1d20'47"
Tara
Vir 14h14'38" -0d41'7"
TARA
Umi 16h36'15" 82d39'18"
Tara
Apu 14h25'35" -74d0'56"
Tara
Psc 1h6'13" 4d16'37"
Tara
Pho 1h41'17" -43d40'40"
Tara
Cam 3h24'2" 56d44'29"
Tara
And 1h28'52" 46d0'31"
TARA
And 2h29'15" 43d59'15"
Tara
And 2h0'23" 42d22'21"
Tara
And 1h40'47" 42d39'33"
Tara
Lyr 18h56'19" 32d47'48"
Tara
And 0h35'39" 36d37'34"
Tara
Crb 15h32'44" 28d56'40"
Tara
Mon 6h39'59" 1d6'1"
TARA
Ori 5h20'46" 0d16'26"
Tara
Ori 5h11'11" 7d38'11"
Tara
Psc 0h18'44" 8d40'45"
Tara
Tau 4h33'5" 21d25'1"
Tara
Leo 11h6'32" 9d17'17"
Tara 221
Umi 12h38'11" 86d30'25"
Tara A. Brady
Tau 5h47'3" 27d27'13"
Tara Akasha Tajiki
Mon 6h24'0" 9d1'21"
Tara and Chris
Leo 9h30'51" 28d13'31"
Tara and Fernando's wish come true!
Cyg 21h14'42" 33d54'31"
Tara and Henry Payne
Uma 11h46'37" 36d31'5"
Tara and Jason
Cyg 21h16'39" 47d3'15"
Tara And Jason
Psc 1h17'57" 28d10'1"
Tara and John
Tau 3h43'42" 12d1'0"
Tara and Joseph
Uma 10h14'46" 50d7'2"
Tara Andrea
Aqr 22h20'55" -22d39'15"
Tara Angel
Crb 15h17'51" 30d54'32"
Tara Angelina Pauley
Cnc 8h54'3" 25d13'37"
Tara Ann Beutler
Umi 14h18'6" 76d7'18"
Tara Ann Luby
Ari 2h27'27" 26d41'37"
Tara Anna Bradshaw
And 23h19'8" 48d33'48"
Tara Ashley Nordquist
Cas 2h0'6" 62d16'13"
Tara Ashley O'Sheal
Umi 15h8'56" 71d32'2"
Tara B. Fagan's Everlasting Star
Gem 7h12'23" 16d49'15"
Tara Bear Manke
Uma 11h57'18" 33d30'55"
Tara Bella
Cnc 8h38'7" 24d50'44"
Tara Bella L. Webb
Cas 1h48'54" 55d2'40"
Tara Bergum
Cas 23h4'57" 56d5'40"
Tara Beth
Sco 15h55'17" -21d1'37"
Tara Bethany Kelly
Vir 14h17'32" 0d11'35"
Tara Bien-Aime Turnbull
Cam 5h13'59" 68d14'32"
Tara Black
Psc 1h22'51" 25d36'1"
Tara Blackshaw Happy 21st 13/04/05
Col 6h2'36" -42d3'55"
Tara C
Cnc 8h45'46" 29d58'38"
Tara Camille
Lmi 10h33'59" 34d1'48"
Tara Carl Star
Lib 15h44'43" -8d47'58"
Tara Catherine Adams
Vir 12h43'33" 3d46'9"
Tara Catherine Shannon
Lib 14h49'24" -3d58'18"
Tara Cathleen Rose McCoy Walchesky
Lib 15h2'14" -2d27'55"
Tara Centauri
Uma 13h9'52" 55d0'42"
Tara Cheri Pullen
Leo 10h19'27" 26d1'24"
Tara Colleen
And 2h20'21" 49d9'20"
Tara Connolly
Uma 13h49'0" 53d54'42"
Tara Consetta
And 1h26'31" 46d51'49"
Tara Cross
Cas 0h0'17" 53d55'26"
Tara Danielle
Gem 7h23'0" 22d47'12"
Tara Dawn
Uma 8h25'46" 64d13'48"
Tara Dawn Ahern
Gem 6h56'57" 18d27'44"
Tara Dawn Bell
Tau 3h38'33" 26d43'7"
Tara Dean
Vir 13h16'57" -12d43'55"
Tara Devine
And 2h18'33" 45d1'55"
Tara Donovan
Lib 15h5'10" -1d36'16"
Tara E. Connors
Lyn 8h15'12" 51d29'53"
Tara Eagan
Leo 10h53'56" 9d52'49"
Tara Elizabeth Enck
Aqr 22h35'58" -2d4'41"
Tara Elizabeth Madden
Lib 15h43'6" -11d42'30"
Tara Ella
Del 20h50'10" 12d42'21"
Tara Fickes
Cyg 20h22'40" 34d13'49"
Tara Firma
And 1h31'58" 41d36'46"
Tara Fournier
Gem 7h7'7" 34d32'0"
Tara Froese
And 0h24'12" 38d0'10"
Tara Gene Brady
Cas 1h11'19" 58d14'45"
Tara Giardino
Cas 2h33'27" 66d54'8"
Tara Gladstone
Crb 16h23'19" 39d29'6"
Tara Grant
Sco 16h12'40" -10d8'58"
Tara Grievo
Lib 15h5'22" -2d2'59"
Tara Hammond
Ari 2h20'29" 14d3'49"
TARA HANNAH
Vir 13h50'22" -3d13'7"
Tara Hegdahl
Sco 16h15'2" -10d16'12"
Tara Herald
Ari 3h1'20" 19d23'17"
Tara Higgins
Cas 23h13'30" 55d30'26"
Tara Honey - Bunny
And 2h34'24" 44d32'43"
Tara Jade
And 0h22'4" 45d47'33"
Tara Jane McLean
Lib 15h39'27" -29d3'26"
Tara Jayn
Aqr 22h0'40" -2d41'12"
Tara Jennifer Kreutzer
Sgr 18h14'44" -20d7'42"
Tara Jill Ross
Leo 11h48'19" 26d6'22"
Tara Jo
And 0h38'37" 41d53'54"
Tara Jo
And 0h40'27" 35d41'23"
Tara K
And 2h25'52" 50d15'58"
Tara Karina
Leo 11h13'22" 20d17'38"
Tara Kathleen
Tau 4h21'37" 26d51'34"
Tara Kay Rabb
Lyr 18h48'16" 37d7'39"
Tara Kay Rogers
Lyn 7h4'51" 56d36'0"
Tara King
Cep 4h27'12" 80d26'34"
Tara Kirchner
Dra 19h48'58" 64d54'50"
Tara L Abeyta (Fragle Rock)
Cyg 19h37'25" 35d25'37"
Tara L. Twine
And 2h17'17" 46d9'3"
TaRa la EsTreLLa
Uma 11h17'3" 66d7'0"
Tara Lackie
Uma 10h59'42" 49d22'29"
Tara LaNae Cole
Gem 6h31'27" 21d19'51"
Tara Large
Umi 14h0'37" 74d21'39"
Tara Lea
Uma 8h47'4" 67d54'16"
Tara Leah Baker
And 0h44'10" 42d38'33"
Tara Lee
And 0h53'6" 40d23'7"
Tara Lee
And 23h52'38" 45d36'15"

Tara Lee
Leo 9h45'24" 29d15'19"
Tara Lee Coker
Cnc 9h7'52" 21d14'16"
Tara Lee Ovitt
Lyn 7h41'25" 53d26'26"
Tara Lee Woodside
Vir 14h47'10" 2d24'36"
Tara Leigh
Sgr 18h10'10" -22d13'6"
Tara Leigh Mathis
Cnc 8h19'2" 23d28'30"
Tara Leigh Richards
Crv 12h27'40" -15d46'4"
Tara Lo
Umi 14h11'58" 78d13'3"
tara l'un amore di i
Cap 21h27'59" -9d51'16"
Tara Lyn Heffernon
Psc 0h47'8" 9d28'9"
Tara Lynn
Vir 13h31'33" 3d49'19"
Tara Lynn
Lmi 9h26'50" 35d4'18"
Tara Lynn
Sco 16h18'3" -13d20'9"
Tara Lynn
Ori 5h19'3" -7d34'58"
Tara Lynn
Uma 8h12'29" 68d27'38"
Tara Lynn and Gregory Robert
Cnc 8h29'11" 11d23'2"
Tara Lynn Bugiel 11-5-1989
Dra 17h52'4" 51d7'7"
Tara Lynn Gustafson
And 2h25'2" 50d26'51"
Tara Lynn Johns
Sco 17h52'10" -42d10'55"
Tara Lynn Kinsella
Tau 3h48'43" 22d21'31"
Tara Lynn My Ray Ray
Leo 10h20'33" 13d17'19"
Tara Lynn Schaust
Lyn 6h28'57" 58d12'0"
Tara Lynn Taylor
Leo 11h31'23" 1d52'22"
Tara Lynne
Dra 18h44'54" 73d16'52"
Tara M. Maier
Cas 23h22'52" 57d35'21"
Tara M. Panzo
Lmi 9h28'39" 33d56'23"
Tara Mannar
And 23h39'49" 46d0'16"
Tara Maria
Cas 1h13'35" 52d57'11"
Tara Maria Gallo
Lyn 6h55'21" 52d41'59"
Tara Marie
Sco 16h31'48" -29d34'18"
Tara Marie
And 0h23'9" 46d36'39"
Tara Marie Allen
Ari 2h13'40" 24d44'55"
Tara Marie Juergens
Cyg 19h48'51" 37d49'25"
Tara Marie Lachapelle
Leo 11h5'19" 16d29'34"
Tara Marie Lusk
Uma 10h57'41" 62d42'4"
Tara Marie Lyons
Vir 12h42'56" 3d44'31"
Tara Marie Miller
Her 17h49'17" 42d2'38"
Tara Marie Mulligan
Tri 2h36'3" 35d1'12"
Tara Marie Woosker Urban
Ari 1h59'42" 14d17'39"
Tara McHugh
Cet 1h29'49" -6d59'8"
Tara McLauchlan
And 2h34'15" 39d37'21"
Tara Meus Divum Incendia
Ari 3h13'13" 15d41'26"
Tara Michelle
Peg 21h57'8" 13d35'57"
Tara Michelle
Tau 4h6'4" 11d33'41"
Tara Michelle
Cas 0h40'32" 47d17'15"
Tara Michelle Newinsky
Sgr 19h6'24" -18d21'56"
Tara Miller
Lib 15h5'54" -5d36'32"
Tara Mirabile
Cam 3h16'9" 58d55'36"
Tara Moe
Sgr 18h20'25" -21d54'2"
Tara Mooney
Leo 9h56'20" 9d52'50"
Tara my Pancake
Leo 9h39'9" 26d29'35"
Tara Mykl McCollough
Uma 11h33'29" 45d41'37"
Tara & Natalie
Cyg 21h31'26" 37d15'9"
Tara Nicole
Lmi 9h45'39" 36d4'27"
Tara Nicole 12031981
And 23h2'37" 48d16'15"
Tara Nicole Costagliola
Lyn 7h5'24" 47d39'25"
Tara Nicole Meeks
Sgr 18h50'9" -21d8'9"
Tara Nicole Summer
And 0h11'14" 30d55'9"
Tara Nicole Todd
Leo 10h51'52" 13d42'9"
Tara O'Callahan
Lib 15h0'22" -17d1'30"
Tara Ortholf
Lyr 18h16'12" 30d25'27"
Tara P
Tau 4h13'59" 22d35'22"
Tara Pandya
Ori 5h33'13" 3d39'35"
Tara Plecinski
Leo 10h24'0" 13d11'14"
Tara R. Sandora
Gem 7h29'28" 20d3'17"
Tara Reane
Uma 11h50'20" 55d27'33"
Tara Reilly
Cam 3h29'1" 56d34'43"
Tara Renae
Sgr 19h8'10" -13d57'22"
Tara Renee
Uma 9h37'29" 52d9'3"
Tara Renée Towers Yaden
Uma 10h8'5" 63d25'57"
Tara Rich
Ari 2h14'11" 23d41'7"
Tara Rogers
And 1h8'12" 44d41'0"
Tara Roisin Otto
Cas 0h20'22" 52d39'37"
Tara Romanella
Psc 0h54'2" 15d18'13"
Tara Rose
Ori 6h22'4" 14d8'1"
Tara Rush
Her 16h8'32" 48d28'49"
Tara Ryan
Leo 11h27'3" -2d54'38"
Tara & Ryan Cellarius
Uma 8h18'6" 68d29'56"
Tara S. Fermaint
Cma 6h27'0" -16d3'45"
Tara Schellhammer
Uma 11h18'54" 62d36'56"
Tara Schultz
And 1h14'22" 35d51'43"
Tara Sharkey
And 23h23'43" 48d44'45"
Tara Shea Shaw Fletcher
Gem 7h51'59" 29d26'52"
Tara T.
Cap 21h39'30" -14d35'19"
Tara Tashjian
Psc 1h1'39" 11d34'37"
Tara Todd
Sco 17h45'41" -36d38'26"
Tara + T-rav = 4 ever
Crb 16h11'4" 31d31'51"
Tara Valentine Clark
Uma 8h53'9" 65d5'51"
Tara Vanderwaal
Cas 0h41'38" 66d4'36"
Tara Vaughn
Aqr 23h33'9" -8d39'6"
Tara Vija Mackay
Sgr 18h17'3" -26d5'17"
Tara Wasikowski
Vir 13h20'54" -17d9'2"
Tara Wayne Melanie Christopher
Tau 4h28'7" 17d23'6"
Tara Wilkin
Psc 1h21'50" 32d11'59"
Tara-Bair Sparkles
Gem 6h31'32" 24d38'28"
Tarabick
And 1h0'16" 38d43'15"
Tara-Carol Sawa
Cru 12h17'2" -55d46'12"
Taracorn
Lib 15h23'2" -26d15'34"
Taradise
Ari 2h5'55" 18d19'20"
TaraFina
Oph 16h55'0" -20d31'26"
Taragh Melwani
Her 18h2'4" 37d24'36"
Tarah Michelle
Leo 9h48'19" 22d55'5"
Tarah Muller
Gem 7h36'25" 22d22'26"
Tarah & Ravi
Per 3h45'47" 41d4'19"
Tarah Santoro
Ori 6h16'7" 14d56'39"
Tarah Taggart
Sgr 17h56'7" -28d29'58"
TarahCortneyMertz10000Kisses
Gem 6h48'20" 33d36'50"
Tarah-Fee
Tau 5h58'40" 24d55'33"
Tarah's Star (The Love of My Life)
Cyg 20h25'10" 35d5'4"
Tara-Joe
Psc 1h0'32" 26d23'27"
Tarak Gandhi
Uma 10h11'26" 50d28'54"
Taraliam
Lib 15h20'26" -4d8'51"
Tara-Lyn
Cam 6h7'29" 63d34'49"
TaraLynn81
Aqr 22h10'0" -2d39'26"
Tara-Lynne Law
Cas 2h35'37" 75d40'47"
Taramattie Ramphal
Lyn 7h41'23" 38d34'6"
Taran Agarwal
Her 18h6'52" 20d47'36"
Taran Dee Bowman
Mon 6h48'47" -0d5'39"
Taraneh Eshaghi
Leo 11h22'34" 22d41'56"
TaraNova
Gem 7h41'18" 34d13'55"
Tara's Beauty
Cap 20h35'54" -13d51'32"
Tara's Eyes 1974
Gem 6h30'51" 25d52'5"
Tara's Guiding Star
Sgr 18h0'36" -24d13'0"
Tara's Shining Star
And 2h5'27" 41d56'23"
Tara's Star
Vir 11h37'33" 4d51'30"
Tara's Starshine
Aqr 22h42'20" -11d56'27"
Tara's Tribute
Leo 10h19'19" 26d13'42"
Tara's Twins
Umi 15h12'0" 73d54'30"
Tarasa BS
Sgr 18h40'47" -28d58'32"
TaraStar1023
Mon 7h13'52" -9d41'14"
tara-sugar
Psc 1h18'0" 25d58'19"
Tarcisio Salvatore Ionta
Umi 14h43'33" 73d35'44"
Tare-Da Way-Gantt
Her 17h2'29" 32d30'20"
Tareia
Lyn 8h8'32" 48d7'54"
Tarek Glenn
Uma 11h52'12" 30d29'11"
Tarek McCarthy
Her 17h19'30" 18d13'54"
Tarek Rahal
Ori 5h27'35" 13d19'53"
Tarek w Tuma
Equ 21h9'46" 11d53'37"
Taren
Lyn 7h49'25" 56d45'22"
Targa 9/11
Peg 21h40'1" 18d39'20"
Target Fashion Sdn Bhd
Ori 5h37'33" 3d50'13"
Tari
Her 17h16'38" 27d28'30"
Tari
Per 3h16'7" 46d14'5"
Tari K. Perri
And 23h48'15" 38d44'44"
Tari Lynn Toppe Friedrich
Psc 0h40'36" 17d44'23"
TaRie
Dra 18h33'59" 51d17'15"
Tarik Toukan
Cnc 8h36'2" 18d42'16"
Tarika Lovegarden
Ori 5h43'54" -2d43'35"
Tarikua Gebereselassie
Vir 12h38'54" -3d13'25"
Tarin
Gem 7h20'3" 14d32'42"
Tarin Lee Balchitis
Cas 0h38'38" 66d36'32"
Tarin "The Star of Heaven"
Uma 9h15'7" 49d31'13"
Tarina
Psc 1h2'1" 26d7'19"
Tariq
Lmi 10h14'21" 31d52'12"
Tarique
Uma 8h49'40" 69d31'57"
Tarique Devon Wilson
Aur 5h45'24" 38d49'49"
Tarja
Gem 7h10'53" 15d22'3"
Tarkas DeFabrizio
Cma 7h14'35" -30d25'12"
TARKSHYA
Gem 6h45'56" 20d57'19"
Tarla
Vir 14h48'1" 6d7'20"
Tarlà Pamela
Peg 21h24'55" 21d27'4"
Tarn (Thomas Marko)
Ari 3h4'8" 14d7'2"
Tarnedvroera
Umi 13h14'9" 70d32'46"
Taro & Hiroko
Psc 0h59'40" 20d38'55"
Taro Morris
Her 18h2'29" 25d1'11"
Tarodi, Andras
Uma 12h5'35" 37d4'23"
Taron Piloyan
Sgr 18h10'22" -26d20'4"
Taro-Stern
Uma 11h50'39" 40d3'20"
Tarren Amalia
Tau 5h50'13" 15d46'46"
Tarri & Denys
Ori 6h12'12" 18d33'56"
Tarricone Family Star
Ori 6h11'32" 15d37'32"
Tarry1030Woowie
Cas 2h21'38" 66d26'39"
Taryl
Lyn 7h47'48" 38d43'58"
Taryn
And 23h14'13" 47d49'4"
Taryn
Leo 11h22'30" 13d37'18"
Taryn
Ari 2h31'38" 26d2'46"
Taryn
Uma 11h50'15" 55d18'59"
Taryn
Leo 11h17'10" -3d50'43"
Taryn
Sgr 17h52'41" -25d8'39"
Taryn A Rosales
Uma 10h45'1" 60d46'25"
Taryn Alexis Beckett
And 23h13'57" 47d33'21"
Taryn Alyssa Saijo
Uma 10h22'1" 69d8'22"
Taryn and Brandin Forever
Pho 23h30'16" -48d9'21"
Taryn and Jesse
Gem 7h36'5" 28d46'10"
Taryn & Ashley - 5th May 2007
Cru 12h39'9" -59d37'7"
Taryn B. Fisher
And 22h58'55" 52d3'34"
Taryn Brooke Lucks
Cam 3h59'32" 65d29'50"
Taryn Coe
And 0h12'31" 25d20'0"
Taryn Dawn Henderson
Per 3h38'34" 46d51'10"
Taryn Elaine
And 23h25'27" 49d16'11"
Taryn Elisabeth
And 0h46'17" 43d17'23"
Taryn Jeanne Komma
Tau 4h36'1" 26d28'33"
Taryn & Joseph Sharp
Cyg 19h52'40" 30d38'57"
Taryn L. Yonkin
And 1h32'35" 43d58'38"
Taryn Mantagas
Psc 0h52'42" 29d9'46"
Taryn Marchelle
Cnc 8h45'18" 12d41'31"
Taryn Maria Vincent
And 0h39'26" 38d20'44"
"Taryn My Love"
Gem 7h18'14" 19d59'12"
Taryn Nicole Steed
And 2h0'41" 37d49'39"
Taryn Perry
Tau 5h6'4" 20d39'57"
Taryn Reed
Sgr 19h21'51" -14d40'0"
Taryn Riley Young
Peg 23h24'0" 17d32'40"
Taryn "Serendipity" - Love Mum x
Cas 0h1'55" 59d55'13"
Taryn Shea Foley
And 23h4'35" 35d31'15"
Taryn Swart
And 1h35'46" 47d51'25"
Tarynn Nicole
Cas 1h25'53" 62d32'47"
Taryn's Little Star TLM3
Sco 16h59'53" -39d38'3"
Tarzan
Psc 22h55'11" 0d16'31"
Tas
Lyn 7h3'7" 44d41'8"
TAS Star
Pho 0h21'10" -51d43'35"
Tasadie Olivia Hall
Lyn 8h33'17" 35d21'16"
Tasaiya Bounton Xaypanya
Cyg 19h40'17" 29d34'29"
tasama.I.L.U.3 X infinity +1
Cnc 8h30'20" 20d10'48"
Tascha Nytes
Sco 16h7'14" -26d9'10"
Tash Maio
And 23h35'53" 49d1'35"
Tasha
Crb 16h15'51" 37d32'21"
Tasha
And 1h31'17" 34d57'45"
Tasha
And 2h9'6" 39d59'10"
Tasha
And 1h42'26" 41d30'38"
Tasha
Vir 13h41'55" 1d53'40"
Tasha
Ari 2h35'8" 22d24'58"
Tasha
Sco 17h3'19" -31d26'18"
Tasha
Aqr 21h5'31" -13d51'22"
Tasha
Sgr 18h12'48" -18d58'30"
Tasha Anne Marie Streib
Uma 8h25'20" 71d26'18"
Tasha Clark
Aqr 22h38'34" 2d13'52"
Tasha Cunnane
Cnc 9h8'56" 9d0'45"
Tasha Flowers
Crb 16h18'31" 38d58'55"
Tasha Habibi Jackson
Aqr 21h45'27" -1d11'40"
Tasha K. Rowley
Uma 9h40'31" 63d23'14"
Tasha Kalanik
Uma 10h55'31" 47d25'59"
Tasha Leigh Ludeman
Lib 15h17'13" -26d3'30"
Tasha Marie
And 1h31'15" 36d13'33"
Tasha Miller
Ari 2h31'33" 25d47'13"
Tasha - MojaJedinaZvezda
Cas 1h55'33" 64d43'6"
Tasha & Peter Bergman
Uma 9h42'49" 54d24'13"
Tasha Rene Kuhn
Tau 4h56'33" 18d51'22"
Tasha Renee
Leo 9h53'54" 24d56'17"
Tasha Stover
Aql 18h51'33" -0d54'20"
Tasha Sue
Uma 11h28'2" 60d19'47"
Tasha Turner
Del 20h49'45" 9d46'24"
Tasha Voss
Cma 7h25'3" -18d33'16"
Tasha Waller
Peg 22h36'44" 11d46'46"
Tasha XXI
Cas 0h52'7" 47d27'50"
Tasha-Kannuuk-Smokey
Cma 6h40'30" -13d59'6"
Tashana Rose
Leo 11h21'7" 11d30'4"
Tashanya Amelié Duncan
Psc 0h55'4" 25d10'54"
Tashaopolis
Uma 10h27'54" 42d30'41"
Tasha's Dreams
Sgr 19h40'13" -15d38'44"
Tasha's Eye
Uma 11h46'4" 54d4'34"
TASHA'S HEART
Sgr 19h6'18" -36d23'38"
Tasha's Star
And 23h16'35" 42d56'56"
Tashi
Uma 11h3'9" 67d40'24"
Tashia
Psc 1h3'48" 3d57'55"
Tashina
Cnc 9h18'46" 26d45'17"
Tashina "My Cloak"
Psc 0h40'36" 5d12'2"
Tashmima Qayyum (Judy)
Lib 14h49'47" -4d18'14"
Tashnia
And 23h14'11" 52d33'20"
Tashua Biber
Cnc 8h40'3" 32d44'49"
Tasia
Lyr 19h7'48" 45d8'33"
Tasia Johnson
Sgr 18h44'21" -16d54'17"
Tasia Leigh
Cap 21h37'32" -12d22'7"
Tasia Maria
Cyg 19h23'51" 47d6'34"
Tasia Nicole Shelley
Uma 12h25'44" 62d28'19"
Tasin Annachiara
Cas 23h4'5" 55d59'14"
Tasiyagnunpa Win
Cam 4h27'20" 66d16'26"
Tasja Sachs
Aqr 23h45'10" -24d16'18"
Tasker Daniel Goldston
Peg 22h5'5" 19d34'48"
Taskien
Sgr 17h53'0" -29d53'14"
Tasmin Eloise Logan
Cru 12h54'3" -58d53'1"
Tasmin Maree Star
Col 5h35'51" -39d0'38"
Tassell K (Victor)
Per 3h25'0" 36d4'58"
Tassie
Lmi 10h14'5" 36d22'46"
Tassilo
Ori 6h1'13" 11d2'25"
TASSILO dein Weg ist mein Weg, Tina
Ori 4h53'38" 11d3'44"
Tassos Boulmetis
Uma 10h28'27" 49d25'54"
Tassy's Twinkle
Sco 17h18'12" -43d24'9"
Tat - 1
Aql 20h13'55" -0d6'33"
Tata
Uma 14h26'54" 58d11'49"
Tata
Peg 0h8'15" 19d53'0"
Tata
Gem 7h25'54" 17d21'32"
"Tata Cinzia"
Lyr 18h53'17" 27d35'46"
Tata Joe-Joseph Macias Martinez
Vir 13h44'57" -5d18'45"
Tata John
Sco 16h47'43" -26d40'56"
Tatana Sterba
Lib 15h24'34" -29d15'18"
TaTanisha Katisha Kelly
Leo 10h11'10" 26d5'54"
Tatanka
Lac 22h52'4" 53d35'45"
Tate
Crb 16h17'8" 29d28'55"
Tate
Lyn 8h11'20" 34d50'35"
Tate Agnew
Peg 21h30'51" 18d44'10"
Tate Alexander Goldberg
Umi 16h19'10" 77d8'47"
Tate Alexander Rathwell
Umi 14h4'30" 74d13'7"
Tate Andrew Degener
Her 18h28'0" 23d38'47"
Tate Matthew Sennett
Aql 20h8'33" 3d33'3"
Tate Michael Hyde
Her 16h25'46" 43d26'51"
Tater McVey
Per 3h22'12" 43d12'26"
Tater's Lone Star
Lmi 10h51'7" 32d49'36"
Tate's Angel
Sco 16h14'40" -13d57'48"
Tatia und Robert Schirrmacher
Lyr 19h6'1" 27d38'2"
Tatian Odesho...*Khayet Chrissy*
Uma 10h50'48" 44d51'8"
Tatiana
Lyn 7h40'36" 43d9'22"
Tatiana
Per 3h31'53" 49d9'39"
Tatiana
Lyn 6h59'8" 51d47'14"
Tatiana
Del 20h24'16" 16d50'38"
Tatiana
Aqr 21h15'18" -12d7'33"
Tatiana
Cap 21h51'44" -8d36'4"
Tatiana
Sgr 18h46'4" -20d59'30"
Tatiana
Sco 16h15'12" -17d30'37"
Tatiana
Cas 1h25'10" 63d47'8"
Tatiana 444
Umi 14h59'30" 78d19'14"
Tatiana Alissa Ingraham
Uma 8h57'34" 50d29'57"
Tatiana "Anais" Gomez
Aqr 23h37'53" -19d33'15"
Tatiana Berejuk
Lyn 7h26'54" 53d3'47"
Tatiana Ceasar
Crb 16h11'39" 37d50'27"
Tatiana DeAlmeida
Cnc 8h24'22" 13d48'32"
Tatiana Echeverry
Ari 2h55'23" 27d20'15"
Tatiana Eichkorn
Per 3h0'13" 43d11'0"
Tatiana et Alexandre
Ori 5h21'38" 6d47'31"
Tatiana Fairy Queen
Cap 20h32'50" -20d51'13"
Tatiana Giraldo
Cas 23h48'46" 61d45'52"
Tatiana Grace of Cassiopeia
Cas 0h17'1" 56d30'26"
Tatiana Guadalupe Vanegas
Vir 14h8'35" -5d19'13"
Tatiana & Helly
Crb 16h3'52" 33d14'18"
Tatiana Kalchuk
Lib 15h3'48" -24d57'17"
Tatiana Kushnarenko
Cnc 8h59'24" 32d15'48"
Tatiana Lanae Silvia
And 2h38'39" 40d35'35"
Tatiana Loveday Capildeo's Star
Umi 13h45'20" 78d43'59"
Tatiana Luca
Uma 9h25'36" 42d51'47"
tatiana m bojczuk
Lyn 7h38'24" 46d28'39"
Tatiana Michelle
Sgr 17h53'2" -29d6'36"
Tatiana Miranda
Vir 13h57'2" -1d47'36"
Tatiana "My Lil Luv"
Cas 1h0'19" 63d20'55"
Tatiana Nicole Frost
Cap 20h23'31" -12d52'22"
Tatiana Ridley Matthew Hogan
Cas 2h15'3" 64d21'1"
Tatiana Rose
Sgr 18h30'4" -16d25'9"
Tatiana Satos
Aqr 23h50'44" -23d55'45"
Tatiana Serova
Psc 1h7'40" 33d9'8"
Tatiana Taraday & Boris Kopylov
Psc 1h11'49" 28d26'8"
Tatiana Volfson
Cam 5h49'36" 73d15'28"
Tatiana Witt-Roper
Cas 2h19'2" 66d45'47"
Tatiana, Fairy Queen
Cas 1h4'3" 63d6'51"
Tatiana's Star
Leo 9h36'39" 12d38'44"
Tatillana Obskety
And 0h29'55" 43d24'22"
Tatin
Cnv 12h48'1" 46d51'58"
Tatina
Uma 10h32'12" 59d9'31"
TATIYANA-NIKIVA GREEN
And 0h21'39" 46d36'0"
Tatiyania Harris-Tati
And 0h27'34" 37d48'35"
Tatjana
Leo 9h53'12" 29d44'22"
Tatjana 08.11.1971
Uma 10h35'37" 39d25'16"
Tatjana & Alexander
Uma 11h9'42" 71d24'50"
Tatjana Bölsterli
Cep 21h48'44" 72d25'27"
Tatjana Ristic
Cas 1h16'34" 53d59'59"
Tatjana & Roger
Ori 6h18'16" 11d1'27"
TATMAN
Cnc 8h24'11" 19d24'10"
Tatman 2001
Vir 12h31'42" 11d37'7"
Tatoe Princess
Lyn 6h57'7" 51d17'38"
Tatooine
Her 17h21'46" 26d51'11"
Tator Baby
Ari 2h29'44" 25d41'3"
Tatsgirl
Cnc 8h4'34" 20d37'56"
Tatsiana Bandarchyk
Cyg 19h53'38" 38d48'27"
Tatsu Todoroki
Aqr 20h52'30" -5d19'19"
Tattooed Mother
Ori 6h13'59" 16d15'16"
Tatty Levenson
Aql 19h50'8" 4d20'47"
Tatum
Vir 11h47'53" 10d16'19"
Tatum
Cyg 19h58'36" 40d3'14"
Tatum Aubrey
Gem 6h47'40" 18d14'40"
Tatum Candace Liverio
And 0h37'2" 30d7'27"
Tatum Elizabeth Gertrude Krelle
And 1h44'12" 38d27'4"
Tatum Emily Powell
Cnc 8h51'8" 11d48'9"
Tatum Fiona Heinle
And 23h40'22" 39d7'42"
Tatum Kathleen Brumble
Ari 3h5'31" 14d46'1"
Tatum Lee Fuller
Lmi 10h49'13" 27d43'13"
Tatum Marie Faust
Sgr 18h54'46" -31d36'4"
Tatum Nicole Dartsch
Cyg 21h39'42" 32d44'28"
Tatum Nicole Edgar
Uma 10h44'55" 68d43'0"
Tatum Rae Utterback
Crb 15h59'8" 39d4'59"
Tatum Williams
Leo 11h28'56" 0d46'46"
Tatum-Leigh
Pyx 8h38'49" -34d42'44"
Tatyana
Lyn 7h21'11" 50d4'48"
Tatyana Belousova Dass
Psc 0h52'41" 27d38'31"
Tatyana Belovodskaya
And 2h20'14" 45d55'23"
Tat'yana Dogileva
Psc 0h25'17" 10d46'42"
Tatyana Isabel Quintanilla
And 0h34'47" 39d9'2"
Tatyana Korolyova
Lib 14h42'35" -11d15'44"
Tatyana Leonov
Uma 10h17'14" 50d49'49"
Tatyana Olyvia Steadman
Cnc 8h45'29" 16d3'12"
Tatyana Orlik
Lib 14h58'54" -16d6'30"
Tatyana Sleiman,"TAT"
And 2h22'5" 45d14'7"
Tatyana Volnikova
Cas 1h39'55" 60d29'20"
Tatyo
Vir 13h10'41" -22d8'46"
Taucan
Cru 12h25'7" -56d44'0"
taucher
Cas 1h15'38" 53d16'31"

"T-Dizzle"
Vir 14h30'58" 3d20'28"
T-Dizzles Banjo
Umi 11h44'44" 85d57'19"
TDKowalczyk 09/02
Uma 11h37'49" 52d53'54"
Te Ador Beloved & Beautiful Soul
Psc 0h38'53" 6d39'23"
Te Adoro
Sgr 20h12'34" -29d20'36"
Te adoro mi chiquitira
Sgr 18h17'43" -35d10'42"
Te Amo
Cyg 19h35'26" 28d0'32"
Te Amo
Uma 9h58'37" 50d23'12"
te amo
Lyr 18h47'31" 38d53'22"
Te Amo
Crb 15h32'56" 39d2'26"
Te Amo
Cyg 21h13'7" 31d43'28"
Te Amo Carmen
And 23h22'54" 47d36'31"
Te amo con todo mi corazon
Cas 1h14'32" 54d48'7"
Te amo forever mi vida
And 2h25'2" 42d10'36"
te amo hmd
Cap 21h39'20" -13d29'5"
Te Amo Israel
Crb 15h47'42" 28d58'53"
Te Amo Mi Ciela
Cyg 20h21'40" 56d21'45"
Te amo muito e para sempre
Pho 0h15'40" -49d15'51"
Te amo siempre Marcella N. Craig
Cap 21h33'10" -16d52'27"
Te Amoaeterno
Sco 16h8'57" -16d14'32"
Te amos
Cyg 21h14'56" 46d31'25"
TE & CC Skipper
Leo 11h6'25" -0d8'26"
Te Dua
Per 4h29'24" 51d20'19"
Te Gam
Umi 13h39'23" 73d19'14"
Te iubesc Traian...
And 2h23'51" 46d7'43"
Te(k)nos vagy?
Uma 9h52'37" 50d11'52"
Te queremos Rosy
Cep 21h53'55" 61d12'0"
te quiero mi amor
Leo 10h19'39" 20d21'59"
Te quiero Reyna!
Ori 5h32'35" 10d11'1"
Te Ruru Po
Col 5h49'18" -33d46'59"
Te Te
Psa 22h6'51" -26d14'10"
Te Whetū o Wiremu
Cru 12h11'51" -62d3'22"
Tea
Boo 14h2'11" 9d26'24"
Téa Antonietti
Cnc 8h56'16" 31d40'50"
Tea Guru & Little Girl <3
Teti
Vir 13h15'52" -21d30'43"
Téa - Little Sparkle in the Sky
Cru 12h3'30" -60d55'21"
tea & pie
Lyr 18h41'21" 26d9'24"
Tea Time
Cmi 7h37'23" 5d5'34"
Téabell
And 0h9'41" 33d20'21"
Teacher Josie Young
Cas 1h58'4" 62d1'38"
Teacher Nina
Tau 5h33'59" 25d51'44"
Teacher Patty
Gem 7h3'22" 18d31'37"
Teacher Renee J. Peccia
Cas 0h46'52" 66d59'49"
Teacher Shelly
Ari 2h53'37" 27d40'0"
"Teacher"—L. Jane Niebergall
Ari 2h45'56" 17d15'28"
Teagan
Ari 1h47'24" 19d37'29"
Teagan
Aqr 22h56'43" -10d0'18"
Teagan A. Abbott
Ori 6h1'12" 9d33'49"
Teagan Ann Burgmann
Cnc 8h24'0" 14d18'45"
Teagan Anna Rudy
And 0h39'51" 24d18'4"
Teagan Autumn
Dra 16h8'26" 63d42'31"
Teagan Davis Kopp
Uma 10h37'53" 57d39'43"
Teagan Delaney Hough
Boo 15h7'55" 33d0'0"
Teagan Elise
Leo 10h1'29" 18d6'38"
Teagan Liam Lawson
Cep 22h27'8" 77d42'13"
Teagan Lukacs
Psc 1h21'28" 11d49'32"
Teagan Marie
Umi 15h1'39" 69d14'59"
Teagan Maxine
Gem 7h14'25" 33d22'23"
Teagan Rae Rutkowski
Uma 14h22'41" 57d50'5"
Teagan's Star
Uma 11h31'11" 57d30'57"
Teagan's Star
Tau 5h59'17" 24d57'39"
Teage Ezard - 25.09.1966
Cru 12h26'16" -61d37'49"
Teaglach
Ari 2h8'55" 21d14'28"
Teague
Vir 12h56'24" 11d33'42"
Teah Georgia Gandossini
Aqr 20h57'37" 1d36'2"
Teal
Gem 6h23'8" 21d16'28"
Teal Cowgirl H.L.W
Psc 1h21'1" 29d27'38"
Teal Marie 21
Lyn 7h27'28" 46d2'59"
Teal The Tool Amy Lynch
Lib 15h28'58" -17d36'31"
Teale
Sco 16h8'50" -29d21'19"
Team Baby Goose
Cru 12h0'3" -62d1'4"
Team Barvandahl
Uma 9h26'14" 65d52'36"
Team BFF <3
Sgr 19h37'15" -14d14'3"
Team Casey
Aql 19h33'27" 10d40'17"
Team D
Ori 4h51'35" 10d49'23"
Team DJ
Pho 2h23'5" -45d33'48"
Team Illinois
Her 18h15'31" 15d2'48"
Team Maverick
Uma 8h54'16" 48d45'54"
TEAM MIG
Lyn 7h5'15" 59d30'52"
Team Shampine
Lyn 8h51'34" 34d52'59"
Team Vinik 20th.
Cyg 20h50'43" 34d0'55"
Team Welberry
Cyg 21h49'4" 53d58'35"
Team Whoregoat
Lac 22h31'39" 50d2'0"
Team-Dolphin.com
Umi 15h1'37" 70d48'56"
teamosiempre
Ori 6h20'49" 9d26'55"
Teana Greenwood
And 2h20'58" 49d51'40"
Teanna Hill
Cam 6h33'33" 68d36'26"
Te'anna Pearson
Umi 16h6'17" 80d49'43"
Teany In Loving Memory
Lyr 18h52'44" 36d48'24"
Tear in the Ocean
Eri 2h34'3" -40d41'42"
teardrop's star
Vir 12h27'38" -7d57'19"
Tearlach
Hya 9h17'54" -0d4'7"
Tears of the Heliades
Sco 17h24'28" -42d58'53"
Teary Lynn Cambron
Sgr 18h39'6" -28d23'13"
Teasa
Cap 20h49'30" -15d21'55"
Teawi Angel
Cnc 8h16'45" 20d0'46"
Teaya Kaye Hamilton
Psc 1h27'24" 27d7'5"
Tebone
Lyn 7h32'20" 52d57'1"
Techsmiths Cyberlan
Peg 23h31'24" 17d56'51"
Tecla Gloria
Cas 0h52'41" 57d28'28"
Tecoda-Faith
Cru 12h1'58" -62d35'5"
Ted
Umi 15h42'31" 74d52'51"
Ted
Leo 9h29'50" 16d51'32"
Ted A. Pennypacker, Jr.
Per 4h6'29" 34d10'47"
Ted and Barbara Gutt
Hya 9h39'43" -21d53'9"
Ted and Carol Nussdorf
Aql 20h10'22" 9d25'12"
Ted and Christy
Lmi 10h52'46" 31d47'36"
Ted and Helen 7-12-1947
Ori 5h35'33" -1d52'36"
Ted and Rachel
Aqr 21h7'58" 0d27'5"
Ted Bastian Jr.
Boo 15h31'3" 42d41'32"
Ted Bruun
Ori 6h9'57" 21d2'0"
Ted Charles Scherer
Cyg 19h45'37" 55d9'12"
Ted & Clara Hazelbaker
Uma 13h19'16" 56d7'14"
Ted Cloutier
Uma 11h33'30" 57d50'34"
Ted D. Ailanjian
Sgr 19h43'28" -14d43'17"
Ted Duncan
Her 17h9'20" 27d15'2"
Ted F. Robinson
Cnc 9h7'51" 15d6'15"
Ted Flaherty
Lib 15h9'32" -21d50'45"
Ted Giles
Uma 11h10'1" 33d33'41"
Ted Gunderson
Sco 16h25'37" -26d38'11"
Ted Hearne The Celestial Wisdom
Gem 6h23'12" 20d57'1"
Ted & Hedy Orden
Sge 19h40'57" 18d45'35"
Ted Hobbs Memorial Star
Uma 10h20'21" 54d24'33"
Ted Hoffmann
Uma 8h37'11" 54d21'9"
Ted Hummerston
Ori 5h32'12" -3d5'17"
Ted J. Larkin Eternal Life
Umi 14h29'14" 75d28'21"
Ted Joynes
Uma 13h5'26" 59d8'43"
Ted K. Encke
Lac 22h43'15" 49d22'31"
Ted Kopec
Uma 13h50'24" 58d25'7"
Ted Lang
Srp 18h27'17" -0d13'43"
Ted M Shema
Boo 14h33'22" 41d13'22"
Ted & Mary Anne "For Good"
Sgr 19h47'28" -14d13'51"
Ted McDonald
Uma 9h19'58" 65d33'51"
Ted McDowell
Lmi 9h49'29" 38d30'13"
Ted Mogel
Lmi 10h44'58" 23d34'45"
Ted Morris ~ Someone Special
Psc 1h8'52" 27d14'41"
Ted Page
Psc 0h34'2" 7d6'41"
Ted *Papa*Kill
Cnc 8h16'2" 22d33'12"
Ted Paradise
Her 18h11'2" 28d38'15"
Ted Punter - Ted's Star
Cep 0h19'16" 74d10'38"
Ted R. Robinson
Vir 12h34'40" -0d37'59"
Ted Robinson
Per 2h19'1" 52d37'34"
Ted & Sally Aldeen
Uma 10h35'44" 64d15'26"
Ted Schoff
Sgr 19h51'8" -13d14'7"
Ted & Shirley
Dra 17h6'51" 55d0'15"
Ted & Shirley Levy
Lyr 18h16'25" 33d41'15"
Ted Spivey
Cnc 8h5'56" 24d27'47"
Ted Sulkowski
Lyn 7h58'12" 35d8'21"
Ted & Wendy Haaf's Star
Lyr 18h46'55" 40d10'17"
Ted Wiard - The Soul Star
Lyn 6h21'12" 60d13'1"
Ted William Grant
Lyn 8h29'53" 34d52'31"
Ted Wilson
Per 2h51'18" 46d12'29"
Ted - "Your Guiding Star"
Cyg 19h42'30" 39d59'56"
Ted, Debi Musgrave Anniversary 1977
Cyg 19h54'5" 37d13'12"
TeddahxStar our love undying
Cnc 8h43'25" 24d26'47"
Tedde Kadison
Aqr 22h55'24" -6d46'35"
Teddi Ann 463
Cap 21h5'28" -16d7'9"
Teddi (Bear)
Uma 12h11'44" 56d43'54"
Teddi Bear
Cnc 8h7'14" 17d32'0"
Teddi Wade
Lyr 19h27'13" 40d25'49"
Teddi8245
Leo 11h16'2" 26d11'44"
Teddie Joe
Gem 7h20'30" 20d54'25"
Teddie & Mike
Dra 18h51'3" 65d31'12"
Teddler's Heart
Ori 5h22'28" 3d4'47"
Teddly Bear
Cnc 8h46'45" 9d54'18"
Teddy
Aql 19h4'50" 9d18'11"
Teddy
Uma 10h59'51" 64d45'29"
Teddy
Mon 7h56'42" -4d51'13"
Teddy
Col 6h25'41" -34d19'25"
Teddy Bear
Uma 13h21'17" 53d5'7"
Teddy Bear
Leo 10h47'18" 19d54'45"
Teddy Bear
Uma 12h1'12" 46d42'20"
Teddy Bear Berman
Uma 9h16'59" 50d15'48"
Teddy Bear Rubio
Uma 10h43'46" 65d10'33"
Teddy Bear & Sunshine
Psc 0h41'43" 7d6'58"
Teddy Bright
Uma 11h48'37" 54d50'21"
Teddy Buttskins
Uma 12h9'41" 59d45'4"
Teddy Cool
Sco 16h11'51" -10d46'24"
Teddy Coughlin 12/17/95
Cyg 20h30'17" 35d58'55"
Teddy Durkin
Umi 14h47'24" 77d19'34"
Teddy Ebersol
Per 3h42'14" 39d2'59"
Teddy Federwitz
Uma 10h34'19" 53d44'27"
Teddy Fidelis Lionheart
Lyn 8h7'38" 35d35'40"
Teddy Haldis
Leo 10h11'15" 26d32'48"
Teddy Krass
Mon 7h41'17" -0d42'0"
Teddy L. Hughes
Her 18h9'1" 23d50'52"
Teddy Long
Per 4h8'2" 33d41'3"
Teddy Love Kumar
Uma 11h37'55" 47d27'59"
Teddy Parks
Uma 11h25'9" 37d11'38"
Teddy & Patricia Stalnaker
Cyg 19h55'18" 39d58'14"
Teddy Paul Miller
Gem 6h51'49" 25d50'23"
Teddy Pooh Bella
Cmi 7h53'45" 11d45'54"
Teddy Seiler
Uma 11h42'59" 59d11'22"
"Teddy" Theodore Coderre Jr.
Cep 21h59'42" 70d19'14"
Teddy Wing
Cap 20h9'38" -9d2'56"
Teddy Zangenberg
Tau 4h41'21" 22d54'35"
Teddy, Aggie & TJ Bartels
Uma 10h16'54" 51d19'56"
TeddyBean
Her 16h34'52" 31d40'22"
Teddybear
Uma 10h27'52" 39d31'37"
Teddy's Special Birthday Star
Leo 11h38'36" 25d22'44"
Teddy's Star
Leo 11h47'45" 22d25'48"
Teddy's Twinkle
Her 18h39'18" 19d55'33"
Ted's Star
Psc 1h32'18" 17d51'14"
Tee
Cnc 9h6'1" 17d35'40"
Tee
Uma 12h28'4" 54d11'46"
Tee Catalano... REACH...
Uma 12h48'20" 59d25'56"
Tee Light
Sgr 18h34'57" -28d5'51"
Tee Wooding
Her 17h26'27" 27d23'0"
Tee1258
Lyn 6h56'6" 50d28'6"
Teebz Brannan
Umi 16h11'56" 80d5'5"
Teedlebug
Boo 14h30'43" 54d19'21"
Teegan Wallace
Gem 6h3'6" 24d48'45"
Teegan Wynn O'Neal
Uma 9h7'29" 49d0'56"
Teejay T. Fountain "07/31/1979"
Leo 9h41'4" 11d6'41"
Teej's Star
Cru 12h2'5" -59d47'45"
Teela
Uma 10h55'41" 71d52'10"
Teela
Mon 6h32'30" 1d45'49"
TeeLa Mae Hagy
Cap 21h35'13" -13d33'23"
Teen America 2006 - Brittany Monico
Cas 0h43'53" 63d51'16"
Teena
Cas 2h32'30" 67d38'49"
Teena
Umi 16h54'52" 75d51'23"
teena
Vir 12h42'16" -11d31'15"
Teena Lee Smith
Aqr 22h9'44" -1d30'48"
Teena Majzoub
Psc 0h37'20" 7d39'50"
Teena Mari
Uma 10h35'15" 59d53'28"
Teena Marie Brizzi
Leo 9h48'17" 13d55'6"
Teenie Weenie Super Star
Lmi 10h42'13" 34d41'58"
Teenie's Twinkle
Tau 5h58'50" 27d45'4"
teenuh
Gem 7h46'18" 26d48'1"
Teeny
Lyn 6h29'11" 54d7'30"
Teeny Tiny Tina
Mon 8h2'23" -0d42'21"
Teesa112882
Uma 8h35'29" 47d1'10"
Teesha
Lyr 18h30'52" 36d6'59"
Teeter
Lyn 8h19'8" 36d24'23"
Tegan
And 0h58'46" 40d27'12"
Tegan Anne Hamilton
Cru 12h25'27" -58d31'48"
Tegan Eileen Healy
Lyn 7h22'1" 56d42'44"
Tegan Hunter
And 2h27'20" 46d25'54"
Tegan Lena Grace
Tau 4h30'43" 26d47'23"
Tegan Rose
And 22h59'59" 51d1'54"
Tegan Ruland
Ori 5h34'8" 1d27'50"
Tegbir Singh Chohan
Cep 21h53'28" 63d31'41"
Tegis Astraea - We Love You
Cru 12h27'4" -61d8'45"
Tegtmeyer, Harald
Uma 12h57'56" 60d35'25"
Tegtmeyer, Jürgen
Ori 6h3'33" 10d55'6"
Tehra
Cyg 21h16'39" 46d10'59"
Tehya Gitte
Cas 0h15'21" 55d22'11"
Tehya Pyawasay
Lmi 10h27'13" 36d15'18"
Teichert, Alfons
Uma 10h18'58" 48d15'23"
Teigan Cameron
And 2h12'29" 50d30'4"
Teigan E. Ash
Ari 3h13'23" 27d34'30"
Teigan Elise Meikle
And 0h43'14" 30d1'54"
Teigan Harris
Cyg 20h59'37" 36d26'48"
Teigan's Star
And 0h47'7" 23d22'13"
Teige
Cru 12h28'1" -60d30'36"
Teiji Keakaokalani Kuni
Dra 17h23'51" 54d29'20"
Teiki La-Rue
Dra 14h42'35" 60d31'50"
Teisha Jade Adams
And 1h29'45" 43d20'53"
Tej
Per 4h43'30" 45d23'5"
Te-jai
Vir 13h15'11" 8d16'8"
Tejal Moonlight
Lyn 7h28'11" 45d58'29"
Tejan Charles Dhirmalani
Sgr 19h28'18" -18d0'0"
Tejota
Ori 5h54'14" 12d53'59"
Teka Paris
Ari 2h35'22" 26d22'5"
tekeow
Psc 1h32'41" 26d21'12"
Tel Star
Cep 21h41'31" 65d8'39"
Tela - Il Straordinaria Bellezza
Mon 6h39'46" 6d28'7"
Telamarine
Sco 16h9'52" -13d36'2"
Telford G. Jones
Per 3h18'25" 43d32'19"
Telia Brooke Green
Her 16h16'45" 12d35'27"
Telisa L. Hochstedler
Cas 3h14'14" 59d18'44"
Telisha Moore
Peg 21h45'42" 24d9'10"
Telita Elizabeth Hunter
Crb 16h21'2" 27d31'24"
Tell me a secret....
Cap 20h44'16" -18d21'40"
(Tellas) Shibby King
Cas 23h34'45" 52d52'19"
Tellina Rain
Uma 10h57'44" 54d7'8"
Telly
Ari 2h42'53" 20d33'38"
TE.MA
Peg 3h26'54" 41d38'49"
Tema Achiam
Aur 6h7'57" 35d52'12"
Tema L. Steele
Vir 12h13'27" 12d3'18"
Tembo Zuri Hewa
Cap 20h47'30" -14d42'27"
Temekia Pelletier
Aqr 22h21'8" -5d3'10"
Temitope le lumineux
Sco 16h10'43" -19d14'40"
Temmycaine
Lyn 8h35'25" 34d1'39"
Tempest Guzman
Lyn 8h19'56" 34d45'26"
Tempest Roselene Elmore
Tau 4h40'15" 20d48'48"
Temple Michell Klevesahl
Cnc 8h48'23" 25d7'9"
Templer 66
Gem 7h21'9" 26d33'35"
Ten
Cas 0h41'59" 61d36'45"
Ten and Forever
Cnc 8h26'43" 30d35'29"
ten Loew, Denis
Uma 8h50'24" 47d48'0"
Ten Wonderful Years
Sco 17h49'38" -40d43'8"
Tena Alison Watson
Cas 1h9'6" 49d36'36"
Tena Dawn
And 23h34'11" 50d8'23"
Tena deLaski
Cas 0h44'21" 55d46'42"
Tena Elizabeth Fulghum
Cap 20h20'40" -12d30'8"
TENAZ
Ori 5h18'6" -2d28'55"
Teneyah
Psc 1h12'29" 23d55'21"
Tenez le premier rôle de Janell
Lib 15h30'51" -13d14'34"
Tengstar
Sge 19h39'44" 18d57'58"
Teni
Psc 1h17'45" 18d9'0"
Tenia "Joey's Angel" Simmons
Uma 8h11'40" 66d53'54"
Tenille Dunning
Lib 15h2'31" -1d58'16"
Tenique "Precious" Jaeger
Cru 12h43'15" -63d7'14"
Tenisha S Howard
Vir 12h55'35" -9d4'11"
Tenneagle
Cnc 8h11'52" 6d49'53"
Tennesse
Cap 20h23'27" -22d58'53"
Tennessee Miracle
Tau 5h8'11" 23d31'33"
Tennessee Wildflower
Leo 11h9'57" 16d12'2"
Tennille Black
Cap 20h26'1" -10d33'43"
Tennille Casand
Umi 16h24'54" 80d32'21"
Tenshi
Uma 12h39'14" 57d24'21"
TENSHI
Sco 16h21'53" -32d56'56"
Tenten
Cnc 8h51'29" 12d51'28"
tenyearsafter010297
Uma 11h49'35" 50d36'26"
Tenzin Jamyang Gawa Ngodup (Jangha)
Cnc 8h46'1" 31d35'55"
TEO
Tau 5h17'22" 20d25'29"
Teo
Vir 13h41'46" 6d2'39"
TEO
Vel 9h22'8" -42d50'57"
Teo Jean Scipio
Vir 14h7'38" -1d30'3"
Teo Servajean
Leo 9h38'28" 20d19'0"
Teodora
Ori 5h12'2" 11d37'5"
Teodora
Cyg 20h1'1" 32d27'24"
Teodora
Cap 21h1'28" -25d35'28"
Teodora "Mish" Ilic
Cas 0h34'44" 75d52'34"
Teodoro and Darlene Lumaban
Aqr 23h37'28" -17d0'37"
Teodoro Gonzales
Cet 3h13'23" -0d14'40"
Teodoro Zaniel Soliz IV
Lyn 8h2'10" 41d8'4"
Teofila P. Yambao
Uma 12h45'0" 58d28'24"
Te'oma Nebulae Praesis
Gem 7h36'17" 24d28'55"
Tepaske Garden of Stellar Delight
Cnc 8h43'11" 15d50'30"
Tepasse, Werner
Ori 5h26'37" 14d3'59"
Tepee
Cmi 7h13'27" 5d54'28"
Tepekau ButSki
Pho 23h54'5" -46d6'24"
Teppo
Ori 5h37'5" -3d52'34"
Tequa Festival Marketplace
Peg 21h29'2" 11d46'40"
Tera Barr
Ori 5h52'6" 7d8'26"
Tera & Danny Bailey
Lyr 19h18'35" 28d38'48"
Tera Krabbenhoft
Cnc 8h45'1" 18d19'33"
Tera L. Shaw
Cap 20h41'31" -15d55'28"
Tera Lee Hendrickson
And 23h49'29" 39d5'58"
Tera Lynn
And 0h47'14" 25d44'47"
Tera Lynn Pinkerton
Sco 17h33'39" -39d10'38"
Tera Pavel
Ori 5h39'21" 3d5'17"
Tera Walsh
Uma 10h49'0" 57d52'19"
Terabithia
Sgr 18h26'34" -22d36'21"
Terah Lynn
And 0h52'22" 38d58'53"
TerahMischel
Ari 2h7'40" 24d44'9"
Teraimana
Umi 13h24'33" 73d11'26"
Teraimateata
Sco 17h17'30" -39d47'44"
Teralon
Cas 0h48'7" 49d5'15"
Terance Michael Haight
Dra 16h36'18" 54d40'45"
Terance Tasker - Terry's Star
Per 3h2'16" 44d29'5"
Terassa Starr
Cam 4h36'28" 59d31'56"
TerBear
Ori 6h10'10" 6d56'52"
Terceira & Doah Reynolds
Cas 0h38'44" 60d3'37"
Tereiti
Uma 11h24'32" 32d23'16"
Terek L. Wilson
Lyr 19h26'51" 40d40'49"
Terena Ortega
Cas 0h58'44" 64d17'19"
Terence A Meredith
Ori 5h12'10" -9d24'34"
Terence Dale Downey
Cep 22h28'17" 65d4'18"
Terence Daniel Lee
Ari 2h38'7" 12d50'16"
Terence & Debbie Turner Forever
Cyg 20h37'30" 52d23'55"
Terence Dragula
Uma 11h9'12" 31d52'47"
Terence Edward Vickery
Cep 20h42'15" 58d53'7"
Terence Gabriel Kelly
Sco 16h14'48" -12d2'45"
Terence George Burditt
Ori 6h17'43" 13d35'30"
Terence Harris
Sgr 18h15'42" -33d38'22"
Terence John Williams
Uma 8h37'11" 64d35'12"
Terence McHugh
Ori 6h13'52" 15d49'50"
Terence S. Hatton
Ori 5h26'55" -7d38'21"
Terence Shannon 17.6.84
Cru 12h37'21" -58d4'21"
Terencio S. Pineda, Proud American
Ori 5h31'25" 2d24'54"
Terenia
Umi 14h36'52" 72d59'18"
Teresa
Umi 14h27'6" 66d24'42"
Teresa
Cas 1h7'16" 66d49'39"
Teresa
Lyn 7h44'52" 53d39'3"
Teresa
Lib 14h58'12" -11d6'11"
Teresa
Sco 16h44'6" -31d53'1"
Teresa
Sco 16h51'44" -26d56'48"
Teresa
Aqr 22h38'57" 0d1'37"
Teresa
Aql 19h17'8" 5d59'23"
Teresa
Tau 4h57'30" 15d59'39"
Teresa
Cnc 8h8'59" 13d19'53"
TERESA
Cnc 8h11'27" 16d35'38"
Teresa
Cnc 8h35'21" 22d42'54"
Teresa
Cnc 8h15'38" 32d13'27"
Teresa
Uma 11h57'58" 40d14'57"

Teresa
Uma 11h28'59" 39d40'26"

Teresa
And 0h53'40" 40d8'22"

Teresa
And 23h20'29" 52d17'47"

Teresa 08&51 07-04-97
Umi 15h25'7" 70d41'52"

Teresa A. Furrow
Cyg 20h56'5" 35d49'11"

Teresa A. Hall
And 0h15'19" 29d0'7"

Teresa A. Hill
Cyg 19h39'14" 33d13'0"

Teresa A. Peacock
Lyr 19h27'23" 37d41'59"

Teresa A Wilson
Uma 9h2'37" 72d1'48"

Teresa A Woman With Eternal Purpose
Tri 2h12'45" 34d26'6"

Teresa Amick
Cas 0h6'26" 59d45'13"

Teresa and Bill Schmidt
Umi 15h12'52" 76d20'57"

Teresa and Meagan
Cnc 9h5'52" 13d34'21"

Teresa And Scott
Umi 15h34'15" 73d52'59"

Teresa and William Whelan
Ser 15h33'41" 17d48'54"

Teresa Andreika
Aqr 23h0'25" -6d43'29"

Teresa Ann
And 23h24'6" 52d25'33"

Teresa Ann
And 0h21'27" 38d36'30"

Teresa Ann Bennett
Ori 5h27'57" 9d58'7"

Teresa Ann Davis
Cyg 21h16'54" 33d58'28"

Teresa Ann Fredwest
Lib 15h4'8" -8d11'25"

Teresa Ann Sedlak
Cas 23h59'9" 49d55'44"

Teresa Ann Stephens
Aqr 22h48'6" -4d20'50"

Teresa Ann Tallarico
And 0h27'28" 42d11'43"

Teresa Ashcraft
Ori 5h25'14" 14d43'13"

Teresa Avila Ferretti
Umi 15h25'34" 85d44'1"

Teresa Barnes
Cra 19h1'57" -40d58'18"

Teresa Bolton
Sco 16h15'16" -16d20'36"

Teresa Booth 2005
And 2h25'0" 47d41'34"

Teresa Bozzini Burns
Crb 15h40'24" 27d14'47"

Teresa Bult
And 0h43'47" 43d53'53"

Teresa Caballero
Cas 0h44'27" 53d11'50"

Teresa Callison
And 1h55'40" 37d9'29"

Teresa Cavanagh
Crb 16h15'45" 33d5'9"

Teresa Ciorciari
Psc 1h38'0" 17d1'44"

Teresa Colatorti
Uma 10h56'42" 41d36'48"

Teresa Coleen Hedrich
Ari 2h35'12" 26d4'15"

Teresa "Cricket" Evelyn Gaillard
Tau 4h5'5" 6d0'11"

Teresa Cunningham
And 23h41'55" 46d7'17"

Teresa Czerska-Bicka
Gem 6h43'17" 33d4'31"

Teresa D. Wilke
Cas 2h50'44" 60d5'46"

Teresa Dale Crowley
Lmi 10h17'14" 36d32'54"

Teresa Dangers Hess
Leo 11h15'16" 25d54'6"

Teresa & David Fordham Wedding Star
Cyg 19h31'52" 51d34'51"

Teresa de Jesus
Com 13h5'27" 27d49'32"

Teresa De Jesus Nunez
Dra 18h49'0" 77d56'57"

Teresa Decher
Tau 5h52'33" 27d33'24"

Teresa Denise Delgado
Psc 0h10'14" 7d10'2"

Teresa Diane
Lib 14h47'58" -5d26'6"

Teresa Doherty
Crb 16h11'32" 35d55'6"

Teresa Ducharme
Aqr 21h10'59" 1d57'13"

Teresa Duque Raley
Uma 12h15'2" 62d49'23"

Teresa Elena Dyer
Vir 12h46'53" 6d6'27"

Teresa Elizabeth Carson Bartley
Cap 20h41'15" -17d48'8"

Teresa Escota Cruz
Gem 7h16'23" 26d11'35"

Teresa Farella
Cas 2h17'12" 73d41'26"

Teresa FitzBrown
Uma 10h16'21" 45d56'5"

Teresa Forever Shining
Dra 17h54'51" 66d16'9"

Teresa Fortuna
And 2h17'5" 47d28'28"

Teresa Girardi Orlando Graziani
And 1h45'40" 36d24'24"

Teresa Goldner
Cas 2h23'18" 64d32'36"

Teresa Gonzalez
Cnc 8h53'9" 14d14'39"

Teresa Grace
Cnc 8h49'46" 29d44'57"

Teresa Hane
Cas 23h10'15" 56d9'39"

Teresa Hanson
Cas 0h36'1" 54d37'46"

Teresa Harper
Mon 6h51'37" -0d6'54"

Teresa & Helena
Uma 13h38'46" 57d5'57"

Teresa Horacek
Cas 23h29'19" 58d7'17"

Teresa Hoss
Lyn 8h43'14" 33d29'57"

Teresa Hunter the Twinkle in My Eye
And 0h18'57" 28d7'21"

Teresa Inez Valdez
Tau 4h14'9" 14d1'12"

Teresa Jean
Ori 6h1'2" 13d42'52"

Teresa Jean
Uma 9h14'30" 50d10'23"

Teresa Jo Russell
Psc 1h23'4" 15d35'17"

Teresa June
Cyg 19h35'3" 34d0'56"

Teresa Kaczmarek
And 23h32'58" 48d7'10"

Teresa Kay
Dra 15h50'3" 56d29'40"

Teresa Kay Hansen-Flood-Long
Crb 15h50'39" 28d30'34"

Teresa Kendregan
Her 18h20'36" 23d47'46"

Teresa Kimball
Cas 1h42'36" 61d12'25"

Teresa L. Duke
Cap 21h45'15" -11d18'31"

Teresa Lancaster
Uma 12h4'13" 52d18'32"

Teresa Lee
And 1h5'38" 38d53'29"

Teresa Leigh Barrett
Leo 11h1'56" 5d1'41"

Teresa loves Daniel forever!
Cyg 20h31'55" 36d30'51"

Teresa Lynette Turton
Leo 11h51'32" 25d39'43"

Teresa Lynn
Lyr 19h7'58" 27d0'8"

Teresa Lynn Codelia
And 2h35'31" 42d23'53"

Teresa Lynn Ford
Sco 16h8'25" -17d39'14"

Teresa Lynn Haynes
Crb 16h5'20" 27d23'1"

Teresa Lynn Krumbholz
Psc 0h43'57" 17d43'39"

Teresa Lynn Schmidt
Psc 22h54'35" 3d49'54"

Teresa Lynn Sherman Martin
Cnc 9h6'55" 27d14'51"

Teresa Lynn Walker
Lyn 6h57'30" 58d22'5"

Teresa M Copeland
Gem 6h35'49" 15d28'33"

Teresa M Jakeway
Psc 0h25'5" 8d53'4"

Teresa M. Vinzani
Leo 11h38'51" 18d9'20"

Teresa M. Watters
And 23h11'36" 43d17'34"

Teresa Macias
Ori 5h36'28" 2d3'0"

Teresa Maria Gargiulo Dight
Aqr 22h31'13" -0d55'27"

Teresa Marie Bower
Uma 9h39'23" 42d16'33"

Teresa Marie Erdman
Lib 15h16'54" -12d41'35"

Teresa Marie Haga
Lyn 7h8'48" 49d6'33"

Teresa Marie Haynes
And 0h16'32" 25d17'48"

Teresa Marie Quinn
Psc 23h7'18" -0d10'6"

Teresa Marie Schmedding
Boo 14h50'43" 24d22'10"

Teresa Marie Sealey
Lyn 8h19'30" 45d33'8"

Teresa Marie Tabeek
And 0h30'27" 32d19'38"

Teresa Marie Zsebenyi
Cnc 8h51'5" 14d40'58"

Teresa Martinez
Cnc 8h37'0" 17d42'56"

Teresa Martinez
Lyn 7h13'1" 48d4'46"

Teresa Mary Konier
Tau 5h8'38" 24d33'6"

Teresa Marzolph
And 2h25'18" 49d40'50"

Teresa Matz
Cyg 21h10'35" 34d51'57"

Teresa Medaglia
Cep 22h31'26" 66d44'37"

Teresa Mentz
Ori 5h55'3" -0d11'36"

Teresa Michelle Bogstad
And 0h57'21" 44d14'40"

Teresa Michelle Moots
And 1h48'6" 37d50'21"

Teresa Moen
Gem 7h41'11" 24d3'19"

Teresa Moore
Cyg 20h9'14" 33d58'40"

Teresa Mustang Moore
Tau 5h50'30" 16d54'29"

Teresa Parton
Cas 0h40'4" 48d12'34"

Teresa Pearson
Ari 3h13'50" 29d22'3"

Teresa Pereira Fernàndez
Her 18h14'52" 19d49'20"

Teresa Perrino & Mike Andrzejewski
Uma 11h49'17" 60d44'53"

Teresa Pitek
Lib 15h6'26" -1d23'8"

Teresa Pollock
Peg 21h56'30" 14d55'11"

Teresa Proto
Ori 6h2'4" 11d29'57"

Teresa Quennell
Cyg 20h4'45" 38d29'13"

Teresa Querida
Lib 14h27'54" -19d10'15"

Teresa Quinlan
And 2h14'26" 41d57'52"

Teresa Rizzuto
Cyg 20h3'5" 56d36'56"

Teresa Rodriguez Gomez
And 1h29'36" 50d0'22"

Teresa Rose Cridlin
Cnc 9h0'7" 11d24'38"

Teresa Rose Killeen
Cas 2h2'28" 59d18'21"

Teresa Scott
Dra 17h25'16" 52d0'2"

Teresa Sheehey "Tess's Star"
Lyn 8h56'10" 42d11'36"

Teresa Shweebird Breed
And 1h22'16" 47d6'15"

Teresa Silva
Lib 15h46'33" -12d0'23"

Teresa Smith
Uma 8h58'31" 67d27'50"

Teresa Snow
Lmi 10h41'27" 30d52'51"

Teresa Splendente
Uma 11h11'54" 39d15'23"

Teresa - Steve Simon
Ori 5h27'37" 6d8'41"

Teresa Stubbs
Dra 18h39'52" 56d5'8"

Teresa Sundermann Cherry
And 0h51'47" 23d35'0"

Teresa Supriano
Uma 11h52'22" 36d0'25"

Teresa Templado Garcia
Psc 1h44'51" 21d12'18"

Teresa The Beautiful
Cyg 20h45'27" 39d24'42"

Teresa (Tree) Christiansen
Uma 11h32'0" 37d8'10"

Teresa Vaughn
Lyn 6h24'44" 61d33'15"

Teresa Velez
Lib 15h2'51" -22d29'39"

Teresa Virginia
Leo 10h12'6" 22d56'16"

Teresa Vladessa Caprio
Vir 12h4'52" 4d32'9"

Teresa W.McBlair
Ari 2h13'14" 25d37'20"

Teresa X04
And 0h21'31" 42d12'24"

Teresa Yuan
Uma 11h18'46" 28d50'24"

Teresarc
Lyn 7h59'34" 42d54'44"

Teresa's Brilliance
Sco 16h11'57" -9d19'2"

Teresa's Marathon Memory
Sco 16h3'43" -12d41'10"

Teresa's Star
Sgr 18h24'6" -33d7'12"

Teresa's Star
Lmi 10h34'21" 31d1'27"

Teresa's Star
Cyg 19h36'43" 29d48'14"

Teresa's Star
Gem 7h33'38" 23d32'43"

Teresa's Wish
Sco 16h13'37" -9d0'38"

Terese Allen
Cnc 8h37'36" 23d27'37"

Terese Sesto
And 0h45'24" 40d18'48"

Teresea Hall
Lyn 7h44'44" 35d51'42"

Teresea Lynn Workman
And 0h58'52" 38d33'59"

Teresina del Carmen
Lep 5h16'59" -14d41'53"

Teresita Horney
Lib 15h2'33" -0d48'53"

Teresita Mejias
Crb 15h29'51" 28d35'55"

Teresita Noe
Lyr 18h40'35" 31d31'54"

Teresita Sanchez
Cas 22h58'58" 54d23'26"

Teressa Ann Piper
Sco 16h9'9" -12d12'12"

Teressa Nunez Summers
Psc 1h8'25" 32d49'57"

Terez
Cnc 8h47'49" 31d27'29"

Terezija
Lac 22h48'18" 51d10'33"

Terezka Treslova
Cap 20h20'29" -16d8'9"

Teri
Mon 8h10'27" -1d2'24"

Teri
Sco 16h17'44" -41d18'21"

Teri
Uma 10h21'1" 48d17'0"

Teri
Lyr 18h45'21" 37d53'48"

Teri
Leo 11h15'44" 13d46'56"

Teri 10/17/1967
Lib 15h45'20" -19d56'2"

Teri Abbott
Cyg 20h12'0" 45d53'51"

Teri Allbaugh
Ari 2h31'24" 25d18'30"

Teri and Robert
Cyg 19h59'48" 39d45'0"

Teri Bateman - My Baby
Cyg 20h16'21" 38d38'37"

Teri Beri
Aqr 22h59'27" -8d32'36"

Teri Brown
Tau 5h23'9" 27d50'9"

Teri Camp
And 1h13'53" 42d8'10"

Teri Christine Netter LTC USAF RET
Cam 5h24'27" 63d1'49"

Teri Coomes
Uma 11h3'27" 57d34'32"

Teri Dee
Vul 20h50'38" 22d21'36"

Teri Denise Sytsma
Leo 10h7'29" 16d58'55"

Teri DeSilva
Cas 23h31'28" 58d52'14"

Teri Dunnegan
Uma 10h56'28" 35d19'52"

Teri Faulkner
Ari 2h55'44" 18d21'56"

Teri Hatch
Cnc 9h20'35" 19d14'45"

Teri Horton's Kiwi Star
Del 20h48'23" 5d19'44"

Teri Jo Carman-Murray
Tau 4h55'10" 16d17'59"

Teri L Breeland
Gem 6h55'46" 22d6'18"

Teri L. Stacher
Vir 13h57'48" -13d33'8"

Teri Lawson
Crb 15h48'54" 37d22'12"

Teri Lee Logsdon
Sco 16h6'0" -14d4'51"

Teri Lindsey
Cam 4h1'1" 59d55'5"

Teri Louise Lybecker
Aqr 22h19'2" -1d6'48"

Teri Lyn Uhr
Lmi 10h2'12" 31d47'14"

Teri Mamola
Cas 0h24'24" 55d20'46"

Teri Marie Peyton
Ori 5h55'40" 11d50'30"

Teri McReynolds
Lyn 8h46'26" 35d48'12"

Teri Michelle Phillips
And 23h49'45" 35d25'30"

Teri Nguyen
And 23h50'2" 40d35'58"

Teri O, Star of My Dreams
And 0h16'41" 27d38'11"

Teri Parente
Cnc 8h6'54" 17d17'49"

Teri Parry
Sco 16h16'2" -10d12'29"

Teri Peck
Cas 1h13'30" 57d1'49"

Teri Ripple
Crb 16h12'12" 38d32'56"

Teri & Shahin's Star
Umi 15h29'1" 72d58'44"

Teri Shaver
Ari 3h7'39" 14d9'19"

Teri Stevens
And 1h33'57" 37d52'23"

Teri The Tiger
Psc 1h17'0" 26d33'54"

Teri Weigel
Psc 1h6'18" 5d56'52"

Teri Williams and Family
Uma 11h30'50" 40d12'58"

Teriann Rose
Sgr 18h19'9" -29d4'54"

Terick Strauch
Uma 9h53'48" 63d48'35"

Teriel Del Rosario Go
Lyn 7h52'28" 35d43'14"

Terilareina
Sge 19h46'56" 16d54'39"

TeriLynn Thomas
Mon 6h49'58" -2d41'4"

Terina
Sco 17h46'42" -41d12'6"

Terina's First Kiss
Uma 9h49'16" 51d57'40"

Teris And Richard
Ori 6h10'13" -3d45'47"

Teri's Endless Dream
Uma 10h22'23" 65d13'32"

Teri's -Heart
Cap 20h36'24" -18d36'39"

Teri's little star
Umi 15h17'55" 73d48'56"

Teri's Star
Peg 23h19'0" 32d26'59"

Teri's Stardust
Lyn 7h17'35" 59d15'16"

Teri's Wishing Star
Cyg 21h31'38" 34d3'5"

Teri-Siam-Mia
Lib 14h43'18" -15d49'22"

Terjen
Cyg 19h35'22" 31d38'10"

Termcat
Her 16h32'44" 17d24'39"

Ternis
Dra 16h38'18" 52d37'55"

TERO
Sco 17h40'20" -37d49'28"

terpsichore
Ari 2h49'58" 27d46'44"

Terra 25-12-2005
Cas 0h48'5" 51d48'54"

Terra Ann
Cap 20h53'54" -25d8'27"

Terra EveLynne
Gem 7h20'23" 15d16'52"

Terra Gallo
Lib 14h46'19" -19d46'48"

Terra Grace Hicks
And 0h9'40" 39d28'32"

Terra Hubble
Uma 11h12'27" 48d56'57"

Terra & Jarred
Sgr 18h36'15" -23d46'17"

Terra Jones
Ari 3h5'25" 25d52'0"

Terra Kirk
Lib 15h16'26" -11d45'48"

Terra Lynn
Uma 8h14'5" 63d41'58"

Terra Meinert
Dra 16h23'14" 52d29'12"

Terra "My Porceliean Doll" Fletcher
Cnc 8h0'4" 13d31'30"

Terra North
Uma 11h22'42" 42d35'40"

Terra Scoles
Psc 0h13'26" 9d51'26"

Terrah
Uma 12h28'16" 55d26'19"

Terrah J. L'Allier
Lyn 7h33'23" 50d52'27"

Terrah Olson
Uma 11h34'14" 62d23'9"

Terra-Lynn "WildeKat" Arroyo
Lyn 7h35'9" 38d22'54"

"Terramarvic"
Cyg 19h47'12" 36d4'19"

Terramort
Dra 14h36'8" 55d33'10"

Terran Gregory Whittingham
Leo 10h54'12" 5d49'18"

Terran Kay
Aql 19h28'2" 10d36'32"

Terrance
Leo 9h32'19" 19d2'6"

Terrance Blount
Her 17h14'7" 19d27'52"

Terrance J. Martin
Uma 10h10'5" 53d57'23"

Terrance Kinney's Star
Dra 17h16'59" 58d39'30"

Terrance L. Buchanan (Terry Lee)
Psc 0h43'29" 18d4'23"

Terrance Michael Praud
Per 3h13'52" 42d30'10"

Terrance Michael-The Guiding Light
Sco 16h44'27" -37d17'42"

Terrance Miller
Dra 20h36'58" 71d31'6"

Terrance Robert Courtice
Cnc 8h46'23" 32d0'33"

Terrance Roy Wilson
Her 18h45'58" 13d50'9"

Terrance Smith
Psc 1h38'52" 26d30'11"

Terrance William West O'Bryan
Psa 22h53'59" -24d51'7"

TerranIV
Uma 10h23'2" 51d49'32"

Terra's Destany Star
Cas 0h23'56" 55d31'27"

Terre Anderson
Cap 20h11'48" -10d10'13"

Terre Lee
Leo 10h59'58" 6d6'41"

Terrell Brown
Ori 5h34'7" -0d14'40"

Terrell Gene Miller
Lib 15h0'42" -22d40'53"

Terrell Johnson
Boo 14h27'26" 50d54'24"

Terrell M. Jones
Aql 19h21'41" -9d57'9"

Terrell Mayes
Sgr 19h32'47" -13d49'10"

Terrell Star
Per 3h38'47" 47d11'31"

Terrellariola
Pho 0h45'18" -53d3'52"

Terren Joe Williams
Ori 5h21'41" 7d15'31"

Terrence
Aur 5h38'34" 41d45'24"

Terrence
Sgr 18h53'30" -16d1'45"

Terrence
Mon 7h45'24" -8d20'35"

Terrence Blaze Sekerak
Sco 16h15'7" -9d20'6"

Terrence Holliman Jr.
Uma 8h35'28" 47d6'14"

Terrence Lee Grewe
Ari 3h12'30" 28d46'51"

Terrence Leslie Moore
Ori 6h0'27" 13d37'18"

Terrence Michael Riley
Cap 20h12'14" -10d8'55"

Terrence Patrick McCann
And 23h15'30" 43d1'30"

Terrence R. Yacap, Jr.
Aql 18h53'3" -0d5'0"

Terrence Wade Snyder
Leo 9h41'45" 18d4'54"

Terresa
Lib 15h30'6" -27d22'8"

Terresa And Francisco's Star
Lib 15h36'0" -18d28'33"

Terri
Sgr 19h35'6" -14d30'24"

Terri
Cma 7h20'45" -17d12'48"

Terri
Cap 20h36'43" -12d47'34"

Terri
Lyr 19h8'37" 42d32'12"

Terri
Uma 11h50'48" 39d44'32"

Terri
And 2h18'59" 39d0'59"

Terri - 2006
And 23h35'44" 35d52'59"

Terri and Ellis
Nor 16h4'33" -43d12'46"

Terri and Presh Together Forever
Cyg 19h59'20" 52d43'8"

Terri Ann
Ori 6h18'42" -0d10'29"

Terri Ann Curren
Cnc 9h5'53" 30d58'15"

Terri Anne Matthews - 3.12.1986
Cru 12h19'18" -57d20'26"

Terri & Athena's Star 09-2003
Gem 7h44'17" 30d58'44"

Terri B.
Mon 7h39'18" -0d47'26"

Terri Bishop Duff
Cam 4h8'57" 56d39'43"

Terri Chayko
Sco 17h41'56" -37d6'43"

Terri Cicchetti
Cas 1h26'41" 57d32'50"

Terri Davault
Boo 14h34'22" 28d33'11"

Terri Delo
Leo 10h58'47" 23d42'26"

Terri Denise Karpavicius
Lyn 6h44'12" 61d5'48"

Terri Dingfield
Uma 12h1'11" 39d43'56"

Terri Donna Kestenbaum Feuerstein
Psc 1h8'56" 26d31'26"

Terri Elyse Chaseley
Per 4h27'30" 36d14'44"

Terri Frend
Ori 5h56'42" -0d19'44"

Terri Furry
Leo 9h34'40" 28d48'44"

Terri Gallen Edersheim
Cyg 21h30'39" 50d35'19"

Terri Galloway
Cyg 21h44'3" 50d18'14"

Terri Garnhart
Cyg 21h42'50" 44d14'34"

Terri Garrett
Ser 15h54'28" -0d3'16"

Terri Green
Gem 6h50'41" 17d5'47"

Terri Harris
Psc 1h34'59" 8d7'11"

Terri J. Lamberti
Aqr 22h32'13" -2d35'50"

Terri Jean Heuer
Vir 12h34'13" 7d8'23"

Terri Jean Williams
Lep 5h20'18" -11d29'59"

Terri Jenkins
Mon 6h49'39" -0d47'31"

Terri & Jimmy O'Hanlon 30 Years
Cyg 21h11'49" 33d45'52"

Terri Jo
And 0h25'26" 44d19'45"

Terri & Kole
Lyn 8h49'15" 44d40'39"

Terri Kurtz
Cas 0h43'22" 53d13'48"

Terri L. Riley
Aqr 22h1'18" -17d1'48"

Terri L Tillis & Jeannine M Cruz
Lyr 19h5'45" 33d4'15"

Terri L. Winterbottom
Tau 4h23'38" 18d33'22"

Terri Lacy
Vir 12h26'2" 8d0'35"

Terri Lee
Cam 4h9'22" 66d20'37"

Terri Lee Childers
Cap 21h41'35" -17d58'37"

Terri Lee Davis
Dra 18h47'11" 50d53'56"

Terri Lee Tennill Johnson
Cap 21h36'58" -13d35'39"

Terri Leigh Harper Cook
Sco 16h10'44" -15d34'45"

Terri Loraine
Dra 19h19'16" 61d21'8"

Terri Louise Cray
Crv 12h30'48" -14d30'47"

Terri Lyn Harrell
And 0h37'16" 40d23'58"

Terri Lynn
Cnv 12h42'15" 39d39'22"

Terri Lynn
And 0h42'7" 26d49'8"

Terri Lynn
Umi 17h33'35" 83d24'45"

Terri Lynn Austin
Cen 13h20'54" -34d50'19"

Terri Lynn Bernstein
Lyn 8h42'53" 46d8'3"

Terri Lynn Clark
Leo 9h24'39" 26d55'32"

Terri Lynn Henley
Aqr 22h47'19" -3d53'53"

Terri Lynn McBroom
Lyr 18h58'36" 27d5'9"

Terri Massy
Cyg 20h59'15" 49d23'43"

Terri McCaffrey
Ori 5h35'44" 3d48'40"

Terri Miller
Lyn 6h36'52" 59d52'43"

Terri Miller / Tupac Shakur
Crb 15h43'47" 27d29'21"

Terri Mott-Esteves
Psc 1h16'26" 28d23'2"

Terri Mouse
Ari 2h37'50" 13d6'12"

Terri Murphy
Com 13h0'55" 17d0'32"

Terri My Boo Happy 50th Birthday
And 23h31'46" 47d46'8"

Terri My Love
Tau 5h58'46" 25d13'26"

Terri Nel
Crb 15h28'28" 30d57'13"

Terri Nicole Turner
Ari 3h27'42" 28d25'20"

Terri O'Shea Kerby
Cap 21h8'56" -25d24'25"

Terri Ozarkiw
Aqr 21h3'59" -8d40'49"

Terri Reynolds Berquist
Peg 22h59'24" 9d47'59"

Terri Rhodes
Leo 11h17'53" 16d53'16"

Terri Roseanne Dyer
Cas 0h18'9" 55d52'0"

Terri Sherretta
Lyr 18h44'9" 38d44'20"

Terri Sims
Gem 7h24'4" 29d13'42"

Terri Smith McNamee
Uma 8h21'26" 70d0'12"

Terri Sue
Tau 5h46'52" 22d27'20"

Terri Teuton
Tau 3h49'43" 18d35'8"

Terri (Tiger) Cummings
Peg 22h47'12" 2d53'41"

Terri Vanessa Smith-Davis
Psc 23h49'54" 7d0'45"

Terri Waite
And 0h12'14" 43d58'48"

Terri White
Crb 15h29'48" 30d45'50"

Terri Williams
Uma 13h54'17" 50d24'58"
Terri Wilson "Sign Chi Do Seed"
Crb 15h34'34" 26d14'57"
Terri Zamaites
Ari 2h49'57" 20d6'20"
Terri, A Guiding Light To Many
Cnc 8h53'24" 24d56'38"
terri130204
Sgr 19h47'37" -13d38'33"
TerriAndrew
Cyg 20h56'19" 31d59'7"
Terrianne
Lib 16h1'36" -8d54'37"
Terri-bear
Aqr 23h53'36" -13d36'33"
Terric & Jenna
Lib 15h8'46" -6d22'5"
Terrie A. James
Uma 11h13'21" 50d25'7"
Terrie Alfieri
Psc 1h18'57" 22d22'47"
Terrie Anne Packard
Tri 1h53'31" 29d23'22"
Terrie Caldwell-Ayre
And 23h44'20" 46d17'10"
Terrie Lee Molina
Cas 0h12'40" 51d10'51"
Terrie Louise Austin
And 23h31'53" 42d6'45"
Terrie Lynn "Jinx" Coates
Cas 0h27'25" 51d28'28"
Terrie Marie Edus
Gem 8h4'42" 32d43'59"
Terrie W. Brady - Twin - 3-3-1955
Psc 1h42'55" 9d24'44"
Terrie Westfall
Gem 6h4'45" 23d24'8"
Terrie's Eternal Gift
Mon 6h51'31" -0d18'59"
Terrific Thomas
Lmi 9h34'32" 38d58'4"
Terrific Tinsey
Sgr 19h22'27" -21d39'45"
TERRILL
Gem 7h18'29" 19d12'47"
Terrill Collier
Gem 7h53'12" 28d20'48"
Terrill Norman Brown
Dra 16h12'9" 52d9'31"
TerriLynne and Keir
Dra 17h12'53" 56d37'16"
TERRINA
Tau 4h12'48" 30d47'42"
terrip
Cnc 9h15'32" 28d19'12"
Terris D. Smith Jr.
Her 17h11'56" 24d51'15"
Terri's Diamond
Crb 15h42'56" 26d50'20"
Terri's Ever ~ Glowing Light
Lyr 19h7'54" 36d3'3"
Terris Hall
Tau 5h34'52" 26d20'1"
Terri's Light
Lyn 7h36'47" 37d8'51"
Terri's Light In The Sky
Cap 20h25'53" -11d16'13"
Terri's Sky Diamond
Uma 11h1'28" 60d6'26"
Terri's Star
Sco 16h13'54" -25d9'26"
Terri's Twinkling Bright Star
Aqr 21h53'25" -1d58'4"
Terri's Wishing Star
Mon 6h49'36" 6d55'46"
Terri-Susan
Gem 7h42'11" 30d40'7"
Terror Lynn Howarth
Ori 6h13'19" 15d56'38"
TerrStan
Ori 6h9'39" 19d11'35"
Terry
Boo 14h51'17" 25d10'19"
Terry
Equ 21h17'29" 7d35'38"
Terry
Mon 6h31'57" 8d20'8"
Terry
Cnc 8h37'28" 13d51'41"
Terry
Aql 20h5'26" 4d39'12"
Terry
Aur 7h6'42" 42d26'11"
Terry
Per 3h39'19" 45d35'6"
Terry
Umi 15h32'39" 79d50'0"
Terry
Cap 20h28'5" -20d20'50"
Terry
Cap 21h33'13" -17d11'42"
Terry
Vir 13h56'45" -16d9'28"
Terry
Cas 1h3'22" 60d52'1"
Terry Alan Dodd
Psc 1h7'35" 29d13'52"
Terry Allen Nash
Uma 11h43'13" 62d15'21"

Terry and Amanda
Cas 1h33'52" 68d19'9"
Terry And Cheryl's Destiny
Sco 16h59'47" -38d20'38"
Terry and Denise
Cyg 21h13'33" 52d51'59"
Terry and Gina
Cyg 20h31'16" 34d53'21"
Terry and Jim Bamberg
Cyg 19h46'37" 32d26'4"
Terry and Joan Ferguson
Uma 10h50'12" 65d58'53"
Terry and John's Star
Her 17h35'49" 46d48'1"
Terry and Kristi
Cyg 20h43'15" 42d32'43"
Terry and Rollie Seidler
Aqr 22h9'36" -0d9'22"
Terry and Victoria
Dra 20h33'11" 69d39'36"
Terry Anderson Vosburgh
Aur 6h34'55" 34d20'18"
Terry Ann Moffett
Uma 10h8'23" 48d55'10"
Terry Ann Rodriguez
Lib 15h23'4" -13d40'28"
Terry & Ann Skey
Cyg 19h47'36" 35d24'13"
Terry Armstrong
Per 3h10'18" 54d19'54"
Terry Arnold
Uma 9h41'5" 48d40'43"
Terry Arthur Overby
Vir 14h9'37" -16d18'27"
Terry Austin
And 0h47'1" 40d26'20"
Terry Bardon
Uma 10h55'12" 40d20'56"
Terry Baugus
Uma 10h2'25" 67d57'7"
Terry & Bea Hummel 40 Anniversary
Cyg 19h25'57" 28d39'56"
Terry & Belinda Waggoner
Cru 12h18'3" -58d25'35"
Terry Bell
Mon 6h54'50" -8d53'41"
Terry Bendel
Cyg 20h49'25" 31d54'38"
Terry (Blue) Combs
Lib 14h53'53" -9d10'48"
Terry Brand
Her 16h32'31" 14d7'13"
Terry "Bright Eyes" Hewett
Aqr 22h29'46" -8d44'41"
Terry Brown
Tau 5h54'24" 23d54'28"
Terry Burgess
Ari 2h59'49" 18d39'1"
Terry Burns of Collierville, TN
Uma 12h5'40" 57d30'4"
Terry Carey
Per 4h14'33" 47d31'46"
Terry Carlton Jordan
Cap 21h57'55" -19d15'3"
Terry Carole Azar
Ari 2h48'58" 18d51'37"
Terry Charette
Sco 16h6'20" -15d25'21"
Terry & Christine Coapstick 25 yrs
Cyg 19h44'40" 33d16'16"
Terry Clark
Umi 19h49'21" 87d13'53"
Terry Coe
Per 2h26'40" 51d24'53"
Terry Connelly
And 23h30'31" 46d35'54"
Terry Contreras
And 0h41'4" 25d40'2"
Terry Cullen
Uma 12h10'10" 59d43'36"
Terry D. Ohlemeier
Aql 19h23'31" 0d4'19"
Terry D. Smith
And 0h43'5" 41d5'9"
Terry D. Vanderschuur
Sco 17h37'7" -41d40'12"
Terry Dean Barone
Psc 0h41'53" 6d39'38"
Terry & Debbie Lanni
Cyg 20h4'3" 38d59'29"
Terry DeFilippo
Lyn 6h25'13" 57d1'20"
Terry Duonola
Ari 3h19'17" 28d56'29"
Terry Dye
Aql 19h40'18" 11d48'47"
Terry E Newman
Her 16h8'22" 23d15'41"
Terry Elaine Clements
Uma 9h25'0" 65d27'58"
Terry Elena Raines
Gem 7h26'17" 20d45'27"
Terry F Lynch
Cap 20h55'50" -20d57'34"
Terry Ferguson
Ori 5h59'6" 17d15'6"
Terry Gene Umfleet Jr.
Cnc 8h32'19" 17d34'50"
Terry Gibson
Dra 18h53'40" 66d37'40"
Terry Glen Adams
Lmi 10h33'4" 36d21'26"

Terry Glenn
Lyn 7h55'29" 46d6'4"
Terry Hildebrand
Per 3h13'21" 31d32'58"
Terry Howerton - Shining Star
Uma 11h28'43" 43d27'42"
Terry J. Barilleaux
Sco 16h18'57" -15d33'27"
Terry J Finestein
Psc 23h16'19" 7d15'31"
Terry J. Fontana
Per 3h34'34" 37d1'13"
Terry J. Foust
Umi 15h21'30" 74d0'53"
Terry J. Foust
Umi 14h41'4" 71d49'6"
Terry J. Pakus
Lib 15h35'25" -23d20'33"
Terry J Ulizzi
Sgr 19h56'13" -33d8'0"
Terry & Karen Wells
Cyg 20h59'43" 30d45'33"
Terry Kay Campbell
Cas 2h44'24" 57d51'46"
Terry&Kaz
Cyg 20h0'10" 32d5'30"
Terry Kemple
Lyn 8h31'41" 39d37'9"
Terry Kilburn
Cep 21h10'52" 66d20'59"
Terry Krause
Gem 6h30'51" 21d8'41"
Terry L Franks
Aql 19h8'38" 1d9'37"
Terry L. Janke
Uma 11h21'20" 54d44'19"
Terry L. Smith
Her 16h35'49" 31d47'47"
Terry L. Tripp
Del 20h40'19" 17d11'59"
TERRY L. VINCENT SR.
Leo 10h28'2" 17d49'21"
Terry L. Woolverton Jr.
Vir 13h54'53" -8d5'35"
Terry + Lauren
Lib 15h26'10" -7d42'38"
Terry Layne
Cam 4h33'47" 58d7'48"
Terry Lee
Lyn 7h12'50" 47d45'4"
Terry Lee Burdick
Gem 6h55'43" 33d23'41"
Terry Lee Dauphinee
Vir 14h18'40" 2d25'49"
Terry Lee Fowler Kavorinos
Psc 1h51'0" 9d49'28"
Terry Lee Hickey
Psc 0h51'4" 16d10'15"
Terry Lee McQuery
Lyr 19h25'14" 37d39'49"
Terry Lee Rossi
Sco 17h38'40" -39d22'53"
Terry Lee Warner, Jr.
Tau 4h8'36" 6d0'20"
Terry Lee's Point of Light
Psc 1h33'33" 16d44'56"
Terry Lynn
Leo 10h33'38" 25d34'10"
Terry Lynn
Crb 16h0'8" 27d39'28"
Terry Lynn
Umi 16h37'53" 78d10'27"
Terry Lynn George
Aqr 22h38'19" -15d4'20"
Terry Lynn Hubbard
Uma 10h53'18" 53d58'38"
Terry Lynn Kemp Cavallo
Leo 9h32'3" 27d2'25"
Terry Lynn Kerley Cooke
Cap 21h29'49" -15d22'25"
Terry Lynn Rogers
Umi 14h6'18" 75d9'40"
Terry Lynn Wilson
Cas 0h4'28" 53d19'31"
Terry M. Miller
Cnc 7h59'11" 12d39'51"
Terry Marie Long
Boo 14h56'40" 26d18'19"
Terry Mark DuSell, Star in Heaven
Uma 11h16'55" 31d3'16"
Terry Mark Smith
Her 16h28'39" 47d36'45"
Terry Marley
Cyg 19h37'28" 37d57'2"
Terry Martinson
Vir 12h10'23" 12d33'22"
Terry May Sr.
Uma 11h49'52" 63d46'14"
Terry Mayer
Her 17h50'11" 27d57'7"
Terry Mesritz
Ari 2h22'30" 18d29'53"
Terry Michael Becker
Sgr 18h1'42" -28d0'7"
Terry Micheal Byrd
Her 16h21'47" 20d51'39"
Terry & Mike's Love Burning Bright
Cyg 20h23'6" 40d22'50"
Terry Mills
Uma 10h0'5" 46d0'42"
Terry Myers
Uma 9h30'47" 49d18'24"

Terry Noss
Per 3h34'59" 44d35'31"
Terry Owens & Howard Hill Jr
Cnc 9h1'47" 15d41'48"
Terry P. Maynard
Uma 11h9'37" 45d6'52"
Terry & Patty Baker
Cyg 20h30'57" 53d40'51"
Terry Pieczko
Oph 17h7'6" -0d3'10"
Terry R. Taber
Lib 15h17'31" -8d27'27"
Terry R. Willis
Ari 2h34'41" 23d36'6"
Terry R Zeller Sr
Ori 5h53'15" 19d32'25"
Terry Raymond Conway
Cnc 8h25'58" 15d36'56"
Terry Remley
Uma 9h25'38" 47d4'45"
Terry Richards
Her 16h42'44" 27d12'26"
Terry Robert Hubbard
Cap 20h28'21" -24d51'11"
Terry S Richardson - A Mothers Love
Leo 11h22'27" -6d8'20"
Terry Sams of Data Tech Comm
Sgr 18h30'4" -16d58'46"
Terry & Sandy O. 4-17-98
Umi 15h22'37" 72d17'35"
Terry Silver
Umi 14h35'36" 75d3'41"
Terry Sinnott
Leo 11h40'59" 23d45'21"
Terry Smith
Gem 7h24'46" 32d45'33"
Terry (Sparky)
Vir 13h13'31" -5d0'50"
Terry & Stacie
Cyg 21h22'55" 45d11'23"
Terry & Steve Kraemer
Cyg 20h50'51" 42d59'13"
Terry "Sunnshinne" Utley
Gem 7h38'46" 32d49'25"
Terry Tayler
Uma 8h31'55" 64d22'56"
Terry Terrell Harrell
Lyn 6h56'49" 59d37'24"
Terry Thayer
Cnc 8h39'47" 32d2'32"
Terry Thompson
And 23h36'54" 47d3'8"
Terry "T.T." Hotaling
Uma 9h26'25" 47d43'4"
Terry W. Goff
Ori 5h49'37" -0d4'22"
Terry W. Jordan
Cyg 19h57'37" 39d12'23"
Terry Ward Perkins 07/30/45
Per 3h32'53" 44d32'50"
Terry Wickenhauser
Uma 11h58'40" 56d8'42"
Terry Willis
Gem 6h49'29" 17d35'27"
Terry Winslow-Navarro
Peg 21h43'5" 21d43'11"
Terry, Carol, & Maddie Hunt
Dra 18h48'1" 51d23'55"
Terry, Mandy, & Myah's Star
Umi 16h45'44" 80d42'6"
Terry, "My Destiny"
Cap 21h27'54" -19d43'36"
Terry, Our Goddess of Hearth & Home
Vir 14h7'20" -16d27'54"
Terrye Wagner
Gem 6h33'52" 18d36'39"
Terryheather
Cen 14h2'37" -48d15'52"
Terryl Lee
Tau 4h53'58" 26d4'40"
Terryl Terrell
Peg 22h20'3" 11d33'6"
TerryLee & Terri Jo Durling
Gem 7h41'56" 32d27'22"
TerryLynn Paluch
Psc 1h8'48" 12d19'52"
Terry-Marie
Per 3h13'18" 43d24'39"
Terry's Place
Uma 10h58'27" 46d8'2"
Terry's Red Planet
Uma 11h45'14" 45d5'0"
Terry's Sanctuary
Uma 13h6'8" 56d14'7"
Terry's Smile
Umi 14h17'5" 77d8'53"
Terry's Star
Uma 14h15'57" 58d14'6"
Terry's Star
Uma 13h42'35" 60d49'42"
Terry's Star
Uma 9h49'4" 70d38'29"
Terry's Star
Cam 4h2'46" 70d14'56"
Terry's Star
Lyn 7h35'2" 38d29'55"
Terry's Star
Per 4h21'42" 35d39'55"

Terry's Star, Forever in our Hearts
Per 3h25'36" 39d50'7"
Tertius Septimus Edward & Kelly
Ori 5h59'50" -1d26'31"
Teruna
Vir 13h12'50" -8d48'21"
Terutsugu Ishida
Leo 10h33'25" 17d0'36"
Tervia I
Dra 11h3'14" 78d17'25"
Teryl G. Tenter
Ori 5h57'59" -0d29'53"
Teryn Jo Lukes
And 1h40'3" 48d59'13"
Terynn Gabriel Brink
Col 6h2'5" -28d25'14"
Teryn's Star
Uma 11h25'31" 46d21'22"
Tesch, Beate
Uma 10h54'14" 43d1'57"
Te'Sean Funchess
Uma 9h46'4" 69d59'15"
Tesh Aynalem
Ari 1h47'3" 19d34'21"
Tesha Ramos
Vir 13h12'24" -21d42'1"
TESKA ALEXANDRA POLESCHUK
Tau 3h43'18" 25d55'16"
Tesla
Her 16h58'39" 32d11'5"
Teslynn Dawn-Marie Smith
Lib 15h49'59" -17d16'17"
Tesómnak-Sándornak
Uma 9h50'24" 53d16'6"
Tesora Adorato
Leo 11h10'56" 1d9'51"
Tesorino Mio
Tau 4h59'31" 19d35'3"
Tesoro
Her 17h12'6" 16d27'48"
Tesoro
Her 18h46'3" 19d1'30"
Tesoro di Domenico
Per 3h36'42" 45d24'47"
Tesoro Escondido
Ori 5h21'39" -8d26'0"
Tesoro Hermosa
Uma 9h32'48" 56d4'7"
Tesoro mio, ti amero' sempre
Sgr 18h12'47" -20d19'41"
Tesoromio
Lyn 7h36'15" 38d31'23"
Tess
Ari 3h0'33" 30d8'37"
Tess
Per 3h10'15" 46d2'53"
" Tess "
Cas 1h8'6" 49d5'50"
Tess
Leo 10h34'0" 22d55'10"
Tess
Tau 5h34'0" 18d32'5"
Tess
Lib 15h44'2" -12d59'55"
Tess Annique
Gem 7h12'23" 26d29'42"
Tess Bassford
Cap 21h1'15" -20d48'40"
Tess Brown
And 2h24'45" 45d51'5"
Tess Cameron McDougal
Leo 10h31'18" 12d41'33"
Tess Catherine
Sco 17h9'46" -30d39'3"
Tess Constance Horgan
And 23h38'22" 41d51'20"
Tess Cowherd
Lyr 18h41'28" 33d48'20"
Tess Craft Johnson
Crb 16h13'42" 34d5'58"
Tess Criswell
And 1h57'48" 45d16'26"
Tess Cuda
Lib 14h47'21" -10d49'22"
Tess Eleanor "Smush" Runyon
Gem 7h36'34" 16d23'23"
Tess Eliza
Lib 14h44'30" -14d24'46"
Tess Elizabeth Hayes 19 02 04
Col 6h3'17" -28d34'33"
Tess "Forever Happiness & Love"
Lib 15h59'52" -9d21'33"
Tess Gabrielle
Sgr 19h29'39" -12d46'27"
Tess Galer
Aqr 22h16'9" -12d59'53"
Tess Gray
Uma 10h30'47" 64d43'42"
Tess Hains
Uma 8h47'15" 71d1'35"
Tess Hill
Lib 15h3'5" -24d3'15"
Tess Kidden Metzelaar
Tau 5h8'40" 19d6'49"
Tess Klatt - Anam Cara
Leo 11h0'30" 18d12'37"
Tess Lena Kraus
Uma 11h33'0" 50d11'49"

Tess Lovely
And 0h21'7" 26d44'3"
Tess Madeleine Lee
Tau 4h27'31" 27d9'58"
Tess Menlove
Ari 3h28'34" 20d57'48"
Tess Messina
Uma 8h47'20" 57d16'21"
Tess Rijen Adams
Sgr 18h14'21" -28d36'48"
Tess Sabiston
And 22h58'26" 47d26'12"
Tess the Filipina Angel
Cap 20h35'40" -26d20'58"
Tess Urrutia
Cyg 20h0'19" 33d15'53"
Tess Victoria Potter
Leo 11h24'23" 1d38'54"
TESSA
Tau 5h50'10" 27d18'46"
Tessa
And 0h8'9" 44d29'41"
Tessa
Cru 12h37'7" -59d25'16"
Tessa
Dra 16h42'35" 66d13'21"
Tessa
Vir 12h33'39" -2d11'40"
Tessa Adams
Sco 16h4'50" -9d3'42"
Tessa Claire Gile
Tau 5h8'3" 19d39'42"
Tessa Claire Mills
And 0h34'52" 42d4'45"
Tessa Derby Graham
Umi 14h59'32" 68d18'38"
Tessa Elaine Waxweiler
And 1h1'37" 44d17'39"
Tessa Everett
Lib 15h27'30" -6d59'9"
Tessa Finelli
Leo 9h41'48" 32d27'48"
Tessa Halbritter
Uma 13h49'33" 60d49'32"
Tessa Jane Arbenz
Vir 12h33'51" 8d44'57"
Tessa & Jason
Uma 11h10'14" 62d40'41"
Tessa Jo Petty
Peg 22h29'22" 4d32'8"
Tessa Maria Castaneda
Uma 9h19'28" 69d5'47"
Tessa Marie Bresnen
And 2h18'14" 45d34'36"
Tessa Marie Kristof
Her 16h7'40" 47d31'16"
Tessa Natalina Fratini
Uma 11h10'40" 69d43'23"
Tessa Nicholle Haynie
Sco 16h5'3" -13d16'36"
Tessa Nicole Paneri
Ori 5h57'42" 9d46'46"
Tessa Noel Leone
Gem 7h11'48" 28d38'0"
Tessa Noriega
And 0h7'30" 41d2'47"
Tessa Paige
Uma 13h46'34" 53d4'50"
Tessa Rose Tigar Cross
And 0h41'8" 37d43'54"
Tessa Russo
Cam 4h6'16" 69d2'12"
Tessa Ruth Berkley
Umi 15h57'58" 70d35'29"
Tessa Sherman
Oph 17h43'37" -0d38'11"
Tessa van den Berg
And 23h17'9" 50d48'34"
Tessa Wood
Cru 12h4'31" -63d48'10"
Tessabel C. Welch
Psc 1h5'27" 31d57'27"
Tessah D. Belcher
Ori 6h4'15" 9d32'19"
Tessah Rebecca Todl
Vir 13h7'34" 11d43'43"
TessalovesJonny
Cep 22h7'8" 61d19'57"
Tessamari Ruth Novak
Gem 7h29'20" 32d6'29"
Tessan
Cap 20h27'5" -10d48'46"
Tessaria
Vir 12h38'49" 7d38'12"
Tessa's Wishing Star
And 1h18'4" 37d22'42"
Tessera Rayne
Sgr 18h50'57" -28d35'48"
Tessie d'Auberville Delattre
And 0h38'36" 27d33'39"
Tessie E. Schaller
Cnc 8h32'47" 6d44'8"
Tessie Tangan
Com 12h4'17" 28d9'29"
Tessie Young
Com 12h20'59" 27d29'17"
Tessina Schenk
Tri 2h37'16" 35d3'39"
Tessius
Ori 6h11'12" 16d30'30"
TessJim1955
Uma 13h51'7" 56d11'50"
Tesslar-B
Uma 9h35'7" 66d8'40"

Tesslynn
Aur 5h34'22" 45d51'23"
TessRose
Crb 16h12'25" 33d8'5"
Test
Uma 10h49'23" 69d33'27"
test
Cas 23h31'24" 56d9'56"
Tesya Maria Yaninas
Dra 18h49'8" 59d38'0"
Tetay, mi amor
Cas 0h27'15" 54d23'32"
Tété & Pablo
Peg 21h56'27" 16d26'58"
Teti
Per 4h20'46" 42d46'2"
Tetsu Nishi
Tau 3h45'29" 3d21'21"
Tetsuko Konishi
And 2h27'8" 44d6'7"
Tetsuo
Leo 10h57'20" 15d59'11"
Teuber, Christine
Ori 5h56'2" 21d42'21"
Teubner, Steffen
Uma 13h7'35" 62d11'46"
TEUFEL
Uma 10h34'59" 61d8'53"
Teun van der Starre
Cep 0h23'38" 71d23'1"
Teunis Jaiden Fluit V
Lib 15h11'25" -6d54'1"
Teuta
Umi 15h52'33" 87d24'17"
Teva
Umi 14h0'1" 75d12'10"
Tevethia
Lmi 10h25'23" 37d26'49"
TewieDot
Peg 21h32'33" 24d58'4"
Tews-Jadan, Renate
Uma 14h15'33" 59d26'57"
Tex
Per 3h18'53" 41d33'5"
Tex Andrew's Forever Shining Light
Umi 14h22'13" 68d31'28"
Tex Farrell
Del 20h38'18" 6d5'23"
Tex Francis
Sco 16h16'35" -11d45'11"
Tex Kidd
Uma 11h29'56" 36d37'51"
TEX RISING CELEBRITY STAR
Psc 1h11'51" 28d44'38"
Tex Winter
Lmi 10h3'41" 37d17'32"
Texan Rattler
Vir 12h48'34" 4d1'53"
TexAnnie 13
And 0h37'39" 31d46'17"
Texas A. Durham
Vir 14h5'4" -16d28'28"
Texas Aggie 12th Man
Tommy Orr
Dra 17h45'22" 54d24'11"
Texas Tea
Boo 14h40'36" 37d14'38"
Texas Tornado
Her 18h54'2" 23d7'46"
TEXASANGEL
Lyr 18h38'50" 32d7'18"
Tex's Little Super Star "Andrea"
Umi 14h16'23" 67d32'16"
Tex's Star of Texas
Cnc 8h38'30" 17d15'1"
Teya J. Evans
Uma 8h47'43" 52d19'57"
Teyanea
Ori 6h1'31" 19d22'46"
T-Eye-Ffanny
Psc 23h50'9" 0d11'40"
Teyen Laughlin Persicke
Ori 4h56'50" 15d35'38"
Tez
Leo 11h36'28" 26d51'17"
Tez
And 2h31'24" 42d11'19"
Tezel
Aur 6h24'8" 41d23'29"
TFB & BAS = Soulmates 4 Eternity
Cru 12h27'8" -60d1'43"
T'F.E
Uma 10h18'48" 41d24'27"
TGCWAngelBrnEyz103093 0
Lib 15h56'56" -20d19'20"
Tgirl123
Cnc 8h43'16" 31d16'55"
TGM
Cnc 9h4'30" 31d13'0"
TGR
Sco 16h6'30" -15d50'52"
T.G.'s Desert Diamond
Oph 17h28'11" 10d3'58"
TGS1799
Ari 2h10'27" 22d34'1"
TGS-2006
Per 3h10'14" 50d50'19"
TGSR Owns You
Leo 11h42'36" 18d50'44"

TGuensler
Crb 15h26'18" 31d46'53"
T.H. and Faye - 50 Years Together
Eri 3h35'11" -7d39'45"
Tha Star of Litsa Tsalimopoulou
And 23h0'49" 47d36'47"
Thabiti
Cyg 19h55'47" 37d36'39"
Thad Pryor
Uma 9h12'58" 57d12'41"
Thaddeus
Cep 3h36'35" 80d46'24"
Thaddeus J. Krolicki, M.D.
Lyn 8h58'50" 41d34'49"
Thaddeus Kapla
Uma 12h38'5" 62d52'52"
Thaddeus Margareth
Cen 13h36'43" -46d10'3"
Thaddeus P. Daszkiewicz
Her 17h32'34" 36d30'59"
Thaddeus Tomasik
Boo 13h36'34" 22d12'21"
Thaddis Bosley Jr
Vir 12h50'18" 5d49'36"
Thai & Heather
Cyg 19h50'40" 47d18'44"
Thaila
Cas 0h34'11" 48d46'36"
Thaim in grabh leat
Lyn 7h48'57" 54d19'4"
Thain
Gem 7h9'34" 33d52'43"
Thais Da Silva
Her 18h37'21" 18d57'23"
Thalassa Elizalde
Dra 18h49'23" 51d39'5"
Thaldia James Vick
Ori 4h52'14" 11d46'43"
Thalia
Cyg 20h21'20" 40d18'48"
Thalia
Lyn 7h18'40" 57d42'3"
Thalia Alyss
And 23h31'40" 41d29'46"
Thalia Burkholder
Uma 14h20'35" 61d8'41"
Thalia Danielle
Dra 19h30'18" 67d45'42"
Thalia Dech
Lib 15h13'38" -19d25'20"
Thalia Halperin
Sco 16h9'53" -28d24'27"
Thalia Jones
And 2h31'54" 43d23'17"
Thalia & Justyn's Light
Cra 18h2'32" -37d11'13"
THALMA #60
Lyn 9h12'19" 34d24'43"
Thalya
Lib 14h52'35" -4d42'39"
Thanarat Klinhom
Uma 11h34'0" 53d17'49"
Thanatopsis
Ant 10h11'13" -40d14'3"
Thane Brennen
Psc 23h23'37" 5d33'14"
Thanh
Cap 20h12'38" -25d43'59"
Thanh D. Mead
Cas 0h26'22" 53d42'35"
Thanh Nguyen Tran
Vir 14h31'4" 2d28'30"
Thanh Tu Nguyen
Lib 15h0'44" -1d19'50"
Thanh-Ca Nguyen
Vir 12h36'45" 8d31'31"
Thania y Armando
Pho 0h14'18" -40d53'8"
Thank Fine Love Happy Rej Nulty Met
Pho 23h34'0" -51d54'7"
Thank God We Found Each Other
Peg 22h35'52" 17d40'59"
Thank You
And 23h0'54" 48d35'3"
Thank you Jack!
Vir 14h41'0" 4d10'41"
Thank you Jack!
Vir 12h32'16" 0d12'44"
Thank You Jenn
Vir 12h48'36" 0d18'25"
Thankyou
Dra 9h48'9" 77d43'30"
Thanos, Max
Ori 5h19'52" 3d16'33"
Thanti Powers
Cas 0h35'46" 77d36'46"
Thanuri
Lib 15h6'51" -13d34'34"
thao
Uma 10h50'52" 44d11'59"
Thao 1980
Aqr 21h6'46" -8d45'27"
Thao Le
Vir 13h59'44" -13d47'6"
Thao Nguyen
Psc 1h21'43" 31d38'44"
Tharon's Star
Ari 2h2'6" 23d42'39"
That and Mai
Cyg 20h32'41" 38d6'12"
"That One!" Brian T. Nesgoda
Cap 20h22'22" -16d18'42"
Thatch
Leo 10h22'5" 22d9'46"
Thatch: A Star at Sixty
Cyg 20h0'5" 43d54'50"
Thatch B.
Ori 6h8'12" 8d58'42"
That's Fred
Ari 3h11'11" 23d34'22"
That's my mom, Lucy Mae
Sgr 17h50'55" -28d45'3"
Thaylah Noelle...My Gift from God
Cyg 21h57'23" 45d48'24"
The 17th Floor
Her 16h35'28" 13d52'25"
the 2 lights
Ori 5h36'31" 9d29'15"
The 28th of Feb.MRA + KJM. forever.
Cas 23h59'15" 54d28'59"
The 2Jz
Sgr 20h22'43" -42d27'34"
The 3:30 Star
Cyg 20h50'28" 44d34'28"
The 30 year Molina Star
Per 3h35'27" 49d38'1"
The 3rd of July
Lib 15h3'29" -9d26'16"
The 5 Points of Love
Dra 17h19'19" 53d22'30"
The A & A Hutton Star
Gem 6h49'12" 26d2'7"
The Aaron Joshua Espiritu Star
Cru 12h44'52" -58d45'33"
The Aaron Weart DUH ! ! STAR
Vir 11h51'39" 8d24'32"
The Abba Star for Dr. Ernesto Pruna
Sgr 18h55'41" -16d5'54"
The Abbess Jeanette Eddy
Uma 9h38'31" 55d52'38"
The "Abbey Isla" Star.
Umi 6h18'33" 89d2'36"
The Abbott All Star
Umi 13h23'26" 71d20'58"
The Abe
Lyn 8h15'19" 54d3'23"
The Accomplices' Hideout
Lib 14h49'58" -2d8'42"
The Achiever
Leo 9h54'30" 7d35'57"
The Adam-Kayleigh Star
Cyg 22h0'7" 52d31'3"
The Adams
Umi 16h21'19" 83d44'54"
THE ADAMS FAMILY
Lyn 7h42'43" 53d9'21"
The Addster
Ari 2h0'1" 20d59'47"
The Adelaide and Tom Star
Dra 15h0'50" 60d49'21"
The Adele Anderson Star
Gem 6h34'53" 20d25'43"
The Adler Lane Star
Uma 8h32'20" 62d47'48"
The Adrianne
Tau 5h33'27" 19d49'40"
The Ahl Star
Cas 0h53'5" 61d39'55"
The Aiken Family
Uma 11h14'48" 48d9'1"
The Ajah Star
Per 2h43'5" 55d43'18"
The Al and Eileen Gottlieb Star
Cyg 19h43'45" 55d51'11"
The Alaeddin's Nov. 3, 2006
Hor 2h50'19" -63d27'44"
The Alan Beeton "Super Star"
Cep 23h55'1" 82d45'4"
The Alan Marriott Star
Umi 16h4'23" 71d2'18"
The Alan Shirley Star
Cru 12h18'51" -56d33'10"
The Alanna Star
Cas 0h16'39" 51d35'51"
The Albano and Irene Ponte Family
Her 16h14'11" 49d25'44"
The Alberta Strage Star
Cas 0h49'58" 47d43'23"
The Ale Way
Cnc 9h5'9" 16d20'45"
The "Alex" Alexandra
Cas 2h1'16" 62d40'58"
The Alex Russell Star
Ori 5h25'30" -9d16'1"
The Alexander
Crb 15h35'10" 30d8'11"
The Alexandra Star
Peg 21h34'7" 24d22'3"
The Alexandria
Sco 17h6'23" -41d25'5"
The Alexi M. Wiemer Star
Uma 9h0'7" 71d29'21"
The Alexis Graham
Sgr 18h8'8" -27d35'53"
The Alexix-Brianna
Tau 4h27'44" 23d0'8"
The Ali G
Ori 5h36'10" -1d25'33"
The Alicja Drewnowska Star
Uma 14h17'10" 56d13'6"
The Alison Gwinn
And 0h11'37" 32d10'36"
The Allan Walter Finlay Star
Uma 13h1'33" 63d10'36"
The Alli
Cap 20h24'27" -10d58'39"
The Alli Hubbard Star
And 23h27'35" 47d7'59"
The Allison Michelle Star
Ari 2h45'55" 27d38'13"
The Allison's Star in Heaven
Psc 1h20'22" 17d40'5"
The Ally-Rock
Lyr 18h31'52" 36d41'23"
The Alter Family Star
Uma 10h40'56" 62d17'55"
The Altman Family
Cap 21h9'11" -19d21'8"
The Alvin Infield "Eagle's Eye"
Sco 17h46'37" -36d58'36"
The Alyssa Maree
Ori 5h55'57" 20d14'57"
The Amanda
And 23h12'19" 47d3'39"
The Amanda Linn
Leo 11h7'50" 9d33'57"
The Amanda Lynne Cisko
And 0h36'41" 32d53'38"
The Amanda Polsen Star
Cru 12h32'2" -62d6'0"
The Amanda Star
Crb 15h48'43" 33d15'55"
The Amante
Cyg 20h38'2" 47d29'21"
"The Amazing Dr. Z" Rich Zelkowitz
Cnc 8h24'59" 31d0'9"
The Amazing Elaine Tonks
Cas 0h25'45" 66d28'2"
The Amazing Geo King, David Wright
Cep 3h13'52" 82d30'3"
The Amazing Kevin
Ara 16h56'52" -48d55'54"
the Amazing light of Ceeta
Vir 14h19'34" 4d23'26"
The Amazing Love Star of Rob & Jeni
Crb 16h6'59" 36d32'40"
The Amazing Marc Anthony Clas
Sgr 18h47'37" -36d10'3"
The Amazing Michael Aaron Fisher
Ori 6h8'18" 18d42'51"
The Amazing Susan Race
Tau 5h59'25" 25d1'3"
The Amber Nicole
Lib 15h24'16" -14d43'42"
The Amber Star
Lib 15h32'36" -9d24'25"
The Amberly Weber
Uma 11h48'13" 37d42'32"
The Amy Ariel
And 23h35'21" 44d41'30"
The Amy & Ashley
And 0h42'0" 39d9'22"
The Amy Lee
Lib 15h25'56" -21d35'2"
The Ana Dolatabadi Star
Vir 13h21'55" 7d18'41"
The Anastos Family Star
Uma 10h24'49" 46d58'10"
The Andrea Nelson
Uma 10h20'6" 55d7'10"
The Andrew Mark
Vir 13h24'31" 12d52'43"
the 'Angel Aurora Australis' star
Psa 22h19'59" -30d30'16"
The Angel Called Heather
Cas 1h20'45" 61d50'41"
The Angel Cristina
Sgr 19h7'15" -22d1'15"
The Angel Kristin
And 2h17'22" 47d35'3"
The Angel Robyn's shining light
Cap 20h48'24" -19d49'41"
The Angel Star of Shyla Akers
Tau 4h27'34" 22d55'13"
The Angel Star Sherry Lynn Tanksley
Leo 9h27'55" 8d38'1"
The Angelic Star of LSJ
And 1h44'9" 36d33'10"
The Angie Star +1
Cnc 8h7'10" 26d41'31"
The Ann Snow Star
And 23h25'17" 46d28'11"
The Ann Sugiegeneva
Cap 20h32'7" -16d12'43"
The Anna Fargher Star
And 2h37'23" 41d24'58"
The Anna Marie Smith
And 0h36'16" 46d18'6"
The Anna Star
And 0h17'48" 29d41'35"
The Anna-AM Memorial Star
Lyn 9h38'55" 40d49'53"
The Annabell Starr
Psc 1h29'17" 16d19'36"
The Annapolis Animal Hospital
And 23h22'58" 43d35'17"
The Anne Burns Star
Per 3h1'26" 51d29'13"
The Anne Marie O'Meara Star
Cyg 19h49'12" 37d36'55"
The Annetta Star
And 23h8'32" 47d21'19"
The Annie Lopez
Gem 7h11'2" 32d25'23"
The Annie Owen Star
Cas 23h2'9" 54d49'31"
The Anniversary Star
Ori 5h37'32" 3d55'45"
The Answer to Each Others Questions
Lib 15h42'4" -14d35'35"
The Anthony "Wildman" DiMario Star
Psc 1h47'35" 5d13'17"
The Antigua Star
Cyg 19h55'53" 40d3'46"
The Apple of My Eye 143
Ori 6h18'32" 9d1'45"
The apple of Percival's eye
Cnc 9h16'17" 22d30'39"
The Apples Will Grow Again Twinkler
Uma 10h3'38" 47d45'48"
The April Star
Tau 3h29'32" 8d44'5"
The Aquemini
Uma 9h24'11" 48d17'31"
The Aquino Seas
Cru 12h18'2" -61d24'17"
The Archie James Jones Star
Gem 6h37'1" 19d43'15"
The Arian Wesley Star
Ori 6h14'53" 15d2'3"
The Ariella Rose
Gem 7h29'35" 25d28'59"
The Arlin Star aka Our Star
Sgr 19h47'43" -13d9'48"
The Art Collins Iggy Pop Star
Per 4h2'50" 43d54'24"
The Art & Drama Therapy Institute
Uma 10h0'56" 69d2'59"
The Art History of Ashley Barretto
Aqr 23h14'55" -22d2'33"
The ART Star
Cen 15h0'52" -34d8'41"
The Arthur & Caite Podaras Star
Vir 12h32'51" 8d40'29"
The Arthur F Neuenhaus Star
Her 16h41'47" 12d8'4"
The Arthur & Maggie Bonner Star
Oph 16h29'48" -0d41'30"
The Artist
Vir 14h2'52" 3d1'48"
The Asa Lover Star
Leo 11h26'53" -3d25'36"
The Asbury Family
Per 3h28'48" 43d32'5"
The Ashlynn Perry Sapphire
Crb 15h52'53" 35d24'42"
The Ast Family Star
Crb 16h21'7" 30d41'34"
The Astounding Light of HP Kennedy
And 22h59'19" 39d46'27"
The Attardo's Sapphire
Uma 8h33'53" 73d1'43"
The Aubrey and Raelee Lynn
And 0h15'46" 26d25'36"
The Auburn Star of Greenbank
Dra 18h54'39" 66d31'36"
The Audra Amanda Shining Star 18
And 0h29'34" 29d21'34"
The Audrey Lee
And 23h31'14" 48d25'42"
The Auntie Ger
And 1h36'6" 40d40'59"
The Auntie Grayce
Ari 3h12'38" 28d10'49"
The Auntie Star
Srp 18h18'47" -0d10'22"
The Aurora De Armas
Lyn 8h59'19" 34d34'26"
The Austin Mitchell
Lyn 7h57'16" 41d40'8"
The Avalon - Tony & Stuart
Cyg 20h33'28" 51d34'35"
The Avena Stelle
Uma 9h25'5" 65d50'38"
The "Awesum" John Prince
Cru 12h33'35" -60d32'3"
The Baars' Star
Ara 17h44'52" -48d6'30"
THE BABES
Tau 5h35'20" 25d2'32"
The Babin Wedding Star
Cyg 21h0'32" 47d1'21"
The Baby
And 23h10'1" 41d6'24"
The Bacharama
Uma 9h20'54" 65d21'7"
The Badger Star for Chris & Katy
Dra 15h16'35" 64d17'25"
The BAH
Cam 4h45'29" 58d30'57"
The Bailey Anniversary Star
Cyg 20h18'51" 37d46'17"
The Bailey John Willoughby Star
Cru 12h24'52" -57d22'5"
The Baileys
Sco 16h51'40" -42d27'18"
The B.A.L Star
Vir 14h34'43" -0d43'18"
The Balzer Family Star
Uma 8h13'49" 67d38'8"
The Banana Sprout 1999
Cam 4h32'47" 75d35'39"
The Barans 1997
Her 17h10'37" 48d48'56"
The Barbara Mardene
Cas 1h37'16" 63d54'33"
The Barbara Star
Lib 14h50'34" -1d29'8"
The Barbé Star
Aqr 21h14'28" -8d54'7"
The Barbick Pauls Family Star
Uma 13h38'43" 53d11'37"
The Barbie Carver Star
Tau 3h30'5" 16d58'21"
The Bare
Uma 10h21'51" 55d13'29"
The Barletta Star
Lib 15h48'12" -4d5'46"
The Barney Thomas Star
Ari 2h43'59" 11d59'8"
The Baron
Vir 13h50'32" 1d49'22"
The Baron Star
Ori 5h25'18" -5d5'29"
The Barony
Cep 23h57'28" 74d19'13"
The Barron Family Star
Uma 10h12'23" 50d13'9"
The Barrow Family Star
Lyn 6h49'12" 53d5'39"
The Barry & Debbie Forever Star
Cru 12h27'41" -60d36'7"
The Barry Phillips Star
Leo 11h40'35" 22d15'11"
The Barthelmes Star
Sge 19h44'24" 18d40'17"
The Bat 97
Uma 11h41'27" 62d29'35"
The Batchstar
And 0h47'1" 34d52'30"
The Bates Brothers
Per 3h21'31" 41d42'35"
The Batterson's Bright Light
Ori 4h48'4" 11d57'29"
The Baumann Boys - Conor and Evan
Uma 9h46'19" 42d34'19"
The BCB Is Cancer Free
Uma 11h55'50" 62d16'28"
"The Beach" Software Galactic HQ
Tau 4h37'10" 20d50'23"
The Beacon of Leadership
Crb 15h33'52" 32d8'51"
The Beahms
Her 17h19'26" 34d6'57"
The Beall Family Star
Psc 1h40'39" 27d40'23"
THE BEANSTER STAR (TRACY ELLIOTT)
Dra 17h35'32" 66d12'53"
The Bear
Ari 2h12'20" 23d28'19"
The Bear Blair Jackson-Steinmetz
Uma 11h20'51" 54d10'9"
The Bear Room
Aqr 23h3'27" -24d14'11"
The Beat
Tau 5h34'24" 25d11'57"
The Beatrice Couto Star
And 2h23'58" 45d55'13"
The Beautiful Jennifer
Lib 15h36'22" -13d19'21"
The Beautiful Abby
Ori 5h43'5" 8d39'59"
The Beautiful Adonna
And 0h16'58" 38d40'44"
The Beautiful Amanda
Leo 9h42'33" 26d52'16"
The Beautiful Amy Anne
Sgr 19h27'4" -14d52'46"
The Beautiful Andrea Marie
Sco 17h2'18" -35d53'49"
The Beautiful Ashley Colegrove
Tau 4h23'21" 22d37'36"
The Beautiful Audrey Lynn
Vir 12h14'15" 3d32'44"
The Beautiful Berlynn
Gem 7h37'14" 33d12'18"
The Beautiful Chanda
Peg 23h42'37" 11d21'46"
The Beautiful Cherie Johnston
Cas 0h21'16" 63d1'59"
The Beautiful Crystal
And 0h34'30" 38d51'47"
The Beautiful Desiree
Sco 17h57'30" -39d14'2"
The Beautiful Diane Fitzgerald
Tri 2h3'48" 33d53'21"
The Beautiful Donnalea
Sgr 18h52'47" -30d51'14"
The Beautiful Ebelise Osorto Star
Sco 16h14'9" -17d38'39"
The Beautiful Heart and Mind
Lyr 18h34'0" 36d27'47"
THE BEAUTIFUL HEIDI
Uma 8h9'11" 60d18'7"
The Beautiful Jade
And 0h2'47" 46d12'28"
The Beautiful Jen
Sco 17h58'11" -30d8'22"
The Beautiful Jennifer Hahn
Ori 5h33'49" 9d56'19"
The Beautiful Jennifer Marie
Psc 1h3'13" 32d38'22"
The beautiful Katie T.
Psc 1h9'39" 13d38'52"
The Beautiful Keli Andrea
Vir 12h31'14" -0d1'58"
The Beautiful Kelly Kathryn
And 2h22'22" 47d21'12"
The Beautiful Kiersten
Lyn 6h54'22" 53d29'40"
The Beautiful Laila
Her 18h55'53" 23d46'1"
The Beautiful Leigh Hargreaves
And 0h21'21" 26d7'54"
The Beautiful Lisa Marie
Ori 5h53'32" 16d4'40"
The Beautiful Lyndsey Sullivan
Cas 23h23'58" 53d11'11"
THE BEAUTIFUL MARIAM
Gem 6h53'21" 18d46'19"
The Beautiful Miranda
Uma 12h26'15" 61d52'23"
The Beautiful Miss Jenna Lambra
And 1h46'4" 50d13'51"
The Beautiful Miss. Kayla Pappas
Cnc 8h39'51" 30d46'0"
The Beautiful Molly's Star
And 0h48'16" 41d12'32"
The Beautiful Monessa Patches
Sco 17h55'53" -36d24'55"
The Beautiful Nikki "Yummy" Heinen
Mon 7h22'18" -7d22'49"
The Beautiful Nikolette
Lib 15h16'24" -10d52'19"
The Beautiful one Jessica
Tau 5h38'30" 26d37'36"
The Beautiful " PELI " Star
Tau 4h10'8" 13d4'24"
The Beautiful Pirate Princess
And 2h9'35" 38d51'21"
The Beautiful Princess Lindley
And 1h17'31" 38d5'32"
The Beautiful Sandra Hamilton
And 23h37'41" 50d17'21"
The Beautiful Sarah
Ari 2h15'59" 23d3'41"
The Beautiful Sarah Michelle
Tau 3h45'28" 6d31'42"
The Beautiful Star Of Emma
And 23h59'12" 43d10'14"
The Beautiful Star of Stephanie
And 0h21'6" 25d36'8"
The Beautiful Star of Taylor Renée
Cas 1h48'38" 61d38'53"
The Beautiful Star Yeliz
Cyg 19h53'39" 36d54'49"
The Beautiful Summer
And 0h34'46" 34d26'14"
The Beautiful Tina Marie Star
Com 12h40'14" 18d5'25"
The Beautiful Tracy Marie
Sgr 18h1'55" -27d41'58"
The Beautiful, Jessica Zebrowski
Leo 10h56'34" -1d38'45"
The Beautiful, Rachel Natale Star
Aqr 22h14'22" 2d14'20"
The beautifully eloquent Vonnie Rea
And 0h32'30" 34d56'11"
The beauty of Alexia
Leo 11h56'53" 20d29'5"
The Beauty of Barbie Fuller
Uma 11h24'4" 47d15'50"
The Beauty of Diana L. Sanchez
And 23h10'54" 47d45'45"
The Beauty of Julie
Psc 1h3'30" 14d21'47"
The Beauty of Leah Marie Easterday
Leo 11h5'46" 21d21'51"
The Beauty of Michael Fitch
Sgr 18h15'29" -15d58'46"
The Beauty of Rachel
Sco 16h21'46" -18d13'44"
the bebitos locos
Eri 3h52'26" -0d30'29"
The Becca
Aqr 23h30'39" -16d44'59"
The Beckster
Leo 9h22'14" 16d23'45"
The Becky Star
And 2h15'11" 45d42'35"
The Beddington Star
Lyr 19h19'5" 34d13'19"
The Bee
Sgr 18h29'20" -16d2'56"
The Beeb Star
Umi 15h0'33" 76d1'45"
The Beef Star
Cam 4h35'3" 66d47'24"
The Beep Star
Cnc 8h9'8" 24d41'29"
The Befordaylight
Sco 16h50'54" -26d24'15"
The Beginning
Ori 6h17'8" 16d29'12"
The Beginning:Kerry & Roger Hohman
Lyn 6h23'26" 57d39'47"
The Beginning - Marc & Kristen
Ori 5h9'6" 15d25'38"
The Beginning of my Love for Her
Uma 10h50'51" 49d22'57"
The beginning of our life together
Cyg 19h51'3" 37d26'21"
The Belinda & Wayne Star
Sco 17h56'22" -41d36'57"
The Bell Star
Cas 0h25'26" 60d43'47"
The Bella-Rose
And 1h2'46" 45d2'47"
The Beloved Cari Wittman
Aql 19h41'13" 11d51'45"
The Beloved Monica Nieto
And 22h59'27" 39d18'10"
The Ben and Sunny Wedding Star
Ori 5h51'58" 9d7'43"
The Ben Jimenez Remembrance Star
Aql 19h31'7" 2d49'40"
The Benjamin Phillip Carroll Star
Umi 14h43'32" 69d28'43"
The Bennett Bay Star
Hya 10h38'44" -27d46'38"
The Benoit-Stormes Legacy
Leo 10h19'47" 25d29'35"
The Benson Bears
Uma 11h33'19" 58d21'50"
The Berk
Uma 13h10'54" 57d8'19"
The Bernabo Star
Vir 13h47'13" 0d14'48"
The Berry Family Star
Dra 16h25'16" 54d7'54"
the BEST couple
Vir 13h25'39" 5d7'45"
The Best Cramp Ever
Cyg 19h47'41" 35d22'42"
The Best Dad, Joe Cox
Cnc 8h19'27" 13d7'58"
The Best Daddy Iraj
Lib 14h25'33" -11d22'38"
The Best Day Ever
Cyg 21h15'27" 30d7'8"
The Best Lil Sis EVER
Lib 15h25'33" -5d18'56"
The Best Mom Ever - Louanne Doyle
Ari 3h9'43" 21d26'55"
The Best Mom in the Galixy,Love You
Cyg 20h0'39" 56d37'1"
The Best Mom In The World
Psc 1h11'23" 27d27'22"

The Best Mom, Shelley Carroll
Dra 19h22'46" 60d53'55"
The Best Moms: Anna and Agnes
Lyn 6h59'49" 53d35'35"
The Best Mother in the Universe
Vul 21h28'30" 26d43'50"
The Best Mother Megie Slamin
Tri 2h4'51" 35d23'3"
The Best Mum Kelli Webb
Vir 12h27'37" 3d11'11"
The Best Proof Of Love Is Trust
Cnc 8h8'58" 20d48'44"
The Best Star
Uma 13h46'54" 53d27'2"
The Best Star for the Best Mom
Cap 21h0'59" -22d22'15"
The Best Wedding Planner
Cyg 19h43'38" 39d56'19"
The Bestest Star Ever
Ari 2h54'45" 26d59'40"
The Bethany Bowen Star
And 1h23'29" 39d46'8"
The Bethany Jane
And 1h16'43" 37d17'7"
The Betty Bowen Star
Lib 15h37'53" -7d22'24"
The Betty & Bud McGowan Star
Uma 8h21'57" 73d1'33"
The Betty Milner Star
Cas 3h15'39" 61d6'51"
The Beutel Family Star
Psc 23h49'32" 6d47'35"
The Bev & Robby Star
Cyg 19h50'52" 36d40'4"
The Bianco Family
Uma 14h6'2" 50d27'41"
The Big A
Vir 13h5'8" 10d17'16"
The Big Al
Uma 11h7'19" 35d4'47"
The Big Boppa
Ori 5h29'43" 2d16'16"
The Big Boyum
Cyg 20h0'25" 54d13'20"
The Big Brandon
Her 18h57'3" 16d57'20"
The Big Brother
Her 18h42'13" 22d2'53"
The Big Bumby Star
Gem 7h4'48" 29d2'22"
The Big Chalupa
Umi 13h10'38" 71d3'57"
The Big Deal
Cyg 20h5'24" 37d29'37"
The Big Digger
Aqr 21h20'5" -7d25'52"
"The Big Doc"
Leo 10h14'29" 17d29'47"
The Big Dupper
Vir 14h36'40" 3d33'16"
The Big E
Uma 8h22'55" 60d53'36"
The Big Easy's BOKF Star
Ari 3h3'48" 14d37'46"
The Big Ess
Cep 21h14'45" 66d52'39"
The "Big G" Star
Ori 6h0'18" -0d19'52"
"The Big Gay Bear" MacPherson
Uma 8h59'16" 59d11'32"
The Big Guy
Umi 14h46'8" 74d52'55"
The Big Guy - Dale B
Crb 15h47'47" 28d14'4"
The Big Gypper
Uma 11h27'33" 45d54'1"
The Big I
Uma 12h17'42" 54d49'1"
The Big Kahuna
Aqr 20h48'53" 0d48'31"
The Big Larry
Cep 20h42'1" 67d39'44"
The Big Mama
Cnc 8h23'48" 11d4'33"
The Big Man
Ari 2h3'8" 19d7'38"
The Big Mommy
Cas 1h31'16" 67d51'48"
The Big Munson
Lib 15h8'59" -11d26'46"
the Big Neubizzle
Uma 11h37'5" 52d42'6"
The Big O
Ori 5h0'36" 5d8'12"
The Big Paw
Uma 9h10'58" 67d40'16"
The Big Question
Uma 10h5'30" 48d36'35"
The Big Red Hen
Uma 10h46'28" 57d8'12"
The Big Star
Aqr 22h53'8" -7d58'31"
The Big Unit
Uma 9h54'3" 44d12'35"
The Big Worm
Lac 22h21'18" 51d29'17"
The Bigelow Star
Cyg 20h6'19" 32d27'52"
The Bigger Dream
Uma 10h36'46" 62d33'3"
The Bill and Linda Farthing Star
Cyg 20h7'14" 42d49'50"
The Bill and Mildred Jackson
Sco 16h17'40" -10d26'4"
The Bill Caler Star
Uma 10h25'29" 68d20'21"
The Bill Camp Star
Uma 14h1'12" 52d59'36"
The Bill Voss Star
Dra 20h19'39" 71d6'40"
The Billy B
Uma 11h20'5" 61d27'53"
The Billy (BIG) Mitchell Star
Leo 10h23'23" 18d15'13"
The Billy "Bull" Tedford Star
Per 3h41'12" 39d1'59"
The Billy Cotter
Cru 11h59'10" -62d50'24"
The Billy & Oscar Star
Ori 5h21'7" 6d26'53"
The Binos
Sco 17h15'42" -44d34'38"
The Birch Family Guiding Star
Uma 11h55'53" 49d29'36"
The Birth of "B"
Cam 4h11'39" 66d58'53"
The Birthday Princess
Leo 10h47'27" 20d2'42"
The Bisset Star
Lyr 18h51'3" 35d57'32"
The Bissonnette Family Star
Cap 20h36'26" -18d9'7"
The Bix
Uma 9h17'38" 67d40'45"
The Black Jaguar
Ori 5h38'8" -4d39'20"
The Black Star of Ireland
Sco 17h28'43" -33d25'25"
The Blade's Star
Cyg 19h44'1" 37d11'56"
The Blake
Her 16h24'38" 22d59'48"
The Blakes
Per 2h53'15" 42d13'56"
The Blanca Chez Shining Star
Sgr 18h39'54" -19d56'21"
The Blay Ruby Anniversary Star
Lyn 7h33'33" 37d58'34"
The Bleasel & Spratt Family Star
Cru 12h38'46" -59d31'59"
The Blevins Star
Uma 12h10'53" 54d45'2"
The Blinding Barwicks- Deb & Bob
Sge 20h2'45" 16d36'35"
The Blizzard of Aaahs
Her 17h20'52" 29d58'27"
The Blondheim Light
Tau 4h3'11" 22d4'0"
The Blue Crow
Aql 19h39'50" 7d40'15"
The Blue Rose
Tau 5h22'29" 27d3'59"
The Bluebird's Enchanted Heart
And 0h41'18" 34d20'16"
The " Bo" Sheree"
Sge 19h48'43" 18d2'56"
The Bob and Cristina Duffer Star
Her 17h21'2" 34d4'46"
The Bob and Marcie Thedinger
Cyg 20h30'53" 46d15'23"
The Bob Grabowski Star
Cyg 20h38'48" 55d22'43"
The Bobbie Schunter Star
Cep 21h47'24" 63d39'7"
The Bobby & Sara Star
Aql 19h8'37" 1d56'40"
The Bob-o 1
Uma 14h5'58" 54d34'6"
The Bodick Family Star
Lyr 18h51'15" 32d32'55"
The Bohlen's of North Carolina
Lyn 6h50'47" 58d33'22"
The Bond Of Star Love
Cnc 9h14'2" 27d52'27"
The Bonn Star
Umi 15h23'10" 67d42'5"
The Bonnie Sue Roland Star
Psc 0h57'34" 6d10'25"
The Bonny Brigadoon
Ari 2h57'28" 19d25'59"
The Boofman
Leo 11h20'56" 22d44'34"
The Boojo Star
Uma 11h1'46" 39d49'52"
The Bordeaux Mayfield Farm
Boo 14h43'56" 35d0'36"
The Boren Family Star
Lib 15h25'46" -15d2'50"
The Borgatti Family Wishing Star
Uma 11h36'49" 46d34'31"
The Boring's, Love, Faith, & Hope
Cyg 19h58'28" 34d27'16"
The Boston Blazer Leslie Gurski
Peg 22h21'35" 33d45'25"
The Boswell Diamond
And 23h19'19" 41d56'5"
The Botha Hanson Together Star
Cyg 19h41'32" 34d8'32"
The Bott-Parr Star
Ori 6h18'41" 13d16'0"
The Bower Family
Uma 10h16'53" 48d25'27"
The Bowlus Star
Umi 16h39'58" 82d54'43"
The Boy Next Door
Her 17h27'2" 46d13'57"
The Boys
Per 3h3'18" 45d34'48"
The Boys
Cnc 9h12'21" 25d18'54"
The Boys Little Twinkle in the Sky
Sgr 18h24'21" -26d17'8"
The Brahney Family Star
Uma 9h34'24" 55d45'55"
The Brandon Asher
Lyr 18h33'21" 32d24'26"
The Brandon Family including Casey
Umi 14h23'57" 71d36'55"
The Brantly Family Star
Lyn 6h35'35" 58d6'10"
The Bratty Princess
Sgr 18h1'18" -16d47'52"
The Brazen Hog / Sandy Smith
Cet 1h6'21" -0d25'51"
The Breda Family
And 1h22'49" 47d29'27"
The Bredehoeft Family
And 23h37'23" 47d26'40"
The Brendan Weatherill Star
Sgr 17h53'36" -29d23'50"
The Brian and Holly
Cyg 20h22'4" 36d14'57"
The Brian Richard Thornby Star
Cru 12h39'31" -58d45'9"
The Brian RusStar
Sgr 19h18'29" -17d8'15"
The Brian Taube Sentinel Star
Lmi 10h5'5" 37d5'47"
The Brian Waters Shining Star
Uma 13h27'54" 54d58'8"
The Brick Yards
Crb 15h55'47" 32d3'56"
The Briggs
Uma 12h40'49" 55d55'50"
The Briggs Family
Lyn 7h50'17" 53d47'2"
The Bright Light in Jean 12/03/73
Uma 11h59'1" 28d55'23"
The Bright Light of DH Nathanson
And 1h7'38" 38d13'56"
The Bright Spot In My Life - Deepti
Cap 21h49'40" -22d53'31"
The Bright Star of Sean and Zachary
Aql 19h12'23" 9d59'58"
The Brightest & Best COE Staff 2006
Uma 11h34'56" 61d30'57"
The Brightest Star in MY Galaxy
Aqr 21h34'0" 1d55'19"
The Brightest Star In The Sky
Dra 19h42'30" 61d1'34"
The Brightest Star, My Mom, Valerie
Lib 14h51'47" -1d28'48"
The Brightest Stars Choose Terry
Psa 22h28'9" -25d4'10"
the brightest thing in my life
Uma 10h17'10" 49d29'47"
The Brigman Effect
Her 17h15'20" 17d59'45"
The Brilliance of Leon Freedman
Leo 10h10'29" 16d52'49"
The Brilliance of Michael R Cross
Cru 12h47'44" -64d20'58"
The Brilliant Perseverence
Uma 11h34'33" 52d30'33"
The Brilliant Shepherd
Sco 16h18'44" -18d36'30"
The Brittany Star
Umi 14h50'21" 67d53'23"
The Brittany Star
Sco 17h45'41" -42d11'33"
The Broad Star
Cas 0h51'26" 69d54'47"
The Brog Star
Cyg 20h43'44" 51d5'16"
The Brown Rose
Aqr 23h26'19" -19d18'47"
The Brown's Lucky Bright Star Light
Pho 23h49'12" -41d30'44"
The Bruce Mavec Family Star
Uma 11h33'14" 33d2'37"
The Bruce Rule Star
Cru 12h48'2" -60d31'41"
The Bruce's 10/10/2004
Uma 11h18'51" 72d19'10"
The Brumme Family
Uma 13h44'56" 60d40'35"
The Bryan & Lauren Scherr Star
Cyg 20h55'18" 50d52'36"
The Bubba & Heather Clem Star
Cap 21h32'43" -16d40'56"
The Bubba Star
Cyg 20h26'19" 31d10'55"
The Buck Star
Lyn 7h11'15" 52d3'46"
The Buckeroo-Tinybo Star
Uma 10h13'53" 50d10'47"
The Buckey Star
Cnc 8h40'59" 8d20'49"
The Buckley Star
Lyn 7h55'53" 49d27'54"
The Buddy & Audrey Keenan Star
Cyg 20h40'3" 45d56'30"
The Buergler Family Star
Aql 20h12'28" 15d41'38"
The buffest star in the universe
Cap 20h26'52" -10d30'49"
The Buggh's
Lib 15h39'16" -14d51'25"
The Buller Diamond
Cyg 21h27'55" 34d33'13"
The Bully Star
Uma 8h41'25" 67d45'58"
The Bunny
Uma 8h41'8" 64d40'52"
The Bunny Star
Lep 5h30'12" -11d7'51"
The Bunyips
Ori 5h32'25" -2d58'10"
The Burchill Family's Star
Uma 9h32'51" 56d13'0"
The Burke Family
Cap 20h27'51" -16d27'49"
The Burketts
Uma 10h12'10" 64d7'35"
The Burkholder Star
Aur 5h47'30" 37d54'18"
The Burozski Family
Per 2h50'7" 48d36'9"
The Busby Family Star
Uma 8h18'57" 67d19'57"
The Butterfly
Cnv 13h13'46" 42d19'11"
The Buzz Star for Dennis W Fromherz
Aur 5h21'35" 30d3'32"
The Bykowski Star
Lyn 7h1'52" 50d57'24"
The Cajun Princess-Donna Massari
And 23h52'40" 32d50'1"
The Cakas
Tau 5h45'26" 15d5'9"
The Cake Club
Uma 8h35'35" 71d2'9"
The Caleb Gene Demey Star
Her 17h14'44" 48d46'51"
The Caleb Green Star
Umi 14h29'9" 82d45'25"
The Calender Clan * kjkss* 1989
And 0h1'7" 33d55'41"
The California Star
Lib 14h43'27" -20d12'53"
The Caliri Family
And 23h8'30" 44d12'51"
The Callee Star
Sco 16h22'33" -32d57'57"
The Cameron Brock Star
Cap 20h28'43" -14d11'17"
The Campman Family
Cyg 19h59'20" 58d57'6"
The Capturer of My Heart
Ori 5h38'9" 4d47'35"
The Cara Michelle
Sgr 18h4'27" -27d55'18"
The Caring Humorous Enchanting Star
Per 3h55'26" 32d27'34"
The Carlene Nazarian Dance Center
Uma 11h44'37" 62d50'50"
The Carleton Family
Cyg 19h59'37" 36d1'57"
The Carmelita
And 0h52'32" 41d49'26"
The Carol Bell Star
Crb 15h47'59" 27d13'48"
The Carol Geary Star - GG's Winking
Leo 9h35'20" 24d40'22"
The Caroline Jack Star
And 1h53'46" 40d1'44"
The Caron Family Star
Ori 6h1'1" -0d36'50"
The Carpenter George Weisbecker
Psc 23h31'9" 1d40'41"
The Carpenters Son
Cyg 20h31'23" 36d41'13"
The Carter Rebecca Clabaugh Star
Umi 14h59'32" 81d39'30"
The Carter's Anniversary Star
Cyg 21h11'46" 42d28'18"
The Cartetwards
Uma 11h36'24" 39d29'34"
The Carver Wedding Star
Cyg 21h39'22" 38d43'58"
* The Casey Daniel Howse Star *
Pho 2h12'45" -41d24'52"
The Casey Star
Lyn 8h11'16" 53d50'53"
The Cassell Family
Peg 22h25'41" 20d17'52"
The Catello Family
And 0h43'26" 43d59'38"
The Catherine and Ivan Star
Cyg 20h31'11" 58d24'26"
The Catherine and Martin Star
Cyg 19h44'36" 51d58'33"
The Catherine Legg Lee Family
Cas 0h35'49" 48d49'40"
The Catrina Roe
Cnc 8h3'6" 20d28'30"
The Catty
Cnc 9h20'4" 18d32'48"
The Caudle Family Star
Cyg 19h44'30" 43d36'47"
The Cavig (Marc Cavigli)
Cep 23h37'22" 78d27'38"
The CBD Star
Sco 17h26'25" -42d53'46"
The celestial body of Kim Holyoake
Tau 3h44'53" 28d36'3"
The Center of My WORLD
Uma 11h37'56" 63d16'13"
The Chad
Pho 0h38'18" -49d50'9"
The Chad McCoy Star
Sgr 18h9'21" -33d59'16"
** THE CHAMP ** c.a.c
Leo 11h21'19" 25d10'36"
The Channon Family Star
Uma 12h46'0" 57d37'55"
The Chantelle Anderson Star
Lyn 7h2'22" 46d40'43"
The Chapman Family
Dra 18h18'56" 76d32'24"
The Chardy Wishing Star
Crb 16h2'3" 38d59'22"
The Charles & Janet Heywood Star
Mon 6h27'56" 4d16'8"
The Charles & Jean Brookman Star
Lyr 18h29'57" 36d42'36"
The Charli Erika Russell
Cap 21h7'30" -25d1'23"
The Charlie Perdue Star
Uma 10h41'16" 60d54'13"
The Charlotte Jean Star
Lyn 6h58'24" 57d51'58"
The Charlotte Michael
Lyn 8h6'23" 58d49'41"
The Chase Family
Uma 11h15'11" 33d0'53"
The Chase Star
Uma 11h2'44" 48d4'50"
The Chaudhary Twins
Uma 10h15'9" 72d35'35"
The Chaz Aardema
Sco 17h55'34" -42d25'1"
The Checel-Lund Star
Uma 11h28'9" 40d44'40"
The Chelsea/Seth Star
Cnc 8h36'6" 8d39'52"
The Chibas Star
Uma 11h58'44" 29d41'51"
The Chicken-Fudge Pie Star
Gem 6h30'2" 18d46'29"
The Chief
Vir 13h19'35" 6d23'50"
The Chief
Psc 23h21'57" 6d30'32"
The Chief
Psc 0h33'37" 9d23'52"
The CHIEF of the sky
Her 18h56'44" 23d11'7"
The Children's Learning Center
Lyn 8h24'1" 38d36'47"
The Childrens Place
Lyn 6h50'8" 52d30'37"
The Chinchilla
Tau 4h17'15" 7d49'47"
The Chincoteague Star
Tau 3h36'38" 16d36'55"
The Chip Wendlandt
Cnv 12h40'36" 36d8'33"
The Chipper
Del 20h38'37" 14d0'35"
The Chloe and Amber Sanders Star
And 23h23'42" 41d19'35"
The Chloe Lobsey Star
Ari 3h13'0" 14d2'59"
The Chloe Mari
Lyn 6h20'57" 57d19'50"
The Choice Star
Ori 6h2'5" 3d52'42"
The Chow
Per 3h24'40" 38d43'7"
The Chris Albers Family
Cas 0h21'52" 62d52'1"
The Chris and Dolores McAlinden
Cyg 20h40'12" 50d1'39"
The Chris and Lisa Golden Star
And 0h46'17" 43d53'16"
The Chris Hickenbottom
Her 17h23'22" 17d37'31"
The Christ
Umi 4h21'29" 89d26'20"
The Christal Soehnlein Star
Cnc 8h49'48" 28d39'36"
The Christiana
Lib 15h20'47" -19d42'32"
The Christina
Ari 3h9'34" 19d59'48"
The Christine M. Schumacker
Tri 1h58'2" 31d2'13"
The Christine Politis Friend Star
And 22h59'23" 44d31'45"
The Christopher J. Faisandier Star
Cru 12h18'52" -62d0'49"
The Christopher Scott Miskell Star
Cru 12h25'39" -58d47'18"
The Christopher Wills Star
Cru 12h48'37" -62d28'49"
The Chubbuck
Uma 11h15'30" 69d11'37"
The Cicchetti Star
Tau 4h5'26" 28d30'4"
The Cindy Star 11-21-99
Lyr 18h51'47" 33d8'38"
The Cinnamon Girl Donna Moore
Gem 7h41'37" 33d0'31"
The C.J. Karman Family Star
Uma 11h21'11" 54d27'39"
The Claire and Pat McAlinden
Cyg 19h35'5" 28d20'25"
The Claire J John 14:6
Uma 9h22'58" 72d44'13"
The Claire Wilson Radiant Star
And 0h13'55" 29d54'31"
The Clairmont-Austin
Crb 16h2'43" 36d16'33"
The Clapp Family
Cyg 21h6'33" 48d23'13"
The Clare and Erwin Raabe Star
Ari 2h40'27" 21d49'21"
The Clare Oughton Star
Cyg 21h47'52" 38d47'40"
The Clarice Star
Lyn 7h51'10" 48d16'28"
The Clark Star
Cas 0h35'48" 52d30'25"
The Clark Star
Uma 13h18'29" 57d48'9"
The Clark's
Per 3h35'59" 47d50'18"
The Clay Family
Umi 15h24'11" 74d44'26"
The Claydon Family Star
Ori 5h1'0" 5d22'36"
The Clementine Star
Uma 8h56'1" 59d43'22"
The Click Five
Lyn 7h30'14" 54d55'33"
The Cliff Gundle Star
Ori 5h58'33" 22d15'33"
The Closest Thing
Cyg 20h20'25" 43d32'17"
The Clubhouse
Uma 13h39'56" 56d21'35"
The Clum Family
Uma 14h4'33" 49d49'9"
The Cockney Juan Kerr
Gem 7h3'25" 29d8'58"
The Cole Perry
Sco 17h49'25" -40d55'25"
The Colleen Milford Star
Cru 12h25'14" -59d6'43"
The Collins' Twins
Lyn 6h47'26" 61d37'46"
The Colonel
Sco 16h15'37" -11d44'7"
The Colorado Serio
Psc 0h51'2" 27d55'46"
The Colton Wedding Star
Cnc 8h41'25" 29d25'30"
The Coluccio Family
Aqr 23h14'42" -10d10'24"
The Comfort Zone
Uma 10h59'48" 35d45'12"
The Commodore Poppy John G Rinklin
Sgr 19h45'0" -13d11'58"
The Compound: Charlene, Lee & Cody
Uma 10h44'30" 58d50'35"
the Comptons
Her 17h19'9" 15d33'4"
The Connie Leet Shining Star
And 1h9'43" 40d23'55"
The Connor James Hudson Star
Ori 6h9'46" 20d42'22"
The Constantine Mateos Star
Lib 15h38'22" -11d2'38"
The Constellation of the Lude's
Sgr 18h4'56" -22d10'35"
The Continental Four
Umi 15h55'47" 77d46'28"
"The Cookie Lady"
Uma 9h5'15" 65d4'54"
The Cooper's Golden Anniversary
Cyg 21h10'30" 31d21'34"
The Coopers Star
Lib 16h0'43" -20d6'11"
The Coppermark Star
Uma 12h3'26" 51d45'52"
The Copus'
Cyg 20h4'52" 57d34'33"
The Corey Hawkins Star
Aql 20h14'14" 2d6'8"
The Corinne Hepburn Star
And 23h42'49" 46d57'28"
The Cornett Family
Dra 10h10'41" 74d56'17"
The Cornish
Cra 18h20'1" -36d58'26"
The Corrado Family
Uma 13h5'43" 57d25'44"
The Corrina Dehn Star
Dra 20h18'0" 70d37'7"
The Cosmic Conor Cuchulainn
Sco 16h16'6" -9d0'50"
The Costa Rican Goddess
Psc 1h19'31" 5d41'27"
The Country Bumpkin
Lyn 8h2'1" 45d24'27"
The Courageous Carmelo
Her 17h40'13" 30d45'5"
The Cousins' Star
Uma 9h36'35" 65d8'34"
The Cowboy Star
Tau 5h40'32" 24d20'14"
The Cox's Star
Uma 9h8'29" 72d57'1"
The Cozy Star of Ann
Vir 14h43'15" 1d4'4"
The CPW Group, Inc.
Tri 1h51'10" 30d22'57"
The Crager Family Star
Uma 10h15'8" 51d18'43"
The Craig Newberry-McLeod Star
Cru 12h41'21" -58d51'50"
The Craigheads Star
Uma 13h56'13" 53d0'5"
The Craven Family
Aql 19h48'16" 6d19'56"
The Crawford Family Star
Per 3h5'33" 52d12'19"
The Creative Endeavor
Gem 7h50'27" 31d53'36"
The Creisstoff Family
Lmi 10h35'40" 37d27'26"
The Crouch Family
Sge 19h26'45" 17d36'15"
The Crowley Crew 18 Month
Umi 15h11'40" 75d45'26"
The Crown of Cunningham
Cra 18h17'39" -40d19'58"
The Cruickshank Family Star
Cyg 20h51'3" 31d18'14"
The Crystal Light
Vir 12h50'30" 6d29'48"
The Cubby Star
Cap 21h41'55" -10d6'37"
The Cummings Family
Her 18h37'0" 20d28'1"
The Curtis Star
Peg 22h54'22" 19d39'32"
The Cuthbert Anniversary Star
Ara 17h11'32" -57d13'25"
The Cynthia Anna Wronko Star
Gem 7h20'57" 21d36'45"
The Cynthia Crist-Stommel
Lib 14h53'31" -11d17'13"

The Cynthia Lee Star
Gem 7h18'59" 19d53'11"
The Cynthia Marie
Ari 2h41'31" 21d53'21"
The Daddy and Garrett Do Star
Lyn 8h26'19" 49d9'12"
The Daddy Star
Ari 2h27'35" 25d42'27"
The Daffodil Girl
Tau 4h9'25" 26d47'25"
The D'Agnes Family Star
And 23h9'11" 49d37'33"
The Dahbura Family
Uma 11h25'20" 62d28'13"
The Dakota
Ari 2h42'55" 27d27'3"
The Dale Pfannenstiel Star
Uma 10h37'23" 44d20'34"
The Dale & Sylvia Floriano Family
Uma 13h28'53" 53d35'37"
The Dallas Erin Louden
Lmi 10h23'21" 34d4'35"
The Dalton-Wilson
Cyg 19h32'20" 31d7'38"
The D.A.M. Star
Lmi 10h33'28" 37d0'51"
The Damien Leith Family Star
Cru 12h26'24" -60d17'4"
The Damm Star
Tri 2h3'13" 32d34'8"
The Dance of the Red-Crowned Cranes
Crb 15h52'38" 32d39'26"
The Dancing Star
Aqr 23h2'13" -5d36'9"
The D'Andrea Family
Psc 1h23'39" 14d3'57"
The Daneth my Love
Dra 17h10'37" 51d52'57"
The Daniel John Star
Cep 20h49'44" 59d20'43"
The Daniel Thomas Aiken Star
Ori 5h33'38" 4d1'37"
The Daniele Montefusco
Per 3h43'20" 45d36'12"
The Danielle Cavendish Star
Cru 12h32'19" -58d0'49"
The Danielle Robert
Uma 12h41'24" 57d49'58"
The Danielle Star
And 2h38'11" 45d29'21"
The Dannie Goree Angel Star
Leo 9h37'39" 17d54'4"
The Danny
Lyn 8h26'31" 43d1'26"
The Dapper Star
Leo 11h44'13" 18d59'15"
The Darcy Elizabeth Rose Star
And 2h35'42" 42d13'49"
The Daris
Her 17h33'13" 47d13'56"
The Dark Angel
Sco 17h35'18" -39d7'31"
"The Dark Scorpio"
Sco 17h56'26" -40d42'31"
The Darling Eva Hajok
Sgr 19h43'5" -14d41'41"
The Darren - Fango - Pearce Star
Cru 12h24'37" -57d53'48"
The Darrington Family Star
Ori 6h19'35" 7d38'11"
The Dave Stauffer Star
Cnc 8h47'18" 21d59'14"
The Dave Zacarias Star
Sgr 20h21'23" -42d56'45"
The David Altmark
Cap 21h43'41" -13d8'27"
The David and Lisa Munro Star
Cru 12h18'19" -59d45'0"
The David and Stacy Craig Star
Lyr 19h20'37" 35d54'58"
The David Green Star
Uma 8h41'13" 52d53'1"
The David & Lindsey Vigor Star
Cyg 19h34'54" 28d35'32"
The David Page Star
Peg 22h32'6" 13d8'6"
The David & Pat Anderson Star
Ari 2h42'13" 20d22'0"
The David Trapnell Star
Her 17h31'51" 31d58'51"
The Davidson Family
Per 2h23'8" 56d12'20"
The Davidson Star
Her 17h16'32" 26d41'30"
The Day Everything Changed
Ari 2h8'1" 25d2'50"
The Day I Found My Star Gemma x
And 0h27'20" 40d30'47"
The 'DAY' of Marlène & Daniele
Umi 13h17'12" 70d3'6"
The Day Our Lives Began Together
Uma 9h31'47" 57d26'16"
The Day Star
Uma 12h3'20" 59d44'15"
The Dazzle of David
Umi 17h5'31" 76d55'31"
The Dazzling Star of Ray Dunn
Cru 12h52'22" -63d3'2"
The DB All-Star
Lyn 8h7'44" 58d40'52"
The De Domenicos
Dra 19h13'17" 64d32'22"
The De Young Sisters
And 0h55'35" 38d37'21"
The Dean Folkerts Family Star
Uma 9h5'27" 57d9'4"
The Deana Kay Jackson Star
Mon 6h48'54" 6d56'42"
The Deb Star
Uma 11h52'46" 54d33'49"
The Deblynn P Star
Uma 10h56'28" 52d7'16"
The Deborah June Cox
Gem 7h2'47" 26d46'4"
The Deborah Star
Cap 21h51'24" -10d43'18"
The Debra Lynne
And 23h16'43" 47d8'13"
The Declan Rourke
Uma 13h49'49" 58d24'41"
The DED Constellation
Dra 19h12'17" 61d15'46"
The DeJesus Children of Baldwin, NY
Uma 8h30'58" 67d57'21"
The DelBuono Star
Cyg 19h31'22" 51d37'30"
The deMaagd Family
Cru 12h14'5" -63d35'38"
The Demant Family Star
And 2h37'40" 44d19'22"
The Dennis Cowan Star
Ori 5h53'48" 20d39'19"
The Dennis Dziobak Memorial Star
Uma 11h48'48" 33d22'10"
The Dennis Raatjes Star
Her 17h23'39" 39d15'14"
The Dentler Star
Del 20h36'1" 15d52'42"
The Devan-ator
Srp 18h40'31" -0d11'26"
The Devils Lady
Uma 8h38'9" 50d18'51"
The Devin and Chrissy Star
Lyn 7h15'55" 48d0'9"
The Devin Star
Leo 11h13'35" -3d54'47"
The Devine Star
Lyn 7h58'38" 45d34'44"
The Devonshire
Boo 14h44'29" 32d17'26"
The Dewdrop
Eri 3h52'18" -20d55'1"
The Di Star
Cas 23h12'10" 58d50'43"
The Diamond Bradley
Psc 1h27'48" 27d45'54"
The Diamond Darroch
Uma 9h58'51" 72d36'1"
The Diamond Family Star
Cru 12h27'25" -58d9'27"
The Diamond Star
Crb 15h52'24" 26d6'27"
The Diamonds - Ken & Sheila Sperring
Cyg 21h3'4" 50d25'18"
The Diana Lynn
And 1h12'4" 42d59'5"
The Diana Shannon Star
Vir 12h43'23" -4d30'29"
The Dianne & Dan Star
Umi 16h23'53" 74d2'54"
THE DIC-E-BOY
Cep 20h52'39" 57d0'53"
The Dick and Carol Boerger Star
Ori 6h17'25" 14d36'19"
The Dick & Doris Woods Family Star
And 1h36'17" 45d37'52"
The Dilworth Star
Uma 9h19'50" 48d13'30"
The Dina Jee
Ori 6h5'13" 6d32'53"
The Disneys! - Derek and Daphne
Cyg 20h29'56" 53d44'24"
The Divine Mother: Lord Shiva's Wife
Umi 4h23'3" 88d47'22"
The Dixie Lee
Uma 11h11'44" 52d7'51"
The Dixie Star
Ori 5h36'8" 9d25'45"
The Dixie Star
Leo 11h18'21" 15d10'48"
The Dixon Star
Vir 12h56'30" 8d49'20"
The D.J. Radiant
Uma 10h18'33" 48d11'14"
The Dlorah Star
Aql 19h7'52" 9d9'22"
The Doaney Dazzler
Tau 4h42'20" 22d33'52"
The Dobson 5
Cru 12h25'15" -60d33'20"
The Docter
Aur 5h4'35" 37d24'50"
The Doctor
Uma 9h22'40" 68d54'50"
The Doctor Martin Patrick Newman Star
Cru 12h35'56" -61d53'35"
The Dodd's
Sge 19h26'10" 18d19'7"
The Dodgson-Duff Star
Cyg 19h43'2" 28d41'14"
The Dolly and Louis Pardi Star
Lib 15h56'54" -12d23'26"
The Dolly Momma
Crb 15h44'56" 36d18'59"
The Dolphis
Lib 15h23'44" -7d43'27"
The Domalewski Star
Umi 14h7'4" 68d7'32"
The Donah Rose Star
Tau 5h44'20" 19d28'23"
The Donald A House
Cyg 19h47'55" 35d15'39"
The Donald Floyd
Per 3h11'30" 54d25'29"
The Donald Hiatte Star
Aur 7h19'56" 41d39'57"
The Donald Joseph Macaulay Star
Leo 9h34'41" 12d49'5"
The Donna Jay
And 0h4'59" 44d2'47"
The Donna Mae/Barker
Uma 13h52'15" 61d27'49"
The Donna & Pat Walsh Truelove Star
Sco 17h39'39" -42d36'53"
The Donna Star
And 0h36'4" 41d52'59"
The Donnafaye Evenstar
Lib 15h17'3" -15d38'21"
The Donnellan Star
Vir 14h15'45" 5d17'44"
The Donnelly Dogs, Now and Forever
Cma 7h24'14" -24d54'46"
The Donovan Family
Uma 10h14'23" 65d4'21"
The Dorato Family Star
Uma 9h38'7" 46d51'1"
The Dork
Lyn 8h21'42" 38d26'37"
The Double D. & R.V. Doss Vincent
Sge 19h43'21" 19d16'37"
"The Double D Salazar"
Tau 4h8'39" 5d54'17"
The Double Taurus
Tau 3h38'49" 5d21'8"
The Dougger
Cam 4h28'3" 68d18'50"
The Douglas J Roberts Star S133592
Pyx 9h0'28" -26d30'0"
The Downing Family Star
Lmi 10h13'33" 29d48'53"
The DP Mouat Star
Cru 12h52'4" -63d0'3"
The Dr. David Dennis Memorial Star
Sgr 19h5'17" -17d31'7"
The Dr. Rodney McCarthy Star
Ori 5h31'49" -0d43'28"
the D.R. Star
Uma 10h55'14" 36d4'20"
The Dr. William C. Dundore Jr. Star
Per 3h43'4" 52d11'22"
"the dragonfly star"
Dra 17h17'18" 69d18'2"
The Dream
Dra 18h31'19" 51d12'50"
The Dream Team July 5, 1997
Aql 19h16'29" 3d42'27"
The Drinkers Star
Cru 12h39'48" -59d39'18"
The Drs. Brittain
Uma 9h54'37" 55d22'23"
The Drusilla & Gareth Star
Cyg 19h56'53" 47d28'32"
The Dubliner
Uma 13h50'31" 51d10'57"
The Duchess
Ori 6h10'55" 21d0'31"
The Duchess of Farnsworth
Ari 3h26'11" 25d59'55"
The Duchin Star
Uma 10h33'8" 50d20'21"
The Dudelbarb Stanbridge
Uma 11h25'34" 58d50'41"
The Dudes!
Cyg 21h15'14" 51d2'57"
The Duff Star
And 23h15'55" 49d53'23"
The Duffle Deluxe
Aur 6h27'26" 35d12'34"
The Duffy Family
Cru 12h18'31" -57d37'27"
The Duke Barrett Star
Uma 11h31'31" 47d54'21"
The Dutchman
Per 3h10'59" 52d5'30"
The Duzey
Uma 11h29'45" 55d16'50"
The Duzman Family Star
Uma 13h42'4" 51d25'27"
The Dye Star
Cam 4h54'34" 67d48'13"
The Dynan Star
Uma 11h41'47" 53d32'55"
The E&R Flowerree Star
Cap 21h52'28" -18d8'43"
The Eadie Pearl
Cyg 21h52'2" 48d22'43"
The Eager Family
Pho 1h0'25" -47d32'28"
The Eagle
Aql 19h4'20" 14d16'46"
The Early Family 8-12-2003
Uma 9h30'39" 55d9'24"
The Earnest Wallis Jones Star
Gem 6h23'28" 19d2'20"
The East Family
Cyg 20h59'37" 31d42'32"
The EASY Star
Cyg 21h45'56" 48d13'25"
The Easy Star
Cap 21h39'25" -9d47'1"
The Eberto Family Star
Her 16h53'12" 17d24'47"
The Echo of Suzie French's Laughter
Lyr 18h42'8" 39d17'26"
The Ed & Marie Vincent Family Star
Gem 6h21'15" 22d21'4"
The Ed Sterba Star
And 2h7'5" 45d5'54"
The Ed Struble Star
Cnc 8h17'56" 10d34'25"
The Eddie Aikau Star
Cep 23h14'11" 80d21'21"
The Edgars'
Cyg 21h28'32" 49d5'10"
The Edith Elizabeth Star
Uma 9h59'1" 42d19'48"
The Edith Shallo Star
Cas 0h38'15" 57d11'42"
The Edward Brennan Guardian Star
Ori 4h50'8" 4d39'3"
The Edwards Family
Dra 18h48'9" 51d6'11"
The Eiffel Tower Star - Mike Vance
Vir 13h15'5" 5d5'37"
The Eileen Bowman Sylwestrzak
Tau 4h41'42" 23d0'10"
The EiTo Star
Ori 6h14'7" 20d39'57"
The Elaine and Tim Star
Dra 19h11'33" 60d33'9"
The Elders
Leo 11h14'3" -0d30'21"
The Eleanor-Jane Christening Star
And 23h2'59" 52d5'56"
The Elegant Tiffany
Sgr 19h31'19" -12d31'0"
The Eleonore Cato Lore Star
Ari 1h58'29" 21d7'43"
The Eli & Josh Graves Star
Uma 11h51'31" 46d59'40"
The Elizabeth
Tau 5h11'8" 27d30'36"
The Elizabeth Jane
And 1h18'28" 45d23'58"
The Elizabeth Knoepfel
Sco 17h36'22" -35d54'33"
The Elizabeth Lewin Star
Cas 0h18'27" 51d27'3"
The Elizabeth Rowe Star
Sco 17h53'41" -30d14'26"
The Elizabeth Star
Aqr 22h51'55" -9d5'15"
The Elizade Family
Uma 9h38'0" 56d33'34"
The Ella
Cra 18h48'59" -40d13'49"
The Ella Rouge Beauty Star
Cru 12h18'51" -59d48'3"
The Elliott Craske Star
Vir 12h45'10" -4d39'33"
The Ellis Family Star
Uma 13h19'3" 58d15'48"
The Ellsworth Star
Dra 18h22'9" 58d20'23"
The Elph
Aqr 22h13'29" 1d35'27"
The Elstob Star
Uma 11h28'37" 46d49'0"
The Elusive Inspiration 7-29-59
Cam 4h0'4" 67d16'19"
The Elvin Creedon Star
Cru 12h29'23" -59d3'20"
The Elzy Family
Cyg 20h9'36" 36d50'4"
The Emancipation of Mr. Cool Ice
Ori 5h25'48" 14d17'7"
The Emanelle Star
Cyg 21h51'47" 52d22'44"
The Emerald Star
Col 6h26'32" -35d7'13"
The Emerton Diamond
Cyg 21h6'17" 47d23'39"
The Emery Star
Uma 9h34'57" 56d1'35"
The Emily & Harry Star - 14 08 2005
Cra 18h33'52" -42d54'44"
The Emily Katherine
Uma 13h51'13" 51d43'15"
The Emma Sue Schuerenberg Star
Vir 14h2'32" -11d33'44"
The Emma Toups 100th Birthday Star
Tau 5h19'46" 17d45'57"
The Empress Alexandra Star
Tau 5h25'30" 26d29'58"
The Empress Yvette
And 0h27'2" 26d23'58"
The Enchanting Ashley Brady
Ori 4h49'12" 13d22'28"
The Enchanting Tiffany E. Gonzales
Lyn 6h49'16" 54d3'47"
The Engel Family
Aql 19h1'53" 1d14'7"
The Enid and Bill Breen Star
Lyr 18h49'17" 30d59'41"
The Enos
Uma 12h40'47" 57d17'31"
The Ephie Lowinger Kangarii
Tau 5h54'19" 25d27'14"
The Erdmann Family Star
Uma 11h49'39" 30d2'39"
The Eric Piepenbrock
Ori 5h41'2" 2d44'34"
The Eric Snyder
Ari 3h14'14" 27d53'34"
The Eric Travis Branch Star
Uma 11h30'5" 39d19'0"
The Erica B Star
Lyn 6h37'35" 56d21'16"
The Ericca & Romeo Wright Star
And 0h52'38" 39d30'35"
The Erik Park Star
Cap 20h57'18" -15d4'28"
The Erika Castro Constellation
And 0h14'7" 47d44'50"
The Erin Marie
Gem 6h18'34" 27d10'29"
The Erin Star
Lyn 6h37'26" 53d58'16"
The Ermine Star
Vir 14h31'2" 7d1'49"
The Ernest Feibelman Star 70
Cam 4h32'31" 69d42'10"
The Errol Marron Star
Cru 12h26'54" -59d43'48"
The Ertlet Star
Crb 15h37'56" 26d56'13"
The Eryn Star
Ori 6h10'42" 9d14'54"
The Esmeralda Star
Cas 0h52'15" 65d15'10"
The Essex Serpent
Ser 15h50'55" 23d23'10"
The Esson Star
Ori 5h39'1" -3d48'17"
The Eternal Chief
Ari 2h39'27" 29d19'25"
The Eternal Colette
Vir 12h57'54" -15d4'21"
The Eternal Flame of Dave & Belinda
Aqr 23h10'44" -15d22'13"
The Eternal Freeda B. Pierce
Gem 6h6'10" 25d32'38"
The Eternal Light Gary Grider
Psc 0h54'28" 10d27'8"
The Eternal Love of Joseph and Lisa
Cyg 19h43'34" 31d45'23"
The Eternal Noelle
Cap 20h59'32" -21d37'40"
The Eternal Star of Mallory
And 0h33'52" 31d41'57"
The Ethan Gary James Jones Star
Umi 16h45'28" 81d22'59"
The Ethel & Jimmy Barrish Star 50
Psc 1h24'36" 25d33'14"
The Ethel Marie
Cas 1h37'55" 61d14'45"
The Ethereal Li Ping Tang Star
Leo 9h46'31" 32d22'27"
The Etheridge Family Star
Ori 5h20'34" -4d46'35"
The Etter Star
Ari 2h35'31" 26d47'42"
The Eugene Piccinotti Star
Ori 5h10'35" -0d36'11"
The Eulah Bright
Uma 11h27'17" 32d52'44"
The Eveland Family
Uma 11h15'58" 33d44'56"
The Evelyn Joan Paytosh Star
Cap 20h31'47" -21d41'52"
The Evelyn May Star
Cas 0h32'56" 60d1'50"
The Ever Dazzling......Ruth Mello!
And 0h27'12" 44d2'4"
"The Ever Lasting", Tash & Thomas
Cyg 20h44'36" 46d57'31"
The Everlasting Star
Umi 14h15'7" 76d40'24"
The Evie Grace Star
And 0h8'55" 29d14'28"
The Ewing Family Star
Aqr 21h50'5" 0d56'19"
The Exceptional Tess
Sex 10h15'40" -0d46'6"
The Exquisite Emily Elaine
And 23h6'39" 46d43'5"
"The Extraordinary" Joel Bloomston
Lyn 7h8'25" 46d59'49"
The Eye of Aaliyah Newson
Psc 1h15'23" 15d49'17"
The Eye of Bonnie
And 0h18'26" 46d29'40"
The Eye of Elayne Marie Phillips
Cyg 21h40'6" 43d50'57"
The Eye of Gabrielle
And 0h31'32" 36d16'40"
The Eye of Goddess Heather
Crb 16h6'12" 29d23'54"
The Eye of Goddess Michele
Lyn 7h33'59" 49d55'13"
The Eye of Maranda
Ori 6h12'26" 20d31'43"
The Eyes of Courtney
Uma 10h26'52" 42d4'19"
The Eyes Of Dannielle
Gem 6h35'5" 21d38'41"
The Eyes of Jennifer
Lib 15h15'2" -7d39'2"
The eyes of Stefanie
Lmi 10h8'58" 33d23'38"
The Fab Anne
And 22h57'42" 51d2'48"
The Fabulous Julie Geeting
Ari 2h42'59" 14d15'37"
The Fabulous Meghan Shepard Star!
And 0h19'41" 31d45'22"
The Fabulous Rebecca Jean Rivers
Cyg 21h34'6" 49d27'22"
The Falcon
Aql 19h33'5" 9d58'2"
The Falcone Family Star
Gem 7h37'44" 23d23'5"
The Falconer
Uma 11h47'56" 40d43'4"
The Falconette
Aql 19h11'16" 14d52'27"
the fam damily
Umi 15h31'46" 72d25'12"
The Family of Compton
Uma 8h48'41" 50d22'54"
The Family of Scott Halterman
Lmi 10h53'27" 33d4'30"
The Family of Stoltenow
Umi 15h15'3" 70d0'53"
The Farara Star
Uma 11h32'14" 56d17'3"
The Fast Track Bradley Boys
Ori 5h55'37" 11d43'37"
The Father And Son Star
Vir 13h55'35" -9d10'14"
The Favre Star
Her 16h55'51" 22d3'54"
The Fawn Star
Sco 16h9'53" -12d6'56"
The Fawntia Shay
Uma 13h19'41" 56d13'52"
The Fay School Teachers 2004-2005
Her 16h42'9" 23d22'40"
The Feehan Star
Uma 8h53'59" 50d46'52"
The FEGAN Star
Gru 21h47'37" -36d44'26"
The Fernandes Star
Uma 9h28'2" 66d0'39"
The Fiechtl Phenomenon
Aqr 22h24'50" -2d55'4"
The Five Fitzs
Lmi 10h39'23" 35d15'43"
The Flame
Cyg 19h51'43" 33d23'17"
The Flips Roxski Star
Ori 5h1'15" 6d1'46"
The Flobby and Pen Star
Cru 12h30'45" -60d46'35"
The Florida Pfeil's
Umi 15h5'38" 71d37'42"
The Flossy Q
Aqr 21h58'45" -1d5'22"
The Fluffy Pink Thing
Cas 23h59'48" 53d34'26"
The Flying Dodo
Dra 12h5'24" 70d24'8"
The Flying French Man
Aql 19h42'56" 15d9'51"
The Foley Six
Uma 10h40'26" 56d30'44"
The Foo Man of God
Sgr 18h54'11" -20d23'7"
The Forever Beauty of LeAnne Rider
Uma 11h36'13" 50d0'16"
The Forever Star
And 1h21'30" 38d9'10"
The Forrest Family Star
Ori 5h43'21" 3d55'43"
The "Forward" Star
Aql 19h51'56" 16d17'5"
The Fountster
Aql 19h36'56" 1d57'13"
The Four C's
Umi 16h20'23" 76d44'15"
The Four Kathleen's
Umi 15h56'24" 74d50'9"
The Four Rogers
Leo 11h21'28" 21d38'49"
The Fourth
Umi 17h40'49" 81d27'38"
The Fourth
Umi 17h35'33" 81d20'8"
The Fox
Vul 19h31'57" 26d44'46"
The Fox 75th Birthday Star
Tau 4h37'21" 19d51'45"
The Fox Star
Gem 6h46'43" 22d27'18"
The Fraggle
Oph 17h9'59" -0d3'1"
The Francis J. Bomher Eternal Star
Uma 11h53'11" 44d23'52"
The Francis-MA Memorial Star
Lyn 7h8'5" 59d6'16"
The Frank & Barbara Scotton
Sge 20h5'24" 19d8'43"
The Frank Thomson Star
Uma 9h38'5" 58d52'49"
The Frankcombe Family Happy Star
Cru 12h2'59" -62d13'23"
The Franklins
Uma 10h19'42" 60d34'48"
The Fred Smith
Psc 1h35'15" 23d28'20"
The Fred V & Travis J Star Eternity
Cyg 21h52'27" 49d40'34"
The Frederick and Betty Bensen Star
Tau 5h57'5" 27d11'17"
The Fredric John Seibert, Sr. Star
Cnc 8h23'37" 32d43'40"
The Fredrick and Mary Yezzi
Leo 11h19'26" 11d29'0"
The Frees Family
Uma 8h27'25" 60d27'50"
The Friend
Lyn 9h5'40" 45d3'27"
The Friends of Ashworth House
Cru 12h30'26" -58d36'17"
The Friendship of Jessica & Crystal
And 0h34'5" 43d55'48"
The Fritz & Alice Star
Lyr 19h2'8" 42d59'50"
The "Fru Fru" Tom Boys
Peg 22h19'49" 7d6'1"
The Fruity Star
Cyg 20h0'49" 53d29'38"
The Fuentes Anniversary Love Star
Uma 11h30'22" 47d43'10"
The Fuggiti's
And 0h31'21" 29d52'9"
The Fuller Family Star
Ori 5h17'8" 0d28'8"
The Fun-Pants Kale Star
Cen 13h29'8" -38d2'2"
The Furmato Family
Uma 10h30'30" 59d16'44"
The Fustigator
And 0h24'49" 27d55'51"
The future holds our destiny
Aqr 23h3'14" -5d53'42"
The Fuzzy Star
Vir 12h14'44" 12d10'51"

The G Star
Her 17h50'55" 15d52'34"
The Gabrielle Rosin
Sco 16h14'10" -9d23'32"
The Gail Whittemore
Cas 1h43'33" 64d15'19"
The Galaxy's Best Mom!!
Tau 3h37'46" 1d6'4"
The Galen and Shirley Shultz Star
Cap 20h11'12" -9d47'39"
The Gallacher/Escorza Star
Cyg 21h22'58" 44d31'27"
The Galle Family
Uma 9h27'24" 60d13'23"
The Galley Sweep
And 0h29'37" 37d58'15"
The Gammon
Tau 4h39'40" 12d19'40"
The Gammy Star
Cas 0h43'13" 58d54'19"
The Gannon's - Anne & Steve
Cnc 8h31'26" 21d29'38"
The (Gard)ian
Aql 19h59'19" -0d28'46"
The Gardener Craig Flowers
Boo 15h23'8" 49d32'57"
The Garlic Family Star
Dra 17h25'17" 69d52'23"
The Garnet Star
Col 6h6'3" -28d27'57"
The Garren Six
Aql 19h48'23" -0d36'8"
The Garrett Wade VanGundy Star
Aql 19h41'30" -3d57'22"
The Garth and Riley Star
Cyg 19h44'33" 30d49'10"
The Gary Donald Higgins Star
Aqr 21h54'18" -8d18'27"
The Gary E
Lyn 7h32'38" 53d39'48"
The Gates of Avalon
Vir 14h15'33" -17d2'45"
The Gavster
Uma 12h17'20" 58d31'37"
The Geezer
Sco 16h6'45" -11d51'26"
The Gehrke's 5 Alarm JLNEM 2004
Uma 10h53'23" 62d36'1"
The Gem
And 2h15'53" 42d29'23"
The Gem of excellent phrasing
Lib 15h11'11" -2d48'25"
The Gene and Nancy Phillips Star
Uma 10h1'25" 50d44'22"
The General
Ari 3h23'20" 26d16'7"
The General
Uma 10h29'37" 69d44'10"
The General Phil Sherman Star 80
Boo 14h37'31" 19d55'5"
The Genie D
Leo 11h32'21" 17d11'37"
The George
Dra 10h1'28" 78d2'7"
The George Carletti
Leo 11h38'50" 27d17'7"
The George Frankfort Star
Cep 0h12'31" 74d5'36"
The George Jabbour Family Star
Uma 9h19'24" 61d12'5"
The "George Robinson" Star, My Hero
Per 4h31'21" 48d23'46"
The George Strait Star
Leo 11h13'4" 16d51'51"
The Georgia Hanton Star
Sco 16h38'54" -45d18'0"
The Georgianna
And 1h49'46" 46d32'28"
The Geraldine Elizabeth
Cas 1h14'30" 62d44'31"
The Geren Family Star
Dra 18h32'41" 55d49'59"
The Germanese Star
Ori 5h53'52" 20d51'49"
The GI Tiger
Crb 15h52'34" 31d29'18"
The Gifford Children
Umi 13h43'57" 76d10'47"
The Giggle Star
Leo 10h10'56" 22d6'19"
The Gigi sister star
Cnc 8h21'49" 29d5'34"
The Gilbert Family
Uma 12h21'51" 54d43'19"
The Giles Family Star
Peg 0h4'6" 18d15'59"
The Gillian Goldberg Star
Cas 0h48'16" 47d46'38"
The Gills Family
Aql 19h48'47" 8d56'25"
The Gina Marie
And 2h34'9" 45d25'2"
The Ginny Star
Cnc 8h25'16" 11d24'24"
The Girlfriends
Vir 14h14'51" -1d15'31"
The Girls- Dyan,Judys,Marla,Terry,Wendy
Cas 1h25'26" 59d19'44"
The Giver
Psc 0h19'6" 0d56'27"
The Giving Star
Pho 23h33'32" -43d0'58"
The G.J. Star
Cas 1h40'36" 61d31'55"
The Glavan Family Star
Lyr 18h45'16" 37d51'58"
The Gleam in Papaw's Eye
Uma 10h10'17" 61d10'42"
The Glennifer Family Star
Uma 10h25'17" 46d37'40"
The Glerum Star
Cap 20h23'8" -23d18'0"
The Glimmer of Mark
Ori 4h50'0" 10d19'5"
The Glimmering Jessika
Cru 12h24'39" -56d41'15"
The Glock
Vir 13h14'52" 6d46'35"
The Gloria Star
Sgr 19h20'31" -19d40'45"
The Glowing Beauty of Lindsey Doyle
And 23h26'34" 41d32'52"
The Goad Star
Cyg 19h54'36" 55d41'5"
The Goat
Cap 21h19'52" -23d25'3"
The Goddess Nanette Robinson
Lyn 7h33'38" 41d39'44"
The Goddess Nicole
Psc 0h56'39" 29d30'42"
The Goddess Safia
Cnc 8h9'26" 28d6'42"
The Goddess Star
Lyn 8h32'46" 55d42'37"
The Godfather
Cas 23h22'12" 58d4'56"
The Godspeed
Uma 9h2'39" 63d42'38"
The Godzilla Star
Vir 12h44'0" 4d59'56"
The Goguen's Happily Ever After
Umi 16h17'46" 79d49'0"
The Golden Asher
Uma 10h30'43" 42d58'16"
The Golden Boy's Star
Lib 15h24'24" -8d18'48"
The Golden Chande Star
Ind 21h59'58" -66d54'57"
The Golden Heart / Jelleyman's
And 23h23'54" 38d45'29"
The Golden Norman Wedding Star
Cyg 20h22'35" 41d37'8"
The Goldwell Love Star
Pyx 8h53'1" -26d16'5"
The Goldyn Hart
Ori 5h17'56" 1d22'48"
The Good Life
Cyg 21h22'36" 50d0'1"
The Good Stuff
Uma 9h42'25" 67d3'37"
The Good...The Bad...The Daughters!
Cas 0h20'1" 60d13'53"
The Goose! Bethan M Williams' Star
Cas 0h7'46" 55d7'59"
The Goose Star
Leo 10h50'41" 11d42'23"
The Gordon Family
Cyg 21h49'41" 41d44'22"
The Gorgeous Babe
Vir 11h47'58" -4d30'5"
The Gorgeous Jamie Mosher
Ori 5h1'21" -0d43'16"
The Gorgeous Paul Martin Star
Her 17h47'53" 20d40'0"
The Gothfather
Cas 0h53'45" 62d58'28"
The Grace Carol Amanda Star
Crb 15h34'23" 30d20'7"
The Grace & Jason
Lyr 19h27'21" 41d59'47"
The Grace Katherine Star - 250804
Vir 12h42'53" 1d49'35"
The Grace Sienna Star
And 0h4'22" 43d57'0"
The Gracious Light of Hugh & Diane
Cyg 21h26'21" 37d29'41"
The Graduate
And 2h26'9" 39d7'29"
The Graham and Ryane Star
Ori 4h50'55" 6d55'32"
The Graham David Salon
Peg 0h5'59" 12d35'24"
The Graham, Dad and Pa Daly Star
Nor 16h20'53" -48d12'21"
The Gramps and Nuna star
Uma 8h17'18" 71d16'26"
The Gran & Grumps Star
Cyg 20h48'16" 38d21'36"
The Gran Star
Gem 7h6'34" 16d15'4"
The Grand Poobah
Per 3h11'57" 42d18'54"
The Grand Roman Monk
Cyg 20h0'27" 41d13'23"
The Grandma Madge Mayo Star
Cas 2h41'51" 57d53'28"
The Grandpa Star, Charles Polaski
Sco 17h57'5" -37d46'30"
The Grandpop Luce Star
Cep 22h42'21" 66d6'11"
The Granny & Poppie Dunlap Star
Uma 10h31'7" 49d52'56"
The Granny's Star
Cas 1h24'33" 52d37'2"
The Grantham
Cap 20h34'47" -9d14'43"
The Grants - Forever Light Our Love
Cru 12h38'45" -58d22'24"
The Grasshopper: Lauren Cross
Uma 10h34'38" 54d12'52"
The Gray Star
Cyg 21h27'13" 39d29'55"
The Gray's
Tri 2h1'59" 34d13'46"
The Great and Powerful VU
Sco 17h55'59" -40d16'16"
THE GREAT AUBEL
Vir 14h37'59" 2d12'47"
The Great Believer
Sco 16h50'53" -33d5'39"
The Great Escape
Cyg 20h34'23" 59d46'10"
The Great Furry Burr
Ori 5h12'10" 0d37'38"
The Great Grandma Tosi Star
Lib 15h46'41" -26d10'7"
The Great Herman Viox
Cnc 8h47'15" 13d30'43"
The Great Jadavas
Mon 6h49'54" 8d28'19"
The Great Joe Hannon
Dra 18h44'16" 50d10'18"
The Great John David Herring,Jr MSW
Cep 23h24'39" 77d5'13"
The Great Kay
Leo 10h25'13" 18d36'29"
The Great Loudini
Umi 16h51'55" 86d18'22"
The Great Lynn Chaloupka
Gem 7h21'48" 27d7'26"
The Great Markie
Lyn 8h56'3" 45d59'29"
The Great & Mighty Pip
Crb 16h14'30" 38d3'36"
The Great One
Sco 16h10'17" -9d2'8"
The Great One
Sgr 18h23'17" -23d3'47"
The Great Pharo Galactica
Ori 5h50'39" 2d55'30"
The Great Poobah
Her 17h40'16" 16d16'4"
The Great Shell
Uma 10h24'46" 62d19'48"
The Great Sir
Vir 12h56'0" -2d0'49"
The Great Smolini
Leo 11h44'9" 25d31'57"
The Great Star Of Hope
Ori 5h40'20" 7d53'7"
The Great Star of Talia and Mommy
Gem 6h59'22" 26d18'53"
The Greatest Chapter
Pho 0h40'0" -42d55'18"
The Greatest Dad
Her 16h40'22" 46d8'53"
The Greatest Dad In The Universe
Sgr 18h17'35" -26d33'28"
The Greatest Dad, Jack W. Novin
Per 2h39'54" 55d6'20"
The Greatest Gift
Cyg 19h47'10" 58d1'29"
The Greatest Grandparents
Ori 5h16'3" 0d47'33"
The Greatest Mom Ever
Leo 11h18'15" 13d1'37"
The Greatest Mom In The World
Sgr 18h10'58" -20d1'11"
The Greatest Mom, Marcelle C. Novin
Per 2h48'9" 56d17'5"
The Greatest Mother
Lib 14h52'10" -3d45'30"
The Greatest Union
Cyg 20h1'32" 32d6'45"
The Greats - R & L Lepicier
Crb 15h37'20" 32d12'13"
The Greedy Crocodile
Dra 20h8'58" 71d21'54"
The Greek Star Light
Lyn 8h30'3" 58d56'22"
The Green Eyed Charlie Evans
Uma 10h49'41" 43d33'46"
The Green Family Star
Cyg 21h46'8" 39d58'30"
The Greenough's Star
Dra 18h43'11" 51d48'37"
The Green's Star - 11.11.2006
Ara 17h58'44" -51d10'14"
The Greg and Erin Star
Tel 18h56'3" -45d41'5"
The Greg Doerr Star
Her 17h34'49" 48d21'42"
The Gregory Star
Psc 0h34'29" 14d25'34"
The Griffins Perch
Sgr 19h15'38" -16d1'5"
The "Grills" (Brittany & Courtney)
Umi 14h26'25" 75d56'39"
The Grissom Star
Ari 2h44'6" 29d25'35"
The Group
Tau 5h29'34" 26d11'45"
The Grrr Villy Star
Ori 5h21'59" 13d37'33"
The Guardian
Cnc 8h43'59" 18d17'8"
The Guardian "Lil' Tony"
Her 17h6'45" 31d52'28"
The Gudge
Uma 10h4'22" 60d0'21"
The Guests of our Wedding
Uma 11h45'2" 56d57'37"
The Guide
Lyn 9h36'4" 40d34'18"
The Guiding Light of Love
Lyr 18h42'16" 35d54'14"
The Guiding Star of Kimmie Johnson
And 0h34'27" 26d49'42"
The Gurwitz Family Star
Her 17h16'46" 14d57'33"
The Gutcho
Vir 14h45'26" 4d11'13"
The Gwyneth Marritt 50th Birthday Star
Cyg 19h41'0" 34d14'11"
The Gypsy Rose
Cap 21h14'22" -15d29'11"
The Hader Twins
Pho 0h28'12" -45d7'46"
The Haley Stevenson Star
Lib 15h14'41" -12d5'51"
The Hall of Tranquility
Leo 11h13'52" 7d35'27"
The Hallahan Star
Uma 13h15'21" 63d3'8"
The Halls
Psc 0h34'27" 3d17'4"
The Hally HB Star
Per 4h43'29" 49d2'5"
The Hammett Star
Del 20h41'7" 11d8'21"
The Handsome Cormac McQuillan
Tau 4h37'53" 21d59'6"
The Handsome One. R. B.
Aql 19h21'23" 0d57'41"
The Handsome Scott
Ari 2h36'48" 14d38'55"
The Hanifin's Star
Lib 15h7'17" -22d28'10"
The Hanlon Star
Dra 17h58'39" 56d13'9"
The Hannah Bean
Umi 14h11'1" 77d41'6"
The Hannah Cate
Uma 9h47'3" 61d45'51"
The Hansen Family Star
Uma 11h27'38" 35d34'17"
The Hanson Family Star
Uma 8h42'30" 60d16'2"
The Happer
Uma 8h52'54" 48d23'38"
The Hardy's Star of Love
Crb 16h16'5" 30d40'54"
The Harley Hunter Star
Uma 8h57'13" 61d36'3"
The Harold and Brenda Uphill Star
Cyg 19h48'41" 56d20'11"
The Harpers
Uma 12h51'50" 58d25'34"
The Harringtons
Aql 20h5'10" 3d46'3"
The Harrison William
Uma 11h47'42" 36d28'44"
The Harry Taylor TATOC Star
Uma 11h17'10" 71d42'48"
The Harry - Tina Dreamcatcher Star
Aqr 22h13'8" -1d54'17"
The Hartman's
And 0h39'48" 39d42'6"
The Hartz Boulos
Uma 12h37'3" 57d0'2"
The Harvey Joseph Sell Star
Cnc 8h36'10" 23d38'7"
The Harvey Star
Uma 10h28'58" 60d34'7"
The Harwood Family Wishing Star 07
Uma 9h11'16" 51d15'37"
The Hassfurther Star
Umi 16h25'46" 77d58'12"
The Hatfield Star
Her 17h20'34" 27d21'20"
The Haugen-Savage Love Star
Crb 16h13'48" 39d0'41"
The Hayden
Her 18h20'55" 28d32'58"
The Hayden Bobkowski
Ori 5h17'6" 0d34'45"
The Hayes Family Star
Uma 11h53'28" 34d50'43"
The Hayley Witherington Star
Cyg 20h11'7" 37d32'14"
The Haylor Family Star
Uma 10h2'51" 59d56'7"
The Hayward Star
Uma 11h31'28" 57d58'58"
The HBR "Eternity" Star
Uma 11h31'38" 56d15'14"
The Headliner
Ori 5h55'31" -0d36'50"
The Healer
Lyn 6h16'33" 54d6'51"
the Heart of a Mother
Tau 3h30'22" 17d7'4"
The Heart of Adonis
Per 4h41'55" 49d10'55"
The Heart of Albert
Per 3h5'58" 56d1'34"
The Heart Of Carolyn
Aqr 23h41'44" -24d39'31"
The Heart Of Deb
Sgr 18h12'13" -20d8'36"
The Heart of Dixie
Vir 12h7'23" 0d44'58"
The Heart of Evelyn
Ari 3h14'21" 28d8'41"
The Heart of Kristen
Leo 10h26'18" 16d20'40"
The Heart of Maine
Ari 3h11'42" 28d9'4"
The Heart of Marquel
Peg 23h25'0" 17d24'16"
The Heart of Melanie
And 23h21'33" 35d22'42"
The Heart of Nicole Oesterreich
Cnc 8h39'10" 24d24'24"
The Heart of Tara
And 2h35'18" 42d47'26"
The Heartogram For Rebecca King
Aqr 22h53'18" -7d25'22"
The Heather "Boo Boo" Baker Star
Vir 13h38'37" 0d5'48"
The Heather Leann
Uma 10h14'55" 51d59'52"
The Heather Mayoros Star
Lyn 8h39'39" 38d53'16"
The Heather Renee Bare
Gem 7h14'3" 20d17'15"
The Heather & Shawn Forever Star
Uma 9h31'56" 42d21'20"
The Heavenly Body of Cherie
Pho 0h44'40" -50d5'30"
The Heavenly Body of James Byrne
Pyx 9h26'32" -29d6'7"
The Heavenly Essence of Stacey Anne
Pho 23h51'47" -53d50'2"
The Heavens Declare
Cyg 20h13'43" 38d53'55"
The Heck Family Star
Cyg 21h53'50" 42d38'55"
The Heidi L. Martin Superstar
Crt 11h32'38" -23d31'20"
The Helen Killian Star of Hope
Psc 1h12'1" 26d33'36"
The Helenator
Lib 15h5'38" -18d56'59"
The Henry James Fox Memorial Star
Cep 22h38'40" 76d41'58"
The Henry Star-Ted & Laura Forever
Psc 1h20'47" 24d11'34"
The Herdman's Lucky Star
Ori 6h0'17" 9d39'28"
The Herman Family
Lyr 18h51'10" 34d4'29"
The Hermarge-Arp Constellation
Per 3h25'50" 43d56'18"
The Hibbard Christmas Star
Ori 6h10'35" 15d28'54"
The Hicks Children
Uma 14h17'54" 55d53'34"
The High Rocks Star
Cyg 19h59'52" 39d1'2"
The Hill Family
Cyg 20h6'7" 33d26'50"
The Hillier Star
Uma 11h16'23" 44d2'12"
The Hiroshi Kokubun Family
Vul 19h1'41" 23d6'42"
The Historian
Psc 23h26'3" 4d18'33"
The Hitchhiker
Per 3h42'14" 34d17'25"
The Hochi Star
Gem 6h48'31" 32d12'20"
The Hodges Star
Cru 12h37'45" -59d8'31"
The Holder-Twomlow Day - 17.06.2006
Cyg 20h40'51" 53d41'26"
The Holiday Star
Uma 11h25'5" 32d36'3"
The Holley's Star
Lyn 8h41'47" 38d31'46"
The Holly Star
Aqr 22h13'32" 0d21'18"
The Holly Star
Cnc 8h59'40" 24d23'12"
The Holmes Star
Per 3h23'7" 46d16'2"
The Holt Family Star
Crb 15h36'43" 35d58'35"
The Holy
Ori 5h36'25" 1d20'10"
The Home Depot Appreciates Bonnie Hill
Sco 16h48'51" -29d5'28"
The Home Depot Salutes Elissa Ouchida
Sco 16h44'32" -24d57'44"
The Home Depot Salutes Exhibits South
Sco 16h14'48" -25d23'34"
The Home Depot Salutes Ionnie McNeill
Sco 16h32'19" -29d44'29"
The Home Depot Salutes Jasmine Lawrence
Sco 16h24'34" -28d46'47"
The Home Depot Salutes Lavelle Industries
Sco 16h14'21" -25d10'34"
The Home Depot Salutes Pink Magazine
Sco 16h34'24" -24d56'42"
The Home Depot Salutes Qualis
Sco 16h18'36" -23d4'13"
The Home Depot Salutes Rockwood Shutters
Sco 16h13'7" -28d5'14"
The Home Depot Salutes Stephanie Brown
Sco 16h25'9" -28d26'14"
The Home Depot Salutes Stratix
Sco 16h41'37" -26d30'47"
The Home Depot Salutes Surewood Forest
Sco 16h12'55" -24d26'22"
The Homen Family Star
Uma 10h19'33" 40d33'22"
The Honorable Donald B. Squires
Her 17h45'2" 40d54'2"
The Honorable Dr. K.E. Spencer
Cap 20h51'18" -20d57'48"
The Honorable Michael Nash
Boo 14h53'16" 21d13'51"
The Honorable Paul F. Harris, Jr.
Lyr 18h30'27" 36d2'10"
The Hoodie-Piglet
Ori 6h13'45" 16d3'5"
The Hope of Our Dreams
Cyg 20h24'2" 47d19'1"
The Hoppies
Cam 4h35'23" 68d49'56"
The HORN star
Lyn 7h42'2" 49d51'56"
The Horns of Bradley
Aql 19h46'58" -0d15'43"
The Howard and Marion Smith Star
Uma 9h43'54" 59d15'45"
The Howell's
Cyg 20h59'23" 32d40'58"
The Howman's Star
Cyg 21h13'21" 43d51'11"
The Huckleberry Star
Gem 6h32'21" 21d56'16"
The Hugga Love Star
Aqr 23h51'47" -10d10'59"
The Hugh G. Hanson
Aqr 23h8'56" -7d29'4"
The Hughes Boys
Boo 14h30'10" 51d58'19"
The Hughes Family Star
Uma 9h20'42" 57d35'4"
The Hume Star of Ednam
Umi 15h55'27" 73d16'51"
The Hunny Bee Star
Gem 6h29'5" 15d30'3"
The Hunny Star
Pho 0h35'26" -41d43'53"
The Hunter
Ori 6h7'25" 5d30'30"
The Hunting Legend Joseph Moccio
Vir 13h4'31" 12d30'9"
The Hurst-Walker Star
Uma 10h29'51" 67d54'0"
The I Love You Mom Star
Vir 13h38'33" -15d28'8"
The Ianator
Her 16h39'29" 3d50'49"
The Icie Bud Light
Uma 9h44'47" 44d50'8"
The Ida and Carl Tolchin Star
Cyg 20h21'45" 52d29'11"
The Iglesias Star
Dra 16h5'16" 52d32'53"
The IJL Diamond
Uma 9h24'15" 55d15'6"
The Ilan Aner Kahanov Star
Uma 10h36'53" 42d14'30"
The Illustrious Foxy Lydia
Aqr 23h3'10" -9d24'32"
The I.M.A. "Garnet Star" Adele Laughton
Cas 0h12'54" 54d7'42"
The Immortal Bond
Dra 16h15'43" 60d18'13"
The Incredible Carmen Severino
Lib 15h38'40" -5d46'37"
The Infinate Beauty of Heather
Ari 3h10'52" 28d46'36"
The Ingrid Ly & Elmar "Käbi" Kibal
Cru 12h43'18" -56d41'58"
The Ingrid Mansfield Star
And 2h16'42" 50d2'49"
The Ingrisano Family
Dra 16h9'25" 68d9'30"
The Inspiring Charles Forrest.
Psc 0h0'1" 1d14'50"
The Investigators
Cam 3h51'39" 64d34'39"
The Iris Doreen Christmas Star
Cas 1h20'7" 50d47'12"
The Irma Isaac Star
Uma 10h50'24" 64d10'15"
The Iron Hammer
Lib 15h16'47" -24d38'22"
The Irremistaverse
Crb 15h28'45" 27d21'37"
The Isaac Family
Vol 7h50'43" -69d46'37"
The Isle of Biff
Her 16h18'20" 44d43'26"
The Isobel Catherine Star
And 23h17'53" 40d29'14"
The Ivey Star
Lib 15h45'50" -23d10'14"
The J Hamilton
Aqr 21h34'17" 0d32'15"
The J & S Dunbar
Peg 22h44'58" 32d48'2"
The J * Star
Dra 16h44'7" 51d59'18"
The J. T. Mestdagh Star
Vir 15h6'41" 5d32'35"
The Jack and Jill Star
Com 13h31'57" 16d44'27"
The Jack and Mary Jane
Aql 19h30'41" 11d46'26"
The Jack Nelson Celestial Centaur
Boo 14h42'39" 14d34'47"
The Jack & Pat
Ori 6h24'2" 16d57'27"
The Jack Swan Star
Cyg 20h17'8" 39d57'54"
The Jack's Star
Her 17h31'15" 36d10'21"
The Jackson Star
Lyn 7h34'57" 37d39'46"
The Jacob and Chelsea Noddin Star
Cyg 21h35'13" 52d53'15"
The Jacob James Christening Star
Per 2h34'19" 54d37'36"
THE JACOBS FAMILY NIGHT LIGHT
Cyg 20h3'27" 40d11'1"
The Jacome Family
Uma 11h53'46" 63d1'24"
The Jake Star
Her 16h36'49" 35d37'56"
The Jake Star
Uma 11h29'4" 32d38'48"
The Jakestar
Ari 2h19'6" 11d27'11"
The James Cain
Sco 16h5'47" -15d57'48"
The James Douglas Harris Star
Aql 19h54'20" -0d2'31"
The James F. Bayne Family Star
Uma 11h21'8" 62d1'45"

The James "Red" Barlow Super Star
Aqr 22h27'12" -20d9'51"
The Jamie Star
Ari 2h22'17" 19d49'48"
The Jamison Star
Tau 4h10'51" 7d51'33"
The Janelle Lanae
Lib 15h49'58" -17d18'50"
The Janet Post Hall Star
Psc 1h9'46" 27d7'36"
The Jani Star
Cap 21h44'19" -19d24'25"
The Janice Miriam
And 23h57'37" 41d52'13"
The Janice Stockard Star
Sco 16h10'42" -17d14'59"
The Janie
And 1h39'32" 37d59'25"
The Jard Star
Cru 12h38'18" -62d18'45"
The Jared Boyer Star
Aur 5h19'55" 31d36'59"
The Jason Dean Star
Leo 10h8'13" 16d0'40"
The Jason & Stefani Star
Uma 9h23'7" 58d0'56"
The Jay Cox
Cap 21h45'11" -19d7'0"
The Jay Miller Star
Cep 1h21'8" 86d51'1"
The Jaybird Star
Tau 4h46'25" 21d4'51"
The Jayne Rafferty Star
Cru 12h10'45" -62d15'31"
The Jazzy Girl
Lyr 18h34'5" 38d30'5"
The JB Star
Aql 19h43'8" -3d54'42"
The JBC Star for Reg & Lynda Crick
Cru 12h13'42" -61d18'46"
The Jean 88
Cru 12h16'8" -58d45'47"
The Jeanes' Friendship Star
Uma 9h33'36" 42d49'25"
The Jeanine LaBrenz Light
Uma 11h54'50" 52d39'55"
The Jeannette Three Star Cluster
Umi 14h19'37" 68d58'58"
The Jeff & Beth Swanton Star
Cyg 20h15'47" 38d11'46"
The Jeff Ghaemaghamy Star
Cru 12h15'38" -58d19'38"
The Jeff Hauser Memorial Star
Aql 19h39'12" 10d15'40"
The Jeff & Karen Relationship Star
Cyg 19h43'32" 48d6'31"
The JeffAnne Star
Uma 10h31'9" 44d45'1"
The Jeffrey Mayho Star
Dra 10h3'25" 78d14'31"
The Jen and Alex Star
Cyg 20h11'28" 37d21'36"
The Jenkins' Future
Uma 11h54'25" 34d56'14"
The Jenna Marie
Cas 1h31'12" 67d26'51"
The Jennabee Twinkle
And 1h43'52" 42d30'46"
The Jenners
Her 16h13'58" 44d22'32"
The Jennifer Lee Jones Star
Cru 12h46'10" -62d37'39"
The Jennifer Marie Weyant Star
Cam 4h12'1" 59d14'8"
The Jennifer Rae
Tau 5h25'59" 27d15'36"
The Jennings Family
Cap 21h25'27" -23d17'17"
The Jenny Bear Star
Umi 15h4'53" 71d45'32"
The Jenny Marie
Crb 16h15'38" 35d49'52"
The Jenny Star
And 1h13'53" 42d52'19"
The Jeremy
Her 17h35'22" 27d39'35"
The Jerry Place Divine Star
Cep 20h36'23" 61d48'10"
The Jerry & Trudy Star
Cyg 20h16'23" 42d17'33"
The Jesse Conn Star
Umi 15h28'22" 71d33'17"
The Jesse Star
Cru 12h18'10" -56d29'35"
The Jessica
And 2h33'54" 43d58'47"
The Jessica Rose Rodriquez Star
Leo 9h54'58" 15d7'47"
The Jett Star
Ori 5h56'30" 21d51'49"
The Jewel of Britny's Heart
Ari 2h38'35" 26d20'16"
The Jill Sengel Star
Lyr 18h32'43" 36d24'23"
The JillEric Star
Ori 6h13'53" 17d7'43"
The Jillian Star
Cnc 8h3'22" 20d24'6"
The Jillian Star
And 2h13'28" 39d20'15"
The Jim and Nellie Diamond Star
Cyg 21h35'38" 46d32'20"
The JM Clark Star
Her 16h47'16" 31d10'23"
The JmsMjk-Eternal
Vir 11h51'51" 9d32'5"
The Joan Lindon
Cru 12h2'49" -61d58'34"
The Joan Marie
Leo 11h4'22" 17d3'50"
The Joanne Duggan Star
Cru 12h19'59" -56d33'16"
The Joanne Lobsey Star
Psc 1h4'49" 10d46'26"
The Joanne Star
Cyg 19h54'30" 37d54'45"
The Jocelyn Warner Star
Vir 12h11'37" 12d36'28"
The Jodi Lyn Hawker
Uma 9h43'47" 67d54'30"
The Jodimere
Gem 7h29'36" 17d0'23"
The Joe and Eileen Gigantino
Cyg 20h0'47" 31d50'20"
The Joey Domenici Star
Ori 5h33'21" 5d39'33"
The John Allan sparkler
Uma 11h59'36" 56d30'13"
The John Delance Thompson Star
Lac 22h29'1" 43d43'8"
The John Grover Harmon
Psc 1h42'11" 17d34'59"
The John Hoekstra Star
Ori 5h51'51" 9d0'8"
The John Imperatore Star
Uma 9h8'56" 65d12'54"
The John Kinkenon Star
Per 4h19'10" 51d41'52"
The John Mintz Star
Boo 15h9'40" 40d11'32"
The John Mulholland Burgess Star
Her 18h24'46" 17d52'25"
The John & Santa Catalano
Aqr 23h40'3" -10d19'37"
The John Scott Palladino Star
Her 17h5'56" 13d7'59"
The John William Granzella Star
Tau 4h38'54" 22d34'42"
The Johnathan Nicole
Sgr 18h27'45" -17d16'2"
The Johnny Boy Star
Cnv 12h47'33" 38d29'29"
The Johnson Family
Uma 11h48'43" 38d16'48"
The Johnson Family Happy Star
Cru 12h3'19" -61d57'23"
The Johnson Family Star
Uma 8h47'17" 56d44'59"
The Jonathan G. Schmidt Family Star
Sco 17h42'13" -40d43'54"
The Jones's Discovery
Uma 12h6'44" 53d36'53"
The Jonny C
Uma 11h37'14" 36d24'21"
The Joob Joob Star
Cam 4h7'36" 53d19'27"
The Jordan Lazaroff
Psc 1h19'22" 32d36'15"
The Jordan Star
Uma 10h12'8" 46d43'40"
The Jordan Star
Her 16h32'38" 28d27'43"
The Joseph Cironi Star
Leo 9h30'31" 10d26'25"
The Joseph Sinclair Brazington Star
Umi 14h25'12" 68d8'5"
The Josh Brennan Star
Per 2h55'11" 45d35'36"
The Josh Fleming Rock Star
Dra 19h41'46" 77d12'7"
The Josh Mayo Star
Lib 15h47'59" -18d32'57"
THE JOSHUA STAR
Aqr 22h45'20" -15d27'48"
The Joshua Star
Aqr 23h26'41" -3d41'48"
The JoshuaSaurus Star (J Rizk)
Lib 15h17'59" -20d30'46"
The Journey of Julie Ann Hansford
Cru 12h16'24" -58d9'22"
The Joyce and Jeff Star 959
Cyg 21h23'13" 36d7'11"
The Joyce Ferriabough Star
And 23h56'7" 41d40'56"
The Joyce Piper Star
Ori 5h9'15" 15d21'2"
The Joyful Dog Star
Cyg 19h37'20" 36d47'0"
The JPS Superstar
Dra 18h47'44" 57d52'21"
The J.R. Star
Cap 21h5'33" -26d30'38"
The Judith Eileen
Cnc 9h11'31" 16d33'50"
" The Judy K "
And 0h35'48" 36d51'11"
The Julia Skye Star
And 23h12'48" 51d3'40"
The Julia Star
Uma 9h48'28" 63d49'31"
The Julie Star
Tau 5h26'33" 22d10'12"
The June Reed Winter Light
Cas 23h30'13" 51d13'40"
The June Star
Lyn 7h26'37" 53d26'43"
The Junkyard Dog
Cma 6h42'52" -16d22'1"
The Juric Family Star
Umi 13h50'32" 76d53'51"
The Jurtin and Curtin Star
Cru 12h22'30" -62d10'1"
The Justin Mayo Star
Sgr 18h55'38" -16d22'2"
The K. C. Three Star Cluster
Umi 14h10'20" 67d45'17"
THE KADDATZ PETS
Sgr 19h10'22" -12d23'19"
The Kade Francis Star
Ori 5h20'18" 9d22'1"
The Kahlert Family
Uma 11h36'22" 48d54'47"
The Kai Hogg Star
Tau 3h30'59" 17d12'35"
The Kaitlin and Jerry
Lyn 7h37'30" 52d55'33"
The Kalle Star
Uma 8h36'21" 62d23'4"
The Kamerzell
Sco 17h52'7" -30d43'45"
The Kanaridis Family Star
Cru 11h59'39" -63d22'4"
The Kane LaSalle Star
Cas 1h40'28" 65d21'28"
The Kara K. Star
And 1h46'2" 49d13'25"
The Karamardian Brothers
Lyn 7h57'4" 36d14'16"
The Karen Mascitelli Sun
Psc 1h39'0" 16d59'10"
The Karr Star
Lyn 7h54'10" 35d6'44"
The Kate and Martha Unity Star
Uma 11h27'52" 53d39'32"
The Kateamus
Umi 14h23'4" 77d1'13"
The Katelyn Marie Wittenborn
Cas 23h37'18" 59d36'2"
The Kathleen Ann MCMXLVII
Cap 21h43'55" -20d23'13"
The Kathleen Priest Star
Cas 0h1'1" 53d31'7"
The Kathleen, MSW
And 1h47'42" 42d56'45"
The Kathryn Spanberger Star
And 23h59'57" 34d44'33"
The Katie and Dave Wedding Star
Cyg 19h29'41" 53d55'12"
The Katie Rose Christening Star
And 0h21'5" 45d2'37"
The Katie & Shawn True Love Star
Crb 15h49'49" 36d11'36"
The Katie Sheehan
And 0h42'2" 43d6'55"
The Katrina Bevelander Star
Sco 16h38'2" -34d15'49"
THE KAZIMIERA-IRA
Ori 5h42'47" 3d11'21"
The Keeper of my heart
Sco 16h12'51" -10d38'0"
The Keith and Jennifer Smith Star
Cyg 21h53'25" 48d27'1"
The Keith Ellaway Star
Uma 11h34'33" 35d55'4"
The Kelley Kristine Star of Love
Uma 13h18'35" 56d56'38"
The Kelly Angels
Uma 8h15'19" 69d48'17"
The Kelly Corey Theory
Lib 15h14'4" -5d34'21"
The Kelly Loves Brian Forever Star
Sgr 19h35'7" -14d10'59"
The Kelly Lynn
Ari 3h27'52" 27d20'10"
The Kels'inator
Uma 11h26'41" 38d50'53"
The Ken Pollard Star
Ori 5h21'36" 7d9'5"
The Kendra Rafie
Cnc 9h12'49" 18d6'6"
The Kennesaw Navy
Sex 10h20'3" 3d47'1"
The Kenneth M. Raabe Star
Lac 22h47'15" 52d25'26"
The Kenny
Uma 9h3'43" 59d5'29"
The Kenny (Raemac) Raeburn Star
Leo 10h38'34" 16d44'13"
The Keough Star
Umi 15h33'23" 73d9'35"
The Keri Star
And 0h15'49" 45d58'25"
The Kern
Uma 12h23'30" 59d56'28"
The Kerry Anne Schultz Star
Lmi 10h31'33" 37d44'52"
The Kev and Catherine Star
Cyg 21h38'32" 50d12'7"
The Kew Star "Daryl William Kew"
Pyx 8h49'53" -29d57'17"
The Khamvongsa Family Star
Psc 0h1'57" -3d16'30"
The Kid
Tau 4h3'59" 23d27'32"
The kid (Elaine J Kaminski)
Psc 1h49'23" 6d0'4"
The Kielkowicz Family Star
Lyn 6h32'20" 57d53'34"
The Kieran O'Donnell Star
Uma 11h20'51" 45d7'48"
The Kiersten-Evan Wedding Star
Cyg 21h48'11" 50d1'49"
" The Kim "
Srp 18h14'42" -0d12'1"
The Kim Dobson Star
Cas 23h37'7" 52d37'30"
The Kim & Simon 1st Anniversary Star
And 23h54'25" 35d7'35"
The Kimberly Ann
And 1h0'7" 34d58'10"
The Kimberly Marie Greenblatt Star
And 0h27'25" 38d52'34"
The Kimberly Princivalli
Lmi 10h10'15" 40d31'59"
The Kimbo
Lib 14h49'7" -13d56'43"
The Kimzey Star
Cap 21h50'50" -12d6'53"
The King Family
Cep 21h43'37" 58d43'36"
The King of Smiles and Laughter
Cep 21h26'0" 61d50'16"
The Kingdom of Jeremy Palmer
Per 3h55'40" 33d4'44"
The Kinnevey Family Star
Uma 14h27'8" 59d17'4"
The Kinzel 12
Ari 2h12'10" 22d58'45"
The Kirad
Cnc 8h2'29" 15d7'52"
The Kiran Oak Star
Leo 11h0'6" -0d18'15"
The Kirk Family
Ori 5h15'15" 5d43'58"
The Kirk Saranchuk Star
Cep 21h10'20" 66d49'28"
"The Kiss" James and Karen
Cyg 21h22'10" 46d37'55"
The K.I.S.S. Star
Cnc 9h4'53" 31d10'40"
The Kites
Aql 18h47'8" 7d40'38"
The Kitty Battles Star
Aqr 20h50'11" -12d23'41"
The Klemme Family
Ori 5h54'48" 12d51'34"
"The Klepper"
And 0h3'57" 33d45'25"
The Kniess Star of David
Uma 11h52'56" 63d48'59"
The Knight Star
Dra 17h43'28" 54d3'51"
The Knight Star
And 0h13'38" 33d7'35"
The Knott Star
Cam 4h4'4" 66d1'44"
The Knudsen Cricket
Lyn 8h25'15" 33d47'50"
The Kolb Family Star
Cam 4h6'12" 66d15'9"
The Komberg Star
And 2h32'31" 49d0'23"
The Kornman Star
Uma 13h37'55" 51d44'38"
The Korzeniowski Star
Lib 14h52'18" -13d59'36"
The Koutris' Anniversary Star
Col 5h23'55" -40d34'6"
The KR Star of Granted Wishes & Dreams
Peg 21h20'41" 18d36'32"
The Krasnow-Nalven Family Star
Ari 3h13'54" 29d40'46"
The Krista Dawn
And 23h14'56" 47d31'16"
The "Kristal" Arellano
Psc 0h45'43" 10d25'56"
The Kristen Mattson Star
Cnc 8h39'43" 15d47'25"
The Kristin Lee
Cap 21h25'58" -18d51'21"
The Kristin, Dan & Gus Star
Umi 15h45'1" 77d54'36"
The Kuch Family
Leo 11h28'31" 15d2'33"
The Kutch and Stanley Star
Cyg 19h41'33" 29d18'34"
The Kuteness
Sgr 18h13'1" -20d47'43"
The Kuzio's Gift of the Lord
Aur 5h54'25" 52d0'43"
The K.W. & V.F. Jones Family Legacy
Leo 10h10'21" 24d32'26"
"The Kydd"
Cap 21h32'9" -15d44'54"
The Kyle Klamerus Family Star
Ori 4h55'27" 10d11'38"
The Kyle S. Grant Star
Per 4h46'12" 43d59'24"
The LA Movie Star
Vir 12h48'6" -8d37'16"
The Lacey Star
Uma 12h57'20" 53d23'31"
The Lachapelle's
Umi 15h54'9" 76d21'59"
The Ladies
Cas 0h55'10" 56d10'38"
The Lady Barrow Star
Cam 3h35'41" 56d18'50"
The Lady Eleanor
Vir 12h8'40" -0d25'0"
The Lady From Cally
Cas 1h14'34" 57d11'3"
The Lady Julia of the House Torres
Leo 10h44'49" 19d50'35"
The Lady Melissa
Ari 2h1'37" 13d31'38"
The Lady Nancy
Cas 0h20'39" 51d18'50"
The Lake Star
Ori 5h40'49" 1d38'35"
The LaMay Anniversary Star
Cyg 20h21'12" 46d38'42"
The Lamp Family Star
Cru 12h24'37" -62d2'49"
The Land of the Mash
Ori 5h22'57" 5d9'28"
The Landing Star
Cyg 19h48'46" 53d47'49"
The Langford-Farrier Guiding Star
Uma 9h1'52" 71d33'57"
The Laning Star
Uma 10h0'54" 57d0'13"
The Lansing Star
Uma 8h34'25" 66d54'29"
The Lansing-Webb Star
Cnc 8h25'42" 25d43'9"
The Lantsberg Supernova, MSV
Uma 13h35'39" 55d10'56"
The Laps Family
And 0h39'32" 32d18'13"
The Large Lady
Uma 13h43'20" 61d12'38"
The Larry Abels Star
Ori 6h4'2" -3d55'24"
The Larry and Barbara Star
Cyg 21h54'44" 52d21'55"
THE LARRY & LINDA
Cyg 19h42'7" 28d13'15"
The Larry & Lyn Elgart Star
Uma 8h20'4" 59d44'10"
The Larry Moy Family Star
Aql 20h11'19" 14d25'42"
The Larsen Family
Uma 8h42'59" 64d48'10"
The last tear of Oksanochka
Ori 5h45'1" 1d52'59"
The Laura Alice Woodman Star
Cyg 21h45'34" 39d29'43"
The Laura Peta
Col 5h49'8" -35d37'31"
The Laurels of Mt. Pleasant 2005
And 0h37'59" 38d18'35"
The Lauren Matt Shining Star
And 23h16'17" 44d42'36"
The Lauren, Lindsey, and Leah Star
Cyg 19h36'46" 29d46'3"
The Laurie Boudreau Star
Gem 7h14'33" 21d40'25"
The Laurie Star
Crb 15h24'37" 29d30'8"
The Laurie Star
Uma 9h12'8" 64d29'44"
The Laz Brothers of Judges Nite
Gem 7h24'16" 26d10'39"
The Lazy River of Sandcastle
Eri 4h55'58" -13d15'14"
The Lebel Family
Uma 8h11'1" 65d48'53"
The Lee Fromson
Psc 1h26'46" 28d17'44"
The Legare' Light
Per 4h50'28" 45d40'14"
The Legend
Cru 12h8'37" -62d16'56"
The Legendary Rene Brunet
Aqr 22h10'14" -0d11'11"
The Leigh Zimmerman Star
Ari 2h44'25" 11d7'37"
The Lemmon Family Star
Uma 11h41'42" 61d53'50"
The Lena Meister
Sco 16h10'16" -10d19'47"
The Leo Family Star
Uma 8h40'15" 51d47'27"
The Leon Olivera Family
Lyn 7h52'3" 41d51'38"
The Leonie Lissing, "Lee-Star"
Cru 12h17'51" -57d33'57"
The Leslie David Fountain Memorial Star
Dra 17h41'33" 65d42'31"
The Leslie John Cox Memorial Star
Aql 19h0'17" -6d18'13"
The Leslie & Nicky 25th Anniv. Star
Leo 9h43'27" 25d42'56"
The Lestage's
Uma 10h21'22" 65d15'1"
The Let's Dance Star
Her 16h33'22" 45d29'17"
The Levine Family
Sge 20h6'31" 17d3'58"
The Lewer's "Nifty Fifty" Gold Star
Uma 11h35'25" 57d39'0"
The Lewis Family
Uma 12h11'47" 54d29'32"
The Light
Cru 12h32'29" -59d40'30"
The Light
Uma 11h22'53" 48d30'32"
The Light from Robyn's Heart
Cnc 8h24'42" 6d45'17"
The Light in Anh's Eyes
Leo 11h10'39" 9d50'54"
The Light in my Life, Clint Powell
Her 17h39'56" 18d49'25"
The Light in Rachel's Eyes
Uma 10h7'30" 48d6'3"
The Light leading our Broken Road
Ori 5h52'42" 20d43'53"
The Light of Adele
Uma 9h38'12" 68d40'56"
The Light of Amanda Seelig
Lyn 8h24'8" 36d9'23"
The Light of Bonnie Jean
Aqr 22h57'17" -8d57'51"
The Light of Cara Carmichael
Cen 13h48'33" -60d0'32"
The Light of Casey C. Colosky
Gem 7h19'25" 19d10'10"
The Light of Chermaine (Champs)
Leo 11h5'32" 9d40'15"
The Light of Daelyne Leschuk-Starks
Cap 21h28'3" -17d19'29"
The light of Dave and Mikey
Cyg 20h16'54" 47d57'11"
The Light of Dawn May 23, 1953
Lyr 18h27'13" 40d42'57"
The Light of Flora Elizabeth
Ori 6h18'27" 16d24'27"
The Light of Frank Thomas
Cru 12h49'30" -63d2'25"
The Light of Gintare
Lyn 8h19'26" 51d3'46"
The Light of Healing
Uma 13h54'13" 50d36'28"
The Light of Heather Marie
Aqr 22h53'43" -6d12'35"
The Light of Jay
Lmi 10h7'37" 30d8'33"
The Light of Jim Liebke
Uma 11h14'38" 47d34'47"
The Light of John
Per 2h40'5" 52d31'13"
The Light of Joy
Sgr 17h50'51" -26d6'32"
The Light of Kata Hay
Com 13h5'40" 24d29'28"
The Light of Katy
Psc 1h17'44" 17d31'59"
The Light Of Linda
Ori 6h20'5" 10d1'31"
The Light of Lisa
Uma 11h18'0" 43d41'45"
The Light of Love and Friendship
Her 18h6'9" 36d33'12"
The Light of Lucy & Topsy
Cma 7h26'0" -19d47'17"
The Light of Mary Martha
Cas 0h22'49" 53d19'27"
The light of Mendenhall
Ser 15h57'30" 4d27'33"
The Light of Michelle
Ari 3h22'40" 28d36'11"
The light of my life - Laura McHenry
Pho 2h18'36" -45d37'21"
The Light of My Life ~ Peter Bos
Leo 11h26'13" 2d14'57"
The light of my Sugarbear
Per 3h33'9" 47d43'13"
The Light of Natnya
Lib 15h9'12" -11d38'1"
The Light of our Love
Psc 0h34'32" 7d42'3"
The light of Pammie
Ori 6h9'4" 16d14'58"
The Light of Rini
Leo 10h11'26" 7d9'41"
The Light of Sarah
Crb 16h11'42" 34d59'12"
The Light of Shining Sisters
Uma 11h0'8" 35d24'44"
The Light Of Tara
Cap 20h47'4" -23d6'44"
The Light within Tey's Eyes
And 1h10'17" 34d41'22"
The 'Like Like' Star
Cyg 20h31'1" 38d32'15"
The Lil Hillbilly xoxo
Lac 22h22'54" 44d40'46"
THE LiL WiNKER
Umi 14h35'20" 74d7'47"
The Lilah & Joe Wedding Star
Cyg 19h46'28" 32d25'15"
The Lilliane AngelStar
Gem 7h25'59" 26d34'15"
The Linda Hill Star
Uma 9h58'20" 60d13'57"
The Linden Tree
Cap 21h47'40" -14d11'53"
The Lindsey
Uma 8h13'55" 65d20'55"
The Lindsey Leigh
Psc 0h48'29" 19d8'18"
The Linzness
Lib 15h14'26" -14d39'19"
The Linzy Juan Fairy Star
Crb 16h20'34" 33d52'48"
The Lion and his Moon Maiden
Per 3h25'53" 46d50'12"
The Lisa D.
And 0h20'30" 46d35'18"
The Lisa Gaye
Cnc 8h25'53" 24d0'46"
The Lisa Marie
Psc 22h52'39" 0d26'41"
The Literary Geniuses
Lyn 7h29'11" 58d19'25"
The Little Bumby Star
Gem 7h13'1" 30d8'48"
The Little Dalton
Cam 5h24'56" 60d2'16"
The Little Dixter
Psc 0h10'49" 6d9'48"
The Little General
Umi 14h19'23" 75d4'15"
The Little Gripper
Cnc 8h59'30" 22d53'33"
The Little Guy
Aql 20h14'45" 5d3'15"
The Little King, Ryan S. Johnson
Cep 22h30'38" 66d26'18"
The Little Lemmon Family Star
Cyg 20h10'2" 30d5'3"
The Little Man
Leo 11h35'37" 25d21'48"
The Little Mis
Lyn 7h3'11" 55d43'53"
The Little Miss Amazing Jones
And 0h43'46" 32d10'8"
The Little Missy
And 0h25'52" 29d14'18"
The Little Mommy Star
Cap 20h47'59" -21d4'47"
The Little Mouse
Umi 15h28'22" 73d23'24"
The Little Odie
Umi 14h43'44" 81d18'57"
The Little Piece
Aql 19h8'44" 15d32'51"
The Little Ponderosa
Cnc 8h40'4" 17d20'6"

The Little Prince
Umi 15h33'51" 72d6'0"
The Little Prince and His Rose
Vul 19h51'13" 28d30'30"
The little Quintin star
Gem 7h8'58" 30d4'49"
The Little Skipper
Gem 7h12'2" 22d41'15"
THE LITTLE YORK TAVERN
Uma 10h19'26" 52d5'34"
The Littlest Guy
Umi 14h19'46" 76d0'8"
The Lizzie Star
Ari 2h0'45" 13d28'3"
The Loar Family Star
Uma 8h51'2" 58d16'19"
The Lobster Claw
Tau 5h33'33" 22d27'35"
The Locey Angels
Uma 12h5'4" 46d14'32"
The Lois and Ron Diener Star 50
Cap 20h24'47" -11d48'58"
The Lois - Louise
Cas 23h33'47" 56d59'28"
The London Apprentice
Uma 9h46'13" 70d12'33"
The Lonesome Star
Umi 15h54'23" 72d27'22"
The Lonestar
Vir 13h50'26" -7d10'56"
The Longest
And 1h49'11" 43d34'6"
The Longman Star
Cas 23h44'10" 54d49'17"
The Long's Star
Uma 13h12'33" 55d41'31"
The Lopez Star
Cyg 21h7'31" 48d55'22"
The Lori Lane
Mon 7h32'19" -0d52'25"
The LoriannT
Gem 6h45'16" 15d1'5"
The Lorraine R.
Cnc 8h1'39" 20d14'14"
The Lorren Star
Tau 4h44'5" 27d3'12"
The Lost Scotsman
Boo 14h18'1" 48d0'34"
The Louis Maire Family
Umi 13h47'34" 71d13'11"
The Louise L. Grant Star
And 2h7'19" 42d26'56"
The Louise Star
And 23h0'57" 47d40'43"
The Louisiana Serio
Lib 15h33'47" -25d17'3"
The Love and Light of Annie
Vir 12h34'32" 2d56'59"
The Love of Art
Uma 9h27'23" 65d10'20"
The Love Of Chris & Chris
Cnc 8h30'32" 28d0'17"
The Love of Danika
Leo 9h26'20" 24d59'21"
The love of Doddy and Maniska
Cap 20h49'57" -25d25'0"
The Love of Elaine and Fred
Cru 12h42'8" -62d59'22"
The Love of Elaine & Fred
Cru 11h56'54" -61d26'17"
The Love of Joan
Vir 13h4'16" 13d19'56"
The Love Of My Life
Cnc 9h2'57" 32d16'46"
The love of my life Allyson Hageman
Uma 8h56'36" 53d54'17"
The love of my life Brittany Eagon
Lib 15h43'14" -5d47'12"
The love of my life given by GOD!
Cyg 20h28'31" 56d20'59"
The Love Of My Life—JENNA
Psc 23h33'49" 0d21'19"
The love of my life Jessica
And 0h38'33" 31d25'54"
The Love Of My Life "J.R."
Cap 21h31'11" -16d14'19"
The Love Of My Life - JULIE LIMERI
Cas 0h36'45" 60d13'9"
The Love of My Life — (L and J)
Crb 16h14'33" 33d18'23"
The Love Of My Life Lauren Krieger
Ori 5h44'20" 6d13'3"
The Love Of My Life Nikki
Sgr 18h10'22" -25d28'13"
The Love Of My Life Stephanie Dana
Lyn 8h32'33" 35d8'41"
The Love of My Life's Star
Leo 9h22'46" 16d18'40"
The Love of Robert & Meredith Ellis
Her 18h50'19" 22d3'11"
The Love of Zack and Amy Sasser
And 1h1'53" 37d17'51"
The Love of Zury and Daniel
And 0h17'57" 40d28'3"
The Love Star
Psc 1h41'22" 5d24'28"
The Love Star
Cyg 20h27'2" 53d30'27"
The Love Star - John & Brittney
Per 3h12'10" 43d41'7"
The Love Star Sue and Dick
Dra 18h9'41" 75d58'56"
The Loveable Light
Cyg 20h23'18" 51d55'18"
The Lovebird
Umi 14h7'30" 69d21'29"
The Lovely Angel
And 1h49'22" 42d25'37"
The Lovely Carol Lyons
Peg 23h30'40" 26d50'59"
The Lovely EmmaLee Danilchuk
And 23h20'22" 42d2'5"
The Lovely Jami H Stueck Star
Psc 23h59'36" 5d9'54"
The Lovely Jené Jack
Cnc 8h38'13" 31d33'3"
The Lovely Jennifer Lynn
And 23h37'35" 46d59'14"
The Lovely Lady
And 1h38'43" 42d26'45"
The Lovely LiLi
Dra 16h29'55" 58d23'12"
The Lovely Lily
Sco 16h9'34" -11d47'53"
The Lovely Linda
And 23h56'50" 35d15'3"
The Lovely Lindsay
Cru 12h13'22" -62d0'23"
The Lovely Lorna
And 1h35'29" 49d14'38"
The Lovely Princess Leslie
And 0h18'22" 40d23'35"
the lover
Cyg 21h11'22" 47d5'7"
The Lover Star!
Cra 18h49'58" -41d44'11"
The Lovers
Apu 15h29'3" -72d25'13"
The Loves of My Life
And 23h8'12" 42d52'55"
The Loviner O'Ryan
Crb 16h22'37" 28d36'3"
The Loving Carmen Leal
Cas 1h34'37" 56d37'45"
The Loving Nini
Ori 5h0'59" 11d3'31"
The Lowery Star
Tau 4h41'12" 4d40'34"
The Lucasfilm PR Rockstars
Her 17h19'35" 27d21'50"
The Luce Family
Uma 9h8'0" 54d25'18"
The Luciana
Her 18h35'32" 19d54'36"
The Luck of Lauren's Love
Sco 16h2'48" -18d1'13"
The Luck Of The Irish
Cnc 8h30'44" 7d0'59"
The Luckiest
Ori 6h10'2" 8d41'16"
The Luckiest
Gem 7h35'7" 29d22'51"
The Luckiest- RJ and Spenser
Ari 2h51'40" 26d30'26"
The Lucky Star of Benjamin D. Levin
Psc 1h51'37" 8d15'24"
The Lucy Mabbott Star
Crb 16h1'6" 32d45'50"
The "Luisa" Shining Star
And 0h42'11" 27d23'43"
The Luke
Sco 17h31'10" -36d54'23"
The Luke Family Star
Vir 13h20'14" 7d31'7"
The Luke McKade
Ari 1h49'58" 17d38'58"
The Luminous Star of David
Ori 5h50'13" 19d36'33"
The Lutz Legacy
Uma 9h12'57" 63d43'39"
The Luvlies
Cru 12h28'6" -57d13'17"
The Lyman Star
Cas 1h38'11" 68d23'22"
The Lynch Star
Uma 9h5'33" 63d41'4"
The Lynn Marie Asermily Star
Uma 11h41'11" 47d51'10"
The Lyon of my life
Lmi 10h35'39" 36d21'58"
The Lyrebird Village Memorial Star
Cru 12h31'3" -59d17'38"
The Lytle of Our Lives
Uma 11h16'40" 47d7'3"
The Lytta Jane
Lib 15h0'45" -15d30'4"
The M&M Star (Monica & Mitchell)
Leo 10h18'44" 21d26'9"
THE MACCARDLES
Leo 11h15'30" 5d22'16"
The Mack Family
Per 3h31'41" 43d0'44"
The Maddison Grace
Cen 13h42'11" -53d29'47"
The Maddox Family Star
Uma 11h24'51" 45d1'7"
The Maddy Leader Star
Her 16h19'14" 19d39'5"
The Madeira Class of 2007
Uma 9h0'33" 55d48'42"
The Madera Family Star
Her 18h35'40" 12d17'48"
The Madi Hubbard Star
And 23h22'33" 48d22'23"
The Madison
Umi 14h22'7" 67d45'13"
The Magas
Leo 11h18'31" 22d31'40"
The Maggie D.
Uma 9h23'7" 43d19'40"
The Maggie Frame Sixty Star
Cas 23h14'51" 55d43'12"
The Maggie Star
Cas 1h35'14" 60d56'15"
The Magic Apple
Uma 8h40'46" 59d47'35"
The magic of MacDonald
Ori 6h13'5" 2d9'33"
The magical "mmh"-star
And 23h53'24" 39d19'6"
The Magill Family Star
Uma 8h51'30" 55d12'57"
" The Magnanimous Martin "
Ori 5h14'6" -0d7'18"
The Magnificent Mildred
And 0h16'19" 28d51'44"
The magnificent beauty of Doris
Crb 15h41'9" 32d29'40"
The Magnificent G-MA
Vir 13h25'56" 12d47'37"
The Magnificent Melissa
Leo 11h9'3" -0d14'34"
The Magnificent Monsignor Hartman
Cep 22h11'9" 53d46'4"
The Magnificent Nimaroff
Per 4h32'50" 40d9'39"
The Magnificent Shelley Ann Batch
And 1h58'50" 46d49'35"
The Magyar Gang
And 23h10'45" 43d51'2"
The Maher Family
Sge 20h4'14" 17d12'35"
The Mail Coach
Aur 6h34'49" 35d13'52"
The Main Momma Catharine
Leo 10h25'14" 6d58'0"
The Majestic Meier
Lyr 18h36'54" 31d19'41"
The Mal and Celia Eternal Love Star
Cyg 21h34'20" 40d38'55"
The Maleys
Mon 6h53'47" -0d14'50"
The Malkin Star-Jim-Lu-Todd-Jamie
Lyn 7h39'4" 36d11'49"
The Mallory Renee
Lyn 8h33'26" 35d36'33"
The Malone Star
Cap 21h14'54" -26d29'28"
the man in the moon
Sgr 19h13'4" -20d42'38"
The Man of My Heart - Alan
Sgr 19h7'44" -16d42'25"
The Mandaling
Ari 3h26'6" 26d20'5"
The Mandie Star
And 0h30'39" 25d27'36"
The Mandy Hewitt Star
Umi 14h51'22" 78d19'5"
The Mandy Jones Star
Cru 12h25'7" -58d43'7"
The Manglardi Family Star
Her 18h45'9" 19d17'4"
The Mann Star
Aql 20h35'41" -0d0'20"
The Mannina
Aql 19h50'23" 10d52'17"
The Mannion Star
Umi 15h29'28" 89d13'44"
The Manors
Uma 11h35'9" 54d5'42"
The Mapes Family Star
And 0h6'36" 39d34'30"
The Marc of Lerona
Cyg 20h44'16" 43d27'11"
The Marcie/Jimmy Starlight Express
Cyg 21h11'55" 47d46'21"
The Margae Ann
Aqr 23h38'8" -19d30'57"
The Maria Acevedo Star
Aql 19h50'25" -0d20'41"
The Maria-Louise Everlasting Star
Cru 12h49'42" -63d54'7"
The Marilyn Star
Psc 1h56'24" 7d11'3"
The Marit
Tau 4h29'14" 21d1'2"
The Mark Andrew Stanfield Star
Psc 1h20'3" 16d7'4"
The Mark Bringle Star
Her 17h5'26" 31d17'56"
The Mark C. Browne
Boo 14h47'58" 22d43'5"
The Mark D. Elmore Pathfinder Star
Lib 15h54'34" -19d53'35"
The Mark Larsh Star
Vir 13h53'51" -22d29'10"
The Mark & Stephie Star
Cyg 19h45'46" 31d43'32"
The Markham Star
Sco 16h36'5" -34d37'24"
The Markoe Family
Umi 15h11'33" 72d21'14"
The Marmas Star
Mon 7h15'14" -0d9'37"
The Marriage of Jim & Beti
Cyg 20h21'39" 46d39'36"
The Marriage of Matthew & Caroline
Cru 12h19'57" -56d54'21"
The Marriage of Sean and Nichole
Cyg 19h44'27" 34d22'39"
The Marry Me Star
Ori 6h3'17" 17d15'18"
The Martha Constellation
Uma 10h40'7" 56d30'5"
The Martin Family Star
Ori 5h46'13" 5d43'34"
The Martin Family Star
Lyn 7h39'35" 38d20'13"
The Martin Johnson Star
Lib 15h16'33" -10d40'0"
The Martin Paisner C.B.E. Star
Cep 23h42'8" 74d38'5"
The Marvin Morris Star
Psc 23h16'48" 7d28'50"
The Marvin Siegel
Psc 1h32'0" 13d18'1"
The Mary B.
Uma 11h20'30" 51d26'43"
The Mary D Gordon
Lyn 8h21'32" 36d21'26"
The Mary Erickson Urban Ministry
Cam 5h58'13" 61d9'33"
The Mary Fran & Carl
Uma 8h11'22" 70d15'51"
The Mary Freida Walker Lineage
Cas 0h36'4" 53d3'26"
THE MARY G
Cyg 20h4'34" 51d13'39"
The Mary Katharine
Leo 11h38'56" 26d16'2"
The Mary Katherine
Uma 12h0'48" 32d5'0"
The Mary Mc Star
Umi 14h11'23" 69d22'19"
The Mary McGregor
Uma 10h43'53" 54d2'29"
The Mary Stanford Star
Oph 16h55'50" -0d57'52"
The Mary & Tom Hunt Star
Umi 15h35'26" 73d3'40"
The Mary-Chaim Star
Cap 21h11'26" -21d34'30"
The Mason & Taylor Howard Star
Peg 22h38'21" 15d55'29"
The Matheson Family Star
Cam 7h2'19" 82d13'13"
The Matriarch
Vir 12h19'20" -8d59'12"
The Matt Davies Supernova
Ari 1h59'14" 19d12'37"
The Matthew Angus
Peg 21h29'17" 16d0'18"
The Matthew J. #2 Grandson
Her 17h30'39" 47d13'49"
The Matylou Pierre Flamini Star
Ori 5h1'43" 9d42'18"
The Maurer 50th Anniversary Star
Cyg 20h22'18" 44d14'37"
The Mavis & Ros Rosenberg Star
Uma 8h56'13" 65d43'20"
The Max Day Star
Uma 9h13'16" 51d11'54"
The Maximum Jazz
Her 18h41'0" 21d28'52"
The May Lover's Star
Umi 14h6'56" 69d50'8"
The Maymi Star
Uma 11h31'5" 38d3'17"
The Mayo Family
Cyg 21h37'14" 40d10'15"
The Mayo Family Star
Cas 1h0'42" 60d47'17"
The Mazziotti Star
Cep 22h14'20" 63d2'19"
The McAlarney Twins - Kate & Kathy
Gem 7h14'28" 26d51'52"
The McCarthy Family
Uma 10h47'30" 62d23'54"
The McClelland Star
Ori 5h32'40" 1d16'31"
The McCormick
Dra 18h44'58" 69d12'29"
The McCullochs' Star
Cyg 21h31'18" 48d54'46"
The McDermott Family Star
Cap 20h38'51" -10d20'10"
The McFardy Wedding Star
Sco 17h31'55" -39d57'33"
The McGrew Star
Men 5h15'55" -70d30'5"
The McKenzie Family
Uma 10h12'46" 44d47'26"
The McKinley Love Star
Cyg 20h46'18" 35d5'28"
The McLellan Family
Umi 16h25'20" 74d40'36"
The McNabb 60th Anniversary Star
Cyg 19h58'44" 46d59'35"
The McNeel of Longview aka Big Love
Lyn 9h25'37" 40d19'19"
The McQueen Family Star
Uma 12h0'24" 52d9'10"
The Meacham Star
Her 17h2'55" 47d14'43"
The Mearacle Star
Umi 15h56'26" 75d57'54"
The Medium Bumby Star
Gem 6h56'57" 27d49'4"
The Medo Star
Uma 8h56'14" 69d34'21"
The Megan Elizabeth
Uma 13h33'53" 55d30'42"
The Megan Star
Cam 3h27'9" 62d15'37"
The Meggie Star
Psc 0h48'12" 11d32'42"
The Meggie Star
Crb 16h9'1" 35d39'35"
The Mehner's Family Star!
Sgr 19h50'12" -14d8'38"
The Mei Mei Star
Vir 14h49'20" 3d37'50"
The Meitus Boys
Her 16h13'1" 48d30'3"
The Melchiore Family Star
Umi 9h25'51" 87d50'3"
The Meli Gutierrez
Uma 10h45'8" 54d58'13"
The Melisa Clevenger
And 0h54'27" 45d1'32"
The Melissa Hardman Angel
Vir 13h38'54" -16d47'29"
The Melissa Love Star
Cyg 20h42'16" 37d17'2"
The Melissa Lynn
Cyg 20h8'50" 38d25'42"
The Melissa System
Vir 13h23'4" -6d59'13"
The Mellio
Uma 11h43'27" 62d42'10"
The Melody Heather Harris Star
Uma 12h46'25" 63d8'22"
The Melvin-Scammell Star
Cas 0h44'7" 50d1'49"
The Memory of Us
Cnc 8h15'31" 20d51'52"
The Mena Star
Uma 11h38'29" 45d10'48"
The Mepham Star
Cnv 13h54'53" 34d55'56"
The Meredith
Cam 7h48'38" 69d12'18"
The Meron
Sgr 18h26'17" -32d41'58"
The Merrifield's
Aur 5h38'25" 33d41'12"
The Merrill a.k.a. JdmJerk EM1
Sco 17h3'23" -38d23'24"
THE META STAR
Lyn 8h56'8" 34d18'50"
The Mia Grace Lynch Christening Star
And 0h13'3" 48d31'1"
The Micah Gross
Tau 5h42'10" 27d25'25"
The Michael Dunne Star
Per 3h33'18" 39d19'19"
The Michael F. Brotzman Family Star
Aql 19h31'36" 6d36'0"
The Michael Londeck Phenomenal Star
Psc 1h25'28" 28d53'43"
The Michael P. Satyshur
Tau 3h53'36" 27d28'25"
The Michael Scott Gordon Star
Lyn 6h48'26" 59d32'42"
The Michael Stewart-Smith Star
Per 2h27'11" 56d7'27"
The Michael Sweet Family Star
Cnc 8h19'4" 22d50'16"
The Michalaks
Uma 8h59'34" 46d31'26"
The Michele Star
Leo 9h39'11" 14d53'35"
The Michelle
Lib 15h51'54" -19d52'39"
The Michelle Rocks Mofos Star
And 1h16'33" 34d44'39"
The Mick George Star
Uma 8h29'3" 62d36'11"
"The Mickey Brick", M T Bricklebank
Per 2h56'7" 51d36'51"
The Mickey Weishaus Family Star
Psc 23h13'5" 5d23'14"
The Midnight Blue Star
Cep 22h4'37" 67d36'16"
The Midnight Glow - Bria's Star
Tri 2h23'11" 32d20'21"
The Mighty BLiu
Pup 8h9'33" -22d18'45"
The Mighty Jairo
Her 17h36'15" 14d48'37"
The Mighty Mayhamn
Psc 1h1'52" 14d24'40"
The Mighty Quinn
Her 18h0'42" 14d47'49"
The Mighty Quinn
Cnc 9h5'46" 32d19'4"
The Mighty Quinn
Umi 14h38'12" 76d16'22"
The Mighty Quinn
Uma 8h21'10" 61d15'20"
The Mighty Rod of Power
Uma 10h44'42" 39d42'14"
The Mihara Star
Mon 7h20'59" -0d19'38"
The Mike D Star
Dra 18h29'28" 51d54'32"
The Mike Stewart (Our Pop) Star
Cep 21h58'16" 65d7'58"
The Mike-a-do
Uma 11h36'49" 33d26'48"
The Miko Star
Ori 5h29'52" 2d9'19"
The Miley Star
Her 16h43'7" 15d4'49"
The Milky Wade
Crb 16h16'35" 36d42'20"
The Miller Family
Uma 11h1'38" 37d19'55"
The Miller Wallace Star
Her 18h28'0" 15d1'9"
The Milligail Storm
Her 17h11'57" 36d52'42"
The Mimi Cherin
Ori 5h59'30" 21d24'14"
The Mina Richler
Psc 0h1'43" 3d22'38"
The Mindy Heald Sparkler
Lyn 7h13'58" 56d46'9"
The Minio
Cap 21h29'52" -20d0'58"
The Ministry of Inside Things
Uma 11h11'44" 67d5'37"
The Miracle of *GRACE*
Psc 23h50'31" -0d29'37"
The Miracle of Maeve
Lyn 7h13'5" 49d36'44"
The Miracle of Suzanne
Sgr 19h9'21" -13d49'27"
The Miranda James
Pho 0h30'18" -42d37'22"
"The Miss Libby"
And 2h37'33" 50d35'20"
The Miss Lisa Marie
Lyn 9h13'21" 37d13'57"
The Miss Peace Carla Renae
Psc 1h17'58" 15d22'52"
The Miss Sara Elizabeth Mies
Cnc 8h24'51" 25d43'22"
The Missing & Mitrovich Family
Eri 2h5'1" -52d55'56"
The Mister
Her 16h47'30" 33d58'21"
The MKW - Magnificently Kind Woman
Cas 0h23'46" 53d41'3"
The Modric Family Star
Uma 12h32'28" 58d31'15"
The Moerke Star
Her 17h20'41" 45d28'37"
The Moira Anne Kitchin Star
Gem 7h16'30" 23d46'57"
The Moishe Barish Star
Lyn 8h0'8" 40d2'6"
The Moldenetzel Star!!!
Car 7h40'39" -51d58'43"
The Molly Anne Donnelly Star
Cyg 21h41'58" 33d35'38"
The Molly Dolly
Sco 16h3'53" -10d7'39"
The Molly Hoctor
Cnc 8h15'1" 9d32'44"
The Molly Qing Jing Star
Sco 16h14'22" -11d5'14"
The Money Family Star
Oph 17h0'2" -0d41'30"
The Mongol (Altan Sadik-Khan)
Psc 0h16'58" -4d51'19"
The Monguin
Cas 0h50'38" 62d46'36"
The Monkey Star
Ori 5h14'47" 12d53'48"
The Monna & The Boop
Cas 0h0'23" 53d50'6"
The Mookie Star
Lyn 7h26'55" 57d44'13"
The Moon
Vir 13h16'6" -2d45'30"
The Moon and the Mermaid
Tau 5h39'26" 19d15'4"
The moon is tilted forever & always
Uma 13h26'15" 55d1'34"
The Morgan & Nellie Robert's Star
Per 3h37'44" 44d35'51"
The Morgan Scott Harper Star
Crb 15h47'50" 27d35'52"
The Morgan Star (Anne Marie Geddes)
Umi 15h57'2" 73d58'46"
The Morgans (Pop) Star
Umi 15h45'57" 77d32'49"
The Morning Star
Uma 11h42'32" 36d21'44"
The Morrill Family Star
Uma 13h37'29" 57d16'0"
The Morris Magnum
Uma 8h45'9" 54d9'30"
The Morrison Family Star
Lyn 8h17'0" 55d22'19"
The Morse's
Tri 2h15'20" 31d55'21"
The Moseley's Sparkle
Dra 19h30'11" 67d42'47"
The Most Beautiful Heart
And 22h58'27" 39d39'32"
The Most Beautiful Seema
And 0h57'11" 23d13'3"
The Most Beautiful Star
And 23h0'21" 47d38'56"
The Most Beautiful Star "Nicole"
Cas 1h38'31" 62d34'43"
The Most Bizzare Awesome Star Ever
Tau 3h55'36" 25d58'34"
The Most BUNtastic Star of Them All
Lib 15h44'52" -6d9'59"
The Most Special Mom in the World
Lyn 7h9'41" 47d10'11"
The Mother Star
Cas 0h1'32" 54d4'42"
The Mother Star
Aqr 22h40'58" -2d42'17"
The Mother, Pam Rothman
Gem 6h59'18" 17d52'34"
The Mounter-Newman Guiding Star
Cyg 19h40'18" 32d10'14"
The Mourning Star (MEJR)
Cam 4h39'46" 66d3'41"
The Mouse
Peg 22h30'28" 8d39'18"
The Mousey Princess Ruth
And 2h31'36" 46d35'37"
ThE M.P. bLiNg
Leo 10h18'37" 15d34'48"
The MPDS Chinchilla
Uma 12h40'2" 56d49'33"
The 'Mr. and Mrs. Click' Star
Tau 4h37'0" 17d30'6"
The Mr. Andrew G Tymon Family
Uma 13h21'31" 54d56'48"
The M.R. Faulconer
Her 16h22'49" 9d42'1"
The Mr. Jim
Aql 19h11'1" 2d52'42"
The Mr & Mrs Hinde Anniversary Star
Per 3h33'40" 37d1'38"
The Mr. Rudy Star
Cnc 8h53'11" 26d43'44"
The Mrs. McGee Star
Cas 0h45'20" 56d37'44"
The Muetze Hellmer
Vir 14h38'16" -0d21'0"
The Muli's
And 0h50'18" 42d52'15"
The Mullens' Star - Nathan, Alexa & Grace
Dra 17h49'0" 58d0'38"

The Mundt Family
Umi 15h9'22" 81d27'30"
The Muñoz Star
Uma 11h21'44" 46d18'44"
The Munson Star
Psc 0h47'26" 6d37'31"
The Mur Man
Uma 10h40'22" 65d33'27"
The Murph
Aqr 22h54'59" -7d16'58"
The Murphy Family
Uma 10h36'3" 62d37'30"
The Murray Girls
Uma 14h18'11" 55d11'50"
The Musical Star Named Marc
Per 3h25'26" 40d42'19"
The Myers Star
Uma 11h27'5" 45d23'38"
The Myra-Gene Star
Crb 15h33'19" 29d52'42"
The Mystic Stranger
Crb 16h21'10" 29d17'57"
The Nana Star
Cas 1h32'34" 62d25'36"
The Nancy and Frank Cantwell Star
Cyg 19h47'25" 53d44'48"
The Nancy Rae
Lib 15h52'45" -4d49'30"
The Nancy Star
Vir 13h51'8" -20d16'51"
The Nanna Star
Lib 15h9'23" -29d52'23"
The Nannee Star
Tau 5h15'26" 27d51'2"
The Nanster
Gem 7h7'32" 32d28'11"
The Naomi Star
And 23h24'54" 49d56'49"
The Napier Family Star
Ari 3h5'6" 11d54'43"
The Natalie & Craig Tschirpig
Cru 12h27'59" -60d41'18"
The Natalie Gosper Star
Cru 12h18'6" -57d20'42"
The Nathan Stonebarger Star
Per 3h31'43" 44d39'9"
The Nauth Star
Del 20h50'56" 7d38'1"
The Navan Mono Star
Dra 20h32'54" 75d22'11"
The Navarro Star
Cyg 20h13'45" 37d12'51"
The Nayrae Star
Aqr 22h30'18" -1d55'16"
The Naz
Cas 1h26'26" 68d9'28"
"The NC Star"
Sge 19h49'59" 17d43'12"
The Neal and Jude Wedding Star
Com 12h43'7" 15d4'31"
The Nebula Doris Weinberg
Cas 23h0'29" 59d1'3"
The Ned and Sarah Phillips Star
Uma 8h40'21" 47d37'30"
The Nedwek
Sgr 18h22'16" -20d35'23"
The Negron Twins
Gem 7h31'22" 30d8'56"
The Nehme Miller Constellation
Tau 3h47'0" 25d5'29"
THE NEMECHEK
Tau 5h11'44" 18d21'54"
The Nemiroff Star
Boo 14h41'54" 52d36'57"
The Nervous Wreck
Cnc 9h5'18" 31d0'38"
The Netty
Lib 15h3'55" -17d47'53"
The Never Ending Fire
Sge 19h59'54" 19d34'44"
The Newman's Star
Gem 7h10'29" 31d10'36"
The Next 50 Years David & Carol
Aql 18h55'40" -0d58'0"
The Neysa & Gerard Cantor Star
Lib 15h54'45" -7d52'5"
The Niblett Radio Program
Uma 11h18'28" 53d57'35"
The Niccole Savanna
Cnc 8h43'27" 18d34'1"
The Nichole Love
And 1h30'13" 45d54'12"
The Nick LaBanca
Cnc 8h48'32" 13d39'12"
The Nickolas Ranger
Cyg 20h13'53" 33d48'22"
The Nickstar
Lyr 18h50'23" 35d52'48"
The Nicola Moore Star
And 2h18'50" 49d56'27"
The Nicole Lansdown
Tau 5h57'12" 24d14'35"
The Nicole Schmitt Star
Psc 1h7'46" 20d43'52"
The Nicolo and UnderDog Show
Crb 16h0'29" 39d31'18"
The Nielsen Family
Uma 12h51'35" 54d10'58"
The Nigel John Hawkins Star
Ori 6h12'55" 15d25'52"
The Nigel Johnson Angel Star
Per 4h22'43" 47d6'0"
The Nikita
Sco 17h23'4" -42d14'54"
The Nikki J
Sco 16h41'51" -31d30'8"
The Nildi Star
Umi 16h42'20" 76d14'22"
The Noble-Wood Wedding Star
Cyg 20h35'46" 34d15'11"
The Noel Crawford Memorial Star
Pup 7h39'38" -27d16'19"
The Noel S. Adrias Jr. Superstar
Lib 15h19'19" -11d22'56"
The Noel Star
Cep 20h55'28" 65d32'51"
The Noof Star
Umi 14h13'50" 75d2'20"
The Noralie Star
Uma 8h37'29" 71d37'16"
The Noreaster Star
Uma 8h57'56" 55d16'52"
The Norm Star
Lyr 18h49'41" 34d46'59"
The Norma Clifton Star
Uma 10h31'39" 65d4'21"
"The Norma J"
Lmi 10h50'22" 28d55'33"
The North Paula
Uma 8h20'55" 65d32'44"
The North Star
Umi 17h2'43" 87d3'56"
The North's 1st Wedding Anniversary
Uma 10h6'39" 49d30'17"
The Norton Family
Lyn 8h10'43" 41d29'13"
The "Nuh" Star for Princess Aja
Cnc 9h1'0" 23d9'44"
The Nydia 9/7
Psc 23h49'22" 6d43'35"
The Nykeisha
Umi 16h31'3" 77d24'57"
The Oakwood Star
Uma 9h47'12" 41d25'31"
The O'Beirne Diamond Star
Cas 23h48'54" 54d49'0"
The O'Braps
Cyg 20h25'20" 59d1'6"
The O'Brien Sisters
And 23h57'59" 42d44'8"
The "OC" Man
Lyn 9h23'16" 41d19'5"
The OC ~ Oscar & Christy Cartaya
Lyn 7h44'33" 56d0'2"
The O'Connor Star
Aql 20h5'47" 5d24'26"
The O'Crotty Star System
Uma 10h7'59" 64d56'25"
The Official Jess Loves Dame Star
Cyg 21h55'8" 52d17'16"
The O'Hanlon Legacy
Ori 5h13'57" 6d31'14"
The Ohio Minto
Leo 9h29'19" 29d16'42"
The Ohtmuellers
Lyn 7h0'48" 47d16'46"
The Old Buzzard's Star
Cru 12h39'32" -59d30'3"
The Old Tire Guy Hoffman
And 0h46'50" 36d2'47"
The OLDE Team Starring PEG & WARREN
Uma 8h17'3" 60d32'1"
The Olivejuice Star
Mon 6h53'11" -0d12'20"
The Oliver Patrick Star
Nor 16h28'13" -59d39'41"
The Olivia and Howard Abel Star
Per 3h38'5" 46d22'1"
The Olivia Katz
Uma 13h43'54" 59d24'12"
The Olivia & Melissa Webb Star
Cru 12h32'27" -58d32'41"
The Ollie Lane Tavern
And 23h58'46" 34d49'41"
The Olmos Family Star
Sge 19h8'47" 18d50'34"
The Olsen's 25th Anniversary Star
Tau 4h24'12" 22d32'50"
The Olson Family
Uma 8h33'38" 62d18'24"
The Olwyn Clarke Star
Cru 12h25'23" -60d2'1"
The Olympian Star
Her 16h26'52" 40d12'40"
The O'Meally Star
Cru 12h34'26" -60d13'52"
The Ondi
Lyn 7h38'43" 35d21'46"
The One
Cyg 21h38'32" 46d10'56"
The One
Vir 11h38'4" -0d26'30"
"the one all other stars wait for"
Ori 5h29'55" -8d0'5"
The One and Only
Sco 16h45'11" -31d34'20"
The One And Only ... Bo Baby
Tau 4h33'48" 3d9'21"
the one and only "Gabi"
Ori 5h33'58" -8d21'33"
The one and only Gianina Felix
Her 17h15'1" 31d7'56"
The One and Only Jayke Milton
Srp 18h9'47" -0d19'32"
The One and Only Katie Troisi
Sco 17h24'21" -38d9'38"
"The One and Only" Mrs. Krista
Cas 0h54'3" 54d35'40"
The One and Only Norwaya Ni
Uma 11h29'23" 59d36'7"
The One and Only Sean Francis McNamee
Per 4h8'44" 41d17'31"
The "One" I Luv
Psc 23h50'24" 5d4'23"
The One & Only Amber "Humbar" Khan
Lib 15h7'38" -22d39'43"
The One That Lightens My World, You
Tri 1h51'27" 31d16'9"
The One The Only Donna Watson
Vir 15h10'1" 1d51'28"
The One Year Star
Tau 3h40'17" 28d1'35"
The Only Mother of Mine
Cas 23h36'34" 53d38'38"
The Only One
Cnc 8h26'50" 32d18'57"
The Only One That Matters to Me
Gem 6h44'8" 17d48'10"
The only Peach in my life
Ori 6h5'18" 13d54'4"
The Only Star In My Entire Sky
Sco 17h26'37" -34d40'44"
The only star in my world... Ryan
Cep 0h14'2" 81d21'50"
The Optimistic Star
Pyx 9h0'4" -27d13'47"
The Oran Koren Star
Ori 5h44'36" 6d59'42"
The Order of the Fleur-de-lis Star
Uma 10h21'44" 69d56'13"
The O'Reilly's
And 23h32'17" 35d57'36"
The Original LuieB
Gem 6h45'46" 32d25'5"
The Original Rockstar "NIC"
Sco 16h17'6" -12d49'56"
The Original Star of Seven
Ori 5h52'3" -8d18'9"
The Orphan
Aur 5h12'45" 42d14'53"
The Osborn Heart
Sco 16h50'28" -33d42'10"
The Osterhout Family
Aql 19h58'15" 5d21'17"
The Ott Family
Her 18h1'47" 36d22'7"
The Outlaw
Boo 15h6'32" 33d47'21"
The Pa and Ka Star
Uma 10h39'13" 43d48'6"
The Pa-chui Alliance
Umi 16h58'18" 78d49'8"
The Pack Forever D-C-P
Umi 14h45'6" 67d57'49"
The Padre
Per 4h41'33" 49d1'9"
The Page and Rex Wedding Star
Gem 7h0'46" 25d17'39"
The Paivikan Star
Her 17h39'51" 37d24'33"
The Palladino Five
Aql 18h53'0" 8d46'30"
The Palmisanos
Ori 6h19'5" 14d2'22"
The Pamela
Gem 6h49'27" 27d4'59"
The Panda Star
Gem 6h26'20" 23d18'22"
The Pandy Spade
Vir 13h21'14" 3d26'21"
The Paris and Brionna Shining Star
Lyr 18h39'36" 30d32'11"
The Parke Jensen
Tau 5h52'53" 26d50'36"
The Parker Baby
Gem 7h2'24" 24d33'10"
The Parkettes' Olympic Star
Her 18h36'31" 13d14'25"
The Parris Star
Ori 6h3'44" 10d39'52"
The Parrott Star
Uma 11h31'39" 59d52'45"
The Pat Moorefield and Silas Henry
Ari 3h4'51" 20d36'19"
The Patafio Family Star
Uma 9h17'40" 52d30'12"
The PatPat
Cnc 8h41'20" 32d50'23"
The Patricia Annie Bryan Star
Vir 13h23'31" 2d5'29"
The Patricia Symes Star
Cas 0h37'48" 61d53'12"
The Pattison-Penney Guiding Star
Cyg 19h46'32" 36d51'33"
The Paul Barbour Star
Cru 12h39'17" -58d10'46"
The Paul & Gloria Jackson Star
Cyg 20h1'53" 56d37'31"
The Paul Howarth Star
Per 4h13'34" 31d49'22"
The Paul & Mary Soulmates Forever
Vir 13h16'16" -18d42'14"
The Paul O'Donnell Star
Tau 5h31'59" 28d19'49"
The Paul Rego Heavenly Body
Cnc 8h39'56" 14d46'15"
The Paul Tsai Star
Lib 15h59'47" -4d50'7"
The Paula Avis
Tau 4h38'29" 17d51'41"
The Paulie
Gem 6h45'31" 25d39'3"
The Pauline Star
Ari 2h54'43" 24d42'27"
The Pauline Wenk Star
Her 17h58'45" 21d45'20"
The Pauliny Star
Uma 11h26'9" 65d34'5"
The Pavlich Star
Cyg 19h49'13" 37d41'32"
The Peace & Wildflowers Star
Sgr 18h33'21" -25d46'15"
The Peaner
Lyr 18h31'5" 33d1'28"
The Pearl's together at last!
Per 3h46'39" 33d42'12"
The Pease Boys
Uma 8h28'1" 61d20'4"
The Pebbles Star
And 1h5'0" 37d16'53"
The Pediatric Connection
Peg 23h41'40" 23d46'20"
The Peekaboo
Lmi 10h7'23" 29d50'55"
The Peety James Bowra Star
Sco 17h37'30" -36d23'52"
The Peggy D
Lyr 18h43'30" 35d6'39"
The Peia Point of Light
Uma 10h10'57" 50d7'47"
The Penguins Pebble
Umi 15h5'8" 75d16'43"
The People's Church
Cyg 20h48'49" 36d24'27"
The Peragallo Star
Lmi 10h32'27" 35d12'29"
The "Percyhauser" Star
Cyg 20h37'44" 60d10'6"
The Pereira Family
Uma 8h49'28" 65d41'25"
The Perez Family
Uma 9h22'6" 53d8'44"
The Perfect Balance
Dra 16h53'10" 65d5'36"
The Perfect Love
And 1h7'1" 41d55'54"
The Perfect Love
Cas 1h27'35" 57d42'42"
the perfect man
Vir 12h36'51" 11d14'51"
The Perfect Mistake
Cyg 20h30'26" 46d31'38"
"The Perfect Star"
Umi 15h45'35" 84d6'57"
The Perla and Giuseppe Wedding Star
Cyg 20h43'47" 39d19'9"
The Permezel Professor
Cru 12h26'39" -61d20'25"
The Perrin's Gemini
Gem 6h54'52" 26d50'19"
The Perry Family
Cyg 20h54'29" 45d39'28"
the Peruvian Queen
Gem 7h31'41" 28d10'5"
The Pete
Cnc 8h12'39" 11d50'36"
The Pete
Psc 0h35'25" 7d21'12"
The Pete and Helen Star
Uma 12h37'55" 58d23'21"
The Pete-Shawn Friendship Star
Umi 14h32'40" 74d7'41"
The Pete Weyman Dot
Uma 10h29'44" 70d10'46"
The Peter Lee Star
Vir 13h36'30" -19d35'14"
The Peter Nordhal Star
Per 4h32'51" 47d40'23"
The Petrie Star (For Les and Carol)
Uma 8h39'28" 49d2'37"
The Peyton
Sgr 20h0'11" -32d15'19"
The Phair Family
Dor 5h32'32" -68d4'53"
The Phenomenal Jody White
Ret 3h49'40" -53d13'50"
The Philip Home Star
Ori 6h20'34" -0d18'33"
The Phillip Introna Star - 11/11/82
Pho 0h9'22" -57d6'44"
The Piece
Aqr 22h16'21" -12d34'14"
The Pierson Star of Friendship
Uma 11h35'23" 46d13'3"
The Piggy Star
Ari 3h11'2" 28d58'5"
The Pikey
Lyn 7h35'15" 45d41'11"
The Pink Diva
Cap 21h49'56" -14d20'44"
The Pinsoneaults
Her 17h25'36" 29d52'53"
The Pipp Family Star
Uma 10h32'56" 64d53'33"
The Pixie Star - Mary Beth
Cap 21h11'7" -21d5'7"
The Planet Lenny
Ori 5h36'26" 3d31'29"
The Playwright David Welsh
Aqr 23h25'46" -24d30'53"
The Pod Star
Cru 12h17'2" -57d55'21"
The Poem
Uma 8h57'2" 50d31'49"
The Pohlad Star
Uma 9h27'32" 66d44'45"
The Polish Guiding Light
Uma 11h47'2" 44d49'6"
The Pollard Family Star
Tau 4h18'10" 4d32'22"
The Polly Alexandria Papp
And 0h42'59" 44d58'49"
The Poodle
Cmi 7h27'1" 9d21'9"
The Poody Star
Uma 10h49'7" 48d42'50"
The Pookie Star
Umi 14h23'23" 66d35'33"
The Pookie Star
Vir 14h11'34" -0d39'13"
The Pooser Star
Uma 10h7'28" 69d22'13"
The Popa Star Harry J Bosch
Sgr 18h53'6" -20d30'44"
The Porter/Thomas Family Star
Uma 8h40'29" 51d32'56"
'The Porter-Maule'
Cru 12h46'17" -59d33'28"
The Potance Family Star
Her 18h0'37" 17d53'39"
The Power of the 3rd SWANKLM JECD2
Ori 5h32'40" -1d43'25"
The Poynter Star
Eri 4h9'4" -33d23'39"
The Prado Star
Uma 11h35'18" 46d39'31"
The Preacher
Gem 7h29'47" 28d15'27"
The Precious
Leo 10h48'19" 6d38'49"
The Precious Sparkle of Ryan
Psa 22h55'10" -30d49'30"
The Presley Star
Vir 13h1'26" 11d51'49"
The Prestano Family Star
Uma 10h39'44" 39d29'25"
The Pretty Light
And 1h52'44" 35d57'27"
The Prewett Family
Sge 19h54'13" 19d20'18"
The Price-Turner Star
Cyg 21h33'13" 40d0'12"
The Pride of Andrew Moulthrop
And 2h13'11" 50d24'50"
The Prince Trevor
Gem 7h6'1" 33d41'44"
The Princess and Prince Star
And 0h45'13" 38d21'0"
The Princess and The Bear
Umi 15h9'58" 73d2'28"
The Princess And The Millionaire
Her 16h28'40" 47d56'15"
The Princess Buffy Star
Leo 11h0'41" 1d11'42"
The Princess Isabel
And 23h5'40" 41d14'10"
The Princess Laurie Star
Cap 21h32'7" -15d52'25"
The Princess Lia
And 1h51'25" 46d32'20"
The Princess Michelle M Star
Ori 5h17'22" -7d20'53"
The Princess Ruth
And 0h2'31" 35d18'50"
The Princess Sophia Claire
Cnc 8h12'35" 7d10'41"
The Princess Star
Aqr 21h52'36" 1d13'19"
The Princess Star
And 1h9'14" 35d42'54"
The Princess Star
And 2h14'11" 41d31'7"
The Princess Star
And 1h16'23" 47d59'8"
The Pritchard Star
Ari 2h59'19" 27d41'2"
The Pritiest Star
Vir 13h26'18" 12d21'59"
The Prof
Cep 0h3'50" 75d4'1"
The Promise
Sgr 18h17'40" -17d42'12"
the promise
Sco 16h6'27" -18d52'18"
The Promise
Sgr 20h17'59" -29d32'8"
the promise
Per 3h42'41" 49d52'34"
The Promise Ken and Dara
Ori 6h19'24" 10d27'58"
The Promise of a Life Together
Cyg 20h19'8" 52d19'14"
The Promotional Elements Team
Uma 11h22'58" 57d21'44"
The Pud
Cas 1h23'34" 67d7'2"
The Puddin' Head Lovers
Psc 0h34'48" 3d38'48"
The Puritt Shining Star
Per 3h7'0" 45d0'10"
The Purple Dragon
Cru 12h39'10" -57d41'1"
The Putman Family
Uma 12h19'59" 55d36'59"
The Pyramid Star of TJJ
Aqr 22h37'45" -1d7'5"
The Qualcomm Jacobs Family Star
Per 3h16'46" 42d39'21"
The Quantum Leap
Vir 13h57'24" -16d7'25"
THE QUEEN
Uma 10h27'40" 62d39'34"
THE QUEEN
Gem 6h53'31" 22d22'1"
The Queen
Gem 7h23'30" 20d5'37"
The Queen Jazzette Kowsh
Cnc 9h13'5" 16d16'43"
The Queen Lisa
Cap 20h23'20" -11d40'42"
The Queen Mum
Tau 4h26'22" 20d53'7"
The Queen~Sherry Dorman Is Our Star
Cas 0h23'31" 57d53'8"
The Queen Wallace
Cas 1h12'21" 66d45'9"
The Queens Arms
Cas 3h12'50" 61d46'46"
The Queens Court
Cas 23h51'38" 50d35'25"
The Quilter's Star
Cnc 8h42'23" 20d7'6"
The "R & O" Diamond
Cyg 20h34'54" 48d25'22"
The R & R Wedding Star
Cyg 21h19'1" 38d22'22"
The R S Darisse
Cet 0h59'2" 2d39'55"
The Rachel Pandolfo Star
Cas 0h43'36" 58d6'52"
The Rachel Valley Star
Uma 9h38'18" 44d36'24"
The Radiance & Laughter of Shane
Per 4h27'19" 44d49'11"
The Radiant Ray of Robinson
Aql 19h30'48" 3d54'12"
The Radiant Swan
Gem 7h3'9" 24d47'26"
The Radney Star
Aqr 22h20'24" -17d5'33"
The Raena Min Star
Vul 20h28'3" 28d18'47"
The Raffins
Lyn 7h40'44" 42d32'41"
The Ragatz Brothers
Per 2h50'32" 53d38'6"
The Rainbow Connection
Gem 7h40'36" 32d59'46"
The Rainbow Man
Cru 12h34'21" -56d8'33"
The Ram
Lyn 6h29'7" 56d52'38"
The Ramon Star: Beyond Compare
Ori 5h21'42" 1d19'0"
"The Ramona Stealth"
Aql 19h30'40" -0d3'33"
The Ramones
Per 3h2'48" 46d37'27"
The Ramshaw Family Star
Cyg 19h57'28" 31d15'25"
The Randia Star
And 23h37'16" 35d48'26"
The Randolph Mar of Woonona Star
Vel 9h15'9" -46d7'13"
The Ranger Star
Gem 6h48'25" 15d55'49"
The Rangers Star Eternal & Timeless
Cyg 20h46'16" 43d47'27"
The Rash Star
Leo 11h2'53" 4d56'32"
The Rasmussen Super Star
Uma 10h57'7" 59d46'31"
"The Raven"
Crb 16h4'43" 31d45'21"
The Raven Beauty
Vir 12h36'18" 7d49'3"
The Ray Family Star
Dra 19h9'40" 69d24'35"
The Ray Hackenstar
Leo 10h17'24" 9d57'26"
The Ray of Hope
Uma 9h34'58" 57d11'13"
The Ray Star
Peg 21h46'8" 14d54'31"
The Raymond Commodore Star
Cru 12h37'56" -64d39'19"
The Raymondo
Psc 0h43'49" 21d14'47"
The Rayna Star
Dra 15h21'6" 55d0'52"
The Rayoo
Uma 13h34'47" 58d33'49"
The Razor Angel Sin
Dra 17h23'9" 51d12'48"
The Razzle Dazzle of Jennie Clare
Cen 13h1'33" -47d16'3"
The Real Boz
Uma 12h17'52" 54d21'1"
The Real "Lau Tak Wah" From Extrude
Cmi 8h1'24" -0d18'28"
The Reason
Uma 14h26'30" 58d12'47"
The Reason
And 0h15'45" 46d12'57"
The Rebecca Ann Star
Tau 4h24'41" 24d40'32"
The Rebecca Faye
Cyg 19h46'48" 35d8'13"
The Rebeccian Supernova
And 0h19'4" 37d14'7"
The Rebekah
Cru 12h4'42" -62d50'56"
The Rebekah Raymond Goddess Star
Leo 9h32'9" 24d59'21"
The Recreationalist
Uma 10h17'3" 44d35'37"
The Reeve Star
Del 20h41'58" 11d6'23"
The Regan DiBacco Star
Uma 14h19'7" 56d47'18"
The Rego Star
Aqr 22h40'47" 0d56'38"
The Reier Family
Uma 12h10'43" 58d38'22"
The Reiss Family Star
Oph 17h56'12" -0d1'31"
The Remote Junkies Rock
Lyr 18h46'57" 36d39'49"
The Renaud Star
Leo 11h14'30" 3d3'58"
The Renee Soboleski
Ari 3h26'55" 20d52'39"
The Renegade
Cap 20h22'53" -13d15'28"
The Rennie Three Star Cluster
Umi 14h14'37" 66d46'32"
The Renzulli Family Star
Cnc 8h35'3" 32d30'16"
The Rev. Edee Chase Fenimore
Aur 5h29'15" 40d38'12"
The Reveley Star
Vir 13h16'48" 6d47'38"
The Reverend Neil G. Thomas, MCCLA
Cyg 19h55'4" 41d9'46"
The Rex Dee Pinegar Star
Cyg 19h48'56" 30d49'52"

The Rex G Kervin Family Star
Lyn 8h51'59" 43d18'38"
The Rex & Polly Star
And 1h45'27" 46d14'28"
The Rhino
Uma 11h27'48" 42d0'14"
The Rhoda Star
And 0h27'31" 42d49'10"
The Rhonda K.
Leo 11h1'37" 21d47'49"
The Rhys Holly Star
Sgr 18h8'35" -25d30'14"
The Rhys & Lucy Elavia Forever Star
Cyg 20h5'41" 31d11'29"
The Richard A. Hubbard
Tau 4h8'42" 27d39'49"
The Richard C. Coleman Family Star
Cyg 19h19'55" 54d8'22"
The Richard & Diane Star
Lmi 10h42'2" 25d9'28"
The Richard E. Norton Jr. Star
Gem 7h10'50" 25d9'48"
The Richard Jacqueline
And 23h15'12" 51d41'39"
The Richard L. Prangley Star
Cap 21h48'52" -11d11'41"
The Richard M Genova Star
Per 3h9'41" 55d57'5"
The Richard Mumby Star
Cru 12h28'55" -59d3'27"
The Richard Ptacin "BearCub"
Ori 5h12'2" 12d11'36"
The Richardson Family Star
Ori 5h56'24" -0d44'13"
The Ricky Pentz Star
Aql 20h2'47" 1d47'6"
The Ring Family Legacy
Ori 5h34'58" -0d22'34"
The Ring Family Star
Cyg 20h35'42" 34d37'19"
The Ringer Star
Uma 10h9'0" 47d12'43"
The rising of the son Sean
Psc 0h18'9" 8d26'39"
The Rising Star Academy of 2005
Uma 11h41'20" 62d26'0"
The Rissanen
Cma 6h51'20" -14d3'13"
The River House Trust
Dra 9h35'20" 77d56'2"
The Rivero Family
Sge 19h54'18" 17d35'3"
The R.J. Dillabough Family Star
Leo 10h17'32" 10d26'40"
The Road King
Per 3h14'25" 43d58'40"
The Rob
Ori 4h54'32" -2d55'15"
The Rob & Emma Star
Cyg 21h34'5" 30d42'11"
The Rob & Heather Star
Cyg 21h29'27" 36d31'37"
The Robbins Star
Uma 8h32'56" 61d22'45"
The Robert Danielson Star
Her 17h46'12" 38d30'49"
The Robert Jillett Star
Cru 12h25'24" -62d6'8"
The Robert John Lepre Star
Lib 16h1'51" -10d53'51"
The Robert L. Emmons Star
Cap 21h17'15" -15d13'9"
The Robert P. Myers Star
Uma 12h28'26" 60d43'31"
The Robert Patriquin Star
Per 3h14'58" 54d20'55"
The Robert Schroeder Star
Per 3h48'10" 42d42'26"
The Robert Wrightson Star
Her 17h50'0" 46d28'34"
The Roberta Patrick Star
Cru 12h25'1" -57d2'27"
The Roberts Family
Sct 18h46'3" -10d6'11"
The Robin and The Wren
Uma 11h47'45" 57d35'34"
The Robin and The Wren
Uma 10h39'19" 57d53'44"
The Robin Carson Star
Lib 14h51'19" -18d26'22"
The Robin Sheehan
Ari 2h52'42" 18d51'49"
The Robinson Francis Family
Cyg 19h45'42" 39d55'29"
The Robinson Rocket
Uma 9h31'7" 64d8'55"
The Robinson's Star of Love Above
Lyr 18h50'45" 33d38'45"
The "Robio" Bartlett
Cam 4h47'48" 62d35'15"

The Robyn Sara Gordon Star
Lyn 6h44'5" 57d26'8"
The Robyne Leigh
Lib 15h9'6" -6d50'24"
The Rock Star
Tau 5h22'36" 21d22'20"
The Rocket
Ori 5h29'8" 3d1'0"
The Rockin' Rolemodels
Uma 10h27'50" 58d21'34"
the rockman star
Cyg 20h20'41" 51d3'44"
The Rodan Superstar Status Achieved
Crb 15h57'59" 31d38'58"
The Roderick Family Star
Sgr 19h15'37" -17d15'26"
the Rodney B.
Lib 15h23'18" -7d37'58"
The Roehrer Borealis
Uma 10h59'46" 36d4'38"
The Roger
Per 4h15'6" 47d59'48"
The Roger "G"
Uma 9h15'8" 61d32'5"
* The Roger Norton Star *
Per 2h53'25" 40d21'27"
The Roger Shaw Star
Per 3h5'54" 41d10'34"
The Rogers & Johnson Family Star
Uma 10h54'43" 37d36'7"
The Rogge Family Twinkling Star
Lyn 7h32'12" 38d13'6"
The Roland and Linda Culberson Star
Leo 10h52'29" 7d28'5"
The Roland Ogoshi Family
Aql 20h20'14" 1d59'52"
The Rolli Family
Sge 20h12'31" 17d48'57"
The Romanian Princess
And 23h23'24" 42d25'30"
The Romero Family shines eternal
Uma 11h22'50" 37d49'46"
The Ron & Margaret Woods Star
Cru 12h19'2" -57d31'50"
The Ron & Roni Star
Cnc 8h28'40" 26d25'17"
The Root of Jesse
Vir 13h33'26" -0d3'47"
The Rootbeer Family
Uma 9h40'46" 55d26'9"
The Roots of Owls Head
Ori 6h16'53" 9d27'59"
The Rori Jones Star
Sco 17h52'55" -36d58'4"
The Rory Bear
Cap 20h20'3" -14d1'4"
The Rory Squires Star
Ari 2h57'18" 26d33'59"
The Rose
Sgr 19h33'50" -17d21'2"
The Rose Of Camelot
Aqr 22h2'15" -13d36'40"
The Rose of Mary
Crb 15h37'58" 39d26'43"
The Rose Star
Lyr 18h49'38" 37d48'16"
The Rosenberg Star
Uma 14h14'45" 62d5'9"
The Roses' Star of Texas
Sco 17h52'25" -35d49'47"
The Rosetta
Gem 6h32'2" 18d46'38"
The Ross Family's Star
Ori 5h1'54" -0d2'37"
The Rostrons
Per 2h36'26" 55d51'20"
The Rothschild Family
Uma 10h3'26" 59d48'51"
The Roy Family with Friendship
Tau 4h33'31" 17d30'5"
The Roy Turner Star Most Awesome
Cyg 21h15'45" 47d10'36"
The Royal June Star
Pho 0h33'50" -53d37'51"
The Royal Oarsman
Uma 9h43'19" 52d57'45"
The Royal Royes
Ara 17h12'15" -51d59'18"
The Royce and Jill Imhoff Star
Sco 16h7'6" -17d47'17"
The Roz
Vir 12h32'3" 5d8'59"
The R.T. Karman Family Star
Uma 13h36'20" 61d13'2"
The Ruben Kaim Star of Dreams
Leo 11h35'9" 14d35'4"
The Rubenstein Family
Uma 11h7'15" 56d23'19"
The Ruby and Bob Ferrie Star
Cyg 21h12'36" 43d43'47"
The Rubye C
Uma 10h5'38" 46d24'15"

The Ruckman Family
Uma 10h11'28" 69d12'2"
The Rudy Widmann
Gem 7h47'7" 32d14'16"
The Rue Star
Tau 4h25'36" 22d52'56"
The Ruediger Star
Uma 9h8'15" 51d15'44"
The Ruffner Family Star
Uma 12h23'23" 59d14'21"
The Rugland Star
Gem 7h30'23" 32d49'47"
The Rupert Baron
Cap 20h23'34" -10d0'31"
The Ruth and Arno Drucker Star 50
Cyg 20h37'53" 55d3'42"
The Ruth McBride Newton Star
Crb 15h24'49" 29d34'28"
The Ruthmanns of Eldoret
Eri 4h23'38" -9d23'5"
The Ryan Boswell Star
Cen 12h9'51" -47d20'49"
The Ryan Family Christmas Star
Uma 10h9'43" 46d28'32"
The Ryan Patrick
Lyn 9h7'32" 40d2'40"
The S and I Wedding Star
Cyg 20h2'9" 44d18'7"
The S. and N. Sanders Family Star
Ret 4h11'47" -65d49'13"
The S. Richard & Marlene Gordon Star
Lib 15h10'15" -21d13'28"
The "S" Star
Sco 16h11'50" -9d12'32"
The Sacred Light of Friendship
Vir 12h31'2" 10d0'27"
The Sadati Family Star
Ari 2h2'40" 27d34'28"
The Sadie Lee Dunkin
And 0h20'1" 31d53'53"
The Sales Family
Uma 10h36'57" 43d18'25"
The Salim and Sally Wedding Star
Cyg 20h46'17" 36d24'53"
The Sam
Vir 12h26'26" 3d44'29"
The Sam Alaeddin Star
Vir 12h1'53" -0d7'29"
The Same Bright Star
Umi 14h26'28" 84d44'22"
The Sami Schwartz
Leo 9h41'16" 26d48'52"
The Sammy Star
Uma 10h20'21" 59d55'53"
The Sampson Family Star
Uma 8h23'46" 66d37'27"
The Samuel Joseph Clabaugh Star
Uma 11h18'52" 62d18'42"
The Sande Ortiz Family Star
Sco 16h49'58" -44d21'14"
The Sanderlins
Vir 12h46'48" 2d29'33"
The Sandlin Family
Cyg 20h38'58" 37d21'46"
The Sandpiper Hotel
Per 2h53'30" 51d45'11"
The Sandy Bob
Uma 11h20'36" 43d39'42"
The Sandy Star
Ori 6h1'7" -2d57'0"
The Sangallo Star
Uma 12h33'14" 57d43'57"
The Santosusso Family
Uma 13h18'2" 56d53'34"
The Sarah and Jim Star <3
Cyg 20h18'3" 44d12'24"
The Sarah B. Muska
Lmi 10h50'51" 27d42'12"
The Sarah & Louise Love Star
And 23h7'12" 36d6'6"
The Sarah Lynn
Vir 13h47'45" -7d18'28"
THE SARAH & MANNY STAR
Aql 20h3'5" 3d49'47"
The Sarah Rebecca
And 23h49'3" 46d6'27"
The Sarah & Thomas Richardson Star
Cyg 20h11'37" 57d16'34"
The Sasser Family Star
Uma 10h42'20" 54d30'41"
The Sawyer Shlomo Star
Uma 13h51'19" 58d21'11"
The Scallywagged Princess
And 1h10'29" 38d32'5"
The Scarlett Star
And 23h14'5" 40d18'26"
The Scarpati Clan
Uma 11h31'22" 33d9'21"
The Schaefgen-Burns Family Star
Uma 10h22'18" 61d42'31"
The Schell Star
Cma 6h41'29" -13d9'20"

The Schmoopie Van Hoosier
Uma 10h10'43" 51d38'16"
The Schools Family Star
Uma 13h54'7" 52d52'9"
The Schuler Star
Gem 7h39'14" 31d46'41"
The Schultz Family
Uma 11h3'4" 64d46'27"
The Schuster's Star
Umi 13h10'18" 76d17'37"
The Schwartz Family
Crb 16h14'23" 27d22'31"
The Schyler-Tiana Star
Col 6h23'57" -34d56'20"
The Scivally Star
Mon 7h18'10" -0d36'57"
The Scott Polglase Star - 25/01/78
Cru 12h27'33" -60d30'38"
The Scottie Decelles "Imp Star"
Ori 5h34'12" -4d17'22"
The Scottsdale Charros
Peg 22h48'53" 25d21'31"
The Scotty B
Dra 10h24'0" 74d2'40"
The Scotty Two Chins Star
Per 4h49'29" 46d30'21"
The Scurby Dog
Cru 12h17'55" -61d58'36"
The Seal Family Star
Aql 19h54'57" -1d39'23"
The Sean Thomas Routledge Star
Cru 12h31'24" -61d58'33"
The Secchia Family Star
Cyg 20h44'9" 43d39'18"
The Second Star to the Right
Uma 9h29'45" 46d18'46"
The Second Star To The Right
Tau 3h38'9" 19d52'7"
The Seiboldt
Peg 22h28'13" 29d12'33"
The Seidenberger Adventure
Lyn 7h43'45" 38d5'19"
The Selfridge Family Star
Vir 13h22'46" 11d46'1"
The Senyei Family
Lyn 7h51'31" 58d5'19"
The Seper Star
Lyn 7h55'58" 38d50'44"
The Seth Adam & Danielle SoRa Star
Vir 13h27'10" 11d15'35"
The Sewell 40th Anniversary
Lib 15h8'39" -8d58'27"
The Sexton Family Star
Cyg 19h45'2" 33d49'56"
The Seydel Family Star
Uma 10h1'17" 69d41'29"
The Shaheen Family Star
Per 3h5'9" 54d44'20"
The Shai Stone Star
Sgr 18h14'47" -17d42'16"
THE SHAKEY STAR
Lib 15h6'17" -1d6'17"
The Shana Berry Sparkle
Crb 15h46'36" 35d0'51"
The Shane Karl Maher Star
Tau 4h10'11" 27d47'56"
The Shannon Jade Burns Star
Cru 12h37'25" -58d52'16"
The shape of bonds
Lyr 18h40'12" 38d9'21"
The Shareen and Jade Star
And 23h43'47" 41d37'47"
The Sharkey's Star
Cru 12h5'35" -61d3'37"
The Shaw Star
Ori 5h12'4" 8d52'31"
The Shawn Boggeman Brilliant Star
Cyg 19h51'23" 36d1'26"
The Shawnee Star
Uma 9h7'8" 48d57'32"
The Shelby Rose
Mon 6h53'0" 3d37'56"
The Shelton-Jimenez-Metz-Montgomery
Dra 19h33'34" 63d37'41"
The Sheppard Flyer
Ori 5h41'43" -2d3'54"
The Sheridans
Uma 8h25'26" 65d36'37"
The Sherry Marie Hylton Star
Cas 0h57'51" 59d19'45"
The Shine of Sheree & Lee
Cru 12h35'54" -57d38'48"
The Shining Annie Mercedes
And 1h55'21" 37d9'34"
The Shining Barb Rebola
Lib 14h24'36" -9d31'50"
The Shining Bazza - Barry Hocking
Ari 1h59'17" 20d41'12"

The Shining Jacquelyn JiMoi
And 2h26'55" 42d18'45"
The Shining Liam Shelley
Vir 13h22'26" 12d38'54"
THE SHINING REALM OF AMY LEE
Leo 10h8'7" 22d56'11"
The Shining Skripman
Uma 8h59'51" 65d52'5"
The Shining Star in My Life: Amy
Sco 16h22'59" -39d33'25"
The Shining Star Jason Landau 4ever
Uma 8h26'7" 65d25'46"
The Shining Star Lucille
Psc 0h56'54" 27d14'8"
The Shining Star of Jacob
Col 5h33'54" -39d8'7"
The Shining Star of Sara
Vir 13h5'41" -20d20'56"
The Shining Star of Sherry L. Brink
And 1h18'23" 45d20'53"
The Shining Yvonne
And 2h11'20" 50d20'11"
The Shirley Michelle Amber Star
Crb 15h38'57" 28d31'37"
The Shirley Star
Cas 1h34'13" 61d12'45"
The Shirley & Trevor Akeroyd Star
Cyg 19h56'31" 45d59'20"
The "Shmily" John C. Ruffin Star
Uma 14h0'57" 49d3'10"
The Shmookm's Bag
Pho 1h40'51" -45d17'28"
The Shnooky Star
Uma 8h29'5" 65d7'36"
The Shooting Kapish
Leo 11h10'22" -1d38'35"
The Shroat Family
Cyg 20h16'46" 46d48'58"
The Shuman's
Mon 6h53'33" -0d56'34"
The Shwu Star
Vir 11h37'47" -5d51'20"
The Sibyl Farson
Sco 17h18'23" -30d39'50"
The Sign of the Cream
Aqr 23h3'25" -13d18'49"
The Sikorsky
Uma 11h33'56" 46d24'15"
The Silky Way
Gem 7h27'22" 30d47'12"
The Silver Cricket
Lep 5h15'59" -11d40'33"
The Silver Fox
Per 4h40'5" 41d4'19"
The Silver Thomases
Cyg 21h35'34" 49d51'17"
The Simmons Family
Uma 11h17'45" 64d19'40"
The Simon Bell Star
Cru 12h44'13" -58d37'21"
The Singer of Connection
Lyr 19h24'38" 41d32'52"
The Sir Matthew Little Star
Aqr 22h6'25" -2d15'22"
The Sir Richard Feachem Star
Ari 3h23'44" 15d39'50"
The Sissy
Ori 6h19'1" 19d34'48"
The Sissy
Lyn 6h58'39" 47d19'20"
The Sister Kathy Nolan Star
Cas 1h43'48" 67d11'17"
The Sister Pat Griffith Star
Cas 1h9'34" 65d23'28"
The Sisters - Olivia & Jane Meyers
Lyr 19h14'45" 27d21'30"
The Skies in your Eyes
Lyn 6h57'25" 60d6'34"
The Skillman and Cuesta Family
Eri 3h51'16" -0d15'55"
The Skinner Star
Uma 11h19'23" 55d26'8"
"The sky was made for me and you"
Uma 9h58'14" 48d1'21"
The Skys the limit
Uma 11h9'20" 54d59'17"
The Sladden Star
Aql 19h43'1" 12d7'25"
The Slagter Family
Sge 19h51'45" 18d33'42"
The Sledzik Family
Per 4h37'43" 42d18'47"
The Sleep Fairy - with love always
Com 13h34'31" 16d45'14"
The Slow Motion Champ
Lib 15h3'24" -18d15'6"
The Slubowski Star
Ori 5h32'48" -0d41'45"
The Small Star
Aur 6h12'46" 54d54'47"

"The Smallest Star in the World"
Uma 8h41'15" 69d55'41"
The Smile of Mary Lou Smith
Uma 8h41'56" 72d30'16"
The Smile of Polly Jo
Sco 17h52'57" -35d52'5"
The Smiler
Gem 6h22'1" 27d24'41"
The Smith's
Aql 19h26'50" 4d50'53"
The Smoochie Star
Del 20h18'20" 9d38'59"
The Snyders
Uma 13h50'16" 53d2'15"
The Soderberg
Uma 11h55'36" 54d1'50"
The Soldow Family Star - DJWBR
Ori 5h17'43" -7d7'56"
The Soliz Du Melle
Umi 15h23'37" 71d49'45"
The Solomon - Wagner Star
Cyg 20h8'31" 59d5'17"
The Son of Stanford
Cep 0h59'7" 87d52'54"
The Sonny Newlove Christening Star
Ori 6h22'54" 15d32'10"
The Sophia Munchousen Star
And 2h31'59" 49d39'20"
The Sophiedog Star
Dra 18h16'9" 78d53'35"
The Soul of Maggie Esparza
Ori 5h46'43" 6d44'58"
The Soulmates, Greg & Gerene
And 23h10'22" 41d49'3"
"THE SOUTH" kdskgkaj
Ori 5h39'14" 11d31'5"
The Space Pam
Cap 20h25'14" -19d25'2"
The Spam Queen
Crb 15h31'20" 31d19'17"
The Sparkle in Donna's Eyes
And 1h15'31" 44d59'50"
The sparkle in your eyes
Leo 11h14'14" 11d49'41"
The Sparkle of William Grogan
Her 17h33'43" 43d51'39"
The Sparkling Brigie Star
Crb 15h45'8" 25d55'47"
The Sparkling Kendall Redline
Uma 10h4'53" 54d15'0"
The Sparkling Star of Marta
Ari 2h45'15" 21d15'58"
The Spear Anniversary Star
Cap 20h35'44" -10d28'32"
The special place of Angie Chavera
Her 15h50'57" 49d6'20"
The Speeding Bullet
Her 16h45'0" 37d33'9"
The Spence Star
Lac 22h21'53" 41d21'59"
The Spielvogel Nebula "5-19-02"
Lyn 7h49'54" 59d40'12"
The Spirit
Uma 11h29'34" 52d53'11"
The Spirit of Amanda
Psc 1h33'58" 23d55'27"
The Spirit of Nathaniel
Psc 23h30'54" 3d32'48"
The Spoon Star
Psc 0h7'14" 2d18'56"
The Spring Break Star
Crb 15h43'49" 26d16'31"
The Springer Family Star
And 23h31'53" 36d28'17"
The Squair Family Star
Uma 10h8'32" 47d2'42"
The Squirrel - Goddess of Lost Keys
Psc 23h5'17" 0d10'3"
The Squirrel Queen
Uma 9h27'31" 71d47'44"
The Stacey Star
Cru 12h26'39" -55d44'44"
The Staff of AMT
Lyn 8h19'48" 45d1'18"
The Stahley Family
Cap 21h52'44" -17d42'27"
The Stanbery Star
Uma 11h33'57" 48d49'23"
The Stanbrook Star
Cyg 19h44'48" 37d42'39"
"The STAR"
Cas 0h37'56" 56d36'19"
The Star
And 0h8'43" 28d48'52"
The Star Bud
Leo 10h12'24" 25d40'6"
The Star Dori Ilana
Gem 6h41'17" 18d24'15"

The Star Elizabeth, God's Promise
Lyn 6h53'38" 53d27'1"
The Star for Kahori
Cnc 8h16'37" 13d38'43"
The Star Formally Known as RIGO
Cru 12h35'58" -59d32'39"
The Star In My Heart, Kelly B-O
Cyg 21h24'31" 35d52'7"
The Star in Our Sky
Ori 5h35'45" 6d8'26"
The Star Juanita
Lmi 10h10'58" 37d59'46"
The Star Light of Sheila
Sco 17h56'40" -30d49'21"
The Star Mary Kay, love of my life
Cma 7h23'17" -13d58'22"
The Star Matthew Benjamin
Uma 9h58'54" 46d8'47"
"The Star - Mr Jack Adlington"
Cyg 19h46'29" 29d26'36"
"The Star - Mrs Mavis Adlington"
Cyg 19h44'7" 27d52'39"
The Star of A. Frank Fallone
Uma 10h22'50" 47d47'9"
The Star of Aaron Thomas David Prest
Umi 14h20'42" 66d17'3"
The Star of Abeokuta
Aql 19h58'34" -5d52'40"
The Star of Addy
Lyr 18h58'26" 26d15'16"
The star of AK
Lib 14h37'50" -19d59'15"
The Star of Alan Norman
Per 4h47'30" 43d53'32"
The Star of Alan Tanner
Uma 11h6'20" 48d15'43"
The Star of Alex & Audrey Rogers
Cyg 19h37'23" 36d38'53"
The star of Alexander Perry
Tau 5h48'3" 14d31'56"
The Star of Alisha & Hamish
Cru 12h18'31" -61d36'25"
The Star of All Mothers
Cnc 9h20'50" 7d8'19"
The star of Alma
Aql 19h20'50" -9d26'9"
The Star of Amanda
Ori 5h59'47" 17d46'57"
The Star of Amy
Cru 12h25'8" -60d34'30"
The Star of Amy Cohen
Leo 9h54'4" 20d19'20"
The Star of André Waszczyszyn
Uma 12h23'13" 53d23'12"
The Star of Andrea Evian Baumwald
Ari 3h28'37" 27d21'28"
The Star of Andrea Lebar
Lib 15h36'41" -4d36'29"
The Star of Anika
Sgr 18h27'10" -16d38'19"
The Star of Anna
Ari 2h6'3" 23d26'6"
The Star of Anna and Rob
Gem 7h42'26" 31d31'12"
The Star of Anna and Thomas DeMers
Dra 10h10'49" 75d32'22"
The Star of Antionette
And 0h2'36" 41d13'12"
The Star of Aquamarine
Col 6h25'0" -35d18'42"
The Star of Arthur Buckel
Her 16h42'38" 5d22'6"
The Star of Ashton Yvonne McColl
And 0h23'48" 42d21'56"
The Star of Athena
Cas 23h34'29" 51d47'18"
The Star Of Autumn
And 0h13'38" 46d5'29"
The Star of Avid
Crb 16h8'29" 33d54'21"
The Star of Ayana with Love
Cyg 19h39'4" 55d7'47"
The STAR of BABE
Psc 23h47'47" 5d51'48"
The Star of Baby
Sco 16h22'21" -28d29'55"
The Star Of Baubles
Sco 16h32'42" -37d51'58"
The star of Bea my love
Cep 23h20'26" 72d31'7"
The Star of Beals
And 23h18'22" 49d3'41"
The Star of Ben McLean
Cma 6h53'45" -15d39'6"
The Star Of Benjamin
Lib 15h46'28" -24d34'29"
The Star of Bergers
Uma 9h17'16" 59d20'57"

The Star of Bernadisius
Crb 15h53'17" 28d54'35"
The Star Of Beverly Tall
Uma 9h35'11" 54d13'34"
The Star of Bina Sella
Cas 1h28'3" 53d57'0"
The Star of Bird
Aql 19h24'19" 13d55'40"
The Star Of Birthe
Pho 0h4'16" -44d32'31"
The Star of Bob Neuman
Aql 19h18'46" -0d39'34"
The Star Of Bonita
Lib 16h0'23" -14d12'25"
The Star of Bonnie
Per 3h42'39" 43d51'4"
THE STAR OF BOO BOO WOODLAND
Cyg 21h40'33" 32d24'58"
The Star of Bradley & Sonya
Cru 12h38'28" -59d45'13"
The Star of Brandon
Ari 3h5'34" 29d52'14"
The Star of Brian Luke
Psc 1h45'35" 17d53'25"
The Star of Brian Redding
Ori 6h11'43" 20d41'34"
The Star of Brigitta
And 23h15'46" 35d17'8"
The Star of Bryce
Cnc 7h59'52" 13d6'35"
The Star of Bubba
Her 17h35'5" 33d20'36"
The Star of Carly Anne Taborell
And 23h17'21" 51d29'32"
The Star of Cate
Leo 11h43'3" 24d39'16"
The Star of Chas
Uma 8h27'26" 66d18'35"
The Star of Chetan Ratilal Patel
Uma 10h49'41" 69d8'55"
The Star of Chris and Vaugn
Her 17h31'34" 46d38'27"
The Star of Chrissie and Jeremy
Eri 4h41'9" -0d54'43"
The Star of Christine
Cyg 21h51'7" 54d42'10"
The Star of Christine & Doug
Cyg 21h52'4" 48d42'50"
The Star of Christy
Cnc 9h17'19" 26d0'56"
The Star of Clare
Cnc 8h52'35" 28d11'2"
The Star of Co
Aqr 20h54'0" -8d39'23"
The Star of Cody Leishman
Cnc 8h44'38" 28d11'33"
The Star of Colin
Cyg 21h17'48" 44d52'50"
The Star of Colin David Jehlen
Tau 5h48'41" 13d56'29"
The Star of Colin Jacob
Aqr 21h35'18" 0d34'14"
The Star of Contessa
Sgr 18h37'28" -26d56'53"
The Star of Craig & Sharon's Wedding
Cru 12h21'22" -57d56'20"
The Star of Cuddles
Uma 12h37'41" 54d48'1"
The Star Of Damara
Uma 10h10'15" 47d25'9"
The Star Of Dana
Cyg 19h32'13" 28d10'21"
The Star of Darrell - Guardian Angel
Her 18h54'1" 24d21'15"
The Star of Dave S
Uma 12h1'20" 61d24'47"
The Star of David
Uma 13h6'42" 56d23'12"
The Star of David
Cep 21h22'43" 58d56'31"
The star of David Byrne
Cru 12h33'23" -57d58'46"
The Star of David Kerr
Uma 12h4'2" 36d51'41"
The Star of David Lessore
Cyg 19h30'54" 31d48'31"
The Star of David Mark Prest
Umi 14h7'3" 65d41'55"
The Star of David of Badge Land
Vir 12h1'52" -0d7'8"
The Star of Debbie
Cyg 21h27'23" 50d34'48"
The Star of Dennis & Sharon Roscher
Cyg 21h19'34" 40d38'55"
The Star of Derisory
Cnc 8h8'31" 11d35'13"
The Star of Diane
And 23h10'12" 36d54'54"
The Star of Diane
And 0h59'38" 42d7'55"

The Star Of Dist. 10 Aux. 2005-2006
Uma 11h45'12" 55d31'34"
The Star of Donna
Cas 0h35'50" 53d22'7"
The Star of Donna Ormes
Sge 20h9'55" 19d32'56"
The Star of Dorothy
And 0h45'39" 36d10'11"
The sTar of Dottie
Umi 10h51'34" 88d55'16"
The Star of Douglas
And 0h20'58" 23d56'33"
The Star of Duke
Lac 22h46'26" 50d27'53"
The Star of Eanna
Ori 5h4'55" 5d21'44"
The Star of Edwina "Edwina's Light"
Cas 0h44'31" 60d49'37"
The Star of Elaine Taylor
And 0h10'9" 32d47'28"
The Star of Ellis Douek
Oph 17h5'32" -0d58'25"
The Star of Emilia Mae Kolar
And 2h34'31" 50d17'0"
The Star of Emily Joanne Guiliante
Sgr 18h0'7" -27d37'21"
The Star of Eric Maylam
Peg 22h18'45" 17d37'20"
The Star of Erin
Cir 15h11'3" -57d53'50"
The Star of Erin & Erik
Cyg 19h45'40" 32d23'30"
The Star of Ethan and Jacob Gray
Aql 19h55'42" 11d59'3"
The Star of Eva
Crb 15h48'8" 37d25'20"
The star of Eva Haroldsen
Vir 12h31'14" 10d17'40"
The Star of Eveline
Cas 23h4'26" 58d44'24"
The Star of Feo
Lib 14h30'26" -10d2'8"
The Star Of Ferrin
Aqr 22h32'45" -1d33'53"
The Star of Five Friends
Uma 8h52'26" 60d38'57"
The Star of Florence Fern 2006
Cru 12h37'28" -60d16'22"
The Star Of Fortunato
Lyn 7h32'0" 48d36'17"
The Star of Fr. Martin Daly s.m.
Boo 14h13'33" 42d10'52"
The Star of Gareth the Great
Uma 11h7'13" 33d51'10"
The Star of Gary John Boulton
Cru 12h50'30" -61d6'53"
The Star of Genelyn Oriel
Cnc 8h44'15" 8d40'24"
The Star of Geo Don
Her 17h10'13" 25d3'40"
The Star of Geoffrey
Tau 3h40'59" 16d11'46"
The Star of George C. Magus
Lib 14h43'21" -12d33'51"
The Star of George & Nora Raymond
Pho 1h22'54" -54d51'57"
The Star of Georgia
And 1h54'39" 42d15'57"
The Star of Ghanem
Uma 11h0'51" 69d55'40"
The Star of Gibo
Psa 22h54'25" -24d52'34"
The Star of Gina
Cnc 8h13'19" 15d50'45"
The Star of Graham Savage
Uma 9h47'2" 59d7'19"
The Star of: Grub
Aql 18h59'25" 8d1'24"
The Star of Grunty
Cru 12h28'49" -60d23'26"
The Star of Guidance
Ori 5h55'27" 6d48'33"
The Star of Hailey Bosaw
And 2h36'16" 45d31'33"
The Star of Hannah
Ori 5h56'41" 18d18'55"
The Star of Hannah H
Cyg 21h45'54" 53d30'33"
The Star of Harry, Clem and Ned Dean.
Cep 2h25'52" 86d42'14"
The Star of Hawgs
Vir 12h37'28" 5d1'19"
The Star of Heather and Nipnop
Cyg 21h53'17" 48d16'25"
The Star of Heike and Otto
Eri 2h12'20" -54d40'12"
The Star Of Ian & Jennifer's Love
Ori 5h3'0" 13d21'40"
The Star of Ingrid Scott
And 0h29'41" 46d5'45"

The Star of Isabella Love Daddy
Lyn 7h10'53" 58d2'14"
The Star of Isabella - Love & Light
Psc 1h52'55" 2d56'34"
The Star of Isaiah Lim
Aql 20h11'10" 14d29'44"
The Star of James W. O'Brien
Sco 17h38'30" -33d57'8"
The Star of Jane
Eri 3h48'4" -37d6'11"
The Star Of Janice
And 2h1'27" 47d12'54"
The Star of Jarin Nicole Pruce
Boo 14h19'26" 46d8'15"
The Star of Jason 03/21/1986
Ori 6h4'37" 17d59'14"
The Star of Jen
Leo 10h30'51" 17d57'22"
The Star of Jen
Sco 16h46'56" -30d22'22"
The Star of Jenna Leanne Prest
Umi 14h27'8" 66d4'23"
The Star of Jennifer
Cap 21h0'58" -19d3'32"
The Star of Jennifer
Cyg 19h32'25" 31d45'3"
The Star of Jeremy Atlan
Ant 9h30'56" -35d44'52"
The Star of Jess
And 1h15'46" 44d58'54"
The Star of Jessica Dee Randall
Crb 15h33'13" 38d43'21"
The Star of Jessica Joy
And 0h47'31" 45d20'7"
The Star of Joanne
And 1h49'3" 46d19'0"
The Star of Joey
Ari 2h21'1" 24d33'27"
The Star of John Luke Radford
Uma 9h8'33" 47d16'12"
The Star of John McCamant
Lyn 7h31'2" 49d49'39"
The Star of John W Walker
Ori 6h12'55" 15d22'1"
The Star of Johnny Tiseo
Vir 13h11'24" -5d16'46"
The Star of Josefina & Lucy
Cyg 21h10'16" 45d7'24"
The Star of Joseph Quinton Shuster
Sco 16h8'50" -16d41'23"
The Star of Joy
Lyn 8h27'12" 35d24'56"
The Star of Judy Marie
Cas 2h12'13" 65d22'55"
The Star of Jules
And 0h9'25" 29d16'56"
The Star of Julie
Tau 4h19'46" 12d43'32"
The Star of Julie Joiner
Ori 5h53'10" 6d41'47"
The Star of Justine
Leo 11h4'47" -0d2'35"
The Star of KAB
Umi 16h39'1" 76d57'34"
The Star of Karen
Vir 13h16'8" 12d44'17"
The Star of Karen
And 23h4'37" 36d43'42"
The Star Of Karen And Adrian
Cyg 20h5'12" 32d43'22"
The Star of Karissa
Leo 11h8'52" 20d16'1"
The Star of Kathy
Gem 7h23'21" 20d36'20"
The Star of Kelsey Legresley
And 1h14'44" 45d42'30"
The Star Of Ken & Elaine Peberdy
Cru 12h12'47" -61d16'41"
The Star of Kerrie
Cru 12h39'31" -57d55'32"
The Star of Kerry
Cnc 9h7'20" 26d5'6"
The Star of Kieron Docerty
Uma 9h42'27" 43d0'4"
The Star of Kirk
Her 16h55'46" 26d38'33"
The Star Of KNR
Uma 11h23'9" 64d33'0"
The Star of Kristie
Aqr 21h8'51" -8d54'27"
The Star of Laura
And 23h22'0" 43d28'25"
The Star Of Leo 1/03/60
Per 3h48'39" 31d46'7"
The Star of Leslie Ann
Ari 2h8'7" 12d10'35"
The Star of Liam Thomas Barber
Cru 12h38'1" -59d39'4"

The Star of Linda Bruner Pruce
Lyr 18h32'31" 36d15'28"
The Star of Lisa
Tau 4h23'27" 13d35'15"
The Star of Liz and Nathan, FRIENDS
Psc 0h32'14" 8d32'7"
The Star of Llanedeyrn
Per 4h9'0" 46d27'49"
The Star of London
Umi 14h34'54" 78d32'7"
The Star Of Lori Ann Harral-Stenson
Tau 5h53'8" 25d12'31"
The Star Of Love
Psc 1h26'0" 10d7'41"
The Star of Love
Cas 23h46'26" 50d26'6"
The star of Love, from up above
Cyg 20h40'47" 43d1'2"
The Star of Luck
Uma 14h10'23" 58d44'2"
The Star of Lukrsz
Uma 9h0'36" 55d0'25"
The Star of Mackenzie Pennington
Umi 15h23'51" 72d0'35"
The Star of Madeline
Uma 11h55'36" 41d26'58"
The Star Of Madonna
And 1h42'33" 43d18'5"
The Star of Magda - 06/11/1975
Sco 17h12'35" -30d28'44"
The Star of Margaret Ashton
And 23h2'10" 40d51'42"
The Star of Marilyn Vallejo
Sco 16h15'25" -14d43'17"
The Star of Markey
Ori 5h30'11" -5d34'13"
The Star of Marty
Tau 4h32'36" 16d56'13"
The Star of Mary and Michelle
Uma 13h46'15" 50d46'23"
The Star of Matt Pickles
Ori 5h9'55" 8d24'6"
The Star of Matthew & Amanda Hanson
Lyn 7h25'53" 59d29'27"
The Star of Maximilian David Burns
Cap 21h18'58" -24d40'15"
The Star of Mazin Osama Quotah
Gem 7h42'10" 34d22'58"
The Star of Melvin Morgan - Mel's Star
Dra 10h13'57" 77d44'45"
The Star of Michael
Eri 3h54'22" -0d56'14"
The Star of Michael
Per 2h50'36" 48d11'55"
The Star of Michael P. Poulin
Aur 5h44'51" 47d46'9"
The Star of Milli Ann Kirwan
And 23h58'25" 39d37'15"
The Star of Mormor
Cas 0h38'44" 54d28'47"
The Star of: Mr. I.B.13 Gunselman
Per 3h46'19" 50d41'43"
The Star of Mrs. Parrish
Cas 0h11'44" 53d40'9"
The Star Of My Daddy Dearest
Sco 16h13'51" -9d55'7"
The Star of My Sky
Ori 5h24'11" 2d51'25"
The Star of Nelson Martinez
Her 18h7'16" 42d9'47"
The Star Of Never Say Never
Uma 10h24'28" 60d25'46"
The Star Of Never Say Never
Uma 12h2'37" 64d22'36"
The Star of Nick and Sharon
Her 16h12'37" 46d53'8"
The Star of Nicole
Tau 4h54'1" 17d32'20"
The Star of Nicole & Joe deBuzna
Sco 17h37'22" -41d24'34"
The Star of Nicole Walsh
Lib 14h32'6" -16d55'34"
The Star of Noelle A. Blank
Cap 21h9'34" -19d2'10"
The Star of Opal
Col 5h24'12" -40d45'52"
The Star of Otto & Joanne Penlington
Cma 6h32'7" -31d32'33"
The Star of Paige Alexis Pruce
Her 18h57'37" 16d2'59"
The Star of Patricia
Crb 15h48'27" 35d58'45"

The Star of Pauline
Cru 11h58'49" -58d17'4"
the star of paulsy
Aqr 21h59'5" 1d44'28"
The Star of Peanut
Lmi 10h51'1" 26d59'24"
The Star of Peggy Hughes
Cas 1h26'58" 54d53'42"
The star of Pete
Per 3h51'15" 32d29'10"
The Star of Princess Suzzii
And 1h29'5" 34d40'21"
The Star of Princess Tase
And 1h10'21" 35d10'57"
The Star of Prudence Leigh Carey
Cru 12h17'37" -63d20'27"
The Star of Rebecca Joanne Hinds
Cap 21h54'1" -17d39'34"
The Star of Recovery
And 0h25'26" 41d56'19"
The Star of Rheana
And 23h22'4" 42d23'57"
The Star Of Rich Onorato
Cnc 8h46'17" 14d3'37"
The Star of Richard
Cep 20h56'1" 59d25'19"
The Star of Richard A. Vale
Lyn 7h30'15" 48d10'20"
The Star of Roanne
Gem 7h2'57" 20d38'10"
The star of Robert Colin Iles
Psa 22h2'48" -27d54'11"
The Star of Robin
Mon 7h32'44" -0d53'48"
The Star of Robin
Leo 10h11'54" 10d57'52"
The Star of Robin and Walter
Sco 16h19'56" -10d0'56"
The Star of Robinette
Ori 5h28'42" 3d49'13"
The Star of Roland Rosin
Her 16h26'38" 11d5'17"
The Star of Ron Britland
Cep 23h40'41" 77d13'41"
The Star of Ron Sursely and Friends
Umi 13h55'20" 75d35'28"
The Star of Rosemarie B. Jarocha
Umi 13h13'4" 75d22'3"
The Star of Rosemary
Uma 11h32'17" 53d47'9"
The Star of Roy Gladwell
Cep 22h54'54" 71d6'26"
The Star of Saint Clare
And 2h23'13" 38d30'17"
The Star of Salvo
Leo 9h46'36" 28d8'28"
The Star of Sam & Harry Lee
Cyg 20h44'59" 55d13'20"
The Star of Sam & Phil
Cyg 20h54'4" 41d0'48"
The Star Of Samantha Tappe
Uma 8h36'14" 46d34'48"
The Star of Samson - Our Hero
Per 4h10'57" 41d15'3"
The Star of Sarah & Devin's Song
Sco 17h53'36" -37d29'2"
The Star of Scott & Bridget Rubicz
Her 16h45'59" 34d32'5"
The Star of Scott Mauriello
Aql 19h23'55" 10d24'30"
The Star of S.D.C. "8-28-87"
Uma 9h7'33" 56d43'1"
The Star of Shannon
Umi 15h25'5" 70d50'4"
The Star of Sharon
Cnc 8h31'10" 28d47'46"
The Star of Sharon
And 23h56'32" 39d34'15"
The Star of Sharon (Natalie Morgan)
Umi 15h58'4" 74d49'55"
The Star of Sonja
Psc 23h34'23" 3d19'21"
The Star of STACEY LYNN MOSKAL
Ori 5h29'55" 2d29'7"
The Star of "Stanislav"
Per 3h11'52" 48d19'25"
The Star of Streetopia
Uma 11h32'39" 55d59'58"
The Star of Sunil & Leena Sheth
Cyg 19h59'45" 56d53'4"
The star of Susan
Sco 16h3'58" -25d59'52"
The Star of Susan - 3rd February 1953.
Aqr 21h7'6" -13d21'7"
The Star of Sweeney
Mon 6h49'12" 8d12'26"
The Star of Sweet Pants
Cam 4h8'0" 66d2'34"

The Star of T C Noji
Lib 15h12'39" -13d24'29"
The Star Of Tabitha
Uma 10h35'39" 62d14'37"
The Star of Tanya
Sco 17h47'8" -32d15'6"
The Star of the Grandfather
Tau 3h47'23" 26d39'54"
The Star of the Stars of my Life
Lyn 7h46'9" 52d54'10"
The Star of Themistoklis Loukas
Per 3h5'48" 47d21'0"
The Star of Ther
Ori 5h52'20" 18d35'26"
The Star of Thomas Joseph Hessman
Cap 20h15'22" -12d17'9"
The Star of Tiffany
Lyr 19h12'39" 26d56'11"
The Star of Tippit
Sco 17h44'54" -42d36'21"
The Star Of Trixter
Lmi 10h7'59" 28d51'4"
The Star of True Love
Cyg 19h37'43" 35d42'31"
The Star Of True Love I Love U Brad
Cyg 20h58'27" 48d15'7"
The Star of Twirly and Yo
Cyg 19h39'37" 33d45'5"
The Star of Ubub
Aqr 23h49'30" -4d2'8"
The Star of Veronica & Nona Oct 2004
Cas 0h12'16" 50d47'12"
The Star of Vivien Harvey
And 0h28'12" 38d4'26"
The Star of Wendy Lee Lowe
Vir 14h36'11" -0d12'29"
The Star of Wine & Beermo
Aqr 23h28'56" -7d49'19"
The Star of Wisdom Andrei G. Stoica
Ori 6h14'43" 21d26'58"
The Star of Yanez
Uma 10h41'57" 44d27'34"
The Star Of Yasemin
Gem 7h37'16" 34d55'7"
The Star of Yukiko Nakano
Vir 13h28'35" -1d5'55"
The Star of Zehranaz
Ari 1h57'47" 14d13'3"
The Star of Zeman
Pho 2h19'17" -41d17'43"
The Star Poogan
Leo 9h32'48" 27d56'57"
The Star Suzie
Lyr 18h45'59" 31d42'22"
The Star that is Carol O'Mahony.
And 23h32'18" 48d2'17"
The Star that Won
Ori 4h51'53" 3d34'31"
The Star Uncle Mark, We love you
Cma 7h9'47" -12d4'53"
The Star Uncle Mike,who we all like
Cma 7h13'45" -12d59'56"
THE STAREEH MOOBEAR STAR
Gem 6h32'56" 21d33'47"
The Steads
Per 4h46'47" 46d41'49"
The Steelers Angel
Uma 12h7'4" 59d8'23"
The Steeley Star
Aql 19h41'11" -0d4'24"
The Steers
Peg 0h3'13" 18d43'56"
The Stella Koal
Crb 15h31'19" 26d43'49"
The Stenger's Star
Per 4h19'46" 32d21'44"
The Steph Star
And 0h45'3" 41d44'38"
The Stephanie Roumeliotis Star
And 1h34'31" 47d52'16"
The Stephanie Star
And 23h36'41" 42d43'50"
The Stephen and May Rees Star
Cyg 21h11'53" 42d59'54"
The Stephen Fox
Vul 20h10'18" 23d37'19"
The Stephen Henchy Star
Per 4h45'35" 39d48'9"
The Stephen Higgins Star
Ori 5h36'48" -1d37'30"
The Stephens Star
Cap 20h53'28" -16d7'28"
The Steve Fi "Fhita Star"
Ori 6h15'30" 14d45'46"
The Steve Giumelli (Stevie G) Star
Cru 12h37'51" -59d11'10"
The Steve Hero Star
Per 4h44'53" 49d23'15"
The Steven Lee
Ari 2h23'16" 25d15'39"

The Steven Ray
Uma 9h51'40" 52d17'24"
The Stevenson, Tew and Pittman Star
Uma 9h47'59" 59d37'12"
The Stever's
Vir 13h8'8" 8d3'33"
The Stewart Family Star
Lyn 8h50'59" 43d47'14"
The Stewart Star
Uma 8h31'27" 65d2'13"
The Stinkerbell
Cnc 8h47'46" 26d6'21"
The Story of Us
Sgr 18h43'55" -32d43'14"
The Story of Us, KT and BG
Tau 3h38'22" 16d41'39"
The Stradley Star
Sge 19h48'49" 17d45'47"
The Strand Star
Cas 1h30'20" 61d34'11"
The Strands
Uma 9h30'54" 42d37'26"
The Strawberry Moment
Cyg 19h58'7" 40d44'17"
The Strefford Star
Cyg 19h31'23" 29d56'33"
The Strength Doctor
Her 17h44'7" 48d15'52"
The Strength of Mom
Lib 14h59'18" -16d33'39"
The Stuart Beardmore-Smith Star
Ser 15h37'53" 22d19'44"
The Stuart Gough Birthday Star
Uma 13h59'0" 50d22'55"
The Stunning Star Rita Ciraco
Lyn 7h49'10" 53d17'30"
The Subhash Baburam Mohan Star
Uma 9h44'4" 59d15'14"
The Sue & George 30th Birthday Star
Uma 8h36'22" 68d57'8"
The Sue Tarolla Star
Aqr 21h9'43" -13d53'11"
The Sullivan Group
Aql 20h14'5" -0d52'16"
The Summer Holly Star
And 1h21'49" 49d41'41"
The Summers' Family
Dra 16h59'44" 57d2'24"
The Sunshine Star
Cas 0h56'17" 63d40'58"
The Sunzeri's
Uma 9h36'18" 51d58'46"
The Supa Star
And 23h29'26" 37d6'34"
The Super G
Uma 9h23'32" 43d23'15"
The Super Shirley
Ori 5h35'50" 13d33'54"
The Super Star Kristina
And 1h31'48" 40d23'13"
The Supper (Super Star)
Cyg 20h27'21" 58d42'10"
The Survivor
Aql 20h7'35" 15d16'56"
The Susan & Adam Star
And 2h9'43" 38d11'29"
The Susan Graham Compass Star
Cir 14h49'9" -64d14'21"
The Susan Star
Aur 7h29'16" 40d52'3"
The Susanna Brooks Star
And 0h28'15" 32d26'3"
The Susie Lowe Star
Aqr 22h1'28" -15d54'14"
The Suzanne
Sco 16h8'58" -14d4'23"
The Suzette
Cyg 20h27'59" 31d44'7"
The Swaz
Tau 4h44'24" 27d25'18"
The Sweet Pea Candy Nelson Star
Aqr 20h52'12" -8d21'24"
The sweetest dream would never do
Cyg 21h8'27" 37d0'24"
The Sweetie Star
Lyr 18h25'54" 33d42'31"
The Swimm Family
Cas 23h21'59" 55d12'20"
The Swiss Miss
And 1h29'33" 47d3'12"
The Sword of Cailin
Ori 5h32'29" -2d18'55"
The Sybil Arnold Scorpio Star
Sco 17h54'58" -37d41'54"
The Sykora Family Star
Uma 8h44'27" 59d22'25"
The Symbol of our love
Aqr 21h35'3" -3d41'38"
The T & C Peterson Star of Devotion
Uma 11h25'17" 63d54'54"
"The Taila Gold Star."
Ori 5h43'42" -9d2'30"

The TAJ
Peg 22h12'45" 33d33'33"
The Tamara Lynne
Mon 6h48'57" -0d7'50"
The Tamiki Levena 1st Birthday Star
Cru 12h42'46" -64d5'2"
The Tammy Maria
Leo 10h7'35" 22d44'37"
The Tammy Weldon Star
Leo 9h24'46" 32d25'33"
The Tan Chun Tai
Psc 0h28'54" 13d34'25"
The Tan's Star
Lyn 6h52'55" 56d52'52"
The Tanya Lee
Crb 15h55'25" 27d2'21"
The TARA Borealis 40
Leo 11h20'43" 16d59'47"
The Tavendale Family Star
Uma 10h58'11" 72d39'0"
The Teagan Mae
And 1h8'35" 42d16'15"
The Teddy Star
Uma 9h34'23" 43d30'16"
The Tenjo Family Wishing Star
Sge 19h44'35" 18d19'57"
The Terry Bailey Superstar
Her 16h58'18" 36d2'53"
the Terry doll
And 0h27'19" 42d34'53"
The Tessa August
And 23h59'51" 41d58'28"
The Texan
Per 3h43'43" 49d45'30"
The Texas Holdem Sweetie
Cnc 8h19'32" 10d50'49"
The Thaeler Wedding Star
Tri 2h9'41" 34d38'20"
The Theodore Family Star
Psc 0h21'14" 18d28'31"
The Third Eye of Allan Armenta
Per 3h36'23" 43d54'28"
The Thirsty Camel - 2/19/05
Ori 5h55'53" 18d5'31"
The Thomas Star, 1st Yr Anniversary
Sge 20h3'16" 17d11'19"
The Thompsons (Roy, Leann, Preston)
Lyn 8h3'46" 39d41'14"
The Three Chicas
Tri 2h12'10" 33d13'38"
The Three Jewel Star
And 2h21'23" 49d14'1"
The Three Musketeers
Psc 0h27'3" 11d35'3"
The Three Peepers
Her 18h36'0" 21d23'43"
The Three Sisters
Cas 0h56'4" 66d58'51"
The Tiffany Star
And 0h42'5" 37d50'43"
The Tim Carnes Star 2005
Cru 12h37'43" -58d11'50"
The Tim Loves Annette Star
Cyg 20h27'24" 58d14'34"
The Timberwolf - Class of 2006
Lup 15h35'32" -38d45'52"
The Timmy & Ashley Love Star
Umi 15h5'50" 73d12'15"
The Timothy Mark Barker Star
Cru 12h8'20" -62d25'22"
The Timothy Thornton Star
Cru 12h38'56" -59d24'52"
The Timsar Star
Cru 12h54'29" -55d55'34"
The Tina L
Uma 10h1'46" 43d9'57"
The Tnrdoz Family's star
Dra 18h54'8" 58d19'14"
The Todd Family
Peg 22h44'5" 23d54'27"
The Tom & John Better Bedding Star
Lmi 9h33'46" 35d18'28"
The Tom & Lisa Eternal Light star
Lyr 18h47'58" 31d9'44"
The Tom Menichino Family Star
Aql 20h9'45" 15d31'49"
The Tom Reilly Star
Aur 5h47'19" 49d4'5"
The Toni Ruth Star
Uma 10h33'50" 55d38'20"
The Tool Box
Leo 9h37'51" 29d15'27"
The Toomey Stella Star
Ori 0h0'36" 13d15'29"
The Topher 1
Tau 4h26'42" 20d29'42"
The Toppings Star
Per 4h30'59" 48d52'5"
The TOTINATOR
Cyg 21h9'45" 45d56'33"
The Tower Plaza
Ari 2h8'58" 20d34'30"
The Train Home
Psc 1h19'49" 26d5'45"
The Traveler
Sco 16h48'20" -36d59'54"
The Trevor " Reg " Mudge Star
Vel 8h58'19" -43d52'29"
The Tri Roble Clan
Crb 16h22'5" 29d5'6"
The Tricia Star
Cap 20h35'56" -21d15'23"
The Trinity: Love Strength Wisdom
Pyx 9h26'31" -29d16'2"
The Trio
Ori 6h9'45" -0d7'55"
The Triple Siso Star
Her 16h43'51" 11d26'21"
The Trish and Bob Love Star
Cyg 20h56'35" 46d24'36"
The Tristan James
Tau 5h31'45" 24d40'5"
The Troll Foldy Roll
Uma 8h57'46" 51d56'9"
The Trott Family Star
Peg 22h34'37" 29d48'19"
The Troy Hyman Family Star
Leo 11h3'58" -4d37'12"
The truth... is (Jus) you could...
Ori 5h1'39" 10d55'37"
The Tun
Crt 11h24'47" -13d51'43"
The Turnip
Uma 11h0'10" 35d19'21"
The Turret
Cyg 19h46'6" 40d52'6"
The Turtledove Star
Aqr 20h52'4" 2d11'56"
The Twinkle in Joanie's Eye
Ori 6h18'40" 11d6'18"
The Twinkle in My Sky
Uma 11h26'12" 62d23'6"
The Twinkle In Our Eyes
Tau 4h39'18" 22d21'4"
The Twinkle in Soni's Eyes
Aqr 21h50'0" -1d45'2"
The Twinkle in Vanessa's Eyes
And 23h15'6" 52d24'30"
The Twinkle in Whitney's eyes
Cas 23h3'20" 56d52'47"
The Twinkle In Your Eye
Tau 3h53'5" 0d27'4"
The Twinkle of Savanah
Sgr 18h19'0" -32d24'29"
The Twinkling Eye of Adrik
Psc 1h25'30" 16d49'24"
The Twinkling Linzo
Dra 16h7'19" 67d39'21"
The Twinkling Smile
Cyg 22h1'5" 53d34'55"
The Two of Us
Lyr 19h23'55" 38d18'38"
The Two of Us - Bill and Kim
Umi 15h40'55" 85d22'20"
The UK
Her 18h12'6" 18d43'13"
The Ultimate 817
Leo 9h44'1" 28d49'52"
The Ultimate Dad Michael Meder
Sgr 18h4'56" -26d10'21"
The Underdog
Ori 5h37'22" -4d40'20"
The Undying Flame of Bronwyn Hewitt
Gem 7h19'0" 19d9'30"
The Unforgettable Catherine Lee
Cru 12h52'50" -60d58'48"
The Union of Angela and Steven
Vir 13h22'5" 12d59'5"
The Union of Jane and Joe Williams
Cas 0h34'41" 52d37'20"
The Union of Karen & Jerry Schude
Cyg 20h44'45" 43d54'28"
The Unselfless Angel Jean
Aqr 22h16'40" -17d5'14"
The Upton Family Star
Ori 5h27'6" 3d43'14"
The 'Us' of RCM III and AGB
Hya 9h20'40" 4d47'5"
The "US" Star
Cap 20h58'25" -21d11'37"
The Van Aalst Family
Dra 19h18'24" 66d51'3"
The Van Winkle
Psc 0h19'5" 9d22'46"
The VanReeth Star
And 1h39'37" 45d13'0"
The Vekony Family Star
Umi 15h49'33" 78d5'50"
The Venerable John Yanek
Cnc 9h20'1" 18d18'56"
The Verburg's Shining Star
Crb 15h20'11" 25d43'19"
The Verinia
Cnc 8h33'41" 25d36'21"
The Vernon Lewis Moore 91 Star
Pup 7h35'58" -27d16'33"
The Very Vivacious Claire Smith
Cas 1h16'27" 54d50'51"
The Vianna Star
Sgr 18h4'47" -28d32'2"
The Vicki Matthews Shining Star
Aqr 22h41'2" -8d7'43"
The Vicki Whalan Star
Col 6h5'28" -29d23'48"
The Vickie Lynn Star
And 23h17'38" 44d28'44"
The Vicky Star
And 0h8'34" 48d19'53"
The Victoria Katherine Guiding Star
Ori 6h9'36" 20d19'23"
The Vikki Turton Star
Cru 12h19'16" -57d17'18"
The Villecco Star To Watch Over You
Per 2h48'59" 53d43'8"
The Vince & Lucy Stuart Star
Lib 16h1'22" -15d51'44"
The Vincent D. Prater Galaxy
Gem 7h14'51" 23d33'50"
The Vincent John Docherty Star
Ori 6h13'16" 6d20'50"
The Vincent Star
Uma 11h55'11" 61d27'27"
The Vinny Vieni Star
Psc 0h33'33" 14d42'39"
The Violet Sims Star
Uma 12h16'59" 57d29'56"
The Virgina D & William T Adams
Cyg 20h7'41" 30d0'21"
The Virgina Rose
Cyg 21h43'18" 35d3'26"
The Virgy Louise
Cmi 8h4'54" -0d11'56"
The Visionary
Umi 15h9'12" 71d22'58"
The Viva Beach Star
Ori 5h40'46" 12d24'46"
The Vladdy Star
Her 17h4'2" 13d13'11"
The W. David Chambless Star
Leo 11h16'29" 0d28'23"
The Wadling Star
Cru 12h40'47" -56d18'1"
The Walawender's
Leo 11h12'12" 2d1'5"
The Wald Star
Mon 6h43'44" 8d48'59"
The Waldman Supernova
Aur 5h57'32" 29d29'33"
The Waldron Star
Uma 11h25'13" 52d40'59"
The Wallace Star
Lyr 19h7'40" 27d11'45"
The Wally Hogan Star
Her 18h8'54" 37d7'7"
The Walston Star
Tau 5h52'55" 14d26'10"
The Waltons
Cyg 19h55'42" 39d36'4"
The Wander Star
Her 18h1'24" 36d27'42"
The Warburtons
Uma 11h16'25" 58d32'51"
The Warren E. Buffett
Vir 13h50'22" -17d26'26"
The Warrior
Uma 10h15'15" 70d18'36"
The Watchtower Star
Uma 13h27'45" 58d54'35"
The Watchus Star; Eric,Dawn&Joey
And 0h42'14" 39d40'45"
The Water Store
Lyn 8h37'12" 38d17'37"
The Waters Family Star
Uma 9h55'4" 52d29'18"
The Way Home
Per 2h43'50" 53d35'24"
The Waycaster Family 4-25-1995
And 2h7'1" 42d56'13"
The Wayne Bennett Star
Cra 18h48'22" -40d21'23"
The Wayne Kamemoto Family Star
Aql 19h22'17" -10d11'55"
The WCG Star
Per 4h35'43" 37d31'59"
The Weave
Mon 6h46'50" 9d26'58"
The Webb Family Star
Ari 2h40'43" 17d26'48"
The Weber Family
Uma 10h39'54" 66d52'41"
The Weber Star
And 0h14'41" 29d45'28"
The Weborgs' Star of Friendship
Uma 10h42'23" 44d49'39"
The Wedding of Gina/Stephane Rossan
Lib 14h54'5" -6d7'4"
The Wedding Star
Sco 16h8'48" -15d39'11"
The Wedding Star
Cyg 20h24'30" 47d18'19"
The Wedding Star - Jenny And Eric
Uma 11h7'12" 59d3'50"
The Weeky Lord's
Lyn 7h40'55" 48d3'34"
The Weird Star
Ori 5h0'55" 9d56'44"
The Weissenborn Star
Psc 0h13'25" 12d4'34"
The Welch Family Star
Lyn 9h15'35" 43d22'1"
The Welu Family
Psc 1h37'56" 10d19'51"
The Wendel Henry Family, Patterson
Uma 12h3'22" 46d53'11"
The Wendlers
Cyg 21h33'18" 30d32'32"
The Weston Star
Per 4h45'35" 42d15'31"
The Wheeler Family
Lyn 7h35'51" 38d11'35"
The White Shadow - My Angel forever
Per 3h19'30" 43d31'49"
The Whitness Star
Sgr 19h8'50" -19d7'27"
The Whitney Love
Leo 10h15'16" 22d56'13"
The Whitney Patti
Ari 3h21'19" 20d17'11"
The WHRSPP
Cas 0h26'29" 53d2'49"
The Wigdor Family Star
Umi 13h32'5" 75d2'27"
The Wilbur
Psa 22h28'46" -25d55'14"
The Wilcoxson Fireball
Uma 8h34'53" 63d55'5"
The "Wild Bill" Evans Star
Psc 1h51'0" 7d10'9"
The Willey StaR
Ori 5h38'7" -0d23'8"
The Willi B.
Leo 10h26'30" 25d30'9"
The William C. Rock Star
Ari 3h12'16" 29d8'24"
The William David Bennett Star
Ori 5h7'17" 15d21'17"
The William Farrell Star
Per 4h30'15" 48d47'24"
"The William G. Donne Star"
Ori 5h30'3" 10d28'26"
The William Gaudreau, Sr. Star
Psc 1h22'40" 30d28'27"
The William Grey
Umi 15h30'55" 75d39'49"
The William James Burnett Star
Cep 22h0'21" 56d42'18"
The William P. Craighead Star
Lyn 7h32'22" 41d41'30"
The William Phillip Cote II Star
Lib 15h2'20" -21d39'35"
The William S. Mann Family Star
Sge 19h54'44" 19d3'53"
The Williams '04
Gem 7h18'22" 14d12'57"
The Williams Kids
Tri 2h13'0" 34d59'16"
The Wilson Campbell Star
Ori 6h0'16" -0d47'54"
The Wilson Family Star
Aql 18h49'8" -0d53'33"
The Wilton Star
Uma 13h3'26" 59d46'47"
The Wind Above Our Wings
Uma 11h10'45" 48d37'13"
"The Wine Merchant" Carol Sue Slatt
Crb 16h7'51" 26d53'52"
The Winifred
Cas 1h34'27" 64d26'51"
The Winifred Sell Star
Cas 1h12'25" 56d28'13"
The Winner Masuyama
Gem 7h12'32" 25d18'43"
The Winner's Name Here !!!!
Lyr 18h49'35" 35d10'8"
The Wise Dragon of Cole
Dra 19h32'1" 72d32'32"
The Wise Man
Per 3h12'13" 46d57'6"
The Wish of Laura...
Ari 3h21'43" 16d16'51"
The Wishing Star
Sge 20h1'45" 17d48'16"
The Wishing Star
Uma 8h33'2" 65d38'20"
The Wishing Star of Lilly Rose Saga
And 23h4'9" 51d48'51"
The Witcombe Star
Per 2h14'9" 51d43'59"
The Wizard Star
Tau 3h49'55" 27d44'51"
The Wizard's Princess
Dra 19h34'10" 64d59'26"
The Wolf's Den
Leo 11h53'44" 27d21'46"
" THE WOLIN STAR "
Tau 4h43'39" 6d34'52"
The Wonder of Mike
Per 4h34'48" 31d56'52"
The Wonderful and Beautiful Olivia
Gem 7h26'8" 32d3'24"
The Wondrous Ella
Lib 15h41'31" -11d48'33"
The Wondrous Kate
Leo 9h26'23" 14d14'58"
The Woobie Star
Cap 20h25'18" -22d35'30"
The Woodbury Hill Family Star
Uma 8h53'7" 57d16'41"
The Woodmann
Her 18h50'22" 21d35'3"
The World
Uma 12h54'36" 62d13'44"
The "World Famous" Deb Merdinger
Cnc 8h38'28" 23d52'49"
The Worlds Best Mom
Umi 13h23'22" 76d13'31"
The Worlds Greatest Mom, Cindy Dick
Leo 11h13'42" 15d20'30"
The Worlds Greatest Mommy
Lib 15h25'49" -8d38'37"
The Wright Star
Her 17h19'22" 33d41'46"
The Wyatt Lovebirds
Col 6h1'28" -33d59'6"
The Yering Star
Cyg 19h21'34" 52d17'47"
The Yerrakondreddygari's
Ara 17h10'35" -54d20'50"
The Yia Yia and Pop Pop Star
Crb 15h56'48" 39d6'41"
"The Young Family"
Mon 7h31'38" -0d32'34"
The Your Auction Family Star
Crb 16h20'43" 38d34'36"
The Youth!
Peg 22h50'49" 3d20'4"
The Yuens': Kui, Hung, Yan
Uma 10h47'25" 39d55'11"
The Yuens': Kui, Hung, Yan
Uma 11h41'23" 44d5'20"
The Yvonne Schär star
Dra 9h38'25" 74d50'50"
THE Z STAR
Psc 1h11'41" 22d3'37"
The "Z" (Zachary Ross Koon)
Aqr 23h46'22" -18d55'46"
The Zaccardo Family
Pho 0h40'2" -41d35'36"
The Zoey Star
Sco 16h3'42" -17d33'11"
Théa
Uma 9h7'40" 56d34'40"
Thea Camara
And 0h44'24" 32d41'9"
Thea Castanho
Gem 7h39'39" 26d41'13"
Thea Chesworth
Ori 5h18'40" 0d22'10"
Thea Fleury
And 23h45'44" 48d38'13"
Thea Grace Lowe
Vir 14h12'45" -17d36'8"
Thea Hanson
Cas 1h24'17" 61d24'51"
Thea Ibbs
Aqr 21h21'44" -7d26'34"
Thea Inez Elizabeth Miranda
Lib 15h49'45" -6d16'57"
Thea Jean LaVonne Marsh
Lib 14h30'8" -15d32'20"
Thea Luna
Cas 1h27'52" 66d47'31"
Thea Pence
Her 18h56'18" 24d8'12"
Théa Rose
Cas 1h30'13" 69d18'29"
Thea Rössler
Uma 9h56'16" 56d26'22"
Thea Sophia Hussein
And 0h46'53" 46d12'32"
TheAlison
Ari 3h6'50" 18d41'11"
Thearus
Lib 15h8'13" -6d55'23"
Theda L. Leggore
Sco 16h5'18" -10d12'39"
Theda's Butterfly
Peg 21h25'19" 17d18'34"
THEDENFORME
Psc 1h25'17" 29d1'14"
"Thee I Love" for Carl & Diane
And 0h44'37" 22d16'33"
TheeManzostar
Lac 22h45'25" 42d35'14"
Theen, Volker
Ori 6h8'21" 8d52'37"
Thegreatbarboo
And 2h17'18" 49d42'9"
Theiß, Björn
Ori 6h10'38" 9d7'7"
Theiss, Jürgen
Uma 9h35'57" 49d57'13"
Theisz Tamás
Gem 7h2'12" 20d28'24"
Thekla & Markos Bastl
Cyg 20h41'31" 37d26'49"
Thelda
Mon 6h33'53" 7d32'44"
Thelen, Manfred
Ori 6h13'34" 6d12'27"
Thelen-Cooper Christmas Star
Cas 1h59'38" 60d46'58"
Thelma
Aqr 22h31'48" 1d31'25"
Thelma
Lyn 8h46'58" 33d53'59"
Thelma and Bob, Forever - "60 yrs."
Psc 1h6'39" 27d12'12"
Thelma and Louise
Leo 11h48'35" 25d10'29"
Thelma Augusta Maria Knopp Kelly
Cas 23h42'30" 54d59'44"
Thelma Bohner
Dra 20h28'41" 73d4'6"
Thelma & Don Coleman 7/10/03
Lyr 18h59'29" 33d43'10"
Thelma Donnelly
Lyr 18h33'14" 28d57'20"
Thelma E. Bledsoe
And 2h5'14" 42d24'31"
Thelma Elliott
Cnc 8h45'0" 18d30'38"
Thelma F Diaz
Uma 8h26'52" 69d43'38"
Thelma Flock
Psa 22h13'24" -27d13'10"
Thelma Frazee
And 23h23'32" 40d17'59"
Thelma Germond Stephens Walters
Cas 1h46'45" 61d12'0"
Thelma Hansine Meidinger
Mon 6h39'14" 7d19'7"
Thelma Helen Marie (Hansen) Dowding
Sco 17h51'42" -36d56'8"
Thelma Hess
Aqr 20h42'17" 0d55'53"
Thelma I. Defer
Sgr 18h41'55" -35d49'58"
Thelma Ilene Wyman
Cnc 9h2'40" 8d32'18"
Thelma Jean Schooley nee Guyton
Crb 16h10'21" 37d0'15"
Thelma Jean Speck
Ori 6h6'53" -0d34'22"
Thelma Kathlyn Arends Eichman
Uma 13h38'4" 48d53'48"
Thelma Kingston
Cas 0h16'39" 63d29'49"
Thelma Lutman
Crb 15h27'37" 29d11'29"
Thelma M. McConchie MOM
Psc 0h36'7" 17d45'44"
Thelma Mae
Lib 15h3'8" -7d1'6"
Thelma Maitland Withers - Thel's Star
Peg 22h32'57" 29d36'7"
Thelma Marie's Angel Star
Lyr 18h32'18" 33d0'5"
Thelma McNally
Vul 20h17'33" 23d41'8"
Thelma Myers Endsley 93rd Birthday
Leo 9h48'15" 27d15'49"
Thelma "Nanny" Shelton
Ori 4h49'31" 10d44'14"
Thelma Orsi Courtas
Cas 1h12'47" 69d1'19"
Thelma Pearl Little Michael
Umi 16h20'4" 77d6'31"
Thelma Randall
Mon 6h20'57" -0d29'50"
Thelma Reeder
Cas 1h22'5" 57d19'13"
Thelma Rothman
Lyn 6h41'54" 56d51'41"
Thelma Scott (Mom)
Sco 16h51'14" -45d35'7"
Thelma Southcomb
Crb 16h22'17" 30d55'9"
Thelma Strong
And 23h10'22" 42d48'41"
Thelma Sue Adado
Vir 13h4'2" -7d57'16"
Thelma Sylvia Willits
Cnc 9h2'35" 15d23'25"
Thelma "Tim" Boudreaux
Aqr 22h51'0" -4d39'36"
Thelma Vergara
Lyn 7h23'0" 59d22'28"
Thelma Weatherford
Lib 15h27'17" -13d34'44"
Thelma's Lucky Star
Ari 2h34'54" 21d3'27"
theluckiestash
Peg 21h33'24" 11d12'2"
THEM Corp.'s Star of Sonic
Ori 6h3'14" 9d32'16"
TheMeetingPlace
Ori 5h12'32" -6d7'16"
Themi Vrantsis
Aqr 21h3'30" -12d59'32"
Themis Lianopoulos
Uma 8h23'10" 61d9'8"
Themla Cheneler
Sgr 18h54'53" -16d29'4"
Theo
Leo 11h10'43" 0d10'17"
Théo
Cnc 8h52'49" 19d28'0"
Théo 27-05-2006
Peg 21h23'7" 17d31'15"
Theo Alexander Gallagher
Umi 16h13'34" 83d53'36"
Theo & AngelFace Hudson
Cyg 19h28'16" 59d24'22"
Theo Anthony Halliday-Dorsi
Per 2h55'9" 46d54'48"
Théo Bombardieri
Cnc 8h8'8" 22d10'5"
Théo Burgener
Cas 23h32'1" 56d13'8"
Theo & Elenore Charalambous
Cru 12h18'56" -57d2'51"
Theo Heal
Cep 22h1'36" 56d42'13"
Theo James Williams
Cep 23h10'25" 84d51'29"
Theo Josef Godehardt
Uma 12h19'48" 57d30'1"
Theo Krebber
Uma 9h28'20" 51d11'14"
Theo Paik
Lyn 7h58'45" 37d13'46"
Theo Povondra
Aqr 21h52'41" -0d50'58"
Theo Ryan
Uma 11h36'17" 37d4'23"
Theo Schamerhorn
Per 3h15'0" 54d34'57"
Theo Schwarz
Her 17h18'39" 16d40'36"
Theo Valentino
Per 2h43'4" 50d52'43"
Theo van Dort
Uma 8h38'45" 51d20'12"
Theo Wysocki
Per 3h11'34" 55d52'6"
Theo Yachik
Per 3h44'50" 38d24'48"
Théo, Luca Ghilardi
Uma 8h22'21" 62d4'57"
Theobroma
Umi 16h53'34" 80d28'19"
Theoderic the Great
Per 4h2'57" 47d53'6"
Theodora
Leo 11h15'54" 5d43'21"
Theodora
Vir 14h22'48" -0d42'27"
THEODORA
Uma 14h17'15" 62d13'22"
Theodora
Sct 18h53'42" -12d24'1"
Theodora - Eternal Love - Matthew
Leo 9h25'37" 10d30'23"
Theodora Freiwald
Uma 12h36'34" 57d35'57"
Theodora Iris Williams
Uma 9h19'39" 43d30'47"
Theodora Leaontaridis
Tau 4h20'30" 24d17'23"
Theodora Mary Klein
Gem 7h30'53" 16d17'16"
Theodora McCaffray
Cyg 20h42'44" 54d24'8"
Theodora Muserlian
Cas 0h52'56" 63d15'0"
Theodora Ruth Sonnichsen
Sco 17h53'53" -43d10'13"
Theodora-Roxana Pfeffermann
Uma 13h57'49" 61d12'8"
Theodora's Love
Uma 9h46'20" 56d49'13"
Theodore
Leo 11h25'52" 16d55'23"
Theodore
And 23h3'46" 52d20'27"

Theodore A Myers
Leo 9h49'49" 23d32'53"
Theodore Alfred Attalla
Cnc 8h14'3" 9d34'30"
Theodore Anthony Hess
Uma 11h44'6" 29d52'23"
Theodore Beauparlant
Aqr 21h40'9" 2d42'5"
Theodore Bibza
Uma 11h58'21" 39d40'22"
Theodore C. Kingsbury
And 22h59'28" 47d21'16"
Theodore C Salanti
Uma 10h47'7" 62d39'25"
Theodore D. Keller
Cmi 7h33'9" 4d58'32"
Theodore Dobson Jr.
Per 3h30'12" 45d32'36"
Theodore E. Draus
Cep 21h35'17" 58d30'17"
Theodore E. Warywoda
Cma 6h32'29" -14d55'12"
Theodore Earl Moses, Jr.
Uma 9h17'41" 52d7'0"
Theodore Edward Ritter
Per 2h39'30" 43d19'31"
Theodore Elefter
Sco 17h56'56" -37d35'51"
Theodore Elias Ornstein
Sgr 19h20'7" -14d1'6"
Theodore & Erna Jackson
Leo 11h28'56" 8d40'54"
Theodore Fulton Stevens II
Umi 16h25'43" 82d55'52"
Theodore G Frick
Lyn 7h47'4" 42d11'42"
Theodore Galileo
Stephens-Kalmar
Umi 16h52'29" 79d0'12"
Theodore Garboden-in
God's Heaven
Lmi 10h55'57" 29d39'31"
Theodore Gene "Tex"
Frederiksen
Leo 10h13'22" 14d42'0"
Theodore George Wynn
Outram
Per 3h16'37" 42d40'19"
Theodore Gordon Berk
Umi 18h19'43" 88d34'34"
Theodore Harris
Sco 17h4'47" -37d15'5"
Theodore Hatch Edgerton
Gem 7h16'20" 24d53'44"
Theodore Hugo Hearns
Cap 21h31'12" -12d32'48"
Theodore J. Hill
Her 17h30'41" 37d46'3"
Theodore Jackson (Troy)
Cnc 8h9'1" 7d51'56"
Theodore James Addesa
Cap 20h17'2" -10d25'38"
Théodore Jetté
Umi 14h21'4" 66d52'50"
Theodore Joseph Brodsky
Umi 16h33'30" 78d56'7"
Theodore Joseph Green
Her 17h1'30" 30d11'11"
Theodore Joseph Hynes
Per 2h35'23" 56d48'51"
Theodore Just
Per 3h16'55" 41d57'32"
Theodore K. Johnson II
Gem 6h43'47" 18d32'11"
Theodore L. C. "Teek"
Dekker
Cru 12h48'24" -58d41'23"
Theodore Lawrence Hirsch
Aqr 22h1'13" -1d3'14"
Theodore Ley Andersen
Tau 5h52'45" 14d9'29"
Theodore Louis Curtner
Her 16h43'20" 48d2'4"
Theodore Ludovissie
Cnc 8h1'51" 16d36'11"
Theodore Michael
Tau 5h29'34" 27d49'35"
Theodore Michael Lialios
Her 18h51'54" 23d22'31"
Theodore Michael Lulis
6/17/1986
Aql 20h3'57" 0d46'20"
Theodore Pappas
Uma 9h13'1" 69d3'55"
Theodore Robert Schmidt
IV
Uma 11h47'48" 62d36'21"
Theodore Roosevelt
Whitehead
Boo 14h55'24" 25d13'1"
Theodore Rozoles
Armstrong Booras
Sco 16h59'48" -38d16'40"
Theodore Salkowski
Cam 4h18'42" 66d35'22"
Theodore Sanford
Sherman
Vul 19h30'7" 24d37'16"
Theodore Sanford
Sherman
Vul 19h42'23" 23d24'53"
Theodore Sanford
Sherman
Vul 19h23'16" 21d57'43"

Theodore "Sunshine"
Uma 10h35'7" 45d47'36"
Theodore "Teddy" Michael
Pinaud
Uma 8h27'55" 62d12'29"
Theodore Thomas Platz
Cap 20h30'31" -26d27'38"
Theodore Victor Mohacsi
Dra 18h14'53" 70d26'0"
Theodore Vincent Lemon
Per 3h20'40" 31d57'23"
Theodore Vincent Lemon
Her 17h48'45" 21d34'24"
Theodore Vinciguerra
Tau 3h57'51" 9d15'44"
Theodore W. Metzger
Sgr 19h1'0" -14d5'30"
Theodore W. Whitten
Ori 6h17'27" 11d6'41"
Theodore William
Margisson Ritter
Lib 15h35'16" -9d24'5"
Theodore Worthy
Demetriades
Ori 5h34'55" 9d7'25"
Theodoros Georgiadis
Uma 8h29'19" 61d52'57"
Theodoros Kyriakopoulos
Dra 16h12'31" 54d25'42"
Theodoros Stamatopoulos
Per 3h47'32" 47d23'58"
Theodorus Sandra
Toemion
Her 18h23'20" 12d31'32"
TheoFania Issari-Curado
Psc 22h58'45" 4d39'18"
Theo-Günter Rötjes
Uma 8h37'43" 54d8'34"
Theola
And 0h16'7" 26d42'17"
TheOne
Tau 4h33'27" 14d22'54"
Theonika
Cnc 8h55'33" 13d50'45"
THEOPHANIA.
Sco 16h40'30" -29d25'13"
Theophania Dyeus Hros
Psc 1h27'24" 26d35'27"
Théophile Fischer
Dra 18h35'17" 64d2'43"
Theopilus
Lib 15h25'5" -13d3'50"
Theopoula Kopsa
Uma 13h44'36" 48d3'53"
Thera E. Pearse
And 2h35'28" 45d40'48"
There and Back
Cru 12h25'52" -61d57'24"
There's Always Hope Gary
"Flash"
Aql 19h27'44" 3d41'11"
There's My Lucky Star!
Uma 13h51'58" 52d3'1"
Theresa
And 23h39'35" 39d5'41"
Theresa
And 0h47'22" 46d14'32"
Theresa
And 23h41'43" 43d45'17"
Theresa
Cnc 9h10'45" 32d23'7"
Theresa
Cyg 20h50'25" 35d6'56"
Theresa
And 0h37'23" 39d12'9"
Theresa
And 1h58'53" 39d11'45"
Theresa
Ari 2h59'16" 12d0'46"
Theresa
Tau 4h24'27" 17d53'55"
Theresa
Gem 6h29'0" 15d53'37"
Theresa
Cap 21h28'0" -17d32'25"
Theresa
Dra 10h16'54" 77d22'33"
Theresa
Cam 7h31'25" 63d31'47"
Theresa
Cam 6h42'38" 64d30'32"
Theresa
Lyn 6h32'42" 59d41'7"
Theresa
Cas 23h15'42" 56d13'19"
Theresa A.
Kazmaier,Musiceducator
Crb 16h10'30" 37d52'19"
Theresa A. Materano
"Tessie's Star"
Psc 1h20'33" 11d17'23"
Theresa A. Morales
Vir 12h58'42" 5d37'9"
Theresa Adhiambo
Catherine Obong
Gem 7h26'53" 34d29'29"
Theresa Amaral
Com 12h48'37" 28d28'2"
Theresa and Earl's Heart
Cyg 20h13'20" 48d45'49"
Theresa Ann Campbell
Lib 15h1'43" -14d20'11"
Theresa Ann Elizabeth
Phinazee
Psc 1h15'50" 6d59'52"

Theresa Ann Gonzales
Fournier
Cnc 9h2'50" 25d15'17"
Theresa Ann Knotts
Lyr 19h13'49" 26d15'12"
Theresa Ann Simmons
Cas 0h56'39" 54d18'37"
Theresa Ann Vavra
And 23h8'56" 41d42'12"
Theresa Ann Yodice
And 0h41'22" 40d17'42"
Theresa Anne Pullara
Gem 7h42'23" 15d47'31"
Theresa Bednarz
Her 16h32'35" 13d41'45"
Theresa Bennett
Sgr 20h16'58" -44d1'44"
Theresa Bray
Sco 16h10'55" -13d27'29"
Theresa Brazil
Lib 14h48'20" -20d7'49"
Theresa Bridget Kehoe
Cap 20h19'28" -11d39'26"
Theresa Brigitte
Cnc 7h58'34" 11d47'36"
Theresa Bueno
Sco 17h32'39" -44d24'47"
Theresa Burke
Sco 17h3'27" -33d30'16"
Theresa C. Romeo
Uma 10h17'54" 63d49'43"
Theresa Cantow August
25, 1937
Lyn 8h55'37" 34d59'11"
Theresa Carol Watassek
Sco 17h52'32" -37d37'39"
Theresa Chapman
Uma 9h36'5" 65d40'15"
Theresa Cilento
Vir 12h27'42" -3d46'41"
Theresa Clark
Psc 1h9'58" 19d23'31"
Theresa Conkle Grasssano
Uma 13h12'5" 52d20'32"
Theresa D. Comin &
Robert G. Comin
Lyn 8h39'47" 44d46'23"
Theresa Della McDougal
Cyg 19h54'30" 33d35'53"
Theresa & Dennis Palmer
Uma 11h23'42" 57d38'33"
Theresa Dutch
Sco 17h53'43" -35d54'16"
Theresa E. Giblin
Cas 0h37'18" 57d44'21"
Theresa E. Santilli
Cas 1h22'12" 52d12'32"
Theresa Eileen Markward
Gem 7h0'59" 15d30'17"
Theresa Eiler
Vir 14h4'52" -15d4'6"
Theresa Elizabeth Althea
Yodice
Uma 12h8'3" 52d30'37"
Theresa Ellen Rivers
Leo 11h1'28" 5d8'22"
Theresa Fairless
Gem 6h54'59" 23d36'19"
Theresa Fife
Crb 15h52'1" 37d44'53"
Theresa Flanagan
And 23h10'0" 37d23'59"
Theresa Foster
Sco 16h46'34" -31d42'25"
Theresa Francis McEvoy
Uma 10h33'6" 48d49'34"
Theresa G. Augustin
Gem 7h20'8" 17d34'59"
Theresa Gada
Sgr 18h3'2" -17d43'0"
Theresa Gail
Uma 8h14'4" 69d34'32"
Theresa Gail
And 1h11'39" 36d59'57"
Theresa & Garry
And 0h16'54" 26d56'38"
Theresa Ginger Smith
Ori 6h9'56" 5d53'40"
Theresa Hage
Lyn 7h54'33" 38d18'21"
Theresa Hayden
Ori 5h37'6" 3d5'28"
Theresa Hayes
Crb 16h9'34" 29d34'19"
Theresa Helena Sinacori
Cnc 7h58'23" 14d36'24"
Theresa Hemmer
And 1h9'30" 36d33'37"
Theresa Hoang Diem Ngo
Sco 17h43'46" -32d53'34"
Theresa - Inspiration
Ori 6h15'54" 10d54'22"
Theresa Jean Renfeldt
Aqr 23h0'46" -16d32'3"
Theresa Jessica Italiano
Cnc 8h20'24" 14d1'17"
Theresa Jewel Koziatek
And 0h59'49" 44d1'5"
Theresa Josephine
Skinzdier
Cas 1h23'22" 67d17'51"
Theresa K. Tran
Gem 7h8'5" 15d14'26"
Theresa & Kai
Cyg 20h0'12" 39d32'43"

Theresa Karrie
Cnc 8h38'5" 32d7'59"
Theresa Kastelan
Gem 6h45'42" 32d47'13"
Theresa Kelly
Cas 1h18'20" 57d22'55"
Theresa Krol-Lissauer
And 0h31'42" 37d30'1"
Theresa L Hirt
Uma 11h7'55" 42d26'14"
Theresa L. LaDolcetta
Per 4h15'12" 45d24'19"
Theresa L. Pendola
Uma 10h29'19" 59d58'36"
Theresa Ladocia Lemmond
Camp
Tau 4h59'45" 23d38'7"
Theresa LaPorta
And 0h1'43" 44d47'56"
Theresa LeJohn
Leo 10h19'4" 26d32'52"
Theresa Leonhart
Sgr 18h43'42" -24d58'40"
Theresa Louise Goodwin
Uma 9h46'11" 47d42'16"
Theresa Lynn (Bubbles)
Brown
Lyn 9h13'5" 34d58'38"
Theresa Lynn Norton
Uma 10h49'31" 71d37'4"
Theresa Lynn Toschlog
And 1h24'53" 35d21'40"
Theresa M. Darmenio
Lyr 19h19'17" 41d18'27"
Theresa M. Ewing
Crb 16h15'20" 37d48'59"
Theresa M. Grossman
Cas 0h0'54" 59d56'43"
Theresa M. Jones
Uma 8h45'51" 48d28'30"
Theresa M. Rinaldi
Ari 2h0'16" 17d59'5"
Theresa M. Schino
Cas 2h31'52" 63d38'16"
Theresa Malloy
Crb 16h15'57" 34d25'17"
Theresa & Manny Rivera
Cyg 21h17'37" 46d35'53"
Theresa Margaret Zeidler
Lib 15h46'27" -25d18'46"
Theresa Marie Allen
Cyg 20h20'46" 48d8'7"
Theresa Marie Dundon
Uma 8h51'26" 50d12'29"
Theresa Marie Garrett
And 0h41'6" 40d30'41"
Theresa Marie Harrison-
Wilson
Gem 7h5'8" 33d28'6"
Theresa Marie Marchioni
Cap 20h14'14" -24d22'32"
Theresa Marie Schindler
Sgr 18h47'29" -21d25'58"
Theresa Marie Simmons
Uma 8h53'34" 48d14'54"
Theresa Marie Wilson
Selkirk
Cnc 9h14'7" 28d0'23"
Theresa
Marie,Mrs.Carter,T.
Her 17h58'19" 26d46'7"
Theresa Mary
Aql 19h17'26" 7d21'49"
Theresa Mattijetz
Psc 0h25'27" 15d19'31"
Theresa Mc Lean
Lyr 18h45'55" 41d9'33"
Theresa McDonald
Ari 2h59'36" 18d17'24"
Theresa Medici & Family
Gem 6h50'19" 34d41'51"
Theresa Meinze
Cyg 21h46'30" 44d9'32"
Theresa Michel
Uma 11h37'59" 48d6'3"
Theresa Michele Norman
And 1h13'8" 41d58'43"
Theresa Morahan Simmons
And 2h33'20" 38d40'41"
Theresa Morgan
Cas 1h43'2" 64d48'50"
Theresa Morgenstern
Cnc 9h11'42" 10d17'52"
Theresa Mulholland
And 1h18'9" 34d14'54"
Theresa- My Angel
Ari 3h13'5" 20d59'41"
Theresa My Love
Lib 15h12'22" -16d38'12"
Theresa *My Queen*
Psc 0h36'4" 18d19'32"
Theresa Nana O'Day
Tau 4h34'35" 17d2'19"
Theresa Nichole
Crb 15h52'58" 33d34'7"
Theresa Nicole Coyle
Tau 4h38'51" 28d21'16"
Theresa Nocco
Mon 7h38'8" -0d56'24"
Theresa Onderko
Cas 1h55'42" 61d49'34"
Theresa P. Wilson
And 1h22'0" 44d40'30"
Theresa Panichi
Psc 1h18'10" 32d10'43"

Theresa Pastore
Ari 3h15'43" 29d7'27"
Theresa Peck (Precious)
And 2h36'44" 39d40'14"
Theresa Perez "My Love"
And 23h58'35" 39d35'55"
Theresa Perron Faucher
And 2h19'0" 46d24'2"
Theresa Pettit
Gem 7h5'36" 33d30'52"
Theresa Piccin
Sgr 18h27'32" -27d3'5"
Theresa Pierce
Lyn 8h46'48" 33d37'6"
Theresa R. Aikman
Aqr 22h37'2" -7d19'9"
Theresa Randall
Vir 13h14'50" 4d44'5"
Theresa Reichert
Aqr 22h8'25" -0d3'5"
Theresa "Rhea" Mennen
Psc 1h7'1" 12d33'48"
Theresa Ribeiro
Uma 13h40'56" 55d29'36"
Theresa Rivera-Schaub
Cnc 8h45'45" 6d51'51"
Theresa Rodela Brown
Leo 9h53'0" 32d49'53"
Theresa Romeo
Uma 10h18'30" 63d52'15"
Theresa Rose
Cas 0h33'47" 63d28'39"
Theresa Rose
Cyg 19h31'42" 32d13'4"
Theresa Rose
Cyg 21h51'16" 50d16'4"
Theresa Rose
Tau 5h44'41" 17d31'22"
Theresa Rose Moya
Tau 5h36'2" 26d1'27"
Theresa Rosenberg
Uma 11h36'36" 31d13'10"
Theresa Ryan
Ari 2h46'45" 27d28'43"
Theresa Sempepos
Gem 7h8'21" 34d17'28"
Theresa Sereda
Aql 20h5'58" -11d35'34"
Theresa Skladony
Cnc 9h2'49" 10d11'48"
Theresa Starling Alvarado
Ari 2h13'38" 24d15'6"
Theresa Stawski
Mon 7h54'6" -3d1'15"
Theresa Steele
Per 3h12'29" 31d28'7"
Theresa Stillwell
Aqr 21h10'48" -14d26'26"
Theresa Sue
Sgr 19h48'55" -22d21'34"
Theresa Tanner
Cas 1h20'32" 61d40'55"
Theresa the Beautiful
Umi 14h23'33" 75d0'36"
Theresa Tollefson
Lyn 7h23'58" 46d31'37"
Theresa "T.T." Elizabeth
Cyg 19h43'23" 28d21'56"
Theresa V. Montalbano
Cas 22h59'10" 53d33'47"
Theresa Varano
Dra 19h3'21" 60d24'46"
Theresa Veitenthal
Lmi 11h0'10" 28d27'30"
Theresa Verchio
Psc 1h19'32" 16d15'51"
Theresa Veronica Watson
Cas 1h15'43" 55d8'32"
Theresa Whaley Halford's
Star
And 2h5'57" 42d35'43"
THERESA ZAWADZKI
Cnc 9h6'13" 15d33'17"
Theresa Ziccardi
Lyn 8h20'31" 38d43'36"
Theresa's
Aql 19h42'10" 11d52'29"
Theresa's Shining LOVE
Ori 6h10'38" 8d37'24"
Therese (ACE of HEARTS)
Tau 3h38'53" 16d27'0"
Therese - AJ's Star
Teacher!
Cas 0h48'15" 56d56'36"
Therese Ann
Cyg 20h14'0" 50d29'53"
Therese Ann Leing
Lyn 7h50'17" 56d49'21"
Therese Bächler
Per 4h46'42" 45d37'35"
Therese Bright
Cyg 20h10'30" 40d37'15"
Therese Donno
Dra 17h14'53" 55d21'36"
Thérèse Dubé
Aur 5h15'6" 41d14'41"
Therese Francesca Gould
Clemente
Cnc 9h3'1" 28d18'29"
Therese Goop
Her 17h4'19" 14d32'4"
Therese Joan
Sgr 19h38'55" -13d40'31"
Therese L. Strauss
And 1h27'59" 48d56'35"

Therese Lacasse
Cas 0h41'29" 63d13'13"
Therese Lynne
Cas 23h1'32" 58d55'8"
Therese Marie G. Squires
Cas 1h29'48" 61d41'31"
Therese & Martin's Love
Star
Her 16h50'33" 33d52'12"
Therese Piñeda " Sign Chi
Do Seed"
Crb 15h54'28" 27d57'53"
Thérèse Shehan
Tau 3h29'30" 20d42'37"
Therese Shepetuk
And 1h29'19" 43d49'38"
Therese & Stephen
Cyg 20h43'44" 42d6'43"
Therese Tiongquico
Aqr 22h49'12" -7d39'5"
Therése & Walt Johnson
Cyg 21h47'23" 43d50'9"
Therese Wittwer
Her 17h3'25" 13d18'23"
Thérèse-Marie Delierre
Cap 20h20'37" -15d57'59"
Theresia Elizabeth Ventrice
Psc 1h23'11" 4d44'16"
Theresia Schaufelberger
Aur 6h33'51" 38d52'4"
Theresia W. Atkinson
Lyn 8h29'30" 38d47'15"
Theresus custodus ad
Willemus
Sco 17h57'29" -31d24'58"
Thériault - Ouellet
Vir 13h18'44" -22d15'16"
Therion Aethyr
Psc 1h38'13" 24d3'24"
Theron and Marilyn Pace
Psc 1h32'8" 16d1'56"
Thersa Ann Nora Perricone
Aqr 23h3'48" -11d13'34"
Thersa Marie
Psc 0h8'8" 3d42'32"
Thesaurus
Crb 15h53'13" 34d4'57"
Thespn
Umi 4h37'56" 89d41'25"
Thétis
Aur 4h51'5" 39d28'19"
TheVampireNeko <3
Cnc 8h22'41" 10d20'44"
Theziri
Cep 23h26'24" 79d43'50"
THFE
Aql 18h56'51" -0d33'53"
THG
Aqr 22h12'35" 0d55'26"
THI
And 23h12'43" 50d37'48"
Thiago Perin
Vir 13h10'6" -19d22'8"
Thia's Shiny Star
Sgr 18h5'29" -27d48'5"
Thibault
Uma 10h3'12" 63d58'38"
THIBAULT 54
Sge 19h55'35" 17d37'14"
Thibault LEMASSON
Aqr 22h1'38" -15d50'15"
Thibault Pierrain
Cnc 8h23'45" 14d20'9"
Thibaut Thommen
Equ 21h13'13" 8d34'16"
Thida
Tau 3h56'10" 24d53'9"
Thiel Stella
Umi 14h49'59" 67d57'4"
Thien Huynh~Peanut's
Smiling Light
Per 3h16'23" 50d15'26"
Thien Ngoc Do
Dra 17h9'58" 52d38'23"
Thienan Kiely Bui
Uma 9h29'57" 53d40'9"
Thiermann, Norbert
Ori 6h18'57" 19d15'23"
Thierry Baeriswyl, forever
Umi 17h9'1" 79d28'25"
Thierry Beaumont
Del 20h18'57" 9d0'26"
Thierry Broch
Uma 13h37'13" 48d9'43"
Thierry Couvignou
And 23h59'53" 36d20'3"
Thierry de Monaco
Cap 20h26'43" -11d52'44"
Thierry Debes
Uma 9h4'39" 56d12'28"
Thierry Derlan
Uma 10h3'47" 49d32'52"
Thierry Derrien
Gem 6h53'31" 25d3'55"
Thierry Desbleds - Forever
Per 4h50'41" 49d31'10"
Thierry et Séverine
Her 17h9'28" 17d28'0"
Thierry Guillaudeux
Per 4h11'37" 34d52'59"
Thierry N. Thompson
Tau 4h27'33" 21d10'3"
Thilie
Uma 12h9'39" 59d37'21"

Thin Air
Lib 15h31'0" -19d56'41"
Thinking About You Always
Tau 5h55'39" 28d24'12"
"Thinking of You"
Lib 15h24'25" -8d38'50"
Thinking of You
Ori 5h40'39" -3d5'38"
Thipparut Chutrakul
Her 17h9'17" 32d19'47"
Thireet Swe
Cas 0h54'49" 50d0'2"
Thirteen Knots
Tau 4h36'18" 28d49'17"
Thirty Eight Theresa
Helmer
Uma 11h26'52" 63d57'0"
ThirtyEight
Lyn 7h59'7" 51d17'11"
This
Cap 20h23'59" -22d6'53"
This I promise you...
Umi 16h58'19" 77d51'25"
This I Promise You
Tau 3h48'44" 28d28'9"
This Is For You Mom
Aqr 22h22'49" -6d23'8"
This Is Not A Star
Gem 7h45'18" 28d56'1"
This Is Only Our
Beginning...
Sgr 18h15'43" -23d26'9"
This is You and Me
Cyg 19h38'15" 53d34'20"
This is your friend Jay
Valento
Lyn 8h0'3" 33d51'39"
This Magic Moment
Eri 4h1'36" 0d3'30"
This One Is For You
Gem 6h47'11" 14d18'33"
This star sends light and
love to Michel
Ori 5h24'24" -0d11'45"
This Year's Love
Sge 20h3'15" 18d24'24"
This Year's Love
Lyr 19h5'5" 41d58'5"
This Year's Love
Cyg 20h18'37" 52d0'23"
THISISBOB
Cap 20h27'38" -23d3'1"
Thissssss MuuccccccH
Uma 11h28'5" 33d16'41"
ThiTa
Cyg 21h32'29" 33d58'15"
THIVIMOWEPA
Uma 11h12'5" 70d51'57"
Thoa Thi Kim Pham
Aqr 22h40'30" -4d29'58"
Thoai-An's Shining Girl
Scout Star
And 0h30'15" 45d12'25"
Thogar
Cru 12h51'27" -60d3'57"
Tholkes
Aql 19h4'8" -0d48'29"
Thom and Rod 9/3/2000
Cas 0h35'35" 50d44'51"
Thom Brady
Umi 16h47'37" 76d24'21"
Thom Lee 12-05-51
Aql 19h30'46" 3d42'35"
Thom Morse
Ori 4h48'11" 11d40'12"
Thom Patterson
Tau 5h7'55" 18d47'30"
Thom & Paul's Wedding
Star
Cyg 19h55'0" 31d45'37"
Thom & Robert
Lmi 10h8'37" 29d57'30"
Thom & Ximena
Cyg 20h53'50" 32d28'58"
Thomas
And 2h30'0" 43d3'57"
Thomas
Aur 6h30'29" 35d11'11"
Thomas
Her 18h13'18" 41d16'37"
Thomas
Per 3h9'17" 54d21'49"
Thomas
Her 18h36'15" 24d8'23"
Thomas
Cnc 8h28'57" 9d56'31"
Thomas
Ori 5h10'39" 11d47'51"
Thomas
Cet 2h32'8" 7d44'59"
Thomas
Cnc 8h15'15" 7d25'36"
Thomas
Ori 5h15'18" -4d28'57"
Thomas
Lib 14h48'54" -0d50'0"
Thomas
Cma 6h43'12" -13d1'59"
Thomas
Cap 20h49'16" -22d19'17"
Thomas
Lib 14h23'6" -10d43'6"
Thomas
Umi 14h39'9" 73d14'56"

Thomas
Uma 12h22'40" 56d6'19"
Thomas
Psc 1h51'21" 7d2'18"
Thomas
Col 5h39'33" -31d11'56"
Thomas & Andrea
Cep 22h52'16" 76d17'40"
Thomas A
Hya 9h10'52" 4d14'1"
Thomas A. Ahern
Ari 3h19'40" 23d35'20"
Thomas A. Basti
Leo 10h12'8" 11d25'19"
Thomas A. Dawes
Cyg 20h50'18" 34d14'22"
Thomas A DeLuca, The Train Man
Psc 0h27'28" 16d52'33"
Thomas A. Felton, Jr.
Umi 15h31'11" 70d51'26"
Thomas A. Lasek
Vir 12h46'3" 8d17'25"
Thomas A. Martin
Uma 10h32'56" 67d12'32"
Thomas A. Roberts
Aql 19h9'20" -11d16'49"
Thomas A. Turk
Ari 2h57'6" 20d7'18"
Thomas ACES (direct to family)
Pyx 9h2'29" -36d54'9"
Thomas Addison
Aqr 21h59'55" -17d8'5"
Thomas & Adelina: Eternity
Cyg 21h27'8" 52d36'8"
Thomas Aebi
Aur 5h57'13" 29d18'31"
Thomas Alan Bastian
Umi 15h36'29" 70d54'34"
Thomas Alan Marsh
Leo 11h56'24" 23d30'51"
Thomas Albert Lianza
Per 4h22'41" 43d25'10"
Thomas Albert Rives 10-24-2004
Psc 23h22'4" 4d32'39"
Thomas Albert Toolen
Her 18h41'10" 15d37'40"
Thomas Alexander DeMaria
Ari 3h23'35" 27d23'26"
Thomas Alexander Elliott
Aqr 22h7'40" -12d59'31"
Thomas Alexander Hirsch
Leo 11h30'34" 23d14'12"
Thomas Alexander Howard
Ori 5h57'51" 17d35'37"
Thomas Allen Brannon
Uma 8h43'38" 69d15'9"
Thomas Allen Dickson
Lyn 7h51'21" 52d58'25"
Thomas Allen Hughes
Cyg 19h51'52" 37d59'52"
Thomas Aloysius Callahan V
Boo 14h58'10" 50d23'0"
Thomas Alva Austin
Gem 8h3'17" 28d9'2"
Thomas & Amy Hastings
Cyg 20h17'40" 33d16'50"
Thomas and Anne Trasser
Crb 16h16'44" 31d51'6"
Thomas and Benjamin
Umi 13h9'7" 87d24'43"
Thomas and Carolyn Krueger 2-2-1963
Psc 0h23'43" 4d46'33"
Thomas and Catherine McRae
Cyg 21h17'11" 41d23'37"
Thomas and Emily
Sge 19h51'45" 18d43'14"
Thomas And Gabriella Nastasi
Her 16h39'53" 30d59'38"
Thomas and Ina Llewellyn
Cnc 8h52'30" 31d26'6"
Thomas and Joseph Benson
Per 4h42'59" 48d7'15"
Thomas and Lauren
Cyg 20h26'13" 45d0'21"
Thomas and Lynne McIntyre
Cyg 19h37'26" 30d21'48"
Thomas and Tiffany Forever
Cyg 20h7'36" 51d34'22"
Thomas and Tina Phillips
Gem 6h42'23" 23d46'50"
Thomas Andrew
Cnc 8h27'38" 13d53'57"
Thomas Andrew Jernigan
Gem 7h15'34" 33d12'5"
Thomas Andrew Reich, Jr.
Her 16h32'31" 36d14'2"
Thomas Andrew Windsor
Ori 5h54'43" 21d18'29"
Thomas Anthony
Tau 5h31'5" 25d46'59"
Thomas Anthony Hunt
Her 18h31'33" 14d21'13"
Thomas Anthony Spagnolo Jr.
Uma 11h14'30" 71d4'42"
Thomas Anthony Sylcox
Per 2h46'3" 50d13'16"
Thomas Archie Sanders Jr.
Ari 2h14'2" 24d11'45"
Thomas Arning
Ori 5h57'17" 18d36'50"
Thomas Arthur Brooke
Cep 0h17'16" 70d24'18"
THOMAS ARTHUR EDWIN HESKETH
Men 5h47'11" -79d19'48"
Thomas Arthur Lomonaco
Aur 5h36'3" 48d10'5"
Thomas Arthur Wessgar
Uma 10h25'25" 42d31'12"
Thomas Ashton O'Rourke
Her 17h36'38" 48d14'44"
Thomas Atkins
Cep 22h46'33" 72d34'31"
Thomas Austin Hulsey
Her 18h9'30" 15d38'21"
Thomas B. Anderson
Her 17h42'26" 36d33'34"
Thomas B. Black
Her 17h37'39" 20d58'44"
Thomas B. Crowe
Per 2h29'30" 55d9'15"
Thomas B. Doyle
Uma 8h47'31" 68d9'7"
Thomas B. Murdock
Cnc 8h39'33" 32d3'10"
Thomas B. Vincent
Her 17h26'52" 35d36'47"
Thomas Battersby-Harford - "Tom's Star"
Umi 16h49'16" 82d55'58"
Thomas & Beatrice Frey
Uma 11h24'24" 48d10'6"
Thomas Bendel Mika
Tau 3h37'44" 28d38'46"
Thomas Benjamin Smith
Her 17h28'10" 32d46'54"
Thomas Bernard Michel
Lib 15h20'2" -27d50'17"
Thomas Berry Dyal
Aql 19h54'1" -0d58'22"
Thomas Blanche
Cep 4h22'25" 83d30'15"
Thomas Bloemers
Uma 9h46'33" 72d34'20"
Thomas Bolter in our Hearts Forever
Ori 5h48'43" 5d54'1"
Thomas Bond
Uma 10h57'2" 66d55'8"
Thomas Bond Jr.'s Jirachi
Ori 6h15'30" 15d35'20"
Thomas Bone
Per 2h51'48" 45d2'56"
Thomas "Boo" Gibson
Aqr 22h42'37" 1d9'9"
Thomas Boo Ragan
Sgr 18h45'58" -27d15'36"
Thomas Brauneck
Uma 10h16'1" 60d25'58"
Thomas Brewer
Uma 11h0'2" 34d52'49"
Thomas Brian O'Neill
Sgr 18h29'48" -34d59'21"
Thomas Brice Chesnutt, Jr.
Her 17h39'10" 18d3'18"
Thomas Bridges
Lib 15h24'53" -13d29'21"
Thomas Brown
Leo 10h16'8" 10d18'57"
Thomas Bruce Shouse
Cap 21h3'57" -22d17'36"
Thomas Bryan Evans
Her 17h14'51" 26d1'8"
Thomas Bryant
Her 16h38'2" 15d19'14"
Thomas Bulls
Cyg 19h28'4" 28d15'50"
Thomas Burke
Cnc 8h3'19" 25d21'53"
Thomas Burns
Uma 8h24'37" 68d23'33"
Thomas C. Bleistein, Jr
Uma 8h49'3" 60d47'18"
Thomas C. Brick
Per 3h4'10" 46d45'22"
Thomas C. Kaufmann, Attorney at Law
Crb 15h52'11" 30d38'41"
Thomas C. Spiegel
Boo 14h45'23" 29d22'5"
Thomas C. Wright / T.C.
Vir 13h25'13" 12d53'23"
Thomas Callaghan
Ori 5h56'47" -2d44'47"
Thomas Caraccio
Per 3h45'5" 48d55'33"
Thomas Carmine Dominic Raimondi
Oph 17h39'45" -0d15'44"
Thomas Cartwright
Dor 5h0'41" -67d23'14"
Thomas Carvell
Cep 3h14'49" 81d33'52"
Thomas Cavanaugh
Uma 10h31'4" 51d9'49"
Thomas Cavuoto
Cep 22h36'42" 71d19'9"
Thomas Charles Hannon
Lib 15h38'39" -7d0'28"
Thomas Charles Warnemuende
Her 17h30'52" 30d48'5"
Thomas Cheney
Boo 14h33'55" 32d55'26"
Thomas Chorba
Tau 5h42'59" 20d10'39"
Thomas + Christel
Uma 10h40'55" 59d42'13"
Thomas Christian Sims
Cnc 8h11'51" 12d20'40"
Thomas Christopher
Her 18h43'41" 13d55'32"
Thomas Christopher
Sco 16h6'55" -10d26'40"
Thomas Christopher Carugati
Uma 9h39'13" 64d37'13"
Thomas Christopher Hughes
Her 16h13'51" 44d1'59"
Thomas Christopher Jackson
Lmi 9h48'52" 36d56'46"
Thomas Christopher Keppler
Gem 7h13'30" 20d55'31"
Thomas Christopher Souza
Vir 12h44'56" 5d6'20"
Thomas Clarence Harrell
Her 18h4'53" 27d54'8"
Thomas Clay Biddinger
Uma 11h31'50" 56d5'23"
Thomas Clayton McQuillian
Dra 20h3'45" 68d1'21"
Thomas Clement Loomis, Jr.
Vir 13h2'26" 11d41'10"
Thomas Clifton Drake
Leo 9h55'44" 32d2'55"
Thomas Cobey III
Umi 16h27'20" 76d53'44"
Thomas Coggins
Ari 2h15'51" 25d24'39"
Thomas Cole
Aql 19h56'17" -0d30'46"
Thomas Cole
Sgr 18h58'11" -29d12'40"
Thomas Colmond Martin
Uma 10h30'59" 48d21'24"
Thomas Connolly
Per 2h22'14" 57d19'20"
Thomas Cool
Lyn 7h37'30" 48d37'4"
Thomas Cosgrave
Uma 12h4'22" 29d21'48"
Thomas Cosgro
Cep 21h38'54" 57d58'34"
Thomas Craig Frisbie Jr.
Cep 23h26'52" 71d52'43"
Thomas Crider, watching from above
Psc 1h32'47" 22d43'34"
Thomas Cullen
Ori 5h21'25" 9d0'10"
Thomas D. Burgess
Cnc 8h51'23" 14d49'51"
Thomas D. Centofanti
Uma 12h33'10" 56d44'14"
Thomas D. Cozza Jr.
Sco 16h57'58" -44d45'11"
Thomas D. Gardner
Boo 15h4'53" 41d38'40"
Thomas D. Guilfoyle, Jr.
Uma 10h39'23" 42d17'56"
Thomas D. Mahaney Sr.
Uma 10h19'46" 46d11'57"
Thomas D. Teeple
Uma 13h43'23" 61d9'34"
Thomas D. Walters Jr.
Uma 11h21'6" 69d38'11"
Thomas Dale Vaught
Per 3h38'31" 40d5'33"
Thomas Daley Riley, IV
Ari 3h8'6" 22d13'44"
Thomas Daniel Jacques
Her 16h31'24" 23d3'43"
Thomas Daniel Sheaffer
Vir 14h34'42" 3d16'5"
Thomas David Corlin
Vir 12h18'12" 11d5'47"
Thomas David Derry
Her 17h31'14" 46d1'24"
Thomas Day Caughman "Tom Tom"
Uma 10h57'6" 60d48'29"
Thomas Dean Newton
Sgr 18h26'1" -32d5'20"
Thomas Decius
Uma 9h32'28" 48d14'30"
Thomas DeHays Home Star
Tau 5h38'13" 24d29'10"
Thomas Delores Jetter
Per 3h49'18" 52d17'33"
Thomas DeMartino
Dra 20h20'14" 64d15'29"
Thomas Dimling Morris
Lib 14h53'3" -3d42'30"
Thomas Donald Sedlak
Uma 8h57'17" 59d39'39"
Thomas Douglas Brice Morton
Her 17h57'12" 48d38'40"
Thomas Douglas Driggers, III
Vir 13h8'59" 13d0'33"
Thomas Duane Clemens
Ori 6h7'17" 14d11'13"
Thomas Ducey
Her 16h36'2" 35d8'28"
Thomas Dupre Hall "Troy"
Boo 14h44'32" 36d23'58"
Thomas Dylan James B.A. QTS
Psc 1h23'39" 2d57'40"
Thomas E. Brubaker
Her 17h25'17" 35d22'23"
Thomas E. Dalessandro
Gem 6h41'28" 16d2'51"
Thomas E. Danyluk
Per 3h12'25" 45d8'23"
Thomas E. Detrick
Cas 0h51'0" 48d45'15"
Thomas E. Earley
Her 17h23'4" 38d26'27"
Thomas E. Finnerty
Uma 11h49'59" 56d35'41"
Thomas E. Lantz, PhD.
Ari 2h48'56" 29d5'21"
Thomas E. Lester II
Boo 15h36'11" 42d7'42"
Thomas E. Mitchell III & Chad Bryan
Ori 5h36'35" 0d39'11"
Thomas E. Mullen
Uma 9h18'23" 69d48'35"
Thomas E. Poore
Sco 16h7'23" -12d28'38"
Thomas E Rausch
Vir 13h48'17" 6d9'38"
Thomas E Reeves"Our Adopted Father"
Sco 16h39'34" -39d34'58"
Thomas E. Reining
Per 3h14'47" 43d33'9"
Thomas E. Simone
Uma 9h17'38" 51d50'34"
Thomas E. Stronstad
Lyn 7h28'53" 49d48'42"
Thomas E. Sutherland
Sgr 18h48'15" -20d39'35"
Thomas E. Tripi
Aqr 22h40'13" -2d5'42"
Thomas E. Tucci
Ari 3h17'21" 27d5'37"
Thomas Eakle
Tau 4h29'19" 21d13'54"
Thomas Earl Lacey
Leo 10h47'39" 14d49'9"
Thomas Earl Ranalla
Aur 5h46'31" 54d38'52"
Thomas Eaton Henderson
Uma 13h21'50" 55d12'47"
Thomas Eddie
Lib 15h54'28" -6d42'30"
Thomas Edward Anderson Bubnowski
Cnc 8h30'35" 25d54'19"
Thomas Edward Baize
Her 17h22'8" 35d42'3"
Thomas Edward Cain, Jr.
Uma 13h24'36" 54d55'38"
Thomas Edward Cain, Sr.
Uma 13h45'40" 49d34'8"
Thomas Edward Crawford
Uma 8h56'54" 47d54'16"
Thomas Edward DeHart
Vir 13h43'42" -5d55'21"
Thomas Edward Doubleday
Cnc 9h4'30" 30d26'50"
Thomas Edward Driscoll, Jr.
Sco 17h24'10" -38d48'35"
Thomas Edward Giffard
Per 4h49'46" 42d30'35"
Thomas Edward Hicks
Tau 4h16'24" 4d23'31"
Thomas Edward Kalkstein
Per 3h17'48" 41d10'5"
Thomas Edward Kerber
Cnc 8h49'46" 14d18'33"
Thomas Edward Mahal
Uma 10h29'39" 46d47'8"
Thomas Edward Murray
Her 18h52'47" 24d8'35"
Thomas Edward Nelson
Uma 8h53'6" 69d39'16"
Thomas Edward O'Neill
Cap 21h53'38" -13d51'32"
Thomas Edward Pickney
Cnc 8h46'55" 18d34'46"
Thomas Edward Richter
Cma 7h2'57" -27d17'50"
Thomas Edward Scanlon
Leo 11h1'52" -5d41'47"
Thomas Edward Shepherd
Lib 14h43'28" -24d36'27"
Thomas Edward Tyrrell
Leo 10h25'48" 26d34'14"
Thomas Edward Vincent
Per 3h11'57" 37d15'14"
Thomas Erdmann
Ori 6h7'45" 18d16'22"
Thomas Erfurt
Ori 6h19'59" 7d40'15"
Thomas Erhardt
Ori 5h56'40" 7d22'39"
Thomas Ervin Lucas
Leo 10h58'28" 15d27'43"
Thomas Esquibel
Tau 4h46'43" 22d20'46"
Thomas et Anastasia
Lib 15h41'38" -23d15'18"
Thomas et son papa Thierry
And 23h56'26" 36d54'57"
Thomas Ethan Clarence Williams
Cen 12h38'36" -51d39'36"
Thomas Eugene Lamping
Uma 10h13'2" 52d6'8"
Thomas F Brown V & Cassanda G Brown
Sco 16h13'58" -36d18'15"
Thomas F. Rygiel
Uma 10h6'26" 57d26'5"
Thomas Ferguson
Aur 7h21'27" 41d14'20"
Thomas Finkin
Uma 10h30'5" 65d4'9"
Thomas Finn Shreiber
Cnc 8h53'32" 31d40'13"
Thomas Flemmig
Uma 11h39'17" 35d32'53"
Thomas Fletcher Bleick
Aql 19h30'50" 13d45'49"
Thomas Foley
Her 18h20'35" 16d8'23"
Thomas Foley
Umi 14h30'15" 75d10'14"
Thomas Fournier
Cep 3h48'14" 81d2'2"
Thomas Francis " Daddy " 10/31/03
Uma 9h14'52" 68d51'30"
Thomas Francis Quinn, III
Cap 20h21'42" -13d35'59"
Thomas Francis Saia
Uma 10h11'9" 56d33'14"
Thomas Francis Xavier O'Neill
Aql 20h37'0" -0d11'32"
Thomas Frank
Cap 20h38'26" -17d38'28"
Thomas Frank Foltz 4/24/1932
Lyn 8h35'59" 34d0'54"
Thomas Frank Innace
Vir 12h45'8" -8d23'5"
Thomas Franke
Cap 20h52'1" -17d44'35"
Thomas Franklin Herring, Jr.
Her 16h22'1" 7d12'14"
Thomas Franklin Lloyd, Jr.
Hya 9h22'59" -0d17'9"
Thomas Fredrick Allen ~ Tommy ~
Tau 4h14'20" 14d6'11"
Thomas Fredrick Clark, II
Cap 20h7'45" -16d47'13"
Thomas Fredrick Glover
Sco 16h56'0" -37d54'30"
Thomas Freese
Sco 17h36'41" -40d38'19"
Thomas Frost
Aur 5h33'14" 43d27'13"
Thomas G. Aye
Per 4h11'16" 50d57'18"
Thomas G. Burris (Beep)
Uma 8h57'44" 52d11'32"
Thomas G. Cauble
Cnc 8h39'32" 20d39'20"
Thomas G LaRusso
Uma 9h4'26" 53d18'41"
Thomas G. Sanders
Per 3h18'14" 44d7'8"
Thomas G. Watson
Aql 19h19'16" -0d15'20"
Thomas G Wright - The Chief's Star
Leo 11h53'8" 21d15'54"
Thomas Gabriel Castellanos
Sco 17h43'4" -42d57'59"
Thomas Geon Ortoleva
Mon 6h41'9" 4d8'42"
Thomas George
Cep 2h3'0" 80d42'49"
Thomas George Cooper
Lac 22h24'20" 48d17'24"
Thomas George Corneillie
Psc 1h59'51" 7d33'50"
Thomas George Damien Thorp
Her 17h30'3" 37d3'32"
Thomas George Henry
Per 4h28'44" 31d30'52"
Thomas George Ketcham
Tau 4h36'24" 20d43'27"
Thomas George Pendry
Aql 19h33'55" 11d36'37"
Thomas Georgiew
Uma 10h4'39" 41d40'45"
Thomas Gerard Brown
Uma 9h2'23" 51d22'18"
Thomas Gerard Niemer Jr
Ori 5h20'23" -0d43'26"
Thomas Gernay
Ari 3h20'39" 18d49'3"
Thomas Gibbs Rowe
Dra 15h54'16" 63d19'34"
Thomas Gitau
Her 18h8'1" 26d56'40"
Thomas Gith
Uma 11h21'22" 34d3'13"
Thomas Glenn Bridges
Uma 12h5'37" 65d7'23"
Thomas Glenn Caraway II
Cam 3h58'11" 65d28'51"
Thomas Graeme Luhm
Gem 7h33'3" 31d28'35"
Thomas Graham Stephenson
Uma 11h52'52" 35d56'17"
Thomas "Grandpa" H. Stoner Sr.
Cyg 20h39'42" 46d21'43"
Thomas "Grandpa Lew" Lewandowski
Ori 5h47'24" 11d56'9"
Thomas Grayson Stone
Her 17h38'11" 32d56'37"
Thomas Grebert
Ori 5h59'9" 7d23'48"
Thomas Greenberg
Ori 6h21'18" 10d26'6"
Thomas Gregory Gervais
Her 18h19'15" 25d46'13"
Thomas Gregory Mahon
Ori 5h16'53" -4d34'0"
Thomas Gregory Myer
Gem 6h48'27" 28d27'48"
Thomas Grütter
Her 18h17'0" 22d15'41"
Thomas Guy Distasio
Lib 15h13'16" -10d43'29"
Thomas Guy Jesse
Pho 0h34'21" -40d21'35"
Thomas H. Baker, Jr.
Cep 22h16'38" 61d2'20"
Thomas H. Cook
Cnc 8h37'39" 24d7'1"
Thomas H. Eckert
Her 17h51'3" 47d25'34"
Thomas H. Gaillard
Sgr 18h55'49" -28d5'2"
Thomas H Hertzog 143
Per 3h38'31" 48d41'43"
Thomas H. Hinterman
And 23h48'59" 45d54'14"
Thomas H. Littler
Vir 11h49'32" -4d14'51"
Thomas H. McGinnis
Per 4h17'22" 40d59'3"
Thomas H. Robinson Jr.
Tau 4h31'57" 28d51'13"
Thomas H. Shaw
And 2h21'17" 45d15'46"
Thomas H. Tyree, Jr.
Lep 5h24'39" -11d3'52"
Thomas H. Watkins
Cyg 20h36'54" 39d56'47"
Thomas H Weiss Family Star
Dra 17h47'17" 62d29'9"
Thomas Hale
Her 16h44'0" 32d52'24"
Thomas Hall
Per 4h48'53" 45d15'37"
Thomas Harold Lunham
Cep 20h43'39" 58d8'7"
Thomas Harry Broffman
Her 17h21'53" 14d21'3"
Thomas Harvey
Ori 6h20'6" 8d13'37"
Thomas Hauke
Uma 9h7'16" 59d50'17"
Thomas Hayes
Psc 1h56'33" 6d50'23"
Thomas Heffernan
Sco 17h7'19" -40d43'59"
Thomas Henderson
Psc 1h41'11" 5d11'33"
Thomas Henry
Cep 21h8'39" 66d7'34"
Thomas Henry
Umi 13h34'51" 86d57'7"
Thomas Henry Baker
Ori 6h11'47" 7d0'4"
Thomas Henry Berry - Grandpap
Psc 1h30'32" 22d14'27"
Thomas Henry James
Ori 6h22'59" 10d29'22"
Thomas Henry Mathews
Cep 21h41'4" 61d45'9"
Thomas Henry Pickels III
Psc 1h34'32" 23d58'4"
Thomas Herbick
Aql 19h51'17" -0d16'54"
Thomas Hoehne
Sgr 19h8'33" -15d19'10"
Thomas Holiday
Her 18h35'46" 18d53'57"
Thomas Holzer
Her 18h42'38" 21d6'22"
Thomas Hospidor
Gem 7h26'28" 25d0'55"
Thomas Howard Barnes
Sco 17h53'30" -32d59'10"
Thomas Hutchinson
Per 2h21'32" 55d2'36"
Thomas Hyclak
Ari 2h45'43" 30d24'23"
Thomas il Farfallino
Dra 20h12'35" 71d27'15"
Thomas Immanuel Hux
Her 17h59'12" 18d36'26"
Thomas Inman
Cnc 8h40'43" 13d37'19"
Thomas Irwedd Vaughan
Dra 19h38'54" 65d2'45"
Thomas J
Ari 2h59'16" 26d21'45"
Thomas J. Acee
Cep 21h44'18" 64d16'51"
Thomas J. Banford
Lib 14h53'12" -1d4'11"
Thomas J. Bartis
Uma 12h40'0" 56d40'50"
Thomas J. Boria
Sco 16h54'57" -32d44'51"
Thomas J. Boyle
Cep 22h3'46" 69d5'57"
Thomas J. Brennan
Aur 6h3'14" 37d30'45"
Thomas J. Butler 60th Birthday Star
Sco 16h14'17" -10d19'4"
Thomas J. Corcoran
Cyg 19h27'59" 53d12'40"
Thomas J. Discordia, Jr.
Lac 22h21'48" 37d32'24"
Thomas J. Ekkers, M.D.
Tau 4h33'3" 7d45'8"
Thomas J. Evans
Aqr 22h18'0" -15d43'44"
Thomas J. Frederick
Gem 7h27'28" 29d21'19"
Thomas J. Gillespie, Sr.
Uma 8h37'0" 66d59'38"
Thomas J. Harrald
Leo 10h44'11" 11d13'23"
Thomas J. Harrington
Psc 0h56'33" 8d33'33"
Thomas J. Kelly
Gem 6h33'50" 17d59'36"
Thomas J. Komray
Aur 5h31'6" 44d43'24"
Thomas J. Kuhn
Ori 5h48'52" 2d43'35"
Thomas J. Kulinski Jr.
Vir 13h17'33" 7d53'55"
Thomas J Langan Sr
Ori 5h25'52" 1d29'25"
Thomas J. Lawless
Aqr 21h24'21" -0d54'28"
Thomas J Maddox Jr
Cap 21h8'24" -26d21'30"
Thomas J. Matthews
Ari 2h7'28" 25d27'10"
Thomas J. McGoff
Her 17h44'48" 48d12'51"
Thomas J Mohr
Uma 11h24'18" 64d13'7"
Thomas J. Moore III
Psc 1h22'20" 22d29'45"
Thomas J Mowbray
Cnc 8h19'11" 27d26'24"
Thomas J Murray
Cep 23h53'47" 86d56'20"
Thomas J. Pappenfus
Uma 13h35'3" 54d46'28"
Thomas J. Schneider
Lib 15h17'43" -5d55'8"
Thomas J. Szczepanowski
Cep 23h59'49" 86d32'38"
Thomas J. (Tom) Burns
Uma 9h1'22" 47d32'56"
Thomas J. "Tom" Derichs
Psc 1h13'46" 3d1'29"
Thomas J. Torrington
Vir 13h11'20" 7d6'32"
Thomas J. Troy Jr.
Ari 2h15'37" 26d12'39"
Thomas J. & Virginia M. Collins
Cyg 21h24'20" 51d10'34"
Thomas Jack Mani
Aur 6h14'34" 29d27'25"
Thomas Jackson Fields
Aur 5h17'16" 37d52'56"
Thomas Jackson Jenkins
Her 18h43'39" 20d25'6"
Thomas Jake Naylor
Cep 22h46'47" 80d31'32"
Thomas Jakob - forever love Manuela
Dra 19h17'24" 65d48'54"
Thomas Jakub McManamen
Dra 16h56'1" 59d4'30"
Thomas James Blackwell III
Tri 1h53'28" 33d54'17"
Thomas James Chute
Lib 15h40'58" -28d33'37"
Thomas James Coyle
Cnc 9h13'21" 10d18'46"
Thomas James Egan
Cma 6h48'18" -20d51'0"
Thomas James Fahey
Cep 22h49'35" 66d42'14"
Thomas James George Stanford
Per 3h59'24" 37d5'19"
Thomas James Hubbs
Vir 13h2'33" -20d37'4"

Thomas James Kelly
Del 20h24'40" 9d40'27"
Thomas James Padova
Her 16h41'42" 20d50'21"
Thomas James Roffe Sr.
Cep 23h12'57" 76d44'55"
Thomas James Ryan
Uma 9h51'19" 58d30'3"
Thomas James Smith
Lib 14h39'39" -24d58'31"
Thomas James Stanley Dadd
Her 16h48'29" 9d17'56"
Thomas James West
Uma 10h12'4" 46d40'16"
Thomas James Wickum My Forever Love
Sgr 18h33'10" -26d46'49"
Thomas James Willmott
Peg 21h50'34" 26d30'47"
Thomas Jay Gunn
Cnc 8h43'42" 29d4'11"
Thomas Jay Smith
Sco 16h57'3" -38d52'44"
Thomas & Jeanie Strong Forever
Aql 18h58'31" 7d25'44"
Thomas & Jeanie Strong Forever
Cyg 21h24'4" 34d11'48"
Thomas Jeffery Woods "aka" Daddy
Leo 10h51'58" 6d7'28"
Thomas Jeffrey Anderson
Gem 7h5'6" 15d14'2"
Thomas Jerald Smith
Her 17h49'14" 28d2'31"
Thomas Jerome Cote
Lib 14h32'18" -18d26'1"
Thomas Jerusalem
Vir 13h28'43" 4d13'59"
Thomas Jesse Overbay
Her 17h10'49" 32d27'21"
Thomas Joachim Lutz
Ori 5h48'48" -5d31'54"
Thomas John Berg
Gem 7h18'51" 22d16'58"
Thomas John Burke
Per 3h4'36" 45d31'47"
Thomas John Clarke
Cnc 8h58'5" 13d42'53"
Thomas John Connell O'Shea
Umi 14h35'55" 75d9'22"
Thomas John Dresner
Uma 11h46'37" 49d33'5"
Thomas John Fitzpatrick
Lyn 8h31'18" 46d42'2"
Thomas John Flanigan
Cep 22h0'40" 73d49'25"
Thomas John Frankowski
Aql 18h46'4" 8d1'11"
Thomas John Lynn
Ori 5h45'11" 7d0'34"
Thomas John McInerney Jr.
Cnc 8h7'42" 19d41'1"
Thomas John McMeekin Jr.
Uma 10h39'31" 51d16'23"
Thomas John Nimphius
Cnc 8h28'3" 32d18'54"
Thomas John Parsler
And 23h10'30" 51d43'8"
Thomas John Schembari
Per 3h10'59" 52d18'12"
Thomas John Taverna
Aqr 23h1'2" -15d12'58"
Thomas John Whiteway
Lib 14h40'44" -12d30'1"
Thomas Jonathan Lawrence Russo
Sco 17h5'55" -37d28'48"
Thomas Josef Tohati, Jr.
Uma 10h27'34" 55d21'4"
Thomas Joseph
Cap 21h8'36" -20d45'13"
Thomas Joseph
Cnc 8h34'59" 8d20'6"
Thomas Joseph Bradley
Uma 11h23'5" 56d4'51"
Thomas Joseph Breault II
Per 4h5'56" 42d59'43"
Thomas Joseph Dahir Sr.
Per 4h11'19" 34d30'13"
Thomas Joseph Doody
Her 16h48'34" 35d50'0"
Thomas Joseph Holobyn
Uma 10h34'52" 70d14'39"
Thomas Joseph Kozaczka
Leo 11h11'38" 16d0'4"
Thomas Joseph Kuczynski
Lmi 10h14'57" 35d51'18"
Thomas Joseph Mascia
Boo 14h39'22" 34d58'0"
Thomas Joseph McLaughlin
Ori 5h29'21" 7d51'28"
Thomas Joseph "Niner Nine" Bowker
Lyn 8h56'42" 35d20'45"
Thomas Joseph Stackhouse
Aur 5h28'16" 46d21'44"
Thomas Joseph Stemkoski
Ari 3h3'52" 18d38'3"
Thomas Joseph Wren
Per 3h43'44" 38d56'15"
Thomas Joyce Manning
Umi 15h28'27" 72d16'39"
Thomas Julian Aldridge
Umi 16h23'0" 77d26'14"
Thomas & Julie
Cyg 21h33'43" 52d13'29"
Thomas Julius Rizzo
Tau 3h48'40" 23d17'58"
Thomas Jung
And 1h1'34" 41d44'7"
Thomas K. Groos
Ori 5h54'17" -0d1'5"
Thomas Kearns
Her 18h12'36" 28d15'48"
Thomas Kelly Williams
Cap 20h26'9" -9d57'38"
Thomas Kenny
Lib 14h57'5" -7d18'35"
Thomas Kent
Aql 19h26'32" 15d20'37"
Thomas Kent Koger
Psc 1h25'55" 32d5'37"
Thomas Kent & Virginia B. Wetherell
Aql 19h41'38" 4d45'35"
Thomas Kilger
Aqr 23h11'47" -4d10'18"
Thomas & Kimberly -MMIV
Her 18h50'46" 23d21'28"
Thomas Kirk
Her 17h49'1" 35d9'12"
Thomas Klas
Psc 1h24'12" 9d58'11"
Thomas Koehler
Uma 9h18'38" 70d14'47"
Thomas Koo
Dra 17h45'50" 61d40'2"
Thomas Koppel: Love&Music
Gem 6h37'39" 18d47'6"
Thomas & Kristin
Uma 9h39'56" 69d44'36"
Thomas L. Hamman
Psc 1h32'10" 18d44'3"
Thomas L. Luelling
Sgr 18h29'53" -16d9'15"
Thomas L. Stueber, Jr.
Aql 19h5'57" 15d1'29"
Thomas Lapp
Lmi 10h43'35" 35d16'53"
Thomas Larry Fett
Cnc 8h47'4" 19d51'24"
Thomas Larry Monroe
Uma 11h55'11" 28d50'29"
Thomas Lawrence Whitman
Her 16h46'10" 27d28'0"
Thomas Lee Carman
Lyn 9h4'9" 41d29'34"
Thomas Lee DeSautel
Cap 20h9'31" -25d13'24"
Thomas Lee Payne, Jr.
Cnc 8h46'20" 14d1'50"
Thomas Lee - Sir Galahad Star
Gem 7h43'20" 34d19'21"
Thomas Lee Thompson
And 0h53'56" 34d44'9"
Thomas Lehner
Lyr 19h20'55" 35d31'57"
Thomas Lenton's Star
Cep 0h2'19" 77d17'8"
Thomas Leon
Per 3h55'30" 34d56'14"
Thomas Leonard Augustin Alexandre
Uma 9h4'51" 68d25'2"
Thomas Leonardo Genovese "Tom Tom"
Psc 0h57'5" 25d58'20"
Thomas LeRoy Strader
Lyn 8h57'34" 33d28'48"
Thomas Lesley Graham
Cra 19h3'58" -39d31'54"
Thomas Leslie Heinecke
And 1h42'20" 45d42'7"
Thomas Levasseur (Love of My Life)
Lib 15h50'3" -11d19'19"
Thomas Lewis Morrison
Per 3h13'20" 52d23'58"
Thomas/Liz
Lyr 18h47'9" 34d6'23"
Thomas Louis Kroeger II
Ari 3h20'50" 27d34'50"
Thomas Lucas
Gem 7h28'44" 26d23'29"
Thomas Luigi Marchesini
Vir 13h23'26" 12d48'1"
Thomas Lynn Fortney
Sco 17h52'52" -38d51'10"
Thomas Lynwood Heimberger
And 1h45'29" 42d49'54"
Thomas Lyons
Her 16h32'59" 5d7'50"
Thomas M. Beckett, Jr.
Aql 19h52'26" 6d24'51"
Thomas M. Denzler
Uma 14h15'56" 60d47'49"
Thomas M. Dietrich
Uma 12h3'27" 62d17'39"
Thomas M. Evans
Ori 4h51'30" -0d47'44"
Thomas M. Gallagher
Aur 6h16'3" 31d43'38"
Thomas M. Heter II
Her 16h41'58" 26d47'6"
Thomas M. Hinkle
Uma 10h0'47" 57d14'38"
Thomas M. McNally
Her 16h10'56" 46d23'31"
Thomas M. Spiro
Per 4h34'19" 44d40'11"
Thomas M. Wasel
Uma 9h25'39" 68d42'43"
Thomas Madiara
Uma 9h28'47" 66d28'37"
Thomas Madiara
Sco 16h39'6" -36d1'16"
Thomas Makios
Uma 10h40'0" 63d4'16"
Thomas Manley
Per 4h22'36" 41d14'50"
Thomas Mantovi
Cep 20h52'51" 59d37'39"
Thomas Mark Tordel, Jr.
Uma 8h47'2" 56d41'5"
Thomas Marquis
Aur 6h16'48" 53d15'51"
Thomas Marshall Payn
Sco 17h58'29" -38d26'49"
Thomas Martin Evans
Cap 20h19'3" -26d52'55"
Thomas Martin Junior Lewins V
Uma 10h10'58" 45d31'12"
Thomas Mathew Indiano
Tau 5h33'6" 25d0'37"
Thomas Mathier
Cas 0h16'18" 65d25'30"
Thomas Matthew Sherwin
Umi 13h55'41" 70d0'22"
Thomas Max Shillito
Per 4h23'50" 35d41'3"
Thomas Max Tigani
Cru 12h50'54" -60d39'55"
Thomas Maximilian Gardner
Aqr 22h5'26" -2d24'53"
Thomas & Max's Christening Star
Per 2h41'33" 52d25'58"
Thomas McAndrews
Her 16h22'6" 6d5'16"
Thomas McAuliffe
Aql 19h13'45" 3d21'30"
Thomas McCoy
Uma 11h44'55" 41d6'16"
Thomas McMahon
Cnc 8h48'41" 31d57'22"
Thomas McNeil Callaghan
Pho 1h8'2" -42d36'16"
Thomas Michael
Lac 22h31'33" 39d14'21"
Thomas Michael
Cnc 8h58'39" 29d13'10"
Thomas Michael Anderson
Dra 16h15'9" 62d52'29"
Thomas Michael Christopher Brockway
Lib 15h6'37" -6d1'2"
Thomas Michael Costa
Leo 10h18'57" 26d18'28"
Thomas Michael Cushard IV
Sct 18h47'30" -11d47'53"
Thomas Michael Johnson Evans
Psc 1h14'31" 28d20'40"
Thomas Michael King
Her 18h49'47" 16d24'21"
Thomas Michael Kinney
Ori 6h8'49" 20d44'51"
Thomas Michael Lucas
Her 17h29'23" 18d0'1"
Thomas Michael McFall
Her 18h8'5" 36d28'33"
Thomas Michael Nicholson
Uma 10h22'22" 51d36'2"
Thomas Michael Pena
Lib 15h5'30" -11d47'25"
Thomas Michael Rogers
Gem 6h46'10" 18d25'58"
Thomas Michael Scott - Scott's Star
Ori 6h19'36" 14d11'52"
Thomas Michael Wagner
Sgr 18h20'43" -32d28'36"
Thomas Miele
Uma 10h17'22" 48d9'29"
Thomas Min-Yao
Sco 16h14'30" -14d46'18"
Thomas Misczuk
Leo 9h39'22" 29d22'27"
Thomas Monroe Jones
Aql 19h36'31" 10d21'52"
Thomas Montgomery Pollard
Cnc 8h39'15" 23d21'27"
Thomas Moreno
Boo 15h6'49" 33d50'43"
Thomas Morgan Terry
Dra 18h48'0" 72d2'24"
Thomas Müller
Uma 12h1'15" 47d46'29"
Thomas Mullins
Srp 18h15'30" -0d1'48"
Thomas Muran
Lmi 10h19'22" 37d21'7"
Thomas Murphy
Dra 17h55'53" 59d20'35"
Thomas Murray Bean Jr.
Leo 10h32'26" 8d9'25"
Thomas "My Won-Tom"
Cyg 19h52'6" 37d22'6"
Thomas N. Farrell
Uma 10h48'15" 41d44'4"
Thomas (Napó) Horvath, M.D.
Aql 20h10'38" 14d32'24"
Thomas Nathan Thompson
Ori 5h31'31" -0d33'13"
Thomas Neale Tweedie
Aur 5h16'26" 30d1'26"
Thomas Neitzel
Uma 11h22'2" 59d13'12"
Thomas Nelson Gorsline
Her 16h52'57" 34d25'57"
Thomas Nestor
Her 17h22'12" 37d47'8"
Thomas Neundlinger
Cas 23h21'30" 55d13'15"
Thomas Niccolò Rinaldi
Peg 23h10'49" 31d6'27"
Thomas Nowacki, Sr.
Cep 21h21'14" 67d42'32"
Thomas Nuzzi
And 0h40'51" 40d37'18"
Thomas Oliver Moore
Cru 12h18'45" -57d53'56"
Thomas Oliver Walker
Ari 3h20'48" 27d2'15"
Thomas Olson
Lyn 7h41'56" 39d15'2"
Thomas Orin Maine
Dra 19h10'13" 64d41'49"
Thomas Osborne Irwin
Her 17h49'53" 43d27'57"
Thomas Owen Bainbridge
Cep 22h2'34" 68d31'1"
Thomas Owen Herrick
Psc 23h52'38" 1d13'56"
Thomas P. Casey, Jr.
Uma 10h6'19" 55d56'57"
Thomas P. Janidas, III
Sgr 17h50'2" -26d10'54"
Thomas P. Jankovic, Jr.
Ori 5h4'50" 8d10'18"
Thomas P. Kelly III
Crb 16h2'57" 32d36'27"
Thomas P. Maggio
Ari 2h56'48" 13d56'21"
Thomas P Obade, MD
Cep 20h48'47" 59d13'21"
Thomas P. Ragukonis
Per 2h23'34" 52d25'16"
Thomas (P Squared) Lovestrand
Per 4h18'31" 31d12'30"
Thomas P Sweeney
Sgr 18h20'55" -27d6'17"
Thomas P. Vassallo
Umi 15h54'34" 83d8'59"
Thomas Pacey
Pho 1h46'22" -42d49'0"
Thomas Patrick Greeran
Vir 13h18'23" -19d54'26"
Thomas Patrick James Fetty
Del 20h53'32" 15d59'47"
Thomas Patrick Kiely
Lyr 18h58'12" 33d17'6"
Thomas Patrick Weinschenk
Sco 16h50'56" -38d34'25"
Thomas Paul
Lib 14h36'53" -8d43'49"
Thomas Paul Pajula- Super Dad
Ori 5h19'6" 7d34'10"
Thomas Paul Porter - My Godson
Umi 16h12'44" 72d24'28"
Thomas Paul Rotheroe
Gem 6h44'40" 14d27'27"
Thomas Paul Strunk
Cam 5h3'7" 57d52'58"
Thomas Paul Weismantel
Her 17h36'44" 47d6'38"
Thomas Paul Whelehan
Ori 6h22'28" 16d5'23"
Thomas Paul Zalucki
Vir 13h18'8" -10d24'17"
Thomas Paul Zillessen
Mon 6h50'43" 10d28'30"
Thomas Pena
Uma 11h6'17" 41d9'46"
Thomas Pentz
Lyn 6h25'3" 55d57'54"
Thomas Peter Gallagher III
Aql 19h24'54" -0d2'47"
Thomas Peter Wilson
Tau 3h45'27" 20d26'26"
Thomas Petriello
Her 18h54'24" 24d3'4"
Thomas Petschek
Per 3h25'36" 51d8'50"
Thomas Pires
Uma 10h53'21" 58d4'41"
Thomas Poignet
Sco 16h41'3" -38d6'26"
Thomas "Pop-Pop" Leonard
Cep 21h18'34" 61d12'45"
Thomas Portmann
Cas 0h39'54" 77d39'19"
Thomas Prettyman
Gem 6h33'18" 27d0'25"
Thomas Q
Per 3h19'52" 48d0'1"
Thomas Q. Slyter
Aql 19h32'1" 5d44'37"
Thomas R Carabine
Uma 9h50'49" 44d34'53"
Thomas R. D. Beals
Her 16h49'21" 33d27'22"
Thomas R. Freeze
Cep 1h54'21" 84d59'36"
Thomas R. Harlow "Love Of Wynona"
Aql 19h37'23" -1d13'58"
Thomas R Moleski
Leo 11h55'41" 21d42'2"
Thomas R. Potter father of Stephen
Uma 11h14'41" 44d51'58"
Thomas R. Stahr
Gem 6h45'44" 18d54'34"
Thomas R. Stezzi
Uma 9h27'12" 48d56'13"
Thomas R. Streicher Jr.
Per 3h47'44" 43d55'53"
Thomas R. Warriner
Psc 1h10'53" 5d25'19"
Thomas R. Watson
Ori 5h17'2" 0d0'45"
Thomas R. Weaver
Her 18h34'58" 15d22'46"
Thomas Radar Gryphon Canadiana
Umi 15h54'55" 74d15'35"
Thomas Ray Buckley
Uma 11h46'36" 38d38'25"
Thomas Ray Scott
Cyg 20h25'34" 37d14'10"
Thomas Ray Solyst
Uma 11h51'13" 52d0'3"
Thomas Raymond Gravel
Boo 14h46'25" 43d7'13"
Thomas Raymond Prim
Psc 1h17'9" 4d49'58"
Thomas Recine
Uma 11h58'11" 47d48'10"
Thomas Reed Tyler
Her 18h17'0" 22d24'47"
Thomas Reindersma
Her 17h24'30" 29d6'50"
Thomas & Reinelde Dobnig
Uma 14h20'49" 62d1'12"
Thomas Reto Ruch
Her 18h10'28" 26d11'39"
Thomas R.H. Haycocks
Tau 4h50'16" 26d59'48"
Thomas Richard
Pyx 9h3'43" -26d39'33"
Thomas Richard Crowdis
Uma 9h16'18" 71d58'12"
Thomas Richard Rodriguez
Her 17h54'11" 15d15'42"
Thomas Richard Taylor, Sr.
Sco 16h56'22" -42d23'58"
Thomas Richard West
Ori 6h1'8" 9d35'55"
Thomas Richartz
Uma 12h3'8" 47d22'27"
Thomas Riedel
Uma 9h55'33" 63d56'31"
Thomas Robert Alexander
Her 18h7'6" 27d21'8"
Thomas Robert Beckerleg
Sco 16h18'11" -11d16'35"
Thomas Robert DiLollo
Cyg 20h53'41" 41d21'12"
Thomas Robert Krieg
Ori 5h26'24" -8d51'2"
Thomas Robert Krupa (Cheyenne)
Per 4h23'17" 33d50'0"
Thomas Robert Raney
Ori 5h23'8" 4d28'18"
Thomas Robert Senff
Psc 1h19'4" 27d41'54"
Thomas Robert Taffe
Leo 11h49'11" 18d37'57"
Thomas Robert Towers
Gem 7h27'3" 17d55'20"
Thomas Robert William Plante
Umi 5h14'41" 88d29'40"
Thomas Robert Wright
Umi 15h29'23" 72d35'45"
Thomas Roberts Bradford
Umi 15h24'20" 76d45'30"
Thomas Roger Wilson
Aqr 20h58'51" -12d6'21"
Thomas Romania Costello 4/20/2001
Tau 5h27'23" 27d11'55"
Thomas Rosenzopf
Uma 8h58'43" 48d23'55"
Thomas Ross Barr
And 23h15'1" 42d40'30"
Thomas Ross Morgan
Uma 9h19'36" 46d27'17"
Thomas Ruhala
Per 3h29'30" 41d40'18"
Thomas Russell
Uma 11h0'30" 49d38'38"
Thomas Russell Bacon III
Uma 13h50'44" 51d10'19"
Thomas & Ruva Dixon's Star
Cru 12h25'54" -60d49'21"
Thomas S. Caycedo
Her 17h10'5" 32d29'58"
Thomas S. Fennell
Aql 19h0'15" -0d36'22"
Thomas S. Hamon
Aql 19h43'57" 12d2'36"
Thomas S. Spital
Ori 5h35'13" -0d32'58"
Thomas Saffer (Uncle Tommy)
Cas 1h27'44" 64d21'57"
Thomas Salamone
Per 3h19'55" 52d4'24"
Thomas & Sally Cykon
Gem 7h36'35" 25d11'54"
Thomas Samuel
Lyn 7h25'15" 49d57'10"
Thomas Samuel Head
Aur 5h18'30" 42d34'8"
Thomas Samuel Rooney
Lib 15h47'38" -6d23'45"
Thomas Samuel Zemanian
Leo 11h21'39" 10d11'42"
Thomas Schirow
Uma 8h40'16" 51d36'24"
Thomas Schläpfer
Cas 23h40'58" 56d8'5"
Thomas Schreck
Aur 5h35'46" 33d21'36"
Thomas Schulz
Uma 11h5'28" 58d45'43"
Thomas Scott Collins Jr.
Aur 5h44'18" 49d9'54"
Thomas Scott Lynch
Uma 10h56'20" 52d5'6"
Thomas Scott Mikus
Per 4h15'35" 46d45'22"
Thomas Scott Mikus
Uma 10h54'33" 40d20'47"
Thomas Scott Pullen
Vir 12h4'50" 11d24'7"
Thomas Seth Johnson
Aur 5h56'49" 42d0'21"
Thomas Shackleton Bright
Vir 13h58'8" -13d31'54"
Thomas Shane
Cru 12h30'11" -63d23'35"
Thomas Shane
Per 2h41'13" 54d55'28"
Thomas Shawn Hennessey Jr.
Uma 8h20'17" 65d44'37"
Thomas Shelby Newgent Sr.
Aql 18h58'7" 15d24'45"
Thomas Shock
Cep 22h25'58" 76d56'25"
Thomas Sidney
Uma 8h27'36" 71d2'55"
Thomas Sidor
Her 16h55'38" 31d55'19"
Thomas Simmons Fenner
Lyr 18h38'17" 39d42'1"
Thomas' smile
Leo 10h58'28" 8d37'43"
THOMAS SMITH
Cen 13h9'38" -50d2'55"
Thomas Soarès
Sgr 18h14'24" -34d15'59"
Thomas Sowell
Her 17h35'27" 36d23'44"
Thomas Spencer's Triangle
Her 17h29'57" 37d2'20"
Thomas Stanley
Gem 7h6'51" 15d51'41"
Thomas Stiles
Sgr 19h2'46" -35d36'28"
Thomas Swift
Her 17h5'4" 19d5'26"
Thomas Sydor
Uma 10h0'20" 57d38'5"
Thomas Tallerico
Sco 16h53'35" -42d26'15"
Thomas " TB" Barden
Sgr 18h14'15" -26d36'33"
Thomas Teegan
Crb 16h16'8" 35d26'33"
Thomas Terrestrial Tribute 100679
Uma 13h15'14" 55d37'0"
Thomas - Texas
Boo 14h42'22" 14d43'23"
Thomas "Theo Phania Secreta" Cosad
Per 3h23'22" 38d2'20"
Thomas "TJ" John Duignan
Uma 9h36'12" 44d51'6"
Thomas Tomlin
Mon 6h45'35" -9d56'44"
Thomas "Tommy" Schneider
Umi 15h34'41" 76d51'20"
"Thomas" Truly my best-friend!
Her 17h44'37" 43d17'5"
Thomas Tyner
And 0h19'52" 25d5'7"
Thomas Ullrich
Ori 5h37'9" -5d10'39"
Thomas und Marianne Klokow
Uma 13h3'15" 58d22'53"
Thomas V. Davis
Dra 18h51'25" 68d31'6"
Thomas V. Murphy Sr.
Lyn 7h29'45" 44d17'37"
Thomas Van Dyke
Uma 9h37'4" 50d7'56"
Thomas & Veronica's Angelic Haven
Ori 6h20'23" -1d7'37"
Thomas & Victoria Finklea
Cyg 19h40'11" 32d48'19"
Thomas Villanueva
Lib 15h20'30" -11d26'54"
Thomas Vincent Fury
Cyg 19h38'27" 34d0'56"
Thomas Vize
Ori 5h51'12" 2d40'41"
Thomas W. Bell
Her 17h39'50" 28d26'16"
Thomas W Coyle
Dra 15h3'37" 60d14'4"
Thomas W. Hall
Lib 14h51'0" -6d35'26"
Thomas W. Heck Jr.
Aqr 22h56'37" -9d47'57"
Thomas W. Hobin
Uma 8h44'49" 46d40'23"
Thomas W. Kephart
Ori 5h48'59" 1d34'54"
Thomas W. Kilcollins IV
Leo 10h18'17" 26d0'31"
Thomas W. Wilson
Aur 5h53'25" 45d48'26"
Thomas Wagner
Her 17h37'19" 21d21'2"
Thomas Wake
Cru 11h58'29" -58d5'49"
Thomas Walker Cowan
Aqr 21h25'9" -2d53'29"
Thomas Walker Fischl
Her 17h16'24" 35d50'17"
Thomas Wallace Burkacki
Dra 11h38'7" 73d34'44"
Thomas Ward Early
Uma 11h15'37" 33d28'37"
Thomas Warne 24/05/1925
Cep 22h23'6" 86d19'19"
Thomas Warren Baldocchi
Cep 0h26'3" 80d50'43"
Thomas Weatherspoon
Ori 4h59'18" 14d48'54"
Thomas Welch (My Shrek)
Cyg 21h8'7" 49d51'0"
Thomas Wellborn Sheppard
Boo 14h56'47" 52d34'51"
Thomas Wesely
Sgr 19h10'10" -14d42'3"
Thomas Westbrook Glass
Umi 13h54'28" 75d9'33"
Thomas Wigginton
Tau 3h37'47" 4d32'29"
Thomas William Broomall "Tommy Boy"
Umi 14h15'2" 77d14'12"
Thomas William Cerretani
Tau 4h19'10" 27d29'21"
Thomas William Chapman
Cyg 20h26'31" 47d29'22"
Thomas William Clark
Cnv 12h44'50" 40d27'18"
Thomas William Crull III
Her 17h22'35" 36d30'30"
Thomas William Dorman
Peg 21h23'17" 17d52'46"
Thomas William Fletcher
Aql 19h2'2" -10d2'40"
Thomas William Garrahy
Cap 20h36'31" -18d26'45"
Thomas William Gee
Del 20h51'33" 8d19'30"
Thomas William Hanson
Her 16h53'23" 36d22'22"
Thomas William Hilton
Per 4h12'3" 42d35'3"
Thomas William Powell
Her 18h6'38" 26d57'0"
Thomas William Rendl
Per 3h49'4" 42d24'36"
Thomas William Smith
Per 4h32'24" 43d33'22"
Thomas Wittig
Uma 9h19'51" 72d50'52"
Thomas Wm. Merritt
Boo 15h20'23" 36d1'9"
Thomas Xavier Post
Oph 17h50'59" 4d12'36"
Thomas Young
And 22h59'16" 40d17'30"
Thomas Young McPherson III
Lib 15h7'50" -6d54'38"
Thomas Z. Petrosky
Vir 12h16'35" 4d36'45"
Thomas Zapata
Lib 15h27'54" -26d58'53"
Thomas Zont
Ori 4h51'37" -0d31'31"

Thomas Zscherp
Uma 9h14'0" 57d44'29"
Thomasina
Del 20h46'11" 14d40'52"
Thomasina Maruca
Uma 9h5'23" 71d13'11"
Thomasine Louise Mecham
Cas 0h25'41" 61d40'46"
Thomasine "TAF" Furey
Umi 13h36'53" 77d47'6"
Thomas-James Nicholas Galt
Uma 11h30'16" 58d6'43"
ThomasJessikaHannah
Uma 10h39'18" 70d41'24"
Thomas-Olivier Blais
Psc 0h4'54" -0d0'5"
Thomas-Paul-Köberle
Uma 14h13'53" 60d38'24"
ThomasRobertLovesElaine AnnForever
Ori 6h7'50" 15d2'31"
Thomas's and Kristen's Love
Cnc 8h48'49" 13d27'17"
Thomas's Light
Vel 10h26'0" -42d53'48"
Thomassini (Thomas Hanson)
Umi 15h57'30" 76d8'0"
Thomas.W
Uma 10h35'59" 69d51'6"
Thomed's True Light
Per 3h45'9" 49d33'4"
Thomi20
Cap 21h2'20" -24d42'31"
Thommoward
Cam 4h1'20" 52d55'29"
thompson
And 0h38'19" 36d4'39"
Thompson and Son Forever
Ori 5h31'3" 10d17'43"
Thompson Elliott 35
Psc 0h35'22" 11d4'9"
Thompson Thompson
Dra 17h32'38" 56d55'6"
Thompson Trinity 1 Deanne Dorae
Uma 11h49'40" 57d59'17"
Thompson Trinity 2 Lindy Leanne
Uma 11h45'39" 56d52'14"
Thompson Trinity 3 Teri Lynn
Uma 11h39'17" 58d50'15"
Thompson's Family Christmas
Lyn 7h12'51" 54d12'16"
Thomry Henas
Gem 6h40'40" 22d30'52"
Thom's Eternal Love
Lmi 10h12'25" 34d43'20"
Thömseli
And 1h9'13" 47d16'26"
Thomson Star
Uma 9h27'22" 45d24'54"
Thomsters
And 2h10'18" 38d30'40"
Thomyell
Dra 9h27'57" 76d50'55"
Thop Fox
Her 17h47'53" 42d58'38"
THOR
Sco 17h53'49" -36d27'5"
Thor Eusner
Umi 16h23'35" 75d15'53"
Thor Leif Walker
Ori 5h43'59" 2d0'13"
Thor Marchio
Tau 5h27'4" 24d35'29"
Thor Michael Jones
Gem 6h48'21" 17d17'11"
Thor Webster
Uma 11h11'59" 61d58'3"
Thora L. Birch
Psc 0h11'44" 11d46'55"
Thore Kilian Johannes
Uma 9h59'24" 46d27'38"
Thoresen/Menage
Sco 17h47'59" -36d52'47"
Thoria
Ari 3h27'57" 20d26'0"
Thorleif's
Cep 22h55'37" 59d29'7"
Thormälen, Klaus
Ori 6h20'20" 10d22'17"
Thorn, Holger
Uma 14h2'56" 51d27'30"
Thorne
Ori 5h52'12" 21d29'49"
Thornton 419
Uma 10h27'46" 64d23'51"
Thornton Curry Jr. & Yvonne Skinner
Uma 10h33'28" 50d17'26"
Thornton Star
Cyg 20h39'16" 45d0'51"
Thorny Rose 11 Carpe Diem
Uma 14h0'57" 51d33'8"
Thorpy
Ari 2h14'53" 24d21'54"
Thor's Dream
Sgr 18h50'40" -16d40'40"
Thor's Star
Sgr 18h27'28" -17d6'39"
Thorsten
Uma 12h50'16" 53d18'45"
Thorsten - für ewig Dein
Uma 13h8'41" 59d0'21"
Thorsten Krian i.I.D. M.
Ori 6h9'45" -1d28'18"
Thorsten Rausch
Ori 6h18'15" 5d49'57"
Thorsten Schlingmann
Uma 10h0'11" 55d2'21"
Thorsten Zander
Uma 8h44'34" 67d41'3"
Thos
Psc 1h14'33" 10d30'18"
Thought 57
Sco 16h51'42" -34d45'39"
THP5215
Cyg 20h5'43" 38d11'19"
Three Boys
Uma 13h53'6" 55d40'42"
Three Crowns
Crb 16h1'24" 32d28'36"
Three Crowns Wave Five
Cep 2h4'22" 82d27'29"
Three Daughters' Star
Uma 11h34'39" 57d44'24"
THREE IS THE MAGIC NUMBER DL,AM,DM
Peg 22h53'17" 19d36'40"
Three Obscure Women
Tri 2h12'17" 33d59'44"
Three Sisters
Cas 1h21'56" 64d26'15"
Thresa Hull
Sgr 18h41'54" -35d48'1"
Thresa Richelle Smith
Ari 2h31'38" 22d2'46"
Threse Bonnan
Uma 8h9'46" 63d40'50"
Threse Marie Robinson
Sgr 18h51'0" -16d3'24"
Thrhureyez
Lyn 7h34'56" 43d21'49"
Through the Darkness
Uma 11h1'22" 45d53'52"
Through The Years, Phil & Judy
Sge 19h47'53" 18d40'39"
Through Tyrone's Eyes
Cap 20h22'39" -13d5'59"
Thu Le
Ari 3h19'31" 27d33'18"
Thu Ly
And 1h1'52" 38d32'54"
thu my
Lyr 18h46'33" 35d50'42"
Thuan Tammy Fadler
Aql 20h21'21" 0d26'14"
Thum, Tina
Uma 11h58'47" 61d30'37"
Thumbelina Linda
Lib 14h58'48" -10d17'37"
Thumbelina Valle
And 0h47'11" 45d52'33"
ThumbyBear
Cma 6h56'42" -30d48'35"
Thumper
Cas 1h38'18" 61d8'35"
Thumper
Cas 23h12'30" 55d43'40"
Thumper
Her 16h23'32" 11d10'30"
Thumper Droopy 637
Ori 5h35'35" 2d37'4"
Thumper's Forbidden Love
Lep 5h45'33" -12d8'44"
Thunder
Lyn 7h46'34" 57d13'59"
Thunder
Cap 20h56'58" -22d44'45"
Thunder and Rain
Tau 4h30'4" 22d11'44"
Thunder Cloud's Silver Lining
Cma 6h43'12" -16d17'7"
Thunderation X
Cep 22h31'41" 74d25'16"
Thunderbird
Aqr 21h29'47" 1d18'50"
Thundering Timber
Her 16h51'32" 17d9'29"
ThunderWear
Per 2h54'34" 44d38'5"
Thunker-Bond Forever
Lyr 18h54'58" 38d59'0"
Thuong Qua Di Linh K. Doan
Mon 6h45'29" -0d11'49"
Thurman
Pup 6h49'24" -43d45'51"
Thurman Williams
Uma 10h41'43" 47d42'51"
Thurmond Swaim, Jr.
Cyg 21h51'52" 52d32'35"
Thursday
Umi 16h58'9" 86d42'12"
Thu-Thuy Phan
Lib 14h32'16" -10d41'23"
Thutmose
Cap 20h20'21" -19d46'34"
Thutrang "Bleu" Nguyen
Aqr 20h39'24" 1d12'53"
Thuy Kieu
Aqr 23h27'2" -19d13'37"
Thuy Linh Nguyen
Tau 3h30'32" 11d50'23"
Thuy Shaw
Sgr 18h29'35" -34d11'16"
Thuy T Le
Uma 8h43'20" 46d42'1"
Thuy Tien
And 0h18'0" 36d56'55"
Thuy Trâm
Uma 11h33'42" 38d43'49"
Thuy-Anh Ha Stovall
Ari 2h22'50" 14d3'24"
Thuyet Duc Tran
Uma 11h45'8" 30d32'12"
THX A LOT!
Ori 5h47'50" 6d23'54"
Thyagi - "Papadus"
Cep 21h24'15" 66d17'24"
Thyra Amy Bielfeldt
Sgr 17h56'22" -17d0'28"
Thyra Joy Pedersen
Ari 2h22'14" 10d47'14"
Thyré
Gem 6h39'26" 27d7'7"
TI
Umi 15h3'14" 67d53'11"
Ti Amo
Ori 5h42'54" -4d53'8"
TI AMO
Her 18h5'39" 18d5'45"
Ti amo.
Uma 11h36'37" 31d16'56"
Ti Amo
And 1h13'2" 46d4'42"
Ti Amo Anyway
Cyg 19h57'27" 30d11'14"
ti amo - bella venusta principessa
Ori 6h10'48" 10d44'1"
Ti Amo Carmen
Her 16h8'10" 49d48'56"
Ti amo James
Cyg 21h22'47" 33d50'52"
"Ti Amo" ...love conquers all...
Cyg 19h35'3" 33d52'49"
Ti Amo Madre
Ari 2h48'42" 21d26'50"
ti amo tanto amore
Del 20h44'33" 6d57'54"
ti coco
Ori 5h41'22" -4d36'23"
Ti Marmotte
Uma 9h25'8" 48d27'18"
Ti Penso Sempre
Her 17h47'40" 37d53'54"
Ti Toite
Per 4h46'19" 48d52'11"
Ti Voglio Bene
Sco 17h47'34" -35d7'32"
Tia
Lib 15h36'56" -21d32'48"
TIA
Lyn 7h27'23" 58d5'50"
Tia
And 23h7'1" 50d45'44"
Tia
And 1h59'10" 39d17'28"
Tia
Leo 11h19'9" 24d8'17"
Tia May 12th, 1988
Cmi 7h37'33" 5d21'52"
Tia Angelique Princess Angel
Cru 12h42'34" -58d11'10"
Tia Davies
And 23h29'35" 41d28'45"
Tia Demuth
Umi 13h32'51" 73d26'39"
Tia Golden
Cnc 8h15'15" 12d47'25"
Tia Hall
Pho 1h6'36" -45d19'10"
Tia J. Brown - Happy First Birthday
Umi 15h2'43" 70d59'19"
Tia ja Tuomas Hirvonen
Uma 9h34'0" 57d6'23"
Tia Manasseh
Gem 6h50'3" 21d26'6"
Tia Marie
Aqr 21h44'38" -6d12'17"
Tia Marie Kelly
Tau 5h58'49" 23d32'58"
Tia Marie Malcom
Cnc 8h51'26" 26d31'30"
Tia - Nana's Shining Star
Leo 9h30'57" 10d33'2"
Tia Patton
Psc 0h57'23" 10d51'13"
Tia Shea
Vul 19h15'7" 25d35'39"
Tia, Glenn & Harry Vivash
Umi 16h18'27" 81d46'46"
Tiago Fernandes
Crb 16h7'9" 35d55'41"
Tiago & Mélanie
Lyr 18h46'4" 33d9'20"
Tiago Monteiro
Leo 9h24'48" 10d5'2"
Tiago Ricardo Hernandez
Boo 15h4'13" 32d40'4"
Tiahna Angel Gopaul
Lyr 19h16'11" 28d48'6"
Tia-Liana143
Umi 15h8'59" 71d40'28"
Tiamo
Aqr 23h43'57" -4d53'53"
Tian Huey
Sgr 19h53'15" -28d54'33"
Tiana
Cas 22h57'39" 57d48'32"
TIANA
And 0h52'47" 38d20'25"
Tiana Alexis
And 0h20'10" 25d8'15"
Tiana Alia
And 23h23'21" 51d16'39"
Tiana (Bumpkin) Newhouse
Vir 13h9'48" 7d59'51"
Tiana J. Darder
Ori 4h49'22" 11d44'18"
Tiana Laila Cannuli
Cra 18h55'10" -39d31'25"
Tiana Lucy Australis 21
Leo 11h15'31" -3d27'28"
Tiana Lyles
Vir 13h14'48" 12d17'50"
Tiana Lynn
Uma 8h24'0" 64d21'59"
Tiana Mariah
And 2h13'19" 45d34'33"
Tiana Over
Tau 5h0'12" 19d16'30"
Tiana Rose Frketic 10 November 2004
Cru 12h37'55" -64d16'29"
Tianah Ku'u mele ho'o heno Balberde
Ari 3h3'25" 18d18'47"
Tiana's Light
And 0h47'20" 36d37'57"
Tiana's Star
Vir 13h40'7" 3d37'5"
Tiane Olortegui
Eri 4h31'12" -25d34'20"
Tianeu's Boon
Sco 16h2'50" -21d43'54"
Tiani Solei Salgado
Sgr 19h11'1" -19d7'24"
Tiann Boogie Young
Mon 7h30'32" -0d49'42"
Tianna
And 23h25'6" 39d11'10"
Tianna Dawn Laws
Cnc 8h49'55" 21d8'17"
Tianna M J Ortiz 02/16/1995
Aqr 22h51'11" -6d0'16"
Tianna M. Williams
Aqr 22h6'2" -1d44'17"
Tianna Marie
Leo 11h39'41" 25d22'37"
TiAnna Marie James DiMartino
And 1h18'38" 41d33'27"
Tiannas (Goobies) Peace
Cap 20h52'43" -26d38'54"
Tianni Fredette
Cir 14h38'14" -64d41'48"
Tianyi
Cas 0h12'40" 53d10'14"
Tiara and John 4ever
Gem 7h9'49" 23d57'2"
Tiara Jenel
And 1h11'16" 35d56'32"
Tiara Lee
Aqr 23h32'49" -12d47'50"
Tiara Lynn Schöni
Cyg 21h3'20" 47d51'19"
Tiara Lynne Dye
Cap 21h57'44" -18d27'39"
Tiara Rose Tekla
And 0h49'20" 44d27'22"
Tiara's Star
Tau 5h6'58" 26d17'2"
Tiarne Lee Williamson
Tau 4h35'32" 3d22'58"
Tiarra
Cnc 8h4'26" 6d59'47"
Tiarra Alexander
Dra 9h28'15" 76d10'48"
Tiavonnie
Sco 16h14'35" -11d57'13"
Tiba Parsadoost
Cap 20h20'41" -11d36'30"
Tibby
Aql 19h36'1" 5d4'24"
Tiberius
Uma 11h57'57" 31d29'32"
Tiberius
Uma 11h50'32" 61d56'18"
Tiberius George Wood
Ori 6h3'44" 12d37'53"
Tibi csillaga, Évi
Sgr 18h47'7" -27d16'48"
Tibor Foki
Uma 10h23'25" 42d41'35"
Ticia 01.26.1947
Cas 0h36'5" 53d23'24"
"Tickety-Boo"
Ori 5h44'32" -4d8'5"
Tickle Me Cell
Ari 2h4'25" 22d29'14"
TicTac
Aqr 22h18'47" 1d30'26"
Ticusomnak: Star of the Desert
Cap 21h50'1" -17d13'59"
T.I.D
Lib 15h12'22" -4d55'6"
Tiddy's Stizzar
Lib 15h0'22" -18d41'3"
Tidwell Mostello
Vir 13h37'15" -4d56'29"
Tiedemann, Luuk Heinrich
Uma 10h44'19" 52d46'12"
Tien
And 1h16'51" 46d52'4"
Tien & Hieu
And 2h29'6" 45d2'15"
Tien & Kenny
Her 18h6'41" 18d33'43"
Tieran Michael
Vir 13h20'51" 13d0'56"
Tiercil
Sgr 18h46'57" -32d28'52"
Tiernan
Peg 23h55'31" 10d43'36"
Tiernan Rourk Primm
Gem 6h58'38" 22d29'20"
Tierney Ann Corbett
Leo 11h49'32" 25d40'41"
Tierney D. Mangus
Cas 0h22'21" 57d6'50"
Tierney Grant
Umi 15h31'26" 73d13'32"
Tierney Jayvonne
Leo 10h31'45" 12d50'24"
Tierney's Star
Gem 7h11'33" 29d56'36"
Tierney's Star
And 0h47'34" 41d3'42"
Tierra
Lyr 18h53'5" 35d2'6"
Tierra
And 0h58'48" 36d40'34"
Tierra Licausi
Cmi 7h43'2" 0d25'7"
Tierra Mia
Leo 10h28'27" 8d2'30"
Tierra Powell
Uma 9h54'45" 51d32'17"
Tiesa Ortez
Del 20h46'4" 13d12'17"
Tietje Panasuk
Vir 13h31'44" -4d56'32"
Tietjen, Kai
Uma 11h36'5" 29d3'1"
Tieu and Jeff Cutbush - Aug 6, 2005
Umi 14h52'32" 81d35'2"
Tieva Caila
Leo 9h54'17" 9d38'21"
Tif Me ko`u aloha
Eri 3h26'31" -21d12'59"
Tifani Le
Uma 10h6'21" 41d57'14"
Tifannie Hartwell
Uma 8h35'22" 59d52'19"
Tiff
Dra 18h45'29" 66d31'24"
Tiff
Col 5h38'35" -30d49'40"
Tiff
And 1h4'14" 42d42'54"
Tiff
Ari 2h16'11" 21d48'42"
tiff
Vir 13h18'50" 5d57'47"
Tiff Beezy
Aqr 22h11'45" -0d20'39"
Tiffa
Pho 1h14'34" -42d41'47"
Tiffanee Dawn Elizabeth Rice
Uma 11h3'43" 64d14'29"
Tiffani Chatel
Sco 17h24'36" -44d38'24"
Tiffani & Clayton Boyd 1423
Ori 5h38'25" 2d46'55"
Tiffani & Ed Boyer
Cyg 19h41'21" 50d41'46"
Tiffani Habel
And 1h1'39" 46d31'42"
Tiffani Henn
Cnc 8h19'54" 29d48'11"
Tiffani Jo
Leo 11h40'42" 25d27'30"
Tiffani Magen
Aqr 20h59'53" -13d5'14"
Tiffani Markey
Lib 15h50'41" -3d53'19"
Tiffanie Anna Jenson
Tau 5h24'34" 18d52'4"
Tiffanie Anne Bokor
Uma 13h38'33" 50d47'40"
Tiffanie Barriere's Star
And 23h13'55" 48d40'54"
Tiffanie Diane
Lib 15h28'56" -13d41'35"
Tiffanie Lynn
Leo 11h57'48" 19d2'3"
Tiffanie Lynn Thibodeaux
Lyn 6h57'43" 46d21'15"
Tiffanie Lynne Melero
Ori 5h58'9" 20d53'32"
Tiffanie Renee Anette Church
Cyg 21h46'18" 41d55'27"
Tiffanie Star
And 2h29'36" 38d52'55"
Tiffaniex
Tau 3h26'41" 0d33'51"
Tiffani's Star
Leo 10h18'15" 17d8'15"
Tiffany
Gem 6h53'21" 15d52'4"
Tiffany
Gem 7h6'7" 18d15'15"
Tiffany
And 0h20'51" 27d28'35"
Tiffany
Ari 2h37'16" 27d18'37"
Tiffany
Leo 9h37'13" 28d30'25"
Tiffany
Vul 20h43'36" 27d44'11"
Tiffany
Gem 6h51'4" 28d17'45"
Tiffany
Tau 5h2'25" 24d30'10"
Tiffany
Tau 3h40'4" 27d8'24"
Tiffany
Tau 4h20'43" 24d34'22"
Tiffany
Ori 5h37'14" 4d19'53"
Tiffany
Psc 0h26'48" 12d25'48"
Tiffany
Ari 2h48'25" 17d3'18"
Tiffany
Ari 2h50'36" 16d53'26"
tiffany
Leo 11h36'54" 9d54'5"
Tiffany
Mon 6h48'32" 7d48'27"
Tiffany
And 1h49'24" 38d33'22"
Tiffany
Crb 15h50'23" 33d10'46"
Tiffany
And 23h21'1" 38d24'32"
Tiffany
Cas 2h11'38" 59d18'27"
Tiffany
Cap 21h43'28" -10d0'10"
Tiffany
Cap 21h38'49" -9d59'10"
Tiffany
Cma 7h21'26" -15d38'33"
Tiffany
Aqr 23h13'54" -11d11'33"
Tiffany
Aqr 22h39'25" -17d15'41"
Tiffany
Cap 21h27'38" -17d7'56"
Tiffany
Lib 15h4'38" -6d58'34"
Tiffany
Aqr 22h41'58" -1d10'56"
Tiffany
Aqr 23h16'58" -6d44'23"
Tiffany
Dra 17h16'37" 53d31'22"
Tiffany 1
Aql 19h34'48" 7d15'10"
Tiffany 188
And 0h21'29" 41d39'22"
Tiffany A.G. Drage
Lyn 8h19'12" 40d40'46"
Tiffany AKA Slimy
Lyn 8h59'8" 38d4'2"
Tiffany Alexandria Tanner
Uma 11h29'34" 58d59'20"
Tiffany Amaral
Uma 11h35'34" 32d32'24"
Tiffany Amber Betar
Aqr 21h12'27" -13d50'54"
Tiffany Amber Waldroupe
Her 17h59'13" 22d9'44"
Tiffany and Anthony Forever
Sge 19h54'32" 17d45'41"
Tiffany and Edwin
Col 6h12'52" -32d32'44"
Tiffany and Talyssa
Cyg 19h34'29" 28d40'34"
Tiffany Ann Brown's Star
Gem 7h19'35" 34d39'11"
Tiffany Ann Cole
Cyg 19h33'54" 29d30'38"
Tiffany Ann Cuevas
And 0h15'31" 27d18'58"
Tiffany Ann Kerschner
Sgr 18h49'53" -16d51'43"
Tiffany Ann Sheridan
Gem 6h50'56" 32d44'29"
Tiffany Anne Babb
Mon 7h50'17" -2d15'16"
Tiffany Anne Verhoef
Dra 17h49'32" 51d7'50"
Tiffany B. Miller
Aqr 21h58'16" 1d5'53"
Tiffany "Baby" Lau
Lib 15h15'2" -24d26'59"
Tiffany Beals
Lyr 18h37'10" 36d4'37"
Tiffany Becher
And 0h52'29" 22d2'8"
Tiffany Berner's Star
Psc 0h25'27" 9d3'20"
Tiffany Board
And 0h38'39" 37d13'16"
TIFFANY BODDIE
Lyr 19h18'42" 29d50'15"
Tiffany Breanna
Leo 10h28'16" 9d1'4"
Tiffany Breanne Bradley
Crb 15h28'42" 28d22'43"
Tiffany Bronson
Ori 5h33'38" -0d51'7"
Tiffany Brooke
Lyn 8h30'52" 56d26'28"
Tiffany & Bryan Forever
Gem 7h49'13" 14d26'15"
Tiffany Burch
Ari 1h53'23" 18d41'18"
Tiffany Cappucci
Ari 2h13'43" 26d17'48"
Tiffany Caruso
Sco 17h43'38" -32d26'11"
Tiffany Casterline
Crb 15h59'38" 34d0'31"
Tiffany Cathleen Moore
Boo 14h51'1" 26d19'15"
Tiffany Celeste
Cyg 19h59'3" 44d11'5"
Tiffany Christen
And 23h21'51" 47d12'37"
Tiffany Christine
Ari 3h8'22" 28d26'31"
Tiffany Clement
Lib 15h7'54" -5d13'31"
Tiffany Cook
Ori 6h9'46" 2d28'54"
Tiffany Cox
And 0h42'29" 35d36'32"
Tiffany Cronister
Lib 15h57'25" -17d55'42"
Tiffany Cruz
Ari 3h15'33" 11d27'24"
Tiffany Curameng
Cyg 19h39'26" 31d49'40"
Tiffany D. Clark
And 23h26'31" 50d8'23"
Tiffany Dalin Ostrander
Uma 10h49'53" 51d37'28"
Tiffany Danielle Mansfield
Tau 4h54'52" 23d11'6"
TIFFANY DAWN
Uma 11h22'36" 36d37'32"
Tiffany Dawn
Leo 9h28'26" 30d35'11"
Tiffany Dawn
Sco 16h49'51" -33d57'43"
Tiffany Dawn Atkins
Tau 4h6'22" 23d59'28"
Tiffany Dawn Davison
And 0h19'20" 25d24'17"
Tiffany Dawn Dobrosky
Sgr 19h13'20" -15d36'54"
Tiffany Dawn Matuszak the Beautiful
Ori 5h36'9" 11d31'25"
Tiffany Dawn Reicosky
Lyr 19h6'53" 45d34'11"
Tiffany Dawn Totten
Del 20h42'42" 15d59'24"
Tiffany Devon Plunkett
Sco 16h6'30" -15d27'40"
Tiffany Dianne Bolchoz Lewis
Peg 23h40'24" 15d20'23"
Tiffany Ding
Leo 10h17'59" 25d52'0"
Tiffany & Dragan
Cma 7h0'30" -16d58'52"
Tiffany Drew "God is with me"
Her 17h24'21" 35d28'20"
Tiffany Dudley's star
Gem 6h36'33" 21d16'13"
Tiffany "Dynamite" Bland
Dra 19h42'26" 68d22'19"
Tiffany E. Noblett
And 0h13'51" 43d37'53"
Tiffany Eden
And 1h30'29" 42d9'32"
Tiffany Elizabeth DeFusco
Lib 14h55'18" -17d29'30"
Tiffany Evans
Uma 10h1'25" 71d10'24"
Tiffany F. Sahm's Shining Star
And 0h17'4" 43d42'44"
Tiffany Fabian Felicidad Tran
And 2h22'12" 49d37'30"
Tiffany Fae
Uma 9h11'7" 51d29'12"
Tiffany Faith
Lib 15h19'33" -22d8'55"
Tiffany Fielder
Leo 10h36'21" 22d56'58"
Tiffany Fletcher
Uma 10h22'18" 54d29'42"
Tiffany Fogler
Leo 11h29'56" 14d17'50"
Tiffany: Forever My Babydoll
Ori 5h39'12" 4d16'11"
Tiffany Furman
Psc 0h42'55" 16d22'26"
Tiffany Gale Barth
Vir 11h53'23" 4d35'34"
Tiffany Germain
Lib 15h11'17" -15d29'33"

Tiffany Grace Castellano
Cas 1h5'32" 75d40'49"
Tiffany Graham
Lyn 8h47'44" 36d40'34"
Tiffany Granath
Cas 23h31'14" 58d4'28"
Tiffany Gwen Correll
Eri 3h11'48" -7d0'46"
Tiffany Hamilton Washam
Psc 0h46'40" 18d44'33"
Tiffany Harry
And 23h29'14" 48d7'17"
Tiffany Harvey
Vir 12h8'58" 10d19'35"
Tiffany Hays
Uma 9h6'1" 56d42'26"
Tiffany Helgeson Kritsings
Ari 2h0'22" 21d3'4"
Tiffany Helm
Cap 20h20'24" -11d26'49"
Tiffany Holbrook
Sco 17h3'10" -39d15'11"
Tiffany (Honey Bun) Chelune
Uma 10h45'13" 62d6'8"
Tiffany Huang
Lib 14h56'15" -10d37'43"
Tiffany Iann
Cap 21h54'28" -22d20'3"
Tiffany J. I
Tau 4h23'23" 6d52'19"
Tiffany Jade Chang
Cas 2h19'52" 72d25'4"
Tiffany Jade O'Hern
And 2h3'9" 42d36'52"
Tiffany Jade Thynne - A Star
Sco 17h9'33" -33d17'13"
Tiffany Jae Tazelaar
Tau 5h37'7" 25d40'59"
Tiffany Jean
Tau 4h39'40" 17d19'29"
Tiffany Jean Pinto
Aqr 22h54'47" -7d45'59"
Tiffany Jeanne
Lyn 7h43'54" 38d32'2"
Tiffany Jo Emry
And 23h44'54" 37d34'57"
Tiffany/ JooReeKim
Lib 15h10'1" -7d7'51"
Tiffany K. Cromwell
Lib 15h3'18" -3d7'42"
Tiffany K. Smith
Leo 10h41'1" 12d57'19"
Tiffany Kayla
Sco 16h11'36" -18d11'40"
Tiffany Kern 2006
Uma 10h33'26" 70d5'57"
Tiffany Kinner
Cnc 8h4'0" 21d20'34"
Tiffany Krista Bonello
And 23h22'1" 52d58'53"
Tiffany L Fogarty Angel In My Heart
Crb 16h6'7" 27d31'51"
TIFFANY LAINE MAY
Cnc 8h42'10" 12d9'18"
TiFFanY Lam aka Su-S
And 0h16'2" 43d13'10"
Tiffany LaRie Brinkman
Leo 9h47'51" 19d4'15"
Tiffany Larissa Westerman
Sco 17h52'3" -39d1'11"
Tiffany Lauren
And 0h44'41" 41d27'39"
Tiffany Leanne
Crb 15h52'54" 26d2'4"
Tiffany Lebron
Sgr 18h6'51" -20d24'8"
Tiffany Lee Brusoe McRae
And 2h7'32" 45d28'15"
Tiffany Lee Ingalls
Per 3h34'18" 44d44'15"
Tiffany Leigh
Uma 9h44'14" 46d0'38"
Tiffany Leigh
Del 20h41'26" 17d25'53"
Tiffany Leigh
Lib 15h8'24" -19d17'2"
Tiffany Leigh Erwin
Aql 19h28'28" 3d8'27"
Tiffany Leigh Frederick
Dra 19h21'57" 73d12'36"
Tiffany Leigh McCrary
Cam 4h45'25" 62d8'45"
Tiffany Leigh Rider
And 23h38'20" 44d31'53"
Tiffany London Prager
Cam 5h3'22" 72d46'18"
Tiffany Lopez
Crb 15h36'25" 33d27'14"
Tiffany Lyn Miller
And 1h50'24" 44d5'30"
Tiffany Lynn
Her 17h53'21" 27d12'38"
Tiffany Lynn
Sco 16h10'19" -17d59'23"
Tiffany Lynn Gerstner
And 1h34'25" 37d56'42"
Tiffany Lynn Hall
And 23h48'7" 36d20'38"
Tiffany Lynn Steele
Tau 4h31'14" 27d27'39"
Tiffany Lynn Stout
Psc 0h32'6" 8d16'30"
Tiffany Lynn Umfleet-Cordaway
Aqr 21h49'25" -7d42'15"
Tiffany Lynn Westmoreland
Her 17h21'25" 18d5'14"
Tiffany Lynn Wichman
Cnc 8h11'26" 12d12'27"
Tiffany Lynn Wicklund
Ori 6h4'6" 20d45'3"
Tiffany Lynn Winters
And 23h31'19" 41d58'45"
Tiffany Lynn Wisor
Sco 16h12'53" -9d2'31"
Tiffany M. Thompson
Gem 6h32'37" 24d18'33"
Tiffany Margeaux
Cas 1h35'30" 67d26'8"
Tiffany Marie
Crb 16h9'6" 27d1'1"
Tiffany Marie
Leo 10h7'45" 24d15'34"
Tiffany Marie
Peg 21h28'1" 5d22'56"
Tiffany Marie Ciranni
Sco 16h22'2" -23d23'3"
Tiffany Marie Congdon
Aql 19h39'19" 14d55'57"
Tiffany Marie Godbee Barksdale
Peg 21h35'49" 13d59'47"
Tiffany Marie Jones
Tau 4h34'35" 6d30'26"
Tiffany Marie Kraus
And 0h29'35" 45d40'44"
Tiffany Marie Lesko
Lib 15h7'50" -27d11'30"
Tiffany Marie Maldonado
Ari 2h18'5" 22d3'24"
Tiffany Marie Smith
Her 17h28'25" 32d15'30"
Tiffany Marie Steelman
Sgr 18h13'23" -20d20'14"
Tiffany Marie Tilenni
Tau 4h28'27" 26d5'41"
Tiffany Marie Valdivia
Sgr 18h34'14" -23d22'52"
Tiffany Marissa Cline
And 23h19'56" 51d49'15"
Tiffany Mccumber
Cnc 9h12'39" 7d18'59"
Tiffany Melissa Brown
Tau 5h28'28" 19d29'13"
Tiffany & Michael
Cyg 21h39'8" 30d20'44"
Tiffany Michelle Dickens
Leo 10h56'9" 11d44'41"
Tiffany Michelle Lehman
Lib 15h3'22" -0d42'23"
Tiffany Milane
Sgr 18h5'17" -27d33'18"
Tiffany my angel
And 22h58'4" 39d47'48"
Tiffany My Princess
And 0h27'54" 29d54'27"
Tiffany N O'Brien
Crb 15h54'35" 27d43'23"
Tiffany Neal
Lyn 9h39'12" 39d56'56"
Tiffany & Nevaeh
Leo 9h42'53" 26d42'49"
Tiffany Ng Loo
Tau 3h48'58" 11d21'38"
Tiffany Nichole Mathesius
Cnc 9h2'0" 7d12'38"
Tiffany Nichole Tyson
Gem 7h21'30" 19d7'35"
Tiffany Nicole
Crb 16h3'41" 26d22'16"
Tiffany Nicole
Uma 10h36'33" 61d39'13"
Tiffany Nicole Davis 19880123
Aqr 22h31'37" 1d43'45"
Tiffany Nicole Elmore
Lib 15h27'45" -8d28'51"
Tiffany Nicole Jackson
Cyg 21h49'11" 38d49'3"
TIFFANY NICOLE SWITZER
Sgr 18h51'24" -31d9'12"
Tiffany Nikkole
Leo 10h9'8" 23d33'13"
Tiffany Noble
Cam 3h18'35" 67d8'29"
Tiffany Paige Casey
Tau 5h54'39" 28d14'34"
Tiffany Pais
Cap 21h41'40" -21d28'55"
Tiffany Parker
Cam 5h39'51" 65d41'20"
Tiffany Pollard
Cap 20h7'29" -18d37'51"
Tiffany Popoli
Uma 11h36'33" 36d24'38"
Tiffany R Cothren
Sgr 18h13'35" -26d12'25"
Tiffany Raber
Umi 15h21'28" 72d21'1"
Tiffany Rachelle
And 0h37'52" 40d52'24"
Tiffany Rae Layne
Cas 23h48'28" 53d19'55"
Tiffany Renae Soza
Leo 11h4'7" 10d49'32"
Tiffany Rene' Clements
Tau 5h13'32" 26d16'59"
Tiffany Renee
Uma 10h32'5" 54d13'16"
Tiffany Roa
Lib 14h48'55" -8d21'52"
Tiffany Robin's Night Diamond
Cas 23h37'4" 52d42'30"
Tiffany Rose
Aqr 22h34'21" -5d20'34"
Tiffany Rose Bourne
And 23h34'15" 42d50'25"
Tiffany Rose Helvie
Cnc 8h51'46" 17d41'17"
Tiffany Rose Lewandowski
Leo 10h19'31" 25d29'36"
Tiffany Rose McDevitt 2004
Gem 7h19'9" 20d21'49"
Tiffany Rose Penn
Tau 3h42'9" 27d8'1"
Tiffany Roseboro
Cam 3h58'15" 54d59'17"
Tiffany Russell
Psc 23h7'48" 5d39'5"
Tiffany Ruth Woodby
Uma 9h18'31" 45d31'9"
Tiffany Saenz, my sole-mate
Tau 5h45'31" 22d7'29"
Tiffany Sankar
Psc 22h53'1" 4d59'10"
Tiffany Saxby aka Jesse's Girl
Vir 12h26'22" 4d11'18"
Tiffany Schmiesing
Uma 13h13'46" 56d16'33"
Tiffany Self
And 2h12'50" 37d31'31"
Tiffany Shea
Gem 6h37'43" 20d58'47"
Tiffany Shea Skinner
Uma 10h4'15" 47d4'45"
Tiffany Sherman
Psc 1h24'57" 24d48'26"
Tiffany Sirwet
Sgr 17h54'10" -28d37'3"
Tiffany Spickard
Sgr 17h55'57" -28d35'11"
Tiffany Star
Tau 3h31'50" 7d27'3"
Tiffany Star
Cyg 19h59'57" 39d36'31"
Tiffany Stella * Valedictorian *
And 2h10'46" 43d35'58"
Tiffany Sue Fannin
Lyn 7h26'3" 50d11'48"
Tiffany Taylor
And 1h43'36" 43d2'25"
Tiffany ~ The Apple Of My Eye
And 23h12'51" 40d9'9"
Tiffany Thompson
Lmi 10h40'29" 31d56'40"
Tiffany Tisdale
And 1h24'45" 48d23'23"
Tiffany Trippiedi (nee) Mares
Cam 7h19'40" 60d17'39"
Tiffany Tsarky
Crt 11h12'42" -16d7'38"
Tiffany Ulatowski
Psc 0h54'21" 14d19'31"
Tiffany Van Every Ashing
And 0h19'35" 33d28'41"
Tiffany VanSciver My Soulmate
Dra 18h1'55" 76d0'37"
Tiffany Vaughan
Leo 11h21'27" 15d22'25"
Tiffany Victoria
Cap 20h48'8" -16d11'29"
Tiffany Westhoelter
Uma 8h49'8" 54d58'24"
Tiffany Wurster's Star
Sgr 19h41'59" -16d39'39"
Tiffany & Yasmin's Piece of Heaven
Cyg 20h15'10" 36d53'32"
Tiffany, will you go out with me?
Gem 6h43'16" 13d46'22"
Tiffany2005
Leo 11h41'46" 25d39'0"
Tiffany's Alexa
Vir 12h49'31" -11d10'2"
Tiffany's Destiny
And 0h22'21" 42d23'39"
Tiffany's Eternal Light
And 23h16'30" 41d2'48"
Tiffany's Eternal Light
Cnc 9h0'14" 7d29'58"
Tiffany's Glow
Psc 0h51'54" 16d44'52"
Tiffanys Light
Lib 14h24'43" -10d52'4"
Tiffany's Light
Dra 15h24'47" 63d18'39"
Tiffany's Pooh Bear 808
Uma 8h18'27" 69d38'13"
Tiffany's Smile
Uma 9h5'22" 60d59'22"
Tiffany's Star
Vir 11h41'26" -0d10'7"
Tiffany's Star
Lib 15h18'22" -26d47'40"
Tiffany's Star
And 2h29'26" 37d23'4"
Tiffeny Aleksick
And 0h53'23" 37d58'58"
Tiffini's Starlight Starbright
Uma 10h35'26" 63d9'5"
Tiffiny Lynn Bordali
And 0h14'22" 28d36'56"
Tiff's Star
Ori 5h55'16" 22d28'53"
Tiffs star
Cyg 19h49'52" 31d39'7"
TiffTiff
Crb 15h29'49" 31d5'58"
Tiffury
And 1h29'13" 48d17'15"
Tiffy
And 0h8'36" 42d48'46"
Tiffy Bear
Uma 10h35'30" 66d19'48"
TiffyD
Uma 11h10'59" 52d20'5"
Tif-Stratus
Cas 23h5'49" 59d6'19"
Tig & Lycan
Col 5h37'22" -36d35'39"
TIGAAN
Leo 10h58'43" 5d40'15"
Tiger
Aql 19h31'49" 11d4'58"
Tiger
Cnc 7h57'42" 13d31'14"
Tiger
Ori 5h54'46" 22d23'54"
Tiger
Per 3h23'19" 46d19'48"
Tiger
Per 4h9'12" 35d34'20"
Tiger
Uma 11h22'10" 59d24'49"
Tiger
Uma 11h50'15" 53d58'22"
Tiger and Slash Super Star
Lyn 7h32'51" 46d17'41"
Tiger Blue
Leo 9h40'25" 27d57'22"
TIGER - Champion Best Friend
Her 16h33'32" 48d7'21"
Tiger Eye
Leo 11h28'22" 14d16'5"
Tiger Eyes
Leo 9h46'11" 26d2'49"
Tiger Gaede
Uma 8h44'11" 54d23'48"
Tiger Lady Noura
Sco 16h58'43" -37d5'11"
Tiger Lilly
Peg 22h37'37" 7d40'34"
Tiger Lily Laura
Leo 10h6'58" 19d19'46"
Tiger Lily Skodnek Coffey
Lyn 6h41'57" 59d6'40"
Tiger Man
Cyg 21h15'49" 31d21'29"
Tiger Tim
Sco 16h23'52" -27d59'12"
Tiger Yen Suhu
Aqr 21h23'28" -9d17'4"
Tiger-Alex
Equ 21h5'22" 3d48'9"
Tigerbaum
Lib 15h14'54" -3d51'38"
Tigerchen
Uma 11h24'52" 35d7'37"
Tigeress
Lib 14h51'33" -3d15'43"
Tigergirl
Cap 20h40'46" -25d35'42"
Tiger-Hubert
Uma 8h49'8" 54d58'24"
Tigerli
Cep 22h35'23" 80d2'31"
Tigerli
And 23h24'45" 37d47'17"
TigerLillie Sunshine Johnson
Cas 23h59'30" 59d57'58"
Tiger-Lily Norah Newbould
Ori 5h31'7" 3d19'49"
Tigeroo
Uma 9h27'5" 62d10'34"
Tiger's Eye
Cen 12h25'52" -50d6'58"
Tiger's Love
Gem 7h21'18" 32d0'9"
Tiger's Sweetheart
Cas 0h26'12" 61d13'38"
Tigger
Uma 11h33'37" 61d25'8"
Tigger
Umi 15h55'13" 75d0'2"
Tigger
Uma 13h36'25" 51d11'53"
Tigger
Tau 4h40'46" 20d46'27"
Tigger
Gem 6h37'27" 16d26'5"
Tigger 393
Lyn 7h6'12" 45d29'43"
Tigger & Scooby Doo
Cyg 21h7'31" 53d18'59"
Tiggerific Jenny
Her 17h39'22" 20d43'15"
Tigger's Diamond
Leo 11h0'58" 15d33'0"
Tigger's Dream
Uma 11h15'30" 71d19'56"
Tigges Too
Cnc 8h18'8" 13d15'21"
TIGGIE
Gem 7h18'36" 25d1'58"
Tiggs Pan
Psc 0h37'54" 11d31'27"
Tiggy
Ori 4h55'31" 10d4'17"
Tiggy
Gem 7h50'39" 14d19'14"
Tight Like Us— E.F.B. & C.R.R.
Vir 13h24'10" 12d24'9"
Tigi & Meiti
Per 3h40'30" 33d19'0"
Tigi & Purzel
Lyr 19h2'22" 42d40'39"
Tigistärn
And 1h44'1" 37d42'18"
Tigon Hunter-Selbrede
Dra 17h47'27" 59d22'13"
Tigray Zander
Ari 2h28'43" 16d19'27"
Tigress
Ori 5h36'50" 7d20'3"
Tigress Mommy
Tau 4h28'33" 22d28'40"
Tigris
Lmi 10h17'32" 29d58'6"
Tigrlily
Lmi 10h29'39" 35d39'18"
Tigyi Katalin
Gem 7h9'4" 25d8'36"
Tihomir i Mileva Ravic
Cru 12h35'51" -58d55'3"
Tiina Amanda
Uma 10h40'5" 56d32'19"
Tiina Purin
Tau 5h24'51" 21d2'8"
Tiiya Sherrie Franklin
Lyn 6h49'35" 51d47'17"
Tijana Dapcevic
Her 17h53'10" 40d9'36"
Tijana Filipova
Gem 6h2'43" 23d14'23"
Tijana Noelle Radojicic 10:10 p.m.
Vir 12h51'42" 1d15'13"
Tijn DeWaal
Aqr 22h19'25" -6d38'41"
Tijuana Hooker
Peg 23h19'46" 31d34'6"
Tika - moja zvezda
Uma 11h25'53" 60d28'46"
Tika Tiger
Uma 10h2'26" 71d56'12"
Tiki
Uma 13h21'25" 55d56'44"
Tiki Cat
Cnc 9h0'35" 8d58'53"
Tiki Dusa
And 1h4'7" 42d39'30"
Tiki Mon
Leo 11h31'8" -0d49'33"
Tikqwah Johnson
Cnc 9h5'35" 30d20'1"
'Til Death
Car 6h26'55" -51d41'7"
TIL THE END!
Cnc 8h5'3" 17d2'35"
'Til the Stars Don't Shine
Cnc 8h41'17" 27d40'55"
TIL YAM
And 1h46'0" 39d7'46"
Tilak Raj Sikri
Leo 11h12'56" -1d54'1"
Tildawn
Sge 19h50'27" 17d46'0"
Tilde
Sco 16h7'34" -16d36'57"
Tilde Helen & Stella Maria Keatley
Aqr 22h55'15" -23d41'57"
Tilika Jean
Ari 2h45'28" 14d42'34"
Till Adrian Kandid
Umi 16h22'54" 77d59'9"
Till Death Do Us Part
Sco 17h29'50" -32d33'21"
Till death due us part
Uma 10h19'21" 54d11'56"
Till the end of time.
Her 16h17'52" 45d18'51"
Till-Aurel Anliker
Uma 9h46'19" 65d6'43"
Tilley Mai
And 0h9'20" 33d19'55"
Tillie
Crb 15h24'44" 28d41'26"
Tillie Leija Reyna
Cnc 8h11'3" 12d8'23"
Tillie McGraw
Cas 0h19'7" 52d2'41"
Tillie Violet Pichowski
And 2h4'24" 46d44'58"
Tillie74
Aqr 20h57'32" -11d31'26"
Tillman
Aql 20h13'38" 14d26'50"
Till's Stern
Srp 18h15'39" -0d1'3"
Tilly
Cas 1h27'45" 65d41'24"
Tilly
And 23h18'47" 40d36'36"
Tilly Goldman
Crb 15h44'5" 27d8'9"
Tilly Lucia Catherine Caine
And 1h56'53" 45d26'15"
Tilly Mae
Cas 23h26'3" 51d55'27"
Tilsammen
Ori 5h36'1" 1d10'54"
Tilulu
Aqr 22h45'34" -9d24'37"
Tiluvu
Lyn 7h12'28" 51d51'44"
Tilyn Grace Hodgkin
And 23h40'46" 42d41'46"
Tim
Cyg 20h28'49" 33d58'15"
Tim
Aqr 22h15'42" 0d53'21"
Tim
Cep 21h13'37" 56d27'11"
Tim Akers
Ori 6h0'49" 17d38'22"
Tim Allen
Aqr 22h51'3" -8d10'13"
Tim & Amber Always
Lyn 7h19'39" 59d0'31"
Tim and Anita's Star
Cyg 20h36'1" 53d22'6"
Tim and Anna
Vir 12h51'9" 11d55'43"
Tim and Barbra McKee
And 0h45'30" 35d54'4"
Tim and Betty Forever
Her 18h1'45" 27d6'4"
Tim and Courtney
Cyg 20h42'33" 36d34'21"
Tim and Holly's star
Lyr 18h49'1" 30d16'2"
Tim and Jamie Star of Love
Leo 9h56'16" 24d0'3"
Tim and Kirsten for Eternity
Cyg 21h43'24" 43d55'36"
Tim and Melissa Carter
Peg 21h54'23" 8d51'59"
Tim and Pam Wellborn
Per 3h16'5" 47d40'15"
Tim and Rainbow's Love Star
Sco 16h6'56" -20d19'23"
Tim and Suz Forever Us
Umi 13h7'0" 69d24'51"
Tim Anderberg and Jessica Mitchell
Uma 8h12'4" 62d8'7"
Tim Archibald
Per 2h14'13" 54d28'7"
Tim Atwood
Ori 6h12'6" 7d20'9"
Tim Atwood
Ori 5h54'39" 22d27'35"
Tim Barrett - GaMPI Shining Star
Dra 18h27'40" 58d1'5"
Tim Barthel
Boo 14h37'3" 27d23'38"
Tim & Becky's Star
Cyg 20h38'39" 46d44'9"
Tim & Belinda Pipher
Vir 13h10'23" -2d44'59"
Tim & Betsy Mallon
Lyn 8h15'9" 54d35'49"
Tim Biber
Ori 5h55'5" 2d39'51"
Tim Biggs
Sgr 19h1'21" -16d5'0"
Tim Bohr
Lyr 18h33'14" 37d45'45"
Tim & Breanna
And 0h4'31" 35d11'2"
Tim Brennan
Sgr 18h20'8" -28d22'41"
Tim Brown
Tau 4h40'28" 19d6'38"
Tim Brühlmann
Dra 17h12'34" 55d1'12"
Tim Bunch
Her 17h18'42" 35d42'58"
Tim Caso
Gem 6h31'17" 17d34'35"
Tim Cerutti
Her 17h41'29" 36d17'37"
Tim Coleman
Gem 6h19'42" 22d23'16"
Tim Coscarelli
Uma 11h45'11" 48d41'23"
Tim Cowell
Uma 11h35'53" 38d31'50"
Tim & Dani's Star
Cnc 8h51'58" 28d44'41"
Tim & Dasha
Cyg 20h33'48" 34d24'25"
Tim David Mackereth
Cep 0h2'1" 68d32'48"
Tim Dierkes
Boo 14h32'45" 51d46'0"
Tim Dougherty
Her 18h13'55" 18d8'57"
Tim Duda
Uma 9h59'1" 50d53'52"
Tim E.
Uma 10h45'33" 55d23'30"
Tim Edward Bond
Uma 10h12'57" 58d55'5"
Tim & Em
Her 18h41'0" 20d5'22"
Tim Flasher
Per 3h17'30" 43d42'17"
Tim Fruge' - Vagabond Love
Cru 12h16'40" -58d21'56"
Tim Hardaway
Dra 20h15'40" 70d50'43"
Tim Henneberry "Timwho"
Leo 10h38'53" 10d47'40"
Tim Holder
Cep 22h58'8" 73d1'50"
Tim Hoover & Jessie Espinoza
Ori 6h15'58" 16d5'36"
Tim & Jackie
Tau 4h13'58" 21d24'2"
Tim & Jaimee, Love Always
Cyg 21h43'31" 41d27'19"
Tim & Jeanne St. John Star
Cyg 20h47'11" 38d10'8"
Tim & Jodie Farrow
Cyg 20h1'39" 30d17'16"
Tim Jonas
Uma 13h0'33" 61d9'59"
Tim Julian
Uma 8h24'26" 66d7'46"
Tim & Julie Allen Forever
Lyr 18h36'27" 38d34'28"
Tim Kasper
Uma 10h35'31" 69d51'14"
Tim Kilpin
Aur 5h29'30" 43d34'32"
Tim & Kim
Cyg 21h12'19" 31d3'48"
Tim & Kim 12/5/01
Cyg 20h44'24" 36d38'30"
Tim Koch
Umi 15h7'57" 71d38'11"
Tim L. Cox
Ori 5h35'3" 3d0'31"
"Tim" Lambert
Uma 11h41'1" 42d52'36"
Tim & Laura
Cyg 19h40'6" 40d33'20"
Tim LeBlanc's Special Star
Uma 10h24'18" 57d37'21"
Tim Leger - 31.05.1999
Uma 9h42'17" 49d26'25"
Tim Lehmann
Peg 0h13'3" 17d9'49"
Tim & Leo Levine
Cyg 21h58'54" 46d43'47"
Tim & Liz Hughes
Lyr 18h31'15" 28d35'10"
Tim & Liz, Together Forever
Cma 6h24'38" -16d33'14"
Tim Logan - Shining Star
Aur 5h20'31" 31d14'12"
Tim Louis
Uma 11h25'57" 60d44'9"
Tim Mable
Ori 6h4'42" 18d30'2"
Tim Manning
Per 3h47'26" 42d40'54"
Tim & Max
Vir 13h21'30" -0d11'9"
Tim Montrenes
Boo 15h30'14" 42d54'28"
Tim Moran
Her 16h11'54" 47d41'42"
Tim & Morgan Baker
Cyg 20h55'17" 32d13'17"
Tim Murray Of Pleiades
Tau 5h19'41" 27d28'32"
Tim Naoufel
Sge 20h3'5" 18d42'16"
Tim Nayar the Magnificent
For 3h31'49" -26d56'43"
Tim - "nunc scio quit sit amor"
Cru 12h52'7" -57d17'52"
Tim Odum
Dra 17h16'22" 60d54'33"
Tim O'Toole
Sgr 19h42'32" -36d19'23"
Tim Pellerin
Umi 15h46'4" 79d22'29"
Tim Pilachowski, My Raisin
Aur 5h44'5" 48d44'7"
Tim Reeves
Vir 11h49'27" 2d31'10"
Tim Reynolds
Leo 11h42'2" 14d46'33"
Tim Richard Barney
Dra 16h29'14" 53d40'27"
Tim Rosander
Per 4h24'16" 35d4'29"
Tim Ryan
Uma 11h37'33" 35d56'30"
Tim Ryan Wahlstrom
And 23h14'25" 48d39'2"
Tim & Sally "Marsh Vegas Love"
Cyg 20h4'51" 30d28'41"
Tim Schwarz
Uma 13h59'4" 59d54'20"
Tim<\@>Shawna Rau
Cnc 8h54'11" 14d20'55"

Tim Sibal
Uma 8h32'32" 67d59'27"
Tim Smith ~ Always In Our Hearts
Her 16h27'51" 7d58'24"
Tim 'Stellar' Smith
Cyg 20h51'39" 36d40'4"
Tim Stinebaker
Aur 5h18'19" 40d32'10"
Tim Stricklin's Shining Star
Apu 17h12'14" -72d35'45"
Tim Tarr
Aur 5h19'31" 42d7'38"
Tim & Tricia
Cyg 21h16'27" 46d57'27"
Tim & Trish Eternal, Devoted, One
Umi 15h22'50" 72d36'16"
Tim - Us and the Moonlight
Ori 5h39'56" -1d23'12"
Tim & Valda Forever
Cyg 21h19'24" 32d39'0"
Tim VanderMolen
Ori 5h58'35" 6d44'18"
Tim Werner
Per 3h25'10" 50d20'23"
Tim Williams
Vir 13h17'59" 6d42'53"
Tim Williams
Tau 5h53'53" 27d11'13"
Tim Wilson
Dra 16h57'51" 63d15'2"
Tim Wolf - 24.01.1981
Aqr 21h37'27" -0d20'43"
Tim Wright
Aql 19h30'7" 1d21'25"
Tim, My Love
Cep 22h36'26" 68d14'19"
Tim, Natalie, and Kiley Oswald
Lac 22h22'46" 43d44'4"
Tim, Sonia & Annie Bruno
Cyg 21h36'58" 44d55'35"
TiMaJaRo
Ori 6h20'8" 19d16'56"
TimAnCar
Per 2h51'15" 39d24'12"
Timara Ellen Klotz
And 2h31'50" 49d22'18"
Timatee
Gem 7h1'55" 17d6'21"
TimB
Aqr 21h47'4" 0d59'6"
Timber
Uma 11h15'27" 36d35'6"
Timber
Umi 15h32'37" 71d40'6"
Timberly Ann Cooke
Cnc 8h15'5" 29d16'19"
Timberly D. Eyssen
Cas 2h41'41" 58d0'15"
Timbo
Her 17h47'35" 36d54'54"
Timbo
Cnc 9h15'9" 25d21'52"
Time
Uma 9h57'55" 46d14'44"
Time after Time
Sco 17h58'38" -38d15'37"
Time is for YOU
Ori 6h14'58" 15d25'2"
Time of Your Life
Tau 3h34'43" 21d11'28"
Time Out at the Gap
Umi 15h27'4" 68d18'54"
Timeless
Ori 5h19'54" 6d13'39"
Timeless Butterfly
Lib 15h59'26" -12d4'39"
Timeless Sharon
Ori 5h51'36" 2d42'26"
Timelord
Tau 5h10'40" 22d12'35"
Timeus
Cas 23h24'6" 56d9'20"
Timian
Ari 3h18'52" 29d18'38"
Timika
And 23h55'36" 45d20'35"
Timika
Cyg 21h15'38" 31d27'26"
Timika Mitchell
Uma 11h27'42" 64d51'6"
Timiko Lashai Watkins
Ari 2h57'10" 21d21'10"
Timious Dorithicus Lovis
Aqr 20h55'53" -13d44'24"
Timithy Hargrove
Her 17h18'58" 20d20'33"
TimJen
Cap 20h26'12" -12d47'25"
TimkoToo61805
Lyr 18h49'28" 34d19'6"
Timmermann, Dietrich
Sco 17h15'18" -32d0'17"
Timmie and Punkie Colasanti
Lyn 6h41'24" 57d46'3"
Timmothy Dale Fowls
Per 3h35'5" 45d45'17"
Timmy
Uma 11h40'46" 35d3'25"
Timmy
Gem 7h5'49" 32d7'10"

Timmy
Uma 9h42'48" 41d47'11"
Timmy
Tau 3h54'42" 24d45'11"
Timmy
Umi 14h33'48" 73d52'37"
Timmy Ali
Uma 9h36'18" 58d8'41"
Timmy and Mikayla Gunn
Cyg 21h32'33" 48d44'56"
Timmy Dee 7-59 6-61
Dra 14h34'35" 57d4'39"
Timmy J. McKenrick Jr.
Cep 20h39'56" 55d55'6"
Timmy John Maki
Cap 20h52'33" -16d58'29"
Timmy Kuszak
And 23h47'50" 42d30'10"
Timmy Leiterman
Lyn 6h44'54" 51d44'51"
Timmy Loves Feyza
Her 16h57'56" 33d18'33"
Timmy pooh
Aur 5h38'16" 41d29'44"
Timmy Quigley
Per 3h16'56" 46d15'13"
Timmy Rhodes
Her 16h24'15" 45d13'26"
Timmy Silkaitis
Ari 2h31'35" 10d41'29"
Timmy & Tiffany
Cas 0h54'43" 54d10'11"
Timmy, Chelsie, Megan & Dina Bean
Uma 8h53'21" 51d19'8"
TimmyLawman
Sgr 18h11'59" -22d35'10"
Timmy's Special Star Friend.
Sco 16h10'27" -9d42'15"
Timo
Lac 22h40'16" 54d5'54"
Timo
Tau 3h32'57" 29d36'37"
Timo and Philip's Wedding Star
Cyg 21h44'39" 31d22'31"
Timo Andreas Schaffner
Lyn 7h7'10" 53d30'40"
Timo Bock
Uma 10h24'37" 55d13'25"
Timo Müller
Ori 5h22'49" -8d52'27"
Timo - My Love Ever After
Pho 0h17'54" -39d26'27"
Timonthy Paul McJilton
Her 16h49'17" 46d32'13"
Timor
For 3h16'37" -26d55'27"
TimOShane
Tri 2h1'33" 33d1'7"
Timothé Blanc
Uma 13h36'27" 51d55'39"
Timothy
Uma 13h49'53" 48d48'36"
Timothy
Per 2h40'8" 50d43'8"
Timothy
Her 16h58'22" 39d39'49"
Timothy
Ori 5h36'28" 4d49'14"
Timothy
Leo 11h37'26" 11d47'18"
Timothy
Ori 6h10'8" 8d8'48"
Timothy 3.23
Aqr 23h15'53" -12d59'14"
Timothy A. Brown
Ori 4h50'44" 5d1'27"
Timothy A. Geraghty
Lib 15h3'37" -11d6'37"
Timothy A Jordan II
Cyg 20h52'36" 48d8'19"
Timothy A. Martin 930
Cnc 9h13'50" 9d34'21"
Timothy A. Pennock
Sco 17h53'26" -39d41'18"
Timothy A. Walko
Per 3h4'17" 53d17'46"
Timothy Adams
Her 17h12'56" 35d3'38"
Timothy & Alexandra Parsons
Cyg 21h13'19" 47d10'22"
Timothy Allen Clark, Sr
Ori 5h31'43" 14d52'36"
Timothy and Debra
And 2h38'31" 45d9'52"
Timothy and Mae Cannon
Uma 10h58'43" 40d44'37"
Timothy and Rebecca's Destiny
Cyg 21h19'53" 52d47'33"
Timothy and Richard Harrison
Gem 6h27'34" 25d25'6"
Timothy and Stella Ring - 40 years
Leo 11h39'6" 20d38'40"
Timothy & Andrea Gibson
Lyr 18h41'10" 38d27'1"
Timothy Andrew Bartlett
Ari 2h35'24" 14d0'37"
Timothy Andrew Carter
Ari 2h28'14" 19d50'48"

Timothy Arnold Jeppesen
Aql 19h37'39" 6d24'18"
Timothy Au
Cru 12h26'8" -56d38'21"
Timothy Austin Tackett
Psc 0h42'6" 10d37'21"
Timothy B. Bitler, Jr "BoBo"
Umi 14h9'45" 71d16'29"
Timothy Beaucage
Lyn 6h41'38" 52d51'48"
Timothy Bilash
Cnc 8h1'12" 18d41'8"
Timothy Brandon
Ori 5h57'17" 17d47'14"
Timothy Brandon Bender
Tau 3h31'34" 4d33'59"
Timothy Breen Hofmann
Lyn 6h41'35" 61d3'23"
Timothy Brent Malcolm
Ori 5h19'42" -7d41'29"
Timothy Brian Duggan
Uma 11h30'6" 57d9'24"
Timothy Brian & Melisa Jean Mode
Sge 19h56'26" 16d42'29"
Timothy Brumbaugh
Aqr 22h13'22" -8d36'5"
Timothy Bungato
Per 3h19'37" 52d26'48"
Timothy Cassetty
Uma 9h55'21" 61d46'53"
Timothy Chenette
Aqr 22h6'29" -2d7'3"
Timothy Christopher VanEsley
Lib 15h3'20" -0d55'16"
Timothy Clay Fielding
Uma 10h48'45" 57d34'44"
Timothy Clifford Goetz
Her 18h45'16" 21d46'17"
Timothy Cole
Aql 19h18'39" 12d37'18"
Timothy Cole Jordan
Psc 1h33'41" 15d36'30"
Timothy Craig Edwards
Aqr 22h11'22" 0d15'23"
Timothy D Hawkes
Sco 17h40'28" -42d53'29"
Timothy Daniel Brady
Uma 10h1'28" 62d28'54"
Timothy Daniel Buras
Leo 10h58'9" 18d9'33"
Timothy Daniel Hughes
Uma 11h27'20" 44d30'6"
Timothy Daniel Padgett Jr.
Aql 19h25'47" 2d58'42"
Timothy Daniel Smith
Ori 5h29'2" 5d57'23"
Timothy David Castree
Leo 10h6'6" 24d43'57"
Timothy David Dolon
Vir 13h19'10" 7d14'37"
Timothy David Howe
Psc 1h6'27" 6d8'29"
Timothy David Lane (Konstantine)
Ori 6h13'57" 12d5'22"
Timothy David Mauger
Aur 5h14'59" 42d24'21"
Timothy David Oriel
Uma 13h31'21" 61d4'6"
Timothy David Wilson
Cyg 20h39'54" 38d10'20"
Timothy DeBoer
Uma 11h15'28" 62d7'38"
Timothy & Denise Matrimonial Star
Cyg 19h57'2" 36d24'40"
Timothy Dewayne Adams
Cap 20h19'47" -14d46'49"
Timothy Downey, Rock Star
Gem 7h30'41" 26d43'15"
Timothy Doyle
Ari 3h19'32" 28d50'58"
Timothy Dunn
Cnc 8h50'13" 21d12'30"
Timothy E. Finer
And 0h24'22" 43d5'8"
Timothy E. McVeigh
Tau 5h27'58" 18d26'43"
Timothy Eads McGrath
Sgr 18h37'36" -18d34'41"
Timothy Early Bell
Psc 0h6'31" 6d45'22"
Timothy Edward Philip Jr.
Aur 5h52'51" 52d35'26"
Timothy Ellis
Tau 4h14'33" 10d41'38"
Timothy Farrell
Tau 4h45'36" 29d8'56"
Timothy Ferris Dillman
Cnc 8h15'18" 16d35'11"
Timothy Flynn Glover, Jr.
Lac 22h57'16" 44d58'22"
Timothy Francis Musgrove
Sgr 18h29'42" -16d31'38"
Timothy Francis Rizzo
Sgr 18h38'13" -18d31'4"
Timothy Frankel
Uma 10h58'36" 34d27'57"
Timothy Franklin Weir, Jr.- My Daddy
Cap 20h13'20" -14d56'13"

Timothy Garcia
Her 17h37'19" 47d29'52"
Timothy Gardner
Psc 22h57'38" -0d1'36"
Timothy Giras
Lup 15h48'3" -31d50'0"
Timothy Gorzka
Sco 16h18'1" -11d44'17"
Timothy H. Waugh
Lyr 18h50'13" 33d59'31"
Timothy Hamilton Wisely
Vir 12h43'53" -8d31'48"
Timothy "Handsome" Wilkins
Aqr 23h53'47" -10d53'6"
Timothy Hayes Donahue
Del 20h37'3" 15d47'59"
Timothy Headley
Cma 6h40'20" -14d10'35"
Timothy Herman's star
Psc 0h52'58" 5d43'6"
Timothy Howard
Cma 6h59'7" -13d32'36"
Timothy Hudson
Cru 11h56'22" -56d0'46"
Timothy Ignatius Duffy
Umi 12h8'44" 89d33'13"
Timothy Ira Moran
Her 18h56'22" 16d17'13"
Timothy J
Sgr 18h18'29" -22d31'34"
Timothy J. Frantz
Aur 5h35'12" 48d52'39"
Timothy J. Hadland
Per 4h28'51" 43d19'41"
Timothy J. Homer Simpson Hudson
Uma 12h42'38" 57d9'59"
Timothy J. Ward 1956
Lep 5h42'45" -11d29'41"
Timothy J. Weerasinghe II
Lib 14h31'26" -13d53'27"
Timothy Jack Nicodemo
Ori 6h12'38" 13d33'2"
Timothy James Coad 1971 - 2005
Cru 12h21'16" -56d43'31"
Timothy James Hermann-Soulmate/Hero
Lep 5h14'22" -25d33'30"
Timothy James Hill
Per 1h42'30" 50d51'43"
Timothy James Mosher
Uma 11h57'3" 37d40'22"
Timothy James Richters "Angel"
Cru 12h25'1" -60d15'40"
Timothy James Rogers
Sco 16h6'13" -22d47'22"
Timothy James Strovas
Per 3h14'8" 53d58'38"
Timothy Jansen
Gem 7h21'6" 16d10'1"
Timothy Jay
Aqr 21h2'13" -7d21'31"
Timothy Jay Wilson
Cep 22h10'24" 58d40'15"
Timothy Jenny
And 0h11'17" 31d47'9"
Timothy Jeremiah Flatt
Lyn 6h44'18" 56d46'55"
Timothy John Dennerll
Ari 3h16'15" 29d12'16"
Timothy John Emanoff
Per 4h16'33" 51d1'35"
Timothy John Jankowski
Gem 7h7'20" 26d6'24"
Timothy John Karyczak
Umi 16h15'23" 77d22'19"
Timothy John Mitchell
Per 3h35'45" 51d41'9"
Timothy John Sandberg
Ori 4h52'53" 3d3'26"
Timothy John Seitz
Uma 11h55'29" 41d11'29"
Timothy John Sullivan
Aqr 22h10'25" 1d8'51"
Timothy John Summerfield
Psc 0h38'48" 14d33'30"
Timothy&Jonetta1Heart1Soul
Dra 18h26'58" 63d22'6"
Timothy Jordan Beahm
Leo 10h51'50" 18d44'42"
Timothy Joshi Fisler
Her 17h30'53" 45d26'7"
Timothy Journay Adams
Per 3h12'55" 53d43'30"
Timothy Karl Kuhn
Sgr 19h32'52" -13d0'5"
Timothy Kee Williams
Aqr 20h46'41" 0d16'45"
Timothy Kevan Taney
Umi 15h14'59" 75d46'43"
Timothy & Kimberly Bogdanovich
Lyr 19h27'45" 37d40'45"
Timothy Kossakowski
Ari 2h31'5" 26d20'10"
Timothy L. Archambo
Lib 14h54'47" -0d45'0"
Timothy L. Cantrell
Sco 16h53'41" -42d35'44"
Timothy L. Fredrick
Leo 11h39'2" 20d40'53"

Timothy L. Ready
Lyn 7h21'33" 52d21'50"
Timothy L. Webb
Her 16h49'1" 38d29'38"
Timothy Leadbetter
Tri 2h9'24" 33d52'32"
Timothy Lee
Sgr 18h2'59" -17d36'14"
Timothy Lee Gurecki
Sgr 18h48'44" -28d30'17"
Timothy Lee Tracey
Gem 6h34'41" 17d47'9"
Timothy Lee Waud
Her 16h20'7" 19d8'48"
Timothy Leon Butler, Jr.
Pho 0h8'28" -43d9'51"
Timothy Leong
Hya 9h40'1" -21d48'30"
Timothy Loftus
Lyr 18h49'11" 40d5'22"
Timothy Lorenzo DeSimone
Lib 15h13'28" -19d27'13"
Timothy & Lynne Gilbert
Cyg 20h17'49" 58d45'1"
Timothy M. Steudle II
Cnc 8h24'6" 22d4'39"
Timothy Madden
Aur 5h38'35" 49d42'29"
Timothy & Marilyn Duncan
Mon 6h34'22" 10d20'13"
Timothy Marshall Rotherham
Per 3h54'36" 38d26'0"
Timothy Matthew Young
Sco 17h19'1" -39d55'18"
Timothy & Meghan Harper
Cyg 21h25'28" 43d44'42"
Timothy Meppem
Cru 12h36'31" -64d13'25"
Timothy Michael Daly
Cen 13h36'47" -40d59'44"
Timothy Michael Fort
Gem 6h21'37" 18d22'53"
Timothy Michael Gagnier
Ari 3h22'16" 24d37'55"
Timothy Michael Harden
Vir 13h35'42" -7d38'40"
Timothy Michael Hegglund
Aql 19h28'56" 2d29'34"
Timothy Michael Hodges
Lac 22h19'17" 51d35'35"
Timothy Michael Hurley
Uma 11h23'13" 43d55'54"
Timothy Michael O'Quin
Psc 0h15'34" -2d44'59"
Timothy Michael Sola
Uma 10h57'47" 63d43'1"
Timothy Morgan
Her 16h31'48" 34d9'53"
Timothy Murphy Martin
Tau 3h36'11" 11d38'48"
Timothy Myles Simpson
Cas 0h24'26" 52d35'56"
Timothy Neal Atkins
Lib 14h58'59" -12d14'49"
Timothy Nein
Ori 5h49'35" 3d39'49"
Timothy Nelson Crozier
Aur 6h28'13" 35d23'11"
Timothy Nicholas
Gem 6h4'12" 22d42'56"
Timothy O'Connell
Ori 5h29'0" 1d29'58"
Timothy O'Donnell Murphy
Uma 9h11'35" 69d4'3"
Timothy O'Katz Levenson
Tau 3h50'5" 25d37'3"
Timothy Olen Lott
Per 2h26'18" 52d24'14"
Timothy "Our Shining Star"
Psc 1h45'59" 8d52'51"
Timothy Patrick Cornish
Sco 17h40'53" -42d1'35"
Timothy Patrick Day
Vir 12h10'56" 8d0'34"
Timothy Patrick Kelly
Cas 2h5'44" 61d48'15"
Timothy Patrick Lynch
Her 17h43'12" 15d34'48"
Timothy Patrick Ryan
Her 17h47'37" 44d38'32"
Timothy Paul
Boo 14h57'48" 39d55'18"
Timothy Paul DeCarli
Tau 3h35'16" 17d25'23"
Timothy Paul Eland, Jr.
Cyg 19h47'52" 38d41'57"
Timothy Paul Majka
Mon 7h22'11" -6d18'21"
Timothy Paul Wampler
Aql 19h49'39" -0d13'46"
Timothy Pauxtis
Psc 0h43'9" 3d14'5"
Timothy & Penny Hughes' Star
Lyr 18h40'46" 38d44'6"
Timothy Peter O'Malley
Lib 15h8'8" -1d20'45"
Timothy Peters
Lib 15h39'19" -17d28'46"
Timothy Poorbaugh
Vir 14h32'34" -0d53'55"
Timothy R. Deshaies
Tau 4h30'2" 28d50'30"

Timothy Richard Clark
Aql 18h59'28" 7d18'53"
Timothy Robert Barwald
Vir 13h17'14" 5d53'56"
Timothy Robert Hill
Del 20h47'41" 15d1'8"
Timothy Robert Masters
Uma 13h30'51" 54d35'24"
Timothy Robert Snow
Ari 2h45'44" 23d24'59"
Timothy Rowe
Aql 20h0'8" 10d59'53"
Timothy Ryan Fabrizio
Cep 21h39'37" 61d18'9"
Timothy Ryan Fletcher
Gem 6h27'20" 19d45'30"
Timothy Ryan Shane
Ori 5h58'10" 4d16'38"
Timothy S. Davis
Leo 11h22'40" 15d21'39"
Timothy S. Sweeney
Tau 4h35'20" 2d21'29"
Timothy Schang
Aql 19h44'39" -0d2'27"
Timothy Schigur
Uma 10h4'6" 43d55'17"
Timothy "Scott" Megale
Lib 14h39'41" -18d28'17"
Timothy Sean Hoctor
Aql 19h56'15" -0d27'25"
Timothy Shelly J.R.
Psc 1h3'27" 30d58'15"
Timothy Shine III
Uma 11h36'18" 37d58'32"
Timothy Simmons
Aqr 22h7'35" -13d9'4"
Timothy Smith
And 1h6'22" 40d51'45"
Timothy Steve Polgar
Uma 11h22'49" 51d56'11"
Timothy Sullivan
Per 2h38'55" 55d57'10"
Timothy Sullivan
Her 17h43'24" 37d35'25"
Timothy & Susan Snider's 28th Year
Psc 23h22'19" 6d28'27"
Timothy Swallow
Cep 23h51'16" 74d54'14"
Timothy T. J. Fields
Aql 19h57'55" 8d9'16"
Timothy Telymonde
Gem 7h16'44" 26d44'9"
Timothy Tesauro
Uma 8h45'18" 69d1'10"
Timothy (Tim) R. Smallwood
Uma 10h40'56" 39d41'33"
Timothy (Timmy) A.K. Leos
Ori 4h53'48" 12d41'51"
Timothy "TJ" Moore
Dra 19h50'7" 70d35'23"
Timothy To Infinity and Beyond
Vir 12h12'45" 12d10'5"
Timothy & Tracy Sayler
Uma 10h4'22" 58d38'35"
Timothy Ulakovits
Aqr 20h55'58" -8d18'22"
Timothy Umscheid
Her 16h13'53" 47d17'38"
Timothy W. Bevis
Aql 19h23'19" 9d56'34"
Timothy W. Brown
Her 17h9'26" 32d9'31"
Timothy W. Burke
Her 17h6'25" 21d42'53"
Timothy W. Wiegel
Per 3h5'48" 52d2'45"
Timothy Walchusky
Lib 14h49'59" -0d55'10"
Timothy Wayne Smith
Ari 3h25'52" 24d10'7"
Timothy Wayne White, Jr.
Psc 0h59'12" 12d29'39"
Timothy Wayne Wickline
Ori 5h54'53" 12d51'20"
Timothy Wilck
Per 3h9'22" 55d0'54"
Timothy William
Per 3h18'25" 41d38'17"
Timothy William Richards
Lib 15h2'24" -16d44'40"
Timothy Williams
Uma 10h26'8" 49d38'51"
Timothy-Jon "T.J." Dean
Cas 0h51'47" 61d39'23"
Timothy's heart, Annarita Aler
Per 3h2'32" 54d46'1"
Timothy's Perpetual Light
Sco 17h28'36" -43d13'54"
Tim's 45
Aqr 22h30'33" 1d37'30"
Tim's Heart Star
Per 2h53'40" 54d33'35"
Tim's Little Piece of Heaven
Aql 19h13'56" 6d41'28"
Tim's Love For Brenda
Umi 15h34'55" 74d25'46"
Tim's Lucky Star
Sgr 19h4'40" -31d10'44"
Tim's Place
Uma 11h4'9" 58d43'32"

Tim's Star
Dra 18h55'59" 68d3'13"
Tim's Star
Gem 7h2'37" 19d0'21"
Tim's Star
Cyg 21h39'45" 31d29'36"
Timshel
Vir 13h25'6" 11d25'8"
TimTam
Cru 12h20'4" -63d7'40"
tin min
Tau 5h45'39" 23d59'39"
Tina
Lmi 10h46'41" 29d8'25"
Tina
And 0h40'10" 24d22'46"
Tina
Psc 1h5'4" 27d58'37"
Tina
Ari 2h57'35" 21d19'48"
Tina
Vir 12h14'25" 2d13'0"
Tina
And 0h47'34" 38d50'10"
Tina
Uma 10h15'41" 44d54'40"
Tina
Uma 11h57'54" 39d40'32"
Tina
Lmi 10h11'15" 40d11'29"
Tina
Her 16h32'1" 36d36'44"
Tina
Lyn 7h56'3" 34d4'27"
Tina
And 23h19'59" 47d22'13"
Tina
Lyn 7h1'19" 51d54'34"
Tina
Sgr 18h54'44" -30d26'41"
Tina
Sgr 18h5'55" -27d39'42"
Tina
Cma 7h7'14" -22d50'47"
Tina
Sco 16h21'4" -32d18'51"
Tina
Sco 17h43'42" -33d22'51"
Tina
Her 17h38'18" 49d31'31"
Tina
Uma 8h18'2" 72d39'20"
Tina
Uma 8h13'37" 63d58'54"
Tina
Lib 14h36'41" -18d36'51"
Tina
Sco 16h5'52" -9d14'36"
Tina
Sgr 19h5'35" -13d13'21"
Tina
Ser 18h20'23" -10d54'52"
Tina
Lib 14h51'21" -0d52'45"
Tina
Lib 14h55'43" -1d0'33"
Tina
Aqr 22h20'16" -0d18'28"
Tina
Ori 5h52'30" -0d13'22"
Tina Alvarado
And 0h19'12" 43d31'17"
Tina Amvon - 08.08.2006.
Cas 23h39'13" 53d18'49"
Tina and Ashley
Ori 5h58'47" 20d52'47"
Tina and Kevin's Wedding Star
Cyg 20h0'48" 53d9'18"
Tina and Nicole Caputo
Tau 5h11'56" 17d52'17"
Tina & Andrej
Uma 9h35'39" 64d9'32"
Tina Ann Weingartner
Sco 16h18'56" -15d48'16"
Tina Arleen Tate
Lyn 8h57'11" 40d54'35"
Tina Arnold
Cas 1h40'53" 64d9'39"
Tina & Arthur
Cyg 20h19'35" 38d7'24"
Tina babydoll Taylor
Peg 22h21'43" 12d24'37"
Tina Barbayanis
Aqr 21h15'5" -10d49'58"
Tina Barrenger (Teenie)
Cru 12h27'51" -60d34'18"
Tina Beier
Ori 5h18'41" 9d17'27"
Tina Bell
Gem 7h41'46" 22d32'3"
Tina Benjamin
Leo 11h35'42" 8d48'11"
Tina Blanchard
Ari 2h7'31" 24d30'41"
Tina Boone
And 0h40'25" 34d44'7"
Tina Brady Wharton
Per 3h47'33" 43d52'17"
Tina Burke
Uma 11h16'30" 56d16'7"
Tina Cadman
Vul 20h58'31" 22d10'39"
Tina Campbella
Uma 10h24'58" 56d37'16"

Tina Ceremello Morgado
And 1h17'7" 38d43'3"
Tina Claridge
Aql 19h56'59" 10d41'28"
Tina Cotton
Uma 11h30'38" 34d58'55"
Tina Culver
Lyn 8h29'58" 57d27'11"
Tina Curiel
Cnc 8h59'39" 8d22'35"
Tina Cutler
Dra 16h37'28" 54d38'25"
Tina D. Eakin
And 1h18'21" 41d53'33"
Tina DeSimone
Tau 4h25'20" 21d38'14"
Tina Desjardins
Tau 5h38'35" 22d54'45"
Tina Dickey of Trillistar Interarts
Cas 1h4'6" 56d57'5"
Tina DiCuccio, Pharm D
Cas 1h52'25" 61d40'11"
Tina Doll
And 1h31'1" 50d36'36"
Tina Donegn
Vir 13h6'31" 11d27'40"
Tina Donnelly
Uma 13h58'51" 52d40'13"
Tina E. Davis
And 0h38'34" 33d51'34"
Tina Elaine Wilson Bratt
Nor 16h19'47" -46d54'3"
Tina & Frank
Ori 6h21'59" 10d26'58"
Tina Fry
Lyr 18h28'32" 37d19'46"
Tina & Gary, Forever
And 23h26'40" 35d42'33"
Tina Giannetta
Cam 4h11'21" 55d26'50"
Tina Glekas
And 2h23'58" 48d2'58"
Tina Goldstein
Uma 9h31'55" 63d5'20"
Tina "Grandma" Embry
Gem 7h42'37" 16d19'37"
Tina Grossnickle
Cas 1h21'33" 63d51'6"
Tina Guttillo
Leo 10h8'48" 15d49'50"
Tina Hainsworth
Cas 0h1'46" 53d46'50"
Tina Hovey
And 1h12'48" 37d2'4"
Tina Hugentobler
Sge 19h56'16" 17d48'4"
Tina Hughes
Cyg 20h52'50" 51d24'45"
Tina & Iliana Watts
And 1h15'24" 45d21'43"
Tina J. Kayl
Cnc 8h55'4" 16d7'6"
Tina James
And 1h9'43" 42d56'46"
Tina Jichi
Cyg 19h54'40" 39d13'48"
Tina Johanni
Psc 1h40'6" 5d55'56"
Tina Joy Craven
Cnc 8h25'34" 29d1'9"
Tina Kay Case
Cap 20h20'11" -25d53'36"
Tina Kay Speas
Cas 1h35'9" 67d46'20"
Tina Khammas
Vel 9h12'35" -43d47'43"
Tina Knott
Leo 11h41'39" 15d34'31"
Tina L. Galloway
Mon 7h0'17" 4d27'46"
Tina L Griffin
Cru 12h45'8" -58d55'4"
Tina L Hampton
Gem 7h9'43" 24d35'59"
Tina LaJuan Pace
And 23h17'56" 48d40'8"
Tina LaPlace
Sgr 18h48'48" -30d11'55"
Tina Larae Jones
Psc 1h6'21" 32d29'24"
Tina Lavon Hembree
Ori 6h20'2" 6d33'4"
Tina Ledger
Cam 4h17'4" 57d11'16"
Tina Leigh
Cyg 20h31'54" 38d57'39"
Tina Lile
Cyg 21h35'52" 39d30'41"
Tina Louise
And 2h13'10" 41d40'7"
Tina Louise
Leo 11h12'8" 3d47'23"
Tina Louise Evans
Sco 16h48'38" -33d16'35"
Tina Louise Gardetto
Uma 11h19'25" 44d29'13"
Tina Louise Nicosia
Uma 10h19'1" 46d45'55"
Tina Louise Wightman
And 0h8'40" 44d29'54"
Tina Love
Ari 3h8'53" 12d30'6"
Tina "Lover" Ruth
And 1h44'56" 42d16'14"

Tina loves Bryan
Tau 5h55'17" 28d2'45"
Tina Lubbers
Tau 5h26'54" 23d21'31"
Tina Lynette
Aqr 23h10'58" -12d2'58"
Tina Lynn
And 2h37'5" 45d16'21"
Tina Lynn Wood
Vir 13h53'39" -11d39'42"
Tina M. Brugger
Com 12h36'0" 27d55'50"
Tina M. Butner
And 0h22'31" 45d52'1"
Tina M. Chaides
Vir 14h4'45" -14d41'20"
Tina M. Helwajian
Cap 20h31'2" -12d58'59"
Tina M. Langner
Gem 7h50'38" 32d10'9"
Tina M. Moore
And 1h9'6" 38d6'20"
Tina M. Plante-Jancef
Uma 11h21'19" 41d4'26"
Tina Mae Reed
And 2h15'7" 46d25'5"
Tina Maria
Tau 4h22'50" 27d42'59"
Tina Maria
Vir 13h20'13" -2d52'6"
Tina Maria Russo - "Cherished Friend"
Cru 12h21'18" -56d44'27"
Tina Marie
Sco 16h24'6" -33d43'57"
Tina Marie
Sgr 18h18'56" -27d49'18"
Tina Marie
Vir 13h49'11" -10d6'14"
Tina Marie
Aqr 23h3'19" -10d54'29"
Tina Marie
Aqr 23h52'34" -11d7'44"
tina marie
Lib 15h11'24" -11d58'21"
Tina Marie
Sco 16h14'35" -16d52'42"
Tina Marie
Uma 8h36'46" 71d44'49"
Tina Marie
Cas 1h43'24" 64d6'36"
Tina Marie
Cas 23h54'21" 53d49'42"
Tina Marie
Ari 3h11'42" 28d41'56"
Tina Marie
Vul 21h7'32" 26d42'42"
Tina Marie
Lyr 18h58'23" 26d29'4"
Tina Marie
And 0h11'54" 26d2'54"
Tina Marie
Psc 0h11'58" 8d14'31"
Tina Marie
Vir 13h1'43" 3d36'38"
Tina Marie
Her 18h29'27" 12d9'26"
Tina Marie
Tau 3h26'35" 18d51'12"
Tina Marie
Cyg 21h46'5" 42d51'47"
Tina Marie
Cas 0h43'22" 58d22'6"
Tina Marie Abraham
Cam 4h33'18" 58d27'45"
Tina Marie Amendolara
Lyn 8h25'10" 55d51'4"
Tina Marie Belcarris
And 0h46'12" 40d59'39"
Tina Marie Bindner
And 0h49'17" 32d3'36"
Tina Marie Braun
Cam 4h45'16" 68d58'31"
Tina Marie Brodeur
Cnc 8h1'28" 25d51'44"
Tina Marie Diesz
Leo 9h52'58" 14d10'24"
Tina Marie Fann
And 1h20'23" 48d57'18"
Tina Marie Hahne
Lyn 7h49'36" 59d27'9"
Tina Marie Johnson
Gem 7h29'58" 35d5'5"
Tina Marie Lache
Uma 8h38'52" 67d52'40"
Tina Marie Littley/ I Love You-Rick
Her 17h35'15" 24d12'57"
Tina Marie Majora
Vir 13h42'4" -6d8'46"
Tina Marie Martinez
Com 12h37'38" 17d32'36"
Tina Marie Miracle
And 2h15'44" 46d51'35"
Tina Marie Moore
Tau 5h58'21" 27d15'45"
Tina Marie Quick
Cas 0h5'16" 59d30'44"
Tina Marie Stalzer
And 23h10'18" 35d39'47"
Tina Marie Stephanoff
Cas 1h25'26" 62d0'32"
Tina Marie Trepus
Lib 14h53'5" -4d43'8"

Tina Marie White
And 1h47'17" 46d45'59"
Tina Marie's Shining Star
And 1h38'23" 48d17'24"
Tina Marina Xenos
Sgr 19h41'22" -36d21'14"
Tina Mason Seetoo
And 1h15'21" 38d27'0"
Tina May Rancourt
Psc 22h54'56" 2d44'3"
Tina Maze
Her 18h56'31" 17d1'2"
Tina & Michael 7-23-05
Ari 3h15'17" 27d42'25"
Tina Michelle Howell
Aqr 21h45'29" -0d51'57"
Tina Michelle Lewis
Sco 16h15'26" -16d23'59"
Tina Mon Amour
Uma 11h30'53" 39d16'49"
Tina Morales
Cnc 8h53'27" 15d59'50"
Tina Morand "SemperFi"
Uma 11h49'20" 31d32'6"
Tina Morrison (Puter Butt)
And 0h32'29" 29d34'58"
Tina "Ms T"
Vir 11h38'23" 3d40'33"
Tina Nicole
And 23h34'0" 43d0'0"
Tina Noel
And 23h20'7" 38d32'52"
Tina Novacek
Lib 15h29'51" -4d55'29"
Tina - Number 19
And 23h57'21" 40d6'39"
Tina P aKa Pinky Star
Cnc 8h32'32" 12d27'55"
Tina Padilla
Aql 19h28'28" 12d26'39"
Tina Pina Pine Diane Didi Do TLW
And 0h27'40" 22d37'13"
Tina Potter
Leo 9h43'30" 7d40'46"
Tina Prevas
Uma 12h58'1" 55d19'6"
Tina Price
Cru 12h25'20" -61d13'19"
Tina Ptacek
Gem 7h0'20" 33d34'15"
Tina & Quinten Forever
Cyg 20h27'46" 56d6'15"
Tina Rakhit
Lep 5h7'27" -12d2'9"
Tina Reeves
And 23h0'49" 42d28'59"
Tina Rene Davila
Crb 15h51'53" 34d34'54"
Tina Rosanna Ottaviano
Cas 1h42'58" 66d47'41"
Tina Salsone - Mia Bella
Cru 12h38'49" -59d8'22"
Tina Sanchez
And 0h47'53" 37d44'5"
Tina Schaible 20
Cas 2h19'9" 69d36'53"
Tina Schilling
Mon 6h38'8" 4d56'55"
Tina Sian
Cas 23h25'27" 56d7'32"
Tina Sierra
Peg 22h26'5" 23d38'55"
Tina Smal
Uma 9h30'18" 45d45'30"
Tina Stonecipher
Leo 11h32'48" 27d7'0"
Tina Stowell
Psc 1h35'33" 15d5'27"
Tina Sunshine
Cnc 8h51'19" 26d16'45"
Tina - The Only Star In My Sky
Cru 12h46'32" -57d34'7"
Tina & Thor Forever
Psc 23h22'39" 1d38'48"
Tina Tombs
Ari 2h9'59" 14d4'53"
Tina Trudeau
And 0h2'59" 45d24'39"
Tina Trump
And 23h44'8" 47d40'33"
Tina Tschudin
Cas 2h8'52" 59d6'49"
Tina "Venus" Becker
Lyn 7h13'20" 57d8'51"
Tina Vigar
Leo 10h11'4" 11d54'27"
Tina von Allmen
Uma 8h27'46" 62d23'29"
Tina Wager - Prihar
Cnc 8h35'37" 22d40'39"
Tina Ward
And 2h19'3" 44d16'39"
Tina White (nee Hawkins)
Umi 14h57'55" 79d37'7"
Tina Wicker
Lyn 8h16'22" 46d0'22"
Tina Wolf - 04.11.1985
Aqr 21h35'46" -0d20'0"
Tina Xuan Vu
Gem 7h0'39" 17d10'35"
Tina York
Psc 0h26'39" 13d51'15"

Tina Yuhasz
Leo 11h14'17" 12d47'50"
Tina, Blue-Eyed Baby
And 0h19'5" 33d11'3"
Tina, Mark, Sam
Uma 10h15'34" 70d21'40"
tinabelle
Psc 23h23'8" -3d18'2"
Tinabobina 215
Per 2h25'50" 52d28'35"
TinaBopPrincessStar
Cnc 8h44'53" 16d10'49"
Tinabot
And 2h7'39" 45d7'58"
Tinah Winnie Robles-Shure
Cam 4h5'44" 68d0'34"
Tina-Luna
Psc 1h23'9" 24d39'16"
Tina-Maria Dort-Trüb
And 2h0'24" 46d6'3"
Tinamarie
Lib 15h11'18" -10d28'58"
TinaMario
Cyg 20h8'23" 46d34'55"
TinaMe
Crb 15h30'11" 27d33'43"
Tina's Dugan
Tau 5h51'37" 12d57'13"
Tina's Joy
Ori 6h15'16" 20d53'50"
Tina's Light
Umi 15h19'44" 74d18'19"
Tina's Smile
Gem 6h26'51" 17d9'22"
Tina's Special Star
Lyn 7h48'45" 48d36'51"
Tina's "Special" Star
Vir 11h50'42" -2d51'48"
Tina's Twinkle
Vir 13h17'11" 5d37'27"
Tina-The Sunflower Star
Sco 16h18'31" -12d26'4"
TinaWright
Leo 10h4'52" 21d6'28"
Tinchen
Uma 10h59'52" 72d36'7"
Tincie My Love For All Eternity
Cnc 8h12'53" 16d17'30"
Tine Agus Oigbir
Ori 6h22'13" 10d49'6"
Tine Hagl
Lup 15h26'53" -35d26'32"
Tine Marie Amundsen
Lib 15h20'57" -3d42'55"
Tine Petit Amour Blond
Sco 17h48'22" -40d49'36"
Tine Van Dyck
Cap 20h48'34" -20d12'29"
Tiner
Cam 4h21'52" 73d28'47"
Tineso
Dra 19h47'26" 62d43'6"
Tinette & Joel Sterling 5/6/1951
Uma 11h35'50" 57d54'59"
Tinette Korionoff
And 2h19'17" 46d7'7"
Ting Ting
Lyn 8h52'4" 45d22'54"
Ting Ting
Cnc 8h49'38" 26d30'54"
Ting-Chun Chao & Mei-Hwa Wang
Uma 11h56'48" 34d24'22"
Tinger
Aqr 22h19'11" 1d7'31"
Ting-I (Yoko) Wu
Lib 15h54'6" -17d42'46"
Ting-Ting
Vir 12h9'56" -4d57'51"
Tini
Uma 10h0'40" 72d38'59"
TINI
Sge 19h53'2" 17d24'22"
Tini & Fello
Lib 14h59'38" -16d48'40"
Tini Thomas
Uma 9h48'9" 55d28'57"
Tini-"Angel of Berlin"-20/08/1970
Ori 5h17'18" 3d18'32"
Tinin Forever The First
Lyr 19h12'18" 45d55'38"
Tininha
Uma 11h25'51" 47d38'13"
Tink
And 23h48'58" 42d41'21"
Tink
Gem 7h9'45" 33d54'26"
Tink
Per 3h25'17" 44d1'13"
Tink
Aqr 21h19'57" 1d57'53"
Tink
Cap 21h38'20" -16d57'13"
Tink
Sco 16h3'8" -9d27'49"
Tink
Cap 20h24'11" -27d35'4"
Tink 'n Dawna's Stellar Pixie Dust
Aqr 22h18'14" -1d17'33"
Tink Toad
Per 3h51'50" 52d10'2"

Tinkabell'e
Cru 12h2'29" -62d15'13"
Tinkaymie92781
Lib 15h30'37" -22d53'55"
Tinkdles
Cyg 21h12'50" 46d38'14"
Tinker
And 1h42'38" 45d5'56"
Tinker Bell
And 0h54'30" 40d15'16"
Tinker Belle Marler
And 23h11'25" 40d37'23"
Tinker Hamilton
Psc 1h17'37" 17d22'45"
"Tinker Tom"
Uma 11h59'38" 40d56'0"
Tinkerbell
And 1h45'54" 43d14'0"
Tinkerbell
And 0h26'16" 36d11'3"
Tinkerbell
Uma 11h45'27" 30d25'32"
Tinkerbell
And 0h34'19" 33d4'58"
Tinkerbell
Vir 11h44'41" -4d9'51"
Tinkerbell
Lib 15h1'2" -10d5'11"
Tinkerbell
Lib 14h36'21" -11d38'0"
TinkerBell
Sgr 19h36'52" -16d18'27"
Tinkerbell
Uma 13h30'57" 57d10'35"
Tinkerbell
Umi 14h16'48" 74d59'44"
Tinkerbell Beloved
Vir 12h41'45" 11d45'0"
Tinkerbell Reichert
Lyn 7h39'9" 58d53'31"
Tinkerbelle
Ari 2h7'24" 25d43'52"
Tinkerbelle
Lyn 7h1'21" 48d37'47"
Tinkerbelle Tami Maurer
Psc 0h37'56" 8d38'37"
Tinkerbell's Delight
Gem 6h46'36" 30d28'47"
Tinkerbell's Jade Powers
And 1h19'32" 42d34'2"
Tinky Bear
Lyr 18h31'50" 28d24'59"
tinman2005
And 0h15'31" 30d16'57"
Tinnelle Mancuso
Cas 0h42'42" 54d32'28"
Tinnirello
Uma 11h55'9" 43d1'11"
Tino
Lyn 7h55'44" 42d58'22"
Tino Stemberga
Cyg 20h2'16" 33d34'37"
Tinuviel
Lyn 7h15'49" 52d43'8"
Tiny
Uma 8h12'27" 59d41'30"
Tiny
Vir 12h57'57" 12d54'40"
Tiny A Lopez
Ari 3h8'32" 12d20'7"
Tiny Amy
Ori 6h0'15" 18d25'11"
Tiny Dancer
Psc 1h28'59" 27d14'17"
Tiny Dancer
Psc 1h11'52" 27d42'25"
Tiny Dancer
Ori 5h23'39" 3d16'13"
Tiny Dancer
Lyr 18h45'54" 39d51'40"
Tiny Dancer
Lib 15h5'31" -17d35'13"
Tiny Dancer
Lib 14h52'52" -0d46'12"
Tiny Dancer
Lib 14h52'27" -3d10'41"
Tiny Dancer
Sco 16h38'49" -29d2'30"
Tiny Jack
Uma 8h46'6" 68d4'42"
Tiny Nazo G-Unit
Aqr 22h37'40" -4d14'9"
Tiny T Loves Biggie D
Cyg 20h2'50" 37d50'47"
Tiny Tina Block
And 23h1'31" 47d17'16"
Tiny Toad
Tau 5h37'9" 25d20'41"
Tiny Tuff
Vir 12h57'9" 8d54'15"
Tiny Twinkling Tammy
Lmi 10h37'3" 31d9'6"
Tinyer Davey
Peg 22h9'10" 33d29'43"
TinyKen
Cap 21h55'19" -18d26'35"
tinylilspec tj
Del 20h40'22" 15d34'37"
Tiny's Light
Cnc 8h47'34" 32d45'5"
Tiny's Star
Dra 18h40'57" 57d32'34"
Tio
Dra 16h43'38" 61d58'33"

Tio Nacho
Uma 9h25'32" 52d44'10"
Tio Vic, Esquire
Aql 20h0'21" -0d53'40"
Tiotcont Agonochtonnie
Aqr 23h54'46" -7d15'21"
Tip Licker and Cracker
Lyn 6h23'37" 54d51'13"
Tiphani Nicole Thomas
Tau 4h17'57" 11d11'53"
Tiphni Allison
Cap 21h49'7" -18d18'59"
Tiplyonak
Psc 0h52'58" 9d49'59"
Tipper
Cmi 7h30'2" 8d35'12"
Tipper Broderick
Cma 6h23'12" -29d23'56"
Tippy
Cma 6h31'55" -30d45'16"
Tiramisu
Ori 6h14'4" 16d14'56"
Tirso and Janell Bozan
Her 18h16'34" 20d18'20"
TIS LIZ
Peg 23h1'56" 17d29'24"
Tisa
Psc 1h5'28" 3d33'6"
Tischhauser
Dra 17h19'11" 57d49'19"
Tisdrey Torres
Psc 1h45'3" 8d11'1"
Tish
Cas 0h28'21" 50d11'59"
Tish is a star
Gem 7h32'7" 16d22'54"
Tish Marie Budig
Cyg 20h5'38" 38d14'11"
Tish the Dish
Uma 11h42'10" 42d1'9"
Tisha
Gem 7h38'51" 21d45'44"
Tisha
Uma 10h43'40" 58d43'22"
Tisha
Aqr 21h48'53" -1d1'3"
Tisha Hitt
Aqr 23h33'48" -7d0'34"
Tisha Lee Casey
And 23h34'59" 35d15'29"
Tisha Mary
Cnc 8h18'22" 14d20'0"
Tisha19781117
Uma 10h37'11" 42d1'35"
Tisha's Star
Aql 19h9'15" -7d25'30"
Tishauna Jean King
Cas 23h23'15" 54d56'6"
Tishmasosie
Umi 16h35'35" 76d20'49"
Tisna Djumena
Lib 14h22'16" -10d32'19"
Tissle's Star
Uma 10h18'47" 63d32'20"
Tit
Leo 9h26'20" 25d49'17"
Tit' Fileuse
Ori 4h58'45" -2d4'53"
Tita Poopita My Little Brown Dog
Cas 23h23'23" 53d33'34"
Tita Sheila
Cas 0h38'45" 61d33'38"
Titan
Her 16h27'11" 48d30'30"
Titan Mercier
Her 17h49'31" 15d45'13"
Titan The Great
Lyn 7h54'7" 42d33'24"
Titania Harris
Lib 15h26'43" -26d19'56"
Titapita
Leo 11h21'31" 5d17'38"
Tita's Shining Star
Gem 6h57'19" 34d27'38"
Titch
Crb 15h44'59" 32d34'24"
'tite moitié
Umi 15h6'21" 84d35'13"
Titi
Umi 15h32'43" 69d44'48"
Titi
Tau 5h31'4" 25d52'26"
Titin
Dra 19h40'6" 64d44'58"
Titine et Jeannot
Aqr 21h8'21" -2d56'49"
Tito
Ori 5h21'54" 14d43'48"
Tito Garcia
Her 17h36'51" 37d22'16"
TITOR
Uma 8h16'52" 65d17'32"
Titouan
And 23h10'31" 47d35'31"
Titouan Leproust
Leo 11h5'23" -5d28'36"
Titoule
Col 6h34'43" -36d34'5"
Titta
Cas 23h30'56" 52d52'38"
Titta Masi
Cam 6h38'59" 67d21'11"
Titten
Cas 0h4'10" 56d24'29"

Titter Sheila
Cnv 12h53'47" 48d0'47"
Titti
Peg 22h20'5" 17d7'1"
Titti
And 0h7'12" 37d35'4"
tittin
Ari 3h6'32" 19d41'51"
Tittle
Uma 10h42'41" 68d29'17"
Titus
Ari 3h17'47" 12d9'15"
Titus
Boo 13h46'33" 16d42'38"
Titus & Adela
Leo 11h44'29" 25d2'50"
Titus E. Money
Car 9h34'43" -60d53'7"
Tiva & Manyi
Uma 11h34'57" 36d28'55"
Tivoli West
Leo 11h3'30" 3d39'11"
Tiyára Essence Wingfield
Cnc 8h37'57" 19d14'10"
Tiz
Tau 4h44'11" 28d39'36"
Tiziana
Ari 3h2'9" 29d39'32"
Tiziana
Crb 16h23'52" 30d6'16"
Tiziana
Lyr 18h44'50" 41d23'36"
Tiziana
Per 1h45'22" 50d52'28"
Tiziana
Lyn 6h53'26" 60d44'50"
Tiziana fiore selvaggio delle Antille 10-10-1962
Cyg 19h52'35" 33d31'32"
Tiziana Li Quadri Cassini
Lyn 8h25'24" 50d51'15"
Tiziana Mattoscio Sullivan
Tau 5h23'44" 28d0'17"
Tiziana Rusconi
Her 18h2'45" 32d12'52"
Tiziano
Lyr 18h36'58" 31d21'38"
TJ
Lyn 8h13'2" 45d18'32"
T.J.
Tau 3h59'5" 27d14'12"
tj
Gem 6h23'37" 21d21'17"
TJ
Ori 4h49'17" 13d10'40"
TJ #1 Dad
Cnc 8h19'10" 11d16'44"
TJ and Cheryl Owens
Sge 19h34'45" 18d52'16"
TJ and Emma Forever
Uma 8h36'35" 68d25'18"
T.J. Bell.
Lmi 10h9'9" 40d6'1"
TJ Bradley
Uma 13h44'20" 52d26'6"
T.J. Docena
Uma 10h1'43" 56d46'46"
TJ Fedder
Her 17h7'6" 27d3'26"
T.J. Flatley
Uma 8h55'24" 57d42'49"
TJ Gardner
Uma 8h37'25" 59d47'30"
TJ Haislet
Vir 12h26'59" -6d39'4"
T.J. Higgins
Tri 2h37'30" 34d45'36"
TJ Huyler
Aql 19h37'52" 1d19'8"
TJ Kropp
Crb 16h21'14" 37d6'38"
T.J. Lindsey
Per 1h46'26" 51d9'42"
"T.J." Lindsey
Umi 14h23'35" 66d31'34"
TJ Mahon
Umi 14h39'19" 75d4'25"
T.J. & Marcia
Cyg 21h54'37" 42d9'25"
T.J. MICAH
Vir 14h38'40" -1d14'42"
TJ Murphy
Uma 9h48'5" 54d19'28"
TJ: Seven Years a Jewel In His Eyes
Uma 10h39'1" 72d38'53"
T.J. Shadle Jr.
Aqr 23h25'51" -7d37'44"
T.J. " Tommy John " Stiern
Sgr 18h56'49" -34d54'11"
TJ Unique, Beautiful mind and soul
Aqr 22h14'5" -1d49'25"
T.J. Woodward
Leo 9h31'59" 28d41'26"
T.J.A.A.D MINCHELLA
Uma 13h33'49" 55d38'11"
TJC/Infinitus
Per 2h20'59" 54d46'45"
tjc1798
Cap 21h58'26" -10d52'18"
T-Jen
Lib 15h3'50" -23d44'16"
Tjenneman
Uma 12h38'29" 58d53'27"

T.J.H. 08-11-90
Psc 22h56'46" 6d41'13"
TJL - Hold You In My Heart Forever
Lib 15h51'9" -10d42'46"
TJ's Angel
Tau 4h49'30" 23d30'21"
TJ's Star
Vir 13h47'22" 6d45'13"
TJ's Star
Aql 19h54'54" -0d28'18"
T.J.'s Star
Uma 9h43'43" 66d34'52"
TJThaisSky
Sgr 19h27'43" -13d50'49"
TJW & OLP
Cap 20h30'30" -14d15'41"
T.J.W.L. Galateo 00-04-71-71 Ohana
And 23h51'28" 42d1'3"
TJZforever
Ari 2h46'52" 30d36'34"
TK
Psc 22h53'19" 3d56'52"
TK Harris
Lyn 7h40'31" 43d5'7"
T-Kahootz
Ind 20h40'45" -46d49'26"
TKC In Hoc Signo Vinces Boby Z
Mon 6h52'19" -0d47'10"
TKDRBS1PARK
Gem 7h17'19" 16d28'42"
TKO & Hermit
Ori 5h20'38" -4d17'18"
TKR ~ AOMM.....FIMH
Lib 15h49'7" -12d10'9"
TK's Magic Wish
Sgr 18h7'53" -24d51'42"
TKWhite
Aqr 20h39'46" -11d28'16"
T.L. Henry
Ori 5h39'57" 5d42'47"
T.L. Porretta
Aql 19h41'54" 13d14'29"
TL Princess Butter Cup
Ori 6h4'31" 6d6'21"
TLA
Psc 1h26'50" 14d10'59"
TLC
Lyn 7h6'30" 45d49'21"
TLC 30
Leo 10h58'16" -4d41'18"
TLC and SGC
Col 5h30'46" -33d54'4"
TLC I'll Be There -Terry
Per 4h10'50" 46d49'33"
TLC - To The Stars & Back -16.12.66
Pho 1h12'21" -41d21'21"
TLC123
Aqr 22h14'9" 1d5'46"
Tlee
Psc 1h25'45" 25d2'6"
Tlloa1-d
Lyn 7h50'38" 38d40'0"
TLorMassey-Allegra-Reese
Tau 4h6'31" 26d18'22"
T.L.S.
Vir 13h46'7" -8d15'8"
TLSJDR
Uma 11h34'20" 49d33'45"
TLT
Vir 13h21'39" -11d38'5"
TLV 07/03/04
Crb 15h49'10" 30d0'46"
TLW02
Mon 6h32'5" -10d49'36"
TM-05³
Uma 9h53'27" 56d55'28"
T-Mack
Cmi 7h23'15" -0d9'32"
TMAKK
Cyg 19h58'27" 58d52'7"
T-Maxine
Uma 9h30'3" 45d36'13"
Tmctar
Cam 4h45'4" 61d13'27"
TMFS
Sco 16h4'0" -25d10'32"
TMG2006
Uma 9h40'14" 63d35'19"
TMH61206
Cas 0h22'19" 53d29'21"
TMLA Class of 2005
Her 17h17'30" 36d8'36"
TMLA's melody
Lyr 18h26'48" 35d2'35"
TMMS Class of 2005
Uma 10h22'50" 49d40'48"
TMMS Class of 2006
Uma 9h14'23" 48d10'35"
TMMS Class of 2007
Uma 11h15'16" 60d2'49"
TMP 9
Lyr 18h27'37" 39d41'3"
TMRienzi III
Cap 20h55'51" -16d58'13"
TMT Music Man
Vir 14h46'7" 0d34'47"
TnA
Eri 3h55'30" -15d30'36"
T'nes Nevaeh Smith
Cas 0h20'14" 53d12'22"

TNF4EVR
Leo 11h13'55" -0d51'0"
TNK07
Lyn 7h52'43" 56d29'13"
TNRSOHN
Sco 16h51'40" -34d31'32"
TNS Partners
Cas 23h48'57" 59d9'53"
TNT
Uma 11h20'20" 56d37'48"
TNT
Aqr 21h2'41" -9d27'52"
TNT
Per 3h36'44" 34d0'21"
TnT
Ori 4h54'11" 2d38'12"
T-N-T
Del 20h31'53" 12d22'6"
TNT Forever
Lyn 7h29'56" 57d16'13"
TnT Star
Psc 0h56'38" 10d45'19"
TNT Theresa and Tom Forever
Cap 20h59'0" -27d9'4"
TNT781989
Tri 1h51'7" 31d27'5"
To a new beginning- to a New Year
Per 2h14'29" 55d33'45"
To: Aja Morrow, Love: Matt Burgardt
Uma 10h9'14" 54d27'16"
T.O. (A.K.A. Prince Charming)
Sgr 18h9'59" -25d52'38"
To Ashley Sikes With Love From CWH
Ari 2h57'17" 24d23'16"
TO BEAR
And 0h35'5" 41d34'12"
To Brooke, Love Kurt
Cnc 9h6'47" 17d14'49"
to Chasity-The Blowers Daughter
Crb 15h33'56" 26d59'55"
To Daddy, with love, Rico and Curly
Per 2h41'28" 51d30'21"
To Eddy with love
Lmi 10h37'59" 30d54'38"
To Eva Rainelle Sprewell, With Love
Cas 1h33'58" 62d6'27"
To Garry For Garry & Rani From Mum
Cru 12h17'42" -56d47'12"
to guide you Home
Ori 5h50'21" 7d2'27"
To have met you again is a dream
Sge 19h50'33" 17d17'55"
To Jenell The Love of My Life
Cyg 21h30'39" 35d25'11"
To Justin With Love
Tau 4h3'40" 25d45'32"
To Kevin, My Love, Forever & Always
Vir 12h37'47" 12d35'37"
To Love, & Beauty. To My Lisa.
Uma 11h5'42" 34d41'52"
~ To Make You Feel My Love ~
Ori 6h5'45" 13d22'52"
To Mary McCauley Love, Mike
Cas 0h34'46" 50d10'27"
T.O. McFarland
Boo 13h48'21" 15d41'51"
To Melissa: I will always love you.
And 23h54'40" 42d10'18"
To My Angel Michelle
Aqr 21h37'59" -0d1'29"
To my baby, Rachel I Love you
Cap 21h26'28" -26d48'17"
To my beloved Ciara
Sgr 18h31'4" -18d9'44"
To my best friend, Mom
And 2h35'7" 43d32'5"
To My Best Friend, With Love
Ori 6h1'30" 19d13'36"
To My Dear Briana
And 2h31'49" 45d43'20"
To my Dear Love Charmaine Pienko
Cnc 8h29'17" 14d11'15"
To my Enlightened Mom
Ari 3h8'25" 27d56'12"
to my hunny chris love dawn lol x x
Per 3h17'14" 52d45'8"
To my husband Mark Lovejoy, I Love U
Cyg 21h37'24" 54d22'50"
To my love Catalina Ervin
Ari 2h6'13" 25d51'28"
To My Love - Charbel Kazzi
Cru 12h48'32" -64d23'45"

To My Love David
Aqr 23h17'11" -11d45'33"
To My Love Forever Bud Strauss
Boo 15h39'23" 49d20'50"
To My Love Priscilla Karam
Cru 12h44'43" -63d22'9"
To My Love, Forever Dody
Leo 9h28'51" 9d59'13"
To My Love, Pete Anson
Dra 18h25'14" 57d55'58"
To My Love, Vance M. Jones
Cap 20h38'18" -10d28'12"
To my shining star Kimberly M Pratt
Tau 4h57'27" 16d1'33"
To my shining star Michelle
And 0h47'21" 41d6'35"
To my star Justin from his WifetoBe
Psc 0h18'27" 5d4'13"
To My Sweetheart Steven
Ari 3h17'30" 28d32'7"
To My Sweety Louise
Cru 12h28'55" -62d59'1"
To My"Sweet Patty" With LOVE Rosa!
Cas 0h47'3" 50d26'15"
To Nana & Papa with all our love
Per 3h15'21" 40d25'35"
To Our Love Forever Beth & Ernest
Lyn 7h53'10" 45d58'17"
To Our Shining Stars~Our Grandkids
Cyg 20h39'0" 36d20'2"
To Regina For Once In a Blue Moon
Vir 13h20'5" 12d3'46"
To Terry, my shining light. Katie.
Sco 16h7'1" -26d14'38"
To The Best Mom - Shula
Lib 14h32'9" -22d51'53"
To The Man That Rules My Stars
Gem 6h40'45" 35d5'38"
To The Moon and Back (CCG)
Boo 14h37'35" 50d28'32"
To the Spirit of Love
Cyg 20h22'10" 40d22'27"
To the star of my Life David Soto
Lib 15h53'44" -11d29'25"
To THIS Star and back
Cyg 20h5'43" 52d52'56"
To Timmothy Happy One Year Love Bec
Sgr 18h36'23" -23d54'0"
To Us Gene & Deavah
And 0h54'9" 41d54'13"
To Venus Chang From Admirer Eric Wu
Cnc 8h10'8" 14d7'48"
To you Jayde, my True Love forever!
Cas 23h8'32" 58d15'44"
Toad
Uma 9h35'15" 43d58'3"
Toadette
Tau 4h3'2" 29d2'57"
Toady
Gem 6h55'27" 25d44'47"
Toady
Cyg 19h35'53" 31d31'52"
Toastywarm
Cas 23h2'5" 53d29'39"
Tobaura
Cam 4h31'48" 73d40'27"
Tobey
Uma 9h16'27" 62d11'8"
Tobey Ann
And 0h27'44" 37d27'45"
Tobey Spitzer
Dra 16h39'48" 53d53'15"
Tobeykins
Sgr 19h6'42" -31d40'28"
Tobi
Cas 2h50'36" 60d25'15"
Tobi Ann Good
Crb 15h31'38" 32d48'41"
Tobi Beth Horgan
Psc 0h42'35" 6d26'43"
Tobi JoDean Kellogg
Uma 9h24'59" 50d26'9"
Tobi Louise
Vir 13h21'7" -10d55'18"
Tobi Moser
Lmi 10h32'25" 30d33'52"
Tobi, mein Stern
Dra 18h20'0" 50d46'30"
Tobias
Psc 0h54'53" 13d10'36"
Tobias
Uma 12h9'20" 54d20'3"
Tobias
Sgr 18h33'1" -31d24'50"
Tobias A.J. Freitas-Rogers
Psc 1h9'6" 24d12'53"
Tobias Alexander Edward
Leo 9h25'56" 6d33'23"

Tobias Alexander Green
Ori 5h3'40" 9d36'33"
Tobias Burke
Agr 21h7'35" -3d7'52"
Tobias Christian Schwabe
Aur 5h42'15" 49d59'5"
Tobias Christopher Cole
And 2h28'46" 43d43'47"
Tobias David Stewart
Per 3h4'2" 47d28'59"
Tobias Fritzius
Aur 6h30'21" 38d43'46"
Tobias George Malone
Aql 20h0'30" 7d50'38"
Tobias & Nadine
Uma 8h39'40" 58d51'9"
Tobias Nolan
Ori 5h38'0" 1d56'7"
Tobias Regner
Uma 9h48'57" 49d1'25"
Tobias S. Mitchell
Her 17h30'53" 46d1'3"
Tobias Schneider
Cam 8h4'25" 61d15'14"
Tobias Schröter
Ori 5h14'4" 8d3'49"
Tobias & Simone
And 0h42'42" 42d4'31"
Tobias Stoulil
Gem 7h37'2" 34d44'49"
Tobias William Frankling
Umi 17h21'26" 86d0'10"
Tobias Zettl
Uma 12h50'9" 56d49'29"
Tobie & Bryan Earley
And 0h19'13" 39d29'12"
Tobie's Pantera
Leo 11h24'11" 18d28'38"
Tobin James Wright
Leo 10h24'48" 13d53'30"
Toby
Cnc 8h24'30" 25d48'4"
Toby
Cnv 13h42'51" 33d29'27"
Toby
Lyn 7h36'23" 49d58'58"
Toby and Brittany
Ori 5h26'48" 2d22'2"
Toby Barnes Hast
Uma 14h23'8" 56d9'50"
Toby Becnel
Srp 18h33'51" -0d14'1"
Toby Benjamin
Aql 19h14'8" 11d28'33"
Toby & Bridget
Ori 5h57'44" -0d41'35"
Toby C Minton
Crv 12h34'28" -13d18'16"
Toby Dutfield
Per 4h40'37" 36d40'14"
Toby Edward Plater
Dra 15h23'54" 64d10'43"
Toby Encarnacion
Cnc 9h0'51" 16d1'56"
Toby Ethan Grabruck
Lmi 10h24'20" 29d38'35"
Toby George Corfield
Her 17h15'21" 21d58'23"
Toby George Seller
Cep 0h21'8" 73d16'8"
Toby Graham Pennington
Dra 19h32'50" 65d0'25"
Toby Jay Lockhart
Cep 22h42'3" 77d26'46"
Toby John
Dra 20h26'18" 67d12'16"
Toby John Moore
Psc 0h25'56" 3d56'38"
Toby Klein
Per 4h10'39" 33d54'39"
Toby Lee O'Brien
Leo 11h27'51" 10d44'25"
Toby Michael Bailey
Lib 15h16'24" -4d23'10"
Toby Moore
Uma 10h58'20" 43d33'5"
Toby My Love
Equ 21h3'9" 3d20'55"
Toby Oesd
And 0h43'0" 36d47'24"
Toby of full heart "12-14-91"
Lyn 6h38'16" 54d16'0"
Toby Ouellette
Gem 6h39'14" 27d10'11"
Toby Quaife
Sgr 19h38'54" -13d42'28"
Toby R. Kiker
Uma 11h46'29" 47d31'9"
Toby Sniezek
Cmi 7h32'3" 2d51'42"
Toby Supernova
Cru 12h33'23" -59d28'48"
Toby "Sweet Puppy" Broniarczyk
Uma 11h56'58" 48d9'35"
Toby Swift-Holdcroft
Psa 23h4'21" -34d16'0"
Toby Watchorn
Umi 16h14'39" 81d56'37"
Toby,Ben & Zoe Friedman 1/28/2004
Per 3h26'23" 43d49'15"
TobyBear
Sgr 19h57'39" -41d3'52"

Tobydog
Cma 6h37'46" -25d3'59"
Toby's Birthday Star 30/12/2004
Cap 20h7'21" -15d21'44"
Toby's Eye
Aqr 23h12'4" -15d56'54"
Toby's Love
Cas 0h4'31" 53d57'44"
Toccoa Cotee Brunson
Uma 10h51'0" 60d19'35"
Tochter Bailey Amber
Uma 11h4'22" 70d49'12"
toCRMsomethingPJVnew
Leo 11h44'1" 20d7'13"
Tod Jay Collins
Uma 9h21'12" 59d15'14"
TodAsh
Sge 19h26'47" 17d43'45"
Today Tommorow & Forever I Love You
Cyg 21h11'43" 43d38'18"
today, tomorrow, forever
Tau 3h51'16" 3d27'31"
Todd
Leo 9h45'49" 28d5'54"
Todd
Per 4h16'53" 51d18'9"
Todd
Cap 20h44'30" -17d28'31"
Todd A. Endsley 10-16-1984
Lyn 6h48'5" 50d57'47"
Todd A. Norton
Ari 2h34'58" 13d7'15"
Todd A. Sanzo
Her 16h44'15" 46d55'30"
Todd A. Totty
Leo 10h38'51" 13d18'42"
Todd Aisenbrey
Sco 17h25'49" -37d52'5"
Todd A.K.A. Mr. Alabama
Sgr 18h23'44" -18d28'7"
Todd Alan Anglin 6
Psc 1h11'59" 25d48'49"
Todd Alan Cox
Boo 15h42'14" 40d45'24"
Todd Alan Paige
Ari 2h39'20" 12d38'21"
Todd Alan Sandler
Her 16h23'42" 43d52'1"
Todd Alfred Lawless
Per 3h20'35" 46d52'45"
Todd and Angela
Umi 15h28'13" 73d16'26"
Todd and Brandye
Cyg 19h41'12" 53d12'26"
Todd and Deidra
And 0h51'9" 42d5'2"
Todd and Emily Bourak
Sgr 18h49'28" -23d14'13"
Todd and Heather's Star
Sgr 19h28'3" -15d18'32"
Todd and Jennifer Doobrow
Ori 6h18'24" 14d54'12"
Todd and Jennifer Olson
Gem 7h9'45" 32d36'58"
Todd and Julie's Star
Cyg 20h41'44" 40d8'24"
Todd and Kiki's
Ari 2h43'49" 25d51'51"
Todd and Kimberly's Eternal Love
Leo 9h23'6" 9d34'32"
Todd and Lisa, Through Eternity
Lyn 9h11'52" 46d10'25"
Todd and Nickole
Uma 9h32'51" 54d41'39"
Todd and Rosemary Moss
Cyg 19h47'39" 31d32'33"
Todd and Shiela Scarpino
Lmi 10h58'21" 27d41'30"
Todd Anderson
Gem 7h27'27" 25d37'23"
Todd Andrew Libman
Her 17h45'13" 47d58'58"
Todd Andrew Raivala
Aur 5h41'18" 47d57'22"
Todd Anthony Gugluizza, US Navy
Tau 4h25'30" 13d15'18"
Todd & Ashley
Sgr 18h10'31" -17d36'34"
Todd Ballantine
Uma 8h14'26" 68d3'51"
Todd Barnum
Leo 11h14'30" 18d33'32"
Todd Barry
Ari 3h13'10" 28d19'12"
Todd Barton Shines Eternally
Ari 3h18'30" 20d0'50"
Todd Beasley
Leo 11h26'39" 20d19'42"
Todd & Bertha McIntyre
Cyg 21h29'53" 54d10'22"
Todd & Beverly Ackley
Cyg 21h52'35" 46d43'10"
Todd & Brenda Schook
Cyg 21h38'53" 49d58'12"
Todd & Carissa
Cyg 21h42'17" 47d45'12"

Todd & Carmela
Cyg 20h0'31" 42d49'51"
Todd Christopher Masterson
Leo 11h17'22" -2d38'49"
Todd Christopher Mensch
Gem 7h13'46" 33d1'25"
Todd Clapp
Uma 11h44'38" 47d48'4"
Todd Daniel Joseph Materazzi
Gem 7h37'30" 24d9'2"
Todd David
Aqr 22h4'25" -13d15'50"
Todd Douglas Hendrix
Aur 6h26'12" 32d56'15"
Todd Douglas Moore
Tri 1h48'29" 34d5'52"
Todd Drew
Ori 5h26'43" 3d24'46"
Todd E. Tomich
Tau 5h59'30" 25d22'9"
Todd Edward Ambourn
Ori 6h5'28" -0d32'16"
Todd Eggas
Aql 19h11'1" 4d43'39"
Todd Elliott
Her 18h56'56" 24d5'59"
Todd Estabrook
Per 3h20'16" 44d45'8"
Todd Evan Zwahl
Tau 4h18'32" 9d16'13"
Todd Fritsche
Dra 18h45'8" 50d22'21"
Todd Fritzler
Cyg 20h18'43" 52d9'2"
Todd: God's green eyed wonder
Ori 5h33'18" 9d12'14"
Todd Graham - Our Little Star
Cru 12h9'51" -62d3'20"
Todd H. Brown
Ori 6h12'35" 15d33'45"
Todd Haman
Cep 23h8'40" 76d31'17"
Todd Hellman
Per 3h20'21" 47d48'34"
Todd Henson
Her 16h40'6" 37d21'27"
Todd "Itty Bitty" Davison
Lib 14h33'3" -16d2'51"
Todd J Baker
Cep 23h50'11" 74d51'47"
Todd John Hendrickson
Gem 6h40'42" 34d46'0"
Todd & Joy Eichorst
Cyg 21h42'22" 55d11'15"
Todd K. Heinle
Lac 22h52'46" 38d12'14"
Todd Kenneth Grundner
Tau 4h36'21" 1d26'59"
Todd & Kimmi
Cyg 20h3'39" 30d28'9"
Todd L. Owens
Her 17h53'58" 43d16'38"
Todd & Laura
Lyr 18h31'4" 39d25'40"
Todd Loenhorst
Cyg 19h54'3" 30d42'30"
Todd Luis Tippin
Lep 5h12'17" -12d33'53"
Todd M. Finch
Ori 6h13'18" 13d45'5"
Todd M. Garis
Lib 15h16'15" -25d50'55"
Todd M. Mendoza - Laura Ringemann
Uma 10h55'30" 53d53'52"
Todd M. Whatley
Dra 16h7'19" 68d34'17"
Todd & Margaret Salyards
Ori 5h57'24" 18d1'39"
Todd & Maria
Cyg 20h16'0" 31d52'46"
Todd Mc Closkey
Ori 5h50'5" 20d38'23"
Todd McClintock "1981-2007"
Lmi 9h33'24" 36d59'24"
Todd Michael
Uma 9h38'37" 44d26'11"
Todd Michael Keblish
Ori 6h4'32" 18d45'57"
Todd Michael Montanaro
Cep 21h26'1" 59d29'0"
Todd Michael Shaw, Jr.
Sco 17h33'27" -32d48'36"
Todd Moen
Ori 5h13'15" -9d55'43"
Todd Montague
Per 3h17'18" 52d13'56"
Todd Morris
Uma 9h29'22" 68d14'23"
Todd Muhlenkamp
Mon 7h31'41" -0d34'54"
Todd my forever love 2496
Lyn 6h24'2" 54d13'31"
Todd "Mynony" Bacchi
Her 16h39'28" 27d49'46"
Todd Nelson Humble
Per 3h9'9" 55d41'13"
Todd & Nicole
Cyg 20h55'20" 40d10'45"

Todd Oprzedek
Lac 22h45'1" 52d23'11"
Todd Ostendorf
Per 3h18'45" 39d35'7"
Todd Patrick Hoffman
Uma 14h28'2" 56d20'5"
Todd Paz Fyffe Jr.
Her 17h59'44" 22d19'17"
Todd Quincy Rickett
Tri 1h56'16" 30d34'28"
Todd R. Schlifstein
Cap 20h24'18" -20d1'56"
Todd R. Spicher
Uma 11h25'51" 31d58'15"
Todd Richard Yurko December 4, 1979
Per 4h42'48" 44d40'35"
Todd Robert Nelson
Leo 11h10'58" 15d5'2"
Todd Roberts
Psc 22h54'10" -0d1'23"
Todd S Petzel
Her 18h45'35" 21d35'40"
Todd Shining
Per 4h24'36" 43d30'18"
Todd Siegal
Lib 14h35'8" -22d29'11"
Todd Smith
Tau 5h44'33" 22d18'59"
Todd & Sonya Eury
Cyg 19h38'48" 53d39'17"
Todd Stonnell
Aqr 23h18'48" -14d29'38"
Todd Stuart Rosemurgy
Her 18h3'24" 21d43'49"
Todd The Toad
Her 18h53'30" 23d40'36"
Todd Tubman
Lac 22h31'5" 49d57'39"
Todd Victor
Psc 0h10'51" 6d3'55"
Todd Ward Hickox
Aql 20h19'30" 3d39'36"
Todd Whiteneck
Vir 12h43'21" 9d3'17"
Todd Wyatt
Boo 15h32'24" 40d1'59"
Todd, Brandy, Chelsea and Cassidy
Umi 13h42'9" 86d10'5"
T.O.D.D.1015 "Sparky"
Lyn 8h4'43" 41d58'17"
Todd-n-Kelli
Sco 16h15'4" -9d47'7"
Todd's Great Ball of Fire
Uma 12h39'48" 57d29'37"
Todds Shining Star
Cnc 8h20'59" 24d6'11"
Todd's Star
Per 4h24'2" 44d44'25"
Todd's Star
Cam 3h59'54" 69d27'53"
Todd's Star
Uma 12h4'46" 63d12'13"
Toddy & Sarah's - Star of Schmoopy
Pho 23h33'53" -43d24'22"
Todge
Cep 21h9'56" 66d23'17"
TO.DI.MA 72
Cma 7h15'48" -31d30'58"
Todor and Carmen Djulvezan
And 1h22'21" 45d29'29"
ToE
Gem 6h35'26" 12d24'4"
Toebadoe
Cnc 9h7'21" 17d34'55"
Toester & Hobbit
Cra 18h5'53" -40d6'54"
TOFI
Uma 10h29'7" 49d49'47"
Togeber Foreber
Cyg 21h45'14" 49d46'39"
Together
Crb 15h24'45" 25d43'30"
Together
Aql 19h12'54" 7d5'57"
together 4ever
Ari 2h24'32" 24d2'6"
together again....forever
And 23h23'17" 42d23'34"
Together Always
And 23h55'51" 47d37'3"
Together Always
Uma 11h13'49" 32d42'29"
Together as 1
Cyg 20h13'26" 38d36'35"
Together as One
Aur 5h48'40" 38d40'44"
Together As One
Lyn 6h51'15" 52d59'27"
Together for Eternity
Ori 5h38'56" -3d59'24"
Together for Eternity, Mark & Donna
Uma 10h16'20" 47d44'7"
Together Forever
Cyg 20h39'5" 47d19'33"
Together Forever
And 23h12'46" 47d56'22"
Together Forever
Cyg 20h19'51" 44d11'44"
Together Forever
Cnc 9h6'40" 14d16'49"

Together Forever
Cap 20h8'32" -18d9'21"
Together Forever & Always
Uma 11h24'46" 58d35'45"
Together Forever Claudia Love Jamie
Vir 13h50'49" -7d38'15"
Together Forever Damiano&Nathalie
Aur 5h17'14" 30d1'33"
Together Forever - Mom & Dad
Lib 16h1'15" -11d46'15"
Together Forever never To Part
Sco 16h8'37" -13d23'57"
Together Forever - Peter and Ann
Ori 5h33'44" -8d34'15"
Together Forever Russ & Agathe Hope
Aqr 23h23'50" -7d12'9"
Together Forever, Glenn & Justine
Psc 1h15'10" 4d38'46"
Together in Paradise
Cyg 21h34'53" 46d49'2"
Together In Spirit 813
Aqr 22h27'0" -0d52'4"
...Together Until The End...
Cyg 21h22'37" 39d44'5"
Tognaccini Francesca
Lyr 18h29'51" 26d15'12"
Tognin Stella Miriam
Ori 5h59'5" 18d43'58"
Tohtonku Sdn Bhd
Ori 5h38'59" 3d52'11"
Toi et moi mon amour
Mon 7h19'46" -4d56'44"
Toi Toi
Peg 21h24'4" 21d40'46"
Tojatikawhma
And 0h31'39" 39d53'34"
Toka Terepo
Cru 12h36'21" -58d25'2"
TOKEN
Sco 16h40'57" -37d12'55"
Token of Our Love- Jesse-Laura
Cyg 20h12'47" 42d14'31"
Tokiwa
Vir 12h21'10" -6d2'24"
Tokla
Sgr 18h28'12" -28d19'46"
Toktam Tayefeh
Leo 9h36'9" 29d58'3"
Tokunbo Ayodele
Lyn 6h53'17" 56d24'15"
Tokuyuki Wakiyama D.C.
NEURO-IT
Peg 23h14'17" 30d25'14"
Toledian Rose IV XXIII
Tau 4h40'13" 22d42'44"
TOLGA 42374
Dra 20h32'22" 73d37'50"
To-L-Jo-S
Uma 9h52'28" 57d2'18"
Tolleranza
Uma 11h39'45" 48d31'20"
Tollie's Cynthia
And 23h18'58" 47d9'39"
Tolly Bledsoe
Aql 20h5'42" 3d23'55"
Tollystar
Ori 6h9'24" 7d16'46"
Tolu19800517
And 1h26'27" 47d13'26"
Tom
Per 2h49'59" 51d14'26"
Tom
Lyn 7h38'41" 48d2'27"
Tom
Uma 9h35'8" 47d15'50"
Tom
Her 16h14'3" 47d47'31"
Tom
Uma 12h5'24" 36d58'5"
Tom
Ori 6h7'34" 10d23'9"
Tom
Ari 2h2'17" 14d26'25"
Tom
Tau 5h22'56" 27d15'26"
Tom
Ari 2h35'21" 27d29'6"
Tom
Uma 13h38'38" 55d52'56"
Tom
Umi 14h47'50" 73d13'20"
Tom
Dra 17h58'57" 79d52'4"
Tom
Aqr 23h47'15" -13d53'0"
Tom
Cap 20h10'22" -13d28'55"
Tom 47
Uma 11h1'49" 48d25'57"
Tom & Agnes - Together Always
Cyg 20h6'54" 35d37'27"
Tom Ainsworth
Leo 9h43'5" 9d35'15"
Tom Allinder
Per 3h23'5" 40d42'31"
Tom and Ann's Love Shining Bright
Vir 12h19'37" 4d4'57"
Tom and Audrey Durkin
Sge 19h42'12" 17d45'5"
Tom and Becky
Cyg 20h2'31" 56d4'4"
Tom And Belle's Desert Star
Umi 14h44'57" 72d27'38"
Tom and Brandi Wurtz
Cyg 20h17'39" 54d45'38"
Tom and Christine De Grazia Star
Cyg 21h10'18" 35d30'28"
Tom and Danielle
Cyg 19h58'53" 37d4'14"
Tom and Diana
Cyg 19h36'47" 37d25'52"
Tom and Eileen Shine As One Forever
Per 2h54'37" 48d55'57"
Tom and Elyse
Cyg 20h23'18" 54d53'2"
Tom and Ginny Bean
Cyg 20h44'11" 44d36'29"
Tom and Haley
Cyg 21h20'41" 44d3'33"
Tom and Irene Elliott's Family Star
Uma 10h35'0" 47d21'5"
Tom and Joanna Always
Crb 15h37'9" 27d10'44"
Tom and Julie Fielder
Uma 10h55'39" 41d51'9"
Tom and Kerry Hewson
Cyg 21h30'14" 46d43'15"
Tom and Kimberly Deverell
Psc 0h9'52" 2d2'55"
Tom and Leah = Soulmates
Cnc 8h40'39" 22d56'49"
Tom and Martha Knight
Ari 2h22'54" 12d13'9"
Tom and Mary: 25 years
Cyg 21h44'12" 38d10'54"
Tom and Mary Spears
Uma 10h35'48" 49d46'58"
Tom and Natalie
Cyg 21h10'18" 42d32'9"
Tom and Rebecca: Together We Fly
Aql 19h46'22" -0d45'40"
Tom and Sarah Forever and Always
Cen 14h15'12" -34d56'16"
Tom and Sherri Dream Star
Sge 19h12'58" 18d1'56"
Tom and Stephanie Forever
Sge 19h23'0" 18d33'44"
Tom and Theresa's Star
Cyg 19h45'57" 53d46'8"
Tom Andersen
Her 16h31'32" 21d0'45"
Tom & Ann Dyckman 50 Years
Ori 6h3'42" 16d31'7"
Tom Anne-Ca Star
Aqr 20h39'54" -6d0'35"
Tom Arbisi
Cep 22h23'18" 58d53'51"
Tom Areton
Cep 23h14'9" 72d2'56"
Tom & Arlene
Tau 4h33'0" 17d4'31"
Tom Arone
Leo 10h7'13" 9d36'8"
Tom Arras
Her 18h12'6" 27d50'34"
Tom Barrell
Ori 5h24'14" 8d54'28"
Tom Bear
Uma 10h42'12" 58d29'44"
Tom Bobi
Aqr 23h3'52" -7d2'4"
Tom Bond
Gem 6h18'16" 22d12'53"
Tom Boster "TAB's Perspicacity"
Lib 14h23'50" -16d56'44"
Tom Bourne "My Tommy"
Aql 19h47'49" 3d52'34"
Tom Bove
Vir 12h37'46" 11d27'31"
Tom Brannon
Equ 21h15'29" 5d51'23"
Tom Brennan
Aqr 22h13'44" 1d33'29"
Tom Brennan
Psc 23h8'30" 6d9'56"
Tom Brosche
Aql 19h42'40" 5d55'12"
Tom Brown
Dor 4h10'45" -55d54'7"
Tom Burns
Dra 20h39'30" 69d44'30"
Tom & Carol Smilowski's Star
Lyr 19h19'14" 29d5'46"
Tom & Carole's Christmas Star
Crb 15h37'29" 34d50'15"
Tom Carruthers A Star Always16.1.25
Cru 12h28'6" -63d43'39"
Tom Cat
Vir 13h24'18" -7d52'55"
Tom Cernock
Uma 8h56'3" 52d20'29"
Tom "Chief" Hamilton
Leo 9h39'52" 27d51'16"
Tom Creel
Boo 14h44'4" 21d6'31"
Tom D. Pettry
Gem 7h22'3" 16d14'10"
Tom "Dad" Hutchinson
Cyg 19h48'44" 39d12'55"
Tom Daddy Merlo
Psc 1h0'2" 28d28'14"
Tom & Darla Swiatek
Cyg 20h8'26" 34d21'20"
Tom Darrow
Cap 21h49'10" -13d15'7"
Tom De Kloe
Vir 12h41'59" 12d56'34"
Tom Deedigan 50th Birthday Star
Per 3h31'44" 50d6'6"
Tom & Della Forever
Cyg 21h31'42" 37d37'44"
Tom & Diana Skaff "6-20-1975"
Cyg 20h48'50" 50d12'45"
Tom & Dianne Robinson Birthday Star
Cyg 19h46'30" 37d34'58"
Tom Doherty
Aqr 22h55'42" -8d4'41"
Tom & Dolly Rockell
Psc 0h23'37" 10d34'25"
Tom Dooley
Boo 14h25'11" 28d40'58"
Tom Dorywalski
Sgr 18h7'3" -22d45'15"
Tom Downing
Uma 11h51'10" 48d51'30"
Tom Doyle
Cyg 20h12'15" 59d1'22"
Tom Drougas
Vir 12h16'55" 12d17'25"
Tom Durand
Cep 22h34'24" 57d21'2"
Tom E. Knobel Star
Cyg 19h47'15" 33d21'57"
Tom Eiden
Vir 13h58'16" -10d39'4"
Tom & Emily
And 23h27'22" 48d23'26"
Tom Evans
Per 3h0'15" 45d37'25"
Tom F. Cone Jr.
Uma 13h59'19" 61d58'25"
Tom Fabietti
Uma 8h24'12" 70d58'22"
Tom & Fay
Dra 9h40'51" 78d5'19"
Tom Feil
Aur 6h11'11" 42d11'30"
Tom & Fern McKeown
Umi 15h7'57" 77d44'9"
Tom Flaherty
Psc 1h9'55" 10d2'50"
Tom & Fran "Love Eternal" 04-15-50
Uma 10h26'2" 48d25'23"
Tom G. Schwan
Ari 2h38'37" 24d53'25"
Tom Galvin
Ari 2h46'21" 26d6'8"
Tom Garrison
Crb 15h43'43" 35d23'34"
Tom Gavin
Per 3h11'26" 42d1'4"
Tom & Gayle Digan
Cyg 21h36'59" 33d15'50"
Tom Gerrard
Leo 10h19'4" 26d20'41"
Tom & Gloria Oki
Ori 5h38'28" -0d24'20"
Tom Goebel
Ari 2h51'10" 24d52'42"
Tom Goodwin
Her 16h40'56" 43d2'6"
Tom Goodwin
Sco 17h54'3" -38d35'5"
Tom Goris, Jr.
Her 16h45'28" 27d51'46"
Tom Grzadzieleski
Ori 6h2'8" 19d41'48"
Tom Gunn
Her 16h59'17" 17d55'47"
Tom Haller, The Rock Star
Lyn 7h43'52" 51d35'49"
Tom Heavey
Gem 7h37'0" 23d10'29"
Tom & Helen
Uma 11h14'41" 57d22'56"
Tom & Helen Hartnett Memorial Star
Cap 20h38'14" -10d34'32"
Tom Hennon
Tau 5h20'42" 18d14'6"
Tom Hingerty
Tau 4h38'32" 20d10'2"
Tom Houser
Scl 0h12'10" -25d17'11"
Tom Houston
Aql 19h36'34" 8d36'40"
Tom Hovanec
Cep 23h48'38" 68d13'47"
Tom Ian Barrie Roberts
Umi 14h51'12" 71d14'50"
Tom J Norton
Uma 9h23'17" 45d43'0"
Tom J Yablonski
Vir 13h20'3" 3d24'55"
Tom James Hitt - Tom's Star
Uma 13h25'14" 54d37'27"
Tom Janosky
Psc 0h38'23" 10d42'51"
Tom & Jen
Gem 7h3'9" 19d41'13"
Tom & Jenna Boley's Dream
Cnc 8h54'4" 25d26'18"
Tom Karpecki
Vir 13h26'56" 8d16'40"
Tom & Kathy
Ori 5h49'25" 11d53'8"
Tom & Katie Dean 7-19-94 & 10-25-92
Cyg 20h20'17" 55d34'31"
Tom Kattelman
Lib 15h4'53" -9d29'27"
Tom Kaye
Vir 12h48'2" -1d16'20"
Tom & Kazuko Jones
Uma 9h23'16" 44d20'5"
Tom Keefer
Dra 14h37'33" 55d52'31"
Tom & Kelley: two peas in a pod
Ari 2h50'2" 20d56'37"
Tom & Kim Richards
Lyr 18h33'12" 46d52'29"
Tom Kissling
Cyg 21h19'52" 45d16'45"
Tom Klauer, Jr.
Vir 14h0'42" -9d19'22"
Tom Klein
Vir 13h5'37" 8d15'5"
Tom & Leisa Mullarkey
Cyg 19h48'54" 31d50'49"
Tom & Linda Szabo
And 1h41'19" 43d28'44"
Tom & Lisha
Lib 14h46'32" -0d41'2"
Tom Little
Gem 7h16'20" 20d4'48"
Tom Locchetta, Sr.
Uma 11h59'41" 60d18'38"
Tom Lodemore
Uma 8h30'21" 62d44'38"
Tom & Lois Greiner
Lyn 8h43'32" 45d33'4"
Tom & Lorna
Cyg 21h55'9" 54d53'42"
Tom Lundstrom Eye in the Sky
Crb 16h11'30" 36d35'16"
Tom & Lynn Cameron
Cyg 21h21'5" 36d8'15"
Tom Mack 10/14/37 - 11/19/03
Uma 12h53'19" 60d42'10"
Tom Mahe
Ari 1h53'30" 13d11'8"
Tom Marchioni
Lib 14h48'21" -5d8'20"
Tom Marinich
Her 16h19'13" 45d19'5"
Tom McCartney
Per 4h16'56" 46d53'13"
Tom McLey Go Huskers Go
Dra 18h54'11" 56d17'34"
Tom Mehari & Tsega Asefaha
Cyg 20h4'45" 55d53'52"
Tom Meredith
Pho 0h33'45" -41d39'51"
Tom Mervine
Dra 19h39'8" 70d40'38"
Tom Mervine
Aql 19h29'27" 10d29'57"
Tom Mills
Ori 6h3'8" 17d59'28"
Tom & Molly's Star
Cnc 8h20'0" 27d32'10"
Tom Moore
Her 16h32'57" 38d57'12"
Tom Morley
Vir 12h48'7" -11d8'48"
Tom Mueller
Vir 12h34'33" -11d38'44"
Tom Murray's Shining Star
Uma 10h13'37" 54d4'53"
Tom - My Long Lost Companion
Lmi 9h59'13" 36d0'54"
Tom n Jerry Pelton Aug 9, 1947
Cas 0h29'13" 57d51'41"
Tom & Nancy 2/13/88
Cyg 19h46'48" 30d29'23"
Tom Newman Eleanor WV 7 Aug 1938
Her 17h10'33" 34d21'18"
Tom Nik
Aur 5h24'53" 40d13'48"
Tom Nitschke
Vir 14h19'16" -0d30'16"
Tom Nixon
Sgr 17h53'27" -29d47'47"
Tom Noah´s Glücksstern
Uma 9h55'40" 52d33'27"
Tom Norviel
Leo 10h4'55" 14d42'32"
Tom O'Connell
Ori 5h35'8" 8d33'24"
Tom Oliva
Dra 17h20'36" 62d51'33"
Tom Paccadolmi
Ari 2h2'20" 18d40'11"
Tom & Pat Dickson
Cyg 21h32'12" 45d11'37"
Tom Patrick Collins 022664
Leo 11h40'13" 17d50'24"
Tom Pavlo
Cep 20h52'25" 59d16'50"
Tom Peltier
Aqr 20h45'8" -2d2'30"
Tom Pomeroy
Uma 11h19'15" 33d54'0"
Tom Powers
Per 3h1'24" 41d4'28"
Tom Puckett
Her 17h14'10" 34d15'11"
Tom & Randy O Hell Yes Pot. 2005
Uma 9h45'2" 49d11'59"
Tom Rioux
Cnc 8h35'52" 7d31'29"
Tom & Robin Matrick
Uma 12h26'21" 53d33'29"
Tom & Rose's Wedding Day Star
Cyg 21h40'19" 46d52'26"
Tom Runzo
Lib 15h40'26" -6d53'32"
Tom & Sally
Aqr 22h40'54" -1d48'6"
Tom & Sandy Pierce 25th Anniversary
Cyg 20h11'49" 42d48'18"
Tom Sawyer
Uma 10h14'32" 51d24'16"
Tom Seyler #44
Gem 6h40'53" 31d28'8"
Tom Shay
Vir 12h32'10" 0d24'11"
Tom Simpson, Sr.
Leo 9h51'9" 24d25'8"
Tom Snider
Pyx 8h49'58" -26d4'56"
Tom & Sonya Fanjoy
Cyg 19h47'13" 53d50'6"
Tom & Stacey Reid
Cyg 21h12'30" 32d32'31"
Tom Stanley Edwards
Lib 15h41'32" -25d0'49"
Tom Stephens
Aqr 21h9'8" -10d12'24"
Tom Stockwell
Aur 5h24'43" 44d47'2"
Tom Sturman
Per 3h40'1" 48d23'45"
Tom&Susan&Sophie
Aur 6h6'44" 30d30'40"
Tom & Susan St. Clair
Gem 7h53'10" 18d5'4"
Tom & Teresa Hamilton
Sge 19h26'44" 18d30'34"
Tom the Most Handsome
Aur 4h49'6" 33d22'56"
Tom Thermus Bentleys Celestial Mark
Cru 12h47'49" -64d11'24"
Tom/Tim/Jude Delaware Bond AllStars
Uma 11h29'3" 62d22'28"
Tom & Tinney Leveridge Forever Love
Ori 5h54'49" 12d44'30"
Tom Tipton
Sgr 18h19'31" -24d43'54"
Tom Tom
Aqr 22h38'54" -23d52'21"
Tom Turchetta
Lmi 9h51'49" 34d15'38"
Tom Vernau
Cnv 13h45'9" 30d1'42"
Tom Vojtko
Per 4h39'16" 39d45'18"
Tom Volpe UR 70
Aql 19h45'23" -11d7'53"
Tom Wagenseller
Uma 10h47'53" 44d57'27"
Tom Wartinger
Ori 5h24'4" 14d25'12"
Tom Wass
Uma 11h24'51" 45d2'44"
Tom Wilkening's Star
Lyn 7h14'13" 53d29'38"
Tom Winters
Aur 5h25'13" 43d39'12"
Tom Wolters
Aur 6h43'54" 48d3'35"
Tom Wrede
Aql 19h42'57" -0d5'20"
Tom Yokomizo
Sgr 18h46'26" -20d40'0"
Tom, Dad, We Love You!
Cnc 8h22'38" 18d39'26"
Tom, danke für dini liebi
Cas 1h53'11" 64d29'9"
Tom, Janet, Sarah, Tim Trankle
Uma 9h14'53" 68d7'31"
Tom,Cindy,Briana & Matthew Bradshaw
Uma 13h38'54" 55d7'20"
Toma - Aoife
Uma 8h52'48" 63d14'20"
Toma Jean
And 0h34'59" 35d25'37"
Tomahawk 65 and CC
Uma 13h36'51" 47d56'8"
Tomakamot
Lyn 6h31'53" 56d17'48"
TomandLuanne
Uma 13h39'55" 57d20'15"
Tomar
Cyg 21h12'9" 46d34'50"
Tomar, Randy, & Jonah Gross
Psc 0h45'55" 15d12'9"
Tomare 10142005
And 2h20'33" 46d43'24"
TOMAS
Uma 9h13'7" 54d2'13"
tomas
Sgr 18h45'27" -27d10'45"
Tomas Canevaro
Aql 19h25'25" 8d8'43"
Tomas Ciaran Boothroyd
Uma 9h23'45" 46d41'6"
Tomas Daniel Bignardi
Boo 14h49'44" 38d26'9"
Tomas Edward Jordan
Umi 15h41'55" 80d12'9"
Tomas & Hripsime Mantecon
Cyg 21h25'31" 51d27'39"
Tomas Joseph Robinson
Cnc 9h0'32" 32d12'43"
Tomas My Inspiration My Only Love
Leo 11h2'43" 16d0'55"
Tomas O'Farrelly - Tom's Star
Her 16h37'56" 16d56'12"
Tomas Vera
Gem 6h49'29" 17d17'29"
Tomasina Trocino
Uma 13h44'22" 57d6'8"
Tomasits Martina "Csucsu"
Cas 0h12'1" 54d37'53"
Tomasz
Per 4h40'22" 40d29'34"
Tomasz Atamanczuk
Cep 22h31'10" 81d21'15"
Tomasz Bak
Ori 6h11'36" 15d5'0"
Tomasz Scislowski
Per 3h18'29" 54d48'28"
Tomaszewski, Christel
Ari 2h37'18" 19d54'37"
Tomato
Vir 13h2'9" -10d5'20"
ToMauro Night
Aql 19h54'2" 6d33'39"
Tomberly
Boo 14h40'32" 27d28'51"
Tombo
Her 18h32'53" 21d2'48"
Tomcat
Lyn 8h3'56" 45d25'40"
Tomcat76
Psc 0h34'40" 16d15'24"
Tomchak 2
Gem 6h27'31" 24d53'49"
Tom-Dog
Lyn 9h0'26" 33d52'12"
Tomek Kurczewski
Ori 5h35'49" -6d12'7"
Tomen
Peg 22h25'27" 4d1'10"
Tomer
Vir 13h36'43" -0d56'14"
Tomes Family
Peg 22h30'47" 24d59'43"
Tomes Theodorelos
Uma 9h48'7" 69d7'48"
TomFridline
Her 17h13'18" 33d33'24"
Tom-Helen-Noone-02-11-1956
Cyg 19h55'4" 35d44'25"
Tomi
Uma 9h0'29" 50d18'29"
Tomi 80, Édesanya legked-vesebb fia
Sco 17h4'51" -30d13'7"
Tomi Lynne
Cap 20h11'39" -13d46'52"
Tomia
Lyn 7h38'17" 52d38'29"
TomIanMary
Per 3h8'1" 41d46'32"
Tomidee
Gem 6h59'31" 28d1'30"
Tomikat
Leo 9h44'6" 28d25'25"
Tominus Wurtzimus
Gem 6h41'34" 26d21'0"
Tomio Taki
Ari 2h11'12" 15d50'32"
Tomislav Presecki
Uma 11h38'27" 45d56'6"
TomKat
Lyn 7h47'54" 48d42'32"
TomKat
Cyg 20h45'16" 37d29'12"
TOMKIN
Leo 10h32'38" 27d25'2"
Tom-Lee-Friedrich
Uma 9h48'7" 68d26'29"
Tomm "Nine" Rizzo
Lib 15h46'29" -9d10'33"
Tommaso
Dra 10h35'23" 74d53'50"
Tommaso Bonanni
Cam 6h47'56" 74d27'57"
Tommaso Galligani
Peg 22h41'40" 9d57'56"
Tommaso Guidetti
Cas 23h28'41" 55d48'35"
Tommaso Innocenti
Aur 5h57'28" 34d57'46"
Tommaso Lotto
Per 3h26'54" 41d38'49"
Tommi Sue
Sgr 19h19'34" -32d36'2"
Tommie Ann
Boo 15h20'3" 43d55'19"
Tommie Furlow
Gem 7h36'57" 21d41'35"
Tommie Lane Wiggins
Uma 9h12'28" 53d18'26"
Tommie Mikulice
Her 17h41'26" 28d52'13"
Tommie Yetter
Ari 3h5'49" 29d13'29"
Tommy
Her 18h55'16" 16d50'35"
Tommy
Cnc 9h19'14" 12d31'56"
Tommy
Her 16h49'35" 9d1'24"
Tommy
Boo 15h3'16" 41d24'9"
Tommy
Uma 9h36'42" 42d54'56"
Tommy
Sco 16h16'31" -12d38'0"
Tommy
Umi 14h24'40" 77d5'21"
Tommy & Aija
Ori 5h29'39" 4d37'36"
Tommy and Anna Corbett
Cyg 21h41'36" 46d12'12"
Tommy and Julez
Cas 1h45'0" 65d37'53"
Tommy and Lauren
Vir 13h47'32" -4d24'41"
Tommy and Marylou Gallagher
Vir 12h57'11" -7d2'26"
Tommy and Megs
Vir 14h2'1" -8d53'22"
Tommy and Morgan July 30,2005
Cap 21h17'25" -18d58'46"
Tommy and Nora Lacey
Uma 9h59'26" 59d35'17"
Tommy and Phylis Deegan
Cyg 20h0'9" 43d23'11"
Tommy and Vivian
Cyg 19h50'23" 55d53'22"
Tommy & Ann's Anniversary Star
Sge 19h36'52" 17d11'31"
Tommy Blackwell, Jr.
Vir 14h43'56" 1d36'15"
Tommy Botsko
Umi 13h30'8" 75d12'58"
Tommy Boy
Per 3h14'5" 45d14'6"
Tommy Boy's Harley
Gem 8h7'13" 30d12'45"
Tommy Burke - Super Dad
Per 4h4'48" 34d15'30"
Tommy Cantwell 9-7-1934
Ori 5h38'9" 7d47'15"
Tommy Chang
Psc 23h18'10" 1d33'23"
Tommy Dale Thompson
Per 4h3'34" 46d54'32"
Tommy Davidson
Aql 19h52'9" 12d36'6"
Tommy Dean Wright
Aql 19h45'54" -0d33'1"
Tommy DeSanto
Leo 10h33'12" 18d49'50"
Tommy Downs
Per 2h11'28" 54d42'41"
Tommy Eldridge
Ori 5h38'46" 11d50'13"
Tommy Elian Dos Santos
Tau 4h10'38" 16d17'10"
Tommy & Ellie Axford
Sge 20h3'13" 18d42'14"
Tommy Ethier
Cep 23h48'52" 68d23'25"
Tommy Eugene Parnell
Gem 6h45'33" 25d22'49"
Tommy Faith
Boo 14h40'38" 14d41'53"
Tommy Fischer
Ori 6h12'4" 15d13'2"
Tommy Floyd Hickman
Gem 7h22'57" 26d35'16"

Tommy Forever
Cyg 21h18'32" 33d9'3"
Tommy G.
Per 3h46'37" 49d26'48"
Tommy G. Wells
Gem 6h47'45" 15d49'3"
Tommy Gallegos
Gem 7h29'11" 26d12'11"
Tommy Gower
Boo 14h47'51" 51d12'57"
Tommy Gray
Aql 19h7'50" 4d36'38"
Tommy & Helen
Cyg 20h2'47" 52d24'38"
Tommy Hurlbert
Her 17h39'45" 41d12'19"
Tommy Jacob Hines - TJ's Star
Umi 15h32'51" 75d2'41"
Tommy James Keyworth
Cep 22h55'31" 59d35'39"
Tommy John
Ori 6h3'38" 18d41'23"
Tommy Joseph Krynicki
Per 3h36'10" 45d12'23"
Tommy Kinch
Ori 6h3'37" 6d47'1"
Tommy Kolos
Sco 16h17'11" -10d38'43"
Tommy Krupske
Ori 6h9'35" -1d2'27"
Tommy Lazar
Sco 17h5'52" -39d24'50"
Tommy Lee Darny
Psc 0h45'8" 18d51'22"
Tommy Lee Phillips
Uma 14h23'48" 59d45'34"
Tommy Lee Thompson
Gem 7h31'21" 30d0'27"
Tommy Little
Cnc 8h16'0" 15d52'12"
Tommy Mac
Sco 17h57'38" -30d52'14"
Tommy Meny
Aql 19h6'25" 7d47'22"
Tommy Monaco
Psc 1h28'10" 10d43'57"
Tommy Mote
Pho 23h48'2" -45d26'10"
Tommy&Nicole
Ori 6h5'36" 18d55'12"
Tommy One-Stick 10-6-44
Lib 15h16'11" -6d4'35"
Tommy P. Anker
Aqr 23h16'41" -9d4'50"
Tommy P. Dalling "I Feel Good" -23-
Ari 2h50'10" 28d8'25"
Tommy "Pepsi" Della Bella
Vir 12h44'54" 12d52'23"
Tommy 'Pops & Gramps' Martin
Cap 20h27'59" -10d43'30"
Tommy Rains
Aqr 22h51'51" -5d3'19"
Tommy Ratliff Sr.
And 1h55'56" 44d15'36"
Tommy Ray Minton
Tau 4h5'52" 20d53'4"
Tommy Rinaldo
Tau 4h10'6" 28d11'52"
Tommy Shines
Aur 5h18'43" 30d43'33"
Tommy Sopwith
Per 4h48'56" 40d50'16"
Tommy "SPYKING" Waits
Her 16h50'34" 38d47'6"
Tommy Star Campanelli
Vir 12h19'1" 11d59'2"
Tommy Stensrud
Her 17h28'23" 33d55'3"
Tommy Stuckeman
Tau 4h13'57" 3d58'43"
Tommy Suhrland
Lyn 9h16'50" 35d59'0"
Tommy Swan
Leo 9h27'3" 8d29'21"
Tommy TC Bryda
Sgr 18h45'30" -17d26'40"
Tommy Tree
Per 3h14'44" 52d41'13"
Tommy Tulip
And 1h35'40" 40d34'24"
Tommy Tune
Cet 0h59'46" -0d14'59"
Tommy Two-Tone
Vir 11h40'52" -3d52'14"
Tommy Vorwerk
Ori 5h22'16" 1d46'21"
Tommy Winston
Per 4h11'23" 47d12'56"
Tommy Z
Lib 15h56'12" -12d58'46"
Tommy, Karson & Austen Trotter
Uma 14h2'50" 59d18'33"
TommyB
Cnc 9h4'55" 32d57'37"
Tommye
Cap 20h30'35" -23d53'43"
TommyLe
Her 18h51'33" 23d42'44"
Tommylee Fraser
Umi 14h47'18" 67d59'20"
Tommy's 48
Sco 16h48'16" -27d10'41"
Tommy's Alyssa
Uma 9h58'20" 47d13'40"
Tommy's Girl - Jenny
And 23h36'40" 47d14'45"
Tommy's love
Gem 6h53'23" 25d21'34"
Tommy's Special Star
Her 16h9'30" 48d15'44"
Tommy's Star
Umi 15h10'4" 80d9'55"
Tommy-Star
Lyn 7h42'8" 47d48'47"
TommyTheKing
Lib 15h11'8" -27d20'23"
Tommy-Thur
Cyg 20h38'46" 42d4'13"
Tommyz
Aql 19h12'53" -7d43'39"
Tom'n'WigStar
Aqr 21h31'54" -0d58'2"
Tomoe et Masahiro
Umi 15h1'21" 79d21'20"
Tomoko
Sco 15h54'52" -25d7'4"
Tomoko
Cyg 19h47'40" 36d29'10"
Tomoko Okamoto Schettler
Cnc 8h40'21" 13d21'42"
Tomoko-Shigeru
Vul 18h59'17" 22d39'46"
Tomomi
Ari 3h28'57" 24d50'24"
Tomomi
Cnc 8h16'32" 14d43'31"
Tomomi Phillips
Lyn 8h2'8" 47d58'41"
Tomos Morse
Pyx 9h22'38" -36d23'54"
Tomoyo
Mon 6h49'54" -0d3'35"
TomRita
Per 3h10'32" 54d11'34"
Tom's 50th
Cep 21h32'48" 63d50'14"
Tom's CosmicTrain Dick-Maggie Cooke
Uma 11h35'4" 58d13'30"
Tom`s & Karin`s Liebesstern
Lib 14h41'29" -18d26'17"
Tom's Light
Tau 4h32'44" 30d6'46"
Tom's Luster
Her 16h43'11" 33d21'45"
Tom's Star
Her 17h44'0" 17d8'0"
Tom's Star
Cra 18h56'45" -40d54'35"
Tom's Window from Heaven
Cru 12h46'25" -61d39'23"
Tom's Wishing Star
Per 2h20'39" 53d38'36"
Tom'sStar
Ori 5h47'25" 3d32'45"
Tomstar
Uma 11h22'13" 61d58'58"
TomTom Loves LouLou
Cyg 20h12'52" 33d31'24"
Tomu
Dra 16h46'48" 54d11'29"
Tomukas Piestys
Lmi 10h59'45" 32d20'16"
Ton Ange pour Toujours
Dra 17h33'25" 60d18'10"
Tona Faye Rock "Sparky"
Leo 11h44'12" 25d59'44"
Tonda Lebl
Ori 6h2'15" 9d23'7"
TONDI
Gem 6h57'6" 14d34'17"
tondo
Ori 5h53'16" -8d18'14"
Tone
Uma 11h34'26" 37d59'9"
Tonee
Cas 0h17'10" 57d22'53"
Tonee Lynn Moffett
Psc 1h23'13" 29d3'24"
Toney
Lib 14h57'18" -0d54'7"
Toney John Winter - TJ's Star
Uma 10h58'19" 62d26'38"
Toney Park
Uma 10h20'29" 58d26'3"
TONFER
Leo 10h28'3" 12d40'30"
Tong Family
Uma 8h53'29" 64d4'58"
Tong Weihong
Her 16h24'30" 20d59'11"
Tong Wing Sze
Sgr 19h9'5" -20d24'37"
Tongie
Lyn 8h51'12" 39d21'26"
Toni
Uma 11h23'3" 36d11'48"
Toni
Cam 3h30'58" 52d58'17"
Toni
Uma 9h5'59" 48d15'52"
Toni
Vul 20h26'42" 28d6'28"
TONI
Uma 8h37'56" 65d10'53"
Toni
Uma 9h15'48" 63d31'53"
Toni
Cam 5h20'21" 62d14'10"
Toni
Uma 10h8'14" 71d5'18"
Toni
Cap 21h56'24" -24d33'54"
Toni Adora Belle
Uma 9h20'17" 44d31'4"
Toni and Jon
Sge 19h33'51" 18d20'47"
Toni Ann DiRe
Gem 6h38'3" 22d30'51"
Toni Ann Fracassi
Uma 9h25'3" 43d41'6"
Toni Annette Griffith
Cnc 8h17'18" 22d55'27"
Toni Austrich
Cap 20h27'11" -25d12'25"
Toni Balzar
Lib 16h1'14" -9d28'34"
Toni Benthusen
Uma 8h19'58" 65d14'10"
Toni Berry
Umi 16h46'8" 75d13'47"
Toni Bienkowski
Cyg 20h34'11" 38d39'14"
Toni Brady
Gem 7h6'30" 34d29'47"
Toni Bufi
Cep 21h59'27" 59d4'47"
Toni Button
Aqr 22h40'56" -0d7'40"
Toni Caro
And 0h18'1" 28d58'40"
Toni Carrion
Cap 21h38'26" -19d0'14"
Toni Cook
Psc 1h6'30" 10d43'44"
Toni Darlene Arellanes
Del 20h23'59" 9d11'52"
Toni De Leston
Cam 3h47'20" 58d33'56"
Toni Ellen Turner
And 23h20'25" 47d36'27"
Toni Emma-Louise Watling
And 1h16'14" 44d5'43"
Toni Fisher
Cas 0h49'56" 53d2'37"
Toni Formichella
Del 20h38'33" 16d3'28"
Toni & Frank
Cyg 21h17'55" 45d12'18"
Toni Frankland
Umi 15h27'0" 80d54'51"
Toni Galofaro
Uma 11h24'49" 61d53'49"
Toni Gradsack
Lmi 10h0'29" 32d1'20"
Toni Hamel
Gem 7h20'44" 31d53'19"
Toni Hart
Lib 15h15'28" -28d0'32"
Toni Ivancoe
Uma 9h32'56" 64d27'52"
Toni & Jackson
Psc 23h54'54" 6d52'9"
Toni Jade Avila
Uma 9h55'26" 57d4'29"
Toni Jayne Battista
Tau 4h26'30" 17d52'49"
Toni Jeanne
Tau 4h6'6" 14d16'52"
Toni Jennings' Early Learning Star
Pho 0h38'19" -51d6'14"
Toni Jessica Fink
Ori 4h47'59" 11d43'33"
Toni Jo
Mon 7h13'2" -5d25'41"
Toni & John DeNoble
Lyr 18h50'36" 46d41'27"
Toni Joki-Schlicht
Uma 11h46'57" 32d52'50"
Toni Kachiri Lewis
Aqr 23h48'14" -14d36'35"
Toni Karp
Uma 11h3'34" 72d11'32"
Toni Kelly
Uma 8h34'21" 65d43'31"
Toni L.
Lyn 7h46'13" 49d42'16"
Toni Lou
Lyn 9h32'22" 40d32'6"
Toni Louise Parker
Umi 13h38'21" 70d4'22"
Toni Lynette Warren
Lib 15h52'51" -9d53'52"
Toni Lynn
Del 20h38'13" 15d37'40"
Toni Lynn Chappell
Leo 10h8'57" 20d44'8"
Toni M. Trujillo
Umi 16h30'37" 77d1'28"
Toni Maria Goodsell
Leo 9h51'18" 22d32'0"
Toni Mariani
Uma 10h43'45" 58d6'9"
Toni Marie
Cas 0h32'47" 61d6'28"
Toni Marie
Leo 11h0'16" -6d34'50"
Toni Marie Fredella
Ari 3h6'47" 29d17'40"
Toni Marie Kovach
Vir 14h21'10" 1d17'51"
Toni Marie Otero
Psc 1h3'32" 11d50'0"
Toni Marie Rios
And 2h9'48" 43d21'7"
Toni Marie's Epiphany
Psc 1h20'9" 15d59'30"
Toni Michelle Parvin
And 0h39'25" 25d21'43"
Toni Mishelle
Vir 13h41'54" 6d39'31"
Toni Molina
Uma 13h33'52" 52d36'14"
Toni (Momma) Busch
Cyg 21h52'20" 44d45'1"
Toni My Love
Tau 4h28'59" 2d34'52"
Toni Nicole
Psc 0h59'42" 18d18'41"
Toni Nicole Copeland
Ori 5h41'38" 5d54'0"
Toni Nicole Goulette
Psc 23h32'6" 4d23'36"
Toni Pearse
Vir 15h7'17" 7d25'3"
Toni Q
Her 16h30'24" 48d0'7"
Toni R. Rivera
Lyn 6h18'40" 54d4'22"
Toni Redmond
Uma 11h56'58" 43d56'39"
Toni Renee Jones
Lyr 18h49'19" 33d45'34"
Toni & Rita forever
Uma 11h43'59" 39d0'23"
Toni Rodgers
Psc 0h25'46" 15d49'58"
Toni Roma
Uma 12h49'39" 54d8'55"
Toni Rosslyn Tatum
Pho 23h43'3" -43d13'8"
Toni Sarno
Leo 9h34'49" 26d52'21"
Toni Scaglione
Leo 11h16'40" 13d40'19"
Toni Snell
Lib 14h35'9" -18d19'29"
Toni & Steffi
Uma 10h38'29" 59d52'57"
Toni Taft Star Daughter
Ari 3h12'52" 28d26'16"
Toni Trubachik
Lyr 18h55'56" 36d4'33"
Toni V. Kelli M. Navy V. Aquila
Gem 6h44'56" 25d17'14"
Toni Zehnder
Per 3h40'2" 42d46'8"
Toni, Chip, Ryan, Matthew
Ori 5h25'38" 14d50'52"
Tonia
Leo 10h18'20" 21d31'46"
Tonia
Lyn 6h51'37" 52d58'31"
Tonia Anne Marie Garcia
And 23h35'45" 42d26'57"
Tonia Davero
Uma 11h20'38" 31d36'12"
Tonia Felty
Sco 16h54'47" -39d42'58"
Tonia Jean Nye
And 1h29'43" 47d29'29"
Tonia Leigh-Ann
Sco 16h13'39" -13d51'16"
Tonia Ranee
Per 3h48'22" 45d21'27"
Toniah and Jim Lorge
Uma 11h11'19" 46d18'15"
ToniAnne's Star
Lyn 7h35'41" 53d18'46"
Tonias Mattsdedt
Uma 10h11'36" 61d19'41"
Tonia's Star
Uma 9h7'0" 56d54'45"
ToNiBo
Uma 9h59'23" 44d38'11"
Tonica Chere Streit
Ari 2h26'31" 14d43'52"
Tonie Lolita
Lyn 6h23'5" 58d52'5"
Tonight with Tommy
Tau 3h49'0" 16d39'37"
Tonik2006(shine forever!!)
Dra 16h39'32" 55d11'52"
Tonika
Cmi 7h20'49" 8d51'37"
ToniMarie Alessandro
Leo 9h43'21" 28d11'37"
ToniMike120106 BoundlessHappiness
Uma 9h2'21" 55d59'45"
Tonino e Concetta
Uma 9h10'8" 54d50'36"
Tonino ed Annamaria
Cas 23h32'38" 53d12'9"
Tonio
Ori 6h18'30" 16d48'55"
Tonique Clay
And 22h59'54" 50d49'0"
ToniRae
Lyn 8h20'52" 35d28'35"
Toni.Rose
Cap 20h10'46" -16d53'7"
Toni's 18th March
Ari 2h55'42" 22d27'25"
Toni's Light In The Sky
Her 18h32'37" 19d54'7"
Toni's Sid
Uma 10h54'38" 36d17'55"
Tonita Sanders Vaughan
Cam 4h16'53" 66d34'34"
TONIX - Antonio Tonini
Cam 7h34'7" 70d46'12"
Tonja
And 0h58'44" 39d15'54"
Tonja Robyn Fugate *Mom*
Psc 0h3'7" 1d31'48"
Tonje
Lyr 18h36'1" 27d14'2"
Tonje
Ori 6h24'8" 14d44'2"
Tonje Dyb
Ari 2h15'29" 12d44'40"
Tonje Persson
And 0h18'21" 46d2'54"
Tonka
Cep 22h58'2" 70d8'19"
Tonks & Tanks
Lyn 6h42'36" 56d37'25"
Tonner My Love
Per 3h12'20" 42d2'53"
Tonnie
Sco 17h39'14" -40d55'33"
Tonniette Lenei
And 0h39'49" 36d34'11"
Tonning
Sgr 20h25'43" -41d52'31"
Tono L. & Daisy C.
Tau 5h13'48" 21d28'8"
TonOpi
Cyg 20h33'42" 52d35'50"
Tonto
Ori 5h53'56" 22d14'7"
Tonto
And 0h59'24" 40d45'59"
Tonton Jean-Jacques
Peg 23h41'24" 21d4'27"
Tonuccio and Baberto
Ori 4h51'24" 11d9'59"
Tony
Gem 6h32'3" 15d47'18"
Tony
Gem 6h43'36" 20d35'8"
Tony
Leo 11h3'51" 21d38'30"
Tony
Lyr 19h19'6" 39d18'6"
Tony
Uma 13h37'25" 50d3'52"
Tony
Cep 20h39'16" 56d11'4"
Tony
Cep 21h38'37" 64d51'25"
Tony
Cap 21h41'37" -8d37'33"
Tony
Sco 17h21'59" -30d42'6"
Tony 1947
Uma 8h38'14" 69d32'58"
Tony Aaron Ellis
Her 17h25'44" 35d4'21"
Tony Alan Meitler
Lyn 8h29'5" 40d13'33"
Tony Alemany Loshuertos
Psc 0h54'4" 16d25'12"
Tony Allyn Wilderspin
Vir 13h40'30" -4d37'16"
Tony Alvarado
Her 17h13'35" 48d3'16"
Tony Alvarez III
Cnc 8h19'19" 10d25'27"
TONY ALVIS
Peg 22h54'53" 25d4'38"
Tony~Always Your Angel~DiAnna
Tau 5h47'21" 17d49'47"
Tony & Amanda
Uma 11h16'4" 35d31'11"
Tony and Amy Bruno 60th Anniversary
And 0h36'17" 39d46'17"
Tony and Barbara Leonardi
Boo 15h11'22" 41d19'39"
Tony and Debbie Courcy
Cyg 20h14'8" 51d54'41"
Tony and Helen, Star Crossed Lovers
Cyg 21h38'16" 35d18'20"
Tony and Jackie
Lmi 9h59'8" 37d25'43"
Tony and Jessi <3
Aur 5h43'50" 37d44'41"
Tony and Judith
Dra 15h16'5" 58d8'21"
Tony and Krista's Special Place
Ori 6h24'22" 17d15'29"
Tony and Louise Middleton
Her 16h13'59" 24d27'55"
Tony and Marion's Home Star
Cas 23h53'21" 53d26'48"
Tony and Mary Wilkins
Crb 16h1'28" 36d9'40"
Tony and Michele Zamora
Lyn 8h46'51" 33d31'19"
Tony and Millie Rentos
Psc 0h34'43" 19d8'4"
Tony and Rach
Uma 8h35'50" 62d51'56"
Tony and Sheila Dean, my parents
Cam 7h53'31" 63d36'38"
Tony Andrew Neeck
Her 18h6'7" 37d9'29"
Tony & Angie
Cyg 20h24'45" 34d15'21"
Tony Arriaga
Her 16h43'8" 25d23'16"
Tony Aydt
Sgr 18h55'41" -15d50'38"
Tony Balestreri-Heather Balestreri
Leo 9h24'39" 9d47'39"
Tony Banks
Uma 11h12'52" 56d13'51"
Tony Barca
Her 16h37'5" 31d7'40"
Tony & Betty Vecchione
Leo 9h47'30" 28d18'42"
Tony Bhawani
Lib 15h38'50" -10d9'30"
Tony Blunden
Lib 15h8'47" -16d21'22"
Tony Brandariz
Psc 0h45'3" 4d20'6"
Tony "Bull" Laudicina
Cyg 21h39'39" 33d5'40"
Tony Caputo
Cnc 8h52'44" 31d17'10"
Tony Caso
Uma 10h11'49" 57d17'43"
Tony Cesaro
Her 16h37'49" 26d34'53"
Tony Chilton
Cep 22h55'28" 83d26'39"
Tony Clark - Blue Star of Reading
Cep 21h24'23" 58d14'33"
Tony Cook "the King"
Sgr 18h10'32" -26d12'44"
Tony D. 2054
Ari 3h8'59" 27d28'22"
Tony & Damon
And 1h6'54" 34d34'37"
Tony & Darci
Cyg 21h9'30" 36d27'9"
Tony Dasher
Tau 4h1'18" 21d15'21"
Tony Davis
Gem 7h6'22" 22d29'52"
Tony Dawson
Per 3h53'48" 34d53'16"
Tony DeNicola
Uma 8h39'58" 65d25'52"
Tony - Donn Edward Lorrich I
Cep 0h36'6" 85d48'54"
Tony Drusilla
Her 16h43'32" 11d43'50"
Tony Duffy
Per 3h9'25" 50d58'37"
Tony E Ebeling
Cap 21h51'34" -18d1'16"
Tony E Rivera Sr.
Tau 4h34'56" 22d1'34"
Tony Eugene Mansfeld
Sco 17h55'59" -41d43'43"
Tony Flora
Her 17h28'51" 16d45'2"
Tony from Farmington
Ari 3h17'10" 29d11'59"
Tony Gardner You'll Always B a Star
Uma 10h28'9" 40d57'51"
Tony Gates
Cyg 21h24'5" 37d15'55"
Tony Geiss
Lyr 19h19'20" 30d37'51"
Tony & Giovanna
Tau 5h31'57" 26d41'36"
Tony Groppetti
Per 3h7'9" 39d18'54"
Tony Haine
Col 6h36'15" -37d26'58"
Tony Hammerton
Per 2h19'2" 54d3'38"
Tony Harris
Cru 12h46'55" -64d37'27"
Tony Hat 99
Aur 5h34'45" 53d28'5"
Tony Hernandez
Vir 15h0'8" 0d35'15"
Tony Heyes
Tau 5h55'38" 25d10'6"
Tony Hiller
Psc 0h56'48" 31d53'3"
Tony Hiu Fung Fung
Cap 21h0'43" -15d18'19"
Tony J. Walker
Her 17h48'6" 38d38'55"
Tony James Carr
Lep 5h30'38" -14d2'52"
Tony Jean Intorre
Uma 12h36'51" 57d36'29"
Tony & Jillian Rodgers
Cyg 19h54'3" 39d0'28"
Tony & Joe's Star
Nor 15h57'0" -54d2'0"
Tony Joseph Artzi
Umi 15h29'39" 84d28'17"
Tony Joseph Francis
Her 17h59'18" 49d2'5"
Tony Jules Fernand Jacky
Vir 15h6'7" 2d50'15"
Tony & Julie Knakal
Cnc 9h7'33" 30d52'34"
Tony K.
Dra 18h31'21" 53d14'43"
Tony Kaplan
Sgr 18h14'15" -31d15'41"
Tony & Karen Brittain, 25 years.
Cyg 19h36'1" 52d59'44"
Tony & Kelly Francisco
Tau 5h14'29" 27d43'25"
Tony Kelly's shining star
Cru 12h20'58" -58d3'22"
Tony Kohn
Cru 12h25'19" -63d23'44"
Tony & Kristen Cappiello
Ori 6h8'1" 17d39'10"
Tony Kuszak
And 1h15'6" 50d37'23"
Tony La Violette
Leo 10h44'7" 14d42'48"
Tony Lee Jones
Boo 15h31'17" 40d32'39"
Tony Leeds
Gem 7h7'36" 32d59'42"
Tony Lucarelli
Her 17h26'40" 47d5'33"
Tony MacDougall
Boo 14h48'51" 48d34'29"
Tony & Marjorie - Everlasting Love
Cap 21h32'24" -14d7'13"
Tony Martella
Umi 13h56'51" 72d18'50"
Tony & Martha Pietromonaco
Cyg 21h43'35" 53d3'49"
Tony & Martha — True Love
Cyg 19h38'6" 32d2'57"
Tony Massey
Ori 6h1'3" 6d2'15"
Tony McClyment II
Leo 11h16'10" 2d35'26"
Tony & Memorie's Star
And 0h33'28" 32d45'44"
Tony Mendonca
Her 17h44'23" 45d53'28"
Tony Michael
Per 3h45'35" 44d50'30"
Tony Michael and Lisa Jane
Aur 7h18'46" 42d45'35"
Tony Michael Wachowicz
Per 3h21'55" 54d12'5"
Tony Miguel Barrera
Her 16h35'58" 17d12'50"
Tony Milatos
Umi 17h18'21" 85d20'36"
Tony Nader CEO
Uma 12h37'52" 55d54'19"
Tony Nassour
Her 18h41'59" 20d16'59"
Tony & Neela Gaddis
Uma 9h55'55" 71d23'58"
Tony Nick
Aql 19h41'9" -0d3'22"
Tony O'Donnell
Gem 7h11'33" 27d51'58"
Tony Olhoff
Com 13h14'50" 20d22'29"
Tony Olivieri
Her 17h45'37" 40d29'8"
Tony Parra
Sco 16h11'8" -11d3'37"
Tony Paul
Tau 5h45'4" 17d51'2"
Tony Peterson
Her 17h9'27" 32d19'25"
Tony Petric
Boo 14h30'47" 42d9'48"
Tony & Phyllis
Cyg 20h49'26" 32d12'6"
Tony Poole
Uma 11h12'16" 40d47'3"
Tony Quincy
Vir 14h18'41" 2d47'35"
Tony & Rachel
Tau 3h32'23" 17d1'20"
Tony Record
Cep 22h43'58" 83d42'35"
Tony Rikli
Tau 3h31'34" 13d46'56"
Tony Robinson
Uma 9h31'19" 44d23'45"
Tony Rotondi
Uma 11h16'59" 44d39'41"
Tony Rubio
Ori 4h52'48" 11d28'11"
Tony S. Diamantoni
Dra 16h47'2" 52d6'58"
Tony "Sargettarius" Weaver
Ari 3h20'31" 16d44'46"
Tony Silicato
Her 17h49'24" 39d6'41"
Tony Smalley
Sco 17h6'11" -44d48'1"

Tony Sperry
Uma 10h15'27" 48d8'45"
Tony Spurgeon
Lib 15h52'30" -18d20'42"
Tony Stewart
Lyn 7h44'13" 35d41'40"
Tony & Sue Ayton
Lyn 7h30'13" 44d27'4"
Tony & Susan's Anniversary Star
Cyg 20h49'24" 46d59'47"
Tony. T
Cep 23h59'35" 67d0'54"
Tony T
Sco 17h5'1" -33d1'45"
Tony ~ The Eternal Star Dancer
Crb 15h45'32" 37d48'41"
Tony "The Turbonator" Fields
Ori 4h50'59" 11d43'16"
Tony Thornton
Per 3h46'56" 43d45'22"
Tony "Tig" Ursini
Cep 23h39'1" 74d18'21"
Tony & Tina's Star
Leo 11h20'31" 0d20'6"
Tony TNT Lugo
Cet 0h58'19" -0d20'58"
Tony(Tonz)
Tau 5h30'35" 27d8'32"
Tony & Tricia Tirado Married24/2/05
Eri 2h8'58" -50d18'28"
Tony & Trixie's piece of heaven....
Cir 15h18'19" -58d4'0"
Tony Ulep
Aur 6h8'55" 39d6'11"
Tony Valdez
Gem 6h47'21" 15d29'9"
Tony Valinski
Cnc 8h30'40" 29d5'11"
Tony Van Shans - with the angels
Her 17h23'21" 16d43'41"
Tony Vincent
Her 18h40'9" 13d38'26"
Tony Walsh
Uma 9h16'4" 53d50'59"
Tony Warner
Ori 6h4'37" 16d34'48"
Tony Wayne Moore
Uma 8h57'29" 47d21'31"
Tony Webster
Uma 11h25'29" 33d11'22"
Tony Wilson
And 23h49'5" 37d0'34"
Tony Yesterday, Today & Tomorrow
Cyg 20h28'28" 49d14'0"
Tony & Yolanda - 11-20-1948
Uma 8h45'13" 66d22'18"
Tony Z
Uma 9h21'31" 60d11'0"
Tony Zvirblis
Her 16h42'1" 21d59'21"
Tony, Our Guardian Angel
Sgr 18h39'58" -31d3'55"
Tony, Tony + Robin Mateka
Sge 19h36'15" 18d14'35"
Tony, Trudy, Vaughn
Ori 6h19'19" 14d39'50"
Tony, Val & Steve - "The Johnstone Star"
Peg 21h31'34" 22d18'32"
Tony,Beth,Bailey,Blaise,Zach Maio
Uma 10h41'59" 53d35'36"
Tony23
Uma 10h53'32" 57d34'12"
Tonya
Cma 6h53'29" -12d21'31"
Tonya
Cap 20h20'17" -16d29'13"
Tonya
Gem 7h39'4" 26d20'42"
Tonya
Cnc 8h2'36" 14d41'28"
Tonya Algerine Browder
Lib 14h56'13" -2d56'49"
Tonya and Angelina
And 23h15'51" 48d0'33"
Tonya and Ariana
And 23h14'14" 48d0'16"
Tonya and Josh's Star
Sco 16h13'43" -21d11'25"
Tonya Ann Dopart
Tau 5h46'54" 22d9'45"
Tonya Autry
And 0h25'41" 31d53'1"
Tonya Bea Stout
Cam 3h17'16" 58d36'31"
Tonya & Benjamin Cramer Dec 18 2004
Cyg 19h36'46" 44d12'19"
Tonya Byford
Crb 15h53'25" 34d23'14"
Tonya Celeste
Sgr 19h21'39" -17d31'50"
Tonya Coon
Lib 15h3'40" -27d8'8"
Tonya Dawn Hogue
Cas 2h14'23" 73d34'3"
Tonya Dieker
Sgr 19h7'31" -15d1'52"
Tonya E Bergmann
Lib 14h58'56" -12d34'39"
Tonya Elise Trichel
Lyr 18h51'59" 32d52'40"
Tonya (Filly) Davis
Lib 15h21'20" -26d21'32"
Tonya Frame
Sco 17h29'23" -37d39'34"
Tonya G Kelley
Cnc 9h14'34" 17d21'8"
Tonya Garmon 34
Psc 1h30'57" 15d10'53"
Tonya G.G. Garrett
Uma 13h8'49" 52d29'23"
Tonya Haas
Mon 6h52'39" -0d59'41"
Tonya I Love You
And 0h35'25" 44d7'3"
Tonya Jennings
Uma 11h27'54" 54d15'2"
Tonya Jordan
Gem 7h0'59" 14d57'20"
Tonya Kaye
Psc 1h25'25" 14d51'52"
Tonya Krolczyk
Cam 4h18'24" 68d11'3"
Tonya L. Brown "Love Always"
Tau 3h28'23" -0d39'34"
Tonya L. Jarvis
Lmi 10h43'35" 31d1'54"
Tonya L. Miller
Psc 23h0'21" -0d37'47"
Tonya Lea
Ori 6h18'53" 10d22'37"
Tonya Lea
Ari 3h17'2" 28d27'3"
Tonya Leigh Hannah
Mon 6h31'18" 7d46'5"
Tonya Luann
Crb 16h10'2" 37d40'18"
Tonya Lynette Deveau
Ori 6h3'9" 6d45'48"
Tonya Marie
Tau 4h57'45" 21d58'58"
Tonya Marie Lafountain
Cam 4h9'16" 59d16'28"
Tonya Miller
Per 3h24'50" 44d29'43"
Tonya Painter
Sgr 18h22'17" -16d50'21"
Tonya Pearson Whaley
Leo 9h25'35" 10d32'51"
Tonya Rae
Lyn 8h13'28" 53d0'17"
Tonya Renee Miracle
Lyr 19h10'44" 45d55'11"
Tonya Rose Dams
Aql 20h6'22" 9d22'59"
Tonya Roy
Psc 1h53'11" 5d50'40"
Tonya Sue
Gem 7h23'42" 31d13'55"
Tonya Sue Myers
Cam 6h55'31" 67d28'25"
Tonya Suzette
Leo 11h21'57" 5d1'12"
Tonya Volk
And 23h25'48" 43d53'45"
Tonya, el corazon en el cielo
Cas 1h34'11" 66d49'48"
Tonya's Orbital
Lmi 9h59'58" 36d56'30"
Tonya's star
And 23h47'46" 44d13'29"
Tonya's Star
Aql 19h51'21" 16d12'14"
TonYeli
Ori 6h11'43" 18d41'51"
TonyLongLouiseHalmkanParaSiempre
Per 3h20'19" 41d53'11"
Tony's Bright Light
Leo 10h34'10" 18d28'38"
Tony's Dream
Gem 6h28'44" 23d26'0"
Tonys Esther
Eri 3h47'16" -0d58'14"
Tony's Rose Andre
Gem 6h50'44" 23d20'59"
Tony's Star
Her 17h29'19" 44d21'36"
Tony's Twins
Umi 15h24'43" 73d28'3"
Tony-Star
Mon 6h44'1" -0d20'34"
TonyTinaMartaMARTINI(FR-ITALIA)
Cam 13h34'22" 78d3'24"
Tonzers and Jiffer
Lyn 7h43'30" 55d56'55"
Too Big Karli
Lyn 7h1'57" 51d2'17"
Too Far To Talk
Psc 0h33'28" 9d30'51"
Too Legit To Quit
Uma 11h4'50" 49d46'43"
Too Pancake's
Leo 9h41'54" 21d49'8"
Too Young
Cma 6h35'37" -16d12'54"
Tooch
Uma 8h33'25" 65d13'0"
Toogie
Sgr 18h0'57" -23d13'33"
tookie
Lyn 7h32'28" 58d13'19"
Toomas Lepvalts
Ori 4h46'42" 14d52'22"
Toombs Star
Sgr 20h17'40" -38d39'7"
Toomy & Titoo
Col 5h46'44" -36d17'14"
Toon
Per 4h26'35" 44d50'3"
Toonces
Lyn 9h13'37" 34d53'49"
Toonette
Lyr 19h19'2" 28d5'5"
Toosh Rupert
Uma 9h12'34" 66d27'35"
TOOSHI
Cap 20h54'16" -23d54'45"
Toot
Leo 10h22'58" 19d10'31"
Tootie
Peg 23h19'41" 12d13'7"
Tootie
Crb 16h20'47" 37d11'22"
Tootie
Cyg 21h34'32" 47d22'59"
Tootie
And 2h36'0" 45d59'49"
Tootie Campana
Aqr 21h55'55" 1d3'50"
Tootie Cooman
Cyg 20h51'31" 35d10'2"
Toots
And 23h41'51" 37d8'40"
Toots
And 23h13'25" 47d11'21"
Toots
Cnc 8h18'51" 17d43'26"
Toots
Sco 17h57'11" -30d18'48"
Toots
Aqr 23h8'13" -12d29'17"
Toots Johnson
Lyr 18h29'50" 37d11'33"
Toots Sagaskie
Tau 3h46'58" 27d25'54"
tootsie
Cnc 8h25'18" 18d34'5"
Tootsie
Tau 5h5'22" 18d48'1"
Tootsie
Ori 4h50'4" -0d37'29"
Tootsie
Vir 13h29'55" -3d32'46"
Tootsie: Loving Mother and Grandmom
Crb 15h40'36" 27d8'48"
Tootsie Pop Jelly Bean
Uma 12h30'45" 57d18'1"
Tootsie Shining
Aqr 22h22'54" 2d16'22"
Tootsie Wootsie
And 23h29'7" 38d42'58"
Tootsie's Debbie
Uma 12h42'38" 58d40'54"
Tootsie's Robin
Uma 13h41'9" 55d20'45"
Tootums
Uma 13h48'31" 56d55'30"
Top Banana
Lib 15h48'10" -11d11'56"
Top Dude
Uma 10h30'7" 45d21'43"
Topa Sumac Mallqui
Her 16h50'36" 33d1'58"
Topalidis, Theodoros
Uma 8h50'38" 49d38'1"
"Topanga"
And 2h10'22" 46d4'17"
Toph
Ari 2h31'2" 19d52'18"
Topher
Lyn 7h24'41" 44d46'54"
Topher
Lib 15h7'9" -5d59'57"
Topher and Sweet
Cyg 20h27'55" 34d18'48"
Topola
Umi 15h55'13" 85d31'46"
Topolina
Ari 2h45'23" 15d47'15"
Topolino
Ori 6h9'20" -3d48'42"
Toppina
Ori 5h53'5" 22d39'15"
Toppymackyhappy 2003.7.5
Gem 7h31'50" 27d50'33"
Topridge
Uma 10h33'0" 65d36'34"
Topsie
Cnc 8h12'30" 32d4'40"
Topsy Petal
Umi 13h54'46" 72d25'38"
Topy (Simona)
Per 4h11'1" 45d10'52"
Tor
Ori 4h52'53" 10d59'30"
Tor Fridlund
Uma 11h1'29" 51d24'34"
Tor Peterson
Her 17h23'32" 15d49'20"
Tor & Tiff Linbo-Hadfield-Terhaar
Tau 3h40'31" 16d31'54"
Tora
Uma 11h48'10" 43d55'32"
Toran and Christina
Crb 15h46'35" 26d29'10"
Torben & Linda's Silver Star
Peg 21h40'9" 23d2'42"
Tordjamn Laurence
Uma 13h41'24" 55d38'34"
Torey Jo
Her 16h44'5" 30d39'6"
Tori
Tau 4h42'57" 23d38'51"
Tori
Tau 3h43'37" 28d36'52"
Tori
Del 20h45'1" 15d24'46"
Tori
Psc 1h27'55" 28d59'57"
Tori
Ari 2h8'56" 22d54'32"
Tori
Leo 10h26'46" 18d38'17"
Tori
Gem 7h24'45" 15d57'18"
Tori
Vir 14h51'11" 3d35'32"
Tori and Nick
Uma 8h26'6" 61d33'22"
Tori Anne Lozanoski
Powell Kubat
Cru 12h17'36" -56d39'56"
Tori Brenner
Ori 5h59'59" 17d16'44"
Tori Brooke
Gem 7h9'47" 34d11'4"
Tori Brooke Smith
Psc 1h11'55" 26d15'24"
Tori Glazner's Star
Cam 4h5'41" 55d31'54"
Tori Grace DelliCarpini
Vir 13h24'40" 6d31'50"
Tori Kerrigan Heard
And 23h45'0" 37d55'7"
Tori Lee Beck
Gem 6h2'58" 24d52'26"
Tori Leigh Hadden
Cap 21h26'56" -9d44'42"
Tori Leigh Johnson
Aql 19h3'28" -0d6'12"
Tori May Wright
Col 5h36'55" -32d16'46"
Tori McReynolds
Aql 19h52'43" 9d0'44"
Tori Montgomery
Cap 21h34'30" -14d36'4"
Tori Paige & Lily-Ann Grace's Star
Col 5h47'30" -35d15'36"
Tori Palmer
And 0h54'58" 38d51'38"
Tori Roland
Cnc 8h53'47" 24d42'50"
Tori Rose
Aqr 22h26'5" -1d24'0"
Tori" star
Uma 11h54'5" 62d24'43"
Tori T
Uma 10h56'55" 44d59'8"
Tori Tague
Cnc 9h6'0" 19d33'25"
Tori Taylor
Lyn 7h34'41" 49d11'6"
Tori the sweet 16th star
Psc 23h51'36" 6d53'48"
Tori Weber
Vir 12h21'11" -5d48'30"
Toria
Tau 4h5'21" 26d49'11"
Torias
Lyn 7h53'13" 34d58'38"
Torie 'Akemi' Okemura
Dra 15h42'8" 56d45'55"
Torie and Billy
Cap 21h34'40" -14d49'3"
Torie Elizabeth Avila
And 0h35'22" 40d49'21"
Torie Leonard
Sco 17h46'51" -37d15'35"
Torin Ceryne
Peg 21h23'8" 17d28'56"
Torin Hamish Buchanan McGregor
Per 4h4'6" 33d29'33"
Torin James
Her 18h33'32" 19d36'8"
Tori's Neverland
Vir 12h4'0" -4d32'17"
Tori's Star
Leo 10h46'54" 9d44'8"
Tori's Star
Cnc 8h51'10" 10d57'46"
Tori's Starfire
Uma 13h26'22" 57d42'30"
Torobiny - The Twelfth of Never Star
Cra 18h23'54" -37d0'19"
Torquemada25/5/04
Peg 22h6'10" 27d16'26"
Torrance
Umi 14h49'23" 89d6'15"
Torre
Sco 16h13'46" -10d6'35"
Torre Bear
Mon 7h33'13" -0d58'11"
Torre Jean November 18, 1996
Crb 15h58'36" 33d23'0"
Torrence Clinton
Ari 2h19'16" 25d11'56"
Torres Dániel
Aqr 22h24'33" -7d22'51"
Torretto
And 1h33'48" 42d25'5"
Torrey
Uma 9h2'53" 68d4'46"
Torrey Brooke Speer
Uma 11h48'9" 32d46'28"
Torrey Carlson
And 1h0'8" 46d11'37"
Torrey Lynn Lackey "TC"
And 23h25'20" 51d53'9"
Torrey Marie
Cas 0h46'52" 57d24'39"
Torri Elissa Martin
Mon 6h30'50" 8d18'1"
Torria Larsen
Vir 13h33'49" -20d49'21"
Torrie
Cam 3h48'29" 74d15'0"
Torsten
Uma 11h53'8" 35d47'25"
Torsten Fischer
Ori 6h20'31" 16d19'11"
Torsten Rohde
Uma 11h49'26" 39d22'41"
Torsten Seger
Ori 6h20'47" 8d16'0"
Torsten Strohmeier
Ori 6h17'14" 15d11'47"
Torsten Teske
Uma 8h42'37" 51d14'59"
Torsten Wolf
Ori 6h7'54" 8d10'36"
Tory
And 1h27'13" 46d26'40"
Tory
Uma 11h33'49" 62d43'53"
'Tory' Anthony Marandos
Her 18h49'15" 21d4'53"
Tory Kirby
Crb 16h3'50" 27d31'50"
Tory Kirby
Mon 6h30'45" 3d19'51"
Tory L. Hoskinson
Uma 9h59'53" 64d49'35"
Tosanna
Aqr 21h38'52" -1d11'6"
Tosca Rosa Guiseppa Zuccalmaglio
Vir 12h12'3" 11d48'45"
Toscan Vallet
Cep 20h36'49" 61d27'59"
Toshie
Uma 11h17'0" 60d44'7"
Toshihiro. N
Cnc 8h7'41" 15d5'54"
Toshikazu
Uma 11h35'39" 44d31'39"
toshiko 80
Cap 20h37'31" -20d53'47"
Toshimi
Lib 15h20'7" -29d4'32"
Toshio & Yoshie
Ari 2h38'6" 18d15'11"
Toshiro Sakai
Ori 4h47'19" 4d23'50"
tosso
Ori 6h5'45" 18d25'41"
Tostephmylove
Lib 15h43'49" -18d30'42"
ToSu05
Uma 12h47'35" 61d19'16"
Tosy
Uma 9h9'7" 49d17'29"
Tószegi Fruzsina
Tau 3h59'31" 4d57'37"
Totally Jan
And 1h0'48" 45d15'4"
Tóth Bálint SzeretetCsillaga
Vir 13h51'9" 0d37'59"
Tóth Ferenc
Leo 10h38'55" 10d48'32"
TÓTH Imi, a MotoGP jövo Bajnoka
Cas 0h53'42" 69d45'58"
Tóth József C7-50
Uma 11h49'16" 45d15'0"
Toti
Ori 5h50'39" 7d8'58"
Toti
Lib 14h22'9" -12d48'39"
Totie
Cnc 8h10'23" 25d29'17"
Tótik József
Cyg 19h37'42" 53d19'25"
Toto
Ari 2h28'49" 18d1'55"
Tou Byrd
Uma 9h45'18" 43d29'41"
Touch
And 0h54'14" 41d51'21"
Touch Ek
Crb 16h4'47" 36d41'37"
Toufic, My Abdo
Cyg 19h32'39" 31d50'53"
Toufic, My ABDO
Cyg 19h37'51" 30d10'5"
Touhami and Giovanna
Aqr 22h23'54" -15d40'54"
Toujours
Umi 12h40'58" 89d14'5"
Toujours
Pyx 8h51'43" -26d23'39"
toujours
Lyr 18h32'13" 33d0'27"
Toujours
Uma 11h30'33" 42d33'17"
Toujours
Cyg 21h46'7" 44d3'53"
Toujours à mon coeur ! Kristine..
Eri 3h50'44" -0d52'25"
Toujours Deschamps
Cyg 20h6'27" 34d36'59"
toujours ensemble
Her 16h25'54" 48d19'46"
Toujours Ensemble
Uma 9h25'42" 64d54'37"
Toujours et a` jamais
Cas 1h29'3" 52d19'10"
Toujours et à Jamais je Vous Aimera
Cyg 21h36'7" 33d30'29"
toujours rappelé
Lyn 7h0'51" 47d10'56"
Toula Elli Marinos
Uma 8h16'31" 69d56'24"
Touquey
Del 20h30'56" 18d25'48"
Toutankhamon Julien Soumis
Cep 21h49'13" 59d41'32"
Tove Brigitte
Umi 15h29'58" 68d59'50"
Tove Johansen
Cyg 21h19'48" 42d34'12"
Tove Joline Christensen
Sgr 19h11'3" -21d42'59"
Tove - Unconditionally Yours
And 23h23'27" 38d6'56"
Tovi
Uma 13h52'5" 62d10'24"
Towers of Trinity
Ori 5h57'50" 9d45'2"
Toya
And 2h27'7" 45d47'26"
Toye's Heavenly Beauty
Cyg 19h42'1" 37d30'42"
Toye's World
Cas 23h24'21" 56d31'10"
TOYETTE LENEE'
Uma 11h45'45" 53d34'15"
Toymakeur
Sco 16h14'42" -13d46'49"
Toyston's Angel Donna
Tau 4h42'22" 23d30'41"
TP Pierce - Gentle Giant
Per 3h4'4" 40d34'45"
TP790614
Gem 6h48'22" 19d52'8"
TPH
Her 16h53'44" 32d32'0"
T.P.S. II
Cam 3h58'9" 69d32'1"
tpsunshine
Sgr 18h28'1" -17d44'30"
TQ (aka Tony Quant)
Per 4h44'26" 45d2'12"
TQ CC
Lyn 7h59'25" 51d18'9"
TQ3Navigant
Cru 12h54'15" -60d24'25"
TQ's Star of Love
Cyg 20h22'53" 34d1'28"
TR Vreeland
Ari 2h6'58" 20d52'15"
T.R. Walters Honorary Star
Umi 17h41'58" 80d0'2"
" T.R.A."
Ori 5h30'33" -0d31'52"
Traa Mitchell
Aql 19h16'24" 16d16'37"
Trabacchin Sandra
Umi 16h1'38" 71d15'2"
Traccia Mia Amore
Leo 9h47'14" 9d47'33"
Trace
Cas 23h23'58" 59d47'56"
Trace Aaron Lemasters
Cnc 8h34'22" 15d57'43"
Trace and Abby
Tau 5h47'17" 18d18'12"
Trace Anthony O'Brien
Lmi 10h42'2" 28d7'16"
TRACE AUSTIN SORRELL
Ari 2h56'19" 28d19'24"
Trace Loren Welker
Vir 13h38'21" 4d15'41"
Trace Mitchell Emig
Her 17h7'17" 31d41'59"
Trace Nathaniel Kipfer
Psc 0h13'48" 2d52'20"
Trace Redmond Hasz
Uma 10h51'9" 51d2'46"
Trace S. Barber
Tau 5h24'52" 21d8'26"
Trace Santos-Barber
Umi 16h22'48" 76d3'50"
Trace Winner
Ori 5h20'28" 4d43'0"
Tracee Marie
Vir 12h23'1" -6d49'38"
Tracee Michelle O'Mary
And 0h19'26" 32d57'17"
Tracee Rowe
Cra 18h44'16" -39d2'4"
Tracelink
Cra 19h1'24" -38d32'55"
Tracer 05
Aql 19h12'23" 1d19'1"
TRACEY
And 23h6'19" 48d52'21"
Tracey
And 2h17'27" 49d51'32"
Tracey
Sgr 18h13'16" -22d33'37"
Tracey
Lib 14h52'26" -0d53'4"
Tracey A. R. Harris
Leo 10h46'28" 16d43'48"
Tracey A. Reeves
Ori 6h20'48" 10d11'57"
Tracey and April Cottrell Family
Sge 19h32'56" 16d44'50"
Tracey and Ryan Hayes
Aqr 23h19'30" -12d33'42"
Tracey and William Hancock
Cyg 20h59'32" 48d26'36"
Tracey Ann
Ori 5h15'58" 7d48'4"
Tracey Ann
Lib 14h51'58" -2d40'59"
Tracey Ann
Uma 13h31'46" 57d24'26"
Tracey Ann Crow
Sgr 19h4'23" -30d59'7"
Tracey Ann Oldfield
Cas 0h49'8" 62d34'14"
Tracey Ann Thomason
And 2h15'18" 41d20'44"
Tracey Arbeznik
Sco 16h6'9" -15d42'49"
Tracey Bean Visconti
Uma 11h57'50" 53d8'33"
Tracey Bennett
Umi 14h36'53" 74d26'35"
Tracey Berical
Cyg 21h9'16" 30d23'8"
Tracey Bromley
And 0h21'38" 38d49'1"
Tracey Brown
Cas 1h46'34" 65d10'0"
Tracey C. Lowenhaupt
Gem 6h23'58" 20d30'3"
Tracey Campbell
Cnc 8h15'7" 22d39'5"
Tracey Carol Hamilton
Gem 7h10'34" 25d54'25"
Tracey Cass
And 2h32'41" 50d1'34"
Tracey Cobb
Leo 11h12'15" -5d45'13"
Tracey Corhn "My Baby"
Per 3h30'21" 46d52'12"
Tracey Currey
Psc 0h18'2" 6d7'5"
Tracey Daley
And 2h22'43" 50d30'46"
Tracey Davis
Uma 11h37'18" 48d49'7"
Tracey Delgrande
Tri 2h22'19" 34d11'1"
Tracey DeMartini
Sco 16h11'23" -16d35'2"
Tracey Drakalyiska
Crb 15h50'14" 29d26'33"
Tracey Elizabeth John
Ari 2h48'32" 14d19'10"
Tracey Emma Weston
Leo 10h52'16" 15d1'57"
Tracey Franklin
Gem 6h50'33" 22d18'40"
Tracey Genet and Pat Greytak
Cyg 21h51'20" 43d46'12"
Tracey Gillis
Cep 21h38'14" 62d27'31"
Tracey Gordon
Cnc 8h10'21" 26d47'28"
Tracey Green
Cas 23h19'4" 58d52'48"
Tracey Grimes
Uma 14h2'44" 50d5'31"
Tracey & Her Five J's Duddie
Psc 1h21'22" 30d33'48"
Tracey Hill-to-Be
Cyg 19h57'39" 57d22'13"
Tracey Jane Martin
And 0h10'22" 44d24'56"
Tracey Jane McVilly
Sco 17h37'26" -41d24'58"
Tracey & Jon Haley
Tau 3h44'50" 24d25'3"
Tracey L. Davis
Umi 16h15'43" 74d55'27"

Tracey L. Gauch Cyg 21h41'10" 53d8'8"
Tracey L. Trybom Uma 8h38'39" 58d10'14"
Tracey Leah Gem 6h49'1" 21d44'22"
Tracey Leanne White Cru 12h27'8" -59d25'54"
Tracey Lee Malaea Tangi Ari 1h58'20" 19d20'53"
Tracey Louise Lyn 8h55'24" 37d22'47"
Tracey Lyn Lib 15h12'24" -9d58'34"
Tracey Lynn Lyn 7h21'29" 57d16'4"
Tracey Lynn Jones Cap 21h45'26" -10d47'3"
Tracey Lynn Quick Per 3h57'33" 32d32'59"
Tracey Lynn Telford Uma 14h10'17" 55d50'15"
Tracey Lynne Cam 5h41'26" 59d50'2"
Tracey M. O'Neil's Star Leo 10h12'51" 12d39'59"
Tracey M. Reed Lmi 10h30'5" 28d37'49"
Tracey Marie Psc 0h45'21" 7d6'12"
Tracey Marrs Scl 23h43'46" -28d6'18"
Tracey McCall Lmi 10h27'46" 33d12'10"
Tracey McClean Boo 14h46'49" 19d8'49"
Tracey McMahon Uma 8h26'22" 70d3'26"
Tracey Monique Pendleton Lmi 10h42'45" 35d58'43"
Tracey Morgan Cas 0h3'20" 57d11'32"
Tracey - Mother of Mboya Cru 12h16'24" -62d38'50"
Tracey My Angel Cru 12h26'0" -61d17'55"
Tracey Nicole Rawlings And 2h9'19" 40d40'22"
Tracey Rachelle Cam 4h8'25" 65d55'40"
Tracey Renee Riddell Crb 16h10'23" 32d30'4"
Tracey Richards Uma 12h14'42" 63d3'19"
Tracey Smith And 1h2'54" 43d40'5"
Tracey Thorne's Smile Lyn 7h44'18" 54d30'6"
Tracey Unterseh, Mom, Wife, Friend Cas 0h18'54" 59d3'57"
Tracey Venn Lyn 7h8'14" 58d2'41"
Tracey Wagganer And 0h9'56" 45d37'7"
Tracey Williams - Forever Smiling Cru 11h59'33" -58d43'56"
Tracey-Bright Mommy Star Cnc 8h8'47" 16d45'42"
TraceyCheyenneChase Sco 17h53'23" -35d47'14"
Tracey.S Gem 7h43'25" 28d35'22"
Tracey's Angel Cru 12h25'24" -60d59'28"
Tracey's Eve And 0h7'9" 29d42'43"
Tracey's Guiding Light Ara 17h18'2" -54d37'53"
Tracey's Love for Simon Pho 23h38'0" -43d27'41"
Tracey's Star Cas 0h35'6" 60d59'40"
Tracey's Star Peg 22h36'22" 9d18'46"
Tracey's Star Ori 5h7'25" 4d59'43"
Tracey's Star of Justice And 0h36'22" 43d50'48"
Traci Gem 7h39'51" 32d58'44"
Traci Peg 21h26'45" 11d58'9"
Traci Sgr 18h25'20" -32d41'24"
Traci and James Vir 13h9'9" -15d54'30"
Traci and Rod Cole Umi 15h22'12" 87d18'9"
Traci Anne Ceccherini Lyn 7h53'59" 50d29'44"
Traci & Brandon VanderVoort Cyg 20h8'23" 36d6'11"
Traci Englutt Lyn 7h55'32" 35d50'44"
Traci Fickle Peg 23h19'53" 33d12'1"
Traci Furr Umi 15h38'32" 73d21'14"
Traci Gayle Preston Leo 10h11'20" 13d52'52"
Traci Haggard Sheckler And 1h37'41" 45d21'27"
Traci Heather Markham Leo 11h17'56" 21d20'38"
Traci J. Mitchem Aqr 22h56'17" -4d55'38"
Traci Johnson Cap 21h40'43" -15d54'26"
Traci Jolene Emry Ori 5h38'54" -1d7'45"
Traci Jones (Beam) Cyg 21h55'4" 48d14'34"
Traci Kirk And 0h37'28" 26d26'13"
Traci Klawes And 1h12'37" 46d29'24"
Traci Kotalik Ari 2h55'40" 26d55'18"
Traci L. Rosnack Ari 3h14'7" 19d10'23"
Traci LaNai Gongol Leo 10h27'49" 12d28'59"
Traci Laura Amadio Vir 13h54'30" -1d0'50"
Traci Lee Koszut Ari 2h50'37" 29d36'6"
Traci Leigh Kelly Leo 11h25'54" -1d0'22"
Traci Lidlow And 23h30'55" 49d23'8"
Traci Link Vir 12h48'43" 0d28'50"
Traci & Louis Umi 16h17'9" 76d37'56"
Traci Lynn Johnson Aqr 22h11'23" 1d52'26"
Traci Lynn Mansir And 2h24'24" 48d26'35"
Traci Lynne Gem 6h46'57" 32d55'35"
Traci Lynne Dolph Hickman Vir 14h3'21" -14d58'20"
TRACI MCELDOWNEY Ori 5h7'37" -1d19'57"
Traci Michelle Cnc 8h50'14" 8d55'33"
Traci Michelle Willis Campanella Uma 11h15'13" 62d12'26"
Traci Muriello Horwitz Ari 2h36'3" 30d37'8"
Traci "My Beautiful Angel" Fonseca Gem 7h16'6" 25d44'49"
Traci Noel Shaw Vir 12h36'22" 3d2'37"
Traci Renee' And 1h12'51" 41d42'5"
Traci Rochelle Kristall And 0h31'57" 28d25'43"
Traci Schellhammer Cas 0h18'40" 60d56'15"
Traci Sisson Uma 10h39'44" 64d39'2"
Traci - The Best Mom In The World. Cas 1h5'47" 49d54'14"
Tracie Cas 1h11'18" 50d38'7"
Tracie And 0h42'47" 37d39'0"
Tracie Tau 4h28'28" 16d56'0"
Tracie Sgr 17h47'23" -16d28'14"
Tracie Cap 21h37'42" -14d49'14"
Tracie Ann Jackson Cap 20h31'49" -9d55'20"
Tracie Bidlake Ori 6h13'6" 15d56'42"
Tracie Brooks Felker 5.0 Cas 0h49'20" 61d32'16"
Tracie D Lewis Vir 12h31'10" -3d21'39"
Tracie Dannielle And 2h5'21" 44d42'13"
Tracie Gunnuscio Sco 16h12'23" -12d18'52"
Tracie & John's Love Star Cnc 8h47'30" 21d12'39"
Tracie L. Loranger Cyg 20h3'47" 46d35'50"
Tracie L Shields Lmi 10h27'33" 34d18'25"
Tracie Lanza Sco 17h20'12" -32d57'24"
Tracie Lee Ann Col 6h35'11" -36d35'3"
Tracie Lewis Cas 2h15'8" 72d57'14"
Tracie Lynn McDonald Tau 4h26'54" 22d44'52"
Tracie Marie Ribitch And 23h8'0" 50d43'47"
Tracie Michelle Christian Lib 15h46'13" -11d55'31"
Tracie Nicole Brown Aqr 21h52'3" -1d25'50"
Tracie Ramsden Cas 23h40'5" 56d4'50"
Tracie S. Jones Tri 2h0'30" 31d52'24"
Tracie Taylor Psc 1h9'15" 12d58'16"
Tracie Thoren Gem 7h39'17" 33d24'1"
Tracie143 Uma 13h9'28" 57d4'45"
Tracie's Light And 0h21'40" 41d14'23"
Tracie's Star Lmi 9h29'7" 33d10'37"
TraciRose Sco 17h19'42" -43d55'41"
Tracy Lyn 8h29'40" 59d35'5"
Tracy Cam 4h13'14" 68d20'22"
TRACY Sco 16h8'54" -14d54'0"
Tracy Sgr 19h6'32" -16d9'51"
Tracy Cyg 21h26'16" 32d41'14"
Tracy Cas 0h4'48" 56d18'19"
Tracy Tau 4h41'6" 17d40'17"
Tracy Aql 19h57'28" 7d44'10"
Tracy Psc 1h30'4" 18d40'42"
Tracy Psc 1h17'2" 15d26'31"
Tracy Ari 2h51'39" 28d52'31"
Tracy Ari 2h13'36" 27d8'42"
Tracy - 5/01/1969 Uma 10h58'20" 54d52'6"
Tracy A. Perkins Psc 0h16'15" 0d22'38"
Tracy Akina, Murata Lyn 7h15'10" 50d13'9"
Tracy Alan St. Clair Cas 1h21'20" 60d50'45"
Tracy Allen And 2h27'34" 41d1'42"
Tracy Alma Miller Tau 4h26'19" 23d56'46"
Tracy - An Angel Of Mercy And 23h38'34" 42d40'59"
Tracy and David Whitefield Cyg 19h55'11" 35d35'58"
Tracy and Erich Ori 5h32'42" -1d33'28"
Tracy Ann And 1h26'25" 40d5'27"
Tracy Ann And 23h14'9" 48d41'49"
Tracy Ann Psc 22h52'35" 4d39'39"
Tracy Ann Breedlove Cyg 21h39'53" 41d57'45"
Tracy Ann Carter Tau 5h59'48" 28d24'17"
Tracy Ann Cooke Del 20h49'42" 9d50'58"
Tracy Ann Jones Cas 0h10'10" 61d20'24"
Tracy Ann Plexico Uma 8h38'0" 70d12'20"
Tracy Ann Smalley Cam 5h9'12" 60d38'22"
Tracy Ann Weston Lac 22h19'27" 46d45'36"
Tracy Ann White Cam 6h19'34" 62d36'58"
Tracy Ann Williams Ari 2h45'2" 24d51'23"
Tracy Apple Moran Vir 12h4'48" -3d52'10"
Tracy Baker Foster and Jesse Foster Dra 16h35'34" 60d0'58"
Tracy Ballin Mon 6h55'4" -6d28'53"
Tracy Bernstein Ori 5h44'50" 0d29'54"
Tracy Blair Meese Leo 10h3'38" 24d22'36"
Tracy Bowe Dra 17h2'58" 65d35'53"
Tracy/Bryan Mack Together Forever Psc 23h26'30" 3d1'16"
Tracy Caldwell Cnc 8h33'16" 7d57'9"
TRACY CAMERON GREGORY Cap 20h14'37" -17d58'14"
Tracy Carrillo Umi 13h37'8" 75d45'58"
Tracy Carroll Crb 16h8'41" 38d33'7"
TRACY CATHERINE Mon 6h53'10" -0d22'22"
Tracy Catherine Donachie Cnc 8h22'24" 26d18'45"
Tracy Chafer Cas 2h0'58" 74d25'10"
Tracy Clark Vir 13h34'52" 5d49'3"
Tracy Cohen Cas 0h22'50" 57d53'24"
Tracy D Greene Her 16h25'45" 43d36'33"
Tracy D. Simpson Lyr 18h58'25" 26d33'41"
Tracy Darling Cam 6h5'59" 68d36'33"
Tracy Don Kelly Cyg 19h49'58" 38d41'33"
Tracy Dowler And 23h58'25" 44d11'12"
Tracy Dumais Crb 15h38'14" 37d58'6"
Tracy E Bonanno Cnc 8h52'37" 26d39'53"
Tracy Elizabeth Booth Cas 0h47'41" 64d54'44"
Tracy Fair And 0h30'13" 45d26'9"
Tracy Falkingham Ori 4h50'6" 3d47'11"
Tracy Ferguson Uma 9h54'22" 42d11'6"
Tracy "Forever My Star" Lib 14h55'30" -5d14'32"
Tracy Fuelleman Tau 3h37'49" 16d6'6"
Tracy Harms Peg 23h30'15" 17d7'50"
Tracy Hedgpeth Peg 22h34'29" 11d40'2"
Tracy Hepler Uma 11h1'57" 43d11'31"
Tracy Holcomb Cap 21h41'37" -15d38'45"
Tracy Holland Cas 23h4'4" 56d31'3"
Tracy Huntington Brown Per 3h10'52" 42d47'46"
Tracy James Uma 10h11'10" 59d41'17"
Tracy Jane Armitage And 2h35'58" 49d6'57"
Tracy Jane Navier Baileche And 23h36'8" 36d23'53"
Tracy Jessop Peg 22h25'8" 13d4'41"
Tracy Joan Abalos Tau 5h25'52" 26d16'56"
Tracy Jones - Tredad And 1h10'0" 42d54'13"
Tracy Kay Gem 6h44'2" 20d11'10"
Tracy & Ken Peg 23h22'6" 31d42'16"
Tracy Knapp Cyg 19h58'28" 43d3'8"
Tracy L Baker Cnc 8h43'21" 23d33'47"
Tracy L. Brown Uma 10h33'18" 51d41'16"
Tracy L. Hover Aqr 22h18'15" -1d5'49"
Tracy La Rae Crb 15h29'5" 25d51'40"
Tracy Lee Lmi 10h31'44" 35d34'45"
Tracy Lee Coddington And 0h42'12" 41d18'37"
Tracy Lee Kochanowski Psc 1h22'27" 30d45'1"
Tracy Leigh Noeske Sco 16h8'31" -17d57'55"
Tracy Long Lyn 7h35'4" 36d47'31"
Tracy Long Psc 0h10'59" 8d20'17"
Tracy Lou Butcher-Haynes And 0h8'50" 44d10'34"
Tracy Lyn Ori 6h15'21" 15d20'7"
Tracy LynDara Nomayer Aql 20h12'40" 2d41'32"
Tracy Lynette Uma 11h24'59" 68d16'11"
Tracy Lynn Cam 3h25'33" 64d15'16"
Tracy Lynn Sgr 19h38'16" -16d49'41"
Tracy Lynn Com 12h34'21" 27d41'43"
Tracy Lynn Ari 3h28'53" 24d18'30"
Tracy Lynn Leo 10h38'21" 15d1'46"
Tracy Lynn Benecke Schumann And 2h12'7" 45d9'13"
Tracy Lynn Elkins Cas 0h2'41" 50d27'32"
Tracy Lynn Mahoney Sgr 18h6'7" -26d28'39"
Tracy Lynn Rodriguez Lib 15h12'39" -19d44'51"
Tracy Lynn Therriault Lib 15h37'29" -25d30'5"
Tracy Lynn - TLC Cas 1h22'40" 57d25'21"
Tracy Lynn Vitters Cas 0h28'46" 55d32'10"
Tracy Lynn Williams And 1h41'21" 38d37'6"
Tracy Lynne Del 20h40'16" 19d35'41"
Tracy Lynne Lyman Uma 11h48'11" 37d26'4"
Tracy M. Stewart And 1h44'3" 44d28'40"
Tracy Madelynn Leo 10h12'38" 16d58'42"
Tracy Maida Cnc 8h16'25" 17d48'38"
Tracy "Mama" Trice Crb 15h45'39" 27d24'37"
Tracy Mariano And 1h40'11" 42d36'2"
Tracy Marie Brown Vir 13h26'2" -5d26'27"
Tracy Marie Ferguson Vir 13h22'15" -9d21'55"
Tracy Marie Gore Cam 4h23'17" 56d56'36"
Tracy Marie Morkunas And 0h43'51" 45d10'12"
Tracy Megson La Veyra - Mommy Sco 16h14'40" -10d30'51"
Tracy Meinke Uma 12h33'16" 58d27'30"
Tracy Meredith Anstey Crb 15h47'44" 36d29'51"
Tracy & Michael Cyg 19h23'30" 53d29'31"
Tracy Michelle Barron 23031975 And 23h17'8" 47d33'23"
Tracy Moore Lyn 7h21'47" 53d20'32"
Tracy Nelson Uma 11h15'36" 43d14'31"
Tracy Nicole Cline -Love of my Life Peg 21h52'9" 26d30'43"
Tracy Nicole Rainey Sco 16h12'44" -17d45'7"
Tracy Ohm Vir 13h40'25" -14d31'3"
Tracy Pearson Starry-Eyed Girl Cyg 20h3'1" 42d22'27"
Tracy Pfeifer & Ira Blaufarb Cyg 20h21'38" 55d5'53"
Tracy Plowman Sgr 18h55'26" -31d6'21"
Tracy Powell Gem 6h48'13" 13d2'34"
Tracy R Neeland Gem 7h2'23" 26d29'24"
Tracy & Raffaele Imperato Cyg 19h43'19" 31d11'22"
Tracy Ramig Cyg 20h55'15" 30d43'38"
Tracy Robinson Per 3h17'13" 45d24'8"
Tracy Romero "Tra-Tra's Star" Cnc 8h2'2" 19d29'13"
Tracy Russell Aql 18h51'6" -1d4'22"
Tracy Rysewyk Cas 0h35'24" 52d26'42"
Tracy S. Huber 11-12-81 Sco 17h56'44" -31d2'55"
Tracy S. Morris Sco 16h45'45" -32d57'7"
Tracy S. Teasley III Her 16h43'52" 25d38'46"
Tracy San Juan Psc 1h3'24" 23d28'40"
Tracy Smith Cyg 19h45'49" 30d4'2"
Tracy Smith Eddy Lyr 19h17'50" 28d18'2"
Tracy Springer Gem 7h2'8" 33d6'34"
Tracy Steven Steed Dra 16h42'50" 65d32'20"
Tracy Stuart Ari 1h56'40" 18d45'46"
Tracy Sue McArn Umi 14h21'29" 65d42'9"
Tracy Susan Lessin Aqr 21h59'20" 0d54'29"
Tracy Suzanne Lyn 7h30'50" 52d36'33"
tracy & tatiana luepnitz star Peg 22h40'54" 11d20'40"
Tracy TBuggy Uma 11h24'41" 45d25'25"
Tracy "The Irradiated Star" Tau 3h41'0" 27d51'32"
Tracy "Tinkerbell" Schultz Lib 15h41'8" -4d55'32"
Tracy Tischner Cap 21h55'21" -17d44'11"
Tracy V. Lopez Lyr 18h54'17" 32d29'22"
Tracy Veliz Ari 2h43'28" 28d3'47"
Tracy Villedrouin - My Best Friend And 0h19'51" 35d23'48"
Tracy Vivian Vir 12h40'8" 12d38'34"
Tracy W. Herold Cas 0h46'15" 48d4'54"
Tracy W. Smith Oph 17h32'46" -0d0'7"
Tracy Ward Crb 15h41'6" 38d48'22"
Tracy Williams Lyn 8h38'22" 34d21'49"
Tracy Worthington 40 Uma 11h45'14" 44d50'27"
Tracy Yvonne West-Robinson Cyg 20h19'31" 50d51'7"
Tracy2006 Cnc 8h28'50" 14d6'45"
Tracy-Anne 9-16 Ori 5h17'20" 6d40'46"
Tracye & Tiffany Aqr 23h37'27" -13d25'37"
Tracy's Ari 2h18'18" 24d44'40"
Tracy's Christmas Star Leo 11h14'43" 16d1'3"
Tracy's Dream Cas 0h56'13" 56d55'42"
Tracy's Light Lib 15h2'11" -0d35'21"
Tracy's place in space Tau 4h33'17" 30d3'34"
Tracy's Prince Brett E. Lanz Cyg 21h30'24" 32d19'54"
Tracy's Rock 07/06/2006 Cnc 9h16'52" 32d9'37"
Tracy's Star Gem 7h15'53" 19d49'56"
Tracy's Star - 101263 Sgr 18h55'19" -17d46'50"
Tracy's Wishing Star Cas 0h10'43" 53d24'19"
Tracy-Shane Cnc 8h24'54" 25d48'32"
Tradd Dra 19h8'19" 64d36'41"
Tradd Philip Spadavecchia Per 3h29'2" 50d48'46"
Trae & Brittany's Amsterdam Sge 20h3'37" 17d50'46"
Traevor James Granger Tau 5h23'5" 20d58'0"
Trafalczyk, Hartmuth Uma 12h54'14" 62d23'38"
Trafe Star And 1h3'39" 41d35'0"
Trager Antrim Brown Aqr 22h54'39" -8d49'31"
TRAILBLAZER Uma 11h45'41" 63d54'50"
Trailer Martin Cam 3h56'41" 66d51'42"
Train Cyg 20h31'4" 56d50'50"
Train Station John Vir 14h15'38" -10d42'9"
Trajedy Sgr 19h29'45" -19d53'38"
Trak Potter Ori 6h4'53" -0d45'14"
Tram Per 3h45'53" 40d33'13"
Tram Kim Nguyen Vul 20h39'2" 23d14'14"
Trami Per 3h46'21" 44d47'54"
TRAMPUS Uma 9h31'50" 42d54'47"
Tran Tau 4h35'14" 15d19'58"
Tran LOVES Hien FOREVER Sge 19h44'37" 16d59'48"
Tran LOVES Viet Sge 19h26'34" 17d27'16"
Tran Ngoc Bao Chau Aqr 23h10'55" -5d56'4"
Tran Phan Umi 15h28'3" 72d56'9"
Tran Thi Minh Ngoc Cnc 8h39'36" 22d50'2"
Tranberry Aqr 22h17'0" 1d8'49"
Trancona Ice Uma 11h46'52" 44d32'40"
Trang Bui Cnc 8h12'19" 26d35'9"
Trang Family Cyg 20h39'42" 44d46'56"
Trang Jan Nguyen Gem 7h31'13" 29d23'24"
Trang Nguyen And 0h13'16" 39d35'40"
Trang Nguyen Tran Per 3h14'59" 47d10'34"
Trang Phan Hya 9h14'50" -0d55'4"
Trang Thuy Doan Leo 11h36'2" 26d34'44"
Trang Tran Ori 5h12'23" 3d2'14"
Tranquil Thunder Uma 13h36'22" 55d29'46"
Tranquility Sgr 18h40'58" -24d59'15"
Tranquillitas Dra 15h42'1" 58d39'47"
Transcendance Leo 10h44'18" 19d31'50"
TransDerm Scop Crb 16h20'39" 30d41'15"
Transehunte Uma 9h25'49" 58d43'31"
Transition Umi 14h51'48" 73d51'54"
Transmooditron de Nebula rx800p Boo 14h56'1" 31d55'35"
Trao Sgr 18h24'30" -32d46'4"
Trapper Hap Cap 20h25'11" -19d50'30"
Trasci's Star And 1h18'11" 42d31'50"
Trasee Cosby Uma 8h42'57" 51d11'41"
TRASH Dra 20h17'48" 65d12'21"
TRAS-STAR Ori 5h30'33" 1d17'48"
"Träumchen" Marita Reichelt Uma 12h58'42" 53d24'2"
Traum-Engel Uma 11h52'53" 37d49'52"
Träumerle Uma 11h49'10" 30d34'23"
Traumfrau Karin Tau 3h54'8" 20d28'17"
Traumhäx Cep 22h55'30" 76d34'47"
traummama Uma 8h39'1" 58d35'33"
Trautsch, Rudolf Uma 14h1'37" 51d19'14"
Travadon Aqr 22h59'4" -7d30'10"
Traveler Umi 13h32'39" 72d31'2"
TRAVELER Cnc 9h9'16" 28d25'7"
Travelle Cnc 8h43'56" 24d1'37"
Travellers Palm Apts. And 2h26'30" 48d58'37"
Travellin' Bob Cep 22h7'21" 69d40'12"
travelngirl And 23h46'40" 46d3'10"
Travie Baby Uma 11h26'1" 41d19'24"
TRAVIS Her 17h22'26" 36d28'4"
Travis Uma 11h1'12" 51d53'31"
Travis Per 2h55'13" 52d7'31"
Travis Boo 14h34'20" 42d8'19"
Travis Tau 4h16'13" 27d49'4"
Travis Her 16h39'59" 29d24'59"
Travis Leo 10h36'48" 17d38'7"
Travis Cnc 9h3'48" 12d18'26"
Travis Uma 8h14'14" 73d2'44"
Travis A. Dechene Per 2h22'7" 51d46'12"
Travis A. Moothart Aqr 22h4'35" -4d39'42"
Travis A. Thomas Aur 6h1'52" 48d8'46"
Travis Aaron Claridge Ari 2h17'16" 11d14'34"
Travis Aaron Galey Boo 14h40'53" 35d24'15"
Travis Abbott Uma 11h37'41" 45d29'5"
Travis & Abby King Cnc 8h52'47" 28d49'7"
Travis Alan Bush Leo 11h51'26" 15d43'50"
Travis Alexander Lippard Gem 7h30'28" 26d39'47"
Travis Allen Merritt Psc 1h1'14" 4d42'8"
Travis and Amanda's Star Tau 4h9'58" 6d11'52"
Travis and Arikia Forever Aqr 22h16'12" 1d43'56"
Travis and Hillary Cnc 8h54'33" 15d38'3"
Travis and Misty Willard's And 0h24'11" 32d18'14"
Travis and Trisha Gibson And 2h29'55" 43d45'19"
Travis and Whitney Cyg 20h13'26" 47d2'47"
Travis Andrew Starkey Cap 21h44'49" -13d15'48"
Travis Anthony Monahan Cap 20h31'13" -15d22'47"
Travis Anthony Moraga Ori 6h19'17" 10d1'21"
Travis Benjamin Eckendorf And 23h30'17" 36d20'45"
Travis+Beth Lyn 7h38'13" 52d4'1"
Travis Blackburn Cru 12h56'31" -55d45'52"
Travis Brad Eckmann Lyr 18h46'20" 37d2'42"
Travis & Brandie Uma 13h44'43" 48d0'14"
Travis Brooks Strong Jr. Vir 15h4'51" 4d39'2"
Travis Bussey Leo 9h39'28" 6d41'34"
Travis C. Blunt Lib 14h52'50" -12d0'39"

Travis Christopher Durkee
Tau 4h24'55" 22d59'51"
Travis C.J. Mills
Per 4h15'59" 52d21'41"
Travis Clay Schuermann
Sgr 19h30'14" -13d18'33"
Travis Clyde Mitchell III
Cep 0h4'46" 79d12'48"
Travis Cobb
Tau 3h40'30" 24d14'20"
Travis & Cody
Vir 12h56'15" 11d52'13"
Travis Combs Chafin
Lib 15h28'52" -8d30'10"
Travis Craig Lewandowski
Leo 10h20'3" 23d24'11"
Travis D. Albert
Her 17h51'30" 34d51'27"
Travis Daigle
Aql 20h9'40" 4d24'21"
Travis Daniel Woody
Lac 22h17'52" 54d14'20"
Travis David Woodliff
Sco 16h39'22" -37d11'28"
Travis & DeLaina
Psc 1h7'41" 4d1'47"
Travis Donald Mitchell
Leo 11h26'27" 9d19'41"
Travis Edward Mysliviec
Sgr 18h14'30" -30d25'3"
Travis Eric Walston
Cnc 8h52'16" 26d30'28"
Travis Guy Tunnell
Uma 9h6'8" 50d36'30"
Travis Guymon & Jackie Brock
Ori 6h6'4" 19d51'14"
Travis Head
Her 18h6'10" 32d9'5"
Travis Hicks
Tau 5h26'37" 26d51'36"
Travis J. Kaikkonen
Umi 13h42'23" 72d33'37"
Travis Jackson Thrash
Uma 8h25'5" 68d18'56"
Travis James Jackson
Her 18h42'49" 19d0'38"
Travis James White
Her 16h34'55" 34d21'37"
Travis John Borrenpohl
Sgr 18h48'23" -29d0'36"
Travis John Gilge
Psc 1h3'21" 27d55'30"
Travis John Rhodes
Vir 14h9'22" -9d45'8"
Travis Jon Loves Cristina Renee
Aqr 22h6'27" 0d34'11"
Travis & Julie
Uma 9h41'3" 54d14'40"
Travis/Kate's Mom~Debora Jane Cook
Uma 13h42'17" 53d4'27"
Travis Kaufman
Psc 0h34'50" 19d38'0"
Travis Kearney
Cap 20h25'42" -20d27'53"
Travis King Marrs
Leo 11h12'15" 5d54'12"
Travis & Krystal Hall
And 0h48'14" 39d27'19"
Travis Kunce
Per 3h48'3" 50d50'20"
Travis L. Mahan
Sco 16h14'39" -21d33'42"
Travis Lane and Samantha Elizabeth
Cyg 21h14'28" 52d40'29"
Travis Lanphere
Cyg 20h24'13" 37d11'11"
Travis & Leanne - 7th February 1997
Cru 12h26'16" -60d31'51"
Travis Lee
Aqr 23h28'36" -23d29'8"
Travis Lee
Cap 21h30'4" -16d33'14"
Travis Leo Colgan
Her 17h25'0" 26d58'18"
Travis & Lori
Vir 11h48'14" 4d25'36"
Travis M. Thornhill
Del 20h42'25" 16d5'22"
Travis McGee
Cyg 20h28'52" 34d50'55"
Travis & Megan Bennett
Cyg 20h57'20" 45d49'22"
Travis & Meghan's Star
Ari 2h33'44" 21d24'21"
Travis Michael
Per 2h43'50" 55d15'34"
Travis Michael Brand
Boo 14h47'6" 33d29'40"
Travis Michael Haines
Uma 13h40'30" 49d37'12"
Travis Mitchell Wilde
Her 16h47'56" 29d52'48"
Travis & Morgan
Lib 15h1'48" -24d57'8"
Travis Mullins
Aqr 20h42'32" -13d19'39"
Travis Neil Miller 8/14/1982
Leo 11h57'39" 21d27'52"
Travis Okula
Cap 20h19'10" -16d1'2"
Travis Parker Looney
Sco 16h10'33" -17d1'57"
Travis Patriquin
Her 17h28'54" 39d19'21"
Travis Pennington
Leo 10h50'57" 15d52'28"
Travis Peterson
Cnc 8h51'9" 30d12'21"
Travis Pronk Hafner
Her 16h35'37" 31d23'5"
Travis Ray Brickman
Her 16h57'42" 32d41'46"
Travis Ray Kiser
Her 16h42'54" 48d58'35"
Travis & Renee - 25/01/2003
Cru 12h15'1" -63d44'1"
Travis Robert Roy
Vir 13h21'46" 6d6'17"
Travis Russell
Umi 15h56'4" 83d39'31"
Travis Ryan Day
Sgr 18h37'45" -19d40'34"
Travis Schnurr
Leo 11h19'19" -5d49'46"
Travis Scott Gund
Cyg 19h56'16" 39d56'0"
Travis Semcken
Tau 5h25'52" 18d44'54"
TRAVIS SHAWN HUTTO
Uma 11h11'57" 41d20'48"
Travis Simpson
Dra 18h57'12" 70d31'4"
Travis Son of Vivian Irving
Sco 17h33'30" -31d54'41"
Travis Soulé
Cen 13h57'58" -62d49'19"
Travis Spencer
Uma 10h4'58" 45d8'58"
Travis Stelzer
Uma 8h57'24" 59d56'53"
Travis Swann
Aqr 21h12'15" 1d14'22"
Travis Taber In Loving Memory
Tau 3h56'55" 0d41'42"
Travis The Guardian Angel
Cnc 8h33'22" 23d6'46"
Travis Todd Coder
Ari 2h14'56" 14d58'50"
Travis "T-Star" King
Aql 19h46'55" -0d36'8"
Travis Underwood
Gem 6h31'43" 27d8'44"
Travis Valenzuela
Her 16h13'6" 17d28'45"
Travis W. O'Keeffe
Cnc 8h58'45" 14d41'47"
Travis W. O'Keeffe
Cnc 8h46'6" 32d6'17"
Travis William May
Her 16h38'31" 38d2'30"
Travis Williams
Uma 8h58'41" 72d32'0"
Travis Xere Womack
Sco 17h56'32" -32d16'35"
TravisJMiller
Leo 11h14'49" 24d27'53"
Travis's Blazing Star Fire
Leo 9h27'11" 27d51'59"
Travlr
Crb 15h30'1" 29d38'52"
Travs
Dra 18h6'44" 78d46'21"
TRAX (Patty & John Traxler's Star)
Sgr 18h30'38" -34d14'43"
T-Ray Prime
Sco 16h45'7" -33d9'45"
Traycco
Dra 20h20'40" 67d54'27"
Traywick's Sirius
Cma 6h55'19" -20d11'25"
tre
Vir 12h6'37" -5d20'57"
Tre' B. Adams
Lyn 6h58'5" 60d36'32"
Tré Michael Bower
Leo 10h10'21" 21d45'33"
Tre Mikel Hall
Uma 9h29'47" 50d58'2"
Trea Davis
Cap 20h46'36" -14d44'5"
Treasure
Psc 23h36'15" 1d20'45"
Treasure Beach Hotel
Umi 14h10'20" 68d20'43"
Tréazell
Oph 17h4'39" -0d44'46"
Trebe
Ori 6h6'7" 13d20'49"
Treber, Magnus
Vir 13h36'5" -7d1'49"
Treca
Dra 14h45'12" 57d30'17"
Tree 51
Tau 5h10'55" 17d58'13"
Treehugger
Uma 8h50'44" 52d51'24"
Treela
Mon 6h42'2" 8d51'56"
Treena Marie Ghi
Cap 20h29'24" -14d45'3"
Treesa Ann
Sgr 18h14'3" -21d57'56"
Treesa Marie Francis
Per 4h43'16" 45d28'44"
Treeva Girl
Lyn 8h17'20" 40d58'33"
TREE-Z
Aql 19h49'16" -0d17'59"
Trefzger, Volker
Uma 9h1'44" 49d43'33"
Treganon
Uma 11h40'18" 37d44'9"
Treiber, Alfred
Ori 5h2'27" 6d4'46"
Trek Guillory
And 0h46'44" 38d1'57"
Trekker Tamerou Asrat, MD
Cnc 9h7'12" 23d29'41"
Trekrbear
Cam 5h7'37" 68d59'52"
Trelawne Allison
Uma 11h25'44" 70d35'57"
Trella Luene
Sco 16h8'4" -15d14'58"
TRELYNN
Psc 0h56'29" 10d59'11"
Tremaine "Trikki" Jones
Gem 7h21'50" 24d46'57"
Tremble
Aqr 22h26'8" -2d32'56"
Tremble and Twitch -TNT
Sco 17h11'53" -44d54'56"
Tremendous
Sco 16h16'32" -14d44'0"
Tremendous
Oph 17h26'23" 7d49'17"
Tremescia Kegler
Lib 15h21'51" -22d58'8"
Trena Miller
Lib 15h42'40" -4d35'2"
Trenda-A Wonderful Mom
Cas 0h34'51" 61d36'40"
Trè-Nique Segar
Aql 19h45'44" -0d7'45"
Trenster
Psc 1h20'14" 18d44'4"
Trent
Cas 0h7'40" 57d3'59"
Trent 9-1215225-251521
Ori 5h22'10" 9d14'0"
Trent Allen Birchard
Her 16h40'44" 34d3'6"
Trent and Jen
Crb 15h52'15" 26d16'14"
Trent Constellation
Leo 10h19'23" 24d53'4"
Trent Constellation
Leo 10h13'18" 17d40'34"
Trent Cornell
Leo 10h23'47" 17d44'18"
Trent Edward Schuh
Aql 19h8'28" 11d56'56"
Trent Jeffery Bulow
Uma 11h1'52" 45d14'16"
Trent Joseph Guidry
Cep 21h23'42" 60d22'10"
Trent Lee Bauer
Aql 19h38'18" 10d4'39"
Trent loves Amy
Cam 4h2'10" 71d20'4"
Trent Markwith
Cma 6h41'5" -12d50'58"
Trent Martinez (Lovato)
Her 18h23'16" 14d0'32"
Trent McKenzie
Aqr 23h44'26" -10d31'32"
Trent Michael Holland
Cru 12h32'42" -55d55'31"
Trent Michael Huff
Uma 10h36'33" 55d17'31"
Trent Michael Manhart
Crb 15h41'37" 29d51'14"
Trent Nikolus Bowman
Dra 18h35'3" 54d46'20"
Trent of Alaska
Tau 5h33'52" 25d11'14"
Trent Owen Williams
Leo 10h55'55" -2d25'48"
Trent Piers Jordan-London's 3 Kings
Ori 5h17'56" 6d55'56"
Trent R. Blanton
Ori 5h2'57" 3d52'59"
Trent Reznor
Aql 19h9'31" -0d5'20"
Trent Rhoton
Her 17h20'57" 49d4'17"
Trent Robert Taylor
Aql 19h14'29" 5d51'22"
Trent Russell Strohecker
Her 18h7'24" 17d27'56"
Trent & Sarah Forever
Cru 12h52'22" -58d15'36"
Trent Taylor
Leo 9h37'2" 21d1'47"
Trent Thomas Dye
Sco 16h28'45" -26d41'18"
Trent Turyna
Ori 6h11'40" 15d12'17"
Trent Willmon
Boo 13h46'6" 16d26'31"
Trentin Duke Jones
Vir 13h59'10" -16d52'15"
Trenton
Ori 5h57'58" -0d40'53"
Trenton and Karen Richinski
Cyg 21h43'3" 44d21'57"
Trenton Auston Lawrence 10-24-1981
Ori 5h53'26" 14d39'24"
Trenton Blain Boyer
Her 16h59'10" 13d51'42"
Trenton "Bubba" Eddy
Tau 3h42'40" 13d8'34"
Trenton Carlson
Sco 16h34'44" -26d44'45"
Trenton Jake Cracraft
Lyn 8h5'33" 35d9'4"
Trenton Joel Miner
Aqr 23h9'30" -10d0'2"
Trenton John
Uma 8h50'39" 53d43'52"
Trenton Mayes
Psc 23h9'46" 0d53'22"
Trenton Michael Braithwaite
Umi 0h19'52" 88d56'58"
Trenton Urbain II
Umi 15h36'23" 71d52'57"
Trentson Patrick Sluiter
Her 17h45'36" 48d10'10"
Trentyn
Cnc 8h48'33" 32d6'30"
Trentzsch, Oliver
Uma 12h4'3" 39d8'28"
Trepczik, Rainer
Uma 8h51'44" 59d3'29"
Trepel, Jörg
Uma 8h39'8" 59d8'19"
Trepplin, Julian Manuel
Ori 5h8'14" 15d49'13"
Tres
Sge 20h2'49" 19d4'29"
Très Belle
Tau 4h27'50" 12d40'7"
Tres Estrellas
Uma 13h24'53" 54d5'16"
Tres Mamas (June~Juanita~Kaye)
Cas 0h43'59" 63d55'13"
Tres Randolph
Ori 6h3'40" 21d6'29"
Tres' Spain
Lyn 8h27'46" 33d47'32"
Tre's Sparkle
Umi 15h32'56" 71d8'37"
Tres Tria Una - Jeff, jody, cheryl
Tri 1h48'42" 31d11'55"
tresa
Cas 1h56'53" 67d53'43"
Tresa Jeanne Palmer
And 23h25'57" 42d40'41"
Tresa M. Keith Best Wife&Mom Ever
Psc 23h59'20" 8d53'4"
Tresa Zumsteg
Cyg 20h3'23" 44d15'50"
"Tresena" Star of Darryl 2005
Uma 11h33'1" 64d29'6"
Treskatsch, Hartmut
Uma 11h56'28" 49d12'9"
trésor
Cas 1h17'59" 71d15'20"
Trésor
Uma 10h28'55" 56d41'14"
Tressa
Sgr 19h42'21" -12d8'34"
Tressa Lynn
And 0h52'25" 39d42'3"
tressa mckirgan
Sgr 18h12'15" -23d34'24"
Tressa VanAnda
Lyr 18h49'54" 43d48'46"
Tressah Grace Marra
Gem 6h47'8" 33d38'49"
Tressarannon
Ari 2h12'45" 22d56'19"
Tressa's Magic Star
And 1h12'37" 39d7'25"
Tretter, Erich
Uma 11h54'29" 33d22'7"
Treu Breeze
Cnv 13h30'46" 36d37'15"
trev
Sgr 20h2'4" -25d50'56"
TrEv & EmiLy
Lib 15h22'47" -29d6'4"
Treva
Aql 19h9'23" 10d45'35"
Treva Grace Hedrick
Tau 4h31'42" 30d13'18"
Treva T.T. Traver
Sgr 19h13'33" -14d17'19"
Trevenen Mitchell Wright
Gem 7h11'1" 31d27'22"
Trever
Ari 3h11'30" 28d53'31"
Trever and Melissa 4Ever
Cyg 20h7'45" 39d57'33"
Trever Peck
And 1h23'33" 48d4'51"
Trevin Anthony Frame
Per 4h43'11" 49d52'42"
Trevin J Sacca
Uma 11h5'24" 39d40'41"
Trevor
Vir 12h50'36" 12d47'39"
Trevor Allan Dickson
Sgr 19h5'37" -35d4'4"
Trevor and Cindy Gatus
Equ 21h7'4" 11d45'18"
Trevor Anne Vienna Sander Nina Webb
Crb 16h8'8" 28d47'10"
Trevor "Birdie" Davis
Lac 22h35'8" 50d7'9"
Trevor Anthony Schuetz
Trevor Battles
Per 3h49'34" 40d37'0"
Trevor & Bhavna
Cen 11h31'20" -37d49'56"
Trevor "Birdie" Davis
Ari 2h30'14" 10d59'15"
Trevor Blackwood
Ari 2h11'19" 23d25'56"
Trevor Blake Crawford
Cyg 20h8'6" 31d34'20"
Trevor Brendan Donovan
Dra 18h38'25" 55d26'42"
Trevor Brick
Cep 22h57'8" 71d37'43"
Trevor Brown
Sco 16h8'16" -18d22'36"
Trevor Bynum
Sco 16h14'10" -16d26'10"
Trevor Carl Johnston
Aur 6h35'57" 28d10'36"
Trevor Christian Boyer
Her 17h15'19" 34d42'29"
Trevor Christopher Welsh
Her 18h27'11" 25d20'19"
Trevor Dagg
Cas 1h57'39" 60d30'6"
Trevor Daniel
Uma 9h32'57" 48d25'30"
Trevor Daniel Davis
Ari 1h47'32" 19d49'28"
Trevor David Billinghurst
Umi 16h13'46" 73d7'7"
Trevor David Stringer
Sgr 18h50'26" -25d13'15"
Trevor Dietrich
Gem 7h8'43" 23d6'24"
Trevor Doster
Ari 3h23'41" 15d35'25"
Trevor Duane Savage
Cyg 19h58'27" 33d8'18"
Trevor Ealy
Her 17h53'30" 14d26'17"
Trevor Eden McGuire
Cnc 8h9'47" 13d43'34"
Trevor Edward
Per 3h11'33" 53d20'49"
Trevor Edward Edmunds
Psc 1h19'15" 31d14'17"
Trevor Edward Taison Williams
Leo 10h41'34" 14d22'0"
Trevor Fabian
Ori 6h2'16" 10d57'27"
Trevor Fox Perez
Gem 7h1'43" 16d49'52"
Trevor Gresko
Uma 9h27'53" 67d28'59"
Trevor Helderman
Psc 1h19'16" 23d17'48"
Trevor Hershberger
Umi 17h29'16" 81d48'55"
Trevor Howard
Lmi 10h6'17" 35d10'32"
Trevor J. Teply
Boo 14h32'11" 29d33'34"
Trevor Jacob Fisch
Per 4h11'9" 45d6'55"
Trevor James Campbell
Cep 22h11'0" 53d53'24"
Trevor James Duarte
Her 17h41'32" 40d5'17"
Trevor James Gabriel Williams
Her 16h32'38" 32d19'9"
Trevor James Lay
Vir 14h28'18" 4d23'10"
Trevor & Jami Wiggins Family Star
Uma 11h27'20" 32d51'58"
Trevor Jason Rund
Umi 15h46'0" 79d48'36"
Trevor Jeffrey Orndoff
Her 17h10'53" 42d37'41"
Trevor John Andrews
Cep 22h5'5" 56d17'36"
Trevor John Elgar Beale
Cru 12h39'11" -64d32'9"
Trevor John Lempesis
Lyn 7h16'24" 58d23'38"
Trevor John Rossi
Vir 12h47'58" 2d19'55"
Trevor Jon Jakaby
Gem 6h40'40" 35d15'0"
Trevor Justin Lyons
Uma 11h11'2" 46d51'10"
Trevor Kent Gose
Cep 20h15'47" 61d9'38"
Trevor Laurence Miller
Ari 1h50'53" 19d6'3"
Trevor Loves Amelia
Vir 13h37'31" 1d59'47"
Trevor Matthew Thomsen
Uma 10h9'58" 62d1'47"
Trevor Matthews
Dra 20h14'0" 64d47'50"
Trevor McKissick
Umi 16h34'6" 78d11'4"
Trevor McLaren
Uma 11h54'27" 43d39'4"
Trevor McNaught
Cep 21h50'59" 67d8'29"
Trevor Michael Richard Logan
Ori 6h7'1" 10d21'30"
Trevor Michael Velez
Sco 16h5'54" -9d39'0"
Trevor Mighty Mars Corsello
Her 18h0'27" 36d41'7"
Trevor Moores: Dad in a million.
Cep 22h42'28" 66d58'52"
Trevor Nichols Quest
Lib 15h31'17" -9d50'32"
Trevor & Nid
Cyg 20h56'40" 32d23'47"
Trevor Patton
Lib 15h29'39" -7d33'29"
Trevor Philip
Per 3h16'53" 49d14'56"
Trevor Piggott
Lmi 10h18'41" 32d13'11"
Trevor & Rachael - Forever In Time
Uma 10h8'26" 51d41'42"
Trevor Richard Allan
Ori 6h19'27" 7d8'49"
Trevor Rill
Ori 6h8'15" 17d47'41"
Trevor Robert Chapman
Cep 20h54'52" 61d3'12"
Trevor Roberts
Cru 11h59'19" -61d20'1"
Trevor Ross Smentek
Vir 13h43'50" 2d39'58"
Trevor Scott Airola
Her 17h14'47" 34d18'15"
Trevor Scott Zankl
Tau 3h47'22" 21d29'28"
Trevor Seth Letendre
Lib 14h50'58" -19d19'41"
Trevor Shaw
Uma 10h31'5" 53d9'6"
Trevor Stewart - 'Stargazer'
Sco 17h3'55" -45d7'13"
Trevor T-Diddy Johnson
Aqr 22h57'41" -6d1'53"
Trevor (TJ) Stutt
Lib 15h3'9" -1d3'45"
Trevor "Trevhead" Kingsley 1.06.78
Cru 12h26'23" -58d55'7"
Trevor Underhay
Uma 11h18'51" 61d23'33"
Trevor W. Colburn
Psc 0h48'33" 11d29'50"
Trevor Walshaw
Uma 8h32'3" 70d47'41"
Trevor William Hudson
Cep 20h43'49" 60d37'22"
Trevor William Karp
Gem 6h48'51" 33d2'0"
Trevor William Obal
Psc 0h59'18" 7d23'31"
Trevor William Schuetze
Cen 14h43'33" -41d50'7"
Trevor Win'E
Lib 15h5'57" -29d26'55"
Trevor's Eternal Light
Ori 6h16'43" 5d57'3"
Trevor's Light
Cnc 8h26'16" 21d34'11"
Trevor's Shining Light
Aql 19h43'30" 7d3'14"
Trevor's Star
Sco 16h53'59" -38d41'33"
T-Rex
Cas 1h33'12" 65d11'46"
T-REX 1
Gem 6h54'32" 34d40'34"
TREXALL
Aql 19h52'47" -0d20'11"
Trey
Ori 4h51'24" 10d28'50"
Trey Anastasio
Lyr 18h46'34" 39d31'47"
Trey Andrew Simmons
Tau 5h58'8" 25d22'14"
Trey Anthony Spurgeon
Ari 3h3'16" 20d0'4"
Trey Beattie
Uma 11h31'32" 43d59'5"
Trey Breaux
Uma 11h12'43" 31d59'15"
Trey Davis Martin
Cyg 19h54'10" 37d45'35"
Trey Dog
Lyn 7h7'32" 47d5'30"
Trey Evan Rachunek
Sco 16h46'56" -32d33'1"
Trey Fetter
Ori 6h0'8" 6d54'12"
Trey Hardesty
Psc 1h7'48" 23d46'38"
Trey & Heather 4/Life
Cyg 20h22'1" 39d0'27"
Trey Hollingsworth
Vir 14h12'11" 2d38'22"
Trey Jacob Adams
Her 17h33'34" 36d57'10"
Trey & Jodi
Uma 11h39'34" 64d37'20"
Trey Knight Holmes
Sco 17h37'39" -36d29'34"
Trey Landon Fletcher
Ori 5h30'52" 5d12'39"
Trey Leger
Boo 15h9'3" 13d46'4"
Trey Logan Huffman
Her 16h48'20" 28d52'39"
Trey Lynn and Gram's Star
Lyn 9h3'33" 42d57'33"
Trey Menefee
Aql 19h19'17" -0d23'46"
Trey Olson Thorner
Cyg 20h17'16" 43d15'20"
Trey Robert Davis
Dra 16h33'39" 52d27'14"
Trey "Sgetties" Ostby
Leo 9h36'59" 11d1'43"
Trey Smith
Cap 21h53'51" -19d36'23"
Trey & Stacey Turner's Shining Love
Aqr 22h11'24" -15d38'4"
Trey Thomas Manhart
Crb 15h50'46" 29d14'36"
Trey Xavier
Cyg 20h35'10" 33d52'11"
Trey92
Vir 11h56'53" 4d1'44"
Treyden DeMar Boggs
Cam 4h57'11" 53d37'56"
Treyla
Lyr 19h6'8" 36d39'58"
treynewbern
Cet 3h15'22" -0d47'48"
treyopolis
Uma 10h27'48" 45d20'48"
Trey's Destiny
Uma 10h24'14" 49d16'52"
Trey's Star -Edward Earl Fisher III
Gem 7h29'45" 25d32'41"
Treyson Clauss
Aql 20h7'56" 5d12'24"
TRF 13
Uma 11h45'30" 31d56'42"
Triaminic
Tri 1h40'43" 29d7'8"
Triana Hill
Sgr 18h46'7" -17d25'39"
Triandra Jesico
Lib 15h0'16" -13d36'53"
Triantafillos Karayannis
Ori 6h11'37" 21d4'49"
Trice
Leo 11h15'56" 21d49'3"
Trice
Psc 0h20'46" 12d25'6"
Tricia
Cnc 8h50'54" 28d12'1"
Tricia
Gem 7h24'0" 33d44'7"
Tricia
And 0h37'58" 42d29'16"
Tricia
Her 16h51'1" 48d4'39"
Tricia
Sco 16h12'32" -10d30'49"
Tricia
Sco 16h42'13" -38d35'46"
Tricia 11
And 1h3'20" 41d3'16"
Tricia 4-12-64
Ari 2h27'50" 18d34'8"
Tricia A. Fariss
Lyr 18h26'8" 32d37'6"
Tricia Ann
Cap 20h25'48" -10d48'3"
Tricia Ann Kownacki
And 23h43'4" 47d56'30"
Tricia Anne
Sgr 19h11'6" -21d56'18"
Tricia Burns Scarpellino
And 1h45'49" 50d4'51"
Tricia Carrabba
Uma 8h54'17" 63d54'41"
Tricia Dawn
Vir 13h23'28" 2d12'11"
Tricia Elizabeth Adams
Ari 2h7'22" 27d14'45"
Tricia Erb
Uma 12h5'32" 40d30'6"
Tricia Guarino
Cap 20h24'53" -9d3'44"
tricia hollowell
Sco 17h56'27" -31d22'49"
Tricia Jaskolski
Aqr 23h51'36" -10d19'59"
Tricia Joyce Foertsch
Sgr 18h49'52" -29d42'47"
Tricia K. Hansen
Cyg 19h33'36" 32d36'27"
Tricia Lyn Bulley
Umi 14h27'37" 81d28'24"
Tricia Lynn Letton
Cas 1h43'42" 66d11'17"
Tricia M Griffin
Psc 1h9'25" 5d11'57"
Tricia Martinez
And 0h16'28" 31d55'39"
Tricia McElwee
Lib 15h21'26" -7d57'19"

Tricia Moore
Ari 2h35'38" 27d44'34"
Tricia Nadelee Ruben Hercules
Her 18h42'6" 16d53'49"
Tricia Neault
Cnc 9h4'31" 29d51'52"
Tricia Nicole Snyder
Lyr 19h24'9" 37d35'13"
Tricia Peck
Crb 16h23'54" 35d29'57"
Tricia & Phil 25th Anniversary Star
Sge 20h17'54" 17d37'13"
Tricia Prettypaul
Mon 7h16'39" -9d0'14"
Tricia Purden
Cas 0h20'27" 47d31'34"
Tricia Randolph
Sco 17h22'14" -38d32'40"
Tricia Remacle
Mon 6h52'21" -0d26'41"
Tricia Richardson
Lyn 7h41'16" 44d48'6"
Tricia Rodenbough
Gem 7h32'9" 33d37'6"
Tricia Roesler
Cam 4h5'23" 53d44'29"
Tricia Rose
Leo 9h29'35" 24d59'51"
Tricia & Tom Seaver - Always
Eri 4h38'18" -0d59'40"
Tricia Weaver 40
Cas 1h14'28" 57d15'48"
Tricia's Blumoon
Aql 19h53'15" 10d18'31"
Tricia's Friendship Star 02.02.1976
Aqr 21h6'4" -12d12'58"
Tricia's Heart
Psa 22h5'16" -27d47'50"
Tricia's Light
Cnc 8h36'2" 17d42'1"
Tricia's Mother
Tau 3h39'19" 17d11'15"
Tricia's Star
Cas 1h28'28" 60d23'14"
Trickett - Lichfield DW1-5 16.8.1962
Ori 5h49'14" -3d11'0"
Trick's Midnight
Tau 3h36'35" 16d7'21"
tricky
Lac 22h19'56" 54d37'27"
Tricky Ricky
Uma 11h9'29" 52d37'10"
Tricky4Nicky
And 23h29'28" 39d27'8"
Trident
Peg 21h21'9" 23d52'2"
Tried and True
Sgr 19h10'50" -34d26'48"
Trigger 19
Ori 5h14'22" 7d22'11"
Triggs Eternal Love
Cyg 21h7'34" 36d54'40"
Triginta
Tri 2h9'51" 33d30'48"
Trihan Ursachi
Ori 5h16'49" -7d34'6"
Triinu
Uma 12h2'44" 60d14'36"
Trike
Crb 16h9'7" 31d3'41"
Trillinderland
And 2h34'8" 40d9'28"
Trilly
Aql 20h13'28" -0d40'2"
Trillz
Gem 7h6'59" 22d12'11"
Trimboli Jean-Pierre
Cen 14h35'28" -64d4'36"
Trina
Ari 2h37'52" 20d21'10"
TRINA
Cas 0h18'17" 50d23'4"
Trina Allen
Ori 5h43'10" 6d8'38"
Trina Amy
Psc 0h10'50" 1d56'28"
Trina B's 16th
And 23h43'46" 46d56'28"
Trina Dawn Fotenos
And 23h54'5" 41d59'10"
Trina Hallows Kedzierski
Uma 10h28'41" 45d46'12"
Trina Hull
Aqr 22h9'38" 0d17'5"
Trina Jane
Gem 6h56'49" 26d18'43"
Trina Joy Terry Owens
Leo 11h6'0" 2d36'47"
Trina Klassa
Lib 14h40'2" -18d6'27"
Trina Martin My Love
Ari 2h48'5" 14d48'38"
Trina May
Aqr 22h52'49" -20d13'1"
Trina Michelle Coday Mar. 15, 2006
Uma 10h54'48" 61d17'23"
Trina Millard
Cnc 7h58'5" 10d57'8"
Trina Price
Lyr 19h12'27" 26d23'36"
Trina Renee
Tau 4h43'40" 23d50'50"
Trina Stephenson
Uma 13h55'33" 61d35'45"
Trina Wawrzyniak
Umi 16h13'15" 76d18'31"
Trinack
Uma 9h27'35" 48d50'1"
Trinadad Rachel Jaime
Uma 11h27'52" 41d21'1"
Trina's Star
Umi 16h13'56" 84d1'27"
Trina's Star
Mon 7h34'31" -0d19'2"
Trinaty Aiko Jones
Psc 1h24'7" 23d51'24"
Trine Andersen
Uma 10h28'14" 44d41'59"
Trine Hurv Nilsen
Leo 10h12'37" 27d49'48"
Trine 'Laekkerhaps' Brønsgaard
And 1h39'40" 42d16'54"
Trine Lise
Psc 1h25'23" 24d39'38"
Trine Nilsen
Sco 16h38'11" -31d45'59"
Trine Olsen
Cyg 20h6'20" 44d54'56"
Trine Olson
Ori 5h55'12" 11d48'22"
Trine Stjerneskinn
Umi 17h11'18" 78d28'27"
TrinellMarie421
Uma 9h43'5" 47d23'56"
Trinh and Lam Nguyen 27th May 2000
Ori 5h0'56" 13d10'57"
Trinh Truong Feb. 22, 1979
Psc 0h34'20" 9d20'26"
Trinh Vu
Cyg 21h30'2" 45d56'46"
Trini
Lyn 8h56'40" 39d48'33"
Trini and Frank
Lyr 18h36'11" 34d34'45"
Trini Ruiz Caballe
Sgr 18h3'18" -17d46'34"
Trini Triggs
Ori 5h8'50" -0d35'13"
Trinidad and Angela Bocanegra
Lib 14h23'53" -9d2'2"
Trinity
Uma 11h38'26" 63d25'18"
Trinity
Dra 18h51'57" 60d47'10"
Trinity
Lyn 8h49'5" 37d52'30"
Trinity
Uma 11h43'26" 49d30'52"
Trinity
Cnc 9h2'51" 20d40'1"
Trinity 5
Gem 6h47'43" 29d11'15"
Trinity A. Armoogam
Cnc 8h2'6" 12d45'22"
Trinity Ann
Aqr 23h24'26" -12d46'30"
Trinity Ann Alya Abdulkhalik
Uma 10h48'58" 41d19'51"
Trinity Grace Butler
Sgr 18h12'49" -34d18'9"
Trinity Jayde - Happy 21st
Cru 12h29'19" -56d9'46"
Trinity Kathleen Koehler
And 0h33'23" 33d8'2"
Trinity Koehler
Sco 16h13'14" -12d14'17"
Trinity L. Harris
Peg 21h50'27" 12d27'1"
Trinity Laura Alexander
Gem 6h34'52" 14d2'15"
Trinity Lee Overholser
Ari 2h1'47" 13d58'20"
Trinity Louise Carlson
And 0h40'46" 32d39'49"
Trinity Lynn Earnhart
Crb 15h42'12" 32d30'38"
Trinity Maria Rodriguez
Sgr 20h13'14" -37d32'25"
Trinity Melcher
Gem 7h6'27" 21d31'40"
Trinity Nicole Boyd
And 0h19'16" 41d18'3"
Trinity Paige Watkins
Uma 11h18'23" 30d26'18"
Trinity Patton Weighman
Tau 4h12'19" 26d34'53"
Trinity Sanford
Aur 6h3'45" 47d10'11"
Trinity Selph
Gem 6h46'33" 31d25'22"
Trinity "Trinnie" Ann
Lib 14h23'42" -16d30'56"
Trinity Viola
Aqr 22h33'3" -7d15'34"
Trinity Yackels
Tau 5h24'58" 25d47'59"
Trinity129
Tri 2h31'44" 30d52'8"
Trinity-2005-Gr6
Ori 5h29'42" -8d18'8"
TrinityBliss
Uma 9h52'6" 63d29'28"
Trinity's Blaze
Sco 16h11'41" -17d44'17"
Trinity's Class of 2005
Uma 11h17'48" 58d59'16"
Trinity's Class of 2006
Uma 11h15'38" 57d25'4"
Trinity's Class of 2007
Uma 11h11'21" 57d56'5"
Trinity's Dream
Tau 5h23'24" 28d5'32"
Trinity's Star
Tri 1h56'53" 33d49'43"
Trinket's Treasure
Uma 11h8'21" 32d30'44"
Trinkett
Uma 9h46'50" 57d9'42"
Trinna Michelle MacIntosh
Cnc 8h51'39" 31d8'51"
Trinxle
Cnc 8h24'1" 8d53'3"
Triple Rodriguez... 3 times the fun
Vir 12h54'8" 1d17'14"
Tripod
Uma 10h10'57" 62d47'19"
Tripodi's Light
Uma 11h8'15" 40d22'26"
Tripp Stanton
Sco 16h49'42" -36d55'34"
Trish
Umi 14h37'8" 72d43'4"
Trish
Uma 11h54'19" 52d38'40"
Trish
Aqr 21h13'11" -10d20'37"
Trish
And 22h59'50" 35d25'9"
Trish
Cyg 20h2'29" 34d50'20"
Trish
Uma 9h50'57" 46d57'45"
Trish
Ari 2h44'42" 14d19'34"
Trish and Bob Together Forever
Umi 15h20'15" 72d7'2"
Trish and David's Star
Crb 15h54'45" 31d17'19"
Trish And Mike Forever
Ari 3h10'58" 20d26'40"
Trish Baker
Leo 11h51'21" 20d45'28"
Trish Bardaro
Gem 6h50'26" 33d55'40"
Trish Beck
Cap 21h32'43" -10d20'57"
Trish Biddle
Vir 14h43'24" 5d8'55"
Trish Cardoza
Cnc 8h54'41" 23d45'37"
Trish Cooper
Cnc 8h16'40" 32d58'13"
Trish Eternally
Psc 1h22'14" 15d31'5"
Trish Hawkins
Uma 11h52'8" 65d33'53"
Trish Hornbeck
Uma 11h15'59" 42d30'46"
Trish Martzolf
Peg 22h48'37" 27d16'38"
Trish McGonagle "A Cloud Away"
Cas 22h59'50" 58d26'38"
Trish McGowen
Tau 4h12'31" 24d37'27"
Trish McHugh
Lyn 8h34'9" 38d34'48"
Trish Mizak
Vir 12h8'45" -6d56'20"
Trish Morlock
Psc 1h21'59" 23d42'41"
Trish - our mother forever shining
Col 6h23'32" -33d47'6"
Trish Pisegna
Cas 23h40'27" 53d47'43"
Trish Schrick
Psc 1h10'5" 32d35'9"
Trish Searfoss
Cas 1h28'45" 68d1'46"
Trish Sowell
Uma 8h12'56" 71d12'47"
Trish Sterling
Ari 2h49'44" 18d43'44"
Trish Telfer Adams - 101256
Cru 12h28'11" -59d53'2"
Trish The Dish
Uma 10h23'30" 47d41'25"
Trish Thuy Trang
Ser 15h20'25" -0d47'28"
Trish Walters
Crb 15h41'5" 26d27'0"
Trish, Kolie's Angel
Per 3h5'14" 34d35'15"
Trisha
Cas 1h27'56" 52d43'48"
Trisha
Lyn 7h44'35" 45d2'51"
Trisha
Psc 1h37'36" 18d29'20"
Trisha_41
Sgr 19h30'48" -14d53'50"
Trisha Ancheta
Vir 14h20'10" 1d29'14"
Trisha Ann
Sco 17h38'59" -40d22'12"
Trisha & Anthony
Uma 8h30'34" 65d3'13"
Trisha Ashby
Aqr 21h12'10" -8d27'48"
Trisha Daly
Lyn 6h54'55" 52d4'27"
Trisha Dori Miller
Cnc 8h44'54" 26d40'7"
Trisha Garner
And 0h51'1" 38d25'44"
Trisha George-Imendorf
Sco 15h53'36" -25d48'9"
Trisha Gordon Zagha
Leo 10h16'4" 17d47'32"
Trisha Hershberger
Vir 12h15'36" 12d40'24"
Trisha Kimberly Palmer
Cas 1h20'17" 55d32'1"
Trisha Konkol
Gem 7h33'20" 20d4'12"
Trisha L. Bright
Leo 10h16'52" 6d40'25"
Trisha Lee
Vir 14h6'22" -15d29'49"
Trisha Lurie
Psc 1h35'21" 9d33'53"
Trisha Lynn
Cas 0h35'8" 60d43'51"
Trisha Lynn Bedford
Ari 3h22'15" 29d58'30"
Trisha Lynn Carmichael
Mon 7h16'48" -0d12'10"
Trisha Lynn Collins
Lib 15h11'46" -18d55'9"
Trisha Lynn Droptiny
Vir 12h58'33" -5d42'38"
Trisha Marie Bohm
Gem 6h21'17" 22d16'59"
Trisha Marie Ciervo
Her 16h30'46" 45d46'18"
Trisha Marie McIntosh
Leo 11h4'35" 15d29'27"
Trisha Michelle
Crb 15h36'25" 29d14'15"
Trisha Miller
And 1h56'15" 45d1'11"
Trisha Nheli Pacho Campos
And 1h56'43" 39d42'30"
Trisha Nicole Ramey
Gem 6h56'1" 15d40'2"
Trisha Okowa Vincent
Cyg 21h36'41" 53d38'8"
Trisha Peldo
Lyn 9h13'5" 39d29'23"
Trisha Perkins
Cap 21h27'37" -20d47'51"
Trisha R.
And 1h26'12" 44d39'24"
Trisha R. Simmons
And 2h27'28" 45d11'57"
Trisha Renae
Leo 11h0'3" 19d5'48"
Trisha Riley
Cap 20h22'32" -10d24'25"
Trisha Stottlemyre
Lyn 7h24'2" 46d2'18"
Trisha the Beautiful
Cnc 9h4'50" 15d11'46"
Trisha V. Mendoza
Cnc 8h36'17" 20d25'22"
Trisha Walker
Tau 3h29'5" 0d6'35"
Trisha You are my world Love Travis
Tau 3h26'7" 19d0'25"
TrishArmand
Aqr 22h56'39" -9d13'58"
Trishas Brian
Lib 15h46'44" -6d22'59"
Trisha's Fire
And 0h42'17" 39d9'33"
Trisha's Star
Cnc 8h0'30" 11d34'50"
Trisha's Star
Psc 23h19'49" 0d38'43"
Trishe
Uma 8h50'25" 48d31'43"
Trishia J
Sgr 19h42'54" -33d45'51"
Trishimal
Ari 2h41'37" 22d0'14"
Trish's Angel
Aqr 22h34'19" -0d24'37"
Trish's Heart
Lib 14h33'10" -14d11'40"
Trish's Legacy - Tedmic & Jasten
Uma 12h57'8" 61d53'56"
Trish's Star Forever Shining
Cra 18h49'44" -40d5'32"
Trish's Sun Shine
Sco 17h39'50" -45d30'27"
Trislynn's Dream
And 0h42'17" 43d49'3"
Trisse & Thomas
Leo 10h36'49" 7d56'31"
Trist
Aur 6h38'2" 34d18'46"
Trista Anne
Cas 1h6'53" 54d23'25"
Trista Beth Livermore
Cap 20h22'54" -15d33'14"
Trista Beth Williams
Del 20h39'43" 15d29'58"
Trista Burnham
Uma 8h59'24" 50d49'42"
Trista Dear
And 0h43'33" 45d19'26"
Trista DeRobbio
Cas 23h26'3" 56d43'45"
Trista Jane Greene
And 23h39'15" 33d7'8"
Trista Joy
Aqr 21h11'55" 1d25'17"
Trista Joyce
Lyn 9h5'0" 36d44'19"
Trista Kareese Ferenz
Vir 13h2'27" 5d10'44"
Trista Kay Dawson
Tau 4h53'50" 24d39'6"
Trista L. Culbreth
Leo 11h24'31" 25d21'4"
Trista Lynn
Cam 3h51'48" 52d58'36"
Trista Lynn
Mon 7h14'11" -3d45'3"
Trista Moore's Little Robin
Leo 9h42'57" 24d57'24"
Trista Moretti
Uma 9h20'21" 65d43'40"
Trista Nicole
And 23h8'37" 45d7'35"
Trista Olivier Lam
Cap 20h45'2" -20d9'22"
Trista Schmidt
Lib 15h18'29" -25d7'42"
Trista Thatcher
And 23h46'22" 47d20'58"
Trista Yackels
Ari 2h51'13" 29d34'20"
TRISTAN
Tau 3h36'36" 28d17'16"
Tristan
Crv 12h28'45" -14d18'24"
Tristan
Dra 15h57'36" 60d6'10"
Tristan
Uma 11h40'59" 62d8'42"
Tristan Andrew Hammond
Her 16h44'32" 15d44'6"
Tristan Bailey Jones
Tau 4h5'42" 12d32'16"
Tristan Charles
Per 2h54'35" 55d41'40"
Tristan Connor
Lib 14h50'59" -1d56'39"
Tristan Drake Edwards
Vir 13h18'31" 3d46'33"
Tristan Eicher - unsere Sonne
Uma 8h46'54" 58d49'49"
Tristan Elaine
Mon 5h58'3" -8d27'56"
Tristan Ferdinand Walter
Ari 2h44'20" 16d40'18"
Tristan Gregory Grimm
Uma 10h27'38" 66d46'41"
Tristan Harris
Umi 17h11'45" 75d59'3"
Tristan Hendrik van Losenoord
Vir 13h42'30" 3d40'49"
Tristan James Robert
Gem 6h25'25" 21d53'3"
Tristan Joel
Gem 7h24'36" 14d30'56"
Tristan John Boyden
Boo 14h13'28" 49d0'23"
Tristan Jordon Todaro
Her 18h33'4" 22d0'37"
Tristan L Bluitt & Jaden L Bluitt
Sgr 19h12'40" -31d13'29"
Tristan Lorenzo
Her 16h50'19" 12d23'9"
Tristan & Mario
Cnc 9h10'22" 17d18'1"
Tristan Michael Katsaros
Cnc 8h23'34" 14d25'51"
Tristan Mikel
Leo 10h40'50" 15d16'54"
Tristan: My Love Forever, Tegan
Pho 2h19'56" -41d2'14"
Tristan Ottet 11.08.2006
Uma 9h20'25" 61d59'23"
Tristan Perry
And 0h25'3" 32d50'40"
Tristan Ralph Ramcharitar
Lib 14h35'6" -10d34'27"
Tristan River Yagley
Uma 12h29'58" 55d2'0"
Tristan Robert Jimenez
Uma 9h14'44" 49d39'48"
Tristan Rock-Stierle
Dra 19h28'38" 68d45'38"
Tristan Scott Stewart
Lib 15h53'15" -14d26'55"
Tristan & Shannen
Cyg 20h21'52" 58d29'33"
Tristan Thomas Schemmerling 9/22/05
Vir 12h36'17" 12d21'47"
Tristan Vaughn Coker
Her 17h3'23" 29d23'28"
Tristan William Nelson
Per 2h41'22" 53d33'57"
Tristan Yuhasz
Lyn 7h53'53" 34d56'55"
Tristano "Vic" DeMauri
Cep 21h37'59" 64d47'25"
Tristan's Star
Peg 22h25'56" 24d34'26"
Tristan's Stella Brilliante
Vir 14h27'50" -0d27'33"
Tristan's wish
Per 3h2'40" 48d47'13"
Tri-Star Mortgage
Lib 15h24'20" -8d18'3"
Trista's Love Genova 1
Leo 11h1'0" 14d11'44"
Tristen Brianne Ellis
And 0h34'56" 36d30'9"
Tristen Gayle (Tigi)
And 1h19'32" 39d20'15"
Tristen Gibbons
Leo 11h55'43" 20d50'51"
Tristen Ray English
Uma 11h35'16" 56d18'23"
Tristen the Great
Ari 2h21'2" 23d12'43"
Tristian & Aidan Peters
Uma 11h38'33" 58d51'3"
Tristian Andrews
Gem 6h4'33" 22d34'18"
Tristian Nathaneal Bean
Aqr 22h34'14" -2d35'34"
Tristin
Gem 7h21'12" 22d57'49"
Tristin Lee Tucker
And 1h38'44" 44d40'29"
Tristin Marie Bergh
Her 17h8'55" 36d37'26"
Tristin Michael Dressler
Lib 15h5'9" -21d14'14"
Tristin Osadche
Leo 11h7'7" 21d25'10"
Tristin Pear
Tra 15h47'41" -62d48'17"
Tristina Renee Martin
Sco 16h5'39" -26d1'22"
Tristine
Uma 8h44'40" 60d45'22"
Tristine
Lyn 8h9'10" 37d32'2"
Tristine's glow
Crb 15h36'39" 27d19'23"
Tristlyn Skylight
Lup 15h3'55" -43d36'35"
Triston Michael
Per 4h30'41" 47d46'30"
Triston Timothy Bulger's Star
Mus 12h46'29" -64d55'3"
Triston's Star
Psc 1h35'52" 21d58'50"
Tristyn Dean Leos
Gem 7h29'53" 19d36'58"
Tristyn Phillips
Cas 0h40'35" 65d34'32"
Tritia J. Manianglung
Uma 10h24'13" 45d51'2"
Triumphant Tessa
Psc 0h45'20" 17d56'48"
Triva Lynn Pino
Gem 6h23'11" 27d24'6"
Trivennia K. Caruso
Uma 10h42'25" 45d28'1"
Trix
Aur 6h30'8" 38d53'9"
Trix Fais
Leo 10h14'28" 22d16'43"
Trixi & Heinrich
Cyg 19h54'16" 39d43'17"
Trixie Ann Smith
Cas 1h24'14" 55d51'29"
Trixie Bell Lou Lou
And 22h59'51" 52d3'21"
Trixie Koontz, Dog
Cma 6h50'15" -13d53'55"
trixmas
Mon 6h44'53" 8d42'2"
Trixy
Uma 9h45'12" 51d2'17"
trna
Gem 7h19'53" 32d14'2"
Trodler, Silvio
Ori 5h20'8" -7d59'27"
Troi Ann Reynolds
Ori 6h4'2" 14d7'27"
Troilus And Cressida CST 2007
Uma 8h42'24" 51d16'11"
Troius Norvigicus
Cmi 7h45'5" 0d29'4"
Trombley
Her 17h24'6" 35d10'49"
Trömi
Uma 9h49'1" 42d31'2"
Tron
Tau 4h24'9" 19d5'45"
Tron Butler
Cas 0h7'40" 54d10'0"
TRON78
Sco 17h40'54" -44d56'5"
Trond Enemo
Uma 10h38'53" 58d46'6"
Tronick's Paradigm
Vir 14h0'37" -2d54'28"
Troonie aka "Chunks"
Lyn 8h54'50" 38d32'9"
TroonLee
Cnc 9h8'55" 32d44'2"
Trooper
Ori 5h53'11" 22d45'27"
Trooper
Cma 6h58'57" -25d51'59"
Trooper's Brave Heart
Cyg 21h26'41" 37d24'27"
T-Roos
Lyn 7h26'38" 53d32'33"
Tropea
Cas 23h31'59" 58d2'20"
Tropez
Uma 11h11'2" 39d36'24"
Trösch, Hans-Ludwig
Ori 6h10'45" 16d20'15"
Trost, Johann
Uma 11h12'22" 47d37'40"
trostyanetskaya
And 1h28'44" 46d10'18"
Trouble
Uma 11h44'23" 50d32'53"
Trouble
Ori 6h15'58" -1d14'52"
Trouble McConnell
Uma 11h25'25" 46d50'26"
Trout Light
Lyr 18h51'42" 32d43'3"
Troutco
Uma 9h28'23" 47d26'15"
Troy
Ari 2h49'41" 29d23'11"
Troy
Boo 14h55'13" 19d17'44"
Troy
Lib 15h15'31" -19d35'28"
Troy
Lib 15h6'28" -14d20'49"
Troy A Hoegner
Vir 12h53'55" -5d18'56"
Troy Agler
Aur 6h14'58" 33d38'44"
Troy Agulla
Cnv 12h48'52" 39d31'1"
Troy Alan Ennis
Aur 5h49'13" 29d40'50"
Troy Allen Adams
Tau 4h0'27" 16d44'43"
Troy Allen Biles
Sco 17h20'7" -32d5'42"
Troy and Jeanne Ray
Ori 5h37'30" -2d43'46"
Troy and Jennifer Polen
Cyg 21h35'13" 51d24'22"
Troy Andrew Savinkoff
Cep 22h29'38" 58d7'21"
Troy Austin
Uma 9h3'54" 69d24'43"
Troy Bondurant
Sge 19h45'46" 18d42'44"
Troy Boy
Cap 21h35'37" -23d0'25"
Troy Bronstein
Aur 5h12'52" 41d54'41"
Troy Burns
Her 17h58'52" 49d20'46"
Troy Campbell Monson
Aur 5h38'58" 49d56'52"
Troy Clifford Zeller
Her 17h32'14" 44d12'30"
Troy D. Major
Per 3h3'37" 48d48'6"
Troy Daniel Gordon
Ori 6h14'16" 15d50'23"
Troy & Darlene Irving
Vir 13h10'46" 7d35'13"
Troy & Dawn
Umi 15h39'40" 73d42'6"
Troy Donovan Matchett
Cnc 8h44'23" 31d28'26"
TROY DORMAN
Leo 9h29'7" 25d25'52"
Troy Du Koppenjan
Peg 22h38'24" 27d29'42"
Troy Du Koppenjan
Tau 5h51'13" 18d0'39"
Troy E. Star
Ari 2h43'32" 25d37'36"
Troy Edward Russiaky
Peg 23h9'4" 27d23'35"
Troy Edwards Blackmon
Per 3h48'4" 44d43'18"
Troy Erick Hunts
Her 18h22'55" 23d18'55"
Troy Evan Patton
Uma 11h33'30" 37d19'19"
Troy Filipou
Sco 16h39'34" -34d43'38"
Troy G. Sirman, "Daddy"
Vir 11h49'25" 5d29'50"
Troy Goodling
Cep 22h32'56" 64d47'55"
Troy Hutnik
Per 3h39'21" 47d29'14"
Troy Jamison Satterley
Per 2h49'17" 36d59'19"
Troy Jeffries
Vir 12h9'38" 12d8'25"

Troy Johnson Mueller
 Her 17h52'0" 24d1'30"
Troy & Katrina Meredith
Wedding Day
 Cru 12h2'31" -61d34'44"
Troy Keasling
 Her 18h29'21" 14d14'35"
Troy L. Tibbetts
 Ari 2h21'0" 14d59'33"
Troy L. Wicker
 Her 17h39'13" 15d12'42"
Troy Land
 Cnc 8h56'44" 13d45'26"
Troy Leslie Petryszyn
 Cru 12h43'40" -57d10'56"
Troy Lucas Reeves
 Aqr 21h40'17" -3d45'4"
Troy M. Kell
 Gem 6h32'10" 13d13'40"
Troy Matthew Thanel's
Shining Star
 Leo 11h23'51" -0d21'58"
Troy Mattingly
 Her 16h49'23" 41d33'32"
Troy Michael Brewer
 Ori 5h53'6" 21d17'42"
Troy Michael Wagaman Jr.
#19
 Uma 10h29'59" 47d2'42"
Troy Moore
 Lyn 9h13'12" 38d2'24"
Troy Olen Dempsey
 Cap 15h50'21" -16d59'52"
Troy Pollett
 Ori 4h51'2" 11d51'36"
Troy Pprice Isaacs C1A4Y3
 Psc 0h42'47" 17d34'20"
Troy Rease
 Uma 9h43'19" 68d1'39"
Troy Robert Pfeffer
 Aur 5h39'54" 38d28'37"
Troy Roth
 And 0h49'57" 44d20'18"
Troy Scott
 Ori 6h9'37" -1d33'6"
Troy Scott Christopher
Pollett
 Aqr 22h35'12" -0d9'34"
Troy Smith
 Crb 15h40'46" 28d26'46"
Troy "Stinky" Gorena
 Sgr 18h28'45" -21d43'3"
Troy Sufferling
 Peg 23h15'57" 17d28'23"
Troy Tarnowski
 Lib 15h58'35" -13d15'59"
Troy & Tasha "Together
Forever"
 Cyg 21h40'2" 52d41'19"
Troy "Tigger" Main
 Her 16h49'35" 32d4'45"
Troy Tuttle
 And 23h40'45" 39d17'56"
Troy William "Little Warrior"
 Her 17h45'14" 47d9'12"
Troy William Ninness
 Psc 0h24'8" 11d41'4"
TroyandKim
 And 0h19'15" 33d9'39"
TroyBren
 Lyr 18h58'34" 28d4'3"
Troy's Shining Star
 Sco 16h4'35" -29d44'10"
Troyster
 Aql 20h37'43" -0d11'51"
tru
 Vir 13h11'56" 1d18'59"
Tru
 Uma 11h46'24" 30d6'40"
Tru Bella
 Psc 1h0'23" 24d51'22"
Tru Bly Domenz
 Vir 12h25'34" 8d12'18"
Truck Fuhrer
 Per 3h19'42" 45d55'47"
Truck2Dan 82
 Per 2h44'46" 54d44'21"
Truckee & Sierra Forever
Together
 Cnv 12h43'52" 36d44'49"
Trudi and Steve Boehler
Forever
 Cyg 19h45'18" 56d6'23"
Trudi Elizabeth
 And 0h5'47" 42d41'24"
TRUDI HAAF
 Umi 14h31'38" 79d9'17"
Trudi Jayne Newton
 And 0h34'42" 27d10'42"
Trudi Lee Markert
 Crb 15h39'1" 36d42'30"
Trudi Leighann Shoemaker
 Aqr 21h26'57" 1d48'33"
Trudi Ming
 Sge 20h19'6" 19d20'7"
Trudi Speirs
 And 1h0'43" 45d1'44"
Trudi, the Best Sister
 Gem 6h43'44" 13d36'0"
Trudicious
 Her 18h12'28" 42d20'21"
Trudie Teague
 Leo 11h36'2" 23d46'10"
Trudine
 Cnc 8h54'31" 26d51'28"

Trudi's Twinkle
 And 0h48'33" 40d41'15"
Trudy
 And 1h40'2" 38d56'36"
Trudy 16
 Lib 15h45'46" -22d48'14"
Trudy Amanda Massey
 Boo 14h26'19" 35d20'4"
Trudy Ann Avis
 And 0h34'18" 46d32'43"
Trudy Cutler
 Cnv 12h37'6" 43d32'42"
Trudy DuMay
 Lib 15h47'25" -6d41'51"
Trudy Gorman
 And 2h15'41" 46d39'38"
Trudy & Greg Gibson
 Cyg 20h50'38" 47d52'21"
Trudy Lauben
 And 2h17'13" 47d6'14"
Trudy LeVarrah Bagley
 Tau 4h39'34" 22d12'44"
Trudy Loretta York
 Lyn 8h44'56" 34d30'36"
Trudy McFarland
 Aqr 23h10'51" -7d43'34"
Trudy O'Connell
 Lyn 7h48'13" 38d44'36"
Trudy Robinson
 Cas 2h10'19" 59d32'6"
Trudy Rose Levine
 Cas 1h0'43" 63d40'40"
Trudy Strombom Harness
 Aqr 22h12'57" -9d23'32"
Trudyangelis Star
 Psc 23h8'42" 7d16'20"
Trudy-anne Victoria Craig
 Cyg 21h32'7" 43d21'22"
Trudy's Evening Star
 Aqr 21h26'11" -0d4'38"
True
 Lep 5h11'34" -11d11'26"
True
 And 23h5'35" 41d48'20"
True
 Cyg 20h1'11" 44d18'21"
True
 Cam 4h10'8" 54d44'55"
True
 Lyn 9h3'40" 33d29'40"
True Blue Romance
 Ori 6h16'2" 10d23'38"
True Companion
 Ori 4h51'49" 4d28'37"
True Definition of Loveable
- Cindy
 Peg 22h49'58" 19d4'57"
True Destiny
 Cyg 20h20'34" 54d25'29"
True Friend
 Cyg 20h22'13" 47d1'38"
True Friends Are Like
Diamonds
 Ari 24h44'27" 26d22'1"
True Friendship
 Uma 11h6'55" 60d2'26"
True Friendship of Liza &
Itai
 Aql 19h47'16" 5d13'13"
True Gage
 Sco 17h57'2" -30d52'6"
True Happiness
 Cap 20h26'12" -11d57'25"
True Joe
 Cyg 20h56'2" 30d17'26"
True Julie Drendel Shine
Effect
 Her 17h12'50" 19d5'51"
True Love
 Leo 11h10'18" 20d11'51"
True Love
 Ari 3h19'42" 27d39'27"
True Love
 Cnc 9h21'4" 26d12'19"
True Love
 Ori 6h20'21" 6d10'47"
True Love
 Tau 4h56'17" 19d12'3"
True Love
 And 0h20'44" 21d56'17"
True Love
 And 1h52'56" 42d48'9"
True Love
 Per 2h56'59" 44d27'46"
True Love
 Tau 5h29'53" 17d13'59"
True Love
 Lyn 7h57'30" 34d16'9"
True Love
 Crb 15h51'17" 36d14'42"
True Love
 Cyg 19h47'41" 31d41'17"
True Love
 Cyg 20h59'35" 47d31'20"
True Love
 Cyg 19h59'17" 49d9'44"
True Love
 Per 2h48'6" 54d9'54"
True Love
 Cyg 21h30'4" 47d43'37"
true love
 And 23h17'16" 44d18'19"
True Love
 Cyg 21h36'42" 44d42'26"

True Love
 Sco 16h10'48" -11d15'23"
True Love
 Uma 10h32'9" 65d9'3"
True Love
 Cyg 20h20'39" 55d19'1"
True Love
 Cyg 20h15'45" 55d8'57"
True Love
 Cyg 19h41'49" 53d34'8"
True Love
 Cyg 21h52'0" 54d1'1"
True Love
 Cas 1h39'18" 63d34'41"
True Love 5/22/05
 Cyg 21h17'8" 43d56'50"
True Love Blooms
 Gem 7h21'17" 30d45'9"
True Love - Cameron &
Christine
 Pho 0h38'39" -49d48'41"
True Love - Daniel
Beardsley
 Cep 20h47'51" 61d22'10"
True Love Forever
 Uma 8h15'0" 62d7'32"
True Love Forever Ryan &
Rachel
 Cyg 19h54'53" 30d0'55"
True Love Is Forever
Theirs
 Lyn 19h26'12" 36d3'0"
True love is found in the
stars.
 Uma 9h25'49" 66d16'27"
True Love Jackie and Ernie
 Sgr 18h3'8" -30d13'38"
True Love ~ John and Gina
7/24/85
 Cyg 21h53'27" 50d20'47"
True Love Kelly & Shawn
 Cyg 21h14'49" 35d28'30"
True Love - Kerry L &
Justin C
 Ind 20h47'10" -45d14'1"
True Love Lives
 Ori 6h13'23" 15d11'36"
True Love - Michael C.N.
Roth
 Uma 11h50'32" 54d46'24"
True Love S&T 05.06.2004
 Uma 9h52'26" 44d41'4"
True Love- *Shauna Rae*
 Mon 6h54'23" -0d16'51"
True Love True Friendship
 Cas 0h56'3" 61d43'27"
True love waits.
 Cyg 21h40'47" 32d6'28"
True Love's Reflection
 Uma 10h57'59" 51d43'25"
True Lovie
 Sge 19h10'3" 19d43'4"
True North
 Her 17h18'9" 17d34'17"
True North
 Ori 6h17'37" 14d28'55"
TRUE139
 Peg 22h32'52" 29d55'20"
Trueheart Love
 Uma 10h15'36" 60d37'46"
Truesdale, Luther &
Heyward
 Uma 11h45'46" 29d46'0"
Truh Sara Pennington
 Aql 19h55'45" -10d18'54"
Truhn, Bodo
 Ori 5h38'53" 15d28'5"
Trula McCarter
 Lyn 9h10'30" 42d31'48"
Trulock's True Light
 Gem 7h13'6" 15d47'27"
Truly Burton
 And 0h13'50" 28d33'6"
Truly Madly Deeply
 Sex 9h47'47" -0d23'4"
Truly Madly Deeply
 Pho 1h13'13" -44d45'39"
Truly Scrumptious
 Ari 2h1'57" 21d4'9"
Truly, Madly, Deeply
 Cnc 8h44'16" 30d56'16"
Truman Barstow
 Uma 9h52'54" 64d3'52"
Truman Camozzi
 Uma 11h45'43" 51d27'49"
Truman Crafts
 Tau 5h29'53" 17d13'59"
Truman Elizabeth
Sherwood
 Sco 16h10'42" -10d51'25"
Truman Jherek Willick
 Lyn 7h31'13" 48d37'54"
Truman Wilcox
 Cnc 8h10'39" 10d44'58"
Trumper
 Lyn 8h11'10" 45d54'57"
Trunk Family Twinkler
 Uma 8h57'37" 69d53'49"
Trust
 Uma 12h55'31" 59d28'38"
Trust
 Cnc 8h42'39" 31d28'24"
Trust Forever
 Cam 4h26'7" 63d6'26"

Trust Obey
 Tau 4h12'50" 28d44'43"
Trust SM&VB
 Lyn 7h33'40" 44d25'36"
Trust-eight Amachika
 Sgr 18h43'3" -19d54'10"
Truth
 Cyg 21h26'6" 53d0'8"
Truth
 Cyg 20h2'30" 33d53'24"
Truth
 Cyg 21h29'37" 46d49'59"
Truth
 Psc 1h7'28" 28d24'20"
Truth and Promise JJE
 Del 20h21'44" 9d22'33"
Truthfully
 Cyg 20h33'24" 37d26'24"
TRWJones
 Psc 1h9'43" 20d26'27"
Trycia
 Col 6h13'39" -39d49'43"
Tryla Nimhawayis
 Uma 9h27'47" 53d2'45"
Trysha
 Lyn 7h29'16" 46d0'12"
Tryst
 Sco 16h38'47" -40d57'41"
Trysta Michelle
 Umi 16h53'15" 84d27'11"
Trystan James
 Sgr 19h16'34" -34d46'53"
Trystin Dennis
Lewandowski
 Aur 5h15'16" 42d51'31"
Trzaska Twilight
 Ori 6h22'54" 14d17'5"
TS - 15
 Del 20h34'25" 13d21'50"
T.S. Bishop
 Uma 9h29'15" 48d32'59"
TS Callie and Paige
 Crb 15h22'32" 28d5'19"
T's Memory
 Aqr 22h4'14" -15d7'58"
ts15
 Cap 21h47'46" -18d4'16"
TS22405
 Sco 17h48'30" -33d26'46"
TS48
 Psc 0h39'3" 7d31'52"
Tsai I Hsuan The Momo
 Uma 9h32'13" 45d36'54"
Tsangaras
 Sco 17h0'58" -38d6'57"
Tschapeller, Anne
 Leo 11h1'32" 20d23'28"
Tschernjakov, Leo
 Ori 5h51'6" 7d2'38"
Tschiby
 Her 16h40'15" 41d25'14"
Tschipsy
 Com 13h14'24" 14d24'33"
Tschunko, Brigitte
 Ori 5h9'31" 15d52'31"
Tschusi
 And 23h6'51" 51d51'6"
Tschynah
 Ori 6h13'46" 3d0'42"
Tsegayesus
 Sgr 19h42'43" -16d46'34"
Tshy Meri Cross
 Cru 12h34'38" -57d48'12"
TSI Telecommuncation
Services Inc.
 Mon 6h27'22" -0d56'33"
Tsiganov, Konstantin
 Uma 9h46'32" 69d26'34"
Tsigóba
 Sgr 18h30'5" -16d43'9"
Tsong-Yuan Ho
 Cap 21h29'25" -22d44'29"
Tsoutsouna
 Mon 6h41'22" -0d52'54"
T.S.P. 27/05/1976
 Cru 12h30'34" -57d46'1"
Tsu Pin & Chia Hui
 Lib 15h21'40" -4d2'22"
Tsubasa
 Aqr 21h57'41" -4d31'41"
Tsubox
 Vir 12h20'41" 6d9'59"
Tsui Suet Ling 143
 Lib 15h49'58" -4d33'17"
Tsuki
 Ori 6h18'18" 16d2'20"
tsukito 0724
 Leo 9h35'40" 8d18'40"
tsuneni Sueshijuu
 Cap 21h43'22" 18d9'19"
Tsvetelina Ivanova
 Gem 6h30'1" 12d52'43"
Tsvety
 Aqr 22h39'10" 1d5'32"
Ttiot MJD 8169
 Dra 19h46'19" 61d44'56"
TTJ-FKA-1111
 Tau 3h38'54" 21d33'57"
TTOSH
 Leo 11h6'20" 15d15'22"
Tu Amor
 Vir 12h36'31" 6d11'39"
Tu es mon unique amour...
 Cas 1h3'52" 53d48'38"

Tu Morena Angel
 Sco 17h52'30" -35d42'18"
Tu Pequeno Pedazo De
Cielo
 Umi 16h42'31" 75d8'29"
Tu sei una stella la mia
stella
 Ari 1h50'32" 18d15'26"
Tu Tu
 Peg 22h52'21" 10d44'5"
Tu Tu
 Lyn 7h58'26" 58d2'53"
Tú Vida, Mein Licht, Our
Star
 Lib 15h33'37" -12d23'59"
Tua Dimora Tra Le Cielo
 Per 3h44'8" 43d56'24"
Tuan H. Nguyen
 Leo 10h53'13" 14d34'53"
Tuba
 Leo 11h28'49" 12d3'39"
Tuba Avcisert
 Aqr 22h22'54" -23d21'43"
Tubby
 Tau 5h19'34" 18d14'34"
Tubby Tumpers
 Cnc 8h47'22" 30d50'32"
tubipimmmm
 And 2h15'29" 48d49'7"
Tubosun Dennis
 And 0h52'4" 37d2'11"
tubs
 Vir 12h38'10" -7d54'6"
Tubsy Wubsy 22-
Cunha/Shea 2004
 Cap 21h33'11" -18d36'54"
TüBü
 Uma 11h21'4" 34d20'38"
Tubular Fiesta
 Cas 0h31'40" 63d1'34"
Tuck and Tillie
 Lyr 18h35'5" 35d41'16"
Tucker
 Cma 7h0'56" -24d51'57"
Tucker Bauman
 Uma 8h12'29" 62d51'23"
Tucker Collins
 Sco 16h8'3" -11d41'31"
Tucker Crawford
 Sgr 18h58'50" -23d25'15"
Tucker Crawford
 Sgr 18h8'47" -26d21'28"
Tucker Crouse Terry
 Her 17h7'33" 32d1'25"
Tucker Debataz
 Aqr 22h7'22" -2d36'48"
Tucker Ferris
 Her 18h41'32" 20d55'4"
Tucker for Eternity
 Ori 5h51'6" 7d2'38"
Tucker Jackson Henry
 Ori 5h45'1" 1d17'11"
Tucker John Czajkowski
 Lib 15h19'25" -13d3'7"
Tucker Johnson-Nieto
 Lib 15h36'43" -19d25'53"
Tucker Lambert
Gustavesen
 Tau 5h30'45" 21d51'32"
Tucker Lawrence Rose
 Boo 14h38'54" 25d29'23"
Tucker Lynch
 Gem 7h0'17" 30d5'31"
Tucker Matthew Green
 Uma 9h41'28" 59d51'6"
Tucker Michael
 Cyg 19h37'44" 28d17'10"
Tucker Monson
 Per 2h25'10" 51d11'28"
Tucker Neil Moyer
 Cnc 8h5'47" 11d23'24"
Tucker Wiley Spraetz
 Umi 15h23'31" 73d37'59"
Tucker Yaguchi
 Uma 11h39'44" 63d12'42"
Tucker Yandell
 Dra 17h0'19" 64d24'49"
Tuckerbear Bailey
 Uma 11h34'29" 46d17'56"
tuckerjohn
 Cma 6h25'3" -29d23'34"
Tudi and John's Everlasting
Love
 Lyn 9h7'55" 37d3'6"
TuEsDaY
 Cnc 8h46'17" 14d50'1"
TUESDAY Jessica Renee
Ingram
 Lmi 9h53'34" 34d3'2"
Tuesday the Cat
 Lyn 7h53'34" 45d11'45"
Tuesday's Dragonfire
 Leo 9h43'17" 28d41'46"
Tuesdee
 Col 5h58'37" -34d37'53"
Tuet
 Tau 5h5'12" 22d30'53"
Tuffe
 Umi 13h37'54" 70d43'41"
Tuffet Sharp
 Psc 1h7'3" 10d21'3"
Tuffi-Taz
 Per 3h28'40" 48d28'31"
Tuffy
 Ari 2h13'57" 25d34'39"

Tug and Chrissy
 Crb 16h6'25" 39d8'31"
Tugce & Tugba, sunshine
sisters
 And 23h3'23" 47d1'47"
Tui Lucidus Stella Semper
 Sco 16h9'52" -10d19'50"
Tuianna
 Vul 19h52'30" 22d50'6"
Tuky
 Leo 9h23'18" 22d25'56"
Tul
 Vir 13h25'10" 8d59'36"
Tula
 And 2h18'25" 42d57'18"
Tulani
 Umi 16h1'51" 71d1'33"
Tulia Zanne
 And 1h36'9" 41d18'38"
Tulia's Diamond
 Uma 10h26'59" 60d28'19"
Tulin Ayse Yuce-Howey
 Crb 16h1'54" 32d45'33"
Tulin Baycan
 Ori 6h18'17" 10d36'51"
Tulio Kay
 Lyr 18h47'11" 39d50'37"
Tulio's Light
 Ori 5h9'35" -0d38'1"
Tullio Castiglioni
 Her 18h8'53" 31d26'48"
Tully
 Cas 0h57'40" 50d25'15"
Tully Stephen
 Ori 6h8'30" 9d19'35"
Tuls
 Leo 11h52'20" 23d58'10"
TULSI
 Ari 3h0'0" 21d57'52"
Tulsi ka Tara
 Per 4h44'12" 47d17'12"
Tuly
 Lib 15h9'54" -5d43'1"
tumaini
 Psa 22h43'11" -26d45'15"
Tumbleweed
 Boo 14h19'43" 17d39'38"
TumTum
 Leo 9h39'3" 29d11'13"
Tum-Tum
 Cap 20h28'53" -9d14'30"
TUNC
 Cnc 7h56'41" 17d50'47"
Tunc Ozer
 Ari 2h54'13" 28d35'31"
Tunca, Alexandra
 Uma 11h36'47" 31d32'40"
Tünde/Hangyalka
 Psc 23h38'38" 6d51'24"
Tünder Ancsin Judit
 Uma 8h13'45" 70d10'5"
Tündérmanó (Korcsok
Tünde)
 Sco 16h48'25" -27d38'27"
TungerinesunVictoria
 Cnc 8h43'38" 31d9'54"
Tunkhannock Field Hockey
Team 2005
 Aql 19h15'21" 7d53'27"
Tunni
 Aqr 21h20'10" 1d19'44"
Tun-Twin
 Lib 15h19'32" -19d36'46"
Tuphph
 Uma 14h21'22" 59d9'50"
Tur nan Callad
 Her 17h20'0" 15d56'58"
Turbo
 Sco 16h11'59" -23d53'57"
Turbo Ferbo
 Cnc 8h53'2" 17d36'50"
Turbo's Star
 Ari 2h16'53" 24d5'45"
" Turi "
 Lyn 8h4'3" 51d54'5"
turi
 Cas 1h9'28" 50d52'2"
Turk
 Umi 16h34'32" 77d16'42"
Turkeev, Jurij
 Ori 6h10'38" 16d25'40"
Turkeyfoot"Danielle"05
 Lyn 8h13'28" 54d8'2"
Turki
 Cap 21h48'15" -19d59'44"
Turkmen Bayraktar
 Her 16h18'58" 25d21'52"
Turnbrook
 Umi 14h43'51" 83d20'34"
Turner
 Uma 10h44'2" 68d41'43"
Turner Hill
 Cnc 9h17'20" 25d49'9"
Turner Lee Wilkerson
 Cep 20h43'13" 61d33'38"
Turner Thomas Prickett II
 Psc 0h23'3" 9d44'38"
Turner's Torch
 Cas 0h1'38" 56d41'28"
Turney-Bozell
 Cyg 21h24'21" 45d38'1"
Turóni Gemma
 Peg 23h42'30" 11d14'5"

Turquoise (Martha
Michaels)
 And 0h31'42" 35d43'36"
Turthalion
 Cyg 21h31'2" 50d20'5"
Turtle
 Gem 6h44'42" 26d43'40"
Turtle
 Ari 3h16'37" 29d56'21"
Turtle 1
 Lib 15h34'38" -10d41'47"
Turtle Beach Resort
 Lmi 9h27'24" 36d27'11"
Turtle Jones
 Per 4h24'1" 43d25'0"
Turtle Star
 Lac 22h26'13" 53d40'43"
Turtle, Kate, Layla, and Big
Guy
 Aur 5h14'45" 40d59'46"
Turtle's Star
 Del 20h33'29" 13d28'16"
TurtlestargalAshleya
 And 1h55'2" 38d36'7"
Turtlesymbiote
 Lyn 6h51'43" 57d27'57"
Tushar Balsara
 Umi 8h26'38" 88d31'1"
Tushie Pie 1999
 Uma 9h35'40" 45d50'31"
Tut Smith
 Cep 23h2'47" 70d21'50"
Tutankhamen's Cavalier
Nichols
 Gem 7h7'55" 25d54'59"
Tutela Validus Patronus
 And 2h24'46" 44d14'52"
TuTi
 Cap 20h38'27" -25d4'46"
Tutti Caringal
 Aqr 22h20'22" -2d59'12"
Tutti Sofia
 And 0h28'9" 45d31'51"
TuttoMotto3.14land
 Gem 7h22'42" 14d24'27"
Tuttuy & Caroline
 Uma 9h38'28" 68d43'34"
Tutu37
 Her 18h52'46" 23d42'18"
Tuula Orvokki Anneli
Paukku
 Umi 15h12'14" 70d16'50"
Tux
 Dra 16h49'57" 68d24'58"
TUySHO
 Uma 11h28'21" 31d1'56"
TV EV
 Aqr 22h46'7" -0d47'7"
Tvffolope Estuce
 Sco 17h0'51" -33d37'32"
TW & Allison West
 Cyg 21h10'28" 46d25'18"
T.W. Veda
 Tau 5h36'36" 27d57'6"
Twaine "Dad" Walters
 Leo 10h14'25" 23d2'11"
Twana & Jamel Always &
Forever
 Aur 5h32'47" 53d4'31"
Twas Brillig
 Lyn 7h53'9" 49d55'50"
Twavis and Sami
 Cyg 20h14'30" 50d28'10"
TWB2005
 Cyg 19h54'39" 36d53'59"
twbcsesw
 Lyn 8h8'34" 41d17'54"
TWBD 2705
 Her 17h52'52" 21d48'57"
Tweedie's Star
 Leo 10h41'2" 8d0'38"
Tweenkle
 Tau 4h37'44" 21d54'7"
Tweet Rubber Star
 Sco 16h13'56" -14d13'51"
Tweetie
 Ari 2h46'11" 13d9'5"
TweetiePie
 Mon 6h54'31" -0d57'9"
"Tweety"
 Cnc 8h44'56" 20d21'24"
Twelfth Year Mimi
 Vir 15h3'2" 3d21'56"
Twenty One
 Gem 7h16'58" 19d44'34"
Twenty One
 Lib 14h56'50" -0d46'2"
Twenty wishes upon your
star...
 Uma 12h32'10" 55d29'21"
Twentybelle
 Cnc 8h51'49" 13d30'16"
Twessa's Twinkle
 Psa 22h45'28" -28d39'1"
Twhirl
 Uma 10h32'23" 62d56'48"
Twice Blessed
 Gem 6h37'23" 23d49'6"
Twiggy
 Leo 11h18'22" 18d15'13"
Twiggy-Lee
 Psc 1h19'37" 14d49'8"
Twiggy's Twinkle
 Tau 4h57'22" 18d57'50"

Twila Case
Cas 1h0'30" 61d16'52"
Twila J. Miles
Cnc 8h46'8" 25d17'12"
Twilight 8-16-03 Joseph &
Dawn
Cyg 19h37'17" 51d47'45"
Twilight Edwards
Crb 15h47'37" 29d4'49"
Twiller Govan McGraw
Mon 7h18'19" -0d24'42"
Twilley #1
Aqr 22h38'45" 1d27'48"
Twilley's Star
Cyg 20h4'10" 59d41'51"
Twilva Lee Meagher
Cnc 8h49'53" 14d44'58"
Twin Bundle - Victoria &
Ryan
Umi 14h39'55" 75d47'23"
Twin J Ewart
Crb 15h41'31" 27d53'25"
Twin Sister: Amy Ann
Ari 2h6'52" 24d51'16"
Twin Souls; Skyler and
Jessica
Sge 20h0'43" 17d42'1"
TWIN STAR
Crb 16h19'15" 35d17'55"
Twink
Cnc 8h54'10" 31d38'18"
Twink
Leo 11h36'11" 16d21'35"
Twink
Gem 6h53'32" 23d47'45"
TWINK
Her 16h47'14" 9d5'59"
Twinkel 091285
Cru 12h26'33" -62d22'20"
Twinker
Uma 8h37'26" 65d19'39"
Twinkie
Uma 9h31'13" 68d26'19"
Twinkle
Umi 15h43'38" 74d36'21"
Twinkle
Umi 14h19'16" 71d21'46"
twinkle
Cas 23h49'16" 54d20'57"
Twinkle
Uma 13h3'16" 55d50'47"
Twinkle
Uma 12h19'53" 57d22'46"
Twinkle
Lyn 7h4'48" 53d15'27"
Twinkle
Umi 14h48'37" 77d6'17"
Twinkle
Vir 12h42'46" -3d10'8"
Twinkle
Aqr 20h48'48" -1d10'4"
Twinkle
Ara 17h35'27" -63d4'2"
Twinkle
Sco 17h9'4" -38d15'24"
Twinkle
Vir 11h40'18" 4d39'32"
twinkle
Crb 15h23'16" 29d8'25"
Twinkle
Psc 1h12'38" 26d21'48"
Twinkle
Lyn 8h4'21" 41d31'51"
Twinkle
And 23h25'40" 38d52'38"
Twinkle A 04-16-1991
Lyn 8h43'38" 36d14'26"
Twinkle Burke
Uma 11h2'56" 45d30'40"
Twinkle Butt
And 0h18'44" 27d29'26"
Twinkle Chuck Chuck's
Star
Umi 16h28'44" 86d36'58"
Twinkle in Dad's Eye
Peg 22h12'42" 5d58'38"
Twinkle In Her Eye
Leo 9h41'32" 27d21'56"
Twinkle Little
Leo 9h45'57" 25d47'30"
Twinkle Little Jaz Star
Ori 6h20'18" -1d51'31"
Twinkle Little Natalie
Crb 15h41'53" 26d33'0"
Twinkle Little Pearl
Vir 12h44'3" 6d25'49"
Twinkle My Little Karyn
Star
Ari 2h25'54" 27d39'54"
Twinkle of Kotoko
Sgr 18h43'21" -20d8'2"
Twinkle of my eye
Aur 5h39'58" 33d16'4"
Twinkle of Timeless Love
Cnc 8h43'23" 19d12'51"
Twinkle *PAIGE* Lights the
Sky
Cnc 8h18'52" 20d17'15"
Twinkle Star
Cru 12h28'36" -60d6'1"
"Twinkle" Thurs B M.O.P.S
'05 - '06
Umi 15h26'3" 71d0'1"
Twinkle Toes
Cas 23h20'29" 55d41'45"
Twinkle Toes
Cap 21h41'45" -23d36'7"
Twinkle Toes
Ari 2h49'28" 13d16'29"
twinkle toes
Psc 0h47'11" 17d42'9"
twinkle twinkle
Ari 2h10'35" 24d32'30"
Twinkle Twinkle
And 0h58'9" 39d53'59"
Twinkle Twinkle Debbie's
Star
Uma 9h13'53" 54d46'35"
Twinkle Twinkle Hannah-
Faye
And 23h32'29" 49d1'39"
Twinkle Twinkle Jessica's
Star
And 0h50'25" 41d48'37"
Twinkle Twinkle Little
Kirstyn
Cru 12h42'54" -60d34'6"
Twinkle Twinkle Little Noel
Her 18h56'11" 16d46'38"
Twinkle Twinkle Little
Sparky
Psc 1h17'6" 10d43'46"
Twinkle Twinkle Tom
Kelly's Star
Cra 18h21'15" -40d55'53"
Twinkle W 04-03-1988
Lmi 10h7'24" 35d51'34"
twinkle, twinkle Fischer's
star....
Tau 4h6'50" 5d13'29"
Twinkle, Twinkle Little Tat
Vir 14h11'33" -5d20'56"
Twinkle, Twinkle, Amy
Starr
And 2h12'31" 45d32'1"
Twinkle, Twinkle, Debbie's
Star
Cas 23h13'46" 59d27'24"
Twinkles
Lyr 18h24'4" 38d27'53"
Twinkles
Lyr 18h42'43" 34d38'58"
Twinkles 11/1/05
Umi 15h9'9" 78d38'52"
Twinkles of Starlight from
Edna Mai
Del 20h37'37" 15d43'31"
Twinklin' Stinkin' Lincoln
Cma 6h37'10" -13d31'3"
Twinkling 50 - Jenny Hager
Cru 12h2'27" -62d58'3"
Twinkling Abigail
And 1h23'16" 50d27'49"
Twinkling Angel
And 2h35'27" 49d20'22"
Twinkling Bridgette
Mon 7h22'18" -0d18'52"
Twinkling Diane:Your Eye
In The Sky
Cap 20h21'8" -19d48'27"
Twinkling for Eternity
Col 5h53'53" -34d23'23"
Twinkling Lawren Jane
Tau 5h25'12" 26d27'55"
Twinkling Star of Kasumi
Sgr 19h2'17" -34d30'43"
Twinkling Tyler
Ori 4h57'27" -0d32'49"
Twinklit
Del 20h21'58" 9d49'21"
Twinks
Sco 16h18'30" -12d23'43"
Twinky!
Leo 10h32'23" 23d28'22"
Twinnie-Twinnie Star
Gem 6h45'12" 31d47'21"
Twister
Cnc 8h33'5" 22d18'56"
Twitchy
And 2h5'38" 37d52'55"
Twitterpated
Cyg 21h14'46" 43d8'54"
Twix
Lyr 18h47'17" 41d54'11"
Two became One
Cnc 8h37'38" 8d8'3"
Two Become One
Leo 11h49'11" 23d10'50"
Two best friends, LaDa
Leo 11h29'42" 27d53'42"
Two Boats
Cyg 19h37'13" 39d27'25"
Two Feathers
Gem 7h28'23" 26d43'39"
Two Hands In Heaven
Cap 20h29'31" -23d0'35"
Two Hearts
Uma 11h56'5" 64d48'44"
Two Hearts
Tau 4h38'41" 13d33'44"
Two Hearts for Eternity
Cyg 21h28'53" 46d39'41"
Two Hearts, One Beat,
Forever
Gem 6h22'6" 27d46'9"
Two Heroes - Derek &
Isaac
Per 4h23'29" 32d15'4"
Two Lights
Ori 5h53'3" 20d30'57"
Two Little Angels
Aqr 21h12'41" -9d17'32"
Two lives as one
Pho 0h26'54" -41d41'32"
Two Lovebirds
Ori 6h6'9" 15d1'37"
Two Lovers
Uma 8h32'52" 61d39'31"
Two More Minutes
Cap 20h36'59" -11d12'10"
Two More Years
Cyg 21h34'43" 35d19'29"
Two Peas
And 23h4'58" 50d51'34"
Two Peas in a Pod
Cyg 19h32'23" 31d43'18"
Two Penny Kenny
Boo 14h52'31" 33d44'25"
Two People, One Love,
Our Star
Her 17h30'35" 16d10'28"
Two Scoops
Uma 8h55'33" 48d38'37"
Two Sisters
Cnc 8h27'2" 6d57'16"
Two Souls
Cyg 21h48'45" 52d31'52"
Two years of Happiness
Uma 10h37'14" 62d54'34"
Two-Gem's
Uma 10h30'3" 44d2'0"
Twomomma Wilson
Leo 11h13'52" -0d35'1"
T.W.W.S.I.
Cyg 21h33'29" 38d49'37"
Twyla
Ori 5h9'52" 5d7'41"
Twyla
Psc 0h28'44" 8d12'43"
Twyla Mae
Cap 20h37'38" -10d56'43"
TXM+G
Per 3h54'36" 52d51'16"
Ty
Per 3h48'56" 40d48'42"
Ty
Ari 3h10'4" 27d3'55"
Ty 77 Lallier
Tau 5h50'40" 27d34'29"
Ty Alexander Carnegie
Uma 10h33'1" 68d7'16"
Ty and Dana's 25th
Anniversary Star
Cyg 20h4'23" 40d12'12"
Ty and Lisa
Cyg 20h13'4" 44d37'5"
Ty Anthony Webster
Tau 4h17'3" 26d41'55"
Ty Arthur Guevara
Boo 14h2'43" 22d33'57"
Ty Austin Witte
Lib 15h3'41" -7d12'20"
Ty Bekaroo
Leo 9h35'42" 10d22'34"
Ty Cameron Watts
Cru 12h18'52" -56d9'2"
Ty Chapman
Uma 12h45'42" 63d10'17"
Ty Charles Bonnett
Her 16h7'38" 45d13'5"
Ty Craig Regis
Sco 17h20'12" -32d57'4"
Ty Daniel Miles
Ori 5h25'12" 13d11'26"
Ty Dennis Paul Tournier
Uma 9h57'23" 59d23'35"
Ty Etheridge
Uma 9h19'59" 49d14'36"
Ty Gabriel
Aqr 22h44'21" 0d50'30"
Ty Heath
Vir 13h24'1" -3d12'3"
Ty Herndon
And 2h35'8" 45d12'13"
Ty Hines Waggoner
Umi 14h43'17" 73d56'42"
Ty James Riccobene
Lyn 8h59'43" 40d28'7"
Ty & Jaye Dubb's Journey
Together.
Cyg 20h0'6" 42d20'16"
Ty & Jenni
Cyg 21h9'50" 46d45'24"
Ty John Cosenzi
Dra 18h53'46" 67d21'33"
Ty Jones Turner
Umi 15h28'46" 76d2'43"
Ty Jurras
Uma 9h33'57" 54d49'34"
Ty Lafe Allie Pippin
Uma 11h43'58" 45d42'52"
Ty Paslay
Lib 15h26'40" -7d19'15"
Ty Phillip Perkins
Lyn 7h44'20" 41d55'40"
Ty Richard Chauvin
Tau 5h55'53" 26d42'32"
Ty Riley Hall
Lyn 7h14'0" 48d34'38"
Ty & Rose Mary Pinchback
Ori 5h35'57" -2d3'12"
Ty & Stephanie's Wedding
Star
Cyg 21h25'42" 45d34'18"
Ty Topol
Cas 1h16'40" 65d26'31"
Ty Ty Our Precious Son
Ori 5h41'42" 7d59'5"
Ty & Vo Forever
Her 18h51'49" 23d25'4"
Ty William Berry
Her 16h57'29" 34d30'19"
Ty Wotherspoon
Cam 4h4'4" 72d45'59"
TYA
Lyn 8h43'1" 44d48'57"
Tyadea
Lyr 18h55'30" 36d26'6"
Tyana Ralston
And 1h10'48" 41d46'42"
Tyanna's First Mommy Day
Mon 7h15'5" -0d30'57"
Tyanne Nicole Campbell
Ari 2h32'33" 20d57'45"
Tyanya & John David
Coomer
Cyg 20h43'58" 37d12'56"
Tyari M. Johnson
Dra 19h43'27" 60d19'46"
Tyauna Ché Charisse
Anthony
Dra 16h27'35" 54d38'6"
Tyazani
And 1h48'34" 36d24'10"
Tyber
Ori 5h52'23" 5d48'33"
Tyboria
Gem 7h53'12" 29d0'45"
Tyche
Ari 3h7'7" 29d31'48"
Tycien Ly
Her 17h8'13" 33d49'38"
Tyco Coleman
Dra 15h52'49" 62d12'57"
Tydooh Stardooh
Vir 13h13'39" 4d15'3"
Tydrew Ligtenberg
Lmi 10h32'12" 38d43'5"
Tye
Sco 16h9'42" -10d14'36"
Tye Joseph Williams
Cap 21h41'37" -19d23'17"
Tye Julian Sutherland
Tau 3h54'39" 26d4'24"
Tye Murphy
Uma 11h34'43" 48d5'59"
Tye Steven Mitnick
Uma 10h55'46" 52d55'37"
Tyecid-28
Pho 1h53'20" -40d3'42"
Tye-dyed sky
Uma 11h45'21" 57d12'4"
Tye-mela'ne
Cyg 20h36'44" 49d26'22"
Tyerinray Bright Grice Love
Star
Cyg 21h38'58" 32d37'11"
TYESE
Sgr 18h58'57" -30d16'36"
Tyesha Walker
Vir 12h1'37" -10d24'2"
Tyger
Lmi 10h27'30" 32d58'47"
Tyger Infinitus
Lyn 7h26'26" 44d46'22"
Tyger & Matt 06/04/03 to
Forever
Gem 7h28'36" 26d3'1"
Tygert Pennington
Lib 15h36'24" -15d56'27"
Tygress Carolyn
Lyn 7h12'28" 55d42'39"
TYJARU
Lyn 7h1'22" 57d51'11"
Tyjon "Ty" Clark
Uma 10h35'7" 64d28'40"
Tykiera Le'Vaunye
Aqr 21h43'53" 1d5'3"
Tyla
Lyn 8h40'20" 43d15'39"
Tyla Jay
And 23h59'16" 41d58'6"
Tyla Jessie Kelly
Sco 17h30'17" -43d34'27"
Tyla Marie Schell
Uma 11h24'15" 57d51'35"
Tyla Megan Fowler
Cnc 8h18'22" 29d50'25"
Tylan Frederick Dean
Her 16h11'4" 47d9'50"
Tyleann
Uma 10h6'13" 44d40'23"
Tyleena Machol
Mon 6h30'51" 8d13'6"
Tylemachos Louggis
Ori 6h3'7" 3d55'53"
Tylene Buck
Ori 5h41'35" -4d47'54"
Tyler
Umi 16h28'33" 77d38'50"
Tyler
Dra 18h48'44" 60d14'18"
Tyler
Uma 10h2'37" 60d36'36"
Tyler
Cet 2h49'55" 7d25'59"
Tyler
Ori 6h16'28" 14d58'17"
Tyler
Ori 6h4'29" 16d29'28"
Tyler
Gem 7h29'46" 26d44'10"
Tyler
Cnc 8h27'23" 29d31'53"
Tyler
Cnc 9h14'42" 24d14'51"
Tyler
Lmi 11h5'52" 26d58'41"
Tyler
Her 17h14'4" 20d25'58"
Tyler 14.11.90
Sco 16h10'11" -15d25'8"
Tyler Aaron Swanson
Her 17h44'14" 46d40'49"
Tyler Aaron Toms
Cap 21h52'45" -11d1'24"
Tyler Alan Folsom
Ori 5h43'18" 3d2'12"
Tyler All-Star 13
Sgr 20h20'3" -35d14'46"
Tyler Amy Love Forever
Sco 16h27'3" -27d6'29"
TYLER AND LEXIE
Sco 16h41'6" -30d4'38"
Tyler and Meghan
Shackelford Star
Ori 6h6'0" 18d32'0"
Tyler and Meredith -
28.08.03
Umi 17h22'10" 76d49'58"
Tyler and Sarah
Mon 7h20'52" -0d47'1"
Tyler And Trish Forever
Lib 14h56'47" -5d57'16"
Tyler Andrew Case
Uma 11h0'25" 60d43'43"
Tyler Andrew Gordon
Her 17h24'15" 48d36'2"
Tyler Andrew Hamvas
Leo 10h7'5" 20d31'11"
Tyler Andrew Jordan
Aqr 22h0'19" -22d19'41"
Tyler Andrew Kennedy
Cap 21h24'30" -27d5'0"
Tyler Andrew Laney
Uma 9h52'28" 60d35'6"
Tyler Andrew Luedtke
Umi 13h37'36" 69d31'13"
Tyler Andrew Swan
Tau 3h40'46" 26d20'15"
Tyler Andrews
Ari 2h19'34" 25d4'9"
Tyler Anthony Duncan
Dra 15h46'19" 53d43'52"
Tyler Anthony Viscardi
Per 3h36'48" 32d49'25"
Tyler Austin Erdman
Per 2h57'54" 48d35'42"
Tyler Austin Reardon
Tau 5h46'19" 18d27'25"
Tyler Austin Ziroe
Lyn 7h54'50" 35d6'4"
Tyler B. Crouch
Aur 5h47'32" 54d2'14"
Tyler Bean
Uma 10h59'44" 41d40'36"
Tyler Benjamin Green
Dra 19h48'45" 63d41'54"
Tyler Benjamin Milliren
Per 2h51'16" 54d9'32"
Tyler Benton
Her 17h31'38" 37d0'56"
Tyler Bibins
Uma 9h41'8" 56d24'38"
Tyler Bienemann
Umi 15h24'34" 72d44'31"
Tyler Blake Eveland
Her 17h32'49" 37d46'19"
Tyler Blake Fabian Lowney
Per 3h11'28" 41d57'0"
Tyler Bland Smith
Her 16h8'20" 22d3'39"
Tyler Braden Dorang
Ori 5h4'58" 4d15'16"
Tyler Braden Dorang
Aql 19h49'38" 5d43'50"
Tyler Brendon Moore
Boo 14h53'14" 22d35'21"
Tyler Bresslin
Dra 17h1'6" 63d22'34"
Tyler Brian Jones
Lib 15h27'56" -22d35'50"
Tyler & Brooke * Feels Like
Home
Uma 9h21'38" 55d53'9"
Tyler Bryant Jensen
Crb 15h55'7" 35d50'9"
Tyler C. Thompson
Aur 5h29'0" 40d38'15"
Tyler Casey Dueker
Ori 5h45'3" 2d22'8"
Tyler Cassetta-Frey
Vir 12h48'16" 7d35'31"
Tyler Charles Montgomery
Vir 11h47'14" 9d13'48"
Tyler Chiou
Uma 13h52'27" 53d37'34"
Tyler Chrisinger A Shining
Star
Boo 14h25'1" 14d5'15"
Tyler Christian
Uma 12h42'27" 56d5'30"
Tyler Christian Nelson
Uma 11h58'44" 61d35'58"
Tyler Christian Parks
Sgr 19h34'2" -28d0'5"
Tyler Christopher
Gem 7h3'9" 23d40'2"
Tyler Christopher Murray
Uma 14h1'58" 54d51'17"
Tyler Christopher Stewart
Lyn 8h29'18" 56d54'17"
Tyler Christopher Wilkins
Aqr 22h44'6" -1d52'42"
Tyler Cirillo
Aqr 22h21'34" -4d6'42"
Tyler Craig Myers
Gem 7h7'0" 27d15'41"
Tyler D. Huss
Uma 11h9'17" 39d29'43"
Tyler Dane Russell
Ori 5h21'19" 5d28'31"
Tyler Daniel
Tau 3h34'38" 3d57'42"
Tyler Daniel Kolb
Cnc 8h34'51" 29d19'47"
Tyler Daniel Parra
Umi 1h31'31" 88d45'2"
Tyler Daniel Sheehan
Leo 11h8'46" 3d7'2"
Tyler Daniel Wise
Her 17h9'2" 16d18'27"
Tyler David Criswell *5/9/84
Tau 5h50'8" 15d27'19"
Tyler Dawain Eaton
Vir 12h9'30" -11d9'7"
Tyler Denise Pritchett
Uma 12h4'4" 57d31'5"
Tyler Derikx
Lyn 8h30'28" 39d37'3"
Tyler Dingman
Ori 5h54'38" -0d40'33"
Tyler Donaldson
Aur 5h50'37" 43d17'55"
Tyler Duelm
Her 17h31'19" 30d45'2"
Tyler Duggins
Sco 16h15'17" -14d42'4"
Tyler Ebner
Her 17h11'45" 30d56'15"
Tyler Edward
Cnc 8h34'12" 17d42'8"
Tyler Erickson
Cma 7h24'46" -22d46'59"
Tyler & Erin Together
Always
Gem 6h32'24" 17d16'3"
Tyler Evan Caparso 10-07-
04, 5:20pm
Umi 15h44'56" 74d54'3"
Tyler Feistl
Cru 12h2'57" -63d21'47"
Tyler Feldman
Aur 5h28'43" 39d28'15"
Tyler Frank Parker
Her 17h7'42" 13d23'42"
Tyler Frank Rosenthal
Her 16h53'18" 19d1'9"
Tyler G. Pittman
Vir 12h51'48" 11d54'11"
Tyler Gary Thomas
Grosvenor Davies
Umi 17h20'21" 83d49'7"
Tyler Goldberg
Uma 9h42'24" 41d56'51"
Tyler Golden
Cha 10h34'6" -76d11'52"
Tyler Gordon Bruyea
Her 16h46'18" 47d22'37"
Tyler Gracy
Umi 14h31'33" 77d26'19"
Tyler Graebner
Lyr 18h53'16" 43d26'54"
Tyler Graham Kennedy
Her 16h25'28" 46d3'39"
Tyler Hansen
Uma 13h54'3" 52d6'3"
Tyler & Hawli's
Sge 19h55'53" 17d35'51"
Tyler Henry Barberio
Aur 5h31'3" 48d32'54"
Tyler Higgins
Vir 12h10'1" 12d0'11"
Tyler Innis Opdyke
Psc 23h30'56" 4d43'50"
Tyler J. Boudreau
Lmi 11h0'11" 30d45'29"
Tyler J. Butterfield
Del 20h38'43" 15d23'35"
Tyler J. Gauthier
Aqr 21h37'22" 1d31'58"
Tyler Jack Kubo
Sgr 18h58'34" -29d14'42"
Tyler Jacob Ballou
Leo 11h29'2" 1d2'16"
Tyler Jacob Brennan
Sco 17h17'56" -43d25'12"
Tyler Jacob Thomas Baer
Sco 17h23'43" -38d10'20"
Tyler Jacob Yousif
Leo 11h20'53" 21d38'26"
Tyler James
Vir 13h5'54" 11d42'56"
Tyler James Bangle
Cnc 8h31'30" 28d49'13"
Tyler James Coburn
Del 20h48'38" 14d51'10"
Tyler James Gruner
Ori 5h31'36" 1d22'28"
Tyler James Gruner
Her 16h47'40" 33d58'31"
Tyler James Hanley
Col 5h50'23" -35d34'36"
Tyler James Harberts
Cam 3h50'32" 68d32'24"
Tyler James Jevaney
Her 16h46'24" 28d24'59"
Tyler James Lomeli
09/21/01
Vir 13h32'6" 3d6'25"
Tyler James Lowe
Per 2h59'33" 55d53'49"
Tyler James Paine
Cnc 8h55'53" 15d27'39"
Tyler James Plauger
Her 16h19'54" 44d38'47"
Tyler James Poncy
Her 17h31'39" 34d31'41"
Tyler James Ramsey
Lmi 10h23'11" 34d18'52"
Tyler James Smith
Dra 18h43'55" 52d19'45"
Tyler James Wall
Her 18h44'47" 12d12'26"
Tyler James Whitman
Psc 1h26'7" 32d35'8"
Tyler James Wren
Dra 15h33'9" 60d23'41"
Tyler Jason Breen
Umi 16h24'23" 76d2'38"
Tyler Jeffrey Swanson
Gem 7h3'53" 19d45'36"
Tyler Jenson Elliott
Psc 0h34'55" 18d27'30"
Tyler Jerome Webb (T.J.)
Aqr 22h32'12" -14d23'47"
Tyler John Bartlett Brudno
Her 17h36'59" 27d10'31"
Tyler John Bock
Ori 6h8'58" 21d7'13"
Tyler John Glover
Ari 3h18'4" 27d16'44"
Tyler John Hornacek
Gem 6h44'34" 16d1'1"
Tyler John Russell
Umi 16h19'17" 78d21'58"
Tyler John Shann
Leo 10h32'11" 13d5'55"
Tyler John Vranick
Cnc 8h48'50" 31d13'0"
Tyler John Watts - The
Little Dolphin
Uma 12h0'9" 56d31'59"
Tyler Jon Copeland
Col 5h47'13" -34d48'30"
Tyler JonEric
Gem 7h52'41" 19d20'43"
Tyler Jordan Williams
Umi 14h52'4" 76d31'9"
Tyler Joseph Carter
Uma 11h12'13" 37d51'26"
Tyler Joseph Charles
McLaughlin
Umi 15h42'8" 73d10'18"
Tyler Joseph Davis
Lyn 7h51'29" 39d17'3"
Tyler Joseph Dunne
Her 18h30'21" 13d49'38"
Tyler Joseph Foster
Her 17h45'2" 46d17'47"
Tyler Joseph Garmon
Ori 4h54'16" -0d6'22"
Tyler Joseph Maar
McClister
Lib 15h42'0" -12d10'25"
Tyler Joseph Segal
Psc 1h7'57" 31d37'21"
Tyler Joseph Shores
Lyn 8h36'35" 34d41'13"
Tyler Junior Shaw
Cma 7h24'16" -29d28'33"
Tyler K 1
Sco 17h22'14" -32d55'46"
Tyler Kai Quinones
Lib 14h57'33" -6d19'45"
Tyler Keir
Ori 6h17'59" 18d57'27"
Tyler Kirby Smith
Sco 17h19'51" -31d16'13"
Tyler Kopp
Her 17h28'46" 14d30'8"
Tyler L. Olswold
Lib 15h10'57" -5d51'38"
Tyler L. Wicker
Sco 16h17'47" -29d57'10"
Tyler Lane Casey
Aql 19h21'35" 0d1'32"
Tyler Lawrence
Vir 12h47'27" -1d9'35"
Tyler Lawrence Carman
Cam 7h17'21" 60d1'31"
Tyler Lee
Sgr 18h15'33" -29d31'49"
Tyler Lee Burgess
Ori 5h48'14" 2d18'1"
Tyler Lee Wright
Psc 1h7'33" 32d46'53"
Tyler Lee Wright
Umi 14h38'11" 73d13'10"
Tyler Leigh Beccari
Aqr 22h11'1" -4d22'38"

Tyler Lewis Thacker
Cep 23h10'0" 79d36'4"

Tyler Liam
Uma 8h35'21" 59d58'54"

Tyler & Lindsey
Cyg 20h28'8" 45d49'44"

Tyler Lord Clark.."Our T.L.C."
Uma 11h13'43" 60d51'12"

Tyler Lorette
Aur 6h28'56" 35d3'55"

Tyler Lowenthal
Aqr 20h55'2" 0d34'2"

Tyler Luedtke
Uma 9h16'23" 66d36'9"

Tyler Luedtke our Hero
Per 3h7'40" 53d40'17"

Tyler Luke Corso
Ari 2h52'43" 26d53'10"

Tyler Luke Dunegan
Sco 16h32'17" -28d5'53"

Tyler M. Riedel
Her 16h21'53" 5d52'42"

Tyler M. Teasdale
Uma 9h39'32" 47d51'3"

Tyler Maddox Young
Tau 4h31'36" 22d30'55"

Tyler Marc Shabinaw
Cnc 8h11'30" 11d11'49"

Tyler Marc Terenzi
Her 17h47'30" 48d31'15"

Tyler Marechal Cerise
Cap 21h12'57" -16d53'32"

Tyler Markey
And 23h34'24" 41d37'55"

Tyler Marshall Jackson
Vir 14h5'48" -6d17'50"

Tyler Matthew Sabodor
Lib 15h10'5" -6d28'21"

Tyler Matthew West
Uma 11h46'9" 50d18'14"

Tyler Mayhew Shipway
Cnc 8h38'26" 9d15'14"

Tyler McNeill Nelson
Per 1h52'4" 48d14'9"

Tyler Megan Cano
Umi 13h30'14" 69d33'4"

Tyler Micai
Psc 23h31'43" 1d37'54"

Tyler Michael
Per 3h30'47" 47d20'43"

Tyler Michael
Uma 10h32'56" 41d34'17"

Tyler Michael Maddix
Umi 14h36'28" 78d50'1"

Tyler Michael Noucas Star
Her 16h24'30" 44d25'43"

Tyler Milton
Leo 10h29'20" 8d37'3"

Tyler (Misselwaithe XVI) Pratt
Tau 5h50'9" 16d57'59"

Tyler Mitchel Robinson
Sco 16h14'26" -12d48'57"

Tyler Mitchell
Ori 5h49'32" 9d21'50"

Tyler Monroe Heckaman
Sgr 19h35'15" -12d4'27"

Tyler Montavon
Lyn 7h53'13" 34d57'22"

Tyler Muchmore
Uma 11h47'19" 38d18'24"

Tyler N. Schook- 2 Pack Nat
Sco 17h27'45" -37d39'15"

Tyler Nathan Thomas
Cmi 7h35'33" 6d0'28"

Tyler Nicole Jones
Cam 7h25'23" 63d55'27"

Tyler Norman George Chapman
Boo 14h52'28" 29d30'23"

Tyler O'Leary
Uma 10h15'29" 54d15'16"

Tyler Paige Weiss
Uma 10h3'23" 60d19'52"

Tyler Philip Roney
Her 16h25'13" 48d36'37"

Tyler Pineda
Lib 15h42'40" -19d39'47"

Tyler Plank (JT)
Aql 19h28'34" -7d46'57"

Tyler Preston Collier
Ori 6h5'42" 10d28'10"

Tyler Preston Spears
Leo 11h4'42" 18d27'0"

Tyler Rainville
Sco 16h6'55" -16d39'39"

Tyler Raven
Her 16h43'21" 44d12'37"

Tyler Ray Norton
Psc 23h40'21" 2d39'50"

Tyler Ray Shreve
Lyn 7h41'42" 38d38'0"

Tyler Ray Steele
Per 3h17'17" 47d23'48"

Tyler Ray Stelter
Per 3h13'53" 44d40'17"

Tyler Raymond
And 23h34'22" 48d41'17"

Tyler Reed Hamish Badger
Uma 12h44'39" 60d34'24"

Tyler Ren Kakiki
Per 3h9'5" 37d2'24"

Tyler & Renee
Tau 3h52'35" 12d8'33"

Tyler Richard Anthony
Tau 3h26'34" -0d27'33"

Tyler Riches
Umi 16h26'48" 73d34'12"

Tyler Rodgers
Uma 9h46'35" 56d54'5"

Tyler Romano
Peg 23h48'38" 18d46'18"

Tyler Royse Greene
Lep 5h32'19" -15d46'30"

Tyler Ryan Cravey
Lmi 10h14'49" 30d19'29"

Tyler Scott Fisher
Boo 14h18'17" 17d16'32"

Tyler Scott Gladieux
Psc 23h46'14" 6d0'25"

Tyler Scott Thacker
Cyg 19h21'29" 29d12'10"

Tyler Shane Smith
Ari 2h46'4" 16d37'1"

Tyler Shines Faithfully
Ori 6h10'45" 16d9'51"

Tyler Small
Crb 15h42'55" 26d18'9"

Tyler Smit
Uma 10h7'1" 66d54'6"

Tyler Smith Ethier
Vir 12h49'58" 3d12'52"

Tyler Snell
Her 17h53'23" 49d23'3"

Tyler SooHoo
Sco 16h8'22" -9d29'18"

Tyler SooHoo
Sco 17h31'31" -33d14'50"

Tyler Stevens
Leo 10h22'48" 25d48'48"

Tyler Stirling Little
Lyr 18h53'32" 43d21'30"

Tyler Stone Rogers
Lyn 6h23'25" 57d40'45"

Tyler Tebori Barth
Aql 19h14'8" 15d51'30"

Tyler Thomas Clay
Uma 9h25'19" 68d28'17"

Tyler Udelhoven (Tinker)
Pho 0h4'41" -45d19'29"

Tyler Uhlich
Ori 5h46'47" 2d23'57"

Tyler Van Hunt
Boo 14h15'45" 24d47'19"

Tyler Varney
Umi 14h43'54" 75d42'0"

Tyler Walters
Uma 11h54'23" 40d58'27"

Tyler Warren Bowman
Ori 5h20'54" 7d19'30"

Tyler Watts
Her 16h50'12" 35d30'41"

Tyler Wayne Mize
Tau 4h31'56" 21d14'33"

Tyler Weber
Her 18h11'25" 22d24'6"

Tyler William Campbell
Ori 6h16'0" 14d45'39"

Tyler William Gilmore
Peg 22h52'19" 16d11'45"

Tyler William Saunders
Aql 19h25'44" 15d35'13"

Tyler William Scott
Her 16h46'15" 37d56'13"

Tyler Wilson
Per 4h8'32" 46d46'10"

Tyler Wong Achuck
Vir 12h11'9" 5d24'15"

Tyler Woodring
Cyg 19h42'24" 54d4'37"

Tyler Younkman
Ori 4h54'3" 5d16'5"

TylerCaitlynGabe
Leo 11h3'25" -6d16'37"

TylerDo
Ori 6h5'9" 0d27'18"

TylerKel
Umi 15h19'13" 79d42'29"

Tyler-man
Psc 0h38'45" 13d51'36"

tyler-marie
Tau 4h28'44" 28d27'50"

Tyler's constellation home-run
Lac 22h32'47" 49d21'0"

Tyler's Dixie Twins All Star 2005
Her 16h59'18" 17d30'6"

Tyler's Ginormous Star
Ori 5h46'31" 7d39'39"

Tyler's Joy
Aql 19h37'46" 3d1'55"

Tyler's Landing TJFDRM8-1-4-7-37
Her 15h52'17" 42d59'33"

Tyler's Night Light
Cas 0h22'18" 53d0'43"

Tyler's Nightlite over Marginal Way
Lyn 8h37'19" 34d42'5"

Tyler's Pinky
Umi 15h0'9" 85d23'51"

Tyler's Sparkle
Umi 13h21'45" 74d37'24"

Tyler's Star
Vir 13h27'22" -7d0'0"

Tyler's Star
Uma 9h10'40" 47d32'20"

Tyler's Taurus
Gem 7h16'48" 24d53'36"

Tyler's Touch (on the Universe)
Ori 5h30'29" -4d40'43"

Tyler's Twilight Twinkle
Per 3h18'34" 44d2'33"

Tylie Bubbles
Cyg 20h29'34" 33d40'36"

Tylier - Precious Gem
Dra 16h25'5" 54d27'8"

Tylor
Umi 14h15'48" 73d31'18"

Tylor Kordzikowski
Sco 16h23'41" -26d45'24"

Tylor Robert Miller
Leo 11h8'4" 6d23'3"

Tylor Robert Worthington
Uma 9h15'44" 50d57'59"

Tymber
Leo 11h26'59" 12d56'53"

Tyme Elizabeth
Crb 15h29'38" 26d31'18"

Tyme Wynde
Uma 10h47'0" 55d33'31"

Tymi
Uma 10h29'2" 42d22'56"

Tyna
Crb 16h3'28" 28d20'11"

Tynan "Piglet" Chau
And 0h34'34" 41d20'5"

tynandd
Tau 4h36'41" 20d16'10"

Tynea Moseley
Cap 21h32'22" -15d50'24"

Tyra
Uma 8h29'17" 60d8'48"

Tyra Ali
Cas 0h55'14" 52d29'51"

Tyra Leah
Cra 18h52'37" -42d36'31"

Tyra Mayes
Sco 17h13'24" -38d51'43"

Tyra Rose Spring
Sco 17h26'20" -38d45'4"

TyraMcKenzieMise
Lib 15h59'49" -18d32'58"

Tyra-My Eternal Love
Uma 9h22'43" 65d13'35"

Tyrannis
Uma 12h8'14" 49d40'17"

Tyra's 36
Aqr 21h6'49" -13d58'57"

Tyree Watson
Oph 17h40'46" -0d33'50"

Tyrel & Desire'
Uma 8h21'9" 63d58'42"

Tyrel Louis Haveman
Aqr 21h20'15" -13d55'30"

Tyrell Johnsrud
Tau 5h6'20" 22d56'58"

Tyrell Keonte Cowling
Cyg 19h57'49" 30d20'42"

Tyrene
Aqr 22h43'31" -3d31'26"

Tyrese Donnell La Maison
Per 2h55'40" 46d57'25"

Tyrese Marcus Cox 4-21-99 - 4-8-06
Uma 8h37'59" 68d34'10"

Tyrin Shane Carré
Uma 9h18'57" 45d20'8"

Tyron
Leo 10h21'56" 12d0'1"

Tyrone
Aur 5h19'46" 31d6'1"

Tyrone
Cam 4h8'33" 71d38'53"

Tyrone Mosby
Uma 9h47'17" 58d7'30"

Tyrone S. Soklaski
Psc 0h52'31" 28d31'22"

tyroun
And 23h21'59" 35d30'5"

Tysan Rose
Uma 10h27'51" 53d54'39"

Tysen Anthony
Lib 15h25'15" -20d47'35"

Tyson
Umi 13h36'19" 76d12'35"

Tyson
Uma 11h32'25" 64d44'27"

Tyson
Psc 23h34'41" 4d50'31"

Tyson and Brittany
Peg 23h24'15" 32d48'59"

Tyson and Callyn
Pho 1h34'8" -44d4'24"

Tyson and Liz
Tau 4h23'0" 13d27'24"

Tyson Frank Sakrison
Sgr 19h46'28" -15d40'33"

Tyson George Wayne Latif
Ari 2h51'1" 27d16'20"

Tyson Kevin McNeese
Cet 1h58'31" -3d16'41"

Tyson Self Scratch: GOD LOVE HIM
Her 18h7'21" 17d51'23"

Tyson W. Sawkins
And 1h32'19" 49d29'35"

Tyson Wersoski
Per 3h17'54" 31d41'25"

Tyson's Birthday Star
Sgr 19h38'23" -15d23'58"

Tyto Alba
Crb 15h47'41" 35d7'18"

Tytus Robert Smith
Cap 21h36'31" -13d32'27"

Ty-Ty Monster
Cmi 7h32'15" 4d25'24"

Tywysoges Sian
And 23h14'55" 51d54'49"

TZ-7-XP-33
Uma 8h19'2" 66d19'34"

Tzeb
Aql 19h13'49" -0d28'7"

Tzima Imirtziadi
Uma 8h54'6" 59d37'58"

Tzipora Maryam
Umi 14h47'48" 72d48'26"

Tzipporah
Vir 13h56'11" -16d30'44"

T-Zone
And 22h58'58" 48d27'17"

Tzvetan Stefanov Lukanov
Aqr 23h6'34" -12d42'27"

U and me
Lib 14h55'11" -17d25'13"

u and me
Cyg 21h27'24" 45d48'48"

U are 4ever in my heart then&always
Cyg 21h23'9" 45d45'13"

U B Parillo
Ori 5h49'46" 11d53'39"

U heart I
Leo 9h44'35" 26d8'24"

U heart I
Leo 9h43'53" 31d23'57"

U. & J. Fankhauser-Cantaluppi
Com 12h48'5" 27d28'55"

U K Sandy Filson
Vir 14h23'39" -4d57'59"

U Kikvo di Lena
Uma 10h34'44" 48d4'18"

U R My Sunshine~"Seth Karl Steiner"
Ori 6h17'19" 10d41'8"

U - Shaka Cat -n- Roo
Cap 20h26'26" -11d55'20"

UAA (M) Sdn Bhd
Ori 5h39'58" 4d16'1"

Ubalda Buraczenski
Uma 11h18'41" 51d29'9"

Uber Manthei
And 1h47'2" 49d42'52"

Uber, Thomas
Sco 16h53'58" -39d3'23"

UBetCha
Cap 20h7'5" -11d6'9"

UBG
Lyn 7h26'25" 44d57'1"

Ubi
Ori 5h46'52" 1d11'7"

UBIARM
Crb 16h8'11" 34d48'5"

UCB Pharma
Per 3h29'22" 45d25'2"

Ucil Gertude Baker
Cam 7h32'21" 62d36'13"

" Ucky "
Uma 9h5'49" 48d30'6"

UCMHS Staff 2005
Uma 13h43'13" 54d30'19"

Uda Cubed
Lyn 8h2'52" 51d16'17"

Udden Cluster
And 0h33'2" 28d2'54"

Udelphimax
Uma 12h12'25" 58d34'57"

Udhe Singh
Vir 14h56'11" 5d19'9"

Udo and Yavaun
Gem 6h52'19" 26d42'34"

Udo Faßbender
Uma 9h51'11" 42d26'59"

Udo Günner
Uma 9h33'40" 57d53'7"

Udo Hassel
Ori 5h40'31" -5d36'22"

UDO KRUPPA (f1JL,vU)
Ori 6h16'50" 19d23'50"

Udo Liebetrau
Uma 13h22'17" 60d41'2"

Udo Zöllner - mein Spatzi
Uma 11h24'49" 30d53'55"

Udyat
Ari 3h12'44" 29d2'46"

UE Sabine
Ori 4h52'25" 11d1'49"

Ueli
Dra 17h42'0" 69d15'55"

UEVOLI
Gem 6h46'18" 14d54'30"

Uff Da
Umi 15h27'4" 73d42'10"

Ufo
And 0h11'16" 46d0'23"

Ufuk
Dra 17h48'13" 62d50'6"

Ugg
Sco 17h47'36" -38d2'28"

U-Glay
Uma 9h37'5" 58d19'27"

Ugly Casanova
Cap 21h24'32" -16d31'15"

UGO TORRES
Her 16h37'43" 8d34'3"

Uhegaxe
Ari 3h17'24" 16d46'20"

Uihlein Hoku
Sgr 19h16'54" -26d45'12"

Uiker, Andreas
Ori 5h36'43" 9d59'51"

Uilani
Peg 22h59'38" 25d6'41"

Uilani9976
Vir 13h30'40" -11d4'39"

Uijfusa04
Uma 10h49'9" 69d37'48"

UK2CA2F4
Vir 13h4'28" -13d42'42"

Ukrit
Sgr 19h6'12" -22d5'59"

UkSidT
Cnc 8h20'42" 19d39'11"

Ulalume
Cas 0h17'45" 61d29'57"

Ulandria
Psc 23h43'44" 3d27'4"

Ulanigida Didatinehi – 'Mama Sue'
Uma 14h5'51" 60d19'9"

Ulf Schindler
Ori 5h0'57" 4d44'6"

Ulf-Thorsten Zierau
Uma 9h27'0" 48d10'55"

ULI Charley Beek
Immortalis Amor
Ori 5h33'51" -0d36'39"

Uli Kappeler
Uma 9h53'34" 47d54'9"

Uli P.
Uma 9h50'4" 64d1'19"

Uli und Anita Happy-Future-Star
Uma 12h9'59" 59d25'45"

Uli & Uschi
Ori 5h17'58" -4d55'50"

Ulis Ilios
Ori 5h16'9" -9d17'29"

Ulises Millan-Munguia
Per 4h26'11" 47d40'53"

Ulka Patel
Aqr 22h12'22" 1d47'

Ull always be the twinkle in my sky
Gem 7h9'56" 34d29'39"

Ull Moros
Dra 17h36'43" 53d31'57"

Ulla Barkey
Cas 2h15'19" 66d26'39"

Ulla Johnson
Aqr 22h11'57" -24d30'26"

Ulla Schoedler
And 0h10'25" 38d3'17"

Ulle Tonne
Lib 14h59'52" -15d10'0"

Ulli
Uma 9h59'1" 44d58'56"

ULLI
Psc 0h51'59" 10d19'10"

Ullrich Weidert
Uma 10h16'51" 45d53'57"

Ulrich
Per 4h49'51" 46d41'50"

Ulrich Bommes
Ori 6h23'21" 15d54'4"

Ulrich Eichstädt
Peg 21h25'16" 23d53'23"

Ulrich Finzler
Uma 9h2'43" 59d55'59"

Ulrich Fricke
Uma 11h44'13" 31d55'14"

Ulrich Kintzinger
Uma 8h12'0" 60d29'45"

Ulrich Kirschstein
Ori 6h15'51" 8d55'12"

Ulrich Krämer
Uma 9h56'52" 61d56'50"

Ulrich Radeck
Uma 9h53'9" 44d6'2"

Ulrike
Aur 6h32'39" 38d40'36"

Ulrike
Uma 9h3'29" 50d49'51"

Ulrike
Uma 12h24'22" 59d5'48"

Ulrike Blank
Uma 9h45'31" 52d34'43"

Ulrike Friedel
Uma 11h8'31" 71d43'15"

Ulrike Hammann
Uma 13h54'35" 54d21'11"

Ulrike Ringleb
Uma 11h38'17" 34d52'12"

Ultimate Angel
Lyr 18h49'50" 44d31'55"

Ultimate Lisafier
Lyn 8h8'48" 33d58'12"

Ultra Hot Cindy Diane Pardy Star
Cap 20h33'25" -14d37'39"

Ultra Running Brian
Uma 13h38'53" 60d6'41"

Ulu, Gülten
Ori 5h28'40" 14d58'59"

Ülviye Emirzade 30.09.1972, 0130 AM
Lib 15h10'48" -6d36'55"

Ulvskjold
Ari 3h12'51" 27d18'35"

Ulysses Mathew Busch
Per 2h37'18" 51d28'2"

Um Set Ramos
Aqr 23h46'28" -4d43'24"

Uma
Vir 14h36'20" 3d14'21"

Uma Eliana Casey
Sgr 19h6'10" -19d4'47"

U-Make-Me-Proud-To-Be-Ur-Daughter!!
Crb 15h41'17" 25d54'49"

Umberto
And 1h54'46" 45d37'56"

Umberto Macri
Cnc 9h6'45" 32d48'4"

Umberto Vittorio Emanuele Zabaroni
Leo 11h44'57" 20d10'38"

Umbleskunk
Cam 4h36'26" 62d36'52"

Umea
Dra 17h35'26" 54d35'36"

Umer Chaudhry
Gem 6h48'54" 32d19'49"

Umesh-Anjali Anand
Umi 16h13'13" 70d21'9"

Umpi Lumpi
Uma 8h34'26" 47d4'49"

Umpqua Bank
Per 4h36'18" 47d13'32"

Umriti Manohar Pawaroo
Cyg 21h8'8" 48d15'42"

UMUT
Uma 8h49'35" 52d51'8"

Umut Daneshpayeh
Lyn 7h47'50" 57d11'44"

Un Amour Eternel
Cas 0h7'19" 58d56'53"

Un Anno Felice
Gem 6h22'53" 18d8'38"

un destino
And 2h22'47" 48d51'24"

Un Gioiello Nel Nielo
Uma 12h20'42" 56d28'22"

Un Jin
Ari 3h16'11" 27d33'2"

Un jour une jeune fille...
Uma 12h6'2" 53d50'57"

Un Kyong Wallace
Umi 16h32'19" 77d48'41"

Un Pequeño Tornado De Tejas
Uma 8h44'32" 68d44'51"

Un symbole de mon amour a vous
Lyr 18h46'16" 36d28'12"

UNA
Cyg 20h43'8" 48d11'37"

Una
Vir 11h52'21" 9d32'22"

UNA
Cap 21h30'9" -13d20'45"

Una amistad y adora para siempre
Leo 9h36'20" 26d41'53"

Una and Baby Beau Simmers
Cap 21h18'12" -24d19'57"

una canzone per te
Peg 22h12'25" 19d29'7"

Una Donagher
Sco 16h11'53" -41d20'48"

Una Ella Fra Le Stelle
Cyg 21h40'55" 30d11'31"

Una Famiglia Bella
Ori 5h32'47" 11d26'12"

Una & Frithjof (Bob) Forsberg
Cru 12h35'41" -58d49'20"

Una Jane Jones, The Eternal Star
Ori 5h54'34" 22d12'49"

Una McLean M.B.E.
Cyg 20h0'14" 43d4'2"

Una O'Moore
Uma 11h22'44" 56d42'45"

Una pro Infinitas - Jonathan/Ashley
Lyr 18h16'54" 34d20'11"

unabuna
Cas 0h11'19" 54d10'8"

Una's Star
Cyg 19h57'8" 43d24'15"

Unbreakable... Phillip & Nani Woods
Cyg 21h51'6" 37d0'59"

UNCHAINED MELODY
Vir 13h50'26" 2d44'13"

Unchained Melody
Cyg 20h30'1" 59d22'3"

Unchained Melody
Umi 15h31'25" 74d23'42"

Unci's Universe
Cap 20h44'30" -25d9'59"

Uncle Al
Per 2h16'41" 56d0'50"

Uncle AL the Kiddies Pal
Leo 11h26'9" 16d44'55"

Uncle Andrew
Umi 16h21'37" 77d6'24"

Uncle Bee
Vir 12h29'53" 10d40'29"

Uncle Benjie
Per 2h23'36" 57d26'49"

Uncle Bill aka Aflac Senior
Psc 1h47'10" 5d49'30"

Uncle Bill Kuni
Aql 19h39'57" 5d57'42"

Uncle Bill Murphy
Per 3h33'6" 47d55'16"

Uncle Billy
Cyg 21h31'19" 42d28'50"

Uncle Bob
Aqr 22h48'51" -19d26'10"

Uncle Bob Special
Sgr 18h45'39" -21d34'35"

Uncle Brian OK See Ya
Her 16h49'33" 35d35'41"

Uncle Bubba
Per 4h11'41" 37d41'51"

Uncle Buddy
Ori 5h24'26" 4d24'34"

Uncle Butch
Her 16h16'15" 13d42'32"

Uncle Butch
Aql 20h8'39" 11d18'22"

Uncle Carl
And 23h17'9" 49d31'3"

Uncle Charlie
Cnc 8h54'5" 15d49'1"

Uncle Dan
Her 16h47'46" 47d32'33"

Uncle Danny
Cnc 9h12'22" 31d47'17"

Uncle Dave
Boo 13h48'32" 10d32'56"

'Uncle Dave' Barnes
Cru 12h39'26" -60d33'16"

Uncle Dave Haber
Vir 13h6'48" 12d43'21"

Uncle Dave Kudron
Aql 19h14'10" 1d59'8"

Uncle Dick
Leo 10h10'25" 25d29'32"

Uncle Don C. Ryan
Her 18h48'22" 24d57'8"

Uncle Don' Star (Don Corley)
Per 2h41'42" 54d58'0"

Uncle Funster's Star for Chapman
Aqr 22h1'14" -14d57'4"

Uncle Funster's Star for Clea
Sgr 20h24'50" -42d33'55"

Uncle Gar
Ori 6h4'45" -3d53'33"

Uncle Gary
Per 3h47'51" 52d15'7"

Uncle Gary Schamberger
Uma 10h4'33" 71d9'45"

Uncle Gavin
Cru 12h24'37" -61d51'51"

Uncle Gene
Ari 2h39'57" 19d58'26"

Uncle Grady
Lib 15h9'3" -3d46'21"

Uncle Greg Uncle Rerry
Gem 7h39'3" 22d34'55"

* Uncle Guy's Star *
Cnc 9h6'34" 18d26'36"

Uncle Harry
Boo 15h20'52" 44d52'46"

Uncle Harry
Umi 13h32'13" 71d33'9"

Uncle Harry, (Our Great Bear)
Uma 10h19'50" 56d16'58"

Uncle Henry
Cep 22h31'11" 77d2'37"

Uncle Henry Stiffel 1925
Uma 10h53'2" 52d32'36"

Uncle Ivan
Lib 15h12'26" -12d45'35"

Uncle Jack
Sgr 18h53'9" -26d10'0"

Uncle Jim
Dra 19h36'45" 69d20'2"

Uncle Jim
Gem 7h20'29" 15d14'27"

Uncle Jimmy
Uma 9h48'39" 42d28'19"

Uncle Jimmy's Star
Her 17h13'23" 18d43'28"

Uncle Jim's Hunting Point
Cnv 12h38'27" 43d0'56"

Uncle Joe
Uma 11h35'53" 30d2'26"

Uncle Joe Insco
Cep 1h9'8" 83d2'30"

Uncle John
Sco 16h33'36" -38d32'47"

Uncle John
Cyg 19h47'7" 35d55'23"

Uncle John Smiling From Heaven
Aur 5h50'18" 51d27'29"

Uncle Johnny
Peg 0h9'33" 13d28'14"

Uncle Justin and Aunt Corinne
Ari 2h2'4" 18d19'7"

Uncle Ken
Uma 9h3'59" 57d38'56"

Uncle Ken Salach's Star
Cnc 9h21'7" 30d17'28"

Uncle Kenny Gold
Aqr 21h38'14" 0d58'10"

Uncle Larry
Tri 2h11'37" 33d14'22"

"Uncle" Liz Goulding
Uma 8h40'8" 66d36'43"

Uncle Lutz
Sgr 18h41'4" -27d14'14"

Uncle Mait
Tau 4h32'15" 30d49'54"

Uncle Mark & Aunt Rhonda
Lib 15h7'37" -11d1'39"

Uncle Marvin
Per 3h51'37" 36d15'35"

Uncle Mike
Per 2h53'29" 52d24'39"

Uncle Mike
Peg 23h47'45" 25d12'13"

Uncle Mikey
Psc 0h47'20" 21d20'17"

Uncle Mikey
Lyn 7h6'56" 61d14'43"

Uncle Mikey
Lib 15h18'46" -28d32'28"

Uncle "Mr." Jeff Beach
Her 17h35'46" 37d13'58"

Uncle Nick
Leo 10h49'52" 14d47'8"

Uncle Nick
Her 17h39'54" 20d2'54"

Uncle Nick Abate, Superstar
Her 18h40'25" 19d46'43"

Uncle Nuni's Star
Aqr 21h51'14" -8d14'27"

Uncle Paul's Big Black Dog
Ari 2h13'23" 15d49'9"

Uncle Pauly
Lib 15h4'30" -1d45'32"

Uncle Peter
Ori 5h45'3" 5d12'28"

Uncle Randy
Vir 12h42'2" 4d55'11"

Uncle Ray
Sgr 18h2'37" -26d15'1"

Uncle Richard
Per 3h42'47" 36d30'38"

Uncle Richard's Place In Space
Vir 14h16'5" -14d59'8"

Uncle Robert
Ori 6h13'23" 13d20'49"

Uncle Rod
Her 17h57'5" 26d53'41"

Uncle Ronnie
Aql 19h4'50" 15d51'20"

Uncle Ronnie
Her 16h49'19" 33d55'44"

Uncle Ror
Cep 2h38'45" 81d15'17"

Uncle Roy Beeler ie Uncle Melonhead
Her 18h19'3" 22d22'38"

Uncle Scott
Her 17h57'55" 26d46'49"

Uncle Stanley
Dra 15h49'36" 51d46'48"

Uncle Steve
Gem 7h15'32" 19d56'5"

Uncle Steve Schopper
Sco 17h56'3" -42d43'16"

Uncle Terry Bache
Psc 0h45'56" 15d39'16"

Uncle Tim
Her 17h20'3" 45d41'9"

Uncle Tom
Per 2h42'20" 55d0'27"

Uncle Tom
And 1h15'24" 35d40'13"

"Uncle Tony" Conte
Peg 23h8'30" 17d20'20"

Uncle Wally
Cep 22h56'21" 78d3'53"

Uncle Wally's Louie
Ori 6h5'11" 19d23'43"

Uncle Warren
Sco 16h41'30" -25d16'36"

Uncle William Read
Cep 21h52'20" 85d0'3"

Uncle Willy
Cnc 8h58'10" 28d8'57"

Uncleboogie3 Charlie
Cap 21h36'5" -9d15'10"

UnCommon
Del 20h30'4" 16d1'4"

Unconditional
Leo 9h30'25" 10d3'39"

Unconditional
Lyn 7h53'33" 33d40'59"

Unconditional "12-26-03"
Lyn 7h16'44" 59d29'53"

unconditional eternal love
Cyg 20h30'41" 53d13'12"

Unconditional Love
Sco 16h42'45" -42d31'36"

Unconditional Love
Lyr 18h27'44" 36d22'30"

Unconditional love - Ryan McDonagh
Vir 12h8'48" 12d43'15"

Undaleigha " Star of my life"
Uma 8h31'53" 69d28'48"

Undeniable
Tau 4h4'31" 15d30'28"

Under Graph
Tau 4h29'9" 23d51'16"

Under The Same Sky TMAJM
Cma 6h46'0" -15d16'53"

Undercover Richie
Her 16h32'16" 6d28'3"

Underground
Ari 1h53'21" 17d26'47"

Undying Love
Cyg 20h5'15" 33d52'13"

Une étoile est née Steve Laporte
Ori 6h16'52" 9d30'16"

une étoile pour mon amour
Lyn 7h55'26" 56d55'17"

U-Neeq Francis
Uma 10h6'1" 57d58'45"

Unforgettable
Cnc 8h7'24" 16d57'59"

Unforgettable
Crb 15h46'36" 39d35'57"

Unforgettable Bliss
Cha 10h36'19" -77d27'47"

Unforgettable -David Lasher Reis-
Aqr 22h34'31" 1d17'44"

Unforgettable Jeanne S. Collins
Mon 6h45'13" 3d6'43"

Unforgettable Love
Sco 17h24'38" -45d25'21"

unforgettable night under the stars
Uma 13h19'41" 56d7'13"

Unforgettable Phyllis
Lyn 7h44'3" 48d42'33"

Unger Tamás
Gem 7h0'25" 20d30'9"

Unger, Celia
Uma 10h1'3" 45d46'41"

unglaubliche Goldstärn
Lac 22h46'48" 52d46'43"

U-N-I
Lib 15h59'55" -18d25'5"

UNI DAKS-13
Peg 22h4'5" 6d57'26"

Uni Q & Ty/Helen Weiler
And 23h15'51" 48d45'59"

Unican
Crb 15h50'58" 37d18'19"

Unico e Bello
Uma 12h7'41" 60d22'6"

Unicorn
Mon 7h34'17" -0d45'47"

Unicorn's Journey
Umi 15h44'37" 75d11'2"

Unie and Fitz
Cyg 19h14'30" 51d59'33"

UNIFI
Umi 14h44'6" 76d14'2"

Unified Western Grocers
Per 2h59'44" 55d34'17"

Unikue
And 23h0'28" 48d45'7"

Unilever (M) Holdings Sdn Bhd
Ori 5h30'30" -0d25'1"

Union Pacific
Uma 11h47'1" 32d36'6"

Unique Edwards
Uma 9h43'51" 69d24'19"

Unique Spirit
Cyg 21h23'8" 38d59'33"

Unique-Beauty
Ari 3h12'20" 15d36'27"

Uniquely Yours Forever.....
Dra 17h54'10" 56d36'28"

Unis
Tri 2h24'46" 30d4'54"

Unita
Her 16h6'8" 49d21'9"

Unity
Cyg 20h8'56" 43d2'31"

Unity
Cyg 20h48'12" 31d38'31"

Unity
Cyg 21h16'13" 34d21'57"

Unity
Vir 13h0'35" 10d9'46"

Unity
Gem 6h35'13" 24d38'0"

UNITY
Lyr 18h27'20" 28d50'43"

Unity
Sgr 19h19'41" -23d21'49"

Unity Of Angela and Joshua
Cnc 9h9'26" 23d29'48"

Unity Serene Sjoblom
Cnc 8h37'53" 17d41'14"

Universal Mom (Janet E. Garland)
Cas 0h30'8" 69d54'49"

Universe and James
Eri 3h46'40" -0d55'3"

unme
And 1h21'18" 42d3'40"

Unmei
Uma 9h43'44" 55d9'35"

Unni Johanne
Umi 15h38'4" 82d12'43"

Unqua's Little People Class of 2005
Her 18h33'13" 16d14'4"

Unreachable Star
Leo 9h22'38" 24d55'42"

Unrivaled Beauty
And 22h59'36" 48d7'20"

Unser Freund Marc Neuenschwander
Her 18h1'34" 25d51'43"

Unser großer Schatz wir lieben Dich
Uma 8h53'33" 61d23'56"

Unser kleiner Engel Julia
Uma 8h45'16" 47d34'36"

Unser Stern
Ori 6h11'14" 9d0'7"

Unser Stern
Uma 10h17'23" 72d6'27"

Unser Sternchen Jen
Uma 8h19'53" 65d3'29"

Unser Wunschbaby
Cas 2h17'10" 70d41'24"

Unstoppable Duo
Ori 6h4'48" 16d34'23"

Unsworth United
Cyg 20h5'27" 31d34'39"

Unterschied
Leo 11h10'5" -1d59'15"

Unti Andinak, az életem értelmének
Uma 9h49'8" 51d14'41"

until forever...
Cyg 19h52'10" 52d35'41"

Until Later
Lib 14h52'44" -4d25'24"

Until we meet again
Uma 11h25'0" 42d2'6"

Untill Forevers Over. LMS Loves DJS
Tau 4h37'35" 13d58'11"

Unto Eternity
Pho 0h34'9" -46d29'27"

Untouchable Shining Love
Ari 3h3'55" 26d24'26"

Unus Dias
And 0h32'41" 45d5'21"

Uomo Universale
Cyg 20h2'8" 32d27'7"

Uoyevoli
Cyg 20h28'26" 33d25'5"

uoyevoli
Cam 4h36'15" 63d41'18"

Uoyevolissor
Ari 1h57'23" 18d51'17"

Up
Sgr 18h36'12" -23d52'35"

Up Where We Belong
Psc 1h22'28" 17d2'57"

Upe
Oph 16h51'51" -0d23'43"

Upendra Wakade
Umi 14h10'31" 73d22'14"

Upenzi
Lyn 7h5'0" 50d0'18"

Upi My Love
Aqr 21h29'59" -5d40'7"

Upmeier, Jochen
Uma 10h32'47" 40d56'3"

Upstar
Umi 16h17'26" 76d31'42"

Upton
Cyg 19h49'36" 38d50'7"

UR SO BEAUTIFUL
Tau 4h36'9" 18d9'44"

U.R. STERN
Lyn 8h51'55" 33d23'58"

Urania 0511
Del 20h40'34" 15d43'49"

Urban, Bernd
Aqr 21h51'32" -1d38'44"

Urbanek
Umi 16h42'5" 76d50'11"

Urho A. Piilo
Cnv 12h32'35" 43d32'47"

Uri e Barbara Rispoli
Ori 6h9'49" 3d41'47"

Uri & Luisana
Cyg 21h11'10" 37d0'34"

Uriah
Per 4h8'29" 45d3'31"

Uriah's Light Shines Forever
Cnc 8h21'19" 13d49'35"

Urielle Guerrero
Tri 2h12'22" 33d44'31"

URIME
Tau 4h24'26" 18d31'54"

URMPFJW2
Sco 16h11'56" -10d21'25"

Uroosa
Cyg 19h38'58" 29d56'57"

UR-r-Skyliter
Uma 10h56'8" 52d27'11"

Urs
Her 18h45'17" 20d11'1"

Urs Battaglia
Lyn 7h35'13" 37d15'1"

Urs Gruber
Ori 5h8'0" 9d8'1"

Urs & Jean-Phillippe
Cas 2h21'27" 68d52'17"

Urs Kohler
Cas 0h56'2" 53d21'17"

Urs & Madeleine
Sct 18h28'38" -15d23'31"

Urs mein Stern, 26.12.1978
Aur 5h57'12" 36d12'29"

Urs - Rahel Schneebeli
Tri 2h32'41" 33d59'50"

Urs Riedi 1969
Per 4h36'40" 39d59'19"

Urs Siegrist
Per 4h7'26" 35d1'39"

Ursa Anna
Tau 3h25'3" -0d28'55"

Ursa Bulgy
Uma 8h20'34" 66d2'53"

Ursa Major Dollina Ruthus
Ori 5h5'50" -1d17'47"

Ursa Margratia Ravenia
Uma 11h11'53" 36d55'20"

Ursa Minor Spinks Star
Umi 14h50'26" 74d12'41"

Ursa Niblet
Psc 1h18'56" 15d14'30"

Ursi Amantes
Uma 10h28'11" 40d18'10"

Ursi & Tinu
Umi 15h0'9" 77d0'59"

ursina
Cam 7h0'50" 69d16'7"

Ursina
Boo 15h31'9" 45d36'20"

Ursina Julia zum 20. Geburtstag
Uma 14h6'20" 56d21'38"

URSODER2ME
Cyg 21h34'8" 47d22'50"

UrSue Math Major
Uma 12h43'41" 57d34'42"

Ursula
Uma 13h47'46" 54d12'7"

Ursula
Umi 15h59'30" 72d15'53"

URSULA
Umi 17h2'44" 79d49'36"

Ursula
Lib 15h46'13" -13d37'13"

Ursula
Cyg 20h2'18" 47d1'57"

Ursula
Cas 0h17'10" 51d26'9"

Ursula
Cyg 19h43'56" 31d20'20"

URSULA
Ari 3h4'59" 22d42'23"

Ursula Arduinna Boynton
Umi 15h11'12" 68d2'2"

Ursula Baker
Uma 9h22'20" 51d16'45"

Ursula Bauer 30.03.1945
Uma 11h30'22" 29d19'57"

Ursula Bertotti
Uma 11h22'19" 44d5'19"

Ursula & Bruno
Ori 5h35'20" 10d1'59"

Ursula Bürgi
Ori 6h0'50" 15d43'50"

Ursula Divanovic
Ori 5h49'29" 9d2'45"

Ursula Eckstein
Lyr 18h29'51" 36d19'39"

Ursula Edinger
Uma 10h38'54" 48d3'47"

Ursula Euler
Tri 1h54'57" 25d55'36"

Ursula & Gilbert (T.F)
Psc 0h1'48" 4d54'35"

Ursula Irene Fleischer
Lib 14h54'59" -10d11'14"

Ursula Irene Schanbacher
Dra 18h17'9" 77d14'44"

Ursula Marie Boehnlein DeMassimo
Aqr 22h43'42" -2d52'46"

Ursula Preisner
Lib 14h48'40" -12d7'3"

Ursula Ring
Uma 11h12'58" 35d25'15"

Ursula & Robert Muhr
Umi 14h53'44" 71d30'21"

Ursula Röbke
Ori 5h55'13" 11d24'14"

Ursula Schmeink
Uma 10h41'59" 70d28'18"

Ursula Schmid
And 0h45'0" 41d49'31"

Ursula Snelius
Sco 16h9'44" -12d25'19"

Ursula Stalder
And 23h3'23" 44d49'56"

Ursula Turner
Uma 10h32'37" 59d7'8"

Ursula und Jürg
And 1h42'14" 38d51'53"

Ursula Zendron-Hess
Cas 0h2'31" 59d1'57"

Ursula-Mischu
Aur 6h37'58" 39d4'48"

Ursula's Heavenly Sparkler
Uma 11h50'51" 55d8'36"

Uruviel Anwamane - Leigh's Star
Lib 15h15'14" -11d39'27"

Urve O Goran O FTJFDSMLHBHMOT
Uma 11h28'2" 59d26'13"

Urvesh & Hinal Patel - United Forever
Cyg 20h14'59" 38d29'21"

Urvi Ghandi
Sco 16h15'43" -21d47'49"

Urwin_Manjit1109
Ari 2h9'19" 27d12'51"

Ury
Lib 15h38'9" -29d18'18"

URYO
Tri 2h7'33" 32d14'22"

" US "
Leo 9h45'18" 27d32'27"

"Us"
Mon 7h34'48" -0d24'21"

Us 2005
Uma 8h40'25" 58d0'17"

US Bank
Per 3h49'45" 39d0'0"

U.S. Blues
Vir 13h49'25" -2d15'10"

"Us" Forever
Cyg 21h8'46" 51d22'55"

U.S. Marine Wacie Sheyne Laabs
Gem 7h3'41" 30d38'13"

U.S. Postal Inspector Caryl Byrd
Crb 15h47'32" 37d43'48"

US TROOPS Protectors of FREEDOM
Ori 5h40'28" 7d32'17"

Usabubu
Cru 12h26'28" -60d29'47"

Usagi
Cap 21h31'17" -12d44'30"

US-AGMG
Gem 6h37'24" 20d58'29"

U.S.C.G. Todd Albert Terrill
Tau 3h40'54" 3d33'20"

Uschi
Ori 5h53'51" 16d22'12"

USCHI
Uma 10h6'43" 48d58'13"

Uschi Born-Schievenbusch
Uma 10h32'11" 42d11'44"

Uschi & Franz Krucker
Cyg 21h26'48" 49d4'16"

Uschi & Franz Krucker
Cas 1h24'26" 68d15'32"

Uschi Meyer-Ottensen
Uma 13h54'13" 50d32'5"

Uschi Stark
Cas 23h31'30" 51d5'59"

Uschi und Herr Sonntag Paradies
Uma 8h22'26" 66d45'15"

Uschi und Lenzi
Scl 23h38'8" -25d13'49"

Uschi Zürner *Gullchen
Uma 10h35'4" 65d42'34"

Usdi
Psc 0h34'8" 12d50'28"

Use Stern für immer
Umi 13h39'25" 70d15'12"

Usha
Crb 15h36'36" 27d14'36"

Usha
And 0h19'33" 45d43'12"

Usha 'The Stellar' Rao
Aqr 23h48'25" -23d34'31"

Usha Wasan
Cas 1h16'8" 66d8'5"

Ushi
Cru 12h18'54" -56d52'21"

Usman Mohammad Awan
Sgr 18h5'32" -27d1'20"

USPA 170919 A-39555 B-25193
Leo 9h22'11" 15d3'18"

Usque Aeternum SEC
Leo 10h35'22" 16d11'12"

Usquequaque Memor
Cyg 21h51'45" 43d49'52"

U.S.S. Oran Jay
Boo 14h40'42" 37d41'36"

U.S.S. Texas Lone Star
Ori 6h13'11" 15d28'3"

UsStuff
Gem 6h8'35" 24d46'59"

Ut Ameris
Com 12h47'3" 26d19'40"

Utaibito S&T Haneboshi
Sgr 18h14'2" -18d58'27"

Ute Bauer-Dörr
Uma 8h28'8" 73d4'31"

Ute Buse
Uma 11h15'28" 42d10'45"

Ute Drescher
Uma 9h30'27" 53d34'30"

Ute Hardt
Uma 8h55'57" 58d48'31"

Ute Ku'uipo Kragl
Uma 11h31'25" 42d30'49"

UTE LENI MARTIN
Cnc 9h7'39" 32d19'45"

Ute Sabine Cozby
Aur 5h9'15" 46d58'31"

Utech, Peer
Uma 13h2'33" 61d23'18"

U-Tellum-Ping...AKA Liz
Gem 6h36'46" 27d23'20"

utopia
And 0h41'29" 43d9'35"

"UTOPIA" The Perfect Place
Uma 10h52'32" 52d59'58"

Utsukushii Chiyoko
Aqr 23h26'15" -11d33'10"

Utsukushiki Kimi no Hoshi
Tau 5h0'29" 26d30'14"

Uttama Narayan
Crb 16h13'16" 35d5'57"

Uttley
Cyg 21h11'59" 32d8'23"

Utu Abe Malae
Sgr 18h31'40" -35d27'5"

Uu!
Aqr 23h39'50" -22d45'23"

U.U.Shevchuk
Tau 3h45'41" 22d41'23"

UVM Valcour
Cyg 20h36'44" 47d29'5"

Uwe
Uma 10h30'4" 41d36'27"

Uwe
Uma 14h4'38" 56d41'31"

Uwe Blechschmidt
Ori 6h14'23" 7d37'15"

Uwe Buttgereit
Uma 11h28'34" 39d51'50"

Uwe Dillschmitter
Ori 6h11'45" 7d56'56"

Uwe Düring
Uma 10h27'12" 49d47'11"

Uwe F. Heiker
Uma 11h0'4" 40d39'8"

Uwe Frenzel
Ori 6h15'32" 8d34'34"

Uwe Hiby
Uma 11h18'46" 33d56'51"

Uwe Hitschfeld
Ori 5h57'34" -2d49'15"

Uwe Kahlert
Uma 8h12'52" 65d37'16"

Uwe Korsten
Uma 9h29'16" 69d42'27"

Uwe Krempasky
Uma 9h19'39" 67d43'14"

Uwe & Leonie
Ori 6h12'57" 8d43'10"

Uwe Martin
Uma 9h20'31" 48d57'25"

Uwe mein Stern
Uma 10h49'25" 47d38'27"

Uwe Pedd 07.06.1975
Uma 11h9'53" 72d18'32"

Uwe Rostock
Uma 9h14'56" 49d56'14"

Uwe Schaback
Ori 5h41'26" -2d10'8"

Uwe Schmidt
Uma 9h30'29" 47d31'51"

Uwe Schubert
Uma 9h33'19" 63d44'45"

Uwe Simmchen
Uma 10h29'10" 39d41'19"

Uwe Starke
Uma 11h44'47" 33d15'43"

Uwe Tittelbach
Uma 11h12'42" 64d44'19"

Uwe Wrage
Uma 11h39'39" 28d38'16"

Uwena
Ori 6h4'39" 21d23'26"

Uwe's & Sandra's Glücksstern
Uma 8h58'32" 53d56'0"

UW-Stout 2006/2007 Graduates
Lyn 7h16'18" 57d56'47"

Uxia
Aqr 23h22'5" -7d28'47"

Uxor Cuius KerryAnn
Uma 10h28'53" 66d24'4"

Uzay proposed to Selin
Cyg 20h22'3" 51d8'15"

Uzi Ovadia
Her 17h32'38" 27d57'26"

Uzma Husain
Cyg 20h3'4" 31d52'24"

V-
Psc 1h19'41" 24d15'34"

V
Tau 5h51'54" 17d54'56"

V
Tau 4h7'34" 10d37'6"

V
Tau 4h9'11" 7d14'2"

V. Aigron
Uma 11h48'5" 42d3'23"

V. Bernice
And 0h25'23" 40d45'29"

V J Conforti SDMF
Uma 10h39'49" 66d16'55"

V. LaVaughn Livingston Artchovia
Aql 19h34'17" -3d35'14"

V. Lee Shaleen
Uma 11h19'18" 36d43'18"

V. Lou Buhl
Vir 12h2'9" 4d22'50"

V. Maxine Ferrara Miles Burnham
Uma 13h49'19" 57d9'47"

V Nadine Smith
Ori 5h39'33" -5d32'25"

V S M
Uma 11h24'5" 57d51'17"

V to the V power
Cyg 19h49'50" 35d38'34"

V2-TOVWL
Lyn 7h35'48" 35d54'6"

V6
Cam 4h21'36" 66d16'41"

Vache
Aur 5h36'35" 49d22'28"

Vaclav Harsa
Uma 14h12'44" 57d32'47"

Vada Rochelle Heron
Cnc 8h32'27" 6d54'8"

Vadah Jean Richey
Uma 9h51'23" 56d26'14"

Vadász Judit
Lib 14h40'29" -21d48'6"

VADDER E.Lu.Sondergeld
Leo 9h30'7" 11d0'44"

Vaddon
Cyg 20h45'42" 35d45'35"

Vadella Worrell
Uma 10h18'18" 57d55'4"

Vaden S. Pitts
Lib 15h53'57" -12d56'22"

Vader Duane Berndt
Lyn 6h19'48" 56d11'6"

Vadim
Cas 1h56'2" 65d9'5"

Vadim Andronov
Ori 5h58'7" -0d32'27"

Vadim Rene
Uma 11h44'52" 61d43'37"

Vadim Solovjov
Ari 1h47'49" 17d45'9"

Vadim W
Cap 20h58'58" -26d4'23"

VAE54
Vir 12h31'31" 6d0'28"

Vágó György és Vágó Györgyné
Cnc 9h7'0" 8d12'21"

Vágó Judit
Sgr 18h28'26" -16d49'41"

Vahagn Setian
Leo 11h15'24" 22d13'6"

Vahan Koshkaryan
Uma 11h29'52" 44d27'31"

Vahida Dzanic
Cep 21h56'9" 72d19'14"

Vaida
Cap 21h41'31" -24d44'35"

Vaillante Umberto Bandini
And 1h36'11" 41d40'31"

Vaishali Dhirajlal Chhatralia
Cyg 21h21'36" 38d33'13"

Vajda Gyöngyvér Morzsi
Uma 10h25'34" 70d4'32"

VAL
Cap 21h6'29" -21d13'11"

Val
Uma 11h40'47" 39d20'26"

Val Aaron Zvinyatskovsky
Ari 2h40'52" 30d41'6"

Val Casale
Vir 13h22'56" -0d12'13"

Val Eastwood - Angel of the North
And 0h11'48" 47d21'51"

Val Guerrier
Lyr 19h4'31" 31d35'15"

Val H. Underwood
Lyn 7h8'51" 54d8'20"

Val Kitty
Lyn 6h41'51" 59d5'37"

Val. Loutsch Marraine Chérie
Ori 4h59'21" -2d13'25"

Val Pezzimenti
Uma 13h18'13" 56d14'45"

Val Powers
Uma 12h37'46" 53d25'39"

Val Ruud
Uma 11h24'11" 34d31'55"

Val The Star
Lac 22h41'1" 50d2'1"

Vala
Lyn 9h5'16" 38d33'4"

Valacia
Lib 14h53'13" -5d43'50"

Valaria J Foster
Lib 15h50'1" -20d20'3"

Valarie Ann
Cnc 9h12'39" 18d28'43"

Valarie Brattin
Psc 23h5'35" 5d2'29"

Valarie Fane
Leo 9h43'47" 28d23'12"

Valarie Ianniello
Cam 6h11'14" 61d25'1"

Valarie L Wunderlich
Vir 13h10'3" -14d55'10"

Valarie Loves Lance Leblang
Cyg 20h20'10" 40d0'37"

Valarie Lynn Tickle
Aqr 22h27'44" -1d48'40"

Valarie Marie Augustine
And 0h28'36" 32d27'46"

Valarie Marie Peterkin
Aqr 23h27'48" -19d6'45"

Valarie/Nana
Mon 6h43'40" 8d52'9"

Valarie's Heavenly Light
Cas 1h24'6" 64d25'47"

Valarie's Star
Leo 9h46'4" 28d10'13"

Valasia - forever my 'Ruby Gem'
Cru 12h36'14" -58d43'34"

VaLBLaZe
Sco 16h5'8" -21d45'0"
Valborg Elisabeth Koldste
Aqr 23h41'28" -10d56'23"
Valcor
Aql 19h41'51" 13d17'12"
Valda Louise Haughton
Vir 12h30'24" 4d35'49"
Valdemar Paul Helt
Sgr 19h17'33" -14d52'4"
Valdis' Star
Aql 19h31'57" 4d25'43"
Vale & Vanni
Lyn 8h51'38" 33d26'10"
Valeia Mae
And 0h29'49" 27d42'37"
Valena
And 1h2'1" 37d44'58"
Valena Burgess
Uma 11h49'36" 52d28'42"
Valenaalice
And 2h31'9" 43d49'57"
Valenbuiz
And 23h19'58" 52d15'30"
Valencia Carol Laniak
Crb 16h3'1" 36d57'38"
Valencia M. Davis
Leo 10h35'38" 13d18'18"
Valencia, Aztec Princess
Leo 9h59'26" 20d28'43"
Valene
Leo 11h50'26" 20d47'47"
Valene L. Smith
Cas 1h39'51" 65d20'45"
Valentin
Crb 16h16'25" 27d14'47"
Valentin
Ori 6h14'55" 1d59'36"
Valentin Nuss
Lib 15h56'44" -17d32'2"
Valentin Train 03.12.2005
Peg 21h57'3" 20d24'15"
Valentin Vrazilov
Uma 9h29'39" 51d24'23"
Valentin Women
Cnc 8h45'14" 12d7'56"
Valentin, Jörg
Leo 10h53'25" 6d17'15"
Valentina
Mon 6h31'22" 6d20'33"
Valentina
Ari 2h0'32" 12d14'20"
Valentina
Ori 6h0'48" 10d54'32"
Valentina
Vir 13h26'3" 9d4'47"
Valentina
And 0h46'46" 27d56'53"
VALENTINA
Cnc 8h16'49" 19d11'31"
Valentina
Her 17h56'56" 24d22'39"
Valentina
Leo 11h35'41" 25d58'49"
Valentina
Cnc 9h13'24" 24d25'58"
Valentina
Lac 22h55'55" 50d9'8"
Valentina
Per 4h43'31" 52d40'1"
Valentina
Her 17h51'53" 40d12'39"
Valentina
Cyg 19h35'51" 30d9'1"
Valentina
Lmi 9h32'54" 34d30'31"
Valentina
Cnc 8h41'54" 32d42'57"
Valentina
Aur 5h23'2" 44d15'32"
Valentina
Ori 5h53'51" -8d22'51"
Valentina
Umi 17h46'23" 80d0'14"
Valentina
Umi 16h59'32" 83d20'39"
Valentina
Cas 23h37'46" 59d30'52"
Valentina Aerts
Uma 11h52'36" 45d58'54"
Valentina Alessi
Uma 11h55'53" 30d34'46"
Valentina Amedea Fulvia
Per 4h13'4" 45d16'25"
Valentina Arevalo
Uma 9h48'38" 45d13'53"
Valentina Athena Lewis
Lib 15h25'43" -15d33'3"
Valentina B
Aqr 21h43'27" -1d4'52"
Valentina Bove
Tau 4h23'18" 12d35'31"
Valentina Cappai
Cyg 21h28'50" 34d37'29"
Valentina Caruso
Cam 4h11'17" 58d18'9"
Valentina Casadio Montanari
Per 2h50'25" 41d27'8"
Valentina Chesi
Uma 9h56'19" 41d30'26"
Valentina Connors
Cas 0h56'29" 59d24'7"
Valentina Dragone
Cas 0h57'33" 57d5'20"
Valentina Esposito
Aql 20h15'0" -0d14'59"
Valentina "Giuggy"
And 22h57'33" 50d8'7"
Valentina Guglielminetti
Her 16h49'51" 8d52'26"
Valentina Kalashnikova
Lyr 18h42'45" 43d57'32"
Valentina Lo Preiato
Umi 13h51'1" 74d31'55"
Valentina Lombardo
Cyg 19h57'4" 33d28'26"
Valentina Marie Lettieri
Cyg 20h52'11" 35d14'38"
Valentina Marie May
Psc 1h26'48" 16d1'0"
Valentina Negro
Peg 23h10'49" 31d6'27"
Valentina NeJame
And 0h18'22" 29d46'23"
Valentina Perugini
Cam 6h25'27" 63d28'47"
Valentina Placido McBride
Tau 3h38'42" 27d57'12"
Valentina Rea
Oph 17h10'57" 2d5'9"
Valentina Restelli
Cas 1h6'11" 54d13'44"
Valentina Sinagra
Ori 6h12'15" 2d58'45"
Valentina's Star
Aur 5h12'43" 46d47'28"
Valentine
Lmi 10h32'40" 38d5'49"
VALENTINE
Tau 4h15'54" 6d42'2"
Valentine
Vir 13h17'50" 1d57'33"
valentine
Leo 11h53'40" 20d35'47"
Valentine
Col 5h6'9" -35d59'45"
Valentine
Cyg 19h25'42" 53d34'43"
Valentine 2004
Cyg 20h10'48" 57d7'2"
Valentine Abhervé
Vir 12h38'38" -7d31'12"
Valentine Angela Little Bear
Umi 15h21'58" 71d4'21"
Valentine Gerrard Carr
Cyg 19h43'29" 30d28'33"
Valentine J. (Chunk) Palo Jr.
Her 17h53'16" 18d25'35"
Valentine J. Palo, Jr.
Uma 10h53'11" 44d14'59"
Valentine Julian Sue
Ori 5h45'18" -8d48'31"
Valentine Amelie Mignogna
Aqr 20h50'25" -7d24'39"
Valentine My Lovable
Aur 5h41'11" 33d44'36"
Valentine R Dunn
Mon 7h24'21" -1d28'55"
Valentine "Sparky" Brown
Cas 1h28'14" 63d40'46"
Valentine's Star
Cyg 20h24'16" 54d42'3"
Valentino
Cep 22h16'56" 66d15'38"
Valentino
Cyg 19h30'30" 28d2'7"
Valentino Family Star
Psc 1h4'44" 3d42'2"
Valentino Rosario
Umi 14h19'50" 78d49'27"
Valentyn
Aqr 23h41'7" -14d7'0"
Valenzuela 2
Uma 11h9'53" 56d42'56"
Valenzuela 5
Sge 19h22'59" 18d12'49"
Valer Capalnas - My Bhudda
Sco 17h9'52" -34d21'33"
Valere Helen Walton
Uma 8h16'11" 67d20'18"
Valeri
Tau 4h40'53" 10d10'4"
Valeri Jean Alston
Lyn 6h58'12" 47d37'32"
Valeri Khan
Tau 3h36'24" 3d55'40"
Valeria
Crb 15h25'11" 30d35'27"
Valeria
Per 4h28'5" 42d18'36"
Valeria
Umi 17h1'41" 79d45'39"
Valeria Aliah
Psc 1h9'54" 23d37'53"
Valeria Andrea Marti
Crt 11h28'35" -17d9'12"
Valeria Baltika D'Costa
Lib 15h30'48" -18d37'58"
Valeria Bryant
Ari 2h51'2" 28d3'1"
Valeria Calles Guerra
Cap 21h16'12" -20d3'0"
Valeria De Francesco
Her 18h24'13" 29d24'11"
Valeria Demont
Her 16h48'47" 41d16'28"
Valeria Johnson
Uma 8h39'31" 50d23'45"
Valeria K. Douglas (Earth Angel)
Cyg 20h53'9" 30d51'36"
Valeria Kunz
Crb 16h13'34" 26d39'57"
Valeria La Camera
Umi 12h32'28" 89d1'41"
Valeria la cigno
Cyg 20h20'51" 56d50'11"
Valeria Langley
Uma 9h26'0" 46d57'19"
Valeria Lesma
Cas 2h8'28" 59d8'34"
Valeria Letizia
Lyr 18h59'43" 45d55'24"
Valeria Massari
Lyn 7h35'38" 37d41'18"
Valeria & Michel
Cas 23h5'10" 56d47'34"
Valeria Scarpa
Cyg 21h32'16" 42d15'40"
Valeria Stella & Altan
And 0h17'0" 46d1'23"
Valeria stellina della mia vita
Umi 14h9'2" 66d41'2"
Valeria Van Soelen
Crb 15h55'28" 32d13'1"
Valeria Vieira Dias
And 23h23'29" 42d8'50"
Valerian
Lib 15h49'30" -19d26'21"
Valerie
Sgr 18h29'23" -17d4'34"
Valerie
Aqr 23h10'7" -16d16'47"
Valerie
Aqr 22h58'18" -7d45'34"
Valerie
Lib 15h17'56" -9d1'23"
Valerie
Sco 16h10'31" -11d7'40"
Valerie
Umi 4h12'35" 88d49'42"
Valerie
Lib 14h48'40" -3d46'4"
Valérie
Uma 8h51'33" 60d13'53"
Valérie
Sgr 18h18'55" -27d27'14"
Valerie
Psc 1h7'42" 3d1'5"
Valerie
Sgr 18h38'52" -32d32'14"
Valerie
And 23h15'52" 43d47'42"
Valérie
Cyg 20h30'27" 41d30'46"
Valérie
Cas 0h46'0" 52d49'51"
Valerie
And 23h13'32" 48d16'58"
Valerie
Cyg 19h44'19" 31d51'32"
Valérie
Lyn 8h16'24" 36d53'54"
Valerie
Gem 6h44'19" 31d28'29"
Valerie
Gem 7h44'49" 31d40'13"
Valerie
Lyn 7h38'55" 37d43'37"
Valerie
And 2h26'16" 42d26'17"
Valerie
And 0h47'9" 39d10'39"
Valerie
Lyr 19h16'56" 29d40'39"
Valerie
And 0h14'33" 29d58'9"
Valerie
Del 20h40'16" 15d52'36"
Valerie
Leo 10h45'24" 15d46'30"
Valérie
Cnc 8h27'3" 15d30'29"
Valerie
Aqr 21h7'43" 1d45'26"
Valerie
Aqr 21h36'6" 2d27'6"
Valerie
Psc 23h45'8" 5d0'47"
Valerie
Psc 1h29'22" 17d43'18"
Valerie
Psc 0h24'41" 15d27'56"
Valerie
Vir 13h26'39" 13d10'27"
Valerie Amber
Ari 3h15'37" 29d37'6"
Valerie Amelia Scott
Cnc 8h14'16" 12d32'24"
Valerie Anais Randall
Ori 5h54'18" -0d6'59"
Valerie and James Cox
Cyg 20h47'19" 33d45'20"
Valerie and Robert Rabinowitz
Leo 11h5'25" 21d20'21"
Valerie Andrews
Leo 9h35'32" 29d53'25"
Valerie Angela LeBlanc
Psc 0h42'27" 10d32'6"
Valerie Ann
Ori 4h48'34" 13d43'18"
Valerie Ann Cegelka
Tau 5h58'10" 24d56'0"
Valerie Ann Egley
Aqr 20h57'11" 1d5'39"
Valerie Ann Etcheverry
Ari 2h15'42" 23d28'36"
Valerie Ann Flores Delgado
Aqr 21h52'15" -2d32'41"
Valerie Ann Gonzales
Dra 19h45'51" 61d1'8"
Valerie Ann Haase
Cap 21h36'30" -22d13'1"
Valerie Ann Hernandez
Lyn 7h42'58" 55d23'49"
Valerie Ann Lawrence
Cas 1h37'59" 62d12'54"
Valerie Ann Stowe
Aqr 21h38'48" 1d21'34"
Valerie Anne Imperial Chappell
And 23h57'43" 39d52'51"
Valerie Anne Maher
And 23h14'50" 42d20'15"
Valerie Anne McGraw
Cyg 19h55'26" 32d17'30"
Valerie Antoinette
And 22h58'11" 48d11'40"
Valerie Arklow Kern
Sgr 19h25'8" -15d54'21"
Valerie Barzetti
Ori 5h32'28" 3d23'2"
Valerie Bene Mater
Ori 5h44'1" 5d29'37"
Valérie Beneventi
Sco 16h15'6" -28d22'41"
Valerie Berrios
Vul 20h37'22" 23d20'38"
Valerie Blondel
Cyg 20h51'4" 34d50'46"
Valerie Blu
Vir 13h57'43" -11d38'6"
Valerie Boucher
Uma 10h37'29" 58d50'53"
Valerie Bronson
Gem 7h9'15" 25d50'5"
Valerie Brown
Ori 4h49'57" 14d47'41"
Valerie Burnier
Umi 17h2'55" 87d47'25"
Valerie C. Gilroy
And 0h15'46" 45d26'50"
Valerie Carney's Amazing Star
Sgr 18h12'37" -20d22'8"
Valerie Ceccherelli Walter
Vir 13h21'54" -6d30'37"
Valérie Chatel
Vir 11h48'59" -0d24'40"
Valerie Chavez
And 1h3'47" 42d4'54"
Valerie Chiarizzi
Uma 9h18'19" 68d2'52"
Valerie Chiyo Jaffee
Sco 16h49'7" -27d24'43"
Valerie Christine
Uma 11h59'7" 61d41'42"
Valerie Clare Allen
And 2h19'33" 46d2'30"
Valerie Collins
Ori 6h18'30" 8d12'4"
Valerie Dawn Burdick
Psc 0h49'8" 12d57'7"
Valerie Diane Carlisle
Ari 3h7'39" 28d36'42"
Valerie E. Bowen
Aqr 22h2'18" -15d0'15"
Valerie E Weisner
And 1h5'36" 39d34'20"
Valerie Eiffes
Cas 1h32'5" 61d15'11"
Valerie Elizabeth Young
Cnc 9h4'54" 31d13'49"
Valerie Ellen
Gem 6h37'0" 20d8'43"
Valerie Erickson
Leo 10h18'46" 6d53'9"
Valérie et Oscar
And 1h9'41" 48d30'23"
Valerie F. Reed
And 2h7'41" 41d46'45"
Valerie Faltas
Cyg 21h30'37" 36d44'15"
Valerie Frances Brady
Umi 15h9'38" 72d6'49"
Valerie Franklin
Cyg 20h0'7" 40d7'51"
Valerie Garcia-Alston
Cnc 9h13'32" 11d32'41"
Valerie Gilbert - Galaxy's Best Mom
Psc 0h42'5" 4d33'57"
Valerie Goodvin
Ari 2h6'6" 13d26'3"
Valerie Grace Dunnivan
Crb 16h20'51" 27d37'57"
Valérie Gravel
And 23h31'0" 39d7'29"
Valerie Gray
Cas 0h0'43" 57d1'30"
Valerie H Legard
And 0h32'9" 44d23'10"
Valerie Hakam Sacay
And 1h17'57" 43d0'25"
Valerie Hartmann
And 2h24'46" 42d36'11"
Valerie Hope Bennett
Leo 10h8'41" 17d26'37"
Valerie Hoskins
Sgr 18h1'59" -30d18'5"
Valerie Irizari
Aqr 22h13'39" -0d29'7"
Valerie Isabel Valentin
Sgr 19h45'15" -14d3'50"
Valerie J. Basile
Ari 2h50'51" 27d58'49"
Valerie J. Bothun
Cet 1h18'44" -21d17'35"
Valerie J. Dow
Cas 1h27'37" 62d15'37"
Valerie J Franklin
Cas 23h10'6" 55d25'22"
Valerie Jean
Tau 4h42'44" 27d12'1"
Valerie Jean
Lyr 18h44'9" 40d59'4"
Valerie Jean 9-4-1952
Uma 11h11'1" 28d51'13"
Valerie Jean Goodreau
And 2h29'22" 39d7'54"
Valerie Jean Niece
Aqr 22h14'15" 1d0'4"
Valerie Jean Paiva
Uma 9h22'39" 55d21'23"
Valerie Jean Simmons
Vir 15h10'14" 2d54'55"
Valerie Jermy
Cas 0h41'19" 62d42'49"
Valerie Jo Brandt
Cnc 8h19'25" 13d42'35"
Valerie & Joseph Callahan 6/23/02
Uma 10h36'44" 45d38'57"
Valerie Joy
Vir 13h18'4" 5d16'34"
Valerie Joy
Crb 15h43'42" 26d40'30"
Valerie Joy McClune
Aqr 20h55'35" -10d38'46"
Valerie Jurries
Cas 0h33'7" 63d32'2"
Valerie Kabouris
Cap 21h16'55" -26d57'47"
Valerie Kagan & Bill Coon
Cyg 20h2'4" 37d13'54"
Valérie Kendall
Psc 23h34'42" 5d50'38"
Valerie Kines
Cas 0h47'42" 47d41'29"
Valerie L. Britton
Tau 4h31'1" 20d47'22"
Valerie L. Kelesyan
Tau 4h7'0" 8d23'18"
Valerie L. Mason
And 2h10'18" 42d38'26"
Valerie L. Morgan
Sgr 19h14'24" -29d22'29"
Valerie L. Phillips
Sco 16h11'16" -16d16'13"
Valerie Lantier
Crb 15h58'37" 26d32'12"
Valerie LaRoque
Cas 2h27'47" 62d0'6"
Valerie Laspino
Sco 17h45'39" -34d2'43"
Valerie Lauren Gale
Cas 23h34'25" 59d38'21"
Valerie Laverty
Crb 15h41'51" 34d5'59"
Valerie Lee Meals Duff "1936-1991"
Leo 9h30'36" 30d32'7"
Valerie Lukens
Cas 22h59'38" 53d32'23"
Valerie Luthwood
Cas 1h58'25" 63d37'36"
Valerie Lynn
Cap 20h35'26" -10d32'15"
Valerie Lynn Buccino
Psc 1h25'37" 27d7'21"
Valerie Lynn Consoer
And 1h41'43" 44d42'51"
Valerie Lynn Fiore
And 23h43'30" 34d44'3"
Valerie Lynn Lawson
And 0h45'37" 36d26'28"
Valerie M. Hill
Uma 13h53'35" 57d1'0"
Valerie "Ma Puce" Grenon
Cas 23h34'49" 55d33'48"
Valerie Mae Smith
And 23h29'56" 48d15'26"
Valérie & Manuel
Uma 12h3'32" 62d17'30"
Valerie Marie Catherine Driscoll
Ari 3h14'44" 23d18'31"
Valerie Marie Hudson-Roberts
Sgr 19h42'12" -16d5'56"
Valerie Martinez
And 23h53'58" 36d29'34"
Valerie Mary (Duckie)
Cru 12h39'50" -60d52'32"
Valerie Mary O'Shea Edworthy
Lyr 18h23'42" 32d25'33"
Valerie & Michael Berger
Eri 4h5'33" -29d55'7"
Valerie Mills
And 0h33'49" 44d5'21"
Valerie Monroe
And 23h35'22" 35d32'38"
Valerie Monroe
Com 12h23'44" 22d15'19"
Valerie Moran
Cas 23h50'46" 58d42'6"
Valerie Morgan
Ari 3h22'28" 25d4'51"
Valerie Morningstar
Com 13h1'46" 17d15'46"
Valerie Nicole
And 0h37'43" 39d40'25"
Valerie Nicole Jones
And 1h34'52" 41d26'29"
Valerie Nicole Scafidi
Lyn 7h23'20" 54d15'6"
Valerie Orsini Loves Jordan Lander
Uma 11h41'47" 38d11'41"
Valerie O'Steen
Aqr 21h10'52" -14d28'55"
Valerie Patricia Gibson
Cnc 8h12'56" 9d57'24"
Valerie Preston
Aqr 21h3'51" -14d22'14"
Valerie Quinn
Aqr 23h23'55" -17d51'8"
Valerie R. Smith
Sgr 17h57'29" -17d4'47"
Valérie Racine
Cap 21h56'58" -8d49'28"
Valérie & Ramez Paradise Island
Psc 1h43'44" 9d42'10"
Valerie Ramos
Aqr 22h30'52" -22d33'28"
Valerie Reno
Uma 10h45'47" 64d52'52"
Valerie Reuter
And 0h22'27" 46d13'59"
Valerie Ricapito
Uma 12h1'48" 33d59'41"
Valerie Rice Enderlin
Sgr 20h15'45" -44d32'21"
Valerie Riggins
Ori 6h17'26" 14d37'22"
Valerie Rose
Psc 1h25'0" 8d6'23"
Valerie Rose
Ori 5h35'54" -1d44'28"
Valerie Rose Shill
Sgr 19h9'8" -18d16'50"
Valerie S.
Sgr 18h45'46" -29d1'5"
Valerie S. Chavez
Psc 1h27'31" 15d9'8"
Valerie Sauber
Cas 1h21'3" 64d58'24"
Valerie Schoenmann
Ari 2h5'1" 19d58'21"
Valerie Shaughnessy
And 0h41'56" 27d58'3"
Valerie Shawhan's Night Light
Sgr 18h25'45" -17d55'40"
Valerie "Spirit" Cavallo
Lyr 19h1'28" 42d52'54"
Valerie Staunton
And 2h8'31" 46d42'1"
Valerie Stayskal
Cyg 20h9'57" 35d6'21"
Valerie Swingler
Ari 2h20'10" 22d2'52"
Valerie & Tim McCulloch* Forever
Sgr 18h15'37" -19d2'41"
Valerie [trans.: Paul's Star]
Sgr 18h51'39" -15d57'9"
Valerie Uprichard
And 23h55'50" 45d33'16"
Valerie Vantrease
Mon 6h55'17" -0d52'31"
Valerie Vega
Ari 2h45'24" 28d0'45"
Valerie W. Hurley
Cyg 20h24'27" 43d59'4"
Valerie Waite
Cnc 8h14'22" 16d19'15"
Valerie Webster
Tau 4h49'18" 19d40'24"
Valerie Williams, My Shining Star
Cap 21h23'10" -23d40'8"
Valerie Wrba
Ant 9h40'25" -25d15'37"
Valérie, 09.01.1981
Mon 6h28'55" -5d1'40"
Valerie, Emma, and Mack
Uma 8h53'21" 70d29'55"
Valerie, Joy & Nadine
And 23h40'18" 41d51'42"
Valerie0214
Uma 10h30'36" 55d54'23"
valerie0302
Psc 1h20'14" 14d34'30"
Valériejolie
Cap 20h15'25" -22d9'22"
Valeriemily
Cnc 8h34'0" 28d47'11"
Valerie's Heaven
Cru 12h35'45" -58d20'51"
Valerie's Home Star
Cnv 12h24'23" 48d55'38"
Valerie's Love
Vir 14h20'57" 1d12'57"
Valerie's Place in Heaven
Sgr 20h4'45" -11d57'12"
Valerie's Star
Ari 2h49'47" 14d28'55"
Valeriestar
Umi 15h28'20" 86d5'7"
Valerine
Uma 9h46'32" 66d5'11"
Valerine Perez
Cap 20h47'54" -22d52'18"
Valerio
Cam 4h31'0" 55d36'49"
Valerio
Cam 4h11'3" 57d5'28"
Valerio
Boo 14h57'10" 50d57'9"
Valerio e Francesca
Lyn 6h47'45" 60d49'39"
Valerio Mattiuz
Cra 18h48'55" -41d36'54"
Valerio Valori
Aur 5h32'42" 32d34'54"
Valerion Majorus
Sco 16h10'10" -17d25'8"
Valeriy Karoglanov
Sco 16h15'51" -13d16'45"
Valeriy L. Makarov
Cam 5h16'6" 68d31'8"
Valery Brandon Mitros
Per 2h54'56" 54d34'42"
Valery Espinoza
Aqr 22h27'58" -0d29'30"
Valery Kulikova
And 1h8'32" 46d0'24"
Valéry Lagassé
Cas 0h30'5" 61d57'45"
Valery Lavrik
Aqr 20h44'38" -12d7'11"
Valery Mints
Uma 9h18'23" 61d14'15"
Valeska
Tau 4h4'5" 6d5'41"
VALETA
Vir 13h19'30" -12d37'40"
Valette
Peg 22h23'11" 19d31'26"
Valgal 31
Crb 15h44'11" 25d43'52"
Valhalla
Ori 5h55'56" 18d45'45"
Vali
Aqr 23h51'40" -4d5'55"
Vali pro semper
Dra 17h25'33" 56d58'35"
VALIA
Lyn 7h34'1" 37d24'28"
Valice
Sgr 19h27'9" -13d21'16"
Valika Anyucika
Lyr 18h16'47" 30d46'7"
Valileo
Eri 3h24'40" -21d42'45"
Valintina
Ori 5h56'48" 21d26'16"
Valiollah
Cnc 8h34'42" 14d38'31"
Valkyrie
Ori 5h59'7" 21d55'16"
Valleen Morris
Sgr 18h9'57" -18d0'18"
Vallen Zukowski
Cap 20h13'50" -24d23'52"
Vallerie
And 0h52'3" 40d53'39"
Vallerie D. Wagner
Ori 5h40'8" 8d55'35"
Vallerie Gabriella Djojoseparto
Leo 11h5'52" -1d14'44"
Vallerie Hamel
Dra 19h14'31" 65d44'8"
Vallery Allord
Lmi 10h30'57" 36d14'40"
Valli Kathleen Davis
Cyg 20h11'26" 57d19'54"
Valli Laub
Uma 13h31'41" 54d7'36"
Valli Rosa
Ori 5h56'55" 11d22'20"
Vallie Robinson
And 1h29'49" 43d3'26"
Vallones' 50th Anniversary
Cnv 13h21'52" 29d14'58"
Vallorrie Ann Clark
Cas 0h38'51" 53d7'54"
Vally 831
Ari 2h21'35" 14d13'9"
Valma
Dra 19h0'39" 52d17'48"
Valma Joan Crow
Tau 3h37'17" 10d13'59"
Valorie A Hennessey & Loved Ones
Cyg 21h44'16" 44d38'47"
Valorie and Tracie
Uma 10h13'5" 64d35'7"
Valorie Annette Murray
And 0h32'45" 33d7'21"
Valorie Nicole
Sgr 19h17'37" -22d17'43"
VALPS
Tau 4h34'55" 22d13'58"

VALRED
Dra 19h38'27" 74d58'49"
Valrie A. Hall
Lmi 9h40'19" 33d18'44"
Val's Love
Uma 8h35'18" 49d50'14"
Val's Rose
Per 4h6'0" 41d47'44"
Val's Star
Cas 0h11'11" 55d3'45"
* Valter da Silva *
Ori 5h34'0" -0d44'56"
ValVessian
Aqr 21h0'16" -10d21'55"
Valya Kukhar
Ori 5h28'45" 7d23'31"
Vályi László
Tau 3h32'38" 10d53'12"
Van
Cap 21h27'47" -24d16'34"
Van Allen Dawes
Sco 17h3'8" -35d41'12"
Van Buskirk's Retreat
Lyn 6h48'45" 53d12'44"
Van C Pressley
And 0h16'19" 26d58'19"
Van Demark
Uma 9h56'26" 70d35'36"
van Drünen, Arnd
Uma 10h49'58" 56d47'11"
van Eymeren, Manfred
Ori 5h40'7" -2d28'30"
Van Farrier
Cam 4h14'59" 66d50'37"
Van & Georgia Stephens
Crb 15h17'4" 27d17'57"
Van & Lee
Umi 16h9'43" 78d36'59"
Van Nguyen
Lib 15h45'36" -5d22'24"
Van Scott Howard
Her 17h18'4" 26d34'24"
Van Slyck
Uma 11h6'16" 59d21'13"
van Stippent, Leo
Ori 6h8'7" 13d25'54"
Van Terrance Drake
Uma 13h47'5" 58d27'59"
Van Thuy Duong
Sgr 18h39'53" -28d24'49"
van Wahden, Adolf
Ori 4h57'16" 15d25'30"
Van Zan
Sco 17h2'11" -39d28'34"
VANA GKOUNTOUVA
Aqr 22h24'12" -14d16'37"
Vana Stamataki
Lib 15h16'17" -16d48'50"
Vana, Sven
Uma 11h30'30" 34d29'40"
Vanadiya
Cam 5h3'9" 57d28'34"
VanAmes
Lac 22h26'27" 49d7'55"
VanAndAubreyForverWritte
nInTheStars
Tau 4h40'34" 24d54'29"
Vance E. Holmes
Aqr 21h18'25" -9d42'59"
Vance Family Star
Uma 10h43'15" 54d57'44"
Vance Hilderman
Aqr 22h22'33" -22d45'50"
Vanda Waller Kenyon
Umi 15h23'33" 68d54'36"
Vandana Sonia Alam
Tau 5h9'42" 25d43'0"
Vandee's Star
Gem 7h20'38" 16d8'36"
Vandenberg's of Jeff Co
Uma 10h30'44" 40d11'6"
Vanderbilt Thompson, Jr. -
3-2-1945
Oph 17h10'41" -0d4'28"
Vanderveer-Gibson
Gem 6h56'24" 28d10'13"
Vandi 68
Cas 0h46'32" 60d44'48"
Vandora
Lyn 8h25'33" 56d49'5"
Vane & Freddy
Peg 22h5'51" 9d57'4"
Vaneese
Sco 16h13'42" -8d42'40"
Vaneeza Rupani
Cap 20h53'5" -26d5'31"
Vanesa Alvarado
And 0h18'55" 31d52'49"
Vanesa Marie
Crb 15h40'44" 27d32'51"
Vanesa Sotomayor
Cnc 7h59'36" 11d41'59"
Vanese
Sco 16h9'31" -14d59'27"
Vanessa
Lib 15h56'16" -11d17'17"
Vanessa
Cap 20h27'12" -12d1'6"
Vanessa
Aqr 23h3'5" -10d50'33"
Vanessa
Lib 15h2'51" -1d10'22"
Vanessa
Vir 14h10'50" -13d10'33"
Vanessa
Aqr 22h27'3" -2d56'18"
Vanessa
Dra 16h56'15" 69d55'5"
Vanessa
Lyn 6h21'59" 60d43'36"
Vanessa
Sco 16h36'26" -31d30'42"
Vanessa
Sco 17h54'52" -36d45'43"
Vanessa
Sgr 18h58'27" -34d9'22"
Vanessa
Leo 10h12'16" 14d10'22"
Vanessa
Vul 20h34'28" 23d47'55"
Vanessa
Vul 19h51'1" 22d47'4"
Vanessa
Cnc 8h26'57" 17d58'26"
Vanessa
And 0h28'1" 28d16'8"
Vanessa
And 2h21'22" 41d25'45"
Vanessa
And 0h43'36" 41d29'37"
Vanessa
And 0h46'16" 41d26'12"
Vanessa
Cyg 20h2'38" 32d18'38"
Vanessa
Her 16h59'24" 30d55'7"
Vanessa
Gem 7h25'12" 33d1'31"
Vanessa
Psc 1h31'11" 33d2'19"
Vanessa
And 23h23'44" 48d11'10"
Vanessa
Cas 23h27'2" 52d28'48"
Vanessa
Uma 10h24'21" 46d35'25"
Vanessa
Crb 16h11'11" 39d21'15"
Vanessa 11/23
Lib 15h57'33" -18d34'23"
Vanessa A Cadden
Ari 3h16'19" 27d32'22"
Vanessa Adam
Cas 0h29'28" 61d28'44"
Vanessa Adofowaa
Debrah-Braxton
Cas 1h5'21" 60d56'33"
Vanessa Agassiz
Sgr 19h42'56" -14d6'6"
Vanessa & Alan Day Star
Ori 5h26'34" -4d30'38"
Vanessa Alejandra
Ari 1h56'5" 22d54'34"
Vanessa Alexandra
Gonzalez
And 0h48'6" 32d33'55"
Vanessa Alfonso
Gem 6h30'1" 14d59'29"
Vanessa Amanda Rose
Gray
Cnc 8h59'54" 14d31'48"
Vanessa And Adam's Star
Sgr 18h49'55" -18d6'45"
Vanessa and Dennis
Forever
Cyg 19h44'31" 44d30'40"
Vanessa and Glenn
Cyg 20h44'11" 45d45'26"
Vanessa Ann Barcello
Leo 9h25'18" 11d13'13"
Vanessa Ann Rogers
Vir 12h38'32" -3d28'51"
Vanessa Ann Viiberg
Her 17h27'4" 24d39'32"
Vanessa Antoinette
Aql 19h57'33" 12d55'20"
Vanessa Aranigo Ceballos
Cam 6h24'5" 65d2'23"
Vanessa Averhoff
Psc 0h26'29" 3d52'53"
Vanessa Baulny
Uma 10h58'2" 65d15'37"
Vanessa bear
Uma 10h23'51" 44d16'16"
Vanessa Beauty
Cas 0h46'16" 65d32'23"
Vanessa Bibz Sebastian
Mon 7h23'0" -5d34'27"
Vanessa Blackman
Aql 19h47'33" 4d55'24"
Vanessa Bolognino
Cyg 19h31'19" 31d54'2"
Vanessa Briggs Mother of
L. Snipes
Uma 11h45'55" 37d23'52"
Vanessa Caitlin Salter
Vir 12h38'35" 2d39'26"
Vanessa Cantin
Umi 15h13'34" 89d6'58"
Vanessa & Cassandra
Nevarez
And 0h24'13" 29d12'5"
Vanessa Celeste
And 23h25'26" 47d8'4"
Vanessa Chan
Leo 11h25'12" 7d13'6"
Vanessa Chapman
Cru 12h33'38" -60d32'8"
Vanessa "Charley"
Crb 16h6'46" 36d31'50"
Vanessa & Chris
Gem 7h18'13" 34d10'31"
Vanessa Claire
And 0h39'8" 22d0'39"
Vanessa Claudia Delgado
Cas 23h50'20" 52d46'12"
Vanessa Cornet
Lib 15h55'48" -17d32'33"
Vanessa Dale Brown
Uma 11h28'2" 38d35'21"
Vanessa D'Auria
And 2h17'13" 49d53'20"
Vanessa Dawn
Uma 10h40'59" 50d21'23"
Vanessa Dawn James
Lib 15h3'49" -26d11'17"
Vanessa Debra Grogan
Cyg 19h45'48" 55d46'0"
Vanessa & Dennis
Gniewosz
And 23h13'11" 51d22'48"
Vanessa Diagnora Villegas
Alfonzo
Psc 1h52'17" 6d13'33"
Vanessa Diana,
18.03.1981
And 23h7'34" 47d26'58"
Vanessa DiNoto
Gem 6h43'47" 20d30'33"
Vanessa & Doug
Psc 1h26'18" 18d15'51"
Vanessa Edwards
Cnc 8h34'7" 27d32'38"
Vanessa Elise Sanchez
Lyn 7h43'39" 42d32'50"
Vanessa Elizabeth Dahl
Psc 1h26'55" 5d48'16"
Vanessa Elizabeth Perry
And 23h33'32" 45d13'15"
Vanessa Elyse
Cnc 8h0'46" 18d42'43"
Vanessa Ennes
And 23h23'23" 46d23'58"
Vanessa Erin Vanhoogen
And 1h33'42" 39d8'10"
Vanessa Fae
Ori 6h19'22" 15d39'39"
Vanessa Faith
Lib 15h7'6" -7d0'0"
Vanessa Farrera
Cap 20h15'42" -20d47'57"
Vanessa Fay Lytton
Uma 13h46'41" 53d29'55"
Vanessa Ferreira
Tau 4h58'22" 22d27'33"
Vanessa Fink
Sgr 19h52'40" -40d21'6"
Vanessa Frazier
Uma 13h6'21" 57d38'28"
Vanessa Furey
Uma 14h18'31" 58d33'54"
Vanessa Furfaro
Umi 17h22'21" 77d49'29"
Vanessa G Garcia
And 23h39'47" 33d49'47"
Vanessa Gabaldon
Psc 1h17'52" 19d31'49"
Vanessa Giancola
Gem 7h47'54" 32d39'45"
Vanessa Gibbs
And 0h38'18" 33d44'5"
Vanessa "Goofy" Garland
Leo 10h12'40" 17d43'16"
Vanessa Hermoso
Carpintero
Lac 22h23'32" 54d53'1"
Vanessa Hernandez
And 1h21'39" 36d40'56"
Vanessa Hernandez
Lyn 7h56'48" 33d15'37"
Vanessa Hetler
Cas 0h57'3" 63d22'24"
Vanessa Hunziker
Cas 23h29'38" 59d59'48"
Vanessa Isabel
Vir 12h40'16" 2d1'31"
Vanessa Ivy Del Rio
Gem 7h0'42" 14d31'20"
Vanessa "J&S" Linton
Lmi 10h45'44" 28d2'44"
Vanessa Jean Thompson
Psc 0h43'37" 18d9'46"
Vanessa Jean Trevino
Boo 15h4'39" 48d6'1"
Vanessa J.M.E.C.
Cleveland
Lyn 6h26'10" 57d42'20"
Vanessa Joyce Michelson
Sgr 18h31'47" -16d39'25"
Vanessa Kay
Lib 14h51'47" -9d47'23"
Vanessa Lee Cannon
Sgr 18h40'8" -30d10'6"
Vanessa Legrand
Cep 21h39'47" 58d7'38"
VANESSA LEIGH HOR-
TON
Aqr 22h17'58" 0d48'7"
Vanessa Leigh Johnson
Vir 13h20'27" -5d41'0"
Vanessa Lespagnol
Cas 23h53'18" 53d55'33"
Vanessa Lopez
Crb 15h50'57" 36d6'53"
Vanessa Loreto
Uma 11h23'12" 58d31'14"
Vanessa Louise
And 0h57'14" 40d50'27"
Vanessa Louise Crawford
Peg 21h58'55" 34d56'27"
Vanessa Lyeve Castillo
Sco 17h40'42" -40d39'7"
Vanessa Lyn Simiola
Sco 17h54'35" -42d39'26"
Vanessa Lyn Weaver
Aqr 22h25'40" -1d44'12"
Vanessa Lynn Dempsey
Ori 6h7'35" 17d44'47"
Vanessa Lynne Bell
Leo 10h44'27" 17d24'39"
Vanessa Macias
Leo 11h2'39" 9d12'41"
Vanessa Macias
Aqr 22h30'5" -2d3'37"
Vanessa Margaret Baden
Psc 23h49'10" 5d31'55"
Vanessa Marie Detwiler
Leo 11h6'33" 16d58'57"
Vanessa Marie Romero
And 0h18'23" 32d1'39"
Vanessa Marie
Vasconcelos
Cru 12h7'28" -61d8'50"
Vanessa Marlene Gomez
Ser 18h35'35" -0d55'10"
Vanessa Marquez
Vir 12h18'37" 3d46'0"
Vanessa Marrero
Lyn 8h44'56" 45d14'10"
Vanessa Marroquin
Cap 21h41'42" -20d7'49"
Vanessa Mei Wong
Vir 12h26'49" 11d57'11"
Vanessa Melanie Vargas
Sco 16h7'8" -10d54'33"
Vanessa meus Afflatus
Sco 16h13'36" -18d36'10"
Vanessa Meyer's Delicious
Star
Lib 15h17'12" -3d53'39"
Vanessa "Mi Amor" Araya-
Trojik
Col 6h25'20" -33d38'29"
Vanessa Michelle
Huntsman
Cam 4h34'52" 77d4'31"
Vanessa Modafferi
Leo 10h58'51" 15d35'43"
Vanessa Molina
Uma 10h40'39" 48d27'12"
Vanessa mon amie pour
l'infini
Tau 5h8'6" 27d55'51"
Vanessa Muhle-pela
Uma 8h57'44" 67d18'40"
Vanessa Muller
Psc 23h36'45" 2d52'59"
Vanessa Mulrenan
Cas 0h34'57" 61d36'12"
Vanessa Najor
Uma 9h20'16" 66d43'24"
Vanessa "Nessie"
Getelman
Cas 1h14'30" 55d52'36"
Vanessa Newton
Uma 10h48'30" 52d51'36"
Vanessa Nicole Butler
Cas 23h30'35" 52d20'49"
Vanessa Nicole Pantano
Lib 15h46'23" -6d40'18"
Vanessa Nicole Sapien
Crb 16h12'21" 26d41'41"
Vanessa Nicole Sukay
Cnc 9h17'13" 12d13'36"
Vanessa Nikole Phillips
And 0h38'31" 27d2'28"
Vanessa Noel
Lyn 6h54'13" 59d59'54"
Vanessa Noelle Show
Aqr 22h36'24" -21d6'40"
Vanessa Nu Nu
Crb 16h9'44" 35d15'56"
Vanessa Papilio
Lyn 7h27'33" 57d25'53"
Vanessa & Pavel Forever
Cyg 20h11'54" 52d55'48"
Vanessa Penelope Hoene
Gem 7h28'4" 32d26'21"
Vanessa Persichetti
Cas 23h4'54" 53d48'21"
Vanessa Peter
Umi 13h33'58" 87d15'52"
Vanessa Petrucci
And 0h18'39" 28d2'38"
Vanessa Priebe
Gem 7h4'4" 31d47'41"
Vanessa Rae
Ari 2h10'51" 23d40'44"
Vanessa Renay
Lmi 10h49'15" 29d28'26"
Vanessa Renee Maestas
Cap 20h56'3" -21d5'40"
Vanessa Rentsch
And 2h12'37" 41d49'35"
Vanessa Reza
Cap 21h31'43" -16d31'41"
Vanessa Rice
Ari 2h47'45" 27d20'27"
Vanessa Rojas
Cas 23h41'2" 52d35'5"
Vanessa Rose Smith
Cas 0h18'55" 59d23'6"
Vanessa Roseann
Castania
And 0h51'1" 36d32'23"
Vanessa Ruiz
Com 13h8'23" 27d13'54"
Vanessa Shining Star
Vicente
And 0h34'9" 29d18'52"
Vanessa Solvay French
Cam 5h51'34" 57d55'1"
Vanessa "Squidgy" Skinner
Peg 21h59'27" 16d21'15"
Vanessa Starr Palm
And 1h52'11" 45d22'27"
Vanessa Szymanski
And 0h31'45" 43d8'6"
Vanessa Tel
Dra 18h56'28" 54d13'52"
Vanessa The 2nd
And 22h58'54" 39d34'59"
Vanessa Torres
And 2h17'32" 46d48'45"
Vanessa Torrijos
Leo 10h3'6" 20d36'45"
Vanessa Travis Nance
Ori 5h37'45" -0d58'23"
Vanessa & Tyrone Bowden
Cyg 19h49'3" 32d55'58"
Vanessa Urena
Aqr 23h3'37" -16d24'14"
Vanessa Valencia
Tau 4h25'6" 10d26'0"
Vanessa Vanlooy
And 0h32'1" 37d16'36"
Vanessa Varignana
Cas 23h33'36" 55d11'0"
Vanessa Venus Attard
Psc 1h33'42" 19d9'48"
Vanessa Vidal
Ari 3h6'58" 28d39'1"
Vanessa Willis
Vir 12h38'21" -9d0'5"
Vanessa Wonderful vb Nhu
Psc 1h12'23" 26d46'56"
Vanessa Wyffels
Lyn 8h58'52" 38d8'9"
Vanessa Yamileth
Cyg 21h47'46" 36d14'50"
Vanessa17
Cas 0h51'19" 57d17'28"
Vanessa17026jse
Tau 4h1'15" 0d28'38"
Vanessa.K.S
Ori 6h9'3" 8d32'24"
Vanessa's Angel
And 23h53'53" 41d56'16"
Vanessa's Dream
Peg 22h46'51" 8d23'33"
Vanessa's Eye in the Sky
Cyg 20h55'40" 40d38'8"
Vanessa's Star
Per 4h7'43" 34d21'53"
Vanessa's Wish
Cap 20h36'12" -22d7'26"
VanessaToriCamiChet#1
And 0h18'39" 30d22'15"
VanEssendelft
Psc 1h5'12" 4d34'33"
Vangel "Peter" Petrovitch
Uma 11h24'2" 31d31'5"
Vangie Balauras
Cas 23h30'9" 57d57'53"
Vangie Dexter My Baby
Girl
Leo 10h30'25" 21d48'49"
Vangie Rodriguez Griffin
Lyn 7h35'14" 49d24'23"
Vangilisti
Uma 11h7'27" 32d11'35"
VanHook's Glory
Gem 7h17'2" 25d45'4"
Vania Marie Lenora
Cnc 9h12'54" 10d59'40"
Vanialena Heardoon
Uma 10h26'46" 64d41'43"
Vaniesessa
Leo 10h11'45" 22d27'48"
Vanilla
Pup 7h51'19" -20d32'55"
Vanilla Bear
Uma 13h32'55" 55d26'15"
Vanilla Cookie in the Sky
Sco 17h9'12" -45d33'26"
Vanilla Gorilla
Cap 21h37'54" -18d49'15"
Vanille Dumont
Psc 1h0'34" 24d4'28"
Vanina Le Gall
Tau 3h24'11" 15d17'15"
Vanita
Cnc 8h46'48" 16d55'49"
Vanita Emavis Little-Hunt
Sco 16h12'25" -10d33'44"
Vanity
Cap 20h43'55" -19d30'9"
Vanity
Peg 22h32'34" 30d37'5"
vanity hacegaba
Tau 4h44'5" 25d38'22"
Vanja
Uma 9h59'36" 43d51'24"
Vanja Petrovic
And 23h40'53" 50d5'49"
Vanja Raznatovic
Her 17h52'51" 24d49'41"
Vanja Sutter
Aur 5h1'46" 46d11'31"
Vanne and Ryan stars for-
ever
Sge 19h50'15" 16d58'22"
Vannessa Dorantes
And 1h5'25" 45d45'48"
Vannessa Urich
Cas 0h44'10" 65d35'42"
Vannie Foris
Cnc 8h9'55" 12d12'11"
Vanny
Cas 0h51'25" 57d21'10"
VANORE
Tau 3h51'40" 14d47'48"
Van's Star
Uma 11h4'50" 35d41'29"
VanShu
Uma 13h8'45" 60d30'55"
Vanthana
Aqr 22h19'10" -1d33'47"
Vanya
Tau 4h45'38" 24d18'50"
Vanya Tarloyan
Gem 7h29'55" 14d57'3"
Vanyel
Leo 9h45'3" 7d40'25"
Vanza Attaviana Lazet
Aqr 21h37'33" -3d38'0"
Vaquero
Tau 3h46'8" 11d25'22"
Var
Cnc 8h10'44" 10d44'5"
Vardaan
Col 5h50'20" -35d3'30"
Varga Izabella
Sgr 18h27'25" -15d58'51"
Vargamania
Tau 5h29'48" 25d0'51"
Vargas-Esquer
Ari 3h5'17" 18d21'51"
"Vargie"
Cep 22h57'38" 65d29'26"
Vári Anita Drága feleségem
Sco 17h37'43" -41d21'3"
Varjú Krisztina
Uma 10h3'46" 59d24'56"
VARL 2K5
Tau 4h39'26" 17d7'53"
Varner
Uma 9h55'50" 58d20'44"
Varonica Lea
Cas 0h16'2" 57d26'22"
Varsha
Vir 12h38'45" -10d48'9"
Vartanush
Ori 4h51'43" 14d52'42"
Vartouhi
Aqr 20h59'29" 1d21'59"
Varvara Popovich
Umi 15h23'18" 71d29'19"
Vasantha Gudiwada
Uma 10h7'28" 49d49'34"
Vasanthakumari
Aqr 22h7'57" 0d22'13"
Vasco Maria Nino
Psc 23h54'55" 0d59'54"
Vashali Shroff
And 23h30'42" 49d6'1"
Vashte Johnson
Uma 11h27'13" 48d15'50"
vasilami
Tau 5h40'10" 21d12'29"
Vasile
Cyg 19h56'23" 38d27'29"
Vasili Hatziris
Oph 17h54'41" -0d25'37"
Vasilia
Uma 10h32'12" 55d6'24"
VASILIKI
Gem 7h38'48" 14d2'23"
Vasiliki
Tau 3h53'5" 26d11'55"
Vasiliki Adamopoulos
Umi 15h58'43" 74d20'45"
Vasiliki Georgoulas
Cnc 9h9'20" 19d13'47"
vasiliki mou
Uma 11h27'44" 54d23'22"
Vasilios Nectarios Bratsis
Uma 14h1'40" 59d52'45"
Vasilios Thornton
Tau 3h44'20" 24d33'31"
Vasilord
Gem 6h6'36" 22d36'55"
Vaso Pantic Memorial Star
Cru 12h21'2" -57d29'57"
Vass Kata 30
Uma 13h19'24" 53d3'12"
Vassallo's Family Star
Uma 11h45'45" 61d50'29"
Vassilios D. Paschalis
Per 4h46'15" 46d47'57"
Vassilis Andronikidis
Uma 9h31'38" 44d2'30"
Vassilis Antzoulis
Uma 10h42'51" 40d40'50"
Vassilis Chimonidis
Uma 8h59'17" 59d1'36"
Vassilis Katsoufis
Uma 11h37'31" 65d0'55"
Vassilis Tsiganos
Uma 9h30'3" 46d21'33"
Vassilis Xenoyiannis
Uma 8h49'55" 70d39'7"
Vasso Theotoka
Uma 10h46'55" 40d3'30"
Vast Com - A.Valdez &
A.Large
Her 16h49'0" 30d55'51"
Vastame
Tau 3h27'1" 13d55'20"
Vater
Ari 2h37'48" 25d28'2"
Vater Bär
Uma 9h6'15" 56d4'3"
Vaughan
Cma 7h26'31" -29d5'0"
Vaughan Clift
Uma 10h44'34" 41d51'40"
Vaughan Thompson - The
Burleigh Star
Cru 12h8'47" -63d31'21"
Vaughan's and Jill's Star
Cyg 21h12'1" 47d1'14"
Vaughn Alexander Ehrgott
Her 17h35'44" 32d17'46"
Vaughn D Renfro Sr. "My
HERO"
Dra 18h49'48" 67d3'17"
Vaughn Dabney
Aqr 20h48'13" -7d35'4"
Vaughn Ezekiel Soltys
Umi 16h45'16" 78d30'40"
Vaughn Forster Myovich
Ari 2h14'58" 24d2'6"
Vaughn Magee Armstrong
Cep 3h47'59" 81d23'24"
Vaughn Mayson Young
Vir 13h22'53" 7d52'30"
Vaughn Premo
Gem 6h5'12" 24d1'34"
Vaughn, We Love You
Forever
Vir 13h31'21" 12d40'14"
Vaughn's Vision
Ari 3h26'5" 22d47'47"
Vauhnius
Dra 13h32'59" 66d40'0"
Vaultz
Her 16h14'27" 11d36'25"
Vazquez-Figueroa Family
Sge 19h34'48" 17d0'17"
Vazzana / Antinucci Sales
Maxumus
Cyg 21h44'21" 41d17'38"
VBN-G5
And 23h2'49" 48d21'20"
VC2 Steve
Cas 0h38'13" 65d19'11"
v-chan
Umi 14h13'47" 72d6'22"
V.C.S. The Butterfly
Ori 6h22'13" 14d4'59"
VCS-02171969
Lyn 8h55'7" 45d47'20"
VeachOrbet
Cam 4h33'42" 54d16'28"
Vearl Everett Ray
Sco 16h16'19" -10d39'58"
Vearnessa Rena'
Leo 9h46'2" 32d42'40"
Véber Andi
Uma 9h42'22" 44d17'4"
Vebri
Cam 4h34'48" 65d13'34"
Veca Meca
Cas 23h41'6" 51d50'3"
Veda
And 0h46'7" 35d21'37"
Veda Gerdes
And 23h1'50" 41d5'33"
Veda JoAnne Kelly
Cnc 9h14'30" 27d58'45"
Vedalia Alicia Lopez
And 0h22'4" 28d33'57"
Vedant
Lyr 18h19'34" 34d31'9"
Veda's Last Date
Gem 6h41'27" 25d39'40"
VedaSiddappa
Gem 7h3'58" 11d40'28"
Vedat Cakova
Sco 17h10'33" -43d18'44"
Vedder10
Cnc 9h6'36" 20d0'43"
Vedere Dominica
Ori 4h55'25" 10d4'8"
Vedette Valerie
Uma 10h2'27" 63d26'12"
Vedrana
Aqr 21h39'6" -1d48'27"
Veebs
Psc 0h39'27" 15d22'22"
Veel Sterkte Mi Amore
Peg 22h50'39" 20d43'43"
"Veena"
Uma 8h38'49" 66d7'0"
Veerco
Cyg 19h41'33" 36d12'47"
Veerle Olieslagers
Tau 3h37'49" 17d50'18"
vega
Lib 15h33'59" -23d53'32"

Vega Ament Psc 1h43'25" 24d14'36"
Vega Beal Cyg 20h50'22" 31d14'54"
Vegas Aqr 21h18'23" 1d56'13"
"Vegas Chance" S&D Vir 14h45'49" 2d36'49"
Vegas In The Spring Vir 14h24'19" 2d48'7"
Vegas Julien-Micheal Ryan Cnc 9h6'5" 27d26'40"
Vegas Moe Sgr 19h6'13" -27d27'57"
Vegas-Vallarta Ari 2h37'48" 26d2'35"
Vehbi Sehirli Uma 14h17'48" 60d8'47"
Veiner Uma 11h29'24" 59d56'43"
Veit, Klaus Uma 11h21'33" 48d47'37"
Vekar Star Lib 15h47'17" -7d12'51"
Velarie M Ruiz Lyn 6h27'37" 61d11'14"
VelaSphere Lyn 7h53'12" 42d41'0"
Velcro Sadighian Cyg 21h42'30" 30d7'29"
Velda Umi 13h51'32" 78d39'38"
Velda's Mom Uma 11h44'7" 40d18'20"
Velena Lib 15h45'21" -17d12'33"
Velga Lidums Brolis Uma 9h1'2" 64d17'46"
Vells Spell Boo 15h9'9" 13d36'31"
Velma Cas 1h5'0" 61d5'49"
Velma and Wayne's Forever Star Pav 20h43'57" -60d20'33"
Velma Ann Menghini Aqr 22h30'13" -0d53'8"
Velma Boykin Lyn 8h50'57" 45d47'23"
Velma Faye Cornwell Ori 5h38'22" 3d13'51"
Velma Hall Per 3h16'26" 41d31'50"
Velma May Lyn 8h31'59" 50d3'11"
Velma Steinman And 0h28'8" 43d58'49"
Velma Tweedie Cas 0h37'47" 63d49'49"
Velta May LaNotte Uma 8h51'26" 68d19'22"
Veltrup, Domenic Ori 5h32'21" 10d31'4"
VELVET Sgr 18h17'37" -22d56'1"
Velvet Puppy Cmi 7h30'34" 5d37'35"
Velveteen Rabbit Gem 6h58'31" 12d44'16"
Vemulakonda Viswanadham Sco 17h40'47" -37d11'16"
Vena Uma 9h6'0" 58d6'32"
Vendethiel Ithilwen Leo 10h11'48" 25d35'21"
Vendetta B Strickland Ari 3h21'47" 28d58'42"
Vendoria Ori 6h6'19" 19d51'18"
Venecia Yesenia Flores And 2h31'20" 49d12'53"
Vener Gem 6h9'56" 26d58'49"
Veneranda Amanda Parras Cas 2h11'18" 59d56'19"
Venesa and Jazmine Cyg 20h36'0" 34d21'53"
Venesio Agustin Flores Cep 22h48'39" 61d3'8"
Venetia "Blossom" Clare Johanna And 22h59'29" 47d21'28"
Venezia Lib 15h21'35" -27d11'38"
Veniamin Muzychenko Tau 5h43'25" 26d13'57"
Venice Earp Per 3h19'17" 46d34'40"
Venice Schoch Gem 6h48'17" 23d4'55"
Venire qui, la mia fede Umi 14h36'23" 74d55'12"
Venita Aqr 22h34'51" -1d26'35"
Venita Lyr 18h55'36" 35d54'26"
Venon H Ison Her 16h16'39" 24d27'6"
VentiSette* Gem 7h8'28" 33d7'48"
Venturi Jordan Peg 23h27'12" 12d21'7"
Venus Crb 15h49'10" 26d54'15"
VENUS And 0h45'13" 31d17'50"
Venus - Buena Ventura Oph 17h33'57" -0d23'23"
Venus Castellano Gem 7h44'56" 25d48'25"
Venus Evette Jones Cas 0h55'35" 50d15'21"
Venus Giavonni Vir 13h22'48" -13d22'47"
Venus Ho Uma 11h18'10" 63d38'1"
Venus Jade Eri 3h41'33" -1d44'12"
Venus Joy Estepa Sco 16h52'32" -44d27'24"
Venus & Kan And 23h19'29" 36d21'56"
Venus von Holger und Karen Uma 9h8'45" 50d4'42"
VenusC Leo 10h39'27" 7d42'1"
VenusHawk Cap 21h44'23" -11d45'30"
Venustas Aql 19h12'53" -0d24'18"
Venustas - Jessi's Star Boo 14h30'57" 52d27'15"
Venustas Leanne Cyg 19h43'25" 40d23'19"
venustas virgo Leo 10h48'30" 17d51'57"
Venustus Astrum Verle Lyn 6h45'49" 53d32'15"
Venustus Unus Aqr 22h28'49" -18d29'55"
Veon Cam 6h35'11" 73d43'37"
Ver Siempre Con Amor Tau 5h53'14" 25d8'39"
VERA Tau 3h36'38" 24d37'28"
Vera Gem 6h49'59" 25d25'56"
Vera Psc 1h27'47" 26d31'44"
Vera Vir 12h36'54" 3d10'13"
Vera Ori 5h9'2" 11d25'6"
VERA Her 16h43'36" 42d29'46"
Vera And 0h44'20" 40d11'57"
Vera Ori 6h3'56" -2d9'16"
Vera 23-04-1986 Ori 5h18'58" 1d38'47"
Vera Achana Cep 23h55'10" 71d9'44"
Vera & Alfred Vlautin Cnc 8h35'59" 24d8'26"
Vera Almgren Uma 11h20'55" 62d42'17"
Vera and Charles Berthalot Uma 8h46'15" 56d52'10"
Vera Ann Baird Gem 7h30'17" 23d35'37"
Vera Beatrice Grace Cross Leo 11h31'29" 21d18'17"
Vera Beth Peak Lyr 19h9'49" 41d34'35"
Vera Botting Lyn 8h37'18" 38d55'57"
Vera Bowling Hutton Ori 5h2'41" -0d44'15"
Vera Capps Oct 19h8'31" -76d10'36"
Vera Christensen And 0h51'4" 41d20'5"
Vera Conway Crb 16h3'14" 28d48'54"
Vera Diane And 2h21'12" 50d32'44"
Vera & Dirk Uma 8h46'3" 72d56'10"
Vera Dubach Her 17h28'46" 32d2'18"
Vera Eagan Crb 15h46'52" 25d43'17"
Vera Ellen Umi 14h52'5" 70d38'42"
Vera Fegley Vir 13h29'19" -6d2'57"
Vera Gunawan Aqr 21h2'31" -11d28'42"
Vera Hairabedian Gem 6h30'28" 26d15'58"
Vera I. Kunnath Mon 6h49'12" 8d21'0"
Vera J Lyn 9h1'41" 43d18'8"
Vera & James Evans Aql 19h3'10" -0d6'51"
Vera Jaunita Reed Cas 23h45'53" 54d54'47"
Vera & Jim Dingman Cyg 20h32'7" 32d57'41"
Vera K. Vieman Cap 21h56'14" -18d7'27"
Vera Kathleen Cornell Crb 16h10'7" 36d36'37"
Vera "Lalie" Price Cas 2h25'51" 65d43'7"
Vera Lanzieri Cam 5h23'23" 68d29'43"
Vera Le Quesne (née Miller) Cyg 20h4'40" 48d24'39"
Vera Lucia da Cunha Umi 17h9'8" 75d19'34"
Vera Lynne Harding Lmi 10h35'30" 34d13'39"
Vera Mae Peg 22h9'17" 30d47'26"
Vera Man Michaelson And 23h10'53" 43d9'57"
Vera Mary Jarvie Psc 1h1'28" 13d50'20"
Vera Mary, Beloved Mum Cas 23h44'37" 54d58'32"
Vera McGuckin Peg 21h53'49" 16d5'8"
Vera Mellody Ari 1h50'9" 24d5'45"
Vera "Mummy" Hanson Cas 23h26'29" 52d57'43"
Vera Nordstärnli And 0h20'58" 46d12'41"
Vera Quinn Mon 6h39'53" 7d7'23"
Vera & Remo Giuntoli Sco 16h54'42" -42d28'31"
Vera Rose And 0h30'36" 43d4'10"
Vera Rose Lankus Uma 11h27'26" 37d37'8"
Vera Sasso Uma 12h6'0" 63d56'31"
Vera Schumacher Uma 14h7'7" 60d51'22"
Vera Toro Cas 0h15'39" 61d29'29"
Vera Troyke Uma 9h54'17" 64d41'11"
"Vera" Veronica L. Noyes Sgr 19h50'23" -17d59'39"
Vera Victorious And 23h16'21" 38d26'58"
Vera W. Gorman Sco 16h6'34" -10d28'48"
Vera-Bett1924-30 Aqr 22h27'31" -6d11'53"
Veraellen Cas 0h1'46" 56d18'3"
veramaebarton Sco 16h43'26" -30d1'23"
Veramona Cmi 7h19'39" -0d8'38"
Verbena Louise Burton Psc 22h52'47" 2d32'30"
Verda Demir Uma 14h0'33" 48d16'22"
Verda Veatrice Lib 15h7'14" -15d47'0"
Verde Occhi Cap 20h10'45" -21d42'14"
Verdel and Betty Golden Anniversary Cyg 20h38'31" 41d22'14"
Verdonna F. Lyn 7h40'26" 41d11'31"
Vere & Sue Cyg 21h20'22" 32d28'59"
Verena Uma 11h51'33" 31d43'6"
Verena Cnc 8h50'53" 31d17'55"
Verena Cas 0h20'59" 50d28'3"
VERENA Uma 9h20'35" 51d49'39"
Verena Uma 10h3'7" 45d26'35"
Verena And 23h13'58" 52d1'24"
Verena (auch genannt Zaubermaus) Cnc 8h47'57" 32d8'33"
Verena Carne And 2h13'16" 43d4'55"
Verena Helen Parent "1914-2006" Aqr 22h24'52" -12d55'54"
Verena Killadt Uma 9h40'4" 65d1'58"
Verena Riedo Crb 15h56'12" 25d43'22"
Verena und Martin Uma 11h56'48" 37d2'21"
Verena Zbinden Dra 18h38'44" 52d46'6"
Verenas vanilla pink sun-bell planet Uma 8h43'1" 59d41'37"
Verenursa Umi 15h25'31" 69d0'54"
Verewyn Cas 0h28'14" 59d31'51"
Vergie Mae Cas 1h8'32" 59d15'24"
Vergiemae Lyn 8h50'33" 40d41'41"
Vergie's Baby 1 of 2 Luv My Brenda Vul 20h40'46" 25d54'26"
Vergißmeinnicht Uma 10h29'30" 63d39'42"
Verica Gligoric And 0h22'11" 25d51'57"
Vericanstvena Cyg 20h32'36" 56d57'16"
v.Erich f. Emilia Silva Bernado Pischetsrieder Uma 10h3'35" 68d30'24"
Veridex Umi 17h7'24" 75d40'37"
Veridian Per 4h3'8" 43d54'48"
Verina Cyg 21h38'45" 39d40'33"
Verinda Willa Dean And 0h30'14" 29d36'49"
Verinia & Steve - Las Vegas -180903 Cru 12h18'48" -57d47'22"
Veritas Mon 6h39'1" 8d59'34"
Veritas Her 17h13'42" 45d54'7"
Veritas 13 Cyg 20h25'58" 40d31'34"
Veritatem Quaerere Peg 22h43'51" 11d37'37"
Verity. Cru 12h56'54" -59d10'58"
Verity Uma 9h11'17" 57d16'32"
Verity Dawn Tau 5h47'16" 13d17'48"
Verity Florence Helen Brown Uma 8h50'19" 55d8'27"
Verity May And 23h49'39" 35d18'26"
Verity Stanistreet - Star of Truth Psc 1h50'54" 2d56'59"
Verla Dyer Brown Aqr 22h51'15" -19d29'10"
Verla K. Haugen Cas 1h19'59" 54d27'45"
Verlean Sco 16h10'57" -17d9'50"
Verlicia Miller And 0h45'55" 41d5'28"
Verlobungsstern Nicole & Christian Cep 22h30'52" 66d31'1"
Vermontious Uma 11h33'58" 61d54'7"
Vermot Peg 22h16'34" 18d48'12"
Vern Vir 13h43'24" 2d46'9"
Vern Uma 9h53'41" 55d39'33"
Vern and Roxanne's Eternal Love Pho 0h31'46" -42d31'14"
Vern F. Perry Tau 4h37'12" 20d5'43"
Vern & Geneva Pine Forever Loved Lyr 18h58'18" 26d56'1"
Vern Hemmingsen Uma 13h35'52" 60d28'24"
Vern Hubbard Per 3h3'40" 46d32'38"
Vern & Linda Howe Cyg 20h6'36" 38d23'47"
Vern Nielson Per 3h11'26" 45d29'52"
Vern Robert Mottern, Sr. Sco 17h17'30" -30d14'39"
Vern Stratton Ari 2h1'7" 22d10'59"
vern1979 Aql 19h17'45" 5d3'41"
Verna Anderson Cas 2h6'50" 59d58'53"
Verna Emma Augusta Weis Krauss-GG Gem 6h47'9" 14d50'59"
Verna June Dunn Aql 20h5'25" 11d40'42"
Verna Lee Lyr 19h2'54" 35d52'58"
Verna Lockhart Leo 10h53'55" 10d42'2"
Verna O'Brien Cas 1h17'46" 61d55'45"
Verna P. Schaeperclaus / Rhoads Uma 12h34'40" 56d36'57"
Verna Reichl-Kalb Cap 21h46'7" -19d46'35"
Verna Rende Cas 23h36'31" 54d43'49"
Verna Zottola Scott Sgr 19h39'1" -14d58'10"
Vernadan Sco 17h37'3" -31d50'18"
Vernal Wilkinson Uma 8h45'48" 60d29'38"
Verna's Retirement Star Cyg 20h52'44" 54d9'33"
Verne Hall Ori 5h23'30" 1d58'41"
Verne Harnish, Young Entrepreneurs Umi 13h34'29" 71d15'59"
Verne Prescott Power Cru 12h46'34" -57d42'45"
Verneda Viola Petty Sco 16h6'31" -17d54'25"
Verner Johnson Aqr 22h57'24" -8d47'31"
Verngully Veronica Uma 11h14'58" 49d52'51"
Vernice Cnc 8h32'44" 9d31'47"
Vernico Aql 19h50'28" -0d54'42"
VERNIKKI Leo 10h28'9" 14d58'14"
Vernley M. Hester Aur 5h56'30" 34d35'16"
Vernli Peg 22h52'33" 8d40'3"
Vernon Tau 3h53'16" 24d21'27"
Vernon and Garda Simpson Cap 21h36'19" -24d13'29"
Vernon and Lucille Schlater Aqr 22h30'36" -2d38'16"
Vernon Annes Uma 11h33'54" 35d37'17"
Vernon Ben Frantzich Uma 14h27'3" 55d39'37"
Vernon Campbell Per 4h48'46" 47d0'39"
Vernon Christie Dra 12h13'18" 73d27'38"
Vernon D Moore (Honey) Vir 11h45'49" 3d0'0"
Vernon Dale Hillard Cap 21h26'26" -21d2'7"
Vernon & Elsa Luber Per 2h44'4" 41d2'16"
Vernon Eugene Rippy And 0h28'59" 32d19'55"
Vernon J. Bolstad Vir 13h24'50" -0d48'0"
Vernon Keith Sco 16h50'45" -35d44'34"
Vernon Keith Wold Sco 17h13'9" -34d39'50"
Vernon L. Castle Cap 21h14'14" -26d36'13"
Vernon LaNeal Austin Per 3h14'34" 46d0'0"
Vernon Lee Tridle Sco 17h7'53" -39d36'35"
Vernon Oliver Cobb Her 16h46'5" 42d51'57"
Vernon Paul Gem 6h50'28" 30d13'47"
Vernon "Popeye" Cockerham, Sr. Ari 2h17'42" 21d31'9"
Vernon R. Fisher Lyn 7h35'34" 53d28'31"
Vernon Taylor Cep 22h54'14" 64d3'21"
Vernon Tejas 143 Psc 0h34'11" 9d2'17"
Vernon Tucker Boo 13h53'50" 12d27'40"
Vernon W. Libby Ari 2h18'13" 23d48'14"
Vernonaloribusgegenschein xxv Lmi 9h28'40" 34d43'59"
vernon's ANGEL Cnc 8h14'56" 22d40'46"
Vero Amor (Ben) Boo 14h6'44" 17d44'51"
Vero Amore Uma 9h1'52" 57d19'52"
VéroAngel Dra 15h1'5" 64d3'36"
Verolga Cnc 8h41'31" 15d8'33"
VeRoN Cyg 19h32'47" 43d53'19"
Verona Aqr 21h41'27" -0d49'45"
Verona M. and Robert H. Stone Lyn 8h46'55" 35d7'41"
Veronia Galena Vir 14h49'14" 2d54'47"
Veronica Vir 14h0'13" 3d51'30"
Veronica Psc 0h10'25" 7d42'7"
Veronica Tau 4h12'11" 16d16'1"
Verónica Ari 2h46'8" 15d23'4"
Veronica Ori 5h55'44" 20d57'35"
Veronica Ari 2h33'57" 26d17'12"
Veronica Gem 6h41'38" 23d10'37"
Veronica Ari 3h9'15" 27d29'30"
Veronica Her 17h17'27" 26d1'20"
Veronica Com 12h48'55" 28d27'42"
Veronica And 0h14'14" 35d11'14"
Veronica Lyr 18h47'47" 34d53'32"
Veronica And 2h31'39" 39d38'52"
Veronica And 23h5'30" 36d43'37"
Veronica Lyr 18h53'17" 37d52'56"
Verónica Cas 0h18'7" 51d55'18"
Veronica Uma 9h7'41" 51d11'29"
Veronica And 23h4'22" 47d4'22"
Veronica Lib 15h3'56" -2d19'23"
Veronica Uma 8h56'15" 68d32'36"
Veronica Lyn 6h43'21" 56d15'15"
Veronica Cas 23h3'46" 53d51'39"
Veronica Pho 1h6'46" -40d59'57"
Veronica Cap 20h39'29" -23d52'44"
Veronica Aqr 23h29'16" -22d32'45"
Veronica ~ 52 ~ Maks Ori 5h47'44" -3d16'11"
Veronica A. Branch Cas 2h13'23" 65d13'56"
Veronica Ale Psc 1h9'57" 23d58'47"
Veronica Alvaro Ori 5h54'41" 17d24'50"
Veronica and Kevin Gann Leo 11h19'42" 12d39'24"
Veronica Andreu And 1h20'13" 45d27'37"
Veronica Ann Garza Leo 11h50'51" 21d3'41"
Veronica Ann Salomone Uma 8h29'3" 68d12'12"
Veronica Ann Saucedo Ari 2h34'17" 24d13'44"
Veronica Ann Spicuzza Ori 5h4'41" 14d13'14"
Veronica Anne "Francine" Cristiano Her 18h52'25" 23d25'43"
Veronica Ashley Beck Lib 14h26'56" -19d41'7"
Veronica Ashley Benitez And 0h42'18" 41d39'29"
Veronica Babci Wojcik Psc 0h26'5" -4d24'16"
Veronica Bailey Stangl Cas 1h35'41" 68d10'5"
Veronica Bedon And 2h26'48" 49d8'6"
Veronica Busse Lyn 6h32'20" 56d25'4"
Veronica Callaway Sgr 18h25'50" -31d40'4"
Veronica Carlin Ori 5h54'1" -0d12'58"
Veronica Carmolinga de Vientos Lib 15h42'2" -6d4'50"
Veronica Catherine Stenberg Leo 9h31'18" 27d40'48"
Veronica Cecilia Novo Sgr 19h4'58" -25d4'33"
Veronica & Chad Forever Ari 2h40'45" 26d11'45"
Veronica Charles Ori 6h6'17" 19d37'38"
Veronica Chavez Gem 7h26'23" 21d58'52"
Veronica Cristal Rocha Tau 4h32'15" 18d20'15"
Veronica Dale Loudermilk Peg 21h44'41" 24d1'3"
Veronica De Andreis Uma 10h36'43" 71d35'17"
Veronica E. Womack Ari 3h11'56" 29d17'53"
Veronica Elizabeth Cra 19h1'3" -39d56'23"
Veronica Fair Cas 2h1'49" 63d56'47"
Veronica Flores Ori 5h12'1" 8d26'27"
Veronica Galvan Cyg 20h44'51" 52d28'7"
Veronica Geneva Umi 14h17'50" 75d39'24"
Veronica Georgeo Cap 20h32'28" -23d32'52"
Veronica Gonzales Gem 7h40'2" 18d26'59"
Veronica Herrera Chavez Tau 4h11'46" 4d59'50"
Veronica Huerta And 2h24'45" 49d14'24"
Veronica Jean Vezzoli Eri 4h18'55" -31d55'28"
VERONICA JO EBY Cnc 9h18'53" 9d5'41"
Veronica JoAnne Carruth Vir 14h34'0" 2d52'16"
Veronica Joy Kriegl Cap 21h3'32" -15d0'2"
Veronica Karpecki Vir 14h46'52" 4d8'21"
Veronica Koala Rossi-Torino, Italy And 0h35'15" 36d9'42"
Veronica Lara Hernandez Cyg 21h40'38" 33d41'23"
Veronica Lee Aqr 23h25'1" -21d18'48"
Veronica Leigh France Cnc 8h37'41" 8d30'17"
Veronica Loizou Lyn 6h47'39" 57d51'19"
Veronica Lynn And 1h29'50" 47d9'0"
Veronica M. Foresta Lyr 18h37'58" 31d36'53"
Veronica "Ma Ma Ma" Himmel Cas 1h38'54" 62d42'30"
Veronica Malinowski Cas 0h46'57" 51d59'59"
Veronica "Mangito" Rodriguez Cyg 20h2'32" 57d45'40"
Veronica Manning Murphy Cas 1h33'28" 65d55'40"
Veronica Maria Canizo Ori 5h39'14" 3d20'42"
Veronica Marie Lyn 7h46'19" 45d32'21"
Veronica Marie Crb 16h13'19" 35d17'50"
Veronica Marie And 1h12'40" 38d30'17"
Veronica McElhinney Lyn 7h19'17" 50d14'50"
Veronica Meleady Gem 6h31'13" 24d54'33"
Veronica Mendoza Aqr 21h48'26" -1d35'27"
Veronica Michele Lyn 8h36'33" 45d32'34"
Veronica Mondany Morelli Umi 16h14'6" 85d4'15"
Veronica Muzurek Cas 1h27'4" 58d0'52"
Veronica My Beautiful Cyg 21h32'43" 37d5'0"
Veronica Nazzaro Cas 0h19'49" 53d23'16"
Veronica O'Hagan Whiting Uma 11h53'12" 54d2'11"
Veronica Olvera Psc 0h45'37" 19d4'46"
Veronica Our Godchild Pho 0h30'0" -42d48'25"
VERONICA PADILLA And 23h59'46" 48d28'7"
Veronica Patricia Ann Griego Crb 15h26'27" 28d26'25"
Veronica Patricia Limon Cnc 9h8'40" 16d3'36"
Veronica Perdue Uma 12h7'13" 53d4'40"
Veronica Pettiford Lib 15h15'39" -9d8'53"
Veronica Proscia Cas 1h26'50" 57d56'35"
Veronica Rita Hennessy Hiler Cas 23h36'31" 53d16'26"
Veronica Rodriguez Lyn 7h18'51" 58d17'17"
Veronica "Ronnie" Quinn Uma 10h38'48" 59d43'22"
Veronica Rose McAdams Lyn 8h48'32" 33d23'20"
Veronica Ruiz Lyn 8h3'48" 51d21'49"
Veronica S Cas 1h23'49" 68d16'48"
Veronica Schembri Cas 1h19'5" 54d35'45"
Veronica Sequeira Aur 5h43'20" 37d32'17"
Veronica Serna And 2h30'12" 43d31'3"
Veronica "Sharon" Glasner Leo 10h58'44" 12d11'49"
Veronica Shurgala Ori 6h17'48" 15d17'56"
Veronica & Stephen Cyg 20h13'5" 53d40'17"
Veronica Sweet Crb 15h36'4" 27d11'28"
Veronica Torres Cap 21h12'16" -17d33'55"
Veronica Urueta Uma 9h28'44" 45d56'54"
Veronica Valencia Vir 13h30'47" -1d19'35"
Veronica Wooten Psc 0h46'50" 19d33'46"
Veronica Zamora Peg 21h34'33" 21d58'33"
Veronica, Juliet Vir 13h24'23" 12d19'30"
Veronica-Angel Baby Ari 3h14'3" 24d7'8"
VeronicaGoergAnner Ari 2h49'51" 18d23'46"
Veronican Leo 11h17'10" 14d49'9"

Veronica's Angel
Vir 12h24'46" 3d49'6"
Veronica's Corner
Cap 20h54'54" -18d34'13"
Veronica's Light
Aqr 20h46'24" -13d37'43"
Veronica's Shining Star
Umi 14h49'13" 80d14'36"
Veronica's Silver Star
Cap 20h31'46" -22d41'7"
Veronica's Snowflake
Gem 6h27'21" 24d40'14"
Veronica's Star
Psc 0h24'19" -1d38'17"
Veronika
Uma 12h38'28" 62d28'15"
Veronika
Gem 7h19'6" 26d21'49"
Veronika
Cam 4h2'5" 55d41'10"
Veronika Bergly of Andenes Norway
Cru 12h18'35" -57d16'4"
Veronika Gritskova
And 0h41'27" 26d42'21"
Veronika Krautschneider
Uma 11h26'17" 72d40'6"
Veronika Marer Foldes
Leo 11h22'4" 7d23'31"
Veronika Puchlova
Psc 1h43'9" 25d12'40"
Veronika V. Zagreda
Sco 17h39'20" -40d51'17"
Véronique
Leo 10h35'28" 20d5'26"
Véronique Alie
Hya 14h53'52" -27d35'56"
Véronique Belley
Uma 14h19'43" 60d34'46"
Véronique Chabert-Jardat
Lib 15h25'47" -5d9'49"
Véronique Champion
Lib 14h35'15" -11d46'6"
Veronique F. Delattre
And 0h37'27" 27d39'30"
Véronique Galloo-Vandamme
Sgr 18h11'10" -19d21'36"
Véronique Landolt
Com 12h43'6" 28d52'29"
Véronique Le Roch
Uma 9h14'58" 51d50'40"
Véronique LECOMTE
Ori 4h47'25" -3d42'40"
Véronique Lefèbvre
Cyg 20h2'6" 52d18'3"
Véronique Lupo
Lyr 18h36'47" 34d29'6"
Véronique Pradeau
Cas 0h3'8" 57d27'4"
Veronique Vijverman
Cnc 9h8'48" 15d12'23"
Veronique Whitaker
Psc 22h58'23" 8d0'17"
Veron's Wish
Del 20h32'38" 16d2'49"
Verquis
Ori 6h0'18" 9d44'25"
Verri Family Star
Ori 6h13'19" 19d15'23"
Verrier
Gem 6h40'44" 34d59'52"
Versi und Min
Uma 9h36'0" 56d8'33"
Vertigon
Psc 0h50'36" 28d55'1"
Verusca & Filippo
Uma 9h2'40" 62d46'1"
Verusca Tarca
Uma 8h43'27" 51d55'50"
Verute Giedraityte-Obydzinska
Gem 8h4'38" 30d33'49"
Verve
Uma 13h49'33" 56d37'40"
Veryl (MOM) Durfey
Cap 21h38'44" -13d45'37"
Verzeih'-mir-Stern für Petra
Uma 13h33'59" 61d32'28"
Vesela's Star
Dra 17h25'2" 57d30'50"
Vesile Demir
Umi 15h25'2" 85d24'20"
Vesi-Vlado
Cas 1h52'37" 64d29'10"
Vesle Andreas
Per 4h31'15" 33d45'31"
Vesna
Her 18h10'10" 42d7'32"
Vesna
Cap 20h55'8" -14d52'21"
Vesna 35 Licina
And 23h23'20" 52d58'13"
Vesna "Boggie" Jovanovic
Leo 11h46'7" 24d27'22"
Vesna Pavlov
Peg 23h39'19" 15d46'51"
Vesna Skocz
Leo 10h12'46" 22d53'25"
Vesta
Tau 6h0'19" 24d28'51"
Vesta
Uma 13h38'47" 56d59'16"
Vesta and Tracy & Emmett
Cyg 21h56'56" 41d28'39"
Vesta Johnson
Lyn 6h52'46" 51d34'32"
Vesta Mae Cochran New
Lib 15h33'8" -21d26'54"
Vesta Renee Heuschkel
Uma 10h5'29" 54d54'45"
Vestally (M) Sdn Bhd
Ori 5h33'14" 4d25'31"
Vester Vestra Vestrum Usquequaque
Uma 10h57'33" 57d24'39"
Vestez
Her 17h31'51" 34d24'50"
Vestfossen
Uma 10h15'55" 63d28'27"
Vestri Olympus
Cnc 8h33'2" 30d6'33"
Vetle
Cam 4h40'32" 66d25'0"
Vetsy
Uma 10h31'56" 47d23'19"
Vetta Mankarios
Lmi 10h33'53" 37d2'45"
Vex Fluxor
Ori 5h54'2" -0d41'34"
Veyond (My Shining Star)
Lyn 6h56'40" 50d42'45"
Veysel ve Güleser
Ori 5h9'26" 15d56'9"
Vezer's Glory'a
Uma 11h57'33" 61d24'16"
Vezilou's Long Overdue Star
Uma 13h38'5" 61d39'25"
VFB09292006
Ori 5h21'4" 1d43'14"
Vhari Edwards
And 1h14'4" 41d47'40"
VHC's Chiang Mai Crystal Diamond
Lyn 8h42'57" 34d53'22"
"VHP" Dragon-77006
Dra 19h10'39" 75d11'58"
Vi - Bill
Per 4h37'21" 36d0'33"
Vi Hamlin
And 1h7'54" 34d8'48"
Vi Isabel McDonald
Cyg 20h32'33" 37d15'34"
Vi - our radiant flower in the sky
Uma 14h22'5" 56d4'28"
VI & Ray Dumais
Leo 11h3'3" 17d26'59"
Viajante Mundial
Lyn 7h20'52" 50d6'44"
Vianca Xenia Davis
And 2h22'52" 45d8'15"
ViAngel
Peg 22h37'19" 34d26'52"
Vianney
And 23h23'8" 38d38'0"
Vianney Carvajal
And 0h49'2" 45d45'27"
VIASYS NeuroCare
Uma 12h43'10" 57d29'30"
Vibeke Amundsen
Lib 15h52'55" -17d47'25"
Vibeke Ring Gerkins Clark
Cnc 8h22'51" 24d5'9"
ViBern Skripata
Cyg 21h15'51" 38d21'37"
Vibrant Marinda
Lib 15h33'23" -19d54'4"
Vic
Uma 10h50'52" 54d52'34"
Vic Amburgy
Boo 15h47'18" 44d39'41"
Vic & Angie Mangone
Ari 2h23'17" 22d8'14"
Vic Australis
Nor 16h27'47" -48d17'59"
Vic Befera
Ari 2h6'52" 26d31'14"
Vic Eilenfield
Her 17h29'36" 31d50'45"
Vic Galef
Per 2h24'27" 51d14'48"
Vic Gregor
Her 17h52'43" 35d23'16"
Vic n Chris
Sex 9h48'13" -0d38'8"
Vic Tomlin
Cnc 8h14'39" 21d19'4"
*Vic*Wanda*Kenzi*Freeman*2006*
Uma 8h51'39" 51d15'17"
Vic Zagaro's Karma
Aqr 23h2'59" -8d18'16"
Vic73
Cap 21h31'41" -9d18'34"
VicA
Ori 6h20'12" 19d46'10"
Vicci Loves Rylle
Ari 2h52'46" 28d10'43"
Vicdu Lodun Of Delton Mi Usaearth
Aql 19h37'31" 12d43'5"
Vicente Barragan Jr.
Gem 6h42'51" 28d7'40"
Vicente Buendia Saez
And 2h10'25" 37d35'41"
Vicente Gabriel Garzo Toro
Cep 0h21'44" 70d0'14"
Vicente Jose Buigues Gardner
Aql 20h34'3" -7d2'42"
Vicente Jr.
Boo 14h50'5" 54d27'25"
Vicente & Kimberly Carmona
Cyg 19h39'21" 34d32'50"
VICENTE MORAN
Her 18h44'29" 20d22'41"
Vicente S. Espinosa
Her 18h19'40" 15d59'19"
Vicheany "Noodle" Samnom
Lib 15h9'30" -8d34'31"
Vicious Creature
Per 3h29'5" 34d32'18"
Vicious Vixen
Gem 6h3'0" 25d41'27"
Vickey-my soul mate
Sgr 19h0'7" -33d48'5"
Vickey's Star
And 23h1'35" 48d35'22"
"Vicki"
And 23h13'51" 42d44'9"
Vicki
Crb 16h5'57" 35d49'53"
Vicki
Gem 7h17'58" 20d29'52"
Vicki
Ari 1h56'35" 19d21'34"
Vicki
Aqr 23h9'52" -6d52'51"
Vicki
Dra 17h5'33" 61d3'40"
Vicki
Uma 10h47'7" 65d50'14"
Vicki - A Slice of Heaven - Kubacki
Cap 21h14'48" -27d9'54"
Vicki & Amber (Mother & Daughter)
Cyg 20h38'3" 53d26'53"
Vicki Anastasia
Cnc 8h51'38" 32d32'58"
Vicki and Caroline Petersen
Cmi 7h37'5" -0d7'8"
Vicki and Don's Wish
Cyg 19h35'49" 31d39'3"
Vicki and Eileen
Cyg 19h54'43" 39d55'41"
Vicki and Gabriel
Aqr 22h43'46" 2d2'28"
Vicki and Riki
Uma 8h51'43" 50d27'47"
Vicki and Robbie
Sgr 19h17'42" -14d16'58"
Vicki & Andy - 22.12.2005
Cyg 20h2'53" 51d26'6"
Vicki Ann
Tau 4h41'53" 23d52'29"
Vicki Atkins
And 0h19'56" 43d23'18"
Vicki Bensel
And 0h49'24" 36d39'52"
Vicki & Bill
Cyg 20h27'29" 39d45'6"
Vicki Brasuell
And 23h36'20" 47d13'8"
Vicki Bryan
And 0h34'20" 33d6'42"
Vicki Carlson my beautiful mother
Cas 0h16'37" 57d32'38"
Vicki Carroll
Del 20h59'8" 14d43'35"
Vicki Clark
Dra 18h22'19" 73d0'59"
Vicki Colledge
Cyg 21h38'33" 47d31'55"
Vicki Cordes
And 0h24'47" 25d49'47"
Vicki Coyle Mother and Friend
Cnc 9h2'56" 20d50'20"
Vicki Cutler
Aqr 21h14'33" -10d25'24"
Vicki D. Lee
Dra 17h35'13" 58d42'19"
Vicki Dunsmore, My Angel
And 2h32'36" 49d44'28"
Vicki Feely
Uma 9h0'10" 47d34'10"
Vicki Fox
Lib 15h8'13" -13d51'36"
Vicki Gordon
Umi 14h57'37" 68d21'5"
Vicki Gray
Ori 4h47'28" -3d38'46"
Vicki Harrison Tolliver
Gem 6h28'26" 24d33'16"
Vicki Harvey
Uma 11h30'59" 28d49'9"
Vicki Hauff
And 23h29'17" 39d22'43"
Vicki & Her Three Suns
Lib 15h28'37" -7d30'46"
Vicki Hoban 7/25/1981
Umi 13h4'55" 75d32'25"
Vicki Imsande
Peg 23h43'31" 26d26'59"
Vicki Jacoba
Tau 5h20'10" 17d27'13"
Vicki Jo
Leo 10h4'24" 26d30'47"
Vicki Jo Staton
Uma 11h30'14" 38d38'17"
Vicki Joan Biehl
Cnc 8h24'4" 10d46'48"
Vicki Johnson Meier Oberleitner
Sco 17h53'22" -36d13'26"
Vicki K
Aqr 23h43'36" -15d42'49"
Vicki Kelley
Lyn 8h16'8" 34d8'34"
Vicki L. Morris
Cnc 8h22'16" 7d15'58"
Vicki Lanell Littlejohn
Aql 20h0'37" 8d48'40"
Vicki Lawrence
Cas 0h29'20" 58d55'2"
Vicki Lee
And 0h57'49" 40d38'7"
Vicki Lee
Lyn 7h32'9" 56d30'52"
Vicki Lee Poff
Cas 1h24'22" 57d36'21"
Vicki Lee Scherer Cooper
And 0h39'22" 40d5'40"
Vicki Love Star
Ori 5h6'34" 8d8'48"
Vicki Lyerly
Gem 6h36'17" 12d39'19"
Vicki Lyn
Ari 2h35'41" 12d43'35"
Vicki Lynn Boyer Douglas
Uma 8h24'4" 66d10'59"
Vicki Lynn Connell
Leo 10h19'51" 21d36'59"
Vicki Lynn Johnson
Ari 3h0'20" 11d35'46"
Vicki Lynn Kinseth Lempke
Ori 6h15'0" 6d56'48"
Vicki Lynn Kunkle
Uma 14h10'48" 62d17'13"
Vicki Lynn Onstad
Cas 0h57'59" 50d14'43"
Vicki Lynn Paro
And 2h6'18" 41d27'51"
Vicki Lynn Richardson
Cyg 21h59'3" 48d41'23"
Vicki Lynn Sessions
Cyg 21h25'24" 39d31'32"
Vicki Lynn's Very Own Star
Vir 13h10'3" 7d38'13"
Vicki Marguerite
Crb 16h21'15" 34d34'26"
Vicki Mayer
Uma 10h32'15" 72d31'40"
Vicki Norton
And 1h54'37" 39d15'28"
Vicki & Paul Reid 40 Years of Love
Umi 15h30'16" 74d34'53"
Vicki Penchosky CMP - GaMPI Star
And 0h29'45" 44d7'20"
Vicki Pino
Psc 23h45'40" 5d16'35"
Vicki S. Boutell
Tau 5h33'37" 19d51'41"
Vicki S. Middagh
Vir 14h6'50" -18d0'57"
Vicki Starr Hedin
Cap 21h42'52" -9d0'42"
Vicki Sue
Sco 17h1'55" -39d13'55"
Vicki Sue Stiles
Uma 9h38'7" 52d9'9"
Vicki: The Star That Guides Me
Lyn 8h10'48" 34d3'4"
Vicki + Tim
Ori 6h10'40" 18d53'41"
Vicki Tiveron
Lmi 10h44'26" 31d1'4"
Vicki Truong Eternal Butterfly
Cas 1h23'3" 68d49'38"
Vicki Turner - msm
Del 20h38'57" 16d40'55"
Vicki Warga
Aqr 23h45'8" -18d43'49"
Vicki Weston
Lyn 7h44'31" 41d33'1"
Vicki Whyman
And 2h29'44" 39d30'20"
Vicki Winner
Vir 13h26'20" 11d36'48"
Vicki Zimmerman
Ori 5h21'12" 8d44'2"
Vickie
Tau 3h46'39" 28d23'35"
Vickie
Gem 7h10'49" 34d1'17"
Vickie
Sgr 18h35'19" -23d55'17"
Vickie
Cas 0h52'56" 61d36'3"
Vickie
Aqr 22h11'35" -13d38'36"
Vickie
Lib 14h48'31" -9d10'4"
Vickie and Dave Wippel
And 1h31'32" 43d52'22"
Vickie And Guy
Sge 19h50'52" 17d40'17"
Vickie Annette
Cam 12h25'9" 77d16'48"
Vickie B. "Til the 12th"
Gem 7h19'39" 16d45'33"
Vickie Bochniak
Cas 1h29'23" 69d1'49"
Vickie Christman
Cap 20h34'59" -22d28'29"
Vickie & Daryll's
Her 18h19'53" 16d55'0"
Vickie Ebeling
Cap 20h8'56" -25d12'5"
Vickie Evers
Cap 21h52'41" -9d40'32"
Vickie F. Turner
And 0h49'59" 43d22'20"
Vickie Fabrizio
Lyn 7h49'52" 35d33'55"
Vickie Golladay Love, Susan
And 23h17'55" 43d36'26"
Vickie "Grammy" Benefield
Uma 10h39'52" 57d35'23"
Vickie Holden
Leo 9h37'47" 12d37'39"
Vickie J. DuBois
Uma 10h15'20" 47d41'34"
Vickie Johnson
And 0h40'11" 27d56'23"
Vickie Joyce
Lyn 7h39'50" 47d42'11"
Vickie Kay Coghan
And 2h11'13" 50d26'12"
Vickie L. Dellinger
And 1h22'51" 44d36'56"
Vickie Lorraine
Sco 17h11'19" -39d24'26"
Vickie Lynn Jones
Mon 6h53'46" -0d10'27"
Vickie Lynn Konkol
Lyr 18h37'16" 31d9'31"
Vickie Lynn Taylor
Lyn 8h48'1" 40d8'22"
Vickie Lynn Work
Mon 6h45'13" 7d16'8"
Vickie Marte
Cas 22h57'49" 53d46'11"
Vickie Murphy
Uma 9h22'21" 70d1'29"
Vickie Nesmith
Cyg 20h29'44" 40d22'49"
Vickie P. Cornejo
Crb 15h25'40" 28d48'13"
Vickie Palumbo Tucker
Gem 7h15'50" 17d35'26"
Vickie Slack GAMPI Shining Star
Aql 19h4'51" -0d19'15"
Vickie Taylor
Aqr 23h40'0" -10d26'37"
VickiJohnChamberlain
Cyg 19h40'17" 28d49'25"
Vicki-Lou & Jerry Goss
Lyr 19h9'49" 43d2'44"
Vickilou Nichols
Mon 7h16'51" -0d17'9"
Vicki-Lynn Gardner
And 2h13'24" 37d55'8"
Vicki-pooh
And 0h12'2" 25d49'11"
Vicki's Beauty
Leo 10h56'23" -4d2'18"
Vicki's Bright Little Stars, EPGRSP
Cas 1h28'11" 59d54'6"
Vicki's First
Ari 3h24'33" 20d31'5"
Vicki's Forty
Tau 4h20'49" 7d42'40"
Vicki's Journey
Ori 5h51'26" 19d12'41"
Vicki's Precious Glow
Lyr 18h31'27" 28d46'6"
Vicki's Star
Cyg 19h52'52" 35d46'42"
Vicki's Star
Gem 7h20'51" 31d12'38"
Vicki's Star
Uma 10h55'53" 62d46'48"
Vicks
And 0h44'7" 30d28'50"
VickTom
Uma 11h10'8" 34d41'56"
Vicktoria Hay
And 23h32'3" 47d49'14"
Vicky
And 2h23'52" 48d47'49"
Vicky A Zimmerman-Harvey(5-11-1958)
Tau 3h53'45" 17d53'56"
Vicky Amaral
Umi 14h21'43" 67d45'27"
Vicky and John 1 love Forever
Lib 15h12'42" -5d39'14"
Vicky and Terry's Guiding Light
Cyg 19h39'33" 31d24'45"
Vicky Ann Ramirez
Vir 13h3'16" -21d57'2"
Vicky bb
Psc 0h46'48" 6d3'36"
Vicky Callahan
And 23h24'58" 48d9'41"
Vicky Caraway
Leo 10h11'12" 12d56'58"
Vicky Caudill
Psc 23h43'39" 3d12'40"
Vicky & Dale
Cyg 20h48'35" 39d7'56"
Vicky Denise Littleton
And 23h34'18" 48d22'56"
Vicky E. Cory
Uma 11h46'31" 47d45'9"
Vicky Enloe Beloved Wife & Mother
Cyg 20h35'49" 42d7'26"
Vicky Foster "We Love You Mawmaw"
Cas 23h40'6" 54d52'50"
Vicky Garrido
Mon 6h41'52" 4d56'41"
Vicky & Geoff Langham
Uma 9h19'27" 62d40'43"
Vicky Harker
And 1h8'8" 45d28'58"
Vicky Jester
Vir 13h23'48" 10d57'55"
Vicky June Goze Smith
Cnc 8h18'36" 8d39'23"
Vicky Kautz
And 0h22'23" 46d37'8"
Vicky Khaov
Sgr 18h9'7" -19d13'30"
Vicky Kossiva
Uma 9h18'58" 42d10'49"
Vicky Lee
Cam 5h50'45" 58d17'53"
Vicky & Lee's Anniversary Star
Cyg 19h39'1" 54d40'18"
Vicky Lynn Bearce
Uma 10h8'10" 67d35'49"
Vicky Lynn Stanger
Cnc 8h31'26" 32d18'12"
Vicky Marie
And 0h32'59" 36d56'46"
Vicky McAree
Lmi 10h1'42" 40d34'56"
Vicky Milner
Vir 13h53'36" 7d13'8"
Vicky Ng
Cyg 19h38'32" 32d20'32"
Vicky Patricia Buckles 24
Sgr 18h44'7" -27d53'29"
Vicky Peace, Coventry DOB 07/03/66
And 2h7'14" 46d17'23"
vicky "peach" willim
Her 18h4'56" 32d22'48"
Vicky Rhind
And 23h58'17" 39d16'42"
Vicky Rhoades
Lyr 18h49'19" 32d36'23"
Vicky + Robert
Cyg 20h56'43" 42d3'30"
Vicky Rose
And 0h18'59" 25d27'1"
Vicky Savas
Crb 15h52'12" 36d43'48"
Vicky Saygnasith
Tau 5h9'19" 21d12'48"
Vicky Sue Roy
Cyg 19h46'59" 33d22'0"
Vicky Swaby
And 2h37'58" 40d49'31"
Vicky USMC
Lyn 9h20'28" 34d20'48"
Vicky Winson
And 1h10'23" 42d14'36"
VickyJo
Cam 6h25'0" 67d38'42"
Vicky's Bright Star
And 2h11'34" 42d26'8"
Vickys Eric
Lyn 9h13'15" 34d55'30"
Vicky's Star
Cas 3h32'6" 77d3'59"
Vicmatlin
Leo 11h46'0" 25d26'57"
Vic's Shining Light
Cyg 20h29'50" 33d48'11"
VicSalEst Dago-Suarez
Aqr 21h5'42" -8d50'41"
Victoire
Cap 20h22'36" -14d17'54"
Victoire
Ori 5h22'5" -8d29'0"
Victoire Bouchayer-Marque
Tau 4h2'53" 26d59'48"
Victoire Joséphine Ghnassia
Tau 3h29'58" 19d36'27"
Victoire,Jeanne,Céleste,Nakri
Uma 11h53'49" 59d35'48"
Victor
Sco 17h38'5" -37d2'40"
Victor
Ori 6h2'19" 13d15'14"
Victor
Cnc 8h47'29" 21d23'1"
Victor
Per 3h47'18" 43d55'46"
Victor
And 23h33'39" 47d20'45"
Victor A. Abrahamsen
Cnc 9h21'18" 11d16'39"
Victor A. Betancourt
Cnc 8h38'29" 23d52'20"
Victor A. Smith
Per 3h31'15" 51d12'33"
Victor Agrest
Ari 3h7'52" 29d2'48"
Victor and Lucille
And 0h19'1" 26d47'6"
Victor and Valerie
Leo 11h17'48" 14d18'34"
Victor Angel Garcia
Boo 13h49'33" 16d24'8"
Victor Anthony Andrade, Jr.
Ori 5h29'52" -2d7'41"
Victor Anthony Garnett
Uma 10h52'26" 52d37'25"
Victor Arnold Hamilton Jr.
Her 16h53'31" 33d47'51"
Victor Benedito
Tau 3h59'47" 0d38'22"
Victor Bodi
Cru 12h12'34" -62d35'10"
Victor Bremson
Aur 6h4'43" 51d46'54"
Victor & Brynn Rangel
Cyg 20h6'3" 35d34'59"
Victor Byrdsong
Cyg 19h52'29" 37d52'8"
Victor C. Buczynski
Uma 10h54'57" 46d23'35"
Victor C Giles
Sgr 18h48'43" -32d25'56"
Victor Carr
Sgr 19h14'44" -26d12'38"
Victor Charles Cicero
Uma 9h20'38" 51d23'52"
Victor Claus Eriksen
Her 18h29'21" 18d55'6"
Victor Clore
Cep 22h33'52" 74d22'30"
Victor "Congero" Baez
Her 17h30'51" 37d26'28"
Victor Cordero
Lyn 7h35'45" 52d12'33"
Victor Cotogno
Ori 5h28'3" 4d31'12"
Victor "CU Grad" Leon
Aqr 20h42'32" -11d21'33"
Victor Daniel Mascorro Ruiz
Her 17h35'23" 33d26'53"
Victor Darth Varma Superstar
Uma 11h0'31" 34d36'45"
Victor & Dawn Schepisi
Cyg 21h28'39" 53d1'39"
Victor Delrue
Lib 15h38'15" -23d14'40"
Victor Diaz (Dada Supreme)
Gem 7h35'33" 16d48'52"
Victor Douangphachanh
Her 16h41'52" 35d10'8"
Victor E. Friesen
Sgr 19h43'43" -14d9'27"
Victor & Elizabeth
Ari 2h8'59" 25d36'59"
Victor Evan Taboadela
Aqr 23h37'42" -7d16'17"
Victor Fainstein
Aql 18h58'5" -0d8'52"
Victor Galfo
Sco 17h6'5" -43d31'20"
Victor George Deacon
Uma 13h49'6" 62d1'30"
Victor H. Solano Jr.
Cyg 20h43'12" 35d45'17"
Victor Hugo Santos
Tau 4h28'20" 16d24'32"
Victor J. D'Errico, Jr.
Vir 14h11'12" -0d14'2"
Victor J. Howe
Ari 2h32'59" 25d45'17"
Victor John Bugariu
Boo 14h43'39" 16d25'50"
Victor "John" Laurie
Cru 11h58'46" -55d43'59"
Victor Joseph
Cep 22h53'35" 72d10'10"
Victor Joseph Moscone
Her 16h24'51" 15d20'41"
Victor & Karen Forever
Lyn 7h43'54" 44d1'5"
Victor King
Uma 11h0'2" 38d19'33"
Victor Kirk
Tau 5h57'34" 27d55'9"
Victor "Little" Garcia
Lyn 7h30'28" 51d25'44"
Victor Luna Castro
Leo 11h34'47" 24d33'12"
Victor M. Farinas
Psc 23h13'5" -2d26'1"
Victor M. Ramos
Ari 2h42'18" 13d27'12"
Victor Manuel Acosta
Cnc 8h3'36" 12d22'0"
Victor Manuel Arroyo
Her 18h10'3" 15d29'19"
Victor Manuel Beltran
Cep 20h55'41" 62d17'50"
Victor Manuel Cruz
Leo 10h18'47" 25d58'48"
Victor & Marie Diaz
Umi 15h24'33" 75d14'24"

Victor mi Amor
Gem 6h17'54" 25d36'0"
Victor mi Amor
Gem 6h17'54" 25d36'0"
Victor Michael Esposito, III
Lib 15h35'26" -11d43'57"
Victor Minnocci
Crb 15h35'6" 38d43'46"
Victor Monroy
Dra 9h30'25" 73d32'2"
Victor Morenovic
Vir 12h0'5" -8d8'34"
Victor Nicholas Sabastiano
Leo 9h29'29" 18d17'21"
Victor O'Brien
Pho 0h38'41" -47d20'42"
Victor P. Gracias
Cyg 20h26'14" 36d43'45"
Victor "PAPA" Podbielski
Her 16h38'40" 38d32'0"
Victor Perry Jacobs
Uma 10h19'44" 59d29'14"
Victor Philip Bilan
Aql 19h39'49" 12d55'35"
Victor Rivara
Ori 5h59'51" 9d45'13"
Victor Ross Hux
Aur 5h41'46" 39d50'27"
Victor S. Hogen Sr., MD
Her 16h35'9" 37d38'3"
Victor S Wilson
Uma 12h52'25" 59d45'15"
Victor & Sabrina Noya, Eternal Love
Umi 17h33'4" 84d26'59"
Victor Sanz
Per 3h16'44" 49d52'41"
Victor Scavelli
Uma 10h19'0" 55d30'14"
Victor Schmeck's Memorial Star
Cep 22h16'53" 60d49'25"
Victor Singleton
Dra 16h8'10" 63d11'43"
Victor Slabodchikov
Vir 13h0'53" 8d32'12"
Victor Smeretskiy
Gem 6h2'10" 23d45'14"
Victor Steven Crea
Aur 5h38'29" 33d27'25"
Victor T. & Kim T.- Soulmates4Life
Leo 9h51'23" 24d33'41"
Victor Tavares Jr. Loving Husband
Per 3h35'47" 34d41'9"
Victor Thomas Gil
Umi 14h22'55" 77d40'44"
Victor Tooker
Lyn 7h57'30" 41d46'6"
Victor W Clark
Ori 5h51'0" 6d31'54"
Victor "W I N D I I- baby" Zubac Isaac
Dra 19h17'8" 65d11'54"
Victor W. Sinopoli
Cep 22h35'36" 64d58'22"
Victor Warren
Her 17h58'28" 49d37'4"
Victor Warren Wernett
Cap 21h17'23" -16d29'1"
Victor Wiedemann
Gem 7h34'56" 27d6'24"
Victor Zalewski, Jr
Her 17h54'39" 27d42'12"
VictorAllison
Cyg 19h39'37" 52d39'42"
Victoria
Cyg 21h42'57" 54d26'46"
Victoria
Lyn 7h5'19" 53d29'33"
Victoria
Uma 10h22'47" 68d20'12"
Victoria
Uma 12h4'17" 62d9'4"
Victoria
Uma 9h34'18" 60d17'50"
Victoria
Cas 1h43'0" 67d29'33"
Victoria
Cap 20h37'17" -17d33'57"
Victoria
Lib 14h26'54" -17d18'4"
Victoria
Lib 15h41'39" -17d9'11"
Victoria
Aqr 21h13'12" -8d6'25"
Victoria
Vir 13h1'13" -13d12'57"
Victoria
Aqr 22h35'48" -2d35'53"
Victoria
Vir 13h42'25" -5d51'10"
Victoria
Lib 14h51'9" -3d48'56"
Victoria
Psc 0h28'56" 4d33'3"
Victoria
Sgr 18h46'52" -29d6'40"
Victoria
And 0h17'15" 33d18'6"
Victoria
Leo 10h19'16" 24d59'56"
Victoria
Cnc 8h30'3" 20d10'52"
Victoria
Psc 1h8'33" 27d43'27"
Victoria
Vir 14h31'33" 2d25'49"
Victoria
Vir 14h14'21" 4d59'10"
Victoria
Psc 22h55'58" 5d8'14"
Victoria
Vir 13h23'52" 11d13'52"
Victoria
Cyg 20h39'58" 51d1'43"
Victoria
Uma 10h7'13" 46d18'57"
Victoria
Cas 1h22'43" 53d20'23"
Victoria
Lyn 7h23'22" 45d33'20"
Victoria
And 0h18'9" 45d6'42"
Victoria
And 23h30'9" 41d15'58"
Victoria
Uma 11h43'3" 38d19'29"
Victoria
Lyn 9h0'13" 44d14'42"
Victoria
Cyg 21h22'13" 33d14'23"
Victoria
And 23h48'43" 33d53'47"
Victoria
And 1h10'18" 38d37'36"
Victoria
And 1h7'16" 39d44'6"
Victoria
And 0h49'32" 34d12'26"
Victoria
Gem 6h46'42" 34d13'49"
Victoria *16th Victory Star*
Lib 14h48'47" -12d13'40"
Victoria 615
Mon 7h31'55" -0d35'50"
Victoria A. Elton
Ari 2h19'46" 25d19'21"
Victoria A. Kirk
Cam 7h58'42" 78d35'55"
Victoria A Ueland
Cnc 8h2'1" 11d49'10"
Victoria Alexandra DeNoto
Cnc 8h18'29" 14d43'38"
Victoria Alexis Bogolubov, "vika"
Ari 2h5'18" 18d55'57"
Victoria Allison Gardner
Tau 5h25'28" 24d38'7"
Victoria Alyse Marrone
Cas 0h18'56" 53d23'24"
Victoria and Brent
Uma 9h22'41" 48d10'38"
Victoria and Haydn Never too much
Cyg 20h4'8" 35d19'40"
Victoria and Nicholas
Gem 7h14'4" 33d47'3"
Victoria and Richard Arden
Cam 4h14'2" 67d40'57"
Victoria And Timothy
Sgr 17h59'5" -28d10'45"
Victoria & Andrey
And 0h29'6" 46d31'34"
Victoria Ann
And 2h24'46" 47d54'11"
Victoria Ann Adkins
Tau 5h55'58" 25d11'15"
Victoria Ann DeLia 4/11/1987
Ari 3h13'54" 19d57'44"
Victoria Ann Garcia
Cap 21h6'6" -20d46'26"
Victoria Ann Gutierrez-Vaughn
Cnc 8h36'19" 24d17'39"
Victoria Ann Harris
Cas 0h28'18" 54d29'1"
Victoria Ann Loomis
And 23h49'22" 41d16'11"
Victoria Ann Quinn
Cma 6h43'14" -14d25'7"
Victoria Ann Santopietro
Vir 13h32'7" -22d7'7"
Victoria Ann Scarpaci
Cas 23h38'51" 56d44'37"
Victoria Ann Shullick
Vir 12h16'16" 6d19'47"
Victoria Ann Spaniol
Sco 17h37'36" -41d22'4"
Victoria Anne Johnstone 31/10/83
Cru 12h40'35" -56d16'20"
Victoria Anne Vila
Cap 21h50'24" -10d22'19"
Victoria Anne Warren
Umi 14h40'6" 83d18'4"
Victoria Ann-Marie
And 23h0'54" 47d31'58"
Victoria Ashleigh Dyson
Aqr 22h50'44" -2d43'3"
Victoria Ashley Sindone
Sco 17h26'45" -40d42'23"
Victoria Ashton
And 1h15'32" 50d20'59"
Victoria Augusta Hoeger
Dra 16h31'33" 61d15'27"
Victoria Bailey
And 23h33'45" 48d10'34"
Victoria Barbra Rahall
Cnc 8h17'51" 8d5'0"
Victoria Bartasek
And 1h40'15" 49d52'9"
Victoria Baudoin
Cyg 20h30'7" 42d35'24"
Victoria & Bill Celebration of Love
Psc 1h58'18" 2d48'5"
Victoria Blais
Ori 5h25'41" -5d31'33"
Victoria "Bobble" Frith
Cyg 19h48'41" 31d24'7"
Victoria Bressano
Cru 12h24'55" -56d14'49"
Victoria Brixey
And 0h29'14" 34d22'14"
Victoria Browne
Cas 1h27'13" 65d25'25"
Victoria C. Lemut
Leo 9h30'32" 10d16'38"
Victoria Campos
Her 17h29'4" 27d56'7"
Victoria Candelaria
Cas 1h20'58" 56d22'37"
Victoria Cara Lee Boucher
Pho 2h23'49" -46d46'42"
Victoria Carnera 4/13/91
Ari 3h11'55" 29d25'1"
Victoria Caroline
Cyg 20h48'56" 49d0'37"
Victoria Caroline Hill
Cas 0h38'47" 57d26'12"
Victoria Catherine Leffler
Lib 15h6'45" -1d13'55"
Victoria Catherine Young
And 23h32'33" 43d53'1"
Victoria Cavallaro
Lmi 10h41'19" 26d46'21"
Victoria Chevalier Prades
Cas 3h31'23" 71d3'12"
Victoria Chun Lee
Uma 11h22'7" 37d26'32"
Victoria Consuelo Abad Sanchez
Crb 15h33'6" 26d44'14"
Victoria Covey
And 0h30'29" 26d51'51"
Victoria Curran
Cas 0h31'25" 61d46'27"
Victoria Curtis - Mummy's Star
And 0h8'52" 29d0'53"
Victoria Danielle Wollum
Lyr 18h26'25" 34d49'8"
Victoria Dawn
Cnc 8h6'45" 11d35'17"
Victoria Dawn Batstone
Dra 10h55'47" 76d27'57"
Victoria Dawn Purvis (nee Haddock)
Lib 14h56'56" -19d38'33"
Victoria Dee
Aqr 23h17'11" -17d21'31"
Victoria Dente
Uma 8h27'51" 70d20'46"
Victoria Dickerson
Mon 6h29'12" 9d4'58"
Victoria Dinkelis
And 1h14'4" 41d39'19"
Victoria Doolittle
Gem 6h48'10" 24d28'21"
Victoria Dutterer
Vir 13h43'13" 4d50'28"
Victoria E. Johnson
Dra 18h31'17" 56d0'43"
Victoria Elena Lisa Zukas
Cam 3h29'21" 57d34'56"
Victoria Elizabeth
Ori 5h38'43" 3d42'48"
Victoria Elizabeth
Aqr 21h38'53" 0d44'16"
Victoria Elizabeth
Vir 12h53'52" 11d15'39"
Victoria Elizabeth
Aqr 22h17'34" -13d13'1"
Victoria Elizabeth
Sco 16h51'20" -38d17'19"
Victoria Elizabeth Brannock
Per 4h46'55" 44d55'54"
Victoria Elizabeth Faith Preiss
Ari 2h51'8" 28d20'57"
Victoria Elizabeth Hirsch
Leo 10h23'31" 22d1'45"
Victoria Elizabeth Miele
Vir 13h4'41" 4d44'19"
Victoria Elizabeth Moore
Lyr 19h14'29" 27d10'35"
Victoria Elizabeth Sellens
Crb 15h29'30" 27d3'37"
Victoria Elizabeth Stanford
Psc 0h50'52" 8d40'49"
Victoria Ella-Jean Guez
Del 20h38'28" 17d7'46"
Victoria Emily
Lib 15h27'35" -15d51'30"
Victoria Emma Long BA (Cantab)
Cas 1h59'7" 61d23'23"
Victoria Eve
Uma 10h26'17" 46d23'33"
Victoria Finney
Uma 11h44'40" 55d51'49"
Victoria Fitchlee (tori)
Tau 5h24'25" 27d5'48"
Victoria & Florence's Star
Cyg 21h25'8" 39d41'17"
Victoria Forever
Lib 14h50'28" -16d29'43"
Victoria Foster
Lib 15h18'18" -28d31'56"
Victoria G. Nazario (Viqui)
Ori 6h14'4" 15d15'10"
Victoria Gail
Vir 14h6'31" -15d26'34"
Victoria & Gary Fairweather
Cyg 21h40'54" 53d4'14"
Victoria Georgopoulos
Cru 12h35'14" -59d42'9"
Victoria Gilligan
Sco 16h16'14" -9d20'28"
Victoria Godkin
Leo 11h40'55" 26d14'39"
Victoria Grace
Crb 15h51'6" 34d25'14"
Victoria Grace Hetherington
Uma 9h34'8" 69d51'1"
Victoria Guay Sieradzki
Aqr 22h29'2" -0d3'36"
Victoria Hillman
Cyg 20h30'24" 37d28'18"
Victoria in Zambia
Ori 5h56'29" 11d23'20"
Victoria Isabella Plaza
Ari 2h29'46" 22d3'26"
Victoria Isabella Power
Cyg 21h49'18" 50d57'51"
Victoria J. Nebel
Gem 6h39'32" 20d26'55"
Victoria James Lambert
Sgr 20h11'34" -39d2'7"
Victoria Jane
Cap 20h35'26" -22d27'29"
Victoria Jane
And 23h56'37" 38d36'2"
Victoria Jane
Cyg 21h13'1" 43d0'14"
Victoria Jane Ford
Cap 20h26'6" -22d53'55"
Victoria Janell Farmer
Tau 3h44'54" 29d41'20"
Victoria Jayne Martin
Cas 23h56'56" 54d24'9"
Victoria Jean
Sco 16h37'0" -31d24'41"
Victoria Jean Ibbetson
Ori 6h8'10" 17d26'14"
Victoria Jean Valone Fowler
Sgr 19h20'46" -41d31'47"
Victoria Joanne Chenevert
Leo 11h52'15" 19d42'34"
Victoria Julia Vanderhoff Angelo
Ser 15h29'21" 19d14'38"
Victoria Katherine Kindred
Mon 6h47'8" -0d38'26"
Victoria Kathryn Large
Umi 16h46'12" 82d44'12"
Victoria Klinkowitz
Tau 4h26'31" 16d57'3"
Victoria Koydl
Sgr 17h55'53" -17d0'48"
Victoria Kravets
Tau 5h48'15" 27d14'54"
Victoria Kwan Wing Kwan
Cap 20h21'36" -21d22'2"
Victoria Laetitia v. S.
Uma 10h30'10" 51d47'2"
Victoria Lakers
Psc 0h54'0" 30d46'32"
Victoria Lambert
Psc 0h49'39" 9d48'45"
Victoria Langan 2nd February 1977
And 23h49'1" 40d16'55"
Victoria Leann Scott
Lyr 19h22'51" 38d7'30"
Victoria Lee Uhl
Lyr 18h32'25" 39d47'25"
Victoria Leigh
Crb 15h36'29" 30d34'39"
Victoria Leigh Bennett
Lib 15h11'53" -28d2'47"
Victoria Leigh Roveto
Aqr 22h55'58" -7d38'2"
Victoria Leona
Cas 23h25'47" 58d14'42"
Victoria Lombardi Jackson
Cap 20h50'31" -18d37'41"
Victoria López Agut
Leo 10h15'8" 24d35'40"
Victoria Loqui March 6, 1949
Psc 1h10'23" 29d10'15"
Victoria Louise
And 1h57'10" 38d2'7"
Victoria Louise
And 23h9'55" 38d26'3"
Victoria Louise Shaw
And 0h16'53" 39d20'25"
Victoria Louise Smith
Ari 3h12'20" 28d23'5"
Victoria Lovejoy
Lyn 6h37'12" 61d43'2"
Victoria Lucas
And 0h24'27" 29d16'40"
Victoria Luisa Rada Daniele
Cap 20h24'1" -16d6'15"
Victoria Lynn
Lib 15h4'21" -5d20'43"
Victoria Lynn
Lyn 6h41'34" 58d18'51"
Victoria Lynn
And 0h14'38" 29d37'5"
Victoria Lynn
Ori 6h7'8" 10d3'8"
Victoria Lynn 73079
Leo 10h43'39" 11d44'20"
Victoria Lynn Adley
Vir 12h21'33" 9d7'18"
Victoria Lynn Belmore
Gem 6h50'23" 19d52'3"
Victoria Lynn Bertness
Cas 1h21'38" 65d4'46"
Victoria Lynn Cody
Gem 7h12'1" 27d39'38"
Victoria Lynn De Leo
Psc 0h29'33" 9d50'53"
Victoria Lynn Evans
Sgr 19h47'47" -39d3'24"
Victoria Lynn "GiGi" Bailey
Vel 8h27'12" -45d5'37"
Victoria Lynn Heim
Lyn 6h29'2" 59d21'48"
Victoria Lynn Keller
Crb 15h34'41" 26d27'39"
Victoria Lynn Kimmel
Sgr 17h51'41" -28d30'14"
Victoria Lynn White
Vir 13h4'24" -17d2'0"
Victoria Lynn Witham-Wilkinson
Com 12h44'15" 23d55'23"
Victoria Lynne
And 0h49'34" 37d51'24"
Victoria M. Hunter
Lyr 19h22'35" 37d40'48"
Victoria M. Muir (27-03-82)
Umi 12h6'13" 88d31'33"
Victoria M. Turco
Lup 15h2'49" -43d50'36"
Victoria Mae Humphrey
Ari 3h18'3" 27d19'57"
Victoria Malia Skyer
And 2h26'11" 46d15'5"
Victoria Margaret Vanderpan
Lyn 9h32'46" 41d18'50"
Victoria Maria Casas
Ari 2h41'19" 14d33'34"
Victoria Maria Howell
Eri 3h42'39" -38d32'56"
Victoria Maria Velocci
And 1h8'12" 41d49'46"
Victoria Marie
Vir 12h9'1" 11d3'30"
Victoria Marie Arteaga
Cmi 7h37'0" -0d2'16"
Victoria Marie Fencl
And 2h12'22" 45d43'30"
Victoria Marie Holland
Psc 0h46'1" 6d33'49"
Victoria Marie Murphy
Hya 9h27'7" 4d28'4"
Victoria Marie Rose Miller
Ant 9h56'30" -38d13'53"
Victoria Marie Spears
Tau 4h7'1" 17d45'31"
Victoria Mary Mackie
Sgr 20h25'5" -42d16'35"
Victoria Mary Payne
Psc 23h17'40" 7d35'47"
Victoria Mase
Gem 6h41'52" 23d13'54"
Victoria Masters
Umi 14h18'19" 75d8'52"
Victoria Meppem
Col 5h59'32" -35d6'50"
Victoria Michele Nieliwocki
And 23h16'5" 44d16'38"
Victoria Moreno
Aqr 22h12'55" -12d46'6"
Victoria Morgan
Cas 0h59'42" 56d27'40"
Victoria Murray
Psc 23h28'5" -3d12'21"
Victoria My Love
And 23h8'7" 40d8'27"
Victoria My Queen
Cas 0h9'20" 55d52'24"
Victoria Nancy
Sco 16h58'43" -34d8'26"
Victoria Naomi Roslan
Aqr 22h11'2" -14d51'11"
Victoria Nickerson
Tau 3h41'10" 4d49'4"
Victoria Nix
Umi 15h11'45" 72d1'54"
Victoria Noel Washington
Cnc 9h5'18" 10d43'45"
Victoria Nohra - My Sister Forever
Cap 21h5'59" -25d15'59"
Victoria Oram
And 23h54'22" 36d4'1"
Victoria Orfelina Gonzalez
Vir 14h23'23" 2d10'53"
Victoria Paige Collins
Leo 9h54'8" 16d15'57"
Victoria Parry
Cap 21h28'30" -23d47'23"
Victoria Pearl
Leo 9h54'6" 7d2'35"
Victoria Pezzarossi
Lmi 9h52'48" 36d50'33"
Victoria Pickersgill
Leo 9h28'51" 24d42'50"
Victoria Pie in the Sky
Sgr 20h13'9" -38d42'22"
Victoria Praeclarus
Psc 0h9'16" 2d13'55"
Victoria Princess
Tau 5h10'51" 18d43'52"
Victoria Pussy Cat
Lyn 7h52'3" 57d11'58"
Victoria Quitevis
Lyn 7h53'49" 59d13'58"
Victoria R. Janik 6/10/20-12/14/02
Cas 1h8'50" 62d52'16"
Victoria Ramshaw
Lyr 18h29'16" 28d48'3"
Victoria Regina
And 0h14'35" 45d40'7"
Victoria Reid Jeter
And 1h25'7" 41d48'43"
Victoria Reidl
Peg 21h49'51" 13d50'15"
Victoria Rej
Cyg 21h40'59" 50d20'12"
Victoria Rene Black
Tau 5h44'9" 23d36'54"
Victoria Renee Brodeur-Sime
Uma 11h19'57" 37d36'17"
Victoria Rose
And 23h54'18" 33d57'12"
Victoria Rose
Ari 2h59'14" 29d2'46"
Victoria Rose
Leo 10h16'14" 27d21'52"
Victoria Rose
Sco 16h15'1" -12d15'17"
Victoria Rose 03152000
Lyn 6h54'41" 51d54'28"
Victoria Rose Fortner
Lyn 8h31'35" 34d45'21"
Victoria Rose Kennedy
Tau 4h35'42" 29d16'27"
Victoria Rose Teegan
Crb 16h21'12" 36d15'1"
Victoria Roshini Guruswamy
Psc 1h30'35" 11d51'26"
Victoria Ross Sayers
Lib 14h49'55" -1d29'57"
Victoria Russell
And 23h10'56" 41d59'48"
Victoria Ruth Joyce
Psc 1h6'24" 28d17'46"
Victoria Ruth Neiman
Hya 9h26'32" -0d38'16"
Victoria Salomone Meisler
Ori 6h25'34" 17d13'13"
Victoria Salute-Marion...A Star Mom
Crb 15h38'12" 28d39'29"
Victoria Sepulveda
Uma 10h24'12" 60d9'53"
Victoria Sergeyvna Rasskazova
Psc 1h44'32" 12d8'48"
Victoria Seyana Jamali
Com 12h50'13" 21d51'57"
Victoria Smart
Uma 11h34'47" 41d44'3"
Victoria Somlo
Leo 10h20'13" 20d4'24"
Victoria St. Mae
Cnc 8h47'24" 8d23'23"
Victoria Tate 59
Tau 5h33'19" 17d26'36"
Victoria Taylor Walker
Sgr 19h48'46" -16d41'33"
Victoria und Jörg
Uma 12h19'4" 60d25'33"
Victoria & Vanessa Ellsworth
Uma 11h52'31" 28d31'9"
Victoria VanLoan
Cru 12h40'25" -56d35'58"
Victoria Viernes
Col 5h50'43" -34d53'44"
Victoria Vittoria
And 23h48'37" 42d9'13"
Victoria Vrablic
Sco 17h3'16" -44d30'37"
Victoria Webb
And 0h37'59" 29d13'54"
Victoria Werner
Lmi 10h49'44" 29d30'18"
Victoria Wing - Ying Wong
Per 4h44'30" 41d1'12"
Victoria Yachiyo Ditaranto
Vir 13h38'19" 5d44'39"
Victoria Yen
Vir 14h15'49" 2d29'44"
Victoria York
Lyr 18h30'39" 36d25'41"
Victoria Young-Selvey
Uma 9h17'1" 64d8'36"
Victoria Zelter
Her 17h55'15" 20d6'7"
VictoriaCCT
And 23h40'38" 48d24'26"
Victoriadam
Sco 17h51'40" -42d35'55"
VictoriaFrakes
Crb 15h53'55" 26d54'5"
VICTORIAJASON
Vir 12h14'20" -1d12'11"
Victoria-Lafite
Cep 21h54'53" 65d2'41"
Victorian Beauty
And 23h50'43" 37d56'44"
VictoriaRose
Sgr 18h58'14" -34d46'17"
Victoria's Aster, Amor e Tria Amita
Ori 6h0'59" 7d13'49"
Victoria's Belly
Crb 15h42'54" 28d0'1"
Victoria's Brick
Cas 0h46'22" 65d16'29"
Victoria's Castle
Cap 21h36'29" -11d50'30"
Victoria's Dream
And 23h50'30" 40d58'23"
Victoria's Grace
Aqr 21h42'39" -0d12'33"
Victoria's Light
Cas 1h26'52" 51d6'41"
Victoria's little piece of heaven
Sgr 18h11'15" -23d17'39"
Victoria's Magic
Cap 20h56'4" -16d40'27"
Victoria's (not so) secret
And 0h41'27" 36d31'9"
Victoria's Personal Planet
Lyn 6h49'26" 55d8'32"
Victoria's Shining Star
Psc 23h25'48" 3d33'18"
Victoria's Special Star
Cap 21h43'0" -23d40'8"
Victoria's "Spot On" Star
Cap 21h47'38" -18d47'2"
Victoria's Star
Uma 9h49'17" 55d50'48"
Victoria's Star
Tau 4h6'54" 5d25'18"
Victoria's Star
Cnc 9h15'4" 13d6'33"
Victoria's Star
Gem 6h54'52" 21d51'1"
Victoria's White Lightning
Cru 12h54'29" -64d5'30"
Victoria's Wish
Peg 22h12'5" 23d18'0"
Victorious Judge Of Beloved Pearls
And 0h51'7" 38d38'0"
Victorius Ballerinus
Cet 1h13'49" -0d53'47"
Victor's 20th Birthday Star!
Vir 14h38'4" -1d30'24"
Victor's and Priscilla's Love Star
Sco 16h4'55" -12d24'48"
Victory
Cap 20h31'30" -16d34'48"
Victory
Cnc 8h32'31" 9d14'36"
Victory Anna
Tri 2h30'38" 31d47'38"
VictoryMan (GES)
Uma 13h1'2" 55d22'27"
Vida
Sgr 18h18'44" -19d23'33"
Vida
Per 2h22'28" 52d10'58"
Vida Mirzadeh
And 1h10'34" 41d23'11"
Vida Radice
Cas 1h30'20" 60d15'0"
Vida Roozen
Her 16h47'5" 22d51'40"
Vida Sax "My Shining Star"
Ori 5h34'52" 6d54'20"
Vida Zoltán
Cas 2h34'49" 65d33'56"
Vidal Fabela-My Eternal Star!
Her 17h21'47" 35d30'25"
Vidal Rodriguez Melguizo
Aur 5h53'40" 55d3'39"
ViDar 1985
Uma 10h36'34" 62d35'34"
Vidko Umek
Uma 11h10'0" 44d47'21"
Vidmantas Nemura
Uma 11h8'5" 40d40'36"
Vidok Batriz
Uma 11h12'46" 60d11'1"
Vidon
Per 3h13'31" 47d27'44"
Vidooch
Uma 11h19'17" 60d31'41"
Vidur Parashar
Her 17h1'14" 30d4'8"
VIDYA
Ari 2h45'11" 14d10'31"
Vidya Marie's Wishing Star
Ari 2h30'36" 24d50'31"
Vielka Jimenez
Crb 15h36'57" 32d13'52"
Viena
Uma 11h43'55" 49d9'11"

Viengdara Khenmy Gem 6h33'9" 16d47'37"
Vienna Carina Cnv 12h42'17" 41d14'56"
Viera Cechova Uma 11h49'24" 45d41'55"
Viera Hauriskova My baby Guinness. Leo 11h5'19" 0d4'43"
Viganò Laura Uma 10h7'41" 48d35'52"
Vigen Grigoryan Trdat Lyn 9h26'49" 40d50'20"
Vigigoodiba Uma 10h0'36" 46d47'21"
Vignan Cru 12h28'22" -61d7'35"
Vihor Victoria And 2h17'37" 47d29'18"
Vijay Lib 15h35'28" -26d17'11"
Vijay Ahire Vir 13h24'20" 13d43'16"
Vijay & Melissa Cas 0h37'53" 53d53'54"
Vijay Singhal Vir 13h39'58" 2d44'7"
Vijayamohan Xavier Palat Gem 7h46'27" 32d16'40"
Vijita Ori 4h52'25" 4d46'34"
Vikaly Cas 23h31'47" 53d4'2"
Vikas Kirin Bobba Leo 10h29'3" 27d6'53"
Viki Jo Olczak Cas 1h45'46" 61d9'21"
Viki Lynn Teafatiller And 23h41'50" 37d5'3"
Viking Uma 10h29'59" 46d42'33"
Vikki And 2h0'22" 45d18'34"
Vikki 116 Cap 20h24'10" -10d14'19"
Vikki Anne Cas 1h1'10" 48d43'55"
Vikki Bennett Lyn 7h22'53" 44d51'38"
Vikki D. Wright Aur 5h18'38" 30d25'42"
Vikki Lefkowitz Cas 0h8'18" 53d31'57"
Vikki Leigh And 1h29'26" 36d38'59"
Vikki LeSuer Lyr 18h30'21" 36d41'26"
Vikki Naomi Jones Psc 1h8'16" 27d22'1"
Vikki P And 23h20'32" 41d39'9"
Vikki Pond Dra 17h6'10" 62d8'53"
Vikki Skok Vir 11h53'22" -4d36'7"
Vikki's Star Sgr 18h11'27" -19d39'18"
Vikky Cam 7h58'45" 61d1'12"
Viktor Vir 13h2'34" 11d58'48"
Viktor Antony Carola Charbonneau Psc 0h30'59" 15d18'13"
Viktor Eigenmann & Andrea Odermatt Tau 5h45'12" 17d36'24"
Viktor Hensel Cam 6h33'31" 65d55'26"
Viktor Kavcic Uma 9h21'5" 41d49'47"
Viktor Klochai Cnc 9h14'42" 13d52'23"
Viktor Matrosov's Lucky Star Lib 15h57'10" -18d2'23"
Viktor Radians Cep 22h59'19" 75d42'42"
Viktor Solin Umi 15h3'5" 73d15'36"
Viktor "Viko" Serralde Cnc 9h17'29" 28d40'21"
Viktor Wallberg Cnc 8h30'46" 13d42'40"
Viktoria Tau 4h30'44" 10d34'0"
Viktoria Bodnar Psc 23h3'24" 7d40'0"
Viktoria Emily Kamenshine Tau 3h43'8" 16d35'26"
Viktoria Faith Hunt Uma 10h57'41" 48d52'57"
Viktoria Helena Ori 6h18'9" 10d43'50"
Viktoria Isabel Uma 10h15'23" 65d2'39"
Viktoria Makarevich Sgr 17h52'58" -25d20'29"
Viktoria Melik Cap 20h31'13" -26d56'2"
Viktoria Michelle Stumpek Tau 5h26'39" 19d4'5"
Viktoria Papagiannis Mon 6h44'37" -0d3'57"
Viktoria Retschmeier 25.01.1988 Uma 8h14'0" 64d46'19"
Viktoria Treneva Ari 3h2'48" 22d32'49"
Viktorija Lyn 9h28'55" 39d26'58"
Viktoriya Cyg 20h10'9" 38d38'1"
Viktoriya Cas 3h13'19" 65d7'21"
Viktoriya Frenkel Vir 14h33'33" 4d5'43"
Vil & Lah Lib 15h8'44" -6d15'29"
Vilero Lib 15h10'14" -9d59'58"
Vilia Co Aqr 23h19'13" -12d17'24"
Vilis Futurus Psc 1h6'34" 13d59'54"
Villa la Renardiere Crb 16h7'57" 36d47'40"
Villa Nova Hotel Lyr 18h18'30" 47d2'26"
Villamil Family Cyg 21h59'5" 45d55'14"
Villanella-Kellemeyer Uma 9h23'8" 44d23'5"
Villian Lopena Eri 3h48'36" -5d20'31"
Vilma Dra 12h48'44" 76d0'43"
Vilma C. Da Silva Crb 15h59'4" 26d36'39"
Vilma K. Pallette Sgr 19h34'46" -18d9'2"
Vilma Madeleine Nuñez Villanueva Cap 20h26'26" -18d58'58"
Vilma Rosa Arroyo-Williamson Lmi 10h24'39" 33d48'8"
Vilma S Tau 4h22'39" 6d59'19"
Vilma Stellar Elliott Cru 12h17'47" -57d3'24"
VILMAI IR PRANUI KRIAUCIUNAMS Cam 4h18'15" 67d1'36"
Vilte-Hope Olson Cnc 8h51'47" 25d49'8"
Vimlesh Mittal Crb 16h13'19" 25d44'12"
Vimone Dra 20h22'48" 63d0'26"
Vin Rosa 50 Psc 1h26'21" 32d1'44"
Vina Dra 16h33'33" 58d25'31"
Vina My Love Umi 15h28'15" 86d9'15"
Vina's Star Vir 12h46'7" 6d40'27"
Vinc Love Cep 23h6'20" 70d50'51"
Vince Uma 9h8'3" 54d17'27"
Vince Ori 6h0'52" -0d43'34"
Vince Boo 15h27'31" 49d18'26"
Vince 33 Cnc 8h27'59" 32d39'57"
Vince and Jen Uma 8h27'59" 60d57'7"
Vince and Mary Beth Mathews Lyr 18h18'0" 39d7'3"
Vince Brianlee Smith Sco 17h12'17" -33d55'10"
Vince Carlyle Hozier Umi 16h20'52" 76d55'30"
Vince & Doris True Love 4ever more! Col 5h40'14" -30d48'28"
Vince - Forever My Shining Star Cru 11h59'25" -62d13'29"
Vince Hesser Her 17h41'36" 44d13'47"
Vince & Janet Stockhausen Uma 9h14'40" 50d42'53"
Vince & Jeanette Gorman Lyr 18h47'31" 31d54'4"
Vince Lamanna Boo 15h11'0" 48d39'49"
Vince Landry Aql 19h39'36" 5d14'59"
Vince Pantano Cnv 12h40'56" 44d30'36"
Vince & Rose Uma 11h12'1" 56d27'13"
Vince Sides Her 17h8'7" 33d12'20"
Vince Spano Ori 5h27'5" 13d55'42"
Vince "sparkie" Ciolino Cra 18h4'38" -37d45'25"
Vince - The Duke - Howe Cru 12h34'15" -59d51'55"
Vince & Tyler Sticca Pho 1h13'0" -48d29'29"
Vince Vella Cru 12h38'40" -61d24'3"
Vince Visone Lib 15h36'34" -17d41'34"
Vince Watkins Her 17h37'43" 35d53'54"
Vince, Christie, & Jessie Joy Tri 2h0'25" 31d31'55"
Vincellie Tau 5h33'3" 20d27'0"
Vincent Her 16h36'37" 23d41'48"
Vincent Leo 11h40'49" 25d33'27"
Vincent Lmi 9h30'59" 34d47'8"
Vincent Cas 1h4'55" 53d58'9"
Vincent Cap 21h29'8" -14d48'21"
Vincent Ori 5h31'27" -0d45'41"
Vincent Uma 10h10'58" 62d20'50"
Vincent A. Proscia Uma 10h9'16" 58d14'13"
Vincent Aaron Gibbs Dor 4h40'35" -58d46'16"
Vincent Alexander Psc 1h12'47" 28d48'8"
Vincent Alexander Kisiel Ori 6h20'5" 10d33'33"
Vincent Alexander Sainato Leo 9h52'52" 16d19'23"
Vincent and Arline Benedetti Eri 3h40'14" -6d1'41"
Vincent and Yolanda Psc 1h4'24" 24d19'58"
Vincent Andreu Perez Vir 13h14'58" 6d19'54"
Vincent Angelo Ari 2h29'57" 19d22'45"
Vincent Anthony Alagna Per 3h52'2" 49d17'19"
Vincent Anthony DeSimpliciis Ori 6h15'37" 16d26'12"
Vincent Anthony & Nicholas Francis Aql 19h42'16" 10d30'33"
Vincent Aversa Cep 22h12'24" 66d3'48"
Vincent Baldassare Her 16h55'47" 32d55'54"
Vincent Benjamin Valter Cyg 20h4'58" 34d37'45"
Vincent Bernard Thompson Ori 5h42'22" 12d56'17"
Vincent Brian Linguanti Cep 21h40'59" 63d47'56"
Vincent Caputo Cnv 13h5'6" 47d3'46"
Vincent Carmine Cap 20h43'45" -26d32'42"
Vincent Carovillano Vir 13h29'24" 12d29'35"
Vincent Charles Sanchez Her 18h14'27" 16d29'8"
Vincent Christopher Spalletta Uma 8h50'34" 47d11'29"
Vincent Couture Cyg 19h54'8" 34d26'13"
Vincent D. Bowlen Vir 12h21'46" 3d33'59"
Vincent D. Mirabito Her 16h23'41" 20d29'58"
Vincent D. Steward "Vin-Vin" Lyn 7h10'15" 58d43'10"
Vincent Daniel Greto Sco 17h5'18" -35d39'23"
Vincent Daniels Aur 5h25'22" 54d2'53"
Vincent Darin Talley Dra 15h50'35" 61d37'52"
Vincent Daunais Tau 5h8'4" 21d2'3"
Vincent Dauria Vir 12h29'39" 3d10'21"
Vincent David Tobey And 0h35'19" 29d14'51"
Vincent DeBenedictis Her 17h17'9" 28d17'40"
Vincent Deleonardo Lib 14h26'24" -17d19'42"
Vincent Dominic Calvanese Umi 14h3'52" 76d23'35"
Vincent Drankwalter Jr. Cap 21h34'49" -13d8'13"
Vincent E. Brooks Jr. Aqr 22h9'57" -0d51'28"
Vincent E. Brought, Jr. Peg 23h40'53" 14d40'44"
ViNceNt EdWarD Lib 15h42'17" -19d58'21"
Vincent Emil DeMarco Per 3h25'36" 52d29'2"
Vincent et Bénédicte Her 17h0'59" 19d25'8"
Vincent Francis Gallo Psc 0h8'30" -4d18'42"
Vincent Frank Gural Umi 14h47'4" 76d48'10"
Vincent Geefei Wong Boo 14h38'6" 34d37'57"
Vincent George-Heidi Arnold's Daddy Sgr 18h25'39" -22d33'52"
Vincent Gerald & Lois Lee Sourbeer Umi 10h31'7" 85d56'52"
Vincent Ginnetti Lib 15h21'15" -4d19'38"
Vincent Giovanni Bandini Her 17h6'15" 32d28'12"
Vincent Grimm Always and forever Per 4h5'10" 37d17'31"
Vincent Hankins Her 16h33'45" 33d50'0"
Vincent Haywood Uma 8h42'27" 48d13'21"
Vincent & Hiromi "Forever & ever" Sge 19h20'6" 16d53'20"
Vincent Hong Uma 10h15'20" 49d26'42"
Vincent Iovine Her 16h44'58" 30d46'45"
Vincent Isaac Pimentel Leo 9h37'33" 13d24'32"
Vincent J Agosta Cep 22h3'29" 61d25'33"
Vincent J. Capece, Jr. Cnc 8h36'30" 31d1'28"
Vincent J. Mannino III Happy 40th. Cas 23h26'13" 59d48'47"
Vincent J. McCartney Gem 6h46'46" 27d20'6"
Vincent J. Rosa Lib 15h5'5" -0d37'53"
Vincent James Caputo Umi 15h51'43" 70d47'39"
Vincent James Hanson Aql 19h39'10" 14d10'27"
Vincent James Russo, Jr. Sco 17h57'42" -31d46'2"
Vincent James Scalio Lib 15h51'40" -18d22'52"
Vincent & Janice 10.26.02 Uma 10h42'55" 46d31'39"
Vincent (Jim) D'Annunzio Per 4h45'42" 41d40'53"
Vincent "Jimmy" Cusmano Gem 7h37'7" 28d3'35"
Vincent Joe Higens Aqr 21h50'57" -1d2'15"
Vincent John Hyland Ari 2h50'3" 29d16'43"
Vincent John Sorrentino Aqr 22h25'15" -12d48'29"
Vincent Joseph DeSantis Per 2h14'56" 56d9'40"
Vincent Joseph Lacanne Vir 14h4'46" 4d9'38"
Vincent Jude Creighton Aql 19h27'50" 7d42'12"
Vincent Koczurik Ari 2h41'2" 27d46'38"
Vincent Landgraf And 1h9'26" 47d31'50"
Vincent Lawrence Lambiase Vir 13h21'42" -0d32'18"
Vincent Legotte Leo 9h42'46" 28d33'35"
Vincent Lesnak Cyg 21h33'16" 47d9'41"
Vincent Llyod Thompson Psc 0h31'17" 3d51'58"
Vincent Louis Bourquin Ori 6h21'17" 13d32'36"
Vincent Louis Kruse Tau 5h38'8" 19d5'1"
Vincent Louis Linder Dra 15h20'18" 60d7'33"
Vincent Lubrano, Sr. Psc 0h9'45" -5d1'32"
Vincent M. DeMartino Uma 10h14'15" 45d32'29"
Vincent M. Kelly Her 18h13'21" 23d27'49"
Vincent M. Patrizi Cep 21h30'3" 63d59'12"
Vincent M. Philippi Aur 5h15'48" 31d34'39"
Vincent Maddox Ari 1h52'58" 18d26'51"
Vincent Marconi Lyn 7h36'10" 38d54'40"
Vincent Martin Magazzolo Per 4h17'37" 51d22'53"
VINCENT MASINO Uma 12h33'45" 52d27'35"
Vincent Matthew Comeaux Aqr 23h12'0" -4d51'13"
Vincent Mazzillo Vel 8h24'28" -47d28'7"
Vincent & Mia Tau 3h41'2" 29d20'47"
Vincent Michael Cap 20h48'42" -16d25'51"
Vincent Michael Litto Her 16h21'19" 41d32'12"
Vincent Michael Paglucci Per 3h17'5" 54d15'47"
Vincent Mirizio Uma 8h44'19" 57d28'31"
Vincent Montalbano Lib 15h17'27" -9d50'27"
Vincent Nappi Sco 17h20'37" -41d41'12"
Vincent Nesi Aqr 23h19'32" -21d16'56"
Vincent Novak, S.J. Leo 9h46'49" 28d16'30"
Vincent Orlando Schiavoni Psc 1h8'49" 28d29'0"
Vincent P. Clark Cap 20h17'25" -14d49'52"
Vincent P. DePonto III Lib 14h45'39" -17d36'4"
Vincent P Masi Ori 5h41'13" 2d16'20"
Vincent P. Rivera "I Love You" ATL Ari 2h5'28" 21d40'58"
Vincent Paul Cannariato Cep 21h33'11" 68d9'29"
Vincent Paul Martorelli Vir 13h55'45" -6d0'11"
Vincent Peter Manganiello Ari 2h56'24" 24d38'41"
Vincent Principe Aur 5h48'18" 49d8'41"
Vincent R. DeLeo Ori 5h47'2" 6d0'1"
Vincent R. Santise Aql 19h20'47" -0d15'36"
Vincent Raya Cap 21h34'38" -15d15'49"
Vincent Reed Her 17h18'46" 37d8'43"
Vincent Robert Federle III Ori 5h52'20" 9d20'11"
Vincent Robert Ferraro Psc 1h21'14" 32d31'17"
Vincent Roberti Uma 10h48'0" 58d14'34"
Vincent Roos And 23h14'37" 52d2'58"
Vincent S. Bennett Uma 9h57'22" 56d6'9"
Vincent Salvatore Conti Aql 18h51'21" -0d33'3"
Vincent Salvatore Minniti Lib 15h45'3" -29d12'30"
Vincent Sante Her 18h23'16" 22d48'16"
Vincent Sciullo Her 17h21'28" 36d14'30"
Vincent Sodd, Camel Jockey Cnc 7h57'48" 12d45'39"
Vincent - Soul Connection 5/30/49 Cep 20h50'3" 64d13'47"
Vincent Sr. Dra 17h56'40" 60d51'31"
Vincent Steven Hall Leo 11h42'40" 15d52'48"
Vincent Stinton Per 3h20'40" 52d41'50"
Vincent Thomas DiPietri Sco 16h9'48" -12d10'59"
Vincent Thomas Munno Boo 14h18'28" 12d52'59"
Vincent Timothy Nauheimer Cnc 8h0'20" 12d50'50"
Vincent Tricarico Ori 5h31'22" 6d49'6"
Vincent "Vinnie" Sperduti Ori 6h15'46" 13d52'31"
Vincent Visoiu Vir 13h20'43" 3d38'28"
Vincent W. Hogg Cap 20h8'59" -10d1'13"
Vincent W. Sparks Cep 22h20'12" 63d35'34"
Vincent Washington Aqr 21h43'8" 2d15'16"
Vincent Young Ori 5h37'22" -0d32'16"
Vincent Zansardino "1960-2004" Gem 7h48'39" 16d40'41"
Vincente Her 17h20'24" 36d12'28"
Vincent's Star Uma 8h38'7" 49d35'57"
Vincenza Ciampi Her 17h55'18" 21d23'32"
Vincenza Isabella Cyg 21h49'48" 38d16'13"
Vincenza's Dream Sgr 18h54'6" -34d21'51"
Vincenzio Her 16h52'56" 30d25'31"
Vincenzo Cyg 20h12'20" 36d57'42"
Vincenzo Crb 15h58'52" 39d7'6"
Vincenzo Her 16h37'48" 29d39'26"
Vincenzo Umi 14h23'45" 69d12'59"
Vincenzo Lyn 6h50'43" 60d45'48"
Vincenzo Uma 9h51'58" 55d14'41"
Vincenzo Albano Lyr 19h18'29" 28d16'19"
Vincenzo And Sophia Ruccolo Umi 14h39'39" 86d53'7"
Vincenzo Colosimo Dor 5h3'34" -66d41'42"
Vincenzo Galati Uma 11h54'27" 29d48'53"
Vincenzo John "Vinny" Milazzo Per 4h36'38" 49d15'24"
Vincenzo Joseph Guarano Aqr 22h14'23" -8d20'29"
Vincenzo Milione "Lil' Guy" Tau 3h37'35" 14d43'2"
Vincenzo & Nancy Star Ori 6h4'24" -0d13'9"
Vincenzo Saverio Sanacore Cam 4h13'50" 69d55'39"
Vincenzo "Vinny" Romeo Cnc 8h51'28" 7d24'44"
Vincenzo-Eva Col 5h46'53" -27d53'56"
Vince's Shining Star Cas 1h27'40" 57d27'32"
VinCin Cyg 19h45'49" 43d44'41"
Vincit Omnia Amor Boo 14h35'28" 53d52'26"
Vinco Zvaigzde Cep 23h12'0" 70d16'48"
Vinculum Unitatis Aeternus Cyg 20h15'34" 54d38'13"
Vincy & Norman Chan Gem 6h46'24" 17d34'56"
vindaloo Cep 23h10'1" 79d34'29"
Viness Love Eternal Cyg 19h15'49" 51d42'46"
Vineyard 208 Dra 18h3'14" 77d48'12"
Vinh Sgr 17h58'35" -19d25'2"
Vinit Nair Aqr 20h59'39" -12d48'19"
VinitaB Lib 15h58'1" -18d4'51"
Vinnie And 1h15'45" 42d6'6"
Vinnie Boy Her 17h54'59" 26d55'0"
VINNIE D'ANDREA Uma 11h17'26" 60d18'27"
Vinnie & Marie "The Honeymooners" Uma 14h16'50" 61d2'2"
Vinnie & Susanne's Star Sge 19h46'49" 16d42'11"
Vinnie Teigen Cnv 13h52'47" 39d25'57"
Vinnie The Kid Her 16h48'8" 32d47'49"
Vinno The Great Leo 11h46'12" 18d51'24"
Vinny And Antonella Cyg 20h13'54" 55d47'32"
Vinny Buzzurro Her 17h37'54" 27d24'43"
Vinny & Caterina's STAR Uma 13h53'16" 48d45'37"
Vinny DeMaio Per 1h47'13" 50d54'24"
Vinny & Nikki Romeo Cyg 21h47'32" 41d50'57"
Vinny "Superman" Nicosia Her 16h18'55" 48d55'53"
Vinny & Susan's Engagement Star Cyg 21h40'42" 36d27'33"
Vino Guarnieri Cyg 19h31'31" 52d40'12"
Vinod Mehta Cep 22h31'39" 62d49'50"
VinRod (Vincent Olmos Rodrigues) Her 16h38'53" 15d37'31"
VINTON Tau 5h45'30" 15d20'3"
Vinton Arthur & Phyllis Mary Corwin Cyg 21h55'42" 52d18'3"
Vinz Leo 9h57'31" 8d5'41"
Vinzi Carcione Cru 12h15'3" -61d53'25"
Viola Cas 23h49'40" 54d1'9"
Viola Psc 1h29'52" 10d14'33"
Viola Leo 11h17'0" 15d27'32"
Viola Cas 0h12'54" 59d32'23"
Viola Lyn 7h38'46" 41d7'58"
Viola A. Barraclough Uma 8h34'33" 59d38'17"
Viola & Abe Jones Uma 11h7'42" 36d11'48"
Viola Arlene Tau 4h19'35" 19d43'15"
Viola Barlotta Cyg 21h56'16" 50d39'1"
Viola Belle Cas 1h33'10" 58d2'7"
Viola Blakley And 0h12'11" 29d47'27"
Viola Britsch Uma 11h2'13" 58d58'10"
Viola Charlotte Cas 1h28'26" 66d55'57"
Viola E. Kimbrew Uma 9h55'42" 57d32'46"
Viola E. M. Simons Vir 12h46'45" 3d16'27"
Viola Esther Mecum Her 18h2'34" 25d1'39"
Viola Fay Wright Ewen And 1h11'16" 37d49'29"
Viola Fern Blanton Sgr 18h45'19" -23d35'58"
Viola G. Allen-Young Cas 1h8'27" 50d43'8"
Viola Grace Peg 22h41'11" 32d11'42"
Viola Harrigan 4-3-1915 11-11-2005 Ari 1h53'39" 18d22'0"
Viola Jerussi Tau 5h47'47" 15d57'20"
Viola Kozlowski Vir 12h50'45" 4d15'49"
Viola L. Husen And 0h13'18" 27d21'52"
Viola Patricia Bachino Sco 16h5'34" -17d11'48"
Viola Petrangeli Boo 15h8'41" 24d48'39"
Viola Purinton Giffin Psc 1h3'59" 8d50'49"
Viola Ronge Vitale Crb 15h39'45" 39d7'15"
Viola Tamberi Cas 0h29'17" 59d28'51"
Viola Walter (Best Mom) Aqr 22h17'14" 0d43'22"
Violesq Ari 3h21'30" 23d47'23"
Violet Lyn 9h12'44" 34d55'32"
Violet A. Giambalvo Gem 7h13'2" 15d18'30"
Violet Ann Moquin 11-19-35 And 1h38'58" 43d46'39"
Violet Beebee Sgr 18h25'55" -22d55'4"
Violet Duffy Vir 12h4'18" 7d37'27"
Violet Elaine Hughes Lib 14h31'29" -13d23'40"
Violet Eudora Koenig Cyg 20h50'40" 33d1'59"
Violet Gloria Millard Sgr 18h16'43" -26d52'25"
Violet Isabella Johnston Pho 0h27'14" -53d52'20"
Violet Johansen And 0h20'42" 27d29'5"
Violet Joy Weathersby Uma 11h20'32" 48d3'35"
Violet Julie Zena Young And 23h5'20" 39d34'36"
Violet Koresian Gem 7h39'18" 31d55'54"
Violet Lena Rose Crb 15h30'59" 29d14'13"
Violet Llolla Ari 2h50'59" 28d4'48"
Violet Lohr Uma 9h7'17" 62d52'8"
Violet Louise Davis Foltz Aqr 23h15'51" -16d17'40"
Violet M. C. Lyn 6h51'47" 53d57'17"
Violet Mabel And 1h11'50" 34d9'10"
Violet Mae Ari 3h20'21" 11d30'57"
Violet Marie Ham Tau 5h38'50" 22d52'57"
Violet Marion Joyce - Joyce's Star Uma 12h58'28" 53d29'13"
Violet May Morgan Cap 20h25'37" -11d25'42"
Violet Noella Gem 7h16'50" 16d10'24"
Violet Picabo Crabtree Cnc 7h56'10" 14d6'59"
Violet Richards And 0h8'46" 42d29'5"
Violet Smith Cnc 8h48'19" 18d6'48"
Violet Tergliafera/Scott Lyn 8h3'7" 59d38'33"
Violet Veloso Lmi 10h15'45" 31d1'56"
Violet "VMD 1975" Umi 16h38'18" 79d45'38"
Violet Wells Cas 23h37'52" 51d38'41"
Violet & William's Diamond Wedding Uma 9h30'31" 60d52'15"
Violeta Cep 22h22'27" 62d38'39"

Violeta B. Warner
Ari 2h34'41" 20d59'24"
Violeta Disla
Gem 7h30'17" 25d52'45"
Violeta Guzman
Lib 15h17'21" -27d59'49"
Violeta Maximus
Sgr 19h7'15" -17d29'35"
Violeta Stoevska
Psc 1h44'55" 16d10'35"
Violeta T. Maier "Moia Lubov"
Cyg 20h28'8" 48d20'56"
Violeth
Tau 3h47'57" 8d30'51"
Violetos Pleskeviciutes
Aql 19h46'53" 14d19'38"
Violet's Star
Cnc 8h12'36" 9d47'26"
Violetta
Cap 20h43'51" -20d40'57"
Violetta Ohliger
And 0h59'25" 39d36'32"
Violetta Simonacci
Leo 11h21'54" 18d49'52"
Violette
And 0h29'27" 42d31'56"
Violette Corneloup
And 1h4'47" 41d55'14"
Vionnie
Peg 22h49'5" 14d56'49"
Viotto Simona
Ori 6h6'48" 20d53'37"
Vippper67
Uma 13h37'27" 57d11'9"
Vips Uppal (Nana)
Cap 21h11'27" -16d4'6"
Viqi Diane Wagner
Per 4h41'11" 44d53'18"
Vir
Dra 10h22'0" 79d18'3"
ViRee
Mon 6h51'40" -0d37'44"
Virender Singh Ahluwalia
Gem 7h16'57" 24d57'5"
Virenia
Uma 9h16'30" 57d15'43"
Virg
Cap 21h15'31" -17d3'22"
Virgens Family
Per 3h8'31" 49d43'10"
Virgetta Rose Johnson
Leo 9h34'11" 31d33'41"
Virgie
Aqr 21h36'10" 2d55'32"
Virgie Armije Longwood
Vir 12h42'54" 10d22'50"
Virgie Opal Dobkins Oct. 8, 1910
And 2h8'3" 44d58'11"
Virgiestar
Mon 6h29'0" 9d17'54"
Virgil Alec Bartkowiak
Vir 13h42'13" -13d51'35"
Virgil B. Graves "Brightest Star"
Umi 14h59'20" 73d12'30"
Virgil Glaser
Uma 8h10'31" 67d23'9"
Virgil Greco
Uma 11h8'36" 68d28'21"
Virgil & Jack Cochran
Cyg 20h19'34" 41d31'10"
Virgil Jones
Uma 8h30'0" 64d37'2"
Virgil Lee Homister
Boo 14h17'8" 35d41'33"
Virgil McCoy
Cyg 20h46'40" 36d53'23"
Virgil R. Archbold
Aqr 21h31'5" 2d28'1"
Virgil Ray Brandenburg
Cap 20h25'13" -11d34'8"
Virgilio Cano
Col 6h5'0" -28d10'27"
Virgil's Bright Light
Leo 11h43'20" 20d32'58"
Virgin Maria Ayala
Sgr 19h34'37" -14d46'8"
Virgin Mary
Leo 9h34'14" 27d56'47"
Virgina Baird
Cap 20h16'18" -19d43'37"
Virgina Kogelschatz
Uma 10h15'9" 49d37'33"
Virgina Nardone
Cas 0h29'53" 62d53'4"
Virgini
Tau 3h45'21" 6d15'21"
Virginia
Psc 23h47'25" 6d39'10"
Virginia
Boo 14h23'35" 14d2'17"
Virginia
Ari 2h52'8" 26d0'30"
Virginia
Cyg 21h39'9" 48d55'26"
Virginia
Per 4h23'58" 42d53'26"
Virginia
And 1h15'6" 37d29'18"
Virginia
Uma 9h10'6" 58d52'24"
Virginia
Cas 1h17'28" 67d40'3"
Virginia
Sct 18h35'27" -14d7'17"
Virginia
Lib 14h33'15" -14d29'46"
Virginia
Aqr 22h26'49" -12d59'58"
Virginia
Cru 12h2'3" -59d33'45"
Virginia 140204
Peg 22h7'54" 10d59'32"
Virginia A. & Marvin B. Holstein
Cyg 20h31'19" 31d30'24"
Virginia Abbott Jackson Stanley
And 22h59'36" 47d22'49"
Virginia Abruzzini
And 2h9'49" 43d1'2"
Virginia Adams Smith
Uma 10h38'57" 46d27'4"
Virginia Amato
And 1h1'31" 39d15'21"
Virginia and Howard Nicholson
Cyg 20h13'13" 43d27'59"
Virginia Ann
Aql 18h51'59" 7d54'42"
Virginia Ann Borton Mosiman
Dra 19h21'9" 61d18'44"
Virginia Anne
Aqr 22h44'46" -23d28'37"
Virginia Anne
Cnc 8h53'53" 9d52'30"
Virginia Anne Hurd
Sgr 19h50'2" -13d1'52"
Virginia Ann's Family Star
Umi 13h12'11" 71d31'24"
Virginia Asia
Cyg 20h12'43" 31d12'22"
Virginia B. Stewart
Umi 17h0'23" 84d24'15"
Virginia Barboza
Crb 16h18'2" 34d58'0"
Virginia Barker Dominic
Uma 11h0'6" 48d2'20"
Virginia Bau
Uma 10h42'31" 45d51'50"
Virginia Beasley
And 0h29'10" 32d29'23"
Virginia Belford
Aqr 22h31'37" -1d49'1"
Virginia Bell
Cnv 12h34'10" 42d25'44"
Virginia Bellm
Uma 8h55'54" 57d43'49"
Virginia Bolling
Lib 15h18'19" -11d34'38"
Virginia Bruhn
Psc 0h19'9" 8d32'38"
Virginia C. (Ballentine) Wallerius
Ari 2h41'28" 21d37'26"
Virginia Caroline Webb
Uma 13h0'49" 53d13'40"
Virginia Carolyn Scott
And 23h39'11" 48d56'44"
Virginia Carty Cummings
And 0h35'33" 45d14'37"
Virginia Cassidy Cole
Uma 12h41'18" 62d1'34"
Virginia Catherine
Vir 14h17'43" -2d39'43"
Virginia Christensen
Lib 15h43'54" -17d2'46"
Virginia Christine 1947
Psc 1h3'1" 32d30'15"
Virginia Craddock
Cap 21h49'56" -13d12'40"
Virginia Curtis
Uma 12h59'59" 58d27'10"
Virginia Dale Sedgwick
And 0h47'3" 42d22'24"
Virginia Darnell
Uma 11h47'59" 58d14'18"
Virginia Dean Bryan Word
Uma 11h16'11" 37d16'35"
Virginia DiGiambattista
Cyg 19h29'4" 54d8'14"
Virginia Dilger
And 1h18'9" 38d9'15"
Virginia Drotar
Uma 11h32'58" 63d11'30"
Virginia Drummond Young
Crb 15h50'3" 28d10'47"
Virginia Dunaway Tyre
And 2h0'42" 39d18'7"
Virginia E. Anzivino
Sco 16h53'56" -41d46'39"
Virginia Edwards
Aqr 22h5'7" -2d27'53"
Virginia Egan
Cyg 20h5'39" 33d18'31"
Virginia Elizabeth
Aql 19h12'43" -0d0'13"
Virginia Elizabeth
Cap 20h38'12" -10d3'37"
Virginia Elizabeth Link
Ari 3h8'3" 29d54'42"
Virginia Ellen
Tau 4h37'45" 7d58'24"
Virginia Ellery Wyatt
Cas 23h58'1" 56d29'14"
Virginia Engelman
Sco 16h4'1" -13d25'58"
Virginia Esmeralda Dinzey
Tau 4h26'40" 17d11'48"
Virginia F. Lewis
Aqr 20h43'34" -13d37'33"
Virginia F. Nejdl
Uma 13h41'47" 52d39'47"
Virginia Farrell
Cas 2h35'18" 67d40'9"
Virginia Fay
Lyn 8h0'48" 55d45'3"
Virginia Fay
Gem 7h11'11" 34d14'35"
Virginia Filkins Brockner
And 23h48'52" 42d42'36"
Virginia Fiorillo
Crb 16h10'30" 35d5'1"
Virginia Florence "Ginny" Pierce
Cnc 8h26'59" 18d52'28"
Virginia Foss
Crb 15h52'55" 26d48'49"
Virginia G. Kabuss
Sgr 18h9'3" -23d9'54"
Virginia Gao
And 2h26'9" 49d35'17"
Virginia "Ginger" Chenier
Dra 19h33'34" 65d49'49"
Virginia "Gini" Anne Laurin
Psc 1h0'42" 13d53'31"
Virginia "Ginni" Kay Sexton
Psc 0h30'29" 4d59'7"
Virginia "Ginnie" Mitchell
And 1h17'54" 50d28'45"
Virginia "Ginny" Lee Bogan
Uma 10h1'3" 48d32'44"
Virginia Gonzalez
Uma 13h53'24" 49d55'31"
Virginia Gorychka
Cyg 20h29'3" 45d46'28"
Virginia Gould
Leo 9h28'46" 19d58'3"
Virginia Grace Matheson
And 2h9'50" 44d24'33"
Virginia Gutierrez
Umi 13h51'52" 76d10'41"
Virginia H.
Uma 8h47'42" 64d41'44"
Virginia H. Mooney
Dra 17h35'19" 54d25'1"
Virginia Harriet Breaks
And 23h23'59" 48d39'15"
Virginia Harrison
And 23h26'27" 47d51'28"
Virginia Hayden
Psc 23h48'3" 0d13'44"
Virginia Henderson
Tau 4h48'39" 18d38'19"
Virginia & Henderson Herod
Cyg 19h45'42" 31d49'42"
Virginia Hoeper Morisette
And 0h38'24" 33d9'14"
Virginia Hope Larson
Cnc 8h53'15" 30d44'52"
Virginia Howard Alice Buster Brown
Crb 15h28'3" 29d37'41"
Virginia Irene Bound
Cyg 21h32'33" 32d46'30"
Virginia Irene Wells
And 0h45'11" 39d39'49"
Virginia Irwin
Ser 15h33'14" 19d44'7"
Virginia is for Buckeye Lovers
Uma 9h23'19" 66d11'16"
Virginia J. Priebe Haux
Cap 20h33'30" -11d58'3"
Virginia Jablonski
Lac 22h46'15" 37d3'37"
Virginia Jane Parkes
Cap 21h38'50" -14d16'0"
Virginia Jean Whelan
Cyg 19h35'50" 28d16'0"
Virginia Johnson Richardson
Uma 8h13'36" 64d47'0"
Virginia & Joseph Girardi
Cyg 20h13'1" 54d4'13"
Virginia & Joseph York
Uma 11h29'7" 58d12'56"
Virginia Katherine Boysen
Vir 13h9'7" 5d49'7"
Virginia Kelly (Mom & Nana)
Uma 11h48'54" 49d52'48"
Virginia Kenney
Lyr 18h30'19" 39d28'19"
Virginia Kralj Rekoske
And 0h24'32" 42d9'17"
Virginia K.W. Lickley
Sco 16h13'39" -19d22'4"
Virginia L. Conner
Psc 1h37'42" 21d14'6"
Virginia L. Martin
Sco 16h20'45" -19d11'21"
Virginia L. "Nana" Studyvin
Cas 1h24'19" 56d12'0"
Virginia L. Phillips
Psc 1h5'13" 11d3'35"
Virginia Lea
Dra 19h0'12" 79d6'51"
Virginia Lee Gatliff
Gem 6h48'5" 20d22'51"
Virginia Lee Powell
Cyg 19h48'12" 36d3'53"
Virginia Leigh
Lyr 18h58'16" 26d15'58"
Virginia Leigh Marshall
Aqr 21h38'15" 0d28'25"
Virginia Lockwood Mahoney Star
Cap 20h21'27" -11d8'25"
Virginia Long-Cloutier
Sgr 18h40'23" -22d10'16"
Virginia Lora
Uma 12h2'12" 30d55'35"
Virginia Lowe, Grandma's Star
Lib 15h8'55" -20d59'15"
Virginia Lyle Capshaw
Psc 0h2'45" 4d8'0"
Virginia M Cobb
Ari 2h19'20" 25d17'37"
Virginia M. Doria
Vir 12h46'28" -0d27'57"
Virginia M. (Gengler) Lee 5/7/1940
Tau 3h49'58" 27d56'55"
Virginia M. Mendoza
Cap 20h40'28" -20d53'44"
Virginia Madewell Cummins
Lib 15h11'16" -5d41'34"
Virginia Mae Atkins Collins
Cas 1h12'25" 54d14'57"
Virginia Mae Lambert Enola
Cap 20h18'40" -10d55'9"
Virginia Mae Richards
Uma 13h38'47" 50d11'12"
Virginia Mae(Sue) Kirkland Robinson
Leo 11h12'27" -5d6'35"
Virginia Malone
Uma 8h34'3" 69d55'8"
Virginia Margo
Gem 7h23'52" 22d10'3"
Virginia Maria Brown
Boo 15h12'38" 48d13'16"
Virginia Marie
Lib 15h13'12" -10d9'35"
Virginia Marie Backes Wehrli
Aqr 22h10'26" 1d7'0"
Virginia Marie Kavall
Ori 5h18'5" -4d21'37"
Virginia Marion Halas McCaskey
Col 5h41'31" -33d1'44"
Virginia Marshall
Vir 13h16'17" -2d42'40"
Virginia Mary
Cyg 19h38'27" 34d14'10"
Virginia Mary Goodman
Lyn 8h2'54" 35d27'38"
Virginia Marz and Richard de Thomas
Umi 15h32'37" 72d47'50"
Virginia Maxine Downing
Cyg 19h37'15" 35d15'59"
Virginia May Severns
Cas 1h21'52" 51d57'13"
Virginia McEntee - The Best Mum
Aqr 23h17'16" -20d52'23"
Virginia Mechnig
And 1h11'29" 45d48'24"
Virginia Morland
Sgr 19h15'12" -22d58'19"
Virginia Mouse
Cas 3h16'34" 65d50'13"
Virginia Mudge
Cnc 8h36'58" 20d31'18"
Virginia Mumma
Lib 15h46'56" -23d21'7"
Virginia Mylar
Psc 0h37'21" 4d21'22"
Virginia Nell Scherbek
Cas 0h16'4" 50d10'25"
Virginia Nutt
Cas 1h15'37" 59d20'32"
Virginia Ogurkicwicz McNally
Leo 11h45'27" 25d58'12"
Virginia Oliver Bloomfield
And 0h49'40" 39d39'7"
Virginia R. St.John Reynolds
Cas 0h15'31" 51d23'5"
Virginia Read
Cyg 20h19'1" 41d21'34"
Virginia Reid
And 23h36'55" 48d25'18"
Virginia Renda and Family
Umi 13h39'33" 75d48'49"
Virginia Rhyne Richardson
Aql 19h1'15" -0d10'5"
Virginia Ritan
Crb 16h8'5" 33d14'29"
Virginia Roe
Cnc 9h8'3" 31d23'42"
Virginia Rose Boyce
Cyg 20h3'30" 58d16'17"
Virginia Rose Deatherage "Ginnie"
Cap 20h23'11" -11d37'58"
Virginia Ruiz
Peg 23h8'5" 31d49'34"
Virginia Ruth
Lib 15h41'49" -11d52'22"
Virginia Ruth Ingram
Sco 17h58'12" -40d17'13"
Virginia Ruth Zunner
Mon 7h17'13" -0d26'3"
Virginia Sheffield McClure
Psc 0h30'49" 6d58'53"
Virginia Shines Forever
Gem 6h41'49" 17d14'48"
Virginia Smee Dwyer
Cas 0h35'43" 57d30'29"
Virginia Smirnova
Tau 5h35'37" 22d30'19"
Virginia Sobolewski
Cas 0h40'31" 51d20'53"
Virginia Sue Nanney King
Sco 16h12'15" -14d54'39"
Virginia Townsley
Cas 23h30'29" 53d13'7"
Virginia Tredway Purnell Littleton
Sco 16h10'21" -24d32'34"
Virginia Trela
Aur 5h44'17" 47d24'47"
Virginia Tye Cockerill
Lyn 7h36'17" 55d2'53"
Virginia & Vincent Hoefling
Cyg 21h36'31" 50d41'29"
Virginia Walker Gaunt
Cas 1h28'50" 60d55'24"
Virginia Weber
Cam 6h37'20" 63d54'16"
Virginia Wyly Johnson
Sco 17h28'28" -34d53'22"
Virginia Yates Young
Psc 1h29'37" 5d12'3"
Virginia Young
Crb 16h13'7" 32d1'30"
Virginia Zarndt
Uma 11h41'30" 38d20'15"
Virginia, (Bear and Bull) Star
Tau 4h31'12" 25d27'26"
Virginia41
Sgr 18h12'36" -20d7'15"
VirginiaBlueyes
Lyn 7h10'28" 57d49'48"
Virginia's Jeanette
Cyg 20h26'56" 39d56'18"
Virginia-Star of Faith Hope & Love
Eri 3h41'33" -37d21'36"
Virginie
Ori 6h7'40" -2d9'19"
Virginie
Uma 12h39'46" 54d1'8"
Virginie
Cyg 21h16'2" 47d0'46"
Virginie
Her 17h17'42" 45d24'9"
Virginie
Vul 20h25'55" 22d18'50"
Virginie
Del 20h18'11" 9d39'40"
Virginie et Sébastien Flament
Lyn 7h30'40" 57d8'57"
Virginie et Serge
Vir 12h14'38" -11d12'0"
Virginie & Franck
Her 17h56'35" 49d58'39"
Virginie Jockers
Cas 0h59'11" 63d37'30"
Virginie Lasmazeres
Psc 1h40'12" 19d6'56"
Virginie Menne
Umi 19h24'56" 87d12'34"
Virginie Papaceit
Del 20h37'13" 7d22'19"
Virginie Pasquino
Peg 23h43'54" 26d4'33"
Virginie Yver
Dra 18h40'57" 57d27'16"
Virginie.Kuhfeld.CoeurdeWill.
Dra 19h33'33" 74d3'22"
Virginy # 1
Uma 9h36'42" 58d36'17"
Virgo Norma Hunton
Cyg 20h18'57" 43d34'44"
Viri Castillo
Cyg 21h20'8" 55d0'52"
Viridea
Boo 14h55'4" 53d39'36"
Viridea
Uma 9h59'21" 58d14'32"
Viridea
Umi 13h42'23" 69d50'42"
Viridea
Uma 8h26'36" 66d28'56"
Viridea
And 23h38'43" 50d28'9"
Viridea
And 23h40'10" 49d19'23"
Viridea
Cyg 20h30'30" 32d53'32"
Viridea
Uma 11h40'0" 36d47'57"
Viridea
Her 18h41'22" 24d24'13"
Viridea
Her 18h41'42" 25d51'40"
Viridea
Boo 14h54'10" 23d53'14"
Viridea
Her 18h32'52" 18d22'34"
Viridea
Boo 14h54'0" 19d11'46"
Virigina S. Chrismer
Uma 11h50'29" 52d0'25"
Virlea White Klingenburg
Cas 1h27'13" 55d48'15"
Virta M Walter
Umi 8h52'47" 86d9'59"
Virtara
Sgr 18h13'30" -26d33'2"
Virtue's Jewel
Dra 17h41'41" 66d28'57"
Virtus
Ori 6h1'26" 10d1'30"
Vi's Joy ~ Violet Palmer
Cas 0h21'1" 55d17'37"
Visa aka Rose Nguyen
Sco 17h2'47" -38d23'39"
Vishvesh Agrawal
Tau 4h34'18" 18d30'7"
Vispar
Sge 19h58'4" 16d29'1"
Visually Brilliant James E. Larkin
Tau 4h5'28" 22d45'18"
Vita Anksh, M.D.
And 0h42'59" 25d18'33"
Vita Chernyakhovsky
Aqr 22h55'56" -7d51'22"
Vita Kotsar
Cap 21h30'47" -10d16'5"
Vitali Allan Price
Leo 11h39'49" 24d35'20"
Vitali Franco VFR
Umi 17h32'3" 84d35'51"
Vitalie Bogorad
Ori 6h1'1" 21d14'10"
Vitality
Sgr 19h12'47" -16d16'3"
Vitaliy Baazov
Cnc 8h10'15" 24d39'5"
VITEKI
Uma 11h11'58" 52d25'5"
Viti
Psc 0h35'7" 7d6'25"
Vito
Cyg 21h38'36" 42d13'48"
Vito
Cnc 8h20'40" 14d19'18"
Vito 7
Dra 17h14'51" 57d55'53"
Vito A.Carelli Jr.
Gem 7h21'53" 22d7'29"
Vito and Bobbi Forever
Cyg 19h36'14" 32d28'19"
Vito Brundia
Cep 22h46'20" 66d33'43"
Vito Cammarota
Leo 11h39'43" 26d12'32"
Vito Carusello
Boo 15h35'36" 42d5'28"
Vito Ferretti
Umi 19h4'54" 87d15'11"
Vito Gambino's Star
Cnc 8h23'30" 12d53'31"
Vito "Horns the Adventurer"
Leo 9h37'34" 6d33'39"
Vito J. Posa
Her 16h11'13" 48d28'57"
Vito & Jeanette Mannone
Uma 11h50'51" 38d56'48"
Vito Mennona
Sco 16h9'24" -17d36'40"
Vito Richard Carriel
Lib 15h6'25" -3d5'24"
Vito Terrence Gallello
Tau 3h34'1" 1d14'34"
Vitoria Edwiges Pinheiro Cardoso
Uma 12h53'20" 58d11'23"
Vitrano's Inferno
Aql 19h46'44" 7d22'5"
Vitro G. Tulli
Cyg 19h59'20" 38d29'15"
Vittoria
Cas 0h29'15" 61d13'19"
Vittoria
Cam 5h31'26" 71d28'49"
Vittoria Alessandra Gallello
Sgr 19h23'46" -23d19'22"
Vittoria e Emma
And 1h59'22" 47d8'35"
Vittoria Elizabeth Ray
Cap 20h38'51" -8d49'26"
Vittoria Littlesnail
Per 4h8'39" 33d55'57"
Vittoria Mallano
And 0h26'48" 40d45'28"
Vittoria Sorice
Uma 11h5'49" 41d50'12"
Vittoria Urzetta
Uma 12h28'17" 60d46'42"
Vittorio
Ori 6h18'14" 9d44'18"
Vittorio
Tau 4h31'25" 28d19'30"
Vittorio Bossi
Uma 10h42'20" 40d48'29"
Vittorio Francesco Tiramani
Uma 11h22'49" 58d29'29"
Vittorio M. Morreale
Dra 15h50'5" 52d8'55"
Vittorio Shelby
Uma 8h35'32" 53d20'29"
Viv
Peg 21h52'23" 18d30'50"
Viv & "Big" Mac
Psc 0h39'30" 7d8'21"
Viv = Mummy V LUM x Keep Shining
Cyg 20h40'51" 48d18'30"
Viva! Access 2005
Crb 15h28'31" 27d21'14"
Viva Carter Steed
Uma 10h36'40" 63d40'40"
Viva Forever
Leo 9h22'32" 8d21'40"
Viva la famiglia Hugenberg!
And 0h44'54" 45d7'53"
Viva la famiglia Vicendese!
And 0h25'33" 45d6'11"
Viva Las Barb
Cyg 20h6'46" 32d1'42"
Vivacious
Crb 15h33'47" 28d12'54"
Vivacious Charmaine Keyser
Pav 17h41'24" -61d3'11"
Vivacious Victoria Bernadette
Gem 6h48'55" 34d12'53"
VivaKC
Lyn 7h42'46" 37d37'51"
Vivan & Donald's Guiding Light
Cyg 19h43'13" 40d16'6"
Vive
Her 18h41'16" 19d52'40"
Vive L'Amore Di Risata
Cyg 20h17'42" 54d25'34"
Viven BB
Per 3h39'26" 38d35'15"
Vivere per sempre
Psc 1h45'35" 7d39'58"
Vivi & Serge
And 23h36'25" 47d56'18"
Vivian
Cyg 19h40'33" 31d58'7"
Vivian
Psc 1h8'35" 11d18'49"
Vivian
Peg 21h45'14" 15d36'23"
Vivian
Sgr 17h54'58" -29d14'38"
Vivian An Chun Ku
Psc 1h3'50" 24d30'30"
Vivian and Jamal forever
Crb 15h32'22" 26d59'42"
Vivian and James
Cyg 20h9'19" 40d58'59"
Vivian Arnone
Gem 7h3'58" 15d56'42"
Vivian B. Alluisi
Uma 11h58'40" 36d59'36"
Vivian C. Roth
Vir 14h58'56" 6d32'46"
Vivian Chu
Lyr 19h19'32" 29d22'33"
Vivian Cirillo - VAVA's
Umi 14h48'17" 74d0'25"
Vivian Cobb
Crb 16h24'43" 30d24'4"
Vivian Costantino
And 1h40'57" 43d32'47"
Vivian Daoud
Col 5h31'58" -37d4'30"
Vivian de Lara Laureano Brukhnova
Ari 2h16'6" 16d50'39"
Vivian Deelite Lopez
And 0h35'12" 45d36'52"
Vivian Diaz
Cyg 21h12'28" 40d40'31"
Vivian Donn
Cnc 9h3'4" 15d3'17"
Vivian E. Tomaskie
Cas 23h17'32" 59d46'42"
Vivian Figueroa
Tau 4h6'44" 27d26'37"
Vivian Franco
Vul 19h27'19" 24d24'11"
Vivian Giordano-LaGreca
Cas 0h4'21" 53d46'9"
Vivian Jane Spuehler
Uma 9h47'9" 64d42'27"
Vivian John Mitchell
Uma 11h9'46" 72d16'28"
Vivian Joyce Taggart
And 1h13'18" 46d39'20"
Vivian K
Boo 14h17'58" 47d18'58"
Vivian Keller
And 23h1'40" 47d42'0"
Vivian Kolz
Tau 5h45'42" 17d3'46"
Vivian Kwan Kit Kit
Sgr 17h52'39" -29d29'52"
Vivian L. McGee
Cas 0h39'46" 61d57'58"
Vivian Lane Hardin
Lyr 18h44'6" 38d40'57"
Vivian Leiva
Sgr 19h8'27" -34d7'14"
Vivian Levenson
Her 18h25'34" 17d16'55"
Vivian Lynn
Tau 5h49'16" 22d49'50"

Vivian Madeline
Cas 1h40'51" 73d37'20"
Vivian Magoolaghan
Aqr 23h36'19" -20d38'7"
Vivian Marie Anderson
Lyn 7h53'40" 56d19'20"
Vivian Marie Blye
Umi 15h8'23" 78d55'32"
Vivian Marie Valadez
Leo 11h27'45" 11d39'28"
Vivian Michael Pahos
Ori 5h13'30" 11d52'40"
Vivian "Nanny" Wooldridge
Lyn 6h27'50" 60d2'43"
Vivian Nguyen Van
Sgr 19h54'16" -24d47'2"
Vivian Noel Mead
Leo 11h54'4" 18d49'58"
Vivian P. Morgan
Cas 1h32'19" 58d0'56"
Vivian Perez
Psc 1h25'18" 15d42'28"
Vivian Plantarich
Aqr 22h34'32" -8d44'14"
Vivian Rivera
Lib 15h3'19" -1d46'12"
Vivian Rose S.S.S. Rick
Cas 1h36'41" 65d42'10"
Vivian S. Clark
Cas 0h49'3" 51d58'52"
Vivian Thai Nguyen
Cnc 8h10'58" 22d1'40"
Vivian Tunnell Woods
Sex 9h42'26" -0d19'55"
Vivian Viola Cox Gingrich
Lyr 18h28'17" 37d6'49"
Vivian Virginia Davis
Psc 1h26'15" 12d50'31"
Vivian Walker Lang
Derouin
Tau 3h46'17" 20d38'54"
Vivian Wang
Dra 18h23'59" 72d36'40"
Vivian Welch Searcy
Vir 13h51'41" 4d41'9"
Vivian Windhorn Star Beta
Sigma Phi
Lyn 8h12'59" 50d21'55"
Viviana
And 1h42'10" 47d55'40"
Viviana
Peg 22h40'47" 27d32'11"
Viviana
Ori 5h49'1" -8d20'21"
Viviana Berrios Rosario
Sco 16h11'34" -22d34'37"
Viviana & Daniel Blumberg-Buzzini
Uma 10h56'4" 71d4'18"
Viviana Diletta Leoncini
"Leo"
And 2h19'49" 37d35'22"
Viviana Elias
Lib 15h20'35" -9d38'35"
Viviana Hebe
Tau 4h37'0" 12d0'1"
Viviana L Varas
Ari 3h27'17" 28d5'30"
Viviana Pearson
Com 12h22'52" 28d5'3"
Viviana Reis
And 1h4'59" 38d57'54"
Viviane * 812
Aqr 22h17'20" -24d32'8"
Viviane Be
Vir 14h21'49" 1d52'40"
VIVIANE FARFALLINA
Uma 9h56'46" 71d35'46"
Vivianne Lapointe
Cas 0h14'58" 62d29'41"
Vivianne Sarah Guildford
Lyn 7h35'28" 44d37'41"
Vivica Danielle Lopez
Lyn 7h40'1" 36d39'26"
Vivien
Col 5h32'55" -41d51'47"
Vivien
Del 20h20'37" 10d9'59"
Vivien Autumn
And 1h10'6" 45d0'23"
Vivien & Denise
Lyr 19h13'4" 29d7'55"
Vivien Helen Schwarz
Lyr 18h58'10" 27d44'6"
Vivien Kate Little
Vir 11h49'47" -0d11'29"
Vivien Sun
Uma 8h34'14" 50d46'7"
Vivienne BPG
Leo 11h12'3" 23d54'30"
Vivienne Dawn
Cyg 20h1'44" 34d55'20"
Vivienne Duffy - Twinnies'
Nan
Cas 2h8'40" 74d46'41"
Vivienne Margaux O'Hara
Lyn 7h38'14" 56d56'6"
Vivienne Rosemary Adams
Cyg 20h6'26" 33d7'11"
Vivienne Thanh Lan Duong
And 1h39'41" 43d7'44"
Vivien's Wish
Dra 15h49'42" 58d57'13"
Vivirenti
Cap 21h11'59" -21d39'16"
Vivi's Star
Cas 0h49'5" 52d11'36"
Viviven
Lac 22h46'27" 54d12'13"
ViViX
Aqr 21h49'24" -3d9'38"
Vivo Le S.I.S.
Cam 4h17'59" 66d48'45"
Vivo Per Lei
Lyn 9h15'29" 34d11'33"
Vivy Yan
Psc 1h10'7" 22d38'52"
Vixana Moonbeam Frisque
And 23h18'26" 48d29'22"
Vixen Goddess
Sco 16h11'0" -9d41'26"
Vix's (Victoria Garcia)
Pho 0h45'1" -49d19'32"
VJ
Aur 5h43'38" 41d3'19"
Vj Pizzonia
Uma 8h32'4" 72d13'12"
Vjeko
Ori 5h29'33" 9d24'7"
VKE62678
Cnc 7h59'19" 9d53'35"
Vlad
Psc 0h17'3" 20d54'35"
Vlad & Luda Forever
Lyr 19h9'27" 44d46'40"
Vlad Matei
Gem 7h3'36" 34d51'13"
Vlada - moja najveca zvezda
Cyg 21h34'51" 35d3'20"
Vladár Tamás
Ari 2h14'19" 14d21'57"
Vladeke
Per 4h31'3" 39d13'34"
Vladimil & Kimberly
Forever
Ori 6h3'13" 18d17'40"
Vladimir Columna
Sgr 19h7'14" -13d36'9"
Vladimir Ermakoff
Uma 13h32'11" 52d59'43"
Vladimir Garcia
Ori 5h8'6" 2d16'28"
Vladimir Kelemko
Uma 11h45'5" 43d0'17"
Vladimir Maslov
Ari 2h36'54" 13d11'2"
Vladimir Osinin
Her 17h20'37" 28d22'46"
Vladimir Perlovich
Aqr 22h3'16" 0d34'17"
Vladimir Primilsky
Sgr 19h44'33" -14d56'40"
Vladimir Purlo
Sgr 18h12'17" -18d28'40"
VLADIMIR RUBIN
Uma 9h45'38" 67d14'56"
Vladimir Shishkin
Aur 5h5'43" 34d49'9"
Vladimir Sigalov
Dra 15h49'32" 60d24'31"
Vladimir und Jana
Uma 10h18'45" 50d6'0"
Vladimir V. V.
Lyr 18h33'0" 37d10'55"
Vladimir Walter Petroff
Lac 22h47'40" 50d3'16"
Vladimira
Leo 11h20'59" 3d25'10"
Vladimira
Uma 8h34'34" 62d38'12"
Vladimiro Reynaudi
And 23h13'17" 44d38'30"
Vladislav
Psc 23h24'56" 7d23'17"
Vladislav Zhynov
Tau 5h22'45" 26d55'10"
Vladka Korytarova
Her 17h12'48" 26d55'29"
Vladochka
Ari 2h13'23" 26d46'56"
Vlasis Theodore Pappas
Her 17h34'1" 32d45'13"
Vlasta
Ori 6h11'56" 14d1'53"
V.L.C. III
Lmi 10h5'6" 29d53'3"
Vlliz
Col 5h48'39" -33d25'14"
Vlodjo
Lib 15h2'14" -1d37'45"
Vlynnbear
Peg 22h34'26" 21d1'43"
Vn.Ch.Kilyka'sMs. Rocky
Mtn Jubilee
Cma 6h40'23" -14d50'24"
Vo Phan Kim Khanh
Cam 4h24'31" 75d37'38"
Vo Thi Hong Van
Sgr 19h25'50" -12d27'3"
Voce e especial a mim,
Sarah
Cas 1h3'16" 64d56'9"
Voelker
Lyr 18h51'18" 43d56'10"
Vogel
Cyg 21h45'56" 35d48'9"
Vogel, Estella Cara
Uma 11h29'22" 40d26'40"
Vogel, Roland
Psc 1h55'45" 5d41'9"
Vogey's Journey
Uma 10h19'27" 59d47'22"
Vogue
Uma 9h30'13" 44d3'20"
Vogue
Vul 19h31'52" 19d58'43"
Vohn Patrick Mosing
Srp 18h32'47" -0d49'31"
voiceovereddiekkerneckeljr.
Aqr 22h7'51" -14d27'36"
Voigt, Holger
Ori 5h57'39" 21d2'7"
Volatilis altus procul
quadraginta
Ari 3h23'0" 18d37'36"
Vole and Volette
Leo 10h14'16" 21d23'5"
Volena Howe
Uma 8h20'50" 64d11'20"
VolimAdi
Cap 20h35'18" -22d40'22"
Volkan Ekinci
Leo 11h58'6" 12d30'37"
Volkard Flaig
Uma 8h23'46" 69d32'57"
Volker
Uma 8h44'17" 57d42'34"
Volker
Ari 2h45'43" 20d18'13"
Volker & Carina
Uma 12h45'18" 53d14'6"
Volker Frank
Uma 11h59'18" 30d44'59"
Volker Ludwig
Uma 10h11'37" 43d37'1"
Volker Schmidt
Uma 9h21'45" 41d46'12"
Volker Schröder
Uma 10h0'4" 55d54'54"
Volker Vortmeier
Uma 9h20'21" 56d12'55"
Volker Winzer
Psc 1h5'18" 3d21'18"
Volker Zimmermann
Uma 9h43'53" 59d43'30"
Völkl, Josef
Ori 6h19'47" 15d35'51"
Völkner, Hans-Joachim
Uma 9h52'9" 45d22'7"
Volkwein, Rainer
Uma 10h6'29" 61d26'23"
Volley Lube Banca Marche
Per 3h3'8" 46d55'14"
Volleygirl Bev
Cnc 8h40'6" 19d1'30"
Volmershausen
Her 16h20'48" 6d21'22"
Volovik Anatoly
Cnc 8h53'19" 6d44'42"
Volper
Gem 7h45'59" 34d31'31"
Voltaire F. Escalona
Sco 16h52'53" -29d25'21"
Voluntarius
Umi 15h57'49" 78d0'7"
Volz, Sabine
Uma 8h51'14" 69d40'7"
von Eysmondt, Thassilo
Ori 6h5'37" 6d4'24"
Von Heaters Destiny
Umi 15h57'8" 86d31'20"
Von Herman Harris
Lib 15h35'15" -16d46'54"
von Nell, Konstanze
Uma 9h32'10" 65d49'20"
von Oertzen, Klaus
Uma 10h35'14" 65d27'41"
Von Patricia Ramirez
Lib 14h53'15" -2d48'37"
Von Ray & Mary Dungan
And 23h16'59" 41d0'43"
von Rössing, Dirk
Uma 8h19'57" 63d49'54"
von Seidlitz, Friedrich-Ernst
Uma 11h3'41" 48d52'33"
von Tengg-Kobligk, Heinz
Uma 9h56'48" 46d30'3"
Vona Jean
Cma 7h18'10" -14d40'34"
Voncile Hovard Bush
Cas 23h38'50" 58d18'51"
Vonda Noe (My Girl)
Gem 7h17'13" 25d16'9"
Vonda's Alpheteous
Guiding light
Ori 6h13'7" 16d53'38"
Vonetta & Krystal
Gatewood
Cyg 19h43'4" 42d36'58"
Vonna P5
Uma 8h37'33" 69d46'30"
Vonnie
And 0h52'39" 45d27'58"
Vonnie
Cmi 7h29'6" 2d53'22"
Vonnie Jack
Vir 12h49'12" 5d42'41"
vonnsche
Uma 8h36'22" 62d42'36"
VonStula
Cma 6h33'25" -15d30'57"
Voodoo Prophet
Psc 23h16'52" 5d1'17"
Voody Voo
Uma 10h41'59" 53d25'51"
Vorndran
Mon 6h32'42" 7d53'46"
Vos, Freddy
Uma 10h40'43" 45d20'25"
Voss
Umi 14h55'55" 67d53'27"
Voti~La
Cyg 20h13'37" 54d33'14"
Voto Su Sopra a Stella
Lib 15h12'20" -6d27'19"
Votum Andreae
Leo 11h2'7" 10d9'36"
vous eclairez mon monde
Ori 5h59'44" 17d59'30"
Vous serez a jamais beau
Psc 1h12'36" 27d27'20"
Vova's & Sonya's Love
Lyr 19h13'23" 27d22'5"
Voveo
Psc 1h19'31" 24d50'11"
Vovo 3.141592
Sco 17h21'55" -42d18'4"
Vovó Lulu
Cnc 8h5'31" 21d13'16"
Voxelectronica
Cnc 8h50'30" 12d8'7"
Voyage
Eri 4h12'40" -0d7'15"
Voz Hermosa
Ori 5h55'58" 17d51'26"
Vozecita de Oro
Vir 12h7'1" -7d45'4"
VP Lubbo 2007
Ori 4h50'52" 13d31'10"
V.Patrick Main
Lyn 7h29'0" 51d55'24"
VPC III
Lyn 7h4'7" 54d59'20"
VPGVVMMAJ
Werner*Rörig
Uma 9h25'46" 64d16'5"
Vrai Amour
Uma 11h9'50" 31d3'26"
Vreni
And 0h48'4" 37d54'27"
Vreni
Umi 15h51'58" 83d10'10"
VRENISTAR
Cas 23h55'43" 53d41'53"
Vrezh Melikyan
Tau 3h49'42" 22d40'39"
Vrlik
Uma 12h48'24" 56d46'22"
Vroni & Andy
Uma 8h51'55" 54d3'0"
Vu Tram
Ari 2h40'28" 13d51'11"
Vukovac, Nadine
Ori 5h55'41" 21d33'13"
Vulpes
Cep 2h25'50" 81d18'59"
Vulpes Vulpes
Vir 13h50'53" 2d19'25"
Vunihar
Cir 15h0'13" -62d54'2"
Vunvi
Ori 5h57'22" 17d16'19"
Vupog
Per 3h23'22" 37d44'46"
V.W.B.
Uma 9h0'15" 68d15'7"
Vy & Anderson
Leo 11h7'17" -0d34'9"
Vy Shinn
Cyg 20h44'20" 42d30'42"
Vyara
And 1h23'12" 34d8'35"
Vyard
Lyn 8h12'17" 54d21'56"
Vylit O' Hara
Cam 3h59'24" 65d6'39"
Vytas Nausedas
Aql 19h56'58" -6d12'52"
Vyvy Truong
Umi 16h42'27" 83d40'52"
W. Attila
Cas 23h28'14" 55d42'32"
W. B. Higgins "A Soldier"
Uma 11h43'18" 54d8'50"
W. B. Sullivant Family
Ori 6h18'37" 5d42'28"
W. Belles, Jr.
Umi 13h28'19" 74d54'3"
W C Compton
Ori 6h3'15" 20d4'13"
W. D. & C. D.
Cyg 20h34'48" 40d59'14"
W. D. S.
Dra 19h26'29" 60d43'58"
W. D. Schellig
Her 18h36'49" 19d19'18"
W. Daemon Hillin
Pho 23h41'12" -41d18'29"
W. Foster Rich
Aur 5h44'58" 42d16'41"
W. George Thompson,
M.D.
Ori 6h19'10" 9d33'46"
W. Glenn Salmons
Her 18h54'35" 24d6'56"
W. Hawley Van Wycke
Uma 8h25'58" 63d28'34"
W. Herbert Goodick
Per 3h2'0" 42d35'6"
W J Gray 80
Sco 16h15'12" -13d10'57"
W. Jean Evans
Aqr 22h21'11" -16d43'13"
W. K. Camp
Psc 1h9'33" 32d48'10"
W. K. Penn III
Aql 19h55'31" 7d25'43"
W. L. Ili
Cas 0h55'21" 62d5'0"
W. Lance Speidell
Per 3h52'49" 34d26'54"
W & MB
Uma 12h48'16" 61d35'5"
W. Ned Croshaw
Uma 9h59'19" 58d1'15"
W P Catalano
Tau 5h22'15" 17d1'27"
W. Peter Maxwell
Vir 14h22'56" 2d56'31"
W. Preston Crook
Lib 14h58'8" -13d39'59"
W. Richard Adams
Ori 5h31'39" -3d13'35"
W Rombalski 01142318
Uma 13h13'34" 56d1'33"
W. S. Wood
Her 18h44'12" 21d33'45"
W. Scot MacDonald
Pho 0h43'35" -40d6'24"
W. Scott Applegate
Uma 13h35'50" 58d59'24"
W. Stephen Roop
Sgr 18h53'12" -18d54'0"
W. Steven
Cap 21h33'36" -16d18'50"
W Struan Robertson
Uma 8h10'56" 66d24'54"
W. Thomas. Gutowski
Her 16h22'17" 47d28'4"
W. & V. TREDER
Crb 15h53'9" 29d15'33"
W. William Rudolph
Her 16h42'15" 42d8'29"
W. Y. M. M.
Del 20h49'52" 9d53'30"
W1MMM
Cyg 20h9'45" 41d53'30"
W4- Daddy Weezer's
Christmas Star
Umi 13h12'35" 73d33'52"
W5
Cas 0h50'45" 75d5'27"
W.A. Jackson Ross &
Benjamin Ross
Sgr 20h4'51" -43d41'17"
Wa Wa
Lib 15h53'59" -17d53'52"
Wa Wa & Allen
Vir 12h27'40" 2d58'19"
wa8oba
Cyg 21h49'34" 53d50'4"
Waaaou
Sgr 18h14'2" -19d22'1"
Wabnig, Heimo
Ori 5h4'56" 6d46'41"
W.A.C.E.M.
Vir 13h7'19" 10d27'5"
Wachowski, Melanie
Uma 8h21'15" 73d6'30"
Wachtmann, Rolf
Gem 6h26'8" 20d11'43"
Wacki Milacki
Col 6h6'23" -35d28'44"
Wacko
Leo 11h3'46" 20d34'41"
WACOBE
Lyr 18h48'59" 43d35'30"
Wadan
Dra 17h8'43" 56d42'43"
Wade
Cep 20h31'23" 60d29'16"
Wade
Aur 5h46'50" 39d13'59"
Wade Aaron
Psc 1h27'33" 27d4'17"
Wade Aaron Zorn
Ori 5h30'3" 1d38'27"
Wade and Jennifer Holley
Uma 11h28'41" 40d48'39"
Wade Andrew Duncan's
Star
Cnc 8h50'48" 28d3'11"
Wade & Billie Wheeler
Cyg 21h33'55" 54d56'15"
Wade Chappele
Per 3h21'23" 37d1'58"
Wade & Chasidy
Tau 3h50'7" 26d7'20"
Wade Douglas Larson
Boo 14h30'44" 17d45'45"
Wade Krambeer
Vir 13h46'47" -7d7'32"
Wade Matthew Fricke
Uma 9h57'49" 60d44'28"
Wade Mickey Nye
Dra 19h5'8" 52d12'57"
Wade Patrick Wyatt
Her 18h56'26" 24d3'45"
Wade Phillips Wedding
Star
Aql 19h43'57" -0d1'25"
Wade S. Baker
Uma 9h21'3" 49d36'8"
Wade Scott Berger
Her 16h31'20" 6d52'23"
Wade, Prince of Stars
Tau 5h10'30" 16d47'53"
Wadenova
Tau 5h13'53" 20d43'29"
Wadih El Manneh
Sco 17h48'46" -42d31'54"
Wadii
Cap 21h7'25" -26d45'32"
Wael
Sgr 17h57'47" -19d36'0"
Wael
Uma 10h39'33" 69d16'10"
Wael R. Al-quqa
Uma 11h42'44" 64d48'52"
Waffles
Uma 12h34'5" 56d32'1"
WAG 232
Tri 2h21'2" 33d15'46"
Waghubinger, Franz
Uma 10h32'58" 57d51'14"
Wagner
Cen 13h44'22" -32d38'20"
Wagner
Her 17h7'30" 31d22'22"
Wagner Triplets
Ari 3h21'29" 27d36'24"
Wagner, Albert
Uma 9h54'17" 47d26'22"
Wagner, Matthias
Ori 5h58'5" 21d32'31"
Wagoner's Wish
Aqr 21h38'32" -0d34'44"
Wagonman102
Cru 12h0'18" -62d49'2"
W.A.G.S.
Vir 12h29'58" 0d13'10"
Wags La Vie est Bonne
Dra 15h8'7" 55d10'45"
Wagschal, Hermann
Uma 11h11'42" 56d26'35"
Wahahuy
Umi 13h18'19" 74d45'42"
Waheeda Heyliger
Uma 13h40'54" 57d4'15"
Wahkan (Sacred)
Sco 16h41'48" -25d57'55"
Wah-Wah
Uma 11h3'20" 62d12'12"
Wai Kin Hong
Uma 11h24'33" 60d15'26"
Wai -Yau and Wendy
Cnc 8h26'8" 18d44'52"
Waichill
Uma 9h29'47" 48d53'39"
Waidhas, Thomas
Uma 11h35'18" 39d31'31"
Waikit ~ Jas's Light in the
Night
Col 5h49'16" -34d12'40"
Wailing Sam
Ori 5h47'32" 7d22'34"
Wainani
Ari 3h11'30" 29d22'23"
wAiT wHaT ... LuV u
Sge 20h1'17" 19d30'32"
Waj & Kay 4 Eva
And 0h14'11" 45d47'28"
Wakana Lu Smith
Ori 5h30'56" 3d46'41"
Wakan'gli Peta
Uma 9h9'5" 49d32'14"
Wake up, wake up, wherever you are
Sco 16h42'52" -28d53'53"
Waker
Cet 0h52'36" -0d35'9"
Waki Maki
Umi 14h13'33" 76d20'54"
Wakita
Uma 9h1'46" 59d16'36"
Wala
Ori 5h38'22" -2d52'26"
Walbert's Zack-n-Bleu
Zena
Cnv 12h32'7" 42d53'22"
Walbert's Zack-n-Bleu Zeus
Cnv 12h37'0" 42d24'20"
Walbie Kolodziej
Leo 10h38'2" 14d8'41"
Wald, Alan
Uma 11h59'21" 51d17'13"
Waldemar Badia Catala
Cen 11h37'26" -48d28'53"
Waldemar Pelz
Uma 11h4'13" 69d56'20"
Walden Pampennamar
Uma 10h11'38" 50d9'47"
Waldo
Lyr 18h47'43" 45d25'47"
Waldo
Uma 12h37'27" 58d3'18"
Waldo Crow
Uma 13h6'30" 63d19'40"
Waleed Monzer
And 2h35'11" 49d13'42"
Walet Wishes
Cyg 20h21'12" 46d58'39"
Walid Nasef
Aur 5h38'34" 41d49'3"
Walin Pena
Sgr 18h40'4" -28d49'11"
Walk With Me
Pho 2h18'11" -40d27'53"
Walker
Cyg 20h27'21" 52d45'25"
Walker Owen Thompson
Umi 15h33'42" 79d6'44"
Walker Roach, Jr.
Uma 10h22'11" 64d20'36"
Walker, Glenn
Cap 20h36'59" -24d17'8"
Walker's "Aonghus"
Cru 12h48'13" -62d2'6"
Walker's Dark Helmet
Ori 5h59'26" 6d54'57"
Walker-Stone
Cyg 19h37'42" 32d16'59"
Walking in Oakwood
Psc 23h35'22" 4d43'30"
Walking Tree
Cam 6h19'28" 78d20'52"
Wallace and Heather
Pickett
Lac 22h25'3" 48d24'58"
Wallace Covington
Mon 7h33'23" -0d37'56"
Wallace Dillon Whiteley
Vir 13h29'43" -7d0'25"
Wallace Eugene Beebe
Cnc 9h19'32" 10d43'55"
Wallace H.J. Chang, M.D.
Tau 4h30'15" 0d35'6"
Wallace Hubbard
Per 3h22'24" 45d57'49"
Wallace (Leif) Nelson
Gem 6h48'20" 22d14'16"
Wallace Main
Aql 20h14'28" -0d2'29"
Wallace & Marion's Star
Cyg 20h6'13" 43d53'13"
Wallace Medralys
Ori 5h17'42" 7d7'41"
Wallace Morgan Kuehl
Vir 14h2'40" -11d37'34"
Wallace Ramirez
Aql 19h46'38" -0d1'59"
Wallace Silver Star
Sco 16h15'43" -16d25'27"
Wallace Wong
Aur 6h14'39" 52d36'42"
Wallaman
Lib 15h46'0" -4d12'45"
Walle
Cyg 20h9'18" 36d54'42"
Wallenpaupack
Crb 15h31'39" 35d34'7"
Walle's World
Cyg 19h36'23" 32d8'47"
Wallisont
Cas 0h8'57" 58d43'22"
Wallsie
Cap 20h30'59" -14d21'36"
Wally
Tri 2h29'3" 31d29'41"
Wally
Cnc 8h58'44" 14d8'55"
Wally "B" Blake
Lib 15h46'58" -16d58'59"
Wally & Ellen Richardson
Forever
Cas 0h20'7" 56d43'43"
Wally & Irene Jester's 60th
Dra 18h41'3" 52d54'29"
Wally Kleinfeldt
Tau 3h59'6" 25d17'35"
Wally & Lina-Stars
Together Forever
Cyg 21h22'24" 30d11'58"
Wally Saylor
Aqr 23h0'23" -8d13'56"
Wally Smith, Beloved
Father
Ori 5h41'28" 1d5'12"
Wally Swiercz
Per 3h34'34" 49d19'44"
Wally's Star
Cen 14h7'50" -61d42'42"
Wally's Strawberry
Spokadelli
Cas 23h29'39" 59d7'40"
Walmar-rukwor 121391
Aql 19h45'24" 14d39'27"
Walmer Lodge Apts.
Lyn 6h50'59" 52d53'59"
Walnut
Lyn 8h49'37" 33d37'1"
Walrus and the Mermaid
Cyg 20h5'33" 45d50'30"
Wals
Cyg 19h52'55" 51d48'27"
Walsh
Aql 19h6'17" 5d32'41"
walsh80
Tau 5h18'54" 25d45'20"
Walt & Chelsea 38
Aql 19h28'58" -0d27'27"
Walt & Donna Myers
Cap 20h54'15" -20d47'29"
Walt Dugan
And 0h36'52" 26d22'0"
Walt Gdowski
Leo 10h30'23" 13d55'45"
Walt Jones (Bama)
Ari 3h6'16" 28d23'38"
Walt & Lisa
Cyg 21h31'0" 41d48'53"

Walt Luszczak
Aur 5h43'37" 36d59'11"
Walt Rich
Per 3h7'8" 54d59'55"
Walt & Steph's Star Phone
Cnc 8h37'20" 21d11'33"
Walt Swarthout
Cnc 8h44'27" 16d28'36"
Walt Weaver's Rising Star
Umi 11h35'29" 87d32'44"
Walter
Sgr 18h47'57" -16d30'9"
Walter
Sct 18h25'21" -15d17'26"
Walter
Aqr 22h31'0" -9d13'47"
Walter
Lyn 8h14'17" 56d58'44"
WALTER
Uma 8h55'8" 56d33'41"
Walter
Uma 13h54'45" 57d34'48"
Walter
Leo 9h47'59" 26d25'4"
Walter
Aqr 22h29'57" 1d46'31"
Walter
Aur 5h53'5" 52d41'19"
Walter
Per 3h30'6" 49d16'7"
Walter A Lee Jr Family Star
Uma 13h47'41" 52d18'48"
Walter A. Martishius
Her 16h37'14" 47d50'4"
Walter A Oprych
Uma 9h49'20" 47d37'13"
Walter & Aida's 60th Anniversary
Uma 11h20'16" 57d22'43"
Walter Albert Dombroski
Her 16h41'35" 14d22'46"
Walter Amundsen
And 0h53'57" 35d45'59"
Walter and Ann Hurley
Sco 16h12'51" -11d20'34"
Walter and Essie Poole
Cyg 20h11'1" 53d55'21"
Walter and Genevieve Kee
Leo 11h29'31" 10d23'34"
Walter and Jean
Aqr 22h26'19" 1d14'9"
Walter & Anne Always
Aqr 22h37'49" -8d13'13"
Walter & Antonia Bebirian
Sgr 19h2'50" -26d48'22"
Walter B. Wriston
Leo 11h21'46" 16d26'29"
Walter Bassett
Srp 18h9'4" -0d18'33"
Walter Baumhoff
Ori 6h13'43" 20d53'1"
Walter Bays
Psc 23h59'26" -3d49'44"
Walter Beeyt
For 3h13'11" -26d24'5"
Walter Blaine Smith
Sgr 19h57'58" -19d56'4"
Walter Blessey, Jr. 4-11-06
Uma 10h23'14" 43d34'49"
Walter C Righter
Per 2h24'12" 55d31'52"
Walter Casabonne-My Grampee's Star
Aql 19h39'49" 5d34'11"
Walter Castracane
Cap 20h20'52" -11d36'57"
Walter Charles Noon
Per 4h30'21" 40d34'12"
Walter Cox
Uma 11h2'15" 46d49'21"
Walter Curphey
Aur 5h34'50" 53d27'15"
Walter D. Galloway
Her 17h17'12" 34d3'5"
Walter Day
Cep 0h34'44" 85d22'26"
Walter & Delphine Brenner *60th*
Cyg 19h56'10" 51d17'37"
Walter Dennis Oliver
Aql 19h56'19" -8d30'26"
Walter Dettmann
Her 16h44'53" 36d32'26"
Walter E. Andersen
Uma 8h50'41" 49d14'54"
Walter E. Graff
Her 17h2'22" 34d14'40"
Walter Edward Dahlem
Cyg 20h35'14" 47d35'29"
Walter Edwards, Jr
Uma 11h21'6" 51d31'26"
Walter Eidson
Her 17h23'0" 25d8'30"
Walter & Eileen Stokowski
Tau 4h40'3" 17d5'0"
Walter F Baron Sr
Lib 15h22'22" -27d52'21"
Walter F. Cummings
Aur 5h48'11" 44d38'2"
Walter F. Smith
Uma 10h28'15" 66d6'58"
Walter F. Terlecki
Lib 15h44'55" -17d40'45"
Walter Family Star
Uma 9h11'48" 50d55'32"
Walter Feuerhahn
Uma 10h3'26" 44d43'25"
Walter Fitzgerald
Cnv 12h40'13" 40d21'25"
Walter Florian Haemmerle, M.D.
Aql 20h20'0" -0d32'10"
Walter "Foxy" Frye's Family Star
Ori 6h19'25" 14d48'38"
Walter Fredrick Kuehne
Leo 10h53'57" 6d34'25"
Walter G. A. P. H. Heidenreich
Uma 12h53'8" 62d10'22"
Walter G. Arias II
Lib 14h59'33" -13d18'28"
Walter G. O'Connell
Cep 0h8'10" 77d53'7"
Walter G. Potter A Shining Star
Her 18h47'12" 20d23'9"
Walter Galarza
Cap 20h40'52" -17d2'22"
Walter George Saleman
Per 3h40'5" 34d5'2"
Walter H. Cardona
Sco 16h9'17" -16d20'43"
Walter H Hubbard II
Leo 11h24'27" 2d6'15"
Walter H. Musser
Uma 10h20'44" 58d54'47"
Walter H. Zaptin
Uma 11h49'10" 43d2'54"
Walter Haefner
Vir 14h39'20" 6d21'19"
Walter Hammon
Boo 14h55'8" 51d36'53"
Walter Harold Elkins Jr.
Cap 20h8'34" -12d23'37"
Walter Holidy Criner
Ori 5h21'40" 1d14'15"
Walter Holiday Criner
Gem 7h20'59" 16d54'45"
Walter Holmich
Lyn 6h49'47" 60d36'9"
Walter J. Lubanty,Jr."You're loved"
Ari 3h14'13" 22d18'14"
Walter J. Pierson
Aur 5h31'57" 47d27'42"
Walter J Pituch
Per 3h50'38" 34d8'39"
Walter J. Rouse
Her 18h22'46" 23d22'30"
Walter James Aitchison
Uma 8h11'20" 70d1'2"
Walter James Lacy
Per 3h31'53" 45d37'30"
Walter James Riess
Ori 5h25'27" 5d11'25"
Walter & Joann Belowsky
Uma 13h39'1" 62d15'57"
Walter John & Erin Shultz
Cyg 20h55'31" 31d58'48"
Walter John "Jack" Writt Jr.
Sco 17h47'35" -42d1'38"
Walter John Walkington 10-07-54
Cyg 21h28'13" 39d34'57"
Walter Kachel
Uma 9h43'26" 58d2'13"
Walter & Kathleen
Cyg 20h20'19" 55d24'33"
Walter Koehler
Uma 8h21'31" 70d10'11"
Walter Krummenacher-Gacita
Boo 14h54'15" 13d35'37"
Walter Kuhn
Sco 16h9'34" -11d18'35"
Walter Kulesza
Uma 8h13'51" 64d22'26"
Walter L. Beyer
Leo 10h15'13" 22d47'10"
Walter L. Mikulski
Tau 5h36'51" 25d26'31"
Walter L. Swanson
Ori 5h23'3" 9d57'57"
Walter Lambuth Thomas
Cep 23h41'37" 79d31'4"
Walter Leslie Brooker
Sco 17h54'32" -39d27'33"
Walter Liebig
Ori 5h51'3" -5d25'11"
Walter Lindsay Varley
Vir 12h19'40" 10d47'18"
Walter Louis
Uma 11h25'1" 33d58'48"
Walter M.
Crb 15h32'27" 36d21'8"
Walter M. Donohue
Psc 1h44'14" 7d17'59"
Walter M. Johnson
Aur 5h51'7" 51d9'45"
Walter Malcolm Crews
Cyg 20h2'58" 38d43'40"
Walter & Marian Martin
Cyg 21h34'56" 51d9'31"
Walter Martin Soltys
Tau 4h9'1" 29d27'29"
Walter Matthew Rasic
Cep 20h25'21" 60d22'53"
Walter McKillop (Mac)
Uma 14h12'32" 60d20'24"
Walter McKinley Adams
Umi 15h17'23" 71d50'42"
Walter Meyns
Uma 8h21'55" 66d52'55"
Walter Montogomery
Cyg 19h42'12" 32d0'25"
Walter Moore
Cap 21h52'52" -16d22'19"
Walter Munn - A Star in Our Eyes
Sgr 18h30'26" -27d41'13"
Walter Napoleon
And 23h1'55" 51d3'53"
Walter & Noreen Meyer
Umi 16h55'41" 76d28'46"
Walter Notheis
Uma 10h56'23" 63d26'43"
Walter Nuyens
Aqr 22h48'41" -17d10'9"
Walter O. Johnston
Uma 10h27'56" 58d35'50"
Walter Olin Brown
Tau 4h18'47" 13d47'13"
Walter Ortega
Ori 5h56'5" 2d20'12"
Walter Patrick Allen
Ari 2h54'9" 21d28'38"
Walter Paul
Cen 11h47'26" -52d8'10"
Walter Pavlo
Cep 21h37'30" 61d39'50"
Walter Pedro
Aur 5h53'39" 52d42'20"
Walter Powell
Cyg 20h12'24" 41d9'23"
Walter Priebe
Leo 11h34'41" 8d17'24"
Walter R. Conner
Aql 19h8'46" 0d0'49"
Walter R. Troynack
Cnc 8h14'29" 15d34'19"
Walter Ray Wingate
Cep 22h17'13" 64d37'10"
Walter Revere Marsh
Aur 5h37'50" 38d41'12"
Walter Rhodes
Cap 21h4'1" -19d53'38"
Walter Richard Peppers
Lib 15h16'23" -27d12'0"
Walter Robert Bowker
Tau 5h17'35" 16d23'8"
Walter Robert Braden
Lyr 18h25'16" 32d43'40"
Walter Royster Jr.
Uma 13h7'37" 62d12'56"
Walter S. Davies
Psc 0h5'57" 7d45'8"
Walter S. Huff, Jr.
Sco 15h53'25" -20d41'11"
Walter Scollard Jr
Cap 20h21'33" -24d54'29"
Walter Scott
Gem 7h1'38" 22d11'7"
Walter Scott Buchholz
Aur 5h52'10" 47d28'54"
Walter Scott "Hero" Star
Cyg 20h34'22" 53d33'2"
Walter Serbin
Uma 8h17'35" 60d15'55"
Walter Stanley Niedewiecki
Uma 11h56'7" 55d26'11"
Walter Stefan Novotny
Cnc 8h10'41" 17d55'17"
Walter Stein
Aql 19h51'48" 6d34'39"
Walter Thibeault
Sgr 18h34'23" -23d37'31"
Walter Vernau
Her 17h40'4" 20d0'8"
Walter Vukcevich, M.D.
And 0h52'47" 40d16'59"
Walter W O'Brien
Tau 5h44'48" 22d18'0"
Walter Wade Frankli
Gem 6h44'35" 26d43'6"
Walter Waite
Aqr 22h34'1" -0d53'20"
Walter Werthman
Ori 5h55'45" 18d0'21"
Walter William Furley
Tau 5h56'46" 25d24'32"
Walter Windheim
Aqr 21h8'2" 1d13'13"
Walter & Winifred
Sco 16h46'41" -27d43'42"
Walter Winston
Leo 11h28'52" 1d51'9"
Walter,Forever I Will
Umi 15h38'41" 69d47'46"
Walter's 60th
Tau 3h47'53" 8d16'40"
Walter's Big Bet
Peg 22h19'47" 10d56'21"
Walter's Essence
Uma 10h56'41" 64d43'56"
Walter's Light
Gem 7h24'0" 24d37'6"
Waltham High School Class of 2013
Cyg 21h37'4" 50d43'2"
Walther Grober
Ori 5h55'32" 1d59'1"
Walther, Barbara
Uma 10h48'10" 41d48'39"
Walther, Dirk
Uma 11h4'40" 70d56'59"
Walton Bertrand Baldwin
Crb 16h7'43" 38d39'43"
Walton Russell "Topper" Winder
Uma 11h28'9" 40d49'27"
Waltraud
Uma 8h13'4" 62d24'17"
Waltraud
Ori 5h1'43" -0d30'22"
Waltraud (*29.9.1939)
Uma 12h51'33" 55d10'1"
Waltraud Karl
Sgr 18h23'52" -23d23'44"
Waltraud Kornherr
Cas 3h0'48" 65d10'58"
Waltraud Lutterbei
Uma 9h46'44" 48d29'34"
Waltraud & Peter Lawrenz
Cam 7h59'51" 72d32'10"
Waltraud Ramser
Uma 8h30'59" 65d8'8"
Waltraud Theilen
Eri 4h25'12" -0d36'51"
Waltraud und Theo
Uma 8h53'43" 49d33'56"
Walt's Birthday Star
Her 18h55'8" 13d41'56"
Walt's Moon Jack
Cep 22h18'1" 69d49'37"
Waltz
Leo 9h55'45" 6d58'19"
Walyn
Gem 6h44'55" 14d2'44"
Wameling, Franz-Hubert
Uma 12h15'9" 60d18'40"
Wamon Ellis Klett
Uma 11h28'10" 63d27'4"
Wampy
Uma 10h35'20" 41d7'50"
Wan Chi Lau
Cnv 12h44'43" 42d10'26"
Wan Shuan
Cyg 19h35'7" 31d27'59"
Wanda
Uma 10h20'29" 48d23'24"
Wanda
Cas 0h27'41" 53d21'28"
Wanda
Leo 9h55'8" 13d53'57"
Wanda
Cas 0h24'38" 61d26'25"
Wanda "8/08/1940"
Leo 11h6'25" 11d34'59"
Wanda Abbott
Cas 1h24'7" 60d57'54"
Wanda and Jim Forever
Cyg 21h15'24" 44d30'32"
Wanda Ann
Leo 9h53'5" 23d22'46"
Wanda Arlene Cain
And 0h44'35" 38d4'47"
Wanda Big Grandma McCorkle
Cap 21h2'53" -18d34'38"
Wanda Bleich "Google Girl"
Ari 1h56'49" 19d39'1"
Wanda Buzzini
Peg 23h29'50" 22d17'40"
Wanda C. Williamson
Lyr 18h51'41" 38d48'31"
Wanda Capece
Ori 4h49'13" 7d9'16"
Wanda D.
Cas 1h30'43" 64d48'19"
Wanda Dee
Aql 19h34'49" 5d29'29"
Wanda Dee Steelman-White
Aqr 22h37'56" -1d15'33"
Wanda Duplessie
Uma 14h5'54" 60d50'35"
Wanda Earlene Johnson Bagwell
Cap 20h34'34" -10d7'6"
Wanda Eason's Star
Cnc 8h6'46" 10d48'6"
Wanda Ellouise Jones
Peg 22h55'6" 30d42'53"
Wanda Fay
Cas 1h14'2" 50d30'17"
Wanda G. Camarota
Cas 23h30'59" 58d12'8"
Wanda G Wheeler
Vir 14h23'59" 5d19'14"
Wanda Gail Taylor
And 1h0'18" 46d29'2"
Wanda Gale DéJesus
Aql 19h31'22" 9d19'56"
Wanda Grace White
Cas 1h22'53" 66d18'48"
Wanda H Bowman Shines Forever
Vir 15h3'49" 3d2'28"
Wanda Howard
Psc 0h25'18" 8d47'20"
Wanda Jean Oliver
Lyn 7h54'43" 48d46'29"
Wanda Jean Smith
Crb 16h3'32" 35d34'8"
Wanda Jewell
Crb 16h1'27" 33d4'39"
Wanda Jo
Leo 10h24'58" 17d10'55"
Wanda Joy Stevens
And 0h34'38" 38d23'23"
Wanda Kay
Lyn 7h0'52" 50d18'29"
Wanda Kay Hogue
Psc 0h30'8" 4d52'57"
Wanda Kaye / Heath's Angel
Vel 9h12'47" -43d42'0"
Wanda Kim Gaddy
Leo 10h12'11" 8d40'26"
Wanda L. Hatfield
Cas 0h15'32" 53d37'12"
Wanda Lafferty
Cas 0h19'51" 59d8'3"
Wanda Laura Sathoff Hughes
Uma 11h34'28" 51d31'21"
Wanda Lee
Aqr 23h16'42" -13d33'44"
Wanda Lee Davis
Tau 5h49'18" 25d49'16"
Wanda Lester
Cyg 21h40'31" 47d16'25"
Wanda Lisa Dulac
Vir 12h36'18" 11d28'16"
Wanda Mack
And 0h52'9" 41d59'0"
Wanda Mae
Uma 9h55'42" 61d59'24"
Wanda Maria
Crb 15h52'55" 36d21'28"
Wanda Meacham
Crb 16h5'50" 36d28'55"
Wanda Metcalf
Crb 16h10'53" 31d32'6"
Wanda "Moo" White
Crb 15h37'30" 28d0'36"
Wanda Mother
Pav 20h25'0" -58d41'18"
Wanda Nemec
Per 4h13'58" 45d12'59"
Wanda Nolan
Crb 15h39'56" 27d35'19"
Wanda Pinello
Sco 17h33'41" -33d39'48"
Wanda Powell
And 0h56'39" 40d37'54"
Wanda Riddle
Tau 5h34'45" 24d22'55"
Wanda Santana
Cap 20h38'49" -11d16'21"
Wanda Smith
Col 6h5'41" -42d38'28"
Wanda Strew
Uma 13h23'7" 63d7'26"
Wanda Sue
Cas 23h8'2" 54d58'18"
Wanda Sue, PPPD
And 1h12'38" 41d10'39"
Wanda Wade Locke
And 22h58'23" 48d20'57"
Wanda Watling Pedersen
Gem 6h36'57" 14d47'29"
Wanda Wyman
And 0h34'50" 27d27'28"
WANDALOO
Vir 13h16'18" 2d22'44"
Wandalyn Ure
Gem 6h32'33" 15d59'49"
Wanda's Spirit
Vir 12h27'58" -9d19'53"
WandasTranquility
Uma 11h6'54" 52d57'25"
WandE
Lyn 8h4'46" 57d4'9"
Wander Rezender
Umi 13h56'44" 70d30'40"
Wandinger, Maximilian
Uma 8h56'47" 56d22'30"
Wang Di
Ari 2h11'43" 22d0'6"
Wang Jingna
Aql 20h5'12" 8d46'31"
Wang Lili
Ari 2h20'16" 11d29'29"
WANG Rui-yu
Uma 11h4'5" 34d31'28"
Wang Xiao Nan
Aqr 22h31'33" -2d0'49"
Wang Yanjing
Aql 19h15'24" -10d20'18"
Wang Yaoqian
Cep 22h5'14" 60d10'59"
Wang, Marco
Uma 13h51'6" 55d7'32"
wanja
Ari 3h3'28" 15d51'47"
Wanna see my monkey?
Ori 4h55'23" 10d20'30"
Wannavetch Uthaiwat
Ari 3h15'53" 21d51'54"
Wanner-Sekella
Gem 7h29'7" 33d43'27"
Want To
Dra 19h50'17" 75d51'12"
Wantana
Lib 15h28'49" -11d45'24"
Wan-ting Rebecca Kung
Ori 5h7'34" 8d58'36"
Wanting & Shulong
Peg 22h36'58" 11d44'46"
Wanza Gayle Samara
Ari 2h35'9" 22d16'3"
Wanza Uilani Falevai
Vir 13h21'10" -11d47'48"
Ward Abbott
And 1h35'9" 48d32'6"
Ward "Bud" Pire
Cap 20h27'35" -15d24'9"
Ward & Joyce Love Star
Sge 19h43'18" 17d9'27"
Ward Sherwood Leum
Uma 11h35'35" 49d54'15"
Warda BB
Ari 2h20'4" 24d58'23"
Warder's Own
Lmi 10h0'46" 32d18'17"
Ward-Harger 7
Uma 10h26'11" 62d20'27"
Wardini33
And 0h14'18" 25d54'37"
Wareeporn Donkasem
Umi 17h15'0" 82d39'30"
Warenetta Letze's Star
Umi 14h6'48" 75d40'13"
Warfrost
And 0h22'47" 25d43'37"
Warm Fuzzies
Per 4h17'54" 48d34'9"
Warm Summer Night
Uma 11h22'18" 62d34'23"
Wärmi
Ori 6h11'25" 9d52'7"
Warmkin Snugglebum
Gem 7h6'10" 24d26'25"
Warmness on the Soul
Peg 22h21'55" 25d17'45"
Warner Hiatt
Her 16h35'46" 34d33'20"
Warner Rubnitz Ferratier
Lyn 8h56'1" 41d25'5"
Warnock's Warlock (Tommy DuRell)
Her 16h35'20" 11d50'36"
Warran Peterson
Aur 5h34'48" 33d24'45"
Warran Thomas Ballantyne
Umi 15h47'13" 81d44'58"
Warren
Her 17h27'36" 24d8'35"
Warren A. Everard
Cra 18h50'28" -38d23'1"
Warren and Alycia Zehner Love Star
Ori 6h0'20" 13d51'0"
Warren and Evelyn
Cyg 21h56'27" 45d9'12"
Warren and MaryAnn Golden
Cyg 20h5'13" 33d25'47"
Warren Andrew Kneissl
Cap 20h20'36" -13d10'20"
Warren Andrew Pawlowski (WAP)
Hya 9h22'6" 5d2'38"
Warren Anthony Murphy
Vir 12h59'58" 2d2'58"
Warren Ashley Maude
Gem 6h48'17" 32d23'12"
Warren Bailey 8/21/24 - 10/13/99
Sge 19h54'13" 18d51'54"
Warren Bayless
Cep 22h23'34" 60d36'26"
Warren Beans Campbell
Lib 15h42'1" -10d2'31"
Warren Belding
Per 4h30'21" 41d4'18"
Warren Blake Williams
Per 3h23'21" 51d26'40"
Warren Bopp
Gem 7h48'5" 32d56'36"
Warren Brown
Aql 20h5'38" 3d42'31"
Warren Bruce Miller
Gem 7h13'0" 24d0'4"
Warren C. Faucher
Dra 17h57'2" 55d8'33"
Warren & Cindy Brandenburg
Cyg 19h39'55" 51d38'51"
Warren Clancy Haslam
Cnc 8h11'28" 9d7'16"
Warren Darrick Epps
Her 17h15'17" 38d34'50"
Warren & Davy Rainey's "Twin Flame"
Cyg 21h12'54" 31d15'12"
Warren Derstine
Lac 22h11'6" 43d7'15"
Warren "Doc" Schnitzer
Uma 8h48'50" 65d31'54"
Warren Edgar Wasson
Cap 20h38'55" -16d10'39"
Warren Einhorn
Boo 14h54'12" 15d1'9"
Warren Ernest Helmreich
Aur 5h46'15" 32d19'48"
Warren Ferlandy III
Cap 20h37'49" -18d27'23"
Warren G. Blank
Uma 10h15'25" 47d13'23"
Warren George Duffield, Jr.
Aql 19h47'8" -0d13'23"
Warren Grant Egger
Tau 4h20'42" 21d27'3"
Warren H. Brown
Sgr 19h9'31" -16d11'47"
Warren Hawley
Uma 14h28'3" 55d31'37"
Warren Henry Whisonant III
Sco 16h19'36" -28d22'15"
Warren Jackson Walker
Cnc 9h17'30" 25d57'59"
Warren Jefferson McCarter
Sco 16h6'49" -18d27'2"
Warren John Brewer (Scooter)
Vir 12h8'3" 7d28'53"
Warren L. Shaw, Sr.
Cep 23h28'46" 86d32'41"
Warren Lee Ouwenga
Per 2h50'21" 53d46'11"
Warren Leroy Taylor
Lib 15h26'4" -15d41'15"
Warren Liem Pham
Ari 2h39'42" 23d24'24"
Warren & Lyn's Wedding Star
Uma 11h30'27" 56d4'35"
Warren & Mary Hegidus
Leo 11h16'38" -1d30'12"
Warren Mattox
Cep 21h7'40" 85d28'38"
Warren Nelson Gorton
Vir 14h38'9" -4d59'11"
Warren Obadiah Gobel
Vir 12h19'51" -0d59'57"
Warren Paul Jowett
Cru 12h34'30" -61d28'34"
Warren Roger Chick
Cnc 8h44'46" 16d14'37"
Warren Salbeck
Uma 11h40'50" 54d49'59"
Warren Sellers
Aqr 22h51'13" -7d31'38"
Warren Shiller
Psc 23h24'13" 7d14'50"
Warren & Stacey's Valentines Love Star
Cyg 20h37'23" 34d58'41"
Warren & Sue Campion
Col 5h53'25" -32d22'38"
Warren Thomas Major
Cap 20h26'46" -14d52'41"
Warren Thomas Russell 11/11/1946
Sco 17h53'13" -30d26'50"
Warren Tomlinson
Cep 3h37'59" 82d24'36"
Warren W. Smith & Julia V. Smith
Umi 14h30'38" 82d46'1"
Warren Wilcox
Ori 6h18'33" 6d31'53"
Warren William Marquardt
Per 4h4'53" 49d37'41"
Warren Zevon
Per 2h15'30" 51d32'1"
Warren, Jill M.K.
Per 4h0'46" 32d4'35"
Warren's 50th. Anniversary 6/30/56
Per 3h28'2" 48d52'7"
Warren's Angel
Ari 2h7'52" 26d1'44"
Warren's Phish 5!
Cnc 8h32'33" 23d35'19"
Warren's Star
Leo 9h56'23" 30d17'20"
Warrior
Pyx 8h28'9" -35d46'54"
Warrior Chief #20
Her 18h45'49" 13d31'40"
Warrior for Life
Lmi 9h33'36" 36d12'39"
Warrior of Light
Lmi 10h7'55" 39d24'43"
Warrior of Love
Leo 10h25'51" 12d46'34"
Warrior Sargent
Uma 10h36'21" 66d55'44"
Warriors As One JB & Kristina
Ori 5h36'48" 2d37'18"
Warsha's Star
Sco 16h43'12" -32d31'50"
Warwick Franks
Cru 12h18'26" -57d8'17"
Warwick II
Cap 21h8'27" -21d45'41"
Warwick Stewart Poon
Ori 5h59'19" -1d25'4"
WAS (Warren Arthur Spencer)
Aur 5h45'0" 49d57'11"
Wasela
Cir 15h19'31" -57d19'12"
Waser Stéphanie
Ori 5h52'29" 6d52'59"
Wash. DC So. Mission Pres. Bretzing
Her 16h52'22" 21d21'40"
Washie
Gem 6h56'48" 21d41'19"
Wasim & Gurpreet
Cyg 22h1'23" 53d53'9"
Wasna Kanoo
Uma 8h27'39" 65d36'50"
Wassem Khalil
Leo 11h17'36" 27d22'32"

Wassim Bouras
Del 20h43'3" 16d18'17"
Wasyl and Marion Kohut
Tau 3h47'51" 9d49'37"
"Watash" William R. Lewis
Ori 6h23'15" 12d18'5"
Watch It Shine
Sgr 18h34'34" -15d59'29"
Watching over Heidi
And 0h18'16" 27d36'42"
Watching Over You Always
Uma 11h26'10" 58d24'39"
Water Boy 27
Per 2h54'25" 56d21'13"
Water Lillies
Umi 13h18'57" 75d46'16"
Waterboy
Cyg 19h59'12" 50d31'28"
Waterdog
Oph 17h9'28" -0d8'55"
waterfire
Uma 9h28'56" 67d13'40"
Watermaster
Uma 10h31'30" 67d7'40"
Watermelon Head
Del 20h42'15" 16d55'30"
Watosh
Uma 9h57'41" 44d21'29"
Watoshi William Wesley Grove
Aur 5h52'36" 53d18'32"
Watson
Ari 2h10'31" 24d24'18"
Watson Gaydos
Ori 6h10'46" 19d47'41"
Watterson~Prime
Lyr 18h42'41" 39d57'21"
Watterson's Halo
Uma 11h15'14" 47d48'22"
WATZ
Tri 1h45'52" 34d51'26"
Waui Lana Star
Lmi 10h38'49" 32d11'51"
Wauneta Casner Lauder
Crb 15h32'28" 29d43'21"
Waunika Y Cofresi Angel Star
Sgr 18h12'32" -19d45'34"
Wavah
Ari 3h16'24" 19d43'11"
Wave Bharam Umi Yi-Dyer
Cyg 20h34'13" 47d22'1"
Waverleigh Elizabeth Jenkins
Mon 6h46'20" -0d3'26"
Wawa Store 118
Sco 17h51'32" -31d56'2"
Wax
Uma 13h40'40" 53d52'9"
WAY
Tau 5h8'20" 24d31'53"
Wayani Taylor
Lyn 8h1'50" 51d44'18"
Wayde Bailey Thorne
Boo 15h31'37" 49d24'45"
Wayland Cooley
Aur 6h2'25" 47d13'22"
Wayland Keith Baldwin
Aql 19h49'55" 11d43'51"
Wayland Marshall West
Ori 5h32'1" 1d29'36"
Waylon
Lyn 7h27'0" 45d49'29"
Waylon Emory Tate Sutton
Psc 0h19'29" 7d56'17"
Waylon Jay Hawley
Ori 5h50'59" 6d39'9"
Waylon Jennings, Outlaw Legend Hero
Per 3h39'53" 48d5'10"
Waylon Roman Biernacki
Eri 4h16'34" -0d59'28"
Waymar
Lyn 7h7'42" 60d29'43"
Waymond Elam
Ori 5h11'2" 6d40'35"
wayne
Aqr 21h40'31" 1d8'38"
Wayne
Cyg 19h55'50" 51d47'52"
Wayne
Uma 10h8'55" 52d53'25"
Wayne
Cep 21h26'26" 82d5'14"
Wayne A. Owens
Ori 5h57'54" 11d17'9"
Wayne A. Spinner
Cep 21h9'11" 62d56'2"
Wayne A. Stillings
Aql 20h0'34" 10d19'29"
Wayne Adams
Per 2h47'57" 50d34'14"
Wayne Addessi
Per 4h24'25" 44d32'55"
Wayne Alan Forester 2/08/1990
Uma 11h24'24" 65d10'16"
Wayne Alton, Always and Forever
Cas 0h11'37" 54d39'3"
Wayne and Andin
Sge 19h37'32" 17d25'20"
Wayne and Danielle Hunter
Uma 14h27'44" 58d12'35"
Wayne and Debie
Cyg 20h30'21" 34d50'26"
Wayne and Dorothy Rankin
Cyg 20h46'6" 36d15'50"
Wayne and Irene Blackburn
Sco 16h7'50" -17d2'50"
Wayne and Janet McGee
And 0h20'28" 45d47'48"
Wayne and Lisa Norton
Crb 15h28'43" 27d52'47"
Wayne and Marian Romberger
Psc 1h33'29" 14d55'3"
Wayne and Merle
Pho 1h23'35" -42d0'49"
Wayne Babler
Her 18h23'12" 21d33'58"
Wayne Ballinger
Gem 7h55'5" 29d2'18"
Wayne Barton Study Center
Uma 11h24'5" 35d2'24"
Wayne Benevides
Uma 11h39'36" 55d12'8"
Wayne Benson
And 1h17'14" 46d6'46"
Wayne "Bo" Peacock
Ari 2h9'35" 21d36'12"
Wayne Bryant Jr.
Uma 10h14'30" 41d27'4"
Wayne Buckley
Her 17h39'32" 37d31'17"
Wayne C. Lay
Uma 9h2'28" 49d11'38"
Wayne C. Williams, Jr.
Per 3h42'40" 46d31'11"
Wayne Cahoon
Cep 21h30'17" 63d46'56"
Wayne&Charlotte Dutch Italy Holland
Uma 8h46'4" 54d24'48"
Wayne Christopher Lewis
Per 4h5'22" 49d39'38"
Wayne Creswick
Cen 13h39'17" -40d52'45"
Wayne D. Bozeman, esq.
Sgr 18h16'14" -19d36'18"
Wayne "Daddy" John Redfern
Per 3h11'27" 48d43'44"
Wayne Douglas Brunt
Psc 1h15'11" 17d47'9"
Wayne Edward Hall
Lib 14h50'8" -7d58'34"
Wayne 'Eternally Ours' Kerrie
Ara 16h56'34" -48d47'41"
Wayne F. Winters
Uma 8h20'45" 71d35'26"
Wayne Fielder
Cep 23h1'31" 70d8'11"
Wayne Flett
Cru 12h51'58" -64d13'49"
Wayne & Fleur for Eternity
Tau 4h1'52" 19d17'57"
Wayne Fuqua
Her 17h6'48" 32d9'20"
Wayne Garrett
Cnc 8h56'50" 19d9'17"
Wayne Good
Aqr 21h37'19" 0d48'15"
Wayne Grabis
Per 2h53'55" 53d41'24"
Wayne Green
Vir 12h55'1" -0d20'57"
Wayne Gregory
Sco 16h50'20" -32d34'22"
Wayne H. Pagani
Cas 0h50'1" 65d32'16"
Wayne H. Purdin *Pappy Poo*
Tau 4h6'16" 6d35'3"
Wayne Hafner
Lib 15h58'21" -12d52'56"
Wayne Hansen
Dra 19h49'39" 61d15'39"
Wayne Hartless
Uma 11h39'12" 51d43'24"
Wayne Heath Godfrey
Cas 0h56'31" 57d25'10"
Wayne Henry Hubbard
Cep 22h31'1" 71d24'42"
Wayne Hironaka
Aqr 23h2'36" -9d44'33"
Wayne Hopkinson
Cep 0h8'8" 75d13'46"
Wayne "Hot Chocolate" McIntosh
Cyg 19h29'31" 30d51'33"
Wayne J. O'Keefe
Dra 17h7'11" 65d47'37"
Wayne Jackson
Crb 16h0'30" 25d48'27"
Wayne James De-Leston
Leo 9h54'24" 23d35'58"
Wayne Jervis
Aql 19h57'14" 0d21'47"
Wayne John Weber
Cep 2h34'7" 82d24'29"
Wayne K. Larson
Cnc 8h32'45" 27d30'28"
Wayne Klunk
Uma 13h28'46" 58d14'33"
Wayne Lawrence Martin
Sco 17h23'32" -41d21'47"
Wayne Lee Hepler
Ori 5h12'33" 5d29'29"
Wayne Leo Loomis
Her 17h22'17" 35d10'30"
Wayne Leonard Warner
Uma 13h4'37" 59d48'21"
Wayne & Leontine
Cyg 20h28'23" 40d42'24"
Wayne Leslie Moore
Ori 5h15'43" 2d56'39"
Wayne Luthor
Umi 16h39'10" 83d52'42"
Wayne M. Squires
Ori 6h22'22" 14d38'56"
Wayne & Maria's Mystical Night Star
Cru 12h36'51" -58d32'12"
Wayne & Martha Mincey
Sge 19h53'51" 18d51'43"
Wayne McDermott
Cap 20h24'41" -26d17'44"
Wayne Mills
Cyg 20h0'21" 41d7'26"
Wayne Morrison
Uma 12h28'26" 62d3'53"
Wayne My Angel
Cyg 21h15'9" 47d28'30"
Wayne Nickle
Lyn 7h38'43" 55d8'19"
Wayne Noseworthy
Dra 18h35'4" 61d23'18"
Wayne P. Stearns
Uma 9h48'14" 65d19'32"
Wayne Peterson
Per 2h39'40" 55d26'27"
Wayne "Pookie" Jung
Per 4h17'53" 40d31'25"
Wayne R Carter
Peg 21h32'40" 7d44'12"
Wayne R. Schneider, MD
Psc 0h50'54" 15d8'27"
Wayne & Regina Mize 6-17-2005
Uma 9h47'56" 45d2'26"
Wayne Richard Wilkes
Dor 5h8'29" -65d48'56"
Wayne Richards
Uma 10h49'34" 65d14'34"
Wayne S. Parker
Psc 0h31'12" 13d32'48"
Wayne Schwartz
Cap 20h54'5" -26d2'27"
Wayne Scot Lukas Major
Aqr 22h36'57" 0d52'43"
Wayne & Sharlene Haven
Ori 4h45'32" 4d43'24"
Wayne Shaw
Aql 20h11'48" 5d51'19"
Wayne Shields
Ori 6h16'56" 13d42'3"
Wayne Simpson
Aqr 22h0'31" -17d24'43"
Wayne Slemp
Dra 17h24'40" 59d0'20"
Wayne Spray
Lib 15h47'5" -16d55'4"
Wayne Stanton
Cnc 8h45'22" 17d25'33"
Wayne "Stargate" Wallace
Ori 5h37'23" 3d13'19"
Wayne Sterling Schoniger
Lib 15h38'8" -7d53'46"
Wayne Thompson
Lib 15h54'15" -20d9'39"
Wayne Trevor Kirby
Uma 13h28'8" 62d2'52"
Wayne Vogel
Lyn 6h32'45" 59d56'1"
Wayne (Waz) Sprunger
Aqr 22h18'13" 1d53'45"
Wayne William Thomas
Vir 14h24'54" -2d24'36"
Wayne Wodach
Uma 11h14'57" 57d13'24"
Wayne Zatarain
Aql 20h0'45" -0d36'47"
Wayne Zeevering
Ari 2h13'17" 23d48'45"
Wayneandtamsin
Cyg 21h58'48" 45d16'50"
Wayneo's star
Cen 13h41'39" -42d20'55"
Wayne's Shining Star
Ori 5h42'10" -2d10'24"
Wayne's Wonder
Ori 6h2'56" 17d35'59"
Wayne-T Bird
Aqr 21h43'46" 2d6'36"
Waynie-Pooh
Aur 6h27'16" 51d36'3"
Waynoka
Dra 18h38'6" 51d6'59"
Waywah
Cyg 20h4'48" 58d55'20"
Wazza's Bit Of Heaven
Cru 12h37'37" -58d9'3"
WB Joseph Fuller Steilacoom #2 6004
Uma 14h13'16" 55d26'49"
Wbyers1
Uma 12h3'12" 48d2'35"
WC Watson III
Vir 13h24'48" -2d54'34"
wCovell
And 2h18'2" 49d19'42"
W.D. Finlayson IV
Psc 1h42'13" 5d46'17"
W.D. Galloway
Lib 15h7'40" -9d13'35"
wdla
Psc 1h0'58" 23d44'1"
WDO-What've u done 4 beer whlslrs?
Tau 5h44'0" 16d16'56"
WDS
Cnc 8h59'54" 14d10'56"
We are blessed with Shari Fay
Umi 14h36'0" 73d43'59"
We Are Eternity
Cap 21h44'36" -14d23'9"
We Are One
Uma 8h53'35" 62d45'32"
- We are so proud of u Mum - Love u xx
Car 10h21'15" -59d45'35"
We are the Star of Love. SoulMates
Cyg 20h58'25" 45d23'19"
We Baby Chickens
Crb 15h20'4" 25d58'29"
We Could Live Like Jack And Sally
Tau 4h10'11" 10d29'10"
We love Mommy Mu!
Leo 10h43'24" 22d46'25"
We Love Our Nana
Lyn 8h6'4" 52d6'17"
We Love Our Parents From Kris & Meg
Cyg 20h22'26" 46d39'48"
We Love You (A.J.& Gabriella)
Cyg 19h45'22" 53d16'4"
We Love You Gramma and Grampa
Uma 14h17'28" 59d49'47"
We Love You Grandma Val
Sgr 19h20'49" -30d21'41"
We Love You Mam
Cep 21h50'33" 68d23'31"
We Love You Mamaw Evelyn M. Green
Leo 11h19'39" 17d39'27"
We love you mom and dad!
Sco 16h8'57" -12d12'42"
We Love You Pap Earl B. Green Sr.
Leo 11h22'7" 16d34'45"
We love you, Pa!
Per 3h44'32" 44d55'16"
We Love You, Peter Juhlin
Cas 1h32'1" 67d40'26"
W.E. Orrick
Del 20h47'19" 19d45'37"
WE: our love will shine forever XXX
Vel 9h26'58" -50d34'43"
we think of you too
Pho 0h48'55" -42d20'29"
We Will Always Be Close Ma, Donna
Ari 2h6'15" 21d34'3"
We will sparkle in the sky forever
Cyg 20h36'40" 34d43'23"
Weahs Love
Tau 4h14'28" 8d43'21"
Wearethecokeclub
Wearethegreatest
Tri 2h31'12" 31d7'53"
Weasel
Cas 0h10'40" 54d53'23"
Weasel
Uma 9h17'19" 71d33'32"
Weather Marie
Aqr 23h6'13" -10d1'36"
WeaverStar
Vir 13h9'19" 3d47'50"
Webb
Gem 7h0'33" 17d21'44"
Webb - Methe
Col 6h1'29" -29d55'27"
Webb Morton
Tau 4h9'8" 7d0'4"
Webber
Leo 11h43'13" 12d52'6"
Webb's Wonder
Uma 10h56'57" 72d38'6"
WebbSeven
And 0h48'1" 31d52'30"
Weber
Per 2h59'29" 54d38'15"
Weber
Leo 11h2'34" 20d35'49"
Weber
Uma 8h35'5" 64d19'21"
Weber, Dominik
Uma 14h5'13" 61d32'13"
Weber, Ilse
Uma 10h26'28" 39d26'35"
Weber, Joachim
Uma 9h55'9" 62d10'6"
Weber, Renate
Uma 8h38'35" 59d27'20"
WEBOFLIFE
Uma 9h30'39" 68d13'28"
WebStar 2426
Uma 11h18'0" 52d11'46"
Webster
Lyn 7h47'35" 49d38'26"
Webster
Uma 8h58'24" 59d27'25"
Wedad Adli Ghdamsi
Ori 5h25'21" 2d39'45"
Wedding Bell Rings
Uma 11h20'26" 58d37'15"
Wedding Day
Cyg 19h32'19" 51d50'15"
Wedding Day Julian and Angie Sear
Pho 0h10'44" -46d22'29"
Wedding Night
Cet 2h15'34" -12d57'1"
Wedigo Darsow
Ori 5h57'57" 17d49'26"
WEDIKA & JEROEN
Per 2h15'32" 54d44'23"
Wednesdai Aprylle
Uma 12h35'17" 55d55'59"
Wednesday's Star
Lib 15h4'16" -13d14'55"
Wee Averrey
Cnc 8h19'51" 20d42'15"
Wee Billy Scott Ferguson
Cmi 7h45'54" 5d59'13"
Wee Ella Mc Aleese
And 1h53'25" 36d1'7"
Wee Gran
Cas 0h35'18" 67d24'33"
Wee Mari Roden - 22 November 1972
Her 18h13'39" 41d33'40"
Wee Sheila
And 1h0'6" 45d55'46"
Wee Stevie Boy
Umi 17h13'9" 83d39'51"
Wee Stewie
Sco 16h49'1" -26d12'29"
Wee William
Cru 12h11'25" -64d9'1"
Weeble
Psc 1h1'12" 9d11'48"
WEEBLE/L.F.W.
Sgr 18h7'38" -23d12'39"
Weeda "Rita" Asaad
Cas 23h33'41" 53d21'5"
Weedi Beeson
Peg 22h20'54" 7d37'9"
WEEDLE
Uma 14h16'17" 61d43'22"
Weedle Weedle
Cnc 8h51'50" 19d24'28"
Weeeee
Cnc 8h26'18" 20d52'50"
Weefish-LSF
Psc 0h41'56" 17d56'24"
Weegi
Lyn 7h23'56" 45d21'12"
Weehar
Crb 16h17'10" 36d29'2"
Weejoe Louis Joseph Paluzzi
Cap 21h57'30" -9d26'53"
Weekley Family Spirit of Spindrift
Uma 11h44'44" 52d31'2"
Weeks & Assoc. Insurance Service
Ori 5h38'52" -0d50'26"
Weeky Weeky Weeky
Cyg 20h41'58" 38d26'47"
Weems Star
Dra 19h19'43" 64d48'47"
Weenie
Lyn 7h12'54" 59d17'57"
Weep not,allow yourself to be tamed
Cyg 21h18'43" 54d26'13"
Weeple
Dra 18h45'11" 54d50'34"
Weesie
Tau 5h33'34" 17d8'22"
Weezee
Lyn 7h56'12" 39d27'13"
Weezie
Per 3h48'54" 49d1'0"
Weezie's Star
Tau 4h27'57" 20d42'41"
Weezybean's Skittlefritz
Uma 11h18'56" 47d55'54"
Wefan Donnchadh Duncan
Umi 14h39'12" 68d15'42"
Wehi 'ia ke alaula no ka pu'uwai
Eri 4h24'4" -31d25'31"
Weibel-Maanum
Lyn 7h7'34" 53d51'0"
Weichbrodt, Frank
Uma 9h37'34" 49d59'36"
Weidenmüller, Jürgen
Uma 10h43'23" 46d24'38"
Weidl, Heinz
Uma 10h33'12" 58d35'5"
Weidmann, Peter
Uma 12h57'17" 60d45'28"
Weigl, Leopold
Uma 11h40'58" 33d18'41"
WeiJun & Jacelyn
And 23h49'5" 41d6'51"
Weil, Melanie
Uma 11h24'11" 62d38'26"
Weil, Peter
Ori 6h19'6" 16d8'50"
Weili
Aqr 22h49'45" -8d39'47"
Wein
Cyg 19h49'50" 46d24'17"
Weinberger, Axel
Uma 10h42'47" 65d36'48"
Weininger, Anastasia
Uma 10h2'31" 52d47'45"
Weiran Zhang
Uma 9h34'27" 58d1'41"
Weirich, Joachim
Ori 5h0'31" 6d9'24"
Weis, Helga
Uma 9h41'43" 66d1'1"
Weise, Beate
Uma 9h37'44" 72d30'36"
WEISS Family Brett Heather Bella
Psc 0h15'19" 7d19'32"
Weiss, Daniela
Uma 8h50'50" 52d28'59"
Weiss, Isabella
Uma 14h23'53" 59d24'39"
Weißmann, Manfred
Ori 5h0'1" -0d42'29"
Weiss-Robbins Engagement
Uma 11h58'38" 52d5'49"
Weiss-Schiele, Irmgard
Uma 11h10'6" 35d51'21"
Weisweiler, Peter
Ori 4h46'27" -0d51'26"
Weisz, Nancy
Uma 14h11'27" 55d41'33"
Weitsch, Sandy
Uma 11h47'11" 63d48'43"
Weitzel, Luke
Ori 6h3'26" 13d42'35"
Wei-Yu
Aqr 23h39'50" -18d55'46"
WEJoye
Sgr 18h21'52" -17d26'53"
Welch
Cap 21h32'0" -16d43'53"
Welch Legacy VI
Lyn 6h45'15" 57d0'30"
Welcome Ruben & Tati
Lyr 18h25'25" 32d21'11"
Welcome William
Umi 4h27'29" 89d31'55"
Weldon Faucett
Lyn 9h7'50" 33d46'58"
Weldon Francis Stephens
Her 17h17'10" 15d26'34"
Weldon Kyle Holland - # 1 Dad
Vir 14h13'17" 3d59'55"
Weldon-Mom-Smith
Lib 14h53'48" -5d58'49"
We'll Always Have LA
Leo 10h12'18" 25d29'33"
Wellenbrock
Uma 10h44'57" 41d58'37"
Wellington
Cmi 7h38'23" -0d5'2"
Wellington Alves
Ori 5h26'32" 2d33'3"
Wells Campbell Peery
Uma 13h52'10" 58d20'18"
Wells Family Star
Uma 13h38'53" 52d49'19"
Wells Paul Johnson
Her 18h9'31" 25d43'48"
Wellwood
Gem 7h25'45" 20d48'34"
W-E-M your Angles
Lib 15h34'0" -19d16'26"
W-E-M your Angles
Lib 15h41'41" -24d35'11"
Wen
Sgr 19h13'38" -16d36'11"
Wen
Vir 14h36'23" -0d42'11"
wEn
Uma 11h44'9" 54d55'19"
wEn
Uma 11h5'31" 38d0'0"
Wen En's Distant Jewel
Cap 21h31'5" -24d5'1"
Wen Ling Chua
Sgr 19h7'38" -22d55'14"
Wen, Shine On Little Sis-Love J
And 22h58'53" 41d0'32"
Wen-Bin Hsieh
Uma 8h43'46" 52d45'29"
Wenceslao Serra Deliz
Aqr 22h16'16" -3d47'42"
Wenda L. Webb
Lib 15h25'47" -4d0'35"
Wenda Mae
Cas 1h40'30" 62d35'55"
Wendel McConnell
Per 3h27'38" 52d1'48"
Wendel, Tanja
Ori 6h12'47" 16d20'49"
Wendelin
Uma 8h35'12" 65d48'41"
Wendell A Williams
Aqr 22h9'35" -2d30'30"
Wendell Banks, Junior
Cnc 8h39'53" 30d20'39"
Wendell D. Johnson
Gem 7h4'18" 33d48'29"
Wendell Harris
Cep 21h29'39" 64d34'6"
Wendell Jeffery Seaver
Gem 7h33'44" 28d0'18"
Wendell Kyyitan
Ori 5h29'1" 2d33'31"
Wendell Mohling
Per 3h46'31" 41d33'51"
Wendell Nathaniel Graham
Gem 6h55'37" 32d57'9"
Wendell W. Jesseman
Dra 19h0'51" 65d19'21"
Wendell W Wenneker, M.D.
Uma 11h28'46" 47d52'26"
Wendell (Wendy Bolin)
Sco 17h32'13" -32d26'35"
Wendella
Per 2h20'34" 51d37'21"
Wendell's Shining Star
Cyg 19h56'12" 43d21'22"
Wendell's Wishing Star
Leo 11h14'16" 13d10'4"
Wendell's World
Uma 9h16'44" 65d1'41"
Wendelyn Louise Wallace
Ari 2h9'19" 23d19'48"
Wendi
Cam 4h6'51" 56d52'4"
Wendi
Cas 1h24'46" 57d51'48"
Wendi
Lib 14h55'54" -10d21'45"
Wendi
Psc 1h19'31" 7d28'42"
Wendi Ann Rode
Cam 6h46'36" 69d58'2"
Wendi Elizabeth Pool
Dra 15h40'13" 58d56'34"
Wendi Fernandez
Leo 10h20'38" 26d16'20"
Wendi Jean
Cas 0h43'57" 57d55'22"
Wendi L. - L&W BD Guru
And 0h7'4" 44d38'56"
Wendi Mae Tackett
Sco 16h6'24" -18d14'16"
Wendi Pedicone
Uma 9h39'59" 68d39'27"
Wendi Renae
Per 3h45'54" 43d47'15"
Wendi Ruggiero
And 2h19'15" 47d40'44"
Wendi Sisson - Tupperware
Lmi 10h32'23" 31d16'35"
Wendi Sue
Lyn 9h5'53" 37d32'18"
Wendi Taklo
Uma 11h9'51" 53d32'54"
Wendi & Wayne
Pho 1h35'32" -51d51'54"
Wendi Woo x
Cyg 20h14'44" 48d3'16"
WendiAnn Sethi
And 23h23'55" 47d17'50"
Wendie Fisher
Leo 9h55'46" 20d57'4"
Wendie Rose Kaminski
Gem 6h52'51" 24d49'25"
Wendiness
Uma 12h0'37" 32d27'48"
Wendi's White Chocolate Wonderland
Tau 5h30'40" 18d38'8"
Wendkunni
Tau 5h29'53" 21d9'57"
Wendling ~ Don & Ruby
Cyg 19h52'46" 39d11'20"
Wendolyn Forbes
And 0h46'45" 28d12'52"
Wendryn and Xander
Cyg 20h32'38" 51d42'31"
Wendy
Cas 0h2'48" 54d20'39"
Wendy
Cam 5h31'12" 56d8'13"
Wendy
And 0h14'1" 35d41'34"
Wendy
And 2h37'40" 40d35'8"
Wendy
And 0h13'7" 44d50'45"
Wendy
And 1h20'59" 38d6'56"
Wendy
Her 18h43'56" 19d18'49"
Wendy
Crb 16h10'11" 26d12'53"
Wendy
Crb 16h10'5" 26d5'17"
Wendy
Aql 20h16'41" 8d26'44"
Wendy
Ori 5h29'51" 6d17'33"
Wendy
Ori 5h14'18" 12d57'24"
Wendy
Cap 20h21'8" -24d19'35"
Wendy
Sgr 17h58'34" -24d34'8"
Wendy
Cas 0h31'37" 62d28'21"

Wendy
Uma 9h9'11" 59d40'39"
Wendy
Uma 8h41'57" 57d5'51"
Wendy
Cyg 19h48'53" 54d41'37"
Wendy
Cas 1h29'40" 63d32'6"
Wendy
Cam 5h41'4" 68d2'19"
Wendy
Lib 15h35'31" -14d31'34"
Wendy
Aqr 21h42'57" -6d30'52"
Wendy
Cma 6h47'10" -14d38'24"
Wendy
Vir 13h43'26" -5d56'52"
Wendy
Leo 11h23'59" -2d23'24"
Wendy
Umi 15h36'45" 78d20'37"
Wendy- 22156
And 0h40'38" 34d26'32"
Wendy A. Brunner
And 2h23'54" 46d59'7"
Wendy Abel
Aqr 20h41'53" -1d8'42"
Wendy Alison Rice-Hughes
And 0h13'20" 29d6'55"
Wendy and Billy
Cyg 20h36'24" 39d4'1"
Wendy and Emma
Cyg 21h13'25" 41d5'36"
Wendy and Gary
Cyg 21h20'51" 37d57'35"
Wendy and Tom Duvall
And 0h49'47" 37d56'46"
Wendy Ann Allen
Cyg 20h33'5" 37d41'4"
Wendy Ann Kumbier
Ari 3h20'25" 29d1'45"
Wendy Ann Powers
And 0h28'20" 26d32'9"
Wendy Ann Weber
Aqr 21h8'36" -10d58'59"
Wendy Anne Conant
Tau 3h52'25" 21d26'22"
Wendy Anne Davidson
Tau 4h43'25" 23d56'16"
Wendy Arthur Special Friend
Cas 0h13'34" 51d44'33"
Wendy Beasley
Crb 15h43'17" 26d33'6"
Wendy Berg
And 1h58'27" 41d39'15"
Wendy Beth Humphries
Psc 1h22'37" 21d47'24"
Wendy Boe
Uma 9h44'24" 60d11'59"
Wendy Brillon
Lyr 18h31'27" 36d38'20"
Wendy Brook Faircloth
Lac 22h39'0" 43d9'49"
Wendy Bryan
Psc 0h51'4" 15d55'12"
Wendy "Bun" Hartman
Cap 20h36'56" -17d38'31"
Wendy C. Johnson
Cas 1h57'5" 72d44'58"
Wendy Carolina Feliz
Aqr 21h10'8" -10d33'13"
Wendy Chace Balch
Uma 8h31'57" 71d3'19"
Wendy Cheng
Lyn 7h54'34" 34d0'51"
Wendy & Chris - 40 Years
Cyg 20h56'39" 37d36'30"
Wendy Cole
Cas 1h48'51" 64d46'17"
Wendy Colleen Kelly
Dra 17h37'29" 67d5'20"
Wendy Coulthurst
Leo 9h43'46" 28d27'10"
Wendy Dae
Uma 8h32'44" 60d18'25"
Wendy Davis
Crb 15h45'57" 32d12'21"
Wendy Dawn Clevenger
Cnc 8h22'49" 13d38'26"
Wendy De Thame
Sco 16h54'55" -41d42'23"
Wendy DeAnn Barnes
Uma 10h15'49" 70d40'7"
Wendy Diane
Leo 11h43'44" 23d10'51"
Wendy Diane Sissons
Ori 5h5'30" 15d58'20"
Wendy Domres
Cep 23h9'50" 59d47'42"
Wendy E. Graham
Vir 14h10'23" 6d23'48"
Wendy Edwards
Cas 0h19'27" 53d53'1"
Wendy Elizabeth Cooper
Peg 21h59'16" 21d49'25"
Wendy Elizabeth Evans
Ori 5h36'14" -1d26'19"
Wendy Elizabeth Millman
And 2h18'35" 38d5'47"
Wendy & Emily
Oph 17h41'47" -0d20'5"
Wendy F. Turcotte
Ari 2h21'49" 26d38'41"
Wendy Francis Valentine Jones
Com 13h1'52" 17d3'39"
Wendy Franklin
Lyr 18h36'30" 34d41'47"
Wendy Franklin
Cas 1h47'34" 61d14'57"
Wendy - friend, mother, angel.
Cyg 20h39'27" 54d13'26"
Wendy Gail
Vir 11h42'42" 8d5'22"
Wendy Gail Reyes
Uma 13h35'17" 58d45'34"
Wendy Geisinger
Vir 12h59'54" -17d7'49"
Wendy Gibson
Leo 11h6'38" -1d53'59"
Wendy Gibson
Lyn 7h10'40" 51d57'32"
Wendy Hall
Tau 4h28'21" 10d38'28"
Wendy Hall - Best Mum Ever
Umi 16h50'22" 83d58'27"
Wendy Helker One
Ori 5h40'59" 4d38'40"
Wendy Higdon Carberry
Cyg 21h9'34" 53d33'1"
Wendy Hollmann
Leo 9h42'2" 16d21'56"
Wendy Inderlin-Maid of Honor
Uma 13h18'3" 56d38'39"
Wendy Ingram
Vir 13h4'26" 12d40'26"
Wendy J.A. Stevenson - 10.04.1962
Ari 2h49'9" 22d9'33"
Wendy Jane
Cas 0h5'2" 56d0'43"
Wendy Jane
And 23h16'3" 47d38'55"
"WENDY JANE" forever shine for us
Cru 12h47'28" -64d24'47"
Wendy Jane Frech Kramer
And 1h39'25" 42d23'45"
Wendy Jarrette Levesque
Oph 17h31'41" -23d5'22"
Wendy Jean Dunbar
Cyg 20h2'22" 39d0'35"
Wendy Jenner
Cru 12h15'8" -57d26'0"
Wendy Jill Stein, Esq.
And 0h34'26" 28d30'30"
Wendy Jo Long
Ori 5h54'54" 22d28'44"
Wendy Jolly
And 2h0'0" 46d25'1"
Wendy K Buck
And 23h46'17" 48d21'49"
Wendy Kaminski
Sco 16h6'56" -16d29'21"
Wendy Karina
Cas 1h44'21" 62d47'6"
Wendy Kate Oates
Uma 11h6'44" 57d47'46"
Wendy Keegan
Sco 17h12'21" -34d29'10"
Wendy & Keith HOPE - Star of Love
Cru 12h57'8" -59d18'3"
Wendy Kwok
Vir 12h19'55" 5d49'14"
Wendy L. and Rick C.
Cyg 20h44'29" 36d12'13"
Wendy L. and Salvador R.
Cnc 9h4'13" 27d44'36"
Wendy L. Bryson
Srp 18h4'25" -0d6'52"
Wendy Lady
Per 2h53'54" 33d21'13"
Wendy Lamontagne
Mon 7h59'7" -0d38'17"
Wendy Lee
Oph 17h38'32" -0d46'48"
Wendy Lee
And 23h58'58" 40d20'40"
Wendy Lee Harwood
Cas 1h26'33" 63d1'12"
Wendy Loo
Lib 15h11'9" -5d58'49"
Wendy Lorene Crowell
Lib 15h26'9" -25d30'38"
Wendy Lori Lynch
Tau 4h37'14" 17d35'58"
Wendy Louise
And 0h47'13" 25d8'9"
Wendy Louise Seamon
Crb 15h47'49" 36d13'38"
Wendy Loves Dale
Col 6h11'43" -35d39'52"
Wendy Lucille Montgomery
Aqr 22h34'49" -16d43'9"
Wendy Lupoli
And 1h48'38" 45d11'41"
Wendy Lynn
And 0h43'58" 36d10'26"
Wendy Lynn
And 2h14'7" 41d29'23"
Wendy Lynn
Crb 15h52'6" 27d24'38"
Wendy Lynn
Dra 11h45'48" 70d18'41"
Wendy Lynn Friedland
Sco 17h54'40" -36d1'47"
Wendy Lynn Killian
Crt 11h30'55" -17d13'49"
Wendy Lynn Maxim
Tau 5h38'30" 25d17'45"
Wendy Lynn Purser
And 23h38'52" 37d28'26"
Wendy Lynne
Cnc 8h43'16" 6d44'30"
Wendy Lynne Berrier
And 1h45'9" 50d35'13"
Wendy M. Larson
Com 13h9'7" 28d59'51"
Wendy M. Wainscott
Uma 9h43'53" 53d24'54"
Wendy Maeline Ward
Aqr 20h39'31" -9d53'49"
Wendy Maria Mendoza M-13
And 1h46'5" 38d9'0"
Wendy Marie Black
Tau 4h12'4" 6d32'4"
Wendy Marie Shay
Tau 4h32'44" 19d8'27"
Wendy Marsh - Little Star
Cma 6h12'31" -26d37'47"
Wendy Martucci Hamm
Sgr 18h25'27" -22d56'7"
Wendy & Matthew Peralta
Umi 15h19'20" 74d26'49"
Wendy McVay Ligon
And 23h20'46" 48d11'29"
wendy & meme
Cyg 19h26'14" 35d43'55"
Wendy Messy Perez
Lib 14h29'6" -9d40'43"
Wendy Michelle Tomko
Psc 22h54'35" 4d4'19"
Wendy Millard
Crb 15h51'41" 32d4'18"
Wendy Mora
Lyn 7h22'28" 52d25'50"
Wendy Mora Gibson
Sco 15h51'17" -20d55'21"
Wendy My Shining Rose
Gem 6h45'0" 34d53'50"
Wendy Nevard
And 1h41'47" 45d41'13"
Wendy Newman Inspiration and Dreams
And 23h2'17" 40d52'36"
Wendy Nicole Washabaugh
Ori 5h57'17" 18d33'12"
Wendy Ona Werkheiser
Per 2h47'11" 53d23'38"
Wendy Osborn
And 0h49'26" 37d9'6"
Wendy Parmelee
Ari 2h37'22" 21d34'36"
Wendy Payne
Lyn 9h37'18" 40d5'59"
Wendy Peaches Trussell
Lyr 18h32'26" 36d29'39"
Wendy Plymire Baker
Uma 13h46'15" 56d21'47"
Wendy "Pooh Bear" McFee
Umi 15h44'0" 75d34'0"
Wendy Reibelt
Cru 12h1'19" -63d56'10"
Wendy Renee` Bennett
And 1h32'43" 42d57'9"
Wendy Renee Knight Farmer
Leo 9h45'1" 23d49'9"
Wendy Rice-Larson
Her 18h6'7" 28d3'0"
Wendy Roberts
And 1h25'4" 36d18'18"
Wendy Rubel Stevens
Cyg 19h44'16" 31d5'17"
Wendy Serringer
Cam 6h35'25" 66d33'46"
Wendy Shairs/ Flemister
Vir 12h9'45" 11d29'41"
Wendy Simpson
Ari 3h20'44" 25d54'17"
Wendy Snyder
And 23h26'33" 48d34'16"
Wendy (starchild) Setright
Lib 14h52'16" -22d14'41"
Wendy Sterns
Sco 17h41'21" -32d17'3"
Wendy Stevenson Howard
Gem 7h12'57" 15d3'22"
Wendy Stracka
Cap 20h27'20" -9d42'31"
Wendy Sue
Cap 21h41'10" -23d0'19"
Wendy Sue
Gem 7h34'50" 21d5'54"
Wendy Sue Bovee
Cam 7h30'56" 64d41'24"
Wendy Sue Leichter
And 2h19'54" 49d8'42"
Wendy Sweetie
Gem 6h48'6" 20d46'9"
Wendy T
Gem 7h42'40" 34d43'45"
Wendy Tang
Lib 15h5'57" -1d16'26"
Wendy Taylor
Vir 12h28'39" 2d18'10"
Wendy Tazartes
Uma 11h25'36" 62d3'17"
Wendy "the guiding Agunzo light"
Sco 17h12'37" -33d33'27"
Wendy the Mother
Her 16h51'28" 37d35'45"
Wendy Topoozian
Uma 9h44'34" 44d13'11"
Wendy Tulk
Ari 3h13'14" 29d44'18"
Wendy Veneda Burgess
Cru 12h2'8" -59d27'11"
Wendy & Victor
Cas 0h45'46" 54d5'47"
Wendy Waddell (Farnum)
Uma 11h7'55" 31d23'10"
Wendy West
Lyn 8h42'48" 34d37'36"
Wendy Wiederholt
Cnc 8h32'48" 26d23'9"
Wendy Wilcox
Aqr 22h39'36" 0d21'2"
Wendy Williams
And 0h56'33" 39d9'48"
Wendy Wilson
Ari 3h20'56" 28d10'38"
Wendy Wolowicz
Ari 3h6'40" 30d8'5"
Wendy Woo
Sgr 18h40'25" -29d40'26"
Wendy Wood Bruns
Sco 16h16'50" -39d26'23"
Wendy Wrenn
Leo 9h54'23" 24d49'4"
Wendy Wu
Per 4h24'37" 41d48'18"
Wendy & Zag
Ori 6h1'52" 17d11'45"
Wendy Zi Gui
Cas 23h45'13" 57d48'32"
Wendy Zoe O'Connor
Cru 12h21'12" -56d42'43"
Wendy0225
Psc 0h31'19" 9d20'46"
Wendylynn Labosh
Sco 16h6'10" -12d22'39"
Wendypooh
Lyr 18h33'26" 35d51'39"
Wendy's 4-26-05 Birthday Star
Tau 4h24'57" 19d49'17"
Wendy's Birthday Sparkle
Cas 23h2'10" 55d0'16"
Wendy's Birthday Star 5/9/43
Vir 13h21'59" -0d47'34"
Wendy's Butterfly
Psc 22h56'43" 7d1'55"
Wendys holding you forever star
And 23h52'44" 39d24'6"
Wendy's Light
Cas 22h58'18" 57d49'55"
Wendy's Love
Cam 7h49'55" 76d7'49"
Wendy's Love Star
Cyg 21h52'51" 50d9'48"
Wendy's Star
Sco 16h14'39" -11d51'43"
Wendy's Wishes
Dor 5h42'53" -68d23'40"
Wendy's Yellow Scientist
Leo 11h7'27" -5d46'18"
Wendysue Raymond
Cap 20h56'40" -15d27'54"
Weninger, Lena
Uma 9h42'19" 47d38'1"
WeniRena
Uma 13h47'51" 51d17'39"
Wenjie Zhu
Leo 11h30'42" 15d9'59"
Wenke
Ari 2h30'11" 10d47'5"
WenKen04
Vir 13h31'11" -1d43'43"
Wenna's star
Cas 1h18'38" 57d24'46"
Wenndy & Elaine, Eternal Soulmates
Umi 15h47'0" 72d43'14"
Wennie
Gem 6h15'4" 23d37'7"
Wenona M Young
And 1h36'14" 41d15'2"
Wensanna
And 0h54'5" 40d15'43"
WenTing §Forever§ ChinHsuan
Cnc 9h17'26" 16d37'52"
Wentworth Miller
Ori 6h16'5" 10d8'44"
Wenzel, Hedwig
Uma 9h19'57" 43d0'17"
Wenzel, Jasna & Michael
Sgr 19h26'50" -30d39'38"
Werdin, Klaudia
Uma 8h19'53" 64d12'2"
Were
Aqr 23h56'22" -11d34'24"
"We're 37"
Tau 4h4'29" 6d1'45"
We're Forever
Aqr 21h35'24" 1d12'20"
We're KINDA a Big Deal
Aqr 22h38'38" -0d26'59"
Wernedi
Uma 11h37'55" 61d56'6"
WERNER
Cnc 9h16'15" 28d40'22"
Werner
Cyg 21h7'35" 47d25'44"
Werner Amon
Uma 11h36'53" 33d24'44"
Werner Brauneck
Ori 6h21'49" 11d2'31"
Werner Dewitz
Ori 5h22'21" -7d28'15"
Werner Engelhardt
Uma 13h29'7" 59d39'39"
Werner Farwick
Uma 12h0'53" 47d45'6"
Werner Fiedler
Uma 13h53'37" 50d28'14"
Werner Floors
Uma 9h0'45" 69d45'56"
Werner H. Gumpertz
Cap 20h30'55" -26d32'38"
Werner Jasper
Ori 6h16'1" 6d55'53"
Werner Jathe
Uma 11h4'7" 38d52'28"
Werner Karl Czepluch
Ori 6h16'49" 9d13'52"
Werner Kühl
Uma 9h12'56" 48d2'32"
Werner Löhlein
Ori 6h17'21" 6d32'36"
Werner Lux
Uma 10h57'48" 72d10'40"
Werner Maximillian Heimberger
And 1h36'45" 41d10'22"
Werner Nüse
Uma 8h39'4" 56d12'56"
Werner Reich
Uma 11h12'22" 71d54'19"
Werner Reuber
Uma 9h58'0" 72d46'45"
Werner Schmidt
Uma 10h16'5" 43d18'36"
Werner Schorr
Uma 9h51'31" 42d25'37"
Werner Schott
Uma 11h21'23" 34d21'24"
Werner Siegenthaler
Uma 8h34'57" 50d29'26"
Werner Simon Schott
Uma 10h16'22" 53d5'3"
Werner Stark
Ori 6h16'16" -0d32'13"
Werner Von Finckenstein
Ori 5h36'48" -2d24'21"
Werner Wendt
Uma 8h19'21" 65d11'37"
Werner Wittwer
Per 3h40'13" 43d47'3"
Werner Ziebold
Ori 5h53'20" 17d41'29"
Werner Zimmermann
Ori 5h55'16" 17d40'6"
Werner, Hartmut
Ori 5h0'48" 6d42'3"
Wernitz, Torsten
Ori 4h57'55" 15d5'56"
Werno, Manfred
Ori 5h52'3" 5d39'38"
Weronika
Uma 11h35'7" 52d38'12"
Werren, Hermann
Uma 11h18'49" 40d55'14"
Wes
Sgr 19h51'13" -12d58'2"
Wes and Jamie Banner
Tau 5h9'7" 21d52'58"
Wes and Kate's Star
Lmi 10h23'57" 32d32'48"
Wes and Mandy
Cyg 20h9'5" 40d52'33"
Wes B. Robinson
Cap 20h16'24" -26d50'49"
Wes Blalock
Vel 8h51'19" -38d40'11"
Wes Castelsky
Cyg 21h57'8" 49d12'16"
Wes Friedman
Lib 15h6'10" -19d21'45"
Wes & Jan Fredrick
Cyg 21h20'59" 45d50'27"
Wes & Jiggs
And 23h17'40" 43d4'6"
Wes Jones
Per 4h19'45" 44d14'57"
Wes Leroy and Ashley Emile
Ori 6h4'4" 21d10'27"
Wes & Pooki's Place
Cyg 19h47'38" 35d56'25"
Wesley
Cnc 8h49'3" 30d55'38"
Wesley
Lyn 6h49'44" 49d57'55"
Wesley
Cnc 8h32'33" 26d9'39"
Wesley
Cap 21h5'14" -16d11'22"
Wesley 9/29/90
Uma 10h43'45" 49d11'34"
Wesley Alan Richmond
Her 18h18'44" 17d13'41"
Wesley Bear
Sco 17h26'50" -41d15'26"
Wesley Bradley & Krianne Spencer
Cyg 19h43'6" 33d49'44"
Wesley Bryan Killen
Gem 6h47'20" 26d45'19"
Wesley Burgin
Leo 11h19'2" 18d54'58"
Wesley Camerson
Uma 9h37'43" 65d21'4"
Wesley Charles Taff
Aqr 22h20'57" -9d51'19"
Wesley Chastain
Aql 19h8'40" 1d56'46"
Wesley Clark
Cap 21h27'48" -8d32'26"
Wesley Clayton Dixon
Tau 4h57'41" 22d12'13"
Wesley Collins
Gem 6h14'11" 26d14'12"
Wesley D. Hanshaw
Lyr 18h47'34" 31d43'41"
Wesley David Ellingson
Sco 16h19'19" -25d10'19"
Wesley Dean Contryman
Ori 5h24'51" 10d50'3"
Wesley E. Wagoner
Cnc 8h41'36" 22d1'44"
Wesley Edward Ott
Psc 0h11'18" 5d55'21"
Wesley Edward Pope
Psc 0h36'3" 14d5'20"
Wesley Eric Norlund
Cap 21h40'16" -14d48'1"
Wesley Evan Herzing
Umi 14h31'2" 73d12'53"
Wesley Francis Nelson
Uma 9h37'36" 66d33'24"
Wesley Fuller
Aur 5h36'42" 33d27'51"
Wesley Gene Liehs
Per 3h30'47" 34d26'44"
Wesley Gifford
Uma 8h38'48" 66d23'25"
Wesley Howard Johnson
Ori 6h3'47" 18d37'24"
Wesley "Jay" Bradshaw
Uma 10h17'38" 48d14'33"
Wesley Jay Walsh
Her 16h38'55" 35d24'32"
Wesley Josef Hein (Wesley)
Umi 16h43'3" 83d33'42"
Wesley Joseph Isam
Lyr 18h30'5" 32d25'44"
Wesley Jr. & Dakota Russell
Uma 11h22'49" 53d37'49"
Wesley K. Camic
Aur 5h48'34" 48d52'2"
Wesley & Kristi
Lyr 19h12'12" 26d28'13"
Wesley P. Holden
Per 3h32'0" 45d25'9"
Wesley Parker
Her 18h6'28" 32d52'30"
Wesley Parkes Hurst
And 23h11'12" 35d23'8"
Wesley Quandt, Jr.
Her 17h0'44" 33d23'3"
Wesley Roberts
Aql 19h55'35" 13d8'20"
Wesley Rutzler
Her 18h9'53" 32d47'55"
Wesley Schulz
Aur 5h22'34" 53d46'30"
Wesley Scott Bannister
Boo 14h23'39" 19d33'57"
Wesley Shawn Sparks
Uma 10h25'15" 56d52'39"
Wesley Steven Young
Boo 14h49'54" 51d12'11"
Wesley Vernon Gorton
Sgr 18h6'18" -26d13'38"
Wesley W. Boyanski
Her 16h18'38" 47d51'17"
Wesley Weinman
Sgr 18h7'26" -25d52'28"
Wesley Wells
Per 3h57'30" 38d16'3"
Wesley William Beal May 27, 1949
Ori 5h34'0" 11d55'49"
Wesley's Eternally II
Ori 5h22'0" 10d55'23"
Wesley's Light in the Galaxy
And 0h11'41" 32d19'18"
Wesley's Star
Ori 6h7'5" 17d26'5"
Wesryn
Cnc 8h46'26" 32d26'22"
Wessely Kravitz
Lib 15h14'39" -11d19'17"
Wesselmann, Werner
Uma 10h31'8" 69d19'15"
Wessels, Werner
Ori 6h16'54" 10d5'29"
Wessu Wells
Ari 2h28'56" 19d25'2"
West 40 SAC Teams
Her 16h42'6" 30d18'41"
West - Belle
Col 5h52'8" -30d27'3"
West Coast Bank
Per 2h46'28" 48d46'10"
West Dakota
Vir 13h46'44" 2d54'36"
West End
Cnc 9h12'20" 16d48'35"
West Haven Community House
Psc 1h41'30" 8d0'30"
West Hope
Aqr 22h9'3" -22d36'2"
West Island Palliative Care Residence
Per 4h20'22" 45d56'57"
West Jefferson High - Class of 1986
Her 16h25'16" 47d30'53"
Westab-04
Cam 3h56'3" 69d30'5"
Westberg
Dra 18h14'40" 57d57'49"
Westberg
Gem 7h19'40" 33d34'3"
Westbound Passage 8/20/88
Umi 15h36'43" 77d36'59"
Westen McConnell Muntain
Umi 13h51'18" 77d3'1"
Westen Presley Evans
Aqr 22h41'4" 0d46'2"
Western Dreams
Uma 12h10'1" 52d39'25"
Westerworld
Sco 16h13'25" -21d17'55"
Westfall
Aql 19h13'56" -7d48'30"
Westfall
Mon 7h18'7" -0d13'19"
Westin Lee Hoyt
Lyn 7h36'2" 56d46'25"
Westley
Psc 23h52'37" 0d3'44"
Westley Hennessee
Cas 0h37'22" 51d44'2"
Westly White
Psc 1h16'18" 7d0'46"
Weston
Cyg 20h24'7" 51d49'35"
Weston
Tau 4h42'38" 17d33'21"
Weston and Amanda Green
Cma 7h11'9" -24d51'40"
Weston Brian
Cap 20h7'44" -9d52'3"
Weston Derek Insley
Per 3h9'52" 53d20'35"
Weston Edward Baxter
Cnc 8h7'21" 15d10'23"
Weston forever in my heart
Lyn 6h52'54" 51d58'36"
Westview Pommies 2007
Pho 1h17'17" -43d12'13"
Wetzig, Stephan
Ori 6h0'32" 10d53'58"
WeUs
Cyg 21h33'49" 51d57'43"
Weylspinor
Uma 9h47'40" 46d41'15"
Weyman Noel "Jay" Jaques
Ori 5h34'18" -0d57'48"
Wezits
Psc 1h25'16" 23d2'36"
W.F.B. III
Uma 8h53'24" 55d8'23"
WGKA Stewart
Gem 6h49'1" 26d53'25"
WGN-LSS
Ori 5h46'59" 7d54'55"
Whakahoanga
Uma 8h44'52" 66d38'22"
Whale with Takumi
Sgr 18h26'33" -32d37'6"
Whamboozle
Per 2h14'41" 56d30'14"
Wharton's Star
Cyg 19h26'30" 53d49'31"
What About Bob?
Her 16h59'41" 14d32'43"
What If
Uma 12h57'4" 61d5'12"
Whataboutbob
Vir 13h8'24" -1d59'53"
'whataboutnow'
Cru 12h1'23" -61d32'52"
W.H.B.My Dad My Hero My Best Friend
Per 3h16'41" 48d34'56"
Wheat Kings
Cas 1h27'15" 59d29'17"
Wheaton's Star
Lyn 8h2'23" 43d46'7"
Wheeler Wood Soos
Psc 1h16'26" 10d39'29"
Wheeler's Eagle
Aql 20h1'29" -0d18'18"
Wheezy's Star
Ori 6h12'55" -1d28'57"
Whejackwellson
Uma 9h36'3" 66d18'45"
When Day is Done
Lyr 18h51'9" 33d35'0"
When I look To The Sky...
Umi 14h35'56" 79d34'30"

When Stars Collide
And 23h40'34" 41d51'18"
When The Stars Align
Uma 8h33'48" 62d35'55"
When Two Stars Collide Like You & I
Cep 23h29'35" 79d10'35"
When you Wish Upon a Star...3/24/05
Vir 12h46'17" 4d21'2"
WHEN YOU WISHED FOR ME...
Per 3h10'16" 46d19'37"
Whenever I Need You
Sco 17h35'45" -35d48'33"
Where all our Wishes Go
Gem 7h14'35" 26d52'6"
Where Brown Meets Green
Lib 14h55'2" -4d40'14"
Where Flowers Grow
Cas 1h14'8" 58d3'23"
Where life & death meet.
WCE & CWM
Vir 14h33'41" 1d25'28"
Where Soulmates Gaze
Ari 3h27'3" 26d35'44"
Where We'll Meet...I Love You
Uma 8h45'35" 61d47'27"
Where's Teddy?
Cnc 7h57'21" 18d7'53"
Wherever in the World xxx
Ori 6h7'48" 16d22'9"
Wherever - Janette With Ali And Boy
Cru 12h18'4" -56d30'26"
Wherever we are...
Uma 9h53'57" 53d59'56"
Wherever We Go - The Walter's Star
Ori 5h5'44" 2d49'25"
Wherever You Are I Am There Too
Sco 17h39'41" -40d24'28"
Whickstar
Vel 9h57'14" -51d34'17"
WhirligigDave
Uma 11h12'14" 28d36'13"
Whirlygigging Frank
Aqr 22h35'13" -0d24'36"
Whiskey Whistler
Cma 7h15'50" -16d9'5"
WhiskyFWL80
Per 2h52'57" 41d43'44"
Whispers In The Breeze
Del 20h33'55" 6d20'7"
Whispers in the dark
Uma 11h40'49" 34d10'7"
Whistle
Umi 16h13'34" 73d36'46"
Whit Kreigh
Lib 15h13'35" -9d24'53"
Whit Lee 1
Sgr 19h11'56" -21d41'7"
Whit P. Whitaker
Sgr 18h31'41" -31d45'28"
Whit Sloan 10
Sco 17h23'29" -30d37'51"
White angel
Cap 20h52'47" -20d9'24"
White Angel
Tau 5h49'11" 21d16'5"
White Bear
Uma 13h24'13" 58d50'14"
White Chocolate
Vir 14h28'20" 3d6'52"
White Cloud LaPresti
Cnc 8h41'52" 13d11'49"
White Feather
Gem 6h57'43" 14d12'41"
White Flower Destiny
Lyn 6h24'48" 56d23'16"
White Island
Tri 1h53'25" 30d49'10"
White Knight
Ari 2h7'12" 23d40'30"
White Oleander
Uma 13h28'6" 56d33'59"
White on Rice
Uma 9h57'13" 57d11'57"
White Pelican
Peg 0h2'17" 21d25'0"
White Rose
Sco 16h58'33" -33d12'17"
White Sand
Cas 1h23'23" 55d40'29"
White Shark
Leo 11h53'28" 22d0'4"
White Star
Mon 7h51'15" -5d27'47"
White-Charles First Anniversary
Ara 17h15'37" -50d49'12"
Whitefield NH Fire and Rescue
Her 17h10'29" 15d38'46"
Whitehead
Per 4h13'39" 45d11'16"
Whitehorn
Equ 21h7'31" 7d54'27"
Whitehouse Junior School
Uma 12h6'52" 49d30'8"
Whitelums Benny
Uma 8h49'50" 50d21'7"
Whiter Galfand
Sgr 19h17'11" -12d30'18"
Whitewing
Per 2h59'38" 32d7'10"
* Whitey *
Uma 10h3'17" 46d36'7"
Whitey J
Uma 10h28'52" 47d7'5"
Whitey Webb World
Uma 8h14'33" 71d4'52"
Whitford Dean Brown
Uma 9h32'53" 57d55'35"
Whitley Marie Smith
Cnc 9h7'55" 31d56'24"
Whitley Simone'
Lib 15h7'20" -13d26'8"
Whitman
Vir 15h3'16" 1d35'52"
Whitman/Williams
Per 4h7'44" 42d49'35"
Whitnagale
Gem 7h27'40" 19d53'8"
Whitnee Brianne
And 2h9'54" 41d16'8"
Whitners
Lib 14h52'10" -3d44'57"
Whitney
Aql 19h41'55" -0d1'52"
Whitney
Lib 15h3'20" -7d35'12"
Whitney
Sco 16h17'29" -14d49'45"
Whitney
Sgr 17h50'6" -18d14'11"
Whitney
Cas 23h32'8" 52d55'45"
Whitney
Lyn 9h6'44" 34d20'43"
Whitney
Psc 0h55'55" 29d36'29"
Whitney
Ari 2h48'34" 16d18'32"
Whitney Adam (Whit) Warren
Uma 9h44'0" 71d55'54"
Whitney Alexander Parsons
Sgr 18h44'53" -29d2'17"
Whitney Alexandria French
Cyg 20h56'45" 48d7'54"
Whitney Allen
Aur 5h51'16" 36d8'12"
Whitney and Dustin, Eternal Love
Gem 7h9'0" 34d2'30"
Whitney and Johnny 7/7/04 1:57
Umi 15h9'59" 74d22'8"
Whitney "Angel" Ann Gibb
Aqr 23h45'0" -16d36'56"
Whitney Ann
Sgr 19h24'15" -15d23'3"
Whitney Ann
And 23h41'55" 44d34'35"
Whitney Ann
Ari 2h16'20" 25d6'30"
Whitney Ann Elizabeth Smith
Sgr 17h49'38" -17d31'51"
Whitney Ann Lucado
And 0h36'11" 29d34'16"
Whitney Anne
Psc 0h48'41" 16d0'47"
Whitney Anne
Cyg 21h35'19" 51d1'39"
Whitney Anne Vanhoose
Aqr 22h25'27" -24d1'45"
Whitney Ashton Whitmer
Leo 10h39'36" 14d9'33"
Whitney B. Kays
Cap 20h59'21" -25d19'25"
Whitney "Babies" Lester
Ori 5h33'14" -1d5'25"
Whitney (Babys) Rose
Lib 14h49'41" -1d59'20"
Whitney (BEbe) & Stephen (Rotag)
Tau 3h45'10" 28d46'49"
Whitney Blair Minyard
Vir 12h35'34" 4d6'46"
Whitney Brice Harrington
Cas 0h7'9" 55d6'58"
Whitney & Bryce
And 1h2'3" 38d45'47"
Whitney Burse
Ori 4h59'43" -2d37'54"
Whitney Bushart
Lyn 8h9'14" 44d21'20"
Whitney C. Passint
Per 3h50'16" 33d22'40"
Whitney Captain Underpants Rupp
Leo 10h36'36" 20d49'1"
Whitney Christine Berger-Whit.
Psc 0h19'45" 7d33'9"
Whitney Claire
Ari 2h22'30" 27d14'47"
Whitney Dale Norsworthy
Cnc 8h21'20" 12d22'50"
Whitney Day Solem
Psc 1h15'58" 7d16'8"
Whitney Dianne Bae
Cas 23h27'19" 54d57'18"
Whitney Drew Hand
Cnc 8h51'21" 28d43'27"
Whitney Edwards
Sco 16h24'22" -32d1'56"
Whitney Eggleston
Vir 11h55'28" 8d24'29"
Whitney Elaine
Lib 15h44'46" -7d41'10"
Whitney Elizabeth
Lib 14h55'27" -8d54'21"
Whitney Elizabeth
Cas 2h32'19" 67d12'27"
Whitney Elizabeth
And 23h33'54" 50d16'35"
Whitney Elizabeth Spradlin
Cap 20h17'28" -10d50'32"
Whitney Ellen Carleton
Snuggle Bear
Leo 10h59'44" 6d28'8"
Whitney Enuis Richoux
Leo 9h50'30" 23d31'37"
Whitney Erica Seabolt
Vir 13h36'35" -0d58'11"
Whitney Gagliardo
Uma 8h27'41" 61d56'57"
Whitney Gayle Dunn
Lyn 7h51'33" 42d33'56"
Whitney Gladney
Leo 10h16'53" 7d26'57"
Whitney Gooding
Crb 16h11'26" 39d36'36"
Whitney Hare
Crb 15h45'59" 32d8'32"
Whitney Hebel
Psc 1h39'41" 7d50'7"
Whitney in space
Lib 15h7'3" -7d30'59"
Whitney Jenais Griffin
Gem 6h43'7" 15d59'2"
Whitney Jo
Lib 14h23'53" -17d44'15"
Whitney & John Strassel
Uma 9h11'13" 60d9'28"
Whitney Kathleen Vigil
And 1h57'9" 45d40'16"
Whitney & Kenneth
And 2h22'45" 46d46'33"
Whitney Lane
Tau 4h10'16" 28d16'34"
Whitney Langsjoen-Hill
Aqr 22h19'52" -23d29'22"
Whitney Large
And 0h39'14" 36d10'46"
Whitney Lee Dean
Uma 11h21'29" 46d36'47"
Whitney Lee Norcross
Vir 12h35'49" -5d45'2"
Whitney Leigh
Sco 16h12'18" -16d52'11"
Whitney Longmire
And 23h18'30" 47d51'9"
Whitney Lynn O'Connell
Tau 3h42'58" 27d22'43"
Whitney Lynn Robinson
Aqr 22h7'29" -11d51'6"
Whitney M. Cowdin
Cnc 9h12'51" 15d25'20"
Whitney Marie
Ori 5h2'46" 10d44'2"
Whitney Marie
Cnc 9h12'20" 10d5'1"
Whitney Marie Young
Leo 11h20'36" 0d54'36"
Whitney Mark Gracie Julian
Uma 13h37'17" 56d37'27"
Whitney Massey
Lyn 7h2'21" 51d18'27"
Whitney Maya Bass
Cyg 20h7'10" 48d7'53"
Whitney Melroy
Crb 16h15'57" 32d46'44"
Whitney Michele Stevenson
Psc 1h22'14" 31d45'31"
Whitney Michelle
Ori 6h17'36" 10d34'0"
Whitney Miller and Ben McComb's
Lyn 7h59'7" 40d12'1"
Whitney - My Binary Star
Ari 2h36'7" 24d17'40"
Whitney Nicole
Cnc 8h29'22" 25d7'54"
Whitney Nicole
Sgr 17h51'4" -25d53'5"
Whitney Perdue
Sgr 20h5'24" -33d12'27"
Whitney Priest
Cam 6h54'3" 65d0'3"
Whitney R. Nelson
Ori 6h5'14" 13d4'8"
Whitney Rae Wilson
Uma 10h5'24" 68d22'0"
Whitney Reeder the love of my life
And 0h17'40" 29d8'37"
Whitney Rene Compton
Sgr 19h3'34" -33d53'17"
Whitney Rheel
Lyn 6h31'37" 55d7'9"
Whitney & Ricky Forever
Ari 2h55'32" 26d41'6"
Whitney Rose
Sco 16h15'5" -10d56'35"
Whitney Rosenblatt
Sgr 18h34'53" -25d21'38"
Whitney S. Turner
Cyg 20h30'24" 41d51'22"
Whitney Saunto Fitzgerald
Her 17h28'19" 16d15'6"
Whitney Schoultz
Cap 21h52'17" -18d53'2"
Whitney Shae Ferguson
Ari 2h48'18" 25d59'53"
Whitney Siehl, Eternally Dedicated
Uma 10h42'33" 55d13'38"
Whitney Sullivan
Lyr 18h36'55" 34d50'16"
Whitney Tatum Henne
Mon 6h45'55" 8d39'28"
Whitney the Darling Angelic Muse
Gem 6h44'16" 15d44'39"
Whitney Tyler
Sgr 18h53'21" -19d2'43"
Whitney Weaver
Cas 1h16'27" 53d27'9"
Whitney Wutzler
Vir 12h23'28" -9d13'17"
Whitney-Lea
Ari 2h42'50" 20d21'40"
Whitney's
Leo 10h11'43" 14d1'46"
Whitney's Sochor Star
Aqr 22h30'29" -2d6'50"
Whitney's Star
Lyn 7h30'8" 54d43'22"
Whitney's Star
Uma 8h55'11" 52d47'26"
Whitney's Star
Cyg 20h0'19" 53d36'7"
Whitney's Star
Uma 8h46'22" 68d29'8"
Whitney's Star
And 0h26'8" 38d15'16"
Whitni the Magnificent
Cyg 20h9'22" 46d22'21"
Whittany
Lib 14h54'37" -4d56'25"
Whitters Star
Uma 11h56'51" 29d19'42"
Whmes
Aql 19h36'41" 6d48'45"
Whoa! Where am I ?
Uma 10h43'31" 69d55'6"
Who-Doo Douglas E. Kaser
Per 3h32'30" 43d57'19"
WhoLeoDad47
Leo 11h26'34" 15d29'36"
WhoLeoMom48
Vir 12h0'29" 7d25'24"
Wholissa
And 23h3'54" 45d23'20"
Whoosh
Her 18h1'46" 18d18'0"
Who's Your Buddy - Doug Robins
Lib 14h41'48" -13d35'17"
Who's Your Daddy
Her 17h21'50" 16d25'23"
Whowolf
Cep 23h28'25" 71d24'19"
WHRSIR1946
Ori 6h0'20" 19d9'39"
W.H."Sid" Killingsworth 100th B-Day
Per 3h43'4" 43d58'53"
Whyte Carter & Isaac Flint
Uma 11h13'44" 28d29'41"
Wibbelmann, Manfred
Cap 20h30'23" -16d55'45"
Wichaxpi Ina
Uma 9h1'28" 58d29'22"
Wick Kok
Uma 8h56'31" 55d42'0"
Wickersham's Wonders
Cma 6h42'15" -13d24'13"
Widderli 'Roger'
Lac 22h43'39" 48d55'35"
Widemann, Fritz
Uma 10h28'52" 64d20'25"
WIEBKE
Lac 22h53'5" 54d19'54"
Wiebke & Mike
Uma 10h57'15" 41d19'3"
Wiedemann, Albert
Uma 10h4'24" 61d39'50"
Wiedner, Tom
Uma 9h43'16" 64d34'35"
Wiegard
Ori 6h0'6" 14d12'21"
Wien Manguerra
Vir 12h13'26" 10d33'14"
Wienold
And 1h34'21" 48d42'49"
Wiesalex
Umi 4h19'2" 89d6'14"
Wiese
Psc 0h16'26" 6d41'55"
Wiese, Eckhardt
Gem 7h17'26" 20d28'8"
Wiese, Marlies
Uma 12h1'13" 28d26'50"
Wiesel, Josef
Uma 12h58'53" 53d32'40"
Wieser Paul
Uma 12h24'4" 58d15'22"
Wiezie
Leo 11h4'24" 22d17'44"
Wifee
Vir 11h45'12" 2d35'21"
Wifey
Leo 10h57'28" 6d25'9"
Wifey
Vir 12h33'1" 13d9'27"
Wifey
Uma 9h44'35" 46d37'8"
Wiggey Anne - "Peppa"
Uma 10h17'50" 50d3'19"
WIGGI, 26.09.2001
Cyg 19h42'12" 29d8'18"
Wiggle
And 0h11'12" 40d2'52"
Wiggles
Lyn 7h29'26" 44d34'11"
Wiggles
Leo 11h19'47" 2d5'17"
Wiggles 2252004
Psc 1h20'9" 16d19'55"
Wiggles and Giggles
Cyg 19h48'49" 29d40'55"
Wiggles Magoo
Uma 8h51'16" 57d9'20"
Wigglesworth Must Be Fate "Parker"
Aqr 22h9'18" -0d4'0"
Wiggy
Aur 7h27'49" 35d22'52"
Wigi Aquilae
Aql 19h26'38" -9d27'35"
Wig-Katona, Susanna
Ari 2h21'2" 17d50'57"
Wijdan
Crb 15h17'14" 25d45'9"
Wikeen Fire
Uma 13h15'58" 52d42'13"
Wikolia
Leo 9h32'43" 28d36'53"
Wikouswitus Pesesum - Mom's Star
Gem 7h13'33" 34d45'50"
Wiktor Kapszewicz
Vir 14h19'22" -20d52'1"
Wiktoria Danko
Ari 2h50'44" 28d40'29"
Wil and Erika 1Year Anniversary
Cyg 20h25'48" 46d47'46"
Wil DiSabitino
Per 3h38'15" 45d30'39"
Wil Henry Yarnold
Ori 5h42'23" -1d34'45"
Wil Hoskovec
Cnc 8h15'43" 15d39'16"
Wil Oosterveld
Lyn 8h15'32" 39d4'23"
WIL RYAN GORDON
Aql 19h33'9" 4d0'42"
Wil Thouvenin
Ori 5h40'5" 2d22'26"
"Wil" - William Hume
Cyg 19h41'38" 29d2'30"
Wilbert Francis Warner
And 0h58'3" 39d44'57"
Wilbert Shepard
Uma 9h4'22" 53d35'0"
Wilbert Snow Elementary School
Uma 11h46'47" 57d27'9"
Wilbert Wayne Morris
Uma 10h39'20" 60d11'22"
Wilbois, Frank
Uma 11h6'8" 64d27'28"
Wilbur
Lyn 7h34'34" 56d31'2"
Wilbur
Lyn 7h50'59" 57d38'7"
Wilbur A. Sweeten, Jr.
Her 16h48'47" 43d23'37"
Wilbur and Mary Joseph
Uma 8h55'24" 53d33'27"
Wilbur Dooley
Umi 14h43'33" 83d7'59"
Wilbur F. Schlette II
Leo 10h16'14" 17d24'26"
Wilbur Harris
Vir 13h3'46" 12d49'56"
Wilbur J. VanDiest
Per 3h50'0" 33d37'26"
Wilbur Mueller
Umi 16h47'33" 80d13'57"
Wilbur Summerson
Ori 5h48'8" -2d23'1"
Wilcox & Pirate
Lyr 18h28'7" 29d52'15"
Wild
Her 16h33'27" 13d28'36"
Wild
Ori 5h24'52" 14d52'20"
Wild Bill
Psc 1h42'56" 12d48'50"
Wild Bill
Aql 19h8'30" 2d10'7"
Wild Bill
Psc 0h21'52" 17d40'39"
Wild Bill
Her 16h14'49" 16d20'29"
Wild Bill
Boo 14h43'55" 37d14'21"
Wild Bill Harvey
Ori 5h58'11" 12d32'50"
Wild Bill Harvey
Ori 5h54'34" 12d3'16"
Wild Bill Menna
Cnc 8h38'25" 10d9'16"
Wild Bill Vickers
Sgr 19h25'40" -30d46'53"
Wild Bill Young
Lyr 19h22'16" 37d45'57"
Wild Bill's Jewell
Cnc 8h14'34" 15d41'54"
Wild Bunch
Aql 19h29'7" 15d44'27"
Wild Flower
Psc 22h56'4" 4d54'11"
wild horses
Peg 22h27'58" 26d14'11"
Wild Man
Her 17h38'2" 31d59'33"
Wild Shep
Dra 16h10'34" 56d13'45"
Wild Woman Bev
Psc 1h53'41" 6d48'3"
Wilda Georgia Sampo
Uma 8h23'29" 63d7'5"
Wilda Janette Tuttle
Leo 10h34'34" 24d10'43"
Wilda Jeanne Holbrook
Tau 5h10'24" 18d28'18"
Wilda Lou Calico
Tau 5h48'28" 19d55'28"
Wilda McCullough Maley
Lyn 7h52'36" 57d10'24"
Wilda Stacey
Sgr 19h4'10" -20d55'10"
Wildcat
Lyn 6h39'48" 58d32'53"
Wilder Springs
Uma 9h46'1" 51d38'4"
Wilder Yasci
Aql 19h46'39" -0d35'17"
Wildgrube, Harald
Uma 11h3'49" 56d24'47"
Wilene Macale
And 0h40'55" 33d56'26"
Wiley Five-O
Cyg 19h47'51" 40d48'2"
wiley-g
Cma 6h28'19" -17d11'7"
Wiley's
Per 3h15'12" 44d21'25"
Wilf Kewley's Star
Uma 9h5'43" 56d49'43"
Wilfred
Sgr 20h6'33" -22d35'19"
Wilfred "Billy" Norman Bentham
Aqr 22h10'12" 1d13'42"
Wilfred C. Carr
Boo 14h46'54" 42d57'3"
Wilfred & Ethel Carey
Lyr 18h48'47" 36d21'1"
Wilfred Feinberg
Gem 7h50'49" 32d41'10"
Wilfred H. "Bill" Gump
Uma 11h19'13" 56d3'30"
Wilfred J. Carr
Uma 12h9'7" 46d44'44"
Wilfred John Ferreira Vazquez
Ori 4h55'8" 3d53'38"
Wilfred Lee
Sco 16h34'39" -26d8'10"
Wilfred "Willy" Lacas
Lyn 7h31'45" 58d3'21"
Wilfrid
And 0h9'5" 39d44'45"
Wilfried Groß
Uma 13h47'34" 51d36'12"
Wilfried Hülsmann
Ori 6h14'13" 8d18'47"
Wilfried Kadler
Uma 8h59'59" 57d0'53"
Wilfried Rütters
Uma 8h24'28" 59d38'15"
Wilfried Virmond
Uma 8h40'29" 56d8'0"
Wilhda J Rios Sepulveda
Uma 11h43'18" 44d43'18"
Wilhelm Garre
Ori 6h17'3" 9d18'20"
Wilhelm Resch
Uma 9h49'56" 56d20'17"
Wilhelm Staab
Uma 9h26'17" 52d31'0"
Wilhelm von Reth
Uma 10h25'17" 56d15'8"
Wilhelm Wolfgang
Uma 9h57'48" 69d49'23"
Wilhelm Zettl
Uma 8h56'11" 64d50'52"
Wilhelmina Clara Pollpeter
Uma 12h0'8" 28d26'1"
Wilhelmina Fluck "Mimi's Star"
Cas 0h29'40" 62d0'32"
Wilhelmina G. Harvey
And 2h36'30" 45d20'17"
Wilhelmina H. Hawkins
Crb 15h47'10" 36d51'51"
Wilhelmina Jaqueline Maria Wells
Psc 1h26'31" 15d34'22"
Wilhelmina Juurlink nee Belt
Psc 1h54'49" 6d47'3"
Wilhelmine Kiesswetter
Uma 11h23'25" 39d27'40"
Wilho C. Kivi
Uma 10h48'13" 53d57'20"
Wil-Hue View
Uma 9h16'24" 71d38'0"
Wiligene Clitus
Ari 2h53'3" 25d29'29"
Wilk
Uma 9h22'24" 45d22'20"
Wilken, Helmut Gerhard
Ori 5h51'54" 7d27'21"
Wilke-Sellers Starchild
Tau 4h18'1" 12d52'54"
Wilkie
Vel 9h31'38" -44d51'52"
Wilko
Ari 2h30'12" 18d56'29"
Wilko Gerlitz
Uma 9h24'47" 53d41'56"
Will
Uma 9h59'24" 55d41'6"
Will
Lib 15h7'29" -11d47'19"
Will
Ari 2h9'20" 16d37'47"
Will
Ari 2h42'49" 22d2'16"
Will
Her 18h1'36" 18d4'6"
Will
Uma 11h4'28" 48d21'40"
Will
Cas 0h21'43" 56d37'37"
Will
Per 3h49'39" 51d11'1"
Will
Boo 14h48'41" 33d36'29"
Will and Jaclyn...True Love
Cyg 19h39'56" 53d29'46"
Will Beria
Her 16h39'38" 8d16'47"
Will Blaha
Uma 9h36'4" 41d35'34"
Will & Chelsea - Star of Hope
Psc 1h20'3" 8d5'44"
Will Clifford
Boo 14h59'33" 50d24'49"
Will Costa
Ari 1h51'18" 12d5'23"
Will Crowley
Her 18h52'51" 22d50'24"
Will David Patton
Lib 14h50'14" -6d5'57"
Will Dupsha
Ari 3h22'35" 24d12'47"
Will F. Moesel
Leo 9h26'16" 25d21'37"
Will Fore
Cep 20h48'56" 60d39'27"
Will Foreman
Per 3h9'6" 41d11'13"
Will Francisco
Vir 12h36'51" -8d31'50"
Will Franklin
Sco 16h52'10" -38d54'9"
Will Fronek
Lyn 6h46'4" 57d39'41"
Will Geller
Cyg 21h14'19" 45d49'22"
Will Gregg-The Light of My Life XAC
Cru 12h23'29" -56d30'26"
Will Henry Kurth
Lib 14h54'1" -2d1'34"
Will Hermes
Cep 21h30'2" 60d22'15"
Will James
Uma 10h19'45" 65d38'38"
Will James Dougal
Umi 13h46'20" 71d14'17"
Will Jones
Lyn 7h22'37" 53d15'4"
Will Jowett
Uma 8h42'12" 65d11'56"
Will & Kelly Hawkins
Ari 2h33'3" 19d7'19"
Will Kesler
Dra 17h39'29" 51d55'6"
Will LaCharite
Per 3h12'20" 54d54'45"
Will Lacy Mary Joe
And 23h8'12" 36d6'25"
Will Landy
Aqr 22h42'19" 1d55'20"
Will Longman
Sco 16h10'6" -11d44'17"
Will Love You Forever, Terry
Cas 0h22'59" 54d10'43"
Will & Marie Barnes
Vir 11h46'5" -4d20'57"
Will Mark Laubach
Cep 20h48'28" 59d30'14"
Will N Bini
Lyn 7h36'8" 40d16'57"
Will Nathaniel Sierleja
Cmi 7h15'30" 10d52'45"
Will & Nicola's Anniversary Star
Per 3h42'29" 41d12'13"
Will Poteet-Berndt
Lyn 8h15'47" 47d58'43"
Will Rees
Lib 15h28'50" -9d58'58"
Will Roberts
Her 15h58'47" 44d42'50"

Will Rogers
Tau 4h28'54" 18d30'38"
Will & Sara Buckman
Cyg 20h23'57" 43d32'32"
Will Sopha
Vir 12h57'3" -19d31'7"
Will Thomas Birthday Star
Cru 12h34'44" -64d36'53"
Will Thomas Naylor
Cep 21h52'37" 63d1'19"
Will Tucker
Cas 0h24'20" 64d26'59"
Will "William" Magliocco
Ori 5h29'17" 14d54'43"
Will you go out with me?
Sco 17h43'25" -37d52'19"
Will you marry me?
Cir 14h12'36" -66d22'10"
Will You Marry Me?
Vir 12h13'28" -2d18'50"
Will You Marry Me?
Tau 4h22'16" 21d45'33"
Will You Marry Me.
Gem 6h27'28" 21d8'1"
Will You Marry Me
Ori 6h5'43" 18d30'20"
Will You Marry Me??
And 1h19'54" 47d29'17"
Will You Marry Me?
Lyr 18h32'38" 37d16'3"
Will You Marry Me
And 0h53'46" 36d58'39"
Will You Marry Me Beautiful?
Cyg 20h47'48" 47d11'35"
Will You Marry Me Cathy?
Lib 15h0'43" -15d19'47"
Will you marry me Chantel
Cru 12h0'58" -60d26'47"
Will You Marry Me Henh?
Leo 11h2'0" 13d17'26"
Will You Marry Me Kristen?
Gem 7h18'0" 24d36'42"
Will you marry me, Victoria
Ari 2h48'45" 29d0'32"
Will Young
Aqr 23h31'53" -16d2'14"
Will. Your light is with us, Always
Uma 11h38'16" 52d50'35"
Will, Heinz
Uma 11h36'1" 29d8'25"
Willa
Ori 4h59'39" -0d3'47"
Willa Dorothy Osborn
Cas 0h30'41" 64d14'10"
Willabee
Ori 5h22'13" 0d13'36"
Willadean Biagi
Lyr 18h45'17" 33d57'25"
Willam Stanley Butler - 01-06-1931
Uma 11h52'22" 34d0'3"
Willamena Barnes
Lmi 10h24'47" 34d20'15"
Willard
Ari 2h12'27" 13d6'54"
Willard Burnett Hicks IV
Her 17h23'36" 38d52'42"
Willard Byrd
Cap 21h29'8" -24d44'45"
Willard Darryl Jersey
Cas 1h34'11" 66d2'36"
Willard Frew Davis
Ori 4h51'35" 1d46'45"
Willard K. King
Lib 15h41'30" -4d50'9"
Willard & Virginia Oplinger & Fam.
Cyg 19h52'20" 32d34'23"
Willbeca
Dra 17h11'54" 56d49'16"
Wille and Popp
Gem 7h36'16" 23d16'11"
Willeke, Julia Marie
Uma 11h34'28" 33d10'45"
Willelmi and Alianora Duglas
Lyn 6h48'52" 54d41'25"
Willem Finn Boxen
Pyx 8h44'9" -27d35'59"
Willem Kuntz & Angela Heurter
Tau 5h39'55" 22d50'55"
Willena
Leo 11h19'7" 0d31'18"
Willene Francis Collier
Cas 1h39'22" 64d54'59"
Willerbadoo
Aur 5h43'13" 54d22'58"
Willershausen, Ingeborg
Uma 11h35'52" 39d46'9"
Willette Barnetta Ward Garrett
Per 3h39'4" 34d12'14"
Willi Büchler
Uma 11h55'1" 32d58'58"
Willi Emil Stein
Uma 9h34'54" 58d57'10"
Willi Knops
Uma 9h42'20" 48d1'56"
Willi Lenk
Uma 10h23'0" 48d37'3"
Willi1950
Lyr 19h19'29" 35d21'56"
WILLIA FAYE STARNES
Cap 20h36'17" -9d39'17"
Willia Susan Donnelley
Lyr 18h28'23" 38d13'0"
William
Per 4h14'0" 52d6'34"
William
Gem 7h39'38" 33d13'58"
William
Per 3h25'1" 40d45'0"
William
Ari 2h35'32" 19d23'14"
William
Leo 9h46'47" 26d54'10"
William 143
Gem 7h3'52" 10d36'10"
William 84
Cep 21h45'12" 61d8'31"
William A. Cervi
Cep 20h44'26" 59d20'51"
William A. Ciaburri Jr.
Sco 17h29'42" -41d12'39"
William A. Cutchin
Her 16h43'44" 46d23'20"
William A. Dailey
Aql 18h52'3" -0d0'31"
William A. Etheridge
Leo 11h7'2" -6d27'7"
William A. Galvin
Per 3h8'59" 49d11'19"
William A. Jordan
Cep 22h41'42" 63d44'51"
William A. Marshall
Cap 20h19'9" -10d18'44"
William A. Myers
Ori 6h21'45" 16d39'51"
William A. & Patricia A. Calderon
Vul 19h26'49" 23d31'4"
William A. Perez
Gem 7h30'26" 33d12'56"
William A. Reilly 63rd 255th Co B
Vir 13h24'47" 13d15'5"
William A Schaaf
Aql 20h2'53" 10d45'28"
William A. Temlak
Sco 16h9'28" -13d51'6"
William A. Thompson
Cep 20h39'57" 64d13'48"
William A. Van Ditto
Lib 15h14'2" -25d37'22"
William A Walsh
Cnc 9h19'2" 26d16'12"
William A. Wawak
Umi 17h22'51" 76d21'36"
William Ace Russell
Cap 20h30'8" -15d18'39"
William Adair LaForge
Tau 4h21'2" 24d58'55"
William Adam Ashurst
Cep 20h38'8" 61d46'54"
William Adams Seale
Per 3h40'5" 38d35'5"
William Adkins-H1A7A2L05Y7N2A06F
Aqr 22h29'35" 2d16'17"
William Alan Blakemore Hays
Vir 15h3'55" 5d11'55"
William Alan Meng
Ori 6h7'42" 9d7'27"
William Alan Weissbrodt
Cap 21h20'42" -24d57'37"
William Alex Mattero
Sco 17h8'34" -31d52'42"
William & Alexander
Aql 19h9'33" -0d3'48"
William Alexander Charouhis
Boo 15h31'12" 40d21'31"
William Alexander Piazzi
Leo 10h23'10" 8d39'7"
William Alexander Procter
Vir 13h6'27" 7d55'4"
William Alexander Skidmore
Leo 10h40'54" 9d48'26"
William Alexander Tally
Uma 9h30'4" 50d7'4"
William Alfred Brown, Jr.
Leo 10h12'35" 25d56'33"
William Allen Copeland III
Per 2h53'13" 41d53'46"
William Allen Gabriel Brewi
Aur 5h34'34" 46d2'29"
William Allen Jarosz, Jr.
Ori 5h53'45" 6d12'32"
William Alvin Hubbard, 03221949
Ari 2h9'9" 25d15'12"
William and Catalina
Sge 19h25'56" 17d1'4"
William and Christine
Cyg 21h18'5" 46d5'32"
William and Chrystal Fischbacher
Cyg 21h5'54" 48d3'51"
William and Jaime Martin Family
Aql 18h55'31" -0d14'43"
William and Justina
Boo 15h8'53" 40d30'55"
William and Karleen Lawrence
Ari 3h22'14" 27d5'21"
William and Kathleen Hinshaw
Ori 5h12'30" 6d51'11"
William and Kathleen McMillian
Cyg 19h45'16" 34d50'36"
William And Katie For True Love
Sgr 19h36'13" -16d1'15"
William and Lorraine Osborne
Cyg 19h33'55" 32d7'29"
William and Margaret Parkhurst
Cyg 21h45'14" 54d35'3"
William and Melissa
Her 18h51'59" 25d50'25"
William and Nellie Logan
Uma 10h39'25" 56d4'54"
William and Patricia Zeile
Peg 21h41'22" 18d17'36"
William and Ramona Hershberger
Ara 17h30'29" -47d53'25"
William and Sarah Camblin
And 2h17'51" 45d38'45"
William and Shelly Eddy
Uma 11h6'55" 65d56'27"
William And Stephanie Carmichael
Ori 5h31'56" 6d0'28"
William and Tommie Jones
Her 17h12'23" 32d51'53"
William and Vickie Lovingood
Her 18h30'9" 25d27'20"
William and Victoria Beers
Ori 4h50'40" 12d58'12"
William and Waneta Smith
Lyn 8h23'16" 35d51'20"
William Andrew
Cep 22h10'47" 60d17'17"
William Andrew Gray
Ori 5h41'59" -3d43'23"
William Andrew Henderson
Lyn 7h45'13" 53d4'31"
William Andrew Karo II
Psc 1h7'8" 32d0'24"
William Andrew Patzig
Leo 11h11'9" -5d27'56"
William Ansell
Umi 14h26'44" 73d22'18"
William Anthony Komma
Ari 2h23'14" 26d8'20"
William Anthony Mac Lopez
Dra 18h30'38" 55d49'6"
William Anthony Pipesh
And 2h2'25" 37d37'52"
William Anthony Stone
Cep 23h40'50" 74d37'50"
William Aronson
Vir 14h5'4" -15d29'11"
William Asher Price
Sgr 19h22'13" -23d36'13"
William & Ashley
Cyg 19h28'54" 57d41'11"
William Ashley Fisher
Per 4h6'38" 43d4'24"
William Atticus Parker
Aql 19h30'16" 8d0'26"
William August Duckworth
Tau 3h50'0" 11d24'5"
William August Weigelt
Ari 3h28'7" 23d3'19"
William Augustus Blount Stewart III
Uma 9h34'30" 60d27'38"
William Avelino
Cap 21h54'22" -21d9'22"
William B. Black, Jr.
Boo 14h50'8" 54d14'4"
William B. Coleman
Aql 19h55'53" -0d26'17"
William B. Kelly
Tau 4h21'13" 3d7'9"
William B: Loving Father and Healer
Umi 14h46'44" 73d39'9"
William B. Peregrine
Gem 6h58'47" 32d23'43"
William Babkiewicz
Cma 6h44'32" -26d19'28"
William Barnes
Her 16h43'41" 11d20'22"
William Barry Lawson
Her 17h42'35" 44d10'15"
William Bates Mattson
Cap 21h32'17" -10d10'20"
William "Beel" " Pappy" Michael
Per 3h15'15" 54d18'47"
William Bekker
Cep 22h24'52" 75d38'3"
William Bell White Howe
Mon 6h54'48" -0d2'47"
William Bennett Aldridge
Ori 6h17'3" 11d33'32"
William Bernard
Vul 20h41'8" 25d1'18"
William Bettencourt
Lib 14h52'28" -2d27'32"
William & Betty Hoffer
Tau 3h48'38" 16d21'3"
William B.Holmstock
Per 3h5'51" 40d29'6"
William (Bill) Andrade
Her 17h4'50" 46d6'43"
William "Bill" Bell
Ori 5h45'44" 1d58'43"
William "Bill" Bell
Ori 6h7'52" 5d40'28"
William "Bill" C. Dresser
Per 2h45'49" 54d19'17"
William (Bill) Fasulo
Psc 0h55'51" 26d6'9"
William Bill Fishman
Lib 15h3'12" -1d43'20"
William (Bill) George Capper
Tau 4h6'4" 4d56'47"
William "Bill" Grimsley
Uma 11h49'1" 57d41'14"
William "Bill" Hornberger Jr.
Per 4h14'52" 36d31'13"
William (Bill) Joy Heitz
Cap 20h32'43" -9d31'14"
William "Bill" Michael Jordan
Ori 5h36'52" -1d16'12"
William "Bill" Nutt
Tau 4h32'10" 12d26'55"
William "Bill" Owens
Ori 5h31'50" 1d19'41"
William "Bill" Paul Harrell
Tau 4h40'2" 0d51'49"
William (Bill) Pemberton
Per 2h19'47" 54d30'8"
William (Bill) Peters
Per 3h37'55" 44d47'35"
William "Bill" Wickham
Per 3h23'23" 51d52'49"
William "Billy" Barteld, Jr.
Vir 13h54'42" 0d27'56"
William (Billy) Benson
Cep 22h0'19" 56d50'55"
William "Billy" DeFeo
Her 16h11'21" 13d46'7"
William "Billy" Depalma
Lib 15h46'14" -29d25'8"
William (Billy) F. England
Cyg 19h37'13" 35d35'35"
William "Billy" Hartley
Per 3h38'43" 41d22'39"
William "Billy" James Murphy
Psc 0h55'39" 32d39'2"
William "Billy" Rishel
Per 4h12'39" 33d22'34"
William "Billy" Sigers
Tau 4h8'25" 16d55'52"
William (Billy) Wolf Jr.
Cep 22h51'40" 79d12'13"
William Birk, Jr. *Billy*
Per 2h24'45" 55d42'0"
William Birk, Jr. *Billy*
Per 2h24'45" 55d42'34"
William Blake Butcher
Cep 22h23'52" 86d39'15"
William Bly Blandini
Her 16h46'6" 28d55'42"
William B.Mattull,Sr.
Uma 12h54'53" 54d46'8"
William Bobrink Hare
Cnc 8h23'43" 6d44'24"
William "Boompa" Hunsburger
Uma 11h27'41" 60d25'22"
William Bottenfield
Crb 15h49'40" 38d51'48"
William Boulton
Per 3h39'1" 49d55'21"
William Bowman
Per 2h50'1" 53d50'44"
William Boyd Woodward Jr.
Per 4h33'50" 47d53'51"
William Brackney
Uma 10h43'46" 72d26'35"
William Bradford Marsh
Cas 23h35'26" 59d56'20"
William Bradley Beckett
Gem 6h52'18" 33d28'57"
William Bradley Blair
Cap 21h10'28" -24d39'0"
William Bradley Jordan XxLostxX
Lib 15h36'33" -22d2'31"
William Bradley Van-Go to the Stars
Per 3h27'8" 44d35'29"
William Brander III
Psc 0h8'28" 0d15'55"
William Brendon Schneider
Uma 8h40'0" 63d34'22"
William Brennan Conrad
Sco 16h10'0" -11d24'32"
William Brent Jeffers III
Leo 10h59'40" -2d34'29"
William Brett Wightman
Cap 21h16'53" -16d28'29"
William Brian Adams
Ori 5h1'3" 13d44'56"
William Brian Hurt
Her 16h38'47" 31d9'58"
William Brian Sneddon
Ser 15h17'22" 13d16'18"
William Brian Willdridge
Leo 9h26'28" 9d34'21"
William Bryant Armas
Cnc 8h19'14" 6d39'25"
William Brynn
Uma 9h2'56" 53d29'6"
William Bryon Chaapel II - Grape
Her 16h15'28" 7d24'17"
William 'Bud' Ellsworth
Lib 14h25'58" -19d27'33"
William Buending
Dra 19h1'3" 64d28'20"
William Burke
Per 3h49'58" 43d42'58"
William Busby
Her 16h6'16" 16d26'39"
William C. Abbott
Ari 2h4'39" 23d0'5"
William C. Akunevicz III
Her 17h53'25" 18d21'28"
William C Culp
Vir 14h55'11" 5d33'22"
William C. Dotson Jr.
Lib 15h51'40" -18d21'10"
William C. Gates
Lib 15h22'3" -8d55'33"
William C. Hinton, Sr.
Ori 5h20'38" 6d50'23"
William C. Hursch
Lyn 8h46'3" 35d17'2"
William C. Lee, Jr.
Uma 8h58'58" 52d7'43"
William C. Onofry
Her 16h57'8" 17d28'55"
William C. Pommerer Sr.
Aql 20h7'15" 0d52'59"
William C. Raspet, Jr.
Her 17h1'5" 31d40'22"
William C. Schnell
Lib 15h9'28" -4d10'2"
William C. Smith
Gem 6h52'35" 16d14'21"
William C. Smith Sr.
Uma 9h32'21" 67d10'1"
William C. Trotter
Uma 9h8'25" 69d58'30"
William C. Valentine
Per 4h10'0" 51d47'4"
William Calderon
Tau 4h25'17" 22d26'16"
William Calkins Senior
Psc 1h14'50" 24d1'37"
William Campbell (Cael) Allen
Lyn 8h56'7" 45d56'41"
William Canlas & Steph Lastoskie
Uma 10h45'3" 43d42'17"
William Carlton Bruce
Cnc 8h31'11" 13d31'14"
William Casey
Cap 20h16'41" -9d15'36"
William Casey Cordell
Cep 22h5'15" 69d27'48"
William Catesby Gilbert
Lib 15h6'26" -3d42'28"
William Charles Arnold
Cyg 21h27'0" 32d44'47"
William Charles Clifford, II
Boo 15h13'50" 45d37'35"
William Charles Gregorio
Lib 15h54'35" -6d34'34"
William Charles Grisack III
Per 3h3'48" 56d24'37"
William Charles Heuter
Her 18h43'43" 20d4'6"
William Charles Stewart
Ari 3h2'51" 12d38'45"
William Christian Beck
Her 18h49'23" 21d22'39"
William Christopher Griggs
Oph 17h39'37" -0d6'51"
William Christopher Monaghan
Uma 10h49'38" 46d37'0"
William Christopher Penhall
And 2h6'31" 45d34'22"
William Christopherson
Cyg 20h22'32" 48d27'21"
William Clark Bennett III
Her 16h34'4" 34d35'48"
William Clark Seneff
Cap 21h49'41" -8d57'19"
William Clifford Parris III
Vir 13h2'28" 4d30'31"
William Cochran Turbiville
Tau 4h23'41" 22d30'58"
William Coe
Ari 2h38'54" 25d45'22"
William Cole
Ari 2h26'28" 27d40'8"
William Cole Gawf
Lmi 10h48'41" 24d41'7"
William Colin Clark
Psc 1h1'51" 4d7'11"
William Colli, Jr.
Crb 16h9'30" 34d57'13"
William Colmenares
Cyg 21h49'5" 45d36'54"
William Conklin Borynack
Cap 21h26'59" -26d39'5"
William Connelly
Cep 0h0'32" 73d32'43"
William Connor
Lib 15h28'15" -17d3'47"
William Constantine Kekeris
Crb 15h33'57" 28d59'13"
William Constitution O'Rights
Crb 15h33'50" 27d55'2"
William Corbin Tanner
Cnc 8h28'52" 23d21'19"
William Corey
Aur 5h46'55" 29d45'8"
william corey
Uma 9h16'6" 54d44'42"
William Cormac McCarthy
Ori 5h58'57" 7d29'18"
William Cotter Demmon
Lyn 6h32'30" 57d27'54"
William Council III
Per 4h50'22" 50d27'5"
William Cree Hicks
Vir 13h28'29" -14d38'57"
William Curtis Bovender
Her 16h47'5" 31d6'40"
William Curtis Jr. Family
Psc 0h54'12" 26d50'11"
William D. Bush II
Ori 5h31'7" 4d49'47"
William D. Carroll
Cyg 20h16'57" 43d17'50"
William D. Cochran
Ori 5h20'5" 8d54'45"
William D. Finnegan
Cep 3h1'28" 84d24'1"
William D. Hadfield
Umi 14h6'52" 74d27'8"
William D. Mercer I
Cap 20h29'17" -10d52'34"
William D. Pinder III
Cnc 8h58'30" 10d39'42"
William D. Popp
Uma 11h46'32" 31d55'47"
William D. Samaha
Tau 5h35'7" 25d51'58"
William D. Smith
Her 17h12'58" 35d52'18"
William D. Weber
Sgr 19h44'15" -15d18'46"
William D. Zeedyk
Aur 5h41'57" 49d1'25"
William D. Zimmerman Jr. USN Ret'd
Lyn 7h49'55" 57d33'3"
William "DAD" Snyder
Cyg 19h37'31" 38d31'53"
William Dagle
Per 3h41'49" 42d50'52"
William Dale Bickerstaff
Sco 16h5'30" -18d57'51"
William Dale Grosh
Her 17h4'20" 29d4'45"
William Dale Speckels
Ari 2h55'47" 26d16'56"
William Dale Wilkinson II
Ari 2h42'50" 26d35'13"
William Daniel Arden III
Tau 5h56'19" 26d26'59"
William Daniel Huff
Cnc 8h38'12" 7d21'28"
William Daniel Rudski
Sgr 18h49'43" -17d25'51"
William David English
Ori 6h10'34" 15d37'35"
William David Fletcher Rogers
Ori 5h32'6" -1d55'25"
William David Galonis Lipscomb
Cap 21h21'36" -16d36'45"
William David Jowett
Uma 14h27'18" 60d31'57"
William David Joy
Uma 8h55'59" 49d5'41"
William David Llewelyn
Uma 9h23'14" 53d32'10"
William David Short Jr.
Cnc 8h49'31" 24d44'3"
William David Sigee
Per 2h36'31" 51d47'4"
William Dean Douglass, Jr.
Aql 19h40'9" 4d53'28"
William Dean Heaton, Jr.
Her 17h16'52" 35d7'40"
William Dean McWhorter
Uma 11h32'52" 51d47'34"
William & Deborah Rittenhouse
Dra 15h42'25" 56d44'30"
William Demerio - 75
Vir 13h12'5" 11d48'38"
William Dennis Boland
Gem 7h30'52" 30d0'36"
William Dennis Spradlin
Uma 12h26'22" 53d58'14"
William Denver Sabree
Vir 12h43'9" 1d12'55"
William DePonte
Per 4h47'22" 42d37'11"
William Derek Jorden
Lib 15h4'11" -7d45'33"
William Dillon Gabriel
Tau 3h31'1" 11d36'21"
William Dobek
Her 17h37'6" 32d11'48"
William Doherty
Per 2h23'43" 56d38'39"
William Donald Balthaser
Ori 5h53'27" 22d41'25"
William Donald Halverstadt
Boo 14h47'1" 19d48'39"
William Donald Heiser
Ori 5h20'17" 6d54'44"
William & Donna Hubbard
Cyg 21h36'59" 46d25'17"
William Douglas
Lyn 7h50'22" 50d52'54"
William Doyle Erard
Uma 13h3'7" 57d1'37"
William "Dude" Wiley
Cep 0h2'9" 77d40'6"
William Duncan Nov 1945-July 1981
Uma 13h28'6" 52d36'32"
William Dwight Lowry
Gem 7h45'59" 25d54'21"
William Dylan O'Hurley
Sgr 20h4'25" -35d35'10"
William E. Alexander
Per 3h35'17" 47d25'34"
William E. and Colette Ricco
Uma 13h14'8" 56d45'50"
William E. and Grace Ricco
Uma 13h59'20" 52d56'18"
William E. and Ruth Schiebel
Cma 6h40'26" -15d59'34"
William E. Baab
Psc 0h19'55" 8d40'49"
William E. Barlow
Per 3h15'21" 51d41'34"
William E. Bartee's 70th B-Day Star
Ori 5h32'36" -0d38'8"
William E. Brenner, Jr.
Psc 0h49'58" 3d57'3"
William E. Brown III
Aql 18h51'27" -0d44'56"
William E. Bushyager
Cnc 8h17'58" 12d2'18"
William E. Calvert
Sco 16h6'47" -17d8'50"
William E. Cardinal, Jr.
Her 17h42'54" 19d12'41"
William E Corbett
Uma 9h16'26" 57d48'58"
William E. Fenton, Jr.
Uma 10h10'28" 67d14'33"
William E. Ferris Sr.
Sco 17h15'5" -43d31'59"
William E. Frawley, Sr.
Dra 16h58'42" 67d37'5"
William E. Hodapp III
Uma 14h17'31" 61d55'41"
William E. Kattmann
Vir 13h46'17" -7d21'33"
William E Kenealy
Cep 20h59'49" 67d30'21"
William E. Marks
Per 3h45'5" 37d51'46"
William E. May
Oph 16h23'5" -0d9'24"
William E. McCraw Sr.
Lyn 7h56'49" 47d33'53"
William E. Muller
Boo 15h14'44" 40d31'45"
William E. Naugle
Per 3h6'15" 51d30'30"
William E. Nelson Jr.
Dra 15h18'24" 58d22'24"
William E. Purcell
Sgr 18h48'17" -18d46'19"
William E. Readel
Uma 10h43'53" 45d49'29"
William E. Scheets
Psc 23h56'8" 0d32'59"
William E. Shiels III
Ori 5h54'58" 21d48'43"
William E. Siburt
Aqr 22h45'33" -2d32'46"
William E. Sowell
Vir 14h33'28" -2d24'29"
William E Stelzer
Vir 13h17'52" 5d24'8"
William E Underwood Jr
Gem 6h51'46" 27d56'37"
William E. Vance
Cep 23h28'59" 78d1'47"
William E. Wall
Per 3h32'21" 34d8'54"
William E. Wolf III
Lyn 6h31'44" 60d0'11"
William E. Wright, Jr.
Per 3h8'24" 55d23'52"
William Earl Hartpence
Uma 12h29'50" 60d59'49"
William Earl Moore
Tau 5h35'35" 21d21'1"
William Edgar Jorgensen Sept. 24, 1947
Her 16h43'1" 28d54'58"
William Edward Blakeley
Per 4h7'41" 32d14'1"
William Edward Ford III
Per 3h25'35" 45d40'56"
William Edward Kirke
Uma 14h17'45" 56d9'58"
William Edward Luzar
Sgr 18h42'24" -33d2'35"

William Edward
MacWhorter
Cap 21h30'3" -16d37'5"
William Edward Necessary
Vir 14h22'27" -0d50'31"
William Edward Smith
Cyg 20h58'9" 47d19'59"
William Edward Stovall
Sco 16h59'21" -42d5'53"
William Edwin Banks
Uma 9h56'33" 56d32'47"
William Eifert
Aql 19h45'48" 7d8'27"
William "El Caballero"
Bowker
Peg 22h29'35" 3d47'7"
William & Ellen Vidro
Aqr 20h51'36" 1d57'2"
William Ellic Cherry
Uma 9h56'17" 56d18'14"
William Ellis - The Love Of
My Life
Tau 3h32'14" 24d39'23"
William Elmer Hayek Jr.
Her 16h9'34" 24d2'44"
William Emery Hapworth
Jr.
Her 17h30'2" 30d15'43"
William Ernest Chilton
Ori 5h56'45" 17d15'29"
William et Emma Campeau
Uma 13h22'39" 55d3'18"
William Eugene Bird
Cyg 20h6'53" 35d18'5"
William Eugene Cureton
Aur 5h50'6" 42d4'21"
William Eugene & Laurie
Ann Sollows
Sge 20h4'20" 17d1'53"
William Eugene Mauck
Tau 5h34'52" 19d32'31"
William Eugene Wood
Ari 1h52'40" 18d41'34"
William & Eula Garnett
Ori 6h3'28" 10d0'47"
William F. Banks
Peg 23h51'44" 10d54'21"
William F. Baxter
Psc 0h42'16" 21d0'37"
William F. Cassin
Psc 23h21'16" 4d6'59"
William F. Cooper
Tau 5h10'3" 22d51'30"
William F. Coyne, Sr
Cep 22h32'51" 75d26'52"
William F. Davis
Uma 13h45'37" 61d54'5"
William F. Elliott Jr.
Ori 5h34'10" 3d58'4"
William F. Lovett
Per 2h12'13" 55d55'37"
William F. Madison
Per 4h13'50" 49d16'47"
William F Matarazzo
Psc 23h33'22" 4d26'56"
William F. Meany
Uma 13h5'41" 62d49'50"
William F. Meyer (The
WFM)
Vir 13h56'32" -18d38'18"
William F. Renz, II
Aur 5h29'40" 41d7'17"
William F. Sharp
Lmi 10h16'23" 38d3'30"
William F Talbott
Cnc 8h41'14" 17d47'48"
William F. Weiss
Ari 3h12'38" 28d18'11"
William F. York, Sr.
Uma 8h46'35" 50d17'9"
William Fairbourne
Armstrong
Aur 6h33'47" 34d35'32"
William Fairchild
Sgr 18h0'48" -18d24'8"
William Fairley's Angels
Uma 11h30'4" 56d1'41"
William Fang
Umi 13h24'36" 69d25'12"
William & Farah Flores
Tau 4h29'56" 17d44'7"
William Fay Vicory
Uma 11h5'3" 38d28'58"
~* William Felipe O'Donnell
*~
Leo 10h14'52" 22d21'0"
William Forest Moody
Gem 7h11'58" 16d0'2"
William Foss Jr.
Sgr 19h11'36" -23d41'55"
William Fountaine Allen
Ori 4h52'41" 2d40'42"
William Frances O'Neil
Gem 7h16'54" 20d50'47"
William Francis
Her 16h22'39" 18d41'29"
William Francis
Uma 13h4'32" 53d47'32"
William Francis Galloway,
III
Lib 14h58'40" -15d39'11"
William Francis Galloway,
III
Lib 15h23'22" -4d7'26"
William Francis McGrath
Uma 12h40'56" 58d28'0"
William Francis Murphy the
Great
Aqr 22h37'50" -2d36'23"
William Francis Ward
Per 4h0'29" 38d55'43"
William Frank Bacon
Uma 11h11'42" 52d44'0"
William Frank Green
Aqr 20h50'58" 1d18'32"
William Frank Hatter
Sco 16h7'3" -26d29'29"
William Frank Leonard
Uma 8h34'49" 70d20'35"
William Frank Maurer
Vir 13h46'37" 4d7'27"
William Frank Wisler
Ori 6h1'0" 18d50'48"
William Frankel
Sgr 20h5'57" -26d23'2"
William Franklin
Her 17h21'20" 39d5'49"
William Franz Webber
Uma 11h0'19" 52d5'37"
William Fred Traber
Ori 6h6'7" 4d35'52"
William Frederick Adams
Cas 23h45'2" 51d48'27"
William Frederick McHenry
Cnc 8h54'5" 32d20'10"
William Fredrick Hess
Uma 11h18'36" 37d16'45"
William Freidreck Halbach
Ori 5h54'23" 6d37'56"
William G. Allenbaugh III
Her 18h47'52" 21d45'17"
William G. Bankhead, Jr.
Ari 2h10'1" 24d9'59"
William G. Corrigan
Gem 7h30'37" 18d54'48"
William G E Clark
Cep 21h31'15" 55d53'5"
William G. Fey
Del 20h37'48" 17d38'27"
William G. Larson
Sgr 17h58'32" -28d44'25"
William G. MacAdam
Com 12h24'50" 28d55'26"
William G. Rorison
Her 17h12'40" 30d4'3"
William G. Tackett
Tau 5h31'19" 26d40'10"
William Gary
Uma 8h53'46" 60d18'53"
William Geary
Cyg 21h34'38" 43d23'5"
William Genader
Per 3h30'20" 51d44'25"
William Gene Hill
Sgr 18h24'55" -23d3'18"
William Geoffrey McMullin
Per 1h45'35" 51d12'10"
William George Aston
Cherry
Per 3h20'15" 40d4'25"
William George Fearn
Uma 10h4'7" 44d29'20"
William George Grant
Hewett
Her 18h33'11" 22d59'34"
William George Kerr
Per 4h25'55" 33d30'28"
William George Louch
Ori 6h15'13" 9d44'11"
William George Mace
Per 4h31'43" 38d13'7"
William George McCarthy
Sgr 18h12'9" -22d52'19"
William George Mullins
Psc 23h29'2" 1d58'23"
William George Roberts
Cru 12h45'33" -57d29'37"
William George Russell
Uma 12h10'16" 49d15'51"
William Glenn Holt
Ari 3h25'43" 27d52'39"
William Glenn Huebner's
Star
Sgr 19h30'36" -38d1'33"
William Glenn Hutchinson
Lyr 19h20'3" 40d19'41"
William Glenn Worsham
Her 17h56'57" 14d54'29"
William "Goat" Klein
Uma 11h51'7" 43d58'2"
William Goble
Her 17h59'14" 42d59'13"
William Gonyo
Uma 10h58'55" 48d50'11"
William Goodman
Uma 12h1'2" 55d50'47"
William Graham Carter
Lib 15h37'38" -6d8'42"
William Grampy Davis
02/09/37
Lyn 8h11'59" 57d29'23"
William "Grandpa" Rusen
Sr.
Gem 6h23'58" 22d45'41"
William Grant Adkins
Aur 6h23'5" 48d5'44"
William Grant Adkins
Sco 16h17'3" -28d57'44"
William Grant Davidson
Ori 5h22'42" 5d10'47"
William Grey
Vir 12h51'28" 6d15'38"
William Grey
Boo 14h47'55" 26d37'28"
William Grover Redden
Per 3h52'45" 37d11'58"
William Grumley Hayes
Ari 3h24'8" 26d52'23"
William Guy Smith
Cap 21h35'50" -16d57'34"
William H. Arthur IV
Sco 16h17'28" -17d33'15"
William H. Barlow
Her 16h53'4" 13d6'38"
William H Berry III
Psc 1h14'54" 18d16'12"
William H. "Billy" Jervis II-
#5
Lib 15h50'37" -12d3'35"
William H. Connor
Her 16h46'37" 47d58'27"
William H. D. Clausen
Cap 21h29'52" -14d17'22"
William H. Dewey
Sco 17h56'27" -30d40'6"
William H. Fraizer, Jr.
Boo 14h13'57" 28d33'9"
William H. Green
Ari 3h16'5" 21d27'30"
William H. Harbaugh
Uma 10h10'56" 56d19'5"
William H. McBride
Tau 3h44'1" 28d25'35"
William H Peacock
Uma 11h18'13" 29d2'25"
William H. Rex
Her 17h18'42" 37d2'12"
William H Richards
Ori 4h52'58" 12d45'11"
William H. Robinson
Uma 11h46'38" 59d28'32"
William H. Stevens
Per 2h47'10" 53d25'55"
William H. Teesdale
Sco 16h11'51" -34d32'7"
William H. White
Aqr 21h6'0" -11d25'43"
William H. White Sr
Per 3h16'27" 51d23'29"
William Hall
Ari 2h3'1" 14d2'42"
William Harold Baldwin
Uma 13h8'18" 56d7'10"
William Harold Carter
Cnc 8h9'36" 22d25'32"
William Harris
Uma 9h5'23" 51d42'51"
William Harris
Uma 11h10'35" 38d31'47"
William Harry Dunn
Sgr 17h46'9" -17d35'43"
William Harry Smith
Cep 23h15'46" 79d33'7"
William Harvey Squally Jr.
Psc 1h2'59" 26d11'51"
William Haselden
Lyr 18h55'36" 32d16'52"
William Hastings Lanza
Dra 17h58'45" 53d6'44"
William Hayden Bystrom
Psc 1h8'49" 31d47'30"
William Hayden Partain
Psc 0h34'24" 13d52'43"
William Heiser
Her 17h29'58" 47d49'38"
William Henning McFarlane
Leo 10h19'33" 17d57'51"
William Henry Bayes-Green
Sgr 20h0'42" -29d21'38"
William Henry Buntin III
Her 18h24'5" 12d17'15"
William Henry "Bushie"
Schafer Jr.
Leo 9h55'32" 24d23'25"
William Henry Clever
Her 16h34'34" 45d39'54"
William Henry Dickson
Aql 20h10'19" -6d9'23"
William Henry Glasser
Per 3h17'39" 51d55'53"
William Henry Scott
Psc 0h56'37" 12d4'44"
William Henry Sewell
Her 17h33'37" 26d9'55"
William Henry Smith IV
Uma 11h19'44" 63d48'21"
William Hestand Baxter
Lib 14h55'9" -0d34'35"
William Higgins
Cnc 9h6'20" 16d1'56"
William Holmes Black
Sco 17h51'11" -42d50'5"
William Horton
Ori 5h18'2" -5d22'25"
William "Howie" Barczak
Cyg 21h41'45" 43d54'19"
William Howland Eddy
Per 3h17'56" 54d35'21"
William Hubert Hurdle, Sr.
Gem 7h21'49" 25d58'9"
William Hugh Bradford
Per 4h41'30" 46d23'50"
William Hunter Taylor
Ori 5h30'38" 6d55'41"
William Idris Aubrey
MacNamara
Hys 0h32'8" -64d32'37"
William Imboden Campbell
Sco 16h18'2" -12d0'49"
William Isaac Phillips
Her 17h23'11" 22d33'47"
William Issac Wolfson
Sgr 18h24'23" -32d24'4"
William J. Anderson
Cyg 20h1'28" 36d56'40"
William J. Bernard
Cap 21h30'22" -17d15'0"
William J. Brkovich
Ori 5h55'32" -0d44'46"
William J. Burns Jr.
Umi 18h49'44" 87d50'25"
William J. Carew
Tau 5h34'53" 24d57'18"
William J. Cooper
Uma 12h2'11" 50d21'19"
William J Corbett
Uma 9h54'56" 52d24'27"
William J. Deming
Ori 5h38'12" 7d12'6"
William J. Dopirak
Cnc 9h1'21" 24d10'54"
William J Entley
"1/29/1955"
Her 17h31'15" 34d2'37"
William J. Esser
Her 17h42'26" 16d0'50"
William J. Felker
Cyg 19h37'41" 35d41'34"
William J. Flaman
Sgr 19h10'6" -16d49'41"
William J. Floyd
Col 5h12'0" -32d19'3"
William J. Forbotnick
Aur 5h47'10" 52d21'57"
William J. Ganus
Aqr 22h35'36" 0d32'37"
William J Gowen
Her 17h39'41" 36d41'18"
William J. Graley
Ari 2h57'43" 21d12'55"
William J. Graley
Ser 18h17'11" -15d55'8"
William J. Green
Leo 11h35'43" 8d39'14"
William J. Hatfield
Her 17h22'45" 35d54'58"
William J. Keller, Jr.
Cap 21h53'19" -13d12'27"
William J. Lober III
Ori 5h36'8" -0d33'17"
William J. McGookin
Uma 8h37'13" 50d52'6"
William J. McPherson
And 0h30'55" 45d21'24"
William J. & Merrilee M.
Rice
Per 4h9'45" 51d4'48"
William J. Northerner
Ori 6h4'17" 17d53'2"
William J. O'Keefe
Dra 15h41'59" 59d18'34"
William J. Ortiz Jr.
Lib 15h2'5" -10d40'22"
William J. Ortiz Sr.
Sco 17h15'38" -43d10'18"
William J. O'Shea
Vir 11h57'51" 6d27'5"
William J R
Cyg 19h26'4" 35d46'45"
William J. Robinson
Boo 14h46'9" 23d31'14"
William J. Rotchford
Cep 21h51'2" 66d30'33"
William J. Shannon
Cap 20h30'26" -9d9'18"
William J. Storer
Ari 3h8'41" 18d24'16"
William J Vetter
Her 18h46'14" 25d54'27"
William J. Wiese
Uma 12h0'21" 42d10'46"
William J Yost
Per 2h25'18" 57d4'35"
William J. Ziegler
Sco 16h31'26" -37d12'9"
William Jack McHenry
Uma 11h2'29" 44d32'25"
William Jack Smith
Ori 5h54'59" 17d15'4"
William Jack Waterhouse
Lmi 10h39'38" 28d31'5"
William Jackson Crosby
Ori 4h49'20" -0d23'48"
William Jacob Reddell
Per 3h29'47" 35d48'43"
William & Jacqueline
Younger
Cyg 19h48'57" 33d23'56"
William James
Aqr 21h58'41" 0d23'1"
William James Armstrong
"11-9-04"
Boo 14h24'43" 24d59'45"
William James Boisoneau
Uma 9h50'53" 69d27'59"
William James Casterton
Cep 3h40'16" 80d50'22"
William James Doty
Psc 0h6'45" 4d21'38"
William James Fulton III
Psc 23h28'38" 4d14'55"
William James Gallagher
Jr.
Lib 15h6'39" -1d6'22"
William James Gately II
Per 2h25'5" 51d58'25"
William James Glover
Gem 7h14'32" 17d23'43"
William James Kelly
Dra 20h9'49" 74d10'36"
William James Kendall
Per 2h11'6" 57d11'17"
William James Kidd
Ori 5h26'20" 2d59'32"
William James Mace
Ori 6h1'22" 15d3'32"
William James Mulligan
Tau 5h15'19" 16d15'12"
William James Pisano "Mr.
P Star"
Per 3h4'5" 55d25'29"
William James Pletcher
Lib 15h46'51" -17d34'14"
William James Sansalone
Uma 14h0'14" 49d33'41"
William James Steffan
Green
Dra 16h18'25" 67d48'31"
William "Jamie" Scobie
Ori 5h39'14" -2d10'50"
William & Janet
Hendrickson
Uma 13h55'18" 53d15'7"
William Jared Maupin
Aur 5h24'0" 40d3'14"
William Jarrett Flowers
Per 2h15'3" 54d18'46"
William Jay Fields
Dra 16h54'45" 66d58'6"
William "Jay" Rector
Aql 19h59'37" 8d57'19"
William&Jean
Boo 13h47'19" 10d2'12"
William Jefferson Clinton
Aql 19h49'48" 6d23'26"
William Jeffrey Eastham's
Baptism
Cru 12h25'31" -56d59'14"
William Jeffrey Lloyd
Cnc 8h51'35" 27d1'55"
William Jeffrey Strouse
Aqr 22h56'54" -23d46'53"
William Jeffrey Waldsmith
Sco 17h26'59" -42d51'48"
William Jeffrey Weidner
Aur 5h56'40" 37d29'47"
William Jenkins
Aqr 22h58'55" -7d36'55"
William J.F. Roll III
Ori 6h3'54" 21d13'53"
William J.M. Wald
Cep 22h29'13" 66d28'11"
William Joel Bly
Cap 21h46'2" -14d16'32"
William John Allan Born 25
Aug 1947
Cru 12h40'42" -59d48'18"
William John Arrigoni
Cnc 8h57'48" 10d46'57"
William John Buehler
Cmi 8h4'44" -0d13'47"
William John Carney
Sco 16h9'52" -14d51'34"
William John Daly
Cep 21h3'56" 57d47'12"
William John Dodsworth
Cep 0h13'22" 71d47'9"
William John Dowse
Per 3h53'38" 32d12'25"
William John Durrant
Leo 10h11'39" 26d52'27"
William John Feickert
Cep 23h22'56" 70d23'59"
William John Gaguski
Her 17h2'1" 14d4'23"
William John & Isobel
Fraser King
Cep 20h47'39" 60d38'25"
William John Larner
Cep 22h52'23" 72d31'54"
William John Majewski
Vir 14h43'38" 5d45'58"
William John Martin
Per 2h25'0" 53d49'23"
William John Muller IV
Uma 9h58'31" 44d24'57"
William John Nape, II
Psc 1h19'56" 32d17'37"
William John Smith
Aqr 21h11'47" -8d57'9"
William John Stevens
Umi 14h49'4" 73d28'2"
William John Walsh
Her 17h41'29" 47d52'19"
William John Wendt
Per 3h38'10" 43d28'55"
William Johnston
Cep 22h21'34" 82d12'58"
William Jon York
Dra 16h45'54" 56d36'40"
William Jonathan Georges
Ari 2h21'15" 18d21'11"
William Jordan Applegate
Her 16h28'11" 42d32'21"
William Joseph Baldwin
Sgr 19h14'0" -16d51'59"
William Joseph Becker
Cep 4h26'45" 82d16'7"
William Joseph Brewer
Psc 1h40'17" 19d33'7"
William Joseph Crowley III
Her 17h22'16" 16d41'42"
William Joseph Fitzpatrick,
Jr.
And 0h46'50" 34d44'24"
William Joseph Gabriel
McSherry
Cap 20h30'34" -13d50'57"
William Joseph Girvan 60th
- Love Owen
Leo 11h20'46" 0d13'56"
William Joseph Kennedy III
Uma 10h56'26" 47d20'35"
William Joseph Loftus 111
Aql 19h45'12" 0d45'13"
William Joseph M
Psc 0h50'35" 15d53'32"
William Joseph Maier
Boo 14h32'50" 19d35'25"
William Joseph McCullough
Jr.
Per 3h21'15" 44d58'26"
William Joseph Meyer
Sgr 18h8'22" -31d15'27"
William Joseph Noah
Boo 14h18'30" 17d39'44"
William Joseph Olsen
Vir 12h49'30" 7d34'46"
William Joseph Pennington
Tau 3h59'35" 25d3'38"
William Joseph Reger
Aql 19h7'6" -0d37'1"
William Joseph Rosser
Ari 2h22'37" 25d56'23"
William Joseph Wilson
Sgr 17h54'43" -17d45'33"
William Joshua Meadows
Tau 4h9'20" 5d12'6"
William Josiah Romero
Peg 22h40'15" 25d53'55"
William Judson Thomason
Uma 11h38'45" 33d13'59"
William Justin
Uma 10h45'13" 47d8'12"
William K Gumpert
Leo 11h33'57" 17d47'46"
William K. Runyeon, M.D.
Uma 8h54'53" 52d11'49"
William K. "Skip" Hodapp
Vir 13h29'10" -20d18'47"
William K. Trowsdal Jr.
4/13/03
Ori 5h36'37" -3d16'55"
William Kaplan
Aur 5h9'1" 35d44'20"
William Katz #1 Equity
Production
Cep 0h6'3" 78d6'5"
William Katzberg
Uma 9h19'2" 72d57'0"
William Kee
Boo 13h40'47" 20d3'17"
William Kelly Turner
Huggett
Per 3h23'38" 34d23'23"
William Kelvin Zaladonis
Lib 14h22'29" -17d19'46"
William Kendrick
Cap 21h51'54" -10d49'37"
William Kennerley 80
Cnv 13h52'19" 32d21'49"
William Kenneth Hansen
Sgr 17h57'54" -22d45'9"
William Kenneth Morgan
Sge 19h56'53" 19d0'9"
William Kerwick
Cep 22h35'14" 78d8'46"
William Kiefer
Psc 23h33'52" 5d17'22"
William & Kim Chappell
Uma 13h33'36" 59d30'31"
William & Kimberly Corder
Dra 14h55'22" 59d0'19"
William Kinzler
Uma 11h17'34" 54d27'12"
William Kirby Koen Family
Aql 19h35'11" -0d4'23"
William Klein
Ori 6h17'52" 10d31'34"
William Koleniak
Cyg 20h25'5" 35d51'20"
William Korach III
Dreamfish
Umi 14h5'36" 74d52'30"
William Kristopher and
Sara Dorothy
Tau 5h12'26" 21d32'57"
William L. Benjamin
12/4/48-2/17/03
Uma 10h37'36" 44d21'15"
William L. Berkeyheiser
Per 2h11'23" 57d1'9"
William L. "Bill" Agambar,
Sr.
Sgr 19h15'47" -15d44'51"
William L. Gunlicks
Leo 10h10'27" 22d40'17"
William L. Hopkins
Her 18h18'14" 18d26'52"
William L. Longe
Dra 19h25'42" 76d30'47"
William L. McKown Jr.
Leo 9h39'37" 27d27'30"
William L. Mize M.D.
Boo 13h54'47" 24d31'21"
William L. Ross II
Her 17h47'47" 15d0'40"
William L. Rugg
Cru 12h22'58" -59d10'28"
William L. Sheets
Her 17h57'16" 49d51'5"
William L. Shoemaker
Uma 11h40'31" 45d49'58"
William L. Smith
Her 16h37'30" 35d35'40"
William L. Steck
Uma 10h49'58" 49d35'27"
William L. Sterrett
Cnc 8h50'12" 17d47'58"
William L. Walker
Uma 13h13'58" 55d52'31"
William Landau
Uma 8h42'8" 58d35'14"
William Landon Fleeman
Gem 7h29'10" 22d19'28"
William Lane
Per 3h28'5" 50d20'15"
William Langdon - Bill's
Star
Uma 8h48'53" 50d55'49"
William Lange Sherrin
Aur 5h27'10" 40d51'46"
William & LaRae Maguire
Forever
Uma 10h27'12" 52d5'50"
William Laurens Pressly
Her 16h18'36" 17d12'16"
William Lawrence & Connie
Severson
Uma 9h7'56" 62d32'51"
William Lechleitner
Cnc 9h7'3" 31d13'24"
William Lee Rosenbaum
Boo 14h36'9" 27d33'37"
William Lee Transier
Cnc 8h57'30" 10d22'56"
William Leo Frank
Aur 7h19'28" 42d9'15"
William Leonardo Zipeto
Umi 2h11'47" 88d43'46"
William "Liam" White
Per 3h49'20" 36d40'53"
William Liberty Wellington
Cyg 20h50'7" 31d10'11"
William & Lillian Roach
Sge 19h42'13" 18d52'40"
William LLoyd Haley
Cnc 7h59'40" 12d39'46"
William Lloyd Whitworth
Her 16h48'29" 35d44'15"
William Lloyd Whitworth, Jr.
Ori 5h38'52" 5d13'53"
William Luke Governali
Gem 6h34'26" 20d10'32"
William Lynch Irving
Ori 6h8'13" 14d58'52"
William M Curran
Ari 3h17'28" 27d9'36"
William M. Katz
Lac 22h44'15" 49d48'0"
William M. Lane
Uma 13h36'54" 52d55'13"
William M. O'Neal
Vir 12h28'7" -5d38'59"
William M. Rue
Her 17h34'8" 46d58'44"
William M. Stringer
Tau 4h42'17" 27d42'51"
William M Walker
Uma 13h54'28" 52d51'45"
William "Mac" Esposito
Peg 23h35'43" 28d3'15"
William MacPherson
Lyn 7h50'6" 57d53'5"
William MacQuire
Per 4h41'16" 46d26'48"
William Malloy Jr.
Uma 8h10'38" 62d54'17"
William Manasse
Cam 4h3'33" 72d10'7"
William Manis
Leo 11h35'56" 2d23'29"
William Manson Gough III
Uma 13h37'27" 50d19'33"
William & Margaret Lange
Cyg 19h43'5" 40d23'53"
William & Margaret Walser
Cyg 19h57'11" 33d23'3"
William & Marian Peters
Crb 15h56'37" 30d28'34"
William Marinelli
Ari 2h47'1" 24d56'59"
William Marius Durani
Ser 18h17'6" -13d23'54"
William Mark Kirkland -
Will's Star
Ori 6h6'36" 10d43'20"
William & Marti Lee
Cas 23h47'43" 57d59'52"
William Martin
Cnc 8h44'21" 10d50'2"

William & Mary
Cyg 19h52'46" 39d31'10"
William Masamichi Fujitaki
Crb 15h41'47" 25d56'2"
William Matthew Butler
Cep 21h30'44" 58d46'48"
William Matthew Rice
Uma 8h51'15" 62d35'52"
William Matthew Stembridge
Her 16h27'17" 20d0'33"
William Matthews Long III
And 0h8'53" 41d30'45"
William Mauger
Cep 23h31'39" 77d20'21"
William & Maureen
Cyg 20h30'44" 39d20'10"
William McCallister
Boo 13h51'49" 11d20'15"
William McCann
Ori 5h48'54" 7d24'57"
William McComb Chamberlin
Per 4h11'56" 34d27'7"
William McDougall Lukens
Uma 11h36'33" 39d53'2"
William McKinley Batts Jr.
Cet 2h14'8" -14d10'40"
William McManus
Uma 8h36'26" 58d18'39"
William McNay
Uma 12h6'15" 58d46'12"
William Messner
Per 3h20'18" 42d11'35"
William Michael
Ori 5h34'10" 4d38'35"
William Michael Black
And 23h27'1" 44d20'54"
William Michael Brown
Cnc 8h49'28" 8d45'50"
William Michael Brown
Cap 21h37'56" -12d29'16"
William Michael Connors
Her 16h50'41" 32d46'54"
William Michael DeMeo
Uma 9h29'39" 44d58'24"
William Michael Fech
Cas 0h14'9" 54d24'35"
William Michael Fox
Her 17h4'39" 42d18'12"
William Michael Gooler
Aql 19h55'8" 13d55'20"
William Michael Johnson
Gem 7h29'16" 26d21'7"
William Michael Malone
Sco 17h39'4" -45d16'27"
William Michael Mora
And 0h34'32" 41d35'5"
William Michael "Sparky" Lawson
Aql 19h3'43" -0d7'58"
William Michael Wenz
Cas 0h53'44" 54d29'44"
William Michael Wilson
Her 16h18'39" 20d24'38"
William Miguel Thoms
Her 18h5'57" 17d32'26"
William Monaghan Jr.
Tau 4h35'17" 6d32'26"
William Monico Dec III
Sco 16h7'32" -21d37'18"
William Monroe Tucker II
Cap 21h5'15" -16d37'4"
William Montanez JR.
Per 3h1'41" 43d48'2"
William Moody III
Ari 2h26'52" 11d7'5"
William Mudd
Cep 22h52'14" 70d55'30"
William Multack
Lib 15h28'43" -8d4'43"
William Murray Penhale
Vir 11h46'39" -0d57'18"
William My Love
Aur 5h45'26" 46d12'32"
William Nagel MacMillan
Vir 12h11'17" -5d6'12"
William Nathaniel Dougherty
Leo 10h52'9" 0d21'59"
William Nation
Uma 11h21'34" 49d21'40"
William Neely Setzer
Umi 13h10'25" 70d11'48"
William Nicholas Jeffrey Hammond
Dra 18h51'29" 71d9'42"
William Nicholas Krohn
Her 18h52'42" 21d22'47"
William Nickerson
Cas 23h46'41" 57d36'6"
William Noll, My Valentine
Cep 22h5'11" 54d2'6"
William O. Jensen Jr.
Ari 2h23'59" 26d45'17"
William O Schuman
Uma 13h48'58" 57d16'59"
William O'Donnell
Uma 8h59'23" 58d15'28"
William Oehler
Ari 2h24'1" 11d19'27"
William of Allagash
Ari 3h18'53" 29d11'59"
William of Wynyard's Wishing Star
Umi 15h25'21" 69d10'22"
William O.J. Herbert
Leo 9h34'33" 7d14'30"
William Oliver Bickhart
Her 18h4'52" 18d0'49"
William Olmsted Dougherty, Jr.
Aql 19h51'57" -0d19'8"
William O'Neill
Per 3h31'57" 51d19'3"
William Orin Airy
Cnc 9h1'1" 6d55'32"
William O'Ryan
Aqr 23h25'10" -10d56'36"
William Osceola, Sr.
Aqr 22h35'56" -0d49'52"
William Osco Keys Jr.
Boo 14h38'43" 37d4'31"
William P Mack
Lib 15h40'44" -10d10'24"
William P. Moore
Tau 5h15'58" 27d47'38"
William P Pully, Jr.
Sco 17h38'53" -34d33'54"
William P. Stough
Cam 3h39'55" 60d31'20"
William Pack
Dra 18h32'39" 69d7'15"
William Pack
Uma 10h42'57" 48d8'28"
William Paladine
Ori 5h24'19" -0d56'20"
William Panzer
Uma 9h55'5" 49d46'15"
William "Pa-Pa Bill" Fahy
Her 17h11'10" 31d52'0"
William Pappas
Lmi 10h33'46" 35d11'53"
William Parsley
Uma 10h54'57" 50d46'42"
William Patrick 21 *xod*
Psc 1h14'53" 8d23'39"
William Patrick Anthony Garcia
Boo 14h36'59" 27d18'22"
William Patrick Diamond
Her 16h40'59" 6d51'1"
William Patrick Sine
Cen 13h25'13" -36d23'34"
William Paul
Umi 14h3'46" 71d16'16"
William Paul
Aqr 22h16'55" 0d11'15"
William Paul Cress
Cyg 21h34'43" 41d11'41"
William Paul Mamounis Jr.
Her 17h52'57" 27d39'32"
William Paul Rose
Dra 20h3'51" 70d43'53"
William Paul Rotheroe
Gem 6h47'51" 18d4'48"
William Paul Sullivan Longmire Star
Cru 12h37'45" -61d54'0"
William Payne Howard
Ari 2h20'31" 26d34'38"
William Pearson
Vir 13h0'29" 5d32'4"
William Peavey, Jr.
Cyg 20h51'13" 41d57'34"
William "Pepa" Dunbar
Aqr 21h20'4" -11d12'57"
William Perez
Ori 5h53'31" 12d30'12"
William Perry Fuller III
Tau 5h53'5" 24d27'6"
William (Pete) Ross
Dra 16h54'50" 74d29'40"
William Peter Garten
Gem 7h26'20" 17d23'56"
William Peter Mader
Cep 21h31'24" 62d9'30"
William Peter Verberne
Del 20h35'21" 4d21'22"
William Peter Walker
Ori 6h23'45" 14d3'56"
William Pfeiffer
Gem 7h2'54" 33d13'26"
William Philip Murphy
Cnc 9h6'16" 15d58'22"
William Poole Miller
Aql 19h13'53" 1d6'8"
William Pop McClain
Umi 4h41'16" 88d37'19"
William "Poppop" Keating
Lib 15h27'43" -18d18'16"
William "Poppy" Becton
Uma 10h57'8" 70d45'39"
William Pops Settan
Aqr 23h0'1" -8d35'6"
William Prentice
Ori 5h25'38" 2d30'1"
William Preston Cole
Lyn 7h43'1" 54d24'36"
William Purton
Cap 21h24'6" -14d39'21"
William R. and Anne A. Taylor
Lyr 18h43'58" 41d1'19"
William R. "Bill" Hoover
Cap 21h26'56" -16d39'18"
William R. Brent
Uma 9h33'53" 72d55'51"
William R. Brown
Umi 14h55'11" 78d32'36"
William R. Carmon
Per 2h54'50" 52d27'58"
William R. Cooke
Uma 10h35'56" 69d21'32"
William R Creighton ~ Feb 11, 2000
Gem 7h44'14" 33d41'24"
William R. "Doc" Palumbo
Aql 20h19'40" 0d48'26"
William R. Doyle
Ori 5h10'9" 5d3'59"
William R. Fair
Ori 5h28'19" 3d39'43"
William R. Garrett
Cap 20h15'24" -19d36'17"
William R. Goss 77
Her 17h54'14" 28d2'15"
William R. Hoffman
Lac 22h56'31" 51d10'4"
William R. Jepson
Cyg 21h44'26" 38d46'11"
William R. & Lillian A. Lincoln
Lyn 7h37'4" 47d50'42"
William R. McCay
Aur 5h32'45" 45d58'19"
William R. McCoy
Cnc 8h17'32" 19d4'9"
William R. Olsen
Cyg 19h57'56" 31d2'38"
William R. Perretti
Vir 12h43'25" 7d41'59"
William R. Ruiz III
Lib 15h49'51" -6d57'12"
William R. Sosa II
Sco 17h24'3" -38d22'48"
William R. Webster Sr.
Cep 21h55'10" 60d10'40"
William R. Wheeler
Ori 5h20'34" 8d46'28"
William R. Wick
Cnc 8h51'42" 30d57'5"
William R. Wuensche
Cnc 8h20'11" 22d54'48"
William R. Wuensche
Cnc 8h19'32" 23d44'2"
William Ralph Bowers
Ari 2h10'48" 23d34'24"
William Ramzey Kholeif
Uma 9h39'14" 46d47'52"
William Randy Redden
Cnc 8h54'18" 22d31'1"
William Ray
Her 17h46'6" 43d30'47"
William Ray Sigmund
Aql 19h9'32" 9d1'6"
William Ray Smith Birthday Star
Cap 21h14'22" -18d7'0"
William Ray Weaverling
Vir 12h9'9" 11d35'23"
William Raymond Miller
Per 3h56'50" 52d42'43"
William Raymond Schell
Leo 10h31'39" 9d51'3"
William Raymond Turner
Aur 6h30'2" 39d4'27"
William Raymond White
Lib 15h10'25" -21d10'31"
William & Rebecca Stanford
Lyr 18h44'24" 34d36'48"
William Reed Davis
Sgr 18h48'18" -16d42'29"
William Reed's Virtus
Uma 11h32'43" 47d8'59"
William Reese
Leo 11h39'51" 24d9'43"
William Richard
Uma 10h31'54" 45d13'47"
William Richard
Aqr 22h1'28" -3d21'51"
William Richard Halopoff
Her 16h37'54" 35d22'23"
William Richard Neild
Gem 6h46'31" 26d16'24"
William Richard Prohinsie
Ari 2h55'21" 28d31'30"
William Richard Sherwood
Lib 15h5'43" -0d46'22"
William Richards
Ori 5h25'47" 2d54'45"
William Richardson
Her 16h12'23" 47d46'14"
William Riley
Ori 5h43'39" 8d46'56"
William Riley Litz
Umi 13h29'24" 73d53'44"
William Rivenburg
Per 4h10'48" 51d55'36"
William Rivera
Aql 19h16'42" 14d24'16"
William Robert Epright
Cnv 12h52'59" 45d21'41"
William Robert Glenn
Gem 6h24'0" 22d54'17"
William Robert Haniford, Jr.
Tau 5h24'57" 26d20'33"
William Robert Knox
Boo 15h7'50" 41d44'11"
William Robert Lloyd (Billy)
Cep 0h5'2" 67d56'9"
William Robert Netro
Aur 6h32'26" 36d6'44"
William Robertson Walsh
Lyn 6h23'24" 55d41'56"
William Roger Bohl
Aur 5h40'18" 47d23'19"
William Rome
Uma 11h33'28" 49d0'25"
William Ron Anderson
Aql 20h0'55" 9d4'39"
William Ronald Johnson, Jr.
Sco 16h11'55" -13d14'5"
William Ross Shook
Vir 13h19'50" -4d34'38"
William Rossiter & Alexa Winner
Aqr 22h35'37" -2d25'36"
William Russo
Boo 14h39'39" 29d20'2"
William Ryan Bovender
Crb 15h47'30" 27d57'38"
William S. Achey
Leo 10h32'10" 26d9'32"
William S. "Billy" Roberts
Per 3h45'35" 50d55'3"
William S Campbell
Leo 9h39'13" 30d59'16"
William S. Chinn
Uma 9h51'13" 46d14'34"
William S. Fernstrom Jr.
Her 17h40'42" 42d36'33"
William S. Fernstrom Sr.
Per 3h33'54" 46d8'32"
William S. Malone -Daddy-
Lmi 10h22'8" 35d23'13"
William S. Slater
Her 17h17'46" 36d4'28"
William S. & Virginia L. Applegate
Cyg 20h56'43" 34d18'29"
William Sampson Franks, Jr.
Men 4h13'36" -81d22'11"
William Sandford Frampton
Her 17h41'37" 16d56'40"
William Sansalone the Lionhearted
Her 17h47'25" 44d59'19"
William Schmitt "Pops"
Per 4h9'32" 31d9'35"
William Schwiemann
Psc 0h47'31" 14d28'36"
William Scott Brown and Family
Tau 4h42'50" 4d22'56"
William Scott Fletcher
Umi 14h4'17" 78d42'18"
William Scott Munro
Per 2h49'14" 57d13'55"
William Scott Peters
Sco 16h3'52" -17d49'1"
William Scott Sellers
Aur 6h30'52" 34d13'32"
William Scott Severn
Sgr 18h16'1" -20d58'37"
William Scott Thompson
Her 17h16'2" 18d13'54"
William Scott Werner
Ori 6h5'20" -0d3'58"
William Scott Wright
Tau 5h34'48" 19d52'50"
William Shayne Noel
Cep 3h41'44" 80d57'0"
William Sherwood
Lib 14h53'5" -19d28'31"
William & Shirley Crawford
Col 5h8'55" -35d42'12"
William Sidney Henderson-Yendrys
Cep 22h31'20" 62d10'10"
William Sidney Moss
Lib 15h6'23" -17d20'51"
William Simonian
Uma 8h35'40" 48d15'54"
William Sisco
Her 18h35'16" 25d11'52"
William "Skip" Boyer
Gem 6h47'14" 13d48'30"
William "Skipper" Roney
Cas 1h28'5" 65d22'13"
William Slade Rhodes
And 0h39'22" 35d19'14"
William Smith
Psc 1h25'3" 29d14'8"
William Smith
Umi 8h53'33" 86d8'52"
William Snyder
Her 17h18'38" 35d46'49"
William & Sondra Long
Cap 21h35'33" -14d1'38"
William Soreca
Her 17h25'12" 32d32'53"
William Spencer Randle, III
Per 4h28'20" 33d45'37"
William Stairiker & Alaina Bradley
Cyg 19h42'32" 55d17'22"
William Stanford Douglass
Umi 15h17'57" 68d44'33"
William Stanley David Hickin
Cru 12h43'21" -61d32'57"
William Stanley Novak, Sr.
Uma 8h28'14" 65d17'47"
William Stanley Powell
Per 3h15'35" 41d29'46"
William Stark
Sgr 17h48'26" -17d37'42"
William Steier, M.D.
Cep 22h37'57" 72d5'15"
William Stephen Burcham
Lib 14h35'51" -17d44'29"
William Stephen Capozzoli
Boo 14h13'56" 34d16'3"
William Stephen Dise
Cyg 20h45'36" 35d38'27"
William Stephen Higgins
Crb 16h15'16" 34d46'43"
William Stephen Weber
Aql 19h50'24" 7d36'25"
William Stuart Rupp
Lib 15h39'45" -5d23'17"
William Summers
Aql 19h42'31" 8d22'12"
William & Susie Ferrante's Star
Leo 10h37'28" 17d46'23"
William & Susie Wiseman
Uma 11h6'29" 42d40'57"
William Sven Larsen, Jr.
Uma 12h2'49" 55d45'46"
William Sweet Tolentino
Aqr 23h37'28" -21d55'54"
William Szigeti
Sgr 19h18'5" -14d42'57"
William T. and Betty B. Chumney
Crb 15h19'52" 26d37'20"
William T. Clary
Aur 6h54'50" 38d31'19"
William T. Condefer, M.D.
Per 3h35'49" 44d13'45"
William T. Guck
Aur 5h45'33" 48d19'51"
William T. King
Cep 21h21'39" 56d58'7"
William T. Pitts
Psc 0h49'42" 17d25'16"
William T. Stephens
Aql 19h50'24" -0d26'38"
William T. Weldon
Vir 12h42'42" 11d6'52"
William Tanto
Uma 11h50'53" 45d14'25"
William Teal Clark
Uma 11h36'20" 44d17'49"
William Ted Millhouse
Lib 15h8'30" -4d44'17"
William Terrelle Goddard
Cnc 8h48'47" 24d56'21"
William "Terry" Carpenter
Psc 1h10'19" 22d38'3"
William Terry & Marthanne Waugh
And 1h17'6" 41d19'24"
William The First
Aqr 22h34'39" -0d58'39"
William Thomas
Sco 17h17'30" -33d7'50"
William Thomas Crosby
Her 18h49'55" 21d40'4"
William Thomas Hansell
Aur 5h35'26" 32d58'37"
William Thomas Jekot
Ori 6h6'4" 14d51'7"
William Thomas Kidd, Jr.
Sge 19h49'32" 17d26'9"
William Thomas Linthicum
Cnc 8h22'40" 12d28'11"
William Thomas Perdue
Sco 17h24'9" -42d51'29"
William Thomas Pope
Lyn 9h17'53" 41d21'1"
William Thomas Reardon
Uma 9h47'26" 44d41'34"
William Thomas Reisen
Uma 13h32'29" 56d35'40"
William Thomas Schoenfeld
Aqr 23h14'3" -9d52'43"
William Thomas Sexton
Sco 16h22'13" -31d11'43"
William Thomas Vaughn Jr.
Ori 5h34'24" 10d6'6"
William Thomas Vidrick
Cep 20h19'37" 61d42'54"
William "Thug Nasty" Quigley
And 1h46'6" 45d56'55"
William Timothy Brooks
Ari 2h59'59" 22d9'29"
William Tinsley
Sgr 17h57'17" -24d35'30"
William Treger, Sr.
Uma 11h32'1" 32d26'47"
William "Trey" Melson III
Her 18h43'19" 19d53'54"
William Trible's Hope
Aur 5h57'24" 29d49'42"
William Troy McCraw
Cap 21h52'38" -17d50'44"
William Turner
Her 17h4'32" 36d46'15"
William 'Ty' Kailey
Cep 21h51'27" 56d8'9"
William Tyler
Ori 5h19'16" 6d14'47"
William Tyler Warren Hartwig
Sco 17h7'16" -37d14'44"
William U. Wise
Cep 22h30'2" 62d57'3"
William Urick
Lib 15h26'35" -8d22'57"
William Van Ditto
Aur 5h48'39" 54d56'29"
William Vassilios Mark Borowski
Cap 21h53'20" -8d32'9"
William Virgil King
Leo 11h43'45" 25d36'20"
William Vito Buffa
Tel 18h15'48" -50d8'1"
William Voigt IV
Tau 3h38'12" 9d42'30"
William W. Bourbeau
Cnc 8h8'49" 16d42'55"
William W. Brown
Uma 11h8'0" 37d44'27"
William W. Fisher, Jr., DMD
Ori 5h58'26" 16d58'33"
William W. Klosin
Aur 5h42'27" 52d52'54"
William W. Taylor
Uma 11h42'47" 58d19'57"
William W. Walters
Leo 11h54'53" 20d39'52"
William Wade Martin
Gem 6h52'13" 22d9'30"
William Waldschmidt
Vir 13h26'7" 12d34'16"
William Walker
Vir 14h47'29" 6d44'39"
William Walker Janek
Per 3h11'39" 45d17'43"
William Walter Green Jr.
Sco 17h23'40" -39d29'1"
William Walter Higgins
Per 2h49'12" 35d54'56"
William Watkins Walker 2/7/25
Aqr 21h52'6" -6d17'46"
William Wayne Lefler
Lyn 7h26'30" 52d26'22"
William & Wendy Sandillo
Cyg 20h58'35" 44d33'8"
William Wesley Bartley
Her 17h42'34" 48d18'14"
William Wesley McCullough Sr.
Lib 14h45'18" -17d14'9"
William West
Her 17h32'7" 37d50'12"
William Wetterau
Ori 6h7'14" -0d37'30"
William Whalen
Aur 6h29'47" 41d18'40"
William Wharton Eckert
Aur 5h53'14" 52d59'45"
William Wheeler Castellano
Cyg 20h18'57" 35d34'25"
William Wierda
Uma 9h52'30" 46d9'4"
William "Wild Bill" Kalbfleisch
Gem 7h45'21" 32d0'27"
William "Wild Bill" Scales
Boo 14h42'39" 28d21'49"
William Wiley Rhodes
And 0h46'47" 44d0'26"
William "Will"
Boo 14h41'50" 53d39'31"
William (Will) Patterson
Ori 6h16'36" 15d31'53"
William "Willy" Rosenthal
Per 3h11'39" 55d41'47"
William Wipperfurth
Ori 4h52'16" 4d30'10"
William Wittich
Tau 5h49'19" 23d27'19"
William Wouwenberg
Cma 6h47'27" -16d24'5"
William Wray
Cnc 7h55'48" 15d28'12"
William Wright
Uma 8h37'29" 50d48'7"
William Xavier Jimenez Marana
Ori 5h7'17" 15d27'54"
William Yngve Rice, II
Aql 19h26'52" 5d30'26"
William Yost
Uma 11h5'10" 55d43'25"
William, Beloved Husband and Father
Uma 8h51'49" 49d17'4"
Williamary
Cyg 19h55'0" 33d8'58"
WilliamGarneau
Lib 14h39'7" -18d13'25"
William-John O'Sullivan
Gem 7h17'47" 24d17'20"
William's Bliss TLS4
Cep 23h14'42" 79d5'51"
William's Ella
Cnc 8h16'42" 22d41'49"
William's Star
Peg 21h48'48" 21d23'13"
William's Star
Cep 21h46'48" 59d32'46"
William's Wish
Leo 11h20'4" 11d25'26"
William's Wishing Star
Her 17h15'59" 27d20'14"
William's Wonder
Per 4h47'20" 45d12'38"
Williams-Fulmer Future
Lac 22h27'7" 43d27'53"
Williamstown Middle/ High School
Uma 9h57'21" 69d26'43"
Willian
Leo 10h30'55" 15d40'7"
WILLIANNE DOCARMO DANIEL
Ari 2h32'45" 11d8'8"
Willibald 60
Uma 10h1'45" 63d8'30"
Willie
Uma 8h34'5" 61d4'19"
Willie
Vir 13h25'30" -12d48'27"
Willie
Boo 14h14'54" 38d53'25"
Willie & Alan
Lmi 10h4'27" 37d13'34"
Willie B. Hill Sr.
Cyg 20h47'36" 37d17'26"
Willie Burns Brightly Morgan
Ari 3h8'4" 27d44'45"
Willie & Candi Schmidt
Cyg 20h33'40" 41d22'52"
Willie Clarence Calvin 3rd aka Bob
Uma 9h5'26" 56d2'39"
Willie D
Cep 22h11'19" 59d30'22"
Willie Dottie Ben Camie Watson
Uma 14h2'57" 56d6'54"
Willie Foster
Per 2h29'30" 51d48'37"
Willie James Crawford Sr
Sgr 18h19'29" -29d13'37"
Willie James Loynes
Her 16h51'5" 33d38'40"
Willie Johnson
Per 2h47'56" 55d24'40"
Willie Jukes
Uma 12h30'4" 59d26'54"
Willie Kunkis
Cap 20h41'53" -17d19'40"
Willie L. Johnson, Jr.
Aql 19h46'19" 3d14'19"
Willie London
Gem 6h1'41" 25d40'28"
Willie Mae
Lib 15h56'41" -12d25'10"
Willie Mae Clark
Aur 5h36'15" 45d19'32"
Willie & Michelle Baby's Day
Per 4h11'28" 34d13'53"
Willie Mungia
Aqr 22h43'3" -16d4'52"
Willie Murrell (Pepper) Lamp
Equ 21h16'47" 9d15'33"
Willie Richardson
Sco 17h41'37" -33d2'11"
Willie Sullivan
Per 4h40'57" 40d33'8"
Willie the Great
Dra 16h22'11" 56d56'46"
Willie VanMeter
Cma 6h46'15" -14d20'21"
Willie Vern Hickman III
Sco 16h34'5" -38d35'41"
Willie's Charger
Cas 0h28'23" 56d3'59"
Willie's Love For Meagan
Uma 8h47'51" 48d31'2"
Willimantic Rain
Lac 22h46'1" 54d9'5"
WilliMatt
Sco 16h17'58" -13d18'6"
Willing Rhys
Uma 8h13'56" 71d47'31"
Willis
Boo 15h21'23" 48d41'54"
Willis 300705
Cyg 20h49'30" 45d24'20"
Willis Boroto
Lyn 6h57'8" 52d14'5"
Willis Clarence Hughes
Boo 14h57'44" 15d50'24"
Willis H. Wamsley
Cap 20h33'42" -16d57'47"
Willis Mattei
Gem 7h30'41" 26d37'3"
WillKel
Ori 6h8'32" 15d5'57"
Willliam Sean Norris AKA Zenergy
Sco 17h18'26" -31d11'18"
Willll!!
Sgr 19h18'38" -19d39'51"
Willma C. Wilhite, Barry W. Wilhite
Cep 22h50'1" 71d56'48"
Willma C. Wilhite, Kim & Bob
Cyg 19h37'45" 29d26'51"
Willma C. Wilhite, Tania AJ & Bobby
Cyg 20h16'48" 50d2'36"

Willma Celeste Wilhite, Jillie Dana
And 23h41'47" 37d31'3"
Willmann, Michael
Uma 12h53'42" 54d5'43"
willnheather Mayer
Lyn 8h4'36" 57d5'12"
Willo
Ser 18h33'20" -0d50'50"
Willoughby
Gem 6h47'53" 25d6'5"
Willoughby Parker
Sgr 18h50'31" -18d59'54"
Willow
Leo 9h40'5" 16d8'53"
"Willow"
Gem 6h49'32" 17d23'1"
Willow
And 1h47'37" 50d26'31"
Willow 1969
Cas 0h17'20" 51d55'19"
Willow Amelie Kwiecien-Hunt
Cyg 19h35'34" 44d54'10"
Willow Aurora
Peg 22h26'37" 4d36'32"
Willow Chavez
Gem 6h58'22" 14d11'15"
Willow Guo Qian Louis
Vir 12h7'50" 12d44'13"
Willow Liadon
Ori 4h52'47" 4d19'54"
Willow Mary Brear
Cas 23h2'37" 53d44'20"
Willow Qiryn Small
Cru 11h57'0" -60d42'2"
Willowby
Crb 15h47'51" 27d44'16"
WillowLark
Lyn 8h32'10" 38d21'25"
Willow's G-PAW
Crb 15h17'31" 28d25'53"
Wills
Dra 20h24'59" 68d57'55"
Wills Exceeding
Per 3h7'0" 48d15'32"
Will's Eyes
Ori 5h46'35" 6d59'53"
Will's Flower Bed
Uma 11h29'9" 63d36'12"
Will's Future
Her 17h7'38" 37d34'51"
Wills Lorraine
Crb 16h7'29" 38d55'7"
Will-V
Her 17h34'47" 33d36'23"
Willy
Uma 8h37'12" 63d5'34"
Willy and Wifey
Col 6h18'58" -33d56'53"
Willy Bertotto MF
Uma 12h33'41" 60d39'56"
Willy Brock-Broido
Leo 9h47'48" 21d35'13"
Willy et Rebecca *Pour la Vie...*
Cyg 21h29'22" 49d9'36"
Willy Guardiola
Vir 12h20'49" -6d47'40"
Willy Hauser
Uma 8h33'25" 63d15'22"
Willy James Lamson March 23, 1991
Uma 12h49'7" 58d5'34"
Willy June 26, 1972 - March 13, 2000
Ori 5h34'49" -1d38'59"
Willy Kissling
Cas 0h46'1" 48d42'19"
Willy Massieu
Apu 15h25'40" -71d13'21"
Willy & Nilly's
Sgr 19h21'54" -14d39'35"
Willy Rodriguez
Ori 5h57'11" 6d18'21"
Willy & Shelby Luther
Leo 9h50'30" 18d30'31"
Willy und Monika Gebert
Lyn 8h52'40" 37d51'43"
Willy & Wetty
Uma 9h35'13" 42d48'32"
Wilma
Cam 4h0'42" 56d43'56"
Wilma
Vir 14h1'28" -5d25'50"
Wilma B. Paronto
Uma 8h22'50" 63d46'15"
Wilma Bird
Lyn 7h57'52" 47d36'17"
Wilma & Branislav's silver love
Lmi 10h21'53" 29d48'37"
Wilma Cameron Gustafson
Tau 5h44'37" 26d29'20"
Wilma Edith Hart
Peg 21h48'24" 11d0'14"
Wilma Eloise Collings
Lyr 19h12'48" 46d3'7"
Wilma G. Barnes
Uma 9h10'56" 46d35'6"
Wilma "Granny" Cartwright
Sgr 18h38'34" -19d52'13"
Wilma Gregory
Cnc 8h33'56" 20d43'9"
Wilma Irene Way Marr Weyand
Cas 1h21'42" 57d32'51"
Wilma Jean Hale
Uma 9h20'6" 43d21'1"
Wilma Jean Manville
And 0h43'51" 41d1'19"
Wilma Jean Milligan
And 22h57'58" 36d25'50"
Wilma L
Lib 15h21'22" -8d51'29"
Wilma Lasser
Uma 11h28'16" 43d49'44"
Wilma M. Jackson
Tau 3h46'47" 14d51'8"
Wilma M. Renzoni
Lyn 7h3'41" 46d26'29"
Wilma Magramo
Tau 5h3'47" 19d32'1"
Wilma Peters
Uma 10h31'24" 55d12'2"
Wilma Reedy(Mommy)
Vir 12h39'54" 6d14'44"
Wilma Roberta Carder Higdon
Lyr 19h20'8" 29d54'33"
Wilma Sassard MacRae
Ori 5h22'25" 13d26'51"
Wilma Soto
Sco 16h43'56" -31d49'0"
Wilma The Wicked Witch
Cas 0h20'44" 61d10'25"
Wilma Vaughan
Cas 0h9'32" 53d26'42"
Wilma Venyce Shore
Lib 15h4'23" -4d24'11"
Wilma Wishon
Ari 3h10'53" 18d38'39"
Wilma, where we go, you are with us
Umi 13h38'17" 86d20'46"
WilMalAnn
Lib 15h47'19" -9d41'49"
Wilma's Everlasting Light
Uma 9h27'52" 42d5'50"
Wilma's Smile
Tau 4h36'3" 21d54'6"
Wilmaviola 2°
Leo 11h50'57" 11d36'48"
Wilmer Nofrey
Uma 10h34'10" 70d39'17"
Wilmer Phillips
Uma 12h10'42" 60d58'58"
Wilmer R Buote
Cap 20h38'14" -15d38'37"
Wilnia Aristilde
Cas 1h46'31" 61d14'29"
Wiloja
Umi 14h58'56" 81d59'40"
Wilothlafoev
Cyg 20h44'15" 34d14'2"
Wils
Dra 19h2'21" 60d27'30"
Wilsaghan
Her 17h48'11" 41d24'3"
Wilson Abitang Olalia
Aur 5h36'13" 47d5'51"
Wilson Alberto Loureiro
Ori 6h15'59" 15d49'46"
Wilson and Carle'
Sco 17h15'17" -39d47'54"
Wilson Annosus
And 23h30'26" 36d2'52"
Wilson Carreras "My Daddy"
Cep 0h15'58" 74d41'58"
Wilson Castillo - Shining Star
Eri 3h53'58" -5d2'39"
Wilson David Clark
Cap 21h4'25" -15d3'1"
Wilson Douglas Fletcher
Sco 16h16'8" -11d43'24"
Wilson Eric Allford
Sgr 19h10'2" -17d13'12"
Wilson Eugene Boyles
Uma 12h37'40" 56d30'35"
Wilson Henry Rath
Her 17h45'11" 27d29'49"
Wilson K.S. Miao
Uma 9h5'11" 46d44'27"
Wilson Lamar 1946
Gem 6h42'7" 34d57'58"
Wilson Palomino
Cyg 20h5'55" 38d10'43"
Wilson Pena
Lib 15h12'47" -15d26'32"
Wilson Sonny Ruckman
Aur 5h48'39" 55d23'59"
Wilson/Stephens
Uma 10h25'26" 69d48'36"
Wilson Sun Britten
Leo 10h1'35" 8d45'35"
Wilson the Goldfish
Uma 11h37'34" 63d52'44"
Wilson Troy Blackmon
Cma 6h54'58" -22d3'18"
Wilson's Star
Lyn 6h17'21" 56d15'47"
Wilton
Ori 5h50'57" 2d5'55"
Wilton and Eleanor Syckes
Cnc 8h11'40" 15d29'53"
Wilton Wiggins
Leo 11h14'33" -0d14'34"
Wilwythyn *50th*
Cyg 19h37'25" 38d51'43"
Wimberly Mattly
Cmi 7h22'57" 3d37'8"
Wimmer, Hans
Uma 11h16'56" 33d56'12"
Wimösterer, Alexander
Ori 6h8'19" -1d45'18"
Wimp
Uma 12h38'21" 62d19'8"
WIMSAM93
Cyg 19h52'4" 32d28'45"
Win and Bubby
Vir 14h11'18" 4d20'25"
Win Smith Taylor - 8 May 2005 1930
And 2h0'7" 39d9'5"
Win White
Cas 0h3'44" 53d26'59"
Winburn & Kenneth Jones
Psc 0h20'12" 15d20'23"
Winchester Tool Collector
Uma 11h16'8" 55d13'50"
Wind Beneath My Wings
Cyg 20h11'22" 39d32'46"
Wind Beneath My Wings
Cnc 9h8'32" 30d46'30"
Wind Swept
Aqr 21h8'52" 0d18'51"
Windcrest's Blue Christmas
Cma 6h46'9" -13d41'34"
Windi J. Miller
Uma 10h32'49" 47d30'14"
Windland Smith Rice
Lyn 8h46'7" 37d20'0"
Windman
Ori 6h18'52" 14d39'31"
Window In the Skies
Her 17h21'37" 35d49'9"
Winds Fire
Ori 6h18'16" 13d51'51"
Windsor Hutto
Uma 10h28'13" 46d39'16"
Windsurf Beach Hotel
Cep 22h47'29" 74d54'40"
Windswept
Tau 4h28'55" 27d44'37"
Windwer's Star
Uma 9h12'8" 60d30'52"
Windy
Sgr 19h56'51" -37d4'15"
Windy
Tau 3h36'58" 16d35'1"
Windy Idnar Jones Eichten
Sco 16h14'15" -12d48'7"
Windy K. Clark
Ori 6h14'1" 15d1'35"
Windy McClendon
Lep 6h0'43" -24d42'2"
Windy Wong
Uma 9h59'46" 62d0'27"
Windy Zopfi
Gem 6h47'26" 26d20'36"
Winfred D. Littlejohn
Per 3h16'30" 34d24'51"
Winfred Eton McAdoo
Ari 2h30'47" 11d9'14"
Winfried Berener
Uma 8h45'41" 54d45'57"
Winfried Hensel
Uma 11h24'23" 30d21'31"
Winfried Rudloff
Uma 9h47'46" 50d31'15"
Winfried Swidersky
Uma 10h46'5" 65d9'23"
Wing
Sgr 19h21'26" -41d18'27"
Wing on Wing
Aql 19h55'23" -0d59'33"
Wing Sze Anita Au-Yeung
Tau 3h58'43" 2d38'47"
Wingfield Jr.
Lib 15h12'14" -19d56'4"
Wingman 85
Psc 1h2'32" 14d28'24"
Wingnut
Umi 16h3'55" 74d25'1"
Wingnut710
Tri 1h53'26" 29d20'12"
Wings
Cma 6h47'31" -17d46'23"
WINGS 2005
And 1h23'1" 45d49'54"
WINGS 2006
Aql 19h44'15" 7d9'27"
Wings 2007
Aql 19h49'55" 12d22'45"
Wings of the Golden Songbird
Lyr 18h45'44" 33d52'0"
Winifred
Dra 15h50'40" 64d40'0"
Winifred Ann Sissy Isetta
Aqr 22h31'37" -4d31'50"
Winifred Anthony
Cru 12h28'10" -59d17'11"
Winifred Bunnell Weig
Ari 2h51'36" 30d40'29"
Winifred Collins
Aqr 22h4'15" -15d23'32"
Winifred Piechoski
Cas 1h3'36" 63d43'58"
Winifred Schachinger-Scott
Umi 14h42'57" 69d4'9"
Winifred T. Dowd
Uma 9h17'57" 64d15'12"
Winifred Tibstra
And 0h40'47" 44d37'9"
Winifred W. Bingham
Crb 16h24'23" 32d49'14"
Wink & Nancy's Love Is Eternal
Cyg 20h3'49" 30d54'9"
Winkelmann-Schmidt, Doris
Uma 10h11'2" 67d37'16"
Winki
Uma 11h28'4" 46d3'46"
Winkiblinki
Uma 11h12'36" 71d25'55"
Winkler, Heidrun
Ori 6h19'45" 11d11'46"
Winkler, Jürgen
Sgr 18h4'14" -27d37'55"
Winky
Tau 4h50'34" 20d35'8"
Winky Coleman
Srp 18h18'1" -0d10'47"
Winky Lea
Lib 15h1'52" -12d42'53"
Winni
Lac 22h4'49" 45d42'11"
Winnie
Cas 0h35'44" 57d11'7"
Winnie
Boo 14h32'4" 52d29'30"
Winnie
Tau 3h28'36" 8d32'58"
Winnie
Vir 14h36'43" -0d22'53"
Winnie Cooper
Ari 3h2'16" 28d52'43"
Winnie de Fatima
Umi 13h42'41" 73d44'31"
Winnie Eng
Leo 10h26'7" 13d23'26"
Winnie Evelyn Heywood - Winnie's Star
And 1h10'39" 46d22'9"
Winnie Gabriel
Lib 15h5'7" -0d33'2"
Winnie Girl
Aqr 23h39'8" -13d46'46"
Winnie (Jia) Wang
Cnc 9h1'6" 16d19'15"
Winnie Kong
Tau 3h26'36" -0d54'16"
Winnie May
And 23h29'9" 49d53'42"
Winnie May Byng
Cru 12h22'21" -57d56'2"
Winnie Sowerby "Nanna" - I Love You
Cas 3h12'38" 61d18'25"
Winnie Van der Woude
Ari 2h34'11" 24d38'46"
Winnie Walker's Star
Cnc 9h4'20" 31d54'19"
Winnie White
Cnc 9h2'53" 23d27'32"
Winnie Wong/Boba's Star
Sgr 17h51'52" -28d46'41"
Winnie W.Y. Siu
Cas 3h28'21" 73d26'26"
Winnie Zhao
Cyg 19h57'47" 35d47'12"
Winnifred Mae Fairbanks
Eri 3h29'24" -6d53'42"
Winnig, Jens
Ori 5h24'39" 10d29'59"
Winning Ticket Cafe
Ori 5h16'16" 8d22'1"
Winnona Campbell
Leo 11h20'34" 23d3'20"
Winns Eternal Love
Sco 16h4'5" -19d19'56"
Winoa
Gem 7h10'30" 26d0'58"
Winona
Lyn 9h10'26" 39d16'2"
Winona Lucretia Pion
Aqr 20h57'15" 1d33'18"
Winona Mandrell
Uma 8h45'4" 53d54'34"
Winona's Song of God
Mon 6h34'34" 0d42'41"
Winona's Star
Lyn 8h11'41" 55d19'21"
Winslow's Star
Uma 8h13'19" 63d40'10"
Winstarr
Uma 13h40'14" 56d0'13"
Winston
Cyg 20h33'31" 55d57'10"
Winston
Umi 14h37'43" 68d9'35"
Winston
Ori 5h26'30" 3d45'58"
Winston
Vir 13h50'11" 2d46'32"
Winston
Cmi 7h27'2" 8d18'56"
Winston
Boo 14h41'9" 21d47'16"
Winston
Cnv 13h38'1" 32d22'10"
Winston 12-12-2004
Cep 3h32'51" 80d53'45"
Winston Alderson
Boo 14h46'41" 44d18'25"
Winston and Pansy Greene
Col 6h13'42" -34d6'7"
Winston and Toya
Cyg 20h45'46" 32d17'9"
Winston & Guiseppa Theuma
Cru 12h27'44" -62d5'11"
Winston McZeal
Tau 4h36'44" 30d25'26"
Winston Terrell Tillman
Ori 5h23'52" 2d39'42"
Winston Walsh Klika
Umi 14h18'42" 76d18'53"
Winston, Stella, Minus & StirFry
Ari 2h7'14" 19d5'56"
Winston's Love
Lyn 9h5'35" 37d29'57"
Winter Ann Runyan
Uma 10h46'38" 64d40'24"
Winter Carol Grant
Psc 0h23'41" 3d29'41"
Winter Darling
Aqr 22h35'21" 1d40'1"
Winter Gray Owl
Psc 0h42'55" 9d39'38"
Winter M.E. Rinehart
Tau 3h47'33" 21d55'52"
Winter Rose
Lyn 8h37'15" 44d45'44"
Winterjess
Uma 11h56'26" 47d17'37"
Winters Eve of Starry Night
Cma 7h7'53" -26d14'42"
Winters Family
Her 16h46'8" 38d52'41"
Winter's Star
Her 18h8'8" 14d31'47"
Winter's Star
Aql 19h42'57" -0d5'11"
Winter's Touch
And 0h13'51" 23d58'59"
Winterz Glow
Tau 4h33'53" 6d47'31"
Winton Dell
Umi 13h26'24" 87d19'20"
winton150905
Vir 13h32'55" -20d59'56"
Winwood Mathews Bowman
Sco 17h24'14" -44d27'7"
Winx
Dra 20h21'1" 79d22'47"
Wipf, Vera
Vir 12h12'53" -5d31'40"
Wip-hun-sis
Lib 15h38'3" -9d30'58"
"Wir alle können mehr" - Coach Klaus Gunkel
Uma 9h18'39" 66d47'55"
WIR ZWEI
Cas 23h31'51" 55d5'51"
Wirbelwind
Lac 22h52'9" 53d53'27"
Wirth, Hermann
Uma 8h52'32" 50d29'39"
Wisam Ismair
Crb 16h3'48" 36d28'14"
Wisconsin Heart Hospital Family
Uma 13h24'30" 56d27'19"
Wisdm
Cnc 7h56'44" 9d51'14"
WISDOM
Leo 10h42'35" 14d16'13"
Wisdom & Strenght "Byron McArthur"
Aql 20h0'6" 14d11'27"
Wise Hero John
Tau 4h32'22" 17d5'37"
Wise Myriam
Mon 6h32'52" 5d37'24"
Wiseguy493
Psc 0h29'0" 8d42'13"
Wiseman
Cyg 20h35'27" 35d7'17"
Wisey Vkolas-Wv Thompson
Cam 3h29'47" 57d3'56"
Wish
Mon 7h33'51" -0d38'53"
Wish Come True
Uma 11h15'34" 46d13'53"
Wish Eternal
Sgr 17h56'37" -29d38'14"
Wish Fame
Dra 19h23'58" 61d34'0"
Wish Star Tatiana Sinkova
Cam 5h44'53" 58d33'31"
wish upon a
Ari 2h26'51" 24d38'33"
Wish upon a Fallon Star
Cma 7h8'7" -28d1'36"
Wish Upon a Wandering Star
Uma 10h2'52" 64d50'37"
Wish Upon Me
Lmi 10h47'48" 32d50'51"
Wish you were here
Uma 9h11'39" 49d44'22"
Wish You Were Here
Uma 13h13'6" 55d36'25"
Wisheart
Cyg 19h47'55" 50d25'15"
Wishes Forever
Ori 5h55'27" -0d36'59"
Wishes of Carole Waki Goddess!
Car 10h10'17" -62d54'22"
Wishing Star
Crb 16h21'17" 38d17'58"
wishing star
Crb 16h8'41" 30d4'48"
Wishon Emily
Leo 10h26'6" 20d46'11"
WishStar A&J
Lib 15h41'25" -16d40'59"
Wiskoski-Weitzel
Cyg 21h51'22" 52d52'52"
Wiskowski
Lyn 7h57'35" 44d38'40"
WISKUP 1
Uma 12h36'51" 57d46'2"
Wisp
Cap 20h42'53" -14d37'6"
wisping fire
Her 17h25'52" 35d36'5"
Wiston Wedding
Cyg 19h30'7" 31d57'8"
Wisty
Cnc 9h6'10" 9d29'13"
Witakae
Dra 15h37'37" 56d17'58"
Witchy Woman
Aqr 22h55'42" -10d26'50"
With Love
Uma 8h42'14" 61d44'35"
With Love for Julia
And 0h49'11" 26d45'51"
With Love to Gamma from Ayden Sirak
Lib 14h23'27" -9d6'52"
With Love To Lisa Always & Forever
Ari 3h16'42" 29d1'26"
With Love to Mama Always & Forever!
Vir 13h11'19" 12d3'4"
With-in Kadens' Reach
Uma 11h16'14" 43d33'46"
Witold i Krystyna
Aqr 22h6'48" -1d53'26"
Witold Zapala
Gem 7h15'13" 16d52'51"
Witonya
And 1h8'18" 35d42'39"
Witte, Klaus
Uma 10h16'10" 61d20'33"
Wittle Bear Star
Lib 15h34'39" -10d6'59"
Wittrim, Oliver
Ori 5h33'1" 10d29'0"
Wittschier, Bernd
Ori 5h54'9" 21d46'35"
Wittyis Maximus
And 0h34'41" 26d15'9"
Witwer's Valentine
Ari 2h38'22" 30d59'41"
Wizard
Leo 11h6'47" 5d14'4"
WIZTIGE LEB
Lib 15h40'5" -17d48'35"
W.J. Byrnes & Co.
Uma 9h56'2" 51d9'53"
WJB's Fantastic 40th
Per 3h17'13" 45d34'49"
WJS 22
Gem 6h59'26" 16d36'9"
WJSNA1124
Lyr 18h39'43" 36d11'5"
W.J.T., MY One
Sgr 18h44'23" -21d48'10"
WKM2005
Cyg 22h0'51" 52d32'15"
W.L. Fuzzy York
Uma 9h34'57" 41d50'19"
W.L. Stoppel
Her 17h55'26" 19d17'4"
WLP-4-TLC
Ori 6h14'17" 6d0'27"
WLS
And 23h29'22" 41d27'53"
Wm Bradford Freese
Her 17h27'38" 38d12'5"
W.M. Fitzpatrick Jr.
Per 3h4'5" 41d5'43"
Wm H Strehle Jr/Hawker 1000 Father
Uma 8h40'4" 58d57'51"
Wm. & Lois English
And 1h44'56" 45d54'38"
Wm. Riley U R My Everlasting Star
Gem 6h49'12" 27d52'34"
Wm. Timothy Lyons
Uma 9h53'51" 56d9'26"
Wo Ai Ni
Sco 16h11'35" -28d17'29"
Wo Ai Ni
Leo 11h5'27" 20d24'29"
Wo ai ni Pengge
And 0h46'11" 32d0'45"
WO9LF Night Sky Watcher
Sgr 19h11'57" -16d39'12"
WOB
Cnc 8h19'39" 19d51'11"
Wobbie
Cyg 21h7'1" 46d16'38"
Wodie Bug
Ori 6h2'39" 21d24'11"
wo-di-ge yo-na 082995
Uma 11h39'19" 28d37'4"
Woe&Weon
Uma 8h36'35" 68d37'2"
Wöhltjen, Nikolaus
Ori 6h11'33" 8d54'18"
Woidt, Nicole
Ori 5h6'56" 7d38'5"
Woitke, Clemens
Uma 8h56'16" 48d5'29"
Wojciech Dawidowski
Uma 11h13'29" 51d7'17"
Wojciech Dawidowski
Cnc 9h4'28" 15d49'1"
Wojciechowski, Marek
Ari 2h19'49" 17d55'2"
Wojcik
Aqr 20h54'0" -12d19'4"
WOJCIK EVA
Sgr 17h49'28" -17d39'51"
Wojtus & Beatcia
Dra 19h44'3" 61d31'8"
Wokpetanka Wayace
Her 16h33'17" 37d8'7"
WOLCNMANNCASSC-VAMD
Ori 5h54'28" 21d47'31"
Wolcott Allen Marsh
Vir 13h42'49" -14d51'27"
Wolcott's Spire
Uma 11h55'11" 36d25'3"
Wolf Brardt
Leo 11h7'6" 27d35'22"
Wolf, Heiko
Ari 1h58'48" 15d5'53"
Wolf, Karl-Friedrich
Uma 11h40'45" 31d23'25"
Wolf, Manfred
Ori 4h56'3" 2d20'38"
Wolf, Vera und Johann Josef
Uma 9h53'55" 63d5'44"
Wolf-Andreas Eichel
Ori 6h20'57" 7d31'36"
Wolfe - 3
Uma 12h0'29" 59d34'22"
Wolfen Forever Bright
Cnv 13h49'8" 30d2'58"
Wolff, Axel Dirk
Uma 9h19'12" 70d7'9"
Wolfgang A. Music
Ari 3h19'21" 28d38'7"
Wolfgang Allehoff
Ori 5h13'41" -6d46'37"
Wolfgang Becker
Uma 11h17'51" 42d38'13"
Wolfgang Behr
Uma 8h54'42" 67d57'36"
Wolfgang Blazejewski Born: Jan 31 1944
Uma 11h32'6" 56d45'42"
Wolfgang Bröhr
Uma 9h20'18" 47d37'3"
Wolfgang Dettmann
Ori 5h54'37" 7d22'28"
Wolfgang Eisermann
Uma 10h59'22" 65d53'48"
Wolfgang Erber, 30.04.1952
Aql 19h4'34" -7d42'49"
Wolfgang Haak
Uma 11h37'12" 32d40'38"
Wolfgang Haehn
Uma 9h42'57" 48d56'58"
Wolfgang Haferkamp
Ori 5h57'41" 18d32'42"
Wolfgang Harth
Uma 8h44'59" 50d37'21"
Wolfgang Jungo
Per 3h43'4" 45d28'19"
Wolfgang Kaufmann
Her 18h22'39" 12d16'50"
Wolfgang Keith
Uma 8h15'5" 60d10'20"
Wolfgang Kortegast
Uma 9h1'37" 53d31'59"
Wolfgang Kraus
Gem 7h8'28" 32d38'46"
Wolfgang Kußmann
Ori 5h55'45" 7d10'25"
Wolfgang Lampka
Uma 11h2'59" 42d53'55"
Wolfgang Martin Baier 50
Ori 5h11'2" 15d33'39"
Wolfgang Mattern
Uma 9h56'35" 56d35'12"
Wolfgang Mayerhofer
Aur 4h58'34" 42d51'25"
Wolfgang Metz
Ori 4h44'39" 5d16'14"
Wolfgang Meyer
Ori 6h2'25" 21d26'4"
Wolfgang Otto August
Crb 16h24'40" 32d36'19"
Wolfgang Pfeiffer
Uma 11h57'40" 32d5'58"
Wolfgang Richter
Ori 5h54'36" 3d9'49"
Wolfgang Rittenbruch
Uma 8h45'49" 53d4'22"
Wolfgang Roth
Ori 6h19'0" 8d17'10"

Xian
Aql 20h13'43" 2d1'6"
Xiang-nian ni
Uma 10h29'16" 66d3'7"
Xiao Fang Fang
Dra 20h19'17" 74d13'59"
Xiao Hai Ning
Vir 13h46'55" -8d47'45"
Xiao Li
Psc 23h28'2" 6d9'54"
Xiao Xia
Dra 17h29'6" 52d22'35"
Xiao Yanzer
Psc 1h19'39" 18d34'22"
xiao zhu loves ta fei zhu
Cyg 21h10'5" 33d41'4"
Xiaohong Yin
Leo 11h13'15" -4d9'1"
XiaoMing Wang
Cyg 20h2'30" 55d37'48"
xiaosa
Sgr 19h5'36" -22d28'54"
Xiaoxia Ke
Lib 14h51'18" -0d41'36"
xiaoxia shi
Uma 9h13'26" 53d55'39"
Xiaoxuan Li
Vir 12h38'2" 0d17'3"
xiii
Lyr 18h42'54" 39d29'31"
XII-My Star in The Making-Malissa
Cnc 9h7'24" 10d6'2"
Ximena Victoria Brook
Cru 12h33'13" -62d45'37"
Ximena's Star
Dra 18h31'5" 58d48'19"
Ximo Llopis Ortiz
Sco 17h15'36" -33d30'27"
Xin
Lyn 6h57'24" 50d18'48"
Xin ai Robinson
Aql 19h19'20" 15d34'1"
Xiomara Alejandra Valencia
Tau 4h51'51" 29d41'48"
Xiomara González
Cnc 8h16'45" 18d35'6"
Xiomara Maria Carrillo Jimenez
And 2h20'32" 38d31'28"
Xiomara Reyes
Sge 20h15'56" 18d34'11"
Xiomara Senior
Sgr 18h50'5" -17d58'58"
Xiomara T. Garza
Mon 7h56'0" -8d59'1"
Xiquena
Uma 9h49'57" 46d17'10"
Xiuhtonal
Aqr 21h58'31" 0d21'21"
Xiuli
Ari 1h54'10" 18d11'53"
xmaslover4ever
Psc 23h52'21" 7d37'3"
XO - Judith Louise Gregory - XO
Uma 11h39'24" 44d20'57"
Xochil Castillo
Dra 19h14'52" 61d26'27"
Xochitl Dominguez Benetton
Leo 10h4'42" 25d31'45"
Xochitl Fernandez
Umi 14h31'59" 77d58'14"
Xochitl Nerina
Tau 4h20'13" 26d30'0"
xoKaraSox
Tau 5h59'55" 23d48'28"
xokarinecdf
Uma 13h39'23" 55d6'11"
Xolotl's Estrella
Aur 5h24'42" 35d54'42"
Xong Lor
Tau 5h39'58" 23d37'17"
XO-Sweet Danny-XO
Her 16h29'56" 45d16'19"
XOXO JD
Ori 5h42'36" 0d39'22"
XOXOWAM!
Gem 6h44'39" 26d23'31"
XOXQODCH42106AMLAR OXO
Lyn 8h32'55" 48d5'42"
Xpert.Line
Cep 0h9'45" 82d22'35"
xplosive0131
Uma 13h14'40" 53d9'55"
X-RAY
Cap 20h29'15" -22d37'43"
Xrisi The Sleepy-Head
Dra 20h24'13" 67d11'55"
Xtal
Aqr 21h12'47" -12d48'0"
Xtiangel
Pho 0h19'24" -40d24'49"
Xtinas star
Aqr 20h40'25" -13d0'8"
xtine
Psc 1h21'4" 13d8'10"
Xtleoid
Cep 21h32'18" 65d47'48"
Xtreme
Dra 16h30'25" 74d22'39"
Xu Yan
Cyg 21h2'40" 36d30'55"
Xuan & Bichha
Sge 19h42'50" 17d44'31"
Xuan Thai
Ori 5h32'50" 5d48'27"
Xusha
Tau 3h56'8" 22d17'57"
XUXA MENEGHEL
Del 20h36'11" 12d32'36"
XuXinru
Vir 12h35'34" 5d13'51"
xx Charlotte xx
And 2h23'40" 39d0'37"
xX Kelly Xx
And 1h8'34" 46d26'23"
xx Marma xx
Cas 23h59'6" 60d10'10"
-XX Star of Denise & Steve XX-
Ara 17h13'35" -57d13'14"
xx The Gorgeous Magdani xx
And 0h18'28" 26d49'21"
XXelaA
Crb 15h16'34" 26d32'26"
xxx Forever Yours xxx
Cyg 21h50'51" 53d47'15"
xxx You Will Never Dream Alone xxx
Pho 0h30'17" -39d30'51"
xxxBeautifulxAngharad21xx x
And 1h38'17" 40d13'38"
xxxConniexxx
Sgr 18h17'17" -29d57'56"
XXXtina, My Lady Lover
Leo 11h23'45" 7d7'31"
xxxx Lance & Vix Lost in Space xxxx
Uma 8h34'39" 53d9'23"
xxxx Lucy's Star xxxx
And 23h46'58" 42d5'6"
Xyon-Picurro
Vir 14h58'20" -0d3'17"
Xzabier Wynn
Ori 5h56'31" 14d45'35"
XZanthia
Lib 14h51'47" -5d30'3"
Xzavian
Ori 4h49'10" 3d39'48"
Y. I. B. S.
Vir 12h35'15" -5d15'44"
Y J Dickert
And 1h3'28" 37d33'34"
Y L Hudson
Cyg 20h5'47" 32d19'19"
Ya Ching Chen
Cyg 21h33'30" 48d59'45"
Ya Lyublyu Tebya
Crb 15h57'58" 27d43'50"
Ya Me Fil
Per 2h50'33" 53d48'22"
Ya tebya lyublyu, Angela
Crb 16h13'9" 37d0'20"
Ya Ya Mother
Cap 21h41'9" -16d57'26"
Ya Ya Sisters
Tau 3h44'54" 23d39'39"
Yaacov Rasier
Cnc 8h19'49" 11d57'5"
Yaakov Bellas
Cnc 8h49'15" 31d11'43"
Yaakov Gluck
Cnc 8h24'1" 10d8'12"
Ya'akov lysa
Leo 9h44'35" 21d15'52"
Yabo
Lmi 10h8'53" 37d36'13"
Yachai
Eri 3h49'14" -2d38'39"
Yacov Mordechai Torgow
Tau 3h40'30" 26d5'36"
Yadi
Ari 3h6'14" 29d28'22"
Yadi
Gem 7h10'59" 33d28'30"
Yadira
Crb 15h31'39" 33d57'22"
Yadira
And 0h36'44" 42d45'9"
Yadira
And 0h14'17" 45d51'16"
Yadira Alarcon
Lyn 7h32'41" 49d47'3"
Yadira Avalos
Tau 5h32'18" 18d15'30"
Yadira Esmeralda Martinez Salazar
Cap 21h17'18" -16d15'26"
Yadira la estrella mas preciosa
Cas 1h22'22" 52d10'50"
Yadira Lucero
Dra 20h11'23" 61d59'12"
Yadira Yarazeth
Crb 15h53'42" 27d39'55"
Yadira's Star
Aqr 23h12'20" -8d58'9"
yadoo (Heidi Morris)
Cma 7h11'48" -15d37'23"
Yadviga
Lyr 19h18'32" 29d46'1"
Yaël
Leo 11h43'18" 21d34'8"
Yael O'Hagan 08092005
Vir 13h55'47" 5d8'40"
Yael Rebish
Aqr 20h45'10" -13d9'59"
Yael Sofia
Umi 15h50'0" 77d33'1"
Yaeli and Trevor Green
Cyg 20h8'8" 30d42'56"
Yaeli's Star
Psc 23h46'43" 6d7'10"
Yaelisa Noelle Flores (Yaya)
Umi 15h5'48" 69d15'54"
Yaelisa Noelle Flores (Yaya)
Lyn 7h36'31" 58d47'25"
Yaëll & Raphaël
Del 20h40'59" 10d52'16"
Yaffa Ariel
Sgr 19h27'50" -12d23'55"
Yagaya Kumar
Leo 10h45'23" 10d13'40"
Yahaira Itzel Ortiz
Uma 8h20'41" 72d31'10"
Yahaira Rivera
Cyg 19h58'54" 46d12'21"
Yahalom
Sgr 19h19'39" -21d26'5"
Yahel "Mi niña Hermosa"
Col 5h47'36" -35d23'12"
Yaholkovsky
Ori 4h46'51" -0d59'30"
Yahoody
Dra 15h41'13" 55d37'53"
Yahtzee
Tau 4h12'47" 29d2'14"
Yahya
Uma 11h40'37" 44d22'11"
Yainel
Her 16h37'2" 29d57'32"
Yair Chaim Aburus
Her 16h19'19" 5d52'15"
Yajaira Little One Pereira
Cnc 9h12'56" 31d30'49"
Yakema Brisman
Cam 4h54'27" 57d55'38"
Yakimus
Tau 3h57'15" 18d30'35"
Yakini Star
Lmi 10h31'52" 35d48'51"
Yakky Paige
Uma 10h10'26" 63d52'26"
Yako
Her 18h41'16" 13d10'55"
Yakowenko
Per 3h29'48" 48d56'47"
Yal 33
Lyr 18h29'38" 36d49'37"
YalanMyLove
Cyg 20h22'15" 54d38'36"
Yalavarthi Prakasa Rao
Leo 9h53'54" 31d6'14"
Yalchin&Melinda Shaban
And 22h58'46" 41d47'25"
Yalda Hakim - The Longest Night
Cnc 8h56'11" 9d34'34"
Yale 1
Lyn 7h55'24" 57d26'20"
Yalietza Sauri
Leo 11h18'19" -5d16'57"
Yalla
Cas 1h12'9" 55d15'56"
Yamagata, Tatsuya
Ori 6h3'19" 10d8'30"
Yama's Twinkling Brightness
Lyn 7h33'8" 44d8'34"
Yamashita Tomohisa
Uma 12h7'52" 63d3'34"
Yamchi
Cap 20h30'16" -11d12'29"
YAMEL
Gem 6h13'3" 22d11'47"
Yamel Yvette Guerra
Umi 15h57'19" 74d57'51"
Y.A.M.F.
Sgr 19h21'34" -12d21'0"
Yamile Saavedra
Vir 12h14'5" -3d15'51"
Yamilet de Filippi
Lib 15h26'21" -6d56'11"
Yamileth
Lyn 6h31'46" 55d35'43"
Yamill Rodriguez
Gem 6h31'10" 13d6'46"
Yamina Star
Cap 21h47'17" -9d24'16"
Yamu
Aqr 22h50'14" -7d3'27"
Yamuna
Vir 12h53'5" 6d42'25"
Yamush Matokush 6
Ori 5h11'8" 7d17'26"
YAN
Psc 0h41'25" 16d40'5"
Yan
Dra 16h53'57" 60d54'58"
Yan Bo Zhang
Sco 16h12'17" -9d4'26"
Yan Bonzon
Lyn 7h37'43" 41d27'28"
YAN DAVID BILLETER
Crb 15h36'13" 25d48'10"
Yan Jie
Ori 5h50'5" 7d27'50"
Yan Lvovitch
Umi 14h27'8" 76d54'32"
Yan Maxwell, Etoile de la Vie
Cyg 21h43'46" 44d46'21"
yAn yAn ylm
Psc 1h43'18" 9d37'57"
YANA
Ari 2h21'12" 26d7'53"
Yana Bender
Cyg 20h38'50" 57d47'39"
Yana Jones
Del 20h48'19" 15d33'47"
Yana Levin
Cnc 8h24'41" 9d55'7"
Yana M. J. Cohen
Gem 7h37'18" 14d53'9"
Yana Mikhaylov
Cap 21h58'24" -13d18'58"
Yana Vass
Lib 14h57'7" -9d57'44"
Yana Zuo
Equ 21h7'31" 5d51'32"
Yana, 15.04.1983
Peg 21h28'4" 8d9'34"
Yanary Anderson
Uma 9h47'12" 45d27'38"
Yanek Riber
Gem 7h43'48" 22d31'3"
Yanel
Tau 4h29'1" 17d48'50"
Yanessa Flores
Sgr 19h19'56" -44d0'58"
Yanet DeLeon
Cam 3h33'3" 55d46'2"
Yanet Meza
Cyg 21h32'26" 43d51'54"
Yanett
Gem 7h28'38" 20d9'44"
YANG
Uma 11h18'37" 34d56'55"
Yang Guang
Cep 22h54'51" 71d31'18"
Yang Jie Sheng
Cyg 20h46'46" 47d29'4"
YanHong Bi
And 2h30'57" 38d22'41"
YANI
Lib 15h36'28" -16d11'55"
Yani Anastasia
Sco 17h52'5" -35d58'48"
Yania Rodriguez
Aqr 22h0'36" -22d24'26"
Yanicke - Major
Uma 10h35'14" 58d59'49"
Yanik Chicoine
Uma 11h25'27" 60d47'45"
yanileidis
Umi 17h29'43" 82d5'45"
yanin
And 0h6'22" 44d42'48"
Yanina Kopyt
Ari 3h9'21" 27d29'11"
Yanina Zborovskaya
Psc 0h39'18" 19d48'32"
Yanira Lakhlani
Lyn 7h2'2" 48d29'8"
Yanira Moreno
Uma 8h40'13" 62d37'2"
Yanira "yummy" Naugle
Peg 23h22'36" 33d10'47"
Yanirah
Tau 4h39'3" 3d8'9"
Yanisstar
Cam 4h39'53" 65d12'46"
Yanjun Jing
Vir 13h28'16" 5d35'40"
Yank
Aql 19h10'44" 14d38'27"
Yankee Boy
Uma 10h40'35" 46d24'23"
Yankee girl
Lyn 6h36'11" 57d19'26"
Yankuen
Sco 17h0'43" -44d41'43"
Yanly
Vir 13h4'42" -10d22'19"
Yann
Lac 22h24'10" 54d7'3"
Yann
Lac 22h43'1" 53d14'51"
Yann
Lmi 9h28'21" 38d20'20"
Yann Daniel
Cyg 21h39'48" 41d25'50"
Yann Granet l'amour de ma vie
Col 5h50'52" -41d48'35"
Yann Hadifé
Sco 17h55'53" -42d41'38"
Yann Richard
Uma 12h40'45" 53d57'10"
Yann Schläfli
Cnv 13h55'19" 36d20'10"
Yanna Caroline Fee
Lib 15h16'27" -7d27'32"
Yanna Machi
Ari 2h45'12" 13d3'49"
YANNI
Leo 11h3'3" 15d47'59"
Yanni
Lyn 7h25'18" 56d33'36"
Yannic
Dra 17h15'36" 65d51'22"
Yannic
Tri 2h36'7" 35d4'26"
Yannic Guillon
Uma 11h45'7" 46d35'54"
Yannick Colot
Uma 11h43'59" 37d23'46"
Yannick Comenge
Her 18h26'37" 18d37'19"
Yannick Gérin
Sgr 19h15'27" -26d3'43"
Yannick Jacquier
And 1h0'5" 33d47'49"
Yannick Kramer
Her 18h42'27" 13d49'28"
Yannick Le Monnier Rabinel
Cep 0h8'21" 76d58'16"
Yannick Loves Aishling 4 Ever
Cnc 8h28'24" 10d35'21"
Yannick Nael
Cnc 8h57'54" 7d37'50"
Yannick Oussalem
Umi 18h7'44" 87d4'18"
Yannick Tschanz
Cas 23h5'59" 58d17'41"
Yannique . Plus que 2 et 4
Tau 4h21'28" 24d27'49"
Yannis Krähenbühl
Uma 11h50'59" 54d48'46"
Yannis Lucas
Uma 12h46'4" 62d54'28"
Yannis Reiner
Ori 6h16'16" 2d13'14"
Yannou
Ori 4h59'51" 10d35'38"
Yannouch
Cap 20h58'25" -15d27'43"
Yanny Nabula
Aql 18h52'53" 6d50'25"
Yanti and Tony
Leo 9h22'23" 24d53'15"
Yanzi
Sgr 19h17'16" -14d57'56"
Yao Lifei
Tau 4h19'59" 17d51'35"
Yao-yao Zhi
And 23h9'44" 48d9'31"
Ya-Ping Chang "Cow Cow"
Uma 9h13'23" 70d3'3"
Yaqout
Lyn 7h26'36" 53d32'9"
YAR
Aqr 21h41'50" 2d31'33"
Yara
And 23h28'28" 50d39'20"
Yara
Uma 13h44'16" 48d10'16"
YARA
Sco 16h22'10" -26d13'44"
Yara Kanoo
Uma 10h3'18" 59d24'58"
Yaravicu
Lyn 9h3'44" 37d57'58"
YARDO
Uma 14h5'51" 56d38'37"
Yards MD
Ari 2h16'15" 13d46'9"
YARELI "mi gordita" LOPEZ
Vir 12h42'3" 4d44'24"
Yarelis Perez
And 22h58'7" 41d42'5"
Yarinette Ayala
Psc 1h10'28" 27d17'37"
Yari's Beauty
Cnc 8h18'49" 12d16'38"
Yarisbeth
Aqr 22h26'52" -2d3'32"
Yaron Sandalon
Uma 9h46'4" 52d46'20"
Yaroslavna Vladimirovna Kadulina
Cam 3h53'5" 64d35'56"
Yaruji
Cnc 8h15'17" 21d55'14"
Yasemin
Ori 4h54'13" -0d3'12"
Yasemin Karahasan
Ari 3h20'15" 16d45'0"
Yasemin Ozumerzifon
Vir 13h8'25" 13d28'59"
Yasemin Stillig 08.10.1982
Uma 11h26'31" 40d18'25"
Yasenia Alderete
Mon 6h56'38" -5d14'37"
Yashima Tutomu
Gem 7h13'40" 24d59'58"
Yasil Arcia
Lib 15h5'29" -27d48'30"
Yasin
Leo 11h18'20" 0d34'52"
Yasin "Deine Insel Ufenau im Himmel"
Ori 5h36'26" -8d46'17"
Yaslyn's Star
Lib 15h11'14" -5d25'54"
Yasmeen Linjawi
Lib 15h16'32" -20d42'27"
Yasmeena
Cam 4h28'15" 66d50'26"
Yasmeena
And 1h2'55" 37d8'45"
Yasmen Simonian
Mon 6h47'58" -6d8'28"
Yasmin
Lyn 6h50'36" 57d20'54"
Yasmin
Cas 1h3'15" 49d40'37"
Yasmin Ayala
Sco 16h7'58" -12d32'8"
Yasmin Bassam
Vir 13h26'22" -7d20'2"
Yasmin C. Spiegel
Cap 20h7'44" -18d17'41"
Yasmin Ghaith
Cas 23h5'8" 56d6'24"
Yasmin Hui
And 23h55'25" 42d2'42"
Yasmin Hunter
Ori 5h17'0" 15d51'27"
Yasmin India Leacock - Yasmin's Star
And 23h25'45" 36d48'50"
Yasmin Leigh
And 23h16'14" 41d49'59"
Yasmin Liliana Ouart
Mon 6h48'1" 6d55'44"
Yasmin Loves Seth
Lib 14h34'38" -14d48'11"
Yasmin Madison Staveley
And 23h53'25" 42d44'56"
Yasmin Olivia Rodgers
Tau 4h36'39" 16d56'45"
Yasmin - our little princess
Leo 11h30'55" 6d21'38"
Yasmin Sara
Cas 0h2'2" 54d34'3"
Yasmin Vieira Braga
Uma 12h45'51" 56d28'39"
Yasmin Watson
Cas 0h19'8" 51d53'32"
Yasmina
Tau 4h26'58" 21d55'44"
Yasmine
Lib 15h48'10" -11d53'8"
Yasmine
Sco 16h15'12" -28d59'39"
Yasmine Azaria Strickland
Crb 15h57'45" 34d0'13"
Yasmine El Mouallem Qannati
Cyg 21h13'52" 55d2'59"
Yasmine Ilana Katherine Klein
Psc 23h45'8" 5d47'49"
Yasmine James
Lyr 18h41'51" 30d57'54"
Yasmine Mueller
Vir 15h7'59" 1d13'54"
Yasmine Star Pertiller
Sgr 18h37'17" -20d43'6"
Yasmine Sulamith
And 2h11'35" 50d30'18"
Yasmine Vickers
And 2h8'17" 39d27'59"
Yasmine Virji
Lib 14h45'2" -22d1'24"
Yasmin's Twinkle
And 2h23'33" 37d35'30"
Yasna / Ron
Cyg 19h41'12" 30d14'48"
Yasseen Al-Nae'eb
Uma 9h52'0" 55d13'46"
Yasser Shahin
Sco 16h55'4" -44d16'29"
Yasso
Cas 0h59'16" 49d55'44"
Yasuhiro & Ako & Tara
Tau 5h50'2" 24d55'0"
Yasuoyrevetahwekimsey
Uma 11h49'6" 38d53'21"
Yasuyo Uchida
Uma 14h9'35" 57d34'18"
Yatin
Lib 15h26'18" -11d16'14"
Yatri
Cas 0h19'14" 65d9'29"
Yau Arumda-un
Lib 15h5'49" -23d53'43"
Yavin4
Aqr 21h6'22" -13d11'56"
Yaxin Duan
Cnc 8h34'16" 15d17'11"
Yay Tara
Cyg 21h25'44" 54d21'5"
Yaya
Tau 5h7'38" 18d28'59"
YAYA
Crb 15h39'55" 30d50'5"
YAYA "1-30-1951"
Lyr 18h44'37" 39d37'32"
" Yaya " Ma. Lourdes A. de Valles
Lyr 18h50'54" 32d13'38"
YaYa The Wise Queen of the Universe
Srp 18h5'21" -0d23'34"
YaYa Vandermolen
Uma 11h13'53" 43d49'41"
Yayoiya
Cyg 19h41'33" 35d28'5"
yaz2005
Sco 16h59'55" -30d16'35"
Yazmin Angelica
Col 5h50'20" -33d24'45"
YAZMIN ANGUIANO
Aqr 21h12'59" -8d42'12"
Yazmin Salangsang
Psc 2h2'25" 4d34'5"
Yazmin & Yozdy
Sgr 18h2'16" -28d35'8"
Yazz
Cnc 8h27'27" 13d6'49"
Ybarra Love For Eternity
Psc 0h36'5" 7d36'32"
Ybot Tropptauri
Uma 10h36'24" 46d31'2"
Ye Olde Piano Tuner - Warren Park
Tau 3h48'10" 26d47'31"
Yea MmmHmm
Uma 11h46'24" 42d3'41"
Yeager Way
Vir 14h23'6" 2d37'19"
Yeager's Star
And 2h38'30" 50d1'9"
Yeaji
Umi 14h24'10" 65d48'21"
Yeanor
Per 3h21'42" 44d10'50"
Year of the Carp
Uma 10h39'33" 53d14'1"
Yeardley Eustis Smith Williams
Psc 0h21'5" 2d46'58"
YeaterNation
Ari 3h12'43" 28d30'42"
Yedinak
Mon 7h33'6" -0d48'43"
Yeekhun Lee
Uma 12h8'0" 62d38'42"
Yefera
Uma 13h50'27" 57d35'27"
Yefim Leo
Gem 7h7'42" 33d16'53"
Yegoiants
Per 3h18'59" 41d20'11"
"Yehonatan"
Ori 6h16'7" 16d39'45"
Yehoshua Sofer
Vir 12h39'24" 7d38'37"
Yehuda Lieb
Tau 4h14'31" 21d10'40"
Yeiry
Gem 6h49'15" 22d15'32"
Yeisika Yazmin McKean 1
And 0h41'26" 38d5'2"
Yeji Lee
Psc 1h21'42" 29d4'30"
Yekara's Star
Leo 10h41'28" 8d21'40"
Yekaterina Bogush
Vir 13h30'47" 9d22'51"
Yekaterina Borzhemskaya
And 1h5'56" 38d30'33"
Yekatrina ivgenivna arbozov (kety)
And 0h48'26" 27d45'29"
Yelena
Gem 6h55'51" 26d9'29"
Yelena
Vir 13h24'34" 13d51'53"
Yelena
Psc 0h34'51" 11d35'50"
Yelena
Cyg 21h46'55" 49d47'41"
Yelena & Boris
Ari 2h43'59" 20d33'46"
Yelena Golshteyn-Orodovskaya
Cap 21h38'28" -15d19'31"
Yelena Gozener
Lib 14h34'33" -9d4'32"
Yelena Kofman
Lyn 6h35'12" 58d4'32"
Yelena Meus Carus
Lyr 18h28'17" 37d12'14"
Yelena Rikhman
Cap 20h57'53" -16d32'51"
YelenaBanshchik
Leo 9h49'21" 9d59'8"
YELEX
Umi 14h13'29" 68d6'18"
Yeliz
Ori 5h53'56" 16d51'33"
Yellia
Per 4h23'39" 35d11'6"
Yellow
Gem 7h27'34" 33d1'2"
Yellow
Ari 3h10'58" 28d31'55"
Yellow
Lyn 6h47'42" 61d17'42"
Yellow Bird Hotel
Cyg 21h11'56" 47d47'46"
Yellow Jellybean and Pink Jellybean
Her 18h10'6" 21d57'5"
Yellow Wings
Tau 4h40'31" 20d18'37"
Yellowbird
Tau 3h32'30" 9d11'52"
Yen
Peg 21h38'41" 7d15'46"
Yenastanunha
And 23h47'26" 48d18'10"
Yency Patricia
Ari 1h54'25" 21d38'59"
Yeneilyn, My Angel
Cyg 20h30'6" 51d49'43"
Yeni
Lyn 8h12'43" 49d51'45"
Yenifer R. Rodriguez
Lyn 7h13'31" 53d51'57"

Yenior
Sco 16h10'21" -11d52'13"
Yennece May
Lyn 8h26'23" 36d58'42"
Yenny Lo Lau
Sgr 18h47'19" -20d20'13"
Yen-Phin & Hon-Kit
Sge 19h38'31" 17d11'44"
Yen's Special Birthday Star
Psc 0h56'2" 8d15'0"
Yenter Twins
Vir 14h38'34" 6d40'48"
Yenushka Gopal -with love-Kumar
Cyg 21h45'45" 44d19'2"
YEOHOO USA
Her 16h14'55" 44d55'19"
Yeon Seon Terry
Ori 6h13'26" 5d43'57"
YEP! Montana Michalek Forever
Cyg 20h42'11" 42d39'7"
Yepo
Cas 0h51'22" 49d9'38"
Yerin
Crb 16h14'21" 27d29'22"
Yes Star
Cap 21h38'39" -17d8'44"
Yesenia
Lib 15h32'27" -19d38'57"
YESENIA
Ari 3h21'27" 28d40'25"
Yesenia
Cnc 8h1'11" 18d18'2"
Yesenia
Tau 4h29'23" 15d15'53"
Yesenia
And 2h27'53" 42d13'5"
Yesenia Almonte
Cyg 19h36'38" 35d19'37"
Yesenia Baez
Lyn 7h56'20" 37d44'36"
Yesenia Carde
Aql 19h32'56" 9d0'51"
Yesenia Colon-Delgado
Leo 11h31'48" 1d58'6"
Yesenia Delgdillo
And 0h42'46" 25d3'19"
Yesenia Muñoz
Her 17h38'13" 23d26'36"
Yesenia Pena
Aqr 23h18'44" -13d33'9"
Yesenia Quintana
Ari 1h55'27" 19d23'32"
Yesenia "Yesi" Amanda Saenz
Cnc 8h46'46" 20d28'19"
Yesh
Sex 9h47'49" -0d27'41"
Yesni
Ori 6h5'32" 20d2'16"
Yessenia
Del 20h36'12" 16d36'48"
Yessenia
Sgr 17h54'50" -16d57'59"
Yessenia Alvarado
Uma 12h26'28" 60d1'43"
Yessenia Ramirez
Crb 15h41'51" 32d2'11"
Yessenia Vasquez
Leo 11h48'1" 25d10'13"
Yessica Estrella Hermosa Del Cielo
Sco 16h43'23" -34d35'23"
Yessina DelValle
Lyn 7h54'38" 51d23'16"
Yesterday-Today-Tommorrow
Uma 8h42'9" 71d19'25"
Yetkin Elaine
Cap 21h41'0" -10d28'6"
Yetta Glassoff
Uma 11h43'46" 59d5'31"
Yeuk Huen Au
Cyg 21h24'54" 46d32'59"
Yevgeniy Shteynman
Lib 15h42'14" -14d30'34"
Yi Du
Sex 10h36'7" -0d25'20"
Yi Ji Hae
Psc 1h53'46" 6d20'5"
Yia Ttavta
Lyn 6h58'9" 47d1'9"
Yia yia'
Uma 9h31'7" 47d38'32"
Yia yia
Uma 8h30'10" 67d1'50"
Yia Yia and George Venderis
Ari 3h6'53" 28d33'31"
Yia Yia's wish
Vir 14h29'3" 2d21'53"
Yianna Assimina
And 1h41'25" 50d7'37"
Yianni
Lyr 18h40'22" 30d8'26"
Yianni H. Skourtis
Her 17h41'9" 19d2'41"
Yiannis Kambouroglou
Aql 19h0'0" -6d18'53"
Yiayia
Cap 20h24'13" -21d25'34"
Yiayia Margarita
Peg 22h48'12" 21d53'39"
Yikes
Uma 10h6'33" 65d0'24"
Yilay
Cam 4h34'28" 74d30'25"
YILDIZ BEDENLIER
Tau 4h2'25" 26d11'38"
Yillian's Star
And 23h25'42" 39d15'4"
Yin
Sco 16h34'55" -29d29'57"
Yin and Yang Lobsters July 24, 1996
Cas 0h48'9" 66d14'18"
Yin Liu
Del 20h38'36" 13d21'33"
Yin Min & Jay Jay
Peg 0h4'40" 16d52'59"
Yin Xia - Angel
Cas 0h10'14" 53d24'29"
Yin-Chu Jou
Tau 4h54'52" 18d52'26"
Ying Chile
Cru 12h24'16" -56d43'49"
Ying Chu
Umi 15h6'49" 69d20'38"
Ying Chu
Umi 15h59'19" 82d28'57"
Ying Chu
Crb 16h21'44" 38d9'40"
Ying & Guo
Cyg 20h24'10" 58d51'26"
ying lek
Cas 0h37'0" 48d58'21"
Ying Yan
Lib 15h30'40" -12d47'58"
Ying-Hsiung Lin
Srp 18h35'20" -0d2'51"
Ying-Yan Njoo
Aqr 22h48'16" -24d23'50"
Yinka Kamson
Leo 10h25'29" 22d20'28"
Yinz Guys Star n'at
Boo 14h35'48" 51d40'0"
Yiota Pavlou
Lyr 19h6'2" 45d25'24"
YIP HO WAI
Her 16h51'32" 25d51'48"
Yisel
Sge 20h5'40" 17d36'2"
Ykükü
Umi 17h21'43" 83d51'55"
Yldsol
Peg 22h33'25" 25d28'17"
Ylenia
Umi 16h24'16" 74d6'40"
Ylenia Venzo
Sco 16h58'18" -42d27'25"
Yliana
Cyg 20h23'12" 38d22'22"
YLJ1977
Sgr 19h37'6" -32d35'22"
Ymer
And 23h39'44" 37d37'28"
Ymm Asad
Uma 9h9'31" 72d3'50"
Y.N. Faulks
Lyn 7h32'45" 55d7'43"
Ynes Depuydt
Crv 12h8'2" -13d50'31"
Ynette Siegelman Hogue
Oph 17h11'45" -0d20'45"
Yngrid
Cnc 8h58'15" 24d7'30"
YnotYelhsa
Lyn 7h56'53" 37d19'44"
Yo Cupcake
Tau 3h45'3" 18d14'48"
Yo - Star of Love & Happiness
Sge 19h45'21" 17d36'53"
Yo Yo
Uma 8h49'31" 52d14'8"
Yoan Avrila
Sco 17h3'46" -41d6'45"
Yoandra Sanchez
Ari 2h9'50" 24d47'38"
Yoann
Umi 13h5'22" 74d21'57"
Yoann et Amandine
Cap 20h22'43" -17d10'33"
Yoanna De Leon
Vir 13h14'23" 10d52'1"
Yoanna My Queen
Cas 1h23'15" 52d14'41"
Yobie
Cyg 19h32'8" 30d6'57"
Yochris
Uma 9h40'23" 58d32'31"
Yoda Levenson
Tau 3h41'47" 25d41'9"
Yodi - Pumpkin
Umi 14h50'3" 72d56'8"
Yoee
Lib 15h57'49" -7d4'18"
Yoey's Stellar Guardian
Vir 12h45'28" -2d48'32"
Yoffie
Per 3h19'14" 46d40'3"
Yogi
Uma 10h52'22" 54d11'40"
Yogi Bear "The Yogster"
Cma 7h24'35" -16d23'15"
Yogi C
Her 18h44'42" 22d12'27"
Yogi Dave Sheppard
And 0h40'5" 26d28'48"
Yogo & Yumika Kanda
And 2h31'30" 39d43'44"
Yohan Chauliac
Psc 0h58'6" 27d48'56"
Yohania T. Santana
Sco 16h51'40" -38d48'46"
YoHillEin NewAndriEnce
Cnv 13h49'9" 34d51'0"
Yoko Macuaga
Ari 2h52'25" 26d17'59"
Yolanda
Gem 6h8'13" 25d27'29"
Yolanda
Gem 7h31'39" 16d47'15"
Yolanda
Psc 1h18'24" 11d30'56"
Yolanda
Cas 0h0'42" 56d17'33"
Yolanda
Eri 4h12'7" -25d38'31"
Yolanda Alvarez Estrada
And 0h43'52" 24d35'1"
Yolanda Anne Bindics
Lib 15h29'40" -5d52'31"
Yolanda Canales Dowdy
Uma 10h45'2" 43d11'29"
Yolanda Crowley
Uma 9h30'24" 50d24'43"
Yolanda Crupper
Lmi 10h31'27" 37d15'19"
Yolanda Hickey
Sgr 17h58'13" -19d18'25"
Yolanda J. Jefferson
Sgr 18h27'9" -25d28'47"
Yolanda L Hunter
Leo 11h6'40" 22d17'35"
Yolanda Lynette Hankins
Gem 6h47'37" 32d39'28"
Yolanda M. Discenzo
Cas 23h40'10" 56d43'31"
Yolanda Maltezos
And 0h16'42" 26d52'43"
Yolanda Montelongo
Sgr 18h12'7" -29d1'50"
Yolanda P. Fryson
Cap 21h2'43" -14d51'27"
Yolanda Plaza Suarez
Her 18h39'28" 21d48'50"
Yolanda "Princess" Groff
And 2h30'59" 47d42'11"
Yolanda R.
And 0h20'52" 26d37'33"
Yolanda Ramirez
Psc 1h36'35" 19d7'4"
Yolanda Roller
Cas 0h58'57" 63d36'15"
Yolanda Unzueta
And 2h17'15" 49d17'19"
Yolanda Vega Cardenas
Cnc 8h38'2" 18d10'24"
Yolanda Villarreal
Cnc 8h15'12" 23d54'42"
Yolanda110660
And 22h58'0" 36d10'39"
Yolande Berthiaume Allard
And 0h11'22" 26d39'10"
Yolande Gadeau
Sgr 18h31'40" -32d55'7"
Yolande Van Niekerk
Pho 0h46'18" -42d50'49"
Yolande's Star
Sco 16h17'31" -19d32'24"
Yolanta
Tau 5h6'50" 26d41'4"
Yola-VSJ
Ori 5h37'48" 3d1'42"
Yole Miceli
Vir 14h17'27" -16d50'10"
Yolène
Uma 9h9'44" 62d53'14"
Yolene Wilson
And 1h13'21" 42d15'0"
Yoli y Alejandro
Cep 22h57'48" 75d30'58"
Yoli Zbinden
Cas 3h26'2" 68d36'32"
Yolly
Tau 4h8'34" 15d20'34"
Yolonda Purkett (Booke)
Vir 13h59'11" -7d46'47"
Yolotl Aztecatl
Cyg 21h54'33" 52d59'15"
Yoly
Ori 6h7'9" 19d5'22"
Yoly Deseo
Crb 16h21'44" 39d36'12"
Yoly Valdes
And 1h0'12" 40d37'39"
Yomari Crespo
Per 2h18'46" 51d34'47"
Yonatan Gozdanker's family
Ori 5h57'56" 21d50'25"
Yonda Elaine Hannaford(Y.E.H.)
Per 2h54'12" 51d0'46"
Yone
Cru 12h0'48" -63d40'46"
Yonga: My Shinging Star
Crb 15h43'34" 28d20'41"
Yonga (My Shining Star)
Tau 5h46'12" 22d15'20"
Yongchun Selen Yan
Psc 0h23'23" 7d42'39"
Yong-The Seat of Resplendence
Leo 9h24'1" 12d15'54"
Yoni's Blazing Star
Dra 18h20'20" 52d36'45"
Yoo, Eui Dong "Caden"
Ori 6h0'12" -0d56'59"
Yoon Hae Kyung
Ori 5h4'24" 4d2'56"
Yoon Lea Pierce
Cap 21h32'55" -11d7'1"
Yoon Mi Won
Cyg 19h47'40" 29d55'29"
Yoon Soh Jung
Leo 9h22'37" 14d53'33"
Yoon-Jean Kang
Vir 13h0'19" -3d46'5"
Yoonmee Choi & Changhak Sohn
Cyg 20h57'41" 33d14'52"
Yordies I'll love you forever
Lyn 8h23'56" 34d0'5"
Yorgos Koniaris
Ori 5h37'47" 8d3'58"
York Kiwanis Club
Uma 8h17'9" 61d7'58"
Yorleny Fallas
Gem 6h8'58" 23d36'34"
Yort Star
Sco 16h34'52" -44d3'32"
Yosef
Leo 11h29'4" 9d6'44"
Yoselyn Perez
Crb 15h38'39" 39d29'45"
Yosemite Tom "3-20-2004"
Lyn 6h55'21" 53d16'38"
Yosephus Filbert Star
Her 17h36'6" 21d1'38"
Yoshi
Sco 16h7'3" -15d27'26"
Yoshi
Sgr 19h26'28" -39d3'51"
Yoshida's Fine Art Gallery
Per 2h50'33" 42d48'53"
yoshie
Vir 14h5'3" 1d12'56"
Yoshie`s shining star
Aqr 22h17'11" -0d31'5"
Yoshiko Villamor
Uma 9h13'28" 68d45'10"
Yoshikuni & Maki
Tau 4h18'47" 25d57'20"
Yoshimasa.H
Vir 13h14'11" -9d42'15"
Yoshimi
Ori 6h16'36" 15d8'29"
Yoshinari & Hisato
Sgr 20h11'16" -34d58'7"
Yoshinori & Mika
Lib 15h20'8" -14d3'44"
Yoshio
Uma 11h31'48" 58d25'39"
Yoshiteru Tamura
Aqr 22h33'7" -2d23'18"
Yosi-Monkey
Cnc 8h13'52" 14d43'1"
Yosip Alexander
And 0h29'48" 34d35'19"
YOSRA
Psc 23h55'18" 1d5'43"
Yosri
Gem 7h6'55" 22d9'31"
Yossi
Aur 5h34'48" 33d25'55"
Yossi and Sharon Adjedj
Cyg 20h45'37" 54d50'55"
You
Cap 21h0'5" -23d32'21"
You
Psc 1h24'47" 21d57'54"
You and Always
Ori 5h9'26" 12d59'12"
You and I...Together Always
Sco 16h11'46" -30d34'39"
You and Me
Uma 11h30'36" 33d8'15"
You and Me
Cyg 21h14'50" 32d31'44"
You and Me
Cyg 20h2'32" 43d50'41"
You and Me
Cyg 21h10'2" 44d8'49"
You and Me
Uma 8h34'13" 50d48'23"
You and Me
Per 4h38'43" 47d23'56"
You and Me
Cyg 20h42'7" 52d15'18"
You and Me
Cyg 20h53'58" 46d23'9"
"You and Me"
Cyg 19h35'13" 51d54'14"
You and Me
Cyg 21h56'2" 48d3'11"
You and Me Against the World
Uma 14h2'59" 53d48'49"
You and me ... c'est pour la vie!
Cap 20h13'55" -26d31'37"
You and Me Forever
Cyg 19h43'56" 34d18'0"
You are a star, John Paul Dera
Cnc 9h3'41" 8d57'7"
You Are And Always Will Be A Star!
Ari 1h55'15" 20d25'10"
You Are Loved CML
Gem 7h34'25" 22d50'36"
You Are Loved(John & Jackie Dubose)
Tau 3h44'19" 6d52'51"
You are my everything.
Per 3h37'5" 34d15'23"
You are my favorite part of the day!
Cas 23h0'28" 56d3'53"
You are my shining star
Sgr 19h23'35" -22d42'58"
You are my shining star Grandma!
Cyg 19h46'23" 36d33'29"
You are my star, Ski!!!
Uma 9h35'47" 44d7'12"
You Are My Sunshine!
Lyn 9h14'26" 34d38'20"
You are my "Sunshine"...
Cyg 20h33'33" 57d6'26"
You Are my Sunshine Craig S. Forand
Cnc 8h39'23" 31d45'48"
You Are My Super Star Michilin Chau
Umi 16h50'13" 79d33'47"
You are the Battlefront Star
Boo 14h49'56" 51d56'2"
You are the best Mom,I Love You!BMC
Psc 23h55'46" 1d54'56"
You Are The Love Of My Life - Joni
Leo 11h32'40" 16d2'55"
You are the One
Cyg 19h27'39" 46d57'25"
You Are The Sparkle In My Life!
Ari 3h12'6" 29d17'32"
You Are The Star That's In My Sky
Lyn 6h18'27" 54d42'35"
You Bet!
Uma 13h9'34" 52d51'8"
You don't have to wory
Sco 16h14'20" -13d18'11"
You fell down to me
Lyn 7h35'34" 48d27'56"
you forever
Cep 20h43'7" 72d12'18"
You Light Up My Life
Cnc 8h42'16" 31d35'6"
You Me ~ Me You
Cas 0h31'52" 64d21'8"
You & Me Together Forever
Sex 9h50'4" -0d26'47"
You never know
Ori 5h54'45" -0d32'47"
YOU ROCK
Aqr 22h45'8" -5d14'30"
"YOU SATISFY MY SOUL"
Aqr 23h11'3" -14d41'38"
You Shine John William Gane
Uma 11h28'17" 58d56'7"
You will always be my man Dan
Sco 17h53'15" -36d48'52"
YouAndMe: Samson's Star
Lyr 18h36'51" 35d56'56"
Youcef
Cas 23h39'42" 56d44'32"
Youcef Zrig 22 janvier 1979
Uma 11h30'4" 57d59'24"
You...Finally
Del 20h42'6" 11d8'15"
Youhna Wang Ayala
Gem 7h30'46" 27d23'37"
Youkam
Sco 16h12'8" -41d28'14"
YOUMEWE777
Sco 16h5'32" -11d30'4"
Youn Ha
Sgr 19h17'25" -16d22'30"
Youna Lola
Cap 20h32'58" -18d14'41"
Youna Lola
Umi 16h19'28" 82d35'1"
Young Adam
Cep 20h43'58" 59d47'6"
Young Choon Bae
Cam 5h28'26" 78d7'0"
Young Forever!
Her 16h23'4" 19d26'20"
Young & Gorgeous Milton
Cnc 9h5'8" 27d31'24"
Young Jerr
Ori 6h12'21" 17d30'8"
Young Love
Ori 6h8'55" 15d26'48"
Young Love
And 23h2'55" 42d46'43"
Young Marianne
And 23h34'19" 47d43'16"
Young Stars
Cyg 19h41'3" 28d44'26"
Young Stud
Pic 5h30'34" -45d7'6"
Young Ye Di Netta
Leo 11h25'46" 12d27'16"
Young Yi
Sgr 19h10'58" -18d41'14"
Youngblood
Umi 15h24'51" 69d5'16"
Youngblood50
Gem 7h9'1" 16d13'11"
Young-Hee Cho
Cas 23h22'50" 53d50'44"
Younghoney
Boo 14h34'59" 38d39'19"
Your 21st and 1st Fathers Day
Per 4h22'52" 34d58'58"
Your a star Karl
Tau 4h30'52" 17d16'1"
Your Angel
Cyg 19h39'52" 36d42'22"
Your Angel Grace
And 23h32'56" 47d43'54"
Your Angel: Sophia Fotopoulos
Cas 1h15'45" 50d43'40"
Your Aziza
Lyr 18h53'43" 35d28'36"
Your Beautiful Dreams
Aqr 22h8'51" -0d31'56"
Your Beauty In the Sky
And 1h1'51" 37d46'52"
"Your chosen name here..."
Per 4h0'29" 32d56'53"
"Your chosen name here..."
Uma 12h10'54" 47d24'28"
Your chosen name Here !!
Ori 5h37'30" 12d0'51"
Your chosen name here!
Umi 14h58'24" 79d7'51"
"Your chosen name here..."
Cas 23h53'10" 59d37'48"
Your Chosen Name Here...
Cas 0h59'20" 60d59'10"
"Your chosen name here..."
Uma 10h4'18" 56d41'6"
"Your chosen name here..."
Umi 13h31'49" 71d15'57"
Your Chosen Name Here!
Umi 16h0'57" 73d22'48"
Your chosen name here!
Equ 21h8'15" 6d22'38"
Your Chosen Name Here - Park School
Aql 20h13'52" 3d35'52"
Your Chosen Name Here! - St Luke's Hospice
Per 4h24'17" 40d33'57"
Your Chosen Star Name Here !!
Peg 21h39'29" 14d29'31"
Your Dani
Uma 12h55'59" 57d54'34"
Your Future
Aqr 23h8'44" -21d48'59"
Your Grandson's Love
Uma 11h26'8" 40d5'54"
Your Guardian Angel
Her 18h30'27" 21d51'14"
Your Kimmarie 12-15-06 430AM
And 1h6'12" 47d15'2"
Your Laughing Girl Loves You x x x
Lib 14h55'44" -2d13'52"
Your Loving Brother
Lib 15h5'17" -15d35'0"
Your Loving Light
Leo 10h51'37" 5d42'29"
Your Loving Wife, Daughter and Son
Uma 11h28'43" 58d49'44"
Your Mam - Watching over you Always
Tau 4h59'28" 27d19'55"
Your Mom
Cap 20h59'8" -17d10'15"
Your My Shining Star Forever... AG
Leo 10h5'22" 23d44'55"
Your Name
Uma 11h43'50" 51d28'24"
Your Past, My Present, Our Future
Ori 5h55'41" 17d47'10"
Your Perfect Lid
Uma 10h36'32" 44d11'47"
Your Place in the Stars Debra Seman
Sgr 18h24'29" -32d25'37"
Your Rights <\@> Work
Cru 11h58'40" -59d55'10"
Your Shining Star
Uma 8h26'59" 63d3'58"
Your Source of Connection - WAH
Cnc 8h26'41" 15d19'5"
Your Star Name Goes Here
Ori 5h20'44" 11d22'46"
your stellar fairy
Gem 6h29'27" 12d13'5"
Your The New Star In My Sky Meg
Uma 9h39'21" 62d27'15"
You're All I want You're All I Need
Cas 1h54'1" 61d42'48"
You're Always Our Star, Mr.Harrison
Tau 3h55'4" 26d44'5"
You're A-Ward
Cyg 21h50'8" 47d2'43"
You're Beautiful
Sco 17h6'40" -34d6'37"
you're my Baby
Crb 16h21'41" 29d37'0"
You're my choice truly madly deeply
Uma 12h19'36" 57d27'28"
You're my forever, stars never die
Cyg 19h36'32" 34d54'20"
"You're my guy"
Cap 21h30'56" -9d53'54"
"You're My Shining Star"
Gem 7h14'0" 25d24'22"
You're My Shining Star John Halbur
Aur 5h26'56" 45d43'42"
You're my star forever, Dennis!
Psc 1h17'20" 29d1'14"
you're my wonderwall
Tau 5h44'57" 22d47'39"
You're Our Shining Star! Ashjen
Ari 2h22'43" 26d51'29"
You're Still The One
Col 5h34'14" -33d58'51"
You're Still The One Honey - Deb
Umi 16h42'29" 77d55'57"
You're The Instrument To My Heart!
Tau 4h54'40" 15d45'14"
You're The One I Love ~Margaret~
Uma 9h6'39" 67d53'10"
Youri Chliaifchtein
Ari 3h26'26" 20d32'35"
Yours
Ori 6h0'57" 17d17'13"
Yours and Mine
Cyg 20h18'49" 43d16'24"
Yours Forever - Laura Elizabeth
Sco 17h26'10" -33d24'35"
Yours Forever, Ghost Orchid
Cyg 19h45'57" 34d7'46"
Yours Truely
Cyg 21h10'32" 43d16'58"
Yours, Anne
Aql 18h59'10" -10d50'27"
Yousef
Lib 15h47'14" -28d31'12"
Yousef
Cnc 9h5'6" 9d7'33"
Yousef Adam Salti
Cep 2h33'9" 80d56'11"
Yousra - The Princess
And 0h7'22" 30d3'9"
Youssef A Saleh
Her 17h11'48" 31d0'29"
Youssef Mouallem
Her 18h28'23" 22d7'34"
Yousseph Slimani
Cam 6h13'53" 74d44'52"
Yovan Bordin Zlicic
Aql 20h10'31" 14d30'53"
Yowyex Star
Uma 9h38'35" 54d5'29"
Yoyikitu
Gem 6h54'57" 17d23'6"
Yoyo
Sco 16h9'37" -15d31'52"
Yoyo
Vel 9h52'58" -43d21'19"
Yoyo Youssef Iskandar
Peg 22h50'19" 21d46'5"
Yrma's Star
Cyg 20h27'0" 45d50'6"
Yrrag
Per 4h6'8" 44d15'54"
YRROS Carlos Immortalis Amor
Ori 5h30'12" 0d13'28"
Y's Dragon Hold
Dra 18h6'41" 57d55'8"
Ysela and Ryan
Cyg 20h16'45" 48d45'58"
Ytram Moolbnrok
Her 16h16'10" 13d29'20"
Yu Hua Kate Chan TW.LA.12.12.1973
Sgr 19h28'8" -41d26'54"
Yu Lee
Uma 11h12'11" 68d42'3"
Yu Ling
Sco 16h37'55" -38d23'37"
Yu & Sylvia
Cnc 8h48'16" 26d6'49"
Yu Yang
Dra 18h16'44" 75d45'56"
Yu Zhang
Ori 5h55'20" 2d47'17"
Yuan Bundschu Childers
Sco 17h36'38" -32d34'15"

Yuan Chen
Sco 16h14'42" -19d10'49"
Yuan Shuyun
Cam 4h33'33" 66d53'46"
Yuc
Lib 15h48'6" -17d35'38"
Yuca
Sgr 19h7'32" -13d43'40"
Yudi
Lib 14h46'30" -10d43'16"
Yudi
Lyn 7h41'57" 55d50'25"
Yuen -The Star of Chinmoy's Galaxy
Sco 16h51'42" -44d16'45"
Yuen Yee Joey Chan
And 0h19'52" 28d56'29"
Yuen Yuen
Lib 14h54'59" -5d43'49"
Yuen's Method
Lyr 18h44'42" 40d45'19"
Yuhae
Vir 14h11'50" -15d37'4"
Yuhki
Lyn 8h33'51" 44d38'20"
Yu-Hsien Judy Hung
Tau 4h25'47" 5d16'11"
Yuka
Psc 1h17'34" 15d18'42"
Yuka Majima
Ari 2h48'44" 12d23'23"
Yuka "Preciosa" Yazaki
Leo 11h22'55" 4d25'3"
Yukako Naito 1941
Ser 16h13'6" -0d20'25"
YUKI
Ari 2h47'21" 16d11'27"
Yuki
Vir 13h4'37" 10d41'11"
Yuki
Cas 1h16'56" 50d50'53"
Yuki and Akihiro Ito
Ori 4h56'56" 15d23'36"
Yuki since 2006
Cnc 8h10'37" 22d25'57"
Yuki Star
And 2h20'0" 45d25'6"
Yuki Tsubasa Kanako Star
Sco 16h59'43" -37d53'32"
Yukie
Aqr 22h24'52" -22d11'41"
Yukihito 3342g
Tau 5h37'38" 21d42'6"
Yukiko
Uma 10h3'16" 55d30'14"
Yukiko Clark
Vir 13h24'45" 13d2'47"
Yukimi
Cnc 8h51'13" 27d57'15"
Yukimi Nagakura
Psc 0h19'59" 15d34'16"
Yukio Ninagawa 70th
Lib 15h31'5" -18d0'4"
Yuki's Star
Cyg 20h5'11" 42d58'14"
Yukko
Aqr 21h30'15" 1d47'44"
Yuko
Aqr 21h19'52" -6d0'36"
Yuko Ito
Cnc 8h53'18" 19d28'0"
Yukon and Hannah
Gem 7h41'48" 33d25'15"
Yukon Dutton
Uma 9h0'45" 64d17'21"
Yukon Jason Cornelius
Ari 2h45'54" 29d44'47"
Yuli
And 23h25'49" 42d8'43"
Yuli Chan
Ari 1h58'37" 19d8'31"
Yulia
Peg 22h7'2" 27d21'47"
Yulia
Umi 15h30'38" 78d33'5"
Yulia & Luca
Uma 9h13'23" 66d17'44"
Yulia Mikhalis Sergeevna
Lyr 18h46'35" 35d15'8"
Yulia Nikulina
Sgr 18h19'29" -27d54'17"
Yulia Sidorova
Ori 5h16'1" 6d21'36"
Yulia Tatchin
Uma 12h17'55" 55d45'53"
Yulian Ramirez
Uma 11h40'42" 33d36'56"
Yuliana Rossi
And 1h26'24" 45d10'34"
Yulichka
Sgr 18h14'56" -26d9'40"
Yulinka
Leo 9h23'10" 10d16'25"
Yuliya
Cru 12h49'9" -60d11'41"
Yuliya Gil
Cnc 8h52'0" 20d46'25"
Yuliya IV
Sco 16h15'35" -16d22'45"
Yuliya Mihailovna
And 0h43'21" 40d59'24"
Yuliya Serhiyivna Pepelyayeva
Sgr 18h0'1" -26d9'2"
Yulkyu Hwang
Cap 21h3'53" -15d55'18"
Yulla
Cas 2h38'51" 66d19'9"
Yulrayz Love
Aql 19h54'3" 15d34'14"
Yulya 1976
Uma 12h36'1" 57d23'44"
Yume
Dra 17h32'2" 51d18'40"
Yumekichi & Akiko no Sato
Vir 12h33'19" -8d6'3"
Yumi Miura
Tau 4h53'12" 19d53'57"
Yumi Nagamatsu
Cnc 8h52'44" 29d49'37"
Yumi Shirai
Ori 5h22'24" -5d19'3"
Yumi Wilson
Tau 5h16'30" 27d25'21"
Yumiko
Tau 5h3'22" 24d51'23"
Yumiko Kimura Smith
Eri 3h37'44" -13d42'6"
Yummy's Star
Leo 11h20'44" -0d45'41"
Yumstar
Aqr 23h49'15" -3d45'55"
Yum-Yum
Cmi 8h0'52" 1d21'30"
Yun Chun Cheng
Leo 11h22'26" 17d7'39"
Yung Su
Ori 5h51'40" -8d22'8"
Yungvirticus
Leo 11h13'44" 16d42'11"
Yunhee Song
Uma 10h59'8" 70d49'18"
Yunona Badaeva
Sgr 18h58'6" -19d9'10"
Yunseon Esther Kim
Aqr 20h41'17" -14d27'21"
Yup Yup
Cap 21h27'14" -9d29'2"
Yurachka
Cam 4h14'51" 72d2'47"
Yuri Khodakov
Cyg 20h31'18" 42d30'22"
Yuri Kwon Clark
Ori 5h42'54" 6d4'56"
Yuri Lopez
Aur 6h10'26" 49d32'5"
Yurie Niikura
Lmi 10h48'35" 35d44'26"
Yurie Treneer Hewitt
Sgr 19h41'39" -13d37'1"
Yurij Lutsenko
Uma 10h27'22" 45d13'47"
Yuriko
And 1h11'4" 45d4'7"
Yuriko Yamashiro Stewart
Psc 0h44'41" 6d47'17"
Yurina chan
Sgr 18h20'40" -31d53'7"
Yuriske Cornus
Tau 4h6'33" 6d22'35"
Yuriy Kolkevich
Cyg 19h38'15" 36d51'39"
yusa
Vir 13h41'23" -11d57'17"
Yusef Mohamed
Ori 5h26'5" 3d40'44"
Yushaa Adam
Per 3h36'32" 45d43'39"
Yusong Rose Lv
Lyr 18h41'53" 34d3'24"
Yusuf Saeed
Lyn 8h51'15" 40d10'38"
Yutaka
Cnc 8h14'14" 21d21'20"
Yutzmann
Cyg 21h30'22" 41d37'14"
yuu.7.7 1981
Cnc 8h38'35" 27d36'38"
Yuuki Matsuzaka
Cap 20h52'12" -19d34'5"
Yuuma
Gem 6h20'21" 25d33'7"
Yuuya Hakoda
Cap 20h29'33" -22d5'30"
yuvaika
Leo 9h56'47" 8d54'52"
Yuvonne Believes
Uma 9h28'21" 68d55'55"
YUYANG my love forever
Cnc 8h4'31" 22d14'0"
YUYU
Sgr 18h38'29" -19d26'27"
Yuyu Lapin
Hya 9h41'58" -21d45'10"
Yuzmany
Gem 6h36'56" 23d10'8"
Yvan Bussières
Ori 5h55'36" -0d46'52"
Yvan & Pamela
Sge 19h58'36" 18d4'10"
Yves & Christine
Tri 2h6'43" 32d50'23"
Yves Durant
Aqr 20h50'57" 0d4'0"
Yves Eicher
Cam 4h47'10" 78d45'13"
Yves para siempre
Per 3h10'56" 45d13'37"
Yves Pouliot
Ori 5h5'9" 14d34'31"
Yves Schwander - Elwe
Uma 9h6'1" 48d36'23"
Yves Tremp
Crb 16h11'40" 35d50'45"
Yvette
And 1h20'24" 49d19'20"
Yvette
And 0h17'41" 26d27'9"
Yvette
Vul 20h26'42" 26d39'43"
Yvette
Lib 16h1'38" -6d20'39"
Yvette
Cas 2h19'4" 65d43'24"
Yvette
Sgr 19h0'4" -28d41'49"
Yvette C Kaufman
Cet 1h22'39" -0d38'18"
Yvette & Carley Casali
Mon 8h1'17" -0d47'42"
Yvette "Chiv"
Crb 15h50'41" 35d50'30"
Yvette Cohen
Aqr 23h2'29" -13d50'37"
Yvette Coppick
Aqr 22h22'41" -19d57'25"
Yvette Darcy
Cas 23h13'5" 59d35'40"
Yvette De Soignie
And 23h35'58" 37d37'36"
Yvette "Enana" Vazquez
Vir 13h28'8" 4d47'46"
Yvette Hernandez
Vir 12h42'49" 3d16'36"
Yvette "Lady Y" White
Cap 21h21'18" -26d40'46"
Yvette Lopez-Lagunas
And 0h19'0" 39d23'2"
Yvette Michele Sanders-Walton
Vul 20h24'1" 27d45'17"
Yvette Morin
Uma 9h22'5" 45d41'8"
Yvette Orjuela
Sco 17h58'35" -38d52'56"
Yvette Salinas
Sco 16h35'29" -26d38'44"
Yvette & Tom Reed
Uma 8h16'29" 65d48'2"
Yvette Torres
Ari 2h42'5" 27d57'52"
Yvette W. Johnson *Sungoddess*
And 2h7'52" 42d53'52"
Yvette Wang-Allen
Gem 7h14'53" 17d12'23"
Yvette Y. Edmond
And 0h30'22" 29d22'8"
Yvette Yvonne Yates
Cap 20h55'12" -22d4'28"
YvetteRalphy
Psc 0h53'20" 33d1'13"
Yvi...für immer!
Umi 14h6'56" 74d37'18"
Yvin
Aur 6h13'31" 31d2'6"
Yvon Cayer
Her 15h53'48" 40d15'14"
Yvon of the Yukon
Her 17h12'31" 27d12'48"
Yvonne
Ori 6h19'42" 10d56'16"
Yvonne
Tau 4h20'18" 17d59'8"
Yvonne
And 23h21'19" 38d44'30"
Yvonne
And 23h40'0" 41d52'44"
Yvonne
Uma 11h5'49" 49d52'2"
Yvonne
Lac 22h3'49" 45d50'45"
Yvonne
Cas 1h1'10" 56d30'24"
Yvonne
Crb 15h52'26" 34d50'45"
Yvonne
Uma 11h22'21" 39d55'9"
Yvonne
And 2h14'58" 39d17'42"
Yvonne
Uma 10h27'58" 69d34'40"
Yvonne
Lyn 6h19'28" 60d0'4"
Yvonne
Cas 1h1'50" 63d7'45"
Yvonne
Uma 8h53'8" 59d14'57"
Yvonne
Cap 20h38'2" -21d55'49"
Yvonne
Umi 16h52'52" 76d11'28"
Yvonne
Vir 12h37'23" -1d9'3"
Yvonne 30
Ari 2h11'55" 12d47'7"
Yvonne 619
Dra 19h6'46" 51d28'51"
Yvonne Alford
And 0h57'28" 35d3'25"
Yvonne Ann Durant
And 0h25'35" 43d34'49"
Yvonne Annette
Sco 17h57'58" -31d26'40"
Yvonne Arnold
Aur 5h13'19" 43d6'51"
Yvonne Bailey - Mummy's Star
And 23h10'6" 43d3'47"
Yvonne Beatrice Brushey
Cas 23h6'57" 58d36'27"
Yvonne Beeckmans Johnson
Lib 15h3'40" -2d11'24"
Yvonne Byrom
Crb 15h29'17" 28d33'49"
Yvonne & Carsten
Cyg 20h30'29" 44d28'6"
Yvonne & Charles Murray-40 Years!!
Cyg 20h52'35" 46d21'26"
Yvonne Collett - we love you pcrjlbt
Cas 0h45'38" 60d19'1"
Yvonne Collins
And 23h18'34" 42d23'45"
Yvonne Courtney
Cas 0h5'28" 55d12'58"
Yvonne Crenshaw
Sgr 19h36'23" -13d57'41"
Yvonne Crispino
Psc 1h19'50" 11d59'13"
Yvonne Dahint
Lmi 9h34'58" 35d19'35"
Yvonne deLima 13.8.1942
Ori 5h23'32" -4d5'16"
Yvonne (Destined to)
Lib 14h52'29" -1d8'18"
Yvonne Doris Banholzer/Meitä
Her 17h49'34" 37d56'31"
Yvonne Duddie
Cnc 8h17'1" 10d49'35"
Yvonne E. Mackie (Beebee)
Uma 13h29'29" 55d11'21"
Yvonne Eaton
And 0h43'56" 39d14'20"
Yvonne Elizabeth Lefebvre
Tau 4h14'23" 28d45'29"
Yvonne Elvira Salas
Lyn 8h5'38" 59d24'2"
Yvonne Evelyn Even
Psc 1h24'9" 17d45'13"
Yvonne Frye
Uma 14h7'3" 56d24'2"
Yvonne Grace & Chris Hahn's Love
Sco 16h46'3" -38d13'41"
Yvonne Hallsworth
Cas 2h31'54" 67d28'26"
Yvonne Jagiello Murray
Cyg 21h53'27" 43d36'24"
Yvonne K. Merz
Aqr 22h36'57" -6d58'48"
Yvonne Karrer
Uma 11h13'45" 67d34'43"
Yvonne Kavanaugh McGookin
Uma 9h32'54" 68d6'56"
Yvonne KM Jenkins
Aql 19h47'1" -0d0'14"
Yvonne Lash
And 23h20'43" 42d17'47"
Yvonne Lowenthal
Peg 21h18'28" 18d11'47"
Yvonne Lujan Rebsch
Mon 6h37'39" 0d15'34"
Yvonne M. Moore
And 0h45'0" 36d37'27"
Yvonne M. & Walter E. Dedrick
Cnc 8h26'45" 16d10'13"
Yvonne Maiorano
Gem 6h57'50" 13d24'1"
Yvonne Manser
Ori 6h15'52" 21d3'15"
Yvonne Marie Teriet
Ari 2h11'52" 11d40'22"
Yvonne Memeger-Caesar
Uma 8h26'35" 62d38'5"
Yvonne Merzlak
Cyg 21h54'14" 52d34'57"
Yvonne Morris
Aqr 23h28'24" -12d37'37"
Yvonne Morrison's Star
Lyr 18h30'51" 39d59'2"
Yvonne my Angel
Uma 9h47'45" 42d20'34"
Yvonne Nevels "Heart, Body & Soul"
Crb 15h29'48" 27d44'38"
Yvonne Nevlin
Cas 23h29'34" 51d58'50"
Yvonne (Nonne) Tucker
Cas 1h27'29" 62d39'6"
Yvonne O'Connell's Star
Dra 18h20'29" 70d32'26"
Yvonne Odette
Lyn 7h35'47" 41d36'34"
Yvonne Pida
Ori 6h20'12" 8d59'7"
Yvonne Poon
And 2h17'31" 46d55'46"
Yvonne Quinn
Cas 3h24'0" 70d20'33"
Yvonne R. Carder Brandenburg
Cas 2h6'8" 59d17'51"
Yvonne R. Hopkins
Mon 7h1'1" -0d48'48"
Yvonne R Peaslee
And 23h22'57" 48d37'53"
Yvonne Raelynn
Psc 1h40'28" 7d24'49"
Yvonne Ramirez
Cas 0h0'59" 56d28'39"
Yvonne Reinette Cleverly-Campbell
Leo 9h26'33" 15d8'47"
Yvonne Richards
Uma 14h0'6" 53d1'47"
Yvonne Rosado Curity
Ori 6h0'44" 0d53'7"
Yvonne Rose Banks
Psc 1h7'43" 26d3'55"
Yvonne Ross
Uma 13h50'34" 53d3'18"
Yvonne S Cabral
Sco 16h4'25" -11d52'45"
Yvonne Schneider
Vir 11h42'24" 3d31'21"
Yvonne Solis
Cas 23h43'49" 58d0'3"
Yvonne Suzette Fulmer
Cas 23h35'7" 52d43'41"
Yvonne White
Lib 15h34'21" -25d27'22"
Yvonne & Wolfgang
Cyg 19h58'19" 33d25'37"
Yvonne's Lasting Star
Ori 6h22'34" 14d34'50"
Yvonne's Star
Ari 2h40'59" 21d1'29"
Yvonne's Star
Mon 8h1'26" -0d24'28"
Yvy Schätzibönzgerli
Her 18h35'14" 19d6'32"
Y.W.A.M.T.W.T.S
Uma 11h19'52" 54d37'50"
Y.Y.S. Astra Bellus
Sco 17h27'40" -43d13'20"
Yzusdnacire
Psc 22h59'58" 7d10'15"
Z
Leo 11h26'22" 0d14'50"
004 Andrachelle Straka
Cnc 9h8'28" 11d35'12"
01
Umi 17h6'53" 80d33'46"
010203
Cru 12h29'22" -61d12'34"
01191991 van Amen
Her 18h6'42" 26d15'47"
020727 Ivan, The Gentle
Aqr 22h43'3" -2d31'46"
02121985 Toni Lauren 21
Col 5h35'39" -34d19'35"
0255 Akkp
Tau 5h41'15" 21d25'19"
04 / 05 varsity basketball
Uma 11h32'25" 57d56'59"
040836 Helen Haruko Yamamoto-Tanaka
And 0h33'51" 38d28'39"
05.09.1955 - Geoffrey Robert Best
Cru 12h41'15" -55d41'24"
058429 Geoffrey Phillip Roberts
Psa 21h35'44" -35d45'18"
06 M-L-G-Z
Ori 6h1'57" 20d5'26"
06071977Polly06072001
Srp 18h27'38" -0d19'26"
06-18-66 cjs 02-02-04 cjn love
And 1h9'6" 45d37'53"
0649 & 25
Vir 13h36'53" 1d11'17"
07-07-07 Wedding-Star
Uma 11h22'6" 39d19'19"
072002-MBE
Leo 11h0'31" 1d59'55"
080704
Cap 20h22'29" -26d35'38"
0 Always
Cyg 20h54'38" 46d45'57"
z' Hundschi vam Bärli
And 22h59'53" 40d13'48"
Z i l d a M i c h e l
Uma 9h17'34" 46d45'18"
Zabe
Ari 3h24'41" 26d51'33"
Zabeen "Sajni" Shah
Leo 11h4'0" 12d13'15"
Zabel, Rolf
Uma 9h37'56" 68d14'46"
Zabflo
And 22h58'50" 40d19'10"
Zabia Racine
Psc 0h39'50" 14d54'55"
Zabidin
Uma 12h45'26" 54d5'40"
Zable Bunny McClurg
Lep 5h24'36" -11d49'39"
Zabrina Gravelle
Uma 13h16'11" 58d30'24"
Zabrinna's Star
And 0h42'35" 22d3'17"
Zac Anthony Malanga
Aqr 22h54'35" -8d24'4"
Zac Copp
Srp 18h23'51" -0d21'7"
Zac Corbin Olsen
Ori 5h11'58" -7d1'23"
Zac Dogg 93
Uma 11h24'26" 42d0'34"
Zac & Ethan Perks 'Forever Shining'
Col 5h13'34" -35d34'57"
Zac Francis Long - 05/07/2006
Ori 5h38'23" -2d43'0"
Zac M. Huffman
Cnc 8h22'44" 10d30'1"
Zac Michael Pietersen-Smith
Cra 18h5'2" -39d6'1"
Zac Miller and Arin Jerauld
Cyg 19h45'27" 42d42'52"
Zac Murtagh Flood
Umi 15h29'50" 70d2'42"
Zac the Great One
Sco 16h16'58" -15d24'40"
Zac Tropf
Ori 5h24'11" 2d58'24"
Zac Trumble
Uma 9h34'2" 46d35'6"
Zac Zomaya
Psc 1h17'4" 32d28'44"
Zacarie James Gonzalez
Sgr 19h9'11" -21d38'0"
Zacchia
Uma 11h11'7" 33d4'17"
Zac-E.Alt
Pho 0h55'53" -53d26'2"
Zach
Mon 7h21'16" -0d36'31"
Zach
Her 16h36'59" 44d9'1"
Zach
Vul 19h30'9" 23d16'58"
Zach and Allie's Forever Star
Uma 11h55'45" 34d8'51"
Zach and Bea
Uma 9h39'45" 47d16'4"
ZACH AND DYLAN
Uma 11h10'32" 50d25'49"
Zach and Eva
Her 16h17'36" 7d11'24"
Zach and Maritza
Cyg 21h51'40" 37d50'51"
Zach and Teenie's 1st Anniversary
Cyg 19h29'46" 51d35'56"
Zach BB Gun Boy Hengel
Tau 4h3'23" 27d48'42"
Zach David Varnam - "Zach's Star"
Per 2h51'22" 42d0'17"
Zach Edward Thomas Oliver
Umi 13h59'32" 72d44'16"
Zach Feldman
Aql 19h25'30" 8d2'55"
Zach Fell
Lyr 18h36'25" 30d47'4"
Zach Glickman
Aur 5h35'27" 33d28'12"
Zach Jasinski
Cap 21h6'41" -25d44'29"
Zach Joseph Martinelli
Sco 16h45'20" -24d54'26"
Zach Kopsell
Vir 13h22'53" 3d56'32"
Zach & Laura's Love Forever Shines
Cyg 19h41'5" 39d16'26"
Zach & Nicole 4-15-05
Cyg 20h16'51" 54d31'9"
Zach & Shannon
Sge 19h33'45" 18d50'9"
zach + simona = better together
Cyg 21h47'31" 45d37'57"
Zach Wingert
Aql 19h52'10" -0d51'40"
Zacharia David Schueler
Cma 7h8'16" -26d48'27"
Zacharia Grant Stover
Aqr 22h9'15" -15d38'36"
Zachariah Daniel Robert Marsh
Cep 3h50'40" 81d54'11"
Zachariah Henry
Gem 6h34'47" 20d19'3"
ZACHARIAH LIVING-STONE
Tau 5h52'29" 13d16'40"
Zachariah Marinos
Cnc 8h39'41" 10d5'29"
Zachariah Onslow Nydes
Boo 15h30'57" 44d57'20"
Zacharia-Paige
Sge 19h26'52" 17d26'47"
Zacharias Gehl Plate
Uma 10h40'9" 57d2'13"
Zacharius
Cnc 8h43'57" 16d13'6"
Zach-a-roni
Sco 16h11'36" -15d1'35"
Zacharoo
Sgr 19h38'18" -16d30'56"
Zachary
Lyn 7h43'33" 52d50'59"
Zachary
Her 18h33'30" 15d39'9"
Zachary
Tau 4h53'0" 26d12'5"
Zachary
Her 17h37'28" 28d16'19"
Zachary
Her 16h41'22" 47d50'39"
Zachary
Uma 10h16'53" 48d43'47"
Zachary A. Mulkins
Sco 16h11'30" -9d42'22"
Zachary A. Schmidt
Ori 5h31'11" 2d27'1"
Zachary Aaron
Ori 6h7'41" 12d54'41"
Zachary Aaron
Her 16h14'30" 45d8'0"
Zachary Aaron
Uma 11h46'3" 35d13'49"
Zachary Adam Glorioso
Psc 1h32'9" 18d18'47"
Zachary Albanese
Lyn 7h42'15" 36d6'33"
Zachary Alexander
Per 3h18'6" 38d5'45"
Zachary Alexander Bossie
Uma 11h56'28" 59d41'0"
Zachary Allan
Ori 6h1'51" 9d49'8"
Zachary Allen Coates
Per 3h50'5" 43d3'41"
Zachary Allen Collins
Her 16h34'51" 13d37'9"
Zachary Allen Groen
Aur 5h41'5" 49d53'37"
Zachary Allen Konik
Her 16h50'37" 14d28'0"
Zachary Allen Langster
Lyn 9h18'45" 38d38'10"
Zachary Allen Sauder
Umi 16h16'57" 76d5'36"
Zachary and Ami
Mon 6h52'13" -0d39'24"
Zachary and Elizabeth's Star
Cyg 19h39'21" 31d46'0"
Zachary and Hannah Together Forever
Col 5h48'1" -29d29'6"
Zachary Andre Zwierko
Cap 20h31'50" -23d47'46"
Zachary Andrew Castellarin
Tau 4h12'49" 8d7'10"
Zachary Andrew Rowland
Gem 7h23'52" 26d21'24"
Zachary Arthur Welch
Umi 16h22'1" 77d45'20"
Zachary Aston Philips
Umi 14h10'27" 66d15'36"
Zachary Atcheson
Cyg 20h21'46" 39d40'2"
Zachary Austin Clardy
Aqr 22h15'0" -16d53'18"
Zachary Ayao Endow
Ori 4h48'10" 13d4'35"
Zachary Barry
Cep 20h26'17" 59d58'20"
Zachary Bastian
Aur 5h47'15" 39d48'37"
Zachary Baxter White
Cnc 8h21'38" 25d35'16"
Zachary Benjaman Thomas
Her 17h6'53" 30d41'0"
Zachary Benjamin Avery
Gem 7h47'15" 27d33'12"
Zachary Benjamin Stein
Uma 11h17'22" 53d12'18"
Zachary Blake Tucker
Ori 6h22'16" 16d26'8"
Zachary Blake Turcotte
Psc 0h44'2" 15d52'34"
Zachary Braff
Ari 2h1'56" 21d15'45"
Zachary Brent Malott
Leo 9h52'9" 17d25'33"
Zachary Brian 13
Sgr 20h3'22" -40d52'28"
Zachary Brody
Cnc 9h15'45" 25d57'17"
Zachary Bruce Cole
Uma 8h30'14" 62d29'22"
Zachary Bryan Gage
Psc 1h11'41" 14d32'16"
Zachary Carl Tapper
Lyr 19h21'40" 38d1'18"
Zachary Carl Tapper
Uma 8h20'18" 61d39'45"
Zachary Charles
Ori 5h33'51" 3d8'13"
Zachary Chasteen
Ori 5h19'46" 3d12'23"
Zachary Christopher Thompson
Uma 11h24'14" 62d9'38"
Zachary Clarence Spiegel
Uma 14h5'52" 61d34'49"
Zachary Cole Ghannam
Aqr 22h4'50" -2d35'44"
Zachary Connor Porter
Leo 9h29'27" 27d40'54"
Zachary Constantine
Dra 18h22'31" 67d31'46"

Zachary Cox
Ori 5h45'45" 3d16'56"

Zachary D. Romeika
Ori 5h56'20" 17d38'9"

Zachary Daniel
Umi 15h0'16" 72d49'32"

Zachary Daniel Ferguson
Uma 10h9'3" 51d37'46"

Zachary Daniel Romeika
Cap 20h29'35" -10d1'6"

Zachary Daniel Spooner
Vir 14h43'35" -4d20'26"

Zachary Daniel-Louis Nelson
Tau 5h13'39" 26d45'7"

Zachary Darosa Dossantos
Boo 15h38'42" 49d6'27"

Zachary David Russo
Her 17h48'3" 46d58'7"

Zachary David Swennes
Uma 9h40'53" 71d41'40"

Zachary David Weltz
Sco 16h5'58" -16d59'50"

Zachary Davis
Uma 11h52'22" 59d3'45"

Zachary Dean Dufort
Uma 13h22'34" 56d44'42"

Zachary DeVouge
Ori 5h58'44" 18d18'8"

Zachary Dylan Auville-Parks
Uma 10h58'4" 52d30'46"

Zachary Edward Marquardt
Leo 11h30'25" 8d29'39"

Zachary Edward Neal
Sgr 17h56'26" -29d15'49"

Zachary Edwards
Lmi 10h16'58" 31d47'22"

Zachary Eric Burdick
Uma 10h30'32" 49d1'24"

Zachary Ess Major
Uma 11h37'27" 63d52'32"

Zachary Ethan Barnes
Her 17h49'37" 37d15'8"

Zachary Evan Noel
Leo 11h54'8" 24d27'8"

Zachary Fausone
Cmi 7h31'34" 7d51'55"

Zachary Fec
Uma 12h13'33" 57d29'47"

Zachary Flowers
Aur 5h52'10" 47d4'45"

Zachary Francis Rusnock
Sgr 18h9'47" -27d53'30"

Zachary Glenn Clemons
Sgr 18h11'43" -18d32'30"

Zachary Glosser
Uma 11h43'47" 37d17'19"

Zachary Grayson Walters
Lmi 10h44'3" 33d8'19"

Zachary Gregory
Cep 22h32'52" 77d28'51"

Zachary Guy
Psc 23h0'39" 5d45'41"

Zachary Harrison
Aqr 22h31'5" -0d36'9"

Zachary Henry Semmens
Per 3h13'37" 43d59'59"

Zachary Hodge
Gem 7h44'45" 24d27'11"

Zachary Holbrook Spruill
Lib 15h22'27" -7d9'19"

Zachary Ian
Uma 11h51'26" 49d56'36"

Zachary Ian Browne
Aqr 20h51'16" 0d49'47"

Zachary Ian Ponce
Per 3h23'40" 33d26'24"

Zachary J. Ramsey ZJR-Reborn
Ori 5h42'56" 8d23'54"

Zachary & Jacqueline Snyder Twins
Tau 4h3'11" 26d33'33"

Zachary James Albert Aichele
Uma 9h33'22" 69d48'28"

Zachary James Belan
Lyn 8h4'38" 34d35'42"

Zachary James Harris
Lib 14h51'1" -2d19'42"

Zachary James Healy
Psc 1h58'35" 3d15'31"

Zachary James Iozzi
Aqr 22h30'32" 2d8'5"

Zachary James Lamprecht
Ori 4h49'32" 10d59'15"

Zachary James Rivera
Cnc 8h3'20" 19d36'1"

Zachary James Shipley
Aqr 21h43'44" -1d53'55"

Zachary James Wadas
Vir 13h7'20" 5d26'30"

Zachary James Weidler
Dra 17h31'45" 56d10'23"

Zachary Jason
Lmi 10h37'53" 35d57'21"

Zachary Jason Voikos
Uma 8h47'9" 53d14'52"

Zachary Jay
Cru 12h35'50" -57d52'19"

Zachary Jefferson McCombs
Her 17h21'49" 25d41'52"

Zachary Joesph Rheault
Leo 10h25'16" 10d52'51"

Zachary John Beshay
Vir 12h53'43" 7d6'6"

Zachary John Clark
Boo 14h36'23" 32d1'52"

Zachary John Franklin Moore
Sco 17h53'5" -41d32'40"

Zachary John Mello
Tau 5h33'32" 17d18'28"

Zachary John Richard Smith
Vir 12h39'53" 7d42'45"

Zachary John Walinder Ryan
Cnc 8h43'7" 22d50'36"

Zachary Jordan
Lib 15h6'32" -26d30'28"

Zachary Joseph Bauer
Sco 16h6'31" -17d4'36"

Zachary Joseph Blaine Wright
Vir 14h5'12" -1d3'59"

Zachary Joseph Daniel Novak
Lyn 7h40'38" 43d59'31"

Zachary Joseph Decker
Aqr 23h15'32" -8d14'20"

Zachary Joseph Escott
Cnc 8h16'22" 15d35'43"

Zachary Joseph Hicks
Her 16h24'16" 44d28'7"

Zachary Joseph Metzger
Ari 2h20'40" 26d37'48"

Zachary Joseph Mullins
Leo 10h44'46" 17d58'14"

Zachary Joshua Joly
Umi 15h30'17" 74d18'12"

Zachary Kent Feuerbach
Her 18h25'47" 12d17'41"

Zachary Kevin Saad-Love You Forever
Leo 10h10'18" 26d26'44"

Zachary Klausner
Aql 19h31'40" 3d18'41"

Zachary Lan Schaffer
Cap 20h26'43" -14d50'52"

Zachary Lee Dobbin
Uma 11h6'11" 69d16'12"

Zachary Lee Garner
Her 17h43'48" 36d40'26"

Zachary Lieberman
Ari 2h28'58" 26d25'9"

Zachary Lohrey
Uma 10h21'15" 43d19'54"

Zachary Lucas
Uma 10h32'10" 43d35'36"

Zachary Martin
Cap 20h56'28" -19d14'22"

Zachary (Master Zach) Morgan
Ari 2h49'22" 29d4'44"

Zachary Mathias Harvanek
Ori 6h18'56" 14d18'11"

Zachary McFadden
Sgr 18h57'5" -24d23'34"

Zachary Mead
Umi 15h40'20" 83d56'55"

Zachary Mealman
Ori 5h54'38" 18d24'48"

Zachary Mellinger
Per 2h15'34" 56d54'37"

Zachary Michael
Per 2h15'53" 55d58'47"

Zachary Michael Alvarez
Cnc 8h26'3" 26d44'9"

Zachary Michael Kaster
Dra 18h17'50" 58d0'59"

Zachary Michael Trissel
Lib 15h47'5" -9d8'7"

Zachary Moore
Boo 14h39'1" 31d50'47"

Zachary Morgan
Dra 10h7'6" 77d59'38"

Zachary "My Chip"
Lib 15h23'29" -8d2'6"

Zachary Nash
Vir 12h36'35" 7d52'30"

Zachary Nathan Gold
Uma 13h53'46" 52d10'33"

Zachary Nathan Prechtel
Peg 22h58'55" 9d53'40"

Zachary Noel Ramos
Sgr 17h52'11" -17d55'52"

Zachary & Norma Hosler
Nor 16h3'28" -43d40'12"

Zachary Paul Wood
Lib 15h4'36" -15d44'58"

Zachary Perkitny
Uma 12h32'33" 62d36'33"

Zachary Perry
Cep 22h42'29" 67d12'38"

Zachary Peter Perkins
And 23h23'51" 52d6'10"

Zachary Philip Brown
Her 16h53'7" 24d46'47"

Zachary Portnoy
Her 17h34'18" 30d11'53"

Zachary Prizant
Sco 16h5'0" -13d27'42"

Zachary R. Kennedy
Uma 10h36'21" 58d37'13"

Zachary R. Purnell
Uma 10h49'13" 46d25'36"

Zachary Rankin
Per 3h54'33" 36d52'13"

Zachary Raymond Hadley
Ori 5h25'58" 14d38'2"

Zachary Reaugh
Per 3h37'9" 40d48'42"

Zachary Richard Savage Hertel
Cep 0h2'16" 76d2'38"

Zachary Robert Kennedy
Psc 0h26'20" -4d29'39"

Zachary Robert Wilson
Dra 17h36'23" 64d45'23"

Zachary Ross Jones
Uma 8h48'47" 67d10'8"

Zachary Roy
Per 3h16'25" 54d9'42"

Zachary Ryan
Gem 7h35'12" 14d12'4"

Zachary Ryan
Uma 11h11'4" 56d9'25"

Zachary Ryan Bryson
Umi 15h36'13" 75d10'29"

Zachary Ryan Cleveland
Dra 9h47'58" 77d55'57"

Zachary Ryan Copeland
Umi 15h17'44" 70d20'3"

Zachary Ryan Germana
Her 16h37'36" 27d40'32"

Zachary Ryan Jassenoff
Ori 5h25'37" 9d57'44"

Zachary Ryan Rich
Tau 3h39'59" 27d48'31"

Zachary Ryan Rockefeller
Ari 2h17'24" 23d23'10"

Zachary Ryan Schoem
Ori 5h55'40" 21d22'10"

Zachary Ryan Tipton
Cnc 8h5'37" 7d23'43"

Zachary Ryan Williams
Uma 9h1'33" 58d38'45"

Zachary Ryan Williams
Ori 6h10'56" -1d34'56"

Zachary S. Woodward
Umi 15h43'23" 71d14'19"

Zachary Schalberg
Lib 15h54'5" -6d4'22"

Zachary Schwarz
Per 3h15'11" 51d52'23"

Zachary Scott Buda
Uma 10h27'21" 51d34'40"

Zachary Scott Farrell
Dra 19h38'49" 68d35'14"

Zachary Scott Goodfellow
Ori 5h21'15" -3d55'58"

Zachary Seltzer
Cyg 20h15'16" 37d23'52"

Zachary Seth Wooten
Cap 20h33'18" -15d46'29"

Zachary Smith
Uma 10h16'54" 44d39'7"

Zachary S.P. Kelley
Ari 3h20'49" 25d18'46"

Zachary Spencer
Aqr 21h44'27" -1d10'6"

Zachary Star
Sco 16h11'30" -10d9'51"

Zachary Steele
Uma 11h45'13" 53d24'21"

Zachary Sterner
Psc 23h48'36" 1d10'0"

Zachary Steven Doak
Per 2h47'49" 53d24'53"

Zachary Steven Meckstroth
Gem 7h9'40" 17d39'48"

Zachary Stone
Uma 10h10'29" 59d16'27"

Zachary Stone Buchanan
Her 17h25'16" 23d57'46"

Zachary Swingle
Leo 11h6'33" 24d32'54"

Zachary T Ebright
Cap 21h38'9" -11d38'24"

Zachary T. Parker
Aur 5h24'21" 43d18'55"

Zachary Taylor Caldwell
Uma 10h14'59" 44d3'54"

Zachary Taylor Kerstetter
Her 17h38'39" 32d51'53"

Zachary Telljohann
Aur 5h38'33" 49d1'9"

Zachary Thade McLaren
Vir 13h5'29" -6d56'39"

Zachary Thomas Beadling
Ori 5h59'55" 19d32'3"

ZACHARY THOMAS KLOHE
Cap 21h42'29" -14d44'48"

Zachary Thomas Warner
Cma 6h43'35" -28d3'29"

Zachary Timothy
Eri 3h55'48" -39d2'54"

Zachary Toskovich
Uma 11h44'49" 59d59'17"

Zachary Tyler Hubbard
Ori 6h3'34" 17d54'0"

Zachary Tyler Marvel
Aqr 20h51'32" 2d25'48"

Zachary Tyler Tebbetts
Her 17h54'12" 18d56'13"

Zachary Vigliotta
Cnv 13h11'11" 48d50'49"

Zachary Vincent
And 23h32'16" 47d47'31"

Zachary Vincent
Per 3h24'45" 45d38'12"

Zachary W. Roth
Ori 5h34'12" -4d35'44"

Zachary Wade Brown
Ori 6h2'56" 14d8'5"

Zachary Wadelon Ford Z-Bo
Uma 11h10'3" 51d53'21"

Zachary Warrin
Ori 4h53'54" 10d46'7"

Zachary William Harris
Umi 14h40'19" 73d49'32"

Zachary William Haug
Psc 0h19'7" 15d26'6"

Zachary Z. Topper
Lib 14h52'26" -2d0'36"

Zachary Zogran
Boo 14h41'2" 24d37'0"

Zachary "Zulu" Andrew Johnson
Ari 3h5'27" 13d9'59"

Zachary Zumsteg
Dra 18h34'27" 60d45'13"

Zachary's Star
Leo 11h1'37" 0d6'14"

Zachary's Star
Her 17h25'51" 44d28'52"

Zach-a-Zooby
Lyn 9h1'59" 37d44'0"

Zachery
Umi 14h34'17" 76d13'13"

Zachery Alan Miles
Ori 6h0'56" 21d12'17"

Zachery Edward Davies
Dra 16h16'35" 53d52'30"

Zachery Feldbauer
Umi 14h34'14" 74d54'15"

Zachery Isaac Allen
Cep 21h8'36" 65d14'43"

Zachery Lee Westerhoff
Aqr 20h54'26" -8d22'43"

Zachery (Mountain Man) Forman
Per 3h18'4" 51d11'32"

Zachery Smith's Papa Bear
Uma 11h12'59" 34d26'8"

Zachery T. Hillman
Lib 14h37'38" -15d52'53"

Zachery Walter Gay
Aqr 23h27'57" -11d2'15"

Zachory "Brave Soldier" Carter
Aql 19h42'42" 7d27'4"

Zachry
Ori 5h45'38" 5d45'10"

Zach's Dream
Uma 9h59'3" 41d30'10"

Zach's Karate Star
Uma 10h28'42" 62d22'26"

ZachsPlace
Uma 8h15'11" 67d37'14"

ZaciStace 91604
Uma 11h56'40" 37d21'1"

Zack
Tau 4h20'19" 28d12'14"

Zack & Alicia Bruscella
Cnc 9h7'12" 27d58'56"

Zack and Michelle's Shining Star
Tau 3h42'34" 27d39'31"

Zack and Sarah Wedding Anniversary
Cyg 19h47'7" 47d36'58"

Zack Attack
Psc 1h14'56" 23d17'10"

Zack Christopher
Psc 0h42'46" 14d44'15"

Zack Jared Lacey Richardson
And 0h16'4" 25d9'43"

Zack Lynch
Cma 7h19'1" -30d26'17"

Zack "Moose" Triplett
Umi 13h29'10" 72d2'47"

Zack Phillips
Vir 14h19'28" 1d8'14"

Zack (Robert) Lynn
Vir 14h15'23" 2d27'6"

Zack Ryan Smith
Ari 2h2'54" 14d48'14"

Zack+Sarah=Love
Cyg 21h43'54" 32d10'8"

Zack the God
Lib 15h30'58" -15d2'14"

Zack Toskovich
Uma 11h39'50" 38d0'57"

Zack W. Watson
Ori 5h23'52" -0d48'19"

Zack Westermann
Per 3h43'20" 42d9'9"

Zack Wilson
Psc 0h29'14" 9d11'52"

Zack Wolfrey Creation
Psc 0h5'36" 2d41'3"

Zackary A. Hixon
Ori 6h11'51" 16d3'55"

Zackary Allen
Sco 16h50'42" -27d16'4"

Zackary Couch
Uma 11h8'42" 49d42'20"

Zackary David Twardowski
Uma 8h58'2" 57d34'10"

Zackary Drucker
Tau 4h43'10" 19d0'40"

Zackary James O'Canna
Uma 9h58'19" 60d10'57"

Zackary John Boucher
Per 3h14'11" 54d26'32"

Zackary John Boucher "1994-2004"
Aqr 22h50'34" -19d22'50"

Zackary Laven
Sgr 18h41'41" -34d41'8"

Zackary Lawrence Broadwater
Lib 14h40'53" -12d29'41"

Zackary Major
Uma 10h19'39" 47d49'45"

Zackary Michael
Aqr 22h16'24" 1d0'44"

Zackary Paul Montognese
Ari 2h8'39" 21d38'14"

Zackary Raymond Baldaramos
Vir 13h16'36" 8d16'3"

Zackary Ryan Metoff
Ori 6h2'7" 19d0'49"

Zackary Scott Hiza
Lyn 6h51'12" 53d5'42"

Zackeroo
Tau 5h15'57" 21d27'33"

Zackery Allan Woods
Ari 3h21'53" 19d10'33"

Zackery James Paul Morgan
Uma 10h6'5" 52d34'28"

Zackrey Larz
Ari 3h20'21" 28d4'15"

Zack's Star
Umi 16h5'52" 84d40'59"

zaclecia
Lep 5h42'32" -13d25'23"

Zac's Mum - Amanda J. Charlesworth
Sco 17h49'13" -34d49'28"

Zadie Elizabeth Stack
Gem 7h40'27" 27d33'58"

ZAFER ÖZDEMIR
Tri 2h36'5" 35d10'34"

Zah Zah
Vir 14h16'11" 4d33'57"

Zahava Tzini Ganchrow
Peg 22h43'21" 17d1'48"

Zahava Wein
Uma 12h14'21" 55d6'36"

Zaheer Ahmed Ellahi
Sgr 18h54'10" -16d31'33"

Zahira
And 1h14'12" 46d50'50"

Zahmel, Günter
Uma 9h44'16" 72d34'34"

Zahra
Gem 7h3'29" 17d14'40"

Zahra Ahmadi
Peg 22h38'8" 12d18'17"

Zahra Bahbahani
Per 3h28'37" 54d33'55"

Zahra Darvishi
Cnc 9h14'18" 10d40'22"

Zahra & Sameer's Eternal Star
Lib 14h50'41" -3d40'7"

Zahryn
Aql 19h50'28" -0d18'54"

Zaia Taylor
Ari 2h5'19" 19d8'26"

Zaid Ghasib
Sco 16h6'14" -12d38'38"

Zaida
Uma 9h1'14" 66d52'56"

Zaida
Sgr 17h54'52" -26d31'30"

Zaida Garcia
Umi 14h39'58" 70d9'14"

Zaida's Star
Com 13h9'3" 28d23'42"

Zaika
Cas 22h58'24" 56d57'48"

Zain
Aqr 23h41'48" -15d29'56"

Zain
Gem 7h6'34" 17d47'13"

Zain
Ari 2h36'42" 20d28'25"

Zain Ryland Cozens
Psc 1h27'25" 24d39'3"

Zain Taelor Fulmer
Cap 20h57'55" -21d55'17"

Zainab Hodroj
Lyn 7h21'13" 57d38'17"

Zainka
Leo 10h17'4" 17d0'3"

Zaira
Ori 6h6'53" 21d8'49"

Zaira
Aur 4h38'21" 30d27'23"

Zak
Ori 6h8'20" 7d26'52"

Zak
Dra 17h53'32" 60d51'2"

Zak Bailey
Per 3h3'28" 45d29'18"

Zak Beastall
Lmi 10h10'51" 27d57'19"

Zak den Hertog - 28.04.2004
Cru 12h54'26" -60d54'24"

Zak Stuart - 04/02/2004
Uma 9h30'31" 41d28'23"

Zak Tiffany
Leo 10h46'54" 19d53'43"

Zak Watts Gustavesen
Psc 1h13'15" 30d48'56"

Zakariah Blye
Uma 9h7'27" 69d55'4"

Zakary
Her 18h24'35" 28d38'10"

Zakary Julian Turner
Umi 16h0'16" 79d34'50"

Zakian's Destiny
Leo 11h24'11" 5d52'51"

Zakiya Kiara Roberts
Sco 16h10'2" -29d33'15"

Zakol
Lib 15h5'10" -4d59'2"

Zak's Star
Umi 15h8'58" 74d29'25"

ZAKSAM
Per 3h4'38" 42d22'18"

Zali Belle
Lyn 8h16'1" 39d50'9"

Zali Q Elias
Aqr 22h57'26" -9d39'5"

Zallares "47"
Tau 5h46'23" 22d52'44"

Zalman Ilishayev
Cnc 9h9'41" 31d17'33"

Zamandra MD
Mon 6h36'30" 10d53'56"

Zamara
Uma 11h53'42" 38d31'56"

Zamboni
Ori 5h45'18" 2d21'6"

Zamgwar
Lyr 18h52'4" 33d17'43"

Zamora's Union
Cyg 21h47'58" 45d46'46"

Zan
Dor 5h3'8" -65d55'50"

Zan Narayan
Cyg 19h16'34" 53d48'58"

Zana Goodhew
Uma 10h3'8" 42d23'18"

Zana June
Leo 11h38'19" 14d34'26"

Zanda Frances Lavilla
Leo 10h18'9" 16d8'41"

Zandboer
Lyn 7h23'54" 45d56'36"

Zander
Her 17h55'6" 49d22'59"

Zander
Tau 4h16'57" 17d53'6"

Zander Allen Matthews
Gem 7h39'36" 17d21'40"

Zander Joe McMinn
Peg 22h38'16" 5d1'53"

Zander Paul Fairweather
Her 18h9'11" 22d15'37"

Zander Vance Kubajak
Uma 10h58'49" 55d2'55"

Zander, Bernd
Ori 6h15'54" 10d33'58"

Zander's Candor
Cyg 20h41'52" 43d8'47"

Zander's Castle In The Sky
Boo 14h31'45" 51d29'5"

Zander's Star
Uma 8h13'2" 60d27'46"

Zandra
Cam 3h37'44" 64d53'54"

Zandria
Cma 6h37'53" -17d25'19"

Zane
Uma 11h35'59" 44d43'58"

Zane
Leo 11h15'33" 16d8'6"

Zane
Boo 13h47'4" 17d22'7"

Zane Aaron
Her 17h31'47" 32d44'51"

Zane Anthony Kenneth Samuels
Uma 9h10'59" 66d44'56"

Zane Anthony Zavindil
Sco 17h39'8" -35d51'19"

Zane Arthur Zeiner
Per 2h59'21" 46d21'50"

Zane Cash Kiser
Cap 21h17'59" -24d23'33"

Zane David
Boo 15h6'27" 35d7'23"

Zane Diaz
Cap 21h44'12" -13d7'16"

Zane Foice Harrel
Her 18h33'22" 21d2'48"

Zane Forrester Northrip
Uma 9h35'9" 57d3'32"

Zane Graywolf 6221995
Cnc 8h20'28" 29d36'26"

Zane Julian Orellana
Cru 12h41'22" -60d41'48"

Zane Reid Piper
Uma 9h9'39" 57d7'16"

Zane Robert Edgar
Cep 22h9'31" 61d4'15"

Zane Russell Treadway
Uma 10h57'37" 61d48'51"

Zane Schnee
Boo 14h34'52" 17d32'40"

Zane's Star
Ori 5h46'26" -4d5'18"

Zanetta Giuseppe Pumukli
Uma 9h47'13" 60d44'1"

Zanette Douglas
Lib 15h24'33" -7d7'2"

Zaney
Sgr 19h50'18" -13d2'42"

Zanibugll
Peg 23h16'56" 33d1'11"

Zanna
Tau 3h40'28" 22d29'7"

Zanna Gamboa
Umi 15h29'24" 67d53'33"

Zanna Karol Laud
Dra 17h28'30" 68d55'10"

Zanne
Umi 17h21'37" 76d42'0"

Zanne Wiskerson
Aqr 22h42'46" -3d22'34"

Zanni
And 1h57'43" 42d12'36"

Zante's proposal!
Cyg 19h37'15" 31d8'0"

Zanzibar
Her 18h40'48" 20d33'50"

Zanzibar
Cep 21h7'14" 67d44'55"

Zanzibar Castle Rock
Cyg 21h45'14" 44d54'20"

Zanzoun
Psc 1h22'22" 28d31'45"

Zanzoun
Uma 9h17'54" 61d28'15"

Zaphea Nasreen
Lib 14h35'57" -8d48'51"

Zapominaja' Ja Nie
Cru 12h55'52" -59d48'40"

Zappel
Umi 14h37'36" 70d24'55"

ZAR
Cam 7h29'1" 67d41'44"

Zar Wardak
Uma 11h1'53" 35d23'18"

Zara
And 0h21'31" 41d41'16"

Zara
Sge 20h16'19" 19d33'40"

Zara
Gem 6h38'33" 15d13'4"

Zara Abgaryan
Sco 16h50'1" -27d23'36"

Zara Abulova
Psc 0h6'14" 1d27'20"

Zara Bhardwaj
And 1h24'46" 41d37'24"

Zara Dee
Peg 21h33'20" 13d9'14"

Zara i Jakov 08.09.2007.
Cyg 19h44'31" 34d41'18"

Zara Layla Quintana
Vir 13h20'29" 13d11'18"

Zara Leah
And 23h1'57" 49d31'33"

Zara Michelle
Lyn 6h28'25" 60d49'48"

Zara & Philip Cairey
Cyg 20h18'3" 50d30'41"

Zara Salih
And 2h1'24" 47d21'10"

Zara Scarlett Verman
Dor 5h8'11" -66d38'2"

Zara Sofia
Peg 22h11'59" 7d11'31"

Zarah Louise
Cru 12h38'22" -60d21'40"

Zarah Star
Ari 2h28'2" 20d3'4"

Zarana B Patel
Psc 1h34'37" 21d24'50"

Zaranika
Aqr 22h15'22" -9d29'18"

Zarea
Sco 17h9'35" -33d31'39"

Zareen Ash Shaikh
Vir 12h48'31" 11d40'53"

ZarehP
Psc 1h46'51" 22d57'56"

ZARI
Cnc 8h37'51" 13d39'56"

Zariat8
Uma 11h43'51" 42d33'16"

Zarifah *Zanelle*
Sco 16h45'34" -32d0'6"

Zarifé
Ori 5h45'57" -2d25'9"

Zarin Noor
Lib 15h36'11" -16d55'37"

Zarina Valle
Mon 7h19'50" -3d49'44"

Zarymo
Tau 3h43'4" 29d4'58"

zasou-souris-petites chaussettes
Dra 17h26'6" 51d11'23"

ZATHRAS69
Lmi 10h26'47" 29d38'22"

Zato! ANA 24.07.1976.
Cyg 20h3'57" 47d19'46"

Zauber-Bär
Cyg 19h38'46" 32d20'56"

Zaubermaus
Ori 6h5'38" 13d38'38"

Zaubermaus Bianca
Cnc 8h10'8" 20d1'42"

zaubermaus verena
Per 2h22'50" 51d17'14"

Zavion Langrin's Shining Star!
Leo 10h19'59" 23d54'42"
Zawierucha-Zack
Ori 6h8'17" 7d53'36"
Zaya
Uma 13h20'45" 55d16'40"
Zayana Nevaeh Damyn
Cnc 8h25'9" 30d47'58"
Zayda
Tau 3h36'19" 16d42'22"
Zayed Bin Saif Al Nehyan
Uma 13h3'5" 59d38'10"
Zayka Marinka
Lyn 9h24'56" 40d10'36"
Zayna Nisreen Ben-Musa
Peg 0h4'41" 17d58'29"
Zaynah
Cas 0h31'54" 53d24'36"
Zaynah Kamran
Hya 9h22'53" -4d37'17"
Zayne Arthur Woods
Cap 20h49'45" -20d52'12"
Zayne DiPietro's Shining Star
Cru 12h35'33" -58d40'3"
Zayne McVey
Her 17h44'29" 37d39'45"
Zaynolaubedean Adnani
Lib 15h8'33" -7d33'16"
Zaypher
Ori 5h31'43" -5d33'27"
Zayra
Mon 6h45'17" -0d56'55"
Zaza
And 23h59'14" 39d49'47"
Zazie Sky
Aqr 22h42'2" -15d42'47"
Zazu
Cyg 20h28'8" 47d30'6"
Zazza
Ori 5h12'23" -6d40'26"
Zbigniew Gerlach
Her 17h19'33" 29d33'38"
Zbigniew Muziol
Ori 6h19'12" 10d53'54"
Zbysiu Koziol
Vir 12h8'30" 13d5'15"
Zdenek Kedroutek
Uma 11h56'37" 33d10'8"
Z.Dove
Sgr 18h3'29" -26d43'2"
Zdravka Kazanlieva
Lib 15h26'13" -9d51'4"
Zdravos
Dra 20h18'20" 68d42'52"
Zdzislaw
Umi 14h17'46" 76d57'48"
Zdzislaw A Nowinski I Love You Kris
Per 2h39'17" 54d34'22"
Zeal
Umi 14h38'57" 78d39'24"
Zeamon
Gem 6h36'36" 22d26'0"
Zeanon 12709 12
Aql 20h2'41" -8d51'55"
Zeb Franklin
Uma 10h36'54" 40d17'18"
Zebbie
Ari 1h54'27" 17d30'10"
Zebedee Judah Clay
Uma 10h6'27" 48d57'54"
Zebediah
Aql 19h45'11" -0d3'56"
Zebo
Per 3h39'16" 33d47'17"
Zebra
Tau 4h37'28" 18d0'55"
Zebra
Gem 7h16'59" 21d4'45"
Zebra
Sco 16h10'13" -11d31'41"
Zebra - Star 12.05.2001
Her 17h30'40" 32d52'12"
Zebrastern
Uma 9h44'58" 69d36'26"
Zebrella
Umi 13h11'10" 73d35'28"
Zebulon Thomas Bishopric Wolfe
Cnc 8h49'32" 13d13'42"
Zebulon Thomas Wolfe
Cnc 9h8'54" 28d5'40"
Zebulun Wyatt Thomas
Cnc 9h5'5" 22d36'48"
Zebuluv
Per 3h43'41" 47d44'41"
Zecca '77
Her 15h55'51" 44d44'19"
Zech, Stefan
Uma 8h55'23" 48d2'20"
Zecharia Benjamin Burdette
Leo 10h37'13" 15d1'3"
Zechariah Isaac Christopher Wright
Cap 21h39'26" -16d9'34"
Zechariah Isaac Christopher Wright
Cap 20h27'3" -15d2'13"
Zechariah Jordan Castner
Her 18h51'41" 24d11'41"
Zechariah Morrison Davidson
Boo 14h46'7" 20d43'55"
Zechariah Taylor Harris
Sgr 18h21'17" -32d17'49"
Zee
Ori 5h27'59" 7d28'32"
Zee Michael Dickey
Umi 16h22'17" 76d5'9"
Zeeboid
Ari 3h27'19" 27d16'35"
Zeedunn
Col 5h40'14" -36d46'45"
Zeek
Her 16h8'29" 45d19'30"
Zeek
Psc 0h56'1" 31d1'1"
Zeemasters
Gem 6h5'3" 24d35'0"
Zeena Cook
Lyn 8h21'19" 46d5'28"
Zeena I. Ghandour
Ori 5h56'49" 21d36'17"
Zeepa Charlie
And 23h39'26" 42d48'45"
Zeero Everlong
Aur 5h5'10" 47d1'18"
Zeeshan
Tau 5h28'7" 17d36'16"
Zeeshan Elahi
Aur 5h56'16" 53d55'46"
Zeff Hanower
Lup 15h20'3" -30d58'59"
Zeffa
Lyr 18h47'46" 43d43'42"
Zeffanie
Uma 10h55'59" 45d51'3"
Zeger Coelst
Per 4h18'26" 44d31'0"
Zehava HF
Ari 2h16'42" 26d4'48"
Zehna
Crb 15h46'45" 32d37'14"
Zehndi's Shining Star
Psc 1h15'53" 31d49'17"
Zehra "Tranmere Rovers" Dergin
Cas 23h7'45" 55d16'5"
Zeidler, Peter Georg
Uma 9h1'21" 52d17'1"
Zeikiel John
Cen 13h37'0" -42d48'56"
Zeina
Uma 13h30'7" 54d33'49"
Zeina Bulgasem
Uma 12h37'52" 60d21'24"
Zeina G. Bejjani
And 23h43'28" 39d12'59"
Zeina Ghorayed
And 1h53'6" 36d37'57"
Zeina & Rami
Mon 7h28'58" -3d41'15"
Zeisel, Alois
Uma 10h40'37" 44d56'46"
Zeisneiss
Vir 12h41'37" 3d40'5"
Zek Mann
Her 17h18'11" 27d11'24"
Zeke
Per 2h16'10" 56d53'25"
Zeke
Pup 8h7'20" -14d15'20"
Zeke Trout
Uma 10h18'16" 44d54'54"
ZEKER
Leo 9h32'56" 31d6'16"
Zekers
Umi 14h35'54" 73d14'18"
Zelah's Star
Gem 7h9'16" 15d46'15"
Zelda Diesta
Cnc 8h40'15" 8d3'22"
Zelda Mae Hammond
Sgr 18h15'5" -19d54'50"
Zelda Swan Sherwood
Gem 7h10'4" 33d6'5"
Zelenyánszki László
Sco 17h40'15" -32d20'57"
Zeleyna Santiago
Aqr 21h23'48" -9d56'4"
Zelina Machevskaya-Iskanderova
Cap 21h27'0" -26d56'50"
Zella
Ori 5h26'19" 0d31'2"
Zella Barnes
Uma 9h6'10" 56d49'57"
Zeller
Uma 9h0'7" 64d45'57"
Zellmer
Oph 17h3'34" -0d32'18"
Zellua
Cas 1h3'30" 60d30'0"
Zelly "n" Phoenix
Uma 10h12'42" 60d38'11"
Zelma Weinberg
Uma 9h33'8" 56d23'3"
Zelta and Lois
Lib 15h10'36" -16d20'27"
Zemfira Valerevna McAllen
Mon 7h36'41" -1d5'4"
ZemoF9
Cnc 8h39'45" 31d41'49"
Zemra
Sgr 18h12'41" -30d14'47"
Zen
Uma 8h53'43" 49d0'9"
Zen Dreams
Uma 11h50'18" 55d28'50"
Zena Johnson Green
And 0h15'33" 28d33'24"
Zena & Ray's Ruby Star
Cyg 19h34'23" 28d8'56"
Zenaida (Nedy) Dolchanczyk
Cap 20h8'37" -14d29'38"
Zencke, Barbara
Uma 8h57'39" 50d59'7"
Zendy
Ari 2h15'27" 11d48'19"
Zeng Zheng
Lyn 7h22'25" 54d53'25"
Zenia A New Beginning
Cnc 8h16'10" 21d8'47"
Zenika Ariadne Ramirez
Cap 21h28'20" -10d31'15"
Zenith-PL
Sgr 18h38'18" -24d24'3"
Zeno
Cas 1h16'56" 72d37'35"
Zeno Valentin
Umi 13h58'39" 71d17'0"
Zenon
Uma 10h15'25" 44d17'4"
Zenon Markides
Uma 9h29'46" 44d25'1"
Zenora
Cyg 19h48'58" 59d9'53"
Zenses, Eva-Maria
Uma 8h51'0" 52d3'21"
Zent, Georg
Uma 8h59'34" 55d47'10"
Zenzetra
Cnc 8h44'40" 31d55'5"
Zephan
And 1h43'25" 42d7'15"
Zephan Sam Lacey-Rousou
Cep 23h1'58" 70d20'3"
Zephram Alexander Dale Thomas
Aqr 22h8'19" -0d2'37"
Zephyr Krapfel
Leo 10h42'19" 6d24'9"
Zer Vang
Gem 6h51'25" 34d2'37"
Zerita Climaco
Peg 22h8'59" 6d46'14"
Zero in, says Mrs. von!
And 0h20'5" 29d9'5"
Zerona Presslie Westbrook Hartzog
Peg 22h9'59" 7d55'59"
Zeroni
Sco 16h52'41" -27d8'41"
Zeropey
Lyn 8h28'39" 45d41'58"
Zerostar28
Sco 17h50'6" -38d52'32"
Zerres, Hans-Willi
Uma 13h45'48" 56d35'36"
Zerviah Mabee Mariano 8.11.1920
Uma 9h21'26" 56d19'50"
Zessie
Peg 23h5'31" 25d24'18"
Zesty's Place
Per 2h24'36" 55d31'0"
Zetland Arms
Cet 3h0'12" 1d2'58"
ZETSUEI
Uma 9h39'3" 45d7'19"
Zettl, Brigitte
Uma 11h35'23" 51d7'53"
Zetty
Sgr 18h14'30" -20d6'14"
Zeughardt, Mike
Gem 7h39'12" 16d4'17"
Zeus
Tau 5h37'0" 25d20'41"
Zeus
Uma 11h14'47" 30d2'59"
Zeus
Uma 9h37'58" 56d20'8"
Zeus Hearsey
Cnc 8h8'26" 32d35'49"
Zeus Walsh
And 0h25'41" 33d12'11"
Zev
Cap 21h50'59" -11d28'45"
Zev 1825
Gem 7h29'51" 34d37'37"
Zev's Wolfstar 65
Uma 11h37'14" 61d21'42"
Zeydula K. Uzbekov
Sgr 19h36'7" -15d28'52"
ZEYL
And 23h16'32" 46d40'18"
Zeynep & Metin
Cnc 9h20'4" 23d51'46"
Zezé Jakurski
Cnc 9h20'10" 10d13'37"
Zhang Li
Uma 10h41'54" 43d13'2"
Zhang Yan
Uma 12h26'16" 57d25'43"
Zhang Yongjian
Lyn 8h3'38" 35d0'8"
Zhanna Pustilnik
Sgr 18h53'55" -31d58'0"
Zhannachka
Ari 3h7'39" 29d34'53"
Zhao Qun Jessica
Cap 20h30'8" -13d50'5"
Zhao Zheng Bo
Lyn 7h5'26" 44d49'45"
Zhen Xing
Sco 16h18'29" -26d4'22"
Zheng Hui
Uma 11h31'26" 64d49'4"
Zheng Xiao Chun & Lin Zhi Peng
Uma 13h38'50" 48d56'35"
Zhenja Harutyunyan
Cap 20h28'9" -26d39'12"
Zhixin Zhao
Gem 7h33'46" 23d49'13"
Zhon & Marissa Burns
Lyr 19h19'3" 29d42'55"
Zhong Lei & Shen Peigao
Ari 2h58'6" 21d44'20"
Zhor
Cap 20h56'54" -23d9'6"
Zhou Jun
Cep 22h26'29" 81d10'19"
Zhou Qi
Cnc 9h1'42" 14d7'36"
Zhou Wei
Aqr 21h38'50" -1d10'34"
Zhou Yan-Chang
Dra 17h26'54" 51d5'50"
Zhu Ying
Vir 11h55'1" 7d26'50"
Zhukova Svetlana Vladimirovna
Psc 1h10'51" 28d9'29"
ZHU-LI
Uma 8h41'8" 46d57'52"
Zhyra
Uma 12h42'40" 55d22'18"
Zi Lin
Vir 14h14'12" -3d7'13"
Zia Giury
Cyg 21h29'29" 36d57'46"
Ziad
Oph 16h34'23" -0d37'56"
Ziad Z. Ahmed
Lyn 6h57'12" 52d18'26"
Zian Jay Robert Loe
Cnc 9h13'29" 27d32'48"
ZiaSheonie
Sge 19h44'37" 18d20'20"
Zico Scallywag
Cheggybums Chester
Lyn 7h28'9" 53d49'48"
Ziegler Star
Uma 9h26'25" 58d11'19"
Ziegler, Kurt
Uma 8h52'35" 53d48'55"
Zieleninka
Ari 2h11'26" 23d9'15"
Zielinski, Christa
Uma 8h31'28" 73d3'41"
Zielke, Jürgen
Uma 8h51'17" 61d22'18"
Zig Gauthier
Gem 7h21'37" 31d52'38"
Zig, my one and only
Uma 12h37'48" 57d53'13"
Zigger
Cyg 20h13'12" 39d38'2"
Ziggie
Ori 5h38'10" -0d18'31"
Ziggy
Cma 6h52'54" -15d19'21"
Ziggy
Uma 8h56'1" 67d54'21"
Ziggy
Per 2h27'48" 52d43'21"
Ziggy
Ori 6h16'11" 11d0'13"
Ziggy Grandpa Brzozowski
Cnc 8h34'18" 16d34'44"
Ziggy O'Brien
And 1h58'21" 47d3'43"
Ziggy's #52
Uma 11h34'59" 65d35'47"
ziggystar 83
Vir 13h53'10" -2d12'19"
Ziggy-Star-Dust
Ori 6h20'42" 5d58'29"
Zildjian
Uma 11h23'53" 45d20'1"
Zilm
Vir 13h22'44" 9d35'16"
Zilya Autumn
Lib 15h58'49" -17d23'42"
Ziment
Umi 16h21'51" 76d1'51"
Zimmer
Uma 10h12'22" 53d10'33"
Zimmer Gunsul Frasca
Per 3h23'48" 43d41'44"
Zimmermann, Adriana
Serio Rodrigues
Ori 5h19'16" 3d17'40"
Zimmermann, Horst
Ori 6h17'28" 9d28'0"
Zimpelman's Beacon
Cam 4h27'3" 66d41'50"
Zimtschtärn
Cep 23h4'58" 79d18'22"
Zina Krivoruk
Psc 1h19'37" 18d49'53"
Zina Noorani
Cap 21h8'31" -19d9'49"
Zina Ruban
Cnc 9h4'1" 19d54'0"
Zind, Michael
Uma 10h42'54" 71d3'23"
Zinfandel and Akita
And 0h58'23" 36d39'44"
Zink, Franz Josef
Uma 12h3'20" 60d16'6"
Zink, Gudrun
Uma 14h7'57" 57d13'25"
Zinnow, Heinz-Jürgen
Uma 8h12'3" 68d50'47"
Zinny Zue's Star
Crb 15h49'33" 29d28'4"
ZINO B. A.
Psc 0h41'57" 7d40'49"
Zino Monti
Cnc 9h18'18" 10d16'8"
Zinochka
Psc 23h47'29" 5d44'19"
Zinovy Goro
Uma 9h33'14" 55d28'55"
Zinur Kurmakaev
Leo 10h17'16" 26d52'35"
Zinzaro
Leo 10h11'37" 27d46'26"
Zio
Lac 22h21'11" 47d15'55"
Ziobro
Uma 9h22'2" 64d49'20"
zioGianfranco
Uma 8h55'31" 58d35'39"
Ziomara Ocasio
Lib 15h20'40" -16d12'48"
Zion
Umi 14h52'38" 74d45'33"
Ziona
Uma 10h56'50" 34d52'32"
Zipfel, Klaus
Uma 13h56'33" 57d19'1"
Zipper, Bernd
Sgr 17h53'57" -29d31'47"
Zippy Nicky
Cnv 13h18'30" 38d29'46"
Zircon Star for Keith Wigglesworth
Peg 21h40'58" 16d0'15"
Zirka Sophia Diane
Umi 15h46'47" 77d29'51"
Zirmarie
Dra 16h55'28" 57d36'13"
Zita & Brian
Cyg 21h16'3" 52d18'54"
Zita Kühnis Keel
Cyg 21h23'54" 50d4'19"
Zitha
Leo 9h55'38" 15d26'5"
Ziv
Cap 20h13'32" -9d12'44"
Ziv Ben-David
Leo 11h55'16" 20d26'30"
Ziv Tzfir Samberg
Lib 15h25'9" -13d48'4"
Zivi Attwood-Cutter
And 23h16'12" 35d28'13"
Zivil
Cnc 8h6'7" 16d23'4"
Zivko & Roland Katavic
Tau 4h30'3" 15d55'1"
Ziya Guan - My spark of radiance
Pho 0h36'36" -54d43'11"
Ziya Irulan
Sgr 18h12'15" -16d32'50"
Ziz
Uma 10h29'59" 67d37'12"
Ziza
Uma 9h41'1" 54d29'45"
Zizou
Lac 22h30'10" 41d38'20"
Zjawiskowa Odwaga
Tau 5h57'59" 24d8'18"
Zlata
Aqr 23h11'8" -20d20'40"
Zlata Fishel
Lib 15h5'16" -8d1'14"
Zlata zvezdica Metoda
Ari 2h58'11" 17d50'30"
Zlatana 22.06.1986.
Cyg 20h3'22" 35d51'13"
Zlati i Yoana
Uma 11h13'37" 60d46'47"
Zlatna Nicki
Cap 20h18'57" -25d11'44"
ZLPPNEVE 80
Umi 15h15'44" 74d19'47"
zman
Tau 4h50'51" 23d56'18"
Zo Lennon Ocean Hatchman
Leo 11h14'23" -0d42'26"
Zoar
Gem 6h56'33" 13d33'46"
ZOBO
Ari 3h20'4" 22d18'18"
Zobo's Star
Ori 6h6'28" 7d4'49"
Zocca Claudia
Umi 15h15'16" 67d59'48"
Zocha Mary
Uma 10h32'4" 63d57'39"
ZochDR042963
Tau 4h26'18" 23d45'51"
Zoe
Peg 22h49'59" 18d18'25"
Zoe
Uma 9h50'18" 61d43'46"
Zoe
Cyg 21h45'42" 53d44'58"
Zoe
Sco 16h10'24" -9d18'32"
ZOE
Cap 20h42'59" -14d41'9"
Zoe 86
Aqr 21h46'29" 1d12'7"
Zoe Alessia Timo
Umi 13h33'49" 86d46'4"
Zöe Alexa
And 0h28'28" 36d47'15"
Zoe Alexandra Garrett
Sgr 19h22'28" -14d58'46"
Zoe Alexis
Lib 15h52'39" -10d12'44"
Zoe Alexis
Umi 14h39'17" 72d35'13"
Zoe Amelie
Uma 11h56'14" 53d53'1"
Zoe Angel Star
And 23h13'31" 41d53'24"
Zoe Ann Sciullo
Cam 4h17'34" 56d10'58"
Zoë Ann Spencer's Christening Star
And 23h37'47" 41d43'36"
Zoe Anna
Umi 15h17'31" 71d2'49"
Zoe Anne
Cap 20h16'32" -12d42'34"
Zoe Anne Gray (My Little Sister)
Lyn 9h21'7" 39d54'31"
Zoe Ariel
Sco 16h34'19" -28d47'8"
Zoe Barfield
Cas 0h51'58" 57d4'4"
Zoe Bertha Eiden
Dra 17h53'1" 52d26'16"
Zoe Camille Johnston
Cas 1h37'32" 65d11'49"
Zoe Chantiles
Ari 2h21'52" 10d35'43"
Zoe Christianna
Uma 9h25'53" 51d19'16"
Zoe Claire Akesson
Cru 11h59'34" -61d50'8"
Zoe Clayton
Cas 23h49'42" 58d58'31"
Zoe Corrin Wesp
Cnc 9h18'56" 17d54'46"
Zoe Dakota Burns
Del 20h40'11" 16d19'18"
Zoe Dawn Smith
Uma 10h49'47" 40d41'22"
Zoe Dimitri
Ori 6h11'4" 16d52'29"
Zoe Dora
And 0h32'28" 26d38'6"
Zoe Dweck
Leo 10h34'16" 16d12'45"
Zoe Elizabeth Fulton
And 0h1'22" 44d20'43"
Zoe Elizabeth Hicks
Cas 23h27'44" 58d3'23"
Zoe Elizabeth Rose
Cnc 8h51'28" 19d27'34"
Zoe Elizabeth Zannie
And 1h33'14" 45d37'56"
Zoe Ellen Regan, 7/8/04
Apu 14h31'30" -73d30'30"
Zoe (Elliot) Cartwright
Per 3h3'57" 47d9'51"
Zoe Emily Kathleen Bartlett
Cnc 8h5'2" 23d32'9"
Zoe Emma
And 23h2'3" 47d28'57"
Zoé et Lola Delizee
Ori 5h18'26" 15d56'37"
Zoe Faith
Vir 12h26'46" 10d33'23"
Zoe Foster
Col 5h59'30" -34d53'50"
Zoe Fredrique Reiter
Uma 11h26'49" 62d22'58"
Zoé Friedlingstein-Hoibian
Uma 11h43'44" 38d36'25"
Zoe Frigla-Elyse
Aqr 20h46'30" -8d56'42"
Zoe G. Nardone
Sco 17h51'45" -40d32'16"
Zoë Gomes
Uma 8h57'47" 70d24'10"
Zoe Grace
Aqr 20h57'23" -1d33'6"
Zoë Grace
And 1h36'32" 47d32'39"
Zoe Gregory-Paul
Aql 19h46'11" -0d40'5"
Zoe Grillo
And 2h11'35" 38d48'42"
Zoe Hagen Castro
Psc 23h49'15" 0d52'51"
Zoë Hardie
Lyn 7h55'2" 38d21'38"
Zoe Isabel Ross Gilligan
Psc 0h49'42" 8d0'35"
Zoe Isabella Baca
Tau 5h43'1" 24d40'49"
Zoe Isabella Hall
Ori 5h8'49" 1d18'0"
Zoë Isabelle
And 0h11'53" 25d18'59"
Zoe_Jane
Cas 0h53'34" 54d53'22"
Zoe Joleigh
Uma 11h0'9" 45d0'41"
Zoe Kiara Harbart
Leo 9h33'39" 28d30'23"
Zoé Latifah
Uma 9h14'40" 50d46'25"
Zoe LeeAnn King
Lep 5h16'41" -10d57'3"
Zoe Leigh Hansford
Cnc 8h18'18" 8d22'5"
Zoe Ling D'Angelo
Cam 4h21'49" 62d51'24"
Zoe Louise Greenfield - Zoe's Star
Cas 23h57'17" 56d46'20"
Zoe Lydiaann
Sgr 19h8'4" -14d24'46"
Zoe M. Bishop #3
Gem 6h43'40" 16d41'45"
Zoe Macwilliams
Psc 23h27'16" 5d30'53"
Zoe Malia Man
Sco 17h54'23" -35d50'3"
Zoe Mann
Vir 13h34'39" -20d59'43"
Zoe Margaret Crippin
Lmi 10h38'6" 35d37'23"
Zoe Margaret Stack
Gem 7h31'10" 31d9'57"
Zoe Margot Kudrnac
Psc 1h22'18" 15d26'46"
Zoe Maria Burrell
And 23h19'54" 47d51'55"
Zoe Marie Podejko
And 1h42'16" 45d26'7"
Zoe Marie Scott
Uma 11h40'49" 57d13'26"
Zoe May Eldridge - Zoe's Star
And 2h33'52" 46d51'50"
Zoe Melina Levy
Sgr 18h40'44" -24d56'51"
Zoe & Mia
Cas 23h44'7" 56d41'42"
Zoe Michelle Rodriguez
Vir 13h6'53" -16d11'53"
Zoe Michelle Schrappe Nobrega
Ari 2h55'19" 14d15'39"
Zoe (My Angel)
Cas 0h16'41" 51d57'10"
Zoe Noel Curry
Lyn 7h53'40" 54d14'47"
Zoe P. Kashner
Leo 9h41'21" 28d18'26"
Zoe Parker
And 22h59'40" 48d28'10"
Zoe Payne
Psc 0h33'58" 9d0'58"
Zoe Pfeiffer
Uma 8h16'24" 59d45'3"
Zoe Ple-Nemo
Cep 4h0'52" 82d20'39"
Zoe Polaris
Psc 1h33'50" 13d10'45"
Zoe Quartey
Lib 16h0'1" -9d19'58"
Zoe Reagan Hubbard
And 2h28'45" 50d6'24"
Zoe Rebecca Bensusan
Crb 16h4'6" 38d38'32"
Zoe Rene Carcia
Cap 20h36'52" -10d17'4"
Zoe Richardson
And 23h36'34" 48d38'55"
Zoe Rowe
Uma 10h15'51" 44d31'10"
Zoe S. Davis
Ori 5h28'13" -7d42'21"
Zoe Sachero
Uma 9h4'24" 49d11'34"
Zoe Shemera Neville
And 23h56'22" 40d34'15"
Zoe Siobhan
Sco 16h41'13" -42d51'46"
Zoe Summer
Peg 21h53'53" 6d35'44"
Zoe' Tatum Hemphill
Tau 4h7'33" 24d2'35"
Zoe Todorov
Aqr 22h26'56" -3d23'47"
Zoe Trears
Peg 0h2'1" 27d13'7"
Zoe Victoria Hogben
Cas 0h42'36" 51d23'40"
Zoë Victoria Oelz
Aqr 22h14'5" 1d3'15"
Zoe Virginia Gironda
Vir 12h10'19" 7d27'14"
Zoé Walther
Uma 10h22'17" 46d31'27"
Zoe Walz
Cas 3h12'9" 64d47'5"
Zoe Wendolowski
Cam 7h41'46" 77d51'34"
Zoe Wilson
Tau 4h33'48" 20d24'28"
Zoe Wood
Lyn 7h51'48" 52d2'46"

Zoe - You light up my life
 Tau 4h20'1" 30d1'35"
Zoei Alexia
 Umi 15h28'12" 71d35'39"
Zoel Pierre André Simon
 Umi 13h9'23" 71d38'50"
Zoellner's Magical Brandy
Brook Star
 Uma 11h16'51" 63d56'53"
zoen voor koen
 Uma 11h28'57" 37d16'22"
ZoeNista
 Vir 13h18'19" 4d11'54"
Zoe's Star
 Lyn 9h1'36" 33d6'16"
Zoe's Star
 Dra 18h41'28" 70d18'40"
Zoe's Wish Upon A Star
 Ari 3h16'21" 31d2'14"
Zoesophia
 Sco 16h41'39" -30d29'9"
Zoey
 Lib 15h2'31" -8d17'51"
Zoey
 Gem 7h11'32" 34d35'27"
Zoey
 Lmi 10h30'57" 34d8'34"
Zoey A. Sternoff
 Umi 14h17'39" 76d17'10"
Zoey Alexandria Martin
 And 0h17'26" 31d36'33"
Zoey Angel Lyra
 Lyr 19h20'1" 37d44'23"
Zoey Bleu
 Cap 20h57'13" -19d37'54"
Zoey Cecilia
 And 0h13'49" 46d13'5"
Zoey Cooper
 Uma 9h28'9" 53d49'33"
Zoey Horan
 Cam 6h43'32" 75d28'54"
Zoey Kaelin
 Mon 7h23'15" -8d9'49"
Zoey Madison Rouwette
 Ari 3h8'34" 26d37'1"
Zoey Montgomery
 Cma 6h18'21" -27d54'33"
Zoey Rayne Lowe
 And 0h9'33" 46d25'41"
Zoey Renee
 And 23h42'37" 45d50'7"
Zoey Tate Delaney
 Lib 15h1'27" -10d33'1"
zoeylilevil
 Dra 19h30'4" 67d16'9"
Zoey's Piece of Heaven
 Lib 14h52'27" -1d9'49"
Zoey's Wish
 Vir 13h50'33" -10d54'17"
Zoey's World
 And 0h33'33" 28d10'1"
Zohar
 Tau 3h41'18" 29d57'12"
ZOHRA Best Mummy in
the Universe!
 Cnc 8h43'30" 11d3'37"
Zohrab Yedalian
 Leo 11h40'19" 20d22'43"
Zoi Psylla
 Uma 9h32'48" 47d21'56"
Zoia
 Sgr 18h23'53" -33d3'32"
Zoian
 Cam 5h26'57" 73d46'44"
Zoila Pino
 Vir 14h21'4" 2d55'27"
Zola
 Cap 20h31'49" -18d57'39"
Zola Eleanor Barnes
Brubacher
 Cas 1h19'46" 54d23'9"
Zoli 40
 Ari 1h47'4" 11d49'52"
Zolina
 Crb 16h6'52" 36d7'18"
Zomax Marketing
 Ori 5h37'25" 1d50'3"
Zombori Ottó
 Tau 3h58'55" 1d47'14"
Zombro Light
 Umi 13h17'40" 72d20'0"
Zomper Family Star
 Ori 5h29'19" 13d28'50"
Zonia
 Cas 1h6'32" 62d13'32"
Zonora Linda Boyce
 Psc 1h26'1" 26d37'53"
Zontaris
 Cas 2h12'7" 63d57'53"
Zonya Lydia
 Mon 6h49'35" -0d22'37"
Zooker
 Uma 13h45'58" 54d18'37"
Zookoa
 And 1h8'39" 35d10'53"
ZOOM
 Her 16h47'34" 9d18'55"
ZOOM
 Vir 14h11'35" -0d4'18"
Zoom Zoom
 Sco 16h16'47" -17d11'18"
Zoomba2007
 Umi 14h43'45" 75d47'44"
Zoomer
 Cam 4h36'28" 64d29'51"

Zoos
 Aql 19h57'32" 8d35'49"
Zoot!
 Leo 10h56'24" -3d39'13"
zora
 Tau 3h51'43" 8d9'59"
Zora
 Cas 0h0'14" 53d33'43"
Zora Antoinette Dukovac
 Umi 15h30'11" 75d34'14"
Zora Joy Kidron
 Per 3h30'53" 42d18'38"
Zora Kasser
 Aql 19h34'35" 10d21'15"
Zoraida
 And 2h16'35" 45d36'48"
Zoraida Colon
 Uma 13h34'0" 58d19'30"
Zoraida Fernandez
 Leo 9h49'10" 24d7'20"
Zoran M Iliev '66 My Star
Was Born
 Ori 5h38'50" -2d4'6"
Zoran Markovic
 Uma 13h43'36" 50d9'34"
Zoran23 *Forever Love
Jeli+Zoki*
 Cas 23h26'10" 52d45'34"
Zorana Rade
 Cas 1h2'44" 53d34'49"
Zoraya
 Lyn 9h10'55" 38d13'9"
Zordon
 Cnc 8h44'47" 16d53'28"
Zoriana Oreste
 Psc 0h54'32" 7d26'56"
Zorisco
 Aql 19h12'25" 5d45'8"
Zorjana Inna
 Gem 7h16'18" 24d3'41"
Zorn , Susy
 Cap 20h43'44" -24d32'44"
Zorro
 Her 16h46'42" 32d53'15"
Zorrodream
 Cap 20h29'56" -11d59'56"
Zoryana
 Vir 12h1'8" -1d43'56"
Zosha Venetta Womack
 Del 20h39'0" 6d22'29"
Zou Xuan
 Psc 1h11'52" 17d29'57"
Zoua Moua
 Lib 15h26'27" -7d46'27"
Zouille
 Vir 13h23'52" -6d42'18"
Zouzou
 Aqr 21h20'43" -7d46'6"
Zoya Beg
 And 23h27'14" 41d57'58"
Zoya Stott
 Cyg 20h15'36" 32d37'25"
Zoya-Max
 Uma 9h8'41" 49d13'31"
Zron
 Psc 1h6'26" 19d17'0"
ZSB21
 Aqr 22h54'44" -8d19'6"
Zsepi Csillaga
 Leo 11h32'38" 10d9'57"
Zsóka & Jan
 Aqr 22h31'46" -20d44'22"
Zsolti
 Uma 11h21'3" 44d41'27"
Zsuzsa
 Uma 10h32'52" 50d20'37"
Zsuzsa & Zsolt
 Cyg 21h59'3" 45d11'36"
Zsuzsanna
 Cap 21h30'41" -17d50'38"
Zsuzsanna Kiss
 Vir 12h40'9" -7d57'24"
Zsuzsi Nyuszi Hercegnom
 Psc 1h2'6" 13d20'57"
ZsuzsikaHusika
 Sco 17h12'52" -43d36'40"
Zsuzsinak az én
Piroskámnak
 Cnc 9h5'29" 23d57'33"
Zsu-Zsuzsi-Zeti
 Uma 11h4'6" 34d21'5"
Zsyla Honda
 Cap 20h31'17" -18d14'56"
Zubair Ahmed Abbasi
 Per 4h15'55" 51d23'44"
Zubin - Forever And A Day
!!
 Cyg 21h42'35" 43d0'18"
Zubler
 Cam 4h13'29" 66d12'15"
Zufriedenheit
 Uma 13h47'48" 62d7'19"
Zufrinia
 Uma 9h22'52" 57d22'24"
Zula Campagna
 Uma 11h52'35" 63d8'14"
Zuleika
 Aqr 20h55'9" -8d17'46"
Zuleima Torres
 Aqr 20h46'6" 0d20'51"
Zulema Alexandra Lemolt
 Gem 7h17'32" 19d13'35"
Zulema Morales Iribe
 Sgr 18h37'24" -24d35'41"

Zulema Nuno
 Ori 6h13'5" 6d14'59"
Zully
 Leo 9h53'27" 28d50'53"
Zully
 Vir 13h23'28" -6d53'55"
Zulma
 Umi 16h51'3" 79d40'33"
Zulma Y. Orellana
 Sgr 18h29'30" -22d35'11"
Zulphia's Jewel
 Umi 14h21'34" 72d47'55"
Zümran
 Uma 11h19'0" 36d13'13"
Zumwalde
 Uma 9h59'23" 46d34'33"
Zunno's Light of Hope
 Uma 11h34'33" 58d57'45"
Zuokas
 Cet 3h13'34" -0d9'2"
zupkowolff3202006
 Uma 11h31'56" 63d31'41"
Zurgglemorph
 Tau 4h45'33" 28d54'31"
Zuri Al-Uqdah
 Umi 16h18'28" 73d18'11"
Zuri Naima
 Vir 14h46'36" 3d52'21"
Zurr, Michael
 Ori 6h7'38" 9d26'14"
Zusammen Ewig
 Uma 9h54'4" 44d34'38"
Zuzana
 Ari 1h57'2" 13d29'1"
Zuzel Manny's Passion
 Sgr 18h59'34" -13d19'54"
Zuzia
 Leo 9h58'41" 9d37'1"
ZUZKA
 Leo 11h14'54" 23d48'3"
Zuzu
 Sgr 18h16'37" -17d13'17"
Zuzu's Petals
 Uma 12h47'29" 58d18'42"
"Zvezda Brajena"
 Lyn 6h59'21" 58d19'2"
Zvezda Eleni i Georga
 Cyg 20h37'42" 61d3'19"
Zvezda Gale
 Peg 21h51'44" 10d59'10"
Zvezda Milojevica
 Per 2h13'55" 52d7'40"
Zvezda Tamara
 And 2h18'44" 47d12'25"
Zvezdelina
 Vir 13h59'20" -12d43'46"
Zvezdica Iva
 Vir 15h8'19" 0d53'35"
Zvi Goren
 Aql 19h19'16" 10d44'6"
Zvyezda
 Uma 9h21'28" 42d39'40"
zwanzig
 Lyr 18h47'34" 39d30'6"
Zwelithini Mabhena
 Tau 3h53'57" 9d1'57"
Zwergi
 Dra 18h59'49" 62d35'27"
ZWI
 Umi 14h24'13" 73d32'29"
Zyanya
 Peg 22h16'56" 7d3'11"
zycie kolega
 Sco 16h44'35" -32d46'52"
Zynique Michelle Madison
 Vir 12h30'10" 10d46'46"
Zypher
 Lyn 7h27'24" 52d26'15"
Zyrus
 Ori 6h4'24" 17d40'59"

The Astronomer

Dr. James J. Rickard received his M.S. and Ph.D. degrees from the University of Maryland. He was a member of a team of radioastronomers who produced the first high resolution maps of neutral hydrogen in the spiral arms of our own Milky Way galaxy. The 300' dish antenna at the National Radio Astronomy Observatory in Green Bank, West Virginia was used to gather the data for his research in spiral structure. His interest in the interstellar gas led him to Palomar Observatory and then to the European Southern Observatory in Chile to use optical telescopes to continue his research. the years in Chile involved research, development of electronic cameras and the introduction of computers for telescope control. He is a member of the American Astronomical Society and the International Astronomical Union.

Dr. Rickard returned to the USA and the small desert community of Borrego Springs to work again as a radioastronomer at the Clark Lake Radio Observatory, a remote research station for the universities of Iowa, Maryland and California. The research projects included studies of the sun, solar wind and scintillations of compact objects. One of the peculiar compact objects was later discovered to be the first *millisecond pulsar*, a neutron star spinning at over 38,000 RPM.

With the closure of Clark Lake Radio Observatory, due to funding cutbacks and consolidation in federally supported research, Dr. Rickard has turned his attention to education projects. He established popular monthly star parties for visitors to enjoy the dark night sky in the Anza-Borrego State Park, is a science writer for the local newspaper, and teaches at San Diego State University. He is an avid solar eclipse and comet chaser; and has led expeditions to Mexico, the Philippines, Brazil and Chile. He volunteers his time an technical talents to aid the Astronomical Foundation and environmental projects in Southern California.

Illustrations

Taken from Sheglov's reproduction of the 1690 Star Atlas by Johannes Hevelius, these lithographs appear to be drawn backwards because Hevelius drew his star atlas as though he was looking down on the celestial sphere. Johannes Hevelius lived from 1611 to 1687. He was born in Danzig, Poland (now known as Gdansk). His studies of the surface and the movements of the Moon laid the foundation for modern lunar topography. Hevelius was one of the first astronomers and although his star atlas was published after his death he is responsible for designating many of the constellations which are still recognized today.